英汉
土木工程大词典

An English-Chinese Civil Engineering Dictionary

中交第四航务工程勘察设计院有限公司

罗新华 主编

人民交通出版社
China Communications Press

内 容 提 要

《英汉土木工程大词典》共收录词条约82万,内容涉及建筑工程、港口航道工程、道路工程、铁路工程、岩土工程、桥梁与隧道工程、市政工程、水利水电工程、城市轨道交通工程、工程机械、建筑材料、健康安全环保(HSE)等相关学科专业方面的词组、短语,可供从事土木工程方面相关工作的广大科技人员参考使用。

图书在版编目(CIP)数据

英汉土木工程大词典 / 罗新华主编. — 北京:人民交通出版社,2014.2
ISBN 978-7-114-11134-1

Ⅰ. ①英… Ⅱ. ①罗… Ⅲ. ①土木工程—词典—英、汉 Ⅳ. ①TU-61

中国版本图书馆 CIP 数据核字(2014)第 011098 号

书 名:	英汉土木工程大词典
著 作 者:	罗新华
责任编辑:	邵 江 杜 琛
出版发行:	人民交通出版社
地 址:	(100011)北京市朝阳区安定门外外馆斜街3号
网 址:	http://www.ccpress.com.cn
销售电话:	(010)59757973
总 经 销:	人民交通出版社发行部
经 销:	各地新华书店
印 刷:	北京市密东印刷有限公司
开 本:	880×1230 1/16
印 张:	233
字 数:	17334 千
版 次:	2014年2月 第1版
印 次:	2014年2月 第1次印刷
书 号:	ISBN 978-7-114-11134-1
印 数:	0001~1500 册
定 价:	868.00 元(上、下两册)

(有印刷、装订质量问题的图书由本社负责调换)

《英汉土木工程大词典》
编 委 会

总 策 划：朱利翔
主 编：罗新华
副 主 编：覃 杰　张丽君
编 委（按姓氏笔画顺序）：
　　　　卢永昌　朱利翔　余巧玲　张丽君　张宏铨　李 刚
　　　　李 虹　李华强　李伟仪　杜 宇　沈力文　肖玉芳
　　　　陈策源　周晓琳　罗 梦　罗新华　徐少鲲　高月珍
　　　　曹培璇　覃 杰　廖建航
审 定：林 鸣　王汝凯
校 审（按姓氏笔画顺序）：
　　　　刘少珍　吕邦来　何康桂　何智敏　张 勇　张欣年
　　　　张冠绍　肖向东　邱创亮　麦若绵　周 野　林 琳
　　　　林吉兆　苗 辉　祝刘文　贾登科　郭玉华　梁 桁
　　　　曾青松　韩凤亭　廖先斌　蔡泽明　薛轩宇
顾 问：蔡长泗　杨孟愚　王将克
电脑录入及排版：罗新华　余树阳　余树敏　张喜顺　张再嘉　张丽琼

《英汉土木工程大词典》
编 辑 组

责任编辑：邵　江　杜　琛

编　　辑：卢　珊　温鹏飞　富砚博　潘艳霞　卢俊丽　丁　遥
　　　　　　李　洁　杜　琛

复　　审：刘　涛　陈志敏　王　霞　杜　琛　王文华　郑蕉林

终　　审：赵冲久　韩　敏　周往莲　田克运　尤晓暐

特邀审稿：李志荣（民族出版社）

序

随着国家"走出去"战略的实施,我国工程建设行业的海外业务也在不断扩展,施工企业承担的海外建设项目种类越来越多,而国内现有的有关土木工程专业方面的英语辞书已经不能适应这种新形势的要求,很有必要编写一部综合性的土木工程方面的英语辞书,以满足广大科技工作者在工作和学习中的需要。

20世纪90年代初,中交第四航务工程勘察设计院罗新华硕士等人即开始了该方面词条的收集工作,并于2000年4月出版了《汉英港湾工程大词典》(人民交通出版社出版)。在此基础上,罗新华等人又参考了大量的国内外相关辞书、现行规范和标准术语,进一步广泛收集了与土木工程相关的专业词汇,涉及建筑工程、港口航道工程、道路与桥梁工程、铁路工程、水利水电工程、岩土工程、建筑材料、环境保护和安全等相关学科专业。经过多年的不懈努力,终于完成了《英汉土木工程大词典》的编纂工作,共计收录土木工程类英语词条近82万条。该书是国内迄今为止最为完整的土木工程方面的英语辞书。

该大词典的出版,将在很大程度上满足从事土木工程方面工作的广大科技工作者阅读和翻译英文的需求。

中国工程院院士:谢世楞

2013年9月于天津

前　言

20世纪80年代初的某个晚上，编者在广州中山大学图书馆晚自修，因学习需要进入外文工具书阅览室借阅有关地质方面的英汉词典。但看到的是，宽敞的书架上只放了少量仅有几百页（16开本）的非土木工程专业词典和两本厚厚的美国人编的韦氏词典（大16开本）。当时编者便感慨于韦氏词典编者的伟大；佩服之余，心生今后也要编出这样大部头综合性权威词典的理想。1990年，编者硕士毕业，其后一直在国内外从事岩土工程勘察类工作，经常接触到工程报告、招投标书的翻译，也经常因找不到可用的相关专业词典而苦恼，于是萌发了编写汉英土木工程方面词典的想法，并在王汝凯、麦土金等专家教授的鼓励下开始了资料搜集工作。

1997年初，编者完成了《汉英港湾工程大词典》的编撰工作，收录词汇约186000条，该书于2000年4月由人民交通出版社出版。在该词典编写过程中，偶然在一本有关土木工程的辞书中看到这样一句话："土木工程学科发展到今天，英汉土木工程词汇估计有三十多万条"。为了这个答案，在好奇心的驱使下，编者凭借兴趣和长期在国外及港澳地区的工作需要，尝试收集，结果发现，在实践中收集的词条远远超过这些。经过悉心整理和筛选，汇集了词条近82万，编写成本书。编者在本书收录词条的过程中，特别注意以下两方面：第一，高校和科研单位需要；第二，土木工程项目招投标、勘察设计以及施工过程中的需要，以期满足土木工程科技工作者的实际需要。

本词典的词条首先源自近年来土木工程专业大类相关出版单位出版的土木工程类词典和专业图书，并在收集时做了大量核查比对工作，力求对每个词条给出最为准确、常用和实用的解释；其次，编者参考了全国科学技术名词审定委员会公布的《土木工程名词》，并根据行业发展近况，对部分内容进行了修订。此外，据编者近二十年在国外和港澳地区参与国际土木工程建设的经验，目前的土木工程勘察设计与施工特别注重健康、安全、环保（HSE）方面问题，因此本词典也收集了这方面的词条，方便广大科技工作者在工作中使用。

在词典编排顺序上，一般有两种方式：一种是先列出基本词，再在基本词下按字母顺序分别列出相关词组；一种是不区分基本词和词组短语，按字母顺序依次排列。本词典采用了后一种方式，这种排序的优势是读者可按字母顺序快速地找到所需要的词条，节省时间，同时这种方式也符合一般科技工作者的查阅习惯。

本词典在编写过程中，先后有一百多位国内外专家、教授、勘察设计大师、博士、硕士、英语专业毕业的大学生、在校大学生、常年奋战在工程项目第一线的工程师、技术人员参与。同时，也得到了交通运输部科技司、中国交通建设集团有限公司、人民交通出版社、中交第四航务工程勘察设计院以及其他兄弟单位的大力支持。本词典的编纂可以说是凝聚了我国从事土木工程工作的广大科技人员的心血，在此表示衷心的感谢！

科技发展迅速,土木工程新词汇不断出现,且随时间的推移某些词的词义也有所变化,热切期望广大读者使用过程中发现问题给予批评指正,也欢迎随时提供该词典没有收集到的相关词条,为以后再版输送新鲜血液。相信有读者的热心参与,该词典将更加完善。

主编:罗新华

2014 年 1 月 13 日于广州

凡　　例

1. 本词典系按英文字母顺序排列,术语单词、复合词和短语一律顺排。对于含有逗号(,)和阿拉伯数字的英文词条,只考虑逗号(,)及阿拉伯数字前面的英文词的排序。

2. 单个英文单词或单条词组在有若干释义时(即一词/词组多义情况),释义间用分号(;)隔开。如:saddle joint 阶形接槎;咬口接头;咬接头;鞍状关节;鞍形接头;鞍(形)接合

3. 括号的用法

(1)( )圆括号

① 释义的补充,如 sacellum（露天的）小型罗马祭坛;

② 可省略的字母,如 sabin(e)。

(2)[ ]方括号

① 可替换前面的单词,如 overhead railway [railroad] 高架铁路;高架铁道;

② 缩略语,如 uninterrupted power supply [UPS] 不间断(供电)电源;不中断电源;

③ 缩略语解释,如 FVT [field vane test] 现场十字板剪切试验;

④ 词源,如 passing point [美]越行站;相会点;

⑤ 单词的复数形式,如 palestra 体育场;希腊或罗马体育训练馆;[复] palestrae;

(3)【 】鱼尾括号,表示专业或学科,如【铁】,表示铁路工程;

(4)〈 〉尖括号,表示词语属性,如〈旧〉,表示旧时用。

4. 专业或学科代号

【数】数学	【港】港口航道工程	【建】房屋建筑工程
【物】物理	【道】道路工程	【航海】航海学
【化】化学化工	【铁】铁路工程	【航空】航空学
【无】无线电子学	【给】给排水	【岩】岩土工程
【地】地质学	【矿】矿物学	【水文】水文学
【测】测量学	【气】气象学	【机】工程机械
【疏】疏浚工程	【计】计算机	【天文】天文学
【救】救捞工程	【声】声学	【植】植物学
【动】动物学	【生】生物学	

目　录

序 …………………………………………………………………………………………… I

前言 ………………………………………………………………………………………… Ⅲ

凡例 ………………………………………………………………………………………… V

正文 …………………………………………………………………………………… **1～1806**

参考文献 ………………………………………………………………………………… 1807

M

maar 小火山口;火口湖;低平火山口
Maas(borehole)compass 马氏罗盘
Maas survey 马氏钻孔偏斜测量
Maastrichtian 麦斯里希特阶【地】
macaboard 马克板(一种木纤维板)
macadam 碎石
macadam aggregate 碎石集料;锁结式集料;锁结式骨料
macadam-aggregate mix 碎石集料混合料;碎石骨料混合料
macadam aggregate type 嵌锁式路面;碎石路面
macadam aggregate type road mix surfacing 碎石骨料的混合路面
macadam base 碎石基底;碎石基床;碎石基层;碎石地基;碎石底层
macadam bed 碎石路基
MacAdam colo(u)r difference equation 麦克亚当色差方程式
macadam foundation 碎石基层;碎石基础
macadamization 建筑碎石路;碎石筑路法;碎石铺路法
macadamize 建筑碎石路;铺筑碎石路;铺碎石
macadamized road 碎石路;铺碎石的路;碎石铺路
macadamized runway 碎石跑道
macadamizing 碎石铺道法
macadam mixing plant 碎石搅拌机;碎石混合料工厂
macadam pavement 砾石路面;碎石路面
macadam road 碎石路面道路;碎石路
macadam road construction method 碎石路施工方法
macadam roller 碎石路碾;碎石压路机
MacAdam's chromaticness discrimination threshold 麦克亚当辨色阈
MacAdam's construction 麦克亚当筑路法;碎石路面筑路法
MacAdam's ellipse 麦克亚当椭圆
MacAdam's limit 麦克亚当极限
macadam spreader 碎石摊铺机;碎石铺路机
macadam sub-structure 碎石下部结构
MacAdam's unit 麦克亚当色差单位
macadam surface 碎石路面;碎石(路)面层
macadam surfacing 碎石路面
macadmix 拌有沥青或其他黏结料的碎石混合料
Macalloy bar 马卡路合金钢筋
MacArthur pedestal pile 麦克亚瑟扩底桩
MacArthur pile 麦克亚瑟桩
Macartney's collet adjuster 麦卡特尼氏内柱调整夹
macasfelt 三毡两油平屋顶
macasphalt 沥青碎石混合料;马克地沥青混合料
macasphalt type pavement 沥青碎石路面;碎石柏油路面
Macassar ebony 望加锡乌木(深褐色纹理装饰硬木)
macaulayite 羟硅铁石
Macaya trough 马卡亚海槽
Macbeth illu minometer 麦克佩斯照度计
MacCaull corrosion tester 麦考尔腐蚀试验机
maccessiable 不能接近的
Macclesfield 麦克莱斯菲尔德
Macclesfield reed count 麦克莱斯菲尔德筘号
Macco side door choke 马可式旁通节流器;马可式旁通油嘴
macdonaldite 莫水硅钙钡石
Macdownel feed water controller 麦克唐纳式给水控制器
Mace 压缩液态毒气
mace bit 冲击回转钻头
Macedelepinary half-shade 马塞德勒皮内半影仪
macedonite 铅钛矿
MA cell diaphragm cell with metal anodes 金属阳极隔膜电解槽
maceral 显微组分;显微煤岩组分;煤显微成分
maceral classification 显微煤岩组分分类
maceral content 显微组分含量
maceral group 显微组分组
maceral variety 显微组分变种
macerate 浸软;浸解;浸化;薄片状填料
macerated fabric 碎布

macerater 废物研磨机;浸化器;筛渣破碎机;切碎机;浸渍器;果子剥离机;纸浆制造机
macerating 浸离
macerating agent 软化剂
maceration 冷浸(渍);浸渍作用;浸软;浸化作用
maceration-disinfection unit 浸渍消毒装置
maceration extract 浸渍液
maceration tank 浸渍槽
macerator 废物研磨机;浸化器;纸浆制造机;果子剥离机;浸渍器;切碎机;筛渣破碎机
maceria 用多种材料粗砌的墙(古罗马)
Mace's patent 梅斯专利(将球形门执手装到轴上的专利方法)
Macewen's triangle 麦丘恩氏三角
macfallite 钙锰帘石
Mac-Girling 由铸石空心砌块建成的房屋
Macglashan tank 边水柜
Macguarie ridge 马阔里海岭
Machang type pottery 马厂类型陶器
mechanically menipulated electrode holder 自动焊焊把
machatschkiite 九水砷钙石
Mach cone 扰动锥
Mach diamond 菱形激波
macherel sky 鱼鳞天
machete 割灌刀;大刀
Mache unit 马谢氏单位(镭射气浓度单位)
Machiakow limestone 马家沟灰岩(中奥陶世)
machicolate 作突出的堞眼于胸墙
machicolated form 枪眼形;堞眼形;凹凸形
machicolation (古堡女儿墙中的)窗孔;(古堡女儿墙中的)枪眼;突堞;堞眼
Machilus oil 楠木油
Machilus thunbergii 红楠;钓樟
machinability 可切削性;可机加工性;机械加工性能;切削性
machinability annealing 改善加工性退火
machinability test 可加工性试验
machinable 可切削的;可加工的
machinable cast-iron 易切削铸铁
machinable durability of layers 膜层的机械牢固度
machinable glass ceramics 可切削微晶玻璃
machinable mica ceramics 可加工云母陶瓷
machine 机床
machine account 台账
machine acoustic(al)efficiency 机声效率
machine address 机器地址
machine addressing 机器编址
machine address instruction 机器地址指令
machine(age)aesthetics 机械形式的审美观点
machine-aided design 计算机辅助设计;机(器辅)助设计
machine-aided retrieval 机助检索
machine-aided training method 机助训练法
machine-aided translation system 机器辅助翻译系统
machine alarm 机器警报信号;机器报警信号
machine and equipment 机器和设备
machine application 机械应用
machine applied mixed plaster 用机器混合抹灰
machine arithmetic 机器运算
machine art 机器艺术
machine assembler 机器装配工
machine assembly 机器组合;机器装配
machine assembly department 修理车间;机器装配车间;维护部分
machine assist 机器助手
machine attendance 机器保养
machine available time 机器可利用时间
machine-banded pipe 机械加箍木管
machine barrel 机筒
machine base 仪器基座;机(底)座
machine bay 机械工段
machine bearing 机器轴承
machine bit 机钻
machine bite 机器进刀
machine bolt 机制螺栓;机械螺栓;机螺栓
machine bored tunnel(l)ing 掘进机隧洞施工

machine brick 机制砖
machine-broken metal 机碎金属
machine-broken stone 机器粉碎石块
machine-building 机械制造
machine-building factory 机械制造厂
machine-building plant 机器制造厂
machine burden 机器负荷
machine burn 机械致伤;机械磨焦
machine capability 机器能力
machine carbine 冲锋枪
machine card 设备卡片
machine casting 机器浇铸;机器铸造;机铸
machine cavern 机械挖成的洞穴
machine center 机器中心
machine chase 机器版框
machine check 自动校核
machine check interrupt 机器校验中断
machine chest 成浆池
machine cleaned screen 机械清除筛网
machine-cleaning method 机械清洁法;机械清除法
machine code 指令表;机械代码;机器(代)码
machine code instruction 机器(代)码指令
machine code level 机器代码级
machine code program(me) 机器代码程序
machine coding 机器编码
machine components 机器部件
machine computation 机器计算
machine computer 机用计算机
machine construction 机器结构
machine controller 电机控制器
machine control unite type 直接结合式
machine cooling system 机械冷却系统
machine cost 机器成本
machine countersink 机用锥形锪钻
machine cover 机器盖
machine cut 截煤机截槽;机械切削
machine-cut peat 机切泥(炭)
machine cutting tool 机床切削工具
machine cutting torch 机械切割吹管
machine cut tooth 机械切削齿
machine cycle 计算周期;机器(工作)周期;机器工作循环
machine cycle status 机器周期状态
machine cylinder method 机械拔筒制板法
machined 机器加工的;机械加工的
machine data card 机器资料卡
machine data transducer 机器数据传送器
machine days in accordance with calendar days 日历台日数
machine days in accordance with working days 制度台日数
machine-dependent feature 依赖于机器的特性
machine dependent language 计算机专用语言
machine development 机器显影
machined flush 加工齐平
machined gate seat 机加工闸门支座
machine direction 纤维方向;纵向;织机方向;机器方向
machined jet-dust sprayer 动力喷粉喷雾机
machine downtime 停机时间
machine downtime analysis 机器停产时间分析
machine drawing 工程机械图;机器作图
machined rib 机制筋;磨光止水肋条
machine drill 钻机;钻床;机械钻井设备
machine drilling 机钻孔;机械钻孔
machine driven automatic telephone switching system 机动制自动电话系统
machined sealing strip 磨光止条;磨光止水条
machined steel billet meeting face 机加工钢坯贴接面
machined surface 已加工面;经机加工的表面;加工面
machined washer 精制垫圈
machined wood 机加工木料
machine efficiency 机器效率
machine engineer 机械工程师
machine equation 运算方程式【计】;机器方程
machine error 机器误差;机器故障;机器错误

machine excavation 机械开挖
machine-extruded terra-cotta 机制琉璃砖
machine face 机械开挖工作面(采矿)
machine factory 机械制造厂
machine fastening 机械紧固
machine feeder 自动续纸装置
machine final-terminal stopping device 终端停车装置
machine finish 机械加工(光)面;机械抛光;机械最后加工;机器修整;机床加工精度
machine-finish allowance 机械加工裕量;机械加工余量;机械加工留量;机械加工公差
machine-finish concrete 机械修整的混凝土;机器磨面混凝土
machine finishing 机械修整;机械抛光;机械精加工;机械光制
machine-finish paper 机制光泽纸
machine fitter 机器装配工
machine floor elevation 主机室地面高程
machine-foaming process (膨胀渣的)机械成型法
machine for ballasting 铺砟机
machine for cleaning shutter boards 百叶窗板清洗机
machine for decorating rim of ceramic ware 陶瓷饰边机
machine for drilling rails 钢轨钻机
machine for edging 磨边机
machine for flexural tensile tests 抗弯抗拉试验机
machine forging 机器锻造
machine for glazing 施釉机
machine for graduating a circle 度盘刻度机
machine for handling glass sheet 玻璃板搬运机械
machine for living in 高层住房;高层住宅
machine for magnetic welding 磁力焊接机
machine for material testing 材料试验机
machine for parquetry work 镶木机
machine for process(ing)tea 茶叶加工机械
machine for slotting templet 模片切缝器
machine for testing torsion 扭矩试验机
machine for turning over wagons 翻车机
machine foundation 机器基础;机座;机器底脚;设备基础
machine frame 机框;机架
machine framing 机架
machine function 机器功能
machine-glazed paper 单缸光泽纸
machine glazing 单面光
machine gouge 机械刨削槽
machine grading 机械平地
machine grinding 机力磨;机动磨削
machine group 机器组
machine guard 机器上保护操作人员的装置;机器防护装置;机器安全罩;机器保护装置
machine-gun microphone 线列传声器
machine hacksaw blade 机用弓锯片;机用锯条
machine hall 车间;主机室;主厂房;机(器)房
machine halt 停机
machine halting idle 停机
machine hammer 机械锤
machine hand 机械手
machine handling 机器处理
machine handling time 刀架进退时间
machine head 主轴箱;床头
machine-hour 台时;机器工时
machine-hour method 机时法;机器工时法
machine-hour rate 机器小时率;机时率
machine hours 机器运转时间;计算机工作时间;运转时间
machine hours at cargo work 装卸机械作业台时
machine hours at work 装卸机械工作台时
machine hours in idleness 装卸机械停工台时
machine-hour method based a manufacturing expenserate 制造费用率的机器工时法
machine-hour standard 机器时数标准
machine house 机器房
machine idle time 机器窝工时间;机器停工时间
machine-imprinted 打机械标记的
machine improvement time 机器装配时间
machine-independent 机器独立程序;独立于机器的
machine-independent command language 独立于机器的命令语言
machine-independent job control language 独立于机器的作业控制语言
machine-independent language 计算机通用语言;独立于机器的语言
machine-independent programming 独立于机器的程序设计
machine information retrieval 机器检索
machine in normal service 常用机器
machine in running order 机器开动状态
machine instruction 计算机指令;机器指令;机器说明书
machine instruction code 机器指令码
machine interface 机器接口
machine jigging 机械淘选;机器跳汰
machine key 机器键
machine key ring 键控铃呼叫;半自动呼叫
machine knife 机刀;车刀
machine-knotted pile carpet 机器打结绒头地毯
machine-laid 机器铺设成的
machine language 计算机语言;机器语言
machine language instruction 机器语言指令
machine language program(me) 机器语言程序
machine laying 机械铺设
machine learning 机器学习
machine lehr 垂直退火炉
machine length 字长【计】;机器长度
machineless 不用机器的
machine level 机器级
machine level instruction 机器级指令
machine level interrupt 机器级中断
machine level language 机器级语言
machine level routine 机器级程序
machine life 机器寿命
machine-limited system 机械限制系统
machine load card 机器负荷卡
machine load chart 机器负荷图
machine loading 机器装载;机器负荷
machine loading problem 机器分配问题
machine-made 机制的
machine-made brick 机制砖
machine-made glass 机制玻璃
machine-made hose 机制胶管
machine-made mortar 机制砂浆;机拌灰浆
machine-made nail 机制钉
machine-made nut 机制螺母
machine-made paper 机制纸
machine-made plaster 机制灰浆;机拌石膏
machine-made tile 机制瓦
machine mason 机加工石工
machine member 机件
machine method of accounting 会计机器整理
machine-mixed concrete 机拌混凝土
machine-mixed mortar 机拌砂浆
machine-mixed plaster 机拌灰浆
machine mixer 搅拌机
machine mixing 机械搅拌;机拌(法);机械混炼;机器拌和
machine mo(u)lding 机压成型(法)
machine-mounted automatic backfiller 汽车式(轮胎式)自动回填机
machine net weight 机器净重
machine offset 机械偏差
machine of normal series 标准型机器
machine oil 机(械)油;机器(润滑)油
machine oil emulsion 机油乳剂
machine operation 机器操作;机械运转
machine operation code 机械操作码
machine operator 机械算子;机器操作员;机工
machine organization 机器结构
machine-oriented language 面向机器的语言
machine oxygen cutting 机械化氧切割
machine parts 机械零件;机器部件
machine parts grinding 机件磨削
machine payload capacity 机械的有效载重能力
machine pin 机用销
machine pistol 冲锋枪
machine plant 机械厂
machine plotting 自动绘图;展点
machine posting 机器过账
machine press 压力机;压床
machine-pressed brick 机制砖
machine-pressed tile 机制瓦
machine press-packed bale 机器紧压包
machine press-packing 机器打包
machine-printed paper 机印墙纸
machine-printed wall-paper 机器印花壁纸
machine processible form 机器可处理形式
machine processing 机器处理
machine program(me) 机器程序
machine proof 开印样;机器打样
machine punched card 机器穿孔卡
machine ram 机床滑枕
machine rammer 机械夯(具);机动夯(具);机动夯;夯实机;夯击机
machine-rate bases 机器运行率法
machine-rate charges 设备分摊费用
machine rating 机械标称功率;机器额定功率
machine-readable 机器可读的
machine-readable catalogue 机器可读目录
machine-readable file 机读文件
machine-readable form 计算机可读格式
machine-readable map 机器可读地图
machine-readable medium 自动数据介体
machine reamer 机用铰刀
machine reel 收带盘
machine regulator 机器调节设备
machine repair 机修
machine repair shop 机(器)修(理)车间
machine retrieval system 机器检索系统
machine rifle 冲锋枪
machine ringing 自动振铃;铃流机振铃
machine riveting 机械紧固;机(器)铆(接)
machine rock drill 钻岩机;机动钻岩机
machine room 机房;机器间
machine routine 机器程序
machinery 机械;机器;落砂设备
machinery-aft ship 艉机船
machinery alarm system 机舱警报系统
machinery and electric(al)products 机电产品
machinery and equipment moving 迁移机械和设备
machinery and equipment yard 机器和设备场地
machinery and tool 机械及工具
machinery arrangement 机械布置
machinery base 机器基础
machinery breakdown 机器损坏
machinery breakdown insurance 机器损坏(保)险
machinery breakdown policy 机械故障保险单
machinery bronze 机用青铜
machinery building 机房
machinery bulkhead 机舱壁
machinery casing 机舱棚;机舱(口)围壁
machinery casting 铸铁机器件;机器铸件
machinery certificate 机器证书
machinery certificate of class 轮机入级证书
machinery charge qualification 机械负荷合格证
machinery classification and inspection record 船机入级记录和检验报告
machinery class notion 机械装入级标志
machinery components 机器零(部)件
machinery computation 机器计算
machinery consequential loss insurance 机械间接损失保险
machinery control 报警及巡检
machinery control center 机舱集控中心
machinery control console 船机控制台
machinery cost 机械费
machinery damage insurance 机器损坏(保)险
machinery efficiency 机械效率
machinery enamel 机械用(磁)漆
machinery environment 机械环境
machinery equipment 机器设备
machinery equipment management of railway construction enterprise 铁路施工企业机械设备管理
machinery equipment renewal 机械设备更新
machinery for processing woodwork 木工加工机械
machinery for spring work 春耕机械
machinery foundation 机器基座;机器基础;机械基础
machinery foundation structure 机器基础结构
machinery in the casthouse 炉前机械
machinery murmur 机鸣杂音
machinery noise 机器噪声;机械噪声
machinery number 主机特征数
machinery of consultation 协商机构
machinery of production 生产设备
machinery oil 机(器)油;机械(润滑)油
machinery pad 机器垫层
machinery performance monitoring 机械工况监测

machinery repairman 机械修配工
machinery repair parts 车辆用备件
machinery retrieval system 机械补救系统
machinery room 机房;机舱
machinery space 机器处所;机舱
machinery space bulkhead 机器处所舱壁;机炉舱壁
machinery specification 机器说明书
machinery station 机器站;机械站
machinery steel 机器钢
machinery survey 机器检验
machinery vibration 机械振动
machines and tools 机具
machines and tools for construction and erection 施工安装用的机械及工具
machine saw 机器锯
machine saw blade 机器锯片
machine scarfing 机动火焰清理
machine screening 机器筛分
machine screw 联结机件用的螺钉;机械螺钉;机(器)螺钉;机器安装螺钉
machine screw with oval countersunk fillister head 半沉头螺钉
machine sculpture 机械雕塑
machine-sensible language 机器可读语言
machine setting 仪器上输入数据;机械镶嵌
machine-shaping 机械成型
machine shear 机械剪
machine shop 修配(车)间;金工车间;机械(工)厂;机械车间;机器房;机器车间;机铆车间;机工车间;活动机器车间
machineshop car 工程车
machineshop truck 工程车
machine shovel 挖土机;机械铲
machine sieving 机器筛分
machine-sized 机选的
machine's life span 机器的使用年限
machine specification 机床说明书
machine spreading 机械撒布;机械摊铺
machine state 机器状态
machine station 机器站
machine status bit 机器状态位
machine status register 机器状态寄存器
machine steel 机器钢;机件钢
machine stitcher 装订机
machine stop 打炉
machine storage 机器存储
machine strap clamp 机械加工稳定工件用压板
machine stress rated lumber 标定应力的木材(经非破坏试验);机械应力等级木材
machines used in service trade 服务业机械
machine sweeping 机器扫路
machine switchgear 机器开头设备
machine switching system 自动交换机
machine-tamped concrete 机夯混凝土
machine tamper 打夯机;机械夯(具);机夯
machine tap 机器丝锥;机用丝锥
machine taper 机用锥度
machine taper pin reamer 机用锥销孔铰刀
machine-team 台班
machine-tending cycle 看管循环期
machine-tending quota 看管定额
machine time 机动时间;切削时间
machine tool 工作母机;工具机;机械工具;机床;母机
machine tool control 机床控制
machine tool control system 机床控制系统
machine tool drive 机床传动
machine tool plant 机床厂
machine tool works 机床厂
machine translation 机器翻译
machine trowel(ing) 机器抹灰;机器抹光;机器镘平
machine tunnel(l)ing 机械化隧洞开挖
machine tunnel(l)ing cost 机械开挖造价
machine under repairs 修理中的机器
machine unit 机器单元;机器单位
machine utilization 设备利用
machine utilization rate 机器使用率
machine variable 机器变量
machine vibration 机器振动
machine vise 机床虎钳
machine washer 平垫圈
machine weight 机器重量
machine welding 机械化焊接;机器焊接

machine welding torch 机械焊接吹管
machine winding 电机绕组
machine with closed circuit ventilation 闭路通风式电机
machine with inherent self-excitation 内在自励磁电机
machine with natural cooling 自然冷却式电机
machine with open circuit ventilation 开路通风式电机
machine word 计算机字
machine work 机械工作;机械加工;切削加工;机械厂
machine-working taps 攻丝机
machining 制造;开动机器;加工;机(械)加工;器加工;切削加工;操作
machining accuracy 精加工精确度;加工精度
machining allowance 制造公差;加工余量;机械加工余量
machining by repeated impact 冲击加工
machining centers 组合加工中心机床
machining condition 切削条件
machining constant 切削常数
machining data bank system 加工数据库系统
machining efficiency 加工效率
machining error 机加工误差
machining of metals 金属切削加工
machining pile 机扩桩
machining precision 加工精度
machining pressure 加工压力
machining property 可切削性;机械加工性能;机械加工性
machining quality 可加工性
machining set-up 机加工装置
machining shop 机械加工车间
machining stock allowance 机制加工裕度
machining stress 加工应力;机械加工残余应力
machining symbol 机械加工符号
machining time 加工时间
machinist 机械工人;机工;机械师
machinist's file 机工锉
machinist's hammer 机工锤
machinist's level 机工水平仪
machinist's scraper 钳工刮刀
machinofacture 机械制造
machmeter 马赫数指示器
Mach nozzle 马赫喷嘴(超音速喷气脱水装置)
Mach number 马赫数
Mach number calibration 按马赫数校准
Mach region wave 弱冲波
Mach's angle 马赫角
Mach's band 马赫带
Mach's cone 马赫锥
Mach's fringe 马赫条纹;马赫对比环
Mach's front 马赫阵面;马赫扰动面
Mach's indicator 马赫数指示器;马赫表
Mach's line 马赫线;马赫波
Mach's metal 马赫镁铝合金
Mach's meter 马赫计;马赫表
Mach's method 逐层切削法(测残余应力)
Mach's number 马赫数
Mach's number component 马赫数分量
Mach's number computer 马赫数计算机
Mach's number controller 马赫数控制器
Mach's number gradient 马赫数陡度
Mach's principle 马赫原理
Mach's reflection 马赫反射
Mach's reflection region 马赫反射区
Mach's region 马赫区
Mach's region wave 马赫冲波
Mach's stem 马赫扰动面;激波前沿
Mach's(stem) effect 马赫效应(反射冲击波叠加效应)
Mach's unit 马赫单位
Mach's wave 马赫波
Macht's metal 马赫特含铁黄铜;铜锌合金
Mach-Zehnder interferometer 马赫—曾德耳干涉仪
macintosh 胶布雨衣
mack 轻薄防水布雨衣;烟囱楦
mackayite 水碲铁矿
Mackenite metal 麦克内特镍铬耐热钢
Mackensen bearing 麦肯森式三油速楔动压轴承
mackensite 针铁绿泥石
Mackenzie's amalgam 麦肯齐汞(合金)
macker 炭质页岩

mackerel scale 鱼鳞
Mackey test 麦克试验
mackinawite 马基诺矿;四方硫铁矿
mackintosh 雨衣;防水(胶)布;胶布雨衣
Mackintosh probe 麦克托什探头
Mackintosh prospector 麦金托什探测仪
Macklers glaze 麦克勒釉
Mack's cement 麦克水泥
Mac-Lane system 麦克莱恩堆集法
Maclast 马克拉斯特(一种木块地板冷凝乳胶)
Maclaurin's expansion 麦克劳林展开
Maclaurin's formula 麦克劳林公式
Maclanrin's series 麦克劳林级数
Maclaurin's spheroid 麦克劳林球体
Maclaurin's theorem 麦克劳林定理
macle 短字晶石;空晶石;八面体双晶;(矿物中的)暗色斑点
Mac-Leod equation 麦克劳德方程
Mac-Leod ga(u)ge 麦克劳德真空regler;麦克劳德压强计;麦克劳德(压力)计;麦氏压力计
Mac-Leod pressure ga(u)ge 麦克劳德低压计
Mac-Leod vacuum ga(u)ge 麦克劳德麦氏真空规
Mac-Leod vacuummeter 麦克劳德真空计
MacMahon packing 麦克马洪填充物
MacMichael degree 麦克米契尔(黏)度
MacMichael viscometer 麦克米切尔黏度计
Maco-block 混凝土砌块机
Maco ga(u)ge (制作轮廓不规则的线脚的)可调样板
Macomb strainer 马科姆系统
maconite 黑蛭石
Maco template 马克样板
macphersite 亚硫碳铅石
Mac Quarrie test for mechanical ability 麦夸里机械能力测验
Macquer's salt 砷酸钾
Macquisten film-flo(a)tation machine 马奎斯顿薄膜浮选机
Macrame cord 马克莱姆线
macramé 流苏花边
macrinite 粗粒体
macrinoid group 粗粒体组
macro 粗视的;宏观的;微距的;近摄
macroaddress table 宏地址表
macroadjustment 宏观调节
macroaggregate 大颗粒
macroaggregate manpower budgets 宏观总人力预算
macroaim of production 宏观生产目的
macroalgae 大型藻类
macroanalysis 总体分析;总量分析;宏观分析;大剂量分析;常量分析
macroanalytic(al) 常量分析的
macroanalyticle balance 常量分析天平
macroanisotropy 宏观各向异性
macroaperture lathe 大孔径车床
macroapproach 宏观(方)法
macroargument 宏变元
macroasperity 宏观粗糙度
macroassembler 宏汇编(程序)
macroautoradiography 宏观自体放射照相术;宏观自动射线摄影术;宏观放射自显影
macroaxis 长轴;大轴
macrobend loss 宏弯曲损耗
macrobenthos 大型底栖生物
macrobiota 大型生物区系
macroblock 宏模块
macrobusiness 总体企业
macrocall 宏调用;宏调入
macrocausality 宏观因果性
macrocavitation 大空穴
macrocavity 宏观缩孔
macrocharacteristic curve 测微特性曲线
macrocheck 宏观分析
macrochemistry 常量化学
macrochronometer 测微时计
macroclastic 粗粒的
macroclastic rock 显晶碎屑岩;粗屑岩
macroclimate 宏观气候;大气候
macroclimatology 大气候学
macrocoacervation 常量凝聚
macrocode 宏代码
macrocoding 宏编码
macrocoefficient 宏观系数

macrocommand 宏命令
macrocommand list 宏命令表
macrocomous 长发的
macrocomponent 大量组分
macroconcentration 常量浓度
macroconstituent 常量成分
macroconsumer 大消费者
macrocontract 总合同;大型合同;大契约
macrocontrol 宏观控制
macrocontrolled market economy 宏观管理的市场经济
macrocorrosion 大量腐蚀;宏观腐蚀
macrocosm 宏观宇宙;宏观世界
macrocosmos 宏观世界;大宇宙
macrocrack(ing) 宏观裂纹;宏观裂缝;宽裂缝;大裂隙
macrocreep 宏观蠕变
macrocross-section 宏观横截面
macrocrystalline 大结晶体的;粗(粒)结晶;粗晶(质);粗晶(质)的
macrocrystalline dolomite 巨晶白云岩
macrocrystalline limestone 巨晶灰岩
macrocrystalline texture 巨晶结构
macrocycle 大旋回
macrocyclic(al) 大环的;长循环的
macrocyclic(al) polyether 大环聚醚
macrocyclic(al) polysiloxane 大环聚硅氧烷
macrocyclic(al) siloxane 大环硅氧烷
macrodata 宏观资料
macrodecision-making 宏观决策
macrodeclaration 宏说明
macrodefect free cement 无宏观缺陷水泥;无大缺陷水泥
macrodefect free construction material 无宏观缺陷结构材料
macrodefinition 宏定义
macrodiagonal 长对角轴
macrodisequilibrium 宏观经济不平衡
macrodispersoid 粗粒分散胶体
macrodistribution pattern 大分布模式
macrodocuments 宏文件;宏观文献
macrodome 大坡面(指晶体);长轴坡面
macrodrum wheel 测微鼓轮
macrodynamic(al) theory 宏观动态理论
macrodynamics 宏观动力学
macroeconometric(al) model 宏观经济计量模型
macroeconometrics 宏观经济计量学
macroeconomic activity 宏观经济活动
macroeconomic analysis 宏观经济分析
macroeconomic coefficient vector 宏观经济系数向量
macroeconomic control 宏观经济控制;宏观调控
macroeconomic control system 宏观调控体制
macroeconomic cost benefit analysis 大量的费用与经济效益分析
macroeconomic dynamics 宏观经济动力学
macroeconomic financing ratio 宏观资金融通率
macroeconomic forecasting 宏观经济预测
macroeconomic instability 宏观经济不稳定性
macroeconomic lever 宏观经济杠杆
macroeconomic liquidity ratio 宏观经济清偿率
macroeconomic loss function 宏观经济损失函数
macroeconomic management 宏观经济管理
macroeconomic management mechanism 宏观经济管理机制
macroeconomic model 宏观经济模型;宏观经济模式
macroeconomic parameter 宏观经济参数
macroeconomic policy 总体经济政策
macroeconomic regulation and control 宏观经济调控
macroeconomic regulatory mechanism 宏观经济管理机制
macroeconomic resource 宏观经济资源
macroeconomics 总体经济学;宏观经济学
macroeconomics performance target 宏观的经济效益指标
macroeconomic statistics 宏观经济统计学
macroeconomic theory 宏观经济理论
macroeconomic variable 宏观经济变量
macroeconomy 总体经济
macroeddy current 宏涡流
macroeffect 宏观效应
macroelastometer 测微弹性仪

macroelement 大量元素;宏组件;宏单元;常量元素
macroemulsification 粗滴乳化(作用)
macroemulsion 粗(滴)乳状液
macroenergy 宏观能量
macroengineering 大工程学;宏观工程学
macroenvironment 宏观环境
macroenvironmental policy 宏观环境政策
macroetch 宏观试片腐蚀
macroetched for examination 宏观浸蚀检验;宏观浸蚀检定;宏观浸蚀检查
macroetching 宏观组织腐蚀;宏观浸蚀;深度浸蚀;粗彩浸蚀
macroetch test 宏观腐蚀试验
macroevolution 宏(观)进化
macroexamination 宏观检验;宏观检定;宏观检查
macroexamination of seismism 宏观地震地质研究
macroexercise 宏观运用
macroexpansion 宏扩展
macroexpression 宏表达式
macroeyepiece 测微目镜
macrofabric 大型组构;粗组构;宏观组构
macroface 巨面
macrofauna 广动物区系
macrofeed 常量馈给
macrofiber 粗视的纤维;长纤维
macrofissure 宏观裂纹;宏观裂缝
macroflora 广植物区系;广布植物区系;大型植物区系
macroflow 宏观流动
macroflow chart 宏(观)流程图
macrofocusing 微距聚焦
macrofold 巨型褶皱;宏观褶皱;大褶皱;大褶曲
macroforecast 宏观预测;总体预测
macrofossil 大化石
macrofractography 宏观断口金相学;断口低倍检验
macrofracture 大破裂
macrofracture mechanics 宏观断裂力学
macrofungi 大型真菌
macrogamy 成体配合
macrogenerating program(me) 宏生成程序;宏产生程序
macrogenerator 宏(功能)生成程序;宏产生程序
macrogeochemistry 宏观地球化学
macrogeography 宏观地理学
macrograin 宏观晶粒;粗晶(粒)
macrograined 粗粒的
macrograph 宏观组织照片;宏观相片;宏观图;低倍照片
macrographic 宏观的
macrographic examination 宏观组织检查
macrographic test 宏观检验
macrography 宏观(照相)术;宏观检查;肉眼检查;低倍照相
macrography method 肉眼观察法
macrogrowth theories 宏观增长理论
macrohabitat 大生态环境;大生境
macrohardness 粗视硬度
macrohardness testing 宏观硬度检验
macroinsertion 大镶嵌
macroinstruction 广义指令【计】;宏指令
macroinvertebrate population 大型无脊椎动物种群
macroion 高分子离子
macroite 微粗粒煤
macrokinetics 宏观动力学
macrolattice 大网格;大晶格
macrolens 低倍照相镜头;大镜头;超近摄影镜头
macrolide 大环内酯
macrolithology 宏观岩性学
macrologic 宏逻辑
macromanagement indicator 宏观管理指标
macromarketing 总体销售;宏观销售学
macromechanics 宏观力学
macromer 大单体
macromere 大裂球;大分裂球
macromeritic 粗粒花岗状;粗晶粒状(的)
macrometeorology 大(范围)气象学;大尺度气象学
macrometer 宏观度量计;测远仪;测远器
macromethod 宏观(方)法
macromixing 大量混合;粗混
macromode 宏观模式
macromodel 宏观模型
macromodular computer 宏组件计算机
macromodule 超小型器件;大模块
macromolecular 高分子的

macromolecular coagulant 高分子凝结剂
macromolecular complex 高分子络合物
macromolecular compound 高分子化合物;大分子化合物
macromolecular dispersion 大分子分散(液)
macromolecular flocculant 高分子絮凝剂
macromolecular polymer 高分子聚合体
macromolecular property 高分子性质
macromolecular resin network 大分子树脂网络
macromolecular solution 高分子溶液
macromolecular waste 高分子废物
macromolecular weight material 高分子量物质
macromolecule 大分子;高分子
macromolecule complex 大分子配合物
macromolecule surfactant 高分子表面活性剂
macromonomer emulsion 大单体乳液
macromotion and micro-motion 大动作和小动作
macroname 宏名字
macronucleus 大核
macronutrient 大量营养物;常量营养元素
macroobservation 宏观观察;宏观观测
macroorganic matter concentration 常量有机物浓度
macroorganism 大型生物
macrooscillograph 常用示波器;标准示波器
macroparameter 宏(观)参数;宏参量
macroparticle 宏观粒子;大粒子
macroperfection 宏观完整性
macroperformance 宏观效益
macropermeability 大孔隙透水性
macrophenomenon 宏观现象
macrophoto 低倍照相
macrophotogrammetry 超远摄影测量
macrophotograph 宏观照相;宏观照片;放大照片
macrophotography 宏观摄影术;放大照相术;放大摄影
macrophyric 大斑晶状
macrophysics 宏观物理学
macrophyte 大型植物
macrophyte re-establishment 大型植物重建
macropinacoid 长轴面
macropipeline 宏流水线
macropipelining 宏流水线操作
macroplankton 大型浮游生物;大型浮游动物
macroplanning 宏观规划
Macropodus opercularis chinensis 中华斗鱼
macropolicy 宏观政策
macropore 大气孔;大孔(隙)
macropore coefficient 大孔隙系数
macroporosity 宏观孔隙;肉眼孔隙;大气孔率;大孔性;大孔隙(度)
macroporous 多孔的;大孔(隙)的
macroporous boron-specific ion-exchange resin 大孔硼特效离子交换树脂
macroporous ion exchange resin 大孔离子交换树脂
macroporous resin 大孔树脂
macroporous silica gel 大孔硅胶
macroporous soil 大孔(性)土;大孔隙土(壤)
macroporous structure 大孔(隙)结构
macroporous volume 大孔隙体积
macroporphyritic 大斑晶状
macroprecipitation 常量沉淀;大降雨
macroprecursor survey 宏观前兆调查
macroprediction 宏观预测
macroprint 宏观贴印
macroprism 长轴柱
macroprocessing 宏处理
macroprocessor 宏处理程序
macroprogram(me) 宏程序
macroprogramming 宏程序设计;宏编程
macropterous 大翅的
macropyramid 长轴锥面
macropyrometer 测微高温计
macroquake 宏震;大震
macroqualitative analysis 常量定性分析
macroregionalization 大区域划分
macroregulatory measures 宏观调节措施
macrorelation 宏观关系
macrorelief 广域地形;宏观地形;大(区)地形;大起伏
macroreticular polymeric adsorbent 大网状高分子吸附剂
macroreticular type ion exchange resin 大孔型

离子交换树脂
macroreticular type resin 大网络树脂;粗网结构树脂
macroretrenchment 宏观紧缩
macrorheology 宏观流变学
macroroutine 宏程序
macros 麦克罗斯碱性耐火制品
macro-sample 常量试样
macroscale 宏观尺度;大规模;大尺度
macroscheme 宏功能方案;宏方案
macroscope fractography 宏观断口分析
macroscopic 巨型的(指地质构造);目视的;宏观的;肉眼识别的;肉眼可见的;低倍放大的;粗大的
macroscopical 目视的
macroscopic analysis 宏观分析
macroscopic anisotropy 宏观各向异性
macroscopic approach 宏观途径;宏观(方)法
macroscopic causality 宏观因果性
macroscopic cavitation 大空穴
macroscopic concept 宏观概念
macroscopic crack(ing) 宏观裂纹;宏观裂缝
macroscopic cross-section 宏观(横)截面
macroscopic deformation 宏观形变
macroscopic deformation property 宏观形变性质
macroscopic dispersion 宏观弥散
macroscopic dynamic(al) state 宏观动态
macroscopic economic evaluation 宏观经济评价
macroscopic effective cross-section 宏观有效截面
macroscopic environment 宏观环境
macroscopic environmental policy 宏观环境政策
macroscopic equilibrium 宏观均衡
macroscopic examination 低倍检验
macroscopic fault motion 宏观断层运动
macroscopic field 宏观场
macroscopic fold 大型褶皱
macroscopic imperfection 宏观缺陷
macroscopic inclusion 宏观包裹体
macroscopic incomes 宏观经济收益
macroscopic influence field 宏观影响场
macroscopic input-output model 宏观输入/输出模式
macroscopic instability 宏观不稳定性
macroscopic intensity 宏观烈度
macroscopic intensity standard 宏观烈度标准
macroscopic interphase boundary 宏观相间边界
macroscopic irregularity 宏观表面缺陷;外观缺陷
macroscopic landform 巨型地貌
macroscopic lithotypes of coal 肉眼煤岩类型
macroscopic magnetization 宏观磁化
macroscopic measurement 宏观测度
macroscopic method 肉眼识别法
macroscopic mixing 宏观混合
macroscopic mode 宏观模式
macroscopic model 宏观模型
macroscopic motion 宏观运动
macroscopic multiplication 宏观倍增
macroscopic noise 宏观噪声
macroscopic observation 宏观观察;宏观观测
macroscopic phenomenon of fluctuation 宏观涨落现象
macroscopic planning 宏观规划
macroscopic precursor 宏观前兆
macroscopic propagation 宏观传播
macroscopic property 宏观特性;热力学性质
macroscopic readjustment and control 宏观调控
macroscopic scale 宏观规模;大型构造尺度
macroscopic scattering cross-section 宏观散射截面
macroscopic seismic phenomenon 宏观地震现象
macroscopic simulation 宏观仿真
macroscopic state 宏观(状)态
macroscopic stress 宏应力
macroscopic structure 宏观构造;大型构造
macroscopic symmetry 宏观对称
macroscopic system 宏观系统
macroscopic test 直观试验;低倍试验
macroscopic turbulence 宏观素流
macroscopic velocity 宏观速度
macroscopic viscosity 宏观黏度
macroscopic void 大孔;大空洞
macroscopy 肉眼检查
macrosection 磨片组织图;宏观磨片;宏观金相试片;宏观断面图;粗视剖面
macrosegregation 宏观偏析;宏观分离

macroseism 强(地)震;大震
macroseismic area 强震区
macroseismic data 强震资料
macroseismic effect 宏观地震效应;强震效应
macroseismic epicenter 宏观震中;强震震中
macroseismic epicenter survey 宏观震中位置调查
macroseismic field 强震场
macroseismic influence field 宏观地震影响场
macroseismic intensity survey 宏观地震烈度调查
macroseismic investigation of strong earthquake 强震宏观调查
macroseismicity 强震活动性
macroseismic observation 宏观地震观测;强震观测
macroseismic phenomenon 宏观地震现象
macroseismic survey 宏观地震调查
macroseismic zone 强震区
macroseismograph 强震仪
macroseism seismology 强震地震学
macroseptum 大隔壁
macroshot 微距摄影
macroshot lens 微距摄影透镜
macroshrinkage 宏观缩孔;宏观收缩
macroskeleton 宏程序纲要
macroslip 宏观滑移
macrosolifluction 大型泥石流
macrosome 核粗粒体;粗粒体
macrosonics 宏观声学
macrospecies 大型种类
macrospecimen 宏观试样
macrospicule 大针状物
macrosporinite 大孢子体
macrostandard data 宏观定额资料
macrostate 宏观(状)态
macrostatistics 宏观统计数据
macrostrain 宏应变;大应变
macrostratigraphy 宏观地层学
macrostreak flaw test 断面缺陷肉眼检验;粗视条痕裂纹检验
macrostress 宏应力;宏观应力
macrostructural test 宏观组织检查
macrostructure 大型结构(物);宏观组织;宏观体制结构;宏观结构(物);宏观构造;低倍组织;大型构造;粗显构造;粗视组织
macrosubroutine library 宏子程序库
macrosubstitution 宏置换
macrosymbol 宏符号
macrosystem 宏观系统
macrosystem theory 巨系统理论
macrotectonics 巨型构造;大构造;大地构造学
macrotexture 宏观纹理
macrothermophyte 高温植物
macrothermophytia 热带植物群落;高温植物群落
macrotidal range 大潮差(强潮差)
macrotime event list 大延迟时间事件表
macrotime variable 宏时变量
macrotrace 宏追踪;宏观探索;宏观查找
macroturbulence 宏观素流;大尺度素动;大尺度湍流
macrotype 大型
macro-urban 大城市
macrovariable 宏观变量;宏变量
macroviscosity 宏观黏性;稠黏度
macrovoid ratio 大孔隙比
macrozone 大区
macrozooplankton community 大型浮游动物群落
macthed filter 匹配比滤波器
macula 太阳黑点;斑点;岩浆房
maculose 斑结状【地】
maculose rock 斑结状岩(石)
maculosus 斑状(云)
Madagascar aquamarine 二色海蓝宝石
Madagascar basin 马达加斯加海盆
Madagascar hook 马达加斯加钩
madder 鲜红色的;鲜红(色);茜红色;茜草
madder lacquer 茜草色喷漆;茜草色清漆
madder lake 茜草红色淀
madder plant 茜草植物
Maddrell Salt 长链高分子量偏磷酸钠
made bill 境外待付票据
made block 组合滑车;组成滑车
made circuit 闭合电路
made course 制定航线;真航向
made eye 绕匝索环
made from pipe 由管子制造

made good 抵付
made good the deficit 弥补亏损;弥补亏空
made ground 填土空地;人工开辟场地;现代沉积;填筑地;填土(地);填方;填(成)地
made land 人造陆地;填土筑地;人工填地;填筑地;填土;填(成)地
made mast 组合桅(杆)
made road 填土路
madescent 微湿的
made-to-measure 特制
made-to-order 特制;定制;按订户要求制作
made-up 制成的;人工制造的
made-up article 坯品
made-up belt 胶带坯品;带坯
made-up cable 成品钢缆(预应力混凝土用)
made-up fuel oil 补充燃料油
made(-up) ground 垫高;人工填地;填成地;填土造地;填土
made-up ground level 建成后地面高程
made-up road 人工道路
made work 人为的工作
madhouse 精神病院;疯人院
madial wall 内侧壁
madisonite 炉渣石
Madison process 麦迪逊氏废物处理过程
Ma-distance test 马氏距离检验
madistor 晶体磁控管;磁控等离子体开关
madocite 麦硫锑铅矿
Madoi block 玛多地块
madrasa(h) 回教学校
madrepore 石珊瑚
madrier 军用厚橡木板
madrusah 回教大学
maelstrom 灾害;破坏性力量;大旋涡;大漩流
maenianum (古罗马剧场或比武场的)观众席
Maentwrogian 门特罗格阶【地】
maerl 藻团粒
Maestrichtian 马斯里奇特阶
maestro 美斯屈罗风
Mae West 海上救生衣
Maf 锚件;木材连接件
mafelsic 镁铁硅质的
mafic 镁铁质(矿物)
mafic index 镁铁指数
mafic mineral 镁铁(质)矿物
mafic rock 镁铁质岩;镁铁质岩石
Mafi portal-lift 马菲吊运车
mafraite 钠闪辉长岩
Maftex 马夫的斯板(一种木纤维板)
mafurite 橄辉钾霞斑岩
magadiite 麦羟硅钠石
magaldrate 氢氧化镁铝
magamatic rock 岩成因类型
magamp 磁性放大器
magaseism 剧震;强震
magauigite 镁辉石
magazine 线圈框架;装drum暗盒;杂志;料斗;卡片箱;卡片输入箱;军火库;货栈;储料容器;爆破材料库
magazine arclamp 复式电极弧光灯
magazine boiler 自动加煤(小)锅炉
magazine camera 自动卷片照相机;图片摄影机;暗盒摄影机
magazine capacity 暗盒容量
magazine case 弹仓
magazine casing 暗盒箱
magazine chamber 底片盒
magazine chute 料斗斜槽;送料溜槽
magazine cock 火药舱船底塞
magazine creel 复式纱架
magazine data card 暗盒数据卡
magazine feed 自动储存送料
magazine feed attachment 自动储存送料附件
magazine for empty sacks 纸袋库
magazine for empty wall 空袋储[贮]库
magazine for valve sacks 自封口纸袋储[贮]库
magazine grinder 库式磨木机;水浆研磨机
magazine loader 自动储存送料装置
magazine loading camera 插入式片盒摄影机
magazine piler 堆垛机
magazine platen 暗匣
magazine rack 杂志架
magazine rail 料斗送料轨道
magazine rifle 带弹匣的步枪
magazine room 杂志(阅览)室

magazine screen 火药舱防风帆布帘
magazine slot 库房储存区间；储存槽
magazine spring 托弹簧
magazine stove 自动加煤炉
magazine table 储[贮]料辊道
magazine tool holder 库房工具架
magazine type automatic lathe 料斗式自动车床
magazine type charge 储存式装料台
magazine type charger 储料进给台
magbasite 硅镁钡石
magbestic 镁质无缝地板材料；菱镁土地面铺料
mag card 磁卡片
magchrome refractory 镁铬耐火材料
Magclad 双镁合金板
mag coke 镁焦
magco mica 防漏碎云母
Magcoustic 隔声人造石（商品名）
magdolite 两次煅烧白云石
magdynamo 永磁直流发电机组；磁石发电机
magdyno 永磁直流发电机组
mage analyzer 图像分析仪
Magellanic stream 麦哲伦流
Magellanic system 麦哲伦系
magenta 洋红；碱性品红；碱性复红；品红色；品红（碱）
magenta contact screen 品红色接触网目板；品红接触网
magenta flower 洋红色花
magenta layer 深红色层
magenta screen 品红网目板
magentic acupress stylus 磁缇针
magentic ag(e)ing 磁性老化
magetostatic field 静磁场
maggie 劣质煤；不纯煤
maghemite 磁赤铁矿
magic chuck 快换夹具
magic code 幻码
magic dart 魔镖
magic eye 光调谐指示管；电眼
magic eye flame control 光电火焰调节器
magic filter 幻式滤波器
magic glass 单面可视
magic guide bush 张缩导套
magic hand 机械手；人造手
magic heat pack 魔力热袋
magic ink 记号油墨；万能笔；标记油墨
magic lantern 映画器；幻灯
magic list 幻表
magic mirror 哈哈镜
magic nucleus 幻核
magic number 幻数
magicore 高频铁粉芯
magic qaure 魔方
magic scanner 光笔显示器
magic square 幻方
magic stone 透蛋白石
magic T 混合接头
magic veil 电视屏遮光罩；显像管框架
mag-ion pump 磁控离子泵
magister （欧洲中世纪的）匠师；制备；（欧洲中世纪的）手艺师
magister of sulfur 硫的制备
magistral 焙烧黄铜矿粉；按配方制的
magistrate 地方法官；地方行政官
magistrate's court 地方法庭
magjolica 花饰陶器
maglev railway 磁浮铁路
maglev system 磁（悬）浮系统
maglev test line 磁浮试验线
maglev train 磁悬浮列车
maglev transport 磁悬浮运输
maglev vehicle 磁悬浮车辆
magma 岩浆；稠液
magma basalt 岩浆玄武岩
magma basin 岩浆储源
magma chamber 岩浆库；岩浆房；岩浆储源
magma chamber reservoir 岩浆房热储
magma density 稠液密度
magma diapir 岩浆底辟
magma distillation 岩浆蒸馏作用
magma facies 岩相
magma filter 稠液过滤器
magma geothermal system 岩浆地热系统
magma-glass-ash 岩浆玻璃灰

magmagranite 岩浆花岗岩
magma igneous 火成岩浆
magma impact genesis of earthquake 岩浆冲击成因
magma intrusion 岩浆侵入
magma intrusion theory 岩浆侵入假说
magma isotope composition 岩浆同位素成分
magma magnesiae 镁乳浆
magma pocket 岩浆储源
magma pump 糊浆泵
magma reservoir 岩浆库；岩浆储源
magma series 岩浆系列
magma theory 岩浆学说
magmatic activity 岩浆活动
magmatic assimilation 岩浆同化（作用）
magmatic body 岩体
magmatic complex 岩浆杂岩
magmatic concentration 岩浆富集
magmatic corrosion 岩浆侵蚀
magmatic cycle 岩浆旋回；岩浆回旋；火成旋回
magmatic differentiation 岩浆分异（作用）
magmatic digestion 岩浆同化（作用）
magmatic emanation 岩浆喷发
magmatic emplacement 岩浆贯入（作用）
magmatic eruption 岩浆喷溢；岩浆喷发
magmatic evolution 岩浆演化
magmatic explosion 岩浆爆发
magmatic formation 岩浆建造
magmatic gas 岩浆气(体)
magmatic heat budget method 岩浆热估算法
magmatic heat source 岩浆热源
magmatic hot fluid 岩浆热流体
magmatic hydrothermal solution 岩浆热液
magmatic injection 岩浆贯入（作用）
magmatic injection deposit 岩浆贯入矿床
magmatic intrusion 岩浆侵入
magmatic mineral 岩浆矿物
magmatic mixing 岩浆混合作用
magmatic(ore) deposit 岩浆矿床
magmatic orefield structure 岩浆矿田构造
magmatic ore-forming process 岩浆成矿作用
magmatic origin 岩浆起源；岩浆成因
magmatic residuum 岩浆残余物
magmatic rock 火成岩；岩浆岩
magmatic rock association 岩浆岩组合
magmatic rock type 岩浆岩类型
magmatic segregation 岩浆分凝（作用）；岩浆分结（作用）；岩浆
magmatic series 岩浆系列
magmatic solution 岩浆溶液
magmatic steam 岩浆蒸汽
magmatic stoping 岩浆升蚀；岩浆顶蚀（作用）
magmatic water 岩浆水
magmatism 岩浆作用；(花岗岩的)岩浆生成论
magmatism age 岩浆活动时代
magmatism feature 岩浆活动特征
magmatism intensity 岩浆活动强度
magmatism markers 岩浆活动标志
magmatism migration 岩浆作用迁移
magmatism pattern 岩浆活动方式
magmatism polarity 岩浆活动极性
magmatism rock assemblage and genetic type 岩浆作用岩石组合和岩石成因类型
magmatism scale 岩浆活动规模
magmatism way 岩浆作用方式
magmatist 岩浆论者
magmatite 岩浆岩
magmatite factor 岩浆因素
magmatogene rock 岩浆成因岩石
magmatogenic condition 岩浆成因条件
magma type 岩浆类型
magmeter 直读式频率计；直读式仪表
magna-check 磁力探伤机
magnadur 铁钡永磁合金
magnafacies 主相；同性相
magnafloc 一种絮凝剂
magnaflux 磁束探伤法；磁力探伤机；磁力探伤法；磁粉探伤法
magnaflux inspection 磁力线探伤；磁力探伤检查
magnaflux inspection method 磁力线探伤法
magnaflux method 磁通量检测法
magnaflux powder 磁性探伤用铁粉
magnaflux steel 航空用高强度钢
magnaflux test 磁力线探伤；磁力探伤检查

magnaflux test method 磁力线探伤法
magnaglo 磁光探伤法
magnal base 十一脚管底
magnalite 铝基铜镍镁合金
magnalium 镁铝(铜)合金
magnal socket 十一脚管座
magnane 镁烷
magnascope 扩大镜
magnasil 镁氧水泥甲板敷料
magnavue card 磁卡片照相胶片存储器
magnechuck 电磁吸盘
magneform 磁力成型
magneforming 磁力成型法
magneform machine 磁力成型机
Magnel anchoring system 马涅尔锚固体系；马氏预应力张拉系统
Magnel-Blaton jack 马涅尔-布莱登千斤顶
Magnel prestressing system 马涅尔预应力张拉系统；马涅尔预应力张拉体系
Magnel system 马氏张拉系统
magnescope 放像机
magnesedin 镁菌素
magnesia 氧化镁；菱镁矿；菱苦土；苦土；镁氧；镁砂
magnesia alba 白镁氧
magnesia-alumina-silica 硅镁铝合金
magnesia-asbestos 镁石棉
magnesia based 镁基的
magnesia based desulfurization[desulphurization] 镁氧脱硫法
magnesia blythite 镁锰榴石
magnesia borosilicate glass 硼硅酸镁玻璃
magnesia brick 镁砖
magnesia cement 菱镁土水泥；高菱镁水泥；镁质水泥；镁氧水泥(强度很高，常用作无缝楼面的黏合剂)
magnesia cement concrete 镁氧水泥混凝土
magnesia ceramics 氧化镁陶瓷；氧化镁瓷；镁氧陶瓷
magnesia-chrome 镁铬合金
magnesia-chrome refractory 镁铬耐火材料
magnesia clinker 重烧苦土
magnesia crucible 氧化镁坩埚
magnesia expansion 氧化镁膨胀
magnesia glass 氧化镁玻璃
magnesia hardness 镁氧硬度；菱镁硬度
magnesia-insulated metal sheathed wire 氧化镁绝缘金属铠装电缆
magnesia insulation 氧化镁绝缘层(用于管线)
magnesia lime 镁氧石灰
magnesia limestone 镁氧石灰岩
magnesia magma 氧化镁悬浮液；镁乳
magnesia marble 含镁大理石
magnesia mica 金云母；镁云母；黑云母
magnesia mixture 镁氧混合剂；镁剂
magnesian 镁质的；含镁的
magnesian alum 镁明矾
magnesian annabergite 含镁镍华
magnesian calcite 镁方解石
magnesian chalk 镁质白垩
magnesian chromite 镁铬铁矿
magnesian clay 镁质黏土
magnesian coefficient 镁质系数
magnesian concrete 镁质混凝土
magnesian dolomitic marble 白云石质大理石；含镁大理石
magnesian ion 镁离子
magnesianite 菱镁矿
magnesian lime 镁(质生)石灰；镁氧石灰
magnesian lime brick 氧化镁石灰砖
magnesian lime paste 镁氧石灰膏；镁质石岩浆
magnesian lime putty 镁质石灰腻子；镁质石灰油灰
magnesian limestone 白云石石灰；含镁石灰岩；镁质石灰岩；镁质灰岩(层)
magnesian marble 镁质大理石
magnesian matt(glaze) 镁质无光釉
magnesian melanterite 含镁水绿矾
magnesian product 镁质制品
magnesian quicklime 镁质生石灰；含镁生石灰
magnesian semi-hydraulic lime 镁氧半水硬石灰；镁质半水石灰
magnesian spar 白云石
magnesia porcelain 镁氧陶瓷；镁质瓷器
magnesia refractory 镁氧耐火材料；镁氧耐火砖
magnesia refractory materials 镁质耐火材料

magnesia rods 镁氧棒
magnesia-silica gel 氧化镁硅胶
magnesia spinel 镁氧尖晶石
magnesia-spinel brick 镁尖晶石耐火砖
magnesia tooth paste 镁氧牙膏
magnesia type expansive cement 氧化镁膨胀水泥
magnesia unsoundness 氧化镁引起的不安定性
magnesia whisker 氧化镁须晶
magnesic 镁的
magnesio anthophyllite 镁直闪石
magnesio arfvedsonite 镁亚铁钠闪石
magnesioastrophyllite 镁星叶石
magnesioaxinite 镁斧石
magnesiocarpholite 纤镁柱石
magnesiochromite 铬镁尖晶石；镁铬铁矿
magnesiochromite ore 镁铬铁矿矿石
magnesioclinoholmquistite 斜镁锂闪石
magnesiocopiapite 镁叶绿矾
magnesio crocidolite 镁钠闪石石棉
magnesiocummingtonite 镁闪石
magnesiodolomite 镁质白云石
magnesioferrite 镁铁矿
magnesiogedrite 镁铝直闪石
magnesiohastingsite 镁绿钙闪石
magnesioholmquistite 镁锂闪石
magnesiohornblende 镁角闪石
magnesiohulsite 黑硼锡镁矿
magnesiokatophorite 镁钛闪石
magnesiomargarite 镁珍珠云母
magnesion hornfels 镁质角岩
magnesion marble-siliceous rock formation 镁质大理岩-硅质岩建造
magnesion sharn 镁质矽卡岩
magnesiopectolite 镁针钠钙石
magnesioriebeckite 镁钠闪石
magnesiosadanaguite 镁砂川闪石
magnesiospinel 镁尖晶石
magnesiotaramite 镁绿闪石
magnesite 菱镁焦油(炉矿炉衬用)；菱镁土；菱镁矿；菱苦土；镁氧水泥甲板敷料；镁砂；海泡石；天然碳酸镁；碳酸镁
magnesite-alumin(i)um refractory 镁铝耐火材料
magnesite bottom 镁砂炉底
magnesite brick 镁砖
magnesite brick lining 镁砂炉衬
magnesite building board 碳酸镁建筑板；菱苦土建筑板
magnesite(building)sheet 镁氧建筑板
magnesite carbon brick 镁碳质砖
magnesite carbon refractory 镁碳质耐火材料
magnesite cement 菱镁土水泥；镁质水泥；镁氧水泥(强度很高,常用作无缝楼面的黏合剂)
magnesite-chrome 镁铬合金
magnesite chrome brick 铬镁砖；镁铬(质)砖；镁铬合金砖
magnesite chrome refractory 镁铬(质)耐火材料
magnesite chromite brick 镁铬砖
magnesite clinker 烧结镁砂
magnesite composition flooring 菱苦土(组合)地面；菱苦土地平
magnesite concrete 菱苦土混凝土
magnesite deposit 菱镁矿矿床
magnesite dolomite brick 镁白云石砖
magnesite dolomite refractory 镁质白云石耐火材料
magnesite flooring 氧化镁铺面；菱苦土地平；菱苦土地面；镁砖地板
magnesite flooring tile 碳酸镁地板砖；菱苦土地板砖；氧化镁铺面砖
magnesite lining 镁砂内衬
magnesite mixture 菱苦土混合物
magnesite ore 菱镁矿矿石
magnesite powder 镁粉
magnesite refractories 镁氧耐火制件
magnesite refractory 镁质耐火材料；镁氧耐火砖
magnesite refractory concrete 镁质耐火混凝土
magnesite refractory product 镁质耐火制品
magnesite rock 碳酸镁岩石；菱苦土岩石；菱镁岩
magnesite sheet 碳酸镁薄板；菱苦土薄板
magnesite slab 菱苦土板
magnesite subfloor 碳酸镁底层地板；菱苦土底层地板
magnesite suspended arch 镁砖吊式炉顶
magnesite tar (矿炉炉衬用的)菱镁焦油

magnesite wheel 菱镁砂轮；镁砂粉砂轮
magnesite-zirconia brick 镁锆砖
magnesium acetate 乙酸镁；醋酸镁
magnesium alkoxide 醇镁
magnesium alloy 镁合金
magnesium alloy sheet 镁合金薄板
magnesium-alumin(i)um alloy 镁铝合金
magnesium-alumin(i)um hydrotalcite 镁铝水滑石
magnesium-alunina brick 镁铝砖
magnesium ammonium chloride 氯化铵镁
magnesium ammonium phosphate 磷酸铵镁
magnesium ammonium phosphate gravitation settling method 磷酸铵镁重力沉淀法
magnesium ammonium phosphate precipitation method 磷酸铵镁沉淀法
magnesium analyzer 镁含量分析仪
magnesium anode 镁阳极
magnesium anode method 镁阳极法
magnesium arsenate 砷酸镁
magnesium azide 叠氮镁
magnesium-base alloy 镁基合金
magnesium base grease 镁基润滑脂
magnesium-bearing formation 含镁建造
magnesium benzoate 苯甲酸镁
magnesium bicarbonate water 重碳酸镁水
magnesium bichromate 重铬酸镁
magnesium borate 硼酸镁
magnesium bronze 镁青铜(用于窗栏杆扶手等)
magnesium carbonate 碳酸镁
magnesium carbonicum leve 轻质碳酸镁
magnesium casting alloy 镁铸(造)合金
magnesium cell 镁电池
magnesium cement 镁质水泥
magnesium chilled iron roll 冷硬球墨铸铁轧辊
magnesium chips 镁屑
magnesium chloride 氯化镁
magnesium chloride process 氯化镁法
magnesium chloride solution 氯化镁溶液
magnesium chloride water 氯化镁型水
magnesium-chlorphoenicite 砷锰镁石
magnesium-chrome brick 镁铬砖
magnesium citrate 柠檬酸镁
magnesium coke 镁焦
magnesium compound 镁混合物
magnesium copper alloy 镁铜合金
magnesium-copper sulphide rectifier 镁硫酸铜整流器
magnesium couple 镁电偶
magnesium dioxide 二氧化镁
magnesium dust 镁粉
magnesium ethoxide 乙醇镁
magnesium ethylate 乙醇镁
magnesium ferrite 镁铁氧体；铁酸镁
magnesium fertilizer 镁肥
magnesium flare 镁(光)照明弹
magnesium fluoride 氟化镁
magnesium fluosilicate 硅氟酸镁；氟硅酸镁
magnesium flux 氟化镁
magnesium globule 镁珠
magnesium group 镁族
magnesium halide 卤化镁
magnesium hardness 镁硬度
magnesium hydrate 氢氧化镁
magnesium hydride 氢化镁；二氢化镁
magnesium hydroxide 氢氧化镁
magnesium hydroxide mixture 镁乳浆
magnesium hypophosphite 次磷酸镁
magnesium hyposulfite 硫代硫酸镁
magnesium ingot 镁锭
magnesium in lime sludge 石灰污泥中的镁
magnesium iodide 碘化镁
magnesium ion 镁离子
magnesium lamp 镁光灯
magnesium light 镁光
magnesium light lamp 镁光灯
magnesium lime 镁(质)石灰
magnesium limestone 白云岩；镁质石灰石；镁质(石)灰岩
magnesium magma 氢氧化镁悬液
magnesium-manganese alloy 镁锰合金
magnesium-manganese battery 镁锰电池
magnesium-manganese-zinc alloy 镁锰锌合金
magnesium metai 金属镁
magnesium metasilicate 硅酸镁

magnesium mica 镁云母
magnesium nitrate 硝酸镁
magnesium nitride 氮化镁；二氮化三镁
magnesium octahedron 镁氧八面体
magnesium oleate 油酸镁
magnesium ore 镁矿石；镁矿
magnesium oxalate 草酸镁
magnesium oxide 氧化镁；苦土
magnesium oxide and calcium carbonate tablets 镁钙片
magnesium oxide cement 氧化镁水泥；菱苦土水泥；镁氧化水泥
magnesium oxide cement concrete 氧化镁水泥混凝土
magnesium oxide process 氧化镁清洗过程
magnesium oxide slurry 氢氧化镁乳液
magnesium oxychloride binder 氯氧化镁胶结料
magnesium oxychloride cement 索勒尔胶结料；镁石水泥；氯氧化镁胶结料；高菱镁水泥；氯氧化镁水泥；菱苦土水泥；镁氯水泥
magnesium oxychloride composition 氯氧化镁组成；氯氧化镁成分
magnesium oxychloride screed topping 氯氧化镁砂浆层顶面
magnesium oxychloride subfloor 氯氧化镁地板垫层
magnesium oxychloride tile floor cover(ing) 氯氧化镁地砖铺面
magnesium oxychloride tile floor(ing)finish 氯氧化镁地板成品
magnesium oxysulfate cement 氧硫酸镁水泥
magnesium perchlorate 高氯酸镁
magnesium permanganate 高锰酸镁
magnesium peroxide 过氧化镁
magnesium phosphate 磷酸镁
magnesium phosphate binder 磷酸镁胶结料
magnesium phosphate cement 磷酸镁水泥
magnesium powder 镁粉
magnesium propoxide 丙醇镁
magnesium-rare earth 稀土镁合金
magnesium recovery 镁的回收
magnesium-reduced 镁还原(的)
magnesium reduction arc furnace 镁还原电弧炉
magnesium revised geothermometer 镁校正法地热温标
magnesium ribbon 镁带
magnesium rod 镁条
magnesium salt 镁盐
magnesium salt attack 镁盐侵蚀
magnesium salt hydrolysis product 镁盐水解物
magnesium salts 镁盐类
magnesium sheet 镁片
magnesium silicate 硅酸镁；镁硅砂
magnesium silicate pigment 硅酸镁颜料
magnesium silicide 硅化镁
magnesium silicofluoride 氟矽化镁；氟硅酸镁
magnesium soap 镁皂
magnesium solution 镁溶液
magnesium stannate 锡酸镁
magnesium stearate 硬脂酸镁
magnesium sulfate 泻盐(硫酸镁结晶)
magnesium sulfate water 硫酸镁水
magnesium sulfide 硫化镁
magnesium sulphate 硫酸镁
magnesium thiosulfate 硫代硫酸镁
magnesium titanate 钛酸镁
magnesium titanate ceramics 钛酸镁陶瓷
magnesium titanate porcelain 钛酸镁瓷
magnesium trisilicate 三硅酸镁
magnesium tungstate 钨酸镁
magnesium ultramafic rock 镁质超镁铁岩
magnesium vapo(u)r 镁蒸气
magnesium wollastonite 镁硅灰石
magnesium zippeite 水镁铀矾
magnesium zirconate 锆酸镁
magnestat 磁调节器
magne-switch 磁(力)开关
magnesyn 磁(电式)自动同步机
magnesyn compass 远距离指示罗经
magnesyn repeater 磁罗经复示器
magnet 吸铁石；磁铁；磁体；磁石
magnet bar code 磁条码；磁棒形码
magnet base 磁性座；磁力座
magnet-bearing 磁性轴承

magnet belt separator 带式磁选机
magnet-blaton (采用高拉力钢丝的)预应力混凝土系统
magnet chamshell crane 磁力自卸吊车
magnet charger 充磁器
magnet chuck 电磁卡盘
magnet clutch 电磁离合器
magnet coil 电磁(铁)线圈;磁化线圈
magnet core antenna 磁芯天线
magnet coupling 磁性联轴器
magnet cover 磁铁罩
magnet crane 磁力起重机
magnet-cushion railway 磁悬浮铁路
magnet erasing 磁(铁)抹音
magnet exciting coil 励磁线圈
magnet feed 磁性传动
magnet filter 磁选过滤器
magnet for ignition 点火磁电机
magnet gap 磁隙;磁极间空隙
magnet grate 磁化篦条
magnet housing 磁铁壳
magnetic 磁性的;磁(化)的
magnetic abnormal detection 地磁异常探测
magnetic acoustic 磁声
magnetic action 磁(性)作用
magnetic adherence 磁附着
magnetic adhesion 磁吸附;磁附着
magnetic after effect 磁后效(应)
magnetic ag(e)ing 磁性时效;磁性陈化
magnetic aging 磁陈化
magnetic air gap 磁路气隙
magnetic aligning 磁取向
magnetic alloy 磁(性)合金
magnetically actuated shutter 磁致激活快门
magnetically ag(e)ing 磁性老化
magnetically combined turbine and generator 磁耦合水轮发电机组
magnetically confined 磁场约束
magnetically confined laser 磁约束激光器;磁聚焦激光器
magnetically confined plasma 磁约束等离子体
magnetically controlled ignition 电磁控制点火
magnetically coupled circuit 磁耦合电路
magnetically focused cascade image intensifier 磁聚焦级联像增强器
magnetically focused image converter tube 磁聚焦变像管
magnetically hard alloy 硬磁合金
magnetically hard material 硬磁性材料;硬磁材料
magnetically levitated high speed transportation system 磁悬浮高速运输系统
magnetically levitated train 磁悬浮列车
magnetically sensitive recording paper 磁敏记录纸
magnetically soft alloy 软磁合金
magnetically soft material 软磁(性)材料
magnetically soft steel 软磁钢
magnetically supported car 磁悬浮车辆
magnetically suspended guided vehicle 磁悬浮引导车辆
magnetically suspended gyroscope 磁悬浮陀螺仪
magnetically suspended rotor 磁悬挂转子
magnetically vibrated screen 电磁振动筛
magnetical stepping motor 磁性步进电机
magnetic amplifier regulator 磁性放大调节器
magnetic amplifier 磁放大器
magnetic amplifier characteristic 磁放大器特性曲线
magnetic amplitude 天体出没磁方位角;磁化曲线幅值
magnetic analog computer 磁模拟计算机
magnetic analysis 磁分析(法)
magnetic analysis inspection 磁力分析探伤
magnetic analyzer 磁分析仪;磁分析器
magnetic anisotropy 磁(学)各向异性
magnetic anisotropy energy 磁各向异性能量
magnetic annealing effect 磁(致)退火效应;磁致冷却效应
magnetic annual change 年磁差;地磁年差;地磁年变量;磁周年变化
magnetic annual variation 年磁差;磁周年差
magnetic anode 磁性阳极
magnetic anomally detector 地磁异常探测仪;地磁异常探测器

magnetic anomaly 地磁异常;磁异常
magnetic anomaly axis chart 磁异常轴线图
magnetic anomaly characteristics 磁异常特征
magnetic anomaly code 磁异常编号
magnetic anomaly intensity 磁异常强度
magnetic anomaly interpretation chart 磁异常推断解释图
magnetic anomaly name 磁异常名称
magnetic anomaly of horizontal loop method 水平线框法相对磁异常剖面平面图
magnetic anomaly of sea-mount 海山磁异常
magnetic anomaly profile on plane 磁异常平面剖面图
magnetic anomaly spectrum 磁异常谱
magnetic anomaly trend 磁异常走向
magnetic anomaly type 磁异常类型
magnetic anomaly width 磁异常宽度
magnetic antenna 磁性天线
magnetic anti-wax apparatus used in well 井下磁力防蜡器
magnetic aperture of the stellarator 仿星器的磁孔径
magnetic arc 磁吹
magnetic arc controller 电磁弧控制器
magnetic arc lamp 磁弧灯
magnetic arc stabilizer 电磁弧稳定器
magnetic armature 磁衔铁
magnetic armature loudspeaker 磁框式扬声器;电磁扬声器
magnetic associative memory 磁相联存储器
magnetic attachment 磁吸引
magnetic attenuator 电磁衰减器;磁性衰减器
magnetic attraction 磁引力;磁吸引
magnetic automatic calculator 磁性自动计算器
magnetic auto-steering gear 磁性自动操舵装置
magnetic auxiliary division rotating wheel 磁辅助分划转轮
magnetic axis 地磁轴;磁轴
magnetic azimuth 磁方位(角)
magnetic azimuth of profile 剖面磁方位
magnetic balance 磁力天平;(探іс用的)磁秤
magnetic ballasting machine 电磁铺砟机
magnetic bar 磁棒
magnetic barrier 磁塞
magnetic basement 磁性基底
magnetic basement depth 磁性基底深度
magnetic basement depth map 磁性基底深度图
magnetic base point 地磁基点
magnetic bases 磁性底座
magnetic base station 地磁基点
magnetic battery 复合磁铁
magnetic bay 磁湾
magnetic beam-switching tube 磁旋管
magnetic bearing 磁方向角;磁性轴承;磁象限角;磁向位;磁推轴承;磁力轴承;磁方位(角)
magnetic belt 磁铁钢带
magnetic belt type conveyer 磁性带式输送机
magnetic biasing 偏磁;磁偏法
magnetic bit extractor 磁力钻头打捞器
magnetic bit holder 磁力钻头夹
magnetic blast breaker 磁性断路器
magnetic blow 磁偏吹;磁力送风
magnetic blow-out 磁吹;磁性灭弧;磁灭弧
magnetic blow-out arrester 磁性熄弧避雷器;磁吹避雷器;磁性灭弧式避雷器
magnetic blow-out circuit breaker 磁性熄弧断路器;磁吹灭弧断路器;磁吹断路器
magnetic blow-out fuse 磁吹熔断器
magnetic blow-out lightning rod 磁性灭弧式避雷器
magnetic blow-out switch 灭磁开关
magnetic blow-out valve type arrester 磁吹阀式避雷器
magnetic body 磁体
magnetic body isodepth chart 磁性体等深度图
magnetic body minimum depth chart 磁性体最小深度图
magnetic bonding 磁力键合
magnetic bottle 磁瓶;磁捕集器
magnetic bow 磁性熄弧
magnetic brake 磁制动器;磁力制动器
magnetic bridge 磁导率电桥;磁桥
magnetic bubble 磁泡
magnetic bubble detection 磁泡检测

magnetic bubble device 磁泡器件
magnetic bubble domain 磁泡畴
magnetic bubble material 泡畴材料;磁泡材料
magnetic bubble memory 磁泡存储器
magnetic bubble storage 磁泡存储器
magnetic capacity 磁导系数
magnetic card 磁(性)卡片;磁卡
magnetic card code 磁卡片代码
magnetic card device 磁卡片装置
magnetic card file 磁卡(片)文件
magnetic card memory 磁卡(片)存储器
magnetic card reader 磁卡阅读器
magnetic card storage 磁卡片存储器
magnetic card system 磁卡系统
magnetic card unit 磁卡片机
magnetic carrier 磁性载体;磁记录载体
magnetic cartridge 磁性拾音器
magnetic catch 磁铁夹;磁门扣
magnetic cell 磁元件;磁(存储)单元
magnetic ceramics 磁性瓷
magnetic change 磁化变化;磁变(化)
magnetic character figure 地磁特性图
magnetic characteristic 磁特性
magnetic character recognition 磁字符识别
magnetic chart 地磁图;磁性图;磁力线图;磁场图
magnetic chronograph 磁计时器
magnetic chuck 电磁卡盘;磁性吸盘;磁性卡盘
magnetic circuit 磁(回)路
magnetic circuit system 磁路系统
magnetic circular dichroism 磁性圆二色散
magnetic cleaner 磁力清选机
magnetic cloud chamber 磁云室
magnetic clutch 电离磁合器;电磁离合器;磁离合器;磁力离合器
magnetic clutch motor 磁力离合器电动机
magnetic coating 磁性涂层;磁层
magnetic coating anchorage 磁层黏牢度
magnetic coating resistance 磁层电阻
magnetic cobber 磁(粗)选机
magnetic cobbering machine 粗粒磁选机
magnetic cobbing 粗粒磁选
magnetic cobbing machine 磁选机
magnetic code 磁(代)码
magnetic code format 磁代码格式;磁编码格式
magnetic code reading wands 磁码读数杆
magnetic coefficient 自差系数;磁化系数
magnetic coil 电磁线圈;磁性线圈
magnetic coincidence spectrometer 符合磁谱仪
magnetic combinational switch 磁组合开关
magnetic comparator 电磁钢材分类仪
magnetic compass 罗盘(仪);罗经;磁针罗盘;磁罗盘(仪);磁经
magnetic compass deviation curve 磁罗经自差曲线图
magnetic compass pilot 磁罗盘驾驶器
magnetic compass table 自差表;磁罗盘偏差表
magnetic compensating alloy 磁补偿合金
magnetic compensator 校正磁铁;磁性补偿器
magnetic component 磁分量
magnetic composition in specimens 标本的磁性构成
magnetic computer 磁性计算机
magnetic concentrate 磁选精矿
magnetic concentration 磁选
magnetic concentrator 磁力选矿机
magnetic conductance 磁导(性)
magnetic conductivity 磁导性;磁导率
magnetic conductor 磁导体
magnetic configuration 磁位形
magnetic confinement 磁约束
magnetic contact boundary line 磁性接触线
magnetic contact interface 磁性接触面
magnetic contactor 磁接触器
magnetic contact relay 磁触点式继电器
magnetic contour 磁力等值线
magnetic control 磁铁控制;磁力控(制)
magnetic control injection electron gun 磁控注入式电子枪
magnetic controller 磁力控制器;磁控制装置;磁控制器
magnetic control relay 磁控继电器
magnetic convergence 磁会聚
magnetic convergence circuit 磁会聚电路
magnetic conveyer 磁力运输分堆机

magnetic coolant separator 磁力冷却液分离机
magnetic cooling 退磁法冷却;磁致冷却;磁性冷却;磁体冷却
magnetic cooling effect 磁致冷却效应
magnetic core 磁(铁)芯
magnetic core access switch 磁芯数据存取开关
magnetic core antenna 磁芯天线
magnetic core array 磁芯阵列;磁芯体
magnetic core circuit 磁芯电路
magnetic core counter 磁芯计算器
magnetic core gate 磁芯门
magnetic core logical circuit 磁芯逻辑电路
magnetic core matrix 磁芯矩阵
magnetic core memory 磁芯存储器
magnetic core memory photometer 磁芯存储计光度计
magnetic core multiplexer 磁芯多路编排器
magnetic core production line 磁芯生产线
magnetic core register 磁芯寄存器
magnetic core sense amplifier 磁芯读出放大器
magnetic core shift register 磁芯移位寄存器
magnetic core shift register storage 磁芯移位寄存器
magnetic core stack 磁芯体
magnetic core storage 磁芯存储器
magnetic core storage unit 磁芯存储器装置
magnetic core switch 磁芯开关
magnetic core switching time 磁芯开关时间
magnetic corrosion-resistant filter 电磁防蚀过滤机
magnetic counter 磁性计数器;磁计数器
magnetic coupling 磁(性)耦合
magnetic coupling flowmeter 磁耦合式流量计;磁耦合式流量表
magnetic course 磁罗盘航向;磁航向
magnetic crack detection 磁力(线)探伤
magnetic crack detector 电磁探伤仪;磁力探伤仪;磁力探伤器
magnetic crane 电磁吊车;磁盘起重机;磁力起重机
magnetic creep 磁蠕动
magnetic creeping 磁滞现象;磁蠕变;磁漂移
magnetic crochet 磁鼻
magnetic crotchet 地磁扰动
magnetic crystal 磁性晶体
magnetic cup 磁荧光屏;磁性杯;磁屏
magnetic Curie temperature 磁致变居里温度
magnetic current 磁流
magnetic current antenna 磁性天线;磁流天线
magnetic current meter 磁力流速仪;磁力流速计
magnetic curve 磁力线
magnetic cut-out 电磁开关
magnetic cutter 磁刻纹头
magnetic cycle 磁化循环
magnetic daily variation 周日磁变;磁周日变化
magnetic damper 磁(性)阻尼器
magnetic damping 磁阻尼
magnetic datum ring 磁性环【岩】
magnetic declinating point 磁偏角测点
magnetic declination 磁偏转;磁偏角;磁偏差;磁差角
magnetic declination map 磁偏角略图
magnetic defect detector 电磁探伤器
magnetic deflecting field 磁偏转场
magnetic deflection 磁(致)偏转;磁偏移;磁致偏
magnetic deflection cathode-ray tube 磁偏转电子束管
magnetic deflection colo(u)r selection 磁偏转分色
magnetic deflection mass spectrometer 磁偏转质谱计
magnetic deflection sensitivity 磁偏转灵敏度
magnetic deflection system 磁偏转系统
magnetic deflection tube 磁偏转管
magnetic deflector 磁偏转板;磁偏角测定器
magnetic deformation 磁性变形
magnetic deformation hydrophone 磁致伸缩水听器
magnetic degree 磁角度
magnetic delay line 磁(性)延迟线
magnetic densimeter 磁密度计
magnetic density 磁场密度
magnetic detection 磁探测
magnetic detector 磁探测器;磁性检波器;磁力探伤机;磁力检波器
magnetic detector for ore deposit 磁力探矿仪

magnetic deviation 自差;罗差;磁偏转;磁偏角;磁偏(差);磁差角
magnetic deviation map 磁偏角略图
magnetic dial ga(u)ge 磁性千分表;磁性度盘式指示器;磁性百分表
magnetic dielectric 磁介质
magnetic difference of potential 磁位差
magnetic differential 磁通差动;磁差(动)
magnetic differential unit 磁差动装置
magnetic diffusivity 磁扩散率
magnetic dip 磁倾仪
magnetic dip angle 磁倾角
magnetic dip needle 磁倾仪
magnetic dipole 磁偶极子
magnetic dipole antenna 磁偶极子天线;磁偶极天线
magnetic dipole field 磁偶极子场
magnetic dipole moment 磁偶极矩
magnetic dipole moment matrix 磁偶极矩矩阵
magnetic dipole radiation 磁偶极子辐射
magnetic dipole transition 磁偶极子跃迁
magnetic direction 磁方向
magnetic direction indicator 磁航向指示器
magnetic discharge welding 电磁(储能)焊
magnetic disc memory 多磁头装置
magnetic discontinuity 磁的间断性
magnetic disorder 磁无序
magnetic dispersion 磁消散;磁色散;磁漏
magnetic displacement 磁位移
magnetic displacement recorder 磁位移记录器
magnetic displacement transducer 磁性位移传感器
magnetic distortion 磁场畸变
magnetic disturbance 地磁扰动;磁扰;磁暴
magnetic diurnal variation 周日磁变;磁周日变化
magnetic document sorter reader 磁性文件分类阅读机;磁性文件分类读出机;磁性字符分类器
magnetic domain 磁域;磁畴
magnetic domain device 磁畴器件
magnetic domain material 磁畴材料
magnetic domain memory 磁畴存储器
magnetic domain nucleation 磁畴成核
magnetic domain structure 磁畴结构
magnetic domain theory 磁畴理论
magnetic domain wall 磁畴壁
magnetic door holder 磁性门扎头
magnetic double refraction 磁(场)致双折射
magnetic double resonance spectrometer 磁双共振谱仪
magnetic doublet 磁偶极子
magnetic doublet radiator 磁致偶极辐射器;磁性偶极辐射器
magnetic drag 磁阻
magnetic drag tachometer 磁感应式转速计
magnetic drain plug 磁性油塞;磁性排出口塞
magnetic drill 磁性卡盘钻
magnetic drill press 电磁钻床
magnetic driven arc 磁驱动电弧
magnetic drive pump 磁力泵
magnetic driver 磁力钉锤
magnetic drum 转筒式磁铁分离器;磁选鼓;磁力滚筒;磁鼓
magnetic drum-type separator 鼓式磁选机;鼓式磁力分离器
magnetic drum unit 磁鼓装置;磁鼓机
magnetic dust collector 磁力除尘器
magnetic dust core 磁性铁粉心
magnetic earphone 电磁(式)耳机
magnetic effect 磁效应
magnetic effete of shop-body 船体磁效应
magnetic electric(al) tachometer 磁电测速仪
magnetic electron lens 磁电子透镜
magnetic electron multiplier 磁电子倍增器
magnetic electron spectrometer 磁电子谱仪
magnetic element 磁(性)元件;磁力元件
magnetic elongation 磁致伸长
magnetic encoded system 磁编码系统
magnetic energy 磁能
magnetic energy density 磁能密度
magnetic energy storage 磁能存储
magnetic energy-storage spot welder 磁能存储式点焊机
magnetic equator 无倾线;地磁赤道;磁赤道
magnetic equipotential 等磁势

magnetic equivalent current system 等效磁电流系;等磁效电流系
magnetic eraser 消磁器;去磁器
magnetic escapement 磁性擒纵机构
magnetic examination 磁性探伤检查;磁化试验
magnetic exchange valve 电磁换向阀
magnetic excitation 磁力激发
magnetic exploration 磁性探矿;磁性勘探;磁力勘探;磁法勘探
magnetic fabric 磁性组构
magnetic false-twist spindle 磁性假捻锭子
magnetic fatigue 磁疲乏
magnetic fault find method 磁性探伤法
magnetic feeder 电磁喂料器
magnetic feeder unit 电磁进料装置
magnetic feed roll 磁力进料器
magnetic-ferrite memory 铁氧存储器
magnetic ferrites 磁性铁氧体
magnetic fibre sensor 光纤磁性传感器
magnetic field 磁场
magnetic field analyzer 磁场分析仪
magnetic field annealing 磁场退火
magnetic field automatic scanning 磁场自动扫描
magnetic field balance 探测用磁秤;探测磁棒;磁力仪;磁场平衡
magnetic field coil 磁场(激励)线圈
magnetic field component 磁场分量
magnetic field configuration 磁场形态
magnetic field control 磁场控制
magnetic field cooling 磁场冷却
magnetic field detector 磁性场探测器
magnetic field discharge 磁场放电
magnetic field equalizer 磁场均衡器;补偿磁圈
magnetic field excitation 磁场励磁
magnetic field intensity 磁场强度
magnetic field interference 磁场干扰
magnetic field measurement 磁场测量
magnetic field pressing 磁场压制成型
magnetic field pressurization conbustion sintering 磁场加压燃烧烧结
magnetic field quenching 磁场淬火
magnetic field resistor 磁场电阻器
magnetic field spectrum analyzer 磁场频谱分析仪
magnetic field stability 磁场稳定度
magnetic field strength 磁场强度
magnetic field strength sensor 磁场强度传感器
magnetic field strength transducer 磁场强度传感器
magnetic field-supported accelerometer 磁悬加速计
magnetic field test equipment 磁力探伤机
magnetic field-tuned laser 磁场调谐激光器
magnetic field vector 磁场矢量
magnetic field weakening control 磁场削弱控制
magnetic field weakening ratio 磁场削弱比
magnetic field winding 磁场绕组
magnetic figure 磁力线图
magnetic filament 磁丝
magnetic film 涂磁胶片;磁(性薄)膜
magnetic film logic device 磁膜逻辑器件
magnetic film memory 磁膜存储器
magnetic film storage 磁膜存储器
magnetic film unit 磁膜单元
magnetic filter 磁(性)过滤器;磁芯过滤器;磁滤机;磁力过滤器
magnetic filtration 磁性过滤
magnetic firing circuit 磁性点火电路
magnetic fixing die 磁力固定模具
magnetic flaw detecting 磁力探伤
magnetic flaw detection 磁性探伤
magnetic flaw detector 磁性探伤器;磁探伤仪;磁力探伤器
magnetic float 磁浮式
magnetic floating valve 电磁浮式阀
magnetic flocculation 磁团絮;磁(力)絮凝
magnetic flocculation separation process 磁絮凝分离法
magnetic flow 磁通;磁流量
magnetic flow meter 磁(性)流量计;磁通计;磁场流量计;磁流测量仪;磁(合式)流量计
magnetic fluid 磁性液体;磁(性)流体
magnetic fluid clutch 磁性液离合器;磁力流体离合器
magnetic fluid control device 磁性流体控制仪

magnetic fluid power plant 磁流体发电厂
magnetic flux 磁性溶剂;磁性焊剂;磁通(量)
magnetic flux arc welding 磁性焊剂电弧焊
magnetic flux density 磁通(量)密度;磁感应密度
magnetic flux detector 磁通探测器
magnetic flux distribution 磁通分布
magnetic flux gas shielded arc welding 磁性焊剂气体保护电弧焊
magnetic flux intensity 磁通密度
magnetic flux leakage 漏磁;磁漏
magnetic flux-linkage 磁键
magnetic flux measurement 磁通测量
magnetic flux meter 磁通(量)计;磁通(量)表
magnetic flux sensor 磁通量传感器
magnetic flux test 磁流试验;磁(通)(线)检验
magnetic flux transducer 磁通量传感器
magnetic focused image intensifier 磁聚焦像增强管;磁聚焦图像增强器
magnetic focusing 磁(致)聚焦
magnetic focusing coil 磁聚焦线圈
magnetic focusing indicator tube 磁聚焦指示管
magnetic force 磁力;磁场强度
magnetic force anomaly 磁力异常
magnetic force driving pump 磁力驱动泵
magnetic force welding 磁力焊接法
magnetic force welding machine 磁力焊接机
magnetic forming 磁力成型法;磁力成型
magnetic forming machine 磁力成型机
magnetic fraction 磁性部分
magnetic freqeuncy multiplier 磁(性)倍频器
magnetic friction 磁摩擦
magnetic friction clutch 磁(力)摩擦离合器
magnetic gap 磁隙
magnetic gate 磁闸;磁门
magnetic ga(u)ge 磁力测微计;磁力测厚仪
magnetic gear 电磁摩擦联轴器;磁力离合器
magnetic gear shift 磁力变速器
magnetic generator 磁发电机
magnetic geophysical method 地球物理磁测法
magnetic grader 磁力分选机
magnetic gradient 磁梯度
magnetic gradiometer 磁力梯度仪
magnetic grid 磁栅;磁格子
magnetic guard 磁性钢丝面罩
magnetic guidance 磁(力)导(向)
magnetic hammer 磁化锤
magnetic hand scanner 磁手动扫描器
magnetic hardness 磁(力)硬度
magnetic hardness comparator 磁性硬度比较器
magnetic head 磁头
magnetic head azimuth 磁头方位角
magnetic head impedance tester 磁头阻抗测试仪
magnetic heading 磁航向
magnetic head material 磁头材料
magnetic head positioning construction 磁头定位机构
magnetic head potting adhesive 磁头灌封胶
magnetic head stack 磁头组
magnetic health ball 磁健身球
magnetic heat treatment 磁场热处理
magnetic heavy method 磁重法
magnetic high 磁力高
magnetic holder 磁性托架
magnetic holding device 磁夹具
magnetic holography 磁(性)全息术
magnetic horizontal gradiometer 水平磁力梯度仪
magnetic horn 磁喇叭
magnetic hum 磁哼声
magnetic hydraulic clutch 磁性液体离合器
magnetic-hydrothermal theory 岩浆热液成矿说
magnetic hysteresis 磁滞
magnetic hysteresis alloy 磁滞合金
magnetic hysteresis cycle 磁滞回线
magnetic hysteresis damping 磁滞阻尼
magnetic hysteresis effect 磁滞效应
magnetic hysteresis loop 磁滞回线
magnetic hysteresis loss 磁滞损失;磁滞损耗
magnetic hysteresis motor 磁滞电动机
magnetic hysteretic angle 磁滞角
magnetic ignition 磁石电机点火法
magnetic impulse counter 磁脉冲计数器
magnetic impulser 磁脉冲传感器
magnetic inclination 磁倾(角)
magnetic indicator 磁性指示器

magnetic induced polarization 磁激发极化法
magnetic induction 磁感(应)
magnetic induction accelerator 磁感应加速器
magnetic induction component 磁感应分量
magnetic induction density 磁感应密度
magnetic induction flow meter 磁感应流量计;磁感应流量表;磁流量计;磁感应式流量表计
magnetic induction gyroscope 磁感应式回转仪
magnetic induction inspection 电磁诱导检测
magnetic induction intensity 磁感应强度
magnetic induction part 磁感应元件
magnetic induction pump 磁感应泵
magnetic inductive capacity 磁导率;磁导系数
magnetic inductivity 磁导率;磁感应率
magnetic inertia 磁惯性
magnetic ink 磁性油墨;磁(性)墨水;磁性检验液
magnetic ink character reader 磁性墨水字符阅读器
magnetic ink character recognition 支票磁性墨水识别码;磁记录字符识别
magnetic ink character recorder 磁性墨水字符记录器
magnetic ink scanner 磁(性)墨水扫描器
magnetic insert 磁垫片
magnetic inspection 磁性探伤;磁力检查
magnetic inspection equipment 磁力探伤设备
magnetic inspection oil 磁力探伤油
magnetic inspection paste 磁力探伤糊剂
magnetic inspection powder 磁力探伤粉
magnetic insulation 磁绝缘
magnetic intensifier 磁图像增强器
magnetic intensity 磁(场)强(度)
magnetic intensity vector 磁场强度矢量
magnetic interaction 磁相互作用
magnetic interaction number 磁相互作用数
magnetic interceptor 磁力分选机
magnetic interface inversion in frequency domain 频率域磁性界面反演
magnetic interface inversion in space domain 空间域磁性界面反演
magnetic interrupter 磁(力)断续器;磁路断续器
magnetic inversion 磁反转
magnetic ion 磁(性)离子
magnetic ion spectrometer 磁离子光谱仪
magnetic iron 磁铁
magnetic iron ore 四氧化三铁;磁铁矿(石)
magnetic iron ore concrete 磁铁矿混凝土
magnetic iron oxide 磁性氧化铁;磁铁氧
magnetic iron oxide epoxy ester anti-rust paint 磁铁环氧酯防锈漆
magnetic iron oxide epoxy precoat primer 磁铁环氧预涂底漆
magnetic iron powder 磁铁粉
magnetic iron pyrite 磁黄铁矿
magnetic isolation method 磁分离法
magnetic isoporic line 地磁图等年变线
magnetic jack 磁力连接器
magnetic joint 磁性接合
magnetic Kerr effect 克尔磁效应
magnetic key relay 磁力继电器
magnetic lag 磁滞(后);磁化滞后;磁惰性
magnetic lag clutch 磁滞离合器
magnetic latitude 等磁倾纬度;磁纬度
magnetic layer 磁(性)层
magnetic leakage 磁漏
magnetic leakage coefficient 磁漏系数
magnetic leakage transformer 磁漏变压器
magnetic ledger 磁性总账
magnetic length 磁体长度
magnetic lens 磁性透镜;磁透镜
magnetic lens spectrometer 磁透镜谱仪;磁透镜分光计
magnetic levitated vehicle 磁垫车
magnetic levitation high speed ground transportation 磁浮式高速交通
magnetic levitation railway 磁浮铁路
magnetic levitation system 磁浮装置;磁浮系统
magnetic levitation technique 磁浮技术
magnetic levitation vehicle 磁浮车(辆)
magnetic lid lifter 电磁起炉盖机;电磁盖机
magnetic lifter 磁力起重机
magnetic lifting device 电磁板直机
magnetic limit 最大磁感应强度
magnetic line 磁力线
magnetic lines of flux 磁力线

magnetic lines of force 磁力线
magnetic link 磁钢棒
magnetic linkage 磁链
magnetic loading 磁负荷
magnetic local anomaly 磁局部异常
magnetic locator 磁性探测器;磁力检测器
magnetic lock 磁锁;磁门
magnetic lock relay 磁锁定继电器
magnetic log(ging) 磁(法)测井
magnetic logic computer 磁逻辑计算机
magnetic loop system 磁回路系统
magnetic loss angle 磁损失角
magnetic loudspeaker 磁性扬声器;电磁扬声器;磁力扬声器
magnetic low 磁力低
magnetic lubricating oil conditioner 磁性滑油再生器
magnetic Mach number 磁马赫数
magnetic mass spectrometer 磁质谱计
magnetic material 磁性材料;磁体材料
magnetic mate rial wet forming hydraulic press 磁材湿式成型液压机
magnetic matrix 磁存储矩阵
magnetic matrix switch 磁(性矩)阵开关;磁膜开关
magnetic matrix transducer 磁矩阵式传感器
magnetic measurement 磁性测量;磁力测量;磁测定
magnetic mechanical 磁机械的
magnetic medium 磁(性)介质
magnetic memory 磁存储器
magnetic memory material 磁性记忆材料
magnetic memory matrix 磁存储器矩阵
magnetic memory plate 磁性存储板
magnetic memory system 磁存储系统
magnetic mercury cut-off device 磁力水银断流器
magnetic mercury switch 磁性水银开关
magnetic merging 磁力线重接
magnetic meridian 地磁子午线;磁子午线
magnetic metal 磁性金属
magnetic metallic glass 磁性金属玻璃
magnetic metal pigment 金属磁性颜料
magnetic method 磁(选)法;磁勘探法
magnetic method of exploration 磁力勘探法
magnetic micrometer 磁力测微计
magnetic microphone 电磁传声器
magnetic microscope 磁(式电子)显微镜
magnetic mine 磁性水雷
magnetic mine detector 磁性探雷器
magnetic mineral 磁性矿物
magnetic mineral in specimens 标本的磁性矿物
magnetic mirror 磁镜
magnetic mirror field 磁镜场
magnetic mixer 磁力混合器
magnetic modulation 磁性调制
magnetic modulator 磁(性)调制器
magnetic moment 磁偶极矩;磁(力)矩
magnetic moment of electron 电子磁矩
magnetic monopole 磁单极子
magnetic motive force 磁动势;磁动力
magnetic motor starter 电动机磁力启动器
magnetic mo(u)lding machine 磁力造型机
magnetic mo(u)lding process 磁型铸造;磁丸造型法
magnetic multiparticle spectrometer 磁性多粒子谱仪
magnetic multipole 磁多极
magnetic multipole field 磁多极场
magnetic multipole radiation 磁多极辐射
magnetic navigation 磁导航
magnetic needle 指南针;磁针
magnetic needle amplitude 磁针摆幅
magnetic needle bearing 磁轴
magnetic needle button 磁针制动钮
magnetic needle compass 磁针罗盘
magnetic needle declination 磁针偏角
magnetic needle holders 磁性持针钳
magnetic needle inclination 磁针倾角
magnetic needle release loss 磁针制动杆
magnetic needle sensibility 磁针灵敏度
magnetic north 磁北(方)
magnetic north line 磁北方线
magnetic north pole 磁北极
magnetic note 磁偏角注记
magnetic nozzle 磁喷管

magnetic observation 磁力观测;磁力测量
magnetic observatory 地磁(观测)台;地磁观测所
magnetic observatory survey 地磁台测量
magnetic octupole moment 磁八极矩
magnetic oil filter 磁性滤油器
magnetic operation(al) amplifier 磁运算放大器
magnetic optical-character reader 磁光字符阅读器
magnetic optic(al) detection method 磁光探伤法
magnetic optic(al) disc 磁光盘
magnetic optic(al) effect 磁光效应
magnetic optics 磁光学
magnetic order 磁(有)序
magnetic ore dressing plant 磁选厂
magnetic orientation 磁性取向;磁定向
magnetic orientation control coil 磁性定向控制线圈
magnetic oscillator 磁振荡器;磁控振子
magnetic oscillograph 电磁示波器;磁整波器
magnetic oxide iron 磁性氧化铁
magnetic oxygen cutting machine 磁力氧气切割机
magnetic oxygen recorder 电磁测氧仪;磁力测氧仪
magnetic padlock 磁锁
magnetic panel 磁接触器盘
magnetic paper 磁性纸带
magnetic parameter 磁性参数
magnetic particle 磁(性颗)粒
magnetic particle analyzer 磁粒子分析仪
magnetic particle clutch 磁粒子离合器;磁粉离合器
magnetic particle detector 电磁探伤机
magnetic particle examination 磁粉探伤
magnetic particle indication 磁粉指示法;磁粉显示法
magnetic particle inspection 磁力探伤;磁粉探伤;磁粉检查
magnetic particle method 磁粉探伤法
magnetic particle method of inspection 磁铁粉检查法
magnetic particle test(ing) 磁粉检测;磁力探伤(试验);磁粉(探伤)试验
magnetic particle testing machine 磁粒子测试机
magnetic parts 磁性零件
magnetic path 磁路
magnetic peeler 加速器磁反射器;磁激发器;磁反射器
magnetic pen 磁记录笔
magnetic pendulum 磁摆
magnetic performance 磁性
magnetic permeability 透磁性;磁导性;磁导系数;磁导率
magnetic permeability curve 磁导率曲线
magnetic permeability of rock 岩石磁导率
magnetic permeameter 磁导仪
magnetic permeance 磁导性
magnetic permitivity 磁导率
magnetic perturbation 磁扰
magnetic phenomenon 磁现象
magnetic photocatalyst 磁载光催化剂
magnetic pickup 电磁(式)拾音器;电磁式拾波器
magnetic pickup tool 磁化采集工具
magnetic pinch effect 磁箍缩效应
magnetic plane characteristic 磁场平面内的方向性
magnetic plant 磁选设备
magnetic plated wire 镀磁线;磁镀线
magnetic plated wire memory 磁镀线存储器
magnetic plate memory 磁板存储器
magnetic plate storage 磁板存储器
magnetic player 磁放声机
magnetic playset 磁性玩具
magnetic plug 磁(性)塞;电磁铁
magnetic point 磁力点;磁化点
magnetic polarity 磁极性
magnetic polarity reversal 磁极性转向
magnetic polarization 内禀磁感应强度;磁极化
magnetic pole 磁极;磁荷
magnetic pole method 磁极探伤法
magnetic pole position 磁极位置
magnetic pole strength 磁极强度
magnetic position converter 磁性位置转换器
magnetic positioner 磁力垛板机
magnetic position sensor 磁定位传感器
magnetic potential 磁位;磁(标量)势
magnetic potential difference 磁势差
magnetic potential gradient 磁位梯度
magnetic potential regulator 磁势调节器

magnetic potentiometer 磁位差计
magnetic powder 磁(性)粉(末)
magnetic powder brake 磁粉制动器
magnetic powder clutch 磁粉离合器
magnetic powder-coated tape 涂磁粉带;磁粉涂敷带
magnetic powder core 压粉铁芯;磁(性铁)粉芯
magnetic powder detection 磁粉探伤检验
magnetic powder flux 磁粉焊剂
magnetic powder indication 磁粉指示
magnetic powder inspection 磁粉探伤
magnetic powder liquid trough 磁粉液槽
magnetic powder pattern 磁粉图样;磁粉图案
magnetic powder seal for rotating shafts 转轴用磁粉密封
magnetic powder test 磁粉检验
magnetic power 磁功率
magnetic power factor 磁功率因数
magnetic precipitation 磁力除尘
magnetic pressing 磁场压制
magnetic pressure 磁压
magnetic pressure transducer 磁性压力传感器
magnetic prime vertical 磁卯酉圈;磁东西圈
magnetic printing 磁性转印
magnetic printing ink 磁性印刷油墨
magnetic prism 磁棱镜
magnetic prism spectrometer 磁棱镜谱仪
magnetic probe 探磁圈;磁探头
magnetic probe extensometer 磁性沉降计
magnetic process 磁选法
magnetic product 磁性制品;磁铁制品
magnetic profile 磁剖面图
magnetic property 磁性
magnetic property of minerals 矿物磁学性质
magnetic prospecting 磁性勘探;磁力勘探法;磁力勘探;磁力测探;磁法找矿;磁法勘探
magnetic pull 磁引力;磁铁吸引力
magnetic pulley 磁性滑轮;磁性滚筒;磁力滚筒;磁滚轮;电磁皮带轮
magnetic pulley separator 磁力分离滚筒;滑轮式磁选机
magnetic pulsation 磁脉动
magnetic pulsator 磁力脉动器
magnetic pulse 磁脉冲
magnetic pump 磁能泵;磁力泵;电磁泵
magnetic pumping 磁抽运
magnetic pumping heating 磁泵加热
magnetic pyrite 磁(性)黄铁矿
magnetic pyrometer 磁高温计
magnetic quadrupole lens 磁四极透镜
magnetic quantum 磁量子
magnetic quantum number 磁量子数
magnetic quenching 磁猝灭;磁吹灭弧
magnetic quench test 磁性淬火试验
magnetic quiet zone 地磁低缓带;磁寂带
magnetic range 校罗经叠标
magnetic reactance 磁抗
magnetic read-write head 磁读写头
magnetic receiver 磁铁座
magnetic recorder 磁性(带)记录器
magnetic record file 磁记录文件
magnetic record(ing) head 记录磁头
magnetic recording wire 磁性钢丝
magnetic rectifier 磁整流器
magnetic refrigerator 磁致冷机;磁制冷机
magnetic regulator 磁力调节器
magnetic relaxation 磁(性)弛豫
magnetic relay 磁(性)继电器
magnetic relay timer 磁性继电器定时器
magnetic reluctance 磁阻
magnetic reluctivity 磁阻率
magnetic remanence 磁顽
magnetic reproducer 磁性再现装置
magnetic reproducing 磁性再现
magnetic reproducing head 复制磁头
magnetic repulsion 磁推斥;磁拒斥
magnetic resistance 磁阻(率)
magnetic resistance semiconductor 磁阻半导体
magnetic resolution 磁分辨
magnetic resonance 磁共振
magnetic resonance accelerator 磁共振加速器
magnetic resonance AC voltage stabilizer 磁共振交流稳压器
magnetic resonance spectroscopy 磁共振波谱学

magnetic response 磁响应;磁反应
magnetic retardation 磁滞
magnetic retentivity 顽磁性
magnetic return path 磁通量回路
magnetic reversal 逆磁化;地磁反向;地磁倒转;磁性反转;磁极倒转事件;磁反向;反(方)向磁化
magnetic Reynolds number 磁雷诺数
magnetic rigidity 磁刚性;磁刚度
magnetic road sweeper 磁力扫路机
magnetic roasting 磁化焙烧(法)
magnetic rock 磁性岩石
magnetic rod 磁棒
magnetic rod memory 磁棒存储器
magnetic rod storage 磁棒存储器
magnetic roll 磁力辊
magnetic rotation 磁转偏光;磁旋(转)
magnetic rotation of polarized light 极化光的磁致旋转
magnetic rotation spectrum 磁致旋光谱
magnetic rotatory power 磁旋光本领
magnetic rougher 粗选磁选机
magnetic rubber 磁性橡胶
magnetics 磁学;磁性元件
magnetic sand 磁铁(矿)砂;磁矿砂
magnetic saturation 磁(性)饱和
magnetic saturation starter 磁饱和启动器
magnetic saturation voltage stabilizer 磁饱和稳压器
magnetic scalar potential 磁标量势
magnetic scale 磁尺
magnetic scanning 磁扫描;磁偏转扫描;磁偏扫描
magnetic scattering 磁散射
magnetic scattering amplitude 磁性散射振幅
magnetic screen 磁屏(幕);磁屏蔽
magnetic screening 磁屏蔽
magnetic screening action 磁屏蔽作用
magnetic seal(ing) 磁性密封
magnetic secular change 地磁长期变化;磁长期变化
magnetic semiconductor 磁性半导体
magnetic sensing 磁敏
magnetic sensing element 地磁敏感元件
magnetic sensor 磁性传感器
magnetic separating 磁选分离
magnetic separating plant 磁力分离设备;磁选设备
magnetic separating roll 磁力分离辊
magnetic separation 磁选;磁性分离;磁铁分离;磁力选矿;磁(力)分离
magnetic separation method 磁力选矿法;磁力分离法
magnetic separation pulley 磁力分离滚筒
magnetic separator 磁选器;磁选机;磁性分离器;磁(铁)分离器;磁力选矿机;磁力分离器
magnetic servo-amplifier 磁性伺服放大器
magnetic shield 磁屏蔽
magnetic sheet 磁性板;磁数据记录纸;磁片
magnetic sheet handler 薄板电磁分送机
magnetic sheet handling unit 薄板电磁分送机构
magnetic sheet piler 电磁垛板机
magnetic shell 磁壳
magnetic shielded gun 磁屏蔽电子枪
magnetic shield(ing) 磁屏蔽
magnetic shift register 磁移位寄存器;磁移位存储器
magnetic shoal 具有磁干扰的浅水区
magnetic shoe 磁瓦
magnetic shunt 磁分路器;磁分路
magnetic shunt alloy 调磁合金;磁分流器合金
magnetic silencer 电磁消声器
magnetic sine plate 磁性正弦规
magnetic slide valve 电磁滑阀
magnetic slot reader 磁槽阅读机;磁槽读出器
magnetic slot-wedge 磁槽楔
magnetic smoke detector 磁性烟感器
magnetic softening 磁性软化
magnetic solar daily variation 地磁太阳日变化
magnetic solar diurnal variation 地磁太阳日变化
magnetic solar lunar variation 地磁太阳月变化
magnetic solenoid 磁螺线管
magnetic sorting bridge 磁性分拣电桥
magnetic sound carrier 磁性音频载波
magnetic sounder 磁测深仪
magnetic sounding 磁测深
magnetic sound recording 磁录音

magnetic sound talkie 磁录式有声电影
magnetic sound track 声音磁迹；磁声道
magnetic south 磁南极
magnetic space 磁场空间
magnetic spacer 磁隔离物
magnetic spark blow-out 磁铁灭火花(器)
magnetic spark plug 电磁火花塞；磁性火花塞
magnetic speaker 电磁扬声器
magnetic spectrograph 磁摄谱仪；磁谱计
magnetic spectrometer 磁谱仪；磁谱计；磁分光计
magnetic spectrum 磁谱
magnetic spell and learn set 磁性拼学玩具
magnetic spin 磁自旋
magnetic splitter 磁分裂器
magnetic spot 磁化点
magnetic stability 磁稳定性
magnetic stabilizer 磁稳定器
magnetic star delta switch 磁力星形三角开关
magnetic starter 磁力起动器
magnetic starting switch 磁力起动开关
magnetic station 电动机控制盘；磁测站
magnetic station mark 磁力点标志
magnetic steel 磁(性)钢
magnetic steel core 铁芯
magnetic steel detector 钢筋磁性探测仪；钢筋磁性检测仪
magnetic steel module 电工磁性钢
magnetic steering 磁导向
magnetic stepping motor 步进电机
magnetic sticking 磁铁黏结
magnetic stick relay 磁性保持继电器
magnetic stirrer 电磁搅拌器；磁(性)搅拌器；磁力搅拌器；磁力搅拌机
magnetic stirring 磁力搅拌；磁场搅动
magnetic stirring apparatus 磁力搅拌器
magnetic stirring bar 磁性搅拌棒；磁力搅棒
magnetic storage 磁存储器
magnetic storage register 磁存储寄存器
magnetic storage system 磁存储系统
magnetic store 磁存储器
magnetic storm 磁暴
magnetic storm monitor 磁暴记录器
magnetic strain 磁应变
magnetic strain energy 磁应变能
magnetic strainer 磁选筛；磁选应变器；磁性筛网；磁性滤网；磁性滤器
magnetic strength 磁强(度)
magnetic stress 磁应力
magnetic stress tensor 磁应力张量
magnetic strip accounting machine 磁条会计计算机
magnetic stripe 磁条；磁片；磁迹
magnetic stripe accounting machine 磁条会计机
magnetic striped ledger 磁条分类账记录纸
magnetic stripe encoder 磁条编码器
magnetic stripe reader 磁条阅读器
magnetic stripe recording 磁条记录
magnetic stripe storage 磁条存储器
magnetic stripe store 磁条存储器
magnetic strip file 磁条文件
magnetic structure 磁路结构
magnetic subband 磁支带
magnetic substance 磁性物质；磁性体；磁性材料
magnetic substorm 磁亚暴
magnetic support 磁座；磁垫
magnetic surface 磁鼓面
magnetic surface memory 磁(表)面存储器
magnetic surface recording 磁表面记录
magnetic surface storage 磁面存储器
magnetic survey 地磁查勘图；地磁测量；磁力调查；磁测
magnetic survey area 磁测区范围
magnetic survey date 磁测时间
magnetic survey in borehole 井中磁测
magnetic surveying 磁力勘测法
magnetic survey map 磁测图件
magnetic survey method 磁测法
magnetic survey precision 磁测精度
magnetic survey under water 水下磁测
magnetic susceptibility 透磁率；磁化(敏感)率
magnetic susceptibility log 磁化率测井
magnetic susceptibility log curve 磁化率测井曲线
magnetic susceptibility logger 磁化率测井仪
magnetic susceptibility meter 磁化率仪

magnetic susceptibility survey 磁化率测量
magnetic susceptibility test 磁感应试验
magnetic susceptometer 磁化感受计
magnetic susceptor 磁感受器
magnetic suspension 磁悬浮；磁力吊架
magnetic suspension spinning rotor ga(u)ge 磁悬浮转子计
magnetic suspension spinning rotor vacuum ga(u)ge 磁悬浮转子真空计
magnetic suspension technique 磁悬挂技术
magnetic suspension train 磁悬浮列车
magnetic suspension zone-melting 磁悬区域熔炼
magnetic sweep amplifier 磁扫描放大器
magnetic sweeper 磁选扫路机(清除铁屑用)；磁铁清扫机；磁力清扫机；磁力清除器
magnetic switch 磁(力)开关
magnetic switch controller 磁开关控制器
magnetic switching system 磁力开关系统
magnetic synchro 磁同步
magnetic system 磁系统
magnetic tachometer 电磁转速表；磁转速计；磁流速计；磁力转数计
magnetic tell tale 磁感式沉降标
magnetic temperature compensating alloy 磁温度补偿合金
magnetic temperature controller 磁性温控器
magnetic temporal variation 磁时变
magnetic test 磁力探伤
magnetic test coil 检测线圈；探测线圈；测磁线圈
magnetic testing 磁性探伤；磁力探伤
magnetic theodolite 地磁经纬仪；磁经纬仪
magnetic thermogensis 岩浆热力作用
magnetic thermometer 磁(石)温度计
magnetic thickness ga(u)ge 磁(性)测厚计；磁力测厚仪
magnetic thickness tester 磁性(镀层厚度)测厚仪
magnetic thin film 磁(性)薄膜；磁膜
magnetic thin film storage 磁(薄)膜存储器
magnetic thrust bearing 磁性推力轴承
magnetic ticket 磁法车票
magnetic time relay 磁延时继电器
magnetic torque 磁(转)矩；磁扭矩
magnetic tracer roller 磁导轮
magnetic track 磁(航)迹(向)；磁道
magnetic track angle 磁迹角
magnetic track brake 磁性轨道制动器
magnetic transducer 磁换能器
magnetic transfer 磁性转印
magnetic transformation 磁性转变
magnetic transformation point 磁性转变点
magnetic transient recorder 磁性瞬变过程记录器
magnetic transition temperature 居里温度；磁性转变温度
magnetic transmission 磁力传动
magnetic transmission variable aperture flow meter 磁力传动可变孔径式流量计
magnetic trap 磁收集器
magnetic travel(l)ing crane 电磁吸盘式移动式起重机
magnetic treatment 磁化处理
magnetic treatment on disinfection 磁化消毒处理
magnetic trigger 磁性起动装置；磁触发器
magnetic trough separator 槽式磁选机
magnetic tube concentrator 磁选机
magnetic tuning system 磁调谐系统
magnetic type speedometer 磁力式速率计；磁力式速度计
magnetic unit 磁单位
magnetic valve 电磁阀；磁性阀；磁力阀
magnetic valve type surge arrester 磁阀式避雷器
magnetic variable 磁变量
magnetic variation 磁(性)变(化)；磁(偏)差
magnetic variation chart 磁偏差海图
magnetic variation curves 磁差曲线
magnetic variometer 磁变仪；磁变计
magnetic vector 磁向量；磁(场)矢量
magnetic vector potential 向量磁位；磁矢势
magnetic vehicle detector 电磁检车器；磁感检测器[道]
magnetic vertical gradiometer 垂直磁力梯度仪
magnetic vibrating feeder 磁振给矿机
magnetic vibration 磁振动
magnetic vibration actuated feeder 电磁振动给料机

magnetic vibrator 磁振动器
magnetic vise 磁性虎钳；磁力虎钳
magnetic viscosity 磁黏滞度；磁黏性(系数)
magnetic voltage stabilizer 电磁式稳压器
magnetic vortex in a super-conductor 超导体中的磁旋涡
magnetic water 磁化水
magnetic water cleaning device 磁水净化器
magnetic water treatment 水的磁性处理法；磁水处理(法)；磁化水处理
magnetic wave 磁波
magnetic wave detector 电磁检波仪；磁性检波器
magnetic wedge 磁性楔
magnetic well 磁笼；磁阱
magnetic wheel 磁性轮
magnetic wind direction 磁极风向
magnetic wire 电磁线；磁性钢丝；磁(导)线
magnetic wire storage 磁线存储器
magnetic worktable 磁性工作台
magnetic yoke 磁轭
magnetic zenith 磁天顶
magnetic zero balance 磁力零平衡器
magnetisability 可磁化性；磁化率
magnetisable 可磁化的
magnetisation 磁化强度
magnetism 磁学；磁性；磁力现象；磁力
magnetism influence of the winch 绞车磁化干扰
magnetite 四氧化三铁锈层；磁(铁)石；磁铁矿
magnetite aggregate 磁铁矿集料；磁铁矿骨料
magnetite black 四氧化三铁；磁性铁黑
magnetite catch 磁铁碰头
magnetite-chlorite oolite 磁铁矿-绿泥石鲕状岩
magnetite chondrodite sharn 磁铁矿粒硅镁石矽卡岩
magnetite concrete 磁铁矿混凝土
magnetite concrete aggregate 磁铁矿混凝土集料；磁铁矿混凝土骨料
magnetite content 磁铁矿含量
magnetite corundolite 磁铁矿刚玉岩
magnetite feeder 磁铁矿给料器
magnetite ore 磁铁矿砂；磁铁矿矿石
magnetite-phyllite 磁铁千枚岩
magnetite quartz schist 磁铁石英片岩
magnetite-rich rock 富磁铁矿岩
magnetite type granite 磁铁矿型花岗岩
magnetitic quartzite 磁铁石英岩
magnetitite 磁铁岩
magnetizability 磁化能力
magnetization 起磁；磁化作用；磁化强度；磁化
magnetization curve 正常磁化曲线；磁化曲线
magnetization cycle 磁化回路
magnetization direction 磁化方向
magnetization intensity 磁化强度
magnetization mechanism 磁化机理
magnetization of transducer 换能器充磁
magnetization reversal 反磁化
magnetization treatment 磁化处理
magnetization vector 磁化矢量
magnetize 磁化
magnetize attraction 磁性力
magnetized area 磁化区域
magnetized drink water cup 磁化杯
magnetized needle 磁针
magnetized spot 磁化点
magnetized-water health-protection cup 磁化水保健杯
magnetized water treatment 磁化水处理
magnetizer 励磁装置；磁化装置；磁化器；磁化机；传磁物；充磁装置
magnetizer water 磁化水
magnetizing 磁化
magnetizing ampere turns 磁化安匝
magnetizing apparatus 磁化器；充磁装置；充磁器；充磁机
magnetizing coil 磁化线圈
magnetizing current 起磁电流；磁化电流
magnetizing current and reactance 磁化电流和电抗
magnetizing direction 磁化方向
magnetizing exciter 励磁器
magnetizing field 励磁场；起磁场；磁化磁场
magnetizing force 起磁力；磁化力
magnetizing inductance 磁化电感
magnetizing inrush 磁化冲量

magnetizing loss 磁化损失
magnetizing of transducer 换能器充磁
magnetizing pattern 磁化图形
magnetizing power 磁化功率
magnetizing roast 磁化焙烧
magnetizing roasting 磁化焙烧(法)
magnetizing rubber sheet 磁性橡胶压片
magnetizing solenoid 磁化螺线管
magnetizing susceptance 励磁电纳;磁化电纳
magnetizing winding 磁化绕组
magnet keeper 永久磁铁;卫铁
magnet-lagging synchronized motor 磁滞同步电动机
magnet limb 凸极
magnet meter 磁通计
magnet needle 磁针
magneto 永磁式的;磁(发)电机;起动磁石电机;手摇磁石发电机;磁石发电机;磁力发电机
magnetoacoustic effect 磁声效应
magnetoacoustics 磁声学
magnetoactive plasm 磁活化等离子体
magnetoadvance control 磁电机点火提前调整
magnetoaerodynamics 磁空气动力学
magneto alternator 永磁同步发电机;永磁交流发电机
magnetoanemometer 磁电风速计
magnetoball bearing 磁电机球轴承
magnetobase 磁电机座
magnetobeamed tube 磁聚束电子管
magneto bell 永磁电铃;磁铁电铃;磁石(式)电铃
magnetoblock 磁电机座
magnetoboot 磁电机保护罩
magnetobracket 磁电机架
magnetobreaker arm 磁电机断电器臂
magnetobreaker arm point 磁电机断电器臂接触点
magnetobreaker points 磁电机断闭点
magnetobrush 磁电机电刷
magnetocall 手摇(磁石)发电机呼叫;磁石发电机呼叫
magnetocaloric 磁致热的
magnetocaloric cooling 磁热效应冷却
magnetocaloric effect 磁致热效应
magnetocentral office 磁石式电话总局
magnetochemistry 磁化学
magnetocircuit 磁电机电路
magnetocoil 磁电机线圈
magnetocollector ring 磁电机集电环
magnetoconductivity 磁导率
magnetocontact breaker 磁电机断路器
magneto coupler 电磁耦合器
magneto coupling 电磁联轴器
magnetocoupling member 磁电机接合器
magnetocrank 磁电机摇把
magnetocrystalline 磁晶体
magnetocrystalline anisotropy 磁晶各向异性
magnetocrystalline anisotropy constant 磁晶各向异性常数
magnetocrystalline anisotropy energy 磁晶各向异性能
magnetodependent sensor 磁敏传感器
magnetodetector 磁(石)检波器
magnetodielectrics 磁介电体
magnetodiode 磁敏二极管
magnetodistributor 磁电机配电盘;磁电机配电器
magnetodrive 磁电机传动
magnetodrive gear 磁电机主动齿轮
magnetodynamics 磁动力学
magnetodynamics system 磁动力系统
magnetodynamo 永磁发电机;高压永磁(发)电机
magnetoelastic 磁致弹性的
magnetoelastic coupling 磁弹性耦合
magnetoelastic energy 磁致弹性能
magnetoelasticity 磁(致)弹性;磁弹效应
magnetoelasticity non-linearity 磁弹性非线性
magnetoelasticity resonance 磁弹性谐振
magnetoelectret 磁驻极体
magnetoelectric(al) 磁电的
magnetoelectric(al) coil 磁电线圈
magnetoelectric(al) collecting ring 磁电集电环
magnetoelectric(al) coupling 磁电耦合器
magnetoelectric(al) distributor 磁电配电器
magnetoelectric(al) drive 磁电驱动
magnetoelectric(al) flowmeter 磁电流量计
magnetoelectric(al) generator 永磁发电机;磁石发电机;磁电(发电)机
magnetoelectric(al) ignition 磁电(机)点火
magnetoelectric(al) machine 永磁电机;磁电式电机
magnetoelectric(al) relay 磁电式继电器
magnetoelectric(al) scope 磁电式示波器
magnetoelectric(al) tachometer 磁电式转速计
magnetoelectricity 电磁(学)
magnetoemission 磁致发射
magnetoencephalograph 磁脑电描记器
magnetoencephalography 磁脑电描记法
magnetoexchange 磁石交换机
magneto exploder 轻便电动爆炸装置
magnetofield scope 磁场示波器
magnetofluid 磁流体
magnetofluid dynamics 磁流体动力学
magnetofluid mechanics 磁流体力学
magnetogalvanic effect 磁场电效应
magnetogasdynamics 磁性气体动力学
magnetogasket 磁电机垫片
magnetogear spring 磁电机齿轮弹簧
magneto-generator 永磁发电机;手摇发电机;磁石发电机
magnetogram 磁强记录图;磁强度图;磁力图
magnetograph 地磁计;磁强记录仪;磁场记录仪
magnetographic(al) inspection 磁性图示探伤;磁性图示检验
magnetographic inspection 磁性检查
magnetography 磁摄影术
magnetogrease 磁电机润滑脂
magnetogyrocompass 磁陀螺;磁力回转罗盘
magnetoholographic memory type 磁全息照相存储方式
magnetohydrodynamic(al) arc jet 磁流体动力电弧火箭发动机
magnetohydrodynamic(al) conversion 磁流体转换
magnetohydrodynamic(al) cooling 磁流体动力冷却
magnetohydrodynamic(al) cycle 磁流体动力循环
magnetohydrodynamic(al) flow 磁流体流
magnetohydrodynamic(al) force 磁流体力
magnetohydrodynamic(al) generator 磁流体发电机;磁流体动力发电机
magnetohydrodynamic(al) lubrication 磁解体动压润滑
magnetohydrodynamic(al) plant 磁流体装置;磁流体动力装置
magnetohydrodynamic(al) (power) generation 磁流体发电
magnetohydrodynamic(al) pumping 磁流体动力抽运;磁流体动力泵浦
magnetohydrodynamic(al) shock wave 磁流体动力学冲击波;磁流体动力冲击波
magnetohydrodynamic(al) turbulence 磁致湍流
magnetohydrodynamic(al) wave 磁流体动力学波;磁流体动力波
magnetohydrodynamics 磁流体动力学;磁流体动力现象;磁场流体动力学
magnetoignition 磁电机点火;磁电点火
magnetoignition system 磁石发电机点火法;磁点火系统
magnetoil menite 磁钛铁矿
magnetoimpulse coupling 磁电机脉冲接头
magnetoinductor 永磁电机;磁石感应器;磁电机
magnetoionic 磁离子
magnetoionic duct 磁离子波导;磁电离波导
magnetoionic theory 磁离子理论
magnetoionic wave component 磁离子波分量
magnetomechanical 磁机械的
magnetomechanical effect 磁力学效应;磁机械效应
magnetomechanical factor 磁力因数
magnetomechanical tuning 磁力学调节
magnetomechanics 磁力学
magnetometer 地磁仪;磁强计;磁力仪;磁力计;测磁仪
magnetometer arrays 磁力仪组合观测
magnetometer detector 磁强检测器
magnetometer method 磁秤法
magnetometer model 磁力仪型号
magnetometer precision 磁力仪精度
magnetometer scale value 磁力仪格值
magnetometer sensitivity 磁力仪灵敏度
magnetometer sensor 磁力仪传感器
magnetometer survey 磁强计测量
magnetometer type 磁力仪类型
magnetometric(al) 测磁的
magnetometric(al) survey 地磁测量
magnetometric(al) titration 磁量滴定法
magnetometry 磁力测定(术);磁测定法;测磁学;测磁强术;测磁法
magneto microphone 电磁式送话器
magnetomotive 磁势;磁动势的
magnetomotive force 磁通势
magnetomotive potential 磁位;磁动势
magnetomotor 磁力发电机
magneton 磁子
magneton number 磁子数
magnetoohmmeter 永磁式欧姆表
magnetooperated oil valve 磁铁操纵的油阀
magnetooptic 磁光的
magneto-optic(al) birefringence 磁光致双折射
magneto-optic(al) bubble device 磁光泡器件
magneto-optic(al) camera 磁光照相机
magneto-optic(al) deflector 磁光偏转器
magneto-optic(al) devices 磁光器件
magneto-optic(al) disk 磁光盘
magneto-optic(al) disk editor 磁光盘编辑器
magneto-optic(al) display 磁光显示
magneto-optic(al) driver 磁光盘驱动器
magneto-optic(al) effect 磁光(力)效应
magneto-optic(al) glass 磁光玻璃
magneto-optic(al) isolator 磁光隔离器
magneto-optic(al) Kerr effect 磁光克尔效应
magneto-optic(al) laser 磁光激光器
magneto-optic(al) material 磁光(学)材料
magneto-optic(al) memory 磁光存储器
magneto-optic(al) memory material 磁光存储材料
magneto-optic(al) memory technique 磁光存储技术
magneto-optic(al) modulation 磁光调制
magneto-optic(al) modulator 磁光调制器
magneto-optic(al) phase modulation 磁光相位调制
magneto-optic(al) phenomenon 磁光现象
magneto-optic(al) playback system 磁光再现系统
magneto-optic(al) resonance 磁光谐振
magneto-optic(al) rotation 磁致旋光
magneto-optic(al) semiconductor laser 磁(光)半导体激光器
magneto-optic(al) semiconductor shutter 磁光半导体快门
magneto-optic(al) shutter 磁光快门;磁光开关
magneto-optic(al) thin film 磁光薄膜
magneto-optic(al) waveguide 磁光波导
magnetooptics 磁光学
magnet opening 磁铁孔
magnet operation 磁铁吸光金属异物术
magnetophone 磁铁膜录音器;磁石扩音器;磁电话筒
magnetophoto-conductivity 磁光导性
magnetophoto-elasticity 磁光弹性
magnetophoto-reflectivity 磁光反射系数
magnetopiezoresistance 磁致压电电阻
magnetoplasma 磁等离子体
magnetoplasmadynamic(al) generator 磁等离子体发电机
magnetoplasmadynamic(al) plant 磁等离子体动力装置
magnetoplasmadynamics 磁等离子体动力学
magnetoplumbite 磁(铁)铅矿
magnet-optic(al) memory system 磁光存储系统
magnetor 磁电机
magnetoresistance 磁阻;磁致电阻
magnetoresistance detector 磁(致)电阻检测器
magnetoresistance displacement transducer 磁致电阻位移换能器
magnetoresistance effect 磁(致)电阻效应
magnetoresistance microwave wattmeter 磁阻微波功率表
magnetoresistance oscillator 磁致电阻振荡器
magnetoresistive amplifier 磁致电阻放大器
magnetoresistive effect 磁阻效应
magnetoresistive head 磁阻磁头
magnetoresistive magnetometer 磁阻效应磁强计
magnetoresistivity 磁致电阻率
magnetoresistor 磁敏电阻器;磁控电阻器;磁电阻

magnetoringer 磁石式振铃器
magnetoscope 验磁器
magnetosensor 磁敏元件
magnetoshaft 磁电机轴
magnetosheath 磁鞘
magnetosiren 磁警报器
magnetosonic wave 磁声波
magnetospark advance 磁电火花提前
magnetospeed indicator 磁石测速计
magnetosphere 磁性层；磁圈
magnetospheric explorer 磁化层探查器
magnetospheric plasma 磁层等离子体
magnetospheric ring corrent 磁层环流
magnetospheric shockwave 磁层冲击波
magnetospheric structure 磁层结构
magnetospheric substorm 磁层亚暴
magnetostatic 静磁的
magnetostatic electron lens 静磁电子透镜
magnetostatic zone 静磁带
magnetostratigraphic polarity unit 磁性地层极性单位
magnetostratigraphy 磁性地层学
magnetostratigraphy method 磁性地层学方法
magnetostriction 磁致伸缩；磁弹性
magnetostriction alloy 磁致伸缩合金
magnetostriction apparatus 磁致伸缩仪器
magnetostriction coefficient 磁致伸缩系数
magnetostriction compass 磁致伸缩罗盘
magnetostriction cookie cutter 磁致伸缩曲奇切割器
magnetostriction cutter 磁致伸缩切割器
magnetostriction depth sounding apparatus 磁致伸缩测深器
magnetostriction drill 磁致伸缩钻(机)
magnetostriction echo sounder 磁致伸缩回声测深器；磁致伸缩测深仪
magnetostriction effect 磁致伸缩效应
magnetostriction filter 磁致伸缩滤波器
magnetostriction loudspeaker 磁致伸缩式扬声器
magnetostriction measurement 磁致伸缩测量
magnetostriction microphone 磁致伸缩传声器
magnetostriction oscillator 磁致伸缩振荡器
magnetostriction phenomenon 磁致伸缩现象
magnetostriction pressure ga(u)ge 磁致伸缩压力计
magnetostriction strain ga(u)ge 磁致伸缩应变仪
magnetostriction transducer 磁致伸缩换能器
magnetostriction transmitter 磁致伸缩发射机
magnetostriction type 磁致伸缩型；磁致伸缩式
magnetostrictive 磁致伸缩的
magnetostrictive acoustic delay line 磁致伸缩声延迟线
magnetostrictive coupling 磁致伸缩耦合
magnetostrictive cutting machine 磁致伸缩切割机
magnetostrictive delay 磁致伸缩延迟
magnetostrictive delay line 磁致伸缩延迟线
magnetostrictive delay line storage 磁致伸缩延迟存储器
magnetostrictive drill 磁致伸缩钻(机)
magnetostrictive drive 磁致伸缩驱动
magnetostrictive effect 磁致伸缩效应
magnetostrictive effect storage 磁致伸缩效应存储器
magnetostrictive hydrophone 磁致伸缩水听器
magnetostrictive loudspeaker 磁致伸缩扬声器
magnetostrictive microphone 磁致伸缩传声器
magnetostrictive oscillation 磁致伸缩振荡
magnetostrictive oscillator 磁致伸缩振荡器
magnetostrictive reaction 磁致伸缩逆效应
magnetostrictive receiver 磁致伸缩接收器
magnetostrictive relay 磁致伸缩式继电器
magnetostrictive resonator 磁致伸缩谐振器
magnetostrictive rod 磁致伸缩棒
magnetostrictive sensor 磁致伸缩传感器
magnetostrictive storage 磁致伸缩存储器
magnetostrictive storage unit 磁致伸缩存储单元
magnetostrictive stress 磁致引起应力；磁致伸缩应力
magnetostrictive transceiver 磁致伸缩式收发机
magnetostrictive transducer 磁致伸缩传感器
magnetostrictive vibration generator 磁致伸缩振动发生器
magnetostrictive vibrator 磁致伸缩振动器
magnetostrictor 磁致伸缩体；磁致伸缩器；磁致伸缩振子
magnetoswitch 磁电机开关
magnetoswitchboard 磁石式交换机
magnetoswitchboard exchange 磁石式交换台
magnetosynchronizer 磁石同步器
magnetosystem 磁石式制；磁电机系统
magnetosystem exchange 磁石式交换机
magnetotail 磁层尾
magnetotelluric method 大地电磁法；大地大地电流法
magnetotelluric sounding 大地电磁测探
magnetotelluric sounding instrument 大地电磁测深仪
magnetothermal effect 磁热效应
magnetothermodynamics 磁性热力学
magnetothermoelectric 磁热电
magnetothermography 磁热摄影术
magnetoturbulence 磁致湍流；磁性湍流
magnetotype ball bearing 磁式滚珠轴承；磁电机用滚珠轴承
magnetoviscous 磁黏性
magneto voltage regulator 感应式电压调整器
magneto winding 永磁绕组
magnetowire 磁电机点火线；磁电机导线
magnet plunger 电磁阀柱塞
magnet pole face 磁极面
magnet powder 磁铁粉
magnet power 磁体功率
magnet-probe extensometer 磁性探头伸缩计
magnet profile 磁铁剖面
magnet pulley 磁性轮
magnet recording medium 磁性载声体
magnet reed switch 磁性舌簧开关
magnet resistor 磁阻器
magnet road sweeper 磁力扫路机
magnetrol 磁放大器
magnetrom pushing 磁控管频率推移
magnetron 电子注开关管；磁控管
magnetron amplifier 磁控型放大管；磁控管放大器
magnetron beam switching tube 磁控注型开关管；磁控电子束开关管
magnetron cavity 磁控管腔谐振器
magnetron frequency 磁控管频率
magnetron heater 磁控管加热器
magnetron mode 磁控管振荡模式
magnetron modulator 磁调制器
magnetron optics 磁控型电子光学系统
magnetron oscillator 磁控管振荡器
magnetron power source 磁控管电源
magnetron pulling 磁控管频率牵引
magnetron seasoning instrument 磁控管老化仪
magnetron sputtering 磁控溅射
magnetron type traveling wave tube 磁控型行波管
magnetron vacuum ga(u)ge 磁控管真空计
magnetropism 向磁性
magnets bell 交流电铃
magnet screen filter 磁筛过滤器
magnet sensing diode 磁敏二极管
magnet separator 磁选机；磁力分离机
magnet set 磁铁组
magnet slab turning crane 磁力翻板坯吊车
magnet spectrograph 磁谱仪
magnet spool 电磁铁线圈
magnet stand 磁性表架
magnet steel 磁钢
magnet steel sheet 矽钢片
magnet stopper 电磁制动器
magnet store 磁存储器
magnet switch 电磁开关
magnettor 谐波型磁放大器；磁调制器
magnet-type uncoiler 电磁直头式开卷机
magnet valve 电磁阀
magnet-wheel 磁轮(选矿用)
magnet winding 永磁绕组；磁铁线圈
magnet wire 磁(性)钢丝；磁(导)线
magnet yoke 磁偏转系统
magniferous 含镁的
magnification 增益；增大；放大比率；放大；倍率
magnification changer 放大倍数变换器
magnification control 放大控制
magnification curve 放大曲线
magnification degree 放大度
magnification eyepiece 放大目镜
magnification factor 放大因素；放大因数；放大系数；放大倍数；倍率系数
magnification factor for amplitude 振幅放大因子
magnification formula 放大率公式
magnification for rapid ground movements 速震动放大
magnification for rapid waves 快速波放大；速波放大
magnification of eyepiece 目镜放大率
magnification range 放大幅度
magnification ratio 扩大率；伸缩比；放大率；放大比
magnified image 放大影像
magnified scale 放大比例(尺)
magnified sweep 放大扫描
magnifier 扩展镜；简单显微镜；放大器；放大镜
magnifier for reading 读数放大镜
magnifier power supply 放大器电源
magnifying coefficient of eccentricity 偏心距增大系数
magnifying glass 简单显微镜；放大镜
magnifying lens 放大透镜
magnifying measuring scale 带有读数放大镜刻度尺
magnifying mirror with ceramic foot 瓷座放大镜
magnifying optics 放大光学系统
magnifying power 增大；扩大能力；放大能力；放大率；放大倍率
magnifying power of eye-piece 目镜放大率
magnifying viewer 放大观察仪
magniotriplite 氟磷铁镁矿
magni-scale 放大比例尺
magnistor 磁开关；磁变管
magnitude 量值；量级；等级；大小
magnitude anomaly 震级异常
magnitude chart 震级图
magnitude comparator 量值比较器
magnitude contour 等幅线；等值线；等高线
magnitude-dependent peak acceleration 依赖于震级的峰值加速度
magnitude determination 震级测定
magnitude distribution 震级分布
magnitude estimate 震级估计
magnitude estimation 震级估算
magnitude for local shock 近震震级；地方震震级
magnitude formula 震级公式
magnitude frequency law 震级频度法则
magnitude frequency relation 震级频度关系
magnitude intensity correlation 震级重现曲线
magnitude intensity relation 震级裂度关系
magnitude interval 震级区间
magnitude method 偏差法试射
magnitude of a complex number 复数的绝对值
magnitude of aggradation 淤积程度
magnitude of a vector 向量的绝对值
magnitude of closed loop resonant peak 闭环共振峰值
magnitude of creep 徐变量
magnitude of crustal stress 地应力大小
magnitude of current 电流量
magnitude of degradation 刷深程度；冲刷程度
magnitude of earthquake 地震震级
magnitude of enlargement 放大倍数
magnitude of imposed force 外力大小
magnitude of interference 过盈量
magnitude of light 光量
magnitude of load 载重量；荷载大小；负荷量
magnitude of loading 负荷量
magnitude of maximum shearing stress 最大剪(切)应力值
magnitude of motion 运动量
magnitude of near earthquake 近震震级
magnitude of operation 业务工作量
magnitude of physical quantity 物理量值
magnitude of power 功率值
magnitude of pump discharge 泵排水量
magnitude of relative value 相对价值量
magnitude of relocation 迁建规模；搬迁工作量
magnitude of scour and fill 冲淤程度
magnitude of self-stress 自应力
magnitude of source force 发震力大小
magnitude of strain 应变量
magnitude of stresses 单位荷载大小；应力值；应力大小
magnitude of traffic flow 交通流(量)；车流量
magnitude of use-value 使用价值量

magnitude of voltage 电压值
magnitude only arithmetic operation 绝对值算术操作
magnitude ratio 星等比;光比
magnitude recurrence curve 震级重现曲线
magnitude residual 震级残差
magnitude scale (of earthquake) 震级标度
magnitude stability 震级稳定性
magnitude statistics 震级统计
magnitude system 星等系统
magnitude threshold 震级限值
magnitude time curve 震级时间曲线
magnochromite 镁铬铁矿
magnoferrichromite 镁铁铬矿
magnoferrite 镁铁矿
magnoferrogahnite 镁铁锌类晶石
magnofranklinite 镁铁锌尖晶石
magnojacobsite 镁锰铁尖晶石
magnolia 木兰属植物
magnolite 碲汞石
magnon 磁振子;磁量子
magnon emission 磁振子发射
magnoniobite 铌镁矿
magnophorite 镁红钠闪石
magnophyric 粗斑状
Magnorite 硅镁耐火砖
Magnox 镁诺克斯合金
magnum 大酒瓶
Magnuminium 镁基合金
Magnuminium alloy 锰铝镁合金;镁基合金
Magnus' bath 马格纳斯浴槽
magnus hitch 鲁班结;天幕结
Magnus number 马格纳斯数
magnussonite 氯砷锰矿;方砷锰矿
Magog belt 马戈格带
magslep 无触点式自整角机;无触点式自动同步机
magslip 同步机;旋转变压器;遥控自动同步机
magslip resolver 无触点同步机解算装置
Maguel 高强度钢丝的张拉锚固法
Maguel post-tensioning 玛格尔(式)后张法
magurasphyllite 镁铀矿
magursilite 镁硅铀矿
Mahabon teak 婆罗洲樟木
mahapengiri oil 香茅油
Mahler alloy 马勒铝硅(活塞)合金
Mahler calorimeter 马勒量热器
mahlstick 搁手用的手杖;支腕杖
mahoganize 将木料油漆成光花心木状;模拟桃花心木状
mahogany 红木;桃花心木;赤褐色
mahogany acid 磺酸;石油磺酸
mahogany birch 山桦
mahogany brown 赤褐色;红木褐色;桃花心木棕;煅黄土颜料
mahogany chair 红木椅
mahogany colo(u)r 红木色料
mahogany-faced 红木贴面的;红木铺面
mahogany flush 红木色潮红
mahogany furniture 红木家具
mahogany lined 红木装饰;红木衬里的
mahogany ore 密铜铁矿
mahogany petroleum sulfonate 石油磺酸盐
mahogany soap 磺酸(钠)皂;石油磺酸皂
mahogany sulfonate 磺酸盐;石油磺酸盐
Mahometan architecture 穆罕默德建筑
maiden field 未开采矿区;未采矿区
maidenhair tree 鸭脚树;公孙树;白果树;银杏(树)
maiden trial 初次试验
maiden trip 处女航;初航
maiden voyage 首(次)航(行);第一航次;初航
maid room 保姆室
maid's changing room 女子更衣室
maifanshite 麦饭石
Maihak strain ga(u)ge 马依哈克声应变计
mailbag 邮袋
mailbag duck 邮袋帆布
mail boat 小型油船;邮政便船;邮船
mail bomb 邮件炸弹
mailbox 信箱区;信箱;信筒;邮箱;邮筒
mailbox system 电子箱系统
mail bus 邮政车
mail carrier 邮船
mail car(t) 邮车;邮政(汽)车
mail catcher 邮包装卸机

mail chute 信件滑槽;滑槽邮筒
mail clerk 邮局办事员
mail cover 邮封
mail cutter 邮政快艇
mail day 邮件截止日
mail declaration 邮件声报单
mail deposit 通信存款
mail drop 邮筒
mailer method 堆积烧炭法
mail exploder 函件分发器
mail-facsimile apparatus 信函传真机
mail filter 邮件过滤器
mail flag 邮件旗
mail gateway 邮件网关
mailgram 邮递电报
mail handling pigment 处理邮件颜料
mailing facility 邮寄设备
mailing list 邮件列表;发送文件清单;发送名单;发函清单
mailing machine 邮件收发机
mail inquiry 通信调查
Maillechort 铜镍锌合金
mail liner 邮船;定期邮船
mail man 邮递员
mail motor truck 邮政汽车
mail order 邮购;函购
mail-order house 邮寄订购商店;邮购商店
mail packet 定期邮船
mail pennant 邮件旗
mail plane 邮航机
mail railcar 邮政轨道车
mail reimbursement 信函求偿
mail remittance 信汇
mail room 邮件室
mail route 邮路
mail route map 邮路地图
mail server 邮件服务器
mail service 邮政业务;邮递业务
mail ship 邮船
mail signal 邮件旗
mail slot 信槽;投信口
mail sorting room 邮件分拣室
mail steamer 邮轮
mail survey 通信调查
mail tariff 邮运运价
mail transfer 信汇;邮汇
mail transfer advice 信汇委托书
mail truck 邮政车
mail van 邮政车辆
mail wagon 厢式邮政车
maim 残废
maimed person 残废人
main 总水管;主要应力;主要的;主输电线;总管(气、水)等
main accelerator 主加速器
main access 出入口
main access road 主要专用线;主要便道
main account 主要账户
main aerodynamic(al) balance 主气体动力平衡
main-aftershock earthquake 主余震型地震
main agent 主要代理商
main air 主风
main air blower 主风机
main air cleaning system 主净化系统
main air duct 主风道
main air filter 主空气过滤器
main air intake 主进气道
main airport 主风口;主要机场
main air reservoir 主储气筒
main air reservoir pipe 总风缸管
main air supply pipe 总风管
main air valve 主风阀
main airway 主风巷;回风巷
main aisle 主走廊
main aisle exposed tiebeam 挑尖梁
main alarm 主报警器;主报警(信号器)
main alloying constituent 主要合金成分
main altar 主要祭坛;主要梯阶
main amplifier 主放大器
main amplifier circuit 主放大器电路
main anchor 主固定支架
main anchor wall 主锚碇墙
main and auxiliaries 主机与辅机
main and auxiliary lobes restraining rate 主副瓣抑制比
main and retrun cam 主回凸轮;强力回程凸轮
main-and-tail haulage 头尾绳运输
main-and-tail system 头尾绳运输系统
main angle 主对角平巷
main anode 主阳极
main application 主要用途
main arch 主拱;大毡
main arch ring 主拱圈
main archway 主拱道
main arcuate tectonic belt 弧形构造带
main arithmetic processor 主运算处理机
main arm 主臂
main armo(u)r 主护面块
main artery 主要干线
main assemblage of spring 弹簧总成
main assembly 总体组装
main assembly hall 大会堂;主会议厅
main assembly line 总装配线;主装配线
main avenue 主要路线
main axis 主轴线;主轴
main axis of building square grid 建筑方格网主轴线
main axis of inertia 主惯性轴
main axis of stress 主应力轴
main azimuth plate 方位盘
main ballast Kingston(valve) 主压载水通海阀
main ballast tank 主压载(水)舱
main bang 主脉冲信号;探索脉冲
main bang suppression 主脉冲信号抑制;直接波抑制
main bar 主筋;主(干)钢筋;受力筋
main barometric(al) level(l)ing 干长气压高程测量
main barrel 主机筒
main base 主基点;控制点;地脚底;底板
main base station 主要基本站
main basic indicator 主要基本指标
main basin 主港池
main basis 主要依据
main battery 主电池
main battery circuit 主电池电路
main beam 最大宽度处横梁;主(射)束;主(横)梁;主波束;船舶最大宽度
main beam efficiency 主波束效率
main beam longitudinal 短梁;舱口纵桁
main beam of a truss frame 大梁
main beam transverse 短横梁
main bearing 主轴承
main bearing babbitting jig 主轴承浇巴氏合金夹具
main bearing boring bar 主轴承镗杆
main bearing boring tool 主轴承镗刀
main bearing cap 主轴承盖
main bearing cap dowel 主轴承盖销钉
main bearing frame 主承载框架
main bearing journal 主轴承轴颈;曲轴主轴颈
main bearing knock 主轴承敲击声
main bearing lining 主轴瓦耐磨合金层
main bearing lubrication 主轴承润滑
main bearing oil pipe 主轴承油管
main bearing oil seal 主轴承油封
main bearing reamer 主轴承铰刀
main bearing reboring device 主轴承精镗削工具
main bearing removal 主轴承拆卸
main bearing shell 主轴承壳
main bearing structure 主要承重结构
main bearing stud 主轴承柱螺栓;主轴承柱
main bearing wrench 主轴承扳手
main bed 主河道;主河床;主河槽
main bed of river 干河床
main bed of stream 主流河槽
main belt 主传送带
main belt conveyer 干线带式运输机
main bilge line (pipe) 舱底污水主管
main blasting lead 主干导爆线
main blast line 总风管
main block 主街坊;主要部分;主体;主街
main blower 主鼓风机
main board 主板
main body 主要部分;主体;正文;机身;筒体;船体;本体
main body bleeder 主体排气孔
main body building 主体建筑
main body cover 主体外壳
main body entrance 主要进口

main body entrance door 主车门
main body frame 主体框架
main body of bridge abutment 桥台主身
main body of oil 油的主要成分
main body of road 路基
main boiler 主锅炉
main boiler survey 主锅炉检测
main bonding jumper 接地母线
main bottom 基座;基岩
main boundary thrust 主边界(逆冲)断层
main box 主轴承(轴)瓦
main brace 主斜撑;主撑
main brake contact 主制动触点
main brake pipe 制动主管
main brake pull-rod attached to the cylinder lever 连接制动缸杠杆的主制动拉杆
main brake shoe 基本铁鞋
main brake solenoid 主制动螺线管
main braking cylinder 主制动缸;制动总泵缸
main braking reservoir 主制动储气缸
main branch connections 主要分支线
main branch(line) 主支线;主矿脉;(河道的)主支流;主分支;主汊(道)
main branch road 主要支路
main brasses 主轴承轴瓦
main breadth line 船舶最大宽度线
main breaker 主开关
main breaker test bench 主断路器试验台
main breakwater 主防波堤
main bridge 主桥
main bronchus 主支气管
main brush 主电刷
main bucket ladder 主斗桥
main budget 预算本体;总预算;主要预算
main building 正屋;主要建筑(物);主楼;主建筑;主厂房
main building bay 主要建筑跨度
main building of cold store 冷库主体建筑
main building vault 主要建筑拱顶
main bulk 总量;主体
main bulkhead 主要间壁;主舱壁
main bulk of fuel 燃料的主要部分
main bulwark rail 主甲板舷墙栏杆
main burner 主燃烧器;主(燃料)喷嘴
main bus-bar 主母线
main bush 主轴衬;底衬环
main cable 总输送电缆;主索;主(电)缆;载重索;干线电缆;承载索
main cable channel 总电缆沟
main cable trench 主电缆沟
main calculation 主计算
main cam 主凸轮
main camshaft 主凸轮轴
main canal 总渠;主水道;干渠
main cannel 主渠
main canopy 主座舱罩;主座舱盖
main carburetor 主化油器
main cargo line 货油系统总管(油轮上)
main carriage 纵刀架
main carriageway 主要行车道;主要车(行)道
main carrier 主载波;基本载波
main carrier cable 主承力索
main case 主机壳
main casing 主壳
main casting 主要铸件
main catenary wire 主承力索【电】
main cat(head)shaft 绞线主轴
main cell 主格仑;主电池
main center 主中心局
main centerline 主中心轴线
main center thrust 中央大断层
main chain 主锚链;主链条;主链索;主链节;主键
main chain liquid crystal polymer 主链液晶聚合物
main chamber 主室;主燃烧室;主舱
main channel 主用波道;主信道;主通路;主通道;主水道;主频道;主槽;主航道;干渠
main channel flow 主槽流量
main channel sensitivity of oriented sampling 定向取样主道灵敏度
main chapel 大礼拜堂
main characteristic 主要特征
main charge 主要药包
main charging belt 主装料运输带
main check-valve 主止回阀

main chimney damper 总烟道闸板
main chimney flue 总烟道
main choir 主唱诗班
main circuit 主油路;主结线;主回路;主环路;主管线;主干线路;主电路;干线;干路
main circuit breaker 主(电路)断路器
main circuit switch 主电路开关
main circulating[circulation]pump 主循环泵
main classes 主类
main classification yard 主要编组场【铁】
main clause 主要条款
main clock 主(时)钟;母钟
main closed loop 主闭环
main clutch 主离合器
main clutch magnet 主离合磁铁
main clutch shifting lever bracket 主离合器拨叉支架
main cock 总旋塞;总(水)龙头;总开关;主旋塞;主龙头;主开关
main coil 主线圈
main coil current 主线圈电流
main coking coal 主焦煤
main collector 集水干管;集水总管;总干管
main colo(u)r 主色
main column 主要支墩;主塔;主力纵队
main column foundation 主柱基础
main combination of load 荷载的主要组合
main combustion chamber 主燃烧室
main command 主命令
main commercial center 主要商业中心
main commodities 主要矿产
main communication center 主通信中心
main comparator 主比较器
main compartment line 主林班线
main compass 主罗盘;主罗经
main compression coil 主压缩线圈
main compression curve 压缩主支曲线
main compression stress 总受压应力
main compressor 主压缩机
main condensate pump 主凝水泵;主凝结泵
main condenser 总冷凝器;主冷凝器
main condition 主要条件;主要情况
main conduit 总管(气、水等);综合管道;综合地下管道;主涵洞
main cone 主锥
main cone driving wheel 主锥形驱动轮
main connecting rod 主连杆
main connection 主连接;干线连接
main console 主控制台
main-consolidation 主固结
main contact 主触头;主触点
main contact group 主触点组
main contactor 主接触器
main contact spring 主触簧片
main content 主要内容
main content of contract 合同的主要内容
main contract 总(承)包合同;主合同
main contract clauses 合同主要条款
main contractor 总承包者;总承包商;总承包人;总包工;总包单位
main control 主控
main control board 总配电盘;主控制盘;主控盘;主操纵台
main control center 主控制中心
main control chain 主控制链
main control console 主调整台;主操纵台
main control drum 主控制滚筒
main control gate 主控闸门
main controller 主控制器;主调节器
main control lever 主操纵手柄
main control lever interlock release 主操纵手柄联锁释放装置
main controlling board 主操作盘
main controlling element 主控制元件
main control loop 主控制回路
main control magnet 主控制磁铁
main control panel 主控板
main control register 主控寄存器
main control room 总控制室;主控制室;主操纵室
main control station 主控(制)站
main control unit 主控制装置;主控(制)盘;主控制部件
main control valve 主配压阀
main conveyer 主输送机

main cooling water pump 总冷却水泵
main core 钢丝绳芯麻绳
main couple 主屋架;主桁架
main coupling 主联轴器
main course 主梳;主航道;主帆;干道;大横帆
main-course ballast 主层石料
main crab 主小车;主起重绞车
main crank 主曲柄
main crank case 主曲柄箱
main crank pin 主曲柄销
main crater 主火山口
main crop 主要作物;主(伐)林木
main cross girder 主横梁
main crosshead 主连杆十字头
main crossing 主要十字路口;主要交叉口;主交叉
main cross section 主截面
main currency 主要货币
main current 主水流;主气流;主(电)流;主电波
main current direction 主流流向
main current line 主流线
main current relay 主电流继电器
main current terminal switch 主电流终端开关
main current trend line 主流倾向线
main curvature 总曲率
main curvature line 总曲率线
main cushion duct 主气垫管道
main cut-off 总熔断器
main cutting edge 主切削刃
main cutting force 主切削分力
main cycle 主循环
main cylinder gasket 主缸垫密片
main dam 主堤
main data 主(要)数据
main data area 主数据区
main deck 主甲板
main deck beam 主甲板中横梁
main deck stringer 主甲板缘
main defecation 主澄清
main deflection 主偏转
main delay line 主延迟线
main delivery pipe 主压力管道;主输送管道;主润滑油管
main delta arm 三角洲主支股;三角洲主汊(河)
main dewatering pump 主排水泵
main diagonal 主斜杆;主斜撑;主对角线
main diagonal rib 主斜肋
main dike 主堤;干堤;大堤
main dimension 主要尺寸
main dimensional of vessel 船舶主要尺度
main dimensions of ship 船舶主要尺度
main direction of wind 主要风向
main discharge 主排出管
main discharge current 主放电电流
main discharge jet 主喷嘴
main discharge jet nut 主喷嘴螺帽
main discharge jet plug 主喷口塞
main discharge pipe 总排水管
main discharge tube 主量孔喷管
main discharge valve 总放水阀
main dispatching center 主调度中心
main dispatch loop 主循环发送
main display console 主显示台
main distillate fraction 主馏分
main distributing board 主配电盘
main distributing center 主配电站
main distributing conveyer 主分配运输机
main distributing frame 总配线架;主配线架
main distributing point 主配电点
main distributing valve 主分配阀
main distribution 主配水器
main distribution board 总配电盘;总配电板;主配电板
main distribution cable 总配电电缆
main distribution frame 主配线架;主配线板
main distribution pipe 总配水管
main distributor 主配水器
main district of supply 主供水区
main ditch 干沟;主排水沟
main divide 主(要)分水岭;主分水线;主分水界;主分界线
main documents of surveying 测绘生产依据的主要文件
main door 正门
main draft 主牵伸

main drag conveyer 主牵引运输机
main drain 干线沟渠;排水干渠;泄水主管;雨水干管;总排水管;主排水管;主排水沟;排水总管;排水干沟
main drainage ditch 总排水沟
main drain system 主泄水系统
main drain trap 总存水弯
main drift 主平巷
main drill frame 汽车式钻机架
main drive 主传动机构;主平巷;主传动装置;主传动设备;主传动
main drive control 主传动操纵手柄
main drive gear 主传动机构;主(传)动齿轮
main drive gear bearing 主(传)动齿轮轴承
main drive gear bearing cap 主动齿轮轴承盖
main drive gear bearing oil slinger 主动齿轮轴承抛油环
main drive gear bearing retainer 主动齿轮轴承护圈
main drive gear wheels 主传动机构齿轮
main drive motor 主(传)动电动机
main drive pinion 主(传)动小齿轮
main driver 主驱动装置
main drive shaft 主动轴
main drive shaft bearing sleeve 主动轴轴承套
main driving axle 主传动轴
main driving screw 主螺杆
main driving shaft 主动轴
main driving wheel 主(传)动轮;大牙轮
main drum 主卷筒
main duct 总风道;主管(指供气供水管);主风道
main earthquake 主震
main earth station 主地球站
main economic indicators 主要经济指标
main effect 主要效应;主效果
main electric(al) circuit diagram 电气主线图
main electric(al) connection diagram 电气主结线图
main electric(al) power supply fuse 总电源熔断丝
main electrode 主电极
main element 主振子
main emergency feeder 主应急馈电线
main end 主要掘进工作面
main energy 主要能源
main engine 主发动机;本务机车【铁】
main engine abnormal 主机失常
main engine control center 主机中央操纵台
main engine control desk 主机控制台
main engine control program(me) 主机操纵程序
main engine maneuvering stand 主机操纵台
main engine operation console 主机操纵台
main engine overhauling crane 主机解体检修用吊车
main engine portion of bollard test 主机系泊试验
main engine portion of trial test 主机航行试验
main engine power 主机功率
main engine remote control system 主机遥控系统
main engine revolution indicator 主机转速表;主机转数表
main entrance 总入口;住宅正大门;主(要)入口;主门口【港】;主进厂通道;正门
main entrance signal 主无线电指标信号
main entry 主要煤巷
main entry address 主入口地址
main entry door 主入口
main entry pillar 主平巷矿柱
main entry point 主入口点
main environments 主要的外界条件
main equipment 主要设备
main exchange 交换总机;电话总局
main exciter 主励磁机
main exhaust 主排气
main exhaust fan 主排风机
main exposure 主曝光
main facade 主要建筑立面;主立面
main facility 主要设备
main failure 电源故障
main fairway 主要通道(油、气等);主航道
main fan 主(扇)风机;主减速机
main farm road 机耕路
main fault 主断层
main fault belt 主断层带
main fault belt zone 主断层带
main feature 主景;主要特点

main feed 主馈线;主馈
main feed check(valve) 主给水止回阀
main feeder 总馈(电)线;主馈电线
main feeding ditch 主灌溉渠;主引水沟渠
main feed line 主给水管路
main feed pipe 主供风管;主给水泵
main feed pump 主给水泵
main feed water pump 主给水泵
main field 基本地磁场
main field focusing 主场聚焦
main field winding 主磁场绕组
main file 主文件;主存储
main filter 主滤器
main fixed station 主固定站
main flood 主涨流;主潮
main floor 主要楼层;主肋板【船】
main floor beam 地板的主梁
main floor unit 主底板总成
main flow 主流
main flow during the peak period 高峰主流向
main flow orifice plate 主流量喷嘴
main flue 总烟道;主烟道;主烟囱
main flue damper 总烟道闸板
main focusing lens 主聚焦透镜
main fold 主褶皱
main forest-road 主林道
main forge shop 总锻工车间
main forward path 主正向通路
main fracture 主破裂
main fram 主林班线
main frame 主车架;大梁;主梁;主配线架;主肋骨;主框架;主(机)架;主机;主车架;中央处理(装置);骨;机架;底盘;低盘;大型机;承力隔框;总配线架;总装配架
main frame computer 主(计算)机;主体计算机
main frame engine 主机
main frame maker 主机制造者
main frame memory 主机存储器
main frame program(me) 主机程序
main frequency 主要频率;电源频率
main frequency furnace 工频炉;工频感应炉
main fuel 主要燃料
main fuel control 主燃料调节器
main fuel filter 主滤油器
main fuel pump 主燃油泵
main fuel spray nozzle 主燃料喷嘴;主喷油嘴
main fuel tank 主油箱
main fuel valve 主燃油阀
main funding bank 主办银行
main furnace building 炼钢炉主厂房
main fuse 总熔丝;主熔断器;主熔断丝
main gallery 主矿脉;(地下工程的)主巷道;主平巷;主廊道;主走廊
main gallery pipe 主轴承润滑油管
main gangway 主运输平巷
main gantry 主吊架
main gap 主间隙
main garden 主屋前庭园
main gas container 主储气器
main gas pipe 主煤气管
main gate 主(要)平巷;主进风道;主气闸;主波门;工作闸门
main gate generator 主门发生器
main gate valve 主配压阀
main gateway 主平巷
main gating pulse 主选通脉冲
main gear 主传动(装置);大齿轮
main gear actuating cylinder 主轮收放作动筒
main gear box 主变速机;主齿轮箱;主变速器
main gear wheel 主大齿轮
main generating set 主发电机组
main generating station 主发电站
main generator 主发电机
main generator breaker 发电机主断路器
main generator hall 主厂房
main generator room 主发电机室;主厂房
main girder 主(大)梁
main gliding surface 主滑面
main glow peak 主发光峰
main governor valve 主调节阀
main grid 主栅
main grizzle brick 白垩黏土软性砖
main ground water 主地下水
main ground water table 主地下水位

main group element 主族元素
main guide 主导轨
main hall 正厅;正殿
main harbo(u)r 主要港湾;主要港口
main harmonic 主谐波
main hatch(way) 主舱口;重点舱口
main haulage 主运输大巷;主要运输
main haulage conveyer 主(要)运输机
main haulage cross-cut 主要(运输)石门
main haulage horizon 主要运输水平
main haulage level 主要运输水平
main haulage road 主要(运输)平巷
main haulage track 主运输道
main haulage tunnel 主要(运输)平硐
main haulageway 主运输道
main header 总集气管
main header casing 顶部罩
main heading 主平巷
main heater 主加热器
main highway 干路;主要公路;公路干线
main highway mixer 混凝土铺路机
main hoist 主起升(机构);主卷扬;主升降装置;主升降机构;主井提升机;主吊钩
main hoisting mechanism 主提升机构
main hold 主(货)舱;重点货舱
main hollow spindle 主空心轴
main hook 主钩
main house 主房;主厂房;(农村房屋的)正房
main hydraulic power plant 主液压系统泵装置
main hydraulic pump 主液压泵
main hydraulic system 主液压系统
main hydrolysis 主水解
main image 主像
main impregnation 主浸
main incoming switch 进线总闸
main index 主要指标
main industry 主要工业
main injection valve 主喷射阀;通海主吸入阀
main inlet control valve 主调节阀
main inlet throttle-stop valve 主气门
main input 主输入
main input into agriculture 农业主要投资
main inspection and repair shed 主检修厂房
main inspection chamber 主检验井
main installation 主设备
main instruction buffer 主指令缓冲器
main insulation 主绝缘
main intake 总入风(道);主进风道
main intermediate frequency amplifier 主中频放大器
main internal memory 主内存储器
main investigation 详勘;主要勘探工作
main irrigation canal 主灌溉渠;灌溉总渠;灌溉主渠;灌溉干渠
main item 主要项目
main jack 盾构千斤顶
main jet 主量孔
main jet spray hole 主量孔喷油孔
main jet tube 主量孔喷油管
main jib 主臂
main job 主要工作;主要工程;主体工程
main joint 主节理;主接头
main joist 主龙骨;主格栅;大龙骨
main joist of suspended ceiling 吊顶主龙骨
main journal 主轴颈
main junction box 主配电箱
main keel 主龙骨
main keelson 中内龙骨
main kiln room 炉子工作区
main knitting station 主成圈区
main ladder (挖土机的)主梯架;主斗桥
mainland 大陆(指本土)
mainland climate 大陆(性)气候
mainland ice 大陆冰盖;大陆冰川
main landing runway 主着陆跑道
main laser 主激光器
main lattice line 基点阵线
main layer of road 路面主层
main laying 干管敷设
main laying cost 管道装设费用
main lead 主引出线;主入口线
main leaf 主片
main leaf of spring 钢板弹簧主(叶)片
main lens 主透镜

main levee 主堤；干堤；大堤
main level 主要水平；主水准器
main level plan 主平巷平面图
main lever 主杠杆
main lifting elevator 主升运器
main lifting pontoon 主抬升浮筒
main light 主灯
main lighting 主(要)照明
main limit switch 主限制开关
main line 总路线；总管道；主(干)线；正线【铁】；高速干道主线；干线；干管；总管(气、水等)
main-line block 牵引索滑车；主索滑车
main-line channel 干线信道
main-line coach 干线客车
main-line code 正线电码化
main-line control 高速干道主线限流；干线控制
main-line driver 干线司机【铁】；干线驾驶员
main-line for passenger train 客运主要线路
main-line load 干线负载
main-line locomotive 干线机车
main-line meter 煤气总表；干线水表
main-line metering 高速干道主线控制
main-line network 干线铁路网
main line of communication 交通干线
main line of sight 主视线
main line of turnout 道岔主线
main-line pipe 干线管；干管
main-line power supply 干线供电
main-line railroad 干线铁路
main-line railway 干线铁路(车站)
main-line railway electrification 干线铁路电气化
main-line section 干线部分
main lines of communication 交通干线
main-line switch 正线道岔
main-line train 干线列车
main living chamber 主舱
main load 主要荷载；主力
main load-bearing structure 主要荷载承重结构
main load-bearing system 主载系统；主承重系统
main load-carrying structure 主要承载结构；主要承重结构
main loaded structure 主要受载结构
main loading case 主要受载情况
main loading level 主要装载水平
main loading or discharging line 主油管
main lobe 主波瓣；主瓣
main-lobe solid angle 主瓣立体角
main locking force 主锁闭力
main longitudinal force 总纵向力
main longitudinal girder 主纵梁
main loss 主要损失
main losses 总管损失
main lubricating oil pump 主润滑油泵
main machine hall 主厂房
main machinery 主要机具
main machinery console 主机操纵台
main machinery controlling mechanism 主机控制机构
main magnetic field 主磁场
main magnetic flux 主磁通
main magnet ring 主磁铁环
main mall 主要林荫道
main manifold 主歧管；主导管
main manufacturing plant 主要制造设备
main map 主图；基本地图
main mark 主要标记；主标志
main marshalling station 主要编组站【铁】
main marshalling yard 主要编组场【铁】
main mast 主桅杆；主桅
main mast fore stay 主桅前支索
main material cost 主要原材料成本
main materials 主要材料
main maximum 主最大(值)
main meal products 主食品
main means of transport 主要运输手段；主手段
main measuring point 主测(量)点
main member 主(要)构件；主要杆件
main memory 主存储器
main memory address 主存地址
main memory priority 主存储器优先数
main memory protection 主存储器保护
main menu 主菜单
Main Meteorologic(al) Office 气象总局
main meter 计量表；主要测量仪器

main metering jet 主量孔；主测油孔
main mill 主轧机
main mill drive motor 轧机主传动电机
main mode 主模
main moment 主力矩
main moment of inertia 主惯性矩
main monitor rack 主监视台
main mooring line 主系索
main motion 主运动
main motor 主电动机
main motor contactor 主电动接触器
main motor storage rack 主电动机置台
main movement 主运动
main muffler 主消声器
main navigable channel 主航道
main navigable watercourse 可通航的主航道
main navigable waterway 主要水道
main navigation channel 主航道
main network 主网路；主网(络)
main nozzle retainer spring 主喷嘴扣紧簧
main number 主要数字
main observation hole 主观测孔
main office 总行；总局；总公司；总办事处；大会办事处
main office expense 主要行政管理费用
main oil conduit 主机油道
main oil distributing passage 主油道
main oil gallery 主油道
main oil pipe 主油管；主润滑油管
main oil pump 主油泵
main oil reservoir 主储油器
main oil tank 主油箱
main opening 中央跨(孔)；主跨(孔)；主要巷道
main operation 主操作
main operation control program(me) 主操作控制程序
main operation house-keeping routine 主操作内务程序
main orbit 主轨道
main orebody 主矿体
main or first infection 主要的或第一次侵染
main oscillation 主振荡
main oscillator 主振荡器
main outfall 主要排泄口
main outlet 主排水口；主出水口
main outlet channel 主排水渠道；主出水槽
main outlet line 主引出线
main outlet of sewer 下水道出水总管
main panel 主控制屏
main parameter 主要参数
main part 主要部分；主体
main partition 主划分；主分区
main parts 主要零(部)件；主要部件
main path 主要途径；主通路
main peak 主峰
main peak maximum 主峰最大值
main pedestal 总轴架
main pedestrian zone 主行人区
main performance 主要性能
main performance index 主要性能指标
main phase 主相
main phase ring current 主相环电流
main picture monitor 主图像监视器
main pier 主墩；主支柱；主桥墩
main pile 主桩
main pin 主销；中心销
main pin brass 主销铜衬
main pin brass cotter set screw 主销铜衬扁销止动螺钉
main pinion 主小齿轮
main pip 主标记
main pipe 干管；总引水管；总管(气、水等)；主管(指供气供水管)；主管道
main pipe line 总管道；主管路；总管系；主管道；干管
main piping 总管道
main piston 主活塞
main piston bushing 主活塞衬套
main piston ring 主活塞涨圈
main pit 主探井；主矿井
main pivot 旋转轴；主销；支承铰链
main plan 规划总图
main plan design 总平面设计
main plane 主平面
main plate 主夹板

main plate enlargement ring 主夹板衬框
main plate nut 主夹板螺母
main platform 基本站台
main plot 主(小)区；整区
main plunger 主活塞
main point 要点；主点
main point of curve 曲线主点
main point of design 设计要点
main points of a cast 案由
main pole 主支柱；主极；主杆；主磁极
main pole core 主磁极铁芯
main pole gap 主磁极间隙
main pole piece 主极靴
main pole winding 主极绕组
main pollution sources 主(要)污染源
main port 主(要)港(口)；枢纽港
main portal 大门；正门
main portion 主要部分
main post 主柱；主杆
main power auxiliaries 主动力系统辅助设备
main power jet 主射流
main power path 主要能流
main power plant 主动力装置
main power source 主电源
main power station 主发电厂
main power substation 主变电站
main power supply 主电源
main power system 主动力系统
main precipitation center[centre] 主要降水中心
main precipitation core 主要降水中心
main press 主压榨
main pressure 主压力
main pressure differential 电力网电压差；电源电压差
main procedure 主过程
main process equipment 主要工艺设备
main processer 主处理机
main processing block 主处理部件
main processor 中央处理机
main process stream 总流程；主流程线
main process wastewater 主处理污水
main production plant 主要生产装置
main products 主要产品
main program(me) 主程序
main program(me) cycle 主程序循环
main program(me) data area 主程序数据区
main program(me) sequence 主要程序步骤；主程序序列
main project 主体工程
main propeller 主推进器
main propelling machinery 主推进装置
main propulsion 主推进
main propulsion gas turbine 主燃气轮机
main propulsion unit 主推进装置
main propulsive load 主推进装置负载
main protection 主保护装置
main pulley 主动皮带轮
main pulley for secondary change-speed 二级变速主皮带轮
main pull rod 主拉杆
main pulse 主脉冲；探测脉冲
main pulse/bubble ratio 主脉冲/气泡比
main pulse referennce group 主基脉冲群
main pump 主泵
main pumping station 总水泵站；总抽水站
main quantum number 主量子数
main rafter 主脊梁；主橡体；主橡
main rail 舷墙扶手；主轨
main railroad[railway] 主要铁道
main railway line 铁路干线
main ram 主锤
main ramp 主跳板【船】
main range circuit 主测距电路
main raw material 主要原(材)料
main ray 主射线
main reaction 主要反应
main receiver 主用接收机
main receiving station 主收报台
main reducing gear 主减速器
main reef 主矿层
main reflectivity 主反射率
main reinforcement 主筋；受力(钢)筋
main reinforcement bar 主配筋；主钢筋
main reinforcing steel 主钢筋

main relay 主继电器
main repair shop 主修车间
main repairs of fixed assets 固定资产大修理
main repeater section 主中继段
main repeater station 总中断站;主中继站
main reservoir 主储[贮]气器;主风缸
main reservoir chamber cap 储气主室盖
main reservoir drain cock 储气主筒放水旋塞
main reservoir outlet cock 储气主筒开放旋塞
main reservoir pipe 总储气管
main reservoir pressure 主风缸压力
main resonance 主谐振
main response 主响应
main retarder 主减速器;主缓行器
main retarder control 主缓行器控制
main retarder position 主缓行器位置
main return 主回风道
main return air current 主回风流
main return airway 主回风道
main rib 主肋
main ride 主林道
main ridge 正脊
main rim 主轮缘
main ring 主环
main riser 总立管;主立管
main river 主要河流;主河;干流
main river channel 主河道;主河槽
main river control station 干流控制站
main road 主要道路;干线公路;干路;干道;交通干线
main rod 主杆
main rod bearing 主杆轴承
main rod brass wedge bolt 主杆铜楔螺栓
main rod guard 主杆护架
main rod strap 主杆套
main rod strap bolt 主杆套螺栓
main roof 主顶板;炉拱顶
main root 主根
main rope 主绳
main rotor 主转子;主旋翼
main-rotor drive 主旋翼传动
main-rotor shaft 主旋叶(传动)轴
main route 主要路线;主要道路;主路由
main routine 主程序
main rudder 主舵
main rudder piece 下舵杆
main runner 主扶手;主滑条;主墙筋;主龙骨;主格栅;主铁水沟;吊顶主龙骨;大龙骨
main running signal 主体信号
main runway 主跑道
mains 主干线;干管网;干道网;馈电线;电源;电力网
main saddle 纵向刀架
mains antenna 电源线天线;照明网天线;主干天线
mains-borne disturbance 电源干扰
mains calculation 管道计算
main scale 主尺;主比例尺
main scale mark 主刻度线;主标度线
main scan(ning) 主扫描
main scavenging pump 主轴气泵;主排气泵;主抽气泵
main schedule 正表
main scheduler 主调度程序
main scheduling routine 主调度程序
main-scheme station 一等(测)点
mains connection set 电源设备
mains cord 电源软线
main sea 开阔海面;外海
main seam 主矿层
main sea water system 主海水系统
main section 干线段;基本段
mains electricity 电源供电;电力网供电
main separating door 主风流隔开风门
main sequence 主序;主星序
main-sequence fitting 主序拟合
main series 主列
main service 总配水管
main service fuse 总火线熔断(器);进户线熔断器
main service panel 主配电盘
main servomotor 主伺服马达;主伺服电动机;主接力器;导叶接力器
main set 主机组
main sewer 总排水管;主下水道;排水总管;排水干管;污水总管;污水主管;污水干管
mains filter 电源滤波器
mains frequency 工业频率

mains frequency coreless induction furnace 工频无芯感应炉
mains frequency coreless induction melting furnace 工频无芯感应熔炼炉
mains frequency indicator 电源频率指示器
main shaft 主轴;主要巷道;主烟囱;主(竖)井;主矿井;透气主井
mainshaft 3rd speed gear 主轴第三速齿轮
mainshaft 3rd speed gear bushing 主轴第三速齿轮套
mainshaft 3rd speed gear roller 主轴第三速齿轮轴承滚针
mainshaft ball bearing 主轴滚珠轴承
mainshaft bearing 主轴轴承
mainshaft bearing cap 主轴轴承盖
mainshaft bearing cap gasket 主轴轴承盖垫密片
mainshaft bearing cover 主轴轴承盖
mainshaft bearing oil seal 主轴轴承油封
mainshaft bearing oil slinger 主轴轴承抛油环
mainshaft bearing retainer 主轴轴承护圈
mainshaft bearing roller 主轴轴承滚柱
mainshaft bushing 主轴衬套
mainshaft cable 主轴电缆
mainshaft cap 主轴盖
mainshaft differential gear 主轴差速齿轮
mainshaft differential spider 主轴差速器十字架
mainshaft differential spider pinion 主轴差速十字架小齿轮
mainshaft driven gear 主轴从动齿轮
mainshaft driven gear bearing 主轴从动齿轮轴承
mainshaft drive pinion 主轴传动小齿轮
mainshaft 5th speed gear 主轴第五挡齿轮
mainshaft gear 主轴齿轮
mainshaft gear bushing 主轴齿轮衬套
mainshaft gland 主轴压盖
mainshaft high speed sliding gear 主轴高速滑动齿轮
mainshaft intermediate gear 主轴中速齿轮
mainshaft key 主轴键
mainshaft low and reverse sliding gear 主轴低速及倒档
mainshaft low speed sliding gear 主轴低速滑动齿轮
mainshaft nut 主轴螺母
mainshaft oil baffle 主轴挡油圈
mainshaft overdrive gear 主轴超速(传动)齿轮
mainshaft overdrive gear bushing 主轴超速齿轮衬套
mainshaft pilot bearing 主轴导向轴承
mainshaft rear bearing 主轴后轴承
mainshaft rear bearing oil seal 主轴后轴承油封
mainshaft roller bearing 主轴滚柱轴承
mainshaft sleeve 主轴套
mainshaft sleeve clamp lever 主轴套筒夹紧手柄
mainshaft sliding gear 主轴滑动齿轮
mainshaft snap ring 主轴开口环
mainshaft speed gear retainer ring 主轴变速齿轮止动环
mainshaft speed gear sleeve 主轴变速齿轮套
mainshaft synchronizer gear 主轴同步齿轮
mainshaft synchronizer gear sleeve 主轴同步齿轮套
mainshaft thrust cap 主轴推压盖
main shear wire 主剪力线
main shield 主屏蔽
main shipping season 航运忙季
main shipping track 主航线
main shock 主震;本震
main shock-after shock type 主(震)余震型
main shock strut 主减震支柱
main shoe 主滑脚
main shoe casting 主滑脚铸件
mains hold 主电源同步
main shop 主车间
main shopping center 主要购物中心
main shrine 主殿
mains hum pattern 交流干扰图像
main shut-down 主停机
main shut-down system 主停机系统
main shut-off handle 主截止手柄
main shut-off valve 总关闭阀
main side 主侧
main sideband 主边带
main signal 主信号(机);主体信号

main signal(l)er 主体信号机
main signal(l)ing equipment 总信号设备
main sill 主底梁
mains immunity 电源抗扰性
main single line diagram 主结线单线图
mains input 电源(电压)输入
main site 主(建筑)工地
mains jack 电源插座
mains junction 管网节点
main slant 主斜井
main slide mass 主滑动体
main slide valve 主滑阀
mains lighting supply 照明电源
main slope 主斜井
main-slope engineer 主斜井卷扬机司机
main medium frequency induction heating equipment 主中频感应加热设备
main soil group 主要土类
main soil type 大土类
main solar semidiurnal tide 太阳半日主潮
main solenoid 主螺线管
mains-operated instrument 交流电动仪表
main sounding line interval 主测深线间距
main source 主源
main source of electric(al) power 主电源
main span 主孔;主跨
main specification 主要规格
main speed 基本速度
main speed reduction box 主减速箱
main spillway 主溢洪道
main spindle box 主轴箱
main spindle hand feed 主轴手动进给
main spindle head 主轴箱
main spindle quill 主轴套管
main split 主分支
main spot 主黑子
mains power supply 电力网供电;系统供电
mains pressure 电源电压
main spring 主泉;主弹簧;击针簧;表弹簧;主发条;发条
mainspring hook 主发条钩;发条钩
mainspring leaf 主弹簧片
mainspring winder 主发条卷线器;发条缠绕器
main spud 主钢桩【疏】
main squeeze 监工头(俚语);工长;领工员
mains ripple 电压脉动
mains side 电源侧
mains socket 电源插座
mains supply 市电电源;电网供电电源
mains switch 电源开关
main stack 主排气管;主通气管
main staff 主要人员
main stage 主级
main stair(case) 主楼梯
main stand 主看台;正面看台
main standpipe 总立管;主立管
main starting air valve 主起动空气阀
main starting valve 主起动阀
main station 主(要台)站;主要观测站
main station peg 主测站标桩
main stay 主桅前支索;台柱(子);大桅牵条;大桅拉索;主拉索;主要支持;主要台柱;主系木
main steam 主蒸汽
main steam and feed water system 主蒸汽与给水系统
main steam header 主蒸汽集管
main steam isolation valve 主蒸汽止回阀
main steam line 主蒸汽管道
main steam pipe 主(蒸)汽管;蒸汽干管
main steam-piping system 主蒸汽管道系统
main steam range 蒸汽总管
main steam system 主蒸汽系统
main steam valve 主(蒸)汽阀
main steel (混凝土中的)主(钢)筋
main steering arm 总转向杆;主转向臂
main steering equipment 主操舵装置
main steering gear 主操舵装置
main stem 河流主道;主要街道;干流;铁路干线
main step bearing 主支承轴承
main steps 主轴承(轴)瓦
main stock 下舵杆
main stop valve 主停气阀;主截止阀
main storage 主存储器
main storage area 主存储区

main storage buffer 主存(储器)缓冲区
main storage controller 主存储器控制装置
main storage data base 主存储器数据库
main storage need 主存储器需要量
main storage read-write channel 主存读写通道
main storage requirement 主存储器需要量
main storage switch 主存储器转换机构
main storage yard 主堆场;主储[贮]藏场
main store 主累加器;基本库存;基本记忆装置
main stor(e)y 主层;主林层
main storgae dump 主存储区转出
main strain 主应变
mains transformer 电源变压器
main stream 主(要河)流;主河;干流
main stream control station 干流控制站
main stream flow 主流
main street 主要街道;主要大街;大街
main street green 主干道绿化
main stress 主应力
main stress line 主应力迹线
main stroke 主冲程;回返闪流
main structure 主要结构;主(体)结构
main structure system 主体结构体系
main studio 主播送室;中心播送室
main sub station 总变电站
main substation equipment 主变电站设备
main suction 吸水总管
main suction foot 主吸口根部
main suit 主水源
main sump 主水舱
mains unit 供电整流器
main supervisory panel 主监视仪表板
main supply 主供水管;主电源;供电干线
main supply conduit 给水干管
main supply duct 总送风道
main supply line 主供油管;供电干线
main supply road 主要补给线路
main supply source 主要供应来源
main supply switch 总开关
main support 主要支墩;主架
main supporting beam 主梁
main supporting surface 主支承面
mains voltage 供电电压;电源电压
mains voltage heating 干线电压加热
mains water 总水管
main sweep 主扫描
main switch 电源开关;主电门;总开关;主开关;主交换机
main switchboard 总电表板;总配电板;总控制板;总开关盘;主配电盘
main switchboard room 主配电盘室;主开关盘室
main switchgear closing 主开关设备合闸
main switching compound 主配电装置
main switchroom 主配电间
main synchronization pulse 主同步脉冲
main system 配水管系统;主系统
main system of the digital dispatching phone 数字调度电话主系统
main tail feather 舵羽
maintainability 运转的可靠性;可维修性;可维护性;耐用性;维持性;维修能力;维护性能;保养性能;保养率;保全性
maintainability design criterion 维修性设计原则
maintainability function 可维修性函数;可维护度函数
maintainability index 保养率指数
maintainable economic welfare amount 可维持的经济福利量
maintainance line 保持系
maintain a schedule 遵守进度表
maintain at grade 保持坡度
maintain commercial integrity 守信用
maintain confidentiality 保守秘密
maintain contact 保持接触
maintained load method 维持荷载法(试桩)
maintained load pile test 桩的维持荷载试验
maintained load test 恒荷载试验;维持荷载法(桩工试验)
maintained mileage of waterway 航道维护里程
maintainer 养路机;养路工;养护工;刮土机;检修人员;维护人员;保养器
maintainer line 保持系
maintainer scraper 养修平整机;养路工
maintaining 维护保养

maintaining account 维持账户
maintaining furnace 保温炉
maintaining heat 维持热
maintaining power 储能器件
maintaining the navigable depth 保持通航水深
maintaining valve 保持阀
maintaining voltage 维持电压
maintain railways 养路
maintain roads 养路
maintain sustainable development 实现可持续发展
maintain the price level 维持价格水平
maintain value 保值
main tank 主箱
main tapping 主要分接头;主抽头
main target of a plan 主要计划指标
main targets of the plan 主要计划目标
main task 主任务
main technical behavio(u)r 主要技术性能
main technical data 主要技术数据
main technical details 主要技术数据
main technical index 主要技术指标
main technical parameter 主要技术参数
main technical requirements of railways 铁路主要技术要求
main technical specifications 主要技术规格
main technical standard of railways 铁路主要技术标准
main techno-economic targets 主要技术经济指标
main tectonic stress field 主构造应力场
main telescope 主镜
main temperature controller 主温度控制器
main temple 主寺庙
main temple hall 大殿
maintenace energy 维修能量
maintenance 养路工作队;养护;固位法;维修保养;维护;维修;保养
maintenanceability 可维修性
maintenance activity 维修活动
maintenance agency 维修机构
maintenance amplitude section 保持振幅剖面
maintenance and after installation services 维修及安装后保养工作
maintenance and construction brigade 修建队
maintenance and increase ratio of state-owned assets value 国有资产保值增值率
maintenance and management system of plant 设备维修管理系统
maintenance and operation 维修和运营;维护与操作
maintenance and operation(al) of program(me) 程序维修和操作
maintenance and overhaul 维修;保养与大修
maintenance and repair budget 维修预算;维检修预算
maintenance and repair cost 维修费用
maintenance and repair parts 维修备件
maintenance and repairs 维修养护;保养与修理;维修与保养;维护与维修
maintenance and repair shop 维修工场
maintenance and service 维修与使用;保养与维修
maintenance and service manual 维修服务手册
maintenance and supervision of aids 航标维护管理
maintenance apron 机修坪
maintenance area 养护区域;养护面积;维修区;保证供应区;养护区(域)
maintenance assistance modules 维护辅助程序
maintenance behavior 维持行为
maintenance bond 保修单;维修(保)证书;维修保函;维修金;保修协议
maintenance budget 养护维修预算;维修预算
maintenance building 养路工房;道班房
maintenance by contractor 承包者维修;承包商维修;承包人维修
maintenance by firm 承包者维修;承包商维修;承包人维修
maintenance car 修理车;工具车;保养车
maintenance care 技术维护;保养管理
maintenance center 维修中心;维修基地;维护中心
maintenance certificate 维修证书
maintenance charges 保养费用;养护维修开支
maintenance chart 维修图表
maintenance clause 维修项目

maintenance cleaner 清洁工;维修清洁工
maintenance coating 维修涂料
maintenance code 维修编号卡
maintenance compound 养护工长;养护领班;维修用复合膏;维修用化合物
maintenance concept 维修保养原则
maintenance contract 维修合同
maintenance control 维修控制
maintenance control center 运营管理中心
maintenance control data 维修控制数据
maintenance control panel 维护控制板
maintenance control report 维修管理报告
maintenance control retry register 维护控制重算寄存器
maintenance control signal 维护控制信号
maintenance cost 维修成本;维修保养费(用);维护费(用);维持消耗
maintenance crew 养护工(作)队;维修组;维修(工作)队
maintenance current 维持电流
maintenance cutout cock 维修用切断阀
maintenance cycle 维修周期
maintenance cycle of rolling stock 机车车辆维修周期
maintenance data 维修数据
maintenance data system 维护数据系统
maintenance deck 维修平台
maintenance department 修缮部;维修部门;维修保养工段
maintenance department expense 维修部费用
maintenance depot 机械维修厂;修配站;分配厂;修理厂;养路机械修配场;养护补给站;检修站;检修厂;维修站;保养厂
maintenance depreciation 维修折旧;维持折旧;大修折旧;大修提成
maintenance depth 维护水深
maintenance design 维修设计
maintenance detailed planning 维修工作细则
maintenance device 维修装置
maintenance dimensions of channel 航道维护尺度
maintenance division 养路工区;养路段;维修分队
maintenance documentation 维修资料文件
maintenance door 检修门
maintenance dose 维持量;维持剂量
maintenance downtime 修复时间;维修时间;检修停机时间;停修时间
maintenance dredging 维护性挖泥;维护(性)疏浚;维持性挖泥
maintenance drill 技术维修训练
maintenance effectiveness 维修有效性
maintenance efficiency 维修效率
maintenance electric(al) monorail hoist 检修用单轨电动葫芦
maintenance electric(al) travel(l)ing crane 检修用电动吊车
maintenance electrician 维修电工
maintenance energy 维持能
maintenance engineer 养护工程师;维修工程师;维护工程师
maintenance engineering 维修工程;维护工程;保养工程
maintenance engineering analysis 维修工程分析
maintenance equipment 维修设备;养路设备;养护设备;维护设备
maintenance establishment 保养机构;保养编制
maintenance evaluation test 保养鉴定试验
maintenance examination 集装箱定期检验
maintenance expenditures 养护开支;维修费(用)
maintenance expense 维修费用;维护开支
maintenance facility 养路设备;维修设备
maintenance factor 使用系数;维修系数(照明设备);维修率;维修系数(隧道照明的)(保养系数
maintenance fee 维修费(用)
maintenance fertilizer 保养肥
maintenance finish 耐用涂料;维护涂料
maintenance fitter 维修钳工
maintenance flat-roof for machine 养路机械作业平台
maintenance footway 维修便道;维修用便道
maintenance force 养路工作队;养护工队;维修养护分队
maintenance foreman 养路领班;维修班长
maintenance frame 维修帧

maintenance-free 不需维护(的);不需养护的;不需维修的
maintenance-free bearing 自润滑轴承
maintenance-free life 耐久年限;无养护年限
maintenance-free operation 不需维修的运行;不需维修的运作
maintenance-free service 不需维修的使用
maintenance-free track 不需经常维护的轨道
maintenance frequency 维修次数
maintenance funds 维修资金
maintenance gang 养护工作队;养路工队;养路工班;养路道组;维修组;维修(工)班
maintenance gate 检修闸门
maintenance handbook 维护手册
maintenance hangar 维修机库;维修厂棚;修理棚(飞机)
maintenance in storage 库存器材的维修
maintenance instruction 技术维护规程;维修细则;维修规程;维护手册;维护保养说明书;保养指南
maintenance interval 技术保养周期;维修期间隔;维修周期;维护时限
maintenance isolating valve 维修用切断阀
maintenance job 维修工作
maintenance kit 修理包;成套维修器材;保养箱
maintenance level 维护水平
maintenance lighting 维修照明
maintenance load 维修工作量;维护荷载
maintenance load test 维持荷载法(试桩)
maintenance location 技术保养与修理站
maintenance locomotive moving kilometers 本务机车走行公里
maintenance log 维修保养记录;保养时程表;保养记录表
maintenance logistics 技术保证;保养勤务
maintenance lorry 维修车
maintenance machinery 养路机械;养护机械
maintenance machinery and equipment 养路机械和设备
maintenance man 保养人员;维修人员;维修工
maintenance management 维修管理
maintenance management system 养路管理系统
maintenance manhole 检修人孔;检修孔
maintenance manipulation 养路工作;养路操作;养护操作
maintenance manpower 维修人力
maintenance manual 养护手册;维修(工作)手册;维修保养手册
maintenance margin 保证金下限
maintenance marketing 维持性行销
maintenance material 维修材料
maintenance material cost 保修材料费
maintenance media 维持液
maintenance method 保养方法
maintenance moving packing ring 活动备修止水圈
maintenance objective 维修目标
maintenance obligation 维修义务
maintenance of buildings 房屋养护
maintenance of class 保持船级
maintenance of container 集装箱的维修保养
maintenance of dike 坝的维修
maintenance of earthworks 土工建筑物的维护
maintenance of equipment 设备维修
maintenance of estuarine channel 河口航道的维修
maintenance of high-speed track 高速铁路线养护
maintenance of machine 机器保养
maintenance of nitrogen 氮(的)保存
maintenance of plant 设备维护保养
maintenance of possession 保全
maintenance of premises and equipment 房屋和设备维修
maintenance of pressure 保持稳定压力;压力保持
maintenance of road 养路
maintenance of roofs 屋面维修
maintenance of sewers (污水的)管道养护
maintenance of soil fertility 维持土壤肥力;地力保持
maintenance of surface 路面养护;面层养护
maintenance of tracks 线路维修;轨道养护维修
maintenance of value 保留金
maintenance of value obligations 保值债务
maintenance of waterway 航道养护
maintenance of way 线路养护;线路维修;养路;路面维修费

maintenance of way account 线路养护账目
maintenance of way and structures 线路桥隧养护
maintenance of way division 工务段
maintenance of way engineer 养路工程师
maintenance of way section 工务(分)段
maintenance of way shop 工务修配厂
maintenance of work 工程维修
maintenance operator 维修作业;维修工
maintenance organization 养护机构;维修组织;维护组织
maintenance outage 检修(性)停机
maintenance outlay 维修费;维修费(用)
maintenance overhaul 经常性维修;日常修理;日常维护;维修检查;大修(理);大检修
maintenance overhead 经常性维修
maintenance package 修理包;维修器材
maintenance paint 维修用漆;维修涂料;保养漆
maintenance painting 维修涂装;维护性油漆
maintenance painting work 维修油漆工作
maintenance panel 维修板
maintenance parts list 维修部件清单
maintenance party 修理队;维修组;维修班
maintenance patching material 养路补充材料
maintenance period 维修养护期;维修周期;检修期;维修期;保养(周)期;保修期(限)(缺陷责任期)
maintenance personnel 养护人员;维修人员;维护人员
maintenance plan 技术维修计划;维修计划
maintenance planning and control system 维修计划和控制系统
maintenance platform 维修工作台
maintenance platoon 维修排
maintenance point 技术保养与修理站;维修预算;维修点;保养站
maintenance practices 保养程序
maintenance prevention 维护设施;防护检修;保养;安全设施;安全措施;维修
maintenance procedure 维修程序;维护程序
maintenance process 养护程序;养路工程;养护方法
maintenance program(me) 维修程序;保养大纲
maintenance program(me) chain 维修程序链
maintenance project 维修计划;保养计划
maintenance proof cycle 维护质量检查周期
maintenance quality of aids-to-navigation 航标维护质量
maintenance quota 养护定额【道】
maintenance rate 维护率
maintenance ratio 维修率
maintenance record 养护记录;维修记录
maintenance register 维修资料登记簿
maintenance regulation 技术维护规程;维修规则;维护规则;保养细则;技术保养细则
maintenance reliability 维护可靠性
maintenance repair 经常维修;日常维修;日常检修
maintenance repair and operating supplies 维修及作业用品
maintenance repair and replacements 维修和更换
maintenance requirement 维修要求;维持需要;养护要求
maintenance reserves 检修备用;维修费准备金
maintenance rig 维修架
maintenance routine 例行维护;维修程序
maintenance safeguard 维护保障措施
maintenance schedule 维修图表;维修(实施)计划;维修进度表
maintenance seating ring 固定备修止水圈
maintenance section 修理小组;养路工区;维修养护工段
maintenance service 修理勤务;维修;维护业务;维护手段
maintenance service manual 维修手册
maintenance shed 保养库;保养车库
maintenance ship 维修工作船
maintenance shop 维修场厂;维修场所;机修车间;维修工段;保修保养车间;维护车间;保养工场
maintenance sidewalk 维修用便道
maintenance siding 检修线
maintenance staff 维修职员;维修人员
maintenance standard order 维修标准规范
maintenance standby time 维护准备时间
maintenance station 巡线站;技术保养与修理站;维修站;维修台

maintenance statistics 维修统计
maintenance status 修理情况;保养状况
maintenance stores 保养器材
maintenance strategy 维修策略;维护策略
maintenance superintendent 养路段长;养路主管工程师【铁】
maintenance supply 维修器材
maintenance support 技术维护;维修支持
maintenance survey 养路测量
maintenance switching 维修轮换
maintenance system of random repair 随时修缮制度
maintenance target value 维修指标值
maintenance task 维修任务书
maintenance task period 维护工作周期
maintenance team 修理组;维修组;维修班;保养组
maintenance technician 修理行业的技术人员;维修技术员
maintenance test 维护试验;维护测试
maintenance test equipment 技术维护试验设备;维护测试设备
maintenance test method 维护测试法
maintenance time 维修时间;维修保养时间
maintenance tolerance 维修允许公差
maintenance tool 修理工具;维修工具;维护工具
maintenance track 检修线
maintenance train 修理列车
maintenance trouble 维修不良
maintenance truck 维修车
maintenance underpinning 维持性托换
maintenance undertaking 承担维修
maintenance unit 维修装置
maintenance vehicle 检修车;维修服务车
maintenance way department 工务处
maintenance with rent 以租养房
maintenance work 养路工作;日常维修(工作);维修作业;维修工作;维护工作;保养工作;维修工程;保养工程
maintenance worker 维修工;保养工
maintenance work shop 维修车间
main tensile reinforcement 主要受拉钢筋
main tensile stress limit 主要应力极限
main terminal 主要枢纽;主端子
main terminal building 主要航站楼;主要港口建筑
main test 主要检验
main test frame 主试验架
main textural coefficient 主要结构系数
main thoroughfare 主要通道;主要大街
main thread 主流中心线;主流中泓线
main throttle 主节气门
main throttle valve 主气门;主节流阀
main through line 干线通行线
main thrust 主力冲击
main tie 主拉杆;联结杆
main tilting axis 主倾斜轴
main title 主标题
main-topmast 主中桅
main tower 主塔
main tow rope 主拖缆
main track 主要轨道;主线;主航线;正线【铁】;正轨;干线;铁路干线
main track line 主干线路
main track of turnout 道岔主线
main traction system 主牵引系统
main traffic 主流交通
main traffic artery 主要交通干道;交通主干线
main traffic flow 总交通流量;主(向)交通流
main traffic station 主通信站
main trajectory 主轨道
main transept 主要建筑翼部
main transformer 主变压器
main transformer station 主变电站;主变电所
main transmission 主传动系;主传动
main transmit and receiver station 主收发站
main transmitter 主发射台
main transmitting station 主发射台
main transverse bulkhead 主横舱壁
main trap 房屋存水弯;总水封;干管存水弯
main traverse 主(要)导线【测】
main traverse line 主导线【测】
main traverse net 主导线网
main treatment 主处理
main triangulation 一等三角测量;主要三角测量;主杆三角测量

main trolley 主吊运车
main truck 主桅冠
main truck highway 主(要)干(线)公路
main trunk circuit 主干线
main trunk duct 总风道
main trunk highway 主要干线公路;主干公路
main trunk line 主要干线管路
main trunk road 主要干线道路;主干路;主(要)干道
main trunnion bearing 中空轴轴承
main truss 主桁架
main tuning 主调谐
main tunnel 主巷道;主(风)洞;主坑道;正隧道
main turbine 主涡轮机
main turret 主炮塔
main underbeam attachment point 车架纵梁联结点
main undercarriage 主起落架
main underpinning 主体托换
main unit 主件;主单元
main vacuum manifold 主真空管道
main value 总值
main valve 总闸门;总阀;主阀
main valve bush 主阀衬套
main valve bushing 主阀衬套
main valve chamber 主阀室
main valve disc 主阀盘
main valve head cap screw 主阀盖有头螺钉
main valve plunger 主阀柱塞
main valve ring 主阀环
main valve spring 主阀簧
main valve stem 主阀杆
main valve stem nut 主阀杆螺母
main vanishing point 主合点;主遁点
main variable 主变量
main vault 主拱顶
main-vault shielding 主厅屏蔽
main vehicle deck 主车辆甲板
main vent 主通气管;通气主管;通气立管;通气干管;通风干管
main vent stack 主通气主管
main vent tube passage plug 主通风管通道塞
main vertical 主要竖杆
main vertical zone 主竖区(消防分隔制)
mainvocal microphone 主声拾音器
main voltage selector 主电压选择钮
main vortex 外涡旋
main walk 主要走道
main wall 主墙
main wall panel 护墙板;墙裙
main warp 主经
main washbox 主洗跳汰机
main washer 主垫圈;大号垫圈
main wash heating phase 主洗加热期
main water 自来水
main water table 主潜水面;基本地下水面
main wave 主波
main weld 主焊缝
main wheel 主操纵轮;主轮
main wheel door actuating cylinder 主轮舱盖作动筒
main-wheel strut 主轮支柱
main wind 主导风向
main wind direction 主风向
main winding 主绕组
main wing 主叶片
main wire bundle 主电线束
main wire feeder 主送线器
main wiring diagram 主接线图
main working plan 主要平面图
main works 主要工程;主体工程
main workshop 总厂
maioliaca 锡釉彩陶(意大利)
maiolica 不透明釉彩陶
maise 玉蜀黍
maisonette 复式住宅;(跨二层的)公寓套房;公寓住宅;小房屋;分租部分
maize 玉蜀黍;玉米
maize borer 玉米螟(虫)
maize grinder 玉米粉碎机
maize oil soap 玉米油皂
maize rains 玉米雨季
Maizewood 玉米板(一种绝缘保温材料)
maize yellow 玉米黄
maizuru tectonic zone 舞鹤带

Majac jet pulverizer 马亚克对喷式气流粉碎机
Majac mill 马亚克磨
majakite 砷镍钯矿
Majestic Babbitt 马耶斯提克轻载巴比合金
majestic babbitt alloy 高级巴比特合金
Majiabin culture pottery 马家浜文化陶器
Majiayao type pottery 马家窑类型陶器
majolica 锡釉彩陶(意大利);乌釉陶器;不透明釉彩陶
majolica colo(u)r 锡釉彩陶用色料
majolica enamel 乌釉搪瓷;陶釉搪瓷
majolica glaze 砖瓦用釉料;锡釉彩陶釉
majolica glazed coat(ing) 陶器上釉层
majolica mosaic (涂有不透明釉的)锦砖;(涂有不透明釉的)马赛克;彩陶镶嵌
majolica mosaic wall tile (涂有不透明釉的)墙砖
majolica tile (文艺复兴时期意大利出产的)花饰陶器;(石灰质的)陶器瓦
majolica wall tile 陶器墙砖
major 主要的;较大
major accident 主要事故;重大事故;大事故
major accident loss 重大事故损失
major advantage 主要优点
major aftershock 主余震
major air cell 主空气室
major airport 主要机场
major angle 主角
majorant 强函数
majorant for integral 强积分
major arc 优弧;主弧;大弧
major arch 主拱;长拱
major assembly 装配主件
major avenue of approach 主要通路
major axis 主轴;长轴;强轴
major axis of building square grid 建筑方格网长轴线
major axis of ellipse 椭圆形长轴
major axis of inclusion 包裹体长轴
major beam 主射束
major bed 主河床
major betterment 主要改善
major blasting 大爆破
major breakdown 大事故;重大阻断
major bridge 大桥
major calorie 大卡
major cause of failure 断裂的主要原因
major character 主要特征
major characteristic of seepage 渗流的主要特征
major charges 主要费用
major chemical company 大型化学公司
major clique 优系
major combination 主要机组;全套
major commodities 主要货物
major community 大群落
major component 主要元件;主要部件
major conjugate angle 优共轭角
major conjugate arc 优共轭弧
major conjugate segment 优共轭弓形
major constituent 主组分;主要成分;常量组分
major construction element 主要建筑单元
major construction project 重点建筑工程计划;重点建设项目;大型建筑工程计划
major contract 主要合同
major control 主要控制
major control break 主要控制断路
major control change 主要控制改变;高位控制变更
major control data 主控数据
major control net 主要控制网
major control point 主要控制点
major country 主要国家
major critical component 主要关键部件
major cross street 主要横街
major cutting 主切削刃
major cycle 主循环;主旋回;大周期;大循环
major dam 主要水坝;主坝;大坝
major data 主要数据
major defect 主要缺陷;重缺陷
major dense line 主致密线
major desire line (根据城市交通流量而标定的)主要要求线;(根据城市交通流量而标定的)主要愿望线
major diameter 大直径;长径
major diameter fit of spline 花键外径配合
major dimension 主要尺寸

major directional desire line 主流倾向线;交通主流倾向线
major disaster 大灾害;大事故
major discharge 排放大户
major dislocation 大断裂
major disposition 主要配置
major distributary 主要配水渠;斗渠
major divide 主分水线;主分水岭;主分水界;主分界线
major division 大格
major drainage basin 大流域
major earthquake 主震;大(地)震
major economy 较大范围的经济
major element 主要元素;大量元素;常量元素
major-element analysis 主要元素分析
major elements in seawater 海水主要元素
major emitting facility 重要排放源
major engineering test item 主要工程试验项目
major engineering works 大型工程
major engine overhaul 发动机总检修
major equatorial axis 赤道长轴
major equatorial radius 赤道长半径
major equipment 主要设备
major examination 主要测试检验
major exit(of interchange) (互通式立体交叉的)主要驶出口
major expenditures 主要支出;主要经费;主要费用
major factor 主要因素
major factor of seepage movement 渗流运动要素
major failure 严重失效;主要故障
major failure mode 主要失效模式
major fault 主断层;重点病害;大断层
major fault zone 主断层带
major fire 大火
major first face 第一前刀面
major first flank 第一后刀面
major flank 刀具主后面
major flood 大洪水
major flood flow 大洪水流量
major fog signal 较优方案
major fold 主褶皱;主褶曲;大褶皱;大褶曲
major forest 主林
major forest produces 主林产物
major forest tree 主林木
major formula of seepage flow 渗流运动主要公式
major fractional 大部分
major fracture 主破裂
major framework 主要控制网
major freight station 主要货运站
major function 优函数;主要功能
major glacial epoch 主冰期
major grid 主格网
major harbo(u)r 大港
major heading 主目标
major highway 主要公路;干线公路
major highway artery 主要公路干线
major hysteresis loop 主磁滞回线
major impurity 主要杂质
major industry 主要行业;主要工业;主要产业;重点工业;大型工业
major ingredient 大料
major inorganism 主要无机物
major inspection 大检查
major insulation 主绝缘
major intrusion 深成侵入(体)
major ion 主要离子
majorite 镁铁榴石
major item 主要成品
major item repair parts list 主要项目修理零件单
major items of expenditures 重大支出项目
majority 择多;多数;大多数;大部分
majority carrier 主要载流子;多数载流(子)
majority decision 多数决定法
majority decision element 多数决定元件;择多判定元素
majority decision gate 多数判定门;多数决(定)门
majority element 多数元件;多数决定门
majority emitter 多数载流子发射极
majority gate 多数(决定)门
majority holding 拥有半数以上股权
majority logic 择多逻辑;多数逻辑;多数决定门;大数逻辑
majority logic circuit 多数逻辑电路
majority logic decoding 择多逻辑解码

majority operation 多数逻辑运算;多数操作
majority organ 多数元件;多数符合元件
majority-owned subsidiary 拥有过半数股权的附属机构
majority principle 多数原理
majorization 优化
majorized function 优化函数
majorized set 优化集
majorized subset 优化子集
majorizing linear mapping 优化线性映射
majorizing sequence 优化序列
majorizing series 优化级数
major jet 主喷嘴
major job 主要工作
major joint 主节理
major key 主键;常用键
major kitchen 大型厨房
major light 强光灯
major limit 大限
major lineaments 主要断裂线
major line-haul train 主要长途货物列车
major load transfer 主荷载转移
major lobe 主波瓣;主瓣
major lobe of direction(al) diagram 方向图主瓣
major loop 主循环;主回路
major lunar diurnal constituent 太阴主要全日分潮
major lunar elliptic(al) semidiurnal constituent 太阴椭率主要半日分潮
major lunar semidiurnal constituent 太阴主要半日分潮
major maintenance availability 主要维修可行性
major materiel category 主要补给品分类
major metal 主要金属
major mineral or elements 主要矿物或元素
major mineral reserves 主要矿物储量
major minor rotary 主要路与次要路环形交叉
major-minor sorting 主副分类
major module 主模块
major movement 主要行车方向
major network 主网(络)
major nodal earth terminal 主要结点地面终端机
major node 主结点
major objective 主要目的;主要目标
major orthogonal steel-plate bridge 主正交钢板桥
major-orthotropic bridge 主正交各向异性桥
major-orthotropic steel-plate bridge 主正交各向异性钢板桥
major-orthotropic deck structure 主正交各向异性桥面结构
major-orthotropic plate 主正交各向异性板
major-orthotropic plate method 主正交各向异性板法
major overhaul 总检修;大修;车辆大修
major parameter 主要参数
major part 大部分;主要零(部)件
major path 主通道
major path satellite earth station 大通道地球站
major peak 主尖锋
major planet 大行星
major plant 主要建筑设备;主要施工设备
major port 主要港口;大港
major positioning coil 主定位线圈
major power supply 主电源
major primary structure 主要原生构造
major principal plane 第一主平面;大主平面
major principal strain 最大主应变;第一主应变;大主应变
major principal stress 最大主应力;第一主应力;大主应力
major principle strain 最大主应变
major principle stress 最大主应力
major produces 主收获
major production areas 主要产区
major products 主要产品
major profile 主要剖面
major project 大型(工程)计划;重点研究课题;重点项目;重点工程;重点方案;骨干工程
major property of liquid 液体的主要性质
major radial 主要辐射线(路)
major radius 长半径
major radius curve 长半径曲线
major radius elbow 长半径弯头
major radius of ellipsoid 椭球长半径
major relay center 大转报中心站

major relay station 主中继站
major relief 大地形
major repair 大修(理)
major repair depreciation 大修折旧
major repair depreciation expense 大修折旧费
major repair depreciation fund 大修折旧基金
major repair depreciation rate 大修折旧率
major repair expenditures 大修支出
major repair of buildings and structures 房建大修
major repair of track 线路大修
major repair plan 大修计划
major repair renewal and reconstruction expense 大修更新改造费
major repair standard 大修定额
major requirement 主要要求
major rhombohedral 大菱面
major ride 主林班线
major river bed 严重;主河道;主河槽;重大河床;高水河床;高水河槽;洪水河床;大河床
major road 主要道路;干路;大路
major route 主要途径
major runway 主跑道
major saving 较大节约
major section 主要段;主件
major semi-axis 长半轴;半长轴;大半轴
major shell 主壳层
major shock 主震
major sized 大型的
major soil types 土壤主要类型
major solar diurnal constituent 太阳主要全日分潮
major solar semidiurnal constituent 太阳主要半日分潮
major spare parts 主要备件
major state 主状态
major state generator 主(状)态发生器
major state logic generator 主状态逻辑发生器
major station 主电站;主车站;主测站;井顶主车场;电话主机
major stationary source 主要固定(污染)源
major stop (乘客很多的)停车站;(公共车辆的)大站
major street 主要街道;大街
major stress 主要应力
major structure 主(体)结构
major structure system 主体结构体系
major stupa(mound) 主神龛塔(土墩)
major subgraph 主子图
major system 主系统
major system kit 成套主系统
major task 主任务
major technique 主要技术
major tenant 主要承租人
major terminal 主要终点站
major terms 常用名词表
major test 重点试验
major through road 直达干线;过境干道
major thrust face 大推力面
major thrust plane 主逆掩断层面
major tope(mound) 主要灌木林(土墩)
major total 总计;总和;主要统计值;合计;大计
major transactions 巨额贸易
major triangulation 一等三角测量;主杆三角测量
major tune-up 全面检修
major tunnel 主干隧道
major unit 主机组
major unit assemblies 装配式主机组;主机组总成
major unit assembly replacement 主部件总成更换
major water treatment plant 主要水处理厂
major wave 主波;长波
major weaving section 主要交通路段
major world currencies 主要世界货币
Majuro atoll 马朱罗环礁
Makalian wet phase 马卡拉潮湿期
Makarov basin 马卡洛夫海盆
Makassar Strait 望加锡海峡
makatite 马水硅钠石
make 开始看见;制作
make a audit 查账
make a bid 投标;议价还价;递价;出价
make a bid for 就……进行投标;投标承包;(拍卖时)出价买
make accounts square 结清(账目)
make a contract 签定合同;签订合同;订立契约;订立合同

make additional appropriations 增拨款项
make additional provision 增拨款项
make a deduction 折减
make a detailed statement of profit and loss 出具详细的损益计算书
make a difference 有差别;很重要;起作用;产生影响;发生影响
make a draft of money 提取款项
make a draft on a bank 向银行提款
make advance 垫款;垫款预付
make a fortune 发财
make a fraudulent application and claim 虚报冒领
make a fresh start 从头开始
make a full appraisal of 充分估计
make a good landfall 准时看到陆地
make a graduated application of the brake 施行制动机的阶段制动
make airfield 工厂机场
make a land 初见陆地【航海】
make a landing 靠岸
make allowance for 留有余地;扣除;考虑到;计及;为……留有裕量;为……留有余地;对……进行修正
make allowance for loss 对损失给予补偿
make a loss 亏损
make a market of 利用……赚钱;推动(发展)……市场
make an accommodation 通融资金
make an advance to somebody 贷款给某人
make an agreement 签订协议
make an allowance of 10 percent for cash payment 现款九折
make an appearance 出现
make an appropriation for 拨款供……之用
make-and-break 接与断;继续器;启闭;通断;闭合断开
make-and-break cam 电路断续凸轮
make-and-break contact 合断触点;通断触头;断续触点;闭开接点;闭开触头
make-and-break current 断续电流
make-and-break device 断续器
make-and-break ignition 继续发火点火;断电式点火
make-and-break ignition system 继续发火系统
make-and-break key 开关
make-and-break mechanism 开闭机构
make-and-break operation cycle 合断操作回合
make-and-break period 接通断开周期
make-and-break terminal 断续电路线接头
make a new record 创纪录
make an inventory 开列清单
make an inventory of goods in a warehouse 盘库
make an inventory of warehouse 清仓查库
make an offer 递盘;出价;发盘;发价;报价
make an overdraft 透支
make a pile 发财
make a port 进港
make a price 开价;讨价
make a progress report on a project 填报工程进度
make a prompt settlement 迅速解决;迅速结算
make a purchase 采购
make a raise 筹数
make a reparation for an injury 赔偿损害
make arrangements for 安排
make arrangement with 洽商
make a screwed joint 作螺旋接合
make a sketch 绘制草图
make a stern board 船向后退
make a suggestion 提出建议
make a tender for 就……进行承包;投标承包
make atonement for 对……赔偿
make bad weather 碰到坏天气;风浪中船身剧烈摇摆
make bad weather of it 经不起大风浪
make-before-break 先接后断;先开后接
make-before-break changeover switch contact 先闭后断换向开关接点
make-before-break contact 先闭后开触点;先接后断触点;先接后断接点,合一断接点
make-before-break switch 先合后断开关;合一断开关
make best possible firm offer 报……最低实盘
make both ends meet 使收支相抵
make-break operation 通断操作
make-break operation cycle 通断操作回合
make buckle and tongue meet 使收支相抵

make-busy 置忙
make-busy jack 闭塞塞孔;闭塞按钮
make-busy key 闭塞电键
make-busy relay 闭塞继电器
make-busy release 闭合释放
make collections 收款
make compensation 赔款;补偿
make connection 闭路连接
make contact 接通;闭合接触(点);接通电路;接通触头;接通触点;闭路触点
make contract 订合同
make correction 更正
make credit inquiry 作资信调查
make daily deliveries 每日发货
make dead 切断
make default 不履行法律义务
make delay 闭合时间延迟
make deliveries 发货
make do with the meagre means at hand 因陋就简
make economic development our central task 以经济建设为中心
make fast 关紧;绑牢;拴牢;系船柱;拴船桩;固定
make first-phase preparations 前期准备工作
make foreign things serve China 洋为中用
make for land 驶近陆地
make full use of limited resources 小型资源
make further investigation 进一步调查
make good 在线等待(完工后可重新开始);整新;修缮;修理;修复;弥补;完成;兑现
make good a loss 弥补损失;赔偿损失
make good as general average 共同海损补偿
make good the deficit 弥补差额
make good use of talent 人才优势
make good weather 碰到好天气
make good weather of it 经得起大风浪
make harbo(u)r 进港停泊;入港停泊
make head 断开
make headway 向前进;前进
make heavy weather 碰到恶劣天气
make hole 进尺【岩】
make impulse 接通电流脉冲;闭合脉冲
make inroads on crop 侵害农作物
make its goods more competitive 增强产品的竞争能力
make land 根据陆地确定方位;接近陆地
make lee way 向下风
make lending and investment commitment 承担贷款和投资
make life easier for the population 方便生活
make line 直线比例尺
make mistakes 出错
make of water 涌水量
make on 接通
make out a bill 开发票
make out a cheque 开支票
make out a draught of 起草
make out a list 开列清单
make out an account 开账
make over 转让
make overall plans 统筹规划
make payment 付款
make payment beforehand 提前付款
make point 接通点
make position 闭合位置;接通位置
make preparations for ploughing and sowing 备耕
make profit 获利
make progress 进展
make provision for 为…做好准备;采取措施
make public the administration of finance 财政公开
make pulse 接通脉冲;闭合脉冲
maker 制造者;制造机构;制造厂(家);制定者;接通器
make readings 取读数
makeready 垫版
make ready activities 准备工作
make ready time 生产准备时间
make regulation 立章程
make remittance 开支票
make restitution 归还
maker instruction 制造厂说明书
maker's certificate 制造厂证书

maker's mark 制造厂标志;制造厂商标
maker's technical documents 厂家技术文件
maker test 制造厂检测;制造厂的测定
make-run 制气周期
maker use 接合器应用
make satisfaction for 赔偿
makeshift 权宜之计;暂时代用品;临时措施;临时的;权宜措施;代用品
makeshift bridge 便桥;临时便桥
makeshift check dam 临时挡水坝
makeshift construction 临时建筑(物)
makeshift device 权宜之计
makeshift dwellings 临时凑合的住处
makeshift handing stage 临时码头
makeshift measure 临时办法
makeshift method 暂行办法
makeshift valve 变向活门;变向阀
make size 复制比例尺
make slurry volume 造浆量
make something of 利用;使更有用;使更完善
make spark 闭合火花
make tender 投标
make the best use of planter 爱惜播种机
make the harbo(u)r 进港
make the investment inefficient 无效投资
make the investment yield results 发挥投资效果
make the land 驶近陆地;初见陆地【航海】
make through (将某一地层)钻穿
make time 闭合时间
make to access 使访问
make to fine limits 精密制造
make to refer to 使关联于
make traffic easier for the people 方便交通
make treaty 立约
make true an engine 调整发动机
make twice as much 翻一翻
make-up 组成;构成;排版;成本加成;补足;编制;修理;配制;提高标价;电泳漆配槽;制作
make-up a deficiency 补偿差额
make-up air 补充空气;补偿空气
make-up air filter 新风过滤器
make-up air heater 补给式热风炉
make-up air unit 送风机组;空气补偿装置;空气补偿单元;补充空气装置
make-up a loss in proportion 按比例摊赔
make-up a purse 筹款
make-up articles 成品
make-up capacity 编组能力
make-up carbon 碳的补充部分;补充碳
make-up carrier gas 配载气;补充载气
make-up cathead 备用套管;备用绞盘;备用锚栓
make-up circuit 补油回路
make-up compressor 新料压缩机;补充压缩机
make-up copying 拼晒
make-up crane 配料起重机
make-up day 清算日
make-up deficit 调头寸
make-up department 造型车间
make-up feed 补(充)给水
make-up float 浮子式补给水调节器
make-up for 弥补
make-up for a loss 弥补损失
make-up for possible shortages with surpluses 以丰补歉
make-up for the cost of production 补偿生产成本
make-up for the deficiency 弥补不足
make-up gas 补充气体
make-up ground 人工填筑土层;填筑地
make-up heat(ing) 补充加热
make-up isobutane 补充异丁烷
make-up job 弥补工作;配制工作
make-up level 垫高地坪
make-up lubricating system 补偿润滑系统
make-up machine 贴合机;成型机
make-up machinery 装配机械
make-up mirror 化妆镜
make-up of charge 配料
make-up of manpower 劳动力构成
make-up of the imports 进口结构
make-up oil 配制油;不足油;补给油;补充润滑油
make-up operation 编组作业
make-up piping 补给管路
make-up projector 投影装置
make-up pump 供水泵;给水泵;补充(水)泵

make-up rail 标准短轨
make-up recover 弥补
make-up reservoir 调节水库;平衡水库;调节水池;平衡水池
make-up room 化妆室
make-up system 补给系统
make-up table 成型操作台
make-up tank 调节油箱;平衡油箱;调水箱;平衡水箱;补给油箱
make-up the charge 配料
make-up the required number 弥补需要的数额;补足所需数额
make-up time 下钻杆(下套管)时间;准备时间;纠错时间;超支时间;补算时间;补救时间;补偿时间
make-up to 补偿
make-up top 化妆台
make-up torque 补偿扭矩;补偿转矩
make-up valve 补偿阀;备用阀
make-up water 加水;配制水;补(充)给水;补偿给水
make-up water deaeration 补给水脱气
make-up water dealkalization 补给水脱碱
make-up water deionization 补给水去离子
make-up water iron removal 补给水除铁
make-up water of circulation system 循环系统补充水
make-up water percentage 补水率
make-up water pipe 补给水管
make-up water pretreatment 补给水预处理
make-up water pump 补给水泵;加水泵
make-up water tank 补水箱;补水池
make-up water treatment 补充水处理
make-up yard 编组场【铁】
make urgent repair on highway and bridge 抢修公路和桥梁
make use of 利用
make use of indigenous materials 就地取材
make use of scrap material 利用废料
make valve 补偿阀
make way 向前移动;航行
make way over the ground 对地移动
make way through the water 对水移动
make zero 归零
makinenite 三方硒镍矿
making 制作;制造
making a connection time 接单根时间
making-breaking point 断续点
making capacity 接通能力;接入容量;闭合容量;闭合能力
making contact 闭合接点
making cost 计算成本
making current 最大冲击电流;接通电流;闭合电流
making curve 闭合曲线
making design elaborately 精心设计
making different price according to distance categories 按距离别的差别运价
making different price according to freight categories 按货种类别的差别运价
making different price according to transport categories 按运输类别的差别运价
making drugs into frostlike powder 制霜
making full use of economic advantages 发挥经济优势原则
making full use of favorable factors 发挥优势原则
making good 粉刷;修整;修理;修复;修补工作;修补工程;研磨;完成;补偿
making good of concrete surface 混凝土表面修整
making instrument 仪器制造
making iron 捻缝凿
making light 近陆灯标;驶近陆地初见灯光;初见灯标
making low maintenance demands 提出低级维修要求
making mallet 捻缝木槌
making maul 捻缝木槌
making-out 放线
making overlapping run 多层焊
making power 遮盖能力
making price 定价
making report 编写报告
making retaining wall 修挡土墙
making the cash 筹集现金
making the land 驶近陆地
making time 闭合时间
making-up 船在干坞中撑垫材;引头子;上炉;包装

making-up block 封头砌块
making-up brick 间隙砖
making-up day 结算日
making-up differences 补进差额
making-up of train 编排列车
making-up or breaking-out tools 拧卸工具
making-up price 结算价格;结算日价;补进价格
making-up room 包装室
making way 正在航行
makintosh 防水胶布
makite 无水碱芒硝
makore 红木(西非)
makroskelic 长腿型
maksoorah (伊斯兰教寺院中围隔的)祈祷空间
Maksutov catadi optric objective 马克苏托夫反折射物镜
Maksutov corrector 马克苏托夫校正板;马克苏托夫改正镜
Maksutov objective 马克苏托夫物镜
Maksutov system 马克苏托夫系统
Maksutov telescope 马克苏托夫望远镜
Malabar kino 玛拉巴古纳树胶
malabsorption 吸收障碍
Malacca cane 马六甲藤
Malacca tin 马六甲锡
malachite 绿青;孔雀石;石绿
malachite green 孔雀绿露光计;孔雀绿;碱性孔雀绿;品绿
malachite green actinometer 孔雀绿光量计
malachite green lake 孔雀绿色淀;孔雀绿沉淀颜料
malachite green-phosphomolybdic heteropolyacid spectrophotometry 孔雀绿-磷钼多酸分光光度法
malachite green spectrophotometry 孔雀绿分光光度法
malachite green toner 孔雀绿色源
malachra fiber 马拉奇拉纤维(黄麻代用品)
malacolite 透辉石;白透辉石
malacolite gneiss 白透辉石片麻岩
malacolo 贝类学
malacon 水锆石;变水锆石
malactic 软化的
maladjusted 调整不良
maladjusted scale 比例失调
maladjustment 调整不善;顺应不良;适应不良;失配;失调;不一致性;不适应;不匹配
maladministration 管理不善
mala fide 恶意地
mala fide possessor 非法占有人
malaise 不适
malaise-Anglaise 币值萎缩
malakograph 软化率计
malalignment 相对位偏;偏心率;不同轴性;不平行性;不成一直线
malanite 马兰矿
Malan loess 马兰黄土
Malan stage 马兰期
Malaprade reaction 高碘酸氧化反应
malaria 瘴气;疟疾
malarial parasite 疟原虫
malate 苹果酸盐;苹果酸
malaxate 捏
malaxation 揉混;拌和;揉捏法
malaxator 搅拌机器;捏合器;碾泥机;黏土拌和器;黏土和机;揉泥机
malaxerman 拌料工
malayaite 马来亚石
Malay Camphor 马来樟脑
Malay floristic subregion 马来植物亚区
Malaysia 马来西亚
malchite 暗煌斑岩;微(晶)闪长岩
malcolmise = malcolmize
malcolmize 不锈钢表面渗氮处理
malcolmizing 氮化热处理;不锈钢表面渗氮
malcombustion 不正常燃烧
malcompression 未压紧
mal-condition 恶劣条件
malconformation 不均衡性
malcrystalline 残晶(的)
malcrystalline structure 残晶结构
maldevelopment 发育不良
maldistribution 分配不均匀;分配不公;分理纹乱;分布失常;分布不良现象;分布不均
maldonite 黑铋金矿

male 凸形的;插入式器件
male adapter 支承环;外螺纹过渡管接头;凸形密封环;凸面接头
male adaptor ring 外螺纹异径管接头
malealdehyde 马来醛
maleamic acid 马来酰胺酸;马来酸-酰胺
maleamide 马来酰胺;顺丁烯二酰胺
male and female 公与母
male and female face 阴阳face;凸凹面
male and fomale 凹凸面
maleate 马来酸(盐)酯;马来酸盐
maleated rosin 马来松香
maleate resin 马来酸酐二酸酯树脂
maleation 马来化
male bend 管包弯头;凸弯(管)
male box 插入箱
male branch tee 外螺纹三通
male building form 阳模
male caryatid 男性女像柱
male coconut 公椰子树
male cone 圆锥形塞;预应力锚塞;阳锥;锚塞;外锥面;外圆锥
male connection 轴套式连接;外螺纹接头
male connector 外螺纹管接头
male contact 刀口触片;插头;插塞接点
male coupling 阳端接头
male cross 阳四十字接头;阳十字;外螺纹十字接头
male die 阳(冲)模;内螺模;凸模;冲头
male dovetail 榫舌
male elbow 外螺纹弯头
male end 外螺纹端
male end of pipe 外螺纹管端;陶管喇叭头
male-female adaptor 公母接头
male-female fittings piece 公母接头
male-female seal contact face 凹凸密封面
male fitting 阳螺纹配件;阳模配合;阳螺纹管接头;外螺纹配件
male flange 凸缘法兰;凸法兰;凸出折边
male force 阳模
male friction cone 阳摩擦锥轮
male ga(u)ge 内测量用量规;塞规;内径规
male half coupling 外止口;凸止口
maleic acid 马来酸
maleic acid ester resin 马来酸酯树脂
maleic acid-modified vinyl copolymer resin 顺丁烯二酸改性乙烯(基)共聚树脂
maleic acid monoamide 马来酰胺酸;马来酸-酰胺
maleic acid resin 马来酸树脂;顺丁烯二酸树脂
maleic acid resin varnish 顺丁烯二酸树脂清漆
maleic amide 马来酰胺
maleic anhydride 马来酐;顺丁烯二(酸)酐
maleic anhydride terpinene adduct 萜品烯-顺(丁烯二酸)酐加成物
maleic anhydride value 马来(酸)酐值
maleic dialdehyde 马来醛
maleic ester resin 马来酯化树脂
maleic hydrazide 马来酰肼
maleic modified rosin 顺丁烯二酸酐改性松香
maleic oil 顺丁烯二酸化油
maleic polymer resin 顺丁烯二酸(型)聚合物树脂
maleic resin 顺丁烯二酸树脂;马来(酸)树脂
maleic resin varnish 马来酸树脂清漆
maleic rosin 马来松香
maleic treated oil 顺丁烯二酸酐化油;顺丁烯二酸改性油
maleimide 马来酰亚胺
maleinized tung oil 马来化桐油
maleinoid form 顺式
male member 凸状部分;凸件
male mo(u)ld form 阳模;上模
male nipple 有外螺纹的管接头;外丝扣管;外螺纹接管
malenite (呈木纹面的)装饰用胶合板
male nozzle 带外螺纹的喷嘴
MaLeod ga(u)ge 麦氏真空计
maleopimarate 马来海松酸酯
maleopimaric acid 马来海松酸
male pin 栓钉
male plug 插销;阳模;模塞;插头
male punch 凸模
male rest room 男休息室;男厕所
male rod 外丝钻杆
male rotor 凸形(螺旋)转子;凸形轮;被包容转子
male screw 阳螺旋;阳螺纹;阳螺钉;柱螺纹;外螺纹;外螺丝
male screw cutting tool 外螺纹车刀
male screw thread 外螺纹
male section 阳模
male spline 外花键
male square 外方头
male surface 被包容面
male tee [T] 外螺纹三通管
male thread 阳螺旋;阳螺纹;公螺纹;外丝扣;外螺纹;外螺丝
male thread nipple 外螺纹接套
male toilet 男厕所
male tool 压入式阳模
male union 管内接头;外螺纹连接管;外螺纹联管节
Maley brake 机电制动器
malformation 异常形成;结构变形;畸形;不正常部分;变形
malformation crystal 残缺晶
malformed 残缺的;畸形的
malformed fish 畸形鱼
malfunction 运行不正常;故障;机能失调;机能不良;误操作;事故;失灵;失调;错误动作;出错;不正常起动;不正常工作;不正常动作
malfunction advisory 故障报告
malfunction(al) alarm 功能故障警报信号;故障报警
malfunction alert 故障报警器
malfunction analysis 故障分析
malfunction analysis detection and recording 故障分析检查与记录
malfunction detection system 故障探测系统;故障检测系统
malfunctioning 功能不良
malfunction interrupt 故障中断;错误中断
malfunction probability 工作失常概率
malfunction routine 故障(检查)程序;定错程序;错误检查程序;查障程序
malfunction train 故障列车
Malia cement 麦地那水泥
Malibu board 马利布冲浪板
malic acid 羟基丁二酸;苹果酸
malic acid anil 苹果酸缩苯胺;苹果酸靛蓝
malic amide 苹果酰胺
malice 恶意
malicious accusation 恶意控告
malicious act 恶意行为
malicious arrest 非法扣押
malicious call tracing 恶意呼叫追踪
malicious damage 恶意破坏造成的损失
malicious damage clause 恶意损害条款
malicious destruction of property 恶意破坏财产
malicious falsehood 恶意欺诈
malicious injuries to property 恶意毁坏财物
malicious motive 恶意动机
malicious prosecution 恶意控告
malignant bacteria 有害细菌
malignant catarrhal fever 恶性卡他热
malignant catarrhal fever virus 恶性卡他热病毒
malignant malaria 瘴疟;恶性疟疾
malignant scarlet fever 恶性猩红热
malignite 暗霞正长岩
Malimo knitting machine 玛里莫缝编机
mall 绿荫步道;林荫(小)路;步行商业街;步行街(道);手用大锤;大木槌
malladrite 氟硅钠石
mallard 野鸭;绿头鸭
mallard falcated teal feather 白银边毛
mallardite 水锰矾;白锰矾
Mallar's constant 马氏常数
malleability 延展性;可展性;可压延性;可锻性;加工性;韧性
malleability test 锻造性试验
malleability wrought iron 展性锻铁
malleabilization 锻化
malleable 有延展性的;可压制的;可轧的;可锻的
malleable anneal 可锻化退火
malleable brass 孟滋合金;芒茨合金
malleable capital 可延展的资本
malleable casting 展性铸件;韧性铸件
malleable cast-iron 可锻铸铁;韧性铸铁;展性铸件;纯铁
malleable cast-iron pipe fitting 玛钢管件
malleable detachable chain 展性活动链
malleable hard iron 韧性铸铁
malleable iron 韧性铁(可锻铸铁);纯铁

malleable iron clip 展性铸铁箍
malleable iron fitting 锻铁配件
malleable iron pipe 展性铸铁管；玛钢管
malleable iron square 活络角尺
malleable iron trap 玛钢存水弯
malleable iron washer 锻铁垫圈
malleableize 韧化(作用)
malleable mild steel 韧性软钢；韧性低碳钢
malleableness 展性
malleable nickel 展性镍
malleable pig 韧性铸铁
malleable (pig) iron 展性铸铁；制造可锻铸铁的生铁；可锻生铁
malleable pintle chain 展性锁链
malleable shot 可锻铸铁铁丸
malleable steel 展性钢；软钢；韧(性)钢
malleable wrought iron 展性锻铁；韧性锻铁
malleablising 韧性退火；韧铁退火
malleablization 韧化(作用)
malleablize 可锻化
malleablizing annealing 可锻化退火
mallear stria 锤纹
malleate 锻；锤薄
malleate trophus 锤形咀嚼器
malleation 锻
mallee 热带密灌丛
mallee scrub 桉树矮木；桉树密灌丛
malleolar surface 咬合面
mallet 木锤；木槌；小槌
Mallet alloy 马雷特黄铜；马雷特合金；黄铜
mallet cutting 槌形插条
malletfinger 锤状指
mallet handle die 锤压式冲模
mallet-headed chisel 砖工圆头凿；圆头凿；槌头凿
mallet perforator 锤式冲孔机
mallet toe 锤状趾
mallet vase 槌形瓶
mall hammer 手锤；手用大槌
mall net 林荫路网(散步用)
Mallory 马洛里合金
Mallory metal 马洛里无锡高强度青铜
Mallory sharton alloy 钛铝锆合金；马洛里-沙顿钛铝锆合金
malloy 镍钼铁超磁导合金
Malloydium 马洛伊迪昂铜镍锌耐蚀合金
Malm 晚侏罗世；上侏罗统
malm 钙质砂土；白垩砖；白垩土
malm brick 灰砂砖；白垩砖
malm grizzled brick 欠烧软砖
Malmquist correction 马姆奎斯特改正
malm rock 灰砂砖；黏土砂岩；黏土砂岩(玛姆砂岩)
malm rubber 软白垩砖
malmstone 黏土砂岩；玛姆岩；灰砂岩；燧石质砂岩(铺路用)；砂岩
malmy 泥灰岩的
malnutrition 营养不良
malobservation 观测误差
malocclusion 咬合不正
malodo(u)r 恶臭
malodo(u)rant 恶臭物
malodo(u)r control 恶臭控制
malodo(u)r counteraction equipment 中和脱臭装置
malodo(u)r gas 恶臭气体
malodo(u)r masking equipment 掩蔽脱臭装置
malodo(u)rous 恶臭的
malodo(u)rous black river water 黑臭河水
malodo(u)rous black river water purification 黑臭河水净化
malodo(u)rous black water body 黑臭水体
malodo(u)rous cargo 有恶臭货物
malodo(u)rous substance 恶臭物质
Malon 马纶聚丙烯腈系纤维
malonaldehydic acid 丙醛酸
malonamic acid 丙酰胺酸
malonamide 丙二酰胺
malonamide nitrile 氰基乙酰胺
malonate 丙二酸酯；丙二酸盐；丙二酸根
malonic acid 丙二酸
malonic anhydride 丙二酸酐
malonic ester 丙二酸酯；丙二酸二乙酯
malonic methyl ester nitrile 氰基乙酸甲酯
malonic mononitrile 氰基乙酸
malonic semi-aldehyde 丙二酸半醛

malonnitrile 丙二腈
malononitrile 丙二腈
malonuric acid 脲羰基乙酸；脲基甲酰乙酸
malonyl 丙二酰基；丙二酰
malonyl urea 丙二酰脲；巴比土酸
maloperation 误操作；失灵；操作不当；不正确运行；不正确维护；不正确操作
Malora 马罗拉仿金属线
Malotte's metal 马洛特合金
malotth 马罗特易熔合金
Malott metal 锡铅铋易熔合金；马罗特易熔合金
malpais 玄武岩区；熔岩区
Malpighian tube 马氏管
malpractice 职业过失
malreduction 复位不良
malrotation 旋转不良
Malsuyama 松山反向期
maitene 马青烯
Maltese cross 马氏机构；马耳他十字架
maltha 沥青柏油胶；软沥青；软地沥青；土沥青；半液质沥青
malthacite 漂布土
malthene 软沥青质；马青烯
malthite 软(地)沥青；土沥青
malthoid 油毛毡；纸胎油毡；低脂油毡
malthoid roofing 油毛毡屋面
malt-house 发酵厂
Malthusian theory 马尔萨斯(人口)理论
Malthusian thesis 马尔萨斯(人口)论
malticolo(u)r plaid shirt 复色长方格衬衫
malting effluent 酿造排放液
malting wastewater 制麦芽糖废水
maltose 麦芽糖
Maluku Sea 马鲁古海
malunion 连接不正；畸形连接
malus (古罗马剧场中支撑帐篷的)杆子
Malus cosine-squared law 马吕斯余弦平方定律
Malus law 马吕斯定律
Malus law of rays 马吕斯光线定律
Malus theorem 马吕斯定理
Malvinoakffric brachiopod realm 南方腕足动物地理区系
malybdite 钼华
mamagement of urban industrial wastewater 城市工业污水管理
mamatele 马马特勒风(西西里岛的西北风)
mamatili 马马特勒风(西西里岛的西北风)
mambrane tension 膜片刚度
mamelon 乳状突；圆穹；小圆丘；圆丘
m-aminophenol 间氨基酚
mammal 哺乳动物
Mammalia 哺乳动物纲
mammary calculus 乳石
mammillary structure 乳房构造【地】
Mammisi temple 法老诞生神庙(古埃及)
mammock 块石；碎块
mammoth 巨型的；马木斯式起重机；巨物
Mammoth antenna 马木斯天线
mammoth blast 露天大爆破；露天爆破；大爆破
mammoth converter 大型转炉
Mammoth cotton 马木斯棉(美国佐治亚产晚熟陆地棉)
Mammoth crane 马木斯式起重机
Mammoth event 马木斯反向事件
mammoth ocean barge 大型海驳
mammoth pump 气动泵；超大型泵
Mammoth reversed polarity subchronzone 马木斯反向极性时间带
Mammoth reversed polarity subzone 马木斯反向极性亚带
Mammoth reversed polarity superzone 马木斯反向极性超带
Mammoth rotor 马木斯旋曝气器
mammoth structure 巨型结构；巨大结构
mammoth tanker 大型油轮；大型油船(4万～5万吨以上)
mammoth tree 猛犸树
mammoth vessel 巨型船舶
Mammut crusher 猛犸型破碎机
mammy chair 运人吊篮
man 配备人员
man-acoustic(al) stone 隔声人造石
manage 经营
manageability 可使用性；可驾驶性；可管理性

manageable variable 易控变量
managed capacity cost 经营能力
managed cost 预测成本
managed fixed cost 预测固定成本
manage discharge monitoring 管理排污监测
management 管理；操纵
management accounting 管理计算；管理会计
management activity 管理活动
management advisory service 管理咨询服务
management agency 管理处
management agent 经营代理商
management analysis 管理分析
management analysis center 管理分析中心
management analysis reporting system 管理分析报告系统
management and administrative rights 经营管理权
management and general expense 营业开支
management and operation 经营管理
management at different levels 分级管理
management at factory level 厂级管理
management audit 内部审计
management by control 控制式管理
management by drive 压制管理
management by exception 例外管理
management by exception technique 例外管理技术
management by objective 目标管理
management by objectives during term of service 任期目标管理制
management by personal supervision 人的监督管理
management by results 依结果管理；目标管理
management by synchronization 同步管理
management committee 管理委员会
management competence 经营才干
management consultant 业务顾问；经营顾问
management consulting 经营诊断
management consulting firm 管理顾问公司
management content 管理信息目录(数据库)
management contract 管理合同；经营合同；企业经营合同
management control 管理控制；经营管理控制；经营管理
management control console 管理控制台
management control level 管理控制级
management control system 管理控制系统
management cost 管理费
management cost accounting 管理成本会计
management cost control 管理成本控制
management cycle 管理周期
management data 管理资料；管理数据
management decision 管理决策
management decision system 管理决策系统
management direct cost 直接管理费
management ecology 管理生态学
management efficiency 经营效率；经营管理效率
management engineer 管理工程师
management engineering 管理工程；经营管理工程学
management environment 管理环境；经营环境
management expense 管理费用
management factors 管理因素
management function 管理机能
management funds volume 经营性资金容量
management game 管理技巧；管理对策；管理策略；经营比赛
management gap 管理差距
management improvement 经营改善运动
management indirect cost 管理部门间接成本
management information 管理信息
management information center 管理情报中心
management information requirement 管理信息要求
management information system 信息控制系统；管理信息系统；经营情报系统
management leve computer 管理级计算机
management level 管理层次
management map 森林经营图
management measures of radioactive medicines and chemical reagents 放射性药物管理办法
management measures of water body pollution sources 水体污染源管理措施

management model 管理模型
management of agricultural estates 农业地产管理
management of a ship 操纵船
management of blueprint 图纸管理
management of coal resources 煤炭资源管理
management of commercial and industrial firms 工商企业管理
management of commercial erterprises 商业企业管理
management of commission shops 信托商店管理
management of construction and installation enterprise 建筑安装企业管理
management of drilling production 钻探生产管理
management of energy consumption 能源消耗管理
management of equipment 设备管理
management of export contracts 出口合同管理
management of factors of scientific research 科研条件管理
management office 管理处
management of hazardous solid waste 危险固体废物管理
management of human difference 人的差异管理
management of import orders 进口订货管理
management of indirect cost 间接成本管理;间接管理费
management of land development 土地开发管理
management of levee 堤防管理
management of navigation channel 航道管理
management of offshore drilling platforms 海上钻探平台管理;海上钻井平台管理
management of participation 参与经营管理
management of polluted ecosystem 污染生态系统管理
management of pollution risk 污染风险管理
management of pollution sources 污染源管理
management of process 进程管理
management of railway transportation cost and expenditures 铁路运输成本费用管理
management of research and development 科研管理
management of research subjects 课题管理
management of river basin 流域管理
management of science and technology 科技管理学
management of scientific research 科学管理学
management of shifting products 搬运管理
management of solid wastes 固体废物管理
management of taxation 征税管理
management of tools and equipment 农业机具的管理
management of trade 行业管理
management of underground minestopes 矿山井下采场管理
management of urban domestic sewage 城市生活污水管理
management of urban industrial wastewater 城市工业废水管理
management of urban non-point source pollution model 城市非点源污染管理模型
management of water body pollution 水体污染源管理
management of water environment 水环境管理
management of water resources 水资源管理
management of water use 用水管理
management of weeds 防除杂草
management of wholesale enterprises 批发企业管理
management of working groups 小组管理
management operating system 管理信息系统;管理经营系统;管理工作系统;管理操作系统;经营管理制度
management organization 管理机构
management organs of forest resources 森林资源管理机构
management package 管理应用程序包
management participation 经营参与;参与管理
management performance 管理成效;经营成效
management plan of gardens 园林管理规划
management plan area 管理计划范围
management plan(ning) 管理计划;管理规划
management planning and control 管理规划及监督
management planning system 管理计划系统

management policy of wastes 废物管理政策
management power 经营权
management principle 经营方针
management procedure 管理过程;管理方法
management procedure of environment(al) impact assessment 环境影响评价管理程序
management process 管理过程
management profit charges 经营盈利额
management quality 管理质量
management quality control 经营质量管理(法)
management rate 整理发货手续费
management record 管理记录
management reform 管理改革
management research 管理研究
management review 管理评审;经营调查
management right 经营权
management schedule 管理计划
management science 管理结构;管理科学;经营科学;经营管理学;经营管理科学
management science model 管理学模型
management scientist 管理科学工作者
management service 管理机构;管理业务;经营管理业务;管理系统;管理服务(处)
management shares 经营股份
management signal 管理信号
management software 管理软件
management specialization 专业化管理;经营专门化
management sponsor 企业(经营)管理
management standard 管理标准
management strategy 管理对策;经营战略
management structure 管理结构
management structure of company legal body 公司法人治理结构
management study 经营管理研究;管理研究
management supervisory system 管理监视系统
management support utility 管理支援应用程序;管理支持实用程序;管理维护实用程序
management system 管理制度;管理系统;管理法
management system engineering 经营管理系统工程
management system for the paid use of land 土地有偿使用管理制度
management system of choosing supplement stock 选择补充库存管理制度
management system of ration stock 定量库存管理制度
management system of regular stock 定期库存管理制度
management system of water resources 水资源管理体制;水资源管理体系
management tactics 经营策略
management technique 管理技术
management theory 经营管理理论
management theory and method research 管理理论与方法研究
management thought 经营思想
management through cost 成本管理
management through figures 图表管理
management through synchronization 同步对照管理
management training 管理训练;经营管理培训
management training software 管理专业训练软件
management volume inventory 森林蓄积清查
management work station 管理工作站
manager 主任;管理者;管理人员;理事;经营者;经理(人);经管人
managerial and administrative expertise 经营管理水平
managerial and administrative personnel 经营管理人才
managerial and operational rationalization 经营管理合理化
managerial approach 管理方法
managerial audit 经营审计
managerial body 管理集团
managerial decision-making 经营决策
managerial economics 管理经济(学);经营经济学
managerial force 经营管理人员
managerial grid 管理方阵
managerialist 管理学家
managerial man model 经营人模型
managerial method 管理方法
managerial report 管理报告;管理报表

managerial report system 管理报告系统;管理报表系统
managerial role 管理的作用;管理的任务
managerial staff 管理人员
managerial system in commercial enterprises 商业企业管理体制
managerial theories of the firm 厂商的管理理论
manager of programming 程序管理员
manager responsibility system 经理负责制
manager responsibility system of a key project under construction 项目经理负责制
manager's discretionary limits 经理自主权限
manager selection 选用经理
manager's room 经理室
managing agent 代管人
managing bank 经办银行
managing director 总经理;总裁;执行董事;常务董事总经理;常务董事
managing expenditures 管理费
managing for results 目标管理
Managing Measures of Cutting and Reforestation of Forest 森林采伐更新管理办法
managing operator 运务经理;委托经纪人
managing owner 经理船舶业务的船东
managing partner 执行业务股东;执行合伙人;经营合伙人;全权合伙人
managing point source 管理点污染源
managing soil resources 管理土壤资源
managing surveyor 检验主任工程师
managing system of enterprise 企业管理制度
managing wastewater 管理废水
managing water environmental capacity 管理水环境容量
managing water quality 管理水质
man amplifier 体力放大器
man and biosphere program(me) 人与生物圈计划
man-and-material hoist 副井提升机
manandonite 硅硼锂铝石;锂硼绿泥石
Manans faience 马兰期锡釉陶器
manasseite 水碳铝镁石;水镁铝石
man-assignment 人力分配
man-assignment schedule 劳动力调配表
manauto 手控-自控;手控-自动
manbarklac 橄榄绿至红褐色硬木(中美洲产)
man cage 载人罐笼;提人罐笼;乘人罐笼
man-capstan 人力绞盘;人力绞磨
man car 人送人员小车;乘人矿车
man-carried 便携的
Manchester 曼彻斯特
Manchester brown 曼彻斯特棕
manchesterism 自由贸易主义
Manchester plate 曼彻斯特极板
Manchester's code 曼彻斯特编码
Manchester sett system 曼彻斯特经密制
Manchester yellow 曼彻斯特黄
Manchurian ash 水曲柳
Manchurian walnut 胡桃楸木
Manchurian fir 辽东冷杉
man-computer communication 人机通信
man-computer dialog 人机对话
man-computer interaction 人机配合设计;人机联系
man-controlled machine 人控机器
man-controlled mobile robot 人控机器人
man conveyer belt 载客运输带
man crazing 路面网状裂纹
mandamus 法院的训令;职务执行令
mandapa (印度教寺院的)大厅或大凉台
mandapam (印度庙前的)前庭
mandarah (埃及及阿拉伯人的)客厅
mandarin 柑橘(树)
mandarin duck 鸳鸯
mandarinoite 水硒铁石
mandate 指定;委托;财产委托
mandate agency 委托代理
mandated project 指令性项目;委托项目
mandator 委托人
mandatory 命令者;命令人;强制性的;委任的;受(委)托人;代理人
mandatory administration 委托管理
mandatory age for retirement 法定退休年龄
mandatory and customary benefits 个人福利和惯例待遇
mandatory annual survey 海上人命公约年度检查
mandatory arbitration 强制仲裁;强制性制裁

mandatory construction schedule 施工控制性进度
mandatory contribution 法定缴款
mandatory control 强制性控制
mandatory expenditures 强制性支付;强制性支出;强制性开支
mandatory filtration 强制过滤
mandatory forbidding sign 禁止标志
mandatory instruction sign 强制性指示标记
mandatory layer 标准层
mandatory level 标准等压面
mandatory member 永久性成员
mandatory membership class 必须属籍类别
mandatory minimum thickness 规定最小厚度
mandatory period 法定期限
mandatory planning 指令性计划
mandatory provisions 约束性条款;强制性规定
mandatory requirement 强制性要求;必须遵循的要求
mandatory restrictions on exports 强制出口管制
mandatory retirement 强迫退休
mandatory road marking 指示路面标志
mandatory rule 强制性守则;强制性规则
mandatory sanction 强制性制裁
mandatory sign 指示标桩
mandatory standard 强制性标准
mandatory treatment requirement 强制处理要求
man-day 工日;人一天;人(工作)日;人工生产日
man-day of legal holiday 公休工日
man-day of strike idleness 罢工荒废的工日(数)
man-day rate 人日单价
man days actual used in loading/discharging 实际装卸工作日数
man days in absence 缺勤工日数
man days of absence from work 缺勤工日
man days of legal holidays 公休工日数
mandelato 白斑红大理岩
Mandelbaum's reaction 丝状反应
Mandel-Cryer effect 曼代尔-克雷尔效应
mandelic acid 孟德立酸
mandir 印度教庙宇
man door 人行小门;进人门
mandorla 圣像周围的光轮;杏圆形装饰板;圣像周围的光晕
man drawn truck 手拉车
mandrel 圆柱木芯块;旋转筒;芯子;芯轴;芯模;芯棒;桩芯;轴棒;紧轴
mandrel bend test 轴棒棒弯曲试验
mandrel coiler 筒式卷取机
mandrel down coiler 辊式地下卷绕机
mandrel drawing 长芯棒管材拉拔
mandrel expand 卷筒胀大
mandrel for bending 弯管用芯棒
mandrel hanger 油管挂
mandrel holder 芯座
mandrel mill 芯棒式无缝管轧机
mandrel payoff reel 卷筒式开卷机
mandrel piler 卷筒式堆板机
mandrel press 芯轴压机;手动压力机
mandrel roll 卷筒式压力棍
mandrels for pipe bending 弯管用的芯杆
mandrel socket 球窝芯轴;套管矫直器
mandrel stripper 脱芯机;拔芯机
mandrel stroke limiter 顶杆行程限位器
mandrel supporter 马架
mandrel test 卷解试验;紧轴压入试验
mandrel uncoiler 卷筒式开卷机
mandrel work 最后加工工作
mandril 车床旋转轴(加工件固定在轴上);芯轴;芯棒;芯子;顶杆;硬木圆棒;桩芯
mandrin 细探针
man-earth system 人一地系统
Manebach law 马尼巴赫定律
manebach twin 底面双晶
manebach twin law 底面双晶律【地】
Manedach law 曼尼巴律
man efficiency 人工效率
manege 马术学校
manengine 坑内升降机(隧道矿井用)
man equivalent 人等量
maneton 曲柄颈;曲臂夹紧螺栓
maneton bolt 拉紧螺栓
maneuver 机动;手法;操作法;操纵
maneuverability 灵敏性;灵活性;可控性;可操纵性;驾驶性能;机动性;(船舶的)操纵性(能)

maneuverability coefficient 操纵性系数
maneuverability criterion 机动性准则
maneuverability index 操纵性指数
maneuverability trial 操纵性试验
maneuverable 可调动的
maneuverable bridge 开合桥
maneuverable policy 机动灵活的政策
maneuverable ship 操作性能良好的船舶
maneuver boat 作业船
maneuver energy 机动能;操纵能
maneuver for locking 进出船闸的操纵
maneuver for man overboard 救助落水人员操纵
maneuvering air compressor 操纵用压气机;操纵用空气压缩机
maneuvering area 掉头区(船舶);操纵区
maneuvering ballistic reentry vehicle 机动弹道再入飞行器
maneuvering basin 回船(港)池;掉头区;船舶转头(港)池;船舶掉头港池
maneuvering board 船舶运动图
Maneuvering Board Manual 船舶运动图手册
maneuvering cab 操纵室
maneuvering capability 操纵能力
maneuvering chain 操纵链
maneuvering control 操纵控制
maneuvering craft 受控航行器
maneuvering dial 操纵转盘
maneuvering engine 操纵发动机
maneuvering gear (舵机的)操纵装置;操舵装置
maneuvering inside harbo(u)r 港内船舶调动
maneuvering lane 驾驶航道
maneuvering light 操纵灯(拖轮)
maneuvering platform 操纵台
maneuvering property (船舶的)操纵性能
maneuvering range 机动范围
maneuvering room 操纵室
maneuvering satellite 机动卫星
maneuvering ship 从动船
maneuvering simulator 操纵模拟器
maneuvering space 车辆行驶净空
maneuvering speed 机动速率
maneuvering stability 机动稳定性
maneuvering system 机动系统
maneuvering telegraph 车钟【船】
maneuvering test 操纵试验
maneuvering trials 运转试验
maneuvering under sail 帆船操纵方法
maneuvering valve 操纵阀
maneuvering winch 机动性绞车
maneuvering with the aid of tug 拖轮协助操纵
maneuver margin 机动限度;机动能力储备
maneuver rate 航向变换率
maneuver winch 操纵绞链
maneuver with tugs 拖轮协助操纵
man-excited vibration 人工激振
mangal 锰铝合金
Mangalite 曼格莱特铝基活塞合金
manganactinolite 锰阳起石
Manganal 曼加那尔高强度耐磨含镍高锰钢;含镍高锰钢
manganalluaudite 锰钠磷铁矿
manganalmandite 锰铁铝榴石
manganandalusite 锰红柱石
manganankerite 锰铁白云石
manganapatite 锰磷灰石
manganate 锰酸盐
manganaxinite 锰斧石
manganbabingtonite 硅锰灰石
manganbelyankinite 铌钛锰石
manganberzeliite 锰砷钙镁矿;黄砷榴石
manganblende 辉锰矿;硫锰矿
manganchlorite 锰绿泥石
mangandiaspore 锰水铝石
mangandisthene 锰硅铝石
mangandolomite 锰白云石
manganeisen 锰铁合金
manganese 锰钢衬板;锰
manganese acetate 乙酸锰;醋酸锰
manganese addition 锰增量
manganese alloy 锰合金
manganese-alumina pink 锰红
manganese-aluminum 锰铝合金
manganese-amphibole 锰闪石
manganese-anorthite 锰钙长石

manganese-autunite 锰铀云母
manganese bacteria 锰细菌
manganese bath 锰浴
manganese-bearing formation 含锰建造
manganese-bearing limestone 含锰灰岩
manganese-bearing shale 含锰页岩
manganese black 软锰矿;锰黑
manganese black pigment 软锰矿颜料;锰黑颜料
manganese blue pigment 锰蓝颜料
manganese borate 硼酸锰
manganese-boron 锰硼合金
manganese boron alloy 锰硼合金
manganese brass 锰黄铜
manganese bronze 锰青铜
manganese bronze catching finger 锰钢制抓卡
manganese brown 锰棕(色)
manganese carbonate 碳酸锰
manganese carbon steel 锰碳钢
manganese casting brass 铸造锰黄铜
manganese cast-iron 锰铸铁
manganese catalyst oxidation process 锰催化氧化工艺
manganese catalyzed ozonation 锰臭氧催化
manganese cell 锰电池
manganese cement 锰水泥
manganese concentrate 锰精砂
manganese concretion 锰质结核
manganese copper 锰铜
manganese copper alloy 锰铜合金
manganese copper-nickel 锰铜镍合金
manganese copper resistance wire 锰铜电阻线
manganese crust 锰结壳
manganese deposit in the weathering crust 风化壳锰矿床
manganese difluoride 二氟化锰
manganese dioxide 软锰矿;二氧化锰
manganese dioxide electrolytic 电解二氧化锰
manganese dioxide scavenging 二氧化锰清除法
manganese disulfide 二硫化锰
manganese drier 锰干料;锰催干剂
manganese dry cell 锰干电池
manganese epidote 锰绿帘石;红帘石
manganese family element 锰族元素
manganese fayalite 锰铁橄榄石
manganese fertilizer 锰肥
manganese flight 锰钢链板
manganese fluoride 氟化锰(二价锰)
manganese fluosilicate 氟硅酸锰
manganese-gehlenite 锰黄长石
manganese glass 锰质玻璃
manganese green 锰酸钡;锰绿
manganese green sand 锰绿砂
manganese group 锰基
manganese halide 卤化锰
manganese hypophosphite 次磷酸锰
manganese in hair 发锰
manganese in stool 粪锰
manganese iodide 碘化锰
manganese ion 锰酸根离子;锰离子
manganese iron 锰铁合金;锰铁
manganese-iron-copper deposit on the modern ocean floor 现代海洋底部铁锰铜矿床
manganese isotope 锰同位素
manganese linoleate 亚油酸锰
manganese-merwinite 锰镁硅钙石
manganese metal 金属锰
manganese mineral 锰矿
manganese module 锰结核
manganese monosulfide 一硫化锰
manganese monoxide 一氧化锰
manganese mordant 锰媒染剂
manganese mud 锰矿渣
manganese muscovite 锰白云母
manganese naphthenate 锰环烷酸盐;环烷酸锰
manganese nitrate 硝酸锰
manganese nitride 二氮化三锰
manganese nodule 锰质结核;锰团块;锰矿瘤
manganese nodule property 锰结核性状
manganese nodule province 锰结核区
manganese nodules crust survey 锰结核壳调查
manganese nodules on the ocean floor 洋底锰结核
manganese oleate 油酸锰
manganese ore 锰矿石;锰

manganese oxalate 草酸锰
manganese oxychloride 氯氧化锰(制无缝地板用)
manganese pavement 锰结皮
manganese pennine 锰叶绿泥石
manganese permutite 锰人造沸石
manganese peroxide 过氧化锰
manganese phosphate 磷酸锰
manganese pigment 锰颜料
manganese poisoning 锰中毒；锰毒害
manganese powder 锰粉
manganese red 锰红
manganese removal 除锰(法)
manganese removal filter 除锰滤池
manganese removal plant 除锰站；除锰装置
manganese resinate 树脂酸锰
manganese rock 锰质岩
manganese rocks 锰质岩类
manganese salt 锰盐
manganese scraper 锰钢耙斗
manganese sediments 锰质沉积物
manganese sesquioxide 三氧化二锰
manganese-shadlunite 硫铜锰矿
Manganese Shale Group 曼加内斯页岩群
manganese shoe 锰钢履块
manganese-silicon 锰硅合金
manganese silicon spring steel 锰硅弹簧钢
manganese silicon steel 锰硅钢
manganese silicon tool steel 锰硅工具钢
manganese soap 锰质皂；锰皂
manganese spar 菱锰矿；蔷薇辉石
manganese spring steel 锰弹簧钢
manganese stain 锰帽
manganese star 锰星
manganese steel 锰钢
manganese steel insert crossing 锰钢镶嵌辙叉
manganese steel pan feeder 锰钢溜槽给料器
manganese steel rail 锰钢钢轨
manganese sulfide 硫化锰
manganese tipped switch 锰钢尖轨转辙器
manganese titanate 钛酸锰
manganese-titanium 锰钛合金
manganese tool steel 锰工具钢
manganese tungstate 钨酸锰矿
manganese vanadium steel 锰钒钢
manganese vanadium tool steel 锰钒工具钢
manganese violet 锰紫
manganese violet pigment 锰质紫色涂料
manganese vitriol 锰胆矾
manganese wedge-bar liner 锰钢制楔形棒条内衬(球磨机)
manganese zeolite 锰沸石
manganese-zinc ferrite 锰锌铁氧体
manganese zinc pyroxene 锰锌辉石
manganese zinc spar 锰菱锌矿
manganeshoeresite 砷镁锰石
manganesian fayalite 锰铁橄榄石
manganesian magnetite 锰磁铁矿
manganesite putty 耐热油泥
manganfayalite 锰铁橄榄石
manganfluorapatite 锰氟磷灰石
manganglauconite 锰海绿石
manganhedenbergite 锰钙铁辉石
manganhumite 硅锰石
manganic acid 锰酸
manganic compound 锰化合物
manganic drier 锰催干剂
manganic fluoride 氟化锰(三价锰)
manganic oxide 三氧化二锰
manganides 锰系元素
manganiferous 含锰的
manganiferous white cast-iron 含锰白口铸铁
manganilmenite 锰钛铁矿
manganin(alloy) 锰镍铜合金；锰铜镍线
manganin manganin alloy 锰镍铜合金
manganin soil ga(u)ge 锰合金电阻压力计
manganin wire 锰铜线
mangani-sicklerite 磷锰锂锰矿
manganismus 锰中毒
manganite 亚锰酸盐；水锰矿
manganjustite 锰黄长石
mangankoninckite 锰针磷铁矿
manganleonite 锰钾镁矾
manganludwigite 锰硼镁矿
manganmagnetite 锰磁铁矿

manganmelanterite 锰绿矾
manganmonticellite 锰钙橄榄石
manganmuscovite 锰白云母
manganneptunite 锰柱星叶石
mangano-anthophyllite 锰直闪石
manganobarium muscovite 锰钡白云母
manganocalcite 锰方解石；含钙菱锰矿
manganochromite 锰铬铁矿
manganocolumbite 铌锰矿
manganoconiosis 锰尘肺；锰尘沉着病
manganoferrogehnite 锰铁锌尖晶石
manganolangbeinite 锰钾镁矾；无水钾锰矾
manganolite 锰质石；锰矿岩；蔷薇辉石
mangano-manganic oxide 四氧化三锰
manganomelane 锰土
manganonatrolite 锰钠沸石
manganopectolite 锰针钠钙石
manganophyllite 锰黑云母
mangan-orthite 锰褐帘石
manganosiderite 锰菱铁矿
manganosite 方锰矿
manganosteenstrupine 锰菱黑稀土矿
manganostibite 锑砷锰石
manganostrengite 锰红磷铁矿
manganotantalite 钽锰矿
manganotapiolite 重钽锰矿
manganotremolite 锰透闪石
manganous chloride 二氯化锰
manganous dichloride 二氯化锰
manganous diiodide 二碘化锰
manganous fluoride 氟化锰(二价锰)
manganous fluosilicate 氟硅酸亚锰
manganous iodide 碘化锰
manganous oxalate 草酸锰
manganous oxide 一氧化锰
manganous silicate 硅酸锰
manganous sulfate test 硫酸亚锰试法
manganous sulfide 一硫化锰；硫化锰
manganoxyapatite 锰氧磷灰石
manganpalygorskite 锰铝坡缕石
manganpectolite 锰针钠钙石
manganpyrite 锰黄铁矿
manganpyrosmalite 锰热臭石
mangansepiolite 锰海泡石
mangansmithsonite 锰菱锌矿
manganuralite 锰纤闪石
manganvesuvianite 锰符山石
manganvoelckerite 锰氧磷灰石
manganvoltaite 锰钾铁矾
manganwollastonite 锰硅灰石
mange 螨病
Mangelinvar 曼格林瓦钴铁镍锰合金；钴铁镍锰合金
manger 链舱底泄水板；饲料槽；船首楼锚链孔后挡水板
manger board 挡水板；船首楼锚链孔后挡水板
mangerite 纹长二长岩
Mangin mirror 曼金镜(消球差的折反射镜)
Mangin primary 曼金主镜
mangle 割碎；刮刀；矷光机；碾压滚轴光泽机；预压合片机；轧液机；轧扎机；辊式板材矫直机；碾压机；碾土机；吊篮运输机
mangle dryer 立式链板干燥器
mangle gear 轧布轮；滚销齿轮
mangle gearing 往复齿条装置
mangler 矷光机
mangling 碾压切削
mango 芒果树
mango gum 芒果胶
Mangol 曼戈尔铝锰合金
Mangonic 锰基锰合金；曼戈尼克镍锰合金
mangrove 红树属植物；红树(林)
mangrove bark 红树皮
mangrove beach 红树林海滩
mangrove coast 红树林海岸
mangrove community 红树林群落
mangrove ecosystem 美洲红树生态系统；红树林生态系统
mangrove marsh 红树林草沼
mangrove marsh deposit 红树林沼泽沉积
mangrove peat 红树林泥炭
mangrove soil 红树林沼泽土
mangrove swamp 红树沼泽；红树林沼泽；红树林沼；红树低洼地
mangrove tannin 红树丹宁酸

mangrove vegetation 红树林植被
mangrove wetland 红树林湿地
manhaden oil 步鱼油
man handle 人工操作
manhandle boring hole 人力推钻钻孔
manhandled 人力推动的
man-handling of materials 人工转运；材料的人工转运；物料人工搬运
Manhattan 曼哈顿商业区
Manhattan distance 曼哈顿距离
man head 平均水头；人孔
manhole 检查井；升降井；窨井；工作人员进出口；工作人孔；检修孔；人孔；人洞；清除口
manhole block 检查井砌块
manhole bottom 检查井底
manhole chamber 人孔(探视)室；进人孔间
manhole coaming 人孔围壁；人孔围板
manhole cover 进人孔盖；检查井盖；人孔盖；探井盖；阴沟盖
manhole covering 人孔盖
manhole cover plate 人孔盖板
manhole cross bar 人孔横杆
manhole dog 门闩；横杆；人孔扣夹；人孔盖压板；人孔撑架；十字架
manhole door 检查门；人孔门
manhole foundation 检查井基础
manhole frame 人孔框架；检查井盖圈；检查井座；检查井框架
manhole frame and cover 进人孔盖架；检查井盖架
manhole grid 人孔格栅
manhole grill(ag)e 人孔格栅
manhole guard 人孔拦栅
manhole head 铸铁盖；进人孔盖架；检查井盖(架)；人孔盖
manhole in tunnel 隧道躲避所
manhole invert 检查井内底
manhole iron steps 检查井铁踏步
manhole junction box 检查孔套管；人孔套管
manhole ladder step 检查井梯阶
manhole masonry(work) 人孔的砖石工程
manhole on end of air duct 风渠挡头板通道门
manhole opening 人孔口
manhole plate 检查孔盖；人孔盖
manhole removal 检查井回填；检查井拆除
manhole ring 窨井圈；人孔圈；人孔加强圈；人孔环
manhole seal 检查井封口
manhole sewer brick 人孔下水道砖
manhole shaft 检查井井筒
manhole size 人孔尺寸
manhole step 检查井踏步；人踏步；人孔踏步
manhole step iron 人孔铁踏步
manhole sump 检查井落底；检查井沉泥槽
manhole wall 探井壁；人孔墙
manhole walls and invert 人孔壁及仰拱
manhole with butt welded flange 对焊法兰人孔；对焊凸缘人孔
manhole with common welded flange 平焊法兰人孔；平焊凸缘人孔
manhole with convex cover 凸形盖人孔
manhole with hanging cover 吊盖人孔
manhole with horizonal hanging cover 水平吊盖人孔
manhole with sight glass 带视镜人孔
manhole with vertical hanging cover 垂直吊盖人孔
man-hour 一人一小时的工作量；工时；人时；人(工)小时
man-hour base 人工小时法
man-hour consumption of drainage 排水工时消耗
man-hour cost 工时成本；人时成本
man-hour distribution task report 人时分配任务报告
man-hour evaluation 人时评价法
man-hour for loading-discharging 装卸作业人时
man-hour in attendance 出勤时间；出勤工时
man-hour in idleness 窝工工时；停工工时；怠工工时(数)
man-hour method 人工小时法；人工时间法
man-hour norm 工时定额
man-hour quota 工时定额
man-hour quota of cargo handling 装卸工时定额
man-hour rate 人时单价；人工小时率
man-hour record 工时记录
man-hour requirement 工时需要；单位产量所需

工时
man hours consumption of hoisting 提升工时消耗
man hours consumption of transport 运输工时消耗
man hours for auxiliary work 辅助作业人时;辅助作业工时
maniac 高速电子数字计算机
Manic 马尼克铜锰镍合金;铜镍合金;铜锰镍合金
manicure lavatory 整容室
manifest 装货清单;载货清单;货单;清单;舱单;仓口单
manifest covariance 明显协变性
manifest demand 实量需求
manifest of cargo 货物清单
manifest of import and export 进出口载货清单
manifold 一字头(螺杆泵);流形;联箱;拷贝;集液总管;集流箱;集流腔;集流管;集合管;集管总管;多种特征的;多种多样的;多支管;多头(导)管;多接口管段;多叉管
manifold air oxidation 支管空气氧化;歧管空气氧化
manifold (branch) 支管;歧管
manifold burner 复式喷灯
manifold center 管汇总站
manifold clamp 管夹(板);歧管夹
manifold cleaning paste 多用洁净剂
manifold condenser 歧管冷凝器;多管冷凝器;多管冷凝管
manifold cooler 多管冷却管
manifold crane 多用起重机
manifold depression 歧管抽空
manifold drying apparatus 歧管干燥器
manifolder 复印机
manifold exhaust 歧管排气;多样式排气;多管式排气
manifold flow 簇流;分支管中流动;分叉管(路)水流
manifold gasket 歧管垫密片
manifold gland 歧管压盖
manifold heat control 歧管加热控制
manifold heater 多管蒸汽炉;多管加热器
manifold heater plate 歧管加热器片
manifold hood 歧管外罩
manifold ignition test 歧管点火试验
manifold inflation 充气歧管
manifolding 歧管装置
manifold injection 进油歧管喷射;进水歧管喷射;进气歧管喷射
manifold of excited state 受激态簇
manifold paper 复印纸;薄打字纸
manifold penstock 多叉压力水管
manifold pipe 多叉管;多支路管
manifold plug 歧管塞
manifold pressure 歧管压力
manifold pressure control 进气管压力调节;歧管压力调节
manifold ring 集流环;集合环
manifold system 支管法分റ系统;多头管系
manifold tissue 多层绝缘纸;复印纸
manifold truck 管汇车
manifold tunnel 多叉隧洞;多叉隧道
manifold vacuum 歧管真空
manifold valve 会流阀;汇流阀;歧管阀;多管阀
manifold wall wetting 歧管壁浸湿
manifold writer 复写器
manikin 人造模型;人体伦琴当量;微小物
manilabiability 可处理性
Manila hemp 吕宋麻
Manil(l)a board 马尼拉纸板;灰底白板纸
Manil(l)a bristol 马尼拉厚纸
Manil(l)a copal 马尼拉树脂
Manil(l)a fibre 马尼拉纤维
Manil(l)a gold 铅黄铜
Manil(l)a hawser 粗麻绳;白棕绳
Manil(l)a hemp 马尼拉麻
Manil(l)a paper 马尼拉牛皮纸;马尼拉纸(电缆绝缘纸)
Manil(l)a Port Manil(l)a 马尼拉港(菲律宾)
Manil(l)a resin 马尼拉树脂
Manil(l)a rope 马尼拉(棕)绳;马尼拉麻丝;粗麻绳;白棕绳;吕宋麻
Manil(l)a rope paper 马尼拉纸绳纸
Manil(l)a trough 马尼拉海槽
Manil(l)a web sling 白棕绳吊索
Manil(l)a wrapping 马尼拉包装纸

man-induced 人工诱发
man-induced cause 人为原因
man-induced change 人为变化
man-induced channel change 人为的河槽变化
man-induced earthquake 人工诱发地震
man-induced erosion 人为侵蚀;人为冲刷;人为冲蚀
man-induced fluctuations 人为波动
man-induced variation 人为变化
man-induced weather modification 人为气候变化;人工影响天气
man-in environment problems 人为环境问题
man-in-space project 人类进入宇宙时代
man in the sea 海中居住
manipuility 可管理性;可处理性;可操作性
manipulate 操作
manipulated clay 重塑黏土
manipulated key 手动键
manipulated range 操纵范围
manipulated soil 重塑土
manipulated variable 控制(变)量;操纵量;被控变量
manipulate unit 操作部件
manipulate variable 操作变量
manipulating key 手动键
manipulation 控制;人工操作;操作(法);操纵
manipulation chart 联锁操作图
manipulation data 操纵数据
manipulation noise 按键噪声
manipulation of algebraic(al) formulas 代数公式的处理
manipulation of generating function 生成函数的处理
manipulation of goods 货物的搬运和操作
manipulation of name data 地名资料处理
manipulation of periodic income 操纵当期利益
manipulation of string 串的处理
manipulation of supply 操纵供应
manipulation platform 操作台
manipulation ratio 焊条焊着率
manipulative compression joint 手工压接
manipulative device 手工器具;手工工具;手工处理法操作装置
manipulative index 操纵索引
manipulative indexing 相关索引;键控检索
manipulative joint 喇叭(管)接头;手工压接
manipulator 控制器;机械手(臂);焊接操作者;焊接操作机;电键;操作者;操作器;操纵装置;操纵型机器人;操纵器
manipulator and observation bell 机械手和观察钟
manipulator arm 操作器机械手;翻science挺杆
manipulator cell 机械手操作室;操作室
manipulator finger 机械手抓手
manipulator for forging 锻造操作机
manipulator joint 机械手关节
manipulator marks 推床痕(轧材缺陷)
manipulator rack 推床齿条
manipulators extraman 机械手
manipulator slide beam 推床的导板
manipulator traversing gear 机械手传动装置
manipulator turning gear 机械手转动机构
manipulatrion of security market 操纵证券市场
manison 乡村山庄
manjak 硬化沥青;纯沥青
manjiroite 锰钠矿
man-land ratio 人-地比率
man-land relationship 人-地关系
manless face 无人工作面;无人工作间
manley-roew equation 曼利-罗方程
manlid 进人工作间;进人孔盖;检查井盖;人孔盖
manlift 载人提升机;载人电梯
manlift elevator 手动升降机
manlift step platform 载人提升机的踏步板
manlift travel 载人提升机的行程
man-load 人员载重
man lock 密封舱;进入气闸;人(行)闸;人孔闸;(气压沉井的)变压舱;工人进出舱室;进人气闸;人员出入闸
man-machine 人员升降机
man-machine analogy 人机模拟
man-machine chart 人与机器作业图
man-machine commission 人机联系
man-machine communication 人机通信(联系)
man-machine communication system 人机通信

系统
man-machine compatibility 人机相容
man-machine complex 人机组合
man-machine control 人机(联合)控制;手控自控联合控制
man-machine control system 人机控制系统
man-machine conversation 人机对话
man-machine dialogue 人机对话
man-machine digital system 人机数字系统
man-machine engineering 人机工程(学)
man-machine harmony 人机协调
man-machine integration 人机联系
man-machine interaction 人机对话
man-machine interactive system 人机交互系统
man-machine interface 人机联系装置;人机联系;人机接口;人机交互界面
man-machine interface workstation 人机接口动作站
man-machine language 人机语言;人机通信语言
man-machine matching 人机匹配
man-machine problem 人机问题
man-machine process chart 人机程序图
man-machine processing system 人机调节系统
man-machine relationship 人机关系
man-machine reliability 人机可靠性
man-machine simulation 人机模拟
man-machine symbiosis 人机共存
man-machine system 人机(联系)系统
man-made 人造的;人为的;人工的
man-made additive 人造胶黏剂;人工添加剂
man-made air pollution 人为空气污染
man-made asbestos 人造石棉
man-made atmospheric contaminant 人为大气污染物
man-made aurora 人造极光
man-made beta radiation 人为贝塔射线
man-made bonding medium 人造黏合剂
man-made building material 人造建筑材料
man-made building product 人造建筑制品
man-made cementing 人造胶结材料
man-made cementing agent 人造胶结剂
man-made climate room 人工气候室
man-made coarse concrete aggregate 人造混凝土粗骨料
man-made concrete aggregate 人造混凝土骨料
man-made constraint 人为约束
man-made construction material 人造建筑材料
man-made construction product 人造建筑制品
man-made contaminant 人为污染物
man-made contamination 人为污染
man-made desert 人造沙漠;人为(形成的)沙漠
man-made diamond 合成金刚石;人造钻石;人造金刚石
man-made disaster 人为灾难
man-made disturbance 人为扰动
man-made earthquake 人工地震
man-made earthwork 人工土方工程
man-made ecosystem 人为生态系统
man-made effect 人为效应
man-made energy source 人工能源
man-made environment 人造环境;人为环境;人工环境
man-made environment(al) contaminant 人为环境污染物
man-made environment(al) pollutant 人为环境污染物
man-made erosion 人为侵蚀;人为冲刷;人为冲蚀;人工冲刷
man-made feature 人为地物;人工地物
man-made fiber 人造纤维;化学纤维
man-made fiber cordage 化学纤维绳
man-made fiber fabric 人造织品
man-made fiber material 人造纤维材料
man-made fine concrete aggregate 人造细骨料混凝土
man-made floating island 人工浮床
man-made forest 人工林
man-made fracture 人造裂缝
man-made geologic(al) hazard 人为地质灾害
man-made geothermal system 人造地热系统
man-made glue 人造胶水
man-made ground 人造地表;人造庭园
man-made groundwater pollution 地下水的人为污染

man-made harbo(u)r 人造港湾;人造港口
man-made influence 人为影响
man-made interference 工业干扰;人为干扰
man-made island 人工岛
man-made isotope 人工同位素
man-made lake 人工湖
man-made land 人造土地;人造地
man-made land subsidence 人为地面沉降
man-made machine 人造机器
man-made marble 人造大理石
man-made masonry unit 人造砖石单元;人造圬工体
man-made mineral vitreous fibre 人造矿物玻璃纤维
man-made mineral wool 人造矿物绒
man-made noise 工业噪声;人为噪声
man-made obstacle 人造障碍物
man-made obstruction 人为障碍
man-made plain 人造平原
man-made plastic 人造塑料
man-made plastic material 人造塑料材料
man-made pollutant 人为污染物
man-made pollution 人为污染
man-made pollution sources 人为污染(来)源
man-made pond 人工池(塘)
man-made pool 人工池(塘)
man-made pozzolanic material 人造火山灰材料
man-made pozz(u)olana 人造火榴火山灰
man-made quartz 人造石英
man-made radio noise 人体感应噪声
man-made rain 人造雨
man-made reservoir 人工水库
man-made resin 人造树脂
man-made resin-based adhesive 人造树脂基胶黏剂;人造树脂类黏结剂
man-made resources 人造资源;人为资源
man-made rubber 人造橡胶
man-made sabotage 人为破坏
man-made seismic hazard 人力震害
man-made shortage of goods 人为的供应紧张
man-made source 人为来源;人工放射源
man-made source of air pollution 人为空气污染源
man-made source of water pollution 水认为污染源
man-made space-craft 人造宇宙飞船;人造空间飞行器
man-made statics 人为静电干扰
man-made stone 人造石
man-made structural material 人造的结构材料
man-made structural product 人造的结构产品;人造建筑制品
man-made sun satellite 人造行星
man-made surface 人为地表;人工地表
man-made system 人工系统
man-made target 人工目标
man-made travertin(e) 人造石灰华;人造凝灰石
man-made vibration 人为振动
man-made wastes 人为废弃物
man-made waterway 人工水道
man-made wood 改良木材;人造木材
man-made wood board 木质人造板
man-made world 人造世界
man-manageable system 人控系统
man-management 劳动力管理;人员管理
man-management analysis 人员管理分析
man minutes manit 每人每分钟所做的工作
man-month rate 人月单价
Manmoth event 马默斯事件
manna gum 多枝桉树
mannal entry 人工输入;手控入口
manned attitude control 人控的姿态控制
manned ballon 载人气球
manned cableway 载人缆道
manned craft 载人飞行器
manned flight 载人飞行
manned level crossing 有人看守道口
manned microsubmersible 载人微型潜水器
manned non-self propelled barge 有人值班的非自航驳
manned orbital space station 载人轨道空间站
manned orbiting laboratory 载人轨道实验室
manned orbit space station 载人轨道航天站
manned pontoon 有人值班的浮驳
manned satellite 载人卫星

manned spacecraft 载人宇宙飞船;载人航天器;载人飞船
manned spacecraft center[centre] 载人宇宙飞船中心
manned submersible 载人潜水器;载人可潜器
manned tethered microsubmersible 载人系缆式微型潜水器
manned undersea station 载人海底站
manned underwater laboratory 载人水下实验室
manned underwater station 载人水下工作站;载人海底站
manned underwater structure 载人水下结构
manned underwater work 载人水下作业;载人海底作业
manned underwater work system 载人水下作业系统;载人水下工作系统
manned vehicle 载人潜水器;载人飞船
manner 样式;方式
mannered 有手法的;有风格的
mannerism 风格主义;手法主义(16世纪末,意大利古典建筑风格)
Mannerist architecture (17世纪意大利的)守旧派建筑;程式化建筑;强调独特风格的建筑;守恒派建筑
Mannerist device 风格主义手法
Mannerist trait 风格主义特征
manner of collecting export proceeds 出口收汇方式
manner of crustal move 地壳运动方式
manner of occurrence of element 元素存在形式
manner of packing 包装方式
Mannesmann effect 曼内斯曼效应(锻件或轧件内)
Mannesmann mill 曼内斯曼斜轧机
Mannesmann piercer 辊式穿轧机;曼内斯曼式穿孔机
Mannesmann piercing mill 斜轧穿孔机;曼内斯曼式穿孔机
Mannesmann piercing roll 曼内斯曼斜辊穿轧机(管材)
Mannesmann process 曼内斯曼制管法(斜辊轧管法);曼内斯曼式轧管法
Mannesmann roll-piercing process 曼内斯曼辊式穿孔法
Mannheim absorption system 曼海姆吸收装置
Mannheim gold 曼海姆(铜锌锡代金合)金
Mannheim proeess 曼海姆法
Mannheim salt-cake furnace 曼海姆制盐饼炉
mannich base 曼里斯碱
manning 人员组成;配备人员
manning cost 工资支出
manning level 人员配备;定员
manning of life boat 艇筏配员
Manning's coefficient of roughness 曼宁粗糙系数
Manning's constant 曼宁常数
Manning's equation 曼宁方程(式)
Manning's formula 曼宁公式
Manning's friction coefficient 曼宁摩擦系数
Manning's nomograph 曼宁图表
Manning's roughness coefficient 曼宁粗糙系数;曼宁糙率系数
Manning's roughness factor 曼宁粗糙系数;曼宁糙率系数
Manning table 人力配备表
mannite 甘露糖醇
mannite fatty acid ester 甘露糖醇脂肪酸酯
mannitol 甘露糖醇
Mann-Kendall test 曼肯德尔检验
mannure 粪肥
manoc 双代森锰
mano-contact 压力感知触点
manocryometer 融解压力计;加压熔点计
manoeuvreability 灵活性;机动性
man-of-war anchorage 军用锚地
man of wealth 富人
manograph 压力记录器;压力计;高速光学示功器;流体压(力)记录器;测压器
manometer 液柱压力计;液体压力计;压强计;压力计(量器);压力表;流体压强计;流体压力计;检压计;测压计
manometer calibration 压力表校准
manometer cock 液体压力计旋塞
manometer flask 测压计瓶
manometer liquid level 气压计液压指示器

manometer liquid level indicator 压力表式液位指示器
manometer method 测压法
manometer pressure 压力计压力;表压(力)
manometer regulator 风压调整器
manometer tap 压力表接口和螺塞
manometer tube 测压管
manometer tube gland 压力计管(压)盖;压力管塞
manometer(tube)method 测压管法
manometric(al) 液体压力计的
manometric(al) bomb 密闭爆发器;测压器;测压弹
manometric(al) capsule 密封增压座舱
manometric(al) delivery head 液体压力计水头
manometric(al) efficiency 压缩效率;测压效率
manometric(al) head 液体压力计水头;测压管水头
manometric(al) liquid 测压液
manometric(al) method 测压法
manometric(al) pressure head 液体压力计压头
manometric(al) suction head 液体压力计吸水头
manometric(al) test 测压试验
manometric(al) thermometer 感压(式)温度计;压力(表)式温度计;压差温度计;温差式温度计;测压温度计
manometry 压力计测量;容积的压力测量;测压术;测压(法)
manool oxide 迈诺氧化物
man operated propulsion 人工操纵推进
manor 宅第;封地;庄园
manor-castle 庄园主的大住宅
manor house 宅第;庄园主住宅
manor place 庄园住宅
manoscope 流体压强计;流(体)压(力)计;气体密度测定仪
manoscopy 气体密度测定术;流压术;气体容量分析;气体密度测定(法)
manostat 压力稳定器;压力调整器;恒压器;稳压器
man overboard 有人落水;人员落水
man overboard light 人员落水信号灯
man-pack 单人可携带的
man-packed rockfill 人工抛石
man-pack laser unit 便携式激光器;背负式激光器
man-pack loran 背包罗兰
man-pack microwave radio 背负式微波台
man-pack television unit 便携式电视装置
man-passing-man 接续前进法
man pipe 总管道
man-portable laser 便携式激光器
manpower 劳动力;人力;人工;定员
manpower allocation 人力分配法
manpower allocation procedure 人力分配法
manpower analysis 人力分析
manpower authorization 定员管理
manpower budget 劳动力预算;人力预算
manpower budgeting 人力预算
manpower capital 人力资本
manpower control 人力管理
manpower cost 人力费用;工资成本;人工费
manpower deficit 人力不足
manpower demand 人力需求
manpower development 人力开发
manpower drain 劳力外流
manpowered boat 人力船
manpower effectiveness 人力效率
manpower input-output analysis 人力投入产出分析;人力输入/输出分析
manpower-intensive 人力密集的
manpower inventory 人力清单
manpower leveling 人力负荷拉平
manpower management 人事管理;人力管理
manpower needed per unit of product 单位产品的必要劳动消耗量
manpower payoff 人力支付
manpower planning 人力规划;人力安排规划
manpower policy 人力政策
manpower quota 劳动定员
manpower rapids-heaving 人力绞滩
manpower ratios 人力方面的比率
manpower relocation 劳动力再配置
manpower report 劳动力报告
manpower resources 人力资源
manpower resources accounting 人力资源会计
manpower roller 人力压路机
manpower scheduling 人力调配
manpower shortage 人力缺乏;人力短缺

manpower surplus 人力过剩
manpower system 人力系统
manpower training 人力训练；人力培训；人才培养；人才培训
manpower utilization 人力利用
manpower utilization survey 人力利用调查
man-productivity 人的生产率
man-pumped well 手压井
man qualities 人的质量
man rack 平板卡车上的工房
man record chart 人工记录表
man responsible for train lighting system 电车员
man-riding 乘人的
man-riding conveyor[conveyer] 乘人输送机
man-riding haulage 人员运送
man rope 扶手索；扶手绳；安全索；舷门索
manrope knot 双握索结
man's activity 人为影响；人为活动；人类活动
mansard 阁楼；折线形屋顶；热风干燥室；复折屋顶
mansard cornice mo(u)lding 折线形屋顶飞檐线条；复折式檐口装饰线脚
mansard dormer window 折线形屋顶窗；复折屋顶窗
mansard flat floor 折线形平屋顶；复折式平板屋顶
mansard gable tile 复折屋顶山墙顶盖瓦
mansard roof 锥台式屋顶；折线形屋顶；双折屋顶；复折(式)屋顶
mansard roof truss 折线形屋顶桁架；复折屋顶桁架；复折屋架
mansard roof with slanted struts 有斜撑的折线形屋顶
mansard tile 复折屋顶折线瓦
mansard truss 折线形桁架
mansard truss roof 折线形屋顶
manse 教士住宅；牧师住宅
Mansfield 红和奶色砂岩
Mansfield copper process 曼斯菲尔德炼铜法
mansfieldite 砷铝石
Mansfield oil gas producer 曼斯菲尔德油气发生器
Mansfield Sandstone 曼斯菲尔德砂岩
man-shaped walking machine 人形步行机
man-shift 人班；一人在一班中所作的工作量；班-工
man-shift quota for cargo handling 装卸工班定额
man-shift rate 装卸工班效率
man-shifts for cargo handling 装卸工班数
Mansholt plan 曼斯霍尔特计划
man's impact on climate 人类对气候的影响；人口影响气候
mansion(house) 府邸；庄园主住宅；官邸；大厦；大楼
man-sized chimney 人可进出的烟筒
mansjoite 氟透辉石
mansonia 灰褐有斑驳的硬木(尼日利亚产)
mantapam 印度寺庙前廊亭
mantel (岩石或土壤的)表层；壁炉架；壁炉台
mantel board 壁炉(架)板
mantel pick-up 壁炉架
mantel piece 壁炉架；壁炉台；壁炉面饰；炉壁面饰
mantel register 安在炉壁中的铸铁盘；壁炉调节板
mantel shelf 壁炉板；壁炉面饰面；壁炉装饰架；壁炉架；壁炉台
mantel tree 壁炉过梁；壁炉楣
man the capstan 绞缆岗位
man the ship 配备船员
mantienneeite 曼廷尼石
man-time 人次
mantissa 假数；尾数；定点部分；浮点系数
mantissa of logarithm 对数的尾数
mantissa part 尾数部分
mantle 罩盖；高炉环梁壳；机套；斗篷；地幔；覆盖物；风化覆盖层
mantle bloc 过桥盖砖
mantle block 桥砖
mantle carrier 煤气灯纱罩托
mantle cavity 套腔
mantle cone 破碎机可动圆锥
mantle convection 地幔对流
mantle-core boundary 核幔界面
mantle-crust mix 地壳地幔混合物
mantle degassing 地幔脱气
mantle-density magma 幔源岩浆
mantle-derived granite 幔源花岗岩
mantled gneiss dome 带盖片麻岩穹隆
mantled pipe 套管

mantle earthquake 地幔地震
mantle effect 地幔效应
mantle fiber 套丝
mantle filter 罩式滤器；罩式滤池
mantle flow 地幔流
mantle fold(ing) 盖层褶皱；套膜褶
mantle friction 表面摩阻；表面摩擦(阻力)
mantle gas 地幔气
mantle-grade 灯罩品级
mantle groove 套膜沟
mantle heat flow 地幔热流
mantle isochron 地幔等时线
mantle lacquer 无光漆
mantle layer 套层
mantle life 可动衬板寿命(破碎机)
mantle metasomatism 地幔交代作用
mantle of lamp 灯罩
mantle of rock 表皮岩；风化岩；表层岩；岩石表层；风化层
mantle of soil 土(壤)覆盖层；土覆盖层；土层被覆；土被；表皮土
mantle of the earth 地幔
mantle of vegetation 绿化覆盖；植被
mantle of waste 风化残积
mantle petrology 地幔岩石学
mantle piece 罩套构件
mantle pipe 套管
mantle plume (羽状地幔流)地幔柱；地幔羽；地幔热柱
mantle region 套膜节
mantle regolith 覆岩；风化岩；表皮土；表层岩
mantle reservoir 地幔库
mantle ring 支圈；垫环
mantle rock 覆岩；风化岩；风化层；表皮土；表层岩石；表(层)岩
mantle segment 套膜节
mantle soil 表土层
mantlet 盾架；(防御工事的)活动保护幕；防盾
mantling 壁炉面饰
manto 席状交代矿床；矿层；平卧状矿床
Mantoux test 曼托试验
man trip 人员运送
man trolley 人控机动小车；载人吊运车；自升式起重小车；驾驶室小车；人控式起重小车；带人小车
man-type flow process chart 人工型流程程序图
manual 说明书；手控的；手工的；手动的；便览
manual AC and DC tungsten-pole argon arc welding machine 手工钨极交直流氩弧焊机
manual acting 人工操作的
manual activity analysis 体力活动分析
manual actuator 手操器
manual address switch 手动地址开关
manual adjusting knob 手调旋钮
manual adjustment 人工调整；手调
manual-advance lever 手拉点火提前杆
manual alarm 手摇警报器
manual alarm button 手动报警按钮
manual alarm system 手摇警报装置；手动报警系统
manual amendment 人工修整；手工修正
manual answering 人工应答
manual application of lubricant 人工润滑；手工润滑
manual applied plaster 手工石膏
manual arc welding 手弧焊；手工电弧堆焊
manual area 人工电话区
manual assembly 手工成型
manual assisted positioning 手动辅助定位
manual auger 人力钻机；手摇钻机
manual-automatic change-over 手动自动变换
manual-automatic relay 手动自动转换继电器
manual-automatic switch 手动和自动(转换)开关
manual-automation 半自动化
manual auto run-up selector 手动自动起动选择器
manual backup 人工接替；手动备用调节装置
manual backup control system 手动备用控制系统
manual backup unit 人工备份
manual banding 人工打捆
manual batcher 人工操作配料(拌和)机；人工配料器
manual blending mixer 手摇搅拌机；手动拌和机
manual block 人工闭锁；手动闭锁
manual block apparatus 人工闭塞装置
manual block signal system 人工闭塞信号系统
manual block system 人工闭塞系统

manual blowing 人工吹制
manual board 人工台
manual bonder 手工接合器
manual book 手册
manual booster 手动升压机
manual boring 人力钻探；人工钻探
manual boring grouted pile 人工挖孔灌注桩
manual brake 手动制动器
manual brake release device 手动制动释放装置
manual burner 人工点火燃烧器
manual butterfly valve 手动蝶阀
manual button 手动按钮
manual calculating unit 手动计算器
manual calculation and interpretation 人工计算和解释
manual call(ing) 人工呼叫；手工呼叫；手摇呼叫
manual capacity regulator 手动能量
manual card 手穿孔卡片
manual casing hanger 手动套管挂
manual catching 机座间的人工递钢
manual central office 人工电话局
manual chain-driven carrier 人工手链小车
manual chain hoist 手动链轮起重机；手动葫芦；神仙葫芦
manual changeover 手摇转向机构
manual changeover procedure 人工转换程序
manual changeover signal 人工转换信号
manual check 手动检查
manual circuit 人工电路；手工电路
manual cleaning device 人工清扫设备
manual clear 人工清除
manual closed control 手动闭路控制
manual closed-loop control system 手动(闭)环控制系统
manual closed-loop process control 手动环程序控制；手动闭环过程控制
manual closing operation 人工接续
manual coating and wrapping 手工涂刷和缠绕
manual code programming 手编程序
manual collator 人工比较装置
manual commercial office 人工商用电话局
manual compensation 手动补偿
manual compilation 手工编程
manual composition 手工排版
manual computation 手算；笔算
manual consolidation 人力固结；人工固结；手工捣实
manual control 人力控制；人工控制；人工操纵；手(工)控(制)；手工操纵；手动控制；手动操纵
manual-control freewheeling 手动单向离合器
manual-controlled block 半自动闭塞
manual controller 人工控制器；手动控制器
manual control lever 操纵手柄
manual control mode 手动控制系统
manual control of cargo valve 手动货油阀
manual control open loop 人工控制开环；手控开环
manual control panel 人工控制板；手动控制板
manual control pitch propeller 手调螺距螺旋桨
manual control screw 手调螺钉
manual control station 手动控制站
manual control system 手动操纵系统
manual correction 手工修改
manual counting 人工计数；手工计数
manual counting board 手录计数板
manual craft 手工艺
manual crane 手动起重机
manual cranking handle 手摇起动手柄
manual cut-out lever 手动停油杆
manual cut(ting) 人工切割；人工裁切；手工裁切；手工切割
manual damper 风挡；手动风门
manual data 手册资料
manual data collection system 人工数据收集系统
manual data input 人工数据输入；手动数据输入
manual datainput equipment 人工数据输入设备
manual data relay center 人工数据中继中心
manual degaussing 人工消磁
manual desk 人工交换台
manual device backup 手动设备备份
manual digital command 手控数字指令
manual digitizer 手控数字转换器；手持数字化器
manual direction finder 手动无线电测向仪；手动测向仪
manual drilling 人力钻探；人工钻探；手工打眼

manual drive 手操作
manual driven duplicator 手动复印机
manual driver 手动驱动器
manual duplicator 手动复制机
manual duster 手摇喷粉机
manual editing 手动编辑
manual editing table 手编数字化器
manual effort 人力
manual electric(al) arc pile up welding 手工电弧堆焊
manual electric(al) arc welding 手工电弧堆焊
manual electro-slag welding 手工电渣焊
manual element 手工基本动作
manual emergency running device 手动应急运转装置
manual enamelling 手工涂搪
manual engraving 手工刻图
manual environment 人工环境
manual equalizer 手动均衡器
manual examination 手动检查
manual excavation 人工挖掘;人工开挖
manual exchange 人工电话局;手动交换
manual exchanger 人工交换机
manual exposure 手控曝光
manual extinguisher 手提式灭火器;手提式灭火机
manual extraction of radar information 雷达信号人工录取
manual feed 人力推进;手动进给
manual feeding of coordinate 人工输入坐标
manual fine tuning 手控微调
manual finishing 手工清理
manual fire alarm 手动火警报警器
manual fire alarm box 手摇火警器
manual fire alarm system 人工火警系统;手摇式火警装置;手动火警系统
manual fire pump 手动消防泵
manual fire station 手控火警报警器
manual firing 手击发
manual fitting 手工调整;手工装配;手工组装;手工配合
manual floating 手工镘平
manual followed digitizer 手扶跟踪数字化仪;手扶跟踪数字化器
manual followed digitizing 手扶跟踪数字化法
manual following 人工跟踪;手控跟踪
manual frame 手控帧
manual function 手动功能
manual gain 手控增益
manual gain control 手动增益控制
manual gas cutting 手工气割
manual gas sampling valve 手动气体进样阀
manual gas shut-off valve 手动煤气关闭阀
manual gas welding 手工气焊
manual ga(u)ging 人工计量
manual gear shifting 手动变速装置
manual gear-shift transmission 手控齿轮换挡变速器
manual globe valve 手动截止阀
manual governor 手控调节器
manual grease station 手动加油站;手动干油站
manual greasing 手动加注润滑脂
manual grouting 人工灌浆
manual handling 人力装卸;人工装卸;人工处理;手工操作
manual haulage 人力推车
manual hill shading 手工晕渲法
manual hill shading technique 手工晕渲法
manual hoist 手动提升机;手动卷绕机
manual hue 手控色调
manual hydraulic pump 手动液压泵
manual hydraulic tray pushcart 手动液压托盘搬运车
manual hydroelectric(al) station 手动水电站
manual-increment key 手控增量键
manual input 人工输入;手动输入
manual input device 人工输入装置;手控输入装置;手动输入装置
manual input generator 人工输入设备
manual input key 人工输入键
manual input register 人工输入寄存器
manual input unit 人工输入装置;人工输入设备
manual insertion 人工放入
manual intercept 人工截断
manual interlocking 人工联锁

manual international exchange 人工国际交换台
manual interpretation 人工解释
manual intervention 人力干预;人工干预
manual intervention capacity 人工应急能力
manual iris 手动光圈光阑
manual jaw crusher 手动颚式破碎机
manual knob 手动旋钮
manual labo(u)r 体力劳动;手工劳动;手工工人;手工
manual laying 人工敷设;人工铺设
manual laying up 手工涂刷和缠绕
manual lens 手控透镜
manual lever 人工操作手柄;手动操作杆
manual loading 人工装载;人工装料;手动加荷
manual load key 人工输入键
manual lofting 手工放样
manual looping 人工造成活套
manual low gear 手动低速挡
manual lubrication 人工润滑
manual lubricator 人工润滑器
manually 手动地
manually angled cutter 手调斜度切割器
manually assisted digitizer 手扶数字化器
manually coating welding electrode 手涂焊条
manually controlled 人力操纵的;手控的
manually controlled fire pump 手动消防泵
manually controlled traffic signal 人(工)控(制)交通信号;手控交通信号
manually controlled work 手控周期
manually feeding fillered 手工充填焊丝
manually guided 人工操纵的;人力引导的
manually guided drag skip 手扶刮土小车
manually guided vibrating concrete finisher 手扶式混凝土振捣修整机
manually held 手控稳住的
manually lighted boiler 手动点火锅炉
manually operated 用手操作的;人力操纵的;人工操作的;手工操作的
manually operated circuit 人工操作线路
manually operated clutch 手控离合器
manually operated counterweight lift 手动平衡重升降机
manually operated digitizer 手扶数字化器
manually operated door 手动启闭门;手动门
manually operated equipment 手工操作设备
manually operated fire alarm 手动火警报警器
manually operated fire resisting door 手控防火门
manually operated furnace 手烧炉
manually operated gas inflatable lifejacket 手动充气救生衣
manually operated gate 手动启闭闸门;手动栏木;手动阀(门)
manually operated machinery 人工操作机器
manually operated mixing 人工搅拌
manually operated pile driver 人工打桩机
manually operated pneumatic tamper 人工操纵的气动夯具
manually operated point 手扳道岔
manually operated press 手动压力机
manually operated rammer 人工操作夯;人工夯
manually operated screen 手动筛网
manually operated screw 手动压力螺钉
manually operated semaphore signal 人工操纵臂板信号
manually operated signal 人工控制交通信号;人工操纵交通信号
manually operated spray system 手动飞溅润滑系统
manually operated switch 手动开关
manually operated turning gear 手动盘车装置
manually operated valve 手动操纵阀;人工操纵阀;手动阀(门)
manually operated vibrating finisher 手动操纵振动精整机
manually operation 手动
manually operation switch 手扳道岔
manually powered draw bending machine 手动卷缠弯曲机
manually propelled mobile scaffold 活动脚手架;手推移动式脚手架;飞模脚手架
manually reset time switch 手工操作的复位时间
manually shifted transmission 人工变速传动
manual main shutoff valve 手动主截流阀
manual maneuvering 人工操纵

manual manipulation 人工操作;手工操作;手(动)操作
manual memory 人工存储器
manual metal-arc welding 手弧焊
manual method 手工方法
manual mixing valve 手工操作的混合阀门
manual mixture control 人工控制配料;手动节气门
manual mode 人工式;手控式;手控方式;手动方式
manual monitored control system 手动监控系统
manual mucking(out) 人工挖除淤泥
manual multiple type switchboard 人工复式交换台
manual number generator 人工输入设备;手控数码发生器
manual of account 会计规程
manual of cost control 成本管理手册;成本控制手册
manual of earthquake observation 地震观测法
manual-off-automatic selector switch 手动-切换-自动选择开关
manual office 人工局
manual of operating instruction 业务规章手册
manual of photographic(al) interpretation 像片判读手册
manual of procedure 操作手册
manual of seismometry 地震观测法
manual of standards 标准手册
manual oil pressure pump 手动压油泵
manual open-loop control 手动开环控制
manual operated 手操纵的
manual operated control 手动控制
manual-operated cutter 手工操作刀具
manual-operated gate 手动操作闸门
manual operation 人力操作;人工操作;手工生产;手工操作;手动操作;单独操纵作业
manual operational requirement 人工操作要求
manual operator 人工接线员
manual overhead crane 手动桥式起重机
manual override control 手控超越控制
manual panel 手操作仪表板
manual placement 人工浇筑;人工填筑
manual plastering 人工操作抹灰;人工操作粉刷
manual plate making 人工制版
manual plotting 手工作图
manual plugging meter 手动堵塞计
manual pneumatic sprayer 手动气力喷雾机
manual polarograph 手工极谱仪
manual pond 人工池(塘)
manual position 人工坐席;手动位置;手操作位置
manual potentiometer 手动式电位计
manual power 手工操作的动力
manual preset control 手动预调控制
manual press 手压机
manual priming 人工灌注;手动启动注水
manual procedure 人工操作程序;手工操作程序
manual processing station 人工操作信息处理中心
manual program(me) 人工程序;手工程序
manual programming 人工编制程序;手工编程序
manual proportional control system 手控按比例操纵系统
manual proportioning 手工称量
manual pump 手压泵;手动泵
manual pumping unit 手动抽油机;手动润滑油站;手动抽水机
manual punch 手控穿孔机
manual quickopening manhole 手摇快开人孔
manual radio direction finder 手动无线电测向仪
manual ram reverse 手动滑枕反向
manual range selector 人工量程选择器
manual rate 规章费率;保险费率
manual rate-aided tracking 人工速度辅助跟踪
manual read 人工读入【计】
manual readout 手动测读仪
manual reaper 人力收割机
manual reaper binder 人力割捆机
manual reconfiguration 人工改变系统结构
manual record 人工记录
manual regulation 人工调整;人工调节;手动调节
manual release 人工解锁;人工缓解;手动释放
manual release button panel 解锁按钮盘
manual release indication 人工解锁表示
manual release without time locking 不限时人工解锁
manual rendering 手工操作的粉刷

manual reset 人工重调;人工复位;手控复位;手动复位
manual-reset adjustment 人工手调复位;手动重调
manual-reset relay 人工复位继电器
manual-reset safety control 手动重调保险控制
manual restoration 人工复原
manual restore 人工复原
manual retrofire 手控制动发动机点火
manual reversing handle 手动换向把
manual ringer 人工振铃器;手摇振铃器
manual ringing 人工振铃;手摇呼叫
manual ringing telephone 人工振铃电话
manual riveting 手工操作的铆接
manual riveting machine 手工(操作)铆接机
manual rock drilling 手工凿岩
manual rod 手动棒
manual rotary(drum)pump 手动回转泵
manual rotary switch 手动旋转开关
manual route release 进路人工解锁
manual route setting 人工设置进路
manual safety switch 手控安全开关
manual sampler 人工取样器;手工取样器;手动取样器
manual sampling 人工取样;手工取样;手工采样
manuals and monographs 手册和专著
manual saturation(colo(u)r)control 手动饱和度调整
manual scram 手动紧急停堆(反应堆)
manual scraper 手扶机械铲
manual scribing 手工刻图
manual search 人工搜索;手动扫掠
manual searching 手工检索
manual selectivity control 人工选择性调整;手动选择性调整
manual selector cam 手控凸轮
manual selector fork 手控拨叉
manual selector switch 手控选择开关
manual semi-automatic welding 手工半自动焊接
manual service 人工业务
manual set 手调
manual setter 手摇装置
manual setting 手工装配;手工调整
manual shaft turning device 手动盘车装置
manual shield 人工防护;人工开挖盾牌;人工开挖盾构
manual shift 手动变速;人工换挡;手动转换
manual shunt 人工分路
manual shut-down 手动停机
manual shutoff valve 手控断油开关
manual sieve shaker 人工筛分振动器
manual signal 人工信号
manual simplex system 手工单工系统
manual simulation 人工模拟;试操作
manual skin 玛瑙皮
manuals of engineering practice 工程实践手册
manual soil auger 入力取土螺钻;手摇土钻
manual spark advance 手控点火提前装置
manual speed adjustment 人工速度调节;手动调速
manual spinning 人工旋压法
manual sprayer 手压式喷雾器;手揿式喷雾器;手动喷雾器
manual spray gun 手动喷枪
manual spray hose 手动喷雾器;手动喷雾机
manual spreading 人工撒布;人工涂敷
manual S/R machine 手动有轨巷道堆垛起重机
manual starter 人力启动器;手工启动器
manual starter switch 手控起动开关
manual starting 人力启动;手工启动;手动起动
manual starting crank 手工起动摇把;人力起动曲柄;手工起动曲柄;人力起动摇把
manual starting switch 手动起动开关
manual station 手动操作机
manual steering 人力转向;手动转向
manual steering effort 手动转向力
manual steering equipment 手动操舵装置
manual steering force 手操纵力;人工转向力
manual stoker 人工加煤机
manual stop valve 手动止动阀
manual storage switch 手控存储开关
manual stripping of membranes 人工剥膜
manual substation 人工分局
manual supervisory control 人工监督装置
manual switch 手控开关;手动开关;手扳道岔
manual switchboard 人工交换台;人工交换机
manual switchgroup 手动组合开关

manual switching 人工切换
manual switchroom 手工操作的配电间
manual-switch storage 手控开关存储器
manual synchronization 手控同步
manual synchronizing 人工同步
manual system 人控系统;人工制;人工系统;手动系统
manual tachoscope 手控转速计
manual take-over drive 手动传动装置
manual takeup 人力拉紧装置
manual target selector 手控目标选择器
manual telegraph transmitter 手控发报机
manual telephone 人工电话
manual telephone center 人工电话中心
manual telephone substation 人工打电话支局
manual telephone system 人工电话系统
manual temperature control 人工温度控制
manual test equipment 人控试验设备
manual tilting device 手动倾倒装置
manual time release 限时人工解锁
manual timing 手动正时
manual toll board 人工长途交换台
manual toning control 手动色调控制
manual tracing 手扶跟踪
manual tracing digitizer 手扶跟踪数字化仪;手扶跟踪数字化器
manual tracking 人工跟踪;手动追踪瞄准;手动跟踪
manual-tracking tach(e)ometer 人工跟踪转速计;手动跟踪转速计
manual traffic counting 人工交通量计数
manual traffic signal 手动交通信号
manual transfer pump 手摇传送泵
manual transmission 手控传输
manual transportation 手工搬运
manual traverse 手摇旋转
manual traversing handle 手摇旋转手柄
manual trigger 人工触发器
manual trip(ping) 手动跳闸
manual tripping device 手动脱扣装置;手动释放装置
manual-tuned loop detector 人工调谐线圈检测器
manual tungsten inert gas welding 手工钨极惰性气体保护焊
manual tuning 人工调谐;手控调谐
manual twist tester 手摇式捻度计
manual type 手动式
manual unit 手动装置
manual update 手工修改
manual valve 手控阀
manual valve positioner 手动阀位控制器;手动阀门定位器
manual voltage regulation 手动调压
manual voltage regulator 人工电压调节器
manual volume control 手控音量控制
manual water blending mixer 手工操作的水搅和机
manual water mixing valve 手工操作的水混合阀门
manual water shower mixer 手工操作的水喷灌混合机
manual weight batcher 手动重量称量器;手动计重器
manual weight carrying 徒手搬运
manual weld 手焊
manual welding 人工焊接;手工焊接
manual welding installation 手工焊接设备
manual welding machine 手工焊机
manual whistle 人力号笛
manual winder 手动卷绕机
manual word generator 人工输入装置
manual work 手动作业;人工;体力劳动
manual worker 体力劳动者;体力工人
manual working 人工工作;手动作业
manual W-pole argon arc welding machine 交流手工钨极氩弧焊机
manual zoom lens 手控可变焦距透镜
manubrium 垂管
manuduction 手法
Manueline architecture 曼纽林建筑形式
Manueline style 曼纽林建筑风格
manufactory 制造厂
manufactory of hazardous articles 危险品制造厂
manufactory of traffic vehicle 交通工具制造厂
manufactuer's mark 工厂商标
manufacturability 可制造性
manufactural 制造的

manufacture 制作;制造(品);生产
manufacture bases 施工依据
manufacture catalog(ue) 产品目录
manufacture complete sets of equipment locally 就地生产
manufactured abrasive 人造磨料
manufactured adhesive 人造黏合剂;人工黏接剂
manufactured aggregate 人造集料;人造骨料
manufactured alumina 工业氧化铝
manufactured article 人工制品
manufactured articles of daily use 日用工业品
manufactured board 成板
manufactured brick chip(ping) 加工砖片
manufactured building 工业建筑;预制(构件)建筑物
manufactured building product 工业制造的建筑产品
manufactured clay brick sand 人造黏土碎砖
manufactured coarse sand 工业人造粗砂
manufactured construction(al) product 工业制造的建筑产品
manufactured construction materials 人造建筑材料
manufactured dowel 成品冠桩
manufacture deficiency 加工缺陷
manufactured fine sand 工业人造细砂
manufactured fuel gas 人造燃气
manufactured gas 人工燃气;人造气体;人造煤气;人工煤气
manufactured glue 工业人造胶水
manufactured goods 制成品
manufactured goods for daily use 日用工业品
manufactured graphite 人造石墨
manufactured head 铆钉头
manufactured housing 工厂预制住房
Manufactured Imports Promotion Organization 进口产品促销组织
manufactured inventories 库存制成品;库存产成品
manufactured lot 已具备施工条件的地块
manufactured lumber 加工木材
manufactured marble 工业人造大理石
manufactured mineral 人造矿物
manufactured parts for assembly 散件组装
manufactured plastics material 工业人造塑料材料
manufactured product 产品;制成品;工业品;工业产品
manufactured resin 人造树脂
manufactured resin-based bonding medium 人造树脂基黏合剂
manufactured resin-based cementing agent 人造树脂基胶合剂
manufactured rubber 人造橡皮
manufactured sand 人造砂;人工砂
manufactured seamless crown 成品无缝冠
manufactured structural product 制造的结构产品
manufactured travertine 人造钙华;人造石灰华
manufactured unit 生产单元
manufactured waste 工业废水
manufacture of brick 砖的制造
manufacture of iron and steel by melting 钢铁熔炼
manufacture of vessel 容器制造
manufacture on a large-scale 大规模生产
manufacture order 生产任务书;生产任务单
manufacture plant 制造厂
manufacture process 制造工艺
manufacturer 制造者;制造商;制造厂(家);生产者;生产厂(家);厂商;产品生产厂
manufacturer instructions 厂商说明书
manufacturer mark 厂商标志
manufacturer's agent 经销商;厂商代理人
manufacturer's bond 生产厂商保证书
manufacturer's brand 制造商厂牌;厂商牌号
manufacturer's brochures 制造厂样本
manufacturer's certificate 厂商的执照;出厂证(书);厂商证明书
manufacturer's certificate of quality 制造商品质证明书
manufacturer's certificate of quantity 制造商数量证明书
manufacturer's data report 制造厂的数据报告
manufacturer's examiner 制造厂检查员
manufacturer's export agent 厂商外销代理商
manufacturer's fabrication drawing 制造厂装配图

manufacturer's guarantee 制造厂保证
manufacturer's instruction 厂家说明书
manufacturer's invoice 厂商发票;厂家发票
manufacturer's mark 制造厂家标记;厂商商标
manufacturer's model number 制造厂型号
manufacturer's name 制造厂名
manufacturer's new orders 制造商的新订货
manufacturer's output insurance 产品保险
manufacturer's partial data report 制造厂的部分资料报告
manufacturer's qualification statement 制造厂家的资格声明
manufacturer's rating 铭牌出力
manufacturer's recommendation 产品使用说明书
manufacturer's sales branches 制造商的分销机构
manufacturer's serial number 制造厂编号
manufacturer's software 厂家软件
manufacturer's specification 产品说明书
manufacturer's stamp 制造厂的印记
manufacturer's standard 制造(厂)标准;厂家标准
manufacturer's/supplier's guarantee certificate 生产厂家/供应商的保证书
manufacturer's technical documents 厂家技术文件
manufacturer's test 厂家的检验;生产厂的试验;厂方试验
manufacturer's trial 制造厂家试验
manufacturer's wire 钢丝制品用钢丝
manufacturer's working curve 制造厂提供的工作曲线
manufacturer trial 工厂试验
manufacture's brand 厂牌
manufacture sewage 工业污水
manufacture sewer 工业污水管
manufacture shop 专业化生产车间
manufacture system 专业化生产系统
manufacture test 成品试验
manufacture tolerance 制造公差
manufacturing 制成;制动;生产;制造的
manufacturing account 在制账户
manufacturing accounting 制造业会计
manufacturing accuracy 制造精度;加工精度
manufacturing activity 加工生产
manufacturing and inspection record 制造与检查记录
manufacturing assembling parts list 制造装配零件表
manufacturing assembly drawing 制造装配图
manufacturing bay 制造区间
manufacturing budget 制造预算;生产预算
manufacturing burden 制造费(用)
manufacturing business 制造业
manufacturing certificate 生产许可证
manufacturing change note 制造更改说明
manufacturing city 生产城市
manufacturing company 制造公司
manufacturing cost 制造费(用);制造成本;造价;生产成本
manufacturing cost budget 制造成本预算
manufacturing costing 制造成本核算
manufacturing crop 经济作物;工业作物
manufacturing cycle 制造周期;生产循环
manufacturing deficiency 制造上的缺陷;加工缺陷
manufacturing department 制造部门
manufacturing dimension 制造尺寸
manufacturing directive 制造命令
manufacturing district 工业区
manufacturing drawing 制造图(纸)
manufacturing engineer 工艺工程师
manufacturing engineering 制造技术;制造工艺;制造工程
manufacturing equipment 制造设备
manufacturing errors 制造误差
manufacturing establishment 制造公司
manufacturing expense 制造费(用)
manufacturing expense budget 制造费用预算
manufacturing expense budget report 制造费用预算报告
manufacturing expense distribution 制造费用分配
manufacturing expense in process 在制品费用
manufacturing expense ledger 制造费用分类账
manufacturing expense order 制造费用通知单
manufacturing expenses-in-process 在制造费用
manufacturing expense variance 制造费用差异

manufacturing facility 生产设备;生产设施
manufacturing failure 制造缺陷
manufacturing feasibility 制造的可能性;生产的可能性
manufacturing firm 制造商行
manufacturing flow chart 制造过程表
manufacturing gas with oil-wastewater 油制气废水
manufacturing ga(u)ge 制造量规
manufacturing hangar 制造厂棚
manufacturing imperfection 制造缺陷
manufacturing industry 制造(工)业;加工工业
manufacturing installation 加工装置;生产设备;制造装置;加工软化
manufacturing lead time 制造所需时间
manufacturing limit 制造极限;加工极限
manufacturing loss 生产亏损
manufacturing machinery 生产机械
manufacturing management 生产管理
manufacturing manual 制造手册
manufacturing method 制造方法;生产方法
manufacturing miller 专用铣床;生产型铣床
manufacturing network monitor 制造网络监视器
manufacturing of diamond bit 金刚石钻头制造
manufacturing operation 制造操作;生产过程
manufacturing order 生产凭单;生产命令
manufacturing-oriented 与制造有关的
manufacturing overhead 生产管理费用;生产间接费用
manufacturing overhead budget 间接制造费用预算
manufacturing overhead expense 制造间接费用;间接制造成本
manufacturing overhead rate 间接制造费用率
manufacturing overhead spending variance 制造费用耗费差异
manufacturing overhead volume variance 制造费用数量差异
manufacturing parameter 生产参数
manufacturing parts list 制造部件清单
manufacturing permit 生产许可证
manufacturing planning change 制造计划改变
manufacturing plan sheet 生产计划表
manufacturing plant 制造厂;生产厂
manufacturing practice guarantee 制造工艺保证
manufacturing procedure 制造工序;制造程序;加工过程;加工工序
manufacturing process 制作过程;制造过程;制造工艺;制造(方)法;加工过程;加工工艺
manufacturing program(me) 制造程序
manufacturing program(me) of a month 月度生产计划
manufacturing purpose 制造目的
manufacturing quality control 制造质量控制
manufacturing rate 生产率
manufacturing regulation 制造条例
manufacturing requirement 生产技术要求
manufacturing research 生产(工)研究
manufacturing schedule 生产进度
manufacturing sewage 工厂污水
manufacturing shop 专业化生产车间;制造车间;生产车间
manufacturing specification request 生产规模要求
manufacturing standards manual 制造标准手册
manufacturing statement 制造成本表
manufacturing statistics 生产统计
manufacturing status 生产现状
manufacturing supervision 生产监督
manufacturing system 专业化生产系统
manufacturing target 工业目标
manufacturing technical directive 制造技术指令
manufacturing technique 制造工艺;生产技术
manufacturing technology 制造技术
manufacturing test procedures 制造试验程序
manufacturing tolerance 制造裕度;制造公差
manufacturing town 工业城市
manufacturing variable 生产变量
manufacturing waste 制造(业)废物;工业废物;生产废液
manufacturing wastewater 制造(业)废水;生产废水
manufacturing works 制造工厂;制造厂
manufacturing yard 预制场
manufacturing yard of concrete block 混凝土预制场

manula discharge button 手动释放钮
manula train control device 手动列车控制装置
manu-marble 人造大理石
manumotive 手推的;手动的
manumotor 手推车
manure 堆肥场
manure cleaner 厩肥清除机
manure ladle 施肥勺
manure pile 粪堆
manure pit 粪坑;粪池;肥料坑
manure rain equipment 厩液喷洒机具
manure receiver 粪水池
manure separator 粪便分离器
manure slurry 肥料淤泥
manure spreader with mounted loader 带装肥器的撒肥车
manure storage 粪坑;粪窖
manure treatment 粪便处理
manure water 粪液
manurial application 施肥
manuring 施肥
manuring field trial 肥料田间试验
manuring irrigation 肥水灌溉
manuscript 原稿;直接指令单;加工图;手稿;底稿
manuscript base 稿图
manuscript display 原图显示
manuscript level data 原始高程数据
manuscript map 稿图
manuscript plane 绘图平面
manuscript submitted for approval 报批稿
manuse dump 粪堆
manus plane 扁平手
manway 井筒梯子格;人行道;进入道;(坑道里的)人行走道;人孔
manway compartment 人行格
manway compartment size 人行格子尺寸
manway cover 升降孔盖;进人孔盖;人孔盖
manway covering 人孔覆盖
manway frame 人孔框架;进人道框架
manway head 人孔顶部
manway junction box 人行道(检查孔)套管
manway landing 人行格梯子平台
manway masonry(work) 人行道(检查孔)的砖石结构
manway ring 人行道(检查孔)环
manway sewer brick 人行道(检查孔)阴沟用砖
manway shaft 人行井
manway shaft size 人行天井尺寸
manway size 人行道尺寸
manway subdrift 人行分段平巷
man winding 人员提升
Manx slates 曼克斯板岩层
many-body 多体的
many-body force 多体力
many body problem 多体问题
many-body theory 多体理论
many-bottom plow 多体犁
many-celled 多室的
many-columned 多柱的
many-contact relay 多触点继电器
many degrees of freedom system 多自由系统;多自由度体系
many dimensions 多维
man-year 一人一年的工作量;人年(劳动量单位,一个人在一年内完成的工作量)
man-year equivalent 一人一年(劳动)当量
many-element laser 多元激光器
many feed intake capacity 各种集食量
many fire retardant chemicals 各种阻火剂
many fissure 多缝隙
many-gabled style 多山墙风格
many-hued 有许多颜色的
many leaved 多叶的
many-lines spectrum 多线光谱
many-many 多对多
many-many relationship 多对多关系
many materials for mulching 多种覆盖用的物质
many of the strains in the collection 收集到的许多品系
many-one 多对一
many-one relationship 一对多关系
many order derivative of gravity potential 重力位各阶导数

many-partical state 多粒子状态
many particle spectrum 多粒子谱
many-particle system 多粒子系统
many practical decisions 许多实际措施
many reserve system 分散准备制
many-server queue 多服务排队
many-sided 多方面的;多边的
many-sided utilization 多种利用;综合利用
many-stage 多极的
many-storied 多层的(房屋)
many-to-many 多对多
many-to-many communication 多对多通信
many-to-many route control system 多对多线路控制系统
many-to-many service 多点服务
many-to-one 多对一(的)
many-to-one function 多对一函数
many-to-one route control system 多对一线路控制系统
many-turn secondary coil 多匝副圈
many-valued 多值的
many valued function 多值函数
many-valued logic 多值逻辑
many-variable system 多变量系统
many-way selection 多路选择
mao bamboo 毛竹
maolica 花饰陶器
Maorian ammonite region 毛利菊石地理大区
map 映射;映像;图;地图;测绘;布局图
map about ore districts and deposit 矿区矿床图件
map accessions list 新到地图目录
map accuracy 地图精度
map accuracy criterion 地图精度标准
map accuracy test 地图精度检验
map address 变换地址
map adjustment 地图接边
map align 图网对准
map analysis 图分析
map appearance 地图整饰
map archive 地图档案
map archiving 地图档案传输
map arrangement 地图配置
map automation 地图制图自动化
map base 底图
map bibliography 地图目录
map board 图板
map body 图面;地图主体
map border 图廓
map boundary 图廓线;图框
map cartographics 制图综合
map case 图囊
map catalogue 地图目录
map catalogue card 地图目录卡
map cataloguing 地图编目
map cataloguing system 地图编目系统
map chart 陆海图
map clarity 地图清晰性
map classification 地图分级
map cleaning 图像洁化
map cloth 地图布
map code 地图符号;地图代码
map collection 地图收集
map colo(u)r standard 地图色标
map communication 地图传输
map comparison unit 地形图重合比较装置
map compilation 图件编制;地图绘制;地图编制;地图编绘
map compiler 地图编绘员
map complexity 地图复杂性
map composition 地图编绘;编图
map configuration 地图图形
map configuration record 地图图形记录
map construction 地图编绘
map content elements 地图要素
map-controlled mosaic 地图控制镶嵌图
map coordinate code 地图坐标网代码
map coordinate network 地图坐标网
map coordinates 地图坐标
map coordinate system 地图坐标系
map copy 地图拷贝;地图复制
map copy preparation 地图复照
map correction 地图审校
map coverage 有图地区
map crack(ing) 地图形裂缝(常为碱性骨料反应的结果);网状裂纹;龟裂;路面网(状)裂(纹);网状开裂;细裂纹
map crazing 龟裂;路面网(状)裂(纹)
map data 地图资料
map data base 地图数据库
map data structure 地图数据结构
map decoration 图廓整饰;地图整饰
map delineation 地图清绘
map delineator 绘图员
map depot 图库
map design 地图设计
map details 地图内容要素
map development 图件绘制;构图
map digitizing 地图数字化
map dimension 图件规格;图幅尺寸
map display 图像显示;地形图显示板;地图显示器
map distance 图上距离;图距(离)
map distribution 地图供应
map distribution point 地图供应点
map documents 地图资料
map drawing 绘图;地图绘制;地图绘图
map duplicate 地图复照品
map edge 图廓
map editing 图件编辑;地图编辑工作;地图编辑
map edition 地图出版
map editor 地图编辑员
map editorial design 地图编辑设计书
map editorial plan 地图编辑计划
map edit program(me) 地图编辑大纲
map electrostatic copier 地图静电复印机
map enlargement 地图放大(法)
map error 地图误差
map evaluation 地图评价
map exercise 填图
map exhibition 地图展览
map expansion 图距扩散
map expression 地图表示法
map extract 地图摘录;地图片段
map eye for map 图上目测
map face design 图面设计
map face position 图面配置
map fair drawing 地图清绘
map filtering technic 图滤波技术
map finish 地图整饰
map fixation 地图定位
map folding 地图折叠
map format 地图开幅
map forming system 成图系统
map for prediction 预测图
map function 地址变换操作;地图用途;变换操作
map generation 绘图
mapgraph 统计地图
map graticule 地图经纬网格
map grid 制图网格;图网;图格;地图方格网
map history 图历表
map holder 图架
Mapica 除草佳
map image 地图照相图;地图复照图
map image data 地图图形数据
map image record 地图图形记录
map in books and periodicals 书刊地图
map index 接图表
map information 地图资料
map information room 地图资料室
map inset 地图插图
map interlinking 接图
map interpretation 读图;地图判读;地图阐释
map intersection of deposits 矿产断面图
map issue 地图发行
map join 图幅接边
map layout 图面配置
maple 槭属;槭树;淡棕色
map legend 图例;地图符号
maple hornbeam 槭鹅耳枥
maple leaves 枫树叶子
maple silkwood (昆士兰产的)类似桃花心木的硬木
maple tree 枫树
Maplewell beds 马波维尔层
map library 图库
mapmaker 制图员
map making 制图;地图制印;地图制图;地图绘制;地图制作;地图测绘学;绘制地图
map-making office 制图局
map-making organization 制图机构
map making process 制图过程
map-making room 制图室
map-making satellite 测图卫星
map-making unit 制图室
map maneuver 图上作业
map manuscript 地理底图
map margin 图廓;图(幅)边(缘)
map matching 地图匹配
map matching guidance 地图重合制导;地图匹配制导
map mathematic model 地图数学模型
map measure 量图
map measurement 量图;图上量表;图上量标
map measurer 计图器;量图仪;量图器;量图工具;曲线仪;图上测距仪;测图仪
map-microfilm reader 地图缩微阅读器
map nadir 图幅天底点;图面底点
map norm dimension 图件正常规格
map number 图件编号;图幅编号字母;地图图号;地图编号
map of active structural system 活动构造图
map of age provinces 年龄省图
map of alumin(i)um plate 铝板图
map of basement structure 基底构造图
map of blasting zone 爆破区图
map of boundary condition with electro analogue 电模拟边界条件图
map of bursting water prediction 突水预测图
map of deflection of vertical 垂线偏差图
map of distribution of active structures and strong earthquake epicenters 活动性构造和强震震中分布图
map of distribution of dominant tectonic systems and epicenters 主要构造体系与震中分布图
map of distribution of mineral deposits 矿产分布图
map of distribution of sand rocks and various deposits 砂岩及各类储集体分布图
map of environment quality evolution of groundwater 地下水环境质量评价图
map of faults 断层图
map of food 粮食分布图
map of freeboard zone 干舷区域图
map of free hand in field 野外工作手图
map of frequency distribution 频率分布图
map of gas saturation in formation water 地层水中气饱和度图
map of geochemistry anomaly 地球化学异常图
map of geophysics anomaly 地球物理异常图
map of geothermal field exploitation 热田开发图
map of geothermal manifestation 地热显示图
map of groundwater 地下水分布图
map of groundwater pollution degree 地下水污染程度图
map of groundwater resource distribution 地下水资源分布图
map of heat flow contours 热流量等值线图
map of hydrochemical division 水化学分区图
map of hydrochemical type 水化学类型图
map of hydrogeologic(al) division 水文地质分区图
map of hydrogeologic(al) parameter division 水文地质参数分区图
map of hydrothermal alteration 水热蚀变图
map of intensity zoning 烈度区划图
map of isochloside contours 氯离子等值线图
map of karst hydrogeology 岩溶水文地质图
map of lithofacies and palaeogeography 岩相古地理图
map of lithology and lithofacies 岩性岩相图
map of mankind 种族分布图
map of mineral deposit 矿床图
map of mineral resources 矿产资源图
map of molecular ratio 克分子量比图
map of monomial hydrogeology characteristic 单项水文地质特征图件
map of natural environmental background 自然环境背景图
map of neotectonic type 新构造类型图
map of ocean floor 洋底地图
map of oil field concession 油田开采权图
map of open pit limit 露天矿采场境界图
map of ore block design 矿块设计图

map of pal(a)eocurrent direction 古水流方向图
map of parted mining fields 部分矿区图
map of pollution hydrogeology 污染水文地质图
map of prediction of seismic calamity 震害预测图
map of primitive data 实际材料图
map of principal stress trajectories 主应力轨迹图
map of prospect engineering arrangement 勘探工程布置图
map of radioactivity hydrochemistry 放射性水化学图
map of recent horizontal movement 现代水平运动图
map of recent vertical movement 现代垂直运动图
map of research level 研究程度图
map of rhythmic(al) thickness 韵律厚度图
map of sample site 取样位置图;采样点位图
map of seismic intensity zoning 地震烈度分区图
map of seismicity 地震区域图;地震活动(分布)图
map of seismic result 地震成果图
map of seismic risk zoning 地震危险区划图
map of sewers 下水道系
map of shear(ing) stress trajectories 剪应力轨迹图
map of silica contours 二氧化硅等值线图
map of single engineering design 单项工程设计图
map of skeleton lines 山脉线略图
map of soil 土壤(分布)图
map of standard format 标准图幅
map of stratigraphic(al) geology 地层地质图
map of structural systems and earthquake epicenters 构造体系与地震震中分布图
map of structural systems and ground hot water 构造体系与地下热水分布图
map of structural systems and ground water 构造体系与地下水分布图
map of submarine geologic(al) structure 海底地质构造图
map of superficial sediment types 底质类型图
map of surveying work arrangement 调查工作实际材料图
map of the world 世界地图
map of time zones 时区图
map of utilization degree of water resources 水资源利用程度图
map of volcanic geology 火山地质图
map of water content of aquifer 含水层富水程度图
map of water table 地下水位(分布)图
mapograph 投影转绘仪
map orientation 地图标定
map orientation of digitizer 数字化仪原图定向
map out 拟订
map output 地图输出
map over 对映像
map overlay 分色透明图
map overlay analysis 地图叠置分析
mappable material 编图资料
mappable unit 图幅
map paper 地纸
map parallel 图横线;透视轴
map pattern 地图图形
map pattern record 地图图形记录
mapped buffer 映像缓冲器;映射缓冲区;变换缓冲器
mapped data 图廓资料;编图资料
mapped memory 映像存储器
mapped physical storage 地址变换物理存储器
mapped storage 映像存储器
mapped surface 有图地区
mapper 制图者;绘图仪;测绘仪;测绘器
map photocopy 地图复照品
map photography 地图复照
mapping 映像;映射;铸铁鼠尼;绘图;填图;地形制图;地图绘制;测图;测绘
mapping accuracy 制图精度;测图精度
mapping address 变换地址
mapping aircraft 航空摄影飞机
mapping air plane 测绘飞机
mapping algorithm 算法
mapping angle 收敛角
mapping area 测绘面积
mapping axis map 映射轴平面图
mapping base point 图根点
mapping block number 变换程序段数

mapping by computer 计算机绘制
mapping by free hand 人工测制
mapping by instrument 仪器测制
mapping by photography 摄制
mapping by remembrance 记忆测图
mapping by survey and record method 测记法成图
mapping camera 地图测绘照相机;制图照相机;测图摄影机;测绘照相机
mapping carriage 测图仪骨架
mapping cartographic(al) topography 地形测图航空摄影
mapping center 制图中心
mapping class 映射类
mapping cone 映射锥
mapping control 图根控制;测图控制
mapping control chain (net) with angle measurements 测角图根锁(网)
mapping control horizontal survey 图根平面位置测量
mapping control level(l)ing 图根水准测量
mapping control point 图根控制点;图根点;地形控制点;测图控制点
mapping control survey 图根测量
mapping control vertical survey 图根高程测量
mapping cylinder 映射柱
mapping data 地图数据
mapping data bank 地图数据库
mapping date 制图日期
mapping degree 映射度;映射度
mapping device 映像装置;映射装置;布局设备;变换装置;变换机构
mapping digitizing data bank 地图数字数据库
mapping drawing 制图
mapping drill hole 填图孔
mapping equation 地图投影方程
mapping equipment 测绘设备
mapping error 映射误差;绘图误差
mapping extension 映射扩充
mapping fault 变换误差;变换故障
mapping feature data bank 地图要素数据库
mapping field 变换字段
mapping film 测图航摄胶片
mapping flight 地图测绘(摄影)飞行
mapping from photograph 像片测图;像片编图
mapping function 映射函数;射影函数;变换功能
mapping grid 绘图格网
mapping height survey 图根高程测量
mapping information 测绘信息
mapping information bank 地图信息库
mapping infrared system 红外(线)测绘系统
mapping instrument 绘图仪器;绘图机;测图仪器
mapping in the reverse direction 反向起绘
mapping jitter 映射抖动
mapping language 映像语言
mapping lens 测图镜头
mapping lens of camera 测图摄影机主镜
mapping method 作图法;制图方法;成图方法
mapping method with transit 经纬仪测绘法
mapping mission 航摄资料分布示意图
mapping mode 变换方式
mapping objective 测图物镜
mapping of a set in another 映入
mapping of a set onto another 映成
mapping of deformed face 变形面填图
mapping of exposure 露头填图
mapping of ocean 海洋测绘
mapping of surface waters 地表水图
mapping organization 测图部门
mapping parameter 制图参数
mapping pen 绘图笔
mapping photography 地形测图摄影
mapping point of geochemistry 地球化学填图点
mapping problem 图上作业;映射问题
mapping radar 测图雷达;测绘雷达
mapping recorded file 图历薄
mapping route length 观测路线长度
mapping route of geochemistry 地球化学填图路线
mapping scale 成图比例尺;测绘比例尺
mapping schedule 制图计划
mapping space 映像空间;映射空间
mapping standards 制图标准
mapping strip 测图航片
mapping survey 测图

mapping system 测图系统
mapping table 变址表;变换表
mapping technique 测图方法
mapping tool 测图工具
mapping transformation 映射变换
mapping traversing 图根导线测量
mapping triangulation 图根三角测量
mapping truck 映射载
mapping type 测绘种类
mapping unit 填图单位
mapping window 变换窗口
mapping work 地图测绘工作
mappist 制图者;制图员
map place 图面位置
map plane 图幅高程基准面
map plotting 填图;地图绘制;地图编制
map plumb point 天底点
map pocket 图囊
map point 图上定位点;图面点
map printing 地图印刷厂;地图印刷
map printing master 地图印刷原版
map printing plate making 地图制版
map process and mapping 制印
map production 地图生产
map program(me) 映像程序
map projection 地图投影(法)
map projection atlas 地图投影图集
map projection grid 地图投影格网
map projection table 地图投影图表
map projection transformation 地图投影变换
map projector 投影制图仪;地图投影仪
map proof checking 地图审校
map protractor 地图量角器
map publication 地图出版
map publishing house 地图出版社
map range 图上距离
map-read 读图
map reading 航图测读;读图;地图说明;地图读数
map-reading amplifier 读图放大镜
map reconnaissance 图上选点
map record 地图资料
map reduction 地图缩小(法)
map reference 地图基准;地图点位号
map reference code 地图坐标网代码
map reference grid 地图格网
map relationship 图幅接合表
map reliability 地图可靠性
map representation 地图表示法
map reproduction 地图复制;地图复照
map reproduction equipment 地图印刷设备
map reproduction plant 地图印刷厂
map reprography 复照
map revision 地图修订;地图修测;地图更新
maps about ore field 矿田图件
map scale 图(件)比例尺;地图比例尺
map scanner 地图扫描仪;地图扫描器
map scriber 刻图员
map sections 地图片段
map separation film 地图分色片
map set 接图表
map sheet 图页;图件;图幅;地图页;地图图幅
map sheeting 地图图幅
map-sheet motion 图幅移动
map showing distribution of pal(a)-eocontinents 古大陆分布图
map showing drifting path of pal(a)-eocontinent 古大陆漂移轨迹图
map showing pal(a)eoatmospheric circulation 古大气环流图
map showing pal(a)eoclimatic zonation 古气候带分布图
map showing pal(a)eocommunity migration 古群落迁移图
map showing pal(a)eolatitude 古纬度图;古海陆分布图
map showing prospective mineralization area 成矿远景区平面图
map showing regional coalfield prediction 区域煤田预测图
map showing regional coal-forming expectation 区域成煤远景图
map showing regional coal-forming prediction 区域成煤预测图
map showing regional mineralization expecta-

tion 区域成矿远景图
map showing regional mineralization prediction 区域成矿预测图
map showing regional research level 区域研究程度图
map showing the degree of geologic-(al) study 地质研究程度图
map showing the relation between structural systems and the distribution of ore 构造体系与矿产分布图
maps in geologic(al) documentation 编录地质图件
map sketch 草图
map source data 编图资料
map specification 地图设计书；地图规范
map specimen 地图样本；地图图样
map splicing table of survey area 测区图幅接图表
map spotting 填图
map square 地图方格
map standard 地图标准图；标准地图
map stock 地图储备
map storage 图库；地图储备
map style 地图类型
map subject 制图对象
map substitution 临时版地图
map subtitle 副图名
map supply 地图供应
map suppression 地图改版
map surface 图面
map-surfaced model 地图贴面立体模型
map surface lettering machine 图面注记机
map surface lettering photoelectronic machine 光电子式图面注记机
map surface lettering photomechanical machine 光机械式图面注记机
map survey 地图测量
map symbol 地形符号；图例；地图图式；地图符号
map table 映像表；变换表
map target indicator 图上标指示器
map test 地图检测
map title 图名；地图名称
map tolerance 地图限差
map tracer 地图描绘器
map trade 地图门市部
map trimming 地图光边
map type 图件种类；地图种类
map unit 图(距)单位
map updating 地图更新
map use 地图用途
map varnishing 地图上光
map vertical 地球表面法线方向；大地水准面法线方向
map with fluorescent colo(u)ring 荧光地图
maqarna 钟乳形饰(伊斯兰教建筑中天花板装饰)
maqsura (清真寺的)祈祷堂
maqsurah 围护空间
maquette 初步设计模型；蜡模型；胶泥模型；房屋模型
maquis 刺丛灌木群落
mar 毁损；小湖；划痕
Maracray 马拉克赖聚酯纤维
mar-ag(e)ing 改变钢的马丁体结构法；高强度热处理；马氏体时效(处理)
mar-ag(e)ing steel 马氏体时效钢；特高强度钢
Marangoni convection 表面张力梯度驱动的对流
Marangoni effect 马兰戈尼效应(因界面张力梯度引起的液流动)
marantic 消耗的
marasmatic 消耗的
marasmic 消耗的
Marb-l-cote 仿大理石涂料；粉状塑料漆(加水涂刷24小时凝结后像大理石)；速干油漆(干后呈大理石面状)
marble 云石；大理岩；大理石；玻璃球
marble aggregate 大理石集料；大理石骨料
marble balda(c)chino 大理石华盖
marble beam 大理石梁
marble bench 大理石长凳
marble block 大理石块；大理石坯料
marble block-cutter 大理石分切机
marble board 仿大理石墙板；大理石花纹纸板
marble boat 石舫
marble building 大理石建筑；大理石建筑物
marble bushing 玻璃球料碗

marble candelabram 大理石烛架
marble carving 大理石石刻品
marble chip(ping)s 大理石切片；大理石渣；大理石屑
marble chute 玻璃球溜槽
marble coating 大理石墙面
marble column 大理石柱
marble cooling conveyer 玻璃球网带冷却机
marble Corinthian column 大理石考林斯式柱
marblecrete 干黏大理石米饰面；大理石质感镶面
marble cross 大理石十字架
marble crypt 大理石地窖
marble cutter 大理石切刀；大理石切割机；大理石采石工
marbled 仿大理石的
marble dam 大理石坝(特指印度拉贾斯坦邦阿拉瓦利山的两座坝)
marble decoration 大理石装饰
marble decorative finish 大理石的外表抛光；大理石的外表精修；大理石的外表装饰
marbled edge 云石边；大理石镶边
marble defects 玻璃球缺疵
marble deposit 大理岩矿床
marbled finish 仿大理石饰面
marbled glass 斑纹玻璃；大理石纹玻璃；云石形玻璃；大理石状玻璃
marble glaze 大理石釉
marbled glazed paper 玳瑁蜡光纸
marbled grain 粗粒面
marble distribution station 配球台
marble dividing panel 大理石隔板
marble door 大理石门
marbled pattern 大理石模样
marble dressing 大理石装饰；大理石修整
marbled texture 大理石纹理
marble dust 云石粉；大理石粉
marble effect 大理石效果
marble enamel 大理石纹搪瓷
marble enrichment 大理石美化
marble facade 大理石建筑物立面
marble-faced precast concrete panel 大理石面预制混凝土板
marble facing 大理石饰面
marble feeding hole 加球孔
marble feeding tube 加球管
marble feed system 玻璃球加料系统
marble figure coating 石纹涂装法；仿大理石(纹)涂装法
marble filter gravel 大理石过滤砾石
marble finish 大理石饰面
marble fireplace 大理石壁炉
marble flag pavement 大理石板铺面；云石板铺面
marble floor(ing) 大理石地板；大理石地面
marble floor(ing) finish 大理石地面装饰
marble flour 大理石粉(末)
marble flower vase 大理石花瓶
marble for cement 水泥用大理岩
marble for construction 建筑用大理岩
marble for decoration 饰面用大理岩
marble for glass 玻璃用大理岩
marble fracture 大理石状断口
marble frieze 大理石檐壁；大理石饰带
marble furnace 球窑
marble garden table and stool 大理石花园桌凳
marble glass 斑纹玻璃
marble glaze 大理石釉
marble glazed paper 玳瑁蜡光纸
marble grader 分球器
marble grain 大理石纹(理)
marble grain mix(ture) 大理石粗粒面配料
marble gravel 大理石砾石
marble gravity feeding 玻璃球重力加球
marble grill(ag)e 大理石格栅
marble hopper 玻璃球料斗
marble imitating granite 仿大理石花岗岩
marble inlay 大理石嵌体；云石镶嵌
marble inspection table 玻璃球试验台
marble intarsia 云石镶嵌装饰；大理石镶嵌装饰；大理石贴面装饰
marbleization 仿大理石花纹
marbleize 仿大理石花纹
marble ized 大理石花纹(装饰)的
marbleized effect 大理石效果
marbleized finish 仿大理石饰面

marbleized lino(leum) 大理石形的油地毡
marbleized texture 大理石纹理
marbleizing 仿大理石纹
marbleizing glaze 大理石釉
marble limestone 大理岩灰岩；大理石灰岩
marble lining 大理石衬层
marble machine 制球机
marble making process 球法拉丝工艺
marble masonry work 大理石圬工工程
marble melt 玻璃球熔体
marble melt process 坩埚法；球法拉丝法
marble mosaic 大理石马赛克；大理石镶嵌艺术
marble ornament 大理石装饰品
marble ornamental finish 大理石装饰的成品
Marble Palace at Leningrad 列宁格勒的大理石皇宫
marble panel(l)ing 大理石镶板；大理石(面)板
marble paper 云色纸；大理石纹纸；大理石花纹纸
marble pattern 大理石模型；大理石花纹
marble pavement 大理石地面；大理石铺面
marble pedestal 大理石基座
marble pillar 华表；大理石柱子
marbles plank 大理石板材
marble plaque 大理石插屏
marble plaster 云石灰泥；云石灰浆；大理石灰泥；大理石灰浆
marble plate 大理石板
marble polishing machine 大理石上抛光机
marble polishing material 大理石上抛光材料
marble powder 大理石粉末；大理石粉
marble powder colo(u)r brick 大理石粉彩色砖
marble process 球法
marble products 大理石制品
marble pulpit 大理石控制台
marble quadriga 大理石四马拖车雕饰
marble quarry 大理石采石场；采大理石场
marble relief 大理石浮雕品
marble rernelted 玻璃球重新熔化
marble resin 人造大理石
marble roof 大理石屋顶
marble roof(ing) slab 大理石屋面板
marble rot 大理石斑腐
marble sand 大理石砂；云石砂
marble sarcophag 大理石棺
marble sawing plant 切大理石厂
marble set 大理石条石
marble setter 大理石安装工具
marble shaft 大理石柱身
marble shot 大理石爆破；大理石炮
marble size 大理石粒度
marble slab 大理石板
marble slab stair tread 大理石石板楼梯踏步
marble soil 大理石状土
marble sphinx 大理石人面狮身
marble status 大理石雕像
marble structure 大理石结构
marble stucco 大理石毛粉饰
marble sunblind 大理石遮阳
marble surface 大理石面层；仿大理石纹表面
marble surfacing 大理石饰面
marble switchboard 大理石配电盘
marble tabernacle 大理石壁龛
marble test 大理石试验
marble texture 大理石纹理；类似大理石花纹
marble throne 大理石宝座
marble tile 大理石砖；大理石花纹瓦；大理石饰面砖
marble top 大理石台面
marble top wash-stand 大理石面洗面台
marble tread 大理石踏板
marble vein 大理石纹(理)
marble veneer 大理石面板；大理石面层；大理石贴面板；大理石板
marble veneer facing 大理石板贴面
marble veneering 大理石板贴面
marble veneer panel 大理石贴面板
marble washer 洗球机
marble white 天然方解石粉
marble window sill 大理石窗台板
marble wood 大理石(纹)木
marble work 大理石工程
marbling 仿石纹饰；形成大理石纹路；大理石饰纹；大理石装饰法；仿大理石纹；仿大理石
marbling brush 大理石饰纹刷
marbling print 云纹印涂；墨法印刷；大理石纹印涂
marbling varnish 大理石纹饰清漆

marblization 大理岩化作用
marblized 大理石花纹的
marblized finish 大理石花纹装饰
marblized soap 大理石花纹皂
marbly 冷淡的;像大理石的
marbly building 大理石建筑物
marbolike 仿大理石花纹的保温板(用于冷藏方面)
Marbolith 马波里板(由菱苦土和沥青乳液拌制而成的一种无缝地板)
marcasite 二硫化铁;白铁矿
marcasite ore 白铁矿矿石
marcasite rock 白铁矿岩
marcasite sulfur 白铁矿硫
marcel disk 槽形盘子
Marcellus shale 马塞卢斯页岩
march 行程;变程
Marchand tube 马献德管(一种 U 形氯化钙管)
marchasite 黄铁矿;白铁矿
march by compass 按方位角行进
March equinox 春分
marches 边界地区居民
Marchetti's flat dilatometer 马氏扁式松胀仪
marching 行变程
marching dune 移动的沙丘;进展的沙丘
marching problem 步进式问题
marching subpulse 漂移次脉冲
marching type 步进式
marchioness 板岩瓦片;石板瓦(22 英寸乘以 12 英寸);长方石板
marchite 顽火透辉岩
march of temperature 温度变程
march stone 标界石
March type reflection wave 马赫型反射
Marcinkiewicz's theorem 马辛寇维茨定理
marcomizing 不锈钢表面渗氮处理
Marconaflo slurry system 马康那弗罗式矿粉液化装卸法;矿粉液化装卸系统
Marconaflo slurry transport 马康那弗罗回收矿泥输送法
Marconi beam aerial 马可尼定向天线
Marconi beam antenna 马可尼定向天线
Marconi coherer 马可尼金属屑检波器
Marconi commutator 马可尼换向器
Marconi detector 马可尼检波器
marconigram 无线电报
Marconi magnetic detector 马可尼磁性检波器
Marconi self-tuning system 马可尼自动调谐系统
Marconi type antenna 马可尼天线
Marconi undamped generator 马可尼等幅波发射机
marcottage 压条法
marcotting layerage 压条法
marcus 大铁锤
Marcus trough conveyer 马尔古斯槽式输送机
Marcy mill 圆筒形棒子球磨机;马西型(球)磨机
mare clausum 领海
mare frigoris 冷海
maregraph 验潮仪
marekanite 珠状流纹玻璃;似曜岩斑状体
mare liberum 公海
mare-like plain 类海平原
maremma 近海沼泽地(意大利)
marenco 马朗科风(意大利马列一带东南风)
mare nostrum 两国或数国共有的海
mareogram 潮汐涨落曲线;潮汐曲线;潮位自记曲线;潮时曲线
mareograph 潮汐自记仪;自动水位计;自记潮位仪;水位记录仪
mareographic 潮汐作用的
mare spumans 泡海
mare's tails 马尼云
marezzo 人造大理石
marezzo marble 金氏水泥人造大理石;人造大理石
marfanil 高磺胺
Marform process 橡皮模压制成型法
margarine 珠脂;人造黄油
margarine factory wastewater 人造奶油厂废水
margarine wastes 人造奶油废水
margaritasite 钒铯铀矿
margarite 珍珠云母;串珠雏晶
margarodite 珠水云母
margarosanite 针硅钙铅石;针硅钙铅矿
Margelef diversity index 马格列夫多样性指数
Margelef index 马格列夫指数

margigram 潮时曲线
margin 限界;裕度;余量;余额;改正力;利差;容许极限;容限;储备量;成本与售价的差额;差距;额;范围;边缘;边沿;边际收益;边际;保证金买卖;保证金购进;保证金比率
marginal 边缘性的;边缘的;边饰的
marginal aberration 边缘像差
marginal absorption 临界吸收;边缘吸收
marginal abyss 边界深渊
marginal access 边缘通道;侧面通道
marginal access road 进入地块的支路
marginal access street 进入地块的小街
marginal account 边缘账户
marginal accretion 边缘增生
marginal activity 边际活动
marginal adjustment 边限调整;边界调整;边际调整
marginal advantage 边际利润
marginal aggregate 边次骨料;边次集料
marginal amplifier 边频放大器
marginal analysis 限界分析;边缘分析;边界分析;边际(效益)分析
marginal angle 图廓角;边缘角
marginal arc 边缘弧
marginal area 边缘地区
marginal aulacogen 陆缘拗拉谷
marginal balance 余额
marginal bank 路堤岸;边(缘)滩;边岸
marginal bar 边缘钢筋;边筋;边部钢筋;护栏;玻璃窗格条
marginal basin 陆缘盆地;边缘盆地
marginal bast 边缘韧皮部
marginal beam 边缘梁
marginal benefit 边缘效益;边际效益
marginal benefit-cost ratio 边缘效益成本比;边际效益费用比;边际利益费用比
marginal benefit curve 边缘效益曲线
marginal berth 顺岸码头;顺岸泊位
marginal bore 边缘钻孔;边孔
marginal borrower 边际借款人
marginal branch 边缘支
marginal break-even analysis 边际损益点分析
marginal bund 堤;边缘堤岸
marginal business 限界营业;边际营业
marginal capacity 备用容量;界限容量
marginal capacity cost 边陆资本
marginal capital coefficient 边际资本占用率;边际资本系数
marginal capital expenditures 边际资本支出
marginal capital-output ratio 边际资本产出比率
marginal category 边缘范畴
marginal change 边际变化
marginal check 边缘检验;边缘检查;边界检验
marginal checking 边缘校验
marginal clause 栏外条款;边际条款
marginal community 边缘群落
marginal component 临界分量
marginal concrete strip finisher 混凝土板边缘修整机
marginal condition 边界条件;边缘条件;边际条件
marginal conditions of optimization 最优化的边际条件
marginal consumer 边际消费者
marginal containerizable cargo 边际集装箱货
marginal continuous distribution 边缘连续分布
marginal contribution 边际贡献
marginal contribution ratio 边际贡献率
marginal convolution 边缘回卷;边缘叶
marginal correction 边缘校正
marginal cost 增支成本;边缘费用;边际费用;边际成本
marginal cost and revenue 边际费用和边际收入
marginal cost breakeven 边际成本无盈亏点
marginal cost curve 边际成本曲线;边际成本曲线
marginal cost equation 边际成本方程式
marginal costing 边际成本计算法;边际成本核算
marginal cost of acquisition 边际所得物成本;边际取得成本
marginal cost of capital 边际资金成本;边际资本成本
marginal cost of information 信息的边标成本;边际信息成本
marginal cost of production 边际生产成本
marginal cost price 边际成本价格;边际成本定价法
marginal cost pricing 边际成本定价

marginal cost vector 边际成本向量
marginal credit 限界信贷;定额信用证;边际信贷
marginal crest 缘峰
marginal crevasse 边缘裂缝;边(缘)冰隙
marginal crop rows 边行作物
marginal current 定限电流
marginal curve 边际曲线
marginal damage cost 边际损害成本
marginal data 图例说明;边注资料
marginal decoration 图外整饰
marginal deep 陆缘海渊;边缘海渊;边缘拗陷
marginal definition 像片边缘清晰度;边缘清晰度
marginal degree of utility 边际效用(程)度
marginal demand 边际需求
marginal demand price 边际需求价格
marginal density 边际密度
marginal deposit 陆相沉积(物);冰缘沉积;边缘沉积
marginal deposit for security 保证金存款
marginal deposit loan ratio 边际存贷率
marginal depression 边缘坳陷
marginal depression type formation 边缘坳陷型沉积建造
marginal discharge 尖端放电;边缘放电
marginal distortion 边缘畸变
marginal distribution 边缘分布;边陆分配;边际分配;边际分布
marginal disturbance 边缘干扰
marginal disutility 边际负效用
marginal ditch 边缘沟;边沟
marginal earthquake 边缘地震
marginal ecology 边际生态学
marginal effect 边缘影响;边缘效应;边际效应
marginal efficiency 边际效率
marginal efficiency of capital 边际资本收益率
marginal efficiency of investment 投资边际效率
marginal efficiency of investment schedule 投资边际效率表
marginal employment coefficient 边际就业系数
marginal error 镜边畸变;边缘误差;边缘错误
marginal escarpment 陆缘陡坡;海岸崖;边缘海崖
marginal expenditures 限界消费;不能再减的费用
marginal expense 边际费用
marginal explosion crater 边缘喷口
marginal facies 边缘相
marginal facies zone 边缘相带
marginal factor cost 边际(生产)要素成本
marginal factor revenue 边际生产要素收益
marginal false quay 顺岸透空式码头
marginal farm area 边缘作物区
marginal fasciole 缘带线
marginal fault 边缘故障;边缘断层
marginal firm 边缘厂商
marginal fissure 边缘裂缝
marginal flood 边缘注水
marginal fluvial-lacustrine bivalve subprovince 边缘河湖相双壳类地理亚区
marginal focus 边缘校验;边缘焦点;边缘检验
marginal fold 缘褶;边缘褶
marginal form 边缘型
marginal fracture 边缘断裂
marginal frequency 边际频数率
marginal frequency function 边际频数函数
marginal friction 路侧摩阻;路边阻力
marginal furrow 边缘沟
marginal gain 边际增益
marginal geomonocline sediment 边缘单斜沉积
marginal geosyncline 陆缘地槽;边缘地槽
marginal geothermal reservoir 边缘热储
marginal grade 最低品位;边界品位;边际品位(矿业)
marginal groove 临界纹
marginal ground 边缘土地
marginal groundwater 边缘地下水
marginal growth 边缘生长
marginal growth fault 边缘生长断层
marginal habitat 边缘生境
marginalia 页边说明
marginal illumination 边缘照明
marginal import quota 边际进口率
marginal incident ray 边缘入射光线
marginal income 边际收益
marginal income chart 边际收益图
marginal income ratio 边际收益(比)率

marginal income statement 边际收益(计算)表
marginal increase allocation method 小幅度递增分配方式
marginal increment 边际增益
marginal information 边缘信息
marginal internal rate of return 边际内部报酬率
marginal invasion 边缘侵占
marginal investment 边陆投资;边际投资
marginal laceration 边缘撕裂
marginal lagoon 沿海潟湖;边缘潟湖
marginal lake 冰缘湖;冰前湖
marginal land 贫瘠土地;低产地;道路边用地;边缘(用)地;边缘陆地;边缘古陆;边角地;边际土地;边际耕地
marginal land acquisition 界地申请(取得路边地权)
marginal land of city 城市边缘土地
marginal lappet 缘垂
marginal layer 缘层;边缘层
marginal lender 边际贷款人
marginal lenslike organ 缘晶形器
marginal lines of chart 海图边线
marginal load capacity 储备供电量
marginal loss 边际亏损
marginal marine deposit 滨海砂矿床
marginal-marine formation 海岸沉积(物)
marginal market 边际市场
marginal marking 边缘标志
marginal material 临界性质的材料;边缘材料
marginal material production 边际物质生产
marginal membrane 边缘膜;边渗膜
marginal metamorphism 边缘变质作用
marginal migmatization 边缘性混合岩化
marginal moraine 终碛;边碛
marginal mo(u)ld 边缘饰条
marginal name 图边注记
marginal need 边际需要
marginal note 旁注;图边注记;边注
marginal nunatak 边缘岛峰
marginal object 图边地物
marginal obstruction 路边障碍物
marginal-ocean basin 边缘洋盆
marginal operation 边缘操作
marginal opportunity cost 边陆机会成本
marginal output 边际产出;击击产出
marginal output or product 边缘产量或产品
marginal-Pacific active area in Eastern China 中国东部滨太平洋活动区
marginal-Pacific deep fracture system 滨太平洋断裂体系
marginal-Pacific geosynclinal fold region 滨太平洋地槽褶皱区
marginal-Pacific migration region 滨太平洋迁移区
marginal-Pacific tectonic domain 滨太平洋构造域
marginal parallel ray 边缘平行光线
marginal physical product 边际物质产品;边际生产实物量;边际产品
marginal placentation 边缘胎座式
marginal plain 冰水沉积平原;边缘平原
marginal plate 缘板
marginal plateau 陆缘高原;陆上高原;边缘台地
marginal plates 侧蜡板
marginal platform 边缘台地
marginal plexus 缘丛
marginal point 图廊点;边缘点
marginal pool 边缘池
marginal pore 缘孔
marginal price 边陆价格;边际价格
marginal pricing 边际定价
marginal private cost 边际私人费用
marginal probability 界限概率;边缘概率;边际概率
marginal probability density 边缘概率密度
marginal probability distribution 边缘概率分布
marginal producer 边际生产者
marginal product 边际(生)产量;边际产品
marginal production 边际生产
marginal production operating cost 边际生产经营成本
marginal production revenue 边际生产收益
marginal productivity 边际生产率;边际生产力
marginal productivity of labor 边际劳动生产率
marginal productivity theory 边际生产率说
marginal product operation 边际产品生产
marginal profit 限界利润;边际利润

marginal profit chart 边际利润图表
marginal profit ratio 边际收益率;边际利润率
marginal profit yield 边际利润收益
marginal project 边际项目
marginal propensity 边际倾向
marginal propensity to consume 边际消费倾向
marginal propensity to import 边际输入倾向;边际进口倾向
marginal propensity to invest 边际投资倾向
marginal propensity to save 边际储存倾向
marginal propensity to spend 边际支出倾向
marginal propensity to tax 边际课税倾向
marginal propensity to withdraw 边际提取倾向;边际漏出倾向
marginal punched card 边缘穿孔卡
marginal quay 沿岸码头;顺岸码头
marginal rate of interest 边际利率
marginal rate of liquidity preference 边际流动偏好率
marginal rate of personal income tax 边际个人所得税率
marginal rate of return over cost 弥补成本后的边际报酬率
marginal rate of substitution 临界置换率;边际替代(比)率;边际代替比例
marginal rate of tax 边际税率
marginal rate of transformation 边际转换率
marginal ratio 边际比率
marginal ray 边缘射线;周边光线;边缘光线;边(部)光(线)
marginal receipt 边际收入
marginal reinforcement 边(缘钢)筋;边部钢筋
marginal relay 定限继电器;边缘继电器
marginal relief 边际减免
marginal relief of income tax 所得税减免界限
marginal rent 边际地租
marginal representation 图廊外整饰
marginal reserve accounting 边际准备金核算
marginal retailer 边际零售商
marginal return 边际报酬
marginal returns 边际利润率
marginal returns to capital 边际资本收益;边际资本报酬
marginal revenue 边际所得总额;边际收益;边际收入
marginal revenue curve 边际收入曲线
marginal revenue product 边际收益产品;边际收入产品;边际生产收益量;边际生产收入
marginal ridge 围脊;边缘嵴
marginal ring basin 陆缘环形盆地
marginal ring depression 边缘环形坳陷
marginal risk 边际风险
marginal rot 边腐
marginal runway 长度仅够的跑道
marginal salt pan 陆缘盐田
marginal scute 缘板
marginal sea 陆缘海;边缘海
marginal sea basin 边缘海盆
marginal sea level movement 边缘海平面运动
marginal seamount 边缘海山
marginal seller 边际卖主
marginal sharpness 边缘清晰度
marginal-slope accumulation 边坡堆积(作用)
marginal social cost 边际社会费用;边际社会成本
marginal social value 临界社会价值
marginal soil 边际土壤
marginal spot 缘斑
marginal spray 行边喷射;行边喷射
marginal stability 临界稳定性;极限稳定性;边界稳定性;边界稳定度;边际稳定性
marginal standard deviation 边际标准差
marginal streets 边缘街道
marginal strip 路缘带
marginal supplier 边际供应者
marginal swell 沿岸涌浪;边缘隆起
marginal tax 边际税收
marginal tax rate 边际税率
marginal test 临界试验;极限试验;边缘试验
marginal texture 边缘结构
marginal thrust 边缘逆断层;边缘冲断层
marginal tile 边缘砖;镶边面砖;边瓦
marginal tile for gables 用于山墙的边缘砖;山墙边瓦
marginal torque moment 边缘扭矩;极限转矩

marginal torsional moment 边缘扭转力矩;极限扭矩
marginal track 最外侧股道;码头前沿铁路;外侧股道
marginal trading 边际贸易;边际交易
marginal transaction 边际交易
marginal trench 陆缘海沟;边缘海沟
marginal trough 边缘坳陷;边缘凹陷
marginal tubercle 缘结节
marginal twisting moment 边缘扭转弯矩
marginal type wharf 堤岸码头;顺岸式码头
marginal unit cost 边际单位成本
marginal uplift 槽缘隆起;边缘隆起
marginal upthrust 边缘逆断层
marginal user cost 边际使用者费用
marginal utility 限界效用;顺岸码头;边限效用;边陆效用;边际效用
marginal utility analysis 边际效用分析
marginal utility degree 边际效用程度
marginal utility per dollar 每美元边际效用
marginal utility school 边际效用学派
marginal utility theory of taxation 课税的边际效用理论
marginal utility theory of value 边际效用价值论
marginal vacuum 容许真空
marginal value 余量值;临界值;边缘值;边陆值;边际(价)值
marginal-value product 边际价值产品
marginal-value productivity 边际价值生产率
marginal value vector 边际价格向量
marginal variation 边际变异
marginal velocity 边际流通速度
marginal verge 路缘带
marginal vibration 极限振动;边振
marginal wage cost 边际工资费用;边际工资成本
marginal wage quota 边际工资率
marginal well 枯竭井;边缘井
marginal wharf 顺码头;堤岸码头;顺岸码头
marginal yield 边际收入
marginalysis 房地产抵押值与卖价差额研究;成本与卖价差额研究
marginal zone 边区;缘带;边缘区;边缘(地)带
marginal zone of island are 岛弧边缘带
marginal zone of typhoon 台风边缘
margin and guaranty 保证金
margin angle 舭角钢
margination texture 蚀边结构
margin bracket 双层底外侧肘板;舭肘板
margin call 补交保证金通知
margin capacity 备用容量
margin check 边缘校验
margin check earthquake 安全极限校核地震
margin clip 船底边夹铁
margin control 边缘控制(器)
margin deposit 押金存款
margin draft 石块琢琢;石面琢边;细凿边框;路缘琢边;石缘琢边
margin draught 石缘琢边
margin for adjustment 调整余量
marging ecotone 渐进式生态过渡带
margin knot 边节
margin light 窗扇边窄玻璃;侧光;边窗
margin line 限界线
margin lug 船底边夹铁
margin money 预收保证金
margin note 栏外注解
margin of continental shelf 大陆架边缘
margin of controllability 操纵度
margin of cultivation 耕作限度;耕作界限
margin of drill 钻头刃带;钻锋圆边
margin of energy 能量储存;能量储备;后备能量
margin of error 最大允许误差;最大容许误差;误差量;误差界限;误差(极)限
margin of fault block 断块边缘
margin of image 像片边缘
margin of lift 升力限度
margin of optic(al) cup 杯缘
margin of picture 像片边缘
margin of power 功率储备
margin of preference 优惠幅度
margin of profit 利润率
margin of restrictive rights 限制性权利边界
margin of safety 强度储蓄;可靠程度;强度储备;保险限度;安全裕度;安全余裕;安全余量;安全限度;安全系数;安全幅度;安全储备;安全边缘;

安全边际
margin of safety in cost-volume-profit analysis 成本-产量-利润分析中的安全边际
margin of safety ratio 安全边际比率
margin of solvency 偿债基金最低标准
margin of speed 速度限度
margin of stability 稳定裕量;稳定裕度;稳定系数;稳定界限
margin of tectonic upwarping 构造隆起边缘
margin of tolerance 公差范围
margin on sales 销售边标
marginoplasty 缘成型术
margin phenomenon 边沿现象
margin plank 木甲板边(缘)板
margin plate 木甲板边(缘)板;双层底边板;边缘板;舭肘板
margin plate bracket 双层底外侧肘板
margin-profit on sale 销售利润边标;销售的边际利润
margin-punched card 边缘穿孔卡片
margin rate 保证金率
margin regulator 边限调整器
margin release key 极限释放键
margin release mechanism 极限释放机构;边缘释放机构
margin requirement 追加保证金;额定保证金;保证金要求
margin rules 保证金规则
margin scale 极限刻度;边缘标度
margin stop 极限挡块
margin stop indicator 极限挡块指示器;边缘停机指示器
margin stop mechanism 极限挡块机构;边缘停机机构
margin stop setting control 极限挡块定位控制器;边缘停机位置控制
margin strip 边木条
margin temple 定坡板(其宽度与突沿突出距离相等)
margin templet 修边模板
margin tile 边瓦
margin trading 保证金信用交易
margin transaction guarantee money 保证金交易中所交存的保证金
margin trowel 抹边抹子;抹边镘刀;卷边线脚抹子;修边抹子;阴角抹子;方头镘板;边角抹子
margin voltage 容限电压
margin wage 边际工资
margo 加厚边缘;塞缘
margode 泥灰板岩
margosa 楝树
margosa oil 苦楝子油
margrain-needles 珠脂针
marguerite type of bellmouth spillway 菊花形喇叭溢洪道
Margules equation 维特-马古列斯方程
marialite 钠柱石
marialitization 钠柱石化
marial rocks 海岩
Mariana basin 马里亚纳海盆
Mariana ocean basin block 马里亚纳海盆巨地块
Mariana subduction zone 马里亚纳俯冲带
Mariana trench 马里亚纳海沟
Mariana-type subduction zone 马里亚纳型俯冲带
marianna 软质石灰岩(产于美国)
maricite 磷铁钠石
maricultural science 海洋养殖学
mariculture 海洋养殖;海水养殖;海产养殖
Marietta miner 马里埃塔连续采煤机
Marietta process 马里埃塔法;石油加压处理法
Marignac salt 马里纳克盐(硫酸锡钾)
marigold 底片敷纸
marigold window 菊花窗;辐射状花格圆窗;圆花窗;玫瑰窗;车轮窗
marigram 潮汐(自记)曲线;潮汐涨落曲线;潮位(自记)曲线
marigraph 潮汐自记仪;自记潮位计;自记验潮仪;水位记录仪
marina 小艇停靠区;小艇基地;小船坞;系船池;游艇港池;港湾基地;海滨广场
marinada 马兰纳达风(西班牙卡塔尼亚一种海风)
marin blanc 白海风(无锋面海风)
marine 航海灯标;航海的;海的;船用(的)
marine 3-D survey 海上三维地震勘探
marine abrasion 浪蚀;海水腐蚀;海蚀

marine abrasion plain 海底冲积平原
marine abrasion terrace 海蚀阶地
marine accident 海损事故;海事;海上(意外)事故
marine accident report 海事报告
marine accumulation 海洋堆积
marine accumulation coast 海积海岸;海积岸
marine accumulation environment 海积环境
marine acoustics 海洋音响学;海洋声学;水声学
marine aerosol 海洋气溶胶
marine aggregate 海洋集料;海洋骨料;海沙;海砂;海(成)集料;海(成)骨料
marine agriculture and animal husbandry 海洋农牧业
marine aided inertial navigation system 航海惯性导航系统;船舶综合惯性系统
marine air control group 海洋大气控制组
marine aircraft 海上航空器;海上飞机;水上飞艇
marine airfield 水上机场
marine air(mass) 海洋气团;海洋空气团
marine airport 海军航空港;海上机场
marine algae 海藻;海洋藻类
marine algology 海洋藻类学
marine alidade 航海用照准仪
marine alloy 船用合金
marine alluvium 海洋冲积层
marine almanac 航海历(书)
marine anchor 船用锚
marine anchor chain 船用锚链
marine and air navigation light 海空两用灯标
marine and estuary hydrologic(al) element 海洋与河口水文要素
marine anemometer 船用风速表
marine aneroid barometer 船用空盒气压表
marine animal 海洋动物
marine animal injury 海洋动物伤害
marine aquaculture 海产养殖
marine aquarium 海水水族馆
marine aquatic biological resources 海洋水生物资源
marine arch 海蚀拱
marine architect 造船工程师;船舶设计师
marine area 海域;水域
marine associated industry 与海洋有关的工业
Marine Association 海事协会;海运协会
marine atmosphere 海上大气;海洋大气层
marine audio visual instruction system 船用视听指挥系统
marine autochthonous organic matter 海洋原生有机质
marine automatic meteorologic(al) observing station 海上自动气象观测站
marine automatic meteorologic(al) oceanographic(al) station 海上自动海洋气象站
marine automatic meteorologic(al) station 海上自动气象站
marine automatic navigation system 船舶自动导航系统
marine automatic telephone set 船用自动电话机
marine autotraverse positioner 船迹推算仪
marine autotrophic organism 海洋自养生物
marine auxiliaries 船用辅(助)机(械)
marine auxiliary boiler 船用辅锅炉
marine bacteria 海洋细菌
marine bacterial corrosion 海洋细菌腐蚀
marine bacterial hose 海洋细菌寄生
marine bacterial pollution 海洋细菌污染
marine bacteriology 海洋细菌学
marine bacteriophage 海噬菌体
marine ballast 海洋集料;海洋骨料;海产碎石;海产集料;海产骨料
marine bank 海洋灰岩滩
marine barometer 航海气压计;船用晴雨表;船用气压表;船用气压表
marine beach 海滩
marine beacon 航海灯;航行岸标灯
marine beacon-light lantern 海上信号灯;海标灯
marine beacon transmitter 船舶导航无线电发射机
marine bed 海相层;海床
marine belt 沿岸海域;领海带
marine bench 海蚀台地
marine benthic community 海洋底生群落;海洋底栖生物群落
marine benthic diatom 海洋底生硅藻;海洋底栖硅藻

marine benthic dinoflagellate 海洋底生甲藻
marine benthic fishes 海洋底栖鱼类
marine benthic organism 海底生物
marine benthos 海底栖生物
marine bill of lading 海运提(货)单
marine binoculars 船用双筒望远镜
marine bioacoustics 海洋生物声学
marine biochemical resource 海洋生物化学资源
marine biochemistry 海洋生物化学
marine biocycle 海洋生物环
marine bioecology 海洋生物生态学
marine biogeochemistry 海洋生物地球化学
marine biogeography 海洋生物地理学
Marine Biological Association 海洋生物学协会(英国)
marine biological chart 海洋生物图
marine biological laboratory 海洋生物实验室
marine biological larva 海洋生物幼体
marine biological pollution 海洋生物污染
marine biological tester 海洋生物测试仪
marine biology 海洋生物学
marine biosphere 海洋生物圈
marine biota 海洋生物群;海洋生物区;海洋生物
marine boiler 船用锅炉
marine borer 海中蛀船虫;海洋钻蛀虫;海洋穿孔动物;海水蛀虫;海生钻木动物;(蛀船的)海虫;船蛀虫
marine boring organism 海上木船钻孔生物
marine borrow area 海上取土区
marine botany 海洋植物学
marine bottom community 海底群落
marine bottom sediment 海底淀积(物);海底沉积(物)
marine boundary delimitation 海上划界
marine-built platform 海成台地;海积台地
marine-built terrace 海成台地;海成阶地;海积阶地
marine bunker facility 海港油库设备;油库设施【港】
marine bunkering service barge 水上加油驳
marine bunkering unit 水上加油装置
marine bureau 海运局
marine business 航运业
marine cable 海底电缆;船用电缆
marine cargo handling gear 船用装卸机械
marine cargo insurance 海洋货物保险;海上运输保险
marine cargo insurance policy 海上货物保险单
marine casualty 海损事故
marine cave 海蚀洞;海底洞
marine celestial navigation 海上天文导航
marine cement 海工水泥;海水构筑物水泥
marine centrifugal fan 船用离心通风机
marine ceramic wool 船用陶瓷棉
marine chart 海图
marine chemical analyzer 海洋化学分析仪
marine chemistry 海洋化学
marine chronometer 航海时计
marine city 海上城市
marine clastics 海相碎屑岩
marine clay 海相黏土;海成黏土
marine cliff 海蚀崖;海边悬崖
marine climate 海洋性气候
marine climatology 海洋气候学
marine coal 船用煤(炭)
marine coastal sample 近海样品
marine coating 船用油漆;船舶涂料;船舶漆
marine coenology 海洋生物群落学
Marine Commission 航运委员会
marine communication 航海无线电通信;海上通信
marine communication satellite 航海通信卫星;海上通信卫星
marine communities 航运界
marine company 航运公司
marine compass 船用罗盘
marine complete product 船舶成套产品
marine compressor 船用水冷压缩机;船用空冷压缩机
marine concrete 海用混凝土;海工混凝土
marine condition 海上情况
marine conservation area 海洋保护区
marine construction 海洋建筑物;船舶结构;海洋工程;水工建筑物
marine construction engineer 海上建筑工程师;船舶结构工程师
marine construction model 水工建筑物模型

marine control point 海洋控制点
marine conveyance 海运;水运
Marine Cooperation 海事协会
marine corps 海军陆战队
marine corrosion 海洋腐蚀;海水腐蚀;海蚀
marine course computer 船舶航向计算机
marine court 海事法庭
marine crane 甲板起重机;船用起重机;船上起重机
marine crossing 水下穿越
marine cultivation 海水养殖
marine culture 海洋养殖;海水养殖
marine current 海流;海洋流
marine custom 海员习惯
marine-cut bench 海台;海蚀台地;浪蚀阶地;海蚀阶地
marine-cut plain 海蚀平原
marine-cut platform 海蚀台地
marine-cut terrace 浪蚀阶地;海蚀阶地;海切阶地
marine cycle 海洋循环;海(水侵)蚀循环;海蚀周期
marine cylinder oil 船用汽缸油
marine data 海洋水文资料;海洋水文注记
marine deck machinery 船用甲板机械
marine delta 海洋三角洲;海相三角洲
marine delta plain 海洋三角洲平原
marine denudation 海水侵蚀作用;海(水剥)蚀
marine department 航运部;海务部(门)
marine deposit 海洋沉积;海洋沉积(物);海相沉积(物);海积物;海成矿床
marine deposition 海洋沉积作用;海相沉积作用
marine deposition coast 海洋堆积海岸;海积(海)岸
marine deposition plain 海积平原
marine deposition platform 海积台地
marine deposition terrace 海积阶地
marine deposit topography 海积地貌
marine depth correction 海深校正
marine depth sounder 海洋回声测深仪
marine-derived nitrogen 从海水中提取的氮
marine development 航运开发
marine development project 航运开发工程
marine diatom 海洋硅藻
marine diesel 船用柴油机
marine diesel engine 船用柴油机
marine diesel engine oil 船用柴油机油
marine diesel engine parts 船用柴油机配件
marine diesel fuel 船用柴油机燃料
marine diesel generating set 船用柴油发电机组
marine diesel lube 船用内燃机润滑油
marine diesel medium 船用中级柴油
marine diesel oil 船用柴油
marine differential analyzer 船用微分分析器
marine direction finder 船用测向器;船舶方向探测器
marine disaster 海事;海上遇险;海难
marine disc centrifugal separator 船用碟式离心分离机
marine discovery well 海上新发现油井
marine disposal area 海上抛泥区
marine disposal of hazardous waste 危险废物的海域处理
marine distiller 船用蒸馏器
marine distress signals 海上遇险信号
marine dock fender 码头碰垫
marine documents 航海文件
marine dredge(r) 海上疏浚工;水上挖泥船;挖泥船
marine dredging 海洋疏浚
marine drift 海流
marine drilling 海洋钻探;海上钻探;海上钻井;海底钻探;海底钻井
marine drilling engineering 海上钻探工程;海上钻井工程
marine drilling platform 海上钻探平台;海上钻井平台
marine drilling system 海上钻探系统;海上钻井系统
marine drilling technology 海上钻探工艺;海上钻井工艺
marine dynamograph 海洋动力自记仪
marine ecologic economics 海洋生态经济学
marine ecology 海洋生态学
marine economic geography 海洋经济地理学
marine economic history 海洋经济史
marine economics 海洋经济学
marine ecosystem 海洋生态系统

marine electric(al) appliance 船用电气设备
marine electric(al) fan 船用电风扇
marine electric(al) machine 船用电机
marine electric(al) meter 船用电表
marine electrochemistry 海洋电化学
marine electromagnetism 海洋电磁学
marine electronic engineering 船用电子工程学
marine electronics 航海电子学;海洋电子学
marine electronic technical unit 船用电子技术装置
marine electrotechnical maintenance and operation 船电维修与管理
marine elevator 卸粮机;吸粮机
marine end-sealed airtight-sliding door 船用端部密封式气密滑动门
marine energy source 海上能源
marine engine 轮机;船用发动机;船舶发动机
marine engineer 轮机员;轮机工程师
marine engineering 造船工程;轮机工程【船】;海运工程;海洋工程(学);海事工程(学);海工工程学;船舶工程(学)
marine engineering geology 海洋工程地质学
Marine Engineering Society in Japan 日本海洋工程学会
marine engineering survey 海洋工程测量
marine engineer's room 轮机员室
marine engine oil 船用机油
marine environment 海洋环境;海相环境
marine environmental buoy system 海洋环境浮标系统
marine environmental capacity 海洋环境容量
marine environmental chart 海洋环境图
marine environmental law 海洋环境法
marine environmental management 海洋环境管理
marine environmental observation 海洋环境观测
marine environmental pollution damage 海洋环境污染损害
marine environmental prediction 海洋环境预报
Marine Environmental Protection Law of the People's Republic of China 中华人民共和国海洋环境保护法
marine environmental quality 海洋环境质量
marine environmental quality assessment 海洋环境质量评价
Marine Environmental Quality Committee 海洋环境质量委员会
marine environmental remote sensing 海洋环境遥感
marine environmental sensing equipment 海洋环境传感器
marine environmental survey craft 海洋环境调查船
Marine Environment and Pollution Control Act 海洋环境和污染控制法
Marine Environment Data and Information Referral System 海洋环境数据和信息查询系统
marine epipelagic organism 海洋上层生物
marine equipment 水上运输工具
marine erosion 浪蚀;海蚀(作用);海浪侵蚀
marine erosion coast 海蚀海岸;海蚀岸
marine erosion environment 海蚀环境
marine-erosion platform 海蚀平台
marine estuarine deposit 入海河口沉积物
marine eumycete 海洋真菌
marine evapo(u)rate 海相蒸发岩
marine evapo(u)rator 船用蒸馏器
marine extension clause 航程延长条款;海上延伸条款
marine facies 海相
marine facies theory 海相生油说
marine facies theory on origin of petroleum 海相生油说
marine facility 海洋设施
marine-fan trap 海底扇圈闭
marine fauna 海洋动物群落;海洋动物区(系)
marine fill 海底材料填充
marine fill reclamation sea 海底材料吹填区
marine fire alarm 船用火警报警器
marine fish 海水鱼
marine fisheries 海洋水产业
marine fisheries biology 海洋渔业生物学
marine fishery 海洋渔业
marine fishery resources 海洋渔业资源
marine fishes 海洋鱼类
marine fish farming 近海养鱼

marine fishing 海洋捕鱼
marine fishing resources 海洋渔业资源
marine flame-proof axial flow ventilator 船用防爆轴流式扇风机
marine flame-proof three phase asynchronous motor 船用防爆三相异步电动机
marine flooding surface 海泛面
marine flora 海洋植物区系
marine flue boiler 船用烧管锅炉
marine fluid mechanics 海洋流体动力学
marine food chain 海洋食物链
marine forecast 海洋预报
marine foreland 海滩地;海角;海岬
marine fouling 船底长满海藻;海洋污浊;海底污着;船底污物;船底污损
marine fouling organism 海洋污着生物;海船附着生物
marine four strokes diesel engine 船用四冲程柴油机
marine freightage 海运费率
marine frequency band 船用无线电通信波段
marine fresh water generator 船用海水淡化装置
marine fuel 船用燃料
marine galvanometer 船用电流计
marine gas cooling reactor 船用气冷反应堆
marine gas seepage 海上天然气苗
marine gas turbine(engine) 船用燃气轮机
marine gear 船用齿轮
marine gearbox 船用齿轮箱
marine generator 船用发电机;船舶发电机
marine geochemical exploration 海洋化探
marine geochemistry 海洋地球化学
marine geodesy 海洋(大地)测量学;海洋测地学
marine geodetic surveying 海洋大地测量
marine geography 海洋地理学
marine geologic(al) process 海洋地质作用
marine geologic(al) survey 海洋地质调查
marine geologic(al) tester 海洋地质测试仪
marine geology 海洋地质学;海洋地质;地质海洋学
marine geomagnetic anomaly 海洋地磁异常
marine geomorphology 海洋地貌学
marine geophysics 海洋地球物理学
marine geotechnology 海洋岩土工程学;海洋土工学
marine glacier 海相冰川
marine glue 耐水胶;木料胶结剂;鱼胶;捻缝胶;船用胶;防水胶
marine glue ca(u)lking 橡胶捻缝
marine governor 轮机调节器
marine grass marsh 海草沼泽
marine gravel 海洋砾石;海相砾石
marine gravimeter 海洋重力仪
marine gravity anomaly 海洋重力异常
marine gravity meter 海洋重力仪
marine gravity prospecting 海洋重力勘探
marine gravity survey 海洋重力测量
marine gray 船用灰漆
marine grey cloth 海洋灰色布
marine growth 船底附生物
marine growth preventing system 船底防附生物系统
marine guidance 海上制导;海面导航
marine gypsum 盐石膏;海水石膏
marine hazard 海洋危险
marine hazardous chemical worker 海运危险化学品工作人员
marine head 海堤
marine heavy metal pollution 海洋重金属污染
marine herb 海洋草本植物;海生草本植物
marine heterophic microorganism 海洋异养微生物
marine histology 海洋组织学
marine hitch 吊铺捆结
marine hospital 海员医院
marine hull insurance 海洋船壳保险
marine humus 海成腐殖质
marine hydrocarbon survey 海洋烃类测量
marine hydrochemistry 海洋水文化学
marine hydrographic(al) surveying 海洋水文测量;海道测量
marine hydrography 海洋水文学;海道测量学
marine hydrological chart 海洋水文图
marine hydrological forecasting 海洋水文预报
marine hydrological station 海洋水文站
marine hydrometeorologic(al) chart 海洋水文气

象图
marine hydrometeorology 海洋水文气象学
marine hydrometry 海洋水文测验
marine hydrosphere 海水圈；海水界
marine ichthy-anatomy 海洋鱼类解剖学
marine ichthy-geography 海洋鱼类地理学
marine ichthy-morphology 海洋鱼类形态学
marine ichthyology 海洋鱼类学
marine ichthy-pathology 海洋鱼类病理学
marine ichthy-systematics 海洋鱼类分类学
marine identification digits 海上识别数字
marine incineration 海上焚烧
marine incinerator 船用焚烧炉
marine index of orderness 海上秩序指数
marine industry 航业业；海水工业；水产业
marine information center[centre] 海事信息中心
marine information service 海上信息业务
marine installation 海岸设施；船用装置；船舶装置
marine insulation 船舶绝缘
marine insurance 航运保险；海(洋)运(输)保险；海损保险；海上保险；水险
marine insurance act 水险法
marine insurance application 海运投保书；水险投保单
marine insurance Co. 海上保险公司
marine insurance company 航运保险公司
marine insurance condition 海洋运输保险条件
marine insurance contract 货物海上保险契约
marine insurance law 海运运输保险法；海上保险法
marine insurance policy 海(洋)运(输)保险单；海上保险单；水运保险单；水险(保)单
marine interests 航海利息；海商利率；航运界
marine invertebrate 海洋无脊椎动物
marine invertebrate zoology 海洋无脊椎动物学
marine investigation 海洋调查
marine laboratory 海洋研究实验室；海洋实验室
marine lake 海源湖
marine lamp 船用灯
marine land 海产养殖场
marine landform 海成地形
marine landform observation apparatus 海洋地貌观测仪器
marine-land purpose motor 船-陆两用电机
marine lantern 航海灯标
marine lantern lens 航海灯标透镜
marine law 海洋法；海事法；海法
marine lead battery 船用铅蓄电池
marine league 水程里格
marine leather 海产皮革
marine leg 驳船起重臂；(船用)卸料吊斗；浮式驳船提升机
marine level(l)ing 海洋水准测量
marine lichen 海岸地衣
marine life 海洋生物
marine life jacket 海上救生衣
marine life-saving 海上救生
marine lifes-aving equipment 海上救生设备
marine lift-saving apparatus 船用救生器材
marine light 航道标志灯；航路标志灯；航海用导航灯；海上导航灯标
marine light diesel oil 船用轻柴油
marine lightering 海上驳运
marine lighting fitting 船用灯具
marine lighting practice 海上照明设施
marine littoral faunal region 浅海动物区
marine living resources 海洋水产资源
marine loading arm 装油管
marine loran coverage 船用罗兰有效区域
marine loss 海损
marine low speed diesel engine 船用低速柴油机
marine low-voltage electric(al) appliance 船用低压电器
marine low-voltage switch 船用低压开关
marine machinery 轮机；船用机械
marine magnetic anomaly 海洋磁力异常
marine magnetic chart 海洋磁力图
marine magnetic field property tester 海洋磁场性质测试仪
marine magnetic prospecting 海洋磁法勘探
marine magnetic survey 海洋磁力调查
marine magnetometer 海洋磁力仪
marine magnetometer profiling 海洋磁力仪剖面测量
marine magnetometry 海洋磁力测量

marine mammal 海洋哺乳动物
marine management center 航运管理中心
marine management science 海洋经营管理学
marine management system 航运管理系统
marine map 海图
marine map format 海图分幅
marine mapping 海图制图学
marine mareograph 海洋验潮仪
marine market 海运市场
marine marl 海成泥灰岩
marine material 海洋物质
marine measurement 海洋检测
marine measuring instrument 海洋测量仪
marine media 水上运输工具
marine mesopelagic organism 海洋中层生物
marine meteorograph 海洋气象记录仪
marine meteorologic(al) chart 海洋气象图
marine meteorologic(al) observation 海洋气候观测
marine meteorology 航海气象学；海洋气象学
marine meterologal prediction 海洋气象预报
marine microalga 海洋微藻
marine microbial biochemistry 海洋微生物生物化学
marine microbial biophysics 海洋微生物生物物理学
marine microbial ecology 海洋微生物生态学
marine microbial morphology 海洋微生物形态学
marine microbial pollution 海洋微生物污染
marine microbial systematics 海洋微生物分类学
marine microbiology 海洋微生物学
marine micropaleontology 海洋微体古生物学
Marine Midland Bank 海事密德兰银行(美国)
marine mineral 海成矿物
marine mineral resources 海洋矿产资源
marine mining 海洋采矿；海底采矿
marine mollusc 船蛆
marine molysmology 海洋污染学
marine monitoring 海洋监测
marine morphology 海洋形态学
marine multiplexed streamer system 海上多道电缆系统
marine natural product 海洋天然产物
marine nature products chemistry 海洋天然产物化学
marine nature reserves 海上自然保护区
marine navigation 海上航行；航海(学)；航海导航；水运；船舶导航
marine navigation aid 海上救护(设施)；海上导航设备；海上航标
marine navigation light 航标灯
marine navigation radar 船用导航雷达
marine navigator 船舶驾驶人员
marine neuston 海洋漂浮生物
marine non-metal cable tier 船用非金属电缆扎带
marine observation satellite 海洋观察卫星
marine observatory 海洋气象台
marine oil 航海油；船用重质油
marine oil fence 海上围油栅
marine oil pollution 海洋油污染
marine oil seepage 海上油苗
marine oil spill 海上溢油；海上漂油
marine oil-water separator 船用油水分离器
marine onlap 海进
marine operation method 海上作业方法
marine operation mode 海上作业方式
marine optic(al) fibre communication system 船用光纤通信系统
marine optic(al) instrument 海洋光学仪器
marine optics 海洋光学
marine organic chemistry 海洋有机化学
marine organism 海洋有机体；海洋生物
marine organism corrosion 海洋生物腐蚀
marine origin 海相成因；海成
marine-originated salt water lake 海源咸水湖
marine outfall 海流出口
marine overlooker 海务监督
marine paint 防水漆；海船油漆；海用漆；船用灰漆；船舶漆
marine paint for boot topping 船舶水线漆
marine paint for bottoms 船底漆
marine paint for superstructure 船舶上层建筑漆
marine paint for topsides 船舷漆
marine painting 船用油漆

marine parade 海上军事演习
marine park 海洋公园；海水养殖场；海上公园；海底公园
marine pathogenic contamination 海洋病原体污染
marine patrol aircraft 海上巡逻飞机
marine pearl feed 海产珍珠养殖
marine pelagic fishes 海洋远洋鱼类
marine pelagic organism 海洋远洋生物
marine peneplain 沿海海蚀准平原；海蚀准平原
marine perils 航海风险
marine Permian 海洋二叠纪【地】；白云石
marine personnel 海运人员；海员
marine pesticide pollution 海洋农药污染
marine petroleum development 海洋石油开发
marine petroleum exploitation 海洋石油开发
marine pharmacognosy 海洋生药学
marine phenomenon 海上现象
marine phosphorescence 海洋磷光
marine physical chemistry 海洋物理化学
marine physicochemistry 海洋物理化学
marine physics 海洋物理(学)
marine physiology 海洋生理学
marine phytobiochemistry 海洋植物生物化学
marine phytobiology 海洋植物生物学
marine phytobiophysics 海洋植物生物物理学
marine phytochemistry 海洋植物化学
marine phytocoenostics 海洋植物群落学
marine phytoecology 海洋植物生态学
marine phytogeography 海洋植物地理学
marine phytomorphology 海洋植物形态学
marine phytoplankton 海洋浮游植物
marine phytotaxonomy 海洋植物分类学
marine pier 突堤码头
marine pile driver 打桩船
marine piling 海上打桩；水上打桩
marine pipeline 海上管线；水下管线；水下管路
marine placer 海成砂矿；海滨砂矿
marine plain 浪蚀平原；海蚀平原；海积平原
marine planation 浪蚀夷平作用
marine planktonic diatom 海洋浮游硅藻
marine plantation 海洋种养业
marine plastics pollution 海洋塑料污染
marine plywood 船舶胶合板；船用胶合板；防水胶合板
marine pneumatic seismic source 海洋压缩空气震源
marine polarimeter 海洋偏振计
marine police 水上警察
marine policy 海洋政策；海上保险单；水险(保)单
marine pollutant 海洋污染物
marine pollution 海洋污染
marine pollution control unit 海洋污染控制装置；海洋污染控制单元
marine pollution from land-based sources 陆源海洋污染；陆地来源的海洋污染
marine pollution investigation 海洋污染调查
marine pollution monitor 海洋污染监测器
marine pollution monitoring 海洋污染监控；海洋污染监测
marine pollution monitoring program-(me) for petroleum 海洋油污染监测计划
marine pollution of organic substance 海洋有机物污染
marine pollution prevention 海洋防污；防止海洋污染
marine pollution prevention law 海洋污染防治法；海洋污染防止法；海洋防污法规
marine port 海港
marine positioning 海上定位
marine power plant 船舶动力装置
marine prairie 海洋草场
marine premium 海运保险费；海洋运输保险费
marine preserve 禁渔区；海洋保护区；海上自然保护区；海上禁捕(渔)区
marine press 海上新闻业务
marine process 海洋冲积过程；海蚀过程；海侵过程
marine production system 海上采油系统
marine products 海货；海产(物)；水产品货；船用产品；海产品；水产
marine products industry 水产工业
marine products institute 水产学院
marine products water 水产用水；海产用水
marine progressive overlap 海侵超覆；海成覆蔽
marine propeller 船舶推进器

marine propulsion 船舶推进装置
marine protection 航运保护(主义);海洋保护
marine protest 海损声明书;海事声明书;海难抗议书
marine proton gradiometer 海洋质子重力梯度仪
marine proton magnetometer 海洋质子磁力仪
marine proton precession magnetometer 海洋质子磁力仪
marine protozoan ecology 海洋原生动物生态学
marine protozoology 海洋原生动物学
marine pulley 船用滑车
marine pump 船用泵
marine qarosol 海上烟雾
mariner 船员;航海者;航海家;海员;水手
marine radar 航海雷达;海用雷达;船用雷达;船舶雷达
marine radar band 船用雷达波段
marine radar camera 航海雷达摄影机
marine radar observer trainer 船舶雷达观测员训练器
marine radiant boiler 船用辐射型锅炉
marine radio 航海用无线电台
marine radioactive pollution 海洋放射性污染
marine radio aids 船用无线电(助航)设备;船舶无线电(助航)设备
marine radio beacon 航海(用)无线电信标;海上无线电信标
marine radio beacon station 海上无线电信标电台
marine radiocom 海上无线电通信
marine radio communication 海上无线电通信
marine radio direction finder 船用无线电测向仪
marine radio receiver 船舶无线电接收机
marine railroad 修船滑道
marine railway 修船滑道;斜架车滑道;缆索式升船滑道;缆车滑道;缆车道;升船滑道;船坞铁路;船排
marine railway cradle (升船用的)缆车
marine railway ferry 海上铁路轮渡;海上火车轮渡
marine railway ferry terminal 海上铁路轮渡码头;海上火车轮渡码头
marine rainbow 海(洋)虹;海上虹
marine reclamation land 围海造田
marine refraction seism 海上折射地震
marine refrigerating plant 船用制冷装置
marine regatta 海上军事演习
marine regression 海退
marine rescue 海上救助
marine rescue and salvage organization 海难救助机构
marine rescue coordination center 海上营救协调中心
marine researching ship 海洋研究船
marine research ship 海洋科研船;海洋科学考察船;海洋考察船
marine research station 海洋科学研究站;海洋考察站
marine resource economics 海洋资源经济学
marine resources 海洋资源;海成资源
marine resources chart 海洋资源图
marine return tube boiler 船用回管锅炉
marine rig 海洋钻机
marine riser 海底取油管
marine riser system 隔水导管系统
marine risks 航海风险;海险;海上风险
marine rock 海成岩石
mariner program(me) 水手计划
mariner's compass 航海罗盘(仪);航海罗经;船用罗盘;船用罗经
mariner's lien 海员留置权
mariner's needle 罗盘针
mariner's ring 圆圈测天仪;测星盘
mariner's splice 锚索插接
mariner type rudder 半平衡半悬舵
marine safety 海上安全
marine safety corps 海上安全队
marine safety information 海上安全资料;海上安全信息
marine safety information system 海上安全信息系统
marine safety law 海上安全法
marine safety valve 船用安全阀
marine salina 海滨盐沼地
marine saline 海边盐沼;海边盐滩
marine salt 海盐
marine salvage 海上救捞

marine salvage service 海上打捞法
marine salvage station 海上救生站
marine sampling 海洋取样
marine sanctuary 海洋自然保护区
marine sand 海沙;海砂
marine sanitation device 船用卫生设备
marine science 海洋科学
marine's compass 船舶用罗盘
marine scraper 船用刮刀
marine screw propeller 船用螺旋桨
marine sea 外海;水域
marine search 海上搜索
marine searchlight 船用探照灯
marine sediment 海洋沉积(物);海相沉积(物);污染沉积(物)
marine sedimentary bauxite deposit 海相沉积铝土矿矿床
marine sedimentary deposit 海相沉积矿床
marine sedimentary diatom 海洋沉积硅藻
marine sedimentary geochemistry 海洋沉积地球化学
marine sedimentary phosphorite deposit 海相沉积磷块岩矿床
marine sedimentary type 海洋沉积型
marine sediment dating method 海洋沉积物年龄测定法
marine sedimentology 海洋沉积学
marine sediment topography 海积地貌
marine seed botany 海洋种子植物学
marine seepage 海底油苗
marine seismic cable 海洋地震电缆
marine seismic exploration 海洋地震勘探
marine seismic profiler 海洋地震剖面仪
marine seismic source 海洋震源
marine seismic survey 海上地震勘探
marine seismograph 海洋地震仪
marine seismometer 海洋地震计;海洋地震表
marine self-purification 海洋自净(化)作用
marine self-purification capacity 海洋自净(化)能力
marine separator 船用分离器
marine service instruments 船用仪器
marine service instruments and implements 航海仪器及器具
marine service motor 船用电动机
marine sessible organism 海洋固着生物
marine sewage treatment plant 船用污水处理装置
marine sextant 航海(用)六分仪
marine shafting 船用轴系
marine shear wave profile 海洋横波剖面
marine shelf 大陆架
marine shell 海贝壳
marine shipping 海上运输;水上运输
marine signature waveform deconvolution 海上特征波形反褶积
marine silicon rectifier 船用硅整流器
marine simulation 航海模拟;海上模拟
marine simulator 航海模拟器
marine simulator experiment 航海模拟器试验
marine site 水工建筑物地点
marine snow 海上飞雪
marine soap 水皂;船用(肥)皂
marine soil 海积土
marine soil mechanics 海洋土力学
marine solonchak 海成盐;滨海盐土
marine sonoprobe 海声呐探测仪;海洋回声探测仪
marine source detect mode 海上震源检测方式
marine source effect 海上震源效应
marine species 海洋品种
marine stack 海蚀岩柱
marine standing 海上位置
marine steam boiler 船用蒸汽锅炉
marine steam power plant 船舶蒸汽动力装置
marine steam turbine 船用汽轮机;船舶蒸汽涡轮机;船舶蒸汽透平
marine steel 耐海水钢材
marine stone borer 穿石贝
marine storage battery 船用蓄电池
marine store 航海用品供应站;航海物料行;船用商品店;船用品商店;船舶用具;船舶物料
marine stratigraphy 海洋地层学
marine structural dynamics 海洋结构动力学
marine structure 海洋结构;海上结构物;海上建筑物;海上构筑物;海工结构;海工建筑(物);海岸结构;水工建筑物
marine structure model 水工建筑物模型
marine succession 海洋生态演替
marine superintendent 海务监督
marine supply industry 海上供应业
marine supply ship 海上补给船
marine survey(ing) 海洋调查;海事测量;船舶测量(学);海洋考察;海底调查
marine surveying and charting 海洋测绘学
marine surveying and mapping 海洋测绘
marine surveyor 验船师;海事检查人;海事调查人员
marine survey position 海洋测量定位
marine survey report 海事调查报告
marine survival craft 海上救生艇
marine swamp 沿海沼泽;海湿地
marine system electrolysis 海洋系统电蚀
marine talus 海下岩屑堆
marine tanker facility 海港油库设备
marine technique 造船技术
marine technological economics 海洋技术经济学
marine technology 海洋技术;海洋工艺;海洋工程学
marine tectonics 海洋构造学
marine telegraph 船舶收发报话机
marine telephone 船用电话
marine telephone set 船用电话机
marine telescope 航海望远镜
marine terminal 港区车站;码头(区);航海站;海运终点站;海上转运站;海上转运码头;水运终点站;水运枢纽;岸边站
marine terminal operator 码头公司
marine terminus building 海运终点客运站
marine terrace 海洋阶地;海岸
marine thematic survey 海洋专题测量
marine thermodynamics 海洋热力学
marine thermology 海洋热学
marine thermometer 海水温度计
marine three-dimensional seismic 海上三维地震
marine tide 海洋潮(汐);海潮
marine timber 海用木材;海洋森林
marine toilet 船用厕所
marine torch 水中手电筒
marine towage 海上拖运费;海上拖驳运输
marine tractor 小拖轮
marine trade 航运业;海上贸易
marine traffic 航运;海上交通;水上运输;水上交通
marine traffic accident 海上交通事故
marine traffic control 海上交通管制;水上交通管理;船舶交通管制
marine traffic control system 水上交通管理系统
marine traffic engineering 海上交通工程;水运工程(学)
marine traffic investigation 海上交通调查
marine traffic observation 海上交通调查
marine traffic survey 海上交通调查
marine traffic system 海上交通系统
marine transfer terminal 中转码头
marine transgression 海侵
marine transmitter 船舶无线电发射机
marine transport(ation) 海(上)运(输);船舶运输
marine transportation engineering 海上运输工程
marine transport business 海运业务
marine tubular boiler 船用火管锅炉
marine turbidimeter 船用浊度计
marine turbine 船舶汽轮机;船用涡轮机;船用透平;船用汽轮机
marine turbine condenser 船用涡轮机的凝汽器
marine type dynamo 船用直流发电机
marine umbrella type hole stopper 船用伞形堵漏器
marine underwriter 海上保险人;海险公司
marine valve 船用阀门
marine varnish 海上建筑物用清漆;海工清漆;船舰用漆
marine vegetation 海洋植物群落
marine vehicle 船舶
marine vertebrate 海洋脊椎动物
marine vertical seismic profile 海上垂直地震剖面
marine vertical seismic profiling 海洋垂直地震剖面
marine virology 海洋病毒学
marine volcanic granulitic copper and zinc-bearing formation 海相火山变粒岩含铜锌建造
marine waste disposal 废物倾海处理;废水倾海处理

marine waste treatment 废水倾海处理
marine water 海水
marine water quality standard 海水质量标准;海水水质标准
Marine Waters Pollution Control and Quality Enhancement Act 海水污染控制及提高水质法
marine water tube boiler 船用水管锅炉
marine wave telemetering system 近海遥测波浪仪
marine weather data 海上气象数据
marine weather observation ship 海洋天气观测船
marine wildlife 海洋野生动物
marine works 海洋工程;海事工程;海工建筑(物)
marine zoobenthos 海洋底栖动物
marine zoogeography 海洋动物地理学
marine zoophysiology 海洋动物生理学
marine zooplankton 海洋浮游生物;海洋浮游动物
marine zootaxology 海洋动物分类学
marine zootaxy 海洋动物分类学
maringal profit 边际利润
maring and guaranty 保证金
maring calls 追加保证金
marining 海淹
marinization 适用船用技术
marinized aircraft gas turbine 航空改装型船用燃气轮机
marinotherapy 海滨治疗;海滨疗养
marionette 木偶剧院;木偶剧场
marionite 水锌矿
Mariotte's law 玻意耳定律
Mariposa formation 马里波萨组
mariposite 铬硅云母
MARISAT[maritime satellite] 航海卫星
MARISAT system 海事卫星通信系统
maritime 航海的;海(洋)的;海上的
maritime accident 海运事故;海上事故
maritime action 海事诉讼
Maritime Administration 航运总署;海事管理局
maritime affair 海运事务;海事
maritime aids to navigation 海区航标
maritime air 海洋空气
maritime air fog 海洋气团雾
maritime air mass 海洋气团;海洋空气团
maritime anti-stranding sonar system 海上探礁声呐系统
maritime application bridge system 海上适用船桥系统
maritime arbitration 海事仲裁
Maritime Arbitration Commission of China 中国海事仲裁委员会
Maritime Arbitration Committee 海事仲裁委员会
maritime arctic air mass 北极海洋气团
maritime assistance 海事援助
Maritime Association 海事协会;海运协会
maritime atlas 海图册
maritime authority 海运当局
maritime automated flowchart analysis system 海事自动化流程分析系统
maritime belt 沿海地带;领航带;领(海)水(域);海滨区
maritime boundary delimitation 海上划界
maritime buoy 海上浮标
maritime buoyage system 海上浮标制式
maritime business 航运业
maritime canal 海轮运河;通海运河
maritime carriage 海上运输
maritime casualty 海事;海上事故
maritime center 海岸中心
maritime claim 海事索赔;海事求偿权;海事请求;海商法上的债权
maritime claimant 海事债权人;海事索赔人
maritime climate 海洋(性)气候
maritime climatology 海洋气候学
maritime coasting trade 海洋沿海贸易
maritime code 海法
maritime commerce 海上贸易
maritime commission 海事委员会
maritime communication satellite organization 海事通信卫星组织
maritime communication subsystem 海事通信子系统
maritime communication system 海事通信系统
maritime communities 航运界
maritime company 航运公司
maritime construction 海洋工程;水工建筑物
maritime container network 海运集装箱网
maritime control area 海上控制区
maritime convention 海运公约
maritime country 沿海国(家)
maritime country code 洋区码
maritime court 海事法庭
maritime customs 海关
Maritime Customs Administration 海关总署
maritime declaration 海事声明;海事报告
maritime disaster 海事;海难
Maritime Disaster Inquiry Agency 海上仲裁厅(日本)
maritime disputes 海事争议
maritime distress 海难;船舶遇难
maritime distress channel 海上遇险信道
maritime engineering 海事工程(学);轮机工程【船】;海洋工程学;海上工程;海工工程学;水运工程(学);水上工程
maritime engineering budget 水运工程预算
maritime engineering design 水运工程设计
maritime engineering estimate 水运工程概算
maritime engineering programming 水运工程规划
maritime engineering rating 水运工程定额
maritime environment 海上环境
Maritime Environmental Protection Committee 海洋环境保护委员会
maritime exchange 航运电台;海运情报机构;海事交易所
maritime fraud 海事欺诈
maritime frequency bands 海用频带
maritime fruit carrier 水果运输船
maritime geotechnics 海洋岩土工程学
maritime glacier 海洋性冰川
maritime hydraulics 海洋水力学
maritime hydrology 海洋水文学
maritime hygiene 港口卫生
maritime identification code 海上识别码
maritime industrial area 临港工业区
maritime industrial development area 港口工业发展区
maritime industrial development zone 港口工业发展区
maritime information center[centre] 海事信息中心
maritime inquiry 海事查询
maritime insurance 航运保险;海洋运输保险;海损保险;海上保险
maritime insurance company 航运保险公司
maritime intercourse 海上往来
maritime interest 航海利益;航运界
maritime jurisdiction 海事管辖权;海事裁判权
maritime law 海洋法;海事法;海(上)法(规);海商法
Maritime Law Association 海商法协会
maritime law committee 海法委员会
maritime lien 海运留置权;海事优先权;海上优先权;海上留置权;抵偿权
maritime lienee 海事优先权人
maritime limit 海域界线;海上界线
maritime limit in general 一般海区界限
maritime losses 海上损失
maritime meteorologic(al) observation 海洋气象观测;海上气象观测
maritime mile 海里
maritime mobile radio-telephone equipment 海上机动无线电话设备;水上行动无线电话设备
maritime mobile satellite 航海移动卫星
maritime mobile satellite service 海事移动卫星服务;卫星水上移动业务
maritime mobile service 海上移动业务;海上移动通信业务;水上行动业务
maritime mobile service identities 海上移动业务标识
maritime mobile telegraphy calling 海上移动电报呼叫
maritime mobile telephone equipment 海上移动电话设备
maritime mortgage 海上抵押权;海商法上的抵押
maritime museum 航运博物馆
maritime nation 海运国家
maritime navigation 航运;海洋航行;海上航行
maritime negligence 海上过失
maritime operating cost 航行成本
maritime people 航海人员;海员
maritime peril 海上遇险;海难;海上危险;海上风险
maritime pine 海松;海岸松
maritime plant 海生植物
maritime polar air mass 极地海洋气团
maritime policy 海洋政策
maritime pollutant 海洋污染物
maritime pollution 海洋污染
maritime pollution claim fund 海洋污染索赔基金
maritime port 海港
maritime ports convention 海港公约
maritime power 制海权;海运国家;海上强国;海军力量;海军大国
maritime property 海上财产
maritime protection 航运保护(主义)
maritime protection work 海上防护建筑物
maritime provinces 沿海省份;沿海各省
maritime quarantine 海港检疫
maritime radar interrogator transponder 海上雷达询问应答器
maritime radio 海上无线电通信
maritime radio beacon 海上无线电信标
maritime radio direction finding station 海上无线电测向站
maritime radio equipment 航海无线电设备
maritime radio navigation land station 海上无线电导航陆地电台
maritime radio navigation mobile station 海上无线电导航移动电台
maritime radio navigation satellite service 卫星水上无线电导航业务
maritime radio navigation service 海上无线电导航业务;海上无线电导航机构
maritime region 海区
maritime rescue sub-centee 海上救助分中心
maritime resources 海洋资源
maritime rights 海洋权
maritime river 滨海河流
maritime river reach 靠近海岸的河段
maritime safety agency 海上安全机构
Maritime Safety Board 海上保安厅(日本);海上安全局;海运安全委员会;海上保安部门
Maritime Safety Committee 海上安全委员会
maritime safety organization 海上安全机构
maritime sanitary devices 船上卫生设备
maritime satellite 航海卫星;海事卫星
maritime satellite communication and navigation 海事卫星通信与导航
maritime satellite system 海事卫星系统
maritime search and rescue 海上搜救
maritime search and rescue plan 海上搜救计划
maritime search and rescue recognition code 海上搜救识别码
maritime section of stream 感潮河段;盛潮河段
Maritime Silk Road 海上丝绸之路
maritime soil mechanics 海洋土力学
maritime source of pollution 海洋污染源
maritime sovereignty 海洋主权
maritime space 海域
maritime state 海上强国
maritime strategy 海洋政策
maritime stream 滨海河流
maritime structure 海洋建筑物;海工结构;水工建筑物
maritime survey(ing) 海洋测量(学)
maritime term 海事术语
maritime terminal operator 码头公司
maritime territory 领海
maritime tort 海上侵权行为
maritime trade 航运业;海外贸易;海上贸易
maritime traffic 海上交通
Maritime Traffic Safety Law of the People's Republic of China 中华人民共和国海上交通安全法
maritime transport(ation) 海上运输;海洋运输;海运
maritime transportation contract 海上运输合同
maritime transportation insurance 海洋运输保险
maritime transport law 海上运输法
maritime transport statistics 海运统计
maritime tropic(al) air(mass) 热带海洋气团
maritime works 海洋工程;海事工程;海工建筑(物);水工建筑物
mariupolite 钠霞正长岩;钠霞正长石
mark 型号标志;钢筋号;加标记;记号;痕迹;牌号;传号;符号;标明;标记;标志

mark 2 gas cooled reactor 第二代气冷反应堆
mark 3 gas cooled reactor 第三代气冷反应堆
mark and garden 园标
mark-and-space 传号和空号
mark and space impulse 传号和空号脉冲
mark at or below ground level 埋石
markboat 标志船
mark card 符号卡片;标记卡片
mark card reader 标记卡(片)读出器
mark card sensing 标记卡读出
mark check 标记检查
mark condition 传号状态
mark contact 传号接点
mark counter 有码计数器
mark counting check 特征计数检验
mark detail 标志大样
mark detection 标志检测
mark-down 减价;减底标价;减低标价;削价
mark-down percentage 标价减成率
mark-down sale 降价销售
marked capacity 额定容量;注册容量;登记容量;额定生产率
marked capacity of car 车辆标重;车辆标记载重
marked check 保证兑现的支票
marked compound 标记化合物
marked cycle 标号循环
marked difference 显著差异;显著差别
marked elevation 标高
marked end 标记端
marked face 木料正面;标记面
marked factor figure 显著因素图
marked finish (混凝土)木质花纹饰面;花纹饰面
marked freight prepaid 注明运费预付
marked loading capacity of car 货车标记载重(量)
marked loading capacity of wagons 货车标重
marked map 战术态势图
marked number of semi-cushioned seats on passenger train 硬座标记定员
marked page reader 标记页面阅读器
marked point 设标点;标志点
marked price 价目;标价
marked rating 注明额定值
marked ratio 变压器标称变比
marked route 有标志路线
marked station 设标点
marked stone 标石
marked symbol 标记符号
marked terminal 标记终结符;标记的终端
marked transfer 标明转让;保证股票转让
marken 软炭质页岩
marker 指示器;指示层;指点标;路标;可变距标;记分器;划线规;打号员;打号器;打印器;标志(器);标示物;标示器;标记器;标记脉冲;标号畜;标号器
marker and link tester 标志器链路测试器
marker antenna 信标天线;指示天线;指点信标天线
marker atom 标记原子
marker beacon 指向标;无线电信号发射台;标志信标
marker beacon indicator 信标指示器
marker bed 指层;标准层;标志地层;标志层(位)
marker bit 标记位
marker bubble 标志泡
marker buoy 指示浮标;位置浮标;(测量用)标志浮标;航向浮标;临时信号浮标;标位浮标
marker buoy release unit 信号浮标释放装置
marker character 标记符
marker circuit 标志电路
marker clamp 标志电平箝位电路
marker clone 基准品系
marker control 标志控制
marker crude 标志原油
marker design 航标设计
marker effect 标记效应
marker enzyme 标记酶
marker for channel lines 航道边线标志灯
marker gate 标志脉冲闸门
marker generator 标记(信号)发生器
marker group 标志群
marker horizon 指示层;标准层;标志地层
markering cost 推销费用
marker lamination 指示层;标准层
marker lamp 识别信号灯;标志灯

marker layer 标准层
marker level 标志电平
marker light 信号灯;信标灯光;灯光信号;标志物灯;标志灯光;标志灯
marker line 信号线路;标志线
marker mixer 标志信号混合器
marker of biogenesis 生物成因标志
marker of biology 生物学标志
marker off 划线工;放料
marker of lithology 岩石学标志
marker of mineralogy 矿物学标志
marker of texture 结构标志
marker oscillator 标记脉冲发生器
marker out 翻样;划线;放样技工
marker peg 标志桩【测】
marker pen 标记笔
marker pip 指向脉冲;校准标志;标记脉冲
marker pip displacement 标记点位移
marker post 向导标(反射式)
marker post for snow clearing 除雪标桩
marker pulse 标志脉冲;信号脉冲;指向脉冲;指示脉冲;传号脉冲;标记脉冲
marker pulse conversion 标志器脉冲转换
marker radio beacon 无线电信号发射台
marker register 时标寄存器;标志寄存器
marker scan(ning) 标记扫描
marker selector 指向选择器;指向脉冲选择器;标志脉冲选择器;指向脉冲选择器
marker sense 标记识别
markers of geochemistry 地球化学标志
markers of sedimentary facies 沉积相标志
markers of unconformity 不整合标志
marker space 分隔区
marker spacing 标志间距;标志间隔
marker sweep generator 扫频标志发生器
marker system 标志信号系统
marker thread 标志线
marker type scale factor 记号类型比例因子
marker velocity 标准层速度
marker zone 指示层
market 商场;定期集市
marketability 销路;适销性;变现性
marketability ratio 股票易销性比降
marketable equity 可让渡权益
marketable goods 易销货物;适销产品
marketable grain 商品粮
marketable industrial crop 有销路的经济作物
marketable price 适销价格
marketable products 适销产品
marketable securities 可转售证券;畅销证券
market abroad 海外市场
market absorption 市场吸收能力
market absorption capacity 市场容量
market administration 市场管理
market advancing 市价正上涨
market allocation 分配市场
market amount 市场金额
market analysis 市场分析
market and marketing system 市场和销售系统
market appraisal standard 市场议价标准
market approach to value 据市场行情估价
market area 市场所在地;市场区
market assessment 市场估价
market audit 市场审核
market behavior 市场行为
market behaviour 市场走势
market bill 市场国库券
market borrowing 市场借债
market brass 商品黄铜
market building 市场房屋
market chains 连锁市场
market channel 市场渠道
market claim 市场索赔
market classification 市场分类
market cleaning price 市场清算价格
market clearing 市场结算
market closure 市场封闭
market combination 市场联合
market competition 市场竞争
market condition 市场情况;商情
market congestion 市场阻塞(指市场挤迫现象)
market control 市场管理
market-creation of pollution rights 污染权;污染权交易市场;市场污染权建立

market cross 市场办公室
market cycle 销售循环
market data approach 按市场资料估价法(指不动产)
market deal 市场交易
market declining 市价正下降
market demand 市场需求
market demand curve 市场需求曲线
market demand price 市场需求价格
market depreciation 市场价格的下跌
market development 市场开发;市场发展
market difference margin 市场价差保证金
market-directed economics 市场导向经济
market-directed economy 市场经济
market discount allowance 市场交易折扣
market discount rate 市场贴现率
market disruption 市场扰乱
market diversification 市场多样化
market dominance 市场操纵
market economy 市场经济
market economy structure 市场经济结构
marketed cement 市售水泥
marketed pollution permit 市场污染许可证
marketeer 市场卖主
market efficiency 市场效率
market equilibrium 市场供需平衡
marketer 销售经理
market exchange rate 市场汇率
market expansion 市场扩张
market expansion promotion 扩大市场推销
market exploration 市场考察
market failure 市场衰退
market feedback 市场调查资料
market finance 市场金融
market fluctuation 市场波动
market for bill exchange 票据交换市场
market forces 市场力量;市场动力
market forecast(ing) 市场预测
market forecasting of mineral commodities 矿产市场预测
market forecast of commercial enterprises 商业企业市场预测
market for executive talent 行政管理人才的就业市场
market for general labor 一般劳动力市场
market game 市场对策;市场博弈
market garden 商品菜园;市场花园;以销售为目的果菜园;以供应市场为目的的菜园;供应市场的菜园
market gardening 商品蔬菜种植业
market gardening base 商品蔬菜栽培基地
market goods locally 就地销售
market grouping 市场分类
market growth 市场发育
market hall 室内市场
market has openings 市场缺口
market house 市场管理所
market identification 市场确定
market index 市场指数
market influence 市场影响
market information 行情;市场信息
marketing 销售商品;运销;交易;上市;销售;经营学;经销学;市场工作
marketing activity 市场活动
marketing agencies 销售部门
marketing anticipating 市场展望
marketing boards 销售局
marketing channel 销售渠道;销货渠道
marketing concept 销售概念;市场观念
marketing controller 销售检核人
marketing cooperative 销售合作社
marketing cost 销售成本;市场营业费
marketing cost analysis 销售成本分析
marketing cost consultant 销售成本顾问
marketing cost control 控制销售成本
marketing cost control by division 控制部门销售成本
marketing cost control by product 控制产品销售成本
marketing cost control by territory 控制地区销售成本
marketing decision variables 销售决策变数
marketing department 销售部门
marketing expense 营销费用;推销费用

marketing expert 市场专家
marketing forecast of passenger traffic 客运市场营销预测
marketing function 市场业务;市场功能
marketing game 销售对策
marketing information system 行销资讯系统;行销信息系统;行销情报系统;销售情报系统
marketing life 市场寿命期
marketing main body 市场主体
marketing management 销售管理
marketing manager 营业部经理
marketing mechanism 销售机构
marketing mix 行销组合;市场经营组合
marketing mix of passenger traffic 客运市场营销组合策略
marketing-mix program(me) 销售组合计划
marketing model 销售模型
marketing myopia 销售短视
marketing of passenger traffic 客运市场营销管理
marketing of railway passenger traffic 旅客运输市场营销
marketing of railway transport enterprise 铁路运输企业市场营销
marketing organizing 销售组织
marketing-oriented business concept 销售化企业观念
marketing plan 销售计划;市场计划
marketing planning 销售计划
marketing policy 销售方针;市场政策
marketing practice 销售惯例
marketing price policies 市场价格政策
marketing principle 市场原则
marketing process 销售过程
marketing profit center 销售利润中心
marketing rate of fattened stock 出栏率
marketing research 销售调查;市场活动调查
marketing segmentation of passenger traffic 客运市场细分
marketing station 商业基地
marketing strategy 销售策略;市场策略
marketing strategy of railway passenger and freight transport 铁路客货营销战略
marketing strategy simulation 市场战略模拟
marketing system 市场体系
marketing technique 销售技术
marketing trade 集市贸易
marketing unit 销售单位
marketing year 销售年度
market instinct 市场本能
market interest rates 市场利率
market intervention 市场干预
market in the interior 内地市场
market investigation 市场调查
market investigation of passenger traffic 客运市场调查
market lead 商品铅
market letter 商情报告书
market level 市场层次
market liquidity 市场流动性
market management 市场管理
market management strategy 市场经营战略
market manipulation 市场操纵
market marketing 市场营销学
market mechanism 市场机制
market model 市场模型
market objective 市场目标
Market of Caesar and Augustus 恺撒与奥古斯塔斯市场
market of discharge permit 排污许可证市场
market off 市场休业
market on close 收市订单
market order 卖出指示;市价单;市场订购;市场订单;市价订购
market-oriented economy 着重市场的经济
market-oriented industry 市场导向工业
market-oriented production 面向市场的生产
market outlook 市场展望
market overseas 海外市场
market overstocked 市场存货过多
market overt 公开市场
market penetration 市场渗入
market-penetration pricing 市场渗透定价
market place 集市;市场商业中心区;市场;商业中心地

market position 市场地位
market potential 市场潜在需求量;市场潜力
market power 市场支配力
market price 牌价;市价;市场价(格)
market price at time of issue method 时价支出法
market price method 市价法(指联合成本的分配)
market price of labo(u)r 市场劳动力价格
market production 市场生产
market profile 市场扫描;市场概况
market prospect 市场展望;市场前景
market purchasing 市场采购
market quality 市场要求的质量
market quality information 市场质量情报
market quotation 牌价;市场行情
market rate 市场利率
market rate loan 市场利率贷款
market rate of discount 市场贴现利率
market rate of exchange 市场汇率
market rates of exchange in effect 实际市场汇率
market readjustment 市场再调节
market realization 市场实现
market regulation 市场调节;市场法规
market rent 市场租金
market report 市场报告
market requirement 市场需要量;市场需求
market research 市场研究;市场调查
market research and analyze sales potential 市场研究与分析销售潜力
market research company 市场调查公司
market risk 市场风险
market road 市场道路
market rose 行情上涨
market sale 市场销售
market saturation 市场饱和
market scale 市场规模
market section of passenger traffic 铁路客运目标市场选择
market segmentation 市场划分;市场分割
market segments 区隔市场
market sensitiveness 市场的敏感性
market sentiment 市场气氛
market separation 市场割据
market-set prices 市场定价
market share 市场占有率;市场份额;市场分配
market share at percentage 销售量占百分比
market share matrix 市场份额矩阵
market share of railway transportation 铁路运输市场份额
market sharing 市场分配;分占市场
market-sharing arrangement 市场分配协定
market shortages 市场紧张
markets in sales areas 销地市场
market size 市场规模;(商品)市场尺寸
market square 市场广场
market stall 街市摊位
market stead 市场
market strategy 市场策略
market structure 市场结构
market supply-and-demand situation 市场供求情况
market supply schedule 市场供给表
market survey 市场调查
market survey made by commercial enterprises 商业企业市场调查
market survey method 市场探测法
market survey procedure 市场调查程序
market target 市场对象
market term 市场名称(木材);市场用语(木材)
market test(ing) 市场测验
market threshold 市场限值
market town 集镇;集市贸易镇;市集
market transaction 市场交易
market transaction cost 市场交易费用
market transparency 市场明朗度
market trend 市场趋势
market unhealthy 市场不正常;市场不稳定
market unsettled 市场不稳定
market valuation 市场估价
market value 市(场价)值;市(场)价(格);时价
market value approach 市场价值评估法
market value method 市场价值法
market value of building 房屋市价
market value of investment securities 投资证券的市值

market wage 市场工资
market weight 出售体重
marketwide risk 系统性风险
market write-down 市价下跌
mark for compass adjustment 罗盘校正浮标
mark form sequence 符号格式序列;标记形成序列
mark for waterway 水道标【航海】
mark frequency 传号频率
mark function 符号功能;标记功能
mark ga(u)ge 划线器
mark hold 符号保持;标记保存;标记保持;固定符号;固定标记;传号保持
Mark-Houwink constant 马克-豪温克常数
marking 印记;印痕;作记号;作标记;加记号;记分;锚链长度标志;认付(支票);打号;标号;标志
marking and spacing intervals 传号和空号区间
marking awl 划针;划线钻
marking awl thicknesser 划线盘
marking axe 号斧
marking bay 钢材打印跨
marking bias 传号偏压;传号偏离;标记偏移
marking board 交易所揭示牌;记分牌;记分板
marking buoy (测量用)标志浮标
marking buoy moorings 标志浮标的锚碇
marking chalk 划线白垩;记号笔
marking circuit 标记电路
marking class 标记类
marking composition 划线构图
marking compound 划线用涂料;校验表面着色用涂剂;涂色剂
marking contact 传号触点;标志触点
marking crayon 标志色笔;标记色笔
marking current 传号电流;符号电流
marking-cutting shop 划线切割车间
marking depth 设标水深
marking device 划线规;压花机划线规;压花机;记录设备;刺点装置;标示器
marking end distortion 标记端失真;传号末端失真
marking equipment 制标志机;标记设备
marking for street traffic 路线标志
marking frequency 标记频率;发号频率;标志脉冲频率
marking ga(u)ge 边线规;平行划线尺;木工划线尺;划线规尺;线勒子;木工划线盘;木工划线规;划线器
marking gun 喷漆枪
marking hammer 号锤;打号锤
marking impulse 标记脉冲
marking ink 打印油墨;打印油;标记墨水
marking iron 烙印铁
marking knife 木工划线刀;划线刀;剥刀
marking machine 打字机;印字机;划线机;打号机
marking material 路面划线材料
marking needle handle file 半椭圆针锉
marking nut oil 印果油
marking of a car 车辆标记
marking of a channel 航道标志
marking of boundaries 标定界线
marking of buoyant apparatus 救生浮具标记
marking of cargo 货物标志
marking of control station 控制点埋石
marking of cylinder 气瓶标志
marking off 划线
marking off pin 划线针
marking off plate 划线板
marking off plater 划线工
marking off table 划线台
marking of lifeboat 救生艇标记
marking of liferaft 救生筏标记
marking of points 测点标志
marking of surveying points 测量点设标
marking of the bank line 海岸线标定
marking of trigonometric(al) point 三角点埋石
marking out 划线;划出(定出工程边界线和距离);立标桩定线【道】;定线;放线;标出;注意
marking out dimension 注明尺寸
marking out of curve 测设曲线
marking out of the track 轨道放线
marking out section 划线区段
marking out table 划线台
marking paint 路标漆;高速公路路标漆;高速公路划线漆;划线漆;标志油漆;标志色漆
marking painting 路标油漆
marking peg 标针

marking percentage 传号百分率
marking pin 测钎;测标钎;测标杆;标钎
marking pipe 作管子标记
marking plate 标记板
marking point 划线针;标志点
marking pole 标杆
marking post 标杆标桩
marking potentials 标志电位
marking press 压痕机;雕印压榨
marking pulse 传号脉冲
marking-recapture method 标记重捕法
marking ring 标注环
marking scale 标度尺
marking scraper 划线器
marking screw 标号螺钉
marking separated 划线分隔;标线分隔
markings for street traffic 路线标志
marking signal 标志信号;标记信号
marking stake 标桩【测】
marking stamp 压印模
marking steel 测钎
marking steel materials 钢材记号
markings transfer after cutting plates 板材切割以后的转移标准
marking stud contact 标志接点
marking system 标志系统
marking tool 划线工具
marking transfer 标记转印(压力容器)
marking unit 打号装置
marking-up 标字
marking-up day 贴水日
marking-up loans 提高贷款利息
marking vessel 标志船
marking wave 记录波;传号波;符号波;标记波
markite 导电性塑料
markle basalt 拉斑橄玄岩
mark length 记号长度
mark match 标记符合
mark matching 标记选配
mark memory 标记存储器
mark number 标号
mark of acceptable product 合格品标志
mark of car 汽车牌号
mark of conformity 合格标志
mark of conformity certification 合格认证标志
mark of deep fracture 深断裂的标志
mark-off 划痕
mark of groundwater pollution 地下水污染标志
mark of manual strangulation 扼痕
mark of origin 原产地标志;进口商品的产地标记
mark of reference 参照符号;参量符号
mark of remote-sensing image 遥感影像标志
mark of safety certification 安全认证标志
mark of unconformity 不整合标志
mark-on 成本外加数;加利(指成本加利变成售价)
mark-on percentage 标价加成率
mark out 划线
Markovian chain 马尔柯夫链
Markovian decision theory 马尔柯夫决策论
Markovian model 马尔柯夫模型
Markovian process 马尔柯夫过程
Markovian renewal process 马尔柯夫更新过程
Markovian variable 马尔柯夫随机变数
Markovian vector approach 马尔柯夫矢量法
Markovitz equation 马科维茨方程
Markov process with stationary increments 具有平稳增量的马尔柯夫过程
Markov vector approach for linear systems 线性系统的马尔柯夫矢量法
Markov vector approach for non-linear systems 非线性系统的马尔柯夫矢量法
Markov vector approach for SDF system 单自由度体系的马尔柯夫矢量法
mark pen 符号记录笔
mark plate 标志板
mark position 标记位置;标识位置
mark post 标杆
mark pulse 标志脉冲
mark radiation 传号发射
mark reader 符号读出器
mark reading 符号读出;标记读出
mark-reading station 标记读出机构
mark register 标记寄存器
marks and deeps 水砣长度记录;水砣长度记号

marks and numbers 麦头和号码
mark scan(ning) 特征扫描;标志扫描;标记扫描
mark scanning documents 标记扫描文件
mark scraper 划线器
mark-seen punching 标记读出穿孔
mark sense 标记读出;读出标记
mark sensed card 符号读出卡片;标记读出卡
mark sense punch 标记读穿;读标记穿孔
mark-sense row 标记读出行
mark sensing 符号读出;标记检测;标记读出;标号读出
mark-sensing card 符号读出卡片;标记读出卡
mark-sensing column 标记读出列
mark-sensing machine 符号读出机
mark-sensing punch 符号读出穿孔机
mark-sensing row 标记读出行
mark-sensing sheet 标记读出表
mark sensor circuit 标记读出电路
mark sheet 特征表;标记纸;标记图表
mark sheet reader 标记卡读出器
marks indistinct 标志不清;标记不清
marks of plate division 板块划分标志
mark space 符号间隔
mark space multiplier 符号区间乘法器
markstone 标石
mark synchronization 标记同步
Mark system 马克制度
mark tally 检数理货员;记录理货(法);标记点数
mark target 加标志于物标
mark the chain cable 锚链标记
mark-to-market 调至市价
mark-to-space ratio 传号空号比;标记占空比
mark-to-space transition 传号空号转换;符号到间距转换
mark track error 标志磁道错误
mark-up 加价;标高价格;增高标价;加利(指成本加利变成售价);标量;提价;提高标价
mark-up inflation 标高价格的通货膨胀
mark-up loans 标高贷款利息
mark-up percentage 提价百分率
mark-up pricing 加成定价法
mark-up rate 提价率
mark variable coefficient 标志变动系数
mark variable extent 标志变动度
mark with the center 定心
mark with trees 原生树标
mark zone 标记区
marl 泥灰土;捆吊床结;泥灰(岩);灰泥
marlaceous 泥灰质的;泥灰岩的
marlaceous facies 泥灰岩相
marlaceous shale 泥灰质页岩
marl-bond macadam/gravel 泥灰结碎石/砾石
marl brick 泥灰岩砖
marl deposit 泥灰岩矿床
Marlene 马兰变性弹力丝
Marles steering gear 球面蜗杆滚轮式转向器
Marlex 马来克司聚乙烯
Marley 马来制品(多种混凝土制品的统称)
Marlie's alloy 马里抗氧化合金;马里耐氧化合金
marline 细索;细绳;油麻绳;双股油麻绳
marline clad wire rope 包麻钢丝绳
marline knot 绳条结
marline spike 穿索针
marling 抹灰泥
marlin spike hitch 绳针结
marlite 硬泥岩;泥灰质岩;泥灰岩
Marlith 马里板(一种刨花板)
marl loam 泥灰质壤土
marl loess 泥灰质黄土
marlm rubber 软白垩砖
Marlon type oscillation method 栗形运条法
Marlowe kiln 马洛隧道窑
marl pellet 泥灰球条;泥灰粉末;泥灰质团粒
marl pit 泥灰岩采掘场
marl rubber 软白垩砖
marl shale 泥灰页岩
marl slate 泥灰石板瓦;泥灰板岩
marl slate stratum 泥灰板岩层
marlstone 泥灰(质)岩;泥灰石
marl transional association 泥灰岩过渡组合
marl turbidity 泥灰混浊度
marly 泥灰质的
marl yarn 夹色纱线
marly bituminous shale 泥灰沥青质页岩

marly clay 泥灰质黏土
marly earth 泥灰土
marly limestone 泥灰(岩)质石灰岩;泥灰石灰石;泥灰灰石
marly loam 泥灰质壤土
marly mud 泥灰质泥
marly sandstone 泥灰(质)砂岩
marly sandy loam 泥灰砂质壤土
marly shale 泥灰质页岩
marly soil 泥灰土;泥灰质土(壤)
marmatite 铁闪锌矿
marmite can 大型保暖容器
marmolite 白蛇纹石
Marmor 马莫阶
marmoraceous 大理石状的;像大理石的
marmorate(d) 带大理石纹的;具有大理石纹的
marmoration 用大理石贴面;大理石饰面;大理石表面装饰
marmoratum 大理石粉胶泥
marmoreal 大理石状的;像大理石的
Marmorene 大理石平板玻璃(一种磨光墙面平板玻璃)
marmoresina (人造的)大理石砌块
Marmorite 大理石平板玻璃(一种不透明浇注并磨光的平板玻璃)
marmoset 怪状丑角像装饰
marokite 黑钙锰矿
maroon 栗(红)色;红褐色
maroon lake 紫红色淀
maroon oxide 氧化紫红;氧化栗红
maroon prunosus 紫红色
maroon toner 紫红色原;紫红调色剂;栗红色料
maroscopic quantity 宏观量
marouflage 马罗裱糊法;粘贴壁画画布
marpol oil annual 船舶防油污染年检
marpol oil interim 船舶防油污染临时检验
marpol oil renewal 船舶防油污染合格证更新
mar proof 耐磨损性;耐擦伤性;无划痕
Marquardt porcelain 马夸特瓷
marquee 营幕;雨罩;大门罩;挑棚;大帐幕
marquee sign 大雨棚上的招牌
marquench 马氏体淬火;等温淬火
marquench and tempering 分级淬火回火处理
marquenchflame hardening 马氏体淬火
marquenching 马氏体等温淬火;马氏体淬炼
marquench rolling 淬火马氏体轧制
marquench stressing 加载分级淬火
marqueterie 镶嵌细工;镶嵌装饰品;嵌木细工
marquetry 镶嵌装饰品;镶嵌细木工;镶嵌工艺;嵌木技术;嵌木制品;嵌木细工
marquise 雨篷;大帐幕;天幕;挑棚;大门罩
marquisette 薄纱罗;窗帘
marquisette curtain 薄幕;(玻璃纤维的)窗帘
marquois scale 平行线尺【测】;三角尺
mar resistance 抗划伤(性);耐擦伤性;耐划痕性;抗划痕性;耐擦伤性
marriage chain 结合链
marriage-gate 配合门
marriage problem 匹配问题
marriage relation 结合关系
married 密切结合的
married couple 双索吊钩吊索系统
married fall 联杆吊货辘缆;双索吊钩吊索系统
married fall system 联杆吊货法;双吊杆联合作业法
married gear 联杆吊货装置
married joint 编接接头;绳索编接接头
married quarters 已婚者住宅
married rope 插接绳
marrite 硫砷银铅矿
marrying 夹牢;墙接;木杆绑接(制脚手架时)
marrying of wooded pole 木杆绑接
marrying wedges 对插楔
Mars 火星
Mars cartography 火星制图学
Mars chart 火星图
Marschet approximation formula 马斯凯特近似公式
Marsden chart 马士登图
Marsden square 马士登方
Marseilles pattern tile 马塞式瓦
Marseilles tile 马塞瓦
marsh 沼泽;湿地(沼泽);湿草甸;草(本)沼泽
marshal(led)yard 编组场【铁】
Marshall aid 马歇尔援助

Marshall compaction hammer 马歇尔试验中的击压锤
Marshall cylindric(al) test 马歇尔圆柱形试验(沥青混凝土的)
Marshall cylindric(al) test-head 马歇尔圆柱体试验头(沥青混凝土的)
Marshall-Edgeworth-Bowley index 马歇尔-埃奇沃思-鲍利指数
Marshall flow value 马歇尔流动值(沥青混凝土的)
Marshallian demand curve 马歇尔需求曲线
marshalling 排列；整理；列车编组
marshalling airfield 集结待发机场
marshalling and disposition 编组调车
marshalling area 列车编组场；编组区
marshalling capacity of switching lead 牵出线改编能力
marshalling-departure track 编发线
marshalling-departure yard 编发场
marshalling equipment 列车调度设备；编组设备
marshalling facility 调车设备
marshalling masonry 规则砌体；砌筑圬工
marshalling plan 集结场计划；货位计划；前方堆场计划；堆存安排计划；(集装箱)堆场计划
marshalling service 编组作业
marshalling siding 调车线【铁】
marshalling station 编组站【铁】；铁路编组站
marshalling system 材料集体制度
marshalling track 调车线【铁】；编组线【铁】
marshalling tractor 列车调度牵引车；编组车头
marshalling vehicle 列车编组调度车；编组列车
marshalling yard 集装箱前方堆场；调配场；调车场；分区车场【铁】；编组车场；编组站【铁】；编排场
Marshall Islands 马歇尔群岛
Marshall K 马歇尔 K 指标
Marshall-lerner condition 马歇尔勒纳条件
Marshall line 马歇尔线
Marshall mix design method 马歇尔混合设计法
Marshall-Olkin distribution 马歇尔—奥尔金分布
Marshall-Palmer distribution 马歇尔—帕尔默分布
Marshall plan 马歇尔计划
Marshall property 用马歇尔法测定的(沥青混凝土)性质
Marshall refiner 马歇尔精制机
Marshall's indicator 马歇尔指示器
marshall springs 座簧
Marshall stability 马歇尔稳定度(沥青混凝土强度的一种指标)
Marshall stability apparatus 马歇尔稳定度仪
Marshall stability test 马歇尔稳定度试验
Marshall stability value 马歇尔稳定度值
Marshall stiffness 马歇尔劲度
Marshall test 马歇尔试验
Marshall test specimen 马歇尔试验用试件；马歇尔试验标本
Marshall-type design 马歇尔式设计
Marshall-type design method 马歇尔式设计法
Marshall valve gear 马歇尔阀装置
marsh buggy 沼泽地用汽车
marsh deposit 沼泽沉积(物)
marsh excavator 沼泽地挖掘机
marsh formation 沼泽层
marsh forming process 沼泽形成
Marsh funnel viscometer 马氏漏斗(型)黏度计
marsh gas 沼气
marsh gas generation 沼气发生
marsh gas power generation 沼气发电
marsh gas tank 沼气池
marsh-grass coast 沼泽草地海岸
marsh herbaceous facies 沼泽草本相
marsh island 沼泽岛
marshite 碘铜矿
marsh land 沼泽湿地；沼泽地区；沼地；沮洳地；沼泽地(带)
marsh marl 沼泽泥灰土砖；沼泽泥灰岩
marsh muck 湿地腐殖质；沼泽腐殖土
marsh ore 沼铁矿
marsh peat 沼泽泥炭；沼泽泥煤
marsh plant 沼泽植物
marsh plough 沼泽地犁
marsh podzol 湿地灰壤；湿地腐殖土
marsh screw amphibian 沼地螺旋推进水陆两用车
marsh sediment 沼泽沉积(物)
marsh sedimentation 沼泽沉积作用

Marsh's test 马希氏试验(检砷或锑)
marsh tire 沼泽地用轮胎
marsh vegetation 沼泽植被
marsh wetland 草沼湿地
marshy 沼泽地的
marshy area 沼泽地
marshy district 沼泽地区
marshy field 水田
marshy grassland 草甸子
marshy ground 沼泽地；沼泽性土(壤)；湿地土壤
marshy land 湿地；泥沼地；沼泽地；烂泥地
marshy line 湖沼地带管道
marshy meadow 沼泽草地
marshy pasture 沼泽草地
marshy podzol 沼泽灰壤
marshy soil 沼泽(性)土(壤)
marshy wasteland 沼泽荒地
marshy weeds 水草
Mars orange 氧化铁橙色
Mars pigment 马氏颜料；合成氧化铁(系)颜料
Mars probe 火星探测器
Mars red 合成氧化铁红
marstraining 马氏体应变；马氏体形变时效；马氏体形变处理
marstressing 马氏体常温加工
marstressing process 马氏体形变热处理
marsturite 硅锰钠钙石
marsupialization 造袋术
Mars yellow 合成氧化铁黄；天然铁黄
mart 商业中心；市场；商场
Martello tower (防御海岸用)圆堡
martemper(ing) 间歇淬火；热浴淬火；马氏体等温淬火；马氏回火
martempering oil 分级淬火用油
martenaging 马氏体时效
martenaging martempering 马氏体回火
martenite 一种合成平炉炉底打结料
Martens densitometer 马腾斯测淀积银密度计
Martens hardness 马氏硬度
Martens' hardness test 马腾斯硬度试验
Martens-Heyn hardness 马腾斯—海因硬度
Martens illuminator 马腾斯照明器
martensite 马丁体；马氏体
martensite ag(e)ing 马氏体时效
martensite cold working 马氏体冷作处理
martensite finish(ing) point 下马氏点；马氏体转变终止温度；马氏体转变终止点
martensite formation 马氏体的形成
martensite lath 马氏体板条
martensite lattice 马氏体晶格
martensite plate 马氏体片
martensite point 马氏体转变点
martensite quenching 马氏体淬火
martensite range 马氏体转变区；马氏体温度范围；马氏体区域；马氏体范围
martensite region 马氏体区域
martensite start(ing) point 马氏体转变开始温度；马氏体开始形成点
martensite steel 马氏体钢
Martensite temperature 马氏体温度
martensite tempering 马氏体回火
martensite transformation 马氏体(型)转变
martensitic 马氏体的
martensitic cast-iron 马氏铸铁；马氏体铸铁
martensitic component 马氏体成分
martensitic hardening 马氏体硬化
martensitic matrix 马氏体基体
martensitic phase transformation 马氏体相变
martensitic range 马氏体转变区
martensitic stainless steel 马氏体不锈钢
martensitic steel 马氏体钢；马氏永磁钢；马丁钢
martensitic structure 马氏体组织；马氏体结构
martensitic transformation 马氏体转变
martensitic-type phase transformation 马氏体式相变化
Martens polarization photometer 马腾斯偏光光度计
Martens sclerometer 马腾斯刮痕硬度计
Martens scratch hardness 马腾斯划痕硬度
Martens spectroscope 马腾斯分光镜
Martens surfac scratching test 马腾斯表面划痕试验
Martens test 马腾斯(耐热)试验(塑料热弯变形)
Martens wedge 马腾斯光楔

martesia 蛀木海虫
marthozite 硒铜铀矿；铜硒铜矿
martial ethiops 磁性氧化铁
Martinel steel 硅锰钢；硅锰结构钢
Martin furnace 平炉
martingale 方式接线
Martin-Heyn hardness 马丁—海因硬度
martinite 板磷钙石
Martin process 平炉法
Martin's anchor 马丁式锚
Martin's cement (含碳酸钾外加剂的)硬石膏胶结料；马丁水泥；马丁胶结料
Martin's density meter 马丁密度计
Martin's diameter 马丁直径
Martin's filter 马丁过滤器
Martin's(hearth) furnace 马丁炉
Martin's(hearth) steel 马丁钢
martinsite 马丁散镁
Martin's quick setting cement 马氏快凝水泥
Martin's scratch hardness tester 马丁刮痕硬度试验机
Martin steel 平炉钢
martite 假象赤铁矿
Martius yellow 马休黄
martyrium 忠烈祠；殉道堂
martyr memorial park 烈士纪念公园
martyrs park 烈士公园
martyry 烈士纪念堂
marume plate 造粒板
marumerizer 球形造粒机；球形造粒
marver 乳光玻璃板
marverer 滚料工
marvering 滚压料泡
marver plate 玻璃滚料板
Marvibond method 滚压叠层法(氯乙烯叠层金属板的)
Marvin sunshine recorder 马文日照计
marworking 形变热处理(法)；奥氏体过冷区加工法
Marx generator 马克斯发生器
Marx rectifier 大气电弧整流器
Marx regeneration process 强碱再生过程
mary entry point 一次进入点
Maryland road test 马里兰道路试验
Maryland verde antique 马里兰古绿石；蛇纹大理石
marzacotto 珐琅用乳油釉
Marzine 苯甲哌嗪
masanite 马山岩
masanophyre 马山斑岩
Masan Port 马山港(韩国)
mascagnite 铵矾
mascaron 漫画化人头像装饰；怪状头饰
mascerating machine 皱片机；片机
mascerator 皱片机
Masci 薜纲
Mascolite 马斯柯利特绝缘材料(一种消振绝缘材料)
mascon 质量密集；质量瘤
maser 量子放大器；脉塞；微波量子放大器；微波激射(器)
maser action 微波激射作用
maser amplifier 微波激射放大器
maser bandwidth 脉塞带宽
maser gain 脉塞增益
masering material 脉塞材料
maser interferometer 脉塞干涉仪；微波激射干涉仪
maser medium 脉塞介质
maser preamplifier 脉塞前置放大器
maser receiver 脉塞接收器
maser relaxation 微波激射弛豫
mash 矿浆；泥浆泵；捣烂
mashan-type phosphate deposit 麻山式磷矿床
mashed 磨碎的
masher 搅糊机；捣粉机
masher for dry and wet material 干湿料两用搅糊机
Masher process 马歇尔钢瓦挂铅青铜法
mash gas 瓦斯
mash hammer 小铁锤；石工短锤；短柄重铁锤；大榔头；大槌
mashing machine 搅糊机
mashrebeeyeh (封闭阿拉伯建筑阳台窗的)精致旋制木隔板；(在大门上挑出的)防卫阳台
mashroom anchor 菇形清沟器；菌形清沟器
mashroom rock 菌状石

mash seam weld(ing) 滚压焊(接);滚压电阻缝焊;压薄滚焊;压薄缝焊
mash stitch welding 多针缝式焊接
mash tun 泥浆桶
mash weld 点压焊;点焊
mash welder 点焊机
masjid 伊斯兰教寺院;清真寺
mask 掩模;遮盖;障板;光刻掩模;面罩;面具;蒙片;屏框;分离子;表征码;避光框;掩蔽
maskable interrupt 可屏蔽中断指令;可屏蔽中断
mask aligner 掩模套准器;掩模光刻机;掩模对准器
mask alignment 掩模重合;掩模校准;掩模对准
maskant 掩蔽体;保护层
mask aperture 掩蔽孔径
mask artwork 蒙绘;蒙片图
mask bit 掩模位;屏蔽位
mask data 分离数据
mask design 掩模设计
mask detection 掩模检查
masked 戴着面罩的
masked bus 屏蔽总线
masked civet 果子狸;花面狸
masked colo(u)r negative film 遮光彩色负片
masked diffusion 掩蔽扩散
masked group 掩蔽团
masked off 掩蔽
masked position 掩蔽的阵地
masked reaction 隐蔽反应
masked set 屏蔽置位
masked state 屏蔽形态
masked valve 屏蔽式阀
masked wheel 带罩棘轮
maskelynite 阒玻长石;熔(料)长石;长石玻璃质
mask epitaxial method 掩模外延法;掩蔽外延方法
mask etch 掩模腐蚀;掩蔽腐蚀
mask evaporation 掩模蒸镀
mask field 屏蔽栏
mask for inhalation 吸入口罩
mask generation system 掩模生成系统
mask generator 屏蔽生成程序
mask holder 掩模架
masking 掩幕;遮拦法;光刻掩模;护面;屏蔽;掩蔽
masking action 掩蔽作用
masking agent 隐匿剂;掩蔽剂;遮蔽剂
masking amplifier 彩色信号比校正放大器
masking angle 遮蔽角
masking aperture 遮蔽孔;拦光孔径
masking blade 遮光叶片
masking by noise 噪声掩蔽
masking compound 遮蔽剂
masking effect 隐蔽效应;掩模效应;掩蔽效应;遮蔽效应
masking effect of sound 声音掩蔽作用
masking film 遮蔽膜片;蒙片
masking frame 遮蔽框架;遮光框
masking jig 掩蔽模
masking lacquer 保护漆
masking level 掩蔽级
masking liquid 涂液
masking material 罩面材料
masking matrix circuit 屏蔽矩阵电路
masking method 掩蔽法
masking method deodorizing 掩蔽法脱臭
masking of odor 恶臭的掩蔽
masking-out 蒙版
masking pad 掩蔽衰减器
masking paint 遮盖漆
masking paper 伪装文件
masking paste 防渗碳涂料
masking period 掩蔽期间
masking piece 遮幕
masking power 遮盖能力;遮盖力
masking ratio 掩蔽比
masking reagent 掩蔽试剂;掩蔽剂
masking sheet 掩片;蒙纸;蒙片
masking sheeting 伪装挡板
masking signal 掩蔽信号
masking sound 掩蔽声
masking tape 遮盖(纸)带;掩模带;遮蔽胶带;胶纸带;防污带
masking technique 遮蔽技术
masking tension 掩蔽电位
masking value 掩蔽值;屏蔽值
mask inspection densimeter 荫罩密度检测仪

mask interrupt enable flag 可屏蔽中断标志
mask layout 掩模设计
maskless 无掩模
maskless process 无掩模法
mask-like face 假面样面容
mask line 分帧线;分格线
mask-making technology 制版工艺
mask manufacture 掩模制造
mask matching 掩模匹配;掩码配准
mask means 掩模法
mask of beaten gold 金箔罩
mask off code 屏蔽码
mask opening 掩模窗孔
mask option 屏蔽选择
mask part 罩框部分
mask pattern 掩模图案;晶架
mask pitch 屏蔽距
mask plate 掩模板
mask potential 掩蔽电位
mask process 掩模过程
mask processing 掩模加工
mask programmable 屏蔽可程序
mask program(me)-negative film 屏蔽可编程序
mask register 选择寄存器;计时寄存器;屏蔽寄存器
mask replication 掩模复制
mask signal generator 遮蔽信号发生器;屏蔽信号发生器
mask stop (在线脚终止端的)怪状人面装饰石
mask target 对准标记
mask transfer 掩模迁移
mask welding 电焊眼罩
maslovite 等轴铋碲铂矿
Masny directive antenna 麦斯尼定向天线
mason 砖石工;泥瓦工;砌石工;石工
mason-architect 砖工建筑师
mason construction 砖石工程;砖砌工程;砌工工程(学);砖石结构
masoned cast stone 人造石
masonette 出租公寓房间
mason flagger 路面铺砌工;地面砌工;砖石铺路
Mason graph 马森图
masonic 砖瓦工的;砌工的;石工的
masonite 硬纤维板;绝缘纤维板;夹布胶木板;马桑尼特纤维板
masonite board 绝缘纤维板;纸柏板;美生板
masonite die 纤维板模具
mason jar 陶瓷瓶
mason layer 砌砖工
mason lodge 石工小屋
mason mark 石工标志
Mason master 麻生电钻(一种电钻)
masonry 砖石建筑;砖石工程;构筑体;砌筑;砌体;砌工;石工程
masonry abutment 砌工桥台
masonry aggregate 砌筑用集料;砌筑用骨料;水洗砂(用于砂浆拌和料的)
masonry anchor (固定门框于墙体的)锚件;砌体锚碇件;砌工锚碇
masonry anchorage 砌工锚具
masonry apartment tower 砖石公寓塔楼
masonry arch 砖石拱;砖砌炉顶;砌筑拱;砌工拱
masonry arch bridge 砌工拱桥;石拱桥
masonry arch dam 浆砌石拱坝
masonry arched bridge 砌筑拱桥;砌工拱桥
masonry arch gravity dam 砌工拱式重力坝
masonry architecture 砌工建筑
masonry archy bridge 砖石拱桥
masonry backfill 砌工回填料
masonry back-up (加固墙壁的)砖石里壁
masonry base 砌体基底
masonry beam 砖石梁;砌工梁
masonry bearing wall 砖石承重墙
masonry bin 砖砌料仓
masonry block 砌块;砌工块
masonry bond 砌工砌合;砌工黏结剂
masonry bonded hollow wall 砌工结合空心砖墙;砖石砌合空心墙
masonry bonding 砌工砌合
masonry bond principle 砌工联结原理
masonry bond type 砌工黏结类型
masonry bracket 砌工托架
masonry brick 砌工砖
masonry brick for chimney shaft 烟囱筒身用砖(砌工)

masonry bridge 砖石桥;砌工桥
masonry bridge pier 砖石桥墩
masonry building 砖石结构房屋;砖石建筑;砖石房屋;砌工建筑(物)
masonry building block 砌工建筑块材
masonry building component 砌工建筑构件
masonry building tile 砌工建筑砖
masonry bulkhead 砖石隔墙;砖砌密闭墙
masonry buttress 砌工支墩
masonry cavity wall 中空砌块墙;中空砌筑墙
masonry cement 泥瓦工水泥;砌筑水泥;砌工(用)水泥
masonry check 砌工节制闸
masonry check dam 砌工拦沙坝;砌工谷坊
masonry check valve 砌工节制闸阀
masonry chimney 砌筑烟囱
masonry cleaner 砌工除垢剂;砌工清洁剂
masonry cleaning 砌工处理;砌工清洁
masonry column 砖石柱;砖柱
masonry component 砌工构件;砌工建筑构件
masonry conduit 砌工渠道
masonry conduit-type sewer 砌工下水道
masonry construction 砌工施工;砌工结构;砌工建筑;砌工工程(学);砖石结构;砌工构造学
masonry contractor 砖石工承包商;砌工承包商
masonry course 砌工砌层
masonry cupola 砌工圆顶
masonry cutting blade 斩砖泥刀;砌工泥刀;砌工用切刀
masonry dam 块砌坝(混凝土块);砌筑坝;砌石坝;砌工坝;石坝
masonry diaphragm 砌工隔膜
masonry dome 砖石穹隆
masonry donjon 砖石城堡主塔
masonry dowel 砌工定位销
masonry drill 砌工钻
masonry duct 砌工管道
masonry dungeon 砌工地牢
masonry dwelling tower 砖石住宅楼
masonry envelope 砌体
masonry facing material 砌工面料
masonry failure test 砌工破坏试验
masonry fastener 砌工接合件;砌工紧固件
masonry fill 砌工填料;砌工填充(料)
masonry-filled 砌工填充料
masonry filler unit 填充砌块;梁间空心砌块;砌工填充单元;支承搭接楼板的砌体
masonry filler wall 砖石填充墙
masonry finish 砖石建筑面漆
masonry fireplace 砖砌炉子
masonry fixing 砌体附件;石工连接件;固定铁件;固定附件
masonry flue 砖砌烟道
masonry flume 砌工水槽
masonry footing wall 砌工基础墙;砌工建基脚墙
masonry foundation 砖石基础;砌工基础
masonry foundation wall 砌工基础墙
masonry gallery 砖石通道;砖石走廊;砌工廊道
masonry granite 砖石工程用花岗岩
masonry gravity dam 砖石重力坝;砌工重力坝
masonry grout(ing) 砌工灌浆;砌筑砂浆;砌筑胶泥
masonry guard (防砂浆流入的)挡板
masonry high flats 砌工高层公寓
masonry infilled frame 砖石填充框架
masonry in-fill panel construction 砖石填充墙板结构
masonry insulation 砌体绝缘材料
masonry joint 砌工接头;灰缝;砌体灰缝;砌缝;砌工砌缝
masonry joints of different colo(u)rs 各种色彩的砖石接头
masonry layer 砌工层
masonry lime 砌筑用石灰;砌筑石灰
masonry limestone 砌筑用石灰石
masonry-lined 有衬砌的;砌工衬砌的
masonry-lined tunnel 砖石衬砌隧道;砌工衬砌隧道
masonry lining 砖石衬砌;砌工衬砌
masonry lining material 砌工衬垫料
masonry lock 砌工插锁;砌工水闸;砌工船闸
masonry mass 砌块;砌工块
masonry material 砌工材料
masonry member 砌工建筑构件
masonry mortar 砖石砌筑灰浆;砌工砂浆;砌筑砂浆;砌土灰浆

masonry nail 砖石钉;砖钉;砌块钉;水泥钉
masonry of hollow units 砌筑空心墙
masonry opening 圬工开口
masonry paint 圬工用漆;砖石用涂料;砖石(建筑)用漆;圬工涂料
masonry painting 圬工涂料
masonry panel 预制砌块墙板;大型砌块;预制砌筑墙板;砌筑墙体
masonry panel wall 非承重圬工外墙;圬工格板墙;装配圬工墙板;轻型圬工墙板(不承重)
masonry partition 圬工隔断
masonry pier 砖石墩;(用混凝土块砌成的)块砌墩;圬工墩
masonry pitching of slope 砌石护坡
masonry plate 支承垫石;基础板;承梁块;座板;支承圬工垫板;圬工板;承梁板
masonry platform 圬工平台
masonry platform temple 圬工平台庙宇;圬工平台圣堂
masonry podium 圬工墩座墙
masonry podium temple 圬工矮墙庙宇
masonry point 圬工灰缝
masonry pointing 砌体勾缝
masonry principle 圬工原理
masonry quay 砌石护码头
masonry quay wall 整体砌筑码头;砌石护岸壁;砌石码头
masonry reinforcement 砖工砌体增强材料;混凝土加钢筋
masonry reinforcing 砖石砌体配筋;砖石砌体增强材料;圬工加筋
masonry reservoir 砖石蓄水池;圬工储[贮]水池
masonry residence tower 圬工住宅塔楼
masonry residential tower 圬工住宅塔楼
masonry retaining wall 圬工挡土墙
masonry revetment 圬工护岸
masonry rubble 毛石砌体
masonry ruins 圬工碎片
masonry sand 圬工用砂
masonry saw 砖石锯;圬工锯;石工用锯
masonry screen 圬工屏幕
masonry seal 圬工防潮密封
masonry sealing agent 圬工防潮密封剂
masonry seepage pit 圬工渗流坑
masonry sewer 砖石污水沟渠;烧砌沟管
masonry shaft 砖石支护井筒
masonry shell 圬工薄壳
masonry shoulder 圬工拱座
masonry silo 圬工筒仓
masonry skin 圬工壳板
masonry stack 砖砌烟囱
masonry-stone 圬工石
masonry-stone culvert 圬工涵洞;砌石涵洞;石涵(洞)
masonry strength 砌体强度
masonry structural system 圬工结构系统
masonry structure 圬工结构;砌石结构;圬工结构
masonry support 砌石支护
masonry tall flats 圬工高层公寓
masonry tie 砌体连系片;砌体拉杆;砌体金属系板
masonry tier 砖层;砖皮;圬工单层砖墙
masonry tile 圬工瓦;石瓦
masonry tomb 圬工墓穴
masonry tool 砌筑用工具;圬工工具;圬工用具
masonry toothing 砌体马牙接头;砌体马牙槎
masonry unit 砌筑单位;砌块;圬工体;圬工构件;圬工(砌筑)单元
masonry vault 砌筑穹隆;圬工穹隆
masonry veneer 圬工用镶饰材料(体);饰面砌体;砖石建筑饰;砌筑面层;贴面砌体
masonry veneer wall construction 贴面砖石结构
masonry wall 砖石墙;砌筑墙体;圬工(边)墙
masonry wall anchor 圬工墙锚锭装置
masonry wall base 圬工墙座
masonry wall beam 圬工墙梁
masonry wall below grade 地下基础墙;地下砖石墙
masonry wall block 圬工墙块
masonry wallboard 圬工墙面板;砌体墙面板
masonry wall breakthrough 圬工技术革新
masonry wall capping brick 圬工墙顶盖砖
masonry wall center line 圬工墙中心线
masonry wall construction 圬工墙结构
masonry wall coping 圬工压顶
masonry wall coping slab 圬工墙压顶板

masonry wall core 圬工墙核心
masonry wall course 圬工墙砌层
masonry wall crack 圬工墙裂纹
masonry wall crossing 圬工墙交叉
masonry wall crown 砖石墙顶拱
masonry wall dissolution 砖石墙裂开
masonry wall element 圬工墙构件
masonry wall enclosure 砌筑砌体围地;砌筑砌体围圈
masonry wall flashing piece 圬工墙披水板构件
masonry wall footing 圬工墙基础
masonry wall head 圬工墙头
masonry wall hollow 圬工墙洞
masonry wall installation 圬工墙的安装
masonry wall junction brick 圬工墙的联结砖
masonry wall line 圬工墙线
masonry wall lining 圬工墙内衬
masonry wall material 圬工墙材料
masonry wall niche 圬工墙龛
masonry wall opening 圬工墙口
masonry wall-pointing machine 圬工墙勾缝机
masonry wall recess 砖石边墙门槽
masonry walls below grade 地下砖石墙;基础墙
masonry wall skin 圬工墙表皮
masonry wall slot 圬工墙裂口
masonry wall stability 圬工墙的稳定性
masonry wall strength 圬工墙强度
masonry wall system 圬工墙体系
masonry wall temporary clamp 砖石墙砌体临时夹具
masonry wall thickness 圬工墙厚度
masonry wall tie 圬工墙拉杆
masonry wall tile 圬工墙砖
masonry wall top 圬工墙顶
masonry wall tower 圬工墙塔
masonry wall wing 圬工墙翼
masonry weir 圬工溢流堰
masonry well foundation 圬工沉井基础;圬工井筒基础
masonry withe 圬工烟囱隔板
masonry work 圬工工程;砖石工程;毛石砌体;砌筑工程
masonry work angle bead 砌体工程墙角护条
masonry work apartment tower 圬工工程公寓塔楼
masonry work arch 圬工工程拱券
masonry work bead 砖石凸圆线脚
masonry work bridge 砖石桥梁
masonry work building 砖石建筑
masonry work chimney 砖石烟囱
masonry work cleaning 砖石工程的清洁处理
masonry work construction type 砖石结构类型
masonry work course 砖石层
masonry work cupola 砖石圆顶
masonry work diaphragm 砖石隔板
masonry work donjon 砖石城堡主塔
masonry work duct 圬工通道;圬工管道
masonry work dungeon 砖石地牢
masonry work dwelling tower 砖石居住塔楼
masonry work facing 圬工饰面
masonry work failure test 砖石工程破坏性试验
masonry work haunching 砖石工程加腋
masonry work joint 砖石接缝
masonry work keep 砖石工程维护;砖石工程管理
masonry work layer 砖石层
masonry work leaf 砖石薄层
masonry work lining 圬工衬砌;砖石衬砌
masonry work metal bead 砖石墙上金属凸圆线脚
masonry work moisture sealing agent 砖石工程防潮密封剂
masonry work of parallepipedal cut stone 平行六面体石块的砖石工程
masonry work of vertical coring clay bricks 竖向挖心黏土砖的砖石工程
masonry work opening 砖石工程开洞
masonry work pier 砖石工程桥墩;砖石工程码头
masonry work reinforcement 砖石工程配筋
masonry work residence tower 砖石工程住宅塔楼
masonry work screen 砖石工程屏幕
masonry work silo 砖石工程筒仓
masonry work skin 砖石工程表层
masonry work strength 砖石工程强度
masonry work structure 砖石工程结构
masonry work style 圬工风格;砖石工程风格
masonry work surfacing 砖石工程表层
masonry work technique 砖石工程技术

masonry work tomb 砖石工程墓穴
masonry work treatment 砖石工程处理
masonry work type 砖石类型
masonry work type of construction 砖石结构类型
masonry work water tower 砖石水塔
masonry work wythe 单层砖石墙厚
mason's adjustable multiple-point suspension scaffold 瓦工用多点可调悬挂式脚手架
mason's adjustable suspension scaffold 瓦工用可调悬挂式脚手架
mason sand 圬工砂
mason's bolster 圬工垫块
mason's brush 圬工刷
mason's chisel 圬工凿
Mason scribe 麦松刻图片
mason's flat-ended chisel 瓦工平头凿;泥瓦工平头凿
mason's float 刮板;镘刀;泥瓦工抹子;抹子;圬工镘刀;圬工镘板;瓦工镘刀
Mason's formula 梅森公式
mason's guild 砖石行会
mason's hammer 圬工锤;砖石锤
mason's hod 砖石工用砂浆桶
mason's hydrated lime 砌筑砌体用熟石灰;砖石(工)用熟石灰;圬工用熟石灰
mason's joint 圬工接缝;砌体砌缝;V形砌缝;灰缝;圬工砌缝
mason's lath 圬工板条
mason's lead 砌砖拉线
mason's level 圬工水平仪;圬工用水准尺;泥瓦工水准仪;泥瓦工水准器;泥瓦工水准尺
mason's lime 建筑石灰;砌筑石灰
mason's line 砖石工准线;泥工线
mason's mark 石工标记
mason's measure 砌块用量测定法
mason's miter 砌体斜接面;仿料角缝的转角石;砌体斜接缝;斜接头;砌体半分接法;圬工斜接面
mason's mortar 砌筑砂浆;圬工砂浆;圬工灰浆
mason's mo(u)ld 石匠型板;泥瓦工用型板
mason's putty 圬工油灰;石工腻子;砌筑灰膏
mason's rule 圬工尺;泥瓦工(用)尺
mason's runway 砌筑脚手架跳板;圬工跳板
mason's sand 建筑用砂
mason's scaffold 砌筑脚手架;圬工脚手架
mason's square 圬工角尺
mason's stop 石工斜接缝;砌体斜接缝;圬工砌体斜接面
Mason's theorem 梅森定理
mason's trap 圬工储水坑;圬工存水坑;圬工存水井
mason's trowel 圬工泥刀
mason's V-joint pointing 墙面V形勾缝
mason's white washing brush 粉刷墙
masonwork 砖石圬工
mas(o)ut 铺路油;重油;黑油
mass 质量;物质;堆;大量
mass ablation rate 质量消融速率;质量烧蚀速率
mass absorption 质量吸收
mass absorption coefficient 质量吸收系数
mass absorption effect 质量吸收效应
mass abundance 质量丰度;质量产额
mass acceleration 质量加速度
mass accounting 容积核算
Massachusetts Institute of Technology 麻省工学院(美国)
Massachusetts method 麻省设计法(一种柔性路面厚度的古典设计方法)
Massachusetts's rule 麻省准则【道】
mass action 质量作用
mass action effect 质量作用效应
mass action law 质量作用定律
mass action principle 质量作用原理
mass addition 质量增加
massage armchair 按摩椅
massage brush 擦身刷
massage room 按摩室;推拿室
massage shower set with soap holder 带皂托的按摩淋浴器
massaging yongquan 擦涌泉
massa intermedia 中间块
mass analyser 质量分析器;质量分析计
mass analysis 质量分析
mass analysis method 质量分析法
mass analyzed ion kinetic energy spectrometry 质量分析离子动能度谱术

mass and rapid transit 快速公共交通
mass appraising 区域评估
mass area ratio 质量面积比;单位面积质量
mass-asphalt 大块(地)沥青
mass assignment 质量数测定
mass attenuation 质量衰减
mass attenuation coefficient 质量衰减系数;质量减弱系数
mass attraction 质量引力;万有引力
mass attraction vertical 质量引力垂线
mass avalanche 大规模崩坍;大规模崩塌;大崩落
mass average value 质量平均值
mass axis 重心轴(线)
Massay formula 梅赛公式
mass balance 质量平衡;平衡重;配重;物质平衡;物料平衡;气态平衡
mass balance calculation 质量平衡计算
mass balance equation 质量平衡方程
mass balance weight 平衡重
mass-basis response 质量响应
mass blasting 大爆破
mass budget 质量收支
mass burning rate 质量燃烧速率;质量燃烧速度
mass calculation 土石方计算;土方(量)计算
mass calculation for earth works 土方计算
mass caving 大冒顶;大崩落
mass cement 普通水泥
mass center 重心;质心;质量中心
mass-centered coordinate system 质心坐标系
mass-centered system 质心坐标系
mass centric coordinates 质心坐标
mass centroid 质量中心
mass characteristic 质量特性
mass/charge ratio 质荷比
mass chromatogram 质(量色)谱图;质量层析谱
mass chromatographic analysis 质量色谱分析
mass chromatography 质量色谱法
mass circulation velocity 质量循环速度
mass client 大量顾客
mass coefficient 质量系数
mass coefficient of reactivity 质量反应性系数
mass collision stopping power 质量碰撞阻止本领
mass colo(u)r 主色;墨色;表光
mass colo(u)ring 原液着色;原料着色
mass communication 大量通信
mass component 质量组分
mass computation 大规模计算
mass concentration 质量浓度
mass concentration factor 质量浓度系数
mass concrete 大体积混凝土;大块混凝土
mass concrete abutment 大块混凝土支座;大体积混凝土桥台;大体积混凝土支座;大块混凝土桥台
mass concrete block 大型混凝土砌块;大型混凝土块(体);大体积混凝土块体
mass concrete-block wall 大型混凝土砌块墙体
mass concrete dam 重力式混凝土坝;大体积混凝土坝;大混凝土坝
mass concrete foundation 大体积混凝土基础;大块混凝土基础
mass concrete invert (干船坞的)大体积混凝土底板
mass concrete overflow dam 重力式混凝土溢流坝
mass concrete pier 大体积混凝土墩;大块混凝土墩
mass concrete pour 大体积混凝土浇注
mass concrete quaywall 大体积混凝土岸壁
mass concrete structure 大体积混凝土建筑;大体积混凝土结构
mass concrete wall 大体积混凝土墙;大体积混凝土岸壁
mass concreting 大体积混凝土浇筑
mass conservation equation 质量守恒方程
mass consumption society 消费社会
mass control 质量控制
mass-controlled oscillator 质量控制振荡器
mass convergence 质量辐合
mass conversion factor 质量换算因子
mass convervation 质量守恒
mass coupling 质量耦联;质量耦合
mass crystallization 大量结晶
mass curing 整体养护(密封容器内绝热养护);密封绝热养护;密闭容器内隔热养护;蒸釜式养护;绝热养护;封闭养护(混凝土件)
mass curve 质量曲线;累积曲线(图);积分曲线图;土积曲线;土方累积曲线
mass curve of storage 蓄水累积曲线
mass cycle 物质循环
mass data 海量数据;大量数据
mass data multiprocessing 大量数据多道处理
mass data processing 海量数据处理
mass decrement 质量减损
mass defect 质量缺陷;质量亏损
mass deficiency 质量亏损;质量不足
mass degradation 质量降解
mass degradation rate factor 质量降解速度因素
mass density 质量密度;集结密度;单位体积质量
mass density of water 水的质量密度
mass detector 质量检测器
mass deviation 质量偏差
mass diagram 质量曲线;累积图;累积曲线(图);径流积分曲线;积分曲线图;土方累积图;土(方)积(累)图;土方叠积图
mass diagram analysis 累积图分析法
mass diagram of runoff 流量累积图;径流累积图
mass diffusion 质量扩散
mass diffusivity 质量扩散率
mass discharge curve 流量累积曲线
mass discrepancy 质量差异
mass discrimination 质量甄别;质量歧视
mass dispersion 质量分散
mass distribution 质量分布
mass distribution function 质量分布函数
mass divergence 质量散度;质量辐散
mass doublet 质量双线
mass drying rate 质量干燥速率
mass eccentricity 偏心
massed glaze 堆釉
massed trials 集聚测验
mass effect 质(量)效应;质量效果
mass ejection 质量抛射
mass-elastic characteristics 质量弹性特性
mass element 成批构件;质量元素
mass emission rate 质量排放速率
mass-energy absorption coefficient 质能吸收系数
mass-energy conservation 质能守恒
mass-energy conversion 质能转换
mass-energy conversion coefficient 质量能量换算系数
mass-energy conversion formula 质能换算公式
mass-energy curve 质能曲线
mass-energy cycle 质能循环
mass-energy equivalence 质能相当性;质能当量
mass-energy equivalence principle 质能相当原理;质能当量原理
mass-energy equivalent 质能当量
mass-energy parabola 质能抛物线
mass-energy relation 质能关系
mass-energy transfer coefficient 质能传递系数;质量能量传递系数
massenfilter 滤质器
mass epidemic 集体流行;大流行
mass equation 质量方程
mass equivalent 质量当量
mass erosion 重力侵蚀;质量顺坡移动;整体侵蚀
masses of gangue 废石块;矸石块
mass etch 粗蚀
mass examination 集体检查;群体检查
mass excavation 大量开挖
mass excess 质量过剩
mass exchange 质量交流;物质交换
mass extinction 动物区系消亡
mass extinction period 集群绝灭期
mass fabrication 大量装配;大量生产;大量制作
mass factor 质量系数
mass file 大存储文件
mass filter 质量过滤器;滤质器
mass float 惯性浮标
mass flow 质量流量;质量流动
mass flow coefficient 质量流量系数
mass flow constant 质量流量常数
mass flow control 质量流控制
mass flow controller 质量流量控制器
mass flow detection 质量流量检测;质量(流量敏感)型检测器
mass flow equation 质量流量方程
mass flowmeter 质量流量计
mass flow-pressure difference relation 质量流量-压差关系
mass flow rate 质量流量;质量(比)流率;物料流速
mass flow rate perturbation 质量流量扰动
mass flow rate sensitive detector 质量流量敏感性检测器
mass flow ratio 质量流量比
mass flow theory 物体流动学说;质量流动学说
mass flux 质量通量
mass force 惯性力;体积力
mass formula 质量公式
mass for stopping grout 止浆岩盘
mass foundation 大块基础;大体积基础
mass fraction 质量分数;质量百分比
mass fragmentography 质量碎片(谱)法
mass freight 大宗货物
mass function 质量函数
mass fusion splicing machine 多路熔接接头机
mass goods 大宗货物
mass hardness 全部过硬
mass haul 大量运输
mass-haul curve 填挖方设计曲线;开挖运输曲线;土方运距曲线;土方搬运设计曲线
mass holograghy 质量全息术
mass housing 成片的住房
massicot 氧化铅;一氧化铅;黄丹;铅黄
Massieu function 马休函数
massif 岩体(深成的);整体;林区;块体;山丘;断层地块;地块
massif nodule 团块状结核
mass-impregnated insulation 整体浸渍绝缘
mass indicator 质量指示器
mass-inductance analogy 质量电感模拟
mass influence coefficient 质量影响系数
massing 块化;集中;密集
massing flower bed 集栽花坛
mass-in-situ concrete wall 现浇大体积混凝土岸壁
mass integral detector 质量累积检测器
mass in working order 装备质量
massive 整块的;块状矿;块状的;巨大的;大块的;大规模(的)
massive abutment 重力式桥墩
massive and rigid structure 大体积刚性结构
massive anode 实心阳极
massive arch 整体拱;实心拱;巨型拱
massive arch dam 重力拱坝;巨型拱坝
massive article 大型制品
massive bedding structure 块状层理构造
massive bitumen 块状沥青
massive block 整块石料;大块石料
massive buttress dam 肋墩坝;大头坝
massive casting 大型铸造;大件铸造
massive cataclastic texture 块状碎裂结构
massive collapse 大块萎陷
massive concrete 重型混凝土;大体积混凝土
massive concrete abutment 整体混凝土桥台;巨型混凝土桥台;整体混凝土支座;巨型混凝土支座
massive concrete dam 巨型混凝土坝
massive concrete foundation 大体积混凝土基础
massive concrete pier 巨型混凝土墩
massive concrete structure 整体混凝土结构;重型混凝土结构
massive construction 整体结构;实体结构;重力式结构
massive construction project 大型工程(项目);大规模建设工程
massive crystalline graphite 块状晶质石墨
massive dam 圬工坝
massive dead 大量死亡
massive design 大量设计;大规模设计;笨重结构
massive dump 大量转储;大量信息转储
massive flood year 大洪水年
massive footing 整体式结构;整体式底脚;块状底脚
massive foundation 大块式基础
massive gypsum 整块石膏;天然石膏
massive-head buttress 大头式支墩
massive-head dam 大头坝
massive layer 大体积浇筑层
massive martensite 大块(状)马氏体
massive masonry wall 重型砖石墙
massive material 大块材料
massive mountain 块状山
massive movement 块状运动
massive object model 重体模型
massive ore 块状矿石
massive Pb-Zn ore 块状铅锌矿石

massive phase 块状相
massive pier 巨大的码头
massive planting 集植
massive plate 大平板
massive pool 块状油藏
massive quay wall 重力式岸壁;巨块岸壁;大体积岸壁;大块体岸壁
massive rock 块状岩;大块岩石
massive section 大体积断面
massive seeding 丛播法
massive structure 重型结构;重力结构;整体结构;块状构造;块体结构;实体结构;大体积结构;大块结构
massive sulfide-type polymetallic deposit 块状硫化物型多金属矿床
massive sulphide-type copper deposit 块状硫化物型铜矿床
massive texture 整体结构;块状组织;块状结构
massive transformation 块状转变
massive type 巨块型
massive wall 重型墙体
massivity 巨块结构;整体性
mass law 质量定律
mass law of sound insulation 隔声的质量定律
mass limit 质量限额
mass load(ing) 质量负载;质量负荷;大量负荷;惯性力;惯性荷载
mass loading curve 质量负荷曲线
mass loading variation 质量负荷变化
mass loss 质量损失;质量亏损
mass loss rate 质量损失率
mass loss segregation 质量层化
mass luminosity curve 质光曲线
mass-luminosity law 质光定律
mass-luminosity-radius relation 质量发光度半径关系;质光半径关系
mass-luminosity ratio 质光比
mass-luminosity relation 质光关系
mass management 集体管理
mass manufacture 批量制造;大批制造;大量制造
mass marker 质量(数)指示器
mass marketing 大规模销售
mass matrix 质量矩阵
mass measurement 质量测量;质量测定
mass media 宣传工具
mass median aerodynamic(al) diameter 质量中位直径
mass median diameter 质量中位直径
mass mement of inertia 质量惯(性)矩
mass memory system 海量存储系统
mass method 大量生产法
mass migration 集体洄游;大规模移民;大规模迁移
mass migratory 大群迁徙
mass mixing ratio 质量混合比
mass mortality 群集死亡;大量死亡
mass motion 整体运动
mass-motion attitude control system 质量移动姿态控制系统
mass motorization 高度汽车化
mass movement 质量运动;整体运动;整体滑动;块体运动;块体移动;集团运动;水团移动
mass movement deposit 整体滑动泥沙
mass movement of sediment 沉积的块体运动
mass number 质量数
mass observation 大量观测
mass of adsorbates 被吸附物质量
mass of adsorbent 吸附剂质量
mass of air 空气质量
mass of atmosphere 大气质量
mass of caprock 盖层质量
mass of concrete 混凝土浆
mass of crust 地壳质量
mass of dam 坝体
mass of data 大量数据
mass of hydrogen atom 氢原子质量
mass of inner core 内核质量
mass of mantle 地幔质量
mass of masonry 大量砖石;圬工块
mass of oceans 海洋质量
mass of outer core 外核质量
mass of pollutant 污染物质量
mass of pollutant remaining 污染物残留质量
mass of profit 利润总额
mass of rock 岩体;岩块

mass of sand 砂堆;砂团
mass of soil 土团
mass of the earth 地球的质量
mass of the electron 电子质量
mass of vibration 参振质量
mass of water 水团;水体
masson pine 马尼松
mass operator 质量算符
mass optic(al) memory 大容量光存储器
mass output 累计出力;累计(产)量
mass parameter 质量参数
mass peak 质量峰
mass percent 质量百分比
mass peristalsis 集团蠕动
mass per unit length of member 构件单位长度质量
mass per unit of power 单元功率的质量
mass per unit volume 单位体积质量
mass planting 集植;群植
mass plate 厚板
mass point 质点
mass polymerization 大块聚合;本体聚合(法)
mass pouring 大量(混凝土)浇注
mass power 单位质量的作用力
mass pressure 质量压力
mass-produced 大量生产的;成批生产的
mass-produced asbestos-cement product 大量生产的石棉水泥制品
mass-produced asbestos-cement ware 大量生产的石棉水泥制品
mass-produced precast concrete ware 批量生产的预制混凝土加工品;大量生产的预浇混凝土制品
mass-produced reinforced-concrete product 大量生产的钢筋混凝土制品
mass-produced sintered ware 批量生产的烧结制品
mass-produced structural units 成批生产构件
mass production 批量生产;大量生产;大规模挖方工程;大规模生产;大规模开挖;成批生产
mass production house 大规模建造住宅
mass-production-scale 大规模生产
mass profile 土方纵剖面图;土方量;挖方与填方纵断面图
mass program(me) 主程序
mass property 整体特性
mass provisioning 大宗供应品
mass purchasing power 群众购买力
mass quaywall 重力式岸壁
mass radiative stopping power 质量辐射阻止本领
mass radiator 物质辐射器
mass radiography 集体X射线检查
mass-radius relation 质量半径关系
mass rainfall curve 雨量累计曲线;雨量累积曲线
mass range 质量范围
mass rapid transit 大众快速交通;大运量快速运输
mass rapid transit railway 高速大量运输铁路;大运量快运铁路;大运量快速交通系统
mass rate of discharge 质量排放流量
mass rate of emission 卸料速度;排放速率;排放(材料)速度
mass rate of flow 质量流量(率);质量流过速率
mass ratio 质量比
mass reactance 质量抗;声质抗
mass recirculation 质量再循环率
mass reflex 总体反射
mass region 质量范围
mass renormalization 质量重整化
mass reproduction 大量复制
mass residue 大量残渣
mass resistivity 质量电阻率;体积电阻率;电阻率
mass resolution 质量分辨率;质量分辨本领
mass resolving power 质量分辨率
mass retaining wall 重力挡土墙;大体积挡土墙
mass river 主河流
mass run 大量生产
mass runoff 累积径流(量)
mass runoff curve 径流累积曲线
mass scale 质量数标度;质量标度
mass scan 质量扫描
mass scattering coefficient 质量散射系数
mass screening 群集筛选
mass screw 平衡螺钉
mass selection 集团选择;混合选择;群选(择)
mass selling 集体销售
mass-sensitive deflector 质量敏感分离器

mass-sensitive quantity 质量灵敏值
mass separation 质量分离
mass separator 质量分离器;同位素分离器
mass sequential insertion 大量顺序插入
mass shell 质量壳
mass-shifting attitude control system 质量移动姿态控制系统
mass shooting 大(体积)爆破;集爆;齐发爆破
mass sliding 整体滑动
mass slip 地块滑动
mass soil movement 大体土体移动
mass solid wall of concrete 整片大体积混凝土岸壁
mass spectral search system 质谱检索系统
mass spectrogram 质谱图;质谱
mass spectrograph 质谱仪
mass spectrographic analysis 质谱分析(法)
mass spectrography 质谱分析(法);质谱法
mass spectrometer 质谱仪;质谱测定计;质量分光计
mass-spectrometer by car 车载质谱仪
mass-spectrometer detector 质谱仪探测器
mass-spectrometer in laboratory 实验室质谱仪
mass-spectrometer leak detector 质谱检漏仪
mass-spectrometer measurement 质谱测量
mass-spectrometer tube 质谱仪管
mass spectrometric(al) analysis 质谱分析(法)
mass-spectrometric(al) computation 质谱分析计算
mass-spectrometric(al) data 质谱数据
mass-spectrometric(al) detection 质谱仪检定;质谱检测
mass-spectrometric(al) differential thermal analysis 质谱差热分析
mass-spectrometric(al) isotope-dilution analysis 质谱同位素稀释分析
mass-spectrometric(al) method 质谱法
mass-spectrometric(al) stable isotope-dilution analysis 质谱稳定同位素稀释分析
mass-spectrometric(al) technique 质谱技术
mass-spectrometric(al) thermal analysis 质谱热分析
mass spectrometry 质谱学;质谱分析(法);质谱法;质谱测量法;质谱测定法
mass spectrometry data center 质谱分析数据中心
mass spectroscope 质谱仪
mass spectroscopy 质谱(学)
mass spectrum 质谱
mass spectrum digitizer 质谱数字转换器
mass spectrum line 质谱线
mass splitting 质谱线劈裂
mass-spring-dashpot combination 质量弹簧缓冲器组合
mass-spring-dashpot system 质量-弹簧-阻尼(器)体系
mass-spring system 质量-弹簧系统;质量弹簧体系
mass stopping power 质量阻止本领
mass storage 海存;大容量存储器
mass storage control 大容量控制
mass storage control system 大容量控制系统
mass storage dump program(me) 大容量存储[贮]器转储[贮]程序
mass storage of gas 大宗储气(季节调峰用的)
mass storage system 大容量存储系统【计】
mass storage system communicator 大容量存储系统通信程序
mass structure 大体积结构
mass surface 质量面
mass-surface ratio 体积表面积比
mass survey 群体调查;普查
mass synchrometer 高频质谱仪;高频质谱计;同步质谱仪;质量分析仪
mass-temperature relation 质量温度关系
mass term 质量项
mass thickness 质量厚度
mass throughput 大流量
mass-to-charge ratio 质荷比
mass-to-electric(al) charge ratio 质荷比
mass-to-light ratio 质光比
mass tone 主色;质色调
mass transfer 质量转移;质量传移;质量传递;物质传递;传质
mass transfer coefficient 液膜传质系数;质量交换系数;传质系数
mass transfer efficiency 传质效率

mass transfer in gas phase 气相传质
mass transfer in liquid phase 液相传质
mass transfer mechanism 物质传递机理
mass transfer rate 质量转移率;传质速率;质量迁移率;质量交换速度
mass transfer reaction 质量转移反应
mass transfer resistance 传质阻力
mass transfer theory 质量转移理论;传质理论
mass transfer valve plate 传质浮阀塔盘
mass transit 公共汽车运输;公共交通;大量运输
mass transit facility 公共交通设施
mass transit passenger turnover 公交乘客周转量
mass transit rail system 大运量轨道交通运输系统
mass transit railway 大运量轨道交通
mass transit service 公共交通公司
mass transit system 大运量客运系统;大运量交通系统
mass transmission 质量传递
mass transport 质量输运;质量输送;质量传递;(波浪的)质点运动;集体运输;水质点运动;水体转移;水体迁移
mass transportation 公共交通;大量运输;大量客运
mass transportation bus service 公交企业;公共汽车交通企业
mass transportation facility 公共交通设施
mass transportation loading zone 公共车辆站台
mass transport mechanism 质量传递机理;物料迁移机理
mass transport of waves 波浪的质量运移
mass transport process 传质过程
mass transport resistance 传质阻力
mass transport velocity 质量输送速度;水质点运动速度
mass trial 大量生产试验
mass-type foundation 整体基础
mass unbalance 质量不平衡
mass unbalance compensator 质量不平衡补偿器
mass unit 主部件;质量单位
mass unit weight 块体容重
mass velocity 质量速度;质量流(动)速(度)
mass-velocity ratio 质量速度比
mass volume 土方累计体积;土方累积体积
mass wall 现浇大体积混凝土岸壁;整体墙
mass wasting 质量销蚀;矿体废物;物质坡移;物质波移
mass waterproofer 大块防水材料
mass yardage distribution 土方分布数量
mass yield 质量产额
mass-yield curve 质量产额曲线
mass-yield distribution 质量产额分布
mast 坚果;桅杆;电杆;船桅
mastaba(h) 石墓室(古埃及);露天石头平台(伊斯兰教建筑)
mastabah tomb (古埃及的一种长方形的)平顶斜坡坟墓;石樟坟墓
mast and sail 桅帆
mast antenna 桅杆式天线;铁塔(式)天线
mast arm 桅杆;灯具悬臂
mast assembly of forklift truck 叉式装卸车门架系统
mastax 破碎器
mast ball 旗杆顶饰;桅顶球
mast band 桅箍
mast bed 木船的桅座板
mast bench 桅座板(小艇)
mast bottom 起重杆支柱底座;吊杆支座底座
mast cap 紧桅箍
mast clamp 桅铰链
mast climbing work platforms 爬杆式工作台架
mast collar 桅圈
mast colo(u)r paint 桅色漆
mast crane 柱形塔式起重机;桅式起重机;桅杆式起重机;桅杆(式)起重机;起重机双脚架
mast derrick 桅式起重机;桅式吊车
masted 装桅
master 主要的;主管的;主导装置;控制者;上位机;底版
master action 主销后倾作用
master address 基本地址
master agreement 主要合同
master air suction valve 主进气阀
master air valve 主空气阀
master airway bill 空运主提单
master alloy 主合金;母合金

master alloying 主合金
master altimeter 校正用高度计
master altitude 校正高度
master amplifier 分子放大器
master and block valve 主油管堵塞阀
master angle 主销后倾角
master angular reference 主角度基准
master answer sheet 主应答表
master antenna 主天线;共用天线
master antenna distribution system 主天线分配系统
master antenna television 共用天线电视
master antenna television system 主天线电视系统;共用天线电视系统
master arm 主动臂
master ball 标准球
master bar 校对棒;标准棒
master batching 母料着色;母料配料
master beam spreader 子母式吊具
master bedroom 船长卧室;主人卧室
master bevel gear 主锥齿轮
master bistable 主控双稳态
master blade 基准刀齿
master blank 样板毛坯
master block 模座
master board 主控制(仪表)板;主插件板
master body 标准样件
master box 城市火警箱
master bracket 主支架
master budget 总预算;主要预算
master builder 营造工长;营造师;建筑师;建筑承包商;施工工长;承包者;承包商;承包人
master bushing 主衬套;主销衬套;端锁衬套
master busy 主线占线
master cam 主凸轮;靠模凸轮
master card 万能卡;原版卡片;主卡片;导卡
master carpenter 木匠工长;木工师傅
master cast 主模(型)
master castor 主销后倾
master catalogue 主目录
master central timing system 主中央计时系统
master certificate 船长证书
master changeover switch 万能转换开关
master change record 主要更改记录
master characteristic(curve) 主特性曲线
master check 校对
master city plan 城市总体规划
master clear(switch) 主清除开关
master clock 主(时)钟;母(时)钟;标准钟
master clock frequency 主时钟频率
master clock-pulse generator 主时钟脉冲发生器
master clock system 母子钟系统
master clutch 主离合器
master clutch brake 主离合器闸
master clutch release yoke 主离合器分离叉叉
master cock 总(水)龙头;总开关;操纵开关
master collet 主套筒
master compass 主罗盘;主罗经
master computer 主计算机;主电子计算机
master connecting-rod 主连杆
master construction schedule plan 施工总进度计划
master container 主集装箱
master contract 总承包;主要合同;主约;总合同;交易主约
master contrast 主对比度
master contrast control 主对比度调整
master contrast controller 对比度主控器
master control 主调整;中央控制;总控制;主(总)控制;主控(台);中心控制;整体控制
master control automation 主控自动化
master control board 主控制台;主控(制)板;中心控制台
master-control center 总控制中心;主控中心
master control circuit 主控线路
master control console 中央控制台;主控制台;主操纵台
master control gate 主控闸门
master control interrupt 主控(制)中断
master controller 总控制器;主控(制)器;主控机;中心控制器
master controller handle 主控手柄
master control lever 主控制杆
master control operator 主控操作技术员

master control panel 主控制板;主控制盘;主控屏
master control program(me) 主控制程序
master control program(me) procedure 主控程序工序
master control room 总控制室;主控室;中央控制室;中心控制室
master control routine 主(控)制程序
master control switch 主控开关
master control unit 主控制器
master cooperation agreement 公共住房计划的总协议
master coordinate 主坐标
master copy 原本;分色清绘底图;标准本
master curve 样板曲线;主曲线;量板;通用曲线
master cylinder 总液压缸;主液压缸;主(油)缸;制动总泵;控制缸
master cylinder body 主汽缸体
master cylinder boot 主缸罩
master cylinder bottom valve 主缸底阀
master cylinder bracket 主缸托架
master cylinder connecting link 主缸活塞推杆
master cylinder cover 主缸盖
master cylinder filler cap 主汽缸加油口盖
master cylinder filler cap gasket 主缸加油口盖垫片
master cylinder head gasket 主缸盖垫片
master cylinder inlet connection 主汽缸输入接头
master cylinder lever 主缸杆
master cylinder lever bushing 主缸杆衬套
master cylinder lever shaft 主缸杆轴
master cylinder outlet connection 主汽缸放出接头
master cylinder outlet fitting gasket 主缸出油接头垫片
master cylinder outlet plug 主缸出口塞
master cylinder piston 主缸活塞
master cylinder piston cup 主缸活塞皮碗
master cylinder piston push rod 主缸活塞推杆
master cylinder piston return spring 主缸活塞回动簧
master cylinder piston secondary cup 主缸活塞副皮碗
master cylinder plug 主缸塞
master cylinder secondary cup 主缸活塞副皮碗
master cylinder supply tank 主缸供给箱
master cylinder valve 主缸阀
master cylinder valve seat 主缸阀座
master cylinder valve spring 主缸阀簧
master data 主(要)数据;基本数据
master data base 基本数据库
master data sheet 主数据表
master deed 主要契约;总契约
master design 总体设计;总设计图
master design drawing 总设计图
master designer 总设计师
master design of waterway project 航道工程总体设计
master dial 主度盘
master die 标准模;标准板牙
master directory 主目录
master disk 主盘
master distance indicator 主航程指示器
master diversion airfield 主备降(机)场
master dividing gear 标准分度齿轮
master drain 干渠;干管
master drawing 样图;标准图
master drive 主传动
master driver 主驱动器;主驱动电动机;主激励器;主传装置;主传动件
master driving program(me) 推算程序
master drum 控制筒
master electrician 电气技师
master equation 主方程
master equipment list 总装备表;主要设备清单
master fader 主衰减器
master fault 主断层
master fiche 缩微胶片(复制用)母片
master file 主资料;主文件;主外存储器
master file directory 主文件目录
master file index 主文件索引
master file inventory 主文件清单
master file tape 主文件带
master file updating program(me) 主文件修改程序
master film 底片胶卷
master film positive 原版正片

master financial budget 财务总预算
master fixture 基准型架
master flip-flop 主触发器
master form 原模;靠模
master format 标准格式
master for spirit duplicating 酒精复写板
master fracture 主断层
master frequency 主频率;基本频率
master frequency meter 总频率计;积量频率计
master gain 主控增益
master gain control 主增益控制
master gas fitter 煤气管道安装技师
master ga(u)ge 主规;校正标准仪器;校验计;标准仪表;标准压力表;标准量计;标准(量)规
master gear 主齿轮;基准齿轮;标准齿轮
master gear checking instrument 标准齿轮校正仪
masterglass negative 玻璃格网板
master grind 研磨(主)色浆
master group 主群
master group bank 主群排
master group modulation 主群调制
master gull(e)y 集水渠;集水(主)沟
master gyrocompass 陀螺罗经主仪;主陀螺罗经
master gyroscope 主陀螺仪
master handle 主操纵杆
masterhead 柱顶
master history file 主过去文件
master index 主索引
master index map 主索引图
master indicator 校准指示器
mastering recording 原始记录
master instruction tape 主指令带;主程序带
master instrument 主控仪表
Master International Frequency List 国际频率总表
Master International Frequency Register 国际频率登记总表
master jaw 卡爪座
master jig 总装装配架
master joint 主(要)节理
master key 总钥锁;总电键;主钥匙;主键;万能钥匙
master-keyed lock 万能钥匙系列锁;总钥匙系列锁;总钥匙能开的锁;总钥锁
master-keying 设计制作万能钥匙型;万能钥匙型
master-keying system 主键控制系统
master layout 标准规划图
master lease 总租约;一揽子租赁;主租借契约;灵活租赁
master level 主水准器
master level control 主电平控制
master lever 主操纵杆
master lever tatchet 主操纵杆棘齿板
master lift control 主升降控制
master line 分水线
master line test console 主线路测试台
master link (吊具的)主连接杆;主联杆;主连杆;履带闭合链节
master list 总清单
master lode 巨矿脉
master magnet valve 主磁力阀
master map 原图;主图;底图
master mariner 船长
master mark feature 主数据表检查器
master mask 母版
master mason 坏工领班;熟练坏工;熟练泥瓦工;石工工长
master matrix 主盘模型
master mechanic 总机械师;技工长;机长;熟练技师
master meter 煤气总表;基准仪表;标准仪表
mastermeter method 标准仪表比较检验法
master mode 主控方式;主方式
master model 母型
master modem 主调制解调器
master monitor 主监视器;主监控器
master mortgage 主要抵押
master motor 主驱动电动机
master mo(u)ld 标准模型;原始模型;一次过渡模;原版;母型
master multivibrator 主控多谐振荡器
master negative 原底片;主底片;金属头版
master network 总网络
master nozzle 校对喷嘴
master of architecture 建筑学硕士
master of computer science 计算机科学硕士
master of engineering 工程硕士;技术科学硕士

master of science 理科硕士
master of the hold 随船理货员
Master of the King's Masons to Windsor Castle 温莎堡皇家建筑工匠长
master of the work 工程的主管人
master operating system 主操作系统
master operation 主(要)工序;主操作
master operational console 主操作控制台
master operations controller 主操作控制装置
master oscillator 主(控)振荡器
master oscillator dower amplifier 主控振荡器功率放大器
master oscillator frequency 主(控)振(荡器)频率
master oscillator power amplifier 主控功率放大器
master-oscillator radar system 主振器控制雷达系统
master oscillator system 主振荡器系统
master output tape 标准幅度带
master overhaul 大修
master-pallet 大托盘
master parallax table 主视差表
master part master device 主件
master parts reference list 主要部件参考表
master pattern 原始模型;主模;金属芯盒;母型;母模;标准模型
master pedestal 主台座
master pendulum 主摆;母摆
master phase 基本相位
master picture monitor 图像主监视器
master piece 样件;杰作
master piece of work 杰作;杰出工作
master pilot 总信号灯;主控驾驶仪
master pilot lamp 总指示灯
master pin 主销;中心立轴
master plan 总图;总体规划(图);总体布置图;总平面(图);总计划;规划总图
master plan for development of city 城市发展的总图;城市发展的总体规划
master planning 城市总体规划
master planning budget 主计划预算
master plan of port 港口总体规划
master plan position indicator 主控平面位置显示器;雷达主平面位置显示器
master plate 靠模板;测平板
master policy 主要保险单
master porter 搬运工(人)(指码头作业工人)
master positive 主盘正片
master power switch 主功率开关
master pressure controller 主压力控制器
master print 拷贝原片
master printer 主打印机
master prism 标准棱镜【测】
master process 主进程
master processor program(me) 主处理程序
master production schedule 总生产进度表
master profile 标准剖面图
master profile template 靠模样板
master program(me) 总(工程)计划;主(要)程序;大纲
master program(me) clock 母程序钟
master program(me) controller 主程序控制器
master program(me) file 主程序文件
master program(me) tape 主程序带
master programming schedule 主程序日程表
master project schedule 控制性工程进度
master pulse 主台脉冲;主(控)脉冲;主导脉冲
master push button panel 主控按钮盘;主按钮板
master rack gear 主齿条(机构)
master radar station 主雷达站
master radio antenna 主无线电天线;无线电主天线
master recession curve 典型退水曲线
master record 主数据;主记录;基本记录
master reference 主基准
master reference machine 主参考机;控制底版参考机
master regulating control valve 主控调节阀
master relay 主控继电器
master remote controller 主遥控器
master reset 主复位
master retarder 主控减速器;主缓行器;主扼流圈
master reticle 掩模原版
master rheostat 主变阻器
master rig 总装配架
master river 主河;干流;主要河流

master rod 主杆
master roller 触轮
master rotor 主滚筒
master route chart 总工艺卡
master routine 主程序;执行程序;管理程序
Master's 2-step exercise test 二阶运动试验
master sample 标样品;标准样品
master scale 标准刻度;标准秤
master schedule 总日程表;总进度计划;主要作业进度表;主要作业表;进度计划总表
master scheduler 主调度程序
master scheduler task 主调度程序任务
master scheduling 总日程表;主调度
master screw 标准螺旋
master-secondary controller 次级主控制器
master sensor 主传感器
master sequencer 主程序装置
master serial number 总编号
master set 校正调整;校对调整
master's expense 船长费用
master sheet 楔座;原因;原文;原图
master shift 主偏移
master side 主动端
master signal 主信号;主控信号
master skew tape 标准扭斜带
master-slave 主从
master slave accumulator 主从累加器
master-slave computer system 主从计算机系统
master-slave configuration 主从配置
master-slave flip-flop 主从触发器
master-slave manipulator 主从操纵器;从动式机械手;仿效机械手
master-slave mode 主从(方)式
master-slave operation 主从运转
master slave relation 主从关系
master-slave spreader 主从式吊具
master-slave synchronization 主从同步
master-slave system 主从系统
master slice 主控薄片
master's licence 船长执照
master source program(me) 主源程序
master space 主宰空间
master spark gap 主控电火花隙
master specification 主要规范
master speed tape 标准速度带
master spring leaf 主弹簧片
master's protest 海难抗议书
master's report 海事报告
master stacker 主堆垛机
master staff instrument 基本路签机
master start button 主控起动按钮
master starting switch 总起动电门
master start-stop button 主控起闭按钮
master state of control processor 控制处理机主态
master station 总站;总机;主站;主控(制)站;主控台;主发射台;主测点
master stock book 材料总登记簿
master straight edge 检查用直度规
master stratum 主要地层
master stream 主要河流;主流;干流
master street plan 街道总图;街道总体规划
master subswitch 主辅助开关;主控分开关;校准用副开关
master suite 主要序列
master supervisory and alarm frame 集中监视警报系统
master supervisory and test console 主监测台
master switch 主(控)开关;总开关;主控寻线机;主转换开关;校准用转换器
master switching 主控切换
master switching center 主转换中心
master synchronization pulse 主同步脉冲
master synchronizer 主同步器;主时钟;主脉冲发生器
master system tape 主系统带
master table of contents 总目录
master tablet instrument 基本路牌机
master tap 标准丝锥;板牙丝锥
master tape 主带;主程序带
master television antenna 主电视天线
master tempering curve 回火基本曲线
master test station 主试验站;主试验台
master timekeeper 主计时装置
master timer 主要定时装置;主要定时开关;主时

钟;主定时器
master timing device 总定时设备
master timing frequency 主定时频率
master timing pulse 主控定时脉冲
master timing station 主定时台
master timing system 主定时系统
master tool 样板工具;基础工具
master tools of industry 工业加工机床
master-to-slave tape drive 主从带式传动
master track link 履带主链节
master track pin 链轨销
master track section 履带总链节
master traffic count 主要交通计数
master traffic plan 主(要)交通设计
master transmitter 主发射机
master trip 主停车装置
master unit 主台;主部件
master valve 总阀;主(导)阀;导阀
master variable 主变量;控制变数
master vector 主矢量
master viscometer 主黏度计;标准毛细管黏度计
master volume 主卷
master warniag signal 主警告信号
master well 主源泉
master wheel 主操纵轮;大齿轮
master work 杰作
master workstation 工作总站
master wrench 总扳手
mastery 精通
master zero key 总置零键
mast extension 天线扩展;伸缩杆;套管式天线杆
mast frame 塔架;杆架
mast ga(u)ge 支柱侧面限界
mast hasp 桅铰链
masthead 杆顶;桅顶;柱顶
masthead aerial 桅顶天线
masthead angle 桅顶(仰)角
masthead band 桅顶箍
masthead height 桅顶高度
masthead light 前桅灯;桅(顶)灯
masthead lookout 桅顶瞭望者
masthead man 桅顶瞭望者
masthead mounted 桅装(雷达收发机装于雷达天线架上)
mast heel 桅根
mast hole 桅孔
mast hoop 桅箍
mast hounds 桅肩
mast house 桅屋;桅室
mastic 胶剂;胶泥;胶合铺料;砂胶;地板蜡
mastic adhesive 玛琋脂胶黏剂
mastic asphalt 沥青胶浆;沥青玛琋脂;沥青胶泥;石油沥青胶浆;石油玛琋脂;地沥青玛琋脂
mastic asphalt beam 沥青玛琋脂涂面的梁
mastic asphalt cauldron 沥青砂胶煮釜;沥青玛琋脂煮釜
mastic asphalt cooker 沥青胶熔制锅;石油沥青胶熔制锅
mastic asphalt finisher 沥青玛琋脂修整机
mastic asphalt floor topping 沥青玛琋脂的楼板面层
mastic asphalt for jointless floor(ing) 无缝地板用的沥青膏
mastic asphalt jointless floor(ing) 沥青无缝地板
mastic asphalt mixer 沥青玛琋脂搅拌机
mastic asphalt mixer transporter 沥青玛琋脂搅拌输送车
mastic asphalt roof(ing) 沥青玛琋脂屋面
mastic asphalt surfacing 沥青玛琋脂胶浆面;沥青砂胶脂铺面;沥青胶(脂)面层;地沥青砂胶面层
masticate 捏扣;撕捏
masticated rubber 人造橡胶
masticator 揉混机;割碎机;立式黏土搅拌机;素炼机;撕捏机
masticatory movement 咬合运动
mastic based on oil 油基脂胶
mastic bed 砂胶层
mastic-bitumen rubber 沥青砂胶橡皮混合料;沥青砂胶脂橡皮混合料
mastic block 胶泥块;砂胶块
mastic cement 水泥玛琋脂;胶脂水泥;胶质水泥;胶黏水泥;玛琋脂水泥;水泥(砂)胶
mastic coating 厚浆涂料
mastic composition 胶的成分
mastic compound 胶黏材料;玛琋脂合成物
mastic cooker 砂胶加热锅;地沥青砂胶加热锅
mastic cord 涂厚浆式绳索;玛琋脂线;玛琋脂绳
mastic cushion 砂胶垫层
mastic filler 玛琋脂填(塞)料;玛琋脂玛琋脂填料;砂胶填料
mastic floor 砂胶地面
mastic flooring 胶脂地面;砂胶铺(地)面
mastic-grouted surfacing 灌玛琋脂路面;沥青砂胶贯入式路面
mastic gum 乳香;胶黏剂;胶黏脂;胶黏水泥;玛琋脂
mastic in drops 成滴的胶
mastic insulation 玛琋脂绝热层;玛琋脂隔热层;玛琋脂保温
mastic joint(ing) 玛琋脂接缝;胶缝;砂胶勾缝;砂胶接缝
mastic joint sealer 胶缝密封剂
mastic macadam 砂胶碎石路
mastic melting boiler 熔化玛琋脂锅炉
mastic melting kettle 玛琋脂熔锅
mastic membrane 胶膜
Masticon 玛琋脂康(一种装配钢窗玻璃的玛琋脂)
mastic pavement 玛琋脂路面;沥青砂胶铺面;沥青砂路面
mastic penetration pavement 玛琋脂贯入式路面
mastic pointing 胶勾缝
mastic process 沥青膏浆表面防水处理方法
mastics 厚涂涂料;乳香
mastic sealer 密封膏;玛琋脂密封剂
mastic seal(ing) 胶泥密封;砂胶密封;密封油膏;玛琋脂密封
mastic shrub 乳香黄连木
mastic spreader 铺胶器;玛琋脂涂布器
mastic system 黏结剂体系
mastic tree 乳香黄连木
mastic trowel 腻子泥刀;胶黏水泥
masting 装桅;船桅总称
mast jacket 转向柱套管
mast ladder 桅梯
mast lamp 桅灯
mast lift truck 门架式叉车
mast light 桅灯
mast mold 桅杆模型
Mastobas 古埃及墓上线结构
mast of an overhead line 架空线桅杆
mast opening 桅孔
mast paint 桅杆漆
mast partner 桅孔加强板
mast pedestal 夹桅板
mast section 桅杆断面
mast shoulder 桅肩
mast stack(er) 立柱吊重式堆垛机
mast step 桅座
mast switch 杆上开关
mast table 桅台;桅杆柱脚
mast tackle 桅杆绞辘
mast thwart 桅座板
mast timber 桅木;桅杆材
mast tower attachment 塔臂配置
mast truck 桅帽;桅冠
mast trunk 夹桅板
mast wedging 桅楔
mast with arm (照明的)灯具悬臂;臂式起重杆;臂式起重机
mast with double cantilever 双悬臂支撑杆
mast with twin cantilevers 双悬臂支撑塔杆
mast yoke 下桅帽
Masuda-Coryell diagram 马苏达-科勒尔图
masur birch 斑纹桦木
masut 铺路油;重油;黑油
masutomilite 锰锂云母
masuyite 橙红铀矿
mat 席子;吸垫;支柱底板;再生顶板;芦席;垫(子);柴排;保护网罩;板坯
matai 浅褐色黑松(新西兰产)
matal deactivator 金属钝化剂
Mataline 钴铜铝铁合金;麦塔林钴铜铝铁合金
matallic film 金属膜
Matanuska wind 马塔牛斯加风
mat-awning 凉棚
mat-backed film 毛背面胶片
mat base 垫层
mat-boat 护岸工程的船(满载柴枝的连串小舟)
Matbro swing forklift 马特劳摆式叉车装载车
match 熏船燃料;相配;竞赛;火柴;核对;匹配;配比
match arc 匹配弧
match assemble welding 装配焊接
match assembly welding 装配焊接
match block 火柴杆坯料
match board 型板;假型板;模板;合榫板;企口板;拼花板;拼合板
match board door 拼板门
match boarding 镶拼花板;铺企口板;拼板(工作);舌槽企口板
match boarding machine 镶板机;企口镶板机
match-box-type of flat slide valve 火柴盒式滑阀
match by map 利用地图行进
match casting 镶合浇制;镶合浇铸
match casting method 镶合浇制法;配合浇制法
match condition 符合条件;匹配条件
match die process 对模法
matched aggregate 配合的集料;配合的骨料;配好的骨料
matched asymtotic expansions 匹配渐近展开式
matched attenuator 匹配衰减器
matched beam 匹配束
matched bearing 成对安装轴承;成对轴承
matched board 拼花板;企口板;拼合板
matched bolometer detector 匹配辐热检波器
matched bolometric(al) detector 匹配式热照射检测器
matched ceiling 企口板平顶;企口板顶棚;镶配天花板;拼板天花
matched cladding fiber[fibre] 匹配包层光纤
matched coil 匹配线圈
matched colo(u)r 调配色
matched column 匹配柱
matched comparison 匹配比较
matched condition 匹配条件
matched conic(al) method 匹配圆锥线法
matched data 匹配数据
matched die mo(u)lding 配套硬模型;对模成型;对模模具
matched edge 检验边;配合边缘
matched filter 匹配滤波器
matched filter detection 匹配滤光片探测
matched filtering 匹配滤波
matched floor 镶配地板;企口地板;拼花木板;拼花地板;拼合地板;配合地板
matched generator 匹配振荡器
matched group 配比组
matched impedance 匹配阻抗
matched impedance antenna 匹配阻抗天线
matched joint 合榫;企口接缝;舌(接)缝;企口接合;舌槽接合
matched lens 匹配透镜
matched line 匹配线
matched load 协调荷载;匹配荷载;匹配负载
matched loading of dangerous goods 危险货物配装
matched loading table of dangerous goods 危险货物配装表
matched loading to separate one consignment from another 隔离配装
matched load power 匹配负载功率
matched load resistance 匹配电阻;匹配负载电阻
matched lumber 企口接材;企口配合木材
matched material 配合的材料
matched metal mo(u)lding 金属对模成型
matched mo(u)ld 合模
matched-mo(u)ld forming 对模模压成型
matched orders 相配订单
matched output 匹配输出
matched pair 配对
matched pairs design 配对设计
matched parting 凸凹分型面;双面型板分型面
matched photo 配对像片
matched picture 配对像片
matched print 对像片
matched pulse intercepting 匹配脉冲拦截
matched receiving antenna 匹配接收天线

matched roof board 企口屋面板
matched samples 相等样品
matched seal 匹配封接；等膨胀接封
matched shingle 配合的屋顶板
matched study 配对调查
matched sub-floor 企口地板的底层
matched survey 配对调查
matched termination 匹配终端器
matched test tube 匹配试管
matched timber 配合的木材
matched transmission line 匹配的传输线；匹配传输线
matched veneer 配合的贴面板；对纹拼板
matched volume 匹配体积
matched wood 配合的木材
matcher 刨床；装配工人；制榫机；匹配器；四面刨
matcher-selector-connector 匹配选择器
matches 引火材料
match exponents 对阶
match gate 同门
match gearing 配合齿轮对
match grinding 配磨自动定时磨削
match hook 双抱钩
matching 选配；拼配；拼板（工作）；配套；配合；对准；对纹拼合；符合比较；核对；匹配；配比
matching accuracy 匹配准确度
matching amplifier 匹配放大器
matching and grooving 企口加工；舌榫加工
matching at the boundary 晶界匹配度
matching autotransformer 匹配自耦变压器
matching bond area 匹配焊接面积
matching box 匹配箱；匹配器；匹配盒
matching button 配色纽扣
matching can 匹配外壳；匹配罐
matching casting 密贴浇铸（法）
matching cell 匹配盒
matching circuit 匹配电路
matching coaxial cable 匹配同轴电缆
matching coefficient 匹配系数
matching components characteristics 匹配元件特性
matching concept 配合观念
matching condition 匹配条件
matching control 自动选配装置
matching convention 配合惯例
matching criterion 匹配准则
matching current 匹配电流
matching delay-line 匹配延迟线
matching device 匹配装置
matching diaphragm 匹配窗口片
matching disc 匹配板
matching draft 匹配模锻斜度
matching element 匹配元件
matching equiment 配套设备
matching error 匹配误差；配搭误差
matching field 配比信息组
matching filtering 匹配滤波
matching flange 接合凸缘
matching gem 拼合石
matching grant 相应赠款
matching hierarchy 匹配阶层
matching hole 销孔；装配孔
matching impedance 匹配（用）阻抗
matching iris 匹配可变光阑
matching item 项目匹配法
matching joint 舌槽接（合）；企口接合；合榫
matching line 配准线
matching load 匹配负载
matching magnet 匹配磁铁
matching method 选配法；配合法
matching method with well side sounding curve 井旁测深曲线对比法
matching network 匹配网络
matching of expansion coefficient 膨胀系数匹配
matching of exponents 对阶
matching of load 负载匹配
matching of marker bed 标志层对比
matching of pulses 脉冲匹配；脉冲刻度校准
matching of sheet edges 地图接边
matching of stages 级间匹配压气机
matching of successions 序列对比；层序对比
matching of tyre 轮胎配合
matching of waveguide 波导匹配
matching operation 匹配运算；配对操作

matching optics 匹配光学
matching parts 配件；配合件
matching pillar 匹配棒
matching plane 开槽刨；企口刨
matching plane badger plane 槽刨
matching plug 耦合元件；匹配插头
matching point 匹配点；平衡工作点；同名点
matching post 匹配棒
matching principle 匹配原理
matching problem 匹配问题
matching pulse 匹配脉冲
matching punch cards 匹配穿孔卡；配对穿孔卡
matching record indicator 匹配记录指示器
matching requirement 装配要求；配合条件
matching resistance network 匹配用电阻网络
matching seam 配色线缝
matching section 匹配节；匹配段
matching sizing 配磨自动定尺寸
matching stitching 配色线缝
matching storage hierarchy 匹配的存储体系
matching structured object 结构化事物的匹配
matching stub 匹配短截线
matching study 配比研究
matching surface 配合面
matching technique 匹配技术
matching transformer 匹配变阻器
matching transition 匹配过渡
matching trim 配合密封面
matching triplet 匹配三合透镜
matching unit 配件；连接器；匹配元件；匹配装置
matching window 匹配窗；调配膜片
matching word 符合字
match joint(ing) 槽舌接口；合榫；舌槽接合；企口缝；凸凹接头
match line 检验边；对正线；对口线
match-lining 覆面板；企口板
match-making machinery 火柴制造机
match mark 模线缝
match marking 配合记号；配合符号；装配标记；装配编号；装配符号
match operation 符合运算
match plane 合笋刨；槽刨；开槽刨；合准刨；边刨
match plate 模板；双面型板；双面模版；对型板
match plate dies 模板铸模
match plate pattern 双面模（板模）
match point 配合点
match sand 假型砂
match sealing 匹配封接
match stand 火柴插
match stock 火柴用材
match strip 接边带
match-terminated line 匹配负载线；负载匹配传输线
match the colo(u)r 配色
match time 匹配时间
match together 配套；重合
match up 相配；配合；适应；适合；符合
match wax 火柴蜡
match wheel 配合轮
matchwood 细木片；碎木
mat coal 暗煤
mat coat 保护层；面层；罩面；无光泽面层
mat-covered pavement 有保护层的路面
mat crystal 席状晶体
mat curing 覆盖养生；覆盖养护
mat cutting 毛面刻花
mat dike 柴排堤坝
mate 啮合部分；伙伴；匹配；配对物；配对零件；搭配
mate bending strength 等同抗弯强度
mate block 成对分割块
mated 成对的
mated gear 成对齿轮
mated pair 配磨
Mateke reversed polarity chron 马特基反向极性时
Mateke reversed polarity chronzone 马特基反向极性时间带
Mateke reversed polarity zone 马特基反向极性带
matelasse 马特拉赛
mat enamel 无光瓷漆；无光泽的瓷釉
mat enamel paint 无光瓷漆
mate of the watch 舱面值日员
material 资料；物资；物料；材料；消光涂料
material ablation 材料烧蚀
material absorption 材料吸收
material account 材料账户

material accountability 物料衡算量
material accounting 材料核算
material accumulation 物质性蓄积
material adjustment 物资调配；材料调剂
material advanced 垫用原料
material age 材料使用期限
material ag(e)ing laboratory 材料老化试验室
material agglomeration 物料结团
material aggressive to concrete 对混凝土有侵蚀性的材料
material air interface 料面；物料与空气的接触面
material allocation scheme 物资分配制度
material analysis 材料分析
material analysis data 材料分析数据；材料分析资料
material and data of investigation 勘察资料与数据
material and energy pool 材料和能源供应源
material and equipment for leak stoppage 堵漏器材
material and labo(u)r 工料
material and labo(u)r analysis 工料分析
material and labour cost analysis 工料分析
material and service 材料及劳务
material and supply 物资与供应
material and work schedule 材料和工作的安排
material application plan 物资申请计划
material appropriation 材料调拨
material arm 材料臂
material assets 有形资产；物资
material assistance 物资援助
material at site 工作地点存料；工地存料
material-at-site account 工地存料账；厂家原料账
material audit 原材料审计；材料审计
material avalanche 物料崩落
material axis 实轴
material balance 物料（进出）平衡
material balance area 物料衡算区
material balance calculation of pollutant 污染物物料衡算
material balance constraint 物料平衡限制条件
material balance equation 物质平衡方程(式)
material balance method 物质平衡法；物料衡算法
material balance of dryer 烘干机物料平衡
material balance relationship 物料平衡关系
material balance report 物料衡算报告
material balance sheet 物资平衡表；材料平衡表
material balancing calculation 物料平衡计算；物料
material bed 材料过滤层；料层
material behavio(u)r 材料性能
material being burned 被煅烧物料
material benefit 福利；物质利益
material bill 材料清单；材料单
material breach 严重违约；严重违背合同；重大违约
material breakage 材料损耗；材料损害
material breakdown 材料击穿
material budget 原材料预算；材料预算
material cage 运料罐笼
material capital 有形资本；物质资本
material car 材料车
material card 材料卡（片）
material catalog(ue) 物资目录
material cause 材料原因
material certificate 材料合格证
material change rate 材料变换率
material characteristic 物料特性；材料特性
material characterization 材料特性表达
material checker 检料员；材料员
material chemical analysis 材料化学分析
material chute 材料滑槽
material civilization 物质文明
material class number 材料类别号码；材料编码
material clerk 材料员
material collapse inside shaft kiln 立窑内塌窑
material collapse inside the kiln 塌窑
material collector 风力选分器；材料收集器
material compatibility 材料相容性；材料可混用性
material component 材料成分
material composing the embankment 填堤材料；筑堤材料
material composite 复合材料
material composition 物质组成
material composition sheet 材料成分分析表
material composition test 材料的化学分析；材料成分试验
material concentration 物质浓度

material conditions of social life 社会物质生活条件
material consistency 物料黏稠
material constant 材料常数
material consumption 原料耗费;原材料耗用;物资消耗;材料消耗(量);材料耗用
material consumption calculation 物料消耗计算
material consumption norm 原材料耗用定额
material consumption quota 物资消耗定额;材料消耗定额
material consumption rate 材料消耗定额
material contract 购料合同
material control 原料管理;材料控制;材料管理
material conveyed by air 气动运料
material corrosion prevention 材料防腐
material cost 原料成本;原材料成本;材料成本;材料费
material cost accounting 原材料费用计算
material cost basis for manufacturing overhead rates 按制造费用率分配的材料成本法
material cost budget 用料成本预算
material cost method 原材料成本法
material cost variance 原材料成本差异
material credit slip 退料单
material cycle 物质循环
material cycle in ecosystem 生态系的物质循环
material cycle model 循环模式
material damage 重大损害;物资损失;物质损失;物质损伤;材料损害;材料破坏
material damping 材料阻尼
material debris chute 废料溜槽;碎石料溜槽
material defect 材料缺陷
material delivered note 材料出库单
material density 材料密度
material derivative 物质导数
material description 材料描述
material design 材料设计
material designation 材料标志
material deviation 重大偏离
material difference 重大的差别
material dispersion 材料色散
material dispersion parameter 材料色散参数
material dispersion wavelength 材料色散波长
material distributing vessel 物料分配小仓
material distribution 原材料分配;物资配送;材料分布
material distribution audit 材料发出业务审计
material distribution box 布线箱
material distribution building 物资调运库
material distribution center 物资调运中心
material distribution cone 布料锥
material distribution map 材料分配图
material distribution sheet 原材料分配单
material ductility 材料延性
material economy 材料经济学
material efficiency variance 材料效率差异
material-efficient design 材料有效设计
material elevator 材料升降机
material engineer 材料工程师
material engineering 材料工程;材料储运工程
material equity 物质权益;实际产权
material estimator 材料估算员
material evidence 物证
material examination and acceptance report 材料验收报告
material exchange 物资交流
material expense 物资费
material factor 材料因素;材料系数
material fallacy 实质错误
material fault 材质不良
material feeder 给料机
material feed(ing) (生产流程的)供料过程;喂料
material flaw 材料缺陷
material flow 原料周转;物资流通;物流;物料流(程)
material for cell lining 电解槽衬里材料
material for electrical systems 电力系统材料
material for energy application 能源材料
material for energy-saving 节能材料
material for energy storage 储能材料
material for integrated optics 集成光学用材料
material for interior work 内部工程用材料
material for scratch work 打底工作用材料
material for services 服务工作用的材料
material for test 试件材料
material for the toner-carrier combination 墨粉载体配合用材料
material fragment 材料碎片
material function 材料函数
material fund 材料资金
material gains 物质利益
material handling 原料处理;货物装卸;物料装卸;物料运输;材料装卸;材料转运;材料供应;材料处理;材料搬运
material handling analysis 材料搬运分析
material handling and treatment 材料的搬运和处理
material-handling arrangement 装卸工艺;装卸安排
material handling crane 运料吊车;运料工程师
material handling device 物料输送装置
material handling device for construction sites 施工现场的物料搬运设备
material-handling elevator 运物升降机
material handling engineer 材料处理工程师
material handling equipment 装卸设备;物料搬运设备;材料装卸设备;材料搬运设备
material handling expenses 原材料处理费用
material handling pump 物料输送泵
material handling system 材料输送系统;材料处理系统
material hoist 运料升降机;起重机;材料升降机
material hose 输料软管(受压混凝土在软管中输送)
material identification 材料鉴别
material imperfection 材料缺陷
material implication 实质蕴含
material in bulk 松散材料;整批材料;散货
material index 材料指数;材料指标
material industry 材料工业
material inlet 给料口;进料口
material in process 在制物料;在制材料
material inspection 材料检查
material instrument of production 物质生产资料
material interest 物质利益
material internal control system audit 材料内部控制制度审计
material in transit 在途材料
material inventory 物资总量;物资盘存;物料总存量;材料总存量;材料盘存;材料库存;材料储备
material inventory budget 库存材料预算
material investigation 材料调查
material is not available 材料现缺
material issuance summary 材料配给一览表
material issuing unit 发料单位
material issuing warehouse 发料仓库
materiality wasting 有形损耗
materialization 质化;物质化
material laden air 带物料的空气
material law 实体法
material layer 料层
material ledger 原材料明细分类账
material level 料位
material level control 料位控制
material level elevation 料仓料位高度
material level in bin 料仓的料位
material level indicator 料位指示器
material lift 物料提升机
material list 材料(清)单;材料(明细)表
material loader 装载机;装料机
material loading factor 物料负荷率
material lock 运料闸门;料封;进料闸(门);材料(进出)闸
material lock discharging tube 料封卸料管
material loss 有形损失;有形损耗;物资损失
material man 仓库管理员;仓库管理人;材料员;材料工;建筑材料供应商;材料供应者
material management 物资管理;材料管理
material manual 材料手册
material mark 材料记录
material mass 物质质量
material measure 实物量具
material mechanics 材料力学
material men's lien 航行修理工料留置权
material misrepresentation 重要(的)错报
material mixed in site 路拌材料;就地拌和材料
material mixed in situ 路拌料
material mix variance 原材料搭配差异
material model 实体模型
material movement plan 货物运输计划;材料运输计划
material mover 运料车
material needle valve 涂料(喷枪)针形阀
material nonlinearity 材料非线性
material number 材料号
material numerical aperture 材料数值孔径
material object method in working drawing budget 施工图预算的实物法
material odo(u)r 材料气味
material of cushion 垫层材料
material of gravel stuffing 填砾材料
material of heat releasing light 热释光材料
material of low dimensions 低维材料
material of marginal quality 临界性质的材料
material of sculpture 雕塑材料
material of survey area 测区已有测绘资料
material of test piece 试板材料
material of track detector 径迹探测器材料
material of wastage 材料损耗
material on hand 现有物资
material on order 已订购物资
material order 材料单
material outlet 排料口;出料口
material over-valuation adjusted 材料高估整理
material parameter 材料参数
material passing the screen 筛下级物料;过筛的材料;过筛物料;筛下产品
material passing the sieve 筛下产品
material percentage method 材料百分比法
material pile 料堆
material piling machine 堆料机
material pit 储料坑
material plan(ning) 物资计划;材料计划
material plastic 塑料
material platform hoist 运料平板升降机
material point 质点
material policy 物资政策
material preparation 材料制备
material price 材料价格
material price for estimation 材料预算价格
material price list 材料价格表
material price standard 材料价格标准
material price variance 原材料价格差异
material processing 材料加工;材料处理
material process(ing) reactor 材料处理反应堆
material procurement 材料采购
material procure specification 材料采购规格
material produce 物料生产
material products 物质产品
material property 有形资产;有形财产;物料性质;材料性能;材料特性
material pumped 泵送物料
material purchase 原材料采购
material purchase account 材料采购账
material purchase contrast 购料单
material purchase cost audit 材料采购成本审计
material purchase order 购料单
material purchase plan audit 材料采购计划审计
material purchase strategy 材料采购策略
material purchasing and stock 材料购买与储存
material quality 材料质量
material quality cost 材料质量成本
material quality grade 材料质量等级
material quality standard 材料质量标准
material quantity analysis 材料数量分析
material quantity standard 材料用量标准;材料数量标准
material quantity variance 原材料用量差异
material radiation effect 材料的辐射效应
material rate 物料运动速度;物料流量
material receipt 原材料接收;收料报告
material receipt note 材料入库单
material receipt sheet 收料单
material received report 材料接收报告单
material receiving region 受料区
material receiving unit 收料单位
material reciprocity 实质互惠
material record 材料记录;材料核算
material recovery 废物利用
material related cost 材料有关成本
material release 材料开放;物资投放
material reliability evaluation 材料可靠性审定

material reliability report 材料可靠性报告
material remnants of war 战争遗留物;战争遗迹
material-removal rate 材料切除率
material report 收料报告
material requested 所需材料
material requirement 原材料需要量;材料需要量;材料要求
material requirement planning 原材料需求计划
material requirements per square meter 单方用料
material requirements planning 库存管理
material requisition 领料单
material requisition journal 领料日记账
material requisition return system 物资领退制度
material requisition summary 发料汇总表
material requisition unit 领料单位
material reserves 材料储备;物资储备
material resources 物质资源;物力资源
material retained on screen 筛上物料
material retained on sieve 筛上物料
material retained on trash rack 格栅截余物
material returned 退料
material returned journal 退料日记账
material returning slip 退料单
material return report 退料报告
material review standards 材料审核标准
material revision 材料修正
material reward 物质奖励
material safety data sheet 材料安全数据表
materials allocation 材料分配
material salvage 物资回收
material sample 材料样品
materials and work schedule 材料和工作日的日程安排
material saving 节约用料
materials being fabricated 制造用料
materials budget control method 材料预算控制法
materials cage 吊盘;吊篮
material scattering 材料散射
material schedule 材料清单;材料表
material scheduling 编制材料消耗表
materials chemistry 材料化学
material science 材料学;材料科学
materials classification 材料分类
materials consumption 材料消耗
materials consumption norm 材料消耗定额
materials consumption rate 材料消耗率
materials control account 材料控制账户
materials control card 材料控制卡
materials cost budget 材料成本预算
materials cost method 材料成本法
materials credit slip 退料单
materials data 材料数据
materials database 材料数据库
materials delivery 材料转移;材料输送
materials dressing 材料修整
material seal 料封
materials earmarked for unified distribution 统配物资
material segregation 材料离析
material selection 材料选择
materials engineering 材料工程
materials engineering manual 材料工程手册
materials feeder 喂料器;给料器;加料装置
materials flow 物流
material shaft 材料运送竖井
materials handling 材料管理;物料供应;材料装卸;物料搬运
materials handling engineer 物料供应工程师;物料搬运工程师
materials handling machinery 物料搬运机;材料运送机械
materials hauling 搬运材料
material shed 料棚
material shortage 材料短缺
materials in stock 库存材料
materials in storage 库存材料
materials inventory 库存原料
materials lock 材料闸(材料及工具转移到不同气压环境时通过的密室)
materials loss rate 材料损失率
materials-mixed in situ 就地拌料;路拌材料;现场拌和材料
materials now available 现有材料
materials of clinker ring 熟料圈的圈料
materials of equipment 设备材料
materials on hand 现有材料;库存材料
material source 材料源
material specification 原材料明细表;材料规格;材料规范书
material specification requirement 材料规格要求
material spending amounts 材料消耗量
materials physics 材料物理学
materials placed under unified distribution 统配物资
materials planning 编制进料计划
materials planning system 材料计划制度
materials preparation 材料准备
materials price variance 材料价格差异
materials price variance ratio 材料价格差异率
materials processing 材料制造;材料加工;材料处理
materials productivity 原材料生产率
materials programming 编制进料计划
materials progressing 进料进度管理
materials purchasing 购买材料
materials purchasing and stock budget 原材料购买存储预算
materials rate 原材料耗用率;材料率
materials received note 收料单
materials received report 入库原材料报告
materials received sheet 收料单
materials received variance 收料差异
materials requirement 材料要求
materials requisition 领料单;材料申请单;材料请领员
materials requisition on quota 限额领料制;限额领料单
materials reradiate 材料再辐射
materials returned in the stockroom 原材料退库
materials returned report 已退原材料报告;已退原材料报表
materials safety data schedule 物料安全数据表
materials shortage 材料短缺;材料缺乏
materials sinking bucket 浇灌井筒材料的料斗
materials specification manual 材料规格手册
materials standard 材料标准
materials standard cost 材料标准成本
materials standardization 材料标准化
materials storage 材料储量;储料场;材料库
materials strength 材料强度
materials summary 材料汇总账
material standard 原材料标准;实物基准;实物标准
material statement 物料表
material statistics 物资统计
material status report 物料现状报告
materials technology 材料工艺学
materials testing institute 材料试验学会
materials testing laboratory 材料试验室
materials testing machine 材料试验机
materials testing reactor 材料试验反应器;材料试验反应堆
material stock 原材料储备;存料;材料储备
material stocking quota 物资储备定额
material storage 储料场;材料库;材料存储量
material store 材料储藏;材料栈房;材料商店;料库;物料储备;材料库
material store audit 材料存储业务审计
material storehouse 材料库
material strength 材料强度
materials turnover 材料周转率
materials type flow process chart 材料型流程序图
material subsidiary ledger 材料卡片
material supplier 材料供应员;材料供应商;材料供应者
material supply 原材料供应;物资供应;材料供应;实物支持
material supply cost 物资供应成本
material supply department 材料供应处
material supply depot 材料供应仓库
material supplying unit turnover 材料供应单位资金周转额
material supply plan 原材料供应计划;物资供应计划
material supply section 材料科
material supply ship 物资供应船
material surge 料涌
material survey 材料调查
materials usage price variance 材料消费价格差异;材料耗用价格差异
materials usage standards 材料耗用标准
materials usage variance 材料耗用差异
materials used 使用材料
materials used in reinforced concrete members 钢筋混凝土材料用量表
materials used summary sheet 材料耗用汇总表
material term 实质性条款
material test(ing) 材料试验
material test(ing) laboratory 材料试验室
material test(ing) machine 材料试验机
material test(ing) report 材料试验报告;材料检验报告
material time 实际时间
material to be conveyed 搬运的材料
material to be cut 切割的材料
material to be loaded 装载的材料
material to be rolled 辊压的材料
material to be washed 冲洗过的材料
material-to-void ratio 料隙比
material trademark 材料牌号
material transfer 物质传递;材料调拨
material transfer by diffusion 扩散传质机理
material transfer by plastic flow 塑性流动传质机理
material transfer mechanism 黏性流动传质机理
material transport 物料运输
material transportation truck 材料运输车
material transporting bridge 桥形装卸机
material transport mechanism 物质迁移机理
material transport plan 材料运输计划
material turnover rate 原材料周转率
material type 材料类型
material under-valuation adjusted 材料低估整理
material unit price 原材料单价;材料单价
material usage 材料用量;材料用法
material usage ratio 材料用量比率
material usage variance 材料用料差异
material used as a soil corrective 校正土壤的物质
material used summary sheet 材料用量总表
material utilization 材料利用率
material valuation 材料估价
material variance 材料差异
material wagon 材料搬运车
material wastage 材料消耗;材料损害
material wastage rate 材料损耗率
material waterproofing 材料防水
material wave 物质波
material well 材料井
material with anisotropic thermal expansion 各向异性热膨胀材料
material world 物质世界
material yard 料场;堆场;存料场;材料堆放场
material yielding 材料回收;物质收获
material yield rate 材料回收率
material yield table 物质收获表
material yield variance 原材料(使用)效率差异;材料产量差异
materiology 材料检验学
maternal adaptation 母体适应
maternal character 母体性状
maternal effect 母体影响;母体效应
maternal environment 母体环境
maternal form 母体类型;母本类型
maternal handicap 母体阻滞
maternal impression 母体印痕;母体感应
maternal influence 母体影响
maternal-line selection 母体选择
maternal messenger 母体信使
maternal plant 母体植株;母树;母本植株;母本
maternal plant plantation 母本园
maternal seedling 母本实生苗
maternal sex determination 母体性决定
maternal toxicity 母体毒性
maternity 产科医院
maternity and child care center 妇幼保健站
maternity and infant healthy institute 妇婴保健院
maternity home 产科医院
maternity hospital 产院;产科医院
maternity ward 产妇病房;产科病房
mate's chair 船员椅
mate's receipt 大副收据
mate table 成对分割表
mat etching 毛面蚀刻

mat-etching paste 毛面酸蚀膏
mat-etching salt 毛面酸蚀盐
mat fender 护舷碰垫;碰垫;绳碰垫
mat finish 光泽消退;消光修整;消光整理;磨砂表面处理
mat finish paint 磨砂表面涂料;消光罩面漆
mat footing 全面基脚;板式基础
mat formation 毡片成型
mat forming 毡片成型
mat forming chamber 成毡室
mat-forming treatment 路面表面处治
mat foundation 整体基础;席形基础;满堂基础;片筏基础;底板基础;柴排基础;浮筏基础;筏(式)基(础);筏板基础;板式基础
mat foundation supported on piles 支承在桩上的席形基础
mat fracture 无光泽断口
mat glass 磨砂玻璃;毛玻璃
mat glaze 无光釉
mat glazed coat(ing) 无光上釉的涂层
mat glazing 无光上釉
mat gloss 平光
mat gloss paint 平光油漆
Math 2-plus 马赫数大于2
math analysis 数学解析
Mathar method 小孔释放法(测应力)
mathematic(al) analysis 数学解析;数学分析;数理分析
mathematic(al) approach 数学近似(法)
mathematic(al) approximation 数学近似(法)
mathematic(al) argument 数学论证
mathematic(al) arrays 数学阵列
mathematic(al) astronomy 数理天文
mathematic(al) axiom 数学公理
mathematic(al) basis 数学基础
mathematic(al) built-in function 数学固有函数
mathematic(al) cartography 数学制图学
mathematic(al) chance 极小机会
mathematic(al) characteristics of geologic(al) body 地质体数学特征
mathematic(al) characterization 数学表征
mathematic(al) check 数字检验;数学校验;数学检验
mathematic(al) climate 数理气候
mathematic(al) compiler 数学编译程序
mathematic(al) completeness of model 模型的数学完备性
mathematic(al) computation 数学计算
mathematic(al) constant 数学常数
mathematic(al) construction 数学作图
mathematic(al) control mode 数学控制方式
mathematic(al) control model 数学控制模型
mathematic(al) curve fitting 数学曲线拟合
mathematic(al) danger 数学上的危险
mathematic(al) decomposition 数学分解
mathematic(al) definition 数学定义
mathematic(al) ecology 数学生态学
mathematic(al) economic model 经济数学模型法
mathematic(al) economics 数理经济学
mathematic(al) equipment 数学计算装置
mathematic(al) expectation 计算寿命;数学(的)期望(值)
mathematic(al) expectation formula 数学期望公式
mathematic(al) expectation value 数学期望值
mathematic(al) expective value 数字期望值
mathematic(al) experience 数学经验
mathematic(al) expression 数字表达式;数学式
mathematic(al) expression of theory 理论的数学表达式
mathematic(al) forecast 数值预报
mathematic(al) form 数学形式
mathematic(al) formalism 数学形式体系
mathematic(al) foundation 数学基础
mathematic(al) function 数学函数
mathematic(al) function library 数学函数库
mathematic(al) function program(me) 数学函数程序
mathematic(al) genetics 数理发生法
mathematic(al) geodesy 数学大地测量学
mathematic(al) geography 数理地理学
mathematic(al) geology 数学地质
mathematic(al) horizon 理想地平
mathematic(al) horizontal adjustment 解析法平面平差

mathematic(al) idealization 数学理想化
mathematic(al) induction 数学推理;数学归纳法
mathematic(al) inference 数学推论
mathematic(al) instrument 数学仪器;数学设备
mathematic(al) inversion 数学的反演
mathematic(al) law 数学定律
mathematic(al) law of probability 数学的或然律
mathematic(al) linguistics 数理语言学
mathematic(al) lofting 数学放样
mathematic(al) logic 数学逻辑;数理逻辑
mathematically convergent system 数学收敛系统
mathematic(al) machine theory 数学机器理论
mathematic(al) manipulation 数字变换;数学运算;数学变换
mathematic(al) mean 数学平均
mathematic(al) meteorology 数理气象学
mathematic(al) method 数学法
mathematic(al) mode 数学模式
mathematic(al) model 描述系统工作的综合方程;数学模型
mathematic(al) model calculation 数值模拟计算
mathematic(al) model experiment 数学模拟实验
mathematic(al) modeling 数学模拟
mathematic(al) modeling map of tectonic stress field 构造应力场数学模拟图
mathematic(al) model of ecology 生态学数学模式
mathematic(al) model of hydraulic 水力学数学模型
mathematic(al) model of map 地图数学模型
mathematic(al) model of pool 油藏数学模型
mathematic(al) model of prediction 预测的数学模型
mathematic(al) model of river 河流数学模型
mathematic(al) model of stream 河流数学模型
mathematic(al) models of continuous systems 连续系统的数学模型
mathematic(al) models of discrete systems 离散系统的数学模型
mathematic(al) models of interaction between system and its environment 系统及其环境相互作用的数学模型
mathematic(al) module 数学模数
mathematic(al) object 数学对象
mathematic(al) operator 数字算符;数学算符
mathematic(al) optimization method 数学优选法
mathematic(al) parameter 数学参数
mathematic(al) pendulum 数学摆
mathematic(al) philosophy 数理哲学
mathematic(al) physics 数学物理学
mathematic(al) physics model 数学物理模型
mathematic(al) planning model 数学规划模型
mathematic(al) point 虚点;理论点
mathematic(al) power 幂;乘方
mathematic(al) prediction 数学预测
mathematic(al) proceeding 数学计算过程
mathematic(al) programming 数学规划;数理规划(法);最优化理论;线性规划;数学程序编制;线性规划最优化理论
mathematic(al) programming systems 数学程序设计系统
mathematic(al) projection 数学投影
mathematic(al) relation 数学关系
mathematic(al) representation 数学表示法
mathematic(al) residues 数学残数
mathematic(al) routine 数学程序
mathematicals 性能数据
mathematic(al) shaping 数学模拟
mathematic(al) ship model 数学船模
mathematic(al) simplification 数学简化
mathematic(al) simulation 数学模拟
mathematic(al) simulation kerogen thermal degradation 干酪根热降解数学模拟法
mathematic(al) simulation of folded structure 褶皱构造的数学模拟
mathematic(al) simulation of geologic(al) process 地质过程数学模拟
mathematic(al) simulation of magmatic process 岩浆过程的数学模拟
mathematic(al) simulation of sedimentary process 沉积过程的数学模拟
mathematic(al) simulation of sedimentary structure 沉积构造的数学模拟
mathematic(al) simulation of supergenesis process 表生作用过程的数学模拟

mathematic(al) software 数值计算用软件
mathematic(al) solution 数学解(答)
mathematic(al) splines 数学样条函数
mathematic(al) statistic(al) analysis 数理统计分析
mathematic(al) statistic(al) method 数理统计分析法
mathematic(al) statistics 数学统计学;数理统计(学)
mathematic(al) statistics of hydrology 水文数理统计
mathematic(al) structure 数学结构
mathematic(al) subroutine 数学(例行)子程序
mathematic(al) surface 数学面
mathematic(al) symbol 数学符号
mathematic(al) system 数学体系
mathematic(al) system theory 数学系统理论
mathematic(al) table 数学用表
mathematic(al) terrain analog(ue) computation 数学地形模拟算法
mathematic(al) terrain model 数学地形模型
mathematic(al) theory 数学理论
mathematic(al) theory of statistics 数理统计学理论
mathematic(al) tractability 便于数学处理
mathematic(al) transformation 数学转换
mathematic(al) treatment 数学解释;数学处理
mathematic(al) treatment of economics 经济学的数学处理
mathematic(al) vertical adjustment 解析法高程平差
mathematic(al) water quality model 水质数学模型
mathematicasis 数学术
mathematician 数学家
mathematicians seamount 数学家海山
mathematicization 数学化
mathematics 数学
mathematics and statistics 数学和统计学
mathematics equation 数学方程
mathematics mean-diameter of silt 泥沙的算术平均粒径
mathematics model of tritium 氚数学模型法
mathematics of computation 计算数学
mathematics of map projections 地图投影数学
mathematics of molecular biology 分子生物数学
mathematics-physics models of groundwater age 地下水年龄的数学物理模型
mathematic transformation of topographic(al) map 地形图数学基础变换
mathematization 数学化
Mathesius metal 马西修斯铅-碱金属合金;马氏铅锶合金(轴承合金)
Matheson joint 马西森接合;承插式接头
Mathias formula 马赛厄斯饱和蒸汽液体密度关系式
mathiasite 钛钾铬石
Mathieu differential equation 马提厄微分方程
Mathieu equation 马提厄方程
Mathieu function 马提厄函数
matildite 硫银铋矿;硫铋银矿
mat impregnator 毡预浸机
matinal 晨风
mating 配套;配对的
mating annular face 环形配合面
mating area 接合面;衬垫面积
mating collar 套接圈
mating die 配合模
mating gear 配对齿轮
mating holes 配合孔
mating mark 配合标记
mating member 配合件
mating partner 配对
mating period 繁殖季节
mating pinion 配对小齿轮
mating profile 配对齿廓
mating reaction 配合作用
mating ring 接合环
mating season 交尾期;交配期;繁殖季节
mating surface 接触面;啮合触面;配合触面;配合表面
mating type 接合型
mat iron 镀铅薄钢板
mat layout 织物敷层
mat layup 织物敷层
matless cell 充气式浮选槽

mat line 制毡生产线
matlockite 氟氯铅矿
mat machine 制毡机
Matobar 马托巴(一种混凝土用焊接编织钢筋)
mat of brush 柴排
mat of bush 灌木柴排;梢排;柴排
mat of reinforced concrete 钢筋混凝土垫层;钢筋混凝土沉排
Matoushan(glacial)stage 马头山冰期
mat packs 铁丝捆绑方木排
mat paint 无光泽漆
mat patch 粗糙斑
mat-pattern 无光泽类型
mat plant 铺地植物
mat porcelain enamel 无光泽搪瓷釉
matraite 三方闪锌矿
matrass 卵形瓶
mat reinforcement 钢丝网;钢筋网;网状加筋
mat reinforcement bender 钢筋网折弯机;钢筋丝折弯机
matric algebra 矩阵代数
matrices 杂矿石;母表
matricon 阵选管
matriees 填质
matrix 阴模;模型;矩阵;间质;基质;基料;基块;母表;模子;模型;铜模;填质;衬质;版型;凹型
matrix adder 矩阵(式)加法器
matrix addition 矩阵加法
matrix algebra 矩阵代数
matrix algebra table 矩阵代数表
matrix alloy 字模合金;基体合金
matrix amplifier 矩阵放大器;换算放大器
matrix analysis 矩阵运算;矩阵分析;矩阵法;基质分析
matrix and cements of carbonate rocks 碳酸盐岩基质和胶结物
matrix and vector program(me) 矩阵及向量程序
matrix approach accounting 矩阵表示法会计
matrix assessment 矩阵评价法
matrix band 型片;成型片
matrix body 胎体【岩】;基质体
matrix bookkeeping 矩阵簿记
matrix brass 模型黄铜
matrix buffer 矩阵缓冲器
matrix calculation 矩阵计算
matrix calculus 矩阵微积分
matrix capture cross section 岩石骨架俘获截面
matrix cathode 阴模式阴极
matrix check 矩阵检验
matrix circuit 矩阵线路
matrix component 小型元件
matrix computation 矩阵计算
matrix concentration 基质浓度
matrix control method 矩阵控制法
matrix correction 矩阵修正
matrix correlation 矩阵相关
matrix cutting machine 铜模雕刻机
matrix cystal 基体晶体
matrix deflation 矩阵收缩
matrix differential equation 矩阵微分方程
matrix displacement method 矩阵位移法
matrix display 矩阵显示
matrix display panel 矩阵显示板
matrix dryer 烘纸型机
matrix effect 基质效应
matrix eigenvalue 矩阵特征值;矩阵固有值;矩阵本征值
matrix element 矩阵元(素);基体元素;母体元素
matrix encoder 矩阵式编码器
matrix equality 矩阵等式
matrix equation 矩阵方程
matrix fabric 基质组构
matrix force method 矩阵力法
matrix form 矩阵组织;矩阵式
matrix formulation 矩阵形成;矩阵表述
matrix formulation of linear programming problem 线性规划问题的矩阵结构
matrix for the correlation coefficient 相关系数矩阵
matrix fraction 矩阵分数
matrix gain control 矩阵放大器增益控制
matrix game 矩阵对策
matrix glass 基质玻璃;母体玻璃
matrix grammar 矩阵文法
matrix granule 基质颗粒

matrix hardness 基体硬度
matrix in block form 分块矩阵
matrixing 矩阵变换;换算
matrixing network 矩阵网络
matrix inverse 反矩阵
matrix inversion 矩阵求逆;矩阵反演;逆矩阵
matrix inversion technique 矩阵求逆方法
matrix iteration 矩阵迭代法
matrix iterative analysis 矩阵迭代分析
matrix language 矩阵语言
matrix loss (路面中)基本集料的磨耗;(路面中)基本骨料的磨耗
matrix management 矩阵管理
matrix manipulation 矩阵操纵
matrix matching 阵列匹配;矩阵匹配
matrix matching method 矩阵比较法
matrix material 结合料材料;增强复合材料
matrix matrix 母质
matrix mechanics 矩阵力学
matrix metal 基体金属
matrix method 矩阵(方)法
matrix method of environment(al) impact 环境影响矩阵评价法
matrix model 矩阵模型
matrix module 矩阵模块
matrix multiplication 矩阵乘积;矩阵乘法
matrix multiplier 矩阵乘数
matrix norm 矩阵模(量);矩阵的模
matrix notation 矩阵记号;矩阵符号表示;矩阵符号
matrix of a linear transformation 线性变换矩阵
matrix of clastic rocks 碎屑岩杂基
matrix of coal 煤基质
matrix of coefficient 系数矩阵
matrix of consequence 结果矩阵
matrix of imported input coefficient 进口投入系数矩阵
matrix of loadings 负荷矩阵
matrix of partial derivatives 偏导矩阵
matrix of reachability 可达性矩阵
matrix of semi-linear transformation 半线性变换矩阵
matrix operation 矩阵运算
matrix organization 矩阵(式)组织
matrix out of register 合模不正
matrix output 矩阵输出
matrix pin board 矩阵插接板
matrix plane 矩阵板
matrix pliers 成型片镊
matrix polymer 母体聚合物
matrix polymerization 复制聚合;仿基料聚合
matrix porosity 基质孔隙度
matrix printer 点阵式打印机
matrix probability and statistic (al) in conference 矩阵概率及统计推断
matrix progression method 矩阵级数法
matrix proofer 矩阵式打样机
matrix pseudo-inverse 矩阵伪逆元
matrix representation 矩阵表示(法)
matrix resin 基料树脂
matrix retainer 成型片固定夹
matrix ring 矩阵环
matrix sampling 矩阵抽样
matrix save value 矩阵保存值
matrix solution 矩阵解法;基质溶液
matrix storage 阵列式存储[贮]器;矩阵存储器
matrix store 矩阵存储器
matrix stress 基岩应力
matrix structural analysis 矩阵结构分析;结构矩阵分析
matrix subtraction 矩阵减(化)法
matrix suction 基质吸力
matrix summation 矩阵求和
matrix-supported 杂基支撑
matrix switch 矩阵开关
matrix symbol 矩阵符号
matrix system 矩阵组
matrix texture 基质状结构
matrix theory 矩阵理论
matrix theory of photoelasticity 光弹性矩阵理论
matrix trace 矩阵轨迹
matrix transformation 矩阵变换
matrix transpose 矩阵转置
matrix tree theorem 矩阵树定理
matrix type 矩阵类型

matrix type structure 矩阵形式的结构
matrix unit 矩阵单元;换算单元
matrix vapo(u)rization technique 基体汽化技术
matrix velocity 基质速度
mat roller 毛面辊
matronics 材电学
mat runway 金属板装配式跑道
mats 席包
mat screen 无光泽银幕
mat seeding 席播
mat self-elevating drilling platform 沉垫自升式钻探平台
mat shed 芦席棚
mat sheet(ing) 衬垫薄片;镀铅薄钢板
mat shelter 凉棚
mat sink 门毯坑
mat sinking 沉排
mat sinking operation 沉排作业
mat sinking work 沉排工作
Matson method 麦逊法
Matson system 集装箱跨运车搬运方式;麦逊公司方式(集装箱运输);(集装箱码头的)跨运车系统
mat stripping 毡条
matsu 日本松
Matsumura hardness meter 松村氏硬度计
Matsumura indentation hardness test 松村氏钢球硬度试验
mat supported jackup 沉垫式自升平台
mat surface 毛面;粗糙(的表)面;照相机无光泽表面;布纹面(相纸)
mat-surface glazing 无光泽表面打光;无光泽表面抛光
mat(t) 无光泽的
mattagamite 斜方碲钴矿
mattamore 地下仓库;地下室
Mattauch-Herzog mass analyzer 马陶—赫绍格质量分析器
Matt Bitu (具有各种颜色的)可洗水溶性沥青漆
matt dip 无光浸洗液
mat(te) 无光泽;退光;冰铜
mat(te) blue 无光蓝(色料)
mat(te) coat 无光罩面;无光面层
mat(te) coating 无光涂层
matted crystal 晶子
matted finish 闷火处理;无光的毛面
matted glass 磨砂玻璃;无光泽玻璃;毛玻璃
matted growth 密集生长
mat(te) dip 无光浸洗液
matted materials 草垫制品
matted mor 编织状粗腐殖质
matted vegetation 铺地植被
mat(te) enamel 无光搪瓷
mat(te) enamel paint 无光瓷漆
mat(te) film 打毛塑料片
mat(te) finish(ing) 无光油漆;麻面;毛面;无光泽面;毛面加工;无光饰面;面漆消光处理;涂无光漆
mat(te) glaze 无光釉
mat(te) kip 毛面浸液
mat(te) paint 无光泽涂料;无光泽油漆
mat(te) paper 无光相纸
mat(te) picture 无光橡片
matter 物质;物品;物料;实物
matter and energy in ecosystem 生态系统中的物质和能量
matter dominated era 物质占优期
matter energy 物质能量
matter era 物质时代;物质期
matter forming a scum 构成水垢的物质
Matterhorn 马塔角
matter in solution 溶液中的物质
matter in suspension 悬浮物
matter of expediency 权宜之计
matter of necessity 必然事件
matter of principle 原则性问题
matter rate on the sieve 筛上物料率
matters in dispute 争议的事实
matters need attention 注意事项
matters needing attention 有关注意事项
matter tensor 物质张量
matter wave 物质波
mattes 地中海硬土浅滩
mat(te) shot 挡摄
mat(te) surface 消光面;漫射面;无光泽(表)面;无光表面;粗面

mat(te)-surfaced glass 毛玻璃;磨砂玻璃
mat(te) tentering table 拉毡台
matteuccite 重钠矾
mat(te) varnish 无光清漆;平光清漆
matt finish 使无光泽;晦饰;无光修饰;无光涂料;无光光洁度
matt finishing 无光整理
matt-finish structural facing unit 无光泽饰面构件;毛面饰面构件
matt-glazed 乌光釉的;无光釉的
matthew walker knot 桶梁结
Matthias rule 马赛厄斯定则
Matthiessen's standard 马希森铜丝电阻标准
matting 炼铣;去光泽;清洗工序;退光(使玻璃、金属等表面无光泽)栅网
matting agent 毛面剂;消光剂;去光剂;除光剂
matting member 编织构件
matting of roots 根系交织
matting operation 消光处理
matting pigment 消光颜料
matting process 毛面加工过程;装饰性花边加工
matting runner 走廊地毡
matting strip 长条的衬垫;无光泽板条
mattisolda 银焊料
mattness (油漆的)退光
mattock 十字镐;掘根斧;鹤嘴镐;鹤嘴锄
mat(t) paint 平光漆;亚光漆;无光涂料
mat(t) plane 毛面
mat(t) prints 无光印花布
mattress 垫板;筏式基础;满堂式基础;褥垫;台垫;埽褥;床垫;防波堤席
mattress antenna 床垫型天线;多层天线
mattress array 多排天线;矩阵式天线阵;横列定向天线阵
mattress ballasting 柴排压沉
mattress cover 床垫罩
mattress covering 埽褥盖面;沉排护面
mattress for bank protection 护岸沉排
mattress for bottom protection 护底沉排
mattress for protection 防护用柴排
mattress foundation 埽褥基础;沉排基础
mattress netting steel wire 钢绞床用钢丝
mattress pole stiffener 沉排加劲杆;柴排加劲杆
mattress protection 埽褥护岸;柴排护岸
mattress reflector 多层反射器
mattress revetment 柴排护坡;埽褥护岸
mattress slope protection 柴排护坡
matt surface 毛面;无光面
matt-surface glass 无光泽玻璃;毛(面)玻璃
matt varnish 无光漆
mat type 编绕式
mat-type artificial antenna 编绕式仿真天线
mat-type cell 充气式浮选槽
mat-type digitizer 格栅数字化器
mat-type filter 垫板型空气过滤器
mat-type insulant 编绕式绝缘;地席型绝缘材料
matulaite 磷钼钙石
Matura diamonds 锡兰锆石
maturate pond 熟塘
maturation 成熟(过程)
maturation diagram of natural gas 天然气成熟度图
maturation evolution stage of organic matter 有机质成熟作用演化阶段
maturation factor 成熟因子
maturation index 成熟指数
maturation of organic matter 有机质熟化作用
maturation of soil 土壤熟化
maturation period 熟化期
maturation pond 熟化塘;熟化池
maturation pool 成熟池
maturation process 熟化法
maturation zone 成熟区
mature 养护成熟;成熟(的)
mature adult 成熟个体
mature city 老城市;定型城市
mature coast 壮年海岸;次生海岸
mature cutting 熟枝插条;成木插条
matured bill 到期汇票
matured cement 硬化的水泥
matured check 到期支票
matured compost 成熟堆肥;腐熟堆肥
matured concrete 硬化混凝土;养护硬化后混凝土;养护后硬化的混凝土;足龄混凝土;经养护硬化(的)混凝土;熟(化)混凝土;成熟混凝土
matured continental margin 成熟型大陆边缘带
matured liabilities unpaid 未付到期负债
matured liability 到期负债
matured note 到期票据
matured principal 到期本金
matured slag 熟渣
mature ecosystem 成熟生态系统
mature face 成熟面
mature forest 成熟林
mature forest stand 成熟林分
mature form 壮年地形
mature karst 壮年溶岩【地】;壮年喀斯特
mature leachate 晚期渗滤液
mature leaf 定型叶
maturely dissected plain 壮年切割平原
mature mountain 壮年山
mature of resilience 回弹系数
mature product 成熟产品
mature proglottid 成熟节片
mature repayment 到期应付金
mature reservoir pool 成熟储存池
mature river 壮年河;均衡河流
mature river valley 壮年河
mature sediment 成熟沉积物;成熟节片
mature shore line 壮年海滨线;壮年滨线
mature sieve elements 成熟筛管成分
mature sieve tube elements 成熟筛管成分
mature soil 显域土;熟土;成熟土
mature source rock 成熟源岩
mature stage 壮年期;成熟期
mature-stage landform 壮年期地貌
mature stream 壮年河;均衡河流
mature substrate 成熟基质
mature system 成熟系统
mature technology 成熟技术
mature topography 壮年地形
mature valley 壮年谷
mature with storage 储存熟化;储存陈化
mature wood 成熟枝插
maturing 烧成(陶瓷);(泥料的)陈化;(混凝土的)硬化;熟化;成熟(作用)
maturing bin 滤灰池;熟化池;熟化箱
maturing check 支票到期
maturing investment 期满投资;到期投资
maturing of concrete 混凝土结硬
maturing of sludge 污泥的熟化;污泥熟化
maturing process 熟化法
maturing range (陶瓷的)烧成温度范围;成熟温度范围
maturing temperature 熟化温度;成熟温度
maturity (混凝土的)成熟;到期
maturity basis 按到期计算
maturity classification 成熟期分类
maturity date 到期日
maturity gap exposure 满期日缺口暴露;期限差距风险
maturity index 成熟指标;成熟度指数
maturity law 成熟度定律
maturity meter 成熟度仪(随时测得新浇混凝土的温度以计算已达到的强度);成熟度计
maturity of cement 水泥的成熟度
maturity period 成熟期
maturity phase 壮年期
maturity stage 成熟阶段
maturity structure of debt 债务到期结构
maturity value 到期值
maturity yield 全期获利率
maturnity pen 下崽牛栏
maturometer 成熟度测定计
Matuyama reversed polarity chron 松山反向极性时
Matuyama reversed polarity chronzone 松山反向极性时间带
Matuyama reversed polarity epoch 松山负极性期
Matuyama reversed polarity zone 松山反向极性带
mat varnish 平光清漆;无光漆
mat vibrated 表面振捣的
mat-vibration 表面振捣
mat vitreous enamel 无光搪瓷釉
mat vitrifiable colo(u)r 无光泽彩料
mat wall paint 无光墙漆
mat well 门垫塘;擦鞋垫池;擦鞋垫凹坑;擦鞋垫坑
mat work 席工;编织工
Mauch Chunk series 毛庄克统【地】
Mauch Chunk bed 毛庄克层【地】
maucherite 砷镍矿
maufite 镍柬纹石
Maui-type well 莫艾型井
maul 大木锤;大(木)槌
maulstick 支手杖
Maumene test 莫梅埃油脂不饱和试验
Maurer constitution diagram 毛雷尔组织图
Maurer equilibrium diagram 毛雷尔组织图
Maurer's nomogram 毛雷尔等方位曲线航用图
Maurer type screw press 毛雷尔型螺旋压榨机
Mauritius hurricane 印度洋热带性低压
Mauritius trough 毛里求斯海槽
mauritzite 蓝黑镁铝石
mausoleum 陵;陵墓;【复】mausoleums 或 mausolea
Mausoleum at Halicarnassos 哈利卡那索斯陵墓
Mausoleum of Augustus 奥古斯塔斯陵墓
Mausoleum of Theoderic at Ravenna 拉韦纳的西奥地里克陵园(美国)
mauve 紫红色;红紫色;苯胺紫;苯胺子
mauve colo(u)r 淡紫色
mauveine 苯胺紫
mauve pale 淡紫;淡红紫色
Mauzin inspection car 法国轨道检测车
mawsonite 硫锡铁铜矿
max aperture 最大光圈
max containing asphalt 普通石油沥青
max distortion 最大畸度差
Maxey-Silberston curve 马克西-西尔伯斯通曲线
max flash temperature index 最高闪点指数
max-flow 最大流
max-flow min-cut theorem 最大流最小截定理
maxi 最大的
Maxicadre container 内陆用集装箱
maxicomputer 大型计算机
maxillo-alveolar index 上齿槽指数
Maxilvry 马克西尔赖含铜奥氏体不锈钢
maximal 最大的
maximal Abelian extension 最大阿贝耳扩张
maximal acceleration 最大加速度
maximal allowance exposure 最大容许照射
maximal allowance irradiation 最大容许照射
maximal allowance irradiation level 最大容许照射水平
maximal breathing capacity 最大通气量
maximal clearance 最大清除率;最大间隙
maximal compatible 极大协调的
maximal compatible set 极大协调集
maximal concentration function 最大集中函数
maximal condition 最大条件
maximal connected set 最大连通集
maximal crack width 最大裂缝宽度
maximal deficiency 最大亏量
maximal density 最大密(实)度
maximal detectable range 最大可测距离
maximal dilatation 最大伸缩率
maximal dose 极量
maximal draw ratio 最大拉伸比
maximal element 极大元
maximal error 最大误差
maximal excretion rate 最大排泄率
maximal exercise 极限运动
maximal exercise test 极量运动试验
maximal false vector 极大伪向量
maximal flow algorithm 最大流量算法
maximal friction 最大摩擦力
maximal hearing threshold 最大可听阈
maximal humidity 最大湿度
maximal ideal 极大理想
maximal illumination 最高照度
maximal inspiratory capacity 最大吸气量
maximal invariant statistic 最大不变统计量
maximal limitation 最大限度
maximally flat 最大平坦
maximally flat bandpass network 最平特性带通网络
maximal member 极大元素;极大元
maximal non-negative boundary space 最大非负边界空间
maximal order (module) 最大序模
maximal outerplannar graph 最大外可平面图
maximal oxygen uptake 氧上限

maximal phase 最大震相;最大相
maximal planar graph 最大可平面图
maximal point 极大点
maximal pressure 最高压
maximal rate 最大速度
maximal ratio combiner 最大比率并合器
maximal reabsorption rate 最大重吸收率
maximal resistance 最大抵抗力
maximal secretory capacity 最大分泌量
maximal separable field 最大可分区域
maximal solution 最大解
maximal term 最大项
maximal tolerable concentration 最大耐受浓度
maximal tolerance dose 最大耐受量
maximal tree 最大树
maximal value 极大值
maximal ventilatory capacity 最大通气量
maximal ventilatory volume 最大通气量
maximal weight 极大权
maxima minima 最大区最小区
maximatily 最大限度
maximax criterion 极大极大准则
maximin 极大极小
maximin approximation 极大极小近似;极大极小逼近
maximin criterion 极大极小准则;极大极小判据
maximin error norm 极大极小误差范数
maxi-mini system 最大小型系统
maximin method 极大极小法
maximin principle 极大极小原理
maximin solution 极大极小解
maximin strategy 极大极小战略;极大极小策略
maximin system 极大极小系统
maximin technique 极大极小技术
maximin test 极大极小判别;极大极小检验
maximin theorem 极大极小值定理
maximization 极限化;极大化
maximization of consumer satisfaction 消费者满足极大化
maximization of living standard 生活素质极高化
maximization of storms 暴雨极大化;暴雨极大
maximization of structural parameters 结构参数极大化
maximization over discrete 离散集合上的最大化
maximization problems 最大化问题
maximize 极限化;极大化
maximize profit 追求最大利润
maximizer 达到极大
maximize the use of resources 使资源利用达到最大限度
maximizing 求最大值;求极大值;达到最大值;达到极大值
maximizing control mode 达到最大值控制方式
maximizing expected profit 最大期望利润
maximizing power output 最大功率输出
maximizing production potential 最高产量
maximizing storage 最大储藏
maximizing storm in place 暴雨原地极大化;暴雨当地放大
maximizing storm in situ 暴雨原地极大化;暴雨当地放大
maximum 最高;最多;最大值;最大;极大(值);【复】maxima
maximum abnormal value 最大异常值
maximum abrasion resistance 最大抗磨力
maximum absorbed dose 最大吸收剂量
maximum absorbed water volume 最大吸着水容量
maximum absorption 最大吸收
maximum absorption wavelength 最大吸收波长
maximum acceleration 最大加速度
maximum acceleration of vibration 振动最大加速度
maximum acceleration response 最大加速度反应
maximum acceptable concentration 最大容许浓度;最大接受浓度;极大允许浓度
maximum acceptable toxicant concentration 最大容许毒物浓度;最大可接受毒物浓度
maximum acceptance angle 最大接受角
maximum accumulation of snow 最大积雪量
maximum accuracy 最大精度
maximum activity 极大活动性
maximum adhesion 最大黏着力
maximum adhesion torque 最大附着扭矩

maximum adiabatic flame temperature 最高绝热火焰温度
maximum admissible axle load 最大允许轴荷重
maximum admissible bar stock diameter 棒料最大许可直径
maximum admissible concentration 最大容许浓度;最大接受浓度
maximum admissible flood 最大可能洪水
maximum admissible load 最大容许荷载
maximum admissible radius of curvature 最大容许弯曲半径
maximum admissible velocity 最大容许速度
maximum admittance frequency 最大导纳频率
maximum admitted diameter 最大许可直径
maximum admitted diameter of work 工件最大许可直径
maximum admitted weight 最大许可重量
maximum admitted weight of work 工件最大许可重量
maximum adsorption capacity 最大吸附容量
maximum adsorption density 最大吸附强度
maximum adsorption quantity 最大吸附量
maximum advance 最大旋回纵距
maximum aggregate size 最大骨料尺寸;骨料最大粒径;集料最大粒径;集料最大尺寸
maximum air concentration 最大空气浓度
maximum allowable biological concentration 最大生物容许浓度
maximum allowable coefficient of friction 最大容许摩擦系数
maximum allowable concentration 最高容许浓度;最大允许浓度;最大容许浓度;可允许的最大浓度
maximum allowable concentration for toxicant 毒物最大容许浓度
maximum allowable concentration of substance 最大容许物质浓度
maximum allowable contamination concentration 最大容许污染物浓度
maximum allowable deflection of tray column 板式塔的最大容许挠度
maximum allowable discharge 最大容许排放量
maximum allowable discharge concentration 最大容许排放质量浓度
maximum allowable dissolved oxygen deficit 最大容许溶解氧缺量
maximum allowable dose 最大容许剂量
maximum allowable drawdown 可利用的最大水位降
maximum allowable ecological discharge 最大容许生态排放量
maximum allowable effect level 最大容许效应水平
maximum allowable error 最大容许误差;误差极限
maximum allowable level 最大允许水平;最大允许高度;最大容许水平
maximum allowable load 最大允许载重
maximum allowable misalignment 最大容许位移
maximum allowable no-effect level 最大容许效应水平
maximum allowable offset 最大允许偏差
maximum allowable operating temperature 最高操作温度;最大允许操作温度
maximum allowable pressure 最高容许压力;最大许可压力;最大容许压力
maximum allowable pressure drop 最大容许压力降
maximum allowable sample size 最大允许试样量;最大试样量;最大容许样本尺寸
maximum allowable speed 允许最高转速
maximum allowable standard 最高容许标准
maximum allowable toxicant concentration 最大容许毒物浓度
maximum allowable velocity 最大许可速度;最大容许速度;最大容许流速
maximum allowable water table decrease 最大允许水位下降值
maximum allowable weight capacity 最大允许抓取重量
maximum allowable working pressure 最高允许工作压力;最大允许工作压力;最大许用工作压力;最大容许工作压力
maximum allowance error 最大容许误差
maximum allowance exposure 最大容许照射
maximum allowance flow velocity 最大容许流速

maximum allowance irradiation 最大容许照射
maximum allowance surface gradient 最大容许水面比降
maximum altitude 最大高度;天体最大高度
maximum amount 最高额;最大量;最大金额
maximum amount limitations 最高额限制
maximum amount of accident indemnity 赔偿限额
maximum amount of fecal coliform pollution 最大大肠菌污染量
maximum amount of operating rigs 最高开动钻机数
maximum amount of oxygen 最高含氧量
maximum amplitude 最大振幅
maximum amplitude filter 最大振幅滤波器
maximum amplitude of variation 最大变幅
maximum amplitude pulse 最大振幅脉冲
maximum amplitude signal 最大幅度信号
maximum and minimum 极大和极小
maximum and minimum proportion of length 极限长度比
maximum and minimum relay 极值继电器
maximum and minimum tariff 最高最低税则
maximum and minimum thermometer 最高最低温度计;最高最低温度表
maximum angle of dumping 最大倾卸角
maximum angle of friction 最大摩擦角
maximum angle of inclination 最大倾角
maximum angular distortion 最大角度变形;角度最大变形
maximum angular frequency deviation 最大角频率偏移
maximum annual flood 最大年洪水量;年最大洪水量
maximum annual flood flow 年最大洪水流量
maximum annual flow 年最大流量
maximum annual hourly traffic volume 年最大小时交通量
maximum annual hourly volume 年度最高小时(交通)量
maximum annual precipitation 最大年降水量
maximum annual royalty 每年最高支付额
maximum annual runoff 最大年径流量
maximum antenna current 最大天线电流
maximum anticipated settlement 最大预估(计)沉降量
maximum apparent error 最大外显误差;最大可视误差;最大表观误差
maximum atmospheric concentration 最高大气浓度;最大大气浓度
maximum authorized speed 最高允许速度
maximum available drawdown 可利用最大洪水位
maximum available gain 最大有效增益;最大可用增益;最大可大增益
maximum available moisture 最大有效水分
maximum available moisture content 最大有效含水量
maximum available power 最大可用功率
maximum available storage 最大有效库容;最大可用存储量
maximum available water 最大有效水分
maximum average power output 最大平均功率输出
maximum average price 最高平均价格
maximum average temperature 最高平均温度
maximum average uncertainty principle 最大平均不确定性原理
maximum averaging time 最大平均时间
maximum axle load 最大轴荷重;最大轴荷载;最大轴负载
maximum azeotropic mixture 最高共沸混合物
maximum backfill force 最大回填力
maximum backfill vector 最大回填力矢量
maximum background equivalent illumination 最大背景等值照度
maximum bandwidth 最大带宽
maximum bar diameter 最大棒料直径
maximum barge train 最大驳船队
maximum base shear 最大基底剪力
maximum base width 最大底宽
maximum beam 最大束流
maximum belt slope 传送带最大坡度
maximum belt tension 传送带最大张力
maximum bending moment 最大弯矩

maximum bending moment diagram 最大弯矩图
maximum bending moment envelope 最大弯矩包络线
maximum bending stress 最大弯曲应力
maximum berthing angle 最大靠泊角
maximum biohydrogen production rate 最大生物制氢率
maximum biosorption capacity 最大生物吸附容量
maximum bit hydraulic horsepower 最大钻头水马力
maximum bit weight 最大钻压
maximum blade width ratio 最大叶宽比
maximum block length 最大信息组长度
maximum blow-distance 最大吹程
maximum boiling-point mixture 最大沸点混合物
maximum bole inclined angle 最大井斜角
maximum bollard pull 最大系柱拖力
maximum boost 最大增压
maximum boost horsepower 最大升压马力;最大增压功率
maximum brake power 最大制动功率
maximum brake torque 最大制动力矩
maximum braking gradient 最大制动坡度;最大制动坡道
maximum breadth 最大宽度;船舶最大宽度
maximum breadth line 船舶最大宽度线
maximum breaker 最高限断路器
maximum breaking capacity 最大断开容量
maximum breaking load 最大断裂负载;最大断裂负荷
maximum brightness 最高亮度
maximum bubble method 最大气泡压力法
maximum bubble pressure method 最大气泡压力法
maximum buried depth of basement 基底的最大埋深
maximum camber 最大弯度
maximum capability operation 最大出力运行
maximum capacity 最高生产能力;最高电容;最大容量;最大能量;最大能力;最大出力
maximum capacity of runway 跑道最大容量
maximum capillarity 最大毛细作用;最大毛细管力
maximum capillary capacity 最大毛细水量;最大毛细吸湿量;最大毛细水量
maximum capillary water capacity 最大毛管水量
maximum cargo stowage ga(u)ge team 货物装载极限
maximum carrier level 最大载波电平
maximum carry capacity 最大负荷容量
maximum carrying capacity 最大负荷(容)量;最大承载能力
maximum carrying capacity in the section in the pairs of trains or numner of trains 区间最大通过能力
maximum cavity pressure 最大腔内压力
maximum chamfer 最大倒棱
maximum change 最大变化
maximum changing rate of hole inclination 最大井斜角变化率
maximum changing rate of hole size 最大井径变化率
maximum charge 最大装药量
maximum charge weight 最大充装重量
maximum charging capacity 最大充电容量
maximum circulation rate of pumps 最大排量
maximum cleanup width 最大清理宽度
maximum clearance 最大限界;最大净空;最大间隙
maximum climbing 最大上升率
maximum coefficient of heat transfer 最大传热系数
maximum coercivity 最大矫顽磁力
maximum collective utility 集合体效用极大化
maximum colo(u)r acuity 最大彩色分辨率
maximum combination 最大结合
maximum combustion pressure 最高燃烧压力
maximum comfortable side friction 最大舒适度的横向摩擦力
maximum compacted dry density 最大压实干密度
maximum compacted dry unit weight 最大压实干容重
maximum compression axis 最大压缩轴
maximum compression deflection 最大压缩变形
maximum computed flood 最大计算洪水(流量);最大估算溢流;最大估算流量;计算最大洪水(量)
maximum computed flow 最大估计流量
maximum concentrated weight per meter 每米最大集中荷载
maximum concentration 最大浓度
maximum concentration goal 最大污染水平目标
maximum concentration level 最大浓度水平
maximum concentration of pollutant 最大污染物浓度
maximum concentration site 最大浓度点
maximum condition 极限条件
maximum consolidation pressure 最大固结压力
maximum consumption 最大用水量;最大消耗量
maximum consumption per hour 最大每小时消耗量
maximum contaminant level 最大污染水平
maximum contaminant limit 最大污染限度
maximum contamination level 最大污染水平
maximum continuous 最大连续
maximum continuous duty 最大连续工作负载
maximum continuous load 最大连续荷载;最大连续负载
maximum continuous output 最高持续出力;最高连续出力;最大连续输出功率;最大连续产量;最大持续出力
maximum continuous rate 最大连续速率
maximum continuous rating 最大连续出水量;最大连续速率;最大连续出力;最大持续功率
maximum continuous revolution 最大连续转速;最大持续转速
maximum continuous service rating 最大持续运行功率
maximum continuous speed 连续运行最大转速
maximum continuous voltage 最大连续电压
maximum contraction 最大收缩度
maximum-contrast 最大对比度
maximum conversion 最高转化率
maximum convexity 最大凸率
maximum cooking pressure 最高蒸煮压力
maximum cooling load 最大冷却负荷
maximum core recovery 最高岩芯收获率
maximum corona 最大日冕
maximum coverage 最大覆盖面
maximum crack width 最大裂缝宽度
maximum credible accident 最大置信事故;最大设想事故;最大可信事故
maximum credible earthquake 最大可信地震
maximum credible ground motion 最大可信地面运动
maximum credible intensity 最大可信烈度
maximum crest 最大幅度
maximum critical value 极大临界值
maximum cross adjustment 最大横向调整表
maximum cross current to operate 作业最大横流
maximum cross-section 最大(横)截面
maximum cross-section coefficient 最大横剖面系数
maximum cross-section passenger volume 最大断面客流量
maximum crushing strength 最大压碎强度
maximum crushing stress 最大压应力
maximum current 最高电流
maximum current capacity 最大电流负荷
maximum current density 最大电流密度
maximum current limitation 最大电流限制
maximum current velocity 最大流速
maximum curvature 最大曲度;最大曲率
maximum curvature for normal crown section 不设超高的最小平曲线半径
maximum cut 最大切削量
maximum cut-off 最大截止频率
maximum cut-out 高限自动开关
maximum cutting capacity 最大切削容量
maximum cutting height 最大挖掘高度
maximum cutting radius at ground level 停机面最大挖掘半径
maximum cut(ting)width 最大挖宽【疏】
maximum cyclic(al)shearing stress 最大循环剪应力
maximum cylinder volume 最大汽缸容积
maximum daily average discharge 最大日平均流量
maximum daily consumption 最高日用水量;最大每日用水量;最高耗量
maximum daily consumption of water 最高日用水量;最大日用水量
maximum daily discharge 日最大流量
maximum daily dose 最大日剂量
maximum daily footage 最高日进尺
maximum daily output 最高日供水量
maximum daily storm rainfall 日最大暴雨量
maximum daily water consumption 最大每日用水量
maximum daily water supply output 最高日供水量
maximum data rate-to-band-width ratio 最高信道频带利用率
maximum daylight 折光棱镜(地下室采光用)
maximum day sendout 最大日供气量
maximum dead weight 最大载重吨位
maximum deckle 最大有效网宽
maximum deflation quantity 最大风蚀量
maximum deflection 最大偏转
maximum deflection angle 最大偏转角
maximum degree of curvature 最大曲度
maximum delivery pressure 最高排出压力
maximum demagnetizing field 最大退磁场
maximum demand 高峰要求(如电力、供水等);预期最大负荷;预期最大水量;最高需量;最大需要量;最大需(求)量
maximum demand indicator 最大需量指示器
maximum demand meter 最大负荷指示器
maximum demand per hour 最大每小时需水量
maximum demand power meter 最大需量功率计
maximum demand recorder 最大需量记录器
maximum-demand tariff 最大需量收费制
maximum demand watt-hour meter 最大需量瓦时计
maximum density 最大密(实)度
maximum density gradation criterion 最大密度级配规范
maximum density grading 最大密度级配
maximum dependable capacity 最大可靠出力
maximum depression 最大俯角
maximum depth 最大深度;最大水深
maximum depth-area-duration data 最大深度—面积—历时资料
maximum depth of exploration 最大勘探深度
maximum depth of frozen ground 最大冻土深度;最大冻结深度
maximum depth of frozen soil 最大冻土深度;最大冻结深度
maximum depth of lot 场地的最大深度
maximum depth of mixed layer 最大混合层厚度
maximum depth of plastic deformation area 塑性变形区最大深度
maximum depth of plastic zone 塑性区最大深度
maximum depth of scour 最大冲刷深度
maximum depth of scour pit 冲刷坑最大深度;冲坑最大深度
maximum depth of water 最大水深
maximum depth of water to operate 作业最大水深
maximum design berthing angle 最大设计靠泊角度
maximum designed shaft horsepower 最大设计轴马力
maximum designing gradient 最大设计坡降
maximum designing velocity 最大设计流速
maximum design pressure 最高设计压力;最大设计压力
maximum design spacing 最大设计距离
maximum design voltage 最高设计电压
maximum design wave height 最大设计波高
maximum detectable range 最大探测距离
maximum deviation 最大偏差
maximum diameter 最大直径
maximum diameter of gear cut 齿轮切削最大直径
maximum diameter of hole 最大孔径
maximum diameter of slotting saw 开槽锯最大直径
maximum diameter of workpiece 工件最大直径
maximum dielectric(al)strength 最大介电强度
maximum die set height 最大装模高度
maximum difference 极大差
maximum differential settlement 最大差异沉降
maximum digging under ground 最大铲掘深度
maximum dilatation 最大膨胀度

maximum dimension 最大尺度;最大尺寸
maximum dimensions of workpiece 工件最大尺寸
maximum direction(al) distortion 方向最大变形
maximum discharge 最大排污量;最大排放量;最大流量;极大流量;洪水流量
maximum discharge of spring 泉的最大流量
maximum discharge per second 最大秒流量;最大每秒流量;每秒最大流量
maximum disinvestment floor 投资收缩的最低限度
maximum displacement 最大位移(量)
maximum displacement response 最大位移反应
maximum display surface size 最大显示面大小
maximum distance 最大距离
maximum distance between centers 最大中心距
maximum distance between manholes 检查井最大间距
maximum distance of face control 最大控顶距离
maximum distance of perceptibility 最大有感距离
maximum distance of shuttle 货叉伸出最大行程
maximum distortion 最大失真
maximum distributed load 最大分布荷载
maximum dive 最大深度潜水
maximum diving depth 最大下潜深度
maximum dominating time 最大支配时间
maximum dose 极量
maximum dose level 最高剂量
maximum draft 最强通风;最大拉力;最大吃水(深度)
maximum drag 最大阻力
maximum draught 最强通风;最大拉力;最大吃水(深度)
maximum drawbar pull 最大牵引力
maximum drawdown 最大水位降深;最大降深
maximum draw ratio 最高拉伸倍率
maximum dredging depth 最大挖深
maximum dredging width 最大挖宽【疏】
maximum drilled diameter 钻孔最大直径
maximum drilling depth 最大钻孔深度
maximum drilling diameter 最大钻孔直径
maximum driving power 最大传动功率
maximum dry density 最大干密度
maximum dry unit weight 最大干容重
maximum dry weight 最大干重
maximum dump clearance 最大卸载高度
maximum dumping angle 最大卸载角
maximum dumping radius 最大卸载半径
maximum duration 最长时间
maximum dust content 最大含尘量
maximum duty of water 最大灌溉定额;最大灌溉率
maximum dynamic(al) head 最大动压头
maximum earth contact pressure 最大接地压力
maximum earthquake 极大地震
maximum earthquake deflection 最大地震变位
maximum earthquake force 最大地震力
maximum ebb 落急(指潮汐)
maximum ebb current 最大落潮流速
maximum ebb strength 最大退潮流(速);最大落潮流(流速)
maximum eccentricity 最大偏心距
maximum economic(al) ascending grade 最大经济升坡
maximum economic(al) plus grade 最大经济升坡
maximum economic(al) recovery 最经济的采收率
maximum economic(al) yield 最高经济收益
maximum effect force 最大有效力
maximum effective range 最大有效射程
maximum effective rate 最大有效率
maximum effective temperature 最高有效温度
maximum effective thickness of source rock 有效母岩最大厚度
maximum efficiency 最高效率;最大生产率;最大效率;最大效果
maximum efficiency curve 最高效率曲线
maximum efficient rate 最大有效采收率
maximum efficient rate of production 最大有效生产率
maximum elastic moment 最大弹性力矩
maximum elevation 最大仰角;最大射角;最大高程;制高点
maximum elevation of terrace 阶地最大高度
maximum elongation 最大伸长率;扯断伸长率
maximum emission concentration 最大排污浓度;最大排放浓度

maximum emission rate 最大排放率
maximum end load of cable 钢绳终端最大荷重
maximum endurance 最大续航时间
maximum energy 最高能量;最大能量
maximum energy transfer 最大转移能量
maximum engine speed 发动机最大转速
maximum engine torque 发动机最大扭矩
maximum entropy 最大熵
maximum entropy deconvolution 最大熵反褶积
maximum entropy method 最大熵法
maximum entropy spectrum estimation 极大熵谱估计
maximum envelope curve 最大包络线;最大包迹线;上包络线
maximum equivalent daily handling quantity 最大等效日操作量
maximum erosion depth 最大冲刷深度
maximum error 最大误差
maximum estimation 最大估计
maximum evapo(u)ration rate 最大蒸发率
maximum evapo(u)rative capacity 最高蒸发量
maximum excess disinfectant index value 最大剩余消毒剂浓度指数值
maximum excess disinfectant level 最大剩余消毒剂水平
maximum excess disinfectant level goal 最大剩余消毒剂水平目标
maximum expectant earthquake 最大期望地震
maximum experienced flood 历史最大洪水;经历最大洪水;实测最大洪水
maximum exposure limits 最大暴露极限
maximum extension 最大延长;最大扩张
maximum extension of shuttle 货叉伸出最大行程
maximum extent 最大程度
maximum extrusion pressure 最大挤压力
maximum failure rate 最高故障率
maximum fan-out 最大扇出数
maximum far earthquake 最大远震
maximum feasible pressure 最大可能的压力
maximum feeding length 最大送料长度
maximum feed stop lever 最大供油量限止杆
maximum fibre stress 最大纤维应力
maximum field capacity 最大田间容量
maximum field moisture capacity 最大野外持水量;最大田间持水量
maximum filamental flow 最大线流速
maximum financial requirement 最高资金需要额
maximum flame height without smoking 最高无烟火焰高度
maximum flat response 最大平坦响应
maximum flexural strength 最大挠曲长度
maximum flight height 最大航高
maximum floatable size 最大可浮粒度
maximum flood 最大涨潮流(速);最大洪水;涨急;急涨
maximum flood control operating level 最高防洪运行水位
maximum flood discharge 最大洪水流量
maximum flood flow 最大洪水流量;最大洪流流量
maximum flood level 最高洪水位
maximum flood line 最高洪水位线
maximum flood peak 最大洪峰
maximum flow 最高流量;最大流;最大(径)流量;极大流
maximum flow capacity 最大过水能力
maximum flow problem 最大流问题;最大流量问题
maximum flow rate 最大流率;最大泵流量
maximum flow temperature 最大流动温度
maximum fluctuation 最大波动
maximum fluidity 最大流动度
maximum flying speed 最大飞行速度
maximum fording depth 最大涉水深度
maximum foreseeable cost 最大可预测价格;最大可预测费用
maximum format 最大幅度
maximum frequency 最大频率
maximum frequency deviation 最大频(率)偏(移)
maximum frequency of picture current 最高图像电流频率
maximum frequency operation 最大频率工作
maximum friction 最大摩擦(力)
maximum frictional force 最大摩擦力
maximum fuel limit screw 最大燃油限制螺钉

maximum full load speed 全荷载时最高速度
maximum gain 最大增益
maximum gas temperature 最高燃气温度
maximum gate power 最大门功率
maximum gate voltage 最大门电压
maximum ga(u)ge (木材或线材的)最大直径;(木材或线材的)最大厚度
maximum generating watt 最大发电量
maximum good field radius 最大工作场半径
maximum governed speed 最大可调速度
maximum gradability 最大爬坡度
maximum grade 最大坡度
maximum gradient 最大梯度;最大坡降;最大坡度;最大陡度
maximum gradient of river 河流最大比降
maximum grain size 最大粒度
maximum granular size 最大粒径
maximum green change 最长绿(灯)换相(感应信号用)
maximum green period 最长绿灯时间
maximum grinding diameter 最大磨削直径
maximum grinding diameter on centers 中心最大磨削直径
maximum grinding efficiency 最大磨矿效率
maximum grinding length 最大磨削长度
maximum grinding travel 最大磨削行程
maximum gross 最大载重量
maximum gross head 最大毛水头
maximum gross mass 最大总质量
maximum gross weight 最大总重(量)
maximum ground concentration 最大地面浓度
maximum ground contact pressure 最大接地压力
maximum growth period 最高生长期
maximum growth rate 最大生长率
maximum growth temperature 最高生长温度
maximum guaranteed capability 最大保证出力
maximum gust 最大阵风
maximum habitat stability 最大生境稳定度
maximum habitat volume 最大生境容量
maximum hardness test 最大硬度试验
maximum head 最大水头;最大压头
maximum headwater 上游最高水位
maximum headwater level 最高上游水位
maximum heat demand 最大需热量
maximum heat flux 最大热通量
maximum heating load 最大热负荷
maximum heat output 最大发热量
maximum heat-release rate 最大放热速率
maximum height 最大高度
maximum height above sea level 最大海拔高度
maximum high discharge 最大流量
maximum high water(level) 最高(高)水位;最大高水位;最高(高)潮位
maximum high water level of neaps 小潮最高潮位
maximum high water springs 大潮最高潮位;大潮最高潮面
maximum hoisting capacity 最大起重量
maximum hoisting height of grab 抓斗最大吊高
maximum hole size 最大井径
maximum horizontal range 最大水平射程
maximum horizontal tensile stress 水平方向最大拉伸应力
maximum horsepower 最大马力
maximum hourly consumption 最大小时耗水量;最大小时用水量;最大小时消耗量
maximum hourly demand 最大小时需量
maximum hourly rainfall 最大小时降雨量
maximum hourly requirement 最大小时需水量
maximum hourly sendout 最大小时供气量
maximum hourly traffic volume 最高小时交通量
maximum hourly volume 最大小时交通量
maximum hourly water consumption 最大小时用水量
maximum hourly water demand 最大小时需水量
maximum humidity 最大湿度
maximum hydraulic head 最大水头
maximum hydraulic horsepower 最大水马力
maximum hydrostatic(al) pressure 最大静水压力
maximum hygroscopicity 最大吸湿水;最大吸湿度
maximum ice thickness to operate 作业最大冰层厚度
maximum impedance frequency 最大阻抗频率
maximum in basin relief 流域最大高差
maximum in peak 峰值中最大值

maximum input level 最大输入电平
maximum input repetition rate 最大输入重复频率
maximum input signal level 最大输入信号电平
maximum input voltage 最高输入电压
maximum instantaneous discharge 最大瞬时流量
maximum instantaneous fuel change 最高瞬间燃料变动率
maximum instantaneous power 最大瞬时功率
maximum instantaneous wind speed 最大瞬时风速
maximum in-station cabling 最长局内电缆
maximum intensity 最大强度;最大烈度
maximum intensity map 最大烈度图
maximum interference 最大过盈;最大干涉
maximum intermittent gross torque (发动机试验时)短时间最大扭矩
maximum internal dimensional of rear body 车厢最大尺寸
maximum intersection angle 最大交会角
maximum inverse peak current 最大反向峰值电流
maximum issue method 最高额发行法
maximum J condition 极大 J 条件
maximum jet force 最大冲击力
maximum jet height 最大喷射高度
maximum jet width 最大喷射宽度
maximum joint gap structurally obtainable 构造轨缝【铁】
maximum junction temperature (晶体管的)最大结温度
maximum justification rate 最大调整率
maximum keying frequency 最高键(控)频(率)
maximum kinetic energy 最大动能
maximum knock-out stroke (压力机的)最大打料行程
maximum lag 最大滞后距
maximum landing weight (飞机)最大降落重量
maximum latency 最大等待时间
maximum lateral amplitude of bridge 桥梁最大横向振幅
maximum length 最大长度
maximum length code 最大长度码
maximum length of bar feeding 棒料最大送料长度
maximum length of workpiece 工件最大长度
maximum length sequence 最大长度序列
maximum lethal concentration 最大致死浓度
maximum lethal dose 最大致死剂量
maximum lethal oxygen concentration 最高致死氧浓度;最大致死氧浓度
maximum liability 最高责任;最大的赔偿责任
maximum lifting capacity 最大起重机
maximum lifting height above ground 最大举升高度
maximum lifting load 最大起重量
maximum light transmission 最大光输
maximum likelihood 最大似然(性);最大似然率;最大可能性;极大似然
maximum likelihood classification 最大似然分类
maximum likelihood criterion 最大似然(率)准则;极大似然准则
maximum likelihood decision rule 最大似然判别法则
maximum likelihood decoding 最大似然解码
maximum likelihood detection 最大似然检测
maximum likelihood errorcorrecting parser 最大似然误差校正剖析程序
maximum likelihood estimate 最大似然估计(法);极大似然估计
maximum likelihood estimator 最大似然估计量
maximum likelihood membership function 最大似然从属度函数
maximum likelihood method 最大似然量法;最大似然(方)法;最大可能量法;极大似然法
maximum likelihood ratio 极大似然比率
maximum-likelihood-ratio algorithm 极大似然比算法
maximum likelihood technique 极大似然法
maximum likelihood test 最大似然检验;最大似然检测
maximum lime 饱和石灰
maximum lime content 石灰极限含量
maximum limit 最高限制;最大限额;最大限度;最大极限
maximum limit deposit velocity 最大极限沉积流速

maximum limit of size 最大极限尺寸
maximum line capacity 线路最大通过能力;最大线路通过能力
maximum live shear 最大活剪力
maximum load available daily 最大有效日输送量
maximum load crown block 天车最大负荷
maximum load current of feeding section 供电臂最大负荷电流
maximum load design 最大载设计法
maximum load indicator 最高负载指示器
maximum load(ing) 最大荷载;最大装载量;最大荷载重;最高负荷;最大载重;最大荷载量;最大负载;最大负载;荷载极限
maximum loading capacity 最大容许载重量
maximum loading hours 最大负荷利用小时数
maximum loading rate (油轮)最大装卸率
maximum load of hook 大钩最大负荷
maximum load of swivel 水龙头最大负荷
maximum load of traveling block 滑车最大负荷
maximum load per axle 最大轴荷重;最大轴荷载
maximum longitudinal distance 最大纵距
maximum longitudinal grade 最大纵坡(度)
maximum longitudinal gradient 最大纵坡(度)
maximum loose 最大松散度
maximum loss expectancy 估计最大损失
maximum lump size 最大块尺寸
maximumly flat filter 最大平坦滤波器
maximum machine voltage 最大运算电压
maximum machining diameter 最大加工直径
maximum machining height 最大加工高度
maximum machining length 最大加工长度
maximum magnetic susceptibility 最大磁化率
maximum magnetizing force 最大磁化力
maximum magnitude 最大允许量
maximum maintenance weight 最大维护重量
maximum matching 极大匹配
maximum material condition 最大实体条件
maximum material principle 最大实体原则
maximum mean diurnal range 最大平均日潮差
maximum mean temperature 最高平均温度;最高平均气温
maximum melting capacity 最大熔化能力
maximum melting rate 最高熔化率
maximum meterage of hole 钻孔最大进尺
maximum-minimum principle 最大最小值原理;极小极大定理
maximum-minimum thermometer 最高最低温度计
maximum mining yield 最大开采(产)量
maximum mixing depth 最大混合深度
maximum-mobility period 激烈期
maximum modulating frequency 最高调制频率
maximum modulation 最大调制度
maximum modulation frequency 最高调制频率
maximum module of gear cut 最大齿轮切削模数
maximum module to be cut 最大切削模数
maximum modulus 最大模数
maximum modulus principle 最大模原理
maximum-modulus theorem 极大模定理
maximum moisture capacity 最大容水量;(土壤)最大持水量
maximum moisture content 最大含水量
maximum molecular moisture(holding)capacity 最大分子吸水量;最大分子容水量;最大分子持水量
maximum molecular specific retention 最大分子持水度
maximum molecular water content 最大分子吸水量
maximum moment 最大弯矩;最大力矩
maximum momentary speed 瞬间最大转速
maximum momentary speed variation 最大瞬时速度变化率;瞬时最大速度变化率
maximum monthly efficiency 最高月效率
maximum monthly footage 最高月进尺
maximum moving dimension 车辆(最大)限界
maximum nail holding power 最大握钉力
maximum navigation discharge 最大通航流量
maximum near earthquake 最大近震
maximum negotiable radius 最大弯道半径
maximum net head 最大净水头
maximum net residential density 最大住房密度
maximum nitrification efficiency 最高消化效率
maximum nitrogen uptake 最大摄氮量

maximum noise level 最大噪声(等)级
maximum non-effect dose 最大无影响剂量
maximum non-effect level 最大无影响量;最大无作用浓度
maximum non-load speed 最大空载转速
maximum non-observed effect concentration 最大无影响浓度
maximum non-observed effect level 最大无觉察到的效应水平
maximum non-observed effect concentration 最大无觉察到的效应浓度
maximum norm 最大模
maximum normal strain criterion 最大正应变理论
maximum normal subgroup 极大正规子群
maximum normal working pressure 最大额定工作压力
maximum-norm estimates 最大模估计量
maximum nozzle velocity 最大喷速
maximum number of balls 最多球数
maximum number of passengers in peak hours 最高聚集人数
maximum number of persons allowed 额定人数
maximum number of revolutions 最高转数
maximum number of rollers 最多滚子数
maximum number of strokes per minute 每分钟最高行数
maximum number of trains per hour in each direction 每小时单向最大列车对数
maximum number of trains per hour in two direction 每小时双向最大列车对数
maximum observed hourly volume 最大实测每小时交通量
maximum observed precipitation 最大观测降水量
maximum occupancy 最多居住人数
maximum of a set 集的极大值
maximum offered line capacity 最大提供的线路容量
maximum offset 最大偏移距;最大偏差
maximum of regular waves occurring on the preliminary tremor 初期微震最大波
maximum one-day rainfall 最大日降雨量
maximum opening diameter without reinforcement 不另行补强的最大孔径
maximum operating depth 最大作业深度
maximum operating efficiency 最大运转(效)率
maximum operating frequency 最高工作频率
maximum operating level 最高工作水位
maximum operating pressure 最大作业压力
maximum operating speed 最高运行速度
maximum operating temperature 最高运行温度
maximum operating voltage 最大工作电压
maximum operating water level 最高运行水位
maximum operating width 最大作业宽度
maximum operation frequency 最大工作频率
maximum operation mode 最大(功率)运行方式
maximum operation pressure 最高使用压力
maximum operation resistance 最大工作阻力
maximum operation temperature 最高使用温度
maximum ordinate 最大纵坐标
maximum output 最高输出量;最高出力;最高产率;最高产量;最大输出(功率);最大流出量;最大产量;峰值输出
maximum output current 最大输出电流
maximum output of the power station 电厂最高出力
maximum output pressure 最高排出压力
maximum output voltage 最高输出电压
maximum outreach 最大伸距;最大跨距
maximum outrigger load 外伸支腿最大压力
maximum outside diameter 最大外径
maximum outside diameter of gear cut 最大齿轮切削外径
maximum overall efficiency 最高总效率;最大总效率
maximum overall length 最大总长度;最大全长
maximum overall profitability 最大总盈利
maximum overall stress in rail 钢轨内部最大总应力
maximum overbreak 最大超挖地段
maximum overflow 最大溢流量
maximum overshoot 最大过调量
maximum overtemperature 最大过热温度
maximum oxide thickness 最厚氧化层厚度
maximum oxygen consumption 最大耗氧量

maximum packet lifetime 包装有效期
maximum packing case size 最大包装箱尺寸
maximum particle diameter 最大粒径;最大粒度
maximum particle size 最大粒径;最大粒度
maximum part weight 最大部件重量
maximum passengers in waiting room 旅客最高聚集人数
maximum passengers per day per direction 每天单方向断面最大客流量
maximum past apparent pressure 最大前期表观压力
maximum pattern area 最大花型范围
maximum payload 标定负荷;最大装载量;最大载重量;最大容许载重量
maximum peak discharge 最大洪峰流量
maximum peak forward current 最大正向峰值电流
maximum peak forward voltage 最大正向峰值电压
maximum peak inverse voltage 最大反向
maximum peak inverse volts 最大反向峰压
maximum peak voltage 最大峰值电压
maximum penetration rate 最大钻速
maximum penetration volume 最大透过容量
maximum percentage modulation 最大调制度
maximum percentage recovery of core 最高岩芯采取率
maximum percent overshoot 最大百分比超调量
maximum permeability 最大渗透率;最大磁导率
maximum permissble revolution 最大容许转数
maximum permissible acceleration 最大允许加速度
maximum permissible accumulated dose 最大容许累积剂量
maximum permissible amount 最大允许量;最大容许量
maximum permissible atmosphere concentration 最大容许大气浓度
maximum permissible body burden 最大允许人体负荷;最大允许机体负荷;最大容许机体浓度
maximum permissible boost pressure 最大容许增压
maximum permissible concentration 最大允许浓度;最大容许浓度
maximum permissible concentration in water 水中最大容许浓度
maximum permissible concentration of unidentified radionuclides in water 水中最大容许未鉴别出的放射性核素浓度;未鉴定放射性核素在水中的最高容许浓度
maximum permissible contamination level 最大容许沾染水平
maximum permissible cumulated dose 最大容许累积剂量
maximum permissible current 最大容许电流
maximum permissible daily intake 最大容许日摄入量
maximum permissible deviation 最大允许偏差;最大容许偏离
maximum permissible displacement 最大容许位移
maximum permissible dosage 最大容许剂量
maximum permissible dose 最高容许量;最大允许剂量;最大容许剂量
maximum permissible dose equivalent 最大容许剂量当量
maximum permissible dose rate 最大容许剂量率
maximum permissible drawbar load 最大允许牵引荷载
maximum permissible dustiness 最大允许含尘量
maximum permissible error 最大容许误差
maximum permissible exposer (放射线)最大容许线量
maximum permissible exposure 最大允许辐照(量);最大允许曝光量;最大容许暴露(量)
maximum permissible flux 最大容许通量
maximum permissible ga(u)ge 最大容许轨距
maximum permissible gross laden weight 最大容许总装载重;最大容许满载总重
maximum permissible indicated speed 最大允许表速
maximum permissible individual concentration 最大容许个体浓度
maximum permissible intake 最大容许摄入量;最大容许摄取量
maximum permissible integral dose 最大容许累积剂量

maximum permissible level 最大容许限度;最大容许水平;最大容许能级;最大可容水平
maximum permissible limit 最大容许限度;最大容许极限
maximum permissible load(ing) 最大容许负载;最大容许荷载
maximum permissible noise level 最大容许噪声电平
maximum permissible porosity indication 最大容许气孔显示
maximum permissible power 最大容许功率
maximum permissible pressure 允许压力;最大容许压力
maximum permissible rate 最大容许采收率
maximum permissible release 最大容许排放
maximum permissible revolution 最大允许转数
maximum permissible service temperature 最大允许使用温度
maximum permissible speed 最高允许速度;最高容许速度;最大允许转速;最大允许速度;最大容许速度
maximum permissible stress 最大许用应力
maximum permissible temperature 最高允许温度
maximum permissible value 最大允许值;最大容许值
maximum permissible velocity 最大允许流速;最大容许速度
maximum permissible weekly dose 最大容许周剂量
maximum permissible weight 最大容许重量
maximum permissive speed 最大允许速度
maximum permitted concentration 最大容许浓度
maximum permitted speed 最高允许速度;最大容许速度
maximum phase 最高期;最大相位
maximum picture frequency 最高帧频;最高图像频率;最大图形信号频率
maximum picture height 最大图像高度
maximum picture width 最大图像宽度
maximum piston rod pull 活塞杆最大拉力
maximum pitch rolled 最大滚丝距
maximum planing height 最大刨削高度
maximum planing length 最大刨削长度
maximum planing width 最大刨削宽度
maximum plant capacity 设备最高生产能力
maximum plastic resistance 最大塑性阻力
maximum plate input 最大板极输入
maximum point on a curve 曲线上的极大点
maximum polarization 最大极化强度
maximum polluted level 最高污染物水平;最大污染程度
maximum polluted level goal 最高污染物水平指标
maximum pore volume 最大孔隙体积
maximum porosity 最大孔隙率;最大孔隙度
maximum porosity in clean rock 纯岩石最大孔隙度
maximum positive moment 最大正弯矩
maximum possible adsorption density 最大可能吸附强度
maximum possible earthquake 最大可能地震
maximum possible flood 最大可能洪水;极限洪水
maximum possible intensity 最大可能烈度
maximum possible magnitude 最大可能震级;最大可能允许量
maximum possible operating factor 最大可能利用率
maximum possible precipitation 极限降雨量;最大可能降雨量;最大可能降水量
maximum possible rate 最大可能速率
maximum possible saturation concentration 最大可能饱和浓度
maximum possible wave 最大可能波浪
maximum potential 最大电位
maximum potential output 可能达到的最高产量
maximum potential strength 最大潜在强度
maximum power 最高功率;最大功率
maximum power altitude 最大动力高度
maximum power dissipation 最大功耗
maximum power of draw works 绞车最大功率
maximum power output 最高输出功率
maximum power rating 最大功率定额
maximum power-transfer theorem 最大功率传输定理
maximum precipitation 最大降雨量;最大降水量

maximum precipitation at a time 最大次降水量
maximum pressure 最高压力;最大压力
maximum pressure arch 最大压力拱
maximum pressure boost 最大增压
maximum pressure governor 最高压力调节器;最大压力调节器
maximum pressure line 最大压力线
maximum pressure of mud pump 泥浆泵最大压力
maximum pressure valve 最大压力阀
maximum price 最高价(格)
maximum principal energy 最大主能量
maximum principal strain 最大主应变
maximum principal strain criterion 最大主应变理论
maximum principal stress 最大主应力
maximum principal stress criterion 最大主应力理论
maximum principal stress value 最大主应力值
maximum principle 极大值原理
maximum probability 最大概率
maximum probable error 最大可能误差
maximum probable flood 最大可能洪水;极限洪水
maximum probable loss 最大可能损失
maximum probable precipitation 最大可能降雨量;最大可能降水量
maximum probable rainfall 最大可能降雨量;最大可能降水量
maximum probable storm 最大可能暴风雨
maximum producible oil index 最高可采油指数
maximum production 最高产量
maximum productive capacity 最大生产能力
maximum productivity 最高生产率
maximum profit 最大利润
maximum program(me) 最高纲领
maximum pull 最大拉力
maximum pump discharge 最大排量
maximum pump pressure 最高泵压
maximum punch length 凸模最大长度
maximum quantity acceptable 最大可接受交货量;可接受最大数量
maximum quantity of precipitation 最大降雨量;最大降水量
maximum quantity of stores 最高库存量
maximum quantity of wind 最大风量
maximum quenching hardness 最大淬火硬度
maximum radar range 雷达最大作用距离
maximum radiation 最高辐射
maximum radius 最大半径
maximum rafter distance 最大椽子距离
maximum rainfall 最大降雨(量);最大降水量
maximum rainfall intensity 最大(降)雨强(度);最大降水强度
maximum range 最大续航距离;最大行程;最大射程;最大量程;最大距离;最大航程;最大范围;最大测距;最大变幅;船舶续航力
maximum range of electromagnetic distance measurement 电磁波测距最大测程
maximum range of ship 船舶最长续航距离
maximum rate 最高率;最高定额;最大强度
maximum rated load 全部最高荷载(包括施工荷载);最大额定荷载
maximum rated sound-power level 最大标定声功率级
maximum rate of fire 最大射速
maximum rate of flow 最大流量
maximum rate of rotary table 转盘最大转速
maximum rating 最高地方税征收额;最大额定值
maximum ratio 最大值比;最大比率
maximum ratio combining 最大比值合并
maximum reach at ground level 停机面最大挖掘半径
maximum receiving efficiency 最高接收效率
maximum-recorded 最高记录的;实测最大的
maximum-recorded flood peak 实测最大洪峰
maximum-recorded flood peak discharge 实测最大洪峰流量
maximum-recorded flow 记录最大流量;实测最大流量
maximum-recorded stage 记录最高水位;实测最高水位
maximum recording attachment 最高需量记录装置
maximum recording level 最高记录磁平
maximum recording thermometer 最高自记温度表

maximum recovery rate 最大开采速度
maximum reduction 最大压下量
maximum reflectivity 最大反射率
maximum reinforcement 最大补强
maximum relative error of closure of traverses 导线最大相对闭合差
maximum relative time interval error 最大相对时间间隔出错
maximum relay 过载继电器;过电流继电器;极限继电器
maximum relieving depth 最大铲齿深度
maximum relieving length 最大铲齿长度
maximum remanent flux 饱和剩磁
maximum-remembering circuit 最大存储电路
maximum reserves 最大储量
maximum reset time 最大复位时间
maximum residual disinfectant level 最大剩余消毒剂水平
maximum residual disinfectant level goal 最大剩余消毒剂水平目标
maximum residue limit 最大残留(限)量
maximum resistance 最高电阻;最大阻力;最大电阻;极限电阻
maximum resistive cutoff frequencies 最高电阻截止频率
maximum resultant ground displacement 最大合成地面位移
maximum retention time 最大保留时间
maximum return 最大收益
maximum return signal 最大返回信号
maximum reverse voltage 最大反向电压
maximum revolution 最大转数
maximum rigidity 最大刚性
maximum rock crushing strength 最大岩石压碎强度
maximum rolling angle of buoy 浮标最大摇摆角
maximum rotary speed 最大转速
maximum rotary torque 最大转矩
maximum rotational speed 水龙头最大转速
maximum rotative speed 最大旋转速度
maximum r. p. m. 每分钟最大转数
maximum running speed 最高运行速度
maximum running torque 最大转矩
maximum runoff 最大径流量;最大径流
maximum runoff rate 最大径流速度
maximum safe air-speed indicator 最大安全空速指示器
maximum safe capacity 最大安全容量;最大安全度
maximum safe concentration 最大安全浓度
maximum safe density 最大安全密度
maximum safe earthquake intensity 最大安全地震烈度
maximum safe firing current 最大安全电流
maximum safe load 最大安全荷载
maximum safe rating 最大安全速度
maximum safe speed 最高安全速度;最大安全速率;最大安全速度
maximum safety factor 最大安全系数
maximum safety setting 最大安全定值
maximum safety temperature 最高容许温度
maximum safe velocity 最高安全速度
maximum safe water-depth 最大安全水深
maximum safe working pressure 最高安全工作压力
maximum sag 最大垂度
maximum sag of cable 线缆最大垂度
maximum saturation 最大饱和度
maximum saturation diving condition 最大饱和潜水条件
maximum sawing or filling thickness of work 工件锯割或锉削最大厚度
maximum scale 最大刻度;最大标度;最大比例(尺)
maximum scale value 最大刻度值;标度终点值
maximum scan 最大扫描
maximum scour depth 最大冲刷深度
maximum scoured line 最大冲刷线
maximum screen luminance ratio 最大屏亮度比率
maximum secondary discharge 最大二次排放量;最大次级排放量
maximum secondly discharge 最大每秒流量
maximum section of cutting tools 刀具最大截面
maximum sediment concentration 最大含沙浓度;最大含沙量

maximum segregation 最大偏析
maximum seismic intensity 最大地震烈度
maximum self-conjugate sub-group 最大自共轭子群
maximum sensitivity 最高灵敏度;最大灵敏度
maximum service life 最长使用寿命;最长使用期限;最长工作寿命
maximum service pressure 最大使用压力
maximum service rating 最大工作功率
maximum service speed 最高运行速度
maximum service temperature 最高使用温度
maximum settling rate 最大沉淀(速)率
maximum shaft power 最大轴功率
maximum sharpness 最大锐度
maximum shear 最大剪力
maximum shearing strength 最大剪(切)强度
maximum shearing stress 最大剪(切)应力;最大切胁强
maximum shear resistance 最大抗剪阻力
maximum shear strain 最大切胁变;最大剪(切)应变
maximum shear strain energy theory 最大剪(切)应变能量理论
maximum shear stress criterion 最大剪(切)应力准则;最大剪(切)应力强度理论
maximum shear stress yield criterion 最大剪(切)应力屈服条件
maximum shear theory 最大剪力理论
maximum shift 最大轮班
maximum shift footage 最高班进尺
maximum shock pressure 最大冲击压力
maximum shop assembly 最大程度工厂组装
maximum signal level 最大信号电平
maximum signal method 最大信号法
maximum signal recognition time 最大信号识别时间
maximum size 最大粒径;最大孔径;最大尺度;最大尺寸
maximum size of aggregate 最大粒径骨料;最大骨料粒径;集料最大粒径;集料最大尺寸
maximum size of plate cut 冲剪板最大尺寸
maximum size pump liner 泵最大缸径
maximum skid-resistance potential (of surface) (路面)最大抗滑势能
maximum slip 最大滑动
maximum slope 最大坡率;最大坡度
maximum snow 最大雪量
maximum snow clearing capacity 最大除雪量
maximum snow clearing distance 最大除距离
maximum snow fall line 最大降雪线
maximum solubility 最大溶解度
maximum solution 最大解
maximum sorption capacity 最大吸附容量
maximum sorption density 最大吸附强度
maximum sound intensity 最大声强
maximum sound level 最大声级
maximum sound pressure 最大声压
maximum sound pressure level 最大声压级
maximum span 最大跨径
maximum span-depth ratio 最大跨深比
maximum spawning area 最大产卵区
maximum specific substrate utilization rate 最大底质利用率
maximum spectral sensitivity 最大光谱灵敏度
maximum speed 最高转速;最高速度;最大速度;全速
maximum speed governor 最高速度调节器
maximum speed mixture 最高燃速混合物
maximum speed of hydraulic motor 液压电机最高转速
maximum speed of revolution 最高转速
maximum speed on cross-country 最大越野行驶速度
maximum speed regulator 最高转数调速器
maximum speed test 最高速度试验
maximum speed-up hierarchy 最大加速层次
maximum spoil-discharge distance 最大排泥距离
maximum spoil-discharge height 最大排泥高度
maximum-stability period 和缓期
maximum stable gain 最大稳定增益
maximum stable tilting angle 最大稳定倾角
maximum stack height 最大烟囱高度
maximum stacking fold 最大叠加次数
maximum stacking height 最大堆码高度

maximum stage 最高水位;最高阶段
maximum standard deviation 最大标准差
maximum starting current 最大起动电流
maximum starting load 最大起重荷载
maximum static deflection 最大静挠度
maximum stationary phase 最大静止期
maximum station interval 最大站间距
maximum steady speed of hydraulic motor 液压电机最高稳定转速
maximum steam working pressure 最大蒸汽工作压力
maximum steepness 最大陡度
maximum stock 最高库存量
maximum stock control 最高存量管理法
maximum storage coefficient 最大存储[贮]系数
maximum storage factor 最大存储[贮]系数
maximum storage level 最大蓄水水位
maximum storage time 最大存储时间
maximum storm rainfall 最大暴雨量
maximum strain 最大应变
maximum strain axis 最大应变轴
maximum strain energy 最大应变能
maximum strain theory 最大应变理论;最大变形理论
maximum strain theory of failure 最大应变破坏理论
maximum stream 最大水流
maximum stream flow 最大河流流量;河道最大流量
maximum strength 最大强度
maximum strength of current 最大流速
maximum stress 最大应力
maximum stress limit 最大应力极限
maximum stress theory 最大应力理论
maximum stroke 最大行程
maximum strokes of pump 泵最高冲程数
maximum stuffing rate 最大塞入率
maximum submerged macrophyte biomass 最大沉水大型植物生物量
maximum subsidence 最大沉陷
maximum subsidence factor 含水层组比储量
maximum subsidence rate 最大沉降速率;最大沉降量
maximum subsidence time 最大沉降时间
maximum sucking height 最大吸程
maximum suction 最大真空度
maximum suction lift 最大吸入高度
maximum suction pressure 最高吸入压力;最大吸入压力
maximum sum of hourly cooling load 逐时冷负荷综合最大值
maximum superelevation 最大超高(度)
maximum superelevation rate 最大超高率
maximum support clearance 最大支承间隙
maximum surface elevation 最高水位
maximum surface runoff 最大地表径流
maximum surge 最গ涌浪;最大浪涌
maximum survival time 最高存活时间;最大存活时间
maximum susceptibility 最高磁化率
maximum suspension time 最大中止时间
maximum sustained yield 最大可持续获水量;最大持续产水量
maximum swell height to operate 作业最大涌高
maximum swelling 最大膨胀
maximum swing diameter of work over-bed 车床面上加工最大直径
maximum swing diameter of work over-carriage 车床在刀架上的最大加工直径
maximum swivel of table 工作台最大回转度数
maximum swivel of tool slide 刀架最大回转度数
maximum system voltage 最高系统电压
maximum table speed 工作台最大速度
maximum take-off weight 最大起飞重量
maximum tariff 最高税率
maximum taxing weight (飞机)最大滑行重量
maximum temperature 最高温度
maximum temperature period 最高蒸压温度期限(指养护期限);最高温度期限;(蒸汽养护混凝土的)恒温期
maximum temperature-rise 极限温升;最高温升;最大温升
maximum temperature variation 最大温度变化量
maximum tensile force of hoisting machine 提

升机最大拉力
maximum tensile stress theory 最大拉应力理论
maximum tension 最大张力;最大拉力
maximum tensional stress 最大张应力
maximum tension stress 最大拉应力
maximum thaw depth 最大解冻深度
maximum theoretic(al) liability 理论上的最高责任
maximum theoretic(al) yield 最大理论流量
maximum thermal discharge 最大热量排放量
maximum thermal gradient 最大温度梯度
maximum thermometer 最高温度计;最高温度表;极大温度计
maximum thickness of aquifer 含水层最大厚度
maximum thickness of bending plate 最大卷板厚度
maximum thickness of work 工件最大厚度
maximum thread diameter 最大螺纹直径
maximum throat depth (点焊机的)夹钳外伸长度
maximum throughput 最大排出量
maximum throughput capacity 最大通过能力
maximum thrust 最大推力
maximum tidal current 最大潮流
maximum tidal current velocity 最大潮流速度;最大潮流流速
maximum tidal flood strength 最大涨潮流速
maximum tide 最高潮
maximum tide level 最高潮位
maximum tide range 最大潮差
maximum tight 最大紧密度
maximum-to-average-power ratio 最大功率与平均功率的比
maximum toggle speed 最大反复速度
maximum tolerable concentration 最大容许浓度
maximum tolerable noise 最大容许噪声
maximum tolerance limit 最大容许限度;最大耐受量
maximum tolerance risk 可容许风险
maximum tolerated dose 最大耐受(剂)量
maximum tonnage capacity 最大通过能力
maximum torque 最大转矩;最大扭矩
maximum torque of motor 电动机最大转矩
maximum torque phase 瞬时力矩最大值
maximum torque spark timing 最大扭矩点火时
maximum torque speed 最大扭矩转速
maximum total resource 最大总资源量
maximum total sag 最大总垂度
maximum total traverse 最大方向转动界
maximum trace amplitude 最大波振幅
maximum track capacity 最大线路通过能力
maximum tracking error 最大跟踪误差
maximum traction truck 最大牵引车架
maximum tractive current of a catenary system 接触网最大牵引电流
maximum tractive current of a overhead catenary system 接触网最大牵引电流
maximum tractive effort 最大牵引力
maximum tractive force 最大牵引力
maximum traffic capacity 最大通过能力
maximum trailed load 最大牵引荷载
maximum transfer 回转初径
maximum transfer rate 最大传送速率
maximum transmission ratio 最大传动比
maximum transmission time 最长传输时间
maximum transmitting efficiency 最高发射效率
maximum transverse section coefficient 最大横剖面系数
maximum travel of cross slide 横刀架最大行程
maximum travel of tool slide 刀架最大行程
maximum travel of turret slide 回转刀架最大行程
maximum travel of vertical slide 立刀架最大行程
maximum treatment efficiency 最大处理效率
maximum turbidity zone 最essibility大混浊区
maximum turbine discharge capacity 水轮机最大过水能力;水轮机最大过流量
maximum turning length 最大车削长度
maximum turn separation point 最大圈距点
maximum tyre load 最大轮胎负载
maximum unambiguous range 最大单值区域
maximum undistorted output 最大无畸变输出功率
maximum undistorted power output 最大无失真输出功率;最大不失真功率输出
maximum unilateralized power gain 最大单向功率增益
maximum unit flow 最大单位流量

maximum unit stress 最大单位应力
maximum upsurge 最大上涌浪
maximum usable frequency 最高可用频率;最大可用频率
maximum usable writing speed 最大有效记录速度
maximum useful discharge 最大使用流量
maximum useful frequency 最大使用频率
maximum useful load 最大可用负载;极限有效荷载
maximum utility 最大效用
maximum utilization 最大负载;最大负荷
maximum utilization factor 最大利用系数
maximum utilized line capacity 最大利用的线路容量
maximum vacuum 最高真空度
maximum valence[valency] 最高价【化】
maximum value 最大值;极大值
maximum value indicator 最高值指示器
maximum value of anomaly 异常极大值
maximum value of log 测井曲线极大值
maximum value of organic carbon 有机碳最大值
maximum value over years 多年最大值
maximum value variogram 最大变差值
maximum vapo(u)r pressure 最大蒸气压(力);最大水汽压(力)
maximum vapo(u)r tension 最大水汽张力
maximum vector 最大向量
maximum vehicle-extension green period 最大延长绿设置
maximum velocity 最高速度;最大速度
maximum velocity locus 最大流速轨迹
maximum velocity of entrance 最大进口速度
maximum velocity of flow 最大流速
maximum velocity ratio 最大速度比
maximum velocity response 最大速度反应
maximum vertical curve radius 最大竖曲线半径
maximum vertical speed 最大垂直速度
maximum vibrating acceleration of surface 地面最大振动加速度
maximum visibility 最大能见度;最大可见度
maximum void ratio 最大孔隙比
maximum voltage 最高电压
maximum volume 最大体积;最大容积
maximum warm-up speed 最大加温转数
maximum water bursting yield 最大突水量
maximum water capacity 最大含水量
maximum water consumption 最高用水量;最大用水量;最大需水量;最大耗水量
maximum water-consumption per capita per day 每人每日最大用水量
maximum water-consumption per head per day 每人每日最大用水量
maximum water daily output 最高日供水量
maximum water demand 最大需水量
maximum water head 最大水头
maximum water-holding capacity 最大含水量;最大持水量;(土壤)最大持水量;最大保水能力;最大保水量
maximum water level 最高水位;最高潮位
maximum water-level drawdown in well 水井最大水位降深
maximum waterline beam 水线最大宽度
maximum water need 最大需水量
maximum water pressure 最大水压力
maximum water rate 最高用水量
maximum water surface elevation 水面最大高程
maximum water yield of mine 矿坑最大涌水量
maximum water yield of mine in rainy season 雨季矿坑最大涌水量
maximum water yield of well 水井最大出水量
maximum water-yield per well 单井最大出水量
maximum wave 最大波浪
maximum wave height 最大浪高;最大波高
maximum wave height to operate 作业最大波高
maximum wave of end portion 尾震最大波
maximum wave of principal shock 主震最大波
maximum weight 最大重量
maximum weight of equipment 设备最大重量
maximum well discharge 井的最大流量
maximum wheel load 最大轮载
maximum width 最大宽度
maximum width cut by single-side planer 单边刨最大刨削宽度
maximum width of cut 最大开挖宽度
maximum width of ditch 最大越壕宽

maximum width of face 表面最大宽度
maximum width of vehicle outside 车辆外侧最大宽度
maximum wind and shear chart 最大风速和风切变图
maximum wind force 最大风力
maximum wind speed 最大风速
maximum wind velocity 最大风速
maximum work 最大功
maximum working height 最大工作高度
maximum working pressure 最高允许压力;最大使用压力;最大工作压力;最大使用压力;最大工作压力
maximum working space 最大工作空间
maximum working temperature 最高工作温度
maximum working tension 最大工作张力
maximum working value 最大工作面积
maximum working voltage 最高工作电压;最大工作电压
maximum workpiece weight 工件最大重量
maximum yearly peak 年最大流量
maximum yield 最大出水量;最大产量;屈服上限
maximum yield of drillhole 钻孔最大出水量
maximum zonal westerlies 最大纬向西风带
maximun reflectance of vitrinite 镜质体最大反射率
maxiorder 巨额订单
maxirolling (锻压机的)预变形轧辊
maxi-taxi 巨型出租车
maxivalence 最高价【化】
maxlite 扯窗机构;扯窗系统
max-min system 库存高低点定量控制系统;库存最大最小定量控制系统
max-min value of organic carbon 有机碳最大最小值
maxmun field capacity 田间最大持水量
max possible compression ratio 最大理论压缩比
max roll(angle) 最大横摇角
maxterm 最大项;极大项;大项
maxterm expansion 最大项展开
maxweld (用于混凝土中)金属编织物
Maxwell 麦克斯韦;麦(磁通量单位)
Maxwell-Betti reciprocal theorem 麦克斯韦—贝蒂互换定理
Maxwell-Boltzmann classical statistics 麦克斯韦—玻耳兹曼经典统计学
Maxwell-Boltzmann distribution 麦克斯韦—玻耳兹曼分布
Maxwell-Boltzmann distribution law 麦克斯韦—玻耳兹曼分布律
Maxwell-Boltzmann's equation 麦克斯韦—玻尔兹曼方程
Maxwell fluid 黏弹流体
Maxwellian character 麦克斯韦特性
Maxwellian distribution 麦克斯韦分布
Maxwellian distribution law 麦克斯韦分布(定)律
Maxwellian energy distribution 麦克斯韦能量分布
Maxwellian equilibrium 麦克斯韦平衡
Maxwellian gas 麦克斯韦气体
Maxwellian velocity distribution 麦克斯韦速度分布
Maxwellian viewing 麦克斯韦观察法
Maxwellian viewing system 麦克斯韦观察系统
Maxwellmeter 麦克斯韦磁通计;麦克斯韦计;磁通计
Maxwell-Mohr's Method 麦克斯韦—摩尔法
Maxwell polygon of forces 麦克斯韦力多边形
Maxwell's and Harrington impact test 麦克斯韦—哈林顿冲击试验
Maxwell's body 麦克斯韦液体;麦克斯韦体
Maxwell's circulating current 麦克斯韦环流
Maxwell's coefficient of diffusion 麦克斯韦扩散系数
Maxwell's colo(u)r triangle 麦克斯韦色三角;麦克斯韦色品图
Maxwell's constant 麦克斯韦常数
Maxwell's cross-section 麦克斯韦截面
Maxwell's diagram 麦克斯韦图
Maxwell's disk 麦克斯韦圆盘;麦克斯韦混色盘
Maxwell's distribution 麦克斯韦分配;麦克斯韦分布
Maxwell's distribution law 麦克斯韦分布(定)律
Maxwell's effect 麦克斯韦效应
Maxwell's electromagnetic theory of light 麦克斯韦光的电磁论
Maxwell's electromagnetic wave theory 麦克斯韦电波理论

Maxwell's equal area rule 麦克斯韦等面积法则
Maxwell's equation 麦克斯韦方程
Maxwell's equation of state 麦克斯韦状态方程
Maxwell's field 麦克斯韦场
Maxwell's field equation 麦克斯韦(电磁)场方程
Maxwell's fluid 黏弹流体;麦克斯韦流体(即黏弹性流体)
Maxwell's force 麦克斯韦力
Maxwell's formula 麦克斯韦公式
Maxwell's fundamental equation 麦克斯韦基本方程
Maxwell's hypothesis 麦克斯韦假说
Maxwell's isotropic(al) viscous medium 麦克斯韦各向同性黏性介质
Maxwell's law 麦克斯韦定律
Maxwell's law of reciprocity 麦克斯韦互等变位定律
Maxwell's law of velocity disribution 麦克斯韦速度分布定律
Maxwell's law reciprocal deflection 麦克斯韦变位互等定律
Maxwell's liquid 麦克斯韦液体
Maxwell's mean free path 麦克斯韦平均自由行程
Maxwell's model 麦克斯韦模型(表示材料流变性质的一种力学模型)
Maxwell's orthogonal rheometer 麦克斯韦正交流变仪
Maxwell's primary 麦克斯韦基色
Maxwell's relation 麦克斯韦关系式
Maxwell's relaxation equation 麦克斯韦松弛方程
Maxwell's relaxation test 麦克斯韦松弛试验
Maxwell's relaxation time 麦克斯韦松弛时间
Maxwell's rheological model 麦克斯韦流变模型
Maxwell's rule 麦克斯韦定则
Maxwell's spectrum 麦克斯韦能谱
Maxwell's spot 麦克斯韦光斑
Maxwell's stress-strain relation for elasto-viscous fluid 麦克斯韦黏—弹性流体的应力应变关系
Maxwell's substance 麦克斯韦体
Maxwell's theorem 麦克斯韦定理;麦克斯韦原理
Maxwell's theorem of reciprocity 麦克斯韦互等定理;麦克斯韦变位互等原理
Maxwell's theory of light 麦克斯韦光学理论
Maxwell's triangle 麦克斯韦三角形;色三角
Maxwell's Wagner mechanism 麦克斯韦—瓦格纳机构
Maxwell-turn 麦克斯韦匝
maxwhite 白革涂饰剂
Maya arch 玛雅(式)拱(门);突拱
Maya architecture 玛雅建筑(中美洲印第安人的建筑);玛雅式建筑
Mayan astronomy 玛雅天文学
mayapis 柳安
Mayari R 迈亚里 R 低合金耐热钢
Maybach impact testing machine 梅氏冲击试验机
maybush 山楂
Maycoustic 隔声用人造石;人造石(隔声用)
mayday 无线电话求救信号
mayen 高山低草原
mayenite 钙铝石
mayer 迈尔(热容量单位)
Mayer curve [M-curve] 迈尔曲线
Mayer problem in calculus of variations 变分法中的迈尔问题
Mayer's condensation theory 迈尔凝结理论
Mayer's fluid 迈尔溶液(氯化银磺化锌溶液)
Mayer's formula 迈尔公式
Mayer's process 迈尔精密冲裁法
Mayer's synthesis 迈尔合成法
May Hill sandstone 梅丘陵砂岩
Maynani board machine 马雅尼制瓦(板)机
Maynani pipe machine 马雅尼制管机
mayonnaise 低温残渣
may self-combustion factor 可能自燃因素
May's graticule 梅斯分度镜
Maysvillian 迈斯维尔阶【地】
May thorn 山楂
maytree 山楂木
mazaedium 黏表层
Mazak alloy 马扎克锌基压铸合金
mazarine 深蓝色
mazarine blue 深蓝色料;深蓝色
Mazatzal geosyncline 马萨兹尔地槽

maze 迷宫;曲折前室;曲径
maze-bright 敏于迷路的
maze domain 迷宫型磁畴
maze-like conduit 迷宫式通道
maze method 迷宫法
maze-running algorithm 迷路走线算法
Mazier triple tube retractor core barrel 美兹三管牵引器岩芯管
Mazlo alloy 镁合金;梅兹洛合金
maz(o)ut 铺路油;重油;黑油
mazy 迷宫式的
Mazza process 马扎抄取制管法
mazzite 针沸石
m-benzenediol 间苯二醇
mbobomkite 硫硝镍铝石
mcallisterite 三方水硼镁石
McColl protective system 麦克科尔保护系统
McCollum terrestrial ammeter 麦克隆姆大地电流指示计
mcconnellite 铜铬矿
McConnell's curve 麦克康奈螺线
McCully gyratory 麦卡利悬轴式旋回破碎机
McDonnel Douglas Aircraft Company 麦克唐纳道格拉斯飞机公司
McGill metal 麦吉尔铝铁青铜;麦吉尔可铸可锻铝青铜
mcgovernite 粒砷硅锰矿
mcguinnessite 麦碳铜镁石
Mcguire-Keogh Act 麦夸尔—基奥法
m-chloroaniline 间氯苯胺
Mchmullen accelerated cabinet 麦克米仑加速试验箱
McIntosh classification 麦克因托希分类
Mckechnie's bronze 麦基奇尼高强度黄铜
mckelveyite 碳钇钡石
Mckinley type curve matching 麦金利图版法
mckinstryite 马硫铜银矿
Mclead vacuum ga(u)ge 麦克劳真空规
Mcleod formula 马克里奥公式(一种设计柔性路面厚度的公式)
Mcluckie gas detector 麦露奇型瓦斯检查器
McMahon packing 金属网鞍形填料;麦克马洪填料
McMarth's formula 麦马马斯公式
McMarth telescope 麦马马斯望远镜
McMichael visco(si)meter 麦克迈克尔黏度计
McMurray inverter 麦克默里变流器
McNally tube 反射速调管
mcnearite 麦砷钠钙石
McNemar's test 麦克尼马尔检验
Mc-Quaid-Ehn test 麦克奎德爱恩晶体粒度试验
m-cresol sulfonphthalein 间甲酚磺酞
M-C-S intensity scale 莫—坎—西烈度表
m-design 模量比法设计
Mdlin transform 梅林变换
meacon(ing) 虚造干扰设备;假象雷达干扰
meader 河曲;河流曲折
meadering stream 大弯度河
Meadfoot beds 米德富特层
meadow 牧草地;平稳区;草甸;草场
meadow and steppe landscape 草原景观
meadow bog 水湿地;低洼沼泽;草甸沼泽;草地沼泽
meadow-boggy soil 草甸沼泽土
meadow burozem 草甸棕壤
meadow chernozem 草甸黑钙土
meadow chestnut soil 草甸粟钙
meadow community 草甸群落
meadow deervetch 草地百脉根
meadow gley soil 草地格列土;草地潜育土
meadow grass rotation 草地轮作
meadow land 草地
meadow-marsh-pond treatment method 草地—沼泽—池塘处理法
meadow moor 水湿地;低位沼泽草地;草甸沼泽;草地沼泽
meadow ore 沼铁矿
meadow peat 草木泥炭;草甸泥炭
meadow podzolic soil 草甸灰化土
meadow sanne soil 草甸盐土
meadow soil 草甸土;草地土壤
meadow solonetzic solonchak 碱化草甸盐土
meadow steppe 草甸草原
meadow swamp deposit 草甸沼泽沉积
meadow sweet 欧洲合叶
meadow tundra 草甸冻原

meadow vegetation 草甸植物
meadowy land 牧场草坪;牧草地
meager 不足的;贫瘠的
meager clay 贫瘠黏土
meager coal 贫煤
meager lean coal 贫瘦煤
meager lime 有杂质的石灰;贫石灰;不纯石灰
meager oat 沙丘野麦
meager profit 微利
meagre 贫瘠的
meagre clay 脊黏土;贫瘠黏土;瘦黏土
meagre lime 不纯石灰;有杂质的石灰;贫石灰
meal 细磨石料;粗粉
meal hours 吃饭时间
meal mixer 混粉器
meal outlet 出粉口
meal push 跑生料
meal ring 生料圈
mealtime 吃饭时间
mealy 粉状的
mealy sand 粉砂土(壤);粉砂
mealy structure 粉状结构;粉末结构
mean 均值;均数;平均的;平均;手段
mean absolute deviation 均值绝对差;平均绝对偏向;平均绝对偏离;平均绝对偏差
mean absolute error 平均绝对误差
mean absolute magnitude 平均绝对星等
mean absolute percent error 平均绝对百分误差
mean absolute value 平均绝对值
mean acceleration 平均加速度
mean accumulation 平均积厚
mean activity 平均活度
mean activity product 平均活度乘积
mean aerodynamic(al) center 平均气动中心
mean alley 陋巷
mean altitude 平均海拔(高度)
mean-altitude photography 中等航高航摄
mean ambient flow vector 平均环境水流流速矢量
mean amount 平均量
mean amplitude of tide (潮的)平均振幅
mean angle 角均数;平均角
mean angle of emergence 平均引出角
mean angular deviation 平均角离差
mean annual 年平均
mean annual abstraction 平均年抽水量
mean annual change 平均周年变化
mean annual concentration 年平均浓度
mean annual discharge 年平均流量;内平均流量;平均年流量
mean annual efficiency 全年平均效率
mean annual erosion 平均年冲刷量
mean annual evapo(u)ration 年平均蒸发量
mean annual flood 年平均洪水流量;内平均洪水量;平均年洪水(流量)
mean annual growth rate 年平均增长率
mean annual growth rate percent 群体每年平均增长百分率
mean annual higher high water 多年平均较高高潮位
mean annual highest water level 多年平均最高水位
mean annual increment 平均年生长量
mean annual lowest water level 多年平均最低水位
mean annual maximum flood 年平均最大洪峰;多年平均最大洪水
mean annual maximum flood discharge 多年平均最大洪水流量
mean annual maximum level 多年平均最高水位
mean annual minimum flood 多年平均最小洪水
mean annual minimum flood discharge 多年平均最小洪水流量
mean annual minimum level 多年平均最低水位
mean annual precipitation 年(平)均降水(量);平均年雨量多年平均降水(量)
mean annual rainfall 年(平)均雨量;平均年降水量;多年平均雨量
mean annual range 年平均较差;多年平均较差;多年平均变幅
mean annual range of temperature 平均温度年差
mean annual rate 每年平均速率
mean annual rate of interest 平均年利率
mean annual runoff 正常径流量;年平均径流(量);内平均径流量;平均年径流;多年平均径流(量)

mean annual sea level 多年平均海(平)面
mean annual sedimentation rate 平均年沉积率
mean annual sediment discharge 平均年输沙量
mean annual sediment yield 平均年泥沙沉积量
mean annual soil loss 平均年土壤流失量
mean annual temperature 年平均温度;年平均气温;内平均温度;多年平均温度;常年平均温度
mean annual temperature range 平均年温差
mean annual water level 年平均水位;平均年水位;多年平均水位
mean anomalistic motion 平均近点运动
mean anomaly 平均近点角;平近点角
mean approximation 平均近似
mean areal depth of precipitation 流域平均雨深;面平均降水深度
mean areal precipitation 面平均雨量;面平均降水量
mean astronomical north pole 平均天文北极
mean astronomic latitude 平天文纬度
mean astronomic meridian plane 平天文子午面
mean atmospheric(al) refraction 平均大气折射
mean availability 平均可用性
mean available discharge 平均可用流量
mean avoiding angle 平均避碰角
mean avoiding method 平均规避法
mean avoiding speed 平均规避速度
mean axis 中轴;平均轴
mean basin height 流域平均高度;流域平均高程
mean basin slope 流域平均坡降;流域平均坡度
mean bearing stress 平均承载应力
mean bed level 平均河底高程
mean blade chord 叶片平均翼弦
mean block anomaly 平均区域异常
mean boundary-layer thickness 平均边界层厚度
mean breadth 平均宽度
mean breakup date 平均解冻日期
mean brightness 平均亮度
mean calorie 平均热值;平均卡路里
mean camber 中曲面
mean camber line 中弧线
mean candle 平均烛光
mean carbon content 平均含碳量
mean carrier 中间载波;平均载波
mean carrier frequency 平均载波频率
mean carrier velocity 平均载气速度
mean center[centre] of the moon 平均月球中心
mean channel height 平均河床高度;平均河床高程
mean chart 平均值等深线图
mean circulation time 平均循环时间
mean circulatory filling pressure 平均循环系充盈压
mean clay 中稠黏土
mean closure 平均闭合差
mean code length 平均电码长度
mean coefficient of elongation 平均延伸系数
mean collisional frequency 平均碰撞频率
mean collision time 平均碰撞时间
mean column pressure 平均柱压
mean complexity constant 平均络合常数
mean composite sample 平均拼合样品
mean composition 平均组成
mean compression[compressive] strength 平均受压强度;平均抗压强度
mean concentration 平均浓度
mean conjugate curve 平均共轭曲线
mean conjugate net 平均共轭网
mean consistency 中稠度
mean consumption 平均用(水)量
mean convergence 平均收敛
mean corpuscular volume 平均微粒子体积
mean-cost curve 平均费用曲线
mean cotidal hour 平均同时潮
mean covariance function 平均协方差函数
mean current 平均流;平均电流
mean curvature 中曲率;平均曲率
mean curvature of level surface 水准面平均曲率
mean cycles between failures 平均无故障次数;损坏周期平均值
meand 老露天矿
mean daily 日平均;日晷仪
mean daily change 平均日变化
mean daily consumption 平均日耗量
mean daily discharge 日平均流量;日平均流量
mean daily dredging quantity 日平均疏浚量
mean daily flow 平均日流量

mean daily humidity 日平均湿度
mean daily maximum temperature 逐日最高温度平均值
mean daily motion 平均日运动
mean daily range 平均日温差
mean daily stage 日平均水位
mean daily temperature 日(常)平均温度;常日平均温度
mean daily traffic 日平均运量;平均日交通量
mean daily water consumption 平均日耗水量
mean date of first snow cover 平均初雪(日)期
mean date of last snow cover 平均终雪(日)期
mean day 平日
mean decade humidity 旬平均湿度
mean decade temperature 旬平均温度
mean declination 平均偏角
mean deficit 平均亏空;平均赤字
mean definition 平均清晰度
mean degree of polymerization 平均聚合作用;平均叠合度
mean delay time 平均缓发时间
mean demand 平均需要量
mean density 平均容重;平均密度
mean departure 平均偏差;平均距
mean deposit number 平均矿床数
mean depth 平均水深
meander 弓形湖;迷宫曲径回文饰;回纹(波形)饰;曲折回纹(饰);曲折(河道);曲径;蜿蜒;蛇曲;羊肠小道;曲流
meander amplitude 曲流幅度;蜿蜒幅度
meander bar 弓弧沙坝;曲湾(内侧)坝;曲流沙坝;曲流边滩
meander belt 河曲(地)带;河道蜿蜒带;河道曲折带;曲流河段;曲流(地)带;蜿蜒带
meander belt length 蜿蜒带长度
meander belt width 蜿蜒带宽度;曲流带宽度
meander bend 河曲
meander bluff 曲流陡岸
meander breadth 河曲曲折宽度;曲折带宽度
meander channel 游荡性水道;弯曲水道;弯曲河槽
meander coefficient 河流弯曲系数;弯曲系数
meander concertina 蛇腹形河弯;蛇腹形曲湾;多重河曲
meander configuration 曲折型
meander core 离堆山;曲流环绕岛
meander crescent 新月形废河湾
meander curve 曲流
meander cut-off 裁弯取直;曲流裁弯
meander deformation 河弯变形
meander element 曲流要素
meandering 羊肠小道;游荡;河道蜿蜒;曲折;曲流
meandering belt motion 胶带扭曲动作
meandering channel 游荡河道;蜿蜒水道;蜿蜒河道;弯曲航道;弯曲河槽
meandering coefficient of traverse 导线曲折系数
meandering motion 蜿蜒运动
meandering movement 迂回运动;游荡运动;曲折移动;蜿蜒运动;曲流移动
meandering of river 河道弯曲
meandering phenomenon 曲流现象
meandering reach 曲流段;蜿蜒河段;弯曲型河段;弯曲河段
meandering river 游荡性河流;游荡型河流;曲流河;蜿蜒(性)河流;弯曲河流;弯曲河;蛇曲河(流);大弯度河
meandering river sedimentation model 曲流河沉积模式
meandering river stretch 蜿蜒河段
meandering stream 游荡性河流;游荡型河流;曲折河;曲流河;蜿蜒(性)河流;弯曲河流;弯曲河;蛇曲河(流)
meandering stream deposit 曲流河沉积
meandering stream sequence 曲流河层序
meandering structure 蜿蜒构造
meandering type 弯曲型
meandering valley 曲流河谷;蜿蜒河谷;蛇曲河谷
meander lake 弓形湖;牛轭湖;曲流湖
meander length 河弯长度;曲曲长度;曲折长度;曲流(带)长度;蜿蜒长度
meander line 折测线;曲流线;曲折线;曲流;蜿蜒线
meander-line travel(l)ing wave tube 曲折线行波管
meander loop 河环;曲流的环状河道;弯曲河环
meander loop cut-off 河套裁弯

meander migration 河弯游荡;河弯迁移;河曲转移;曲流游荡;曲流迁移
meander neck 河弯颈;曲流颈
meander niche 曲流壁龛
meander pattern 河弯形式
meander plain 曲流平原
meander platform 河曲平面形态
meander radius 曲流半径
meander ratio 河道曲率;曲折率;曲流率;曲流比;蜿蜒比;弯曲系数;弯曲比
meander river-bed 曲流型河床
meander scar 曲流痕
meander scroll 曲流迂回扇;曲流内侧坝;曲流湖;蜿蜒河湾
meander spur 曲流山嘴
meander system 河曲系统
meander terrace 曲流阶地
meander theory 河曲理论
meander types 曲流类型
meander wave 蜿蜒波
meander wavelength 曲流波长
meander width 河弯宽度;河弯宽度;曲折宽度;曲折带宽度;曲流宽度;蜿蜒宽度
meander wriggling 河曲蠕动
meander zone 曲流地带
mean detection limit 平均检出限
mean deviation 离均差;均差;平均自差;平均漂移;平均偏移;平均偏差;平均离差;平均差(值)
mean deviation in daily rate 平均日偏差
mean deviation of daily rate 平均日偏差
mean diameter 平均直径;平均粒径
mean difference 均差;平均离差;平均概率偏差;平均差分;平均差(值)
mean difference of elevation 平均高(程)差
mean difference parameter 平均差参数
mean dimension 平均尺度
mean discharge 平均流量
mean discharge modulus 平均流量模数;平均流量模量
mean discharge ratio 平均流量比
mean dispersion 中间色散;平均色散
mean displacement 平均位移
mean dissipation 平均耗散
mean distance 平均距离
mean distance between failures 平均故障行程;平均无故障距离
mean diurnal change 平均周日变化
mean diurnal high-water inequality 日潮高潮不等;平均日高潮不等(值)
mean diurnal low-water inequality 平均日低潮不等(值);平均日潮低潮不等
mean diurnal motion 平均周日运动;平均日运动
mean diurnal variation 平均周日变化
mean divergence 平均散度;平均扩散度
mean division method 平分法
mean dosage 平均剂量
mean down time 平均空闲时间;平均故障时间;平均不工作时间
mean downtime 平均故障间隔期
mean draft 平均吃水(深度)
mean draught 平均吃水(深度)
mean draught of vessel 船舶平均吃水
mean drift 平均漂移
mean dry weather flow 平均旱天流量
mean dry year 一般枯水年;平均枯水年
mean duration curve 平均历时曲线
mean duration of ebb 平均落潮历时
mean duration of failure 平均故障时间;平均故障历时
mean duration of flood 平均涨潮历时
mean dust concentration 烟道气粉尘;粉尘平均浓度
mean dust concentration in stack gas 烟道气粉尘平均浓度
mean dwelling 陋宅
mean dynamic(al) amplification factor 平均动力放大系数
mean dynamic(al) head 平均动压头
mean earth ellipsoid 平均地球椭球
mean echo balance return loss 平均回波平衡返回损耗
mean ecliptic 平黄道
mean ecliptic sun 平黄道太阳
mean effective diameter 平均有效直径

mean effective horsepower 平均有效马力
mean effectiveness factor 平均效率因素
mean effective pitch 均效螺距;平均有效螺纹;平均有效螺距
mean effective power 均方根功率;平均有效功率
mean effective pressure 均效压力;平均有效压力
mean effective stress 平均有效应力
mean effective temperature 平均有效温度
mean effective value 平均有效值
mean effective viscosity 平均有效黏度
mean effect range median quotient method 平均效应区间中值商法
mean efficiency 平均效率
mean efficiency factor 平均有效系数
mean effort 平均作用力
mean electron concentration 平均电子浓度
mean element (取景器的)中镜片
mean elevation 平均高程
mean elevation of basin 流域平均高程
mean elongation 平均延伸量
mean end area 平均端面积
mean end area formula 平均端面积公式
mean end area method 平均端面积(计算)法
mean epoch 平均时纪;平均历元
mean equator 平赤道
mean equatorial moment 平均赤道惯性距
mean equatorial sun 赤道平(均)太阳
mean equinoctial springs 平均春分大潮
mean equinox of date 平春分点
meaner 方法
mean ergodic 平均遍历的
mean ergodic matrix 平均遍历矩阵
mean ergodic semi-group 平均遍历半群
mean ergodic theorem 平均遍历定理
meaner of execution 实施方法
mean erosion(al) depth 平均冲刷深度
mean erosion(al) velocity 平均冲刷速度
mean error 中误差
mean error ellipse 中误差椭圆
mean error of angle 角度中误差
mean error of direction 方向中误差
mean error of prior weight 先验权中误差
mean establishment 平均朔望潮高潮间隙;平均朔望;平均潮候时差
mean evolute 中点渐屈线
mean extreme high water 平均大潮极高潮
mean extreme high water springs 平均大潮极高潮面
mean extreme low water 平均大潮极低潮
mean extreme value 平均极值
mean failure rate 平均故障率
mean fall velocity 平均沉落速度;(泥沙的)平均沉降速度
mean fiber 中层纤维
mean fiber strength 纤维平均强度
mean filling pressure of circulatory system 循环系平均充盈压
mean flood level in past years 历年平均洪水位
mean flow 平均流量
mean flow at outgoing flow 落潮平均流量;退潮平均流量
mean flow direction in section 断面平均流向
mean flow duration curve 平均流量历时曲线
mean flow rate 平均流率
mean flow velocity 平均流速
mean fluctuation power 平均涨落功率值
mean fracture load 平均断裂荷载
mean free-air anomaly 平均空间异常
mean free collision time 平均自由碰撞时间
mean free error time 平均无故障时间
mean free path 平均自由行程;平均自由程;平均自由通路
mean free time 平均自由时间;平均自由时
mean freeze-up date 平均封冻日期
mean freezing index 平均冻结指数;平均冰冻指数
mean frequency 中间频率;平均频率
mean fuel conversion ratio 平均燃料转换比
mean function 平均值函数
mean future time 平均预期寿命
mean geometrical pitch 几何平均螺距
mean geometric(al) distance 几何平均距离
mean geothermal gradient 平均地温梯度
mean girth 平均干围
mean grade 平均比降

mean gradient 平均坡降;平均坡度;平均比率;平均比降
mean grading 平均粒径级配
mean grading method 平均级配方法
mean grain diameter 平均粒径
mean grain firn 中粒雪
mean grain size 平均粒径;平均粒度;平均颗粒直径;平均颗粒粒径
mean gravity 平均重力
mean gravity anomaly 平均重力异常
mean ground elevation 平均地面高程
mean ground level 平均地面高程
mean ground surface 平均地面高程
mean groundwater velocity 地下水平均流速
mean half-tide level 平均半潮面
mean haul 平均运距
mean haul distance 平均运距
mean heat capacity 平均热容(量)
mean heat flow density 平均热流密度
mean height 平均高度;平均高
mean height difference 平均高(程)差
mean height map 平均高度图
mean height of burst 炸点平均高度
mean height of partial high waves 部分大波平均波高
mean height of survey area 测区平均高程面
mean height of the tide 平均潮高
mean hemi-spheric(al) candle power 平均半球面烛光
mean hemi-spheric(al) intensity 半球面平均烛光;平均半球面亮度
mean higher high water 平均高高潮位;平均(较)高高水位;平均(较)高高潮(面)
mean higher high-water lunitidal interval 平均月潮高高潮间隙
mean higher high water springs 平均大潮高高潮(面)
mean higher low water 平均高低潮
mean highest discharge 平均最大流量
mean highest high water 平均最高高水位;平均最高高潮面
mean highest water level 平均最高水位
mean high of ordinary spring tide 大潮平均高潮位;大潮平均高潮面
mean high range 平均高潮差
mean high tide 平均高潮位;平均高潮
mean high water interval 平均高潮间隙
mean high water(level) 平均高潮(面);平均高水位;平均高潮位;多年平均高水位
mean high water line 平均高水位线;平均高潮线
mean high water lunitidal interval 平均高潮(月潮)间隙;平均月潮高潮间隙
mean high water neaps 平均小潮高潮位;平均小潮高潮面;小潮平均高潮位;小潮平均高潮面
mean high water of ordinary neap tide 平均小潮高潮位;平均小潮高潮面
mean high water quadrature 方照平均高潮
mean high water springs 平均大潮高潮面;大潮平均高潮位;大潮平均高潮面;平均大潮高潮位
mean high water spring tides 大潮平均高水位;平均大潮高潮位
mean high water tides 平均高潮水位
mean horizontal candle power 平均水平烛光;平均平面烛光
mean horizontal indensity 平均平面光强
mean horizontal sound speed 平均水平声速
mean hourly output of handling machinery 装卸机械平均台时产量
mean hours 平均小时
mean hydraulic depth 平均水力深度
mean hydraulic radius 平均水力半径
mean ice edge 平均冰界
mean igneous rock 平均火成岩
mean impulse indicator 平均脉冲指示器
mean incidence 平均冲角
mean inclination 平均倾角;平均坡降;平均坡度
mean increment 平均年生长量
mean indicated pressure 平均指示压力;平均图示压力
mean indicator 平均指示器
mean infiltration 平均渗率;平均渗透量;平均渗入率;平均渗入量
mean information content 平均信息量
meaningless figures 虚假数字

mean instantaneous frequency 平均瞬时频率
mean intensity 平均强度
mean interdiurnal change 平均日际变化
mean interdiurnal variation 平均日际变化;平均日变化
mean interval between successive generations 两代人平均间隔
mean in vertical 沿垂线平均;垂线平均
mean ionization 平均电离
mean ionization energy 平均电离能
mean isopleth 平均等值线
mean isopleth map 平均值等深线图
mean kinetic energy 平均动能
mean lake depth 平均湖深
mean lake level 平均湖面
mean land level 大陆平均面;大陆平均高度
mean latitude 平纬;平均纬度;平黄纬
mean length 平均长度
mean length of flow mark 流痕平均长度
mean length of life 平均寿命
mean length of overland flow 平均坡面漫流长度
mean length of turn 匝的平均长度
mean lethal dose 中间致死量
mean level 平均水平(面);平均水面;平均海面;平均高度;平均电平
mean-level auto-gain control 平均电平式自动增益控制
mean level detection system 平均电平检测系统
mean level of spring low tide 大潮平均低潮位;大潮平均低潮面
mean level of the sea 平均海平面;平均海面
mean level of wave 波浪中分面
mean life 平均(使用)寿命;平均使用年限;平均年限
mean life rate 平均寿命率
mean lift 平均升举力
mean lift force 平均升举力
mean light curve 平均光度曲线
mean limit 平均极限
mean line 中线;中心线;平均线;平分线;顶线;等分线
mean linear diameter 平均线性直径
mean linear range 平均直线射程
mean linear successive difference 平均线性连续差
mean line system 中线系统
mean load 平均荷载;平均负载;平均负荷
mean load density 平均装填密度
mean logarithmic decrement 平均对数衰减率
mean logarithmic energy loss 平均对数能量损失
mean log residual 平均对数差余
mean longitude 平经(度)
mean longitudinal error 平均距离误差
mean longitudinal resistivity method 平均纵向电阻率法
mean long-term yield 多年平均产量
mean loss ratio 平均损失比
mean lot elevation 场地平均标高
mean lower high water 平均低高潮
mean lower limit 平均下限
mean lower low tide 平均低低潮(面)
mean lower low water 平均较低低水位;平均低潮(面)
mean lower low-water lunitidal interval 平均月潮低低潮间隙
mean lower low water springs 平均大潮低低潮(面)
mean lowest discharge 平均最小流量
mean lowest water level 平均最低水位;平均最低潮位;平均最低潮面
mean low flow 平均枯水流量
mean low tide 平均低潮位;平均低潮面;平均低潮
mean low water 平均低潮位;平均低潮面;平均低潮
mean low water flow 平均枯水流量
mean low water full and change 平均低潮间隙
mean low water interval 平均低潮间隙
mean low water level 平均低水位;平均低潮面
mean low water lunitidal interval 平均低潮月潮间隙;平均月潮低潮间隙
mean low water mark 平均低潮水位
mean low water neaps 小潮平均低潮位;小潮平均低潮面;小潮低潮
mean low water neap tides 小潮平均低潮位;平均小潮低潮位;平均小潮低潮面
mean low water of ordinary neap tide 小潮平均低潮位;小潮平均低潮面;平均小潮低潮位;平均

小潮低潮面
mean low water of ordinary spring tide 大潮平均低潮位;大潮平均低潮面
mean low water springs 平均低潮位;平均大潮低潮(面);大潮平均低潮位;大潮平均低潮面
mean low water spring tides 大潮平均低潮位;平均大潮低潮位
mean luminance 平均亮度
mean lunar semidiurnal constituent 平均月潮半日分潮
mean lunar semidiurnal tide 平均太阴半日潮
mean magnetizing curve 平均磁化曲线
mean mass range 平均质量射程
mean matrix 平均矩阵
mean maximum 平均最大值;平均极大值
mean measured value 平均量测值
mean meridian 平子午线;平均子午线
mean meridional circulation 平均经向环流
mean mesh 平均筛目
mean method 折中办法
mean-metric(al) 平均度量的
mean midnight 平子夜
mean mileage 平均里程
mean minimum 平均最小值;平均极小值
mean minimum flow 平均枯水流量
mean modulation rate 平均调制速率
mean modulus of deformation 平均变形模量
mean modulus of error 平均波动模数(交通量)
mean molar quantity 平均摩尔数量
mean molecular weight 平均分子量
mean molecule 平均分子
mean monthly air temperature 月平均气温
mean monthly discharge 月平均流量
mean monthly highest tidal level 月平均最高潮位
mean monthly highest water level 月平均最高水位;朔望平均高潮位
mean monthly lowest water level 月平均最低水位;朔望平均低潮位
mean monthly maximum temperature 月平均最高温度;月平均最高气温
mean monthly minimum temperature 月平均最低温度;月平均最低气温
mean monthly stage 月平均水位
mean moon 平月亮
mean motion 平均运动;平均行度
mean movement 平均运动
mean navigation period 平均通航期
mean neap 平均小潮
mean neap range 小潮平均潮差;小潮潮差;平均小潮差
mean neap rise 小潮平均潮升;小潮小潮升
mean neap rise of high tide 小潮小潮高潮升
mean neap rise of low tide 小潮小潮低潮升
mean neap tide 平均小潮
mean net head 平均净水头
mean neutron lifetime 中子平均寿命
mean night flow 平均夜流量
mean nine-month flow 平均低水流量
mean noon 平正午;平午
mean normal deformation 平均法向形变;平均法向变形
mean normal gravity 平均正常重力
mean normal stress 平均正应力;平均垂直应力
mean number of collisions 平均碰撞数
mean obliquity 平均黄赤交角
mean obliquity of ecliptic 平黄赤交角
mean observatory 平均天文台
mean occurrence rate 平均发生率
mean ocean floor 平均海底
mean of a fuzzy event 模糊事件均值
mean of a probability distribution 概率分布均值
mean of binomial distribution 二项分布平均值
mean of conditional simulation 条件模拟均值
mean of high stages 平均高水位
mean of high water stages 平均洪水位
mean of low stages 平均低水位
mean of low water stages 平均枯水位
mean of mean 平均值的平均;平均的平均
mean of non-conditional simulation 非条件模拟均值
mean of normal water levels 平均常水位
mean of observation 观测值中数;实测平均值
mean of population 总体平均值;总体(平)均数;人口平均值

mean of tide 平均潮高
mean of valve loop curve 平均调节阀组合特性曲线
mean of variance 方差平均
mean operating time between failures 故障间平均运行时间
mean operation time 平均运算时间
mean orbit 平均轨道
mean orbit radius 平均轨道半径
mean ordinary water level 平均水位
mean orthogonal reflectivity 平均正交反射率
mean output 平均效率;平均产量
mean parallax 平均视差
mean parameter 平均参数;参数平均值
mean particle diameter 颗粒平均粒径;平均粒径;平均颗粒直径;平均颗粒粒径
mean paste 中级研磨膏
mean penetration of pile 桩的平均贯入度
mean penetration rate 平均钻速
mean percent error 平均百分误差
mean performance 平均表现
mean period 平周期
mean periodic functions 平均周期函数
mean period of storage 库场平均堆存期
mean per unit 每单位(平)均值
mean picture brightness 图像平均亮度
mean picture level 图像平均电平
mean piston speed 平均活塞速度
mean pitch 平均螺距
mean place 平位置;平均位置
mean point of group 点群平均点
mean pole 平极
mean pole of rotational axis 自转轴北极
mean pole of the epoch 历元平极
mean pole position 平均磁极位置
mean pore pressure 平均孔隙压力
mean pore size 平均孔径
mean position 平位置;平均位置
mean potential water power 平均水力蕴藏量
mean power 平均功率
mean precipitation 平均雨量;平均降水量
mean precipitation intensity 平均降雨强度
mean prediction 均值预测
mean pressure 平均压力
mean pressure grade line 平均压力梯度线
mean price 平均价(格)
mean principal stress 中间主应力;平均主应力
mean probable error 平均概率误差;平均概率偏差
mean profit 平均利润
mean profit rate 平均利润率
mean proportion(al) 比例中项
mean pulse profile 脉冲平均轮廓
mean pulse time 平均脉冲时间
mean p-valent function 平均p叶函数
mean quadratic error 均方误差;均方差;二次平均误差
mean quality 平均质量
mean quantity of precipitation 平均降水量
mean radial error 平均径向误差
mean radiant temperature 平均辐射温度
mean radiation intensity 平均辐射强度
mean radioactivity meter 平均放射性测量计
mean radius 平均半径
mean radius of curvature 平均曲率半径
mean radius of the earth 地球平均半径
mean rainfall 平均(降)雨量;平均降水量
mean rainfall intensity 平均降雨强度
mean range 极差率均值;平均差(值);平均幅度;平均变(化)幅(度)
mean range of tide 平均潮差
mean rank 平均等级
mean rate of irrigation 平均灌溉率
mean ratio 平均比值
mean recession curve 平均退水曲线
mean recurrence time 平均重现时间;平均回转时间
mean referred pressure 平均参考压力
mean reflectance 平均反射能力
mean reflectivity of vitrinite 镜质体平均反射率
mean reflectivity 平均反射率
mean refraction 平均折射;平均蒙气差;平均大气折光差
mean refractive index 平均折射率
mean refuelling rate 平均换料率
mean region anomaly 平均区域异常

mean relative deviation 平均相对偏差
mean relative fluctuation 平均相对偏差;平均相对波动
mean relative humidity 平均相对湿度
mean repair time 检修平均周期;平均检修时间;平均修复时间
mean reserves of deposit 矿床平均矿石量
mean residence time 平均残留期
mean residual 平均残差
mean resistance 平均阻力
mean resource of unit volume 单位体积平均资源量
mean response spectrum curve 平均反应谱曲线
mean return air temperature 平均回风温度
mean return time 平均重现时间
mean return water temperature 平均回水温度
mean rise 平均潮位升高
mean rise interval 平均涨潮间隙;平均潮升间隙
mean rise of high tide 平均高潮升
mean rise of low tide 平均低潮升
mean rise of tide 平均潮升
mean rise of water 水位上升平均高度
mean river flow 平均河流流量;平均河川流量
mean river level 江河平均水位;平均河水位;平均河面高程;平均河面
mean rotation 平均旋度
mean roundness 平均圆度
mean runoff 平均径流量
mean runoff rate 平均径流率
means 收入;办法
mean sample 平均样品;平均样本
mean sample tree 中央木
mean sampling 平均取样
mean scale 平均比例尺
means clustering analysis 平均值聚类分析
means depreciation kilometer 公共交通工具折旧里程
means depreciation time limit 公共交通工具折旧期限
mean sea depth 平均海深
mean sea level 平均海(平)面;平均海平面高程;平均海水面
mean sea level datum 平均海水面基点;平均海平面基点
mean sea level of Huanghai 黄海平均海平面
mean sea level of the Yellow Sea 黄海平均海平面
mean sea level pressure 平均海平面气压
mean sea level surface 平均海水面
mean seasonal river 平均季节性河流
mean seasonal runoff 平均季节性径流
mean sediment concentration 泥沙平均浓度;平均泥沙浓度;平均含沙量(指油、水、空气中含沙量)
mean sediment concentration in-section 断面平均含沙量
mean sediment concentration on a vertical 垂线平均含沙量
mean sediment discharge 平均输沙率
mean sediment size 泥沙平均粒度
mean seeking time 平均查找时间
means-ends analysis 手段与目的分析
mean service life 平均使用年限
mean settlement velocity (泥沙的)平均沉降速度
mean shading 平均明暗度
mean shear strain 平均剪应变
mean shear stress 平均剪应力
mean shift 平均位移
mean shipping weight 平均装载重量
mean side length 平均边长
mean sidereal day 平均恒星日
mean sidereal time 平均恒星时;平恒星时
mean silt charge 平均含沙量(指油、水、空气中含沙量)
mean size 平均大小;平均尺度
mean sky 平均天空
mean slope 平均坡降;平均坡度;平均比降
mean smoothed spectrum 平均平滑谱
means of communication 交通工具;通信工具
means of compensation with compensating drum 平衡卷筒补偿法
means of compensation with compensating pulley 平衡滑轮补偿法
means of compensation with principle of ellipsograph 椭圆规原理补偿法

means of compensation with pulley block 滑轮组补偿法
means of consumption 消费资料
means of conveyance 交通运输工具;传送工具;财产转让方式
means of defence 防御手段
means of delivery 运载方法;输送工具;输送方法
means of demonstration 论证方法
means of determination 测定手段
means of development 发展资料
means of dispatch 疏运工具
means of driving 传动工具;传动手段;驱动工具;驱动手段;传动方法;驱动方法
means of egress 疏散设施
means of enjoyment 享受资料
means of erection 安装方法;安装方式;安装的设备
means of escape 消防安全通道;消防安全措施;应急离机手段;疏散设施;安全疏散路线
means of exchange 交换手段
means of execution for project 工程实施方法
means of financing 融资手段
means of hydraulic fracturing 水力破碎法
means of injecting slurry 注水泥方法
means of labo(u)r 劳动资料
means of light signalling 光信号工具
means of line communication 有线通信工具
means of livelihood 生活资料
means of living 生活资料
means of one's satisfaction 满足需要资料
means of payment 支付手段;支付方式
means of production 生产资料;生产手段;生产工具
means of production in category 1 一类物资
means of production in category 2 二类物资
means of relief 补救措施
means of remote sensing 遥感方式
means of representation 描绘手段
means of security 保全措施
means of slinging 起吊工具
means of subsistence 消费资料;生活资料
means of transport(ation) 运输方式;运输工具;装运方式
mean soil loss 平均土壤流失量
mean soil temperature 平均土温
mean solar day 平太阳日;平均太阳日
mean solar hour 平太阳小时
mean solar second 平太阳秒
mean solar semidiurnal constituent 平均太阳半日分潮
mean solar semidiurnal tide 平均太阳半日潮
mean solar time 平太阳时;平均太阳时
mean solar year 平太阳年;平均太阳年
mean sound absorption coefficient 平均吸声系数;平均声吸系数
mean sounding velocity 平均声速
mean sound pressure 平均声压
mean sound pressure level 平均声压级
mean spacing of joint walls 节理平均壁距
mean specific budget 平均比收支
mean specific gravity 平均比重;平均相对密度
mean specific heat 平均比热
mean specific pressure 平均比压
mean speed 平均速率;平均速度;平均车速
mean sphere level 平均地表面;地面平均高度
mean spheric(al) candle power 平均球面烛光
mean spiral angle 中心螺旋角
mean spring high water 大潮平均高潮位;大潮平均高潮面
mean spring low water 平均大潮低潮(面)
mean spring range 平均大潮差;大潮(平均)潮差
mean spring rise 平均大潮(潮)升;大潮平均潮升
mean spring rise of high tide 平均大潮高潮升
mean spring rise of low tide 平均大潮低潮升
mean springs 平均大潮
mean spring tide 平均大潮
mean square 均方;平均平方;均方离散度
mean square band width 均方带宽
mean square canonial correlation coefficient 均方典型相关系数
mean square consecutive fluctuation estimator 均方连续波动估计量
mean square contingency coefficient 均方列联系数
mean square continuity 均方连续性
mean square departure 均方偏差

mean-squared error optimization 均方差的最优化
mean square deviation 均方偏移;均方(偏)差;均方离差
mean square displacement 均方位移;均方偏移
mean square distance 均方距离
mean square distance error 距离均方误差
mean square distribution 均方分布
mean square error 中误差;均方(根)误差;均方差
mean square error criterion 均方误差准则
mean square error efficiency 均方误差效率
mean square error inefficiency 均方误差低效率
mean-square error in magnetic survey 磁测均方误差
mean square error norm 均方误差范数
mean square error of angle observation 测角中误差
mean square error of a point 定位中误差
mean square error of azimuth 方位角中误差
mean square error of azimuthal angle 方位角中误差
mean square error of base station gravity 基点网重力值均方误差
mean square error of coordinate 坐标中(均方)误差
mean square error of density measurement 密度测量均方差
mean square error of displacement for normal direction of position line 位置线法线方向位移中误差
mean square error of distance measurement 测距中误差
mean square error of elevation measurement 高程测量均方差
mean square error of height 高程中误差
mean square error of height of the weakest point 最弱点高程中误差
mean square error of horizontal position 平面位置中误差
mean square error of initial azimuth 起始方位角中误差
mean square error of point 点位中误差
mean square error of position of the weakest position 最弱点点位中误差
mean square error of reading data 读数中误差
mean square error of side length 边长中误差
mean square error of the weakest side 最弱边中误差
mean square error of unit weight 单位权中误差
mean square error per kilometer 每公里测量中误差
mean square integral 均方积分
mean square line 均方线
mean square modulus 均方模
mean square moment 均方弯矩
mean square prediction error 均方预测误差
mean square regression 均方回归
mean square regression plane 均方回归平面
mean square response 均方反应
mean square root 均方根
mean square scatter 均方分散度
mean square spectral density 均方普密度
mean square successive difference test 均方递差检验
mean square value 均方值
mean square velocity 均方速度
means square contingency 均方列联
mean standard deviation 平均标准偏差;均数标准差
mean station method 平均站点法
means to prevent freezing 防冻方法;防冻措施
mean storm sediment discharge 平均暴雨输沙率
mean strain 平均应变
mean stream flow 平均河流流量;平均河川流量
mean stream velocity 平均流速
mean strength 平均强度
mean stress 平均应力;八面体法向应力
mean stripping ratio 平均剥采比
mean subareal velocity (两测流垂线间的)部分断面平均流速
mean summer level 夏季平均水位
mean sun 平太阳
mean supply air temperature 平均送风温度
mean supply water temperature 平均给水温度
mean survival time 平均生存期;平均存活时间

mean synodic lunar month 平朔望月;平均朔望潮
mean tank method 平均秩法
mean temperature 平均温度
mean temperature difference 平均温差
mean temporal velocity gradient 平均瞬时速度梯度
mean tensile strength 平均受拉强度
mean tension 平均张力
mean term 中项;中间项;内项
mean term proportion 比例中项
mean terrestrial pole 平均地极
mean thawing index 平均融化指数
mean thermal transmittance 平均传热系数
mean thickness 平均厚度
mean thickness of aquifer 含水层平均厚度
mean threshold shear stress 平均临界剪应力
mean tidal current 平均潮流
mean tidal curve 平均潮位曲线
mean tidal flow 平均潮流
mean tidal level 平均潮位
mean tidal range 平均潮差
mean tide 中潮;平均潮汐;平均半潮水位;半潮
mean tide level 平均潮位;平均潮面;中潮面;平均潮(汐)水位;平均半潮水位
mean tide level rise 平均潮位升高
mean tide range 平均潮差
mean time 平时;平均时间
mean time before failure 修理间隔平均时间;故障前平均时间
mean time between arrivals 两次到达之间的平均间隔时间
mean time between calls 平均呼叫时间
mean time between defects 故障平均间隔时间
mean time between failures 平均无故障时间;平均失效间隔;平均稳定时间;平均故障(间隔)时间;平均故障间隔时间;失效故障平均间隔期
mean time between maintenance 平均维修时间(间隔);平均维护时间(间隔);平均保养(间隔)时间;维修平均间隔时间;平均保养间隔时间
mean time between overhaul 大修平均间隔时间
mean time between repairs 平均修理时间间隔
mean time between stops 平均停机间隔时间
mean time clock 平时钟
mean time of no failure operation 平均无故障工作时间
mean time of true noon 真午平时
meantime screw 调时螺钉
mean time to catastrophic failure 大故障前平均时间
mean time to detect 检测平均时间
mean-time-to double failure 平均无故障工作时间
mean time to fail 失效平均时间
mean time to failure 故障前平均时间;平均无故障时间;平均故障间隔时间;平均初次出故障时间;平均故障间隔期
mean time to repair 平均修理时间;平均修复(间隔)时间;平均(故障)修理时间
mean time to restore 平均故障修理时间
mean-time type time signal 平时式时号
mean tonnage 平均吨位
mean top wood height 平均最大木高
mean trajectory 平均轨道
mean transmission loss 平均隔声量
mean triangle misclosure 三角形平均闭合差
mean trip length 平均旅程
mean tropic(al) range 平均回归潮差
mean tropic(al) year 平回归年
mean turn 平均匝数
mean turning moment 平均转动矩
mean type 平均型
mean unbiased estimator 平均无偏估计量
mean unit hydrograph 平均单位(水文)过程线
mean upper limit 平均上限
mean up time 平均正常运行时间;平均(可)工作时间
mean value 平均值;中间值;平均数
mean value control system 平均值调节器
mean value function 均值函数;平均值函数
mean value method 平均值法
mean value of observation 观测平均值
mean value of tracing oscillation 曲线振动均数
mean value process 平均值法
mean value theorem 中值定理;均值定理;平均值原理;平均值定理;第一中值定理

mean variability 平均变异性;平均变化率
mean-variance criterion 均方差判据
mean variation of daily rate 平均日变差
mean variation of rates 平均日变差
mean variogram 平均变差函数
mean vector 均值向量
mean velocity 平均速度;平均流速
mean velocity agitation 平均骚动速度
mean velocity curve 平均速度曲线;平均流速曲线
mean velocity gradient 平均速度梯度
mean velocity in section 断面平均流速;断面平均风速
mean velocity in vertical 垂线平均流速
mean velocity of molecular 分子平均速率
mean velocity of vertical line 垂线平均流速
mean velocity on a vertical 垂线平均流速
mean velocity point 平均流速点
mean velocity position 平均速度点;平均流速点
mean volume of particle 颗粒的平均体积
mean volume-surface diameter of particle 颗粒体积—表面积平均(直)径
mean wall thickness 平均壁厚
mean water 中水位;平水
mean water depth 平均水深
mean water depth in section 断面平均水深
mean water flow 平均流量
mean water level 平均水位;平均水平(面);常水位
mean water neaps 平均小潮高潮高
mean water regulation 中水治理;中水位整治
mean water river bed 中水位河床
mean water stage 平均水位
mean water training 中水治导;中水导治
mean wave 平均波浪;平均波
mean wave direction 平均波向
mean wave drift 平均波浪漂移力
mean wave height 平均波高
mean wavelength 平均波长
mean wave period 平均波(浪)周期
mean weight 平均重(量)
mean width 平均宽度
mean width ratio 平均宽度比
mean wind 和风
mean wind speed 平均风速
mean wind velocity 平均风速
mean winter level 冬季平均水位
mean winter temperature 冬季平均温度
mean yaw 平均偏航角
mean year 平年;平均年
mean yearly discharge 年平均流量
mean yearly loading 年平均负荷(容)量
mean yearly loading capacity 年平均负荷(容)量
mean yearly rainfall 平均年降水量
mean yearly water level 年平均水位
mean yield 平均收入;平均产量
measeta 陆台
measles 斑点(图像)
measurability 可测性
measurability attribute 可度量的属性
measurable 可计量的;可测量的
measurable benefit 可计量的收益
measurable characteristic 可度量特征
measurable discharge 可测流量
measurable distance 一定距离
measurable function 可测函数
measurable funds effect 可衡量的资金效果
measurable future 不久的将来
measurable quantity 可测量
measurable set 可测集
measurable space 可测空间
measurable stochastic process 可测随机过程
measurable utility 可计量的效用
measurable value 可测值
measurand 被测数;被测量;被测对象
measurand transmitter 被测量发射机
measuration 量测;测量学;测定法
Measuray 带材厚度X光测微计
measure 丈量;量器;量度;计施,测度;办法
measure against erosion 防侵蚀措施;防冲刷措施
measure algebra 测度代数
measure analysis 容量分析(法)
measure and a half 半边线脚
measured 有规则的;整齐的;量过的
measured shovel packing 垫砂起道
measured altitude 测量高度

measured amount 实测数量;被测数量
measured backspace 精确回退
measured concentration 实测浓度
measured consideration 适度的报酬
measured coordinates 量度坐标
measured course line 测速线
measured curve 实测曲线
measured data 实测资料;实测数据;测量数据;测定数据;测得数据
measured data map of exploration 勘探区实际材料图
measured data map of ore deposits 矿床实际材料图
measured data map of ore district 矿区实际材料图
measured day rate 已测日工资率
measured day work 计时日工;基准工作量
measured density 实测密度
measured deviation 测量偏差
measured discharge 实测流(量)
measured distance 船速校验距离
measured distance mark 测速标
measured drawing 实测图;测绘图
measured fall 实测落差
measured feedback 实测反馈
measured flood 实测洪水
measured flood discharge 实测洪水流量
measured flow velocity 实测流速
measured gas 已测煤气量
measured geologic(al) map 测定地质图
measured highest stage 实测最高水位
measured hole 测量孔
measured income 已评定的收入
measure distance 船速校验距离
measured load 实测荷载;测出荷载;实测含沙量
measured longitudinal profile 实测纵剖面;实测纵断面
measured lowest discharge 实测最小流量
measured lowest stage 实测最低水位
measured lubrication 计量润滑
measured method 丈量法;测量学方法
measured mile 实测英里;船速校验线;测速标间距;标准海里
measured mile beacon 标准里程标杆
measured mile buoy 标准里程浮标
measured mile course 测速线
measured mile mark 测速标
measured mile trial 短航程试验;船速校验试航
measured of line of sight distance 照准线距离测量
measured of water prevention on surface 地面防水措施
measured ore 可靠储量;开拓储量
measured overshoot 实测过调量
measured points 实测点据
measured power 实测功率
measured pressure 已测压力
measured profile 实测纵剖面;实测(纵)断面
measured quantity 测定量;被测量
measured rate 按次计费率(电位)
measured reserves 开拓储量;确定储量;查明储量;测定储量
measured resources 测定资源
measured runoff 实测径流
measured sediment discharge 实测输沙率
measured service 计次制通话业务;计次制
measured signal 测定信号
measured specific gravity 实测比重;实测相对密度
measured stage 观测水位;实测水位
measured stress 实测应力
measured value 观测值;量测值;实测数值;测量值;测定值
measured variable 被测变量
measured water depth 实测水深
measured water equivalent of snow 实测雪水当量
measured water yield mine 矿坑实测涌水量
measured width of goods 货物的实测宽度
measured wind 实测风
measure equation 测量公式
measure expansion 体积膨胀
measure expenditures by income 量入为出
measure for disinfection 消毒措施
measure function 测度函数
measure hopper 定量斗

measure-kernel 测度核
measurement 丈量;量度;量测;计量;度量;尺码;尺度;尺寸;测量;测定
measurement accuracy 量测精度
measurement adjustment 测量平差
measurement amplifier 测量放大器
measurement and control system 测量与控制系统
measurement and cost reimbursements contracts 检测分期付款合同和成本加酬金合同
measurement and identification 测量和鉴定
measurement and laboratory test equipment 计量化验设备
measurement and payment 定额付款(制度)
measurement and weight list 容积重量表
measurement apparatus 测量装置
measurement area method 测量区域法
measurement arrangement 量测器具
measurement at fixed hours 定时观测
measurement basis 以容积计(费)
measurement bias 测量偏性
measurement bolt 测量螺杆;测量螺栓
measurement by leakage field 漏磁测量
measurement by magnification 放大测量
measurement by repetition 复测法
measurement by volume cubing 按体积计量
measurement by weight 按重量计
measurement capability 测量能力
measurement capacity 容积量
measurement cargo 丈量船货;量积货;量尺货;容积货(物);轻(泡)货;体积货物;大量船货;按体积计量的货物
measurement category 测量范围
measurement chamber 测量室
measurement chopper 测量限制器
measurement circuit 量测线路;测量回路;测量线路
measurement compartment 试验车测试间
measurement component 测量分量;测量部件
measurement condition 测量条件
measurement contract 检测分期付款合同
measurement control processor 计量控制处理机
measurement conversion table 丈量换算表
measurement correction 测量校正
measurement cross-section 测流断面
measurement data 计量资料;计量数据;测量数据
measurement device 量测装置;测量装置;测量设备
measurement distance 测量距离
measurement endpoint 测定终点
measurement equipment 测量设备
measurement error 量测误差;测量误差;测定误差
measurement freight 计容运费;按体积计算运费
measurement goods 丈量船货;体积货物;按体积计量的货物
measurement hydrophone 测量水听器
measurement in radian 弧度
measurement interval 测量区间
measurement items 测量项目
measurement length 测量尺度
measurement mark 测标
measurement method of environment-(al) pollution 环境污染测量法
measurement microphone 测量传声器
measurement mineral content 测量矿物含量
measurement mineral graininess 测量矿物颗粒
measurement mode 测量模式
measurement network of regional SO_2 地区二氧化硫污染测量网络
measurement of accuracy 测量精(确)度
measurement of acoustic(al) quantity 声学量测量
measurement of advection 平流测量
measurement of altitude 高程测量;高程测定
measurement of angle 测角
measurement of angles in all combination 全组合角度测量法
measurement of area 面积测量
measurement of area distribution 面积归一法;面积分布测定法
measurement of available soil phosphorus 测定土壤中有效磷
measurement of azimuth 方位角观测
measurement of azimuth on map 方位角量测
measurement of bearing 方向角观测
measurement of bleeding 泌水测量(测泌水速率及总泌水量)
measurement of channels 脉度

measurement of chemical composition of water 水化学组分测定;水化学成分测定
measurement of consistency 稠度的测定
measurement of contaminant 污染物测定
measurement of coordinate 坐标量测
measurement of crystofabric 结晶纤维测量
measurement of current velocity 测流速
measurement of dam's behavior 坝的性能测定
measurement of daylight 日光测定
measurement of depth 深度测量;测深
measurement of difference in elevation 高差测量
measurement of diffusion 扩散测定
measurement of discharge 流量测;测流量
measurement of discontinuity 不连续性测量
measurement of distorsion 变形测量
measurement of dust concentration 粉尘浓度测量
measurement of elongation by strain 应变伸长测定
measurement of environment(al) pollution 环境污染测量
measurement of evapo(u)ration 蒸发量测定;蒸发观测
measurement of exhaust smoke concentration 排烟浓度测量
measurement of exhaust smoke concentration from diesel automobiles 柴油汽车排烟浓度的测量
measurement of fill 吹填方的测定
measurement of flow rate 流量测定
measurement of flow through 径流测定
measurement of function 功能测量
measurement of gas phase inclusion pressure 气相包体压力测定
measurement of gloss 光泽测定
measurement of gradient 斜度测量
measurement of horizontal angle 水平角测量
measurement of horizontal distance 水平距离测量
measurement of humidity 湿度测定
measurement of length 测定长度
measurement of level 高程测量;水平测量
measurement of light 光度测量
measurement of line of sight distance 视线距离测量
measurement of luminous flux 光流测量
measurement of magnetic field 磁场测量
measurement of material level 料位测量
measurement of minerogenetic solution concentration 成矿溶液浓度测定
measurement of minerogenetie solution 成矿溶液浓度测定
measurement of models 模型量测
measurement of molecular property 分子性质测定
measurement of muffler performance 消声器测量
measurement of noise source 噪声源测量
measurement of non-linear distortion 非直线性畸变测量
measurement of odor 恶臭测定
measurement of oxide layer thickness 氧化层厚度测试
measurement of parcels 地块测量
measurement of performance 业务完成的评价;业务活动的评价
measurement of photographic 像片量测
measurement of physical parameter of natural gas 天然气物理参数的测量
measurement of plate thickness 板厚测定
measurement of power 功率测定;测功
measurement of precipitation 降水量测定
measurement of pressure 压力计量;压力测量
measurement of profile 纵断面测量
measurement of quantities 工程量计算;量方;量的测定;计量
measurement of radioactivity 辐射(性)测量;放射(性)测定
measurement of rainfall 降雨量测定
measurement of rate flow 流速测量
measurement of rate of flow 流率测定;流量测定
measurement of reflection-factor 反射系数测量
measurement of refractive power of lens 镜片屈光力鉴定
measurement of rock pressure 岩石压力测量
measurement of run off 径流测量
measurement of sample concentration 样品浓度测量

measurement of sample thickness 样品厚度测量
measurement of ship 船舶丈量
measurement of size aerosol 气溶胶粒径测定
measurement of skidding resistance 抗滑性能测量;抗滑能力测量
measurement of soot and dust concentration 烟尘浓度测量;烟尘浓度测定
measurement of sound insulation 隔声测量
measurement of specific surface 比面积测量
measurement of strata movement 地层错动测量(巷道)
measurement of stresses in the earth's crust 地应力测量
measurement of stress lining 衬砌应力测量
measurement of surface subsidence 地表下沉测量
measurement of tension 拉力测量
measurement of tension of bolts 锚杆轴力测量
measurement of the polysaccharide content 多糖含量的测定
measurement of tilt 斜度测量
measurement of track distance 股道距离测量
measurement of vertical angle 竖直角测量
measurement of volume on map 体积量测
measurement of water adsorption 吸水量测定
measurement of water pollution 水污染测定
measurement of water quantities 水量的测量
measurement of wreck dimension 探摸丈量
measurement pattern 测量模式
measurement plane 测量平面
measurement quantity 测定量
measurement range 测程;量程范围;量程;量测范围;测量范围;测定范围
measurement range of laser range finder 激光测距机测程
measurement rate 容积吨运费率
measurement-recording system 测量记录系统
measurement reproducibility 测量重复性
measurement result 测量结果
measurement runoff 实测径流
measurement science 计测学
measurement service 计量业务
measurement signal 测量信号
measurements in digital 数字化测图
measurement site 测试地点
measurements of the stereo-model 模型量测
measurement standard 测定标准
measurement subsystem 测量子系统
measurement system 度量制;测量系统
measurement technique 量测技术;测量技术
measurement test 测定试验;测试
measurement theory 计测理论
measurement time 测量时间
measurement tolerance 测量公差
measurement ton method 容积吨法
measurement ton(nage) 体积吨;尺码吨;丈量吨(位);载货容积吨数;载货体积单位;容积吨
measurement transducer 量测传感器;量测变换器;测量传感器
measurement transformer 测量变压器
measurement translator 测试变换器
measurement unit 度量单位;测量装置
measurement unit of plan 平面单位尺寸
measurement update 观测值修正;校正观测值;测量校正
measurement updating 观测值更新;测量校正;观测值修正
measurement value of nature base number 自然底数测定值
measurement wind 测风
measurement window 测量窗
measurement with drilling inclinometer 随钻测斜仪
measure of absolute variability 绝对变异测度
measure of accuracy 精度
measure of a pointset 点集的测度
measure of association 相联度量
measure of capacity 容量度量;容量
measure of central tendency 集中趋势度量
measure of characteristics 特征量度
measure of consumption 消费的量度
measure of curvature 曲率感度;曲率
measure of damages 损害程度的测定;损失赔偿估量;损害赔偿估量
measure of discontinuity 不连续(的)测度

measure of dispersion 离散度;离差的度量;变异度
measure of economic welfare 经济福利量;经济福利尺度
measure of effectiveness 效果的衡量
measure of excess 超额度量
measure of fixed sand 固沙措施
measure of fuzziness 模糊性测度
measure of goodness of fit 拟合优度度量
measure of indemnity 赔偿限度;赔偿标准;保险赔偿标准
measure of information 信息量度
measure of kurtosis 峰度测度
measure of length 长度量算;长度量测
measure of linear correlation 线性相关度量
measure of location 定位测量
measure of obstructing sand 阻沙措施
measure of performance 性能指数;性能指标
measure of precision 精确量度;精确度测度;精确度;精密度测量;精密程度
measure of relative variability 相对变异度测度
measure of relative variation 相对变异度量
measure of resistance 阻力测定
measure of risk aversion 厌恶风险的度量
measure of saving energy 节能措施
measure of skewness 偏斜度;偏态度量;偏度(测度)
measure of spread 分布测定
measure of surface 表面度量
measure of translating sand 输沙措施
measure of value 价值尺度;衡量价值;定价
measure of variation 变异的度量;变异测度
measure precision 精密度
measure preserving transformation 保测变换
measurer 计量表(指仪表);测量员;测量元件;测量仪器;测量仪表;测量器
measures and weights 权度;度量衡
measure scale 分度尺;比例尺
measures for conserving and improving water quality 防护改善水质措施
measures for pollution control 控制污染措施
measures of artificial recharge 人工补给方法
measures of dispersion 离差测定
measures of dust prevention 防尘措施
measures of effectiveness 效率量度
measures of indemnity 赔偿程度
measures of mud-stone flow treatment 泥石流防治措施
measures of prevention 预防措施
measures of protecting and administering ground water pollution 防治地下水污染的措施
measures of protection 防护措施
measures of silting reduction 减淤措施
measures of waste liquid disposal 污水处置措施
measures of water prevention 防水措施
measures of water prevention under the mine 井下防水措施
measures of wind-sand treatment 风沙防治措施
measure space 测度空间
measure tank 计量槽
measure tape 带尺
measure theory 测量理论
measure unit 量度单位
measure up to standard 符合标准
measure up to the contracted standard 达到合同规定标准
measure-while-drilling time 中途测试时间
measure zero 零测度
measuring 检测;计量误差;度量;测试
measuring accuracy 量测精度;计量精度;测量准确度;测量精(确)度
measuring altitude 测定高度
measuring amplifier 量测放大器;测量(用)放大器
measuring analysis 测量分析
measuring and controlling system 量测监控系统
measuring and cutting tool 量具刀具
measuring and identifying cyclic(al) patterns 循环类型测定与识别
measuring and testing device 测试装置
measuring and testing instrument 测试仪表
measuring and testing technique 测试技术
measuring and transmitting bridge 测量传输桥
measuring aperture 测量口径;测量孔径
measuring apparatus 量测仪(器);量测设备;计量装置;定量装置;定量给料器;测试仪器;测量

装置;测量仪器;测量设备;测量工具
measuring apparatus for solidification point 凝固点测定装置
measuring appliance 测量设备;测量仪器;计量装置;测量器具;仪表;量具
measuring arrangement 量测器具
measuring balance 天平;秤
measuring bar 量杆;量尺;测杆
measuring base 测量基线
measuring basis 测量基准
measuring beam 测量光束
measuring bellow 测量波纹管
measuring bin 量箱;量(料)斗;量仓;计量(料)斗;定量装矿器
measuring block 量块
measuring boat 测船
measuring bolt 量测用螺栓;量测锚杆;锚栓测点;测试锚杆
measuring bottle 计量烧瓶
measuring box 量箱;量水箱;计量箱
measuring bridge 量电桥;量测电桥;测量用电桥;测量电桥
measuring buoy 测验浮标;测流浮子;测流浮标
measuring buret 量液滴定管
measuring by conveyor[conveyer] belt 用运输皮带测量(运输量)
measuring by repetition 测回[测];复测法
measuring by sight 目测
measuring by volume 容积量测法;以体积计
measuring cableway 水文测流缆道
measuring capacity 测量容量
measuring car 测车
measuring cart 度量的手推车
measuring case depth for steel 钢(的)表层硬化深度测定法
measuring cell 测量元件;测量池
measuring chain 丈量链条;测量链;测链(长66英尺,1英尺=0.3048米)
measuring chamber 计量室
measuring channel 量水槽;测流槽
measuring chemical traits 测定化学特性
measuring circuit 量测电路;测量电路
measuring coil 测量线圈
measuring column 水银柱
measuring comparator 测量补偿器;测量比较仪
measuring compressor 计ة空气压缩机;测量用空气压缩机
measuring condition 测量状态;测量条件
measuring cone 测量锥体
measuring container 测量筒;测量用筒;计量容器
measuring conveyer 计量输送机
measuring conveyor belt scale 用运输带作衡器的秤
measuring current 测试电流
measuring cuttent 测量电流
measuring cycle 测量周期
measuring cylinder 量筒;量杯
measuring dam 溢流式量水堰;量水堰;(有测流堰的)量水坝
measuring data 测试数据
measuring day 收尺日
measuring day work 标准日工作量
measuring deck 测量仪器舱
measuring device 仪表;量具;量测装置;量测仪(器);计量装置;测量装置;测量仪器
measuring diaphragm 测量孔板
measuring diode 测量用二极管
measuring drum 计量转筒;测量鼓筒;读数鼓
measuring duct 测量管道
measuring electret microphone 测量用驻极体传声器
measuring electric(al) firing element instrument 电爆元件测试仪
measuring electrode 测量电极
measuring element 量测元件;测量元件
measuring elevation 测定标高;测定高程
measuring equipment 量测装置;测量仪(器);量测设备;测量设备
measuring error 量测误差;测试误差;测量误差
measuring eyepiece 量测目镜;测量目镜
measuring faucet 量水龙头
measuring feeder 称量喂料机
measuring flask 量瓶;计量烧瓶;容量瓶
measuring flume 量水槽;测流槽;测量(水)槽

measuring force 测量力
measuring frame (混凝土骨料计量用的)无底量斗;量料框;配料箱
measuring frequency 量测时间间隔;量测频率;测量频率
measuring frequency meter 测量用频率计
measuring ga(u)ge 量具车间;量规;测试表(指仪表);测量规;测厚计
measuring glass 量筒;量杯;玻璃量杯
measuring glassware 测量用玻璃器皿
measuring grid 量测络网;测量格片;方格测片
measuring head 探头;测试头;测量头
measuring hole 测量孔
measuring hopper 计量漏斗;量计斗;配料器;定量斗
measuring hydrophone 测量水声器
measuring hygrometer 测量用湿度计
measuring implement 量具
measuring impulse 测量脉冲
measuring impulse generator 测量脉冲发生器
measuring in catenary 悬链线测距
measuring index 度量指标
measuring in space 立体测量
measuring installation 测量设备;架设测量仪器点;测站;测量装置
measuring instrument 量具;量测仪(器);度量仪器;测量仪器;测量仪表;测量器具
measuring instrument and technique 测量仪器及技术
measuring instrumentation 测量用仪表
measuring instrument engineering 测量仪器工程学
measuring instrument for vibration 振动测试仪器
measuring interference flow 实测干扰流量
measuring jar for rain ga(u)ge (雨量器的)量雨瓶
measuring jaw 量爪;测爪
measuring jet 针阀调节喷嘴;计量喷嘴
measuring junction 高温接点;测量结
measuring key 测试电键
measuring lag 测量滞后;测量数值的指示滞后
measuring line 观测剖面线;测线;测绳;测量线
measuring lock 量测阀
measuring longitudinal and lateral dip angle 测量纵横向倾角
measuring machine 量度仪;测量机;测长机
measuring magnifier 测量放大镜
measuring mark 丈量标;测标
measuring-mark mirror 测标(反射)镜
measuring-mark plate 测标板
measuring means 量度工具;测量方法
measuring mechanism 测量机构
measuring method 丈量法;测量方法
measuring method of isotope tracer in single well 单井同位素示踪测定法
measuring method of occurrence element of rock in outcrop 露头产状测定方法
measuring microscope 量度显微镜;测(量显)微镜
measuring modulator 测量调制器
measuring nozzle 量水喷嘴;测量管嘴
measuring of angle in sets 测角法测角
measuring of human resources 人力资源的衡量
measuring of hydrophysical property of rock 岩石水理性质测定
measuring of smoke density 烟气浓度测量
measuring of structural element 构造要素测量
measuring of water pollution 水污染测定
measuring orifice 测量用孔口;测量孔板
measuring oscillator 测量振荡器
measuring oscilloscope 测量示波器
measuring out 测量
measuring pachymeter 测厚计
measuring pad 测量极板
measuring panel 测量仪表板
measuring peg 测桩【测】;测量标志
measuring period 测量周期
measuring phototube 测量用光电管
measuring pin 量针(测量油位用)
measuring pipet(te) 计量吸管;刻度移液管;带刻度吸管;量液(移液)管
measuring plane 测量平面
measuring plant 检测车间
measuring plate 量测网板
measuring platform 观测台

measuring pliers 测量钳
measuring plug 测量标志
measuring pocket 给料量斗;量矿器;配料槽
measuring point 量测点;计量起点;测量点;测量部位;测点
measuring point anomaly 测点异常
measuring point coordinates 测点坐标
measuring point density 测点密度
measuring point error 测点误差
measuring point index 测点索引
measuring points of vertical line 垂线测点
measuring position 测量位置
measuring potential 测量电位
measuring potentiometer 测定电位计
measuring precision after adjustment 调差后观测精度
measuring precision before adjustment 调差前观测精度
measuring pressure 测试压力;测量压力
measuring pressure basis 测压基准
measuring principle 测量原理
measuring prism 测量棱镜;测距仪棱镜
measuring probe 测量探针;测量探头
measuring process 测量过程
measuring procision 测量精(确)度
measuring profile 量测断面;测量断面
measuring projector 轮廓投影仪;测量投影仪
measuring pump 量液泵;计量泵
measuring range 量程;量测范围;测试范围;测量量程;测量范围;测定范围
measuring reach 量测河段;测流河段
measuring reel 卷尺
measuring resistance 测量(用)电阻;标准电阻
measuring resolution 测试清晰度;测试分辨率
measuring result 量测结果;测量结果
measuring reticle 测量分划板
measuring robot 测量机器人
measuring rod 测量棒;测量杆;液位量杆;量杆;测杆
measuring rod of economic performance 经营成果指标
measuring roller 测鼓
measuring room 计量室
measuring rope 测索;测绳
measuring rope regulator 测绳调节器
measuring rule 量尺;测尺
measuring scale 刻度尺;比例尺;量尺;测量用刻度;测量尺;标尺;比例直尺
measuring screw 测量螺旋;量测螺旋;测微螺钉
measuring section 观测断面;观测段;量测断面;试验段;测验断面;测流断面;测量段
measuring sediment range 测含沙量断面
measuring selector 测量用选线开关
measuring sequence 测量顺序
measuring set 测量仪表组;测量设备;测量成套设备
measuring signal 测试信号;测量信号
measuring silo 测量圆仓;测量筒仓
measuring siphon tube 测量虹吸管
measuring size range 测定粒度范围
measuring slide wire 测量用滑线电阻
measuring slit 测量用狭缝
measuring soil moisture 测定土壤湿度
measuring spanner for force moment 测力矩扳手
measuring spectroscope 波长分光镜
measuring spindle 量测轴
measuring spot 测量地点
measuring staff 测杆
measuring stake 测桩【测】
measuring star 测量星号
measuring stereoscope 量测立体镜;测量立体镜
measuring stick 测杆;量尺;测量杖
measuring stress of steel support 钢支撑应力测量
measuring symbol 测标
measuring system 计量系统;度量系统;测量系统
measuring table 测量平台
measuring tank 量油箱;量桶;量水箱;测箱
measuring tape 皮尺;卷尺;测带;测尺
measuring technique 量测技术;计量技术;测量技术;测量方法
measuring technique of pore structure 孔隙结构参数测定方法
measuring telescope 测量望远镜
measuring temperature basis 测温基准
measuring terminal 测量端子

measuring the potential of hydrogen 测量氢潜力
measuring thermistor 测量用热变电阻器
measuring thickness micrometer 测厚千分尺
measuring thickness of oill-water transition zone 实测油水过渡带厚度
measuring tool 量度仪器;测量工具
measuring tour 野外测量
measuring tower 计量塔
measuring transducer 测量传感器
measuring transductor 测量用饱和电抗器
measuring transformer 测量用变压器
measuring translator 测量变换器
measuring triangle 轨缝尺
measuring trough conveyer 配料传送带;配料传送槽
measuring tube 量筒
measuring type of instrument 测试方式
measuring uncertainty 测量不确定性;测量不确定度
measuring unit 计量单元;量测装置;计量单位;测量装置;测量元件;测量单位
measuring updating 测量更正
measuring valve 手动定量给进阀
measuring vertical 测点垂线
measuring vessel 量器;测深船;测量船
measuring viewfinder 测量用瞄准器
measuring voltage 测量电压
measuring water 测水流量
measuring waveguide 测量波导管
measuring wedge 量测楔;测量(光)楔
measuring weigh gear 称重装置
measuring weir 量水堰;测流堰
measuring weir formula 测流堰计算式
measuring weir with linear discharge 用线性排量关系计算测流堰
measuring well 观察井;观测井;量测井;测量井
measuring wheel 测轮
measuring worm conveyor[conveyer] 计量用蜗杆运输机
measurment of film thickness 膜厚测量
meat-and-fish preparation room 鱼肉加工间
meat by-product 肉类加工副产品
meat cannery 肉类罐头厂
meat carrier 肉类装运船
meat cooling plant 肉类冷藏库
meat curing 肉类加工
meat elevator 冻肉提升机
meat, fish and poultry processing wastes 肉、鱼、家禽加工废物
meat freezing 肉类冷冻
meat industry wastes 肉类加工废物
meat industry wastewater 肉类加工废水
meat packer 推助公共车辆乘客上车的人
meat packing 肉类加工
meat packing industry 肉类包装工业
meat packing plant 肉(类)联合(加工)厂;屠宰加工厂;肉类加工场
meat packing waste 肉食包装废料
meat plant 肉类加工厂
meat processing 肉类加工
meat processing byproduct 肉类加工厂副产品
meat processing factory 肉类(联合)加工厂
meat processing wastewater 肉类加工废水
meat product industry 肉类加工工业
meat ship 肉类装运船
meat ware factory 肉食加工厂
meat work 肉类加工厂
meaty cotton 纯净棉
mecalix 切削力测定仪
mecarta 胶木
Mecasol cone 密加索尔圆锥(测定泥浆稠度的仪器)
mech 机械师;技工;机械工人
mechanic 机修工;技术员;技工;机械员;机械师;机(械)工(人)
mechanical 机械絮凝
mechanical abrasion 机械性磨损
mechanical accident 机械事故
mechanical accident time 机械事故时间
mechanical actuated 机械传动的
mechanical adhesion 机械黏性附力;机械黏结;机械黏接;机械黏附
mechanical adjustment 机械调整
mechanical admittance 力导纳;机械导纳

mechanical admixture 机械杂质;机械混入物
mechanical advantage 节省倍力;机械优点;机械效益;机械利益;滑车功率
mechanical aeration 机械通气;机械曝气(法);机械充气;机械掺气
mechanical aeration basin 机械曝气池
mechanical aeration system 机械曝气系统
mechanical aerator 机械曝气装置;机械曝气器
mechanical aerator in activated sludge process 活性污泥法机械
mechanical affixment 机械辅(助)件
mechanical agitating clarifier 机械搅拌澄清池
mechanical agitation 机械搅动;机械搅拌;机械拌和
mechanical agitation flo(a)-tation cell 机械搅拌式浮选机
mechanical agitation pneumatic flo(a)-tation machine 机械搅拌充气式浮选机
mechanical agitator 机械搅拌器
mechanical air classifier 机械空气分级机;机械粗粉分离器
mechanical air flo(a)tation machine 压气机械搅拌式浮选机
mechanical air separator 机械空气选粉机
mechanical air supply 机械送风
mechanical air supply system 机械送风系统
mechanical algebraic operation 机械化代数运算
mechanical alloying 机械合金化
mechanical amusement device 机械式载人运移娱乐设备
mechanical analog(ue) 机械模拟
mechanical analog(ue) computer 机械模拟计算机
mechanical analog(ue) instrument 机械模拟测图仪
mechanical analog(ue) stereoplotter 机械模拟立体测图仪
mechanical analogy 机械相似法;机械模拟
mechanical analogy of city 城市机械模拟
mechanical analysis 过筛分析;力学分析;颗粒(级配)分析;机械分析
mechanical analysis curve 粒径(分析)曲线;力学分析曲线;颗粒分析曲线;机械分析曲线
mechanical analysis of sand 砂粒(机械)分析;砂的粒径分析;砂的机械分析
mechanical anchor 机械锚杆
mechanical anchorage 机械锚具
mechanical and electric(al) area 机电面积
mechanical and electric(al) equipment 机电设备
mechanical anemometer 机械式风速仪
mechanical appliance 机械设备;机械用量
mechanical application 机械作业
mechanical aptitude test 力学性能试验
mechanical arm 机械手(臂);机械臂
mechanical arm template 机械支臂模片
mechanical arrangement 机械化布置
mechanical ashed producer 机械清灰煤气发生炉
mechanical ashing 机械除灰
mechanical atomization 机械雾化
mechanical atomizer 机械雾化器;机械喷雾器
mechanical atomizer burner 机械雾化燃烧器;机械喷灯
mechanical atomizing oil burner 机械雾化油料燃烧器
mechanical attraction 机械引力
mechanical auger 机械螺旋钻土器;机械麻花钻;机械螺旋钻
mechanical automatic lubrication 机械强制润滑;机械加压润滑
mechanical axis 力轴;机械轴
mechanical back-to-back test 机械连接式效率试验
mechanical balance 机械平衡
mechanical ballast cleaner 道床清筛机
mechanical ballast sieve and cleaning team 机械清筛工队
mechanical ballast tamper 机械式道砟捣固机
mechanical bandspread 机械波段扩展
mechanical barrier 机械障碍物
mechanical barrow 机械式手推车
mechanical bearing 轴承
mechanical bearing cursor 雷达测方位透明板;方位游标
mechanical behavio(u)r 机械性能;力学性能;机械特性
mechanical behavio(u)r under high temperature 高温力学性能

mechanical belt weigher 机械皮带秤
mechanical bias 机械偏置
mechanical bilge block 机械舱墩;活动舱墩
mechanical birefringence 受力双折射
mechanical blending 机械混合;机械倒库
mechanical blocking 机动区截制【铁】
mechanical blower 机械式鼓风机
mechanical blowpipe 自动焊(割)炬
mechanical boat davit 提出式艇架
mechanical bolt 螺钉
mechanical bond(ing) 机械连接;机械结合;咬合作用;咬合握裹力;机械砌合
mechanical booster 机械升压器
mechanical bound water 机械结合水
mechanical brains 人工脑
mechanical brake 机械(式)制动器
mechanical breakdown 机械性损伤;机械性损坏;机械事故;机械故障
mechanical breakdown classification 机械事故分类
mechanical bruise resistance 抗机械损伤性
mechanical buffing 机械抛光
mechanical bulging die 机械胀形模
mechanical burn 机械摩擦灼焦
mechanical burner 高压燃烧器;机械燃烧器
mechanical CAD 机械计算机辅助设计
mechanical caging 机械锁定
mechanical calculator 机械计算器
mechanical carburetor 机力汽化油器
mechanical car park 机械设施停车场
mechanical cascade machine 机械搅拌泻流式浮选机;机械搅拌梯流式浮选机
mechanical casting extractor 铸件顶出器
mechanical cell 机拌池
mechanical center 机械中心
mechanical centering 机械置中
mechanical central interlocking 机械集中联锁
mechanical change of urban population 城市人口机械变化
mechanical character 力学性能;机械性能
mechanical characteristic 力学特性;力学性能;机械特性
mechanical characteristic of motor 电动机机械特性
mechanical character of structural plane 结构面力学性质控矿
mechanical charge out rates 机械台班费
mechanical charging 机械化装药
mechanical checkout 机械校正
mechanical chimney 机械抽风烟囱
mechanical chopper 机械斩光盘;机械斩波器;机械切碎机;机械调制盘
mechanical circulation 机械循环(作用)
mechanical clarification 机械澄清(法)
mechanical clarifier 机械澄清池
mechanical classification 机械分类
mechanical classifier 机械分粒器;机械分级机
mechanical cleaned rack 机械清除式格栅
mechanical cleaner 机械清洁器;机械清洁剂
mechanical cleaning 机械选矿;机械清洗;机械清除(法);机械精选
mechanical cleaning device 机械清除设备
mechanical cleaning off 机械除尘
mechanical cleaning of sewers 机械清通管道
mechanical clearance 机械间隙
mechanical clock 机械钟
mechanical close 机械封闭
mechanical clutch 机械(式)离合器
mechanical coaling station 机械化加煤站
mechanical collector 机械集流器;机械除尘器
mechanical colo(u)rant dispenser 自动混色机;自动调色机
mechanical combinator 机械协联机构
mechanical compaction 机械压实
mechanical comparator 机械比较仪;机械比较器;机械比长仪
mechanical compass inclinometer 罗盘重锤测斜仪
mechanical compilation 机械编图
mechanical completion 功能完工
mechanical compliance 力顺;机械顺从性
mechanical compoment of soil 土壤结构组成
mechanical component 机械构件;机械部件;机器构件
mechanical composition 力学构成;机械组成;机

械成分
mechanical composting 机械堆肥
mechanical compression refrigeration system 机械压缩制冷系统
mechanical computer 机械计算机
mechanical computing system 机械计算系统
mechanical conditioning 机械调节
mechanical conditioning process 机械调节法;机械调节处理
mechanical connection 机械连接
mechanical connector 导线连接钳
mechanical constraint 机械约束
mechanical construction 机械作图;机械施工;机械结构;机械化施工
mechanical contrivance 机械装置
mechanical control 机械控制;工程措施(水土保持)
mechanical controlled magazine 机动料斗
mechanical control system 机械控制系统;机械操纵系统
mechanical conveying 机械输送
mechanical conveying system 机械输送系统
mechanical cooling 鼓风冷却;机械冷却
mechanical cooling bed 机动冷床
mechanical cooling tower 机械冷却塔
mechanical copying attachment 机械仿形附件
mechanical core 建筑设备管线预制件;机械中心
mechanical core wall 机械型堤坝芯墙;机械型岩芯墙
mechanical correlation 机械相关
mechanical corrosion 机械溶蚀;机械腐蚀
mechanical counter 机械计数器
mechanical counter wheel 机械数字轮
mechanical couple 机械连接
mechanical couplers 车钩机械部分
mechanical coupling 机械性耦联;机械式联轴器;机械耦合
mechanical covering 机械覆盖
mechanical cow 机械牛
mechanical crust breaking 机械打壳
mechanical current meter 机械传动流速仪
mechanical cutting 机械切削;机械切割
mechanical cycling 机械循环
mechanical damage 硬伤;机械损伤
mechanical damper 机械阻尼器;机械减振器
mechanical damping 机械阻尼
mechanical davit 机械吊艇柱
mechanical declinometer 机械倾斜仪
mechanical de-enamelling 机械除瓷
mechanical deflector 机械偏转器
mechanical deformation 机械变形
mechanical deformation stage 机械变形阶段
mechanical deformer 机械式变形器
mechanical degradation 机械性破碎
mechanical dehydration 机械脱水
mechanical densification 机械压实法
mechanical department 机务处【铁】
mechanical depolarization 机械退极化;机械去极化
mechanical deposit 机械沉积(物);动力沉积(物)
mechanical deposition 机械沉积作用
mechanical derusting 机械除锈
mechanical descaling 机械除锈
mechanical deslimer 机械脱泥器
mechanical desludging 机械除泥
mechanical deterioration 机械性损伤;机械性损坏
mechanical device 机械装置;机械仪表
mechanical dewatering 机械脱水;机械降水
mechanical dewatering of sludge 污泥机械脱水
mechanical dew point 机械露点
mechanical differential analyzer 机械微分分析机;机械式微分分析器
mechanical differentiation 机械分异作用
mechanical differentiation way 机械分异方式
mechanical digger 挖掘机
mechanical directly driven vibration generator 机械(直接驱动)振动台
mechanical discharge 机械排水
mechanical discharge aid 机械助卸装置
mechanical discipline 设备专业
mechanical disintegration 机械粉碎
mechanical dispersivity 机械弥散系数
mechanical displacement pouring 置换浇注
mechanical displacement seismometer 机械位移地震计
mechanical dissemination 机械传布
mechanical distortion correcting device 畸变机械校正装置
mechanical dividing heads 机械分度头
mechanical dotter graver 机械刻点仪
mechanical dowel and tie bar installer 机械式销榫传力杆支座
mechanical dowel installer 销棒装置
mechanical downtime 事故停工时间
mechanical draft 机械通风;机动气流
mechanical draft burner 机械送风燃烧炉
mechanical draft cooling tower 机械通风冷却塔
mechanical draft dry cooling tower 机械通风干法冷却塔
mechanical drafting 机械制图
mechanical draft water cooling tower 机动气流水冷(却)塔;机械通风水冷却塔
mechanical drainage 机械排水
mechanical draught 机械通风;机械通气;机械排烟;机械抽风
mechanical drawing 设备布置图;工程图;机械制图;机械图(样);机械绘图;设备图(纸)
mechanical drawing from a bushing 漏板拉丝法
mechanical dredge(r) 机械挖泥船;铲斗(式)挖泥船
mechanical dredging 机械清淤
mechanical drier 干燥机
mechanical drift 机械性固定偏差;机械漂移
mechanical drift indicator 机械式倾角计;机械测斜器
mechanical drill(ing) 机钻
mechanical drive 机械传动装置;机械传动
mechanical drive system 机械驱动系统
mechanical drive turbine 机械驱动用透平
mechanical driving ratio 机械传动比
mechanical dryer 干燥机
mechanical drying 机械烘干
mechanical durability 力学耐久性
mechanical dust collector 机械除尘器
mechanical dust extractor 机械除尘器
mechanical dust separation 机械除尘
mechanical earth auger 螺旋钻土机;挖土螺钻;挖穴螺旋
mechanical earthwork 土方机械施工
mechanical effect 机械效能
mechanical effects of ultra-sound 超声波机械效应
mechanical effects of curvature 弯道的机械影响
mechanical efficiency 机械效率
mechanical electric(al) heat-flow meter 机械电气化热量表
mechanical electroterminal 机电终端
mechanical element 机械元件;机件
mechanical elevator 机械提升机;提升机;(矿料)升运机
mechanical emptying 机械排空
mechanical endurance 机械寿命;机械疲劳强度
mechanical energy 机械能
mechanical engineer 机械工程师;设备工程师
mechanical engineering 机械制造业;机械工程(学);设备工程学
mechanical engineering installation 机械工程安装
mechanical enlargement 机械放大
mechanical equation of state 力学状态方程
mechanical equilibrium 机械平衡
mechanical equilibrium condition 力学平衡条件
mechanical equipment 机械设备
mechanical equipment installation work 机械设备安装工程
mechanical equipment room 机械设备室;机械设备间;机房
mechanical equipment schedule 机械设备清单;机械设备表
mechanical equipment station 机械设备站
mechanical equivalent 功当量
mechanical equivalent of heat 热功当量;热的机械当量
mechanical equivalent of light 光当量;光的机械当量(单位:瓦/流明)
mechanical erosion 刻蚀;动力冲刷
mechanical excavation 机械挖掘;机械开挖
mechanical excavator shovel 挖掘机
mechanical excitation 机械激发;机械共振
mechanical execution 机械施工;机械化施工
mechanical exhaust 机械排气;机械排风
mechanical exhaust system 机械排气系统;机械排风系统
mechanical exhaust system of ventilation 机械排气通风系统
mechanical expression 压榨
mechanical exterior plastering 机械的外部粉刷
mechanical exterior rendering 机械的外部抹灰
mechanical extraction 机械抽风
mechanical extractor 机械拔桩机;机械抽提机;机械抽风机
mechanical fabric 工业用织物
mechanical face seal 机械面密封
mechanical factor of merit 机械优值
mechanical failure 机械失效;机械破坏;机械故障;设备机械故障失效
mechanical fastener 机械接合件
mechanical fatigue 机械疲劳
mechanical feedback 机械反馈
mechanical feeder 给料机
mechanical feel device 机械感触装置
mechanical fender 机械化护舷
mechanical fiducial marks 机械准标;机械框标
mechanical filter 机械滤(波)器;机械过滤器
mechanical filter respirator 机械过滤呼吸器
mechanical filtration 机械过滤(作用)
mechanical fine-grader 细磨机
mechanical finger 机械手
mechanical finisher 机械抛光机;整修机
mechanical finish(ing) 机械精加工;机械整修;机械抛光
mechanical fired boiler 机械燃烧锅炉
mechanical fitting 机械属具
mechanical fitting worhshop 机钳间
mechanical fixture 机械属具
mechanical flexure 机械弯曲
mechanical float 机动镘板(混凝土路面);整平机(整修水泥混凝土路用);机械抹灰(板);机镘;机动抹灰板;抹灰板;镘平机
mechanical flo(a)tation cell 机械浮选槽
mechanical float barge lift 机械式驳船升船机;浮筒式升船机
mechanical flocculation 机械絮凝作用
mechanical flocculator 机械絮凝器
mechanical floor 设备层
mechanical flow diagram 机械流程图
mechanical flyball-type governor 飞球式机械调速器
mechanical foam 机械泡沫
mechanical foamed plastics 机械发泡塑料
mechanical foam extinguisher 机械泡沫灭火机
mechanical force 机械力
mechanical form block 机械的成型模块
mechanical forming 机械成型
mechanical fragility 机械性脆性
mechanical fragmentation of rock 岩石机械破碎
mechanical friction 机械摩擦(力)
mechanical friction coefficient 机械摩擦系数
mechanical friction loss 机械摩擦损失
mechanical fuel pump 机械燃油泵
mechanical fuse 机械保护器
mechanical gadget 机械配件
mechanical gantry 机架
mechanical garage 机械化车库
mechanical garbage grinder 机械的垃圾磨碎机
mechanical generator 机力发电机
mechanical getting 机械化开采
mechanical gradation 机械级配分析;机械(化)筛分;机械化筛分选;机械(化)分级
mechanical grading 机械筛分;机械颗粒分级;机械分级
mechanical grain blocking 机械颗粒堵塞
mechanical grain size analysis 机械粒径分析
mechanical gravity marshalling yard 机械化驼峰编组场【铁】
mechanical grease lubricator 机械滑脂润滑器
mechanical grinding wheel forming equipment 机械式磨轮成型设备
mechanical gripper feed 机械夹持爪式送料
mechanical grit arrester 机械除尘器
mechanical hammer 机动锤
mechanical handling 机械装载;机械运输;机械(化)装卸;机械操作;机械操纵
mechanical handling industry 机械操作搬运工业

mechanical handling method 机械装卸法;机械搬运法
mechanical hand press 机械手动压力机
mechanical hand pressing pliers 手动压接钳
mechanical hanger 机械式悬挂器
mechanical hardening 机械硬化
mechanical haulage 机械运输方式
mechanical heading machine 联合掘进机;掘进联合机
mechanical high pressure method 机械高压法
mechanical highway equipment 筑路机械
mechanical hoe 动力锄
mechanical hot water heating system 机械热水供暖系统
mechanical hump 机械化驼峰【铁】
mechanical hump yard 机械化驼峰编组场【铁】
mechanical hydraulic control 机械液压式控制
mechanical hydraulic deep drawing 机械液压联合深拉延
mechanical hydraulic governor 机械液压调速器
mechanical hydraulic servomotor 机械液压接力器
mechanical hysteresis 机械滞后(现象);力学滞后
mechanical hysteresis effect 机械迟滞作用
mechanical hysteresis loss 机械吸震能力
mechanical impact screen 机械式冲击筛
mechanical impact strength 机械碰撞强度
mechanical impedance 力阻抗;力学阻抗(作用);机械阻抗
mechanical impurity 机械杂质;机械混合物
mechanical incomplete combustion 机械不完全燃烧
mechanical index 机械指数
mechanical industry 机械工业
mechanical injection 机械喷射
mechanical injury 机械性损伤
mechanical inspection 机械性能试验力;力学性能检测;力学性能试验
mechanical installation 机械装置
mechanical instrument 机械仪表
mechanical integration 仪器积分法
mechanical integrator 机械积分器
mechanical intelligence 机械智能
mechanical interlock 机械联锁装置
mechanical interlocking 机械联锁
mechanical interlocking cabin 机械闭锁工作间
mechanical interlocking function 机械联锁功能
mechanical interlocking machine 机械联锁机
mechanical intersection 机械交会
mechanical isolation 机械隔离
mechanical jack 机械(式)千斤顶;机力起重器;机动千斤顶;机动起重机
mechanical jamming 机械干扰
mechanical joining 机械方法连接
mechanical joint 机械连接;管道法兰螺栓连接;铰接;机械(连)接头;外线端钮;接线夹;管道凸缘螺栓连接
mechanical joint accessories 机械接头附件
mechanical joint and bell connecting piece 机械接头与钟口连接件
mechanical joint and flange connecting piece 机械接头与法兰连接件;机械接头与凸缘连接件
mechanical joint and plain end connecting piece 机械接头与平插连接件
mechanical joint cast-iron bend 机械接头铸铁弯头
mechanical joint cast-iron concentric-(al) reducer 机械接头铸铁同心大小头
mechanical joint cast-iron cross 机械接头铸铁四通
mechanical joint cast-iron eccentric reducer 机械接头铸铁偏心大小头
mechanical joint cast-iron tee 机械接头铸铁三通
mechanical joint fitting 机械接头配件
mechanical joint gasket 机械接头垫圈
mechanical joint(ing) 机械接头
mechanical key 机械性锁接
mechanical key groove 机械键槽
mechanical keying for plastering 为粉刷面而作的机械嵌缝
mechanical kinematics 机械运动学
mechanical kinetic energy 机械动能
mechanical kinetics 机械动力学
mechanical language 机械语言
mechanical life 机械寿命
mechanical lift 机械式升降机;机械升力;升降机
mechanical lift dock 机械提升(式)船坞

mechanical lift lock 机械单级船闸
mechanical limit 机械极限
mechanical line 力学线路
mechanical linkage 连杆机构;机械联动
mechanical liquid limit device 液限仪;流限仪
mechanical loader 装载机;机械式装载机
mechanical loading 机械装载;机械装卸
mechanical loading system 机械式加载系统
mechanical lock 机械锁
mechanical locker relay 机械闭锁继电器
mechanical locking 机械锁定;机械锁闭
mechanical logic control 机械逻辑控制
mechanical looper 机械传动的活套挑
mechanical loss 机械损失
mechanical lubrication 机械润滑;强制润滑
mechanical lubricator 机械润滑器;机力润滑器
mechanically actuated 机械起动的
mechanically actuated power meter 机械驱动式功率计
mechanically aerated cell 机械充气式浮选机
mechanically aerated stabilization process 机械曝气稳定法
mechanically anchored bolt 机械锚固式锚杆
mechanically ashed producer 机械除尘发生炉
mechanically capped steel 机械封顶钢
mechanically circulated hot water heating 机械循环热水供暖
mechanically clamped tool 机械夹固车刀
mechanically cleaned bar screen 机械清理格网;机械清理格栅
mechanically cleaned rack 机械清洗格栅;机械清除的格栅
mechanically cleaned screen 机械清污滤网;机械清污拦污栅;机械清理筛;机械清除筛
mechanically cleansing screen 机械清除筛
mechanically deposited sedimentary rock 机械沉积形成沉积岩
mechanically despun antenna 机械消旋天线
mechanically driven 机械传动的
mechanically driven flour grinder 机动磨粉机
mechanically driven interrupter 机械传动装置断续器
mechanically driven optic(al)reflector 机械传动光反射器
mechanically driven pump 机械传动泵;机动泵
mechanically excavated material 机械挖出的物料
mechanically fastened roofing 机械固定屋面
mechanically fired boiler 机械化燃烧锅炉
mechanically foamed plastic 机械法发泡塑料
mechanically formed sedimentary rock 机械作用形成的沉积岩
mechanically held water 机械结合水
mechanically laid 机械砌砖;机械铺轨;机械压平
mechanically laminated 机械层积构件;机械层积材
mechanically lubricated axle-box 轴箱机械润滑
mechanically lubricated axle-box bearing 轴箱轴承的机械润滑
mechanically mixed disinfectant 机械混合消毒剂
mechanically mixed disinfection technology 机械混合消毒技术
mechanically mixed flux 机械混合焊剂
mechanically mixer 机械混合器
mechanically mixing device 机械混合设备
mechanically operated 机械操作的;机动的;机械控制的;机械操纵的
mechanically operated cutter 机械操作的切割刀具
mechanically operated float 机动镘刀;机动浮球;机动浮筒
mechanically operated nozzle 机动喷嘴
mechanically operated tamping bar 机械捣棒
mechanically operated trowel 机动抹子;机动泥刀
mechanically polished slice 机械抛光片
mechanically polished substrate 机械抛光衬底
mechanically propelled 机械推进的
mechanically propelled lifeboat 机械推进救生艇
mechanically propelled vehicle 机动车辆
mechanically rated lumber 机械评定木材等级(木材应力等级机械地压印上 E 或 F 标志)
mechanically refrigerated container 机械式冷藏集装箱
mechanically refrigerated train 机械冷藏列车
mechanically refrigerated wagon 机械式冷藏车
mechanically resistant material 硬质材料
mechanically scanned guidance system 机械仿形导向系统
mechanically strength 机械强度
mechanically thickener 机械浓缩器;机械浓缩池
mechanically timed relay 机械定时继电器
mechanically tuned tube 机械调谐管
mechanically ventilated 机械通风的
mechanically vibrated screen 机械振动筛
mechanical magnetometer 机械化磁力仪
mechanical maintenance 机械保养
mechanical management of respiration 机械呼吸管理
mechanical manipulation 机械操作
mechanical manometer 自动测压计;机械压力计
mechanical mapping 机械作图
mechanical masking 机械掩蔽法
mechanical mass 力学质量
mechanical means 机械法
mechanical measurement 力学测量
mechanical metallurgy 机械冶金学;机械冶金
mechanical metamorphism 机械变质(作用);动力变质(作用)
mechanical meter 机械仪表
mechanical method 力学方法
mechanical migration of population 人口机械迁移
mechanical mine 机械水雷
mechanical misalignment 机械不同轴性
mechanical mixed tank 机械混合槽
mechanical mixer 机械混合机;机械混合器
mechanical mixing 机械搅拌;机械拌和
mechanical mixing device 机械混合设备
mechanical mixing flocculating tank 机械搅拌絮凝池
mechanical mixture 机械拌和(物);机械搅拌(物);机拌混合料
mechanical model 力学模型
mechanical modulator 机械调制器
mechanical mole 联合掘进机;隧道全断面掘进机
mechanical movement 机械运动
mechanical mucker 装岩机
mechanical mucking 机械装岩
mechanical mule 机械骡
mechanical multiplier 机械乘法器
mechanicalness 机械性
mechanicalness opening 机械孔径
mechanical newsprint-mill white water 机制新闻纸厂白水
mechanical noise 机械噪声
mechanical ohm 力学欧姆;力欧姆
mechanical oil valve 机械式油阀
mechanical operated semaphore signal 机械臂板信号机
mechanical operation 机械运转;机械操作
mechanical operator 开窗器;机械操纵工
mechanical optic(al) acceleration 机械光学式加速器
mechanical optic(al) comparator 机械光学比较仪
mechanical optic(al) deflector 机械光学偏转器
mechanical optic(al) seismograph 机械光学地震仪
mechanical optic(al) strain ga(u)ge 力学—光学应变仪;机械光学应变仪
mechanical optic(al) switch 机械光开关
mechanical oscillator 机械振荡器
mechanical oscillograph 直接记录器
mechanical output 机械功率
mechanical paiting work 机动油漆工程
mechanical pantograph 机械缩放仪
mechanical paper tape reader 机械纸带输入机
mechanical parallel circuit 机械并联回路
mechanical parts 机械构件;机械零件
mechanical passivity 机械钝性
mechanical penetration 机械穿透
mechanical penthouse 屋顶机房
mechanical performance of material 材料力学性能
mechanical perfusion 机械灌注法
mechanical phase shifter 机械移相器
mechanical pick 凿碎机;机凿;机(械)镐
mechanical picker 机械筛;机械挑选机;机械分选机
mechanical pick-up 机械拾音器
mechanical pilar 机动堆垛机
mechanical pilot 自动操纵器
mechanical pipe 机械缩孔
mechanical pipe joint 管道机械接口;管道的机动接头;机械管道接头;机械管道接合

mechanical piping 机械管桶
mechanical pit cutter-loader 机械化青储[贮]窖用切碎装载机
mechanical pitting 麻点
mechanical plant 机械设备
mechanical planter 植种机
mechanical plant for construction 施工机械设备
mechanical plastering 机械抹灰
mechanical platform barge lift 机械式平台驳船升船机；平台式驳船升降机
mechanical plating 机械喷镀；机械镀敷；机械辅助电镀
mechanical plotter 机械绘图仪；机械绘图机
mechanical plotting 机械作图；机械制图；机械绘图
mechanical polishing 机械磨光
mechanical positioner 托架支承
mechanical position readouts 机械位置读数器
mechanical power 机械功率；机械动力
mechanical power house 机动室；动力站（指水、汽、压缩空气）
mechanical power lift 机械式自动起落机构
mechanical power lift clutch 机力起落离合器
mechanical precipitation 机械力除尘器；机械除尘
mechanical precision instrument 机械精密测图仪
mechanical pre-cutting method 机械预切割法
mechanical predictor 机械预报器
mechanical press 机械压机
mechanical pressing 机压成型（法）
mechanical pressure atomizer 机械压力雾化器
mechanical principle 机械原理
mechanical process 机械过程
mechanical producer 机械发生炉
mechanical programmer 机械程序设计
mechanical projection 机械投影
mechanical property 力学性质；力学性能；力学特性
mechanical property of material 材料力学性质
mechanical property of rock 岩石的力学性质
mechanical property of soil 土的力学性质
mechanical property test(ing) 力学性能试验
mechanical proportional 缩放杠杆
mechanical puddling 振动捣实
mechanical pulp board 机械的纸浆板；磨木浆制纸板
mechanical pulp-free printing paper 道林纸
mechanical pulsing 机械脉冲
mechanical pump 机械泵
mechanical pusher ram 机械堆料机
mechanical pyrometer 机械高温计
mechanical Q-switch 机械 Q 开关
mechanical quality factor 机品质因素
mechanical quantity 力学量
mechanical quarry shovel 挖掘机
mechanical rabble 机械搅拌器
mechanical rabbling 机械搅拌
mechanical radical triangulation 机械辐射三角测量
mechanical rake 机械清污耙；机械耙；机动耙；耙路机
mechanical raked 机械耙除的
mechanical raked fine screen 机械除渣格网
mechanical raking system 机械耙动系统
mechanical rammer 机械夯（具）；机动夯
mechanical rapids-heaving 机械绞滩
mechanical rapids-warping 机械绞滩
mechanical rapper 机械振打锤
mechanical raster scanning type system 光栅型机械扫描系统
mechanical reactance 力抗；机械反作用力
mechanical reaction tank 机械反应池
mechanical realization 机械构造
mechanical reaming 机械铰孔
mechanical rearranged stage 机械重排阶段
mechanical receiver 机械接收机
mechanical recording 机械记录
mechanical recording current meter 机械自记海流计
mechanical recording element 机械记录元件
mechanical recovery 力学回复
mechanical rectifier 机械整流器
mechanical re-cycling plant 机械分类厂
mechanical reduction 机械缩小转绘
mechanical reduction factor 机械原因缩减系数
mechanical reduction gear 齿轮减速装置
mechanical refining 加工退火；机械调质
mechanical refrigerating system 机械制冷系统

mechanical refrigeration 机械制冷；机力冷冻
mechanical refrigerator container 机械式制冷柜
mechanical refuse grinder 机械垃圾磨碎机
mechanical register 机械配准；机械记录器
mechanical registering 机械记录
mechanical relationship 力学关系
mechanical relay drop 机械式继电器释放
mechanical releasing mechanism 机械开门机构
mechanical reliability 机械可靠性
mechanical reliability report 机械可靠性报告
mechanical removal of solids from settling tank 沉淀池固体的机械清除
mechanical repairing department 机械修理车间
mechanical (repair) shop 机修车间
mechanical resistance 机械力阻；力阻；机械（摩）阻力
mechanical resonance 机械共振
mechanical resonator 机械谐振器
mechanical reversibility 机械可逆性
mechanical rig 机械钻机
mechanical roaster 机械搅拌焙烧炉
mechanical robustness testing 机械强度试验
mechanical rock bolt 机械式锚杆
mechanical roller compaction method 机械碾压法
mechanical roof extractor 屋顶机械抽气机
mechanical room 机械设备间；设备用房（如通风房、电梯间、密闭水箱、水泵间等）
mechanical roughening pump 机械初步抽空泵
mechanical rubbish grinder 机械垃圾磨碎机
mechanical rust removing 机械除锈
mechanicals 工业制品
mechanical safety 机械保险装置
mechanical sampling 机械取样；机械抽样；机械采样
mechanical sand control 机械固砂
mechanical sander 机械砂光机
mechanical saw 机锯；机械锯
mechanical scale 机械秤
mechanical scanner 机械扫描器
mechanical scanning 机械扫描
mechanical scour 机械冲刷
mechanical scraping and cutting 机械搅刮法
mechanical screeding 机械整平；机械刮平
mechanical screw batch charge 螺旋加料器；螺旋加料机
mechanical seal 机械密封；端面密封
mechanical sealing 机械封接
mechanical seal spare parts 机械密封备件
mechanical seal specification 机械密封规格
mechanical sectioning 机械截面图
mechanical sediment 机械沉积物
mechanical sedimentary deposit 机械沉积矿床
mechanical sedimentary differentiation 机械沉积分异作用
mechanical sedimentation 机械沉积
mechanical seismograph 机械地震仪
mechanical self starter 机力自动起动机
mechanical separation 机械分离法；机械分离
mechanical separator 机械选粉机；机械分离器
mechanical series circuit 机械串联回路
mechanical servo 机械随动系统
mechanical setting 机械镶嵌
mechanical sewage treatment 污水机械（法）处理
mechanical shaft kiln 机械化立窑；机立窑
mechanical shaker 振动机；机械振动器
mechanical sharpening machine 锻钎机
mechanical shear connector 机械剪切接合器
mechanical sheeting 机器用薄片
mechanical shield 机械式盾构；机械开挖盾构；机械化盾构
mechanical shock 机械振动；机械冲击
mechanical shock condition 力学振动条件
mechanical shock resistance 抗机械冲击性
mechanical shovel 电铲；机械铲；挖土机
mechanical shutter 机械快门
mechanical side block 机械舣墩
mechanical side guides 机械化侧导板
mechanical side shore 机械撑柱
mechanical sieve 机动筛
mechanical sieve analysis 机械筛分
mechanical sieve shaker 振动筛的激振器；摇筛机；试验筛机械振动器
mechanical sifter 机动筛
mechanical signal 机械信号；机力信号

mechanical similarity 力学相似性
mechanical single seal 单端面密封
mechanical slewing derrick 机械摆动单杆装置
mechanical slicer 刨床
mechanical slide block 活动舣墩
mechanical slipway 机械化滑道
mechanical sludge dewatering 机械污泥脱水；污泥机械脱水
mechanical sludge drying bed 机械化污泥干化场
mechanical sludge thickener 机械污泥浓缩池
mechanical smoke control 机械控烟
mechanical smoke discharge 机械排烟
mechanical smoke exhaust system 机械排烟系统
mechanical snow plow 排雪机
mechanical soil cultivation 土壤机械化耕作
mechanical soil stabilization 土的机械加固法；土的机械稳定法
mechanical sources 机械源
mechanical sowing 机播
mechanical spalling 机械（性）剥落；机械崩裂
mechanical spot displacement 光点的机械位移
mechanical spot misalignment 光点的机械位移
mechanical sprayer 机械喷雾器
mechanical spraying 机械喷射
mechanical spreader （混凝土）分布机；摊铺机；撒布机
mechanical spreading 机械撒布
mechanical spreading paver 机动摊铺机；机动铺路机；摊铺机
mechanical spring 弹簧
mechanical spring and flyweight governor 飞锤调速器
mechanical stability 机械稳定性；机械安定性
mechanical stabilization 机械稳定；机械加固
mechanical stabilization for soil 机械加固土
mechanical stabilized base 机械稳定基层
mechanical stabilizer 机械稳定器
mechanical stacker 机械垛材机
mechanical stage 机械台
mechanical stage measure method 机械台测量法
mechanical stamping 机械冲压
mechanical stamping device 机动打印装置
mechanical starter 机动起动机
mechanical starting time 机组起动时间；机组加速时间
mechanical start up 机械设备试运转
mechanical state 力学状态
mechanical stereoviewing 机械化立体观察
mechanical stiffness 力劲
mechanical stimulus 机械刺激物
mechanical stipple 机械点刻
mechanical stirrer 机械搅拌器；机械搅拌机；机器搅拌机
mechanical stirring 机械搅拌
mechanical stirring clarifier 机械搅拌澄清池
mechanical stoker 自动加煤机；加煤机；机械加煤机；机动炉排；机动加煤机；机动加料机
mechanical stoking 机械加煤
mechanical stoppage 机械故障
mechanical stops 货叉超行程停止器；货叉超行程挡块
mechanical storage 机械存储器
mechanical stowing 机械充填
mechanical stowing with slingers 用抛掷机进行装填
mechanical strain 机械应变
mechanical strainer 机械隔滤器
mechanical strain ga(u)ge 机械（式）应变仪
mechanical strain ga(u)ge extensometer 机械式应变引伸仪
mechanical straining 机械隔滤
mechanical strain meter 机械应变仪
mechanical strake 机械洗矿槽
mechanical strength 力学强度；机械强度
mechanical stress 机械应力
mechanical stress relieving 机械方法消除应力
mechanical stress relieving method 机械应力消除法
mechanical stretching 机械张拉
mechanical stroboscope 机械示速器；机械闪光仪；机械频闪观测器
mechanical suberosion 机械潜蚀作用
mechanical subgrader 路基整修机；路基修整机；平地机

mechanical surface aeration 机械表面曝气
mechanical sweeper 机动帚;扫路机;扫路机
mechanical sweeping 机械排气;机械刮除;机械清扫
mechanical system 力学系统;力学体系;机械装置;机械系统;机工系
mechanical tamped 机械振捣的
mechanical tamper 机械路夯;机械砸道器;机械夯(具);机动夯;夯击机
mechanical tamping 机械夯实;机械捣固
mechanical tape puncher 机械纸带穿孔机
mechanical tapping machine 机械放液机
mechanical telemetry 机械遥测
mechanical temperature rise 机械温升
mechanical template plot 模板机械制图
mechanical tension 机械张拉
mechanical tensioning method 机械张拉法
mechanical test(ing) 力学试验;机械(性能)试验
mechanical testing apparatus 力测仪器
mechanical test of structure 结构力学试验
mechanical test screen 动力试验筛
mechanical theorem proving 定理的机器证明
mechanical thinning 机械疏伐;机械定距疏伐
mechanical threshold 机械阈
mechanical tidal predictor 机械潮汐预报器
mechanical tilt 机械倾角
mechanical timer 机械计时器;机械定时(器)
mechanical tissues 机械组织
mechanical tooling of concrete 混凝土机械压印(图案花纹)
mechanical torque converter 机械变扭器
mechanical tower 机械(干燥)塔
mechanical track-maintenance train 机械化轨道保养列车
mechanical translation dictionary 机器翻译词典
mechanical translation machine 翻译机
mechanical transmission 机械性传播;机械传动;机械传播
mechanical transmission of power 动力的机械传送
mechanical transmission system 机械传输系统
mechanical transmitter 机械传播者
mechanical transport 机械输送;机械迁移;机械化运输;机动运输;机动车交通;汽车交通
mechanical trauma 机械性创伤;机械伤
mechanical treatment 机械加工;机械处理
mechanical treatment process 机械处理法
mechanical trench plow 犁式挖沟机
mechanical trimmer 平舱机
mechanical trip 机械释放装置
mechanical trowel 机械抹子;抹面机;抹灰机
mechanical tube-type cleaner 管式清洁机
mechanical tuning 机械调谐
mechanical tunnel(l)ing 机械法施工
mechanical turbulence 机械湍流
mechanical twin 机械双晶;机械孪晶;塑变双晶
mechanical-type dust collector 机械除尘器;机械集尘器
mechanical-type dust removal 机械除尘
mechanical-type exhauster 机械式排气风机
mechanical-type inflator 机械式轮胎打气泵
mechanical-type manometer 机械式压力表
mechanical-type unit 机械单位
mechanical union 机械活接头
mechanical unit 水暖电设备的装配部件;机械部件;力学单位;机械设备装置
mechanical vacuum system 机械真空系统
mechanical valve prosthesis 机械瓣膜
mechanical variable speed drive 机械变速拖动装置
mechanical vector 机械(性)媒介物
mechanical vehicle 机动车
mechanical ventilating 机械通风
mechanical ventilating exhaust system 机械排风系统
mechanical ventilating supply system 机械送风系统
mechanical ventilating system 机械通风系统
mechanical ventilation 机械通风;强制通风
mechanical verification 机械核对
mechanical vibrating feeder 机械振动给矿机
mechanical vibration 机械振动
mechanical vibration analysis 机械振动分析
mechanical vignetting 机械渐晕

mechanical viscosity 机械黏性;机械黏度
mechanical warm air heating plant 机械热风供热站;机械热风供热厂
mechanical washing machine 机械洗舱机
mechanical waste grinder 机械垃圾磨碎机
mechanical water bar (向内开启门窗底边的)挡水杆;(向内开启门窗底边的)挡水条;(向内开启门窗底边的)挡风条;(向内开启门窗底边的)挡风杆
mechanical water lift 机械式提水器
mechanical wear 机械损耗;机械磨损;机械磨耗
mechanical weathering 机械风化(作用);动力风化
mechanical weigh-bridge 机械地磅
mechanical welding 机械(化)焊接
mechanical wood 机械用材
mechanical wood pulp 机械木浆
mechanical work 机械功率
mechanical working 机械加工
mechanical working property 加工性能;机械加工性能
mechanical works 机械工程
mechanical workshop waste 机械废水厂
mechanical wrench 机动扳手
mechanical yielding prop 机械让压支柱
mechanical zero 机械零点;机工零点
mechanic attenuating of glass fibre 玻璃纤维机械控制工艺
mechanic deepened gamma ray survey 机械加深伽马测量
mechanichal respirator 机械呼吸器
mechanician 机械技术人员;技工;机械师;机(械)工
mechanic induced noise 机械感应噪音
mechanic industry 机械制造工业
mechanic of continua 连续体力学
mechanic of failure by piping 接管引起的力学损坏
mechanic of fracture 断裂力学
mechanic of set(ting) 凝固力学
mechanics 力学;机械学;机理
mechanic's and materialman's lien 人工及材料费用留置权
mechanics experimental equipment 力学实验设备
mechanic('s) level 机工水平仪
mechanic's lien 技工留置权(建筑施工中的工资和费用扣押权);留置权;扣押权
mechanic's lien surety bond (由保险公司出具的)人工及材料担保书
mechanics of bulk materials handling 散装材料起重运输机械
mechanics of closed die forging 闭模锻造力学
mechanics of communication 通信技术
mechanics of discontinuum 非连续体力学;非连续介质力学
mechanics of elasticity 弹性力学
mechanics of elasto-plasticity 弹塑性力学
mechanics of frozen soil 冻土力学
mechanics of jointed rock 节理岩石力学
mechanics of landslide 滑坡力学
mechanics of machinery 机械力学
mechanics of management 管理结构
mechanics of materials 材料力学
mechanics of microcirculation 微循环力学
mechanics of mixture 混合体力学
mechanics of particles 质点力学
mechanics of plasticity 塑性力学
mechanics of rock blasting 岩石爆破力学
mechanics of rock bursting 岩石爆破力学
mechanics of rock cutting 岩石切割力学
mechanics of rock masses 岩体力学
mechanics of structure 结构力学
mechanics of super-plasticity 超塑性力学
mechanics of the foreign exchange market 外汇市场机制
mechanics of treatment system 处理系统动力学
mechanics of turbulence 湍流力学
mechanics of viscous fluids 黏性流体力学
mechanics or material lien 机械材料的留置权
mechanic's scaffold(ing) 装配工用的脚手架
mechanics workshop 机工车间
mechanic telescopic spreader 机械伸缩吊具
mechanisation of maintenance 维修保养机械化
mechanism 机制;机芯;机械构造;机械;机理;机构学;机构;河流自净机理
mechanism conversion reform in management 经营转换机制改革

mechanism for closing 闭路机理
mechanism loader 装载机
mechanism of a checking device 检查校正机构
mechanism of action 作用机理
mechanism of adhesion 黏附机理
mechanism of a measuring device 测量机构
mechanism of biofilm formation 生物膜生成机理
mechanism of cavitation 空蚀机理;汽蚀机理
mechanism of change 变革方式
mechanism of chemical corrosion 化学腐蚀机理
mechanism of combustion 燃烧机理;燃烧过程
mechanism of compensation 补偿途径;补偿方式
mechanism of continental rift originesis 大陆裂谷成因机制
mechanism of corrosion 腐蚀机理
mechanism of creep 蠕变机理;徐变机理
mechanism of decompressor 减压机构
mechanism of diaphragm 光阑结构
mechanism of earthquake 地震机理
mechanism of earthquake foci 震源机制
mechanism of economic activities 经济活动机制
mechanism of electrolysis 电解机理
mechanism of failure 破损机理
mechanism of feedback 反馈的工具
mechanism of formation 形成机理
mechanism of formation of spectrochemical smog 光化学烟雾形成机理
mechanism of fracture 破坏机制;破坏机理;破坏机构;断裂机理
mechanism of hardening 淬火机理
mechanism of hydrodynamic(al) dispersion 水动力弥散机理
mechanism of ionic conduction 离子导电机制
mechanism of isostatic compensation 均衡补偿机制
mechanism of mass transfer 传质机理
mechanism of ore formation 成矿机制;成矿机理
mechanism of poisoning 中毒机理(树脂)
mechanism of polymerization 聚合历程
mechanism of production 产生机理
mechanism of pulping 成浆机理
mechanism of rain formation 雨形成机理
mechanism of reactions 反应历程;反应机制;反应机理
mechanism of reaeration 复氧机理
mechanism of retention 保留机制
mechanism of riverbed evolution 河床演变机理
mechanism of roasting 熔烧机理
mechanism of rock-forming 成岩机制
mechanism of sediment transport 输沙机理
mechanism of selection through competition 竞争淘汰机制
mechanism of self-lubrication 自动润滑机理
mechanism of self-purification 自净机理;自净化机理
mechanism of silt deposit 淤积的形成机制;泥沙沉积机理
mechanism of soil liquefaction 土液化机制
mechanism of solid state creep 固态蠕变机制
mechanism of super-plastic deformation 超塑性变形机制
mechanism of transmission 传播机制;传播机理
mechanism study 机理研究
mechanism support 机构支承架
mechanism with small machines 小型设备机械化;小型机器机械化
mechanist 技工;技匠;机械师;机工
mechanistic organization structure 机械性组织结构
mechanization 机械化
mechanization in weed control 机械化控制杂草
mechanization of building operation 施工机械化
mechanization of cargo-handling 装卸作业机械化;装卸工作机械化
mechanization of construction 施工机械化
mechanization of equation 方程式的机械编排
mechanization of maintenance 养路机械化;维修(保养)机械化
mechanization system of cargo handling 装卸机械化系统
mechanization with large equipment 大型设备机械化
mechanized accountant 计算装置(机械);机械计算装置

mechanized bulk system 机械化散装系统
mechanized car park 机械化停车场
mechanized classification 自动化分类
mechanized classification yard 机械化编组场
mechanized coal mining 煤矿机采型
mechanized construction 机械化施工;施工机械化
mechanized construction company 机械化施工公司
mechanized data base 机械化数据库
mechanized data processing system 机械数据处理系统
mechanized dewpoint meter 露点记录器
mechanized dipping 机械浸渍涂搪
mechanized equipment 机械化设备
mechanized farming 机械化耕作
mechanized feed lot 机械化饲料分配器
mechanized fleet 机械化船队
mechanized freight-handling facility 货物装卸机械化设备
mechanized hotbed 机械化温床
mechanized isostatic pressing 自动等静压制
mechanized ladle 机动铁水包
mechanized longhole drilling 机械化深孔凿岩
mechanized lorry 自卸大卡车;具有装卸设备的载货汽车;起重汽车;载货汽车
mechanized magazine 机械化仓库
mechanized marshalling yard 机械化编组场
mechanized mill 机械化轧机
mechanized operation of building 机械化施工
mechanized packing yard 机器操作包装场
mechanized parking garage 机械化停车库
mechanized parking garage equipment 机械化停车库设备
mechanized press line 机械化冲压生产线
mechanized shaft kiln 机立窑
mechanized shaft kiln with pendulum roller 摆辊式机立窑
mechanized shaftmouth platform 井口机械化平台
mechanized shield 机械化盾构
mechanized stowage 机械理仓
mechanized track inspection 机械化轨道检查
mechanized vertical kiln 机械化立窑
mechanized working gang 机械化养路队
mechanized work-platform of maintenance 养路机械作业平台
mechanizing circuit 机械计算电路
mechanocaloric 机械制热
mechanocaloric effect 机械(致)热效应
mechanochemical degradation 机械化学降解
mechanochemical reaction 机械化学反应
mechanochemical reduction 机械化学还原
mechanochemical treatment 机械化学处理
mechanochemistry 力学化学;机械化学
mechano-electronic transducer 机电变换器
mechanoenergetic effect 机械能效应
mechanograph 模制品
mechanography 机械复制法;模制法;办公自动化
mechanomotive force 交变机械力均方根值;机动力
mechanoreceptor 机械感受器
mechanostriction 力致伸缩;机致伸缩
mechatronics 机电仪一体化;机电学
mechinable carbide 可切削硬质合金
mechine ringing 机械振铃
mechining test 机械切削试验
Mechlenburg equation 麦克林伯格方程
meconism 阿片中毒
mecysteine 半胱甲酯
medallion 团花(地毯图案形式);圆形装饰;圆形浮雕;圆雕饰板
medallion gutta 圆锥饰
medallion mo(u)lding 圆雕饰线脚
Medang 几种性质不同的木材(马尼拉产的,用于建筑、家具等)
Medart straightener 梅达尔特(型)矫直机
meddle 干预
media 介质;媒质
media characteristics 介质特性
media charging 装研磨体
media coherence disturbances 媒质相干干扰
media compatibility 介质兼容
mediacy 中间状态;媒介
mediad end 向中端
media design 路中央分隔带设计
media disturbance phase 介质扰动阶段

Mediaeval architecture 中世纪建筑;中古时代的建筑
Mediaeval city wall 中古时代的城墙
Mediaeval road pattern 中世纪道路模式
Mediaeval town 中世纪城市
media filter 介质滤器
media grading 研磨介质的级配
medial 中层的;内侧的
medial allegation 中间混合法
medial alternative law 中间择一律
medial angle 内侧角
medial area 中域
medial basal branch 内侧底支
medial basal segment 内侧底段
medial branch 内侧支
medial camber line 中弧线
medial cell 中室
medial complex 中央复合体
medial divider 中央分隔带
medial eminence 内侧隆起
medial extremity 内侧端
medial friction 交会阻力;交会摩阻
medial groove 中沟
medial head 内侧头
medial humidity 湿度中值
medial island 中央分隔岛;中央分车岛【道】
medial law 中间律
medial leg 内侧脚
medial lip 内侧唇
medially 向中线
medial margin 内侧缘
medial moraine 中碛
medial moraine bar 中碛堤
medial opening 中等张开的
medial part 内侧部
medial plate 中板
medial quasi-group 中间拟群
medial rough discontinuity 中等粗糙的结构面
medial saddle 中鞍
medial section 中央截口
medial segment 内侧段
medial strip 中央分隔带;中央分隔岛
medial surface 内面;内侧面
medial temperature 温度中值
medial type 中间型
media migration 介质迁移
median 中(位)数;中间值;正中(的)
median and small scales 中小比例尺
median band 中带
median bank 河口沙洲
median bar 中条
median barrier 路中(分隔带)护栏;中央分隔带护栏
median basin 中间型盆地
median bed material size 底质中值粒径;底质粒径中值;废砂粒径中值
median bed sediment size 底质中值粒径;废砂粒径中值
median carina 中隆线
median cell 中室
median chart 中间管理图
median chorisis 中央劈开
median class 中位(数)组;中数所在组
median claw 中爪
median cleavage plane 中裂面
median concentration 中值浓度;中位浓度
median crack 中间裂纹
median design 路中分隔带设计
median deviation 中位离差;中位差
median diameter 中值直径(分成重量相同的两部分的直径);中值粒径;中值粒度;中位(数)粒径;中央直径;中径;中间粒径;等重径;等量径;等径
median direction 中位方向
median discharge 中值流量;中位流量;中水位流量;常年流量
median divider 中央分车带;中央分隔带
median dose 中等剂量
median effective concentration 半有影响浓度;半数有效浓度
median effective dose 有效剂量中值;半数有效量;半数有效剂量
median effective time 半数有效时间
median eminence 正中隆起
median energy 平均能量

median error 中间误差
median error norm 中间误差范数
median fiber[fibre] 中间纤维
median filtering 中值滤波
median flow year 平水年;中水年
median fold 中隆褶皱
median furrow 中沟
median game theory 中线对策论
median groove 中沟
median guard bar 中间带护栏
median gutter 中沟
median humidity 平均湿度
median index number 中位指数
median island 中央分车岛【道】
median lamella 中隔壁
median lane 中央分隔带车道;分隔带车道
median lethal concentration 中数致死浓度;致死中浓度;半致死浓度
median lethal dosage 中间致死量
median lethal dose 半致死剂量;半(数)致死量
median lethal time 半致死时间;半数致死时间
median level 中值电平
median line 中线;中间线;正中线
median longitudinal carina 中纵隆线
medianly zygomorphic 左右两侧对称的
median magnitude 中位星等
median mass 中央地块;中间地块
median massif 中间地块
median method 中值方法
median monthly runoff 中值月径流
median of a triangle 三角形中线
median of sound level 噪声声级的中间值;噪声中值
median opening 分隔带开口;中央分隔带开口
median orbital elements 平均轨道根数
median particle diameter 中值粒径;中值粒度;中数粒径;中等粒径
median particle size 中值粒径;中值粒度;中数粒径
median period 中位周期;平均周期
median photographic(al) magnitude 中位照相星等
median plane 中央面
median-plane correction 中平面校正
median-plane field 中平面场
median-plane focusing 中平面内聚集
median-plane location 中平面位置
median-plane orbit 中平面轨道
median planting 分隔带绿化
median plate 中板
median point 中点;重心
median pollution 中度污染
median random variable 中值随机变量
median raphe 正中缝
median ridge 中脊
median rift 中间裂谷
median rise 中间隆起
median section 正中切面
median sector 中分脉
median sediment size 泥沙中值粒径;泥沙中值粒径
median separator 中央分隔带;中间分隔(带)
median septum 中板
median size 中值粒径;中值粒度;中等大小;平均粒径
median space 中区
median speed 中位地点车速;中间车速
median stage 常水位
median stream flow 中值流量;河流中值流量
median strip 中间分车带;中央分隔带;中间分隔(带)
median strip planting 中央分隔带栽植
median surface 中平面;中间面
mediant 中间数(字)
median tarim upwarping region 中部隆起区
median temperature digestive treatment 中温消毒法
median tolerance level 中耐受水平;中间忍受水平;半数耐受水平
median tolerance limit 中值容许浓度;中耐受限度;平均耐受限;半数生存界限;半数耐受量;半耐受极限
median tolerant limit 耐药中浓度
median toxic dose 半数中毒量
medianum colpi 沟中线
median unbias(s)ed estimator 中位数无偏估计量

median unbias(s)edness 中位数无偏性
median valley 中间沟谷；中谷
median value 中央值；中值
median wall 中央壁
median water level 中水位；常水位
median water period 中水期
median water stage 中水位
median year 中水年；中常年；平均年
media-sieve embankment 介质筛护岸
mediastinography 纵隔 X 线照相术
mediastinum 纵隔
media-stored format 存储在媒体上的格式
media supported 滤材支承
mediate 硅钒锰石
mediate contour 中间等高线；基本等高线；首曲线
mediate counter gear 中速齿轮副
mediate curve 首曲线
mediated electrochemical oxidation 媒介电化学氧化
mediation 调停；调解
mediation abode 调解处
mediation application function 协调应用功能
mediation decision 调解书
mediation device 协调设备
mediation function 协调功能
mediation of disputes 调解纠纷
mediation of environment(al) dispute 环境纠纷的调解
mediator 媒质；媒剂；调解者；调解人
media velocity 过滤速度；过滤风速
Medicago sativa 苜蓿
Medicaid 医疗补助计划
medical aid station 救护所
medical air lock 医疗用气压室；医疗用气压舱
medical and dental expense 医疗费
medical and health work 医疗卫生工作
medical apparatus and instruments 医疗器械
medical apparatus and instruments factory 医疗器械厂
medical appliance 医疗器材
medical assessment 医学评价
medical assistance 医疗援助；医疗护理
medical assistant's school 医士学校
medical atomic reactor 医用原子反应堆
medical attention 医疗护理
medical bag 医用包
medical bandage 医用绷带
medical benefits 医疗津贴
medical benefits fund 医疗保险基金
medical boat 医疗船
medical building 医疗建筑
medical cabinet 医用橱(柜)
medical care 医疗保健
medical-care worker 卫生保健员
medical case 医用箱
medical center 医疗中心
medical certificate 医疗诊断书；医生证明书；健康证明(书)
medical chamber 气压(调节)治疗室
medical charges 医药费
medical chart cabinet 医疗图表柜
medical check-up 医学检查；健康检查
medical check-up in turn 挨个儿检查身体
medical control 医疗管理；医疗监督
medical corps 医疗队
medical cost 医药费
medical coverage 医疗保险
medical cupboard 药品柜
medical department 医务部门
medical examination 体格检查
medical examination on the spot surveillance 就地诊验
medical expense 医药费
medical expensededuction 纳税时的医疗费用扣除
medical expenseinsurance 医疗费保险
medical exposure 医疗照射
medical facility 医疗设备(施)
medical faculty 医学院
medical geology 医学地质学；医疗地质学
medical gloves 医用手套
medical history 病历
medical inspection 检疫
medical installation 医疗设备(施)
medical institution 医疗机构

medical insurance 医疗保险
medical insurance premiums 医疗保险费
medical kit 保健箱
medical kit for home visit 出诊箱
medical launch 医务艇
medical lock 医用气闭室(用于治疗减压病的特殊房间)；医疗闸(沉箱)；递物筒
medical locker 医药柜
medical microscope 医用显微镜
medical officer 军医
medical optic(al) instrument 医疗光学仪器
medical record 医疗记录
medical record room 病历室
medical recuperate environment 医疗休养环境
medical room 医务室
medical scales 海船备用药物器材标准
medical security 医疗保障
medical security of ventilative diving 通风式潜水医务保障
medical service 医疗服务；医疗费
medical services of the armed force 军队卫生
medical spring 医疗泉
medical station 医疗站
medical stone 麦饭石
medical subsidies 医疗补贴
medical team 医疗队
medical telemeter 医用遥测计
medical treatment 医疗
medical treatment and sanitation building 医疗卫生建筑
medical vacuum pump 医用真空泵
medical vehicle 医疗车
medical ward 内科病房
medicamentarius 药店
medicament form 剂型
Medicare plan 医疗保障方案
medication 投药法
medichair 医疗椅
medicinal extract 膏剂
medicinal glass 药用玻璃
medicinal granules 冲剂
medicinal herb 药材
medicinal materials 药材
medicinal oil 药用油
medicinal spring 医疗(矿)泉
medicinal tolerance 耐药量
medicine box 药品箱
medicine cabinet 药箱；药品箱
medicine chest 药品箱；药品柜
medicine examination at the port 口岸药检
medicinerea 中间灰质
Medici porcelain 麦迪西软瓷
medico mineral 药用矿物
Medieval architecture 中古时代建筑
medieval maximum 中世纪极大期
mediiphyric 显微斑晶的
medimarimeter 平均海面记录器；平均海面测定仪
Medina cement 麦地那水泥
medina deceleration lane 中央减速车道
medina quartz 中温石英
Medina quartzite 麦地那石英岩
mediocommissure 中连合
mediophyric 中斑晶的
Medio-Pleistocene 中更新世
mediosilicic 中硅质的
medio-twist 中等捻度
meditation chapel 沉思小教堂
mediterranean 中间地槽
Mediterranean agriculture 地中海式农业
Mediterranean architecture 地中海建筑
Mediterranean belt 地中海带
Mediterranean bivalve province 地中海双壳类地理区
Mediterranean climate 温带冬季气候；地中海气候
Mediterranean ecosystem 地中海生态系统
Mediterranean faunal region 地中海动物区
Mediterranean foraminifera realm 地中海有孔虫地理区系
Mediterranean front 地中海锋
Mediterranean gangway 船尾舷梯
Mediterranean geosyncline 地中海地槽
Mediterranean-Himalayan seismotectonic zone 地中海—喜马拉雅地震构造带
Mediterranean karst 地中海岩溶

Mediterranean ladder 船尾舷梯
Mediterranean moor 船首向洋面抛锚
Mediterranean regional oil combating center 地中海区清除油污中心
Mediterranean salinity crisis event 地中海盐度危机事件
Mediterranean Sea 地中海
Mediterranean-south Asian belt 地中海—南亚地震带
Mediterranean suite 地中海岩组；地中海岩套
Mediterranean tectonic zone 地中海构造带
Mediterranean type geosynclines 地中海型地槽
medium 中央的；中等的；介质；媒质；媒介；平均的；培养基；存储媒体；存储介质；【复】media
medium abrasion furnace black 中耐磨炉黑
medium abrasive rocks 中等研磨性岩石
medium accelerator 中速促进剂
medium accelerator dosage 中等量促进剂
medium access memory 中速存取存储器
medium accuracy inertial navigation system 中等精度惯性导航系统
medium acidic volcanic assemblage 中酸性火山喷发组合
medium activity liquor 中放射性废液
medium air bubble 中等气泡
medium air bubble aeration 中气泡曝气
medium alkali glass 中碱玻璃
medium alkali glass fibre 中碱玻璃纤维
medium alloy 中合金
medium alloy steel 中合金钢
medium altitude aircraft 中空飞机
medium altitude communication satellite 中高度通信卫星
medium altitude satellite 中高度卫星
medium and long range hydrologic forecast 中长期水文预报
medium and long-term plan of railway construction enterprise 铁路施工企业中长期经营计划
medium and small scale enterprises 中小型企业
medium and small sized unit 中小型装置
medium angle boundary 中等角度晶界
medium angle thrust 中等角度冲断层
medium-baked brick 半烧成砖；适度烧成的砖；适度烘成砖
medium-baked tile 适度烧成的瓦；中等烧过的瓦
medium banded coal 中条带状煤
medium band photometry 中带测光
medium band width 中带宽度
medium bank 中等倾斜
medium barium crown glass 钡冕玻璃
medium base 中等基础
medium basic eruptive 中基性喷出岩
medium bed 中厚层；中间层
medium bedded 中厚层状；中厚层的
medium bedded orebody 中层矿体
medium bedded soil 中厚层土
medium bedded structure 中层状构造
medium block 中型砌块
medium boiler 中等锅炉
medium boiling solvent 中沸点溶剂
medium boulder 中巨砾
medium break(ing) 中裂的(乳化沥青)；中度裂化
medium breaking dispersion 半稳定沥青分散液
medium breaking emulsified asphalt 中裂乳化沥青
medium breaking emulsion 中裂乳液；半稳定乳液
medium brick 中砖；中级砖
medium bridge 中桥
medium brightness 中等明度
medium bubble aeration system 中气泡曝气系统
medium burned brick 标准烧成砖；中温烧成砖；标准成砖
medium burned refractory ware 中等焙烧的耐火瓷器
medium caliber 中口径
medium calorific value coal 中等发热量煤
medium calorific value gas 中等发热量燃气
medium capacity 中等容量
medium capacity plant 中型电站；中容量电站
medium capacity storage 中容量存储器
medium car 中型轻便汽车
medium carbon steel 中碳钢
medium casting 中铸件

medium cast-iron 中铸铁
medium cell 中型电解槽;中室
medium cement castable 普通水泥浇注料
medium centrum 次中心质
medium check 中度裂缝
medium chip(ping)s 中等小块;中等碎片
medium chrome yellow 中铬黄
medium clastics 中粒碎屑
medium clay 普通黏土;中黏土
medium clay(ey) loam 中黏壤土;中亚黏土
medium cleaning 中净化;中间净化;中级净化
medium cleavage 中等解理
medium close shot 中近景镜头
medium cloud 中层云
medium coarse sand 中等粗砂;中粗砂
medium coarse texture 中—粗粒结构
medium collapsible 中等湿陷
medium colo(u)r 中色
medium compact sand 中密度砂
medium complexity site 中等复杂场地
medium concentration 中等浓度
medium concrete 普通混凝土
medium consistency 中等稠度
medium constant 介质常数
medium construction item 中型建设项目
medium construction project 中型建设项目
medium contour 中间等高线
medium contrast 中反差;中等反差
medium control 中等治理
medium cross-section 中断面
medium crushing 中碎;中级压碎
medium crystal 中等晶体
medium crystalline dolomite 中晶白云岩
medium crystalline limestone 中晶灰岩
medium crystalline texture 中晶结构
medium curing 中速凝固的;中级处理的;中凝的(沥青等)
medium curing asphalt 中凝沥青;中等稀释沥青
medium curing cutback asphalt 中凝轻制(地)沥青
medium curing liquid asphalt 中凝液体石油沥青;中凝液体(地)沥青
medium current 中等强度电流
medium deep drill hole 中深钻孔
medium deep well 中深井
medium dense 中密(的)
medium density 介质密度
medium density baler 中密度捡拾压捆机
medium density building fiberboard 中密度建筑纤维板
medium density fiberboard 中密度纤维板
medium density fiberboard wastewater 中密度纤维板废水
medium density flaxbased board 中密度亚麻板
medium density hardboard 中密度硬质纤维板;中密度纤维板
medium density hay press 中等密度干草压捆机
medium density overlay 中密度贴面(胶合板)
medium density particle board 中密度刨花板
medium density particle wood 中密度碎料板
medium density plywood 中密度贴面胶合板
medium density polyethylene 中密度聚乙烯
medium density polyethylene plastics 中密度聚乙烯塑料
medium density wood chipboard 中密度碎木板
medium depth 中埋
medium depth(bore) hole 中深孔
medium depth rig 中深型钻机
medium diameter 中值直径;中值粒径;中值粒度;中间直径
medium diamond-point knurling rolls 中金刚钻压花滚轮
medium dimension 中等尺寸
medium dip 中倾斜
medium dipping bed 中倾斜岩层
medium discharge 中流量
medium dissemination 中粒浸染
medium distance 中程
medium distance aeroplane 中程飞机
medium distance aids 中距离无线电导航仪
medium draft 中绵
medium drill 中型凿岩机
medium drying 中速干燥的(指油脂)
medium duty 中批量生产;中型的;正常工作情况下
medium duty engine 中型发动机

medium duty fireclay brick 中级(黏土质)耐火砖
medium duty flexible(swing) door 中型双开弹簧门
medium duty scaffold 中型脚手架;中负荷脚手架
medium duty truck 中型载货汽车
medium effect 介质效应
medium effective concentration 中值有效浓度
medium effective dose 中值有效剂量
medium efficiency air filter 中效空气过滤器
medium egg-plant screw indicating lamp bulb 中螺口茄形指示灯泡
medium emery powder 中级金刚砂粉
medium emulsion 中等乳浊液
medium energy 中等能量
medium energy accelerator 中能加速器
medium energy cyclotron 中能回旋加速器
medium energy type dyes 中能型染料
medium error 中误差
medium exchange 交换媒介(货币、支票等)
medium fast plate 中等感光速度硬片【测】
medium fast scan 中速扫描
medium fast sweep 中速扫描
medium fiber 中间纤维
medium field earthquake motion 中场地震运动
medium filtered cylinder oil 中级过滤汽缸油
medium fine grinding 中细粉磨
medium fine texture 中细粒结构
medium finger spacing cutter 中距护刃器切割器
medium finish(ing) 中级精加工;中等光洁度
medium first-year ice 当年中冰(厚度70~120厘米)
medium fissured 中等裂隙化
medium fit 中级配合
medium floe 中冰块
medium floor 中地板
medium flow rate paper 中流速纸
medium force fit 中级压(紧)配合;中等压入配合
medium format 适中的格式
medium fraction of oil 原油中的中馏分
medium fracture 中冰隙
medium frequency 中频率;中频
medium frequency amplifier 中频放大器
medium frequency band 中频频带
medium frequency changer 中频变换器
medium-frequency concrete finisher 混凝土中频修整机
medium frequency dielectric(al) separator 中频介电分离仪
medium frequency induction furnace 中频感应电炉
medium frequency loss 中频损耗
medium frequency oscillator 中频振荡
medium frequency response 中频率响应
medium frequency signal 中频信号
medium-frequency sound insulation 中频率的隔声
medium gear 中速齿轮
medium glaze 中温釉
medium grab bucket 中型抓斗
medium grade 中级品位
medium graded bituminous concrete pavement 中粒沥青混凝土路面
medium grade glass fibre 中级玻璃纤维
medium grade metamorphosed 中变质的
medium grade ore 中富矿石
medium gradient 中等坡度
medium grain 中密度纹理;中粒;中等纹理;中等磨粒;中等颗粒;中等晶粒
medium-grained 中粒(度)的;中颗粒的;中等粒径的;中等粒度的
medium-grained granite 中粒花岗岩
medium-grained gravel 中等大小的砾石
medium-grained grinding stone 中粗磨石
medium-grained limestone 中粒灰岩
medium-grained material 中等粒状的材料
medium-grained meander belt deposit 中粒曲流带沉积
medium-grained particle 中等粒状的微粒
medium-grained sand 中粒状的砂子;中(粒)砂
medium-grained sandstone 中等粒状的砂岩
medium-grained sandy gravel 中等粒状的砂砾
medium-grained silt 中粒径粉土
medium-grained snow 中粒雪
medium-grained structure 中等粒状结构

medium-grained texture 中粒结构
medium granular 中粒状的;中粒(度)的
medium granular crystalloblastic texture 中粒变晶结构
medium granular psamitic texture 中粒砂状结构
medium granular rocks 中粒岩石
medium granular sand 中砂;中颗粒砂
medium gravel 中卵石;中砾石
medium gravel and fine gravel 中粒砾石和细粒砾石(美国标准25.4~9.52毫米为中粒,9.52~2.00毫米为细粒);中砾和细砾
medium gravity 中等比重
medium gray[grey] 中灰;中等灰色
medium grinding 中磨;中级磨碎;中等粒度研磨
medium grinding compartment 中间卸料卸料磨的回粉仓;中间粉磨仓
medium ground 中硬岩层
medium hair grease 中级黏度润滑脂
medium hall of metro station 地铁中间站厅
medium harboard 半硬纤维板
medium hard 中(等)硬度(的)
medium hardboard 中密度硬质纤维板
medium-hard coal-tar pitch 半硬质煤焦油柏油;半硬质煤焦油沥青
medium hardening 中速硬化
medium hard formation 中硬地层
medium hardness 中等硬度
medium hard rock 中硬岩石
medium hard steel 中硬钢;中等硬度钢
medium hard white coat 中硬度白灰罩面
medium haul aeroplane 中程飞机
medium head 中水头
medium head(hydraulic) turbine 中水头水轮机
medium head plant 中水头电站
medium-head range 中等压头范围
medium head scheme 中水头电站
medium heavy-duty 中等工况
medium heavy-duty class 中等载重等级
medium heavy fuel 中级重燃油
medium heavy lathe 中型车床
medium heavy loaded circuit 中等加载电路
medium heavy loading 中等加载
medium heavy oil 中—重质石油
medium heavy piece of cargo 中等重件货物
medium heavy rail 次重轨
medium heavy traffic 中(等)密度交通
medium height grass 中高草
medium height mountain 中高山(绝对高度3500~5000米)
medium hewn squares 中方材
medium high frequency 中高频
medium high frequency direction finder 中高频测向仪
medium high frequency wave 中短波
medium high mountain 中高山(绝对高度3500~5000米)
medium-high partition 中等高度隔墙;中等高度隔板;中等高度间壁
medium high solid 中高固体分
medium hydraulic project 中型水利工程
medium hydrophilicity 中等亲水的
medium ice content frozen soil 多冰冻土
medium ice-field 中冰源
medium ice floe 中浮冰块;中冰盘;中冰块
medium impedance 中间阻抗
medium infrared 中红外
medium inlet 介质入口
medium insulation 中级绝缘
medium intensity 中等强度
medium intermittent duty 中间歇负载;中间歇负载
medium investigation induction log 中感应测井
medium investigation induction log curve 中感应测井曲线
medium irrigated emulsion 中等含水量乳化液
medium knot 中等树节
medium landslide 中型滑坡
medium lead crystal glass 中铅玻璃
medium leaded brass 中铅铜合金
medium length(bore) hole 中深孔
medium length fibre 中长纤维
medium-length haul 中等托运长度
medium lethal radiation dose 半致死辐射剂量
medium level 中值电平
medium level cloud 中云

medium-level radioactive waste 中等程度的放射性废弃物
medium level waste 中放废物
medium level work 中等放射性操作
medium lever radioactive waste liquor 中放射性废液
medium lever radioactive wastewater 中放射性废水
medium lift lock 中水头船闸
medium lift pump 中扬程(水)泵;中压泵
medium line 中线
medium line of triangle 三角形中线
medium liquid asphalt 中凝液体(地)沥青
medium lived 中等耐用的;中等使用寿命的;中等耐久的
medium loader 中型装载机
medium loading district 中等负荷区
medium loam 中壤土
medium long shot 中远景镜
medium lorry 中型(载重)货车;中等载重货车
medium-loss fiber 媒质损耗光纤
medium lot producer 中型企业
medium low mountain 中低山(绝对高度500~1000米)
medium maintenance 中修
medium maintenance group 中修组
medium maintenance of track 轨道中修
medium manganese steel 中锰钢
medium media 次介质
medium mesh 中筛
medium metamorphism 中级变质作用
medium mill 介质磨
medium moist underground water area 中等湿度潜水区
medium-moisture silage 半湿青储[贮]料
medium molecular weight 中等分子量
medium mountain 中山(绝对高度1000~3500米)
medium mull 中腐熟腐殖质
medium objective 中期目标
medium of advertisement 广告手段
medium of circulation 流通手段;流通媒介
medium of exchange 流通媒介;交易媒介
medium of heat transmission 热媒
medium of international payment 国际支付手段
medium of payment 支付手段
medium of sound level 声级中值
medium of varying refractive index 折射率可变介质
medium oil 中黏度(油类);中油(度)
medium oil alkyd 中油度醇酸树脂
medium oil alkyd resin 中黏度醇酸树脂
medium oil paint 中油性油漆;中油度油漆
medium oil resin 中油度树脂
medium oil varnish 中油度清漆;中性油清漆
medium operating condition 中等使用条件
medium orebody 中等矿体
medium overhaul 中修
medium oxygen consumption water 耗氧量中等的水
medium paint media 漆料
medium paving sett 中等大小铺路块石
medium pebble 中型石砾
medium performance sealant 中等功能密封膏
medium permeability 中等渗透率;中渗透度
medium persistance phosphor 中等余辉的磷光体
medium phosphor brake-shoe 中磷闸瓦
medium pickup car 中型货重(汽)车
medium pitch 中等倾斜
medium plain emery wheel 中砂轮
medium plant 中型设备
medium plasticity 中塑性
medium plastic soil 中等塑性土
medium plate 中厚钢板;中(厚)板
medium plate mill 中板轧机;中板厂
medium point 重心;中点
medium porosity 介质孔隙度
medium position 中间位置
medium powered car 中等功率汽车
medium power objective 中(等)倍(数)物镜
medium power reactor 中等功率堆
medium pressure 中压;中等压力
medium pressure absorber 中压吸收塔
medium pressure accumulator 中压蓄力器
medium pressure acetylene generator 中压乙炔发生器
medium pressure ammonia absorber 中压氨吸收塔
medium pressure blowpipe 中压喷焊器
medium pressure boiler 中压锅炉
medium pressure burner 中压燃烧器
medium pressure engine 中压发动机
medium pressure mercury lamp 中压汞灯
medium pressure metamorphism 中压变质作用
medium pressure oil absorption process 中压油吸收过程
medium pressure pneumatic conveyer 中压气流式输送器
medium pressure pump 中压泵
medium pressure regional metamorphic facies series 中压区域变质相系
medium pressure regional metamorphism 中压区域变质作用
medium pressure rubber sheet 中压胶板
medium pressure sprinkler 中压洒水车;中压喷洒器;中压喷淋器
medium pressure steam 中压水蒸气
medium pressure steam system 中压蒸汽系统
medium pressure steam turbine 中压汽轮机
medium pressure tank 中压槽
medium pressure tunnel 中压隧洞;中压隧道
medium pressure type 中压型
medium pressure tire 中压轮胎
medium pressurevessel 中压容器
medium price 中等价格
medium processing channel black 中等加工槽黑
medium quality 中等货
medium quartz spectrograph 媒质石英摄谱仪
medium range 中期的;中距离;中程(的)
medium range earthquake prediction 中期地震预报
medium range forecast 中期(天气)预报
medium range plane 中程飞机
medium range positioning system 中程定位系统
medium range system 中程导航系统
medium rank 中级
medium rank anthracite 中级无烟煤
medium rank bitumite 中级烟煤
medium rate of exchange 中间汇率
medium ratio of modulus 中等模量比岩石
medium reflectivity 中间反射率
medium reinforced 有中等配筋率的;中等配筋(率)的
medium relief 中等地势
medium repair 中修
medium reservoir 中砂岩储集层
medium resolution infrared radiometer 中分辨率红外辐仪;中分辨率红外辐射计
medium resolution picture transmission 中分辨率图像传输
medium resolution spectrometer 中等分辨率光谱仪
medium return 中等回波
medium ring compound 中环化合物
medium rise block 中层街坊;中等高层街坊
medium rise building 中等高层建筑;中层建筑;中等高度建筑
medium road oil 中质铺路油
medium rock 中性岩
medium rocks 中性岩类
medium rotten water 中等腐败性水
medium rough toothed cutter 中等粗齿的切刀
medium rubble fill 中等毛石填充
medium salinity 中等含盐量
medium salinity water 中盐度水;中度盐水
medium salted fish 中盐鱼
medium salting 中盐渍
medium sand 中砂;中粒砂
medium sand filling 回填中砂
medium sandstone 中砂岩
medium sandy clast 中砂屑
medium sandy clay 中砂黏土
medium saturation 饱和度中等
medium scale 中型规模;中型的;中间尺度;中(等)规模的;中(等)比例尺
medium scale chart 中等比例尺海图
medium scale computer 中型计算机
medium scale cyclone 中尺度气旋
medium scale geologic(al) survey-(ing) 中比例尺地质调查;中比例尺地质测量
medium scale integrated circuit 中规模集成电路
medium scale integration 中等规模集成电路
medium scale map 中比例尺图
medium scale of bursting water 中型突水
medium scale test 中试
medium scan 中等长度扫描
medium screen 中筛;中等屏幕
medium screening 中级筛分;中级筛选
medium section 中型材
medium section mill 中型型钢轧机
medium sensitivity 中度敏感
medium septum 中隔壁
medium set(ting) 中裂的(乳化沥青);中凝(的)
medium setting(asphaltic) emulsion 中裂乳液
medium setting emulsified asphalt 中裂乳化沥青
medium setting time 中凝时间
medium shade 中色(调);中等色
medium shaft 中间竖井
medium sheet 中板材;中片材
medium short wave 中短波
medium silt 中粒粉(砂);中粉(砂)土
medium silt and fine silt 中粒粉砂和细粒粉砂;中细粉砂
medium sinus 中槽
medium size 中型;中等尺寸
medium size bearing 中型轴承
medium size coal mine 中型煤矿
medium sized 中型的;中号的;中规模的
medium sized aggregate 中颗粒集料;中等颗粒骨料
medium sized city 中等城市
medium sized coal 中块煤
medium sized computer 中型计算机
medium sized format 中等尺寸的版本;中等规模格式
medium sized green house 中型温室
medium sized mosaic tile 尺寸中等的锦砖;中等大小的彩色镶嵌瓷砖
medium sized motor 中型电动机
medium sized plot test 中型小区试验
medium sized quarry 中型露天矿
medium sized tanker 中型油轮(常指6万~12万吨级的)
medium sized tractor 中型拖拉机
medium size enterprise 中等企业
medium size log 中原木
medium size ore deposit 中型矿床
medium size panel 中型板材
medium size sheet stacker 中片堆垛机
medium size turbine generator 中型汽轮发电机
medium sizing 中间筛(分);中级筛选;中级筛分
medium slaking (石灰)中等程度的潮解;(石灰)中等程度的消化
medium slaking lime 中化石灰;中消石灰
medium smooth circular-cut file 中等平滑圆锉刀
medium smooth file 中细锉
medium sodium water 中度钠质水
medium soft 中等软度的;中软的;半硬的
medium soft pitch 中软沥青;中度软焦油脂;半软沥青
medium solder 软焊料
medium solid 重介质固体;加重质
medium soluble salt 中溶盐
medium soluble salt test 中溶盐试验
medium-span bridge 中跨桥梁
medium-span roof 中跨度屋盖;中跨度屋顶
medium speed 中速
medium speed computer 中速计算机
medium speed diesel 中速柴油机
medium speed distance relay 中速远距继电器
medium speed elevator 中速电梯
medium speed emulsion 中速乳胶
medium speed engine 中速发动机;中速柴油机
medium speed lift 中速电梯
medium speed memory 中速存储器
medium speed ship 中速船(舶)
medium speed storage 中速存储器
medium speed V-engine 中速V形发动机
medium speed vessel 中速船(舶)
medium square 中枋;中方格;中等方木材
medium stain 中等变色
medium stained sapwood 中等变色边材
medium standard frequency 标准中频

medium staple 中绒
medium staple fibre 中长纤维
medium steel 中碳钢;中硬钢
medium steep seam 中倾斜矿层
medium stiff clay 中硬黏土
medium stone 中碎石
medium stone bit 中粒金刚石钻头;中号金刚石钻头
medium strain 中等应变
medium straw 中型塑料细管
medium-strength steel wire 中强钢丝
medium strip 中间分车带;中央分隔带;中间分隔(带)
medium strong mo(u)lding sand 半肥砂
medium structure 中粒结构;中等(粒度)结构
medium suspension 黏结剂的悬浮
medium sweep elbow 中等弯曲半径弯头
medium swelling-shrinkage soil 中等胀缩性土
medium tar 中质柏油;中质焦油(沥青)
medium temperature 中等温度;介质温度
medium temperature chimney 中温烟囱
medium temperature digestive treatment 中温消化处理
medium temperature fuel cell 中温油箱;中温燃料电池
medium temperature grease 中温润滑脂
medium temperature incinerator 中温焚化炉
medium temperature sensor in pipeline 管道内介质温度传感器
medium temperature setting adhesive 中等温度凝结胶黏剂
medium temperature water-heating system 中温暖系统
medium temper steel 中等回火钢
medium tenacity 中等强度
medium tension bushing (变压器的)中压套管
medium term 中项;中期(的)
medium term and longterm credit 中长期信贷
medium term and long term earthquake prediction 地震中长期预报
medium term bonds 中期债券
medium term capital 中期资本
medium term credit 中期信贷
medium term earthquake prediction 地震中期预报
medium term Eurocredit market 中期欧洲贷款市场
medium term forecast 中期预测
medium term loan 中期放款;中期贷款
medium term parking 中等时间停车;中等时间存车
medium term schedule 中程调度
medium texture 中型结构
medium textured loam 中亚黏土
medium texture soil 中等质地土壤
medium thermal black 中粒子热裂炉黑;中等热裂黑
medium thermal carbon black 中粒热裂法炭黑
medium thick bedded texture 中厚层状结构
medium thick coal seam 中厚煤层
medium thickness coal seam 中厚煤层
medium thickness seam 中厚矿层
medium throat width 中等喉道宽度;平均喉道宽度
medium tidal environment 中潮汐环境
medium tide curve 中潮曲线
medium-to-air interface 介质空气界面
medium tolerance limit 中耐受限度
medium to-low temperature brine 中低温热卤水
medium tone 半音度
medium tonnage mine 中型矿山
medium torn grain 中级裂纹
medium to-short range prediction 中短期预报
medium tractor 中等功率拖拉机
medium traffic 中量交通;中等量交通
medium traffic density line 中等运输密度线路
medium transmission line 中长输电线路
medium transportation truck company 中型运输汽车连
medium truck 中等载货汽车;中型载货汽车
medium type 中间型
medium type layout 中式布置(机电)
medium type truck 中型载货汽车;中型载重(汽)车
medium vacuum 中真空
medium valley 中间沟谷
medium value 中值;平均值
medium value of fracture width 裂缝宽度中值
medium value of pore 孔隙中值

medium value of pore throat 喉道中值
medium value reflectivity 中值反射率
medium viscosity 中等黏滞性;中黏度
medium volatile 中挥发性物
medium volatile fuel 中等挥发性燃料
medium volatile pituminous coal 中度挥发烟煤
medium voltage 中压;中(等)电压
medium voltage bushing (变压器的)中压套管
medium voltage line 中压线路
medium voltage network 中压供电网;中压电力网;中压(电力)系统
medium voltage switchgear 中压开关装置;中压开关柜
medium voltage system 中压系统
medium voltage unit 中压元件
medium voltage winding 中压绕组
medium volume nozzle 中流量喷嘴
medium volume spraying 中容量喷雾
medium volume swelling 中体积溶胀
medium water depth density 中等水深密度
medium water level 中水位
medium water stage 中水位
medium wave 中波
medium wave antenna 中波天线
medium wave band 中波(波)段
medium wave condition 中浪海况
medium wave receiver 中波接收机
medium wave transmitter 中波发射机
medium weathering 中风化
medium weight 中重的
medium weight breed 中型品种
medium weight load 中等负载;中等荷载
medium weight truck 中型载重(汽)车
medium wide strip mill 中型宽带材轧机
medium width steel strip 中等宽度带钢
medium year 中常年
medium yearly runoff 中值年径流
medius 中指
medley 杂拼物
medomin 庚巴比妥
medulla 骨髓;木髓(拉丁语)
medullary 髓心的
medullary ray 髓线;木纹;木射线
medusa 水母
medusa white 美杜莎白(美国白硅酸盐水泥)
Medvedey intensity scale 麦德维地震烈度表
meehanite 密烘铸铁
meehanite cast-iron 密烘铸铁
meehanite metal 密烘铸铁;变性铸铁
meerschalminite 铝海泡石
meerschaum 海泡石
meet 交切;满足;会合;会船;如期付款;如期偿付
meet a bill 准备支付到期票据;支付到期的票据
meet a cassation 满足上诉
meet a condition 满足条件
meet a criterion 符合标准
meet a demand 满足需要
meet an objective 达到目的
meet commitments 履行承诺
meet engagement 清偿债务
meeter 迎面来的船
meet expenses 偿付开支
meeting 汇合点;交汇点
meeting a rudder 压舵
meeting beam 近光
meeting documents 会议文件
meeting face 止水面(坞口);贴接面
meeting hall 会议厅
meeting house 教堂;会堂
meeting liability 履行债务
meeting of trains 列车交会
meeting pile 接合桩
meeting place 交会处;会船处;会场
meeting point 集合点;会车地点;合流点;交(汇)点
meeting post 斜接柱;闸门立柱;(人字门上的)接合柱;交会竖杆;碰头门梃
meeting rail 交会横挡;滑动窗框碰头横档(关窗时上下靠在一起的横档);横档;(窗的)碰头横档;提拉窗横档
meeting report 会议报告
meeting room 会议厅;会议室;会议交接班室
meeting situation 相遇局面;会船局面
meeting station 会让站
meeting stiles 碰头边梃;中门梃;双开门门梃

meeting strip 窗扇盖缝条
meeting the quality standards projects 全优工程
meeting to adjust shortages and surpluses 调剂会议
meetion 相遇
meet one's expenses 使收支相抵
meet one's liabilities 偿还债务
meet the challenge 满足要求
meet the necessity 符合
meet the needs 满足需要
meet the obligation 履行义务
meet the requirement 满足要求;满足需要
meet the requirement of specifications 适应规范的要求;符合规范要求
meet the serious challenge 面对严峻挑战
meet the specification 满足技术条件;合乎规格;符合规格
meet with stresses 承受应力
meg 小型绝缘试验器
mega 十进制数的百万;百万
mega-ampere 兆安
megabacterium 巨型细菌;【复】megabactera
mega bar 兆巴
megabit 兆比特;百万位;百万二进制位;百万比特
megabit memory 兆位存储器;兆位
megabits/second 兆比特/秒
megabit storage 兆位存储器;兆位
megabreccia 巨型角砾岩
megabusiness 超巨大企业
megabyte 百万字节;兆字节
megacalorie 兆卡
megaclast 粗碎屑
megacorporation 特大企业
megacryst 大晶体;大晶
megacrystalline 大晶的
mega cubic(al)meter 兆立方米
megacurie 兆居里;百万居里
megacycle 兆周;巨旋回
megacycles per second 兆周/秒;兆赫;每秒兆周
megacyclothem 巨旋回层;巨旋层
mega-depression tectonics 巨型坳陷构造
megadyne temperature 位温
mega electron-volt 兆电子伏特
megaerg 百万尔格
megafacies 大相【地】
megafamily 大杂院;多户杂居住所
megaflop 百万次浮点运算
megafog 警雾扩音器;浓雾报警器
megafossil 大型化石;大化石
mega gallons per day 兆加仑每日
megagea 巨陆【地】
megagon 多角形
megagrained 巨砾的;粗粒的
megagram 百万克
megagramme force 兆克力
megagramme force meter 兆克力米
megahenry 兆亨(利)
megahertz 兆周每秒;兆赫;百万赫
megahydroproject 大型水利工程
megajoule 兆焦(耳)
megaliter 兆升
megaliters per day 兆升每日
megalith 巨石
Megalithic 巨石器阶;巨石的
Megalithic Age 巨石器时代
megalithic architecture 巨石建筑
megalithic burial chamber 巨石墓室
megalithic masonry 密封砌筑巨石圬工;巨石圬工
megalithic monument 大石纪念碑;巨石纪念碑
megalith masonry 巨石砌体;大块石砌体
megalitre 兆升
megalitres per day 兆升每日
megalo-earthquake 极大地震
megalograph 显微图形放大装置;显微图形放大器
megaloplankton 巨型浮游生物
mega(lo)polis 特大城市;大型工业城镇;人口高度集中的大城市;巨型城市;都市组群;大都市连绵区;大城市地区;超级城市
mega(lo)politan 特大城市(的)居民;特大城市的人
megalopolitan corridor 巨大城市走廊
megalosphere 大球壳
megalosyndactyly 巨并指
megamere 大分裂球
megameter 摇表;兆欧表;兆米;高阻表;千公里;大

公里(等于1000公里)
megampere 兆安;百万安
meganewton 兆牛顿
megaohm 兆欧(姆)
megapark 特大公园
megaparsec 百万秒视差;百万秒差距
Megaperm 梅格珀姆铁镍锰高磁导率合金
megaphanerophyte 大乔木
megaphenocryst 大斑晶
megaphone 传声器;喇叭(筒);扩音器;传声筒
megaphyric 大斑晶的
megaplate 巨板块;大板块【地】
megapoise 兆泊(黏滞度单位);百万泊
megapolis 特大都市;巨型城市
megapore 大孔隙的
Megapyr 梅格派尔铁铬铝电阻丝合金;铁铝铬电阻丝合金
megarad 兆拉德
megarelief 宏观地形;大起伏;大地形
mega-rhythm 大韵律
megaripple 巨波纹;大波痕
megaron (古希腊建筑的)中央大厅
megascope 肉眼识别;粗视显微镜;放大镜
megascopic 肉眼可见的;粗视的
megascopic anthraxylon 肉眼镜煤
megascopic fusain 肉眼丝炭
megascopic method 宏观(方)法
megasedimentology 宏观沉积学
megaseism 剧烈地震;强震;伟震;大(地)震
mega sequence 大层序
mega signal(l)ing 元信令
megasome 核粗粒体
megastage 大阶段
megastere 兆立方米
megastructural 巨大建筑的
megastructure 巨型建筑;复合体建筑;巨型建筑物(聚合体);庞大建筑物;特级大厦;超级建筑
megasweep 摇频振荡器
megatanker 百万吨级油轮;超级油轮
megatectonics 巨型构造;宏观构造学;大构造;大地构造学
mega-temperature 高温(带)
megatexture 巨观纹理
megatherm 高温;热带植物
megathermal climate 高温气候;热带雨林气候
megathermal period 高温期
megathermal type 高温植物型
megathermal zone 热带雨林带;高温带
megathermophyte 高温植物
megatherm plant 高温植物
megaton 百万吨
megatonnage 百万吨级
megatron 盘封管
megavar 百万乏
megavarve 粗纹泥
megaversity 超级大学
megavolt 兆伏;兆达因;百万伏
megavolt ampere 千千伏安;兆伏安;百万伏安
megawatt 兆瓦;千千瓦;百万瓦
megawatt day 兆瓦日
megawatt hour 兆瓦时;千度;百万瓦时
megayear 百万年
megeresin 白瑞香脂
megger 兆欧计;摇表;高阻计;高阻表;绝缘试验器;迈格表;测兆欧计;测高阻计
Megger earth tester 梅格尔土壤电阻测定器
Meggers lamp 高频电源水银灯
megger test 绝缘测试
MeGill alloy 麦吉尔可铸可锻铝青铜
megilp 油画载色剂;油画载色体
megnetron 磁控电子管
megohm 百万欧姆
megohm box 高阻箱
megohm bridge 高阻电桥
megohmit(e) 绝缘物质
megohmmeter 高阻计
megrims 眩晕症
mehrab 指向麦加的祈祷壁龛
Meigen's reaction 迈埃金反应
meinbranous structure 膜相结构
meiobar 低压区;低压等值线(低于1000毫巴)
meiobenthos 小型底栖生物
meionite 钙柱石
meiosis 减数分裂

Meissen green 迈森绿
Meissen porcelain 迈森瓷器
Meissner effect 迈斯纳效应
Meissner method 迈斯纳法
Meissner-Ochsen feld effect 迈斯纳效应
Meissner oscillator 迈斯纳振荡器
meixnerite 羟镁铝石
Meiyer form bow 梅亚型船首
Meiyer form ship 梅亚型船
Meiyer's sransform 梅亚变换
Meizer unit 民泽单位【地】
meizoseismal area 最强地震区;极区区;强震区
meizoseismic 最强震度力的
MEK double rubs 甲乙酮往复擦拭
Meker burner 梅克尔灯
mekometer 光学测距仪;光学精密测距仪;测距器
mel 美【声】;蜂蜜
melabasalt 暗色玄武岩
melacarbonate 暗碳酸岩
melaconite 黑铜矿;土黑铜矿
meladiabase 暗色辉绿岩
meladiorite 暗色闪长岩
melagabbro 暗色辉长岩
melamac 蜜醛塑料
melamine 蜜胺树脂;蜜胺
melamine alkyd resin paint 三聚氰胺醇酸树脂涂料
melamine decorative laminate 三聚氰胺装饰板
melamine foam 三聚氰胺泡沫
melamine-formaldehyde 三聚氰胺甲醛树脂;蜜胺树脂
melamine-formaldehyde adhesive 蜜胺甲醛黏胶剂
melamine-formaldehyde glue 三聚氰胺甲醛胶;蜜胺甲醛胶
melamine-formaldehyde paper 三聚氰胺甲醛纸
melamine-formaldehyde plastics 三聚氰胺甲醛塑料
melamine-formaldehyde resin 蜜胺甲醛树脂;三聚氰胺(甲醛)树脂
melamine glue 蜜胺胶;三聚氰胺树脂胶黏剂
melamine impregnated 浸渍三聚氰胺的
melamine laminate 蜜胺层压板;三聚氰胺层压板
melamine phosphate 磷酸三聚氰胺酯
melamine plastic board 三聚氰胺塑料(面)板
melamine plastics 蜜胺塑料
melamine resin 蜜胺树脂;三聚氰胺树脂
melamine resin adhesive 密胺树脂胶;三聚氰胺树脂黏合剂;三聚氰胺树脂胶
melamine resin decorative plastic film 三聚氰胺树脂塑性装潢层
melamine resin glue 蜜胺合成树脂胶
melamine resin plastics 三聚氰胺树脂塑料
melamine resin varnish 蜜胺树脂清漆;三聚氰胺清漆
melamine-surfaced 表面涂三聚氰胺的;三聚氰胺表面涂层
melamine-surfaced chipboard 三聚氰胺贴面碎木胶合板;三聚氰胺贴面碎木刨花板
melamine-surfaced decorative laminate 三聚氰胺表面处理装饰层压板
melamine-urea resin 蜜胺—脲醛树脂
melamine varnish resin 三聚氰胺清漆树脂
melamino-formaldehyde 蜜胺甲醛;三聚氰胺甲醛
melaminoplast 三聚氰胺塑料;蜜胺塑料
melaminoplastic board 三聚氰胺塑料(面)板
melane 铁镁矿物;暗色矿物
melanedema 炭尘肺
melange 糜滥石;混杂岩;混杂体;混杂堆积(物)
melange accumulation 混杂岩堆积
melange accumulation association 混杂岩堆积组合
melaniline 蜜苯胺
melanite 灰黑榴石;黑榴石
melanite microsyenite 黑榴石微晶正长岩
melanite syenite 黑榴石正长岩
melano-alteration 暗色蚀变
melanoandesite 深色安山岩
melanocerite 黑稀土矿
melanocrate 暗色岩
melanocratic 暗色的
melanocratic mineral 暗色矿物
melanocratic rock 深色岩;暗色岩
melanodiorite 暗色闪长岩
melanogabbro 暗色辉长岩
melanogabbro norite 暗色辉长苏长岩

melanohornblende gabbro 暗色角闪辉长岩
melanoid 人造黑素
melanomicrodiorite 暗色微晶闪长岩
melano-olivine gabbro 暗色橄榄辉长岩
melano-olivine norite 暗色橄榄苏长岩
melanophlogite 硫方英石;黑方石英
melanoresinite 黑色素树脂体
melanorite 暗色苏长岩
melanorite gabbro 暗色苏长辉长岩
melanoscope 红外(线)镜
melanosome 暗色体
melanostibite 黑锑锰矿
melanosyenite 暗色正长岩
melanotekite 硅铅铁矿
melanothallite 黑氯铜矿
melanotrichia 毛黑变
melanotroctolitic norite 暗色橄榄苏长岩
melanotroctolite 暗色橄长岩
melanovanadite 黑钒钙矿
melanovandite 黑钙钒矿
melanterite 水绿矾
melanuric acid 异氰脲酰亚胺
melaphyre 蚀变玄武岩;暗玢岩
melasyenite 暗色正长岩
melatope 光轴点
Melbourne Port 墨尔本港(澳大利亚)
Melbourn rock 墨尔本岩
meldometer 高温温度计;熔点测定计
melilite 黄长石;镁硅钙石;高炉结晶渣
melilite basalt 黄长石玄武岩
melilite-bearing lamprophyre 含黄长石煌斑岩
melilite-bearing ultrabasic rock 含黄长石超基性岩
melilitite 黄长岩
melilitite-bearing ultrabasic extrusive rock 含黄长石基性喷出岩
melilitite group 黄长岩类
melilitolite 黄长石岩
melilitolite group 黄长岩岩类
meliorate 修正
melioration 改善;改进;土壤改良
meliphane 蜜黄长石
meliphanite 蜜黄长石
melissane 三十烷
melissate 三十酸;蜂花酸盐
melissic acid 三十烷酸;蜂花酸
melissyl 三十烷基
melissyl alcohol 三十烷醇;蜂花醇
melitin(e) 波状热菌素
melitin test 波状热菌素试验
melitite leucitite 黄长石白榴岩
melkovite 磷钼钙铁矿
mellate 苯六甲酸盐
mellite 蜜蜡石
mellitic acid 苯六羧酸;苯六甲酸
mellitic acid trianhydride 苯六(羧)酸三酐
mellitoxin 蜂膏
mellophanic acid 苯偏四甲酸
Mellotte's alloy 梅洛特铋锡铅易熔合金
mellow 松软
mellow consistency 松软结持
mellow consisting 松软结持
mellow earth 海绵土;熟土
mellowing 柔软处理;柔和
mellow loam 松散垆姆;松壤土
mellowness 松软性
mellow soil 松透性土;松软土(壤);熟土;疏松土(壤)
mellow-soil field 海绵田
mellow tilth 耕耘土壤
mellozing 金属喷涂
melmak 麦尔马克塑料
Melni alloy 梅尔尼镍基耐热合金;梅尔尼合金
melocol 三聚氰胺
melon crop 瓜类作物
melonite 碲镍矿
melonjosephite 磷铁钙石
melons 额外股利
melon seed file 瓜子形锉刀
melon-shaped dome 瓜形穹顶;瓜形圆屋顶
melosira 直链藻
Melotone 米洛通(一种隔声挡板)
melphalan 曼尔法伦;苯丙氨酸氮芥
Melrose pump 梅耳罗斯泵
melt 熔体;熔解

meltability 熔度;熔融性
meltable 熔度;可熔性;可熔化的
melt-additive brightener 熔体增白剂
meltage 熔解物;熔解量
meltallizing 熔射
melt back 反复熔炼
melt backing 焊接垫板;焊剂托板
melt-back process 熔化过程
melt blend 熔体混合物
melt-blown 熔毁;熔化
melt-blown fabric 熔喷法非织造织物
melt-bonding process 熔融黏结工艺
melt-bonding soil 热胶黏土
melt canal 熔体通道
melt-casted 熔铸的
melt catalyst 熔媒剂
melt cement 熔化水泥
melt coating 熔体涂覆;熔融涂装
melt coating traffic paint 热熔型道路划线;热喷涂道路划线漆
melt colo(u)ration technique 熔体着色技术
melt compounding 熔融复合
melt crystallization 熔体结晶(作用)
melt down 回炉;熔化金属;变卖财产;熔毁
melt-down time 熔化时间
melted addition 熔体加入剂
melted asphalt 摊铺(地)沥青;熔化沥青
melted cement 热熔融性胶黏剂;热熔性胶黏剂;熔融水泥
melted gangue 熔融矿渣
melted-glass lining 熔融玻璃镶衬
melted ingot 熔制锭
melted iron 铁水
melted pool 熔池
melted slag 熔融矿渣
melted snow 融雪
melted snow runoff 融雪径流
melted temperature 融化温度
melted water runoff 熔化水径流
melted zone 熔化带
melteigite 霓霞钠辉岩
melt elasticity 熔体弹性
melten weld pool 焊接熔池
melt equilibrium 熔融平衡
melter 熔铸工;炉前工(人);熔窑;熔炉;熔炼炉;熔化器;熔化工
melter for joint sealer 填缝料熔化器
melter temperature control 熔化池温度控制
melt extractor 熔料分料器
melt extrusion 熔体挤塑;熔体挤出;熔体挤塑;熔融挤出
melt-flow 熔融流
melt flow index 熔体流动指数
melt flow instability 熔体流动不稳定性
melt fluidity (塑料等的)熔融流动性
melt fracture 熔体破坏;熔裂
melt grid 熔栅
melt growth 熔体生长;熔融生长
melthene 石油质
melt hopper 熔体料斗
melt impregnation 熔体浸渍
melt inclusion 熔融包裹体
melt index 熔体指数;熔融指数;熔化指标;熔点指标
melt indexer 熔体指数测定仪
melting 熔解;融化;熔化的
melting ablation 熔化烧蚀
melting and blowing process 熔融喷吹法
melting and fusing centrifuge 熔化离心机
melting and refreezing 熔化再凝
melting area 解冻区;熔化面积
melting band 融化带
melting bat 熔池
melting boiler 熔化锅炉
melting capacity 熔化能力;熔化率;熔化量
melting chamber 熔油罐;熔化室;熔化池
melting characteristic 熔化特性
melting characteristic of an electrode 焊条熔化性
melting charge 熔化料
melting coefficient 熔化系数
melting coke 铸造焦炭
melting compartment 熔化室
melting compound 熔化混合物
melting condition 熔化状态;熔化条件;熔化状况
melting cone 熔锥;测温锥;测热熔锥

melting crucible 熔炼坩埚
melting cup 熔化坩埚
melting cup assembly 熔化坩埚装置
melting current 熔化电流
melting curve 固液平衡线;熔化曲线
melting decladding 熔融去壳
melting decomposition method 熔融分解法
melting diagram 熔化图;熔度图
melting diffusion 熔融扩散
melting duct 熔池导管
melting efficiency 熔融效率;熔化效率;熔敷效率
melting end 熔化端部;熔化部
melting expansion 熔融膨胀
melting furnace 熔窑;熔炼炉;熔化炉
melting furnace for joint sealer 填缝料熔化炉
melting furnace shop 熔制车间
melting grid 熔栅
melting grid spiral 熔栅盘管
melting heat 熔解热;熔化热
melting ice 融冰
melting-in 熔挂法(一般指滑动轴承上减摩合金的浇挂法)
melting interface 熔化分界面
melting kettle 熔化釜;熔化锅;熔炼坩埚
melting ladle 熔勺
melting level 融化高度
melting loss 熔炼损耗;熔化损失;配合料挥发率;烧损
melting method for the design of foundation in frozen soil 冻土地基容许融化法设计
melting of ice 消融冰
melting operation 熔化操作
melting-out temperature 熔开温度
melting period 解冻期;融化期;熔化期;融雪期
melting platform 熔化台
melting point 软化点;熔(化)点;融(化)点;熔(化)温度
melting point apparatus 熔点测定器;熔点测点器
melting point bar 熔点块;测温块
melting point curve 熔点曲线
melting point depression 熔点降低
melting point diagram 熔点图
melting point indicator 熔点指示器
melting point of paraffin(e) wax 石蜡的熔点
melting point test 沥青熔点试验;熔点测定
melting point tube 熔点(测定)管
melting pot 坩埚;熔料坩埚;熔化锅;熔化罐
melting practice 熔炼操作
melting process 熔化法;熔化过程
melting quantity 熔化量
melting quartz 熔融石英
melting range 熔化范围
melting rate 熔化速率;熔化率
melting ratio 熔化比;铁焦比
melting receptacle 熔炉
melting refining process 熔融精制过程
melting-rock-and-earth method 融化土石法(建筑隧道用)
melting season 融雪期;解冻期;融化期
melting section 熔化区;熔化带
melting shop teeming bay 炼钢车间的铸锭跨
melting solid 熔化体
melting speed 熔化速度
melting stage 熔化阶段
melting state 熔化状态;熔化状态
melting stock 熔料
melting stock column 熔化料柱
melting tank 熔化池
melting tank backup lining 熔化池壁外帮砖
melting temperature 解链温度;熔融点;熔解温度;熔化温度;熔点
melting temperature curve 熔化温度曲线
melting temperature transducer 熔化温度传送器
melting test 熔点测定
melting through weld 熔透焊缝
melting time 熔融时间;熔化时间
melting unit 熔化设备;熔化单元
melting visocity 熔融黏度
melting with partial oxidation 部分氧化熔炼
melting zone 熔化区;熔化带;熔化层;熔化部
melt kettle 熔炉
melt loading 熔注装药;熔融装料
melt mixer for powder coating 粉末涂料熔化混合器

melt number 熔号
melt-off 熔脱
melt-off pore 熔化孔
melt-off rate 熔化率
melt-out 熔融
melt per hour 每小时熔量
melt period 融雪期
melt pigmentation 熔体着色法
melt pit 地坑
melt polycondensation 熔融缩聚
melt polymerization 熔体聚合(作用)
melt pool 熔体潭
melt potential 可能融雪量
melt pressure controller 熔化压力控制器
melt pressure transducer 熔化压力传送器
melt-pulling method 熔融淬火法;熔体拉伸法
melt-quenching method 熔化淬火法;熔融淬火法
melt-quench process 熔化淬火过程
melt rate 融雪(速)率;融化速率
melt refining 熔融精制
melt release system 熔释体系
melt residence time 熔体逗留时间
melt run 熔合线
melt settlement 融雪沉陷;融沉
meltshop 熔炼车间
melt sinking 熔陷
melt slippage 熔体滑动
melt snow and ice 冰雪融化水
melt spinning 熔体纺丝;熔融旋压法
melt spinning by extrusion method 熔体挤压纺丝法
melt spinning process 熔体旋凝工艺
melt-stoichiometry 熔体计量;熔体化学计量学
melt swell 熔体突然膨胀
melt tank 熔化槽
melt thru weld 熔透焊缝
melt-transesterification 熔融酯基转移(作用)
melt transfer printing 熔融转移印花
melt treatment process 熔体处理法
melt uniformity 熔体均匀性
melt urea pump 熔融尿素泵
melt viscometer 熔体黏度计
melt viscosity 熔体黏度;熔化黏滞度;熔解黏性
melt water 融(化)水;熔融液;熔化液;冰雪融化水;冰水
melt-water deposits 融水沉积
member 构件;杠件;杆件;机件;会员;成员;成件
member axis 杆件轴线
member buckling 构件受屈曲;构件受弯曲
member by member 逐项
member center line 构件的中心线
member company 分公司
member condition 杆件状态;杆件条件;成分条件
member connection 构件连接;杆件连接
member cross-section 构件截面
member ductility 构件延性
member field 构件场
member force 构件受力
member in bending 受弯构件
member in compression 压杆;受压杆件;受压构件
member in flexure 受弯构件;受挠杆件;受拉构件
membering bending 受弯构件
membering compression 受压构件
membering shear 受剪杆件
membering tension 受拉构件
member in shear 受剪构件;受剪杆件
member in tension 受拉构件;受拉杆件
member in torsion 受扭构件;受扭杆件
member length 构件长度
member moment 构件力矩;杆件弯矩
member name 构件名称
member number 杆件编号
member of a consortium 联合体成员
member of certification system 认证体系的成员
member of driving crew 驾驶乘务组成员【铁】
member of equation 方程式项
member of full standing 正式会员
member of train crew 列车乘务组成员
member property change point mode 杆件质变节点
member record 从记录;成员记录
membership 会籍;成员资格;成员关系;成员
membership function 隶属函数;成员关系函数
membership grade 隶属度

membership groups 团体成员小组
membership in professional organization 参与职业组织
membership of subcommittees 分委员会成员身份
membership of technical committees 技术委员会成员身份
membership problem 成员(资格)问题
membership table 成员表
member size 构件大小;构件尺寸
member slope 构件斜度
members' lounge 会员休息室
members of take-apart 拆卸人数
member splice node 杆件拼接节点
member state 杆件状态
member stress 构件应力;杆应力
member structure 杆系结构
member subject to buckling 受屈曲构件;受弯曲构件
member system 构件系(统)
member transformation 构件变换
member transposition 构件变换;杆件互换
member type 杆件类型
membrane 膜片式风箱;表层;防渗护面;隔板;膜片;膜;薄膜
membrane absorption 膜吸收
membrane absorption system 膜吸收系统
membrane action 薄膜作用
membrane aerated biofilm reactor 膜曝气生物膜反应器
membrane aeration 膜曝气
membrane aeration bioreactor 膜曝气生物反应器
membrane air stripping 膜气提法
membrane anaerobic reactor 膜厌氧反应器;膜序批间歇式反应器
membrane anaerobic reactor system 膜厌氧反应器系统
membrane analog(y) 薄膜模拟;弹性薄膜模拟法;薄膜比拟法
membrane analysis 薄膜分析
membrane-assited electroylsis 膜辅助电解
membrane assymmetry 膜的不对称性
membrane-attached biofilm 附着膜生物膜
membrane backed precast panel 防渗层衬底预制墙板;防渗层衬底预制面板
membrane balance 膜渗平衡
membrane barrier 隔膜;隔水膜;隔气膜
membrane-based selective separation 膜的选择性分离
membrane biofilm reactor 膜生物膜反应器
membrane biologic(al) reactor 膜生物反应器
membrane biologic(al) reactor fouling 膜生物反应器污着
membrane biologic(al) reactor system 膜生物反应器系统
membrane biologic(al) technology 膜生物技术
membrane biology 膜生物学
membrane bioreactor 膜生物反应器
membrane bioreactor fouling 膜生物反应器污着
membrane bioreactor system 膜生物反应器系统
membrane boundary layer 膜界层
membrane capacitance 膜电容器
membrane capacity 膜容量
membrane-case manometer 膜盒压力计
membrane charge 膜电荷
membrane chemistry 膜化学
membrane chromatography 膜色谱
membrane cisterna 膜池
membrane cleaning 膜清洗
membrane compliance 橡皮膜顺变性
membrane compound 薄膜养护液;薄膜养护剂
membrane concrete curing 混凝土薄膜养护法
membrane concrete curing agent 混凝土薄膜养护剂
membrane conductance 膜电导
membrane contactor 膜接触器
membrane container 薄膜容器
membrane contractor's pump 隔膜式工程用泵
membrane control of anodic side 负极半透膜控制
membrane correction shaft test 皮膜校正轴试验
membrane coupled activated sludge process 膜耦合活性污泥法
membrane coupled activated sludge system 膜耦合活性污泥系统
membrane curing (混凝土的)薄膜养护;薄膜防火;薄膜养生
membrane curing compound 薄膜养护化合物;薄膜养生剂;薄膜养护剂
membrane curing for concrete 混凝土薄膜养护
membrane curing solution 薄膜养护液;薄膜养护剂
membrane current 膜电流
membrane degradation 膜退化
membrane demineralization process 膜法脱盐法
membrane densimeter 薄膜密度计;薄膜比重计
membrane desalting 膜脱盐法
membrane detector 膜式探测器
membrane displacement 膜位移
membrane distillation 膜蒸馏
membrane duct 膜导管
membrane effect 薄膜效应
membrane efficiency 膜效率
membrane eigenfunction 膜本征函数
membrane eigenvalue 膜本征值
membrane eigenvalue problem 膜本征值问题
membrane electric(al) current 膜电流
membrane electrode 膜电极;模电极
membrane electrolysis 隔膜电解(法);膜电解
membrane electrophoresis 膜电泳
membrane element 膜元(件)
membrane-enveloped soil layer 薄膜封闭的土层
membrane envelope 包膜
membrane equation 膜方程;薄膜方程
membrane equilibrium 膜渗平衡;膜平衡
membrane equipment 薄膜设备
membrane filter 膜(式)过滤器;膜滤器;薄膜(过)滤器
membrane filter chamber 膜滤器室
membrane filter count 膜滤器计数
membrane filter index 膜滤器指标
membrane-filter procedure 薄膜过滤法
membrane filter standard plate count 膜滤器标准板式计数
membrane filter technique 膜滤器技术;膜滤法
membrane filtration 膜滤(法);薄膜过滤
membrane filtration activated sludge process 膜滤活性污泥法
membrane filtration characteristic 膜滤特性
membrane filtration culture method 薄滤膜培养法
membrane filtration index 膜滤指标
membrane filtration method 滤膜法
membrane filtration technique 膜滤技术
membrane fireproofing 薄膜防火;防火薄膜;隔板防火;防火隔板
membrane flashing 薄膜闪光;薄膜披水层;薄膜泛水层;膜泛水
membrane flashing piece 薄膜闪光片
membrane fluidity 膜流体性
membrane flux 膜通量
membrane-foaming compound 泡膜剂
membrane foil pressure sensitive keyboard 轻触式薄膜键盘
membrane force 薄膜受力;薄膜力
membrane-forming curing compound 成膜养护液;成膜养护剂
membrane-forming type bond braker 形成膜状的黏合剂制动器
membrane foulant 膜污着物
membrane fouling 膜污着
membrane fouling control 膜污着控制
membrane fouling layer 膜污着层
membrane fouling mechanism 膜污着机理
membrane fouling resistance 膜污着阻力
membrane fusion 膜融合
membrane granulosa 颗粒膜
membrane grouting 帷幕灌浆
membrane grouting process 帷幕灌浆法;薄膜灌浆法
membrane-immersed process 淹没式膜工艺
membrane inductance 膜电感
membrane integration technique 膜一体化技术
membrane ion-exchange plant 薄膜离子交换设备
membrane lipid phase transition 膜脂质相变
membrane logic 膜片式逻辑
membrane manometer 膜式压力计;膜测压力计
membrane material 膜材料
membrane method 膜法(离子交换);薄膜法
membrane method of water-proofing 隔膜防水法;薄膜防水法;防水隔膜法
membrane model 膜模型;薄膜模型
membrane module 膜组件
membrane of electrodialysis 电渗析膜
membrane of water-proofing 薄膜防水;防水薄膜
membrane penetration effect 橡皮膜嵌入效应(三轴试验)
membrane permeability 膜透率;膜(通)透性;膜(的)渗透性
membrane poisoning 膜中毒
membrane pollution 膜污染
membrane potential 膜(片)电位;膜电势
membrane pressure ga(u)ge 薄膜压力计
membrane process 膜法(离子交换);薄膜法
membrane pump 隔膜泵;膜(式)泵;薄膜泵
membrane reaction curve 膜反应曲线
membrane reactor 膜反应器
membrane receptor 膜受体
membrane recycling hypothesis 膜再循环假说
membrane regulating device 油毡调节装置
membrane resistance 膜电阻;膜阻(力);薄膜电阻
membrane resonance 膜共振
membrane roofing 薄膜屋面;卷材屋面
membrane rotation 薄膜旋转
membrane salinometer 薄膜盐度计
membrane science 膜科学;薄膜科学
membrane screen printing machine 薄膜丝网印花机
membrane seal 膜封闭
membrane sealed 膜封闭的;薄膜止水(的);膜封闭的
membrane sealed soil layer 薄膜封闭的土层
membrane seal mechanical pump 膜密封式机械泵
membrane seepage prevention 薄膜防渗
membrane selectivity 膜选择性
membrane separating process 膜分离法
membrane separation 膜分离
membrane separation activated sludge process 膜分离活性污泥法
membrane separation bioreactor 膜分离生物反应器
membrane separation device 膜分离装置
membrane separation efficiency 膜的分离效率
membrane separation process 膜分离工艺;膜分离法;薄膜分离法
membrane separation system 膜分离系统
membrane separation technique 膜分离技术
membrane separator 隔膜分离器;膜分离器
membrane shear(ing) force 薄膜剪力;薄膜切力;薄膜剪切力
membrane shell theory 薄膜理论
membrane spectroscophy 膜谱学
membrane stabilization 膜稳定作用;薄膜稳定作用;薄膜稳定(土)法
membrane stabilized soil road 薄膜稳定土路
membrane stabilizing action 膜稳定作用
membrane stack 膜堆
membrane stack of electrodeionization 电去电离膜堆
membrane state of stress 应力的膜态
membrane steady-state flux 膜稳态通量
membrane strain 膜应变
membrane strength 薄膜强度
membrane stress 膜应力;薄膜应力
membrane structure 膜结构;薄膜结构
membrane surface 薄膜表面;膜面
membrane surfaced 薄膜面层的
membrane syringe filter 膜吸设滤器
membrane system 薄膜叠层系统;隔膜法
membrane tank 膜式舱柜;薄膜储[贮]罐;薄壁储罐
membrane technology 膜技术
membrane technology for wastewater disinfection 废水消毒膜技术
membrane technology of water disinfection 水消毒膜技术
membrane tension 膜片张力;薄膜张力
membrane theorem 薄膜理论(一种薄壳设计的理论)
membrane theory 膜学说;薄膜理论(一种薄壳设计的理论)
membrane theory of conduction 膜传导理论
membrane theory of plasticity 可塑性薄膜理论
membrane theory of shell 壳体薄膜理论
membrane translation 薄膜位移
membrane transport 膜运载
membrane treatment for water-proof-(ing) 薄膜

防水处理
membrane type absorber 膜式吸收塔;膜式吸收器
membrane type fabric support 膜式织物支撑
membrane type manometer 膜式压力计
membrane type pressure ga(u)ge 薄膜式压力盒
membrane ultrafiltration 薄膜超滤法
membrane ultrafilter 薄膜超滤器
membrane valve 薄膜阀;膜片阀
membrane vesicle transport 膜泡传输
membrane wall 膜式水冷壁
membrane waterproofing 薄膜防水;膜防水;防水膜(层)
membrane white rot fungus bioreactor 白腐真菌膜生物反应器
membrane yield 膜壁屈服
membraniform 膜状的;膜样的
membranization 膜化
membranoid 膜样的
membranoid substance 膜状物
membranometer 薄膜压力计
membranous 膜状的;膜(质)的;薄膜的;被膜状
membranous desquamation 鳞片样脱屑;表皮剥脱
membranous disk 圆形膜盘
membranous layer 膜层
membranous part 膜部
membranous pump 隔膜泵
membranous semicircular canal 膜半规管
membranous spiral lamina 膜螺旋板
membranous type 膜型
membranous wall 膜壁
membranous water-proofing 薄膜防水;薄膜防火
membron 功能膜子
Memel timber 米蒙尔木(一种红色硬木)
memery 存储器
memister 可调电阻存储器
memistor 电解存储器;存储电阻器
memnescope 瞬变示波器
memoir 学术论文;学术报告;研究报告
Memoir Grassland Research Institute 草地研究所专刊
memometer 记忆测验器
memomotion 控时摄影;慢速摄影
memomotion photography 延时慢速摄影;慢速摄影(术)
memomotion stucy 慢速摄影研究
memorabilia 大事记
memorability 记忆率
memorandum 摘要;备忘录;便函
memorandum account 备查账户
memorandum book 备查簿
memorandum clause 附注条款;备忘条款
memorandum copy of bill of lading 提单抄本
memorandum dial 记录拨盘;备忘拨盘
memorandum entry 备查记录
memorandum ledger 备查分类账
memorandum of agreement 协定书
memorandum of an association 公司章程;组织大纲
memorandum of complaint 民事起诉书
memorandum of deposit 保证金附加条款
memorandum of exchange 开列其计算方式
memorandum of procedure 施工程序摘要说明书
memorandum of satisfaction 清偿债务备忘录
memorandum of understanding 备忘录;理解备忘录
memorandum record 备查记录
memorandum trade 备忘录贸易
memorandum value 非正式定价;备忘录价格;备忘价值
memorial 仪式;纪念物;纪念的
memorial arch 纪念(性)拱门;拱形纪念门;(纪念性的)拱门
memorial architecture 纪念性建筑
memorial archway 牌坊
memorial basilica 纪念型大庙;纪念型大教堂
memorial building 纪念建筑
memorial chapel 纪念小教堂
memorial church 纪念教堂
memorial column 纪念柱
memorial cross 纪念十字架
memorial figure 纪念雕像
Memrial forest 纪念林
memorial gateway 牌坊
memorial hall 纪念厅;纪念堂;纪念馆

memorial monument 纪念碑
memorial museum 纪念馆
memorial park 公墓陵园;公墓公园;陵园;纪念公园
memorial plaque 纪念碑;纪念牌
memorial rose 光叶蔷薇
memorial stone 纪念石;纪念碑
memorial tablet 纪念碑
memorial temple 祠堂
memorial tower 纪念塔
memoric instruction 记忆指令
memorise 信息存储
memorize 内存化
memorizer 存储器
memorizing meter 记录仪(表)
memory access 存取(器);存储器取数;存储(器)访问;存储器存取
memory access conflict 存储器访问冲突;存储器存取冲突
memory access control 存储器的存取控制;存储器存取控制
memory access fault 存储器存取故障;存储访问故障
memory access port 存储器存取口
memory access protection 存储器访问保护;存储器存取保护
memory access rate 存储器存取速度
memory access time 存储器取数时间;存储器存取时间
memory accumulator 存储累加器
memory acknowledge signal 存储器应答信号
memory action 存储作用
memory address 存储(器)地址
memory address assignment 存储器地址分配
memory address bus 存储器地址总线
memory address counter 存储器地址计数器
memory address driver 存储地址驱动器
memory addressing 存储器寻址
memory address line 存储器地址线
memory address pointer 存储器地址指针;存储器地址指示器
memory address register 存储(器)地址寄存器
memory address selector 存储器地址选择器
memory address space 存储器地址空间
memory aid 记忆辅助工具
memory-aided profile scanning 辅助记忆断面扫描
memory allocation 内存分配;存储器配置;存储(器)分配
memory allocation subroutine 存储分配子程序
memory allocator 存储分配器;存储分配程序
memory and device control unit 存储器和外设控制部件
memory area 存储区(域)
memory arrangement 存储排列
memory array 存储器阵列
memory assignment map 存储器分配图
memory-autoincrementing operation 存储器自动增量操作
memory bank 存储库
memory block 存储区(域);存储块
memory board 存储板
memory body 存取体
memory-bound 受存储器限制的
memory bounds register 存储器界限寄存器
memory buffer 存储缓冲区;存储缓冲器
memory buffer register 存储(器)缓冲寄存器
memory bus 存储器总线;存储器总线
memory busy signal 存储忙碌信号
memory button 记忆键;存储器按钮
memory byte format 存储字节格式
memory byte location 存储字节单元
memory cache 超高速缓冲存存储系统
memory capacitance 记忆;记忆量;存储量
memory capacitor 存储电容器;存储电容
memory capacity 记忆容量;记忆能力;存储(器)容量;存储能力;存储量
memory card 记忆卡片;存储器插件板
memory cell 记忆元件;存储元件;存储(器)单元
memory cell array 存储单元阵列
memory change 存储更换
memory check program(me) 存储检查程序
memory chip 存储器片
memory circuit 记忆电路;存储电路
memory clamp 存储钳位
memory clear key 存储器清除键

memory clear subroutine 存储器清除子程序
memory code 记忆码;存储码
memory compaction 存储器压缩;存储紧缩
memory configuration 存储格局
memory construction 存储器结构
memory contention 存储争夺;存储冲突
memory content 存储内容
memory contents dump 存储内容转储
memory control 记忆控制
memory control logic 存储器控制逻辑
memory control unit 存储器控制部件
memory core 存储磁芯;磁芯
memory core tester 存储磁芯试验器
memory counter 记忆计数器
memory cycle 周期时间;存储(器)周期
memory cycle time 存取周期时间;存储周期时间;存储循环时间
memory data bus 存储(器)数据总线
memory data ready 存储数据准备
memory data register 存储(器)数据寄存器
memory density 存储密度
memory depth 存储器深度
memory device 记忆设备;存储装置;存储器(件)
memory diagnostic 存储器诊断(程序)
memory disc 存储盘
memory drum 存储(磁)鼓
memory dump 信息转储;清除打印;存储(器)转储;存储器清除打印;存储器清除
memory dump routine 存储转储例行程序;存储(器)转储程序
memory effect 记忆效应;存储效应
memory element 记忆单元;存储元件;存储器元素
memory enter 存储器入口
memory error 记忆误差;存储错误
memory exchange 存数互换;存数变换;存储互换(装置)
memory exchanger 存数互换装置
memory exerciser 存储演算器
memory expansion 存储器扩展
memory extender 存储(器)扩展器
memory fetch 取数据;从存储器取数据
memory field 内存区
memory fill 存储器填充
memory function 记忆功能;存储作用;存储功能
memory gradient method 记忆梯度法
memory guard 存储保护
memory guard error 存储器保护错误
memory hardware 存储器硬件
memory hierarchy 存储器分级体系;存储层次;分级存储器系统;分级存储结构
memory idle 存储器闲置
memory image 存储图像
memory immediate instruction 存储器立即指令
memory indexing 存储变址
memory indicator 记忆指示灯;记忆选合接受指示灯
memory inhibit 存储器禁用
memory in key 存储器输入键
memory in metal 金属存储器
memory input bus 存储器输入总线
memory instruction data 存储器指令数据
memory interface 存储接口
memory interface protocol 存储器接口约定
memory interleave system 存储器交错存取方式
memory interleaving 存储器交叉存取
memory latency time 存储器等待时间
memory layout 存储配置
memory limit 存储容量极限;存储量极限
memory limit register 存储限制寄存器
memory loader 存储器输入程序
memory load key 存储输入键
memory location 存储器位置;存储(器)单元
memory locking 存储器封锁;存储器闭锁
memory lockout 存储封锁
memory lockout register 存储界限寄存器
memory machine 记忆机;存储器
memory management 存储器管理;存储管理
memory management exception 存储管理异常
memory management hardware 存储器管理硬件
memory management policy 存储管理策略
memory management register 存储管理寄存器
memory management status 存储器管理状态
memory management unit 内存管理单元;存储管理装置
memory map 存储映像(表);存储器分配形式;存

储器变换
memory-map list 存储变换表
memory-mapped device 存储器映像设备
memory-mapped interface 存储器映像接口
memory-mapped video 映像内存显示
memory mapping 存储(器)映像;存储(器)交换
memory mapping device 存储变换机构
memory mapping enable 存储变换赋能(位)
memory margin 记忆余量;存储余量;存储器容限
memory match 存储器匹配
memory material 记忆材料
memory matrix 存储矩阵
memory minus key 减存键
memory modify 存储交换改
memory module 存储组件;存储体;存储器模件;存储器模数;存储(器)模块;存储模(式)
memory multiplex data link 存储器多路数据链
memory organization 存储器组织
memory oriented system 面向存储器系统
memory or search lock 记忆或寻像锁定
memory output bus 存储器输出总线
memory over-lay 存储(器)重叠;存储器覆盖
memory package interface 存储部件接口
memory page 存储页面;存储页
memory parity 存储器奇偶性
memory parity error 存储器奇偶错
memory parity generator 存储器奇偶发生器
memory parity interrupt 存储(器)奇偶中断
memory partitioning 存储区分
memory patch 存储插入程序
memory peak 记忆峰
memory period 记忆时间;存储周期;存储期间
memory plane 存储板
memory plus key 加存键
memory point 记忆点;存储点
memory pool 存储器组合
memory power 存储器能力
memory print 存储转储;存储印刷;存储打印
memory printout 存储器打印输出
memory priority 存储优先(权)
memory program(me) 存储器程序
memory property 存储性能;存储能力
memory protect 存储器保护
memory protection 存储器保护;存储保护
memory protection feature 存储保护功能
memory protection key 存储保护键
memory protection option 存储保护选件
memory protection register 存储保护寄存器
memory protection scheme 存储器保护方案
memory rate 记忆速率;存储速率
memory-read 读存储器;存储(器)读出
memory-read cycle 读存储器周期
memory ready 存储准备;存储器就绪;存储器备用
memory reference 存储访问
memory reference address 存储器转移地址;存储器访问地址
memory reference instruction 存储器引用指令;存储器取数指令;存储器访问指令;访问存储器指令
memory reference order 存储访问指令
memory reference subroutine 存储器访问子程序
memory refresh 存储器再生
memory refresh feature 存储器再生功能
memory refresh register 存储更新寄存器
memory register 存储(器)寄存器
memory relay 存储式继电器
memory relocation 存储器的重新分配;存储单元的重新分配
memory request 存储器请求
memory request attribute 存储器请求属性
memory request counter 存储请求计数器
memory resident 存储器驻留
memory resident program(me) 常驻程序
memory resource 存储器资源
memory response 存储响应
memory scan 存储器检索
memory scanning 存储扫描
memory scan option 存储器扫描选择
memory scheduler 存储器调度程序
memory scheduling 存储调度
memory scope 存储式(同步)示波器
memory search 存储器检索
memory search routine 存储查找程序
memory-segmentation control 存储分段控制

memory sense amplifier 存储器读出放大器
memory shared multiprocessor 共享存储器的多级处理机
memory share system 存储器共用系统
memory sharing 存储器共享
memory size 内存大小;存储器体积;存储(器)容量;存储量
memory skewing scheme 存储斜移方案
memory space 存储量;存储空间
memory speed 存储速率
memory stack 存储栈;存储体
memory stitcher 电子缝纫机
memory storage 存储[贮]器;存储装置
memory swapping 存储交换
memory synchroscope 存储式同步示波器
memory system 记忆系统;存储系统
memory system architecture 存储系统结构
memory tape 记忆带
memory test 记忆调查法
memory test system software 存储器检测系统软件
memory time 暂记时间;记忆时间;存储时间
memory time-delay relay 存储延时继电器
memory time-slice method 存储时分方式;存储器时段方式
memory timing generator 存储定时产生器
memory-to-memory transfer 存储器间转移
memory tracking 记忆跟踪
memory transfer 转储;存储转移
memory transfer rate 存储传送速率
memory transparency 存储器透明性
memory trap condition 存储器陷阱条件;存储器俘获状态
memorytron 记忆管;存储管
memory tube 记忆管;存储管
memory tuning 存储调整
memory unit 记忆装置;存储装置
memory update 存储器更新
memory utilization 存储规划
memory velocity compensator 存储速度补偿器
memory violation 存储违章
memory width 存储器宽度
memory window 存储窗
memory windowing 存储器窗口
memory word 存储(器)字
memory write 存储(器)写入
memory zone 存储区(域)
memoscope 记忆(管)示波器;存储管式示波器
memotron 存储管
memo update 扼要更新
Mempening 马来西亚橡木
menaccanite 含钛火山灰砂岩;钛铁矿;碎屑钛铁矿
menagerie 动物展览
menakanite 钛铁砂
Menapian cold epoch 曼奈普寒冷期【地】
Menapian glacial stage 曼奈普冰期【地】
Menard pressuremeter 梅纳旁压仪
Menbranipora 膜孔苔虫
menbutone 孟布酮
mend 修补
mendaden oil 鲱油
mended crack 修补裂纹
Mendeleev's periodic system 门捷列夫周期系
mendelevium 钔
mendeleyevite 钙铌钛铀矿
Mendel's first law 分异定律
Mendelyeev's chart 门捷列夫周期表
Mendelyeev's law 门捷列夫周期律
Mendelyeev's periodic table 门捷列夫周期表
mender 修理工;修边工;报废板材
mender piler 次品板堆垛机
mendery 修理店
Mendez shales 门德斯页岩层
Mendhein kiln 孟海窑
mendicant order church 托钵僧团教堂
mending 修布;修补;订正;补版
mending bagging 补包黄麻布
mending clay 黏结用泥
mending cotton 缝补棉线
mending material 补炉料
mending plate 连接盖板;加固板
mending slip 修补用泥浆
mending stitch 缝线针脚
mending vessel 修理(工作)船
mendipite 白氯铅矿

Mendocino fracture zone 门多西诺断裂带
Mendola dolomite 门多拉白云石(晚寒武世)
mendozite 纤钠明矾;钠明矾;水钠铝矾
mend the service 修补受损的索具
meneghinite 斜方辉锑铅矿;辉锑铅矿
Menelaus theorem 米尼劳斯定理
Menger's theorem 门格尔定理
mengwacke 蒙瓦克岩
menhir (史前的)粗石巨柱;粗糙巨柱
menilite 硅乳石
meniscus 月牙饰;新月形(物);新月饰;新月形饰;弯月面;弯液面;凸凹透镜;半月板;板根;凹凸透镜;【复】menisci 或 meniscuses
meniscus astrograph 弯月形天体摄影仪
meniscus camera 新月形照相机;弯月形摄影机
meniscus cement texture 新月形胶结物结构
meniscus correction 弯月面校正
meniscus corrector 弯月形改正镜
meniscus curvature 月牙形曲率;弯月面曲度;透镜曲度
meniscus dune 新月形沙丘
meniscus effect 新月效应
meniscus heliograph 弯月形透镜太阳照相仪
meniscus lens 凹凸透镜
meniscus of liquid 弯液面
meniscus photographic(al) telescope 弯月形望远镜
meniscus photoheliograph 弯月形太阳照相仪
meniscus shape 弯月形
meniscus shaped lens 弯月形透镜
meniscus shaped lens system 弯月形透镜系统
meniscus sign 半月征
meniscus telescope 弯月形(透镜)望远镜
meniscus transit 弯月形透镜中星仪
menisus-Schmidt telescope 弯月镜施密特望远镜
menology 月历
mensa 圣坛的台面
mensal 每月的
mensao 史前纪念石碑
men's changing room 男更衣室;男衣帽间
men's cubicle hotels 隔仓式男旅舍
mensole 拱顶石
men's room 男卫生间;(公共建筑中的)男盥洗室
menstruation 赤龙
menstrum 溶媒
mensurable 可度量
mensural 关于度量的
mensuration 量度;求积法
mensuration by parts 分块求积法
mental asylum 精神病院
mental hospital 精神病院
mental institution 精神病研究治疗学院
mental map 环境感知图
mental model 构思模型
mental nursing home 精神病疗养院
mented survey point 埋石控制点;标志点
mentel 斗篷
menthane 甲基异丙基环己烷
Mentha spicata 留兰香
menthene 薄荷烯
menthol 薄荷脑
menthol crystal 薄荷脑
menthone 薄荷酮
men toilet 男盥洗室;男厕所
mentor coach 教练车
mentor method 蒙导法
mentor plant 蒙导植株
mentzelia sop 刺莲花
menu 项目单;功能选择单;菜单【计】
menuals of engineering practice 工程实践手册
menu bar 菜单条
menu card 特征码清单
menu card holder 菜单卡片夹
menu creation 菜单建立
menu display 菜单显示
menu generation system 菜单生成系统
menu keypad 菜单键
menu mode 菜单方式
menu output 菜单输出
menu parameter 菜单参数
menu processing 菜单操作
menu processor 菜单处理器
menu prompt 菜单提示
menu section 菜单段

menu selection 菜单选择
menu selection procedure 菜单选择过程
menus form of air photogrammetric-(al) survey 航测数据清单
menu technique 特征码清单法
men working sign 前方施工标志
Meny oscillator 梅尼振荡器
Menzies hydroseparator 孟席斯型新型水力分选机
Meotian 迈欧特阶【地】
mep formation 毛粒构成
mephenesin 甲苯丙醇
mephentermine 甲苯丙胺;甲苯丙胺
mephetedrine 甲苯丁胺
Meral alloy 梅拉铝合金
Meramec Group 梅拉梅克群【地】
Meramecian 梅拉梅克群【地】
meranti 红杉木;梅兰蒂木
merblastic 部分分裂的
merbromin(e) 红汞
mercadium orange 硫化汞橙
Mercalli intensity scale 麦卡利地震烈度表
Mercalli scale 麦卡利地震烈度表;麦氏地震烈度表
mercallite 重钾矾
mercantile 贸易的
mercantile agency 商业代理机构
mercantile building 商业大楼
mercantile center 贸易中心;商业中心
mercantile credit 贸易信用;商业信用
mercantile firm 商店
mercantile fleet 商船队
mercantile inquiry agency 商业信用咨询所
mercantile marine 商船队
mercantile occupancy 商业用户;商务占用
mercantile paper 商业票据
mercantile system 商业制度
Mercanton normal polarity hyperchron 梅坎顿正向极性巨时
Mercanton normal polarity hyperchronzone 梅坎顿正向极性巨时间带
Mercanton reversed polarity hyperzone 梅坎顿反向极性巨带
mercaptan 硫醇
mercaptan absorber 硫醇吸收塔
mercaptide 硫醇盐
mercaptobenzothiazole 快热粉
mercary protoiodide 碘化亚汞
mercast 冷冻水银法;水银模铸造
mercast pattern 水银模;冰冻水银模
mercast process 冷冻水银模铸造;冰冻水银模铸造
Mercator(al) bearing 恒向线方位;墨卡托方位(角)
Mercator chart 渐长海图;墨卡托海图;墨卡托航用图;赤约圆柱形投影海图
Mercator course 墨卡托航向;恒向线航向
Mercator direction 墨卡托方向;墨卡托方位;恒向线方向
Mercator sailing 墨卡托航迹计算法;墨卡托航行法;恒向线航迹计算法
Mercator's mile 墨卡托海里
Mercator's projection 渐长投影;墨卡托投影
Mercator's projection chart 墨卡托投影海图;墨卡托投影地图
Mercator's projection map 墨卡托投影地图
Mercator track 墨卡托航线;恒向线航线
Mercator-type graph 墨卡托型图
mercerisation = mercerization
mercerization 碱化;丝光(作用)
mercerize 丝光处理
mercerized cellulose 丝光纤维素
mercerized cotton 府绸;丝光棉布
mercerized finish 丝光整理
mercerized sewing thread 丝光缝纫线
mercerized yarn 丝光纱线
mercerizing 碱化;丝光(处理)
mercerizing assistant 丝光助剂
mercerizing machine 丝光处理机
mercery 绸缎店;布店;绸缎类;布类
merceya latifolia 苔藓
merchandise 商业;商品
merchandise account 商品项目
merchandise balance 进出口商品平衡表;贸易差额;商品差额
merchandise broker 商品经纪人
merchandise buying budget 商品采购预算

merchandise car 货车
merchandise control 商品管理
merchandise cost 商品成本
merchandise credit slip 退货清单
merchandise declaration 货物申报单
merchandise discount 商品折扣
merchandise expense 商品费用
merchandise in transit 转口商品
merchandise inventory 商品存货
merchandise mark 商品标志;商标
merchandise marks act 商标法
merchandise money 商品货币
merchandise on hand 现货;库存商品
merchandise outward on consignment 出口寄销商品
merchandise priced below cost 亏损商品
merchandise procurement cost 商品采购成本
merchandise purchases 进货
merchandise purchases account 进货账户
merchandise sales 销货
merchandise sales account 销货账
merchandise selling budget 商品销售预算
merchandise shipped on instalment sales 分期收款发出商品
merchandise subject to controlled purchases 控购商品
merchandise trade 商品交易
merchandise transfer 商品转移
merchandise turnover 商品周转
merchandise valuation 商品估价
merchandising enterprise 商业企业
merchandising location 销售场所
merchandising plan 商品化计划
merchandising salesman 推销员
merchandising turnover 商品周转;商品回转率
merchandize 商业
merchant 贸易商;商业的;商人
merchantability 可销性;适销性
merchantability specification 商品材规格
merchantable (美国木材的)商品等级
merchantable bole 商品干材
merchantable fuel 商品燃料
merchantable gasoline 商品汽油
merchantable height 商用材高;商品材高度
merchantable ore 商品矿石
merchantable product 商品
merchantable quality 可销售品质
merchantable timber 商品木材
merchantable title 可出售的产权
merchantable volume 商品材积
merchant bank 商人银行
merchant bar 商品铁条;劣质熟铁;多种用途型钢;小型轧材
merchant bar iron 商品条钢;不合(规)格的铁条
merchant bar mill 条钢轧机
merchant builder 营造商
merchant capital 商业资本
merchant copper 商品铜
merchant cruiser 改装巡洋舰
merchant fleet 商船队
merchant furnace 工厂熔炼炉
merchant haulage 货主内陆运输;货方接运
Merchantile Marine Office 商船局
merchantile usages 商业习惯
merchanting trade 三角贸易
merchant iron 优质熟铁;商品铁
merchant man 商船船员
merchantman trading vessel 商船
merchant marine 商船队
Merchant Marine Act 海运法(美国);商船法;海商法
Merchant Marine Department 商船局;海运局
merchant marine officer 商船高级船员
merchant mill 条钢轧机
merchant navy 商船队
Merchant Navy Establishment Administration 船员配备管理局
merchant navy officers pension fund 船员养老基金
Merchant Navy Training Board 船员培训局
merchant pipe 商品管(材)
merchant prince 富商
merchant rate 商业汇价
merchant roll 辊式中碎机

merchant's credit 商人信用证
merchant sdock 商用船坞
merchant seaman 商船水手;商船海员;商船船员
Merchants Group 招商局集团
merchant shape 商品钢材
merchant ship 民用船;商船
merchant ship flag 商船旗
merchant shipping 商船运输
Merchant shipping act 英国商船法案;商船法;运法;海商法;商船运输法规;商船条例
merchant shipping law 海商法
merchant shipping notes 商船航运通告
merchant ship reactor 商船用反应堆
Merchant Ship Search and Rescue Manual 商船搜索救助手册
merchant steamer 商船
merchant steel 商品条钢;商品钢
merchant vessel 商船
merchant whole sale 批发商
merchant wire 钢丝制品
merchrome 异色异构结晶
merchromize 水银铬化
Merco bronze 默科(铅)青铜
mercoid 水银转换开关
mercoid switch 水银开关
mercolized wax 面蜡
Mercoloy 铜镍锌耐蚀合金
mercurate 汞化
mercurating agent 汞化剂
mercuration 汞化作用;加汞作用
mercurial 汞制剂;汞的;含汞的;水银的
mercurial barometer 水银气压计;水银气压表
mercurial exchange rates 变动不定的兑换率
mercurial fungicide 汞系防霉剂
mercurial funicide 汞杀菌剂
mercurial horn ore 汞膏
mercurial intensification 水银加厚法
mercurialism 汞中毒;水银中毒
mercurialization 汞剂化
mercurial lamp metal-film resistor 汞灯金属膜电阻器
mercurial level 水银水准仪
mercurial mud 含汞盐泥
mercurial ointment 汞软膏
mercurial pendulum 水银摆
mercurial pesticide 汞制农药
mercurial poisoning 汞中毒;水银中毒
mercurial slimicide 汞杀黏菌剂
mercurial thermometer 汞温度计;水银温度计;水银温度表
mercurialtype differential pressure recorder 水银式压差记录器
mercuric 汞的;二价汞的
mercuric acetate 醋酸汞
mercuric arsenate 砷酸汞
mercuric atoxylate 氨基苯胂酸汞
mercuric ballast 高汞镇流器
mercuric barium iodide 碘化汞钡
mercuric benzoate 苯甲酸汞
mercuric bichromate 重铬酸汞
mercuric bulb 高汞灯泡
mercuric chloride 氯化高汞;升汞;二氯化汞
mercuric chlorite 氯化汞
mercuric complex 汞络合物
mercuric compound 汞化合物
mercuric cyanate 雷酸汞
mercuric cyanide 氰化汞
mercuric fluoride 氟化汞
mercuric fulminate 雷酸汞
mercuric hydroxide 氢氧化汞
mercuric iodide 碘化汞;二碘化汞
mercuric iodide X-ray detector 碘化汞X射线探测器
mercuric isocyanate 异氰酸汞
mercuric nitrate 硝酸汞
mercuric nitrate titration 硝酸汞滴定法
mercuric nitrate volume method 硝酸汞容量法
mercuric oleate 油酸汞
mercuric oxalate 草酸汞
mercuric oxide 氧化汞
mercuric oxide method 氧化汞法
mercuric para-chlorbenzoate 对氯苯甲酸汞
mercuric perchlorate 高氯酸汞
mercuric pressure ga(u)ge 水银压力计;水银压

力表
mercuric stearate 硬脂酸汞
mercuric subsulphate 碱式硫酸汞
mercuric sulfate 硫酸汞
mercuric sulfide 朱砂；硫化汞
mercuric thiocyanate 硫氰酸汞
mercuride 汞化物
mercurimetric(al) determination 汞滴定法
mercurimetric(al) titration 汞量滴定法
mercurimetry 汞液滴定法；汞量法；汞量测量；汞滴定法
mercurochrome 红汞
mercuro-iodo-formaline 汞醛碘
mercuro-iodo-formaline concentration method 汞醛碘浓集法
mercuro-iodo-formaline precipitation method 汞醛碘沉淀法
mercurous acetate 乙酸亚汞；醋酸亚汞
mercurous barometer 汞液大气压力计
mercurous chloride 氯化亚汞
mercurous chloride prism polarizer 氯化亚汞棱镜起偏(振)器
mercurous chromate 铬酸亚汞(绿色色料)
mercurous discharge lamp 汞银放电灯；汞弧灯
mercurous iodide 碘化亚汞
mercurous iodine lamp 碘化亚汞灯
mercurous nitrate 硝酸亚汞
mercurous oxide 氧化亚汞
mercurous salts 亚汞盐类
Mercury 水星
mercury 自然汞；水银；捕汞器
mercury absolute pressure 水银柱压力
mercury absorption 汞吸收
mercury accumulation 汞累积；汞积累
mercury acetate 乙酸亚汞
mercury air pump 汞(气)泵；水银气泵；水银扩散泵
mercury analyzer[analyser] 测汞仪
mercury analyzer type 测汞仪类型
mercury arc 汞弧
mercury-arc converter 汞弧变换器
mercury-arc lamp 汞弧灯；水银(弧光)灯
mercury-arc oscillator 汞弧振荡器
mercury-arc power converter 汞弧功率变换器
mercury-arc rectifier 汞弧整流器；水银电弧整流器
mercury arrester 水银避雷器
mercury arseniate 砷酸汞
mercury ballistic of gyrocompass 陀螺罗经的水银稳定器
mercury barometer 水银(柱)气压表；水银气压计
mercury bath 水银槽
mercury bath collector 水银集流器(电测扭矩用)
mercury battery 水银电池
mercury-bearing 含汞的
mercury-bearing content 含汞量
mercury-bearing formation 含汞建造
mercury-bearing fume 含汞烟气
mercury-bearing salt mud 含汞盐泥
mercury-bearing wastewater 含汞废水
mercury benzoate 苯甲酸汞
mercury bichloride 氯化高汞；升汞；二氯化汞
mercury blend 晨砂
mercury bob 水银垂球(测垂直度用)
mercury boiler 汞锅炉；水银蒸煮器；水银锅炉
mercury break(er) 水银断路器
mercury-cadmium lithopone 汞镉钡黄
mercury-cadmium pigment 硫化汞镉颜料
mercury cathode 汞阴极
mercury cathode cell 汞阴极电(解)池
mercury cathode separation 汞阴极分离法
mercury cell 汞极电池；汞干电池；汞电解池；水银槽
mercury chloride 氯汞矿
mercury circuit breaker 汞断路器
mercury cleaner 水银清洗设备
mercury coherer 汞检波器
mercury column 汞柱；水银柱
mercury column type 汞柱式
mercury complex 汞的络合物
mercury concentration 汞浓度
mercury condensation pump 汞气冷凝泵
mercury condenser 水银冷凝器
mercury connection 水银连接(环形水银开关)
mercury contact 汞电钥；水银开关；水银接点
mercury contact relay 汞接点继电器；水银触点继电器

mercury-contact switch 水银接点开关
mercury-containing sludge 含汞污泥
mercury-containing sludge treatment 含汞污泥处理
mercury-containing wastewater 含汞废水
mercury contaminant 汞污染物
mercury contaminated 汞污染的
mercury contamination 汞污染
mercury content 汞含量；含汞量
mercury controller 水银蒸煮器
mercury converter 水银蒸煮器；水银换流器
mercury cooled valve 水银冷却气门
mercury cosmetic 氨基汞化氯
mercury coulometer 汞极电量计
mercury cup connection 水银杯连接
mercury cut-off 水银开关
mercury cycle 汞循环
mercury dashpot 水银缓冲器
mercury delay line 汞延迟线；水银延迟线
mercury-depleted brine 脱汞盐水
mercury deposition 汞沉积
mercury detector 水银检波器；测汞仪
mercury diethide 二乙基汞
mercury diethyl 二乙基汞
mercury differential thermometer 水银示差温度计
mercury diffusion pump 汞扩散泵；水银扩散泵
mercury dimethide 二甲基汞
mercury diphenide 二苯基汞
mercury direct current transformer 水银直流变换器
mercury discharge lamp 汞弧灯；水银放电灯；水银避电器
mercury distant-reading thermometer 远距离水银温度计
mercury dropping electrode 汞滴电极
mercury electrode 汞电极；水银电极
mercury-electrolyte transducer 水银电解质换能器
mercury electrolytic cell 水银电解槽
mercury electrolytic meter 水银电解电度计
mercury-filled 水银充满式
mercury filled thermometer 水银温度计
mercury film electrode 汞膜电极
mercury film relay 水银薄膜继电器
mercury float type manomer 浮子式水银压计
mercury free ship-bottom paint 无汞船底漆
mercury freezing test 水银冷冻试验
mercury frequency changer 水银频率变换器；水银变频器
mercury fulminate 雷酸汞；雷汞
mercury fume 含汞烟气
mercury gas discharge lamp 水银气体放电管
mercury ga(u)ge 汞压计；水银压力计；水银压力表；测压计
mercury gilding 镀汞材料；镀汞术
mercury globule 汞珠
mercury halogen compound 卤素汞化合物
mercury handling 汞的处理
mercury high-pressure lamp 高压汞灯
mercury horizon 水银地平；水银盘地平
mercury hypersensitizing 水银超增感
mercury indicator 水银指示器(测量油舱位用)
mercury-in-glass thermometer 水银温度计；玻璃(管)水银温度计
mercury injection method 压汞法
mercury-in-steel thermometer 钢壳水银温度计；钢管水银温度计
mercury intensification 水银加厚法
mercury interrupter 水银断续器
mercury intrusion analysis 吸汞分析；压汞分析
mercury intrusion method 压汞入法
mercury intrusion porosimeter 汞注入孔隙计
mercury inverter 水银换流器；水银反向换流器
mercury iodine lamp 碘汞灯
mercury ion 汞离子
mercury isotope lamp 水银同位素灯
mercury jet rectifier 汞流整流器
mercury lamp 汞灯
mercury laser 汞激光器；水银激光器
mercury level 汞水平仪
mercury line 汞线；汞谱线
mercury manometer 水银压力计；水银压力表
mercury memory 汞存储器；水银延迟线；水银存储器
mercury memory storage 水银存储器

mercury metabolism 汞代谢
mercury meter 水银电度表
mercury methide 二甲基汞
mercury method 水银法
mercury methyl 二甲基汞
mercury motor meter 水银电动式仪表；水银电动式安培小时表
mercury motor type 水银电动机式
mercury naphthenate 环烷酸汞
mercury nitride 二氯化三汞
mercury-oil pressure cell 水银-油压力盒
mercury ointment 汞软膏
mercury oleate 油酸汞
mercury ore 汞矿(石)；水银矿石
mercury oxide 氧化汞；汞氧化物
mercury oxide cell 汞干电池
mercury oxide red 红色氧化汞
mercury penetration method 水银压入法；水银渗透法(测气孔率)
mercury peroxide 过氧化汞
mercury persulfate 硫酸汞
mercury phenide 二苯基汞
mercury phenyl 二苯基汞
mercury plunger relay 水银插棒式继电器
mercury poisoning 汞中毒；汞毒
mercury pollutant 汞污染物
mercury pollution 汞污染
mercury pond 水银槽
mercury pool 汞池；水银阴极；水银池
mercury pool cathode 汞池阴极；水银槽阴极
mercury pool mirror 垂准镜
mercury porosimeter 压汞仪；汞压测孔仪
mercury porosimetering 汞压入法
mercury porosimetry 水银孔率法(用水银测孔率的方法)
mercury power rectifier 大功率水银整流器
mercury pressure 水银柱高度
mercury pressure porosimeter 水银测孔仪；汞压孔隙计
mercury process electrolysis 汞法电解
mercury program(me) 水星计划
mercury protoacetate 乙酸亚汞
mercury protochloride 氯化亚汞
mercury pulser 水银脉冲发生器
mercury pump 汞泵；水银扩散泵；水银泵
mercury purifying tunnel 纯化汞用漏斗
mercury recording thermometer 水银记录式温度计
mercury rectifier 水银整流器
mercury reflection 镜面反射
mercury reflector 水银反射器
mercury regulating unit 水银调节装置
mercury relay 水银继电器
mercury-relay pulse generator 水银继电器脉冲发生器
mercury remains 汞残渣
mercury removing reagent 除汞剂
mercury replacement 汞取代
mercury reservoir 水银容器
mercury residue 汞残留物
mercury residue in fish 鱼体残留汞；鱼的汞残留量
mercury-resistant 抗汞的；耐汞的
mercury-resistant bacteria 耐汞菌
mercury resistor 水银电阻器
mercury resonance radiation 水银共振辐射
mercury-rich sediment 富汞沉积物
mercury salt 汞盐
mercury seal 汞封(口)；水银(密)封
mercury-sealed 汞封的
mercury seal pot 水银密封釜
mercury settlement ga(u)ge 水银沉降计
mercury spectral lamp 汞光谱灯；水银光谱灯
mercury spectrum 水银光谱
mercury spectrum lamp 汞灯
mercury spot 水银阴极辉点
mercury standard 水银标准
mercury stearate 硬脂酸汞
mercury storage 汞存储器；水银延迟线；水银存储装置；水银存储器
mercury sulfate 硫酸汞
mercury sulfide red 红色硫化汞
mercury super-pressure lamp 超高压水银灯
mercury survey by air pump 抽气测汞法
mercury survey by car 车载测汞法

mercury switch 汞开关;水银开关
mercury switch interlock 水银开关联锁装置
mercury tank 汞槽;水银槽
mercury tank rectifier 汞槽整流器
mercury teleswitch 水银遥控开关
mercury telethermometer 遥测水银温度计
mercury test 水银试验(测应力腐蚀)
mercury thermometer 水银温度计
mercury thermoregulators 水银温度调节器
mercury thermostat 汞控恒温器;水银恒温器
mercury timing valve 水银延时装置
mercury trap 汞捕集器;捕汞器
mercury trough 水银槽
mercury tube 汞弧整流管
mercury-tungsten lamp 水银钨丝灯
mercury turbine 汞汽轮机;水银涡轮机;水银汽轮机
mercury type temperature relay 水银温度继电器
mercury vacuum ga(u)ge 水银真空计
mercury vacuum pump 水银真空泵
mercury valve 水银阀
mercury vapormeter 测汞仪
mercury vapo(u)r 汞蒸气;汞气
mercury vapo(u)r air pump 水银扩散泵
mercury vapo(u)r analyzer[analyser] 汞蒸气分析仪
mercury vapo(u)r arc lamp 汞气弧灯
mercury vapo(u)r detector 汞蒸气检验器
mercury vapo(u)r diffusion pump 汞气扩散泵;水银扩散泵
mercury vapo(u)r discharge lamp 汞蒸气放电灯;水银蒸气放电灯
mercury vapo(u)r jet pump 汞气喷射
mercury vapo(u)r lamp 汞(气)灯;水银(蒸气)灯
mercury vapo(u)r lantern 水银灯泡
mercury vapo(u)r light 汞气灯;水银灯
mercury vapo(u)r meter 汞蒸气(测量)仪;汞蒸气测定仪;汞蒸气测定器
mercury vapo(u)r pump 汞(蒸)气泵;水银蒸气泵
mercury vapo(u)r rectifier 汞蒸气整流器
mercury vapo(u)r rectifier locomotive 汞弧整流式机车
mercury vapo(u)r rectifier tube 汞气整流管
mercury vapo(u)r survey 汞蒸气测量
mercury vapo(u)r tube 汞气管;水银充气整流管
mercury-wetted contact 水银触点
mercury-wetted reed switch 水银湿簧片开关
mercury-wetted relay 水银接点继电器;湿簧继电器
mercury with chalk 汞白垩
mercury xenon-arc lamp 汞氙灯
mercury-zinc cyanide 氰化锌汞
mere 海湾;浅湖;三角湾;池
mere international technological transfer 简单国际技术转移
merely positive subdefinite 仅次正定
merely pseudo-convex 仅伪凸
merely quasiconvex 仅拟凸
merely supported end 自由支撑端
merenskyite 碲钯矿
merestone 界石
mereurous sulfate 硫酸亚汞
Merevale shales 梅雷瓦尔页岩层
merganser 秋沙鸭
merge 归并;合流;并接
merge application 合并处理
merge command 归并命令;并合指令
merge control 交汇控制
merged 合并的
merge detector 交汇检验器
merged relief map 分层设色地势图
merged transistor logic 集成注入逻辑
merge exchange 归并交换法
merge exchange sort 合并交换分类
merge into 并入
merge line 分割线;合并线
mergency closing device 备用关闭装置
merge network 归并网络
merge node 合并节点
merge order 合并阶
merge point 交合点
merger 兼并;合并;吞并;并入;合且量
merger diagram 状态合并图
merger of title 权利汇合;合并权利
merger system 联合制
merge sort 归并排序;归并分类;合并分类

merge sorting 归并分类
merge way 合并通路
merging 合流;汇合;熔合;数据并合
merging angle 交汇角
merging area (车辆的)交会区段
merging cluster 聚类合并
merging conflict 交汇点;合流冲突点
merging distance (车辆的)交织距离
merging end 汇合端;合流端
merging intersection 汇合交叉口
merging lane 合流车道
merging maneuver area 汇合行驶区
merging of files 文件归并
merging phase 合并状态
merging program(me) 合并程序
merging road 交汇道路
merging routing 合并程序
merging space (车辆的)交织区段
merging star 并合星
merging traffic 汇合交通;合流交通
Merian's formula 梅立恩公式
mericarp 分果片
mericlinal chimera 边缘嵌合体
meridian 子午圈;经(度)线
meridianal 子午线的
meridianal arc 子午线弧
meridianal arc measurement 子午线弧度测量
meridianal arc method 子午线弧法
meridianal component 子午分量
meridianal convergence 子午线收敛角
meridianal differential of latitude 不同纬度的子午线差
meridianal ellipse 子午线椭圆
meridianal field stress 径向像场应力
meridianal flow 子午流
meridianal height 子午高度角
meridianai interval 子午线间距
meridianal parts 纬度渐长率
meridianal plane 子午面
meridianal projection of the sphere 子午面投影
meridianal section 子午截面
meridian altitude 子午(线)高度;子午圈高度;中天高度
meridian altitude below pole 下中天高度
meridianal zone 子午带
meridianal zone of grid system 坐标网子午带
meridian angle 时角;子午(圈)角;子午角距
meridian angle difference 时角差;子午(圈)角差
meridian arc method 子午线弧法
meridian astrometry 子午天体测量学
meridian astronometry 子午天体测量学
meridian astronomy 子午天文学
meridian catalogue 子午星表
meridian circle 子午仪;子午圈;子午环
meridian circulation 子午圈环流
meridian convergence 子午线收敛角
meridian coordinate 子午线坐标系
meridian day 日界线通过日
meridian distance 纬度
meridian extension 经线延长线
meridian finder 子午线测定仪
Meridian Gate 故宫(北京);午门
meridian instrument 子午仪;中星仪
meridian interferometer 中星仪式干涉仪
meridian line 子午线;经线;南北线
meridian mark 子午线标;子午标
meridian observation 子午观测;中天观测
meridian of an observer 测者子午圈
meridian of the earth 地面子午线;地面子午圈
Meridian parts for spheroid 椭球图纬度渐长率
meridian passage 中天
meridian photometer 子午光度计;中天光度计
meridian plane 子午面;经圈面
meridian plane of Greenwich 格林尼治子午面
meridian projection 子午线投影
meridian quadrant 子午线四分仪
meridian reflection 子午线反射
meridian rib (圆屋顶的)子午肋
meridian rib of a dome 圆顶子午肋
meridian ring 子午刻度环
meridian ring globe 子午环地球仪
meridian sailing 子午线航行
meridian section 子午截面
meridian seeking torque 指向力矩

meridian spacing 经差
meridian stress 经线方向应力;(穹中的)垂直应力
meridian telescope 子午望远镜
meridian transit 子午仪;中天
meridianus 子午线
meridian wire 子午丝
meridian zenith distance 子午(圈)天顶距;中天顶距
meridiem 正午
meridional bending 子午线弯曲(度)
meridional cell 经向环流
meridional circulation 经向环流
meridional contour 子午线剖面轮廓
meridional difference 纬差
meridional direction 子午线方向
meridional distance 子午线距离
meridional fibers 经线纤维
meridional flow 经向气流
meridional focal line 子午焦线
meridional focus 主焦点
meridional force 经向力
meridional gradient 经向梯度
meridional index of circulation 经向环流指数
meridional intensity 子午线方向强度
meridional parts 渐长纬度;渐长区间
meridional plane 子午线断面
meridional ray 子午面光线
meridional scale 经线尺度
meridional stress 经向应力
meridional stress resultant 经向应力的合力
meridional structural system 经向构造体系
meridional syzygy tide 朔望潮;大潮
meridional tangent 经向正切;经向切线
meridional tangential ray 子午正切光线
meridional tectonic system 经向构造体系
meridional type 经线型
meridional velocity distribution 轴面速度分布
meridional wind 经向风
meridional X-ray diffractogram 子午线的X射线衍射图
meridonal transport 经向输送
merism 节构造
merismite 斑杂混合岩
merisoprol 放射性汞丙醇
meristele 分体中柱
meristic variation 部分变异
merit 荣誉;优点;功劳;功绩;价值
merit factor 质量优良程度
merit function 优质函数
merit grade 质量水准
merit increase 考核加薪
merit increase pay 考绩提资
merit number 价值指数
merit pay plan 记功付酬计划
merit point system 优选打分制
merit rating 质量评价;质量评定;质量鉴定;考绩;人事考核
merit rating plan 考绩制度;考绩征税办法
merits and demerits 优缺点;优点和缺点
merits and drawbacks 优缺点
merits and faults 优缺点
merit shop 自由雇佣企业
merits rating 优点评价
merlinoite 麦钾沸石
merlon 城齿;雉堞;城垛齿
mermaid 美人鱼(城市雕塑)
meroblastic cleavage 部分分裂
mero cave 半洞
merocrystalline 半结晶的
merohdrism 缺面体
merohedral 缺面的
merohedral form 缺面形
merokarst 半岩溶
meromictic lake 局部混合湖;分层湖;不完全混合湖;不完全对流湖;半混合湖;半对流湖
meromixis 部分融合;(湖水)半混合状态;半混合层
meron 后基节
meroplankton 暂时性浮游生物;季节(性)浮游生物
meros (最古希腊建筑中的)竖条纹饰两槽间的平面;排档间距;三槽板间平面
merosymmetric(al) 缺面的
merotomize 裂成几块;分成几部分
merotomy 分成几部分
Merovingian architecture 墨洛温王朝建筑

Merrick weightometer 梅里克称重计
merrihueite 陨铁大隅石;陨硅钾铁石
Merrill-Crowe process 梅里尔—克劳法
merrillite 陨磷钙钠石;磷钙钠石
merry dancers 北极光
merry-go-ground 旋转木马
merry-go-round 钢水罐回转架
merry-go-round equipment 循环式预应力张拉设备;"走马灯"式预应力钢丝连续张拉设备
merry-go-round for baggage 在回转架上递送行李包裹;"走马灯"式展示行李包裹
merry-go-round garage 旋转式升降机停车库
merry-go-round train 环行列车
merry-go-round windmill 转塔式风力发动机
Mersenne's law 梅逊定律
Mersey yellow coal 塔斯曼油页岩;塔斯曼煤
Mershon diagram 梅尔生电压调整图
mersida 纸背饰面壁柱
mersolite 苯基水杨酸汞
Merte surface 默脱面(复杂三棱镜的胶合面)
mertieite 砷锑钯矿
mert index 价值指标
Merulius lacrymans 干朽;皱孔菌
mer weight 基体量
merwinite 默硅镁钙石;默硅镁;镁硅钙石
Merze-Price protection system 梅兹—普赖斯保护系统
merzlota 冻土
mesa 桌子山;平顶山;台面;台地;方山
mesa-butte 平顶孤丘
mesaconic(al) acid 中康酸;甲基反丁烯二酸
mesa device 台面型器件
mesa diode 台面二极管
mesa etch 台面腐蚀
mesal 中线的
mesal penellipse 侧缺环
mesa mask 台面掩模
mesa plain 方山平台
mesarch sere 中生演替系列
mesa structure 台式结构
mesa surface laser 台式激光器
mesa technology 台面工艺
mesaulos (古希腊住宅中连接餐厅和女宾室的)过道及过道门
meseta 小方山;高原
mesexine 中表层
mesh 栅网;网眼;网络;网或金属网;网格;啮合;筛目
mesh adjustment 啮合调正;啮合校正;啮合调整
mesh analysis 网筛分析;网络分析;网孔分析法;筛析;筛(分)分析
mesh and plaster wall 钢丝网粉刷墙
mesh aperture 筛孔
mesh bag 网袋
mesh band 钢丝型表带
mesh-belt 织网运输带;织带;钢丝网运输带
mesh-belt conveyer 织带式输送器;网带式输送器
mesh belt drier (单板用)网带式干燥机
mesh-belt kiln 网带窑
mesh bipolar transistor 网状双极晶
mesh blasting net 网格式爆破网
mesh boundary 网格边界
mesh cathode 网状阴极
mesh circuit 回路;网孔电路
mesh cloth 筛网
mesh configuration 网形
mesh-connected circuit 网接电路
mesh connection 网状联结;网状连接;网状结线;三角形接法
mesh connexion 网状接线
mesh coordinates 格网坐标;网格坐标
mesh core 网格门芯
mesh-core door 蜂窝空心胶合板门
mesh current 网络电流
mesh cycle 啮合周期
mesh distortion 格网畸变;网格畸变
mesh division 划分网格(有限元法);分格
mesh duct 网状管道
meshed 有孔的;网状的;啮合的
meshed anode 网状阳极
meshed gate structure 网状栅极结构
meshed gears 啮合的齿轮
meshed network 节点网络;网状网络
meshed wire 金属丝网
mesh emitter 网状发射极

mesh equation 网格方程
meshes of tectonic network 构造网眼
mesh fabric 网眼织物
mesh facility 网眼装置
mesh filter 网式滤清器;网式(过)滤器;筛网过滤器
mesh-form earthing device 接地网;网形接地器
mesh fraction 筛分粒度率;筛分粒(度)级
mesh ga(u)ge 网目板;网目测量仪;筛号;网目号;网目规;网目规
mesh generator 网格发生器
mesh grid 网状栅;编织网
meshing 结网;啮合
meshing bevel gear 啮合锥齿轮
meshing condition 啮合条件
meshing depth 啮合深度
meshing engagement 啮合
meshing frame 啮合检查结构
meshing gear 啮合齿轮
meshing interference 啮合干涉
meshing needle 结网梭
meshing point 啮合点
meshingtester 啮合检查仪
meshing zone 啮合区
mesh ionization chamber 网状电极电离室
mesh knot 网眼结
mesh lapping 网布搭接
mesh laying jumbo 网式隧道运渣车
mesh length 网孔长度
mesh length of river 河网长度
mesh lines 格网线;网格线
mesh marks 筛号
mesh method 网络法
mesh multiplier 网络倍增器
mesh network 环状管网;网状网络
mesh node 网孔结点
mesh number 网眼号;网(目)号;筛号
mesh of filter 过滤器孔眼
mesh of wires 导线网眼
mesh of wrapped net 包网网眼
mesh opening 网孔;筛目;筛孔
mesh opening width 网孔宽度
mesh-operated network 网式操作运行的网络
mesh or grid pattern roller 网式碾压机
mesh panel 网状墙板
mesh parameter 网格参数
mesh partition 钢丝网隔墙;钢丝网隔断;金属网隔断
mesh per inch 每英寸(筛网)孔(眼)数
mesh per square inch 每平方英寸孔数
mesh point 网点;网格点
mesh pot 网孔盆
mesh range 筛目范围
meshrebeeyeh (封闭阿拉伯建筑阳台窗的)精致旋制木隔板;(在大门上挑出的)防卫阳台;挑台窗
mesh reef 网状礁
mesh region 网区区域;网格区域
mesh-reinforced 钢丝网加筋的;网状钢筋的;网状加强筋
mesh-reinforced masonry 网状配筋砌体
mesh-reinforced shotcrete 钢丝网喷射混凝土
mesh reinforcement 钢丝网;钢筋网;网状加筋;网状钢筋
mesh Reynolds number 网格雷诺数
mesh-rod placement 铺网孔筋
mesh roller 网格路面碾
mesh scale 筛系列;筛分级别;网格距;网格尺度
mesh screen 网孔;过滤器;滤网;网筛;网格篦;筛子;筛网;筛目
mesh segment 网眼边
mesh series 筛孔系列;网眼系列
mesh sewage screen 污水网筛
mesh side cutter 组合侧铣刀
mesh sieve 网式网筛
mesh sinker 钢筋网沉放器(路面)
mesh size 粒度;网眼大小;网目尺寸;网孔尺寸;筛眼孔径;筛眼(尺寸)
mesh spacing 格网间隔;网格步长
mesh-star connection 三角星形接线法
mesh strainer 网式(过)滤器
mesh structure 格状构造
mesh tarpaulin 网状盖布
mesh test 筛分试验
mesh texture 网状结构;网眼结构
mesh type air outlet 网式风口
mesh-type face of the shield 网格式盾构

mesh-type oil filter 网式滤油器
mesh voltage 多相制线电压
mesh weld 凸点焊
mesh width 筛格尺寸;网孔宽度;网格宽度
mesh width of filter net 滤网网眼宽度
mesh wire 金属网线;网线
mesh work 网状结构工作;网络;网状物
meshwork structure 网眼结构
meshy 网状的
meshy filter 网式过滤器
meshy surface 网状表面
mesial 正中的
mesial margin 近中缘
mesial rare earth 中稀土
mesic 中生的
mesic atom 介(子)原子
mesioversion 近中向位
mesite 中性岩
mesitol 三甲苯酚
mesitylene 均三甲苯
mesityl oxide 异亚丙基丙酮
Mesnager hinge 曼斯纳格铰链
Mesnager impact test 曼斯纳格冲击试验
Mesnager notch 冲击试验用的槽口;曼斯纳格缺口
mesne assignment 中间转让
mesne lord 中间业主
mesne profit 中间收益
Meso-American architecture 中美洲建筑
mesoanalysis 中分析
mesobacteria 中型细菌
mesobar 中间气压
mesobenthos 中深海底的
mesobiota 中型生物群
mesocardia 中位心
Meso Cathaysian 中华夏系【地】
mesochronous 平均同步的
mesoclimate 中(期)气候;中尺度气候;局部气候;地方气候
mesoclimatology 中尺度气候学
mesocline 中度倾斜坡
mesocole 湿生动物
mesocolloid 近胶体;介胶体
mesoconch 中醒型
mesocormophyta 中生茎叶植物
mesocosm 中宇宙
mesocrate 中色岩
mesocratic 中色的
mesocratic rock 中色岩
mesocrystalline 中晶(质)的
mesocrystalline particle 中晶颗粒【地】
mesocumulate texture 中堆积结构
mesocycle 中旋回
Mesocyclops 中华水蚤
mesodermal type of gravitative tectonic 中皮型重力构造
mesodialyte 中性石
meso-environmental policy 中观环境政策
mesofauna 中型区系
mesogen 中介体
mesogene 中源;中深成的
mesogenetic fabric 深成组构
mesogeosyncline 中间地槽
mesoglea 中胶层
mesogloea 中胶层
mesogneiss 中带片麻岩
mesograined 中粒(度)的
mesohaline 中盐性;中盐度(的)
mesohigh 中(尺度)高压
Mesoid 中古
Mesoide 中生代褶皱带;中生代构造带
mesoiden 中生代山脉
mesoid tectonic system 中古构造体系
meso-impsonite 中脆沥青
mesoionic 介离子
mesokarst 半喀斯特
mesokurtic distribution 常峰态分布
mesokurtosis 正态峰;常态峰
meso-layer 中间层
meso-layer drift net 中层流网
mesolimnion 中湖;湖水中层
mesolimnion zone 水层温度差异区
mesolite 中沸石
Mesolith 中石器时代的
Mesolithic 中石器时代

Mesolithic age 中石器时代
Mesolithic period 中石器时期
meso-low 中尺度低压;中低压【气】
mesolyte 中介电解质
mesome 中干
mesomer 内消旋体
mesomeric 中介的
mesomeric effect 中介效应
mesomeric state 中介态
mesomerism 中介(现象);缓变异构(现象);稳变异构
mesometamorphism 中变质(作用)
mesometeorology 中尺度气象学
meso method 半微量法
mesomorphic phase 介晶相
mesomorphic soil 自成土
mesomorphic state 介晶态
mesomorphism 中性结构;介晶现象
mesomorphous 介晶的
mesomorphy 中性结构的特性;体育型体质
meson 重电子(即介子);介子
mesoneritic 中近海带
meson-meson effect 介子相互作用
mesonorm 中带标准矿物
meson resonance 介子共振态
meson scattering 介子散射
mesopause 中(间)层顶
mesopeak 中(间)层最高温度点
mesopegmatophyre 中色伟晶斑岩
mesopelagic(al) 中远洋的;中深海(2000～4000米水深的)
mesopelagic(al) fauna 海洋中层动物区系;海洋中层动物群
mesopelagic(al) fishes 中层鱼类
mesopelagic(al) plankton 半浮游生物
mesopelagic(al) zone 中层带
mesophase 中间阶段
mesophase pitch 中间相沥青
mesophile 中温微生物;嗜温生物
mesophilic 嗜温的
mesophilic and thermophilic co-phase anaerobic digestion 中温和高温同相厌氧氧化
mesophilic bacteria 中温菌;嗜温性细菌
mesophilic digestion 中温消化;嗜温消化
mesophilic methane fermentation 中温甲烷发酵
mesophilic process 中温过程
mesophilic range 中温(消化)范围;嗜温消化范围
mesophilic sludge digestion 污泥中温消化
mesophilic zone 中温带
mesophillic 中温的
mesophillic bacteria 中温细菌
mesophillic digestion 中温发酵
mesophilllic microorganism 中温微生物
mesophilous bacteria 喜温细菌;中温细菌
mesophotic algae 中光度藻类
mesophreatophyte 中间地下水湿生植物
mesophyte 中生代植物
mesophytic environment 中生植物环境
mesophytic forest 中生林
mesophytic formation 中生植物群系
mesophytic habitat 中生生境
mesophytism 中生性
mesopic spectral luminous efficiency 中间视觉光谱光视效率
mesopic vision 中间视觉;薄暮视觉
mesopiestic water 中承压水
mesoplankton 中型浮游生物;中生浮游生物;中层浮游生物
mesoplate 中间板块;中板块【地】
meso-pollution 中度污染
mesopore 中孔隙;间隙孔
meso-position 中位
Mesopotamian architecture 美索不达米亚建筑
mesoprosopic 中面型的
mesoprosopy 中面型
mesopsammon 沙隙生物
mesorelief 中地形
meso rock 中带岩
meso sample 小量试样;半微量试样;半微量称样
mesosaprobia 半污水生物
mesosaprobic(al) 中污染的
mesosaprobic(al) belt 中等污染带
mesosaprobic(al) river 中等有机污染河流
mesosaprobic(al) stream 中等有机污染河流

mesosaprobic(al) zone 中污(生物)带;中腐水性地区;中腐生带;中度污染带;中等污染区;中等污染带;中等腐水性地区;半污水生物带;半腐生生物带
mesosaprobic(al) zone indicator 中污带指示种
mesosaprobium 中污腐生物
mesoscale 中涡漩;中等规模;中(等)尺度
mesoscale diffusion 中尺度扩散
mesoscale eddy 中涡漩;中尺度旋涡;中尺度涡动
mesoscale high 中尺度高压
mesoscale low 中尺度低压
mesoscale model 中间比例模型
mesoscale system 中尺度系统
mesoscope 中视镜
mesoscopic fold 小型褶皱
mesoscopic scale 中型构造尺度
mesoscopic structure 中型构造
mesosiderite 中陨铁;中(石)铁陨石
mesosilexite 电气英石岩
mesosilicate 焦硅酸盐
mesosilicic acid 中硅酸
mesoslope 中坡(度)
mesosome 中体
mesosphere 下地幔;中圈;中间层;中带;中大气圈;中层;散逸层
mesospheric ozone 中间层臭氧
mesostasis 最后充填物
mesostroma 中基质
mesosyncline 陆间地槽
mesosystem 中尺度(天气)系统
mesotectonics 中构造
meso texture 中观结构
mesotherm 中温植物
mesothermal 中深热液的;深成中温热液的
mesothermal climate 中温气候
mesothermal deposit 中温热液矿床;中深热液矿床
mesothermal process 中深热液作用
mesothermal type 中温植物型
mesothermal vein 中深热液矿脉
mesothermic climate 中温气候
mesothermophilous 适暖带
mesothermophytia 温带植物群落
mesothyrid 中孔型
mesotidal range 中潮差
mesotill 中等风化冰碛
mesotourmalite 中电气岩
mesotron 重电子(即介子)
mesotrophic 半自养的
mesotrophic lake 中营养湖
mesotrophic mire 中营养泥沼
mesotrophic peat 中滋育泥炭
mesotrophic river system 中营养河系
mesotrophic state 中间营养状态;半自养状态
mesotrophic swamp 中滋育沼泽
mesotrophic water body 中营养水体
mesotrophy 中间型营养
mesotype 中型;钙沸石类;中沸石;钠沸石;中色的
mesotype dynamic(al) sounding 中型动力触探
mesowax 帆布防水材料
mesoxalic acid 中草酸;丙酮二酸
Mesozoic 中生代的
Mesozoic basin 中生代盆地
Mesozoic crude 中生代原油
Mesozoic era 中生代
Mesozoic erathem 中生界
Mesozoic group 中生界
Mesozoic ocean 中生代海洋
mesozone 中深(变质)带;中间区;中(变质)带
mess 公共食堂
message accuracy 消息的准确度
message acknowledgment 消息答复
message area 信息输出区
message authentication code 信息证实代码
message block 信息组;信息块
message blocking 信息编组
message buffer(ing) 信息缓冲
message buoy 救生浮标
message capacity 通信容量
message center[centre] 通信中心
message coding 信息编码
message communication 信息通信
message communication function 消息通信功能
message composer 信息编排器
message control 信息控制

message control block 信息控制块
message control program(me) 信息控制程序
message control task 信息控制任务
message data rate 信息传递速率;数据传递速率
message data set 信息数据集
message delay 信息延迟
message digit 信息(数)位;消息位
message display 信息显示
message display console 信息显示控制台
message editing 信息编辑
message edit tasking 消息编辑作业
message ending 消息尾
message exchange 信息交换
message exchanger 信息交换装置
message feedback 信息反馈
message flow 信息流
message format 消息格式
message group 信息组
message handler 信息处理器
message handling system 电函处理系统
message header 消息首部;消息标题
message indicator 信息指示器;消息指示符
message input/output 远距离信息输入/输出
message interpolation 信息内插法;信息插入;消息内插
message journalizing function 信息日志功能
message length 信息长度
message level 信息电平
message memory 信息存储器
message modulation 信息调制
message monitoring task 消息监控作业
message numbering 消息编号
message-oriented 面向消息的
message panel 信息显示板
message parity 信息奇偶性
message passing 消息传递
message polling 信息轮询
message polynomial 信息多项式
message precedence 消息优先等级
message priority 信息优先级
message processing distribution system 信息处理分配系统
message processing program(me) 信息处理程序
message process tasking 消息处理作业
message protection 消息防护
message queue 信息排队;信息列队;信息队列
message queue data set 信息队列数据集
message queueing 信息排队
message rate 消息率
message reference 信息访问
message register 信息寄存器
message retrieval 报文检索
message routing 信息指定通道;信息通路(选择);信息发送
message routing code 信息路径选择码
message sample 信息取样
message section 信息段
message segment 信息段
message sink 信息接收器;信宿
message source 信息源;消息源
message stream mode 信息输出方式
message structure 信息结构
message switching 信息转接;信息转换
message switching center 电文交换中心;报文交换中心;信息转接中心
message switching computer 信息转接计算机
message switching network 信息交换网(络);信息转接网络
message switching system 信息交换系统;信息转接系统
message switching technique 信息交换技术
message text 信息正文;报文正文
message throughput 信息总工作量
message-timed release 通话限时释放
message time stamping 信息记时;发信记时
message to noise ratio 信息噪声比
message transfer agent 电函传送代理
message transfer time 信息传送时间
message transmission rate 数据传输率
message volume 通信业务量
messaging application programming interface 通信应用程序接口
mess allowance 伙食津贴
mess canteen 食堂

mess cloth 餐巾
mess deck 住舱甲板【船】;寝居甲板
messenger 邮递员;引缆索;引(缆)绳;引火绳;钻孔取样器;拖索;通信员;电缆吊绳;传令器;承力索
messenger bottle 漂流瓶;浮标瓶
messenger cable 吊线(缆);吊绳;承力吊索;悬缆线
messenger call 传呼
messenger calling system 传呼电话系统
messenger chain 驱动链
messenger clamp 吊线缆夹
messenger winch 输索绞盘
messenger wire 承力索;悬缆(线);吊(缆)线;承力吊索
messenger wire and contact wire without balance 无补偿接触网
messenger wire clamp 吊线夹
messenger wire connecting clamp 承力索接头线夹
messenger wire for main line 正线承力索
messer sheet-pile method 梅赛尔施工法
mess gear 餐具
mess hall 职工食堂;职工饭堂;职工饭厅;内部食堂;食堂;餐具室;饭厅
mess house 食堂;职工食堂;职工饭堂
Messier Catalogue 梅西叶星云团表
Messier's catalog 梅西叶星表
Messier's number 梅西叶号数
Messina cycle 墨西拿造山旋回
Messing cone press 双锥辊挤浆机
mess kid 餐具
messmate 桉木
messmotor 积分电动机
messograph screen 雕刻网目板
Messo interglacial stage 米索间冰期【地】
Messo interstade 米索间冰阶【地】
messosilexite 电气石英岩
mess room 内部食堂;食堂;职工饭堂
messuage 住宅及屋基;住宅及基地;住宅;宅院
mess utensils 餐具
Mesta bearing 梅斯塔油膜轴承
mesulfamide 美磺胺
mesure analysis 量测分析
metaadolerite 变粗玄岩
meta-alginite 高煤化藻类体
metaaluminite 变矾石
metaalunogen 变毛矾石
meta-aminotoluene 间甲苯胺
metaandesite 变安山岩
meta-ankoleite 变钾铀云母
meta-anthracite 高煤化无烟煤;高级碳化无烟煤
meta-anthropology 超验人类学
meta-argillite 变质厚层泥岩
meta-arkose 变长石砂岩
meta-autunite 变钙铀云母
metab 金属底座
metabariscite 变磷铝石
meta-basalt 变玄武岩
metabasite 准基性岩;变质晶;变质基性岩
metabentonite 准斑脱岩;变斑脱岩
metabiosis 后生现象;同生作用;半共生
meta-bituminite 高煤化沥青体
metabituminous coal 高煤化(无)烟煤;超烟煤
meta-boghead coal 高煤化藻煤
metabograph 新陈代谢记录器
metabolic nucleus 静止核
metabolic pathway analysis 代谢途径分析
metabolic pollutant 新陈代谢污染物
metabolism 新陈代谢说
metabolism of pesticides 农药的代谢
metabolism zone of organic mater 有机质代谢作用带
metabolites of varying potency 不同活性的代谢产物
metabond 环氧树脂类黏结剂;环氧树脂类黏合剂
metaboric acid 偏硼酸
metaborite 偏硼石
metabreccia 变质角砾岩
metacalciouranoite 变钙铀矿
metacannel coal 高煤化烛煤
metacartography 比较制图学
metacenter 稳心;稳定中心;定倾中心
metacentre = metacenter
metacenter curve 稳心曲线
metacenter for longitudinal inclination 纵稳定(中)心
metacentre for transverse inclination 横稳心
metacentre height of gyrosphere 陀螺球稳心高度
metacentric 稳心的
metacentric arm 稳052力臂
metacentric diagram 稳心图;稳心曲线(图)
metacentric height 稳定中心高度;重稳距;稳心高度;定倾中心高度
metacentric height above center of buoyancy 稳心距浮心高度
metacentric involute 稳心轨迹曲线
metacentric radius 稳心半径;定倾半径
metacetaldehyde 变乙醛
metachemistry 超级化学
metachloral parachloral 聚三氯乙酸
metachoresis 变位
metachromasia 因光异色现象;变色现象;变色反应
metachromasia metachromate 异染物
metachromasy 异染性;变色反应性
metachromatic filter 变色滤光器
metachromatism 因光异色现象;异染性
metachrome blue 偏铬蓝
metachrome bordeaux 偏铬枣红
metachrome olive brown 偏铬橄榄棕
metachrome red 偏铬红
metachrome yellow 偏铬黄
metachromia 变色反应
metachromic acid 偏铬酸
metachrosis 变色机能
metacinnabar 黑辰砂矿
metacinnabarite 黑辰砂矿
metacinnavar 黑长砂
meta-clarite 高煤化微亮煤
metaclase 次生劈理岩
meta-cleavage 偏位裂解
metacolizing 铸铁表面渗铝法
metacolloidal mineral 变胶体矿物
metacolloidal structure 变胶状构造
meta-compound 间位化合物
metacone 后尖
metaconglomerate 变质砾岩
meta-cresol 间甲酚
metacrylic acid ester 甲基丙烯酸酯
metacryptozoite 后期潜隐体
metacryst 次生晶;变晶
metacrystal 变晶的;变斑晶
metacrystallic rock 变晶岩
metacrystal texture 交代晶体结构
metacryst inclusion 变晶包体
metadacite 变英安岩
metadelrioite 变水钒钙钙矿
meta-derivative 间位衍生物
metadiabase 变辉绿岩
metadiagenesis 外成作用
metadiazine 间二氮杂苯
metadiorite 变闪长岩
metadromic progression 奔走步态
metadurain 变质暗煤
metadurite 高煤化微暗煤
metadyne 交磁放大机;微场扩流发电机
metadyne generator 交磁旋转扩大机;微场扩流发电机;放大机发生器
meta-element 母体元素
meta-exinite 高煤化壳质组
meta-exsudatinite 高煤化渗出沥青体;变渗出沥青体
metafilter 层滤器;层滤机
metafiltration 层滤
metaform 金属模板
metaformaldehyde 变甲醛
metagabbro 变(质)辉长岩
metagabbroid 变质辉长岩类
metagalaxy 宇宙
metagame theory 变化博弈论
metage 称量
metagelatin 变性明胶
metagenesis 后成作用;世代交替;深变质作用;沉积变质作用
metagenetic gas 变生阶段成因气
metagenetics 复成沉积物
metagenetic stage 变生作用阶段;变成作用阶段
metagenic 变替的
metagenic deposit 复成沉积
metaglygh 变质印痕
metagneiss 变片麻岩
metagreywacke 变质杂砂岩
metahalloysite 准埃洛石;变埃洛石;扁管状高岭土
metaheinrichite 变钡砷铀云母
metahewettite 变针钒钙石
metahohmannite 变褐铁矾
metaigneous rock 变火成岩
metainstruction 中间指令
metakahlerite 变铁砷铀云母
metakaolin 准高岭土;偏高岭土
metakaolinite 偏高岭石;二水高岭石
metakeratophyre 变角斑岩
metakinesis 中期分裂
metakliny 基团位变
metakoritnigite 变水红砷锌石
metal 金属;米托(一种铝漆)
metal abrasive material 金属研磨材料
metal-absorbent char 金属吸附炭
metal abundance 金属丰度
metal acetylide 乙炔基金属;金属的乙炔化物
metal acoustic(al) ceiling 金属吸声板吊顶
metal additive 金属添加剂
metal adhesive 金属胶合剂;金属黏合剂;金属黏结剂
metal age 金属时代
metal aggregate 金属集料
metal-air battery 空气去极化电池
metal alcoholate 金属醇化物;醇淦
metal alkoxide 金属烷氧化物;金属醇盐
metal allowance 加工余量
metal alloy 金属合金
metal analysis 金属分析
metal anchor 金属拉杆;金属支撑物;金属锚碇
metal anchorage 金属锚头;金属锚具;金属锚固件
metal angle 金属边角料;金属角
metal angle bead 金属边角小珠;金属边角小球;金属包角
metal angle cutter 金属角切刀
metal anode 金属阳极
metal-anodicoxide 金属阳极氧化物
metal antenna 金属天线
metal anti-reflection coating 金属减反射膜
metal aperture mask 金属孔掩模
metal-arc cutting 金属(极)电弧切割
metal-arc gas-shielded welding 金属极气保护焊;气保护金属极电弧焊
metal arch casing 金属拱套
metal-arc inert-gas welding 惰性气体保护金属极电弧焊
metal-arc welding 金属电弧焊(接);金属极电弧焊
metal armo(u)ring 金属铠装
metal aryl 芳基金属
metal arylide 芳基金属
metalastic mounting 金属弹性减振器;金属弹性减振垫
metalastik 金属橡胶结合法
metal astragal 镶窗玻璃金属条;金属半圆线脚;金属半圆联珠饰
metalate 使金属化
metalation 金属化作用
metalaumontite 黄浊沸石
metal awning-type window 金属遮蓬形窗;金属遮篷式窗
metal back 金属壳;金属衬垫
metal-backed seal 金属背衬密封环
metal-backed tape 金属背衬胶带
metal backing 金属壳衬垫;金属壳背;金属敷层;金属垫衬;金属垫板
metal back tube 金属敷层显像管
metal backup 金属挡块
metal baluster 金属栏杆柱
metal bar 金属(撑)条;金属(拉)杆;金属条;金属棒
metal base 金属基体;金属管基;金属底座
metal base clip 金属基层卡
metal base plate 金属底座工作台;金属基座;金属底板
metal bas-relief 金属浮雕
metal bath 金属浴;焊接熔池
metal batten 金属狭板;金属板条
metal bead 金属压条
metal beam 金属梁
metal-bearing 合金轴承;含(有)金属的
metal-bearing sand 矿砂
metal-bearing solid 含金属固体
metal bellows 金属伸缩软管;金属膜盒;金属波纹管

metal belt fastener 金属带扣
metal-binding ligand 金属接合配位体
metal-binding plate 金属连接板
metal blade 金属扇叶
metal bladed drag 刮板式刮路机;金属刃刮路机
metal block 铁滑车
metal boat 金属盘
metal body 金属车身;坯胎
metal bond 金属黏结;金属结合剂
metal-bonded brush 金属丝刷;金属丝装成的刷子
metal bonded wheel 金属结合砂轮
metal bonding 金属胶接;金属结合;金属焊接;金属连接
metal-Braun tube 金属显像管
metal break out 铸型裂口;跑水
metal breakout error 漏箱
metal bridge 金属桥梁
metal bridging 马钉;金属桥接
metal bronze 古铜;(铜锡合金的)青铜
metal brush 钢丝刷
metal builders fitting 金属建筑零配件;建筑金属设备;建筑五金
metal builders hardware 建筑金属器具;建筑金属零件;建筑金属元件;建筑金属附件;建筑金属构件
metal building 金属房屋;金属建筑(物)
metal building material 金属建筑材料
metal building panel 建筑用五金板材
metal build-up 金属制成品
metal buried welding 埋渣焊
metal button-sewing machine 缝金属扣机
metal caisson (天花板的)金属凹槽;金属沉箱
metal caisson type ceiling 金属凹槽型天花板
metal cap flashing 金属帽盖泛水
metal capping 金属帽盖的
metal capping strip 金属套环
metal carbide 金属碳化物
metal-carbide rock drill 碳钢钻头
metal carbon 金属相碳
metal carbonate solid 金属碳酸盐固体
metal carbonyl 羰络金属
metal card memory 金属卡片存储器
metal-cased brick 金属夹圈砖;铁皮砖;铁壳砖;包金属砖
metal-cased pile 具有钢壳的现浇混凝土桩;现浇混凝土钢壳桩
metal-cased refractory 铁皮镶套耐火材料
metal casement 铝合金窗;钢窗;金属窗扇;合页金属窗
metal-casement putty 金属窗扇油灰
metal-casement window 金属竖铰链窗
metal casing 金属压条;金属壳;金属护板
metal cassette 金属铠装暗箱;金属铠装暗盒
metal cassette type ceiling 金属铠装暗盒式天花板
metal casting 金属铸件;金属外壳
metal catcher 金属杂质分离器
metal cation 金属阳离子
metal cavity joint 埋铁(接)缝
metal ceiling 金属顶棚;金属天花板
metal ceiling panel 金属天花板框架
metal ceramic 金属包瓷;金属陶瓷
metal ceramic agglomerate 金属陶瓷烧结制品
metal ceramic coating 金属陶瓷涂层
metal ceramic combinations 金属陶瓷制品
metal ceramic composite material 金属陶瓷复合材料
metal ceramic filter 金属陶瓷过滤器
metal ceramic friction material 金属陶瓷摩擦材料
metal ceramic-plastic composite 金属陶瓷塑料复合材料
metal ceramics 金属陶瓷学
metal ceramic seal 金属陶瓷封接
metal channel 金属管路;金属通道;槽钢
metal charge 金属料
metal chelating agent 金属络合剂
metal chemistry 金属化学
metal chimney 金属烟囱
metal chip(ping)s 金属屑
metal-clad 装甲的;金属保护的;铁壳的;有金属包层的
metal-clad building 金属外皮建筑物
metal-clad cable 铠装电缆
metal cladding 金属敷镀;金属覆盖;金属包层(法)
metal cladding sheet 金属包层片材;金属包层薄板
metal-clad door 包金属门;金属外包门

metal-clad fire door 包金属防火门;金属包皮防火门
metal-clad laminated wood door 金属包皮拼合木门
metal-clad motor 铠装电动机;金属加固电动机
metal-clad optic(al) waveguide 金属包层光学波导
metal-clad plywood 金属贴面胶合板
metal-clad station 金属铠装变电站
metal-clad waveguide 金属包层波导
metal-clad window 包金属的窗
metal cleaner 金属清洗剂;金属净化剂
metal-cleansing solution 金属清洁液
metal clip 金属夹
metal closure coating 金属盖涂料
metal cloth 金属丝筛布
metal-coated glass fiber 镀金属玻璃纤维
metal-coated paper 敷镀金属的纸
metal coating 金属涂料;金属镀层(法);金属涂层;金属镀面;金属层敷涂;金属保护;电镀金属
metal coating surface treatment 金属镀层表面处理
metal coffer 金属围堰;金属平顶镶板
metal coffer type ceiling 金属镶板型天花板
metal-colloidal colo(u)ration 金属胶体着色
metal colo(u)ring 金属着色
metal column 金属柱
metal combustion chamber 金属燃烧室
metal compact 金属坯块
metal comparater 磁力金属比测仪
metal complex 金属配位染料;金属配合物
metal complexation 金属络合
metal complexation reaction 金属络合反应
metal concentration 金属浓度
metal conditioner 金属调理剂;金属(表面)处理剂
metal conduit 金属导管
metal cone 锥形金属壳
metal connecting fitting of high grade furniture 高档家具五金连接件
metal connection plate 金属连接板
metal connector 金属插头座;金属接头;金属连接器
metal construction 金属建筑;金属结构
metal construction(al) material 金属结构材料;金属建筑材料
metal container 金属容器;金属集装箱
metal cope 金属顶层;金属顶盖
metal coping 金属压顶
metal core 金属核;金属芯片;金属芯子
metal cored carbon 金属芯碳棒
metal core glass fibre 金属芯玻璃纤维
metal corner bead 墙角金属线脚;金属墙角护条;金属护角
metal corner plate 金属角(撑)板
metal cornice flashing piece 金属挑檐泛水;金属挑檐挡水板;金属飞檐泛水;金属飞檐挡水板
metal counter ceiling 金属吊平顶
metal counter flashing 金属帽盖泛水
metal cover 金属盖;金属包皮
metal covered 外包金属的
metal-covered door 金属外包门;包金属门
metal-covered window 金属外包窗
metal covering 金属罩;金属包皮;包金属皮
metal cover strip 罩金属带状物;盖金属带状物;罩金属板条;盖金属板条
metal cradle 金属插座;金属支架;金属托架
metal cramp 夹缝石用铁夹;金属夹;扒锔(用以连接、固定相邻石块);(砌筑用的)石块铁夹
metal creep 金属蠕变
metal crib 金属框笼
metal crown 金属冠
metal crystal 金属晶体
metal cupola 金属化铁炉;金属冲天炉;(立式的)金属圆顶炉;金属穹顶;金属圆顶
metal curb 金属边饰;金属路缘
metal current 玻璃液流
metal curtain wall 金属幕墙
metal curtain wall panel 金属幕墙(镶)板
metal cut saw 金工用锯
metal cutting 金属切削
metal-cutting capacity 金属切削能力
metal-cutting machine 金属切削机床;金属切削机
metal-cutting machine tool 金属切削机床
metal-cutting oil 金属切削油
metal cylinder 金属圆柱筒;金属柱筒
metal deactivator 金属惰化剂
metal deck 白铁皮平屋顶;金属(薄)板平屋顶;金属(铺)板

metal deck floor 金属楼板;金属(铺)板楼盖;金属铺板
metal defect 玻璃(内部)缺陷
metal defect detection 金属探伤
metal degradation 金属降解
metal degreasing 金属脱脂;清除金属油渍
metaldehyde 聚乙醛;四聚乙醛
metal deposit 金属沉积;熔敷金属
metal depth 玻璃液深度
metal detector 金属探测器
metal detector/separator 金属检测/分离器
metal diaphragm pump 金属隔膜泵
metal-dielectric(al) filter 金介质滤光片
metal-dielectric(al) interference filter 金属介质干涉滤光片
metal-dielectric(al) interferometry 金属介质干涉量度学
metal dip brazing 金属浴钎焊
metal dish 金属盘
metal dissolution 金属溶解
metal distribution ratio 金属分布比
metal door 金属门
metal door factory 金属门工厂
metal door frame 金属门框
metal draft pipe 金属尼水管
metal drain tile 金属排水管
metal drill 金属钻
metal drip 金属滴水【建】
metal drop 金属熔滴
metal droplet 金属熔滴
metal droplet detachment 金属熔滴的分离
metal dross 金属浮渣
metal drum 铁桶
metal dust 金属粉末;金属粉尘
metal dye 金属染料
metal eaves gutter 金属檐沟
metaled gasket 金属垫
metal-edged gypsum plank 金属镶边石膏板
metal edged mirror 金属镶边镜子
metal edge element 金属卷带周边镜清元件
metal edge sealed glazing unit 空气夹层玻璃(用于保温隔热);金属密封镶边的玻璃窗组件
metal electrode 金属电极;金属电弧焊条
metal electro-sparking work 金属电火花加工法
metal eliminator 金属杂物分离装置;金属杂物分离机
metal-encased brick 铁壳砖
metal-enclosed 铁壳的
metal-enclosed apparatus 金属封闭型电器
metal-enclosed switchgear 金属封闭开关柜
metal endurance 金属耐久力;金属耐久性;金属耐疲劳度
metal enrichment 金属增富
metalepsis 取代作用
metalepsy 取代作用
metaler 钣金工
metaleucite 蚀变白榴石
metal expansion strip 金属伸缩片
metal extraction efficiency 金属提取率
metal extrusion 金属轧制
metal fabrication 金属预制件;金属结构
metal facade 金属立面;金属外观
metal-faced 金属贴面的
metal-faced board 金属贴面板
metal-faced door 金属面门;金属贴面门
metal-faced joinery 包金属面的木制件;金属面木作
metal-faced plywood 金属贴面胶合板
metal-faced rubber 金属包面橡胶;金属贴面橡胶
metal-faced window 金属包面窗;金属贴面窗
metal facing 金属贴面工艺;金属贴面法;金属饰面
metal fall 金属提取率
metal fastening 金属紧固件;金属固定件
metal fatigue 金属疲劳
metal fence picket 金属围篱尖柱
metal fence post 金属围篱支墩;金属围篱支柱;金属围篱支柱
metal fence stake 金属围篱竖管;金属围篱柱;金属围篱杆;金属围篱柱
metal fever 金属热
metal fibre-reinforced composite 金属纤维增强复合材料
metal filament lamp 金属白炽灯;金属白热灯
metal filings 金属锯屑;金属锉屑;金属填料
metal film 金属膜

metal film filter 金属薄膜滤光器
metal-film resistor 金属膜电阻器
metal filter elements 滤清器金属滤片
metal finish 金属饰面
metal finishing 金属抛光;金属精整;金属表面处理
metal finishing industry 金属抛光工业
metal finishing waste 金属抛光废料;金属表面处理废料
metal fish plate 金属接轨(夹)板;金属接合(夹)板;金属鱼尾(夹)板
metal fittings 小五金
metal fixing 金属固定件
metal flake 金属片状粉末;金属小片;金属鳞片;金属薄片;金属旗杆
metal flashing 金属泻水板;金属挡水板;金属泛水
metal flashing shingle 金属披水板;金属泛水板
metal flexible conduit 金属挠性导管;金属蛇管;金属软管
metal floor decking 金属楼板;楼面金属覆盖板;金属楼面
metal flume 金属水槽;金属渡槽
metal focal plane shutter 钢片快门
metal fog 金属雾
metal foil flashing piece 金属箔泛水薄膜片;金属箔泻水薄膜片
metal foil heater 金属箔加热器
metal-foil heat treating method 金属箔热处理法;包装热处理法
metal foil insert(ion) 金属箔垫片;金属箔衬垫;金属箔插入物
metal foil layer 金属箔层;金属箔夹层;金属箔垫片
metal foil paper 金属箔纸
metal foil resistor 薄式平面电阻器
metal foil surfaced asphalt felt 金属箔覆面油毡
metal folding door 金属折(叠)门
metal for hammering 锻打金属;锻造金属;锤击金属
metal forming 金属成型
metal forming equipment 锻压设备
metal forming machinery 锻压机械
metal form(work) 钢模(板);(混凝土的)金属模板
metal fractionary pump 金属分馏真空泵
metal frame 轻钢龙骨;金属框架
metal frame building 金属框架建筑物
metal framed partition(wall) 金属框架分隔墙
metal framed window 有金属框的窗;金属框(架)窗
metal framery 金属门窗结构
metal frame work 金属框架
metal-free 不含金属的
metal-free phthalocyanine 五金属酞菁颜料
metal frieze 金属雕带(古典檐梁装饰);金属壁缘;金属中楣
metal fume fever 金属烟尘热
metal furnace 金属加热炉
metal furring 钉金属薄板条;刮去金属水垢
metal furring channel 金属槽形龙骨
metal gasket 金属填密片;金属衬垫
metal gate 金属栅
metal gathering 金属爆炸差厚成型法
metal gauze 钢丝网;金属(丝)网
metal girder 金属桁架
metal glazing bar 装配玻璃用金属条;金属上釉杆;装配玻璃用金属杆;金属上釉棒
metal-grade 金属级品位
metal-graphite brush 金属石墨电刷
metal grating 金属格栅;金属光栅;金属算子
metal grating filter 金属栅栏滤光片
metal gridcoating 金属栅状涂层
metal grid floor(ing) 铺设金属格栅地板;铺设金属网格地板;铺设金属格子地板;金属网格楼板
metal grill(e) 金属格栅;金属格子窗
metal grit 棱角形碎金属;金属屑
metal grommet 帆布金属眼环
metal guard 金属护角
metal gusset plate 金属加强板;金属角板;金属结点板;金属连接板;金属节点板
metal gutter 金属天沟;金属檐沟;金属沟槽;金属边沟
metal hack saw 金属弓锯
metal halide 金属卤化物
metal halide lamp 金属卤素灯;金属卤化物灯
metal handle 金属提手;金属拉手;金属把手;金属摇柄;金属曲柄;金属手柄

metal handrail 金属扶手
metal hanger 金属梁托;金属支架;金属吊钩;金属挂钩;金属钩子
metal hardener 金属硬化成分;金属淬火剂;金属硬化剂
metal hardening process 金属硬化过程(法)
metal head frame 金属井架
metal heat exchanger 金属材料换热器
metal hinge 金属活动关节;金属活页;金属铰接;金属铰链
metal holder 金属架
metal hose 金属挠性导管;金属蛇(形)管;金属软管;柔性管
metal hydroxide solid 金属氢氧化物固体
metalic contamination 金属掺杂物
metaliding 金属沉积法
metalignitous coal 高煤化褐煤;超褐煤
metalikon 金属喷镀法
metalikon process 金属喷镀法
metalimnion 中层湖水;温跃层;变相湖沼;变温层
metal impurity 金属杂质
metal inclusion 金属夹杂物
metal indicator 金属指示剂
metal-inert-gas 金属焊条惰性气体
metal-inert-gas arc welding 惰性气体金属电弧焊;金属极惰性气体保护焊
metal-inert-gas cutting 金属惰性气体切割
metal-inert-gas welding 金属惰气焊;惰性气体保护焊;金属焊条惰性气体保护焊;熔化极惰性气体保护焊
metal insert(ion) 金属衬垫;金属插入物
metal inspection 金属探伤;金属检验
metal insulator 金属绝缘体
metal-insulator-semiconductor laser 金属绝缘体-半导体激光器
metal interference filter 金属干涉滤光片
metal ion buffer 金属离子缓冲
metal ion coagulant 金属离子混凝剂
metal-ion concentration 金属离子浓度
metal ion dopant 金属离子掺杂物
meta-liptinite 高煤化类脂体
metalirchheimerite 变钴砷铀云母
metalized lamp bulb 喷镀金属的灯泡
metal jacket 金属隔层;金属外壳;金属套管;金属外套
metal jacket gasket 钢包石棉垫片;金属包(石棉)垫片;铁包石棉垫片;包金属垫片
metal jet (爆炸焊的)金属喷流
metal-joint filler 金属填缝料
metal-joint plate 金属连接板
metal-joint strip 金属连接片;金属连接条;金属连接带
metal knee bracket plate 金属拐弯托座板
metal laden food material 含金属的食品原料
metallargist 冶金工作者
metal lath 抹灰金属网;钢丝网板条;金属(拉)网
metal lath and plaster 钢丝网抹灰;钢板网抹灰
metal lath and plaster ceiling 金属网抹灰顶棚;金属网粉刷天棚;金属网刷平顶
metal lath clip 金属拉网卡
metal lathing 钉金属肋板条;金属肋板条;钢板(板)网;墁墙金属板条;钢板网;钉钢丝网
metallation 金属取代
metal lattice 金属晶格
metal lattice work 金属格构工
metal leaf 金属薄箔;金属箔
metalled 碎石铺路的;金属的
metalled road 碎石铺面道路;碎石路
metal lens antenna 金属透镜天线
metaller 钣金工
metallergy 变型变态反应性
metal level 金属液面;玻璃液面
metallic 金属的
metallic absorbing layer 金属吸收膜
metallic absorption 金属吸收
metallic-additive lamp 金属卤素灯;金属卤化物灯
metallic aggregate 金属集料;金属骨料
metallic-aggregate mortar 金属集料砂浆;金属骨料灰浆;金属黏结灰浆
metallic air recuperator 金属的空气换热器
metallic aluminium 金属铝
metallic antimony 金属锑
metallic arc welding 金属弧焊
metallic area of wire rope 钢丝绳有效金属断面

metallic automotive coating 汽车闪光涂料
metallic band tape 金属卷尺
metallic bath surface 金属液面
metallic bead 金属粒;金属夹砂
metallic behavior 金属性能
metallic bellows ga(u)ge 金属膜盒压力计
metallic bond 金属键
metallic brown 金属褐色
metallic brush 金属刷
metallic cable 钢丝索
metallic carbide 金属碳化物
metallic carbon 金属质碳
metallic catalyst 金属催化剂
metallic cathode 金属阴极
metallic cement 矿渣硅酸盐水泥;金属硬化物;金属水泥;金属粉腻子
metallic cementation 喷镀金属;渗金属
metallic cementation process 渗金属法
metallic channel 导线槽
metallic character 金属特性
metallic chloride 金属氯化物
metallic chlorinate waterproofing agent 金属氯化物防水剂
metallic clad diamond 金属衣金刚石
metallic clothing mounting machine 金属针布包卷机
metallic clothing unwinding machine 金属锯条倒条机
metallic clothing welder 金属针布焊接器
metallic coating 金属外层;金属涂敷;金属闪光涂层;金属覆盖层
metallic coating by hot-dipping process 热浸镀金属法
metallic colo(u)r 金属色(料)
metallic commodities 金属矿产
metallic component 金属成分
metallic compound 金属化合物
metallic conductivity 金属导电性
metallic conductor 金属导体
metallic conduit 金属管
metallic contaminant 金属污染物
metallic contamination 金属污染
metallic content 金属含量
metallic continuity 金属的连续性
metallic copper 金属铜
metallic corrosion 金属腐蚀
metallic corrugate pipe 金属波纹管
metallic creep 金属蠕变
metallic crystal lattice 金属晶格
metallic diamond 金属性金刚石
metallic diaphragm 金属膜片
metallic drift tube 金属壳漂移管
metallic dropper 金属滴管
metallic dust 金属粉尘
metallic earth 金属接地
metallic electrode 金属焊条;金属电极
metallic element 金属元素
metallic element content 金属元素含量
metallic fabric 金属丝布
metallic fiber 纤维金属材料;金属纤维
metallic filament 金属灯丝
metallic film 金属薄膜
metallic filter 金属过滤器
metallic finish 金属表面精加工
metallic finish coat 金属闪光漆;金属光泽面漆
metallic fittings 五金件
metallic flake pigment 片状金属颜料
metallic foil 金属箔
metallic form 金属模板
metallic fume 金属烟雾;金属烟
metallic gasket 金属垫片
metallic glass 金属玻璃
metallic glaze 金属釉
metallic graphite brush 金属碳刷
metallic grey 银灰色的
metallic grid coating 金属栅状膜
metallic grit 钢砂;金属砂粒
metallic helium 金属氦
metallic hollow O-ring for selfseal 自紧式金属空心O形环
metallic hollow O-rings of non-self-energizing 非自紧式金属空心O形环
metallic hollow self-energizing O-rings 自紧式金属空心O形环

metallic inclusion 金属夹杂物
metallic ink 金粉油墨
metallic insulation 金属质绝热隔声材料
metallic insulator 金属绝缘子
metallic ion 金属离子
metallic iron 金属铁
metallicity 金属性;金属特征;金属丰度
metallicity parameter 金属性参数
metallic joiner 金属接缝条
metallic joint 金属接缝
metallic lacquer 金属亮漆
metallic lath(ing) 金属挂瓦条
metallic lead 铅粉漆
metallic lead frame 金属引线架
metallic lead printer 红丹(又名红铅)
metallic lifeboat 金属救生艇
metallic line star 金属线星
metallic link belt 金属链带
metallic liquid 金属防水液(一种防水液)
metallic luster 金属光泽
metallic luster paint 金属光亮漆
metallic material 金属材料
metallic materials warehouse 金属材料库
metallic matrix 金属基体
metallic membrane 金属膜片
metallic membrane plating 镀膜
metallic mercury 金属汞
metallic mineral 金属矿物;金属矿石
metallic mineral deposit 金属矿床
metallic mineral products 金属矿物产品
metallic mineral resources 金属矿产资源
metallic mirror 金属反射镜
metallic network 金属网络
metallic-nonmetallic O-ring seal 金属-非金属双O形环密封
metallic ore 金属矿石;金属矿产
metallic packing 金属垫片;金属填料
metallic packing ring 金属密封环
metallic paint 金属粉涂料;金属光泽涂料;金属漆;金属涂料
metallic paper 金属涂布纸;金属箔
metallic phase 金相
metallic pigment 金属颜料
metallic pigmented paint 金属闪光漆
metallic pipe 金属管
metallic piping 金属管
metallic piping fitting 管道装配工作;金属接管零件;金属管子配件
metallic plane 金属刨子
metallic poison 金属性毒物
metallic pollutant 金属污染物
metallic pollution 金属污染
metallic powder 金属粉末
metallic prominence 金属日珥
metallic pyrometer 金属高温计
metallic recuperator 金属换热器
metallic reducing agent 金属还原剂
metallic reduction of fluorides 氟化物金属热还原
metallic reflection 金属反射
metallic reflective coating 金属反射膜
metallic reflector 金属反射器
metallic resinate 金属树脂酸盐
metallic return 金属回线
metallic return circuit 金属回路
metallic rheostat 金属线变阻器
metallic ring 金属环
metallic ring filter 金属环过滤器
metallics 金属物质;金属粒子
metallic salt 金属盐类
metallic scrap 废金属
metallic screen 金属筛
metallic seal(ing) 金属密封(材料)
metallic sheath 金属套
metallic sheave 金属滑车
metallic shell 金属壳
metallic shop 金属加工间
metallic short circuit 金属性短路
metallic sieve 金属筛网
metallic silicon 金属硅
metallic soap 金属皂;分型皂
metallic soap formation 金属皂的形成
metallic soap stabilizer 金属皂(类)稳定剂
metallics of charge 装料中的金属物质
metallic sound 金属音;金属声

metallic spatula 金属刮刀
metallic spiral core wire rope 带金属螺旋式钢丝绳
metallic spray coating 喷涂金属膜
metallic sprayed coat(ing) 喷涂金属膜
metallic spraying 金属喷涂
metallic spring gravimeter 金属弹簧重力仪
metallic stabilizer 金属稳定剂
metallic standard 金属本位
metallic standard system 金属本位制
metallic star 金属星
metallic stearate 金属硬脂酸盐
metallic stuffing box 金属填料箱
metallic substance 金属物质
metallic substrate 金属坯胎;金属基片
metallic tank 金属水槽;金属槽
metallic tape 钢皮尺;钢(卷)尺;金属卷尺;金属带
metallic target 金属靶
metallic taste 有金属味
metallic template 金属样板
metallic thermocouple 金属热电偶
metallic thermometer 金属温度表;金属温度计
metallic tile 金属瓦片
metallic tone quality 金属音色
metallic toxicity 金属毒性
metallic tube 金属管
metallic-tube arch-type heating exchanger 金属管拱形换热器
metallic tubing 金属管
metallic valve stem packing 阀杆金属填密
metallic water-piping system 金属管线供水系统
metallic waterproofing 铁粉防水材料
metallic waveguide 金属波导管
metallic work 金属加工间
metallic zinc 金属锌
metallic zinc coating 锌粉涂层;金属锌镀层
metallic zinc pigment 金属锌颜料
metallide 金属与金属化合物
metal lifeboat 金属救生艇
metalliferous 含金属的
metalliferous material 金属物料
metalliferous mining 含金属采矿;含金属矿业
metalliferous ore resources 金属矿产资源
metal lift-off technology 金属剥离工艺
metallike 似金属(的)
metallikon 金属喷涂;喷镀金属;喷镀法
Metalline 梅达林铜钴铝合金;玻璃液面线;金属性质;含金属盐;含金属
metal line attack 液面线侵蚀
metal line control 玻璃液面控制
metal line corrosion 料液面侵蚀
metal line cut 液面线侵蚀;料液面侵蚀
metal-lined 金属衬里(的)
metal line stability 玻璃液面稳定性
metal(l)ing 碎石料;碎石层;盖以金属;筑路碎石;铺(路)碎石;喷镀金属;道砟
metal(l)ing clause 磨耗条款;镀铁件磨耗条款
metal lining 金属衬层;金属衬板;金属衬里;铁壳板;镀覆金属
metal l insert(ion) 金属垫片
metallio-oxide electrode 氧化金属电极
metal lipping 金属镶边
metallisation 喷涂金属粉;敷金属
metallised plastics 镀金属塑料
metallization 喷镀金属法;硬化;金属涂敷法;金属喷镀;金属化;喷镀;成矿作用;敷金属法
metallization apparatus 金属喷镀器
metallization of plastics 塑料涂敷金属
metallization pattern 金属化图形;金属化互连图
metallize 金属喷涂;金属化;涂金属
metallized brush 金属石墨电刷
metallized capacitor 镀金属电容器
metallized carbon 金属渗碳
metallized coating 金属闪光涂层;金属光泽涂料
metallized contact 敷金属接头
metallized film resistor 涂金属膜电阻
metallized finish 金属闪光面漆
metallized foils 镀金属箔
metallized glass 喷镀金属(膜)玻璃;镀金属玻璃;敷金属玻璃
metallized glass fiber 镀金属玻璃纤维
metallized lamp 镀金属灯泡
metallized lamp bulb 敷金属灯泡
metallized optic(al) fiber 镀金属光学纤维
metallized paper 镀金属纸

metallized paper capacitor 镀金属纸质电容器
metallized paper capacitor for elecetrical fan 镀金属纸介电风扇电容器
metallized paper condenser 镀金属纸质电容器
metallized plastic capacitor 镀金属塑料电容器
metallized plastics 镀金属塑料;掺金属塑料
metallized plastic tape 金属化塑料带
metallized plywood 强化胶合板
metallized resistor 镀金属电阻器
metallized screen 金属膜荧光屏;金属背荧光屏
metallized slurry blasting 金属浆爆破
metallized strip 敷金属带
metallized titanium dioxide 包覆二氧化钛
metallizer 金属粉末;喷镀金属器
metallizing 涂敷金属层;真空镀膜烫金法;金属化涂层;喷涂金属;喷镀金属(法);喷镀;闪金属本泽涂层;镀金属;包镀金属
metallizing temperature 金属涂覆温度;金属化温度
metalllic thermometer 金属温度表
metal locator 金属探寻器
metallocene 金属茂;茂金属
metalloceramics 喷镀金属陶瓷
metallochrome 金属色;金属面着色
metallochromic indicator 金属指示剂
metal lock fastener 金属扣件;金属卡固件
metallocyanide 氰化金属
metallogenesis 成矿作用
metallogenetic 成矿的
metallogenetic region 成矿区域
metallogenic 成矿的
metallogenic belt 成矿带
metallogenic belt of Tathyan porphyry copper deposits 特提斯斑岩铜矿成矿带
metallogenic belt of the middle to lower reaches of the Changjiang river 长江中下游成矿带
metallogenic elements 成矿元素
metallogenic epoch 成矿时代
metallogenic formation 成矿建造
metallogenic formation Diwa-type 地洼型成矿建造
metallogenic formation platform type 地台型成矿建造
metallogenic hypothesis 成矿假说
metallogenic map 金属成矿地图;成矿预测图;成矿规律图
metallogenic model 成矿模式
metallogenic prognosis 成矿预测
metallogenic province 成矿省;成矿区
metallogenic specificity 成矿专属性
metallogenic tectono-geochemistry 成矿构造地球化学
metallogeny 区域成矿学
metallogeny of deep lineaments 深部构造线成矿说
metallograph 金相照片;金相仪;金相显微照相机;金相显微镜;金属版
metallographic(al) 金相(学)的
metallographic(al) analysis 金相分析
metallographic(al) etching 金相侵蚀
metallographic(al) examination 金相试验;金相检验;金相断面磨片
metallographic(al) laboratory 金相实验室
metallographic(al) microscope 金相显微镜
metallographic(al) specimen 金相试样
metallographic(al) test 金相试验
metallographic province 成矿区
metallography 金相学
metalloid 准金属;类金属;非金属
metalloid element 类金属元素
metalloid-organic compound 类金属有机化合物
metalloid phase 似金属相
metalloid pollution 类金属污染
metalloid substance 非金属物质
metalloid toxicant 类金属毒物
metallometer 金属试验器
metallo-metric(al) survey 金属量测量
metallomicroscope 金相显微镜
metallomicroscopy 金相显微术
metallorganics 金属有机物(质)
metalloscope 冶金显微镜;金相显微镜
metalloscopy 金属反应检查法
metallosiloxane polymer 金属硅氧烷聚合物
metallostatic 金属静力学的
metallostatic pressure 金属静压力
metallothermic reduction 金属热还原

metallothionein 金属硫因
metallothiopeptide 金属硫肽
metallotrophy 金属移变作用
metal louver 金属百叶
metal low relief 金属浅浮雕
metallurgic(al) accountability 冶金衡算
metallurgic(al) addition agent 冶金添加剂
metallurgic(al) blast cupola 冶金冲天炉
metallurgic(al) bonding 冶金结合
metallurgic(al) calculation 冶金计算
metallurgic(al) cement 熔渣硅酸盐水泥;熔渣水泥;冶金(熔渣)水泥;矿渣水泥
metallurgic(al) cement rendering 矿渣硅酸盐水泥粉刷;矿渣硅酸盐水泥抹灰
metallurgic(al) certificate 冶金专业证明书
metallurgic(al) chemistry 冶金化学
metallurgic(al) coal 炼焦煤;冶金(用)煤
metallurgic(al) coke 冶金焦(炭)
metallurgic(al) crude 冶金粗料
metallurgic(al) dust 冶金粉尘
metallurgic(al) engineer 冶金工程师
metallurgic(al) engineering 冶金工程
metallurgic(al) equipment 冶金设备
metallurgic(al) extraction 冶金提取(率)
metallurgic(al) feature 冶金特征
metallurgic(al) fuel 冶金燃料
metallurgic(al) fume 冶金烟气
metallurgic(al) fume emission 冶金尘雾排放
metallurgic(al) furnace 冶金炉
metallurgic(al) furnace emission 冶金炉排放物
metallurgic(al) heat treatment 冶金热处理
metallurgic(al) industry 冶金工业
metallurgic(al) instability 冶金不稳定性
metallurgic(al) machinery 冶金机械
metallurgic(al) magnesite ore 冶金菱镁矿矿石
metallurgic(al) microscope 金相显微镜;冶金显微镜
metallurgic(al) phase diagram 金相图
metallurgic(al) plant 冶炼厂;冶金(工)厂
metallurgic(al) process 冶金过程
metallurgic(al) product 冶金产品
metallurgic(al) product analysis 冶金产品分析
metallurgic(al) slag 冶金炉渣;冶金废渣
metallurgic(al) structure 金相组织
metallurgic(al) technology 金属工艺学
metallurgic(al) test 冶金试验
metallurgic(al) thermodynamics 冶金热力学
metallurgic(al) waste 冶金废料
metallurgic(al) waste gases 冶金厂废气
metallurgic(al) wastewater 冶金(工业)废水
metallurgic(al) working of ores 矿石冶炼工作
metallurgic(al) works 冶金工厂
metallurgist 冶金学者;冶金学家
metallurgy 冶金(学);冶金术
metallurgy cost 冶炼成本
metallurgy extraction 冶金提取(率)
metallurgy furnace 冶金炉
metallurgy of alumin(i)um 铝冶金
metallurgy of light(weight)metals 轻金属冶金学
metallurgy of magnesium 镁冶金
metallurgy of nickel 镍冶金
metallurgy plant building 冶炼厂建设
metallurgy product 冶金产品
metallursical goal 选矿指标
metal lustre 金属光泽
metal machining liquid 金属切削液
metal making resistance 耐金属划痕性
metal marking 陶瓷釉刀痕
metal mask 金属掩模
metal masonry wall flashing piece 圬工墙体的金属防雨泛水板;圬工墙体的金属防雨披水板
metal masonry wall tie 圬工墙体的金属连接件;圬工墙体的金属拉杆
metal mast 空心钢桅;铆接板桅
metal master 金属主盘;金属母盘
metal material 金属材料
metal matrix 金属模版;金属基体
metal matrix composite 金属基复合材料
metal media 金属滤板
metal melting works 金属熔炼厂
metal membrane 金属膜
metal mesh 金属(丝)网;金属挂网
metal mesh container 金属网集装箱
metal mesh facing 金属网面层

metal-metal eutectic alloy 金属—金属共晶合金
metal milling machine 金属铣床
metal mine seismic prospecting 金属矿地震勘探
metal mirror 金属镜
metal mixer 金属混合炉
metal model 金属模型
metal-modified 金属改性的
metal-modified oxide 金属改性氧化物
metal modifier 金属修饰剂
metal monocoque 金属壳机身
metal morphology 金属形态学
metal mother 金属母模
metal mo(u)ld 金属模子;金属模型;金属铸模
metal mo(u)ld casting 硬模铸造
metal mo(u)lding 金属线脚;明装的金属电缆管道
metal mo(u)ld panel 金属定型模板
metal-mounted board 金属裱板
metal-mounted paper 裱糊纸
metal nailing plug 受钉金属块;金属插塞;金属钉条
metal naphthenate 环烷酸金属盐
metal negative 金属主盘;金属母盘;金属底片
metal newel(post) 楼梯金属栏杆柱
metal-nonmetal eutectic alloy 金属—非金属共晶合金
metal octagonalring closure 金属八角垫密封
metalodevite 变锌砷铀云母
metalogic 元逻辑
metal-on-glass plate 敷金属玻璃板
metal-organic compound 金属有机化合物
metal-organic 金属有机物(质)
metal-organic complex 金属有机络合物
metal-organic compound 金属有机化合物
metal-organic sulfonate 磺酸金属盐
metal O-ring and nonmetallic O-ring seal 金属—非金属双 O 形环密封
metal ornament 金属饰件
metal oval-ring closure 金属椭圆环密封
metal overlaid 金属贴面
metal oxide 金属氧化物
metal oxide catalyst 金属氧化物催化剂
metal oxide crucible 金属氧化坩埚
metal oxide electrode 金属氧化物电极
metal oxide film 金属氧化膜
metal-oxide-hydroxide mud 金属氧化—氢氧化物泥
metal oxide precipitate 金属氧化物沉淀物
metal oxide semiconductor 金属氧化物半导体
metal oxide semiconductor integrated circuit 金属氧化物半导体集成电路
metal oxide surge arrester without gaps 无间隙金属氧化物避雷器
metal packing 金属填料
metal paint(ing) 金属涂漆
metal pair 金属偶
metal pan 金属底壳;金属盘;金属箱;金属盒;金属模
metal panel 金属镶板;金属护墙板
metal panel ceiling 金属镶板天花板;金属板顶棚;金属板顶板
metal panel(l)ing of body 车体金属墙板
metal panels for concrete form 金属模型板
metal particle 金属微粒
metal particle in urban air 城市空气中的金属微粒
metal partition 金属间壁;金属墙;金属隔断;金属隔板
metal partition base 金属隔墙抹灰骨架
metal parts 金属零件;金属部件
metal path 碎石小路
metal pattern 金属铸模;金属模型
metal pattern shop 金属车间
metal pencil 焊条
metal penetration 金属渗入;机械黏砂
metal phase 金属相
metal phenyl stearates 苯基硬脂酸金属盐
metal photomask 金属光掩模
metal physics 金属物理学
metal picket 金属围篱桩
metal pickling 金属浸酸;金属酸洗
metal pickling plant 金属酸洗厂
metal picture rail 金属挂镜线
metal pieces 金属件
metal pile shoe 金属桩靴
metal pipe staging 金属管脚手台
metal plaque 金属装饰牌
metalplastic crown 金属塑料联合冠

metal plate 金属板
metal plate cover(ing) 金属板罩
metal plate culvert 金属板涵洞
metal plate Erichsen indentation test 金属板杯突试验
metal plate girder 金属板大梁
metal plate guard rail 铁板护栏
metal plate lens 金属片透镜;金属板透镜
metal plate lining 金属板衬里
metal plate lithography 金属平板印刷术
metal plate rigidity 金属板刚度
metal plate sheathed door 包金属板门
metal plate stiffness 金属板劲度
metal plate surfacing 金属板铺面;金属板表面加工;金属板贴面
metal plate suspension roof(ing) 用金属板作悬挂屋顶材料
metal plating 金属电铸;金属电镀
metal plating process wastewater 金属电镀污水;金属电镀过程废水
metal plating wastewater 电镀(工业)废水
metal plug 金属丝刷
metal-ply truck tire 钢丝载重轮胎
metal pointing 制尖
metal poisoning 金属中毒
metal polish 金属抛光剂;金属擦光剂;擦铜水;擦铜剂
metal-polysilicate complex coagulant 金属聚硅酸盐络合物混凝剂
metal-poor brine 贫金属卤水
metal porcelain crown 金属陶瓷联合冠
metal positive 金属正片;金属正版;母版;第一模盘
metal post 金属支撑;金属桩;金属杆;金属柱
metal powder 金属粉
metal powder cutting 金属粉末切割
metal powder grain 金属粉粒
metal powder magnet 粉冶磁铁
metal powder method 金属粉末(喷涂)法
metal powder oxygen cutting 金属粉末氧割
metal powder pigment 金属粉颜料
metal powder press 金属粉末压机
metal powder system 金属粉末喷涂系统
metal power cutting 金属粉切割
metal precipitation 金属沉淀
metal precipitation reaction 金属沉淀反应
metal-precipitator 金属沉淀剂
metal pretreating primer 金属预处理底漆
metal pretreatment 金属预处理
metal primer 金属用底漆
metal priming 刷金属底漆
metal principal post 金属主支撑;金属主桩;金属主杆;金属主柱
metal probe 金属探针
metal processing 金属加工
metal processing and finishing 金属加工和修饰;金属加工和表面处理
metal processing industry 金属加工工业
metal production 金属生产量
metal production plant 冶炼厂
metal products 金属制品
metal profile 金属轮廓;金属断面;金属外形;金属截面
metal property 金属特性
metal protection 金属保护措施;金属保护设备;金属保护装置;金属防护
metal pulverization 金属粉末化
metal-purifying process 金属精炼法
metal purity 金属纯度
metal quality 金属质量
metal raceway 金属电缆管道;金属线槽
metal radial panel 金属采暖板;金属辐射板
metal radial panel heating 金属辐射板采暖
metal radiator shield 金属散热器罩;金属暖气罩
metal rail 铁栏杆;金属栏杆
metal railing 装金属栏杆
metal reactivity 金属反应性
metal recovery 金属回收
metal recovery rate 金属回收率
metal recovery unit 金属回收设备
metal rectifier welding set 整流焊机
metal refinement 金属精炼
metal reflective coating 金属反射涂层
metal reflectivity standard R value 金属反射率标准 R 值

metal-reinforced composition roofing 含金属加强的组合屋面(材料)
metal-reinforced hose 金属加固软管
metal-reinforced ready roofing 金属加强的预制屋面材料
metal reinforcement 金属丝增强
metal remains 金属残渣
metal removal 金属切除量;除金属
metal removal factor 金属切削率
metal removal from edges of plates 由加工板边去除金属
metal removal rate 金属切削率;切削率
metal reservoir 金属液池
metal revolving door 铁栓门;金属转门
metal ribbon source 金属带状光源
metal-rich brine 富金属卤水
metal ridge capping 金属脊帽
metal ring resistance 耐金属划痕性
metal road 碎石路(面)
metal rod 金属杆;金属棒;金属条
metal roller shutter 金属卷升百叶门;金属卷升百叶窗;金属辊式百叶窗
metal rolling grill(e) 金属滚压栅栏;金属滚压网格
metal rolling shutter 金属卷升百叶门;金属卷升百叶窗;金属辊式百叶窗
metal roof cladding 金属屋顶的表面覆盖
metal roof cover(ing) 金属屋顶面;金属屋顶覆盖层
metal-roofed building 金属屋顶建筑物
metal roofing 金属屋面;铺金属屋面
metal roof-light 金属天窗;金属屋顶窗
metal roof sheathing 金属屋面板(条)
metal roof tile 金属板瓦
metal runner 钢制主龙骨;金属主龙骨
metal runout 跑水;漏铁水;漏箱
metal runway 金属(道面)跑道;金属装配式跑道
metal rust 金属锈
metal rust proofing 金属防锈处理
metals 五金
metal sandwich 金属填料夹层结构
metal sandwich wall 金属隔心墙
metal sanitary cove 金属凹(嵌)条
metal sash 金属窗框;金属窗扇
metal sash bar 金属窗芯子;金属窗锁闩
metal sash block 装钢窗用混凝土砌块;金属窗框砌块
metal sash putty 金属窗用腻子;金属窗用油灰
metal sash window 金属框窗
metal sash work 金属窗框工程
metal saw 金属锯;钢锯
metal saw of coarse teeth 粗齿的锯片铣刀
metal saw blade 钢锯片
metal sawing machine 金属锯床
metal scaffolding 金属脚手架
metal scrap 金属屑
metal screen 金属网印;金属丝筛网;金属纱窗;金属屏;金属滤网
metal screw 金属螺钉
metal sealing strips 金属止水片
metal section 金属剖面;金属截面
metal separator 金属分离器
metal shape 金属造型;金属型材;金属毛坯
metal shaping equipment 金属轧制设备
metal shears 金属剪
metal-sheathed 金属包皮的
metal-sheathed cable 金属包皮电缆
metal-sheathed mineral-insulated cable 金属皮绝缘电缆
metal sheathing 金属覆盖层;金属包皮;金属铠装;金属护套
metal sheathing tendon 钢筋束金属包皮
metal sheet 铁皮;金属片;金属板
metal sheet caisson 金属板潜水钟;金属板沉箱
metal sheet cassette 金属板盒;金属板箱
metal sheet cladding 金属外挂板
metal sheet coffer 金属隔声板;金属天花板镶板
metal sheet cover(ing) 装金属板盖;装金属板套;装金属板壳;金属板罩
metal sheeting 金属薄板;金属鱼尾板;金属挡板
metal sheet panel 金属片镶板;金属片护墙板
metal sheet pipe 钢板桩
metal sheet roof cladding 金属屋顶的表面覆盖
metal sheet roof cover(ing) 金属板屋顶面;金属板屋顶覆盖层
metal sheet roof(ing) 铺金属片屋面材料

metal sheet roof sheathing 金属片屋面板(条)
metal sheet waffle 金属板蜂窝;金属板方格
metal shell cover 端盖
metal-shielded wire 金属屏蔽线;金属包覆线
metal shingles 金属屋面板;金属屋顶板;金属技工;金属鱼鳞板
metal shirting 金属踢脚板
metal shop 金属车间;金工车间;船体车间
metal-shrinkage allowance 金属收缩留量
metal shutter 卷帘式铁门;金属百叶窗;金属鱼尾板
metal shuttering 金属模壳;装卷帘式铁门;装金属百叶窗;金属模板
metal siding 金属挡板;金属壁板;金属墙板
metal silencer 金属消声器
metal-silicate mud 金属硅酸盐泥
metal silo 金属竖井;金属仓库;金属地窖
metal sink 金属容污器;金属洗涤盆
metal site hut 工地金属棚屋
metal skin 金属皮
metal skylight 金属天窗
metal slat 金属板条
metal sleeve 金属套筒;金属套管
metal-slitting saw 金属开槽锯;切缝铣刀
metal smelting 金属冶炼
metal-smelting and refining 金属冶炼和精制
metal-smelting furnace 金属熔炼炉
metal soap 金属皂
metal solubility 金属溶性
metal sorption 金属吸附
metalsorter 金属鉴别仪
metal spacer 金属隔片
metal spinning 金属旋压
metal spiral stair(case) 金属螺旋楼梯
metal splasher 金属泼散器
metal splash flat bar 散水条
metal-sprayed 金属喷涂过的
metal spray gun 金属喷枪
metal spraying 金属喷涂;金属熔融喷涂;金属喷镀;喷镀法
metal spraying system 金属粉末喷涂系统
metal spray plating 金属喷涂
metal spring isolator 弹簧隔振器
metal stair(case) 金属楼梯
metal stake 金属竖柱;金属竖管;金属杆
metalster 金属膜电阻
metal strap 扁钢;金属带(条)
metal strength of the electrolyte 电解质溶液中的金属浓度
metal strip 金属带;金属(板)条
metal strip for expansion joint 伸缩缝中的(止水)金属片
metal strip stiffener 金属带加强件
metal structural cladding 金属围护结构;金属外挂板
metal structural material 金属结构材料
metal structure 金属结构
metal structure plant 金属结构厂
metal structure works 金属结构厂
metal strut 金属支柱;金属压杆
metal stud 金属墙柱;金属螺栓;金属壁骨;金属龙骨;金属立筋
metal stud curtain wall 金属立筋隔板墙;金属龙骨幕墙
metal stud partition 金属立筋隔墙;金属龙骨隔墙
metal stud partition wall 金属立筋分隔墙
metal substitution refining process 金属替换精炼法
metal-sulfide mud 金属硫化物泥
metal sulphide concentrates 精选金属硫化物
metal support 金属支承;金属支架
metal surface 碎石路面
metal-sur faced board 金属包面板;金属贴面板
metal-surfaced door 金属包面门
metal-surfaced window 金属包面窗
metal surface precipitation 金属表层沉积
metal surface treatment 金属表面处理
metal surface treatment waste 金属表面处理废料
metal surfacing 铺碎石路面;金属贴面
metal surround 金属围栏
metal-tank mercury-arc rectifier 金属槽汞弧整流器
metal tape 钢卷尺
metal temperature determination 金属温度确定
metal template 金属样板;金属模片
metal tender 熔制工;熔化工

metal thermometer 金属温度计
metal threshold 金属门槛
metal thru-wall flashing 金属穿墙披水板
metal tie 钢系杆;金属系杆;金属系杆;钢枕
metal tie rod 金属系杆
metal tile 金属砖片;金属瓦
metal timber connector 金属木材连接器
metal-tired vehicle 铁轮车辆
metal-to-glass bond 金属与玻璃黏结
metal-to-metal bond 金属和金属连接
metal-to-metal fit 无过盈与间隙配合
metal-to-metal seal(ing) 金属与金属密封
metal-to-metal water stop 金属与金属间的水密封
metal-to-metal wheel 有金属轮圈的轮子
metal toner 金属盐调色剂
metal tongue depressor 金属压舌板
metal transfer 金属过渡
metal transfer characteristics 金属过渡特性
metal transfer rate 金属过渡率
metal trap 杂铁捕集
metal tray 金属托盘;金属碟
metal trim 金属线脚;金属贴脸;金属镶边;金属饰;金属包角;金属门框
metal-tube rotameter 金属管转子流量计
metal tube staging 金属管脚手台
metal tungsten powder 金属钨粉
metal turning 金属车削
metal tying 金属捆绑;金属接结
metal type composition 机器排版
metal-tyred vehicle 铁轮车辆
metal unit 金属制的设备
metal upper/lower 上下轴承
metal valley 金属檐沟;金属天沟;金属沟槽;铅皮天沟
metal vapo(u)r lamp 金属蒸气放电管;金属蒸气灯
metal vapo(u)r vacuum arc 金属蒸气真空弧
metal waffle 金属蜂窝;金属方格
metal waffle type ceiling 金属方格形天花板
metal wall 金属内壁;金属(围)墙
metal wall panel 金属护墙板;金属墙板条;金属墙镶条;金属墙镶板
metal wall plug 金属墙上插头;金属墙栓;金属墙上灯座
metal wall tie 系墙铁;金属墙(腰)箍;墙面金属连接件
metal waste 金属废物
metal-water-jet pump 金属喷水抽气泵
metal waterstop 金属水密封;金属止水器
metal welding 金属焊接
metal window 钢窗;金属窗
metal window frame 钢窗框;金属窗框
metal window sash 钢窗扇;金属窗扇
metal window shutter 金属百叶窗
metal window sill 金属窗台板;金属窗盘
metal wire suture 金属缝线
metal wood 金属胶合板
metal wool 铁回丝;钢丝绒;金属绒;金属纤维
metal work 金属制造;金属制品;金工;金属(构)件;金工车间
metal working 金属加工;金工
metal working compound 金属加工混合料
metal working fluid 金属加工液
metal working industry 金属加工工业
metal working machinery 金属加工机械;金工机械
metal working shop 金工车间;机修车间;钳工车间
metal working tool 金属加工工具
metal workshop 金属加工间;金工车间
metal wrapping 金属包装;金属包封
metal yield 金属成品率
metal zinc pigment 金属锌颜料
metal zinc reduction 金属锌还原法
metamagnetic transformation 变磁性转变
metamagnetism 变磁性
metamarketing 曲线推销
metamathematics 元数学;超数学
metamelaphyre 变暗玢岩
metamer 位变异构体;条件等色光;条件等色
metameric 位变异构的
metameric colo(u)r 异谱同色;条件等色
metameric colo(u)r match 条件配色
metameric colo(u)r stimuli 同色异谱刺激
metameric match 条件等色
metameric objects 条件等色物体
metameric state 位变异构态

metameride 位变异构体;位变异构态
metamerism 光学条件配色;位变异构现象;位变异构;条件等色;分节现象;分节
metamerism index 条件等色指数
metamers 不同光谱能量分布的同色光
metamic 铬氧化铝金属陶瓷;铬铝陶瓷金属
metamict 混胶状;变生非晶质
metamict mineral 变生矿物;变晶矿物
metamorphic 改变结构的;变质(作用)的
metamorphic age 变质年龄
metamorphic amylum production waste-water 改性淀粉生产废水
metamorphic assemblage 变质杂岩;变质矿物组合
metamorphic aureole 接触变质带;变质晕;变质圈
metamorphic basal conglomerate 变质底砾岩
metamorphic belt 变质带
metamorphic belt of coal 煤变质带
metamorphic breccia 变质角砾岩
metamorphic carbonate formation 变质碳酸盐建造
metamorphic complex 变质杂岩(体)
metamorphic concrete 变性混凝土
metamorphic core complex 变质核杂岩体
metamorphic crystalline rock 变核岩
metamorphic crystallization 变质结晶作用
metamorphic cycle 变质旋回
metamorphic-deformational event 变质—变形事件
metamorphic degassing 变质脱气作用
metamorphic deposit 变质矿床
metamorphic differentiation 变质分异(作用)
metamorphic differentiation way 变质分异作用方式
metamorphic diffusion 变质扩散作用
metamorphic discontinuity 变质结构面
metamorphic event 变质事件
metamorphic facies 变质相
metamorphic facies group 变质相组
metamorphic facies series 变质相系(列)
metamorphic fluid 变质流体
metamorphic formation 变质建造
metamorphic formation of geosynclinal type 地槽型成矿建造
metamorphic geology 变质地质学
metamorphic glutenite formation 变质砂砾岩建造
metamorphic grade 变质级(别);变质程度
metamorphic gradient 变质梯度
metamorphic granite 变质花岗石
metamorphic granite gold-bearing formation 变质花岗岩含金建造
metamorphic greywacke-metamorphic basite formation of glaucophane type 蓝闪石型变硬砂岩—变基性岩建造
metamorphic heat source 变质热源
metamorphic homogenization 变质均匀化
metamorphic-hydrothermal solution 变质热液
metamorphic intermediate-acid porphyritic copper bearing formation 变质中酸性斑岩含铜建造
metamorphic iron-bearing formation 变质含铁建造
metamorphic limestone 变质灰岩
metamorphic map 变质岩地图
metamorphic mechanism 变质机理
metamorphic metallization 变质成矿作用
metamorphic mineral 变质矿物
metamorphic mineral association 变质矿物组合
metamorphic ophiolitic suite 变蛇绿岩套
metamorphic ore deposit 变质矿床
metamorphic overprint 变质叠加
metamorphic petrology 变质岩石学
metamorphic phase 变质相;变质幕
metamorphic phase boundary 变晶相界
metamorphic plot 变质土地块
metamorphic polarity 变质极性
metamorphic pseudo conglomerate 变质假砾岩
metamorphic rank 变质级别;变质程度
metamorphic rank of coal 煤变质程度
metamorphic reaction 变质反应
metamorphic recrystallization 变质再结晶
metamorphic reservoir 变质岩储
metamorphic rock 变质岩;变成岩
metamorphic rock formation 变质岩建造
metamorphic rock gas 变质岩气
metamorphic rock reservoir 变质岩储集层

metamorphic rock type 变质岩类型
metamorphic schist 变质片岩
metamorphic sediment 变质沉积物
metamorphic sequence 变质顺序
metamorphic soil 变质土
metamorphic stage of coal 煤变质阶段
metamorphic structure 变质构造
metamorphic subfacies 变质亚相
metamorphic terrain 变质地层
metamorphic terrane 变质地体
metamorphic texture 变质结构
metamorphic type 变质类型
metamorphic type and genesis of metamorphism 变质作用的类型和变质
metamorphic water 变质水
metamorphic weathered paleo-crust 变质古风化壳
metamorphic zone 反应带;变质带
metamorphic zoning 变质分带
metamorphide 变质褶皱带
metamorphism 岩石变质作用;变质作用;变质;变形现象
metamorphism factor 变质作用因素
metamorphism feature 变质作用特征
metamorphism markers 变质作用标志
metamorphism migration 变质作用迁移
metamorphism of coal 煤变质作用
metamorphism of natural water 天然水变质
metamorphism polarity 变质作用极性
metamorphism zone 变质带
metamorphite 变质岩
metamorphogenic condition 变质成因条件
metamorphogenic deposit 变质成矿床
metamorphosed argillaceous limestone 变质含土石灰石
metamorphosed basement 变质基底
metamorphosed granite 变质花岗岩
metamorphosed limestone 变质石灰岩
metamorphosed (ore) deposit 受变质矿床
metamorphosed rock 变质岩
metamorphosed schist 受变质片岩
metamorphosed sedimentary ore deposit 沉积变质矿床
metamorphosed stratiform copper deposit 变质岩层状铜面;变质岩层状铜矿床
metamorphosis 变质;变态
metamorphosis of value 价值转化
metamorphous 变质的
metaneutrophil 中性异染的
metanic 暗色的
metanil yellow 酸性间胺黄
meta-nitroaniline 间—硝基苯胺
metanovacekite 变镁砷铀云母
meta-orientation 间位定向
meta-orkose 变质长石砂岩
metapegmatite 变伟晶岩
metapelite 变泥质岩
metaperidotite 变橄榄岩
metaperiodic acid 偏高碘酸
metaphase 分裂中期
metaphor 隐喻;比喻
metaphyllite 变千枚岩
metaphysical painting 无形抽象画
Metaphyta 后生植物
metapicrite 变苦橄岩
meta-platform 准地台【地】
metapole 等角点
meta-porphyropsins 变视赤质
meta-position 间位
metaprogram (me) 亚程序
metaquartzite 变(质)石英岩
metaranquilite 准水硅钙铀石
metarhodopsin 变视紫红;变视玫红质
metarhyolite 准流纹岩;变流纹岩
metaripple 不对称波痕
metarossite 准水钒钙石;变水钒钙石
metasandbergerite 准钡砷铀云母
metasandstone 变质砂岩
metasanidine 准透长石
metascarbroite 准碳酸铅矿
metaschoderite 准水磷钡铝石;变水磷钒铝石
metaschoepite 准柱铀矿;变柱铀矿
metascope 红外(线)指示器;红外(线)显示器;红外(线)望远镜;红外(线)探测器
metasediment 变质沉积物;变沉积岩

metasedimentary rock 变质沉积岩
metaseds 变质沉积岩
metasequoia 水杉
metashale 变质页岩
metasideronatrite 准纤纳铁矾;变纤钠铁矾
metasilicate 硅酸盐;偏硅酸盐类;偏硅盐;偏硅酸岩
metasilicic acid 偏硅酸
meta-sociology 超验社会学
metasomatic alteration 交代蚀变作用
metasomatic anti-perthitic texture 交代反条纹(长石)结构
metasomatic anti-zonal texture 交代反环带结构
metasomatic corrosion texture 交代蚕蚀结构
metasomatic edulcoration-border texture 交代净边结构
metasomatic lead-zinc deposit carbonate rocks 碳酸盐岩交代铅锌矿床
metasomatic metamorphism 交代变质作用
metasomatic mineral deposit 交代矿床
metasomatic mosaic-border texture 交代镶边结构
metasomatic myrmekic texture 交代蠕英结构
metasomatic ore deposit 交代矿床
metasomatic perforated texture 交代穿孔结构
metasomatic perthitic texture 交代条纹结构;交代条纹长石结构
metasomatic porphyritic texture 交代斑状结构
metasomatic pseudo-morph texture 交代假象结构
metasomatic relict texture 交代残余结构【地】;交代残留结构【地】
metasomatic ring texture 交代环状结构
metasomatic rock 交代岩
metasomatic texture 交代结构
metasomatic vein 交代脉
metasomatic zonal texture 交代环带结构
metasomatism 交代变质作用
metasomatism way 交代作用方式
metasomatite 交代岩
metasome 交代矿物;脉体;出溶矿物
metaspilite 变细碧岩
metastability 亚稳度;亚稳定性
metastable 亚稳(定)的;准稳定的;介稳度;似稳态的
metastable decomposition 亚稳分解
metastable defect 亚稳缺陷
metastable diagram 亚稳定平衡图
metastable dolomitic calcite 亚稳态白云方解石
metastable energy level 亚稳态能级
metastable equation 亚稳态方程
metastable equilibrium 亚稳(定)平衡;暂时稳定平衡
metastable growth 亚稳相生长
metastable helium magnetometer 亚稳氦磁强计;准稳态氦气式磁力仪
metastable intermediate 亚稳中间体
metastable level 亚稳电平
metastable limit 亚稳极限
metastable memory element 亚稳态存储元件
metastable mineral 准稳(定)矿物
metastable nucleus 亚稳核
metastable peak 亚稳态峰;亚稳峰
metastable phase 亚稳相
metastable phase diagram 亚稳相图
metastable population 亚稳态粒子数
metastable range 亚稳区
metastable region 亚稳区;亚稳境;介稳区
metastable soil 亚稳定土
metastable solution 亚稳溶液
metastable state 亚稳(状)态;准稳定状态;介稳状态
metastable-state jump 亚稳态跃迁
metastable-state transition 亚稳态跃迁
metastable structure 亚稳定结构
metastable supersaturation 亚稳过饱和
metastable transition 亚稳过渡
metastable type 亚稳定型
metastasis 移位变化;同ід蜕变;同质衰变;地壳侧向移动
metastasy 水平均衡调查;地壳侧向移动;地壳侧向均衡调整;侧向均衡调整
metastate 亚态
metastatic abscess 流注
metastibnite 胶辉锑矿
metastrengite 准红磷铁矿
metastructure 亚结构;介结构;次显微组织
metastudtite 变水丝铀矿

meta-substitution 间位取代作用
metasulfite 焦亚硫酸盐
metasymbol 元符号
metasynchronism 亚同步
metatarsal guard 脚面罩
metatect 变熔体
metatectite 交代深成岩
metatexis 带状混合作用;分异深熔作用
metatexis way 分异深熔作用方式
metathenardite 六方无水芒硝
metatheorem 元定理
metatheory 元理论
metathesis 置换作用;复分解
metathetical reaction 复分解反应
metathetical salts 复分解盐
metatholeiite 变拉斑玄武岩
metathomsonite 准杆沸石
metatitanic acid 钛酸
meta-toluidine 间甲苯胺
metatome 齿饰中两个齿间的空间
metatorbernite 准铜轴云母;变铜铀云母
metatriplite 准磷铁锰矿
metatrophic 嗜有机质的
metatrophic bacteria 腐生菌
metatrophy 腐生营养
metatuberculosis 变型结核
metatuff 变凝灰岩
metatyyuyamunite 变钒钙铀矿
meta-uranocircite 变钡铀云母
metauranopilite 变铀矾
metauranospinite 变钙砷铀云母
metavanadic acid 偏钒酸
metavandendrisscheite 变橙黄铀矿
meta-vanmeersscheite 变万磷铀矿
metavanuralite 变钒铝铀矿
metavariscite 准磷铝石
metavauxite 变蓝磷铝铁矿
metavivianite 三斜磷铁矿
metavolcanic rock 变质火山岩
metavolcanics 变质火山岩
metavolcanite 变(质)火山岩
metavolcanite formation of laumonite type 浊沸石型变火山岩建造
metavoltine 变绿钾铁矾;变钾铁矾
metaxite 硬纤蛇纹石
meta-xylene 间二甲苯
metayage system 分成制
metazellerite 变碳钙铀矿
metazeunerite 变碲砷铜铀云母;变翠砷铜铀矿
Metcolising 铸铁喷铝法
mete 界石;边界
meteor 陨星;流星
meteor crate 陨石坑
meteoric dust 流星尘
meteoric iron 陨铁
meteoric material 流星物质
meteoric matter 陨石物质;流星物质
meteoric minerals 陨石矿物
meteoric origin water 大气成因水
meteoric seism 陨致地震
meteoric shower 陨石雨;流星雨
meteoric spring 大气降水泉
meteoric stone 陨石
meteoric stream 陨石群
meteoric swarm 陨石群
meteoric water 雨水;气象水;天落水;大气水
meteoric-water hydro-dynamism 雨水流体动力作用
meteoric-water isotope composition 大气降水同位素成分
meteorite 陨石
meteorite ages 陨石年龄
meteorite crater 陨星坑
meteorite-impact structure 陨石撞击构造
meteorite isotope composition 陨石同位素成分
meteorite shower 陨石雨
meteoritic abundance 陨石中
meteoritic crater lake 陨石坑湖
meteoritics 陨石学
meteoritic sample 陨石样品
meteor-nature satellite 流星—自然卫星
meteorogram 气象图解;气象(记录)图
meteorograph 气象自记仪;气象记录仪;气象记录器;气象计

meteoroid 陨星群;流星体
meteorolite 陨石
meteorologic(al) acoustics 气象声学
meteorologic(al) aids service 气象辅助业务
meteorologic(al) aircraft 气象观测飞机
meteorologic(al) analysis 气象分析
meteorologic(al) anomaly 气象异常
meteorologic(al) automatic reporting station 气象自动报告站
meteorologic(al) balloon 气象探测气球
meteorologic(al) broadcast 气象广播
meteorologic(al) bulletin 气象公报
meteorologic(al) buoy 气象浮标
meteorologic(al) bureau 气象局
meteorologic(al) calamity 气象灾害
meteorologic(al) calculus 气象计算
meteorologic(al) chart 气象要素图;气象图
meteorologic(al) codes 气象电码
meteorologic(al) condition 气象状态;气象条件
meteorologic(al) constituent 气象分潮
meteorologic(al) correction 气象改正数;气象改正量
meteorologic(al) data 气象资料;气象数据
meteorologic(al) data center 气象资料中心
meteorologic(al) data series 气象资料系列
meteorologic(al) datum plane 气象基准平面
meteorologic(al) depression 低气压
meteorologic(al) depression area 低气压区
meteorologic(al) disaster 气象灾害
meteorologic(al) dispersion 气象色散
meteorologic(al) district 气象区(域)
meteorologic(al) diversity scenery 气象景观
meteorologic(al) dynamics 气象动力学
meteorologic(al) earthquake prediction 气象方法预报地震
meteorologic(al) effect 气象影响;气象效应
meteorologic(al) element 气象要素
meteorologic(al) elements chart 气象要素图
meteorologic(al) experiment 气象试验
meteorologic(al) extreme 气象的极端情况
meteorologic(al) factor 气象因子;气象因素
meteorologic(al) flag 气象旗号
meteorologic(al) humidity 气象湿度
meteorologic(al) index of production 气候生产指数
meteorologic(al) influence 气象影响
meteorologic(al) information 气象资料;气象信息;气象情报
meteorologic(al) information network and display 气象情报接收网;天气情报网与显示
meteorologic(al) instrument 气象仪(器)
meteorologic(al) instrumentation 气象仪器设备;气象测量仪表
meteorologic(al) laser radar 激光气象雷达
meteorologic(al) limit 极端气象条件
meteorologic(al) log book 气象日记簿
meteorologic(al) map 气象(要素)图
meteorologic(al) message 气象通报;气象电报
meteorologic(al) minimum 气象极小值
meteorologic(al) model 气象模型
meteorologic(al) multi-instrument 综合气象仪器
meteorologic(al) network 气象台站网;气象监测网
meteorologic(al) noise 气象噪声
meteorologic(al) observation 气象观测
meteorologic(al) observation network 气象观测站网
meteorologic(al) observation ship 气象观测船
meteorologic(al) observation station 气象观测站
meteorologic(al) observatory 气象台;气象观测台
meteorologic(al) observer 气象观测员
meteorologic(al) office 气象站
meteorologic(al) optic(al) range 气象光学视距;气象光学距离
meteorologic(al) optics 气象光学
meteorologic(al) parameter 气象参数
meteorologic(al) phenomenon 气象现象;气象
meteorologic(al) post 气象哨
meteorologic(al) proverbs 气象谚语
meteorologic(al) radar 气象雷达
meteorologic(al) radar station 气象雷达站
meteorologic(al) radio telegram 气象无线电报
meteorologic(al) reconnaissance flight 气象侦察飞行
meteorologic(al) record 气象记录

meteorologic(al) region 气象区(域)
meteorologic(al) report 天气预报;气象报告
meteorologic(al) representative error 气象代表性错误
meteorologic(al) rocket 气象探测火箭;气象火箭
meteorologic(al) rocket network 气象火箭网
meteorologic(al) route 气象航线
meteorologic(al) routing 气象选线
meteorologic(al) satellite 气象卫星
meteorologic(al) satellite image 气象卫星图像
meteorologic(al) service 气象业务;气象服务
meteorologic(al) service centre-east 东风气象中心
meteorologic(al) service station 气象服务站
meteorologic(al) ship 气象船
meteorologic(al) sign 气象符号
meteorologic(al) signal 气象信号
meteorologic(al) simulation 气象模拟
meteorologic(al) situation 天气形势;天气情况
meteorologic(al) society 气象学会
meteorologic(al) sounding 气象探测
meteorologic(al) standard range 气象标准视距
meteorologic(al) station 气象站;气象观测站;测候站;测候所
meteorologic(al) storm 气象风暴
meteorologic(al) summary 气象报告
meteorologic(al) survey 气象测量
meteorologic(al) symbol 气象符号
meteorologic(al) table 气象表
meteorologic(al) telecommunication network 气象电传通信网
meteorologic(al) telegraph 气象电报
meteorologic(al) teleprinter network 气象电传打字通信网
meteorologic(al) tide 气象潮汐;气象潮(气压变化所引起的潮升变化)
meteorologic(al) tide sheet 气象潮汐表
meteorologic(al) tide table 气象潮汐表
meteorologic(al) tower 气象塔
meteorologic(al) tower layer 气象塔层
meteorologic(al) transmission 气象传送
meteorologic(al) visibility 大气能见度
meteorologic(al) visual range 气象视距;气象视程
meteorologic(al) visual range table 气象视程表
meteorologic(al) warning 气象警告;气象警报
meteorologic(al) watch office 气象监视台
meteorologic(al) water 大气水;气象水
meteorologic(al) wind tunnel 气象风洞
meteorologic(al) working chart 气象(分析)作业图
meteorologic(al) yearbook 气象年鉴
meteorologist 气象学家;气象工作者
meteorology 气象
meteorology aerology 气象学
meteorology and hydrologic(al) data 气象水文数据
meteoropathy 气候病
meteorotropism 气候影响性反应;气候向性
meteosat 气象卫星
meteosat dual-channel radiometer 气象卫星双信道辐射计
meteotron 造云器
meter 仪表;公尺;量度;计器;测量计;表(头)(指仪表)
meter adjusting device 电度表的调整装置
meterage 量表使用费;计量;测定
meterage per drill working-month 钻月效率;钻月进尺
meterage per rig-hour 台时效率【岩】
meterage per rig-month 台月效率【岩】;台月进尺
meterage per rig-shift 台班效率【岩】;台班进尺
meterage per rig-year 台年进尺【岩】
meter-ampere 米安(培)
meter arrangement 仪器安排;仪器布置
meter base 电度表底座
meter board 仪器表
meter box 电表箱;水表箱
meter box ring and cover 水表盒圈和盖
meter branch 仪表支线
meter bridge 滑线电桥
meter cabinet 仪表室;计数器室
meter calibration 仪表标定;计量器校准
meter calibration tank 校准槽
meter-candle 米·烛光;米烛
meter-candle-second 米烛秒
meter coil 表头线圈

meter comparator 米制比长仪
meter compartment 仪表柜;仪表箱;仪表间;仪表室
meter component 末波分量
meter constant 仪表常数;校正常数;计数器常数
meter control 表阀
meter correcting factor 仪表校正因数
meter cover 电度表盖
meter diagram 仪表图
meter dial 仪器刻度盘;仪表(刻)度盘;计量器标度盘;表头刻度
meter display 仪表指示
metered billing 按表计价
metered concrete 计量混凝土;定量混凝土
metered connection 有计量连接
metered consumption 计量消耗量;表计(用)水量
metered error 计量误差
metered fare system 计程票制
metered flow 计量流量
metered flow of oil 配量油流
metered gate 计量漏斗门
metered-leak method 计量器计量漏率法
metered lubrication 计量润滑
metered orifice 计量孔
metered service 装水表用户
metered station 计量站
metered system 计量系统;计量装置
metered valve 量阀
metered water use 表计用水量
meter enclosure 仪表罩;仪表箱;仪表盒子;仪表封装;仪表包装
meter error 仪表误差
meter factor 仪表校正因数
meter feeder 饲料计量分送器
meter flume 测流槽
meter for liquid 液体仪表
meter frame 电度表基架
meter full-scale 最大量程
meter gate 量水门;斗门
meter ga(u)ge 量水仪;米轨铁路;米尺
meter ga(u)ge line 米轨线路
meter ga(u)ge railway 米轨铁路
meter ga(u)ge track 米轨线路
meter gear train 水表齿轮系
meter glass 刻度烧杯
meter hand 仪表指针
meter horsepower 米制马力
meter housing 仪表壳
meter in 进油节流
meter-in circuit 入口节流式回路
metering 量测;记录;计量;测量
metering auger 计量螺旋;配量螺旋
metering band 计量卷尺
metering block 配量阀组
metering chamber 按量分配室
metering characteristics of nozzle 喷管流量特性
metering circuit 仪表电路;测量电路
metering contact 测量触点
metering control 定量调节器
metering device 量斗;量具;量配装置;量器;计量装置;计量(仪)器;计量仪表;配量装置;定量给料器;测量装置
metering equipment 量度仪器;测试仪器;测量装置;测量仪表
metering error 测量误差
metering flume 量水槽
metering function 限流作用
metering gallery 量测廊道
metering hole 限流孔;计量孔;定径(校准)孔
metering hopper 计量料仓
metering-in control system 进口控制系统
metering installation 计量装置
metering jet 限油喷嘴;限流嘴;测油孔
metering jet plug 测油孔塞
metering line 计测控制线
metering needle 调节针
metering nipple 测量螺纹接管
metering nozzle 限流喷管;计量管嘴;定流油嘴;定径喷嘴;测流管嘴
metering of heat supply system 热计量
metering orifice 测流口
metering-out control system 出线控制系统;出口节流调速式
metering performance 计量性能
metering pin 控流阀;阀门活塞;定径销

metering pin valve 调节针阀
metering plug 计量塞;计量孔塞
metering point 量测点
metering practice 流量测定工作
metering profile 量测断面
metering pulse 计费脉冲
metering pump 测量泵;限量泵;计量泵
metering rate 限流率
metering relay 计数级继电器
metering rod 油尺;液位量杆
metering screw 计量螺旋绞刀;计量螺旋
metering section 量测断面;计量剖面;计量段;测流断面
metering signal 限流信号
metering station 原油计量站
metering switch board 仪表盘
metering system 记录制;限流系统;量测装置;计量装置;计量系统;测试装置;测量系统
metering tape 带尺;卷尺
metering valve 计量阀
metering wheel 草捆推长度控制轮
meter installation 成套煤气表
meter key 滑线电键;计数器电键
meter kilogram 米·千克
meter kilogram second ampere system 米·千克·秒·安制
meter kilogram second system 米·千克·秒制;公制
meter ladle 定量浇包
meter lamp 仪表指示灯;记录器指示灯
meter leather 仪表皮革(无漏孔性)
meter loss 仪表显示损耗;表头损耗
meterman 仪表调整者
meter mercury column 米汞柱(压强单位)
meter movement 仪表的测量机构
meter multiplier 仪表量程倍增器;测量仪表扩程器
meter needle 仪表指针
meter niche 仪表储藏室;仪表壁龛
meter noise 仪表噪声
meter number 仪表编号
meter oil 仪表油;计量器用油
meter out 回油节流
meter-out circuit 出口节流式回路
meter-out system 出口节流系统
meter-panel 仪表屏;仪表板;仪表盘
meter parking 计时停车
meterpercent 米百分值
meter pit 藏仪表的地下温室;仪表地窖;水表坑
meter point method 计点法
meter protection circuit 仪表用保护电路;表头保护电路
meter prover 校表气罐
meterproving tank 校准箱
meter-pump 计数泵
meter rack 仪表架
meter range 表头量程
meter rate 水表收费率;电表计费率
meter reading 仪表读数;计数器读数;表读数
meter regulator 计量器
meter relay 记录继电器
meter repair shop 水表修理工场
meter rod 高精度水准尺;精密水准尺;米制水平标尺;(流速仪的)吊杆;测深竿;测深杆
meter room 仪表间
meter rule 米尺
meter run 仪表长度
meters and valves 计和阀
meter scale 表头刻度仪表标度;表头刻度
meter scale switch 量程开关
meter screw 公制螺纹;公制螺钉
meter seal 煤气表铅封
meter/second 米/秒(速度单位)
meter sense 表头读数
meter sensitivity 仪表灵敏度;计算灵敏度
meter series 公制系列
meter set 成套煤气表;仪器调节
meters feet conversion 米英尺换算
meter sill 人字门槛
meter slip 计数偏慢
meter spanner for force moment 表式测力矩扳手
meters per second 每秒公尺数
meter square 平方米
meter station 仪表站
meter stick 米尺

meter stop 记录中止;表阀
meter suspension cable 流速仪吊索
meter switch 仪表转换开关;仪表开关
meter system 记录制
meter testing bench 计量器测试装置
meter testing machine 仪表测试机械
meter-ton-second system 米·吨·秒单位制
meter-ton-second units 米·吨·秒单位制
meter transformer 仪表用变压器;电表用互感器
meter-type 米字型
meter-type relay 仪表型继电器;仪表式继电器;电流式继电器
meter unit 表头部件
meter voltage 计器电压
meter water column 米水柱(压力单位)
meter water ga(u)ge 米制水表;米水柱(压力单位)
meter wave radar 米波雷达
meter wheel 测量轮
meter wire 计数器引线
meter zero adjustment circuit 表头零位调整电路
metes and bounds (丈量的)地块边界;(土地的)四止尺寸;边界;土地边界(线)
metes and bounds survey 地块界线测量法
methacrolein 异丁烯醛
methacrylate 异丁烯酸;聚甲基丙烯酸乙酯;甲基丙酸酯
methacrylate polymer 甲基丙烯酸聚合物
methacrylic acid 异丁烯酸;甲基丙烯酸
methacrylic ester 异丁烯酸酯
methacrylic plastic 甲基丙烯塑料
methacrylonitrile 异丁烯腈;甲基丙烯腈
metham 威百亩
methamidophos 甲胺磷
methanation 甲烷化
methanation of wastes 废物沼气化
methane 沼气;甲烷;瓦斯
methane accumulation 瓦斯积聚
methane bacteria 产甲烷菌
methane blower 瓦斯泄出
methane burning equipment 甲烷燃烧设备
methane concentration 甲烷浓度
methane determination 沼气测定
methanedicarbonic acid 丙二酸
methane dioxide 二氧甲烷
methane drainage 甲烷排放;沼气排放;排放瓦斯
methane drainage pipe 甲烷排放管;沼气排放管
methane emission 沼气泄出
methane fermentation 甲烷发酵
methane fuelled engine 沼气发动机
methane-generating pit 沼气池
methane grid 甲烷栅网;沼气栅网;甲烷格栅;沼气格栅
methane layering 瓦斯分层
methane liberation 沼气放出
methane monitoring system 瓦斯监控系统
methane outburst 沼气喷出
methane phase 甲烷相
methane phone 沼气报警铃
methane pocket 沼气包
methane recorder 瓦斯记录器
methane series 烷系
methane tank 沼气池
methane terminal 甲烷终端站;沼气终端站
methane tester 沼气测定器
methane water 含甲烷水
methanite 密桑奈特炸药
methanization 甲烷化
methanogenic organism 产甲烷生物
methanol 甲醇;木酒精;木醇
methanol poisoning 甲醇中毒
methanol solubility test 甲醇溶解试验
methanol solution 甲醇溶液
methanol wastewater 甲醇废水
methanometer 沼气检定器
methanophone 瓦斯警报器
metharmosis 后期成岩固结作用
methazolamide 醋甲唑胺
methenamine 乌洛托品
methenyl-次甲基
methine and polymethine colo(u)ring matter 次甲基和多次甲基染料
methiocarb 灭梭威
methiodal sodium 碘甲磺钠

methiodide 甲碘化合物
methoamidophos pesticide 甲胺磷农药
methoamidophos pesticide wastewater 甲胺磷农药废水
method 梯温法;方法
method air sampling 空气采样法
method analysis 方法研究
method by hour angle of Polaris 北极星任意时角法
method by intersection of three-directions 三点交会法【测】
method by inversion 逆点法
method by naked eye 目测标定法
method by trial 试算法;试合法
method by volume 材积法
method-dependent 随方法变化的
method detection limit 方法检出限
method engineer 方法工程师
method expert 工艺专家
method for analysis of water quality 水质分析法
method for arbitration 判优法
method for assessment 评定方法
method for cooling 冷却法
method for determination of ammonia in exhaust gas 排气中氨测定法
method for determination of carbon dioxide in exhaust gas 排气中一氧化碳测定法
method for determination of chlorine in exhaust gas 排气中氯测定法
method for determination of hydrogen cyamide in exhaust gas 排气中氰氢酸测定法
method for determination of oxides of nitrogen in exhaust gas 排气中氮氧化物测定法
method for determination of total sulfur oxides and sulphur dioxide in fuel gas 燃料气中硫氧化物和二氧化硫测定法
method for horizontal mapping control survey 图根平面位置测量方法
method for hydrogen ion concentration 氢离子浓度测定法
method for measurement of noise level 噪声级测量法
method for measuring 测定的方法
method for repairs and component replacement 部件修理法
methodical error 方法误差
method improvement 方法的改进
method in optimal control 最优控制法
method library 方法库
method man 工艺员
method of absorption coefficient section 吸收系数剖面法
method of access 存取法;访问法
method of actual stress 有效应力法
method of addition 增量法
method of adjustment 平差(方)法;配合法
method of aeration 充气法;掺气法
method of aerophotogrammetry 航测成图方法
method of age determination 年龄测定方法
method of a given deviation 给定偏差法
method of agreement 契合法
method of airborne fix 航空定位方法
method of air supply 空气补充法
method of alignment 视准线法
method of analysis 分析方法
method of analysis of variance 方差分析法
method of analytic(al) calculation 解析计算法
method of analytic(al) smoothing 解析修匀法
method of anchorage 锚碇(方)法
method of anchoring 支撑方法;锚碇方法;固定方法;锚固方法
method of angle observation in all combinations 全组合测角法
method of antithetic variable 对偶变量法
method of application 使用方法;施用方式;施工方法;申请方法
method of approach 研究方法;逐渐趋近法;逐步逼近法;逼近法
method of approximation 近似法
method of arc measurement 弧线法
method of artificial rainfall experiment 人工降雨试验法
method of assumed pressure 假定压力法
method of attachment 连接方法

method of attack 溶解方法
method of attributes 属性检验法
method of auxiliary intersection point 副交点法
method of auxiliary plane 辅助平面法
method of average 平均法;统计(方)法
method of average ratios to trend 趋势平均比率法
method of averaging observation 平均观察法
method of axial ratio measurement 轴比测量法
method of backward intersection 后方交会法
method of balance element determination 均衡要素确定方法
method of bandage 包扎法
method of batch treatment 分批处理法
method of bearing 格栅座法
method of bearing and distance 方位距离法
method of bearings 坐标方位角法
method of big aperture well 大井法
method of bioassay 生物试验法
method of biologic(al) association 生物组合法
method of bisection 对分法
method of bisectors 二等分线法
method of blasting 爆破方法;爆破法
method of borehole strain measurement 钻孔应变测量法
method of boring 镗孔法
method of bottom sampling 底质取样方法
method of braking 制动方式
method of calculating depreciation 折旧计算方法
method of calculating depreciation charges 折旧费计算方法
method of calculating lighting 照明计算法
method of calculation 计算方法;计算步骤
method of calculation by assumed unit weight 假定容重计算法
method of calculation reserves 储量计算方法
method of calibration 校验方法
method of careening 倾斜法
method of cargo handling 操作方法(指装卸)
method of carrier distillation 载体分馏法
method of cartographic(al) representation 地图表示法
method of center charging 中心装料法
method of center of rotation 旋转中心法
method of central projection 中心投影法
method of chaining by horizontal steps 分段水平丈量法
method of characteristic 特性线法
method of characteristic curves 特征线法
method of charge in well 钻孔充电法
method of checker packing 格子体排列方式
method of chemical treatment 化学处理法
method of chord deflection distance 弦线偏距法
method of circle average 圆周平均法
method of closed contour 闭合等值线法
method of closure 截流方法
method of clustering 聚类分析方法
method of clustering center[centre] 聚合中心法
method of coils with ruled surfaces 可控表面线圈法
method of coincidence 重合读数法;符合法
method of column analogy 柱比法
method of combination 合成法;组合法
method of compacting wagon kilometres at halt of the length of district 区段里程折半计算法
method of compaction 压实方法
method of comparative statics 比较静态方法
method of comparision spectra 比较光谱法
method of compensation 补整法;补偿法
method of computing wagon kilometers actually run 实际里程计算法
method of concreting 混凝土浇筑方法
method of condensation 压缩法
method of conduit tracing test 连通试验方法
method of coning and quartering 堆锥四分法
method of constant deviation 恒偏法
method of construction 建造方法;施工方法
method of continuous beams on elastic supports 弹性支承连续梁法
method of continuous exponential smoothing 连续指数修匀法
method of control 控制方法
method of controlling current capital 流动资金控制法

method of control sample 控制试样法
method of conversion factor 换算因数法
method of correcting landslides 滑坡防治法
method of correction 校正法
method of correlates 联系数法;相关法
method of correlative sediments 相关沉积法
method of corresponding altitudes 等高法
method of cost 成本计算方法
method of cost accounting 成本计算方法
method of course and angle 航向角法
method of course and timing 航向计时法
method of critical levels 临界水平法
method of cross bearings 方位角交会法
method of curing 养护方法
method of curve fitting 曲线拟合法;适线法;适点法
method of curve realignment 曲线整正法
method of data reference 数据引用方法
method of deduction 演绎法;推论(方)法
method of deflection 挠度法
method of deflection angles 偏角法
method of deformation 角变位移法
method of degassing 脱气法
method of deoxidation 脱氧(方)法
method of depreciation 折旧方法
method of descending 递降法
method of description 描述法
method of design 设计方法
method of detached coefficients 分离系数法
method of detection 探伤方法
method of determination 鉴定方法;测定方法
method of determination by Zinger star-pair 津格尔星对测时法
method of determination of the exploratory grid 勘探网度确定方法
method of determination project number 确定工程数量方法
method of determining yield by area 面积法
method of determining yield by volume 材积法
method of development 展开法
method of dewatering with compressed air 压气排水打捞法
method of difference 可变差异法;差异法;差分法
method of diffusion analogy 扩散模拟法;扩散类比(法);扩散比拟(法)
method of diffusion index 扩散指数法
method of digging 开挖方法
method of dilution 稀释法
method of dimensions 维量法;因次法
method of direct combustion decomposition 直接燃烧分解法
method of direct cooling 直接冷却法
method of direct deflections 直接偏转法
method of direction(al) observation 方向观测法
method of direction observation in rounds 全圆方向观测法
method of direct plummet observation 正锤线(观测)法
method of discharge measurement 流量测验方法;测流方法
method of discharging 泄流方法
method of dismantling and lifting in section 解体打捞法
method of dispersion experiment in laboratory 室内弥散试验
method of dispersion test 弥散试验方法
method of dispersion test in field 野外弥散试验
method of distance measurement 边长测定方法
method of distance measurement by phase 相位法测距;相位测距法
method of distance measurement by pulse 脉冲测距法
method of distributing points 布点法
method of distribution 分配法
method of distribution of sewage 布水方式
method of diverless underwater intervention 无潜水员水下装调方法
method of diversion 引水方法;导流方法
method of divided bar 分割棒法
method of division into groups 分组平差法;分组法
method of docking 靠泊方法;入坞方法
method of dogging 锁定方法(闸门)
method of double differences 重差法
method of double refraction 双折射法

method of double sight 复观法;复觇法
method of drag-free satellite 无阻力卫星法
method of drilling 钻孔法
method of dry dosing 干投法
method of dumping 卸料方法;倾倒(方)法
method of earthquake coefficient 地震系数法
method of economic analysis 经济分析方法
method of economic calculation 经济核算方法
method of effective stress 有效应力法
method of elastic center 弹性中心法
method of elastic weights 弹性荷载法
method of electric(al) blast 电爆破法
method of electric(al) prospecting section 自然电场法;电测剖面法
method of electrodialysis dewatering 电渗排水法
method of electrostatic analogy 静电比拟法
method of eliminating block effects 消除区组效应法
method of elimination 消元法;消去法;淘汰法
method of engineering geologic(al) studies 工程地质研究方法
method of entasis 收分法
method of equal coefficient 等系数法
method of equal effect 等效法
method of equal-weight substitution 等权代替法
method of equivalent stiffness 等效刚度法
method of erection 装配方法;架设方法;安装方法
method of estimation 估计法
method of evaluating dynamic(al) load rating 额定动负荷计算方法
method of evapo(u)ration 蒸发法
method of excavation 开挖方法
method of excavation work 掘进方法
method of excess 超量法
method of excessive pressure 余压法
method of exchange of members 构件交换法;构杆交换法;杆件交换法;杆件代替法
method of execution 执行方法;施工方法
method of exhaustion 穷举法
method of experimental value 经验取值法
method of exploration 勘探方法
method of explosion 爆炸方法;爆破(方)法
method of extensional abstraction 外延抽象方法
method of extrapolation 外延法;外推法;外插法
method of fabrication 制作方法;装配方法
method of facing 贴面法;修饰表面法;表面加工方法
method of false position 错位法;插根法;试位法
method of feeding 供料方法;给矿方法;馈电(方)法;喂料方法
method of fictitious loads 虚(拟)荷载法;虚构荷载法;假设荷载法
method of field work of gravity survey 重力测量工作方法
method of financing 筹资方法
method of finite difference 有限差分法;差分法
method of finite element 有限单元法
method of finite increment 有限增量法
method of fire protection 防火方法
method of fixed points 定点(方)法
method of flashing 防雨法;急骤蒸发法;闪光法
method of flow field analysis 流场分析法
method of flow interception from cross-section under ground river 暗河断面截流法
method of fluxions 微积分法
method of forward intersection 前方交会法
method of fractional steps 分步法
method of fraction level(l)ing 分段水准测量法
method of framing 成帧法
method of frost protection 防霜法
method of Gaussian elimination 高斯消去法
method of geochemistry 地球化学法
method of geographic(al) comparison 地理比较法
method of graphic(al) analysis 图解分析法
method of graphic(al) rating scale 图解评价量表法
method of gravity measurement 重力测量法
method of grid 网格法
method of grounding 接地方法
method of grouping 分组法
method of guided wave factor 导波系数法
method of hammer ampoule 锤击安瓶法
method of heat budget 热平衡法
method of heat conductivity measurement 热导率测量方法
method of heating 加热法
method of heavy mineral association 重砂矿物组合法
method of heavy minerals 重砂找矿法
method of high frequency photoconductivity 高频光电导法
method of homologous line pair 均称线对法
method of horizontal control survey 平面控制测量方法
method of hydraulic pressure fracture 水力致裂法;水力压裂法
method of hydrochemistry composition classifications 水化学成分分类法
method of hydrogeologic(al) mapping 水文地质测绘法
method of hydrograph separation 水文分割法
method of identification 证实法;鉴定法
method of image 源像法;映像法;镜像法;反影法
method of image interpretation keys 图像解译样片法
method of impregnation 浸制法(木材防腐)
method of inbred-variety cross 顶交法
method of independent image-pairs 独立像对法
method of independent model 独立模型法
method of index fossil 标准化石法
method of indicating the centerline of turnout 道岔中心表示法
method of indirect elimination 间接消失法
method of induction 归纳法;感应法;推理法
method of ingredient 配矿方法
method of initiation 起爆方法
method of inspection 检验法;视察法
method of installation 安装方法
method of integral water-proofing 总体防水法
method of interior point 内分点法
method of interpolation 插入法;内插(值)法;插值法
method of intersection 相交法;交会法;交叉法;基线法
method of inverse plummet observation 倒垂线法
method of inversion 反演法
method of inverted beam and girder floor 倒楼盖法
method of investigation 勘测方法;调查方法
method of isocline 等倾法
method of isotope 同位素法
method of iteration 迭代法
method of J-integral J 积分法
method of joint e center determination 联合震中测定法
method of jointing 节点法
method of joints 结点法;节点法
method of karyotype analysis 核型分析方法
method of laser alignment 激光准直法
method of lattice 网格法
method of leading marks and angle 导标—角法
method of leading variable 引入变量法
method of least squares 最小平方法;最小二乘法
method of least work 最小功法
method of limit equilibrium 极限平衡法
method of limits 极限法
method of lineament analysis 线性体分析方法
method of linearization 线性化法
method of line intersection 方向线交会法
method of link relatives 链环指数法
method of lithologic(al) comparison 岩性比较法
method of litho-stratigraphy 岩石地层学方法
method of load classification number 荷载分级数法
method of local ventilation 局部通风方式
method of logarithmic additament 对数附加数法
method of log-normal distribution 对数正态分布法
method of long chord 长弦法【测】
method of long probe 长探头法
method of magnetic survey 磁测方法
method of magnetism measurement 磁性测定方法
method of manufacture 制造方法
method of marker bed 标志层法
method of mathematic(al) simulation 数学模拟的方法
method of maximum likelihood 最佳相似法;极大似然法;极大拟然法
method of mean 消防安全措施
method of mean thickness 平均厚度法
method of mean value 均值法
method of measurement 丈量法;量测方法;测量方法
method of measuring dust content in stack gas 烟道气中粉尘含量测量法
method of measuring potential 测电位法
method of metallurgy 冶金方法
method of middle area 中面积法
method of middle ordinates 中距法
method of middling content per cent 中煤含量法
method of mineral comparison 矿物对比法
method of miniature thermal probe 微探针法
method of minimum potential 最小势能原理法
method of minimum squares 最小平方法;最小二乘法
method of mirror 镜像法
method of mixed acid reduction 混合酸还原法
method of mixture 混合法
method of moment distribution 力矩分配法;力矩分布法;弯矩分配法;弯矩分布法
method of moments 矩量法;动差法;力矩法;矩法
method of morphologic(al) survey 地貌调查方法
method of mo(u)lding 模塑法
method of moving arithmetic(al) average 移动算术平均法
method of moving averages 移动式平均法;流动式平均法;滑动式平均法
method of moving geometric(al) average 移动几何平均法
method of mucking 出渣方法
method of multidimensional field 多维场法
method of nodal displacement 节点移位法
method of non-destructive examination 无损检验法
method of non-destructive inspection 无损探伤法;无损检验法
method of nuclear recoil 核反冲法
method of nuclear technology seeking water 核技术找水方法
method of number index 指数法
method of numeric(al) calculation 数值计算法
method of observation 观测方法
method of observation sets 测回法
method of obtaining sample and equipment (collecting method of sample) 取样方法与设备
method of one probe spreading resistance 单探针扩展电阻法
method of opening 开拓方法;开窗方式
method of operation 运转方法;运行方式;作业方法;工作规范;工作方法;经营方法;操作方法
method of opposability 冲日法
method of optic(al) shaft plumbing 竖井光学垂准法
method of optimal control 最优控制法
method of order reducing 降阶法
method of ore-forming temperature determination for gas-liquid inclusion 气液包裹体成矿温度测定法
method of ore prospecting by use of extracting remotely sensed minerals product information 遥感矿产信息提取找矿方法
method of ore prospecting using remotely sensed characteristic spectral information 遥感特征光谱信息找矿方法
method of overhanging lifting 驮拍打捞法
method of overlapping maps 重叠图法
method of pairing arrangement 对比法排列
method of pal(a)eo-bioecology 古生态学法
method of pal(a)eoclimate determination 古气候测定方法
method of pal(a)eomagnetite 古地磁法
method of panel points 节点法
method of parallel sections 平行断面法
method of parallel straight curve 平行直线法
method of path difference of one meter 一米程差法
method of payment 支付方式
method of payment for construction work 建筑工程的付款方法
method of penetration 透孔法
method of percentile 百分位数法

method of persistent graph 持续曲线法;持久曲线法
method of perturbation 扰动法;摄动法
method of photometric(al) interpolation 光度插入法
method of plane mean value 平面平均值法
method of plate constants 底片常数法
method of plate orientation elements 底片方位元素法
method of plotting position 定位法;定点(方)法
method of point rating 评分法
method of polar coordinates 极坐标法
method of polynomial fitting 多项式拟合法
method of population enumeration 群体计数法;群体查点法
method of portrayal 描绘法
method of position line by altitude 天体高度定位线法
method of powder preparation 粉末制备法
method of precipitation 沉淀法
method of prediction 预测方法;预报法
method of presentation 描绘法;显示法;表演法;表示法
method of prevent floods by water control 治水措施
method of price of last purchase 最终购入价格法
method of primary treatment of sewage 一级污水处理方法
method of procedure 程序法
method of producing a new tree 生产新树的方法
method of projection 投影法
method of proportional computation 比例计算法
method of proportioning 配合法
method of purification 净化法;提纯方法
method of pyramid 角锥法
method of quadrature 求面积法;求积分法
method of quality control 质量控制方法
method of quantitative analysis 定量分析(方)法
method of quarrying 采石法
method of quartering 四分取样法;四分法
method of radiation 射线(方)法;射出法;放射法
method of rainfall-runoff 雨量径流法
method of raising in steps 逐步打捞法
method of raising with salvage pontoons 浮筒打捞法
method of random sampling 任意抽样法;随机抽样法
method of reciprocal bearing 对向方位角法
method of rectification in zones 分带纠正法
method of redress 补救法
method of reduction of dimensions 降维法
method of redundant reaction 赘余反力法;冗力法;超静定反力法
method of regional geophysical survey 区域地球物理调查法
method of reiteration 反复迭代法;反复逼近法
method of relaxation 松弛法
method of remote sensing image interpretation 遥感图像解译法
method of repeated smoothing 反复修匀法
method of repetition 复测法
method of representation 表示法
method of resection 后方交会法
method of resection by sextant 六分仪后方交会法
method of residuals 剩余法;残数法;残差法
method of residues 剩余法;扣除法
method of resultant 联解法;合力法
method of resultant stress 总应力法
method of reverse osmosis 反渗透法
method of right plummet observation 正锤线(观测)法
method of rigid plate 刚性板法
method of river diversion 导流方法
method of rock-forming temperature determination for melt inclusion 熔融包裹体成岩温度测定法
method of rock fragmentation 岩石破碎方法
method of rotation 转角法;转动法
method of round 全圆观测法
method of salvaging a stranded vessel with frameworks 构造式打捞法
method of sample enlargement 样本扩大法
method of sample interest 单利计算法
method of sampling 取样方法;采样方法

method of satellite gravity gradient 卫星重力梯度法
method of satellite radio altimetry 卫星天线电测法
method of saturation 饱和(方)法
method of sculpture 浸蚀法;雕刻法;雕塑法
method of secondary grouping 二次分组法
method of sections 计桁架内力的截面法;断面法;截面法;桁架断面分析法
method of sedimentary rhythm 沉积韵律法
method of seismic coefficient 地震系数法
method of seismic prospecting 地震勘探法
method of seismic wave refraction 地震波折射法
method of selected points 选点法
method of separate joint displacement 分离节点位移法;个别节点位移法;可分节点位移法;节点移位法
method of separation metal ion 金属离子分离法
method of separation of variables 变数分离法;分离变量法
method of sequential sorting 序列分类法
method of settlement 解决办法
method of settling accounts 结算方式
method of shielding factor 屏蔽系数法
method of shift 变换法
method of side charging 侧边装料法
method of sieving 筛分法
method of slices 条分法;分片法;分块法;分割法
method of slip circle analysis 圆弧滑动分析法
method of slope deflection 转角变位法;角变法;角变位移法
method of slope deformation 倾度变位法
method of small angle measurement 小角法
method of small parameter 小参数法
method of smelting 熔炼法
method of smoke screen tracing 烟幕示踪法
method of smooth curve 平滑曲线法
method of solving equation 解方程法
method of sounding 测深法
method of sound level measurement 声级测量法
method of specific drawdown within funnel 漏斗单位下降值法
method of specific gravity 比重法;相对密度法
method of specific heat 比热法
method of specific loss pressure 比压损耗法
method of spring attenuation equation 泉流量衰减方程法
method of spring spurt 弹簧喷射法
method of stable isotopic analysis 稳定同位素分析方法
method of stationary phase 驻相法
method of statistic(al) analysis 统计分析(方)法
method of statistic(al) survey 统计调查法
method of steady flow calculation 稳定流计算法
method of steady graph 固定曲线法
method of steady pumping test 稳定流抽水试验法
method of steepest descent 最快下降法;最陡下降法
method of stepwise elastic analysis 逐次弹性分析法
method of storage attenuation 储[贮]存衰减法
method of streamers 飘带法
method of stripping 露天开采法;脱模法;剥落法
method of structural style 构造样式法
method of study 研究方法
method of substitution 置换法;替代法;代入法;代换法
method of successive approximation 渐近法;逐次近似法;逐次渐近法;逐步近似法;逐步渐近法;逐步逼近法
method of successive comparison 逐次比较法;逐步比较法
method of successive conjugates 逐次共轭法
method of successive corrections 逐步校正法;连续校正法
method of successive displacement 逐次变换法;逐次移位法;逐次替换法;逐步置换法
method of successive elimination 逐次消去法
method of successive interval 毗连区间法
method of successive itration 逐次代入法
method of successive substitution 逐次代换法
method of summation 总和法;求和法
method of superposition 叠合法;叠加法;叠置法
method of sweeping out 括去法
method of symmetric(al) coordinates 对称坐标法

method of synthesis 综合法
method of tagged 标记方法
method of taking samples 采样法;取样法
method of tangent(ial) offsets 切线支距法
method of telemetering 遥测法
method of tension wire alignment 引张线法
method of terrain point identification 地面点辨认法
method of test for resistance to penetration 贯入阻力试验法;插入阻力测定法
method of testing 试验法;抽查法
method of testing materials 材料试验方法
method of testing shaft alignment 假轴找中法
method of the truss joints 桁架节点分析法
method of three-direction 三方向法
method of three-moment 三力矩法
method of three-planes 三平面法
method of three-standard sample 三标准试样法
method of three-tripods 三联脚架法
method of tillage 耕作方法
method of tilting screw 倾斜螺旋法
method of timbering 支撑方法
method of time determination by star transit 恒星中天测距法
method of time determination by two stars in equal altitude 双星等高测时法
method of time determination by Zinger star-pair 金尔星对测时法;东西星等高测时法
method of time measurement 时间计量方法
method of time standard 标准工时法
method of total karst cutoff 岩溶截流总合法
method of tracer 示踪剂法
method of tracer dilution in single well 单井示踪剂稀释法
method of tracer in wells 多井示踪剂法
method of tracer pumping test 指示剂抽水试验法
method of tracer putting in 示踪剂投放方法
method of tracing test 示踪试验法
method of transformed section 折算截面法
method of transient pumping test 非稳定流抽水试验法
method of transition matrices 转换矩阵法
method of transit projection 经纬仪投点法
method of transliteration 转写法
method of traversing 导线法
method of trial 试探法
method of trials and errors 逐步逼近法;试探和误差法;试探法;试算法;试凑法
method of triangular superposition 三角形叠加法
method of truncation 舍项法
method of tunnel(l)ing 隧道掘进方法
method of two-stage least squares 两段最小平方法
method of unconstrained optimization 非约束最佳法
method of underhand square setting 下向方框支架安装法
method of undetermined coefficients 未定系数法;待定系数法
method of undetermined multipliers 待定乘法
method of unifying the rock materials 石料统一法
method of unifying the rock veins 纹理统一法
method of unsteady flow calculation 非稳定流计算法
method of variable difference 变差法
method of variation of coordinates 坐标变换法
method of variation of parameters 参数变分法
method of verification 验证方法
method of virtual displacement 虚位移法
method of virtual work 虚功法
method of water infiltration test 渗水试验方法
method of water infiltration with dual ring 双环渗水法
method of water infiltration with single ring 单环渗水法
method of water-level transfer 水位传递试验法
method of water-level variation-precipitation ratio 水位变幅-降雨量比值法
method of water rammer 水撼法
method of water recharge blocked off 断源法
method of water resources calculation 水资源计算方法
method of water sealing 止水方法
method of water yield prediction 矿井涌水量预

测方法
method of weight analysis 重量分析法
method of weighted mean 加权平均法
method of weighted residuals 加权余数法;加权余量法;加权残数法;加权残差法;权余法
method of weighting 加权法;权重法
method of well cleaning 洗井方法
method of well uniform distribution 平均布井法
method of work 功法
methodological 方法论问题
methodological analysis 方法论的分析
methodological framework 方法论体系
methodology 一整套方法;研究法;操作法;分类研究法;方法学;方法论
methodology for recovery 回收工艺
methods analysis chart 方法分析图
methods analysis specification 方法分析说明
methods department 方法部门
methods employed 使用方法
methods engineer 工艺工程师
methods engineering 工艺设计;工艺工程;方法工程
Methods Engineering Council 方法工程协会(美国)
methods engineering department 方法工程部门
methods for handling imports 进口产品的处理办法
methods for reserve calculations 储量计算方法
methods of air sampling and analysis 空气采样与分析方法
methods of computation of the volume of goods traffic 计算运量方法
methods of determination 鉴定方法
methods of determining depth 确定断层深度手段
methods of estimating soil organic phosphate 测定土壤的有机磷方法
methods of evaluation 评价方法
methods of fabrication 制造方法
methods of geochemical gas survey 气体测量的方法
methods of geologic(al) observation point 地质点定位方法
methods of installation 设置方法
methods of measurement 计量方法
methods of production 生产方法
methods of programming 计划编制方法
methods of regional geologic(al) survey 区域地质调查方法
methods of soil analysis 土壤分析方法
methods of soil tillage 土壤耕作方法
methods of strata correlation 地层对比方法
methods of structural survey 构造调查方法
methods of threshing 脱粒方法
methods simplification 方法简单化
method standard 方法标准
methods-time standard 方法—时间标准
method study 方法研究;工艺研究;操作方法分析
method-time data 方法—时间数据
method time measurement 操作方法时间测量;操作方法的时间测定
method with hand 徒手法
method with normal distribution 正态分布法
methoxone 二甲四氯
methoxy butyl acetate 乙酸甲氧基丁酯
methoxy ethyl acetoxystearate 乙酰氧基硬脂酸甲氧基酯
methoxyethylmercury silicate 灭菌硅
methoxyhexanone 甲氧基己酮
methoxyl value 甲氧基值
methoxymethyl-ethoxymethyl benzoguanamine 甲乙混合醚化苯代三聚氰二胺
methoxytrimethylsilane 甲氧基三甲基(甲)硅烷
methydopa 甲基多巴
methyl abietate 松香酸甲酯;枞酸甲酯
methylacetal 醛缩
methyl acetate 乙酸甲酯
methyl acetate poisoning 甲基醋酸盐中毒
methyl acetoacetate 乙酰乙酸甲酯
methyl acetophenone methyl tolyl ketone 甲基苯乙酮
methyl acetylene 丙炔
methyl acetyl ricinoleate 乙酰(基)蓖麻酸甲酯
methylacrylamide 甲基丙烯酰胺
methyl acrylate 丙烯酸(甲)酯
methylacrylic acid 甲基丙烯酸

methylaffinity 甲基亲合性
methylal 甲缩醛;甲醛缩二甲醇
methyl alcohol 甲醇;木醇
methylamine 甲胺
methylamine manufacturing wastewater 甲胺生产废水
methyl aminoacetate 氨基乙酸甲酸
methyl amyl acetate 乙酸甲(基)戊酯
methyl amyl ketone 甲戊酮;甲基戊基甲酮
methyl anthracene 甲基蒽
methyl anthranilate 邻氨基苯甲酸甲酯;氨茴酸甲酯
methylated methylol melamine 甲基化羟甲基三聚氰胺
methylated spirit 含甲醇酒精;变性酒精;甲基化酒精
methylation 甲基化作用
methylation/demethylation ratio 甲基化/脱甲基化比
methylation of mercury 汞的甲基化作用
methylbenzene 甲苯
methyl benzene sulfonate 苯磺酸甲酯
methyl benzilate 二苯乙醇酸甲酯
methyl benzoate 苯甲酸甲酯
methylbenzoic acid 甲苯(甲)酸
methyl benzoylacetate 苯甲酰乙酸甲酯
methyl benzoylsalicylate 苯甲酰水杨酸甲酯
methyl benzylketone 苯基丙酮
methyl blue 甲基蓝
methyl borate 硼酸(三)甲酯
methylbutyl benzene phthalate 邻苯二甲酸甲基丁苯
methyl butyrate 丁酸甲酯
methyl caprate 癸酸甲酯
methyl caproate 己酸甲酯
methyl carbamate 氨基甲酸甲酯
methyl-carbithionic acid 二硫代乙酸
methyl carbitol acetate 甲基卡必醇醋酸酯
methyl carbonate 碳酸甲酯
methyl cellosolve 甲基溶纤剂
methyl cellosolve acetate 甲基溶甲基三纤剂醋酸酯;乙二醇-甲醚乙酸酯
methyl cellulose 甲基纤维素
methyl cellulose glue 甲级纤维素胶
methyl chloride 氯甲烷;氯代甲烷;甲基氯
methyl chloride wastewater 氯甲烷污水
methyl chlorofluoride 二氟二氯甲烷
methyl chlorosilane 甲基氯硅烷
methylcholine iodide 碘甲胆碱
methyl cyanoacetate 氰乙酸甲酯
methyl cyclohexane 甲基环己烷
methyl cyclohexanone 甲基环己酮
methylcyclopentane 甲基环戊烷
methyl decanoate 癸酸甲酯
methyl derivative 甲级衍生物
methyl dichloroacetate 二氯乙酸甲酯
methyl diethano lamine 甲基二乙醇胺
methyldiphenyl 甲基二苯
methylene blue 亚甲蓝;美蓝;次甲蓝;次甲基蓝
methylene blue absorption 吸蓝量
methylene blueeosin staining method 美蓝伊红染色法
methylene blue reduction test 美蓝还原试验
methylene blue resin method 美蓝树脂法
methylene blue test 甲基蓝试验
methylene bromide 二溴甲烷
methylenecyclopentadiene 富烯
methylene dianiline 二苯氨基甲烷
methylene diiodide 二碘甲烷
methylene diphenyl-4 亚甲基二苯基二异氰酸酯
methylene halide 二卤甲烷
methylene iodide 二碘甲烷
methylene removal of stress 亚甲基消除力
methylene stress relieving 亚甲基应力缓解
methyl ether 二甲醚
methyl ethyl ether 甲基乙基醚
methyl ethyl ketone 甲基乙基酮;丁酮
methyl ethyl ketone peroxide 过氧化甲乙酮;过氧化丁酮
methyl ethyl ketoxime 甲乙酮肟
methyl ethyl toxime 甲乙基甲酮肟
methyl fluoride 氟甲基
methyl furoate 糠酸甲酯
methyl glycol 甲基乙二醇
methyl gravitometer 甲基比重计

methylguanidine 甲胍
methyl heptanoate 庚酸甲酯
methyl hexalin acetate 甲基环己醇乙酸酯
methyl hydrite 甲烷
methyl hydrogen siloxane 甲基氢硅氧烷
methyl hypochlorite 次氯酸甲酯
methylic chloride 氯代甲烷
methylimine 甲亚胺
methyl iodide 碘代甲烷
methyl isoamyl ketone 甲基异戊基甲酮
methyl isobutyl carbinol acetate 乙酸甲基异丁基甲酯
methyl isocyanate 异氰酸甲酯
methyl malonate 丙二酸(二)甲酯
methylmalonic acid 异丁二酸;甲基丙二酸
methyl mercuric iodide 碘化钾汞
methylmercury 甲基汞
methylmercury chloride 氯化甲基汞
methyl mercury cyanide 氰化甲汞
methylmetal-crylates 丙烯酸酯类
methyl methacrylate 异丁烯甲酸酯;甲基丙烯酸甲酯;有机玻璃
methyl methacrylate butadiene styrene copolymer 甲基丙烯酸甲酯-丁二烯
methyl n-amyl ketone 甲基-正戊基甲酮
methyl n-butprate 丁酸甲酯
methyl n-caprate 癸酸甲酯
methylol 羟甲基
methyl oleate 油酸甲酯
methylol urea 耐火的浸渍化合物;木材耐潮;羟甲基脲
methyl orange 甲基橙
methyl-orange alkalinity 甲基橙碱度
methyl-orange wastewater 甲基橙废水
methyl ortho-silicate 原硅酸甲酯
methyl-oxalacetic ester 草酰乙酸酯
methyl oxitol acetate 甲氧基乙酸乙酯
methyl p-aminophenol sulfate 硫酸甲基对氨基苯酚
methyl para-aminobenzoate 对氨基苯甲酸甲酯
methyl para-cresol 对甲酚甲醚
methyl para-hydroxybenzoate 对羟基苯甲酸甲酯
methyl-parathion 甲基对硫磷
methyl phenanthrene 甲基菲
methylphenidate 苯哌啶醋酸甲酯
methylphenol 甲基苯酚;甲酚
methyl-phenoxide 苯甲醚
methyl phenyl acetate 苯乙酸甲酯
methyl phthalyl ethyl glycolate 邻苯二甲酸(甲酯)乙醇酸乙酯
methyl p-methoxybenzoate 对甲氧基苯甲酸甲酯
methyl p-methyl benzoate 对甲基苯甲酸甲酯
methyl p-nitrobenzoate 对硝基苯甲酸甲酯
methyl polysiloxane 聚甲基硅氧烷
methyl prooenyl ketone 甲基丙烯基甲酮
methyl propionate 丙酸甲酯
methyl propyl ketone 甲基丙基甲酮;甲丙酮
methyl proxitol 丙二醇甲醚
methyl proxitol acetate 丙二醇甲醚乙酸酯
methyl pyruvate 丙酮酸甲酯
methylquinoline 甲基喹啉
methyl radical 甲基
methyl red 甲基红
methyl red test 甲基红试验
methyl red wastewater 甲基红废水
methyl ricinoleate 蓖麻醇酸甲酯
methylsilane 甲基硅烷
methyl silicate 硅酸甲酯
methyl silicone 聚甲基硅氧烷;甲基有机硅油
methyl silicone fluid 甲基硅酮液
methyl silicone gum 甲基硅酮胶
methyl silicone resin 聚甲基硅酮树脂
methyl stearate 硬脂酸甲酯
methyl styrene 乙烯基甲苯;甲基苯乙烯;对甲苯乙烯
methyl sulfate 硫酸二甲酯
methyl sulfide 二甲硫
methylthioninium chloride 美蓝
methyl vinyl ether 甲基乙烯基醚
methyl violet 甲基紫;红光碱性紫
methyl violet lake 甲基紫色淀
methyl violet toner 甲基紫调色剂
methymercury chloride 甲汞化氯
methyne 次甲基

meticulous 过细的
meticulous construction 精心施工
meticulous design 精心设计
meticulous inspection 严格检查
meticulous plan all expenditures 精打细算
metlbond 环氧树脂类及无机黏合剂;金属黏合工艺
met-L-check 金属染色探裂法
Metlex plug 梅特列克斯插销
metoche 檐下齿形装饰间的空间;齿饰中两个齿间的空间
Metonic cycle 月龄周期;麦东周期【给】;朔望日期变化
metonymy 换喻
metope 排档间饰;两饰柱间的壁;三槽板间平面;三槽板间的面石
metopic 额侧的
metoplankton 阶段浮游生物
metosomatism 交代作用
metraster 曝光表
metre = meter
metre-ampere 米·安(培)
metre case 仪表外壳
metrechon 双电子枪存储管
metre dial 仪表刻度盘
metre full scale 刻度范围
metre ga(u)ge 米轨;米轨铁路
metre-ga(u)ge track 轨距量仪
metre-kilogram-second system 米千克秒制
metre module 公制模数
metre run 跑米
metre scale 仪表刻度
metrewave radar 米波雷达
metric(al) 公制的;米制的;度量的;度规
metric(al) addressing 矩阵寻址
metric(al) and inch vernier calipers 公制和英制游标卡尺
metric(al) atmosphere 公制大气压
metric(al) calibration 测量校正
metric(al) camera 量测摄影机
metric(al) carat 公制克拉;克拉(宝石重量单位,1克拉=200毫克)
metric(al) centner 公担(1公担=100公斤)
metric(al) character 计量性状
metric(al) characteristic 量测性能
metric(al) charts 米制海图
metric(al) convention 米制公约;米制换算
metric(al) count 公制支数
metric(al) data 测量数据;米制数据
metric(al) data system 计量数据系统
metric(al) density 度量密度
metric(al) diameter tape 公制直径尺
metric(al) dimension 米制尺寸;公制尺寸
metric(al) fidelity 量测保真度
metric(al) fine thread 公制细(牙)螺纹
metric(al) form 度量形式
metric(al) ga(u)ge 米制尺;米尺
metric(al) geometry 度量几何学
metric(al) grain 米制格令
metric(al) horsepower 公制马力(1公制马力=0.986马力);米制马力;米马力
metric(al) hundred weight 公担(1公担=100公斤)
metric(al) information 尺度信息;可度量信息;计量信息;测量信息
metric(al) instrument 计量仪器;量具
metric(al) line 毫米
metrically dese 度量的稠密
metric(al) measure 公制单位;米制度量
metric(al) modular unit 米制模数单位
metric(al) nut 公制螺母
metric(al) product 度量积
metric(al) property 度量性质
metric(al) rating 鉴别
metric(al) relation 度量关系
metric(al) scale 公制比例尺;米制标度;米制比例尺;米尺;米制尺(度)
metric(al) screw 公制螺旋;公制螺(丝)钉;公制螺杆;米制螺纹
metric(al) screw pitch ga(u)ge 国际螺距规;米制螺距规
metric(al) screw thread 公制粗牙螺纹
metric(al) series 公制系列
metric(al) size 公制尺寸
metric(al) space 度量空间

metric(al) system 公制;公制单位制;米制
metric(al) system of measurement 米制计量度衡
metric(al) taper 公制锥度
metric(al) tensor 度量张量;度规张量
metric(al) thread 公制螺纹;米制螺纹
metric(al) thread gearing 米制螺纹齿轮装置
metric(al) thread tap 公制螺纹丝锥
metric(al) ton 公吨;米制吨;千公斤
metric(al) ton per year 年吨
metric(al) tons per hour 每小时公吨
metric(al) transformation 度量变换
metric(al) transitivity 度规传递性
metric(al) unit 公制单位;米制单位
metric(al) updating 测量校正
metric(al) value 折合公制
metric(al) wave 米波
metrication 公制化;米制化
metric ga(u)ge 米制规
metrics 计量学;测量学
metric-technical unit of mass 质量的米制技术单位
metrizable 可度量
metrizable space 可度量化空间
metrization 度量化
metro 地(下)铁(道);都市地下铁道
metro administration 地铁管理
metro alignment 地铁线形
metro area 大都市圈;大都市区
metro corporation 地铁总公司
metro engineering traction 地铁工程牵引
metro ga(u)ges 地铁限界
metrograph 汽车速度计
metrolac 胶乳比重计
metroliner 都市间高速列车
metrological 度量衡学的
metrological and inspecting work 计量检验工作
metrological evaluation 计量学评定
metrological examination 计量学检验
metrological moisture 计量水分
metrologic instrument 计量仪器
metrologist 度量衡学家;计量学家
metrology 计量制;度量衡学;计量学;度量衡制;测量学
metrology equipment 测量仪器
metrology grating disk 计量图光栅
metrology microscope 测量显微镜
metron 密特勒恩(计量信息的单位)
metro network 轨道交通网;地下铁道网
metro noise 地铁噪声
metronome 节拍器
metrophic bacteria 后生营养细菌
metrophotography 摄影测量学;测量摄影学
metropolis 文化商业中心;中心地;巨大城市;首府;首都;大都市;大都会;大城市
metropolitan area 首都地区;大都市圈;大都市区;大城市市中心区;大城市市区;大城市区域;大城市地区
metropolitan area network 城域网
metropolitan area population 大城市区人口
metropolitan area rapid transit 大城市地区快速交通系统
metropolitan axes 都市轴心
metropolitan city 大都市
metropolitan district 首都行政区;大城市市中心区;大城市市区;大城市区域
metropolitan mass transit railway terminal 都市高速大量运输铁路枢纽
metropolitan primacy 大都会首市现象
metropolitan railroad 市内铁道
metropolitan railroad station 市内铁道车站;市内地下铁道车站
metropolitan railroad track 市内地下铁道钢轨;市内铁道钢轨
metropolitan rail transit system 大城市轨道交通系统
metropolitan railway 城市铁道;大城市铁路;市内铁道;城市铁路
metropolitan railway track 市内铁道钢轨
metropolitan region 大城市地区;大都市圈;大都市区
metropolitan sewer district 城市下水道区
metropolitan shopping center 大都市购物中心
metropolitan station 都市电台
metropolitan underground railway 都市地下铁道
metropolitan waste(water)control 城市污水控制

metro route 地铁线路
metroscope 测长机
metro's noise 地下铁道噪声
metro-station 地铁车站
metrostaxis 漏下
metro's waste 地下铁道废弃物;地铁废弃物
metro tunnel 地铁隧道
metroum 量器;表;计;公尺;米
Mettex plug 螺旋嵌固金属塞
met-xylylene diamine 间位亚二甲苯基二胺
me-tyl-wood 镶贴金属板(防水)
Meudon white 麦东白
Meuse valley incident 马斯河谷烟雾事件
meutral wedge filter 中性光劈式滤光片
meutriere (堡垒上的)射孔
mews 小巷;马厩改建的住房;马车屋;马厩;(设马车房的)马店
mexan 甲氨胺
Mexican architecture 墨西哥建筑
Mexican back-hoe 鹤嘴锄;墨西哥式十字镐
Mexican crusher 墨西哥大铁锤
Mexican geosyncline 墨西哥地槽
Mexican Gulf Coast geosyncline 墨西哥湾地槽
Mexican linaloe oil 沉香木油
Mexican oiticica oil 墨西哥产巴西果油
Mexican onyx 墨西哥缟玛瑙;条纹大理岩
Mexican speed wrench 墨西哥钳子;钳子;扳钳
Mexico asphalt 墨西哥柏油;墨西哥沥青
Mexico Gulf Coastal basin 墨西哥湾岸盆地
Meyer hardness 迈氏硬度;迈尔硬度
meyerhofferite 三斜硼钙石
meyerhofferite water reducer 三斜硼钙石减水剂
Meyerhof method 梅耶霍夫法
Meyerhof's formula 梅耶霍夫公式
Meyer pump 梅耶(往复)泵
Meyer's absorber 梅耶吸器
Meyer's bar 梅耶棒
Meyer's condensation theory 梅耶凝结理论
Meyer's formula 梅耶公式
Meyer's hardness 梅耶硬度
Meyer's hockey stick incision 弯形切口
Meyer's index 梅耶指数
Meyer's law 梅耶酯化定律
Meyer's problem 梅耶问题
Meyer's pump 梅耶泵
Meyer's synthesis 梅耶合成法
Meyer's test 梅耶试验
Meyer's tube 梅耶二氧化碳吸管
meymacite 水氧钨矿
meymechite 玻基纯橄岩
mezogeosyncline 陆中地槽
mezzanine 夹楼层;夹层楼面;地铁中间厅
mezzanine floor 夹层楼面;夹层(尤指介于一层与二层之间的)
mezzanine stor(e)y 夹层层;夹层楼面
mezzanine window 夹楼层窗子
Mezzera counter acting open width washing machine 梅氏逆流平洗机
mezzograph 网目图形
mezzo relievo 中浮雕;半浮雕
mezzotint 网线金属版
Mg-cation exchange capacity 镁离子交换容量
Mg-Fr ratio 镁铁比
M-glass fiber 高模量玻璃纤维
Mg[megagram] 兆克
Mg-Mn dried battery 镁锰干电池
Mg-point 变形软化温度;变形点
mgriite 莫砷硒铜矿
Mg-Si ratio 镁硅比
MGT 通过轨道总重
mhometer 欧姆计
miacalcic ampoule 密钙息注射液
miagite 球状闪长岩
mia-mia 临时棚屋(澳洲)
mianthite 深熔暗包体;暗色包体(深熔岩中)
miargyrite 辉锑银矿
miarolitic 晶洞状
miarolitic cavity 晶洞
miarolyte 晶洞花岗岩
miarophitic structure 晶洞构造
miascite 云霞正长岩
miaskitic nepheline syenite 云霞正长岩型霞石正长岩
miasma 瘴毒

M-l bus 存储器输入总线
mica 云母
mica alba 云母石
mica alvikite 云母方解石碳酸岩
mica andesite 云母安山岩
mica band heater 云母带加热器
mica-basalt 云母玄武岩
mica-bearing arenite 含云母砂屑岩
mica-bearing arkose 含云母长石砂岩
mica-bearing graywacke 含云母杂砂岩
mica-bearing quartz sandstone 含云母石英砂岩
mica-bearing wacke 含云母瓦克岩
mica beforsite 云母镁云碳酸岩
mica body 云母小体
Micabond 迈卡邦德绝缘材料
mica book 书页云母；云母片
mica box 云母箱
mica capacitor 云母电容器
micaceous 云母状的；云母土；含云母的
micaceous clay 似云母黏土
micaceous gneiss 云母片麻岩
micaceous granite 云母花岗岩
micaceous hematite 云母(状)赤铁矿
micaceous iron ore 云母铁矿；云母赤铁矿
micaceous iron oxide 云母氧化铁；云母铁矿
micaceous iron oxide anticorrosive primer 云铁防锈底漆
micaceous iron oxide paint 云母铁矿涂料；云母铁矿油漆
micaceous porphyry 云母斑岩状紫色彩饰
micaceous quartzite 云母石英岩
micaceous sand 云母砂
micaceous sandstone 云母砂岩
micaceous schist 云母片岩
micaceous shale 云母页岩
micaceous structure 云母构造
micaceous tile 云母瓦
micacite 云母片岩
mica clay 云母黏土；低品位瓷土
mica cleavage 云母裂缝；云母劈理(面)
mica clip 云母绝缘夹
mica cloth 云母片；云母布；云母箔
mica collar 云母圈
mica condenser 云母电容器
mica content 云母含量
mica content in aggregate 骨料云母含量
mica deposit 云母矿床
mica detector 云母探测器
mica diaphragm 云母膜片
mica disc 云母盘
micadon 云母电容器
mica dust 云母碎屑；云母粉末
mica end-window counter 云母端窗计数管
mica flake 云母碎片；云母片；云母薄片
mica-flake insulation 云母薄片绝缘
mica flap 铰支的云母板；云母页岩；云母页片
mica flap valve 云母片截门；云母页片阀
mica flour 云母粉(状物质)
mica foil 云母箔
micafolium 云母箔；胶合云母箔
micafolium shellac paper 虫胶云母纸
mica fragments 云母碎片
mica gneiss 云母片麻岩
mica granular poprhyrite 云母安山岩粉岩
mica group 云母类
mica heating unit 云母加热单元
mica insert 云母插片
mica in sheet 云母板；云母薄片
mica insulation 云母绝缘
mica insulator 云母绝缘子
mica kimberlite 云母金伯利岩
mica lamination 云母片
mica lamprophyre 云母煌斑岩
mica law 云母律
micalex 云母石；云母玻璃
mica mat 云母垫
mica mo(u)lded insulator 云母模制绝缘物
mica nepheline melilitite 云霞黄长岩
micanite 云母(塑料)板；绝缘云母板；胶合云母板；人造云母；层压云母板；层合云母板
micanite paper 云母纸
micanite pipe 云母管
micanite sleeve 云母板套管
mica of commerce 贸易云母；商用云母

mica packing plate 衬垫云母板
mica paper 云母纸
mica parition 云母分隔片
mica particle 云母颗粒
mica peridotite 云母橄榄岩
micaphyre 云母斑岩
mica picrite 云母苦橄岩
mica pig 云母晶块；合成云母晶块
mica pigment 云母颜料
mica plate 云母板
mica pneumoconiosis 云母尘肺
mica-porphyrite 云母玢岩
mica porphyry 云母斑岩
mica powder 云母粉
mica pyroxene melilitite 黑云辉石黄长岩
mica pyroxenite 云母辉石岩
mica rauhaugite 云母白云(石)碳酸岩
mica resistance 云母电阻
micarex 云母石；云母玻璃
mica-rich domain 富云母域
mica ring 云母环
micarta 胶质板；胶合云母纸板；米卡他绝缘板；层状酚醛塑料；云母板
micarta board 胶合板；装饰板；米卡塔胶纸板
Micarta insulating ring 层状酚醛塑料绝缘环
mica schist 云母片岩
mica schist porphyry 云母片麻斑岩
mica scrap 碎云母
mica segment 云母片
mica separating screen 云母分选筛
mica shale 云母页岩
mica sheet 云母片
mica-sheet insulation 云母片绝缘
mica silicosis 云母矽肺
mica slate 云母片岩；云母板岩
mica sovite 云母黑云碳酸岩
mica spacer 云母垫片
mica spark(ing) plug 云母火花塞
mica specks 云母状斑点
mica splitting 云母分裂；剥片云母；剥制云母
mica strip 云母条
mica-syenite 云母正长岩
mica talc 云母滑石
mica tape 云母带
mica taper ring 锥形云母环
mica test-plate 云母检查板
micatization 云母化
mica trap 云母煌斑岩
mica tube 云母管
mica undercutter 云母绝缘片修整刀
mica washer 云母垫圈
mica window 云母窗
micell 晶子；微胞团
micella(e) 胶团；胶粒；分子团
micellar angle 分子团角
micellar catalysis 胶团催化剂
micellar colloid 胶束胶体
micellar electrokinetic capillary chromatography 胶束电动毛细管色谱法
micellar enhanced ultrafiltration 胶束强化超滤
micellar-polymer flooding 胶束—聚合物驱动
micellar solution 胶束溶液
micellar structure 胶束结构
micellar surfactant solution 胶团表面活性剂溶液
micelle 胶束；胶粒；微(胞)团；微胞；分子团
micellerization 胶粒化
micelle solubilization spectrophotometric(al) 胶束增溶分光光度法
micelle structure 胶粒结构
micelle weight 胶团量
micellization-demicellization process 胶态化—去胶态化法
Michael addition reaction 迈克尔加成反应
Michaelis constant 米凯利斯常数
Michaelis Mast pressure pile 米凯利斯·马斯特压力灌注桩
Michaelis's rhomboid 米凯利斯菱形区
Michaelson actinograph 迈克尔逊日射计
michaelsonite 复硅锆钡矿
Michelia oil 白兰花油
michelison bimetallic pyrheliometer 迈克尔逊双金属直接日射计
Michell-Banki turbine 米歇尔—班克水轮机；双击式水轮机

Michell's bearing 米歇尔式轴承
Michell's parameter 米歇尔参数
Michell thrust bearing 楔块式推力轴承；米歇尔推力轴承；倾斜瓦块式推力轴承
Michell type thrust bearing 米歇尔型推力轴承
Michell's clips 米歇尔小夹
Michelson echelon 迈克尔逊阶梯光栅
Michelson-Gale ring interferometer 迈克尔逊—盖尔环形干涉仪
Michelson interference microscope 迈克尔逊干涉显微镜
Michelson microinterferometer 迈克尔逊微干涉仪
Michelson-Morley experiment 迈克尔逊莫雷试验
Michelson rotating mirror 迈克尔逊旋转镜
Michelson's interferometer 迈克尔逊干涉仪
Michelson stellar interferometer 迈克尔逊星体干涉仪；迈克尔逊恒星干涉仪
Michelson type cadmium lamp 迈克尔逊型镉电池灯；迈克尔逊型镉灯
Michelson type laser 迈克尔逊型激光器
Michel-type bearing 米歇尔轴承
michenerite 等轴铋碲钯矿
Michigan cut 密执安式平行龟裂陶槽；平行空炮眼掏槽法
Michigan tripod 钻孔设备三脚架
mickey 雷达手
mickle 软黏土；大量的
mickle hammer 碎石锤；大锤
miconcave 微凹表玻璃
micoquille 薄曲面玻璃
micor-rotation 微观转动
micrergy 显微操作术
micrfile 缩微档案
micrinite 微粒体；碎片体煤岩
micrinite groundmass 微粒体基质
micrinoid group 微粒体组
micrite 泥晶灰岩；微晶灰岩；微晶
micrite enlargement 泥晶加大作用
micrite envelope 泥晶套
micrite envelope fabric 泥晶套组构
micrite fabric 泥晶组构
micrite matrix 微晶基质
micrither mometric(al) measurements 显微量热测定
micritic dolomite 微晶白云岩
micritic limestone 泥晶灰岩；微晶灰岩
micritilization 微晶化作用；泥晶化作用
micro 微型；缩微；百万分之一
microabsorption 微量吸收
microaccounting 微观会计
microacoustic(al) system 微声系统
microactive switch 微动开关
microacupunoture 微针术
microadding 微量添加
microaddition 微量加入物
microadjuster 精调装置；精密调节器；微量调节器
microadjustment 微量调整
microadmixture 微掺和物
microadsorption 微吸附
microadsorption column 微吸附柱
microadsorption detector 微(型)吸附监测器
microaeraophilic 微需氧的
micro-aeration reactor 微曝气反应器
micro-aerobe 微好氧菌
micro-aerobic 微好氧的
microaerobion 微需氧菌
microaerobiology 微需氧微生物；微需气细菌
microaerophilic bacteria 微好氧菌；微需氧细菌
microaerophilic test 微需氧试验
microafter-shock 微余震；微系震
microagglutination test 微量凝集试验
microaggregate 小颗粒聚集体；微团粒；微团聚体；微集合体
microalignment telescope 微调对中望远镜
microalloy 微合金
microalloy diffusion type 微合金扩散型
microalloying 微合金化
microalloy technology 微合金工艺
microammeter 微安计；微安表；微安(安培的百万分之一)
microampere 微安(安培的百万分之一)
microampere circuit 微安电路
microamperemeter 微安计；微安表
microampere sensibility 微安灵敏度

microanaerobic bacteria 微厌氧性细菌
microanalysis 显微分析;微量分析
microanalytic(al) balance 微量分析天平
microanalytic(al) chemistry 微量分析化学
microanalytic(al) method 微量分析法
microanalytic(al) reagent 微量分析试剂
microanalytic(al) standards 微量分析标准
microanalyze 显微分析
microanalyzer 微量分析器
microanalyzer mass spectrometer 微量分析质谱仪;微量分析质谱计
microand semimicroviscometer 微量和半微量黏度计
microangiography 微动脉造影术
microanisotropy 微观各向异性
microanticipatory crash sensor 微波预测碰撞传感
microaphanitic 显微隐晶质
microaphanitic texture 显微隐晶质结构
microapplication manual 微应用手册
microarchitecture 微(型)体系结构
microarea chemical analysis 微区化学成分分析
microarea diffraction 微区衍射
micro-aeration 微曝气
micro-argon detector 微氩检测器
microarm 细臂
microassembler 微汇编程序
microassembly 微组合;微系集
microassociation 小群丛
microatoll 微环礁
microaulacogen 微型裂陷
microautograph 显微放射自显影照相
microautoradiogram 显微放射自显影相片
microautoradio graph 微射线自动照相机
microautoradiography 微射角自动摄影术;显微放射自显影术
microbalance 测微天平;微量天平
microball-mill 微粉球磨机
microballoons 微球
microballoon sphere 微球
microbar 微巴;巴列(气压单位)
microbarn 微靶
microbarogram 微压记录图;微气压记录图
microbarograph 微气压计录器;微气压计
microbarometer 精测气压计;微(气)压表
microbase 微机座
micro-basification 微碱化
microbasin 微平原
microbasis for macroeconomic management 宏观经济管理的微观基础
microbasis for macroeconomic regulation 宏观经济管理的微观基础
microbathymetry 精密测深(法)
microbattery 微电池
microbe 细菌;微生物
microbe adsorption capacity 微生物吸附容量
microbeaker 微烧杯
microbeam 微束;微光束
microbeam irradiation 微(光)束照射
microbeam X-ray analyzer 微束X射线分析器
microbedding 微层理
microbe degradation 微生物降解
microbe density 微生物密度
microbenthos 微型海底生物
microbe satellite 伴随细菌
microbial activity 微生物活性
microbial adhesion to hydrocarbon 微生物黏着碳烃化合物
microbial assimilation 微生物同化
microbial association 微生物结合
microbial attack 微生物侵蚀
microbial bacterial population 微生物菌群
microbial bioassay 微生物生物测定
microbial biochemistry 微生物生物化学
microbial biodiversity 微生物生物多样性
microbial biomass 微生物生物物质;微生物量
microbial biomass carbon 微生物生物物质碳
microbial biomass nitrogen 微生物生物物质氮
microbial biomass phosphorus 微生物生物物质磷
microbial biosensor 微生物生物传感器
microbial breakdown of pesticide 农药(的)微生物降解
microbial characteristic 微生物特征
microbial closed ecosystem 微生物闭锁生态系
microbial cometabolism 微生物共代谢

microbial community diversity 微生物群落多样性
microbial community structure 微生物群落结构
microbial concentration 微生物浓度
microbial contaminant 微生物污染物
microbial contamination 微生物污染
microbial contamination of water 水的微生物污染
microbial control 微生物防治
microbial conversion 微生物引起的转化作用
microbial conversion of mercury compound 汞化合物的微生物转化
microbial decolo(u)rization 微生物脱色
microbial decomposition 细菌分解(作用);微生物分解(作用)
microbial decomposition process 微生物分解工艺
microbial decomposition system 微生物分解系统
microbial degradation 微生物降解
microbial dephosphorus 微生物除磷
microbial deterioration 微生物引起的变质
microbial dissociation 微生物离异
microbial diversity 微生物多样性
microbial ecology 微生物生态学
microbial ecosystem 微生物生态系(统)
microbial film 微生物膜
microbial flocculant 微生物絮凝剂
microbial fuel cell technology 微生物燃料电池技术
microbial growth 微生物生长
microbial growth curve 微生物生长曲线
microbial herbicide 微生物除草剂
microbial hydrolysis 微生物水解
microbial imbalance 微生物不均衡
microbial insecticide 微生物杀虫剂
microbial kinetics 微生物动力学
microbial leaching 微生物浸出
microbially available phosphorus 微生物可用磷
microbial material 微生物性原料
microbial metabolism 微生物代谢(作用)
microbial metabolism activity 微生物代谢活性
microbial metallurgy 微生物冶金学
microbial methylation 微生物甲基化(作用)
microbial modification 微生物变体
microbial nitrogen removal process 微生物脱氮工艺
microbial numbers 微生物数
microbial parameter 微生物参数
microbial pesticide 微生物农药
microbial population 微生物群系;微生物群体
microbial process 微生物过程
microbial proliferation 微生物繁衍
microbial reaction 微生物反应
microbial reduction 微生物还原
microbial remediation 微生物修复
microbial resources 微生物资源
microbial sludge 活性污泥;微生物污泥
microbial standard 微生物标准
microbial technology 微生物技术
microbial toxin 微生物毒素
microbial transformation 微生物的转化
microbial wastewater treatment 小型藻类废水处理
microbial water quality 细菌性水质
microbian 微生物
microbibliography 缩微目录
microbicide 杀菌剂
microbic insecticide 微生物农药
microbioassay 微生物测定(法)
microbiochemcial reaction 微生物化学反应
microbiodegradation 微生物降解
microbiofacies 微体生物相
microbiologic(al) analysis 微生物(学)分析
microbiologic(al) anomaly 微生物异常
microbiologic(al) antagonism 微生物对抗作用
microbiologic(al) assay 微生物学测定;微生物鉴定
microbiologic(al) characteristic 微生物特性
microbiologic(al) component 微生物成分
microbiologic(al) concentration in water 水中微生物的浓度
microbiologic(al) condition 微生物条件
microbiologic(al) contaminant 微生物污染
microbiologic(al) contamination 微生物污染
microbiologic(al) control 菌藻处理;微生物控制
microbiologic(al) corrosion 微生物腐蚀
microbiologic(al) cultivation 微生物培养
microbiologic(al) cycle 微生物循环
microbiologic(al) degradation 微生物降解
microbiologic(al) fouling 微生物污着

microbiologic(al) indices 微生物学指标
microbiologic(al) industry 微生物学工业;微生物工业
microbiologic(al) infallibility 微生物的确实性
microbiologic(al) insecticide 微生物杀虫剂
microbiologic(al) leaching 微生物浸出
microbiologic(al) measurement 微生物测定(法)
microbiologic(al) method 微生物学方法
microbiologic(al) oxidation 微生物引起的氧化作用;微生物氧化(作用)
microbiologic(al) phosphorus 微生物磷
microbiologic(al) pollutant 微生物污染物
microbiologic(al) pollution 微生物污染
microbiologic(al) pollution of atmosphere 大气微生物污染
microbiologic(al) population 微生物群落
microbiologic(al) process 微生物工艺
microbiologic(al) products 微生物产物;微生物产品
microbiologic(al) property 微生物特性
microbiologic(al) reaction 微生物反应
microbiologic(al) research 微生物学研究
Microbiologic(al) Resources Centre for Developing Countries 发展中国家微生物资源中心
microbiologic(al) specification 微生物属性
microbiologic(al) survey 微生物测定
microbiologic(al) treatment 微生物处理
microbiologist 微生物学家
microbiology 微生物学
microbiology of resource development 资源开发微生物学
microbiology of waste recycling 废物再生微生物学
microbion 微生物
microbiophagy 微生物噬菌作用
microbiophotometer 微生物浊度计
microbiorite 微晶闪长岩
microbioscope 微生物显微镜
microbiosis 微生物病
microbiostratigraphy 微体生物地层学
microbiota 微生物相;微生物群;微生物区系;微生物区;微生物丛
microbiotest 微生物测定(法)
microbit 微型钻头;微位
microbit drilling rig 微型钻头钻机
microbit drilling test 微型钻头凿岩试验
microbivorous 溶解微生物的
microblasting drilling 微爆凿岩
microbody 微体
microbonding 微焊
microbook fiche 缩微图书胶片
microbore 精调刀头
microbore heating system 微管采暖系统;小管径集中式采暖系统
microbore(lock type) unit 精密微调镗刀头;精调刀头
microbore system 微管采暖系统
microboring head 精密镗刀头;微调刀头
microboulder 微粒
microbreccia 显微角砾岩;微(晶)角砾岩
microbridge 微桥
microBrownian diffusion 微观布朗扩散
microBrownian movement 微观布朗运动
micro-bubble pure oxygen aeration technique 微气泡纯氧曝气技术
microbulking 微膨化
microbundle 微丛
microburet(te) 微量滴管
microburner 小型本生灯;微焰灯;微型燃烧器
microbus 小型客车;微型公共汽车
microcal(l)ipers 测微器;千分卡(尺);测微计;千分尺
microcalorie 微卡
microcalorimeter 微热量计
microcalorimetric(al) method 微量热法
microcalorimetry 微量热法
microcamera 微型照相机;缩微照相机
microcapacitor 微电容器
microcapillary 微细管
microcapillary pore 微毛细管孔隙
microcapilliarity 微毛细管(作用)
microcapsule 微胶囊
microcard 缩影胶片卡;缩微卡(片)
microcartridge 微调镗刀

microcast process 精密铸造法
micro-cataclasite 碎粒岩
microcatalytic reactor 微型催化反应器;微观催化反应器
microcationic titration 微量阳离子滴定法
microcator 指针测微计
microcausality 微观因果性
microcausality condition 微观因果条件
microcell 微量试池;微极化池;微电池对
microcellular rubber 微孔橡胶
microcellular structure 微孔构造
microcellular texture 微孔结构
microcentrifugal tube 微型离心管
microcentrum 中心粒团
microchannel 微通道
microchannel inverter 微通道变换器
microchannel plate 微通道板
microcharacter 显微划痕硬度计
microcharacter hardness 显微划痕硬度
microcharacteristic 微观特征
microchecker 微米校验台
microchemical analysis 显微化学分析;微量化学分析;微化分析
microchemical apparatus 微量化学仪器
microchemical balance 微量化学天平
microchemical pollutant 微(量)化学污染物
microchemical pollution 微(量)化学污染
microchemical test 微(量)化学试验
microchemistry 微量化学
microchip 微片
microchip computer 微片计算机
microchip technology 微片技术
microchromatography 微量色谱(法)
microchromatoplate 微型色谱板
microchronometer 微动计时器;精密计时器;测微计时表;测微分计;分秒表
microci chip 微电路片
microci computer 微循环计算机
microci isolation 微电路隔离
microcilium 微纤毛
microcircuit 微(型)电路;超小型电路
microcircuitry 微型电路学
microcircular technique 微量循环技术
microcirculation 微循环
microcirculation dysfunction 微循环功能障碍
microcirculatory disturbance 微循障碍
microcirculatory perfusion 微循环灌流量
microcirculatory system 微循环系统
microci stencil 微电路模板
microcite-bearing nepeline syenite 含微斜长石霞石正长岩
microci technique 微型线路技术
microclarain 显微亮煤
microclast 微屑
microclastic 细屑质的;微细屑状
microclastic rock 微细屑岩
microclastic texture 微屑结构
microcleanliness 显微清洁度;显微纯洁(度)
microclimate 小气候;微(小)气候;地方气候
microclimate data 小气候资料
microclimate effect 微气候影响
microclimate measurement 小气候测定
microclimatic effect 小气候效应
microclimatic factor 小气候因素
microclimatic heat island 小气候热岛
microclimatic survey 小气候测量;微气候测量
microclimatology 小气候学;小尺度气候学;局部气候学;微(小)气候学;地方气候学
microcline 微斜长石
microcline perthite 微斜纹长石
microcline syenite 微斜长石正长岩
microcline twin law 微斜长石双晶律
microclinite 微斜正长岩
microclinization 微斜长石化
microclock 微型钟
microcoacervation 微凝聚
Micrococus foetidus 恶臭细球菌
microcode 微指令;微代码;微程序;微操作码;编制微码
microcoded controller 微编码控制器
microcoded instruction 微编码指令
microcode instruction set 微代码指令系统
microcode instruction set component 微代码指令系统成分

microcoding 微程序设计;微编码
microcoenosium 小群落
microcollar 微动轴环
microcolony 小集落
microcolorimeter 微量比色计
microcolorimetric(al) determination 微量比色测定
microcolorimetry 微量比色法
microcolo(u)r reaction 微量颜色反应
microcolumn 微型柱
microcommand 微命令
microcommunity 小群体;小群落
microcomparator 微比长计;测微比较仪
microcomponent 微型元件;微量组分
microcomposer 分解者
microcomputer 微(型)计算机;微型电子计算机
microcomputer-aided engineering 微计算机辅助工程
microcomputer application design 微计算机应用系统设计
microcomputer-based image processing system 微计算机图像处理系统
microcomputer control drill 微计算机控制钻机
microcomputer-controlled data collection device 微计算机控制数据采集装置
microcomputer-controlled recording spectrophotometer 微计算机控制自记分光光度计
microcomputer control system 微计算机控制系统
microcomputer detecting control system 微计算机检测控制系统
microcomputer development system 微计算机开发系统
microcomputer execution cycle 微计算机执行周期
microcomputer interfacing kit 微计算机接口套件
microcomputer kit assembler 微计算机配套汇编程序
microcomputer network 小型计算机网络
microcomputer prototyping system 微计算机样机系统
microcomputer system 微(计算)机系统
microcomputer teaching equipment 微计算机教学设备
microcomputer technology 微型电子计算机技术
microcomputer terminal 微(计算)机终端
microconcentrator 微型浓缩器
microconcrete 微型混凝土;微粒混凝土;微晶混凝土;缩小比例混凝土
microcone penetration test 微锥针入度试验
microcone penetrometer 微锥针入度计
microconformation 微构象
microconjugate 微观共轭
microconsistometer 微型稠度计
microconsole 微控制台
microconstituent 微量成分
microconsumer 分解者
microcontaminant 微(量)污染物
microcontent 微量
microcontinent 海底高原;微大陆
microcontinental block 微型陆块
microcontinuum approach 微观连续介质法
microcontinuum model 微观连续介质模型
microcontrolled modem 微控制调制解调器
microcontrolled seeder 微控制播种机
microcontrolled terminal 微控制终端
microcontroller 微控制器
microcontroller architecture 微控制器结构
microcontroller capability 微控制器能力
microcontroller external input signal 微控制器外部输入信号
microcontroller interface 微控制器接口
microcontroller production system 微控制器生产系统
microcopy 显微照片;缩影印刷品;缩微复制图;缩微复制品;缩microcopying 缩影复制
microcoquina 细介壳灰岩;微贝壳灰岩
microcorrosion 显微腐蚀
microcosmic 微观的
microcosmic salt 磷酸氢铵钠
microcosm(os) 微观宇宙;小宇宙;微观世界;实验生态系统
microcoulomb 微库(仑)
microcoulometer 微库仑计
microcoulometric(al) detector 微量检测器;微库仑检测器

microcoulometric(al) gas chromatography 微量库仑气相色谱(法);微电量气相色谱法
microcoulometric(al) method 微库仑法
microcoulometric(al) titration 微量库仑滴定;微电量滴定法
microcoulometry 微库仑滴定分析法;微电量分析法
microcrack 细微裂纹;细微裂缝;显微裂隙;显微裂缝;微(细)裂纹;微细龟裂;微裂(缝);微观裂缝
microcracking 微裂纹;显微裂纹;细微裂缝;微疵点;发丝裂缝
microcrack(ing) in martensite 马氏体中的显微裂纹
microcrack toughening 微裂韧化
microcrazing 显微裂纹
microcreep 显微蠕变;微(观)蠕变
microcrith 微克立司(一个氢原子重)
microcross-bedding 微交错纹层
microcross section detector 微切面检测器;微截面检测器
microcryostat 微冰冻箱
microcryptocrystalline 显微隐晶质
microcryptocrystalline texture 显微隐晶质结构
microcrystal 微晶(体)
microcrystalline 微晶质;微晶搪瓷;微晶的
microcrystalline alumina 微晶氧化铝
microcrystalline cellulose 微晶纤维素
microcrystalline dispersion 微晶分散
microcrystalline film 微晶薄膜
microcrystalline fused alumina 微晶刚玉
microcrystalline limestone 微晶灰岩
microcrystalline mica 微晶云母
microcrystalline ooze 微晶软泥
microcrystalline paraffin 微晶石蜡
microcrystalline paste 微晶浆体
microcrystalline powder 微晶粉末
microcrystalline silica 微晶二氧化硅
microcrystalline structure 微晶构造
microcrystalline texture 显微晶质结构;微晶(体)结构
microcrystalline wax 微晶体蜡;微晶蜡
microcrystallite 微晶
microcrystallization 微粒结晶;微晶结晶
microcrystalloblastic texture 显微变晶结构
microcrystallography 显微晶体学;显微结晶学;微晶学
microcrystal wax 微晶蜡
microculture 显微培养;微型培养皿
microcurie 微居里
microcurrent meter 微型流速仪
microcutting pliers 微型钢丝钳
micro-cyanide 微量氰化物
microcycle 微周期;微循环
microcycle time 微周期时间
microcyclone 小气涡
microcylinder 微柱体
microdamage 显微损伤
microdefect 微缺陷;微观缺陷
microdelta 微三角洲
microdensitometer 显微(光)密度计;显微光度计;微量(光)密度计;测微密度计
microdensitometering 微密度测定
microdensitometer parameter 微密度仪参数
microdensitometer tracing 显微密度计测定线
microdensitometry 显微密度测定;微密度测定法
microdetection 微量测定
microdetector 灵敏电流计;微量测定器
microdetermination 显微测量;微量法;微量测量;微量测定
micro-Devel test 微型狄法尔试验【道】;小型狄法尔试验【道】
microdevice 微装置
microdiabase 微晶辉绿岩
microdiablastic texture 显微筛状变晶结构
microdiagnosis 微诊断
microdiagnostics 微诊断(法)
microdial 精确分度盘;精密标度盘;精密刻度盘;精密分度盘;精密标度盘
microdialysis 微量渗析
micro-diamond 金刚石微粉
microdiecast 精密压铸
microdifferential thermal analyzer 显微差热分析仪
microdiffraction 微衍射
microdiffusion 微量扩散法;微扩散
microdiffusion analyser 微量扩散分析仪

microdilatancy 微膨胀
microdilatometer 微膨胀计
microdiorite 微闪长岩
microdispersoid 微粒分散胶体
microdisplacement meter 微位移测量仪
microdist 精密测距仪
microdistancer 精密测距仪;微波测距仪
microdistillation 微量蒸馏
microdistillation flask 微量蒸馏烧瓶
microdistillation tube 微量蒸馏烧管
microdistribution 微区分布;微观分布
microd ocumentation 缩微文献
microdosage 微小剂量;微量
microdose 微小剂量
microdosimetry 微剂量学;微剂量测定法
microdosis 微小剂量;微量
microdot 微点照片
microdot strain ga(u)ge 微粒形应变片
microdrawing 微型图
microdrill(ing) 微型穿孔;微型打孔
microdrive 微动
microdroplet status 微滴状态
microdrum 微分筒;测微鼓轮
microdrying tube 微量干燥管
microdurain 显微暗煤
microdurometer 薄样硬度计;薄材硬度计
microd ynamic feature 微动态特征
microdynamometer 微型测力计
microdyne scrubber 微达因除尘器
microdyn tester 微型强力试验仪
microearthquake 微震;微小地震
microearthquake activity 微震活动
microearthquake array 微震台阵
microearthquake data 微震资料
microearthquake forecasting 微震预测;微震预报
microearthquake movement 微震运动
microearthquake network 微震台网
microearthquake observation 微震观测
microearthquake zoning 微震区划
microecology 微生态学
microeconomic model 微观经济模型
microeconomic policy 微观经济政策
microeconomic returns 微观效益
microeconomics 微观经济学
microeconomic theory 微观经济理论
microeconomy 微观经济
microecosystem 微小生态系;微生态系(统)
microecosystem test 微生态系统试验
microeffect 显微效应;微观效应;微观效益
microelectrode 微型电极;微电极
microelectrode amplifier 微电极放大器
microelectrode log 微电极测井
microelectrode log curve 微电极测井曲线
microelectrode log plot 微电极测井图
microelectrodialysis 微量电渗析
micro-electrolysis 微(量)电解
micro-electrolysis disinfection 微电解消毒
micro-electrolysis oxidation 微电解氧化
micro-electrolysis reaction 微电解反应
micro-electrolysis sterilization 微电解杀菌
micro-electrolysis technology 微电解技术
micro-electrolysis-upflow anaerobic sludge bed-contact bio-oxidation process 微电解—升流厌氧污泥床—接触生物氧化
micro-electrolysis wastewater treatment 微电解废水处理
microelectrolytic determination 微量电解测定
microelectronic circuit 超小型电子电路
microelectronic device 微(型)电子器件
microelectronic element 微电子元件
microelectronic material 微电子材料
microelectronic modular assembly 微电子微型组件装置
microelectronic radar array 微电子学雷达相控阵
microelectronics 微电子学
microelectronic technique 微电子技术
microelectrophoresis 微量电泳法;显微电泳(法);微观电泳
microelectrophoresis apparatus 微量电泳仪
microelectrophoretic cell 微型电泳池
microelectrophoretic technique 微电泳技术
microelement 痕量元素;微型组件;微型元件;微量元素
microelement analysis 微量元素分析

microelement fertilizer 微量元素肥料
microelement grinder 微粒磨机
microelement wafer 微型元件片
microemulsion 微乳液;微乳剂;微滴乳状液
microemulsion copolymerization 微滴乳液共聚合(作用)
microemulsion membrane 微乳液膜
microenvironment 小环境;微小环境;微(观)环境
microenvironment policy 微观环境政策
microerg 微尔格
microetch 微刻蚀;微刻
microetching 显微侵蚀
microevolution 微进化;微观进化
microexamination 显微检验;显微检查;显微观察;微观研究;微观检验
microexplosion 微爆
microfabric 微组构
microfacies 显微相;微相【地】
microfacsimile 缩微相传真
microfarad 微法(拉)
microfarad meter 微法计
microfauna 微型区系;微动物群;微动物区系
microfeature 微型要素
microfeed 微量进给
microfeeder 微(量)加料器
microfiber 微(细)纤维;超细玻璃棉
microfibre = microfiber
microfiber grease 微纤维润滑脂
microfibril 微纤丝
microfibrillar angle 微纤丝角
microfibrillar structure 微纤结构
microfibrous texture 显微纤维结构
microfiche 显微照相卡片;缩微制品;缩微平片;缩微卡片;缩微胶片
microfiche viewer 缩微卡片观察器
microfield 微指令段
microfilament 微丝
microfile 高分辨度缩微胶卷
microfilm 缩微制品;缩微影片
microfilming 缩小摄影
microfilm terminal 缩微片终端
microfilter 微型滤波器;微滤器;微滤机;微过滤器
microfilter separation-activated sludge method 微滤机分离—活性污泥法
microfilter tube 微量过滤管
micro filtrate tubular membrane 微滤管式膜
microfiltration 微滤;微量过滤
microfiltration fouling 微滤污管
microfiltration membrane 微滤膜
microfiltration membrane process 微滤膜法
microfine 细微的
microfinishing 精密磨削;精滚光
microfission 微裂变
microfissure 显微裂隙;细小裂纹;微裂隙;微裂纹
microfitration membrane process 微量过滤薄膜法
microflame ionization detector 微火焰电离检测器
microflare 微耀斑;微闪光;微喷发
microflash 高强度瞬时光源
microflat bottom beaker 微量平底烧杯
microflat bottom flask 微量平底烧瓶
microflaw (钢材的)发裂纹;显微裂纹;微裂缝
microflip-flop 微型触发电路
microfloat 微浮沉子
microflo(a)tation 微量浮选;微浮选(法)
microflo(a)tation cell 微量浮选试验机;微量浮选试验槽
microfloc 微细絮凝物;微絮凝物
microflocculation 微絮凝
microflocculation contacting filtration 微絮凝接触过滤
microflocculation deep bed direct filtration 微絮凝—深床直接过滤
microflocculation-didrect filtration-biological activated carbon process 微絮凝—直接过滤—生物活性炭工艺
microflocculation fiber 微絮凝纤维
microflocculation filtration 微絮凝过滤
microflocculation sand flow filtration 微絮凝流砂过滤
microflora 微植物群;微生物群落;微生物区系
micro-flow cell 微流量电池
microflow point 显微流动点
microfluctuation 微涨落;微观起伏;微观脉动
microfluid 微流体

microfluorescence analysis 显微荧光分析
microfluorometer 显微荧光计;测微荧光计
microfluorometric(al)determination 显微荧光测定
microfluorometry 微量荧光分析法
microfluorophotometry 显微荧光光度法
microfluoroscope 显微荧光镜
microflute 微槽
microfluxion 微流结构
microfoam 微泡沫(体)
microfocus 微聚焦
microfocused log 微聚焦测井
microfocused log curve 微聚焦测井曲线
microfocused resistivity curve in dip log 微聚焦电阻率曲线
microfocus(s)ing 显微测焦法
microfog lubrication 微量喷雾润滑
microfolding 微型褶皱
microforge 显微拉制仪
microfork 显微操作叉
microform (文件等的)缩微形式;缩微印刷品;缩微过程;缩微复制品;缩微版本;小型物像
microform display device 微型显示装置
microform reader-printer 缩微胶片阅读复印器
microform rubber 微海绵橡胶
microfossil 微体化石;微化石
microfossil zone 微体化石带
micro-fouling 微污着
microfractionating column 微量分馏柱
microfractionating tube 微量分馏烧管
microfractography 显微断口谱学
microfracture 微细裂缝;微破裂;微裂(缝)
microfracturegraphy 显微断谱学;显微断口检验法;显微断口分析
microfracture system 微裂系统
micro-friction welder 低摩擦力焊机
microfungus 微小真菌
microfusain 显微丝炭
microfuse 微型保险丝
microfusible point 显微熔点
microfusion 微量熔化
microfusion method 微熔鉴定法
microfuze 微型保险丝
microgal 微伽
microgalvanometer 微量检流表
microgap 微间隙
microgap switch 微间隙开关
microgap welding 微间隙焊接
micro-gas analysis in inclusion 包体微量气体分析
microgasburner 微型煤气灯
microgasuss 微高斯
microgel 微凝胶体;微粒凝胶
microgel grease 微粒凝胶润滑脂
microgel particle 凝胶微粒
microgenerator 微型发电机
microgeographic race 小地理品种
microgeography 微(观)地理学
microgeomorphology 微地貌学
microglass 盖玻片;显微镜用载片
microglass bead 微珠玻璃
microglass electrode measuring chain 微量玻璃电极测链
microgoniometer 显微测角仪
micrograined 微粒的
micrograined limestone 微粒灰岩
micrograined texture 显微粒状结构;微粒结构
micrograin hard alloy 微晶粒硬质合金
microgram balance 微克天平
microgram 百万分之一克;微克;显微图;微观图;显微照片
microgram per liter 微克每升
microgranite 细花岗岩;微(晶)花岗岩
microgranitic 微晶粒状;微花岗状
microgranodiorite 微晶花岗闪长岩
microgranular 微晶粒状
micrograph 显微照片;显微图;显微放大器;微画器;微观照片;微动描记器;缩微照片
micrographic 显微摄影业;显微的;微文象状【地】;低倍照相的
micrographic(al)examination 显微检验;显微检查
micrographic(al) texture 显微文象结构;微文象结构
micrographics 缩微技术
micrographic test 显微镜试验
micrography 显微制图;显微照片(术);显微镜描

记法;显微绘图术;缩微摄影
microgravity 微重力
microgrid 微网栅模型法;微细网眼;微网格
microgroove 密纹
microguide 微型导管
microhabitat 小生境;微(观)生境
microhardness 显微硬度(值);微硬率;微观硬度
microhardness determination method 显微硬度测定法
microhardness head 显微硬度锥头
microhardness instrument 显微硬度计
microhardness ratio of composite to unimpregnated body 浸渍与非浸渍材料的微硬度比
microhardness scale 显微硬度计
microhardness test 显微硬度试验
microhardness tester 显微硬度计
microhardness testing 微观硬度测量
microhardness value 微观硬度值
microhardometer 显微硬度计
microH circular radio navigation 微 H 圆形无线电导航法
microheater 微型加热器
microheight ga(u)ge 高度千分尺
microhenry 微亨(利)
microheterogeneity 微观不均一性
microhistology 显微组织学
microhm 微欧(姆)(欧姆的百万分之一)
microhmmeter 微欧(姆)计
microhologram 显微全息照片;(显)微全息图
microholograph 微型全息照相
microholography 显微全息照相术;显微全息摄影;微(型)全息照相术
microhomogenisation 微量匀浆法
microhomogenizer 微量匀浆器
microhoner 微型珩磨头
microhsteresis effect 微观滞后效应
microhydra 小水螅
microhydraulic station 小型水力发电厂
microhydraulic turbine 微型水轮机
microhysteresis effect 微观滞后效应
microijolite 微晶霓霞岩
microimage 缩微影像
microimage analyzer 微影分析器
microimage data 显微数据(录在胶片上的数据);微像数据;缩像数据
micro-image storage 缩微存储[贮]
microimpurity 微量杂质
microinch 微英寸;百万分之一英寸
microinching 微动
microincineration 微量灰化
microinclusion 显微包裹体;微量含有物
microindentation 微刻痕
microindentation hardness testing 微量压痕硬度试验
microindication 微量指示
microindicator 指针测微器;米尼表;微指示器;纯杠杆式比较仪;测微指示器
microinhomogeneity 显微不均匀性;微观不均匀性
microinjury 微小损伤
microirregularity 显微不平度
microinsertion 微镶嵌
microinstability 微观不稳定性
microinstruction 微指令
microinstruction cycle 微指令周期
microinstruction debug 微指令排错
microinstruction field 微指令字段
microinstruction length 微指令字长
microinstruction register 微指令寄存器
microinstruction storage 微指令存储器
microinstrument 显微器具
microinterferometer 显微干涉仪
microinterrupt 微中断
microionics 微离子学
microionic sensor 微离子传感器
microirradiation 微束照射;微量照射
microisoelectric(al) focusing 微量等电聚焦
micro-jet roll coater 气力辊涂机
microkernel technology 微内核技术
microkymatotherapy 微波透热法
microlaikte 微晶歪碱正长岩
microlamp 显微镜用灯
microland bridge transport 微陆桥运输
microlandscape 微观景
microlandschaft 小景观;微景观

microlateral log 微梯度测井
microlateral log curve 微梯度测井曲线
microlaterolog 微侧向测井
microlaterolog curve 微侧向测井曲线
micro-lathe 钟表台式车床
microlayer 细层;微(表)层
microlens 微摄镜
microlesion 小损害
microleucosyenite 微晶英碱正长岩;微晶浅色正长岩
microlevel 微级水准仪;微级水平仪
microlevel ga(u)ge 精密水位计;微水位计
microlevel interrupt 微级中断
microlevel interrupt capability 微级中断能力
microlimnigraph 精密湖泊水位计
micro-liquid-liquid extraction 微量液—液萃取
microlite 细晶石含量;细晶石;细晶刚玉;微晶
microliter 微升
microliter syringe 微升注射器
microlith 微块;微晶
Microlithic 细石器时代的(中石器阶)
microlithic culture 细石器文化
microlithofacies 显微岩相
microlithography 缩微平版印刷术
microlithology 显微岩性学;显微岩石学;微岩性学
microlithon 微劈石
microlithotype classification 显微煤岩类型分类
microlithotypes of coal 显微煤岩类型
microlitic gabbro 微晶辉长岩
microlitic structure 微晶构造
microlitic texture 微晶结构
microlock 微波锁定
microlog 微(电极)测井
micrologic 微逻辑
micrologic adder 微逻辑加法器
micrologic element 微逻辑元件
micrologic timing circuit 微逻辑定时电路
micrologic unit 微逻辑单元
micrology 显微(科)学;镜检学
microlubricant 微润滑剂
microlug 球化率快速测定试棒
microlustre method 微光泽测定法
microlux 微勒克斯
micromachine 微级机;微电机
micromachinery 微型机械
micromachining 显微机械加工;微量切削加工
micromag 微放大器
micromagnetic survey 微磁测量
micromagnetometer 微磁磁力仪
micromanipulation 显微操作;显微操纵;精密控制;微操作
micromanipulator 小型机械手;显微镜操作设备;显微检验设备;显微操作器;显微操纵器;精密控制器;微操作设备
micromanometer 精密压力表;精测流体压力计;微压计;微压力表;微压表;微量测压计;测微压力计
micromap 微型地图
micromap colo(u)r separation 微型地图分色片
micromap generator 缩微地图绘图机
micromap generator system 微型地图绘制系统
micromap master 缩微地图原片
micromap start and stop control 缩微地图起止控制
micromap transport device 缩微地图传输装置
micromark 微测标
micromark stereo point marking instrument 微测标立体刺点仪
micromatic setting 微调装置
micromation 微型化;缩微化
micromatrix 微矩阵
micromatrix approach 微矩阵法
micromatted roller 微粒磨砂滚筒机
micromechanics 微观力学
micromechanism 微观机制
micromedia 微介质
micromelasyenite 微晶暗色正长岩
micromelteigite 微晶霓霞钠辉岩
micromelting point apparatus 显微熔点测定器
micromercurialism 微量汞中毒
micromeritic 微晶粒状;粉末状的
micromeritics 微晶粒学;粉体学
micromerograph 粒度分析仪;粒度分析器;微粒测

定仪
micromerograph analysis 气析微粒分析
micromesh 微型小网眼;微孔(网);微孔筛(网)
micromesh sieve 细目筛
micrometallic filter 微孔金属过滤器
micrometeor 微流量
micrometeorite 陨石微粒;微陨石
micrometeorogram 微气象图
micrometeorograph 微气象计
micrometeorologic(al) evaluation 小气候评定
micrometeorology 小气象学;局部地区气象学;微气象学
micrometer 显微量尺;千分尺;微米(千分之一毫米,符号为μm);测微器;测微计;测微表;测距器;分厘卡
micrometer adjusting screw 微调螺钉
micrometer adjustment 测微计调整
micrometer cal(l)ipers 千分卡规;螺旋测径器;千分测径尺;千分卡(尺);千分尺
micrometer collar 千分尺套圈
micrometer compensator 测微补偿器
micrometer constant 测微器常数
micrometer cutting depth adjustment 千分表切削深度调整
micrometer depth ga(u)ge 深度千分卡规;深度规测微器;千分深测规;深度千分尺
micrometer dial 千分刻度盘;微米刻度盘;测微仪;测微表
micrometer dial ga(u)ge 测微计刻度盘规
micrometer-driven tuning mechanism 千分卡微调机构
micrometer drum 鼓轮;测微鼓
micrometer eyepiece 测微目镜
micrometer feed 细进给;测微计进给
micrometer for inside measuring 内径千分(卡)尺
micrometer ga(u)ge (光学的)小角度测定仪;千分卡尺;测微表;测微器;千分表;测微计;测微规;测丝规
micrometer head 千分卡头;测微头;测微计头;测微鼓
micrometer hook ga(u)ge 钩形精密水位计
micrometering 微测
micrometer instrument 光学经纬仪;微米经纬仪
micrometer knob 测微旋扭
micrometer measuring rod 测微量杆
micrometer mechanism 测微机构
micrometer method 测微器法;测微法
micrometer microscope 微观显微镜;微微显微镜
micrometer microscope theodolite 测微显微镜经纬仪
micrometer microscope transit 测微显微镜经纬仪
micrometer microscopy 测微显微镜检查
micrometer mount 测微器支架
micrometer ocular 接目测微计;测微目镜
micrometer of mercury 微米汞柱
micrometer reading 测微器读数
micrometer regulator 微测快慢针调整器
micrometer rods 长杆千分尺
micrometer run 测微器行差
micrometer scale 测微尺
micrometer screw 千分丝杠;微调螺钉;测微螺杆
micrometer screw cal(l)ipers 测微螺旋卡尺;测微螺旋测微器
micrometer screw ga(u)ge 螺旋千分尺;螺旋测微器;螺旋测微计
micrometer sextant 测微六分仪
micrometer slide calipers 测微滑动卡尺
micrometer space 千分尺能伸缩的隔套
micrometer spindle 微米轴
micrometer stand 千分尺座;千分尺架;测微计座
micrometer stop 千分定位器;千分尺定位器
micrometer theodolite 测微(器)经纬仪
micrometer thimble 测微鼓
micrometer tooth rest 微米支齿点(磨工具用)
micrometer transit 测微器经纬仪
micrometer turn 测微器周值
micrometer unit 显微单位
micrometer value 测微刻度值
micrometer with dial ga(u)ge 带表千分尺
micrometer with dial indicator 千分表;带千分表的千分尺
micrometer with vernier 游标千分尺;游标测微计
micromethod 微量法
micrometric(al) 测微的

micrometric(al) measurement 精密测定
micrometric(al) microscope 微分显微镜
micrometric(al) reader 测微读数器
micrometric(al) regulating screw 微调螺钉
micrometric(al) scale 测微尺
micrometric(al) screw 微测螺旋;微测螺杆
micrometric(al) timing adjuster 精密定时调整器
micrometry 显微测量;测微数量;测微法
micromicro 毫纤;皮可;微微
micromicroammeter 微微安培计
micromicroampere 微微安(培)
micromicrocurie 微微居里
micromicrocurie per liter 微微居里每升
micromicrodiffraction 微微衍射
micromicrofarad 皮法;微微法(拉)
micromicrogram 微微克
micromicron 皮(可)米;百万分之一微米
micromicron 微微米(1 微微米 = 10⁻¹² 米)
micromilled clay 微粉碎黏土
micromillimeter 微毫米;纳米
micromillimetre = micromillimeter
micromineralogy 显微矿物学
microminiature 微小型;超小型
microminiature circuit 微型电路
microminiature diode 超小型二极管
microminiature modul 微型线路
microminiature packaged circuit 超小型装配电路
microminiature relay 微型继电器
microminiature technique 超小型技术
microminiature tube 微型管;超小型电子管
microminiature welder 微型焊机
microminiaturization 微型化;微型产品;超小型化
microminiaturized circuitry 微小型化电路
micromist 微粉
micromite ore 细晶石矿石
micromizer 胶体磨;雾化器
micromodul 超小型组件
micromodule 微型组件
micromodule electronics 微型组件电子学
micromodule equipment 微型设备
micromodule pack 微型组件封装
micromodule package 微型组件;微型封装件
micromodule stable 微型组件
micromodule technique 微型组件技术
micromolar 微克分子
micromolar concentration 微克分子浓度;微摩尔级浓度
micromolar factor 微分子因素
micromole 微摩尔
micromolecule 小分子
micromonzonnite 微晶二长岩
micromorphic continua 微观形态连续体
micromorphological analysis 微形态分析
micromorphology 微观形态学
micromosaic 细微马赛克
micromotion 分解动作
micromotion study 细(微)动作研究
micromotion unit 微动装置
micromotor 微电机;微型马达;微型电(动)机
micromo(u)lding series 微模型系列
micromount 显微载片
micromovement control system 微移动控制系统
micron 微米(千分之一毫米,符号为μm)
Micronaire 马隆尼气流式纤维细度测试仪
micronaire scale 马克隆尼值标尺
micron dimension 微米量级
microne 微粒
microneedle 显微(操作)针;微针状体
microneedle holder 微型持针钳
micronekton 微型自游泳生物
micronephelometer 微量散射浊度计
micronetwork 微小网
microneutralization 微量中和
microneutralization test 微量中和试验
micron filter (网孔尺寸以微米计的)微孔过滤器
micron gear 小模数齿轮
micron hob 小模数滚刀
micronics 超精密无线电工程
microniser 气动高速冲击磨;粉碎机
micronize 微粒化;使成为微小粒子
micronized clay 微粒黏土
micronized mica 微粒云母
micronized talc 微细滑石
micronizer 声速喷射无粉机;气动高速冲击磨;喷射式磨机;微粒化设备;微(粉)磨机;(超)微粉碎机
micronizing 微粉化
micron micrometer 精密千分尺;千分尺
micron mill 微粉磨机
micron order 精密级
micronormal 微电位曲线
micronormal log 微电位测井
micronormal log curve 微电位测井曲线
micron separator 微粉分离机
micron sprayer 弥雾机
micronucleus 小核
micronucleus test 微核试验
micron ultra-shifter 超微粉碎装置
micronutrient 微量营养
micronutrient level 微量元素含量
micronut tightener 微型螺母套紧器
micron wave 微米波
microobject 显微样品
microobjective 小型物镜;显微(镜)物镜
microocclusion 微粒夹持
microocean 微大洋
microohm 微欧(姆)(欧姆的百万分之一)
micro-Omega 微欧米加
micro-opaque 缩微卡片
microoperation 微操作
microoptic(al) level 光学测微水准器
microoptic(al) theodolite 光学测微经纬仪
microorder 微命令
microorganic contaminant 微量有机污染物
microorganic decomposition 微生物分解(作用)
microorganic pollutant 微量有机污染物
microorganics 微量有机物
microorganism 微生物
microorganism accumulation 微生物累积
microorganism adsorption 微生物吸附
microorganism biologic oxygen demand test 微生物生化需氧量试验
microorganism control 微生物控制
micro-organism corrosion 生物发霉
microorganism degradation 微生物降解
microorganism determination 微生物测定(法)
microorganism dispersant 微生物分散剂
micro-organism electric cell 微生物电池
microorganism film 微生物膜
microorganism growing adjustment agent 微生物生长调节剂
microorganism identification 微生物鉴定
microorganism intrinsic metabolism 微生物内源代谢
microorganism ratio 微生物比
microorganism sorption 微生物吸附
microorganism synthetic metabolism 微生物合成代谢
microorganism treatment 微生物处理
microorgranic contaminant 微生物污染物
microoscillation 微观波动
microoscillator 微型振荡器
microoscillograph 显微示波器;超小型示波器
microosmometer 微渗压计
micro-overconsolidated clay 微超固结黏土
microoxidation technique 微量氧化技术
microoxygenic upflow sludge bed 微氧升流污泥床
microoxygenic upflow sludge bed reactor 微氧升流污泥床反应器
microoxygenic upflow sludge blanket 微氧升流污泥床
micropacked column 微(型)填充柱
micropal(a)eontology 古微生物学;微体古生物学
micropaleontology analysis 微体古生物分析
micropane 玻璃缩微片
micropantograph 缩微片缩放仪
microparasite 微寄生物
microparticle counter 显微粒子计数器
microparticle support 微粒状载体
microparticulate packing 微粒填充
microparticulate stationary phase 微粒固定相
micropedology 微土壤学
micropegmatite 微文象岩【地】
micropegmatitic structure 显微伟晶结构
micropellet texture 微球粒结构
micropenetration 微针入度
micropenetrometer 微针入度计
microperfection 微观完整性
microperforated absorber 微穿孔吸声结构
microperforated panel 微穿孔板
microperforated panel acoustic(al) construction 微穿孔板吸声结构
microperfusion technique 微灌注术
microperthite 微纹长石
micropetrographic(al) index 微岩相指标
microphase theory 微相理论
microphenomenon 微观现象
microphilic 狭温微生物
microphilic bacteria 狭温细菌;微需气细菌
microphone 扩声器;机内话筒;麦克风;话筒;送话器;传声器
microphone amplifier 传声(器)放大器
microphone arm 传声器(支)架
microphone assembly 传声器组件
microphone boom 送话器架;传声器架
microphone capsule 传声器炭精盒
microphone carbon powder 传声器用碳精粉
microphone noise 颤噪噪声
microphone response 传声器响应
microphone sensor 微声传感器
microphone singing 传声器振鸣
microphone transducer 传声器换能器
microphone transformer 传声器变压器
microphone transmitter 送话器
microphonic 颤噪(声);颤噪效应的
microphonic bar 颤噪效应条纹
microphonic contact 颤动接触
microphonics 图像微音效应;颤噪效应;颤噪声
microphonic tube 颤噪管
microphonism 传声效应;颤噪效应;颤噪声
microphony 颤噪效应;颤噪声
microphony trouble 颤噪故障
microphoto densitometer 显微光密度计;微光密度计
microphoto density 显微光密度
microphotoelectric(al) photometer 微光电光度计
microphotogram 显微照相图;显微光谱图;分光光度图
microphotogrammetry 显微摄影测量
microphotograph 小型照片;显微照相;显微照片;显微镜摄像;微观照片;缩微照片;缩微片;缩微胶卷
microphotographic(al) camera 显微摄影机
microphotography 小型照相术;显微照相术;显微摄影术;缩微照相术;缩微摄影
microphotolithographic technique 显微光刻技术
microphotometer 显微镜光度计;显微光度计;偏光显微镜;微光度计;测微光度计
microphotometer comparator 显微光度计比较器;测微光度计比较仪
microphotometer method 显微分光光度计法
microphotometer tracing 测微光度计扫描图
microphotometry 显微光度术
micro-photosizer 光透式粒度测定仪
microphthalmoscope 小型眼底镜
microphyric 微斑状
microphysics 微粒物理学;微观物理学
microphytes 微植物
micropicnometer 微量比重瓶
micropile 小型桩
micropipe 显微缩孔;微管;晶间缩孔
micropipette 微型移液吸管;微吸管;微量吸(移)管
micro-Pitot tube 微型毕托管
microplankton 小型浮游生物
microplasma arc welding 微束等离子弧焊
microplastic yielding 微塑性屈服
microplastometer 微量塑性计
microplate 小板块;微型板块;微板(块)
microplethysmography 微差体积描记法
microplex 微粉分粒机
microplotter 微型绘图机
micropluviometer 微雨量器
micropoikilitic texture 微嵌晶结构
micropoise 微泊
micropolar 微极
micropolar analyser 测微偏振分析仪
micropolar boundary layer flow 微极边界层流
micropolar fluid 微极性流体
micropolarimeter 测微偏振计
micropolariscope 偏(振)光显微镜;测微偏振镜
micropolarograph 显微极谱仪
micropole diffusor 小孔消声器

micropolitan 微型城市
micropollutant 微(量)污染物
micropollutant removal 除微污染物
micro-polluted and eutrophic water treatment 微污染富营养化水处理
micro-polluted drinking water sources 微污染饮用水源
micro-polluted drinking water treatment 微污染饮用水处理
micro-polluted raw water 微污染原水
micro-polluted source water 微污染水源水
micro-polluted source water treatment 微污染水源水处理
micro-polluted surface water 微污染地表水
micro-polluted water 微污染水
micro-polluted water body 微污染水体
micro-polluted waters 微污染水体
micro-polluted water source 微污染水源
micropollution 微污染;微量污染
micropopulation 微型生物群;微生物群(体)
micropore (介壳虫中的)小蜡孔;小孔;细微孔隙;微孔隙;微孔的
micropore muffler 小孔消声器
microporosity 多微孔性;显微疏松;微气孔率;微孔(性);微孔隙度
microporous 微孔性的
microporous alumina membrane 微孔氧化铝膜
microporous and dense inorganic membrane 微孔密集无机膜
microporous barrier 微孔膜
microporous ceramic filter 微孔陶瓷滤器
microporous ceramic medium 微孔陶瓷介质
microporous ceramics 微孔陶瓷
microporous ebonite 微孔硬质胶;微孔橡胶
microporous film 微孔膜
microporous filter 微孔过滤器
microporous filtering film 微孔过滤膜
microporous filter method 微孔滤池法
microporous filtration 微孔过滤
microporous hollow fiber 微孔中空纤维
microporous membrane 微孔膜;超微孔(隔)膜
microporous membrane technique 微孔膜技术
microporous membrane technology 微孔膜技术
microporous plastics 微孔塑料
microporous polymer 微孔聚合物;微孔浆合物
microporous polyvinylene difluoride membrane 微孔聚偏二氟乙烯膜
microporous polyvinylene fluoride membrane 微孔聚偏氟乙烯膜
microporous rubber 微孔橡皮;微孔泡沫胶
microporous tube 微孔管
microporous tube filtration 微孔管过滤
microporphyritic 微斑状
microporphyritic texture 微斑结构
microposition 微定值
micropositioner 微位移器;微动台
micropot 微(型)电位计
micropowder 微粉;超细粉
micropower 微小功率;微功率
micro-power detector for pulsed laser 脉冲激光器微功率探测仪
microprecipitation 超细沉淀;微量沉淀
micro-pressure-ga(u)ge 微压计;千分压力表;千分压力计
microprint 缩微照片;缩微印刷品;缩微复制片
microprism 微型棱镜
microprobe 显微探针;微探针;电子探针
microprobe analysis 显微电子探针分析;微探针分析;微区分析
microprobe spectrometry 微探针度谱术
microprobing 微区探查
microprocedure 微量过程
microprocessing 显微加工;显微处理
microprocessing unit 微处理机
microprocessor 微(型)处理器;微(型)处理机;微波处理机
microprocessor assembler simulator 微处理器汇编模拟程序
microprocessor auto error correction 微处理机自动错误校正
microprocessor-based programmer 采用微处理机的程序编制机
microprocessor cache memory 微处理器超高速缓冲存储器

microprocessor chip 微处理器芯片
microprocessor code assembler 微处理器代码汇编程序
microprocessor compiler 微处理器编译程序
microprocessor control 微处理机控制
microprocessor control description 微机控制装置说明
microprocessor control system 微(处理)机控制系统
microprocessor debugging procedure 微处理机调试程序
microprocessor development system 微处理机开发系统
microprocessor electronics 微处理机电子学
microprocessor instrument 微处理器控制的仪器
microprocessor interlocking 微机联锁
microprocessor language assembler 微处理器语言汇编程序
microprocessor master clock 微处理机主时钟
microprocessor protector 微处理机保护装置
microprocessor RAM interface 微处理机随机存取存储器接口
microprocessor ROM programmer 微处理器ROM写入程序
microprocessor terminal 微处理器终端;微处理机终端
microproduction 缩微复制
microprofile 微断面
microprofilometer 表面光洁度轮廓仪;显微轮廓仪
microprogram(me) 微程序
microprogrammable processor 微程序控制的处理机
microprogrammed computer 微程序计算机
microprogrammed controller 微程序控制器
microprogram(me) development 微程序开发
microprogrammer 微程序设计员
microprogramming 微程序设计
microprogramming language 微程序设计语言
microprojection 显微映像;显微投影仪;显微投影(法)
microprojection apparatus 显微投影器
microprojector 显微映像器;显微投影仪;射微镜;显微放映机
micropublication 缩微出版
micropulsation 微脉动
micropulsation field 微脉动场
micropulsation frequency 微脉动频率
micropulsation noise 微脉动噪声
micropulsation sensor 微脉动探测设备
micropulsation whistler 微脉动哨声
micropulse 微脉冲
micropulser 矩形脉冲发生器;微脉冲发生器
micropulverizer 微粒磨碎机
micropulverizing 微粉碎
micropump 微型泵;电渗激发器
micropunch 微穿孔
micropunch plate muffler 微穿孔板消声器
micropuncture 微穿刺
micropycnometer 微型比重瓶
micropyrometer 小型高温计;显微高温计;精测高温计;微型高温计
microquake 微震
microradian 微弧度
microradiogram 显微射线照片;显微放射照片
microradiographic 显微放射照相的
microradiography 显微射线照相术;显微放射自显影;显微放射自显术;显微放射照相术;显微放射显影术;显微X射线照相术;X射线显微照相术
microradiography unit 显微射线照相机
microradiology 显微X射线法
microradiometer 微辐射计;显微辐射计
microrange photogrammetry 微距摄影测量学
microranger 小型光电测距仪;微型红外测距仪
microray 微射线;微波
microreaction 显微反应;微量反应
microreaction technique 微量反应技术
microreactor 微型反应器
microreader 微(照)片阅读器
micro-reciprocal degree 微倒度
microrefractometer 显微折射计
microrefractometry 显微折射率测量法;显微法折射率测量
microregion 微型地区
microregionalization 小区划化

microregional plotting 小区域区划
microrelief 小起伏;小地形;微(域)地形;微起伏;微量代换;微地貌
microrepiraometer 微量呼吸器
microreproduction 缩微片复制
microresistance welding 显微电阻焊
microresistivity survey 微电阻测量
microresistor 微电阻器
microrespirometry 微量呼吸测定法
microreticular resin 微孔径树脂
microretractor 微型拉钩
microrheology 微(观)流变学
microrhythm 小节律
microribbon connector 微矩形插头座
microriscosity 微黏度
microbot 微型机械手;微型机器人
microroughness 微粗糙度
microround bottom flask 微量圆底烧瓶
microrule 微型尺
microrunning 慢速运转
microrutherford 微卢(1微卢 = 10⁻⁶卢,放射性单位)
microsample 微量试样;微量称样
microsampling 微量采样
microscale 小规模;小单位;小尺度;微型比例尺;微量;微尺度
microscale crossbedding 微型交错层理
microscale distribution 微尺度分布
microscale landform 微型地貌
microscale of turbulence 紊流微尺度
microscale zone refining apparatus 微量区域提纯设备
microscan 显微扫描
microscanning 细光栅扫描;微扫描
microscanning instrument 微扫描仪器
microschlieren 显微纹影仪
microscler 长形的
microscope 显微镜
microscope adapter 显微镜转接器
microscope base 显微镜座;镜座
microscope camera 显微(镜)照相机
microscope carrier 显微镜载物台;载台
microscope count method 显微镜计数法
microscope cover glass 显微镜载体玻璃
microscope draw tube 显微镜抽筒
microscope electron 电子显微镜
microscope electrophoresis cell 显微镜电泳池
microscope for multiviewing 多人观察显微镜
microscope for research 科研用显微镜
microscope illuminator 显微镜照明装置
microscope micrometer 显微镜测微器;显微测微计
microscope photometer 显微镜光度计
microscope sizing 显微镜粒度法
microscope slide 显微镜载(片)片;显微镜载玻结;显微镜用玻片;磨砂(边)显微镜载玻片
microscope slide glass 显微镜用盖玻片
microscope slide single concave ground edges 单凹磨砂边显微镜载玻片
microscope stage 显微镜载物台;显微镜载片台;载台;镜台
microscope stand 镜座
microscope tube 显微镜筒
microscope turn table 显微镜旋转台
microscope with line 刻线放大镜
microscope work 显微镜操作
microscopic 显微镜的;高倍放大的;微观的
microscopic acidity constant 微观酸度常数
microscopic agglutination 镜检凝集
microscopical 显微镜的
microscopical analysis 显微镜分析法
microscopical chemical method 显微化学法
microscopical examination of water 水的显微镜检验
microscopic algae 微型藻;微生藻类
microscopically 显微镜下
microscopical slide 载玻片
microscopic analysis 显微镜分析法;显微(镜)分析;微观分析
microscopic anisotropy 微观各向异性
microscopic approach 微观方法
microscopic bacterial count 显微镜数菌法
microscopic capillary 微观毛细管
microscopic chemistry 显微化学
microscopic constant 微观常数

microscopic count 显微镜计算;显微镜计数
microscopic counting 显微镜计数
microscopic crystal 显微结晶
microscopic dispersion 微观弥散
microscopic dust particles 显微尘粒
microscopic economic evaluation 微观经济评价
microscopic effective cross section 微观有效截面
microscopic electrophoresis 显微电泳
microscopic examination 显微镜检验;显微镜检查;显微镜观察;金相试验;微观检验
microscopic fatigue crack 显微疲劳裂纹
microscopic field 显微(镜)视野;微观场
microscopic fissure 微裂隙
microscopic fold 显微褶皱
microscopic food chain 小型食物链
microscopic identification 显微(镜)鉴定
microscopic inclusion 微观包裹体
microscopic incomes 微观经济效益
microscopic instability 微观不稳定性
microscopic lifetime 真实寿命;微观寿命
microscopic method 显微镜法;微观方法
microscopic model 微观模型
microscopic objective 显微镜物镜
microscopic observation 微观观察
microscopic ocular 显微镜目镜
microscopic particle 微观粒子
microscopic pile theory 微观反应堆理论
microscopic precursor 微观前兆
microscopic projector 显微投影仪
microscopic reading 显微镜读数
microscopic reversibility 微观可逆性
microscopic sand 粉砂
microscopic scale 显微构造尺度
microscopic size 显微粒径
microscopic sizing 测微法
microscopic spectrophotometry 显微分光光度法
microscopic stage 显微镜载物台
microscopic state 微观态
microscopic stress 显微应力
microscopic structure 显微组织;显微构造;微型构造;微观构造
microscopic stubs 极微的突起物
microscopic system 显微系统
microscopic television 显微镜电视
microscopic test 显微镜试验;金相试验
microscopic texture 显微结构
microscopic texture examination 微观组织检查
microscopic theory 微观理论
microscopic unevenness 微观不均匀性
microscopic unit 显微单位
microscopic view 微观
microscopic void 显微孔隙;显微空隙;微型空穴
microscopic warming table 显微镜加温台
microscopic Widal agglutination test 显微威达凝集试验
microscopist 显微镜学家
microscopy 显微术;显微镜检查(法);镜检
microscosm pollution control 微观污染控制
microscosm pollution monitoring 微观污染监测
microscratch 微痕
microscreen 微孔筛网;微(孔)滤网
microscreening 微筛选;微筛分
microscrew 微型螺钉;测微螺旋;测微螺钉
microsecond 微秒;百万分之一秒
microsecond impulse 微秒脉冲
microsecond pulse generator 微秒脉冲发生器
microsecond switch 微秒开关
microsection 显微切片;显微切面;显微磨片;显微断面;显微薄片;金相切片;磨片;微观磨片
microseepage 微孔过滤筛;微渗液;微过滤;微渗(漏)
microsegregation 显微偏析;微观偏析
microseism 微震;微地震
microseismic activity 微震活动
microseismic data 脉动数据
microseismic epicenter 微观震中
microseismic forecasting 微震预测;微震预报
microseismic instrument 顶板岩层微震听测仪
microseismicity 微震活动性
microseismic monitoring network 微震监测网
microseismic noise 脉动噪声
microseismic noise spectrum 脉动噪声谱
microseismic observation 脉动观测
microseismic peak 微震峰

microseismic storms 脉动暴
microseismic zoning 微震区划
microseismogram log 微地震测井
microseismograph 微震计;微震仪;微(型)地震仪;微动计;地脉动仪
microseismology 微震学
microseismometer 微震计;地脉动计
microseism spectrum 微震谱
microsensor 微型敏感元件;微传感器
microsequence 微层序
microsere 小演替系列;微小演替系列
microservo actualor 微型伺服执行机构
microset presetting machine 精密预调仪
microshear deformation 微剪切变形
microshrinkage 显微缩孔;显微收缩;微观缩孔
microshrink-hole 显微缩孔
microsilica 硅粉
microsize 微小尺寸
microsize grade 微粒级
micro-slide 载物台
microslide 显微镜(载)玻片
microsnap ga(u)ge 手提式卡规
microsoftware 微软件
microsolid 微粒
microsolifluction 微泥流
microsolubleness 显微可溶性
microsome 微粒体
microsome fraction 微粒体部分
microsonde 微电极系
microsound scope 小型测振仪
microspaced sonde log 微极距测井
microspar 微亮晶
microspar fabric 微亮晶组构
microsparite 微亮晶灰岩
microspec function 微专用功能;特定微功能
microspecific gravity method 显微比重法;显微相对密度法
microspecimen 磨片;微观磨片
microspectrofluorimeter 显微荧光分光计
microspectrofluorometry 显微荧光分光度(测定)法;微量荧光光度法
microspectrograph 显微摄谱仪
microspectrography 显微(镜)光谱描记法
microspectrometry 显微光谱术;显微测谱术
microspectrophotometer 显微分光光度计;微量分光光度计
microspectrophotometry 显微分光光度
microspectroscope 显微分光镜
microspectroscopy 显微分光谱学;显微分光镜检查;微观谱学;微波光谱学
microsphere 精密圆球;微球体
microspheric 微球形的
microspheric(al)focused log 微球形聚焦测井
microspherically focused log curve 微球形聚焦测井曲线
microspherulitic texture 微球粒结构
microspindle 千分螺杆;千分尺轴
Microspora stagnorum 池生微胞藻
microsporinite 小孢子体
microspot 微黑子
microspot welder 微型点焊机
microspun 超细纺丝
microspur 微距
microstainer 微滤机
microstandard 微标准
microstat 显微镜(载物)台
microstate 微态;微观(状)态
microstatistics 微统计学
microstep 微步
microstepping motor 微型步进电动机
microstereocomparator 小型立体坐标量测仪
microstilb 微照提(表面亮度单位)
microstirrer 微量搅拌器
microstome 小口
microstone 细粒度油石
microstoning 超精加工
microstorage 微存储器
microstrain 微应变
microstrainer 微滤机;微滤器;微孔滤网
microstraining 微量应变
microstraining filtration 微滤
microstrand 小植物群落
microstratification 微层理
microstratified structure 显微层状构造

microstratigraphical 微观地层学的
microstratigraphical switch 微观地层(分析)仪
microstratigraphy 微观地层学
microstream 微液流
microstrength 微观强度
microstress 显微应力;微(观)应力
microstretch expander 微伸扩幅装置
microstretching 微伸
microstrip 微带;微波传送带;微波传输带;带状传输线
microstrip line 微带线
microstrips 缩微长条
microstroke 微动行程
microstructrue of powder particles 粉粒显微构造
microstructure 显微组织;显微构造;微(型)结构;微(观)构造
microstructure and fabric 显微构造和组构
microstructure manipulation 微结构变换
microstructure of clinker 熟料显微结构
microstructure of coal 煤显微结构
microstructure of metal 金属显微组织
microstructure strain 微结构应变
microstylolite 微缝合线【地】
microsublimation 微量升华
microsubmersible 微型潜水器
microsubmersible category 微型潜水器类型
microsuccession 微小演替
microswitch 精密开关;微型开关;微动开关;微动电门
microsyenite 微正长石
microsyn 精密自动同步机;微动同步器
microsyringe 微量(调节)注射器
microsystem 微小系统;微系统
microsystem organization 微系统结构
microtacticity 微观规整性
microtasimeter 微压计
microtaxonomy 微分类学
microtear 微裂纹
microtechique 精密技术
microtechnic 显微技术;精密技术
microtechnique 显微技术;微量技术
microtechnology 精密工艺
microtectonics 显微构造学;微观构造学
microtectono-geochemistry 微构造地球化学
microtelevision 微型电视
microtensile strength 显微拉力强度;显微抗拉强度
microtensiometer 测微张力计
microterminal 微终端
microtest 显微检验;精密试验
microtest plate 微量培养板
microtext 缩微资料;缩微文本;缩微版
microtexture 显微结构;微纹理;微观组织;微观结构
microtexture test 微观组织试验
microtherm 低温植物;低温
microthermal 低温的
microthermal analysis 微热分析
microthermal analyzer 微热分析仪
microthermal climate 低温气候
microthermal type 低温植物型
microthermistor 微热敏电阻
microthermodynamics 微热力学
microthermometer 精(密)温度计
micro-thin 微薄的
microthin-layer chromatography 微(量)薄层色谱法
microthroat with connecting micropore 微孔微喉型
microthrust 微推力
microtidal range 小潮差
microtime scale 时间微比尺
microtite-bearing metagranite deposit 含细晶石交代蚀变花岗岩矿床
microtiter complement fixation test 微滴补体结合试验
microtiter plate 微量滴定板
microtitration 微滴定法;微量滴定
microtitrimetry 微量滴定法
microtome 显微切片机;金相切片机;切片刀;切断器;切(薄)片机;超薄切片机;薄样切片机
microtome knife 切片机刀片
microtome knife holder 切片刀刀架
microtome method 切片法(测表面浓度)
microtome section 切片
microtomy 切片法(测表面浓度)

microtonometer 微测压计;二氧化碳分压测定仪
microtopography 小起伏;小地形;微起伏;微观地形学;微地形;微地貌;表面细微结构学
microtoroid 微型环芯
microtorr 微托(1微托 = 10⁻⁶托,压强单位)
microtorsion balance 微量扭力天平
microtrace analysis 微痕量分析
microtransistor 微晶体管
microtraps 微筛
microtremor 脉动;微震;微力震动;地面脉动
microtremor behavior 脉动性状
microtremor data 微震数据
microtremor excited vibration 脉动激振
microtremor instrument 脉动仪
microtremor measurement 脉动测量
microtremor process 脉动过程
microtremor signal 脉动信号
microtremor spectrum 脉动谱
microtriangulation 小三角测量
microtron 电子回旋加速器
microtrotrondhjemite 微晶更长花岗岩
microtsunami station 微海啸观测站
microtube 微管
microtublue organizing center 微管组织中心
microtubular membrane 微管膜
microtubular vesicle 微管泡
microtubule 细管;微小管;微细管道
microtubule assembly 微管组装
microtubule disassembly 微管拆散
microtubule fasolculus 微管束
microtubule system 微管系统
micro-tunnel 小直径隧道;小断面隧道
micro-tunnel(l)er 小直径隧道掘进机;小断面隧道掘进机
microturbulence 微湍(流态的一种);微(观)湍流
microturbulence theory 微湍流理论
microturbulent motion 微(观)湍动
microtwin 显微孪晶
microunit 微单位
microup 超小型电子管
microvalve 微量阀
microvariation 小变化;微变化
microvariometer 微型变感器
microveneer 微薄板
microvesicle 微泡
microvesicular texture 显微气孔结构
microvibraograph 微振动计
microvibrating shifter 微粉振动筛
microvibration 微小振动
microvibrograph 微震仪;微震计
microviscometer 微黏度计;测微黏度仪
microviscosimeter 微(量)黏度计
microviscosity 稀黏度;微粒度计
microvision 微波观察仪
microvitrain 小镜煤;显微镜煤
microvoid coalescence 显微空穴聚结
microvoid coalescence fracture 微孔集聚型断裂
microvoid compression stage 微裂隙压密阶段
microvoid hiding 隐匿微孔
microvoids 微孔
microvolt 微伏(特);百万分之一伏
microvolter 交流微伏计;微伏计
microvolt(o)meter 微伏(特)计;微电位计
microvolt per meter 微伏每米
microvolume method 显微体积法
microvolumetric(al)flask 微量容量瓶
microvolumetry 微容量计数法
microvortex 微涡
microwash slope 微洗刷坡
microwatt 微瓦
microwatt electronics 微瓦功率电子学
microwattmeter 微瓦计
microwave 微波固化;微波测距;微波
microwave absorbent material 微波吸收材料
microwave absorption 微波吸收
microwave absorption moisture gage 微波测定仪
microwave absorption spectroscopy 微波吸收频谱学
microwave accelerator 微波加速器
microwave acoustics 微波声学
microwave activation 微波活化
microwave aerial 微波天线
microwave amplification by stimulated emission of radiation 微波激射

microwave analysis 微波分析
microwave anechoic cell 微波消声室
microwave antenna 微波天线
microwave assisted extraction 微波辅助萃取
microwave assisted oxidation process 微波辅助氧化法
microwave attenuation ceramics 微波衰减陶瓷
microwave attenuator 微波衰减器
microwave avalanche diode 微波雪崩二极管
microwave background 微波背景
microwave background radiation 微波背景辐射
microwave beacon 微波无线电导航标
microwave brightness temperature 微波亮度温度
microwave burst 微波暴
microwave ceramics 微波陶瓷
microwave channel 微波信道
microwave circuit 微波电路
microwave communication 微波通信
microwave communication equipment 微波通信设备
microwave communication network 微波通信网
microwave communication system 微波通信系统
microwave component 微波分量
microwave course beacon 微波导标
microwave curing 微波养护
microwave cyclotron 微波回旋加速器
microwave damage 微波损伤
microwave decomposition 微波分解
microwave decomposition of toxic 毒物微波分解
microwave degradation 微波降解
microwave desulfurization 微波脱硫
microwave detector 微波检测器;微波检波器
microwave device 微波元件
microwave diagnostic apparatus 微波诊断仪
microwave diagnostics 微波诊断
microwave diagnostic technique 微波诊断技术
microwave diathermy 微波透热法
microwave digestion 微波消解
microwave discharge detector 微波放电检测器
microwave discriminator 微波鉴频器
microwave distance measuring instrument 微波测距仪
microwave distance measuring system 微波测距仪
microwave doppler speed sensor 微波多普勒效应速度传感器
microwave drill 微波钻机
microwave dryer 微波干燥器
microwave drying 微波干燥
microwave early warning 微波预警雷达;微波预警
microwave electron cyclotron 微波电子回旋加速器
microwave emission detector 微波发射检测器
microwave emissivity 微波发射率
microwave energy 微波能量
microwave engineering 微波工程学
microwave extraction 微波萃取
microwave ferrite 微波铁氧体;微波铁氧化体
microwave field 微波场
microwave filter 微波滤波器
microwave firing 微波烧成
microwave frequency 微波频率
microwave frequency spectrum 微波频谱
microwave gyrator 微波回转器
microwave heating 微波加热
microwave heating method 微波加热法
microwave heliograph 微波日像仪
microwave hologram radar 微波全息雷达
microwave holography 微波全摄影
microwave hybrid integrated circuit 微波混合集成电路
microwave illumination 微波照射
microwave image 微波图像
microwave induced catalyst 微波诱导催化剂
microwave induced oxidation process 微波诱导氧化工艺
microwave injury 微波损伤
microwave inspection 微波检验
microwave inspection of highway 微波检查公路
microwave integrated circuit 微波集成电路
microwave interferometer 微波干涉仪
microwave interferometry 微波干涉量度法
microwave label(l)ing 微波标记
microwave landing system 微波着陆系统
microwave line 微波谱线

microwave link 微波中继器;微波链
microwave maser 微波脉塞
microwave measurement 微波测量
microwave method 微波法
microwave modulator 微波调制器
microwave moisture apparatus 微波含水量仪
microwave moisture meter 微波测湿仪
microwave network 微波网络;微波通信网
microwave noise 微波噪声
microwave optics 微波光学
microwave oven 微波炉;微波烤箱;微波烘炉(快速测定土壤含水量)
microwave oven gasketing 微波炉垫布
microwave pasteurization 微波巴氏灭菌法
microwave phase equalizer 微波相位均衡器
microwave plasma 微波等离子体
microwave plasma chemical vapo(u)r deposition 微波等离子体气相化学沉积
microwave plasma deposition 微波等离子体气相沉积
microwave plasma gun 微波等离子体枪
microwave position fixing system 微波定位系统
microwave powered plasma light source 微波动力等离子体光源
microwave probe 微波探头
microwave processing unit 微波处理机
microwave propagation 微波传播
microwave protection unit 微波保护装置
microwave quantum amplifier 微波量子放大器
microwave radar 微波雷达
microwave radiation 微波辐射
microwave radio meter 微波辐射仪;微波辐射计
microwave radio metric(al) navigation system 微波无线电导航系统
microwave radio metry 微波辐射测量
microwave radio relay communication system 微波无线电中继通信系统
microwave rangefinder 微波测距仪
microwave ranging 微波测距
microwave ranging traverse 微波测距导线
microwave receiver 微波接收器;微波接收机
microwave reflectometer 微波反射计
microwave refractometer 微波折射计
microwave region 微波区;微波段
microwave relay 微波中继(器)
microwave relay communication 微波中继通信
microwave relay communication system 微波中继通信系统
microwave relay station 微波中继站
microwave remote sensing 微波遥感
microwave remote sensing of atmosphere 微波大气遥感
microwave remote sensor 微波遥感器
microwave repeater 微波中继器
microwave resonance 微波共振
microwave resonance cavity 空腔共振器
microwave resonant cavity 微波空腔谐振器
microwave rotational spectroscopy 微波旋转光谱学
microwave scanner 微波扫描仪
microwave scanning radiometer 微波扫描辐射仪
microwave scanning radiometry 微波扫描辐射测量
microwave scatterometer 微波散射仪;微波散射计
microwave sensor 微波遥感器;微波传感器
microwave shielding eyeglass 微波防护眼镜
microwave sintering 微波烧结
microwave solid state maser 固体微波激射器
microwave sounder 微波探测器
microwave sounder unit 微波探测装置
microwave spectrometer 微波频谱仪;微波分光计
microwave spectroscope 微波分光镜
microwave spectroscopy 微波频谱学;微波频谱法;微波(光)谱学;微波谱学
microwave spectrum 微波谱;微波波谱
microwave station 微波站
microwave strip 微波传送带
microwave stripline 微波带状传输线
microwave surface resistance 微波表面电阻
microwave switching tubes 微波开关管
microwave synthesis 微波合成
microwave system 微波系统
microwave technique 微波技术
microwave telephone 微波电话

microwave terminal equipment 微波终端设备
microwave terminal station 微波终端站
microwave transmission 微波传送
microwave transmission circuit 微波传输电路
microwave transmission line 微波传输线
microwave transmission tower 微波传输塔
microwave transmission unit 微波传输设备
microwave transmitter 微波发射机
microwave transmitting mixer 微波发送混频器
microwave trisponder 微波三应答定位仪
microwave tube 微波管
microwave ultrasonics 微波超声学
microwave ultrasonic wave 微波超声波
microwave ultrasound 微波超声
microwave wavemeter 微波波长计
microwave wind field scatterometer 微波风场散射计
microwave window 微波窗口
microwax 微晶蜡
microweather 小天气;小尺度天气
microweighing 微量称量
microwelding 显微焊接;微型焊接;微件焊接;微焊
microwire 超精细磁线
microwire cutter 微型钢丝钳
microyield strength 微屈服强度
microzonation 小区划分
microzone electron diffraction 微区电子衍射
microzone relfining 微区精炼
microzoning effect of earthquake 地震小区划效应
microzoning of earthquake 地震小区(域)划(分)
microzoning parameter of earthquake 地震小区划参数
microzonning 小区域划分
microzoon 微(型)动物
microzooplankton 小型浮游动物;微型浮游动物
microzyme 微胶粒
micrurgical technique 显微操作术
micrurgy 显微放大技术;显微操作术
mictite 混染岩
mictosealy texture 显微磷片结构
mi-cuit silk 半脱胶丝
micyophysical property 显微物理性质
midalkalite 霞石正长岩类
mid-America trench 中美洲海沟
mid-anchor clamp 中心锚结线夹
mid-anchor clamp for busbar 汇流排中心锚结线夹
mid-anchor clamp for catenary wire 承力索中心锚结线夹
mid-anchor clamp for contact wire 接触线中心锚结线夹
mid and high river level 中高水位
mid and long range hydrologic forecast 中长期水文预报
mid and long term hydrology forecast 中长期水文预报
mid-ash coal 中灰煤
mid-Atlantic belt 中大西洋带
mid-Atlantic ridge 大西洋中脊;中大西洋海岭;大西洋中央海岭
mid-Atlantic rift valley 中大西洋裂谷
midautumn 仲秋
mid-band frequency 频带中心频率;波段中心频率;中波频率
mid band impurity 中心带杂质
mid-band streaking 中频拖尾
midbandwidth 中心带宽
midbarrel 中间馏分
mid-batch recovery 包传输过程中恢复
mid-bay bar 海湾中间的沙洲;湾中沙洲;湾中沙滩;湾中坝
mid-bay spit 湾中沙嘴
midbin 中间仓
mid-block 中间街区;道口区间
mid-block bus stop 区段中间的公共汽车站
mid-blue 淡蓝的
mid-board 中间纸板;中(隔)墙;间壁
mid-body 中部船体;中(间)体
mid-body of ship 船体中段
mid-body of vessel 船体中段
mid caking coal 中黏结性煤
midcapacity 中等容量
mid-channel 河流中部;河道中流;中央航道;中流;中航道;航道中央;航道中心;水路的中段
mid-channel bar 心滩;江心沙洲;河中沙滩;河心沙洲;江心洲;江心滩
mid-channel buoy 中央航道浮标;航道中心浮标;航道中线浮标
mid-channel depth 航中水深
mid-channel mark 航槽中心标志;中央标
mid-channel surface slope measurement 河心(水面)比降测量
mid-circle 中点
mid class calorific value coal 中等发热量煤
mid-class mark 中组值
mid-coil 半线圈
mid-continental 中大陆的
mid-continental region 中心区
mid-continent-Andean foraminigeral realm 中大陆—安第斯有孔虫地理区系
mid-contour 中值等高线
midcourse 中途
midcourse control 中航程控制
midcourse guidance 航程引导
midcourse navigation 中间过程导航
mid crack edge 中等裂缝边
mid-cut 中针距
mid-cut mark 中线标志
midday 中午;正午;日中
midday break 午间休息
mid deck 中层甲板
mid-deep floe system 中深水流系统
midden 灰坑;垃圾坑;粪坑;垃圾堆
mid-depth 半深
mid-depth concentration 中深浓度;中深含沙量
mid-Devonian extinction 中泥盆世绝灭【地】
mid-diameter 中央直径;平均直径
middle 中央(的);中冒头;中间;中等的
middle abrasive formation 中研磨性地层
Middle Ages 中世纪
middle Alpine geosyncline 中间尔卑斯期地槽
middle America seaway 中美海路
middle and long term credit 中长期信用
middle and long term investment 中长期投资
middle and long term plans 中长期计划
middle and low-pressure vessel 中低压容器
middle and short term special loan 中短期专项贷款
middle aquifer 中等含水层
middle Asia Mongolian migration region 中亚蒙古迁移区
middle atmosphere 中层大气
middle axle 中轴
middle axle differential case 中桥差速器壳
middle band 中矿带
middle-banded structure 中条带状结构
middle bar 心滩
middle bearing 中(间)轴承
middle bench 中部台阶
middle block 龙骨墩
middle body 中央部分;船中平行体
middle box 中间砂箱
middle-bracket 中间等级(的)
middlebreaker sweep bottom 双壁开沟铲式犁体
middle buoy 中洲浮标
middle Caledonian geosyncline 中加里东期地槽
Middle Cambrian 中寒武世【地】
Middle Cambrian series 中寒武统
middle capacity of hump 中能力驼峰
middle carboniferous series 中石炭统
middle carpenter file 中木工锉
middle case 中间壳
middle casing 中间壳体
middle city 中等城市
middle class goods 中等货
middle class residential zone 中等阶层住宅区
middle clay(ey) loam 中亚黏土
middle clinoid process 鞍中突
middle cloud 中层云
middle coating 中涂;中间涂层;中层涂料
middle colo(u)r 中间色调
middle compressible soil 中压缩性土
middle contrast 影像反差适中;中等反差
middle corona 中冕
middle corridor 内廊
middle corridor type 内廊式
middle corridor type dwelling house 内廊式住宅
middle course 中游河段;中游;中流
Middle Cretaceous 中白垩世
middle cross section 中央横剖面
middle crusher 中碎机
middle crushing 中碎
middle cut 中间掏槽
middle-cut cutter 中割型切割器
middle-cut file 中纹锉
middle-cut rasp 中等锉刀
middle-day plant 中日照植物
middle deck 中层甲板
middle-deep level 中深层次
Middle Devonian 中泥盆世【地】
Middle Devonian series 中泥盆统
middle diagenesis 中期成岩作用
middle distance race 中距离跑
middle distance spray 中距离喷药
middle distillate 中间馏分;照明灯油
middle double cut file 中双纹锉
middle draft 船中(部)吃水
middle dry strength 干强度中等
middle ductility 韧性中等
middle ear barotrauma 中耳气压伤
middle earthquake 中震
Middle East 中东
middle elevator 中矿提升机
middle emulsion 中层乳剂
middle entry 中间项
middle error 二次误差
middle evolute 中渐屈线
middle exploration area 勘探程度中等区
middle family of cloud 中云族
middle fan 中扇
middle fan deposit 中扇沉积
middle field layer 中草层
middle file 中锉
middle filling machine 中切机
middle finger 中指
middle flask 中箱
middle flat file 中扁锉
middle flue damper 中间烟道闸板
middle forest 中林
middle fraction 中间馏分
middle frame 中部肋骨
middle gas field 中气田
middle gasket 中部衬
middle gear 死点位置
middle girder 中主梁;中间梁
middle gouge 弧口中凿
middle ground 折中观点;水道边浅滩;航道中浅滩;中沙;河中潜洲;潮间浅滩;分蘖沙坝
middle ground buoy 左右通航浮标;中洲浮标;中间地面标志;航道中浅滩浮标
middle ground mark 左右通航标
middle ground shoal 心滩;河中潜洲
middle half 中间的二分之一
middle half-round file 中半圆锉
middle hanger 中部吊杆
middle-heat calorie gas 中热值煤气
middle-height discharge tunnel 中高闸料通道
middle herbaceous layer 中草本层
middle-high cost clean production option 中高费用的清洁生产方案
middle hopper 中间料斗
middle housing 中小房
middle-income group 中等收入组别
middle infrared 中红外
middle infrared spectroscopy 中程红外光谱学
middle infrared spectrum 中红外光谱
middle interpretability 可判程度中等
Middle Jurassic 中侏罗世【地】
Middle Jurassic series 中侏罗统【地】
middle-keyed picture 色调适中图像
middle knife file 中刀锉
middle lane 中间车道
middle latitude 真中分纬度;中(分)纬度
middle-latitude belt 中纬度地带
middle-latitude climate 中纬度气候
middle-latitude cyclone 中纬度气旋
middle-latitude desert 中纬度荒漠
middle-latitude method 中纬度法
middle-latitude region 中纬度(地)区
middle latitude sailing 中纬度航迹计算法;中纬度航行法
Middle-latitude zone 中纬度地带
middle layer 中间层

middle leaf 中部叶
middle level 中等；中层
middle level overflow gate 中位溢流门
middle limb 中翼
middle line 中心线；中线
middle line awning stanchion 中幕中央撑住
middle line keelson 中内龙骨
middle line pillar 中线支柱
middle loam 中亚黏土
middle locator 中距导航台；中间定位器
middle lock wall 中间闸墙
middle longitude 平分经度
middle-long slope 中长坡
middle machine oil 中质机械油
middle Magdalena basin 中马格达莱纳盆地
middleman 岩层中夹层；中间人；中介人；中间商；经纪人；夹石层；调解人
middle management 中层管理
middleman's fee 佣金
middle marker 中指点标；中指标点（机场）；中点指标
middle marker beacon 中线指示信标
middle mast 中桅（五桅船的第三桅）
middle mineralized water 中等矿化
Middle Mississippian 中密西西比世
middle moraine 中碛
middle mountain 中山区
Middle Muschelkalk 无水石膏族
middle night watcher 值半夜班的人
middle nodule 中间节
middle-of-chain 链的中间
middle of curve 曲线中点
middle of stroke 中间行程
middle-of-the-road class （小客车等的）中等尺寸级
middle of water quantity 水量中等
middle oil 中油；中间馏分油；中级油
middle oil distillate 中间馏分
middle oil field 中油田
middle oil paint 中油性油漆；中油度油漆
middle ordinate 中距；曲线正矢
middle ordinates method 正矢法
Middle Ordovician 中奥陶世
Middle Ordovician series 中奥陶统
middle part 中部地区；中部
Middle Pennsylvania 中宾夕法尼亚世
middle permeable stratum 中等透水层
Middle Permian 中二迭世
middle phase of planning 中间设计阶段；中间阶段设计
middle piece 连接段
middle plane 中面（壳体）
middle plate 中板
middle pleistocene 中更新世
middle point 中点
Middle Pointed Style 中期（英格兰）哥特式
middle pole 中间杆
middle post 桁架中柱
middle post leaf 有中梃的门窗扇；中等柱叶形泥刀
middle-pressure reactor 中压反应器
middle pressure rubber asbestos 中压橡胶石棉板
middle-pressure steam turbine 中压汽轮机
middle-pressure turbine 中压涡轮机
middle price 中间价格
middle prop 中间立柱
middle proterozoic subera 中元古代
middle proterozoic suberathem 中元古界
middle punch 中穿孔
middle purlin(e) 中间桁条
middle quality 质量一般
middle rail 中间栏杆；（安装门锁的）横档；中冒头；（门的）中（横）档
middle-range 中程的
middle-range plan 中期计划
middle-range positioning system 中程定位系统
middle rasp 中等锉
middle rate 中间价
middle reach 中游河段；中段；中游；河道中游
middle relief 中等浮雕；中等松弛；中等卸载；中等缓解；中间浮雕
middle ring 中环；中等环
middle roll 中辊
middle rolling car 中行车
middle rough 中等粗糙
middle round file 中圆锉

middle rudder 中舵
middle runnings 中间馏分
middles 中矿
middle salinization 中度盐渍化
middle sample 中间试样
middle scale 中型的
middle-scale landform 中型地貌
middle school 中学
middle seam 中间矿层
middle-season rock mass 中等的岩体
middle section 中截面
middles grade 中矿品位
middle shaft 中间轴
middle sharpness 各种标志清晰度中等；清晰度中等
middle shift 中班（下午四点到半夜）
middle shore 中间斜支撑
Middle Silurian 中志留世
Middle Silurian series 中志留统
middle size 中等尺寸
middle-sized 中型的；中等尺寸的
middle-sized grains 中等大小的粒度；中等大小的颗粒
middle-sized gravel 中等大小的卵石；中等大小的石子
middle-sized gravelly sand 中等大小的砾砂
middle-sized particles 中等大小的微粒；中等大小的粒子
middle-sized sandstone 中等大小的砂岩
middle-sized sandy gravel 中等颗粒的砂质砾石
middle-sized silt 中等颗粒的淤沙；中等颗粒的泥沙
middle-sized water source 中型水源地
middle span 中跨
middle span of bridge 桥梁中跨
middle square file 中方锉
middle square method 平方取中法；中平方法
middle stage wastewater 中段废水
middle standing pillar 中（间）立柱
middle stand oil 中级熟油
middle-steep slope 中陡坡
middle stile 中竖樘；门中梃
middle stoa 中间拱廊；中间柱廊
Middle Stone Age 中石器时代
middle stratosphere 平流层中部
middle-strength explosive 中等威力炸药
middle strength rock 中等强度的岩石
middle strip 中条板；中间（地）带；跨中板板；中跨板带；无梁板结构中间板带
middle subtropical zone 中亚热带
middle surface 中面（壳体）；中面；中部面层
middle tank air 中槽空气（容器中层的蒸汽空气空间）；容器中层的蒸汽
middle temperate zone 中温带
middle temperature digective treatment 中温消化处理
middle-temperature error 平均温度误差
middle term 中项；内项
middle-term material plan 中期物资计划
middle-term plan 中期计划
middle term pollution 中期污染
middle third 中三分；中间三等分；中间的三分之一；三等分的中部一等分
middle third point 三分点
middle third rule 三等分法（则）；三分点准则
middle thread 中泓
middle thread float 中泓浮标
middle throat with connecting large pore 粗孔中喉型
middle throat with connecting small pore 细孔中喉型
middle-time solution 中间时刻解
middle transport volume 中运量
middle transverse section 中央横剖面
middle triangular file 中三角锉
Middle Triassic epoch 中三迭世
Middle Triassic series 中三迭统
middle tropical zone 中热带
middle trunk 中干
middle twill 中等斜纹
middle type 中间型
middle ultraviolet 中紫外区
middle velocity 中速
middle ventilation 中央式通风
middle vessel 中间船舶；中间导管；中间容器
middle vessel pier （教堂的）中堂的支柱

middle vessel range of columns （教堂的）中堂的柱列间的间距
middle vessel row of columns 中堂的柱列；（教堂的）中堂的列柱排列
middle vessel vault （教堂的）中堂的拱顶
middle vessel vault bay （教堂）中堂的拱顶跨度
middle vessel window （教堂）中堂窗
middle voltage transmission line 中压输电线路
middle wall （双线船闸）中墙；隔墙
middle watch 夜值；午夜值班；半夜班
middle water 中层水；层间水
middle water reuse 中水回用
middle watery overflow zone 中等富水溢出带
middle way test 中途测试
middle Weichselian glacial epoch 中魏克塞尔冰期
middle weight smooth wheel roller 中型平碾
middle wheel 中间轮
middle Wisconsin 中维斯康辛
middle work 中作业
Middle Yanshanian subcycle 中燕山亚旋回
middling 中选；中煤；第二流的；中级的；中矿；中级品；中级矿砂；中等的
middling board 中段挺紧的挡土板
middling content 中煤含量
middling content per cent graduation 中煤含量百分比分级
middling flour 粗石粉
middling particle 中矿粒
middling regrinding circuit 中矿再磨回路
middlings 中煤；中级材
middling sample 中煤样
middling sampling 中矿采样
middlings band 中矿带
middlings elevator 中矿提升机
middling yield 中等产量指标
middling zone 中矿带
mid-door 中间出车台
middy 见习生
Mideast 中东
mideddy 中涡旋
mid-European sandstone facies 中欧砂岩相
mid-extreme tide 极中潮位（最高和最低潮位的中途潮位）；正中潮位
mid-fan 中部冲积扇；中扇
mid-fan deposit 扇中段沉积
mid-fan of alluvial fan 冲积扇扇中
mid-feather 纵隔板支承的隔墙；烟道纵向分隔墙；承板；挡板；隔板；中间墙壁；烟道隔墙；纵向隔墙；吊窗隔条；中间隔墙；风火隔墙
midfeed system 中分式系统
midfellow 中间壁
mid-focal length 中焦距
mid forceps 中位钳
mid-frequency 中心频率
mid-frequency noise 中频噪声
mid-gap line 禁带中间线
mid-gear 中间齿轮；死点位置
midget 袖珍的；小型焊枪；小型的；小尺寸的；微型物；微型焊炬；微型的
midget bulb 小型灯泡
midget car 微型汽车；微型小客车
midget circuit tester 小型电路试验器
midget concrete mixer 微型混凝土搅拌机；微型混凝土拌和机
midget condenser 小型电容器
midget construction crane 微型工程起重机
midget current meter 微型流速仪
midget duster 手动风箱式喷粉器
midget excavator 微型挖掘机
midget impinger 小型冲击式采样吸收管；微型冲击式检尘器
midget implinger 微型撞击式检尘器
midget inductor 小型电感器
midget microphone 微型传声器
midget motor 小型电动机；微型电动机
midget pile driver 微型打桩机
midget plant 小型装置；小型工厂
midget polyunit 小型叠合装置
midget receiver 小型接收机
midget relay 小型继电器
midget roller 微型滚筒；微型压路机
midget self-propelled roller 微型自行式滚筒；微型自行式压路机
midget sheeps-foot roller 微型羊蹄滚筒；微型羊

足压路机
midget socket 小型套节
midget tester 小型万用表
midget tractor 小型拖拉机;微型拖拉机
midget trailer 微型拖车;微型挂车
midget tropical storm 小型热带风暴
midget truck 微型载货汽车
midget vibrating plate 微型振动切入板;微型振动片
midget water power station 小水电站
mid-girth 中央干围
mid-gley soil 中位潜育土
mid-grade metamorphic bituminous 中变质烟煤
mid-gradient method 中间梯度法
mid-gradient profiling curve 中间梯度曲线
midgrass 中高草
mid-hard rock 次坚石
mid-height 半高
midheight-deck bridge 中承(式)桥
mid high calorific value coal 中高发热量煤
mid-Himalaya old land 中喜马拉雅古陆
mid-Himalaya tectonic knot 中喜马拉雅构造结
mid humic acids coal 中腐殖酸煤
Midhurst white 灰砂砖;硅酸钙砖
midi-bus 中型公共汽车;中型出租汽车
midified groundwater 变质地下水
mid-Indian basin 中印度洋海盆
mid-Indian ocean ridge 印度洋中央海岭;印度洋中脊
mid-infrared interferometric (al) spectrometer 中红外干涉光谱仪
mid-intertidal 潮间带内的
midlake 湖中(心)的
midland 远离海洋的;中心地带;中部地区;内陆(的);内地
mid-late carboniferous transgression 中晚石炭世海浸
mid-late Jurassic climatic zonation 中晚侏罗世气候分带
mid latitude 中分纬度;(墨卡托投影的)等比例纬线
mid latitude atmosphere 中纬度大气
mid latitude sailing 中纬度航迹计算法
mid layer correction 中间层校正
mid length draft 船中剖面处吃水
mid lethal dose 中等致死剂量
mid-level rapids 中水急滩
mid-life update 中期改造
mid-lift link (拖拉机半悬挂系统的)中间提升杆
mid-lift linkage 半悬链系
mid-line 中线
midline air dose 中线空气剂量
midline exposure 中线照射量
mid-littoral 潮间带以上的
midloc 玻璃窗扣件;门式窗插销
mid-montane forest 中山地森林
mid-month settlement 月中结算
midmorning 上午的中段时间
midmost 正中
mid-mounted delivery shaken sicve digger 中间悬挂抖动输送筛式挖掘机
mid-mounted frame 中间挂架
mid-mounted mower 中间悬挂式割草机
mid-mounted reversible plough 中间悬挂式双向犁
mid-mounted tool 悬挂在拖拉机中部的农机具;半悬式农具
mid-mounted toolbar 中部悬挂机具架
mid-mounting 中间悬挂
mid-negative relief 中负突起
midnight 子夜;零时;午夜;半夜
midnight blue 午夜蓝
midnight dumper 废物偷倒人;废物倾倒者;偷倒人;倾倒者;午夜倾倒者
midnight noise 午夜噪声;深夜噪声
midnight sun 夜半太阳;子夜太阳
midnoon 正午;日中
midnormal 垂直平分线
Mid-north Tianshan marine trough 中北天山海槽
mid-ocean 外洋;远洋;洋中;公海
mid-oceanic bottom 洋底山岭
mid-oceanic canyon 洋中峡谷
mid-oceanic channel 洋内中水道
mid-oceanic dynamics experiment 洋中动力学研究计划
mid-oceanic islands 洋中群岛
mid-oceanic ridge 洋中脊;大洋中脊;中央海岭;中央海脊;大洋中央海岭
Mid oceanic ridge of Atlantic Ocean 南部大洋中脊
mid-oceanic rift 洋中裂谷;大洋中脊裂谷;裂谷
mid-oceanic rise 洋中隆;大洋中脊
midohm alloy 铜镍合金
mid-ordinate 中央纵距;中纵坐标
mid-ordinate method 中距法;中央纵距法
mid-ordinate of chord 弦中距
mid-Pacific basin 中太平洋海盆
mid-Pacific Ocean 中太平洋
mid-Pacific Ocean basin block 中太平洋海盆巨地块
mid-Pacific project 中太平洋调查计划
mid-Pacific rise 太平洋中隆
mid-parallel 中纬线
mid-part 中箱
midperpendicular 中垂线
mid-phase grounding 相中点接地
mid-phosphorus coal 中磷煤
midpiece of complement 补体中段
mid-plane 中平面;中分面
mid-plate 中板块【地】
mid-plate earthquake mechanism 板块中地震机制
mid-plate melting anomaly 板内熔融异常
mid-plate volcanism 板块间火山作用
mid Pleistocene epoch 中更新世
mid Pleistocene series 中更新统
midpoint 中性点;中心值;中点;正中时刻;平均点
midpoint anchor 中心锚结
midpoint circle 中点圆;中点
midpoint connection 中点连接
midpoint convex functional 中点凸泛函【数】
midpoint crossing 中点交叉;区间交叉(口)
midpoint line 中点线
midpoint method 中点法
midpoint of curve 曲线中点
midpoint of slope 斜坡的中点
midpoint of span 跨距中点
midpoint pole 中心锚柱
midpoint protective system 中点保护系统
midpoint rule 中点法则
midpoint tap 中心抽头
midpoint value 中点值
mid-position 中间位置
mid-position contact 中位触点
midpotential 等(电)位的
mid-Qilian island chain 中祁连岛链
midrail 中间(安全)扶手;(在护轨及站台之间的)护栏
mid-raised dwelling 多层住宅(2～7层)
mid-range 中距;中点值;中列数
mid-range estimate 中期预测
mid-range forecasting 中期预报
mid-range frequency 中频率;中频
mid-range frequency amplifier 中频放大器
mid-range frequency signal 中频信号
mid-range lateral acceleration handling maneuver 中等横向加速转向运动
mid-range performance (发动机的)中等运转性能
mid-range speed 中间速度
mid region stretch 中间灰度扩张(图像处理中)
mid reserve box 中储棉箱
mid-rise building 中层建筑;多层建筑(物)
mid-river 河流中游
mid rudder 中舵
mid-score 中间值
midsea 外海
mid seaming 中缝
mid-section 正中切开;中间截面;半节
mid-section impedance 半节阻抗
mid-section method 中断面积法
mid-separate surface seal 中分面密封
mid-series 串中剖;半串联
mid-series derived filter 半串联导出式滤波器
mid-series section 半串联节
mid-shaft 中间水平
midship 船中央;船中(部);船中部
midship area coefficient 船中剖面系数
midship beam 舯宽;船体中部主甲板横梁;船体中部宽度
midship bend 船中平行体
midship bulkhead 舯部舱壁
midship draft 船中(部)吃水
midship engine 中置式发动机
midship floor 船体中部肋板
midship frame 中部肋骨;船中部框架;船体中部肋骨
midship guy 中稳索
midship line 船中中线
midship log 船中计程仪
midshipman (船上的)见习生
midshipman's hitch 半钩结
midship mounting 中间支承
midship profile 总中剖面【船】;船体中段断面图
midships 在中线上
midship section 横中剖面【船】;船中横剖面;船中横截面;船中横断面;船体中段断面图
midship sectional area 船体中段断面积
midship section area 船中剖面面积
midship section area coefficient 船中剖面面积系数
midship section coefficient 船中横截面饱满系数;船中横断面饱满系数
midship section plan 舯剖面图
midship engine 中置发动机
midship shaft bearing 中间轴轴承
midship shaft bearing cover 中间轴轴承盖
midship spoke 上舵轮柄
midship superstructure 船中部上层建筑
mid-shunt 半并联
mid-shunt image impedance 并中剖镜像阻抗
mid-shunt section 半并联节
mid-side nodes 边界中间节点
mid-span 档距中间;跨中;跨距中点;中跨;中孔
mid-span deflection 中跨挠曲;中跨挠度
mid-span load 跨中荷载
mid-span mast 中心柱
mid-span moment 跨中弯矩;中(间)跨弯矩
mid-span reinforcement 跨中钢筋;中跨钢筋
mid-span section 翼展中央断面;中跨断面
mid specification quality 中等质量
mid-speed centrifuge 中速离心机
mid-square generator 中平方生成程序;平方取中生成程序
mid-square method 平方取中法;中平方法
midst 正中
mid stone paint 中石色油漆
midstream 中游;中流;中泓;中点水流(源与出口的中点);河(流)中心;河道中游;河道中流;水流中心线
midstream bar 江心洲;江心滩
midstream channel 河心航道
midstream depth 中泓水深
midstream transfer 水上过驳
mid-strength coal 中强度煤
mid-stroke 中间行程
mid-sulfur coal 中硫煤
midsummer 仲夏
Mid Taklimakan tectonic segment 中塔克拉玛干构造段
mid-tap 中心抽头;二道螺丝攻
mid-term production-programming forecast 中期生产规划预测
mid thermal stability coal 热稳定性中等煤
mid-Tianshan isthmus 中天山地峡
mid-timber diameter 用材中央直径
mid-timber girth 中央杆干围;用材中央干围
midtones 中间影调
midtown 城市中区
mid-travel 中间行程;中间位置(指开关位置)
mid-value 中值;中间值
mid velocity 中等流速
mid-wale corduroy 中条灯芯绒
mid-wall column 承墙柱;墙壁柱
mid-wall shaft 承墙柱
mid washability 中等可选
midwatch 午夜值班;半夜班
midwater 中层水
midwater feeder 中层摄食者;水下摄食者
midwater fishery 外海渔业
mid-water fishes 中层鱼类
midwater level 中水位
midwater mass 中层水团
midwater trap net 中层固定网;中层定置网;水下定置网
midwater vehicle 中层水潜水器
midwater zone 中层水区
mid-wave 中间波;中波
midway 半路;中途
midway atoll 中途岛环礁

midway between tracks 线间距
midway deflection 中间挠度
midway return coach 回转车
midway return operation 中途折返;中途返回
midway station 错车站
midwifery school 助产学校
midwinder 仲冬
midwing 中置翼
midwinter 隆冬
midwood 密林
mid-workings 中间采区;中间平巷
mid-year population 年中人口
mid zone draft 中区牵伸
Miebach high efficiency flash welding machine 米巴赫高效闪光对焊机
miersite 黄碘银矿
Mie scattering coefficient 米氏散射系数
Miesian architecture 密斯风格建筑
Miesian style 密斯风格【建】
Mie's scattering 米氏散射
Mie's scattering function 米氏函数
Mie's scattering intensity 米氏强度
Mie's scattering laser radar 米氏激光雷达
Mie's scattering parameter 米氏参数
Mie's theory 米氏理论
migakite 三宅岩
migma 混合岩浆
migmatic complex 混合杂岩
migmatite 混合岩
migmatite dome 混合岩穹隆
migmatitic gneissic structure 混合片麻状构造
migmatitic granite 混合花岗岩
migmatitic granodiorite 混合花岗闪长岩
migmatitic striped structure 混合条带状构造
migmatitic structure 混合岩构造
migmatization 混合(岩化)作用
migmatization way 混合岩化方式
migmatized metamorphic rock 混合岩化变质岩
mignonette 灰绿色
Migra iron 米格拉生铁
migrant 移栖动物;移居者;洄游鱼;迁移植物;迁移人口
migrant bird 候鸟
migrant labo(u)r 流动工人;流动劳动力;外籍劳工
migrant labo(u)r housing 流动劳工住房
migrant labo(u)r system 流动劳工制
migrant worker 农业季节工人
migrate 迁移
migrating 迁移
migrating bar 游移沙洲;移动沙洲
migrating colo(u)r code interval velocity section 偏移彩色编码速度剖面
migrating dune 游移沙丘;游动沙丘;移动沙丘
migrating meander 游移河弯;游荡性曲流;游荡(性)河弯
migrating property 色移性
migrating reach 游荡河段
migrating sand wave 移动沙波
migrating wave 游荡波;迁移波
migration 运移;移栖;移居;移动;洄游;迁移;飘移
migration agent 迁移动力
migrational direction of materials 物质迁移方向
migrational mass flux 迁移质量通量
migration aptitude 迁移倾向
migration area 移民区;迁移区
migration behavio(u)r 迁徙行为
migration bird 迁徙鸟
migration characteristics 耐转移性
migration circle 迁移范围
migration coefficient 迁移系数
migration constant 迁移常数
migration current 迁移电流
migration depth section 偏移深度剖面
migration distance 迁移距离;动程
migration distance, time and depth 运移距离时间和深度
migration elements 迁移元素
migration energy 移动能;迁移能
migration enhanced epitaxy 表面迁移增强外延
migration enhancement epitaxy 迁移增强外延
migration fishes 洄游性鱼类
migration form 洄游型;迁移型
migration from high population density area to low one 均民

migration in direction of ocean 向洋迁移
migration inhibition factor 移动抑制因子;迁移抑制因数
migration inhibition test 移动抑制试验
migration in space time domain 空间—时间域偏移
migration intensity 迁移强度
migration lake 迁徙湖
migration length 迁移长度
migration loss 迁移损失
migration mechanism 运移机理;迁移机理
migration model 运移模式
migration model of continuous oil phase 连续油相运移模式
migration model of gas phase diffusion 气相扩散运移模式
migration model of gas phase solution 气相溶液运移模式
migration model of kerogen network 干酪根网络运移模式
migration model of pore center network 孔隙中心网络运移模式
migration model of thermal water upward from deep basin 深盆地热水垂直运移模式
migration of divide 分水岭移动;分水界迁移
migration of earthquake 地震迁移
migration of eels 鳗的洄游;鳗的迁移
migration of elements 元素迁移作用
migration of gas 气体的运移
migration of meander 游荡河弯;曲流游荡;曲流迁移
migration of oil 石油的运移
migration of plasticizer 增塑剂渗移;增塑剂迁移
migration of population 人口迁移
migrationofrainfall 雨区移动
migration of river 河流迁移;河流迁徙;河流改道;河道迁移
migration of seismic focus 震源位置移动
migration of valley 河谷迁移
migration origin 迁移起点
migration pattern 洄游路线
migration potential 移动电位;移动电势;电泳电位;电泳电势
migration probability 迁移概率
migration rate 迁移速度;迁移率
migration rate constant 迁移速率常数
migration rate equation 迁移速度公式
migration rate theory 迁移速率理论
migration stack 偏移叠加
migration stack section 偏移叠加剖面
migration test 迁移试验
migration tool 移植工具
migration toward atlantic 向大西洋方向迁移
migration toward pacific 向太平洋方向迁移
migration tube 迁移管
migration velocity 移动速度;洄游速度;迁动速度;漂移速度;偏移速度;徙动速度
migration velocity of particle 粒子驱进速度
migration volume 卷转移;迁移卷
migrator 移居者
migratory 移栖的;移居的;流动的;洄游的
migratory anticyclone 移动性高气压;移动性高(气)压;移动性反气旋
migratory aptitude 迁移性
migratory behavio(u)r 迁移行为
migratory bird 旅鸟;候鸟
migratory bird observation station 候鸟观察站
migratory birds 候鸟类
migratory community 游移群落;迁移群落
migratory direction 迁移方向
migratory dune 流动沙丘;游动沙丘
migratory fishes 洄游(性)鱼类
migratory group 迁移群
migratory highway 迁移途径
migratory instinct 迁移本能
migratory labo(u)r 流动工人
migratory movement 迁移活动
migratory permutation 可移排列
migratory route 迁移路线
migratory species 移栖(物)种;迁移(物)种;洄游类
migratory stain 移附污染
migratory sypminihole 单臭小洞
migratory system 迁移系统
migratory wave 游荡波

migratory worker 季节性工人
miharaite 硫铋铅铁铜矿;三原岩
mihrab 壁龛(清真寺院面向麦加的那道墙内的)
mike 麦克风;送话器;传声器
mike boom 传声器柱
mike cord 送话器软线;传声器(软)线
mike noise 传声器噪声
mike stand 送话器支架
mike stew 传声器噪声
mike technique 传声器技术
miking 千分尺测量
Mikrolit 一种陶瓷刀具
Mikrolux 杠杆式光学比较仪
Mikrotast 组合杠杆式可调比较仪
Mikro-tester 米克洛硬度试验机
mil 密耳;毫英寸;千分之一英寸;千分角
Milan Cathedral Church 米兰大教堂;米兰主教堂
milarite 整柱石;铍钙大隅石
milch cattle 奶牛;乳牛【动】
milch cow 奶牛;乳牛【动】
Milchell's screw anchor 螺旋形锚
mild 软性的;柔软的
mild abrasive 柔和磨料
mild agitation 轻微搅动
mild alkaline solution 微碱性溶液
mild alloy 软合金
mild-carbon steel strip 低碳钢带;低碳带钢
mild carburizer 缓和渗碳剂;软渗碳剂
mild carburizing 缓和渗碳;软渗碳
mild channel 平缓河道
mild clay 砂质黏土;垆姆黏土;瘦黏土;亚黏土;软(质)黏土;柔性黏土;柔和黏土
mild climate 温和气候
mild cracking 轻度裂化
mild dehydration 轻度失水
mild detonating fuse 缓爆导火索
mild disinfectant 软性消毒剂
mild drawn wire 软拉钢丝
mild-duty operating condition 轻型运转状态
mildew 发霉
mildew and rot 霉烂
mildewcide 杀霉菌剂;防霉剂
mildewed surface 霉变表面
mildew prevention 防霉
mildew preventive 防霉剂
mildew-proof 防霉;不发霉的
mildew-proof latex paint 防霉乳胶漆
mildew-proof paint 防霉漆
mildew proof paper 防霉纸
mildew resistant paint 防霉涂料;防霉漆
mild extreme pressure lubricant 中度极压润滑剂
mild fever 微热
mild fire 文火
mild form 轻型
mild heat stroke 轻度中暑
mild humus 温性腐殖质
mild intoxication 轻度中毒
mild iron 软性铁;软铁
mildly alkaline soil 轻度碱性土壤
mildly arid region 轻旱境;轻度干旱地区
mildly corrosive 轻度腐蚀的
mild moderate dose 中等剂量
mild mortar 非水硬性石灰砂浆;(由石灰和砂组成的)无水砂灰浆
mild oxidation 轻度氧化
mild pollution 轻污染;轻度污染
mild sand 亚砂土;瘦(型)砂
mild sheet steel 软钢板
mild side slope 平缓边坡
mild slope 缓坡;缓比降;平缓比降
mild steel 软钢;低碳钢
mild steel angle 低碳角钢
mild steel arc welding electrode 低碳钢焊条
mild steel bar 低碳钢钢筋
mild steel channel 槽钢
mild steel checkered plate 花纹软钢片;花纹(软)钢板
mild-steel domestic boiler 软钢家用锅炉
mild steel equal angle 等边角钢
mild steel expanded sheets 钢板网
mild steel fire box 软钢板火箱
mild steel flat 低碳扁钢
mild steel hexagonal bolts 六角螺丝梗
mild steel hexagonal bolts and nuts 六角螺钉口

mild steel hollow section 低碳钢空心型材
mild steel I-beam 低碳工字钢
mild steel ingot 低碳钢锭
mild steel pipe 低碳钢管
mild steel plate 软钢板
mild steel reinforcement 软钢钢筋（含钢量 0.12%~0.25%）
mild steel reinforcing 低碳钢增强
mild steel reinforcing bar 低碳钢钢筋
mild steel shank 软钢手柄
mild steel sheet 软钢皮；低碳钢薄板
mild steel structural tube 低碳结构钢管
mild steel welding rod 低碳钢焊丝
mild steel window 低碳钢窗
mild steel wire 低碳钢丝
mild type 轻型
mild water 温水
mild wear 微缓磨损
mild weather 温和天气
mile 英里；哩
mileage 覆盖厚度；一层涂层厚度；按英里支付的旅费；按英里计算的运费；行车里程；英里里程；里数；里程
mileage allowance 以里程计算的差旅费
mileage chart 里程图；航程海图；道路图
mileage counter 里程计；里程表
mileage dial 里程碑
mileage indicator 里程指示器；航程指示仪；航程指示器
mileage number 密西西比河里程数
mileage of aids-to-navigation installed 设标里程
mileage of shipping line 航线里程
mileage of train runs 列车运行里程
mileage open to traffic 通航里程
mileage per car 每车平均英里程
mileage per gallon 每加仑英里程
mileage per quart 每四分之一加仑英里程
mileage post 里程碑
mileage scale rate 随着里程递远递减的运价
mileage sensor 里程计；里程表
mileage stone 里程碑
mileage table 里程表
mileage tester 里程试验机；燃料消耗里程测试计
mile/hour 英里/小时
mile of standard cable 标准电缆英里
mileometer 路码表；里程计；里程表
mile per hour 英里每小时
mile post 英里标；里程标；船速检验线标柱；里程碑；英制里程碑
miles of travel 行程英里数
miles on course 路程英里数；某航向上的航程
miles operated 行驶里程
miles per gallon 加仑英里
miles per hour 每小时英里数
milestone 立里程碑；英里标；重要事件；重大事件；里程标；里程碑；程标
milestone budget 里程碑预算
milestone payment 分期付款
milestone scheduling 里程碑排程
milestone state 里程碑状态
milestoning 测量打桩
mile table 航程表
mil-foot 密耳·英尺
Milford pink granite 美国麻省粉红色花岗岩
mil formula 密位公式
miliary pillar （罗马公路上的）里程碑
mil-inch 千分之一英寸
miliolite 糜棱岩
military academy 军校；军事学院
military airfield 军用飞机场
military airplane 军用飞机
military architecture 军事建筑
military base 军事据点；军事基地
military boundary 防界层
military brick architecture 砖砌军事建筑
military bridge 军用桥（梁）
military building 军用房屋；军用建筑
military building construction 军用房屋营造；军用建筑工程；军用房屋构造
military camp 兵营
military cartography 军事制图学
military chart 军用海图；军用地形图
military city map 军用城市地图
military civil engineering work 军事建筑工程
military command system 军事指挥系统
military construction 军事修建；军用飞机场建造
military district 军区
military diving 军事潜水
military effect 军事影响
military engineering 军用工程；军事工程(学)
military engineering geology 军事工程地质学
military engineering psychology 军事工程心理
military engineering survey 军事工程测量
military engineering surveying 军事工程测量学
military exercise zone 军事演习区
military function 军事功能
military game 军事策略
military gas-mask 军用防毒面具
military geodesy and cartography 军事测绘
military geodetic control point 军用控制点
military geography 军事地理(学)
military geology 军事地质学
military goods 军用物资
military grid 军用坐标网
military grid reference 军用参考坐标
military harbo(u)r 军港
military high-precision plotter 军用高精度测图仪
military hospital 陆军医院；军(事)医院
military industry 军需工业；军事工业
military infrared equipment 军用红外装置
military installation 军用设施；军事设施
military institute 军事学院
military large scale mapping 军用大比例尺测图
military level 军用水准仪
military map 军用地图
military map series 军用地图系列
military material 军事物资
military medical academy 军医学校
military medical college 军医大学
military medical school 军医学校
military meteorology 军事气象学
military model 军事模型
military motor lorry 军用卡车
military museum 军事博物馆
military oceanography 军事海洋学
military optic(al) instrument 军用光学仪器
military platform 军用站台
military port 军港
military post 军用哨所
military practice area mark 军事演习区标
military project 军事工程
military project survey 军事工程测量
military railroad 军用铁路
military railway 军用铁路；军用铁道
military reconnaissance 军用勘察
military reserve decrease 削减军需储备
military reserve increase 增加军需储备
military road 军用公路；军用道路
military road survey 军用道路测量
military ruggedized tube 军用强化管；军用耐震管；军用可靠管
military saddle 军用鞍
military school 军校
military sketch 军用地形略图
military specification 军用规格
military specification design 按军用规范设计
military standard 军用标准
military statistics 军事统计
military stores 军需品
military stronghold 军事据地；军事要塞
military surgeon 军医
military survey 军用测量
military surveying and mapping 军事测绘
military system engineering 军事系统工程
military telescope 军用望远镜
military terminal 军用码头；军港
military topographic(al) data 军事测绘资料
military topographic(al) equipment 军事测绘装备
military topographic(al) service 军事测绘勤务
military topography 军事地形学
military tower 军事碉堡
military toxicant 军用毒物
military traffic controller 军事运输调度员
military train 军用列车
military transport charges paid after arrival 军运后付运费
military transport charges receivable 应收军运后付运费
military tunnel 军用隧道
military upper traffic control area 军事高空交通管制区
military vehicle 军用车辆
military waste 军事废物
military wheel load 军队轮载
military zone 军用地带
milk 乳状物
milk agitator 牛奶搅拌器
milk and cream pump 牛奶与奶油泵
milk bar 乳吧间；牛奶房；牛奶铺
milk-bottle bobbin 奶瓶形筒子
milk bottle washer 奶瓶清洗机
milk can conveyer 奶桶输送机
milk cattle 奶牛
milk churn 大奶桶
milk coagulant 凝乳剂
milk cooler 牛奶冷却器
milk cooling equipment 牛奶冷却设备
milk cooling unit 牛奶冷却装置
milk dispenser 牛奶装瓶机
milker 挤奶装置；分别充电发电机
milk flow meter 牛奶流量计
milk glass 乳白玻璃；乳白色玻璃
milk-glass scale 毛玻璃标度盘
milk goat 奶山羊
milk house 牛奶房
milkiness 乳浊；乳状；乳白色；清漆漆膜发白；漆膜发雾
milking bay 挤乳框；挤乳台
milking booster 补充充电的升压电机
milking generator 电池充电用低压直流电机；分别充电发电机
milking machine pump 挤奶装置泵
milking parlour 挤乳休息室
milking stage 灌浆阶段
milking system 挤奶作业系统
milk jug 牛奶壶
milk lifter 输奶泵
milk lorry 牛奶运输车
milkness 乳白色
milk of bismuth 铋乳
milk of lime 石灰乳；石灰泥浆；石灰浆
milk of lime coat 石灰浆涂层；石灰浆面层；石灰浆粉刷层
milk of lime feeder 石灰浆投加量
milk of magnesia 镁乳
milk opal 乳蛋白石
milk parlour 挤奶厅；挤乳厅
milk plant 牛奶厂；乳品厂
milk-processing waste 牛奶加工厂
milk production equipment 牛奶加工设备
milk product waste 乳制品废物
milk product wastewater 乳制品废水；乳品加工废水
milk pump 奶泵
milk sap 乳状汁；乳状液
milk scale 乳白度
milk scale buret(te) 乳白刻度滴管
milk spots 乳状斑；乳色斑
milk-stone 白燧石；乳石
milk the street 操纵(有价)证券从中牟利
milk-transfer system 输奶系统
milk waste 制奶废水
milkweed butterfly 斑蝶
milk well 乳井
milk white 乳白色
milky 乳白色的
milky crust 融化壳
milky glass 乳白玻璃；玻璃瓷
milky glaze 乳白釉
milky quartz 乳白石英；乳(色)石英
milky sea 沉淀性(乳白色)发光；海面磷光
milky surface 乳白镀层
milky surface of glass 乳白面玻璃
milky water 浑浊水
Milky Way 银河
milky weather 乳白天气；乳白天空
milky-white 乳白色；乳白
milky-yellow 乳黄色
mill 再加工车间；工厂；厘(千分之一美元,只用于统计)；木材加工工厂；磨(炼)机；磨坊；成型车间
Milla 米拉硬木(印度产)
millability 可轧性；可铣性

millable 可轧的;可铣的
mill addition 磨添加物;球磨机添加物料
mill addition method 磨加法
mill aisle 轧机跨
mill approach table 轧机输入辊道
mill apron 返料带
mill atmosphere 磨机气流
mill attendant 看磨工
mill auxiliaries 轧机的附属设备
mill ball 研磨球
mill bar 轧材
mill barrel 磨碎机筒;磨矿机滚筒
mill base 漆浆
mill batch 磨机配料
mill bay 轧机跨
mill bed plate 底盘
mill bench 选矿凳
mill bent 磨坊的立柱屋架;厂房立柱屋架;厂房的主柱屋架
mill blank 夹层纸板
mill board 马粪纸;碾压板;麻丝板;书皮纸板;厚柏纸
millboard roofing material 屋面麻丝板
mill building 磨房;磨坊式厂房;厂房
mill building bent 厂房排架
mill cake 磨饼
mill casing 磨机外壳
mill certificate 钢厂合格证;钢厂材料证明书
mill certified 企业标准
mill charge 磨内荷载;磨机装料
mill charge monitoring 磨机喂料监控
mill chats 选矿厂废石;采石场废料
mill cinder 轧屑
mill circuit 磨机循环;粉磨流程
mill coal 工厂用煤
mill-coated 工厂绝缘的
mill-coated pipe 厂涂涂层管子
mill coil 热轧窄带钢;盘圆(细钢筋或钢丝)
mill condition 轧制规范
mill construction 半防火结构(楼层木梁厚度不小于6英寸);工厂建筑;耐火木结构;耐火构造;厂房结构
mill control by permeability fineness meter 透气细度仪控制磨机
mill control expert system for grinding process 粉磨工艺的磨机控制专家系统
mill cost 工厂生产费
millcourse 磨槽
mill cull 下脚料;工厂下脚料;等外材
mill cylinder 磨机筒体
mill dam 小坝;磨坊(水)坝;水闸
mill data 轧机参数
mill delivery table 轧机输出辊道
mill dell 楞台
mill diameter inside liners 磨机内径
mill diaphragm partition 磨机隔仓板
mill discharge 磨机排料
mill disintegrator grinder 细碎研磨机
mill drive 磨机传动
mill dust 粉尘
mille 商用单位数(相当于10个×120个);千量材尺
milleb wood 铣成的木材
milled 研磨的;铣成的;滚出齿槽的;磨光的
mille dam 矿井防水墙
milled bit 铣齿钻头
milled bolt 铣制螺栓
milled border 研磨边缘
milled clay 磨细的白土;漂白土
milled cloth 毡绒织物
milled edge 铣成边
milled edge thumb screw 铣边翼形螺钉
milled end 正方端
milled fair face 加工的光面
mill edge 热轧缘边
milled glass 磨光玻璃
milled glass fiber 磨碎玻璃纤维
milled granite 粉碎的花岗岩
milled head 滚压头
milled lead 铅板;碾压铅板;铅皮;碾压铅皮
milled natural stone 修琢加工过的石块
milled nut 铣制螺母;滚花螺母
milled plate 轧制板
milled product 轧制成品;铣成的产品
milled refuse 压实废物;粉磨垃圾
milled ring 滚花环

milled screw 滚花头螺钉
milled sheet 轧制板
milled sheet lead 薄铅片
milled soap 研压块皂
milled stonework 修琢加工过的石方工程;修琢加工过的石料
milled thin section 磨光薄膜
milled tooth 铣齿
milled wood lignin 磨木木素
mille face 采矿工作面
mille map 百分点地图
millenary 千周年纪念
mill end 磨头
mill end trunnion 磨端轴颈;磨端中空轴
mill engraving machine 钢芯轧纹机;钢芯雕刻机
millennium 千年期;一千年;千周年纪念
mill equipment 粉碎厂设备
miller 铣床;看磨工
miller center 铣床顶尖
Miller classification of rock 密勒岩石分类
Miller effect 密勒效应
miller hand 铣工
Miller index 密勒指数
Miller integrator 密勒积分器
millerite 针镍矿
Miller machine 密勒全自动辘轳机
Miller radiator 密勒发射天线
Miller sawtooth generator 密勒锯齿波产生器
Miller square code 密勒平方码
Miller's symbol 密勒符号
millesimal 千分之一;千分
millet 小米
millet in husk 谷子
millet-seed sandstone 粟粒状砂岩
mill exhauster 排粉机
Millexpert 粉磨专家系统(产品名)
mill fan 排粉机
mill feed 磨机给料
mill feed bin 磨头仓;磨机给料仓
mill feed control by air separator 空气选粉机控制磨机喂料
mill feed control by separator and acoustics 选粉机和声响信号控制磨机喂料
mill feed control by weigh(ing)circulating load 称量循环负荷控制磨机喂料
mill feed control with electric(al)ear 电耳控制磨机喂料
mill feeder 磨机喂料机;磨机给料器
mill feed hopper 磨机喂料斗;磨头仓
mill feed material 入磨物料;粉磨物料
mill feed size 入磨粒度
mill file 粗纹锉刀;单纹平锉;锉;扁锉
mill finish 磨光;轧光;机械加工光洁度;轧制光(洁)度
mill finished sheet 精轧板;机磨光片
mill finish printing paper 粗道林纸
mill finish sheet 精轧板
millfleurs garden tapestry 什锦香精花园挂毯
mill floor 车间地面;厂房楼面;厂房地面
mill flour 磨面
millfore glass 千花玻璃
mill for rolling sections 型钢轧机
mill friction 轧机部件内的摩擦
mill furnace 加热炉
mill furnace cinder 轧钢屑;加热炉渣
mill-gate price 出厂价(格)
mill general layout 轧机设备的总布置
mill grease 磨机用润滑脂
mill-hand 研磨工人
mill hardening 轧制(余热)淬火
mill head 铣头;磨头
mill head assay 入选原矿分析
mill headed screw 滚花头螺钉
mill-head elevator 磨矿机卸料端提升机
mill hole 露天放矿漏斗;放矿口
mill housing 轧机机架
mill housing cover 粉碎室上盖
mill housing window 轧机牌坊窗口
milliammeter 毫安(培)计;毫安表
milliampere 毫安培;毫安
milliampere man 弱电工程师
milliampere meter 毫安表;毫安(培)计
milliampere-second 毫安秒;微安(培)秒
milliard 十亿(英国);十万万

milliarium 里程碑;距离单位(等于1.48公里)
milliarium aureum 金质路标柱(在罗马帝国马路的终点)
milliatom 毫克原子
millibar 毫巴(测气压的单位)
millibar-barometer 毫巴气压表
millibarn 毫靶
millibarometer 毫巴气压表;毫巴气压计
millibar scale 毫巴标尺
millicoulomb 毫库仑
millicurie 毫居里
millicurie-destroyed 衰变毫居
millicycle 毫赫
millidarcy 毫达西
milli equivalent 毫(克)当量
milli-equivalent method 毫克当量法
millifarad 毫法(拉)(电容单位)
milligal 毫伽
milligamma 毫微克
milligamms per liter 毫微克每升
milligauss 毫高斯
milligram 公丝;毫克
Milligramage 放射性照射量(每小时每毫克镭)
milligram-atom 毫克原子
milligram-hour 放射性照射量(每小时每毫克镭)
milligram-ion 毫克离子
milligram/liter 毫克每升
milligrammes per liter 每升毫克数
milligram-molecule 毫克分子
milligram radium equivalence 毫克镭当量
milligrams loss per square decimeter per day 每日每平方分米毫克损失
milligrams per liter 毫克每升
millihenry 毫亨
millihertz 毫赫
milli Hg 毫米汞柱
milli-inch 密耳·英寸
milli-jansky 毫央斯基
Millikan electrometer 密立根静电计
Millikan oil-drop experiment 密立根油滴试验
Millikan rays 超X射线
milliken conductor 分割导体
millilambda 毫微升
millilambert 毫朗伯
milliliter 毫升
millilux 毫勒克斯
milli-mass-unit 毫质量单位
millimeter 公厘;毫米
millimeter for nominal size 公称直径以毫米表示
millimeter of mercury 毫米水银柱高;毫米汞柱
millimeter of water 毫米水柱
millimeter paper 米格纸(俗称);毫秒格纸
millimeter squared paper 毫米方格纸
millimeter wave 毫米波
millimeter wave communication 毫米波通信
millimeter wave generator 毫米波振荡器
millimeter wave radar 毫米波雷达
millimeter wave satellite communication 毫米波卫星通信
millimeter wire ga(u)ge 毫米线规
millimetric(al)wave 毫米波
millimicro 毫微
millimicrodolomitization 毫微白云岩化
millimicrofarad 毫微法拉;纳法
millimicron 纳米;毫微米
millimicrosecond 纳秒;毫微秒
millimole 毫分子;毫分子量
millinery felt 女帽毡
millinery store 妇女用品商店
milling 铣削加工;铣削;铣(法);研磨;轧制;碾磨
milling action 磨碎作用
milling aids 研磨助剂;粉碎助剂
milling angle 铣角;分度头心轴偏角
milling arbor 铣刀杆
milling arbour 铣刀轴
milling attachment 铣削装置;铣削附件
milling automatic 自动铣削
milling back flute 铣背槽
milling capacity 铣机生产能力;磨机产量
milling characters 磨粉品质
milling cone groove 铣锥形槽
milling cutter 铣刀
milling cutter arbor 铣刀轴
milling cutter diameter 铣刀直径

milling cutter grinding machine 铣刀磨床
milling cutter shank 铣刀柄
milling cutter sharpening machine 铣刀修磨机
milling cutter spindle 铣刀杆
milling cutter tooth form 铣刀齿形
milling depth 铣削深度;铣切深度
milling equipment 研磨设备;研磨机(械);磨碎设备
milling excavator 滚切式挖掘机
milling fixture 铣削夹具;铣刀夹具
milling flat 铣平面
milling flutes 铣槽
milling gear teeth 铣齿轮齿
milling grade 可精选的矿石
milling grinding 细碎;磨矿
milling head 铣刀头
milling jig 铣床夹具
milling loss 磨粉损耗
milling machine 铣床;研磨机;磨削机;切削机
milling machine accessory 铣床附件
milling machine arbor 铣刀轴
milling machine dog 铣床轧头
milling machine hood 铣床抽风罩
milling machine operation 铣削操作
milling machine operator 铣工
milling machine parts 铣床零件
milling machine shop 铣工车间
milling machine speed dial 铣床转速刻度盘
milling machine spindle 铣床主轴
milling machine table 铣床工作台
milling machine upright 铣床立柱
milling machine with universal turret head machine 万向转塔铣床
milling mark 铣削刀痕
milling medium 磨矿介质
milling method 磨矿法
milling off 研光;铣去
milling of grooves 铣槽
milling of limestone 石灰石的处理;石灰石的粉磨
milling of ore 选矿;矿石处理
milling operation 选矿作业;铣削操作
milling orange 磨橙
milling ore 需选矿石;二级矿石
milling ore grade 入选矿石品位
milling out a bridge plug 钻桥塞
milling pit 磨矿机
milling pitch 滚花节距
milling plant 粉磨设备;粉磨车间
milling prophase operation 选矿前作业
milling pulverization 磨碎
milling ring 磨削环;铣环
milling shoe 铣加工导座;磨鞋
milling slide 铣削刀架;铣齿刀架
milling straight flute 铣直槽
milling tap 铣锥;套铣筒
milling time 铣削时间
Millington reverberation formula 米林顿混响公式
milling-tool 铣刀
milling unit head 铣削动力头
milling work 铣工工作
mill inlet 磨机进料口
millinormal 毫当量的
mill interstand tension 轧机机架间张力
milliohm 毫欧(姆)
milliohmmeter 毫欧表
million 百万
million barrels per day 百万桶每日
million electron-volt 兆电子伏特
million gallons/year 百万加仑/年
million gross tones 通过轨道总重
million instructions per second 每秒百万条指令;百万条指令每秒
million years 百万年
milliosmol 毫渗克分子
millipore 微孔的
Millipore aeration 微孔曝气
millipore chamber 微孔扩散盒
millipore filter 微孔滤膜;微孔(过)滤器
millipore filtration method 微孔筛方法
Millipore membrane bioreactor 微孔膜生物反应器
millipore membrane filter technique 微孔薄膜过滤法
millirad 毫拉德
milliradian 毫弧度
milliroentgen 毫伦琴

milliroentgnometer 毫伦琴计
milliscope 金属液温度报警器
millisecond 毫秒;千分之一秒
millisecond blasting 毫秒爆破
millisecond delay 极短时延雷管
millisecond delay blasting 毫秒延时爆破;毫秒延发爆破;毫秒(延)迟爆破
millisecond delay detonator 毫秒延时雷管;毫秒延发雷管;毫秒延迟雷管
millisecond delay electric(al) detonator 毫秒级的延缓电雷管;毫秒延时电雷管;毫秒延发电雷管;毫秒延迟电雷管;微差迟发电雷管
millisecond delay shotfiring 毫秒延时爆破;毫秒延发爆破;毫秒延迟发爆破
millisecond detonator 毫秒雷管
millisecond electric(al) blasting cap 毫秒电爆雷管
millisecond meter 毫秒计;毫秒表
millisecond priming 微差起爆
millisecond pulsar 毫秒脉冲星
millisecond round 毫秒迟发炮眼组
millisecond wavelength scanning spectrometer 毫秒波长扫描分光计
millisite 水磷铝碱石
millisomole 毫渗量
millite 莫来石
millite-based composite 莫来石基复合材料
millite block 莫来石砌块
millite brick 莫来石砖
millite ceramics 莫来石陶瓷
millite-cordierite refractory 莫来石—堇青石耐火材料
millite fiber 莫来石纤维
millite porcelain 莫来瓷
millite refractory 莫来石耐火材料
millite whiteware 莫来石瓷
millitization 莫来石化
millitsmol 毫渗透压克分子
millivolt 毫伏(特);千分之一伏
millivolt ammeter 毫伏电流表;毫伏安培计
millivoltampere 毫伏安
millivoltmeter 毫伏计;毫伏表
millivoltmeter pyrometer 毫伏计式高温计
millivoltmeter regulator 毫伏计调节器
milliwatt 千分之一瓦;毫瓦(特)
milliwattmeter 毫瓦计
mill knife 磨刀
mill language 轧制专业用语
mill layout 轧机布置
mill lead 铅片
mill length 轧制长度;管子出厂长度
mill level control 磨机料位控制
mill limit 轧制公差
mill liner 磨矿机衬里;磨机衬里;磨机衬板
mill liner wear 磨机衬板磨损
mill lining 磨矿机衬里;磨衬
mill load control system 磨机负荷控制系统
mill load(ing) 磨机负荷;轧机负荷
millman 选矿工
mill market 制造者市场
mill-mixed 厂拌的
mill-mixed gypsum concrete 预拌石膏混凝土
mill-mixer 粉碎搅拌机;粉末混合机
mill mixing 辊筒混合
mill mixture 预拌料
mill motor 磨坊用电动机
mill of edge runner type 双辊碾磨机;研磨机
mill off 铣掉
Millon's base 氨络双氧化汞
mill opening 轧辊开度
mill operator 轧钢工
millosevichite 紫铁铝矾
mill outlet 磨机卸料口
mill over meal bin 装在储[贮]粉箱上的磨粉机
mill pack 叠板;叠轧板材;单张薄钢板
mill-painted 车间油漆的;工厂油漆的
mill partition 磨机隔仓板
mill pebble 磨砾(砾磨机用)
mill personnel 轧机操作人员
mill pinion 齿轮座的齿轮轴
mill pond 储[贮]木场;磨坊(储[贮])水池
mill power draft 磨机需用功率
mill practice 工厂的标准生产方法和程序
mill price 厂价
mill primary hopper 磨头仓

mill-primed 工厂打底漆的
mill primer 工厂底漆
mill processing 加工加工
mill product 工厂产品;轧制成品;磨机产出物
mill production 轧机生产能力
mill pulpit 轧机操纵室
mill quenching 轧制余热淬火
mill race 磨坊引水槽;磨坊进水道;磨槽;水车进水槽
mill rigs 缩呢折痕
mill roll 轧辊
mill roller 轧钢工
mill rolling practice 轧制实践
mill roll opening 滚隙
mill roll scale 轧制铁鳞
mill room 辗料间;碾料间
mill run 未分(等)级的;出材量;水车用水流
mill run alumin(i)um plate 砂目铝板
mill run-in table 机前输入辊道
mill-run mortar 机制砂浆
mill run-out table 机后输出辊道
mill sampling 选矿取样
mill saw 框锯
mill scale 热轧氧化皮;锈皮;氧化皮;轧(制)铁鳞;轧制鳞片;轧屑;轧钢鳞皮;黑皮;热处理鳞状结构热;铁氧化皮;碎锈铁片
Mill's colo(u)rimeter 米勒比色汁
mill scrap 工厂废品
mill screw 轧机压下螺钉
Mills cross bottom mapping sonar 精密海底地图测绘声呐
mill separator 机座横梁;粗粉分离器
mills error 千分误差
mill setting 轧机调整
mill shears 轧制剪切机
mill sheet 制造工艺规程;钢厂检验单资料;材料成分分析表
mill shell 粉碎机壳罩;磨机外壳
mill shoe 轧机鞋
millsite 选矿厂厂址
mill slurries 磨机矿浆
mill solution 工艺溶液
mill speed 轧制速度;磨机转速
mill spindle 轧机连接轴;竖轴
mill spring 轧机机座的弹势;轧机弹跳
mill springing 轧机弹跳
Mills-Reincke phenomenon 米尔斯—内克现象【给】
mill stand 轧机机座
mill stand cap 轧机机座盖
mill star 星形铁
millstone 细砂质磨石;磨石;磨盘石;粉碎器
millstone cutter 磨盘上的刻槽
millstone grit 磨石(粗)砾砂;磨石
mill streaks 轧制条痕
mill stream (工矿的)引水渠;引水道;磨机流量
mill surface 磨碎面
mill table 轧机辊道
mill tackle 轧机吊索工具
mill tail 磨坊放水槽;水车尾水渠;水车出水槽
mill tailings 工场废渣;选矿厂尾矿
mill tap 轧制铁鳞;轧机铁鳞
mill teeth 磨齿
mill to death 重复碾磨
mill tooth 臼齿
mill train 轧机机组;轧钢机组;轧道
mill trammel 连接磨机的滚筒筛;磨机滚筒筛
mill-turn 铣车床
mill-type classifier 现场用分级机
mill-type cylinder 大型油缸
mill-type lamp 防震灯泡;耐震灯泡
mill-type timber construction 抗震型木结构
mill under-thickness tolerance 出厂厚度负公差
mill under tolerance 钢厂负公差
mill vent air 磨机通风
mill ventilation 磨机通风
mill water 选矿用水;选厂用水
mill-weir 水堰
mill weld 工厂熔焊;工厂焊接
mill wheel 磨坊水轮
mill white 细磨白漆;厂房用白色涂料
mill with flat hammers and slotted grill 带平面锤和栅式格筛的磨机
mill with horizontal meal mixer 带水平式粗粉混合器的磨粉机

millwork 工厂预制(木构)件;细木工制品;预制木构件;工厂机械的安装;磨光工作;水磨机械;定制木作
millwright 设备安装工;装配工人;磨坊安装工;水磨匠;水车工;设备的安装修理工
milneb 代森环
Milne-Shaw seismograph 米尔恩—肖氏地震仪
milori blue 米洛丽蓝(一种铁亚氰酸盐,铁蓝的别名)
milori blue pigment 米洛丽蓝颜料
milori green 米洛丽绿
milori green pigment 铬绿颜料
miloschite 铬铝英石
milpa 栽培地
milpa system 烧林垦地法
milrule 弦线量角器;密位尺
milscale 密位刻度尺;千分度盘;千分尺
milt-cell framed tube structure 框筒群结构
miltiple cropping systems 多熟耕植物
miltiple-furrow 多犁体的
mil unit of angular measure 密位
milunit of thickness 密耳(等于千分之一英寸)
Mima mound 小土墩
mimbar 穆斯林布道坛;(伊斯兰教寺院的)讲经台;清真寺讲经坛
Mimella 拟态贝
mimeo 油印品
mimeograph 油印品;油印机;滚筒油印机
mimeograph machine 油印机
mimeograph room 油印机室
mimesite 粒玄岩
mimetic 拟态的;拟晶的;模拟的;后变晶的
mimetic crystallization 拟晶结晶【地】;后构造结晶;后变结晶;似晶结晶
mimetic diagram 模拟图
mimetic recrystallization 拟态重结晶
mimetic tectonite 拟构造岩
mimetite 砷铅石;砷铅矿
MIM Holdings Ltd. 芒特艾萨矿业控股公司(澳大利亚)
mimical 模拟的
mimic board 模拟屏;模拟盘
mimic board driver 模拟盘驱动器
mimic board interface 模拟屏接口
mimic bus 模拟母线;模拟电路
mimic channel 模拟电路
mimic colo(u)ration 拟色
mimic colo(u)ring 模拟色;保护色
mimic diagram 站场模型图;站场模拟图;模拟(线路)图;模拟现场图;仿真图
mimic-disconnecting switch 模拟断路器;模拟断路开关
mimic imput signal 模拟输入信号
mimic mutant 模拟突变体
mimic panel 模拟显示屏;模拟屏
mimic panel interface 模拟屏接口
mimic photosynthesis 模拟光合
mimicry 拟态;模仿;仿制品
mimic system panel 模拟系统板
mimic transmission line 模拟输电线
mimiteller 小型出纳机
minability 可开采性
minable boundary 可采边界线
minable thickness 可采厚度
minable thickness of coal 煤层可(开)采厚度
minable width 可采宽度
Mina Jebel Ali 阿里山港(阿拉伯联合酋长国)
minal 纯矿物
Minalith 木材防腐剂
Minalpha 米纳尔法铜锰镍合金;锰铜标准电阻丝合金
Minamata bay 水俣湾
Minamata disease 水俣病
Minamata disease control 水俣病防治
Minamata disease incident 水俣病事件
minamiite 钙钠明矾石
minar 角楼;望楼;灯塔;印度纪念塔
Mina Rashid Port 拉希德港(阿拉伯联合酋长国)
minaret 伊斯兰教寺院光塔;清真寺光塔;叫拜楼;〈旧〉回教(寺院)尖塔
minaret-like 仿伊斯兰教寺院光塔的
Minargent alloy 明纺金特铜镍合金
Minas Gerais type iron deposit 米纳斯吉拉斯型铁矿床

minasragrite 钒矾
minbar 敏拜楼(清真寺内的讲坛);清真寺讲经坛
mince 剁肉
minchery 女修道院
minch house 休息所;小旅店;路边客店
mincing knife 剁刀
mincing machine 切碎机;剁碎机
min-cut 极小截;极小割
Mindel glacial epoch 民德冰期
Mindel-Riss Ice Period 民德—里斯冰期
Mindel-Riss interglacial stage 民德—里斯间冰期
Mindel(stage) 民德阶【地】
minder 照料人员;看管人员;看守人
mindingite 铜水钴矿
Mindlin theory 明德林理论
mind your helm 注意操舵
mine 矿山;开采;水雷;地雷坑
mineability 可采性
mine acid drainage 矿山酸性排水
mine acid wastewater 矿山酸性废水
mine adit 平硐
mine age 矿山服务年限
mine age of planning 设计生产年限【地】
mine air 矿井空气
mine air analyzer 矿内空气分析器
mine analyst 矿山安全技术工程师
mine anchor 雷锚
mine anomaly 矿异常
mine area 雷区;水雷区
mine area tax 矿产税
mine are construction 矿区建设
mine atmosphere 矿坑大气
mine backfilling 矿坑回填
mine barrage 水雷阵;水雷障碍
mine basic geology map 矿山基本地质图
mine block 矿井区段
mine blower 矿用压风机
mine capacity 矿山生产能力
mine capital construction 矿山基建
mine car 运矿车;矿车;斗(翻)车
mine car cleaner 矿车清扫器
mine car cleaning plant 矿车清洗装置;矿车清洗车间
mine car loader 装车机
mine car winding 矿车提升
mine chamber 药室;炮眼
mine characteristic 矿井通风阻力特性
mine characteristic curve 矿井通风特性曲线
mine clearance 清除水雷;扫雷
mine compass 矿山罗盘仪;矿山罗盘
mine compass dialing 矿山罗盘仪测量
mine compound 采矿工业场地
mine detecting sonar 探雷声呐
mine detector 探矿器;金属探测器;探雷器
mine development 矿山开发;矿井开拓工程
mine development cost 矿山开发费
mine development program(me) 矿山开拓方案
mine dewatering 矿坑排水;矿山排水
mine dial 矿用罗盘
mined-out area 采空区
mine drainage 矿山排水;矿井排水
mine drainage contaminant 矿井排水污染物;矿(山)水污染物
mine drainage gas 矿井气
mine drainage water 矿井废水
mine dredge(r) 采矿船
mine dressing 选矿
mine driver 矿车驾驶员
mine dump 矸石场;矿山排土场;矿山矸石堆;排矸场;废石场
mine dump/old working 废矿堆/旧工程
mine dumps 矿渣废料堆成地
mine dump tip 废渣堆
mine dust 矿尘
mine dustproof 矿山防尘
mine-electrocian 矿井电工
mine end 矿山结束
mine engineering 矿山工程
mine engineering survey 矿山工程测量
mine enterprise 矿山企业
mine enterprise economic benefit 矿山企业经济效益
mine enterprise morphology 矿山企业构成
mine enterprise planning 矿山企业设计

mine evaluation 矿山评价
mine exploitation 矿山开发
mine fan 矿井风机
mine fan signal system 矿山扇风机信号系统
mine field 矿田;井田;水雷区;布雷区
mine field exploration 井田勘探
mine field structure 矿区构造
mine-fill 充填材料
mine filling 矿山回填
mine fills storing and mixing shaft 注砂井
mine fire 矿井火灾
mine fire truck 矿山消防车
mine gallery 平巷
mine gas 矿井气
mine gas distribution map 矿山气体分布
Mine Gel 矿山爆胶
mine haulage 矿山运输
minehillite 水硅铝钙钾石
mine hoist 矿用升降机;矿井提升机
mine-hoist chain 矿井提升链;矿井起重链
mine hoisting 矿井提升
mine housing estate 矿工居住区
mine hunter 扫雷艇
mine inflow 矿坑涌水量
mine information 矿山资料
mine jumbo 矿用钻车
mine laser geodimeter 矿用激光测距仪
mine layer 布雷舰
mine layer submarine 布雷潜水艇
mine-laying aircraft 布雷飞机
mine laying boat 布雷艇
mine level 矿山水准仪;开采水平;采掘水平
mine level(l)ing 矿山水准测量
mine life 矿山开采年限
minelite 采矿炸药
mine locator 探矿仪;地雷探测器
mine locomotive 矿车机车;矿用机车;矿山机车;(采矿用)机车
mine-lot tax 矿产税
mine manager 矿山经理
mine map 矿山图
mine-mechanism 矿井机工
mine mill 竖井式磨煤机
mine mouth 坑口;矿坑口;矿井口
mine-mouth power plant 矿山电厂;坑口动力厂;坑口电站;坑口(火力)发电厂;矿口电厂
mine-mouth price 矿口价格
mine neutralization vehicle 灭雷器具
mine occupation or severance tax 矿山占用或采掘税
mine organization 矿业组织;矿业机构
mine output 矿山产量
minepentate 米奈喷酯
mine planning 矿山设计
mine planning program(me) asseverated 矿山设计方案选择
mine plan of geology 矿区平面地质图
mine-planting equipment 布雷设备
mine pollution 矿山污染;矿毒污染
mine pore (矿井的)井口
mine production 矿山生产
mine prop 坑木;矿柱
mine prospecting reserves 探矿准备金
mine pump 矿井水泵;矿用泵
mine quarry 矿山采石;矿石
mine quarrying 地下矿开采
miner 矿工;开采机;超深耕犁;采矿工
mine radio telephone system 矿用无线电话系统
mine railroad 矿山铁路
mine railway 矿山铁路
mineral 矿物;无机的;长英质矿物
mineral acid 矿物酸;无机酸
mineral acidity 矿物酸度
mineral admixture 矿物外加剂;矿物掺合剂
mineral adsorbent 无机吸附剂
mineral ages 矿物年龄
mineral aggregate 矿质集料;矿质骨料;矿物集料;矿物集合体;矿料
mineral aggregate bin 矿物骨料仓;矿物骨料料斗
mineral aggregate filling 矿物骨料的填充料;矿物集合体填充料
mineral aggregate mix(ture) 矿物骨料混合物
mineral amber 矿物琥珀
mineral analysis 矿物分析;无机物分析

mineral ash 矿物灰
mineral asphalt 矿质沥青
mineral assay 矿物分析
mineral assemblage 矿物组合;矿物共生体
mineral association 矿物资源;矿物共生(体)
mineral bath 矿泉浴
mineral belt 矿物带;含矿(地)带
mineral beneficiation 选矿
mineral binder 矿物类结料;矿物黏结剂
mineral binder bond 无机黏结剂
mineral black 矿质黑颜料;矿物黑;黑颜料(石墨);黑色氧化铁;石墨
mineral blue 矿物蓝;锰蓝
mineral board 矿物矿纤维板;矿物板
mineral bond 矿物类黏结剂
mineral-bonded board 矿物黏合板
mineral bound 矿物类黏结的;矿物类黏合的
mineral broken 矿物错断
mineral building material 矿物类建筑材料
mineral bunker 矿仓
mineral caoutchouc 天然沥青;弹性沥青
mineral car 矿车
mineral carbon 石墨
mineral cementation trap 胶结圈闭
mineral cement trap 矿物交代圈闭
mineral charcoal 矿物炭;乌煤;天然木炭
mineral chemistry 矿物化学
mineral classification 矿物类别
mineral-CO₂ carbon isotope fractionation 矿物二氧化碳同位素分馏
mineral coating 矿物涂层
mineral coke 天然焦
mineral colloid 无机胶态;无机胶质;无机胶体;矿物胶体
mineral colo(u)r 矿物颜色
mineral combination 矿物集合
mineral commodities supply and demand analysis 矿产品供需分析
mineral commodity indicator 矿产品指标
mineral commodity prospecting to diminish in strength 矿产普查削弱
mineral commodity prospecting to increase in strength 矿产普查增强
mineral commodity strategic stockpile 矿产品战略储备
mineral component of nodule 结核矿物成分
mineral component of ore 矿石矿物成分
mineral composition 矿物成分;矿物组成
mineral composition core 矿物成分的门芯
mineral composition determination of rock 岩石矿物成分鉴定
mineral composition of ore 矿石的矿物组成
mineral composition of soil 土的矿物成分
mineral compound 无机化合物
mineral concentration 选矿
mineral concentration of lake water 湖水矿化度
mineral concrete aggregate 混凝土矿物骨料
mineral constituent 矿质成分;矿物组分;矿物成分;矿物组成
mineral construction(al)aggregate 矿物建筑集料
mineral construction(al)material 矿物建筑材料
mineral content 矿物组成;矿物含量
mineral content of water 水的矿化度
mineral cord 无机纤维;矿物纤维
mineral cotton 矿(渣)棉
mineral crystal 无机晶体;天然水晶;天然晶体
mineral cutting oil 矿物切削油
mineral data base 矿物数据库
mineral deposit 矿物沉淀;矿床;矿藏;泉华;矿物储量
mineral deposit by cavity filling 充填矿床
mineral deposit by mechanical sedimentation 机械沉积矿床
mineral deposit by weathering process 风化矿床
mineral deposit development 矿床开拓
mineral deposit dewatering and water prevention in mine 矿床疏干与矿井防水
mineral deposit evaluation 矿床评价
mineral deposit exploration package 矿床勘探程序包
mineral deposit exploration to diminish in strength 矿床勘察削弱
mineral deposit exploration to increase in strength 矿床勘察增强

mineral deposit-hydrogeology 矿床水文地质
mineral deposit of direct inundation 直接进水的矿床
mineral deposit of fissure inundation 裂隙充水矿床
mineral deposit of indirect inundation from apical layer 顶板间接进水的矿床
mineral deposit of indirect inundation from bottom layer 底板间接进水的矿床
mineral deposit of karst-fissure inundation 溶蚀裂隙充水矿床
mineral deposit of karst flooding 大水岩溶矿床
mineral deposit of karst grotto 溶洞充水矿床
mineral deposit of karst inundation 岩溶充水矿床
mineral deposit of pore inundation 孔隙充水矿床
mineral deposit of underground river inundation 暗河充水矿床
mineral deposit prepared 矿床采准
mineral deposit stoping 矿床回采
mineral disintegration 矿物分解
mineral district 矿区
mineral dressing 矿物撒料;富集;选矿
mineral dressing flow 选矿流程
mineral dressing machine 选矿机
mineral dressing plant 选矿厂
mineral dust 细集料;矿物性粉尘;矿粉;矿尘;(混凝土用的)细粒料;细骨料
mineral earth oil 矿物油;矿物燃料
mineral engineering 采矿工程
mineral enrichment composition 矿物富集作用
mineral exploration 矿床勘探;探矿
mineral exploration right 探矿权
mineral extraction 采矿
mineral facies 矿物相
mineral facing 矿物贴面(材料)
mineral fat 地蜡
mineral fertilizer 矿质肥料;无机肥料
mineral fiber 无机纤维;矿物纤维
mineral fiber acoustic(al)tile 矿棉吸声砖;矿纤维吸声瓷砖
mineral fiber board 无机纤维板;矿(物)纤(维)板
mineral fiber board acoustic(al)ceiling 矿物纤维吸声天花板
mineral fiber brick 矿质纤维砖
mineral fiber cloth 矿物纤维布
mineral fiber insulating material 矿物纤维(热)绝缘材料
mineral fiber loose-fill thermal insulation 矿物纤维松散保温材料
mineral fiber mat 矿物纤维衬垫;矿物纤维毡
mineral fiber pad 矿质纤维衬垫
mineral fiber pipe section 矿物纤维管道截面
mineral fiber slab 矿物纤维板材
mineral fiber sound-deadening board 矿物纤维隔声板
mineral fiber tile 矿棉瓦;矿棉瓦;矿物纤维瓷砖
mineral fiber type acoustic(al)tile 矿物棉吸声板
mineral-filled asphalt 矿质填粉沥青;填充细矿料沥青;掺矿质填料的地沥青;细砂沥青;掺填细料地沥青
mineral-filled material 矿物质填料
mineral filler 石屑;细集料;矿质填(充)料;矿物质填料;矿物填(充)料;石粉
mineral fines 矿粉
mineral flax 石棉毡
mineral flour 矿粉
mineral flux 石棉丝
mineral food 矿质食物
mineral for insulation material 绝缘矿产材料
mineral for pigment 颜料矿物
mineral fuel 矿质燃料;矿物燃料
mineral generation 矿物世代
mineral geothermometer 矿物地热温标
mineral glue 矿胶
mineral grain 矿物颗粒
mineral grain hardness determination 矿物颗粒硬度测定
mineral granule 绿豆砂;矿物粒(料)
mineral granule finish 绿豆砂面层
mineral granules 矿质颗粒
mineral graphite 黑色氧化铁
mineral green 石绿;铜碳酸盐;矿物绿;孔雀石绿;铁铬绿;碳酸铜绿;舍雷绿
mineral grey 矿物灰

mineral ground 矿物质底层;矿物质基层;矿物基体
mineral hardness factor 矿石硬度系数
mineral hydroxyapatite 无机氢氧化物磷灰石
mineral identification 矿物鉴定
mineralige horizon depth 矿化层位深度
mineralige occurrence 矿化产状
mineral inclusion 矿物杂质
mineral indicators(indicator mineral)矿物指标
mineral industry 矿业;采掘工业
mineral inorganic substance 无机物
mineral-insulated cable 矿质绝缘电缆;无机绝缘电缆
mineral-insulated copper-covered cable 矿质绝缘铜皮电缆
mineral insulation 矿质(热)绝缘;非金属绝热隔声材料
mineral inversion 矿物转化
mineraliser 撒料机
mineral isotope composition 矿物同位素成分
mineralization 供给介质(法);矿化作用;含矿性;成矿作用
mineralization belt 矿化带
mineralization belt division 成矿带分区
mineralization coefficient 矿化系数
mineralization cycle 成矿旋回
mineralization degree 矿化度
mineralization degree of groundwater 地下水矿化度
mineralization epoch 矿化期
mineralization feature 成矿作用特征
mineralization feature of intrusive body 侵入体矿化特征
mineralization graft 矿化接枝
mineralization intensity 矿化强度
mineralization intensity index 矿化强度指数
mineralization liquor 成矿溶液
mineralization migration 成矿作用迁移
mineralization of groundwater 地下水矿化
mineralization of nitrogen 氮素矿化
mineralization of thermal brine 热卤水成矿作用
mineralization of water 水的矿化
mineralization pattern 矿化形式
mineralization period 矿化期
mineralization rate 矿化率
mineralization region division 成矿区分区
mineralization stage 矿化阶段
mineralization type 成矿类型
mineralization zone 矿化带
mineralize 矿化
mineralized bubble 矿化气泡;带矿粒的气泡
mineralized carbon 矿化碳棒
mineralized cell 矿化细胞
mineralized cell treatment 矿化细胞法处理
mineralized control 成矿控制
mineralized degree of groundwater 地下水矿化度
mineralized groundwater 矿化地下水
mineralized intrusive body 含矿岩体
mineralized limestone 矿化灰岩
mineralized point 矿化点
mineralized water 矿化水
mineralize peak 矿化峰
mineralizer 探矿者;矿化物;矿化剂;撒料机
mineralizer agent 矿化剂
mineralizing agent 矿化剂
mineral jelly 矿脂;矿物胶冻;凡士林
mineral kind 矿产种类
mineral kinds of soil 土的矿物种类
mineral kingdom 矿物界
mineral-laden bubbles 矿化泡沫
mineral lake 铬酸锡玻璃;锡铬红
mineral lard oil 矿物猪油
mineral limestone 无机碳酸钙
mineral line 矿区铁路线
mineral lineation 矿物线理
mineral load 矿物质(流水挟带的);无机物质
minerall-surfaced felt 砂面油毡
mineral lubricant 矿物润滑剂
mineral manure 无机肥料
mineral map 矿产图
mineral material 矿物原料
mineral material for artware 工艺美术原料矿物
mineral material for chemical industry 化工原料矿物
mineral-matrix pair 矿物—基质对

mineral matter 矿物质;无机物质
mineral matter in the coal 煤中矿物质
mineral maturity 矿物成熟度
mineral membrane 无机膜
mineral mining permit 采矿许可证
mineral monument 矿区永久界石
mineral nitrogen 无机氮
mineral nitrogen-containing substance 含无机氮物质
mineral nutrition 矿质营养;无机营养
mineral occurrence 矿点
mineral occurrence and deposit cell 矿点矿床单元
mineral occurrence inspection 矿点检查
mineral of leached and oxidized zone 淋滤氧化带矿物
mineral of rare metals 稀有金属矿物
mineral of secondary enrichment zone 次生富集带矿物
mineralogenetic epoch 成矿时代
mineralogenetic province 成矿区
mineralogic(al) analysis 矿相分析;矿物分析
mineralogic(al) assemblage 矿物共生体
mineralogic(al) character of particle 颗粒矿物特性
mineralogic(al) composition 矿物成分
mineralogic(al) composition of aggregate 集料的矿物组成;集料的矿物成分
mineralogic(al) constituent 矿物成分
mineralogic(al) constitution 矿物组成
mineralogic(al) description 矿物学描述
mineralogic(al) indication 矿物学标志
mineralogic(al) microscope 矿物显微镜
mineralogic(al) phase 矿相
mineralogic(al) phase rule 矿物相律
mineralogic(al) piezometer 矿物压力计
mineralogic(al) property 矿物特性
mineralogic(al) survey 矿山测量
mineralogic(al) temperature 矿物温标温度
mineralogic(al) thermometer 矿物温度计
mineralogist 矿物学家
mineralographic(al) microscope 矿相镜
mineralography 矿相学
mineralogy 矿物学
mineralogy article 矿物学论著
mineralogy data 矿物学资料
mineralogy of ore-deposits 矿床矿物学
mineraloid 准矿物;胶质矿物;似矿物
mineral oil 矿物油;石油
mineral oil concentrate 石油浓缩物
mineral oil refining plant 炼油厂
mineral orange 橘色矿石;氧化铅橙
mineral paint 矿质油漆;矿物颜料
mineral paint commodiy 矿物涂料矿产
mineral pair 矿物对
mineral pair carbon isotope fractionation 矿物对碳同位素分馏
mineral pair hydrogen isotope fractionation 矿物对氢同位素分馏
mineral pair oxygen isotope fractionation 矿物对氧同位素分馏
mineral paragenesis 矿物共生
mineral paragenetic sequence 矿物生成顺序
mineral particles 矿物颗粒
mineral phase 矿物相
mineral phase change 矿物相变
mineral physics 矿物物理学
mineral pigment 矿物颜料;无机颜料
mineral pitch 矿质沥青;矿物(硬)沥青;地沥青;柏油
mineral pollution 矿物污染;无机物污染
mineral pool 矿泉池
mineral powder 矿物粉末;矿粉
mineral powder plasticizer 矿粉增塑剂
mineral preparation engineer 选矿工程师
mineral processing engineer 选矿工程师
mineral processing flowsheet 选矿流程
mineral products 矿产
mineral property 矿业权
mineral property lease 矿山设备租赁
mineral prospecting 探矿;找矿;矿产勘探
mineral prospection 矿藏勘探
mineral pulp 矿浆
mineral purple 氧化铁
mineral railroad 矿区铁路;矿区铁道
mineral railway 矿区铁路;矿区铁道

mineral raw material 矿物颜料
mineral raw material commodities for medical use 药用原料矿产
mineral red 矿石红
mineral reserve forms 矿产储量表
mineral reserves 矿物储量;矿储藏量;矿产储量;矿产储藏
mineral reserves data base 矿床数据库;矿产储量数据库
mineral reserves dynamic 矿产储量动态
mineral reserves sheet 矿产储量表
mineral resin 矿质树脂;矿物树脂
mineral resources 矿物资源;矿产资源;矿藏
mineral resources appraisal data base 矿产资源评价数据库
mineral resources appraisal package 矿产资源评价程序包
mineral resources conservation 矿物资源保护;矿产资源保护
mineral resources law 矿产资源法
Mineral Resources Law of the People's Republic of China 中华人民共和国矿产资源法
mineral resources map 矿产资源图
mineral resources quantitative prediction and appraisal 矿产资源定量预测及评价
mineral resources survey(ing) 矿产资源调查
mineral resources tax 矿产资源税
mineral resources tax rate 矿产资源税率
mineral rights 矿产权;地下产权;矿权;采矿权
mineral rock 矿物岩
mineral rubber 矿物胶;黏稠的沥青;沥青胶块
mineral sampling 矿物采样
mineral sand 矿砂
mineral seal oil 重质煤油;重质灯油
mineral selection extractions 矿物选择性萃取
mineral separation 选矿;矿物分离
mineral separation method 矿物分离法
mineral separation process 选矿方法
mineral separation station 选矿站
mineral separation weighing method 矿物分离称重法
mineral separative and metallurgic(al) problems unsettled 选冶问题未解决
mineral sequence 共生次序
mineral skeleton 矿井架;矿物骨架
mineral sludge 矿质污泥;无机污泥
minerals of meteorites 陨石分类
mineral soil 矿质土(壤);生土
mineral solvent for paints 油漆用矿物质溶剂
mineral specimen 矿物标本
mineral spirit 松香水;石油溶剂油
mineral spirit tolerance 石油溶剂油容忍度
mineral spring 矿泉
mineral sprinkling material 撒布料
mineral stabilizer 矿物稳定剂
mineral stratum 矿层
mineral streak 矿物条纹;矿质条纹;天然变色
mineral structural material 矿物性结构材料
mineral substance 矿物质
mineral substrate 涂矿物质的底层;涂矿物质的基层;矿物基体
mineral suite 矿物系列
mineral sulfate 无机硫酸盐
mineral surface 绿豆砂面层;铺撒小粒石面层
mineral-surfaced asphalt felt 绿豆砂面油毡;撒砂沥青油毡
mineral-surfaced asphalt sheet 矿物粒料覆面沥青油毡
mineral-surfaced bitumen felt 矿物集料覆面沥青油毛毡;绿豆砂面油毡
mineral-surfaced roll roofing 矿料面层卷材屋面;矿物粒料覆面沥青屋面卷材
mineral surfacing 矿物贴面材料
mineral survey 矿山测量;矿区测量;矿产调查
mineral tallow 地蜡;矿蜡;伟晶蜡石
mineral tar 矿质焦油;软沥青;软地沥青;土沥青;(质)焦油
mineral thinner 矿物稀释剂;无机稀释剂
mineral trass 矿质火山灰
mineral turpentine 矿油精;矿(物)质松节油;汽缸油;石油溶剂
mineral turps 石油溶剂油

mineral varnish 石漆
mineral vein 矿脉
mineral violet 矿物紫
mineral void 矿质填料孔隙;矿质填料孔
mineral wagon 矿石车
mineral washer 洗矿工;洗矿机
mineral waste 矿物废料
mineral waste residue 矿渣
mineral wastewater 无机废水;矿物废水
mineral water 矿质水;矿水;矿泉水
mineral-water hydrogen isotope fractionation 矿物—水氢同位素分馏
mineral-water oxygen isotope fractionation 矿物—水氧同位素分馏
mineral wax 矿物蜡;地蜡
mineral wealth 矿产财富;矿物财富;矿产资源
mineral well 矿质水井;矿泉井
mineral white 重晶石;结晶石膏;石膏微粉;石膏;粉磨石膏
mineral white primer 碳酸钙油性填孔料
mineral winning 矿物采运
mineral wool 矿质棉;矿渣绒;矿渣棉;矿(物)棉;石纤维
mineral wool acoustic(al) tile 矿物棉吸声板;矿棉管吸声板
mineral wool belt 矿物棉带
mineral wool blanket 矿棉毡
mineral wool board 矿渣绒板;玻璃棉板;矿棉板
mineral wool decorative panel 矿(物)棉装饰板
mineral wool dust 矿渣棉粉尘
mineral wool felt 玻璃棉毡;矿(物)棉毡
mineral wool fiber tile 矿棉纤维瓦
mineral wool insulating material 矿棉(热)绝缘材料
mineral wool insulation 矿棉(热)绝缘
mineral wool panel 矿棉片;矿棉板
mineral wool paper 矿(物)棉纸
mineral wool paper asphalt sheet 矿棉纸油毡
mineral wool pipe insulating section 矿棉管道保温套
mineral wool pipe insulation 矿棉管道保温套
mineral wool pipe section 矿(物)棉套管
mineral wool quilt 矿棉毡
mineral wool rope 矿物棉绳
mineral wool sheet 矿棉薄片;矿棉薄板
mineral wool slab 矿物棉板
mineral wool sound absorbing sheet 矿物棉吸声板
mineral wool strip 矿棉带;矿棉板条
mineral wool tissue 矿棉纸
mineral wool twine 矿物棉绳
mineral yellow 氯氧化铅;矿物黄;碱式氯氧化铅黄
mineral zone 矿物带
mineral zoning 矿物分带
mine refuse 矿渣;矿场废物
mine refuse impoundment 尾矿坝
mine removal 水雷扫清;地雷扫清
mine rescue work 矿山救援工作;矿山抢救工作;矿井救援工作;矿井抢救工作
mine reserves 矿山储量
mine resistance 矿井通风阻力
mine rig 矿用钻具;矿(山)用钻机
mine rights 采矿权
mine risk 水雷区危险
minerloid 类矿物
mine roadway 矿山巷道
mine rock 矿石;矿物岩
minerogenic 成矿的
minerography 矿相学
miner's anemia 矿工贫血症
miner's asthma 矿工气喘病
miner's compass 矿山罗盘
miner's dip needle 矿用磁倾仪
miner's dwelling 矿工寓所;矿工住宅;矿工住处
miner's estate 矿工新村
miner's hand lamp 手提自给矿工灯
miner's horn 矿工牛角
Miner's hypothesis 迈因纳假设
miner's inch 水流速度单位(经过1平方英寸洞口的流水速度,用于美国西部);矿井涌水率单位;矿工英寸
miner's lamp 矿工灯
Miner's linear damage theory 迈因纳线性损伤理论
miner's lung 矿工肺

miner's phthisis 矿工尘肺病
miner's powder 黑色火药
miner's self-rescuer 矿工自救器
miner's truck 矿车
miner's wagon 矿车
miner tar 软质天然地沥青
minertial motor 小惯量电动机
mine run 原矿
mine run coal 原煤
mine run rock 矿坑碎石；未筛(分)石料
miner village 矿工村
miner winding engine 矿井提升机；矿井卷扬机
minery 矿区
mine safety supervision 矿山安全监察
mine salt 岩盐；矿盐；无机盐
mine scale 矿区规模
mine section 矿山断面图；开采区；采(矿)区
mine separation plant 石料筛分厂
mine shaft 矿井竖井
mine shaft drilling 井筒钻进
mine shaft liner 矿山竖井衬里；矿山竖井支护
mine shaft lining (矿井的)井筒支护；井筒衬砌；井壁
mine signal system 矿山铁路信号
mine site 矿区；矿场
mine smoke 矿山烟雾
mine staff 矿区水准尺
mine stone 煤矿废石料
mine sump 矿井水仓
mine supervision 采矿技术监督
mine support section 矿山支撑型材
mine surface survey 矿山地面测量
mine survey(ing) 矿山测量
mine surveying instrument 矿物测量仪器
mine sweeper 扫雷艇；扫雷舰
mine sweep(ing) 扫雷；扫海
mine sweep(ing) area 扫雷区
mine sweep(ing) boat 扫雷艇
mine sweep(ing) float 扫雷浮标
mine sweep(ing) gear 扫雷具
mine sweep(ing) light 扫雷灯
mine tailings 尾矿渣；尾矿砂
mine tailings dam 尾矿坝
Minet aluminum process 迈尼特制铝法
mine tax 矿产税
mine technical inspection 采矿技术监督
mine tenor of asbestos ore 石棉矿石矿山品位
mine timber 坑木；坑柱
minetisite 砷铅矿
mine total head 矿井通风总压头
mine totality geologic(al) map 矿山总体性地质图
mine track 雷轨
mine tractor 矿用拖拉机
mine transport building 矿山运输建设
mine treatment flow 矿山加工流程
minette 云煌岩；云煌石
minette-felsite 云煌霏细岩
mine tunnel 矿山平巷；坑道；平巷
mine tunnel(l)ing method 矿山法(指隧道开挖)
mine unwatering 矿山排水
mine velocity head 矿井通风速度压头
mine ventilating fan 矿用扇风机
mine ventilation 矿井通风
mine ventilation station 矿山通风站
mine ventilation system 矿井通风系统
mine ventilator 矿用风机
mine warfare ship 布扫雷船
mine waste 矿山废物；采矿尾砂
mine waste sand 采矿尾料
mine wastewater 矿山废水；矿井废水
mine wastewater treatment 矿山废水处理
mine water 矿坑水；矿(井)水
mine water hammer 矿山水锤
mine water treatment 矿山水处理
mine wool 矿物棉；矿棉
mine wool belt 矿棉带
mine wool rope 矿棉绳
mine wool slab 矿棉板
mine wool twine 矿棉绳
mine working 矿山工作面；采矿作业；矿山巷道；矿山井巷；矿场
mine yard 矿场
mingle 交织；使混合；掺和
mingled tile 混合瓦

mingle-mingle 大杂烩
Ming-style furniture of hardwood 明式家具
Ming three-colo(u)red ware 明三彩器
Ming Tomb of Emperor 十三陵
minguetite 黑铁绿泥石
minguzzite 草酸铁钾石
mini 微型的；缩影；缩型
mini amplifier 小型放大器
mini-amplifier modulator 小型放大器调制器
mini-arc orbit determination 微弧轨道测定
mini-area 小区
mini-ash 少沉淀的；少灰的
miniative 小型的
miniature 微型；缩影；超缩微品
miniature base 小型管底
miniature bayonet base 小型卡口灯座
miniature bearing 微型轴承；超小型轴承
miniature bearing race 微型轴承座圈
miniature bulb 小灯泡
miniature camera 小型照相机；小型摄影机
miniature camera lens 小照相机物镜
miniature cap 小型管帽
miniature circuit 小型电路
miniature circuit breaker 小型断电器；袖珍式断路器；微型断路器
miniature clad 小型包层
miniature clock case 小型钟壳
miniature component 小型元件；超小型元件
miniature counter 小型计数管
miniature current meter 小型流速仪；微型流速仪；微型海流计；微量电流计
miniature device 微型器件
miniature earthquake 小规模地震；人工地震
miniature Edison screw-up 小型爱迪生式螺旋灯头
miniature edition 袖珍版；特小型版
miniature electromagnetic clutch 微型电磁离合器
miniature electronic autocollimator 微型电子准直仪
miniature electrostatic gyro 小型静电陀螺
miniature end-plate potential 微终板电位
miniature fermenter 微型发酵罐
miniature film 微型胶片
miniature gears 小齿轮
miniature glass tube machine 小型玻壳管制造机
miniature golf course 小型高尔夫球场
miniature hydrocyclone 小型水力旋风(分粒)器
miniature image orthicon 微型超正析像管
miniature inertial gyro 小型惯性陀螺
miniature inertial navigation system 微型惯性导航系统
miniature integrating gyro 小型积分陀螺
miniature ionization chamber 微型电离室
miniature lamp 小型灯泡；指示灯；微型灯
miniature lampholder 小型灯座
miniature lamp-socket 小型灯座
miniature lancet(window) 锐顶小窗
miniature landscape 盆景
miniature military submersible 小型军用潜水器
miniature mixer-settler 微型混合澄清槽
miniature motor 小型电动机；微型电(动)机
miniature motor-car 微型汽车
miniature negative 微型底片
miniature negative photography 小型底片摄影
miniature nuclear battery 小型核电池
miniature open flume 小型试验水槽
miniature oscilloscope 微型示波器
miniature photomultiplier 小型光电倍增管
miniature plate potential 微小终板电位
miniature power unit 小型动力装置
miniature pressure cell 小型压力传感器
miniature probe 微型探针
miniature projector 小像幅投影仪
miniature radio receiver 小型接收机
miniature relay 小型继电器；微型继电器
miniature retarder 小型缓冲器
miniature rotor 整个埋于主夹板内的小型自动锤
miniature scenery 缩景
miniature set 小型装置
miniature smoke chart 小型烟雾图
miniature socket 微型套筒
miniature submersible 小型潜水器
miniature switch 微型开关
miniature table 小型摇床
miniature television 微型电视

miniature thickness ga(u)ge 袖珍测厚仪
miniature thyratron 微型闸流管
miniature-tonnage mine 小型矿山
miniature tractor 小型拖拉机
miniature train 模型列车
miniature transformer 小型变压器
miniature transmitter-receiver 小型发报机—接收机
miniature trees and rockery 盆景
miniature tube 小型管；微型管
miniature TV camera 微型电视摄像机
miniature type 小型
miniaturization 微型化；小型化
miniaturization system 微型化系统
miniaturize 使微型化
miniaturized circuit 小型(化)电路
miniaturized circuit element 微型电路元件
miniaturized component 小型元件
miniaturized recorder 小型化记录仪
miniaturized system 小型化系统
mini-automated device 微型自动化装置
minibar 小型条信号
mini-bore central heating 小孔集中加热
mini-bore heating system 微腔供热系统
minibul 小型推土机
minibulk 小型散装
minibulk carrier 小型散装(货)船
mini-bulker 小型散货轮；小型散货船；小型散装(货)船
minibus 小型客车；小型交通车；小型公共汽车；小客车；面包车
minicab 微型出租汽车
minicam 小型照相机
minicamera 微型照相机
minicar 小型汽车；微型汽车
minicard 小型卡片；缩微胶片系统
minicatolog 微型产品目录
minicell 微型电池；微细胞
mini cement industry 小水泥工业
mini cement plant 小型水泥厂
minicollector 小型收集器
mini column 微型柱
minicom 小(型)电感比较仪
minicomponent 小型元件；微型元件
minicomputer 微型计算机；通用微型计算机；小型(电子)计算机
minicomputer communication processor 小型计算机的通信处理机
minicomputer controlled terminal 小型计算机控制的终端
minicomputer digitizing system 小型计算机数字化系统
minicomputer display 微型计算机显示器
minicomputer input and output 小型计算机输入输出设备
minicomputer instruction 小型计算机指令
minicomputer operating system 小型计算机操作系统
minicomputer peripherals 小型计算机的外围设备
minicomputer software 小型计算机软件
minicomputer system 小型计算机系统
minicomputer terminal controller 小型计算机终端控制器
mini cross-section 特小断面
minicrystal 小型晶体；微型晶体
minicrystal diffusion 微晶扩散
minicrystal diffusion method 微晶扩散法
mini-devaluation 小幅度贬值
mini-dredge(r) 小型挖泥船
miniemulator 小型仿真程序
minification 微小；缩小(率)；缩小尺寸
minification rate 缩微率
minifier 缩小镜
miniflow 微流
mini-foam 微泡沫(体)
mini-four-electrode array method 露头法
mini fullfacer 小型全断面隧洞掘进机
minifundium 小庄园
minify 削减；缩小尺寸；缩减
minigroove 密纹
minihead air sampler 小型空气取样器
minihost 小型主机
mini hydro 小水电
mini hydro turbine 小水轮机；微型水轮机
minikin 微小的东西

mini-land bridge 小大陆桥
mini-land bridge service (集装箱)小型陆桥运输
mini-land bridge service system 小型陆桥运销
miniliform 念珠状的
miniload system 小件储[贮]存系统
mini lorry 小型运货汽车
minilot 超小型电磁旋转开关
mini lube 小型滑脂加注器
minim 量滴(美制液量单位)
minimal 最小的;最简的;最低的;极小;极微的
minimal access coding 最小存取编码;最快存取编码
minimal access program(me) 最快存取程序
minimal access programming 最快取数程序设计
minimal area 群落最小面积
minimal art 抽象派艺术
minimal audible intensity 最低可听强度
minimal automaton 最小自动机
minimal average dosage for mutation 最小平均致突变剂量
minimal capital required by law 法定最低资本额
minimal chain 最小链
minimal change group 轻微变化型
minimal closed extension 最小闭开拓
minimal complete class 最小完备类;极小完全类
minimal complete class of decision rule 决策规则的极小完全类
minimal condition 最小条件
minimal configuration 最小配置
minimal contact angle 最小接触角
minimal convex polygon 最小凸多角形
minimal-cost network-flow 最小成本网络流程
minimal cost path 最小代格路径
minimal current gradient 最小电流梯度
minimal daily requirement 一日必需量
minimal detectable activity 最低检出放射性强度
minimal detectable concentration 最小检测浓度
minimal detectable limit 可检出的最低限值
minimal dose 最小量
minimal down time 最短停产时间
minimal dyadic expansion 最小并矢展开
minimal effective concentration 最小有影响浓度
minimal effective dose 最小作用量;最小有效(剂)量;最低有效量
minimal effect level 最小作用水平
minimal element 极小元
minimal equation 极小方程
minimal face 极小面
minimal fee 最小费用
minimal flight path 最省时飞行路线
minimal flow pump 最小流量泵
minimal form 极小形式
minimal geodesic 极小大地线
minimal identifiable odo(u)r 最低可嗅度
minimal illumination 最小照度;最低照度
minimal index 最小指数
minimal inhibiting dose 最小抑制剂量
minimal inhibitory concentration 最小抑制浓度
minimal inhibitory dose 最小抑制剂量
minimality 最小性;极小性
minimalization 极小化;取极小值
mini-mall 小的步行商业街
minimal-latency coding 最快取数编码;最短等待时间编码
minimal-latency program(me) 最快存取程序
minimal-latency programming 最快存取程序设计
minimal-latency routine 最快取数程序
minimal-latency subroutine 最快存取子程序
minimal lattice 微小网
minimal lethal oxygen concentration 最小致死氧浓度
minimal mean number of inquiry 最小平均询问次数
minimal medium 最低限度培养基;基本培养基
minimal member 极小元素;极小元
minimal national standard 最低国家标准
minimal normal subgroup 最小正规子群
minimal path 最小路径;最短路径;最短程
minimal path method 最短路径法
minimal percentage of reinforcement 最低配筋率
minimal perturbation 极小扰动
minimal polynomial 最小多项式;极小多项式
minimal positive integral solution 最小正整数解
minimal pressure 最低压(力)

minimal product-of-sums 极小和之积
minimal reacting dose 最小反应量
minimal realization 最小实现
minimal reflector 最小反射体
minimal residual air 最少余气
minimal resistance 最小抵抗力
minimal safe distance 最小安全距离
minimal set 极小集
minimal set of generators 生成元的极小集
minimal solution 最小解
minimal sparking voltage resistance 最低耐火花电压
minimal standard 最低标准
minimal sufficient statistic 最小充分统计量
minimal sum-of-products 极小积之和
minimal surface 极小曲面
minimal temperature 最低温差
minimal time control 最短时间控制
minimal toxic dose 最小毒性剂量
minimal tree 极小树
minimal true vector 极小真向量
minimal value 最小值;极小值
minimal variety 极小簇
minimal vector 极小向量
minimal visual angle 最小视角
minimal wave length 最短波长
mini-master batch 最少母料
minimax 极小化最大;极大中的极小;鞍点
minimax approach 极小极大方法
minimax approximation 极小极大近似;极小极大逼近;极大极小逼近
minimax criterion 小中取大法则;极小极大准则;极小极大判据;极大极小准则
minimax design 极限设计
minimax detection theory 极小极大检测理论
minimax estimate for binomial distribution 二项分布的最小最大估计
minimax estimate of location parameter 局部参数的最小最大估计
minimax estimation 极小极大估计
minimax estimator 极小极大估计量
minimax inequality 极小极大不等式
mini-maxi regret 最小的最大失误
minimax loss 极小极大损失
minimax method 极大极小法
minimax negative utility 极小极大负效用
minimax principle 最小最大原则;极小极大原理
minimax rule 极小极大规则
minimax search in game trees 博弈树(中)极大极小搜索
minimax solution 极值解;极小极大解
minimax strategy 极小极大策略
minimax system 最大最小系统;极小极大系统
minimax technique 极小极大技术;极大极小法
minimax theorem 极小极大定理
mini medium cement mill 微介质水泥磨
minimeter 气流负压仪;指针测微计;千分比较仪;测微计
mini-mixer-settler 微型混合澄清槽
minimization 极小化;求最小(参数)值;求极小值
minimization of band width 带宽最小化
minimization of Boolean function 布尔函数的最小化
minimization of the sum of absolute value 绝对值中的最小值
minimization problem 最小化问题
minimization process 极小化过程;求极小值法
minimize 最小限度;减到最少;极小化
minimized friction 最小摩擦
minimize social impact 最小社会影响
minimizing 求最小参数值
minimizing chart 缩图
minimizing control mode 最小化控制方式
minimizing diameter therapy 减径矫治法
minimizing energy method 最小能量法
minimizing sequence 极小化序列
minimizing transport cost 节约运输费用
minimo 缩印版
minimodem 小型调制解调器
mini-motor-home 小型旅宿车
minimotorway 窄公路
minimower 微型割草机
minimum 最小(值);最低值;最低限度;最低点;最低(的);极小

minimum absorbed water volume 最大分子水溶量
minimum acceptable current 最小允许电流
minimum acceptable discharge 最小允许量;最小容许排水量
minimum acceptable flow 规定流量;最小容许流量;最小允许流量
minimum acceptable gap 最小可接受空档
minimum acceptable reliability 最低容许可靠性
minimum acceptable standard 可接受的最低标准
minimum access 最快取数
minimum access code 最快存取码
minimum access coding 最快存取编码
minimum access programming 最快存取编程;最小存取程序设计
minimum access routine 最快取数程序
minimum access time 最速存取时间;最快存取时间
minimum accounting rate 最低结算费率
minimum acquisition flux density 探测目标最小通量密度
minimum action range 最小作用距离
minimum adjustable table speed 工作台最小调整速度
minimum admittance frequency 最小导纳频率
minimum advance for best torque 最大扭矩的最小点火提前角
minimum air requirement 最小需要风量
minimum allowable dissolved oxygen 最小容许溶解氧(量)
minimum allowable dissolved oxygen concentration 最小容许溶解氧
minimum allowable headway 列车最小容许间隔
minimum allowable radius 最小允许半径
minimum allowable reliability 最低容许可靠性
minimum altitude 最低高度
minimum amount 最小量;最小金额
minimum amount of insurance 保险的最低数额
minimum amount of interim certificate 临时支付证书的最小限额
minimum amount of third party insurance 第三方保险金的最低金额
minimum and deposit premium 最小预付保险费
minimum angle of elevation 最小仰角
minimum annual flood 最小年洪水(量);多年平均最小洪水(量)
minimum annual precipitation 最小年降水量
minimum annual royalty 每年最低支付额
minimum annual runoff 最小年径流量
minimum anomaly value 最小异常值
minimum antenna height 最低天线高度
minimum area 极小区域
minimum audibility 最小可听度
minimum audible field 最小可闻区域;最小可闻场
minimum audible pressure 最小听压
minimum audible threshold 最低可听阈
minimum average 极小平均值
minimum average B configuration 最小平均B位形
minimum-B configuration 非零最小B场位形
minimum bearing capacity 最小承载力
minimum bearing property 最小承载性能
minimum bend(ing) radius 最小弯曲半径;最小挠曲半径
minimum bill of lading 最低运费提单
minimum bill of lading charges 最低提单费
minimum block length 最小信息组长度
minimum blowing current 最小熔断电流
minimum boiling point azeotrope 最小共沸混合物
minimum braking speed 最低制动速度
minimum breakdown voltage 最低击穿电压
minimum breaking load of a rope 钢丝绳的最小破断荷载
minimum brightness 最小亮度
minimum burden 最低抵抗线
minimum buried depth of basement 基底的最小埋深
minimum bus 小型公共汽车
minimum capacity 最小起点电容;最小能力;最小容量
minimum capsizing moment 最小倾翻力矩
minimum carrier level 最低载波电平
minimum cash balance 最低现金余额
minimum cement content 最低水泥含量;水泥最小用量
minimum charges 最低费用;起码收费

minimum charged distance 起码里程
minimum charge quantum 起码收费额
minimum chemical potential energy 最小化学势能
minimum circumscribed circle method 最小外接圆法
minimum clearance 最小限界;最小净空;最小间隙
minimum clear floor space 最小净面积
minimum clearing 清零
minimum clock frequency 最小时钟频率
minimum clock-pulse duration 最小时钟脉冲持续时间
minimum cognoscibile 最小辨视阈
minimum colo(u)r acuity 最小彩色分辨度
minimum command interval 最小指令间隔
minimum compactive effort 最小压实力
minimum compression strength 最小抗压强度
minimum compressive strength 最小抗压强度
minimum concentration 最小浓度
minimum concrete cover 最小混凝土保护层
minimum condition 最低条件
minimum conduction time 最小传导时间
minimum consumption 最低消费量
minimum content 最低含量
minimum continued speed 最低持续速度
minimum contour size 最小外形尺寸
minimum control 最小控制
minimum convex set 最小凸集
minimum-cost estimate 最小代价函数估计
minimum-cost estimating 估计最低成本
minimum-cost rule 最低成本法
minimum-cost structure 造价最低结构
minimum creepage length of insulator 绝缘子最小爬距
minimum critical heat flux ratio 最小临界热通量比
minimum critical mass 最小临界质量
minimum-critical-mass experiment 最小临界质量试验
minimum cross-section 最小断面
minimum cube strength 最小立方体受压强度
minimum curb offset 最小缘石支距
minimum current 低值电流
minimum current circuit breaker 最小电流断路器
minimum current relay 低值电流继电器
minimum current velocity 最小流速;最低流速
minimum curve radius 最小曲线半径
minimum cut length 最小切削长度
minimum cutout 欠电流自动断流器
minimum cut(ting) width 最小挖宽【疏】
minimum daily discharge 日最小流量
minimum-damage-fire-control theory 最小损失防火理论
minimum day sendout 最小日供气量
minimum deflection 最小偏转;最小偏斜;最小偏差;最小挠曲度
minimum deflection method 最小偏差法
minimum delay 最小延迟
minimum delay circuit 最小延迟电路
minimum delay coding 最小延迟编码
minimum delay condition 最小延迟条件
minimum delay programming 最小延迟的程序设计
minimum demand 最小需要量
minimum density 最小密度
minimum deposit 最低预付押金
minimum deposit premium 最低预付的保险费
minimum depth 最小深度;最小水深
minimum depth of pipe 管道最小埋深
minimum depth of socket 最小套节深度
minimum depth of water to operate 作业最小水深
minimum design 极小设计
minimum designing gradient 最小设计坡降
minimum designing velocity 最小设计流速
minimum design weight 最小设计重量
minimum desired rate of return 预期最低收益率;最低预期报酬率
minimum detectable activity 最小可检出放射性强度;最小检出(放射)活性;最小检测能力
minimum detectable amount 最小检出量;最小检测量
minimum detectable brightness 最小可检测亮度
minimum detectable concentration 最小检测浓度
minimum detectable difference 最小可检测差
minimum detectable echo level 最小检测回声级
minimum detectable flux density 最小可检流量密度
minimum detectable level 最小检测水平
minimum detectable limit 最低检出量
minimum detectable power 最小检测能力
minimum detectable quantity 最小检测量
minimum detectable range 最小探测距离
minimum detectable signal 最小可检测信号;最小检测信号;临界信号
minimum detectable signal power 最小可测信号功率
minimum detectable temperature 最小可检测温度
minimum detectable temperature difference 最小可探测温差;最小检测温差
minimum detectable threshold odor concentration 最小可检合臭味值
minimum detectable value 最小检测值
minimum detection amount 最小检出量;最小检测量
minimum detection level 最小检出水平;
minimum detection limit 最小检出限;最小检测范围
minimum detection quantity 最小检测量
minimum deviation 最小偏向角;最小偏差
minimum deviation grating 最小偏向光栅
minimum deviation law 最小偏差角法则
minimum deviation method 最小偏差法
minimum diameter 最小直径
minimum diameter of thread 螺纹最小直径
minimum diameter vessel requiring manhole 要求人孔的最小直径容器
minimum dimension 最小尺寸
minimum discernible signal 最小可辨别信号
minimum discharge 最小排放量;最小流量;最低流量
minimum discharge of spring 泉的最小流量
minimum dissolved oxygen concentration 最小溶解氧浓度;最低溶解氧浓度
minimum distance 最小距离;最小间距;最短距离
minimum distance angle 最小求距角
minimum distance between centers of tracks 最小线间距
minimum distance classification 最小距离分类
minimum distance classifier 最小距离分类器
minimum distance code 最小距离码;最短距离码;最小间隔码
minimum distance criterion 最小距离准则
minimum distance decision rule 最小距离决策规则
minimum-distance decoding 最短距离译码
minimum distance error-correcting 最小距离误差校正
minimum distance for sunlight 日照间距
minimum distance method 最小距离法
minimum-distance-to-mean algorithm 极小距离平均算法
minimum down time 最短的停机时间
minimum drag coefficient 最小阻力系数
minimum drawdown level 最小消落水位
minimum dredging depth 最小挖深
minimum dredging width 最小挖宽【疏】
minimum-driving-point function 最小策动点函数
minimum dry density 最小干容重
minimum dry weather flow 最小枯季径流;最小旱天流量;最(小)枯(水)流量
minimum duration 最小历时;最短时间;最短历时
minimum dynamic(al) condition 最小动力条件
minimum earnings level 最低工资水平
minimum earthquake 最小地震
minimum ebb 最小退潮流;最小落潮流(速度);半周日低低潮
minimum echo 最小回波
minimum economical descending grade 最小经济坡降
minimum economic thickness 最小工业厚度
minimum effective concentration 最小有影响浓度
minimum effective dose 最小有效剂量;最低有效剂量
minimum effective height 最小有效高度
minimum effective liquid rate 最小有效流量
minimum effective temperature 最低有效温度
minimum effective thickness of source rock 有效母岩最小厚度
minimum effeminor constituent of the atmosphere 大气的微量组分;大气的次要成分
minimum elastic collapsing strength 最小弹性抗挤强度
minimum elastic-plastic collapsing strength 最小弹塑性抗挤强度
minimum elevation 最小仰角;最小射角
minimum elevation of terrace 阶地最小高度
minimum embedment depth (桩的)最小入土深度
minimum embedment length (桩的)最小入土长度
minimum energy gradient 最小能量梯度
minimum energy line 最小能量线
minimum energy method 最小能量法
minimum energy principle 最小能量原理
minimum energy ratio principle 最小能量比原理
minimum energy theory 最小能量理论
minimum enroute instrument altitude 航途最小仪表高度
minimum entropy 最小熵
minimum entropy deconvolution 最小熵反褶积
minimum envelope curve 下外包线;下包络线;最小包络(曲)线
minimum environment(al) discharge 最小环境流量
minimum environment standard 最低环境标准
minimum error 最小误差
minimum error conic(al) projection 最小变形圆锥投影图
minimum error decoding 最小误差译码
minimum excavation line (隧洞工程中的)最小爆破断面;最小开挖线
minimum excitation potential 最低激发电位
minimum exposure time 最短曝光时间
minimum face 最小面
minimum factor 最小因素
minimum factor of safety 最小安全系数
minimum failure strength 最小破坏强度;最小破坏应变
minimum fall 最小坡降
minimum fee 最小费用
minimum fetch 最小风区;最小对岸距离;最小吹送距离
minimum field capacity 最小田间持水量
minimum fill 最小装入量
minimum fill height 最小填土高度
minimum fill height of subgrade 路基最小填筑高度
minimum film-forming temperature 最低成膜温度
minimum firing current 最小点火电流;最低起爆电流;最低发火电流
minimum-fission warhead 最小裂变弹头
minimum fixed structure 建筑最小限界
minimum flight altitude 最低飞行高度
minimum flood 最小涨潮速度;最小涨潮流(速度)
minimum floor area requirement 最小楼板面积要求
minimum flow 枯水流量;最小流量;最低流量;枯水位
minimum-flow control 最小流量控制
minimum fluidized voidage 最小流态化
minimum form 最简形式
minimum freeboard 最小干舷(高度)
minimum freeboard draught 最小干舷吃水
minimum freight 最小运费;最低运费
minimum freight rate 最低运费率
minimum fresh air requirement 最小新风量
minimum fuel limiter 低限燃烧限制器
minimum fuel limit screw 最小燃油限止螺钉
minimum function 最低函数
minimum funding standard 资金最低标准
minimum ga(u)ge 最小厚度
minimum girth 最小干围
minimum good field radius 最小工作场半径
minimum grade 最小坡度;最低坡度
minimum grade of ore 可采矿石的最低品位
minimum gradient 最小坡度
minimum gradient of river 河流最小比降
minimum gravitational potential energy 最小重力势能
minimum green period 最小绿灯时间;最少绿灯时间;最短绿灯时间
minimum grinding diameter 最小磨削直径
minimum ground clearance 最小离地距离;最小离地间隙
minimum growth temperature 最低生长温度

minimum habitable room height 最小可住房间高度
minimum habitable room size 最低可住房间大小
minimum habitat 最低生境要求
minimum head 最小水头
minimum head loss 最小水头损失
minimum headway 最小车头时距;最小车间距;列车容许最小时间间隔
minimum heat 最小热量
minimum height above sea level 最小海拔
minimum height of fill 最小填土高度
minimum height tree 最小层次树
minimum hop 最小跳步
minimum hop path 最小跳步通路
minimum horizontal curve length 最小平曲线长度
minimum horizontal curve radius 最小水平曲线半径
minimum horsepower 最小马力
minimum hotel-motel units 最小汽车旅馆单元
minimum house size requirement 最小住房尺寸要求
minimum housing code 最低住房法规
minimum housing standard 最低住房标准
minimum human infectious dose 最小人体有害剂量
minimum ignition energy 最低着火能
minimum illumination 最低照度
minimum impedance frequency 最小阻抗频率
minimum impedance relay 低阻抗继电器
minimum initial green period 最短初始绿灯时间
minimum insolation hours in winter 冬季最小日照数
minimum installation allowance 最小安装调节距离
minimum instream flow 最小河道流量
minimum instream flow algorithm 最小河道流量算法
minimum intensity 最小强度
minimum interevent time 最短事件内时间
minimum intersecting angle 走道最小宽度(室内车辆直角拐弯时)
minimum intersection angle 最小交会角
minimum interval 最小间隔
minimum inventory investment 最低库存量投资
minimum ionizing speed 最小致电离速率;最小电离速度
minimum irrigation rate 最小灌注速率
minimum joint strength 最小焊接强度;最小连接强度
minimum-lag shaping 最小延迟整形
minimum landscaped open space 最少绿化空地
minimum latency 最快存取
minimum latency code 最快存取码
minimum latency coding 最小等数时间编码;最快存取编码
minimum latency programming 最小等数时间程序设计
minimum latency routine 最快存取程序
minimum law 最小定律
minimum lease payment 最低租赁支出
minimum least squares velocity filter 最小二乘法速度滤波
minimum legal age of employment 法定最低就业年龄
minimum lending rate 最低借款利率
minimum length 最小长度
minimum length cut 最小切削长度
minimum length of grade section 最小坡段长度
minimum length of intermediate straight line 夹直线的最小长度
minimum length of transition curve 缓和曲线最小长度
minimum length of vertical curve 最小竖曲线长度
minimum lethal concentration 最小致死浓度;最低致死浓度
minimum lethal dose 最小致死剂量;最低致死剂量
minimum lethal oxygen concentration 最低致死氧浓度
minimum life 最短寿命
minimum life section 危险截面
minimum light-intensity 最小需光度
minimum limit 最小限度;最小极限
minimum limit of size 最小极限尺寸
minimum line capacity 线路最小通过能力;最小线路通过能力

minimum liquid-gas ratio 最小液气比
minimum liquidity 最低清偿能力
minimum liquid rate 最小液流量
minimum live load 最小工作负荷;最小动力负荷;最小活荷载
minimum living standard security system 最低生活保障制度
minimum load 最小荷载;最小负荷;最低负载
minimum load operation 最低负荷运行
minimum longitudinal grade 最小纵坡
minimum longitudinal gradient 最小纵坡
minimum-loss attenuator 最小损耗衰减器
minimum-loss matching 最小损耗匹配
minimum-loss pad 最小损耗衰减器
minimum lot 最小地块
minimum maintained flow 最小维持流量
minimum manufacturing quantity 最少制造量
minimum maximum basis 最小最大法
minimum maximum property 极小极大值
minimum meter 小量程仪表
minimum meterage of hole 钻孔最小进尺
minimum minable thickness 最低可采厚度
minimum minable width 最小可采宽度
minimum modulus 最小模
minimum modulus principle 最小模原理
minimum moisture content 最小含水量
minimum mortality 最低死亡率
minimum navigable depth 最小通航水深
minimum navigation depth 最小通航水深
minimum network 最稀站网
minimum night ratio 最小夜间比
minimum noise 最小噪声
minimum non-passing sight distance 安全停车视距
minimum normal operational level 最低正常运行水位
minimum nozzle area 喷管通道最小截面
minimum number of chargeable words 最低计费字数
minimum number of pipe threads for connections 管螺纹连接的最少螺纹牙数
minimum number of theoretical stage 最小理论级数
minimum observed adverse effect level 最小可见损害作用水平
minimum octane rating 最低辛烷值
minimum offset 最小偏移距
minimum operating current 最小工作电流
minimum operating time 最小运转时间
minimum operating width 最小作业宽度
minimum operation mode 最小运行方式
minimum output 最低出力
minimum oxide thickness 最薄氧化层厚度
minimum oxygen layer 最小含氧层
minimum oxygen requirement 最小需氧量
minimum passing sight distance 最小超车视距;安全越车视距
minimum path cover 最小路径覆盖
minimum path-length algorithm 最短通路算法
minimum pause 最小间隙
minimum peak width 最小峰宽
minimum pendulum 最短摆
minimum percentage by weight passing 最小重量百分率
minimum perceptible brightness difference 最小视觉亮度差;最小可辨亮度差
minimum perceptible chromaticity difference 最小可辨色(度)差
minimum perceptible difference 最小可辨差量
minimum perceptible noise 最小可辨噪声
minimum performance criterion for brake system 制动系统最低性能准则
minimum period 最小期限;极限时间
minimum permeability 最小渗透率
minimum permissible content 最小允许含量
minimum permissible ga(u)ge 最小容许轨距
minimum permissible metal temperature 最低许用金属温度
minimum permissible temperature 最低允许温度
minimum permissible velocity 最小允许流速
minimum phase 最小相位
minimum phase angle 最小相角
minimum phase filter 最小相位滤波器
minimum phase operator 最小相位因子

minimum phase shift keying 最小相移键控
minimum phase-shift network 最小相移网络
minimum phase shift system 最小相移系统
minimum phase system 最小相位系统
minimum photographic(al) distance 最近拍摄距离
minimum plastic collapsing strength 最小塑性抗挤强度
minimum play 最小游隙
minimum point 极小点
minimum point on a curve 曲线上的极小点
minimum pollution 最低污染
minimum pool level 最低库水位;水库最低水位
minimum population 最低人口
minimum porosity 最小孔隙率;最小孔隙度
minimum potential energy 最小势能
minimum potential water power 最小水力蕴藏量
minimum pour point 倾点下限
minimum power consumption 最小功耗
minimum precipitation 最低降雨量
minimum pressure 最低压(力)
minimum pressure governor 最小压力稳定器
minimum prestressing 最小预应力
minimum pretightening sealing load 最小预紧密封比压
minimum price 最低限价;最低价(格)
minimum price fluctuation 最低波动价位
minimum price movement 最低波动价位
minimum priming charge 最小起爆药量
minimum principal stress 最小主应力
minimum principal stress value 最小主应力值
minimum principle 最小值原理
minimum principle in investment decision 投资决策的最小风险原则
minimum problem 极小(值)问题
minimum profit 最低利润
minimum property 极小性
minimum property standards 房地产最低标准
minimum protective layer in full-scale 最小安全防护层
minimum pulse 最小脉冲时间
minimum pulse duration 最小脉冲持续时间
minimum quadrant elevation 最低射角
minimum quadrat area 最小样方面积
minimum quantity acceptable 可接受最小数量
minimum quantity of organic carbon 有机碳最小含量
minimum quantity of stores 最低库存量;库存最低量;存料最低量
minimum quantity per order 最低订货量
minimum radar range 雷达作用最小距离
minimum radius 最小半径
minimum radius of curve 最小曲线半径
minimum radius of horizontal curve 最小平曲线半径
minimum rafter spacing 最小椽子间距;最小椽子间隔
minimum range 最小范围;最小(作用)距离;最小射程;最小量程;最近距离
minimum range adjustment 最小距离调节
minimum range potentiometer 最小量程分压器
minimum range selection 最小距离选择
minimum ranging distance 最小测程
minimum rate 最小速率;最低率
minimum rate of return 最低收益率
minimum rate of vacancy condensation 最低空位凝聚速率
minimum rating 最小状态
minimum ratio of reinforcement 最小配筋率
minimum reaction dose 最低反应剂量
minimum readable temperature differential 最小可辨温差
minimum reception 最小信号接收
minimum recorded 记录最低的;实测最低的
minimum recorded depth 记录最小水深
minimum recorded discharge 实测最小流量
minimum recorded stage 记录最低水位;实测最低水位
minimum redundancy 最小多余度码
minimum-redundancy array 最小重复天线阵
minimum reflectance of vitrinite 镜质体最小反射率
minimum reflectivity 最小反射率
minimum reflux 最小回流
minimum reflux cascade 最小回流级联

minimum reflux ratio 最小回流比
minimum relay 低值继电器;低载继电器
minimum reliable yield 最小的可靠产量
minimum required concentration 最小要求浓度
minimum required radiation power 最低要求辐射功率
minimum required receiving field intensity 最低要求接收电场强度
minimum requirement 最低要求;最低需要(量)
minimum reserves 最低储备金
minimum reserved rate of return 最低预定收益率
minimum reserve quota 最低储备定额
minimum resistance 最小电阻
minimum resistance of heat transfer 最小传热阻
minimum resolution 最小分辨限度;最小分辨率
minimum resolvable angle 最小可分辨角
minimum response 最小响应
minimum response concentration 最小反应浓度;最低响应浓度;最感应浓度;最低反应浓度
minimum retention 最低自留额
minimum return flow 最小回流
minimum revolution 最小转数
minimum risk concentration 最小风险浓度
minimum risk estimation 最小风险估计
minimum rivet spacing 最小铆钉(间)距
minimum rolling angle of buoy 浮标最小干弦
minimum roof slope 最小屋顶坡度
minimum room height 房间最低高度
minimum running current 最低工作电流
minimum running interval 最小行车间隔
minimum running interval on ordinary hours 平峰小时最小行车间隔
minimum running interval on peak hours 高峰小时最小行车间隔
minimum runoff 最小径流(量);最低径流量
minimum safe altitude 最低安全高度
minimum safe distance 最小安全距离
minimum safe thickness in theory 理论最小安全厚度
minimum safe thickness of aquifuge 隔水层最小安全厚度
minimum safety factor 最小安全系数
minimum safe(ty) radius 最小安全半径
minimum safety stability 最小安全稳定性
minimum sale 最低销售额
minimum sampling frequency 最低脉冲调制频率;最低抽样频率
minimum scale 最小刻度
minimum scale value 最小刻度值;标度始点值
minimum scientific satellite 最小科研卫星
minimum seat spacing 最小座距
minimum seismic area 最小地震区
minimum selling price 最低售价
minimum separation 最小间隔
minimum service head 最小服务头
minimum service range 最小服务距离;最短通达距离
minimum settlement 最小沉降量
minimum sewer size 最小管道管径
minimum shift keying 最小转换键控;最小频移键控
minimum shut height (模具的)最小闭合高度
minimum sight distance 最短视距
minimum sight triangle 最小视距三角形
minimum signal level 最低信号电平
minimum signal method 最小信号法
minimum signal recognition time 最低信号识别时间
minimum size 最小尺寸
minimum-sized station 最小型站
minimum size of coarse aggregate 粗集料的最小尺寸;粗骨料的最小尺寸
minimum size of manhole 人孔最小尺寸
minimum size of pump liner 泵最小缸径
minimum size problem 最小尺寸问题
minimum slope 最小坡度
minimum soil cover thickness 最小覆土厚度
minimum space headway 极限车头间距
minimum spacing 最小间距;最小间隔
minimum spanning tree 最小生成树
minimum specification 最低规格
minimum speed 最小速度;最低速度
minimum speed of hydraulic motor 液压电机最低转速
minimum speed of wind for carrying sand 起动风速
minimum-squared-error approximation 最小方差近似
minimum stage 最低水位
minimum stage cascade 最少级的级联
minimum stage ga(u)ge 最低(水位)水尺
minimum standard 最低标准
minimum standard of living 最低生活水平
minimum standards of occupancy 最低居住标准
minimum station interval 最小站间隔
minimum steady speed of hydraulic motor 液压电机最低稳定转速
minimum steel ratio 最小配筋率
minimum stock method 最低存量法
minimum strain 最小应变
minimum strain axis 最小应变轴
minimum streamflow 最小径流(量)
minimum stream power 最小水流功率;最小河流功率
minimum stress 最小应力
minimum subscription 最低额应缴股本
minimum sunlighting spacing 最小日照间距
minimum superelevation 最小超高度
minimum supply 最低供应量
minimum survival time 最低存活时间
minimum takeup allowance 最小张紧调节距离
minimum tariff 最低税率
minimum taxable ceiling 最低应税限度
minimum taxable income 最低应税所得额
minimum tax on reference item 优惠项目最低税额
minimum temperature 最低温度
minimum temperature difference 最低温差
minimum temperature of cold starting 最低冷态起动温度
minimum tental 最低租金
minimum terrain clearance altitude 最低离地高度
minimum test duration 最短试验时间
minimum thermal resistance 低限总热阻
minimum thermometer 最低(值)温度计;最低温度表
minimum thickness 最小厚度
minimum thickness of aquifer 含水层最小厚度
minimum thickness of masonry bearing walls 砖石承重墙的最小厚度
minimum thickness of neck 缩颈的最小直径
minimum thickness of plate 板材的最小厚度
minimum thickness of skirt 直边的最小厚度
minimum thickness of tunnel(l)ing 最小衬砌厚度
minimum thrust 最小推力
minimum tie set 最小通信连接量
minimum tillage 少耕法
minimum tillage machine 最少耕作法机具
minimum-time control 最短时间控制
minimum-time headway 极限车头时距
minimum-time lag 最低时延
minimum-time path 最小时程;最速降线;最短时程;捷线;最捷径
minimum-time path tree 最短时间路径树;捷径树
minimum toggle frequency 最低触发频率
minimum torque phase 瞬时力矩最小值
minimum total Gibbs energy 最小吉布斯能总量
minimum total resource 最小总资源量
minimum toxic dose 最小毒性剂量
minimum track capacity 最小线路通过能力
minimum transmission time 最短传输时间
minimum turning circle 最小转弯圆圈;最小旋回圈;最小回转圆
minimum turning curve 最小回转曲线
minimum turning inner radius 内侧轮胎最小转弯半径;外侧车轮最小转弯半径
minimum turning path 最小转弯轨迹
minimum turning radius 最小转弯半径;最小回转半径;汽车最小转弯半径
minimum two-part prepayment meter 最低量两部式预付电度表
minimum unit cost 最低单位费用;最低单位成本
minimum useful range of leading marks 导标最近作用距离
minimum useful signal 最小可用信号;最低有效信号
minimum value 最小值;极小值
minimum value aperture 极小孔径;最小孔径
minimum value of log 测井曲线极小值
minimum value of organic carbon 有机碳最小值
minimum value over years 多年最小值
minimum variance 最小方差
minimum variance criterion 最小方差(判别)准则
minimum variance estimate 最小方差估值;最小方差估计;极小方差估计
minimum variance quantizing 最小方差量化
minimum velocity 最小流速;最低流速
minimum vertical curve radius 最小竖曲线半径
minimum visibility 最小能见度;最低能见度
minimum-void mix 最小孔隙混合料
minimum void ratio 最小孔隙比
minimum voltage 最小电压;最低电压
minimum volume 最小体积
minimum wage 最低工资;法定最低限度工资
minimum wall thickness 最小墙厚度;最小壁厚
minimum water consumption 最小耗水量;最低用水量
minimum water head 最小水头
minimum water-holding capacity 最小持水量
minimum water level 最低水位
minimum water pressure 最小水压力
minimum water quality criterion 水质最低标准
minimum water quality discharge 低限水质排放
minimum water requirement 最小需水量;最低需水量
minimum water stage 最低水位
minimum watt amplifier 最小功率放大器
minimum wavelength 最小波长
minimum weather 最劣天气
minimum weight 最小重量;起码重量
minimum weight design 最轻(重量)设计
minimum weight of equipment 设备最小重量
minimum weight routing 最小权值路由选择
minimum wetting rate 最小润湿速度
minimum width 最小宽度;最小行宽
minimum window size 窗的最小尺寸
minimum work 最小功
minimum working current 最小工作电流;最小动作电流
minimum working excitation 最小动作励磁
minimum working probit 最小工作概率值
minimum X^2 最小卡方
minimum yield collapsing strength 最小屈服抗挤强度
minimum yield point 最小拐点;最小屈服点
minimus 小生物
mining 开矿;采矿
mining academy 矿业学会;矿业学院
mining accounting 矿业会计;采矿会计
Mining Act 矿业法;采矿法
mining activity 采矿活动;采矿工程;采掘作业
mining advancing 前进式回采
mining amount 采矿量
mining and quarrying commodity 采矿业商品
mining aneroid 矿用无液气压计;矿井空盒气压计
mining area 矿山;矿区;采区;采矿区;采矿场;布雷区
mining area tax 矿区税
mining art 采矿工艺学
mining authority 矿业管理;采矿管理局
mining barge 矿用驳船
mining base regions 矿产开采基地
mining blasting explosive 矿井爆破
mining block method 分段崩落采矿法
mining by areas 分区开采
mining by stages 分期开采
mining capacity 采矿能力
mining car 矿车
mining code 采矿规范;采矿法规
mining compass 矿山罗盘
mining complex 采矿联合企业
mining condition 开采条件
mining contamination 采矿污染
mining core tube 岩芯管
mining cost 采矿成本
mining cost of groundwater 地下水开采成本
mining damage 矿山事故;开采损害
mining depression 矿业萧条
mining depth 开采深度
mining dial 矿用罗盘
mining dilution ratio 采矿贫化率
mining district 矿区;采矿区
mining downing velocity of year 年开采下降速度
mining dredge(r) 采矿船

mining effect 地下爆炸效力;爆炸效力
mining electric(al) locomotive 矿用电机车
mining engineer 采矿工程师
mining engineering 采矿工程(学)
mining engineering plan 采掘工程平面图
mining environment 采矿环境
mining excavation 矿山巷道;矿山掘进
mining excavator 采矿挖掘机
mining explosive 矿用炸药
mining feasibility report 开采可行性报告
mining field 采矿场
mining field subdivision map 矿区图
mining geodesy 矿山测量学
mining geologist 矿山地质师;采矿地质师
mining geology 矿山地质(学);采矿地质(学)
mining geology phase 矿山地质阶段
mining geology strengthen 加强矿山地质工作
mining geophysics 矿山地球物理学;采矿地球物理学
mining ground 布雷区
mining ground horizontal control survey 矿区地面平面控制测量
mining in areas 分区开采
mining inclinometer 矿山测斜仪
mining-induced earthquake 矿山诱发地震
mining industrial city 矿业城市
mining industry 矿业;采矿(工)业;采掘工业
mining in installments 分期开采
mining intensity method 开采强度法
mining intensity of groundwater 地下水的开采强度
mining ladder 露天矿场的梯子;矿井中的梯子
mining ladder rung 矿井梯子的踏板;矿井梯子的横木
mining lamp 矿灯
mining law 矿业法规;采矿法;矿业法律
mining lease 开采租约
mining locomotive 矿用机车
mining loss 采矿损失
mining losses ratio 采矿损失率
mining machine 采矿机;采掘机
mining machinery 矿山机械;采矿机械
mining machinery and equipment 采矿机械和设备
mining machinery plant 矿山机械厂
mining method 矿山法(指隧道开挖);采矿方法
mining method of tunnel construction 矿山法隧道施工
mining microscope 矿山显微镜
mining modulus analogy 开采模数法
mining motor 矿用电动机
mining nuisance 矿山公害
mining of clay 黏土的开采;黏土的挖掘
mining of groundwater 地下水开采
mining operation 采矿作业
mining pick 采矿用镐
mining plans 矿山采掘计划
mining plow 超深松犁
mining power 采矿动力
mining practice 开采方法;采矿实践
mining problem 开采问题
mining product 矿业产品
mining property 矿业财产
mining prospecting 探矿
mining pumping method 开采抽水法
mining quantity 采矿量
mining rate 开采速度
mining recovery 采矿回收率
mining region 开采区(域);矿山开采区域
mining regulation 采矿规程
mining rent 采矿租金
mining research 开发技术研究
mining reserves 开采储量
mining retreating 后退式开采法
mining right 矿产权
mining rope 矿用钢丝绳
mining royalty 矿区使用费
mining safety 采矿安全
mining section 采区
mining sequence 开采程序
mining shaft 矿井
mining shaft and open pit 采矿井场
mining shield 掘进护盾
mining shovel 矿用挖掘机;采矿(型)电铲;采掘电铲

mining siding 矿山专用线
mining solid waste 矿业固体废物
mining speed 开采速度
mining stage 开采阶段
mining stock 矿业股票
mining subsidence 矿穴沉陷;矿区场陷;矿坑下陷;矿坑坍陷;开采沉陷;采空坍陷;采空塌陷
mining subsidence survey 矿山沉降水准测量
mining survey 矿山测量;矿区测量
mining surveying 矿山测量学
mining surveyor 矿山测量员
mining system 坑道工事
mining tax 矿业税
mining technique 采矿技术
mining technologic(al) condition of mineral deposits 矿床开采技术条件
mining technology 采矿工艺学
mining technology for marine mineral resources 海洋矿物资源采矿技术研究
mining-test method 开采试验法
mining theodolite 矿山经纬仪
mining thrived 矿业开发兴旺
mining town 矿业城镇
mining transit 矿用经纬仪;矿山经纬仪
mining traverse 矿坑导线
mining tunnel 矿山隧道;矿井隧道;暗挖隧道
mining unit 开采单位
mining waste rock 采矿废石
mining waste 采矿废(弃)物
mining waste water 采矿废水
mining water use 采矿用水
mining width 可采矿床的最小厚度
mining winch 矿用绞车
mining work 矿井工作;采矿工作;采掘
mining yield 开采量
mining yield decrease 减少开采量
mininoise cable 低噪声电缆
minion 直梗;中梃;粗砂、碎石、道砟的俗称
minipad 小垫片
mini-panel 遥控板
mini pickup 小型轻便货车
mini-pile 小型桩;微型桩
mini-plant (试生产用的)小型设备;中间工厂
miniplug 小型插头
minipore 小孔
miniprint 缩印品
mini-processor 微型处理机;小型处理机
minipump 小型真空泵
mini-ranger 小型测距仪
mini-rapid scan electron microscope 小型快速扫描电子显微镜
minirecession 小衰退
mini rotary kiln 小型回转窑
mini-roundabout 微型环交【道】
mini-scanning method 小扫描法
mini-ship 小型船舶
minisize 微型
mini-slump 微坍落度
minispread 小排列
mini steel plant 小型钢铁厂
Minister for Communications 交通部部长
ministerial standard 部(颁)标准
Minister of Power 电力部长
Minister of Transport 运输部长
mini-straw 微型塑料细管
ministry and committee grade 部委级
ministry-issuing standard 部颁标准
Ministry of Agriculture and Forestry 农林部
Ministry of Communications 交通部
Ministry of Health 卫生部
Ministry of Housing 住房建设部
Ministry of Posts and Tele Communication 邮电部
Ministry of Public Works 市政工程部;公共工程部
Ministry of Railways 铁道部
Ministry of Railways Traffic Management Center 铁道部运输指挥中心
Ministry of Transport 运输部
Ministry of Urban and Rural Construction and Environmental Protection 城乡建设环境保护部
Ministry of Water Conservancy and Electric(al) Power 水利电力部
Ministry of Water Resources 水利部
Ministry of Water Resources and Electric(al) Power 水利电力部
Ministry of Works 建筑工程部
mini-supercomputer 小矩型计算机
minisurvey 小型调查
miniswitch reed matrix 小型接线器
minitelevision 袖珍电视机
mini-test 小型试验
minitexture meter 手推式构造深度仪
minitillage system 少耕制
minitrack 无线电跟踪装置;卫星跟踪系统;电子跟踪系统
minitrack radio 微型跟踪无线电
minitrack system 微型人造卫星跟踪系统;干涉仪卫星跟踪系统;微型跟踪系统
mini-tractor 微型拖拉机
minitrain 小型列车
minitram 微型电车;小型电车
minitransistor 小型晶体管;微型晶体管
Minitron 米尼管
minitube 袖珍电视机
minituner 小型调谐器
minitype 小型;微型
mini-type loading plate test 小型承载板试验
minium 四氧化三铅;红铅粉;铅丹;丹铅
minium-based building mastic 以铅丹为基底的胶黏涂料;以铅丹为基底的胶黏水泥
mini-vessel 小型船舶
miniwatt 微小功率;小功率
miniwatt amplifier 小功率放大器
miniwatt station 小功率电台
minizaton of uncertainty 把不确定性减少到最低限度
minizone 小区间
Minkovski geometry 明可夫斯基几何学
Minkovski inequality 明可夫斯基不等式
Minkowski electrodynamics 明可夫斯基电动力学
Minkowski-Leontief matrix 明可夫斯基—列昂吉夫矩阵
Minkowski matrix 明可夫斯基矩阵
Minkowski method 明可夫斯基氏法
Minkowski metrics 明可夫斯基度规
Minkowski norm 明可夫斯基范数
Minkowski space-time 明可夫斯基时空
Minkowski world 明可夫斯基世界
min-max technique 极小极大法
min-max theorem 极小极大定理
min-min value of organic carbon 有机碳最小的最小值
minmum energy 最低能量
Minnesota black 明尼苏达黑色花岗石(美国)
Minnesota engineering analogies test 明尼苏达工程类比测试
minnesotaite 铁滑石
Minnesota Mankato 明尼苏达曼卡托石灰岩(美国)
Minnesota rate of manipulation test 明尼苏达操纵速度试验
Minnesota stone 明尼苏达花岗岩(美国)
minnow net 细眼网
minnow net machine 编网机
minnow seine 小鱼围网;密眼围网
Minoan architecture 米诺斯建筑(古希腊)
Minofar 餐具锡合金
Minofor Alloy 明诺福合金
Minol 迈纳尔炸药
minolith 木材防火防腐剂
minophyric 细斑状
minor accident 小事故
minor adjustment 小调整
minor aids-to-navigation 小型航标
minor air cell 涡流室
minor and low level rapids 中枯水急滩
minor angle 次角
minor angle method 小角法
minorante 劣函数
minor arc 主弧;劣弧;副弧
minor arch 局部拱;短跨拱;辅助拱
minor arterial road 次干道
minor auxiliary circle 小辅圆
minor axis 弱轴;短轴;短径
minor axis of building square grid 建筑方格网短轴线
minor axis of inclusion 包裹体短轴
minor base 稀有碱基

minor bed 中水位河床;中枯水位河床
minor bend 微弯波导;短弯头
minor betterment 次要改进;小改进;局部改造;局部改善;次要改善
minor biome 小生物群落
minor blasting 二次爆破
minor breach of contract 轻微违约
minor bridge 小桥
minor caliber 小口径
minor change 小换工;小更动;小修改;小变更
minor chord 小三和弦
minor circulation 小型环流
minor community 小生物群落;小群落
minor constituent 微量组分;微量成分;次要组分;次要成分
minor control 低等控制;次要控制
minor control change 次要控制改变
minor control data 次要控制数据
minor control framework 低等控制网
minor control plot 辐射三角测量
minor control point 低等控制点
minor crack 细微裂纹;细微裂缝
minor crane 小型起重机
minor critical speed 次临界转数
minor cycle 小周期;小循环
minor cycle counter 短周期计数器
minor cycle pulse generator 小周期脉冲发生器
minor dam 小坝;次要水坝;副坝
minor damage 轻微破坏
minor decision 次要决策
minor defect 轻(微)缺陷
minor details 次要细节;细节
minor determinant 子行列式
minor development 采准
minor diagonal 次对角线
minor diameter 小直径
minor diameter fit 内径配合
minor diameter fit of spline 花键内径配合
minor diameter of screw 螺钉最小直径
minor disaster 小灾害;小事故
minor division 小格
minor earthquake 小地震
minor element 微量元素;少量元素;次要元素
minor element analysis 次要元素分析
minor enterprise 中小企业
minor equipment 次要设备
minor estimate 较低的工程估算
minor exchange 电话支局;分交换机
minor failure 小事故;轻度失效;非重要故障
minor fault 小断层
minor feature 微地貌
minor first flank 第一副后刀面
minor flank 刀具副后面
minor fog signal 较小雾号
minor fold 小褶皱
minor forest products 林副产品
minor function 下函数
minor gas measurement 微量气体测定
minor geologic(al) details 次要地质细节
minor geosyncline 小地槽
minor grade 较小坡度;次要坡度
minor groove 小沟
minor head 分项项目
minor inconsistence 小量不一致
minor ingredient 小料
minor inorganics 微量无机成分
minor inspection 小检查
minor insulation 局部绝缘;次绝缘
minor intrusions 小型侵入体
minor isotopes safeguard techniques 次要同位素安全监督技术
minority 少数
minority carrier 少数载流子
minority carrier emitter 少数载流子发射极
minority control 少数股控制
minority emitter 少数载流子发射极
minority group 少数群
minority holding 少数股权
minority impurity 少数杂质
minority interest 少数权益;少数股权
minority joint-venture 少股合资
minority language 少数民族语言
minority nationalities commerce 民族商业
minority stockholders 小股东;少数股权股东

minor karst features 幼年溶岩地形
minor light 小型灯标;较小灯标
minor line 次要线路
minor loading 次要荷载
minor loading condition 次要荷载情况
minor lobe 后瓣;旁波瓣;旁瓣;副瓣
minor loop 小回路;小环路;局部回路;内环;副环
minor-loop feedback 局部反馈
minor-loop feedback compensation 并联补偿法
minor-loop feedback compensation network 并联校正网络
minor loss 次要损失
minor losses in pipe 管道局部阻力损失
minor maintenance 日常保养
minor meander 小河弯;次要河弯
minor metal 次要金属
minor mine disaster 次要矿山事故
minor mode 小调式
minor nutrient 次要肥料
minor of a determinant 子行列式
minor office 支局;分局
minor-operating theatre 小手术室;小工作室;小操作室
minor overhaul 小修(理);工程小修
minor package unit 小型装配单元
minor parameter 次要参数
minor path 劣路线;次要途径;次要路径
minor place of (outdoor) assembly 小型集会场所;小型室外集合场所
minor planet 小行星
minor pool 浅滩
minor port surcharge 偏僻港增额;偏僻港附加费
minor principal plane 小主平面
minor principal strain 小主应变
minor principal stress 小主应力;最小主应力;次要主应力
minor principal stress plane 小主应力面
minor produce 副收获
minor product 副产物
minor radial 次要辐射(路)线
minor radius 短半径
minor relay station 小中继站;中继站;小转运站;小转发站
minor repair 小修(理)
minor residential layout 住宅小区布置
minor rhombohedral face 小菱面
minor ride 区划副线
minor river 支流;次要河流
minor river bed 小河床;中枯水位河床
minor road 小路;支路;次要公路;次要道路
minor road junction 次要道路枢纽;次要道路交叉口
minor segment 劣弧段
minor seismic area 小地震区
minor seismic belt 小震带
minor semi-axis 短半轴
minor sequence 小层序
minor shock 小(地)震;副震
minor sized 小尺寸的
minor spiral 小螺旋
minor station 乘客不多的停车站;小站
minor stop 小站
minor strain 最小应变
minor stream 支流;次要河流
minor stress 最小应力
minor structure 小工程项目;附属结构;附属房屋;小型构造;子结构;次要结构;次要构筑物
minor structure scale 小型构造尺度
minor subdivision 小区
minor swimming pool 小游泳池
minor switch 小型选择器;小型开关
minor third 小三度
minor thrust face 小推力面
minor total 次总计;小计
minor treatment 局部处理
minor triangulation 小三角测量;低等三角测量
minor triangulation point 小三角点
minor trochoidal axis 旋轮线短轴
minor trough 小型低压槽;小槽;副地槽
minor tune up 小改革;小调整;部分改造
minor use of mineral commodities 矿产品次要用途
minor use of mineral commodities at percentage 次要用量占百分比
minor use tonnage 次要用量

minor value 痕迹量
minor watershed 小流域
Minovar 米诺瓦低膨胀高镍铸铁
Minovar metal 米诺瓦合金
minoxidil 长压定
minrecordite 碳锌钙石
minster (修道院的)礼拜堂;大教堂
minsteryard 教堂寺院
minstrel gallery 教堂中最高楼座;大厅内楼座;室内小眺台
mint 制币厂;造币厂
mintage 印记
minter stereo system 明特立体声系统
mint fine bars 纯金块
mint-mark 刻印
mint oil 薄荷油
Minton dryer 真空干燥装置
mint parity 法定平价
mint par of exchange 法定平价
mint par value of coin 法定货币价
min-tree finding algorithm 求最小树算法
Mintrop wave 明特洛甫波
mint weight 标准重量(货币)
minuano 米牛阿诺风
minuend 被减数
minuend address operand 被减数地址操作数
minuend length operand 被减数长度操作数
minuend string 被减数串
minus 零下;减;负号
minus #00 material 通过00号筛的细料
minus angle 俯角
minus blue 缺蓝;减蓝(色)
minus-blue filter 去蓝滤光片
minus caster 主销负后倾角;负后倾角
minus-cement porosity 无胶结孔隙度
minus charge 负电荷
minus colo(u)r method 减色法
minus correction 凹面修正;负数修正
minuscule 小写字母
minus current connection to earthing rail 负回流接地轨
minus declination 负赤纬
minus deflection 减偏斜
minus deviation 负偏差
minus earth 阴极接地
minus effect 不利影响;副作用;反效果;不良效果
minus electricity 负电
minus flip-flop 减触发器
minus g 负过载
minus gate voltage 负栅压
minus grade 降坡;下坡(道)
minus involute (齿轮的)负的渐开线
minus lap 负余面
minus lens 负透镜
minus line 负线
minus material 低劣材料;次品
minus mesh 筛下料;筛下(产品)
minus mineral 负矿物
minus ooid 负鲕
minus plate 阴极板
minus pole 负极
minus punch 补孔
minus quantity 负数
minus screw 一字槽头螺钉;左旋螺纹
minus side 负端
minus sieve 筛下
minus sight 前视
minus sign 减号;负号
minus size of a screen 筛下级物料尺寸(即小于筛孔的物料尺寸)
minus skewed histogram 负偏态直方图
minus sounding 负深度
minus strain 负宗【植】
minus strand 负链
minus symbol 减符号;负符号
minus terminal 负号端子
minus thread 逆螺纹;负螺纹
minus tolerance 负差
minus tree 负号树
minus zero 游标零分划
minus zone 负数区;负区;负号位
minute 细微的;微小的;纪要;会谈纪要
minute addition 微量添加料
minute adjustment 精密调节

minute angle 微小角度
minute associate 伴生矿物
minute bonder 制动蹄摩擦片快速黏结器
minute book 会议记录簿
minute break 瞬时断流
minute bubbles 小气泡
minute crack 细裂缝;发状裂缝
minute finish 精致修饰;抛光至镜面光泽
minute folding 细褶皱
minute freezing 微冻结
minute gun 分炮(船舶遇险时每分钟发一次的号炮)
minute hand 分针
minute interstice 微小孔隙
minute jump clock 分跳钟
minute mark 分标
minuteness 精密
minuten theodolite 分级经纬仪
minute nut 分轮齿轴
minute of angle 分(角度单位)
minute of arc 弧分
minute of the equator 赤道上每分距离
minute output 每分输出量
minute pinion 分针小齿轮;分轮齿轴
minute projections 极微小突出部分;摩擦面的粗糙度
minute quantity 极少量
minutes of meeting 会议记录
minutes of pre-tender meeting 标前会议记录
minute structure 细微构造
minute totalizer 累积计分器
minute trace survey with truck 汽车微迹法
minute ventilation 每分通气量
minute volume 分钟量
minute volume divider 低量分压器
minute wave type 微波型
minute wheel 跨轮;分针轮
Minvar 镍铬铸铁;敏瓦尔低膨胀系数合金铸铁
minverite 钠长角闪辉绿岩
min xia satin brocade 明霞缎
minyulite 水磷铝钾石;水磷钾铝石
Miocene clay 中新世粘土
Miocene epoch 中新世【地】
Miocene series 中新统【地】
miogeoclinal association 冒地斜组合
miogeocline 冒地斜
miogeosynclinal prism 冒地槽棱柱体
miogeosyncline 冒地槽【地】
miogeosyncline type formation 冒地槽型沉积建造
miomagmatic zone 冒地槽【地】
miorotome knife rack 切片刀刀架
miospore 冒孢子
miostagmin reaction 阿斯科利氏反应
miothermic period 温和期(指间冰期)
Mipolam 麦波郎(塑料专用名)
mipor 微孔的
Mipora 米波拉保温材料
mipor rubber 微孔橡胶
mipor scheider 微孔橡胶隔膜
Mira 米拉铜铅合金;米拉铜基合金
Mira alloy 米拉(铜基)合金
mirabilite 芒硝
Mirabilite alloy 米拉比来铝镍合金
mirabilite cutting 芒硝侵蚀
mirabilite lake 芒硝湖
mirabilite rock 芒硝岩
miracle 自然界奇迹
miracle rice 奇迹米
miraculous image 神像;圣像
Miraculoy 米拉丘洛依耐高压铸造合金
miraculum (防腐漆中的)红丹填料
mirador (西班牙建筑中的)守望处;屋顶亭子;角塔;凸(肚)窗;凉廊
mirage 幻像;海市蜃楼;蜃景
Miralite 米拉赖特耐蚀铝合金
Miramant 米拉合金
Mira metal 米拉耐蚀合金;米拉耐蚀铜合金
Miraplast 高密度聚乙烯合成纸
mirbane oil 硝基苯
mircrometer ga(u)ge 测微器
mire 照准标;沼泽;泥沼;泥塘;泥潭;污泥;梯形目标;标点
mired 迈尔德;微度;微倒度
mire down in mud 陷入泥潭
mire-drum 天然盐水
Mired value 迈尔德值
mire layer 淤泥层
mire vegetation 沼泽植被
miriness 泥泞
miro 带褐色软木(新西兰产)
mirocrack toughening 微裂增韧
mirror 镜子;鸟翼上的翼镜;反映
mirror adjustment knob 反射镜调制钮
mirror and scale method 镜尺法
mirror antenna 反射式天线
mirror aperture 反射镜孔径
mirror-apparatus 收缩测定器
mirror arc 带反射镜弧光灯
mirror arc lamp 反射(镜)弧光灯
mirror autocollimator 反射式自准直仪
mirror backed fluorescent screen 背面反射式荧光屏
mirror ball 小型球面反射镜
mirror bar 反光镜调节杆
mirror black glaze 乌金釉
mirror blank 镜毛坯
mirror cabinet 带镜立柜
mirror camera 反射摄影机
mirror cell 镜室
mirror coating 反射镜(涂)膜;反射镜涂层;反射镜镀层
mirror collector 反射镜集热器
mirror collimater 反射式准直仪
mirror compass 反光式罗盘;反光罗盘仪
mirror concentrator 反射镜聚光器
mirror condenser 反射(镜式)聚光器;反射镜式聚光灯
mirror curvature 镜曲率;反射镜曲率
mirror dehumidifier 镜面降湿;镜面去湿
mirror-deposit work 镜面镀银加工
mirror disk 镜盘
mirror dislocation 镜面位错
mirror drive cable 反射镜驱动钢丝绳
mirror drive clutch 镜式传动离合器
mirror drum scanner 镜鼓扫描器
mirrored cabinet 镜柜
mirrored chopper 反射镜调整器
mirror edge 镜缘
mirror effect 镜像效应
mirror electrodynameter 反射镜式电测功率计
mirror electrodynamometer 镜式力测电流计;镜式功率计;镜式电测力计
mirror element 镜像元
mirror erecting system 反射镜正像系统
mirror extensimeter [extensometer] 镜示伸长计;反光延伸仪;镜式伸长计;反射镜式伸长计;反光应变计;镜面伸长计
mirror finish 光洁表面;镜面抛光;镜面加工;镜面光洁度
mirror finished 高度磨光的;高度抛光的;镜面磨光
mirror finished surface 镜面加工面
mirror-finish grinding wheel 镜面磨削砂轮
mirror finishing 镜面精加工光洁度
mirror foci 镜焦点
mirror galvanometer 镜式检流计;镜示检流计;反射式检流计;反射镜电流
mirror geometry 磁塞形态;磁镜形态
mirror glass 镜(子)玻璃;镜面玻璃
mirror glazing quality 高质量平板玻璃
mirror grating 镜栅
mirror grinding 镜面磨削
mirror housing 反射镜外罩
mirror image 镜像;反射镜图像
mirror image function 镜像函数
mirror image language 镜像语言
mirror image measurement 镜像测量
mirror image reversal 镜像反转
mirror image switch 镜像开关
mirror instability 磁镜不稳定性
mirror instrument 转镜式仪表;镜示仪表
mirror interference 镜面干涉
mirror interferometer 镜式干涉仪;反射镜干涉仪
mirror inversion 镜像
mirror iron 镜铁
mirror lens 反射(镜)透镜;反射镜式透镜
mirror lens optics 反射透镜光学
mirror-lens system 反射透镜系统;折反射系统
mirror-like tongue 镜面舌
mirror luster 镜面光泽度
mirror machine 磁镜装置
mirror making 制镜
mirror matrix 镜面矩阵
mirror matrix target 反射镜矩阵靶
mirror-metal electrode 金属反射镜电极
mirror microscope 反射镜式电子显微镜
mirror monochromator 反射镜式单色仪
mirror mount 反射镜固定架
mirror nephoscope 测云镜;反射镜式测云仪
mirror nuclei 镜像核
mirror-optical system 反射镜光学系统
mirror optics 反射镜光学
mirror oscillograph 镜式示波器
mirror-parallelism 镜平行度
mirror-phone 磁录音机
mirror plane 镜平面;反射平面;反光镜平面
mirror plating 镜面涂敷
mirror polish(ing) 镜面抛光
mirror pool 镜池;倒影池
mirror principle 镜子原理
mirror projector 镜面探照灯;镜面聚光灯;反射探照灯;反射放映机
mirror quality 镀银质量
mirror ramp 反射镜斜面
mirror ratio 磁镜镜比;磁镜比
mirror reader 反光读数镜
mirror reading 镜示读数;镜测读数
mirror reflection 规则反射;镜面反射;单向反射
mirror reflection fluorescent screen 金属背荧光屏
mirror reflector 镜式反射器;反射镜反射器;反射镜;反光镜
mirror reflex camera 反射镜摄像机;反光照相机
mirror reversed negative 反阴片
mirror scale 镜度盘;镜面标度
mirror scan 反射扫描
mirror scanning mechanism 镜扫描装置
mirror scattering 镜面散射
mirror screw 镶镜用螺钉;镜面螺钉;镜螺旋
mirror set 反光镜装置
mirror sextant 分射镜六分仪;反光六分仪;反光镜六分仪
mirror shot 反射镜合成摄影
mirror shutter 镜面快门;镜面反射式遮光器
mirror-slide 镜面拉门
mirror spotlight 反射聚光灯
mirror square 反光直角器
mirror stereoscope 反射立体镜;反光立体镜
mirror stone 白云石;白云母
mirror surface 镜面;发光面
mirror surface machine 镜面磨削机流
mirror sweep 反射扫描
mirror symmetry 镜像对称;镜面对称;面对称
mirror system 反射镜系统
mirror telescope 反射(式)望远镜;发光望远镜
mirror teletereoscope 反射式立体望远镜
mirror tilt 反射镜倾斜
mirror-tilting mechanism 镜倾斜装置;倾斜镜装置
mirror transit circle 镜式子午仪
mirror transit instrument 地平中星仪
mirror trap 镜俘获
mirror-trough lighting 镜面照明槽;槽形镜面照明器
mirror-type electron telescope 反射镜式电子显微镜
mirror wave 镜波
mirror wave-beam guide 镜式光导管
mirror wave-beam structure 镜式波束结构
mirror wheel 镜轮;镜鼓
mirror with glass foot 玻璃座架镜
mirror writing 倒写
miry 淤泥的;沾满泥的;泥沼的;泥泞的
miry soil 淤泥土
mirzapur 默扎珀地毯
misadministration 管理失当
misaim 灯光不正确投射;失准
misalignment 未对准;安装不成直线
misalign 不重合
misaligned contact point 错开接触点
misalignment 线向不正;直线不重合度;校正不当;偏心度;调整不当;失准值;失调准;定线不准;错位;不重合(度);不同轴度;不对准;不对中;非准直
misalignment drift 安装误差漂移
misalignment of wheels 车轮不平行

misalignment stress 装配应力
misalignment voltage 失调电压
misapplication 应用不当;不正确使用
misapplication of law 法律的误用
misappropriation 私吞
misappropriation or diversion of financial investment 挤占、挪用财政投资
misarrangement 错误布置;布置不当;不正确配置;安排不当
misassembly 不正确装配
miscalculate 估计错误;失算
miscalculation 计算误差;计算错误;误算;算错
miscella 含油(脂)溶剂;溶剂混合油
miscellaneous 杂项的;混杂的
miscellaneous accessories 杂件
miscellaneous account 杂项账户
miscellaneous and contingencies 杂项及意外费用;杂费及预备费
miscellaneous assets 杂项资产
miscellaneous backfill 各种回填;其他回填;多种回填
miscellaneous business 杂务
miscellaneous cargo 杂货
miscellaneous cereals 杂粮;杂谷
miscellaneous charges 各种费用价目表
miscellaneous charge book 运价手册;运价表
miscellaneous charge of goods transport 货运杂费
miscellaneous charge order 杂费支付通知
miscellaneous correlation 混杂相关
miscellaneous declaration 杂项说明
miscellaneous division 总务科
miscellaneous expenditures 杂支费;辅助支出
miscellaneous expenses 杂项开支;杂费;其他开支
miscellaneous fill 杂填土
miscellaneous fill landslide 杂填土滑坡
miscellaneous fishes 杂鱼类
miscellaneous function 辅助功能
miscellaneous general expense 杂项管理费
miscellaneous goods 小产品;杂货
miscellaneous income 辅助收入
miscellaneous interference 混杂干扰
miscellaneous investment 各项投资
miscellaneous items 杂项
miscellaneous load 杂项荷载;零星载荷;混杂载重;混杂荷载
miscellaneous loss 杂项损失
miscellaneous machinery 杂用机械
miscellaneous observation 混杂观测
miscellaneous operation 杂操作;其他操作
miscellaneous parts 其他部件;杂件;各种各样的部件
miscellaneous payments 杂项支付
miscellaneous power 杂项用电
miscellaneous profit 杂项利润
miscellaneous provisions 杂项规定;其他有关事项;附则
miscellaneous pump 混杂泵
miscellaneous repairs 分散修理
miscellaneous reservoir 杂类储集层
miscellaneous retail stores 杂货零售商店
miscellaneous revenue 杂项营业收入;杂项收入;其他收入
miscellaneous service charges 综合服务费
miscellaneous small parts 各种小零件
miscellaneous sources of pollution 杂污染源
miscellaneous subsidy 杂项津贴;杂项补助;杂项补贴
miscellaneous tapping machine 各种形式攻丝机
miscellaneous tax 杂税
miscellaneous ternary refractory 三元混合耐火材料
miscellaneous test 多项试验
miscellaneous testing 多项试验
miscellaneous tools 杂项工具
miscellaneous tress 杂木
miscellaneous vessel 杂用船
miscellaneous waste glaze 杂色釉
miscellaneous wiring 杂项敷线
mischance 故障
mischief 故障
mischmetal 铈镧稀土合金;铈钚合金;稀土金属混合物
Mischrome 铁铬系不锈钢
mischzinn 锡铅锑合金

miscibility 可混(合)性;可拌和性;混溶性;溶混性;掺混性
miscibility gap 混溶隙;混溶间隙;溶混性间隔;不相混溶区
miscibility test 混和性试验
miscibility test in linseed oil 亚麻仁油混溶性试验;亚麻仁油混合性试验
miscibility with water 水溶性
miscibility with water test 水溶性试验
miscible 可混合的;能溶合的
miscible displacement 混相驱油
miscible displacement method 混相驱油方法
miscible displacement process 混溶置换工艺
miscible fluid 可混合流体
miscible liquid 可溶混液体
miscible phase 混合相
miscible technique 混溶工艺
misclassification 错(误)分类;分类错误;分错
misclosure 闭合导线误差[测]
misclosure of round 归零差;半测回归零差;半测回闭合差
Misco 米斯科镍铬铁系耐热耐蚀合金
miscolo(u)r 着色不当
Misco metal 米斯科合金
misconception 误解;错误概念
misconduct 错误行为;处理失当;处理不当
misconnect 误接
misconnection 错误连接;错接
misconstruction 盖错(建筑、房屋);建错
misconvergence 会聚失调;收敛失效;收敛不足;失聚
misconvergence of beams 束发散
miscount 计数错误;误算;算错;错误计数;错算
miscut 切割错误
misdate 误记日期;错误数据
misdeclaration 误报
misdelivery 错误交货
misdescription 误记
misdirected investment 方向错误的投资
misdirected ion 错向离子
misdivision haploid 错分单倍体
misdraft 意外牵伸
mise-a-la-masse 充电法
mise-a-la-masse method in borehole curve 井中充电法测井曲线
mise en scene 舞台装置
misenite 纤重钾矾
miser 钻探机;钻湿土用大型钻头;锥钻头;凿井机;管状提泥钻头
miserere 唱诗班正折叠式座位;椅子靠背板
misericord(e) 折叠座位;折椅;椅背突板;(教堂唱诗班座位下的)活动托板;小斋堂;(修道院的)纪律放宽室
miserilite 硅铈钙钾石
misering 钻探;管状提泥钻头钻进
miserite 钠硬硅钙石
misery index 悲惨指数
Mises cylinder 米赛斯圆柱面
Mises yield criterion 米赛斯屈服准则
misfeed 误传送;传送失效
misfire 不着火;瞎炮;拒爆;盲炮;不发火(发动机或者炮);不点火
misfire charge 残留炸药
misfired detonation 拒爆
misfired hole 瞎炮孔;炸药残孔;未爆眼
misfire handling 瞎炮处理
misfiring 瞎跑;引火不良;走火;不点火
misfit 错配;不吻合;不符值;配合不良
misfit dislocation 错配位错
misfit river 遗迹河;河流与河谷不相称的河;弱水河;水量不足的河流;废河;不相称河(流);不适应河
misfit stream 水量不足的河流;不相称河(流);不适应河
misfocused image 散焦像
misfocusing 散焦
misfurnish 脱浆
misgather 偏装(集装箱)
misgotten 用不正当手段取得的
misgrind 研磨不正确
mishandling 误操作;违反运行规程;处置失当;处理不当;不正确运转
mishandling failure 误操作故障
mishap 灾祸;机械事故;事故

mishima 镶嵌装饰法;三岛法
Mishima magnet steel 铝镍强磁钢
mishmash 混杂物
misidentification 错误判断;错误判读;错误辨识
misinterpret 曲解;误判断
misinterpretation 误译;错误判读
misinterpretation rate 错判率
misjudge 误判断
misjudgement 判断错误
misjudgement failure 从属故障
miskeyite 似蛇纹石
mislanding 误卸
misleading declaration 申报失实
misleading indications 易引起误解的标记
mislevel 未安平
mislinement 不对中
mislocation defect 误置缺陷
mismachine 不正确加工
mismachined 不正确加工的
mismachining tolerance 加工误差
mismanagement 处置失当
mismatch 不协调;不良合缝;配合不当;误配;失谐;失配;失调;错箱;错位;错缝;(模锻件上下部间的)错差;不相配
mismatch condition 失配条件;不匹配条件;不符合条件
mismatch contact ratio 失配重叠系数
mismatched 错缝的;对缝不良的;不对称的(相邻两板);配错的
mismatched duals 不相配的成双轮胎
mismatched generator 失配信号发生器
mismatched line 不拟合曲线
mismatched material 误配材料
mismatched plywood 合缝不良胶合板
mismatch error 配合误差;失配误差
mismatch floating rate note 不对称浮动利率本票
mismatch in core 型芯错位
mismatch indicator 失配指示器;不匹配指示器
mismatching 不重合;不匹配
mismatching factor 失配因数
mismatching loss 失配损耗
mismatch in mo(u)ld 铸型错箱
mismatch lumber 边角不整齐的木材
mismatch parameter 失配参数
mismatch slotted line 失配开槽线
mismate 配合不当;错配合
misoperation 异常运行;误动作;误操(动)作;操作失误
misorientation 取向错误;失向;错向;错误取向
misphasing 相移
mispickel 砷黄铁矿;毒砂
misplace 误置
misplaced outcrop 移位露头
misplaced size 晶粒错位;颗粒错位
misplaced winding 偏位线圈;配置错误的线圈;失位线圈
misplacement 误放;错位
misplacement of the reinforcing 钢筋错位
misplacing of stencil 脱版
misplug 插插;插塞错插
mis-pressing 不压针
misprocurement 错误的采购
misproportion 不成比例
misquotation 误引
misreading 读错;错误读数
misrecording 记录错误
misregister 指示不准确;记录不准确
misregistration 记录错误;配准不良;位置不正;对版不准;重合失调
misregistration correction 图像重合失调校正
misregistry 图像重合失调
misrepresentation 虚报;误报;歪曲;错误表述;不实的告知
misrun 滞流;浇铸不满;浇不足
miss an oil sand 无效油砂
miss-cropping 下料重量超差
miss-distance 脱靶距离;差距
miss-distance bias 系统引导偏差
miss-distance indicator 脱靶量指示器
missed approach climb 复飞爬高
missed discount 折扣失效;漏记折扣
missed heat 不合格熔炼炉次
missed hole 失效炮眼;瞎跑;瞎炮孔;未爆眼
missed pulse 漏失脉冲;失误脉冲

missed round 拒爆炮眼组
missed synchronization 漏同步
misses 漏涂区;漏涂点
missetting 墙纸图案错位
miss-fire shot 不爆发炮眼;瞎跑眼
misshape 弄歪曲
misshaped particle 不良颗粒
misshapen particle 长或扁形颗粒
missible base 导弹基地
missible site 导弹发射场
missile 照明弹;投射器;发射物
missile barrier 投射物屏障
missile cruiser 导弹巡洋舰
missile guidance set 制导设备
missile hangar 导弹库
missile oxidant waste treatment vehicle 导弹氧化剂废水处理车
missile photogrammetry 导弹摄影测量学
missile range 导弹靶区;导弹射程
missile-range-instrumentation radar 靶场测量雷达
missile submarine 导弹潜水艇
missile target data 导弹目标数据
missile-track radar 火箭跟踪雷达
missile wing 弹翼
missing 故障
missing address marker 漏地址标记
missing argument 缺变元
missing a train 漏乘
missing bit 失位元
missing cargo 遗失货物
missing code 遗漏码
missing data 漏失数据;缺项
missing design 遗漏设计;设计不全
missing detail 碎部遗漏
missing end 缺径;断头
missing error 遗漏误差;遗漏错误;疏忽错误
missing factor 缺因子
missing goods 遗失货物
missing interruption handler 遗漏中断处理程序
missing level 失误电平
missing line 遗漏划线
missing link 连接链环;可拆链环
missing list 遗失货物目录清单
missing locking 漏锁闭
missing mass 短缺质量
missing of ignition 不着火
missing order 缺序
missing page interrupt 缺页中断
missing page interruption 缺页中断
missing pick 缺纬
missing plot 缺区
missing precipitation data 缺测降水资料
missing pulse 漏脉冲;脉冲遗漏
missing release 漏解锁
missing sample 缺失样品
missing segment interrupt 缺段中断
missing ship 去向不明的船舶;失踪船(舶)
missing stratigraphy 缺失的地层
missing triangle 损缺三角形
missing value 遗漏值;漏测值;欠测值
missing variable 缺变元
mission 任务;使命;代表团
mission allowance 出差津贴
mission analysis 任务分析
mission architecture 教会建筑
missionary 推销员
missionary school 教会学校
mission assignment 任务委托书
mission capability 完成任务的能力
mission cloth 粗厚方平织物
mission control center 任务控制中心
mission data card 任务数据卡
mission duration 任务期限
mission flow diagram 任务流程图
mission net 大网眼粗窗帘布
mission-oriented network 面向任务的网络
mission roofing tile 西班牙屋顶瓦
mission statement 任务书
mission success rate 成功率
mission tile 阴阳瓦;筒瓦;西班牙式圆瓦;拱形瓦;半圆形截面瓦
Mississippi abyssal fan 密西西比深海扇
Mississippian period 密西比纪【地】
Mississippi delta 密西西比三角洲
Mississippi river gulf outlet 密西西比河墨西哥湾人工水道
Mississippi river-type buoy 密西西比河型浮标
Mississippi-valley-type lead-zinc deposit 密西西比河谷式铅锌矿床
missive 公文;公函
Missourian 密苏里统【地】
Missouri design method 密苏里州柔性路面设计法(美国)
Missouri marble 密苏里州大理石
Missouri red granite 密苏里州红色花岗岩
missourite 白榴橄辉岩
misspecified explanatory variable 错定的解释变量
misspecified model 错定的模型
misspent sum 滥用款项
miss probability 漏报概率;失误概率
miss punching 错位冲孔
miss ratio 比例失调
miss shothole 瞎炮孔;拒爆炮孔;未爆眼
misstatement 错报
mist 细雾;烟雾;合剂;轻雾(霭);雾气;四级能见度;薄雾;雹
mistake 过失;人为错误;误差;错误;出错;差错
mistake error 过失误差
mistaken judgement 错误判断
mistake of facts 误解事物
mistake of law 法律错误
mistally 理货错误
mist atomizer 除雾装置;除雾器
mist bed 喷雾苗床
mist belt 雾带
mist blower 鼓风弥雾器;喷雾器
mist catcher 捕雾器
mist coat 喷涂层;喷涂面层;薄雾状(流平罩光)涂层
mist cooling 喷雾冷却;喷雾降温
mist Cottrell precipitator 科特雷尔烟雾排除器
mist duster 弥雾喷粉机
mist eliminator 烟雾消除器;除雾器;捕沫器
mister 弥雾机
mister head 喷雾器喷头
mistermination 终ँ 端失配
mist extractor 湿气分离器
mist fan motor 喷雾吹风电动机
mist filter 过滤式雾滴分离器
mist flow 雾状流;雾流
mist forest 云雾林
mist generator 弥雾发生器
mist grey 雾灰色
mis-tie at intersection 交叉点
mistiness 云翳
misting 飞雾
mist interval 生雾温差
mist light-signal 雾灯光信号
mist lubrication 喷雾润滑
mist of chromic acid 铬酸雾
mist projector 雾用探照灯
mistrack 失去跟踪
mistral 密史脱拉风;法国南部海洋的凛冽北风
mistranslation 错译
mistregistration 记录失真
mist retting 喷雾沤
mistrimmed 偏离配平位置的
mist separation equipment 雾滴分离装置
mist separator 集湿器;湿气分离器
mist spray 喷雾
mist spray damping machine 喷雾给湿机
mist sprayer 喷雾器;弥雾机
mist-spraying cooling fan 喷雾(冷却)风机
mist sprinkler 弥雾机;喷雾喷灌器
mist trap 捕雾器
mistune 失谐
mistuned circuit 失调电路
mistura magnesii hydroxidi 氢氧化镁合剂
mistura oleobalsamica 油香树脂合剂
misty 有轻雾的;湿雹,薄雾的
misty horizon 模糊地平线
mistying 不正确捆扎
Misubishi Heavy Industries 三菱重工业株式会社
misunderstanding 误解;误会
misuse failure 误操作失效;超常使用故障
mis-use land 使用不当的土地
misuse of symbol 象征误用
misylite 叶绿矾
mitchboard 桅搁架
Mitchell's screw anchor 密歇尔螺旋锚(杆)
Mitchell's thrust bearing 密歇尔止推轴承;密歇尔推力轴承
mite 螨
mitella 臂吊带
miter 斜拼缝;斜边;阳角接
miter angle 斜接角
miter arch 三角形拱;斜角拱;山墙三角形檐饰(法国);斜接拱
miter bar 斜接柱
miter bearing 斜接支承;人字形支承
miter bend 斜弯头
miter bend bonder 斜面弯管
miter bevel both sides 双面斜边
miter bevel gear 等径锥齿轮传动
miter block 斜锯架
miter board 刨斜角木模型板
miter box 锯木成斜接缝的工具;斜锯导引的工具;轴锯箱;45°角尺;辅锯箱
miter brad 斜角钉;斜接缝角钉
miter cap 僧帽装饰顶;楼梯扶手柱头饰;楼梯扶手起柱垫子
miter clamp 斜角黏结夹具;斜卡;斜(接)夹
miter column 斜接柱
miter cut 斜截头;斜切锯
miter cut ring 斜口环
miter cutting 斜截头;斜锯切;斜切割
miter dovetail 暗燕尾榫;斜楔榫;合角榫
miter dovetailing 暗燕尾榫接
miter drain 斜接排水沟;V形排水槽
miter drainage V形排水槽
mitered 成45°角斜接的
mitered-and-cut string 明楼梯梁(梯级竖板与搁板斜接的)
mitered bat 斜角半砖
mitered bend 斜接弯管
mitered closer 斜角堵头砖;斜角接砖
mitered decoration 斜角装帧
mitered elbow 斜接弯管
mitered hip 斜屋脊
mitered inlet 水道斜坡进口
mitered jib 斜接缝三角帆
mitered joint 斜角接头;斜角榫;斜角接缝
mitered knee 斜接弯头
mitered sheet 斜角铁板
mitered sill 人字闸门
mitered stringer 榫接踏板楼梯梁
mitered type gate 人字闸门
mitered valley 斜接瓦屋面沟;斜(接)天沟
miter elbow 斜接弯头
miter end 斜接端;门扇斜接端
miter framing 斜接框架
miter gate 人字形闸门;人字门
miter gate leaf 人字门门叶
miter gate strut 人字门推拉杆
miter ga(u)ge 斜接规;定角规;定角规;半斜接
miter gear 转角方向装置;斜齿轮;锥齿轮;等径(直角)锥齿轮;等径锥齿轮
miter half 半斜接;斜半砖;半斜度
mitering 斜接
mitering gate 人字闸门
mitering machine 斜角切割机;斜面研磨机
mitering post 斜接柱;接合柱
mitering sill 人字闸槛
mitering type 人字式
miter joint 45°斜削接头;斜接接头;斜(削)接缝;(活塞环)斜切口;斜面接头;斜面接合;斜面缝;斜接口;斜角连接;斜角接头
miter joint with spline 抽心斜接
miter knee (楼梯扶手)弯曲斜接;(楼梯扶手)斜弯头
miter line 斜接线;斜角缝线
miter lock gate 人字式船闸闸门
miter pier 斜面墩
miter plane 斜刨;斜刨
miter plane miter cutting plane 斜割刨
miter post 斜接柱;接合柱;人字闸门斜接柱;闸门对口端柱;船闸斜接柱
miter post of lock 船闸斜接柱
miter return 大陡度斜面
miter rod 45°斜角铁板(抹灰工用);斜角杆;斜角棒
miter saw 手锯;斜榫锯;夹背锯;横切锯
miter-saw cut 斜锯导引的工具
miter-sawing hoard 斜锯导引的工具

miter shoot 刨斜角木模型板
miter sill 人字(闸)槛;人字门门槛(船闸);斜梁;船闸门槛
miter square 斜角规;斜角尺
miter templet 斜向凿割带刷的模板
miter timber 斜口木
miter to miter 上下闸门距离
miter-type gate 船坞人字闸门
miter valve 锥阀;斜面阀;斜接阀;座面阀;锥形座面阀;锥形活门(阀)
miter wall 斜墙;人字墙
miter weld 斜接角焊缝;平顶角焊缝
miter welding 斜接头焊接
miter wheel gear 等径锥齿轮装置
miter wheel gearing 等径锥齿轮装置
miticide 杀螨剂
mitigate 减免;调节冷热
mitigate corrosion 防腐
mitigate damages 减轻损失
mitigated ether 缓和醚
mitigated silver nitrate 缓和硝酸银
mitigated smallpox 变轻天花
mitigating damage 减轻震害
mitigation 减轻,和缓;水的软化
mitigation measure 减轻灾害措施;减排措施;减免措施;缓解措施;补救措施
mitigation of damages 减轻损失;赔偿的减少
mitigation of flood 减洪;洪水调节;分洪
mitigation of seismic hazard 减轻地震灾害
mitis 可锻铁;可锻铸件
Mitis green 巴黎绿
mitis metal 可锻铸件;可锻铁铸件
mitochrome 线色素
mitogenetic radiation 分生辐射
mitogenetic ray 分生射线
mitoplast 丝状体
mitoschisis 丝状分裂
mitotic activity 核分裂能
mitotic apparatus 分裂装置
mitotic figure 分裂像
mitrate 钟状的;钟形的
mitre = miter
mitre angle 斜接角
mitre arch 三角形铰接拱;人字拱;三角拱;山墙三角形檐饰(法国)
mitre bearing 斜接支承
mitre bend 斜弯头
mitre bend bondstone 斜面弯管
mitre bevel (玻璃的)斜角
mitre bevel wheel 等径锥齿轮
mitre block 斜砌块;斜锯架;斜锯辅助块
mitre board 斜锯槽;斜锯板
mitre-box 轴锯箱;45°角尺(木工用)
mitre-box saw 截斜角木抖箱锯;截45°木抖箱锯
mitre brad 波纹扣片;波纹扣件;斜接波纹连接片
mitre cap 楼梯扶手柱头饰
mitre-cut piston ring 斜接式活塞环
mitred-and-cut string 隅角侧木;45°角斜锯小梁;上缘踏步形楼梯梁
mitred border 斜角边
mitred cap 斜接帽;斜接帽盖
mitred closer 切角镶边砖
mitred core 斜接式铁芯
mitred drainage 鲱骨形地下渗水系统;人字形地下渗水系统;鲱骨形地下排水系统;人字形地下排水系统
mitred gate 人字闸门;混合水门
mitred hip 斜脊梁;斜接戗角
mitred joint 斜接接缝
mitred knee 斜接弯头;斜接扶手弯折处;45°角扶手弯头
mitre dovetail 合角接榫;暗燕尾榫;暗马牙榫;斜楔榫;斜接燕尾榫;暗楔榫
mitre dovetailing 暗燕尾榫接
mitred post 斜切柱;斜接柱
mitred sill 人字窗槛;斜切门槛;人字门槛;斜切窗槛
mitred valley 斜接瓦屋面沟;斜接天沟
mitre elbow 斜接弯头
mitre end 门扇斜接端
mitre fillet weld 角焊
mitre folding 斜接褶皱
mitre gate 人字门;船坞人字闸门
mitre ga(u)ge 斜接规
mitre gear 等径锥齿轮;等径直角锥齿轮

mitre grinding machine 斜面研磨机
mitre half 斜转砖;半斜度
mitre joint 斜角连接;斜面接合;斜削接头;斜面接头;斜接缝;斜向接头;合角接
mitre jointing 斜面连接
mitre line 斜接线
mitre lock gate 人字式船闸闸门
mitre machine 切斜角机
mitre pier 斜面墩
mitre plane 斜割刨
mitre post of lock 船闸斜接柱
mitre rod 斜角棒
mitre saw 夹背锯;横切锯;斜切锯;手锯
mitre shoot 线脚支架;斜角槽;刨斜角槽导板
mitre sill 斜梁;人字门;人字门槛;船闸门槛
mitre square 斜接正方形;角曲尺;斜角尺
mitre stern rampway 尾斜跳板
mitre templet 模板;样规
mitre to mitre 上下闸门距离
mitre valve 斜面阀;斜接阀;锥形活门;锥形阀(门)
mitre wall 斜墙;人字墙
mitre welding 斜接焊接
mitridatite 斜磷钙铁矿
mitriform 钟状
mitring machine 斜面修整机;斜割机;切边机;成角切削机床
mitron 交叉指型电压调谐磁控管;米管
mitscherlichite 氯钾铜矿
Mitsubishi fluidized calcinator suspension preheater 三菱流态层分解炉悬浮预热器
mitten 连指手套
mitten money 冬季引航补贴金
mix 混合;拌和物(混凝土的);拌和料
mixability 混合程度
mix-and-settle extractor 混合沉降萃取器;混合沉淀槽
mix and stir 拌和
mix batch concrete 手拌混凝土
mix batching 混合料配料
mix-bed 混合床
mix-bed exchanger 混合床树脂交换器
mix calculation 配料比计算
mix cement 复合水泥
mix component 混合组分
mix composition 混合结构;混合组成
mix consistency 混合黏性;混合稠度;拌和稠度
mix control 拌和控制
mix crystal 混合晶体;共晶体
mix design 混合料成分设计;混合料配合比设计;配料设计;配合比设计
mix design of concrete 混凝土配合比设计
mixed 拌和的;混合的
mixed account 混合账户
mixed acid-base constant 混示酸碱平衡常数;混合酸碱平衡常数
mixed adhesive 混合黏结剂
mixed air controller 混合空气控制器
mixed algae 混合藻
mixed alkali effect 混合碱效应
mixed amine 混合胺
mixed amplifier 混合放大器
mixed aniline point 混合苯胺点
mixed arbitration 混合仲裁
mixed arch 混合拱
mixed avalanche 混合崩坍
mixed bacteria 混合菌
mixed bacterial bacteria 混合菌性菌
mixed barites grains 混合一向粒度的重晶石
mixed base 混合料基层;混合底子
mixed base asphalt 混合基沥青
mixed base course 拌和式基层
mixed base crude oil 混合基原油
mixed-base crude petroleum 混合基原油
mixed base notation 混合基数记数法
mixed-base petroleum 混合基石油
mixed basin 混合池
mixed batch 配合料
mixed batch capacity 混合配料重度;成批拌和能力
mixed bed 混合床;混合
mixed bed demineralization 混合床除盐
mixed bed demineralizer 混合床除盐器;混合床去除矿质器
mixed bedding 复合层理

mixed bed exchanger 混合床树脂交换器;混床离子交换器
mixed bed filter 混合床过滤器
mixed bed filtration 混合床过滤
mixed bed ion exchange 混合床离子交换
mixed bed system 混(合)床系统
mixed benzene 混合苯
mixed biochemic(al)treatment 混合生化处理
mixed bituminous 拌和沥青路
mixed bituminous macadam 沥青碎石混合料(路);拌和式沥青碎石(路)
mixed bituminous macadam pavement 拌和式沥青混凝土碎石路(面)
mixed bituminous road 沥青混合料路;拌和式沥青路
mixed boat 客货轮;客货船
mixed border 混合花境
mixed boundary 混合边界
mixed boundary condition 混合边界条件
mixed boundary value 混合边值
mixed boundary value problem 混合边(界)值问题
mixed brewery wastewater 啤酒厂混合废水
mixed brine 混合型卤水
mixed broad-lived deciduous and evergreen forest 常绿落叶阔叶混交林
mixed cargo 杂货;混载货物;混合(船)货
mixed cargo ship 件杂货船
mixed cement 混合水泥
mixed chambers of commerce 贸易协会
mixed chemical industry wastewater 混合化工污水
mixed chromatogram 混合色谱
mixed circulation 混合循环
mixed closure method 混合法截流
mixed cloth 混纺布
mixed cloud 混合云;冰水混合云
mixed coal 混煤
mixed coal sample 混合煤样
mixed cock 混合水龙头
mixed cold 冷拌和
mixed colo(u)r 混合色
mixed colo(u)r method 混色法
mixed colo(u)r plate 混色板
mixed compaction facies 混合压实相
mixed compaction zone 混合压实带
mixed concrete 搅拌好的混凝土;拌(制)好的混凝土;拌和混凝土
mixed conductor 混合导体
mixed constant 混合常数
mixed construction 混合结构;混合建筑;混合构造
mixed coordinate map 混合坐标图
mixed coordinates 混合坐标
mixed copper ore 混合铜矿石
mixed corrosion 混合溶蚀
mixed corrosion exotic material 外来物的混合溶蚀
mixed corrosion of saturated solution 饱和液的混合溶蚀
mixed crystal 固溶体;晶体混合物;混晶;混合晶体
mixed culture 混合培养
mixed current 混合潮流
mixed cut 混合掏槽
mixed cycle 混合循环
mixed cycle engine 混合循环发动机
mixed data 混合数据
mixed deciduous forest 落叶混交林
mixed decimal 带小数的数
mixed department 综合部门
mixed design 混合设计
mixed development area 混合建筑区;混合发展区
mixed dike 复成岩墙;复成岩脉
mixed dislocation 混合位错
mixed diurnal tide 混合周日潮;混合日期潮
mixed drive 复合驱动
mixed dust 混合性粉尘
mixed duties 混合税
mixed dye 拼混染料
mixed economy 混合经济
mixed electrode 混合电极
mixed electrolyte solution 混合电解质溶液
mixed electroplating wastewater 电镀混合废水
mixed element 混合元
mixed-en-route 在运输途中拌和的
mixed epihmnion 混合变温水层
mixed explosion 瓦斯和煤尘混合爆炸

mixed explosive 混合炸药
mixed exponent 带分数指数
mixed extension of a game 对策的混合扩充
mixed fabric 混纺织物
mixed face 隧道中同时平行掘进的面；混合界面；混合表面
mixed face tunnel 混合面隧道
mixed farming 综合经营；混合农业经营
mixed fatty acid 混合脂肪酸
mixed faucet 冷热水（混合）龙头
mixed feed 混合喂料；配合饲料
mixed fertilizer 混合肥料
mixed financing 混合资金融通
mixed finite element formulation 混合有限元公式
mixed firing 混合燃烧
mixed flask culture method 混合瓶式培养法
mixed flat 混合坪
mixed flat deposit 混合坪沉积
mixed flow 乱流；跨音速流动；（混）合流
mixed flow blower 离心轴流式压气机；混流式鼓风机
mixed flow centrifugal compressor 混流式离心压气机；混流式离心压缩机
mixed flow centrifugal pumps 混流离心泵
mixed flow combined aeration-setting tank 混合流合建式曝气沉淀池
mixed flow compressor 混流式压缩机
mixed flow fan 混流式扇风机
mixed flow Francis turbine 混流式法兰西斯水轮机
mixed flow impeller 混流式叶轮
mixed flow propeller pump 混流式（螺）旋桨泵
mixed flow pump 螺旋离心泵；混流式（水）泵；混流泵
mixed flow pump turbine 混流式水泵水轮机
mixed flow pump with open impeller 开式叶轮混流泵
mixed flow reactor 混流反应器
mixed flow sewage pump 混流式污水泵
mixed flow turbine 混流式涡轮机；混流式透平；混流式水轮机；混流式汽轮机；混流式水轮机
mixed flow variable pitch turbine 斜流式水轮机
mixed flow volute pump 蜗壳混流泵
mixed flow water turbine 混流水轮机
mixed flow wheel 混流式水轮（机）
mixed forest 混交林；混合林（带）
mixed format 混合格式
mixed forme base 混合版基
mixed fraction 带分数
mixed fuel burning 混合燃烧
mixed fuel burning ratio 燃料混烧率
mixed function 混合功能
mixed garden-wall bond 园墙交替砌合；三顺一丁砌墙法
mixed gas 掺混气；混合气（体）；混合气；混合煤气
mixed gas producer 混合煤气发生炉
mixed ga(u)ge turnout 套线道岔
mixed glue 混合胶；混成胶
mixed goods 杂货；混合织物
mixed grain 混合晶粒；混合粒度；混合纹理；粗细不均的颗粒
mixed-grained 多种粒径混合的；多粒的；多种粒径的；多种粒度的
mixed-grained lumber 混杂木纹的木材
mixed-grained of various of grain-size 颗粒大小混杂的
mixed-grained sand 多种粒径的混合砂
mixed granules 混合砂粒
mixed graph 混合图
mixed gravel 混合石子；混合砾石
mixed ground 混合围岩
mixed hair and fibre polishing wheel 发纤维混合抛光轮
mixed high frequency signal 高频混合信号
mixed high signal 高频混合信号
mixed high system 高频混合制
mixed holding company 混合股份公司
mixed holding fastenings 半分开式扣件
mixed hot 热拌和
mixed housing development 住房联合开发；多类型住房
mixed hydrocarbon 混合烃
mixed hydrothermal alternative 水火电混合替代方案
mixed indication 混合指示剂

mixed-in-place 路拌的；工地拌和（的）；就地拌和的；现场搅拌；现场拌和；工地拌和；就地搅拌
mixed-in-place bituminous surfacing 路拌沥青路面；就地拌沥青路面
mixed-in-place cement 工地拌和的水泥
mixed-in-place construction 工地拌和施工；路拌法；就地拌和施工
mixed-in-place construction of road 就地道路施工焦油；就地搅拌筑路
mixed-in-place course 就地拌和路面层
mixed-in-place method 现场拌和法；现拌法；路拌法；现场搅拌法；工地拌和法；就地拌；就地拌和法
mixed-in-place pile 就地搅拌桩；就地搅拌混凝土桩；拌和桩
mixed-in-place surface 路拌混合料面层；路拌混合料路面
mixed-in-place surface treatment 就地拌和表面处治
mixed in route 在运输途中拌和的
mixed integer linear programming 混合整数线性规划
mixed integer programming 混合整数规划
mixed-in-transit 路拌（混凝土）；运送搅拌（混凝土）；运输搅拌；在运输过程中拌和的；混凝土运送搅拌；途中搅拌；途中拌和；自动机搅拌
mixed ion exchanger 混合离子交换器
mixed joint 混合接头
mixed karst 混合型岩溶
mixed land use 混合土地利用
mixed layer 混合层
mixed layer depth 混合层厚度
mixed layer height 混合层高度
mixed layer mineral 混层矿物；互交层矿物
mixed layer sound channel 混合层声道
mixed layer structure 混合层结构
mixed lead ore 混合铅矿石
mixed ligand complexes 混合配位化合物
mixed light 混合光；混合灯光
mixed light illuminator 混光灯具
mixed lighting 混合照明；混合采光
mixed light lamp 杂光灯；混合灯；各色各样光的灯
mixed limestone 混合石灰石
mixed linear programming 混合线性规划
mixed liquid suspended solids 混合液悬浮固体
mixed liquid volatile suspended solids 混合液挥发悬浮固体
mixed liquor 混合液
mixed liquor sample 混合液样本；混合液水样
mixed liquor suspended solids 混合液中的悬浮固体（有机物和活性污泥）
mixed liquor suspended solid in activated sludge 活性污泥中悬浮固体含量
mixed liquor total suspended solids 混合液总悬浮固体
mixed liquor volatile suspended sludge 混合液挥发性悬浮污泥
mixed liquor volatile suspended solids 混合液挥发悬浮固体
mixed load 混合负载；混合荷载；混合负荷
mixed logic 正负逻辑
mixed macadam 混合碎石；碎石混合料；拌和式碎石路；拌和式沥青路
mixed macadam road 混合碎石路
mixed magma 岩浆混合作用
mixed mark 标记混杂
mixed mass-service system 混合服务系统
mixed material 混合料
mixed material storage bin 混合料储料斗；混合料仓
mixed material storage hopper 混合料储料斗
mixed matte 混合冰铜
mixed-media filter 混合介质过滤器；混合滤料滤池
mixed media filtration 混合介质过滤（污水处理）
mixed melting point 混合熔点
mixed metal oxide pigment 共晶复合型颜料；晶格颜料；金属氧化物混相颜料
mixed metal salt/t-amine accelerator 金属盐/叔胺混合促进剂
mixed method 混合法
mixed microbe 混合菌
mixed middle coal 混中块煤
mixed model 混合模型
mixed models of modern water 年轻水混入模型

mixed-mode stress 混合型应力
mixed Mo ore 混合钼矿石
mixed needle-broad-leaf forest 针阔叶混交林
mixedness 混合度
mixed nickel ore 混合镍矿石
mixed nucleus 混合核【气】
mixed number 带分数
mixed or combination container roll-on/roll-off ship 混合式或组合式水平装卸集装箱船
mixed ore 混合矿石
mixed origin 混合成因的
mixed packing column 混合填充柱
mixed paint 混合涂料；调和漆
mixed passenger and freight station 客货运站【铁】
mixed pastoral farming area 半农半牧区
mixed Pb-Zn ore 混合铅锌矿石
mixed peat 混合泥炭
mixed-phase flow 混流
mixed pigment 调和油漆；调和涂料
mixed plankton 混合浮游生物
mixed plant 混合式电站；拌和站
mixed planting 混合种植
mixed plaster 混合灰浆；混合灰泥
mixed plasticizer 混合增塑剂
mixed policy 混合保险
mixed pollution sources 混合污染源
mixed population 混合总体；混合种群；混合群体
mixed population of microorganism 混合微生物群体
mixed population screening 混合总体筛分
mixed Portland cement 混合的普通水泥；混合硅酸盐水泥；混合波特兰水泥；复合硅酸盐水泥
mixed-powder 混合粉末；混合炸药
mixed pressure steam turbine 混压式汽轮机
mixed-pressure turbine 混压涡轮机；双压透平
mixed product 混合乘积
mixed proportion 混合比例
mixed pumped-storage plant 混合式抽水蓄能电站
mixed pumping 混合取水
mixed pumping test 混合抽水试验
mixed pump with shrouded impeller 闭式叶轮混流泵
mixed queue system 混合制排队系统
mixed quota surplus treaty 比率溢额混合契约
mixed rags 成型；组合；装配
mixed rain and snow 雨夹雪；混雪雨
mixed reflection 混合反射；发散反射
mixed resin bed 混合树脂床
mixed rinse water 混合漂洗水
mixed rock 混合岩
mixed rotor winding 混合转子绕组
mixed salt 复盐
mixed sample 混合样品；混合样（本）
mixed sampling 混合采样
mixed sand 混合砂
mixed sand-and-shingle spit 砂砾混合喷嘴
mixed Sb ore 混合锑矿石
mixed sea 大小波浪混杂的海
mixed sea and land risk 水陆联合保险
mixed sedimentary rock 混合水成岩；混合沉积岩
mixed-sheet structure 交替层结构
mixed sludge 混合污泥
mixed small coal 混末煤
mixed soil 杂填土；混杂土；混合土（壤）
mixed solar diurnal tide 混合太阳日潮
mixed solid phase 混合固相
mixed solvent 混合溶剂
mixed specific gravity method 混合比重法；混合相对密度法
mixed stain 混合斑
mixed stand 混合林
mixed stand oil 混合亚麻仁油；混合厚油；混合熟油
mixed-state cloud 冰水混合云；混合态云
mixed station dialing 混合拨号
mixed stone conglomerate 混合岩砾岩
mixed-strand bar 交叉股线棒
mixed structure 混合结构
mixed stuff 混合织品；混合填料
mixed style 混合式
mixed style road system 混合式道路系统
mixed substation 混合变电所
mixed subtropic(al) rain forest 亚热带混交雨林
mixed superheater 混流式过热器

mixed superlattice 混合型超晶格
mixed support 混合支架
mixed surface solid approach 混合面固体方法
mixed system 混合系统
mixed system of training 混合导流系统
mixed tar 混合柏油;混合沥青;混合焦油
mixed team 混合施工队
mixed term 混合项
mixed tidal current 混合型潮流
mixed tidal harbo(u)r 混合潮港
mixed tidal harbo(u)r of irregular diurnal tide 不正规日潮混合港
mixed tidal harbo(u)r of irregular semi-diurnal tide 不正规半日潮混合港
mixed tide 混合潮
mixed timber 混堆木材
mixed toxicity 混合毒性
mixed toxicity index 混合毒性指数
mixed toxicity test 混合毒性试验
mixed traction and step-down substation 牵引降压混合变电所
mixed traffic 客货混运【铁】;客货混流【铁】;混运;混向运行;混合交通
mixed traffic line 混运线路
mixed traffic volume 混合交通量
mixed train 客货混合列车;混合列车
mixed training system 混合式导治设施;混合导流系统
mixed transaction 混合交易
mixed tube 混合波管;混波管
mixed twill 复合斜纹
mixed-type container ship 混载型集装箱船
mixed-type face 混合型工作面
mixed-type freight yard 混合式货场
mixed-type goods yard 混合式货场
mixed-type groundwater 混合类型水
mixed-type highway 混合行驶公路
mixed-type of tide 混合潮型
mixed-type platform 混合式站台
mixed types of construction 混合构造;混合型建筑物
mixed-type source rock 混合型源岩
mixed unit train 客货混合列车
mixed-use area 综合区
mixed-use building 多用途房屋
mixed-use center 多功能中心
mixed-use development 混合使用发展用地;多用途建设
mixed-use property 多用途建筑用地;多用途房屋土地
mixed-use zone 多用途区
mixed-use zoning 混合使用分区;多用途区划
mixed value of zero shift 零漂的混合值
mixed-vendor environment 不同厂家的设备配置
mixed ventilation 径向—轴向通风
mixed wastewater 混合废水
mixed water spring 混合水泉
mixed wave 混合浪
mixed xylene 混合二甲苯
mixed zinc ore 混合锌矿石
mixed zone 混合地带
mixed zone of smectite-illite 蒙脱石—伊利石混合带
Mixee 米克斯粉末混合度测量仪;粉末混合度测定器
mix en route 路拌(混凝土);混凝土运送搅拌;途中搅拌;途中拌和
mixer 搅拌机;混频器;混合机;混炼机;混合装置;混合龙头;混合机;混合阀;搅拌机
mixer action 混频器作用;混合物作用
mixer-aerator 混合曝气机
mixer-agitator tank 搅拌池;拌和机拌缸
mixer amplification by variable reactance 可变电抗混频放大;脉代;低噪声微波放大
mixer amplifier 混频放大器
mixer arm 搅拌机桨叶;混合器桨臂
mixer blade 混合刀片;搅拌机叶片
mixer capacity 搅拌机容量
mixer car 搅拌车(混凝土)
mixer charging 搅拌机喂料
mixer circuit 混频电路
mixer control 搅拌机控制
mixer controller 混频控制器
mixer crystal 混频晶体
mixer cube 倾翻式方筒搅拌机

mixer designation 搅拌机命名法
mixer diode 混频二极管
mixer driver 搅拌机司机;搅拌机传动装置
mixer-drive shovel 手推刮铲
mixer drum 搅拌机
mixer duplexer 混频双工器
mixer efficiency 搅拌机效率
mixer engine 搅拌机的发动机
mixer filter 混频器滤波器
mixer-first detector 混频管
mixer for road construction 筑路用搅拌机
mixer frame 搅拌机机架
mixer-gradation unit 级配控制的搅拌设备
mixer-granulator 混合制料(碎石)机
mixer hopper 搅拌机加料斗;搅拌桶;拌和箱
mixer liner segment 搅拌衬筒板
mixer-lorry 移动式搅拌机;移动式混凝土拌和车;移动式拌和机;混凝土搅拌车;混凝土搅拌(汽)车;混凝土拌和(汽)车;汽车搅拌机;搅拌车(混凝土)
mixer-mill 混磨机
mixer noise 混频器噪声
mixer of continuous method 连续拌和机
mixer operator 搅拌机操作工
mixer outlet temperature 出机口温度(混凝土)
mixer paddle 搅拌桨叶
mixer pan 盘式拌和机
mixer performance test 搅拌性能试验
mixer plant 拌和楼(混凝土);拌和工厂;拌和厂(混凝土)
mixer protection agent 搅拌机保护剂
mixer scale 搅拌机(配置的)秤
mixer-settler 混合沉降器;混合澄清槽
mixer-settler extractor 混合沉降萃取器
mixer-settler type plant 混合沉降器萃取设备
mixer shaft 搅拌机轴;搅拌轴
mixer size 搅拌机容积
mixer skip 搅拌机装料斗;混凝土拌和机装料斗
mixer stage 混频级
mixer time 搅拌时间
mixer trestle 搅拌机机架
mixer truck 搅拌卡车(车上设搅拌设备);混凝土搅拌运料车
mixer tube 混频管;混波管
mixer-type hot metal car 搅拌式钢水车;搅拌式铁水车
mixer-type truck 搅拌车
mixer valve 混频管
mixer vehicle 搅拌车(混凝土)
mixer with conic(al) screw 锥形螺旋搅拌机
mixer with propeller 螺旋桨式搅拌机
mix-flow compressor 混流式压缩机
mixflow type gearing-up mono-blower 齿轮增速式单级混流鼓风机
mix formula(tion) 混合物配方
mixing 混合;拌和;搅混;搅拌;混料
mixing action 搅拌作用;混合作用
mixing agitator 混合用搅拌器
mixing aid 操作助剂;混炼助剂
mixing air distributor 搅拌空气分配器
mixing amplifier 混频放大器
mixing and reaction chamber 混合反应室
mixing angle 搅拌角
mixing arm 搅拌叶片;搅拌轮叶;搅拌机轮叶;搅拌机桨叶;搅拌杆;搅拌叶;搅拌臂
mixing at site 现场搅拌;现场拌和;工地拌和;就地拌和
mixing basin 搅拌池;混合器;混合池;拌和池
mixing bed 拌和堆场
mixing bed method 混合料堆法
mixing blade 混合刀片;搅拌叶(片)
mixing board 拌灰板
mixing box section 混合段
mixing bunker 搅拌机槽;拌和槽
mixing cement 混合水泥
mixing chamber 搅拌室;搅拌间;混合室;拌和室
mixing chamber silo 混合室库
mixing channel 混合沟渠;混合槽
mixing charging 搅拌时的加料
mixing circuit 混合电路
mixing coat 拌合式面层;混合面层;复合面层
mixing coefficient 混合系数
mixing concentration 拌和浓度
mixing cooling 混合冷却

mixing cycle 搅拌周数;搅拌周期;搅拌循环;混炼周期;混合周期;混合顺序;拌和周期
mixing depth 混合深度
mixing device 搅拌装置;混合装置;混合设备
mixing device of gas 煤气发生炉混合装置
mixing distance of river pollutant 河流污染物混合距离;河流污染物混合长度
mixing distribution 混合分配;混合分布
mixing distributor 混合布水器;混合分配器
mixing drag 搅拌刮铲
mixing dressed soil for dilution 客土混层稀释
mixing drum 混凝土搅拌筒;混凝土拌和机;搅拌筒;搅拌鼓;混合筒;拌和筒;拌和鼓
mixing effect 混炼效果;混合作用;拌和作用
mixing efficiency 搅拌效率
mixing element 搅拌工具;拌和工具
mixing equipment 搅拌设备;混合设备;拌和设备
mixing error 配料误差
mixing faucet 混合水龙头;混合龙头
mixing filter 混频滤波器
mixing filtering material 混合滤料
mixing flow 混合流动
mixing-flow pump 混流泵
mixing fog 混合雾
mixing head 混频头
mixing heat exchanger 混合式换热器
mixing height 混合高度
mixing hexode 六极混频管
mixing homogeneity 混合均匀性
mixing impeller 搅拌机叶轮
mixing index 混合指数
mixing in estuary 河口混合
mixing-in-place 就地拌和;工地拌和;工地搅拌
mixing-in-place method 现场拌和法;现场拌法;工地拌和法;就地拌和法
mixing-in-situ 就地拌和
mixing installation 搅拌装置
mixing intensity 混合强度
mixing-in-transit 在运输过程中拌和(混凝土)
mixing joint 混合厚度
mixing lacquers 配色调和喷漆
mixing ladle 搅拌铲;混合勺;混铁罐
mixing layer 混合层
mixing leach 搅拌沥青
mixing length 混合长度
mixing length theory 混合长度理论
mixing machine 搅拌机;混合机;调和器;拌和机
mixing machinery 搅拌机械;拌和机械
mixing method 拌和法;混合法
mixing method mixed-in-place piles 深层搅拌法
mixing mill 拌和机;两辊磨;开放式混炼机;混砂机;混合碾磨机
mixing motion 拌和动作
mixing nozzle 混合喷嘴
mixing of cement 水泥和水搅拌
mixing of coal sample 煤样的掺和
mixing of concrete 混凝土拌和;搅拌混凝土
mixing of groundwater 地下水混合作用
mixing of lake water 湖水混合
mixing of primary pigment colo(u)rs 混色;色混合;原色颜料的混合方法;原色颜料的混合过程
mixing operation 操作作业;搅拌作业
mixing pad 混频器
mixing paddle 搅拌机桨;搅拌桨;搅拌机叶片;搅拌机叶轮;拌和机叶片
mixing pan 盘式拌料机
mixing pan grinder 混合碾磨机
mixing pan mill 混合盘磨
mixing parameter 拌和参数
mixing phase 混波相位
mixing phenomenon 混合现象
mixing-placing train 工地搅拌列车(用于浇筑混凝土隧道)
mixing plant 搅拌设备;搅拌工场;搅拌工厂;搅拌厂;混凝土拌和站;混合装置;拌和料拌和厂;拌和站;拌和设备;拌和楼(混凝土);拌和厂(混凝土);拌和工厂
mixing-plate 调和板
mixing platform 搅拌台;拌和(平)台
mixing plough 混和犁
mixing pool model 混合池模型
mixing Portland cement 混合硅酸盐水泥
mixing position 混合位置
mixing process 搅拌操作;搅拌过程;混合操作;混合过程

mixing proportion 混合比例
mixing proportion by volume 按体积配料
mixing proportion by weight 按重量配料
mixing proportioning 混合比;配合比
mixing pump 混合泵
mixing rate 混合速率
mixing ratio 混合比;配合比
mixing ratio by weight 重量配合比
mixing ratio indicator 混合比指示器
mixing ratio sheet 配料单
mix ingredient 拌和料组分;拌和成分
mixing rod 搅拌棒
mixing roll 研磨机;两辊混炼机
mixing room 配料房;拌和间
mixing rule 混合律
mixing scale 混合比例
mixing screw 螺旋搅拌器;螺旋搅拌机;螺旋混合器;螺旋混合机;混合螺杆;拌和螺旋输送器
mixing screw conveyer 混合螺旋输送器
mixing sequence 混合顺序;混合工序;混合程序
mixing shaft 搅拌轴
mixing silo 混合库
mixing site 搅拌场所;搅拌地点;混合场所;混合地点
mixing space 混合室
mixing speed 搅拌速度;拌和速度
mixing sphere 混合球
mixing stage 混频级
mixing star (强制式搅拌机的)搅拌星形架
mixing system 混合系统
mixing system of heat(ing) 混合(式)加热系统
mixing table 搅拌台;拌和台
mixing tank 冷热水调和箱;混合池冷热调和箱;混合池;混合槽;拌缸
mixing tap in lavatory 洗脸盆混合龙头
mixing temperature 搅拌温度;混合温度
mixing time 搅拌时间;混топ时间;混合时间;拌和时间
mixing timer 搅拌定时器
mixing tower 拌和塔
mixing traces within common reflection area element 共反射面元的混波道数
mixing train 混合列车(客货同列)
mixing truck 搅拌运料车;(混凝土)搅拌车
mixing unit 搅拌设备;搅拌机;混频管;混频部件;成套拌和设备;拌和装置;拌和站;(混凝土)拌和厂
mixing valve 混合阀
mixing varnish 调和漆;配漆用漆液;调和清漆
mixing vat 混料桶
mixing velocity 混合速度
mixing vessel 搅拌器;调浆桶;混合容器
mixing waste 混合废料
mixing wastewater 混合污水
mixing water 冷热调和水;搅拌(用)水;(混凝土的)拌和(用)水
mixing water heating 混合水加热;搅拌用水加热
mixing-water requirement 拌和需水量
mixing water test 拌和用水化验
mixing wave 混波
mixing(wave) coefficient 混波系数
mixing well 混合(水)井
mixing worm 搅拌机螺杆
mixing worn 混合螺旋输送机
mixing zone 混合区
mixing zone model 混合带模型
mixing zone of river 河流混合区;河流混合带
mixing zone of stream 河流混合区;河流混合带
mix-in-place 就地拌和;路拌
mix-in-place machine 土壤拌和机;现拌机;现场搅拌机
mix-in-place mixing plant 现场搅拌厂;工地搅拌厂
mix-in-transit 在运输途中拌和的
mix-in-travel plant 移动式搅拌厂
mixite 砷铋铜矿
mix line 混频线
mix muller 混砂机
mixoclarite 混合微粒煤
mixoclarodurite 微混合亮暗煤
mixodurite 微暗煤
mixoduroclarite 微混合暗亮煤
MKW mill 偏八辊轧机
mix of concrete 混凝土拌和料
mix of styles 混合风格;风格的混合
mixohaline water 半咸水
mixolimnion 混成层;环行层

mix(o)meter 搅拌计时器;拌和计时器
mixotroph 混养生物;混合营养生物
mixotrophic 半腐生
mixotrophic aerobe 兼养微生物
mixotrophic lake 混合营养湖
mixotrophic series 混溶序
mix plate injector 带混合板的喷雾器
mix preparation (混凝土)拌和物配制;(混凝土)拌和料配制
mix proportion 混凝土配合比;混合比;配合比;拌和料配合比
mix proportion by absolute volume 绝对体积配合比;绝对体积计算法
mix proportion by loose volume 现场松散体积配合比;松散体积配合比
mix proportion by weight 混合重量配合比
mix proportion(ing) calculation 配料计算
mix ratio 混合比;配合比
mix seal 混合料封层;拌和料封层
mix selection mechanism 混合料挑选机理;混合料挑选机构
mix selector 选料器;搅拌方式选择器;配料者;配料器
Mixtec architecture 米斯特克建筑
mixtinite 混合微粒体
mixtion 调料;媒介剂
mixtite 杂岩;混杂岩
mixtum 混杂沉积物
mixture 混合料;配料;混合物;混合剂;混合;合剂;拌和料
mixture adjustment 混合物调制;混合物用量调节;混合料配合比调整;混合物质量调节
mixture calculation 配料计算
mixture component 混合物组元;混合物组成;混合物成分
mixture composition 混合成分;混合物组分;混合物组成
mixture conditions of hydrogeology 水文地质条件复杂
mixture control 混合料配合比调整
mixture control assembly 混合比调节装置
mixture hopper elevator 混合料斗式升送机
mixture length 混合流程
mixture making 配料计算;配料
mixture mass adjustment 混合物质量调节
mixture method 混合法
mixture-method lubrication 混合法润滑
mixture of boric acid and bleaching powder 硼酸漂白粉
mixture of colo(u)rs 色混合
mixture of distribution 分布的混合
mixture of Indian alfalfa and red clover 印度苜蓿和红三叶草的混合料
mixture of ores and fluxes 矿砂熔剂混合物
mixture of sand silt and clay 砂粉砂
mixture of trace metals 微量金属合剂
mixture optimum 最适宜的混合
mixture output at maximum height and distance 最大排高排距时的泥浆排量
mixture placing 摊铺混合料;拌和料浇筑
mixture ratio 配合比;配比
mixture-ratio control 混合比调节
mixture risk assessment method 混合风险评价法
mixture sedimentation device 泥浆沉淀装置
mixture sowing 混播
mixture specifications 混合料的配料规范;配料规范
mixture strength 混合物强度
mixture toxicity 混合毒性
mixture toxicity index 混合毒性指数
mixture turbine 混合涡轮(机)
mixture type of reservoir 混合式水库
mixture velocity indicator 泥浆流速指示仪
mix water 拌和用水
mizen rigging 后桅支索
mizer 锥钻头
mizzenmast 后桅
mizzle 混合轻雾(毛毛雨与雾同时出现)
mizzonite 针柱石;钠钙柱石
MKW mill 偏八辊轧机
ML-test 维持荷载法(桩工试验)
mmat sinkage (放置蹭鞋垫的)下陷口
Mn carbonate ore 碳酸盐锰矿石
mnemonic 记忆存储器;帮助记忆

mnemonic address 助记地址
mnemonic address code 助记地址码
mnemonical code 记助码
mnemonic code 助记码;记忆码
mnemonic mark 辅助号
mnemonic operation code 助记操作码
mnemonic symbol 记忆码
Mn-Fe ratio 锰铁比
Mn oxide ore 氧化物锰矿石
Mn-rich (Co/Ni-poor) Mn nodule 富锰(贫钴镍)型锰结核
Mn sulfide ore 硫化物锰矿石
Mn(V) peroxo complex 锰(V)过氧络合物
mo 冰积粉土或岩粉;粒径0.02~0.1毫米的细砂
moar 泊
moar elevation of boiling point 摩尔沸点上升
moat 火口壕;护城河;海壕;堑;城壕;冰川壕;冰川沟
moating 围以水沟;流沙井壁黏土背带
moattock 鹤嘴锄
Mo-bearing copper ore 钼铜矿石
moben paper 毛边纸
mobilcrete pump 汽车混凝土泵
mobile 流动(的);活动的;发动机
mobile accelerator 移动式加速器
mobile acetylene generator 可移式乙炔发生器
mobile air 流动空气
mobile air compressor 移动式空(气)压(缩)机;拖车式空(气)压(缩)机
mobile analyzer 移动式分析器
mobile antenna 移动式天线
mobile anticyclone 移动性高(气)压;移动性反气旋
mobile autodoffer 小车式自动落纱机
mobile barge 机动驳(船)
mobile barrage 活动堰
mobile batch drier 移动式间歇干燥机
mobile bed 游荡性河床;移动床(层);脉动床层;跳汰床层;动床;不稳定河床
mobile bed engineering model 动床河工模型
mobile bed gas adsorber 流动气体吸附器
mobile bed model 动床模型
mobile bed model test(ing) 动床模型试验
mobile bed river 动床河流
mobile bed stream 游荡性河流;动床河流
mobile belt 活动带
mobile belt conveyer 移动式皮带输送机;移动式带式输送机
mobile belt trough elevator 移动式斗带升运机;移动式斗带升运器
mobile bitumen(ite) road-making plant 移动式沥青筑路设备
mobile bituminous mixing plant 移动式沥青混合料设备
mobile block 动滑轮;动滑车
mobile boring island 移动式镗孔岛;移动式钻孔平台
mobile boring platform 移动式钻台
mobile breaker 移动式破碎机(用于旧路面破碎翻新);可动式粉碎机
mobile building 活动建筑
mobile cable-operated excavator 汽车式钢丝操纵的挖掘机
mobile cage 吊罐
mobile camera 便携式电视摄像机
mobile channel 游荡性河槽;动床;不稳定河床;不稳定河槽;不稳定航道
mobile-charging machine 无轨装料机
mobile checkout 流动检查
mobile checkout and maintenance 流动检查与维修
mobile coal belt 移动式输煤带
mobile communication 移动通信
mobile communication network 移动通信网
mobile compactor 移动式压实工具;移动式夯具
mobile component 活动组分
mobile compressor 移动式压风机
mobile concrete placer 移动式混凝土浇注机
mobile concrete pump 混凝土泵车
mobile concrete pump with boom 带布料杆的混凝土泵车
mobile container crane 高架集装箱轮胎起重机
mobile conveyor[conveyer] 自行式运输机
mobile conveying belt 移动式传送带
mobile cover filter 移动罩滤池
mobile crane 汽车式起重机;移动式起重机;移动

mobile crane with articulated 铰接臂流动式起重机
mobile crane with box section 箱形臂流动式起重机
mobile crane with lattice jib 桁架臂流动式起重机
mobile crew 流动作业队;流动工作班
mobile crusher 移动式粉碎机;移动式轧碎机;移动式破碎机
mobile crushing plant 移动式破碎设备;移动式破碎机组;移动式破碎厂
mobile crushing plant with hydraulic walking mechanism 带液压行走机构的移动式破碎机设备
mobile crustal zone 地壳活动带
mobile data qcquisition system 可携式数据采集系统
mobile design 活动装置设计;汽车设计
mobile device 移动装置
mobile diaphragm liquid manure pump 移动式薄膜厩液泵
mobile digital computer 移动式数字计算机;可移动式数字计算机
mobile diving unit 可移动式潜水装置
mobile drill 自行式钻机;机动钻
mobile drilling island 移动式钻孔岛
mobile drilling machine 移动式钻机
mobile drilling platform 移动式钻(孔)台;移动钻孔平台;可移动式钻井平台
mobile drill(ing) rig 可移动式钻机;移动式钻机
mobile earth station 移动式地面站;移动地球站
mobile electric(al) crane 移动式电动起重机
mobile electric(al) generator unit 发电机车
mobile electrode 移动式电极
mobile element 活动元素;活动元件
mobile element absorption coefficient 活动元素吸收系数
mobile elevating conveyer 移动式上向运输机
mobile emergency winding equipment 移动式紧急提升设备
mobile emissions test laboratory 排气有害物流动试验室
mobile equilibrium 流动平衡;动态平衡
mobile equipment 移动式设备;自行设备
mobile excavator 移动式挖掘机;移动式挖土机
mobile extrusion plant 活动挤压厂
mobile field office 活动工地办公室
mobile filling 自流充填(法);流动充填;从上自重流下充填法
mobile flat belt elevator 移动式平带升运机
mobile foam generator 移动泡沫混合发生器
mobile forehearth 可称式前炉
mobile form 滑模
mobile frame 移动式门(窗)框;可移式框架
mobile froth apparatus 大型泡沫灭火机
mobile fund 可动基金
mobile gamma irradiator 移动式伽马辐照器
mobile gang 流动工作班;流动工班
mobile gas turbine 移动式燃气轮机
mobile gate 活动(栅)门
mobile generating equipment 移动式发电设备
mobile generating set 移动式发电设备;车载式发电设备;发电车
mobile generator 移动式发电机
mobile generator set 移动发电机组
mobile generator unit 活动发电机组
mobile geothermal-power station 移动式地热电站
mobile grease 滑润脂;铅皂润滑;铅皂润滑脂;低黏度润滑脂
mobile hammer mill 锤式移动粉碎机
mobile handling equipment 机动式搬运设备
mobile-hearth furnace 炉底移动式电炉
mobile heightfinder 活动测高计
mobile high-pressure air stand 移动式高压空气供应台
mobile hoist 移动式卷扬机;移动式绞车;移动式吊车;活动吊车
mobile home 活动旅游家庭车;拖车式住宅;由汽车拉的活动房屋;住宿拖车;活动住房;汽车住房
mobile home accessory building 活动住宅附属建筑
mobile home community 活动住房社区;活动房屋社区
mobile home court 活动住宅泊地;活动房屋泊地

mobile home development 活动住所;活动房屋建设
mobile home dwelling 活动住宅;活动住所
mobile home loan insurance 活动性住房贷款;活动房屋贷款保险
mobile home lot 活动住宅泊地;活动房屋泊地
mobile home pad 活动住宅泊地;活动房屋泊位
mobile home park 活动房屋集中基地
mobile home park service 活动住宅泊地服务设施
mobile home park street 活动住宅泊地小道
mobile home service equipment 旅游居住车服务设施
mobile home space 旅游居住(车)所需面积;活动屋单栋用地
mobile home stand 活动住房地;活动房屋场地
mobile home subdivision 活动住房土地分块
mobile hopper 可移式漏斗;可移式料斗
mobile house 活动住房;活动房屋
mobile housing development 活动住房开发区
mobile hydraulic crane 移动式液压起重机;移动式液压吊车
mobile hydraulic jack 移动式液压千斤顶;移动式水力千斤顶;移动式水力起重机
mobile impeller impact breaker 移动式转子冲击破碎机
mobile industrial handling equipment 移动工业搬运设备;移动式工业装卸设备
mobile irradiator 移动式辐照器
mobile jack 移动式千斤顶
mobile jib crane 移动式旋转起重机;移动式转臂起重机
mobile job crane 工地用移动式吊车
mobile jumbo 自行台车;机动钻车
mobile laboratory 移动式实验室;移动实验室;流动实验室
mobile level of new tectogenesis 新构造运动活跃程度
mobile lifting jacks 移动式驾车机
mobile liquid 易流动液体;流性液体
mobile liquid phase 流动液相
mobile load 活载;易变荷载;活动荷载;动载
mobile loader 移动式装载机;自行装载机;自行装车机;自行式装载机;流动装载机
mobile loader vehicle 移动式装填车
mobile loading machine 自行装载机
mobile loading spout 移动式装料嘴
mobile lubrication equipment 移动式润滑设备
mobile lubrication truck 润滑油加油车
mobile machinery load 流动机械荷载
mobile machine shop 活动机器车间
mobile maintenance and support equipment 移动式维修保养设备
mobile maintenance plant 移动保养设备
mobile maintenance unit 磁悬浮维护单元;磁悬浮移动式维修车辆
mobile mechanical crane 移动式机械起重机
mobile mechanical shovel 移动式机动铲
mobile metamorphic zone 变质活动带
mobile mill 移动式制材厂
mobile mixer 可移式搅拌机
mobile model 活动模型
mobile model trailer type 拖挂式活动模型
mobile moisture 游离水分
mobile monitoring 流动监测
mobile monitoring station 流动监测站
mobile monitoring station for water pollution 水污染流动监测站
mobile mounting unit 移动式台座
mobile observation 流动观测
mobile observation bell 移动观察钟(潜水装置)
mobile observation system 流动观测系统
mobile observation unit 移动观测装置
mobile office 移动办公室
mobile oil 发动机油;流性机油;机油;润滑油;车用汽油
mobile oil dyke system 流动性油坝系统
mobile oil testing equipment 车用机油试验器
mobile operation 流动作业
mobile operation process 流动作业法
mobile optic(al) tracking unit 机动光学跟踪装置
mobile parking meter 活动的汽车停放收费计
mobile parts store 汽车零件库;汽车零件商店
mobile party 流动工作班
mobile phase 移动相;流动相

mobile phase ion chromatography 移动相位离子色谱法
mobile phone 移动(式)电话;手机
mobile pilot plant equipment 活动中试设备
mobile plant 移动式工厂;活动装置
mobile platform 游动式平台;移动式平台;可移动式平台;活动地台
mobile platform type formation 活动地台型沉积建造
mobile pollution source 移动(性)污染源;流动污染源
mobile pollution source model 移动污染源模式
mobile population 流动人口
mobile power station 活动电站
mobile power unit 移动式动力装置
mobile primary crushing plant 移动式初级破碎厂
mobile property 动产
mobile quay 活动码头
mobile radio 移动无线电通信;移动式无线电通信
mobile radio service 移动无线电业务
mobile radio station 移动无线电电台
mobile radio system 车辆无线电系统
mobile radiotelephone private line 移动式无线电话专用线路
mobile radio teletype station 移动式无线电传通信电台
mobile radio unit 移动式无线电台;流动无线电台
mobile reactor 移动式反应堆
mobile receiver 轻便接收机;便携式接收机
mobile relay rack bay 插拔式继电器盘
mobile repair shop 修理车;流动修理站;活动修理站
mobile repair truck 流动修理车
mobile residence 移动住房
mobile residence park 移动住房停车场
mobile restaurant 移动式餐馆
mobile riverbed 不稳定河床
mobile robot 移动式遥控装置;移动式机器人
mobile rope carrier 自行式支索器
mobile rotary crane 移动式旋转起重机
mobile rotary drum drier 移动式转筒干燥机
mobile sampler 活动抽样装置
mobile sand 流沙(也称流砂)
mobile sand mill 移动式混砂机
mobile-satellite 移动卫星
mobile satellite service 移动卫星业务
mobile saw-mill 移动式制材厂
mobile scaffold 移动式脚手架;机动施工架;活动脚手架
mobile scatter communication equipment 移动式散射通信设备
mobile screening and crushing plant 移动式破碎筛分厂
mobile self-supporting hoist 移动式自支承升降机
mobile service 移动业务;移动电台通信
mobile ship station 移动船舶站
mobile shop 修理工程车;流动售货商店;流动(修理)车间;流动商店
mobile site cabin 移动式工房
mobile site handling unit 工地多用搬运机
mobile site office 工地活动办公室
mobile slewing crane 移动式旋转起重机
mobile source 移动性污染源;流动源
mobile source emission 移动性污染源排污;流动源排放
mobile source model 移动性污染源模型
mobile source of air pollution 移动空气污染源
mobile source of pollution 移动性污染源
mobile source of water pollution 移动性水污染源
mobile space heater 移动式对流加热器;携带式液化石油气采暖炉
mobile stage 移动式平台;可移式平台
mobile staging (船坞内的)活动脚手架
mobile station 行动吊车;行动电台;移动台(电波设备);移动式电台;流动站;流动台(站)
mobile stone crushing and screening plant 碎石和筛选移动作业车
mobile structure 活动建筑物
mobile studio 移动播音室
mobile subcriber 移动用户
mobile submersible 可移动操作潜水器
mobile substation 移动变电站
mobile switching cabinet 移动式开关柜
mobile switching center 移动交换中心
mobile switching network 移动计算网络

mobile system 流动系统;机载系统
mobile systems equipment 可动系统设备
mobile tank 油槽车;液罐车
mobile target tracking system 活动物标跟踪系统
mobile team 流动工作班
mobile telephone 移动电话
mobile telephone switching office 移动电话交换局
mobile telephone system 移动式电话系统
mobile television 便携电视
mobile television unit 便携式电视装置
mobile telmetering station 移动遥测台
mobile termination 移动终端
mobile tester 流动性试验器
mobile test stand 移动式试验台架
mobile tire inflation unit 移动式轮胎充气装置
mobile tower crane 移动式塔式起重机;轮胎式塔式起重机
mobile tracking 移动式跟踪
mobile tracking mount 随动跟踪装置
mobile tracking scanner 移动式跟踪扫描器
mobile tracking unit 移动式跟踪装置
mobile transformer 移动式变压器;车式变压器
mobile transmitter 移动式发射机
mobile treadway bridge 轻型机械化桥
mobile TV receiver 汽车电视接收机
mobile tyre inflation unit 轮胎活动打气装置
mobile unit 移动式架空装置;流动设备;活动装置
mobile-unit truck 电视车;可移装置车
mobile vacuum cleaner 移动式真空除尘器
mobile vehicle 移动车
mobile version 汽车形式;活动装置形式
mobile wagon 自行矿车
mobile warping frame 移动式分条整经机
mobile water 自由水;流动水
mobile water distilling plant 蒸馏水装置车
mobile welding equipment 移动式焊接设备
mobile welding set 移动式焊接机组
mobile wharf 活动码头
mobile winder 移动式提升机
mobile work platform 移动式工作平台
mobile work shop 流动(修理)车间;修配车;修理工程车;移动式修配间
mobile X-ray machine 移动式 X 射线机
mobile X-ray unit 移动式 X 射线机
mobil home park 活动房屋园区
mobilideserta 流沙荒漠(群落)
mobilite 照明车(工地用)
mobility 移动性;移动率;流度;流动性;灵活性;可(移)动性;机械导纳;机动性;活动率;活动度;人口流动;漂移迁移率;尚度;变动率
mobility constant 迁移常数
mobility control agent 流动度控制剂
mobility housing 方便行动住房
mobility number 可动性指数
mobility of concrete 混凝土的流动性
mobility of elements 元素活动性
mobility of oxygen ion 氧离子迁移率
mobility ratio 迁移率;流度比;迁移系数
mobility safety index 流动安全指数
mobility scale 变动范围
mobility table 变动率表
mobility test 流动性试验(混凝土的)
mobility type analogy 导纳型类比
mobilisation = mobilization
mobilizable 可移动的
mobilizaion of financial resources 财力的动用
mobilization 活动化;动员;调遣
mobilization advance 动员预付款
mobilization cost 动员费用;工地设施费用;调遣费
mobilization expense 开办费
mobilization of domestic resources 国内资金的筹措
mobilization of dredge(r) 挖泥船调遣
mobilization time 调遣时间
mobilized angle of friction 动致摩擦角
mobilize domestic and foreign capital 筹借国内外资金
mobilized plane 动用面
mobiloader 流动装载机
mobiloil 流性油
mobilometer 流动度计;尚度计
mobirepair(work)shop 流动修理车间;流动修理站
Mobire transformation method 摩必尔转化法
Mobitz type II heart block 模氏 II 型

Mobius band 茂比乌斯带
Mobius function 茂比乌斯函数
Mobius transformation 茂比乌斯变换
Moboron 香柏木(尼日利亚产)
mobot 移动式遥控装置;流动机器人
M-O bus 存储器输出总线
mobus 大容车
Moby Dick balloon 高层等高(空)探测气球
mocca 蛛网线迹
moccasin flower 杓兰属【植】
mocha 深褐色;柔软羊皮手套
mock 制造模型;船体弯曲处外板型模
mock arcade 模拟有拱的长形房屋;模拟拱廊
mock architecture 模拟建筑
mock bomb 模型弹
mockery 冒牌
mocket 餐巾
mocketer 餐巾
mock fission neutron spectrum 模拟裂变中子谱
mock fog 假雾
Mock gold 铂铜合金
mock gravitational force 伪引力
mocking up 做样盒
mock lead 闪锌矿
mock leno weave 充纱罗
mockmain 木棉
mock moon 近幻月;幻月
mock ore 闪锌矿
mock pendulum 假摆;模拟摆
Mock platina 高锌黄铜
mock rafter 仿椽;假椽
mock sun 远幻日;幻日
mock sun ring 幻日环
mock-up 按比例实物模型;全尺寸实物模型;模装;实物大模型;大模型;制造样机;制造模型;假雷达
mock-up car 改装车
mock-up experiment 模拟试验
mock-up reactor 模拟反应堆
mock-up test 模型试验;大模型试验;足尺模型试验;模拟试验;大比例尺模型试验
moctezumite 碲铅铀矿
Mo-Cu ratio 钼铜比
mod 取模函数
modacrylic 改性聚丙烯腈纤维;变性腈纶
modacrylic fibre 变性聚丙烯腈纤维
modal 最常见的;模态的;模态
modal analysis 模态分析;模式分析
modal analysis method 振型分析法;振型叠加法
modal analysis of lineament 线性体模式分析
modal analysis program(me) 模态分析程序
modal balancing 振型平衡
modal balancing theory 振型平衡理论;模平衡理论
modal class 众数组
modal column 模态列
modal combination 模态组合
modal component 振型分量
modal concentration 重量克分子浓度
modal controls 模态控制
modal damping ratio 模态阻尼比
modal decomposition technique 振型分解法
modal diameter 众数粒径
modal direction 蜂向
modal dispersion 模式弥散
modal-domain fiber optic(al) sensor 模域光纤传感器
modal frequency 众数频率
modal group 密集组
modality 模态;程式
modality of payment 支付方式
modal line 波结线
modal logic 模态逻辑学;模态逻辑
modal loss 模式损耗
modal matrix 振型矩阵;模态矩阵
modal noise 模拟噪声
modal operator 模态运算子
modal parameter 模态参数
modal phase dispersion 模式相位弥散
modal point 节点
modal response 振型反应;模态反应;动态反应;动力反应
modal speed 最多车速
modal split 交通方式划分;定型的交通分流;交通工具分流;交通工具分类
modal stiffness 模态刚度

modal superposition 振型叠加;模态叠加
modal system 模态系统
modal value 最常见值;众数
modal vector 振型矢量;模态矢量
modal velocity 振型速度
modderite 砷钴矿
mode 样式;众值;众数【给】;格式;类型;模型;式样;(岩石的)实际矿物成分;方式;波模
mode address 方式地址
mode amplitude 振型幅值;模振幅
mode analysis 振型分析
mode average 众值平均
mode beating 模拍频
mode bit 状态位;模式位;方式位
mode bit wire 方式位线
mode centre cluster 集群法
mode change 作用方式变化;模式变换;模式变化
mode change-over disturbance 模转化干扰
mode chart 模式图;波形图
mode combination 振型组合
mode compatibility 方式兼容性
mode competition 模式竞争;波型竞争
mode configuration 模式结构
mode confinement 模限制
mode control 状态控制;模(式)控制
mode-controller 可控模
mode control register 状态控制寄存器
mode control word 状态控制字
mode conversion 模转换;模式变换;波型转换
mode conversion interference 模转换干涉
mode converter 模变换器
mode coupled laser 模式耦合激光器
mode coupler 模式耦合器
mode coupling 振型耦联;振型耦合;模态耦合;模式耦合;模间耦合;波形耦合
mode crossing 波模交叉
mode cross section 模截面
mode damping matrix 振型阻尼矩阵
mode declaration pattern specification 模式说明
mode-definition 模式定义
mode degeneracy 模式简并;波模简并度
mode density 模密度
mode direction handle 行车方向控制手柄
mode-discriminating interferometer 模式鉴别干涉仪
mode dispersion 模式色散;模色散
mode distortion 模失真
mode envelope 模式轮廓;模迹;波模包迹
mode excitation 模激发
mode exciter 模式激励器;波型激励器
mode expansion(method) 振型展开法
mode factor 振型系数
mode fiber 光模纤
mode field 模场;方式字段
mode field concentricity 模场同轴度
mode field diameter 模场直径
mode filter 振荡型滤波器;滤模器;波形滤波器;波模滤波器
mode frequency separation 振荡型分离;波形频率间隔
mode function 模式函数;模函数
mode group 模系
mode hopping 模跳跃;模式跳越
mode identifier 模式标识符
mode indication 模式指示
mode instruction 方式指令
mode instruction format 方式指令格式
mode jump 模式跳变;跳模
model 型机;型号;样机;木模;模样;模型;模式;宏程序纲要;典型;典范;波形
model A benchmark 甲型水准标石
model acceptance 模型验收
model acceptance test 模型验收试验
model adaptive control 模型自适应控制
model adjustment 模型调整
model adjustment technique 模型调整法
model adsorbent 准备吸附剂
model agreement 协议规格
model aircraft 模型飞机
model algorithm 模型算法
model analysis 模型分析;模拟分析(法)
model and mechanism of change 变化的模式与机理
model and prototype 模型和真型

model approximation 模型趋近
model area 模型面积;模型范围
model atmosphere 模式大气;大气模型
model atmosphere re analysis 模型大气分析
mode launching 模发射
model-based vision 模型视觉
model base(line) 模型基线;投影基线
model basing 模型试验台
model B benchmark 乙型水准标石
model behavior 模型行为
model blasting 模型爆破
model boundary 模型边线;模型边界
model builder 制作模型者
model building 模型建造;制作模型;建模;模型制作;模型制造;模型结构;模型建立
model building code 典型建筑法规
model by-laws 法律模式
model calibration 模型校准;模型率定曲线;模型率定;模式校准
model cam 样板凸轮
model car 模型车
model casing 模型箱
model change 产品变化更新;产品更新;型号改变;产品变化
model channel 模型河槽;模拟河槽
model checking 模型校验
model chromosphere 按模型色球模型
model city 典型城市;样板城市;文明城市
model clause 标准条款
model clinker 模拟熟料
model code 样板法规;典型法规
model coefficient 建模型系数
model coil 模型线圈
model complete 模型完全性
model completion 模型完全化
model complexity 模型复杂性
model compound 模型化合物
model computer 积木式计算机
model concrete shell 标准混凝土壳体
model configuration 模拟结构
model conformation 模型验证
model connection 模型连接
model consistent 模型一致的
model constraint 模型约束;模型限制
model construction 模型制作
model contract 范本合同;标准合同
model control 模型控制
model coordinates 模型坐标
model coordinate system 模型坐标系统
model corrector 模型校正器
model corridor facility 模拟走廊设备
model data 模型数据
model date 模式年龄
model deformation 模型变形
model design 模型设计
model designation 型号表示法
model design method 模型设计法
model development and refinement 模型建立与改进
model deviation 模式偏差
model device for the model 模型校正装置
model discrimination 模型识别
model distance 模型距离
model distortion 模型扭曲
model dwelling 样板住宅;典型住宅
model earthquake motion 模型地震运动;模型地裂运动
model ecosystem 模型生态系统;模式生态系统
modeled-in-placed 工地模制的
modeled reach 制模河段
model efficiency 模型效率
model element 模型元件
model equation 模型方程
model error 模型误差
model estimation 模型估计
model expansion 模态展式
model-expansion method 模式展开法
model experiment 模型试验;模拟试验
model experiment method 模型试验法
model experiment with a rotating sphere 旋转球体模拟试验
model experiment with clay 泥巴试验
model fabrication 模型制作
model fabrication technique 模型制作技术

model fiber 模型纤维
model filament 模型丝
model file 模型文件;模型记录组
model fill 模型堤坝;堤坝模型
model fitting 模型拟合
model flow 模型流动
model fluid 模拟液体
model flume 模化流道;模化水槽;模型水槽
model footing test 模型基础试验
model for discrete search 离散搜索模型
model for drift ccmpensation 漂移补偿模型
model formation 建立模型
model formulation 模型组成;模型建立;模型构成;模型表述
model framework chart 模型框图
model garden 模型园
model hardener 模型硬化剂
model head-pond 模型前池
model height 模型高
model home 模型住房;示范性住房
model house 样板住宅;样板房
model industrial city 模范工业城
model information 模型信息;模型数据
model investigation 模型研究;模型(试验)研究
model launching 新型号投产
model law 相拟定律;模型律;模型定律
modellde relationship 按模型构造的关系
modelled sandblast 喷砂雕刻
model lens 模型镜头
model(l)er 黏土模型制作者;塑造者;制作模型者
model(l)er's clay 塑家泥;橡皮泥;制模黏土;模型黏土
model library 模型库
model(l)ing 模型制作;制作模型(方法);造型(术);模型制造;模型建立;模(型)化;制造模型
model(l)ing bar 比例杆
model(l)ing burner 造型用煤气喷灯
model(l)ing by mechanism 机理建模
model(l)ing clay 制模黏土;做模型用黏土;雕塑黏土
model(l)ing criterion 模拟准则
model(l)ing effect 立体感效果
model(l)ing gradation 模拟级配
model(l)ing light 立体感灯光;造型光
model(l)ing method 标准方法
model(l)ing methodology 模型建立的方法论
model(l)ing of sewerage system 建立排水工程系统模型
model(l)ing of the soil nitrogen cycle 土壤中氮循环模拟
model(l)ing paste 模型用浆
model(l)ing phase 建模阶段;模型建立阶段
model(l)ing plaster 造型用灰泥
model(l)ing process 模拟程序;模型建立的过程
model(l)ing program(me) continuous system 模拟程序连续系统
model(l)ing rule 模拟准则
model(l)ing seismic inverse 模拟地震反演
model(l)ing shop 木模车间;制模车间
model(l)ing technique 模拟试验技术;模型制作技术;模型术;模型试验技术;模型技术
model(l)ing test(ing) 模型试(验)
model(l)ing test technique 模拟试验技术
model(l)ing tool 制模木工具
model(l)ing transformation 建模变换;模型变换
model(l)ing verification 模型验证
model(l)ing with air 空气模拟
model(l)ing with conductive paper 导电纸模拟
model(l)ing with soil tank 土槽模拟
model(l)ing with thin later of water 薄水层模型
model(l)ing with water tank 水槽模拟
model liquid 模化流体
model listing 模型排列
Modell number 莫氏耐磨性指数
model logic 模拟逻辑
modelly 模特儿似的
model machine 样机
model magnet 模型磁铁
model maker 模型工
model marriage 模型拼接
model matching 模型匹配
model material 模型砂;模型材料
model measurement 模型测试;模型测量
model membrane 膜模型

model metal 模型金属
model method 分型估值法;式样法;模拟法;样板法;模型法
model methodology 模式研究法
model modifying mode 模型修改方式
model motor 作为试验样品的电动机;模型电动机
model mo(u)ld 原始模型
model municipal traffic ordinance 标准城市交通法规
model network 模拟网络
model notation 模型的记法;模型(的)表示法
model number 型号
model objective 模型目的;模型目标
mode lock 振荡型同步;波型同步
mode locked 锁模
mode-locked dual polarization operation 锁模双偏振运转
mode-locked frequency-doubled operation 锁模倍频运转
mode-locked laser 锁模激光器
mode-locked state 锁模态
mode-locked train 模式同步组列;锁模列
modelocker 锁模器
mode locking 模锁定;模式锁定;锁模;波模锁定
mode-locking by interactive loss modulation 损耗调制锁模
mode-locking by saturable dye 饱和吸收染料锁模
mode-locking effect 锁模效应
mode-locking frequency doubler 锁模倍频器
mode-locking technique 锁模技术
model of acidification of groundwater in catchment 流域地下水酸化模型
model of aerial camera 航摄仪型号
model of a quadric 二次曲面模型
model of atmosphere 大气模式
model of carbon migration 碳迁移模式
model of chemical dilution 化学稀释模式
model of communication system 通信系统的模型
model of compensation 补偿模型
model of concrete armo(u)r unit 混凝土护面块体模型
model of construction 建筑方式
model of decision making by administrative staff 行政人员决策模式
model of deposition 沉积模式
model of distributed parameter system 分布参数系统模型
model of economic order quantity 经济订购批量模式
model of estimated experience 经验估计模式
model of expected value 期望值模式
model of fissuration 裂隙模型;缝隙模型
model of gear 齿轮模型
model of hydrogeologic(al) calculation 水文地质计算模型
model of input-output in investment 投资投入产出模型
model of input-output in labo(u)r consumption 劳动消耗投入产出模型
model of integrated parameter system 集中参数系统模型
model of isostasy 地壳均衡模型
model of isotopes mixed-exchanged equilibriums 同位素混合—交换平衡模式
model of man 人的模型
model of metropolis 城市模式
model of mixed-exchanged isotopes 同位素混合交换模式
model of mixing isotopes 同位素混合模式
model of mobile bed 动床模型
model of ocean circulation 大洋环流模式
model of population dynamic(al) projection 人口动态预测模型
model of porphyrite iron deposit 玢岩铁矿床模式
model of porphyry copper deposit 斑岩铜矿床模式
model of porphyry molybdenum deposit 斑岩钼矿床模式
model of preservation 保存范例
model of propellant 推进剂牌号
model of pump 水泵型号
model of qualitative choice 品质选择模型
model of quality assurance 质量保证模式
model of quantity migration 质量迁移模式

model of river 河道模型
model of river engineering 河工模型
model of river water quality control 河流水质控制模型
model of statistic(al) table 统计表式
model of structure 结构模型
model of symbolic logic 符号逻辑模型
model of terrestrial camera 地面摄影机型号
model of tidal river and canal 潮汐河道和运河模型
model of tungsten-tin-beryllium-niobium-tantalum and rate earth deposit 钨锡铍铌钽及稀土矿床模式
model of water environmental quality 水环境质量模型
model of water quality 水质模拟
model of water quality management 水质管理模型
model of water quantity 水量模型
model of water quantity management 水量管理模型
model of water resources calculation 水资源计算模型
mode one crack[opening or tensile crack] Ⅰ型裂纹(张开型或拉伸型裂纹)
model order 模型的阶
model output diagnosis 模式输出分析
model output statistics 模式输出统计
model parameter 模型参数
model photo 示范图片;示范照片
model photonuclear reaction 模型光核反应
model photosphere 光球模型
model pile 模型桥墩;模型桩
model pile group 模型桩群
model plan 模型计划
model plane 模型平面
model plant 制模工厂;模型厂
model plaster 模型石膏
model plastic 模型塑料
model point 模拟点
model pond 模型池塘
model prediction 模型预测法
model preparation room 模型准备室
model problem 模型问题
model-prototype comparison test 模型与原型比较试验
model-prototype relationship 模型与原型关系
model radiation belt 模型辐射带
model railway 模型铁路
model range 模型距离;模型范围
model rating curve 模型率定曲线
model raw meal 模拟生料
model recognition 模式辨认
model reference adaptive control system 模型基准自适应控制系统
model reference system 典型参考系统
model refinement 模型改进
model reliability 模型可靠性;模型可靠度
model response 模型频率特性
model Reynolds number 模型雷诺数
model rigging 模型装配
model river 模型河流;典型河流
model room 样板间;模型室;模型房;模型车间
model runner 模型转轮
model sampling 模型抽样
model sand 模型砂
model scale 模型缩尺;模型卷尺;模型比(例)尺
model scale effect 模型缩尺效应
model scanning 模式辐描
model scope 模型范围
model seismology 模型地震学
model select 模型选择
model selection criterion 模型选择准则
model selector 模选择器
model select program(me) 模型选择程序;模式选择程式
model set 模型组;模型
model shape 振型形式
model ship 船模
model shop 模型(制造)车间
model shot 模型技巧摄影
model similarity 模型相似性
model similarity law 模型相似律
models of application layer 应用层模型
model soil 模拟污垢

model solar atmosphere 太阳大气模型
model space 造型空间;模型空间
model spacing 模型距离
model specification 模型说明书;模型设定
model split 模型分割;定型交通分流
model stability 模型稳定性
model standard 模型标准
model statement 模型语句
model stator 模拟定子
model stocklist 流行品存货单
model stock plan 标准库存量制度
model stream 模型河流;典型河流
model structural analysis 模型结构分析
model structure 模型结构
model study 模型(试验)研究;典型调查
model substance 模型物质
model support 模型架
models with a sill 有基台模型
models without a sill 无基台模型
model symbol 模型符号
model system 模型系统;模态系统
model tank 模型试验池
model tectonics 模型构造地质学;模型大地构造学
model test 模型试验;模拟试验;水力型试验
model test equipment 模型试验设备
model test for local scour 局部冲刷模型试验
model testing 模型检验
model testing basin 模型试验池
model test method 模型试验法
model test of bridge 桥梁模型试验
model test of cavitation 汽蚀模型试验
model test of open channel diversion 明渠导流模型
model test of similar material 相似材料模拟试验
model theory 模型论;模型理论
model tillage test 模拟耕作试验
model top view 模型顶视图
model tormulation 模型表达
model tree 标准木
model trimmer 模型修整器
model turbine 模型水轮机
model turning machine 仿型修坯机;仿形修坯机
model type 模型类型
model value 模式数值
model violation 模型违例
model warpage 模型扭曲
model water turbine 模型水轮机
model with all sediment 全沙模型
model year 新产品年度;年型
model-year variation 新产品年度差异
modem 调制解调器;反调制装置
mode matching lens 模匹配透镜
modem chip 调制解调(器)芯片;单片调制解调器
modem conversion function 调制解调器转换功能
mode memory 模式存储器
modem hardware 调制解调器硬件
modem interface 调制解调器接口
mode mixer 模式混合器
modem multiplexer diagnostics 调制解调器多路转换器诊断
modem quick turnaround 快速周转调制解调器
modem section 调制段
modem sharing unit 调制解调器共享装置
modem stage 调制级
modena 深紫色
mode name 方式名
mode number 模数
mode of action 作用方式
mode of anti-plane slide 撕裂式扩展
mode of arc transfer 电弧过渡形式
mode of binomial distribution 二项分布方式
mode of blasting 爆破模式;爆破方式;爆破布置
mode of combined rotor 组合转子的振动波型
mode of composition 构图方式
mode of connection 联结方式;接线方式
mode of construction 构造方式
mode of decay 衰变方式
mode of deformation 变形种类
mode of denudation 剥蚀方式
mode of deposition 沉积条件;沉积方式
mode of distribution 分配方式;分布模式;分布的众数
mode of element occurrence 元素存在形式
mode of entry 侵入方式

mode of establishing horizontal control 平面控制网布设形式
mode of exogenic process 外力地质作用方式
mode of failure 破坏模式;破坏方式
mode of financing 资金筹集方式
mode of free vibration 自振振型;自振模态
mode of illumination 照明方式
mode of instrument 仪器型号
mode of laser 激光模
mode of magnetization 磁化方式
mode of metal transfer 金属过渡方式
mode of motion 运动方式
mode of occurrence 埋藏条件;存在形式;产状【地】;产生条件
mode of opening 开拓方式
mode of operation 运营模式;运行方式(状况);工作状态;工作方式;处理方式;操作方式
mode of optic(al) cavity 光学共振腔模
mode of oscillation 振荡方式
mode of payment 支付方式
mode of photography 摄影方式
mode of plane-slide 滑移式扩展
mode of priority 优先方式
mode of production 生产方式
mode of production measurement 产量测量方式
mode of propagation 传码形式;传播方式
mode of random variables 随机变量的众数
mode of relative displacement 相对位移方式
mode of removal 迁移形式
mode of resonance 谐振模式;共振模
mode of satisfaction 赔偿方式
mode of signal indication 显示方式
mode of speed regulation 调速方式
mode of tension 张开式开展
mode of texturing 变形方式
mode of traction 牵引方式
mode of trade 贸易方式
mode of transmission 传播方式
mode of transport 输运方式
mode of transportation 运输方式;运输方法;交通运输手段
mode of transportation passing through the dam 过坝运输方式
mode of treatment 森林经营法
mode of unified construction of houses 统建方式
mode of vibration 振型;振动模型;振动方式
mode of weathering 风化作用方式
mode oscillation 模式振荡
mode-participation factor 振型参与系数
mode partition noise 模分配噪声
mode pattern 模样花式;振荡型图;模图样;模式图样;模式花样;模式;波型图样
mode perturbation 模式微扰
mode pulling 模式牵引
mode pulling effect 模式牵引效应;波模牵引效应
mode purity 模式纯度;波型纯净度;波模纯度
mode pushing 模推离
moder 脉冲编码装置;酸性腐泥;编码装置
mode radius 模半径
moderate 适度的;中度;中等的;平均的;温和的
moderate accelerator 中速促进剂
moderate activity 中等活动
moderate adsorption 中等吸附
moderate agitation 缓慢搅拌
moderate attack 中等侵蚀
moderate availability 中度有效性
moderate average-length swell 中中涌
moderate breeze 和风;四级风
moderate casualty agent 中级伤害剂
moderate cave 中洞
moderate cementation 中等胶结
moderate change test project 适中改变试验计划
moderate climate 温和气候
moderate condition 中等条件
moderate consumption 适度消费
moderate cost 收费低廉
moderate crack 中度开裂
moderate cracking 中度裂化
moderate current 中等水流;中常水流
moderate damage 中等破坏
moderate damping 中等阻尼
moderated core 慢化堆芯
moderate decrepitation 爆裂中
moderate deep well oil well cement 中深井油井

水泥
moderate demand 适度需求;适当的要求
moderate dense 中密(的)
moderate density line 中等运输密度线路
moderate directivity antenna 弱方向性天线
moderated kernel 慢化核
moderated neutron 慢化中子
moderated nuclear reactor 慢化核反应堆
moderate dose 适度剂量
moderated radiation 慢化辐射
moderated reactor 慢化反应堆
moderate drilling 中硬岩层钻进
moderate drying 中庸干燥度
moderate dust 中等灰尘
moderate duty 中等负荷
moderate-duty compressor 中型压缩机
moderate-duty floor(ing) 中等地面;中等楼面
moderate-duty service 一般条件下操作;中等负荷运行
moderate earthquake 中强地震;中地震
moderate effluent 中等污水
moderate energy 中等能量
moderate energy coast 中等能量海岸;中能海岸
moderate energy neutron reactor 中能中子反应堆
moderate erosion 中度侵蚀
moderate fading 中度衰落
moderate failure 中等破坏
moderate fever 中等热
moderate fines 中等细粒
moderate fire hazard 中等火灾危险
moderate flood 中等洪水
moderate-flow year 中水年;平水年
moderate fog 中雾;中等的雾;中常雾;适度的雾;三级能见度
moderate gale 疾风;七级风
moderate grade 中坡(度);中等坡度
moderate ground shaking 中强地震动
moderate hard water 中等硬水
moderate-hazard industrial buildings 中等防火工业建筑物
moderate head 中(等)水头
moderate head plant 中水头电站
moderate heat-of-hardening 中热(水泥)
moderate heat of hydration and sulphate-resistant cement 中等水化热抗硫酸盐水泥;中热抗硫酸盐水泥
moderate heat of hydration cement 中等水化热水泥;中热水泥
moderate heat oil well cement 中热油井水泥
moderate-heat Portland cement 中热硅酸盐水泥;中热波特兰水泥
moderate-income group 中等收入组别
moderate-income housing 中等收入者住宅
moderate injury level 中等伤害碰撞强度
moderate intensity 中等强度
moderate layer landslide 中层滑坡
moderate-length stop 中等距离停车站
moderate lift lock 中水头船闸
moderate loop optimization 适度循环优化
moderately 适度地
moderately acid loamy soil 中等酸性土壤
moderately active 中等活动的
moderately brackish 中等盐渍化的;中等含盐的
moderately coarse 中等粒度的
moderately coarse texture 中等粗质土
moderately fine texture 中等细质土
moderately firm ground 中等硬度土
moderately hydraulic lime 中等水硬石灰
moderately large deformation 中等大形变
moderately persistent gas 半持久性毒气
moderately plunging fold 中等倾伏褶皱
moderately polluted water 中等污染水
moderately populated density 中量人口密度
moderately rapid permeability 常速渗透(度)
moderately robust 硬度适当
moderately salified soil 中盐渍土
moderately slow permeability 稍慢渗透
moderately soluble 中等溶度的
moderately soluble salt test 中溶盐试验
moderately stable 适当稳定的
moderately swelling-shrinkage foundation 中等胀缩地基
moderately turbid 中浊
moderately volatile fuel 中等挥发性燃料

moderately weak rock 中等软弱岩石
moderately weathered 弱风化的
moderately weathered rock 弱风化岩石
moderately well drained 排水中等的
moderate management 适度管理
moderate moist swamp 中湿沼泽
moderate moisture 适中的温度
moderate odor 中等臭味
moderate opening of canopy 郁闭疏开
moderate operating condition 中等使用条件;中等操作条件;中等运转条件;中等工作条件
moderate oven 中温炉
moderate oxidation 中度氧化
moderate physical labor 中等体力劳动
moderate plant 适度植物
moderate price 适度价格
moderate protective duty 适度保护关税
moderate pruning 适度修剪
moderater 减速剂
moderate rain 中雨;和雨
moderate rainfall 中雨量;适中的雨量
moderate range 中等幅度
moderate refrigerating 中度冷冻
moderate relief 中等起伏(地势)
moderate resistance mineral 中等稳定矿物
moderate reverberation 调制反响(测量吸声能力的一种方法)
moderate salinity reservoir 中盐度热储
moderate scale management 适度规模经营
moderate sea 中浪(三级风浪);中常波;中波(海浪三级);四级浪
moderate sea deposit 半深海沉积(物)
moderate shock 中等地震
moderate short swell 短中涌
moderate-size(d) 中型的
moderate size earthquake 中等地震;中强地震
moderate-sorted 中等分选的
moderate-sorted sand 中分选的砂
moderate speed 中速;缓速;平均速度
moderate-speed lubrication 中等速度润滑
moderate spirality sector 中等螺旋扇块
moderate strength 中等强度
moderate strong earthquake 中强地震;中等地震
moderate sulfate-resistant oil well cement 中抗硫油井水泥
moderate sulfate resisting cement 中等抗硫酸盐水泥
moderate sulphate resisting cement 中等抗硫酸盐水泥
moderate supersonic speed 中超音速
moderate supply 适当的供给
moderate swell 一般长涌;中(级)涌浪;四级涌;短涌
moderate swell short 三级涌
moderate tectonic activity 中等构造活动
moderate temperature 中等温度;温和热度
moderate temperature water reservoir 中温热水储
moderate testing level 中等试验水平
moderate thinning 适度疏伐
moderate-to-strong earthquake motion 中强地震运动
moderate traffic 中量交通;中等交通量
moderate velocity 中速
moderate visibility 中等能见度(能见度范围为4~7公里)
moderate water 中硬水
moderate weathering 中风化;中等风化的
moderate wind 中度风;和风;四级风
moderate zone 温带
moderating 慢化的
moderating detector 慢化探测器
moderating effect 慢化效应
moderating efficiency 慢化效率
moderating material 减速剂;慢化材料
moderating power 慢化能力
moderating process 减速过程;慢化过程
moderating property 慢化特性
moderating ratio 减速系数;减速比;慢化系数;慢化比
moderating reflector 慢化反射层
moderating sphere 慢化球
moderation 减速;适中
moderation age theory 慢化年龄理论
moderation length 慢化长度

moderation of business cycles 商业周期的缓和
moderation of neutrons 中子慢化
moderation velocity 慢化速度
moderative method 慢化法
moderator 减速剂;慢化剂;缓和器;缓和剂
moderator assembly 慢化剂组件
moderator band 节制索
moderator capture 慢化剂俘获
moderator circulation system 慢化剂循环系统
moderator control 减速剂控制
moderator coolant 减速冷却剂;慢化冷却剂
moderator-density fluctuations 慢化剂密度起伏
moderator dumping safety mechanism 慢化剂排放安全机构
moderator heat-exchanger 慢化剂换热器
moderator lattice 慢化剂栅格
moderator overflow 慢化剂溢流
moderator purification 慢化剂净化
moderator structure 慢化剂砌体
moderator-to-fuel ratio 慢化剂与燃料比
mode reconversion 波模再变换;振型再变换
mode register 方式寄存器
mode rejection 模式抑制
mode repulsion 模态互斥
mode response 振型反应
modern 现代的
modern agriculture 现代农业
modern and contemporary architecture 现代与当代建筑
modern architecture 新派建筑;现代(化)建筑;近代建筑
modern art 现代艺术;近代艺术
modern Baroque style 近代巴洛克式园林
modern bridge 现代桥梁
modern bridge building technique 现代桥梁建筑技术
modern business administration 近代企业管理
modern campground 现代化野营场地
modern car 现代汽车
modern ceramics 现代陶瓷
modern city 近代城市
modern civilized illness 现代文明病
modern communications technology 现代通信技术
modern computer network technologies 计算机网络新技术
modern computing method 现代计算方法
modern connecter 新型连接件;环接件
modern control equipment 新型控制设备
modern control theory 现代控制理论
modern conveniences (生活方面的)现代化工具;(住房的)现代化设备
modern crust movement 现代地壳运动
modern data entry techniques 现代数据输入技术
modern deposit 新淤积泥沙;现代沉积物
modern digital computer 现代数字计算机
modern digital system 现代数字系统
modern econometrics 现代经济计量学
modern economic management 现代经济管理
modern economics 现代经济学
modern electrical equipment 现代电气设备
modern English architecture 现代英格兰建筑
modern enterprise management 现代企业管理
modern equipment 现代化设备
modern face 现代体
modern gardens in China 中国近代公园
modern general equilibrium theory 现代一般平衡理论
modern Georgian architecture 现代乔治式建筑
modern history 近代史
modern hump yard 现代化驼峰编组场
modern industry 现代工业
modernism 西班牙卡达兰的新艺术形式;现代派;现代式;现代方法
modernistic 现代风格的
modernist school 现代派
modernity 现代性;现代风格
modernization 现代化
modernization management 现代化管理
modernization of accounting 会计现代化
modernization of agriculture 农业现代化
modernization of farming 农业现代化
modernization of railway 铁路现代化
modernization of railway statistic(al) means 铁

路统计手段现代化
modernization of railway transport statistics 铁路运输统计现代化
modernization plan 现代化计划
modernization program(me) 现代化计划
modernized large industry 现代化大工业
modernized technological processes 现代化工艺过程
modern karst 现代岩溶
modern kernal estimate 现代核估计
modern lead 现代铅
modern liquid chromatography 近代液相色谱法
modern management 现代管理
modern motor spirit 现代车用汽油
modern mo(u)ld and core making process 现代型模及型芯制造法
modern national defense 现代国防
modernness 现代性
modern painting 现代绘画
modern pigment 现代颜料
modern plate 现代板块
Modern Power Engineering 现代动力与工程(期刊)
modern refinery 现代化炼油厂
modern science and technology 现代科学技术
modern sectors 现代部门
modern sediment 新淤积泥沙
modern signal processing technique 现代信号处理技术
modern specialized large-scale production 现代的专业化大生产
modern statistics 现代统计学
modern style 现代派式样;现代样式;现代风格
modern style faience 现代风格瓷器
modern system 建筑高烟囱的现代系统;现代高烟囱施工法
modern system(s) theory 现代系统理论
modern taxation system 现代税收制度
modern technology 现代技术
modern terrestrial lead 现代地球铅
modern theory of uneven growth 现代非均衡增长理论
modern times 近代
modern tracing technique 近代示踪技术
modern transport 现代化运输
modern transport sector 现代运输部门
modern type 现代类型
modern version 现代化的型式;新型
modern welfare economy 现代福利经济
mode scrambler 搅模器
mode select 方式选择
mode selection 模选择
mode selection by Fabry-Perot etalon 法布里—珀罗标准具选模
mode selection by Fresnel's number of resonator 菲涅尔数选模
mode selection switch 工作状态选择开关
mode selectivity 模选择性
mode selector 选模器;状态选择器;模(式)选择器;方式选择器;波形选择器
mode selector switch 工件状态选择开关
mode separation 振荡型频差;模式分隔;模的分离;波形分离;波模频率间隔
mode sequence 波型序
mode(shape) 模式;形式;自振型;自振模态;振型
mode shape vector 振型矢量
mode shear coefficient 振型剪切系数
mode shift 振荡型转移;模移
mode ship 模漏
mode simulator 模式模拟器
modes of business operation 经营方式
modes of record in geologic(al) observation point 地质点记录方式
modes of sale 销售方式
mode spacing 模间隔
mode specifier 方式说明符;方式分类符
mode spectrum 模(式)频谱;模谱;波模频谱
mode standard 标准模式
mode step 模阶
mode stop 波型光栏
modest price increase 适当涨价
modest price reduction 适当降价
mode structure 模结构
modesty panel (桌前或台前的)遮腿挡板

mode superposition 模态叠加;振型叠加
mode superposition method 振型叠加法
mode suppresion technique 模抑制技术
mode suppression 模抑制
mode suppression technique 波模抑制技术
mode switch 模转换器;方式开关;波模转换开关
mode switching 波形转换开关
mode symbol 模式符号
mode synthesis method 模式综合法
mode table 方式表
mode top 模式中心
mode transducer 模变换器
mode transfer function 模式传递函数
mode transformation 模变换;波形变换
mode transformer 模变换器;波形变换器
mode voltage 模电压
mode volume 模体积;模容量
modifiable 可减轻的
modificating pureline method 改良纯系法
modification 修饰(作用);修改(方案);革新改造;饰变;变质处理
modification address 修改地址
modification by program-self 程序自修改
modification coefficient 变位系数
modification factor 改正系数
modification kit 改型工具;成套改装器材;附件改型工具;附加器
modification loop 调整循环;数据改变环;改址环
modification of airmass 气团变性
modification of bids 投标文件修改
modification of contract 修改合同
modification of goals 目标的修正
modification of nature 改造自然
modification of objectives 目标的修正
modification of orders 指令修改;指令变化
modification of program(me) 程序修改
modification of tank 池子改装
modification of tenders 投标文件修改
modification of the microclimate 改变小气候
modification of the transverse Mercator projection 改进的横轴墨卡托投影
modification polyconic(al) projection 改良多圆锥投影
modification work 改造工程;改建工程;整修工程
modificator 变质剂
modificatory air mass 变性气团
modified 改良的;改进的
modified AASHTO compaction test 改进的AASHTO击实试验
modified accrual basis 修正权责发生制
modified acetate 变性醋酯纤维
modified acryl fibre 变性聚丙烯腈纤维
modified-acrylic adhesive 改性丙烯酸黏合剂
modified-acrylic fibre 改性丙烯酸系纤维
modified activated carbon 改性活性炭
modified activated carbon fiber 改性活性炭纤维
modified activated sludge process 变型活性污泥法;改良活性污泥法;改进型活性污泥法;改进的活性污泥法
modified address 修改地址
modified aeration 改良(型)曝气(法);改进(型)曝气(法);改进通气法
modified aeration process 改良曝气(法);改良活性污泥法
modified A-frame 变A型架
modified alkyd resin 改性的醇酸树脂;改性醇酸稠
modified alloy 改良合金;变质合金
modified asphalt 改良(地)沥青;改性(地)沥青
modified asphalt grinder 改性沥青研磨机
modified asphalt membrane 改性沥青油毡
modified asphalt sheet 改性沥青油毡
modified Atkinson formula 艾金森简化公式
modified attapulgite clay 改性凹棒石黏土
modified austempering 改良式等温淬火
modified Bauer-Vogle process 改良型鲍尔—沃尔盖法
modified bearing ratio 修正的承载比;改正承载比
modified bentonite 改性膨润土
modified Bessel function 修正贝塞耳函数
modified beveled edge 改型斜削边
modified binary code 改进的二进制代码
modified binomial distribution 修正的二项分布
modified bitumen 改性沥青
modified block 改良方块

modified blow count 修正的贯击数
modified Bouguer anomaly 修正的布格异常
modified California bearing ratio 修正加州承载比
modified Cassegrain antenna 改造型卡塞格伦天线
modified cast-iron 变质铸铁
modified catonic flocculant 改性阳离子型絮凝剂
modified cavitation number 修正空化数
modified cellulose 改性纤维素
modified cellulose fibre 变性纤维素纤维
modified cellulose wastewater 改性纤维废水
modified cement 中热水泥;改性水泥;改良水泥
modified charge balance equation 准电荷平衡方程
modified chemical vapo(u)r deposition process 改性化学气相沉积工艺
modified clay 改性黏土
modified coal tar pitch 改质沥青
modified coefficient 修正系数
modified complement 修正补码;变形补码
modified conic(al) projection 准圆锥投影;改良圆锥投影
modified constant-voltage charge 改进恒定电压充电
modified contact ratio 修正总重合度
modified continuous filament 变形长丝
modified continuous layering 改良连续压条
modified control limit 修正后的控制界限
modified correction 改正刚度
modified Coulomb field 变形库仑场
modified cube 改良方块;人工方块
modified cube quay wall 异形方块码头
modified curve 修改水泵曲线
modified cyclic activated sludge system process 改进型循环式活性污泥系统
modified cylindric(al) projection 准圆柱投影;改良圆柱投影
modified densification process 改性致密工艺
modified depth 修正深度
modified derrick system 改良型吊杆装前方式
modified design 修正的设计;改变标准设计
modified design seismic coefficient method 修正地震影响系数法
modified diaphragm 改性隔膜
modified diatomite 改性硅藻土
modified digit encoding 改进型两位编码
modified distribution method 修正分配法;改良调配法
modified drying oil 改性干性油
modified eccentric compression method 修正偏心受压法
modified Einstein method 修正的爱因斯坦法
modified ellipse equation 修正椭圆公式
modified English bond 改进的英国式交叉砌合
modified English mounting 改进英国式装置
modified epoxy 改性环氧树脂
modified equation 转换后的方程
modified Euler method 改进的欧拉法
modified expanded granular sludge bed 改良型膨胀颗粒污泥床
modified exponential curve 修正的指数曲线
modified fan system 改良扇形体系
modified ferric alginate 改性藻酸铁
modified Fibonacci search 改进黄金分割搜索
modified fiber 变性纤维
modified filter media 改性过滤介质
modified flocculant 改性絮凝剂
modified fly-ash 改性粉煤灰
modified form 修改型;改良型
modified formula 修正公式
modified frequency modulation 改进型调频制
modified frequency modulation system 改进调频制
modified frequency table 修正的频数表
modified Freundlich model 修正弗兰德利希模型
modified Galerkin vector 修正盖勒金矢量
modified Galitzin seismograph 改进的伽利津地震仪
modified gear 变位机构
modified Hankel functions 修正汉克尔函数
modified Hooker's law 修正的虎克定律
modified horseshoe 修正的马蹄型
modified hypercrosslinked polymeric adsorbent 改性超交联高分子吸附剂
modified I beam 改进工字梁
modified impedance 变形阻抗

modified impedance relay 修正阻抗继电器;变形阻抗继电器
modified index 修正指数
modified index of refraction 修正折射指数
modified involute gear 渐开线修正齿轮
modified isochronal test 修改等时法测试
modified Julian date 简化儒略日期
modified Julian day 修正的儒略日
modified Lambert conformal chart 改良兰伯特投影地图;改进的兰伯特正形海图
modified Lambert conformal projection 改良兰伯特正形投影;改进的兰伯特正形投影
modified length 计算长度
modified letter 变形字体
modified lignosulfonate 改性木质素磺酸盐
modified lignosulfonate flocculation 改性木质素磺酸盐絮凝
modified line 修正线;变更线
modified linear polyethylene 变性支链聚乙烯
modified liquid 修正液
modified loess 次生黄土
modified log odds 修正对数成败优势比
modified longwall system 变相长壁开采法
modified loose lay 修改松铺(法);修改干砌(法)
modified mass 修正质量
modified mean 修正平均值;修改平均数
modified mean square successive difference 修正均方连续差
modified Mercalli intensity scale 修正的麦卡利地震烈度表
modified Mercalli scale 修订的麦卡利地震烈度表
modified method of Hassan and Singh 哈森和辛格修正法
modified-modified frequency modulation 改进的改进调频制
modified natural resin 改性天然树脂
modified Newton method 改进牛顿法
modified normal distribution 修正的正态分布
modified objective function 修正的目标函数
modified oleoresinous paint 改性油性树脂漆
modified Ostwald's visco(si)meter 改良的奥斯特瓦尔德黏度计
modified Parshall flume 改良的巴歇尔量水槽
modified paste of pineneedle 改性松针浸膏
modified pedigree method of selection 改良的系谱选择法
modified pendulum gyroscope 摆修正陀螺仪
modified performance level 修正性能等级
modified phenol formaldehyde resin 改性酚醛树脂
modified phthalic resin 改性邻苯二甲酸醇酸树脂
modified plastics 改性塑料
modified polyamine fibre 改性聚酰胺纤维
modified polycarbonate 改性聚碳酸酯
modified polyconic(al)projection 改良多圆锥投影
modified polyepoxysulfosuccinic acid 改性聚环氧磺酸基琥珀酸
modified polyimide 改性聚酰亚胺
modified polymerized ferric sulfate 改性聚合硫酸铁
modified polyphenylene oxide 改性聚苯醚
modified polypropylene biofilm carrier 改性聚丙烯生物膜载体
modified Portland cement 改性硅酸盐水泥;改性波特兰水泥;改良的波特兰水泥
modified precision approach radar 改进型精密进场雷达
modified prismoidal method 修正的平截头棱锥体法
modified Proctor 修正普罗克托密度试验法
modified Proctor density test 改进的普罗克托(土壤)击实试验;改进的普罗克托(土壤)密实度试验
modified Proctor's compaction test 修正普罗克托击实试验
modified Proctor test 重型击实试验;修正击实试验
modified product 变形产品
modified radiation 变频辐射
modified radiation of wavelength 波长已变辐射
modified rayon 变性黏胶;变性人造丝
modified rectorite complex flocculant 改性累托石复合絮凝剂
modified refractive index 修正折射指数
modified regressor 修正的回归自变量

modified resin 改性树脂;改良树脂
modified Reynolds number 修正雷诺数
modified roll mechanism 差动机构
modified rosin 改性松香
modified rosin sizing agent 改性松香施胶剂
modified rubber 改良橡胶;改性橡胶
modified scheme 修改方案
modified scissors truss 改良剪式桁架;改良(的)斜撑桁架;剪刀撑桁架
modified seasonal index 修正的季节指数
modified separation ratio 改良的分离比值
modified sequencing batch reactor 改良型序批间歇式反应器
modified short-bond 修正短路棒
modified silicate cement 改良硅酸盐水泥
modified silicone resin 改性有机硅树脂
modified silicone sealant 改性硅酮密封膏
modified silumir 改性硅铝明合金
modified simplex method 修正的单纯方法
modified sludge aeration 改良污泥曝气
modified smallpox 变形天花
modified soda 改性苏打
modified soil 改良土
modified soil pile 改性土桩
modified solid wood 变性的实心木;改良的实心木
modified stair(case)wave 变形阶梯波
modified staple fibre 变性短纤维
modified starch 改性淀粉
modified starch flocculant 改性淀粉絮凝剂
modified starch-polyamine complex 改性淀粉—聚胺复合物
modified starch production wastewater 改性淀粉生产废水
modified-stiffness 改正刚度;修正劲度
modified straw 改性秸秆
modified straw-aluminum salt combined flocculant 改性秸秆—铝盐复合絮凝剂
modified successive over relaxation 修正逐次超松弛法
modified swimming pool 改良游泳池
modified systematic sampling 修正等距抽样
modified system of flexible exchange rate 修正浮动汇率制
modified system of normal equations 修正的正规方程组
modified temperature difference factor 温差修正系数
modified test 改进试验
modified tetanus 局限性强直
modified titanium dioxide 改性二氧化钛
modified treatment 调质处理
modified trenchstack silo 半地面青储[贮]窖
modified two-bath process 变型二浴法
modified umbrella type 半伞式
modified uniform liability system 修正统一责任制
modified University of Capetoen process 改进型开普敦大学工艺
modified urea-formaldehyde resin quanternary ammonium salt 改性脲甲醛树脂季铵盐
modified value 修改值;修正值
modified Van Dorn water sampler 改良范多恩采水样器
modified velocity 修正速度;修正流速;改正速度;改正流速;调整流速
modified version of fixed model 修正的固定模型变形
modified wood 改性木材;变性木材
modified wool fiber 改性羊毛纤维
modified zeolite 改性沸石
modified zeolitized fly-ash 改性粉煤灰沸石
modifier 修饰因子;修改量;变址量;改良剂;改变装置;调整剂;调剂;促进剂;变址数;变性剂;变换装置
modifier area 修改区域
modifier command 变址命令
modifier register 变址寄存器;变数寄存器
modifier spark advance 发火提早装置
modifier storage 变址数存储器;变址寄存器;变址存储器
modify 修改;更改
modify address 变址
modifying agent 改良剂;改性剂;变性剂;质量改善剂;调整剂
modifying asphalt 改性沥青

modifying factor 修饰因子
modifying oxide 调整氧化物
modifying processing 修饰性处理
modify one's registration 变更登记
modify price 调整价格
modillion 檐口托饰;飞檐托;托饰;飞檐下悬臂石
modinature 建筑线脚的排列和分布;建筑模型的排列和分布
moding 模的;模变
modliboyite 云橄黄煌岩
modulability 调节能力
modular 组件化;制成标准组件的;积木的;模数的;模块化的;模块(的);模的;微型组件组成
modular air handling unit 组合式空气调节机组
modular algebra 模数代数;模代数
modular all-relay interlocking 组闸式电气集中联锁
modular angle 模角
modular architecture 积木结构;模块结构
modular arithmetic 模运算
modular automated container handling 积木式集装箱自动装卸;模数化集装箱自动装卸
modular automated container handling system 定型自动化集装箱装卸系统
modular automatic container handling 模数化自动化集装箱装卸
modular automatic container handling module 模数化组件
modular automatic container handling portainer 模数化自动化集装箱装卸桥
modular azimuth positioning system 模方位角定位系统
modular block 组闸;组匣
modular brick 模数尺寸砖;模砖;模数砖
modular building unit 模数(化)砌块;标准建筑单元;标准化砌块
modular CAD system 模数CAD系统
modular check 模数校验;模数检验
modular clay brick 符合模数尺寸的黏土砖
modular component 模数单元;模数化部件
modular computer 模块化计算机
modular concept 积木式设计原理;积木式概念;模块式概念
modular connection 标准接头
modular connector 组合式接插件
modular constant 模常数
modular construction 用标准构件组装;模数结构;积木式结构;单元组合结构;构件组合结构;定型结构;单元结构;组件结构;组合式结构;模数化施工(构造);模数化构造;模块结构;部件结构
modular construction method 模式化施工法;单元结构法;标准组件结构法
modular construction unit 模数化构件
modular control 模块控制
modular converter 转换模块;模块化转换器
modular coordinating 模数协调法;模数协调
modular coordination 模数制;模数协调(法);模数配合;标准量测系统
modular crane 模数化起重机;标准构件起重机
modular curtain wall 定型悬墙
modular decomposition 模块分解
modular design 组件结构;典型结构;模数法设计;组合式信号机构;组合设计;积木式设计;积木化设计;模数设计;模块化设计;定型(部件组合)设计;典型设计
modular design method 模数法设计;模数设计法;定型设计法
modular deviation 标准偏离;标准偏差
modular dimension 模数量纲;模数尺寸
modular drafting 标准制图
modular draghead 模块耙头
modular duplex milling machine 复式组合机床
modular effect 模量效应
modular electronics 标准组件电子设备
modular element 模数化部件
modular expansion 模块扩充
modular field 模域
modular form 模形式
modular format 标准形式;标准规格;标准格式
modular function 模数函数;模函数
modular functional 模泛函
modular function theory 模函数论
modular furniture 组合家具;模式化家具
modular gas turbine power plant 组装式燃气轮

机装置
modular grid 模数网格
modular group 模群
modular home 预制模件住宅
modular horizontal floor mill 卧式落地组合铣床
modular house 模数制房屋
modular housing 标准住房;积木式房屋;模数制住房;模数制房屋
modular housing part 定型房屋构件
modular ideal 模理想
modular inequality 模不等式
modular instrument system 仪器调节系统;积木式仪器系统
modular integrated utility system 模块式综合公用事业系统
modular invariant 模不变式
modularity 积木性;模型性;模块性;模块化;模块度;模件性;模化程度;调制率
modularization 积木化;模数化;模块化
modularized 模块化的
modularized circuit 模块化电路
modularized computer 积木式计算机
modularized hardware 模块化硬件
modularized program(me) 模块化程序
modularized system 模数制;模量制
modular jack 模数插口;标准组合插口
modular laboratory computer 模块化实验室计算机
modular lattice 模格
modular law 模律
modular length 定型长度;定型尺度
modular line (模量参考系统中的)一条线
modular machine components 组合机床的组件
modular majorant 模强函数
modular manifolding systems 组件式歧管系统
modular masonry unit 砌块模数单位;模数砌块;模数砌筑单位
modular matrix 模矩阵
modular measure system 标准化程度的设备;定型设备;标准系统;模数制
modular member 模数构件
modular method 模数法;标准化方法
modular micro-program(me) machine 模块化微程序机
modular multilevel sampler 模块式多层采样器
modular multispectral scanner 模式多谱段扫描仪
modular-n adder 模数 n 加法器
modular operator 模算子
modular opticaltest instrument for fiber 模块化光纤光学测试仪
modular organization 积木式结构;模块(式)结构
modular plane 模数平面
modular planning grid 模数规划格
modular platform 组装平台
modular point (模数参考系统中的)一个点
modular processing system 模块处理系统
modular program(me) 定型程序
modular programming 模数程序设计;模块(化)程序设计
modular proportion 柱半径倍数比;模数比
modular pulse generator 模块化脉冲信号发生器
modular rack 组闸架;组匣架
modular radar 模式式雷达
modular ratio 比模量;模量比;弹性模量比;弹率比
modular-ratio design 模量比法设计
modular ratio of concrete and steel 混凝土和钢筋的弹性模量比
modular redundancy 模块冗余度
modular reference grid 模格;标准参考格栅;标准参考框格
modular reference system 标准参考系统;模量参考制;模数参考制
modular representation 模(数)表示
modular resolution 模分解
modular ring 模环
modular satellite 模块式卫星
modular semi-ordered linear spaces 模半序线性空间
modular size 模数大小
modular sofa 组合沙发
modular space 模数空间;模空间
modular-stack-type wall system 模数叠合式墙体(系统)
modular structure 积木结构;模数结构;模块(化)结构

modular structure microcomputer 模块结构式微型计算机
modular substitution 模变换
modular supercomputer 模块化巨型(计算)机
modular supercomputer architecture 模块化巨型机体系结构
modular surface 模面
modular surveillance radar 模块式监视雷达
modular system 模数制;组装系统;积木系统;积木方式;模数系统;模块体制;模块体系;模块(化)系统;单元组合系统;单元化设计;标准系统
modular system program(me) 积木式系统程序
modular technologic(al) line 示范工艺线
modular theory 模理论
modular tile 标准瓦
modular tooling system 模块式工具系统
modular tube 束筒;成束筒
modular type kiln 组装式窑;装配式窑
modular unit 模数化砌块;模数化构件;可互换标准件;模数化构件;模件;标准化砌块;多分格池
modular water treatment plant 组装式给水处理厂
modular width 标准宽度
modular zone 模数带
modulate 调制
modulated amplifier 已调放大器;受调放大器;被调放大器
modulated amplifier valve 被调制放大管
modulated carrier 调制载波
modulated continuous wave 已调连续波;已调等幅波
modulated control 模数化控制
modulated current 已调(制)电流
modulated dimension 模数化尺寸
modulated exposure 调整曝光
modulated light 调制光
modulated light beam 调制光束
modulated light source 调制光源
modulated oscillator 受调振荡器
modulated pulse amplifier 脉冲调制放大器;被调制的脉冲放大器
modulated quantity 调制量
modulated receiver 调制接收机
modulated signal 已调制信号
modulated sinusoid 已调正弦波
modulated spectrum 调制光谱
modulated subcarrier technique 调制副载波法
modulated time signal 调制时号
modulated voltage 已调电压
modulated wave 已调(制)波;调制波;调节波;调幅波
modulated wave amplifier 已调波放大器;波调制放大器
modulating action 调制作用
modulating amplifier by variable reactance 可变电抗调制放大器;变抗调制放大器
modulating audio frequency 调制音频
modulating burner 可调节燃烧器
modulating characteristic 调制特性
modulating choke 调幅扼流圈
modulating coefficient 调变系数
modulating control 调幅控制
modulating control system 调制控制系统
modulating current 调制电流
modulating equipment 调制设备
modulating function 调制函数
modulating grid 调制栅
modulating oscillator 调制振荡器
modulating pedal 控制踏板;操纵踏板
modulating pulse 调制脉冲
modulating signal 调制信号
modulating thermostat 调制式恒温器
modulating tube 调制管
modulating valve 调制管;调幅阀
modulating wave 调制波
modulation 确定古典柱型比例;调制;调幅
modulation alarm 调制警告
modulation amplifier 调制放大器
modulation balance 调制平衡
modulation band 调制频带
modulation bandwidth 调制带宽
modulation broadening 调制加宽
modulation capacity 调制能力
modulation code 调制码
modulation coherence 相干调制

modulation coil 调制线圈
modulation collimator 调制准直器
modulation component 调制分量
modulation condition 调制状态
modulation controller 调制控制器
modulation converter 调制变换器
modulation crest 调制峰值
modulation crystal 调制晶体
modulation defocus(s)ing 调制散焦
modulation degree 调制度
modulation demodulation system 调制解调方式
modulation depth 调制深度;调制度
modulation deviation 调制位移
modulation device 调制设备
modulation distortion 调制失真;调制畸变
modulation distortion characteristic 调制失真特性
modulation divider tube 分频管
modulation efficiency 调制效率
modulation efficiency factor 调制效率系数
modulation element 调制元件
modulation eliminator 解调器;调制消除器
modulation envelope 调制包络线
modulation error 调制误差
modulation factor 调制因素;调制因数;调制系数;调制度;调幅度
modulation factor meter 调制系数测试计
modulation frequency 调制频率
modulation frequency feedback 调制频率反馈
modulation frequency limitation 调频极限
modulation frequency ratio 调制频率与载频的比
modulation frequency response 调制频率响应
modulation function 调制函数
modulation gap 调制间隙
modulation generator 调制振荡器;调制发生器
modulation hum 调制哼声
modulation index 调制指数
modulation indicator 调制指示器
modulation interferometer 调制(型)干涉仪
modulation law 调制定律
modulation light source 调制光源
modulation loss 调制损耗
modulation measurement 调制度测量
modulation meter 调制计;调制度器;调制度测量仪;调制度测定器
modulation method 调制法
modulation mode 调制方式
modulation monitor 调制监视器
modulation noise 调制噪声
modulation-noise improvement 调制噪声的改进
modulation of accelerating voltage 加速电压调制
modulation of beam 束流调制
modulation of beam intensity 射线强度的调制
modulation of continuous wave 连续调制
modulation of fringes 条纹调制度
modulation of laser 激光调制
modulation parameter 调制参数
modulation pattern 调制图形
modulation percentage 以百分比计的调制深度;调制百分比
modulation period 调制周期
modulation point 调制点
modulation polarography 调制极谱法
modulation power 调制功率
modulation product 调制积;调制分量
modulation range 调制范围
modulation rate 调制(速)率
modulation ratio 调制比
modulation rise 调制度增高
modulation scheme 调制方案
modulation signal 调制信号
modulation spectroscopy 调制光谱学
modulation spectrum 调制光谱
modulation standard 调制标准
modulation suppression 调制抑制
modulation system 调制方式
modulation technique 调制技术
modulation theorem 调制定理
modulation-tone factor 调制声因数
modulation transfer factor 调制传递系数
modulation transfer function 调制转换函数;调制传递函数
modulation transformer 调制变压器
modulation tube 调制管
modulation type 调制类型

modulation type compensation 调制型补偿
modulation type multiplier 调制型乘法器;调制式乘法器
modulation type switched transponder 调制变换转发器
modulation wave 调制波
modulation waveform 调制波形
modulation width 调制宽度
modulation with a fixed reference 固定参考调制;带固定基准的调制
modulator 调制器;调整器;调节器;调幅器
modulator approach 调制方法
modulator band electric(al) system out 调幅器带通滤波器输出
modulator band filter 调幅器带通滤波器
modulator band filter in 调幅器带通滤波器输入
modulator band filter out 调幅带通滤波器输出
modulator cabinet 调制器箱
modulator carrier 调制器载频
modulator condenser 调制器电容器
modulator control signal 调制控制信号
modulator crystal 调制器晶体
modulator-demodulator(unit) 调制解调器
modulator divider 调制器分频器
modulator driver 调制激励器
modulator element 调制器元件
modulator firing 调制器启动
modulator function 调制器作用
modulator glow tube 调制发光管
modulator grid 调制栅
modulator linearity 调制器线性
modulator multiplier 调制倍频器
modulator of source 源调制器
modulator-oscillator 调制器振荡器
modulator output voltage 调制器输出电压
modulator sensitivity 调制器灵敏度
modulator transmitter receiver 调制器—发射机—接收机
modulator tube 调制管
modulator valve 调制管
modulator voltage 调制器电压
modulatory 调制的
module 因数;组件;流量单位;机芯;模数;模式;模(块);定量装置;程序块
module attribute 模块属性
module board 模块板
module body 模块体
module-built block 标准大楼;按标准模数建造的街坊
module check 模块检查;按模检查
module circuit board 模块电路板
module complex library 模块复合库
module content 模块内容
module declaration 模块说明
module diameter 组件直径
module dissipation 组件功耗;模块功耗
module extender board 模块扩展板
module feature 模特性
module fixture 组件夹具
module frame 模帧
module homomorphism 模同态
module information 组件信息
module interaction 模块交互作用
module machine 模块化机器
module matrix 组件矩阵
module method 模数法;标准尺寸法
module-milling cutter 模数铣刀
module name 模块名
module object 模对象
module of a family of curves 曲线族的模
module of coboundaries 上边缘模
module of cycles 循环模
module of elasticity 弹性模数;弹性模量
module of gear 齿轮模数
module of homomorphisms 同态模
module of program(me) 程序组件
module of quotients 商模
module of resilience 回能模数;回弹系数
module of rigidity 刚性模量;刚性模数;刚性系数
module of telemetry equipment 遥测仪器舱
module oil 脱模油
module production 模块化生产
module specification 模块指明
module structure 模块构造

module tester 组件测试机
module testing 组件测试
module test program(me) 组件测试程序
module test set 模件试验装置
module theory 模理论
module training 模件培训法
module treatment unit 处理单元组件
module units 组装单元
module width 模件宽度
modulo 按模计算
modulo arithmetic 模运算
modulo check 模式校验
modulometer 调制表;调制计
modulo N check 按模数 N 校验
modulo-nine's checking 模九检验
modulo symbol 模符号
modulus 模(数);模量;[复]moduli
modulus at rapid deformation 速形变模量
modulus decay 衰减系数
modulus in elasticity in bending 曲弹性模量
modulus in shear 剪切模数;剪切模量;剪力模数
modulus in tension 拉伸模数
modulus method of groundwater runoff 地下径流模数法
modulus of a congruence 同余式的模
modulus of attenuation 衰减模数
modulus of bending 弯曲模量
modulus of bulk elasticity 体积弹性模数;体积弹性模量
modulus of complex number 复数模(量);复数的模数
modulus of compressibility 压缩性模数;压缩模量
modulus of compression 压缩模量;抗压弹性模数
modulus of congruence 同余模
modulus of continuity 连续性的模数;连续模
modulus of conversion 转换系
modulus of creep 蠕变模量;徐变模量
modulus of decay 衰减模量;衰变模量
modulus of deformation 形变模量;变形模量
modulus of dilatation 膨胀模量
modulus of discharge 流量模数;流量模量
modulus of distribution 分布模数
modulus of dowel bar reaction 传力杆反力模量
modulus of dowel bar support 传力杆支承模量
modulus of drainage 排涝模数
modulus of elastic compression 弹性压缩模量
modulus of elasticity 杨氏模量;弹性系数;弹性模数;弹性模量
modulus of elasticity after effect 后效弹性模量
modulus of elasticity due to shear 受剪弹性模量
modulus of elasticity in bending 弯曲弹性模量
modulus of elasticity in comperssion 压缩弹性模量
modulus of elasticity in compression perpendicular to grain 横纹抗压弹性模量
modulus of elasticity in direct stress 拉伸弹性模数
modulus of elasticity in shear 抗剪弹性模量;剪切弹性模量;剪切弹性模量
modulus of elasticity in static bending 抗弯弹性模量
modulus of elasticity in tension 拉伸弹性模量;抗拉弹性模数
modulus of elasticity of bulk 容积弹性模数;容积弹性模量
modulus of elasticity of wood 木材弹性模量
modulus of elliptic(al) function 椭圆函数的模
modulus of elliptic(al) integral 椭圆积分的模
modulus of elongation 延伸率;拉伸模量;伸长模量
modulus of erosion 侵蚀模数
modulus of fineness 细度模数;细度模量
modulus of flexibility 挠性模数
modulus of flexure 曲挠模数
modulus of flow 流量模数;流量模量
modulus of foundation 基础模数;基础模量
modulus of foundation support 地基支承模量
modulus of granularity 粒度模数
modulus of incompressibility 非压缩性模量;不可压缩性模数;不可压缩(性)模量
modulus of inertia 惯性模数;惯性模量
modulus of instantaneous elasticity 瞬时弹性模量
modulus of linear deformation 线性变形模量;线变(形)模量
modulus of logarithm 对数模量;对数的模

modulus of longitudinal elasticity 纵向弹性模量;纵弹性模数
modulus of non-convexity 非凸性模
modulus of periodicity 周期的模
modulus of plasticity 塑性模量
modulus of precision 精度模
modulus of pressing 压制模制数
modulus of reaction 反应模数;反力模量
modulus of reaction of soil 土壤反力模数;土壤弹性均压系数;土的弹性均压模量
modulus of regularity 正则模
modulus of resilience 回弹系数;回跳模数;回能模量;回弹模量
modulus of resilience test 回弹模量试验
modulus of resistance 阻抗模数;抗力模量;电阻系数
modulus of rigidity 刚性系数;刚性模数;刚性模量;抗剪模量
modulus of road metal 道砟模量值
modulus of runoff 径流模数
modulus of rupture 弯折模量;极限强度;挠曲强度;挠折模量;破裂系数;弯曲极限强度;断裂模数;断裂模量
modulus of rupture in bending 抗折断模量;弯曲破坏系数;弯曲破坏模量;弯曲断裂模量;弯裂模量
modulus of rupture in torsion 扭转断裂模量;扭转的破坏模量;扭裂模量
modulus of section 剖面模量;截面模量;截面模量;剖面模数;断面系数;断面模数;断面模量
modulus of sediment yield 产沙模数
modulus of shear 剪切模量
modulus of shear deformation 剪切变形模量;剪切模量
modulus of shearing 抗压模数;抗压模量;剪断模数
modulus of shear resilience 剪切回弹模量;剪切冲击忍度;剪力回弹模量
modulus of silt discharge 输沙量模数
modulus of sliding movement 滑动模数
modulus of soil compression 土的压缩模量
modulus of soil deformation 土的变形模量
modulus of soil dynamic(al) elasticity 土的动弹性模量
modulus of soil elasticity 土的弹性模量
modulus of soil reaction 土反力模数
modulus of soil static elasticity 土的静弹性模量
modulus of specific compression 单位压缩模数
modulus of subgrade 基床模量
modulus of subgrade reaction 地基反力模量;地基反力系数;基床反力模量
modulus of subgrade reaction for pavement design (路面设计的)地基土反应模量
modulus of torsion 扭转模量;扭转弹性模数;扭弯模量;扭剪(切)模量
modulus of torsion(al) elasticity 扭转弹性模量
modulus of torsion(al) shear 扭切模量;扭剪(切)模量
modulus of toughness 韧性模量;韧度模量
modulus of transverse elasticity 刚性模量
modulus of traverse elasticity 横弹性模数
modulus of volume 体积弹性模量;体积模量
modulus of volume change 体积变化模量;体变模量
modulus of volume deformation 体积变形模量
modulus of volume elasticity 体积弹性模数
modulus of volume expansion 体胀模量;体积膨胀模量
modulus of volumetric(al) elasticity 容积弹性模数;容积弹性模量;体积弹性模量;体积弹性模量
modulus of water glass 水玻璃模数
modulus principle 模原理
modulus ratio 模数比;模量比
modulus system 模数制
modulus-temperature curve 模量温度曲线
modulus tester 模数测定仪
modulus theory 模理论
modulus-to-density ratio 模量密度比
modulus-versus-temperature curve 模量温度曲线
modulus-weight ratio 模量重量比
modumite 霓辉斜长岩
modus 方法
modus operandi 做法;运用法
modus ponens 演绎推理
modus vivendi 暂行协定;妥协;权宜之计

moela 向斜状抬升高原
moellen 油鞣余物
moellon 碎石工（填塞墙缝）；圬工墙内的填料；乱石砌体；油鞣废油；乱石圬工
Moentjang tina 桐油中毒
mofette 碳酸喷气孔
moffrasite 水锑铅矿
mog 磨细筛目
mogas 车用汽油
Mogden formula 毛登公式
Mogen David 六角星（犹太教的六芒星形标志）；大卫王之星（犹太教的六芒星形标志）
Mogenson screen 莫根森筛
Moghrebin architecture 北摩尔族建筑
Moghrebin minaret 北摩尔族伊斯兰教尖塔
moghul architecture 莫卧儿建筑（印巴建筑）
mogote 小块灌丛；密生灌丛；灰岩残丘；单笔孤峰
mogul base 大型灯座；大型电子管底座；大电灯座
mogul cap 大型灯头
mogul lamp-holder 大型灯座
mogullizer 真空浸渗设备
mohair 安哥拉山羊毛
Mohammedan architecture 伊斯兰教建筑；穆罕默德建筑；〈旧〉回教建筑
Mohammedan blue[Chinese blue] 回青（陶瓷色料）
Mohammedan calendar 伊斯兰教历；回历
Mohaupt effect 聚能效应
mohavite 八面硼砂
mohaw-algodonite 杂砷铜矿
Mohawk babbitt alloy 莫霍克巴比特合金
mohite 穆硫锡铜矿
Mohlaman index 莫尔曼指数
mohn effect 雾号声音异常传播
Moho boundary 莫霍面边界
Moho contour map 莫霍面等深线图
Moho depth 莫霍面深度
Moho-discontinuity 莫霍面等深线
Mohole 莫霍钻；莫霍（面）钻探；超深钻
Mohole drilling 超深钻井
Mohorovicic discontinuity 莫霍面；莫霍洛维奇不连续面；莫霍间断面
Mohorovicic discontinuity hole 超深钻；莫霍钻；莫霍面钻探
Moho's discontinuity 莫霍不连续面
Moho surface 莫霍面
Mohr-Caqout envelope 莫尔—卡柯脱包络线
Mohr-Coulomb criterion 莫尔—库仑准则
Mohr-Coulomb equation 莫尔—库仑方程
Mohr-Coulomb failure criterion 莫尔—库仑破坏准则
Mohr-Coulomb failure envelope 莫尔—库仑滑落包络线
Mohr-Coulomb law 莫尔—库仑定律
Mohr-Coulomb model 莫尔—库仑模型
Mohr-Coulomb shear strength parameter 莫尔—库仑剪切强度参数
Mohr-Coulomb soil model 莫尔—库仑土模型
Mohr-Coulomb strength envelope 莫尔—库仑强度包线
Mohr-Coulomb theory 莫尔—库仑理论
Mohr-Coulomb yield criterion 莫尔—库仑屈服准则
mohrite 六水铵铁矾
Mohr's balance 莫尔比重天平
Mohr's bending curve 莫尔弯曲曲线；莫尔变曲线
Mohr's bleaching process 莫尔练漂法
Mohr(s)circle 莫尔圆；莫尔图；莫尔应力圆；破坏包(络)线
Mohr's circle analysis method 莫尔圆分析法
Mohr's circle applicability 莫尔圆的适应范围
Mohr's circle diagram 莫尔应力圆图
Mohr's circle of strain 应变莫尔圆
Mohr's circle of stress 莫尔应力圆
Mohr's clamp 弹簧夹；莫尔夹
Mohr's condenser 莫尔冷凝器
Mohr's correction method 莫尔校正法
Mohr's cubic centimeter 莫尔立方厘米；莫尔毫升
Mohr's diagram 莫尔圆图；莫尔图
Mohr's envelope 莫尔包(络)线；莫尔包络面；莫尔包迹线；强度包(络)线
Mohr's failure envelope 莫尔破坏包络线
Mohr's graphical solution 莫尔图解法
Mohr's graphic representation 莫尔图示法
Mohr's litre 莫尔升（约为1.002公升）

Mohr's method 莫尔法（用硝酸银滴定测定卤素离子的方法）
Mohr's pinchcock 莫尔弹簧夹
Mohr's pinchcock clamp 莫尔弹簧夹
Mohr's rupture diagram 莫尔破坏图解
Mohr's rupture envelope 莫尔破裂包络线；莫尔破坏包络线
Mohr's salt 硫酸亚铁铵
Mohr's scale 莫尔硬度仪；莫尔硬度计
Mohr's scale number 莫氏标度值
Mohr's scale of hardness 莫氏硬度标；莫氏硬度值
Mohr's strain circle 莫尔应变圆
Mohr's strength theory 莫尔强度理论
Mohr's stress circle 莫尔应力圆
Mohr's stress diagram 莫尔应力圆
Mohr's theory 莫尔理论
Mohr's theory of failure 莫尔破坏理论
Mohr's theory of strength 莫尔强度理论
Mohr's titration 莫尔滴定法
Mohr yield criterion 莫尔屈服准则
Moh's hard ga(u)ge 莫氏硬度计
Moh's hardness 莫氏硬度
Moh's hardness scale 莫氏硬度计；莫氏（相对）硬度标
Moh's number 莫氏硬度值
Moh's scale 莫氏分级；莫氏标度；莫氏硬度标
Moh's scale number 莫氏硬度值；莫氏强度值
Moh's scale of hardness 莫氏硬度计；莫氏硬度（分级）表；莫氏硬度（标）；莫氏硬度分度法
Moh's strength theory 莫氏强度理论
Moh's taper 莫氏锥度
moiety 一部分；一半；分子的某部分；半股；半个
moil 杆头废料（残留在成型工具上的玻璃）；十字镐；做苦工
Moiré batten 波纹筘座
Moiré calender 木纹轧光机；波纹轧光机
Moiré effect 莫尔干涉效应
Moiré fringe 莫尔条纹；莫尔干涉条纹；波动光栅条纹；云纹条纹
Moiré fringe measuring system 莫尔干涉条纹测量系统
Moiré fringe technique 莫尔条纹法
Moiré grille 莫尔格子
Moiré holography 莫尔全息术
Moiré-holography method 莫尔全息照相法
Moiré linoleum 云斑油地毡；波纹油地毡
Moiré method 云纹(方)法
Moiré pattern 莫尔图形；莫尔干涉条纹图样；水纹图样；波纹图形；波纹
Moiré ronde 木纹波纹绸
Moiré technique 莫尔技术；莫尔法（用硝酸银滴定测定卤素离子的方法）
Moiré topograph 莫尔条纹图
Moiré topography 莫尔条纹测量；莫尔地形测量学
Moissan furnace 莫桑电炉
moissanite 碳硅石
moist 湿的；多雨的；潮湿的；潮的
moist adiabat 湿绝缘；湿绝热线；湿绝热；饱和绝热线
moist adiabatic change 湿绝热变化
moist adiabatic lapse rate 湿绝热直减率
moist adiabatic process 湿绝热过程
moist adiabatic temperature difference 湿绝热温度差
moist after-treatment（混凝土）湿处理；（混凝土）湿养护
moist air 湿气；潮湿空气；湿空气
moist air cabinet 保湿箱
moist air mass 湿气团
moist-air pump 湿空气泵
moist area underground water area 潮湿地区潜水区
moist ash free basis 恒湿无厌基
moist cabinet 湿治室
moist calendering 湿压光
moist catalysis 湿催化
moist chamber 雾室；湿治室；湿气室；保湿室
moist climate 湿润气候
moist closet 雾室；湿气室；保湿室
moist colo(u)rs 水彩画颜料
moist continental climate 湿润大陆性气候
moist cure 湿养生；（混凝土）湿养护
moist-cured 湿润处理的；保持湿润的；湿养护的（混凝土）

moist-cured concrete 湿治混凝土；湿养护混凝土
moist curing 湿治；雾室养护；湿养生；（混凝土）湿养护；湿润养生；湿润养护；湿润处理
moist-curing of concrete 混凝土湿养护
moist detector 凝露检测器
moist earth 湿土
moisten 沾湿；弄湿
moistened 弄湿的
moistened batch 湿配料
moistened with gasoline 用汽油润湿
moistener 加湿器；润湿器；湿润器
moistener pad 润湿器垫板
moistening 增湿(作用)；加湿；浸湿；润湿
moistening agent 增湿剂；加湿剂
moistening and mixing machine 湿拌机
moistening apparatus 增湿器；加湿器
moistening installation 增湿装置；给湿装置；加湿装置
moistening machine 给湿机；增湿机
moistening of air 空气加湿
moistening of mixture 燃烧混合物的润湿
moistening pressure roller 加湿加压滚筒；湿压力滚子
moistening roller 加湿滚筒；潮湿滚子
moist felt 耐火纤维湿毡
moist gradient 湿度梯度
moist heat 湿热
moist heat sterilization 湿热灭菌法
moist heat sterilizer 湿热灭菌法
moist heat transfer 湿热转换器
moist meadow 湿草甸
moist meadow black soil 湿草甸黑土
moist mineral matter free basis 恒湿无矿物质基
moist mixing 湿搅拌；湿混合；湿拌和
moistness 湿度；水分
moistograph 湿度仪
moist region 湿润（地）区
moist rodding 湿捣法
moist room 雾室；湿治室；湿养室；湿气室
moist room dampproofing 湿室防潮；湿气室的防潮；保湿室的防潮
moist sand 潮湿砂
moist screening 湿筛分
moist snow 湿雪
moist soil 滋润的土地
moist steam 湿(蒸)汽；潮湿蒸汽；饱和水蒸气
moist subtropic(al)climate 亚热带湿润气候
moist tamping 湿击法；湿捣法
moist tongue 湿舌
moist tropical forest 潮湿热带林
moisture 水汽；水分；湿气；潮湿水分；潮湿；潮气
moisture absorber 吸湿器
moisture absorption 吸水；吸潮；吸湿力；吸湿；水分吸收；湿气吸收；湿度吸收；潮气吸收
moisture absorption capacity 吸湿容量
moisture absorption test 湿度吸收试验
moisture-absorptivity 收湿性
moisture adjustrnent 水分修正；水分调整；湿度调整
moisture adjustment factor 水汽改正因子；水分修正系数
moisture ag(e)ing 湿度陈化
moisture air method 湿气干燥法
moisture analysis 水分分析
moisture analyzer 湿度分析器
moisture and ash-free 干燥无灰的
moisture and ash-free basis 无水无灰基准
moisture and flash over voltage curve 湿度对跳火电压特性曲线
moisture and fungus proof 防潮防霉
moisture apparatus 测湿器；湿度计
moisture as charged 装炉时的水分
moisture attraction 水分引力
moisture balance 水分平衡；湿度平衡
moisture barrier 隔汽层；防水层；隔水体；隔汽层；隔潮层；防潮层
moisture barrier material 阻水材料
moisture-bearing 含水的
moisture-box 水分测定箱
moisture can 含水量盒；土盒
moisture capacity 含水量；含水率；水分；容水度；湿容量；持水量；湿度
moisture-catcher 去湿装置；去湿槽；水分分离器
moisture cell 湿气盒；小湿室
moisture chamber 湿气室；保湿室

moisture charge 水汽输入;水汽含量;水分输入;水分含量
moisture coefficient 湿润系数
moisture-colding capacity 最高内在水分
moisture combined ratio 加水燃比;水分结合比;湿度结合比
moisture compensation 水分补偿
moisture compensator 含水量补偿装置
moisture condition 含水状态;雨雪量
moisture conservation 水分保持
moisture constant 水分常数
moisture conten meter 测湿计
moisture content 含水量;含湿量;含湿度;水汽含量;水分(含量);湿量;湿(的)含量;湿度
moisture content alarm 含水量警报器
moisture content by volume 体积含量率
moisture content control 水分控制;湿度控制
moisture content in crude oil 原油含水量
moisture content limit 界限含水量
moisture content measuring apparatus 含湿量测定仪
moisture content meter 湿度计;测湿器
moisture content of air 空气湿含量
moisture content of subgrade 路基含水量
moisture content of the atmosphere 大气中的水汽含量
moisture content of wood 木材含水率;木材含水量
moisture content profile 土壤深度与含水量的关系;土壤含水曲线
moisture-content ratio 重量含水率;湿度比
moisture content tester 测湿计
moisture control 湿度调节;含水量控制;水分控制;湿度控制
moisture corrosion test 湿度腐蚀试验
moisture cover 湿封皮;湿套;潮湿覆盖层
moisture crazing 吸湿开裂
moisture-cured 湿态固化
moisture cured polyurethane coating 湿固化氨基甲酸酯涂料
moisture-cured urethane 湿固化氨基甲酸酯;湿气固化聚氨酯
moisture curing 湿固化;潮气固化
moisture curve 湿度(变化)曲线
moisture cycle 水循环
moisture damage 湿损坏;潮气损坏
moisture deficiency 水分不足;湿度差
moisture deficit 水汽缺乏
moisture degree 湿度
moisture density 缺少水分;湿密度;湿量密度
moisture-density control 湿密度控制
moisture-density curve 击实曲线;含水量—密度曲线;湿度—密度曲线
moisture-density ga(u)ge 含水量—密度计
moisture-density nuclear ga(u)ge 含水量—密度同位素测定仪;同位素含量—密度测定仪
moisture-density relation 含水量—密度关系
moisture-density relationship 湿度—密度关系
moisture-density test 击实试验;含水量—密度(关系)试验;湿度—密度试验
moisture determination 含水量测定;湿度测定
moisture determination instrument 含水量测定器
moisture difference stress 湿差应力
moisture-distribution 水分分布;含水量分布
moisture-dry unit weight curve 含水量—干容重曲线
moisture eliminator 干燥器;去潮器;脱湿器
moisture entrance 透入湿气
moisture entry 湿入口;湿通路;湿引入
moisture equilibrium 水分平衡;湿度平衡
moisture equilibrium at dry side 吸湿平衡
moisture equivalent 含水当量;湿度当量;持水当量
moisture excess 余湿
moisture-excluding efficiency 湿气排除效率
moisture expansion 吸湿膨胀;湿胀;湿膨胀;潮湿膨胀
moisture expansion of bricks 砖的湿膨胀
moisture expeller 水分压榨机
moisture factor 湿润因素
moisture field 湿度场
moisture field deficiency 田间水分不足
moisture figure 水分数值
moisture film 水膜;湿膜
moisture film cohesion 水膜黏聚力;湿膜黏聚力;薄膜水凝聚力;薄膜水内聚力

moisture flux 涡流通量;水汽通量
moisture-free 干的;脱湿;不含水分的;除湿的;干燥的;不受潮的;不潮的
moisture-free basis 干燥基;无水基
moisture-free material 干材
moisture free weight 无水重量
moisture-friction stress (混凝土路面板因湿度变化而发生的)湿度—摩阻应力
moisture gain 散湿量
moisture gain from appliance and equipment 设备散湿量
moisture gain from occupant 人体散热量
moisture ga(u)ge 含水量测定器;含水量测定计;湿度计
moisture gradient 含水率梯度;水分梯度;湿气梯度;湿含量梯度;湿度梯度
moisture-hardening varnish 湿固化漆
moisture holding 保水
moisture-holding capacity 含水量;水容量;湿度;潮度;保水能力;持水能力;土壤持水量;保水量
moisture hydrodynamic(al)pressure 水动力压力
moisture impermeability 不透湿性
moisture index 含水指数;含水量指数;土壤含水指数;土的含水指数;水分指数;湿润指数;阿太堡极限指标;阿氏极限指标
moisture indicator 湿度计;水分指示计
moisture inflow 水汽入流(量)
moisture ingress 水分侵入
moisture integrator 湿度积分器
moisture in the analysis coal sample 分析水分
moisture in the soil 墒
moisture inversion 湿度逆增
moisture-laden 含水分的;饱水的
moisture-laden aggregate 饱水集料;饱水骨料
moisture-laden air 潮湿空气;饱和空气
moisture-laden chemical absorbent 化学吸湿剂
moisture-laden gas 含水煤气;含水的煤气
moisture limit 潮湿限度
moisture loss 水分效应;水分损失
moisture maximization 水汽极大化;水汽放大;水汽调整
moisture measurement 湿度测量
moisture-measuring apparatus 测湿仪
moisture meter 湿度计;水量计;水分测定仪;湿度仪;湿度测量计;测湿计
moisture migration 水分渗移;(混凝土)湿度传移;(混凝土)湿度转移;水分迁移;潮气迁移
moisture monitor 潮气检测器
moisture movement 湿度(造成的)胀缩;湿致胀缩;水分移动;水分变化;湿度胀缩;湿度移动;湿度变形
moisture observation in aeration zone 饱气带水分观测
moisture of absorption 吸湿性
moisture of air 空气湿度
moisture of capillary bond disruption 毛管连系破裂含量
moisture of the soil 土壤含水量
moisture only depth meter 地下湿度探测计
moisture penetration 透湿现象;水分渗透;水分渗入(深)度
moisture percentage 含水量百分数;含水量百分比
moisture permeability 透湿性;透湿性;渗湿性;渗潮性
moisture permeability index 水分渗透指数
moisture permeance 透湿系数
moisture pickup (土的)吸水量;带水
moisture potential 湿度势
moisture prevention 防湿;防潮措施
moisture probe 同位素含水量探测仪
moisture profile 含水量纵剖面图;湿度廓线
moisture-proof 水稳定的;不透水的;耐湿的;抗湿的;耐湿性;防湿的;防潮(湿)的
moisture-proof adhesive 防水胶黏剂;防潮胶黏剂
moisture-proof bag-maker 防潮纸袋机
moisture-proof board 防潮纸板
moisture-proof carton 防潮纸盒
moisture-proof coating 防潮涂料
moisture-proof envelope 防潮封套
moisture proofing 防湿处理;防潮;防水处理
moisture-proof lampholder 防潮灯座
moisture-proof lamp-socket 防潮灯座
moisture proofness 防湿性

moisture-proof paper 防潮纸
moisture-proof roof(ing)material 防潮屋面材料
moisture-proof roofing sheet(ing)防潮屋面板材;防潮屋面片材
moisture-proof tile 防潮砖
moisture protection 湿度防护;防潮
moisture protection plaster 防湿涂料
moisture quantity 湿量
moisturer 湿度器
moisture ratio 含水比
moisture recorder 湿度记录仪
moisture regain 回潮;吸湿性
moisture regime(n)湿度范围;湿度动态;含水状态
moisture regime(n)in aeration zone 饱气带湿度动态
moisture regulating function 水分调节作用
moisture-removal 除湿
moisture-removal device 去湿装置
moisture repellent 憎水的;抗湿的;防潮的
moisture resistance 湿阻;抗潮湿(性);耐湿性;耐潮性;防湿性;防潮性(能)
moisture-resistant 水稳定的;耐湿的;防潮的;抗湿的
moisture-resistant adhesive 耐水黏合剂;耐潮胶黏剂
moisture-resistant board 防潮石膏板
moisture-resistant chipboard 防潮刨花板
moisture-resistant fiberboard 防潮纤维板
moisture-resistant grade 防潮等级
moisture-resistant gypsum wallboard 防潮石膏墙板
moisture-resistant insulating material 防潮绝缘材料
moisture-resistant insulation 防潮隔层;防潮绝缘
moisture-resistant outer covering 防潮包裹层
moisture-resisting 抗湿的
moisture-resisting agent 防潮剂
moisture-resisting lacquer 防潮(油)漆
moisture-retaining property 保水性
moisture retardant 防潮剂;抗湿剂
moisture retention 吸湿性;保水(性)
moisture retention curve 持水曲线
moisture-retentive 吸湿的;保湿的
moisture-retentive chemicals 吸湿剂
moisture room 雾室;湿工室;保湿室
moisture sample 水分试样
moisture sampling 水分取样
moisture seal 湿密封;防潮密封
moisture section 含水量试片
moisture sensing probe 潮湿传感器
moisture sensitivity 对湿气的灵敏度
moisture sensor 湿敏元件
moisture separator 去湿器;气水分离器;水分分离器
moisture-set ink 湿固着油墨
moisture setting 风干处理;潮致凝固
moisture soil 潮土
moisture-solid relationship 含水量—密实度关系;水分—固体关系
moisture source 水汽源(地);水分来源;湿源
moisture stabilization 湿度稳定土法
moisture status 水分现状
moisture stop 防湿;防潮;阻隔潮湿;防潮密封
moisture storage 水分储存器
moisture storage and use 水分储[贮]藏及利用
moisture storage capacity 含水量;水分储存能力;储水量
moisture stress 水分应力(土内)
moisture suction 水分吸力
moisture supply 水汽供应;水分供给;湿源
moisture system 潮湿系统
moisture teller 水分快速测定仪;水分测定仪
moisture-temperature condition of subgrade 路基水温状况
moisture-temperature index 温湿指数
moisture tension 水分张力
moisture tester 湿度试验器;湿度计
moisture test(ing)湿度试验;含水量试验;透湿试验;水分试验
moisture-tight 不透水的;防湿的;防潮的
moisture tightness test 防潮试验
moisture transport 水分输送
moisture trap 气水分离器;脱湿器;水分收集器;水分捕集器;疏水器;除湿器;除湿器;防潮
moisture vapo(u)r permeability 透湿气性

moisture vapo(u)r sorption 潮气的吸附
moisture vapo(u)r transmission 湿气渗透
moisture vapo(u)r transmission rate 湿气透射率
moisture volume percentage 容积含水率
moisture warping 湿度变化引起的翘曲
moisture-warping stress 湿度—翘曲应力
moisture weight percentage 重量含水量;干重含水率
mokador 餐巾
mokaya oil 顶束毛榈油
mol 摩尔
mola 磨齿
molal average boiling point 摩尔平均沸点
molal concentration 摩尔浓度
molal configurational entropy 摩尔构型熵
molal heat content 摩尔热函数
molality 重量摩尔浓度;质量摩尔浓度;重摩(浓度)
molal quantity 摩尔数量
molal solution 摩尔溶液
molal specific heat 摩尔比热
molal weight 摩尔量
molar 臼齿;体积摩尔浓度
molar absorbance 摩尔光吸收
molar absorbancy index 摩尔吸光系数
molar absorption coefficient 摩尔吸光系数
molar absorptivity 摩尔吸收系数;摩尔吸收率;摩尔吸光系数
molar abundance 摩尔丰度
molar amount 摩尔量
molar charge 摩尔电荷
molar cohesion energy 摩尔内聚能
molar concentration 质量摩尔浓度;摩尔浓度;容积摩尔浓度;体积摩尔浓度
molar concentration of oxygen 氧的质量摩尔浓度
molar conductance 摩尔电导
molar conductivity 摩尔电导率
molar decadic extinction coefficient 十进摩尔消光系数
molar depression constant 摩尔冰点下降常数
molar depression of freezing point 摩尔凝固点降低
molar dispersion 摩尔色散
molar electrochemical energy 摩尔电化学能
molar electrochemical energy of electrons 电子摩尔电化学能
molar electrochemical potential energy 摩尔电化学势能
molar elevation constant 摩尔沸点升高常数
molar elevation of boiling 摩尔沸点升高
molar enthalpy 摩尔焓
molar enthalpy of reaction 反应摩尔焓
molar entropy 摩尔熵
molar extinction coefficient 摩尔消光系数
molar fraction 摩尔(分)数
molar free energy 摩尔自由能量
molar gas constant 摩尔气体常数
molar Gibbs energy 摩尔吉布斯能量
molar Gibbs energy of electrons 电子摩尔吉布斯能量
molar Gibbs energy of reaction 反应摩尔吉布斯能量
molar halogen concentration 卤素质量摩尔浓度
molar heat capacity 摩尔热容
molar heat content 摩尔热函;摩尔量热函数
molar internal energy 摩尔内能
molarity 摩尔浓度;体积摩尔浓度
molar mass 摩尔质量
molar number method 摩尔数法
molar olfactometry 摩尔臭气度
molar olfactory 摩尔臭味单位
molar optic(al) absorbance 摩尔光吸收
molar parachor 摩尔等张比容
molar polarization 摩尔极化度
molar property 摩尔性质
molar ratio 摩尔浓度;摩尔比
molar rotation 摩尔旋光度
molar solubility 摩尔溶解度
molar solution 摩尔溶液;体积摩尔浓度
molar specific heat 摩尔比热
molar surface energy 摩尔表面能
molar susceptibility 摩尔磁化率
molar teeth 臼齿
molar thermodynamic(al) energy 摩尔热力学能量
molar unit 摩尔单位
molar volume 摩尔体积
molar volume of reaction 反应摩尔体积
molar weight 摩尔量
molasse 磨拉石
molasse clastic basin 磨拉石碎屑盆地
molasse facies 磨拉石相
molasse formation 磨砾层;磨拉石建造【地】
molasse molecular cohesion 磨砾层分子内聚力
molasses association 磨拉石组合
molasses sandstone 磨拉石砂岩
molasses slop 糖蜜废液;糖浆废液
molasses tanker 糖蜜运输船;糖浆运输船
molasse-type sediment 磨拉石型沉积
molassoid formation 类磨拉石建造
molave 黄褐硬木(亦称菲律宾柚木)
mol-chlonic compound 分子氯化合物
mol concentration 质量摩尔浓度
moldavite 莫尔道熔融石;黑地蜡;绿玻陨石(产于波希米亚)
moldboard 型板;模板;推土机上犁板;样板
molder 成型工;制模工;造型工
moldovite 暗绿玻璃
molds reuseability 模型可再生性
mole 港口外堤;摩尔;平巷联合掘进机;突码头;突堤;隧道掘进机;堤道码头;防波堤
mole attachment 暗沟塑孔附加装置
mole ball 挖沟机拖球;暗沟塑孔球
mole-basis response 摩尔响应
mole blade 开排水沟刀具
mole channel 鼠道排水管;地下水道
mole coefficient 摩尔系数
molecular 分子的
molecular absorption band 分子吸收带
molecular absorption spectroscopy 分子吸收谱学
molecular absorption spectrum 分子吸收光谱
molecular accoustics 分子声学
molecular action 分子作用
molecular activity 分子活性
molecular addition compound 分子加成化合物
molecular adhesion 分子黏附;分子附着力
molecular adsorption 分子吸附
molecular aerodynamics 分子空气动力学
molecular aggregate 分子聚集
molecular airborne intercept radar 分子机载截听雷达
molecular amplifier 分子放大器
molecular amplitude 分子幅度
molecular anthropology 分子人类学
molecular apparatus 分子蒸馏器
molecular arrangement 分子排列
molecular aspect 分子概念
molecular association 分子缔合现象;分子缔合
molecular astronomy 分子天文学
molecular asymmetry 分子不对称性
molecular attraction 分子引力;分子吸引
molecular attraction constant 分子吸引常数
molecular axis 分子轴
molecular band 分子带
molecular bandpass amplifier 分子带通放大器
molecular beam 分子束
molecular beam apparatus 分子束装置
molecular beam magnetic resonance 分子射线磁共振
molecular beam maser 分子辐射脉泽
molecular beam spectrometer 分子束谱仪
molecular biochemistry 分子生物化学
molecular biological technique 分子生物学技术
molecular biology 分子生物学
molecular biophysics 分子生物物理学
molecular birefringence 分子致双折射
molecular bond 分子键
molecular catalysis 分子催化学
molecular centrifugal distortion 分子离心变形
molecular chain action 分子键作用
molecular chain length 分子链长
molecular channel 分子通道
molecular circuit 分子电路
molecular circuitry 分子电路学
molecular clock 分子钟
molecular cloning 分子扩增
molecular cluster 分子簇
molecular cohesion 分子黏聚力
molecular collision 分子碰撞
molecular collision dynamics 分子碰撞动力学
molecular colloid 分子胶体
molecular complex 分子配合物
molecular composition 分子组成
molecular concentration 分子浓度
molecular conductance 分子电导;分子态气导;分子流导
molecular conductivity 克分子电导率;分子电导率;分子导电系数;分子传导性
molecular conductor 分子导体
molecular configuration 分子构型
molecular conformation 分子形态;分子构象
molecular conjugation 分子共轭
molecular connectivity 分子连接性;分子接合度
molecular constitution 分子结构
molecular crystal 分子式结晶;分子晶体
molecular crystal lattice 分子晶格
molecular current 分子电流
molecular current hypothesis 分子电流假说
molecular death 分子死亡
molecular density 分子密度
molecular depression of freezing point 克分子凝固点降低
molecular design 分子设计
molecular diamagnetism 分子抗磁性
molecular diameter 分子直径
molecular differential weight distribution curve 分子量微分重均分布曲线
molecular diffusion 分子扩散;分子分散
molecular diffusion coefficient 分子护散系数
molecular diffusion rate 分子扩散速率
molecular diffusivity 分子扩散性;分子扩散系数
molecular dilution 分子稀释度
molecular dimension 分子大小
molecular dimer 分子二聚物
molecular dipole 分子偶极子
molecular disorder 分子无序
molecular disperse solution 分子分散溶液
molecular dispersion 分子色散;分子弥散;分子扩散;分子分散体
molecular dispersivity 分子分散性
molecular dispersoid 分子色散体;分子弥散体
molecular distillation 分子蒸馏
molecular distortion 分子畸变
molecular distribution 分子分布
molecular drag ga(u)ge 分子阻压规;分子阻尼真空规
molecular drag pump 分子拖曳泵
molecular dynamics 分子动力学
molecular eddy current 分子涡流
molecular effect 分子效应
molecular effusion separator 分子扩散分离器
molecular electronics 分子电子学
molecular electronic technique 分子电子技术
molecular electron microscope 分子电子显微镜
molecular electro-optics 分子电(子)光学
molecular elevation 分子沸点升高
molecular elevation of boiling-point 克分子沸点升高
molecular emission cavity analysis 分子发射空穴分析
molecular emission spectroscopy 分子发射光谱学
molecular energy 分子能
molecular energy distribution 分子能量分布
molecular energy level 分子能级
molecular engineering 分子工程
molecular engram 分子印迹
molecular environment 分子环境
molecular equation 分子反应式
molecular equilibrium 分子平衡
molecular evolution 分子进化
molecular excitation 分子激发
molecular exclusion chromatography 凝胶色谱(法);凝胶过滤色谱(法);凝胶过滤层析;分子排阻色谱法;分子排阻层析
molecular extinction coefficient 分子消光系数
molecular field 分子场
molecular field theory 外斯理论
molecular filter 膜滤器;分子(过)滤器
molecular fit 分子契合
molecular flow 分子流
molecular flow of gas 气体分子流
molecular fluorescence 分子荧光
molecular flux 分子通量

molecular force 分子力
molecular force field 分子力场
molecular formation 分子形成
molecular formula 分子式
molecular fragment 分子碎片
molecular freedom 分子自由度
molecular free path 分子自由程
molecular gas 分子气体
molecular ga(u)ge 分子压力计
molecular geometry 分子几何形状
molecular grating 分子晶格
molecular group 分子团
molecular heat 分子热
molecular heat capacity 摩尔热容;摩尔热函;分子热容
molecular hypothesis 分子假说
molecular integrated device 分子集成组件
molecular interaction 分子间(相)互作用
molecular ion 分子离子
molecular ionization potential 分子电离电位
molecular ion peak 分子离子峰
molecularity 分子状态;分子性(状)
molecularity of reaction 反应分子数
molecular jostling 分子冲撞
molecular kinetic energy 分子动能
molecular kinetic equation 分子运动方程式
molecular kinetics 分子动力学
molecular kinetic theory 分子运动学说
molecular laser 分子激光器
molecular laser spectroscopy 分子激光光谱学
molecular lattice 分子晶格
molecular lattice distortion 分子晶格畸变
molecular layer 分子层
molecular lead 分子铅
molecular lesion 分子损害
molecular level 分子水平
molecular line 分子线
molecular linkage 分子键
molecular luminescence 分子发光
molecularly imprinted ion exchange resin 分子印迹离子交换树脂
molecularly imprinted membrane 分子印迹膜
molecularly imprinted polymer 分子印迹聚合物
molecularly imprinted polymer membrane 分子印迹高分子膜
molecularly imprinted polymer sensor 分子印迹聚合物传感器
molecularly imprinted synthetic receptor 分子印迹合成受体
molecularly imprinting 分子印迹
molecularly imprinting technology 分子印迹技术
molecular magnet 分子磁铁
molecular magnetic rotation 分子磁致旋光
molecular mass 分子质量;分子物质
molecular mechanics 分子力学
molecular mechanism 分子机制
molecular melting 分子熔解
molecular memory 分子记忆
molecular microwave spectrometer 分子微波分光计
molecular model 分子模型
molecular modification 分子改造
molecular moisture capacity 分子吸湿量;分子含水量
molecular motion 分子运动
molecular movement 分子运动
molecular multi-electron spectroscopy 分子多电子光谱学
molecular multiplier 分子放大器;分子倍增器
molecular network 分子网络
molecular number 分子序数
molecular number in the unit cell 晶胞中分子数计算
molecular optics 分子光学
molecular orbital 分子规道方法;分子规道
molecular orbital model 分子轨道模型
molecular orbital theory 分子规道理论
molecular orientation 分子定向
molecular oscillator 分子振荡器
molecular oxygen 分子氧
molecular packing fraction 分子紧束分数
molecular paleontology 分子古生物学
molecular paramagnetism 分子顺磁性
molecular parameter of steranes 甾烷的分子参数
molecular parameter of terpanes 萜烷的分子参数
molecular percent 分子比
molecular photochemistry 分子光化学
molecular photoionization 分子光致电离
molecular physics 分子物理学
molecular polarizability 分子极化度
molecular polarization 分子极化(作用)
molecular property 分子性质
molecular proportion 分子比
molecular proposition 分子命题
molecular pump 分子泵
molecular radiation biology 分子辐射生物学
molecular radiobiology 分子放射生物学
molecular radius 分子半径;分锥半径
molecular ratio 分子比率;克分子量比
molecular ray 分子射线
molecular reaction 分子反应
molecular reaction kinetics 分子反应动力学
molecular reactivity 分子反应性
molecular rearrangement 分子重排(作用)
molecular reflection 分子反射
molecular reflection coefficient 分子反射系数
molecular refraction 分子折射度;分子折射
molecular refractivity 分子折射率;分子折射差率
molecular relaxation 分子弛豫
molecular repulsion 分子排斥
molecular rheology 分子流变学
molecular rotary power 分子旋光本领
molecular rotation 分子转动;分子旋转;分子旋光度
molecular rotation spectrum 分子转动光谱
molecular rotatory power 分子旋光度;分子旋光本领
molecular scattering 分子散射
molecular scission 分子断键
molecular screen 分子筛
molecular self-reinforce polymer 分子自增强聚合物
molecular separator 分子分离器
molecular sieve 分子筛
molecular sieve chromatography 凝胶过滤色谱(法);凝胶过滤层析
molecular sieve desiccant 分子筛干燥剂
molecular sieve drying 分子筛干燥
molecular sieve filtration 分子筛过滤法
molecular size 分子大小
molecular slip 分子滑动
molecular solution 克分子溶液;分子溶液
molecular specific heat 摩尔比热;分子比热
molecular spectroscopic(al) analysis 分子光谱分析
molecular spectroscopy 分子光谱学;分子波谱学
molecular spectrum 分子光谱
molecular speed measurement 分子速率测量
molecular state 分子状态
molecular statement 分子命题
molecular still 分子蒸馏器
molecular stopping power 分子阻止本领
molecular stratigraphy 分子地层学
molecular structural formula 分子结构式
molecular structure 分子结构
molecular structure amplitude 分子结构振幅
molecular structure of coal 煤(的)分子结构
molecular structurology 分子结构学
molecular stuffing 分子填料
molecular stuffing process 分子填料法
molecular surface energy 分子表面能
molecular symmetry 分子对称性
molecular temperature boundary layer 分子温度边界层
molecular tension 分子张力
molecular theoretical energy 分子理论能量
molecular theoretical surface tension 分子理论表面张力
molecular theory of magnetism 磁分子理论
molecular thermal diffusivity 分子热扩散系数
molecular tie 分子联结
molecular topology 分子拓扑
molecular transformation 分子变化
molecular transport 分子输运
molecular transposition 分子重排
molecular type analysis 分子类型分析
molecular vacuum ga(u)ge 分子真空计
molecular vacuum pump 分子真空泵
molecular vehicle 分子载体
molecular velocity 分子速度
molecular vibration 分子振动
molecular vibration-rotation spectrum 分子振—转光谱
molecular vibration spectrum 分子振动光谱
molecular viscosity 分子黏滞度;分子黏度
molecular volume 摩尔体积;分子体积
molecular water 分子水
molecular weight 分子重量;分子量
molecular weight cut-off 截留分子量
molecular weight determination apparatus 分子重量测定仪
molecular weight distribution 分子量分布
molecular weight distribution of organics 有机物分子量分布
molecular weight estimation apparatus 分子量测定器
molecular weight exclude limit 分子量排斥极限
molecular weight thermometer 分子量温度计
molecule 分子
molecule bind 分子结合
molecule change quantum state 分子变化量子状态
molecule lead 分子铅
molecule physics 分子物理学
molecules 分子学
molecule separator 分子分离器
molecule stated contaminant 分子态污染物
moleculo-dosiology 分子剂量学
mole drain 暗渠;地下排水沟;刚性黏土管沟;盲沟;排水暗沟;鼠道式地下排水道
mole drainage 暗管排水;地下排水工程;地下鼠洞式排水;地下穿洞排水;开沟排水;鼠道(式)排水(沟);暗沟排水
mole drainage works 地下排水工程
moleelectronics 分子电子学
mole fraction 克分子分数;摩尔分数
mole gas constant 摩尔气体常数
mole head 突堤外端;堤头(防波堤)
mole heat 摩尔比热
mole jetty 港堤
mole mining 全断面掘进采矿法
mole number 摩尔数
mole of stoichiometric(al) reaction 化学计量反应摩尔
mole per cent 摩尔百分数
mole pipe drainage 暗管排水
mole plough 开沟犁;挖沟犁;暗沟(塑孔)犁
mole-plow 开沟犁
mole press 暗沟塑孔器
moler 硅藻泥岩;矿石碎磨
mole ratio 克分子比;摩尔数比
mole ratio method 摩尔比(率)法
moler block 硅藻土块;硅藻土砖
moler brick 硅藻土砖
moler cement 硅藻土水泥
moler earth 硅藻土
mole-root (防波堤的)堤根(段);防波堤堤根
mole shaft (防波堤的)堤干
mole sieve drier 分子筛干燥器
mole-skin 软毛皮
moles per unit volume 单位体积摩尔
Mole thermopile 摩尔热电偶
moletron 分子加速器
moletronics 分子电子学
MO level ship MO 级船
mole volume 摩尔体积
mole with vertical face 直壁防波堤
Mole wrench 摩尔扳手
molfraction 克分子数;摩尔分数
moling 铺排水管;挖排水沟(用挖沟犁)
moling machine 制模机;隧道全断面掘进机;暗沟塑孔机
molion 分子离子
Molitor precise level(l)ing rod 莫利特精密水准尺
molking fault 压制件缺陷
Moll checker 椭圆格子砖;摩尔格子砖
Moller homogenizing process 莫勒均匀化法
mollerising 钢铁表面热浸铝法
mollerize 钢渗铝化处理
mollerizing 钢铁热浸铝法;钢铁表面热浸铝法
moll heat capacity 摩尔热容
mollic epipedon 松软表土层
mollielast 软弹性体

mollient 缓和剂
Mollier chart 摩尔图
mollifier 软化器；软化剂
mollifying 使软化
molliplast 软塑性体
mollisol 流动土层；活动层；软土；溶化土层；松软土（壤）
mollite 天蓝石
Moll thermopile 摩尔热电偶；摩尔温差电堆
molluscicide 软体动物清除剂
mollusks faunal province 软体动物地理区
mollusk shell 软体动物壳
Mollweide projection 摩尔维特投影
molly（埋在塑料及混凝土中的）膨胀螺栓座；（钢丝缆中的）散断股；螺栓销；螺栓插座
mollybdenosis 慢性钼中毒
molochite 煅烧高岭土；烧瓷土
Molodensky method 莫洛金斯基法
Molodensky theory 莫洛金斯基理论
Moloka fracture zone 莫洛凯破裂带
molozonide 分子臭氧化物
mol percentage 克分子百分数
mol ratio 摩尔比
mol stoichiometric(al) reaction 摩尔化学计量反应
molten 熔融的；熔化的
molten alloy 熔融合金
molten alumina 熔融氧化铝
molten alumin(i)um 液态铝；熔融铝
molten alumin(i)um filtration fabric 熔融铝水过滤布
molten basalt 熔融玄武岩
molten bath 熔体；熔池
molten-bath arc welding 熔池电弧焊；熔池焊
molten-bath gasification 熔浴法气化
molten-bismuth cooling 熔化铋冷却
molten bitumen 熔化石油沥青
molten bullion 熔融粗铅；粗铅熔体；粗金属锭熔体
molten cabonate process 融碳酸盐法
molten carbonate 碳酸盐融烧
molten-cast brick 熔铸砖
molten-cast refractory 熔铸耐火砖；熔铸耐火材料；熔融浇注耐火材料
molten caustic 熔融烧碱
molten cement 熔融水泥
molten charge 热装料
molten condition 熔融状态
molten copper 熔融铜
molten dust 熔融粉尘
molten electrolyte 熔化电解质
molten electrolyte cell 熔盐电解液电池
molten filler 热灌填(缝)料；热贯填(缝)料
molten fluoride fuel 熔融氟化物燃料
molten glass 熔融态玻璃；熔化玻璃；玻璃液
molten glass level 玻璃熔体液面高度
molten glass stirrer 玻璃液搅拌器
molten heat-transfer salt 熔盐热载体；熔盐传热剂；传热熔融盐
molten iron 熔融铁
molten iron blast furnace slag 化铁鼓风炉炉渣
molten lead 熔铅
molten lead quench 熔铅淬火
molten magma 液态岩浆；岩浆；熔岩浆；熔化岩浆
molten mass 熔体；熔融物；熔化物
molten material 熔料
molten matte 熔化冰铜
molten metal 已熔金属；金属液；熔融金属；熔化金属
molten metal bath 金属液槽
molten metal decladding 熔融金属去壳
molten metal dip coating 熔融金属浸涂
molten metal dy(e)ing 熔态金属染色
molten metal flame spraying 熔融金属火焰喷涂
molten metal flowmeter 熔融金属流量计
molten metal level indicator 熔融金属液位指示器
molten metal pump 金属液泵
molten mixture 熔融混合物
molten pad 熔融减摩垫
molten particle deposition 熔化颗粒沉积
molten pig 熔融铁
molten polymer 熔融聚合物
molten pool 熔池
molten rate of batch 生料熔成率
molten rock 熔融岩石
molten-rock casting 铸石
molten salt 熔盐

molten salt bath 熔融盐槽
molten salt bath treatment 熔盐浴处理
molten salt breeder reactor 熔盐增殖(反应)堆
molten salt chronopotentiometry 熔盐计时电位滴定法
molten salt circulating pump 熔盐循环泵
molten salt container 熔盐电解槽
molten salt convertor 熔盐转换反应堆
molten salt coolant 熔盐冷却剂
molten salt cracker 熔盐裂解炉
molten salt cracking 熔盐炉裂解
molten salt electrochemistry 熔盐电化学
molten salt electrocutting 熔盐电切削
molten salt electromigration 熔盐电迁移
molten salt extraction 熔盐萃取
molten salt fuel 熔盐燃料
molten salt growth 熔盐生长(法)
molten salt heater 熔盐加热器
molten salt method 熔盐法
molten salt pyrolysis 熔盐裂解
molten salt reactor 熔盐反应堆
molten salt reactor experiment 熔盐反应堆试验
molten salts 熔融盐类
molten salt sampler 熔盐取样器
molten salt voltammetry 熔盐伏安法
molten silicate 熔融硅酸盐
molten slag 液态熔渣；熔渣
molten slag electrolysis refining 熔融电解精炼
molten solder 熔融钎料；熔融焊料；软焊料；熔焊料
molten solvent 熔化溶剂
molten spelter tank 锌熔池
molten state 熔融状态；熔化状态
molten steel 钢液；钢水
molten surface 熔液表面
molten test sample 熔液试样；熔体试样
molten weld metal 熔融焊接金属
molten zinc 熔态锌；熔融锌
molten zone 熔区
molten zone moving mechanism 熔区移动机构
moluranite 多水钼铀矿
molybdate 钼酸盐
molybdate glass 钼酸盐玻璃
molybdate orange 钼橘红；钼铬橙
molybdate orange pigment 钼橙红颜料；钼橘红颜料
molybdate red 钼橘红；钼铬红
molybdate red epoxy primer 钼铬红环氧底漆
molybdate red pigment 钼铬红颜料
molybdena 氧化钼
molybdenic 钼的
molybdenic acid 钼酸
molybdenire ore 辉钼矿矿石
molybdenite 辉钼矿
molybdenit-wolframite-quartz vein formation 辉钼矿石英脉建造
molybdenum alloy 钼合金
molybdenum alloy resistance furnace 钼合金电阻炉
molybdenum analyzer 钼含量分析仪
molybdenum aperture 钼光阑
molybdenum bar 钼棒
molybdenum bar furnace 钼棒炉
molybdenum bearing austenitic steel 含钼奥氏体钢
molybdenum blue 钼蓝
molybdenum blue colo(u)rimetry 钼蓝比色法
molybdenum blue condensation-organic solvent dissociation spectrophotometry 钼蓝凝聚—有机溶剂溶解分光光度法
molybdenum blue-derivative spectrophotometry 钼蓝—导数分光光度法
molybdenum blue reaction 钼蓝反应
molybdenum boride 硼化钼
molybdenum carbide 碳化钼
molybdenum carbide ceramics 碳化钼陶瓷
molybdenum-chrome alloy steel 钼铬合金钢
molybdenum-chrome steel 钼铬合金钢
molybdenum concntrate 钼精矿
molybdenum-containing wastewater 含钼废水
molybdenum-copper 钼铜合金
molybdenum dichloride 二氯化钼
molybdenum die 钼质模具
molybdenum dioxide 二氧化钼
molybdenum dioxysulfide 一硫二氧化钼

molybdenum disilicide 二硅化钼
molybdenum disilicon resistor 二硅化钼电阻器
molybdenum disulphide 二氧化钼
molybdenum disulphide lubricant 二硫化钼润滑剂
molybdenum electrode 钼电极
molybdenum feed retainer 钼隔料网
molybdenum fertilizer 钼肥
molybdenum filament 钼丝
molybdenum foil 钼箔
molybdenum-foil diaphram ga(u)ge 钼箔薄膜压力计
molybdenum glass 钼玻璃；含钼玻璃
molybdenum-group glass 钼组玻璃
molybdenum heater 钼加热器
molybdenum hemipentoxide 五氧化二钼
molybdenum high speed steel 钼高速钢
molybdenum hydroxide 氢氧化钼
molybdenum lake 钼色淀
molybdenum-liner 钼衬里
molybdenum materials 钼材
molybdenum minerals 钼矿
molybdenum monoxide 一氧化钼
molybdenum nitride 氮化钼
molybdenum ore 钼砂；钼矿(石)
molybdenum oxide 氧化钼
molybdenum oxide fibre 氧化钼纤维
molybdenum-permalloy 钼镍铁导磁合金
molybdenum plate 钼板
molybdenum pollution 钼污染
molybdenum powder 钼粉
molybdenum-quartz vein 辉钼矿石英脉
molybdenum red 钼铬红
molybdenum reflector 钼反射器
molybdenum rod 钼杆；钼棒
molybdenum sesquioxide 三氧化二钼
molybdenum shield 钼屏
molybdenum silicide 硅化钼
molybdenum silicide ceramics 硅化钼陶瓷
molybdenum silicide furnace 二硅化钼电炉
molybdenum single crystal 钼单晶
molybdenum solder glass 钼焊接玻璃
molybdenum steel 钼钢
molybdenum supporting grid 钼承料网
molybdenum target 钼靶
molybdenum tetrachloride 四氯化钼
molybdenum trihydroxide 氢氧化钼
molybdenum trioxide 三氧化钼
molybdenum-vanadium steel 钼钒钢
molybdenum wire 钼线；钼丝
molybdenum wire furnace 钼丝炉
molybdenum wound furnace 钼丝炉
molybdic acid 钼酸
molybdic bromide 三溴化钼
molybdic hydroxide 氢氧化钼
molybdic ocher 水钼铁化矿
molybdoantimony anti-molybdenum blue method 钼锑抗钼蓝法
molybdoantimony anti-spectrophotometry 钼锑抗分光光度法
molybdoformacite 钼砷铜铅石
molybdomenite 白硒铅石
molybdophyllite 钼镁铅矿
molybdous bromide 二溴化钼
molydbate-reactive phosphorus 钼酸盐活化磷
Molykote 二硫化钼润滑剂
moly orange 钼铬橙
molysite 铁盐
mom and pop 家庭式的
mom and pop store 家庭式商店；夫妻店
Mo-mat 钼垫板
Mombasa gum 蒙巴萨鞘刃脂
Mombasa Harbo(u)r 蒙巴萨港(肯尼亚)
moment 动差；力矩
moment about point of support 线支点处力矩；支点处的力矩；对支点的力矩
moment about point support 支点弯矩
moment about the mean 中心力矩；中心矩
moment about the origin 原点矩
moment acting on the flange for gasket seating 法兰预紧弯矩；凸缘预紧弯矩
momental 惯量的
momental ellipse 惯量椭圆
momental ellipsoid 惯量椭球
moment allowance 许可力矩；允许力矩

momental vector 力矩矢量
moment area 力矩图面积;挠矩面积;弯矩面积
moment area method 力矩(图)面积法;弯矩面积法
moment arm 力矩臂;巨臂;矩臂
momentary action 瞬间作用
momentary connection 快速连接;瞬时接通
momentary-contact push button 瞬时按钮开关;瞬时接触按钮
momentary contact switch 瞬时接触开关
momentary discharge 瞬时流量
momentary efficiency 瞬时效率
momentary excess current 瞬时过电流
momentary fall of pressure 瞬时压力下降;瞬时压(力)降(落)
momentary fluctuation 瞬时波动
momentary high power effort 瞬时功率增加
momentary interruption 瞬时断路
momentary load 瞬时荷载;瞬时负载;瞬时负荷
momentary maximum allowable concentration 一次最高容许浓度
momentary minimum discharge 瞬时最小流量
momentary output 瞬时输出;瞬时容量;短时生产率
momentary overload(ing) 瞬时超载;瞬间超载;短时间过载
momentary peak discharge 瞬时洪峰流量
momentary pollution 瞬时污染
momentary pollution source 瞬时污染源
momentary power 瞬时功率
momentary rate 瞬时速率
momentary rise of pressure 瞬时增压;瞬时压力升高
momentary speed droop 暂态转速系数
momentary speed variation 暂态转速变化
momentary state 暂时状态;瞬态
momentary state analysis 瞬态分析
momentary value 瞬时值
momentary water-level 瞬时水位
momentary water stage 瞬时水位
moment at base 底部力矩
moment at column 作用于柱的力矩
moment at extremity of rod 杆端弯矩
moment at fixed end 固端弯矩;固定力矩;定端弯矩
moment at head 顶端弯矩
moment at support 支点弯矩;支承挠矩
moment axis 力矩轴;挠矩轴;弯矩轴
moment balance 力矩平衡
moment-balance method 力矩平衡法
moment bending chart 弯矩荷载图
moment buckling 力矩引起的屈曲
moment capacity 力矩载量;力矩容量
moment center 力矩中心;弯矩中心
moment coefficient 转矩系数;力矩系数;矩系数;弯矩系数;动差系数
moment coefficient of combination 系统力矩系数
moment compressing failure 弯压破坏
moment condition 力矩(平衡)条件
moment connection 刚性连接;刚结点;连续结点;抗弯联结
moment couple 偶矩
moment coverage of reinforcement 钢筋抵抗矩包络图
moment-curvature characteristic 弯矩曲率特性曲线;弯矩角变位特性曲线
moment-curvature diagram 力矩与曲率关系图;弯矩曲率图
moment-curvature law 弯矩曲率定律
moment-curvature relationship 力矩与曲率关系图;弯矩曲率关系
moment curve 力矩图;力矩曲线;挠矩曲线;弯矩曲线
moment-curve slope 力矩曲线斜率
moment derivative coefficient 力矩系数导数
moment-diagram 力矩图;弯矩图
moment distribution 弯矩分配;弯矩分布
moment distribution method 力矩分配法;分布法;弯矩分配法;弯矩分布法
moment due to elevator deflection 偏转升降舵产生的力矩
moment due to lift force 升力扭转力矩
moment due to vortices 旋涡产生的力矩
moment due to yawing 偏航引起的力矩
moment envelope 力矩包(络)线;弯矩包(络)线
moment equation 力矩方程(式);弯矩方程(式)
moment equilibrium 力矩平衡;弯矩平衡

moment estimator 矩估计量
moment flexure 挠曲受力筋;挠矩;弯矩
moment for designing flange 法兰计算弯矩;凸缘计算弯矩
moment-free 无力矩
moment-free tank 无力矩油罐
moment function 矩函数
moment generating 矩量生成
moment generating function 矩量生成函数;矩量母函数
moment gradient 弯矩梯度
moment iteration method 力矩迭代法
moment in extreme span 端跨中弯矩
moment-influence factor 力矩影响系数;弯矩影响系数
moment-influence line 力矩影响线
moment in roll 倾侧力矩
moment in yaw 偏航力矩
moment limit(ing) device 力矩限制器
moment load 弯矩荷载
moment load chart 弯矩荷载图
moment magnifier factor 弯矩增大系数
moment matrix 矩量矩阵
moment method 力矩法
moment of a family of curves 曲线族的矩
moment of area 面积矩
moment of area of transverse 横剖面积力矩
moment of area of water plane 水平面面积力矩
moment of a vector 向量矩
moment of bending 弯曲力矩;弯矩
moment of buoyancy 浮力矩
moment of couple 力偶矩
moment of deflection 挠曲受力筋;挠曲力矩;挠矩;弯矩
moment of deformation 形变力矩;变形力矩
moment of dipole 偶极子矩
moment of distribution 分布(力)矩
moment of elasticity 弹性力矩
moment of exposure 曝光瞬间
moment of flexion 弯曲矩
moment of flexure 弯曲力矩;弯矩
moment of flow 液流力矩
moment of force 力矩
moment of forces tending to capsize 倾斜力矩
moment of friction 摩擦力矩
moment of gyration 转动力矩;转动惯量;惯性矩;惯性动量;回转力矩
moment of impulse 冲量矩;冲击力矩
moment of inertia 转动惯量;惯性力矩;惯(性)矩
moment of inertia about equatorial axis 绕赤道轴转动惯量
moment of inertia about polar axis 绕极轴转动惯量
moment of inertia in yaw 偏航惯性矩
moment of inertia method 惯性矩法
moment of inertia of atmosphere 大气转动惯量
moment of inertia of core 核心转动惯量
moment of inertia of cross-section 横截面惯性矩
moment of inertia of gyrorotor 陀螺转子转动惯量
moment of load(ing) 加载力矩;负载矩
moment of magnet 磁矩
moment of mass 质量矩
moment of momentum 角动量;动量矩
moment of momentum equation 动量矩方程
moment of momentum theorem 动量矩定理
moment of overturning 倾覆力矩
moment of overturning resistance 抗倾力矩
moment of precession 进动力矩
moment of probability distribution 概率分布矩量
moment of resistance 阻力矩;抗力矩;内力矩;抵抗力矩
moment of resultant wind force coefficient 合成风力矩系数
moment of rotation 旋转力矩
moment of rupture 破损力矩;破坏力矩;断裂力矩
moment of shearing 剪断力矩
moment of span 跨中弯矩;跨内弯矩;跨矩
moment of sparking 点火瞬间
moment of spin 自旋矩
moment of stability 稳性力矩;稳定力矩
moment of stiffness 劲矩
moment of torque 扭矩
moment of torsion 扭转力矩;扭矩
moment of twist(ing) 扭转力矩

moment of velocity 动量矩
moment of volume 体积矩
moment of wind pressure 风压力矩
moment on body 作用在机体上力矩
moment operator 力矩运算子;力矩算符
momentous 重大的
moment pattern 弯矩图式
moment peak 力矩峰值
moment plate 加强板
moment pole 旋转中心;矩极
moment ratio 矩比
moment redistribution 力矩再分配;力矩再分布;弯矩重分配;弯矩重分布;弯矩再分配
moment reinforcement 抗弯钢筋
moment resisting connection 抗弯联结;抗弯结点
moment resisting frame 抗弯框架
moment resisting space frame 抗弯空间框架;空间刚架
moment restriction 矩量限制
moment-rotation-hysteresis loop 弯矩转角滞后回线
moment-rotation relationship 弯矩转角关系
moment scale 力矩比例尺
moment sequence 矩序列
moment shear ratio 弯矩剪力比
moments method 矩量法
moment spectrum 矩谱
moment splice 力矩拼接
moment strength 力矩强度
moment table 力矩表;测力矩台
moment taken about the point of fixation 固定力矩
moment tensor 力矩张量
moment test 力矩试验;挠矩试验;弯矩试验
moment to alter trim one inch 每英寸吃水差力矩
moment to change trimone centimeter 每厘米吃水差校力矩
moment to change trim per centimeter 每厘米吃水差平力矩
moment transmitting joint 固接结点
moment triangle 力矩三角形
momentum 动量;冲力
momentum coefficient 动量系数
momentum correction coefficient 动量修正系数
momentum correction factor 动量修正系数
momentum curve 动量曲线
momentum density 动量密度
momentum diffusion 动量扩散
momentum discontinuity 动量骤变
momentum drag 动量阻力
momentum effect 冲力作用;冲力效应
momentum energy vector 动能矢量
momentum equation 动量方程(式)
momentum exchange 动量交换
momentum exchange coefficient 动力传递系数
momentum flow 动量流
momentum flow vector 动流量矢量
momentum flux 动量通量
momentum force coefficient 动量力系数
momentum grade 动能坡度;动量坡度;动量坡度;动力坡道;动力坡度;冲力坡度
momentum gradient 动能坡度;动量梯度
momentum integral boundary layer equation 边界层动量积分方程
momentum integral method 动量积分法
momentum interchange 动量交换
momentum law 动量定律
momentum loss thickness 动量损失厚度
momentum method 动量法
momentum moment 动量矩
momentum of cyclic(al) motion 周期动量
momentum of development 发展的态势
momentum of electron 电子动量
momentum of the jet 喷流动量
momentum operator 动量算符
momentum plume 动力烟云
momentum principle 动量原理
momentum range 信号间隔
momentum redistribution 动量再分布
momentum segregation 动量分离法
momentum space 动量空间
momentum spectrum 动量谱
momentum starter 动量启动器
momentum survey 动量测量

momentum theorem 动量定理
momentum theory 动量原理;动量理论
momentum thickness 动量厚度
momentum thickness of boundary layer 边界层的动量厚度
momentum transfer 动量转移;动量转换;动量交换;动量传递;动量变换
momentum transfer coefficient 动力传递系数
momentum transfer method 冲量法;动量转移法
momentum transfer theory 动量转换理论;动量传递原理
momentum transport 动量输移;动量输送
momentum-transport hypothesis 动量输运假设
momentum-type mass flowmeter 动量式质量流量计
momentum wave function 动量波函数
moment vector 力矩矢量
monacid 一元酸
monadic Boolean operation 一元布尔运算;一目布尔运算
monadic Boolean operator 一元布尔运算符
monadic formula 单值公式
monadic indicant 一元指示符
monadic indication 一元指示
monadic operand 一元运算对象;一元操作数
monadic operation 一元运算;一目运算;单值运算;单值操作
monadic operator 一元(运)算符;一目算子;单值运算符;单一算符
monadnock 残山;残丘;残留山丘
monalbite 蒙钠长石;单斜钠长石
monarch 灰尘与气流分离器
monastery 僧侣住房;修道院(男);寺院
monastery cemetery 寺院墓地
monastery garden 寺院庭园;寺院园林
Monastery of Pure Compassion 净慈寺(杭州)
monastic architecture 寺院建筑
monastic building 寺院房屋
monastic cell 僧室;禅房;寺院小室
monastic cemetery 寺院墓地
monastic choir 寺院歌唱班席位;寺院唱诗班
monastic church 修道院教堂
monastic community 寺院社区;寺院群落
monastic hall 寺院会堂;寺院大厅
monastic house 寺院居住场所;寺院住宅
monastic kitchen 寺院厨房
monastic library 寺院藏经楼;寺院藏书楼
monastic quire 寺院歌唱班席位;寺院唱诗班
monastic recreation room 寺院文娱室;寺院娱乐室;寺院休息室
monastral blue 单星蓝;颜料酞青
monastral green 单星绿
monatomic gas 单原子气体
monaural 单耳的
monaural sound 单耳传声
monaxon 单轴骨针
monazite 独居石含量;独居石
monazite ore 独居石矿石
monazite sand 磷铈镧矿砂;独居石矿砂
moncheite 碲铂矿
monchiquite 沸煌岩;方沸磺煌岩
Moncrief's formula 蒙里克夫公式(用于设计低碳钢柱子)
Mondarine china 东北产瓷器
Mond gas 半水煤气;蒙德煤气
mondhaldeite 白榴闪辉斑岩
Mond process 蒙德法
Monel 镍铜锰铁合金;蒙乃尔合金
Monel flashing piece 蒙乃尔泻水板;蒙乃尔防雨板
Monel-lined 蒙乃尔合金衬里
Monel metal 高强度耐蚀镍铜合金;抗蚀金属;蒙乃尔(铜镍)合金;蒙乃尔高强度耐蚀镍合金;蒙乃尔白铜;镍铜锰铁合金
monetary aggregate 货币总额指标
monetary assets 货币资产
monetary benefit 货币效益
monetary capital 货币资金
monetary correction 财政改善措施
monetary crisis 货币危机
monetary facility 金融机构
monetary fund 货币资金
monetary funds of commercial enterprises 商业企业货币资金
monetary gain or loss 货币损益

monetary inflation 通货膨胀
monetary issue 货币发行
monetary liability 货币债务
monetary limitation 金额限制
monetary limit of liability 赔偿金限额
monetary loss 金钱损失
monetary loss on non-current monetary assets 非流动货币资产的货币亏损
monetary payments 货币支付
monetary payoff 货币报酬
monetary penalty 罚款处分
monetary policy 货币政策
monetary reform 币值改革
monetary saving 节约金钱
monetary stability 稳定币值
monetary system 货币制度;财政制度
monetary telephone 投币式公用电话机
monetary term 以货币表示的
monetary unit 货币单位
monetary value 经济价值;货币价值
monetite 三斜磷钙石
monetization of credit 信用货币化
money 金钱;货币;财帛
money advanced 垫款
money and valuables 财宝
money appropriated 已指定用途的款项
money asset 货币资产
money at call 通知拆放;随时可收回的借款;短期贷款
money at call and short term 可短期内收回的款项
money at long 长期贷款
money award 奖金
money-back guarantee 保证退款;退款保证
moneybags 财东
money bill 财政法案
money box 钱柜
money broker 代办短期借款经纪人;兑换商
money capital 货币资本
money capital audit 货币资金审计
money changer 兑换商
money circulate 货币流通
money claim 货币索取权;债权
money cost 货币成本
money cowrie 贝币
money crop 经济作物;商品作物
money dealer 兑换商
money debt 货币债务
money down 现付;现款;现金
money due 应付款
money economy 货币经济
money equivalent 现金等价物
money exchange's licence 外币兑换业执照
money for new product development 新产品试制费
money house 钱庄
money income 现金收入
money in trust 信托资金
money invested 股本;投资
money-lender 放债者
money-lending business 贷款业务;借贷业
money locker 钱柜
money-making proposition 赚钱计划
moneyman 投资者;财政专家;金融专家
money market 金融市场;货币市场
money market mutual funds 货币市场互助基金
money market rate 短期资金市场利率
money mobilization 资金筹措
money no object 不计报酬
money of account 记账货币
money of small denominations 小额货币
money on account 账上使用
money on call 暂借款;通知拆放;随时可收回的贷款
money on the wing 流动中资金
money order 汇票;汇款单
money order payable to bearer 凭票即付票人
money oriented housing allocation 住房分配货币化
money out of hand 无现款
money raising method 筹款方法
money rate 利息;利率
money received in advance 预收款
money refund 货币退回
money remaining 余款
money remittance 汇兑

money returned 退款
money returned of current term 本期货币回收
money rewards 现金酬劳
money sitting 静止中的资金
money spinner 依靠投机或高利贷发财的人
money stock 货币流通量;现金股票
money supply 货币供应量;纸币供给量;投放通货
money-time value 货币的时间价值
money value adjustment accounting 币值调整会计
money wage 现金工资
Monge's theorem 蒙奇定理
Mongol architecture 蒙古(式)建筑
Mongolia arc 蒙古弧形构造带
Mongolian linden 白皮椴
Mongolian oak 柞树;蒙古栎
Mongolian yurt 蒙古包
monial (门窗上的)直桱;框架立柱;门框竖桱;门窗的竖框
monic 首一
monica 机尾警戒雷达
Monic method 随机搜索法
monic polynomial 首一多项式;首项系数为一的多项式
Monimax 蒙尼马克斯软磁合金
monimolimnion 永滞层;(湖的)深部滞水层
monimolite 绿锑铅矿;丝锑铅矿
monism 一元论
monistic economic structure 一元经济结构
monistic normal distribution 一元正态分布
monitered automatic control system 监督自动控制系统
monitor 指示器;高压水枪;控制装置;监听设备;监听器;监视(器);监视盘;监控(器);监护(器);监督;监测装置;监测(器);通风顶
monitor algorithm 监督程序算法
monitor and control panel 监视器和控制面板
monitor array 监测网系
monitor bore 监测孔
monitor call 监督程序调用
monitor command 监督指令
monitor concept 监督程序概念
monitor console 监视控制台;监控台
monitor console routine 监督控制台例行程序
monitor control 监视控制;监督程序控制
monitor control dump 监督程序控制转储;监察控制转储
monitor counter 检验计数器
monitor curve 监视曲线
monitor deflector 水枪转向装置
monitor desk 监视台
monitor disc[disk] 检验器盘
monitor display 监视显示
monitored control system 监控系统;监督控制系统;监察控制系统
monitored drilling 仪表化钻进
monitored instruction 监视用指令
monitored mode 监督方式;监督状态
monitor elbow 监测器弯头
monitor filming 监视器屏幕摄影
monitor foil 箔检验器
monitor groundwater pollution 监测地下水污染
monitor groundwater quality 监测地下水质
monitor head 监听放音头;水枪压头
monitor hole 监测孔
monitoring 监视;监测;监听;监护;剂量测定;施工监督
monitoring aerial 监听天线;监控天线
monitoring amplifier 监听放大器;监测放大器
monitoring amplifier 监听用放大器
monitoring and controlling facility 监控设备
monitoring and information system 监视和智能系统
monitoring and intelligence system 监视和智能系统
monitoring and maitenance vehicle 检测维修车
monitoring audiometer 监听听力计
monitoring booth 监听箱;监听亭
monitoring building settlement 建筑物沉降观测
monitoring car 监测车
monitoring chart 监视图
monitoring circuit 监听电路;监视电路;监控电路
monitoring coil 检音线圈
monitoring compliance 监测符合

monitoring control 监控
monitoring control system 监控系统
monitoring cost 控制成本
monitoring coupon 监测试片【给】
monitoring criterion 监测基准
monitoring data 监测资料；监测数据
monitoring desk 质量监测台；监听台
monitoring device 监控装置；报警装置；监视设备
monitoring element 监控元件；输入信号转换器
monitoring equipment 监视设备；监控装置；监控设备；监测设备
monitoring equipment in tunnel 隧道监控设备
monitoring feedback 操纵反馈
monitoring frequency 监测频率；监测次数
monitoring groundwater pollution 监测地下水污染
monitoring heat exchange 监控换热器
monitoring indicator 监控指示器
monitoring information 监控信息
monitoring installation 监控装置；监测设备
monitoring instrument 控制仪表；监控仪器；监测仪器；监测仪表
monitoring key 监听键；监视键
monitoring laser 监控激光器
monitoring leakage 监视漏斗
monitoring measurement 监控流量测
monitoring method 监测方法
monitoring method of active fault 活断层监测方法
monitoring method of lichen moss 地衣苔藓监测法
monitoring net(work) 监测网
monitoring of atmospheric pollution 大气污染监测
monitoring of automobile exhaust emission 汽车排气监测
monitoring of coal mine safety 煤矿安全生产监测
monitoring of corrosion 腐蚀监控
monitoring of deformation of rock mass of slope 斜坡岩体变形观测
monitoring of deformation of surrounding rock of tunnel 洞室围岩变形观测
monitoring of dicharge 排污监测
monitoring of displacement of dam 坝体位移观测
monitoring of fouling 污垢监控
monitoring of hygiene 卫生监控
monitoring of induced earthquake of reservoir 水库诱发地震监测
monitoring of kiln 看火
monitoring of landslide moving 滑坡动态观测
monitoring of land subsidence 地面下沉观测
monitoring of mobile fault 活动断层监测
monitoring of poison 毒物监测
monitoring of process variables 工艺过程参数监测
monitoring of surcharge settlement 堆载沉降监测
monitoring of surrounding rock pressure 洞室围岩压力观测
monitoring of the workplace 工作场所监测
monitoring of uplift pressure on the dam 坝基扬压力观测
monitoring of urban environment 城市环境监测
monitoring of working environment 工作环境监测
monitoring panel 监视屏；监视盘
monitoring parameter 监测参数
monitoring picture 监视图像
monitoring plan 监视计划；监测计划
monitoring plant 监测植物
monitoring point 监测点；检测点；监视点
monitoring point for level 水准监测点
monitoring printer 监视打印机；监控印字机；监控复印机
monitoring procedure 监督程序
monitoring process 监视进程
monitoring profile 监测断面
monitoring program(me) 监测规划
monitoring project 监测项目
monitoring receiver 监控接收机
monitoring record 监测记录
monitoring recorder 监测记录设备；监控记录器
monitoring relay 监视继电器
monitoring report 监测报告
monitoring sample 监控样
monitoring scheme 监测布置
monitoring section 监测断面
monitoring service 监听业务
monitoring ship 监测船

monitoring signal 监控信号
monitoring site 监测点
monitoring standard 监测标准
monitoring standard method 监测标准方法
monitoring station 监控站；监视站；监控台；监测站；监测台
monitoring station description 监测站说明
monitoring station of environment (al) atmospheric pollution 环境大气污染监测站
monitoring store 监测信号存储器
monitoring system 监视系统；监听系统；监测系统
monitoring technique 监控技术
monitoring television 监测电视
monitoring terms of groundwater pollution 地下水污染监测项目
monitoring test 监测试验
monitoring test paper holder 检测试纸夹
monitoring thief 监测盗水
monitoring tower 监测塔
monitoring treatment efficiency 监测处理效率
monitoring unit 监护系统
monitoring well 检查井；检测井；监测井
monitoring well of saturated zone 饱和区检测井
monitor interrupt 监督程序中断
monitor kinescope 监视显像管
monitor loudspeaker 监听喇叭
monitor message 监控信息
monitor mode 监控方式
monitor net 监测网系
monitor operating system 监视操作系统；监督操作系统
monitor out 监听输出；监控输出
monitor overlay 监督重叠
monitor photoelectric(al) cell 监控光电管
monitor position 监视席位
monitor printer 监视器打印机；监控程序用打印机
monitor procedure 监督过程
monitor program(me) 管理程序；监视项目；监控程序；监督程序
monitor record 监视记录；监测记录
monitor roof 采光屋顶；带天窗屋顶；横向天窗屋顶；顶窗；采光通风的部分屋顶；采光
monitor room 监听室
monitor routine 执行程序；监控程序；监督程序
monitor sample 监控样品；监控样
monitorscope 监护听诊
monitor screen 监视屏
monitor select 监视选择
monitor signal 监听信号；监视信号
monitor skylight 日照监测
monitor stack area 监督程序栈区
monitor state 监督状态
monitor system 执行系统；监护系统；操作系统
monitor system components 操作系统组成部分
monitor task 监视任务
monitor terminal 监督终端
monitor top 横向天窗屋顶；气楼；顶窗
monitor well 监测井
monitor working area 监督程序工作区
monitor zone adapter module 探测模块
monitron 放射监视器
monk 天窗窗扇；僧侣；修道士；墨层过厚
monk bond 每皮两顺一丁砌砖法；两顺一丁砌法；二顺一丁砌砖法
monkery 寺院；修道院
monkey 心轴；渣口；活动扳手；猴子；起重机小车；通风巷道；树脂胶化批量；打桩(落)锤；锤头
monkey block 起重小滑车
monkey blood 喷洒剂；养护剂
monkey boat 小型半甲板驳船；船坞内用的一种小艇
monkey bolt 锤头螺钉
monkey bridge 天桥
monkey-chatter interference 邻信道干扰
monkey cooler 渣口冷却器
monkey drift 小探巷；导向巷道；通风联络斜巷
monkey driver 卷扬式打桩机；锤式打桩机
monkey driver engine 锤式打桩机
monkey driving engine 锤式打桩机
monkey dung 将牛粪、石灰、砂子和牛毛涂在帆布上作为圆管热绝缘材料
monkey engine 卷扬式打桩机；锤式打桩机；打桩机
monkey face 眼板；三眼环；三角(眼)铁；三角眼板
monkey fist 撇缆头结
monkey forecastle vessel 低首楼甲板船

monkey hammer 落锤
monkey heading 窄小巷道
monkey hook 桩头钩；桩锤钩
monkey island 罗盘甲板
monkey jack 打桩起重器；打桩起重机；拔树起重器
monkey jacket 渣口冷却套
monkey ladder 树杆制矿用梯子
monkey of pile driver 打桩机桩锤
monkey pipe 锚链管
monkey pole 三脚架吊杆(俚语)；(简便起重的)吊杆；拔杆
monkey poop 低船尾楼
monkey puzzle 南洋杉
monkey rail 后甲板栏杆
monkey roll crusher 小型辊式破碎机
monkey rolls 小型辊式破碎机
monkey slide 架空索道(俚语)；缆车(俚语)；缆索吊车
monkey spanner 活旋钳；万能螺旋扳手；螺丝扳手；活(动)扳手
monkey spar 小横桁；短杆；尺寸不够的圆材
monkey's tail 卷尾状扶手端；扶手上的涡卷花饰；扶手卷尾端
monkey stay 前桅与烟筒间支索
monkey-tail 扶手卷尾端；卷尾状扶手端
monkey-tail bolt 长柄门插销；卷尾状扶手端；长插销；卷尾插销
monkey wheel 小滑轮；单滑轮车
monkey winch 手摇小绞车；手摇绞车
monkey wrench 螺丝扳手；管子钳；活旋钳；活(动)扳头
monk's choir 寺院唱诗班；寺院歌唱班席位
monk's cloth 粗厚方平织物
monk's park 浴缸石；巴思石(最好的一种)
monlateral fall 单侧面斜坡
monlayer cell culture 单层细胞培养
monme 日本重量单位
monmouthite 绿质霞石正长岩
Monnoyer system 曼诺耶施工法(一种高烟囱施工法，由预制八边形混凝土块体分段砌筑而成)
monoaccelearator 单加速器
monoacetate 单醋酸盐
monoacetin 一醋精；甘油一乙酸酯；单乙酸甘油酯
monoacid 一元酸
monoacid base 一酸(价)碱；一价碱
monoacidic base 一元碱
monoalkyl ammonium chloride 氯化烷基铵
monoalphabetic cipher 单表密码
monoammonium phosphate 磷酸一铵
monoamplifier 单端放大器
mono-anchor mooring 单锚系泊
monoarsenide 一砷化铀
monoatomic gas 单原子气体
monoatomic metal 单原子金属
monoatomic semiconductor 单质半导体
monoaxial 单轴的
monoaxial interference figure 一轴晶干涉图
monoazo dye 单偶氮染料
monoband coil 单频带线圈
monobar conveyer 单链运输机
monobasic acid 一元酸；一碱(价)酸
monobasic acid ester 一元酸酯
monobasic calcium 磷酸二氢钙
monobasic calcium phosphate 一代磷酸钙
monobasic lead chromate coated silica 碱式铬酸铅包二氧化硅；包核碱式铬酸铅
monobasic sodium phosphate 一代磷酸钠
mono bath 单浴；单浴显影
mono-bed 单床
mono-bed and single cycle moving 单塔单周期移动床【给】
mono-bed ion exchanger 单床离子交换器
monobel 硝酸甘油；硝化甘油；单贝尔
monoblade scutching machine 单刀式打麻机
monobloc 整块
monobloc design 整体(机组)设计
monobloc engine 铸成整体的发动机
monobloc forging 整锻
monoblock 整体汽缸座；整块大砖；整块；壳体；单元机组；单块
monoblock cast(ing) 整体铸造；整块铸造
monoblock drill steel 整体钎子
monoblock engine 整体汽缸发动机；单排发动机
monoblock forged vessel 单层整体锻造式容器

monoblock prestressed concrete tie 整体式预应力混凝土轨枕
monoblock pump 直联泵；整装泵；单体泵
monoblock rotor 整体转子
monoblock unit 整块组件
monobloc type drill rod 整体式钻杆
monobloc wheel 整体车轮
monoboard computer 单片计算机
monobox crane 箱形单梁起重机；单箱体式起重机
monobrid 单片组装法
monobrid circuit 单片混合电路
monobromacetanilid 一溴乙酰苯胺
monobromcamphor 一溴樟脑
monobromide 一溴化物
monobromination 一溴化
monobromo 一溴
monobromo-benzene 一溴代苯
monobromo-ester 一溴代酯
monobromo-ether 一溴醚
monobromonaphthalene immersion system 一溴化萘浸液系统
monobromophenol 一溴酚
monobucket 单斗
monobucket excavator 单斗挖土机；单斗挖掘机
monobucket loader 单斗装载机
monobucket shovel 单斗挖掘机
mono-buoy mooring 单浮标停泊；单浮筒系泊；单点系泊
mono-buoy mooring system 单浮筒系泊系统
mono-buoy terminal 单点系泊码头
monobutyltin 一丁基锡
monocable 架空索道；单线缆索；单索架空索道
monocable aerial ropeway 架空单缆索道
monocable bridge 单索桥
monocable crane 单缆索起重机
monocable grab 单索抓斗
monocable ropeway 单线架空索道；单索索道；单缆架空索道
monocalcium aluminate hydrate 水化铝酸一钙
monocalcium phosphate 磷酸一钙
monocalcium silicate 硅酸一钙
monocaltural farming 单一经营
monocarbide 一碳化物
monocardian 单心的
Monocast process 树脂砂衬离心铸管法
monocell 单电池
monocellular 单格的
monocellular foam 单体泡沫
monocentric bundle 单心光束
monocentric eyepiece 单心目镜
monocentric ocular 单心目镜
monochlorethane 一氯乙烷
monochloride 一氯化物
monochlor(in)ate 一氯化
monochloroacetate 一氯乙酸盐
monochloroacetic acid 一氯乙酸；一氯醋酸
monochloroacetone 一氯丙酮
monochloroamine 一氯胺
monochlorobenzene 一氯苯
monochlorobenzol 一氯苯
monochlorodifluoromethane 二氟一氯甲烷
monochlorophenol 单氯酚
monochloropropionic acid sodium salt 单氯丙酸钠盐
monochlorosilane 一氯甲硅烷
monochlorphenol 一氯酚
monochord 单弦杆
monochroic 单色光的
monochromasia 全色盲
monochromasy 全色盲
monochromatic 单色的；单色的
monochromatic aberration 单色像差
monochromatic absorption 单色吸收
monochromatic absorptivity 单色吸收能力
monochromatic amplitude 单色光变幅；单色变幅
monochromatic anaglyph 单色立体图
monochromatic chart 单色图
monochromatic colo(u)rimeter 单色色度计；单色比色计
monochromatic colo(u)r scheme 单色配色法
monochromatic corona 单色日冕
monochromatic edition 单色版
monochromatic film 单色胶片
monochromatic filter 双折射滤光器；单色滤光器；单色滤光片；单色滤光镜
monochromatic flux 单色流量
monochromatic fringe 单色边纹
monochromatic harmony 单色协调
monochromatic heliograph 单色日光仪
monochromatic illumination 单色光照明
monochromatic illuminator 单色照明器
monochromatic image 太阳单色像；单色像
monochromatic interference 单色干涉
monochromatic ion beam 单色离子束
monochromaticity 单色性
monochromatic light 单色光；单光
monochromatic map 单色地图
monochromatic neutron beam 单能中子束
monochromatic photograph 单色像片
monochromatic photoimage 单色影像
monochromatic print 单色印样；单色翻印
monochromatic printing 单色印刷
monochromatic pyrometer 单色高温计
monochromatic radiation 单色辐射
monochromatic radiative equilibrium 单色辐射平衡
monochromatic rendering 单色渲染
monochromatic scale 单色刻度盘
monochromatic scheme 单色色调
monochromatic sensibility 单色灵敏度
monochromatic shading 单色晕渲法
monochromatic shading map 单色晕渲地图
monochromatic signal 单色信号
monochromatic source 单色光源
monochromatic specification 单色表示；单色标志
monochromatic spectrum 单色光谱
monochromatic stimulus 单色光刺激
monochromatic temperature scale 单色温标
monochromatic time slices stack chart 等时切片分色叠加图
monochromatic transparency 单色透明正片
monochromatic wave 单色波
monochromatic X-ray 单色X射线
monochromatic X-ray line 单色X光谱线
monochromating crystal 单色晶体
monochromating crystal spectrometer 单色晶体谱仪
monochromating spectrometer 单色分光计
monochromation illumination 单色光照明
monochromatism 全色盲
monochromatization 单色化
monochromatizing 成单色
monochromator 单色仪；单色器；单色光镜；单能化器；分光仪
monochromator in visible and ultraviolet 可见光区与紫外区单色仪
monochrome 黑色影片；黑色图像；单色照片；单色画；单色
monochrome band 单色波段
monochrome channel 单色信道；单色图像信号信道
monochrome decoration 单色装饰
monochrome display 单色显示
monochrome film 单色胶卷
monochrome glaze 单色釉
monochrome graphics 单色图形
monochrome grey scale 黑白亮度等级
monochrome image 单色图像
monochrome information 亮度信息
monochrome monitor 单色监视器
monochrome multipersistence screen 单色多余辉穿透屏
monochrome picture 黑白图像
monochrome presentation 单色显像；单色素描
monochrome press 单色印刷机
monochrome receiver 黑白电视接收机
monochrome receiver set 单色接收机
monochrome recording 单色信号记录
monochrome scale 黑白标度
monochrome signal 黑白信号；单色信号
monochrome television 黑白电视
monochrometer 单能化器；单色光镜；单色仪；单色计
monochrome test pattern 黑白测试图
monochrome voltage 亮度信号电压；单色信号电压
monochromic 各向同等吸光的
monochromical signal 黑白信号
monochromic television 黑白电视
monochromism 光的各向同等吸收
monochrom painting 单色调配；单色绘画
monocircle goniometer 单圆测角仪
monocle 单片眼镜
monocleid(e) 单锁柜
monoclimax 单元演替顶极
monoclimax theory 单顶极学说
monoclinal 单斜的
monoclinal analbite 单斜钠长石
monoclinal block 单斜断块
monoclinal block faulting structure 单斜断块构造
monoclinal coast 单斜海岸
monoclinal fabric 单斜组构
monoclinal fault 单斜断层
monoclinal flexural zone 单斜挠褶带
monoclinal flexure 单斜挠曲
monoclinal fold 单斜褶皱
monoclinal form 单斜晶形
monoclinal mountain 单斜山；单面山
monoclinal oil-gas field 单斜油气田
monoclinal pyramid 单斜锥
monoclinal ridge 单斜脊
monoclinal rising flood wave 单斜上升洪水波
monoclinal rising wave 单斜上升波
monoclinal river 单斜河
monoclinal scarp 单斜崖
monoclinal slope 单斜坡
monoclinal spring 单斜泉
monoclinal strata of confined water 承压水斜地
monoclinal stratum 单斜岩层；单斜地层；单斜层
monoclinal stream 单斜河
monoclinal structure 单斜构造
monoclinal system 单斜晶系
monoclinal texture 单斜结构
monoclinal valley 单斜谷
monocline 单斜
monocline fold 单斜褶皱
monocline layer 单斜层
monocline pool 单斜油藏
monocline zone 单斜带
monoclinic 单斜晶的；单斜的
monoclinic angle domain 单斜角域
monoclinic clinoprism 单斜斜轴柱
monoclinic crystal 单斜结晶
monoclinic fabric 单斜组构
monoclinic feldspar 单斜长石
monoclinic hemipinacoid 单斜半轴面
monoclinic hemiprism 单斜半柱面
monoclinic hydrous zirconium oxide 单斜含氮氧化锆
monoclinic lattice 单斜点阵
monoclinic lead chromate 单斜晶型铬酸铅
monoclinic prism 单斜柱晶；单斜棱晶
monoclinic sulfur 单斜晶硫
monoclinic symmetry 单斜对称
monoclinic system 单斜晶系
monoclonal 单斜层的
monoclone planting 单一天性系植区
monocode 单码
monocoil 单线圈(的)；单线管
monocolo(u)r 单色
monocolo(u)r H$_2$S analyzer 单色硫化氢分析器
monocomponent conglomerate 单成分砾岩
monocomponent fibre 单组分纤维
mono-conduction 单向传导
monocontrol 单一控制；单一调节
mono coordinate comparator 单片坐标量测仪
monocoque 硬壳式构造；单壳机身；硬壳式机身
monocoque body 无骨架式车身；单壳体车身；硬壳机身
monocoque coach body 无骨架式车体
monocoque construction 硬壳(式)结构；壳体式结构；壳体施工
monocoque container 薄壳式集装箱
monocoque type 单壳式
monocord 单塞绳；单软线
monocord switch board 单塞绳交换机
monocord system trunk board 单塞绳式中继台
monocord venetian blind 单绳软百叶窗
monocotyledon 单子叶植物
monocrepid 单基片
monocresyl glyceryl ether diacetate 甘油一甲苯醚二乙酸酯；甲苯醚甘油二乙酸酯
monocrotophos toxicity 久效磷毒性
monocrystal 单晶体；单晶

monocrystal fibers 单晶光纤
monocrystalline 单晶(质)的;单晶型
monocrystalline alumina 单晶矾土
monocrystalline analysis 单晶分析
monocrystalline filament 单晶长丝
monocrystalline furnace 单晶炉
monocrystalline fused alumina 单晶刚玉
monocrystalline gallium 单晶镓
monocrystalline germanium 单晶锗
monocrystalline ingot 单晶锭块
monocrystalline particle 单晶粉粒
monocrystalline photoconductor 单晶光导体
monocrystalline silicon 单晶硅
monocrystalline texture 单晶结构
monocular 单筒望远镜;单筒;单目望远镜;单目
monocular analysis 单目分析
monocular camera 单目摄影机
monocular hand level 手水准(仪);单筒手提水准仪
monocular inclined tube microscope 单筒斜管显微镜
monocular instrument 单镜仪器
monocular microscope 单目显微镜
monocular observation 单目观察;单镜观测
monocular range finder 单目测距仪
monocular straight tube microscope 单筒直管显微镜
monocular tube biologic microscope 单筒生物显微镜
monocular tube eyepiece 单筒目镜
monocular tube microscope 单筒显微镜
monocular vision 单眼视觉;单眼观测
monoculture 单一经营
monoculture system 单种栽培系统
monocycle 独(立)轮车;单周期;单(循)环;单独轮车
monocycle geosyncline 单旋回地槽
monocycle landform 单循环地形
monocyclic(al) 单周期的;单(循)环的;单旋回的;单轮的
monocyclic(al) land form 单循环地形;单旋回地层
monocyclic(al) orogensis 单相造山运动
monocyclic(al) terpene 单环萜(烯)
monocyclic(al) aromatic hydrocarbon 单环芳香烃
monocyclic(al) method 单周期法
monocyclic(al) orogenesis 单旋回造山运动
monocyclic(al) ring 单环
monocyclic(al)-start induction motor 单相感应电动机
monocycloalkane 单环烷烃
monocycloaromatic 单环芳烃
monocylic 单一间隔期的
monocylic(al) start induction motor 单周期起动感应电动机
monodentate adsorbate 单配位基吸附质
Monod equation 莫诺方程
Monod formula 莫诺公式
mono-digital comparator 单片数字坐标量测仪
monodiplopia 单眼复视
monodirectional 单向的
monodisperse 单分散
monodispersed aerosol 单分散气溶胶
monodispersed pollutant 单色散污染物;单分散污染物
monodisperse system 单分散系
monodispersity 单分散性
monodrill 单粒播种机
monodrome 单值
monodrome function 单值函数
monodromy 单值性
monodromy theorem 单值性定理
mono-drum winch 单卷筒卷扬机
monoelectrode log 记录电流测井
monoelectrode log curve 记录电流测井曲线
monoenergetic 单能量的;单能的
monoenergetic cross-section 单能截面;单色截面
monoenergetic gamma rays 单能伽马射线
monoenergetic radiation 单能辐射
monoenergetic spectrum 单能能谱
monoergic 单能量的
monoethanol amine 一乙醇胺
monoethanolamine oleate 单乙醇胺油酸盐
monoether 单醚
monofier 振荡放大器
monofilament 单纤维;单丝

monofilar variometer 单丝变感器
monofil-coner 单丝锥形筒子络筒机
monofile monovolume 单文件单卷;单卷单文件
monofile multivolume 多卷单文件;单文件多卷
monofilm 单膜;单层膜
Monofilter 莫诺菲尔特丙纶土建布
monofired 一次烧成
monofluoracetate 单氟乙酸盐
monofocal 单聚焦的
monoform base sheet 合成屋面片材;单型组合屋面材料
monoformer 光电单函数发生器
Monofrax 高密度耐火材料;电熔高铝砖(商品名)
monofrequency radiation 单频辐射
monofrequent 单频率
monofrequent radiation 单频辐射
monofuel propulsion 单一燃料推进系统
monofunctional 单函数的;单功能的
monofunctional exchanger 单功能离子交换剂
monogamy 一一对应;一对一
monogen 单价元素
monogene rock 单成岩
monogenetic 单色的;单成因的;单成
monogenetic dye 单色染料
monogenetic gravel 单成砾石
monogenetic rock 单成岩
monogenetic volcano 单成火山
monogenic 单成因的
monogenic rock 匀质岩;单成岩
monogeosyncline 单地槽
monogirder bridge 单主梁桥架
monoglyceride 单酸甘油酯;单甘油酯
monoglyceride process 单甘油酯法
monoglyceride stage 醇解阶段
monograin 流质烈性炸药
monogram 带字母图案的玻璃器皿
monograph 专著;专题著作;专题文章;专题论文;记压器;示功器
monographic(al) demonstration 专题论正
monographic(al) study 专题研究
monographist 专题论文的作者
monogroove stereo 单纹槽立体声
monogyro compass 单转子陀螺罗经
monohalogenated phenol 一卤代苯酚
mono-hearth refuse incinerator 单炉式焚化炉
monohedral 单面的
monohedron 单面体
mono-hull ship 单体船
monohydrate 一水化合物;单水化合物
monohydrate bauxite ore 一水铝石型铝土矿
monohydrate crystal 含一分子结晶水晶体
monohydrate dolomitic lime 单水白云化石灰
monohydrated sodium carbonate 一水碳酸钠
monohydric acid 一元酸
monohydric alcohol 一羟醇
monohydric phenol 一元酚;一羟(基)酚
monohydrocalcite 单水碳钙石
monoid 类群;独异点
monoid-closure 独异点闭包
monoindication obstruction signal 遮断信号(机)
monoindication obstruction signal button 遮断信号按钮
monoindication repeating signal(l)er 复示信号机
monoiod(in)ate 一碘化
monoiodo-acetic acid 一碘醋酸;一碘乙酸
monoiodo-ester 一碘酯
mono-iodotyrosine 一碘酪氨酸
monojet 单体喷雾口
monoketone 一元酮
monolayer 单层;单分子吸附层;单分子层
monolayer adsorption 单分子层吸附
monolayer capacity 单分子层吸附容量
monolayered vessel 单层容器
monolayer material 单层材料
monolayer of particles 单层粉粒
monolever 单手柄
mono-lever control 单手柄操纵;单杆操纵;单臂操纵
monolever switch 单柄四向交替开关;单手柄十字形开关
monolingual legend 一种文字图例
monolinolein 甘油亚麻酸酯
monolinuron 绿谷隆

monolite maroon 莫诺赖特栗红
monolith 孤赏石;原状土样;正石柱;整体砌块;整(体)料;整块石料;整块石体;整段土壤剖面;块体混凝土;磐石;独石柱;独石(碑);单块;沉井;坝块;(两条伸缩缝的)坝段
monolith caisson construction 整体沉箱结构
monolith foundation 沉井基础
monolithic 整体式的;整体浇筑的;龟甲网衬里;独块巨石的;单块(的)
monolithically assembled floor panel 整块材料组装的地板
monolithically cast wall 整体浇灌墙
monolithically concreted 整体浇的
monolithically poured 整体浇注的(混凝土);整体浇的
monolithic arch 整体式拱
monolithic beam 整体式梁
monolithic beam without ballast and sleeper 整孔无砟无枕预应力混凝土梁
monolithic blast protection structure 整体式抗爆结构
monolithic block 大型砌块;巨型砌块;整体坝块;巨型块石;大块体
monolithic body 整体
monolithic boundary layer capacitor 独石晶界层电容器
monolithic brick pavement 整体式炼砖路面;整体式炼砖结构
monolithic caisson 整体式沉箱
monolithic capacitor 单体电容器
monolithic case 整体外壳
monolithic casting concrete 整体浇筑混凝土
monolithic ceramic capacitor 独石陶瓷电容器
monolithic chamber floor 整体式闸底
monolithic circuit 单片电路;单块电路
monolithic column 整体式柱
monolithic column shaft 整体柱身
monolithic computer 单片计算机
monolithic computing system 整体计算系统
monolithic concrete 整块混凝土;整体浇筑混凝土;整体(式)混凝土;块体混凝土;大块混凝土
monolithic concrete bed 整体道床
monolithic concrete construction 整体(式)混凝土结构;整体(式)混凝土建筑
monolithic concrete encasement 整体混凝土围护
monolithic concrete lining 整体混凝土支护;整体混凝土衬砌;整体混凝土衬护
monolithic concrete panel wall 整体混凝土板墙
monolithic concrete sleeper 整体式混凝土轨枕
monolithic concrete structure 整体混凝土建筑物
monolithic concrete tie 整体式混凝土轨枕
monolithic construction 整体(式)结构;整体(式)建筑;整体(式)构造;整体施工
monolithic construction method 整体建筑法
monolithic crystal monochromator 单晶体单色仪
monolithic curb 整体式路缘
monolithic dam 整体式坝
monolithic device 单块器件
monolithic display 单片显示器
monolithic filter 单片滤波器
monolithic finish 整体修整;整体修饰;整体饰面
monolithic fireproof floor 整体式防火地板
monolithic floor 整体浇灌楼板;整体底板;整浇楼板
monolithic floor surface 整体面层
monolithic footing 整体底脚;整体基础;独石基脚
monolithic foundation 整体式基础
monolithic grillage 整体式格床
monolithic head 单块磁头
monolithic hybrid circuit 单片混合电路
monolithic integrated circle 单块集成电路
monolithic integrated circuit 整体集成电路;单片(式)集成电路
monolithic integrated operational amplifier 单块集成运算放大器
monolithic integrated receiver 单片集成接收机
monolithic joint 整体连接;(填充后形成的)整体式缝
monolithic laser 单片激光器;单块激光器
monolithic layout 单块电路设计
monolithic light detector 单块光探测器
monolithic lining 整体式衬砌;整体炉衬;整体(浇注)衬砌;整体衬里
monolithic lysimeter 原状土测渗计
monolithic masonry 整体圬工

monolithic masonry lock chamber 整体式圬工闸室
monolithic masonry wall 整体式圬工墙
monolithic material 单块材料;整块材料
monolithic memory 单片存储器
monolithic monument 独石碑
monolithic mo(u)ld 整体铸型
monolithic multi-layer wall 多层整体墙
monolithic nonmetallic fireproof floor 整体非金属防火地板
monolithic operational amplifier 单片运算放大器
monolithic patch 整体修饰
monolithic photodiode 单片光电二极管
monolithic piezoelectric(al) ceramic transformer 独石压电陶瓷变压器
monolithic pillar 整体墩;整体支柱;整体式柱
monolithic pouring 整块灌注
monolithic power 单块功率
monolithic power devices 单块功率器件
monolithic pretensioning 整体预张拉
monolithic processor 单片处理机
monolithic refractory 现浇无接缝耐火材料;整体耐火材料
monolithic reinforced concrete 整体式钢筋混凝土
monolithic reinforced concrete structure 现浇式钢筋混凝土结构;整体式钢筋混凝土结构
monolithic reinforced concrete wall 整体钢筋混凝土墙
monolithic reinforcement cement pontoon 整体式水泥砼船
monolithic retaining wall 整体式挡土墙
monolithic roof 整体屋顶
monolithic roofing 刚性防水屋面
monolithic roof structure 整体屋顶结构
monolithic screed 整体式找平层
monolithic ship 整体结构船;单体结构船
monolithic slab 整块板
monolithic slab and foundation wall 整体式板和基础墙
monolithic solid rood 整块实体屋顶
monolithic stability 整体稳定性;整体稳定
monolithic storage 单片存储[贮]器;单块存储器
monolithic structural system 整体结构系统
monolithic structure 整体式结构;整体式构筑物;独石结构;单岩构造
monolithic substrate 单片衬底;单块衬底
monolithic surface treatment 整体饰面;整体面层处理;整体表面处理
monolithic system 单片系统
monolithic system technology 单片系统技术;单片系统工艺
monolithic technology 单片技术;单片工艺;单块工艺
monolithic terrazzo 现浇水磨石;磨石整体罩面;整体水磨石;整块水磨石
monolithic topping 整体面层
monolithic track bed 整体道床
monolithic type wharf 整体式码头
monolithic unit 整体单元
monolithic wall 整体墙;整体式(挡土)墙
monolith joint 坝段接缝
monolith lysimeter 原状土样的溶度测定仪;坝段测量计
monolith quay 整体式码头
monolith quay wall 整体式码头岸壁
monolith type dock wall 整体式船坞墙
mono lock 单体锁;独件锁;单锁
monomaceral microlithotype 单组分显微煤岩类型
monomaran 单体船
monomark 商品符合;注册标记
monomast crane 单立柱桅杆式起重机
mono-medium filter 单层介质滤池
mono-medium sand filter 单层介质砂
monomer 单体;单基物;单分子物体
monomer content 单体含量
monomer geologic(al) map 单体性地质图
monomeric compound 单体化合物
monomeric plasticizer 低分子量增塑剂;单体增塑剂
monomeric unit 链节
monomer impregnated concrete 单体浸渍混凝土
monomer loading 单体浸渍率
monomerous antenna 单节触角
monomer-polymer 单体聚合物
monomer ratio 单体配比
monomer reactivity 单体活泼度

monometallic 单金属的;单金属
monometallic balance 单金属摆轮
monometallic standard 单本位制
monomethylamine 一甲胺;一甲基胺;甲胺
monomethylolurea 一羟甲基脲;脲基甲醇
monomethyl oxalate 草酸一甲酯
monometric(al) equivalent 压力表当量
monomial 单项式
monomial evaluation of enterprise performance 企业效绩单项评价
monomial expression 单项式
monomial matrices 单项矩阵
monomial representation 单项表现
monomict 单矿物碎屑;单矿沉积岩的
monomict breccia 单矿物碎屑角砾岩
monomictic lake 冬回水湖;单对流(混合)湖
monomict rock 单矿物碎屑岩
monomineral 单矿物
monomineral analysis 单矿物分析
monomineralic 单矿物碎屑的;单矿物的
monomineralic breccia 单矿物碎屑角砾岩
monomineralic clay 单矿物黏土
monomineralic claystone 单矿物黏土岩
monomineralic rock 单矿物岩
monomode 单模
monomode and light pack 单模束管式
monomode coupler 单模耦合器
monomode optic(al) fiber 单模光纤
monomode optic(al) waveguide 单模光波导
monomolecular 单层分子
monomolecular adsorption 单分子(层)吸附
monomolecular film 单分子层
monomolecular layer 单分子(吸附)层
monomolecular reaction 单分子反应
monomolecular surface film 单分子表面膜
mono-mooring 单点系泊
monomorph 单形;单晶物
monomorphic 单形的
monomorphism 单一同态;单同态
monomorphous 单形的;单晶形的
monomotor 单发动机;单电动机
Monongahela Series 孟农加希拉统【地】
mononitraniline 一硝基苯胺
mononitrate 一硝酸盐
mononitration 一硝基化
mononitride 一氮化合物
mononitro-benzene 一硝基苯
mononitro-compound 一硝基化合物
mononitro-derivative 一硝基衍生物
mononitromethane 一硝基甲烷(喷气燃料)
mononitronaphthalene 一硝基萘(油料的去荧光剂)
mononitrophenol 一硝基酚
mononitrotoluene 一硝基甲苯
mononuclear 单核
mononuclear aluminum 单核铝
mononuclear complex 单核络合物
mono-objective binocular microscope 单物镜双筒显微镜
monooleate 一油酸
monoolein 一油清;甘油油酸酯
mon(o)oxide 一氧化物
monooxirane 单环氧基环烷
monopack 三色片片
monopalmitin 甘油棕榈酸酯
monopaste 显定合一浆
monophagous 单变的
monophase 单相(的)
monophase asynchronous motor 单相异步电动机
monophase current 单相电流
monophase equilibrium 单相平衡
monophase field 单相区
monophase metamorphism 单相变质作用
monophase microinstruction 单步微指令
monophase system 单相系统
monophase transformer 单相变压器
monophasic 单相的
monophasic orogenic cycle 单相造山循环
monophone 送受话器;收发话器
monophonic 单声道;单路的
monophonic recorder 单声道录音机;单音的
monophonic sound 单通道扩声
monophony 单声道
monophotogrammetry 单像摄影测量学

monophyletic 单源的
monophyletical evolution 单系列演化
monophyletic theory 一元论
monophyletist 一元论者
monophyodont 不换性齿系
monopinch 单收缩
mono-pitched roof 单坡屋顶
Monoplacophora 单板类
monoplanar 单平面的
monoplane 单翼机;单平面
monoplanetic 单游的
mono-plate processor 单板机
monoplayer 单放机
monoplotting 单像测图
monoplunger pump 分配式油泵
monopod 单柱(指系船装置)
monopodial inflorescence 单轴花序
monopod ship loader 单腿装船机
monopolar cell 单极式电解槽;单极电解槽
monopolar direct current dynamo 单极直流发电机
monopolar induction type relay 单极感应式继电器
monopolar type 单极式
monopole 对中柱;单极子;单极;磁单极子
monopole antenna 单极天线;鞭状天线
monopole automatic flame cutter 光学自动追迹气割机
monopole automatic gas cutter 自动光学曲线追踪气割机
monopole cell 单极式电解槽
monopole double throw switch 单刀双掷开关
monopole generator 单极发电机
monopole mass spectrometer 单极质谱仪
monopole oil 磺化蓖麻油;土耳其红油(别名)
monopole single throw switch 单刀单掷开关
monopole source 单极子源
monopolies 垄断财团
monopolistic 专利的
monopolistic competition 垄断性竞争
monopolistic financial group 垄断财团
monopolistic market 垄断市场
monopolization 专利;垄断;获得专利权
monopolize 取得垄断
monopolized commodities 专卖货物
monopolized market 市场垄断
monopol soap 蓖麻油皂
monopoly 专卖权;专利权;专利品
monopoly capital 独占资本
monopoly clause 独家经营条款
monopoly contract 独家经营契约
monopoly control over its income and expenditures 统收统支
monopoly of sale 包销
monopoly price 垄断价格
monopoly sale 专卖
monoprogrammed 单道程序的
monopropellant 单元推进剂
monopropylamine 单丙胺
monopteral 单列柱的;拱顶圣殿;拱顶庙宇;圆形柱(廊)建筑
monopteral architecture 圆形外柱廊式建筑
monopteral temple 排柱寺院;排柱圆庙
monopteron 排柱圆屋;圆形柱(廊)建筑;【复】monoptera
monopulse 单脉冲
monopulse antenna 单脉冲天线
monopulse autotracking 单脉冲自动跟踪
monopulse comparator 单脉冲比较器
monopulse optic(al) receiver 单脉冲光接收器
monopulse radar 单脉冲雷达
monopulse sensor 单脉冲传感器
monopulse tracking 单脉冲跟踪
Mono pump 莫诺泵
monorail 独轨铁路;单线铁路;单频道;单轨线路;单轨索道;单轨梁;单轨
monorail bridge 单轨铁路桥(梁)
monorail car 单轨矿车;单轨车
monorail chain block 单轨链滑车
monorail charger 单轨装料机;单轨加料机
monorail concreting skip 单轨混凝土浇灌车
monorail conveyor[conveyer] 单轨输送机;单轨吊运器
monorail crane 单轨起重机;单轨吊
monorail crane truck 单轨起重车
monorail grab crane 单轨抓斗起重机

monorail hoist 单轨起重机;单轨吊车;单轨电动滑车
monorail ladle 单轨吊包
monorail loop system 环形单轨系统
monorail motor crab 单轨电动起车绞车;单轨电动起重机
monorail motor hoist 单轨电动绞车
monorail opened system 开式单轨系统
monorail over-head feed carrier 单轨悬吊式饲料分送车
monorail station 单轨车站;单轨铁路车站
monorail system 单轨运输系统;单轨调度系统
monorail-tramway 单索道
monorail transporter equipment 单轨运输设备
monorail transport system 单轨运输系统
monorail trolley 单轨小车
monorail way 单轨铁路;单轨铁道
monorail weight car 单轨称量车
monoreactant 单元燃料;单一反应物
monoreactive glyceride 单活性基甘油酯
monorefringent 单折射的
monoreturn valve 止回阀
monorobot 小型会计机
mono-rotor 整体转子
mono-runway aerodrome 单跑道机场
monosalient pole 单显磁极;单凸极
monoscience 专门学科;单项学科
monoscope 单像管
monoscope camera 单像管摄像机
monoscope equipment 单像管设备
monoscope tube 单像管
monoscopic image measurement 单张像片测量
monoscopic quantitative monophotogrammetry 单像管定量单像摄影测量学
monoscopic radar interpretation 单像雷达判读
monoscopic viewing 单目镜观察
monoseaplane 单翼水上飞机
monoseeding 单粒播种
monosemantic 单义的
monosemous 单义的
monoseptic culture 单腐化培养
mono signal 单信号(的)
monosiphonous 单管的
monosize distribution 单一粒度分布
monospace 单间隔
monospaced text 单空格文本
monospar 独立柱式;单桅杆;单梁
monospindle 单轴
monospindle boring machine 单轴镗床
monosplines 单项仿样函数
monostable 单稳态的
monostable blocking oscillator 单稳态间歇振荡器;单稳多谐振荡器
monostable circle 单触发电路
monostable circuit 单稳(态)电路
monostable element 单稳元件
monostable flip-flop 单稳态触发器
monostable multivibrator 单稳(态)多谐振荡器
monostable trigger 单稳态触发器
monostable trigger-action circuit 单稳态触发电路
monostable trigger circuit 单稳触发电器
monostack crane 单轨堆垛起重机
monostatic configuration 单元组态
monostatic pulse radar 单稳态脉冲雷达
monostatic radar 单站雷达;单基地雷达
monostatic relay 单稳态继电器
monostatic sonar 收发合置声呐
monostichous 单列的
mono-strand cable 单股钢丝绳;单股钢(丝)索
monostratum 单一地层;单层
monostyle 单柱式(建筑)
monosulfide 一硫化物
monosulphate 单硫酸盐
monosulphide 一硫化物
monosymmetric(al) 单轴对称的;两侧对称的
monosymmetric(al) dispersion of interference colo(u)rs 干涉色单对称分散
monosymmetry 两侧对称;单轴对称
monotectic 偏析的;偏共晶
monotectic alloy 偏共晶合金
monotectic equilibrium diagram 偏共晶平衡图
monothermite 单热石
monotint 单色(的);单色画
monotone 单色调;单调性
monotone convergence 单调收敛

monotone convergence theorem 单调收敛定理
monotone decreasing 单调(递)减
monotone decreasing function 单调递减函数
monotone function 单调函数
monotone increasing 单调增;单调递增
monotone increasing function 单调递增函数
monotone multiple decision problem 单调决策多维问题
monotone non-decreasing function 单调非减函数
monotone non-increasing function 单调非增函数
monotone order 单调次序
monotone sequence 单调序列
monotone system 单调系统
monotone variable 单调变量
monotonic 单调性;单调的
monotonically increasing function 单调上升函数
monotonically non-decreasing function 单调非减函数
monotonically non-increasing function 单调非增函数
monotonic analysis 单调分析
monotonic convergence 单值收敛
monotonic decreasing quantity 单调减量
monotonic function 单调函数
monotonic functional 单调泛函
monotonic increasing quantity 单调增量
monotonicity 单调性
monotonic loading 简单负荷;单冲荷载
monotonic model 单调模型
monotonic model of the first kind 第一类单调宇宙模型
monotonic operation 单调运算
monotonic operator 单调算子
monotonic quantity 单调量
monotonic quiet 单调的
monotonic sequence 单调序列
monotonic system of algebra 单调代数系
monotonic transformation 单调变换
monotonic utility 单调效用
monotonic variable 单调变数
monotonous loading 单调加载
monotony 单调(性)
monotower crane 塔吊;塔式起重机;单塔起重机
monotower derrick 塔(式)桅(杆)起重机;单塔起重机
monotrack recorder 单踪记录仪
monotractor 单轮驱动拖拉机
monotrailer 单轴挂车
monotriglyph 单三槽板柱间布置;单排档
monotriglyphic 柱间为一个三槽板的(陶立克式)
monotron 直越式速调管;莫诺特龙硬度检验仪;莫诺特龙速调管;单像管
monotron hardness test 莫诺特龙硬度试验
monotron indentation hardness test 莫氏刻痕硬度试验;莫诺特龙刻痕硬度试验
monotrophic 单变的
monotropic 单变性的
monotropic transition 不可逆过渡
monotropic function 单值函数
monotropic inversion 单变性转变
monotropic transformation 单向转变
monotropy 单变性;单变现象
monotropy coefficient 单变系数
monotube 单管
monotube boiler 直流锅炉;单管(式)锅炉
monotube column 单管柱
monotube pile 独管桩;单管桩
monotube pole 单管杆柱
monotube post 单管柱
monotube type boiler 单管式直流锅炉
monotubing 单管
Monotype 莫诺铸排机;自动排浇机;单型;单版画制作法
monotype printing 单型印刷
monotypic genus 单模属
monounsaturate resin 单一不饱和树脂
monouranate 一铀酸盐
monovacancy 单空位
monovalent 单价离子
monovalent base 一价碱
monovalent ion 一价离子
monovariant 单变量的;单变状态的;单变的
monovariant state 单变状态
monovariant system 单变体系

monovinyl acetylene 一乙烯(基)乙炔;丁烯炔
monovolume 单卷
monowheel 单轮
monowheel chassis 单轮起落架
monowheel rotary ditcher 单轮旋转式开沟机
monox 氧化硅
monozone dam 均质土坝
monozygotic 单合子的
monpavet 单色铺面层;本色沥青瓦片;单色沥青瓦片
Monray roof tile 蒙瑞屋面瓦
Monroan(stage) 门罗阶【地】
Monsanto resonance elastometer 蒙桑托共振弹性计
Monsel's solution 次硫酸铁溶液
monsmedite 水钾铊矾
monsoon affected country 受到季风影响的国家
monsoon air 季风气流;季风气候
monsoonal circulation 季风环流
monsoonal climate 季风气候
monsoon and trade-wind littoral climate 滨海季风—信风气候
monsoon broadleaf evergreen forest 季风常绿阔叶林
monsoon circulation 季风环流
monsoon climate 季风气候
monsoon climatology 季风气候学
monsoon cluster 季风云团
monsoon current 季风海流
monsoon depression 季风低(气)压
monsoon drift 季风漂流;季风流
monsoon flow 季节风期流量
monsoon fog 季风雾
monsoon forest 季雨林;季风林
monsoon index 季风指数
monsoon low 季节性低气压;季风低(气)压
monsoon ocean current 季风洋流
monsoon precipitation 季风降雨(量);季风降水(量)
monsoon rain 季风雨
monsoon rainfall 季风降雨(量);季风降水(量)
monsoon region 季风区
monsoon season 季风季节
monsoon surge 季风潮
monsoon troush 季风口低压槽
monsoon weather 季风天气
monsoon wind 季风
monster buoy 大型观测浮标
monster wave 狂浪
monstrance 圣餐饼陈列器
monstrometer 坑道钻孔方向计
monstrous sum 巨氯
montage 镶嵌;镜头组接;剪辑画面;剪辑;蒙太奇;拼贴;拼接
montage amplifier 图像剪辑放大器
Montague profile 半重力抛物线形斜坡
montaine mossy forest 山地苔藓林
Montanan 蒙塔纳统【地】
montance thicket 山地灌丛
montane forest 高山树林
montanic acid 褐煤酸
montanite 碲铋华
montan pitch 褐煤硬沥青
montant 窗梃;门梃;构架式竖杆;竖杆
montan wax 蒙丹蜡;褐煤蜡;粗褐煤蜡
montan-wax tar 褐煤蜡焦油;地蜡焦油
montasite 纤铁闪石
Montauk building 蒙陶克大楼
montbeliard 厚床单
montbrayite 亮碲金矿
montdorite 芒云母
Monte-Carlo analysis 随机搜索法;蒙特卡罗分析法
Monte-Carlo approach 蒙特卡罗法
Monte-Carlo calculation 蒙特卡罗计算
Monte-Carlo exercise 蒙特卡罗实验
Monte-Carlo experiment 蒙特卡罗试验;蒙特卡罗实验
Monte-Carlo forecast 蒙特卡罗预测
Monte-Carlo generator 随机数产生程序
Monte-Carlo investigation 蒙特卡罗实验
Monte-Carlo method 统计试验法;蒙特卡罗法
Monte-Carlo process 蒙特卡罗法
Monte-Carlo sampling distribution 蒙特卡罗抽样分布
Monte-Carlo simulation 蒙特卡罗模拟(法)

Monte-Carlo simulation method 蒙特卡罗模拟法
Monte-Carlo study 蒙特卡罗研究
Monte-Carlo technique 蒙特卡罗技术;蒙特卡罗法
monte community 有刺短灌木丛
Montee caisson 蒙蒂沉箱
Montegal 蒙蒂盖尔铝合金;镁硅铝合金
Montegal alloy 蒙蒂盖尔合金
montejus 压气升液器;蛋形升液器
Montenegro test 蒙特尼格罗试验
monteponite 方镉石
monteregianite 硅碱钇石
Monterey shale 蒙特雷页岩
Monterey Spanish architecture 蒙特雷西班牙式建筑
Montes Alpides 阿尔卑斯山脉
Montes Altai 阿尔泰山脉
Montes Apennines 亚平宁山脉
Montes Carpatus 喀尔巴阡山脉
Montes Caucasus 高加索山脉
Montes cordillera 科迪勒拉山脉
Montes haemus 海姆司山脉
Montes harbinger 前驱山脉
Montes Pyrenean 比利牛斯山脉
montes recti 直列山脉
Montes riphaeus 里菲山脉
Montes rook 卢克山脉
Montes spitzbergensis 斯皮兹柏金西斯山脉
Montes teneriffe 泰纳里夫山脉
montgolfier 热空气气球
montgomeryite 磷铝镁钙石
month 月份
month cam 月凸轮
month counter 月计数器
month-end delivery 月末交付
month-end fund 月末资金
month-end payment 月末支付
month footage 月进尺
month for cyclic(al) dominance 循环波动支配月数
month for cyclic(al) dominance curve 循环波动支配月曲线
Monthier's blue 蒙提蓝
monthly 每月的;按月的
monthly accounting 按月结算
monthly allotments 每月分摊数
monthly allowance 每月津贴
monthly amplitude 月振幅;月较差;月变幅
monthly average 月平均(值)
monthly average daily traffic 月平均日交通量
monthly average discharge 月平均流量
monthly average rainfall 月平均降雨量
monthly average stage 月平均水位
monthly average temperature 月平均温度
monthly average traffic 月平均交通量
monthly balance sheet 月资产负债表
monthly capacity 月生产能力
monthly certificate measurement 按月结算
monthly closing entries 每月结账记录;按月结算分录
monthly consumption 每月消费量
monthly consumption use 月耗水量
monthly data 每月数字
monthly delivery 每月交货量
monthly depreciation reserves 月折旧回收
monthly detailed schedule 月明细表
monthly discharge 月平均流量
monthly distance hydrograph 月流量过程线
monthly distribution of precipitation 月降水量分布
monthly earnings 月工资
monthly efficiency 台月效率
monthly error 月误差
monthly estimate 月预算
monthly evaporation discharge 月蒸发量
month ly financial statement 月财务报表
monthly flood 月最大洪水量;月最大(洪峰)流量;月洪水
monthly flow duration curve 月流量历时曲线
monthly freight transport plan 月度货物运输计划
monthly goods transport plan 月度货物运输计划
monthly index 月指数
monthly index chart 每月指数图表
monthly individual payroll sheet 每月个人工资计算表
monthly inspection 月度检查
monthly installment 按月摊付;按月付款

monthly interest 月利息
monthly interim payment certificate 月中付款证书
monthly investment plan 月投资计划
monthly irregular hours worked 每月不规则劳动时数
monthly load curve 月负载曲线
monthly load duration curve 月负荷历时曲线
monthly load factor 月负载率
monthly magazine 月报
monthly maintenance 月修
monthly maximum 月最大
monthly maximum load 月最高负载
monthly mean 月平均(值)
monthly mean absolute humidity 月平均绝对湿度
monthly mean air temperature 月平均气温
monthly mean discharge 月平均流量
monthly mean rainfall 月平均降雨量
monthly mean relative humidity 月平均相对湿度
monthly mean runoff 月平均径流量
monthly mean sea level 月平均海平面
monthly mean sediment concentration 月平均含沙量
monthly mean sediment discharge 月平均输沙量
monthly mean stage 月平均水位
monthly mean temperature 月平均温度;月平均气温
monthly mean value 月均值
monthly mean value of atmospheric pressure 月平均大气压力值
monthly minimum 月最小
monthly minimum temperature 月最低温度
monthly nutation 周月章动
monthly order 每月订货
monthly output 每月产量
monthly pay 月薪
monthly payment 按月支付;每月付款;按月付款
monthly payment certificate 月付款证书
monthly payment loan 每月还款的借款
monthly peak load 月峰荷
monthly precipitation 月降水量
monthly premium 每月保险费
monthly process report 每月进度报告
monthly production report 每月生产报告
monthly progress 月进度
monthly regulating reservoir 月调节水库
monthly report 月末报告书;月度报告;月报
monthly report of cash income and expense 现金收支月报
monthly report of stopping leakage and keeping income 堵漏保收月报
monthly report of transport revenue 运输进款月报
monthly returns 月报表
monthly rose 月季花
monthly run 每月运转
monthly salary 月薪
monthly schedule of fixed charges 固定费用月明细表
monthly settlement 按月结算
monthly settlement report 月结算报告
monthly statement 月统计表;月付款申请;月度结算;月度计算;月报表
monthly statement of account 每月对账单
monthly store balance sheet 材料月结平衡表
monthly tables 月报表
monthly traffic pattern 交通量月变化图
monthly transport plan 月度运输计划
monthly unbalance coefficient of cargo handled at the port 月吞吐量不平衡系数
monthly unbalance factor 月不均系数
monthly variation 月(度)变化;逐月变化;交通量月变化图
monthly wages 月薪;月工资
monthly wagon requisition form 月度要车计划表
monthly water discharge 月径流量
monthly withholding tax 每月预提所得税
month of dried spring 泉水干枯月份
month of the phases 太阴月
months after due date 到期以后的月数
month star-wheel 月星轮
month to month 逐月
month to month tenancy 按月租赁
Montian 蒙泰阶【地】
monticellite 钙镁橄石
monticule 小山岗;小丘;小火山(丘);尖峰

montmartite 灰性石膏
montmorillonite 胶岭石;蒙脱石;微晶高岭土;微晶高岭石;蒙脱土
montmorillonite clay 蒙脱石黏土;蒙脱石黏土;活性蒙脱土
montmorillonite content 蒙脱石含量
montmorillonite mudstone 蒙脱石泥岩
montmorillonitic clay 蒙脱黏土
montmorillonoid 蒙脱石;细黏土;蒙脱岩
Montpellier yellow 蒙特佩利厄黄
Mont Qomolangma 珠穆朗玛峰
montre 装饰音管
montroseite 黑铁钒矿
montroydite 橙红石;橙汞矿
Mont sopris series digital logging system 蒙特公司系列数控测井仪
monument 界碑;纪念性建筑物;纪念物;纪念塔;标石
monumental arch 纪念拱
monumental architecture 纪念性建筑
monumental brass 黄铜纪念碑
monumental building 纪念建筑;纪念性建筑物
monumental chapel 纪念小教堂
monumental church 纪念教堂
monumental column 纪念柱
monumental courtyyard 纪念庭院
monumental effect 纪念效应
monumental entrance 纪念性入门口
monumental facade 有纪念意义的建筑物立面
monumental gateway 神殿入口;纪念性通廊;纪念性门口;纪念牌坊
monumental height 纪念性高地
monumental inscription 碑文
monumentality 纪念作品;纪念物;纪念性
monumental mason 精雕石工
monumental material 纪念资料
monumental plaza 纪念性广场
monumental portal 纪念门座
monumental ruin 纪念物遗址;纪念物废墟
monumental sculpture 纪念雕塑
monumental stairway 纪念楼梯
monumental structure 纪念塔构造;纪念塔建筑;纪念碑构造;纪念碑建筑
monumental style 纪念风格
monumental tomb 纪念墓
monumental window 雄伟的窗;庄重的窗
monumentation 埋石;埋标石
monument construction and records 施工记录纪念碑
monumented benchmark 埋石水准点
monumented boundary peg 永久界桩
monumented control point 埋石控制点
monumented cross-section 标定断面
monumented point 埋石点;设标点
monumented station 埋石点
monumented survey point 埋石点;设标点
monument for birds 鸟类纪念碑
monument mark 标志桩;标石;标点;界碑;纪念碑
monument of liberty 自由之碑
monument of limestone 石灰石纪念碑
monument of Lysikrates at Athens 雅典的列雪格拉德文体纪念碑
monument point 标志石
monument sacred to the memory of martyrs 烈士纪念碑
monument survey 测量标石;测量标志
monument tie 界碑系桩;标石系桩
monument to martyrs 烈士纪念碑
Monument to the People's Heroes 人民英雄纪念碑
monzodiorite 二长闪长岩
monzodiorite porphyrite 二长闪长玢岩
monzonite 碱铁钙硅铝酸盐;二长岩
monzonite group 二长岩类
monzonite porphyry 二长斑岩
monzonitic 二长岩的
monzonitic granite 二长花岗岩
monzonitic texture 二长岩结构;二长结构
monzosyenite 二长正长岩
mooihoekite 褐硫铁铜矿
moon 太阴
moon age 月球年龄
Moon atlas 月面图集
moonbeam (打桩架底下的)向上弯梁;月光束;上

弯梁
moonbow 月虹
moon-chalk 月色白垩
moon clock 月钟
moon compass orientation 月亮罗盘定向
moon culmination 月中天;月亮中天
moon dog 幻月
moon down 月落
moon-earth transportation system 月地间运输系统
moon enters umbra 本影食始
Mooney rhomb 门尼菱形镜
Mooney shearing-disk viscometer 门尼剪切圆盘黏度计
Mooney unit 门尼单位
Mooney visco(si)meter 门尼黏度计
Mooney viscosity 门尼黏度
moon-face 月样圆面容
moon gate 月亮门;(中国建筑中的)月洞门
mooning 发黄
moon knife 月牙刀
moon leaves umbra 本影食终
moonlet 小月亮
moonlight 月光
moonlight gasoline 月光汽油
moonlighting 兼职
moon map 月面图
moon mapping 月球测图
moon milke 月奶石
moonmist 月雾色
moonmobile 月球车
moon month 阴历月份
moonpath 水波中月影
moon phase 月相
moon pillar 月柱
moonpool 船井(海洋钻探船)
moon port 月球火箭发射站;月球火箭发射场
moonquake 月震
moonquake monitor 月震检收器
moon rise 月出
moonrise and moonset tables 月亮出没时间表
moon's age 月龄
moon's attraction 月球引力
moonscope 人造地球卫星望远镜
moon scraper 月球取样器
moon's disc 月轮
moon seismograph 月震仪
moon set 月没
moon-shaped face 满月脸;月样圆面容
moon-shaped feed opening 弯月形进料口(旋回破碎机)
moon-shaped object 月形物
moon sheered 首尾上翘的船型
moonshine 月光
moon's horizontal parallax 月球地平视差
moon's horizontal semidiameter 月球地平
moon shot 月球探测器
moon's motion 月离
moon's orbit 月球轨道;太阴轨道
moon's path 白道
moonstone 月长石;无光白釉;冰花石
moonstone glass 月长石状玻璃
moon tidal component 太阴分潮
moon tides 月球潮汐
moontracking 月球跟踪
moon transit 月中天
moon watch 月球观察;人卫监视
moonwatcher 卫星观察者
moonwatch telescope 广角镜
moon work 月龄装置
moor 系留;沼泽;高沼;矿脉富集部分;泥炭沼泽;双锚泊;湿原
mooraboolite 钠沸石
moorafe 停泊场
moorage 系泊处;系泊作业;停泊费;停泊处
moor alongside 靠泊;靠岸系泊
moor anchors ahead 船首双锚泊
moorburn 烧荒
moor coal 松散泥煤;易碎沼煤;易碎褐煤;沼煤
Moore-Camhbell kiln 莫-肯二氏电隧道窑
Moore chain 锚链
moored 系泊的
moored alongside 靠拢停泊
moored array 锚系基阵

moored boat 系泊小船
moored mine 系留水雷
moored monitoring system 系留监测系统
moored ship 系泊船泊
moored sonobuoy 锚系声呐浮标
moored suction dredge(r) 直吸式挖泥船
moored suction type 锚碇吸扬式
moored surveillance system 系留(式)监视系统
moored vessel 系泊船泊
Moore filter 莫尔过滤器;莫尔型多叶真空过滤器
mooreite 锌镁矾;锰镁锌矾
Moore lamp 莫尔灯
Moore machine 莫尔机
Moore-Penrose inverse 莫尔—彭罗斯逆变换
Moore's bridge 莫尔桥(支承在过梁上)
Moor filter 莫尔滤机
moor fore and aft 首尾系泊
moor forming process 泥炭沼泽形成过程
moorgae 系泊动作
moor grass 酸沼草原
moor head and stern 艏艉系泊;首尾系泊
moorhouseite 水镍钴矾
mooring 系留;系船处;系泊(费);锚泊;停泊(指系泊作业)
mooring accessories 系缆铁件;系泊设施;系船设备
mooring anchor 系船锚
mooring anchorage 系泊锚位
mooring area 锚泊区(域);系泊地;泊地;泊船区
mooring arrangement 系缆设备;系船设施;系船设备;系泊设备
mooring barge 系船驳(船);趸船
mooring basin 系船港池;系泊港池;系船港池;锚泊区(域);锚(泊)地;停泊区;船舶作业水域;泊(船)地;泊船池
mooring berth 系泊地;锚地
mooring bitt 系缆桩;系缆柱;系船柱
mooring block 锚碇块;浮筒沉坠;浮筒沉附
mooring boat 绞滩接头船;带缆艇
mooring bollard 系缆柱;双柱系缆柱;双柱系船桩
mooring bolt 系船栓
mooring boom 靠船臂架;锚碇臂杆
mooring bridle 系在浮筒上的链或短链(供系泊时使用)
mooring buoy 系船浮筒;系船浮标;系泊浮筒
mooring by the head and stern 首尾系泊
mooring cable 系泊链;系泊缆(索)
mooring caisson 系泊沉箱
mooring capstan 系泊绞盘
mooring chain 系船锚链;锚链
mooring charges 停泊税
mooring chock 导缆钩
mooring cleat 系缆耳;碇栓
mooring clump 锚碇沉块;浮筒沉坠;浮筒沉附
mooring craft 系缆艇;运锚小艇
mooring design 锚泊装置设计
mooring device 停泊装置
mooring dolphin 系缆泊位;系缆浮标;系船桩柱;系留椿;系缆桩;系船墩;靠船墩;系船簇桩
mooring drag 活动锚
mooring equipment 系船设施;系船设备;锚泊设备
mooring facility 系船设施;系船设备;泊船设施
mooring fast 系船缆;系泊缆(索)
mooring fiber rope 系船纤维缆
mooring fittings 系缆铁件;系泊设备
mooring force 系泊力;系船力
mooring fore and aft 首尾系泊
mooring gear 系船设施;系泊设备;系泊装置
mooring geometry 锚泊装置外形
mooring ground tackle 碇泊装置
mooring guy 系留索;系船索
mooring harbo(u)r 锚泊港
mooring hardware 系缆铁件
mooring harness 系泊装具
mooring hawser 系留索;系船索;系船缆(索);系泊缆(索);大缆
mooring head down 船首向外靠岸(即船首背向港口门)
mooring head in 船首向内靠岸;船首朝港内的系泊
mooring head out 船首向外靠岸(即船首背向港口门);船首朝港外的系泊
mooring hole 导缆孔
mooring hook 系缆钩
mooring island 系泊岛;岛式系泊地
mooring lighter 港内运送长期系泊设备的小船;港

口运送长期系泊设备的小船
mooring line 系船缆绳;系船缆;系泊缆(索)
mooring line characteristic 锚链性能
mooring load 系船荷载;系泊荷载
mooring maneuver 靠泊调动
mooring mast 系留塔;系留杆
mooring mat 护缆垫
mooring network 系泊观测网
mooring operation 锚泊作业
mooring pattern 系泊模式;锚泊形式
mooring peg 系泊用桩桩
mooring pendant 连接系船浮筒与浮筒底链的小链
mooring pier 系船码头
mooring pile 系船桩;锚碇桩;码头系缆桩
mooring piling 系船桩
mooring pipe 导缆管
mooring place 系泊码头;泊地;泊船所
mooring platform 系泊平台
mooring point 系泊荷点;系泊点;停泊点
mooring point structure 系泊荷点结构物
mooring port 锚泊港;导缆孔
mooring post 系泊柱;系船柱;码头系缆桩;带缆桩;带缆柱
mooring pull 系船拉力
mooring rack 靠船用排桩(运河口设备)
mooring right 系泊权
mooring ring 系船环;浮筒系船环
mooring rod 锚碇拉杆
mooring rope 系船索;系泊缆(索);系泊缆(索)
moorings 停泊处
mooring screw 螺旋形锚碇块(体);螺旋形锚;螺旋式锚;锚碇螺旋
mooring shackle 系船卸扣;系船卡环;系船钩环;双链转环
mooring sinker 锚碇沉块;浮筒沉坠;浮筒沉附
mooring space 锚泊地
mooring speed 靠码头速度;靠泊速度
mooring spud 带锚桩;带缆桩
mooring stall 浮台;锚这
mooring structure 系泊结构物;系船建筑(物)
mooring stump 系泊桩
mooring swivel 系船卸扣;系船转环;双锚锁环;双链转环
mooring system 系泊方式
mooring tackle 系泊索具
mooring telegraph 起锚传令钟
mooring terminal 系船码头
mooring tower 系留塔;系船塔架
mooring tractor 船坞牵引机
mooring trial 系泊试运转;系泊试验;系泊试车
mooring type at anchorage 锚泊方式
mooring type in anchorage 锚地停泊方式
mooring watch 早班
mooring wharf 系船码头;顺岸码头
mooring winch 系泊绞车;绞缆机
mooring wire 系留索;系船索;系船钢绳
mooring wire rope 锚用钢丝绳;锚索
mooring with two anchors 双锚系泊;双锚锚泊
Moorish arabesque 摩尔式阿拉伯花饰
Moorish arch 马蹄(形)拱;摩尔式拱;马蹄形拱(圈)
Moorish architecture 摩尔式建筑
Moorish capital 摩尔式柱顶;摩尔式柱头
Moorish cupola 摩尔式穹顶;摩尔式圆屋顶
Moorish dome 摩尔式穹隆
Moorish horseshoe arch 摩尔式马蹄形发券;摩尔式马蹄形拱
Moorish multifoiled arch 摩尔式多叶形拱
Moorish palace (中古时代北非或西班牙的)摩尔式风格的大型公共建筑;摩尔式宫殿
Moorish style architecture 摩尔族建筑;摩尔式建筑
moorland 泥炭沼地;沼泽地;高沼地;高沼草原;荒野
moorland soil 沼泽土(壤);沼(泽)地土壤
moor light 停泊灯
moor peat 沼煤;高沼泥炭;泥沼土;高沼泥;泥煤
moor-salt 海盐
Moor's filter 莫尔过滤器
moor soil 沼泽土(壤);泽地土壤
moorstone 锚石;花岗岩;花岗石(质孤石)
moor with an open hawse 迎风抛八字锚
moor with stern way 后退抛双锚法
moor with two anchors 抛双锚
Moorwood machine 浸镀锡机
moory 荒野的;原野的;沼地的;多沼泽的

mooscape 月面景色
mop 布轮;抛光轮;拖布;拖把;挡渣器;擦光辊
mopboard 踢脚板;护栏板;壁脚板
mopboard component 踢脚板构件
mopboard member 踢脚板成件;踢脚板构件
mopboard radiator 踢脚板暖气管;踢脚板暖气片;踢脚板散热器
mopboard unit 踢脚板构件
mop-cat 扫油双体船
mop-catamaran 扫油双体船
moped 装有小发动机的自行车;轻型摩托车胎
moped track 机动自行车专用道
mopen 摹本缎
mope pole 圆木杆;支承管道杆;支撑管道杆
Mo-perm alloy 钼波马合金
Mo-perminvar alloy 钼波民瓦尔合金
Mopin system 穆宾体制(指钢结构和混凝土结构的大生产方式)
mop oil 家具油;地板油
mopping 用拖把涂沥青;拖刷;刷浆
mopping coat 热刷层
mopping method 刷浆法
mopping of hot asphalt 热柏油涂刷;热沥青涂刷
mopping up 清理火场
mopping up time 清理火场时间
mop plate 踢脚板;门脚板;踢板
mop polishing 毡轮抛光;毛刷擦亮
Moppon 莫邦材料
mop rendering 拖把涂刷
mop sink 拖把槽;拖布池;墩布池
mopstick 半圆形(木)扶手;圆形扶手;拖把柄;墩布把
mopstick handrail 平底圆形栏杆扶手;圆形扶手
mor 酸性有机质;森林土壤;粗腐殖土;森林粗腐殖质;粗腐殖质
Mora 矮柳条凳或搁脚凳;莫拉(杂色硬木,产于圭亚那)
moraesite 水磷铍石
morainal 冰碛的
morainal accumulation 冰碛积集
morainal apron 冰水沉积平原
morainal chip(ping)s 冰碛碎石片;冰碛石屑
morainal dam 冰碛坝
morainal dam lake 冰碛湖
morainal delta 冰碛三角洲;冰接三角洲
morainal deposit 冰碛沉积
morainal gravel 冰碛砂砾;冰碛石子;冰碛砾石
morainal plain 冰水沉积平原
morainal sand 冰碛砂
moraine 冰碛土;冰堆石;冰碛(层);(熔岩流表面的)火山碎屑;土石礁;石堆
moraine bar 冰碛砂洲;冰碛埂
moraine breccia 冰碛角砾岩
moraine clay 冰碛黏土
moraine-dammed lake 冰碛堰塞湖
moraine debris 冰碛物
moraine fan 冰碛扇
moraine hill 冰碛丘陵
moraine kame 冰碛阜
moraine lake 冰碛湖
moraine loop 冰碛环
moraine of advance 前进冰碛
moraine soil 冰碛土
moraine terrace 冰碛地坪;冰碛门廊;冰碛阶地;冰成阶地
morainic 冰碛的
morainic dam 冰碛坝
morainic delta 冰碛三角洲
morainic lake 冰碛湖
morainic soil 冰碛土
morainic soil layer 冰碛土层
morainic topography 冰碛地形
moral depreciation 无形损失;无形损耗
moralization tax as pollution abatement tool 缓减税法作为除污手段
moral person 法人
morass 沼泽;泥沼(地);泥塘
morass ore 褐铁矿
moratoria 延期偿还
moratorium 延期付款(命令);依法给债务人的延期偿付权;延期偿付期间;延期偿付(权);延缓偿付期;缓期偿债;欠款的延缓偿付期;停止兑现;暂停活动;[复]moratoria
moratory law 延期偿付法

Morava school 摩拉瓦学校
morbidity(rate) 发病率
morbidity statistics 发病率统计
morbidity survey 疾病调查
morcellation 分碎
mordant 金属腐蚀剂;金箔黏合剂;媒染剂;媒染的;化学清洗剂;染料媒介剂;酸洗剂
mordant based on turpentine 以松节油为基料的金属蚀剂
mordant colo(u)r 媒染染料
mordant dye 媒染染料
mordant for staining wood 腐蚀性的木染料
mordant in dy(e)ing 媒染剂
mordanting 媒染
mordanting assistant 媒染助剂
mordant rouge 媒染剂红液;媒染红
mordant yellow 媒染黄
mordenite 丝光沸石;发光沸石
moreacol 柬漆酚
more active element 易带出的元素
more advanced generations 更高世代材料
moreauite 水磷铝铀云母
more-centered arch 多中心拱;多心圆拱
more cold water 很冷的水
more complete 较完整
more confidential 机密
more developed country 较发达国家
more efficient fixed assets 高效能的固定资产
more favorable signal 较大允许信号
more furrow irrigation 进一步沟灌
more gentle descending gradient 较缓下坡【铁】
Morehouse mill 摩尔豪斯研磨机
more investigation by means of experimment 通过试验做更多的调查
more land 更多耕地
more light black 淡黑
more light gray 淡灰
more loss less gain 得不偿失
more loss than gain 得不偿失
morencite 绿脱石
morenosite 碧矾
more or less clause 超过或不足条款
more or less open 半公开的
more or less shipowner's option 船东有增减选择权
more or less terms 数量增减范围条款
more-out-than-in method 出多入少法
more pay for more work 多劳多得
more population 密生苗
more replenishing soil moisture reserves 提高补给土壤的储水量
more restrictive signal 较大限制信号
mores 生态类群
more sanitary water 相当卫生的水
more seedlings 密株
Moresque 马蹄形拱;摩尔式建筑
more than they can afford 超过支付能力
more volatile component 易挥发成分;轻组分
Morgagni's columns 摩尔干格尼柱
morganite 锰绿柱石
Morgan-Marshall test 摩一马公司试验法
Morgenstern method of slope stability analysis 摩根斯坦边坡稳定性分析法
Morgoil bearing 铝锡合金轴承
morgue 太平间;陈尸所;资料室;参考图书室
Morie fringe 摩尔条纹
Morie frings technique 摩尔条纹技术
Morie osmolarity 摩尔渗透压浓度
Morinda citrifoira oleum 水冬瓜油
morinite 红磷钠矿;水氟磷铝钙石
Morin's tables 摩林氏表(列有各种静止角和摩擦系数)
morioplasty 补缺术
mor layer 粗腐殖质层
Mormon sandstone 莫尔蒙砂岩
morning amplitude 晨角距
morning and evening tides 潮汐
morning calm 晨静
morning dress 常礼服
morning fog 晨雾
morning glory 牵牛花
morning-glory column 牵牛花式柱;顶有扩大托座的尖细空心混凝土柱;喇叭顶的柱
morning-glory shaft spillway 喇叭形竖井溢洪道

morning-glory spillway 直井式溢洪道;喇叭形溢洪道;喇叭口溢汽道;喇叭口溢洪道
morning group 早晨星组
morning peak hours 早晨高峰期;早高峰
morning peak traffic 早晨高峰交通量
morning room 早餐室
morning rush hours 早高峰
morning sight 上午测天
morning star 启明星;晨星
morning tide 早潮
morning twilight 曙光
Morocco faced tile 摩洛哥面砖
morocco pocket case 绘图器袖珍皮盒
morolic acid 摸绕酸
Moro reflex 紧抱反射
morozevicite 硫锗铅矿
morphallaxis 变形再生
morphic assemblage 变质杂岩体
morphodifferentiation 形态分化;形态变化
morphogenesis 形态形成;形态发生;地貌形成作用;地貌发生;地貌成因
morphogenetic 形态形成的;形态成因的;地貌发生的;地貌成因的
morphogenetic force 地貌发生力
morphogenetic region 地貌成因区
morphogenic 形态成因的;地貌成因的
morphographical method 地貌形态图法
morpholine 对氧氮己环
morphologic(al) 形态学(上)的;地形的
morphologic(al) age 地貌年龄
morphologic(al) analysis 构成分析法;结构分析术;地形分析;地貌分析;形态分析
morphologic(al) area 地貌区
morphologic(al) astronomy 形态天文学
morphologic(al) basin 地貌盆地
morphologic(al) change 地形变化;形态变异;形态变化
morphologic(al) character(istic) 形态特征
morphologic(al) characters 形态性状
morphologic(al) classification 形态学分类;形态分类
morphologic(al) classification of mountain 山地形态分类
morphologic(al) classification of trop 圈闭形态分类
morphologic(al) complexities of orebodies 矿体形态复杂程度
morphologic(al) correlation 形态相关
morphologic(al) cycle 地貌旋回
morphologic(al) data 地貌资料
morphologic(al) description 地貌形态描述
morphologic(al) details 地貌细部;地貌碎部
morphologic(al) differences 形态差异
morphologic(al) differentiation 形状分比;形态分化;形态变化
morphologic(al) elements of erosion(al) groove 侵蚀沟形态要素
morphologic(al) error 形态误差
morphologic(al) evolution 地形演变
morphologic(al) examination 形态学检查
morphologic(al) feature 地貌特征
morphologic(al) identification 形态鉴定
morphologic(al) index 形态标志
morphologic(al) map 地势图
morphologic(al) mapping language 形态制图语言
morphologic(al) mineralogy 形态矿物学
morphologic(al) observation 形态观察
morphologic(al) process 地形演化过程;地形变化过程;形态演变过程
morphologic(al) profile 地貌剖面图
morphologic(al) property 形态特性
morphologic(al) region 地貌区
morphologic(al) research 形态研究法
morphologic(al) stability 形态稳定性
morphologic(al) structure 形态结构
morphologic(al) survey 地貌测量
morphologic(al) unit 地貌单元;地貌面
morphologic(al) variation of orebody 矿体形态变化
morphology 形态学;形貌学
morphology classify of sinter 泉华的形态分类
morphology factor 形态因素
morphology of martensite 马氏体形态学
morphology of minerals 矿物的形态

morphology of orebody 矿体形态
morphology of river 河流形态(学)
morphology of river system 河系形态
morphology of the wool fiber 毛纤维形态学
morphometric(al) parameter 形态测量参数;流域形态测量参数
morphometry 形态测量(学);形态测定法;地貌形态量测;地貌测量
morphophysics 形态物理学
morphopoiesis 形态形成
morphosis 形态结构;形态构成
morphostratigraphic(al) unit 地貌地层单位
morphostructural line 地貌结构线
morphostructure 构造地貌
morphostructure marker 构造地貌标志
morphotectonic analysis 地貌构造分析
morphotrophy 变晶性
morphotropism 准同形性;变形性
morphotropy 变形性;变晶影响;形变
Morris chair 大安乐椅;莫里斯式安乐椅
morro 沿岸弧丘
Morse code 莫氏电码;莫尔斯码
Morse code light 莫尔斯信号灯
Morse conical reamer 莫氏圆锥铰刀
Morse distribution 莫尔斯分布
Morse drill 莫尔斯钻
Morse equation 莫尔斯方程式
Morse ga(u)ge 莫氏量规
Morse's chart 莫氏蒸汽图
Morse's cone 莫氏圆锥
Morse signal light 莫尔斯信号灯
Morse standard 莫尔斯标准
Morse standard taper 莫氏标准锥度
Morse standard taper plug ga(u)ge 莫氏标准锥度(塞)规
Morse's taper 莫氏锥度
Morse taper bush with tang 带扁尾莫氏圆锥衬套
Morse taper drill sleeve 莫氏锥度钻套
Morse tapered hole 莫氏锥形孔
Morse taper finishing reamer 莫氏锥形精加工铰刀
Morse taper ga(u)ge 莫氏锥度规
Morse taper reamer 莫氏锥形铰刀;莫氏锥度铰刀
Morse taper roughing reamer 莫氏锥形粗加工铰刀
Morse taper shank drill 莫氏锥柄钻
Morse taper shank twist drill 莫氏锥柄麻花钻头
Morse taper sleeve 莫氏锥度套筒
Morse tapper ga(u)ge 莫氏锥度量规
Morse twist drill 莫氏麻花钻
mort 弓形湖
mortality 死亡数;死亡率;死亡量;必死性
mortality coefficient 死亡率系数
mortality curve 死亡率曲线
mortality factor 磨损系数
mortality multiple regression analysis 死亡率多元回归分析
mortality product 致死指数;致死积
mortality rate 死亡率
mortality ratio 人寿保险死亡比率;死亡率比
mortar 研钵;岩钵;臼;胶砂;水泥;灰浆;乳钵;砌砖浆;砂浆
mortar additive 砂浆添加剂;砂浆添加料;灰浆添加剂;灰浆添加料;灰浆附加剂
mortar admixture 砂浆混合料;灰浆或水泥砂浆掺合料;灰浆混合料;砂浆外加剂;砂浆掺合料
mortar aerated with foam 加气灰浆;泡沫砂浆
mortar agent 灰浆附加剂;灰浆添加剂
mortar aggregate 灰浆骨料;砂浆用砂;灰浆集料;砂浆集料;灰浆骨料
mortar-aggregate bond 砂浆集料结合力;砂浆骨料结合力
mortar-aggregate bond strength 砂浆集料结合强度;砂浆骨料结合强度
mortar aid 灰浆辅助料
mortar anchor 砂浆锚杆
mortar and rocket apparatus 抛绳设备
mortar anti-freezing admixture 砂浆防冻剂
mortar bar 砂浆试棒
mortar bar method 砂浆试棒法
mortar bar test 砂浆棒试验
mortar base 砂浆基础;砂浆底座;灰浆基础;灰浆底座
mortar base course 灰浆勒脚层;灰浆底层;灰浆基层
mortar basin 化灰池

mortar bed 坐浆;灰浆层;化石灰池;化灰池;碎斑沉积;砂浆垫层
mortar bedding 砂浆打底;铺砂浆垫层
mortar-bed method 灰砂层法;砂浆层法
mortar-board 镘板;灰(浆)板;砂浆托板;托灰板
mortar bolt 砂浆锚杆
mortar bond 灰浆砌筑;灰浆砌合
mortar-bound 涂有灰浆的;灰浆黏结的
mortar-bound macadam 灰浆结碎石面
mortar-bound surface 砂浆结碎石面层;砂浆结碎石路面
mortar box 研钵;灰浆池;灰槽;拌灰槽
mortar bricking 砂浆砌筑
mortar briquet(te)s 八字形砂胶试体;砂浆模制试件
mortar-calked joint 灰浆嵌缝;砂浆勾缝
mortar cement 灰砂水泥;灰浆胶结料
mortar class 灰浆类;砂浆类
mortar coating 砂浆涂层
mortar coating machine 砂浆涂复机
mortar composition 砂浆成分
mortar consistency 灰浆坚实度;灰浆稠度
mortar consistency tester 灰浆稠度仪
mortar consistometer 砂浆稠度计
mortar content 灰浆含量
mortar course 灰浆面层;灰浆覆盖层
mortar cover 灰浆盖面;灰浆罩;灰浆抹面
mortar covering 灰浆覆盖层;砂浆覆盖层;灰浆涂层;砂浆涂层
mortar coveying pump 灰浆输送泵
mortar creep 灰浆蠕变;灰浆塑性蠕变;灰浆塑性变形
mortar cube 灰浆压块;砂浆立方试块
mortar-cube test 砂浆立方体试验;砂浆立方块试验;砂浆标号试验;砂浆试块试验
mortar cylinder 灰浆圆筒;灰浆圆柱块
mortar dab 轻敷灰浆;涂饰;锤琢;砂浆涂抹
mortar densifier 灰浆稠化器
mortar densifying agent 灰浆稠化剂
mortar dropping 灰浆喷涂;灰浆喷洒
mortared joint 砂浆接缝
mortar fabrication 灰浆生产;灰浆制造;灰浆拌和
mortar fabrication installation 砂浆制备设备;灰浆制备设备
mortar factory 灰浆工厂
mortar fill(ing) 填充灰浆;灰浆填补;灰浆嵌入;灰浆勾缝
mortar fixative 砂浆固定液
mortar for(clay) brickwork 砖砌用砂浆
mortar for coating 抹面砂浆
mortar for pounding 捣碎用研钵
mortar fraction 灰浆成分的百分含量
mortar from chamot(te) 火泥灰浆
mortar from trass 火山灰砂浆
mortar ga(u)ging water 计量灰浆混合水量
mortar grade 砂浆标号
mortar grouting 灌浆材料;灌浆;水泥砂浆灌浆
mortar grouting method 压浆方法
mortar gun 灰浆喷枪
mortar gunite method 喷浆法
mortar handling 灰浆运送;灰浆输送
mortar hod 灰砂斗;砂浆桶;灰浆桶
mortar ingredient 灰浆成分;灰浆配料
mortar injecting machine 灰浆喷射机
mortar injection 砂浆注射;灰浆喷射
mortar integral waterpellent admix(ture) 灰浆总体防水掺合剂;灰浆稠化剂
mortar integral waterproof(ing) agent 灰浆稠化剂;灰浆总体防水剂(指不透水性)
mortar intrusion 压浆;灌注砂浆;灰浆侵入(岩层)
mortar joint 砂浆接缝;灰浆接缝;灰缝
mortar joint pipe 灰浆接管
mortar joint reinforcement 灰缝加固
mortar joints for pipes 管道接缝-水泥砂浆
mortar lava 碎斑熔岩
mortar layer 砂浆敷面;灰浆层
mortarless 无灰浆的;干砌的
mortarless wall 干砌墙
mortar level(l)ing 砂浆找平
mortar lining 灰浆衬砌;砂浆衬里
mortar lump 灰浆团
mortar machine with trowel 带抹子砂浆机
mortar-making property test 成浆性质试验
mortar masonry(work) 灰浆砌墙工作;灰浆圬工;浆砌圬工
mortar material 砂浆配料;砂浆材料;灰浆材料
mortar mill 研磨机;臼研机;碾磨机;灰浆搅拌机;灰浆拌和机;砂浆料粉碎机;砂浆搅拌站;砂浆搅拌机;砂浆搅拌厂;砂浆拌和机;砂浆拌和厂;捣矿机
mortar mix 砂浆;灰浆;砂浆配合比
mortar mixer 胶砂搅拌机;灰浆搅拌机;灰浆混合机;灰浆拌和机;砂浆搅拌机;砂浆拌和机
mortar mixing aid 灰浆混合助剂;灰浆搅拌助剂
mortar mixing machine 砂浆搅拌机;灰浆搅拌机;灰浆混合机
mortar mixing plant 灰浆拌和厂;灰浆搅拌厂;砂浆搅拌站;砂浆拌和厂;砂浆搅拌设备;砂浆拌和设备
mortar mix(ing) ratio 灰浆混合比;灰浆配比;砂浆配合比
mortar mixture 砂浆混合料;胶砂混合料;砂浆混凝土
mortar of harshness 干硬性砂浆;干硬性灰浆
mortar pan 灰浆桶;灰浆盘;灰浆底座
mortar paving 灰浆胶砌的(块料)路面;砂浆路面;砂浆铺砌
mortar percentage 灰浆组分百分比
mortar ping 灰浆塞
mortar pitching 浆砌块石
mortar plant 砂浆工厂;灰浆工厂
mortar plasticizer 灰浆增塑剂;砂浆塑化剂
mortar plasticizing agent 灰浆增韧剂;灰浆塑化剂;灰浆增塑剂
mortar plug 灰浆塞子;灰浆塞
mortar pocket 灰浆蜂窝
mortar preparation 灰浆制备
mortar prism 灰浆棱柱体
mortar pump 灰浆泵;砂浆泵
mortar pumping and air placing machine 灰浆泵送和气力喷注机
mortar pumping tube 灰浆泵送管
mortar reducing 拉稀薄灰浆;减少灰浆
mortar replacement method 砂浆置换法
mortar resistance 耐灰浆性
mortar rubble 浆砌片石
mortar rubble foundation 浆砌片石基础
mortar rubble masonry 浆砌片石圬工;浆砌片石砌体;浆砌毛石圬工;浆砌毛石砌体;浆砌块石(圬土)
mortar rubble masonry pavement 浆砌片石护坡;浆砌毛石护坡
mortar rubble retaining wall 挡浆墙
mortar sand 建筑用砂;灰浆用砂
mortar screed(for plastering) 灰浆刮板;灰浆样板
mortar seal(ing) 灰浆密封;灰浆止水;灰浆密封;灰浆封层
mortar setting 胶砂凝结;灰浆凝固;砂浆硬化
mortar skin 灰浆皮;灰浆表面;灰浆薄层
mortar slab 砂浆板
mortar sledge 灰浆刮片;灰浆铲
mortar specimen 砂浆样品;灰浆样品;灰浆试块
mortar splashing 灰浆喷溅(法);灰浆喷射
mortar sprayer 灰浆喷射器;灰浆喷涂机
mortar spraying machine 喷砂浆机
mortar spray machine 灰浆喷涂机
mortar spreader 灰浆摊铺机;灰浆撒铺机
mortar stain 砂浆着色剂;砂浆污点;砂浆污斑;灰浆着色剂;灰浆污点;灰浆污斑;灰浆斑点
mortar stone 硬化路面;硬化泥浆;硬化灰浆;砂浆石
mortar stratification tester 砂浆分层度测定仪
mortar strength 灰浆强度;砂浆强度
mortar strip 灰浆冲筋
mortar structure 胶结状构造;碎斑构造
mortar supply hose 供砂浆软管
mortar tank 灰浆罐;灰浆池
mortar texture 胶结状构造;碎斑结构
mortar top 灰浆抹面;灰浆抹面
mortar transport 灰浆运送;灰浆运输;灰浆输送
mortar tray 砌筑铺灰板
mortar trough 灰浆槽;砂浆槽
mortar void 砂浆空隙
mortar-void method 灰浆孔隙法(混凝土);砂浆空隙法;砂浆空隙配料法
mortar-void ratio 灰浆孔隙比;灰浆空隙比;砂浆空隙比
mortar-void test 砂浆空隙试验
mortar wall(ing) 浆砌墙

mortar waterproofer 灰浆防水工;砂浆防水工;砂浆防水层;灰浆防水层;(用于干拌的)灰浆防水粉
mortar water-reducing agent 砂浆减水剂;灰浆减水剂
mortar wetting agent 砂浆可湿剂;砂浆润湿剂;砂浆浸润剂;灰浆增湿剂;灰浆可湿剂;灰浆润湿剂;灰浆浸润剂;灰浆增润剂
mortar with metallic aggregate 耐磨骨料砂浆
mortar workability agent 砂浆塑化剂;灰浆增塑剂
mortar works 灰浆工厂;砂浆工厂;灰浆工程;灰浆工作;砂浆工程;砂浆工作
morted ware 水渍器皿
mortgage 抵押
mortgage a house 押房
mortgage amortization schedule 抵押贷款分期偿付计划表
mortgage assets 抵押资产
mortgage assignment 抵押转让
mortgage assumption 抵押承担
mortgage-backed securities 有抵押保证的证券
mortgage bank 抵押贷款银行
mortgage banker 抵押贷款经理人
mortgage banking 银行抵押业务
mortgage bond 抵押债券;抵押单
mortgage broker 抵押贷款经纪人
mortgage commitment 抵押贷款承诺;抵押贷款承担
mortgage company 抵押公司
mortgage contract 抵押合同
mortgage contracts with warrants 具保证的抵押合同
mortgage credit 抵押信贷
mortgaged assets 抵押资产
mortgaged contract with risks 风险抵押承包
mortgage debt 抵押债务;抵押负债
mortgage deed 抵押契约;押契
mortgage discount 抵押折价
mortgaged property 被抵押的财产
mortgagee 受押人;承受抵押者;承受抵押人
mortgage financing 抵押贷款资金融通
mortgage goods 押货
mortgage guaranty insurance 抵押保险
mortgage insurance commitment 抵押保险承担
mortgage insurance premium 抵押保险费
mortgage investment company 抵押投资公司
mortgage liability 抵押负债
mortgage lien 扣押权
mortgage life insurance 抵押贷款人寿保险
mortgage loan 抵押贷款
mortgage loan disclosure statement 抵押贷款表述文件
mortgage money 抵押资金
mortgage moratorium 抵押延期偿付权;抵押延期偿付期
mortgage note 抵押票据
mortgage on real estate 房地产抵押
mortgage open-end 开放抵押;可加抵押
mortgage payment 应付押款
mortgage pool 联合抵押
mortgage protofolio 若干抵押贷款集合;抵押卷宗
mortgage purchase 抵押买卖
mortgager 出押人
mortgage receivable 应收抵押
mortgage reduction certificate 抵押余额证明书
mortgage release price 抵押贷款清偿数
mortgage relief 抵押契约解除
mortgage requirement 抵押贷款的分期还款额
mortgage revenue bond 抵押收益债券
mortgage securities pool 联合抵押证券
mortgage servicing 抵押贷款服务
mortgage take-back 抵押收回
mortgage warehousing 抵押囤积
mortgaging out (房地产的)全值抵押
mortgagor 出抵人;抵押者;抵押人;出押人
mortgagor's insurable interest 出押人保险利益
mortice 榫孔;榫接合
mortice and tenon joint 镶榫接头;镶榫接合
mortice chisel 榫凿
morticed cremone bolt 暗通天插销
morticed espagnolette bolt 暗通天插销
mortice ga(u)ge 开槽用划线器;开槽规;榫规;槽用分划器
mortice latch 闩;插锁
mortice lock 片锁

mortice lock with handle 执手插锁
mortice lock with knot 执手插锁
mortice lock with lever handle 执柄门锁
morticing 凿榫
mortic lock 暗锁
mortiferous 致死的
mortise 开榫;榫眼;榫孔;榫接合;榫槽
mortise and tenon 镶榫;阴阳榫;公母榫;榫头榫眼接合;雌雄榫
mortise-and-tenon joint 凹凸榫接合;公母榫接合;凸凹榫接头;雌雄榫接合;镶榫接头;镶榫接合
mortise astragal 镶入式半圆件
mortise axe 榫孔斧
mortise block 链滑轮
mortise bolt 门(窗)暗插销;暗插销
mortise chisel 榫孔凿;榫齿;榫(眼)凿
mortised astragal 榫眼腰线;镶入式半圆件
mortised block 镶轮滑车
mortised cremone bolt 暗通天插销
mortised espagnolette bolt 暗通天插销
mortise ga(u)ge 可调量规;划榫器;画榫线具;划榫线具;榫规
mortise hole 榫眼
mortise joint 榫接头;榫接;榫齿接合;插锁
mortise latch 槛内锁;插锁;执手插锁
mortise lock 暗锁;插锁
mortise machine 开榫机;凿榫(眼)机
mortise of cap block 斗口
mortise pin 榫销
mortise preparation 凿眼;榫眼开凿
mortiser 凿榫(眼)机;链插床;开榫机
mortise-type lock 嵌入式锁;暗锁
mortise wheel 嵌齿轮
mortising 暗榫眼;凿榫
mortising bit 榫眼钻头;榫眼凿
mortising machine 凿眼机;凿榫机;制榫机;开榫机
mortising slot machine 钻榫眼机;凿槽机
mortlake 弓形湖;牛轭湖;死水湖;废河道湖
mortlake deposit 牛轭湖沉积
mortlake facies 牛轭湖相
mortmain 土地的死守保有
mortuary 验尸所;埋葬的;停尸房;太平间;殡仪馆
mortuary basilica 殡仪馆内长方形会堂;公墓中的教堂
mortuary block 停棺处;尸体陈列处;墓穴
mortuary chapel 殡仪馆小教堂;公墓中的小礼拜堂
mortuary church 公墓教堂
mortuary cult 殡葬仪式;葬礼仪式
mortuary house 殡仪馆
mortuary monument 墓碑
mortuary temple (祭帝王伟人的)庙宇
mortuary urn 骨灰罐
moruro 紫褐色硬木(中美洲产)
Morus alba 桑树
morvan 准平原面交切
mos 生态种群
mosaic 镶嵌(性);镶嵌细工;镶嵌式构造;镶嵌花样;镶嵌光电元件;感光特性镶嵌幕;硫化锡黄粉;马赛克;镶嵌细工;嵌镶(式);嵌镶的;陶瓷锦砖;地砖
mosaic air photo 镶航空像片
mosaic area 镶嵌面
mosaic artist 马赛克能手;马赛克艺术家;镶嵌能手;镶嵌艺术家
mosaic asphalt 预制镶嵌地沥青
mosaic assembly 像片镶嵌
mosaic bed 镶嵌花坛
mosaic binding 镶嵌装订
mosaic block 镶嵌块;嵌镶块
mosaic building block 拼件;镶嵌块件
mosaic cathode 嵌镶阴极
mosaic cementation 镶嵌胶结
mosaic church 马赛克教堂
mosaic circuit diagram 镶嵌式电路图
mosaic clad 马赛克覆盖层;镶嵌覆盖层;镶嵌式外壳
mosaic clay tile 马赛克瓷砖;马赛克黏土地砖;彩色拼装黏土瓦
mosaic coating 镶嵌层
mosaic community 镶嵌群落
mosaic cover(ing) 马赛克覆盖;镶嵌式外包装
mosaic cracking 龟裂
mosaic crystal 镶嵌晶体;嵌镶晶体
mosaic culture 模纹种植
mosaic cupola 镶嵌小圆顶;马赛克圆顶;镶嵌式圆屋顶

mosaic cutter 嵌镶切削工
mosaic decoration 马赛克装潢;马赛克装饰;镶嵌装饰
mosaic decorative finish 镶嵌装饰性整修;马赛克装饰性整修
mosaic development 嵌合式发育
mosaic diagram 镶嵌图
mosaic dome 马赛克穹隆;镶嵌穹隆
mosaic dominance 嵌合显性
mosaic electrode 镶嵌电极
mosaic endosperm 镶嵌胚乳
mosaic enrichment 马赛克装饰;镶嵌装饰
mosaic exterior finish 镶嵌外表最后加工;马赛克外观装饰
mosaic external finish 马赛克外观装饰
mosaic facade 镶马赛克的建筑物的立面;镶马赛克的建筑物的正面;镶嵌门面
mosaic facing 马赛克饰面
mosaic figure 镶嵌图
mosaic film 嵌镶式照相底片
mosaic fingers 镶花地板;马赛克地板;镶嵌细木工地板
mosaic fissure 镶嵌缝隙
mosaic floor(ing) 马赛克(楼)地面;铺设锦砖地面;马赛克楼板;锦砖楼板;锦砖地面;镶嵌地板
mosaic flower bed 图案花坛
mosaic frame 镶嵌框架
mosaic function 马赛克功能
mosaic glass 镶嵌玻璃;锦玻璃;花玻璃;嵌镶玻璃;玻璃马赛克
mosaic gold 装饰用黄铜;铜锌合金;彩色金;仿金铜箔
mosaic gold alloy 嵌镶金合金
mosaic granoblastic texture 镶嵌粒状变晶结构
mosaic imaging 镶嵌成像;嵌镶成像
mosaic infrared system 镶嵌式红外系统;嵌镶式红外系统
mosaic intergranular pore 镶嵌粒间孔隙
mosaicism 镶嵌性;镶嵌现象
mosaicist 嵌镶设计师
mosaicker 像片镶嵌仪;镶嵌者
mosaicking board 镶嵌图板
mosaic lace 镶嵌花边
mosaic laying 铺马赛克
mosaic map 像片镶嵌图
mosaic masonry work 镶嵌石块圬工工程
mosaic mount 镶图板
mosaic mural 马赛克壁画
mosaic of leaded glass 铅条玻璃嵌镶
mosaic ornamental feature 镶嵌马赛克装饰特色;镶嵌马赛克装饰特征
mosaic outdoor finish 镶嵌室外表面的修整;马赛克外观装饰
mosaic panel 镶嵌面板;镶嵌框架;镶嵌板
mosaic parquet panel 木镶板;镶嵌拼花板
mosaic parquetry 镶花地板;马赛克剧院乐池;镶嵌地板;镶木地板
mosaic pattern 镶嵌模样;镶嵌图案;镶嵌形式
mosaic pattern of the crust 地壳镶嵌格局
mosaic pavement 马赛克铺面;嵌镶式地面;拼花铺面;拼花地面
mosaic paving sett 马赛克铺地小方石;镶嵌小方石;小方石拼花路面
mosaic photocathode 嵌镶光电阴极
mosaic photographic(al) film 嵌镶式照相底片
mosaic photography 嵌镶图摄影学;重叠航摄
mosaic photo series 镶嵌像片排系列
mosaic photo strip 镶航空像片
mosaic piece 镶嵌部件;用于镶嵌的小块
mosaic plate 嵌镶板
mosaic process 嵌合突起
mosaic random rubble masonry 镶嵌任意块圬工
mosaic screen 感光嵌镶屏;感光嵌镶幕;镶嵌屏幕;感光镶嵌幕
mosaic screen film 镶嵌屏薄膜
mosaic shape 镶嵌状
mosaicsit 嵌镶制作工
mosaic stair(case) 镶嵌楼梯;镶嵌式楼梯间
mosaic structure 镶嵌构造;镶嵌地块;晶体镶嵌结构;嵌镶结构
mosaic surface 嵌镶表面;感光嵌镶屏;马赛克饰面;拼花面;马赛克(贴)面;镶嵌层;嵌镶贴面
mosaic target 嵌镶靶

mosaic terrazzo 马赛克磨光石；水磨石马赛克；马赛克水磨石
mosaic tessera 镶嵌用小块
mosaic texture 镶嵌结构
mosaic theory 镶嵌说
mosaic tile 嵌镶瓷砖；锦砖；马赛克瓷砖；彩色瓷砖
mosaic tile surface 锦砖面层
mosaic tiling 马赛克镶面
mosaic type 嵌合型
mosaic vault 镶嵌拱顶；镶嵌式圆屋顶
mosaic wall tile 马赛克墙砖
mosaic window 镶嵌砖；镶嵌玻璃窗
mosaic wood 镶嵌用材
mosaic woodwork 镶嵌细木工
mosaic wool work 镶嵌式绒线刺绣
mosaic work 镶嵌细工；镶嵌工程
mosai-lined 镶嵌衬里的
mosandrite 褐硅铈石；褐硅铈矿；层硅铈钛矿
Moschcowitz's test 莫斯科维茨氏试验
moschellandsbergite 银汞矿；丙银汞膏
Moscicki condenser 莫斯西基电容器
Moseley law 莫塞莱定律
Moser equation 莫泽方程
mosesite 黄氮汞矿
mosit-heat sterilization 蒸汽消毒
Moslay's law 莫斯利定律
Moslem architecture 回教建筑；穆斯林建筑
Moslem tomb 穆斯林墓地
mo-slewing system 瞬时回转方式
mosophase 中间相
mosor 残山；残丘
mososeries 弦月形
mosque 清真寺；伊斯兰教寺院
Mosque and Mausoleum of Sultan Hasan at Cairo 开罗的苏丹哈桑礼拜寺
mosque arch 清真寺拱；马蹄形圆拱；马蹄拱
mosque architecture 伊斯兰教建筑；〈旧〉回教寺院建筑
Mosque of Al-Aqsa at Jerusalem 耶路撒冷阿格沙清真寺
Mosque of Ibn Tulun at Cairo 开罗的伊木图伦清真寺
mosquito boat 鱼雷快艇；鱼雷舰；快艇；驱潜艇
mosquito bomb 灭蚊弹
mosquitocide 灭蚊药；灭蚊剂
mosquito-craft 鱼雷快艇；驱潜艇
mosquito fever 疟疾
mosquito fleet 快艇舰队；快艇队
mosquito killer 灭蚊器
mosquito-net decorative pendant 蚊帐垂饰
mosquito screen 纱窗；防蚊绳纱
moss 青苔(绿藻类)；苔藓；地衣
moss animal 苔藓动物
Moss bar 莫斯氏加强混凝土粗筋
Mossbauer effect 穆斯堡尔效应
Mossbauer gravity meter 穆斯堡尔重力计
Mossbauer nuclear 穆斯堡尔核
Mossbauer parameter 穆斯堡尔参数
Mossbauer spectrogram 穆斯堡尔谱图
Mossbauer spectrometer 穆斯堡尔谱仪
Mossbauer spectroscope 穆斯堡尔谱仪
Mossbauer spectroscopy 穆斯堡尔谱分析；穆斯堡尔谱法；穆斯堡尔光谱
Mossbauer spectrum 穆斯堡尔(能)谱
moss bog 苔藓沼泽
Moss-Burstein effect 莫斯—布尔斯坦效应
moss copper 苔纹铜
moss-covered 长满青苔的
moss crepe 纱绉；苔绒绉
mossery 苔藓地
moss forest 温带雨林
moss fringe 嵌条；绒缀条
moss gold 苔状金
moss hag 泥炭采后的废坑
Moss hardness 莫斯硬度
moss heath 石楠荒地
moss-land 泥炭沼泽
moss layer 泥炭层；苔藓层
mosslichen vegetation 藓类地衣植物
mosslike 似苔的
moss moor 高地酸泽；苔藓沼泽
moss peat 高沼泥炭；高位泥炭
moss rubber 微孔泡沫胶
moss sedge peat 苔芦泥炭

moss silver 苔状银
moss sinter 苔纹泉华
moss starch 苔淀粉
mosstone 苔藓色
mossy 苔状的；苔似的；苔藓状
mossy cotton 多绒棉
mossy green 青苔色；苔绿色
mossy zinc 海绵状锌
most active element 最易带出的元素
most advantageous section 最佳断面
most advantage velocity 最佳流速
most assumed accident 最大假设事故
most cold water 非常冷的水
most common metal 最普通金属
most common scale 最常用比例尺
most confidential 绝密
most critical path 最关键路线
most dangerous slip circle 最危险(的)滑动圆
most dense pile 最紧密堆积
most destructive 最有破坏性
most difficultly selfcombustion 不自燃
most division has ceased 多数分裂已停止
most economical grade 最经济坡度
most economical range 最经济航程
most economical section 最经济断面
most economic conductor 最经济的导线
most economic load 最经济负载
most efficient load 最大效率时的负载；经济负荷
Mostem restaurant 清真食堂
most favo(u)rable direction 最有利方向
most favo(u)rable distribution of weight 最适当的权分配；最佳权分配
most favo(u)rable route 最顺进路
most favo(u)rable signal 最大允许信号
most favo(u)red licence clause 最惠特许条款
most-favo(u)red nation 最(优)惠国
most favo(u)red nation clause 最(优)惠国条款
most favo(u)red nation tariff 最惠国税率
most favo(u)red nation treatment 最(优)惠国待遇
most favo(u)red nation treatment clause 最惠国待遇条款
most frequency water level 最常水位；出现频率最高的水位
most frequency water stage 出现频率最高的水位
most frequency wind direction 最常见风向
most frequent size 最可几大小
most frequent water level 最常水位
most frequent wind direction 最多风向；最常见风向
most general unifier 最一般的合一
most imperfect 极不完全
most important variable rule 最重要变量规则
most likely errors 非常可能存在的错误
most likely time estimate 最可能时间估计
mostly 主要地；多半
most obvious 极明显的
most of the commonly grown varieties 多数普通栽培品种
most perfect 极完全
most perfect cleavage 最完全解理【地】
most permeable system 高渗透系统
most possible number 最大可能数
most powerful 最大功效的
most powerful test 最大功效检定
most probability position 最大概率位置
most probable 最可能的；最可几的
most probable correction function 最或然修正函数；最或然改正函数
most probable distribution 最大可能分布
most probable end-to-end distance 最可几端距
most probable error 最可能误差；最概然误差；最(大)或然误差
most probable extreme value 最可能极值
most probable number 最可能数；最近似值
most probable number index 最可能数指数
most probable number method 最可能数法
most probable number value 最可能数值
most probable position 最可能位置；最或然船位
most probable quantity 最大可能数量
most probable selling price 最有可能的售价
most probable speed 最概然速率
most probable value 最小或是值；最可能值；最可几值；最概然值；最(大)或然值
most probable velocity 最可几速度；最概然速度

most reactive condition 最大后备反应性状态
most restrictive signal 最大限制信号
most serious 极严重的
most severe injury degree 最大伤害度
most short 严重不足；资源急缺
most significant bit 最有效比特；最高有效位；最高位；最大有效位
most significant digit 最高有效位(数字)；最高有效数；最大有效位
most significant end 最高端
most significant position 最高数位
most smooth discontinuity 极光滑的结构面
most stringent level alpha test 最紧迫阿尔发水平检定
most stringent tests 最紧迫检验
most suitable field intensity 最适应的电场强度
most turbid 极浊
most turbulent flow stage 最汹水位
most turbulent stage of rapids/hazardous passage 急、险滩最汹水位
most-unfavorable-condition test 最不利情况试验
most unsanitary water 极不卫生的水
most-up-to-date science and technology 最新科学技术成就
most working hydrostatic pressure 工作最大静压
most worthy 最值得(的)；最有价值(的)
mota 黏土
Mota metal 锡基高强度轴承合金
motanic acid 二十九酸
motar proportioning 灰浆配合(比)
mote 微尘；尘埃；小缺点
motel 公路旅馆；路旁旅社；汽车旅馆
motel of water travel 驾船游客旅馆
motel-tourist cabins 公路两旁的旅馆；汽车游客旅馆
moth 蛀虫
moth ball 卫生球；封存
mothball fleet 后备舰队
moth balling 防腐
mothball ship 闲置船舶
moth-eaten 陈旧的；过时的；虫蛀了的
moth-eaten alopecia 虫蛀样秃发
mother 第一模盘
mother aircraft 运载飞机；母机
mother alloy 母合金
mother and baby unit 医院产妇科；母子单元
mother block 主程序段
mother board 主板；母插件；母板
mother-child relationship 母子关系
mother city 母城(城市规划用语)；核心城(市)
mother clock 母钟；母时钟
mother conveyer 主运输机；主输送机
mother crystal 母晶(体)
mother current 主流；干流；母(电)流；本流
mother earth 地面；大地
mother geosyncline 母地槽
mother glass 基样玻璃；样品玻璃；母体玻璃
motherham 丝炭；乌煤；天然木炭；矿物炭
motherland program(me) 国土规划
mother liquid 母液
mother liquor 母液
mother liquor occlusion 母液包藏
mother liquor tank 母液罐；母液槽
mother lode 母脉
mother machine 工作母机
mother map 底图
mother materials 母质
mother metal 母材；母金属；基体金属
mother nuclide 母核
mother nut 主螺母
mother of coal 丝煤；煤母；丝炭；乌煤；天然木炭；矿物炭
mother ofIthyme 百里香
mother-of-pearl 珍珠母；螺钿
mother-of-pearl clouds 珠母云
mother-of-pearl glass 珠光玻璃
mother-of-pearl lustre 珠光光泽
mother-of-pearl pigment 云母珠光颜料
mother of salt 盐母
mother plate 样板；模板；母模
mother reactor 母型反应堆
mother reef 母礁
mother rock 源岩；母岩；围岩
mother rod 主连杆
mother seed 原种

mother set 母版
mother shaker 主振动运输机;主摇动式运输机
mother ship 母舰;母船;航空母舰
mother's mark 母斑
mother stock 母分
mother stream 干流
mother substance 母体
mother town 母城(城市规划用语)
mother-type fishery 母船式渔业
mother vat 母液瓮
mother vessel 母船
mother whaler 捕鲸母船
moth patch 褐色斑
mothproof 不蛀的;防蛀的
mothproofer 防蛀剂;防蠹剂
mothproof finish 防蠹加工
moth proofing 防蛀性
mothproofing agent 防蛀剂
mothproofing paper 防蛀纸
mothproofing preparation 防蛀制剂
mothproof knitwear 防蛀针织品
mothproof screen cage 防蛾网室
moth repellency 抗蛀性
motif 主题;建筑主题;花样;设计主题
motif of architecture 建筑(的)主题;建筑主旨;建筑特色
motif of crystal structure 晶体结构基型
Motif Palladio 帕拉第奥建筑构图特色
motile 能动的
motility 能动性;能动力
Moti Masjid at Agra 阿格拉珍珠清真寺(位于印度北部的一个城市)
motion 议案;移动;位移;动作
motion aberration 运动象差
motion accelerometer 三分量加速度仪;三分量加速度计
motion aiding 助动的
motional 动的
motional admittance 动态导纳
motional electromotive force 动生电动势
motional feedback 动圈反馈
motional feedback amplifier 动态反馈放大器
motional feedback circuit 动反馈电路
motional induction 动生电磁感应
motional resistance 运动阻力
motional vision 动体视力
motional waveguide joint 活动波导连接
motion analysis 运河分析;动作分析
motion and time study 动作和时间研究
motion arm 操作臂
motion balance pressure sensor 动作平衡式压力传感器
motion block 动块
motion characteristic 运动特性
motion characteristic of flowing surface water 地面流水运动特征
motion characteristic of glacier 冰川运动特征
motion characteristic of lake water 湖水运动特征
motion characteristic of ocean current 洋流运动特征
motion characteristic of tide 潮汐运动特征
motion characteristic of turbidity current 曲流运动特征
motion characteristic of wave 破浪运动特征
motion characteristic of wind 风的运动特征
motion-compensation 影像位移的补偿
motion cycle 运动循环
motion-dependent resistance force 随动而变的阻力
motion detector circuit 动作检测电路
motion driver 斜坡道管理工
motion equation 运动方程
motion equation for uniform flow 均匀流运动方程
motion for compressible media 可压缩流体的流动
motion indicator 速度指示器
motion in one dimension 一维运动
motion law 运动定律
motionless 固定;静止的;不动的
motionless mixer 固定搅拌机;静态混合器
motion lever 移动杆
motion limited to small disturbances 限于微扰动运动
motion limiter 工作运动限位器
motion link 运动链系;导向装置

motion model 运动模型
motion mode of lake water 湖水运动方式
motion mode of surface runoff 地面流水运动方式
motion of Earth poles 地基移动
motion of elements 动作要素
motion of ground 地面运动
motion of membrane 膜片运动
motion of poles 极移
motion of progressive wave 前进波运动
motion of sea 海面状态
motion of the ground 地动
motion of the oscillator 振荡器的固有运动
motion of translation 线性运动;平移运动
motion of uniform velocity 匀速运动;等速运动
motion pattern 流线图(谱)
motion picture experts group 运动图像专家组
motion picture film processing plant 电影洗印厂
motion picture house 电影放映室
motion picture projection room 电影放映室
motion-picture screen 电影银幕;银幕
motion picture theater 电影院
motion progressive wave 前进波动
motion-promoting force 驱动力
motion range 活动范围
motion recorder 运动参数记录器
motion register 运转寄存器
motion resistance 动态电阻
motion-resistance-force characteristic 运动阻力特性
motion resistance of mine car 矿车运行阻力
motion resistance statistic (al) evaluation 运动阻力的统计评价
motion round wing 绕机翼的环流
motion sickness 晕动;晕船;晕车(病)
motion simulation 运动模拟
motion simulator 运动模拟器
motion space 运动空间;动作空间
motion spectrum 运动谱
motion study 动作研究
motion time analysis 运动时间分析;工时分析
motion valve 节气阀
motion with variable vel ocity 变速运动
motion work (机构的)运动机件
motion worksgear train 传动机构
motion work wheel 主动轮;传动轮
motique 流动商店;流动售货车
motivation 激励;动机;促动
motivation research 动机研究;动机分析
motivator 激励手段;舵;操纵面;操纵机构
motive axle 主动轴
motive column 矿井风压空气柱
motive cycle 工作循环
motive force 动力;发动力
motive power 原动力;原动机;机车;驱动功率;动力
motive-power battery 牵引型电池
motive power load 动力负载;动力负荷
motive power machine 动力机械
motive power unit 动力机组;发动机单元;发动机组
motive steam 工作蒸汽
motive unit 牵引车
motobloc 拉床;拉丝机;拉拔机
moto-bug 机动小货车;动力车
motocade 汽车队伍
motofacient 促动的
motometer 转数计
motomixer 混凝土搅拌(汽)车;混凝土拌和(汽)车
motopaver 自动铺路机;自动铺砌机;自动铺料机
motor 马达;电动机;发动机
motorable 可通行汽车的
motor activity test 运动能力试验;活动能力试验
motor-age city 汽车时代城市
motor alternator 电动交流发电机(组);电动(机)发电机组
motorama 新车展览
motor-ambulance 救护车
motor amplifier 电机放大器
motor analyzer 发动机试验台;发动机试验机
motor-armature 电动机电枢
Motora's ship stabilizer 船舶防摇器
motor-assisted pedal cycle 机动脚踏车
motor atomizer 动力弥雾机
motor auger 机力螺旋钻
motor automatic relay 电机自动继电器
motor bark remover 机动剥树皮机

motor barrow 马达驱动的手推车
motor base(frame) 电动机座;电机底座;电动机基础
motor base pin 电动机座销轴
motor battery 电动机电池
motor bearing 电动机轴承;电机轴承
motor bed-plate 电动机机座
motorbi-cycle 摩托车
motorbicycle lead-acid storage battery 摩托车用铅酸蓄电池
motor bike 机动脚踏两用车;摩托车
motor block 电动滑车
motor-blower 电动鼓风机
motor board 电动机配电盘
motor boat 汽艇;小汽艇;轮船码头;机(动)艇;摩托艇;汽船;机动船
motorboating 汽艇声;低声频振荡
motor body 发动机壳体
motor bogie 自动转向架;动力转向架
motor-booster 电动升压机
motorborne 汽车运送的;汽车运输的;汽车拖运的
motor boss 汽车调度员
motor brake magnet 电动机闸磁铁
motor branch circuit 电动机馈电支路;电动机分支电路
motor brass 电动机黄铜
motor brass alloy 电机黄铜合金
motor brush 电机电刷;电动机刷
motor-bug 机动小车
motor buggy 机动小车
motor bus 公共汽车;大轿车;大(型)客车
motorcab 出租汽车
motor cabinet 电机座
motor cable 电动机电缆
motor cable connection 电动机连线
motor cable winch 自动升降机;自动卷扬机;自动绞车;汽车绞盘
motorcade 汽车长列;汽车行列
motor capacity 电动机容量
motor car 小汽车;机动轨道车;汽车;动车;电(力)动车;小轿车
motor caravan 汽车式住宅
motor carbon 电机碳刷
motor carbon brush 电动机碳刷
motorcar brake 自动车闸
motorcar ferry 汽车轮渡
motorcar fitter 汽车修配工
motor cargo boat 机动货船
motorcar industry 汽车制造业
motorcar insurance 汽车保险
motorcar interference 汽车(发动机)干扰
motorcar jack 汽车千斤顶
motorcar lacquer 汽车蜡克漆
motorcar liability insurance 汽车责任保险
motor carriageway 机动车道
motor carrier 公路运输业者;传送机
motor carrier noise 机动车噪声
motorcars on hire 出租汽车
motor cart 机动轻便运货车
motorcar with driver cab 带驾驶室的动车
motorcar with driving stand 带驾驶台的动车
motor case 电动机壳;发动机壳体
motor casing(frame) 电动机壳;电机壳
motor cavity 电机座位
motor chamber 发动机燃烧室
motor chamber diameter 发动机燃烧室直径
motor characteristic 电动机特性(曲线)
motor circuit 动力电路
motor-circuit switch 电动机馈线开关
Motor City 汽车城(美国底特律的俚语)
motor coach 长途汽车;公共汽车;客车电动机;动车;电动车;长途公共客车
motor coach route 长途公共汽车线
motor coach train 动车列车;电动旅客列车
motor coaster 沿海汽艇
motor collapsible ladder 救火车的伸缩梯
motor combination 电动机的组合
motor commutator 电动机整流子
motor compartment 电机舱
motor complement 动力配套;电动补充
motor-compression refrigerator 电动机压缩式冰箱
motor compressor 电动压缩机;电动压气机
motor console 发动机试验操纵台
motor constant 电动机常数

motor control 电动机(电子)控制;电动机操纵
motor controller 电动机控制器
motor control panel 电动机控制盘
motor control relay 电动机控制继电器
motor converter 单枢换流器;电动机一发电机组;电动变流机;整流电动机;电动传送机
motor cooling 电动机冷却
motor cooling jacket 发动机冷却套
motor coupling 电动机联轴器
motor court 汽车游客旅馆;汽车旅馆;停车场
motorcrane 汽车起重机
motor current transducer 电机电流变换器
motor current-transformer 电动变流器
motor cut-out switch 电动机停机开关
motor cuts out 发动机停车
motorcycle 机动脚踏车;机器脚踏车;机车;摩托车
motorcycle-racing arena 摩托车比赛场
motorcycle sidecar 机器脚踏车边车
motor damper 电动机减振器;电动机消声器
motor depot 汽车站;汽车场
motor direct coupled pump 直接联动泵
motor dory 汽船;摩托艇
motor drawn 机械拖带的,(机动的);汽车牵引的
motor drill 内燃凿岩机;摩托凿岩机;手电钻
motor drill and breaker 电钻和破碎器
motor drive 马达传动;电机驱动;电动驱动;电动机拖动
motor drive asphalt pump 电动沥青泵
motor-drive lathe 电动车床
motor-driven 机械牵引的;汽车牵引的;电动的;内燃机牵引的;马达驱动的;电动机拖动的;电(动)机驱动的
motor-driven barge 机动驳(船)
motor-driven (blade) grader 自动平地机;机动平路机;机动平地机
motor-driven blower 电风扇;电动鼓风机
motor-driven boat 机动船
motor-driven cable-winch 电动电缆绞车
motor-driven carrier 电动小车
motor-driven control valve 电动机拖动控制阀
motor-driven distributor 电动分配器
motor-driven feed pump 电动给水泵
motor-driven grab 电动抓斗
motor-driven hoist 电动绞车
motor-driven interrupter 电动断续器
motor-driven layer radiographic X-ray apparatus 电动断层X射线机
motor-driven miniature pump set 电动微型水泵机组
motor-driven mower 电动剪草机
motor-driven psychrometer 电动型通风干湿计
motor-driven pump 电动机泵
motor-driven saw 电锯
motor-driven seed cleaner 电机驱动种子清选机
motor-driven sludge excavator 电动挖泥机
motor-driven slush pump 电动泥浆泵
motor-driven starter 电动起动机;电动机起动器
motor-driven swing hammer 电动摆锤
motor-driven switch 电动机驱动开关
motor-driven switch-group 机动组合开关
motor-driven turbine pump 电动涡轮泵
motor-driven valve 电动机拖动阀
motor-driven vehicle 机动车
motor-driven welding machine 电动机拖动式焊机;电动焊机;电动旋转式焊机
motor-drive oil lifter 电动油压升降机
motor-drive shaft 马达轴;电动机驱动轴
motor-drive shaft and pinion 电动机传动轴和小齿轮
motor-drive type 电动机传动型
motor driving time relay 电动机式时间继电器
motor-drome 汽车场;汽车比赛场;汽车试验场
motor duct 高架箱形桥
motor dynamo 电动直流发电机;电动发电机;电动直流发电机;电动发电机组
motor dynamometer 电动机功率计
motor dynamo unit 电动直流发电机组
motor eccentricity 电机偏心率
motor effect 电动机效应
motor efficiency 电动机效率
motor element 电动机元件
motor enclosure 电机壳
motor end closure 发动机喷口盖
motorette 绝缘寿命试验模型

motor excitation 电动机励磁
motor exciting current 电动机励磁电流
motor fan 电扇;电动鼓风机
motor fault 电动机缺陷
motor ferry 汽船轮渡
motor-field 电机磁场
motor-field control 电机励磁控制;电动机磁场控制
motor-field failure relay 电动机磁场故障继电器
motor fire brigade vehicle 消防车;救火车
motor fishing vessel 机动渔船
motor flusher 洒水车;洒水车;车辆冲洗装置
motor for boat 船用发动机
motor for kicker and doffer 抖动器和滚筒用电动机
motor for wood working 木工电动机
motor foundation 电动机基础
motor frame 电机座架;电动机架
motor frame through bolt 电动机长螺栓
motor freight depot 货运汽车站;货运汽车场
motor freight terminal 货运汽车(终端)站
motor fuel 发动机燃料
motor fuel additive 发动机燃料添加剂
motor fuel constituent 发动机燃料组成
motor fuel service station 机动车燃料加油站
motor fuse 电动机熔断器
motor gain 电动机增益
motor garage 汽车修理厂;汽车间;汽车库
motor gasoline 动力汽油
motor generator 电动发电机
motor-generator arc welder 电动发电机式直流弧焊机;电动发电弧焊机
motor-generator locomotive 电动发电机机车
motor generator panel 电动发电机控制板
motor-generator set 电动机组;整流机;电动发电机组
motor-generator welder 电动发电电焊机
motor-generator welding unit 电动直流发电焊接设备
motor glider 电动滑翔机
motor grab 电动抓斗
motor grader 自动平地机;自动平地机;机动平路机;机动平地机;平路机;平地机
motor gravity 自行式起重机
motor group 电动机组
motor head 发动机头;发动机前端
motor head drill 动力头钻机
motor head reverse circulation drill 动力头反循环钻机
motor highway 机动车道
motor hoist 电动起重机;电动起重机;电动提升机;电(动)葫芦
motor home 活动住宅;含住房的汽车;住宿汽车
motor hotel 汽车饭店;汽车游客旅馆
motor hull insurance 汽车车身保险
motor inclosure 电动机壳
motoring friction 空转摩擦力
motoring map 公路(交通)图;公路旅行路线图
motoring offence 无执照开车罪
motoring ring test 带载试验
motoring run 发动机试运转
motoring test 空转试验
motoring torque 运动力矩
motor inn 汽车旅游旅馆;多层汽车游客旅馆;有停车场的路旁旅馆;汽车游客旅馆;汽车饭店
motor inspection and repair workshop 电机检修间
motor integrating meter 感应式电度表
motor interrupter 电动断续器
motor-in-wheel 车轮电动机
motorised 装有发动机的
motorised roll 电机传动辊
motorist 乘汽车者;汽车驾驶员
motorization 机械化;机动化;摩托化
motorization rate 摩托化比率
motorized 装有发动机的
motorized bogie 机车驱动转向架
motorized caravan 敞篷汽车
motorized cart 自行装置
motorized cart track 机动车展赛场
motorized copying camera 电动复制照相机
motorized damper 电动风阀
motorized direct shear apparatus 电动直剪仪
motorized door 电动门
motorized flat wagon 机动平车
motorized grader 机动平路机;机动平地机;平地机

motorized home 旅游居住汽车
motorized inspection trolley 机动检查电车;机动检查缆车
motorized isolator 电动隔离开关
motorized knapsack 背负式机动喷雾器
motorized knapsack mistblower 背负式机动鼓风喷雾器
motorized level(l)ing 摩托化水准测量
motorized non-return valve 电动止回阀
motorized pan and tilt head 电动驱动的云台
motorized panning head 电动驱动的摇头
motorized pivoted distributor 电动立轴式回转分料溜槽
motorized ramp 自动梯
motorized reducer 带电动机的减速器
motorized road grader 自动平地机;机动平路机
motorized scraper 机动铲运机;机动铲运机;铲运机
motorized shaker 机械振动器
motorized snowplow 摩托化扫雪机
motorized solar control blinds 电动日光控制百叶窗
motorized speed reducer 电动减速器
motorized tar spreader 自动式喷洒沥青机
motorized three-way valve 电动三通阀
motorized traffic 机动车交通
motorized traffic congestion 机动车堵塞;机动车交通拥挤
motorized traverser dolly 电动旋转移动台车
motorized tuning control 电动机驱动的调谐系统
motorized valve 电动(调节)阀;传动阀
motorized wheel drive 机动轮驱动;马达轮驱动
motorized winch batten 电动吊杆
motorized zoom lens 马达驱动的可变焦距镜头
motor launch 小汽艇;舰载大艇;机艇;机动小艇;汽艇
motorless 无动力的
motor licence plate method 汽车牌照法
motor lifeboat 机动救生艇;救生汽艇
motor lighter 机动驳(船)
motor line 电动机系列
motor load 电动机负载
motor load control 电动机负载控制
motor load indicator 电动机荷载指示器
motor lodge 汽车游客旅馆
motor lorry 运货汽车;载重汽车;摩托艇
motor machinery 电动机械
motor magnet 电动电磁铁
motor maintainer 自动养路机
motormaker 电机厂
motor man 机匠工;司机;电机操作工;加油工
motorman's cab(in) 驾驶室;司机室
motor manufacturer 电动机制造者
motor meter 感应式电表;电动机型仪表;电动机式电度表;电磁作用式仪表;电动机型积算表;运动力计;汽车仪表
motor method 发动机开车法;发动机法
motor mix 乙基液
motor-mounted bicycle 机动脚踏车
motor-mounted mixer 汽车式搅拌机
motor-mount pump 马达泵
motor mount ring 发动机安装环
motor mower 自动割草机;机动割草机
motor mower with binder attachment 机动青草割捆机;带打捆装置的动力割草机
motor mower with center drive 中央驱动式动力割草机
motor mower with side drive 侧驱式动力割草机
motor nozzle 发动机喷管
motor off switch 电动机切断开关
motor oil 机油;电机用油;车用机油;电动机(润滑)油
motor oil wax 中间蜡
motor omnibus 公共汽车
motor on-off switch 电动机启停开关
motor-operated 电动的;电动机带动的
motor-operated cam type flash welder 电动凸轮式闪光焊机
motor-operated horn 电动喇叭
motor-operated potentiometer 电动机操作的电位器
motor operated rheostat 电动操作变阻器
motor-operated shutter 机动阀门
motor-operated switch 电动开关;电动断路器
motor operated valve 电动阀(门)

motor-operated welder 电动加压焊机
motor operational valve 电动阀(门)
motor-output 电动机出力
motor overload protection unit 电动机超负荷保护装置
motor panel 电动机配电盘
motor park 汽车停车场
motor passenger ship liner 远洋客轮
motor patrol 路面平整机;巡逻车(道路养护用)
motor pattern 运动模式
motor-paver 自动铺路机;电动铺路机
motor performance 电动机性能
motor petrol 车用汽油
motor pinion 电动机小齿轮
motor pitch 电动机节距
motorplane 动力飞机
motor plough 汽车犁;自走犁;机动犁
motor plow 机动犁
motor pool 汽车集中调度场;停车场;汽车库;汽车调度场;车场
motor power (output) 电(动)机功率;发动机推力
motor press 机动压力机
motor-protecting switch 电动保护开关
motor protection against overheat 马达过热保护;电动机过热保护
motor protection relay 电动机保护继电器
motor pulley 电动机皮带轮
motor-pump 电泵;电动泵;马达泵;电动泵
motor-pumped well 机井;机泵井
motor pump works 电泵厂
motor rear end plate 电动机后端盖
motor reducer 马达降速器;电机减速机;电动机减速器
motor reducing gear 电机减速机;电动机减速器
motor reduction unit 降速电动机
motor reel 电动机轴
motor repair 汽车修理(车)间
motor repair shop 汽车修配厂;汽车修理厂
motor rescue boat 机动救助艇
motor road 公路;机动车路;汽车道;汽车路
motor roller 机碾;单独传动辊道;机动压路机;机动路碾
motor room 电机室
motor rotor 电动机转子
motor rotor tester 电机转子试验装置
motor sailing vessel 机帆船
motor sailor 机帆船
motor saw 动力锯
motorscooter 小型机车;低座小摩托车
motor scraper 自行式铲运机;自动铲运机;机动铲运机;电动铲运机
motor scythe 机动割草机
motor-selector 电动选择器
motor service 公共交通;汽车服务;汽车运输;汽车交通
motor shaft 电动机轴;电机轴
motor shell 电动机壳
motor ship 机动船;内燃机船;发动机推进飞行器;发动机船;汽船
motor ship engine 轮机
motor side 电动机侧
motor siren 电动警笛;电动警报器;电笛
motor slide rails 电动机导轨
motor smack 机帆船
motor specification 电动机规格
motor speed 电动机转速
motor speed control 电动机转速控制
motor speed controller 发动机转速调节器
motor speed sensor 电动机速度传感器
motor speed switch 转子速度加减键(陀螺罗经用)
motor spirit 汽油;车用汽油;发动机燃料
motor spring 汽车弹簧
motor-sprinkler 自动喷淋器
motor-sprinkler system 自动喷水灭火系统
motor sprocket 电动机链轮
motor squadron 汽车队
motorstair 自动扶梯;自动楼梯
motorstair shutter 自动扶梯护板;自动扶梯鱼尾板
motor starter 钻孔器;电动启动机;电动启动器;电动(机)起动器;发动机启动器
motor starting and control equipment 电动机起动控制设备
motor starting characteristic 发动机起动特性

motor starting current 电动机起动电流
motor-starting reactor 发动机起动反应器
motor starting relay 电动机起动继电器
motor starting rheostat 电动机起动变阻器
motor starting switch 电动机起动开关
motor stator 电动机定子
motor steering 汽车转向
motor step 电动机距
motor stirrer 电动搅拌器
motor stoppage 停车
motor sub-assembly 电动机组件
motor support bearing 电动机支承轴承
motor suspension 电动机悬置
motor sweeper 机动扫帚;清扫车;扫路机
motor switch 电动开关;电动机开关
motor switch oil 电动机灭油;电气开关油
motor synchronizing 自同步
motor tanker 内燃机油船
motor tank trunk 油槽车
motor terminal voltage 电动机端子电压
motor test stand 电机试验台;发动机试验台
motor test vehicle 试验发动机用的样机
motor third party liability insurance 汽车第三者责任险
motor threshing machine 动力脱谷机
motor throat diameter 发动机喷管喉道直径
motortilter 自动倾卸汽车
motor time constant 电动机时间常数;发动机时间常数
motor timer 电动机驱动计时器
motor time relay 电动式时间继电器
motor torpedo boat 鱼雷舰
motor torque 电动机转矩
motor-torque generator 一种同步驱动发电机
motor totally enclosed 全封密风冷式电动机
motor tractor 牵引车
motor tractor for semitrailer 用于半挂车的牵引车
motor-tractor industry 汽车拖拉机工业
motor traffic 机动车交通;汽车交通
motor traffic tunnel 汽车隧道;汽车交通隧道
motor train set 电动列车组
motor train unit 动车组
motor transmission shaft 电机传动轴
motor transport 汽车运输;电机输送
motor transport battalion 汽车运输营
motor transport conversion and computation chart 汽车运输换算和计算表
motor transport depot 汽车运输器材库
motor transport distance and time conversion chart 汽车运输里程时间换算图表
motor transport facility 汽车运输设备
motor transport maintenance 汽车运输保养
motor treak 电动机驱动的断续器
motor trend 汽车的发展趋向
motor tricycle 三轮摩托车
motor trolley 轨道摩托车;轨道车
motor truck 运货汽车;载重汽车;载重卡车;载货汽车
motor-truck concrete mixer 混凝土搅拌车;汽车式搅拌机
motor-truck mixer 混凝土搅拌输送车
motor-truck oscillator 电动机调谐振荡器
motor-truck supply 载重汽车补给
motor-truck terminal 载重汽车站;货运汽车(终端)站
motor-trunk road 汽车干道
motor tug 拖轮
motor tune-up 发动机调整
motor turning 汽车转向
motor type 电动机型号
motor type coder 电动机式发码器
motor type insulator 电动机型绝缘子
motor type relay 电动(机型)继电器;电动机式继电器
motor uniselector 机动选择器
motor unit 运动单位;电动机组
motor valve 马达阀
motor van 机动运货车
motor vehicle 机动车(辆);机车动车;汽车;动力车
motor vehicle accessories 机动车配件;汽车配件
motor vehicle accident 汽车事故
motor vehicle care 汽车维修
motor vehicle chassis 汽车底盘
motor vehicle combination 连接车;机动车组

motor vehicle driver 汽车驾驶员
motor vehicle emission 机动车排放;机动车辆排放物
motor vehicle emission standard 机动车辆废气排放标准
motor vehicle equipment 汽车装备
motor vehicle exhaust 汽车尾气排放
motor vehicle exhaust emission 机动车废气排放
motor vehicle hangar 机动车停车间;机动车停车库;汽车停车间;汽车停车库
motor vehicle non-traffic accident 路外汽车事故
motor vehicle passenger insurance 汽车乘客险
motor vehicle plant 汽车制造厂
motor vehicle pollution 机动车污染
motor vehicle prohibited sign 禁止机动车通行标志
motor vehicle repair shop 汽车修理工厂;汽车修理车间
motor vehicle service station 汽车维修站;汽车加油站
motor vehicle storage 汽车库
motor vehicle traffic 汽车交通
motor vehicle use 汽车使用
motor vehicle use study 汽车使用调查
motor vertical 立装电动机
motor vessel 机动船;内燃机船
motor voltage transducer 电机电压变换器
motor wagon 小型运货汽车;电动货车
motor water cap 车辆冲洗装置
motor water car 洗车机;洒水车
motorway 高速公路;公路;高速道路;快速车道;机动车(行)道;汽车专用路;汽车路;汽车道(路)
motorway aesthetics 公路美学
motorway bridge 公路桥梁;高速公路桥
motorway church 公路旁教堂
motorway concrete 公路混凝土施工现场;公路混凝土车行道;公路混凝土
motorway construction 公路施工;公路建造
motorway cross-sectional profile 公路横截面图;公路横断面
motorway emergency communication system 公路应急通信系统
motorway emergency telephone 公路紧急电话;公路事故电话
motorway entrance 公路入口
motorway lighting 公路照明
motorway location 公路定线
motorway network 公路网;高速公路网
motorway pavement 公路路面;高速公路铺面
motorway planing 公路平整
motorway plantation 公路植树
motorway restaurant 公路路边餐厅;公路饭店
motorway ring 公路环形道
motorway runoff 公路径流
motorway section 公路截面
motorway surfacing 公路路面
motorway suspension bridge 公路悬索桥;公路吊桥
motorway sweeper 公路扫路机
motorway traffic 公路交通
motorway traffic lane 公路交通车道
motorway traffic sign 公路交通标志
motorway tunnel 公路隧洞;公路隧道
motorway viaduct 公路高架桥
motor winch 机动绞盘;电动起货机;电动绞车
motor windlass 电动起锚机
motor wire brush 电动钢丝刷
motor wiring 电动机布线
motor with change cable pairs of poles 变极电机
motor with combined ventilation 带风扇电机
motor with compound characteristic 复励特性电动机
motor with reciprocating movement 往复运动电动机;反转电动机
motor with reciprocation 反转电动机
motor with self-excitation 自激电动机
motor with series characteristic 串励特性电动机
motor with shunt characteristic 并励特性电动机
motor with water cooling 水冷式电动机
motor works 汽车制造厂
motor wrench 管子钳
motor yacht 机动游艇
mototilter 倾卸汽车
moto-vibro screen 机械振动筛
Motransformation 奥氏体-马氏体转变开始温度

motte (城堡外的)陡土堤;(上建城堡的)高地;(城堡的)土墩
motte-and-bailey 土墩和外廊
motte castle 丘陵城堡
motte ditch 丘陵沟渠
motte-top castle 圆锥墩形顶
mottle 花纹刷;杂斑纹;花斑纹(涂饰法);斑纹;斑点
mottle cast-iron 杂品铸铁;麻口铁
mottled 杂色的
mottled bamboo 斑竹
mottled cast-iron 马口铁;麻口铸铁
mottled clay 杂色黏土
mottled discolo(u)ration 起斑点的褪色
mottled effect 斑迹现象;斑点效应;斑点效果
mottled enamel 斑釉;斑点搪瓷
mottled glaze 斑点釉
mottled glazed coat(ing) 斑釉涂层
mottled grain 斑点木纹
mottled ore 斑杂状矿石
mottled osseous amber 斑骨状琥珀
mottled patterns in tempered glass plates 在钢化玻璃上的应力斑;钢化玻璃应力斑
mottled(pig)iron 杂色铁;马口(生)铁;斑驳生铁;麻口(生)铁
mottled plate 有杂斑模的镀锡钢板
mottled reddish soil 具有斑块的红色土壤
mottled rot 斑腐
mottled sandstone 杂色砂岩
mottled schist 斑驳结晶片岩;斑点页岩;片麻岩
mottled structure 斑杂状构造
mottled subclay 杂色亚黏土
mottled surface 有斑点的表面
mottled tile 有花斑纹的瓦
mottled wood 花色木料;杂色木料
mottled zone 斑纹带
mottle effect 斑点效果;斑点效应
mottle leaf 花斑叶
mottler 云石纹刷
mottling 制毛面;去光源;色斑;斑点涂法;麻口化;麻点;涂面斑点;成斑作用;斑状花釉装饰;斑影;斑点状阴影;斑驳
mottling brush 杂色刷子
mottling zone 土壤斑驳层
mottramite 钒铜铅矿
motukoreaite 硫碳铝镁石
moucharaby 挑台窗;挑窗台
mouchette 花格窗;火焰状饰;剑形花格
moudle 模件
moulage 印模术;印模材料
mo(u)ld 塑胶;发霉(化石);造型模板;耕作土;模(子);模筛样板;模盘;模具;霉菌;成型模具;造型;模塑
mo(u)ldability 可塑性;可模性;造型性能;模压性能;模压加工性;成型性
mo(u)ldability controller 型砂水分控制器
mo(u)ldable material 耐火捣筑料
mo(u)ldable refractory 可以模塑的材料;可塑性耐火材料;浇注耐火材料
mo(u)ld addition 钢锭模添加剂
mo(u)ld a form 霉菌;线脚;铸型;模子
mo(u)ld alignment 模具配对;模具对准
mo(u)ld and die copying machine 型模靠模机床
mo(u)ld and die milling machine 模具铣床
mo(u)ld assembling and closing 铸型合箱
mo(u)ld assembly 铸型装配;合型
mo(u)ld base line 型基线
mo(u)ld bed 砂床
mo(u)ld blacking 砂型涂料
mo(u)ld blade 模型板
mo(u)ld blocks for slope protection 坡框式砌块
mo(u)ld blower 吹风机
mo(u)ld board 成型纤维板;型板;样板;造型板;犁壁;模板;平板;推土板;底板
mo(u)ld board bracket 犁壁翼撑杆
mo(u)ld board extension 犁壁延长板
mo(u)ld board kicker 犁翼
mo(u)ld boardless plough 无壁犁
mo(u)ld board plough 模板整型犁;板犁;平地犁
mo(u)ld board plow 铧犁;铧式犁
mo(u)ld board reinforcement 犁壁加强筋
mo(u)ld board ridger 犁壁式起垄器;起垄犁
mo(u)ld board stay 犁壁撑杆
mo(u)ld board type scraper (圆盘犁的)犁壁刮土板

mo(u)ld bottom half 下型
mo(u)ld box 铸型箱;模壳
mo(u)ld brick 成型砖
mo(u)ld buggy 钢锭模车
mo(u)ld case 模型壳;模壳
mo(u)ldcase fused knife switch 模型箱闸刀开关
mo(u)ld casing 模衬
mo(u)ld casting 成型铸件
mo(u)ld cavity 阴模;铸型腔;模腔;模巢;凹模
mo(u)ld change 成型变化;换模
mo(u)ld charge 装模料
mo(u)ld clamping(capacity) 合模力
mo(u)ld clamp mechanism 合模机构
mo(u)ld clamp ram 合模活塞
mo(u)ld clamps 马蹄夹;卡具;模型紧固夹;铸型紧固夹;砂箱夹
mo(u)ld cleaner 模具清洁剂
mo(u)ld cleaning compound 清模剂
mo(u)ld closer 铸型装配机;合箱装置;合箱机
mo(u)ld closing 铸型装配
mo(u)ld cloth 模具布
mo(u)ld coating material 钢锭模涂料
mo(u)ld coating of plastics 塑料的模塑涂层
mo(u)ld compound 制模用泥子;模型混合料
mo(u)ld construction 模板施工;模型建筑;模型构造;制模;模制结构
mo(u)ld conveyer 铸型输送器
mo(u)ld cooling 模型冷却
mo(u)ld cope 盖箱;上半型
mo(u)ld core 铸造泥芯;模芯
mo(u)ld cover 腐殖质层
mo(u)ld cream emulsion 乳液脱模剂
mo(u)ld cure 模型硫化;模型硫化
mo(u)ld dimension 铸型尺寸
mo(u)ld drag 下半型;下半型;底箱
mo(u)ld drier 铸型烘干炉
mo(u)ld drying 铸型烘干
mo(u)lded 模制的;模塑的
mo(u)lded aerial photograph 标有军事目标的航片
mo(u)lded and flat panel 平面边线修饰;平面边线镶饰
mo(u)lded articles 模制品;模制件;模制零件;模制构件
mo(u)lded asbestos 模制石棉
mo(u)lded bakelife case 塑料盒;塑料电木盒
mo(u)lded base 模制底板;模型基线;模型基础;模制基础
mo(u)lded beam 型宽
mo(u)lded blank 压型坯料;成型坯件;成型毛坯
mo(u)lded block 成型大砖
mo(u)lded block fuel element 模制块状颗粒燃料元件
mo(u)lded board 模制纸板;模压制板;模型板
mo(u)lded breadth(of vessel) 型宽;船型宽(度)
mo(u)lded brick 模制砖
mo(u)lded casting 模铸
mo(u)lded cathode 模制阴极;模压阴极
mo(u)lded chamfer 模制棱角;模制斜面;模制倒角
mo(u)lded chime lap joint 模型互搭接头;塑口互搭接头
mo(u)lded coal 模制煤;煤砖
mo(u)lded composition 模塑合成物
mo(u)lded concrete 模制混凝土
mo(u)lded concrete ashlar 模制混凝土块
mo(u)lded concrete block 模制混凝土块;模制混凝土板
mo(u)lded core 模压铁(粉)芯
mo(u)lded cornice 模制挑檐
mo(u)lded decoration 模印浮雕装饰法
mo(u)lded depth 型深;船型高(度)
mo(u)lded depth(of vessel) 型深
mo(u)lded dimension 造型尺寸;模型尺寸
mo(u)lded displacement 型排水量
mo(u)lded draft 型吃水
mo(u)lded draught 型吃水
mo(u)lded edge 线脚边缘
mo(u)lded element 模塑元件
mo(u)lded expanded pearlite block 模制膨胀珍珠岩砌块
mo(u)lded fabric 模压织物
mo(u)lded-fabric bearing 模制纤维轴承

mo(u)lded facing 热模压制表面镶片
mo(u)lded ferrite 模制粉末铁氧体
mo(u)lded floor 模制楼板
mo(u)lded foam plastics 模压泡沫塑料
mo(u)lded form 型模
mo(u)lded fuel element 模制颗粒燃料元件
mo(u)lded fuel sphere 模制颗粒燃料球
mo(u)lded glass 模制玻璃;模压玻璃
mo(u)lded glass block 模制玻璃块
mo(u)lded goods 模制品
mo(u)lded gutter 模制檐槽;模制天沟
mo(u)lded head(extrusion head) 铸模头
mo(u)lded hull surface 船体型(表)面
mo(u)lded-in 嵌入模内的
mo(u)lded-in-place 就地制模;现场模制
mo(u)lded-in-place pile 就地灌注桩;现场模制桩;现场浇制桩
mo(u)lded insulating material 模制绝缘材料
mo(u)lded insulation 隔热型材;模制塑料绝缘;模制绝缘材料;模制隔热制品;定型绝缘材料
mo(u)lded insulation product 隔热材料制品
mo(u)lded insulator 模制绝缘子;模制隔热材料
mo(u)lded-intake belt course 镶进线饰带层
mo(u)lded laminate 模制层压材料
mo(u)lded laminated tube 模制层压管材
mo(u)lded lens 模制透镜;模压透镜
mo(u)lded lens blank 模压透镜毛坯
mo(u)lded line 型上线;船型基线
mo(u)lded liner 模压橡胶内套
mo(u)lded lines 型线图
mo(u)lded masonry 模制圬工
mo(u)lded matching door 镶板门
mo(u)lded mica 模制云母;人造云母
mo(u)lded net 净纯模塑
mo(u)lded-on type 模制式
mo(u)lded panel 边饰镶板;缘饰镶板
mo(u)lded part 模制件;模压制品;模压部件
mo(u)ld ed peat 压制泥炭
mo(u)lded piece 铸造零件;模制件;模压制品
mo(u)lded pipe insulation 模制的管道保温材料
mo(u)lded plastic 模压塑料;模制塑料
mo(u)lded plastic compound 模塑化合物
mo(u)lded plastic light fitting 模压塑料照明灯具
mo(u)lded plastic pirn 塑料纤管
mo(u)lded plastic skylight 模制塑料天窗
mo(u)lded plumbing unit 定型模制卫生设备;模制管工设备
mo(u)lded plywood 缘饰胶合板;模压胶合板;成型胶合板;曲面胶合板
mo(u)lded polyethylene foam flo(a)-tation 模制聚乙烯泡沫浮选
mo(u)lded powder 模造黑药
mo(u)lded product 模制品;成型产品
mo(u)lded projecting course 台口线;挑砖饰线
mo(u)lded PVC gasketed sewer fitting 模制聚氯乙烯衬垫污水管配件
mo(u)lded relay 模制继电器
mo(u)lded resin 模制树脂;模制树脂
mo(u)lded rod 模制棒;模压棒
mo(u)lded rubber 模铸橡胶
mo(u)lded section 冷弯型钢
mo(u)lded solid catalyst 模制固体催化剂
mo(u)lded specimen 模制试件;模制试样;成型试件
mo(u)lded specimen of structure 结构扰动的试样
mo(u)lded steel 模铸钢
mo(u)lded suitcase 模压箱
mo(u)lded tablet 膜印片;模制片
mo(u)lded timber 模压材
mo(u)lded tube 模制内胎;模制管
mo(u)lded waffle slab 模壳楼板
mo(u)lded width 型宽
mo(u)lded window sill 模制窗台
mo(u)lded work 模制品;花在模制线脚上的人工;线脚工作;模塑品
mo(u)ld efficiency 模塑效率
mo(u)ld emptier 脱模机构
mo(u)ld-encased curing 密封模内养护
mo(u)lden press 自动压机
mo(u)lder 薄板坯;毛轧板;毛轧机;开坯切机;制模工;造型者;模塑者;铸工;造型机;造型工;成型铣床
mo(u)ldered wood 朽木
mo(u)ldering 造型;腐败作用

mo(u)ldering ruins 废墟
mo(u)ld erosion 铸型冲蚀
mo(u)lder's blacking bag 铅粉袋
mo(u)lder's brad 造型通气针;砂型钉
mo(u)lder's bulb 挤水器
mo(u)lder's hammer 造型锤
mo(u)lder's peel 铸工铲凿;造型铁铲子
mo(u)lder tool 造型工具
mo(u)ldery 造型车间
mo(u)ld explosive 模制炸药
mo(u)ld face 铸型表面
mo(u)ld facing 铸型涂料
mo(u)ld filling ability 铸型充填性
mo(u)ld fitter 铸型合箱工;合箱工
mo(u)ld form 模型型腔;模型型;模型内腔
mo(u)ld formation 模子成型
mo(u)ld formed bottle 模制瓶
mo(u)ld for precast work 预浇模型;预制构件模型
mo(u)ld fracture tester 砂型抗裂试验仪
mo(u)ld ga(u)ge 模子量规
mo(u)ld growth test 长霉试验
mo(u)ld half 半型
mo(u)ld hanger 模具悬杆
mo(u)ld hardness 铸型硬度
mo(u)ld hardness tester 铸头硬度试验器;砂型硬度计
mo(u)ld-hinge pin 模型铰链销
mo(u)ld-holder 模夹
mo(u)ld hold time 闭模时间
mo(u)ld humus 黑钙土
mo(u)ldic porosity 印模孔隙
mo(u)ld impression 模穴;模型型腔;模型内腔
mo(u)ld in a snap flask 无箱造型
mo(u)lding 凹凸线脚;线脚;线角;装饰线条;装饰板条;模型制品;装饰线脚;装饰嵌线;铸造物;模制;嵌条;成型铣削;成型;翻动;造型;模塑
mo(u)lding and shaping machine 线条和成型机
mo(u)lding apparatus 造型设备
mo(u)lding article 模型制品
mo(u)lding base 踢脚线顶部线脚
mo(u)lding bed 胎模
mo(u)lding board 造型底板;样板;型板;翻箱板;造型平板;模板
mo(u)lding book 放样尺寸说明书
mo(u)lding box 型箱;模箱;翻砂箱
mo(u)lding box closer 合箱机
mo(u)lding chamber 模腔
mo(u)lding composition 模制组成
mo(u)lding compound 造型混合料;模塑料
mo(u)lding cutter 嵌条切割器;线脚切割器;成型刀具
mo(u)lding cycle 制模周期;造型周期;模塑周期
mo(u)lding defect 模型缺陷
mo(u)lding department 造型工段;成型工段
mo(u)lding departure 造型工段
mo(u)lding-die 压模;塑型模
mo(u)lding draw-wire machine 成丝机
mo(u)lding epoxy resin 成型环氧树脂
mo(u)lding equipment 造型设备
mo(u)lding fault 模制件缺陷
mo(u)lding fiber board 模制纤维板
mo(u)lding fillet 小圆饰线
mo(u)lding flash 型箱
mo(u)lding floor 翻砂车间
mo(u)lding frame 模铸架
mo(u)lding gravel 特粗砂
mo(u)lding grease 造型油脂;造型润滑油
mo(u)lding gypsum 装饰线条用熟石膏
mo(u)lding in cores 组芯造型
mo(u)lding index 模压系数;模塑指数
mo(u)lding ink 造型(用)涂料
mo(u)lding lath 压模板条
mo(u)lding line 造型(生产)线
mo(u)lding loam 造型黏土
mo(u)lding loft 放样间
mo(u)lding machine 线脚压制机;线条机;线脚压割机;线模机;铸模机;造型机;模塑成型机;切模机;成型模
mo(u)lding machine with down sand frame 下压式造型机
mo(u)lding material 线脚材料;嵌条材料;造型材料;模制材料;模型材料;模塑材料;成型材料
mo(u)lding mat machine 打纸模机
mo(u)lding medium 造型材料

mo(u)lding method 造型法
mo(u)lding micanite 模压云母板;塑型云母板
mo(u)lding mica sheet 模制云母纸
mo(u)lding mixture 型料;造型混合料
mo(u)lding moisture content 成型湿度;塑性湿度;成型含水量
mo(u)lding of model 模型制造
mo(u)lding ogee S 形(装饰)线脚;葱形饰
mo(u)lding oil 模油
mo(u)lding operation 造型操作
mo(u)lding paper 模制纸
mo(u)lding pattern 线脚饰件样本
mo(u)lding pin 铸型用销钉
mo(u)lding pit 造型地坑
mo(u)lding plane 线(脚)刨;嵌条刨;线条刨
mo(u)lding planer 造型刨线脚刀;线脚刨床;样板刨
mo(u)lding plaster 造型涂层;装饰线脚用灰泥;线脚灰泥;造型石膏
mo(u)lding plastics 成型塑料;模压塑料
mo(u)lding plate 造型底板
mo(u)lding powder 铸模粉末;模塑粉料;塑料粉
mo(u)lding powder blends 模塑粉合料
mo(u)lding practice 造型法
mo(u)lding press 模压机;压型机;压模机;模制机;模塑压机
mo(u)lding pressure 造型压力;模制压力;模塑成型压力;成型压力
mo(u)lding process 成型工艺;压印过程;模制工序
mo(u)lding room 造型工段
mo(u)lding sand 型砂
mo(u)lding sand binder 型砂黏结剂
mo(u)lding sand composition 型砂成分
mo(u)lding sand hopper 砂斗
mo(u)lding sand mixture 型砂混合料
mo(u)lding sand moisture 型砂湿度
mo(u)lding-sand preparation 型砂配制
mo(u)lding-sand property 型砂性能
mo(u)lding scraper 造型刮板
mo(u)lding shop 造型工段;造型车间
mo(u)lding shrinkage 模制件收缩量
mo(u)lding spade 造型砂铲
mo(u)lding stock 制线脚针叶材
mo(u)lding surface 型面;陶形曲面
mo(u)lding temperature 模压成型温度
mo(u)lding texturing 造型法变形
mo(u)lding time 成型时间
mo(u)lding tolerance 模制公差
mo(u)lding track 造型线
mo(u)lding turntable 造型转台
mo(u)lding unit 饰线单元;造型部件;造型机组
mo(u)lding water content 塑性湿度;成型含水量
mo(u)ld inhibiting pigment 防霉颜料
mo(u)ld injection press 立体地图压印机
mouldinng box trunnion 砂箱轴
mo(u)ld in place 就地灌注
mo(u)ld insert 模件插入件;模具嵌件
mo(u)ld insulation 模制绝缘
mo(u)ld in the pit 地坑造型
mo(u)ld jacket 型套;砂型套箱
mo(u)ld joint 分型面
mo(u)ld kernel 模芯
mo(u)ldless forming 无模成型
mo(u)ldless technique for hot pressing 无模热压技术
mo(u)ld life 铸型寿命;模型寿命
mo(u)ld lifting 铸型起模
mo(u)ld line (模)型线;饰线;模线
mo(u)ld liner 模板衬
mo(u)ld lining 坩埚衬里;模衬
mo(u)ld locking force 锁模力
mo(u)ld loft 仓库;(工厂的)统楼层;(讲堂的)楼厢;阁楼;放样间
mo(u)ld lofter 放样员;放样工(人)
mo(u)ld loft floor 放样台
mo(u)ld lofting 船体放样
mo(u)ld loftsman 放样员;放样工(人)
mo(u)ld lubricant 造型润滑剂;隔离剂;模型用润滑剂;模型润滑剂;脱模剂;分型粉;分离剂;防黏剂
mo(u)ld machine 模塑机
mo(u)ld-made paper 模制纸
mo(u)ld-making 造型
mo(u)ldman 铸模工
mo(u)ld manufacture 模型制造
mo(u)ld mark 制模工;模具痕迹;模缝线;模缝条痕;模型接合缝;模型飞边;模痕
mo(u)ld matrix cathode 模制阴极
mo(u)ld oil 隔离剂;模型油;模板油;脱模油;脱模剂
mo(u)ld on end 垂直分型造型
mo(u)ld opening 铸型浇注口;模型浇口
mo(u)ld paint 模型油漆
mo(u)ld parting agent 分模剂;脱模剂
mo(u)ld parting force 开模力
mo(u)ld parting line 开模线;分型线
mo(u)ld permeability 铸型透气性
mo(u)ld pig iron 型铁;铸铁块
mo(u)ld pit 造型地坑
mo(u)ld plaster 模型用石膏;模型石膏
mo(u)ld plate 线脚模板;模型板
mo(u)ld plunger 阳模
mo(u)ld plunger force plug 凸模
mo(u)ld portion 模型的一部分
mo(u)ld positioner 压模定位器
mo(u)ld preparation 整模
mo(u)ld preparation bay 钢锭模准备跨
mo(u)ld press 压模机
mo(u)ld pressing 旋压成型;模压成型;模压
mo(u)ld preventive 防霉剂
mo(u)ldproof 不发霉的
mo(u)ld-proofing 防霉
mo(u)ld-proofing coating 防霉涂料
mo(u)ld-proofing wash 铸模涂料
mo(u)ld rain 梅雨
mo(u)ld release 隔离剂;离型剂;模型润滑剂;脱模(剂);分型粉;分离剂;防黏剂
mo(u)ld release agent 松模剂;脱模剂;下模剂
mo(u)ld release oil 脱模油;模型油
mo(u)ld-repair shop 修模车间
mo(u)ld resistor 模制电阻器
mouldrite 脲醛塑料
mo(u)ld rollover unit 翻箱装置
mo(u)ld scale 模鳞
mo(u)ld seam 模缝;合缝线印
mo(u)ld separator 脱模剂
mo(u)ld setting-and-stripping bay 钢锭模准备和脱锭跨
mo(u)ld shaft 模轴
mo(u)ld shift 错箱
mo(u)ld shooter 射型机
mo(u)ld shot 铸造铅丸
mo(u)ld shrinkage 成型收缩;模压收缩;模型收缩;模内收缩;脱模前的收缩;脱模后的收缩
mo(u)ld spacer block 模具定位块
mo(u)ld split 模组分
mo(u)lds stockyard 模型堆场
mo(u)ld stability 铸型稳定性
mo(u)ld steel 型模钢;模具钢
mo(u)ld sticking 黏膜
mo(u)ld stone 模制门框石;门边石;窗边石;铸石;刻有线脚的石头;门窗边石
mo(u)ld strength tester 砂型表面强度试验仪
mo(u)ld stripping 铸型起模;拆模(板)
mo(u)ld surface 模型表面;模面
mo(u)ld test 霉菌试验
mo(u)ld top half 上型
mo(u)ld train 模组(熔模铸造用)
mo(u)ld transfer 铸型输送器
mo(u)ld turnover 翻箱
mo(u)ld turnover rate 模具周转率
mo(u)ld unloading 卸模;脱模;出模
mo(u)ld veining 铸型裂纹
mo(u)ld vent 成型透气孔
mo(u)ld venting 扎出气孔
mo(u)ld vibration 模具振动
mo(u)ld vibrator 模型振动器;模板振动器;外部振动器
mo(u)ld void tester 砂型透气性测定仪
mo(u)ld vulcanization 模塑硫化
mo(u)ld wall 铸型壁
mo(u)ld wash 造型涂料;脱模涂料
mo(u)ld wash mixer 涂料搅拌机
mo(u)ld weight 压重;压铁
mo(u)ld yard 整模间
mo(u)ldy odour 霉腥味
moulin 冰臼麻工艺;冰臼;冰川壶穴
moulinet 扇间;风扇制动
mounanaite 羟钒铁铅矿
mounce 公制两
mound 丘陵;土丘;(螺旋桨等引起的)水面升高

防液堤;宝球纹
mound breakwater 斜坡式防波堤;抛石防波堤;堆石分水堤;堆石防波堤
mound breakwater with laying stones armo(u)r 砌石护面堤
mound breakwater with wall 混合式防波堤
mound castle 护堤船楼;护堤城堡
mound crown 堆石堤顶
mound drain 堆石排水;盲沟
mounded tank 有防波堤的储罐;防液堤保护的储罐;半埋式罐
mounded wall 压筋车体墙板
mounding 埋土法;堆土法
mounding of gravel 砾石堆筑
mound layer 压条
mound layerage 堆土压条法
mound layering 堆土压条
mound of earth 土堤;岗陵;小丘;土墩;土堆
mound planting 丘状栽植
mound spring 岗陵泉;冈陵泉
mound storage 堆藏
mound-type breaker 堆石防波堤
mound-type breakwater 斜坡式防波堤;倾斜式防波堤
mount 固定架;架
mountable 可安装的
mountable curb 斜式(路)缘石
mountain 山;大山
mountain observation 高山观察
mountain air 山地空气
mountain airfield 山地机场
mountain and valley breezes 山谷风
mountain and valley region location 山区河谷定线;山区河谷地区选线
mountain and valley winds 山谷风
mountain aneroid 山地空盒气压计
mountain apron 山麓冲积平原
mountain arc 山弧
mountain area 高山区
mountain ash 浅褐色橡木(澳大利亚产);野花楸树;欧洲花楸树
mountain barometer 高山气候表;山地气压计
mountain barrier 高山屏障
mountain beech 山地假山毛榉
mountain belt 山岳(地)带;山脉
mountain blue 石青
mountain bog 山地沼泽
mountain breeze 下降风;重力风;山风
mountain bridge 山区桥梁
mountain brook 溪涧
mountain-building 山脉形成;造山运动;造山的
mountain building period 造山期
mountain butter 铁明矾
mountain cable railway 山区缆索铁道
mountain chain 山系;山脉;山链
mountain channel 装配线;山区河道;山区航道
mountain city 山城
mountain climate 高山气候;山岳气候;山区气候;山地气候
mountain coast 山岳海岸;山地海岸;陡岸
mountain cork 石棉;坡缕石;山软木;淡石棉
mountain creek 山溪;山涧
mountain creep 滑坡;坍发;(地层结构上的)塌方;山蠕动;山崩;崩塌
mountain crest 山脊
mountain crossing 越岭垭口
mountain crystal 岩晶;水晶
mountain daylight time 山区日光时间
mountain deposit 水晶矿床
mountain driving 山区行车
mountain ecosystem 山地生态系统
mountaineering 登山运动
mountaineering rope 登山绳索
mountain effect 山地效应
mountain farming 山地农业
mountain flax 净化纤维;石棉;石麻
mountain flesh 挠性石棉
mountain flour 硅藻石;石粉
mountain folding 造山运动;山岳形成作用
mountain foot 山麓;山脚
mountain forest 高山树林
mountain forest soil 山地森林土
mountain forest terrain of torrid zone 热带山岳林地

mountain formed by folding 褶皱作用形成的山
mountain formed by plateau-forming movements 海底高原运动形成的山
mountain formed of disrupted folds 断裂褶皱形成的山
mountain formed of fauls 断裂形成的山
mountain front 山前
mountain-gap wind 山口风
mountain glacier 高山冰川;山岳冰川;山地冰川
mountain glacier of Alpine 阿尔卑斯山岳冰川
mountain goat 声呐定位器;声呐
mountain green 孔雀石;矿山绿;天然孔雀石绿;石绿
mountain group 山群
mountain gulch 山区峡谷;山地冲沟
mountain highway 山区公路
mountain house 山庄
mountain hypoxia 高山缺氧
mountain ice-cap 平顶冰川
mountain intake 山麓取水口
mountainite 无铝沸石;水针硅钙石
mountain lake 高山湖(泊);山地湖(泊)
mountain land 山地草原;山地
mountain landscape 高山景观;山区景观
mountain landslip 高山滑坡
mountain leather 山柔皮;铁镁坡缕石;坡缕石;石棉
mountain lift 山区缆索铁道;山区钢缆铁道;登山索道;登山缆车
mountain lime 坚硬石灰;山地石灰
mountain limestone 石炭纪灰岩;坚硬石灰岩;密实石灰岩
mountain line 山岭线
mountain location 山区定线
mountain-making 造山的;造山作用
mountain-making force 造山力
mountain-making movement 造山运动
mountain-making process 造山运动过程
mountain meadow soil 山地草甸地
mountain meadow-steppe soil 山地草甸草原土
mountain meal 石粉
mountain meteorology 高山气象学
mountain milk 软方解石
mountain of accumulation 积高山
mountain of circumdenudation 环蚀山
mountain of dislocation 断层山
mountain of erosion 侵蚀的山;剥蚀的山
mountain orogenesis 造山运动
mountainous 多山的
mountainous area 山区;高山区;山地地区;山地;多山地区
mountainous area with woods 山林地
mountainous channel 山区航道
mountainous country 山区地带;山岭(地)区;山地;多山国家;多山地区
mountainous diffraction 山地衍射
mountainous district 山区
mountainous ground 山地
mountainous region 山区;山岳地区;山岭(地)区
mountainous region landform 山地地貌
mountainous regions of tropics 热带山区
mountainous river 山区河流
mountainous river regulation 山溪性河道整治;山区河道整治
mountainous sea 巨浪;怒涛
mountainous stream 山区河流
mountainous house terrain 山区;山岭(地)区;山地
mountainous territory 多山地区
mountainous topography 山区地形;山岭地形;多山地形
mountainous tunnel 山岭隧道
mountainous watershed 多山流域
mountainous waterway 山区航道
mountain pass 山路;山口;隘路;隘口
mountain pass area 山口区域
mountain pass highway 山口公路
mountain pass saddle back 山垭口
mountain peak 山峰
mountain peat 山地泥炭
mountain pediment 麓原;山麓侵蚀平原;山间碛原
mountain pepper 山胡椒
mountain pine 山松
mountain plant 高山植物
mountain precipitation 山地降雨量
mountain railway 山区铁路

mountain railway location 山区铁路定线
mountain rain 山雨
mountain range 山系;山脉
mountain red earth 山地红壤
mountain region 山岭区
mountain reservoir 山区水库
mountain ridge 山岭;山脊
mountain river 山溪
mountain road 山区道路;山路
mountain road location 山区道路定线
mountain root 山麓;山根
mountain rope hoist 山区用绳索卷扬机
mountain route 山区线路;山区路线;山岭线
mountain sand 山砂
mountains and waters painting 山水画
mountain savana 山地稀树草原
mountain scarp 悬崖
mountain sickness 高原病;高山病
mountain side 山腰上半挖半填;山腰;山坡
mountain slide 滑坡;山崩;地滑
mountain slip 滑坡;地滑
mountain slope 山坡;山腰;山坡
mountain soap 皂石
mountain soil 山地土壤
mountains perennially covered with snow 终年积雪的高山
mountain spring 山泉
mountain standard time 山区标准时
mountain station 高山站;山地(气象)站
mountain storm 山区暴雨
mountain stream 溪涧;荒溪;山溪;山涧
mountain stream regulation 山溪整治
mountain stream training 山溪整治
mountain surface mine 山坡露天采场
mountain system 山系
mountain tarn 冰斗湖;山上小湖
mountain terrain 山岭区
mountain test 山区试验
mountain test course 山区试验路线
mountain thickness 山体厚度
mountain tile 山形瓦
mountain top 山顶
mountain topography 山岳地形
mountain top removal 全剥开采法
mountain torrent 山区洪水;山溪;山区急流;山洪
mountain torrent improvement 山溪整治
mountain town 山城
mountain tract 上游河段;山区段;山地河段
mountain trench 山沟
mountain tundra 高山冻原;山地冻原
mountain tunnel 山岭隧道;穿山隧道
mountain type 山型
mountain uplift 山脊
mountain valley 山谷
mountain vegetation 高山植被
mountain villa 山庄
mountain village 山村
mountain waste 荒山(地);山地岩屑
mountain watershed 山区流域
mountain water tank 山上给水池
mountain waterway 山区航道
mountain wave 地形波
mountain white oak 蓝栎
mountain white pine 爱达荷白松;美国山地白松;山白松
mountain wind 下降风;重力风;山风
mountain wood 镁坡缕石;铁石棉;不灰木
mountain wool 矿棉
mount a map 原图拼贴
mount attribute 安装属性
mount a volume 安装文件卷
mount detector 爬跨检查器
mounted 安装式的
mounted 24-disc harrow 24 片悬挂圆盘耙
mounted alternate plow 悬挂式双向犁
mounted cylinder 分置式油缸
mounted drill 架式凿岩机;架式风钻
mounted earth scoop 悬挂式挖土铲斗
mounted elastically 装配弹簧的
mounted equipment 悬挂装置
mounted general purpose plow 悬挂式通用犁
mounted in a vehicle 装在车上的
mounted lens 装配透镜
mounted load carrier 悬挂式装载车

mounted map 裱板图
mounted model 悬挂型
mounted on a skid frame 滑橇机架安装
mounted on swinging jib machine 摇臂钻床
mounted on wheels 配有车轮的
mounted paper 裱纸;裱糊图纸
mounted plate 固定板
mounted plough 悬挂式犁
mounted plow 悬挂犁
mounted-plow wheels 悬挂犁犁轮
mounted point 活套销轴;活动销
mounted position 安装地点
mounted spares 装配备件;安装备件
mounted split pattern 单面模板
mounted sprayer for walking tractor 手扶拖拉机配装式喷雾机
mounted toolbar 悬挂式机具架
mounted wheel 磨头
mounter 镶嵌工;裱装工;装配工;安装工
Mountford's paint 硅酸盐防火漆
mounting (系统的) 座架;装片;镜头组接;登上;乘骑;衬托纸;衬托板;封片;安装
mounting and dismounting device for flexible pipes 软管拆装机
mounting and dismounting device for traction motor 牵引电机解体组装设备
mounting bar 钻架支架;(凿岩机的) 安装支柱
mounting base 托盘;底座;安装基座;安装机座
mounting block 装配架;安装台;安装部件
mounting board 裱糊板
mounting bolt 装配螺栓;安装螺栓
mounting bracket 装配架;安装托架;安装支架;固定架;安装隅撑
mounting card 安装卡
mounting cardboard 衬纸板
mounting cement 装配胶
mounting ceramics 装置陶瓷
mounting channel 装配导轨;固定槽;固定通道
mounting circle 装架环
mounting clip 装配夹;安装夹
mounting column 钻架;安装支柱
mounting cost 安装费(用)
mounting coulmn 安装柱(钻)机
mounting diagram 装配图(样);安装(接线)图
mounting dimension 安装尺寸
mounting distance 装配距离
mounting drawing 装配图(样)
mounting error 安全误差
mounting fittings 装配附件
mounting flange 安装用法兰;安装用凸缘
mounting flowline for bearings of shaft box 轴箱轴承组装流水线
mounting gap 安装间隙
mounting height 装置高度;安装高度
mounting holder 装配支架
mounting hole 安装孔
mounting horn 安装喇叭
mounting jig 安装夹具;安装工具
mounting lock washer 装配锁紧垫圈
mounting lug 装配用悬架
mounting medium 固定介质;封固剂
mounting nut 装配螺母
mounting of grating 光栅装置法
mounting of optics 光学部件的装配
mounting of telescope 望远镜装置
mounting of the crane head (附着式塔式起重机的) 塔帽的安装
mounting pad 安装垫片
mounting panel 支持板;安装板
mounting photograph 裱板片
mounting pipe 管柱
mounting plate 安装平台;座板;装配平台;装配板;固定台;固定板;紧固板;上托板;安装机;安装板
mounting platform 安装平台
mounting position 安装位置;安装地点
mounting pulley 拉线轮
mounting rabbet 安装接榫
mounting rack 装拆架;工作台;机架;安装支柱;安装架
mounting ring 装配架;安装环
mountings 配件;附加装置
mounting screw 安装螺钉
mounting shaft 安装(竖)井

mounting spindle 安装主轴;安装机轴
mounting step 登车阶梯;蹬马踏步
mounting stud lock nut 装置柱螺栓防松螺母
mounting surface 安装面
mounting system 支架系统
mounting technique 显微镜试样制片技术
mounting test 吸收度试验
mounting tissue 裱糊薄纸
mounting torque 安装扭矩
mounting tripod 安装三脚架
mounting trunnion 吊挂耳轴;安装轴颈
mounting unit 安装部件
mountkeihite 英特克石
mount meal 硅藻土
mount of the picture 像片镶嵌
mount pin 安装销
Mount Rose snow sampler 蒙特罗斯型雪取样器
mount slab 山块
Mount Sorrel 淡红褐色花岗岩(英国莱斯特郡产)
mount the copies 原图拼贴
Mouray's solder 锌基铝铜焊料;穆雷铝焊料
mourite 紫钼铀矿
mourner 哀像
mourning cypress 柏木
mourning hall 灵堂
mourolite 碳草酸钙石
mouse 狭窄的出入口;光笔;窥探;鼠标(器)
mouse ahead 钻小直径超前孔
mouse algorithm 窥探算法;窥视算法
mouse beacon 可控指向标
mouse-colo(u)r 灰褐色;鼠色
mouse hole 圆形钻(俚语);小房间壁橱;进气通道;鼠洞
mouse jar 小动物缸
mouse killer 灭鼠器
mouse mill 静电动机;超小型静电起电机
mouseproof 防鼠的
mouse proof construction 防鼠构造
mouse roller 匀墨辊;附加传墨辊
mouse station 指挥台
mouse-trap 鼠笼式打捞器;捕鼠笼;灭鼠器;捕鼠器;捕鼠夹
mousing 起重钩上的安全索;扣件(俚语);吊钩的封口闩;起重钩防滑配件
mousing hook 小绳缠扎钩口的钩;封口的吊钩
mousing shackle 制滑卡环;吊钩止滑卸扣
Mousquet 默奥斯奎地毯
mousseline 透明平纹薄织物
mouth 排出口;收口部分
mouth bar 江口沙洲;河口沙洲;河口拦门沙
mouth-bar liable to change 易变拦门沙
mouth-bar-type deltaic deposit 口坝型三角洲沉积
mouth blowing 人工吹制
mouthful 抓斗抓取量;满口
mouthful of grab 抓斗的抓取量
mouth gag 开口器
mouthing 套口;承口;漏斗形开口
mouth mask 口罩
mouth of a river 河口
mouth of a stream 溪流口
mouth of blast pipe 吹风管嘴
mouth of distributory 汊口
mouth of entrance 入口截面
mouth of harbo(u)r 港口口门
mouth of hook 钩口
mouth of inlet 进水口
mouth of pipe 管口
mouth of plane 刨削口
mouth of pot 坩埚口
mouth of river 河口
mouth of shackle 卸扣开口
mouth of shears 剪口;冲剪口
mouth of spring 泉眼
mouth of the(bore)hole 洞口;孔口
mouth of tongs 钳口
mouth of well 井口
mouthpiece 接口圈;套口;口承;吸口;管嘴;管口;管子接口;管(接)头;口罩;管(接);受话筒
mouthpiece for musical instrument 管弦乐吹嘴
mouthpiece of a jet pipe 喷射管管嘴;喷射管嘴
mouth plug 锚链孔塞
mouth tool 瓶颈加工工具
moutonnee hummock 圆冰丘
movability 可动性;活动性

movable 可拆卸的;移动式的;活动的
movable air compressor 移动式空(气)压(缩)机
movable alumina-fusing furnace 活动式刚玉冶炼炉
movable anchor 活动锚
movable anode 可动阳极
movable antenna 移动式天线
movable anti-static floor board 防静电活动地板
movable arm 动臂
movable arm loading unit 动臂式装油设备
movable arm rotating shaft 动臂转轴
movable arm valve stem 动֣阀杆
movable barrage 活动堰坝;活动拦河坝
movable bathtub 移动式浴盆;移动式浴缸
movable bearing 可动轴承;活动支座;活动支承
movable bearing support 活动支座
movable bed 移动河床;活动河床;活动床【给】;动床;不定河床
movable-bed engineering model 动床河工模型
movable-bed model 活动河床模型;动床模型
movable-bed model of river engineering 动床河工模型
movable-bed model test 动床模型试验
movable-bed model with bed load 推移质动床模型试验
movable-bed plotter 动床绘图仪
movable-bed river model 动床河工模型
movable-bed routing in open channel 明渠动床演算
movable-bed scale model 活动缩尺河床模型
movable-bed stream 游荡性河流;动床河流
movable-bed with suspended load 悬移质动床测量
movable belt conveyer 移动式胶带输送机
movable bilge block 机械舭墩;活动舭墩
movable blade 可动叶片;活动桨叶;动叶片
movable blade turbine 可调叶片水轮机;可调桨式水轮机;动叶片式水轮机
movable block 移动滑车;活动滑车;动滑轮;动滑车
movable boom 活动吊杆
movable breaker-contact 活动触点
movable breaker disc 断电器活动托架
movable bridge 开启桥;开合桥;活动(式)桥;滑动桥
movable bridge system 活动桥方式
movable bridging approach 活动引桥
movable brush type polyphase series motor 活动电刷式多相串激电动机
movable bullet 可调节尾喷管锥体
movable car body support 移动式车体支座
movable cell guide post 移动式箱格导柱【船】
movable center 可移动顶尖;弹性顶尖
movable centre points crossing 可动心轨道岔
movable chute 活动溜槽
movable clamp 可动夹钳
movable coil 可变线圈;可动线圈
movable compressor 移动式压缩机
movable computer 移动式计算机
movable connection 活动连接
movable contact 活动触点;滑动接触头;滑动触点
movable core 可动铁芯
movable core transformer 可动铁芯变压器
movable coupling 活动联轴器
movable cover 活动盖
movable crane 移动式起重机;移动式吊车;桥式吊(车)
movable crest device 坝顶活动调节设备
movable crossing 可动辙叉
movable crusher 移动式扎碎机;移动式破碎机
movable crusher and screen classifier 移动式破碎及筛分设备
movable dam 活动坝
movable davit 可移动吊杆
movable deck 可动平台
movable device 移动式装置
movable distributor 移动式喷洒机;移动式配电器;移动式分线盒;移动式分配器;活动配水器;活动布水器
movable distributor of biologic(al) filter 生物滤池活动固定布水器
movable divertor hopper 移动式分料斗;移动换向漏斗
movable dock 活动码头;活动船坞
movable dust collector 移动式集尘装置

movable dwelling 活动房屋
movable eccentric 可动偏心轮
movable ecliptic 移动黄道
movable electro-static prevention floor board 防静电活动地板
movable element 活动元件
movable end 可动端
movable equipment 可移动设备;活动装置
movable estate 动产
movable exchange rate 浮动费率
movable eye bolt 带眼螺栓
movable falsework 移动式脚手架
movable fence 活动栅栏
movable fender system 活动防撞系统
movable filling arrangement 移动式罐装设备
movable fit 动配合;松配合;可动配合
movable fixture 可移动装置
movable floor 活动地板
movable floor board 活动木地板
movable flower bed 活动花坛
movable form(work) 活动模板;滑动模板;重复使用模板
movable frog 可动心轨
movable furnace 活动炉
movable gangway 活动跳板;活动坡台
movable gantry 移动式高架起重机;活动起重台架
movable gate 活动闸门
movable glass wall 可动玻璃墙
movable go-devil platform 移动式清管平台
movable grate 活动炉栅;活动格栅
movable grizzly 移动式格筛
movable guard 活动防护网
movable hair 移动丝
movable head 可动磁头
movable head stock 活动车床头
movable hood backwashing filter 移动罩反冲滤池
movable hood filter 移动罩滤池
movable horizontal louver shading device 水平活动遮阳板
movable index 活动指标
movable installation 移动装置
movable insurance 动产保险
movable jaw 活爪;活动卡瓦;活动颚板
movable jet 可偏转喷射流
movable joint 活节;活动接缝
movable joint for bridges 桥梁伸缩接缝;桥梁伸缩接头
movable joint of discharge pipeline 排泥管活动接头
movable lever davit 可移动吊杆;活动吊杆
movable lifting table 移动式升降平台
movable load 移动荷载;活载(荷);活动荷载;动(荷)载
movable magnetic powder detector 移动式磁粉探伤机
movable mark 可动刻度标志;可动刻度;可动标记
movable mask 可动掩模
movable mast 移动式桅杆;可折桅;可倒的桅
movable material 松土
movable metal ceiling 金属活动天棚
movable monitoring 流动监测
movable monitoring station 流动监测站
movable mounting 活动装置
movable nozzle 活动喷嘴;活动喷管
movable oil plot 可动油图
movable part 活动部分
movable partition(wall) 可移动隔墙;活动隔墙;活动隔断
movable pig platform 移动式清管平台
movable pin shaft sleeve 活销轴套
movable pipe clamp 管钳
movable platen 动型板
movable platform 活动式运输站台;活动月台;可动平台;活动平台
movable platform construction 活动水平模板施工
movable-point crossing 可动心轨辙叉
movable-point frog 活动辙叉;可动心轨辙叉
movable pole 活动标杆
movable propeller blade 可动螺旋桨叶
movable propeller turbine 轴流转桨式水轮机
movable property 动产税;动产
movable property in trust 信托动产
movable pulley 动滑轮;动滑车
movable purifying device for paint sprayer room 移动罩式喷漆室喷雾净化装置
movable rack 活动栅格;活动拦污栅;活动格栅
movable radial drilling machine 车式摇臂钻床
movable ramp 活动坡台
movable rate 移动汇率
movable reflector 指标镜;动镜
movable reticule 移动丝
movable river bed 移动河床
movable rocker bearing 活动摇轮支承
movable roof working platform 移动式车顶工作平台
movable sash 活动式窗扇
movable scaffold(ing) 活动脚手架;移动式脚手架
movable scale 可动刻度盘
movable scraper platform 移动式清管平台
movable screen 活动筛(网)
movable seat 活动座位
movable shaft 转动轴;活动轴
movable shower 可移式淋浴器
movable shuttering 活动模壳;活动模板;活动百叶窗材料
movable-sieve jig 动筛跳汰机
movable-sieve type washbox 动筛式跳汰机
movable signal 移动信号
movable singular point 可移异点;可去奇点
movable sleeve 活动套筒
movable sluice 活动水闸;活动闸门
movable span 可动桥跨;活动桥跨;活动跨;可伸缩桥跨
movable spheric(al) brass cup 可移动球形铜杯
movable spline 活动曲线规
movable spray aerator 活动喷洒曝气机
movable sprayer 活动喷雾器
movable sprinkler 活动喷洒器;活动喷淋器;移动式喷灌机
movable stabilizer 活动安定面
movable stage 活动舞台
movable stand 活动看台
movable suction tube 活动式吸管
movable support 跟刀架;可动支架;活动支座;活动支架;活动支承;随动中心架;随动刀架;动刀架
movable table 活动工作台
movable tail weir 移动式尾水堰;活动尾水堰
movable tangential bearing 活动的切向承载力
movable tangential rocker bearing 可移动的切向摆动支座;可移动的切向摆动支承
movable target 活动靶标
movable telescope 活动望远镜
movable trestle 活动栈桥
movable type mucosa 可动黏膜
movable vane 可动叶片;动叶
movable-vane type turbine 动叶式水轮机
movable vertical fin shading device 垂直活动遮阳板
movable vessel 移动式容器
movable voltage testing stand 移动式耐压试验台
movable weir 活动(堰)坝;活动堰;临时堰
movable weir mouth 活动堰口
movable winch 移动绞车
movable wing frog 可动翼轨(式)辙叉
movable wire 动丝
movable working table 移动式中间工作平台
moveable connection 活动连接
moveable form 活动模板;重复使用的模板
moveable glass wall 活动玻璃墙;可移式玻璃墙
moveable muddy bed 活动淤泥质床
moveable opening 活动孔
moveable reference 活动标
moveable shower 活动淋浴装置;活动式莲蓬头
moveable sprayer 活动式喷射器
moved scene 变动现场
moved ton-kilometer 运行吨公里
move function 移动函数
moveing weighted average forecasting 移动加权平均预测法
move instruction 传送指令
move mapping 动作映射
movement 移动;胀缩;动作
movement aberration 运动像差
movement accommodation factor 位移适应因数
movement after image measuring instrument 运动残像测定仪
movement against the current of traffic 反向运行【铁】
movement area 行动地区
movement by saltation 跃移
movement capability 位移能力
movement capacity 运输能力
movement controller 动作控制器
movement direction of glide mass 滑体滑移方向
movement direction of nappe 推覆体运移方向
movement documents 转移文件
movement forecast 动态预报
movement formula of groundwater in river-canal 河渠地下水运动公式
movement ga(u)ge 测动仪
movement in a curve 弧线运动
movement in a curved line 曲线运动
movement in depth 景深移动
movement in masonry 砌体胀缩变形;砌体伸缩
movement in net liquid fund 流动资产净值变动
movement in price 价格变动
movement in the market 市场动态
movement joint 活动(接)缝;施工缝;变形缝;伸缩缝
movement of armo(u)r layer 护面层位移
movement of bed material 河底沙移动;河底质移动
movement of current assets 流动资产动态
movement of detritus 岩屑移动
movement of earth 运土;土体移动
movement of earth's crust 地壳运动
movement of equilibrium lever 平衡杠杆运动
movement of floating debris 悬移质运动;悬浮质运动
movement of fund 资金移动
movement of glacier 冰川运动
movement of goods 货物搬运;商品流动
movement of goods and passengers 客运运输
movement of goods and persons 客货运转
movement of hazardous material 有害物的输送
movement of ice 流冰;冰凌运动
movement of mud 淤泥移动;泥浆移动
movement of planting trees 植树运动
movement of population 人口流动
movement of quotation 开价趋势
movement of reverse order 反向运行【铁】
movement of river sediment 河流泥沙移动
movement of river silt 河流泥沙移动
movement of rotation 旋转运动
movement of sand-wave 沙浪运动
movement of sediment 泥沙运动
movement of seismic fault 地震断裂的活动
movement of silt 泥沙运动
movement of soil particles 土粒运动
movement of stratum 岩层移动
movement of switch points 尖轨运程
movement of trains 列车运行
movement of valve 移动阀
movement of water in soil 土中水分移动
movement of water into soil 水分运动到土壤内
movement on roads 沿道路行进
movement pickup 位移传感器
movement picture 运动图形;运动图像
movement picture experts group 运动图像专家组
movement pile 位移边桩
movement plan 运输计划
movement protection detachment 运输掩护分队
movement system 转运系统
movement towards equalization of value 价值平均化运动
movement type 新构造运动类型
movement way of dune 沙丘移动方式
movement weight 重锤
movement weight cable 重锤绳
movement weight cartridge 重锤筒
movement weight chain 重锤链
movement weight cord 重锤绳
movement weight hook 重锤钩
movement weight mass 重锤块
movement weight pulley 重锤滑轮
movement weight rod 重锤杆
move mode 传送方式
move-off position 撤离原位
move order 物料搬运单
moveout curve 时差曲线
mover 搬场工人;原动机;推进器;提议人;发动机
mover for material 电动运料车
mover stair(case) 自动楼梯

moves of rig 钻机搬迁
move statement 传送语句
move switch 浮放道岔
move to check-count sampling 搬动检验
move to point-of-use 搬到使用地点
move-up 前进法
movie 电影院
movie film factory 胶片厂
movie house 电影院
movie palace 电影大厦;电影宫
movie studio 电影制片厂
movie theatre 电影院
moving 移动;活动的
moving angle 移动角
moving apart 移离
moving apron 活动护板
moving arcing contact 动电弧触头
moving-area sampling 运动面积取样
moving armature 动衔铁
moving-armature loudspeaker 动舌簧式扬声器;动片式扬声器
moving arm disc 活动臂盘
moving assembly method 移动装配法
moving average 移动平均值;移动平均(数);流动平均(数);滑动平均(数);动态平均值
moving average adjustment 移动平均调整
moving average analysis 移动(式)平均分析
moving average cost method 移动平均成本法
moving average forecast method 移动平均预测法
moving average method 移动平均值法;移动平均法;滑动平均法;平均流动法
moving average model 移动平均(数)模型;流动平均(数)模型;滑动平均(数)模型
moving average process 移动平均过程;流动平均过程;滑动平均过程
moving average value figure 移动平均值图
moving axis 活动轴;动轴
moving axle 动轴
moving ball type visco(si)meter 动球式黏度计
moving band conveyer 移动带式输送机;移动带式传送机
moving bar grizzly 转动棒条筛;动柱栅筛
moving-based navigation(al) aids 动基导航设备
moving bathtub 移动式浴盆;移动式浴缸
moving bed 移动床(层);(排种装置等的)活动底;动床
moving bed activated carbon contactor 移动床活性炭接触器
moving bed adsorber 移动床(超)吸附器
moving bed adsorption 移动床吸附剂
moving bed adsorption column 移动床吸附柱
moving bed biofilm reactor 移动床生物膜反应器
moving bed contacting 移动床接触
moving bed filter 移动床过滤器;动床滤器;活动床过滤器
moving bed filtration 移动床过滤
moving bed gasification 移动床气化
moving bed ion-exchange 移动床离子交换
moving bed ion-exchange equipment 移动床离子交换设备
moving bed process 移动床过程;移动床法
moving bed reactor 动态床反应器
moving bed reduction 动态床还原
moving bed reforming 移动床重整
moving bed sand filter 移动床沙滤池
moving bed scrubber 移动洗涤床
moving bed sequencing batch biofilm reactor 序批式移动床生物膜反应器
moving belt 运送带
moving belt contactor 动带接触器
moving between lanes 变换车道
moving blade 转动叶片;可动剪刃;回转叶片;动叶片
moving blade shutter 动片快门
moving blade stage 转动叶片级
moving block 移动滑车;动滑车
moving block system 移动闭塞系统
moving boat discharge measurement 动船流测(量)
moving boat method 动船法
moving bolster 移动式工作台
moving bolster press 可移动工作台式压力机
moving boundary 运动边界;移动界面(指牵伸变速点)

moving boundary electrophoresis 移动界面电泳;界面电泳
moving boundary method 移动界面法
moving boundary problem 移动边界问题
moving bridge 活动桥
moving bridge crane 移动式桥式起重机;移动桥架式起重机
moving brush 活动电刷
moving budget 移动预算
moving budgeting 滚动预算
moving cable system 循环索系
moving capacity 携沙能力
moving car method 流动车测定法
moving chaingrate retort 移动链篦碳化炉
moving cistern mercurial barometer 动槽水银气压表
moving coil 可转线圈;可动线圈;旋转线圈;动圈
moving-coil actuator 动圈式螺线管;动圈式激振器;动圈式传动装置
moving-coil alternate current arc welder 动圈式交流弧焊机
moving-coil ammeter 动圈式电流表;动圈式安培计;磁电式安培计
moving-coil galvanometer 动圈式检流计;动圈式电流计
moving-coil instrument 动圈式仪器;动圈式仪表;磁电式仪表
moving-coil loudspeaker 电动扬声器
moving-coil meter 动圈式仪表;动圈式电表
moving-coil microphone 动圈式传声器;电动式传声器;电动传声器
moving-coil mirror galvanometer 动圈式镜示电流计
moving-coil pickup 动圈式拾振器;动圈式拾声器;电动拾声器
moving-coil receiver 动圈式受话器
moving-coil regulator 动圈式调压器;动圈式调节器
moving-coil relay 动圈式继电器;动圈继电器
moving-coil seismograph 动圈式地震仪
moving-coil seismometer 动圈式地震计
moving-coil speaker 动圈式扬声器
moving-coil transducer 动圈式传感器
moving-coil transformer 动圈式变压器
moving-coil type 动圈式;电动式;活动线圈型
moving-coil type ammeter 动圈式安培计
moving-coil type galvanometer 动圈式检流计
moving-coil type instrument 动圈式仪表
moving-coil type microphone 动圈式传声器
moving-coil type oscillograph 动圈式示波器
moving-coil type over-current relay 动圈式过流继电器
moving-coil type torque motor 动圈式转矩马达
moving-coil type voltmeter 动圈式伏特计
moving-coil voltmeter 动圈式伏特计
moving-coil wattmeter 电动式瓦特计
moving-conductor electromagnetic pendulum seismograph 动导体电磁摆地震仪
moving-conductor electromagnetic seismograph 动导体电磁地震仪
moving-conductor loudspeaker 动导体扬声器
moving-conductor microphone 动圈式传声器;动导体传声器
moving-conductor seismograph 动导体地震仪
moving-conductor transducer seismograph 动导体换能器地震仪
moving constraint 动约束
moving contact 滑动接点;滑动触点;动触头;动触点
moving contact assembly 动触点点组
moving contact member 可动触点部件
moving contact piece terminal 动触头端子
moving conveyer 移动式输送带
moving coordinates 活动坐标
moving coordinate system 运动坐标系统
moving cost 迁移成本;搬迁费
moving crane 行走吊车
moving current 动流
moving curtain air filter 自动卷绕式空气过滤器
moving curtain filter 活动帷幕式过滤器
moving density of train 列车运行密度
moving diaphragm element 膜片型元件
moving die half 动型
moving die plate 移动模板
moving dislocation model 移动断裂模型
moving dune 流动沙丘;游动沙丘

moving-envelope kiln 移动式封套窑
moving equilibrium 移动均衡
moving expense 迁移费用
moving expense allowance 迁移费补贴
moving field 移动场;动磁场
moving-fire kiln 火位移动窑;环窑
moving-fire zone 移动烧火区
moving-flap liquid meter 活瓣式流量计
moving floor body 活动底车厢
moving-floor conveyer 活底式输送器
moving fluidized bed photoreactor 移动流化床光致反应器
moving force 原动力;动力
moving form(work) 移动式模壳;移动模板;活动模板;活动模壳;活模
moving function 动作机能
moving gaseous medium 气流介质
moving grid 可动铅栅;活动栅条;活动网格;活动格网
moving grizzly 活动栅筛
moving ground excitation 强烈地震
moving hair 移动丝
moving head 可移动头;可动头
moving hydraulic jump 移动式水跃
moving ice 活动冰
moving in and out cars 取送调车
moving in a rig 移动钻机
moving indemnity 搬迁费
moving-iron 动铁式
moving-iron ammeter 动铁式电流表
moving-iron instrument 电磁式仪表
moving-iron logometer 动铁式比率表
moving-iron meter 动铁式仪表
moving-iron osillograph 压电式示波器
moving-iron relay 动铁式继电器
moving-iron type ammeter 动铁式安培计
moving-iron type instrument 动铁式仪表
moving-iron type loud-speaker 动铁式扬声器
moving-iron type voltmeter 动铁式伏特计
moving-iron voltmeter 动铁式电压表
moving jaw 活动颚板
moving light spot 动光点
moving liquid out of well bore with water mixed with air 混气水排液
moving load 活载;行载;行动负荷;移动荷载;移动负载;活动荷载;平复式荷载;动荷(载)
moving loading 活负载;活动荷载
moving-load test 动载试验;移动荷载试验
moving-magnet 可动磁铁;动磁铁式
moving-magnet galvanometer 动磁式检流计;动磁式电流计;动磁工检流计
moving-magnetic type 动磁式
moving-magnet instrument 动磁式仪表
moving magnet magnetometer 动磁式磁强计
moving magnet type instrument 动磁铁式仪表
moving-magnet voltmeter 动磁铁式电压表
moving map display 活动地图显示器
moving mean 移动平均值
moving merge control 移动交汇控制
moving model 移动模型
moving moraine 移动碛
moving mo(u)ld casting machine 动模铸锭机
moving mo(u)ld casting process 移动模铸造法
moving object 运河目标
moving observer method 随车观测法;浮动车观测法
moving observer technique 移动观察技术
moving partition wall 活动隔墙;活动隔断
moving parts 活动(部)件;运动(部)件
moving passenger conveyor[conveyer] (输送式)自动步道
moving pavement 移动路面;可移动路面;活动人行道
moving pavement public transport 活动式路面公共运输
moving pedestrian 自动(式)人行道
moving-pen 移动笔
moving people to other places for land reclamation 移民垦荒
moving plan 可变统计划
moving plate 移动模板;动型板
moving platen document copying machine 移动台板式文件复印机;动板式文件复制机
moving plates 动片

moving plates of variable condenser 可变电容器转动片
moving platform 活动(平)台;移动式装卸台;自动搬运台;活动搬运台
moving-plug method of cementing 上塞法注水泥
moving point 动点
moving point load 移动点荷载
moving points 动的道岔
moving poise 滑板;滑块;移动平衡重块;滑动平衡重
moving power 原动力
moving rainstorm 移动暴雨
moving ramp 电动斜道;活动走道;移动坑线;自动坡道;活动坡道
moving ramp landing 电动斜道的终端平台
moving range 移动范围
moving-ripple cross lamellar structure 移动沙纹交错纹理构造
moving roller path 滚道
moving runner 工作轮
moving sand debris 移动沙闸
moving sash 活动式窗扇
moving scaffold 移动脚手架
moving scale 游标尺;移动标尺;活动度盘;活动标度尺;滑尺
moving screen 运动筛;移动屏;活动筛(网);活动格网;测流屏
moving screen concentrator 移动筛脱器
moving screen concentrator system 活动滤网浓缩器系统
moving seasonal variation 移动季节变动
moving sediment 运动泥沙;移动泥沙;活动泥沙
moving shuttering 活动模壳
moving side walk 自动人行道;自动行人道;活动人行道
moving slowly 慢慢移动
moving spare system 移动(式)备用方式【电】
moving speed 活动速度
moving spider 转动辐
moving stage 变动阶段
moving stair(case) 活动楼梯;自动升降梯;自动楼梯;自动扶梯;可变扶梯
moving stairway 活动扶梯;活动楼梯;自动楼梯;自动扶梯
moving stay follow-rest 跟刀架
moving stop 移动挡板
moving structure 结构物迁移
moving submarine haven 潜艇活动锚泊地
moving surface ship haven 水面船舶活动锚泊地
moving system 动作系统
moving tackle 动绞辘
moving targegt identification 动目标识别
moving target 活动靶标
moving target detection system 移动物标探测系统
moving target indicator 移动式目标指示器;活动目标指示器;活动目标显示器
moving target search plan 动目标搜索方案
moving thread 移动丝
moving time 搬迁工时
moving total 移动总量
moving traffic 行驶车流;行车交通;车流
moving traffic lane 行车行列;通车车道
moving trihedral 动标三面形
movingue 尼日利亚椴木
moving uniform load 均布动荷载
moving-up 前移
moving urban pollution source 移动(性)城市污染源;城市移动污染源
moving vane 活动叶片;活动轮叶;活动导叶;动叶片;动瓣
moving-vane instrument 动叶片式仪表
moving vector 动向量
moving vehicle method 随车测速法;浮动车观测法
moving walk 自动行人道;电动走道;(输送式)自动步道
moving walk landing 电动走道的终端平台
moving walk threshold plate 自动人行带的槛板
moving walkway 自动走廊
moving wall 活动壁板;活动隔墙
moving waters 动荡水域
moving-water sedimentation 动水沉淀
moving wave 行波
moving weighted average method 移动加权平均法

moving weight stabiliser 重力移动消摆装置
moving window 滑动窗口
moving window analysis 动窗分析
moving window Fourier spectrum 动窗傅立叶谱
moving window technique 动窗法
moving wire 移动丝
moviola 音像同步装置
mow 禾草堆积处;禾堆;干草堆;麦秸堆
mow conveyer 干草输送器
mower 割草(工)人;割草机;剪草人
mower backshaft 割草机后轴
mower-crusher 割草压扁机
mower cutter 割草机切割器
mower cutterbar attachment 割草机切割器
mower cutterbar release 割草机切割器松脱装置
mower guard 割草机切割器护刃器
mower knife 割草机割刀
mower knife grinder 割草机刀刃磨石
mower lift 割草机起落机构
mower section 割草机动刀片
mower-shredder 割草切碎机
mower sickle 割草机切割器割刀
mowing apparatus 割草装置
mowing-drive 割草机传动装置
mowing machine 割草机
mowing machine oiler 割草机注油器
mowing of verge turf 割除路边杂草;割除檐口杂草
Mowlem 摩勒姆式体系建筑
mowrah oil 阔叶赤铁树油
moxa 灼烙剂;艾绒
moxa cauterization 艾灼法
moxa cautery 加热急冷矫正法
moxa cone 艾炷(用艾绒压制而成的锥状物)
moxa roll 艾条
moya 泥熔岩;火山泥;泥岩
moyle 鹤嘴锄;十字镐
moyno pump 螺杆泵
Mozambique basin 莫桑比克海盆
Mozambique channel 莫桑比克海峡
Mozambique current 莫桑比克海流
Mozarabic architecture 摩莎拉布式建筑(9世纪以后建的西班牙建筑)
m-phenylenediamine 间苯二胺
m-phthalic acid 间二羧基苯;间苯二甲酸
mpororoite 水钨铁铝矿
MR adhesive 耐水黏合剂;耐水胶黏剂
M-roof 双山头屋顶;双山墙屋顶
mroseite 碳碲钙石
MR scanner 磁共振扫描器
Mrs. Murphy 小规模租屋经营者
M-series bit M 系列钻头
M-shaped fold M 形褶皱
M signal 中间信号;单色信号
MSK intensity scale 麦—施—卡烈度表
M-station 主控制站
Ms transformation 奥氏体—马氏体转变开始
M-system critical excitation voltage of elements 元素的 M 系列临界激发电压
M-tooth bit 形齿钻头
M-tooth saw M 形齿锯
mu 市亩(等于十五分之一公顷)
mucago 黏液
muceque 胶结泥砂
much branched 多枝的
much-branched adventitious roots 分枝多的不定根
muchematein 明矾苏木精染液
much gas escape 大量气体逸出
much-weathered 强烈风化的;深受气候影响的
mucialga of arabic gummy 阿拉伯胶浆
mucic acid 黏酸;半乳糖二酸
mucilage 黏胶;胶水;胶浆剂;黏质物
mucilaginous 黏性的
mucilaginous fiber 胶质纤维;黏质纤维
mucilaginous gum 植物粘胶
muck 淤泥;软泥;弃渣;污泥;渣土;腐泥;废土石;熟铁扁条;腐殖土
muck bank 弃渣场
muck bar 碾条;压条;熟铁初轧扁条;熟铁扁条
muck bar piling 束熟铁扁条
muck bin 垃圾箱;储渣仓
muck-blasting operation 崩落岩矿的爆破作业
muck-bog soil 腐泥沼泽土
muck boss 装岩工长

muck bottom 淤泥底;软泥底
muck-bottom lake 淤底湖
muck box 装渣箱
muck bucket 挖污泥斗;挖泥斗;出渣斗
muck cake 泥饼
muck cake production 泥饼产率
muck car 废石列车;土斗列车;出渣斗车;运泥车;矸石车;泥车;土斗车;出渣车
muck conveyer 废石运输机;石渣传送机;石渣传送带
muck disposal 弃渣场
muck ditch 基沟;污水沟;污染沟
muck ditch of levee 堤基沟
mucker 淤泥运泥机;装渣机;装岩机;装岩工;软土挖运机;挖泥船工人;土方工;出渣机
mucker belt 装岩机传送带
muckerism 装岩机特性
mucker operator 装岩机司机
muck excavation 腐土开挖
muck flat 沼泽平原;腐泥平原
muck fork 石渣抓斗
muck foundation 淤泥地基;泥炭地基
muck garden 沼泽地菜园
muck haulage 出土运程(土方工程);运渣;运土;出渣运程
mucking 清理坍方;钢筋调整(浇注混凝土时);吊桶;装岩;装渣;弃渣处理;挖土;出渣
mucking and haulage 除渣
mucking arrangement 装岩方式
mucking bucket 岩石吊桶
mucking car 出渣车
mucking equipment 装岩设备;装岩机
mucking gismo 吉斯莫型装载万能采掘机
mucking loader 装渣机
mucking machine 装渣机(械);装岩机;挖泥机;出渣机
mucking machine drawpoint 装矿机放矿点
mucking out 出渣作业
mucking pan 装载溜槽;装岩溜槽
mucking plate 装岩用垫板
mucking platform 出渣平台
mucking ratio 排土率
mucking slot 装岩檐口
mucking time 装岩时间;出渣时间
mucking tools 螺旋钻(取土样用);吊桶;铲斗(土方工程)
mucking unit 装载机;装岩机;抓岩机
muckite 小粒黄色树脂;小粒黄色琥珀
muck kibble 淤泥吊桶;矿石吊桶;出渣吊桶;出渣斗
muckland 淤泥地;腐殖土;沼泽地
muckle 大型钻机;大型凿岩机
muckle-gypsum rock 泥膏岩
muck loader 隧道挖运机械;装矿机;装岩机
muck loading 装渣;装矿(作业);装渣
muck loading platform 装渣平台
muck lock 材料闸
muck-marsh soil 腐泥沼泽土
muck mill 熟铁扁条粗轧机
muck out 挖除软土
muck pile 矿堆;矿石堆;熟铁扁条束;废渣堆
muck raise 岩石溜井
muck rake 粪耙
muck removal 出渣
muck roll 熟铁扁条轧辊
muck-rolling mill 熟铁扁条轧机
muck sample 淤岩样品;岩样
muck shaft 矿岩竖井
muck shifter 废渣装运机
muck shifting 污泥装运;软泥移动;污泥移动;土方搬运
muck sinking bucket 出渣斗;挖污泥斗
muck soil 软泥;污泥;污土;垃圾土;淤泥土;腐殖(质)土;腐泥土
muck-spreading attachment 撒厩肥附加装置
muck stage 淤泥层;淤泥段
muck surface 淤泥表面
muck train 矿岩列车;运渣列车
muck trench 堤基防渗槽
muck truck 运渣卡车
muck wagon 矿岩矿车
mucky clay 淤泥质黏土
mucky loam 淤泥亚黏土
mucky soil 淤泥质土(壤)

muco-cellulose 黏纤维素
muconic acid 己二烯二酸
mucosa 黏液
mucose 黏的
mucosin 黏精
mucosity 黏稠性;黏滑性
mucous membrane 黏膜
muculent 黏质的
mucuri 褐色硬木(中美洲产的)
mucus 黏液
mud 稀泥;淤泥(浆);滤泥;烂泥;矿泥;污泥
mud ability 泥地行驶能力
mud accumulation 淤泥累积;淤积;灌浆;淤泥
mud acid 泥酸;无机泥酸
mud additive 泥浆添料
mud analysis log 泥浆分析记录
mud analysis logging 泥浆分析测井
mud and snow tire 泥泞路和雪路用轮胎
mud and straw brick 草泥砖
mud apron 叶子板;挡泥板
mud auger 土钻;土层螺旋钻(具)
mud avalanche 泥石流;泥崩
mud avalanche aqueduct 泥石流渡槽
mud avalanche cave 泥石流明洞
mud avalanche ditch 泥石流排头沟
mud avalanche retaining dike[dyke] 拦挡坝(泥石流)
mud avalanche retarding field 停淤场(泥石流)
mud baffle 挡泥板;泥浆挡板
mud balance 泥浆天平
mud ball 泥球;贴砂砾泥球
mudbank 泥洲;泥滩;泥堆
mud bar (河海口的)淤积沙洲;泥坝
mud barge 泥驳
mud barrel 泥浆抽筒(冲击钻进用);掏泥筒
mud base 泥浆基;泥浆底子
mud basin 泥地;泥池
mud bath 泥浴
mud beach (河海口的)泥滩
mud bearing 含泥质
mud-bearing calcareous dolomite 含泥钙质白云岩
mud-bearing dolomitic limestone 含泥云灰岩
mud-bearing limestone 含泥灰岩
mud bed 淤泥质河床
mud belt 淤泥带;泥沉积带
mud berth 坐泥泊位;泥锚位
mud bin 泥浆储[贮]藏箱
mud bit 菱形钻头;黏土层钻头;软地层钻头;土层钻头;匙形钻
mud blanket 泥毡层
mud blasting 炸泥;泥盖(爆破)法;泥封爆破(法);冒泥;覆土爆破(法)
mud boat 载泥船;泥驳;挖泥船
mud body 泥浆基体
mud boil 泥涌;泥土翻腾;泥浆翻腾;涌土
mud boulder 泥砾
mud box 夹心板桩堤;泥箱;泥浆箱;板桩夹心子堤(加高堤顶方法之一)
mud breaker 碎泥刀
mud brick 泥土砖;泥砖
mud bridges (孔壁上的)泥饼沉积
mud bucket 取粉管;挖泥斗
mud cake 滤饼;滤饼;泥皮
mud cake correction 泥饼影响校正
mud cake density 泥饼密度
mud cake growth (孔壁上的)泥皮形成
mud cake resistivity 泥皮电阻率;泥饼电阻率
mud caking formation 造浆岩层(指在钻进过程中可以形成泥浆的岩层)
mud calcite-bearing dolomite 含泥含方解石白云岩
mudcap 封泥;覆盖土爆破(法)
mudcap blasting 覆土装药爆破
mudcap method 裸露药包二次爆破法;覆土爆破法
mudcapping 松动爆破;爆破孔盖泥土;泥盖;糊炮爆破;糊炮;外部装药爆破(法);封泥
mudcapping method 泥盖(爆破)法
mudcap process 覆土爆破法;封泥爆破
mud-cell dike[dyke] 蜂房状泥室堤
mud chamber 泥浆池;排泥室;沉淀室
mud chemist 泥浆配制员;泥浆化学师
mud chute 溜泥槽
mud circulating system 泥浆循环系统
mud circulation (旋钻技术中的)泥浆循环

mud circulation-gallons per minute 每分钟泥浆循环加仑数
mud clast 泥屑
mud clastic texture 泥屑结构
mud cleaner 除泥器
mud cleaning device 泥浆净化设备
mud cleaning machine 清泥机
mud collar 泥圈;泥环;带流通阀的加重钻杆
mud collector 集泥器;污水过滤器;污水过滤池
mud column 泥浆柱
mud conditioner 淤泥调质剂;泥浆调质剂;泥浆处理剂
mud conditioning 泥浆调节;泥浆处理;泥浆性能调整
mud cone 泥锥;泥丘;泥火山(锥);沸泥锥
mud content 含泥量
mud content in aggregate 骨料含泥量
mud control 泥浆处理
mud cooling tower 泥浆冷却塔(地热钻进时用)
mud correction 泥浆校正
mud cost 泥浆费;泥浆成本
mud cover (轴流泵上的)挡泥罩
mud crack 泥裂(痕);泥裂缝
mud cracking 龟裂;大龟裂
mud crack polygon 泥多边形土
mud cross 泥浆四通阀;泥浆交叉
mud crusher 碎泥刀
mud cup 泥浆杯
mud cutter 碎泥刀
mud dam 防泥石流坝
mud daub 修补裂缝;泥铺裂缝;泥补裂缝
mudded off 泥封的
mudded up 泥浆堵钻
mud density 泥浆密度
mud desander 泥浆筛;泥浆去砂器;泥浆除砂器
mud design 泥浆设计
mud diapir 泥底辟
mud diapir barrier 泥底辟遮挡
mud diapire 泥穿刺
mud diapir trap 泥刺穿圈闭
mud diffuser 泥浆喷射器
muddiness 泥泞
mudding 泥浆造壁;泥浆堵漏;泥封
mudding action 泥浆造壁作用
mudding up 引头子;上炉
mud discharge line 泥浆排卸管道
mud ditch 泥水(沟)槽;泥浆池
mud ditch sampling 泥浆槽取样
mud dock 挖泥船坞
mud dolomite-bearing limestone 含泥含云灰岩
mud drag 疏浚机;刮泥机;挖泥机;挖泥船
mud draghead 淤泥耙头
mud drape 泥盖
mud dredge bulb 挖泥机
mud dredge(r) 刮泥机;挖泥船;疏浚机
mud drift 淤泥漂移
mud drilling 泥浆钻进;泥浆冲洗钻进
mud drum 炉底锅筒;泥箱;泥包;排泥筒;排泥管
muddy 泥泞的
muddy ballast 泥浆覆盖的道砟浆
muddy beach 泥质海滩
muddy bottom 泥底
muddy clay 淤泥质黏土
muddy coast 淤泥质海岸
muddy creek 泥水溪;混水溪
muddy estuary 淤泥质河口
muddy gravel 泥砾沉积;泥砾层
muddy ground 淤泥地;淤泥底
muddy ice 含有泥沙贝壳石等的冰块
muddying 翻浆
muddy intercalation 泥质夹层
muddy land 泥地
muddy loam 淤泥质亚黏土
muddy logging 泥浆录井
muddy plain coast 淤泥质平原海岸
muddy print 模糊印刷
muddy river 混浊河流
muddy road 泥泞道路
muddy soil 淤泥质土(壤);淤泥(类)土;泥浆(状)土;粉土
muddy stream 混浊河流
muddy terrain 泥泞地带
muddy wastewater 泥状废水;多泥废水
muddy water 泥水;混浊水;浑浊水

Mudejar architecture 马德加建筑
mud end of the pump 泥浆泵动力端
mud engine 潜入式水力穿孔器
mud engineer 泥浆工程师
mud eruption 泥水喷发
mud-field 泥地
mud filling 淤泥填塞;泥浆淤积;淤泥沉积;淤积
mud filter 滤泥器
mud filtrate 泥浆滤液
mud filtrate resistivity 泥浆滤液电阻率
mudfish 泥鱼
mudflat 淤泥滩;泥滩;泥地;泥坪;河口浅滩;海滨滩;海滨泥地;潮泥滩
mud flat coast 泥坪海岸
mud-flat community 淤泥滩群落
mud flat deposit 泥坪沉积
mudflow 泥石流;泥流(物);泥流构造痕迹火山泥流;混浊水流
mud flowage deposit 泥流沉积(物)
mudflow creep 泥流蠕动
mudflow deposit 泥流沉积(物)
mudflow fan 泥石流冲积扇
mudflow fill indicator 灌泥浆指示器
mudflow front 泥石流前锋
mudflow gully 泥石流冲沟
mudflow levee 泥流堤
mudflow protection 泥石流防治
mudflow soil 泥流土
mudflow terrace 泥流阶地
mud flue 泥浆通道
mud fluid 泥浆
mud fluid displacement 更换泥浆
mud fluid yield 造浆率
mud flume 泥浆侧流槽;泥浆池;泥浆槽
mud flush 泥浆冲洗;钻孔用浆(俚语)
mud flush boring 泥浆洗孔钻进;泥浆冲洗钻进
mud flush drilling 泥浆钻探;泥浆洗孔钻进;泥浆冲洗钻进
mud flush method of sinking 泥浆冲洗钻井法
mud flush system 泥浆冲洗法;泥浆循环系统
mud flush test 钻泥浆冲洗试验
mud foreshore 淤泥岸;泥滩
mud-forming clay 造浆黏土
mud ga(u)ge 泥浆尺
mud geyser 间歇泥喷泉;泥水间歇泉;泥喷泉
mud ground 淤泥土
mud grouting 灌泥浆;黏土灌浆;泥浆灌注;泥浆灌浆作业
mud guard 叶子板;挡泥板
mudguard lamp bulb 挡泥板灯泡
mudguard lamp glass 挡泥板灯玻璃
mudguard lamp rubber pad 挡泥板灯橡皮垫
mudguard stay 挡泥板撑条
mud gully(trap) 泥浆集流沟
mud gun 泥枪;注泥枪;泥浆搅拌装置;炮泥枪
mud hog 隧道掘进工;摆锤式碎土机;摆锤粉碎机组
mud hog pump 泥浆泵;砂泵
mud hold 泥舱
mud hole 泥浆钻孔;排泥孔;出泥孔;淤泥清理孔
mud hopper 泥浆漏斗
mud hose 泥浆软管
mud house 泥浆房
mud hut 泥浆棚屋
mud in 注泥浆下套管法
mud injection 泥浆注浆;泥浆灌注
mud island 泥滩
mud jack 压浆;注浆泵;(填充混凝土路面下空隙用的)压浆泵
mud jacking 压(泥)浆;压浆法;喷射泥浆
mud jack method 压浆(方)法
mud jack pump 压浆泵
mud kibble 挖泥吊桶
mud laden fluid 泥浆体;泥浆(比重大于1.2)
mud lark 清沟工人
mud launder 泥浆沉淀槽
mud lava 泥流火山碎屑;泥熔岩
mud layer 淤泥层;泥浆层
mud lighter 运泥船;泥驳;挖泥船
mud line 泥线;泥浆管线;泥浆管路;泥浆管道;水土交界线
mudline suspension system 泥线悬挂系统
mud lining 结泥饼
mud liquid phase 泥浆液相
mud log(ging) 泥水测井记录;井中泥浆电阻率测

mud loss 泥浆漏失
mud loss instrument 泥浆漏失测定器
mud lubrication 泥浆防喷
mud lubricator 泥浆压井器;泥浆均压器
mud lump 淤泥滩;淤泥层;泥滩;泥丘;泥火山
mud making formation 造浆岩层(指在钻进过程中可以形成泥浆的岩层)
mud manifolding 泥浆管线设备
mud mat 擦泥毡
mud material 泥浆材料
mud meter 泥浆浓度计;含泥率计
mud minnow 泥荫鱼
mud mixer 碾泥机;黏土拌和器;黏土拌和机;泥浆搅拌机
mud mixing appliance 泥浆混合器
mud mortar 泥砂浆;泥灰(砂)浆
mud movement 淤泥运动
mud oack 泥浆泵
mud off 泥封
mud of foreshore 泥前滨
mud oil 钻井采油
mud outlet 排泥口
mud pan 灰槽;污垢沉淀池
mud peat 烂泥炭
mud percolation 泥浆渗透
mud performer 泥地高通过性车辆
mud pile 泥浆桩
mud pipe 取粉管;排泥管(道)
mud piping 排泥管
mud piston 泥浆泵活塞
mud pit 泥浆坑;泥浆池
mud plant 泥浆站;泥浆厂
mud plasticity 泥浆可塑性
mud plug 泥箱管道开关
mud polygon 泥质多边形土;泥多边形土
mud port 泥浆口;泥浆孔;泥浆槽
mud pot 泥喷泉;泥浆罐
mud precipitation 泥质沉淀
mud press 滤泥机;泥浆压滤机
mud pressure 泥浆压力
mud pressure indicator 泥浆压力计
mud pressure shield 泥浆加压盾构
mud program(me) 泥浆使用说明书;泥浆规格
mud propelled turbine 泥浆传动涡轮机
mud property control 泥浆性能处理
mud property ratio 泥浆性能指数
mud puddle 泥潭
mud pump 吸泥泵;泥浆泵;抽泥机;抽泥泵
mud pump hose 泥浆泵软管
mud-pumping 翻泥冒泥;冒泥;唧泥作用
mud-pumping condition 冒泥情况
mud-pumping of track 轨道翻浆冒泥
mud pump pulse 泥浆泵压力波动
mud pump shock pressure 泥浆泵振动压力
mud pump trailer 泥浆泵拖车
mud pump valve 泥浆泵阀;泥泵阀
mud rain 泥雨
mud rake 泥耙;挖泥耙
mud ratio 含泥率
mud reclamation 泥浆再生
mud record 泥浆记录
mud relief valve 泥浆泵安全阀
mud removal 除泥浆;排放泥浆
mud replacement by blasting 爆破排淤法;爆破挤淤法
mud residue 泥渣;残泥
mud resistivity 泥浆电阻率
mud resistivity log 泥浆电阻率测井
mud resistivity log curve 泥浆电阻率测井曲线
mud resistivity log plot 泥浆电阻率测井曲线图
mud resistivity tester 泥浆电阻率测定器
mud retaining shed 拦泥库
mud return ditch 泥浆槽
mud rheology 泥浆流变学
mud ring 接线盒盖;料浆圈;泥浆圈;容器盖(俚语)
mud rock 泥状岩;泥岩;泥石
mud-rock flow 泥石流
mud room 泥鞋室
mud run 泥浆突然涌入
mud-sand pressure 泥沙压力
mudsaver bucket 泥浆护罩;泥浆防溅盒
mud saw 磨料锯

mud scale 泥浆比重计
mud scow 运泥船;(取海泥的)泥浆漏斗;泥驳船;泥驳;平底运泥船;平底泥驳;方驳式泥浆箱
mud scraper 泥工;刮泥机;刮泥刀;砖瓦工(俚语)
mud screen 泥浆筛
mud scum 浮泥;翻浆冒泥
mud seam 流沙灰层;泥缝;泥夹层;泥缝
mud separator 泥浆分离器
mud settler 泥浆沉降器;沉渣机;沉泥机
mud settling pit 泥浆池;泥浆沉淀池
mud settling pump 泥浆沉淀泵
mud shaker 泥浆筛
mud shale 泥页岩
mud sheath 孔壁泥皮;井壁泥皮
mud sheet 泥藏
mud shell 泥壳
mud shield 泥水加压盾构
mud shield of output shaft 输出轴防泥护罩
mud shoal 泥质浅滩
mud shovel 泥铲
mud sill 下槛;门槛;排架座木;底梁;底基木;底基;卧木
mud silting 淤泥沉积
mud slab 湿软土上的底脚;混凝土垫层;底板
mud slinger 向前出口的混凝土搅拌机;除泥罩;特殊混凝土搅拌机(俚语)
mud slip 泥浆清孔底
mud sluice 泥浆槽
mud slush 稀泥浆
mud smeller 地质化学家
mud socket 淤泥筒;泥浆抽筒
mud soil 泥浆土壤
mud solid phase 泥浆固相
mud solution 泥浆
mud-spate 泥流
mud spoil 泥涌
mud spout 喷泥浆
mud spring 泥泉
mud stability 泥浆稳定性
mud-star 泥海星
mud still 泥浆沉降器
mudstone 泥(板)岩;泥屑灰岩;泥石;灰泥石灰岩
mudstone flow investigation 泥石流调查
mudstone for cement burden 水泥配料用泥岩
mud stream 泥浆液流;泥流
mud stump 积泥池;沉泥池
mud suction hose 泥浆吸入软管
mud suction tank 吸泥浆罐
mud sump 集泥井;积泥坑;泥浆池;泥浆沉淀池;泥池;沉泥池
mud-supported 泥支撑
mud-supported fabric 泥支撑组构
mud surface 淤泥表面
mud swivel 泥浆钻进用水龙头;泥浆回转头
mud system 泥浆循环系统;泥浆冲洗法
mud tank 泥浆箱;泥浆桶;泥浆罐
mud tank area 泥浆区面积
mud thickener 泥浆增稠剂
mud thickness contour map 淤泥厚度等值线图
mud thinner 泥浆稀释剂
mud transport 淤泥移动
mud trap 沉泥井;沉泥池
mud treating equipment 泥浆处理设备
mud trough 泥浆池;泥浆沟;泥浆槽
mud tube 取粉管;沉淀管
mud tubular bit 管形提泥钻头
mud type 泥浆类型
mudulation effect 调制效应
mud-up 泥封
mud valve 泥浆泵排出阀;泥阀;排泥阀
mud viscosity 泥浆黏度
mud-volcanic activity 泥火山活动
mud-volcanic gas 泥火山气
mud volcano 泥火山
mud volume 泥浆流量
mud wall 干打垒;夯实土墙;土墙
mud wall-cake 孔壁上的泥饼;井壁上的泥饼
mud water hydrometer 泥浆比重计
mud wave 泥滑;泥波;泥板波
mud weight 泥浆重量;泥浆比重
mud weight balance 泥浆比重计秤
mud weight indicator 泥浆比重计
mud well 泥浆钻孔
mud work 泥藏

mu-factor 放大系数
muff 保温套;绝热套;套筒;衬套
muff chuck 套筒卡盘
muff clamp chuck 套筒卡盘
muff concrete 套筒混凝土
muff coupling 套管式接合;套接;筒接头;套筒联轴器;套筒接合;套管连接
muff coupling joint 筒形万向联轴器
muffet 屋面石板
muff heater 保温加热器
muffity 屋面石板
muff joint 套筒连接;套筒接头;套管连接
muffle 窑炉的隔焰耐火罩;玻璃灯罩
muffle carrier 玻璃灯罩架
muffle coating 消声涂层
muffle colo(u)r 隔焰窑烧成的颜料
muffled glass 遮光玻璃
muffled sheet 导光灯罩架;涟漪型玻璃
muffle furnace 膛火炉;回热炉;隔焰炉;高温炉;马弗炉;套炉
muffle kiln 隔焰窑;马弗窑
muffle lehr 隔焰式退火窑
muffler 消声器;减声器;灭声器
muffler bracket 消声器支架
muffler clamp 减声器夹
muffler cut-out 消声器隔断器
muffler cutout valve 消声器切断阀
muffler device 消声装置
muffle resistance furnace 茂福式加热用电阻炉
muffler explosion 消声器反火;消声器爆声
muffler flange 减声器凸缘
muffler inlet 消声器进口
muffler inlet pipe clamp 减声器进气管夹
muffle roaster 闭式烤炉
muffler outer pipe 减声器外管
muffler outlet pipe bracket 消声器出气管托架
muffler outlet piper 消声器出气管
muffler piling 消声打桩(法)
muffler plate 消声板;减声板
muffler repair jacket 消声器修理套
muffler roaster 套式焙烧炉
muffler section 消声段
muffler shell 消声器外壳
muffler strap 消声器带条
muffler support 减声器支架
muffler tail pipe 消声器排气管
muffler tail pipe clamp 消声器尾管夹
muffler tail pipe support 消声器尾管支架
muffler tail pipe support clip 消声器尾管夹
muffling 消声;深孔爆破
mug 带柄啤酒杯
mugearite 橄榄粗安岩
mugearite glass 橄榄粗安玻璃
mugger 沼泽鳄
muggy 湿热
muggy monsoon 湿热季候风
muggy weather 闷热天气
Mughal architecture 莫卧儿建筑(印度)
mugho pine 矮生山松
mugwort stalk door curtain 艾杆门帘
Muhammadam architecture 穆罕默德建筑;〈旧〉回教建筑;伊斯兰教建筑
mu-H curve 磁导率磁场强度曲线
muirite 硅钛钡石;羟硅钡石
muistanden 山墙上缘砖砌成排齿饰型
muitigrid tube 多栅极管
muitiple-die press 多冲模冲床
mukhinite 钒帘石
mulatto 黄褐色的
mulay saw 去角锯
mulberry 深紫红色;桑树
mulberry field 桑园;桑田
Mulberry harbo(u)r 摩尔布里港(用预制块成的人工港);预制码头;预制港口结构
Mulberry invasion harbo(u)r 摩尔布里登陆港("二战"时期盟军登陆港)
mulberry orchard 桑园
mulch 林地植被;地面覆盖物;覆盖物;覆盖料(保护地面用的木屑、砾石、纸等);覆盖;腐土
mulch cover(ing) 落地层;枯枝落叶层
mulched ground 覆盖地;被覆盖土壤
mulched land 被覆盖土壤
mulcher 土壤覆盖机;表土疏松机
mulch farming 覆盖农作制;覆盖耕作;覆盖防旱

农作制
mulching 铺覆盖料;铺覆盖层;地面覆盖(层)
mulching-machine 表土疏松机
mulching material 地面覆盖物;覆盖物
mulching paper 覆盖纸
mulching soil 覆盖土
mulch layer 土壤覆盖层
mulch method 护路面法;覆盖法;护树根法
mulch planter 带覆盖薄膜的播种机
mulch saturant 覆盖料的浸制剂
mulch spreader 覆盖物分散机;切碎植物撒布机
mulch treatment 覆盖处理
mulde 向斜层;舟状槽;凹地
muldex 复接分接器
mule 小型电动机车;上坡牵引车
mule cart 轻型牵引机
mule-foot 单边磨损(钻头)
mule head 摇杆端部
mule-head hanger 悬挂抽油杆到摇杆端部的装置
Muler-Breslau's principle 米勒-白氏原理
mule's foot 钢绳结
mule shoe 定向靴【岩】
mule track 驮运路
mule trail 大车道;骡马车道
mule travel(l)er 爬行(式)起重机
muley saw 去角锯;去角机
muliple-core 多芯的
mulitple ignition 多火花点火
mulitple impulse weld 多脉冲焊缝
mulitple metering relay 复式计数继电器
mulitple pump circuit 多泵回路
mull 黑泥土;细腐殖质;熟腐殖质;腐熟腐殖质
mullah bridge 沟壑桥
mull-dressing 软布敷料
mullen 板材抗酸性能试验
muller 研磨轮;研磨杵;研磨;辊子;碾轮;平底乳钵
muller and plate 研磨机和刮板
Muller-Breslau's principle 马勒-布雷斯劳原理;变位线原理
muller crusher 辊式破碎机;滚轮破碎机
mullerite 软绿脱石
Muller mixer 马勒混合粉碎机;悬轮混合机;研磨搅拌机;轮碾式混料机;湿式悬轮混料机
Muller model 马勒模型
muller plow 混砂机刮板
Müller's glass 玻璃蛋白石
mullet 测厚卡尺
mulleting 测量板厚
mulling 碾磨;混砂;湿式和干式混合破碎法
mulling agent 研糊剂
mulling machine 碾碎机
mulling ratio 混合比
mullion 中梃;中竖框(门窗);立梃;门窗扇中梃;窗中梃;窗门直梃
mullion anchor 窗梃锚固件
mullion cover 窗梃面板
mullioned 有直梃的
mullion stiffener 窗梃加劲件
mullion structure 栅状构造;窗梃构造
mullion-type curtain wall 直梃式幕墙
mullion window 有竖框窗
mullite 多铝红柱石;富铝红柱石(耐高温材料)
mullite block 莫来石大砖
mullite brick 高铝红柱石砖;莫来石砖;富铝红柱石砖
mullite fire brick 莫来石质耐火砖;富铝红柱石耐火砖
mullite porcelain 莫来石(质)瓷
mullite refractory 富铝红柱石耐火材料;莫来石质耐火材料;多铝红柱石耐火材料
mullite refractory material 富铝红柱石耐火材料
mullite white ware 莫来石瓷
mullitization 莫来石化
mull layer 成熟腐殖质层
mull leaf beds 腐叶层
mullock 隧洞弃土;采石废石堆;矿山废土;矿山废石;废料
mullocker 废石装运工
mullocking 装岩
mull on 窗门的竖框;窗门的直梃
mull technique 黏结法
mulseal 乳化沥青喷剂
mulser 乳化机
mulsh paper 园圃纸
multiaccelerator 多重加速器

multiaccess 多用户访问;多通道;多路访问;多路存取;多处访问
multiaccess computer 多用户计算机;多通道计算机
multiaccess computing 多路存取计算
multiaccess controller 多路存取控制器;多存取控制器
multiaccess network 多路存取网络
multiaccess receiver 多通道接收机
multiaccess system 多存取系统;多处访问系统
multiaction computer 多作用计算机
multiactivity 多活动
multiactivity cruise ship 多功能旅游船
multiadapter 多用适配器;多用附加器
multiaddress 多重地址;多址;多地址(的)
multiaddress calling 多(地)址呼叫;多重寻址
multiaddress code 多地址(代)码
multiaddress computer 多地址计算机
multiaddressing 多重寻址
multiaddress instruction 多地址指令
multiaddress instruction code 多地址指令码;多地址编码
multiaddress message 多地址信息
multiaddress order code 多地址指令码
multi-aisle 多中堂;多通道;多侧廊
multialkali photocathode 多碱光电阴极
multialternative evaluation 多方案评价
multialternatives 多方案
multiamplifier 多级放大器
multianalysis 多重分析;多方面分析;全面分析
multianchorage system 多层锚碇系统
multi-anchored wall 梯式加筋锚碇墙
multianemometer 多点测定风速仪
multiangle glossmeter 多角度光泽计
multiangular 多角的
multiangular bar 多角形钢;多角形棒材;多边形棒
multiangular block 多角形街坊;多角形块;多边形建筑
multiangular bond 多边连接;多边结合;多边形圬工连接
multiangular building 多边形房屋;多边形建筑
multiangular cavity block 多边形空心块
multiangular choir 多边形歌唱班席位
multiangular church 多边形教堂
multiangular column 多边形柱
multiangular cupola 多边形圆顶
multiangular diminutive tower 多边形小塔楼;多边形尖塔
multiangular dome 多边形穹隆
multiangular donjon 多边形城堡主楼
multiangular drum 多边形鼓筒
multiangular dungeon 城楼上防御的塔;多边形地牢
multiangular frame 多边形桁架;多边形骨架;多边形框架
multiangular hall-choir 多边形会堂圣坛-歌唱班席位
multiangular keep 多边联系;多边接触;多边形城堡主楼
multiangular masonry bond 多边形圬工连接;多边形圬工结合
multiangular ornament 多边形装饰
multiangular plate 多边形平板;多边形板材
multiangular pot 多边形箱;多边形槽;多边形筒;多边形罐
multiangular quire 多边形歌唱班席位
multiangular rod 多边形钢条;多边形棒材
multiangular roof 多边形屋顶
multiangular slab 多边形扁钢;多边形厚板
multiangular small tower 多边形小塔楼
multiangular termination 多边形终端
multiangular tower 多边形塔
multiangular truss 多边形桁架
multiangular tunnel curing chamber 折线式养护窑
multiangular turret 多边形塔楼
multiangular vault 多边形拱顶
multiannual average 多年平均值
multiannual storage 多年平均库容
multiannual storage capacity 多年平均库容量
multiannual storage plant 多年调节(库容)电站
multianode 多阳极的
multianode mercury arc rectifier 多阳极汞弧整流器
multianode rectifier 多阳极整流器
multianode tube 多阳极管

multianvil sliding system device 多压砧滑移式装置
multianvil sliding system ultra high pressure device 多压砧滑移式超高压装置
multianvil sliding vessel 多压砧滑块式装置
multianvil system ultra high pressure device 多压砧式超高压装置
multiaperture 多孔
multiapertured 多孔的
multiaperture device 多孔器件;多孔磁芯
multiaperture photohead 多光孔曝光头
multiaperture reluctance switch 多孔磁阻开关
multiaquifer formation 多含水层构造
multiaquifer well 多含水层井
multiarch 连拱
multiarch analysis 连拱分析
multiarch bridge 多孔拱桥;连拱桥
multiarch dam 多拱坝;连拱坝
multiarched 多拱形的;多连拱的
multiarchitecture machine 多体系结构计算机
multiarch tunnel 连拱隧道
multiar circuit 多向振幅比较电路
multiarc welder 多相电极弧焊机
multiarc welding 多弧焊
multiarm caliper device 多臂井径仪
multiarray 多路排列
multiarticulated 多头连接的;多点铰接的
multiaspect 多面;多方面;多标号
multiaspect indexing 多形式索引;多信息组编目法
multiassociative memory 多相联存储器
multiatage axial flow compressor 多级轴流式压气机
multiaterally harmonized standard 多边协调标准
multiattribute assessment 多属性评价;多品质评价
multiattribute utility function 多属性效用函数
multi-automatic vertical spindle automatic chucking machine 多工位卡盘立式自动车床
multiaxial 多向的
multiaxial creep 多轴向徐变(混凝土)
multiaxial fabric 多轴织物
multiaxial filament winding 多轴纤维缠绕
multiaxial filament winding machine 多轴纤维缠绕机
multiaxial knitted glass fabric 多轴缝编玻纤织物
multiaxial multianvil ultra pressure and high temperature device 多轴多压砧式超高压高温设备
multiaxial roving fabric 多轴向无捻粗纱布
multiaxial stress 多轴应力;多轴向应力;多向应力
multiaxial stress condition 多轴应力状态
multiaxial test 多轴试验(三向部等压)
multiaxis turning center 多轴车床中心
multiaxle car 多轴车
multiaxle expression 多轴车
multiaxle lorry 多轴运货汽车
multiaxles 多轴
multi-axle steering 多桥操纵方向;多桥转向
multi-axle trailer 多轴挂车;多轴拖车
multi-axle vibrater 多轴振动筛
multibag filter 复式袋滤器
multibalcony type auditorium 多楼座观众
multiball type mandrel 多球式芯棒
multiband 多谱段;多频带;多波段
multiband airphoto 多波段摄影航片
multiband antenna 多频带天线
multiband camera 多谱段摄影机;多波段摄像机
multiband colo(u)r indices 多色测光色指数
multiband interpretation method 多波段解译法
multiband remote sensing 多波段遥感
multiband spatial resolution 多谱段空间分辨率
multiband spectral analysis 多波段光谱分析
multiband spectral camera 多谱段照相机
multiband spectrometer 多谱段分光计
multiband television antenna 多频道电视天线
multibank 多排的
multibank engine 多排式发动机
multibanking system 多存储体系统
multibank strip chart recorder 多通道长图记录仪
multibar linkage 多杆机构
multibarrel 多管式的
multibarrel system 多腔系统;多管系统;多筒系统
multibarrel winch 多滚筒绞车
multibarrier 多重屏蔽
multibase 多层片基

multibasic channel 多基通道
multibatch processing 多批处理
multibate delta 多趾三角洲
multi-bay frame 多跨度桁架;多跨度构架;多跨度框架;多跨度屋架
multibeacon 三重调制信标点;多信标
multibeam 多束光;多波束
multibeam amplifier tube 多电子束放大管
multibeam antenna 多束波天线
multibeam echo sounder 多波束回声测深仪
multibeam holography 多光束全息术
multibeam interferometer 多光束干涉仪
multibeam resonator 多光束谐振腔
multibeam sonar 多射束声呐;多波束声呐
multibeam sounding 多波束测深
multibeam ultrasonic flow meter 多束超声流量计
multibearing 多支座的
multibed 多段床;多层床
multibed demineralizer 多床除盐装置
multibed system 多级系统;多床系统
multibench blasting 多台阶爆破;多阶段爆破;多级爆破
multibit latching 多位封锁
multibit momentary 多位瞬时
multibit shift 多位移位
multiblade 多刃的;多片(的)
multiblade blower 多叶片送风机
multiblade damper 多叶减振器;多叶气流调节器;多叶消声器
multibladed bit 多刃(式)钻头
multiblade(d) circular grab 多颚圆形抓斗
multiblade electric(al) fan 多叶电扇
multiblade fan 多叶片风机;多叶电扇
multiblade grab 多刃式抓斗;多刃式抓扬机;多瓣式抓斗
multiblade pocket knife 多用小刀
multi block 复滑车;多滑轮的
multiblock machine 多卷筒拉丝机
multibode positive displacement mud motor drill 多头螺杆钻具
multi body structure 多体结构
multibody system 多体系统
multiboom jumbo 多臂钻车
multibore channel 多孔隧道
multibottom directional hole 多孔底定向孔
multibottom settling tank 多底沉淀池
multibranch 多路转移
multibranch fitting 多分支配;多路装配
multibranch network 复式网络
multibreak 多重开关;多断点的
multibreak circuit breaker 多重开关断路器
multibreak switch 多断点开关
multibreak switch or circuit-breaker 多重开关
multibridge intersection 多层立(体)交(叉)
multibridge over one river 一河多桥
multibucket 多斗式
multibucket appliance 多斗用具;多斗设备;多斗装置
multibucket continuous thickener 多斗连续式浓缩池
multibucket dredge(r) 链斗式挖泥船;多斗采砂船;多斗挖泥机
multibucket elevator dredge(r) 多斗式挖泥机;链斗式挖泥船
multibucket excavator 多斗挖掘机;多斗(式)挖土机
multibucket feeder 多斗喂料机
multi bucket-ladder dredger 链斗式挖泥船;多斗梯式挖泥船
multibucket loader 多斗装载机
multibucket system 多斗系统
multibucket trench digger 链斗式挖沟机;多斗挖掘机;多斗(式)挖沟机
multi bucket trencher 链斗式挖沟机;多斗式挖沟机
multibucket trenching plant 链斗式挖沟机
multi-building development 多幢住房设计;多幢住房开发
multi-building project 多幢住房方案;多幢住房规划
multi-building scheme 多幢住房布置平面图;多幢住房布置示意图;多幢住房布置草图
multibulb 多灯泡的
multibulb rectifier 多管臂整流器
multibuoy mooring system 多浮标系泊系统

multiburner cutting machine 多头切割机
multiburst 多波群
multiburst generator 多频(率)脉冲发生器;多脉冲群发生器;多波群脉冲发生器
multiburst signal 多波群信号
multibus 多总线网络标准
multibus compatible 多总线兼容的
multibus control 多总线控制
multibus module 多总线模块
multibus multiprocessor 多总线多处理机
multibyte 多字节
multibyte data 多字节数据
multi-cable stayed bridge 密索斜拉桥
multi-cable system 密索体系;多索体系
multicalcium tablet 多种钙片
multicamera installation 多镜摄影装置
multicamerate 多室的
multican 多分管的
multicapacity control system 多容量控制系统
multicarbide 多元碳化物;复合碳化物
multicarity 多腔;多槽
multicarrier distance measurement 多载波测距
multicarrier transmitter 多载波发射机
multicast 多信道广播;多播
multicast backbone 多播主干网
multicast communication 多目的传送;多点传输
multicategory system 多范畴系统
multicavity 多共振器的
multicavity container 多腔容器
multicavity die 多孔模
multicavity klystron 多腔速调管
multicavity magnetron 多腔磁控管
multicell 多室的;多环流
multicell battery 并联蓄电池组
multicell battery mo(u)ld 多格成组模(预制混凝土用)
multicell box girder 多室箱梁
multicell box girder bridge 多室箱梁桥
multicell culverts 多孔涵管;多孔涵洞
multicell dust collector 多格仓吸尘器
multicell-framed tube structure 框筒群结构;圈筒群结构
multicell heater 多室加热炉
multicell plenum craft 多气室气垫船;多气室充气艇
multicell storm 多单体风暴
multicell type 多箱式
multicellular 多网格的;多格的
multicellular aquatic animal 多细胞水生动物
multicellular bridge decks 多格桥面(板)
multicellular collector 多格集尘器
multicellular electrometer 多室静电计;复室静电计
multicellular electrostatic voltmeter 复室静电伏特计
multicellular glass 多孔玻璃;泡沫玻璃
multicellular hollow section 多格空心截面
multicellular horn 多管喇叭;多格喇叭
multicellular mechanical precipitator 多管式机械除尘器
multicellular metal foil 多孔泡沫金属箔
multicellular metal foil insulator 多孔泡沫金属箔(热)绝缘体
multicellular pump 多室唧筒;多室泵
multicellular static voltmeter 复室静电伏特计
multicellular straight through cyclone 多室直通式旋风分离器
multicellular system 多格系统
multicellular voltmeter 复室静电伏特计
multicentreed 多心的
multicentred arch 多心拱;三心拱
multicentred arch bridge 多心拱桥
multichain 多链的
multichamber 多室;多层
multichambered 多室的
multichamber kiln 多室窑
multichamber locks 多室船闸;多级船闸
multichamber vessel 多室容器;多承压室容器
multichannel 多槽的;多信道;多通道蓄热室;多(通)道;多频道;多路的
multichannel access 多道存取
multichannel amplifier 多路放大器
multichannel analyzer 脉冲高度分析仪;脉冲高度分析器;多路分析器;多(通)道分析器
multichannel and dual identification operation 多信道双识别码操作

multichannel anemometer 多通道风速仪
multichannel applicative signal analyzer 多通道实用信号分析仪
multichannel biochemical analyzer 多道生化分析仪
multichannel bulk encryption unit 多路总体加密机
multichannel bundle cable 多通道束式光缆
multichannel burner 多通道燃烧器;多通道喷（煤）管
multichannel car kiln 多通道(窑)车式隧道窑
multichannel carrier 多路载波装置
multichannel carrier-frequency telephone terminal 多路载频电话终端机
multichannel carrier path 多路载波通路
multichannel carrier telephone 多路载波电话
multichannel coherence filter 多道相干滤波器
multichannel combustible gas detector 多通道可燃气体探测器
multichannel communication 多路通信
multichannel communication program(me) 多通道通信程序
multichannel correlator 多通道相关器;多路相关器
multichannel data 多道数据
multichannel data recorder 多道数据记录器
multichannel demultiplexed receiver 多信道去复用接收机
multichannel digital record system 多通道数字记录系统
multichannel digital signal 多通道数字信号
multichannel discriminator 多道甄别器;多道鉴频器
multichannel distributary estuary 多汊河口
multichannel editing streamer 多路编排电缆
multichannel electric(al) signals comparator 多通道电信号比较仪
multichannel electrocardiograph 多通道心电图仪
multichannel electromagnetic oscillator 多道电磁波示波器
multichannel field-effect transistor 多沟道场效应管
multichannel filtering 多道滤波
multichannel flame spectrophotometer 多通道火焰分光光度计
multichannel gamma-ray spectrometer 多道伽马射线谱仪
multichannel general bio-signal detector 多道通用生物信号检测仪
multichannel grating demultiplexer 多通道光栅解复用器
multichannel head 多路磁头
multichannel HF communication equipment 多路高频通信设备
multichannel indication 多通道指示;多路指示
multichannel loading 多路负载
multichannel locks 多级船闸
multichannel logging recorder 多信道记录器
multichannel main receiver 多信道接收主机
multichannel microwave telecommunication equipment 多路微波通信设备
multichannel mixer 多路混合器
multichannel modulator 多路调制器
multichannel monitoring system 多通道监测系统
multichannel multiplier 多通道乘法器
multichannel multipoint distribution service 多路多点分配业务
multichannel ocean colo(u)r sensor 多通道海洋彩色遥感器
multichannel optic(al) analyzer 多通道光学分析仪;多通道光学分析器
multichannel optic(al) cable stranding machine 多路光缆绞制机
multichannel optoanalyzer 多通道光学分析仪;多通道光学分析器
multichannel oscillator 多道振荡器
multichannel oscillograph 多线示波器;多路示波器;多道示波器
multichannel particle analyzer 多通道粒子分析仪
multichannel physiological monitoring equipment 多路生理监视设备
multichannel predictive deconvolution 多道预测反褶积
multichannel processing 多道处理
multichannel pulse peak analyzer 多通道脉冲峰值分析仪
multichannel Raman spectrometer 多道拉曼光

谱仪
multichannel readout mass spectrometer 多通道读数质谱计
multichannel receiver 多信道接收机;多通道接收机
multichannel recorder 多通道记录仪;多(通)道记录器;多路记录仪
multichannel recorder chart 多通道记录纸
multichannel recording 多道记录
multichannel recording head 多路记录磁头
multichannel recording oscillograph 多路(记录)示波器
multichannel redio transceiver 多路无线电收发讯机
multichannel roller kiln 多通道辊道窑
multichannel seismic instrument 多道地震仪
multichannel seismic profiler 多道地震剖面
multichannel seismic survey 多道地震法调查
multichannel seismograph 多道地震仪
multichannel sensing capacity 多通道灵敏度
multichannel simultaneous X-ray spectrometer 多通道同时式 X 射线光谱仪
multichannel single-fiber cable 多通道单芯光缆
multichannel slab kiln 多通道推板窑;多孔窑
multichannel slip ring 多槽集流环;多槽滑环
multichannel spectrograph 多通道摄谱仪
multichannel spectrometer 多道光谱仪;多道分光计
multichannel stripchart recorder 多道带状图记录仪
multichannel system 多通路制;多通道系统
multichannel telemeter 多信道遥测计
multichannel telephone system 多路电话系统
multichannel television 多信道电视;多频道电视
multichannel temperature controller 多通道温度控制器
multichannel time analyzer 多通道时间分析仪
multichannel transmission 多路传输
multichannel transmitter 多路发射机
multichannel voice frequency 多路话音频率
multicharge 混合装药;多种炸药
multichip 多片状;多片
multichip circuit 多片电路
multichip hybrid circuit 多片混合电路
multichip integrated circuit 多(芯)片集成电路
multichip interface 多片接口
multichip microcircuit 多片微型电路
multichip microcomputer 多片微型计算机
multichromatic 多色的
multichromatic spectrophotometry 多色分光光度计
multichrome multipersistence screen 多色多余辉穿透屏
multichrome penetration display tube 电压穿透多色管
multichrome press 多色印刷机
multicipital 多头的
multicircle wirewound potentiator 多圈线绕电位器
multicircuit 多电路的
multicircuit hydraulic system 多回路液压系统
multicircuit relay 多路继电器
multicircuit switch 多路转换开关
multicircuit winding 多匝绕组;多环向缠绕
multicity sewage treatment 多城市污水处理
multiclass body 混合级车体
multiclass estimation 多类的估计
multiclass passenger ship 多等集客船
multiclass sender 万用记录器
multiclass ship 多等集客船
multiclone 多聚尘机;多管旋流器;多管式旋流除尘器;多管式旋风除尘器
multiclone collector 多管集尘器;多管旋风收尘器
multiclone dust collector 多管集尘器;多管旋风除尘器
multicoat 多层(指涂层)
multicoated lens 多层加膜处理镜片
multicoating 多层镀膜
multicoat lacquer 多道喷涂漆;多层喷漆
multicode state assignment 多码状态分配
multicoil feed heater 多盘管式给水加热器
multicolinearity 多重共线性
multicollar thrust bearing 多环式推力轴承
multicollector electron tube 多集电极电子管
multicollimator 多投影准直仪
multicolo(u)r 多色(的);彩色
multicolo(u)r brick 五彩砖;多色砖

multicolo(u)r cloth 多色织物
multicolo(u)r container screen-printing machine 多色容器网板印刷机
multicolo(u)r copy multicolo(u)r reproduction 多色复制
multicolo(u)r detector 多色探测器
multicolo(u)r digital plotter 多色数字绘图仪
multicolo(u)red 杂色的
multicolo(u)red brick 多色建筑砖
multicolo(u)red clay 杂色黏土
multicolo(u)red effect 多色效应
multicolo(u)red filter 多色滤光器
multicolo(u)red finish 多色饰面
multicolo(u)red subclay 杂色亚黏土
multicolo(u)red system 多色系统
multicolo(u)r effect 彩色效应
multicolo(u)r finish 多色饰面;彩色饰面;多色罩面;多彩涂层;复色面漆
multicolo(u)r graphic adaptor[adapter] 多彩色图形适配器
multicolo(u)r hologram 多色全息照相;多色全息图
multicolo(u)r holography 多色全息照相术;多色全息摄影
multicolo(u)r image 多色像
multicolo(u)r imaging 多色成像
multicolo(u)r intaglio rotary printing press 多色凹版轮转印刷机
multicolo(u)r lacquer 复色喷漆
multicolo(u)r laser 多色激光器
multicolo(u)r light beam 多色光束
multicolo(u)r lithography 多色平印术
multicolo(u)r map 多色地图
multicolo(u)r map reprography 多色地图复照
multicolo(u)r novelty finish 多彩美术漆
multicolo(u)r offset press 多色胶印机
multicolo(u)r original 彩色原图
multicolo(u)r paint 多彩(色)涂料;多彩漆;复色漆
multicolo(u)r paint finish 多彩色涂料饰面
multicolo(u)r pattern 彩色图案
multicolo(u)r photometry 多色光学;多色测光
multicolo(u)r press 多色印刷机
multicolo(u)r printer 多色打印机
multicolo(u)r printing 多色印刷;彩色印刷
multicolo(u)r printing machine 多色印刷机;彩色印刷机
multicolo(u)r recorder 多色自动记录器
multicolo(u)r rotary offset press 多色轮转胶印机
multicolo(u)r rotogravure press 多色凹印机
multicolo(u)r sheet-fed offset press 多色单张纸胶印机
multicolo(u)r sheet-fed press 多色单张纸胶印机
multicolo(u)r symmetry 多色对称
multicolo(u)r twist cable 多色股线
multicolumn foundation 多柱基础
multicolumn hydraulic press 多柱式液压计
multicolumn instrument 多柱仪器
multicolumn operation 多柱操作
multicombination meter 多用途复合仪表
multicombustion chamber 多管燃烧室;多个燃烧室
multicombustion chamber heater 多室立式炉
multicommodity flow 多重物资流
multicompartment 多室
multicompartment bed 多层床
multicompartment building 多分隔房屋
multicompartment cell 多室浮选机;多槽式浮选机
multicompartment electrodialysis 多室电渗析
multicompartment electrodialyzer 多室电渗析器
multicompartment mill 多室磨机;多仓磨(机)
multicompartment septic tank 多层腐化池【给】
multicompartment silo 多室料斗;多室筒仓;多室料仓
multicompartment system 多格系统
multicompartment tank furnace 多室池窑
multicompartment washer 多室洗涤器
multicompletion well 多层合采井
multicomponent 多组分(的);多元的;多分量的;多成分的
multicomponent admixture 多成分掺合料;复合外渗剂;复合料;复合掺合料
multicomponent adsorption equilibrium 多组分平衡
multicomponent analysis 多项分析
multicomponent architecture 多部件体系结构

multicomponent biological system 多成分生物学系统
multicomponent blance 多分力天平
multicomponent blending bed 多组分预均化堆场;多组分混合料堆
multicomponent die 组合模(具)
multicomponent distillation 多元蒸馏
multicomponent effect 多分量效应
multicomponent fertilizer 多种混合肥料
multicomponent fiber 多组分复合纤维;多组分纤维
multicomponent film fibre 多组分薄膜纤维
multicomponent flow 多组分流动
multicomponent gas 多组分气体
multicomponent gas mixture 多元气体混合物
multicomponent glass optic(al) fiber 多组分玻璃光学纤维;多组分玻璃光纤
multicomponent lacquer 多组分亮漆;多组分涂漆;多组分蜡克
multicomponent non-aqueous phase liquid 多组分非水相流体
multicomponent pesticide 多组分农药
multicomponent phase 多成分相
multicomponent pigment 复合颜料
multicomponent polymer 多组分聚合物
multicomponent polymer fiber 多组分共聚物纤维
multicomponent samples analysis 多组分样品分析
multicomponent sealant 多组分密封膏
multicomponent seismic record 多分量地震记录
multicomponent slag 多组分炉渣
multicomponent spectrum 多组元频谱
multicomponent structure 多组元结构;多部件结构
multicomponent system 多组(元)系;多组分系统;多组分体系;多元组
multicomponent varnish 多组分清漆;多组分(罩)光漆;多组分凡立水
multicompound bronze 多金属青铜
multicompound mixture 多成分混合物
multicompression 多级压缩
multicomputer 多计算机
multicomputer communication system 多计算机通信系统
multicomputer mangement system 多计算机管理系统
multicomputer system 多计算机系统;多处理器系统
multicomputing 多值计算
multicomputing unit 多运算器
multiconcentric 多层同心的
multiconcentric arrangement type winding 多层同心式绕组
multiconductor 多触点
multiconductor cable 多芯电缆
multiconductor plug 多线插头;多触点插头
multicone 多管式旋风集尘器
multicone angle cocurrent decanter centrifuge 多锥角并流型沉降式离心机
multicone mixer 多锥型混合机
multi-connector 复式连接器
multiconstituent fiber 多成分复合纤维
multicontact auxilliary relay 多触点辅助继电器
multicontact regulator 多触点调节器
multicontact relay 多接点继电器;多触点继电器
multicontact switch 多触点开关
multicontrol refrigerator freezer 多控式冷藏冷冻箱
multicord 复丝
multiorder 多笔尖记录器
multicore 多芯的
multicore clay brick 多芯黏土砖
multicore cord 多线电绳
multicored brick 多孔砖
multicored clay brick 多孔黏土砖
multicored forging press 多向模锻压力机
multicore flex 多芯软电线
multicore magnetic memory 多磁芯磁存储器
multicore magnetic storage 多磁芯存储器
multicore panel 多芯镶板;多髓嵌板;多层预制板
multicore submarine cable 多芯水下电缆
multicotree 多重余树
multicoupler 多路耦合器
multi-course 多方面;多航线;多路线;多层
multicrank 多曲柄
multicriterion decision making 多标准决策
multicrystal 多晶体
multicrystal autofluoroscope 多晶体自动荧光镜

multicrystal focusing spectrograph 多晶聚焦摄谱仪
multicrystalline diamond 多晶体金刚石
multicrystallizing method 大量结晶法
multicrystal silicon 多晶硅
multiculture 多种养殖
multicurrent silo 多点流库
multicurrent traction unit 多种电流牵引单元
multicustomer group operation 多组用户公用系统
multicut 多刀的
multicutter 多刀;复式铰刀
multicutter lathe 多刀车床
multicutter shears 多刀剪切机
multicutter suction dredge(r) 多铰刀型绞吸挖泥船
multicutting bar 多刀镗杆
multicycle 多循回;多循环;多旋回;多(时段)周期
multicycle feeding 多周期馈送
multicycle intrusions 重复侵入【地】
multicycle ore deposit 迭生矿床
multicycle sort 多周期分类
multicyclic 多循环的
multicyclic orogenesis 多旋回造山运动
multicyclone 旋风分离器组;多级旋风预热器;多级旋风分离器;多管旋风分离器;多管(式)旋除尘器;多管除尘机;复式旋风吸尘器;复式旋风分离器
multicyclone collector 旋流除尘器组
multicyclone device 多管式旋风装置
multicyclone dust collector 多级旋风除尘器;多管式旋流除尘器;多管式旋风集尘器;多管式旋风除尘器;复式旋风式附尘器
multicyclone heat exchanger 多级旋风(筒)热交换器
multicyclone sepaeator 多管式旋风分离器
multicylinder 多汽缸的
multicylinder ball mill 多筒式球磨机
multicylinder compressor 多缸压缩机
multicylinder engine 多缸发动机
multicylinder hydraulic press 多缸式液压机
multicylinder long mesh paper machine 长网多缸造纸机
multicylinder pump 多缸泵
multicylinder steam engine 多(汽)缸蒸汽机
multicylinder turbine 多(汽)缸汽轮机
multidata base 多数据库
multidata entry system 多数据输入系统
multidata set volume 多数据集卷宗
multi-date 不同日期的
multidate interpretation method 多日期解译法
multidate photography 重复摄影
multidaylight press 多柱压力机;多层压(力)机
multidecision 多级判决;多级决策
multidecision game 多步判决对策
multideck 多层
multideck bridge 多层(桥面的)桥
multideck cage 多层罐笼
multideck clarifiers 多层澄清器
multidecked ship 多甲板船
multidecked transit shed (码头的)多层前方仓库
multideck partitioned wagon 多层间隔车
multideck screen 多层栅枷;多层筛
multideck shield 棚架式盾构
multideck sinking platform 多层凿井吊盘
multideck table 多层摇床
multideck vibrating separator 多层振动分选机
multidegree-of-freedom 多自由度
multidegree-of-freedom system 多自由度系统;多自由度体系
multidentate 多配位基
multidesign ribber 多花式罗纹机
multidevelopment method 多次展开法
multidiameter 阶梯轴
multidiameter boring cutter 多直径镗孔铣刀;多径镗孔铰刀
multidiameter expansion reamer 多径膨胀式铰刀
multidiameter reamer 多径铰刀
multidie continuous wire drawing machine 多模连续式拉丝机
multidie machine 多模拉丝机;多次拉丝机
multidiesel 多种柴油机
multidie wire-drawing machine 多模拉丝机
multidigit 多位
multidigitato-pinnate 多指状羽状的
multidigit shifting 多位移位

multidimensional 多维的;多面的
multidimensional active carbon electrode 多维活性炭电极
multidimensional allocation 多维分配;多维分布
multidimensional analysis 多维分析法
multidimensional architecture 多维体系结构
multidimensional array 多维数组
multidimensional array element name 多维数组元素名
multidimensional assignment problem 多维分配问题
multidimensional code 多维码
multidimensional consequence 多维结果
multidimensional derivative 多维导数
multidimensional dynamic(al) system 多维动力系
multidimensional fine angle goniometer 多维精密定角仪
multidimensional gas chromatography 多维气相色谱法
multidimensional language 多维语言
multidimensional maximization 多维极大化
multidimensional normal distribution 多维正态分布
multidimensional optimal control 多维最优控制
multidimensional outcome variable 多维输出变量
multidimensional probability distribution 多维概率分布
multidimensional region 多维区域
multidimensional space 多维空间
multidimensional statement 多维命题
multidimensional subscript 多维下标
multidimensional system 多维系统
multidimensional tree 多维树
multidimensional trellis coded modem 多维格子码调制解调器
multidimensional utility 多维效用
multidipole plasma Langmuir probe 多偶极等离子体朗缪尔探测仪
multidirectional 多(方)向的
multidirectional fabric 多向织物
multidirectional hot working 多向热加工
multidirectional irregular (random) waves 多向不规则(随机)波
multidirectional laminates 多向层压板
multidirectional loading multianvil ultra high pressure and high temperature device 多向加载多压砧式超高压高温装置
multidirectional martensite 多向马氏体
multidirectional photography 多方向摄影
multidirectional shear 多向剪切
multidirectional table test 多向振动台试验
multidirectional vibration 多向振动
multidirectional weave 多向编织
multidisc 多片
multidisc clutch 多片(式)离合器;多盘离合器
multidisc cross cutter 多盘式横锯
multidisc electromagnetic clutch 多片电磁离合器
multidisc flexible coupling 多盘弹性离合器;多盘弹性接头
multidisc friction clutch 多片摩擦离合器;多盘摩擦离合器
multidisciplinary 多(种)学科的
multidisciplinary analysis method 多学科分析法
multidisciplinary group 跨专门小组
multidisciplinary study 多(种)学科研究
multidisc mower 多盘式割草机
multidisk = multidisc
multidisplinary analysis 多科学分析
multidisplinary approach 多科性研究法
multidisplinary study 多科性研究
multidivergence laser 多分支激光器
multidomain network 多畴网络
multidoor refrigerator 多门冰箱
multidoped material 多掺杂材料
multidraught machine 多次拉拔机
multidraw 多点取样
multidrift 多导坑开挖施工法
multidrift method 多导流法;多导坑开挖法
multidrill 多轴钻
multidrill head 多轴钻床主轴管
multidrill head machine 多轴钻机
multidrilling machine 多轴钻机
multidrill trenching machine 多头钻挖槽机
multidrive 多重激励

multidriving technology 多机牵引技术
multidrop 多站线;多站通信网络;多点
multidrop circuit 多点线路
multidrop communication network 多站通信网络
multidrop connection 多点连接;多点连线
multidrop line 多支线;多站线路;多点(共线)线路
multidrop linkage 多点通道
multidrop loop controller 多支循环控制器
multidrop network 多点网络
multi-duplicating machine 连晒机
multiduty 多用途的
multiduty pliers 多用钳
multiechelon inventory model 多阶层库存模型
multiecho 颤动回声;多重回声
multieffect 多效应的
multieffect compressor 多效压缩机
multieffect distillator 多效蒸馏器
multieffect evapo(u)rator 多效蒸发器
multieffect incineration 多效焚化
multieffect multistage flash distillation process 多效多级闪蒸法
multieffect multistage(flash) vapo(u)rizer 多效多级闪蒸蒸发器
multieffect vacuum distillation process 多效真空式蒸馏法
multieffect vacuum evapo(u)rator 多效真空蒸发器
multieffect vapo(u)r compression 多效蒸汽压缩
multieffect water distillatory 多效蒸馏水机
multielectric(al) spot welder 多头点焊机
multielectrode 多极(电)电极;多极电极
multielectrode counting tube 多电极计数管
multielectrode mercury arc rectifier 多极汞弧整流器
multielectrode spot welder 多头点焊机;多极点焊机
multielectrode tube 多极管
multielectrode tube converter 多极管变频器
multielectrode valve 多极管
multielectrode voltage stabilizing tube 多极稳压管
multielement 多元素;多元的
multielemental activation analysis 多元素活化分析
multielemental analysis 多元素分析
multielement anomaly map 多元素异常图
multielement anomaly resolution map 多元素异常剖折
multielement array 多元天线阵
multielement beam 多元梁
multielement earthquake 多摆line地震计
multielement infrared detector 多元件红外探测器
multielement interpretation system 多元素解释系统
multielement member 预制多单元构件
multielement oscillograph 万用示波器;多元示波器
multielement parasitic array 多元无源天线阵
multielement photoconductor 多元光电导体
multielement prestressing 多构件预应力;预应力拼装
multielement raster scan 多元光栅扫描
multielement scanner 多元扫描装置
multielement silicon cell 多元硅光电元件
multiechelon survey 多元素测量
multielement vacuum tube 多元真空管
multielefin 多位数字显示器
multiemitter output 多发射极输出
multiend reeling machine 多条缫丝机
multiengine 多曲柄式发动机
multiengined 多发动机的
multiengined aeroplane 多发动机飞机
multienhancement 多重增强
multientry 多路入口
multiexchange system 多局系统
multiexhaust 多排汽口
multi-exhaust blade 复排汽叶片
multiexhaust turbine 多排汽口汽轮机
multiexpansion 多次膨胀式
multiexposure X-ray powder camera 多次曝光 X 射线粉末摄影机
multiextent file 多范围文件
multi-factor experiment 复因子试验
multifamily 多家庭;多户住房;多户
multifamily apartment house 多住户公寓
multifamily building 多户房屋;多户建筑物;多户住房

multifamily development 多户住房建设
multifamily dwelling 合住住宅;多家合住的楼房
multifamily habitation 多户住宅;多户居住
multifamily housing 多户住房
multifamily housing coinsurance 多户住房共同保险
multifamily project 多户住房建设项目
multifamily rental housing 供出租的多户住房
multifamily rental project 供出租的多户住房建设项目
multifamily residence 多户住所
multifamily sales project 供出售的多户住房建设项目
multifarious 多种多样的
multifeature telephone 多功能电话机
multifee 重复收费
multifeed 多点供油的
multifeed bushing 多点供料套
multifeed lubricator 多点润滑器
multifeed oiler 多点润滑器;多点加油器
multifeed parabolic reflector antenna 多馈源抛物面反射器天线
multifeed rib machine 多路罗纹机
multifee metering 复式计费记录计
multifee registration 复式计费记录
multifiber 复型纤维
multifiber array 毛细管整列
multifiber cable 多纤光缆
multifiber cable connector 多芯光缆连接器
multifiber connector 多芯光纤连接器
multifiber loose tube optic(al)fibre 多纤松套光纤
multifiber optic(al)cable 多芯电缆
multifiber optic(al)submarine cable 多芯海底光缆
multifiber ribbon 多路纤维带
multifiber rod 硬质光学纤维棒
multifier splicer 多纤接头机
multifil 复丝
multifilament 多纤维丝;多丝纤维;多丝;复丝
multifilament lamp 多灯丝白炽灯
multifilament product 复丝产品
multifilament tyre yarn 复丝轮胎帘线
multifilament yarn 复丝
multifile 多文件
multifile reel 多文件卷
multifile search 多文件检索;多文件查找
multifile sorting 多文件排序;多文件分类
multifile volume 多文件卷(宗)
multifiltration 多层过滤
multifinger 多触点的
multifinger contactor 多接点接触器;多触点开关;多触点接触器
multi-fired boiler 多燃料锅炉
multifired power plant 多燃料电厂
multiflame blow pipe 多焰喷管
multiflame burner 多焰燃烧器;多焰焊炬
multiflame pot furnace 多焰坩埚窑
multiflame torch 多焰焊炬
multiflex galvanometer 多次反射电流计;复式检流计
multiflight stair(case) 多梯段楼梯
multifloor building 多层住宅;多层房屋;多层建筑(物)
multifloor car park 多层停车房;多层停车处;多层停车场
multifloor factory block 多层工厂建筑
multifloor garage 多层修车间;多层(停)车库
multifloor hotel 多层旅馆
multifloor library 多层图书馆
multifloor span escalator 多层跨升降梯
multifloor system composed of three hinge(d) frame 三铰接框架组成的多层结构系统
multiflow 复流
multiflow heater 多流式加热器
multiflow mixer 多料流搅拌机
multiflow silo 多股流料库
multi-flue 多路烟道
multifluidized bed adsorption column 多层流化床吸附柱
multiflute drill 多槽钻
multiflying punch 多针穿孔
multifocal 多焦点的
multifocal lens 变焦透镜
multifocus 多电极聚焦
multifocus lens 多焦点透镜

multifocus viewfinder 万能取景器
multifoil 繁叶饰的;多叶饰的
multifoiled arch 多叶(式)拱;多孔叶拱;多瓣拱;复叶形拱券
multifoil heat insulating coating 多层箔隔热涂层
multifold 多倍的
multifolded yarn 多股线
multifolding door 折叠门;多折扇门
multifold towel cabinet 多折毛巾柜
multifont 多字体(型)
multiforce system 多力张拉系统
multiform 多种的;多形的
multiform abacus (多形式的圆柱顶上的)冠板
multiformat camera 多规格摄影机
multiformational anomaly 多建造晕
multiformation cave 混合成因洞
multiformed 多种形式的
multiform function 多值函数
multiformity 多形性
multiform layer 多形层
multiform padstone 多形状的承梁垫石;多形状垫石
multiform raised table 多形状的升高垫板;多形状的升高台板
multifrac 多种破碎
multiframe 多帧;多框架的
multifrequency 多频(率)的
multifrequency alternator 多频(率)交流发电机
multifrequency code signal(l)ing 多频(率)编码信号方式
multifrequency excitation 多频(率)扰动
multifrequency fluid oscillator 多频(率)射流振荡器
multifrequency generator 多频(率)振荡器;多频(率)发生器;多频(率)发电机
multifrequency keying 多频(率)选择;多频(率)键控
multifrequency microwave radiometer 多频(率)微波辐射计
multifrequency motor-generator 多频(率)电动发电机
multifrequency nature 多种特性
multifrequency pulse 多频(率)脉冲
multifrequency push-button set 宽频带按钮机;多频(率)按钮组合
multifrequency radar 多频(率)雷达
multifrequency receiver 多频带接收器;多频(率)接收器;多频(率)接收机
multifrequency remote control system 多频(率)遥控
multifrequency shift demodulator 多频(率)移频解调器
multifrequency signal 多频(率)信号
multifrequency signal(l)ing 多频(率)信令;多频(率)信号方式
multifrequency signalling system 多频(率)信号制
multifrequency sinusoid 多谐波正弦振荡
multifrequency sounder 多频(率)探鱼仪
multifrequency system 多频(率)制
multifrequency terminal 多频(率)终端
multifrequency transmission 多频(率)发射
multifrequency transmitter 多频(率)发射机
multifrequency vibration 复频振动
multifuel burner 多种燃料燃烧器;多燃料复合喷燃器
multifuel diesel 多种燃料柴油机
multifuel engine 多元燃料发动机;多(种)燃料发动机
multifuel firing 多种燃料燃烧
multifunction 多功能
multifunctional 多官能的;多功能的
multifunctional array 多功能阵列
multifunctional automobile communication system 多功能汽车通信系统
multifunctional box-type solar cooker 多功能箱式太阳能灶
multifunctional chromatograph 多用色谱仪
multifunctional concept 多功能思想;多功能概念
multifunctional counter 多功能计数器
multifunctional die processing machine 多功能模具加工机
multifunctional drilling and tap holder 多用钻孔攻丝夹头
multifunctional drinking-water purifier 多功能饮水净化器

multifunctional dual ion beam processing machine 多功能双离子束加工机
multifunctional ecological engineering 多功能生态工程
multifunctional fluid control apparatus 多功能液流控制装置
multifunctional group 多官能团
multifunctional hall 多功能厅
multifunctional integrated circuit 多功能集成电路
multifunctionality 多功能性
multifunctional mobile combination belt conveyer 多功能移动组合式皮带输送机
multifunctional operating system 多功能操作系统
multifunctional polyurethane solution 多功能聚氨酯溶液
multifunctional room 多功能室;多功能工厂;多功能车间;多功能房间
multifunctional sensitive ceramics 多功能敏感陶瓷
multifunctional service boat 多功能工作船
multifunctional shadowless lamp 多功能无影灯
multifunctional spotlight cooker 多功能聚光灶
multifunctional support vessel 多功能支持船
multifunction architecture 多功能建筑
multifunction array radar 多性能阵列雷达;多功能相(控)阵雷达;多功能天线阵雷达
multifunction automatic fire alarm 多功能火灾自动报警器
multifunction card 多功能卡
multifunction catalyst 多功能催化剂
multifunction center[centre]area 多功能中心区
multifunction city 多功能城市
multifunction cold drink machine 多功能冷饮机
multifunction component 多用途部件
multifunction control valve 多功能控制阀
multifunction converter 多功能转换器
multifunction corrosion test 多功能腐蚀试验
multifunction digital voltage meter 多功能数字电压表
multifunction display 多功能显示器
multifunction display and control system 多功能显示与控制系统
multifunction graphics 多功能图形
multifunction hall 多功能厅
multifunction health inspecting apparatus 多功能人体监护仪
multifunction inorganic flocculant 多功能无机絮凝剂
multifunction laminating machine 多功能层压机
multifunction lamp 多功能灯
multifunction machine tool 多功能机床
multifunction milling holder 多用铣床夹头
multifunction multimeter 多功能万用表
multifunction oxygen therapy monitor 多功能输氧监测仪
multifunction phase shift signal source 多用移相信号源
multifunction phone 多功能电话机
multifunction plotting head 多功能绘图头
multifunction printer 多功能打印机
multifunction processor 多功能部件处理机;多操作部件处理机
multifunction reader 多功能阅读机;多功能输入机
multifunction reservoir 多功能水库
multifunction robbery protection alarm 多功能防盗报警器
multifunction routine polarograph 多功能常规测量极谱仪
multifunction sensor 多功能传感器
multifunction-simulator for rail-bound public transport 公共轨道交通运输系统多功能模拟器
multifunction solenoid valve 多用途电磁阀
multifunction station 多功能站
multifunction system 多功能系统
multifunction telephone analyzer[analyser] 多功能电话分析仪
multifunction TV set 多功能电视机
multifunction ultrasonic scanner 多功能超声扫描仪
multifunction ventilator 多功能通风机
multifunction walking stick 多功能拐杖
multifunction watch 多功能手表
multifunction water treatment reagent 多功能水处理剂
multifunction wax 多功能蜡

multifunction wrist-instrument 多功能电子表
multifuse igniter 多信管点火器
multigable building 多坡屋顶建筑；复坡屋顶建筑
multigang switch 多联开关
multigang variable condenser 多联可变电容器
multigap 多隙的
multigap arrester 多隙(式)避雷器
multigap discharger 多隙(式)放电器
multigap head 多隙磁头
multigap sparkplug 多隙火花塞
multi-garage 多层停车场；多层停车库
multigas and multirange infrared gas analyzer 多气体多量程红外气体分析器
multigas burner 多种气体燃烧器
multiga(u)ge 多用检测计；复试测量仪；多用(量)测仪表；多用规；复用测量仪
multiga(u)ge projector 多用放映机
multigirder bridge 多梁桥；多梁式桥
multiglazing unit 多层玻璃窗
multigraded oil 稠化润滑油
multigrade oil 多级通用润滑油
multigrain structure 多粒结构
multigranular charge 多粒炸料
multigraph 多重图；重图
multigreaser 多点润滑器
multigrid 多栅(的)；多层格栅
multigrid detection 多栅检波
multigrip system 多夹具系统
multigrog 多用途耐火材料
multigroove friction clutch 多槽摩擦离合器
multigroup calculation 多组计算
multigroup diffusion 多群扩散
multigroup method 多群法
multigroup model 多群模型
multigroup theory 多群理论
multigun cathode ray tube 多枪阴极射线管
multigun oscilloscope tube 多枪示波器
multigun tube 多枪管
multiharmonic function 多重调和函数
multiharmonograph 多谐记录仪
multihead 多磁头；多传感头
multihead automatic arc-welding machine 多弧自动电焊机；多头自动电弧焊机
multihead automatic welder 多弧自动焊机
multihead automatic welding machine 多头自动焊机
multiheading gradient 多机坡度
multihead welder 多头焊机
multihearth 多层炉
multihearth dryer 多膛干燥炉
multihearth furnace 多膛炉
multihearth incineration 多膛焚化
multihearth incinerator 多膛焚化炉
multihearth refuse incinerator 多炉膛式垃圾焚化炉
multihearth roasting 多膛焙烧
multihole 多孔的
multihole brick 多孔砖
multihole directional coupler 多孔定向耦合器
multihole drilling 多筒钻井
multihole extrusion 多孔挤压
multihole lance 多孔枪
multihole washing bottle 多孔式洗瓶
multihole wire drawing machine 多孔拉丝机
multihoming 多归属
multihop 多次反射
multihopper horizontal-flow sedimentation tank 多斗平流式沉淀池
multihopper radial-flow sedimentation tank 多斗辐流式沉淀池
multihop transmission 多次反射传输
multihost 多主局
multihost operation 多主机操作
multi-household compound 杂院
multihulled ship 多体船
multiimage 多重图像；重复图像；分裂影像
multiimage enhancement 多重影像增强
multiimpression die 多腔模
multiimpression mo(u)ld 多腔模
multiimpule order method 多脉冲序列法
multi-index 复指数
multiindustry company 多国产业公司；工业的多种经营公司
multiinjector 多喷嘴
multiinput 多端输入

multiinput switch 多输入开关
multiinterpreter system 多解释器系统
multiionic system 多元离子系
multiion mass spectrometer 多离子质谱仪
multijaw grab 多爪抓斗；多颚板抓斗
multijet 多喷嘴的；多喷嘴
multijet blowpipe 多焰喷焊器
multijet burner 多焰焊炬
multijet condenser 多喷嘴冷凝器；多喷口凝汽器
multijet dryer 多孔喷射干燥器
multijet ejector condenser 多喷嘴冷凝器
multijet element 多喷嘴管
multijet nozzle 多孔喷嘴
multijet spray 多喷口
multijet turbine 多射流(冲击)式水轮机；多喷嘴(冲击式)水轮机
multijet-type burner 多嘴喷燃器
multijet weaving machine 多喷嘴喷气织机
multijib coalcutter 多截盘截煤机
multijob 多重工作；多道作业
multijobbing 多作业操作；多重作业
multijob operation 多(重)作业操作；多任务操作；多道作业(操作)
multijob scheduling 多作业调度
multijoint system 多铰链传动系
multijunction 多结点
multijunction structure 多结结构
multikey 多键
multikey shaft 多键轴
multikeyway 多键槽
multikeyway shaft 多键槽轴
multiknife chipper 多刀片削片机
multiknife disk chipper 多刀圆盘削片机
multiknife turning machine 多刀修坯机
multilacunar node 多叶隙节
multilaminate 多层的
multilaminated glass 多层夹层玻璃
multiland drill and reamer 多棱钻铰刀
multilane 多车道
multilane bridge 多车道桥
multilane divided road 多车道分隔行驶的道路
multilane facility 多车道设施
multilane highway 多车道公路
multilane pavement 多车道路面
multilane road 多车道道路
multilane tunnel 多车道隧道
multilated 残缺不全；被扰乱
multilateral 多国参加的；多方面的；多边的
multilateral agreement 多边合约
multilateral aid 多边援助
multilateral arrangement 多边协议
multilateral arrangement with trade liberalization 放宽贸易的多边安排
multilateral composition 多镜归正镶嵌
multilateral cooperation 多边合作
multilateral lending 多边借贷
multilateral market group 跨国市场集团
multilateral per capita quantity comparisons 人均量的多项比较
multilateral settlement 多边结算
multilateral tariff treaty 多边税收协定
multilateral technical assistance 多边技术援助
multilateral trade 多边贸易
multilateral treaty 多边条约
multilateration 多边测量
multilaterative comparator 多向坐标量测仪
multilayer 多层
multilayer adhesive 多层胶黏剂
multilayer adsorption 多层吸附
multilayer bit 多层钻头
multilayer blocking filter 多层截止滤光片
multilayer board 多层(印制)板；多层印刷板
multilayer cable drum 多层绕索鼓筒
multilayer capacitor 多层电容器
multilayer ceramics 多层陶瓷
multilayer classification 多重判别分类
multilayer coated eye-glasses 多层镀膜眼镜片
multilayer coating 多层涂膜；多层涂敷；多层镀层
multilayer coil 多层线圈
multilayer colo(u)r negative imaging 多层彩色负片成像
multilayer colo(u)r positive imaging 多层彩色正片成像
multilayer consolidation 多层固结

multilayer construction 多层路面；多层结构；多层构造
multilayer contact material 多层电接触器材
multilayer control structure 多层控制结构
multilayer copolymer 多层共聚物
multilayer cylinder 多层圆筒
multilayer dielectric(al) coating 多层介质镀膜
multilayer dielectric(al) film 多层介质膜
multilayer dielectric(al) filter 多层介质滤光器
multilayer dielectric(al) plane reflector 多层介质平面反射镜
multilayer dielectric(al) reflective coating 多层介质反射膜
multilayer dielectric(al) reflector 多层介质反射器
multilayer dimple-plank isolating oil pool 多层波纹板式隔油器
multilayered 多层的
multilayered aquifer 多重含水层；多层含水层
multilayered board 多层板；多层板材
multilayered construction 多层结构
multilayered elastic medium 多层弹性介质
multilayered elastic theory 多层体弹性理论
multilayered fluid 多层流体
multilayered insulant 多层绝缘材料
multilayered intaking 分层取水
multilayered liquid containment system 液体多层料密封系统
multilayered medium 多层介质
multilayered weld(ing) 多层焊(接)
multilayer field 多油层油田
multilayer film 多层胶片；多层薄膜
multilayer film circuit 多层薄膜电路
multilayer film interconnection circuit 多层薄膜互连电路
multilayer filter 多层滤光器；多层滤光片；多层滤池；多层过滤器
multilayer floating sampler 多层漂浮式采样器
multilayer foam pipe 多层泡沫芯管
multilayer gas-discharge display panel 多层气体放电显示板
multilayer glass 多层玻璃
multilayer heater 多层灯丝
multilayer het-erostructure microcell 多层异构微电池
multilayer high pressure vessel 多层高压容器
multilayer infiltration media system 多层渗滤介质系统
multilayer insulation 多层绝缘；复层绝缘
multilayer interchange 多层互通式立交
multilayer interconnection 多层互连；多层布线
multilayer interface theory of crystal growth 多层界面晶体生长理论
multilayer interference filter 多层干涉滤光片
multilayer ionic device 多层离子装置
multilayer karst cave 多层溶洞
multilayer laminated ceramics 多层陶瓷基片
multilayer laminated glass 多层夹层玻璃
multilayer low-temperature insulation 多层低温绝热
multilayer material 多层材料
multilayer metallization 多层布线
multilayer mode 多层模式
multilayer model 多层模型
multilayer monolithic wall 多层整体墙；多层整块石墙
multilayer network 多层薄膜网路
multilayer packaging 多层组装
multilayer paper board 多层纸板
multilayer piezoelectric(al) ceramic transformer 多层压电陶瓷变压器
multilayer piezoelectric(al) device 多层压电器件
multilayer precision printed board 多层精密印制板
multilayer prestressed reinforcement 多层预应力加筋；多层预应力钢筋
multilayer printed circuit board 多层印制板
multilayer printed wiring board 多层印刷电路板；多层印刷布线板
multilayer radial thickener 多层辐射式浓缩池
multilayer relaying excavating process 多层接力开挖法
multilayer sampler 多层采样器
multilayer sampling 多层采样
multilayers blast against structure 成层式抗爆结构
multilayer screen 多层屏蔽

multilayer soap film lead stearate crystal 多层皂膜硬脂酸铅晶体
multilayer soap film spectroscopic crystal 多层皂膜分光晶体
multilayer stenter 多层拉幅机
multilayer stock 多层滤光器
multilayer structure pavement 多层结构路面
multilayer substrate 多层基片
multilayer system 多层系统；多层体系
multilayer thermal press 多层热压机
multilayer type insulations 多层不同隔热材料
multilayer weld 多层焊缝
multilayer welding 多层焊接
multilayer welding method 多层焊接法
multilayer winding 多层绕组
multilead 多引线；多引入线的
multileaf damper 多叶减振器；多叶气流调节器；多叶消声器；多叶风阀
multileaf door 多扇式门；多翼式门；多叶门
multileaf mechanical mucker 多爪式抓岩机
multileaf up and over door 多页提升向上开启门
multileaved spring 多片式钢板弹簧
multileaving 多点传送
multileaving support 多点传送支持
multilegged armo(u)r block 多足护面块；多脚护面块体
multilegged portal frame 多支腿移动式框架；多支腿移动式构架
multilegged sling 复式多支腿吊索；多脚桁架；多脚吊索
multilegged structure 多支腿构筑物
multileg mooring system 多浮筒系泊系统；多点系泊系统；多点系泊设施
multilegs intersection 多路交叉(口)；复式交叉；多条道路交叉
Multilene 玛尔蒂纶聚烯烃纤维
multilength 多倍长度的
multilength arithmetic 多倍长度运算
multilength number 多倍长数
multilength working 多倍长工作单元
multilens 多物镜的
multilens colo(u)r light signal(l)er 透镜式色灯信号机
multilevel 多平面；多用水平尺；多级；多层的(房屋)
multilevel action 多位作用；多级作用
multilevel address 多级地址
multilevel addressing 多级寻址；多级定址；多级编址
multilevel air sampling 多级
multilevel approach 多层引桥法；多重引桥法；多级法
multilevel auditorium 多层观众厅
multilevel carpet 多层(绒头)地毯
multilevel contact oxidation system 多级接触氧化系统
multilevel control 多级控制
multilevel control system 分级控制系统；多级控制系统
multilevel design 多层设计
multilevel diagnostic software 多级诊断软件
multilevel difference equation 多层差分方程
multilevel distribution system 多级分布系统
multilevel dryer 多层干燥器；多层干燥机
multileveled mountain landform 山地多层地形
multilevel elevator 多级高度升运器
multilevel evaluation method 多层次评价方法
multilevel fall back structure 多级后退结构
multilevel filter 多层滤料滤池
multilevel filtration 多层滤料过滤
multilevel groundwater sampling well 多层地下水采样井
multilevel guyed tower 多级拉索塔
multilevel hierarchical structure 多层递阶结构
multilevel hierarchy 多层体系；多级结构
multilevel high concentration and auto-concentrating activated sludge technology 多级高浓度自浓缩活性污泥法
multilevel high tonnage press 多级高吨位压机
multilevel hoisting installation 多水平提升设备；多级提升设备
multilevel index 多级索引
multilevel indirect addressing 多重间接寻址；多级间接寻址
multilevel intake 多层进水口；分层取水式进水口

multilevel interchange 多层(立体)交叉
multilevel interconnection 多层互连
multilevel interrupt 多级中断
multilevel intersection 多层交叉口
multilevel junction 多层立(体)交(叉)
multilevel laser 多能级激光器
multilevel logical circuit 多级逻辑电路
multilevel media 多层滤料
multilevel memory system 多级存储系统
multilevel modeling structure 多级建筑结构
multilevel-multiobjective structure 多级目标结构
multilevel operation 多级操作
multilevel outlet shaft spillway 多层进水竖井(式)溢洪道；多层出水竖井式溢洪道
multilevel priority interrupt 多级优先权中断
multilevel processing 多级处理
multilevel pulse code digital sum 多级脉冲数字和
multilevel reasoning system 多级推理系统
multilevel road system 多层道路系统
multilevel sampler 多层采样器
multilevel sampling device 多层采样器
multilevel signal 多级信号
multilevel stabilization 多级镇定
multilevel storage machine 多级存储机
multilevel structure 多层结构；多层构筑物
multilevel subroutine 多级子程序
multilevel system 多级系统
multilevel town 多层城镇；多级城镇
multilevel transmission 多电平传输
multilevel utilization 多级利用
multilevel vectored interrupt logic 多级向量中断逻辑
multilift 多层模板
multilift holder 多层储气罐
multiline 复式线路；多线
multilinear 多线性的
multilinear algebra 多重线性代数
multilinear form 多重线性型
multilinear tariff 复式税率
multilinear utility function 多线性效用函数
multiline control 多线路控制
multiline function 多行函数
multiline laser 多谱线激光器
multiline maxtures 多品系混合体
multiline message 多路信息
multiline production 多线生产法；分类生产法
multiline railway 多线铁路
multiline section 多线路区段
multiline text 多行文本
multilingual edition 多种语言混合版
multilingual lettering 多种语言对照注记
multilink 多链接；多级链路
multilinked list 多链接表
multilinked structure 多重链接结构
multilist 多重表(格)；多目表；多表列；多表
multilist file 多重表文件
multilist organization 多表组织
multilist processor 多道程序处理机；多种程序处理机
multiload 多负载
multilobate delta 多叶三角洲
multilobe 多瓣的；多叶的
multilobe tracery 多叶瓣形花格窗
multilocation microcirculation microscope 多部位微循环显微镜
multilocular 多腔的
multiloop 多环；多匝；多重环路；多回路的
multiloop control 多环路控制
multiloop control system 多回路调节系统；多回路(分级)控制系统
multiloop feedback 多回路反馈
multiloop motor 多回路电动机
multiloop regulating system 多回路调节系统
multiloop servo system 多回路随动系统；多环路伺服系统
multiloop superheater 蛇形管过热器
multiloop system 多(环)路系统
multimachine 多机
multimachine assignment 单人多机操作
multimachine group 复式机器群
multimachine operation 多机操作
multimachine system 多电机系统
multimagnetic components 多种磁性成分
multimagnifier 多倍率放大器

multimask 多层彩色蒙片
multimass 多质量
multimedia 多层滤料；多媒体；多种方式的；多种手段的；多媒介
multimedia authoring system 多媒体写作系统；多媒体编著系统
multimediuma bed 多滤料滤床
multimedia card 多媒体卡
multimedia chip 多媒体芯片
multimedia communication 多媒体通信
multimedia communication cable 多媒介通信电缆
multimedia courseware 多媒体课程软件
multimedia data 多媒体数据
multimedia database manage system 多媒体数据库系统
multimedia database system 多媒体数据库系统
multimedia device 多媒体设备；多媒体装置
multimedia extension technique 多媒体扩展技术
multimedia file input/output 多媒体文件输入输出
multimedia file server 多媒体文件服务器
multimedia filing system 多媒体文件系统
multimedia filter 多层滤料滤池
multimedia filtration 多层介质过滤
multimedia format 多媒体格式
multimedia hardware 多媒体硬件
multimedia information system 多媒体信息系统
multimedia interface 多媒体接口
multimedia interface design 多媒体接口设计
multimedia monitoring station 多媒介监测站
multimedia network 多媒体网络
multimedia network system 多媒体网络系统
multimedia personal computer 多媒体个人计算机
multimedia photogram 多介质摄影像片
multimedia sand filter 多层介质砂滤池
multimedia sand filtration 多层介质砂滤
multimedia service 多媒体业务
multimedia shaping environment 多媒体造型环境
multimedia software 多媒体软件
multimedia stratified filter bed 多滤料分层滤床
multimedia system 多媒体系统
multimedia technology 多媒体技术
multimedia terminal 多媒体终端
multimedia unstratified filter bed 混合滤料不分层滤床
multimember 多构件的
multimembrane 多层薄膜
multimer 多体
multimetal deep-etch plate 多层金属平凹板
multimetallic Bi ore 多金属铋矿石
multimetallic cinnabar ore 多金属辰砂矿石
multimetallic copper ore 多金属铜矿石
multimetallic lead ore 多金属铅矿石
multimetallic Mo ore 多金属钼矿石
multimetallic zinc ore 多金属锌矿石
multimeter 复用表；万用表；万能电表；通用量测仪；多用途计量器；多用表；多量程(测量)仪表；伏欧毫安表
multimetering 多次测量；多点测量；复式记录；复式计数
multimetering instrument 多点测量仪
multimicrocomputer 多微计算机
multimicroelectrode 微电极组
multimicroprocessor 多微处理机
multimicroprocessor master-master system 多微处理机主主系统
multimicroprocessor master-slave system 多微处理机主从系统
multimill 双碾盘连续混砂机
multimillion 数百万
multiminiprocessor network 多小型处理机网络
multimirror telescope 多镜头望远镜
multimission aerodrome 多用途机场
multimission altitude determination and autonomous navigation 多用途高度确定与自导
multimission underwater remotely operated vehicle 多功能水下遥控深潜器
multimixer 组合式搅拌器
multimodal 多峰
multimodal distribution 多种模态分布；多重模态分布；多峰分布
multimodal transport 复合运送；联运
multimodal transport bill of lading 复合运输提单
multimodal transport contract 复合运送契约
multimode 多(状)态(的)；多模(的)；多方式

multimode behavio(u)r 多模性能
multimode cavity 多模谐振腔
multimode control 复式控制
multimode direct-view storage tube 多态直视存储管
multimode disturbance 多模干扰
multimode feed 多模馈电器
multimode fiber 多模纤维；多模光纤
multimode fibre-optic microbend sensor 多模光纤微弯传感器
multimode filter 多模滤光器
multimode flow 多型物质流
multimode glass fiber 多模玻璃纤维
multimodel 多种模式；多模式
multimode laser 多模激光器
multimode laser diode 多模激光二极管
multimode laser oscillator 激光多模振荡器
multimodel transport 多种方式运输
multimode-multihorn feed 多模多喇叭馈电器
multimode optic(al)cable 多模光缆
multimode optic(al)fiber 多模光(导)纤(维)
multimode optic(al)fiber hydrophone 多模光纤水听器
multimode output 多模输出信号
multimode propagation 多模传播
multimode radar 多方式雷达
multimode silica fiber 二氧化硅多模光纤
multimode system 多模系统
multimode transit 多种形式联运
multimode wavelength-division demultiplexer 多模波分去复用器
multimode wavelength-division multiplexer 多模波分多路复用器
multimodulus 扩大模数
multimolecular dust 多分子吸附
multimolecular layer 多分子层
multimomlecular film 多分子膜
multimotor 多电动机的
multimotored aircraft 多发动机飞机
multi-mull(er) 双碾盘混砂机
multi-multi fiber 二次复丝
multi-mu tube 可变放大系数管
multi-mu valve 可变放大系数管
multinational accounting 跨国公司会计
multinational accounting firm 跨国会计事务所；跨国会计公司
multinational bank 跨国银行
multinational company 跨国公司；多国公司
multinational corporation 跨国籍企业；跨国公司；多国公司
multinational enterprise 跨国企业
multinational firm 跨国商号；跨国企业
multinational operation 跨国经营
multinatural ecological treatment technology 多自然生态处理技术
multinest 多重嵌套；多嵌套
multinitriding 多段氮化法；多次氮化
multinodal 多节的
multinodal seiche 多节点假潮；多节(点)波动
multi-node 多节
multinomial 多项式
multinomial coefficient 多项式系数
multinomial distribution 多项分布；多维正态分布
multinomial expansion coefficient 多项式展开系数
multinomial expansion coefficient method 多项式展开系数法
multinomial index method assessment 多相指数法评价
multinomial theorem 多项式定理
multinomial trials 多项试验序列
multinozzle 多喷嘴；多口喷枪
multinozzle cutter 多嘴火焰切割器；多嘴割炬
multinozzle engine 多喷管发动机
multinozzle turbine 多喷嘴水轮机
multinuclear city 多中心城市
multinuclear mode 多核(心)模式
multinuclear model 多核(心)模型
multinuclear nodule 多核锰结核
multi-nucleus city 多中心城市
multiobjective 多目标
multiobjective analysis 多目标分析
multiobjective berth 多用途泊位
multiobjective decision 多目标决策
multiobjective decision making model 多目标决策模型
multiobjective dynamic programming 多目标动态规划
multiobjective planning 多目标规划
multiobjective programming 多目标规划
multiobjective structure 多目标结构
multiobjective system 多用途系统
multiobjective water quality management 多目标水质管理
multiobjective water quality management model 多目标水质管理模型
multioffice 多局制
multioffice exchange 多局制交换局
multiopening press 多层压机
multioperand 多操作数
multioperand adder 多操作数加法器
multioperating mode automatic conversion 工况自动转换
multioperating mode control system 多工况控制系统
multioperation 多运算；多操作
multioperation capability 多操作能力
multioperator arc welder 多人电弧焊机
multioperator welding machine 多站(电)焊机
multioperator welding set 多站焊机；多站电焊机
multiorder 多阶的
multiorder xi-type structures 多级多字型构造
multiorifice 多孔板(的)
multiorifice nozzle 多孔喷嘴
multioscillator 多谐振荡器
multioutlet 多引线
multioutlet assembly 多引线接口；多孔插座
multioutlet television coupler 多引线电视耦合器
multipack 一包多件的商品包装
multipacket message 多组信息
multipacting 次级发射倍增
multipactor 高速微波功率开关；次级电子倍增效应
multipactor breakdown 次级电子倍增击穿
multipactor effect 次级电子谐振效应
multi paint shaker 涂料分散(性)试验机
multiparameter 多参数；多参量
multiparameter adjustment algorithm 多参数调整算法
multiparameter air quality analyzer 多参数空气质量分析仪
multiparameter assessment of water quality 多参数水质评价
multiparameter control 多参数控制
multiparameter detector 多参数检测器
multiparameter estimation of water quality model 水质模型多参数估值
multiparameter estimation of water quality model by gradient searching 水质模型多参数梯度搜索法估值
multiparameter estimation of water quality model by network searching 水质模型多参数网络搜索法估值
multiparameter integrated decision analysis model 多参数综合决策分析模型
multiparameter random process 多参数随机过程
multiparameter regulation 多参数调整
multiparameter self-optimizing system 多参数自寻最佳系统
multiparameter trophic state indices 多参数营养态参数
multiparameter water quality kit 多参数水质成套检测器
multiparameter water quality (monitoring) instrument 多参数水质监测仪(器)；多参量水质监测仪(器)
multiparasitization 多寄生
multipart 多部件的；几部分组成的
multipart bearing 弓形轴承；组合轴承
multipartite door 多成多部分的门
multipartite vault 多成多部分的拱顶
multipartition 多分区
multipartition support 多分段支持
multipart sealant 多组分密封膏
multiparty line 公用线；合用线；同线
multipass 多回路；多道的；多程
multipass aeration tank 多廊道曝气池
multipassage kiln 多通道窑
multipassage mass spectrometer 多路质谱计
multipass airlay dryer 多路气流式烘干机
multipass algorithm 多遍算法
multipass and baffled heat exchanger 多程及折流式换热器
multipass compiler 多趟编译程序；多次编译程序
multipass condenser 多程冷凝器
multipass distribution 多次通过区熔的杂质分布
multipass dryer 多次通过式干燥机；多程干燥器
multipass exchanger 多程换热器
multipass flow 多程流量
multipass heater 多程加热器
multipass holographic interferometer 多光路全息干涉仪
multipass interferometry 光倍增干涉法
multipass kiln 多孔窑
multipass laser 多通激光器
multipass method (of construction) 多次行程(筑路)法
multipass mix-in-place work 多次通过就地拌和稳定土的工作
multipass modulator 多通调制器
multipass regenerator 多程蓄热室
multipass scanning 多遍扫描
multipass scanning algorithm 多遍扫描算法
multipass soil stabilizer[stabiliser] 多次通过稳定土拌和机
multipass sort 多路分类；多类数据分类程序；多次扫描分类；多次排序；多次分类；多遍排序；多遍分类
multipass sorting program(me) 多遍分类程序
multipass spectrometer 多光路光谱仪
multipass tubular heater 多程管式加热器
multipass type mixer 多次通过(作业的)拌和机
multipass weld 多道焊缝
multipass welding 多道焊(接)；叠层焊接
multipass zone refining 多次通过区域精炼
multipath 多途径(的)；多路径；多路(的)；多流程
multipath approach electronic computer 多路进场电子计算机
multipath burning device 多通道燃烧装置；多通道燃烧器
multipath cancellation 多径抵消
multipath communication 多路通信
multipath core 多(通)路磁芯
multipath delay 重复信号延迟
multipath effect 多路效应；多径效应
multipath error 多路径反射误差；多径误差
multipath propagation 多路径传播
multipath reduction factor 多径缩减因子
multipath signal 多程信号
multipath simulator 多路模拟器
multipath system 多通道系统
multipath transmission 多路输送
multipeak anomaly 多峰异常
multipeaked mountain 山峦
multipeel bucket 多瓣式抓斗
multipen plotter 多笔绘图仪
multipen recorder 多笔记录仪
multipen turret 多笔回转头
multiperiodic(al)activity 多期活动
multiperiod model 多时期模型
multiperson decision making 多人决策
multipetticoat insulator 多裙式绝缘子
multiphase 多相(的)
multiphase alloy 多相合金
multiphase ceramics 多相陶瓷；复相陶瓷
multiphase circuit 多相电路
multiphase composite ceramics 多相复合陶瓷
multiphase current 多相电流
multiphase diversion 分期导流
multiphase equilibrium experiment 复相平衡试验
multiphase eutectic 多相低共熔混合物
multiphase flow 多元相流；多元流；多相流(动)
multiphase generator 多相发电机
multiphase health screening 多项健康筛检(法)
multiphase inclusion 多相包裹体
multiphase inverter 多相换流器
multiphase load module 多相输入模块
multiphase material 多相材料
multiphase mixture 多相混合物
multiphase motor 多相电动机
multiphase multicomponent model 多相多组分模型
multiphase photocatalysis degradation 多相光催化降解

multiphase polymer material 多相聚合物材料
multiphase project 多期工程项目
multiphaser 多相发电机
multiphase sampling 多阶段抽样(法)
multiphase solidification 多相凝固
multiphase surface 多相表面
multiphase system 多相系(统);多相(体)系
multiphase timing signal 多相调时信号
multiphonic stereophony 多路立体声
multiphotocoupler 多重光电耦合器
multiphoton scattering 多光子散射
multipick insertion 多段引纬
multi-piered bridge 多墩桥
multipin plug 多路插头
multipipe system 多管系统
multiplace glider 多座滑翔机
multiplane 复翼飞机
multiplane-multispeed balancing 多平面多转速平衡法
multiplaner 多功能刨床;多功用刨床
multiplate 多片
multiplate angle valve 复片角阀
multiplate clutch 多片(式)离合器
multiplate condenser 多片电容器
multiplate disc brake 多片盘闸
multiplate friction clutch 多片摩擦离合器;多盘摩擦离合器
multiplate glass 安全玻璃;防弹玻璃;多层玻璃
multiplaten 多层的
multiplaten press 多层压力机
multiplate steering clutch 多片式转向离合器
multiplate vibrator 多片式振动器;多板式振动器
multiplayered structure 多层结构
multiplayer elastic theory 多层弹性理论
multiplayer laminated glass 多层夹层玻璃
multiplaying circuit 乘法电路
multiple 多重的;多元的;多数的;多倍坯料;倍(的);复(合)的;倍数
multiple absorption 重复吸收
multiple access 多址连接;多路接入;多路访问;多路存取;多处访问
multiple access computer 多用户计算机;多路存取计算机
multiple access coupler 多重存取耦合器
multiple access retrieval system 多路存取检索系统
multiple access virtual machine 多路访问虚拟机;多存取虚拟机
multiple accumulating registers 多次累积寄存器
multiple accumulator 多重累加器
multiple action 多重作用;重合动作
multiple action controller 多作业控制器
multiple action press 多效压机
multiple action scrubber 多用洗涤器
multiple activity process chart 多样活动过程图
multiple address 多址;多地址
multiple address code 多地址(代)码
multiple address computer 多地址计算机
multiple address instruction 多地址指令
multiple address message 多重地址报文;多地址信息;多地址消息
multiple address space 多地址空间
multiple address space partition 多地址空间区域;多地址空间分区
multiple adsorbent 多重吸附剂
multiple-aisle 多侧廊;多通道
multiple alternative decision 多择一判定
multiple alternative detection 多择检测
multiple amplifier 多路放大器;多级放大器
multiple analysis 多重分析
multiple annual ring 复年轮
multiple anodes cell 多阳极电解槽
multiple antenna 多单元天线
multiple anvil 多面顶
multiple anvil type 多顶锤滑动式
multiple aperture core 多孔磁芯
multiple appearance 复现
multiple apron dryer 多帘式烘干机
multipipe arc 多重弧;多弧
multiple arc circuit 并联弧光电路
multiple arch 多拱;连拱
multiple arch bridge 连拱桥
multiple arched dam 多孔坝;连拱坝
multiple arch retaining wall 连拱式挡土墙
multiple arch tunnel 连拱隧道

multiple arch type buttressed dam 连拱式支墩坝
multiple arc lamp 并联弧光灯
multiple arc method 复弧法
multiple arc unit 多弧焊接机
multiple arc weld 多弧焊缝
multiple arc welder 多弧电焊机
multiple arc welding 多弧焊
multiple arc welding plant 多弧焊接机
multiple arithmetic 多重运算;多路运算;复式运算
multiple arithmetic unit 多(重)运算器
multiple arm relay 多接点继电器
multiple aspect indexing 相关标引;多方位索引
multiple aspect searching 多面查找
multiple association 复关联
multiple attachment support 多连接支持
multiple attribute decision making 多属性决策
multiple attribute retrieval 多属性检索
multiple attribute transformation 多属性变换
multiple attributive classification 多属性分类
multiple band 多波段
multiple band mapping system 多波段成像系统
multiple bandpass filter 多通带滤波器
multiple band receiver 多波段接收机
multiple band remote sensing 多波段遥感
multiple band scanning system 多波段扫描系统
multiple band spectrum transformation 多波段频谱变换
multiple band system 多波段系统
multiple banking 提供全面服务的银行业务
multiple bar bell jar 多棒钟形烧结炉
multiple barrel shell 多波筒壳
multiple barrel structure 多圆筒结构
multiple barrier 多重屏蔽
multiple barrier concept 多级屏障思想
multiple barrier treatment 多道处理
multiple base altitude 多基线测高法
multiple batcher 多次称量的配料器;复式配料斗
multiple bay 多开间;多跨
multiple bay frame 多跨排架;多跨框架;多跨构架
multiple bay frame as spandrel beam 起托梁作用的多跨构架
multiple bay portal 多跨门架;多跨框架;多跨刚架
multiple bay portal frame 多跨可移式框架;多跨可移式构架;多跨移动式框架;多跨移动式构架
multiple bay prismoidal roof 多跨棱柱体屋顶
multiple bay scrubber 多格洗涤塔
multiple beam 多光束;复光束
multiple beam antenna 多波束天线
multiple beam flame planing machine 多焰嘴火焰刨削机
multiple beam fringe 多光束干涉纹
multiple beam holography 多光束全息术
multiple beam interferometer 多光束干涉仪
multiple beam interferometry 多光束干涉(法)
multiple beam lateral-shear interferometer 多光束横向切变干涉仪
multiple beam radar 多波束雷达
multiple beam radiation 多射束辐射
multiple beam scale 多杠杆磅秤
multiple beam torus antenna 多波束环形天线
multiple bearing 多次定位
multiple bed 多层床;复合层
multiple bed filter 多床滤池
multiple beginning activity 多项初始活动;多项初始工作
multiple bell pier 复钟式桥墩
multiple belt 多层皮带
multiple benches 多层工作面
multiple blade 多叶片(的);多刀的
multiple blade circular saw 多刀圆锯
multiple blade drag 多刃刮路机;多刀刮路机
multiple blade grader 多刃平地机;多刀平路机;多刀平地机
multiple blade maintainer 多刀养路机
multiple blade mixing drag 多刃拌和刮路机;多刀拌和刮路机
multiple blasting 多炮爆破
multiple blowpipe machine 多割嘴气割机
multiple body problem 多体问题
multiple borer 多轴钻孔机;多轴穿孔机
multiple bottom reflection 海底多次反射;多次海底反射
multiple bowl overshot 多级打捞器
multiple box culvert 多孔箱涵

multiple box section 多孔箱式结构
multiple box siphon 复式箱形虹吸管
multiple branching construction 多分支结构
multiple branch manifold 多支路歧管
multiple break 多重断开
multiple breakdown 多因素停机故障
multiple break switch 多断式开关
multiple-bridge 群桥;多臂电桥
multiple-bridge intersection 复式立交桥;群桥交叉
multiple bucket appliance 多斗设备
multiple bucket dredge(r) 多斗式挖泥船
multiple bucket excavator 多斗挖掘机
multiple bucket ladder 复式斗杆
multiple budget 复式预算;复合预算制度
multiple buffer 多缓冲区
multiple bundle cable 多束光缆
multiple-buoy mooring 多浮筒系泊(法);多点系泊(法);多浮标锚泊;多浮标锚碇
multiple burner 多头燃烧器
multiple bus and multiport system 多总线与多入口系统
multiple bus architecture 多总线结构
multiple bus control processor 多总线控制处理机
multiple bus system 多总线系统
multiple byte instruction 多字节指令
multiple-cable 多股电缆;复电缆
multiple cage motor 多鼠笼电动机
multiple cage rotor 多鼠笼型转子
multiple call transmission 多重通话传输
multiple camber 多段曲面
multiple camera 多角摄像机
multiple camera system 多相机系统
multiple cam timer 多凸轮定时器
multiple capacitor 多联电容器
multiple carbide alloy 多元碳化物合金
multiple carbide ceramics 复式碳化物陶瓷
multiple carbide hard metal 多元碳化物硬质合金
multiple card record 多重卡片记录
multiple cargo vessel 多用途船(舶)
multiple cartridges 多级滤片
multiple casement window 多扇窗;多层窗扇
multiple casing 多缸式
multiple casting 劈模铸造;多模齐铸法
multiple cavity 多腔的
multiple cavity block 多孔砌块
multiple cavity casting die 多型腔压型
multiple cavity klystron 多腔速调管
multiple cavity mo(u)ld 多型腔铸型;多腔模;多巢模
multiple cell flo(a)tation machine 多槽浮选机
multiple centre joint 多条中线缝
multiple chain 复式链
multiple chain drag conveyor[conveyer] 多链牵引式输送机
multiple chamber compound mill 多仓管磨
multiple chamber furnace 多室式炉
multiple chamber incinerator 多室焚化炉
multiple chamber locks 多厢船闸
multiple-chamber mill 多仓磨(机)
multiple chamber refuse incinerator 多室垃圾焚化炉
multiple channel 多信道;多通道;复式河槽
multiple channel analyzer 多路分析器
multiple channel buffer 多通道缓冲器
multiple channel river 多汊河流
multiple channel stream 多汊河流
multiple character set 多字号字符组;复式字符组
multiple check 多重校验;多路查看
multiple chimney 复式烟囱
multiple choice question 选择问答法
multiple chord 复弦杆
multiple circuit 多重环路;复杂分路;复式电路;复接电路;并联电路;倍增电路;倍增电极
multiple circular transition curve 多圆弧缓和曲线
multiple clarifier 多层沉淀池
multiple classifying analysis 多重分类分析
multiple cleaning test 多次清洁试验
multiple cloth 多层织物
multiple clutch 复式离合器
multiple coat 多层(指涂层);复层
multiple coat application 进行多次涂刷
multiple coating eye-glass 多模眼镜片
multiple coat shceme 多层涂层格式
multiple coil 多极螺旋

multiple coincidence 多(次)符合
multiple collet winder 多筒拉丝机
multiple column 多柱式
multiple column bent 多柱式排架
multiple combinations 重复组合
multiple combustor 多室燃烧室
multiple community 复合群落
multiple compactor 多头夯土机
multiple comparison 多重比较
multiple comparisons of proportions 比例的多重比较
multiple compartment sinks 分格式洗涤槽
multiple completion 多层完井
multiple completion packer 多级封隔器
multiple completion well 多层合采井
multiple component 多组分
multiple-component external spray nozzle 外混合型多组分喷嘴
multiple component units 组合零件；配套无线电零件
multiple compressed tablet 复压片
multiple compressor 复式压缩机
multiple computer 多媒体计算机；多功能计算机
multiple computer operation 多计算机操作
multiple computer system 多机系统；多层次计算机系统
multiple conductor 分裂导线
multiple cone bit 多牙轮钻头
multiple cone pelletizer 多锥式造球机
multiple-connected battery 并联蓄电池(组)
multiple connection 并联连接；复接(线)；并列连接；并联接法
multiple connector 多路流程图符号；多路连接器；多路插头
multiple console 多话务台操作
multiple console support 多重控制台配套；多控制台支持；多操作台支持
multiple constraint 多约束；多约束
multiple contact 多触头的
multiple contact detector 多触点检波器
multiple contact high-pressure mo(u)lding machine 多触头高压造型机
multiple contact relay 多接点继电器
multiple contact switch 多接点开关；多触点开关；选择器开关
multiple context 多重正文
multiple control 复杂调节；复式控制；复合管理；并列调节
multiple conveyer dryer 多级传送带烘干机
multiple coordinate symbol 多重坐标代号
multiple coordination 多重协调
multiple copy averaging 多拷贝平均(法)
multiple copy control 多厚度复制控制器；多副本控制
multiple cord 多芯软线
multiple cord tyre 多层外胎
multiple-core 多心的
multiple-core box 组合芯盒
multiple-core cable 多股绳；多芯电缆
multiple-core casting 组芯铸造
multiple correlation 多重相关；复(合)相关
multiple correlation analysis 多重相关分析
multiple correlation coefficient 多重相关系数；多元相关系数；多相关系数；复相关系数
multiple correlation method 多元相关法
multiple correlation ratio 多元相关比
multiple cost system 复合(式)成本制(度)
multiple counting 多次计数
multiple course construction 多层结构
multiple course construction stage 多层结构平台；多层式建筑步骤
multiple course surface treatment 多层式表面处治
multiple covariance 多重方差
multiple coverage 多次覆盖
multiple covered type 多层掩盖型
multiple covering layout 多次覆盖观测系统
multiple criterion 多准则
multiple criteria decision-making 多准则(目标)决策
multiple-crop index 复种指数
multiple cropping 多熟种植；复种
multiple-cropping area 复种面积
multiple-cropping index 复种指数
multiple cross 多交系

multiple cross glide 复交叉滑移
multiple crossing 多路交叉(口)；复式交叉
multiple-crystal 多晶体
multiple crystal X-ray spectrometer 多晶体X射线光谱仪
multiple cuase of death 多种死因
multiple currency standard 复合通货制度
multiple current generator 多电流发电机
multiple current hypothesis 多支流假说
multiple curvilinear 多重多曲线
multiple cut 多刀切削
multiple cutter 复式铰刀
multiple cutting edge countersink 多刃锥口钻
multiple cutting-edge tool 多刃刀具
multiple cut trench excavator 多刃挖沟机；多刀挖沟机
multiple cycle 多旋回
multiple cyclone 组合式旋风分离器；多管式旋风除尘器；复式旋风吸尘器
multiple cyclone in parallel 并联旋风分离器
multiple cyclone separator 多管式旋风分离器
multiple cylinder 复式汽缸
multiple cylinder engine 多汽缸发动机
multipled 并联的
multiple data channel 多数据通道
multiple daylight press 多层压机
multiple decade counter 多级十进位计数管
multiple decay 多重衰变
multiple decision 多重判定
multiple decision method 多重决策方法
multiple decision problem 多维决策问题
multiple deck 多层的；多筛面(筛分机)
multiple deck cage 多层畜笼
multiple decker 多层装置
multiple deck oven 多层炉
multiple deck station 多层车站
multiple declaration 多重说明；多种说明
multiple degree of freedom 多自由度
multiple design 多重设计
multiple detector 多重探测器
multiple device 复合器件
multiple diaphragm ga(u)ge 多膜片压力计；多模片压力表
multiple-die 复锻模
multiple-die medium heavy wire-drawing machine 多模中粗拉丝机；多次中粗拉丝机
multiple-die press 连续冲床；复式压机
multiple diesel plant 多发动机装置
multiple die tubing machine 多模管压机
multiple digit decimal adder 多数位的十进制加法器
multiple dike[dyke] 复合岩墙
multiple dilution ratio 多次贫化率
multiple dimensioned double head wrench 多尺度双头扳手
multiple diode 复式二极管
multiple direction migration 多向迁移
multiple direction shaft 万向轴
multiple directory 多重目录表
multiple disc 多盘
multiple-disc air-operated brake 多盘式气动制动器
multiple-disc brake 多盘闸；多盘式制动器
multiple-disc clutch 多片(式)离合器；多盘离合器
multiple discharge 多道闪电
multiple discriminant analysis 多元判别分析；复识别分析
multiple disc sampling apparatus 多盘式采样器；多盘采样装置
multiple disintegration 多重衰变；倍速蜕变
multiple disjoint decomposition 多重不相交分解
multiple distribution 并联配线
multiple distribution system 并联配电制
multiple domain network 多畴网络
multiple dome dam 穹顶连拱坝；双曲连拱坝；多穹顶坝；连穹坝
multiple dosing tanks 复配投配池
multiple drawbench 多线式拉拔机
multiple drift 周壁导坑；多层导坑；复式导坑
multiple drift method 多导流法；多导坑开挖法
multiple drill 多轴钻机；多轴钻床
multiple drill head 多轴钻主轴箱
multiple drilling 密集钻进；多孔钻法；丛式钻进
multiple drilling machine 多轴钻机；多头钻机
multiple drill press 多轴钻机

multiple drill rig 多机式钻车
multiple drill shaft sinker 多钻凿井机
multiple drive 多机驱动
multiple drop check dam 多级谷坊
multiple drum conveyor[conveyer] 多传动轮胶带输送机
multiple drum winch 多卷筒绞车；复鼓式卷扬机
multiple duct conduit 多孔管道
multiple dump processing 多重转储处理
multiple dwelling 多户住宅；多单元住宅
multiple-dwelling building 多种住宅建筑；多种套间建筑
multiple-dwelling system 多用户送冷风系统；复式寓所系统
multiple earth 多点接地
multiple earthing 重复接地
multiple earthquake 多发性地震；复地震
multiple earth system 多重接地制
multiple eaves 重檐(指两层之间加假屋檐)
multiple eaves roof 重檐屋顶
multiple echoes 多重回声；多次回声；多次打回；复式回波
multiple effect 多重效应；多重效果；多效应
multiple effect distillation 多效蒸馏
multiple effect distillator 多效蒸馏器
multiple effect evapo(u)ration 多效蒸发
multiple effect evapo(u)rator 多(重)效蒸发器
multiple effect flash evapo(u)rator 多效闪蒸器
multiple effect incineration 多效焚化
multiple effective 多效的
multiple effect refrigerator 多效制冷机；多效冷却器；多效冷冻机
multiple eigenvalue 多重特征值
multiple electric motor drive 多电动机驱动
multiple electrode 多级焊条
multiple electrode furance 多极电炉
multiple electrode spot welding machine 多极点焊机
multiple electrode submerged arc weld 多电极埋弧焊
multiple electrode welding 多焊条焊接
multiple electrometer 多量程静电计；复式静电计
multiple electron gun zone refining 多电子枪区域精炼
multiple electronic spectrum scanning radiometer 多光谱电子扫描辐射计
multiple element coating 多元素镀层
multiple element cooler 多元冷却器
multiple elutriation 多次淘洗
multiple emitter transistor 多发射极晶体管
multiple end thread 多头线
multiple end user 多端用户
multiple end yarn 多头纱
multiple engine 多发动机装置
multiple entrances to subroutine 对子程序的多个入口
multiple entry 多入口
multiple entry-exit permit 多次进出许可
multiple entry pallet 多向叉入托盘
multiple entry system 多输入系统；多平巷开拓系统
multiple entry volume table 多元材积表
multiple epidermis 复表皮层
multiple error 多(重)误差；多级误差
multiple-error-correcting code 多误差校正码；多差校正代码
multiple-error-stochastic model 复合误差随机模型
multiple evapo(u)rator 多效蒸发器
multiple event flood 多峰洪水
multiple exchange rate 多种汇率
multiple excitation 多励(磁)；多次激发；复励磁；复激
multiple exhaust 分流排气
multiple exits from subroutine 从子程序的多个出口
multiple expansion 多级膨胀
multiple expansion engine 多胀式发动机；多级膨胀发动机
multiple expansion of credit 信用多倍扩大
multiple expansion steam engine 多胀式蒸汽机
multiple expansion turbine 多胀式涡轮机；多胀式汽轮机
multiple explicit route 多显式路由；多显式路径
multiple explosion 多点爆破
multiple exposure 多次曝光；分次曝光；反复接触

multiple exposure hologram 多次曝光全息图
multiple extraction 多级提取
multiple extrema 多重极值
multiple fabric 多层织物
multiple factor experiment 复因子试验
multiple factorial experiment 复因子试验
multiple factors 多对因子；多因子
multiple factors method 复合因素法
multiple factors stratigraphic(al) trap 多因素地层圈闭
multiple factors structural trap 多因素构造圈闭
multiple factors trap 多因素圈闭
multiple false echo 多次反射回波
multiple family 多户住宅；多家庭住宅
multiple family dwelling 多户住宅
multiple family house 多户住宅
multiple fan deposit 叠覆扇沉积
multiple faults 组合断层；阶状断层；阶梯形断层；叠断层；复断层；多故障
multiple feed 多路进线；多口进料；多级馈电；多电源供电
multiple feeder 复馈电线
multiple-feed network 复馈电力网
multiple-feed system 复馈电系统
multiple fender piles 丛桩缓冲设备
multiple fiber 复合丝；多束纤维
multiple fiber cable 多纤光缆
multiple fibre structure 多缕结构
multiple field generator 多绕组发电机
multiple film 多重片
multiple filter 多重滤波器；多节滤波器；多层过滤器；复式滤池
multiple filter plate 多层过滤板式分布器
multiple filtration 多层过滤
multiple firing 齐爆；多发爆破；成组点火
multiple flighted screw 多导程螺杆
multiple floor station 多层车站
multiple flow state 多种流态
multiple flue chimney 多烟道烟囱
multiple flue smoke stack 多烟道烟囱
multiple flue stack 多烟道烟囱
multiple fold 复式褶皱
multiple folded plate roof 复式折板屋顶
multiple folding rule 多折尺；多摺尺
multiple fork type blade root 多叉形叶根
multiple fracture treatment (岩层的)多次压裂处理
multiple fracturing 多次压裂；分层压裂
multiple frame 接线架；多跨框架；多开间结构；复式框架
multiple frame argon-ion laser 多幅氩离子激光器
multiple frame saw mill 多框锯制材厂
multiple freight 复式运费账单
multiple frequency 多重频率；倍频
multiple frequency very low frequency technique 多频(率)甚低频技术
multiple frequency vibration 多倍频振动
multiple fruit 复果
multiple function 复官能
multiple function alarm controller 多功能报警控制器
multiple function bone milling cutter 多功能骨铣刀
multiple function drinking-water cleaner 多功能饮水净化器
multiple function water-treating agent 多功能水处理剂
multiple galaxy 多重星系
multiple gap lightning arrester 多隙(式)避雷器；复隙式避雷器
multiple gas analyzer 多气体分析器
multiple gate 多浇口浇注系统
multiple gated spillway 多闸(门)溢洪道
multiple gate nozzle 多口喷嘴
multiple gating 多直浇口浇注系统
multiple generation area 多代区
multiple generator 倍数发生器
multiple geophone 组合地震检波器
multiple glass 防弹玻璃；夹层玻璃
multiple glazed unit 多层玻璃窗门
multiple glazed window 多层玻璃窗
multiple glazing 多层(窗)玻璃
multiple glazing glass 多层中空玻璃
multiple glazing unit 多层上光单元；多层玻璃窗(门)

multiple grade cooler 多箅床冷却器
multiple graphs 多图
multiple grating 复式光栅
multiple gravel layers 多卵石层
multiple grid 多栅
multiple groove bearing 多油槽轴承
multiple grooved gathering shoe 复槽集束器
multiple grooved pulley 多槽轮
multiple grooved shear 多槽剪切机
multiple grooved wheel 多槽轮
multiple groove sheave 多槽皮带轮
multiple harmonic current 多谐波电流
multiple harmonic function 多调和函数
multiple head broaching machine 复式拉床
multiple head bruner 多头燃烧器
multiple heading 多工作面掘进
multiple head machine 多头成型机
multiple head sander 多头砂光机
multiple head unit 多磁头装置
multiple hearth dryer-incinerator 多床干燥-焚化炉
multiple hearth furnace 多膛焙烧炉；多层炉膛反射炉；旧砂再生炉
multiple hearth incinerator 多炉膛焚烧炉
multiple hearth reactor 多室反应器
multiple hearth sludge incinerator 多炉膛废物焚烧炉
multiple-heating coil 复式加热蛇形管
multiple hoistway 复式提升井
multiple hole 多孔
multiple hole core 多孔磁芯
multiple hole die 多孔模
multiple hologram 多重全息图
multiple hop 多次反射
multiple hop transmission 多次跳跃传输；多次反射传输
multiple hulled ship 多体船
multiple hydraulic pump 多层水力泵
multiple hydrophobic adsorbate 多种疏水吸附质
multiple idler belt scale 多托辊胶带秤
multiple image 多重图像；叠影
multiple imagery 多重图像
multiple image system for depot 车辆段监控复视系统
multiple imaging 多重影像
multiple impact tester 多冲式试验机
multiple impeller pump 多级叶轮泵
multiple impression die 多模槽锻模
multiple impression forging 多型槽模锻
multiple impulse 多脉冲
multiple impulse welder 多脉冲接触焊机
multiple impulse welding 多脉冲接触焊；多脉冲焊
multiple increment test 多次增量试验
multiple indeterminate 多次超静定的
multiple index coupler 复杂折射率耦合器
multiple indexing addressing 多重变址型寻址
multiple index method 多折射率法
multiple indirect addressing 多重间接寻址；多重间接编址
multiple induction loop 多次归纳循环
multiple inequality coefficient 复不等系数
multiple ingredient ore 多次配矿
multiple inputs 多端输入
multiple input system 多输入系统
multiple input terminal equipment 多端输入终端设备
multiple installation 复式装置
multiple instruction 多指令
multiple instruction flow 多指令流
multiple instruction multiple data stream 多指令多数据流
multiple instruction-single data stream 多指令单数据流
multiple instruction-single data stream system 多指令单数据流系统
multiple instruction stream 多指令流
multiple integral 多重积分；重积分
multiple integration 重积分法
multiple interdependency 相互间多重依存关系
multiple interferometer 复式干涉仪
multiple internal reflectance 多重内反射
multiple internal reflection face-pumped laser 多路内部反射表面泵浦激光器
multiple interpolation 多重插值
multiple interrupt 多(重)中断

multiple interrupt system 多重中断系统；多级中断系统
multiple intersection 立体交叉；多线交叉；多路交叉(口)；复式交叉
multiple intrusions 重复侵入【地】
multiple ion detector 多离子探测器
multiple-ioned 多次电离的
multiple-ionized 多次电离的
multiple isomorphism 同态；多重同构
multiple jack 复式插孔
multiple jet 多喷嘴
multiple jet burner 多焰燃烧器；多焰焊枪
multiple jet carbureter 多嘴汽化油器
multiple jet nozzle 多孔喷嘴
multiple jib machine 多截盘截煤机
multiple jig 多钻头钻模
multiple job processing 多(道)作业处理
multiple joint (湖、海面的)多节点；复式缝
multiple junction 多线铁路枢纽
multiple keys 花键
multiple label 多重标号
multiple laccolith 叠岩床
multiple lake zone 多湖地区
multiple lamp holder 复式灯座
multiple lane 多车道
multiple lane bridge 多车道桥
multiple lane highway 多车道公路
multiple lane road 多车道路面；多车道道路
multiple lane tunnel 多车道隧道
multiple language 复式语言
multiple lattice 多重格体
multiple latticed truss 多腹杆桁架
multiple latticework 多重格构；复式格构
multiple layer adhesive 多层黏合剂；复层胶
multiple layer composition roof 多层接合屋顶
multiple layered 多层(次)的
multiple layered composition roofing 多层覆盖屋顶；多层材料组成的屋顶
multiple layered fabric for roofing 覆盖屋顶的多层纤维
multiple layered insulation 多层(热)绝缘
multiple layered monolithic wall 多层整体式墙
multiple layered panel 多层板
multiple layered pavement 多层路面
multiple layered pavement system 多层路面体系
multiple layered ready roofing 多层预制屋顶；多层现成屋顶
multiple layered sandwich radome 多层结构天线罩
multiple layered seal(ing) 多层密封
multiple layered system 多层体系
multiple layered tile 多层瓷砖；多层瓦(片)
multiple layered waterproofing 多层防水
multiple layer fabric for roofing 多层屋顶结构
multiple layer metallization 多层金属化
multiple layer mode 多层模式
multiple layer model 多层模型
multiple layer monolithic wall 多层独居墙
multiple layer panel 多层框
multiple layer structure 路面多层结构
multiple layer Z-buffer 多层Z-缓冲区
multiple leader 复顶芽；多顶枝
multiple leaf 多叶饰；多叶型
multiple leaf gate 多叶闸门
multiple leaf spring 多片簧
multiple leaf valve 多簧片阀
multiple leg intersection 多岔交叉
multiple length 多倍长；倍尺
multiple length number 多倍长数
multiple length numeral 多倍长数
multiple lens 多镜头的
multiple lens aerial camera 多镜头航空摄影仪；多镜头航空摄影机
multiple lens camera 多透镜照相机；多镜头摄影机
multiple lens camera assembly 多镜头摄影机组
multiple lens photograph 多镜视准管法
multiple level 多能级
multiple level combinational network 多级组合网络
multiple level line 复测水准路线
multiple level mining 多水平开采(法)
multiple level station 多层车站
multiple level subroutine call 多级子程序调用
multiple lever system 复式杠杆系统

multiple lift 多层的(房屋);多层;多式的
multiple lift construction 多层施工(法);多层铺筑路路面;路面多层施工(法);分层浇筑施工法;分层浇注施工法;分层浇灌施工法
multiple lift macadam 多层碎石路
multiple lift packing 散装
multiple lift surface treatment 多层(式)表面处治
multiple lift treatment 多层处理
multiple light signal 多灯信号机
multiple light signal(l)er 透镜式色灯信号机
multiple line 多重线;复式线
multiple linear correlation 多重线性相关;多元线性相关;复线性相关
multiple linearization storm-drain routing 多重线性暴雨排水演算法
multiple linear regression 多重线性回归;多元线性回归;复线性回归
multiple linear regression model 多元线性回归模型
multiple linear regression technique 多重线性回归技术
multiple line cutter 多头刻线仪
multiple line structure 多线结构
multiple line track 多线铁路
multiple liquid phase 多液相
multiple list 多重表
multiple listing service 房地产商合营业务
multiple living quarter 公寓
multiple load module processing 多装入模块处理
multiple lobe deposit 叠覆扇舌沉积
multiple lock 复式船闸;多厢船闸;多级船闸
multiple locomotive traction 多机牵引
multiple loop 多闭合环路
multiple loop demodulator 多环路解调器
multiple loop detector 多环框式检测器
multiple loop servomechanism 多回路伺服系统
multiple loop system 母线系统
multiple louver 复式百叶窗
multiple lumped-mass system 多自由度系统;多自由度体系
multiple machine-head 多切削头
multiple manometer 组合压力计;联合压力计;多量程压力表
multiple Markov chain 多重马尔可夫链
multiple masonry wall 多层圬工墙;多层砌筑墙体
multiple-mass system 多质点系统;多体系
multiple match 多重符合
multiple match resolution 多重符合分解
multiple material (scale) batcher 多种物料的配料称量器
multiple meaning 多义性
multiple meaning association 多重含意结合
multiple media photogrammetry 多介质摄影测量学
multiple medium flow model 多重介质渗流模型
multiple member structure 复式交叉结构;复式构件结构
multiple mercury column manometer 多级水银柱压力计
multiple metamorphism 多次变质作用
multiple meteor 多重流星
multiple meter 万用电表
multiple metering 多次计数;多次测量;复式读数
multiple microorganism 多样微生物
multiple modulation 多重调制;多级调制;复调制
multiple module access 多重模块访问;多级存取
multiple money 倍数货币
multiple motor 多电动机
multiple motor all-electric crane 多发动机驱动的起重机
multiple motor all-electric shovel 多发动机驱动的正铲挖掘机
multiple motor unit 多发动机装置
multiple mo(u)ld(ing) 叠箱铸型;多模;多次模塑法;多件石膏模;多件模
multiple mo(u)ld machine 多模成型机
multiple mounted cam 复合凸轮
multiple navigation lock 多航道船闸
multiple nib scriber 多头刻图仪
multiple normal correlation 多重正态相关;复正态相关
multiple nozzle 多喷嘴
multiple nozzle pump 多喷嘴扩散泵
multiple nucleated theory 多中心理论

multiple nuclei theory 多核心理论
multiple object function 多重目标函数
multiple objective 多目标
multiple objective decision 多目标决策
multiple objective decision making model 多目标决策模型
multiple objective function 多目标函数
multiple objective optimization theory 多目标优化理论
multiple objective parameters estimation 多目标参数估算
multiple objective planning 多目标规划
multiple objective planning and evaluation 多目标规划评价
multiple objective structure 多目标结构
multiple observation 多次观测
multiple of direct personnel expense (按人工、开支加权的)收费估算法
multiple of drift 漂移倍数
multiple of object shadow 地物阴影倍数
multiple on-line 多路联机
multiple opening culvert 多孔涵洞;多开口式涵洞
multiple operated window 驱动开启窗
multiple operation 平行工作;多机同时操作;并联运行
multiple operator arc welding machine 复式弧焊机
multiple operator welding-unit 多重操作焊接单元
multiple order 多级的
multiple order pole 多阶极点
multiple or duplicate vessels 多个或两个相同的容器
multiple ore deposit 叠生矿床
multiple oriented berth 多方位码头;多方位泊位
multiple oriented quay 多方位码头
multiple oriented wharf 多方位码头
multiple orifice meter 多孔测量计
multiple orifice nozzle 多孔喷嘴
multiple oscillograph 多射线示波器
multiple outlet 多管出水口;复式排水口;多口排水;多点出水口
multiple outliers 复奇异点
multiple output change flow table 多输出变化流程表
multiple output circuit 多输出电路;输出端连接电路
multiple output combinational network 多输出组合网络
multiple output network 多端输出网络
multiple output prime implicant 多端输出质蕴涵项
multiple output switching function 多输出开关函数
multiple output system 多输出系统
multiple over-band separator 多次磁铁带上分离机
multiple overlap 多重重叠
multiple pair 复接线对
multiple panel door 格子式镶板门;多层拼花板门;多层镶门板
multiple panel type air filter 多镶板式空气过滤器
multiple pane unit 多框格玻璃窗;多框格玻璃板
multiple parallel-tubing string 多根平行油管
multiple parallel winding 复并励绕组
multiple parameter method 多参数法
multiple part core box 多分型面芯盒
multiple part form 多联表格
multiple part gravity die 多分型面的金属型
multiple partition support 多分区支持
multiple part mo(u)ld 多箱铸型
multiple part pattern 多开边模
multiple pass 多次通过;多程;多路
multiple pass coat 多道涂布;多层涂装
multiple passing grating 多次通过光栅
multiple pass prism monochromator 多通棱镜单色仪
multiple pass rotary mixer 多行程转轴(式)拌和机
multiple pass stabilizer 多程稳定机;多程土壤稳定机;多层土壤稳定机
multiple pass weld 多路焊接;多道焊缝
multiple pass welding 多道焊
multiple path 多途径(的)
multiple path communication 多路通信
multiple pathcoupler 多路耦合器
multiple pattern 复式图形
multiple peak 多重峰
multiple peak correlation function 多峰值相关函数

multiple peak hydrograph 多峰式水文过程线;多峰值过程线
multiple perforation 复穿孔
multiple pharmacy 混合制药废水
multiple phase 多相位
multiple phase correlation 多相位对比
multiple phase flow 多相流(动)
multiple-phase generator 多相发电机
multiple photo analytic block adjustment program(me) 多重像片解析区域平差程序
multiple photogrammetric(al) camera 多镜测量摄影机
multiple photograph orientation instrument 多点照相钻孔定位仪
multiple piercing 多孔冲孔
multiple piezometer 多头测压管
multiple piles 群桩
multiple pin 多刀式
multiple-pin-hole 复针孔
multiple-pinhole camera 复针孔照相机
multiple pin mixer 多叶片式拌和机;多刀式拌和机
multiple pin plug 多路插头;多脚插头
multiple pintle rudder 多支承舵
multiple pipe inverted siphon 多管式倒虹吸管
multiple piston pump 多活塞泵;多缸泵
multiple piston rotary pump 多活塞回转泵;多缸回转泵
multiple plane 多重平面
multiple plane projection 多面投影
multiple plankton sampler 复式浮游生物采样器
multiple plate 多层板的;多层板
multiple plate brake 多板闸制动器;多盘闸
multiple plate clutch 多片(式)离合器
multiple plate expansion joint 多片式膨胀节;多片式伸缩缝
multiple plate flange 多层板翼缘;多层板翼缘
multiple plate sampler 多盘式采样器
multiple plough 复式犁
multiple plug 复式插头
multiple plunger press 多柱塞压力机
multiple plunger pump 多柱塞泵
multiple ply canvas 加料帆布
multiple plywood 多层胶合板
multiple point 多重点;多点的;多点
multiple point borehole extensometer 钻孔多点位移计;钻孔多点伸长计
multiple point displacometer 多点位移计
multiple point extensometer 多点应变计;多点位移计
multiple point loading 多点装料;多点填料;多点加料;多点集中加载
multiple point pencil 多点铅笔
multiple point recorder 多点记录器
multiple point recording potentiometer 多点式自记电位计;多点记录式电位计
multiple points measurement 多点测量
multiple point source dispersion model 多点污染源扩散模型;多点污染源
multiple point tool 多刃刀具
multiple pole switch 多级开关
multiple pollution sources 复合污染源
multiple polyaluminum sulfate 复合聚合硫酸铝
multiple pontoon landing stage 多浮筒式码头平台;多浮筒式简易码头;多浮筒式趸船码头;多浮筒式登岸栈桥;多浮筒登陆栈桥
multiple portal frame 复式龙门框架
multiple port tank 多小炉池窑
multiple position borehole extensometer 钻孔多点位移计;钻孔多点伸长仪
multiple position circuit 并席电路
multiple position shift 多位移位
multiple power source 通用电源
multiple precipitation process 多次沉淀法
multiple precision 多倍字长精度;多倍精度
multiple precision arithmetic 多精度算术;多倍精度运算
multiple precision current transformer 多精度变流器
multiple press 多效压机
multiple pressure pad die 带多个压料块的模具
multiple pressure technic 多压技术
multiple pricing 多种价格
multiple priming 并联起爆
multiple priority level 多优先级

multiple prismoid roof 复棱柱式屋顶
multiple process 并联法
multiple processing 多重处理
multiple processor operating system 多处理机操作系统
multiple processor system 多处理机系(统)
multiple-producing factory 综合性工厂
multiple production 多次操作生产;重复发生
multiple program(ming) 多重程序;多道程序(设计);多程序设计;并行编程
multiple project analysis 多目标分析
multiple projected profile 并列断面片组
multiple projection welding 多点凸焊;多层凸焊
multiple proportion 倍比
multiple pulse coding 多脉冲编码
multiple pulse detector 多脉冲探测器
multiple pulse linear predictive coding 多脉冲线性预测编码
multiple pulse problem 多脉冲振荡问题
multiple pulse spectrometer 多脉冲谱仪
multiple pumping log 多孔抽水记录
multiple pumping test 多孔抽水试验
multiple punch 多冲头冲床;多种穿孔机;多滑块压力机;复式冲模
multiple punching 多孔;多(次)穿孔;复穿孔
multiple punching machine 多头冲床
multiple punch press 多冲杆压机
multiple-purpose carrier 多用途车辆底盘
multiple-purpose cold store 多用途冷库
multiple-purpose dam 多目标坝
multiple purpose development 综合开发;多目标开发
multiple-purpose dock 多用途码头
multiple-purpose land development 土地综合开发
multiple-purpose lathe 多用机床
multiple-purpose machine 多用途机械;多用机械
multiple-purpose platform 多用途平台
multiple-purpose port 多用途港口
multiple-purpose project 综合开发计划;综合开发工程;多效用工程;多目标计划;多目标工程;综合利用工程
multiple-purpose reservoir 综合利用水库;多效用水库;多目标水库;通用蓄水池
multiple-purpose river basin development 多用途河流流域开发
multiple-purpose room 多用途房间
multiple-purpose stackyard 多用途堆场
multiple-purpose structure 多用途结构;多用途建筑物
multiple-purpose submersible 多用途潜水器
multiple-purpose support vessel 多用途支援船
multiple-purpose survey 多目标测量
multiple-purpose terminal 通用装卸货物码头
multiple-purpose tester 万能试验机;万用测试器;伏欧毫安表
multiple-purpose use 多目标利用
multiple-purpose water utilization 多目标水利资源利用
multiple quasar 多重类星体
multiple queue 多路排队
multiple radio source 多重射电源
multiple ram forging 多向模锻
multiple ram forging method 多压头锻造法
multiple ram press 多头压床
multiple random variables 多元随机变量;多维随机变量
multiple range 多量程的;复叠标
multiple rate 多丁率
multiple rate of exchange system 复汇率制
multiple rating-curve 复式关系曲线;复式水位-流量关系曲线;复式率定曲线
multiple reading 复音
multiple reclosing breaker 多次重合闸断路器
multiple record card 多重记录卡片
multiple recorder 多项记录仪
multiple rectifier 复合电路整流器
multiple redefinition of data 数据多次重新定义
multiple redshift 多重红移
multiple reflection 多次反射;多次反复
multiple reflection echo 多次反射回波
multiple reflections section 多次波剖面
multiple regeneration 多次再生
multiple register 复合记录器
multiple registering 复式记录

multiple registration 复式记录
multiple regressing 多重回归
multiple regression 多重回归(分析);多元回归;多次回归;复回归
multiple regression analysis 多重回归分析;多次回归分析;复(合)回归分析
multiple regression equation 多重回归方程
multiple regression model 多重回归模型
multiple regression technique 多重回归技术;多重回归法
multiple request 多重(中断)请求
multiple resistance welding 多点电阻焊接;复式电阻焊
multiple resonance 多重共振;复谐振
multiple response 多重响应
multiple restart 多次再起动
multiple return path 多重返回路径
multiple rhombic antenna 菱形天线网
multiple ribbed bridge 多肋式桥梁
multiple ribbed pillar 多肋柱
multiple ribless shell 无肋多层薄壳;复式无肋壳
multiple rib pillar 多肋墩;多肋柱
multiple ring valve 多环阀
multiple riveting 多行铆接
multiple riveting die 多点铆接模具
multiple rivet lap 多行铆钉搭接
multiple rod extensometer 多杆伸缩仪
multiple roll 多辊;多滚筒
multiple roll crusher 多辊破碎机;多滚筒式轧碎机;多滚筒(式)碎石机
multiple roller bearing for bridge 桥梁辊轴支座
multiple roller bit 多牙轮钻头
multiple roller cone machine 多牙轮(式)钻井机
multiple roll mill 多辊磨机
multiple room building 多室房屋;多房间建筑
multiple room dwelling unit 多室套房单元;多房间居住单元
multiple room heating 多房间采暖;多室采暖
multiple roots 多重根;重根;复根
multiple rotary screen 多层旋转筛
multiple rotate instruction 多次旋转指令
multiple rotor 多转轴
multiple rotor mixer 多转轴拌和机
multiple route 多路径
multiple routing 多路由选择;多路径选择
multiple row 多行的
multiple row blasting 多行(炮孔)爆破;多排(炮眼)爆破
multiple row plot 多行区
multiple row rivet 多行铆钉
multiple-row seeder 多行播种机
multiple row shot 多排(炮眼)爆破
multiple rows of cantilever bearing pile 多排悬壁承重桩
multiple runner 多转轮;复式转子;复式转轮;复式叶轮;复式导向滑轮
multiple safety valve 多重安全阀
multiple sample 多重抽样
multiple sample control chart 多样控制图
multiple sample counter 多试样计数管
multiple sample plan 多次选择方案
multiple sampling 多次取样法;多(次)抽样(法)
multiple sampling inspection 多次抽样检查
multiple sampling test 多次取样检验
multiple sash shingle strip 多条木瓦条;多条木瓦挂条;多带木瓦压条;多带木瓦挂条;多框带状板
multiple sash window 多扇窗;多框(格)窗
multiple scale batcher 多等级配料计量器
multiple scaled instrument 多刻度仪器
multiple scale system 多刻度系(统)
multiple scan interferometer 多扫描干涉仪
multiple scanning 多重扫描;多带扫描;重复扫描;复扫描
multiple scattering 多重散射;多次散射
multiple-schedule tariff 复式税率
multiple scraper 多台式铲运机
multiple screen 多层筛
multiple screw 多纹螺旋
multiple screw conveyer 多头螺旋输送机
multiple-screw extruder 多螺杆挤出机
multiple screw hull 多推进器船体
multiple screw ship 多螺旋桨船
multiple sea-sampler 多筒海水采样器
multiple seated valve 多级座阀

multiple section train 多组列车
multiple segment die 拼合模
multiple seismic event 多发性地震
multiple seismometer 组合地震计;复式地震仪
multiple selection 多重选择;多次选择
multiple selector 复式选择器
multiple selector valve 多位置换向阀
multiple separate locks 分离式多级船闸
multiple separation locks 分开布置的多级船闸
multiple sequence operation 多序操作
multiple sequential access method 多顺序存取法
multiple serial camera 多镜连续摄影机
multiple series 多重级数;多路串联;复联
multiple series capacitor 并串联电容器组;串并联电容器组
multiple series connection 并串联;混联
multiple service pipe 多户用户管
multiple sexuality 多性现象
multiple shaft ship type 多轴船型
multiple shaft turbine 多轴透平
multiple shank beam (松土机的)多齿横梁;(松土机的)多耙横梁
multiple shank ripper 多耙松土机;多齿松土器;多齿松土机
multiple shearing machine 复式剪床
multiple shear rivet 多(面)剪铆钉
multiple shear rivet joint 多剪铆钉接头
multiple-sheave block 多轮滑车
multiple shed insulator 多裙形绝缘子
multiple sheet grab 多片抓斗
multiple shifting 多班工作
multiple shifts 多班制
multiple shock 多发性地震;复地震
multiple shooting 齐放
multiple shop 连锁商店;联号商店
multiple short delay blasting 成组瞬时迟发爆破
multiple shot blasting 多发起爆
multiple shot blasting unit 多发起爆装置
multiple shot contact watch 多点测斜仪钟表
multiple shot film 多点测斜仪胶片
multiple shothole 组合爆破孔;多井爆炸
multiple-shot instrument 多点测斜仪
multiple-shot survey 多点测量;多点测斜
multiple-shot tool 多点测斜仪
multiple shot underwater camera 多次拍摄水下摄影机
multiple shovel 多斗挖土机;多斗挖掘机
multiple sided dowel pin 多边销钉
multiple signal 多次信号;复式信号
multiple sill 叠岩床
multiple simultaneous blasting 多眼同时爆破;成组炮眼同时起爆
multiple single plant selection 多次单株选择
multiple sintering 多次烧结
multiple size aggregate 多(种)粒径集料;多(种)粒径骨料;多径级集料
multiple slab anode 多层板式阳极
multiple slide press 多滑块压力机
multiple slip 多滑移
multiple slope 复式边坡
multiple slot coiler 多钳口卷取机
multiple socket 复式插座
multiple sound-insulating construction 多重隔声构造
multiple sound insulation 多层隔声
multiple sound insulator 多层隔声装置
multiple sound track 多重声道;多声道
multiple source 多源;复合源;多光源
multiple source diffusion model 多源扩散模式
multiple source method 多光源法
multiple source model 多源模式
multiple sources diffusion model 复合污染源扩散模型
multiple span 多跨度的
multiple span bridge 多跨桥;多孔桥
multiple span continuous truss 多跨连续桁架
multiple span frame 连续构架;多跨构架;多跨框架
multiple-span gabled frame 人字形的多跨构架;多跨三角形框架;多跨人字形框架
multiple span girder 多跨大梁;多跨梁
multiple-span structure 多跨结构
multiple span suspension bridge 多跨吊桥
multiple spark camera 多火花摄影机
multiple spark system 多火花系统

multiple spectral bathymetry 多波段测深法
multiple spectral echo sounder 多波段回声测深仪
multiple spectral image 多波段图像
multiple spectral scanning system 多波段扫描系统
multiple spectral sonar 多波段声呐
multiple speed gearbox 多级变速器;多挡变速器
multiple speed motor 多速(度)电动机;变速电动机
multiple spindle 多轴
multiple spindle automatic bar machine 多轴自动棒料车床
multiple spindle automatic chucking machine 多轴自动夹头车床
multiple spindle automatic cycle lathe 多轴自动循环车床
multiple spindle automatic lathe 多轴自动车床
multiple spindle automatic machine 多轴自动车床
multiple spindle bar automatic lathe 多轴半自动车床
multiple spindle bar automatics 多轴棒料自动车床
multiple spindle boring machine 多轴镗床
multiple spindle broaching machine 多轴拉床
multiple spindle chucking automatics 多轴夹盘自动车床
multiple spindle drill 多轴钻机;多轴钻床
multiple spindle drill chuck 多轴钻床卡盘
multiple spindle drill head 多轴钻削头
multiple spindle drilling machine 多轴钻机;多轴钻床
multiple spindle drill unit 多轴钻削头装置
multiple spindle heads 多轴头
multiple spindle honing machine 多轴珩磨机
multiple spindle lathe 多轴车床
multiple spindle milling machine 多轴铣床
multiple spindle semi-automatic lathe 多轴半自动车床
multiple spindle vertical internal honing machine 多轴立式内圆珩磨机
multiple spindle vertical lathe 多轴立式车床
multiple spindle wood lathe 多轴木工车床
multiple spline 花键
multiple spline hub 花键轮毂
multiple spline shaft 花键轴
multiple splitting coalseam 多分裂煤层
multiple spot 多点的
multiple spot beam antenna 多点波束天线
multiple spot scanning 多光点扫描
multiple spot welder 多点焊接机;多点点焊机
multiple spot welding 多(点)点焊
multiple spot welding machine 多点点焊机
multiple spring 复式弹簧
multiple stability 多重稳定性
multiple stable-state 多稳态的
multiple stable-state storage 多稳态存储器
multiple stack 多次叠加
multiple stacks and queues 多个栈和队列
multiple stage 多级(的)
multiple stage air compressor 多级式空气压缩机
multiple stage amplifier 多级放大器
multiple-stage cementing 多级注水泥
multiple stage cementing collar 多管注水泥撞箍
multiple-stage comminution 多级粉碎;多级破碎
multiple-stage crushing 多级破碎
multiple-stage drainage 多级排水
multiple-stage dust collection 多级收尘
multiple-stage impinge(ment) separator 多级冲击除尘器
multiple stage intensifier 多级式增压器
multiple stage packer cementing collar 多级封隔注水泥接箍
multiple stage press 多级压(榨)机
multiple stage sampling 多阶段抽样(法)
multiple stage separating column 多级分馏塔
multiple stage separator 多级分离器
multiple-stage sludge digestion 多级污泥消化;多酸污泥消化
multiple-stage triaxial compression test 多级三轴压缩试验
multiple-stage triaxial test 多级三轴试验
multiple stage triaxial test 多级三轴试验
multiple stage washer 多段洗涤机
multiple stand rolling mill 多机座轧钢机
multiple start 多头的
multiple start capability 多次起动能力
multiple start screw thread 多头螺纹

multiple station dial index machine 多面度盘式回转机床
multiple stays 多拉索
multiple steam heat-supply network 多管蒸汽热网
multiple stem budding 多干芽接树
multiple stem system 多干修枝法
multiple step 多级的
multiple step bit 多台阶钻头;多阶梯钻头
multiple step format 分段式;分步式
multiple step form of income statement 分段收益表;分步式收益表
multiple step income statement 分部收益表
multiple step pattern 多阶段型
multiple step type 多级式
multiple stereomodel 复合立体模型
multiple stops 多次停车
multiple stor(e)y 多层(指楼层)
multiple stor(e)y dwellings 多层住宅楼
multiple stor(e)y furnace 多层炉
multiple stor(e)y structure 多层结构
multiple stor(e)y trestle 多层高架桥
multiple stor(e)y warehouse 多层货栈;多层仓库
multiple storied building 多层建筑(物)
multiple straight pipe injector 多直管式喷射器
multiple strand 多线的
multiple strand cable 多股钢绞线;多股钢绞索
multiple-strand caster 多流连铸机
multiple-strand casting machine 多流连铸机
multiple strand chain 多级链
multiple strand conveyer 多根输送机
multiple-strand jacking device 多级千斤顶机具
multiple strand twist block 多线扭转线材轧机机组
multiple stratification 多层现象
multiple string cementing 多管注水泥
multiple string completion 多条绳索的完成;多股绳索的完成
multiple string processing 多串处理
multiple strip processing 机组作业线上的多条带材精整
multiple structure 多桥式;多层结构
multiple-structure interchange 多桥式立(体)交(叉);多层立(体)交(叉);复式交叉
multiple subroutine level 多子程序级
multiple substitution enciphering system 多次代换编码系统
multiple subsystem 多子系统
multiple subsystem simulator 多子系统模拟程序
multiple suppression 压制多次波
multiple surface 多面的
multiple surface treatments 多层式表面处理
multiple surge tank 复式调压井;复式调浆
multiple suspension grab 多悬挂抓斗
multiple sway frame 多侧移(动)框架;复式侧移框架
multiple swing check valve 多行程止回阀
multiple-switch 复联开关
multiple-switch board 复式交换机;复接式人工交换机
multiple-switch controller 多开关控制器
multiple-switch starter 复式开关起动器;复式电门起动器
multiple system 多路系统;并联式
multiple system operator 多系统操作器
multiplet 相重项;多重态;多重(谱)线;多重结构
multiple tail 多重彗尾
multiple tank 多储罐
multiple target 多目标
multiple target detection 多目标探测
multiple target generator 多目标发生器
multiple target system 多种目标测距系统
multiple target tracking 多目标跟踪
multiple tariff 复式税率
multiple tariff meter 多种计价计算器;多费制电度表
multiple tarring 铺浇多层焦油沥青
multiple task management 多任务管理
multiple task operating system 多任务操作系统
multiple taxation 复税制
multiple tax system 复合税制
multiplet component 多重线子成分
multiple telecommunication 多路通信
multiple telegraphy 多路电报
multiple temperature 多次回火
multiple tempering 多次回火
multiple terminal 多终端

multiple terminal access 多终端存取
multiple terminal access feature 多终端存取特点
multiple terminal manager 多终端管理程序
multiple thermocline 复斜温层
multiple thread 多(头)螺纹
multiple threaded bolts 多头螺钉
multiple threaded screw 多头螺钉;多纹螺旋
multiple thread hob 多头螺纹滚刀
multiple thread mill 多头螺纹铣刀
multiple thread worm 多头蜗杆;多头螺纹螺杆
multiple throats 多流液洞
multiple throttle valve 复式节流阀
multiple tidal zone 多潮地区
multiple tide staff 排式水尺;水尺组
multiple tier 多排;多层
multiplet intensity rules 多重线强度定则
multiple tire roller 多轮压路机
multiple tool block 多刀刀架
multiple tooling 多刀切削
multiple-tool lathe 多刀车床
multiple-tool slide 复式刀架滑台
multiple torch machine 多炬气割机
multiple track 多线轨道;多(磁)道
multiple track error 多磁道错误
multiple tracking radar 多路跟踪雷达
multiple track line 多线铁路
multiple trait 多性状选择
multiple transfer automatic presses 复式连续自动工作压床
multiple transform 多重变换
multiple transition time decomposition 多次转换时间分解
multiple transmission 复合式传动
multiple tray 多层浅盘;多层沉淀池
multiple tray aerator 多盘曝气
multiple tray clarifier 多层澄清池
multiple tropopause 复对流层顶
multiple truss 复式桁架
multiplet structure 多重线结构
multiple tubeb burner 多管燃烧器
multiple tube column 多管塔
multiple tube confirmation process 多管证实法
multiple tube cooler 多筒式冷却器
multiple tubed burner 多焰燃烧器;多焰焊枪
multiple tube hydrogen furnace 多管式氢还原炉
multiple tube method 多管法
multiple tube-type press 管式污泥脱水机
multiple tubing string 多根平行油管
multiple tuned antenna 多调谐天线;双调谐天线
multiple tuner 复式调谐器
multiple turn 多匝的
multiple turn cam 复旋凸轮
multiple turn cylindrical cam 多筒形凸轮
multiple turning operation 多道车削程序
multiple twin 聚片双晶;双绞;多重双晶;复对的
multiple twin cable 扭股四芯电缆;多对电缆
multiple twin quad 双股四芯电缆;复式双四芯线组
multiple tyre(d) roller 多轮胎压路机;多轮式压路机
multiple unit 多车厢的;联列的;复合的;多单元的;并联机组
multiple unit cable 复合电缆
multiple unit control 多元控制
multiple unit controller 多元控制器
multiple unit dwelling 多单元住房
multiplet unit interurban train 多节车组市郊列车
multiple unit knitting machine 多节式针织机
multiple unit locomotive operation 多机牵引
multiple unit of measurement 倍数计量单位
multiple unit running 多机重联运行;多机牵引
multiple unit steerable antenna 复合菱形天线;可变方向的菱形天线;可控式多菱形天线
multiple unit suburban train 多列市郊列车;多组市郊列车
multiple unit traction 多机牵引
multiple unit train 多机组列车;多机牵引的列车;多动力单元列车;多车厢列车;动车列车
multiple unit tube 复合管
multiple unit type retarder 多组件式缓行器
multiple unit valve 多路阀
multiple use 多用(途);综合利用;多目标利用
multiple use and joint development 多种使用和联合开发
multiple use audits 满足多种需求的审计
multiple use barn 多种用途畜舍

multiple use bonding adhesive 多用途胶合剂；多用途结合剂；多用途黏合剂
multiple use building 多用途大楼；多用途建筑物；多用途房屋
multiple use cementing agent 多用途水泥结合剂；多用途黏结剂；多用途胶结剂
multiple use door 多用途门
multiple use forestry 多种经营林业
multiple use high rise building 多用途高楼；多用途高层房屋；多用途高层建筑
multiple use liquid coating material 多用途涂敷材料；多用途液态材料
multiple use lock 多用途栓；多用途门；多用途锁
multiple use package 复用包装
multiple use reservoir 多目标水库；多用途水库
multiple use resources 综合利用资源；多用途资源
multiple user storage equipment 多用户存储［贮］设备
multiple use stadium 多用途(的露天大)体育场；多用途的露天大运动场
multiple use stage 多用途场所；多用途站台；多用途舞台
multiple utility 多应用
multiple utility program(me) 多(重)应用程序
multiple utilization 综合利用
multiple utilization of thermal wastewater 热废水综合利用
multiple value 多重值
multiple value clause 多定额条款
multiple valued 多值的
multiple valued function 多值函数
multiple valued logic 多值逻辑
multiple valve 多级管
multiple valve unit 多阀组
multiple vane diffuser 多导向叶片式扩散器
multiple V-belt 多条三角皮带
multiple V-belt drive 多条三角皮带传动；多根三角皮带传动
multiple vibratory compactor 多头振动压实机；多头振动夯实机
multiple virtual storage 多重虚拟存储系统；多虚拟存储器；多虚存
multiple volcano 重叠火山
multiple voltage system 多电压配电制
multiple wait 多重等待
multiple wall 复合墙
multiple wash sink 多组盥洗盆；多组盥洗槽
multiple water ga(u)ge 排式水尺
multiple water sampler 复式取水样器；复式采(水)样器
multiple watershed method 多段并列流域法
multiple wave 多次波
multiple wave filter 复式滤波器
multiple wavelength 多波长
multiple wavelength method 多波长法
multiple wavelength monochromator 多波长单色仪
multiple wavelength nephelometer 多波长浊度计；多波长浊度表
multiple wave reflection 波的多次反射
multiple way cable duct 多路电缆进线通道；多路电缆进线管道；电缆槽
multiple way cock 多路旋塞；多通道开关；多通道旋塞
multiple way directional control valve 多路换向控制阀
multiple way switch 多路开关
multiple way system 多线制；多腹杆体系
multiple weaving 多车道紧密交会
multiple web 多肋式
multiple-web systems 多腹杆系统
multiple weight batcher 多组重量配料器
multiple weight shaft plumbing 多荷重竖井投点
multiple well completion 多井钻探
multiple well system 井群系统；群井系统
multiple well system pumping test 多孔抽水试验
multiple wheel grinding machine 多砂轮磨床
multiple-wheel load 多轮荷载；复轮荷载
multiple-wheel roller 多轮路碾；多轮压路机
multiple wick oiler 多芯加油器；多点润滑器
multiple winding 多筒绕丝；复绕组
multiple winding transformer 多卷变压器
multiple-window opener 开窗器
multiple-window operator 连续窗开关器

multiple-wing door 多翼门；多扇门
multiple wire multiple power submerged-arc welding 多丝埋弧焊
multiple-wire submerged-arc welding 多丝埋弧焊
multiple withe 多烟道隔墙；纵向多道墙；烟囱内多层隔墙厚度；多层砖墙厚度
multiple wound armature 简单并绕组电枢
multiple wound glass filament 多筒卷绕玻璃丝
multiple wound glass staple yarn 多筒卷绕定长玻璃纤维丝
multiplex 多路转输；多路传送；多路信道；多路编排；多路转换；多路(复用)；多道传输；多倍仪；多倍(投影)测图仪
multiplex adapter 多路转接器
multiplex aeroprojector 多倍航测投影制图仪
multiplex aero-triangulation 多倍仪空中三角测量
multiplex aggregate bit rate 多路传输加密位速度
multiplex broadcasting 多路广播
multiplex bus 多路传输总线
multiplex control 多倍仪加密控制
multiplex data terminal 多路数据终端；多路复用数据终端
multiplex demodulator 多路通信解调器
multiplex development 复杂展线
multiplexed address and data bus 多路传送地址与数据总线
multiplexed bus structure 多路转换总线结构
multiplexed bus system 多路转换的总线系统
multiplexed hologram 多重全息片
multiplexed information and computing system 多路信息计算机系统
multiplexed laser 多路复用激光器
multiplexed message processor 多路信息处理机
multiplexed operation 多路分时操作；多路操作
multiplexed sequence 复合序列
multiplexed system 复合系统
multiplexed voice communications 多路语言通信
multiplexel circuit 复用线路
multiplexer 信号倍增器；多路转接器；多路控制器；多路开关(选择器)；多路复用(转接)器；多路调制器；多路传送；多路传输；复接器；复合信号装置；倍增器
multiplexer and demultiplexer 多路转换器与多路分配器
multiplexer channel 多路转接(器)通道；多路转换通路；多路通路；多路通道；多路复用信道；多路复用通道
multiplexer digital transducer 多路数字传感器
multiplexer-filter 复用滤光器
multiplexer input 多路转换输入
multiplexer mode 多路转换器方式
multiplexer section adaptation 复用段适配
multiplexer section protection 复用段保护
multiplexer section termination 复用段终端
multiplexer set 多路复用设备
multiplexer simulator 多路转接器模拟程序
multiplexer switch 多路复用器开关
multiplexer switch circuit 多路转换器开关电路
multiplex extension 多倍仪加密
multiplex fast printer 多路快速打印机
multiplex filter 多路复用滤波器
multiplex heat treatment 反复热处理
multiplexing 多路转换；多路复用；多路转接；多路(化)；多路调制；复用；标度放大；倍增；倍加
multiplexing equipment 多路复用设备
multiplexing fiber-optic sensor 多路复用光纤传感器
multiplexing gate 多路门
multiplexing interface 多路复用接口
multiplexing of processors 处理机的多路传输
multiplexing unit 多路复用器
multiplex lap winding 复叠绕组
multiplex link 多路通道；多路复用链路
multiplex machine 多路复用装置
multiplex map compiler 多倍仪测图员
multiplex mapping 多倍制图仪制图；多倍仪测图
multiplex mapping equipment 多倍投影测图仪
multiplex method 多倍制图仪测图法
multiplex mixer 复式搅拌机
multiplex mode 多路转接方式；多路转换方式；多路工作方式；多路复用方式；多路传送方式
multiplex model 多倍仪立体模型
multiplex modulator 多路通信调制器
multiplex module 多通道组装模件

multiplex operation 多路工作；多路复用操作
multiplexor 多路转接器；多路转换器；多路复用器
multiplexor channel operation 多路工作方式
multiplexor terminal unit 多路转接器终端部件
multiplex plot 多倍仪测图
multiplex printing 多道印刷
multiplex projector 多倍仪投影仪
multiplex reduction printer 缩微仪
multiplex safety glass 多层安全玻璃
multiplex spectrometer 多路光谱仪
multiplex switch 多路传输转换开关；多次转换开关
multiplex system 多路传输系统
multiplex terminal 多路终端
multiplex thread 多头螺纹
multiplex time division system 时间分割多路通信制
multiplex transmission 多路传输
multiplex transmitter 多路(通信)发射机
multiplex triangulation 多倍仪三角测量
multiplex wave winding 复波绕组
multiplex weave 复式编织
multiplex yarn 复合长丝
multiple X-Y recorder 多对变量记录器
multiple yarn 多胶纱
multiple yield table 复式收获表
multiple zone 多油层
multiple zone completion 多层完井
multiple zone refining 多熔区区域精炼
multiplicable 可增加；可加倍的；可乘的
multiplicand 被乘数
multiplicand address operand 被乘数地址操作数
multiplicand divisor register 被乘数除数寄存器
multiplicand length operand 被乘数长度操作数
multiplicand register 被乘数寄存器
multiplicand string 被乘数串
multiplicate flower 重瓣花
multiplication 相乘；增殖；乘法；倍增；倍加
multiplication by constant 乘以常数
multiplication by variable 乘以变数
multiplication circuit 乘法电路
multiplication comparator 乘法比较器
multiplication constant 增殖系数；乘法常数；倍增系数；倍增常数
multiplication cross 交叉相乘；叉乘
multiplication factor 增殖因素；乘数；乘法因数；倍增因数；倍增系数
multiplication microscope 高倍显微镜
multiplication nursery 增殖范围
multiplication of complex numbers 复数乘积
multiplication of matrices 矩阵的乘法
multiplication of polynomial 多项式的乘法
multiplication of series 级数相乘；级数乘法
multiplication of the brake gear ratio 制动倍率
multiplication operand 乘法操作数
multiplication operator 乘法算子
multiplication potentiality 增殖潜力
multiplication process 分支过程
multiplication rate 增殖速率
multiplication routine 乘法程序
multiplication seasonal factors 倍增季节因素
multiplication shift 乘法移位
multiplication table 九九表；乘法表
multiplication theorem 乘法定理
multiplication time 乘法时间
multiplicative array 乘积阵
multiplicative axiom 乘法公理
multiplicative breakdown 倍增击穿
multiplicative channel 相乘信道
multiplicative cipher 乘法密码
multiplicative congruential method 乘同余法
multiplicative error term 倍增误差项
multiplicative function 可乘函数；乘性函数
multiplicative halo 累乘晕
multiplicative inverse 乘法逆元
multiplicative mixing 相乘混频
multiplicative Mobius inversion 乘积形式莫比乌斯反演
multiplicative noise 相乘噪声
multiplicative process 倍增过程
multiplicative ratio 累乘比值
multiplicative reconstruction method 图像倍增恢复法
multiplicative stage 增殖阶段
multiplicator 增益器；求积器；乘数；乘法器；倍增器

multiplicatrix 倍积
multiplicatrix curve 倍积曲线
multiplicity 相重性;多重性;多重数;多样性;多根数;重复度;复合
multiplicity of image 映像的多重性;景象的多重性;图像的多重性
multiplicity of root 根的阶
multiplied yarn 多股线
multiplier 因子;增益器;扩程器;乘子;乘效;乘数;乘式;乘积放大倍数;乘法器;放大器;倍增器;倍增管
multiplier amplifier 倍增放大器
multiplier bit 乘数位
multiplier detector 倍增管探测器
multiplier digit 因子数字
multiplier effect 收入增殖作用;乘数效应
multiplier factor 乘数因式
multiplier field 乘数字段
multiplier function 乘子函数
multiplier kinescope 倍增式显像管;倍增器显像管
multiplier orthicon 倍增式超正析像管;倍增器正析像管
multiplier penalty function 乘子罚函数
multiplier phototube 光电倍增管
multiplier-quotient register 乘数商数寄存器;乘商寄存器
multiplier register 乘数寄存器;乘数法则
multiplier resistor 扩程器
multiplier rule 乘数寄存器
multiplier sensitivity 倍增器灵敏度
multiplier speed 倍速器
multiplier system 倍增系统
multiplier traveling-wave photodiode 倍增行波光电二极管
multiplier tube 倍增管
multiplier word 乘数字
multiplunger pump 多柱塞泵
multiply 多层作法
multiply bellows 多层波纹管
multiply cloth 多层织物
multiply column by column 逐列相乘
multiply connected body 多连通体;复连通体
multiply connected domain 复连通域
multiply connected region 多连通域;多连通区域
multiply connected track circuit 并联式轨道电路【铁】
multiply construction 多层胶合板构造;多层构造;多层薄板构造;叠层构造;多层胶合板结构
multiply defined symbol 多重定义符号
multiply deformed belts of low and medium metamorphic grade 中浅变质的多期变形带
multiply-divide instruction 乘除指令
multiply-divide package 乘除软件包;乘除程序包
multiply field 乘法字段
multiply gilding 多层镀法
multiply glass 多层玻璃
multiply glazing 多层玻璃(窗);多层上光
multiplying arrangement 放大装置;放大系统;放大设备
multiplying constant 乘常常数;常倍数;倍增常数
multiplying factor 放大率;倍数;倍率;倍加系数
multiplying ga(u)ge 放大规
multiplying gear 增速传动装置
multiplying glass 放大镜
multiplying lever 放大杠杆
multiplying lever balance 倍数杠杆天平
multiplying manometer 倍示压力计
multiplying operator 乘法运算子;乘法算子
multiplying power 放大倍率;乘率
multiplying power of pulley block 滑轮组倍率
multiplying signal 乘法信号;乘法脉冲信号
multiplying system 放大系统
multiply laminate 多层板
multiply operation 乘法运算;乘法操作
multiply paper bag 多层纸袋
multiply periodic 多周期的
multiply periodic(al) function 复周期函数
multiply periodic(al) motion 多重周期运动
multiply plot 繁殖区
multiply plywood 多层胶合板;多层夹板
multiply reentrant winding 复式闭路绕组
multiply statement 乘法语句
multiply the channels of circulation 增加流通渠道
multiply tire[tyre] 多层外胎
multiply wood 多层(胶合)板

multipoint 多位置(的);多点(式)
multipoint breaker 多点断路器
multipoint channel 多点通道
multipoint circuit 多点(通信)线路;分支回路
multipoint CO_2 meter 多测点二氧化碳测定仪
multipoint communication 多点通信
multipoint configuration 多点结构
multipoint connection 多点连接;多点连法;多点接法
multipoint earthing 多点接地
multipoint earthing tester 多点接地检查仪
multipointed grab 多爪抓岩机
multipoint fire and single-channel receive 多点放炮单道接收
multipoint indicator 多点指示器
multipoint interface 多点接口
multipoint interpercolation 多点插入法
multipoint line 多点线路;复点线
multipoint link 多点链路
multipoint long chart temperature recorder 多点长图式温度记录仪
multipoint method(of stream ga(u)ging) 多点(测流)法
multipoint mooring 多点系泊(法)
multipoint mooring system 多点系泊设施
multipoint network 多点网络
multipoint priming 多点起爆
multipoint prober 多针探器
multipoint recorder 多点式记录计;多点记录仪;多点记录器
multipoint rotational visco(si)meter 多点旋转黏度计
multipoint-scanner 多点扫描器
multipoint screwdriver 多头螺钉起子
multipoint segment 多点段
multipoints mooring system 多点系泊系统;多点系泊设施
multipoint spot welder 多点头焊机;多点点焊机
multipoint strip chart temperature record 多点条幅式温度记录仪
multipoint supporting method 多点支持点法
multipoint switch 多接点开关
multipoint system 多点系统;多点方式
multipoint temperature indicator 多点温度指示仪
multipoint tool 多刃刀具
multipoint ultrasonic visco(si)meter 多测点超声黏度计
multipoint water heater 多龙头热水器
multipoint welder 多点电焊机;多点头焊机
multipolar 多极的;多极
multipolar coordinates 多极坐标
multipolar dynamo 多极电机
multipolar generator 多极发电机
multipolarization photography 多偏振摄影
multipolar machine 多极电机
multipolar synchro 多极自整角机;多极同步机
multipolar tube 多极管
multipole 多极的
multipole configuration 多极位形
multipole emission 多极发射
multipole field 多极场
multipole magnet 多极磁铁
multipole moments 多极矩
multipole radiation 多极辐射
multipole switch 多极开关;多端开关;多刀开关
multipole transition 多极跃迁
multipollutant 多种污染物
multiporous acrylic fibre 多细孔性丙烯腈系纤维
multiporous polyvinylidene fluoride membrane 聚偏氟乙烯多孔膜
multiporphyritic texture 多斑结构
multiport 多端口
multiport arbitrator 多端口仲裁器
multiport arrangement 多孔排列
multiport burner 多火孔燃烧器;多孔烧嘴
multiport coupler 多端口耦合器
multiport drainage system 多孔排水系统
multiported valve 多通阀;多口阀;多端口阀
multiport intake 多支进水口;多孔进水口
multiport memory 多(进出)口存储器;多端(口)存储器
multiport memory interface 多端口存储器接口
multiport memory system 多端口存储器系统
multiport modem 多端口调制解调器

multiport storage interface 多端口存储器接口
multiport system 多通路方式
multiport valve 多进口阀
multiposition 多位置(的)
multiposition action 多位作用
multiposition connector 多位连接器
multiposition cylinder 多位汽缸
multiposition lathe 多位车床
multiposition mechanical pump 多位机械泵
multiposition regulator 多位式调节器
multiposition relay 多位继电器
multiposition switch 多触点开关
multiposition valve 多位阀
multiprecision 多倍精度
multiprecision arithmetic 多(倍)精度运算
multipressure stage 多压级
multipressure vessel 多重压力容器
multiprinter 多用印字机
multipriority 多优先数
multipriority level 多优先级
multipriority queue 多优先权队列;多优先级排队;多级优先排队
multiprobe 多探针;多探头的;多功能探针
multiprobe surface chemical analyzer 多探头表面化学分析仪
multiprocessed image 多方法处理图像
multiprocessing 多重处理;多路处理;多机处理;多工位加工工艺;多道处理
multiprocessing function 多重处理功能
multiprocessing machine 多用伐木联合机
multiprocessing mass data 多处理大量数据
multiprocessing organization 多处理机组织
multiprocessing recovery 多(道)处理恢复
multiprocessing system 多重处理系统;多处理机系(统);多处理机
multiprocessor 多重处理机;多重处理程序计算机;多(路)处理机;多计算机;多处理器
multiprocessor communication link 多处理机通信连接
multiprocessor interleaving 多处理机交错存取;多处理机交错;多处理机交叉
multiprocessor operating system 多处理机操作系统
multiprocessor running 多道处理运行
multiprocessor system 多(重)处理机系统;多处理机系(统)
multiprocessor system interface 多处理机系统接口
multiprocess production 多程序生产
multiproduct inventory model 多产品存储[贮]模型
multiprogram(me) 多(道)程序
multiprogram(me) computer 多程序计算机
multiprogrammed computation 多道程序计算
multiprogrammed computer 多道程序(控制)计算机
multiprogrammed information retrieval 多道程序控制情报检索
multiprogrammed system 多道程序系统
multiprogrammed time-sharing system 多道程序分时系统
multiprogramming 多重运程;多(道)程序(设计)
multiprogramming computer 多道程序计算机
multiprogramming efficiency 多道程序设计效率
multiprogramming executive 多重程序执行程序;多道程序执行(部件)
multiprogramming executive control 多道程序执行控制
multiprogramming executive operating system 多道程序执行的操作系统
multiprogramming memory protect 多道程序存储保护
multiprogramming priority 多道程序(设计)优先权
multiprogramming real-time executive system 多道程序实时执行系统
multiprogramming sequencing 多道程序的定序
multiprogramming system 多道程序(设计)系统
multiprogramming with a fixed number of tasks 固定任务数多道程序设计系统
multipropellant 多元推进剂
multipulse 多脉冲
multipulse linear proditive coding 多脉冲线性预测编码
multipulse mode chain 多脉冲型链
multipulse transmission 多脉冲发射

multipulse welding 多脉冲焊接
multipunch press 多冲头压力机
multipurpose 通用的;多(种)用途的;多效剂;多目的;多官能的;多功能的
multipurpose adhesive 多用黏结剂
multipurpose analysis 综合分析
multipurpose architecture 多功能建筑
multipurpose automatic data analysis machine 多功能自动数据分析机
multipurpose automatic wrench 多用自动扳手
multipurpose base 综合基地
multipurpose bench vice 多用台虎钳
multipurpose berth 综合性泊位;多用途泊位
multipurpose block 多用途一组建筑物;多用途街坊;多用途地区
multipurpose breed 多用品种
multipurpose bucket 通用挖斗;多用途挖斗;多用途铲斗;多用吊斗
multipurpose building 多用途建筑
multipurpose building block 多用途建筑街坊
multipurpose bulk berth 多用途散散泊位
multipurpose bulk carrier 多用途散货船
multipurpose bulk terminal 多用途散码头
multipurpose cadastre 多用途地籍图
multipurpose cargo carrier 通用货轮
multipurpose cargo ship 多用途货轮
multipurpose carrier 多用途运输车;多用途船(舶)
multipurpose center 多用途中心;多功能中心
multipurpose chemical processing plant 多种生产的化工厂
multipurpose clearing solution 多功能清洗液
multipurpose coating plant 生产多功能的(路面)面层材料工厂;通用铺料厂
multipurpose cold storage 综合性冷库
multipurpose compound 多用混合料
multipurpose computer 通用计算机
multipurpose computer aided design system 多用途计算机辅助设计系统
multipurpose conveyer 多用途输送器
multipurpose core shooter 多用射芯机;多功能射芯机
multipurpose cutter 多用刀
multipurpose dam 综合利用坝;多目标坝
multipurpose development 综合开发;多目标开发
multipurpose disk mill 多用盘式粉碎机
multipurpose door 多用途门
multipurpose double variation refractometer 多用双变折射仪
multipurpose dredge(r) 多功能挖泥船
multipurpose dry chemical fire extinguisher 多用干式化学药品灭火器
multipurpose dryer 多用途干燥器;多用途干燥机;通用干燥剂
multipurpose dustmeasuring instrument 多功能测尘仪
multipurpose electrician screwdriver 多用电工螺钉起子
multipurpose electro acupunctur apparatus 多用途电针仪
multipurpose equipment 多种用途设备
multipurpose evaluating method 综合评价(方)法
multipurpose excavator 多用挖掘机
multipurpose extinguisher 多功能灭火器
multipurpose farm vehicle 多用途农用车
multipurpose feed preparation machine 多用饲料调制机
multipurpose fiber test set 多用途光纤测试仪
multipurpose fishing vessel 多用途渔船
multipurpose floating pontoon 多用途浮筒
multipurpose freighter 多用运输机;多用途货船
multipurpose furnace 多用炉黑;多能炉黑
multipurpose grease 万能润滑脂
multipurpose grinding machine 多用磨床
multipurpose hall 多用途办公大楼;多用途餐厅;多用途大(门)厅;多用途会堂;多功能大厅
multipurpose hammer 多用锤
multipurpose hand saw 多用手锯
multipurpose hardness tester 多用硬度检测计
multipurpose hydraulic press 多用液压机
multipurpose inhibitor 多功能抑制剂;通用防锈剂
multipurpose instrument 全能仪器;万用仪表;多用仪器;通用工具
multipurpose instrument tool 万能工具

multipurpose inter-row cultivator 通用行间中耕机
multipurpose investigation 综合考察
multipurpose knife 多用刀
multipurpose ladder 多用梯
multipurpose lamp 多功能灯
multipurpose level 多用水平尺
multipurpose lo-lo ship 多用途吊装船
multipurpose lubricant 万能润滑油;多用润滑油
multipurpose lubricating grease 多用途牛油;多用途润滑油
multipurpose machine 多功能工具机;多功能车床
multipurpose magnetic analytical meter 多用磁性分析仪
multipurpose magnetic force separator 多用磁力分离仪
multipurpose meter 多用途测试仪表
multipurpose micrometer ga(u)ge 多用千分尺
multipurpose mini machine 多功能小型工具机
multipurpose oceangoing salvage ship 多功能救捞船
multipurpose oil 多用途油类;多用途油料
multipurpose oil skimming system 多效撇油沫系统
multipurpose optimization system 多目标优化系统
multipurpose optimization theory 多目标优化理论
multipurpose oscilloscope 多用示波器
multipurpose pile frame 多功能桩架
multipurpose pipeline pig 多用途清管器
multipurpose plan 多用计划
multipurpose plankton sampler 多用途生物采样器
multipurpose plant 多用途设备
multipurpose plastic adhesive 多用塑料胶黏剂
multipurpose platform 混合式车站
multipurpose pliers 多用钳
multipurpose primer 多用途底漆
multipurpose project 综合利用项目;综合开发工程;多目标计划;多目标工程
multipurpose projectile 多用途弹
multipurpose punching and shearing machine 多功能冲剪机
multipurpose quay 综合性码头
multipurpose quick change collet chuck 多功能快换夹头
multipurpose radar 多用雷达
multipurpose reactor 多用途电抗器;多用反应堆
multipurpose refractory 多用途耐火材料
multipurpose refrigerator 综合性冷库
multipurpose register 多功能寄存器
multipurpose research vessel 综合调研船
multipurpose research vessel ship 综合调研船
multipurpose reservoir 多用途水库;多目标水库;综合利用水库
multipurpose retractor 多用牵开器
multipurpose rice milling machine 多用碾米机
multipurpose robot 多用途机械手;多用途机器人
multipurpose ro-ro ship 多用途滚装船
multipurpose sail 多用途帆
multipurpose scanning microscope 多用途扫描显微镜
multipurpose scheme 多用途方案;多用途设计;多用途草图;多用途计划;多目标计划
multipurpose screwdriver 多用螺钉起子
multipurpose screwdriver set 多用套装螺钉起子
multipurpose scriber 多功能划线器
multipurpose shearing machine 万能剪切机
multipurpose shears 多用剪
multipurpose ship 多用船(舶)
multipurpose signal generator 多用信号发生器
multipurpose simulator 多用途仿真器
multipurpose space 多功能空间
multipurpose spraying system 多种用途喷灌系统
multipurpose stage 多用途场所;多用途站台;多用途舞台
multipurpose steel door 多用途钢门
multipurpose stripper 多用剥皮器
multipurpose stripping pliers 多用剥线钳
multipurpose structure 多功能结构
multipurpose submersible 多用途潜水器
multipurpose survey 多目标测量
multipurpose system 多用途系统
multipurpose tap 多用丝锥
multipurpose terminal 综合性码头;多用途码头
multipurpose test equipment 多用试验设备
multipurpose thermal imager 多用热像仪

multipurpose tile 多用途瓦;多用途瓷砖
multipurpose timber-harvesting machine 多用伐木联合机
multipurpose tractor 万能拖拉机
multipurpose trademark machine 多用商标机
multipurpose trailer chassis 多用拖车底盘
multipurpose triaxial test apparatus 多能三轴仪
multipurpose trip 多目标旅程
multipurpose tube 多用途电子管
multipurpose type pile frame 多能桩架
multipurpose use 多用途;综合利用
multipurpose use of "the three wastes" 三废综合利用
multipurpose use of resources 资源综合利用
multipurpose use of three types of wastes 三废综合利用
multipurpose use of wastes 废物综合利用
multipurpose use of water 水的综合利用
multipurpose use profit-sharing 综合利用利润提成
multipurpose utilization of thermal water 热水综合利用
multipurpose utilization of water resources 水资源综合利用
multipurpose vehicle 多用途车辆
multipurpose vessel 多用途船(舶);多用船
multipurpose vice 多用虎钳
multipurpose vise grip pliers 多用大力钳
multipurpose wastewater reclamation 废水综合利用
multipurpose watch 多用途手表
multipurpose water control project 多目标水利枢纽;综合利用水利工程
multipurpose water impoundment 多用蓄水池
multipurpose water utilization 水(力)资源综合利用;水利资源综合利用;多目标水资源利用
multipurpose wheel loader 多用途轮式装载机
multipurpose wood working machine tool 多用木工机床
multiput and multioutput light cable 多进多出光缆
multip-well pumping unit 多井共用抽油机
multipylon cable stayed bridge 多塔斜拉桥
multiquantum well 多量子阱
multiqueue 多路排队
multiqueuing system 多队列系统
multiradix computer 多基数计算机
multiradix index 多基数指数
multiram forging 多向锻
multiram hydraulic sheet pile driver 多缸液压打板桩机
multirange 多量程的;多波段的
multirange channel 多量程测量电路
multirange indicating instrument 多量程指示仪表
multirange instrument 多量程仪器;多量程(测量)仪表
multirange meter 多量程仪表;多量程仪
multirange receiver 多波段接收机
multirange test meter 多量程万用表;程测计
multirange turbidity-suspended solids monitor 多量程悬浮固体浊度监测器;多量程浊度悬浮颗粒监测仪
multirate digital filter 多级数字滤波器
multiread feeding 多读馈送
multiread head 多读出头
multireasonable check 多次合理校验
multireduction gear unit 多级减速机
multiredundant structure 多次超静定结构
multireel file 多卷文件
multireflected sound 混响声;多次回声
multireflection 多次反射
multireflection lighting fixture 多层反射灯具
multireflection star coupler 多反射星形耦合器
multireflection technique 多次反射技术
multireflection theory 多次反射理论
multireflector 多反射器
multireflector optic(al) resonator 多反射镜光学共振器
multireflex klystron 多反射速调管
multirefracting crystal 多折射晶体
multiregime speed-concentration model 多段型车速-密度模型
multiregime speed-density model 多态速度-密度模型
multiregion 多范围;多种区域;多区带的

multiregion gun 多区电子枪
multiregion operation 多区域运行;多区操作
multiregion reactor 多区反应堆
multiregion reflector 多区反射层
multireinforcement 多层加强
multiresonant circuit 多谐电路
multirevolved view 多次旋转视图
multiribbed floor 多肋楼板;多肋地板
multiribbed grinding wheel 多齿磨轮
multiribbed plate 多肋板;密肋板
multiribbed structure 多肋结构
multiring 多环
multiring file 多环形文件
multiring structure 多环构造
multirole 多重作用的;多用途(的)
multirole light transport aircraft 多用途轻型运输机
multiroll bar-straightening machine 多辊式棒材矫直机
multiroll bearing 滚针轴承
multiroll beater-refiner 多刀辊打浆精研机
multi-roll bridle 多辊张紧装置
multi-roll crusher 多辊破碎机
multiroller centrifugal process 多辊离心法
multiroller centrifuge 多辊离心机
multiroller vertical machine 多辊立式延压机
multiroll flattener 多辊矫直机
multiroll mill 多辊轧机;多辊轧(钢)机;多辊磨机
multiroll press 多辊压滤机
multiroll sizer 多辊格栅(筛分用)
multiroll sludge dewatering press 多辊污泥脱水压滤机
multiroll straightener 多辊矫直机
multiroll straightening machine 多辊式矫直机
multiroll toilet tissue dispenser 多轴手纸配出器
multiroll withdrawal machine 多辊拉坯机
multirope friction hoist 多绳摩擦式提升机
multirope friction winder 多绳摩擦卷筒
multirope skid gear 多钢绳拖拉机
multirope winder 多绳提升机
multirope winding 多绳卷扬;多绳卷绕
multirow 多排的;多列的;多行的
multirow bearing 多列轴承
multirow core 多排孔
multirow-crop rotary cultivator 多行作物旋转中耕机
multirow radial engine 多排星型发动机
multirow radical thrust ball bearing 多列向心推力球轴承
multirow roller bearing 多列滚柱轴承
multi-rubber-tire roller 多胶轮压路机
multirun 多道
multirunning 多道运行;多(道)程序设计
multirunway airport 多跑道机场
multirun weld 多道焊缝
multisample liquid scintillation spectrometer 多试样液体闪烁谱仪
multisample survey 多样品测量
multisampling 多次抽样(法);多抽样(法)
multiscale 多刻度;各种比例尺的;通用换算
multiscale mapping 多种比例尺测图
multiscale ohmmeter 多刻度欧姆表
multiscaler 万能定标器;通用换算线路;通用定标器
multiscale rule 多刻度计算尺
multiscan ultrasonic diagnostic equipment 超声波断层诊断仪
multiscope 简易绘图仪
multiscreen 多层筛
multiscreen graphic 多屏图形
multiscrew extrusion machine 多螺杆挤出机
multiseasonal storage reservoir 多季调节水库
multiseater 多座(飞)机
multisecond 多秒钟的
multisection 多段
multisectional floating dock 多段式浮船坞
multisection filter 多节滤波器
multisection pile 多节桩
multisection power feed 多段馈电
multisection vehicle 多节车辆
multisection vertical door 分段提升门
multisection vertical lift door 多段吊门
multisectoral operation 跨行业经营
multisegment 多节的;多段的
multisegmental baffle 多弓形折流板
multisegment magnetron 多腔磁控管

multiselector 复式选择器
multisensibility 多敏感性
multisequencing 多序列;多机工作
multisequential system 多时序系统
multiseriate 多列的
multiseriate ray 多粒射线
multiservice oil 通用润滑剂;通用机油
multiset 多重集
multishaft arrangement 多轴钻机
multishaft mixer 多轴混合机
multishaft(rotor)gas turbine 多轴燃气轮机
multishare system 多共用系统
multisheave 复滑车组
multisheave block 多饼滑车;复滑车;多轮滑车
multi-sheet drawing 多页图
multishell condenser 组合式冷凝器;多壳式冷凝器
multishell insulator 多裙边绝缘子
multishift 多班制的
multishift operation 多班制作业
multishift work 多班工作(制)
multiship maneuvering simulator 多船操纵模拟器
multishock 激波系
multishoed vibratory compactor 多蹄式振动压实机
multishot exploder 多点爆炸剂;多点雷管
multi shot firing 成组爆破
multishot measurement 多点测斜
multi-site adsorption 多点吸附
multisize 多种大小;多种尺寸
multi-skip dumper 多箕斗卸器;多箕斗卸载机
multislectrode 多丝
multisleeve dust separator 多袋式除尘器
multislide forming machine 多滑块成型机
multislot 多槽
multislot magnetron 多槽磁控管
multislot winding 多槽绕组
multisocket 多孔插口
multisolute system 多溶质系统
multisource distribution box 多电源配电盘
multisource of ore-forming material 成矿物质多来源
multisource water problem 多水源给水问题
multispan 多跨的
multispan beam 连续梁;多跨梁;多跨连续梁
multispan bridge 多跨桥梁
multispan building 多跨房屋建筑
multispan cable-stayed bridge 多跨桥斜拉桥
multispan cable supported bridge 多跨缆索吊桥
multispan elastic rotor system 多跨弹性轴系
multispan girder 多跨栅格
multispan girder bridge 多跨梁桥
multispan portal 多跨刚架
multispan rigid frame 多跨钢框架
multispan skyline 多跨式架空索道
multispan structure 多跨结构
multispan trussed roof structure 多跨屋架结构
multispar 多梁
multispecies dynamic(al)model 多种类数量变动模式
multispecies habitat protection 多物种生境保护
multispecies mature ecosystem 多物种成熟生态系统
multispecimen 多试件
multispecimen testing machine 多试样试验机
multispectral 多谱的
multispectral approach 多谱段法
multispectral camera 多谱段摄影机;多光谱照相机;多光谱摄影机
multispectral camera film viewer 多波段相机胶片观察器
multispectral camera system 多谱段摄影系统
multispectral classification 多光谱分类法
multispectral composite imagery 多光谱合成图像
multispectral data 多谱段资料
multispectral data system 多谱段数据系统
multispectral image 多谱段影像
multispectral imagery 多谱段图像;多谱影像;多频(率)谱影像
multispectral line scanner 光谱线扫描仪;多光谱行扫描仪
multispectral optoelectronic imaging spectrometer 多谱光电图像光谱仪
multispectral photography 多光谱线照相;多谱段摄影;多光谱摄影

multispectral photometer 多谱段光度计
multispectral point scanner 光谱点扫描仪
multispectral projector 多谱段投影仪
multispectral remote sensing 多谱段遥感;多波段遥感
multispectral remote sensing technique 多光谱遥测技术
multispectral remote sensing technology 多光谱遥测技术
multispectral satellite photo(graph) 多光谱卫星像片
multispectral scan 多光谱扫描
multispectral scanner 多谱段扫描仪;多频谱扫描器;多光谱扫描仪;多光谱扫描器
multispectral scanning camera 多谱段扫描摄影机
multispectral scanning system 多谱段扫描系统
multispectral sensor 多谱段传感器
multispectral survey 多光谱测量
multispectral viewer 多光谱观察器
multispectrum scanner 多光谱扫描仪;多光谱扫描器
multispectrum signal 多谱信号
multispeed 多级变速的;多种速度的;多速的
multispeed clock feature 多速时钟特点;多速率时钟特性
multispeed control action 多速控制作用
multispeed drive 多级变速传动
multispeed drum switch 多速鼓轮开关
multispeed floating action 多速无静差作用
multispeed motor 多速(度)电动机;变速电动机
multispeed reduced gear 多级变速器
multispeed rheometer 变速流变仪;多速流变仪
multispeed supercharger 多级转速增压器
multispeed transmission 多级变速
multisphere 多弧
multisphere tank 多弧球形储[贮]罐
multispheric(al)vessel 多级球形容器
multispigot classifier 多室分级机;多塞栓式分级机
multispindle 多轴的
multispindle automatic lathe 多轴自动车床
multispindle automatic machine 多轴自动车床
multispindle bar 多轴杆
multispindle bar automatic machine 多轴棒料自动车床
multispindle chucking automatic machine 多轴卡盘自动机床
multispindle drilling machine 排钻;多头床式;多轴钻孔机;多轴钻床
multispindle semi automatic lathe 多轴半自动车床
multi-spline hobbing machine 花键轴辊铣床
multi-spout in-line packer 平排多嘴包装机;固定式包装机
multispout packing machine 多嘴包装机
multistability 多重稳定性
multistable 多稳态的
multistable circuit 多稳态电路
multistable configuration 多稳电路
multistable system 多稳系统
multistage 分阶段进行;多阶段;多级(的)
multistage absorption 多级吸收
multistage action turbine 多级式冲击水轮机;多级式冲击汽轮机;多级式冲击透平;多级式冲击涡轮机
multistage activated sludge treatment 多级活性污泥处理
multistage aeration 多级曝气
multistage aeration system 多级曝气系统
multistage amplification 多级放大
multistage amplifier 级联放大器;多级放大器
multistage analysis 多阶段分析
multistage anvil 增压块
multistage biochemical treatment 多级生化处理
multistage biological oxidation column 多级生物氧化塔
multistage biological process 多级生物处理法
multistage bleaching 多级漂白法
multistage blower 多级鼓风机
multistage burner 多级燃烧器
multistage caisson 多级套筒式沉井
multistage centrifugal pump 多级(式)离心泵
multistage centrifugal ventilator 多级离心式风机
multistage circuit 多阶电路
multistage cleaning 多段精选
multistage column 多级柱

multistage compression 多级压缩
multistage compressor 多级压缩机;多级压气机
multistage construction 多级施工
multistage control 多级控制
multistage convertor 多级转换器
multistage cooking process 多级蒸煮法
multistage crystallization 多级结晶
multistage crystallizer 多级结晶器
multistage Curtis turbine 多级高速级式汽轮机
multistage cylinder 多级(液压)缸
multistage deaerator 多级除气器
multistage decision process 多阶段判定过程;多阶段决策过程;多级判决过程;多级决策过程
multistaged evapo(u)rator 多效蒸发器
multistage dewaterer 多级降水机
multistage distillation 多级蒸馏
multistage driving 分级掘进
multistaged turbine 多级式涡轮机
multistage effeciency 多级效率
multistage elutriation 多级淘洗
multistage evapo(u)rate 多级蒸发
multistage evapo(u)rator 多级蒸发器
multistage explosion effect 缓爆效果
multistage extraction 多级提取;多级抽提
multistage extraction cascade 多级级联萃取装置
multistage fan 多级风扇
multistage filter 多级滤池;多级过滤器
multistage filtration 多段过滤
multistage flash distillation process 多级急骤蒸馏法
multistage flash distiller 多级闪蒸蒸馏装置
multistage flash evapo(u)ration 多级闪蒸
multistage flash evapo(u)ration method 多级闪蒸法
multistage flight locks 连续梯级船闸
multistage fluidized bed drier[dryer] 多层流化床干燥器
multistage fluidized bed reactor 多级流化床反应器
multistage fluidized calciner 多层流化煅烧炉
multistage furnace 多段炉
multistage game 多阶段对策
multistage gas turbine 多级汽轮机
multistage grinding 多段研磨;分段磨矿
multistage high-rate filter 多级高负荷生物滤池
multistage ignition 多段点火;分段点火
multistage impeller pump 多级式叶轮泵
multistage impulse generator 多级脉冲发生器
multistage impulse steam turbine 多级冲动式汽轮机
multistage impulse turbine 多级式冲击水轮机;多级式冲动汽轮机;多级式冲击透平;多级式冲击涡轮机;多级冲动式汽轮机
multistage internal circulation anaerobic reactor 多级内循环厌氧反应器
multistage interpretation method 多级解译法
multistage liquid-liquid extraction 多级液液提取
multistage metallogenesis 多阶段成矿
multistage method 多步法
multistage mixed flow pump 多级混流式水泵
multistage multianvil technique 多级多压砧技术
multistage multianvil vessel 多级多压砧容器
multistage multieffect flash method 多级多效闪蒸法
multistage multiple element ejector 多级多联喷射器
multistage network parameter 多级网络参数
multistage open-circuit grinding 多级式开环磨矿(工艺过程为研磨-分级-送出,与闭环工艺不同)
multistage optimization 分级最佳化
multistage ore dressing 多段选矿法
multistage oxide lagoon 多级氧化塘
multistage paddle agitator 多级桨式搅拌器
multistage phase 多级相
multistage power turbine 多级动力汽轮机
multistage precipitator 多级电滤波器;多级电除尘器
multistage press 多级压制;多级压力机
multistage process 多级加工
multistage production and scheduling 多阶段生产与调度
multistage pump 多级式水泵;多级抽水机;多级泵
multistage push-type centrifugal 多级推挤式离心机
multistage ram 多级油缸;多段冲头

multistage reactor 多级反应器
multistage recirculation system 多级回流系统
multistage rectification 分带纠正
multistage recuperator 多级换热器
multistage reduction 多级减速装置;多级粉碎
multistage reversing flow bioreactor 多级往复水流生物反应器
multistage rocket 多级火箭
multistage sampler 多级式取样器;多级式采样器
multistage sampling 多阶段收样;多阶段抽样(法);多级取样(法);多级抽样(法);多级采样
multistage self-priming vortex pump 多级自吸旋涡泵
multistage-separation 多级分离
multistage settling tank 多级澄清池;多级沉淀池
multistage shift 多级变换
multistage shredding 多级研碎
multistage single element ejector 多级单联喷射器
multistage stabilization pond 多级稳定塘
multistage steam turbine 多级汽轮机
multistage storage pump 多级蓄能水泵
multistage stressing 多级应力张拉;多阶段张拉;多阶段施加应力;多级应力;多级施加应力
multistage supercharger 多级增压器
multi stage switch of cutter drive 铰刀驱动多级开关
multistage system 多级管网系统;多级水库系统
multistage thinfilm scan generator 多级薄膜扫描发生器
multistage transmitter 多级发射机
multistage turbine 多级叶轮机;多级涡轮(机);多级透平;多级式涡轮机;多级汽轮机
multistage upflow filter system 多级上流过滤系统
multistage volute pump 多级螺旋泵
multistage washing 多级洗涤
multistage water concentration 多级水选
multistage water-cooled compressor 多级水冷压缩机
multistage water power development 多级水电开发(工程)
multistage water power stations 多级水电站
multistage well point system 多级井点系统
multistaging 增加级数
multistandard 多标准
multistand strip mill 多机座带钢轧机
multistar 满天星
multistar equal-altitude method 多星等高法
multistart screw thread 多头螺纹
multistart thread 多头螺纹
multistart worm 多头蜗杆;多头螺纹螺杆
multistate variable 多态变量
multistatic radar 多元静态雷达
multistation 多站;多台站
multistation communication network 多站通信网络
multistation configuration 多终端结构;复台结构
multistation die layout 连续模多工位模具布置
multistationery set 多用文具盒
multistation head 多笔位绘图头
multistation mo(u)lding machine 多工位造型机
multistation photography 多站摄影
multistation toolholder 多笔位绘图头
multistator watt-hour meter 多静子瓦时计
multistep 多步的;阶梯式;多级的
multistep action 多位作用;多级作用;多步作用
multistep concept 多层概念
multistep cone pulley 多级塔轮
multistep control 多步控制
multistep control servomechanism 多级控制的伺服机构
multistep detritus-intercepting dam 多级拦石坝
multistep formula 多步公式
multistep ignition 分级点火
multistep locks 多级船闸
multistep method 多步法
multistepped hole 多级阶梯形内孔
multistepped type quay 多级式码头
multistep production 多级生产
multistep task 多步任务
multistep valve 多级活门
multistone bit 细粒金刚石钻头
multistorage system 水库系统;水库群;多级蓄水系统
multistor(e)y 多层的(房屋);多层(楼)

multistor(e)y basement 多层地下室
multistor(e)y bent pier 多层排架桥墩
multistor(e)y brick building 多层砖房
multistor(e)y brick structure 多层砖结构
multistor(e)y building 多层建筑(物);多层房屋(建筑)
multistor(e)y car park(ing) 多层停车场;多层车库
multistor(e)y catalytic reactor 多层催化反应器
multistor(e)y cell 多层堆放格仓
multistor(e)y city 多层城市(指建筑在高低不同的地面上的城市)
multistor(e)y cold store 多层冷库
multistor(e)y curtain wall system 多层悬墙体系
multistor(e)y dryer 多层干燥器;多层干燥机
multistor(e)y dwelling 多层住宅(2~7层)
multistor(e)yed pavilion 楼阁
multistor(e)y factory 多层厂房;分层出租厂房
multistor(e)y factory building 多层工厂建筑
multistor(e)y frame 多层框架(结构)
multistor(e)y framed building 多层框架建筑
multistor(e)y frame pier 多层刚架桥墩
multistor(e)y garage 多层停车库
multistor(e)y hotel 多层旅馆
multistor(e)y parking 立式停车场
multistor(e)y portal structure 多层隧道门结构;多层桥门结构;多层门架结构
multistor(e)y sand body 纵叠砂体
multistor(e)y sands 纵叠砂
multistor(e)y settling tank 多层沉淀池;多层沉淀槽
multistor(e)y structure 多层建筑(物);多层结构(物)
multistor(e)y villa 多层别墅
multistoried 多层的(房屋)
multistoried apartment 多层公寓
multistoried bamboo pavilion 竹楼
multistoried building 高层建筑(物);高层房屋;多层房屋(建筑)
multistoried factory building 多层厂房
multistoried garage 多层车库
multistoried rigid frame 多层刚性框架
multistoried sedimentation tank 多层沉淀池;多层沉淀槽
multistoried storehouse 多层仓库
multistoried transit shed (码头的)多层前方仓库
multistrand 多股的
multistrand cold-tube rolling mill 多根管材冷拉拔机
multistrand continuous pickle line 多条连续酸洗作业线
multistrand crop and cobble flying shears 多线式端头废料飞剪
multistrand mill 多线式轧机
multistrand rolling 多线轧制
multistrand rope 多股索;多股钢丝绳
multistrand roving 多股无捻粗纱
multistrand stay 多束股拉索
multistrand steel cable 多股钢缆
multistrand steel prestressing cable 多股预应力钢缆
multistrand tendon 钢绞线束
multistrand tendon beam 多股钢束梁
multistrand wire rope 多股钢丝绳
multistream 多管的;多股流的
multistream heater 多路加热炉
multistream profiles 一机拉挤多根型材
multistream system 多流系统
multi-stress 综合因子
multistressing 多级应力
multistrip coupler 多带耦合器
multistrip laser 多带激光器
multistrip shearing machine 多条剪切机
multi-structure 复合结构
multistylus echo sounder 多笔回披探鱼仪
multisupported rudder 多支承舵
multi-symbol disc 符号盘
multisystem 多系统
multisystem environment 多系统环境
multisystem galvanometer 多制式检流计
multisystem locomotive 多制式(电力)机车
multisystem mode 多系统方式
multisystem network 多系统网络;复用系统网络
multisystem network facility 多系统网络设施;多系统网络设备

multisystem operation 多重系统操作
multitandem valve 组合阀;多路阀
multitap 多抽头的;多插头的;转接插座;多插头插座
multitask 多(重)任务
multitask function 多任务功能
multitasking 多(重)任务处理;多任务处理;多任务
multitasking built-infunction 多任务内部函数
multitask link control 多任务连接控制
multitask monitor 多任务监督程序
multitask operation 多(重)任务操作
multitask program(me) 多任务程序
multitask system 多任务系统
multitemporal 不同时间的
multitemporal analysis 多时相分析
multitemporal image 多时相图像
multitendon beam 多钢丝索(预应力)梁
multiterminal 多终端;多端的
multiterminal addressing 多终端编址
multiterminal interface 多终端接口
multiterminal network 多端网络
multiterminal spark array 多端电火花组合
multithematic 多主题的
multithematic picture 多主题图像
multithematic representation 多主题图像
multithread 多线索;多头处理;多头;多流
multithread application program(me) 多执行路线应用程序;多流应用程序
multithreading 多执行路线(操作);多流方式
multi-thread processing 多线处理
multithread test 多流测试
multithread worm 多头蜗杆
multithroated 多路口的
multithrow switch 多掷开关
multitier conveyer drier 多层输送带式干燥机
multitier stack 多层书架
multitime flash lamp 多次闪光灯
multitired tractor 多轮拖拉机
multitle 外层油漆
multitone 多频(率)声
multitone circuit 多音电路
multitone etching 多种深度结合的蚀刻
multitone transmission 多频(率)信号传输
multitool 多刀工具
multitool block 多刀刀架
multitool cutting 多刀切削
multitool head 多刀刀具头
multitool holder 多刀刀架
multitool lathe 多刀车床
multitool machining 多刀切削加工
multitool semiautomatic lathe 多刀半自动车床
multitool turning head 多刀转动刀架
multitooth coupling 多齿联轴器
multitoothed cutter 多齿铣刀
multitooth tool 多齿刀具
multitrace oscilloscope 多线示波器;多迹示波器
multitrace seismograph 多扫描地震仪
multitrack area 多股道地段
multitrack crossing 多线路交叉
multitrack function 多道功能;多磁道功能
multitrack head 多道磁头
multitrack line 多线铁路
multitrack operation 多磁道操作
multitrack railway 多线铁路;复线铁路
multitrack recording system 多道记录系统
multitrack section 多线路区段
multitransformer spot welder 多变压器式点焊机
multitray aerator 多槽曝气池
multitray thickener 多层浓缩机;多层浓缩槽
multitree 多重树
multitriaxial tensile test 多轴向拉伸试验
multitube 多管的
multitube circuit 多管电路
multitube cooler 多筒冷却器;多管冷却机;多管冷凝器
multitube cyclone 多管旋流器
multitube dust cleaner 多管除尘器
multitube dust collector 多管集尘器
multitube elutriator 多管式淘析器
multitube evapo(u)rator 多管蒸发器
multitube heat exchanger 多管式换热器
multitube furnace 多管炉
multitube knock back type condenser 多管反冲式冷凝器
multitube manometer 多管式压力表;多管式测压计

multitube packing machine 多嘴包装机
multitube pressure ga(u)ge 多管压力计
multituberculate 多尖齿
multitube revolving drier 多管回转式干燥机
multitube supported suspended structure 多筒悬挂结构
multitubular 多管道的
multitubular boiler 多管锅炉;多管道锅炉
multitubular collector 多管式收集器
multitubular condenser 多管(式)冷凝器
multitubular counterflow-water-heater 多管式逆流热水器
multitubular gas condenser 多管气体冷凝器
multitubular heat exchanger 多管式换热器;多管热交换器
multitubular radiator 多管状散热器
multitubular reactor 多管反应器
multitubular slab for cable 电缆用多管板;多孔电缆槽
multitubuler boiler 焰管锅炉
multitude 组
multitudinous 大批量的
multituning antenna 多路调谐天线
multiturn 多匝的;多转的;多圈的;多螺线的
multiturn current transformer 多匝电流互感器
multiturning head 多位刀架
multiturn potentiometer 多匝电位计
multitwister 复捻机
multitype highway 多功能公路
multiunder reamed pile 多节扩底桩
multiunit 多机组;多单元的;多部件;复合单位
multiunit control valve 多单元控制阀
multiunit drill 多组排种器播种机
multiunitized cargo 多吊成组货物
multiunit machine 多部件计算机
multiunit message 多单元消息
multiunit plant 多机组电站
multiunit power transmitting tube 多单元功率发射管
multiunit reservoir system 多层热储地热系统
multiunit tank car train 多节罐车组合列车
multiunit train 多机组列车
multiunit tube 复合管
multiunit wall 厚墙;空心墙;多重墙;带空气夹层的墙
multiusage furniture 多用家具
multiuse 多功能;多种用途的;多用
multiuse architecture 多功能建筑
multiuse auditorium 多功能礼堂;多用途观众厅
multiuse bit 多用途钻头;多用钻头;多次使用钎头
multiuse building 综合楼;多用途建筑物;多用房屋;多用途大楼;多功能建筑物
multiuse cross vice 多用十字钳
multiuse drill 多功能钻机
multiuse equipment 多用仪器
multiuse grinding machine 多用磨床
multiuse hammer 多用锤
multiuse hatchet 多用斧
multiuse oscillograph 多用示波器
multiuser 多用户的;多用户
multiuser berth 公用码头;公用泊位;多用户泊位
multiuser capability 多用户功能
multiuser card 多用户卡
multiuser channel 多用户信道
multiuse real-time operation system 多用实时操作系统
multiuser engine 多用燃料发动机
multiuse reservoir 多用途水库
multiuser installation 多用户设备
multiuse room 多用途房间
multiuser operating system 多用户操作系统
multiuser pipeline 多用户管道
multiuser system 多用户系统
multiuser terminal 多用户码头
multiuse test paper 多用试纸
multivalent 多价的
multivalent cation 多价阳离子
multivalent function 多叶函数
multivalent metal 多价金属
multivalent metal compound 多价铁化合物
multivalley semiconductor 多能谷半导体
multivalued 多值的
multivalued decision 多值选择
multivalued dependence [dependency] 多值依赖;多值相关性
multivalued displacement 多值位移
multivalue dependency 多值相关
multivalued function 多值函数
multivalued logic 多值逻辑
multivalued switching theory 多值开关(理)论
multivalue logic simulation 多值逻辑模拟
multivalue method 多值法
multivalve 多管的
multivalve filler 多阀灌装机
multivalve turbine 喷嘴调节汽轮机
multivane 多翼片的
multivane electric(al) fan 多叶电扇;多叶风扇
multivane rotary compressor 多滑片回转式压缩机
multivariable 多变量;多变数的
multivariable control 多相关量的调节;多变量控制
multivariable control system 多变量控制系统
multivariable discriminant analysis 多元判别分析
multivariable feedback control system 多变量反馈控制系统
multivariable function generator 多变量函数发生器
multivariable interval estimate 多元区间估计
multivariable prediction 多变量预测法
multivariable statistic(al) analysis 多元统计分析
multivariable statistic(al) model 多变量统计模型
multivariable system 多变量系统
multivariant 多元的;多方案的
multivariant eutectic 多变复杂共晶合金
multivariate 多变量;多变的
multivariate analysis 多元分析;多变数分析;多变量分析
multivariated analysis 多变分析量
multivariate decision situation 多变决策形势
multivariate distribution 多元分布;多维分布;多(维)变量分布
multivariate equilibrium 多变平衡
multivariate factor analysis 多元因素分析
multivariate function generator 多变量函数发生器
multivariate Gaussian distribution 多元高斯分布
multivariate interpolation 多元内插;多元插值;多变数插值
multivariate linear regression 多元线性回归
multivariate linear regression model 多元线性回归模型
multivariate normal distribution 多元正态分布
multivariate population 多变数总体
multivariate probability density function 多元概率密度函数
multivariate program(me) 多元程序
multivariate random process 多变量随机过程
multivariate regression 多元回归;多变量回归
multivariate regression analysis 多元回归分析
multivariate regression equation 多元回归方程
multivariate sample 多变元样本
multivariate simulation 多元模拟
multivariate statistic(al) analysis package 多元统计分析程序包
multivat machine 多圆网造纸机
multivator 多用万能自动测试仪;复式变换器
multivector 多重矢量;多重向量
multi-vehicle accident 车辆碰撞事故
multivelocity 多速的
multi-Venturi scrubber 多元文丘里洗气槽;多级文丘里洗气器
multiversity 综合大学
multivertor 复式变换
multivessel system 多容器系
multivibrator 多谐振动器;多阶振动器
multivibrator circuit 多谐振荡器电路
multivibrator frequency modulator 多谐振荡器式调频器
multivibrator modulator 多谐振荡器式调制器
multiviewports 多视屏口
multiviscosity 稠化
multiviscosity oil 稠化油(料)
multivoltage accelerator 倍压加速器
multivoltage control 电压变化调节;多电压控制
multivoltage generator 多电压发电机
multivoltine 多窝的;多化的
multivoltmeter 多量程伏特计
multivolume 多卷的
multivolume file 多卷文件;多卷文档
multivolume multifile 多卷多文件

multivortex mechanical collector 多管式(机械)除尘器
multiwafered switch 多片开关
multiwall 有多层外皮的;多层
multiwall cement bag 有多层外皮的水泥袋;多层水泥袋
multiwall corrugated board 多层瓦楞纸板
multiwalled vessel 多层式容器
multiwall kraft 多层牛皮纸
multiwall paper bag 多层纸袋
multiwall sack 多层纸袋
multiwash 多层洗涤(砂石等)
multiwash collector 多层洗涤收尘器
multiwash scrubber 多段洗净器;多段洗涤器
multiwave 多波的
multiwave interaction 多波相互作用
multiwavelength distance meter 多波长测距仪
multiwavelength laser 多波长激光
multiwavelength linear regression 多波长线性回归
multiwave length nephelometer 多波长浊度计;多波长浊度表
multiwave-particle interaction 多波粒子相互作用
multiway 多方向的;多种方法的;多条道路的;多路的
multiway classification 多向分组;多向分类
multiway duct 多路管道
multiway intersection 复合交叉口;多路交叉(口);复式交叉
multiway junction 多路交叉(口)
multiway service pipe system 双向供水
multiway socket 多脚插座
multiway stop of intersection 交叉口多路停车
multiway switch 多路转换开关;多路开关
multiway turntable 多轨道转车盘
multiway type 多主轴式
multiway valve 多通阀
multiwell gas-lift system 多层合采井气举系统
multiwell platform 群井平台;多井平台
multiwell production system 多井采油系统
multiwheel 多轮的
multiwheeled drive 多轮传动
multiwheeled trailer 多轮挂车;多轮拖车
multiwheeler 多轴汽车;多轮汽车
multiwheeler roller 多轴碾压机
multiwheel load 多轮荷载
multiwheel roller 多轮压路机;多轮碾(压机)
multiwheel seam welder 多辊式滚焊机
multiwindow 多窗口
multiwindow user interface 多窗口用户界面
multiwire 多线的
multiwire antenna 多振子天线
multiwire branch circuit 多路电路;多分支电路
multiwire counter 多线计数管
multiwire helical strand 多丝螺旋钢绞线
multiwire tendon 钢丝束
multiwire-triatic antenna 三角形多束天线
multiwire wiring board 多线路配线板
multiword 多字
multiword addition 多字加
multiword direct addressing 多字直接寻址
multiword immediate addressing 多字立即寻址
multiword indexed addressing 多字变址寻址
multiword instruction 多字指令
multiword relative address mode 多字相对寻址方式
multiyear ice 多年冰
multizonal 多区的
multizonal rectification 分带纠正
multizone 多区(域)
multizone cooler 多区冷却器
multizone flow regulating 分层配注
multizone high-speed distance relay 分段限时高速距离继电器;分域高速远距继电器
multizone reheating furnace 多段式加热炉
multizone relay 分段限时继电器
multizone reservoir 多层油气藏;多层油层
multizone supervision 分层管理
multizone system 多区系统
multizone test 多层合试
mult-nave 多中堂的
mume flower 梅花
mu-metal 镍铁高磁导率合金;锰游合金
mu-metron 微米测微表
mummification 干化
mummy 古埃及木乃伊;褐色氧化铁粉

mum permissible concentration 默认容许浓度
mundic 黄铁矿
mundite 穆磷铝铀矿
mundrabillaite 蒙磷钙铵石
mung(bean) 绿豆
mung bean improvement 绿豆改良
Mungoose metal 曼古司(铜镍锌)合金
Mungo reversed polarity subzone 蒙哥反向极性亚带
municipal 市政的;城市的
municipal accounting 市政会计
municipal administration 市政;市政管理
municipal architecture 城市建筑
municipal area 市区
municipal block 市内街道
municipal bond 市政公债
municipal boundary 市区界
municipal budget 市政预算
municipal building 市政建筑(物);市政大厦;市政大楼;市政厅
municipal bus 城市公共汽车;市区公共汽车
municipal by law 城市法规
municipal center 市政中心;市区中心
municipal civil engineering 市政工程
municipal compliance monitoring 市政合质监视
municipal compost 城市堆肥
municipal configuration 城市构造形式
municipal construction 城市建设
municipal construction financing 城市建设资金
municipal cooperation 市自治机构
municipal destructor 城市垃圾焚烧炉;城市废料焚化炉
municipal development 市区开拓;城市发展
municipal development committee 城市发展委员会
municipal discharge amount 城市排污量
municipal drainage 城市排水
municipal drainage system 城市排水系统
municipal drinking water source 城市饮用水源
municipal drinking water supply 城市饮用工水
municipal effluent 城市污水;城市排放污染物;城市废水
municipal engineer 市政工程师
municipal engineering 市政工程(学);城市工程(学)
municipal engineering facility 市政工程设施
municipal engineering survey 市政工程测量
municipal environment 城市环境
municipal expressway 城市高速公路
municipal extension 市区扩展
municipal facility 城市设施;公用设施
municipal fire alarm box 城市火灾警报箱
municipal fire protection requirement 城市防火条件
municipal forest 市区树林
municipal fund 市政基金
municipal garden 街心花园
municipal government 市政府
municipal graphic information system 市政图形信息系统
municipal hall 市政厅
municipal harbo(u)r engineering 城市港口工程
municipal heating 城市供热
municipal heat-supply 城市供热
municipal highway 都市公路;城市公路;城市道路
municipal hotel 市区旅馆
municipal housing management bureau 市房管局
municipal incineration 城市垃圾焚烧
municipal incineration plant 城市废料焚化炉
municipal industrial waste treatment 城市工业废水处理
municipal infrastructure 市政基础设施
municipal intercepting sewer 城市截流(污水)沟管;城市下水道
municipality 自治市;自治区;直辖市;市政府;市政当局
municipality of centrally administered 直辖市
municipality of provincially administered 省辖市
municipal library 市政图书馆
municipal living place 城市生活区
municipal loan 市政公债
municipally administered county 市辖县
municipally affiliated county 市管县
municipally owned housing 房管部门住房

municipal management 市政管理
municipal market 批发市场
municipal museum 城市博物馆
municipal noise 城市噪声
municipal office 市政厅
municipal ordinance 城市条例;城市环境保护条例;城市环境保护法令
municipal orphanage 城市孤儿院
municipal park 城市公园
municipal planner 城市规划者
municipal planning 城市规划
municipal planning act 城市规划条例
municipal planning institute 城市规划研究院
municipal planning study 城市规划研究
municipal planning system 城市规划系统
municipal planning theory 城市规划理论
municipal polluted waters 城市污染水体
municipal pollution 城市污染
municipal port 城市港口
municipal port engineering 城市港口工程
municipal power plant 公用发电厂;市营发电厂
municipal power supply system 公用电力系统;城市供电系统
municipal program(me) 城市规划
municipal project 市政项目
municipal rainfall pipe system 城市雨水管道系统
municipal refuse 市政垃圾;城市垃圾
municipal refuse compost 城市垃圾堆肥
municipal refuse disposal 城市垃圾处置
municipal refuse incineration 城市垃圾焚化
municipal refuse incinerator 城市垃圾焚化炉
municipal refuse management 城市生活垃圾管理
municipal reuse 市政再用
municipal road 市区道路
municipal rubbish 市政垃圾
municipal sanitary 城市卫生
municipal sanitary engineering 城市卫生工程
municipal sanitary landfill 市政卫生掩埋场
municipal sanitation 城市环境卫生;市政卫生;城市卫生
municipal sauna bath 市区桑那浴室
municipal service 公用设施;市政业务;市政服务
municipal service tunnel 市政隧道
municipal sewage 城市污水;市政排水工程;城市排水工程
municipal sewage plant 城市污水处理厂
municipal sewage treatment 城市污水处理
municipal sewage treatment plant effluent 城市污水处理厂出水
municipal sewage treatment system 城市污水处理系统
municipal sewerage 市政排水工程;城市污水工程;城市排水系统;城市排水工程
municipal sewerage system 城市下水道系统
municipal sewer pipe 城市污水管
municipal sewers 城市污水管;城市下水道;都市下水道
municipal sludge 城市污泥
municipal solid waste 市政固体废料;城市废弃物;城市固体废物
municipal solid waste incinerator 城市固体废物焚化炉
municipal station 市区火车站
municipal supervision 市政监督
municipal surface 市区道路铺设
municipal surfacing 城市道路铺面
municipal swimming pool 城市游泳池;市区游泳池
municipal tap water 城市自来水
municipal tax 市政税
municipal theatre 市政剧院
municipal transport(ation) 城市运输;城市交通;城市公共交通
municipal use of water 城市用水
municipal ward 政府区
municipal waste 城市垃圾;城市污水;城市废物;城市废水
municipal waste compost 城市废物堆肥
municipal waste disposal 城市废物处置
municipal waste leachate 城市废物渗滤液
municipal waste treatment 城市废物处理
municipal wastewater 城市污水;城市废水
municipal wastewater effluent 城市污水处理后出水
municipal wastewater flow 城市污水流

municipal wastewater sludge 城市污水污泥
municipal wastewater treatment plant 城市废水处理厂
municipal wastewater treatment plant effluent 城市污水处理厂出水
municipal wastewater treatment trend 城市污水处理发展趋势
municipal water demand 城市需水量
municipal water department 城市供水管理局
municipal water distribution network 城市配水网
municipal water distribution system 城市用水系统;城市配水系统;城市配水网
municipal water distribution system concentration 城市配水网密度
municipal water district 城市供水区
municipal water drainage 城市排水
municipal water facility 城市供水设施
municipal water pollution 城市水(系)污染
municipal water quality standard 城市水质标准
municipal water supply 城市自来水供应;城市供水;城市给水
municipal water supply quality standard 城市供水水质标准
municipal water supply system 城市自来水
municipal water system 城市给水系统;城市给水体系
municipal water treatment 城市水处理
municipal water use 城市用水
municipal works 市政工程
municipal works bureau 市政工程局
municipal zone 政府区
muniment 记录;档案;证书;契约
muniments house 贵重品库;文件保险房库;文件库;档案馆
muniments of title 财产所有权证明
muniments room 贵重品室;服饰品室;文件室;文件库;档案室
muninga 暗褐色斑驳的硬木(南非产)
munion 中梃
munirite 穆水钒钠石
muniting 窗格条
munition 军需品
munitions factory 兵工厂
munition ship 军火补给船
munjack 硬化沥青
munnion 中梃;窗门的直梃;竖框;立梃
munnion window 直梃窗;竖框窗
munsell 孟塞尔云母
Munsell chroma 孟塞尔色品;孟塞尔彩度;色品
Munsell chroma circle 孟塞尔表色系统彩度环
Munsell chroma number 孟塞尔表色系统彩度值
Munsell colo(u)r notation 孟塞尔颜色表示法
Munsell colo(u)r scale 孟塞尔色标
Munsell colo(u)r solid 孟塞尔色立体
Munsell colo(u)r standard 孟塞尔彩色标准
Munsell colo(u)r system 孟塞尔表色系统;孟塞尔颜色体系;孟塞尔色素;孟塞尔彩色制;孟寒尔颜色系列;孟塞尔色表坐标系统
Munsell colo(u)r tree 孟塞尔表色系统色树
Munsell curve 孟塞尔曲线
Munsell hue 孟塞尔(颜色)色调;孟塞尔色(彩)
Munsell renotanon system 孟塞尔新表色系统
Munsell scale 孟塞尔标度
Munsell system 孟塞尔体系
Munsell value 孟塞尔值;孟塞尔明度;孟塞尔色度值
Munsell value function 孟塞尔明度函数
muntin 门中梃;门扇中梃;窗梃;窗格条
muntin bar 门窗中梃;窗芯子;窗格条
munting 门格条;门中梃
muntin window 直梃窗
Muntz brass 孟滋黄铜
Muntz metal 孟滋(锌铜)合金;孟滋黄铜;熟铜
muqarnas 钟乳形细工;立体挑出蜂窝状装饰
mural 墙壁上的;壁画
mural arch 壁拱
mural background 墙壁背景
mural circle 墙仪
mural column 壁柱
mural decoration 壁饰
mural enrichment 壁装饰
mural hanging 墙上悬吊物
mural in mosaic 马赛克壁画
mural monument 壁碑
mural mosaic 壁上镶嵌

mural ornamentation 壁上装饰
mural painting 墙壁油画;壁画
mural precipice 峭壁
mural quadrant 墙象限仪;测墙壁象限仪
mural tablet 壁碑
muranese 图案玻璃;花纹玻璃
murasakite 红帘石石英片岩
murataite 钛锌钠矿
murder clause 删去的条款
murderer 钩棍
Murderian 穆德尔阶【地】
murdochite 黑铅铜矿
murgatroyed belt 瓶壁热应力集中带
muriate 氯化物
muriate of ammonia 氯化铵
muriatic acid 盐酸(旧名)
muriform 壁砖状的
murine typhus 刚果红色热
muritan 捕灭鼠
Murite 石膏粉饰
murkiness of air 空气朦胧阴沉
murky water 混水
murmanite 硅钛钠石;水硅钛钠石
muromontite 钇褐帘石
Murphy 折床;(不用时折叠在壁橱内的)隐壁床
Murphy bed 隐壁床;橱式折床
Murray fracture zone 默里破裂带
Murray loop test 默里环路定位试验
Murray's circuit testing method 默里环线试验法
murrey 桑椹色;紫红色
murunskite 穆硫铁铜钾矿
muschelkalk 壳灰岩
Muschelkalk series 壳灰岩统(三叠纪)【地】
Muschelkalk stage 壳灰岩阶【地】
muscovite 钾云母;白云母
muscovite-biotite granite 二云花岗岩
muscovite deposit 白云母矿床
muscovite fluorite greisen 白云母萤石云英岩
muscovite glass 白云母玻璃
muscovite granite 白云母花岗岩;白云花岗岩
muscovite greisen 云母云英岩
muscovite mining 白云母采矿
muscovite pegmatite 白云母伟晶岩
muscovite quartz greisen 云母石英云英岩
muscovite quartzite 白云母石英岩
muscovite-quartz rock 白云母石英岩
muscovite-quartz schist 白云母石英片岩
muscovite rock 白云母岩
muscovite schist 白云母片岩
muscovite topaz greisen 云母黄玉云英岩
muscovite tourmaline greisen 云母电气石云英岩
muscovy glass 白云母玻璃
museum 博物馆
museum fatigue 博物馆疲劳症
Museum of Anthropology 人类学博物馆
Museum of Antiquities 古物陈列室
Museum of Archaeology 考古博物馆
Museum of Chinese History 中国历史博物馆
Museum of Natural History 自然博物馆
Museum of Natural Science 自然科学博物馆
Museum of Science and Technology 科技博物馆
Museum of the Chinese Revolution 中国革命博物馆
museum piece 老古董;艺术珍品;博物馆珍藏品
museum science 博物馆学
museum zoology 博物馆动物学
mush 烂泥;噪声;泥煤;碎冰群
mush area 不良接收区
mush coil 散下线圈
Mushet('s) steel 马歇特钨钢
mushing error 干扰误差;颤噪误差
Mushlims' canteen 清真食堂
Mushlin pulpit 清真寺讲经坛
mushrabeyed work 细格花窗(伊斯兰教建筑);细木工花格挑窗
mushrabiya (封闭阿拉伯建筑阳台窗的)精致旋制的木隔板;在大门挑出的防卫阳台
mushroom 扩流锥;蘑菇状(的);蘑菇
mushroom anchor 菌形清污器;圆锚;菇形清污器;菌形锚;蘑菇形锚;伞形锚
mushroom antenna 蘑菇形天线;伞状天线
mushroom bit 杯形车刀
mushroom button 菌形按钮
mushroom cloud 蘑菇状(烟)云

mushroom construction 菌形结构;蘑菇形结构;蘑菇式结构;无梁楼板结构;板柱结构
mushroomed head buttress dam 大头支墩坝
mushroom excavation 蘑菇形开挖法
mushroom floor 环辐式楼板
mushroom floor slab 楼板材
mushroom follower 曲面随动件
mushroom hair style 蘑菇式
mushroom head 蘑菇形螺钉头;蘑菇形柱头;(支撑无梁板的)锥形头
mushroom-head bolt 蘑菇头形螺栓
mushroom-head buttress dam 大头坝;蕈头式支墩坝
mushroom-head column 蘑菇形柱
mushroom-head rivet 大平头铆钉
mushroom-head square neck bolt 蘑菇头方颈螺栓
mushroom hothouse 蘑菇栽培室
mushroom ice 蘑菇冰
mushroom insulator 宽裙式绝缘子;蘑菇式绝缘子
mushroom mixer 蕈状混合器
mushroom mo(u)ld 蘑菇形模型;伞状模具
mushroom nail 蘑菇钉
mushroom pavilion 蘑菇亭
mushroom reinforcement 环辐式排列的钢筋;环辐钢筋
mushroom rivet 扁圆头铆钉
mushroom road 蘑菇钉路
mushroom rock 蘑菇石
mushroom roof 伞形屋顶
mushroom-shaped superposed folds 蘑菇形叠加褶皱
mushroom shell 伞形壳(体)
mushroom slab 伞形柱顶;带蘑菇形柱头的无梁楼板
mushroom slab construction 蘑菇形楼板建筑;无梁楼板结构
mushroom spindle 阀杆
mushroom steamer line 不定期船
mushroom system 环辐式体系
mushroom town 新兴城市
mushroom-type shell 蘑菇形壳(体)
mushroom-type tunnel(l)ing method 蘑菇形开挖法;蘑菇形先拱后墙法
mushroom valve 锥形阀(门);菇形阀;菌形阀
mushroom ventilator 锥形通风筒;蘑菇形通风筒
mushroom workers' lung 蘑菇工人肺
mush winding 散下绕组;分布绕组
mushwound coil 散绕线圈
mushy 多孔隙的
mushy concrete 流态混凝土;浆状混凝土
mushy consistency 软练;流态稠度
mushy consistency of concrete 混凝土流态稠度
mushy freezing 糊状凝固
mushy road 松软道路
mushy stage 固液态;糨糊状态
mushy winding 软绕组
mushy wool 干草毛
music 藓类植物
musical arc 音弧
musical clock 音乐钟
musical echo 音乐回声
musical frequency 音乐频率
musical interval 音程
musical quality 音色;音质
musical saw 锯琴
musical scale 音阶
musical waters 水声悦耳的泉水;水声悦耳的流水
music atmosphere cast 环境谐音
music book 乐谱
music hall 音乐会场;音乐厅
music library 音乐图书馆
music line 传音线路
music-note type rubber seal 音符式橡皮止水
music rack 谱台
music room 音乐室
music studio 音乐室;音乐播音室
music wire 弦线;琴弦钢丝
musiga (东非产)绿心硬木
Musily silver 穆西利特殊型锡铋汞合金
musivum 彩色小玻璃块马赛克(罗马)
musjid 清真寺
musk 暗灰黄棕色
Muskat intercept 莫斯凯特截距
Muskat method 莫斯凯特法

musk deer 香獐子;麝
muskeg 稀淤泥;稀泥炭,沼泽泥土;林木沼泽;泥泽土;泥炭沼(泽);苔沼土
muskeg stream 沼地河流
muskeg swamp 泥沼地
Muskingum method 莫斯金干法
Muskingum routing method 莫斯金干洪水演进法
muskmuskeg stream 麝香
muskobads 莫斯科贝地毯
muskoxite 水铁镁石
musk xylene 二甲苯麝香
musk xylol 二甲苯麝香
Muslim architecture 穆斯林建筑;伊斯兰教建筑
Muslim dome 清真寺穹顶
Muslims' canteen 清真食堂;回民食堂
Muslim school 回教学校
muslin 府绸;薄纱织物;软棉布;平纹细棉布
muslin bag 棉布隔膜袋
muslin ceiling 平纹顶棚
muslin-delaine 细薄平纹毛织物;棉布
muslin scrim (油地毡的)背衬织物
mussel 蛤贝;贝壳类
mussel bed 贝壳层
mussel bind 贝壳页岩
mussel gold 淡菜金
mussel processing waste 贝类加工废物
mussel processing wastewater 贝类加工废水
mussite 透辉石
mustard 芥子气;混凝土(俚语)
mustard-gas poisoning 芥子气中毒
mustard mixer 混凝土搅拌机
mustard oil 芥子油
mustargen 氮芥
muster list 应变部署表;集合部署表
muster roll 应变部署表;集合部署表
muster station 应变部署表;集合部署表
must not-touch 不能接触
must-operate value 必作值
must-release value 必释值
must run energy 强制电能;必需电能
must run output 强制出力;必需出力
mutabilite 高氧烯烃沥青
mutability 易变;可突变性;突变性能;突变性
mutable 多变的
mutafacience 突变加强
mutagen 诱变剂;突变剂
mutagencity 诱变性
mutagenesis 突变形成;突变发生
mutagenic agent 诱变因素;诱变剂
mutagenic carcinogen 诱变致癌物
mutagenic components 突变成分
mutagenic effect 诱变效应;变异效应
mutagenic hazard 突变性危害
mutagenicity of water 水的突变性
mutamerism 变旋构现象
mutant 突变型;突变体
mutant bacteria 突变体细菌
mutant bacterial product 变性细菌产物
mutant character 突变性状
mutant phenotype 突变表型
mutant site 突变位点
mutarotation 旋光改变作用;变旋光(作用)
mutation 突变子
mutational equilibrium 突变平衡
mutational load 突变负荷
mutation fixation 突变固定
mutation frequency 突变频率
mutation index 突变指数
mutation of water depth 水深突变
mutation of water-level 水位突变
mutation pressure 突变压力
mutation rate 突变率
mutation stress 突变应力
mutation test 突变试验
mutation theory 突变理论
mute 静噪;无声片
mute antenna 假天线
mute control 静噪控制
muted colo(u)rs 暗色
muted stripe 暗色条纹
Mutegun 铁水口堵眼机;堵眼机
Mutemp 马坦普整磁用铁镍合金;铁镍合金
mute pulley 导向皮带轮
mute value 置切值;剔野值

muthmannite 碲金银矿
mutilated check 破损支票
mutilated cornice 断损的檐口
mutilated pediment 断损的山头
mutilated profile 割切剖面
mutilate driver 截断河;裁短河
mutilated river 被袭夺河
mutilation 残毁
mutilative traffic 破坏路面的交通
muting 噪声抑制;静噪调谐键;无噪声的
muting circuit 镇静电路;噪声抑制电路
muting sensitive 低灵敏的
muting sensitivity 低灵敏度
muting valve 无噪(声)管
mutograph 拍摄
muton 突变子;突变单位
mutton tallow 羊脂
mutual 相互的;交互
mutual acceptance 相互承兑
mutual action 相互作用
mutual advantage 互惠
mutual affinity 亲和力;相互吸引
mutual agency 相互代理
mutual agreement 互相同意;双方协定
mutual-aid construction 相互支援的建设
mutual alignment 相互位置;相互定线
mutual anchorage 钢筋搭接;搭接(钢筋混凝土内的钢筋的)
mutual assent 相互同意
mutual assimilation 相互自化作用
mutual assurance 互保险
mutual attraction 相互引力;交互引力;互相引力
mutual benefit 互利
mutual benefit building society 互利的房屋辅助协会
mutual branch 共分支
Mutual Broadcasting System 相互广播公司(美国)
mutual calibration 互相校准
mutual capacity 交互容量
mutual characteristic 屏栅特性
mutual check 相互核对
mutual check of financial discipline 财政纪律检查
mutual circuit 互通电路;双向通话电路
mutual coagulation 相互渗透
mutual coherence 互相干性
mutual coherence function 互相干函数
mutual-complementing code 互补码
mutual complements in economy 经济互补
mutual conductance 跨导;互导
mutual constraint of variational form 变分形式的相互约束
mutual contact zone 交互接触带
mutual cooperation 相互合作
mutual correlation 互相关
mutual couple factor 互耦合因数
mutual coupling factor 互耦合系数
mutual credit facility 互惠信贷
mutual credits 互惠信贷
mutual currency account 相互通货账户
mutual currency holding 相互通货持有
mutual current account 相互往来账户
mutual debt 相互债务
mutual dependence 相互依存
mutual deposit premium 相互存款保险公司
mutual effect 相互影响;相互作用
mutual exchange reaction 互换反应
mutual exclusion 相互排斥;互斥现象;互斥
mutual exclusion rule 互斥规则
mutual exclusive call 互斥调用
mutual flux 互感磁通;气隙磁通
mutual force 相互作用力
mutual fund 共同基金;互相资金
mutual fund investment 共有资金投资
mutual gift 相互赠与
mutual housing 互助住房协会
mutual impedance 互阻抗
mutual indemnification agreement 相互补偿约定
mutual inductance 互感系数;互感
mutual inductance attenuator 互感衰减器
mutual inductance particle velocimeter 互感粒子测速仪
mutual-induction 互感应;相互诱导;互感
mutual-induction protector 电磁铁线圈的屏蔽罩

mutual inductive coupling 互感耦合
mutual inductor 互感线圈;互感器
mutual information 交互信息;互通信息
mutual inhibition competition 相互抑制竞争
mutual insurance 相互保险;互相保险
Mutual Insurance Association 船东互保协会;相互保险协会
mutual insurance society 相互保险社
mutual interest 双方的利益
mutual interference 相互干扰;互相干扰
mutual interference of piles 桩的相互干扰
mutual interference of wells 井的相互干扰
mutual interlocking gear 互锁机构
mutual intersecting 互切
mutual investment company 合股投资公司;共同投资公司
mutualism 共生关系;互生;互利共生;互惠共生(现象)
mutualist 共生生物;共栖生物
mutualistic effect 互惠作用
mutualistic symbiosis 互惠共生
mutuality 相关
mutual loan society 联合信贷公司
mutually affecting analysis method 相互影响分析法
mutually compatible 相互协调的
mutually coupled coils 互相耦合线圈
mutually equilateral polygons 互等边多角形
mutually exclusive 互斥的;互不相同的;互不相交的;不相容的
mutually exclusive alternatives 互斥方案
mutually exclusive event 互不相容事件
mutually exclusive project 互斥项目
mutually independent random variables 相互独立的随机变量
mutually non-soluble liquid 不互溶液体
mutually perpendicular axis 正交轴(线)
mutually soluble liquids 互溶液体
mutually synchronized network 互同步网
mutually synchronized system 相互同步系统
mutual modulation 互调制
mutual mortgage insurance fund 连环抵押保险基金
mutual movement 相互运动
mutual overtaking 相互超车
mutual passing 相互超车
mutual perpendicular 相互垂直(的);互相垂直(的)
mutual pole of alike point 同点共极
mutual property 共有资产
mutual reactor 耦合电抗器
mutual recission 双方解约
mutual relationship 相关系
mutual rotation 相对转角
mutual savings bank 互助储蓄银行
mutual self-help program(me) 互助自建计划
mutual serving credit 相互透支额度
mutual share transfer 相互转股
mutual solubility 互溶性;互溶度
mutual solution 共溶
mutual solvent 互溶剂
mutual surge impedance 冲击互阻抗
mutual swing credit 相互透支
mutual synchronization 互同步
mutual trading credit 相互贸易信用
mutual understanding 相互理解
mutual visibility 对向通视
mutual water 共同用水组织
mutual water company 互助自来水公司
mutule 托檐石;檐底托板
mutule cornice 有托块的飞檐;有托板飞檐
mu-tuning 动铁调谐;磁性调谐
muzzle 炮口;吻突
muzzle anchor 挽绑锚链
muzzle bell 套口
muzzler 迎面强风;强送风
muzzle velocity 喷口速度;炮口速度;出口速度
mvule 褐色硬木(产于西非)
M-wire 计数器引线
m-xylene 间二甲苯
mycalex 压黏云母块;云母玻璃
mycelium 菌丝体
Mycenaean architecture 美锡尼建筑
mycenoid 小菇状的

mycetocole 栖菌动物
myckle 软黏土
mycobacteria 分支细菌
mycobacterium 分支细菌
mycobacterium marinum 海鱼分支杆菌
mycobiont 地衣共生菌
mycocide 杀霉菌剂
mycoderm 菌膜
mycolic acid 分支菌酸
mycolic acid-containing actinomycetes 含分支菌酸放线菌
mycology 真菌学
mycolytic bacteria 溶真菌细菌
mycophasma 类菌质体
mycophobic 嫌菌的
mycrodyne process 挤压法拉丝
Mydrim limestone 米德里姆灰岩
Mydrim Shales 米德里姆页岩
Myer-Prausnitx adsorbed solution theory 迈尔-普劳斯尼茨吸附溶液理论
Myer's flood scale 迈尔洪水等级
Myer's rating 迈尔率定法
Mykroy 米克罗依(绝缘材料)
mylar 聚酯胶片；涤纶薄膜
mylar capacitor 聚酯树脂电容器
Mylar diaphragm 聚酯薄膜
mylar-faced acoustic(al) tile 聚酯薄膜面吸声板
Mylar film 聚酯薄膜
Mylar plastic 密拉聚酯薄膜
mylius test 碘曙红试验
mylone 棉隆
mylonite 糜棱岩
mylonite gneiss 糜棱片麻岩

mylonite series 糜棱岩系列
mylonite zone 糜棱岩带
mylonitic 糜棱状
mylonitic coal 糜棱煤
mylonitic loosen texture 糜棱化散体结构
mylonitic structure 糜棱构造
mylonitic texture 糜棱结构
mylonitization 糜棱岩化(作用)；糜棱化
mylonitization way 糜棱岩化方式
mylonitized rock 糜棱化岩
mylonization 糜棱岩化(作用)
Mylor series 美乐统
mynchery 女修道院
myodil 碘苯酯；碘苯十一酸乙酯
myostagmin 减滴质
myotatic reflex 牵张反射
myrceane 月桂烷
myrcene 香叶烯；月桂烯
myrcenone 月桂烯酮
myrcia 月桂叶
myrcia oil 香叶油
myriabit 万位
myriabit memory 万位存储器
myriabit storage 万位存储器
myriad 万
myriagram 万克
myrialiter 万升；万公升
myriameter 万米；万公尺
myriametric(al) 万公尺的
myriametric(al) wave 超长波；万米波
myriapoda 多足纲
myriastere 万立方米
myriastere per year 万立方米每年

myriatonne per day 万吨每日
myriatonne per year 万吨每年
myriawatt 万瓦
myrica oil 月桂油
myringo-camera 鼓膜照相机
myringoscope 鼓膜镜
myriopoda 多足纲
myriorama 万景画
myrisc acid 肉豆蔻酸
myristic acid 十四烷酸
myristic oil 肉豆蔻油
myrmecoclepty 蚁客共生
myrmecodomatia 植物蚁巢
myrmecologist 蚁学家
myrmecophile 蚁冢动物
myrmecophilous plant 好蚁植物
myrmekite 蠕状石
myrmekitic 蠕状的
myrmekitic structure 蠕状构造
myrmekitic texture 蠕状结构；蠕虫状结构
myrol 硝酸甲脂和酒精的混合物
myrtan extract 桉树栲胶
myrtle 桃木；铁籽；桃金娘
myrtle burl 月桂树
myrtle wax 香桃木蜡；桃金娘蜡
myrtus 桃木；铁籽；桃金娘
mystery 神秘的事物；白金、锡和铜合金；手艺
mytilotoxin 蛤贝毒
Mytilus 壳菜(属)
Mytilus edulis 贝壳类；紫贻贝
Mytton flags 米顿板层
myvaseal 迈瓦西尔高真空绝缘材料
myxobacteria 黏细菌

N

No. 2 ev. apo(u)rator 二效蒸发器
nab 门锁槽;水下礁;山嘴;暗礁
nabaphite 氟磷钙钠石;磷钙钠石
Nabarro-Herring creep 纳巴罗—赫林蠕变;扩散蠕变
nabby 苏格兰东岸渔帆船
nabe 居住区内电影院
nabivnoy ice 多层冰;重叠冰;筏冰
nabla 劈形算符;微分算符;倒三角算子;倒三角形（算子）
nablock 圆结核体岩石
nacaphite 磷钙钠石
nacarat 鲜橘红(色)
Na-cation exchange capacity 钠离子交换容量
nacelle 短舱;吊舱;发动机舱
NA-cement 低碱水泥
n acetyl isatin 乙酰靛红
nacre 珍珠层;珠母贝;珍珠质
nacreous cloud 珠母云
nacreous glass 珠光玻璃
nacreous luster 珍珠光泽
nacreous pigment 珠光颜料;珠光颜色
nacreous varnish 珍珠光清漆
nacreous wall 珠光壁
nacrite 珍珠陶土;珍珠石;珍珠结构
Nada 纳达抗变色铜合金
Nada alloy 纳达铜合金
Nadai's sandheap analogy 那达砂堆模拟法(求塑性状态扭转剪应力的方法,薄膜模拟法的延伸)
Nadal's formula 脱轨公式
Nadanhada Early Mesozoic subduction zone 那丹哈达早中生代俯冲带
Nadanhada Eugeosynclinal fold belt 那丹哈达优地槽褶皱带
Nadanhada marine trough 那丹哈达海槽
nadel 针状突起;针形突起
nadic acid 降冰片烯二酸
nadic anhydride 降冰片烯二酸酐
nadic methyl anhydride 甲基降冰片烯二酸酐
nadir 天底;最低点;最底点
nadir angle 天底角;底点角
nadir distance 天底距
nadir observation 天底观测
nadir-photograph 垂直摄影像片
nadir plummet 底点对点器
nadir point 像底点;天底点;最低点
nadir-point method 天底点法
nadir-point plot 底点辐射三角测量
nadir point triangulation 天底点三角测量
nadir radial line 天底点辐射线;底点辐射线
nadir reading 天底读数;底点读数
nadorite 氯氧锑铅矿
naegite 苗木石
Naeser process 内塞雾化铁粉生产法
nafalaratite 氟磷铝钙石
nagana 纳加那(非洲锥虫病)
nagana red 纳加那红
Nagaoka's coefficient 长冈系数
nagarot 纳加那红
nagashimalite 硼硅钡钒石
nagatelite 磷褐帘石
Nagele's obliquity 内格累氏倾斜
Nagele's rule 内格累氏规律
nagelfluh 泥砾岩层;泥砾岩(杂色块状砾石)
nagelschmidtite 叠磷硅钙石
nager 手持凿岩机
Nagler type turbine 内格勒式水轮机
nagyagite 叶碲矿;叶碲金矿
nahcolite 苏打石
nahlock 圆块
Nansen zinc process 南森炼锌法
naihead 钉头饰
nail 固定钉;钉子;纳尔(长度单位)
nailability 可钉性能;受钉性
nailable 能受钉的;可受钉的;可打钉的
nailable base 钉牢的地基;钉牢的底座;钉牢的基础
nailable block 可受钉砌块;受钉块

nailable brick 可受钉的砖块;受钉砖
nailable castlite 可受钉的混凝土板
nailable concrete 可受钉(钉的)混凝土;受钉混凝土
nailable material 可钉材料
nailable plate 受钉板
nailable precast concrete plank 可受钉预制混凝土板
nailable strip 可受钉条
nailable tile 可钉花砖;可钉瓦
nail bar 撬棍;起钉棍
nail-base fiberboard sheathing 钉接纤维夹衬板;钉基纤维板衬
nail bearing 针形轴承
nail-bed function 钉板函数
nail box 铁钉箱
nail care preparation 保护指甲制剂
nail carving 钉雕
nail catcher 钉扣
nail claw 拔钉钳
nail clinch 弯铆钉;弯脚钉
nail concrete 受钉混凝土;可钉混凝土
nail-connected girder bridge 钉板梁桥
nailcrete 可钉混凝土;受钉混凝土
nail cutter grinder 磨刀机
nail dog 勾头钉
nail dowel 销钉
nail down 钉住;钉牢
nail drag 钉板刮器
nail drawer 拔钉钳;拔钉器
nail drift 冲孔器
nailed beam 钉成的桁条;钉成的横梁;钉成的梁;钉梁
nailed connection 钉固连接;钉结合
nailed construction 钉成的建筑物;钉成的结构;钉结构
nailed construction method 钉接合构造法
nailed girder 钉成的桁梁;钉成的横梁
nailed-glued truss 胶针结合桁架
nailed joint 钉连接;钉成的组件;钉成的关节;钉成的接头;钉结合;钉接合;钉固接头
nailed plate girder 木钉板梁;钉板梁
nailed roof 钉接屋顶
nailed roof truss 钉成的构架工程;钉成的屋顶桁架
nailed truss 钉固桁梁;钉结合桁架;钉接桁架;钉结桁架
nailed trussed rafter 钉合屋架
nailed wooden girder 钉结合板梁;钉合木板梁
nailed wood girder bridge 木钉板梁桥
nailed wooden plate girder 钉接木板梁;钉合木板梁
nailed wood rib arch bridge 木钉板肋拱桥
nail enamel 指甲油
nailer 射钉机;受钉木条;受钉块钉;敲钉工人;制钉工人;钉钉板
nailer joist 钉有木条的钢格栅
nailery 制钉厂
nail extracting forceps 拔指甲钳
nail extractor 起钉钳;起钉器;拔钉钳;拔钉器
nail eye 钉眼
nail firing tool 射钉枪
nail fixing 钉固定
nail float 划毛抹子;针式浮子;带钉抹子
nail for prepared roofpaper 用于铺屋面油毡的铁钉
nail for reed 芳草用钉;钉薄片用的铁钉
nail-glued roof truss 铁钉与胶结的屋顶桁架
nail-glued truss 胶钉结合桁架
nail gun 射钉机
nail hammer 拔钉锤;鱼尾锤;羊角榔头;羊角锤;钉锤
nail hardness 指甲硬度
nail head 钉头;钉头状装饰;钉帽;钉头(形)饰
nail-head bonder 钉头式接合器
nail-head(ed) 钉头形的
nail-headed bond(ing) 球焊;钉头式接合(法);钉头(式)焊;球焊
nail-head(ed) mo(u)lding 楔形线脚(诺曼底式建筑特征);方形钉头饰的线脚;楔形钉头装饰线脚;钉头线脚;四叶花(装)饰;钉头(形)饰

nail-headed rusting 钉头状锈蚀
nail-head scratch 丁字头擦痕
nail-head spar 钉头石
nail-head type welding 钉头式焊接
nail holding 握钉性;受钉性
nail-holding ability 握钉力
nail-holding force 钉楔裹力
nail-holding power 握钉力
nail hole 钉孔;钉眼
nailing 打钉;插钉术
nailing anchor (将门框固定于板墙筋的)钉入锚件;木立筋锚件
nailing base 钉基板;钉底板
nailing batten 钉挂瓦条;钉狭板;钉板条;(用来挂瓦或压缝的)可钉板条
nailing block 木栓;木砖;可受钉块;受钉木块;可钉砌块
nailing brick 可受钉的砖;受钉砖
nailing channel 受钉槽钢;钉床;钉槽
nailing concrete 可钉混凝土;受钉混凝土
nailing ground 受钉基板;可钉地面
nailing joist 钉有木条的钢格栅
nailing machine 制钉机;钉接机;钉钉机
nailing marker 送钉定位板
nailing pattern 制钉模
nailing plug 钉条;受钉木塞;钉木塞;可受钉木塞;受钉块
nailing property 可钉性
nailing property of concrete 混凝土的可钉性
nailing strip 可钉条;受钉(嵌)条;钉条
nailing strip bracket 可钉板条托架;钉条地板梁托
nailing wall 土钉墙
nailless 没敲钉的
nail(making) machine 制钉机
nail material 制钉材料
nail nippers 钉钳;起钉钳;拔钉钳
nail of small ga(u)ge 小号钉
nail picker 检钉器
nail plate 钉接板;钉板
nail plate girder 钉结合板梁
nail point 钉尖
nail popping 钉凸
nail-pressure 押切法
nail puller 撬棍;取钉锤;起钉钳;起钉器;冲钉器;拔钉钳子;拔钉器
nail puller with bent claw 用弯爪起钉的拔钉器
nail puller with straight claw 用直爪起钉的拔钉器
nail punch 敲钉穿孔器;钉形冲头;钉冲压机;钉冲头;冲钉器
nails barrel clamp 钉子桶夹
nail set 用钉冲孔;压钉杆;冲钉器
nailsick board 因钉孔太多而失去强度的木板
nailsick boat 因钉太多而漏水的船
nail smiths chisel 制钉用凿;钉头模凿
nails of heavy ga(u)ge 大号钉
nail spike 道钉
nail staining 铁钉污染
nail up 钉牢
nail wire 制钉钢丝;制钉(铁)丝
nail withdrawal 握钉力
nail wooden girder bridge 钉板梁木桥
Nain Nucleus 奈恩陆核
Na-ion exchanger 钠离子交换器
Nairn sandstone 奈恩砂岩
Nairobi 内罗毕
NaI scintillation detector 碘化钠闪烁探测器
naive model 单纯模型
Na-K alloy 钠钾合金
nakaseite 辉锑铜银铅矿
nakauriite 水碳硫矾
Na-K-cooled reactor 钠钾冷却堆
naked 裸露的
naked bud 裸芽
naked chip 未封装的片子
naked contract 无赔偿合同;无担保契约;无担保合同;无偿契约

naked debenture 无担保债券
naked displacement 裸船排水量
naked electrode 裸焊条
naked eye 肉眼
naked-eye examination 肉眼检查
naked eye navigation 目测航行
naked eye observation 肉眼观测
naked feet 赤脚
naked fire 裸火;明火头;活火头
naked flame 明火;无遮盖火焰
naked(-flame) lamp 明火灯
naked-flame locomotive 无瓦斯矿山用机车
naked-flame mine 明火灯矿井
naked floor 裸露地板;透光地板
naked flooring 大梁格栅;未铺面层的地板;裸楼层;未铺面层;未铺楼板
naked frame 裸构架
naked hole 裸眼
naked hull 光船(体);裸船体
naked karst 裸露岩溶;裸裸喀斯特
naked light 无遮盖光线;开放光线;明火;无遮光线;无罩灯
naked molten metal 裸露金属液
naked position 暴露的部位
naked radiator 无保护罩散热器
naked reactor 裸反应堆
naked singularity 裸奇点;露奇点
naked substrate 露底
naked trust 无担保信托
naked truth 真相
naked viewing 肉眼观察
naked wall 光秃墙;光面墙;未抹灰板条墙
naked warrant 纯保证凭证
naked wire 裸线
nakhlite 透辉无球粒陨石;透橄无球粒陨石
Nakhodka Port 纳霍德卡港(俄罗斯)
nakrite 珍珠陶土
Nakshatra 月站
Nal detector 碘化钠探测器
naled 二溴磷
n-alkanal 正构醛
n-alkane 正烷烃
n-alkanol 正构醇
namatium 溪河群落
nambulite 硅锰钠锂石
Namco chaser 组合板牙平板梳刀
name 名字;名称
name address server 名地址服务器
name and number of diapiric structure 底辟名称和编号
name(author) entry 著者款目
name block 图签;(图纸右下角的)图标;名块
name board 船名板;招牌;站名牌
name board of ship 船名牌
name brand 名牌货;名牌商品
name call 名字调用
name card 名片
name carrying capability of a map 地图注记容量
name catalog 人名地名目录
name classification 注记分类
name code 名代码;名称码
name condition 名字条件
name constant 名常数
named 命名的
name day 结算日;让渡日
named bill of lading 记名提货单;记名提单
named cargo container 按货物命名的集装箱
named common 公用命名;命名公用块
named common area 有名公用区;命名公用区
named consignee 指定的收货人
named constant 命名常数
named departure point 指定发货地点
named destination 有名目标;指定目的地
named entity 命名实体
named file 命名文件
named insured 提名受保人
named output file 命名输出文件
named peril(s) 指定保险
named policy 指名保险证券;指定船名保单;记有船名的海运保单;确定保险单
named port 指定港(口)
named port of destination 目的港名称
named port of shipment 指定装船港
named principal 显名委托人
named program(me) module 命名程序模块
named source file 命名源文件
named system 命名系统
named user program(me) 命名用户程序
named vessel 记名货船
name entry 名入口;名表目名称登录【计】
name field 名字域;名字栏
name form 命名式;地名译名规则
name guide card 人名导卡
name in a name space 名空间的名
name index 人名索引
name information management 地名资料管理
nameless height 无名高地
name merchandise 名牌商品
name of abyssal fan 深海扇名称
name of abyssal mineral resources 深海矿物资源名称
name of abyssal plain 深海平原名称
name of account 账户名称;会计科目
name of adjacent tectonic region 相邻构造单元名称
name of ancient seaway 古海路名称
name of an object 客体的名称
name of archipelago 群岛名称
name of article 货名;品名
name of bar 沙坝名称
name of basin 海盆名称
name of beach 海滩名称
name of beach ridge near shore barrier 滨岸堤名称
name of buyer or payee 抬头
name of canyon 峡谷名称
name of chain(net) 锁网名
name of chronostratigraphic(al) unit 年代地层单位名称
name of coal seam 煤层名称
name of coal seam group 煤组名称
name of coast 海岸名称
name of coastal terrace 海岸阶地名称
name of coastline 海岸线名称
name of coast zone 海岸带名称
name of commodity 品名;商品名称
name of commodity and specification 货物名称和规格
name of commodity in English 英文品名
name of continental ice-sheet 大陆冰盖名称
name of continental shelf 大陆架名称
name of continental slope 大陆坡名称
name of current 海流名称
name of deep 海渊名称
name of deep fracture 深断裂名称
name of delta 三角洲名称
name of district 地域名称
name of divisional map 图幅名称
name of documents 文献名称
name of drowned valley 溺谷名称
name of estuary 河口名称
name of explosive 炸药名称
name of fault zone 断层带名称
name of geophysical prospecting in borehole 井中物探方法名称
name of fold 褶皱名称
name of fracture zone 破裂带名称
name of gliding tectonics 滑动构造名称
name of graphs 图件名称
name of groundwater level line 地下水位线名称
name of groundwater regime 地下水动态名称
name of gulf 海湾名称
name of guyot 平顶海山名称
name of impact structure 撞击构造名称
name of instrument 仪器名(称)
name of intertidal zone 潮间带名称
name of investigation work 已有调查报告名称
name of island 岛屿名称
name of item 项目名称
name of lake 湖泊名称
name of leveling line 水准路线名称
name of lithostratigraphic(al) unit 岩石地层单位名称
name of log chart 测井图件名称
name of log method 测井方法名称
name of main documents 依据主要文件名称
name of manufacture 制造厂名
name of marketing surveying report 市场调查报告名称
name of material 资料名称;物料名称
name of members 构件名称
name of mine 矿山名称
name of mineral commodities 产品名称;矿产名称
name of mineral resources 矿种名称
name of mud 泥浆名
name of niche 洞室名称
name of pal(a)eoclimate zone 古气候带名称
name of pal(a)eocontinent 古大陆名称
name of pal(a)eo-ocean 古大洋名称
name of pal(a)eoshoreline 古海岸线名称
name of part 品名
name of place of production 产地名称
name of plate 铭牌
name of plateau 海台名称
name of platform 浅台名称
name of project 工程名称
name of publishing house 出版社名称
name of pump 水泵名称
name of recent ocean 现代大洋名称
name of regime observation station 动态观测站名称
name of regional tectonic unit 区域构造单元名称
name of ridge 海岭名称
name of rise 海隆名称
name of river 河流名称
name of road 公路名称
name of rocks 岩石名称
name of sample 样品名称
name of sea 海名称
name of seamount 海山名称
name of seamount chain 海山链名称
name of sediments 沉积物名称
name of shoal 浅滩名称
name of soil formation 土层名称
name of soil layer 土层名称
name of spring 泉水名称
name of spring precipitation 泉水沉淀物名称
name of station 火车站名称
name of steamer 航道名称
name of strait 海峡名称
name of strata 地层名字
name of stream 河流名称
name of survey area 测区名称
name of surveyed region 工作区名称
name of tectonic layer 构造层名称
name of tectonic movement 构造运动名称
name of testing soil sample 试验土样名称
name of the volcanic cluster or zone 火山群或火山带名称
name of trench 海沟名称
name of trough 海槽名称
name of unit 单位名称
name of variables 变量的名字
name of vessel 船名
name of volcanic cones 火山锥名称
name of water resources law 水资源法的名称
name of well logging equipment 测井仪器名称
name plate 铭牌;名牌(门上刻写姓名的牌子);设备商标;设备出厂证;设备标牌;商标;船名板;厂名牌;报刊名
name-plate capacity 铭牌额定值
name-plate current 铭牌电流
name-plate information 铭牌数据
name-plate of manufacturing plant 制造厂铭牌
name-plate rating 铭牌额定值;铭牌出力
name-plate set 铭牌位置
name policy 定名法规
name resolution 名称转换
names 复合名
names authority 命名者
name scope 名字作用域;名字域
name server 名服务程序
names for tectonic movements recognized commonly used in abroad 国外地壳运动
names of volcanic cones 火山锥名称
name space 名字空间
name-spelling authority 地名拼写准则
names trace 注记透写图
name table 名字表
name unknown 船名不详
namibite 纳米铜铋钒矿
naming 命名;定名

naming convention 命名约定
naming of aids-to-navigation 航标命名
naming pailou 点景牌楼
naming rule 赋名规则
namogoniogram 自差曲线图
Namurian 那慕尔阶【地】
n-amyl acetate 醋酸正戊酯
nancy receiver 红外(线)接收机
Nandan trench 南丹海沟
nandina domestica 南天竹
nandrolone phenylpropionate 苯丙酸去甲睾酮；苯丙酸诺龙
Nandus 南鲈属
nanigation 树木矮化法
nanism 矮态
nanization 人工矮化
nankeen 南京土布；本色布；棕黄色
nankin yellow 铁黄
nanlingite 南岭石
nanmu 楠木
nannocyte 微胞
nannofossil 超微化石
nannofossil ooze 超微化石软泥
nannofossil zone 超微化石带
nannoplankton 微小浮游生物；超微浮游动物
nan(n)o-relief 微地形；微起伏
nannostone 超微化石岩
nano 纳；毫微
nano adsorbent 纳米吸附剂
nano-adsorption material 纳米吸附剂
nanoammeter 纳安计；毫微安计
nanoampere 毫微安(培)
nano-anatase TiO2 纳米锐钛型二氧化钛
nanoceramics 纳米陶瓷
nanochemistry 纳米化学
nanocircuit 超小型集成电路
nano-composite 纳米级复合材料
nanocomposite material 纳米复合材料
nanocomposite membrane 纳米复合膜
nanocrystal 纳米晶体
nanocrystalline 纳米晶体的
nanocrystalline ceramics 纳米晶陶瓷
nanocrystalline composite material 纳米晶复合材料
nanocurie 纤居；毫微居里
nano-dosimetry 超微剂量学
nano droplet 纳米液滴
nanofarad 纳法；毫微法拉
nanofiber membrane polycaprolactone 聚己内酯纳米纤维
nanofiltration 纳滤
nanofiltration ceramic membrane 纳滤陶瓷膜
nanofiltration membrane 纳滤膜
nanofiltration membrane broaden 纳滤膜加宽
nanofiltration membrane separation 纳滤膜分离
nanofiltration membrane surface charge 纳滤表面电荷
nanofiltration process 纳滤工艺
nanogram 纤克；毫微克
nanohenry 纳亨；毫微亨利
nanoid enamel 薄釉质
nano-kaolin 纳米高岭土
nanomachine 毫微级计算机
nanometer[nm] 纳米；纤米；毫微米
nanometer club 纳米棒
nanometer resolution 纳米分辨率
nanometer scale crystal 纳米级晶体
nanometer scale function material 纳米级功能材料
nanometer-size modification coagulant 纳米改性混凝剂
nanometer titanium dioxide photoelectrocatalytic degradation 纳米二氧化钛光电催化降解
nanometer zinc oxide 纳米氧化锌
nanometric alumin(i)um oxide ceramic membrane 纳米氧化铝陶瓷膜
nano-particle 纳米颗粒
nanoperm 纳普姆(材料透气率单位，纳普姆 = 10^{-9}普姆)
nanophanerophyte 矮高位芽小灌木
nanophase 纳米相
nanophase ceramics 纳米陶瓷
nanophase materials 纳米材料
nanophotocatalytic oxidation 纳米光催化氧化
nanoplankton 微型浮游生物；微小浮游生物

nanoplankton organic matter 微型浮游生物有机质
nanoplankton plant 微型浮游植物
nanoponge adsorbent media 纳米海绵状吸附介质
nanoporous polymer 纳米多孔聚合物
nanoprogram(me) 纳程序
nanoprogram(me) level 毫微程序级
nanoprogramming 毫微程序设计
nanoscale zero-valent iron 纳米零价铁
nanoscope 纳秒示波器；毫微秒示波器；毫微秒超高频示波器；超高频示波器
nanosecond 纳秒(1 纳秒 = 10^{-9}秒)；纤秒；纳秒；毫微秒
nanosecond fluorescence spectroscopy 毫微秒荧光光谱学
nanosecond image converter streak camera 毫微秒变像条纹相机
nanosecond logic 毫微秒逻辑
nanosecond pulse 纳秒脉冲
nanosecond pulse height analyzer 毫微秒脉冲振幅分析仪
nanosecond pulse oscillator 毫微秒脉冲振荡器
nanosecond rise time 纳秒上升时间
nanosecond time-resolved spectroscopy 纳秒时间分辨率光谱学
nanosized silica-polyvinyl alcohol composite ultrafiltration membrane 纳米二氧化硅—聚乙烯醇复合超滤膜
nano-structure 纳米结构
nano titanium dioxide 纳米二氧化钛
nano titanium dioxide catalysis 纳米二氧化钛催化剂
nano titanium dioxide photocatalyst 纳米二氧化钛光催化剂
nano titanium dioxide photocatalytic oxidation 纳米二氧化钛光催化氧化
nanotorr 毫微托
nanotrace analysis 纳痕量分析
nanovolt 毫微伏特
nanovoltmeter 毫微伏特表
nanowatt 纳瓦(1 纳瓦 = 10^{-9}瓦特)；毫微瓦(特)
nanowatt electronics 毫微瓦功率电子学
nano zeolite molecular sieve 纳米沸石分子筛
nano zeolite molecular sieve adsorption 纳米沸石分子筛吸附
Nansen bottle 海水标本瓶；南森瓶；南森集水器；南森采水瓶
Nansen cast 采水测温
nantokite 铜盐
nanuwite 钠铜锌矾
NaOH solution circulating pump 碱液循环泵
naos 古希腊或罗马庙宇的中央部分
napalm 胶化汽油；凝汽油剂；凝固汽油(剂)
napalm tank 凝固汽油箱
napery 抹布；餐巾
napex 项部
naphoite 磷氢钠石
Na-photocell 钠光电池
naphtha 挥发性油；轻汽油；石油精；石脑油；粗汽油
naphthacene 丁省
naphtha column 石脑油塔
naphtha cut 石脑油馏分
naphtha furnace 石脑油炉
naphtha gas 石脑油气
naphtha gas reversion 石脑油气回馏过程
naphtha insolubles 汽油不溶物
naphthalene acetamide 萘乙酸胺
naphthalene and ammonia washing water 洗萘氨水
naphthalene ball 萘球；萘丸；卫生球
naphthalene base 原油
naphthalene black 萘黑
naphthalene carboxylic acid 萘羧酸
naphthalene content 萘含量
naphthalene derivative 萘衍生物
naphthalene diamine 萘二胺
naphthalene dichloride 二氯代萘
naphthalene fracture 萘状断口
naphthalene hydrocarbons in jet fuel 喷气式燃料中萘系烃含量
naphthalene oxidizing bacteria 萘分解菌
naphthalene removal 脱萘
naphthalene resin 萘醛树脂
naphthalene sulfonic acid 萘磺酸
naphthalene tetrachloride 四氯代萘

naphthanl pigment 萘甲酰苯胺颜料
naphtha polyforming 石脑油聚合重整
naphtha reformer 石脑油重整器
naphtha reformer stripper 石脑油重整的汽提塔
naphtha residue 石脑油脚；石脑油残渣
naphtha solvent 石脑溶剂
naphtha-xylene equivalent 石脑油-二甲苯当量
naphthein 石脑油英
naphthenate 环烷酸盐
naphthenate of lead 环烷酸铅
naphthenate solution 环烷酸盐溶液(木材防腐剂)
naphthene 环烷属烃；环烷
naphthene base 环烷基
naphthene-base crude 环烷基粗石油
naphthene-base crude petroleum 环烷基原油
naphthene base lubrication oil 环烷基润滑油
naphthene series 环烷族(环石蜡族)；环烷系
naphthene soaps 环烷皂
naphthenic 环烷的
naphthenic acid 环烷酸；环酸
naphthenic base 环烷基
naphthenic base crude oil 环烷基原油
naphthenic circle 环烷环
naphthenic crude 环烷基原油
naphthenic oil 环烷油
naphthenic residual 环烷质残油
naphthenic residual oil 沥青质残油；环烷残油
naphthenic soap 环烷酸皂
naphthenic solvent 环烷溶剂
naphthenone 环烷酮
naphthine 伟晶蜡石
naphthionic acid 对氨基萘磺酸
naphthisotriazine 三氮杂菲
naphthlacetic acid 萘乙酸
naphthochrome green 萘铬绿
naphthoflavon 萘黄酮
naphthoic acid 萘酸
naphthol 萘酚
naphtholate 萘酚盐
naphthol benzoate 苯甲酸萘酯
naphthol blue black 萘酚蓝黑
naphthol dyestuff 萘酚染料
naphtholism 萘酚中毒
naphtholithe 沥青页岩
naphthol lake colo(u)r 萘酚色淀颜料
naphthology 石油科学
naphthols pigment 萘酚类颜料
naphtholum 萘酚
naphthol yellow 萘酚黄
naphthophenanthrene 二苯并蒽
naphthoquinone 萘醌
naphthotriazines 三氮杂蒽
naphthoxazine 吩恶嗪
naphthyl 萘基
naphthylamine 萘胺
naphthylamine black 萘胺黑
naphthylamine orange 萘胺橙
naphthylamine poisoning 萘胺中毒
naphthylamine sulfonic acid 氨基萘磺酸
naphyhacene 并四苯
napier 奈培【物】
Napier diagram 纳皮尔自差曲线图
Napierian base 自然对数的底；讷皮尔对数底；纳氏底(自然对数的底)
napierian base 自然对数底；纳皮尔对数底；纳氏底
Napierian log 纳皮尔对数
Napierian logarithm 自然对数；纳氏对数；纳皮尔对数
Napierian rule 纳皮尔法则
Napier's analogies 内皮尔类比；纳皮尔相似式；纳皮尔对偶式
Napier's compasses 多用圆规
Napier's deltaic two stroke engine 纳皮尔三角式双冲程柴油机
Napier's Deviation table 纳皮尔罗经自差表
Napier's differential screw steering gear 纳皮尔差动螺旋操舵装置
Napier's equation 纳皮尔氏公式
Napier's log(arithm) 纳氏对数
Napier's rules 内皮尔法则；纳皮尔法则
napiform root 球根
napkin 餐巾；布巾
napkin pattern 饰有布褶的雕刻板；布褶纹装饰

napkin ring 套餐巾用的小环
napkin tissues 餐巾薄纸
Naples jar 显微载片染色缸
Naples yellow 那不勒斯黄；锑酸铅
Napoleon green 拿破仑绿（1000度高温色料）
napoleonite 球状闪长岩；球状辉长岩
Napoleon marble 褐色带纹的法国大理石
Napoleonville(stage) 拿破仑维尔阶【地】
nap pattern 毛结花纹
nappe 溢流水舌；溢过堰顶薄层水流；漫流水层；熔岩流；推覆体；水舌；餐巾；覆盖层
nappe acquifer 蓄水层
nappe face 水舌面
nappe inlier 构造内窗层
nappe interrupter 水舌掺和齿坎
nappe-like structure 似推覆构造
nappe of weir 过堰水舌
nappe outlier 孤残推复体；飞来峰
nappe profile 溢流水舌截面
napper 拉绒机
nappe separation 水舌脱离；水舌分离；射流分离（现象）；表面分离
nappe-shaped crest of spillway 流线型溢流堰顶；水舌型溢洪道顶
nappe-shaped crest profile 流线型峰值分布
nappe structure 推覆构造
nappe structure belts 推覆构造带
nappe structure covered type 推覆构造掩盖型
nappe tectonics 推覆构造
napping 起绒；拉毛；推覆作用；刷布
nappy 菜盆
nap roller 石印转印墨辊
naptholithe 沥青质页岩；沥青页岩
nararno 一种远程无线电导航系统
narbonnais 纳尔旁风
narcissus 水仙属
narcotics pollution 毒品污染
nard 甘松油脂；甘松
nardil 苯乙肼
nargusta 青褐色硬木（产于巴西）
nari 钙积层；纳里钙质层
narig 斑驳硬木（从黄色到棕色的，产于菲律宾）
Narite 纳丽特铝青铜合金
Narite(alloy) 纳яти特合金
Narizian(stage) 那里兹阶【地】
Narm tape 纳姆合成树脂黏结剂
narra 紫檀（木）
narration 解说词；分录说明
narrative 解释文
narrative appraisal(report) 评估报告
narrative form 叙述式报表
narrative form of profit and loss statement 叙述式损益表
narrative form of statement 文字报告式报表
narrow-aisle fork truck 狭窄通道用的叉车
narrow angle 狭角
narrow angle acquisition 窄角搜索
narrow angle camera 窄角摄影机
narrow angle coordinator 窄角坐标方位仪；窄角配位仪
narrow angle diffusion 深照型漫射
narrow angle lens 窄角透镜；窄角镜头
narrow angle lighting fittings 小角度照明配件；深照型照明器
narrow back 室内敷线工（俚语）；电工（俚语）
narrow band 小差距汇率幅度；狭小汇率幅度；窄频带；窄带；狭窄
narrow band absorption 窄频带吸收
narrow band amplifier 窄频带放大器；窄带放大器
narrow band analyser 窄频带分析器
narrow band antenna 窄带天线
narrow band axis 窄通带轴；狭通带轴
narrow band barrage jamming 窄带阻塞干扰
narrow band blocking filter 窄带闭塞滤波器
narrow band channel 窄频带信道；窄带信道
narrow band charactascope 窄频带特征观测设备；窄频带特性观测设备；窄频带特性观测器
narrow band characteristic 窄频带特征
narrow band circuit 窄频带电路
narrow band colo(u)r-difference demodulator 窄带色差解调器
narrow band communication system 窄带通信系统
narrow band controller 窄范围控制器

narrow band crystal filter 窄带晶体滤波器
narrow band digital filtering 窄带数字滤波
narrow band direct printing 窄带直接印字
narrow band filter 窄带滤光器；窄带滤光片；窄带滤波器；窄带滤波片；窄带段滤光片
narrow band frequency 窄带频率
narrow band frequency modulation 窄带调频
narrow band frequency response 窄带频率响应
narrow band integrated serviceable digital network 窄带综合业务数字网
narrow band interference 窄带干扰
narrow band interference filter 窄带干涉
narrow band level 窄带电平
narrow band link 窄带通信线路
narrow band-long-wave omnidirectional range 窄带长波全向无线电信标
narrow band magnitude 窄波段星等
narrow band noise 窄带噪声
narrow band noise factor 窄带噪声因数
narrow band noise measurement 窄带噪声测量
narrow band pass filter 窄带通滤光片；窄带逼近滤波器；狭通带滤波器
narrow band path 窄带通路
narrow band photometry 窄带测光
narrow band ping 窄带脉冲信号
narrow band proportional control 窄范围比例调节
narrow band pyrometer 窄带高温计
narrow band radio frequency channel 窄带射频频道
narrow band random process 窄带随机过程
narrow band random response 窄带随机响应
narrow band random vibration 窄带随机振动
narrow band random vibration wave 窄带随机振动波
narrow band receiver 窄带接收机
narrow band reception 窄带接收
narrow band saw 窄带锯
narrow band-selective receiver 窄带选择接收机
narrow band-selective voltmeter 窄带选择性伏特计
narrow band signal 窄频带信号
narrow band spectrum 窄带谱
narrow band subscriber terminal 窄带用户终端
narrow band system 窄频带体系；窄带制；窄带（宽）体系
narrow band telemetry 窄带遥测
narrow band transmission 窄频带传输；窄带传输
narrow band transmission line 窄带传输线
narrow band vibration 窄带振动
narrow bandwave analyser 窄频带波分析仪
narrow band width 窄波段宽度
narrow band width interpolation 窄带宽内插法；窄带宽内插
narrow base 窄基底
narrow-base photodiode 薄基底光电二极管
narrow base terrace 窄底阶地
narrow base tower 窄基底杆塔
narrow beam 细束；窄射束
narrow beam absorption 窄束吸收
narrow beam absorption coefficient 窄束吸收系数
narrow beam antenna 窄束天线
narrow beam attenuation 窄束减弱
narrow beam bathymeter 窄束水深测量器
narrow beam echo sounder 狭束回声测深仪；窄束声回声测深仪
narrow beam echo sounding 窄束声回声测深
narrow beam geometry 窄束几何
narrow beam radar 窄波束雷达
narrow-beam radiation 狭束辐射
narrow beam sound source 窄束声源
narrow beam sound wave 窄束声波
narrow boards 窄板
narrow boiling cut 窄馏分
narrow boiling range fraction 窄馏分
narrow boiling range product 窄沸程产品
narrow-bore column 细孔径柱；窄孔柱
narrow-bore tube 小直径管
narrow braid 狭幅饰带
narrow breakwater 窄防波堤
narrow breakwater opening 窄防波堤口门
narrow bridge sign 狭桥标志；窄桥标志
narrow carpet 狭幅地毯；窄幅地毯
narrow channel 狭窄水道；狭窄航道；水道；窄河槽；窄波束

narrow cloth 狭幅织物
narrow crooked fairway 弯窄航道
narrow cut filter 窄带滤色片；窄带截止滤器；高吸收滤光镜
narrow cut petroleum fractions 石油窄馏分
narrow cut share 窄切犁铧
narrow deviation frequency modulated transmitter 窄频偏调频发射机
narrow dimension 窄尺寸
narrow down 变窄
narrow duck 狭帆布
narrowed-neck(ed) bottle 细颈瓶
narrowed portion of duct 风管的狭窄部分
narrowed visual field 视野狭小
narrow-end-up mo(u)ld 上小下大钢锭模
narrow exchange rate margin 小差距汇率幅度
narrow fabric 窄幅织物
narrow fabric skirt 缘饰
narrow face 窄工作面
narrow face flange 窄面法兰；窄面凸缘
narrow face ring 窄面环
narrow field railway 工地轻便窄轨铁路
narrow fire box 窄式火箱
narrow fissure 狭窄裂纹；窄缝
narrow flame 舌焰
narrow-flanged 狭突缘的，狭凸缘的；狭法兰的
narrow flat dump car 窄平底自动倾覆车
narrow flat pliers 窄平钳
narrow fraction 窄馏分
narrow frame screen 窄框纱窗；窄框屏幕
narrow frequency band 窄频带
narrow frequency modulation 窄带调频
narrow front 狭窄正面
narrow gap 狭窄通路
narrow gap helix 窄缝螺旋线
narrow gap one-side automatic welding 窄间隙单面自动电弧焊
narrow gap one side welding 窄间隙单面焊
narrow gap spark chamber 窄隙火花室；窄隙火花室
narrow gap welding 狭间隙焊；窄间隙焊接
narrow gate 狭闸门；窄闸门；窄连通脉冲门；窄门电路
narrow gate amplifier 窄选脉冲放大器
narrow gate circuit 窄选通电路；窄门电路；窄电闸电路
narrow gate multivibrator 窄门多谐振荡器
narrow gate pulse 窄选通脉冲
narrow-gate range potentiometer 精确距离电位计
narrow-ga(u)ge 窄胶片；窄轨（距）；狭轨铁路；狭轨距
narrow-ga(u)ge dump car 窄轨废料车；窄轨倾斜车
narrow ga(u)ge film 窄规胶片
narrow ga(u)ge flat dump car 窄轨平板自卸车
narrow ga(u)ge lighting 窄轨灯光设备
narrow ga(u)ge railroad 窄轨铁路；轻便铁路；轻便铁道
narrow ga(u)ge railway 窄轨铁路；窄轨距轨道；轻便铁路；轻便铁道
narrow ga(u)ge railway bridge 窄轨铁路桥
narrow ga(u)ge steam loco(motive) 窄轨蒸汽机车
narrow-ga(u)ge surface haulage 地面窄轨运输
narrow-ga(u)ge track 窄轨铁路；窄轨（道）
narrow-ga(u)ge track system 窄轨道系统
narrow goods weaving 织带
narrow gorge 狭谷；嶂谷
narrow gradation 均匀级配
narrow guide 窄导轨
narrow hook rule 窄钩尺
narrowing 狭缩；缩窄；缩小；收缩；断面收缩；变窄
narrowing attachment 收针装置
narrowing circuit 脉冲锐化电路；变窄脉冲电路
narrowing combing 收针板
narrowing movement 退缩运动
narrowing of direction(al) pattern 方向图变窄
narrowing of ga(u)ge 轨距缩小
narrowing of pulse 脉冲压缩
narrowing of the ring 环的收缩
narrowing picker 挑针器
narrowing point 套针
narrowing ribbon 缩边
narrowing rod 收针杆

narrowing toward each end 两边小
narrow jaw pipe wrench 窄牙管钳
narrow kerf core bit 薄壁取芯钻头
narrow ladies 14 英寸×7 英寸板条
narrow laydays 缩短受载期
narrow-leaved 狭叶的
narrow light door 带窄玻璃窗的门;带观察窗门
narrow line array 窄线组合
narrow line emission 窄带发射
narrow line source 锐线光源
narrow line spectrum 窄线光谱
narrow linewidth 窄线宽
narrow locomotive 窄轨机车;窄轨火车头
narrow London 砌墙的泥刀
narrow margin 薄利
narrow margin arrangement 小差距汇率幅度安排
narrow market 交易量少的市场
narrow mesh 细筛眼;狭筛眼
narrow meshed 小孔的;带小孔的
narrow mouth container 小口瓶
narrow-mouth(ed) 细口的;窄口的
narrow-mouth(ed) bottle 细口瓶
narrow-mouth(ed) flask 细口烧瓶
narrow mouth reagent bottle 小口试剂瓶
narrow-narrow gate 超窄选通脉冲
narrow-mouth(ed) bottle 细颈瓶
narrow neck press-blow process 小口瓶压一吹成型法
narrow neck ware 细颈器皿
narrow opening 狭孔;窄巷道
narrow orifice 狭口
narrow outlay 销路呆滞
narrow pass 狭窄水道;狭道;窄路
narrow passage 海峡航道;隧道;窄孔道
narrow path impeller 窄流道叶轮
narrow pennant 长三角旗
narrow period band 窄周期
narrow pillar file 窄齐头平锉
narrow place 窄工作面
narrow plume model 窄烟云模式
narrow price gap 缩小差价
narrow pulse generator 窄脉冲发生器
narrow pulse sampling 窄脉冲取样
narrow rail bogie 窄轨矿车;窄轨四轮转向架;窄轨台车
narrow railroad 窄轨铁路(美国)
narrow railway 窄轨铁道
narrow railway track 窄轨铁路线
narrow range cut 窄留分
narrow resonance 窄共振
narrow resonance approximation 窄共振近似
narrow ringed 密纹木料;窄年轮的木材
narrow-ringed timber 细纹木材;密年轮木材;密纹木(材)
narrow row 窄行的
narrow row drilling 窄行播种
narrow row seeder 窄行播种机
narrow row sowing 窄行播种
narrow rule 狭轨;窄尺
narrows 峡口;山峡
narrow scale 窄尺;条尺
narrow scan 窄扫描
narrow-sector recorder 狭扇面记录仪
narrow(sense) currency 狭义货币
narrow(sense) marketing 狭义市场
narrow sheeting 狭幅床单布
narrow shift system 窄移频系统
narrow shoe 窄闸瓦
narrow shrink 缩小
narrow side 门与门框相碰的一面
narrow slip beam 窄条梁
narrow slit 窄缝
narrow slot 窄缝
narrows of river 河流峡谷
narrow sound beam 窄声束
narrow spectrum insecticide 专效性杀虫剂
narrow spectrum pesticide 专效性农药
narrow-spiral glide 小半径盘旋下降
narrow stall 窄矿房
narrow steam locomotive 窄轨蒸汽机车
narrow steel track 小钢轨便道
narrow street 狭车行道;窄街道
narrow strip 窄板材
narrow tape 狭带

narrow tee 小口径三通
narrow the gap 缩短差距
narrow topology 狭义拓扑学
narrow track 窄轨;箭翎线[铁]
narrow track landing gear 窄轮距起落架
narrow track tractor 窄轮距拖拉机;窄履带式拖拉机
narrow T rest 狭丁字刀刀架(木工车床)
narrow tube 小直径管
narrow type bearing 窄系列滚动轴承
narrow upright tree 金字塔形的树
narrow U stair(case) 双跑平行楼梯(指无梯井的双跑平行楼梯)
narrow wale 狭棱条
narrow waters 窄窄水域;海海峡;窄狭水域
narrow waterway 狭窄水道
narrow whip 长三角旗
narrow white line 白而窄的线条
narrow width tractor 窄型拖拉机
narrow winding trail 羊肠小道
narrow workings 窄工作面
narrow zone 狭生长带
narrow-zoned 狭年轮的
narsarsukite 短柱石
narthex 门廊;前厅;古教堂前廊;(教堂的)前廊
Nasa 纳塞(导电玻璃)
nasal atomizer 鼻喷雾器
nasal height 鼻高
nasal part 鼻部
nasal prognathism 中面角
nasal rating 气味分级;按气味评价;按气味分级
nasal septum 鼻柱
nasal tip height 鼻尖高
nasal tip salient 鼻尖突
nasat 导航卫星
nascent alumin(i)um 新生态铝;初生态铝
nascent continental margin 初生大陆边缘
nascent cyclone 初生气旋
nascent hydrocarbon 初生烃
nascent hydrogen 初生态氢;初生氢
nascent ocean 初生洋盆
nascent oil 初生石油
nascent oxygen 新生氧;新生态氧;初生氧;初生态氧
nascent state 新生态;初生态
nascent water 新生水
n'aschi 东北风
Nash compressor 液封压缩机
Nash pump 液封型真空泵;纳希液封型真空泵;纳氏泵;纳计泵
Nashs instantaneous unit hydrograph 纳希瞬时单位过程线
Nash's theorem 纳希定理
Nash-Williams formula 纳希-威廉公式
nasinite 奈硼钠石
nasion 鼻根点
nasledovite 硫碳铅锰铝石
Nasmith pile-driver 内史密斯打桩机;汽锤打桩机
Nasmyth focus 内氏焦点
nasograph 鼻测量计
nasonite 氯硅钙铅矿
nastinic acid 分支菌脂酸
nastrophite 水磷钠锶石
nasturan 方铀矿
nasty place 险恶地方
nasty smell 难闻(的)气味
nasus incurvus 鞍状鼻
natality(rate) 出生率
natality table 出生率表
natant 浮在水上的
natant hydrophyte 漂浮水生植物
natatores 游禽类;水禽类
natatorial bird 游禽
natatorium 游泳馆;室内游泳池;[复]natatoria
National Academy of Sciences 国立科学院;国家科学院
national administration on import and export of endangered species 国家濒危物种进出口管理机构
National Aerometric Data Bank 国家大气数据库
National Aeronautics and Space Administration 美国国家航空和宇宙航行管理局
National Agricultural Exhibition Center 国家农业展览馆
National Agrinomical Research Center 国立农艺研究中心
national aid 国家补助
National Air Sampling network 全国空气取样网
National Air Surveillance Network 国家大气监测网;全国大气监视网
national ambient air quality objective 国家环境空气质量目标
national ambient air quality standard 全国环境空气质量标准;国家环境空气质量标准
national annual emissions of air pullutants 全国每年空气污染物排放量
National Apartment Association 全国公寓协会
national archaeology 民族考古学
national architecture 民族建筑(学)
National Archives on Rainfall 国家雨量记录档案馆
national area 民族集居地区
National Art Gallery 国家美术展览馆
national assets 国家资产
National Association Architectural Metal Manufacturers 全国建筑金属生产者协会
National Association for Community Development 全国社区发展协会
National Association of Assessing Officers 全国财产估价者协会
National Association of Cooperate Real Estate Executives 全国房地产经理联合协会
National Association of Corrosion Engineers 美国国家腐蚀工程师协会
National Association of Cost Accountants 全国成本会计师协会
National Association of Counties 全国各县联络协会
National Association of Home Builders 全国房屋建造商协会
National Association of Home Manufacturers 全国工业化房屋建造业者协会
National Association of Housing and Redevelopment Officials 全国住房及开发商协会
National Association of Housing Cooperatives 全国住房信用社协会
National Association of Independent Fee Appraisers 全国房地产独立收费评估员协会
National Association of Manufacturers 全国制造商协会(美国)
National Association of Master Masons 全国熟练泥瓦工协会(英国)
National Association of Mutual Savings Banks 全国互助储蓄银行协会
National Association of Pension Funds 全国养老基金协会
National Association of Real Estate Investment Trusts 全国房地产投资信托协会
National Association of Real Estate License Law Officials 全国房地产特许律师协会
National Association of Regional Estate Brokers 全国房地产经纪人协会
national atlas 国家地图集
national autonomous areas 民族自治地方
national balance sheet 国家资产负债表
national bank 国家银行
National Bank of Iran 伊朗国家银行
National Bank of Switzerland 瑞士国家银行
national base map 国家基本(地)图
National Board of Fire Under Writer's Label 国家防火部标记
national body 国家团体
national bond 国家公债
national boundary 国境线;国界;国家边界
National Brass foundry Association 全国黄铜铸造协会
national budget 国家总预算;国家预算
national budgeting 国家预算编制
national budget system 国家预算制度
National Building Code 国家建筑法规;国家建筑规范
National Bureau of Economic Research 国家经济研究局
National Bureau of Legal Metrology 国家法制计量局
national bureau of standard NBS 标准
National Bureau of Standards 国家标准局(美国)
national call 国内呼叫
national capital 民族资本
national capital market 国内资本市场

national capital region 国家首都特区
national census 人口普查
National Center for Atmospheric Research 国家大气研究中心
national center 全国电话总局
national certificate 国家合格证书
national channel 国家航道;国道(指航道);全国波道
national character of vessel 船舶国籍
national characters 国家特有字符;国家标准符号
national character set 国际字符组
national city bank 国家城市银行
National Clay Pipe Institute 全国陶土管研究所
National Coal Board 煤炭总局;国家煤炭管理局
National Coarse Thread 美制粗牙螺纹
National Coil Coaters Association 美国全国卷材涂装工作者协会
national commission on water quality 国家水质委员会
National Committee against Discrimination in Housing 全国反对住房歧视委员会
National Committee on Radiation Protection and Measurement 国家辐射防护与测量委员会
national composite indices 全国综合指数
National Composition Association 国际排印工人协会
national comprehensive development plan 国土综合开发计划
national comprehensive equilibrium 全国综合平衡
national conservation program(me) 国家保护计划
national control pollutant discharge elimination system 国家控制污染物排放制度
national control survey net 国家控制测量网
national coordinate system 国家坐标系(统);国家大地坐标系
national corporation 国营公司
national council for technologic(al) awards 国家技术奖委员会
National Crushed Stone Association pavement design method 美国碎石协会路面设计法
national currency 本国货币
national currency per US dollar 每一美元兑换本国货币数
national customs 国民习俗
National Dairy Research Institute 国立乳业研究所
national data management system 国家资料管理系统
national debt 国债
National Decorating Products Association 全国装饰品协会
national defence 国防
national defence economy 国防经济
national defence highway 国防公路
national defence industry 国防工业
national defence (packaging) standard 防卫厅(包装)规格
national discharge standard 国家排放标准
national distance dialing 国内长途拨号
national dividend 国民所得
national drawing 国画;民族画
national ecology 民族生态学
national economical development target 国家经济发展目标
national economic and social development planning 国民经济和社会发展计划
national economic appraisal 国民经济评价
national economic benefit 国民经济效益
National Economic Commission 国家经济委员会
national economic complex 民族经济综合体
national economic development 发展民族经济
national economic evaluation 国民经济评价
national economic plan 国民经济计划
national economics 民族经济学
national economy 国民经济
National Effluent Toxicity Assessment Center of USEPA 美国环境保护局国家排污物毒性评价中心
National Electrical Code 国家电力规范
National Electrical Safety Code 国家电力安全规范
national emblem 国徽
National Emission Data System 全国排放数据系统
national emission standards for hazardous air pollutants 全国空气有害污染物排放标准
national enterprise 国营企业
National Environmental Action Plans 国家环境行动计划
National Environment(al) Policy Act 国家环境政策法
National Environmental Protection Administration 国家环境保护总局
National Environmental Protection Agency 国家环境保护局
National Environmental Research Centre 国家环境研究中心
National Environment(al) Satellite Center 全国环境卫星中心
National Environmental Satellite Service 国家环境卫星局
National Environmental Secretariat 环境秘书处
national environment of a species 物种的自然环境
national environment policy act 国家环境保护法规
national environment protection planning 国家环境保护规划
national estate 国有地产
national farm historic 国家农业历史
national farming economy 国民农业经济
national feature 民族特色
national features of architecture 建筑的民族特征
National Federation of Building Trades Operatives 全国建筑工人联合会
national finance 国家财政
national fineness 国定精度
national fine thread 美制细牙螺纹
National Fire Protection Association 全国防火协会(美国)
National Fire Protection Association of the USA 美国国家防火学会
national first grade gravity base station 国家一等重力点
national flag 国旗;国家旗帜
national flower 国花
national focal point 国家协调中心
national forest 国有森林;国有林;国家森林
National forest park 国家森林公园
national form 民族型式;民族形式
national-geodetic nets 国家大地(测量)网
National Geologic(al) Survey 国家地质局
national gravity base station 国家重力基准点
national gravity base station network 国家重力基本网
national grid 国家测量网;国家网;国家经纬网
national grid square 国家方格网;国家统一坐标网
national group 国家集团
national harbo(u)r 国家港口
National Harbo(u)r Board 国家港口管理局;港务管理总局
National Harbo(u)r Management Bureau 国家港口管理局
National Height Datum 1985 1985年国家高程基准
national highway 国有公路;国道(指公路);全国性公路
national highway system 国道系统
national historic(al) preservation act 国家文物保护法
national history 民族史
national holiday 国家法定假日
National Home Ownership Foundation 全国居者有其屋基金会(美国)
National Housing Agency 美国住宅建设厅
National Housing and Economic Development Law Project 全国住房及经济发展法律服务中心
National Housing Bank 国家住房银行
national housing policy 国家住房政策
National Housing Rehabilitation Association 全国住房更新协会
national hurricane center 国家飓风中心
national hurricane severe storms forecast center 国家猛烈风暴预报中心
national hydrological benchmark network 国家水文基准点网
national income 国民收入
national income per capita 人均国民收入
national industrial corporations 全国性工业公司
National Industrial Pollution Control Council 全国工业污染控制委员会
national industry 民族工业
national information center 国家情报中心
national information infrastructure 国家信息基础结构
national information network 全国情报网
national information system 国家信息系统;国家情报系统
national inheritance 民族遗产
national inquiry 国内查询
National Institute for Occupational Safety and Health 国立职业安全卫生研究所
National Institute of Environmental Health Science 国家环境卫生科学研究所;国家环境卫生科学研究院
National Institute of Farm and Land Brokers 国家农业土地经纪人协会
National Institute of Legal Metrology 国家法制计量研究所
national insurance 国民保险;国家保险
national integrated wastewater discharge standard 国家综合排污标准
national interest 国家利益
national interim primary and secondary standard 国家暂行一级和二级标准
National Interim Primary Drinking Water Regulations 国家一级饮用水暂行法令
National Interim Primary Drinking Water Standards of USEPA 美国环境保护局国家一级饮用水暂行标准
National Interim Primary Water Regulations 国家暂行初级饮用水基本规程条例
national investment fund 国家投资资金
National Iranian Oil Company 伊朗国家石油公司
nationalistic form 民族形式
nationalistic style 民族风格
nationalities affairs 民族事务
nationality 国籍;民族学;民族性;部落
nationality carpet 民族地毯
nationality mark 船舶国籍符号
nationality products enterprises 民族贸易企业
nationality township(s) 民族乡
nationalization 国有化
nationalization of insurance 保险国有化
nationalization of land 土地国有化
nationalization of railways 铁路国有化
nationalized bank 国有化银行
nationalized industries 国有化产业
nationalized undertaking 国营企业
National Joint Council on Materials Handling 货物装卸委员会
national jurisdiction 国家管辖权
national key scientific research project 国家重点科研项目
national land fund 国家土地基金
national land planning 国土规划
national land satellite 国土卫星
national laws 国法
National League of Cities 全国城市联盟
National Leased Housing Association 全国租赁住房协会
national legislation 国内立法
national level 全国水平
national level(l)ing network 国家水准网
National Lime Association 全国石灰协会(美国)
National Limestone Association 全国石灰石协会(美国)
Nationally Designated Eco-Demonstration Region 国家级生态示范区
nationally designated garden city 国家级园林城市
national map accuracy standard 国家地图精度标准
national mapping program(me) 国家制图计划
national marine service ship 国家商船船队
national maritime board 国家海运局;国家海运部门
National Maritime Bureau 国家海事局
National Maritime Electronics Association 国家航运电子协会
national market 民族市场;全国市场
national measurement system 国家计量系统
National Meteorologic(al) Center 全国气象中心
national minorities 少数民族
national mobile highway source standards 国家公路移动污染源标准
national mobilization 全国总动员
national monument 国家保护文物;民族纪念碑
national name authority 国家地名机构
national network 国内网络

national network of seismograph 国家地震台网
National Nuclear Safety Bureau 国家核安全局
national number 国内号码
national numbering plan 国内编号计划
national observation net 全国性观测网
National Oceanic and Atmospheric Administration 海洋及大气管理署;国家海洋和大气管理署
national oceanic satellite system 国家海洋卫星系统
National Oceanographic(al) Data Center 全国海洋资料中心;国家海洋资料中心
national ornamental 民族装饰;民族图案
national output 全国产出量
national-owned fleet 国有船队
National Paint and Coatings Association 美国全国油漆与涂料协会
National Paint Varnish and Lacquer Association 美国全国油漆协会
national paper 船舶国籍证书
national park 国家公园
national park map 国家公园图
national patent 国家专利
National Patent Commission 全国专利权委员会(美国)
national plan 国家计划
national policy 国家政策;国策
national policy for environmental protection 环境保护方针
National Pollutant Discharge Elimination Permit System in Federal Clean Water Act 美国联邦清洁水法国家污染物排放消除许可证制度
National Pollutant Discharge Elimination System of USEPA Clean Water Act 美国环境政治法清洁水法国家污染物排放消除体系
national pollutant discharge standards 国家污染物排放标准
National Pollution Control Foundation 国家污染控制基金会
national population census 全国人口普查
National Ports Authority 港务管理总局
national ports council 国家港务委员会
National Precast Concrete Association 全国预制混凝土协会(美国)
national preferences 民族嗜好
national pretreatment standard 国家废水处理标准
national primary ambient quality standard 国家一级环境空气质量标准
National Primary Drinking Water Regulations of USEPA 美国环境保护局国家一级饮用水法令
national product 国家产值;国民生产;国民产值
national program(me) 国家计划
national project 国家项目
national purse 国库
national quality product 全国优秀产品
National Radiation Protection Committee 国家辐射防护委员会
national railroad 国有铁路;国有铁道;国家铁路
national railway 国有铁路;国有铁道;国家铁路
national railway network 国家铁路网
national reactor test site 国家反应堆试验场
National Ready Mixed Concrete Association 全国商品混凝土协会(美国)
National Realty Committee 全国房地产委员会
national regulation 国家规范
national report 国家报告
national requirements 国家要求;国家特有要求
national requirements feature 国家要求特点;国家特有要求功能部件;国家特有要求功能
National Research Council 全国研究理事会;国家科学研究委员会
national resource committee 国家资源委员会
national resources 国家资源;民族资源
National Resources Planning Board 全国资源计划委员会
national revenue 国税
National Revised Primary Drinking Water Regulations of USEPA 美国环境保护局国家一级饮用水修订法
national road 国家公路;国道(指公路)
national road network 全国道路网
national romantic style 民族浪漫风格
National Rural Housing Coalition 全国农村住房联盟
National Safety Council 国家安全委员会
National Sand and Gravel Association 全国砂石协会(美国)
National Sanitation Foundation 全国卫生基金会
National Savings and Loan League 全国储蓄及贷款联盟
national scale 全国规模
national school 国立学校
National Science Foundation 全国科学基金会;国家科学基金会
national seclusion 闭关自守
national secondary drinking water regulation 国家二级饮用水条例
National Secondary Drinking Water Regulations of USEPA 美国环境保护局国家二级饮用水法令
National Securities Exchange 全国证券交易所(美国)
national service 国内业务
national service of legal metrology 国家法制计量部门
National Society of Real Estate Appraisers 全国房地产评估员学会
national sovereignty 国家主权
national standard 国家标准
national standard arbor 国家标准轴
national standard freight container 国家标准集装箱
national standard hose coupling 国家标准水龙带接头
national standardization 国家标准化
national standardization plan 国家标准化计划
national standard map 国家标准地图
National Standard of PR China 中国国家标准
National Standard Reference Data 国家标准参考数据
national standards body 国家标准机构
national standard thread 国家标准螺纹
national stream quality accounting network 国家河流质量统计网
national structure of urban population 城市人口民族构成
national style 民族形式;民族风格
national subsidies 国家津贴
national supergrid 国家超级电力网;国家超高压电网
national tape 国家标准锥度
national tariff 国定税率;固定税率;法定税率
national tax 国税
National Technical Committee for Standardization 全国标准化技术委员会
national tools monument 国家农具文物
national topographical series 国家地形图系列
national tradition 民族传统
national treasure 国宝
national treasure building 国家保存的文物建筑
national treasury 国库
national treatment 国家待遇
national treaty 国家条约
national triangulation point 国家三角点
national triangulation system 国家三角网
national trunk highway 国家干线公路(国道)
national trunk road network 国道网
national trust 国家信托
national uniform price 全国统一价格
national university 国立大学
national utility 国家公用事业
national vertical datum 国家高程起算面
National Vertical Datum 1985 1985 年国家高程基准
National Waste Policy Act of United States 美国国家废料政策法令
national water council 国家水委员会
national water policy 国家水政策
National Water Pollution Control Act of United States 美国国家水污染控制法令
National Water Quality Commission of United States 美国国家水质委员会
National Water Quality Laboratory of United States Geologic(al) Survey 美国地质调查局水质实验室
national water quality network 国家水质网络
National Water Quality Surveillance System of USEPA 美国环境保护局国家水质监视系统
national water resources 国家水资源
National Water Resources Committee of United States 美国国家水资源委员会
national waters 国有水域
national waterway 国家航道;国道(指航道)
National Water Well Association 全国水井协会
national welfare and people's livelihood 国计民生
national well-being 全民福利
nationlity 民族
Nation Market System 全国证券市场体系
nation safety net service 国家安全网络
nation science 民族学
nation-state 单一民族国家
nationwide 在全国范围内;全国性的;全范围内的
nation-wide sample survey 全国性抽样调查
nation-wide scientific information system 全国科技情报系统
natirvik 风搅雪
natisite 氧硅钛钠石
native 自然的;天然的;本国的;本地人;本地的
native air cooling 自然通风冷却;空气冷却
native alloy 天然合金
native alum 天然明矾
native amalgam 天然汞齐
native and subsidiary products 农副产品
native aquatic plant 天然水生植物
native arsenic 天然砷
native asphalt 天然石油沥青;天然沥青;天然地沥青
native asphaltic pyrobitumen 天然地沥青质焦沥青
native attachment 本机连接;本机附件
native bank 钱庄
native barium sulphate 天然硫酸钡
native beach profile 海滩原地形(剖面);岸滩原地形
native bismuth 天然铋
native bitumen 天然沥青
native blockness 天然块度
native boulder 天然大砾石
native breed 地方品种
native cellulose 天然纤维素
native cinnabar 天然朱砂
native cloth 土布
native compiler 本机编译程序
native compound 天然化合物
native copper 自然铜
native country 国内
native disease 地方病
native effect 自然效应
native element 天然元素
native estate 天然农圃
native fiber 天然纤维
native fish habitat 天然鱼类生境
native floatability 固有可浮性;天然可浮性
native gas 原有天然气
native gold 原金
native goods 土产(品)
native graphite 天然石墨
native grass 天然牧草
native green 天然绿
native groundwater 原生地下水;自然地下水;天然地下水;初始地下水
native habitat 原产地;自然生境
native hydrocarbons 天然碳氢化合物
native image 本机映像
native industry 地方工业
native interest 当地利息
native iron 天然铁
native land 祖国;本土
native language 本机语言;本地语言
native lignin 天然木素
native magnet 天然磁铁
native manganic hydrate 天然锰水化物;天然锰的水合物
native map 本土地图
native meadow 天然草地
native mercury 天然汞
native metal 天然金属
native mineral wax 天然矿蜡
native minium 天然铅丹;天然红铅;天然丹
native mode 本机方式
native mud 天然泥浆
native paraffin 天然石蜡;天然蜡
native pasture 天然牧场;天然草地
native pitch 天然沥青
native plant 乡土植物
native platinum 天然铂
native predator 天敌

native products 土特产品;土产(品);天然产品
native product shop 土产商店
native Pt-Pd ore 自然铂钯矿矿石
native red oxide of lead 天然红的铅氧化物;天然铅丹
native rock 原生岩
native sanctuary 自然禁猎区;自然禁猎地
native silver 天然银
native soda 泡碱
native soil 残积土
native species 乡土种;当地种;本土种;本地种;本地品种
native state 天然状态;天然态
native stress 天然应力
native style 本地作风;本地风格;当地风格
native style blast furnace 土高炉
native substrate 同质衬底;天然衬底
native sulfate of barium 天然硫酸钡
native sulfur 天然硫
native sulfur ore 天然硫矿石
native sulphur 自硫
native sulphur deposit 天然硫矿床
native system demand 本系统需求量
native uranium 天然铀
native variety 地方品种
native vegetation 乡土植被;原生植被;天然植被;地方植被
native vermillon 天然辰砂
native water 原地水;岩层水;同生水;天然水
native white lead 天然白铅
NATM 新奥法
natramblygonite 钠磷锂铝石
natric horizon 碱土层
natrii diatrizoas 泛影钠
natrikalite 杂钠钾盐
natrite 钠碳石
natrium amalgam 钠汞齐
natrium arsenicum 砷酸钠
natrium biboricum 硼砂
natrium brine 钠盐水
natrium carbonicum 无水碳酸钠
natrium citricum 柠檬酸钠
natrium cooled valve 钠冷却气门
natrium hydrocitricum 柠檬酸氢钠
natrium iodide 碘化钠
natrium lamp 钠蒸汽灯;钠光灯
natrium lead 含钠铅合金
natrium oleinicum 油酸钠
natrium tetradecylsulfuricum 十四烃基硫酸钠
natrium uranospinite 钠钙砷铀矿
natrium wolframate 钨酸钠
natroalunite 钠明矾石
natroapophyllite 斜钠鱼眼石
natroautunite 钠钙铀云母
natrobistantite 钠细晶石
natrocalcite 钠方解石
natrochalcite 钠铜矾
natrodavyne 钠钾霞石
natrodavynite 钠钾霞石
natrodine 碘钠石
natrodufrenite 钠绿鳞高铁石
natrohitchcockite 钠磷铝铅矿
natrojarosite 钠铁矾
natrolite 钠沸石
natrometer 测钠计
natromontebrasite 钠磷锂铝石;羟磷铝钠石
natron 含水苏打;泡碱;天然碱;碳酸钠
natronbiotite 钠黑云母
natron calc 碱石灰
natroncalk 烧碱石灰
natroncancrinite 钠钙霞石
natron catapliite 多纳锆石
natroninobite 钠铌矿
natronite 碳酸钠
natron lake 高碱湖;碱湖;泡碱湖;苏打湖
natron saltpeter 钠硝石
natroopal 钠蛋白石
natrophilite 磷钠锰矿
natrophosphate 水磷钠石
natrosilite 硅钠石
natrotantite 钠钽矿
natroxonotlite 钠硬硅钙石
nat-rubber 天然橡胶
natrum 天然碱

natte 花篮状编织花饰;席纹柱
nattier blue 淡蓝色(的)
natty 干净的;整洁的
natural 自然的;固有的;天然的
natural γ-ray background 天然伽马射线背景
natural abode 习生地
natural abradant 天然磨料;天然磨蚀剂;天然研磨料(如金刚砂)
natural abrasive 天然研磨料(如金刚砂);天然磨料
natural abundance 自然丰度;天然丰度
natural acanthus leaf 莨苕叶形装饰;未加工叶板;天然叶板
natural acceleration 固有加速度
natural accretion 自然淤积
natural acid 天然酸
natural acoustic(al) impedance 固有声阻抗
natural activity 天然放射性
natural adaptation 自然适应
natural adhesive 天然胶合剂;天然黏结剂;天然黏合剂;天然胶黏剂
natural admittance 固有导纳;特性导纳
natural adobe clay 天然灰质黏土
natural adsorbing agent 天然吸附剂
natural aeration 自然曝气
natural aeration pond 自然曝气池
natural affection 亲缘关系
natural afforestation 天然造林法
natural ageing 自然时效;天然时效;自然老化;自然陈化
natural agent 自然力
natural agglutination valence 自然凝集价
natural agglutinin 自然凝集素
natural aggregate 自然团聚体;天然骨料;天然团块;天然聚合体;天然集料;天然骨料
natural aggregate concrete 天然集料混凝土;天然骨料混凝土
natural aging 自然时效
natural air 自然空气
natural air circulation 自然空气循环;空气自然循环
natural air cooling 自然通风冷却;自然空气冷却
natural air cooling system 自然空气冷却方式
natural air crossing 天然风桥
natural air drying 自然风干(法)
natural air inflow 自然进气量
natural air pollution 天然空气污染
natural air void 空隙
natural alignment diagram 自然列线图
natural allowance 自然容许度;自然公差
natural alloy 天然合金
natural alloy iron 天然合金铁
natural alumina 天然氧化铝
natural alumin(i)um and baked enamel finish 普通铝及烤漆饰面
natural aluminosilicate mineral 天然硅酸铝无机物
natural amelioration of pollution 污染自然净化
natural anaerobic digestion 自然厌氧消化
natural angle of repose 自然休止角;天然休止角;天然坡息角;天然安息角
natural angle of slope 自然倾斜角;自然坡角;天然休止角;天然倾斜角;(松散体的)天然坡角
natural angular frequency 固有角频率
natural angular velocity 固有角速度
natural anhydrite 天然无水石膏;天然硬石膏
natural anomaly 自然现象异常
natural antenna frequency 固有天线频率
natural aquatic humic substance 天然水生腐殖质
natural aquatic system 天然水体系;天然水生生态系统
natural arch 天然平衡拱;自然拱;天然拱
natural area 原始区;自然区;自然发展的地区
natural arts 天然技艺
natural asphalt 天然(地)沥青
natural asphalt colo(u)r 天然沥青色的;自然沥青色的
natural asphalt tile 天然沥青瓦片
natural aspirated diesel engine 自然吸气式柴油机
natural aspiration 自然吸气
natural assimilating power 自然净化能力
natural atmosphere 天然大气
natural atmospheric dispersion 天然大气分散胶体
natural atmospheric dispersoid 天然大气分散胶体
natural attenuation quantity of noise 噪声自然衰减量
natural attenuation 自然衰减;固有衰减

natural attenuation of recharge source 补给源自然衰减
natural background 自然本底;自然背景;天然本底
natural background analysis 自然本底分析
natural background level 自然本底值;自然背景值
natural background radiation 自然本底辐射;天然本底辐射
natural background radiation equivalent activity 天然本底放射性当量活度
natural balance 自然平衡;天然均衡
natural bank 天然岸坡
natural bar 天然砂洲;天然沙洲;天然沙滩;天然沙坝;天然浅滩;天然潜洲
natural barium sulfate 天然硫酸钡
natural barrier 天然障碍
natural base 自然对数的底;自然地基;天然碱;天然地基;生物碱
natural basin 自然流域
natural bathing water standards 天然浴水标准
natural bauxite cement 天然高铝水泥;天然矾土水泥
natural beach 天然沙滩;天然湖滩;天然海滩
natural beauty 自然美
natural bed 天然岩石表面;天然基层;天然岩层;自然床层;自然层;天然河床;天然底盘;天然层
natural belt 自然地带
natural bentonite 天然膨润土
natural binary code 普通二进制代码
natural biogeochemical cycling 自然生物地球化学循环
natural biologic(al) method 自然生物法
natural biologic(al) treatment 天然生物处理
natural biologic(al) treatment method 自然生物处理法
natural biologic(al) waste(water) process 自然生物污水处理法
natural biotic area 自然生物区
natural biotic community 自然生物群落
natural birefringence 自然双折射
natural bit 自然比特
natural bitumen 天然沥青
natural bituminized sandstone 天然沥青砂岩
natural blanket 天然铺盖
natural bleaching 天然漂白
natural bond sand 天然黏结砂
natural-born 天生的
natural boundary 自然边界
natural boundary condition 自然边界条件
natural boundary line of ore body 矿体自然边界线
natural boundary survey 自然边界测量
natural bow 静挠度
natural breadth 自然宽度
natural bridge 天生桥;天然桥
natural brine 天然卤水;天然盐水;天然咸水
natural bristle brushes 天然鬃刷
natural buffer 天然缓冲
natural building sand 天然建筑用砂
natural building stone 天然建筑石料
natural business year 自然营业年度;自然年度
natural butimen 天然沥青
natural calamity 自然灾害;天灾
natural calcium silicate 天然硅酸钙
natural capacitance 固有电容
natural capacity of the environment to pollutant 环境污染物自然容量
natural carbon 天然炭
natural catalyst 天然催化剂
natural cause 自然因素
natural cavern 天然洞室
natural cavity 天然洞穴
natural cellulose 天然纤维素
natural cement 自然水泥;天然黏合剂;天然水泥
natural cementing agent 天然黏结剂;天然胶结剂
natural cement rock 天然水泥灰岩
natural chalk 天然粉笔;天然白垩
natural change 天然变化
natural channel 自然河槽;天然河道;天然航道
natural channel control 河槽自然调节;河槽的自然调节;天然河槽卡口段
natural channel control section 天然河槽控制断面
natural characteristic water temperature 自然特征水温
natural charcoal 天然木炭
natural chasm 天堑

natural chemical concentration 自然化学浓度
natural chlorine 生物产生的氯
natural cinnabar 天然朱砂
natural circuit 自然循环
natural circular frequency 自然周频率;自然圆弧频率;固有圆频率
natural circulation 自然循环;自然环流
natural circulation boiler 自然循环锅炉
natural circulation cooling 自然循环冷却
natural circulation cooling system 自然循环冷却系统
natural circulation evapo(u)rator 自然循环蒸发器
natural circulation heating 自然循环供暖
natural circulation kiln 自然循环干燥窑
natural circulation reactor 自然循环反应堆
natural circulation water cooling 自然循环的水冷却
natural clarification 自然澄清;天然澄清
natural classification 自然分类法;自然分类;按性质分类
natural classification system 自然分类系统
natural clay 天然黏土;天然白土
natural clear varnish 天然纯清漆
natural cleft 自然裂缝;天然裂口;天然裂缝
natural coagulant 天然混凝剂
natural coagulation 自然絮凝;自然凝固
natural coal 天然焦
natural coarse aggregate 天然粗集料;天然粗骨料
natural coarse sand 天然粗砂
natural coating 加固橡胶保护层
natural coke 天然焦
natural colloid 天然胶体
natural colo(u)r 彩色;自然色;固有色;天然色;本色
natural colo(u)red 自然色的;天然色的
natural colo(u)red earthenware 天然彩色陶瓷
natural colo(u)ring matter 天然色素;天然色料
natural colo(u)r photography 天然色照相法;天然彩色摄影
natural colo(u)r reception 天然色接收
natural colo(u)r-relief presentation 自然光照地貌立体表示法
natural community 自然群落
natural commutation 自然换向
natural complex 自然综合体
natural concentration 自然浓度
natural concrete aggregate 天然混凝土骨料;天然混凝土集料
natural concrete heat 天然混凝土比热;天然混凝土释放热
natural condensated oil 天然凝析油
natural condition 自然条件;天然条件;真实条件
natural conditioning 自然温湿度调节
natural condition test 自然条件试验;天然腐蚀试验
natural conservation 自然保障;自然环境保护;自然保护
natural conservation area 自然保护区
natural conservation campaign 自然保护运动
natural conservation education 自然保护教育
natural conservation law 自然保护法
natural conservation policy 自然保护政策
natural conservation zone 自然保护区
natural consistency 天然稠度
natural consistency of soil 土的天然稠度
natural consistency test 天然稠度试验
natural construction material 天然建筑材料
natural containment 天然防护层(如水、土);天然保护层(如水、土)
natural control 自然控制;自然防治
natural convection 自然对流;自由对流
natural convection air-cooled condenser 自然对流空气冷却式冷凝器
natural convection air cooler 自然对流空气冷却器;自然对流风冷器
natural convection cooling 自然对流冷却
natural convection heat transfer 自然对流换热
natural convection recirculation 自然对流再循环
natural convector 自然对流放热器
natural convolution 天然转曲
natural cooling 自然冷却;对流冷却
natural cooling pond 自然冷却池
natural cooling water 自然冷却水
natural coordinates 自然坐标
natural copal 天然快干树脂;天然树脂;天然硬树脂(胶)
natural copper 天然铜
natural cork 软木塞;天然软木
natural cork insulating board 天然热绝缘软木板
natural correlation 自相关
natural correspondence 自然对应
natural cosecant 余割真数
natural cosine 余弦真数
natural cotangent 余切真数
natural count 自然计数
natural course of river 河道天然走向;河道天然行径
natural cover 自然覆盖(植)物;天然掩蔽;天然覆盖物
natural crack 自然裂纹
natural cracking catalyst 天然裂化催化剂
natural crook timber 天然曲材
natural crystal 天然矿石;天然晶体
natural curbstone 天然阶沿石;天然路缘石;天然侧石;天然路边石
natural curing 自然养护;自然干燥;自然发酵
natural current 自然流;自然电流;中性线电流;地电流
natural-current method 自然电流勘探法
natural cutoff 自然裁弯(段);天然裁弯段
natural cut-off channel 河槽自然裁直
natural cyanobacterial bloom 天然蓝细菌水华
natural cycle 自然循环;自然环流
natural cycle system 自然循环系统
natural damping 自然阻尼;固有阻尼
natural day 太阳日;常用日
natural(day)light 自然光
natural decay 自然腐化
natural decay of organic materials 有机物质自然腐化
natural decay resistance 耐自燃腐朽性
natural defect 自然缺陷;自然缺陷
natural defence 自然防护
natural degradation 自然退化;自然刷深;自然降解;自然分解
natural density 天然重度
natural deposit 天然沉积物
natural depreciation 自然折旧
natural depression 自然贬值;自然折旧;地面沉陷;天然洼地;天然凹陷
natural depth 天然水深;天然深度
natural desiccate 自然干燥
natural detector 矿石检波器
natural deterioration 自然恶化
natural detrition 自然风化破碎作用
natural development area 自然发展地区
natural diamond 天然金刚石
natural diamond bit 天然金刚石钻头
natural dike 天然堤;冲积堤
natural diminution 自然减损
natural disaster 自然灾害;天灾
natural disaster reduction 减少自然灾害
natural disaster warning system 自然灾害预告系统
natural discharge 日常流量;天然排泄;天然流量
natural disintegration 天然衰变
natural dislysis 自然渗析
natural dispersion 自然色散
natural divide of groundwater 地下水天然分水岭
natural division 自然区划;自然划分
natural draft 自然通风;自然抽风
natural draft boiler 自然通风锅炉;自然抽风锅炉
natural draft burner 自然通风燃烧器
natural draft chimney 自然抽风烟囱;自然抽气的烟囱
natural draft cooling tower 自然通风冷水塔;自然通风冷却塔
natural draft drier 自然通风干燥机
natural draft gas burner 自然通风气体燃烧器
natural draft kiln 自然循环干燥窑
natural draft machine 自然通风电机
natural draft pit furnace 自然通风地坑炉
natural draft transformer 自然风冷变压器
natural draft ventilation 天然通风;自然通风
natural drainage 自然排水;天然排水
natural drainage-channel 自然排水沟
natural drainage flow 自然水流
natural draught 自然通风;自然干燥
natural draught burner 大气式燃烧器
natural draught chimney 自然通风烟囱
natural draught cooling tower 自然通风冷却塔
natural draught indirect contact cooling tower 自然通风间接触式冷却塔
natural draught ventilation unit 自然通风组件;自然通风单元
natural draw ratio 自然拉伸比
natural dry(ing) 自然干燥
natural drying out 自然风干;自然弄干;自然疏干
natural dry unit weight 天然干容重;天然干重度
natural durability 天生持久性
natural durability test 自然耐久性试验
natural dust 自然尘土;天然粉尘
natural dyestuff 天然颜料;天然染料
natural earth 天然土
natural earthed body 自然接地体
natural earth potential 自然地电势
natural earthquake 天然地震
natural earth surface 天然土路面;天然土表面
natural ecology 自然生态学
natural economy 自然经济
natural ecosystem 自然生态系(统)
natural elastic limit 自然弹性限度;自然弹性极限
natural electric(al) current 自然电流
natural electric(al) field 自然电场
natural electric(al) field method 自然电场法
natural electric(al) potenti(al) instrument 自然电位仪
natural electrode 天然石墨电极
natural electromagnetic energy 自然电磁能量
natural electromagnetic field fluctuation 自然电磁场起伏
natural electromagnetic phenomenon 自然电磁现象
natural element 自然元素;自然要素;天然元素
natural elimination 自然消除;自然淘汰
natural embedding 自然嵌入
natural emulsion 天然乳化液;天然乳剂;天然乳液;天然乳胶
natural endemicity 地方的特性
natural endowment 天然条件
natural enemy 自然敌害;天敌
natural energy cycle 自然能量循环
natural energy source 自然能源
natural environment 自然环境;天然环境
natural environmental anomaly 自然环境异常
natural environmental carcinogen 自然环境致癌物
Natural Environment(al) Conservation Act 自然环境保护法(令)
natural environmental element 自然环境要素
natural environmental structure 自然环境结构
natural environment deterioration 自然环境恶化
natural environment influence 自然环境影响
natural environment preservation area 自然环境保护区
natural environment rehabilitation 自然环境复原;恢复自然环境
natural environment stress 自然环境压力
natural equilibrium 自然平衡
natural equivalence 自然等价;固有等级
natural erosion 自然侵蚀;自然冲刷
natural error 自然误差;天然误差
natural escape 自然流道;天然溢洪道(小水库);天然逸水口
natural etched figure 天然蚀像
natural eutrophication 自然富营养化
natural evapo(u)ration 自然蒸发
natural everglades area 天然沼泽区
natural excitation 自然励磁
natural exhaust 自然排风
natural exhaust system 自然排风系统
natural expenses 正常费用
natural exposure 天然曝晒;天然露头
natural extender 天然增量剂;天然体质颜料
natural extension 自然扩张
natural face of stone 石料天然面
natural face stone 天然护面石;天然饰面石
natural factor 自然因素
natural factors of mortality 死亡率的自然因素
natural fall 天然落差
natural farming 自然农业
natural fat 天然脂肪
natural fault 天然断层
natural feature 自然形态;自然地貌;天然地形;自然地形;自然地理要素;天然地物;地形;地势
natural feature name 自然要素名称

natural feature of terrain 地貌
natural feed(ing) 重力送料
natural ferrite 天然铁氧体
natural fertilizer 自然肥料;天然肥料
natural fiber 天然纤维
natural field method 天然场法
natural filtration 自然过滤(法)
natural filtration rate 自然过滤速率
natural fine aggregate 天然细集料;天然细骨料
natural fine grain 天然细晶粒;天然细颗粒
natural finish 本色处理;天然涂料;天然表面层色;本色饰面;自然罩面;显真饰面;天然装饰;天然饰面材料;防护漆处理
natural finish tile 天然色釉面砖;不上釉的瓷砖
natural fitness 自然界的合理性
natural flag(stone) 天然薄层砂岩;天然石板;天然铺路石板
natural flank pressure 初始围眼压力
natural flow 自然流量;未调节水流;未调节的流量;天然水流;天然径流
natural flowering hedge 自然式花篱
natural flow field 天然流场
natural flow hot water 自然流动式热水
natural flow station 自流水力发电站;自然发电站;自流发电站;水利发电站
natural-flow theory 天然流理论
natural fluorescence 自然荧光;天然荧光
natural flushing 自然冲洗
natural flux 自然通量;自然熔剂
natural foliage 天然簇叶;天然叶饰;自然叶饰
natural food 自然食品;天然食物
natural forage basis 天然饲料地
natural forces 自然力
natural forest 自然森林;天然森林;天然林
natural forest protection project 天然林保护工程
natural form 天然形状原貌
natural formation 自然建造
natural fortification 天然防御工事
natural foundation 自然基础;天然基础;天然地基
natural fracture system 自然断裂系统
natural frame 自然标架;天然弯曲的肋骨木材
natural freezing method 天然冻结法
natural frequency 自振频率;自然频率;固有频率;天然频率
natural frequency conveyer 摇臂式输送机
natural frequency of a foundation 基础固有频率
natural frequency of structure or member 结构物或构件的自然频率
natural frequency of tube 管子固有频率
natural frequency of vibration 自振频率
natural fuel 天然燃料
natural function generator 自然函数发生器;解析函数发生器;解析函数编辑程序
natural gamma ray logging 自然伽马射线测井
natural garden 天然花园
natural garden style 自然式园林
natural gas 石油气;自然气体;天然气
natural gas accumulation 天然气聚集
natural gas-based channel black 天然气槽黑
natural gas burner 天然气燃烧器;天然煤气燃烧器
natural gas chemical industry 天然气化工
natural gas chemical plant 天然气化工厂
natural gas compressor 天然气压缩机
natural gas condensate 天然气凝析油;天然气凝析液
natural gas deposit 天然气储[贮]藏
natural gas deposit of dissolved-in-water type 水溶型天然气矿床
natural gas engine 天然气发动机
natural gas exploration 天然气勘探
natural gas extraction 天然气开采
natural gas field 天然气(气)田
natural gas gasoline 天然气体汽油
natural gas gathering 天然气的收集
natural gas grid 天然气管网
natural gas industry 天然气工业
natural gas in igneous 岩浆岩中天然气
natural gas in marine 海洋中天然气
natural gas in marine sediment 海洋沉积物中的天然气
natural gas in metamorphic rocks 变质岩中天然气
natural gas in solution 溶液中的天然气
natural gas line 天然气管线
natural gas liquid 天然气液化;天然气液态产物

natural gas locomotive 天然气机车
natural gas luquefaction 天然气液化
natural gas occurrence 天然气产状
natural gas odorant 天然气增味剂
natural gasoline 无添加剂的汽油;天然汽油
natural gasoline plant 天然汽油厂
natural gasoline stabilizer 天然汽油稳定塔
natural gas policy act 天然气政策法
natural gas power station 天然气动力站
natural gas processing 天然气加工;天然气处理法;天然煤气的处理方法
Natural Gas Processors Association 天然气处理者协会(美国)
natural gas production rate 天然气产量
natural gas pyrolysis 天然气裂解
natural gas refining 天然气提纯
natural gas resource 天然气资源
natural gas rich in oil vapo(u)rs 富有石油气的天然气
natural gas sand 天然气砂岩
natural gas seepage 天然气苗
natural gas show 气显示
natural gas spring 天然气泉
natural gas storage 天然气储存;天然煤气的储藏
natural gas system 天然气系统
natural gas terminal 天然气终端
natural gas transportation 天然气输送
natural gas well 天然气井
natural gas with helium 含氦天然气
natural gas with high carbon dioxide 高二氧化碳天然气
natural gas with high helium 高氦天然气
natural gas with high nitrogen 高氮天然气
natural gas with high sour 高硫化氢天然气
natural gas with low nitrogen 低氮天然气
natural gemstone 天然宝石
natural generation of electric(al) power 天然地热电站
natural geologic(al) factor 自然地质因素
natural geologic(al) force 天然地质力
natural geologic(al) hazard 自然地质灾害
natural geologic(al) process 自然地质作用
natural geomorphology 风景地貌
natural geomorphy 自然地貌
natural geothermal reservoirs 天然的热储[贮]藏
natural geothermal system 天然地热系统
natural geothermal well 天然地热井
natural glass 火山玻璃;天然玻璃
natural glaze 天然原料配成的釉
natural glue 天然黏合剂;天然胶黏剂;天然胶;动植物胶
natural gradation 自然分次
natural grade (原始地表的)自然标高;自然高程;自然坡度;自然地平;天然地平
natural grade line 天然地平线
natural gradient 自然坡度
natural grading 天然级配
natural grain 天然粒面
natural grain finish 露木纹油漆
natural granary 天然粮仓
natural granite block 天然花岗石荒料
natural granule 天然团粒
natural graphite 天然石墨
natural graphite brush 天然石墨电刷
natural graphite powder 天然石墨粉
natural grass 杂草
natural grass(land) 天然草地;天然牧场
natural gravel 天然砾石
natural gravel material 天然砾材
natural gravel-sand mix(ture) 天然砾石和砂的混合物
natural gray 天然灰色
natural gray yarn 天然本色毛纱
natural grazing ground 天然牧场
natural grease 天然脂(重质原油)
natural grindstone 天然磨石
natural ground 原地面;自然地面;自然地层;天然地面;天然地基
natural ground airfield 天然土机场
natural ground level 天然地面高程;天然地面标高;原地面高程;原地面标高
natural groundwater level 天然地下水位
natural groundwater pollution 地下水的自然污染;地下水的天然污染

natural groundwater storage 地下水天然储量
natural group 自然类群
natural growth rate of urban population 城市人口自然增长率
natural gum 天然树胶;天然胶树胶;天然胶
natural gypsum 天然石膏
natural habitat 自然生境
natural half-width 自然半宽
natural harbo(u)r 自然港(口);港湾;天然港口;天然港(湾)
natural hard aggregate concrete tile 天然硬集料混凝土瓦;天然硬骨料混凝土瓦
natural hardness 自然硬度;天然硬度
natural haven 天然避风港
natural haversine formula 半正矢自然函数公式
natural hazard 自然灾害;天灾
natural hazard anomaly 自然灾害异常
natural hazard plan 自然风险保护计划
natural head 自流压头;天然水头
natural heat 比热;内在热;固有热
natural heat dissipation 自然散热
natural heat flow 天然热流
natural heating 自然采暖;自然加热;自然发热
natural heat output 天然热流量
natural heat release light curve 自然热释光曲线
natural heat resource 天然热资源
natural heritage 天然遗传性
natural high polymer 天然高聚物
natural history 自然史
natural history museum 自然历史博物馆
natural homomorphism 自然同态
natural horizon 视地平
natural humidity 自然湿度;固有湿度
natural hydraulic mortar 天然的水凝砂浆
natural hydrocarbon 天然碳氢化合物
natural hydrograph 自然水文曲线
natural hydrologic hazards 自然水文灾害
natural ice 天然冰
natural ignition engine 自燃着火发动机
natural illumination 昼光照明;天然光照明;天然采光
natural illumination factor 自然采光照度系数;天然采光系数
natural immersion 自然浸入
natural impedance 固有阻抗;特征阻抗;特性阻抗
natural impregnation 自然浸渍
natural increase 自然增长;天然增殖
natural increase of population 人口自然增长
natural increase of value 自然增值
natural increasing 自然递增
natural increasing process 自然递增过程
natural indigo 天然靛青
natural inductance 固有电感
natural infection focus 自然疫源地
natural inflow 天然入流;天然来水
natural injection 自然内射
natural inner product 自然内积
natural inorganic pigment 天然无机颜料
natural insecticide 天然杀虫剂
natural interest rate 自然利率
natural interference 自然干扰
natural interlock 自然联锁
natural involution 自然对合
natural iron 天然铁
natural iron oxide 天然氧化铁
naturalism 自然主义(指建筑、艺术品)
natural isomorphisms 自然同构
natural isotope 天然同位素
natural isotopic abundance 天然同位素丰度
natural isotopic composition 天然同位素组成
naturalistic design 自然式设计;自然设计
naturalistic form 博物学的方式;自然主义形态;自然主义形式
naturalistic garden 自然景致园;自然式庭园
naturalistic treatment 自然主义处理
naturalization 归化;入籍
naturalized 归化的
naturalized plant 归化植物
natural key 自然分类特征
natural knee 天然弯曲的肘材
natural laboratory 自然实验室
natural landscape 自然风景;自然景色;自然景观;天然景色;天然景观;天然风景
natural landslide 自然滑坡

natural landslip 自然滑坡
natural land subsidence 自然地面沉降
natural lane 天然通道
natural latex 天然胶乳
natural latex rubber 天然胶乳橡胶
natural law 自然规律；自然法则
natural law generator 解析函数发生器
natural layer 自然层；天然层
natural leaching 自然浸出
natural leakage 自漏
natural length 固有长度
natural levee 自然堤；天然堤；冲积堤；滨河床砂坝
natural levee deposit of delta 三角洲天然堤沉积
natural level of water 天然水位
natural life-span 自然寿命
natural light 天然光
natural lighting 自然照明；自然光照；自然采光；天然照明；天然采光；采光
natural light source 天然光源
natural lightweight aggregate 天然轻骨料；天然轻集料
natural lime 天然石灰
natural lime-silica cement 天然石灰硅砂水泥
natural limestone 天然石灰石
natural limit 自然极限
natural linear mapping 自然线性映射
natural line broadening 自然致宽
natural line of resistance 天然抵抗线
natural line shape 谱线自然形状
natural line width 自然线宽；固有线宽
natural liquid fuel 天然液体燃料
natural load 自然荷载；天然含沙量
natural load-transmitting arch 天然卸载拱
natural lodestone 天然磁石
natural logarithm 自然对数
natural look 自然外观
natural loss 自然损耗
natural loss of weight during transit 途中自然减重
naturally air-cooled engine 自然空冷式发动机
naturally aspirated engine 自然吸气式发动机；自然抽气机；非增压式发动机
naturally bonded mo(u)lding sand 天然型砂
naturally bonded sand 天然黏土砂
naturally developed slope 自然形成的边坡
naturally hard steel 天然硬钢；高碳合金钢
naturally hard water 天然硬水
naturally lighted 天然采光的；自然光的
naturally lighted from one side 单侧采光的
naturally occurring isotope 自然界存在的同位素
naturally occurring radio active substance 天然放射性物质
naturally occurring radio activity 天然来源的放射性
naturally occurring repository 天然资源储藏所
naturally occurring substance 天然物质
naturally occurring uranium 天然铀
naturally refractory construction material 天然耐火建筑材料
naturally regulated river 自然调节河(流)
naturally regulated stream 自然调节河(流)
naturally seasoned 自然干燥的
natural macromolecular flocculant 天然大分子絮凝剂
natural magnet 天然磁铁
natural magnetism 天然磁性
natural maintenance 自然养护
natural man-made rubber 合成天然橡胶
natural marble 天然大理石
natural marble block 天然大理石荒料
natural margin 天然界限
natural material 自然材料；天然材料
natural mating 本交
natural matrix 固有矩阵
natural matter 天然物质
natural maturity 自然成熟
natural meadow 天然草地
natural measure 自然测度
natural mechanical property 自然机械性
natural medium 自然介质
natural medium sand 天然中等粒径砂
natural membranes 天然膜
natural metal 自然金属
natural metric 自然度量
natural mica 天然云母

natural migration of population 人口自然迁移
natural mineral aggregate 天然矿物材料；天然矿物骨料；天然矿物集料
natural minerals 天然无机物质
natural mixture of asphalt 天然石油沥青混合物；天然沥青混合物
natural mode 自然振型；自然态；自然模态
natural model 自然模型
natural mode of vibration 固有振型；固有振动方式
natural mode shape 固有振荡频率
natural moire 天然波纹
natural moisture 天然含水量；自然湿度
natural moisture content 天然湿度；天然含水率；天然含水量
natural moisture of aeration zone 包气带天然湿度
natural monument 自然纪念碑；自然标志；天然纪念物；天然标石
natural mortality rate 自然死亡率
natural mo(u)lding sand 天然型砂
natural mud 自然泥浆；天然泥浆
natural mud surface 天然泥面线
natural multiple 自然倍数
natural multiplication 自然乘法
natural muscovite 天然白云母
natural naturalized plant 自然归化植物
natural (navigable) waterway 天然航道；天然通航水道
naturalness 自然；天然性
natural neutron 天然中子
natural noise 自然噪声；天然噪声
natural non-tectonic movement 自然的非构造运动
natural norm 自然模
natural number 自然数
natural number designate 自然序数编号法
natural obstacle 天然障碍
natural occurrence 天然产状
natural oil 天然矿物油；天然油；天然石油
natural oil circulation 自然油循环
natural oil cooling 自然油冷却
natural order 自然顺序
natural ordering 自然顺序关系
natural ore ingredient 原矿石配矿
natural organic acids 天然有机酸
natural organic chelating agent 天然有机螯合剂
natural organic chemistry 天然有机化学
natural organic compound 天然有机化合物
natural organic dye(stuff) 天然有机染料
natural organic fertilizer 天然有机肥
natural organic matter 天然有机物
natural organic matter removal 除天然有机物
natural organic pigment 天然有机颜料
natural organic pollutant 天然有机污染物
natural organic polymeric flocculant 天然有机高分子絮凝剂
natural organism 天然有机物
natural oscillation 自振；自由摆动；固有震荡；固有振荡；基本振荡；本征振荡
natural oscillation frequency 固有振荡频率
natural oscillation of the pantograph frame 受电弓框架自然振荡
natural outcrop 天然露头
natural outdoor weathering test 天气自然作用试验；风蚀暴露试验；室外自然老化试验；自然露天风化试验
natural outlet 天然出口
natural oxidation inhibitor 天然抗氧剂
natural oxidation method 自然氧化法
natural oxide of iron 天然氧化铁
natural oxygen resources 自然氧资源
natural ozone 天然臭氧
natural pace 原步法
natural paraffin wax 天然石蜡
natural parameter 特性参数
natural parent 天然放射系母体
natural park 自然公园；天然公园
natural pasture 天然牧场
natural pattern 实样模；实物模
natural paving flag(stone) 天然铺路砂岩薄板；天然铺路板石
natural paving sett 天然铺路小方石块
natural period 自然周期；固有周期
natural period of swing 摆动固有周期
natural period of vibration 自振周期；固有振动周期
natural permeability 自然渗透性；天然渗透率

natural person 自然人
natural petroleum gas 石油天然气
natural phenomenon 自然现象
natural philosophy 自然哲学
natural pigment 矿质颜料；天然颜料；天然色素
natural pit-run gravel 天然料坑砂砾石
natural planning 自然规划
natural plant height 草高
natural plasma 天然等离子体
natural plastic 天然塑料
natural plastic material 天然塑料性材料
natural point 天然地物点
natural pollination 自然传粉
natural pollutant 自然污染物；自然发生的污染物；天然污染物
natural pollution 自然污染；天然污染
natural (pollution) source 自然污染源；天然污染源
natural (pollution) source waters 自然污染源水域
natural polymer 天然聚合物；天然高分子
natural polymeric flocculant 天然高分子絮凝剂
natural population 天然人口；自然群体；自然群落
natural population growth 人口自然增长
natural port 自然港(口)；天然港口；天然港(湾)
natural positive cone 自然正锥
natural potential 自然位能；自然势能；自然电位；重力势能
natural potential method 自然电位探测法
natural pozz(u)olan(a) 天然混合水泥(波特兰水泥和火山灰的混合物)；天然火山灰
natural pozz(u)olanic material 天然火山灰材料
natural premium 正常奖金
natural pressure cycle 自然压力过程
natural prevention and treatment 天然防治
natural preventive function 天然防护功能
natural price 自然价格
natural privilege 天然特权
natural process 自然过程；天然过程
natural product 自然产物；天然产物；天然产品
natural productiveness 自然生产力
natural profile 自然剖面；平均坡度
natural projection 自然投影
natural property of commodity 商品的自然属性
natural proportion 自然比例
natural proportional limit 自然比例极限
natural propping 自然支撑采矿法
natural protected landscapes area 自然风景保护区
natural protection area 自然保护区
natural pumice 天然轻(浮)石；天然浮石
natural pumice concrete 天然浮石混凝土
natural purification 自然净化(作用)；自净作用；污水自净作用
natural purification capacity 自然净化能力
natural purification characteristic 自然净化特性；自净化特性
natural purification of wastewater by water microorganism 水体微生物污水自然净化
natural purification process 自然净化过程
natural puzzolan(a) 天然火山灰
natural quarrying 自然崩落；(岩石的)天然分解
natural quartz 天然石英
natural radiation 自然辐射；本底辐射
natural radiation belt 天然辐射带
natural radiation environment 天然辐射环境；天然放射性环境
natural radioactive background value 天然放射性背景值
natural radioactive decay 放射性自然衰变
natural radioactive element 天然放射性元素
natural radioactive gas 天然放射性气体
natural radioactive isotope 天然放射性同位素
natural radioactive nucleus 天然放射性核
natural radioactive series 天然放射系
natural radioactivity 自然放射性；自然放射现象；天然放射性
natural radioactivity background 天然放射性本底
natural rain 自然雨
natural range 自然分布范围；天然叠标
natural rate of growth 自然增长率
natural rate of interest 自然利率；正常利率
natural rate of wages 自然工资率
natural reach 天然河段
natural reaeration 自然再充气
natural recharge 降雨补给；天然补给(量)
natural rectifier 矿石检波器

natural reference frame 自然参考架
natural refractory 天然耐火材料
natural refreshment forest 天然休养林
natural regain 自然吸湿率
natural regeneration 自然更新；天然更新
natural regeneration by seeds 天然播种再生
natural regime of groundwater 地下水天然动态
natural region 自然区域；自然区
natural regional environment 自然区域环境
natural regionalization 自然区划；自然分区
natural regulation 自然调节
natural remanent magnetization 固有剩余磁化；天然剩余磁化强度
natural renewal of labo(u)r power 劳动力自然更新速度
natural rent of land 自然地租
natural redox titration 自然氧化还原滴定法
natural reproduction 自然更新
natural reserves 天然储藏量；自然保护区
natural reserve area 自然保护区
natural reservoir 天然水库；天然储库
natural reservoir fluid 天然热储流体
natural resin 天然树脂
natural resin adhesive 天然树脂胶合剂；天然树脂黏结剂；天然树脂黏合剂
natural resin-based concrete curing agent 天然树脂基养护剂
natural resin-based mastic 天然树脂基腻子；天然树脂基玛琋脂
natural resin bonding 天然树脂黏合
natural resin glue 天然树脂胶黏剂；天然树脂胶水
natural resin mastic 天然树脂胶黏水泥
natural resin modified phenol resin 天然树脂改性酚醛树脂
natural resin paint 天然树脂漆
natural resin varnish 天然树脂清漆
natural resistance 天然抗病性；自然阻力；自然阻抗；自然抵抗力
natural resonance 自然共振；固有谐振
natural resonant frequency 固有共振频率
natural resource ecosystem 自然资源生态系统
natural resource endowment 天赋自然资源
natural resource protection 自然资源保护
natural resources 自然资源；自然资源；天然资源；天然财富
natural resources conservation 自然资源保护
natural resources consumption 自然资源消耗
natural resources defense council 自然资源保护委员会
natural resources distribution 自然资源分布
natural resources management 自然资源管理
natural resources of coastal waters 沿海自然资源
natural resources of groundwater 地下水天然资源
natural resources planning 自然资源规划
natural response 必然反应
natural restriction 自然限制
natural retting 天然浸渍
natural reuse 自然再用
natural rigidity 自然稳定性；自然硬度；自然刚性；固有刚度
natural risk 自然风险；正常风险
natural river 天然河流
natural river course 河道天然走向；河道天然行径
natural river deposit 天然河道沉积(物)
natural river engineering model 自然河工模型
natural road 天然土路；天然路
natural roadbed 天然路基
natural rock 天然岩石；天然石料
natural rock aggregate 天然岩石集料；天然岩石骨料
natural rock asphalt 自然岩沥青；天然岩沥青
natural rock asphalt mortar 天然岩沥青(乳)液
natural rock concrete 天然岩石混凝土
natural rock-fill dike 天然堆石堤；天然填石堤
natural rock jointing 岩体自然节理
natural rod 天然圆条；天然棒；天然杆；自然标
natural rolling 自然横摇
natural roughness 天然糙率
natural rubber 自然橡胶；天然橡胶
natural rubber adhesive 天然橡胶胶黏剂；天然胶黏合剂
natural rubber emulsion 天然橡胶乳化剂；天然橡胶乳化液

natural rubber emulsion paint 天然橡胶乳化涂料；天然橡胶乳化油漆
natural rubber hydrocarbon 天然橡胶烃
natural rubber powder 天然橡胶粉末
natural running temperature 正常运转温度
natural runoff 天然径流
natural runoff field 天然径流场
natural sag 自然垂度
natural salt 天然盐
natural sand 自然砂；天然砂
natural sand fine aggregate 天然砂细集料；天然砂细骨料
natural satellite 天然卫星
natural scale 天然尺寸；自然(级)数；直径比率；普通比例尺；固有容积；原尺度；原比(例)尺；自然比例尺；正态比例；基准比例尺；天然比例；天然比尺；数字比例尺；实物大小；非缩尺
natural scale model 天然比例模型
natural scenery 天然景色；天然景观
natural scenic area (未开发的)自然景物区
natural science 自然科学
natural scour 自然演变冲刷
natural screen 天然遮障
natural sea-level 自然海平面
natural seam (岩石中的)天然夹缝
natural seasonal phenomenon 自然季节现象
natural seasoned lumber 风干木材；自然干燥木材
natural-seasoned wood 天然干燥材
natural seasoning 自然时效；自然干燥法；自然干燥；自然缝干法；自然风干(法)
natural seawater 天然海水
natural sebum 天然皮脂
natural secant 正割真数【数】
natural sediment 天然沉积物
natural sedimentation 自然沉淀
natural sedimentation method 自然沉淀法
natural sedimentation tank 自然沉淀池
natural selection 自然选择；自然淘汰；天演；天然选择；天然淘汰
natural selection theory 自然选择说
natural self-cleaning effect 自然自净效力
natural self-purification 自然自净作用
natural self-purification capacity 自净能力；天然自净能力
natural self-supporting arch 自然平衡拱；天然拱
natural separation 自然分离
natural sequence 自然顺序
natural series 天然系
natural sett 天然小方石(块)；天然铺石
natural shelter 自然屏障
natural sheltered harbo(u)r 天然屏蔽港
natural sheltered port 天然屏蔽港
natural shrinkage 自然收缩
natural sienna 天然浓黄土；天然赭土(一种矿物颜料)
natural silk 天然丝；真丝；蚕丝
natural sine 正弦真数
natural sink 自然降低
natural size 原来尺寸；原物尺寸；原大小；原尺寸；自然尺寸；天然尺寸；实物大小；实际尺寸
natural skyline 自然轮廓；地平线
natural slaking 自然消解
natural slate 天然石板(瓦)
natural slop angle above water surface 水上坡角
natural slope 原地面坡度；自然坡度；天然坡(度)；天然边坡
natural smoke control 自然控烟
natural smoke discharge 自然排烟
natural snow cover 天然雪盖；天然积雪
natural sod 天然草皮
natural soft deposit 天然软土层
natural soft porcelain 天然软瓷
natural soil 原生土(壤)；自然土壤；天然土壤；天然土
natural soil deposit 自然堆积土
natural soil drainage 土壤自然排水；土壤天然排水
natural soil road 天然土路
natural soil stratum 天然土层
natural solid fuel 天然固体燃料
natural solute 天然溶质
natural sonar 天然声呐
natural source 天然震源；天然源
natural sources of air pollution 天然空气污染源；空气污染的天然来源
natural sources of water pollution 水污染自然源；水天然污染源

natural spawning place 天然产卵场
natural spiral turn 右螺旋
natural split 天然分隔；天然分离；天然分开；天然分裂；天然裂口；天然裂缝
natural splitting 自然风流分支；自然分风
natural spreading rate 固有涂布率
natural stability 自然平衡状态；自然稳定(性)；固有安定度
natural stability limit 自然稳定极限；天然稳定极限
natural state 自然状态；天然状态
natural steam 自然蒸汽；天然蒸汽
natural-steam field 天然蒸汽田
natural steam power plant 地热力发电站；地热发电厂
natural steam power station 地热发电站
natural steatite 天然皂石
natural steel 天然硬度钢；天然钢；初生硬度钢
natural step length 自然步长
natural stiffness 天然硬度；天然坚硬性；天然刚度；固有刚度
natural stock 自然资源
natural stone 琢石；天然石料；天然石(材)
natural stone altar 天然石头祭坛
natural stone(angle)quoin 天然隅石块
natural stone arch 天然石拱
natural stone arch bridge 天生石拱桥
natural stone bench 天然石阶；天然石凳
natural stone block 天然大块石
natural stone-block paving 天然石块铺面；天然石块铺砌
natural stone capping 天然石头覆盖物
natural stone column 天然石柱
natural stone coping 天然石顶
natural stone corbel 天然石托；天然石撑
natural stone course 天然石层
natural stone dressed ready for building 已修琢成的建筑天然石料
natural stone dressing machine 天然石料琢面机
natural stone drilling machine 钻天然石机
natural stone facing 抛石护坡；抛石护面
natural stone foundation 天然石基础
natural stone frieze 天然石雕带；天然石壁缘
natural stone keep 天然石城堡
natural stone layer 天然石材铺砌工；天然石层
natural stone mate figure 天然成双石像；天然配对石像
natural stone matting 天然石垫层；天然石垫块；天然石垫子
natural stone order 天然石柱型(尤指古典建筑的柱型)
natural stone paving sett 天然小方石
natural stone polishing machine 天然石料擦光机；天然石料抛光机；天然石面抛光机
natural stone pulpit (教堂中的)石头布道坛
natural stone rubbing machine 天然石磨机
natural stone sculpture 天然石雕塑品；天然石雕刻品
natural stone sidewalk paving flag 人行道铺面用天然石板
natural stone splitter 天然石料裂开器
natural stone table altar 天然石板祭桌
natural stone texture 天然石纹理
natural stone threshold 天然石槛
natural stone tile 天然石瓦
natural stonework 天然石方工程
natural stop 自然光圈
natural storage 天然水库
natural straightening of channel 河槽自然裁直
natural strain 自然应变；固有应变
natural strained well 天然滤水井
natural strainer well 天然过滤器井
natural stream 天然河流
natural(stream)channel 天然河槽；天然河床
natural strength 自然强度
natural stress 固有应力；自然应力；初始应力
natural stress observation 地应力观测
natural stress relief 自然应力消除
natural structure 自然结构
natural style 自然式
natural subgrade 天然地基
natural subirrigation 天然地下渗灌；天然地下灌溉
natural subsoil 天然地基
natural succession 自然演替；天然演替

natural sulphur 天然硫
natural suppression 天然抑制
natural surfaced road 天然路面道路
natural surfaced runway 土跑道
natural surface level 原地面高程;天然地面高程
natural surface line 原地面线;天然地面线
natural synoptic(al) period 自然天气周期
natural synoptic(al) region 自然天气区
natural synoptic(al) season 自然天气季节
natural system 自然系统;自然分类系统;自然分类法;排水系统;自然体系;分类系统
natural system of coordinates 自然坐标系
natural tailwater 天然尾水
natural tangent 正切真数
natural tar 天然沥青
natural target 自然目标
natural termination 自然终止
natural terrain obstacles 天然障碍
natural thinning 自然稀疏
natural tide 自然潮;真潮
natural time of oscillation 固有振荡时间
natural time of viscoelasticity 黏弹性特征时间
natural-tinted fabric 天然本色织物
natural tolerance 天然容许度;天然公差
natural torsional frequency 自由扭振频率
natural tracer 天然示踪物;天然示踪剂
natural trail 天然小径;天然小道
natural transformation 自然变换
natural transient stability limit 自然暂态稳定极限
natural transition curve 自然缓和曲线
natural transmission 自然传输
natural trans-polyisoprene 天然反式聚异戊二烯
natural trass 天然火山灰
natural trigonometric(al) function 三角函数(的)真数
natural tunnel 地下河道
natural tunnel wind 自然隧道风
natural twist 天然转曲
natural type 煤的自然类型
natural type of ore 矿石自然类型
natural ultramarine 天然青蓝
natural unit 自然单位
natural unit weight 天然容重;天然重度
natural unit weight of soil 土的天然容重;土的天然重度
natural uranium 天然铀
natural-uranium fission chamber 天然铀裂变室
natural-uranium reactor 天然铀原子反应堆;天然铀反应堆;天然铀堆
natural varnish resin 天然清漆树脂
natural vaseline 天然凡士林
natural vegetation 自然植被;天然植被
natural velocity 自然流速
natural ventilating dry kiln 自然通风干燥窑
natural ventilating pressure 自然通风压力
natural ventilating system 自然通风系统
natural ventilation 自然通风
natural ventilation pressure 自然通风压力
natural ventilator 自然通风器
natural vermillion 天然朱;天然银朱
natural vibration 自振;自由振动;自然振动;固有振动
natural vibration frequency 自振频率;固有振动频率
natural vibration period 自振周期
natural vibration period of bridge 桥梁自振周期
natural void ratio 天然孔隙比
natural wage 实物工资
natural wastage 自然损耗
natural water 泉水;天然水
natural water body 自然水体
natural water categories 天然水类型
natural water content 天然湿度;天然含水量
natural water course 天然水道;天然河道;自然水道;河道天然走向;河道天然行径
natural water cycle 自然界水循环
natural water drive 自然水力驱动
natural water level 自然水位;日常水位
natural water line 天然水面线
natural water pollution 自然水污染
natural water quality 天然(水)水质
natural(water) resources 天然水力资源;天然水(利)资源
natural waters 自然水体;天然水域

natural water sample 自然水样
natural water shed 天然流域;天然分水线;天然分水岭
natural water source 天然水源
natural water system 自然水系
natural water table 天然水位;天然潜水面;天然地下水位;天然地下水面
natural water temperature 自然水温
natural water temperature region 自然水温区
natural water way 河道天然走向;河道天然行径;天然水系;天然水道
natural wave 固有波;自然波
natural wavelength 固有波长
natural wax 天然蜡
natural wealth 自然资源;自然财富;天然财富
natural wear and tear 自然磨损
natural weathering 自然老化;自然风化;天然老化
natural weather phenomenon 自然天气现象
natural weight 湿重
natural welding 自然焊接
natural well 自然产油井;天然水井;天然井
natural wetland 天然湿地
natural wetland treatment system 天然湿地处理系统
natural wetness 天然湿度
natural whiteness 自然白度
natural whiting 天然碳酸钙
natural width 自然宽度;固有宽度
natural width of energy level 能级的自然宽度
natural wind 自然风;天然风
natural wood colo(u)r 木材本色;天然木色
natural wood resin 天然树脂
natural woods 自然森林;天然森林
natural world 自然界
natural year 自然年;民用年;回归年;分至年
natural zeolite 天然沸石
natural zeolite medium 天然沸石介质
natural zone 自然地带;天然带
natural zoological garden 天然动物园
nature 性质;自然;本质;本性
Nature and Environmental Protection Ministry of France 法国自然与环境保护部
nature balance 生态平衡
nature conservation 自然环境保护;自然保护;自然保持
Nature Conservation Council 自然保护委员会
Nature Conservation Society of Japan 日本自然保护协会
nature-divisions for highway 公路自然区划
nature energy 天然能量
nature gas pipe 天然气管
nature hazard 天险
nature heat releasing light method 天然热释光法
nature laboratory 自然实验室
nature lean angle 自然倾斜角
nature load 河流携砂量
nature of allowance account 备抵账户性质
nature of bank soil 岸土性质
nature of beam 束流性质
nature of beauty 美的自然性
nature of circuit indicator 电路性质标志信号
nature of coal 煤质
nature of coast 海岸性质
nature of distortion 失真性质
nature of flow 流动性能
nature of friction 摩擦性质;摩擦性能
nature of grains 颗粒状态;粒径性质;颗粒性质
nature of ground 岩层性质;围岩的性质
nature of invention 发明的性质
nature of load 荷载性质
nature of materials of rock and soil 岩土的材料属性
nature of minerogenic tectonics 成矿构造性质
nature of navigation obstruction 碍航性质
nature of oil 石油性质
nature of ore 矿石性质
nature of particles 粒纹;颗粒性质
nature of petroleum 石油的属性
nature of schedule of consumer demand 消费者需求表列的性质
nature of soap 皂性
nature of soil 土质;土壤性质;土壤特性;土的性质
nature of sources 来源的性质
nature of standard sample 标准样性质

nature of stress in rock mass 岩体应力特性
nature of the bottom 水底特性
nature of the ground 地层性质
nature of the stream bed 河床性质
nature of works 工程性质
nature oilstone 天然油石
nature protection area 自然保护区
nature regeneration 天然更新
nature reserves 天然蕴藏量;天然储量;自然保护区;自然庇护区;自然保留地
nature reserve management 自然保护区管理
nature sanctuary 自然禁猎区;自然禁猎地
nature shelter 自然保护
nature shock 天然地震
nature slope angle under water 水下坡角
nature-society system 自然社会系统
nature study 自然研究
nature trail 自然公园观赏路;通往自然景区的小径
nature valley 天然谷
nature ventilation in biologic(al) filter 生物滤池自然通风
nature water level before pumping 抽水前的天然水位
Naudet's barometer 丝锁气压计
naugahyde 瑙加海德(乙烯树脂涂面织物,作为家具和墙布)
naught line 零位线
naught point two 零点二
nauheim bath 热碳酸盐水浴
naujakasite 硅铝铁钠石
naumachia (古罗马人观赏海战表演的)场所
naumachy (古罗马人观赏海战表演的)场所
naumannite 硒银矿
Naumene number 诺曼值
naupathia 晕船
naurite 圆层磷灰石
naurodite 似蓝闪石
nauropemeter 倾斜表
nauruite 圆层磷灰石
Nauru Phosphate Corp 瑙鲁磷酸盐公司
nausea 恶心
nausea marina 航海性恶心
nausea navalis 航海性恶心
Nauta mixer 诺塔混合器;诺塔(固体)混合机
nautical almanac 航海天文历;航海年历;航海历(书)
nautical assessor 海事鉴定人;海事公估人
nautical astronomy 航海天文学
nautical chain 航海测链
nautical chart 航海图;海图
nautical chart for special-purpose 专用航海图
nautical charting 海图制图
nautical chronometer 航海天文钟;航海时计
nautical college 航海学院
nautical day 航用日
nautical depth 适航水深
nautical distance 航海距离;航程
nautical ephemeris 航海天文历;航海年历;航海历(书)
nautical equipment 航海用具
nautical geography 海道地理
nautical harbo(u)r chart 海港图
nautical hydrography 海道测量学
nautical instrument 航行仪表;航行设备;航海仪器
nautical map 海图
nautical mark 航行标志;航标
nautical measure 航海长度单位;海程长度
nautical mile 海里
nautical morning light 航用晨光时
nautical navigation 航海学
nautical planisphere 航海用天球平面图
nautical receiving set 水声接收器
nautical satellite 航海卫星
nautical scale 海图比例尺
nautical scurvy 船员坏血病
nautical sextant 航海六分仪
Nautical Society of Japan 日本航海学会
nautical stars 航用恒星
nautical survey 海洋水文测量;海道测量
nautical surveying 海道测量
nautical surveyor 海道测量员;水道测量者;水道测量员
nautical tables 航海(用)表
nautical term 航海用语;航海术语

nautical time 船用时
nautical twilight 航海曙暮光;航海晨昏蒙影
nautics 航海术
nautiloid faunal province 鹦鹉螺动物地理区
nautophone 高音雾笛;雾信号器;电动雾笛;电动高音雾号
navaglide 飞机盲目着陆系统
navaglobe 远程无线电导航系统
navaglobe beacon station 远程无线电导航系统航标站
navaglobe system 远程无线电导航制
navaho 超音速巡航导弹
navajoite 三水钒矿
Navajo rug 纳瓦霍小地毯
Navajo sandstone 纳瓦霍砂岩
naval 海军的;船用
naval academy 海军军官学校
naval aeronautics 海上航空学
naval air station 海军一级航空站;海军航空兵基地
naval anchorage 海军锚地
naval architect 造船技师;造船工程师;海军造船师
naval architecture 造船学;造船工程(学)
naval arsenal 海军军械库;海军兵工厂
naval atlas 海图集
naval auxiliary 海军辅助舰艇
naval base 海军基地
naval blockade 海上封锁;海军封锁
naval blue 海军蓝
naval brass 海军黄铜
naval bronze 海军青铜;海军黄铜
naval chart 军事海图
naval communication 海军通信
naval construction 造船;舰艇建造;船舶构造
naval constructor 海军造船师
naval diving medicine 海军潜水医学
naval dockyard 海军船坞
naval electronic warfare simulator 海军电子作战模拟器
naval engineer 造船工程师;海军轮机员
naval engineering 造船工程(学)
naval harbo(u)r 军港;海军港
naval hydrodynamics 海洋流体动力学
naval hygienics 海军卫生学
naval maneuvers 海军演习
naval meteorology 海洋气象学
naval operating base 海军作战基地;海军行动基地
naval optoelectronic director 舰用光电指挥仪
naval pipe 锚链管
naval pitch 黑色松脂
naval port 海军港口;军港
naval power 海军力量
naval quay 海军码头
naval reactor 舰艇用反应堆;船用反应堆
naval research laboratory 海军研究实验室
naval reserves 海军后备队
naval ship 军用船
naval ships and boats 舰艇
naval shipyard 海军船厂
naval shore 海岸
naval stores 海军补给品;松脂制品;松脂类原料;树脂产品;船用品
naval tank 船模(型)试验池
naval target 海上目标
naval technician 造船技师
naval telescope 单筒望远镜
naval vessel 军舰;海军舰艇;舰艇
naval warfare 海战
naval wharf 海军码头
naval wireless service 海军无线电台
naval yard 造船厂;海军工厂;海军船坞
naval yard crane 海军工厂船坞起重机;造船厂船坞起重机
navamander 编码通信系统
navar 导航雷达;雷达导航控制系统
Navarroan(stage) 纳瓦罗阶【地】
navar-screen system 导航屏幕系统
navascope 机载雷达示位器
Navascreen 导航屏幕
navaspectrum 导航谱
nave 教堂正厅;中厅;中殿;轮毂;教堂的中部;建筑中间广场;铁路车站中间广场
nave aisle 建筑物中央大厅通道;教堂中殿的通道;教堂中部的通道
nave aisle bay 建筑物中央大厅的耳房;建筑物中央大厅的侧房;教堂中部的耳房;教堂中部的侧房;厢堂桁幅
nave aisle gallery 建筑物中央大厅的走廊;教堂中部的走廊;厢堂廊台
nave aisleless 无通道的建筑物中央大厅;无通道的教堂中部
nave aisle passage 建筑物中央大厅通道;教堂中部通道
nave aisle pier 建筑物中央大厅的通道墙墩;教堂中部的通道墙墩
nave aisle vault 建筑物中央大厅的通道拱顶;教堂中部的通道拱顶
nave aisle wall 建筑物中央大厅的墙;教堂中部通道墙
nave aisle wall window 建筑物中央大厅通道墙上的窗;教堂中部通道墙上的窗
nave arcade 中堂连拱廊;(教堂主厅的)侧拱廊
nave bay 建筑物中央大厅耳堂;教堂中部耳堂;建筑物中央大厅厢堂;教堂中部厢堂
nave ceiling 建筑物中央大厅的天花板;教堂中部的天花板
nave collar 毂环
navel 中心点;中央
nave range of columns 建筑物中央大厅的柱间间距;教堂中部的柱间间距
nave vault 建筑物中央大厅的拱顶;教堂中部的拱顶
nave window 建筑物中央大厅的窗;教堂中部的窗
navicert 准运证;特准运照
navicular 船形的
Navier equation 纳维方程
Navier's hypothesis 纳维假定
Navier's law 纳维定则
Navier's theorem 纳维定理
Navier-Stokes equation 纳维—斯托克斯方程;黏性流体方程;黏性流体方程;不可压缩黏性流体运动方程
Navier-Stokes shock structure 纳维—斯托克斯激波结构
navigability 可操纵性;可通航性;可航性;适于航行;适航性【船】
navigable 可通航的;可航行的
navigable airspace 可航空域
navigable aqueduct 通航渡槽
navigable bridge opening 通航桥孔
navigable by sea-going vessel 可通行海轮航道
navigable canal 通行运河;通航河;通航渠道
navigable canal bridge 通航渡槽
navigable channel 航道;航槽;通航渠道;通航河槽;通航水道
navigable(channel)depth 航道水深;通航水深
navigable channel mark 航道标
navigable clearance 通航净空
navigable clear height 通航净高
navigable clear with 通航净宽
navigable condition 通航情况;适航状态
navigable current velocity 通航流速
navigable dam 可通航坝
navigable depth 通航深度
navigable depth for launching 船舶下水水深
navigable dimension 通航尺度
navigable flume 通航渡槽
navigable for sea-going vessels 可供海轮航行的
navigable headwaters 可通航的上游段
navigable inland channel 内河航道
navigable internal waterways in use 内河通航水域
navigable lock 通行船闸
navigable low water channel 枯水通航河道;枯水通航河槽
navigable movable dam 通航活动坝
navigable net span 通航净跨
navigable net width 通航净宽
navigable pass 狭窄通航水道;航道;通行水道;通航峡口
navigable pass(age) 通航水道
navigable pass sedimentation 航道淤积
navigable period 通航期
navigable power canal 通航发电渠道;发电渠道
navigable reach 通航河段
navigable river 通航河流;通航河道
navigable semi-circle 可航半圆
navigable span 可航宽度;通行跨度;通航桥跨
navigable stage 通航水位
navigable stage control of canal 运河通航水位控制
navigable stage frequency 通航保证率
navigable stream 通航河流
navigable stretch 通航河段
navigable water 通行水道;通航水道
navigable watercourse 通航水道;通航水路
navigable water level 通航水位
navigable waters 可通航的水道;通航水域;适航水域
navigable waterway 通航水道;通航渠道;通航河道;通航水路
navigate 驾驶(船);漫游;导航
navigate by sounding 按测深航行
navigating 航行的
navigating allowance 航行津贴
navigating apprentice 驾驶实习员
navigating bridge 驾驶桥楼;航行驾驶台
navigating instrument 航海仪器;导航仪
navigating mark 助航标志
navigating mate 驾驶员【船】
navigating officer 驾驶员(船);船舶航行负责人
navigating on water surface 在水面航行
navigating sextant 航用六分仪;船舶导航六分仪
navigating sight 导航望远镜
navigating telescope 导航望远镜
navigating zone 航行区域;航区
navigation 领航;航运;航行;航进;航海学;航海术;航海;通航;航行术;导航
navigation acts 航海条例
navigation aid facility 助航设施;助航设备
navigation aids 航标;导航系统;助航设施;助航设备;导航辅助设备;海上航标;助航设备
navigational 与航海有关的;航海的
navigational accelerometer 领航加速表
navigational access method 导航存取法
navigational accuracy 导航精度
navigational aid chart 导航设备图
navigational aid in wind finding 导航测风
navigational aids 航标;导航系统;助航设施;助航设备;导航辅助设备;海上航标;助航设备
navigational and guidance 导航与制导
navigational and intercommunication equipment 导航仪及船内通信设备
navigational aspects 有关航行方面的问题
navigational astronomy 航海天文学
navigational buoy 导航浮标
navigational by earth references 地球基准导航
navigational by electronic aids 电子助导
navigational by pilotage 地标导航
navigational calculator 导航计算器
navigational canal 航道
navigational channel 航槽
navigational chart 导航图;海图
navigational chart datum 航行基准面
navigational clearance 通航净空
navigational clock 导航钟
navigational command and control system 导航指挥与控制系统
navigational computation device 导航计算装置
navigational computer 导航计算机;船用计算机
navigational computer unit 导航计算装置
navigational control console 导航操纵台
navigational control simulator 导航控制模拟器
navigational control station 导航控制站
navigational coordinates 导航坐标
navigational course 航道
navigational danger 碍航物
navigational data 导航数据
navigational data display unit 导航数据显示装置
navigational datum 航行基准面
navigational difficulties 航行上的困难
navigational digit computer 导航数字计算机
navigational dike 导航堤
navigational doppler radar 导航多普勒雷达
navigational echo sounder 导航回声测深仪
navigational equipment 导航设备
navigational error 导航误差
navigational facility 导航设施
navigational flame float 导航火炬浮标
navigational gyrocompass 导航回转罗盘
navigational hazard 航行危险物;航天事故
navigational indicator 导航指示器
navigational instrument 导航仪器;导航仪表;航海仪器
navigational lamp 航行灯
navigational light 航行灯;航标灯;导航灯

navigational light buoy 航标灯浮标
navigational light controller 航向灯控制器
navigational lighting 航标灯；航行灯；航行照明
navigational lighting practice 航行灯实地应用
navigational lock 船闸
navigation allowance 航行津贴
navigational map computer 导航图计算机
navigational map-matching 导航地图匹配
navigational mark 航道标(志)；航标
navigational microfilm projector 导航显微胶卷放映机；航海用显微胶卷放映器
navigational observation system 导航监测系统
navigational obstructer 碍航物
navigational obstructions 航行障碍物
navigational parameter 导航参数
navigational pause 断航
navigational picture 导航图
navigational planet 导航行星；航用行星
navigational plot 导航航绘
navigational plotter 航用测绘板
navigational radar 导航雷达
navigational radar equipment 导航雷达装置
navigational radar station 导航雷达站
navigational radio beacon 无线电航标
navigational radio facility 导航无线电装置
navigational range finder 航用测距仪
navigational receiver 导航接收机
navigational requirement 通航要求
navigational route database 导航数据库
navigational route type 导航型
navigational safety system 导航安全系统
navigational satellite 导航卫星；导航人造卫星；航用人造卫星
navigational satellite position system 导航卫星定位系统
navigational satellite receiver 导航卫星接收装置
navigational satellite system 导航卫星系统
navigational sensor 导航传感器
navigational signal 航标；导航信号
navigational smoke float 导航发烟浮标
navigational sonar 导航声呐
navigational specialist 导航专家
navigational speed 航速【船】
navigational star catalog 导航星表
navigational star catalogue 航用恒星表
navigational star number 导航星号
navigational stars 航用恒星
navigational station 导航台
navigational station location survey 导航台定位测量
navigational system 导航系统
navigational system with time and range 导航星
navigational tables 航用表
navigational time scale 导航时标
navigational transmitter 航用发射机
navigational treaty 通航条约
navigational triangle 导航三角形；航用三角形；航海三角形
navigational unit 导航设备
navigational warning 航行警告；航海警告
navigational warning for Australia waters 澳大利亚海域航海警告
navigational warning for Indian Ocean 印度洋航海警告
navigational warning for South Atlantic 南大西洋航行警告
navigational water level 航海水位
navigational window 导航视窗
navigation and positioning at sea 海上导航定位
navigation approach 进闸航道；进坞航道
navigation approach channel 引航道
navigation area 航行区域；航区
navigation area survey 适航区测量
navigation authority 航行管理机构；航务管理机构
navigation benefits 航运效益；通航效益
navigation blockage 封航
navigation bounty 航海奖励金
navigation bridge simulation 驾驶室模拟
navigation bridge wing 船舶驾驶台侧翼
navigation by aid of satellite 卫星导航
navigation by recognition 识别导航
navigation by space reference 天体导航
navigation by triangulation 三角测量导航法
navigation canal 运河；通行运河；通航运河

navigation certificate 航行证书
navigation chamber 升船箱
navigation channel 航道；通航水道；通航渠道；通航河道；船道
navigation channel at low water 枯水航道
navigation channel change 航道变迁
navigation channel chart 航道图
navigation channel dredging 航道疏浚
navigation channel in lake area 湖区航道
navigation channel of estuary 河口航道
navigation(channel) project 航道工程
navigation channel sedimentation 航道淤积
navigation channel variation 航道变迁
navigation chart 航行图
navigation clearance 航行净空
navigation clock 航海钟
navigation-coal 蒸汽煤；锅炉煤；汽锅用煤
navigation condition 航运状态；通航条件；航行条件
navigation congress 航运会议
navigation construction = navigable structure 通航建筑物
navigation control 航行管制
navigation dam 航运坝；通行坝；通航坝
navigation depth 航深
navigation discharge 通航流量
navigation dome 天文观测窗；天文观测舱
navigation during construction period 施工期通航
navigation effect 通航影响
navigation entrance channel 进港航道
navigation facility 通航设施
navigation fix 领航坐标
navigation flame float 航行灯浮标；通航灯浮标
navigation flare 航行照明弹
navigation gear 航空设备；航海设备
navigation head 航运终点；水陆转运站；水陆转运点
navigation hydrojunction 航运枢纽
navigation impact 通航影响
navigation information 领航资料【船】
navigation instrument 助航仪表；助航设施；助航设备；航行仪表；航行设备
navigation interests 航运界
navigation junction 航运枢纽
navigation kerosene 航海煤油
navigation lamp 号灯；航行信号灯
navigation lamps indicator 航行灯指示器
navigation landing craft 领航登陆艇
navigation lane 航路【航海】
navigation law 航海法规；航行法规
navigation light 助航灯；号灯；航行信号灯
navigation lights board 航行灯指示板
navigation lights indicator 航行灯指示器
navigation line 导航线；航向线
navigation lock 通行船闸；船闸
navigation lock capacity 船闸通过能力
navigation low-water channel 枯水通航河道；枯水通航河槽
navigation mark 助航标志；航行标志
navigation monitor 航行监测仪
navigation net height 通航净高
navigation notes and cautions 航行注意及警告事项
navigation obstacle 航行障碍物
navigation-obstructing bridge 碍航桥梁
navigation-obstructing dam and lock 碍航闸坝
navigation-obstructing flow pattern 碍航流态
navigation-obstructing reef 碍航礁石
navigation-obstructing river bend 碍航河弯
navigation of aerial photography 航摄领航
navigation of sailing vessels impeded 帆船航行有障碍
navigation of sailing vessels prevented only 只有帆船避免航行
navigation opening 港口；开合跨；通航桥孔；通航口门；通船桥孔
navigation pass 狭窄航道；通航峡口；通航水道；通航渠道
navigation pause 航行中断；停航
navigation period 通航期
navigation pool 航运水位
navigation radar 航用雷达；船用雷达
navigation requirement 通航要求
navigation reservoir 航运水库；通航水库

navigation risk 航行险
navigation rules 航行规则
navigation safety communications 航行安全通信
navigation school 航海学校
navigation season 航行季节
navigation shapes 航行号型
navigation signal 航行号志；航行标志
navigation skill 航海技能
navigation space 驾驶中舱室
navigation span 可航宽度；航道间隔；可通航宽度；通航桥孔
navigation standard 通航标准
navigation standard of inland waterway 内河通航标准
navigation steps 航运梯级；通航梯级
navigation strip 航线【航海】
navigation strip survey 航线测量【航海】
navigation structure 过船建筑物；通航建筑物
navigation subsidy 航海补助金
navigation system design 航行系统设计
navigation tables 航海表
navigation toll 通航费(用)；通过费
navigation/traffic control 导航和交通管制
navigation triangle 航海三角形
navigation tunnel 通航隧洞；通航隧道
navigation warning 航行警告
navigation watch 航海值班；航行值班
navigation water course 水道
navigation water ga(u)ge 航行水尺
navigation water level 通航水位
navigation way 航路【航海】
navigator 领水员；领航员；领航仪；驾驶员【船】；航海者；航海家；导航员；导航仪；船舶驾驶员；船舶航行负责人
navigator watch 领航表
navigraph 领航表
navistar 导航卫星
navistar global positioning system 导航星全球定位系统
navistar system 导航卫星定位系统
navite 蛇纹粒玄斑岩
navsat 导航卫星
navvy 挖土机；挖泥机；挖掘机
navvy bar 手推土车
navvy barrow 运土手推车；土车
navvy curve 土主曲线
navvy pick 鹤嘴镐；挖土镐；十字镐
navvy work 土方工程
navy 藏青色
navy bill 海军军票(英国)
navy blue 藏青色；海军蓝；天蓝(色)；深蓝色
navy bronze 海军青铜
navy coating 海军呢
Navy Department 海军部(美国)
navy dock-yard 海军造船厂
navy exchange 海军基地福利商店
navy gasoline 海军汽油
navy hospital 海军医院
navy loudspeaker 舰用扬声器
navy medicine 海军医学
navy method 平板负载试验
navy motor 船用电动机
navy navigation satellite system 海军导航卫星系统(美国)
Navy Observatory 美国海军天文台
Navy Oceanographic(al) and Meteorological Automatic Device(NOMAD) 海军海洋气象自动观测装置(美国)
navy pick 土镐
navy radar 海军雷达
navy receiver 海军用接收机；船用接收机
navy serge 海军哔叽
navy socket 海军用电子管座
navy special fuel oil 舰船专用燃料
navy standard shackle 海军标准链钳扣
navy stockless anchor 海军无杆锚
navy tear test 海军撕裂试验；海军拉裂试验
navy type anchor 海军式锚
navy yard 海军造船厂；海军船厂
Naxas emery 天然刚玉
Naylorite 那洛剂(一种硅酸，用作水泥的硬化剂和防水剂)
nay stone 垫底石
Nazaca plate 纳新卡板块

naze 岬；海角
Nb_2O_2-Bi_2O_3-ZnO Ceramics 铌-铋-锌陶瓷
Nb-eschynite ore 铌易解石矿石
NBS-A time scale 国家标准局原子时标
NBS unit 美国国家标准局色差单位
Nb-Ta ores 铌钽矿
n-butyl 正丁基
n-butylamine 正丁胺
n-butyl ether 正丁醚
n-butyric acid 丁酸
n-capric acid 正癸酸
n-caproic anhydride 正己酸酐
NC-case hardening 渗氮碳硬化处理
NCO content 异氰酸基含量
NCO index 异氰酸基指数
ncomorphism 新生变形作用
nconicotine 阿那巴辛
Nc steel straight-cutting machine 数控调直（钢筋）切断机
n-cyclohexylthio-phthelimade 防焦剂
Nd-doped phosphate glass laser 磷酸盐钕玻璃激光器
n-decoic acid 癸酸
n-decyl n-octyl phthalate 邻苯二甲酸正癸基正辛酯
Nd-glass 钕玻璃
Nd-in glass 钕玻璃
Nd laser 钕激光器
neaf 分片
neagtive quantity 负量
neanic 青年期的
neanic stage 青春期（指河流）
neanic valley 青年期河谷；青春期河谷
neap 小潮；最低潮
neaped 大潮时搁浅
neap(ed)timber 困河材
neap high water 小潮高潮位；小潮高潮面；平均小潮高潮（面）；平均小潮高潮高
neapite 霞磷岩
neap low water 小潮低潮高；平均小潮低潮面
Neapolitan yellow 内亚波利顿黄（锑酸铅黄色彩料）
neap range 小潮涨落差；小潮潮差；平均小潮差
neap rate 小潮流速
neap rise 小潮升；小潮高度；小潮潮升
neap season 小潮期
neap sequence 小潮层序
neap tidal current 小潮潮流
neap tide 小潮；象限潮；弦潮；最低潮；低潮；潮水低落
neap velocity 小潮流速
near and far visual difference 远近视差
near autocollimation arrangement 近似自动准直装置
near bottom current 近底层流；洞底流；底层流
near bottom layer 沿底层；近低层
near bottom temperature 近低层温度
near bottom velocity 近底流速
near bottom waters 近低层水体
near bottom water sampler 近底层水采样器
near(by)earthquake 近震
nearby electric(al)supply 近区供电
near by frequency 邻近频率
near by interference 近区干扰
nearby material 临近材料；当地材料
nearby residential area 附近居民点
nearbys 近期货（指定交收月份最近的期货合约）
near(by)shock 近震
near by surface water 近地表水体
nearby view 近景
near capacity condition 接近极限能量情况
near cash assets 近似现金的资产；准现金资产
near circular orbit 近圆轨道
near collinearity 近共线性
near contact binary star 接近相接双星
near critical 近临界的
near delivery 近期交货
near diffraction-limited mirror 近衍射极限反射镜
near distance 近距
near drop-in substitute 几乎随手可得的代用品；很容易换用的代用品
near earth environment 近地环境
near earth orbit 近地轨道
near earthquake 近震

near earthquake instrument 近震仪
near earth satellite 近地卫星
near earth space 近地空间
near edge emission line 近边缘发射线
near end crosstalk 近端串音
near end interference 近端干扰
near end moment 近端力矩
nearest approach 最近的逼近
nearest neighboring sequence analysis 最近邻顺序分析
nearest neighbors 最近邻
nearest neighbo(u)r 最近邻
nearest neighbo(u)r analysis 最近邻分析（法）
nearest neighbo(u)r approximation 最近邻近似法
nearest neighbo(u)r classification 最近邻域分类
nearest neighbour decision rule 最近邻次策规则
nearest neighbo(u)r energy 最近邻能
nearest neighbo(u)r frequency 最近邻频率
nearest neighbo(u)r interaction 最近邻相互作用
nearest neighbo(u)r pattern classification 最近邻域模式分类法
near estuarial area 近河口区
near extreme multicollinearity 近似极端多重共线性
near feasibility 近可行性
near feasible point 近可行点
near field 近场；拟域
near field accelerogram 近场加速度图
near field analyzer 近场分析器
near field annular-output pattern 近场环状输出图
near field backscatter 近场后向散射
near field beam 近场波束
near field data 近场数据
near field diffraction pattern 近场衍射图
near field earthquake motion 近场地震运动
near field energy 近场能量
near field flow 近区水流
near field fringes 近场条纹
near field holography 近场全息术
near field method 近场法
near field noise 近场噪声
near field particle velocity 近场质点速度
near field pattern 近场图样；近场方向图
near field pattern of laser 激光近场图
near field phenomenon 近场现象
near field photograph 近场照片
near field plume temperature value 近区烟缕水温度值
near field region 近场区
near field river temperature 近区河水温度
near field seismology 近场地震学
near field testing 近场测试
near finished 接近合格的
near focussing device 近距离调焦装置
near forward 接近向前散射
near future 近期
near ga(u)ge hole 井眼直径与钻头直径相近的孔；井眼直径与钻头直径相近的井
near horizontal occurrence 近水平
near infrared 近红外
near infrared band 近边外区；近红外波段
near infrared communication 近红外通信
near infrared communications system 近红外通信系统
near infrared detector 近红外探测器
near infrared image 近红外图像
near infrared interferometer 近红外干涉仪
near infrared laser 近红外激光器
near infrared rays 近红外线
near infrared region 近红外区
near infrared spectrum 近红外光谱
near infrared window 近红外窗口
near-in sidelobe 内旁瓣
nearlake channel 滨湖航道
near lethal dose 近死剂量
near letter-quality 近似铅字质量
near-level grade 水平线
near limit 近距极限
near limit of depth of field 景深前限
near linear dependency 近似线性相关
near liquid assets 近期兑现资产
nearly best linear estimator 近似最佳线性估计量
nearly erect 几乎直立
near lying mode 近模

nearly parabolic orbit 近抛物线轨道
nearly unavailable 几乎无效
near mesh 接近筛孔尺寸的
near mesh bed 接近筛孔的物料床层
near-mesh grains 接近筛孔尺寸的颗粒；分界粒度颗粒
near mesh particles 近筛孔(尺寸的)颗粒；难筛颗粒
near money 近似货币
near monochromatic radiation 近单色辐射
near-natural uranium reactor 近天然铀反应堆；稍加浓铀堆
near neighbor interconnection 近邻互连
nearness to market 邻近市场
near net shape forging 近无余量锻作
near net shape forming 近净形成型
near object 近距物体
near optimization 近似最优
near optimum solution 近优解；近似最佳解
near order 类似指令
near point 近点
near point correction 近点校正
near point of vision 视力近点
near point reaction 近点反应
near polar 近极的
near polar sun-synchronous orbit 太阳同步近极轨道
near-potable water 准饮用水
near proof 接近证实
near range 近侧距
near real time 近实时
near reflex 近反射
near regime 接近稳定的(水文情况)；接近平衡的
near sea fishery 近海鱼类
near sea shipping line 近洋航线
near section 劲缩断面
near shaft point for orientation 定向近井点
nearshore 岸前浅水地带；临滨；近海的；近滨；近岸的；近岸；前滩
nearshore circulation 近岸环流
nearshore current 沿岸流；海岸流；近岸流；海滨流
nearshore current system 近岸流系
nearshore deposit 近岸沉积；临滨沉积；近滨沉积
nearshore ecosystem 近海生态系统
nearshore environment 近岸环境
nearshore environmental analog(ue)prediction system 近岸环境模拟预报系统
nearshore flow 近岸流
near shore ice 近岸冰
near shore marine environment 近岸海洋环境
nearshore oceanography 近岸海洋学
nearshore region 近岸水域
nearshore sediment 沿海沉积物
nearshore shoal 近岸浅滩
nearshore structure 沿海结构物；近岸结构
nearshore waters 近岸水域
nearshore wave 近岸波
nearshore zone 沿岸带；近岸带；近滨带
near shot-point trace number 近炮点道号
nearside 近边；近侧
nearside divider 分禾器
nearside lane 近车道；外侧车道
nearside signal 外侧车道信号
nearside tank 左油箱
near solar space 近日空间
near sonic 近音速
near sonic drag 跨音速阻力
near sonic ridge line 近声速分界线
near sonic speed 近音速
near source 近震源
near source response spectrum 近源反应谱
near source wave 近震波；近海波(浪)
near spherical 近似球形的；类球状
near stars 邻近星
near surface current 近岸表层流
near surface deposit 近地表矿床
near surface effect 近地表效应
near surface excavation 近地表开挖
near surface flaw 近表面缺陷
near surface layer 附面层；近表面层
near surface of earthquake 近地表地震
near surface path 表层声径
near surface sampling 浅层采样
near surface seismic event 近地表地震
near surface ship 近水面船；半潜船

near surface temperature 近岸表层温度场
near surface vehicle 近水面艇
near surface wave 近表面波
near term 近期(的)
near term decision 近期决策
near term prospect 近期展望
near term target 近期目标
near the foot of a mountain 在山脚下附近
near the lower end of the canal 在渠道下游终点附近
near thermal reactor 近热反应堆
near to-surface-test 近表面试验
near trace gather 近迹合成图;近道选排图
near ultraviolet 近紫外
near ultraviolet band 近紫外区;近紫外波段
near ultraviolet lithography 近紫外光刻
near ultraviolet ray 近紫外线
near ultraviolet resist 近紫外抗蚀剂
near ultraviolet zone 近紫外区
near vane 近目瞄准孔
near vertical photograph 近垂直摄影像片
near vertical photography 近垂直航空摄影
near vertical picture 近垂直摄影像片
near vertical transparency 近垂直摄影透明片
near viewing 近距观察
near vision 近视;近距视觉
near water 近海
near zero release 近零排放
near zone 近区
neat 整齐的
neat area 净面积
neat cement 非水化状态的水硬性水泥;净水泥;纯水泥
neat cement binding course 素水泥浆砂浆结合层
neat cement grout 净水泥浆;素水泥浆;纯水泥浆
neat cement mortar 净水泥浆;净水泥砂浆;水泥净浆;纯水泥浆
neat cement paste 净水泥软膏;净水泥浆;水泥净浆;纯水泥浆;纯水泥浆体
neat cement slurry 净水泥浆
neat gas burner 光焰燃烧器;扩散式燃烧器;天然气燃烧器
neat house 清洁房舍(指牛棚、马厩);牛棚
neat lime 净石灰;纯石灰
neat line 红线;墙面交接线;结构切削线;建筑线;开挖边线;施工限制线;细线;准线;净开挖线;内图廓线;墙面准线;图表边线
neat line gradation 内图廓分度线
neatline mark 内图廓注记
neat line(of wall)(墙面)交接线
neat-line system 地图分幅系统
neat model 有效模型
neat oil 净油;不掺水油
neat paste 纯水泥膏;净水泥浆;纯水泥浆
neat plaster 净浆;素灰;纯水泥浆;净灰浆;纯灰浆
neat Portland cement 纯硅酸盐水泥;纯波特兰水泥
neat profit 净利润;纯收益;纯利(润)
neat resin 纯树脂
neatron probe 中子探测器
neat seam 密缝;严丝合缝
neatsfoot oil 牛蹄油
neat size 光洁尺寸(木器和细木工件);净尺寸;完成尺寸;修正后尺寸
neat slurry 纯水泥浆
neat soap 皂粒
neat work 清水砖墙;地脚以上砖工;细活;清水砖墙勾缝均工;清水墙勾缝均工;砌砖
nebelfrost 雨凇(下降时呈液态,着地后就冰冻)
Nebelwerfer 烟雾放射器
Nebraska bed 内布拉斯加层【地】
Nebraska Engineering Society 内布拉斯加州工程学会(美国)
Nebraskan drift 内布拉斯加冰碛
Nebraskan glacial epoch 内布拉斯加冰期
Nebraskan glacial stage 内布拉斯加冰期
nebula 星云;喷雾剂;雾气
nebula covering the pupil 膜入水轮
nebula mo(u)lding 星云线脚
nebular 星云状;云的
nebular hypothesis 星云假说
nebular lines 星云谱线
nebular red shift 星云红移
nebular spectrograph 星云摄谱仪
nebular theory 星形说
nebular transitions 星云跃迁

nebule 诺尔曼式建筑装饰;波边饰
nebule mo(u)lding 云雾状装饰线脚;波边饰;有齿形边线的线脚
nebulite 星云岩
nebulitic 星云状的
nebulitic migmatite 阴影混合岩
nebulization 喷雾作用
nebulization efficiency 喷雾效率;雾化效率
nebulization interference 雾化干扰
nebulizer 喷雾器;喷洒器
nebulizer drop 雾滴
nebulosus 薄幕状云
nebulous 星云状;星云的
nebulous cluster 伴云星团
nebulous mo(u)lding 云状线脚;云状饰;云雾状装饰线脚
nebulous oven Jun glaze 浑炉钧釉
nebuly corbel table 云状挑台
nebuly mo(u)lding 波边饰
necessary 必需品
necessary accessories 配套器材
necessary and sufficient condition 充要条件;充分必要条件
necessary bandwidth 必要带宽;必需带宽
necessary condition 必要条件;必备条件
necessary equipment 必要设备
necessary expenses 必要经费;必需费用;保存费
necessary fund 所需经费
necessary height of hump 需要峰高【铁】
necessary labo(u)r 必要劳动
necessary labo(u)r time 必要劳动时间
necessary means of livelihood 必要生活资料
necessary minimum profit 必要最低利润
necessary operation 必要操作
necessary preconstruction approvals 施工执照;施工认可书
necessary preconstruction permits 施工执照
necessary product 必要产品
necessary protection ratio 必要保护率
necessary supplement 必要补充
necessities of life 生活必需品
necessity 必要性;必然性
neck 狭窄海岬;岩颈;管状矿脉;管颈;炉颈;地峡;地颈;柱颈
neck bearing 中空轴承;中间轴承;弯颈轴承
neck break 辊颈折断
neckbreaking speed 危险速率
neck brick 斜面砖;锁砖
neck bush 内衬套
neckbush in stuffing box 填料箱内套
neck channel 山狭沟槽;狭窄槽;窄河槽
neck collar 轴颈环;轴承环
neck current 狭窄急流;束狭急流
neck cut-off 曲流裁直;狭颈裁弯;曲流裁弯
neck down 形成细颈;(凹)缩;断面收缩
neck-down riser 易割冒口
necked-and-rough turned roll 全加工轧辊
necked bolt 弯头插销;柱颈螺栓;长颈螺栓
necked-down chamber 收敛室
necked-down core 颈缩芯片
necked-down section 收缩断面
necked-in 向内弯曲(边缘)
necked in operation 割缺口
necked out 向外弯曲(边缘);向外弯曲边
necked roll 光制辊颈轧辊
neckerchief 领巾;围巾
neck grease briquette 辊颈润滑油试块;轴颈润滑油试块
neck gutter 屋顶天沟
neck-in 内缩量;缩进扭弯内缘
necking 柱颈(线饰);瓶颈;缩径;缩颈现象;断层收缩;缩口
necking bit 切槽车刀
necking coefficient 缩口系数
necking down 缩口;缩颈;收颈;颈缩
necking-down action 断面收缩作用;颈缩作用
necking down of inclusion 卡脖子包裹体
necking-groove 剐槽;切槽;开槽;颈槽
necking in 缩颈旋回
necking in operation 割缺口
necking in tension 拉伸缩颈
necking phenomenon 缩颈现象
necking tool 开槽刀;切槽工具
necking waist 缩颈

neck journal 轴颈
neck journal bearing 轴颈轴承
neck mo(u)ld(ing) 柱颈线脚;柱颈线饰;领饰;瓶颈模
neck of hook 钩柄
neck of land 地峡;狭窄地带
neck of shaft 井颈
neck of tube 管颈
neck out 向外弯曲;扭弯内缘
neck reach 窄颈段
neck ring mo(u)ld 口模
neck shadow 管颈阴影
necrology 死亡统计
necroparasite 死物寄生菌
necropolis 死城;废弃的城镇;城市大公墓(古代);墓地(大型的)
nectariferous spur 蜜距
nectonic benthon 水底游动植物
need complementarity 需求的互补
need creation 需要的创造
need for filter 反滤层要求
need for relatedness 相互关系和谐的需要
need hierarchy 需要层次
needle 针状物;针状体;针;横支撑木;折页顶针
needle aberration 磁针偏差
needle acacia 针叶相思树
needle annunciator 指针示号器
needle antimony 粗锑
needle apparatus 维卡仪(用于水泥稠度试验)
needle armature 指针枢纽
needle back 针背
needle bar 针床;喷电针支撑杆
needle bar cam lever 针床凸轮杆
needle bar crank 针杆曲柄
needle bar stroke 针杆行程
needle bar thread guard 针杆拦线板
needle bar wrench 针床扳手
needle bath 淋浴;沐池;针式喷浴
needle beam 临时支承梁;横撑木;临时托梁;小托梁;支托小梁;针形梁;针梁;臂梁(托梁基础用)
needle beam method 针梁法
needle(beam)scaffold 横梁支托的脚手架;托梁脚手架;轻型(平台)脚手架
needle-beam underpinning 小梁托撑基础;用小梁支撑基础
needle beard 针钩
needle bearing 针式轴承;滚针轴承
needle bearing bushing 滚针轴承衬套
needle bearing with closed end 一端封闭的滚针轴承
needle bed 针床
needle board 提花机
needle brake 打捆针
needle bronze 针青铜
needle bush 针叶木;滚针套
needle cast 落叶病
needle chamber 针场电离室
needle clearance 针阀间隙
needle clearing cam 退圈三角
needle clutch 打捆针传动离合器
needle coal 纤维状煤;针状煤
needle controlled weir 针阀控制堰
needle control lever 针控制杆
needle control rod 针阀控制杆
needle control valve 针形阀;针孔阀;针阀
needle conveying 行针
needle counter 针状计数管
needle crystal 针状石油焦;针状晶体;针形结晶
needle dam 针坝;栅条坝;横栅活坝;叠梁坝;针板坝
needled carpet 针扎地毯
needle deflection 指针偏斜
needle density 用贯入法测土的密度;针入密(实)度;针测密实度;贯入法测土的密度
needled glass fiber mat 针刺玻璃纤维毡
needle die grinding machine 模孔针磨机
needle die polishing machine 模孔针磨机
needle divider 分规器
needled mat 针刺毡
needled mat machine 针刺毡机
needle drawing 刻图
needle drift chute 散木流放槽
needle drive arm 打捆针驱动臂
needle drive cam 打捆针驱动凸轮
needle(d)steel 加硼硬化亚共析钢

needle electrode 针形电极;针电极
needle electrometer 针式电位计
needle etching 针刻
needle file 针座;针锉
needle fir 辽东冷杉
needle flame burner 针状火焰燃烧器
needle force 针压力
needle fracture 针状断口
needle galvanometer 磁针电流计
needle gap 针隙
needle gap lightning 针隙避雷器
needle gate 针形闸(门);打捆台穿针口
needle ga(u)ge 针状水尺;针尺;水位测针
needle glazing 窄缝镶嵌玻璃
needle grining machine 模孔针磨机
needle guide 打捆针导向器
needle gun 针枪;梅花电极;撞针枪
needle handle 针柄
needle holder 粗针持针钳;测针夹持器
needle hole 针孔
needle hole test 针孔试验(分散性土的试验)
needle hook 针钩
needle ice 针状冰;针冰;水内冰;(河流的)底冰;冰屑
needle implant 探针插入
needle indicator 指针指示器;指针式指示器
needle ingot shaft 针座轴
needle insulator 针式绝缘子
needle ironstone 针铁石;针铁矿
needle jack 挺针片
needle jet 针阀调节喷嘴
needle juniper 杜松
needle karst 峰林
needle knuckle bearing 滚针关节轴承
needle lac 针状紫胶;针状虫胶
needle land 针槽壁
needle leaf 针状叶
needle-leaved evergreen forest 针叶常绿林
needle leaved forest 针叶林;松柏林
needle leaved plant 针叶植物
needle-leaved tree 针叶树
needle leaved wood 针叶木
needle lift 针阀升程
needle lifter 刻针杆;起针三角
needle lifting cam 起针三角
needle-like 针状的
needle-like bainite 针状贝氏体
needle(-like)coke 针状焦
needle-like leaves of pine 松树外状叶
needle-like pigment 针状颜料
needle lock cam 针座三角
needle lubrication 针孔润滑法
needle lubricator 针孔润滑器;针形润滑器;针孔油枪;针孔润滑器
needle machine 针刺机
needle mesh 针状筛孔;针孔筛目
needle nose pliers 尖嘴钳
needle nozzle 针形喷嘴;针形阀;针式喷嘴;针式喷管
needle number 针孔筛目
needle of instrument 仪器指针
needle opening pressure 针阀开启压力
needle-operated punched card 用针选取的穿孔卡片
needle ore 针硫铋铅矿;针硫铋铅矿
needle paper 包针纸
needle passage 针道
needle pattern 针形图案;尖峰状
needle peak 尖峰
needle peak load 尖峰荷载;尖峰负荷
needle pellet 撞针头
needle pen 绘图笔
needle penetration test 针式贯入法试验;针入法试验
needle penetrometer 针入仪(检验水泥凝结时间用);针式贯入仪;针式贯入器;稠度仪;稠度计
needle pick-up 针式拾音器;磁针式拾音器
needle piston 针形活塞
needle piston guide 针形活塞导管
needle pitman 打捆针机构连杆
needle pivot 指针轴
needle plasma arc 微型等离子弧
needle pliering 针钳
needle plug 针形柱塞;针形塞
needle-plug valve 针阀

needle point 针尖
needlepoint compliance 针端柔性
needle pointed stylus 刻针
needle point file set 成套尖头锉
needle point grinding 针尖研磨
needle point method 针刺法
needle point spark gap 针尖火花隙
needle powder 针状粉末
needle pricker 炮眼针
needle probe 探针型探头
needle punching 针刺
needle pushing-up device 推针装置
needle raising cam 起针三角
needle recuperator 针形换热器
needle regulating nozzle 针式调节喷嘴
needle remover 拔栅器;拔销器
needle rest 打捆针限止器
needle ring 针座环
needle roll 针辊
needle roller 滚针的
needle roller bearing 针状滚柱轴承;滚针轴承
needle roller bearing with variation 游隙可调的滚针轴承
needle roller thrust bearing 推力滚针轴承
needle roller with ball end 球头滚针
needle root 针根
needle scaffold 悬臂支托脚手架;单排脚手架;悬臂脚手架;挂脚手架;横架支托脚手架;横架挑出脚手架;横撑木脚手架
needle seat angle 针阀座角
needle seating 针阀座
needle segment 扇形板
needle selection 选针
needle servomotor 针阀伺服马达
needle shaft 针状轴
needle-shaped 针状的
needle-shaped cathode 针形阴极
needle-shaped crystal structure 针状晶体结构
needle-shaped particle 针状磁粉;针状颗粒
needle-shaped structure 针状构造
needle shoring 斜支撑;斜撑
needle sleeve 针形套筒;滚针套
needle-slot screen 细缝筛
needle spark gap 针火花隙
needle spire 针状尖顶;细尖塔
needle stand 针座
needle steel 针状组织钢
needle stem 针杆;针阀杆;喷针杆
needlestone 钙沸石;网状金红石;金红针水晶;针状结晶岩石
needle support 针状支座;横撑
needle test 针孔试验(分散性土的试验)
needle test point 针状测试点
needle thrust bearing 滚针推力轴承
needle timber 柱脚撑木
needle-tin 针状晶体锡
needle tip 针尖
needle track 针道
needle tracking 针迹
needle transmission 行针
needle traverse 罗盘仪导线;罗盘导线
needle tray 针槽
needle trough 针槽
needle tube[tubing] 针形管;针管
needle tungsten carbide bit 针状硬合金钻头
needle type injector 针阀式喷油器
needle type jib 杆式副起重臂
needle type martensite 针状马氏体
needle(type)valve 针形阀;油针;针塞;锥形阀;针状阀;针孔阀;针阀
needle valve of economizer 省油针阀(汽化器内)
needle valve seat 针状活门座;针阀座
needle valve seat hermetic test device 针阀座气密度试验器
needle valve shock absorber 针阀减振器
needle valve spring 针阀簧
needle valve stroke 针阀冲程
needle vibrator 针状振荡器;针形振荡器;插入式振捣器
needle vibrator with handle 带手柄的内部振动器
needle wall 针槽壁
needle wax 针状蜡
needle wax crystals 针状蜡结晶
needle weir 针堰;针排堰;木栅堰;栅条堰;叠梁堰

needle wire 制针用钢丝
needle with hook 钩针
needlework 刺绣(业);缝纫业;木构架与填充墙的组合结构
needle zinc white 针状锌白
needling 固墙撑木;针刺;横支撑木;横梁支托;横撑支托;横撑木(支托)
needling agent 含硼铁合金
needling instrument 针具
needling machine 针刺机
needling to wall 墙身横撑
needling work 横支撑工程
needly 如针状的
need-ranking theory 需要分等级理论
needs and opinions of consumers 消费者的意见和要求
needs-satisfying products 满足需要的产品
Neef's hammer 内夫氏锤(电流启闭锤)
Neel point 奈耳点
Neel's theory 奈耳理论
Neel temperature 尼尔温度;奈耳温度(反铁磁物质最高温度)
Neel wall 奈耳壁
nefud 沙丘荒漠
negacyclic(al)code 负循环码
Negafel process 尼加费尔防缩法
negate 求负;变符号
negated simple condition 否定的简单条件
negate instruction 求反指令
negate novelty 否定新颖性
negater 倒换器;反相元件
negater circuit 倒相电路
negate routine 求反程序
negater spring 反旋弹簧
negation 否定
negative 消极的;照相底片;片;负像;负的;否定保证
negative-a bolometer 负电阻温度系数测辐射热计
negative absorbing material 负吸收材料
negative absorption 负吸收(作用)
negative absorption coefficient 负吸收系数
negative acceleration 减速度;负加速度;负荷加速度
negative acceleration phase 负性加速期;负加速相
negative acknowledge 负应答
negative acknowledge character 否认记号;否定信号
negative acknowledgement 否定信号
negative adsorption 反吸附(作用)
negative after-image 负余像;负余留形象;负后象;负残影
negative after-potential 负后电位
negative aggregate transparency 集料花斑印
negative air pressure 空气负压;负气压
negative album 底片套;底片薄
negative allowance 允许负差;负留量
negative altitude 低度;负高;俯角;伏角
negative amplitude attenuation 负值衰减系数
negative amplitude attenuation constant 负振幅衰减常数
negative and positive pressure surges 负压和正压涌浪
negative angle 负角;俯角
negative angle of attack 负冲角
negative angle of sight 俯角
negative anomaly 负异常
negative answer 否定的回答
negative appliance 挖方机械
negative approach 反证法
negative arc 负弧
negative area 底片幅面;负异常区;负相区;凹下部分
negative area of influence 负的影响面积
negative armature 反极性衔铁
negative artesian head 负涌泉水头;负承压水头
negative artesian well 负涌泉;负自流井;负承压水井
negative assets 负资产;负债
negative assets account 负资产账户;负资产科目
negative association 负相联
negative assurance 消极保证
negative attitude 消极态度
negative axis 底片轴
negative back tipped bit 负斜镶钻头
negative balance 负平衡;负均衡;负差额
negative balance of sodium 钠负平衡

negative base number 负基数
negative-bath 底片浴槽
negative battery plate 蓄电池阴极板
negative bending 反向挠曲;反向弯曲;负弯曲
negative bending moment 反向弯矩;负弯矩
negative benefit 负效益
negative bias 负偏压
negative binary zero 负二进制零
negative binomial distribution 负二项分布
negative binomial distribution simulation method 负二项分布拟合法
negative birefringence 负双折射
negative booster 制动器;降压器;降压(电)机;减压器;减压机
negative boosting transformer 减压变压器
negative brush 负电刷
negative buoyancy 下沉力;负浮力
negative buoyancy coating 防浮包裹层(包裹在水下管线上的水泥或钢筋水泥层);加重包裹层
negative buoyancy jet 负浮射流
negative burner 阴极燃烧器
negative busbar 阴极母线;负母线
negative camber 负曲率;内曲面
negative camber of wheel 车轮负外倾角
negative capacitance amplifier 负输入电容放大器
negative capital 负资本
negative carbon 弧光灯碳头
negative carbonium ion 负碳离子
negative carrier 底片盒
negative carry 负进位;负持有
negative cash flow 负向现金流动
negative caster 转向销负倾角;主销后倾角
negative catalysis 负催化作用;负催化
negative catalyst 阻化剂;负催化剂;防老(化)剂
negative catalyzer 阻化剂
negative center 反顶尖
negative cerium anomaly 负铈异常
negative character 负光性
negative characteristic 下降的特性曲线;负特性
negative charge 负电荷
negative charge group 负电荷基
negative chemical ionization technique 负化学电离技术
negative chronotropic action 负性频率作用
negative circuit 反相电路
negative circuit interlock 负电路联锁装置
negative clamping 负向钳位
negative clause 负子句
negative clipping 负向削波
negative clock 负时钟信号
negative colloid 阴性胶体;阴电荷胶体;负胶体
negative colo(u)r 冷色
negative colo(u)r film 彩色负片
negative compression 负压缩
negative concept 负概念
negative conditioned reflex 阴性条件反射
negative conductance 负电导
negative confining bed 隔水底板;逆向赋压层;负封闭层;负承压底层
negative confirmation 否定询证法;否定式函证;反面证实
negative contact 阴极接头;负电接头
negative contrast 底片反差;负片反差;负反差
negative contribution margin 负贡献毛益
negative control 消极性管制;负控制
negative control thyratron 负控制闸流管
negative copy 负片
negative copying method 底片晒印法
negative correction 底片修改;负校正
negative correlation 负相关
negative correlation coefficient 负相关系数
negative Cotton effect 负考顿效应
negative counting 负脉冲计数
negative credit 负信用
negative creep 负蠕变
negative crystal 负晶形;负晶体
negative current 负电流
negative current rail 负向流轨
negative cutoff 突出地面截水墙;倒截止墙
negative cutoff grid voltage 负截止栅压
negative cutter 阴像刻图员
negative cutting angle 负切削角;负切角
negative cycle 负循环
negative damping 负阻尼

negative damping factor 负阻尼系数
negative definite 负定
negative definite form 负定型
negative definite function 负定函数
negative definite matrices 负定矩阵
negative definite matrix 负定矩阵
negative deflection 负挠度
negative delta 喇叭口湾;三角湾
negative demand 负需求
negative density 负片密度
negative density difference 负密度差
negative developer 反转显影剂
negative deviation 负偏差
negative diazo process 阴图重氮法
negative die 下模;阴模;底模
negative difference 负差
negative difference rate 负差率
negative differential 负差率
negative diopter 负视度
negative direction 逆向;反方向
negative dislocation 负位错
negative dispersion 负色散
negative displacemenet 负位移
negative displacement pump 抽吸泵;负排量泵
negative distortion 枕形失真;枕形畸变;负失真;负畸变
negative disturbance 负扰动
negative drill hole 未见矿钻孔
negative drying 底片晾干
negative drying hanger 底片晾干架
negative echo 负回波
negative edge 负沿
negative effects 消极效果;负效果
negative electricity 负电
negative electrode 阴极;阴电极;负极板;负极;负电极
negative electron 阴电子;负电子
negative element 阴性元素;负元素;负向地貌要素;凹下部分
negative elongation 负伸长;负拉长;缩短;负延性
negative energy 负能量
negative energy state 负能态
negative energy wave 负能波
negative engraving 阴像刻图(测);阴图版
negative entry 负输入;负号输入
negative envelope feedback 包络负反馈
negative environmental effect 负面环境效应
negative environmental impact 负面环境影响
negative error 负误差
negative europium anomaly 负铕异常
negative evidence 反证;反面证据
negative expansion 负膨胀
negative exponent 负指数
negative exponential decay 负指数式衰减
negative exponential distribution 负指数分布
negative exponential law 指数衰减律;负指数定律
negative export 负出口
negative externalities 不利的外部因素
negative eyepiece 负目镜
negative factor 消极因素;负因素;负因数;不利因素
negative factorial multinomial distribution 负阶乘多项式分布
negative feedback 负相馈;负回授;负反馈
negative feedback amplifier 负回输放大器
negative feedback circuit 负反馈电路
negative feedback control 负反馈控制
negative feedback coupling 负反馈耦合
negative feedback factor 负反馈因数;负反馈系数
negative feedback laser 负反馈激光器
negative feedback link 负反馈链
negative feedback loop 负反馈环
negative feedback self-oscilation 负反馈自激
negative feedback system 负回馈系统
negative feeder 阴极馈电线
negative feeding system 负馈系统
negative film 底片
negative filter 负滤光片
negative-filter type infrared gas analyzer 负滤波式红外气体分析器
negative flag 负数标记;否定旗
negative float 负浮动
negative form 负形
negative format 底片幅面
negative forming 阴模成型

negative form of sea floor 海底负地形
negative for screening 挂网阴片
negative frequency 负频率
negative frequency component 负频分量
negative friction 负摩擦(力)
negative friction of pile 桩负摩擦力;下拉荷载
negative friction pile 负摩擦桩
negative function 负函数
negativeg 负过载
negative ga(u)ge pressure 负表压
negative geothermal anomaly 负地热异常
negative geothermal gradient 负地热梯度
negative geotropism 负向地性
negative ghost image 负重像
negative glow 阴极辉光;负辉光
negative glow lamp 辉光放电管;负辉灯;辉光放电灯
negative glow region 负辉光区
negative going 负向的
negative going edge 下降沿;负沿
negative going pulse 负脉冲(信号)
negative-going reflected pulse 负向反射脉冲
negative going sawtooth wave 负向锯齿波
negative-going signal 负向信号
negative goodwill 负商誉
negative gradient 负梯度
negative gravity 负引力
negative grid bias 负栅偏压
negative-grid generator 负栅振荡器;负栅发生器
negative-grid oscillator 负栅振荡器
negative grid super high frequency tube 负栅超高频管
negative grid thyratron 负栅极闸流管
negative grid tube 负栅管
negative group 负性基
negative g tolerance 负重力加速度容限;负过载容限
negative hardening 软化淬火处理;韧化淬火处理;低温退火
negative hardness 负硬度
negative hard water 负硬质水
negative head 负压水头;负水头
negative heterograde profile 负异常梯度溶解氧廓线
negative holder 底片盒
negative hologram 负全息图
negative hydrogen ion 负氢离子
negative hydrostatic(al) stress 负静水应力
negative hypergeometric distribution 负超几何分布
negative image 阴像;负像
negative immittance converter 负导抗转换器
negative impedance 负阻抗
negative impedance amplifier 负阻抗放大器
negative impedance circuit 负阻抗电路
negative impedance converter 负阻抗变换器
negative impedance region 负阻抗区域
negative impedance repeater 负阻抗中继器
negative impulse 负脉冲(信号)
negative incidence 负倾角
negative income tax 负值所得税;负项所得税;负所得税
negative incoordination 负失调
negative increment 负增量
negative index 负指数
negative indication 负号指示;负号
negative indicator 负指示符
negative induction 负诱导
negative input 负投入
negative integer 负整数
negative interaction 负相互作用
negative interest 逆利率;负利息
negative interesting point 反交点
negative interference 负干扰
negative investment 停止投资;负投资
negative ion 阴离子;负离子
negative ion active agent 阴离子型活性剂
negative ion chemical ionization 负离子化学电离
negative ion densitometer 负离子浓度测定器
negative ion formation 负离子形成
negative ion generator 负离子发生器
negative ion laser 负离子激光器
negative ion source 负离子源
negative ion vacancy 负离子空位

negative justification 负码速调整
negative lag 负滞后
negative landform 负地形
negative layer 负片乳胶层
negative leap second 负闰秒
negative lens 负透镜
negative lift 负升力
negative light modulation 负亮度调制
negative line 负线
negative linear correlation 负线性相关
negative line feed 反向行进
negative list 不准进口商品单
negative load 负载;负水载;负荷载
negative logic 负逻辑
negative loop 负值部分
negatively absorbing material 负吸收材料
negatively charged 带负电荷的
negatively charged ion 阴离子;带负电荷离子;负电荷离子
negatively charged species 带负电荷物质
negatively correlated 负相关的
negatively progressive flow 反向行进水流
negatively skewed 向右倾斜的;负偏的
negatively sloped trade-off 负向倾斜的抉择
negatively sloping yield curve 负收益曲线
negative magnetic anomaly area 负磁异常区
negative magnetoresistance 负磁致电阻
negative magnitude 负震级
negative mantle friction 负摩擦力(桩工);负表面摩擦(力)
negative map 阴像地图;底片图
negative margin 底片边缘
negative mass effect 多重效应
negative mass instability 负质量不稳定性
negative matrix 负矩阵
negative meniscus 负液面;负弯液面
negative meniscus lens 负弯月透镜
negative metacentric height 负稳心高度
negative method 底片法
negative modulation 负极性调制;负调制
negative moment 负弯矩;负力矩
negative moment reinforcement 承受负力矩的钢筋;负弯矩钢筋;负力矩钢筋
negative mortgage security clause 不再抵押条款
negative mosaic 负片镶嵌图
negative mo(u)ld 阴模
negative mounting 负片镶嵌
negative movement 负向运动;负相运动;负位移
negative movement of sea level 海面下降
negative multinomial distribution 负多项式分布;负多项分布
negative nitrogen balance 负氮平衡
negative nodal point 负节点
negative non-linear correlation 负非线性相关
negative norm 负模
negative number 负数
negative-number representation 负数表示法
negative ocular 负目镜
negative output 负输出量;负产出
negative ovality ring 负椭圆环
negative oxygen ion generator 负氧离子发生器
negative painting 反绘彩
negative paper 底片纸
negative parallax 负视差
negative-parameter device 负参量器件
negative pattern 底片图案
negative peak 最大负值;倒峰;负峰(值);反峰
negative penetrability 负穿透性
negative perforation 底片片孔
negative permeable torque converter 负可透变扭器
negative phase 负性期
negative phase-contrast 负相衬
negative(phase)relay 反相继电器;负向继电器;反向继电器
negative-phase sequence 负相序
negative-phase sequence component 反相(序)(列)分量
negative-phase sequence component ammeter 反相序分量电流表
negative-phase sequence current 反相电流;反相序电流
negative-phase sequence current relay 反相序电流继电器

negative-phase sequence impedance 反相序阻抗
negative-phase sequence relay 负相序继电器;反相序继电器
negative-phase wave 负相波
negative photodiode coupler 负型光电二极管耦合器
negative photoresist 负性光致抗蚀剂;负型光刻胶
negative phototropism 负向光性;背光性
negative picture 底片;负像;负片
negative picture phase 负图像相位
negative plane 负平面
negative plate 阴极板;底片;负极板
negative plate azimuth 底片方位角
negative plate library 底片库
negative plate making 阴像制版
negative plate size 底片尺寸
negative pledge 负抵押
negative pledge clause 消极保证条款;反面保证条款
negative pledging clause 不抵押保证条款
negative polarity 负极性
negative polarity welding 反极性焊接
negative polarity zone 阴压区
negative pole 阴极;负极;负磁极
negative pole cabinet 负极柜
negative polling limit 否定探询极限
negative polynomial distribution 负多项分布
negative population growth 人口负增长;负人口增长
negative pore pressure 负孔压力
negative pore water pressure 毛细孔水压力;孔隙水压力
negative-positive process 负正照相工艺;负正法
negative post adapter 负极接线柱接线头
negative potential 阴电势;负电位;负电势
negative potential gradient 负势梯度
negative power 负光焦度
negative practice 消退性实践法
negative pressure 真空;负压强;负压力;负压
negative pressure burning 负压煅烧
negative pressure control bottle 负压调节瓶
negative pressure ga(u)ge 负压表
negative pressure leak-stopper 负压防漏器
negative pressure meter 负压计
negative pressure sieving analyzer 负压筛析仪
negative pressure slider 负压磁头浮动块
negative pressure valve 排气阀;负压阀;负压活门
negative pressure ventilation 负压通风
negative pressure wave 负向压力波
negative price scissors 逆向剪刀差
negative principal plane 负主平面
negative principal point 负主面点;负主点
negative profit 亏损
negative proof 反面证据
negative proposition 否定命题
negative prospecting indication 成矿不利标志
negative proton 反质子
negative pulse 负脉冲(信号)
negative pulse stuffing 负脉冲塞
negative rack 底片架
negative radiation 负辐射
negative rail 负轨道
negative rain 带负电雨
negative rake(angle) 负倾角;负切削角;负前角
negative ray 阴射线
negative reaction 阴性反应;负反应;负反力
negative refraction 负折射
negative regulation 负压稳定
negative reinforcement 压力钢筋;负弯矩钢筋;负钢筋
negative reinforcer 负强化因素
negative reinforcing 负弯矩钢筋
negative relationship 负关系
negative relativity 负相关性
negative release wave 负释放波;负放水波
negative relief 负向地貌;负突起;负地形
negative rent 负房租(即免付房租,并得水电、煤气补贴)
negative residual 负余差
negative resist 负性光蚀剂
negative resistance 负阻;负电阻
negative resistance amplifier 负阻放大器
negative resistance bridge 测量负电阻的电桥
negative resistance characteristic 负阻特性

negative resistance device 负阻装置;负电阻器件
negative resistance magnetron 负阻磁控管
negative resistance oscillator 负阻振荡器
negative resistance region 负电阻区
negative resistance relay 负阻继电器
negative resistance repeater 负电阻中继器
negative resistance storage 负阻抗存储器
negative resistance tube oscillator 负阻管振荡器
negative response 否定应答;否定响应
negative restoring force 负回复力
negative reticulum 负网
negative return 负回流
negative return rail 负回流轨
negative rhythm 反韵律
negative rigidity 负柔度;负刚度
negative ripple 负脉冲(信号)
negative root 负根
negative rotation 负转动
negative saving 负储蓄
negative scanning 负片扫描
negative scotoma 负性暗点
negative segregation 反偏析
negative self-selection 否定的自我选择
negative semi-definite function 半负定函数
negative semi-definite matrix 半负定矩阵;负半定矩阵
negative sequence 逆序;负序
negative sequence current 负序电流
negative sequence field impedance 负序磁场阻抗
negative sequence impedance 负序阻抗
negative sequence power 负序功率
negative sequence reactance 负序电抗
negative sequence resistance 负序电阻
negative serial correlation 负序列相关
negative setup 减水;负增水
negative shaft resistance 负阻力
negative shear 负剪力
negative shorelin 上升滨线
negative shoreline 上升岸线;负性海岸线;负海滨线
negative shrinkage 膨胀;负收缩
negative shrink hole 负收缩孔
negative shunt conductance 负分流电导
negative side rake angle 负旁锋刀面角
negative sign 减号;负号
negative signal temperature 负信号温度
negative signed rank 负符号等级
negative size 负片尺寸
negative skew distribution 负偏斜分布
negative skewed distribution 负偏态分布
negative skewed lognormal distribution 负偏态对数正常分布
negative skewness 负斜对称;负向偏斜度;负向偏态;负偏斜度;负偏态
negative skin friction 负摩阻力;负摩擦力(桩工);负表面摩擦(力);表面负摩阻;表面负摩擦
negative skin friction along the pile 桩的负摩擦力;桩侧表面负摩擦力
negative skin friction of piles 桩的表面负摩擦力
negative slip 负转差率;负滑脱;负滑距
negative slope 负斜率;负斜率;负坡(度);反坡(度);反比降
negative sloping yield curve 负斜度收益率曲线
negative sol 阴电荷溶胶
negative source 负源
negative spheric(al)aberration 负球差
negative spike wave 负性棘波
negative spiral striation 负螺纹
negative spool holder 底片卷轴支架
negative spread 负差额
negative spring camber 钢板负垂弯度
negative stability 负稳定性;负稳定法
negative stable matrix 负稳定矩阵
negative stager 负斜罩
negative staining 负染色;负染法
negative standard polarity 标准负极性
negative stay 交叉斜撑;负拉索
negative step 负台阶
negative stiffness 负劲度;负刚度
negative storing 底片储藏
negative strain 负应变
negative strap 负极板连接条
negative streaking 图像负拖尾;白拖黑
negative stress 负应力
negative strike 否定性罢工

negative strip 负重力异常带;负速铸坯
negative stuffing 负的速度匹配
negative suction 吸入负压
negative superelevation 反超高
negative supply 正极接地的电流
negative-supply regulation 负压稳定
negative surface charge 负表面电荷
negative surge 减水;潮升不足;负涌浪;负波
negative sweep 前伸角
negative synergism 反协同效应
negative synthetic(al) detergent 阴离子合成洗涤剂
negative take 负结果
negative taper 倒起模斜度
negative tappet 单向挺杆
negative tax 负数税收;负赋税
negative taxable profits 负的应税利润
negative taxis 负趋性
negative tax plan 退税计划;负税收计划;负税计划
negative temperature 零下温度;负温度;负温;负荷温度
negative temperature coefficient 负温度系数
negative temperature coefficient ceramics 负温度系数陶瓷
negative temperature gradient 负温度梯度
negative temperature laser 负温度激光器
negative temperature material 负温材料
negative terminal 负线端;负极接线柱;负极端子;负极;负端子
negative test 负结果;负检验;反试验
negative thixotropy 抗摇溶现象;反摇溶现象;反触变性;负触变性
negative thread 阴螺纹
negative throat nozzle 渐缩喷嘴
negative thrust 反推力
negative thrust propeller 可逆式螺旋桨
negative titling 底片注记
negative tolerance 负公差
negative topographic(al) unit 负地形单元
negative-to-positive development 反转显影
negative torsion 反扭转
negative torsional moment 负扭矩
negative-transconductance oscillator 负跨导振荡器
negative transfer 负迁移
negative transition 负跃迁
negative transmission 负调制传送
negative transparency 透明负片
negative triggering pulse 负触发脉冲
negative tropism 负向性
negative type photoresist 负型光致抗蚀剂
negative urbanization 消极性城市化
negative utility 反效用
negative valency 负价
negative value 负值
negative valve 负压阀
negative variance 负差异
negative varnish 底片清漆
negative vertical angle 俯角
negative vibration isolation 消极隔振
negative voltage 负电压
negative voltage feedback 电压负反馈
negative vote 反对票
negative wage tax 负工资税
negative water-hammer gradient 负水锤梯度
negative wave 空号波;减水;水面泄降;水面下降;负波
negative well 渗漏井;泄水井;吸水井;灌水井;排水井;渗水井;倒渗井;负井
negative wire 电源负极引线
negative work 负功
negative-working photoresist 负性感光胶
negative writing 底片记录
negative yield curve 负收益率曲线
negative zone 负波带
negative zone plate 负波带片
negativity 负性
negativity condition 否定性条件
negatoscope 底片观察盒
negatron 阴电子;双极板负阻管;负电子
negatron decay 负电子衰变
negentropy 负平均信息量
negion 阴离子
neglected discount 放弃折扣
neglected loading 漏装
neglected plot 荒芜地段
neglect of duty 失职
neglect of economic performance 不讲经济效果
negligence 过失;粗心大意
negligence clause 免责条款;疏忽条款
negligence of duty 失职
negligence of navigation 航行疏忽
negligent accident 责任事故
negligible 可以忽略的;可忽视的;可不计的;不重要的
negligible effect 可忽略影响
negligible environmental impact 可忽略环境影响
negligible error 可忽略误差
negligible probability 可忽略概率
negligible quantity 可忽略(的)量
negligible rate 可忽略速率
negligible residue 微量残渣;可忽视残留
negotiability 流通能力;可转让性;可流通性
negotiable 可转让的大额定期存款单;可协商的
negotiable amount 可议付的金额
negotiable bill 可流通票据
negotiable bill of lading 可转让提单;可流通提单
negotiable certificate of deposits 可转让定期存单;可流通定期存单;可流通仓单
negotiable certificate of time deposit 可转让定期存款证
negotiable check 转让支票;流通支票;可流通支票;可流通来人支票
negotiable credit instrument 可转让证券
negotiable document 可转让证券;可转让文件;可转让单据;可流通单据
negotiable draft 汇票
negotiable instrument 流通证券;流通票据;可转让证书;可转让票据;可过户证券;票据;可流通证券;可流通票据
negotiable letter of credit 可流通信用证
negotiable multiple transport documents 可转让复合运送单据
negotiable note 可流通票据;可流通期票
negotiable note tax 有价证券税
negotiable order of withdrawal 限定取款额度的可转让账户
negotiable order of withdrawal account 可转让提款指令账户
negotiable original copy 议付正本
negotiable paper 可转让票据;可流通票据
negotiable path 可通行的小径
negotiable security 可流通证券;可转让证券
negotiable warehouse receipts 可转让仓单;可流通仓库收据
negotiate 转让;会谈;让购
negotiate business 洽谈业务
negotiated amount 议付金额
negotiated bidding 协商招标;议标;谈判招标
negotiated bills 议付票据
negotiated contract 协商合同;议定发包合同;议标合同;议标;通过磋商成交的合同;特定承包合同
negotiated market 协商市场;议价市场
negotiated offering 出售证券协议
negotiated price 协议价格;协商价格;议价
negotiated quota 协议配额
negotiated tendering 议标
negotiated terms 协商条款
negotiate purchase 议价收购
negotiating bank 议付(银)行;押汇银行;让购银行;让购汇票银行
negotiating bids 议标
negotiating condition 议价条件
negotiating curve 通过曲线
negotiating date 议付期限;押汇有效期限
negotiating machinery 谈判机构
negotiating resistance 坡度阻力;谈判阻力
negotiating tenders 议标
negotiating transaction 议付业务
negotiation 协商;押汇;让购;谈判;商讨
negotiation charges 协商费(用);押汇手续费;让渡手续费
negotiation commission 协商手续费
negotiation credit 让购信用证
negotiation of bids 议标
negotiation of contract terms 谈判合同条款
negotiation of foreign bills 议付外国汇票
negotiation phase 选择标商阶段;协商阶段(承包工程合同谈判);议标阶段
negotiation price 议价
negotiation room 谈判室
negotiations under contract 合同谈判
negotiation under reserves 保结押汇
negotiator 协商者;商议者;交涉者;谈判者;谈判人
negotiator role 谈判者角色
negrohead (压缩的)烟砖;劣等树胶
Negro's phenomenon 齿轮现象
neighborhood effects 相邻效应
neighborhood section of city 郊区
neighborite 氟镁钠石
neighbo(u)rhood 邻里;邻近
neighbo(u)rhood averaging 相邻点平均化
neighbo(u)rhood center 邻里中心;小区中心
neighbo(u)rhood commercial district 邻里商业区
neighbo(u)rhood committee 居委会;居民委员会
neighbo(u)rhood community 邻里社区
neighbo(u)rhood conservation 邻里保护
neighbo(u)rhood density 邻里密度
neighbo(u)rhood dwelling 居住邻里
neighbo(u)rhood effect 邻域作用
neighbo(u)rhood enterprise 街办集体企业
neighbo(u)rhood environmental evaluation and decision system 毗邻环境评价及判断系统
neighbo(u)rhood factory 街道工厂
neighbo(u)rhood group 居民小组;邻里住宅群
neighbo(u)rhood improvement scheme 街道改善计划;社区改善计划
neighbo(u)rhood index 接图表;图幅接合索引(表)
neighbo(u)rhood industry 街道工业
neighbo(u)rhood life cycle 街区生命周期
neighbo(u)rhood model 邻域模型
neighbo(u)rhood noise 周围环境噪声;毗邻环境噪声
neighbo(u)rhood of a curve 曲线邻域
neighbo(u)rhood of a point 点的邻域
neighbo(u)rhood park 邻里公园;街坊花园
neighbo(u)rhood participation 邻里参与
neighbo(u)rhood residential density 邻里住宅密度
neighbo(u)rhood shopping center 邻里购物中心
neighbo(u)rhood shopping district 邻里购货区;邻里商业点;邻里购物区
neighbo(u)rhood unit 邻里单位;住宅区
neighbo(u)rhood unit area 住宅单位面积;住宅单元面积
neighbo(u)rhood workshop 街道工厂
neighbo(u)ring 邻
neighbo(u)ring angle 接近角
neighbo(u)ring aquifer 相邻含水层
neighbo(u)ring areas 邻域
neighbo(u)ring block 邻近大厦;邻近大楼;邻近街区
neighbo(u)ring country 邻国
neighbo(u)ring earthquake 邻近地震
neighbo(u)ring edge 邻近边缘
neighbo(u)ring face 邻近面
neighbo(u)ring flight lines 相邻航(空)线
neighbo(u)ring layer 邻层
neighbo(u)ring line 邻线
neighbo(u)ring line distance 相邻测线线距
neighbo(u)ring opening 相邻跨度
neighbo(u)ring owner 邻近业主
neighbo(u)ring pile 邻近的桩;邻桩
neighbo(u)ring property 邻近产业;邻居财产;相邻地界线
neighbo(u)ring rock 围岩
neighbo(u)ring room 邻室;邻居房间
neighbo(u)ring sheet 相邻图幅;邻接图幅
neighbo(u)ring shot point gap 相邻炮点间距
neighbo(u)ring span 邻孔距;相邻跨度
neighbo(u)ring track 邻线;相邻线
neighbo(u)ring unit 邻里单位
neighbo(u)ring village 邻村
neighbo(u)ring wall 邻居墙壁;邻墙
neighbo(u)rwise 邻向
Neill method 尼耳氏气压分析法
Neil's parabola 尼尔抛物线
nek 山路
nekal 二丁基萘磺酸钠
nekoite 涩水硅钙石
nekrasovite 硫钒锡铜矿

nekton 游泳动物
nektonic redator 自游捕食者
nekton net 自游动物网
nekton organism 游泳动物；自由生物
Nelaton's catheter 软导管
nelenite 砷热臭石
neli-arc welding 氢电弧焊
Nelle system 维勒方式
nelsonite 钛铁磷灰岩
Nelson method 螺栓电弧压力钎焊
Nelson's cell 纳尔逊电池
Nelson's diaphragm cell 纳尔逊隔膜电解槽
Nelson's rollers system 纳尔逊滚筒系统
neltnerite 黑硅钙锰石
Nemafume 二溴氯丙烷（熏蒸杀虫剂）
Nemag 奈迈（负有效质量放大器与振荡器）
Nemagon 二溴氯丙烷（熏蒸杀虫剂）
nemaline 纤维状的；纤维状；纤维质
nemalite 纤水滑石
Nemaset 二溴氯丙烷（熏蒸杀虫剂）
nematath 断块海岭
nematic 向列相；丝状
nematic alignment 向列调整
nematic compound 向列化合物
nematic liquid crystal 向列型液晶；丝状液晶
nematic mesophase 向列型中间相
nematic phase 向列相
nematic state 向列态
nematic structure 向列型结构；向列结构
nematoblastic 纤状变晶质的
nematoblastic texture 纤状变晶结构；纤维变晶结构
nematophyte 丝状植物
nemeous 线形的
nemo 室外广播
nemoral 小树林的
nenadkevichite 硅钛铌钠矿
neo-abiogenesis 无生源新说
Neo-Academic Formalism 新学院派形式主义
Neo-Algonquian megacycle 新元古巨旋回【地】
neoanthropic 新人的
neoantique architecture 新古典建筑
Neoarctic region 新北区
Neo-Austrian school 新奥地利学派
Neo-Baroque 新巴洛克式
Neo-Baroque architecture 新巴洛克式建筑
neoblastic 新组织的
neo-brutalism 新粗野主义
Neo-Byzantine architecture 新拜占廷建筑
neocalcium 乙烯丙酸钙
Neo-Cambrian period 晚寒武纪【地】
Neo-Cambridge school 新剑桥学派
Neocapnodium tanakae 田中新煤炱
Neocaridina denticulate sinensis 中华新米虾
neocatastrophism 新灾变论
neocategory 新范畴
Neocathaysian 新华夏系；新华夏式【地】
Neocathaysian system 新华夏构造体系【地】
Neocathaysian tectonic system 新华夏系构造体系
Neocene 晚第三纪【地】
neocentromere 新着丝点
neoceramic glass 微晶玻璃
neo-classic architecture 新古典（式）建筑；新古典主义建筑
Neoclassical growth theory 新古典增长理论
neoclassical production function 新古典学派生产函数
neoclassical school 新古典学派
neoclassical synthesis 新古典综合派
neoclassicism 新古典主义
neocomer 新生事物
Neocomian(stage) 尼欧克姆阶【地】
neo-computing element 新计算元素
neo-concave area 新凹陷
Neo-construction 新构成主义
neocrystallization 新结晶作用
neocuproin 新式铜灵
neo-depression 新坳陷
neodrenal 纽转纳
neodymium 钕
neodymium acetate 乙酸钕
neodymium acetyl acetonate 乙酰乙酸钕
neodymium boride 硼化钕
neodymium-chromium glass 钕铬玻璃

neodymium-copper glass 钕铜玻璃
neodymium crystal laser 钕晶体激光器
neodymium-doped 掺钕
neodymium-doped calcium tungstate laser 掺钕钨酸钙激光器
neodymium-doped fiber 掺钕光纤
neodymium-doped fiber laser 掺钕光纤激光器
neodymium-doped fluoride fiber 掺钕氟化物光纤
neodymium doped glass 掺钕玻璃
neodymium-doped glass laser 掺钕玻璃激光器
neodymium-doped lanthanum beryllate 掺钕铍酸镧
neodymium-doped laser glass 掺钕激光玻璃
neodymium-doped silicate oxyapatite 掺钕硅酸氧化磷灰石
neodymium-doped yttrium aluminium garnet laser 掺钕钇铝石榴石激光器
neodymium-doped yttrium-aluminum garnet 掺钕钇铝石榴石
neodymium-doped yttrium-aluminum garnet rod 掺钕钇铝石榴石棒
neodymium fluoride 氟化钕
neodymium glass 含钕玻璃；钕玻璃
neodymium glass laser 钕玻璃激光器
neodymium hydroxide 氢氧化钕
neodymium laser 钕激光器
neodymium laser illuminater 钕激光照明器
neodymium laser oscillator 钕激光振荡器
neodymium laser radiation 钕激光辐射
neodymium molybdate 钼酸钕
neodymium ores 钕矿
neodymium oxide 三氧化二钕
neodymium sesquioxide 三氧化二钕
neodymium-strontium isotope correlation 钕-锶同位素相关性
neodymium-strontium mantle vector 钕-锶地幔向量
neodymium ultraphosphate 过磷酸钕
neodymium-ultraphosphate crystal 过磷酸体晶体
neodymium uranate 铀酸钕
neodymium-vanadium glass 钕钒玻璃
neoeffusives 新喷出岩
neo-epoxy resin 新型环氧树脂
neo-fault 新断层
neofield 新域
neoformation 新矿物生成作用
Neoformative 新生的
Neogaic megacycle 新地巨旋回【地】
Neogaikum 新地旋回【地】
Neogen 一种镍黄铜
Neogene 晚第三系【地】
Neogence system 上第三系
neogene climatic zonation 新第三纪气候分带
Neogene(period) 新第三纪；晚第三纪【地】
neogene planation surface 新第三纪夷平面
neogenesis 新矿物生成作用
Neogene system 新第三系【地】
neogenic 新生的
neogenic mineral 新生矿物
neoglaciation 新冰川作用
neo-Gothic 新哥特式
neo-Gothic architecture 新哥特式建筑
neo-Grec 新希腊式（1840 年在法国开创的建筑形式）
Neo-Greek 新希腊式；希腊复兴式；希腊复古式
neohetramine 新海棕拉明
neohexane 新己烷
neo-historicism 新历史主义
neoholotype 新全型
neoid 放射螺线
neoid deep 近期拗陷
neoid depression in the south marginal of Himalaya 喜马拉雅南缘近期拗陷
neoid tectonic system 近古构造体系
neo-impressionism 新印象派
Neokal 尼奥卡尔（阴离子渗透剂）
neokaolin(e) 新高岭石
neokinetic 新运动区的
neolan blue 宜和兰蓝
neolan colo(u)r 宜和兰染料
neolan green 宜和兰绿
neolan pink 宜和兰桃红
neolan red 宜和兰红
neo-liberty 意大利的新艺术复兴（1945 年后）；新自由派风格的
Neolite 耐欧来特（耐磨合成橡胶化合物）；新石
neolith 新石器
Neolithic 新石器时代的
Neolithic Age 新石器时代
Neolithic culture 新石器时代文化
Neolithic period 新石器时期
neo-loess 新黄土
neologistics cost 非运销成本
neolyn resin 马来改性醇酸树脂
neomagma 新生岩浆
Neomagnal 铝镁锌耐蚀合金
neomanoscope 观照片镜
neomask 新蒙片
neo-mediaeval 新中世纪的；新中古时代的
neo-mediaeval architecture 新中世纪式建筑
neomesselite 新磷钙铁矿
neometamorphism 初变质作用
neomineralization 新矿化；新成矿作用
neo-modern 更现代化
neomorph 新形成体
neomorphism 新形体形成
neomorphosis 新变态
neon 霓虹灯广告
Neonal 涅昂纳尔炸药；布特萨
Neonalium 涅昂纳铝铜合金
neon arc lamp 氖弧光灯；氖弧灯；热阴极氖灯
neon argon luminous tube 氖氩发光管
neon argon luminous tube light 氖氩发光管灯
neonatal 新生的
neonatal line 初生线
neon beacon 霓虹灯标
neon bulb 霓虹灯；氖灯管
neon computing element 氖气计算元件
neon diode counter 氖二极管计数器
neon-electroscope 测电笔
neon filled tube 氖气管
neon flash tube 氖气闪光管
neon gas 氖气
neon glim lamp 氖管
neon glow 氖辉光
neon glow lamp 氖辉光灯管
neon-grid screen 氖栅屏
neon-hydrogen bubble chamber 氖氢气泡室
neon indicator 氖管指示器；氖灯指示器
neon lamp 霓虹灯；氖管；氖灯
neon light 霓虹灯；氖光灯；氖灯
neon lighting 氖管照明
neon light regulator 氖灯方向指示器
neon oscillator 氖管振荡器
neon-photoconductor 氖管光电导体
neon pipe 氖管
neon pulse trigger 氖灯脉冲触发器
neon sign 霓虹招牌；霓虹灯广告；氖灯信号
neon-sign installation 氖管广告牌装置
neon spark tester 氖管火花试验器
neon stabilizer 氖管稳定器
neon-street 霓虹灯街
neon strip lighting 霓虹灯
neon tester 氖测电笔
neon timing light tester 定时霓虹灯测器器
neon tube 氖光灯；氖管
neon-tube electrode 氖管电极
neon-tube installation 氖管装置
neon-tube lamp 氖管灯
neon-tube light 霓虹灯管灯；氖灯管
neon-tube peak-voltmeter 氖管峰值电压表
neon-tube rectifier 氖管整流器
neon-tube transformer 氖管变压器
neon-tube type wave meter 霓虹灯式波高计
neon tuning-indicator 氖示谐管
neon type spark plug tester 氖光式火花塞试验器
neon voltage indicator 氖电压指示器
neon voltage regulator 氖管稳压器
neon voltmeter 氖伏特计
neon-xenon gas laser 氖氙气体激光器
neonychium 保护垫
neo-objectivity 新客观性
neo-orthodoxy 新创正统观念
Neopaleozoic 晚古生代
neopelex 十二烷基苯磺酸钠
neopentane 新戊烷；季戊烷
neopentyl glycol 新戊二醇
neophyte 生手

neophytic operation 新手操作
neoplasticism 新塑造主义;新塑型主义
neoplasty 修补术;造形术
neoprene 氯丁橡胶
neoprene adhesive 氯丁橡胶胶黏剂
neoprene base 氯丁基
neoprene bearing 氯丁橡胶支承架;氯丁橡胶支承面;氯丁橡胶轴承;氯丁橡胶支座;氯丁橡胶(桥面)支座
neoprene bearing pad 氯丁橡胶轴承瓦
neoprene channel 氯丁橡胶槽垫;氯丁橡胶管道
neoprene coated fabric 氯丁橡胶涂层织物
neoprene coated nylon fabric 氯丁橡胶涂层尼龙织物
neoprene coating 氯丁橡胶涂料
neoprene compressible seal 氯丁橡胶压缩密封
neoprene compression seal 氯丁橡胶压缩填料
neoprene control strips 橡胶控制条;氯丁橡胶伸缩缝条
neoprene decking 氯丁橡胶盖层
neoprene elastometric roofing sheet 氯丁橡胶防水卷材
neoprene foam 氯丁橡胶泡沫
neoprene foam rubber 合成泡沫橡胶
neoprene gasket 氯丁橡胶垫圈;氯丁橡胶填塞物;氯丁橡胶垫片;氯丁橡胶垫带
neoprene glue 氯丁橡胶釉料;氯丁橡胶胶
neoprene gum 聚氯合成橡胶
neoprene hanger 氯丁橡胶托架;氯丁橡胶吊钩
neoprene insert(ion) 氯丁橡胶衬垫;氯丁橡胶插片
neoprene jacket 氯丁橡胶罩
neoprene joint seal 氯丁橡胶(桥面压力缝)填封料
neoprene joint strip 氯丁橡胶接缝条
neoprene latex 氯丁橡胶乳液;氯丁胶乳;聚氯丁橡胶浆
neoprene pad 橡胶支座(桥梁)
neoprene paint 氯丁橡胶涂料;氯丁橡胶(色)漆
neoprene pot bearing 氯丁橡胶筒形轴承;氯丁橡胶盆形轴承
neoprene preformed gasket 氯丁预制橡胶垫片;预制氯丁橡胶垫片
neoprene ring 浊度计
neoprene roof 氯丁橡胶屋面;氯丁橡胶屋面片材
neoprene roofing 氯丁橡胶屋面
neoprene rubber 氯丁橡胶;氯丁橡胶胶黏剂
neoprene rubber adhesive 氯丁橡胶胶黏剂
neoprene rubber bridge bearing pad 氯丁橡胶的支承梁垫(预应力混凝土结构或钢结构桥梁的);氯丁橡胶桥梁支承梁垫
neoprene rubber sleeve 氯丁橡胶套筒
neoprene seal 人造橡胶密封
neoprene sealing gasket 人造橡胶密封垫片;氯丁橡胶密封垫片;氯丁橡胶密封填料
neoprene seal(ing) ring 氯丁橡胶密封圈;橡胶止水环
neoprene section expansion joint 氯丁分段膨胀接头;橡胶伸缩接头
neoprene sheathed cable 氯丁橡胶包皮电缆
neoprene sheet 氯丁橡胶薄板
neoprene sheet expansion joint 氯丁片状膨胀接头
neoprene sleeve 氯丁橡胶套
neoprene spring 合成橡胶弹簧
neoprene strip 氯丁橡胶板条
neoprene structural gasket 氯丁橡胶大块垫料
neoprene synthetic(al) rubber 氯丁合成橡胶
neoprene vibration pad 氯丁橡胶防振垫;氯丁橡胶防震垫
neoprene washer 橡胶衬垫;氯丁橡胶垫圈
neoprene waterproofing 氯丁橡胶防水
neoprene waterstop 氯丁橡胶隔水器;氯丁橡胶水密封;氯丁橡胶止水
neoprene weather baffle 氯丁橡胶挡风片
neoprontosil 偶氮磺酰胺
neo-purism 新纯洁主义
neopurpurite 异磷铁锰矿
neoquipenyl 伯氨喹啉
neo-rationalism 新理性主义
neo-realism 新现实主义;新写实主义
neo-Renaissance architecture 新文艺复兴式建筑
Neo-Romanesque 新罗马(建筑)风格
Neo-Romanticism 新浪漫主义
neosalvarsan 新胂凡纳明;九-四
neosepta 一种离子交换膜
neosolidizit cement 新索利迪契特水泥(一种混合硅酸盐水泥)
neosome 新成体;新成本
neostratotype 新层型
neostron 一种频闪观测管
neotectonic active region 新构造活动区
neotectonic analytical map 新构造分析图
neotectonic elements 新构造单元
neotectonic frame【地】新构造骨架
neotectonic inheritance 新构造运动的继承性
neotectonic map 新构造图件;新构造图
neotectonic movement 新构造运动
neotectonic movement division 新构造运动区
neo-tectonics 近代大地构造学
neotectonics 新构造运动;新构造学;新构造
neotectonic stabile region 新构造稳定区
neotectonic types 新构造类型
neotectonic zoning 新构造区划图
neotocite 水锰辉石
neotron 充气式脉冲发生管
neotropical 新热带区的
neotropical disjunction 新热带间断分布
neotropical floral realm 新热带植物地理区系
neotropical region 新热带植物区;新热带区
neotropical zoogeographic region 新热带动物地理区
neotype 新型的;新型;新模式
neovitalism 新活力论
neovolcanic rock 新火山岩
neovolcanism 新火山作用
Neozoic 上第三系【地】
Neozoic era 新生代【地】
Neozone 苯基苯基胺衍生物
Nepa polarity superchron 尼帕极性超时
Nepa polarity superchronzone 尼帕极性超时间带
Nepa polarity superzone 尼帕极性超带
neper[np] 奈培【物】
nepermeter 奈培表
nepernepheline 霞石
nephanalysis 云层分析;卫星云图
neph chart 云层分析图
nephcurve 云系分界线
nephelauxetic parameter 电子云重排参数
nepheline 霞石
nepheline basalt 霞石玄武岩;橄榄霞岩
nepheline diorite 霞石闪长岩
nepheline(eleolitic)porphyry 霞石光斑岩
nepheline leucite analcime basanite 霞石碧玄岩;霞石白榴石方沸石碧玄岩
nepheline leucite analcime tephrite 霞石白榴石方沸石碱玄岩
nepheline leucite nosean sodalite hauynophyre 霞石白榴石黝方石方钠石兰方斑岩
nepheline melilitite 霞石黄长岩
nepheline microsyenite 微晶霞石正长岩
nepheline monchiquite 霞石方沸碱煌岩
nepheline monzonite 霞石二长岩;霞石二长碱
nepheline nosean sodalite leucitophyre 霞石黝方石方钠石白榴岩
nepheline phonolite 霞石响岩
nepheline sodalite microsyenite 霞石方钠石微晶正长岩
nepheline sodalite syenite 霞石方钠石正长岩
nepheline syenite 霞石正长岩
nepheline syenite pegmatite 霞石正长伟晶岩
nepheline tephrite 霞石碱玄岩
nepheline worm 霞石条纹
nephelinite 霞石岩
nephelinite-bearing ultrabasic extrusive rock 含霞石超基性喷出岩
nephelinite-leucitite group 霞石岩—白榴岩类
nephelinitic texture 霞石岩结构
nephelinization 霞石化
nephelite 霞石
nephelium 云翳
nepheloid zone 雾状带
nephelometer 悬浮体散射仪;云量计;烟雾计;浊度仪;浊度计;能见度测定计;能见度测定表;散射浊度计;比浊计
nephelometer turbidity 浊度计浊度
nephelometric 浊度的
nephelometric analysis 浊度分析
nephelometric end-point detection 浊度法终点检测
nephelometric method 比浊法
nephelometric turbidity unit 散射(浊)度(单位);测浊度的浊度单位
nephelometry 悬浮体散射术;浊度测定法;散射测浊法
nephelormetry 混浊度测定法
nepheloscope 云室;成云器
nepherite 软玉
nephline syenite group 霞石正长岩类
nephogram 云图
nephology 云学
nephometer 量云器;量云计
nephoscope 云速计;测云仪;测云器;反射式测云器
nephritoid 温肌纹石
nephsystem 云系
nep meter 棉结测试仪
nepouite 镍绿泥石;镍利蛇纹石
nep pattern 毛结花纹
neps 毛粒
nep tester 棉结检验仪
nep testing machine 棉结检验机
Neptune 海王星
Neptune's sheep 白涌头
neptunian 水成论者
neptunianism 水成论
neptunian theory 水成论
neptunic dike[dyke] 水成岩墙
neptunic rock 海成岩;水成岩
neptunism 水成论
neptunist 水成者;水成论者
neptunite 柱星叶石
neptunium 镎
neptunium boride 硼化镎
neptunium oxide-sulfide 氧硫化镎
neptunium protoxide-oxide 五氧化二镎
neptunium tetrachloride 四氯化镎
nepuite 硅镁镍矿
neral 橙花醛
nerchinskite 多水高岭石;埃洛石
Nereid Monument 海王卫二纪念碑;海神纪念碑
nereites 似沙蚕迹(遗迹化石)
Nerex explosive 内里克斯炸药
neritic 近岸的
neritic area 浅海地区
neritic bottom 浅海底
neritic circulation 浅海环流
neritic community 浅海水层群落
neritic deposit 浅海沉积(物)
neritic deposition 浅海沉积作用
neritic environment 浅海环境;大陆架环境
neritic facies 近岸相;浅海相
neritic fauna 浅海动物区系
neritic Fe/Mn nodule 浅海铁锰结核
neritic formation 浅海建造
neritic marine basin 浅海盆地
neritic plankton 近岸浮游生物;浅海浮游生物
neritic province 近岸区;浅海区
neritic region 浅海区
neritic sediment 浅海沉积(物)
neritic shelf 浅海陆架
neritic shelf facies 浅海陆棚相
neritic system 浅海体系
neritic zone 近海带;近岸带;浅海区;浅海带;大陆架
neritic zooplankton 近岸浮游动物;浅海浮游动物
neritopelagic community 浅海水层群落
Nernst-Einstein relation 能斯脱爱因斯坦关系式(扩散定律)
Nernstian response 能斯脱响应
Nernst-Lindemann calorimeter 能斯脱—林德曼热量计
Nernst-Planck equation 能斯脱—普朗克方程
Nernst's approximation formula 能斯脱近似公式
Nernst's body 能斯脱高温发热体
Nernst's distribution law 能斯脱分配定律;能斯脱分布定律
Nernst's effect 能斯脱效应
Nernst's equation 能斯脱公式;能斯脱方程
Nernst's fuel cell 能斯脱燃料电池
Nernst's heat theorem 能斯脱热定理
Nernst's theorem 能斯脱理论;能斯脱定理
Nernst's theory 能斯脱学说;能斯脱理论
Nernst's unit 能斯脱流量单位
Nernst-Thomson rule 能斯脱-汤姆逊规则
Nernst zero of potential 电势的能斯脱零值
nerol 橙花醇

nerolidol 橙花叔醇
nerolidyl pyrophosphate 橙花叔醇焦磷酸
Nertalic method 耗处氩弧焊法
nerritic fauna 近岸动物区系
nerve damage 神经损伤
nerveless 无力的
nerve of a covering 覆盖网
nerves 交叉肋饰(穹窿面上伸出的肋)
nerviness 回缩性
nervous earth 遭受地震地带；地震地带；地壳震动部分
nervure 交叉肋形装饰；交叉侧肋
nervy rubber 回缩性能良好的橡胶
neryl 橙花基
neryl acetate 橙花醇乙酸酯
nesa 奈塞(透明导电模)；透明导电膜
nesacoat 氧化锡薄膜电阻
nesa coating 透明导电膜
nesa glass 奈塞玻璃(一种透明导电薄膜半导体玻璃)
Nesbitt method 纳氏粒铁直接冶炼法
nesh 热脆的
nesosilicate 岛状硅酸盐
nesosilicate mineral 岛状结构硅酸盐矿物
Nesper sampler 奈斯伯(底沙)采样器
nesquehinite 碳氢镁石
nesquehonite 三水菱镁矿
ness 岬；海角
Nesslerization 奈斯勒比色法；等浓(度)比色法
nesslerize 比色处理
Nessler's colo(u)r comparison tube 奈斯勒氏比色管
Nessler's reagent 奈斯勒试剂比色法；奈斯勒试剂
Nessler's reagent photometry 奈斯勒试剂光度法
Nessler's reagent spectrophotometry 奈斯勒试剂分光光度法
Nessler tube 奈斯勒比色管；奈氏比色管
nest 塔式齿轮；数据套；塞孔座；多联齿轮；叠合；叠放；簇状造材；簇状的；巢；嵌套
nestable pipe 套管
nest actuator 定位圈致动器
nest ball type mandel 成束内球形芯棒
nested 内装的；窝形的
nested barges 套装驳
nested block 嵌套块；嵌套分程序
nested boats 叠置舢板
nested caldera 巢状破火山口
nested classification 成套分组；成套分类
nested crater 巢状火山口
nested critical section 嵌套临界区
nested design of experiment 因子套设计
nested domain 域套；区域套
nested double reduction gear 巢式二级减速齿轮
nested family 成套类
nested file transfer format 嵌套文件传送格式
nested hypothesis 套假设
nested intervals 区间套
nested level 套合级数
nested level of conditional simulation 条件模拟套合级数
nested level of non-conditional simulation 非条件模拟套合级数
nested loop 嵌套循环
nested loop mode 嵌套循环方式
nested macro call 嵌套宏调用
nested macro definition 嵌套宏定义
nested monitor 嵌套监督程序
nested monitor call 嵌套监督程序调用
nested mo(u)ld 多叠压模
nested observation wells 观测井群
nested procedure 嵌套过程；嵌套程序
nested procedure structure 嵌套程序结构
nested recursion 嵌套递归式
nested region 区域套
nested sampling 套抽样
nested scope 嵌套作用域
nested sequence of intervals 区间套序列；区间套
nested sets 集合套
nested structure 嵌套结构
nested subroutine 嵌套子程序
nested subroutine linkage 嵌套子程序链
nested surface water monitoring network 套式地表水监测网
nested tables 套儿；套桌

nested-type reduction gear 巢式减速齿轮箱
nested volcanic cone 巢状火山锥
nested wood(en)silo 木制巢窝式储粮箱
nest egg 储备金
nest epiphyte 巢状附生植物
nester 试验筛组的单位筛
nestification 嵌套；叠加
nesting 构成嵌套；嵌叠；无交错突出的；套用；成套；嵌套
nesting level 嵌套级；嵌套层
nesting loop 嵌套循环
nesting of blocks 分程序的嵌套
nesting site 营巢区；筑巢区
nesting unit 嵌套单元(拉合尔花边机梳栉的)
nesting work of strip 冲片排列法
nestion 嵌置
nestle box 鸟舍
nest-like ore body 鸡窝状矿体
nest of boiler tube 炉管束
nest of saws 组合锯条；成套锯
nest of saws with wood(en)handle 木柄套锯
nest of screens 一套筛子；组合筛；粗筛组
nest of sieves 一组细筛；一套筛子；套筛
nest of smoke pipes 烟管群；烟管族
nest of springs 弹簧座
nest of symbols 符号的嵌套
nest of tables 套几
nest of tubes 管组；管群
nest plate 模穴套板；套片
nest spring 双重螺旋弹簧；复式盘簧
nest type blank locator 坯料定位窝
net 面网；无线电网；网络；通信网；纯净的；纯净
net ablation 净消蚀(量)
net absolutely 绝对净值
net absorption 净吸收量
net account due 到期付款净额
net account receivable 应收账款净额；应收账净额
net accumulation 净积累
net adjustment 管网平差；净平差；网平差
net advance 预付款净额
net advantage 净利益
net aerial production 净空气生产量
net after provision for losses 扣除损失后回收坏账净额
net aggregate 净额总计
net amortization charges 摊销费用净额
net amount 细数；有效储量；净总值；净值；净额
net amount of non-operating profit and loss 营业外损益净额
net amount of payment 实际支付数额
net amount paid 实际支付数额
net amplitude 净振幅；合成振幅
net and gross 净额与毛额
net and rope making yard 绳网区
net annual ablation 年净消融量(冰川)
net annual accumulation 年净堆积量(冰川)
net annual value 年净值
net apartment unit area 公寓净面积；公寓使用面积
net aperture area 净采光面积
net approximate-method 纯近似法
net area 净面积
net areal 净面积的
net array 渔网阵
net arrived sound value 完好货物到达时净值
net arrester barrier 网式制动装置(飞机着陆)
net arrived sound value 到达时完好净值
net arrived value 到达时净值
net assets 净资产；实际资产
net assets value 资产净值；净资产值
net assets value per share 每股净资产值
net assimilation 净同化
net assimilation rate 净同化率
net attenuation 净衰减量
net available head 净有效水头
net avails 贴现净得
net back price 净价；出厂价(格)
net balance 净余额；净差额
net balance of trade 贸易净差额
net barter terms of trade 纯易货贸易条件
net before taxes 税前收入
net benefit 净效益；净利益
net benefit flow 净效益流量
net benefit in present value 净收益现值

net board 围网吊货板
net bonded debt 债券负债净额
net book value 账面净值；净账面值；纯账面价值
net borrowed 借入净额
net borrowing 借款净额
net boundary 网格边界
net braking ratio 净制动率
net budget 净整预算；纯预算
net budget imbalance 净收支不平衡
net budgeting 净预算
net buoyancy 净浮力
net buoyancy of gas 气的净浮力
net buoyancy of oil 油的净浮力
net business formation 净企业形成个数；每月企业净增数
net business profits tax 纯营业收入税
net business savings 净企业储蓄
net cage 网箱
net call sign 通信网呼号
net calorific power 净热值；净发热值；净发热量；低位发热值；低位发热量；低热值
net calorific value 净发热值；净发热量；低位发热值；低位发热量；低热值
net canvas 刺绣十字布
net capacity 净载重量；净载货吨位；净容量；净容积
net capacity useful deadweight 净载重量
net capital 资本净额；资本额；净资本
net capital employed 企业运用的净资本
net capital gains 资本净收益
net capital to debt ratio 净资本对债务比率
net carbon dioxide assimilation 净二氧化碳同化
net carbon dioxide exchange 净二氧化碳交换
net cash 现金净额；现款支付；现金净值；净现金；基本金额
net cash flow 现金净流量；净现金流量
net cash investment 净现金投资
net cash price 净现金价
net cement grout 纯水泥浆
net change 净变化；净变(交易所当日与前一日收盘价之差)
net change method 变动净值法
net charge 净电荷
net chargeable income 应纳税的纯人息；应课税收入净额
net charter 净租赁(指租船)；净租船(船方除航运外，不负担一切港务装卸费用)；净运费租船契约
net charter terms 净运费租船条款
net citizen 网民
net clearance length 净空长度
net coke ratio 净焦比
net community productivity 净群落生产量；群落净生产力
net concentration 净浓度
net concessionary loan and grant 优惠贷款和赠款的净值
net confining pressure 净侧限压力；净封闭压力
net consumptive requirement 净耗水量
net consumptive use 作物净耗水量
net consumptive use of water 净消耗用水
net contribution 净分摊额；纯贡献
net control station 通信网管理电台
net correlation 净相关
net correlation coefficient 净相关系数
net cost 成本净额；净花费；净成本；实价
net cost account 净成本账户；进价的净价账户；进货净价账
net cost benefit 净费用效益
net count of spectrometer 能谱仪净计数
net credit 贷款净额
net credit period 付清赊帐期限
net cross-section(al)area 净断面(面)积；净截面(面)积；净横断面面积
net cumulative allocation of special drawing rights 特别提款权累计分配净额
net current 净向流
net current assets 流动资产净值；净流动资产
net current capital 净流动资金
net cut 净挖方(量)；净开挖量 净开挖
net cycle time 总工作时间；工作循环时间
net damaged value 残损净价
net debt 负债净额；借记净余额
net deficiency 亏损净额；净缺电量；净亏损额；净亏(损)；纯亏损
net deficit for the period 本期净损失

net demand 纯需求
net density 净密度
net density filter 纯光强过滤镜
net density of residential structures 居住建筑净密度
net dependable capacity 净可靠能力
net deposition 净沉积量
net depth meter 网深仪
net-depth telemeter 网深遥测仪;网位仪
netdevice 网络设备
net direct investment abroad 国外直接投资净额
net disbursement 净支出
net disbursement besides business 营业外净支出
net discharge 净发量
net displacement 总位移;净位移
net distance 净距(离)
net domestic product 国内生产净值;国内净生产值
net donator 净污染者
net doping 净掺杂
net dredge(r) 网式挖泥机
net drilling days 纯钻进日数
net drilling rate 净凿岩速度;净钻率
net drilling time 纯钻进时间
net dryer 网带式烘燥机
net dry weight 净实重;净干重
net duty 净用水率
net duty of water 净灌溉水;灌溉水定额;净用水率;净用水量;净吸水率;净灌溉率
net dwelling area 住房使用面积
net earned surplus 营业盈余净额
net earning 收益净值;净利;净赚;净所得;净税收;净收益;净收入;纯收益;纯利(润);净盈利
net economic saving 节约净值
net economic welfare 经济净福利
net economy curves 纯经济曲线
N-E tectonic system 北东构造系
net effect 净效应;基本效应
net effective head 净有效水头
net effective interest rate 净有效报酬率
net effective pore 纯有效孔隙
net effective rainfall 有效雨量;净雨量
net effective yield 净有效报酬率
net efficiency 净效率
net effort 净力;净作用力
net electric(al) generation 净发电量
net electric(al) power 净电功率
net energy 净能量;净能
net energy analysis 净能量分析
net energy for gain 增重净能
net energy for maintenance 维持净能
net energy for production 生产净能
net energy ratio 净能比
net error 错误净值
net estate 房产净面积;财产净额
net evapo(u)ration rate 净蒸发率
net exchange capacity of soil 土壤净交换量
net exchange position 净外汇余额情况;净外汇部位
net excitation 净励磁
net expected gain 期待净收益
net expenditures 纯支出
net export 净出口;出口净额
net exporter 净出口国
net export value 出口净值
net exposure 净暴露
net exposure position 净暴露部位
net extraction 净抽气量
net feed-drive power 净功率
net fill 净填方(量);纯填方
net filter 网滤器
net fixed assets 固定资产净额
net fixed capital formation 固定资本构成净值
net fleet 渔网阵
net floor area 楼层净面积;净建筑面积;建筑面积;地板净面积
net floor space 楼层净面积;楼板净面积;地板净面积
net flow 流动净额;净流量
net flow of revenue in budget 预算的收入净额
net flux 净通量
net force 净力
net foreign balance 净国际清算平衡值
net foreign exchange gains 净外汇效益
net foreign investment 净国外投资;外国投资净额;纯国外投资

net forward gain 往测净得
net foundation pressure 基底净压力
net freeboard 净出水高度;净超高,净(坝顶)超高,坝顶净超高;净干弦
net free reserves 净自由准备金;净自由准备额
net freight 净运费
net freight contract 净运费契约
net freight insurance 净运费保险
net freight ton-kilometers 货物净重吨公里
net fuel 现存燃料
net function 网格函数
net gain 净增重;净增益,净利益;净收益(额)
net gas 干汽;干气
net generation 净发电量
net ground lease 未经改善土地的租赁
net growth 净增长量
net hauler 起网机
net-hauling ship 起网船
net head 有效水头;有效落差;净扬程;净水头
net heat 有效热;净热;净耗热量
net heat consumption 净热耗
net heating power 净热值
net heating value 净热值;有效热值;净发热值;净发热量
net heat rate 净耗热率
Netherlandish flexible pavement design method 荷兰柔性路面设计法
Netherlands-west Germany basin 荷兰-西德盆地
net high-polymer 网状高分子
net hoisting time 纯提升时间
net horizontal pressure 净水平压力
net horsepower 有效马力;净马力
n-ethyl acetanilide 乙替乙酰替苯
net import 净进口;纯进口量
net importer 净进口国
net importer of energy supply 净能源输入国
net import value 进口净值
net income 净所得;净税收;净收益;净收入量;净收入;纯收益;纯利(润)
net income after depreciation 折旧后净收入
net income after dividends 扣除股利后净收益
net income after income taxes 扣缴所得税后的净收益;扣除所得税后净收益
net income after interest charges 计息后净收益
net income after tax 扣税后净收益;纳税后净收入;净利
net income and loss 净损益
net income approach 净收益估值法
net income before depreciation 未扣除折旧的净收益;折旧前净收入
net income before income tax 扣缴税前净收益
net income before interest charges 计息前净收益
net income before tax 缴税前净收益;纳税前净收益(纳税指纳所得税)
net income corrections 净利核正
net income distortion 净利计算不实
net income distribution to affiliated units 分配给联营单位利润
net income from operations 营业净收益;经营净收入
net income of society 社会纯收入
net income ratio 净收益比率;净收益率;净收入率
net income ratio of society 社会纯收入收益率
net income to net sales 净收入与销货净额比率
net increase 净增加
net increment of national economy 国民收入净增
net increment ratio of national economy 国民净增值收益率
net incurrence of liabilities 发生的负债净额
net indebtedness 净负债;负债净额
net indirect taxes 净间接税
net inflow 入超净额
net inflow volume 净入流(总)量;净流入量;净来水(总)量
net interception loss 净截留损失(量)
net interest 净息;净利息;净利
net inventory 净存货
net investment 净投资(额);投资净额
net investment income 投资收入净额;投资净收益
net investment required 新投资净额
net invoice price 发票净价;发票价格净额
net irradiance 净辐射辐照度
net irrigation requirement 净灌溉水量;田间灌水定额

net issues 净发行额
netizen 网民
netlayer 布网船
net leasable area 可出租净面积
net lease 净租约(房产税、保险、维修费等均由承租人负担);净租赁(即承租人担负一切税捐费用的租赁约)
net lending 净贷款
net length of crest 堰顶净长
net liability 负债净额
net lift 净扬程
net lifting head of pumps 水泵净扬程
net-like pattern 网状图案
net like structure 网状构造
net line 1.净自留责任网;2.建筑线;红线;施工限制线
net liquid assets 流动资产净额
net liquidity balance 净清偿差额
net littoral drift 沿沿岸漂沙
net load 有效载重;管网负荷;净荷载;净负载;净负荷;实际负载
net load capacity 净装载能力
net load increment 净增荷载
net loading intensity 净加荷强度;净荷载强度
net loading strength 净荷载强度
net load per vehicle 每车净载量
net long position 净多头部位
net long ton 净英吨
net longwave irradiance 净长波辐照度
net loss 亏损净额;净损失;净损(耗);净亏(损);净亏(损);通信网损耗;纯损失;纯损耗
net loss of current term 本期净损失
net loss of heat 热量的净损失
net making machine 编网机
net margin 净利
net marginal value of social output 边际社会纯产值
net markdown 标价净降低数
net mark-up 净加价;标价净提高数
net mark-up percentage 净增价百分率
net mass withdrawal from field 热田的净质量输出
net measure 净量
net migration 净移民量;净迁移(人口)
net mixing water 净拌和水
net moment 净弯矩
net monitor 网位监视仪
net national farm product 国民农业净产值
net national product 国民生产净值
net negative charge 净负电荷
net negative surface charge 净负表面电荷
net ness 净收益度
net-net weight 纯净重;净净重
net news 网络新闻组
net night flow 夜间净流量
net non-cash 非基本金额;净非现金
net obligation 净负债;负债净额
net of administrative expenses 除去管理费的净值
net of amortization 摊提净值
net of canals and ditches 沟渠网
net of control points 控制点网
net of copper wire 铜丝网
net of curves 曲线网(格)
net of depreciation 净折旧额
net official flow 官方净(资产)流量
net of groundwater 地下水网
net of iron wire 铁丝网
net of reserves 扣除准备金后净额
net of tax 扣除税款后净额;除税净额
net of tax income 纳税后净收入;税后净收入
net of triangulation 三角网
net of water courses 水网
net oil analyser 纯油分析仪
net operating assets 可供营运的资产净额
net operating earnings 净营业收益;经营纯益额
net operating hour 净工时
net operating income 扣除经营费用后的收入;净经营收入;纯经营收益
net operating income approach 净营业收益估值法
net operating loss 经营纯损额
net operating profit 营业净利
net operating time 纯运转时间
net output 有效出水量;净输出功率;净输出(量);净出力;净产量;实际产量

net output at terminal 出线端净输出力
net output value 净产值;纯产值
net output value of railway transport enterprise 铁路运输业的净值
net pattern 网格形
net pay 工资净额;实付工资
net payoff 付清贷款本息
net pay zone 纯油层
net peak discharge 净高峰流量
net peak flow 净高峰流量;净峰态流量
net personalty 动产净额
net photo size 净像片尺寸
net plane 点阵平面
net plankton 小浮游生物
net point 网格点
net-point method 网点法
net polluter 净污染者
net population density 人口净密度
net position 净部位
net position risk 净部位风险
net positive head 净正吸升水头
net positive suction head 净正吸水头(水泵);有效吸入高度(水泵);真空吸出高度;净吸水头;汽蚀余量
net positive suction head test 真吸出高程试验;汽蚀试验
net power 净功率;净动力
net precipitation 净降落量;穿降水;净降雨量;净降水量
net premium 保险净额;净贴水;净保费;奖金净额;纯保险费
net premium rate 净费率
net present requirement 净现值法
net present value 净现值;净当前值;纯现值
net present value method 现净值法;净现值法
net present value model 净现值模型
net present value of income from investment 投资收入净现值;投资净收入现值
net present value ratio 净现值指数
net present worth 净现值;纯现价
net pressure 净压(力)
net pressure fluctuation 脉动净压力
net pressure head 净压头
net price 正价;净值;净价
net price method 净价法
net primary aerial production 净第一性气体生产
net primary production 净植物性生产;净第一性生产
net primary productivity 净初级生产力
net proceeds 净所得;净税人;净税收;净收益;净收入;净收款额;净得;卖出实收价
net proceeds from sales 销货净收入
net product 生产净额;净生产额;纯产品
net production 净产量
net production value 净产值
net productivity 净生产率
net products 净产品
net profit 净利(润);纯收益;净利(润);净益(额)
net profit after depreciation 折旧后净利润;扣除折旧后净利
net profit after income tax 扣除所得税后净利(润)
net profit after tax 税后净利(润)
net profit after taxation 付税后纯利
net profit and loss account 净损益计算
net profit before depreciation 折旧前净利润
net profit before taxes 纳税前净利润
net profit earned 已得纯利;赚得纯利
net profit for the current year 本年纯利润
net profit for the year 本年度净利;本年度纯利
net profit from operation 营业净利;业务净利
net profit margin 纯边际利润
net profit of current period 本期净利
net profit of the term 本期净利;本期纯利
net profit on sales 销货净利;销货纯利
net profit or loss 纯损益
net profit or loss section 纯损益部分
net profit quota 净利润分配额
net profit rate 净利润率
net profit rate to sales 销售利润率
net profit ratio 净益比率
net profit ratio of registered capital 资本净利润率
net profit to net worth ratio 纯益净值比率
net profit to sales ratio 纯益销售额比率

net project benefits 工程净效益
net project cost 建设项目净造价
net property income 净财产收入
net protecting coating 净保护层
net public landing 政府放款净额
net pump suction head 水泵有效吸入扬程;水泵净吸入高度;泵的净吸压头;泵的净水头
net purchase 进料净额
net purchase of land and intangible assets 土地和无形资产的净购买额
net purchases 净购入;进货净额
net pyranometer 净总日射表
net pyrgeometer 净地球辐射表
net pyrradiometer 净全辐射表
net quantity of heat 全部热量
net quick assets 净速流动资产额
net rack cross section 拦污栅净断面
net radiance 净辐射强度;净辐射率
net radiation 净辐射
net radiation method 纯辐射法
net radiometer 净辐射仪;净辐射计
net rainfall 有效雨量;净降雨量;净降水量;超渗雨量
net rate 净速率
net rate of attendance 纯出勤率
net rate of tax 净税率
net reaction 净反应
net realizable value 可变现净值;净变现价值
net realizable value method 可实现净值法;可变信托;净变现价值法
net receipt 实收;纯收入
net recharge 纯灌量
net refrigerating effect 净制冷量;实际制冷效果
net region 网格区域
net registered ton 净吨;登记净吨
net register(ed) tonnage 净登记吨位;净注册吨位;净吨位;登记净吨位
NE-trending structure 北东向构造带
net rentable area 出租净面积
net rental 净租费;净租金
net reproduction rate 净再生产率
net reproductive ratio 净再生产比率
net reserves 净储[贮]藏量
net reservoir evapo(u)ration 水库净蒸发量
net residential area 居住用地净面积;居住面积;净居住面积
net residential density 居住净密度
net resident population density 居住人口净密度
net resistance 纯电抗
net resource depletion 资源净损耗
net result 最终结果;净结果
net retained lines clause 净自留额部分条款
net retention 净载流量;净吸收量;净保持量
net retention time 净停留时间
net retention volume 净停留体积;净保留体积
net return 纯收益;纯收入
net reveiver 净接受者
net revenue 净所得;净税收;净收入量;净收入;纯收入
net revenue account 营业净收入
net revenue before deductions 扣除前净收益
net ring 系网环
net roller 滚筒式起网机
net room area 房间净面积
net safe yield of well 井孔净保证出水量
net sale contract 净价包销合同
net sales 销货净额;净销价
net sales area 净营业面积
net sales-value method 销货净值法
net salvage 净残值(指海难后的残货值)
net sandstone 纯砂岩
net savings 净储蓄
net scale 净尺
net scour 净冲刷量
net section 有效截面;净截面;净断面
net sectional area 净断面面积;有效截面(面)积
net sedimentation 净沉积量
net seismic line 网络状地震测线
net settlement 净沉降;净付款;结算净额;结算付款净额
net-shaped 网状的
net-shaped crack 路面网(状)裂(纹);网裂
net-shaped structure 网状构造
net shift 最大相对变位;总移距;总位移
net shipping weight 装船净重量

net shortfall in resources 资金净差额
net short position 净空头部位
net short-term gains 短期净收益
net shortwave irradiance 净短波辐照度
net sieve 网筛
net single premium 一次纯保费
net site area 住房基地净面积;净工地面积;净占地面积
net size 净尺寸
net-size forging 精密锻件
net sling 网兜;吊货网(兜)
net slip 总滑距;总断距【地】;总变位
net social output value 社会净产值
net solar radiation 净太阳辐射
netsonde 网位仪
net spacing 净间距
net spacing between strip hole 条孔垂向净距
net spendable 净花费值
net spoilage cost 废品净成本
net station heat rate 电站净热耗
net stock 净存置;净存量
net stock of capital 剩余固定资本
net storm rain 净暴雨量
net structure 网络构造;网络组织
net subhead 分收支净额
net supply interval 地面径流形成期间;地表径流形成期
net surface charge 净表面电荷
net surplus 净盈余;纯利(润)
net survey method 测网法
net tangible assets 有形资产净值
net tare 净皮重
net taxable assets 应纳税净资产值
netted-disseminated Au ore 细脉浸染状金矿石
netted-disseminated ore 细脉浸染状矿石
netted-disseminated Pb-Zn ore 细脉浸染状铅锌矿石
netted structure 网状构造
netted texture 网状结构;网格状影纹
netted texture clay 蠕虫状黏土
netted turf 网状泥炭;网状菌丛;网状草皮
net terminal value 净终端价值
net terms 净租条件(指租船)
net terms of trade 净交易条件
net terrestrial radiation 净地面辐射
net thermal efficiency 净热效率
net time 加工时间;低定额;纯工作时间
net time on bottom 纯机械井底钻进时间
netting 张网捕鱼;扎钢筋网;挂网;钢筋网;净额交易;进行联络;结网;网状物
netting analysis 网格分析
netting by novation 债务重新编网
netting dam 钢丝网坝
netting fabric 网格组织
netting gear 网具
netting gear mend yard 网具修理场
netting index 网数
netting knot 网眼结
netting lath(ing) 网格板条;板条网架
netting number 网数
netting theory 网格理论
netting twine 网线
netting wire 网钢丝
nettle 小麻绳
Nettlestone bed 内特斯通层
nettle stuff 细拈绳
net ton 净吨;美吨(1美吨=2000磅);短吨
net ton kilometer 净吨公理
net ton kilometers per kilometer of line 净重货运密度
net ton kilometers per loaded car kilometers 重车平均净重量
net ton kilometer per train kilometre 平均列车净重;平均列车净吨
net tonnage 注册吨位;净装载吨数,净重吨位;净载容量;净吨位;净登记吨位
net torque 净力矩
net total irradiance 净全辐照度
net tracery 网状窗花格;网式窗花格
net tractive effort 净牵引力
net trading profit 贸易净利;销货纯利
net traffic 净运输量;净运量
net transport 净输送
net turbine power 水轮机净出力

net turnover 净职工流动率
net turnover formation 净资本构成
net turnover rate 净周转率
netty 网状的
net type bed bottom 网式床屉
net type filter 网式滤清器
net unilateral transfer 净单方转让
net unsaturated gain 净非饱和增益
net utility 净效用;纯收益
net value 净值
net value added 净增值
net value clause 净值条款
net value of capital invested 投资净值
net value of fixed assets 固定资产净值
net value of industrial output 工业净产值
net value of non-operating revenue and expenditures 营业外收支净额
net vault 网格拱顶;网状拱顶
net vault meshes 网状拱顶网目
net-veined leaf 网状叶
net volume 有效容积;净容积
net volume of earthwork 实土方量
net water cement ratio 净水灰比
net water consumption 净用水量
net water requirement 净需水量
net wealth tax 净财产税
net wear 净磨耗
net weight 净重;实物净重
net weight of equipment 设备净重
net weight terms 净重条件
net work 净功;纯功
network 网络;线路【电】;站网;局部地区网;网状物;网格结构;网系
network access 网络存取
network access processor 网络存取处理机
network address 网络地址
network address unit 网络地址单位
network adjustment 测量网平差
network admittance 网络导纳
network analog(ue) 网络模拟
network-analogue method 网络比拟法
network analogy 网络模拟
network analyser 网络分析器
net work analysis 网路分析;网络分析;网格分析
network analysis method 网络分析法
network analysis program(me) 网络分析程序
network analysis team 网路分析小组
network analysis technique 网络分析技术
network analysis theory 网络分析理论
network analyzer 网络分析器;网路分析器;电路模拟机
network architecture 网络体系结构
network assessment group 网路判定小组
network awareness 网络识别
network bill of lading 网状提单
network breaking 断网
network buffer 网络缓冲器
network calculator 网络计算器
network card 网卡
network chain 网链
network characteristic 网络特性
network characteristic at real frequency 网络的实频特性线
network chart 线路图;网线图;网络图;网路图
network circuit 多分支电路
network code 网络码
network communication circuit 网络通信线路
network commutator 网状换向器;网络换向器
network component 电路的组成部分
network computer 网络计算机
network computing environment 网络计算环境
network computing industry 网络计算工业
network conductor 网络导线;网络导体
network configuration 网络配置;网络结构
network connection 网络接线
network connection string 网络连接字符串
network constant 网络常数
network contactor 电力网接触器
network control 网络控制;电力网控制
network controlboard 电力网控制盘
network control center 网络控制中心
network control center equipment 网络控制装置
network control module 网络控制模块
network control program(me) 网络控制程序
network control station 网路运输控制站
network control unit 网络控制装置;网络控制器
network coordination center 网络协调中心
network coordination station 网络协调站
network cupola 网格圆屋顶;网状圆屋顶
network data 网络数据
network database 网络数控库
network data base management system 网状数据库管理系统
network data model 网络数据模块
network data translator 网络数据翻译程序
network data unit 网络数据单位
network decision 网络决策
network decoration 网眼装帧
network density 线路网密度;站网密度;网络密度;网格密度
network deposit 网状矿床
network design 站网设计;网络设计;电力网设计
network detection capability 台网探测能力
network determinant 网络行列式
network device 网络设备
network diagnosis 网络诊断
network diagram 控制分布图;网络网
network diaphragm 网状光阑
network dispatching automation system 网络调度自动化系统
network dispatching center 网络调度中心
network distribution equipment 网络分配设备
network distributor 网络分配器
network domain 网络管区
network dome 网状穹隆
network drainage 网状水系
network drill 网络演习
network duality 网络的对偶性
net-worked dome 网架圆顶
network element 网络元件;网络单元
network environment operating system 网用操作系统
network equation 网络方程
network exchange unit 网络交换器
network fat 网状油脂
network fault 电力网故障
network FAX 网络传真
network feeder 电力网络的馈电线
network fiber 网络纤维
network field 网络场
network file system 网络文件系统
network filter 网络滤波器
network-fixed site monitoring method 网格定点测量法
network flooding technique 网络满益法
network flow 网络流
network flow routine 网络流程程序
network flow theory 网络流量理论
network for location 定位图
network former 网络形成物
network-forming cation 成网阳离子
network forming element 网络形成元素
network forming ion 网络形成离子;成网离子
network forming oxide 网络形成氧化物
network for the explicit method 显式网格
network for the implicit method 隐式网格
network frequency 网络频率
network function 网络函数
network geometry 网络几何(学)
network glass 网格状玻璃
network host 网络主机
network hum 网络哼声
network hypothesis 网络假说
network impedance 网络阻抗
network in-dialing 网络拨入
network information center 网络信息中心
network information service 情报网络服务社
network information system 网络信息系统
networking 联网;连网;网状结合;网状交联
net working assets 营运资产净额;净营运资产
net working capital 净周转资金;净营运资金
net working capital ratio 净运用资金率
net working capital turnover 净营运资本周转率
net working polymer 网状聚合物
network initial condition 网络初始条件
network intelligence 网络智能
network interconnection 网络互连
network interface 网络接口
network interface adapter 网络传输卡
network interface card 网络接口卡;网络传输卡
network interface program(me) 网络接口程序
network interface unit 网络接口单元
network job processing 网络作业处理
network layer relay 网络层转接
network layout 电力网布置;布网方案
network line 网络线路
network line length 线网长度
network linking up 管网平差
network load 电网荷载;电网负荷
network load analysis 网络负载分析
network lock-up 网络死锁
network loss 电力网损失
network maintenance 网络维护
network maintenance signal 网络维护信号
network management 网络管理;网管
network management point 网路管理点
network-management signal 网路管理信号
network management unit 网络管理部件
network map 电力网图
network map of calculation area 计算区部分网格图
network master relay 功率方向继电器
network matching 网络匹配
network matrix 网络矩阵
network measurements 控制网测量
network media card 网络传输卡
network method of environmental impact assessment 网络评价法
network middle plane 网络状中间夹层;网络状的中间面
network mixing 网络混合
network model 网络模型
network modifier 网络调节剂;网络改性剂;网络调整氧化物;网络外体氧化物;网络修整器;网络修饰物;网络修改剂;网络改良器;网络调整物;网络调整体
network modifying ion 网络外体离子;变网离子
network modifying oxide 网络外体氧化物;网络体外氧化物
network monitoring 网络监视
network name 网状名字
network node 网络节点
network node interface 网络节点接口
network of accessory frame 区划网
network of burrows 穴洞网
network of canals 河道网络交叉;水渠网
network of cement particles 水泥颗粒分布
network of chains 三角锁
network of communication 通信网
network of communication lines 交通网
network of compass traverse 罗盘仪导线网
network of computer 计算机网(络)
network of conduit-type sewers 管状排水系统;下水管道系统
network of control 控制网【测】
network of coordinates 坐标网
network of cracks 网状裂缝;微裂缝系统
network of drains 排水管网
network of fault 断层网络
network of fixed points 控制点网
network of flight strip 航线网【航空】
network of image sources 像源网络
network of instruments 仪表网络
network of kerogen 干酪根网络
network of lightning conduct 避雷网
network of maritime engineering 水运工程网络
network of meshes 筛网
network of monitoring and surveillance 监视网
network of national and regional focal points 国家和区域主要测站网;国家和区域网点网
network of nodes and arcs 节点和弧的网络
network of pipe lines 管网;管道网
network of pipes 管网
network of piping 管道网络;管网
network of piping lines 管网
network of primary station 一等网
network of quadrilaterals 四边形网
network of railroad 铁路网
network of railway 铁路网
network of rivers 河流网
network of roots 根系;根丛
network of sailing strip 航线网【航海】

network of samples 样本网络
network of sampling stations 采样站网;取样站网
network of sewers 污水管(道)网
network of sewer system 管道网状系统
network of slip lines 滑移线网络
network of sound phase-shift 声相移网络
network of square 方格网
network of stations 站网;气象网;台站网;台网
network of strong motion accelerograph 强震加速度仪台网
network of surveillance stations 监测站网
network of system lines 系统线路网络
network of trading establishments 商业网
network of urban motorways 城市汽车道网络
network of vein 脉网
network operating system 网络操作系统
network operation center 网络运行中心;网络操作中心
network operation center 网络运行中心
network operations control center 网络作业控制中心
network operator 网络算子
network operator command 网络操作员命令
network operators maintenance channel 网络运行者维护通路
network ore 网脉状矿石
network-oriented computer system 面向网络的计算机系统
network parameter 网络参数
network patching 补网
network path 网络通路
network performance objective 网络性能目标
network phasing relay 网络定相继电器
network physical unit 网络物理部件
network plan 布网方案
network planning 网络分析;站网规划;网络规划(法);网路规则;统筹法;电网规划
network planning model 网络计划模型
network planning technique 网络计划技术
network plan of arrow 双代号网络计划
network plan of node 单代号网络计划
network point 网络点;网点
network polymer 体型聚合物
network primary distribution system 电力网一次侧配电系统
network processing unit 网络处理器
network processor 网络处理机
network program(me) 网络程序;网络计划
network programming 网络规划(法);网络程序设计
network programming of maritime engineering 水运工程网络规划
network protection 电力网保护
network protector 网状保护装置;网络保护装置
network protocol 网络协议
network protocol data unit 网络协议数据单元
network quality 网络质量
network radio center 无线电网中心
network readiness testing 网络备用试验
network refractory (product) 网络式耐火材料
network register 网络记发器
network relay 网络中继器;电网继电器;电力网继电器
network resource 网络资源
network schedule 网络进度表
network secondary distribution system 电力网二次侧配电系统
network security 网络保密;网络安全性
network sensitivity 网络灵敏度
network service 网络通信;网络服务;网道运输
network service access point 网络服务接入
network service protocol 网络服务协议
network servicer 网络服务程序
network simulation 网络模拟;网络仿真程序
network simulator 网格模拟器
network sinter 网络状泉华
network site 网络位置
network software 网络软件
network solution 网络解法;网络法的解
network source 电网电源
network speed 道路网车速
network stability 电力网稳定度
network stabilization 网络稳定性
network stand-alone system 独立网络系统

network state 网络状态
network station 网路站
network strategy 网络策略
network structure 网形结构;网脉状构造;网络结构;网状结构;网状构造
network structure resin 网状结构树脂
network subtransmission 网络状二次输电
network supply 电力网供电;电力网电源
network supports 网状支架
network synthesis 网络综合
network system 网组系统;网络系统;电网系统
network technique 网络技术
network terminating unit 网络终端装置
network texture 网格状结构
network theorem 网络定理
network theory 网络理论
network throughput 网络流通量
network time adjust 网络时间校准
network timing 网络定时
network tinted glass 网格颜色玻璃
network topology 网络位相学;网络拓扑;网络布局
network topology figure 网络布局图
network traffic 网络信息流通
network transfer function 网络传递函数
network transformation 网络变换;网格化变换
network transformer 网络变压器
network transmission 网络传输
network-type analog(ue) computer 网络型模拟计算机
network-type computer 网络型计算机
network type of throats with connecting pores 孔喉网络类型
network vein 网状脉
network virtual terminal 网路虚终端
network voltage 电力网电压
network-wide coordination entity 网络范围协调实体
network with earth-connected neutral 中性点接地网络
network with insulated neutral 中性点绝缘网络
network with junction points 结点网
net worth 净值
net worth accounts 净值账户
net worth tax 财产净值税
net worth to debts ratio 资本净值与负债比率;负债资本率;负债净值率
net worth to fixed capital ratio 资本净值与固定资本比率
net worth turnover ratio 净值周转率
net yield 净收益率;净收益;净产值
net yield coefficient 净产量系数
Neuberg blue 纽伯格蓝
Neuberg chalk 纽伯格白垩
Neumann-Kopp rule 诺埃曼-柯普法则
Neumann's boundary condition 诺埃曼边界条件
Neumann's formula 诺埃曼公式
Neumann's function 诺埃曼函数
Neumann's principle 诺埃曼原理
Neumann's problem 诺埃曼问题
Neumann well function 纽曼井函数
neuphor 富马松香胶
neural network 神经网络
neurenteric canal 布郎氏管
neurogram 印象
neurolite 纤蜡石
neuronal 二乙基溴乙酰胺
neurosurgery 神经外科
neusilber 铜镍锌合金
neustic 漂浮生物的
neuston 漂浮(微)生物;水表层漂浮生物
neutercane 变性热带气旋
neutralization of odor 消除恶臭;恶臭中和
neutral 中性的;中和的
neutral absorbent 中性吸收剂
neutral absorber 中性吸收器
neutral absorption 中性吸收
neutral-acidic lava 中酸性熔岩
neutral activating method 中性活化法
neutral adjustment 中性点调整
neutral ag(e)ing 中性老化
neutral air mass 中性气团
neutral angle 中性角(轧制用语);临界角
neutral-angle air intake 中和角进气口
neutral area 中立区;不带电区

neutral armature 中性衔铁;不带电衔铁
neutral atmosphere 中性气氛;中性大气
neutral atmosphere density 中性大气密度
neutral atmosphere model 中性大气模式
neutral atmosphere temperature 中性大气温度
neutralator 中线补偿器
neutral attitude 中间态度
neutral autotransformer 中性点接地自耦变压器
neutral axis 中性轴线;中性轴;中性层;中心轴;中和轴(线)
neutral axis of rail 钢轨中性轴
neutral background 不带电背景
neutral balance 随遇平衡
neutral beam 中性束
neutral beam injection 中性束注入
neutral beam injection heating 中性束注入加热
neutral beam source 中性束源
neutral beam splitter 中性分光镜
neutral blue 中性蓝
neutral body 中性体;中立体
neutral bottoms 优质油渣
neutral brick 耐火砖
neutral buffered potassium iodide method 中性缓冲碘化钾溶液法
neutral buoyancy 中和浮力;零浮力
neutral buoyant mine 中浮雷
neutral burning 中性燃烧
neutral-burning curve 中性燃烧曲线
neutral bus 中性母线
neutral bus-bar 中性汇流排
neutral carrier membrane electrode 中性载体膜电极
neutral character 中性性状
neutral circuit 中性线路
neutral clay 中性白土
neutral coast 中性海岸;中性岸
neutral coil 中性线圈
neutral colloid 中性胶体
neutral colo(u)r 灰色;不鲜明的颜色;中性颜色;中性色;中和色;无彩色
neutral colo(u)r filter 中性滤色镜;中性滤光片
neutral combustion 定压燃烧
neutral compensator 中性点补偿器;中线补偿器
neutral complex 中性络合物
neutral compound 中性化合物
neutral condition 中间状态;中性条件;中心条件;中立状态
neutral conductor 中性线;中性导线;中性导体;中线[电];带同心中性导线的电力电缆
neutral conjugation 中性共轭
neutral contact 中性接触
neutral-controlled plant 中间控制设备
neutral country 中立国
neutral cross-section 中和截面
neutral current 中性水流;中性流;中性电流
neutral current interaction 中性流相互作用
neutral current relay 中性线电流继电器;中线电流继电器
neutral current sheet 中性电流片
neutral curve 中性曲线
neutral cyclone 中性气旋;变性气旋
neutral decay product 中性衰变产物
neutral density disc 中灰滤光片;中性滤光片
neutral density filter 中性密度滤光片;中性滤色镜;中性滤光器;中性滤光片;中性滤光镜;中性灰滤光片;中灰滤光片
neutral density glass 中性玻璃
neutral density wedge 中性密度光楔
neutral depth 中和水深;正常水深;等流速水深
neutral desulfating process 中和脱硫酸法
neutral detergent 中性洗涤剂;中性去污剂
neutral detergent fiber 中性去污剂纤维
neutral detergent soluble 中性去污(剂)溶解物
neutral diffuser 中性扩散体
neutral direct-current telegraph system 单流电报制
neutral displacement transformer 中性点位移变压器
neutral dissolved species 溶解中性物种
neutral distillation 中性蒸馏
neutral drag instability 中性拖曳不稳定性
neutral dyes 中性染料
neutral earthing 中线接地;中点接地
neutral earthing compensator 接地线圈

neutral earthing resistor 中性接地电阻器
neutral electrode 中性电极
neutral element 中间元素;零元(素)
neutral end 中性端
neutral equilibrium 中性平衡;随遇平衡
neutral estuary 中性河口
neutral fat 中性脂肪;中性油脂
neutral feeder 中性馈线
neutral fiber 中性纤维
neutral field line 中性场线
neutral filler 中性填料
neutral filter 中性密度滤光片;中性滤光器;中性滤光片
neutral filter glass 中性玻璃滤光片
neutral filter working standard 中性滤光片工作标准
neutral filtration 中性滤波
neutral fire brick 中性耐火砖
neutral flag 中立国(国)旗
neutral flame 中性焰;中性火焰;中心焰
neutral flow plane 中性面;中和面
neutral flux 中性熔剂;中性溶剂;中性焊剂
neutral fold 中性褶皱
neutral fragment 中性碎片
neutral gas 中性气体
neutral gear 中间齿轮
neutral glass 中性玻璃
neutral granulation 中性成粒作用
neutral gray 中性灰色
neutral green 中性绿
neutral grey glass 中性灰色滤光玻璃
neutral ground(ing) 中性接地;中性点接地
neutral grounding resistor 中性线接地电阻器
neutral ground resistor 中性接地电阻器
neutral harbo(u)r 中立港
neutral impurity 中性杂质
neutral index 通用主题索引
neutral ion injector 中性离子注入器
neutral ionized stream 中性电离流
neutralisation test 中性试验
neutraliser 中性化器
neutralism 中性现象
neutrality 中性;临界稳定
neutrality condition 中性状况;中性条件
neutrality equation 中性方程
neutrality of glass 玻璃中性
neutralization 中性化;中和作用;中和法;抵消
neutralization adjustment 中和调整
neutralization agent 中和剂
neutralization chamber 中和槽
neutralization coagulation 中和混凝
neutralization coil 中和线圈
neutralization curve 中和曲线
neutralization equivalent 中和当量
neutralization heat 中和热
neutralization indicator 酸碱指示剂
neutralization level 平衡水准面
neutralization method 中和法
neutralization number 中和数
neutralization point 中和点
neutralization pond 中和池
neutralization procedure 中和操作规程
neutralization process 中和法
neutralization ratio 中和比率
neutralization reaction 中和反应
neutralization reagent 中和试剂
neutralization settling 中和沉降
neutralization tank 中和池;中和槽
neutralization test 中和试验
neutralization titration 酸碱滴定(法)
neutralization treatment of wastewater 废水中和处理(法)
neutralization value 中和值
neutralization wastewater treatment 中和废水处理
neutralization with lime 石灰中和处理
neutralize 制压;使中立化
neutralized amplifier 中和放大器
neutralized area 中立区
neutralized buffer 中和缓冲器
neutralized colo(u)r 中和色
neutralized products 中和产品
neutralized radio-frequency amplifier 中和射频放大器
neutralized radio-frequency stage 中和射频级

neutralized receiver 中和接收机
neutralized series motor 补偿串励电动机
neutralized sludge pump 中和污泥泵
neutralized sulfite liquor 中和亚硫酸盐液;中和硫酸盐废液
neutralized surface method 中和面法
neutralized system 中和系统
neutralized track 压制跟踪
neutralizer 中和器;中和剂;平衡器
neutralizing 平衡;抵消;中和处理;中和的;中和
neutralizing agent 中和剂
neutralizing aperture 中和孔径
neutralizing bath 中和槽
neutralizing buoyant 中和浮力
neutralizing canal 中和通道
neutralizing cell 中性化室
neutralizing charge 中和电荷
neutralizing chemicals 中和化学药品
neutralizing coil 中和线圈
neutralizing device 中和器及设备
neutralizing effect 中和作用;中和效应
neutralizing filter 中和滤光;中和滤光器
neutralizing indicator 中和指示器
neutralizing-membrane filtration process 中和膜滤法
neutralizing overacidity 中和过酸度
neutralizing power of lens 透镜的中性焦强
neutralizing precipitation 中和沉淀
neutralizing precipitation tank 中和沉淀池
neutralizing resistance 抵消电阻
neutralizing settling 中和沉降
neutralizing tank 中和罐;中和槽
neutralizing tool 中和工具
neutralizing tower 中和塔
neutralizing transformer 中和变压器
neutralizing treatment 中和处理;碱洗涤
neutralizing value 中和值
neutralizing water tank 中和水池
neutralizing well 中和槽
neutralizing wire 中和线
neutralizing with clay 白土中和
neutralizing zone 中和区;中和段
neutral keying 中性键控
neutral layer 中和层
neutral leach(ing) 中性浸出
neutral lead 中性点引出线
neutral lepton current 中性轻子流
neutral lift line 中和升力线
neutral line 中性线;中和线;磁针中央线
neutral line discharge theory 中性线放电理论
neutral line of main channel 主航道中心线
neutral lining 中性炉衬
neutral liquid 中性液体
neutral loading 中性线加载
neutral lubricating oil 中性润滑油
neutral lubricating stock 中性润滑油配料
neutrally buoyant 悬浮(在水中)的
neutrally buoyant float 中性浮子
neutral main 中性线;中性干线
neutral margin 零压面
neutral mass spectrometer 中性粒子质谱仪
neutral material 中性材料
neutral matter 中性物
neutral medium 中性介质
neutral metal 中性金属
neutral micelle 中性胶束
neutral mineral 中性矿物
neutral mode 中性振荡模式
neutral model 中性类模型
neutral molecule 中性分子
neutral mutation 中性突变
neutral oil 中性油;脱蜡石蜡油
neutral operation 中性工作
neutral ore 中性矿石
neutral organic compound 中性有机化合物
neutral over-current relay 中性过电流继电器;中性超流继电器
neutral oxidation 中性氧化作用
neutral packing 中性包装
neutral pair 中性偶
neutral peroformance curve 空档特性曲线
neutral pH 中性 pH 值
neutral phase 中线相位
neutral pigment 中性颜料

neutral plane 中性(平)面;中和平面;中和面
neutral point 重力分界;中性点;中和点;双曲点;鞍形区
neutral point displacement 中性点位移(星形接线)
neutral point earthing 中性点接地
neutral point earthing resistance 中性点接地地阻
neutral point grounding 中性点接地
neutral point method 中和点法
neutral point no-earthing system 中性点不接地系统
neutral point solid ground 中性点直接接地
neutral point theory 中性点理论
neutral point transformer 中性点变压器
neutral-polar relay 组合继电器
neutral pole 中性极
neutral port 中立国港口
neutral position 中立位置;中和位置;空挡(位置)
neutral position contact 中位触点
neutral position of shuttle 货叉零位
neutral potassium bromide 中性溴化钾
neutral potassium iodide 中性碘化钾
neutral potential 中性电位
neutral pressure 消除压力;中性压力;中和压力;孔隙水压(力)
neutral pressure level 中和界
neutral process desulfurization 中和法脱硫
neutral rail 中性轨
neutral rate of interest 不赢不亏的利息率
neutral reaction 中性反应;中和反应
neutral reactor 中性线接地电抗器
neutral reagent 中性试剂
neutral red 中性红
neutral red iodide 碘中性红
neutral refractory(material) 中性耐火材料;中间耐火材料
neutral region 中性区
neutral relay 中性线继电器;中性继电器;极化继电器;无极继电器
neutral relay with heavy-duty contacts 无极加强接点继电器
neutral resin 中性树脂
neutral resistance 中性电阻
neutral resonance 中性共振
neutral return path 中性线回路
neutral rock 中性岩
neutral rocks 中性岩类
neutral rope 平衡绳
neutral rudder 中立位置方向舵
neutral salt 中性盐
neutral scattering 中性散射
neutral section 分相装置
neutral section insulator 分相绝缘器
neutral series 中性级数
neutral sheet 中性片
neutral sheet reconnection 中性片重连
neutral ship 中立国船舶
neutral shoreline 中性滨线
neutral size 中性胶料
neutral slag 中性渣;中性炉渣;中性矿渣
neutral sludge pump 中和污泥泵
neutral soda 中性苏打
neutral soil 中性土(壤)
neutral soil plant 中性土植物
neutral solute 中性溶质
neutral solution 中性溶液
neutral spectrometer 中性粒子谱仪
neutral speed 平衡速率
neutral spring 中性泉
neutral stability 中性稳定(性);中性稳定(度);随遇稳定
neutral stability curve 中稳态曲线
neutral stain 中性染料
neutral state 中性状态
neutral steel 中性钢
neutral steer 操纵杆的中间位置
neutral step filter 中性阶梯减光板
neutral step weakener 中性阶梯减光板
neutral step wedge 中性阶跃劈
neutral stick 中立操纵杆
neutral stress 中性应力;中性压力;中和应力;对消应力
neutral sulphite pulp 中性亚硫酸盐纸浆
neutral sulphite semi-chemical 中性亚硫酸盐半

化学的
neutral surface 中性面;中面(壳体);中和面
neutral surface species 中性表面物种
neutral suspended substance 中性悬浮物质
neutral switch 中性线开关
neutral synthetic washing agent 中性合成洗涤剂
neutral temperature 中立温度
neutral terminal 中性线端
neutral tint 中间色;不鲜明色
neutral-tinted glass 中性有色玻璃
neutral-tint filter 中性滤光镜
neutral-tone 中性灰(指图像);中和色调
neutral toner 中性调色剂
neutral-tongue relay 中位舌簧继电器;中簧继电器
neutral track circuit 无极轨道电路
neutral transmission 中性传输;单向电流传送;单流式传送;单流传输
neutral type 中性型;中性类型
neutral unit 信息的自然对数单位(等于1.443二进制单位)
neutral vacancy 中性空位
neutral vessel 中立国船舶
neutral washing agent 中性洗涤剂
neutral wash solution 中性洗液
neutral water 中性水
neutral water glass 中性水玻璃
neutral waters 中立国水域;中立国领海
neutral water treatment 中性水处理
neutral wave 稳定波
neutral wavelength 中性波长
neutral wedge 消光楔;中性劈;中性光楔
neutral wetting 中性润湿
neutral wind shear 中性风切变
neutral wire 中性线;零线【电】
neutral zone (道路的)中心岛地带;中性区;中立地带;中间区;中和区;中和界;中和段;无作用区;不灵敏区
neutral zone of road 道路中心岛地带
neutrator 中性点补偿器
neutrino 中微子;微中子
neutrino astronomy 中微子天文学
neutrino detector 中微子探测器
neutrino horns 中微子透镜
neutrino telescope 中微子望远镜
neutroceptor 中性感受器
neutrodyne 中和式高频调谐放大器;中和接收法;中和法;衡消接收法
neutrodyne circuit 平衡路;平差电路
neutrodyne receiver 中和接受机;中和接收机
neutrodyne reception 中和接受法
neutrography 中子照相法
neutron 中子
neutron absorbent concrete 中子吸收混凝土
neutron absorber 中子吸收剂
neutron-absorbing glass 中子吸收玻璃
neutron-absorbing material 中子吸收材料;中子吸收剂
neutron absorption 中子吸收(作用)
neutron absorption cross section 中子吸收截面
neutron accident dosimetry 事故中子剂量测定法
neutron activation 中子活化(法)
neutron activation analysis 中子活化分析
neutron activation energy 中子激活能
neutron activation log 中子活化测井
neutron activation log curve 中子活化测井曲线
neutron activation method 中子活化法
neutron activation technique 中子活化技术
neutron age 费米年龄
neutron angular current 中子角流量
neutron angular density 中子角密度
neutron angular flux 中子角通量
neutron atomic mass 中子原子质量
neutron attenuation 中子衰减
neutron backscatter 中子反向散射装置
neutron balance 中子平衡
neutron balance equation 中子平衡方程
neutron balance sheet 中子平衡表
neutron beam 中子束
neutron beam attenuation 中子束衰减
neutron beam collimation 中子束准直
neutron binding 中子结合
neutron binding energy 中子结合能
neutron bombardment 中子冲击
neutron camera 中子照相机

neutron capture 中子俘获
neutron capture cross section 中子俘获截面
neutron chain reaction 中子链式反应
neutron chain reactor 中子链式反应堆;中子链式堆
neutron chopper 中子选择器
neutron collimator 中子准直仪
neutron collision 中子碰撞
neutron collision radius 中子碰撞半径
neutron conservation principle 中子守恒原理
neutron converter 中子转换器
neutron cooling coefficient 中子冷却系数
neutron counter 中子计算器
neutron cross section 中子截面
neutron crystallography 中子结晶学
neutron curing 中子固化
neutron current 中子流
neutron current density 中子流密度
neutron current density vector 中子流密度矢量
neutron curtain 中子流切断屏
neutron cycle 中子循环
neutron damage 中子损伤
neutron death 中子俘获
neutron decay 中子衰变
neutron-deficient isotope 缺中子同位素
neutron density 中子密度
neutron density disadvantage factors 中子密度不利因子
neutron density fluctuation 中子密度起伏
neutron densuty distribution 中子密度分布
neutron depolarization effect 中子退极化效应
neutron-detecting phosphor 中子闪烁体
neutron detection 中子探测
neutron detection reaction 中子探测反应
neutron detection system 中子探测系统
neutron detector 中子探测仪;中子探测器
neutron diffraction 中子衍射
neutron diffraction analysis 中子衍射分析
neutron diffraction apparatus 中子衍射器
neutron diffraction camera 中子衍射照相机
neutron diffraction pattern 中子衍射图
neutron diffraction study 中子衍射研究
neutron diffract meter 中子衍射仪;中子衍射计
neutron diffractometer 中子衍射仪
neutron diffusion 中子扩散
neutron diffusion coefficient 中子扩散系数
neutron diffusion cooling phenomenon 中子扩散冷却现象
neutron diffusion current 中子扩散流
neutron diffusion equation 中子扩散方程
neutron dispersion curve 中子色散曲线
neutron dose glass 中子剂量玻璃
neutron effective lifetime 中子有效期
neutron element 中子元件
neutron emergency dosimeter 中子事故剂量计
neutron emerging 中子射出
neutron energy 中子能量
neutron energy groups 中子能群
neutron energy spectrum 中子能谱
neutron energy spread 中子能量分散
neutron-epithermal log 中子—超热中子测井
neutron-epithermal neutron log curve 中子—超热中子测井曲线
neutron escape 中子漏泄
neutron eurrent vector 中子流矢量
neutron event 中子事件
neutron excitation of the nucleus 原子核的中子激发
neutron exposure 中子照射
neutron-exposure chamber 中子照射室
neutron-exposure facility 中子照射装置
neutron film badge 中子胶片剂量计
neutron-fissionable 中子作用下可裂变的
neutron fission detector 中子裂变探测器
neutron fission-scintillation detector 中子裂变闪烁探测器
neutron flashtube 中子闪光管
neutron fluctuation 中子涨落
neutron flux 中子通量
neutron flux converter 中子通量转换器
neutron flux density 中子流密度
neutron flux density indicator 中子通量密度指示器
neutron flux density meter 中子通量密度计
neutron flux density scanning assembly 中子通量密度扫描装置

neutron flux desity monitor 中子通量密度监测器
neutron flux detector 中子通量探测器
neutron flux flattening 中子通量展平
neutron flux intensity 中子通量强度
neutron flux level 中子通量水平
neutron flux modifying device 中子通量修正设备
neutron flux monitor 中子通量监测器
neutron flux pattern 中子通量图
neutron flux sensor 中子通量敏感元件
neutron formation rate 中子生成率
neutron form factor 中子形状因子
neutron-fragment angular correlation 中子裂变碎片角关联
neutron-gamma log curve 中子伽马测井曲线
neutron-gamma log(ging) 中子伽马测井
neutron-gamma well log(ging) 中子势阱测井
neutron graphy 中子照相术
neutron hall 中子厅
neutron hardening 中子硬化
neutron hazard 中子危害性
neutron howitzer 中子准直器
neutronic detector 中子探测器
neutronic noise 中子噪声
neutronic reactor 中子反应堆
neutron image intensifier tube 中子象增强管
neutron-induced 中子诱发的
neutron-induced damage 中子引起损伤
neutron-induced process 中子诱发过程
neutron-induced radioisotope 中子诱发放射性同位素
neutron induced reaction 中子诱发反应
neutron-induced thermal stress 中子引起的热应力
neutron inelastic scattering reaction 中子非弹性散射反应
neutron injury 中子损伤
neutron instrument 中子仪器
neutron instrumentation 中子仪器;中子检测仪表
neutron interaction 中子相互作用
neutron inventory 中子总数
neutron-irradiated 受中子辐照的
neutron irradiation 中子照射;中子辐照
neutron irradiation monitor 中子辐射监测器
neutron leakage 中子泄漏;中子漏泄
neutron leakage spectrum 中子泄漏谱
neutron level 中子能级
neutron level spacing 中子能级间距
neutron life cycle 中子寿命循环
neutron life time 中子寿命
neutron life time log 中子寿命测井
neutron lifetime log curve 中子寿命测井曲线
neutron log calibration value 中子测井刻度值
neutron log calibrator 中子测井刻度器
neutron log curve 中子测井曲线
neutron log(ging) 中子测井(记录);中子量测;中子探测法;中子探测;中子势阱测井
neutron logging method 中子探测法
neutron logging system 中子测井系统
neutron log porosity of wet clay 湿黏土的中子孔隙度
neutron mass 中子质量
neutron matter 中子物质
neutron measurement 中子量测
neutron-moderation moisture meter 中子慢化湿度计
neutron moderator 中子慢化剂
neutron moisture ga(u)ge 中子水分计;中子水分测定仪;中子湿度计
neutron moisture meter 中子含水量测定仪;中子测水分仪;中子水分仪;中子湿度计
neutron moisture probe 中子水分探头
neutron monitor 中子监测器
neutron monitoring 中子监测
neutron multiplication factor 倍增系数;倍增
neutron-neutron log 中子—中子测井
neutron-neutron log curve 中子—中子测井曲线
neutron number 中子数
neutron number density 中子数密度
neutronography 中子照相法
neutron parameter 中子参数
neutron period 中子通量时间常数
neutron photography 中子照相
neutron poor isotope 缺中子同位素
neutron population 中子总数
neutron porosity 中子孔隙度

neutron porosity of shale 泥质的中子测井孔隙度
neutron port 中子束孔
neutron probe 中子探测仪
neutron producer 中子源
neutron production rate 中子产生率
neutron productivity 中子产额
neutron proof glass 防中子玻璃
neutron proton collision 中子质子碰撞
neutron-proton correlation effect 中子质子关联效应
neutron-proton diffusion 中子质子扩散
neutron-proton exchange forces 中子质子交换力
neutron-proton interaction 中子质子相互作用
neutron-proton mass difference 中子质子质量差
neutron-proton ratio 中子质子比
neutron-proton scattering 中子质子散射
neutron-proton system 中子质子系统
neutron pulse 中子脉冲
neutron radiation analysis 中子辐射分析
neutron radiation damage 中子辐射损伤
neutron radiograph 中子射线照相
neutron radiographic inspection 中子射线照相检查
neutron radiography 中子探伤法;中子射线照相法;中子放射线照相术
neutron-ray protected glass 防中子射线玻璃
neutron reactor 中子反应堆
neutron reemission 中子再反射
neutron-reflecting mirror 中子反射镜
neutron reflection 中子反射
neutron reflector 中子反射层
neutron-removal cross-section 中子移出截面
neutron reproduction 中子再生
neutron response 中子响应
neutron rest mass 中子静止质量
neutron-rich nuclide 多中子核素
neutron scattering 中子散射
neutron scattering method 中子散射法;(测土壤湿度的)中子扩散法
neutron scattering moisture meter 中子散射湿度表
neutron scattering technique 中子散射技术
neutron scintillation counter 中子闪烁计数器
neutron scintillator 中子闪烁器
neutron-sensing element 中子敏感元件
neutron sensitive 中子敏感
neutron sensitive chamber 中子灵敏室
neutron sensitive coating 中子灵敏层
neutron sensor 中子探测器;中子传感器
neutron shell 中子壳层
neutron shield(ing) 中子屏蔽;中子防护屏
neutron shield paint 中子屏蔽用涂料
neutron shield tank 中子防护箱
neutron shutter 中子束断续器
neutron signal 中子信号
neutron single-particle energy 中子单粒子能量
neutron soil moisture ga(u)ge 中子土壤湿度计;中子土壤湿度计量器
neutron soil moisture meter 中子土壤水分仪;中子土壤水分计
neutron source 中子源
neutron source box 中子源柜
neutron source calibration 中子源刻度
neutron source density 中子源密度
neutron source holder 中子源夹持器
neutron spectrograph[spectrometer] 中子谱仪
neutron spectrometry 中子测谱术
neutron spectroscopy 中子谱学;中子测谱术
neutron spectrum 中子谱;中子能谱
neutron spectrum hardening 中子谱硬化
neutron spectrum intensity 中子能谱强度
neutron spectrum measurement 中子谱测量
neutron speed 中子速率
neutron streaming 中子束流;中子流
neutron streaming rate 中子束流率
neutron stripping 中子削裂
neutron superfluidity 中子超流性
neutron temperature 中子温度
neutron thermalization 中子热化
neutron thermalization theory 中子热化理论
neutron-thermal neutron log 中子—热中子测井
neutron-thermal neutron log curve 中子—热中子测井曲线
neutron thermometer 中子温度计

neutron thermopile 中子热电堆;中子温差电堆
neutron tight 不透中子的
neutron-to-gamma sensitivity ratio 中子伽马灵敏度比
neutron topography 中子形貌术
neutron transconductance 中子跨导
neutron transfer reaction 中子转移反应
neutron transport 中子输运
neutron velocity selector 中子速度选择器
neutron velocity spectrometer 中子速度谱仪
neutron wall 中子防护墙
neutron wave 中子波
neutron wavelength 中子波长
neutron well logging 中子势阱测井;中子测井
neutron width 中子宽度
neutron yield 中子产额
neutropause 中性层顶
neutrophil(e) 嗜中性的
neutrophil(e) granule 嗜中性粒
neutrophilia 嗜中性
neutrophilic 嗜中性的
neutrophilic granule 嗜中性颗粒
neutrophilous 嗜中性的
neutrosonic receiver 中和式超外差接收机
neutrosphere 中性层
neutrovision 中子视
Neuwieder green 纽维德尔绿
Neva 内瓦(聚酰胺纤维)
Nevada 窗用特种透视玻璃
Nevadian orogeny 内华达造山运动
Nevadic orogeny 内华达造山运动
nevadite 斑流岩
Nevastone 涅瓦斯通(厨房设备的商品名称,如洗涤盆等)
neve 粒雪;冰川雪
neve basin 粒雪盆(地)
neve field 永久雪原;永久积雪原;粒雪原
neve glacier 粒雪冰川
neve ice 永久积雪冰;粒雪冰
nevellator 拉线上的水准仪
never blocked 不可分块的
never call 不调用
never failing 绝不间断
never-frozen soil 不冻土
never indebted 无负债
never lay flat 不可平放
never-never system 分期付款制
never quote two price 真不二价
never slip 不打滑
neve snow 永久积雪;粒雪;万年雪
nevyanskite 亮铱锇矿;天然铱锇合金
new account 新立账户;新开账户
new account application 开立账户申请书;开立新账户申请书
new acquisitions 新到图书
new additional duties 新增关税
new additional reserve 新增储量
new adjustment 新调整
Newage vibrating screen 内韦季摆筛
Newall system 纽沃尔制;纽尔制(公差配合基孔制)
new and renewable sources of energy 新能源和可再生能源
new area 新领域
Newark series 纽瓦克系;内瓦克统【地】
new arrival 新来的人;新到货(物)
new assets 新增资产
new Austrian tunnel(l)ing method 新奥地利隧道法【道】;新奥法
new balance 新余额
New-Baroque 新巴洛克
newberyite 镁磷石;水磷镁石
NewBide 纽比特(硬质合金)
new block 新街坊;新区段
new blood 新鲜血液
new blue 新蓝
new boldite 铁闪锌矿
new bond 新债券
newborn 新生的
newborn thing 新生事物
Newbourn crag 纽伯恩砂质泥灰岩
new Britain trench 新不列颠海沟
New Brutalism 新粗野主义;粗野主义【建】
newbuild 重建;新造;新建

new building 新建筑
new-building completed 新完工建筑物
newbuilt 重建的;新造的;新建的;新建的
newbuilt house 新建房屋
newbuilt port 新建港口
new business venture group 新企业经营集团
new calcination with boiler 带余热锅炉的新烧成法
new candle 新烛光;新国际烛光(单位);新国际光单位;坎德拉(发光强度单位)
new capital issue 新发行的资本股票
new car delivery service 新车交接
new cargo handling berth 新货物装卸泊位
Newcastle bed 纽卡斯尔层【地】
Newcastle coal 纽卡斯尔煤
Newcastle kiln 纽卡斯尔窑
new-ceram 新型陶瓷
new ceramic nanofiltration membrane 新型陶瓷纳滤膜
new ceramics 新型陶瓷;新陶瓷;先进陶瓷
new charter 新租船契约
new charter party 新租船契约方
new classification number 新分类号
new colo(u)rs 新彩
new combined 3R(carbon, nitrogen and phosphorus removal) bioreactor 新型一体化3R(除碳、氮脱磷)生物反应器
Newcomb operator 纽康算符
Newcomb theory 纽康理论
newcomer 新生事物
new constructed railway 新建铁路
new construction 新结构;新建工程;新工程
new construction activity 新项目施工
new construction cleanup 新施工管道清扫
new construction item 新建项目
new construction material 新建筑材料
new core bit 新岩芯钻头
new data flag 新数据标帜
new deal 新政策
new decibel 新分贝
new design 新型设计;新式样
new design control 新型设计控制
new developed area 新开发区
new development plan 新发展计划
new diamond 新金刚石
new dimension 新尺寸
new dry process cement production 新型干法水泥生产
New Dutch mooring buoy 新荷兰式系船浮标
new dwelling construction started 新建住宅开工数
new economic policy 新经济政策
new economic system 新经济体制
new edition 新版
new edition chart 新版海图
new edition of chart 海图新版
new effective temperature 新有效温度
new efficacy 新增效益
newel 拐弯角柱;楼梯栏杆柱;扶手转向端
newel cap 柱顶(脚)饰;楼梯栏杆柱顶饰
newel collar 楼梯柱的颈圈;柱饰的细圆部分
newel drop 楼梯栏杆柱垂饰;望柱吊饰;螺旋楼梯中柱柱脚下垂装饰;楼梯扶手转角柱柱脚下垂装饰
newel joint 望柱接头;中柱接头;楼梯扶手处中柱或转弯角柱的接头
newel post 螺旋楼梯中柱;望柱;旋梯中柱;螺旋型楼梯中柱;楼梯扶手转角柱;楼梯端柱;端柱
newel stair(case) 有角柱的楼梯;具有中柱的螺旋楼梯;环楼楼梯;盘旋扶梯
newel tube 楼梯栏杆主管柱;螺旋楼梯管柱;旋梯轴管;轴管
newel type spiral staircase 中柱式(螺)旋(楼)梯;有中柱的盘旋扶梯
newel wall 层间双折楼梯隔墙;楼梯隔墙
New Empiricism 新经验主义(指北欧斯堪的纳维亚人的20世纪40年代的建筑风格)
New English colonial 新英格兰式(美国初期建筑);美国初期新英格兰式建筑
New English colonial architecture 新英格兰殖民式建筑
New English method 新英格兰砌砖法
new entrants to the labo(u)r force 新增就业人员
new environmentalism (1945年后美国出现的)新环境主义
new epoch 新纪元

new era 新纪元
new era socket 绳公承窝
new erection 新建工程
new face 技术不熟练者;新手;新产品;新颖的
new fallen 新落的(雪);新伐的(木材)
new-fashioned 新式的
new field of technical activity 技术活动的新领域
new fixed assets 新增固定资产
new fold 新褶皱
New Formalism 新形式主义(建筑)
new formation 新形成
new for old 新换旧;以新换旧
new for old clause 新换旧条款
newfound 新发现的
Newfoundland 纽芬兰
new freedom type 新自由型
new funds 新资金
new generation 新一代
new generation AutoCAD system 新一代 AutoCAD 系统
new global tectonics 新全球构造(学)
new grain 新颗粒
new grant lot 新批地段
New Grecian type 新希腊式
New-Greek architecture 新希腊式建筑
new growth of soil 土壤新生体
new guard house at Berlin 设在柏林的新守卫室
new Guinea trough 新几内亚海槽
new habitat 新住处;新的栖息地;新场所境
new hand 新手;生手
New Hebrides plate 新赫布里底板块
new high 新的最高价位
new ice 初始冰
new incorporation 新公司
new industrial city 新兴工业城市
new industrial district 新工业区
new industrial state 新工业国
new industry area 新兴工业区
new industry ceramics 新型工业陶瓷
new input queue 新输入排队;新输入队列
new international economic order 新国际经济秩序
new international economy order 国际经济新秩序
new international normal gravity formula 新国际正常重力公式
new international round 新国际回合
new investment 新投资
New Jersey retort process 竖罐提炼法
new jobs credit 新增雇员税收减免
newkirkite 水锰矿
new-laid 新铺的;新浇的
new-laid line 新建铁路
new-laid pipe 新敷设管道
new-laid tube 新敷设管道
newlandite 顽火榴辉岩
newland lake 新台升地面湖
new light source of high transform efficiency 高效新光源
new lime liquor 新灰水
new line 移行【计】;回车换行
new line character 回车换行字符
new line construction 新线建设
new line key 换行键
new lines of production 新生产部门
new liter 新升(1 新升 = 10^{-3} 立方米)
new location 新位置
new look 新面貌;最新样式
new-loopforming system 新型成圈方法
new low 新的最低价位
Newloy 牛洛铜镍合金
newly 新近地
newly-accumulated alluvium 新冲积层
newly-accumulated sediment 新淤积泥沙
newly broken 新垦荒地
newly brushed main 新刷清干管
newly-built 新造的;新建的
newly-built construction 新建工程
newly-built port 新建港口
newly-built port dredging 新港区浚挖
newly-built project 新建项目
newly-built railway 新建铁路
newly-built railway construction 铁路新线建设
newly-commenced project 新开工(程)项目
newly-cut channel 新挖河槽
newly-developed technique 新发展的技能

newly developing 新兴城市
newly developing material production departments 新兴物质生产部门
newly discovered 新发现的
newly emerging forces 新兴力量
newly established enterprise 初创企业
newly-formed bank 新成河岸
newly-formed ice 初期冰;新成冰;新冰;初期岸冰
newly-frozen ice 新冰;初期冰;初期岸冰
newly-increased accumulation funds 新增积累资金
newly-increased consumption funds 新增消费资金
newly-increased purchasing power 新增购买力
newly-increased value 新增价值
newly industrialization country and region 新兴工业化国家和地区
newly industrialized country 新兴工业(化)国家
newly-installed 新设置的
newly laid 新铺的;新浇筑的
newly laid concrete 新浇混凝土;新铺混凝土
newly laid main 新设干管
newly laid track 新铺线路;新铺轨道
newly-located road 新定线道路
newly made 新造的
newly-made panel 新树立的标牌;新嵌板
newly-made products 新制成的产品;新产品
newly-made slab 新板;新浇水泥板;新制平板
Newlyn datum 纽林零点
newly-oiled 刚浸过油的
newly-placed concrete 新铺混凝土;新浇混凝土
newly rising city 新兴城市
newly scraped main 新刮清干管
new made 新做的;新形成的;新建的
new management 变更经营
Newman limestone 纽曼灰岩
Newman's rule of six 纽曼六位规则
Newman's tables 纽曼表(各种土壤的安全抗阻力表)
Newmark's analysis method for steady state vibration 稳定振动的纽马克分析法
Newmark's chart 纽马克图
Newmark's(influence)chart 纽马克感应图
Newmark's method for fundamental mode 基本振型的纽马克分析法
Newmark's method for steady state vibration 稳态振动的纽马克分析法
new marsh 新生沼泽
new material 新型材料;新材料
new methylene blue 新亚甲蓝
new mine 新建
new mine field discovery rate 新矿产地发现率
new mode 新样式
new model 新型号;改造
new money 新货币
new money loan 新货币借款
new moon 新月;朔
new moon pool 月池
new moon spring tide 新月大潮
new motor vehicle 新式机动车;新式汽车
new mown hay 新割干草
new northway method 弹性橡胶新切口法
new objectivity 新客观性
new obligation(al)authority 新全责拨款授权(美国);新专款授权
new oil 新油
new order 新订货单
new orders for durable goods 耐用品新订货量
new or finished 最后高程(土建项目)
New Orleans Port 新奥尔良港(美国)
new paragraph 新段落
new parts 新零件;新部件
new pattern 新型;新花式
new period of vigorous economic growth 经济振兴时期
new plant and equipment expenditures 新增机械设备支出
new plasticity chart 新塑性图
new ploughed field 新翻地
new position area 新阵地地域
new print 修订版;增印
new process 新工艺
new process equipment 新型技术装备
new product development 新产品开发;新产品发展

new product development plan 新产品发展计划
new product development process 新产品发展过程
new product division 新产品车间
new product evaluation 新产品估价
new product introduction plan 新产品推介方案
new production reactor 新生产堆
new productive capacity 新增生产能力
new productive system 新生产方式
new-product monopoly 新产品专利
new products 新产品
new product trial production fund 新产品试制基金
new protectionism 新保护主义
new publication 最新出版物
new record 新记录
new red sandstone 新红砂岩;斑砂岩
new regionalism 美国新地方主义(建筑)
new register book of shipping 新船名录
new resources 新资源
new rich 暴发户
new route 新路由
Newry 纽里花岗石(产于爱尔兰纽里地区)
news agency 新闻社;通讯社
new sand 新砂
news board 新闻报纸板;布告牌
news bogus paper 废新闻纸
news briefing 新闻简报
news briefs 简讯
new science and technology 新科技
new science of management decision 新的管理决策科学
new scientific and technological undertakings 新兴科技领域
news cinema 新闻电影院
news dispatch 通讯报道
news in brief 简讯
news kiosk 报摊;报亭
new snow 新雪;初雪
new soil 处女地;生荒地
new soil coefficient 新土力学系数
new soil parameter 新土力学参数
Newson's boring method 纽森钻井法
new source performance standard 新污染源执行标准
newspaper 新闻纸
newspaper baling press 报纸捆包机
newspaper clip 报纸夹
newspaper display stand 报纸阅览架
newspaper file 报纸夹
newspaper office 报馆;报社
newspaper press 报版轮转机
newspaper room 阅报室;报刊阅览室
newspapers and periodicals 报刊
newspaper stand 报架
newsprint 新闻纸;白报纸
newsprinting paper 白报纸
news release conference 新闻发布会
newsroom 报刊阅览室
news server 新闻服务器
news service 通讯社
news stall 书报亭
news-stand 书报亭;报亭
news summary 新闻简报
news theater 新闻电影院
new stock 新股份
New Stone Age 新石器时代
new style 新型号;新型;新历;新风格
new style pattern 新样式
new suction plate apparatus 新式吸收板仪器
new super-fluid phases 超流新相
new survey 新测
new suspension preheater 新式悬浮预热器
new suspension preheater with flash calciner 新型悬浮预热器快速分解炉
new sync 新同步
new systematics 新分类学
new system of high technology program(me) leadership 高技术计划领导新体制
Newt 牛特(英国运动黏度单位)
newt 水晰
new technique 新技术
new technique exploration expenses 新技术开发研究费
new technique of borrowing 新借款办法;新的借

款办法
new technologic(al) development 新技术开发
new technologic(al) revolution 新技术革命
new tectonic system 新构造体系
new terms 新条款
Newton activation analysi 牛顿活化分析
Newton-Cotes formula 牛顿—柯特斯公式
Newtonian 牛顿的
Newtonian attraction 牛顿引力
Newtonian behavior 牛顿行为
Newtonian capacity 牛顿容量
Newtonian-Cassegrain telescope 牛顿—卡塞格仑望远镜
Newtonian constant of gravitation 牛顿万有引力常数
Newtonian cooling 牛顿定理冷却；牛顿定理冷却
Newtonian cooling law 牛顿冷却定律
Newtonian cosmology 牛顿宇宙论
Newtonian flow 牛顿流动
Newtonian flow equation for tackmeter 黏性计的牛顿流动方程；黏性计的牛顿流动方程
Newtonian flow property 牛顿流动性
Newtonian fluid 牛顿流体
Newtonian focus 牛顿焦点
Newtonian gravitational constant 牛顿引力常数
Newtonian heating 按牛顿定理加热
Newtonian inner capacity 牛顿内容量
Newtonianism 牛顿性
Newtonian limiting viscosity 牛顿极限黏度
Newtonian liquid 牛顿液体；非黏性液体
Newtonian lubricating layer 牛顿润滑层
Newtonian material 牛顿材料
Newtonian mechanics 牛顿力学
Newtonian outer capacity 牛顿外容量
Newtonian potential 牛顿位势
Newtonian potential function 牛顿位势函数
Newtonian reference frame 牛顿参考坐标系
Newtonian reflector 牛顿反射式望远镜
Newtonian region 牛顿区
Newtonian shear viscosity 牛顿剪切黏度
Newtonian speed of sound 牛顿声速
Newtonian telescope 牛顿望远镜
Newtonian time 牛顿时间；牛顿时
Newtonian universal of gravity 重力常数
Newtonian velocity 牛顿速度
Newtonian viscosity 牛顿黏度
newtonite 防水护墙板；钾明矾石；明矾石；块矾石
newtonium 原元素
Newton-Laphson algorithm 牛顿—拉夫逊算法
Newton-Laphson formula 牛顿—拉夫逊公式
Newton-Laphson iteration 牛顿—拉夫逊迭代法
Newton-Laphson method 牛顿—拉夫逊法（潮流计算）
Newton liquid 简单液体（切变率与切应力成正比）
Newton's alloy 牛顿易熔合金；牛顿合金
Newton's approximation 牛顿近似法
Newton's backward interpolation formula 牛顿向后插值公式
Newton's behavior 牛顿特性
Newton's concept 牛顿概念
Newton's cooling law 牛顿冷却定律
Newton's diagram 牛顿图解
Newton's disc 牛顿氏色盘（七色盘）；牛顿色盘
Newton-second 牛顿秒
Newton's equation 牛顿方程
Newton's equation of motion 牛顿运动方程
Newton's finite difference method 牛顿有限差分法
Newton's first law 牛顿第一定律
Newton's formula 牛顿公式
Newton's forward interpolation formula 牛顿向前插值公式；牛顿前向插值公式
Newton's hypothesis 牛顿假说
Newton's interpolation 牛顿插值
Newton's interpolation formula 牛顿插值公式
Newton's iteration method 牛顿迭代法
Newton('s) law 牛顿万有引力定律；万有引力定律
Newton's law of cooling 牛顿冷却定律
Newton's law of gravitation 牛顿万有引力定律
Newton's law of gravity 牛顿万有引力定律
Newton's law of motion 牛顿运动定律
Newton's law of resistance 牛顿阻力定律
Newton's law of universal gravitation 牛顿万有引力定律

Newton's law of viscous flow 牛顿黏性流动定律
Newton's lens equation 牛顿透镜公式
Newton's liquid 牛顿液体
Newton's liquid model 牛顿液体模型
Newton's metal 牛顿合金
Newton's meter 牛顿米
Newton's method 牛顿法
Newton's method of approximation 牛顿近似计算法
Newton's model 牛顿模型
Newton's number 牛顿数
Newton's parallelogram 牛顿平行四边形
Newton's reflecting telescope 牛顿式反射望远镜
Newton's refraction 牛顿折射
Newton's relation 牛顿关系式
Newton's ring 光圈；牛顿圈；牛顿环
Newton's second law 牛顿第二定律
Newton's serpentine 牛顿蛇形线
Newton's square-root method 牛顿平方根法
Newton's system 牛顿制（黏度单位）
Newton's telescope 牛顿式望远镜
Newton's theory of collision 牛顿碰撞理论
Newton's theory of gravitation 牛顿引力理论
Newton's theory of lift 牛顿升力理论
Newton's theory of light 光的微粒理论
Newton's third law 牛顿第三定律
Newton's three-eighths rule 牛顿八分之三法则
Newton's viewfinder 牛顿式指示器
Newton's viscosity 牛顿黏滞度
Newton's viscous body 牛顿粘滞体
new town 卫星城市；新市镇；新城镇；新城
new-town-in-town 新城内城镇
new transmission plan 新传输方案
new transportation system 新交通系统
new trunk 新干线
new type bucket wheel dredge(r) 新型斗轮挖泥船
new type optical glass 新品种光学玻璃
new ultra positive feed 新型超积极喂纱装置
new value 新值
new variable 新变量
new venture group 新事业开发群
new view of investment 投资新观点
new village 新村庄
new volcanism 新火山作用
new wastewater management project 新建废水管理工程
new water 新水；用人工方法引来的水
new wheel 新车轮
new white pigment 新白颜料
new wood 原木
newwork dredging 新巷道开挖；新港区开挖；基建性挖泥；基建性疏浚
newwork engine 新型发动机；新式发动机
new workers' housing estate 工人新村
new work item proposal 新工作项目建议
new world 新域；新世界
new yellow pigment 新黄颜料
New Yoke level(l)ing rod 纽约水准尺
New Yoke Port 纽约港
New Yoke rod 纽约测杆（划有细线的水准杆）
New Zealand 新西兰
New Zealand Alpine transform fault 新西兰阿尔卑斯转换断层
New Zealand bivalve subprovince 新西兰双壳类地理亚区
New Zealand geosyncline 新西兰地槽
New Zealand greenstone 新西兰绿软玉
Nexa fly-trap 灭蝇磷
next best rule 其次最佳法规
next closing of account 下次决算
next constellation 下一卫星（GPS用语）
next decade 后十年
next door 隔壁房间
next-door neighbour 隔壁邻居
nextel 陶瓷纤维
next higher priority group 较高优先群
next lower priority group 次低优先群
next most significant word 次高有效字
next move 下一动作
next of kin 近亲
next record 下一记录
next spring 来年春季
next state 次状态；次态
next statement 下一语句【计】

next-to-central bay 次间
next to oxygen 次于氧
next week 下星期
next year 次年
next year advantage from project 新投资发生后一年内之利益
nexus 连系；连杆；网络
neyite 针硫铋铜铅矿
Neyman allocation 内曼配置
Neyman model 内曼模型
Neyman-Pearson theory 内曼—皮尔逊理论
Neyman sampling 内曼抽样
Neyman-Scott model 内曼—斯科特模型
Neyrpic current meter 奈尔皮克流速仪
Neyrpic sediment sampler 奈尔皮克采样器
Ney's projection 改进的兰伯特正形投影
Ney's projection chart 改进的兰伯特正形海图
névé snow 永久冰雪
N-formyl sarcolysine 氮甲
N-girder N式格构大梁
NH_3 yield 氨收率
NH_4-catio exchange capacity 铵离子交换容量
N-hammer N型回弹仪
n-hexane 正己烷
n-hexyl 正己基
n-hexyl acetate 乙酸正己酯
n-hexyl alcohol 正己醇
n-hexyl carbitol 正己基卡必醇
n-hexyl cellosolve 正己基溶纤剂
niacin 烟酸
Niag 尼阿格兰黄铜
Niagaran 尼亚加拉统【地】
niahite 冰磷锰铵石
Ni-Al gyromagnetic ferrite 镍—铝旋磁铁氧体
Nian ware 年窑
niaye 沼泽洼地
nib 楔尖劈；尖头；尖端；绘图笔尖；后爪；突梭；突边；飞角
nibbed 装了尖头的
nibbed bolt 尖头螺栓
nibbed saggar 内部有突齿的匣体
nibbed sagger 内在凸齿的匣钵
nibbed spring leaf 槽形弹簧片
nibbed tile 背面有挂脚的瓦；尖头瓦；突边瓦；平瓦
nibbing plank 木甲板边（缘）板
nibble 啮咬；四位组；分段冲截；半字节
nibbler 毛坯下料机；分层破碎机；步冲轮廓机
nibbler shears 缺陷切除剪
nibbling 步冲轮廓法
nibbling machine 冲剪切机；步冲轮廓机
nibbling method 渐进法；步冲法
nibbling shear 振动剪；冲型剪；分段剪切
Ni-bearing serpentinite 含镍蛇纹岩
nib-forming 尖头成型
nib-forming die 尖头成型模
nib grade 塄平刮尺；吊顶压条
nib guide 凸导尺；吊顶压条；突边导板；突边尺
nib pen drafting 清绘
nibrucite 镍水镁石
nib rule 吊顶压条
nib strake 船首尾水道内列板
Nicalloy 尼卡洛伊合金
nicarbing 碳氮共渗；气体软氮化法；气体表面硬化法；渗碳氮化
Nicaro process 尼加罗炼镍法
Nicasil 镍硅镀层
niccolite 红砷镍矿
Ni-Cd accumulator 镍镉蓄电池
nice taste 清凉可口
nicety 精密；精细
nicety of grading 分类精度
niche 雪坑；小生境；龛；适当的位置；适当场所；生态位；洞窟；壁龛
niche baldaquin 壁龛龛室
niche breadth 生态位宽度
niche-buttressed 用壁龛支撑的；带凹龛条槽支墩的
niche canopy 壁龛顶篷；壁龛华盖
niche containing the cult statue 内装祭礼用塑像的壁龛
niched column 壁龛柱
niche diversification 小生境多样化
niche for a statue of Buddha 佛龛
niche for emergency telephone place 紧急电话亭洞室

niche for surveillance equipment 监视器配套设备洞
niche generalization 生态位泛化
niche overlap 生态位重叠
niche pediment 壁龛三角楣饰;壁龛三角饰
niche separation 生态位分离
niches for emergency equipment 应急设备壁龛
niches for fire-fighting equipment 消防设备洞
niche specialization 生态位特化
niche width 生态物宽度;壁龛宽度
Nichols curve 尼科尔斯曲线
Nichols radiometer 尼科尔辐射计
Nichols sludge incinerator 尼科尔斯淤泥焚化炉
nichrome(alloy)镍铬合金;镍克罗
nichrome alumino-copper thermocouple 镍铬铝铜热电偶
nichrome alumino-nickel couple 镍铬镍铝热电偶
nichrome alumino-nickel thermocouple 镍铬镍铝热电偶
nichrome bead 镍铬合金珠
nichrome coating 镍铬合金镀层
nichrome film 镍铬合金片
nichrome fuse 镍铬合金保险丝
nichrome heat element 镍铬发热元件
nichrome heating coil 镍铬丝加热线圈
nichrome heating spiral 镍铬发热线圈
nichrome powder 镍铬合金粉末
nichrome resistance 镍铬电阻
nichrome resistance wire 镍铬电阻线
nichrome steel 镍铬耐热钢;镍铬钢
nichrome wire 镍铬合金丝
nichromite 镍铬铁矿
nichrosi 镍铬硅合金
nichrosilal 镍铬硅铸铁
nichtware 非商品
nick 刻痕;断口
nick action 交咬作用
nickalloy 镍铁合金
nick and break test 同质性试验;均匀性试验
nick bend test 刻槽弯曲试验;刻槽挠曲试验;缺口弯曲试验;切槽弯曲试验
nick-break test 凹槽冲击试验;切口断裂试验;切口冲击试验;凹槽冲撞试验;缺口冲击试验;切口试验;双缺口破断试验
nicked 刻痕的
nicked-clad steel tank 镀镍钢水池
nicked column 刻槽柱
nicked-copper-chrome cast iron 镍铜铬铸铁
nicked fracture test 刻槽弯曲断裂试验;断口试验
nicked silver 镍银
nicked silver alloy 镍银合金
nicked teeth milling cutter 切齿铣刀
nicked tooth 刻齿痕
nickel[Ni]镍;自然镍
nickel acetate 乙酸镍
nickel acetylacetonate 乙酸丙酮镍
nickelage 镀镍
nickel alloy 镍合金;白铜
nickel alloy chill roll 冷硬镍铸铁轧辊
nickel alloy grain roll 麻口镍铸铁轧辊
nickel alloy pot 镍合金罐
nickel alloy steel 镍合金钢
nickel alumide 镍衣铝粉粒
nickel aluminate 镍铝尖晶石
nickel aluminide 铝化镍
nickel alumin(i)um alloy 镍铝合金
nickel-alumin(i)um bronze 镍铝青铜
nickel-alumin(i)um coating 镍铝镀层
nickel alumin(i)um-oxide coating 镍氧化铝敷层
nickelalumite 镍矾石
nickel amalgam 镍汞合金
nickel anode 镍阳极
nickel antimonide 锑化镍
nickel antimony glance 镍锑辉矿
nickel arsenate 砷酸镍
nickel arsenide 砷化镍
nickel arsenide structure 砷化镍结构
nickel arsenite 亚砷酸盐;镍华
nickel asbolane 镍钴土
nickelate 镍酸盐
nickel azo yellow 镍络偶氮黄
nickel bare welding filler 镍焊丝
nickel-barium alloys 镍钡合金
nickel base 镍基底

nickel-base alloy 镍基合金
nickel-base alloy bare welding filler metal 镍基合金焊丝
nickel-base alloy covered electrode 镍基合金焊条
nickel-base heat-resisting superalloy 镍基超耐热合金
nickel bath 镀镍槽
nickel bead 镍珠
nickel-beryllium alloy 镍铍合金
nickelbischofite 水氯镍石
nickel black 镍黑
nickel block 镍块
nickelbloedite 钠镍矾
nickel bloom 镍华
nickel borate 硼酸镍
nickel boride 硼化镍
nickel brass 镍铜锌合金;铜镍锌合金
nickel brazing 镍铜钎焊
nickel-bronze 镍青铜
nickel bronze alloy 镍青铜合金
nickel-cabrerite 镁镍华
nickel-cadmium accumulator 镍镉蓄电池
nickel-cadmium battery 镍镉电池;镉镍蓄电池
nickel carbide 一碳化三镍
nickel-carbon alloy 碳镍合金
nickel carbonate 碳酸镍
nickel-carbon couple 镍碳热电偶
nickel carbonyl 羰基镍
nickel cast iron 镍铸铁
nickel cast steel 含镍铸钢
nickel catalyst 镍催化剂
nickel catalyzator 镍催化剂
nickel cathode 镍阴极
nickel cell 镍镉电池
nickel-cemented carbide 镍黏结硬质合金
nickel-cemented tantalum carbide 镍结碳化钽硬质合金
nickel-cemented titanium carbide 镍结碳化钛硬质合金
nickel-cemented tungsten carbide 镍钨硬质合金;镍结碳化钨硬质合金
nickel cermet 镍金属陶瓷
nickel cermet coating 镍金属陶瓷镀层
nickel chloride 氯化镍
nickel-chlorite 镍绿泥石
nickel-chrome 镍铬合金
nickel-chrome alloy steel 镍铬合金钢
nickel-chrome bronze 镍铬青铜
nickel-chrome glow strip 镍铬预热片
nickel-chrome steel 镍铬钢
nickel-chrome wire 镍铬丝
nickel-chromium alloy 镍铬合金
nickel-chromium cast iron 镍铬铸铁
nickel-chromium-iron 镍铬铁
nickel-chromium-iron alloy 镍铬铁合金
nickel-chromium manganese steel 镍铬锰钢
nickel-chromium molybdenum steels 镍铬钼钢
nickel-chromium-nickel silicon thermocouple 镍铬镍硅热电偶
nickel-chromium-nickel thermocouple 镍铬镍热电偶
nickel-chromium resistor 镍铬电阻器
nickel-chromium stainless steel 镍铬不锈钢
nickel-chromium steel 镍铬钢
nickel-chromium thin film 镍铬薄膜
nickel-chromium triangle 镍铬
nickel-clad 钢板包镍板
nickel-clad copper 镀镍铜;包镍铜
nickel-clad iron 包镍铁
nickel-clad steel 镀镍钢
nickel-clad steel plate 镀镍钢板;包镍钢板
nickel coat 包镍层;镀镍
nickel-coated alumin(i)um oxide powder 包镍氧化铝粉;包镍氧化铝粉
nickel-coated graphite powder 镍包石墨粉;包镍石墨粉
nickel-coated wire 镀镍线
nickel coating 镀镍层
nickel-cobalt(alloy)镍钴合金
nickel-cobalt-iron matte 镍钴铁流
nickel-cobalt matte 镍钴锍
nickel-cobalt plating 镀镍钴合金
nickel-containing electroplating waste-water 含镍电镀废水

nickel-containing iron alloy 含镍铁合金
nickel-containing wastewater 含镍废水
nickel contamination 镍污染
nickel converting 镍吹炼
nickel-copper(alloy)镍铜合金;铜镍合金
nickel-copper chromium cast iron 镍铜铬铸铁
nickel-copper ferrite 镍铜铁氧体
nickel-copper matte 铜镍锍
nickel-copper ratio 镍铜比
nickel core 镍芯
nickel covered welding eleetrode 镍药皮焊条
nickel cyanide 氰化镍
nickel dam 镍层
nickel delay line 镍滞延线;镍延迟线
nickel dibutyl dithiocarbamate 二丁基二氨荒酸镍
nickel dipping 镍被膜(搪瓷)
nickeled steel plate 镀镍钢板
nickel electro 镀镍电铸版
nickel electrode 镍电极
nickel electrolyte purification 镍电解液净化
nickel equivalent 镍当量
nickelex 光泽镀镍法
nickel ferrite 铁镍尖晶石;镍铁体;镍铁
nickel filter net 镍滤网
nickel fittings 镍制配件
nickel flail stirrer 镍制连枷状搅拌器
nickel flashing 镍被膜(搪瓷)
nickel fluoride 氟化镍
nickel foil 镍箔
nickel glance 辉砷镍矿
nickel green 镍绿
nickel-hexahydrite 斜镍矾
nickel-hydrogen dattery 镍氢蓄电池
nickel hydroxide 氢氧化镍
nickelic 高镍的
nickelic acetate 乙酸高镍;醋酸高镍
nickelic compound 高镍化合物
nickelic hydroxide 氢氧化高镍
nickelic oxide 三氧化二镍
nickeliferous 含镍
nickeliferous silicate deposit in weathering crust type 风化壳型硅酸镍矿床
Nickelin 尼克林高阻合金;铜镍锌合金
nickeline 锡基密封合金;铜镍锰合金;红砷镍矿
nickeline resistance 镍克林电阻
nickeline wire 镍合金线
nickeling 镀镍
nickel iodide 碘化镍
nickel ion 镍离子
nickel-iron 镍铁
nickel iron alkaline battery 碱性镍铁蓄电池
nickel-iron alkaline cell 镍铁碱性蓄电池
nickel-iron alkaline storage battery 镍铁碱性蓄电池
nickel-iron alloy 镍铁合金
nickel-iron battery 镍铁电池;镍铁电池组;爱迪生蓄电池
nickel-iron bimetal 镍铁双金属
nickel-iron cell 镍铁电池
nickel-iron matte 镍铁锍
nickel-iron roll 镍铸铁轧辊
nickel-iron storage battery 镍铁蓄电池
nickel-isomerization 镍催化异构化法
nickelizing 镍的电处理;镀镍
nickel layer 镍层
nickel-leaded bronze 含铅镍青铜;镍铅铜
nickelled material 镀镍材料
nickel-ligand system 镍配合基体系
nickel-lined 衬镍的
nickel-lined condenser 衬镍冷凝器
nickel-linnaeite 镍硫钴矿
nickel luster 镍光泽彩
nickel-magnetite 镍磁铁矿
nickel-manganese 镍锰合金
nickel-manganese bronze 镍锰青铜
nickel-manganese cast steel 镍锰铸钢
nickel matte 镍锍;镍冰铜
nickel matte converting 镍锍吹炼
nickel matte refining 镍锍精炼
nickel metallurgy 炼镍
nickel minerals 镍矿
nickel molybdate 钼酸镍
nickel-molybdenum alloy 镍钼合金
nickel-molybdenum iron 镍钼铁

nickel-molybdenum steel 镍钼钢
nickel-molybdenum thermocouple 镍钼温差电偶
nickel monoxide 一氧化镍
nickel nitrate 硝酸镍
nickel nitride 二氮化三镍
nickel nutrition 镍营养
nickel ocher 镍华
nickeloid 镍劳特合金
nickel ore 镍矿石;镍矿
nickelous 镍的;二价镍
nickelous acetate 乙酸镍
nickelous ammine 低镍氨络合物
nickelous arsenate 砷酸镍
nickelous borate 硼酸镍
nickelous carbonate 碳酸镍
nickelous compound 正镍化合物
nickelous cyanide 氰化镍
nickelous fluosilicate 氟硅酸镍
nickelous hydroxide 氢氧化镍
nickelous hypophosphite 次磷酸镍
nickelous iodide 碘化镍
nickelous-nickelic oxide 四氧化三镍
nickelous nitrate 硝酸镍
nickelous oxide 氧化镍
nickelous perchlorate 高氯酸镍
nickelous selenate 硒酸镍
nickelous selenide 硒化镍
nickelous sulfate 硫酸镍;酸硫镍
nickel oxalate 草酸镍
Nickeloy 尼克劳合金
nickel paper 镍纸
nickel pellet 镍珠
nickel phosphor bronze 镍磷青铜
nickel pickling 镍被膜(搪瓷)
nickel plate 镀镍
nickel-plated 镀镍的
nickel-plated double line iron screw hood 镀镍双线螺旋帽
nickel-plated double line iron screw hook 镀镍双线螺旋钩
nickel-plated iron screw eye 镀镍铁螺丝眼
nickel-plated iron screw hook 镀镍铁螺旋钩
nickel-plated iron square hook 镀镍铁方形钩
nickel-plated mild steel round head machine screw 镀镍低碳钢圆头机器螺钉
nickel-plated mild steel round head wood screw 镀镍低碳钢圆头木螺丝
nickel-plated plastic handle for folding umbrella 镀镍塑料折伞柄
nickel-plated PVC flexible hose 镀镍聚氯乙烯软管
nickel-plated sheet 镀镍板
nickel-plated steel self tapping screw 镀镍钢制自攻螺丝
nickel-plated steel wire 镀镍钢丝
nickel-plating 镀镍
nickel-plating bath 镀镍浴槽
nickel-plating brightener 镀镍光亮剂
nickel poisoning 镍中毒
nickel porphyrin 紫四环镍矿;镍卟啉
nickel potassium cyanide 氰化镍钾
nickel powder 镍粉
nickel print 镍印法
nickel product 镍产物
nickel protoxide 一氧化镍
nickel radiation screen 镍辐射屏
nickel reagent 镍试剂
nickel reduction autoclave 镍还原高压釜
nickel refining 镍精炼
nickel-resist 耐蚀镍的
nickel-resist cast iron 耐蚀高镍铸铁;高镍耐热铸铁
nickel-resist ring insert 高镍耐蚀铸铁镶环座
nickel rondelle 镍丸
nickel rutile yellow 钛镍黄
nickel salt 镍盐
nickel-saturated liquor 镍饱和溶液
nickel scratch 镍条刮擦;镍刮
nickel screen 镍网;镍屏
nickel sesquioxide 三氧化二镍
nickel shot 镍丸;镍粒
nickel silicon alloy coating 镍硅合金镀覆层
nickel silicon carbide 镍硅碳化物
nickel silicon carbide coated housing 镍碳化硅镀盖缸壁
nickel silicon-carbide coating 镍碳化硅镀层

nickel silver 铜镍锌合金;德银(一种镍铜锌合金);白铜银;铜锌合金
nickel-silver metal 德银(一种镍铜锌合金)
nickel-skutterudite 镍方钴矿
nickel-smaltite 镍砷钴矿
nickel smelting 炼镍
nickel-Span alloy 尼斯盘铁镍合金
nickel stainless steel 不锈镍钢
nickel stannate 锡酸镍
nickel steel 镍钢
nickel steel gas welding rod 镍钢气焊条
nickel steel wire 镍钢线
nickel strap 镍带
nickel-strap bolometer 镍带测辐射热计
nickel subsulfide 碱式硫化镍
nickel sulfamic acid 氨基磺酸镍
nickel sulfate 硫酸镍
nickel sulfate dip 浸硫酸镍
nickel sulfide 硫化镍
nickel sulfide deposit of sedimentary type 沉积型镍硫化物矿床
nickel sulphate 硫酸镍
nickel superoxide 四氧化镍
nickel-tensilorin 镍铸铁
nickeltin 镍锡合金
nickel titanate yellow 钛酸镍黄;钛镍黄
nickel titanate yellow pigment 钛酸镍黄色颜料
nickel-to-cobalt ratio 镍钴比
nickel tungsten 镍钨合金
nickel tungsten-carbide coating 镍碳化钨涂层;镍碳化钨敷层
nickel-vanadium cast iron 镍钒铸铁
nickel-vanadium steel 镍钒钢
nickel vitriol 碧矾
nickel wastewater 含镍废水
nickel white iron 镍白口铁
nickel wire line 镍导线
nickel yellow 镍黄
nickel-zinc 镍锌铁氧体
nickel-zinc cell 镍锌电池
nickel-zinc ferrite 镍锌铁氧体
nickel zippeite 水镍铀矾
nickel-β-kerolite 镍贝塔-蜡蛇纹石
nicker 刻槽器;中心钻;凿子;槽凿
Nickerson colorimeter 尼克森色度计
nickhydroxide 软镍矿
nicking and breaking 刻痕折断法
nicking file 开槽锉
nickings 焦屑;煤屑
nicking tool 刻刀
Nickoline 尼阔林镍合金
nick-point 泥克点;侵蚀交叉点;裂点
nickpoint in gully 冲沟裂点
Nicla 尼克拉黄铜
Niclad 包镍钢板;钢板包镍板
Niclausse boiler 尼克劳斯型锅炉
Nicloy 尼克洛依镍钢
Nicobar fever 尼科巴热
nicochrome 镍铬合金
nicochrome steel 镍铬合金钢
nicofer 镍可铁
Nico hydrostroke feeder 尼科牌液压往复式给矿机
Nicol crossed 正交尼科耳棱镜
nicolite 铜镍合金
Nicoloid 尼科耳乳化油漆
Nicol polarizing prism 尼科耳偏振棱镜
Nicol('s) prism 尼科耳棱镜;偏光镜
Nicol's radiometer 尼科耳辐射计
Nico metal 尼科铜镍合金
Niconmetal 耐蚀铜镍合金
nicopyrite 硫镍铁矿;镍黄铁矿
nicorros(alloy) 尼科罗斯(合金)
Nicotiana bigelovi 印度烟草
nicotinamide 烟酰胺
nicotine 烟碱
nicotinic acid 烟酸
nicotinic acid amide 烟酰胺
nicotyrin(e) 二烯烟碱
nicoulin 臭树树脂
Nicrite 尼克利特镍铬合金
Nicrobraz 镍铬焊料合金
nicrosilal 镍铬硅合金;镍铬硅铸铁
nicrosoable pesticides 可卤硝化农药
Ni/Cu/Mn-rich Mn nodule 富镍铜锰型锰结核

Ni/Cu-rich Mn nodule 富镍铜型锰结核
nidge 琢
nidg(g)ed 粗琢石块
nidg(g)ed ashlar 尖凿琢面方石;剁斧琢石
nidiselite 二硒镍矿
nidus vespa 蜂房
niello 黑金;黑金镶饰品
nielloed 发黑了的
Niemann's test 尼曼试验
Niemann's triangle 尼曼三角
Nier-Johnson mass analyser 尼尔-约翰逊质量分析器
Nier's mass spectrometer 尼尔质谱计
nife 镍铁(带)
Nife accumulator 镍铁蓄电池
Nife cell 镍铁电池
nifesima 镍铁硅镁带
nifesphere 镍铁层
niff 难闻(的)气味
NiFi hypothesis 镍铁地核假说
nifontovite 尼硼钙石;尼钙石
nig 琢
niger 皂脚
Niger abyssal fan 尼日尔深海扇
Niger delta 尼日尔三角洲
Nigerian mahogany 尼日利亚红木;尼日利亚桃花心木
nigerite 尼日利亚石
Niger triple junction 尼日尔三向联结构造
nigged ashlar 尖凿琢面方石;粗琢石
nigger (套在钳子扳手上的)加长套筒;长把手;翻楞机
nigger-brown 深棕色
nigger head 系索柱;系船柱;蛮石;劣质橡胶;绞车滚筒;(提运货的)转轴头;安全盖;拖缆柱;丁字头;(钻机上的)安全卷筒
nigger heaven 剧院最高层楼座(俚语)
nigging 用尖锤琢石
niggliite 六方锡铂矿
nigh 近的
nigh side of cut 路堑傍山一边
night airglow 夜天光
night alarm 夜铃
night alarm circuit 夜铃装置
night-alarm key 夜铃电键
night-alarm switch 夜间信号开关
night arc 夜间弧;夜弧
night bell 夜铃
night bell key 夜铃电键
night binocular(s) 夜视望远镜
night bolt 夜销;弹簧插销;保险插销
night bombing target map 夜间轰炸目标地图
night brightness 夜天亮度
night call 夜间呼叫
night camera 夜间摄影机
night capacity 街道夜间(交通)容量;夜间生产能力;夜间容量
night cart 粪车
night charging 夜间增压;夜间充电;夜间送气;夜间加热
night club 夜间俱乐部;夜总会
night construction noise 夜间施工噪声
night cooling 夜间冷却
night deposit 夜间存款窗口
night depositing 夜间存款业务
night dial 夜规
night differential 夜班津贴
night digestion tank 化粪池
night dip 夜间低谷
night direction-finder 夜间探向设备
night driverscope 夜间驾驶仪
night driving 夜间行车
night duty 夜班
night duty allowance 夜班津贴;夜班费
night duty room 夜班室
night error 夜间误差
nightfall 黄昏
night flow 夜间流量
night-flying chart 夜航图
night-flying equipment 夜航设备
night frequency 夜间通信频率
night gang 夜班
night glasses 夜用望远镜;夜望镜
night glow(emission) 夜天光;夜辉

night-glow of the sky 夜空辉光
night-glow spectrum 夜辉光谱
night header joint 日夜班交接
night hospital 夜间医院
night illumination 夜间发光；夜间照明
night infrared sensor 夜用红外传感器
night interception 夜间截击
night key 弹簧锁钥匙；弹子门锁的钥匙；夜铃电键
night latch 弹簧门锁；保险插销；弹簧锁
night life buoy 带自亮浮灯的救生艇
night light(ing) 通夜照明灯；夜光；夜间照明；晚间照明
night lighting system 夜间照明系统
night load 夜间负载
night lock 圆筒销子锁
night market 夜市
night mark(ing) 夜间标志
night mission 夜间任务
night noise 夜间噪声
night observation 夜间观察
night observation device 夜视设备
night obstruction marking 夜间障碍标志
night-only tariff 夜间税则；夜间税率；夜间使用费
night operation 夜间作业；夜间操作
night order book 夜航命令簿
night pair 夜班矿工
night periscope 夜间潜望镜
night photography 夜间摄影
night plane guard 夜间飞机救护舰
night population 居住人口；夜间人口
night position 夜班台
night power 夜间动力；夜间功率
night range 夜间通信距离
night rate 半夜费率
night repair 夜间修理
night safe 夜间安全；保险箱
night scene 夜景
night school 夜校
night service 夜晚业务；夜间服务
night-service connection 夜间接续
night-service desk 夜间集中台
night setback 夜间降温；夜间沉降
night shift 夜班工人(总称)；夜班；晚班
night shift allowance 夜班费
night shot 夜间摄影
night sight 夜视仪
night sight distance 夜间视距
night signal 夜间信号；夜间航标；夜航信号
night signal(l)ing 夜间信号通信
night sky 夜天空；夜天
night sky background 夜空背景
night sky camera 夜空照相机
night sky light 夜天光；夜辉；气辉
night sky luminescence 夜天光；夜辉
night sky radiation 夜天辐射；夜空辐射
night sky spectrograph 夜天摄谱仪
night sky spectrum 夜天光谱
night soil 粪便；人粪尿
night soil pipe 粪管
night soil purification 化粪池
night soil sludge 人粪尿污泥
night soil treatment 人粪尿处理；粪便处理
night soil treatment plant 人粪尿处理厂
night soisuper 粪便
night stand 床头柜
nightstool 马桶
night storage heater 夜间蓄热式(电)加热器；夜间储[贮]热(水)器
night surveillance 夜间监视
night table 床头柜；床头(小)桌
night-television 微光摄像电视
night tide 夜潮
night-time fog 夜间雾；夜雾
night-time load(ed) 晚间负荷；夜间负荷
night-time ozone profile 夜间臭氧垂直方向分布
night-time population 夜间人口；居住人口
night-time population density 夜间人口密度
night-time seeing 夜间视宁度
night-time visibility 夜间能见度
night-time vision device 夜视仪
night-time vision system 夜视系统
night-time visual range 夜视距离
night traffic 夜间交通
night traffic accident 夜间交通事故

night transparency 夜天透明度；夜空透明度
night value 夜间值
night vent 通风口；腰头气窗；通风窗；小格通风扇；通气格窗；通风管；通风格；小气窗；透气亮子
night view 夜景
night viewing 夜视
night viewing device 夜间观测仪
night viewing pocket scope 夜视袖珍观察镜
night visibility 夜间能见度
night visibility plan 夜视计划
night visibility requirements 夜间行车能见度要求
night vision 夜视觉；夜视；夜间视觉；暗视
night vision binoculars 夜视双目镜
night vision component 夜视组件
night vision device 夜视器件
night vision goggles 夜视眼镜
night vision instrument 夜视仪器；夜视仪
night vision marine sextant 夜视航海六分仪
night vision objective 夜视物镜
night vision sight 夜视瞄准器
night vision telescope 夜视望远镜
night vision viewer 夜视仪
night visual range 夜间能见距离
night watch 夜班
night watchman 夜间看守者；更夫；守夜者
night wind 夜风
night window 夜视窗
night work 夜间作业；夜间工作；夜工；弹子门锁机件
night-work allowance 夜班津贴
nigrescence 变黑
nigrescence time of lamp 灯泡的变黑时间
nigrify 使变黑
nigrine 铁金红石
nigrite 氮沥青
nigritite 煤化沥青
nigritude 黑色之物
nigrometer 墨度计；黑度计
nigrosin 苯胺黑
nigrosine 油溶黑；苯胺黑
nigrosine spirit soluble 醇溶尼格辛
Ni-hard 镍铬耐磨白口铁；含镍铸铁
Ni-hard grinding ball 镍冷硬磨球
Ni-hard lining 硬质镍合金衬板
Ni-hard martensitic white cast-iron 硬镍马氏体白口铸铁
nihil dicit 无答辩
nihilism 虚无主义
Nihowan sediment 泥河弯沉降(更新世)
niigataminamata disease 新泻水俣病(日本)
ni-isopropyl succinate 琥珀酸二异丙酯
Nikasol 醋酸乙烯酯
nikel silicon carbide surface 镍碳化硅缸壁型面
Nikkei stock average futures 股票指数期货(日本)
nikkel oil 月桂油
Nikopor manganese deposit 尼科波尔锰矿床
nilas 冰壳；暗冰
nil dicit 无答辩
nil-ductility(transition) temperature 无延性转变温度；无塑性转变温度
Nile blue 尼罗蓝(稍带绿色的)；耐尔蓝
Nile green 尼罗绿(浅淡的蓝绿色)
Nile mud brick 泥河泥砖
nil factor 零因子
nil interference 无干扰
nill 铁屑
nilmanifold 幂零流形
nil method 零位法
nilo 镍铬低膨胀系数合金
nilometer 尼罗河水位仪；水位计；水位表
Nilos belt fastening nip 尼洛斯皮带扣；尼洛斯皮带接头夹
niloscope 尼罗河水位仪；水位计
nilpotent 幂零
nilpotent group 幂零群
nilpotent matrix 幂零矩阵
nilpotent operator 幂零算子
nil segment 零线段
nil symbol 零符号
nilvar 尼尔瓦合金
nil visibility 不通视
nimbostratus 乱层云
nimbostratus cloud 雨层云
nimbus 雨云

nimbus-cumuliform 积云状雨
nimbus data handling system 雨云数据处理系统
nimbus satellite 雨云卫星
nimite 镍绿泥石
Nimol 尼莫尔(铸铁)
nimonic 尼莫尼克(合金)
Nine-Dragon Screen 天坛九龙壁；九龙壁【建】
nine edge 九边
nine-inch equivalent 标准砖体积重量
nine-lens camera 九镜头摄影机
nine-light indicator 九灯风向风速计；九灯指示器
nine nines 九个"9"(表示半导体纯度，即99.9999999%)
ninepin block 系缆柱滚筒
ninepins 九柱戏
nine-point approximation 九点逼近
nine-point conic 九点二次曲线
nine-point formula 九点公式
nine points moving average 九点移动平均
nine-point steel pad 九齿钢锤
nines check 模九检验
nine's complement 十进制反码
nine's complement representation 十进制反码表示法
nine spot pattern 九点布孔网
nine-stage design 九段设计
nine-tenths 十分之九
nine-track compatibility 九道兼容
ninety degree base elbow 九十度管座肘管
ninety degree bend 九十度弯头
ninety degree long radius elbow 九十度大半径肘管
ninety-east ridge 东经九十度海岭
ninety-nine year lease 九十九年租约(一种对未开发土地的长期租约)
Ningpo lac 金漆
Ningpo varnish 金漆
ningyoite 人形石；水磷铀矿
niningerite 陨硫镁铁锰石；硫镁矿
ninon 薄绸；尼龙绸
ninor feature 小地形
ninscattered radiation 非散射辐射
niobate 铌酸盐
niobate ceramics 铌酸盐陶瓷
niobate crystal 铌酸盐晶体
niobate system ceramics 铌酸盐系陶瓷
niobate system piezoelectric(al) ceramics 铌酸盐系压电陶瓷
niobe oil 尼哦油；苯甲酸乙酯
niobic acid 铌酸
niobite 铌铁矿
niobite ore 铌铁矿矿石
niobite-tantalite ore 铌铁矿—钽铁矿矿石
niobium[Nb] 铌
niobium anatase 含铌锐钛矿
niobium and tantalum-bearing pegmatite deposit 铌钽伟晶岩矿床
niobium and tantalum deposit in altered granite 铌钽交代蚀变花岗岩矿
niobium-base superconducting material 铌基质超导材料
niobium boride 硼化铌
niobium carbide ceramics 碳化铌陶瓷
niobium die 铌制模具(热压用)
niobium nitride 氮化铌
niobium ores 铌矿
niobium oxide 氧化铌
niobium pentoxide 五氧化二铌
niobium silicide 硅化铌
niobium-tantalum ore 铌钽矿石
nioboaeschynite 铌易解石
niobophyllite 铌叶石；铌星叶石
nioboxide 铌石
niobpyrochlore 烧绿石
Niobrara limestone 尼奥博拉拉灰岩
niobyl 铌氧基
niocalite 黄硅铌钙石
niohydroxide 水铌钽石
Nionel 尼沃内耳耐蚀合金
nioro 铜金镍合金
niose distortion 噪声扭变
niostan 锡酸铌尼奥斯坦(超导材料)
nip 小崖【地】；滚距；河底切槽；海岸低悬崖；绳索扭结；绳索急弯处；砭

nip action (钢丝绳股中钢丝的)交咬作用
nipa house 棕榈叶屋面之简易房
nipalgin 对羟苯甲脂
nip angle 咬入角;啮角
nip drum 压滚
nip dyeing 面轧染色
nip guard 压区挡板
Nipher screen 内夫尼尔罩
Nipher shield 内夫尼尔罩
Nipkow disk 尼普科夫圆盘
nip off 剪断
Nipoon Credit Bank 日本债券信用银行
nipped 被夹住的
nipper frame 钳板座架
nipper pliers 尖嘴钳
nippers 剪钳;钳子;吊装夹具(吊装时以夹石块或砖块等);吊石夹钳;拔钳;各种夹具;玻璃切割刀上撬边的钳子
nipping 夹住;搬钻;元宝桦;压平;切边;船舶冰封
nipping fork 垫叉
nipping machine 压平机
nipple 螺纹接套;螺纹接口(管);螺纹接管;螺管接头;螺纹管接口;龙头;缆扎绳;缆夹;径管接头;加油嘴;加油咀;内接头;喷火嘴;喷灯嘴;喷灯喷嘴;外丝接头;丝头;短接管;短管接;放风门(热水供暖系统中释放空气的装置)
nipple bent pipe 弯曲接管
nipple chaser 材料员(钻机上的)
nipple chuck 夹紧卡盘
nippled sheet 有乳头突起状的片材
nipple for auxiliary steam connecting 副气管螺纹接套
nipple for fittings 螺纹接口(管);螺纹接合管
nipple joint 管接头接合;螺纹接头;螺纹接管
nipple nut 锥端接管螺母
nipple plate 凸纹钢板
nipple roll lower 下部压紧胶辊
nipple roll upper 上部压紧胶辊
nippler pilers 尖嘴钳;克丝钳;手钳;切钳;剪钳
nipple union 外螺纹联管节
nipple up 装配设备
nippling lever 钳杆
Nippoki Heavy Industries 日本海重工业株式会社
Nippon Air Cargo Clearance System 日本航空货物通关资料处理制度
Nippon Electric (al) Company 日本电气公司
Nipponese classical architecture (timber structure) 日本古典建筑(木结构)
Nippon Fudosan Bank 日本不动产银行
Nippon Hakuyohin Kentei Kyokai 日本船舶用品检定协会
Nippon Kaiji 日本船级社
Nippon Kaiji Kentei Kyokai 日本海事检定协会
Nippon Kaiji Kyokai 日本检定协会
Nippon Kokan 日本钢管公司
Nippon Kokan Kabushiki kaiska 日本钢管株式会社
Nippon Press Center 日本新闻中心
nip roller 光泽辊;辊式碾压机;夹膜辊
nip rolls 展平辊;光泽辊
N. I. Protnikov classification N. I. 普洛特尼科夫分类
Ni-resist 耐蚀高镍铸铁;镍铬铬耐蚀铸铁
Ni-resist cast iron 不锈镍铸铁
Ni-rich (Co-poor) Mn nodule 富镍(贫钴)型锰结核
Ni-rich Mn nodule 富镍型锰结核
nisbite 斜方锑镍矿
Ni silicate ore of residuum type 风化壳硅酸镍矿石
nisiloy 镍硅
nis matte 镍锍
Ni-Span 镍铬钛铁定弹性系数合金
Nissan Kizai Co./Ltd. 日产机械材料公司
Nissem hut 尼森式小(活动)屋
Nissl bodies 虎斑质
nissonite 水磷镁铜石
nit 尼特(亮度单位)
nital 硝酸酒精溶液
Nitaline 尼塔玻璃
Nitemper (ing) 气体渗氮处理;气体软氮化处理;气体软氮化;氮回火
Ni-tensilorin 特种铸铁
Ni-tensyl iron 含镍高强度铸铁
niteoglycerin substitute 硝化甘油代用品
niter 硝酸钾;硝石;消石;钾硝石
niter ball 硝石球

niter cake 硝饼
niter oven 硝石炉
nitery 夜总会
Ni/Ti-rich Mn nodule 富镍钛型锰结核
nito 料位
nitopilot 料位指示器
nitosol 强风化粘磐土
nitra-amine 硝基胺
nitra-lamp 充气灯泡
nitralising 硝酸钠溶液浸渍净化法;硝酸钠溶液浸渍处理法
nitralloy 渗氮合金;氮化合金;氮化钢
nitramine 硝基胺
nitrammite 铵硝石
nitramon 安全炸药;硝铵炸药;尼特拉蒙炸药
nitranilide 苯基重氮酸
nitrapyrin 氯定
nitrate 硝酸酯;硝酸盐;硝酸根
nitrate and chloride concentration 硝酸盐和氯化物浓度
nitrate and sulphate reduction 硝酸盐与硫酸盐还原
nitrate bacteria 硝酸盐细菌;硝酸细菌
nitrate-base film 硝酸基胶片
nitrate bed 硝石矿床
nitrate concentration 硝酸盐浓度
nitrate cooling 硝酸盐冷却
nitrated 硝化了的
nitrated alcohol 硝化酒精
nitrated asphalt 硝化地沥青;硝化沥青
nitrated asphaltite 硝化地沥青石
nitrated cellulose 硝化纤维素
nitrated cement 硝化纤维素胶泥
nitrated coal-tar 硝化煤焦油(沥青)
nitrate deposit 硝酸岩矿床
nitrated oil 硝化油
nitrate dope 硝基蒙皮漆;硝基航空涂料
nitrated paper 硝化纸
nitrated polyglycerin 硝化聚甘油
nitrated steel 氮化钢
nitrate fertilizer 氮肥
nitrate film 硝酸纤维素胶片
nitrate film base 硝酸胶片基
nitrate green 硝酸绿;硝基绿
nitrate ion 硝酸根离子
nitrate mineral 硝酸盐矿物
nitrate-negative bacillus 硝酸盐阴性杆菌
nitrate nitrogen 硝态氮;硝酸盐氮
nitrate nitrogen load 硝态氮负荷
nitrate of baryta 硝酸钡
nitrate of lime 硝酸钙
nitrate of potash 硝酸钾
nitrate of potassium 硝酸钾
nitrate of soda 硝酸钠
nitrate paint 硝基漆
nitrate plant 硝酸盐植物
nitrate poisoning 硝酸盐中毒
nitrate pollution 硝酸盐污染
nitrate radical 硝酸根
nitrate-reducing bacteria 硝酸盐还原细菌
nitrate reductase 硝酸还原酶
nitrate reduction 硝酸盐还原作用;硝酸盐还原法
nitrate regeneration 硝酸盐再生(作用)
nitrate removal 除硝酸盐
nitrate run-off 流失硝酸盐
nitrate superphosphate 硝酸化过磷酸
nitrate varnish 硝酸清漆
nitrate wastewater 硝酸盐废水
nitratine 钠硝石
nitrating 硝化
nitrating acid 硝化酸
nitrating agent 硝化剂
nitrating centrifuge 硝化离心机
nitrating pot 硝化釜
nitrating separator 硝化离析器
nitrating test 硝化试验
nitration 硝化作用;硝化
nitration acid heat test 硝酸加热试验
nitration benzol 硝化级苯
nitration case 渗氮层;氮化层
nitration grade benzene 硝化级苯
nitration grade products 硝化用产品
nitration grade toluene 硝化级甲苯
nitration grade xylene 硝化级二甲苯

nitration mixture 硝酸和硫酸的混酸
nitration reaction 硝化反应
nitrato 硝酸基
nitrato complex 硝酸基络合物
nitrator 硝化器
nitrator-separator 硝化分离器
nitre 硝石
nitregon cushion pressure 氮气垫压力
nitriability 氮化性
nitriacidium ion 硝酸合氢离子
nitriary 硝化床
nitric 硝石的
nitric acid 硝酸;硝镪水
nitric acid absorber 硝酸吸收塔
nitric acid burn 硝酸烧伤
nitric acid fibre 硝酸纤维
nitric acid fuming 发烟硝酸
nitric acid oxidation solution 硝酸盐氧化溶液
nitric acid pulp 硝酸法浆粕
nitric acid quenching 硝酸淬火
nitric-acid test 硝酸试验
nitric acid unit 硝酸发动机
nitric anhydride 硝酐
nitric hydrate 水合硝酸
nitric oxide 一氧化氮;氧化氮;氮氧化物
nitric oxide detector 氧化氮检定器
nitric plant 硝酸厂
nitridation 渗氮
nitride 氮化物
nitride bond containing silicon oxynitride 含氮氧化硅的氮化物键
nitride bonded silicon carbide 氮化物黏结的碳化硅
nitride ceramics 氮化物陶瓷
nitrided case 渗氮层
nitrided cast iron 氮化铸铁
nitrided iron 氮化铁
nitrided steel 氮化钢
nitrided structure 氮化组织
nitride hardening 氮化淬火
nitride nuclear fuel 氮化物核燃料
nitride passivation 氮化物钝化法
nitride refractory 氮化物耐火材料
nitride refractory material 含氮耐火材料
nitride refractory product 氮化物耐火材料
nitride surface 氮化表面;渗氮表面
nitriding 渗氮硬化;渗氮;氮化
nitriding box 氮化箱
nitriding cycle 氮化循环
nitriding electric (al) furnace 氮化电炉
nitriding furnace 氮化炉
nitriding layer 渗氮层
nitriding medium 渗氮剂
nitriding process 硝化方法;氮化方法;渗氮工艺过程
nitriding steel 渗氮钢;氮化钢
nitriding structure 渗氮组织
nitriding temperature 氮化温度;硝化温度;渗氮温度
nitridize 氮化
nitridizing agent 氮化剂
nitrification 硝化(作用)
nitrification ability 硝化能力
nitrification-anaerobic ammonia oxidation 硝化厌氧氨氧化
nitrification denitrification biological excess phosphorus removal 硝化反硝化生物去除剩余磷
nitrification-denitrification intensity 硝化反硝化作用强度
nitrification inhibitor 硝化抑制剂
nitrification in soil 土壤消化作用
nitrification liquor recycling 硝化液回流
nitrification process 硝化过程;硝化工艺;硝化法
nitrification rate 硝化速率
nitrification rate constant 硝化速率常数
nitrification reaction 硝化反应
nitrifier 硝化细菌
nitrify 硝化
nitrifying 硝化的
nitrifying bacteria 硝化细菌
nitrifying bacterium activity 硝化菌活性
nitrifying biofilm 硝化生物膜
nitrifying biofilter 硝化生物滤池
nitrifying capacity 硝化强度;硝化能力
nitrifying filter 硝化滤池

nitrifying organism 硝化细菌
nitrifying power 硝化力
nitrifying rate 硝化速率
nitrifying sludge 硝化污泥
nitrifying trickling filter 硝化滴滤池
nitrigenation oven 氮化炉
nitril-attack bacteria 腈分解菌
nitrile 腈
nitrile-butadiene rubber 丁腈橡胶;丁二烯丙烯腈共聚橡胶
nitrile glue 腈胶
nitrile nitrogen 亚硝酸盐氮
nitrile of phenylglycine 苯氨基乙腈
nitrile phenolic adhesive 丁腈酚醛树脂黏合剂
nitrile-phenolic rubber 腈—酚醛橡胶
nitrile resin 腈树脂
nitrile rubber 腈橡胶;丁腈橡胶
nitrile rubber adhesive 腈橡胶黏剂;丁腈橡胶黏剂
nitrile rubber glue 腈橡胶胶水
nitrile silicone rubber 腈硅橡胶
nitrilo- 次氮基
nitrilodiacetic monopropionic acid 次氮基二乙酸一丙酸
nitrilotriacetate 次氮基三乙酸酯
nitrilotriacetic acid 氨川三乙酸;次氮基三乙酸
nitrilotriacetic acid disodium salt 氨三乙酸二钠
nitrilotriacetic acid trisodium salt 氨三乙酸三钠
nitrit 二硝散
nitrite 亚硝酸盐;亚硝酸根
nitrite accumulation 亚硝酸盐累积
nitrite amine 亚硝胺
nitrite elastomer 硝化橡胶
nitrite nitrogen 亚硝态氮;亚硝酸盐氮
nitrizing 渗氮的
nitro 硝基
nitroacetyl cellulose 硝酸乙酸纤维素;硝酸醋酸纤维素
nitroalcohol 硝基醇
nitro-alloy 渗氮合金
nitroamino-compound 硝胺化合物
nitroaniline 硝基苯胺
nitroaromatic 硝基芳香化合物
nitroaromatic acid 硝基芳酸
nitrobacteria 硝化细菌;硝化菌
nitrobacterium 硝化菌
nitrobarite 钡硝石
nitrobenzaldehyde 硝基苯甲醛
nitrobenzamide 硝基苯甲酰胺
nitrobenzanilide 硝基苯甲酰苯胺
nitrobenzene 硝基苯
nitrobenzene degradation 硝基苯降解
nitrobenzene poisoning 硝基苯中毒
nitrobenzene wastewater 硝基苯废水
nitrobenzene wastewater treatment 硝基苯废水处理
nitrobenzidine 硝基联苯胺
nitrobenzoate 硝基苯甲酸盐
nitrobenzoic acid 硝基苯甲酸
nitrobenzol 硝基苯
nitrobenzonitrile 硝基苯甲腈
nitrobenzophenone 硝基苯基
nitrobenzoquinone 硝基苯醌
nitrobid 硝酸甘油
nitroblue tetrazolium 氮蓝四唑
nitrobutane 硝基丁烷
nitrocalcite 钙硝石;水钙硝石
nitrocarbamate 硝氨基甲酸酯
nitrocarbol 硝基甲烷
nitrocarburizing 渗碳氮化
nitrocellulose 硝酸纤维素;硝基纤维素;硝化纤维素;硝化纤维;硝化棉
nitrocellulose adhesive 硝化纤维素黏合剂
nitrocellulose chip 硝化纤维(素)漆片
nitrocellulose cotton gelling number 胶数
nitrocellulose dope 硝化纤维素涂料
nitrocellulose film 硝化纤维胶片
nitrocellulose flooring 硝化纤维素地板
nitrocellulose for lacquer 漆用硝基纤维素
nitrocellulose glue 硝化纤维胶
nitrocellulose lacquer 硝化纤维素漆;硝基清漆;硝化纤维素漆;硝化纤维素喷漆
nitrocellulose lacquer resistance 耐硝基漆性
nitrocellulose paint 硝化纤维漆;硝基漆;硝基涂料;硝化纤维素油漆;硝化纤维素涂料
nitro cellulose paint enamel 快干瓷漆
nitrocellulose powder 硝化纤维素火药
nitrocellulose process 硝化纤维法
nitrocellulose product 硝基产品
nitrocellulose silk 硝化纤维丝
nitrocellulose stopper 硝化纤维素塞子;硝基塞
nitrocellulose surface coat 硝化纤维素表面涂层
nitrocellulose varnish 硝化纤维素清漆
nitrocementation 碳氮共渗
nitro-chloraniline red 硝基氯苯胺红
nitro-chlorobenzene 硝基氯苯
nitro-compound 硝基化合物
nitrocotton 硝化纤维素;硝化棉
nitrocresol 硝基甲酚
nitrocumene 硝基枯烯
nitrocyclohexane 硝基环己烷
nitrocyclohexane process 硝基环己烷法
nitro-dextrin 硝化糊精
nitrodimethylaniline 硝基二甲基苯胺
nitrodiphenylamine 硝基二苯胺
nitrodiphenylmethane 硝基二苯基甲烷
nitrodope 硝基清漆;硝化涂料
nitro dyestuff 硝基染料
nitroenamel 硝基瓷漆
nitroethane 硝基乙烷
nitroexplosive 硝化炸药;硝化火药
nitrofatty acid 硝基脂肪酸
nitrofen 除草醚
Nitrofluor process 硝基氟过程
nitrogelatine 硝化胶质炸药;硝化明胶炸药;硝化甘油炸药
nitrogelation 明胶炸药
nitrogen 氮气
nitrogen-absorption 氮吸附
nitrogen absorption method 氮气吸附法
nitrogen acid anhydride 五氧化二氮
nitrogen adsorption 氮吸附
nitrogen analyzer 氮分析器
nitrogen and phosphorous compound 氮磷化合物
nitrogen and phosphorous detector 氮磷检测器
nitrogen and phosphorous pollution 氮磷污染
nitrogen and phosphorous release 氮磷释放
nitrogen apparatus 定氮装置
nitrogen application 氮肥撒施机
nitrogen-arc welding 氮弧焊
nitrogen assimilation 氮元素同化作用;氮同化作用
nitrogenated oil 氮化油
nitrogenation 氮化作用
nitrogen atmosphere 氮气保护气氛
nitrogen atom 氮原子
nitrogen availability 氮有效性
nitrogen balance 氮平衡
nitrogen balance sheet 氮素平衡表
nitrogen blanket 氮气层
nitrogen blanketing valve 充氮管
nitrogen bottle 氮化瓶
nitrogen budgets 氮的平衡
nitrogen bulb 定氮球管
nitrogen-burst 氮爆搅动
nitrogen carrier 氮载体
nitrogen cascade 氮容器
nitrogen case hardening 表面渗氮淬火
nitrogen ceramics 氮化物陶瓷
nitrogen charging apparatus 氮充电装置;氮充电器
nitrogen charging equipment 氮气补充装置
nitrogen chlorides 氮的氯化物
nitrogen complex 氮收回
nitrogen compound 含氮化合物;氮化合物;氮的化合物
nitrogen compressor 氮气压缩机;氮化压缩机
nitrogen concentration 含氮浓度;氮浓度
nitrogen concentration in atmosphere 大气中氮的浓度
nitrogen-containing alloy 含氮合金
nitrogen-containing base 含氮的碱
nitrogen-containing compound 含氮化合物
nitrogen content 含氮量
nitrogen-cooling circuit 氮气冷却回路
nitrogen cycle 氮循环;氮的循环
nitrogen cylinder 氮气瓶
nitrogen deficiency 缺氮
nitrogen desorption curve 脱氮曲线
nitrogen determination apparatus 定氮仪
nitrogen determinator 氮气测定器
nitrogen dioxide 二氧化氮
nitrogen dioxide index 二氧化氮指数
nitrogen distribution 氮素散布;氮的分布
nitrogen equilibrium 氮平衡
nitrogen factory 氮气工厂
nitrogen fertilizer 氮肥
nitrogen fertilizer factory 氮肥厂
nitrogen fertilizer industrial wastewater 氮肥工业废水
nitrogen fertilizer pollution potential index 潜在氮肥污染指数
nitrogen-filled 充氮的
nitrogen-filled lamp 充氮灯
nitrogen fixation 固氮作用;氮素固定;氮气固定作用;氮的固定
nitrogen fixing bacteria 固氮菌
nitrogen-fixing vegetation 固氮植物
nitrogen flush 吹氮处理
nitrogen gains and losses 氮的获得和损失
nitrogen gas container 氮气容器
nitrogen gas cylinder 氮气容器
nitrogen gas sealed valve 氮气密封阀
nitrogen gas tank pressurizing unit 氮气罐压力装置
nitrogen gas thermometer 氮气温度计
nitrogen generation station 制氮站
nitrogen hardening 钢表面氮硬化处理;渗氮硬化
nitrogen hunger 氮不足
nitrogen hydrogen compound 氮氢化合物
nitrogen isotope 氮同位素
nitrogen-lag 氮迟滞
nitrogen lamp 氮气灯
nitrogen laser 氮激光器
nitrogen liberation 氮素释放
nitrogen lime 氮化石灰
nitrogen loading 氮负荷
nitrogen loading rate 氮负荷率
nitrogen loss 脱氮;氮素损失
nitrogen loving plant 喜氮植物
nitrogen management 氮的管理
nitrogen-methane zone 氮气沼气
nitrogen micro-determination apparatus 氮微量定量器
nitrogen mineralization-immobilization 氮素矿化固定过程
nitrogen monoxide 一氧化二氮
nitrogen mustard hydrochloride 氮芥盐酸盐
nitrogen narcosis 氮麻醉
nitrogenous 含氮的
nitrogenous base 氮碱
nitrogenous biochemical oxygen demand 含氮生化需氧量
nitrogenous compound 含氮化合物
nitrogenous content 含氮量
nitrogenous effluent 含氮废水
nitrogenous generator 氮发生器;氮保护气发生器
nitrogenous spring 氮泉
nitrogenous substance 含氮物质
nitrogenous thermal water 含氮热水
nitrogenous waste 含氮废物
nitrogenous wastewater 含氮废水
nitrogenous water 含氮水
nitrogen oxide 一氧化氮;氧化氮;氮氧化物
nitrogen oxide air pollution 氮氧化物大气污染
nitrogen oxide concentration 氮化氮浓度
nitrogen oxides control 氮氧化物控制
nitrogen oxides poisoning 氮氧化物中毒
Nitrogen Oxides Protocol 长程越界空气污染公约的氮氧化物议定书
nitrogen oxides reservoir 氮氧化物吸收库
nitrogen oxides sink 氮氧化物吸收
nitrogen-oxygen diving 氮氧潜水
nitrogen-oxygen saturation diving 氮氧饱和潜水
nitrogen-oxygen saturation technique 氮肥一氧饱和潜水技术
nitrogen pentoxide 五氧化二氮
nitrogen peroxide 过氧化氮
nitrogen pollution 氮污染
nitrogen pressure 氮气瓶压力
nitrogen purging of tank 用氮气扫罐
nitrogen radical 含氮自由基
nitrogen removal 脱氮作用
nitrogen removal from wastewater by denitrifi-

nitrogen removal from wastewater by nitri-denitrification 污水反消化脱氮处理；废水反硝化脱氮
nitrogen removal from wastewater by nitri-denitrification 污水硝化—反硝化脱氮处理
nitrogen removal from wastewater by nitrification 污水硝化脱氮处理；废水硝化脱氮；废水硝化—反硝化脱氮
nitrogen requirement 需氮量
nitrogen retention 氮储[贮]留
nitrogen-sealed 氮气密封的；氮气密封
nitrogen-sealed transformer 氮气密封变压器；氮气封闭变压器；充氮变压器
nitrogen sequence 氮星序
nitrogen servicing cart 氮气服务车
nitrogen solution 氮溶液
nitrogen solution boom 氮肥液喷洒管
nitrogen source 氮源(物质)
nitrogen starved plot 土壤缺氮
nitrogen station 氮气站
nitrogen studies 氮的研究
nitrogen substance 含氮物质
nitrogen supply 氮气供应；氮供应
nitrogen supply unit 供氮设备
nitrogen tetr(a)oxide 四氧化二氮
nitrogen tetroxide resistant coating 耐四氧化二氮涂料
nitrogen transfers 氮转移
nitrogen transformation 氮转化
nitrogen transformation and cycle 氮元素转化与循环
nitrogen transport processes 氮的转移过程
nitrogen treatment 氮化处理
nitrogen trifluoride 三氟化氮
nitrogen trioxide 三氧化二氮
nitrogen valve 氮气阀
nitrogen water 含氮水
nitroglauberite 钠矾硝石
nitroglycerin 硝酸甘油；硝化甘油
nitroglycerine-amide powder 硝化甘油酰胺炸药
nitroglycerine dynamite 硝化甘油炸药
nitroglycerine explosive 硝化甘油炸药
nitroglycerine powder 硝化甘油火药
nitroglycerine solution 硝酸甘油溶液
nitroglycerol 硝化甘油；硝酸甘油
nitroglycol 硝化甘醇；乙二醇二硝酸酯
nitroglyn 硝酸甘油
nitro-grade benzene 硝化级苯
nitrograph 氮气浓度含量测定器
nitro-group 硝基
nitro-group wastewater 硝基废水
nitro-humic acid 硝基腐植酸
nitroic acid 水合硝酸
nitroisophthalic acid 硝基间苯二甲酸
nitrokalite 硝石
nitrolan 硝酸甘油
nitrolar 硝酸甘油
nitrolignin 硝化木素
nitrolime 氰氨化钙
nitrolingual 硝酸甘油
nitromagnesite 镁硝石；水镁硝石
nitromel 硝酸甘油
nitrometal 硝基金属
nitrometer 尼特计；氮量计；氮测定器；测氮管
nitromethane 硝基甲烷
nitromethane-oxygen mixture 硝基甲烷氧气混合气
nitro-methylaniline 硝基甲苯胺
Nitromonas 硝酸菌
nitro-muriatic acid 硝基盐酸；王水
nitron 硝酸试剂
Nitrone 奈特龙炸药
nitrong 硝酸甘油
nitro-oil 硝化油
nitro-paper 硝化纸
nitroparaffin 硝基石蜡；硝基烷烃
nitrophenetol 硝基乙醚
nitrophenol 硝基酚
nitrophenolate 硝基苯酚盐
nitrophenolic dye 硝基酚染料
nitrophenol wastewater 硝基苯酚废水
nitrophenyl-acetic acid 硝基苯乙酸
nitrophenylamine 硝基苯胺
nitrophenylene diamine 硝基苯二胺
nitrophile 喜氮植物
nitrophoska 硝化酸磷酸钾
nitrophosphate 硝化磷酸盐

nitropigment 硝基系颜料
nitroplatinate 硝基铂酸根
nitropo 硝酸钾钠
nitropolymer 硝基聚合物
nitropowder 硝化火药；硝化炸药
nitropropane 硝基丙烷
nitrorectal 硝酸甘油
nitroretard 硝酸甘油
nitrosaccharose 硝化蔗糖
nitrose 硝酸类(硝酸及亚硝酸的统称)
nitrosilal 尼丑西拉尔镍硅铸铁
nitrosilk 硝化纤维人造丝；硝化丝
nitrosomonas 亚硝酸细菌；亚硝化菌
nitroso pigment 亚硝基系颜料
nitrostarch explosive 硝化淀粉炸药
nitrostat 硝酸甘油
nitrostyrolene 硝基苯乙烯
nitrosulfonic acid 硝基磺酸
nitrotartaric acid 硝基酒石酸
nitrotoluene 硝基甲苯
nitrotoluidine 硝基甲苯胺
nitrourethane 硝氨基甲酸乙酯
nitrous acid 亚硝酸
nitrous bumes 亚硝酸红色烟雾
nitrous fume 亚硝酸烟雾；硝烟
nitrous nitrogen concentration 亚硝氮浓度
nitrous oxide 一氧化二氮；氧化亚氮
nitrox saturation diving 氮氧饱和潜水
nitrozell retard 硝酸甘油
nitrum 消石；天然碱
niva felsite 云母霏细岩
nival 终年积雪(地区)的；多雪的
nival belt 雪带
nival erosion 雪水冲刷
nival flora 雪线以上植物
nival surface 雪面
nival zone 雪带
nivation 雪蚀作用；雪蚀；霜蚀
nivation cirque 雪蚀凹地；雪坑
nivation glacier 雪蚀冰川；雪堤冰川
nivation hollow 雪蚀坑；雪蚀凹地
nivation ridge 冬岩屑堆脊
nivation type glacier 雪蚀型冰川
niveau 水平仪
nivelling plummet 测平垂球
nivenite 钇铀矿
niveous 雪白的
nivometer 雪量计；量雪器；量雪计
nivometric ga(u)ge 雪量雪器；量雪计
nivotester 料位测试器
nixie decoder 数码管译码器
nixie display 数字显示器；数码显示器
nixie display indicator 数字显示指示器；数码管
nixie driver 数字管驱动器
nixie light 数码管
nixie readout 数码管读出装置
nixie tube 数码管
nnavigation(al)star 导航星
N-nitrosodimethylamine 二甲基亚硝胺
n-nonyl acetate 乙酸正壬酯
n nozzle frame 喷嘴座
n-n'reaction 中子非弹性散射反应
N. N. Yelmarkoff diagram 耶尔马可夫图解法；耶尔马可夫图解
N. N. Yelmarkoff method 耶尔马可夫法
No. 1 第一号
NOAA satellite 诺阿卫星
no acceptance 拒绝付
no-account 无价值的
no-address computer 无地址计算机
no-address instruction 无地址指令；不寻址指令
no-address instruction format 无地址指令格式
no admittance 不准入内
no admittance area 非游览区
no admittance except on business 非公莫入
no adverse environmental effect 无害环境影响
no advice 无报单；不正式通知
no amount insurance 无保额保险单
nob 雕球饰；球形小柄
no-bake 自硬的
no bake binder 冷硬黏结剂
no-bake core 自硬砂芯
no-bake sand 自硬砂；自干砂
nobbing 压挤熟铁块；熟铁锤炼

nobble 粗琢
Nobel blastometer 诺贝尔检验器
Nobel elutriator (粒度分析用)诺贝尔淘析器
Nobelist 诺贝尔奖获得者
nobelium 锘
Nobelman 诺贝尔奖金获得者
Nobel oil 诺贝尔油
Nobel Prize 诺贝尔奖金
Nobel's explosive 硝酸甘油
nobilite 针碲矿
noble 贵重的
noble antique(style) 高贵的古代风格；古典风格
noble fir 壮丽冷杉；大冷杉
noble gas 稀有气体；惰性气体
nobleiite 诺硼钙石
nobleite 四水硼钙石
noble metal 贵金属
noble metal analysis 贵金属分析
noble metal catalyst 贵金属催化剂
noble metals 贵金属元素
noble metal thermocouple 贵金属温差电偶；贵金属热电偶
noble method 诺布尔法
Noblen 丙烯均聚物
noble potential 贵电势；高电势
noble serpentine 贵蛇纹石
noble sight 壮丽景色
noble spinel 贵尖晶石
noble stone 宝石
no-bond prestressing 无握裹的预应力；无黏结预应力
no bond resonance 无键共振
no-bond tensioning 无黏结张拉；无黏结张力
no bottom found 未能探测到海底
no bottom sounding 不到水深值
no box type valve body 无填料函式阀体
no-break power system 保证供应电力系统
no-break power unit 不间断电源设备
no-break standby 无中断备用
nobular migmatitic granite 阴影状混合花岗岩
nobular migmatitic granodiorite 阴影状混合花岗闪长岩
no caking coal 无黏结性煤
no carbon paper 无碳复写纸
no carbon required paper 无碳复写纸
no-cargo condition 压载状态；空载状态
no cargo on board 空货船；船上无货
no-carry 无进位
no case to answer 无需答辩
no-cell 无单元
no-cement castable 无水泥浇注料
nocerine 针六方石
nocerite 针六方石
no charges 免费(的)
no-charge machine fault time 不收费的机器故障时间
no-charge non-machine-fault time 不收费的非机器故障时间；非机器故障免费时间
nockbush 节流衬套
no claim bonus 无赔款折扣
no claim discount 无赔款折扣
no-coat chip 基体刀片
no-code aspect 无电码示象
no colo(u)r or smell 无色无臭
no-colo(u)r zero-voltage level 无色零电平
no combined sewer flow 非混合污水流
no-commercial value 无商业价值
no commission until paid 付讫前无佣金
no conflict 无抵触
no connection 无连接
no constant boiling mixture 非恒沸液
no-contact pickup 无触点检波器
no-continuous discontinuity 不连续结构面
no-core braid 无芯编带
no-core reactor 空心扼流圈；无铁芯扼流圈
no correction 无须更正；无更正
no correlation 不相关
no cost 免费的
no-count 未计数；无价值的
no crack edge 无绽边
no credit 现款交易
no credit charge 赊欠免言
no-creep type baffle 非蠕爬型挡板
noctec 水合氯醛

noctilucence 夜光
noctilucent clouds 夜光云
noctilucent train 夜光余迹
noctirsor 暗视器
nocto television 红外线电视
noctovision 红外(线)电视;暗视觉
noctovisor 红外(线)摄像机;红外线望远镜
nocturnal 夜间发生的;天体观测仪
nocturnal arc 夜间弧;夜弧光;地平下弧
nocturnal cooling 夜冷却;夜间冷却
nocturnal inversion 夜逆温;夜间逆温
nocturnal measurement 夜间测量
nocturnal migration 夜迁移;夜间移栖;夜间迁移
nocturnal ozone profile 夜间臭氧垂直方向分布
nocturnal penetration range 夜间可见距离
nocturnal radiation 夜间辐射;红外(线)辐射
nocturnal thunderstorm 夜间雷暴
nocturnal timer 夜间计时仪
nocturne 夜景画
n-octyl acetate 醋酸正辛酯
n-octyl isodecyl phthalate 邻苯二甲酸正辛异癸酯
n-octyl mercaptan 正辛基硫醇
nocuous gas 有毒气体
no-current protection 无电流保护
no customary deductions clause 无惯例扣减条款
no-cut round 不掏槽炮眼组
nod 点头
noda 节圆
nodal analysis 节点分析(法);节点法;波节分析
nodal area 结区
nodal axis 节线轴;节轴
nodal carriage 测节轨运器
nodal circle 波节圆
nodal city 中心市
nodal connection 节点连接
nodal cross-section 节截面
nodal cylinder 波节圆柱体
nodal degree of freedom 节点自由度
nodal diagram 波节图
nodal diameter 波节直径
nodal diaphragm 节隔膜
nodal displacement 结点位移;节点移动;节点位移
nodal drive 节驱动;节传动
nodal equation 节方程;波方程;波节点方程
nodal equation of Galerhin method 伽辽金法节点方程
nodal equation of variation method 变分法节点方程
nodal force 节点力;波节力
nodal gearing 节点齿轮装置
nodal increment 截距;节距
nodalizer 波节显示器
nodal line 结点线;节线;节点线;交点线;无潮线;波节线
nodal load 结点荷载;节点荷载
nodal method 节点法
nodal method of analysis 节点电压法
nodal motion 交点线位移
nodal of seiche 驻波节数;静振波节数;假潮节数
nodal parameter 节点参数
nodal pattern 振动节型
nodal period 交点周期
nodal plane 振动节面;节平面;节面;波节平面
nodal point 结点;节点;无潮点
nodal-point keying 波节点键控
nodal point of emergence 后节点
nodal point of incidence 前节点
nodal point of mesh 网格结点
nodal point of vibration 振动节点
nodal points 高斯点
nodal point separation 节点距;节点分离
nodal processor 节点处理器
nodal region 核心区
nodal seal resonator 波节封接谐振器
nodal section 节面
nodal slide 测节器
nodal support 节点支承
nodal surface 节面
nodal system development 节点系统发展
nodal type 波节式(用于流态)
nodal wedge type cutting 楔形节插条
nodal well 节点井
nodal zone 节带;漂沙改向带;波节区节带

no-darkroom print 非暗室晒印
nodding 点头振动
nodding technique 点头技术
nodding tree 倾斜树
node 零速点;零点;结点;节(点);交点;桁架节点;分支点
node admittance determinant 节点导纳行列式
node analysis 节点分析
node base 节点基
node-beam-column 梁柱节点
node branch 节点分支;节点的分支
node carrier 节子式承马;节套式支索器
no-decompression dive 不减压潜水
no-decompression diving 不减压潜水
no-decompression excursion 不减压向上巡潜
no-decompression excursion limit 不减压向上巡潜极限
node current 节点电流
node cycle 交点周期;交点图
node diagram 节点图
node-disjoint subgraph 节点离子图
node earliest start time 结点的最早开始时刻
node equation 节点方程
node equilibrium 节点平衡
node expression 节点表达式
node factor 交点因素
no defect 合格;无缺陷的
node holding moment 受矩节点
node identifier 节点标识符
node initialization block 节点初始化块
node latest finish time 结点的最迟完成时刻
no-delay base 立即接通制
no-delay gate 快速快门
node line 交点线
node-link relationship 道路连接点交通流量分配率
no delivery 未到货
node-locus 结点轨迹;节点轨迹
node matrix equation 节点矩阵
node mechanism 结点结构
node method 节点电位法
node mobility 结点迁移率
node moment 结点矩
node name 节点名
Noden-bretteuneau 木材电处理防腐法
node-node transmission procedure 节点传输过程
node number 节点号码
node of a curve 曲线的结点
node of emission 像节点
node of fault 断层节
node of orbit 轨道节点;轨道交点
node of oscillation 振荡节点
node of rail network 铁路网联轨点
node of tide 潮波节
node of wave 波节
node operator 节点运行程序;节点算符;节点操作符
node-pair method 结点电压法;导纳法
node-pair transformation 结点偶变换
node partition 节点划分
node point 结点;节点
node point displacement 节点位移
node point equilibrium 节点力平衡
node point holding moment 节点约束力矩
node point load 节点荷重;节点荷载
node point loading 节点加载
node point mechanism 节点机构
node point mobility 节点活动性;节点移动性
node point movement 节点位移;节点活动;节点移动
node point plate 节点板
node point rotation 节点旋转;节点转动
node point rotation angle 节点转动角度;节点旋转角度
node point strength 节点极限;节点强度
node point trajectory 节点轨迹;节点轨迹
node point translation 节点直线运动;节点平移
node point velocity 节点运动速度
node processor 节点处理机
node repair 节点修复
node rope 节索
node rotation angle 节点旋转角
node-shift method 角度偏移法;波节移法
nodes in main axis 主轴上的节数
nodes in vibration 振动中的节点

node splitting 节点分裂
node state 节点状态
node strength 结点强度
node structure 节点结构
node switching 节点转接
node table 节点表
no detected toxic concentration 无觉察到的毒性浓度
node time 节点时间
node trajectory 节点轨迹
node type 节点型;节点类型
node variable 节点变量
node voltage 波节电压
nodical 交点的
nodical month 交点月
nodical period 节点周期
nodical year 交点年
nodischarge operation 零流量运行;无水运行(关闭导叶后)
nodischarge pump head 水泵零流量扬程;水泵关死扬程
nodischarge pump input 水泵零流量输入功率
nodischarge state 全闭状态
no discount 不折扣;不贴现
nodon rectifier 铝铅电解整流器
nodon-valve 电解铅铝整流器
nodose 有节的
nodose antenna 多节触须
no-draft close-tolerance die 固定式高精度模具
no-draft export 无外汇出口;不结汇输出
no-draft import 无外汇进口;不结汇输入
no-draw press 无牵引压榨
no dressing 不修整的
no-drift nozzle 防漂流喷嘴
no-drip nozzle 无滴口喷嘴
nodual point 光学中心节点
nodular 结节的;结核状;节状的
nodular anhydrite 瘤状硬石膏
nodular band 节状绑扎;节状条纹
nodular cast iron 球墨生铁;球墨铸铁
nodular cementite 粒状渗碳体
nodular chert 结核状燧石
nodular end cell 节状端壁
nodular fireclay 块状耐火黏土;团块状黏土;团球状耐火土;团块耐火黏土
nodular graphite 粒状石墨;球状石墨
nodular graphite cast-iron 球墨铸铁;可锻铸铁
nodular graphitic structure 球墨组织
nodular growth phase 球状成长相
nodularity 成球性
nodularization 球化
nodularizer 球化剂
nodular phosphoraite 结核状磷块岩
nodular powder 粒状粉料
nodular precipitation 球状沉淀
nodular shale 块状页岩
nodular sinter 结核状泉华
nodular structure 块瘤结构
nodular troostite 细珠光体
nodulated blowing wool 不规则喷吹玻璃棉
nodulation 结核
nodule 小结;小节点;根瘤;料球;料粒;结核体;结核;不规则的球
nodule in sediment 沉积物中结核
nodule limestone 瘤状灰岩
nodule of clinker 熟料球
nodule of rust 锈结核
nodule size 球粒尺寸
nodulization 结粒
nodulize 烧结
nodulized feed 入窑料球;喂入的料球
nodulized raw meal 生料球
nodulizer 制粒机;造粒剂;球化机;成团机
nodulizing 制粒;团矿;成球;成粒
nodulizing agent 球化剂
nodulizing capacity 成球能力
nodulizing drum 成球筒
nodulizing plant 成球设备
nodulizing region 成球区
nodulizing strength 成球强度
nodulizing zone 成球带
nodum 植被单位;植被抽象单位
no echo area 无回声区
no effect concentration 无效应浓度

no effective footage 无效进尺
no-effect level 无作用水平;无影响剂量
Noel 诺埃尔木块地板
no eluviation without porosity 没有孔隙性就没有溶提作用
no entrance 禁止入内
no-entry sign 禁止车辆驶入标志
no exchange surrender(ed) 不结汇;免结汇
no-excitation contactor 无激励接触器
no-excitation detection relay 无激励检测继电器
no-failure life 无故障使用寿命;无故障的工作期限
no-failure operation 无故障工作
nofalling losses 未采下损失
nofalling losses ratio 未采下损失率;非开采损失率
no fault found 没有发现错误
no-fault insurance 不追究责任的保险
no-field track 无磁场径迹(指云室或气泡室中)
no-fines aggregate 无细集料;无细骨料
no-fines concrete 大孔混凝土;无砂混凝土;无细集料混凝土;无细料混凝土;无细骨料混凝土;无砂大孔隙混凝土
no-fines lightweight concrete wall 无砂孔轻质混凝土墙
no-fire area 射击安全区
no-fire line 射击安全线
no fishing period 禁渔区
no fissure 无缝隙;无裂缝
no fix 无船位;不定位
no fix A 多普勒数据不足不能定位
no fix B 固定数据不全不能定位
no fix data 数据不收敛不能定位
no fixed time 无固定时间
no-flash die forging 无飞边模锻
no-float 配合良好;调好的
no-flow cell 无流式压力盒
no-flow condition 无流动状态
no-flow shutoff head 无流关闭压头
no-fly zone 禁飞区
nofractured earth movement 无破裂性地壳运动
no freezing 不冻结
no funds 无存款
no-fuse panel 无熔丝配电盘
no-fuse switch 无熔丝开关;无保险丝开关
nog 支柱垫板;木栓;木垛支架;木垛;磨盘肋
no-gas welding 非气焊
noggin 小杯子
nogging (间墙柱间的)水平联系木条;壁砖;木架砖壁;木架砖梁;填墙木砖;填充的砖石砌体;木架砖壁作业;木架壁砖;横筋;砌木砖
nogging piece 立柱间的水平加固木;填墙木砖;横撑木;砖壁木架;横筋
nogging strip 埋在砖柱间的加固薄片
no-go-ga(u)ge 不过(端量)规;不过丝扣规;不通过量规
no-go nipple 不过端短节;终端短接
no-go thread ga(u)ge 螺纹验规止端;螺纹验规不过端
nog plate 磨板
no-grind gate 不磨浇道
nogueria oil 蜡烛果油;石栗果油
no hang-up 立即接续制通信
no-heat adhesive 不加热黏胶剂;不加热黏性物质;冷黏结剂
no-hinged 无活动关节的;无铰链的;无铰接的;无铰的
no-hinged arch 无铰接(发)券;无铰接拱;无铰拱
no-hinged beam 无铰接梁;固端梁
no-hinged column 无铰接柱;固端支柱
no-hinged frame 无铰接框架;无铰构架
no-hinged girder 无铰接横梁;无铰接大梁;无铰大梁
no-hinged support 无铰接立柱;无铰接支撑;无铰接支座;固定支座
no-hoisting drill bit 不提钻钻头
no-home record 无原位记录
no hook 勿用手钩;不可用钩
no horn 禁止鸣笛
no host 同层
no-hub pipe 无承口管
no-humping car 禁溜车【铁】
no-humping car storage 禁溜车停留线【铁】
noice reducing antenna system 降低噪声天线系统
Noil 诺伊尔青铜
noil cloth 棉绸

noil poplin 棉绸
no interest 免付利息
noise 噪声;噪声;声音
noise abatement 减噪声;消音;消声;噪声抑制器;噪声抑制;噪声控制;噪声降低;噪声减除;噪声的削减;隔音;降噪量;减声
noise abatement act 噪声控制法
noise abatement climb 噪声限制上升
noise abatement device 消声器
noise abatement equipment 消声装置
noise abatement procedure 噪声限制程序
noise abatement wall 隔声墙
noise absorber 噪声吸收器
noise-absorbing 吸音
noise-absorbing ceiling 吸声顶板
noise-absorbing circuit 噪声吸收电路
noise-absorbing elements 消音元件
noise absorption 吸音;吸声料;噪声吸收
noise absorption mat 消声垫
noise-adding receiver 噪声注入接收机
noise-air pollution 噪声一空气污染
noise alleviation 噪声的衰减
noise ambiguity 噪声不定性
noise AM jamming 噪声调幅干扰
noise amplification 噪声放大
noise amplifier 噪声放大器
noise amplifier circuit 噪声放大电路
noise amplitude 噪声振幅
noise analysis 噪声分析;干扰波研究
noise analyzer[analyser] 噪声分析器
noise and deception jammer 双工作方式干扰机
noise and number index 噪声量指数;噪声及次数指标;噪声和数值指数
noise and vibration control 噪声与振动控制
noise and vibration dampening 隔音与减震
noise assessment 噪声评价
noise attenuation 噪声衰减
noise audiogram 噪声听力图;噪声波形图;声掩蔽听力图
noise autocorrelation 噪声自相关
noise background 声底数值;背景噪声;噪声本底;噪声背景;本底噪声
noise baffle 隔声屏障
noise balancing circuit 噪声衡消电路
noise-balancing system 噪声衡消制
noise band 噪声频带
noise bandwidth 噪声带宽
noise barrier 阻声屏;隔音栅;隔声板;吸声板;噪声障;噪声墙;隔声屏障
noise behind the signal 调制噪声
noise blanker 消灭噪声装置;噪声熄灭装置
noise blanking 噪声消隐
noise bounce 噪声跳动;噪声起伏
noise burst 噪声脉冲串
noise by explosion 爆炸噪声
noise cables 有噪声电缆
noise calender 波纹轧光机
noise cancel adjustment 噪声消除调整
noise canceling microphone 抗噪声传声器
noise cancellation 噪声消除
noise canceller 噪声消除器
noise-cancelling amplifier 噪声消除放大器
noise cancel switch 噪声消除开关
noise channel 噪声通路
noise channel theorem 噪声信道定理
noise characteristic 噪声特性
noise circumstance 噪声环境
noise clipper 噪声限制器
noise coefficient 噪声系数
noise communication channel 有噪声的通信道
noise compensation 噪声校正;噪声补偿(费)
noise compensation method 噪声补偿法
noise component 噪声分量
noise constant 噪声常数
noise contours 噪声等值线
noise control 噪声控制;防噪
noise control acoustics 噪声控制学
noise control criterion 噪声控制准则
noise control curtain 控制噪声幕帘
noise control curve 噪声评价 NC 曲线
noise control in city planning 城市防噪声规划
noise control law 噪声控制法;噪声管制法
noise control legislation 噪声控制立法;噪声管制立法

noise control program(me) 噪声控制规划;防噪声规划
noise control regulations 噪声管理条例
noise control system 噪声控制系统
noise correlation system 噪声相关系统
noise count 噪声计数
noise countermeasure 消声措施
noise cover 噪声封闭器(桩工)
noise cover soundproof enclosure 噪声封闭器;封闭器
noise criterion 噪声容许极限;噪声标准(值);噪声判据;噪声基准
noise criterion 噪声标准
noise-criterion allowance curve 噪声容许标准曲线
noise-criterion curve 噪声标准曲线;噪声基准曲线;噪声评价曲线
noise-criterion number 噪声评价值
noise criterion value 噪声标准值
noise critical area 噪声监界区
noise crusader 噪声防治工作者
noise current 噪声电流
noise current generator 噪声电流发生器
noise curve 噪声曲线;噪声曲线
noise damage 噪声致害
noise-damping 消音;消声
noise-deadened wheel 消音盘;消声盘
noise deadener 消音器;消声器
noise deadening 消音;吸音;吸声;隔声
noise deadening foam 隔音泡沫材料
noise degeneration 噪声衰减
noise density 噪声密度
noise detecting circuit 噪声检测电路
noise detector 噪声探测器
noise development 噪声产生;噪声形式;噪声发展
noise digit 噪声数位
noise discharge tube 噪声放电管
noise disease 噪声病
noise distortion 噪声失真;噪声畸变
noise distribution 噪声分布
noise disturbance 噪声干扰
noise dither 噪声抖动
noise dose 噪声强度;噪声量;噪声剂量
noise dosimeter 噪声测量计
noise effect 噪声影响
noise elimination 消音;消声;消除噪声;噪声抑制;噪声消除;静噪
noise elimination measure 消声措施
noise eliminator 噪声抑制器;噪声消除器
noise emission 噪声发射
noise emission standards for products 产品噪声发射标准
noise emission test 噪声发射测试
noise energy 噪声能量
noise environment 噪声环境
noise equipment 有声器材
noise equivalent 噪声等值;噪声当量;等效杂波
noise equivalent angle 噪声当量角
noise equivalent bandwidth 噪声等效带宽
noise equivalent circuit 噪声等效电路
noise equivalent flux 噪声等效通量
noise equivalent flux density 噪声等效通量密度
noise equivalent input 噪声等效输入
noise equivalent intensity 噪声等效强度
noise equivalent irradiance 噪声等效辐照度
noise equivalent line 噪声等值线
noise equivalent neutron source 噪声等效中子源
noise equivalent pass-band 噪声等效通带
noise equivalent power 噪声等效功率
noise equivalent power density 噪声等效功率密度
noise equivalent resistance 噪声等效电阻
noise equivalent signal 噪声等值信号
noise equivalent source 噪声等效源
noise-equivalent target temperature difference 噪声等效目标温差
noise equivalent temperature 噪声等效温度
noise equivalent voltage 噪声等效电压
noise event meter 噪声计
noise-excitation 噪声激发
noise exposure 噪声接触;噪声暴露;曝噪
noise exposure forecast 噪声接触预报;噪声暴露预测;曝噪预报
noise exposure meter 噪声暴露计
noise exposure monitor 噪声暴露监控计
noise exposure rating 噪声暴露评价

noise exposure time 噪声音响时间;噪声暴露时间
noise factor 噪声因素;噪声因数;噪声系数
noise-factor measurement 噪声系数测量
noise-factor nomograph 噪声系数列线图
noise-feedback coder 噪声反馈编码器
noise field 噪声场
noise-field intensity 噪声场强
noise-field intensity measurement equipment 噪声场强测量仪
noise field strength 射频噪声场强度
noise figure 噪声指数;噪声系数;噪声图
noise figure characteristics 噪声因数特性
noise-figure indicator 噪声因数指示器
noise-figure measurement 噪声指数测量
noise-figure measuring set for microwave equipment 微波噪声系数测试仪
noise-figure meter 噪声指数计
noise filter 噪声滤波器;静噪滤波器
noise filtering 噪声过滤
noise flashes 噪声闪烁
noise fluctuation 噪声起伏
noise footprint 噪声印迹
noise-free 无噪声的
noise-free bearing 低噪声轴承
noise-free channel 无噪声信道
noise-free demodulation of laser 无噪激光解调
noise-free equivalent amplifier 等效无杂波放大器
noise-free light 无噪声光
noise-free operation 无噪声工作
noise-free picture 无噪声图像
noise-free receiver 无噪声接收机;低噪声接收机
noise-free seismogram 无干扰地震图
noise-free signal 无噪声信号
noise-frequency response 噪声频率响应
noise fringe 噪扰带
noise from clutter 杂波噪声
noise from heating system 采暖系统噪声
noise function 噪声函数
noise gate 噪声门限
noise-gated automatic gain controller 噪声门限式自动增益控制
noise-gated sync separator 噪声选通同步分离器
noise generation 噪声的产生
noise generator 噪声发生器
noise glitch 噪声假信号
noise grade 噪声等级
noise hazard 噪声危害
noise hindrance 噪声干扰
noise identification 噪声识别
noise immission 噪声照射
noise-immune 抗噪声;抗扰
noise immunity 抗噪声性;抗扰声度;抗声扰性;抗声扰度;抗扰性;抗扰度;抗干扰度
noise immunization 消噪抗扰
noise impact 噪声冲击
noise impact index 噪声冲击指数
noise improvement factor 噪声改善系数;噪声改进因数
noise impulse 噪声脉冲
noise index 噪声指数;噪声指标
noise-induced deafness 噪声性耳聋
noise induced temporary threshold shift 噪声性永久性阈移
noise-inducing material 致噪声材料
noise-inducing vibration 噪声导生振动
noise induction 噪声感应
noise inhibit monostable 噪声抑制单稳态触发电路
noise injury 噪声损伤
noise in natural 自然界噪声
noise in school 学校噪声
noise inside the railroad car 有轨电车内噪声
noise in signal 信号中的噪声
noise in sources 能源噪声
noise insulation 噪声隔离;隔声
noise insulation class 噪声隔声分类;噪声隔声等级
noise insulation factor 噪声隔声率;噪声隔离因素;隔噪声因数;隔声系数
noise insulation glass 防噪声玻璃
noise-insulation installation 隔音装备
noise insulator factor 隔噪声因数;隔音度;隔声系数
noise intensity 噪声强度
noise-intensity contour 噪声强度曲线
noise interference with speech 谈话噪声干扰
noise inversion 噪声抑制

noise inverter 噪声抑制器
noise investigation 噪声查验
noise isolation class 隔音等级
noise isolation unit 噪声隔离器
noise isoline 噪声等值线
noise jammer 噪声干扰机
noise jamming 噪声干扰
noise killer 静音器;消音装置;噪声抑制器;噪声消除装置;噪声消除器;噪声限制器;噪声吸收器
noise klystron 噪声速调管
noise lamp 噪声灯
noise lamp ignitor 噪声灯引燃器
noiseless 无噪声的;无声的
noiseless action 无声操作
noiseless amplifier 无噪声放大器
noiseless arc lamp 无声弧光灯
noiseless channel 无干扰波道
noiseless circuit 无噪声电路
noiseless drive 无声传动
noiseless motor 无噪声电动机
noiseless pavement 无噪声路面
noiseless phase sensitive amplifier 无噪声相敏放大器
noiseless piling technique 无噪声打桩技术
noiseless pulse 无噪声脉冲
noiseless reactor 无噪声反应堆
noiseless run(ning) 无声运动;无噪声运转
noiseless shift(ing) 无声换挡
noiseless slit 无噪声狭缝
noiseless speed 无噪声航速
noise level 噪声水平;噪声强度;噪声量;噪声级(别);噪声度;噪声电平;噪声等级;噪度级;干扰电平
noise-level acceptance test 噪声级验收试验
noise level indicator 噪声电平指示器
noise level measurement 噪声电平测量
noise level meter 噪声计
noise level monitor 噪声电平监测器
noise level test 噪声级试验
noise-like function 噪声样函数
noise limit 噪声限度
noise limitations 噪声限制
noise limited condition 噪声限度情况
noise-limited detector 噪声限制检测器
noise-limited range 噪声限制范围
noise limited receiver 噪声限制接收机
noise limited sensitivity 噪声抑制灵敏度
noise limiter 噪声限制器;杂波抑制器;消噪器
noise limiter circuit 噪声限制电路
noise limiting 噪声抑制
noise limiting circuit 噪声抑制电路
noise loading 噪声负载
noise loading method 噪声负载法
noise loading ratio 噪声负载比
noise loading test 噪声负载试验
noise loading test set 噪声负载测试设备
noise load ratio 噪声负荷比
noise lower limit 噪声下限
noise maintenance component 噪声维护部件
noisemaker 噪声发生器
noise management 噪声管理
noise margin 噪声容限;噪声极限;噪声边限
noise masking level 噪声掩蔽级
noise measurement 噪声检测;噪声测量;噪声测定
noise-measurement technique 噪声测定技术
noise-measuring 噪声测定;噪声测量
noise measuring buoy 噪声测量浮标
noise-measuring equipment 噪声测量设备
noise-measuring instrument 噪声测量仪器
noise-measuring meter 噪声计
noise measuring set 噪声测量仪;噪声测量仪
noise-measuring system 噪声测试装置
noise-metallic 噪声金属导线的
noise meter 噪声计;噪声表;噪声测试器;噪声测试计
noisemeter 噪声计
noise method calibration 噪声法校准
noise mode 噪声态
noise modulated 噪声调制的
noise modulated jammer 杂波调制干扰机
noise-modulated jamming 噪声调制干扰
noise modulation 噪声调制
noise modulator 杂波调制器
noise monitor 噪声检测器

noise monitoring 噪声监控;噪声监测
noise monitoring and control 噪声监控
noise-monitoring equipment 噪声监测仪
noise monitoring unit 噪声监控部门
noise monitor junction 噪声监测接头
noise muffler 噪声衰减器
noiseness 噪声性质;嘈杂
noise network 噪声网络
noise nuisance 噪声干扰;噪声损害;噪声危害;噪声公害
noise objective 噪声标限
noise of aeroplane 飞机噪声
noise of background 背景噪声
noise of blower and compressor 鼓风机压缩机噪声
noise of compressor 空气压缩机噪声
noise of household appliance 家用电器噪声
noise of household electric(al) appliances 家用电器噪声
noise of laser ranger 激光测距仪噪声
noise of social activity 社会生活噪声
noise operation 噪声运算
noise origin 噪声源
noise output 噪声输出(量)
noise pattern 噪声图
noise peak 噪声峰值;噪声巅值
noise-peak clipper 噪声巅值限制器
noise peak limiter 噪声巅值限制器
noise phase modulation jamming 噪声调相干扰
noise photoelectron 噪声光电子
noise photoelectronics 噪声光电学
noise physics 噪声物理学
noise pickup 噪声拾取
noise pickup coil 噪声拾波线圈
noise pollution 噪声污染;噪声骚扰
noise pollution forecasting 噪声污染预测
noise pollution level 噪声骚扰级;噪声污染级
noise-pollution(level)meter 噪声公害测定仪
noise pollution monitoring 噪声污染监测
noise potential 噪声势
noise power 噪声功率
noise power comparator 噪声功率比较仪
noise power level 噪声功率级
noise power measurement 噪声功率测量
noise power ratio 噪声功率比
noise power spectrum 噪声功率谱
noise precaution 噪声预防;防止噪声
noise prevention 噪声防止
noise prevention arrangement 防噪布局
noise probability density 噪声概率密度
noise problem 噪声问题
noise processing circuit 噪声处理电路
noise-producing equipment 噪声发生装置
noise-producing material 产生噪声材料
noise-producing source 杂音(发生)源;干扰发生源;噪声发生源
noise-producing surface 有噪声路面;发噪声地面
noise-proof 抗噪;防杂音;隔音的;抗扰的
noise-proof feature 抗干扰特性;抗噪特性;防噪性能
noise-proof green space 防音绿地
noise propagation 噪声传播
noise propagation in atmosphere 大气中的噪声传播
noise protection 噪声防护
noise protector 噪声防护器
noise quencher 噪声消除器
noise quieting 静止噪声法
noise quieting sensitivity 噪声抑制灵敏度;静噪灵敏度
noise radar 噪声雷达
noise radiation 噪声辐射
noise radiation pattern 噪声辐射图
noise range 噪声范围;噪声区;噪声区
noise ranging 噪声测距
noise ranging sonar 噪声测距声呐
noise rating 噪声评级;噪声分级;噪声评价
noise rating curve 噪声评级曲线;噪声分级曲线;噪声评价 NR 曲线
noise rating number 噪声(评)级数;噪声检定值;噪声分级值;噪声额定值;噪声值数;噪声评价数;噪声评价试验;噪声评级数;噪声检定值
noise rating parameter 噪声评价参数
noise ratio 噪声比率

noise record(ing) 噪声记录
noise rectifier 噪声整流器
noise-reduced 减噪声的
noise-reducer nozzle 消音式喷管
noise reducing splitter 吸声体
noise reduction 减噪(声);消音量;噪声降低;噪声减低;噪声控制;噪声减少;降噪(量);减声
noise reduction by absorption 吸声降噪
noise reduction by building 建筑降噪
noise reduction by foliage 绿化降噪
noise reduction coefficient 减噪(声)系数;噪声消除系数;噪声降低系数;降噪系数;降低噪声系数
noise reduction cushion 减声垫
noise reduction measure 降低噪声措施
noise reduction system 降噪系统
noise regulations 噪声限制规则
noise-rejected tube 噪声抑制管
noise rejection 噪声抑制
noise remover 噪声抑止器;反干扰机
noise research 噪声研究
noise resistance 噪声电阻;噪声阻力;噪声等效电阻
noise response 噪声响应
noise room 噪声室
noise rules 噪声水平标准
noise screen(ing) 挡噪屏;噪声屏蔽
noise selection 噪声选择
noise sensitivity 噪声灵敏度
noise sensor 噪声探头
noise shaping network 噪声频谱成型网络
noise shelter 噪声隔绝套
noise shield 隔噪罩;防噪罩
noise shielding 噪声屏蔽
noise sideband 噪声边带
noise signal 噪声信号
noise(signal)ratio 噪声比;信噪比
noise silencer 噪声消声器;噪声抑制器;噪声消除器;噪声限制器;噪(抑制)器;减噪(抑制)器
noise simulator 噪声模拟装置;噪声模拟器
noise slot mesauring frequency 噪声槽测试频率
noise snubber 消声器
noise sound analyzer 噪声分析器
noise source 干扰源;杂音源;噪声源
noise source coefficient 噪声源系数
noise source identification 噪声源识别
noise source power measurement 噪声源功率测量;噪声源规律测量
noise speckle 噪声斑点
noise spectrum 噪声谱;噪声频谱
noise spike 噪声尖峰;噪声光脉冲
noise spot 噪声斑点
noise squelch 噪声消除器
noise squelch oscillator 镇噪振荡器
noise stability 噪声稳定率;噪声稳定度
noise standard 噪声容许标准;噪声标准
noise standard of mine environment 矿山环境噪声标准
noise-stop device 防噪装置
noise stopping 噪声抑制;静噪
noise stopping machine 噪声抑制器
noise storm 噪暴
noise strength standard 噪声强度标准
noise stretching 噪声扩展
noise suicide circuit 噪声自灭电路;噪声抑制电路;抗噪声电路
noise suppresser 消音器;噪声抑制器;消声器
noise suppressing 噪声抑制
noise-suppressing device 噪声抑制装置
noise suppression 噪声抑制;噪声消除
noise suppression capacitor 噪声抑制电容器
noise suppression circuit 噪声抑制电路
noise suppression control 噪声抑制控制
noise suppression control tube 静噪控制管
noise suppression device 噪声抑制设备
noise suppression equipment 消杂音设备
noise suppression network 噪声抑制网络
noise suppression resistor 噪声抑制电阻
noise suppression tower 噪声抑制塔
noise suppressor 消音器;消声器;噪声抑制器;噪声消除装置
noise suppressor effect 噪声抑制器效应
noise survey 噪声监测;噪声图测绘;噪声调查
noise susceptibility 噪声敏感性
noise switching 噪声倒换
noise technique 噪声技术

noise temperature 噪声温度
noise-temperature ratio 噪声温度比;噪温比
noise test 噪声试验
noise tester 噪声测试器;噪声测试计
noise testing 噪声测试
noise testing instrument 噪声试验仪
noise thermometer 噪声温度计
noise threshold 噪声阈
noise tolerance 噪声容许量;噪声耐量
noise track 噪声轨迹
noise tracking 噪声跟踪
noise transmission 噪声传输;噪声传播;噪声传递
noise transmission impairment 噪声造成的传输质量降低量;噪声传输损害;电路杂音引起传输质量减损
noise transmitter 噪声发送器
noise trap 噪声抑制器;声阱;反干扰装置
noise tube 噪声管
noise type 干扰波类型
noise unit 噪声单位
noise use 噪声利用
noise variance 噪声离散
noise vector 噪声向量;噪声矢量
noise-velocity law 声强速度定律;噪声速度定律
noise voltage 噪声电压
noise voltage generator 噪声电压发生器
noise wave 杂波
noise weighting 噪声加权
noise whistlers 噪哨
noise whitening filter 噪声白化滤波器
noise wind tunnel 噪声风洞
noise word 噪声码字
noise zoning 噪声分区
noisiness 噪声特性;噪声强度;噪声量;噪度
noisy blacks 干扰黑斑
noisy channel 有噪声信道;噪声信道
noisy channel theorem 噪声信道定理
noisy circuit 有噪声电路
noisy communication channel 噪声信道
noisy digit 噪声数字;噪声数位;嘈杂数位
noisy generator 有杂运转发电机
noisy mode 有噪声方式;噪声方式;杂波形式;嘈杂状态
noisy reproduction 噪声重发;干扰图像重现
noisy running 有噪声运转
nolanite 铁钒矿
no-lead automobile fuel 无铅汽车燃料
no-leak 不漏泄;不漏气
no-leakage 无泄漏
nolecule 微点
no left turn 禁止左转(弯)
nolenin 甘油三亚麻酸酯
nolines 全部(线路)占线
no-live load 空载
No. Lloyd 劳埃德船级社编号
No. Lloyd's 劳氏船级社编号
no-load 空车开路;空载;空负苛;免付佣金的;无负载的
no-load air pressure 空载风压
no-load characteristic 空转特性(曲线);空载特性(曲线);空压特性;开路特性;无载特性(曲线)
no-load condition 空载条件
no-load consumption 空载消耗
no-load control 空载控制;无负荷控制
no-load current 空载电流
no-load curve 空转特性(曲线)
no-load cut-out 无负荷切断器
no-load diesel relay 柴油机空载继电器
no-loaded line 空载线(路)
no-load excitation 空载励磁
no-load field voltage 无载励磁电压
no-load flow 无载流量
no-load friction 无荷载摩擦
no-load fund 投资人不用支付佣金的互相信托投资基金
no-load heat consumption 空运转热消耗;无载热耗
no-load heat duty 空载热负荷
no-load indication 无载指示
no-load jet 无荷载喷嘴
no-load loss 空转损失;空载损失;空载损耗;开路损失;无功损耗
no-load method 无载方法
no-load nozzle 空转喷嘴;无载喷嘴

no-load opening 空载开度
no-load operation 空载运行;无载运转;空载运行;无功运行
no-load position 无负荷
no-load power consumed by transformer 变压器无载功率值
no-load power consumption 空载能耗;空载功率消耗
no-load power requirement 空载功率消耗量
no-load relay 无载继电器
no-load release 无载跳闸
no-load release action 无载释放动作;无负载释放动作
no-load release magnet 无载释放电磁铁
no-load resistance 无载电阻
no-load run(ning) 空转;空载运转;空载运行;无负载运转
no load running test 无载运转试验
no load-sharing between substations 变电所间不能负载复用
no-load shear strength 空载剪切强度(指结构支座移动或温度变化时产生的剪切强度);无载剪力强度
no-load short-circuit method 无载短路法
no-load speed 空载速度;空行程速度;空车速度;无负载速度;同步速率
no-load speed changing 空载变速
no-load speed governor 空载调速器
no-load starting 空载起动;无载起动
no-load starting electro-pneumatic valve 空载起动电气阀
no-load state 无载状态
no-load switch 空载断路器
no-load switching-in 无载合闸
no-load switching off 无载切断
no-load tap-changer of transformer 变压器无载分接头开关
no-load terminal voltage 无载端电压
no-load test 空载试验;空载实验;无负荷试验
no-load test run 空载试运转
no load time 无载时间
no-load valve 无载阀门;零载阀门
no-load voltage ratio 空载电压比;变比
no-load voltage regulator 无载调压器
no-load voltage regulator and tap 无载调压抽头
no-load work 空载运行
no-load zone 无载区
no longer 不再
no longer valid 不再有效
no loop white water system 无循环白水系统
no/lose cost clean production option 无/低费清洁生产方案
no loss line 无损耗线
no lump concrete 零坍落度混凝土
nomad 航用自动气象台
nomadic herding 游牧业
nomadic people 牧民
nomadism 流浪生活;游牧生活;漫游
Nomag 非磁性高电阻合金铸铁
nomal fold 对称褶曲
no man's land 两个舱口之间的空间;无人区
no-man's land 无人地带
no margin allowed 金额须全数收取
no mark 无商标;无标记;无标志
Nomarski microscope 诺马尔斯基显微镜
no-match 非配合
nomenclative 命名的
nomenclature 专门用语;专门术语;专门名词;专门技术名词;命名法;命名;名词汇编;名称;术语表;术语;分幅编号
nomenclature of goods 货物名称表
nomenclature of strata 地层命名
nomenclature plate 型号铭牌;技术数据;标牌
nomenclature principle 命名原则
nomen undum 无记述名
no message 不报
nomex 高熔点芳香族聚酰胺
nominal 公称(的);名义的;额定(的);标称(的)
nominal account 虚账户;名义账户;名义账目;非实物账
nominal accuracy 标称精度
nominal accuracy of electromagnetic distance measurement 电磁波测距标称精度
nominal addendum 标称齿顶高

nominal allowed price 名义作价
nominal altitude 标称高度
nominal amount 名义账面额;名义金额;面额
nominal annual rate 名义年利率
nominal aperture 名义孔径;标称口径
nominal aperture width 标称孔宽
nominal area 公称截面积;标称面积
nominal area of screen 标称筛面面积
nominal asbestos cement corrugated shingle (sheet) 石棉水泥波瓦标准张
nominal asbestos cement sheet 石棉水泥瓦标准张
nominal assets 名义资产
nominal audit 名义审计
nominal average power 标定平均功率
nominal balances 名义余额
nominal band 标称频带;标波段
nominal bandwidth 标称带宽
nominal batch 额定批量
nominal bearing pressure 额定轴承压力;标称承载力
nominal blade setting 标称架叶安装角
nominal bore 公称孔径;公称管径;标称内径
nominal breadth 标称宽度
nominal breaking strength 名义上的断裂强度;标称断裂强度
nominal calorific capacity 标称发热能力
nominal candle power 标称烛光
nominal capacity 公称容量;公称容积;名义能力;额定容量;标称生产能力;标称容量;标称能力
nominal capacity of holding weight 公称载重量
nominal capital 名义资本
nominal cash balance 名义现金余额
nominal character dimension 标称字符尺寸
nominal characteristic impedance 标称特性阻抗
nominal circuit voltage 电路额定电压;额定电路电压
nominal colo(u)r 标称色
nominal composition 公称成分;名义成分;标称成分
nominal compression ratio 额定压缩比;额定压缩率
nominal compression stress 标称压缩应力
nominal condition 额定条件
nominal consumption 额定耗量
nominal contract price 名义约定价格;名义合同价格
nominal cost 名义费用;名义成本
nominal cross-section 名义截面;公称截面(积);额定截面积
nominal crude price 标定油价
nominal cube strength 公称立方强度;额定立方强度
nominal current 标称电流
nominal curve 额定曲线
nominal customs duties 名义关税
nominal cut-off frequency 额定截止频率
nominal damage 小额损失赔偿;小额损害赔偿;象征性损失;象征性赔偿;名义损失;名义上赔偿
nominal data 名义型数据;标定数据
nominal decline rate 标准递减率
nominal defendant 名义被告
nominal deflection 名义转折角
nominal density 名义密度
nominal deposit 名义存款
nominal depth 名义深度;额定深度
nominal design value 额定设计值
nominal deviation 公称偏差;名义偏向;标称偏差
nominal diameter 虚直径;规定直径;公称直径;名义直径;毛直径;通称直径;标称直径
nominal diameter of pipes 管材的公称直径;管道公称直径
nominal diameter of tubing 油管通称直径
nominal dimension 图注尺寸;名义尺寸;毛尺寸;额定尺寸;标称尺寸;公称尺
nominal dimension capacity 标称尺寸大小;标称尺寸容量
nominal dimension of masonry 砌体的名义尺寸
nominal dimension of pipeline 管道标称尺寸
nominal dimension range 名义尺寸范围;标称尺寸房屋;公称尺寸范围
nominal discount rate 名义贴现率
nominal distance 标称距离
nominal dollar 名义美元
nominal ductility 额定延性值

nominal earning 名义收益
nominal effective exchange rate 名义有效汇率;名义实际汇率
nominal element 非实物要素
nominal elevation 名义高度;额定标高;标称高程
nominal endurance limit 标称持久极限
nominal error 公称误差;名义误差;标称误差
nominal exchange 名义上的汇兑;名义汇票
nominal exchange rate 名义汇率
nominal expenditures 营业支出;名义支出;名义上的营业支出
nominal expulsion 开除留用
nominal face velocity 额定迎面风速
nominal failure rate 正常损坏率
nominal filter fineness 公称过滤精度
nominal flow capacity 名义流量
nominal flow limit 标称流量极限
nominal flow rate 公称流量
nominal focal ratio 标称焦距比
nominal focal surface 名义焦面
nominal fracture strain 标称破裂应变;标称开裂应变
nominal fracture stress 标称破裂应力
nominal frequency 额定频率;标称频率
nominal frequency range 额定频率范围
nominal gas capacity 标称气量
nominal gas rate 额定用气量;公称用气量
nominal generating head 额定发电水头
nominal grading curve 公称整坡曲线;名义整坡曲线;标称(颗粒)级配曲线
nominal grain size 名义颗粒度;标称颗粒粒度
nominal ground pressure 标称地面压力
nominal group technique 极小组技术;名义群体技术
nominal heating output 标称供热功率;名义输出功率;名义供热输出量
nominal height 名义层高;名义高度;标定高度
nominal holding 名义上持股
nominal horse power 名义功率;公称马力;额定马力;名义马力;额定功率;标准马力;标称马力;标定功率
nominal hours 名义工作时间;名义工时
nominal hydraulic residue 名义水力停留时间
nominal hydraulic retention time 名义水力停留时间
nominal hygiene air requirement 标定卫生空气需要量
nominal impedance 额定阻抗;标称阻抗
nominal income 名义所得;名义收入;名目所得
nominal income and actual expenditures 财政上的虚收实支
nominal input voltage 额定输入电压
nominal inside diameter 通称内径;标称内径
nominal insulation voltage 额定绝缘电压
nominal interest 名义利息
nominal interest rate 名义利率
nominal internal diameter 标称内径
nominal internal width 标称内宽
nominalism 命名主义
nominal justification rate 标称调整率
nominal ledger 名义账项总账
nominal length 名义长度;标称长度
nominal length of liquid level ga(u)ge 液面计公称长度
nominal liability 名义上的负债;名义负债
nominal life 公称寿命
nominal lifetime 名义寿命期
nominal lifting capacity 公称升船能力;标称提升能力
nominal light flux 额定光通量
nominal line 公称行;标称行;公称线;标称线
nominal line pitch 标称行距
nominal line width 标称行宽;表准行距;标准行距
nominal load 名义荷载;额定装载量;额定载重;额定负载;额定负荷;标称荷载;标称负载
nominal load bearing capacity 名义载重量;额定承载力
nominal load capacity 额定载重量
nominal loading capacity 额定荷载
nominal loading payload 额定荷载
nominal main voltage 额定供电电压
nominal margin 标定运行范围;标称容限
nominal mass 公称质量

nominal maximum plant capacity 标定最大工厂生产能力;标定工厂最大生产能力
nominal maximum reduction ratio 最大公称破碎比;额定最大破碎比
nominal maximum size of aggregate 骨料标称最大粒径;集料标称最大粒径
nominal mean frequency 标准中心频率
nominal mean power 名义功率
nominal measure 名义尺寸;名义尺度;毛尺寸;额定尺度
nominal measurement 公称尺寸
nominal meter of asbestos-cement pipe 石棉水泥管标准米
nominal mix 名义配合比;标称配合
nominal mix proportioning 体积配合比(混凝土的)
nominal mix proportions 体积配合比(混凝土的)
nominal mix ratio 标称配合比
nominal net head 额定净水头
nominal ocular hazard area 标称眼睛受害区域
nominal ocular hazard distance 标称眼睛受害距离
nominal operating pressure 公称工作压力
nominal operation 额定操作
nominal orbit 预期轨道
nominal output 定额产量;额定输出;额定出力;额定产量;标称输出(功率);标称功率;标称产量
nominal output voltage 额定输出电压
nominal outside diameter 标称外径
nominal owners 名义上的所有者
nominal par 名义票面值;名义票面价格;票面价格
nominal parameter 额定参数
nominal particle-diameter 标称粒径
nominal particle-size 公称粒子大小;标称粒子大小;公称颗粒大小;标称颗粒大小;标称粒径
nominal partner 名义股东;名义合伙人
nominal party 名义上的当事人
nominal perimeter 公称周长
nominal pipe diameter 额定管径
nominal pitch 公称螺距;公称节距;标称螺距
nominal pitch line 标称节线
nominal pitch ratio 公称螺距比;标称螺距比
nominal population 名义总体
nominal pore size 公称孔径
nominal power 公称功率;额定功率;标称容量;标称功率
nominal power consumption 额定电力消耗
nominal premium 象征性保费;暂时不收保费
nominal pressure 标定压力;公称压力;额定压力;标称压力
nominal pressure grade 公称压力等级
nominal price 虚价;正价;名义价格;定价
nominal productive head 额定发电水头
nominal profit 虚利;名义利润
nominal pull-in torque 名义牵入转矩;标称牵入转矩;牵入同步力矩
nominal quotation 虚牌价;名义牌价;名义报价
nominal range 标称距离;标称范围
nominal range of a light 额定灯光射程
nominal range of light 额定光达距离;标称灯光射程
nominal range of use 额定使用范围
nominal rate 名义利率;名义汇率;名义汇价;名义税率
nominal rated capacity 铭牌名义容量;铭牌额定容量;标称容量
nominal rated power 标称设计功率
nominal rated speed 额定速率;额定速度
nominal rate of exchange 名义汇率;名义汇价
nominal rate of rise 标称上升率
nominal rate of the spring 钢板弹簧额定刚性系数
nominal rating 公称出力;标称规格;标准规格;电机铭牌值;额定值;标称出力;标称定额
nominal resistance 标称电阻
nominal resolution 额定分辨率
nominal retention period 名义停留时间
nominal revenues 财政虚收
nominal reverberation time 额定混响时间
nominal revolution 名义转数
nominal ridigity 名义刚性;公称刚性;标称刚性;标称刚度;公称刚度;名义刚度
nominal risk area 名义危险区
nominal scale 主比例尺;类别尺度;额定值;标称比例尺
nominal scanning line width 标称扫描行宽

nominal screen aperture 公称筛孔;标称筛孔
nominal screen size 标称筛分粒度
nominal sealing line 名义密封线
nominal section 模拟传输线段;标称部分
nominal service condition 额定使用条件
nominal shear(ing)stress 标称剪应力;额定剪应力
nominal shipper 名义托运人
nominal short-circuit voltage 额定短路电压
nominal short-circuit capacity 额定短路容量
nominal significance level 名义显著性水平
nominal size 公称细度;公称粒径;公称尺寸;名义尺寸;毛尺寸;标称尺寸
nominal size of mineral aggregate 矿物集料的标称细度;矿物骨料的标称细度
nominal size of pipes 管材的公称尺寸
nominal size reinforcement 钢筋标称尺寸
nominal slip 公称滑距
nominal speed 公称速度;名义车速;常见正常车速;额定速率;额定速度;标称速度
nominal standard 公称标准;额定标准
nominal steel 构造筋;副筋;非受力钢筋
nominal stiffness 公称劲度;标称劲度;名义劲度;额定劲度
nominal strain 公称应变;名义应变
nominal strength 公称强度;名义强度;额定强度;标称强度
nominal stress 公称应力;名义应力;标称应力
nominal stuffing rate 标称塞入率
nominal supply voltage 额定供电电压
nominal system voltage 标称系统电压
nominal tariff 名义税
nominal tariff rate 名义关税率
nominal tax 名义上的税收
nominal temperature 标称温度
nominal tensile stress 名义拉(伸)应力;标称拉应力
nominal terms 名义条款
nominal thickness 名义厚度;虚厚度;公称厚度;额定厚度;标称厚度
nominal thickness lumber 标称厚度木材
nominal throughput 额定输送能力
nominal time 标称时刻
nominal torque 额定转矩
nominal torsional strength 名义抗扭强度
nominal transaction 名义交易
nominal transformer ratio 标称变压比;标称变换系数
nominal transmittance 标称透射率
nominal tube wall thickness 标称管壁厚
nominal twist 公称捻度
nominal uninterrupted output 额定连续运转产量
nominal upper and lower cutoff frequencies 名义上下限截止频率
nominal value 公称值;名义值;票面价值;额定值;标称值
nominal value of shock pulse 冲击脉冲名义值
nominal variables 名义可变因素
nominal velocity 标称速度
nominal voltage 公称电压;额定电压;标定电压;标称电压
nominal volume 面值体积;额定容积;标称体积;标称容积
nominal wage bill 名义工资总值;名义工资额
nominal wages 以金钱为准的工资;名义工资
nominal wages indexes 名义工资指数
nominal wall thickness 公称壁厚
nominal water-cement ratio 名义水灰比
nominal wavelength 标称波长
nominal weight 公称重量;面值重量;额定重(量);标值重量;标称重量
nominal weight of drill pipe 钻杆名义重量
nominal white signal 额定白信号;标称白电平信号
nominal width 名义幅度
nominal withstand voltage 标称耐压值
nominal working area 标称工作面积
nominal working hours 名义工作时间
nominal working pressure 公称工作压力
nominal working stroke 公称压力行程
nominal work week 名义工作周时间
nominal yardage 公称支数
nominal yield 名义收益(率);名义获利;名义产量
nominal yield stress 名义屈服应力
nominal zero point 标称零点
nominate 推荐

nominated agency 指定代理
nominated compacted thickness 标准压实厚度
nominated contractor 指定承包者;指定承包商;指定承包人
nominated design level 标准设计标高;标准设计高程
nominated design profile 标准设计断面
nominated sub-contract 指定分包合同
nominated sub-contractor 指定分包商;指定分包人
nominated supplier 指定供应商
nomination 任命(权);提名
nomination of arbitrator 指定仲裁人
nominator 分母
nominee 被指定人;被提名人
nominee holding 名义上持股
nominees 被提名者
nomogenic migmatite 均质混合岩
nomogram 尺解图;列线图解;列线图;计算图(表);图解图样;算图;诺谟图
nomogram 线示图
nomograph 诺谟图;列线图解;列线图;图解图样
nomograph alignment chart 解表;诺谟图;图算法
nomograph(chart) 算图
nomographic(al)chart 列线图;诺谟图
nomographic(al)method 算图法
nomographic(al)methods of computation 列线图计算法
nomography 线解法;列线图解术;列线图解法;列线图解;计算图表学;诺模图法;图算法
no money down 买主不付定金全部靠贷款的房地产交易
no more credit 不增加信用贷款
nomotron 开关电子管;计数管
nom rock ratio 名义摇臂比
non-abrasive finishing 非研磨加工
non-abrasive material 非磨蚀性材料
non-abrasive quality 非磨蚀性
non-abrasive rocks 非磨蚀性岩类
non-absorbency 不吸水性
non-absorbent 无吸收能力的;非吸收剂;不吸收性的;不吸收的
non-absorbent finish 不吸收性处理
non-absorbent material 非吸收(性)材料
non-absorbent medium 不吸收性介质
non-absorbent surface 不吸水性的表面;无吸水性的表面
non-absorbing 不吸收的
non-absorbing coating 非吸收膜;非吸收镀层
non-absorbing film 非吸收膜
non-absorbing medium 非吸收(性)介质
non-absorbing reflector 不吸收反射体
non-accelerating flow 非加速流
non-acceptable item 不能报销项目
non-acceptance 不接收;不验收;拒绝接收;不认付;不接受承兑;不承兑;拒绝兑付
non-acceptance usance letter of credit 不承兑远期使用证
non-accepting state 非接受状态
non-accessible flat roof 不上人平屋顶
non-accessible internal register 不可取数内部寄存器
non-accessible lot 禁区
non-accessible roof 不通行屋顶
non-accessibe traffic 非进入交通
non-accessible trench 不通行地沟
non-accrual assets 非应计资产
non-accumulated sinking fund 非积累偿债基金
nonachlorobiphenyl 九氯联苯
non-acid 非酸性的;不含酸类的
non-acid fast 非抗酸性
non-acid gases 非酸性气体
non-acidic 无酸的
non-acidics 非酸性物
non-acid oil 非酸性油
non-acid wash 非酸洗;非酸蚀
non-acosane 二十九烷
non-acosanol 二十九醇
non-actinic glass 非光化玻璃
non-action message 非活动信息;不要干预信息
non-activated 未活化的
non-activated stock 未活化的胶料
non-activated thinner 非活性稀释剂
non-active 无活动的;稳定的
non-active diluent 非活性稀释剂

non-active part of vane 叶片的失效部分
non-active redundancy 非工作储[贮]备
non-adaptability to machine 不适于机械
non-adaptive investigation delay policy 不适合调查延迟策略
non-adaptive search plan 非适应搜索方案
non-add 非加;非附加
non-additive 无添加剂的;非求和的
non-additive effect 非加电效应
non-additive oil 无添加剂润滑油
non-additivity 非相加性;非加和性的;不可加性
non-addressable memory 不可寻址存储器
non-address memory 无地址存储器
nonadecane 十九烷
nonadhering rust 浮锈
non-adhesive 无黏着力的;无黏性的
non-adiabatic 非绝热的
non-adiabatic change 非绝热变化
non-adiabatic cooling 非绝热冷却
non-adiabatic correction 非绝热校正
non-adiabatic heating 非绝热增温;非绝热供暖
non-adiabatic humidifier 非绝热加湿器
non-adiabatic nozzle 非绝热喷管
non-adiabatic process 非绝热过程
non-adiabatic pulsation 非绝热脉动
non-adiabatic reactor 非绝热反应器
non-adiabatic rectification 非绝热精馏
non-adiabatic transition 非绝热变化
non-adiabatic wind profile 非绝热风速廓线
non-adjustable 不可调整的;不可调的
non-adjustable cam 不可调凸轮
non-adjustable conductor rail support 不可调导轨支撑
non-adjustable orifice 不可调节流孔
non-adjustable restrictor 固定节流孔;不可调节流阀
non-adjustable rubber bushing 不可调整的橡胶衬套
non-adjustable signal 固定标尺
non-adjustable simplex pull rod 固定式单拉杆
non-adjustable wrench 固定扳手
non-admitted assets 未核准资产;非受保资产
non-adsorbable 不吸附的
non-adsorbable fraction 不可吸附百分率
non-adsorptive support 非吸附性载体
non-aerated area 非充气面积
non-aerated burner 光焰燃烧器;扩散式燃烧器
non-aerated flame 无空气火焰
non-aerated flow 不掺气水流
non-aerated flow region 非复氧水流区
non-aeration time 非复氧时间
non-affiliated company 非联系公司
non-affine 非均匀的
non-affine deformation 非仿射变形;不均匀变形
non-ag(e)ing 经久的;未时效;未失效的;非老化的;不老化的
non-ag(e)ing steel 无时效钢
nonagenary 九十进制
non-agency station 无人管理的车站
nonagesimal 黄道地平最高点
non-agglomerating 不黏结的
non-aggregated dye 不聚集染料
non-aggressive 非侵蚀性的
non-aging treatment 无时效处理
non-agitating truck 不搅拌车;无搅拌装置的混凝土运料车;途中不搅拌混凝土运送车
non-agitating unit 装载已搅拌好混凝土的卡车;无搅拌装置的混凝土运料车;不搅拌车
non-agitator truck 无搅拌装置的混凝土运料车
nonagon 九边形
non-agricultural activity 非农业生产(活动)
non-agricultural employment 非农业雇佣
non-agricultural nation 非农业国家
non-agricultural season 农闲季节
non-agricultural sector 非农业部门
non-agrobased industry 非农基地工业
non-air blasting process 非空气喷沙处理法
non-air-entrained concrete 无加气剂混凝土;非气混凝土;非加气混凝土;不加气混凝土
non-air-entraining cement 非引气水泥
non-alcoholic beverages 非酒精性饮料
non-algal biogenus matter 非藻生物类物质
non-alignment diagram 非列线图
non-alkali glass 无碱玻璃

non-alkali glass fiber 无碱玻璃纤维
non-alkaline hardness 非碱性硬度
non-alkaline scale 非碱性垢
non-alloyable metal 不能成合金的金属
non-all-phase operation 非全相运行(状态)
non-alphameric sign 非字母数字符号
non-alternant hydrocarbon 非更迭烃
non-ambiguity 单值性
non-ambulatory disability 非移动性损伤
non-amino nitrogen 非氨基氮
non-analytic(al) function 非分析函数
non-analytic(al) job evaluation 非分解式作业评价
nonanedioic acid 壬二酸
non-animated picture 静止图像
non-anionic polyeletrolyte 非阴离子型聚电解质
nonanol 壬醇
non-anticipating 非预测的
non-anticipating channel 非预测信道
non-anticipative 非预报的;不是预先考虑的
nonanticipatory system 因果系统
non-apatite inorganic phosphorus 非磷灰石有机磷
non-appropriation fund 禁止挪拨基金
non-aqueous 无水的;非水的
non-aqueous adhesive 非水溶液(型)黏合剂
non-aqueous catalytic thermal titration 非水催化热滴定
non-aqueous colloid 非水胶体
non-aqueous dispersion 非水分散体
non-aqueous dispersion acrylic coating 非水分散丙烯酸涂料
non-aqueous dispersion resin 非水分散树脂
non-aqueous dispersion thermosetting acrylics 非水分散型热固性丙烯酸树脂
non-aqueous electrochemistry 非水溶液电化学
non-aqueous fluid 不含水流体
non-aqueous indicator 非水指示剂
non-aqueous leaching 非水溶液浸出
non-aqueous medium 非水介质
non-aqueous phase 非水相
non-aqueous phase organic liquid 非水相有机液体
non-aqueous polyurethane 非水溶性聚氨脂
non-aqueous pyroprocessing plant 无水高温处理厂
non-aqueous solution 非水溶液
non-aqueous solution plating 非水溶液电镀;非水溶液镀层
non-aqueous solvent 非水溶剂
non-aqueous titration 非水溶液滴定;非水溶剂滴定法;非水滴定(法)
non-Arab financial and monetary institutions 非阿拉伯的金融货币机构
non-arc downwarped extracontinental basin 非弧型陆外下陷盆地
non-arithmetic iperation 非算术运算
non-arithmetic shift 循环移位;非算术移位
non-arithmetic statement 非算术语句
non-arm's length allocation 非公平独立核算的分摊
non-aromatic 非芳香性的
non-aromatic hydrocarbon 非芳香烃
non-artesian aquifer 无压含水层;非自流含水层
non-artesian groundwater 非自流地下水;非承压地下水
non-artesian springwater 非自流泉水
non-artesian water 非自流水
non-artesian well 非自流井
non-articulated arch 无铰拱
non-articulated loco motive 非关节机车
non articulated streamline train of light weight 自重型非关节式流线型机车
non-asphalt(ic) 非沥青(质)的;非地沥青的
non-asphaltic base oil 非沥青基石油;非地沥青基石油
non-asphaltic petroleum 无地沥青石油;非沥青基石油;非地沥青基石油
non-asphaltic pyrobitumen 非沥青质焦性沥青;非地沥青质焦性沥青
non-asphaltic road oil 非沥青道路油;非沥青(铺)路油;非地沥青(铺)路油
non-assented 非同意的
non-assisted access 非加速存取
non-assisted urban renewal project 无资助城市更新项目

non-associated liquid 非缔合液体
non-associated natural gas 不缔合天然气;非伴生天然气
non-associated signalling 非对应信号方式
non-associating 不缔合的
non-association cable 非标准电缆
non-associative flow rule 非关联流动法则;不相适应流动法则
non-associative theory of plasticity 非相关塑性理论
nonatomic measure space 无原子测度空间
non-atomized water 非雾化水
non-attainment area (空气污染)不合格区;超标区
non-attendance 旷工
non-attendant service 不出席工作;不到桌旁服务的;不出席服务;无人服务(商店)
non-attended light 自动灯
non-attended remotely controlled power station 无人值班的遥控电站
non-attended station 无人值守站;无人值守台;无人(观)测站
non-attended substation 无人值班的变电所;无人变电所
non-attenuating 无衰减
non-attenuating forward wave 无衰减前向波
non-attenuating wave 无衰减波;等波幅;非衰减波
non-audit client 非审计客户
non-audit services 非审计服务
non-auto 非机动车
non-autocorrelated random disturbance 非自相关随机扰动
non-autocorrelation 非自相关
non-autographic signature 非亲笔签字
non-automatic 非自动的
non-automatic block district 非自动闭塞区段
non-automatic extraction turbine 非自动调节抽气式汽轮机
non-automatic rectifier 非自动纠正仪
non-automatic sprinkler system 非自动喷水灭火系统;非自动洒水灭火系统
non-automatic switching 非自动转换
non-automatic system 非自动系统
non-automatic tripping 非自动跳闸
non-automobile 非机动车
non-autonomous system 非自主系统;非自控系统
non-auto-oriented 面向非机动车的;非机动车为主的
non-autoregression 非自回归
non-available 无供应
non-available goods 无现货
non-axial 非轴向
non-axial force 非轴向力
non-axial geocentric dipole field 非轴向地心偶极子场
non-axiality 不同轴性
non-axial mode 轴外模
non-axial stress 非轴向应力
non-axial trolley 旁接滑轮
non-axisymmetric(al) 非轴对称的
non-axisymmetric(al) configuration 非轴对称组态
non-axisymmetric(al) element 非轴对称构件;非轴对称单元;非轴对称部件
non-azeotropic mixture refrigerant 非共沸溶液制冷机
non-back fund-raising institutions 非银行融资机构
non-backing process 无背衬焊接法
non-baking coal 不结焦煤
non-balanced pump 不平衡泵
non-balanced valve 不平衡阀;不平衡的阀
non-banded coal 非条带状煤;非条纹煤
non-banker's credit 非银行信用证
non-bank financial intermediaries 非银行金融中介机构
non-banking concerns 非银行机构
non-banking financial institutions 非银行金融机构
non-banking institutions 非银行机构
non-bank private sector 非银行私人部门
nonbaric diving system 常压潜水系统
non-barrier-coastal sequence 无障壁海岸层序
non-base variable 非基本变量
non-base wash 非碱洗
non-basic industry 非基础工业
non-basic port 非基本港口
non-basic sector 非基础部门

non-basic variable 非基本变量
non-bearing 不承重的
non-bearing block 非承重砌块
non-bearing brick 非承重砖
non-bearing concrete 非承重混凝土
non-bearing exterior wall 非承重外墙
non-bearing facade 非承重立面
non-bearing fixing 非承重固定件
non-bearing floor block 非承重地板块
non-bearing masonry wall 非承重砌体墙
non-bearing partition 非承重隔墙;非承重隔断;不载重墙壁
non-bearing structure 非承重结构;非承载结构;不载重结构
non-bearing surface 非支承面
non-bearing tillering 无效分蘖
non-bearing tree 不结实树
non-bearing wall 非承重墙;不载重墙
non beneficiaries 非受益者
non-bent-up reinforcing bar 非上弯的钢筋;未弯起的钢筋
non-Bessemer iron 非贝氏铁
non-biding clause 非约束性条款
non-binary code 非二进制代码
non-binary switching theory 多值开关(理)论
non-biodegradability 无生物降解性;不能生物降解;不可生物降解性
non-biodegradable 无生物降解能力的;不可生物降解的
non-biodegradable chlorinated hydrocarbon solvent 不可生物降解的氯代烃类化合物溶剂
non-biodegradable material 不能生物降解物质;不可生物降解物质
non-biodegradable organic pollutant 不可生物降解有机污染物
non-biodegradable organic substance 非生物降解有机物
non-biodegradable residue 不能生物降解残留物;不可生物降解残留物
non-biodegradable substance 非生物降解物质;不易生物降解的物质;生物不能降解的物质
non-biogenic deposit 非生物沉积
non-biological decomposition 非生物分解
non-biological method 非生物法
non bio-remediation 非生物修复
non-bituminous 非沥青质的;非沥青的;非地沥青的
non-bituminous material 非沥青质材料
non-bituminous treatment 非沥青质处治;非沥青质处理
non-black body 非黑体
non-black body radiation 非黑体辐射
non-blank 非空的
non-blank symbol 非空符号
non-bleeding 不渗色
non-bleeding grease 不离浆的润滑油
non-bleeding pigment 不渗色颜料
non-blended cement 纯水泥
non-blind cloth 防推筛布
non-blinding screen 不堵塞筛面
non-bloated 无胀性的
non-bloating 不膨胀的
non-blocking 无阻塞
non-blocking access 不中断通路
non-blocking coating 不粘连涂层
non-blocking network 无阻塞网络
non-blocking switch 非封锁转接设备
non-blooming 无荧光的
non-blowing 非涌水的;非鼓风的
non-blushing thinner 防潮剂;防白药水
non bogie coach 无转向架客车
non-boiling spring 非沸泉
non-bond 无联结
non-bonded joint 非胶结节点;非胶结榫;非胶结点;无灰缝砌缝
non-bonded prestressed reinforcement 不黏着预应力钢筋;无黏结预应力钢筋
non-bonded tendon 不黏着力筋
non-bonding 非键合
non-bonnet 无压盖
non-boosting operation 不增压越站
non-bore-safe 非膛内保险
non-bore-type tsunami 非怒潮型海啸
non-borrowed reserves 非借入储备
non-breakable bond 不易破裂的黏合;不易破裂

non-breakable glass 不破碎玻璃;刚化玻璃;刚性玻璃
non-breaking length 固定区长度
non-breaking wave 非破碎波;不碎波;不间断波
non-breaking zone 固定区(域)
non-break oil 无沉淀物的油
non-breathing 不通气
non-bridging contact 非桥接接点
non-bronze blue 非青铜蓝;无铜光铁蓝
non-bronze blue pigment 非青铜蓝颜料
non-browning glass 防辐射玻璃;不变棕色玻璃(辐射后);不变色玻璃;不暗化玻璃
non-browning micro objective 不发黄显微物镜
non-budgetary unit 非预算单位
non-buffered case 非缓冲情况
non-building project 非建筑项目
non-built up area 非围建地区;不建满房屋的地区
non-bulk liner 非散货班轮
non-buoyant circular turbulent jet 非浮力圆形紊动射流
non-burning 不燃烧的
non-business 非企业
non-business accountancy 非商业会计
non-business days 非营业天数
non-business expenditures 营业外支出
non-business expenditures audit 营业外支出审计
non-business expenditures statement 营业外支出表
non-business expense 非营业费用
non-business income 营业外收入
non-business itemized deductions 非营业扣除项目
non-business property 非营业资产;非营业财产
non-business revolving fund 非营业周转性基金;非营业循环基金
non-busy condition 非忙状态
non-busy waiting strategy 非忙碌等待策略
non-caked 不黏结的
non-caking 不结块的
non-caking black 不结块的炭黑
non-caking coal 非黏结煤;不粘煤;不黏结性煤;不黏结块的煤;不结焦煤
non-calcareous alluvial soil 非石灰性冲积土
non-calcareous cement 无钙水泥;无钙胶结料
non-calcic brown soil 非石灰性棕壤;非钙质棕色土
non-callable 不可赎回
non-Canadian ship compliance certificate 非加拿大船舶合格证书
non-cancellable lease 不可取消的租约;不可解除租约;不可废止租赁
non-canning antenna 固定天线
non-capillary 无毛细的
non-capillary porosity 非毛细孔隙
non-capillary water 非毛细管水
non-captive berth 非专用泊位
non-captive container 非限定集装箱
non-captive quay 非专用码头
non-captive terminal 非专用码头;公用码头
non-captive wharf 非专用码头
non-carbonaceous matter 非碳水化合物质
non-carbonate buffer system 非碳酸盐缓冲系
non-carbonate hardness 非碳酸盐硬度
non-carbon black filler 非炭黑填料
non-carcinogen health risk 非致癌健康风险
non-carcinogen pollutant 非致癌污染物
non-carrier multimodal transport operator 非承运人的联运人
non-cartridge double mechanical seal 非盒式双端面机械密封
non-cartridge single mechanical seal 非盒式单端面机械密封
non-carved waterspout 不掏槽的排水口;不掏槽的水落管
non-cash 非现金(的)
non-cash acquisition of subsidiary stock 以非现金购得子公司股本
non-cash assets 非现金资产
non-cash charges 非现金支出;非现金费用
non-cash circulation 非现金流通
non-cash credit 非现金信用贷款
non-cash documents 非现金证券;非现金单据
non-cash entries 非现金项目
non-cash expenses 非现金支出的费用
non-cash input 非现金投入
non-cash issuance of stock 非现金发行的股票
non-cash items 非现金项目
non-cash outlay 非现金支出;非现金取得的资产
non-cash settlement 非现金结算
non-casing rope percussion-grab drill 无套管绳索冲抓钻机
non-catalysis method 非催化法
non-catalytic car 非催化剂净化废气汽车
non-catalytic process 非催化过程
non-cataracting ball mill 非瀑落式球磨机
non-causative fault 非发震断裂;非发震断层;非地震断层
non-cavitation flow 无汽蚀水流
non-cellulosic fiber 非纤维素纤维
non-cellulosic material 非纤维质材料
non-cementation 无胶结
non-cemented 非黏接的
non-cementious matrix 非水泥类基体材料
non-census 非普查
non-centering telemotor 非定心式遥控电动机
non-central chi-square distribution 非中心 X_2 分布
non-central confidence interval 非中心置信区间
non-central conic 无心二次曲线
non-central distribution 非中心分布
non-central F distribution 非中心 F 分布
non-central field of gravitation 非中心引力场
non-central force 非中心力
non-centrality parameter 非中心性参数;非中心参数
non-centralize control system 非集中控制方式
non-centralized communication system 非集中式传输系统
non-centralized interlocking 非集中联锁
non-central load 偏心荷重;偏心荷载;偏心负载
noncentral quadric 无心二次曲面
non-central t distribution 非中心 t 分布
non-central Wishart distribution 非中心威沙特分布
non-central X^2 distribution 非中心 X^2 分布
non-centrosymmetrical 非中心对称的
non-centrosymmetric(al)crystal 非中心对称晶体
non-centrosymmetric(al) projection 非中心对称投影
non-ceramic tile 非瓷砖;非陶瓷面砖;非陶瓷瓦
non-certified 企业标准的
non-chalking 非白垩处理;不起霜;非白垩化
non-chalking synthetic(al)resin paint 耐白垩化合成树脂涂料
non-changeover system 非转换系统
non-characteristic curve 非特征曲线;非特性曲线
non-characteristic product 非特征产品
non-chargeable component 非应付费用部分
non-check point 无检核点
non-check position 无检核点
non-check station 无检核站
non-chemical environmental factor 非化学环境因素
non-chemical means of pesticide control 非化学农药控制方法
non-chemical system 非化学方法
non-chloride accelerator 无氯早强剂
non-chloride antifreezing admixture 无氯防冻剂
non-chlorinated insecticide 无氯杀虫剂
non-chlorine-retentive 不吸氯的
non-chlorodisinfectant 无氯型消毒剂
non-chokeable 不堵塞的
non-choking jaw plate 不堵塞颚板
non-choking nozzle 不堵塞喷嘴
non-chromatic 无色彩的
non-chromatographic identification 非色谱法鉴定
non-chromogenic bacteria 非产色菌
non-circle serial correlation 非循环序列相关
non-circuit edge 非回路边
non-circular 非圆形的
non-circular analysis 非圆弧滑动分析;非圆弧分析法
non-circular cylindric(al) structure 非圆形的筒体结构
non-circular head 非圆形封头
non-circularity 不圆度
non-circularity of cladding 包层不圆度
non-circularity of reference surface 基准面不圆度
non-circular pulley 不圆滑车
non-circular shaft 非圆形轴
non-circular statistic 非循环统计量
non-circular symbol 非循环符号;不再现符号
non-circular vessel 非圆形容器
non-circulating lubrication 非循环润滑作用
non-circulating lubrication(method)非循环润滑法
non-circulating water 静水;非循环水;非流动水
non-circulatory 非循环的
non-circulatory flow 非循环水流;非环状流;非环流
non-clad fiber 无包层纤维
non-clad optic(al)fiber 无包层光纤
non-claim 未在规定时间内提出(索赔)要求
non-clashing gear set 同步器变速箱;常啮合齿轮变速器
non-classical 非经典的
non-classical carbonium ion 非经典碳正离子
non-classical shell problem 非经典薄壳问题
non-clastic 非碎屑的
non-clastic sediment 非碎屑沉积
non-clastic sedimentary rock 非碎屑水成岩;非碎屑沉积岩
non-clathrate 密筛孔状
non-clay block 非黏土块
non-clay drilling fluid 无黏土钻进冲洗液
non-clay mineral 非黏土矿物
non-clear directive variation 不明显方向变化
non-clogging 不堵塞的
non-clogging centrifugal pump 不阻塞离心泵
non-clogging impeller 非堵塞叶轮;不阻塞的叶轮
non-clogging pump 污水泵;不淤塞水泵
non-clogging screen 不会堵塞的筛
non-clog hammer mill 防堵锤式破碎机
non-clog type of pump 不淤塞水泵;污水泵;不阻塞水泵
non-closed system 非封闭系统
non-coal-bearing strata 非含煤地层
non-coastal state 无海岸国家;非沿海国家
non-coaxial 非同轴的
non-coaxial progressive deformation 非共轴递进变形
non-coded graphic 非编码图形
non-coded graphics 非编码图形学
non-coded information 非编码信息
non-code system 非号码制统计货车停留时间计算法
non-coherence array 非相干组合
non-coherent 非黏性的;不附着的;松散的;不凝聚的;无黏聚力的;非相干的;不黏结的;不凝聚;不共格的
non-coherent carrier 非相干载波
non-coherent carrier noise 非相干载波噪声
non-coherent detection 非相干探测
non-coherent Fourier-optical imaging system 非相干傅立叶光学成像系统
non-coherent infrared detector 非相干红外探测器
non-coherent infrared receiver 非相干红外接收器
non-coherent integration 非相干积分
non-coherent light-ranging 非相干光测距
non-coherent optic(al)carrier 非相干光学载波
non-coherent optic(al)computer 非相干光学计算机
non-coherent optic(al)detector 非相干光探测器;非相干光检测器
non-coherent optic(al)processing system 非相干光处理系统
non-coherent radiation 非相干辐射
non-coherent radiator 非相干辐射器
non-coherent re-emission 非相干再发射
non-coherent rotation 非一致转动
non-coherent scattering 不相干散射
non-coherent soil 无黏聚力土壤
non-coherent synchrotron radiation 非相速辐射
non-coherent twin boundary 非相干李晶间界
non-cohesive 无黏性的;无黏聚性的;无黏聚力的;非黏性的;非黏结性的;不黏结的;不附着的
non-cohesive alluvium 无黏性冲积层
non-cohesive fluid 无黏性流体
non-cohesive gel 不黏胶
non-cohesive material 松散材料;非黏性材料;无黏性材料
non-cohesive sediment 无黏性泥沙
non-cohesive soil 松散岩土;无黏性土;无黏结性土;砂性土;非黏性土(壤)
non-cohesive structure 非黏着结构
non-coincidence 不重合

non-coincident demand 非同时需量
non-coking 不结焦
non-coking coal 不结焦的煤;干烷煤;非炼焦煤
non-colinear point 非共线点
non-collapsible container 固定型集装箱;非折叠式集装箱
non-collapsible loess 非湿陷性黄土
non-collapsing soil 非湿化性土
non-collection analysis 非聚集分析
non-collinearly phase-matched 非线相位匹配的
non-collinear vectors 非共线矢量
non-collision plate boundary system 非碰撞板块边界系统
non-colloidal silt 非胶性粉砂
non-colo(u)r 原色(的)
non-colo(u)r-sensitized 非色敏化的
non-combustibility 非燃性
non-combustible 防火的;不可燃的;非燃物;非烧物;不燃物;不燃(的)
non-combustible blind fence 不易燃烧的封闭围墙
non-combustible building 防火建筑;不易燃烧的建筑物
non-combustible component 不燃体
non-combustible construction 非燃性构造;防火结构;不燃烧结构;不燃结构
non-combustible constructional material 不燃性建筑材料;耐火建筑材料
non-combustible fabric 不燃性织物
non-combustible floor 不易燃地板
non-combustible floor with suspended ceiling 有吊平顶的不易燃楼层
non-combustible ground sign 防火地面标志;不燃烧地面标志
non-combustible insulation material 不燃绝缘材料
non-combustible lining 不燃支架
non-combustible material 不燃性材料;非燃性材料;不燃性材料;不燃物质;不燃(烧)材料
non-combustible matter 不燃物质
non-combustible membrane structure 耐火薄膜结构
non-combustible mo(u)lded asbestos cement panel 防火型石棉水泥板
non-combustible refuse 不燃性垃圾
non-combustible screen fence 不易燃烧的网式篱笆
non-combustible slab floor 不易燃楼板
non-combustible stair(case) 不易燃楼梯(间)
non-combustible substance 不燃烧物质;不燃物
non-combustible sulfur 不可燃硫
non-combustible uncovered floor 不燃的无覆盖地板;裸露的防火地板
non-comforming article 不合格品
non-command language input 非命令语言输入
non-command terminal 非命令型终端
non-commercial 非商业性的;非贸易
non-commercial account 非贸易账户
non-commercial agreement 非商业性协议
non-commercial call 非商用通话
non-commercial contracts 非商事合同
non-commercial energy 非商业性能源
non-commercial enterprise 非营利事业
non-commercial expenses 非商业储备
non-commercial fuel 非商品燃料
non-commercial recreation 非营业性游乐场
non-commercial reserves 非商业储备;非工业性储量
non-commercial thinning 非商业性疏伐
non-commercial traffic 无货车交通;非商业性交通
non-commercial well 无经济价值油井;非工业性产油井;非商业性油井
non-commissioned ship 非现役船
non-commodity 非商品
non-commodity exchange of products 非商品的产量交换
non-commodity flow 非商品流量
non-commodity sale 非商品销售
non-common section 非共同段
non-communicating layer 不连通层
non-communicating pore 不连通孔隙
non-community water supply 非公共给水
non-commutability 不可对易性
non-commutative law 非交换律
non-commutative ring 非可换环
non-comparability 非可比性

non-comparable 非可比数据
non-comparable data 无相比性资料;非可比资料;不可比数据
non-comparable elements 非可比元素
non-comparable factors 不可比因素
non-compatibility 不兼容性;非兼容性
non-compatible waste 非压实性废物
non-compelled signalling 非互控信号方式
non-compensable 不能补偿的
non-compensated acceleration 未被平衡的加速度
non-compensated basin 非补偿盆地
non-compensated centrifugal acceleration 未被平衡的离心加速度
non-compensated centrifugal force 未被平衡的离心力
non-compensated magnetic vehicle detector 非补偿式电磁车辆检测器
non-compensated pump 不平衡泵
non-compensated value date 到期履交割日
non-compensated valve 不平衡阀
non-competing adsorbate 非竞争吸附质
non-competing groups 非竞争集团
non-competing retailers 非竞争零售商
non-competition agreement 非竞争协议
non-competition clause 不竞争条款
non-competitive 非竞争性的;不具竞争性
non-competitive adsorption 非竞争性吸附
non-competitive bid(ding) 非竞争性招标;无竞争能力的报价;非竞争性投标;无竞争的出价
non-competitive inhibition 非竞争性抑制
non-competitive price 无竞争力的价格;非公开招标的价格
non-competitive system 非竞争性系统
non-competitive traffic 无竞争能力的交通业务
non-complemented form 非余式
non-completion 不能如期完工
non-compliance 不遵守;不顺从;不履行;不服从
non-compliance of tax law 不遵守税法
non-compliance penalty 违章罚款
non-compliant technology 不符合环保要求的工艺
non-compostable refuse 不能堆肥的垃圾
non-compressed asbestos-cement panel 未加压的石棉—水泥板
non-compressed fiberboard 未受压的纤维板;非压缩纤维板
non-compressible soil 不可压缩性土
non-computational 非计算式的
non-concealed assets 非隐匿资产
non-concentrating solar photocatalytic oxidation reactor 非聚日光光催化氧化反应器
non-concentricity 不同心度
non-concessional official lending 非优惠性官方贷款;非减让性官方贷款
non-concordant cable 非吻合索;非同位钢索;不吻合素
non-concordant tendons 非吻合钢筋束
non-concrete surface 混凝土墙面;混凝土粗面
non-concurrent forces 非集中一点的力;非共点力
non-concurrent insurance 非同时保险
non-concurrent point 非共点
non-concurrent prospective study 非同时性前瞻研究
non-concussive tap 非触发式接头
non-condensable 不凝结的
non-condensable gas 非凝性气体;不凝性气体;不凝气;不可凝气体
non-condensable gas purger 不凝性气体净化器
non-condensed aromatic 非稠相芳烃
non-condensed polycycloalkane 未稠合的多环环烷烃
non-condensible gas 不凝气体
non-condensing 无冷凝的;不凝结;不凝的
non-condensing automatic extraction turbine 背压调节抽气式汽轮机
non-condensing bleeder turbine 背压式抽气汽轮机
non-condensing engine 排汽蒸汽机;不凝气性蒸汽机;排汽蒸汽机
non-condensing roofing 不凝汽屋面(材料)
non-condensing steam engine 不凝结蒸汽机;背压式蒸汽机
non-condensing turbine 不凝气式汽轮机
non-conditional simulation 非条件模拟
non-conducting 不导电的;不传热的;绝缘的

non-conducting material 不导电性材料
non-conducting medium 非导电媒质
non-conducting nose cone 不导热头锥
non-conducting state 非传导态
non-conducting voltage 截止电压
non-conduction interval 不导通间隔
non-conductive 非导体的
non-conductive area 不导电区
non-conductive coating 绝缘涂层
non-conductive extinguishing agent 不导电灭火剂
non-conductive mineral 非导体矿物
non-conductive toner 不导电型显影剂
non-conductivity 非延性
non-conductor 不良导体;绝缘材料;电介质;非导体;绝缘体
non-conferece carrier 非工会成员的运输船
non-conference 非运费同盟
non-conference line vessel 非同盟定期船
non-conference operator 非同盟业者
non-confined 无侧限
non-confined compression test 无侧限压缩强度试验;无侧限抗压强度试验
non-confined concrete 无侧限混凝土
non-confirmed L/C 不保兑信用证
non-confocal 非共焦
non-confocal resonator 非共焦共振腔
non-conform 不符合
non-conformable 非整合的;不整合的
non-conformal 非保角的
non-conformally graded aggregate 不匀级配集料;不匀级配骨料
non-conformal surface 异曲表面
non-conformance 与规范不符
non-conforming 不合格;不符合;不符规范;非相容;不适当使用
non-conforming imputation 不一致的分配
non-conforming load 不一致负载
non-conforming method 非相容方法
non-conforming shape 非整合形状
non-conforming shape factor 非保形因素
non-conforming use 违章使用;不符合规定使用;不符规定使用
non-conforming work 不合格工程;不符合同工程;不合契约的工作
non-conformity 非整合;不整合;异岩性不整合
non-conformity of sand 砂的不均匀性
non-conformity of the pivots 轴颈不一致;轴颈不等
non-conformity spring 不整合泉
non-congealable oil 抗冻油;防冻油;不冻结油
non-congealing paste 不冻膏
non-coincidence of shipping documents and luggage or parcel 票货不符
non-conjugated fatty acid 非共轭脂肪酸
non-connected storage 非连接存储(器)
non-connection 不接
non-conservation 不守恒(性)
non-conservation of energy 能量不守恒
non-conservative concentration 非恒定浓度
non-conservative creep 非保守蠕变
non-conservative design 非保守设计
non-conservative force 非保守力
non-conservative motion 非保守运动
non-conservative pollutant 非恒定污染物;非持久性污染物
non-conservative problem 不守恒问题
non-conservative process 非守恒过程
non-conservative system 非守恒系统
non-conservative waste effluent 非保持性废水
non-conservative water quality 非恒定水质
non-conserve waste effluent 非保存性废水
non-conserve waste gas 非保存性废气
non-consolidated deposit 未固结沉积物
non-consolidated sediment 未固结沉积物
non-consolidated track 未压实轨道
non-constant 不恒定
non-constant cost 非固定成本;非固定费用
non-constant disturbance variances 非常数扰动方差
non-constant significance level 非常数显著性水平
non-constant sum game 非常数和对策
non-constant variability 非常数变异性
non-constant-velocity universal joint 不等速的万向节

non-construction enterprise 非建筑企业
non-construction establishment 非建筑企业基层单位
non-construction unit 非建筑单位
non-constructive cement 低标号水泥；杂用水泥
non-consumable arc-melting process 非自耗电极电弧熔炼法
non-consumable electrode 非自耗电极；不熔化电极
non-consumable ingot 非自耗弧熔锭
non-consumable melting 非自耗电极熔炼
non-consumably melted method 非自耗熔炼法
non-consumble 非自耗的
non-consumption 非消费
non-consumption expenditures 非消费支出
non-consumptive 非消耗性的
non-consumptive use 非消耗性用途
non-consumptive use of water 非消耗性用水
non-contact 非接触的；不接触
non-contact attemperator 表面式减温器
non-contact face 不接触面
non-contacting fiber 不接触光纤
non-contacting laser device 非接触测量仪器激光装置
non-contacting pickup 无触点检波器
non-contacting piston 扼流活塞
non-contacting plunger 扼流活塞
non-contacting single point infrared sensor 非接触单点红外线传感器
non-contacting thickness ga(u)ge 无触点厚度计
non-contact instrument 非接触仪器
non-contact longitudinal recording 无触点纵向记录
non-contact magnetic recording 非接触式磁记录；非接触式磁记录
non-contact measurement 无接触量测法；无接触测量法
non-contact measuring device 非接触测量仪器
non-contact micrometer 非接触型千分尺
non-contact nuclear densimeter 不接触式核子密度计
non-contact operation 非接触运行
non-contact piston 非接触活塞
non-contact plot 非接触绘图
non-contact recording 无接触式记录；非接触式记录；无触点记录
non-contact rediation infrared hygrometer 不接触式红外辐射湿度计
non-contact scanning detector 非接触式扫描跟踪器
non-contact seal 非接触型密封；非接触式密封
non-contact sealed ball bearing 非接触式密封球轴承
non-contact surface contour analyzer 不接触式表面轮廓分析仪
non-contact thermometer 不接触式温度计
non-contact thickness ga(u)ge 非接触式测厚仪
non-contact track measuring system 非接触测量法
non-contact type 非接触型
non-contact vibration monitor 不接触式振动监测器
non-containerizable cargo 非集装箱货
non-contaminated atmosphere 未污染的大气
non-contamination automobile 无污染汽车
non-contamination pesticide 无污染农药
non-contemporaneous deposit 非同期沉积；不同期沉积
non-contemporaneous error 非同时发生的误差
non-contemporary track 非同时径迹
non-contentious 非争论性的
non-contest clause 非争议条款；非争议性条款
non-contiguous data 非邻接数据
non-contiguous item 独立项；非邻接项；非邻接数据项；非连续项
non-contiguous page 非邻接页
non-contiguous working-storage 独立工作存储器
non-contingent event 偶然事件
non-continuity 非连续性
non-continuous 不连续的
non-continuous footing 非连续基础
non-continuous footing analysis 非连续基础分析
non-continuous rolling mill 非连续式轧机
non-continuous series 非连续序列；非连续数列

non-contraband 非违禁品
non-contraband goods 非禁运品货物
non-contracting goods 非禁运品货物
non-contracting party 非缔约方
non-contract rate 非契约议价；非契约运费率
non-contractual claims 非契约索赔；非合同索赔；非合同的索赔
non-contractual documents 非契约性文件
non-contractual income 非契约收入；非合同收入；非合同的收益
non-contractual liability 非契约性责任；非合同性责任
non-contractual obligation 非合同上的责任或义务
non-contributing area 非汇水面积；非产流区；不产流区；不产流面积；闭流区
non-controllable 不可支配的
non-controllable cost 非控制成本；非可控成本；不可控制成本；不可控成本
non-controlled area 非控制区
non-controlled emergency 不可控制的紧急制动
non-controlled rectifier 非稳压整流器
non-controlled variable 非控制变量
non-control system 非控制系统；无控制方式
non-convective lake 无对流湖
non-conventional approaches 非常规手段
non-conventional bearing 非常规支撑
non-conventional colo(u)r photography 特殊彩色摄影
non-conventional design 非常规设计
non-conventional designed structure 非常规设计结构
non-conventional energy 特殊能源；非常规能源
non-conventional machining 特种加工
non-conventional ocean resources 非常规海洋资源
non-conventional oil 非常规石油
non-conventional photography 特殊摄影
non-conventional type 异型
non-convergent 非收敛的
non-convergent series 非收敛级数
non-convertiable currency 非自由兑换货币
non-convertible bonds 不可转换的债券
non-convertible paint 非转化型漆
non-convex programming 非凸规划
non-convex quadratic programming 非凸二次规划
non-cooperative 非合作的；不合作的
non-cooperative constant sum game 非合作常和对策
non-cooperative equilibrium 非合作性均衡
non-cooperative game 非合作对策
non-cooperative game-theoretic model 非合作对策论模型
non-cooperative n-person game 非合作 n 人对策
non-cooperative solution 非合作均衡；非合作解
non-cooperative target 非协作目标；非合作目标
non-cooperative Walras equilibria 非合作瓦尔拉斯平衡
non-coplanar 非交于一个平面的；非共面的
non-coplanar forces 空间力；非共面力；异面力；非共面力系
non-core bit 不取芯钻头
non-core(bore)hole 不取芯钻孔
non-cored footage 无岩芯进尺
non-core drilling 无岩芯钻进；不取芯钻进
non-coring bit 不取芯钻头
non-coring drilling 不取芯钻进
non-coring type diamond bit 无器芯式金刚钻头
non-corporeal property 非实体财产
non-correlated variables 非相关变量
non-correlation 非相关；不相关
non-correspondent bank 非往来银行；非通汇银行
non-corresponding control 无响应控制
non-corresponding microparameter 非对应微观参数
non-corresponding parameters 非对应参数
non-corrodibility 抗腐蚀能力；防腐性；不蚀腐性
non-corrodible 非腐蚀性的；抗蚀的
non-corroding 抗腐蚀的；不锈的；不腐蚀；不蚀
non-corrosion alloy 不锈合金；抗蚀合金
non-corrosion metal 不锈金属
non-corrosion water 非腐蚀性水
non-corrosive 抗腐蚀；非腐蚀性的；无腐蚀性；不锈的；不腐蚀的
non-corrosive alloy 抗蚀合金；防蚀合金；不蚀合金

non-corrosive alloy coating 防蚀合金涂层
non-corrosive flux 无腐蚀性焊剂；不腐蚀钎焊剂
non-corrosive gas 非腐蚀性气体
non-corrosive grease 无腐蚀润滑；不蚀滑脂
non-corrosive material 不锈材料
non-corrosive metal 防蚀金属；防腐金属；不锈金属
non-corrosiveness 无腐蚀性
non-corrosive pipe 不锈钢管
non-corrosive steel 不腐蚀钢；不锈钢；抗蚀钢
non-corrosivity 无腐蚀性
non-cost items 非成本项目
non-countable 不可数
non-counting keyboard 不能调行键盘
non-coupled axle 附轴
non-covalent bond 非共价键
non-coverable timber 不回收的木支撑；不回收的撑垫木材
non-cracking concrete 不开裂混凝土
non-crank pump 无曲柄泵
non-creasing fabric 不皱布
non-creasing finish 无皱整理；防皱整理
non-creeping material 非徐变材料
non-criteria 非标准的
non-critical 非临界的
non-critical activity 非关键活动
non-critical item 非判定项目
non-criticality 非临界性
non-critical microoperation 非临界微操作；非关键微操作
non-critical moiety 非临界部分
non-critical phase matching 非临界相位匹配
non-critical race 非危险竞争
non-critical water quality checkpoint 非临界水质检验点
non-critical wind velocity 非临界风速
non-crop area 非耕作区
non-crop land 非耕地；非产粮地
non-crossing correlation rule 不交叉对比原则
non-crossing rule 不相交规则；不交叉规则
non-cross-linked polymer 非交联聚合物
non-crustic test(ing) 非破坏性试验
non-crystal 非晶体的
non-crystal body 非晶体
non-crystalline 非晶质的；非晶性的；非晶的；非结晶的
non-crystalline cement texture 非晶质胶结物结构
non-crystalline clay mineral 非晶质黏土矿物
non-crystalline electrode 非晶性电极
non-crystalline graphite 非晶质石墨；非晶性石墨
non-crystalline material 非晶材料；非结晶材料
non-crystalline oscillator 非晶体振荡器
non-crystalline phosphate 非晶性磷酸盐
non-crystalline silica 非结晶二氧化硅
non-crystalline solid 非晶态固体
non-crystalline structure 非晶结构
non-crystallizable 不结晶
non-crystallized 不结晶的
non-crystal oscillator 无晶体振荡器
non-crystal plane slip 非结晶面滑移
non-crystal semiconductor 非晶半导体
non-cubic 非立方系的
non-cubic crystal 非立方晶体
non-cultivable land 不可耕地
non-culturable state 不可培养状态
non-cumulative basis 非累积制
non-cumulative clause 不积算条款
non-cumulative curve 差动曲线
non-cumulative letter of credit 不可累积使用的信用证
non-cumulative provisions 非累积条款
non-cumulative quantity discount 非累计数量折扣
non-curl backing 防卷曲层
non-curl backing layer 不卷曲背涂层；防卷曲支持层
non-curl layer 不卷曲层
non-current 非本期的
non-current assets 非流动资产
non-current congestion 非常态拥塞
non-current field 无电流场
non-current liability 非流动负债
non-current method 非流动方法
non-current receivables 非流动性应收款
non-cushioned 无缓冲的
non-cutting movement 非切削运动

non-cutting shaping 无切削成形
non-cutting stroke 非切削行程
non-cutting time 辅助时间
non-cyanide plating 无氰电镀
non-cycle photophosphorylation 非环式光合磷酸化
non-cyclic 非循环的
non-cyclic(al) 非周期性的
non-cyclic(al) code 非循环码
non-cyclic(al) dependence relation 非周期相关性关系
non-cyclic(al) disturbance 非周期扰动
non-cyclic(al) feeding 非周期馈送
non-cyclic(al) iterative method 非循环迭代法
non-cyclic(al) machine 非周期性电机
non-cyclic(al) phosphorylation 非循环磷酸化;非环式磷酸化
non-cyclic(al) recovery 非周期恢复
non-cyclic(al) variation 非周期性变化;非周期变化
non-cyclotron-resonance 非回旋共振
non-cylindrical fold 非圆柱(状)褶皱
non-cylindrical support 非圆柱形架
non-cylindrical surface 非圆柱形面
non-damaging earthquake 非破坏性地震
non-damping vibration 无阻尼振动
non-data input 非数据输入
non-data input/output operation 非数据输入输出操作
non-data-set clocking 非数传机同步;非数传机时钟
non-dazzling lighting 不耀眼照明
non-debiteuse drawing glass 无槽引上玻璃
non-decimal 非十进位的
non-decimal base 非十进制基数
non-decimal system 非十进制数系
non-decision-making 无法决策;未决策
non-decomposable 不可分解的
non-decomposable matrix 不可分解矩阵
non-decreasing function 非减函数;不减函数
non-decreasing random function 非减随机函数
non-decremental conduction 不减衰传导
non-dedicated control channel 非专用控制信道
non-deductibility 不可扣除性
non-deductible contributions 不能扣除的捐献缴款
non-deductible franchise 不扣减免赔额
non-deductible medical expense 不能扣除的医疗费
non-deductible moving expense 不能扣除的迁移费
non-deductive medical expense 不能扣除的医疗费
non-defect 无缺陷
non-defective unit 优良品;合格品
non-deflecting 不挠曲的
non-deflecting foundation 刚性基础
non-deformability 抗变形能力
non-deformable 不变形的
non-deforming steel 不变形钢
non-degeneracy 非退化性
non-degenerate 非退化的
non-degenerate amplifier 非简并放大器
non-degenerate Boolean algebra 非退化布尔代数
non-degenerate circle 非退化圆
non-degenerate continuous distribution 非退化连续分布
non-degenerate critical point 非退化临界点
non-degenerate curve 非退化曲线;常态曲线
non-degenerate device 非简并器件
non-degenerate four wave mixing 非简并四波混频机
non-degenerate gas 非简并气体
non-degenerate junction 非简并结
non-degenerate level 非简并能级
non-degenerate linear system 非退化线性系统
non-degenerate parametric amplifier 非退化参变放大器
non-degenerate parametric oscillator 非简并参量振荡器
non-degenerative basic feasible solution 非退化基本可行解
non-degenerative type 非退化形
non-degnerate normal distribution 非退化正态分布
non-degradable 不可降解的
non-degradable biochemicals 不可降解的生化物;不可降解生化物质

non-degradable organic compound 不可降解有机化合物
non-degradable pollutant 不可降解污染物;非降解性传染物
non-degradable substance 难降解物质;不可降解物质
non-degradable waste 不可降解废物;不可降解废料
non-degraded surfactant 未降解的表面活性剂
non-delay 无时效的;不延发
non-delayed emergency braking 不延时的紧急制动
nondelay fuse 瞬发引信
non-deletable file 不能删除的文件
non-deletion mode 非删除方式【计】
non-delimiter 非定义符;非定界符
non-deliquescent 不潮解的
non-deliverable articles 无法交付物品
non-deliverable goods 无法交付货物
non-delivery 欠交;无法投递;未交货;未交付;提货不着
non-delivery and shortage risk 不交货及短少险
non-delivery of entire package 整件提货不着
non-denominational number system 不名数系
non-dense index 非稠密索引
non-dense set 非密集
non-denumerable 不可数的
non-depolarizer 非去极化剂
non-deposit flow 无沉降水流;无沉积水流
non-depositing velocity 不淤积速度;不淤(积)流速
non-depositional unconformity 非沉积不整合;准整合
non-depreciable property 非折旧资产
non-derailing position 非脱轨位置
non-derogatory 非损的;非减的
non-designated area 未选用面积;待定用地
non-destrductive read memory 非破坏性存储器
non-destructibility 不损坏性
non-destructible test method 非破坏性试验法
non-destructing check 不破坏性检验
non-destructive 无损的;非破坏性的;非破坏性
non-destructive add instruction 非破坏性加指令
non-destructive addition 不破坏加法
non-destructive analysis 非破坏性分析
non-destructive and instruction 非破坏性与指令
non-destructive assay 不破坏(试样)的验定
non-destructive breakdown 非破坏性击穿
non-destructive characterisation 无损表征
non-destructive compare instruction 非破坏性比较指令
non-destructive cursor 非破坏性光标
non-destructive detection technique 无损探伤技术
non-destructive detector 无损检测器;非破坏性检测器
non-destructive earthquake 非破坏性地震
non-destructive evaluation 无损探伤;无损评价;无损评估;非破坏性评估
non-destructive examination 非破坏性检查;无损检验;无损检测
non-destructive(flaw) detection 无损探伤
non-destructive inspection 无损探测;无损探伤;无损检验;无损检查;非破坏性检验;非破坏性检查
non-destructive inspection machine 无损探伤机
non-destructive inspection method 无损探伤法
non-destructive instruction 非破坏性指令
non-destructive investigation of tunnel 隧道非破损探查
non-destructive material test 非破坏性材料试验;无损材料试验
non-destructive measurement 非破坏性测量法
non-destructive measuring 不破损测量法
non-destructive memory 非破坏性存储器
non-destructive method for estimating 利用非破坏性试验方法估测
non-destructive reading 非破坏性读数
non-destructive readout 无损读出;非破坏性读出;不破坏读出
non-destructive readout optic(al) memory 无损读出光存储器
non-destructive storage 非破坏性存储器;不破坏存储器
non-destructive technique 无损技术
non-destructive test by resonance 共振法非破损

检验
non-destructive test by ultrasonic pulse 超声波脉冲非破坏检验
non-destructive test(ing) 无损探伤;无损(探伤)试验;无(破)损试验;不破坏试验;非破损试验;无损检测;无损检验;非破坏(性)试验;非破坏性检验;非破坏性测试;耐损试验;不破坏试验
non-destructive testing equipment 无损检验设备
non-destructive testing equipment works 无损检测设备厂
non-destructive testing method 无损检验法
non-destructive testing of concrete 非破损性混凝土试验;混凝土非破坏性试验
non-destructive testing technique 无损探伤术;无损探伤技术
non-destructive test method 非破坏性试验法;不损坏试验方法
non-destructive test rebound hammer 回弹仪非破损检验
non-destructive traffic 非破坏性交通
non-detachable 不可拆开的
non-detectability 不可检测性
non-detectable 不能检测的;不可检测的
non-detection 未检出
non-detergent 不变质的
non-deteriorating restoring force 不退化恢复力
non-determinancy 不确定性;非确定性
non-determinate pushdown automaton 非确定下推自动机
non-determination 非测定性
non-deterministic 非确定的
non-deterministic algorithm 不确定算法
non-deterministic analysis 非定值分析
non-deterministic automaton 非判定性自动机;不确定自动机;不确定性自动机
non-deterministic automator 非决定性自动机
non-deterministic finite automaton 非确定的有限自动机;不确定性有限自动机;不确定的有穷自动机
non-deterministic flow chart 非确定框图
non-deterministic parsing 非确定的分析
non-deterministic program(me) 不确定程序
non-deterministic programming 非确定性程序设计
non-deterministic stack automaton 不确定的堆栈自动机
non-deterministic system 非确定性系统
non-deterministic Turing machine 非确定性图灵机;不确定(的)图灵机
non-detonating combustion 无爆燃烧
non-detonating fuel 不爆震燃料
non-detrital 非碎屑的
non-developable 非可展的;不可展的
non-developable ruled surface 非可展直纹曲线
non-development 不发育
non-deviated absorption 无折射吸收;非折射吸收
non-diagonal 非对角(线)的
non-diagonal elements 非对角线元素
non-diagram line 图表中没有的线
non-dialyzed ion 不渗析离子
non-diamond carbon growth region 非金刚石生长区
non-diamond core drill 非金刚石取芯钻机
non-diastrophic 非构造的;非地壳运动的
non-diathermic 非透热的
non-diesel engine 汽化器式发动机
non-differentiable 不可微分的
non-differential cost 无差别成本
non-differential function 不可微分函数
non-differential plateau climate 无分异高原气候
non-differential-pressure penetration speed 无压差钻速
non-differentiated intrusive body 无分异岩体
non-differentiated products 没有差别的产品
non-diffuse contrast phase coherent illumination 非漫射相干光照明
non-diffusible calcium 不扩散性钙
non-digit character 非数字符
non-dilational vein 非扩张脉
non-dimension 无因次;无量纲
non-dimensional 无单位的;无因次的;无量纲的;无尺寸的
non-dimensional analysis 无量纲分析
non-dimensional coefficient 无因次系数;无量纲系数

non-dimensional factor 无量纲因子；无量纲因素
non-dimensional frequency 无量纲频率
non-dimensional hydrograph 无量纲过程线
non-dimensional number 无因次量；无量纲数；无尺度数
non-dimensional parameter 无因次参数；无维参数；无量纲参数
non-dimensional quantity 无因次量；无维量；无量纲量；无尺度量
non-dimensional ratio 无量纲比
non-dimensional stability 无维稳定性
non-dimensional stress 无量纲应力
non-dimensional time function 无量纲时间函数
non-dimensional unit 无因次单位；无尺寸单位
non-dimensional unit hydrograph 无因次单位过程线；无维单位过程线；无量纲单位过程线
non-dimensional variable 无量纲变数
non-dipole field 非偶极子场
non-directed graph 非向图
non-directional 无方向(性)的；非直线的；非定向的；不定向(的)；无向的
non-directional antenna 不定向天线；全向天线
non-directional beacon 非定向标灯；全向信标
non-directional beacon light 不定向信标(灯)；全方向信标(灯)；全方向信号灯
non-directional counter 非定向计数管；不定向计数器
non-directional current protection 非方向电流保护装置
non-directional detector 非定向性探测器；无向检测器
non-directional echo sounding 非定向回声测深法
non-directional microphone 无向传声器
non-directional radio beacon 导航台；全向信标；全向无线电指标台；全向无线电信标；无方向性无线电信标
non-directional sound source 无方位声源
non-direct traffic 非直达运输
non-dischargeability 不可拔换性
non-discharging permeameter 不流水渗透计
non-disconnecting fuse 不能隔离的熔断器
non-discountable bill 非贴现汇票；不能贴现票据；不可贴现汇票
non-discrete valuation 非离散赋值
non-discretionary 不可任意支配的
non-discretionary trust 固定性投资信托；非全权信托
non-discriminatory 无差别待遇；非歧视性
non-discriminatory government procurement 非歧视性政府采购
non-disjunction 不离开；不分离现象；不分开现象
non-disjunction mosaic 不离开嵌合体
non-dispersed mud 不分散泥浆
non-dispersed system 不分散体系
non-dispersed variable 非分散变量
non-disperse X-ray analysis 非分散性X射线分析
non-dispersive 非分散的；非色散的；非扩散的
non-dispersive analysis 非色散分析
non-dispersive infrared absorption method 非分散红外吸收法
non-dispersive infrared analyzer 非分散红外分析仪
non-dispersive infrared CO tester 非色散红外一氧化碳测试器
non-dispersive infrared method 非分散红外法
non-dispersive infrared photometry 非分散红外光度法
non-dispersive linear system 非频散线性系统
non-dispersive medium 非色散媒质
non-dispersive spectrometry 非色散光谱法
non-dispersive ultrasonic delay line 非色散超声延迟线
non-dispersive ultraviolet analyzer 非扩散型紫外线分析仪；非分散紫外紫外分析仪
non-dispersive ultraviolet gas analyzer 非分散紫外气体分析仪
non-dispersive wave 不散波
non-dispersive X-ray fluorescence 非色散X射线荧光
non-dispersive X-ray spectrometer 非色散X射线光谱计
non-displacement condition 非排水状态
non-displacement pile 非挤土桩；钻孔桩；钻孔灌注桩；非挤压桩；非沉管桩；不排土桩

non-displacement ship 非排水型船
non-dissipative interaction 非耗散相互作用
non-dissipative muffler 反作用消声器
non-dissipative network 无耗网络
non-dissipative stub 无耗散短线
non-dissipative system 非耗散系统
non-dissolution trace 无溶蚀现象
non-dissolved organic carbon 非溶解有机碳
non-dissolved phosphorus 非溶解磷
non-distillate oil 渣油
non-distinct image 不清晰像；不清晰图(影)像
non-distortion correcting instrument 无畸变校正仪
non-distortion hologram 无畸变全息图
non-disturbance agreement 承租人不受扰动的协议
non-disturbance control 无扰控制
non-disturbed motion 无摄运动
non-divergent flow 非辐散流
non-diversion 不分道；不分流；不转移
non-dog type of pump 不淤塞水泵
non-dollar countries 非美元国家
non-domestic block 非住宅区
non-domestic building 非居住房屋；非住宅楼
non-domestic construction 非民用设计；非民用结构；非民用建筑
non-domestic development 非国内发展
non-domestic origin 非国内原产；非本国原产
non-domestic sector 非国内地段；非国内部分
non-domestic water supply 非生活用水
non-dominant complement 非线性补体
non-dominant hemisphere 非优势半球
non-dominated solution 非控解
non-donor solvent 非给予体溶剂
non-draft room 不通气房间
non-draining compound 不流动绝缘剂(用于浸渍高压电缆绝缘纸)；稠性绝缘剂(用于浸渍高压电缆绝缘纸)
non-drilling time 非钻进时间
non-drinking water 非饮用水
non-drip alkyd paint 不流挂醇酸树脂漆
non-drip paint 不滴淌涂料；不流挂漆
non-dripping oil paint 不滴油漆
non-driven wheel 非从动轮
non-driving cab 非操纵司机室
non-driving track 非驱动履带
non-dropping tungsten filament 不坠钨丝
non-drying 非干性的
non-drying cargo 非干(性)货(物)
non-drying oil 非干性油；不干性油
non-drying sealant 不干性密封剂
non-dry oil modified alkyd resin 不干性油改性醇酸树脂
non-ductile 非延性的
non-ductile fracture 无塑性破坏；无塑性断裂
non-ductive nature 非可延性质
non-dummy variable 非虚拟变量
non-dumping certificate 非倾销证明书
non-durable 不耐用物品；消耗(物)品；非耐用品；非耐用货物
non-durable consumer goods 非耐用消费品
non-durable goods 消耗物品；消耗品；非耐用商品；非耐用品；非耐用货物；非耐火物品；不耐用商品
non-dusting 无粉尘的；不扬尘的；无灰尘的
non-dusting brush 耐磨电刷
non-dusting charge 无尘装料
non-dusting granules 无粉尘性颗粒
non-dusty mine 无尘矿山
non-dutiable goods 无税货物
non-dwelling facility 非居住设施
non-dyeing fiber 不染性纤维
non-earning space 非载货舱位；非运舱位
non-earthquake-resistive building 非抗震房屋
non-earthquake tectonic movement 非地震构造运动
non-ecclesiastical 非宗教纪念碑
non-ecclesiastical architecture 非宗教结构；非宗教造；非宗教性建筑
non-ecclesiastical building 非宗教(性)建筑物
non-ecclesiastical Gothic style 非基督教哥特式
non-ecclesiastical structure 非宗教建筑物；非宗教结构；非宗教构造
non-echo chamber 消声室
non-echo spectrum 消声谱

non-econometric simulation model 非经济计量模拟模型
non-economic 非经济性的
non-economic activity 非经济活动
non-economic compulsion 非经济强制
non-economic gas field 无经济价值气田
non-economic oil field 无经济价值油田
non-economic oil-gas bed 非经济价值油气层
non-economic variable 非经济变量
non-eddy flow 无涡流
non-edible fish 非食用鱼
non-edible part 不可食部分
no need to change lubricant 不需换润滑油
non-effective 无效力的；不起作用的
non-effective storage 无效库容；死库容
non-efflorescence 不起霜；不开花；非风化物
no-negative effect 无负面效应
no-negative environmental impact 无负面环境影响
none gloss varnish 平光清漆
none heated sludge digestion tank 无加温污泥消化槽
non-elastic 无伸缩性的；非弹性的
non-elastic behavio(u)r 非弹性性能；非弹性性状
non-elastic body 非弹性体
non-elastic buckling 非弹性压曲
non-elastic closure 非弹性闭合
non-elastic collision 无弹性碰撞；非线性碰撞
non-elastic deflection 非弹性挠曲
non-elastic deformation 非弹性变形
non-elastic effect 非弹性效应
non-elasticity 无伸缩性；非弹性
non-elastic range 非弹性范围
non-elastic strain 非弹性应变
non-el detonator 非电起爆器；非电起爆雷管
non-electric 非电的
non-electric(al) blast ignition system 非电起爆点火系统
non-electric(al) current meter 非电式海流计
non-electric(al) delay blasting cap 非电延时爆炸雷管；非电起爆迟发雷管
non-electric(al) delayed detonator 非电延期雷管
non-electric(al) detonating system 非电引爆系统
non-electric(al) detonator 非电起爆器；非电起爆雷管
non-electric(al) initiation 非电起爆
non-electric(al) initiation device 非电起爆装置
non-electric(al) misfire 非电爆拒爆
non-electrified line 非电化铁路
non-electrified railway 非电化铁路
non-electrified section 非电气化区间
non-electrolyte 非电解质；不电离质
non-electrolyty 非电解质
non-electromagnetic chuck 非电磁吸盘
non-electromagnetic molecule 非电磁分子
non-elementary reaction 非元反应
non-elevating road mixer 不能升起的道路拌和机
non-eligible bank bills 非合格银行汇票；不合格银行汇票
non-eligible commercial paper 不合格商业票据
non-embedded 非埋藏的；露天的
non-emitting sole 非发射底极
non-empty finite set 非空有限集
non-empty finite string 非空有限串
non-empty model 非空模型
non-empty queue 非空排队
non-empty sequence 非空序列
non-empty subset 非空子集
non-emulsion-coated face 无乳剂面
non-emulsion-coated side 无乳剂面
non-end-anchored tendon 两端不锚着的预应力筋腱；非端锚力筋
non-energy use 非能源消费
non-engineered structure 非正规设计结构
non-eninge aerodynamic noise 非发动机空气动力噪声
non-enumerative algorithm 不可枚举算法
non-epitaxial 非外延的
non-equality 不等式
non-equalized truck 不等均衡梁的转向架
non-equalizing wage differentials 非均衡工资差别
non-equidistant limit 非等距界限
non-equilateral mine 不等翼矿山
non-equilateral triangle 不等边三角形

non-equilibrium 非平衡态;非平衡;非均衡
non-equilibrium behavio(u)r 非平衡特性
non-equilibrium carrier 非平衡载流子
non-equilibrium chemical excitation 非平衡化学激发
non-equilibrium condition 非平衡状态;非平衡状况;非平衡条件
non-equilibrium constant 非平衡常数
non-equilibrium cooling 不平衡冷却
non-equilibrium defect 非平衡缺陷
non-equilibrium distillation 非平衡蒸馏
non-equilibrium distribution 非平衡分布
non-equilibrium effect 非平衡效应
non-equilibrium equation 不平衡方程
non-equilibrium expansion 非平衡膨胀
non-equilibrium flow 非平衡流;非定常流;不平衡流
non-equilibrium formula 不平衡公式
non-equilibrium freezing 非平衡凝固
non-equilibrium isotopic exchange method 非平衡同位素交换法
non-equilibrium material 非平衡材料
non-equilibrium microstructrue 非平衡显微组织
non-equilibrium mode distribution 非稳态模式分布
non-equilibrium model 非平衡模型
non-equilibrium phase 非平衡相
non-equilibrium plasma 非平衡等离子体
non-equilibrium plasma reactor 不平衡等离子反应器
non-equilibrium plasmon 非平衡态等离激元
non-equilibrium profile 非均衡剖面
non-equilibrium pumping test 非平衡抽水试验;不平衡抽水试验
non-equilibrium reaction kinetics 非平衡态反应动力学
non-equilibrium site 不平衡位置
non-equilibrium state 非平衡状态;非平衡态;不平衡状态
non-equilibrium system 非平衡系统
non-equilibrium technique 非平衡技术
non-equilibrium temperature 非平衡温度
non-equilibrium thermodynamical function 非平衡热力学函数
non-equilibrium thermodynamics 非平衡态热力学;不平衡过程热力学
non-equilibrium writing 不平衡书写;不平衡记录
non-equivalence 非等价
non-equivalence gate 异门【计】
non-equivalence operation 异运算;异操作
non-equivalent control group 非等效控制群
non-equivalent element 反重合元件
non-equivalent symmetry element 非等效对称元
non-equivalent-to-element 反重合元件
non-erasable medium 不可清除介质;不可擦媒体
non-erasable memory 只读存储器;不可擦存储器
non-erasable optical disc 不可擦光盘
non-erasable stack automaton 不可擦堆栈自动机
non-erasable storage 非可擦存储器;不可抹去存储器;不可擦存储器
non-erasing stack automaton 非抹除的堆栈自动机
non-ergodic process 非各态经历过程;非遍历性过程
non-erodible bank 不易冲刷的河岸;不冲蚀河岸
non-erodible bed 非冲刷性河床;不冲刷河床
non-erodible channel 非冲刷性渠道;非冲刷性河槽;不冲刷河槽
non-erodible fractions of soil 土壤不易受侵蚀部分
non-erodible velocity 不冲刷流速
non-eroding and non-silting velocity 不冲不淤流速
non-eroding velocity 不冲刷流速
non-erosive 非侵蚀性的
non-erosive bed 不冲刷河床
non-erosive blasting 软粒喷砂;无损喷砂
non-erosive velocity 不冲刷流速
non-error system 无误差制
non-escaping key 非换码键
non-essential 非本质的
non-essential auxiliaries 次要属件;次要辅件
non-essential clause 非主要条款
non-essential provision 非基本规定
non-essential record 非必要案卷
non-essential repair parts 二级备用零件
non-essential resistance 外部电阻

non-essentials 非必需品
non-essential singularity 非本性奇点
non-essential stipulation 非主要条款
non-esterified fatty acid 游离脂肪酸
nonet 九重线
non-etching degreaser 非侵蚀性脱脂剂
none tree 失配树
non-Euclidean geometry 非欧里得几何(学)
non-Euclidean length 非欧里得长度
non-Euclidean space 非欧里得空间
non-euploid 非整倍体
non-evapo(u)rable water 非蒸发水
nonex 铅硼玻璃
non-exchange 不交换
non-exchange commercium 非交易物
non-exchange sodium 非交换性钠
non-excitation state 非激发态
non-excited synchronous motor 反应式同步电动机
non-exciting regulation transformer 非励磁调压变压器
non-excludability 非排他性
non-exclusive 非独家的
non-exclusive and non-transferable license 非专卖不可转让许可证
non-exclusive jurisdiction 非专属性管辖权
non-exclusive license 非独占许可
non-exclusive license contract 普通许可证合同
non-exclusive surveyor 委托检查员
non-excusable delays 不可原谅的拖期
non-executable program(me) 非执行程序
non-executable statement 不可执行语句
non-executive director 非主管董事;非执行董事;非常务董事
non-existent code 非法代码;不存在码
non-existent code check 非法码校验
non-existent memory 非法存储器【计】
non-expanding exit 非扩散排气喷嘴
non-expanding nozzle 非扩散喷管
non-expansion 非膨胀杆
non-expansion cycle 不膨胀循环
non-expansion rod 不胀缩杆
non-expansion steel 不膨胀钢
non-expansive condition 无湿胀状态;土的无湿胀状态
non-expansive condition of soil 土的非湿胀;土的无湿胀状态
non-expansive fusion caking 不膨胀熔融黏结
non-expansive soil 非膨胀性土
non-expendable 可回收的;可长期使用的;非消耗性的
non-expendable equipment 非消耗性设备
non-expendable fund 不可动用基金;不动本基金
non-expendable item 非消耗品
non-expenditures disbursement 非开支性付款
non-expendituring cost 非支出成本
non-experimental data 非实验性数据
non-experimental data model 非实验性数据模型
non-experimental model-building 非实验建模
non-experimental observation 非实验观测值
non-experimental study 非实验性研究
non-explosion hydrothermal eruption 非爆炸性水热喷发
non-explosive 无爆炸性;不爆炸的
non-explosive agent 无焰火药;防爆剂;安全炸药
non-explosive light 防爆灯
non-explosive source 非炸药震源
non-expondable 多次使用的
non-exponential 非指数的
nonexportation 禁止出口
non-exposed area 航摄死角(区);非曝光区
non-extractable 不可萃取
non-extreme pressure lubricant 非极限压力润滑剂
non-extruding joint filler 不挤凸的填缝料;不凸出的填缝
non-faceted-non-faceted eutectic alloy 非小面化—非小面化共晶合金
non-faceted solid-liquid interface 非小面化固—溶分界面
non-factor services 非要素服务
non-fading 不褪色(的)
non-fail-safe unit 非故障安全部件
non-failure 无损坏
non-failure operation time 无故障运行时间

non-family corporation 非家族公司
non-family household 非家庭户
non-farm land 非耕种土地
non-farmsector 非农业部门
non-fatal accident 非死亡事故
non-fatal error 非致命错误;非严重错误
non-fatal trauma 非致命伤
non-feasible region 非可行区域
non-feasible solution 不可行解
non-feasible state 不可行状态;非可行状态
non-fecal drainage 杂废排水;杂废水
non-felting property 不毡合性
non-ferromagnetic 非铁磁性
non-ferromagnetic material 非铁磁材料
non-ferrous 有色金属的;不含铁的;非铁的
non-ferrous alloy 有色金属合金;有色合金;非铁合金;不含铁合金
non-ferrous alloy electrode 非铁合金焊条
non-ferrous casting 有色金属铸件;非铁铸件
non-ferrous electrode 有色金属焊条
non-ferrous emulsion 非离子乳液
non-ferrous industry 有色金属工业
non-ferrous material 非铁材料;有色金属材料
non-ferrous metal 有色金属;非铁金属
non-ferrous metal analysis 有色金属分析
non-ferrous metal commodity 有色金属矿产
non-ferrous metallic slag 有色金属渣
non-ferrous metallurgical industry wastewater 有色工业废水
non-ferrous metallurgy 非铁冶金学;有色冶金学
non-ferrous metal metallurgy 有色金属冶炼;有色金属冶金学
non-ferrous metal refinery 有色金属冶炼工业
non-ferrous metal refinery industry 有色金属冶金工业
non-ferrous metal rolling mill 有色金属轧机
non-ferrous metal semi-round head rivet 半圆头有色金属铆钉
non-ferrous metal smelting work 有色金属冶炼厂
non-ferrous metal waste water 有色金属废水
non-ferrous nail 非铁钉
non-ferrous smelting industry 非金属熔炼工业
non-ferrous vessel 有色金属容器
non-fibered 无纤维状的;非纤维状的
non-fibered board 非纤维板
non-fibred 无纤维状的;非纤维状的
non-fibrillated fiber 非原纤化纤维
non-fibrous grease 非纤维组织的润滑脂
non-figurative art 非图形表示的艺术
non-figurative management 整体结构管理
non-file-structure device 非文件结构设备
non-filling 未充填
non-filterable 不可过滤的
non-filterable residue 非滤过性残渣;非过滤性残渣
non-filterable solid 非过滤性固体;不可过滤固体
non-filtrable matter 非过滤物(质)
non-filtrable residue 悬浮残渣
non-filtration residue 不可过滤残渣
non-filtration water treatment plant 非过滤水处理装置
non-financial business 非金融企业
non-financial corporation 非金融公司
non-financial intangible assets 非金融无形资产
non-fines concrete 无砂混凝土
non-finite state model 非有限状态模式
non-fireable 不可锻的
non-fired construction 非防火建筑
non-fireproof 非防火的
non-fireproof construction 非防火建筑;不防火建筑
non-fireproof metal construction 不防火金属构造
non-firm offer 非稳固报价
non-firm power 非恒供电;特殊电力;备用功率;备用电力
non-fission 不裂变的
non-fission capture 非裂变俘获
non-fissured 基本无裂隙
non-fissured clay 无裂隙黏土
non-fixed block system 非固定阻塞制
non-fixed wholesale 流动批发
non-flame 非火焰
non-flame atomic absorption 无火焰原子吸收(法)
non-flame atomic absorption analysis 无火焰原子吸收分析

non-flame blasting 无火焰爆发；无焰爆破
non-flame conveyor belt 防燃运输带
non-flame fiber 不燃性纤维
non-flame film 安全胶片
non-flame property 不燃性的；不燃性
non-flammable 无焰的；不易燃的；不燃的
non-flammable coating 不燃性涂层
non-flammable film 不燃性胶片
non-flammable fluid 不燃液体
non-flammable fuel 无焰燃料
non-flammable material 难燃烧体；难燃材料；非燃性材料
non-flammable paint 耐火漆；不燃性涂料
non-flammable resin 不燃性树脂；不燃树脂
non-flaredfitting 非扩口式管接头
non-floating arithmetic sign 非浮点算术符号
non-floating rail 固定栏杆；固定拉杆；固定护栏
non-floating type compass 非浮式罗盘
non-flocculated 非絮凝的；不絮凝的
non-flooding feeder 防溢定量给料器
non-flood season 非汛期；非洪水期；非洪水季节
non-flowering plants 不开花植物
non-flowing artesian well 承压井；非自喷井；非自流井；非溢出自流井
non-flowing character 不流动性
non-flowing water 静水
non-flowing well 非自流井
non-flowing zone test 非自喷层测试
non-flow length 非流动方向的长度
non-fluctuating 非脉动的；非波动的
non-fluid friction 非流体摩擦
non-fluid oil 原质机油；厚质机油
non-fluorescent colo(u)r 非荧光色
non-fluorescent radiation transition 非荧光辐射跃迁
non-fluorescing electrolyte 不发荧光的电解液
non-flush rail fastening 不密贴钢轨扣件
non-foaming 不起泡沫的；不起泡的
non-foaming layer 非泡末层
non-foaming test 不起泡沫试验
non-focusable energy 不可聚焦能量
non-focusing 非聚焦的
non-focusing collector 不聚焦集热器
non-focusing neutron spectrometer 非聚焦中子谱仪
non-fogging coating 不结雾涂层
non-follow-up 非随动
non-follow-up controller 自动手动转换柄；非随动转换柄；非随动控制器
non-follow-up type steering gear control 非随动式操舵装置的控制
non-food department 非食品部门
non-food raw materials 非食品原料
non-forced circulatory lubrication 非压力环流润滑
non-forfeiture condition 不能没有的条件
non-format 非格式
non-fossil fuel 非矿物燃料
non-fouling 不污浊的；不污的
non-foundamental mode flow table 非基本模式流程表
non-fractional crystallization 非分离结晶作用
non-fractionation sedimentation method 不分割沉降法
non-free-flow junction 不畅流交叉口
non-freehold estate 非自由拥有财产
non-freezable lubricating oil 不冻润滑油
non-freeze sprinkler system 不冻喷水系统
non-freezing 耐冻的；不冻的；耐寒的
non-freezing brine 不凝结盐溶液
non-freezing estuary 不冻河口
non-freezing explosive 不冻炸药
non-freezing flush hydrant 不冻路平式消防栓（不露出地面）
non-freezing lubricant 不凝固润滑剂
non-freezing lubricating oil 不冻润滑油
non-freezing mixture 不冻溶液；不冻混合物
non-freezing port 不冻港
non-freezing post hydrant 不冻柱式消防栓
non-freezing river 非封冻河流；不冻河流
non-freezing solution 不冻溶液
non-freezing stream 非封冻河流；不冻河流
non-friction guide 非摩擦导轨
non-friction hinge 无摩擦铰（链）
non-frontal precipitation 非锋面降水

non-frost-active 抗霜的；不结霜的
non-frost-active(clay)brick 抗霜砖；不结霜的砖；抗冻砖
non-frosted water content 未冻(结)含水量
non-frost heaving 不冻胀翻浆的；不冻胀；不翻浆的
non-frosting 不局部脱色
non-frost period 无霜期
non-frost-susceptibility 霜冻敏感性
non-frost-susceptible material 对霜冻不敏感的材料；不冻材料
non-fuel commodities 非燃料矿产
non-fulfilment 不履行
non-full functional dependency 非完全函数相关性
non-fully halogenated carbon compound 非全卤化氯氟碳化合物
non-functional 不起作用的；无功能的
non-functional building 不住人建筑物
non-functional characteristics 非功能特性
non-functional packages software 非函数软件包
non-functional side 非功能侧
non-functioning 无机能的；无功能的
non-functioning protein 无功能蛋白质
non-fundamental breach 非重大违约；非根本性违约
non-fund assets 非资金资产
non-fungible thing 不可代替之物
non-fuse breaker 无熔线断路器
non-fusibility 抗熔性；不溶性
non-fusible 不熔化的；不熔的
non-fusible developer 不熔型显影剂
non-fusible treatment 不熔化处理
non-fusing type 不熔型
non-fusion 没有焊透
non-fusion caking 不熔融黏结
non-gap 非空挡
non-gas nonflux process 无气体无焊剂焊接法
non-gas-shielded arch welding 无气体保护的电弧焊
non-Gaussian chain 非高斯链
non-Gaussian light 非高斯光
non-Gaussian peak 拖尾峰
non-Gaussian solute concentration distribution 非高斯溶质浓度分布
non-geared drive 非齿轮传动
non-generating period 非发电时期
non-geometric(al)condition 强制条件
non-geothermal area 非地热区
non-geothermal tectonic zone 非地热构造带
non-glacial 非冰川的
non-glare 防反射的；无眩光的；防眩的
non-glare glass 防反射玻璃；防眩玻璃
non-glass-forming metallic oxide 非玻璃成型用金属氧化物
non-glassy refractory 非玻璃质耐火材料
non-glassy state 非玻璃态
non-gloss 无光泽的
non-glossy paint 无光漆
non-glossy surface 无光表面
non-glued insulated joint 非胶接绝缘接头
non-goods services 非商品服务业
non-governmental audit 民间审计
non-governmental credit 民间信用
non-governmental economic organization 非政府经济团体
non-governmental organization 非政府组织
Non-governmental Organization Environment Liaison Board 非政府组织环境联络委员会
Non-governmental Professional Association 非政府专业协会
non-governmental trade 民间贸易
non-governmental trade agreement 民间贸易协定
non-governmental transaction 非政府交易
non-government trade mission 民间贸易代表团
non-graded 不均粒的；分选差的
non-graded highway 等外(级)公路
non-graded mix 非级配混合料
non-graded sediment 非级配泥沙；非级配沉积物；不均粒沉积(物)；不均粒沉积
non-graded slag filling 非级配矿渣填料
non-graded system 不分级别
non-grain-raising stain 不起毛刺的染色剂；木纹平整着色剂；木纹平整着色剂
non-grain sizing 无粮上浆
non-granular 非颗粒状的

non-graphical character 非图形字符
non-graphical data 非图解数据
non-graphic control panel 非模拟仪表盘
non-gravitating 无重力的
non-gravitational force 非引力作用
non-gravitational motion 非引力运动
non-gravitational perturbation 非引力摄动
non-gravity spring 上升泉；非自流泉；非重力泉
non-gray body 非灰体
non-greenhouse gas 不产生温室效应气体
non-grey atmosphere 非灰大气
non-grounded 不接地的
non-ground(ed) neutral system 不接地中线系统；不接地中线制
non-group code 非群码
non-guaranteed fund 无保证基金
non-guarded crossing 无人看守道口
non-hairing moisture-curable composition 不裂湿固化组分
non-halating 无晕光的
non-hangup base 立即接通制；立即服务方式
non-hardenable 不能淬硬；不可淬硬
non-hardened 非硬式防护的；未淬火的；未硬化的；非淬火的；非硬化的
non-hardening 非防护措施；非硬化的
non-hardy plant 不耐寒植物
non-harmful 无害的
non-harmonic 非调和的；非谐波的
non-harmonic analysis 潮汐非调和分析
non-harmonic constant 非调和常数
non-harmonic constant of tide 潮汐非调和常数
non-harmonic method 非调和法
non-harmonic ratio 非调和比
non-harmonic tidal constant 非调和潮汐常数
non-harmonic tide 非调和潮
non-hazardous 无危险的；安全的
non-hazardous area 安全区；非危险区
non-hazardous oilfield 无害油田废物
non-hazardous waste 无害废物
non-heated sludge digestion tank 不加热污泥消化池
non-heating lead 非加热引线
non-heating region 非采暖地区
non-heat-treatable 不可热处理的
non-heat treated 未经热处理的
non-heat treated alloy 非热处理合金
non-heaving soil 非隆起土；非冻胀土
non-hermetic 不气密的；不密闭的
non-hiding pigment 非遮盖性颜料
non-holographic 非全息
non-holographic imaging 非全息成像
non-holographic magnetooptic memory 非全息照相磁光存储器
non-holomic constant 不完整约束
non-holomic system 非完整系
non-holonomic constraint 不完整约束；不完全约束
non-holonomic system 不完整系
non-homing 不归位
non-homing finder 不归位式导线机；不归零式导线机
non-homing position type 非归位式
non-homing switch 不归位化键
non-homing turning system 不归位调整系统
non-homing type line switch 不归位寻线机
non-homing type rotary switch 不归零位；不归位旋转机键
non-homocentricity 非共心性
non-homogeneity 非同质性；非均质性；不均匀性；不均一性
non-homogeneity slide 不均匀滑移
non-homogeneity strain 不均匀应变
non-homogeneity stress 不均匀应力
non-homogeneous 多相的；非同质；非均质的；非均匀的；不齐的；不均匀的
non-homogeneous boundary condition 非齐次边界条件
non-homogeneous boundary value problem 非齐次边值问题
non-homogeneous bulk flattening 不均匀立体压扁
non-homogeneous chains 非齐次链
non-homogeneous coordinates 非齐次坐标
non-homogeneous deformation 不均匀变形
non-homogeneous differential equation 非齐次微分方程

non-homogeneous dilatancy 非均匀膨胀
non-homogeneous discovery rate 非齐次发现率
non-homogeneous distribution 不均匀分布
non-homogeneous distribution load 不均匀分布荷载
non-homogeneous enamelling 不均匀涂搪
non-homogeneous equation 非齐次方程
non-homogeneous fissured rock 非均匀裂隙岩石
non-homogeneous linear differential equation 非齐次线性微分方程式
non-homogeneous linear equation 非齐次线性方程
non-homogeneous medium 不均匀介质;非均匀介质
non-homogeneous non-stationary time series 非齐次非平稳时间序列
non-homogeneous plane wave 不均匀平面波
non-homogeneous plastic body 非均质塑料车身;非均质塑料车厢;非均质塑性体
non-homogeneous Poisson process 非均匀泊松过程
non-homogeneous population 非同质总体
non-homogeneous problem 非齐次问题
non-homogeneous region 非齐次区域
non-homogeneous search space 非齐次搜索空间
non-homogeneous soil 非均质土
non-homogeneous state of stress 应力的非均质状态;非均质应力状态
non-homogeneous strain 非均质应变
non-homogeneous stress 非均质应力;不均匀应力
non-homogeneous system 非均匀系
non-homogeneous universe 非均匀宇宙
non-Hookeian interaction 非虎克式相互作用
non-hopper-door barge 满底泥驳
non-horizontal simultaneous 斜同步
non-horizontal simultaneous with receiver upper 发低斜同步
non-horizontal simultaneous with transmitter upper 发高斜同步
non-housing building 非住房区段;非住房街坊;非住房大楼;非住宅楼
non-housing construction 非住房设计;非住房建筑;非住房构造;非住房结构
non-housing development 非住宅开发;非住宅群
non-housing sector 非住宅区段
non-hub pipe 无承口管
non-human productive resources 非人力的生产资源
non-human wealth 非人力财富
non-humic substance 非腐植物质
non-hydraulic 非液力的;非水硬性的;非水力(学)的
non-hydraulic binding material 非水硬性胶凝材料
non-hydraulic cement 气硬性水泥;气硬性胶结材料;气硬水泥;非水硬性水泥;非水硬性胶结料
non-hydraulic cement mortar 非水化泥灰浆
non-hydraulic lime 非水硬性石灰;非水化石灰;高钙石灰
non-hydraulic lime mortar 非水硬性石灰砂浆
non-hydraulic mortar 非水硬性砂浆
non-hydraulic mortar cementing agent 非水硬性砂浆胶凝剂
non-hydraulic threshold weight 无水马力门限钻压
non-hydrocarbon gas 非烃气
non-hydrogen oxide acetic acid methanogen 非氧化氢乙酸产甲烷菌
non-hydrophobic organic chemicals 非疏水有机化合物
non-hydrostatic(al) consolidation 不等向固结;非静水压力固结
non-hygroscopic 不受潮的;不吸湿的;不吸潮的
non-hygroscopic dust 不吸湿粉尘
non-hygroscopic insulation 防潮绝缘
non-hygroscopic material 防潮材料
non-hygroscopic powder 不吸潮火药
non-hygroscopic propellant 不吸湿推进剂
non-hypergolic 非相混自燃式的
non-icing carburetor 防冰汽化器
non-ideal adsorbed solution theory 非理想吸附溶液理论
non-ideal behavior 非理想变化过程
non-ideal elastic body 非理想弹性体
non-ideal filter 非理想滤波器
non-ideal flow 非理想流量
non-ideal fluid 非理想流体

non-ideal gas 非理想气体
non-ideal solution 非理想溶液
non-identical 不恒等的
non-identical distribution 非恒等分布
non-identifiability 不可识别性
non-identified coefficient 非识别系数
non-identity 不同一性
non-identity operation 不同运算
non-ignitable 不燃的;耐火的;防火的
non-ignited 未经灼烧的
non-ignitibility 非燃性;不可燃(烧)性
non-ignition 不着火
non-image area 空白版面
non-imaging sensor 非成像遥感器
non-impact drill 非冲击式钻机
non-impact printer 无压印刷机
non-impact vibrator 无冲击式振动器
non-impinging injector 非撞击式射流喷头;非碰撞射流式喷头
non-implementation of contract 合同的不履行
non-importation 禁止进口
non-income base for tax discrimination 纳税差别的非收入基准
non-income-tax multiplier 非收入税乘数
non-increasing function 非增函数;非递增函数;不增函数
non-incremental cost 非增量成本
non-independent accounting unit 非独立核算单位
non-independent construction unit 非独立建筑单位
non-independent event 非独立事件
non-indexed command 非变址命令
non-indicating controller 无分划调节器
non-indigenous 非本地的
non-individual body 连续体
non-induce-earthquake fracture 非发震断裂
non-inductive circuit 无电感电路
non-inductive load 无感荷载
non-inductive shunt 无感分流器
non-inductive winding 无感绕组
non-industrial activity 非产业活动
non-industrial business operation 非工业性事业经营
non-industrial establishment 非产业基层单位
non-industrial oil-gas bed 非工业油气层
non-industrial injury 非工伤事故
non-industrial service 非产业服务
non-industrial service charges 非工业性劳务支出
non-industrial undertaking 非工业性事业
non-industrial worker 非产业工人
non-inert impurity 活性杂质
non-infected tissue 未侵染组织
non-infection disease 非传染病
non-infective disease 非传染病
non-inferior solution 非劣解
non-infiltration from surface 无地表渗入
non-inflammability 不易燃性;不燃性;不可燃(烧)性
non-inflammable 不燃的;不可燃(烧)的
non-inflammable gas 不燃烧气体
non-inflammable insulator 不燃烧绝缘体
non-inflammable oil 非燃性油
non-inflationary wage increase 非通货膨胀的工资增长
non-information firm 非信息企业
non-informative prior distribution 不提供信息的先验分布
non-inhibitable interrupt 不可禁止中断
non-inhibit interrupt 非屏蔽中断
non-injector 不吸引喷射器
non-injurious examination 非损伤性检查
non-injury accident 无伤亡事故
non-input/output split 非输入输出空间分离
non insolution brush 无绝缘电刷
non-instal ment 一次付款;非分期付款的;一次性偿还的
non-instalment credit 一次还清的信贷;一次性还贷信贷;非分期偿还的信贷;非分期偿还的贷款
non-instalment debit or credit side 非分期偿还的借方或贷方
non-institute clause 非学会条款
non-institutional population 非集体户人口
non-instrument 非流通工具
non-instrument runway 不用仪器的滑行道;非仪表跑道

non-insulated line hanger 非绝缘吊弦
non-insulated overlap 非绝缘锚段开关
non-insulated overlapped section 非绝缘搭接区
non-insulated siding 非隔热墙壁
non-insulated surface 非绝缘面
non-insulated switch 非绝缘开关
non-insulated transition mast 非绝缘转换柱
non-insurable risk 非保风险;不能保险的风险
non-integrable 不可积分的
non-integrable component 非集成元件
non-integrable equation 不可积分方程
non-integral 非整体
non-integral quantity 非整数量
non-integral slot winding 非整数槽绕组
non-integrated structure 非综合结构;非整体结构
non-integrated tow 非组合式驳船队
non-integrating 不累计的
non-intellectual factor 非智力因素
non-intelligent 非智能
non-intelligent terminal 非智能终端
non-interacting 无相互作用;不相互使用;不互相响应
non-interacting binary 非互扰双重星系
non-interacting close binary 非互扰邻近双重星系
non-interacting condition 自律性条件
non-interacting control 自身控制;自身调整;非相互作用控制;非互交影响控制;非互相影响控制;不相关联控制;不互相影响控制
non-interacting control system 非交互式控制系统
non-interacting ligand 非相互作用配位体
non-interactive 非交互的
non-interchangeability 不可互换性
non-interchangeable 不能互换的
non-intercommunicating pore 不连通孔;不连通的孔
non-interconnected electrically coordination 非连接电力协调
non-interconnected local controller 非连接局部控制器
non-interconnected systems 非互连系统
non-interestbearing negotiable order of withdrawal 无息可转让提款通知书
non-interest-bearing note 无息票据
non-interest-bearing securities 无息证券
non-interference 无干扰;不相互干扰;不干涉;不干扰
non-interference point 无干扰点
non-interference temperature 无干扰温度
non-interference to navigation 不碍航行
non-interlaced raster 逐行扫描光栅
non-interlace raster scanning 非隔行光栅扫描
non-interlocked area 非联锁区
non-interlocked block 非连锁闭塞
non-interlocked derail 非联锁脱轨器
non-interlocked level 非联锁握柄
non-interlocked point 非联锁道岔
non-interlocked switch 非联锁道岔
non-interlocking station 非连锁站
non-interlocking zone 非联锁区
non-intermittent camera 连续照相机
non-interruptable instruction 非可中断指令;不间断指令
non-interruptible supply 不间断供给
non-intersect 不相交
non-intersecting 不相交的
non-intersecting planes 不相交平面
non-intersection accident 非交叉口的(交通)事故
non-intersection speed 路段车速
non-intersection speed controller 非交叉口速率控制器
non-intervention 不干涉
non-intervisible 不通视的
non-intervisible point 互不通视点
non-intervisible station 互不通视点
non-intrinsically non-linear 非固有的非线性
non-intrusion 没有入侵
non-invasive ultrasonic technique 非侵超声波技术
non-inventoriable cost 不归属产品成本
non-inventory assets 非存货性资产
non-invertible knot 不易散扭结
non-inverting amplifier 同相放大器;非反相放大器
non-inverting buffer 不反相缓冲器
non-inverting input 正相输入;同相输入;不倒相

输入
non-inverting parametric device 不倒相参量器件
non-investment property 非投资性财产
non-ion active agent 非离子型活性剂
non-ionic 非离子物质;非离子的
non-ionic absorbent 非离子型吸着剂
non-ionic ammonia 非离子型氨;非离子氨
non-ionic compound 非离子化合物
non-ionic detergent 非离子型洗涤剂;非离子型去垢剂;非离子洗涤剂;非离子型去污剂
non-ionic emulsifier 非离子型乳化剂;非离子型乳化器
non-ionic organic compound 非离子型化合物
non-ionic organic contaminant 非离子型有机污染物
non-ionic polyacrylamide 非离子型聚丙烯酰胺
non-ionic polymer 非离子型聚合物
non-ionic reaction 非离子反应
non-ionic sorbent 非离子吸着剂
non-ionic surface active agent 非离子型表面活性剂
non-ionic surfactant 非离子型表面活性剂
non-ionic surfactant solution 非离子型表面活性剂溶液
non-ionization polyacrylamide 非离子型聚丙乙烯酰胺
non-ionizing 不电离的
non-ionizing particle 不致电离粒子
non-ionizing radiation 非离子型辐射;非电离辐射
non-ionizing radiation damage 非电离辐射损伤
non-ionizing solvent 不电离溶剂
non-irrigated 未灌溉的;非灌溉的
non-irrigated agriculture 非灌溉耕作
non-irrigated land 非灌溉土壤
non-irrigation canal system 非灌溉渠道系统
non-isentropic 不等熵的
non-isentropic flow 非等熵流
non-isochromous 不规则的
non-isochronous 不等时的
non-isoelastic 非各向等弹性的;非等弹性的
non-isolated essential singularity 非孤立本性奇点
non-isometric 非等距离的
non-isometric linear array 不等距线性排列
non-isometric lines 非等距线
non-isometry grain 非等量度颗粒
non-isoplanatic 不等晕的【物】
non-isothermal 不等温的
non-isothermal decomposition reaction 非等温分解反应
non-isothermal flow 非等温流
non-isothermal growth 非等温生长
non-isothermality 非等温性
non-isothermal jet 非等温射流
non-isothermal surface 非等温面
non-isothermal viscoelastic fluid 非等温黏弹流体
non-isotropic 各向异性的;非均质的;非各向同性的
non-isotropic elasticity 各向异性弹性;非各相同性弹性
non-isotropic mineral reflectivity 非均质矿物反射率
non-issuance of stock certificates 不发行股票
nonius 游标;游尺
nonius zero 游标零点
non-jacking end 死端;非张拉端
non-judicial foreclosure sale 不经法庭而直接出售的抵押品
non-justifiable dispute 不应由法院解决的纠纷
non-knocking condition 正常燃烧条件
non-knocking explosion 正常燃烧
non-lacquer metallized paper capacitor 不浸漆金属膜纸介电容器
non-laminar flow 非层流
non-lasing light emitter 非相干光源
non-lead-covered cable 无铅包电缆
non-leaded swelling 无荷载膨胀
non-leaded swelling rate 无荷载膨胀率
non-leading axle 非主动轴
non-leafing alumin(i)um 非中展型铝粉;非漂浮型铝粉
non-leakage probability 不泄漏几率;不漏失几率;不漏失概率
non-leakage probability of slowing down 慢化非漏失概率
non-leak detector 非泄漏检测器
non-leak memory 永久存储器
non-leaky recharge 无越流补给
non-ledger assets 秘密资产;账外资产
non-lethal 不致死的
non-lethal dose 非致死剂量;不致死药量
non-lettering map 暗射(地)图
non-level 非水平的
non-life insurance 非人寿保险
non-lifting injector 不吸引喷射器
non-lift mixing action 不上翻的搅拌作用;强制(浆叶)式搅拌
non-light splitting infrared gas analyzer 不分光红外气体分析器
non-light tight 透光的
non-linear 非直线型的;非直线的;非线性(的)
non-linear absorption 非线性吸收
non-linear action 非线性作用
non-linear aerodynamic(al) characteristic 非线性空气动力特性
non-linear aerodynamics 非线性空气动力学
non-linear amplifier 非线性放大器
non-linear amplifying medium 非线性放大媒质
non-linear analysis 非线性分析
non-linear anomaly 非线性异常
non-linear behavio(u)r 土的非线性性质;非线性状态;非线性性状;非线性性质;非线性形态
non-linear behavio(u)r analysis 非线性状态分析
non-linear birefringence 非线性双折射
non-linear bivariate regression 非线性二元回归
non-linear body 非线性体
non-linear boundary value problem 非线性边值问题
non-linear breakeven chart 非线性盈亏平衡图
non-linear breakeven model 非线性保本模型
non-linear capacitor 非线性电容器
non-linear car-following model 非线性随车模型
non-linear characteristic 非线性特性(曲线)
non-linear circuit 非线性电路
non-linear circuit analysis 非线性电路分析
non-linear code 非线性码
non-linear coefficient 非直线系数
non-linear coefficient of saw-tooth wave 锯齿波的非线性系数
non-linear coil 非线性线圈
non-linear computation 非线性计算
non-linear computing element 非线性计算元件;非线性计算器
non-linear condition 非线性条件
non-linear conductance 非线性电导
non-linear conductor 非线性导体
non-linear conservative system 非线性保守系统
non-linear constraint 非线性约束
non-linear continuous programming 非线性连续规划
non-linear continuum mechanics 非线性连续力学;非线性连续介质力学
non-linear control 非线性控制
non-linear controller 非线性控制器
non-linear control theory 非线性控制理论
non-linear correction 非线性改正
non-linear correlation 非直线相关;非线性相关
non-linear coupled differential equation of motion 非线性耦联微分运动方程
non-linear coupled-wave equation 非线性耦合波方程
non-linear coupler 非线性耦合器
non-linear coupling 非线性耦联;非线性耦合
non-linear crosstalk 非线性串扰
non-linear crystal 非线性晶体
non-linear damping 非线性阻尼
non-linear data 非线性数据
non-linear deformation 非线性形变;非线性变形
non-linear detection 非线性检波
non-linear dielectric 非线性介质;非线性电介质
non-linear dielectric(al) polarization 非线性电介质极化
non-linear difference equation 非线性差分方程
non-linear differential equation 非线性微分方程
non-linear diffraction 非线性衍射
non-linear digital modulation 非线性数字调制
non-linear direction(al) coupler 非线性定向耦合器
non-linear discretization 非线性离散化
non-linear discriminant 非线性判别
non-linear discriminate analysis 非线性判别分析
non-linear dispersion relation 非线性色散关系
non-linear dispersive fiber 非线性色散光纤
non-linear distortion 非线性失真;非线性扭曲;非线性畸变
non-linear distortion coefficient 非线性畸变系数
non-linear distortion factor 非线性失真系数;非线性畸变系数
non-linear distortion of scan(ning) 非线性畸变扫描
non-linear distribution 非线性分配;非线性分布
non-linear drift 非线性漂移
non-linear duality 非线性对偶性
non-linear dynamic(al) multiplier 非线性动态乘数
non-linear dynamic(al) property of soil 土非线性动力学特性
non-linear dynamic(al) system 非线性动态系统;非线性动力系统
non-linear dynamics 非线性动力学
non-linear earthquake response 非线性地震反应
non-linear echo sounder 非线性回波测深仪
non-linear econometrics model 非线性计量经济学模型
non-linear econometric techniques 非线性经济计量方法
non-linear effect 非线性效应
non-linear elastic behavio(u)r 非线性弹性性质
non-linear elasticity 非线性弹性
non-linear elastic material 非线性弹性材料
non-linear elastic model 非线性弹性模型
non-linear elastic theory 非线性弹性理论
non-linear electro-dynamics 非线性电动力学
non-linear electromagnetic medium 非线性电磁媒质
non-linear electronic susceptibility 非线性电子磁化率
non-linear electrooptical crystal 非线性电光晶体
non-linear element 非线性元件
non-linear energy transfer 非线性能量转移
non-linear equation 非线性方程
non-linear equations 非线性方程组
non-linear estimation 非线性估计
non-linear estimation routine 非线性估计程序
non-linear estimator 非线性估计量
non-linear expanding 非线性扩展
non-linear factor 非直线系数
non-linear feedback 非线性反馈
non-linear feedback control system 非线性反馈控制系统
non-linear field discharge resistor 非线性消磁电阻器;非线性磁场放电阻器
non-linear field theory 非线性场论
non-linear filter 非线性滤波器
non-linear filtering 非线性滤波
non-linear finite element analysis 非线性有限元分析
non-linear finite element program(me) 非线性有限元程序
non-linear flow 非线性流
non-linear forced resonant response 非线性强迫共振反应
non-linear forward drift 非线性正向漂移
non-linear foundation 非线性基础
non-linear fractional function 非线性分式函数
non-linear fracture mechanics 非线性断裂力学
non-linear function 非线性函数
non-linear functional analysis 非线性泛函分析
non-linear gain characteristic 非线性增益特性
non-linear generator 非线性发生器
non-linear geostatistics 非线性地质统计学
non-linear governing equation 非线性控制方程
non-linear gravity wave 非线性重力波
non-linear harmonic distortion 非线性谐波失真系数
non-linear harmonic wave 非线性谐波
non-linear hyperbolic 非线性双曲线
non-linear hyperbolic system 非线性双曲型组
non-linear hysteric system 非线性滞回系统;非线性滞后系统
non-linear hysteric vibration 非线性滞回振动;非线性滞后振动
non-linear impedance 非线性阻抗
non-linear inductance 非线性电感
non-linear influence coefficient 非线性影响系数
non-linear infrared spectrograph 非线性红外摄谱仪

non-linear interaction 非线性相互作用
non-linear interconnection matrix 非线性互连矩阵
non-linear interpolation 非线性内插
non-linear isotherm 非线性等温线
non-linear iterative 非线性迭代法
non-linearity 非线性
non-linearity in parameters 参数的非线性
non-linearized 非线性化的
non-linear kernel 非线性核
non-linear kinetic model 非线性动力模型
non-linear kinetics 非线性动力学
non-linear law 非线性定律
non-linear least square 非线性最小平方；非线性最小二乘法
non-linear least square estimate 非线性最小二乘估计
non-linear least squares estimation 非线性最小平方估计
non-linear least squares estimator 非线性最小平方估计量
non-linear least squares method 非线性最小平方法
non-linearly elastic structure 非线性弹性结构
non-linearly related variables 非线性相关变量
non-linear mapping 非线性映射
non-linear mapping analysis 非线性映射分析
non-linear material 非线性材料
non-linear mechanics 非线性力学
non-linear medium 非线性媒质
non-linear mode 非线性模式
non-linear model 非线性模型
non-linear motion 非线性运动
non-linear network 非线性网络
non-linear network theory 非线性网络理论
non-linear New-tonian fluid 广义牛顿流体
non-linear non-ideal chromatography 非线性非理想色谱法
non-linear non-separable problem 非线性不可分问题
non-linear operation 非线性运算
non-linear optic(al) coefficient 非线性光学系数
non-linear optic(al) crystal 非线性光学晶体
non-linear optic(al) effect 非线性光学效应
non-linear optic(al) material 非线性光学材料
non-linear optic(al) phenomenon 非线性光学现象
non-linear optic(al) processing 非线性光学处理
non-linear optic(al) response 非线性光学响应
non-linear optic(al) susceptibility 非线性光极化率
non-linear optics 非线性光学
non-linear optimal control 非线性最优控制
non-linear optimization 非线性(最)优化；非线性最佳化
non-linear optimum design 非线性优化设计
non-linear organic material 非线性有机材料
non-linear oscillation 非(直)线性振荡
non-linear oscillator 非线性振荡器
non-linear parameter 非线性参数
non-linear parametric function 非线性参数函数
non-linear partial differential equation 非线性偏微分方程
non-linear perturbation 非线性摄动
non-linear phase conjugation 非线性相位共轭
non-linear phase-matching effect 非线性相位匹配效应
non-linear phase shift 非线性相移
non-linear phenomenon 非线性现象
non-linear photomixing 非线性光混频
non-linear planning 非线性规划
non-linear plasma 非线性等离子体
non-linear plastic theory 非线性塑性理论
non-linear polarization 非线性偏振；非线性极化
non-linear polarization field 非线性极化场
non-linear polymer 非线性高聚物
non-linear potentiometer 非线性电位计
non-linear power amplifier 非线性功率放大器
non-linear prediction 非线性预测
non-linear problem 非线性问题
non-linear process 非直线性过程
non-linear processing 非线性处理
non-linear program(me) 非线性程序
non-linear programming 非线性规划；非线性程序
non-linear programming method 非线性规划法
non-linear programming problem 非线性规划问题
non-linear propagation 非线性传播

non-linear quantizer 非线性分层器
non-linear quantum effect 非线性量子效应
non-linear radiative deactivation 非线性辐射激活(作用)
non-linear random theory 非线性随机理论
non-linear reactance 非线性电抗
non-linear reactance amplifier 非线性电抗放大器
non-linear recording 非线性记录
non-linear reflection 非线性反射
non-linear refraction 非线性折射
non-linear region 非线性区
non-linear regression 非线(性)回归
non-linear regression analysis 非线性回归分析
non-linear regression equation of bi-variable 二元曲线回归方程；二元非线性回归方程
non-linear regression equation of one variable 一元非线性回归方程
non-linear regression model 非线性回归模型
non-linear regression technique 非线性回归法
non-linear relation 非线性关系
non-linear relationship 非线性关系
non-linear reservoir 非线性水库
non-linear resistance regulator 非线性电阻调节器
non-linear resistor 非线性电阻(片)
non-linear resistor type surge arrester 非线性电阻式电涌放电器
non-linear resonance 非线性共振
non-linear response 非线性响应(曲线)；非线性特性(曲线)；非线性反应
non-linear restoring force characteristic 非线性恢复力特征；非线性恢复力特性
non-linear restriction 非线性约束
non-linear rolling 非线性滚动运动
non-linear scale 非线性标度；非线性标度尺
non-linear scanning 非线性扫描
non-linear seismic analysis 非线性地震分析
non-linear servomechanism 非线性伺服机构
non-linear shift register 非线性移位寄存器
non-linear smoothing 非线性修匀；非线性平滑；非线性光滑
non-linear spatial filtering 非线性空间滤波
non-linear spectral density 非线性谱密度
non-linear spectroscopy 非线性光谱法
non-linear spring 非线性弹簧
non-linear stability 非线性稳度；非线性稳定性
non-linear stabilization 非线性稳定化
non-linear standard error of estimate 非线性估计标准误差
non-linear stiffness 非线性刚度
non-linear strain 非线性应变
non-linear stress-strain behavio(u)r 非线性应力—应变形状；非线性应力—应变性质
non-linear structural analysis 非线性结构分析
non-linear structural behavio(u)r 非线性结构性能
non-linear structural dynamics 非线性结构动力学
non-linear structure 非直线型结构
non-linear susceptibility 非线性磁化率
non-linear system 非线性系统
non-linear system analysis 非线性系统分析
non-linear systems incorporation 非直线系统公司
non-linear systems of algebra 非线性代数系统
non-linear system stability 非线性系统的稳定性
non-linear taper 非线性电阻分布特性
non-linear technique 非线性法
non-linear term 非线性项
non-linear theory 非线性理论
non-linear theory of elasticity 非线性弹性理论
non-linear tide 非线性潮
non-linear time base 非线性时基
non-linear time-varying model 非线性时变模型
non-linear tolerance 非线性失真容限
non-linear transfer 非线性转移
non-linear transformation 非线性变换
non-linear trend 非直线趋势；非线性趋势
non-linear tsunami wave 非线性海啸波
non-linear unit-hydrography theory 非线性单位过程线理论
non-linear variation 非线性变化
non-linear vibration 非线性振动
non-linear viscoelasticity 非线性黏弹性
non-linear viscous damping 非线性黏滞阻尼；非线性黏滞衰减
non-linear wave 非线性波
non-linear wave particle resonant effect 非线性波—粒共振效应
non-linear wound potentiometer 线绕非线性电位器
non-liner 非班轮的
non-liquid assets 非流动资产
non-liquidation basis 非清算基础
non-liquid claims 非流动债权
non-liquidity 非流动性
non-liquid ratio 不能变现的比率
non-liquid water 非液水
non-living 非生命的
non-living biomass 无生命物质
non-living resource 非生物资源
non-living substance 非生活物质
non-living thing 非生物
non-living tissue 无生命组织
non-loadable character set 不可装载字符集
non-load-bearing 非承重的；不承重的
non-load-bearing brick 非承重砖(块)
non-load-bearing clay 非承重黏土
non-load-bearing clay hollow tile 非承重黏土空心砖
non-load-bearing clay tile 非承重黏土砖
non-load-bearing concrete 非承重混凝土；不受载混凝土
non-load-bearing facade 非承重立面；非承重的建筑正面
non-load-bearing floor block 非承重地面；非承重地板块
non-load-bearing floor brick 非承重地坪砖；非承重地面砖
non-load-bearing hollow brick 非承重空心砖
non-load-bearing panel 非承重护墙板；非承重镶板
non-load-bearing partition 非承重隔墙
non-load-bearing partition(wall) 非承重隔墙；非承重隔断
non-load-bearing porous brick 非承重空心砖
non-load-bearing structure 非承重结构
non-load-bearing tile 非承重瓦；无负荷面砖
non-load-bearing wall 非承重墙；非承压墙
non-loaded container spreader 空箱吊具(集装箱)
non-loaded shear test 抗切试验
non-loading bearing 非承重
non-loading bearing tile 非承重砖
non-loading bearing wall 非承重墙
non-loading run 无负载运转
non-load voltage 无负载电压
non-local 非区域的；非局部的
non-local entry 非本地输入
non-local field theory 非向域场论
non-local identifier 非局部标识符
non-localing bearing 浮动轴承
non-local interactions 非向域相互作用
non-locality 非定域性
non-localizability 不可定位性
non-localizable 不可定位的
non-localized 非局限的；非定域
non-localized damage 不能确定地点的损坏
non-localized electron 非定域电子
non-localized fringe 非定域条纹
non-localized state 非定域态
non-local thermodynamic(al) equilibrium 非局部热动平衡
non-local variable 非局部变量
non-locating bearing 游动轴承；不定位轴承
non-locking 非锁定
non-locking button 自复式按钮
non-locking escape 非锁定转义；不封锁换码
non-locking key 非锁定按钮
non-locking press button 自动还原按钮；反跳按钮
non-locking push button 自复式按钮
non-locking relay 非锁定继电器
non-locking relay or key 自动还原继电器
non-locking shift character 非锁定转义字符；非锁定大小写转换字符；不封锁移位符
non-lock type vertical barge lift 不连锁的垂直升驳船机
non-lodging 不倒伏
non-lodging veriety 不倒伏品种
non-log centrifugal pump 不堵塞离心泵
non-log type of pump 不堵塞水泵
non-looping property 不圈结性
non-loss 无损耗

non-lubricated air compressor 无油润滑空压机
non-lubricated compressor 无油润滑压缩机；无润滑（式）压缩机
non-lubricated gas compressor 无油润滑气体压缩机
non-lubricating 不需润滑的
non-luminance 不发光
non-luminous 无光的；不发光的
non-luminous body 非自发光体；不发光体
non-luminous flame 无光(火)焰；不发光火焰
non-luminous gas 无焰气体
non-luminous sign 无光标志
non-lustrous finish 无光泽涂料；无光泽饰面(材料)
non-lustrous glazed tile 无光釉面砖
non-machined 未经机械加工的；未车光的
non-magmatic water 非岩浆水
non-magnetic 无磁性的；非磁性的
non-magnetic alloy 非磁性合金
non-magnetic body 非磁体
non-magnetic cast iron 无磁性铸铁
non-magnetic compass 非磁罗盘
non-magnetic drill collar 无磁性钻头卡圈；非磁性钻铤
non-magnetic fraction 非磁性部分
non-magnetic ionization ga(u)ge 非磁性金属材料
non-magnetic isomer 无磁性异构体
non-magnetic laboratory 非磁性实验室
non-magnetic mass spectrometer 非磁式质谱计
non-magnetic material 非磁性材料
non-magnetic metal 非磁性金属
non-magnetic mineral 非磁性矿物
non-magnetic particle 非磁化粒子
non-magnetic recording medium 非磁记录媒体；非磁记录介质
non-magnetic rod 非磁性钻杆
non-magnetics 非磁性物
non-magnetic shim 止钉；防磁钉
non-magnetic steel 非磁性钢；无磁性钢
non-magnetic sub 非磁性接头
non-magnetic substance 非磁性物质
non-magnetic vessel 无磁性船
non-magnetic watch 防磁表
non-magnetometry 非磁力测定
non-maintenance time 非维修时间
non-mammade sources 非人为污染源
non-management job 非管理性工作
non-man control 无人控制的
non-mandatory target 非指令性指标
non-maneuverable 非机动的
non-manipulative compression joint 非手工压接
non-manipulative joint 非手工接头
non-manmade pollution sources 非人为污染源
non-manual workers 非体力劳动者
non-manufacturing business 非制造业企业
non-manufacturing cost 非制造成本；非生产费用
non-manufacturing industry 非制造业
non-manufacturing industry product 非制造业产品
non-manufacturing population 非工业人口
non-mapping mode 非变换方式
non-marginal adjustment 非边际调整
non-marine 非海成的
non-marine bacteria 非海洋细菌
non-marine deposit 非海相沉积
non-marine micro-organism 非海洋性微生物
non-marine reservoir rock 非海相储集岩
non-maritime 非海成的
non-maritime leg 非海运远程
non-market borrowing 非市场借债
non-market-economy country 非市场经济国家
non-marketed product 非市场产品
non-market transaction cost 非市场交易费用
non-Markovian process 非马尔夫过程
non-maskable 不可屏蔽的
non-maskable interrupt 非屏蔽中断；不可屏蔽中断
non-maskable interrupt request 不可屏蔽中断请求
non-masking interrupt 非屏蔽中断
non-matched 不相应的；不匹配的
non-matched data 非匹配数据；不匹配数据
non-matching 不匹配
non-match sealing 非匹配封接
non-material capital 非物质资本
non-material goods 非物质财物

non-material input 非物质投入
non-material loss 无形损失
non-material product 非物质产品
non-material service 非物质服务
non-material sphere 非物质领域
non-material strategy 非物质策略
non-mathematical program(me) 非数学程序
non-matting 不消光的；有光泽的；无席垫的；非装饰性花边
non-Maxwellian tail 非麦克斯韦尼部
non-measurable flow 不可测流
non-measurable set 非可测集
non-mechanical 非碎屑的；非机械的
non-mechanical classifier 非机械分级机
non-mechanical noise 非机械噪声
non-mechanical vehicle 非机动车
non-mechanized 非机械化的
non-mechanized coal mining 煤矿非机采型
non-mechanized hump 非机械化驼峰【铁】
non-mechanized hump yard 非机械化驼峰站【铁】
non-mechanized vehicles 非机动工具的运输业
non-melt 非熔化
non-meltable 不熔化的
non-melt-settlement 不融沉
non-member rate 非会员运费率
non-member state 非会员国
non-merchandise 非商品
non-mercurial fungicide 无汞防霉剂
non-mercuric 非汞的
non-mercuric fungicide 非汞杀菌剂
non-mercuric preparation 非汞制剂
non-mercuric preservative 非汞防腐剂
non-metal 非金属
non-metal analysis 非金属分析
non-metal heat exchanger 非金属材料换热器
non-metallic 非金属的
non-metallic additive 非金属添加剂；非金属加入料
non-metallic air channel 非金属风道
non-metallic air duct 非金属风道
non-metallic alloy ingredient 非金属合金组分
non-metallic bearing 非金属轴承
non-metallic binder 非金属黏结剂
non-metallic cable 非金属电缆
non-metallic chemical mechinery works 非金属化工机械厂
non-metallic coating 非金属涂层；非金属覆层
non-metallic commodities 非金属矿产
non-metallic compound 非金属化合物
non-metallic crown 非金属皂冠
non-metallic element 非金属构件；非金属元素
non-metallic enamel 五金属光泽瓷漆
non-metallic fiber-reinforced ceramics 非金属纤维加固陶瓷
non-metallic film 非金属膜
non-metallic fire-proof building material 非金属防火建筑材料
non-metallic fire-proof uncovered floor 非金属防火无面层地板
non-metallic fluxing hole 非金属熔洞
non-metallic friction materials 非金属摩擦材料
non-metallic fungicide 非金属杀菌剂
non-metallic fusion point 非金属软化点
non-metallic gasket 非金属垫片
non-metallic gear 非金属齿轮
non-metallic impurity 非金属杂质
non-metallic inclusion 非金属夹杂物
non-metallic ingredient 非金属组分
non-metallic inorganic coating 非金属无机覆盖层
non-metallic insulation 非金属绝热隔声材料
non-metallic ion 非金属离子
non-metallic lubricant 非金属润滑剂
non-metallic luster 非金属光泽
non-metallic material 非金属材料
non-metallic matrix composite 非金属基复合材料
non-metallic mineral 非金属无机物；非金属矿物；非金属矿
non-metallic mineral deposit 非金属矿床
non-metallic mineral products 非金属矿产品货物
non-metallic mineral resources 非金属矿产资源
non-metallic mo(u)ld material 非金属制模材料
non-metallic ore 非金属矿石；非金属矿产
non-metallic paint 非金属漆
non-metallic pipe[piping] 非金属管道；非金属管

non-metallic products 非金属产品
non-metallic reduction agent 非金属还原剂
non-metallic resistor 非金属电阻器
non-metallic resistor furnace 非金属电阻炉
non-metallics 非金属物质
non-metallic sheath 非金属包皮
non-metallic sheathed cable 非金属铠装电缆；非金属护皮电缆；非金属包皮电缆
non-metallic smelling industry 非金属熔炼工业
non-metallic substance 非金属物质
non-metallic tank 非金属罐
non-metallic tube 非金属管
non-metallic tubing 非金属管
non-metallic type optic(al) cable 非金属型光缆
non-metallic vessel 非金属容器
non-metallic waterproof cable 非金属防水电缆
non-metallic waterproof material 非金属防水材料
non-metallic waterstop 非金属止水片；非金属止水带
non-metalliferous 非金属的
non-metalliferous ore sources 非金属矿产资源
non-metal material 非金属材料；非金属物质
non-metal material detector 非金属材料检测仪
non-metal-non-metal eutectic alloy 非金属一非金属共晶合金
non-metal powder 非金属粉末
non-metal ultrasonic detector 非金属超声波探测仪
non-metering 不计量的
non-metering relay 计数器切断继电器
non-methane compound 非甲烷化合物
non-methane hydrocarbon 非甲烷烃；非甲烷碳氢化合物
non-methane products 非甲烷产品
non-methane volatile organic compound 非甲烷有机化合物
non-methanogen bacteria 非产甲烷菌
non-methanogenic phase 非甲烷生成相
non-metric 非公制的
non-metric camera 非量测摄影机
non-micellar solution 非胶束溶液
non-microbiofilm organism 非微生物膜生物
non-migrating plasticizer 非迁移型增塑剂
non-migrator permutation 非迁移排列
non-migratory 不迁移的；不洄游的
non-military architecture 非军用建筑；非军事建筑
non-mineral adsorbent 非无机吸附剂
non-mineral aggregate 非矿物集料；非矿物骨料
non-mineral anomaly 非矿物异常
non-minimum balance plan 非最少余额计划
non-minimum phase system 非最小相系统
non-miscible phase 不溶混相
non-mixing bowl pollutant 非活性污染物；不起化学反应的污染物
non-mode coupled operation 非模耦合运转
non-modulation system 非调制方式
non-moisture-absorbing dust 不吸湿尘土；不吸湿粉尘
non-moisture proof 不防潮
non-moisture transferring membrane 水分传递阻止膜
non-monetary account 非货币性账户
non-monetary assets 非货币资产
non-monetary assets and liabilities 非货币资产和负债
non-monetary consideration in capital investment decisions 在基本建设投资决策中的非货币因素
non-monetary contribution 非货币支持
non-monetary data 非价值量数据
non-monetary items 非货币性项目
non-monetary liability 非货币债务
non-monetary statistics 非价值量统计资料
non-monetary transaction 非货币交易
non-monetary unit 非货币计量单位
non-monetary values 非金钱上的价值
non-monochromatic emission 非单色放射
non-monochromatic field 非单色场
non-monochromatic radiation 非单色放射辐射
non-monotonic utility function 非单调效用函数
non-monsoon period 无季风期
non-monumented benchmark 不埋石水准点
non-monumented boundary peg 临时界桩；未埋石界桩
non-motile 无动力的

non-motorized traffic 非机动交通
non-motorized traffic lane 非机动车道
non-motorized vehicle 非机动车
non-motorized vehicle lane 非机动车道;慢车道
non-movable oil 不可动油
non-moving stair(case) 非移动式楼梯
non-multative 非破坏性的
non-multative traffic 不破坏路面的交通
non-multiple exchange 单式交换机
non-multiple switchboard 简式交换机;单式交换机
non-multipulse chain 非多脉冲链
non-nailable 钉不进的;不可打钉的
non-nailable deck 不能钉钉子的板面
non-natural 人造的;非天然的
non-navigable 不通航的;不能通航的
non-navigable branch 非通航汊道
non-navigable dam 不通船坝;无船闸水坝;不通航坝;不可通航的坝
non-navigable reach 不通航河段
non-navigable river 不通航河流
non-navigable stream 不通航河流
non-navigational chart 航海参考图
non-negative combination 非负组合
non-negative condition 非负性条件
non-negative definite 非负定
non-negative definite matrix 非负定矩阵
non-negative form 非负形式
non-negative function 非负函数
non-negative integer 非负整数
non-negative integral form 非负积分形式
non-negative matrix 非负矩阵
non-negative number 非负数
non-negative operator 非负算子
non-negative quadrant 非负象限
non-negative restriction 非负限制
non-negative type 非负型
non-negative value 非负值
non-negative variable 非负变量
non-negativity condition 非负性条件
non-negativity constraint 非负性的约束条件
non-negativity restriction 非负(性)约束
non-negotiable 不准转让的;不可转让(的);不可流通的
non-negotiable bills of lading 副本提单;提单副本
non-negotiable cheque 非流通支票
non-negotiable copy 非流通抄本
non-negotiable copy of bill of lading 非流通提单副本;不能提货的提单副本
non-negotiable endorsement 不可转让的背书票据
non-nested family 非成套类
non-nested hypothesis 非成套假设
non-nested loop mode 非嵌套循环方式
non-nested model 非成套模型
non-network control program(me) station 非网络控制程序站
non-neutral contact 非中性接触
non-neutral tax 非中性税收
non-newable 非更新的
non-Newtonian behavio(u)r 非牛顿性状
non-Newtonian flow 非线性黏性流;非牛顿流
non-Newtonian flow measurement 非牛顿流体流量测量
non-Newtonian fluid 非线性黏性流体;非牛顿流体
non-Newtonian fluid flow 非牛顿流体流
non-Newtonianism 非牛顿性
non-Newtonianity 非牛顿黏度
non-Newtonian liquid 非牛顿液体
non-Newtonian suspension 非牛顿悬浮体
non-nitrogen resin 不含氮树脂
non-nitroglycerine explosive 不含硝化甘油炸药;非硝化甘油炸药
non-noble metal 非贵金属的
non-nodal well 非节点井
non-normal 非正规的;不正规的
non-normal assumption 非正态假定
non-normal classical linear regression model 非正态古典线性回归模型
non-normal continuous distribution 非正态连续分布
non-normal data 非正态数据
non-normal distribution 非正态分布
non-normal equation 非模方程
non-normal error term distribution 非正态误差项分布

non-normality 非正态性;不垂直
non-normal population 非正态总体
non-nuclear 非核的
non-nuisance technique 无公害技术
non-null class 非空类
non-null hypothesis 非虚无假设;非零假设
non-numeric(al) 非数值(的)
non-numeric(al) application 非数值应用
non-numeric(al) character 非数字字符;非数字记号
non-numeric(al) coding 非数值编码
non-numeric(al) comparison 非数字比较;非数值比较
non-numeric(al) data processing 非数值数据处理
non-numeric(al) factor 非数值因素
non-numeric(al) information 非数字信息
non-numeric(al) item 非数字项;非数值项
non-numeric(al) literal 非数字文字;非数值文字
non-numeric(al) operation 非数字运算;非数字操作
non-numeric(al) problem 非数值问题
non-numertic processing 非数值处理
nonobjectivism 抽象主义
non-observable-adverse-effect level 无可见有害作用水平
non-observable-effect level 无可见作用水平
non-observable variable 不可观测的变量
non-observance of terms 不遵守条款
non-observed-adverse-effect level 无可见损害作用水平
non-observed-effect level 无可见作用水平
non-obstacle facility 无障碍设施
non-obviously seasonal change range 季节变化不显著带
non-occupational noise exposure 非职业性噪声接触
non-occupied track 空闲线路
non-official 非正式的;非官方的
non-official language 非官方语言
non-official market 非官方市场
non-official sign 非正式标志
non-ohmic contact 非欧姆接触
non-ohmic resistor 非线性电阻器
non-oil base 非油基的
non-oil export 非石油出口
non-oil-exporting developing country 非石油输出的发展中国家
non-oil impreguated prefilter 非油浸式粗滤器
non-oil lubrication 非油润滑
non-oily 非油滑的;非油质的;非油性的;不含油的
non-opaque pigment 体质颜料
non-open harbo(u)r 不开放港口;不开放港埠
non-opening bridge 非开合桥;无孔桥
non-opening die 非开合模
non-opening mo(u)ld 非开合模
non-opening side scuttle 固定舷窗
non-open port 非开放港(口);不开放港口;不开放港埠
non-open type 永闭式
non-operable instruction 非操作指令
non-operated railroad 不通车铁路
non-operated railway 不通车铁路
non-operate value 不吸起值;不工作值
non-operating 停运的;非营业性的;非业务性的;非生产性的
non-operating account of income and expenses 非营业收益及费用账户
non-operating ampere turn 不吸安匝
non-operating assets 非运营资产;非营业资产
non-operating capital 非经营性净资产
non-operating charges 非生产开支
non-operating company 非自营公司
non-operating cost 非营业成本
non-operating current 不吸动电流;不工作电流
non-operating earnings 营业外收益
non-operating expenses 营业外费用;业外费用;非营业费用;非业务费用
non-operating expense audit 营业外支出审计;营业外支出表
non-operating holding company 非营业的控股公司
non-operating income 营业外收入;营业外收入;非营业收入
non-operating income and expense account 非营业收益及费用账户

non-operating income audit 营业外收入审计
non-operating interest 非营业权益;非生产权益
non-operating item 非营业项目;营业外项目
non-operating loss 非生产性损失;非生产性损耗
non-operating luffing 调整性变幅;非工作性变幅
non-opera ting outlay 非营业性支出
non-operating profit 营业外利润
non-operating profit and loss 营业外损益
non-operating property 非营业财产
non-operating revenue 营业外收入;非营业收入;非营业纯收入
non-operating revenue and expenditures 营业外收支
non-operating station 非营业站
non-operating working time 非生产性工作时间
non-operational 非业务性的
non-operational cost 非业务费用
non-operation instruction 空操作指令;无操作指令
non-operation shock 非工作振动
non-operative position 不工作位置
non-operator 非操作人员
non-operculate 无盖的
non-optical propagation 封闭通道传播
non-optimal output coupling 非最佳输出耦合
non-optimal value 非最佳值
non-optional 没有选择的
non-orbital 无轨道的
non-orgenic 非构造的
non-orientable 不能定向的;不可定向的
non-orientable dipole moment 不可取向偶极矩
non-orientable manifold 非定向流形
non-oriented communication network 无向通信网
non-oriented electric(al) steel 无取向电工钢;无定向电炉钢
non-oriented silicon steel 无方向性硅钢片
non-oriented silicon steel sheet 不定向硅钢板
non-orienting force effect 非保向力效应
non-originating content 非本国含量
non-orographic region 非山岳地区;非山区
non-orographix lifting 非地形抬升作用
non-orthogonal 非正交的
non-orthogonal axes 非正交轴
non-orthogonal data 非正交性数据
non-orthogonal frame 非正交框架
non-orthogonality 非正交性
non-orthogonal system 非正交系统
non-oscillating 不振荡的
non-oscillating state 非振荡状态
non-oscillatory 非振荡的;不摆动的
non-oscillatory discharge 非振荡放电
non-overflow 非溢流
non-overflow dam 非溢流坝;不溢流坝;不淹丁坝
non-overflow dam section 非溢流坝段
non-overflow entry 非溢出项
non-overflow entry position 非溢出项位置
non-overflow groin 非淹没式丁坝;不淹丁坝;不过水丁坝
non-overflow section 非溢流区段;非溢流段
non-overlap mode 非重叠方式
non-overlapping 不重叠的
non-overlapping code 不重叠密码
non-overlapping intervals 非重叠区间
non-overlapping line approximation 非重叠线近似
non-overlapping spatial multiplexing 非重叠空间多路传输
non-overlapping winding 非重叠绕组
non-overloading type 不超载式
non-owner 非所有人
non-ownership 非股权式
nonox 醇醛萘胺
non-oxidation 无氧化
non-oxidative deamination 非氧化性脱氨(作用)
non-oxide 非氧化物
non-oxide ceramics 非氧化物陶瓷
non-oxided lubricant 非氧化润滑剂
non-oxide glass 非氧化物玻璃
non-oxide glass fiber 非氧化物玻璃纤维
non-oxidizability 抗氧化性;不可氧化性
non-oxidizable 不可氧化的
non-oxidizable alloy 不氧化合金
non-oxidizing atmosphere 无氧化气氛
non-oxygen flux 无氧焊剂
non-packed cargo 非包装货
non-packet mode terminal 非分组方式终端

non-pageable area 不可分页区
non-pageable dynamic(al) area 不可调页动态区
non-pageable partition 非可分页面部分；不可调页分区
non-pageable region 不可调页区
non-paging mode 非分片记录方式
non-paired terrace 不对称阶地
non-paraffinous oil 不含蜡石油
non-parallel banks method 不平行边坡法
non-parallel-chord truss 上下弦非平行的桁架
non-parallel crank four bar linkage 不平行四杆回转链系
non-parallelism 不平行度
non-parallell-chord truss 上下弦非平行桁架
non-parallel train working diagram 非平行运行图
non-parametric 非参数的
non-parametric Bayesian estimation 非参数贝叶斯估计
non-parametric comparison 非参数比较
non-parametric confidence interval 非参数置信区间
non-parametric estimation 非参数估计
non-parametric hypothesis 非参数假设
non-parametric induction 非参数归纳
non-parametric inference 非参数推断
non-parametric method 非参数方法
non-parametric model 非参数模型
non-parametric one-sample model 非参数单样本模型
non-parametric sequential analysis 非参数序贯分析
non-parametric statistic(al) method 非参数统计（方）法
non-parametric statistics 无母数统计；非参数统计；非参变量性统计
non-parametric technique 非参数方法
non-parametric test 无母数测验；非参数检验；非参数测定
non-parametric tolerance interval 非参数容许区间
non-parametric tolerance limit 非参数容许限
non-parametric two-sample model 非参数双样本模型
non-parasitic 非寄生的
non-paraxial imaging 非旁轴成像
non-parcel-bulk shipment 非整船散装货
non-pareil 六点间隔
non-parent 非亲本
non-participating member 非正式成员
non-participating preferred stock 不参与分红的优先股
non-particulate grout 非颗粒化灌浆
non-passage of signal 信号闭塞
non-passenger ship 非客轮
non-passing 禁止通行的
non-passing line 禁越车线
non-passing minimum sight distance 最短停车视距
non-passing sight distance 停车视距；不超车视距；安全停车视距
non-passing zone 禁止超车区；不准越车区段；不通行区域
non-pathogenic 不致病的
non-pathogenic bacteria 非致病细菌
non-pathogenic organism 非致病生物(体)
non-paying 不合算的
non-paying weight 非有效载重
non-payment 停止支付；不付款
non-peak hours 平时
non-pecuniary goal 非金钱目的
non-pelagic limestone 非深海灰岩
non-penerative 非贯穿的
non-penetrating 非贯穿的；不透明的
non-penetration 未焊透
non-penetrative 非透入性的
non-penetrative element 非透入性要素
non-penetrative fabric 非透入性组构
non-pensionable service 不计养恤金的服务期间
non-pensionable supplements 不计养恤金的补助费
non-perennial canal 非常年使用运河；非常年使用渠道
non-perfect elastic collision 非完全弹性碰撞
non-perfect fluid 非完全流体；非理想流体；不完全流体

non-perforated 未打过孔的
non-performance 不执行；不履行
non-performance of contract 合同的不履行
non-performer 没有成就的人
non-periodic 非周期(性)的
non-periodic(al) calculation of unit expenditures 不定期单位支出计算
non-periodic(al) comet 非周期彗星
non-periodic(al) component 非周期分量
non-periodic(al) cost computation 非定期成本计算
non-periodic(al) current 非周期性气流；非周期性电流；非周期流
non-periodic(al) disturbance 非周期扰动
non-periodic(al) excitation 非周期激振
non-periodic(al) function 非周期函数
non-periodic(al) phenomenon 非周期现象
non-periodic(al) process 非周期性过程
non-periodic(al) production 非周期性生产
non-periodic(al) profit and loss 不定期损益
non-periodic(al) repair 非定期维修；不定期维修
non-periodic(al) report 不定期报告；非定期报告
non-periodic(al) statistic(al) report 不定期统计报表
non-periodic(al) variable 非周期变量
non-periodic(al) variation 非周期性变化；非周期变化
non-periodic(al) wave 非周期波
non-periodic inspection 不定期检验
non-periodicity 非周期性
non-perishable 不易腐烂的
non-perishable cargo 不易腐烂的货物；不易腐败的货物
non-permanent construction 非永久性建筑
non-permanent deformation 弹性变形；非永久性变形
non-permanent foreground 非固定前台
non-permanent residents 非固定居民
non-permanent strain 弹性应变
non-permeable soil 非渗水土
non-permeable soil subgrade 非渗透土路基
non-permeable stringer 不渗透夹层
non-perpetual inventory 非经常性库存
non-persistent 不安定的
non-persistent pesticide 非持久性农药；无残留农药
non-persistent screen 无余辉荧光屏
non-persistent war gas 非持久性毒气
non-personal selling 非人员销售
non-perspective 非透视的
non-perspective azimuthal projection 非透视方位投影
non-petroleum base 非石油基
non-petroleum source 非石油来源
non-phase-inverting 不反相的
non-phase-match 非相位匹配
non-phenolic tanning agent 非酚类鞣剂
non-phosphorus scale and corrosion inhibitor 无磷阻垢缓蚀剂
non-photographic(al) base 印版空白部分
non-photographic(al) material 无银感光材料
non-photographic(al) photogrammetry 非光学成像摄影测量学
non-photographic(al) system 非照相系统
non-phototographic sensor 非热影传感器
non-physical assets 非实物资产
non-physical capital 非质性资本
non-physical loss 无形损失
non-physical primary 假想基色
non-physical wear 无形耗损；无形损耗
non-phytoplankton oxygen consumption 非浮游植物耗氧量
non-pickup 不吸动值
non-pick-up time 不粘时间
non-pigmented 非染色的；未着色的；无色素的；不变色的
non-pipelined processor 非流水线式处理机
non-piping steel 无缩孔钢
non-placement rate 非安置率
non-planar 非平面的
non-planar frame 空间框架；非平面框架
non-planar graph 非平面图
non-planar network 不同平面网络；不同平面四端网络
non-plane motion 曲面运动

non-planeness measurement 不平度测量
non-planned repair 非计划修理
non-plaster mo(u)ld 非石膏质模型
non-plastic 无塑性的；非塑性的；非可塑性的
non-plastic clay 无塑性黏土
non-plastic concrete 干硬性混凝土；非塑性混凝土
non-plastic fracture 脆性破裂
non-plasticity 无塑性
non-plastic material 非塑性物料；非塑性材料
non-plastic powder 无塑性粉末
non-plastic silt 无塑性粉土
non-plastic soil 无塑性土；非塑性土(壤)
non-plastic stage 非塑性阶段
non-platinum bushing furnace 非铂板护丝炉
non-plunging fold 非倾伏褶曲
non-poisonous material 无毒物质
non-poisonous or less-poisonous electroplating 无毒或低毒电镀
non-point agriculture pollution source 非点农业污染源
non-pointed nose 钝头部
non-point pollution 非点源污染；不定点污染
non-point pollution potential 潜在非点污染
non-point pollution source 面污染源；非点污染源
non-point source 非点源
non-point source assessment 非点污染源评估
non-point source discharge 非点污染源排放
non-point source groundwater contamination 非点污染源地下水污染
non-point source inputs as conservative 非点污染源恒定输入
non-point source inputs as non-convervative 非点污染源非恒定输入
non-point source load 非点污染源负荷
non-point source model 非点源模型
non-point source of pollution 非点污染源
non-point source of urban pollution 城市污染非点源
non-point source pollutant 非点源污染物
non-point source pollutant output 非点污染源污染物输出
non-point source pollutant residence time 非点源污染物滞留时间
non-point source pollutant transport modeling 非点源污染物输移模拟
non-point source pollution 非点源污染；非点污染源水污染；不定源污染
non-point source pollution control 非点污染源控制
non-point source pollution control strategy 非点源污染控制对策
non-point source pollution model 非点源污染模型
non-point source pollution model of water body 水体非点源污染模型
non-point source pollution potential 潜在非点源污染；非点污染源势
non-point source pollution risk assessment 非点源污染风险评估
non-point source simulation model 非点污染源模拟模型
non-point source survey 非点污染源调查
non-point source water quality and quantity modeling 非点污染源水质水量模拟
non-Poissonian model 非泊松模型
non-polar 非极性的
non-polar action 非极性作用
non-polar adsorption 非极性吸附
non-polar agent 非极性溶剂
non-polar binding 无极性结合
non-polar biogenic substance 非极性生物剂
non-polar compound 无极性化合物；非极性化合物
non-polar covalent bond 非极性共价键
non-polar crystal 非极性晶体
non-polar dissociation 非极性离解作用
non-polarity 无极性
non-polarizable 不偏振的；不可极化的
non-polarizable electrode 不可极化电极
non-polarized 非极化的
non-polarized electrode 乏极化电极
non-polarized light 非偏振光
non-polarized relay 非极化断电器
non-polarized return-to-zero 非极性归零制；无极性归零记录方式
non-polarized return-to-zero recording 非极性归

零记录(法)
non-polarizing electrode 非极化电极
non-polar liquid 无极液体;非极性液体
non-polar molecule 无极分子;非极性分子
non-polar optic(al)scattering 非偏振光散射
non-polar organic compound 非极性有机化合物
non-polar organic reaction 非极性有机反应
non-polar organic substance 非极性有机剂
non-polar pesticide 非极性农药
non-polar phase 非极性相
non-polar pigment 非极性颜料
non-polar polymer 非极性高聚物
non-polar sample 非极性试样
non-polar solvent 非极性溶剂
non-polar stationary phase 非极性固定相
non-polar support 非极性载体
non-polar symmetry axis 非极性对称轴
non-polar variation of latitude 非极性纬度变化
non-polishing 不易磨光
non-polishing aggregate 耐磨集料;耐磨骨料;不易磨光集料
non-polluted industrial wastewater 清洁工业废水;无污染工业废水;生产废水
non-polluted technology 无污染工艺
non-polluting automobile 无污染汽车
non-polluting coating 无污染涂层
non-polluting energy sources 无污染能源
non-polluting fuel 无污染燃料
non-polluting or pollution-reducing process 无污染或少污染工艺
non-pollution 无污染
non-pollutive technology 非污染技术
non-pored timber 针叶树材
non-pored wood 无孔材
non-porosity 无孔性
non-porous 无孔的;非孔隙的
non-porous article 无孔制品
non-porous bearing 无孔轴承
non-porous film 无孔膜
non-porous particle 无孔颗粒
non-porous solid 无孔隙固体
non-porous surface 无孔表面;非多孔表面
non-porous wood 无孔材
non-positive compressor 离心压缩机
non-positive drive 力的闭环传动
non-positive metric 非正度规
non-positive operator 非正算子
non-potable mineralized water 非饮用矿泉水
non-potable reclaimed water 中水
non-potable reuse 非饮用性再用
non-potable water 杂用水;非饮用水;不宜饮用水;不合饮用水
non-potable water for agricultural purposes 农用(非饮用)水
non-potable water for industrial purposes 工业(非饮用)水
non-potable water network 非饮用水网
non-potable water pumping station 非饮用水泵站
non-potable water reuse 非饮用水回用
non-potential carbon of kerogen in source rock 母岩干酪根中无潜力的碳
non-potential motion 无势运动
non-power 不做功
non-power driven vessel 非机动船
non-power driving ship 非动力船
non-powered 非机动的
non-powered ship 无动力船
non-power vessel 非动力船
non-predetermined 非预先决定的
non-predicative process 不可预计过程
non-predominant traffic direction 非主要运输方向
non-preemptive algorithm 非抢先算法
non-preemptivepriority 非优先顺序排队法
non-preemptive priority model 非优先次序模型
non-preemptive scheduling 非抢先调度
non-preferential duty 非优惠关税
non-preformed 松散的
non-preformed wire rope 非预搓丝股的钢索
non-prejudicial disclosure 非损害性的公开
non-premium cement 廉价水泥
non-presetting control 非调整控制;非程序控制
non-pressed air method 不压缩空气法
non-pressed sintering diamond reaming shell 无压烧结金刚石扩孔器

non-pressure charging 非增压式供气
non-pressure culvert 无压力式涵洞
non-pressure cure 无压硫化
non-pressure(d)drainage 无压排水;重力排水
non-pressure drain 无压排水
non-pressure drainage works 无压排水工程
non-pressure electric(al)water heater 低压式电热水器
non-pressure flow 无压流
non-pressure method of drilling 减压法钻进
non-pressure parts 不受压部件
non-pressure pipe 无压管道;无压管
non-pressure process 常压法;非压力处理法(木材防腐处理)
non-pressure regulating roller leveller 不调压式多辊矫直机
non-pressure thermit weld 不加压热剂焊缝
non-pressure thermit welding 不加压铝热剂焊接;热剂铸焊;常压铝热法焊接
non-pressure treating method 非加压防腐处理法
non-pressure treatment (木材防腐的)无压力处理(法);无压处理;常压处理(法);非加压处理
non-pressure treatments for timber 木材无压防腐处理
non-pressure tunnel 渠流隧洞;无压隧洞;无压隧道
non-pressure welding 不加压焊接
non-pressurized 常压的;不加压的
non-pressurized body 非密封壳体
non-pressurized interior 透气舱
non-pressurized solar water heater 不加压太阳能热水器
non-prestressed bar reinforcement 无预应力的粗钢筋
non-prestressed reinforcement 非预应力钢筋
non-price competition 价格外竞争;非价格竞争
non-primary economic activity 非主要经济活动
non-primary product 非主要产品
non-prime attribute 非主属性
non-priming cooler 不带液冷却器
non-primitive 非初级的
non-principal test 非主要试验
non-principal trial 非主要试验
non-print area 空白版面
non-priority interrupt 非优先(级)中断
non-priority objective 非优先目标
non-prismatic 变截面的
non-prismatic beam 变截面梁
non-prismatic channel 非棱柱体渠道;变截面河槽
non-prismatic continuous beam 变截面连续梁
non-prismatic cross-section 变截面
non-prismatic member 变截面构件
non-privileged mode 非特权方式;非特惠方式
non-privileged program(me) 非特惠程序
non-privileged instruction 非特许指令
non-probabilistic systems of equations 非概率性方程组
non-probability sample 非概率样本;非概率样品
non-probability sampling 非概率抽样
non-procedural language 非过程语言
non-process language 非过程语言
non-producing well 非生产井
non-production cost 非制造成本
non-production time 非生产时间
non-productive 非生产性的;非生产性;非生产的
non-productive accumulation 非生产性积累;非生产积累
non-productive asset 非生产性资产
non-productive branch 非生产部门
non-productive building 非生产性建筑
non-productive capital 非生产资本;非生产性资本
non-productive capital construction 非生产性的基本建设
non-productive construction 非生产性建设
non-productive construction investment 非生产性建设投资
non-productive consumption 非生产性消费
non-productive department 非(物质)生产部门
non-productive enterprise 非生产企业
non-productive expenditures 非生产性支出;非生产性开支;消费性支出
non-productive expenses 非生产性费用;非生产性费用
non-productive fixed assets 非生产用固定资产
non-productive formation 非生产建造;非生产层

non-productive funds 非生产性资金
non-productive heat consumption 空载热耗
non-productive investment 非生产性投资
non-productive labo(u)r 非生产性劳动;非生产劳动;非生产工人
non-productive operation 辅助操作;非生产操作;辅助工作
non-productive personnel 非生产人员
non-productive poll 辅助轮询;辅助查询
non-productive projects in capital construction 非生产性基本建设
non-productive season 非生产季节
non-productive sector 非生产部门
non-productive service 非生产性行业
non-productive services for daily life 非生产性生活服务
non-productive sphere 非生产领域
non-productive task 辅助任务
non-productive thinning 非生产性疏伐
non-productive time 非有效工时;非生产时间
non-productive wage 非生产(性)工资
non-product output 非产品产出
non-professional 外行人;非职业化的
non-professional diver 非职业潜水员
non-profit 非营利的
non-profit accounting 非营利会计
non-profit accounting equation 以不计营利为基础的核算方程式
non-profit association 非营利机构
non-profit book-keeping 非营利簿记
non-profit company 非营利公司
non-profit corporation 非营利性公司;非营利法人
non-profit enterprise 非营利企业
non-profit institution 非营利机构
non-profit making association 非营利性社团
non-profit making company 非营利性公司
non-profit making corporation 非营利公司
non-profit making organization 非营利组织
non-profit organization 非营利组织(机构);非营利机构
non-programmed decision 非定型化决策;非程序判定;非程序化的决策
non-programmed decision making 非定型决策
non-programmed halt 非程序停机
non-programmed lending 非项目贷款
non-programmed problem-solving 非定型化问题的解决
non-programming user 非程序设计用户
non-projected illumination 非投影照明
non-project lending 非项目贷款;非项目出租
non-project loan 无项目贷款;非项目贷款
non-project type assistance 非计划型援助
non-proliferation 禁止扩散
non-propagating crack 不扩展裂纹
non-propagating fatigue crack 不扩展疲劳裂纹;无扩展疲劳裂缝
non-propelled 非自航的;非自动推进的
non-propelled barge 非自航式驳船
non-propelled vessel 非自航船
non-propelling dredge(r) 非自航(式)挖泥机;非自航(式)挖泥船;非自动推进式挖泥船
non-propelling suction dredge(r) 非自航吸扬式挖船
non-proportional test 不成比例试验
non-proportional viscous damping 非正比黏性阻尼
non-proprietary environmental protection technology 非产权环境保护技术
non-proprietary name 无产权名
non-protected beach 无防护海滩
non-protected riverbed 无防护的河床
non-protecting duty 非保护关税
non-protective 无防护的
non-protective platform 无防护站台
non-protein nitrogen 非蛋白氮
non-protestable bill 无追索权汇票
non-proton flare 非质子耀斑
non-protonic solvent 非质子性溶剂
non-public company 非公营公司
non-public offering 不公开报价
non-public works 非公用工厂
non-pudded soil 非紧实土壤
non-pullted industrial wastewater 无污染的生产废水

non-pulverulent residue 非粉状残渣;不脆残渣
non-pumpers 无唧泥现象(混凝土板)
non-purgeable dissolved organic carbon 不可清除溶解有机碳
non-purgeable organic carbon 不可吹除有机碳
non-purgeable organic halide 不可清除有机卤化物
non-purgeable total organic carbon 不可吹除总有机碳
non-purified wastewater 未净化污水
non-quadratic loss function 非二次损失函数
non-quantifiable factor 非量化因素;非定量因素;不可定量因素
non-quantified statement 非量化命题
non-quantitative 非计量性的
non-quantitative description 非数量描述
non-quantitative objective 非定量目标
non-quantitative variable 非数量变量
non-quantum mechanics 非量子力学
non-quarter-wave stack 非四分之一波长膜系
non-quartz silica 非石英二氧化硅
non-quaternary 非季铵盐的
non-quota circulating fund 非定额流动资金
non-quota current fund 非定额流动资金
non-quota hours 非定额时间
non-quoted 非引用
non-radiable 不透(放)射线的
non-radial 非径向的
non-radial oscillation 非径向振荡
non-radiating 不辐射(的)
non-radiation field 不辐射场
non-radiative 不辐射
non-radiative decay process 无辐射衰变过程
non-radiative generation 无辐射产生
non-radiative jump 非辐射跃迁
non-radiative loss 非辐射损耗
non-radiative process 无辐射过程;非辐射过程
non-radiative recombination 非辐射复合
non-radiative resonance 无辐射共振
non-radiative target 非辐射目标
non-radiative transition 无辐射跃迁;非辐射跃迁
non-radioactive component 非放射性组分
non-radioactive electron capture detector 非放射性电子捕获检测器
non-radioactive test 非放射性试验
non-radioactivity air survey method 非放射性气体法
non-radiogenic isotope 非放射成因同位素
non-railway 非铁路
non-random 非随机
non-random access 非随机访问;非随机存取;非任意抽取
non-random comparison group methods 非随机比较分组法
non-random damage 非随机性损害
non-random deviation 非随机偏离
non-random distribution 非随机分布
non-random index 非随机指数
non-random injury 非随机性损害
non-randomized decision function 非随机化判定函数
non-randomized estimators 非随机化估计量
non-randomized risk set 非随机化风险集
non-randomized strategy 非随机化策略
non-randomized test 非随机化检验
non-random probability sample 非随机概率样本
non-random sample 非随机样本
non-random segregation 不随机分离
non-random structure 有序结构
non-random value 非随机值
non-random variable 非随机变量
non-rational function 非有理函数
non-rattling 减震
non-reacting component 不起化学反应化合物
non-reaction phase 不反应相
non-reactive 惰性的;无反馈的;无反应的;非电抗性的
non-reactive aggregate 惰性集料;惰性骨料
non-reactive black 非活性炭黑
non-reactive circuit 无抗电路
non-reactive contaminant 非活性污染物
non-reactive diluent 非反应性稀释剂
non-reactive gas 非反应气体
non-reactive impedance 无抗阻抗
non-reactive load 无抗负载;电阻性负载

non-reactive phase 不反应相
non-reactive resistance 无电抗电阻
non-reactive solute transport 非活性溶质运移
non-reactive voltamperes 无无功伏安
non-reactivity 惰性
non-readout 未检出;未读出
non-ready program(me) 等待状态程序
non-real-time processing 非实时处理
non-recessed mounting 非凹入设置
non-reciprocal 非互惠性;非互易的
non-reciprocal circuit 不可逆电路
non-reciprocal coupling 非倒易耦合
non-reciprocal device 非互易器件
non-reciprocal network 不可逆网络
non-reciprocal observation 单向观测
non-reciprocal parametric amplifier 非互易参量放大器
non-reciprocal preference 非互惠性优惠
non-reciprocal preferential treatment 非互惠性优惠待遇
non-reciprocal reinsurance 无分回业务分保
non-reciprocal treatment 非互惠待遇
non-reciprocity 非互易性;非倒易性
non-reciprocity concept 非互惠概念
non-reclosing pressure relief device 不重关的卸压装置
non-recognition 不识别
non-recoil 无反冲
non-recoiling fender 非回弹性护舷
non-recording device 非自记装置
non-recording ga(u)ge 非自记式仪表;非自记水位计;非自记水尺;非记录水位标尺
non-recording instrument 非记录式仪器
non-recording rain ga(u)ge 非自记式雨量器;非自记雨量计
non-recording spectrophotometer 非自记分光光度计
non-recording stream ga(u)ge 非记录式流速仪
non-recording surface 非记录面
non-recourse 无追索权
non-recourse debt 无追索权的欠款;无追索权的借款
non-recourse finance 无追索资金融通
non-recourse financing 无追索权资金筹措
non-recourse note 无追索权的借据
non-recoverable 不能挽回的;不能回收的;不可回收的;不可恢复的
non-recoverable deflection 不可逆变形
non-recoverable elongation 不可恢复的伸长
non-recoverable error 不可改正的错误
non-recoverable program(me) error 不可改正的程序错误
non-recoverable timbering 无法回收的(木)支撑;不予回收(的)材料;不能回收的木支撑
non-recoverable transaction 不可恢复事项;不可恢复的事务处理
non-recovery process 无回收法
non-rectifying contact 非整流接触
non-rectifying electrode 不整流电极
non-rectifying inspection 非伸长检验
non-rectifying junction 不整流结
non-rectilinear flow 非直线流
non-recuperative system 无回热系统
non-recuperative turbine 非再热式涡轮机
non-recurrent charges 临时性费用
non-recurrent cost 临时费用;偶生费用
non-recurrent expenditures 临时支出
non-recurrent expenses 偶生费用
non-recurrent inquiry 非定期调查
non-recurrent item 非经营项目;非经常项目
non-recurrent profit 非经常利润
non-recurrent receipts 非经常收入
non-recurrent revenue 临时收入
non-recurrent waste 偶然废料;非经常废料
non-recurring charges 一次性费用
non-recurring cost 偶生费用;偶生成本;非经常成本
non-recurring expenses 一次性费用
non-recurring gains 偶生盈利;偶生利益;非经常收入
non-recurring gains and loss 非经常损益
non-recurring income 偶然收益;非经常收入;一次性收入
non-recurring item 非经常(性)项目
non-recurring maintenance 非经常性维修

non-recurring profit 非经常利润
non-recurring profit and loss 偶生损益;偶然损益;一次性损益
non-recurring repair 非经常性维修
non-recurring sources of income 非经常性收益的来源;一次性收入来源
non-recurring transitional cost 临时过渡性费用
non-recurring upkeep 非经常性维修
non-recursive 非递归的
non-recursive digital filter 非循环数字滤波
non-recursive system 非递归系统
non-recyclable material 不能重复使用的物料
non-recycling control relay 双向控制继电器
non-redundant network 非重复网络;非冗余网络
non-redundant system 非冗余系统
non-reef-building coral 非造礁珊瑚
non-reentrant code 非重入代码
non-reentrant format 非重入结构形式
non-referee method 非仲裁法;非鉴定法
non-reference level species 非参考水平物种
non-reference state concentration 非参考态浓度
non-referred method 非仲裁法;非鉴定法
non-reflecting 不反射的
non-reflecting cavity 无反射空间
non-reflecting coating 无反射涂层;非反射膜;非反射镀层;不反射涂膜
non-reflecting film 非反射膜;不反射膜
non-reflecting finish 表面无光泽的精整加工
non-reflecting glass 漫散射玻璃;无反射玻璃;非反光玻璃
non-reflecting shopwindow 无反射橱窗
non-reflecting surface 非反射面;不反射面
non-reflecting termination 匹配终端负载
non-reflection 无反射
non-reflection attenuation 固有衰减
non-reflection coating 无反光层
non-reflection glass 不反射玻璃
non-reflection surface 不反射面
non-reflective ink 非反射墨水
non-reflective material 非反光材料
non-reflective mirror 非反射镜
non-reflective pavement marker 非反光路面标志
non-reflexive 非自反的
non-reflexive relation 非自反关系
non-reflexivity 非自反性
non-refrigerated 无冷却的
non-refueling duration 不加燃料时间
non-refuelling 不加油
non-refundable 不予偿还的;不退款的
non-refundable bonds 不可调换债券
non-refundable payment 不退的付款
non-regeneratability 不可更新化
non-regenerative compound motor 非再生复激电动机
non-regenerative laser amplifier 非再生式激光放大器;非再生激光放大器
non-regenerative laser oscillator 非再生式激光振荡器
non-regenerative oven 非再生炉;非再生护
non-regime channel 不稳定河槽;冲淤不平衡河槽
non-register controlling selection 直接选择
non-registered 非注册的
non-registering 不记录的
non-regular 非正规的
non-regular building 非正规建筑物
non-regular diurnal tide 非正规日周潮
non-regular estimator 非正则估计量;非规估计量
non-regular income 不属经常固定收入
non-regular semi-diurnal tide 非正规半日周潮
non-regular set 非正规集
non-regulated discharge 未调节的流量
non-regulated flow 未调节流量;无调节水流
non-regulated transformer 不可调变压器
non-reheat turbine 非再热式汽轮机
non-reimbursable assistance 无偿援助
non-reimbursable fund 不用偿还基金
non-reimbursible fund 无需偿还基金;不必偿还基金
non-reinforced 非增强的;无钢筋的
non-reinforced asphalt sheet 无胎油毡
non-reinforced brick masonry 无筋砖砌体
non-reinforced concrete 无筋混凝土;素混凝土
non-reinforced concrete pipe 无筋混凝土管
non-reinforced liquid coating encapsulation prod-

uct 非增强液体涂料包封产品
non-reinforced masonry 无筋砌体
non-reinforced pavement 柔性路面；非刚性路面；无筋混凝土路面；素混凝土路面
non-reinforced slab 无筋混凝土板；素混凝土路面板
non-reinforced spread foundation 无筋扩展基础
non-rejection region 非否定区域
non-relativistic 非相对论性的
non-relativistic approximation 非相对论性近似
non-relativistic cosmology 非相对论性宇宙学
non-relativistic kinematics 非相对论性运动学
non-relativistic mechanics 非相对论性力学
non-relativistic particle 非相对性粒子；非相对论性粒子
non-relativistic quantum mechanics 非相对论性量子力学
non-relevant benefits 不相关效益
non-relevant cost 不相关成本
non-relevant documents 非相关资料；非关联文件
non-relevant failure 非关联失效
non-relief 无突起
non-relocatable phase 不可再定位阶段；不可浮动阶段
non-removable 固定的；不可拆卸的
non-removable discontinuity 非可去不连续点
non-renewable 不可再生的；不可恢复的
non-renewable energy resources 不可再生能源；不可更新能源资源
non-renewable fuse 不能重复使用的保险丝；不能更新的熔断器
non-renewable fuse unit 不能更新熔断器
non-renewable mineral 不可再生矿物
non-renewable natural resources 不能更新的自然资源
non-renewable resources 非再生资源；不能更新的资源；不可再生资源；不可恢复的资源；不可更新的资源；不可复用资源
non-renewable source 不可再生能源
non-rental area 非租赁地区；非租赁区
non-repairable component 不可修部件
non-repeatability 不可重复性
non-repeated key 非重复键
non-repeated sampling 非重复样本；不重复抽样
non-repeating design 不重复花纹
non-repetitive 非重复的
non-repetitive manufacturing process 非重复的制造程序
non-repetitive peak reverse voltage 非重复性反向峰值电压；反向不重复峰值电压
non-repetitive sequence 非重复次序
non-repetitive transaction 非循环性业务
non-replaceable unit 不可更换的部件
non-replacement problem 不置换问题
non-representative 无代表性的；非典型的
non-repressed brick 活砌砖；不再压砖
non-reproducible assets 非再生资产；不能再生产的资产
non-reproducible tangible assets 不能再生产的有形资产
non-reproducing code 非复制码
non-reset timer 不复位定时装置
non-resident 非本州居民；非驻留的；非常驻
non-resident account 非居民账户
non-resident error procedure 非常驻误差过程
non-residential account 非居民账户
non-residential block 非居住街坊；非居住区段
non-residential building 非居住房屋；非居住用建筑物；非居住建筑物
non-residential construction 非居住用构造；非居住用结构；非居住用建筑；非宅建筑
non-residential development zone 非宅发展区
non-residential facility 非居住设施
non-residential fixed investment 住宅以外的固定投资；非住房固定资产投资；非居住用固定投资额
non-residential sector 非住宅区段
non-residential structure 非居住用建筑物
non-residential structure investment 非居住用建筑投资；非住房建筑投资
non-residential usable floor area 不能用于居住的楼面
non-resident portion 非常驻部分
non-resident program(me) 非常驻程序
non-resident software 非驻留软件

non-resident unit 非常住单位
non-residual deformation 弹性变形
non-residual strain 弹性应变
non-residue 非剩余
non-resistance 无电阻
non-resistant 不耐久的
non-resistant to earthquake 未经抗震设计的
non-resonance 非共振
non-resonant 非谐振的；非共振的
non-resonant antenna 非共振天线
non-resonant circuit 非谐振电路
non-resonant diffusion acceleration 非共振扩散加速
non-resonant diffusion heating 非共振扩散加热
non-resonant emission 非共振发射
non-resonant energy 非共振能量
non-resonant feeder 非谐振馈线
non-resonant line 无谐振线；非共振线
non-resonant method 非共振法
non-resonant modulator 非谐振调制器
non-resonant oscillator 非共振振荡器
non-resonant particle 非共振粒子
non-resonant photodissociation 非振光解
non-resonant process 非共振过程
non-resonant system 非共振系统
non-response 无反应
non-responsible accident 非责任事故
non-responsive bid 没有作出反应的投标
non-restoring 不复原
non-restoring division 不恢复除法
non-restoring method 非复原法；不恢复法
non-restraint 无约束的
non-restricted preference 无限制优惠
non-retentive alloy 软磁性合金
non-retractable 不能伸缩的
non-retractable awning 不能收缩的遮篷
non-retractable fins 固定鳍
non-return 止回
non returnable container 一次用包装
non-returnable packing 不能再使用的包装
non-return finger 单向安全装置
non-return finger device 非反向安全装置
non-return flap 止回瓣
non-return-flow wind tunnel 非回流式风洞
non-return foot valve 止回底阀
non-return handle 不返手柄
non-return-reference recording 不归零记录法
non-return safety valve 单向安全阀
non-return state 不返回状态
non-return-to-reference 不归零制
non-return-to-reference recording 非归零记录化；不归基准记录
non-return-to-zero 不归零制；不归零制；不归零(的)
non-return-to-zero change-on-one 逢一变化不归零制
non-return-to-zero change recording 不归零变化记录方式
non-return-to-zero code 非归零码
non-return-to-zero coding 不归零编码
non-return-to-zero indicating 标识不归零制
non-return-to-zero inverse 翻转不归零制
non-return-to-zero inverted recording 不归零遇一翻转记录；不归零反相记录
non-return-to-zero level 电平不归零制
non-return-to-zero logic 不归零逻辑
non-return-to-zero mark recording 不归零(制)标记记录；不归零符号记录方式
non-return-to-zero method 不归零法
non-return-to-zero operation 不归零操作
non-return-to-zero pulse 不归零脉冲
non-return-to-zero recorder 不归零制记录器
non-return-to-zero recording 不归零(制)记录(方式)
non-return-to-zero representation 不归零表示
non-return trap 止回汽水阀
non-return valve 止回阀；逆止阀；单向活门；单向阀；不回阀
non-reusable 不可重用的；不可重复使用
non-reusable medium 不可重用型存储媒体
non-reusable packing 不能再使用的包装
non-reusable program(me) 不可重用程序
non-reusable routine 不可重用的例行程序；不可重(复使)用程序

non-revenue 非税款收入
non-revenue passenger 免费乘客
non-revenue receipts 课税外收入；非税款收入；额外收入
non-reverberant 无混响的；非回响的
non-reverberant chamber 非混响腔
non-reverberant room 非混响室；无回响房间
non-reversed 非反像的
non-reversibility 不可逆性
non-reversible 不能反转的；不可转换的；不可逆的
non-reversible clutch 不可逆离合器
non-reversible control 不可逆控制系统
non-reversible crimp 不可逆卷曲
non-reversible deformation 永久变形；不可逆形变；不可逆变形
non-reversible electric(al) drive 不可逆电气传动
non-reversible engine 不可逆转发动机
non-reversible frequency changer 不可逆变频器
non-reversible hammer crusher 不可逆锤式破碎机
non-reversible hammer mill 非逆转锤式破碎机；不可逆锤式破碎机；不可逆锤式粉碎机
non-reversible laydays 不合并计算碇泊期限
non-reversible laytime 不合并计算碇泊期限
non-reversible machine 不可逆式电机
non-reversible movement 不可逆变化
non-reversible pallet 不可翻转的托盘；不可调换式托盘
non-reversible process 不可逆工程
non-reversible reaction 不可逆反应
non-reversible steering gear 不可逆转向装置
non-reversing 单向的
non-reversing mill 不可逆式轧机
non-reversing motor 不反转马达
non-reversing rolling mill 不可逆式轧钢机
non-reversing strip-foil cold rolling mill 不可逆带箔冷轧机
non-reversing tap 直通丝锥
non-revivalist 非重要新出版者；非重新上演者
non-revolving 非回旋的；不可倒转的
non-revolving crane 不可回旋起重机
non-revolving credit 非周转信用证；非循环信用证；不循环信用证
non-revolving door 非旋转门
non-revolving letter of credit 非循环信用证；不循环信用证
non-riding anchor 惰锚
non-riding cable 惰链
non-Riemannian geometry 非黎曼几何(学)
non-rigid 柔性的；非硬性的；非刚性的；非刚性
non-rigid air-ship 软式飞艇
non-rigid base 非刚性基层
non-rigid carriage way 非刚性车行道；非刚性车道
non-rigid connection 非刚性节点
non-rigid coupling device 非刚性联轴器；非刚性连接装置
non-rigid floor 柔性地板；非刚性地板；非刚性楼板
non-rigidity 非刚性
non-rigid joint 非刚性节点
non-rigid joint covering 非刚性盖缝物
non-rigid material 非刚性材料
non-rigid molecule 非刚性分子
non-rigid pavement 非刚性路面
non-rigid plastics 软塑料；非硬质塑料；非刚性塑料
non-rigid relieving platform supported on bearing piles 无刚性承台
non-rigid road 非刚性路
non-rigid road base 非刚性路基
non-rigid steel arch 柔性钢拱；非刚性(发)券；非刚性钢拱
non-rigid system 非刚性体系
non-rigorous adjustment 近似平差
non-riparian 不在水边的
non-rising 不升起的
non-rising stem valve 壳装带螺旋机构的阀；不升杆式阀
non-risk after discharge 卸货后不负责
non-risk till on board 装船前不担保
non-risk till waterborne 装船前不保险
non-rival consumption 非竞争性的消费
non road oil 非沥青质铺路油；非地沥青质铺路油
non-rock 非岩质
non-rotary press 固定模台压机
non-rotating 不回转的
non-rotating cable 不转动钢丝绳；不转动电缆；不

旋转钢丝绳
non-rotating origin 无旋转原点
non-rotating rope 不会打扭的绳;不旋转钢丝绳
non-rotating stabilizer 非旋转稳定器;不旋转稳定器
non-rotational 非转动的
non-rotational strain 无旋应变
non-routine functions 非常规功能
non-rubber component 非胶组分
non-rubber substance 非胶物质
non-rubbing surface 非摩擦面
non-rust 不锈
non-rusting 抗蚀的;防锈的;不锈的
non-rusting solution 防锈溶剂;防锈剂
non-rust steel 不锈钢
non-saccharine 非糖质
non-safety circuit 非安全电路
non-safety lock 非安全锁
non-safety relay 弹力式继电器;非安全型继电器
non-sag sealant 不下垂密封剂;非下垂密封膏;黏稠密封剂;不下垂密封膏
non-sales mode 非销售方式;非出售方式
non-salient 不凸出的
non-salient pole 隐极
non-salient pole alternator 隐极同步发电机;非凸极式同步发电机
non-salient pole machine 隐极机
non-salient pole synchronous generator 隐极同步发电机
non-saline-alkali s soil 非盐碱土
non-saline sodic soil 非盐钠质土
non-sample information 非样本信息
non-sampling 非抽样的
non-sampling error 非抽样误差
non-sampling risk 非抽样风险
non-saponifiable 不能皂化的
non-saponifying 不皂化的
non-saturable 不饱和的
non-saturable reactor 不饱和电抗器
non-saturated 非饱和的
non-saturated air 未饱和空气
non-saturated region 非饱和区
non-saturated switching circuit 非饱和式开关器
non-saturated void count 未饱和空隙数量
non-saturating logic 非饱和逻辑
non-saturation dive 不饱和潜水
non-saturation diving 非饱和潜水;不饱和潜水
non-saturation recording 非饱和记录
non-saturation site 不饱和位置
non-saturation soil 不饱和土(壤)
non-saturation steam 不饱和蒸汽
non-saturation type circuit 非饱和式电路
non-scalar covariance matrix 非纯量协方差矩阵
non-scalar disturbance covariance matrix 非纯量扰动协方差矩阵
non-scalar variance covariance matrix 非纯量方差协方差矩阵
non-scale 无氧化皮的;不起皮
non-scaling 不起皮的;不结垢(的)
non-scaling steel 不起皮钢
non-scanning antenna 非转动天线;不转动天线
non-scanning direction(al) finder 非搜索测向器
non-scanning display 非扫描显示
non-scattering gas 非分散性气体
non-scattering medium 非散射媒质
non-scheduled maintenance 非正规维修;不定期维修
non-scheduled maintenance time 非规定维修时间
non-scheduled repayment 未安排的信贷偿还
non-scheduled service 不定期航线
non-scouring 不脱土
non-scouring velocity 不冲刷速度;不冲刷流速
non-scratch and rubproof ink 抗划耐磨油墨
non-scratch ink 耐擦油墨
non-screened aggregate 未筛(分)骨料
non-scrubbable 不能擦净的;不能擦洗的
non-sealed fluid coupling 开式液压离合器
non-searchable information 不可查信息
non-sedimentable solid 不可沉淀固体
non-sedimentary environment 非沉积环境
non-seepage calculation model 非渗流计算模型
non-segmented computer 非段式计算机
non-segmented gas laser tube 不分段气体激光管
non-segmented machine 非段式计算机
non-segmented mode 非区段方式

non-segmented processor 非段式处理机
non-segregated filling 无离析装料
non-segregated reservoir 无重力分异的油藏
non-segregated structure 无偏析组织
non-segregation alloy 无偏析合金
non-seismic area 非地震区
non-seismic design 非抗震设计
non-seismic fault motion 非地震断层运动
non-seismic lake 无震湖
non-seismic region 无震区;非地震区;不震区
non-seismic vibration 非地震振动
non-seismological 非地震的
non-selective 非选择性
non-selective absorbent 中性滤光片;非选择性滤光片
non-selective absorber 中性滤光片
non-selective absorption 非选择吸收
non-selective detector 非选择性检测器
non-selective entrainers 非选择溶剂
non-selective herbicide 无选择性除莠剂;无选择性除草剂
non-selective mining 非选择性开采法
non-selective party line 非选择合用线
non-selective phase 非选择性相
non-selective pneumatic detector 非选择性滤光片气动探测器
non-selective radiator 非选择性辐射体;非选择辐射体;灰色体
non-selective solvent 非选择溶剂
non-selective trapping 滥设陷阱诱捕
non-self-cleaning 非自清的
non-self-clearing bit 非自洁式牙轮钻头
non-self embedding 非自嵌入的
non-self locking fastener 非自动锁紧固件
non-self-luminous colo(u)r 非自发光色
non-self-luminous object 非自发光物体
non-self-luminous surface 非自发光面
non-self-maintained discharge 非自持放电
non-self-priming pump 非自引液泵;非自吸泵
non-self-propelled 非自身推进的;非自航的
non-self-propelled barge 非自航驳(船)
non-self-propelled cutter suction dredge(r) 非自航绞吸式挖泥船
non-self-propelling 非自航的
non-self-propelling barge 非自航驳(船)
non-self-propelling dredge(r) 非自航式挖泥船
non-self quenching 非自猝灭的
non-self-weight collapse 非自重湿陷
non-self-weight collapsible loess 非自重湿陷性黄土
nonsenge correlation 无意义相关
non-sensation 无感
non-sensitive 不敏感的;非机密的
non-sensitive clay 非敏感性黏土;非灵敏性黏土
non-separable 不可分离的
non-separable directed graph 非可分有向图
non-separable graph 不可分图
non-separable suspended particles 不可分离的悬浮颗粒
non-separated code 非分离码
non-separated rail fastener 不分开式扣件
non-separating zipper 不分离拉链
non-septate 无间隔的
non-sequence 不连续
non-sequenced acknow-ledgement 非顺序(肯定)回答
non-sequenced command 非顺序命令
non-sequent fold(ing) 不连续褶皱;限制褶皱
non-sequential 非时序
non-sequential computer 非顺序型计算机
non-sequential operation 非时序操作
non-sequential scanning 无序扫描;非顺序扫描
non-sequential stochastic programming 无顺序随机规划;非顺序随机规划
non-series-parallel 非串并联的
non-serviceable 在故障状态的;不能使用的
non-serviceable car 非运用车
non-service invention 非职务发明
non-service time 非运营时间
non-setting compound 不凝固的混合物
non-setting glazing compound 不凝固镶玻璃材料
non-setting minium 不沉淀的红丹
non-setting red lead 不沉淀的红铅
non-settleable matter 不易沉降物质;不沉物质

non-settleable solids 不沉(淀)固体;不易沉降固体;不可沉淀固体;非沉降性固体(颗粒)
non-settling 不沉淀的;不沉陷的;不沉降的
non-settling red lead 不沉淀红丹
non-settling solids 不沉固体
non-setup job 非准备任务
non-shared control unit 非共享控制器;非共享控制部件;非公用控制器;非公用控制部件
non-shared port 单用户入口;非公用入口
non-shatterable glass 抗震玻璃;不碎玻璃;安全玻璃
non-shattering 不碎的;不易脆的
non-shattering crop 不落粒作物
non-shattering glass 不易碎玻璃;不碎玻璃
non-sheathed 无外壳的;无护罩的
non-sheathed explosive 无包皮炸药
non shielded arc welding 无保护电弧焊
non-shielded cathode 非屏蔽式阴极
non-shift hours 无工班时间
non-shipping period 停航期
non-shock chilling 无振动冷却
nonshock fluid pressure 水压试验压力
non-shorted output 非短路输出
non-shorting 非短接的
nonshorting contact switch 无短路接触开关
non-shrink 不收缩;防缩;抗缩
non-shrinkable 不收缩的;防缩的
non-shrinking 不缩;不收缩(的)
non-shrinkage aggregate 不收缩集料;不收缩骨料
non-shrinkage cement 无收缩水泥
non-shrinkage concrete 无收缩混凝土
non-shrinkage grout 不收缩水泥浆;不收缩灌缝
non-shrinkage steel 无变形钢;不收缩钢
non-shrink cement 抗缩水泥
non-shrink grouting compound 不收缩灰浆合成物
non-shrinking 抗缩的;不收缩的
non-shrinking aggregate 不收缩集料;不收缩骨料
non-shrinking cement 不收缩水泥;无收缩水泥
non-shrinking concrete 无收缩混凝土;不收缩混凝土
non-shrinking rapid hardening Portland cement 无收缩快凝硅酸盐水泥;无收缩快凝波特兰水泥
non-shrinking steel 抗变形钢
non-shrink mortar 无收缩砂浆
non-sight distance 安全停车视距
non sign 无光标志
non-signaled shunting movement 非信号指挥的调车运行
non-signal indication light 非信号区表示灯
non-signers' clause 非签约人条款
non-significant 不足道的;不显著的
non-significant interaction 非显著性交互影响
non-silicate series 非硅酸盐矿物
non-siliceous sand 非硅质砂
non-silting channel 不淤渠道
non-silting velocity 不淤积速度;不淤(积)流速
non-silvered 非银盐的
non-silver image 非银盐图像
non-silver photography 非银盐摄影术
non-silver unit 无银装置
non-simple curve segment 非简单曲线段
non-simultaneous interferometry 非同时干涉量度术
non-simultaneous observation 异时观测;不同时观测
non-simultaneous prestressing 不同时后张法(预应力);单根张拉预应力;非同时(张拉)预应力;不同时预加应力
non-simultaneous transmission 非同时传输
non-sine 非正弦
non-sine wave 非正弦波
non-singular 非奇异的;非奇异;满秩
non-singular code 非奇异码
non-singular conic 非奇异二次曲线
non-singular corrspondence 非奇异对应
non-singular covariance matrix 非奇异协方差矩阵
non-singular curve 非奇异曲线
non-singular distribution 非退化分布
non-singular interaction 非奇异相互作用
non-singularity 非奇异性
non-singular linear transformation 满秩线性变换
non-singular matrix 满秩矩阵;非退化矩阵;非奇(异)矩阵
non-singular network 非退化网络

non-singular operator 非奇异算子;非奇异算符
non-singular quadric 非奇异二次曲面
non-singular solution 非奇异解
non-singular transformation 非奇异变换
non-sinking support 稳固支承
non-sinusoidal 非正弦的
non-sinusoidal current 非正弦电流
non-sinusoidal curve 非正弦曲线
non-sinusoidal deformation 非正弦形变
non-sinusoidal oscillation 非正弦振动
non-sinusoidal voltage 非正弦电压
non-sinusoidal wave 非正弦波
non-sinusoidal wave form 非正弦波形
non-siphon 非虹吸
non-siphon trap 非虹吸存水弯
nonsked 不定期飞机;不定期航班
non-skeleton phase 非骨架相
non-skid 不打滑;防滑;不滑的
non-skid agent 防滑剂
non-skid braking system 无滑制动系统
non-skid brick 防滑砖
non-skid carpet 防滑地毯
non-skid chain 防滑链
non-skid coating 防滑涂层
non-skid concrete surface 混凝土(抗滑)糙面
non-skid cover 防滑盖
non-skid deck covering 甲板防滑敷面
non-skid deck paint 甲板防滑漆
non-skid depth 防滑深度
non-skid device 防滑装置;防滑设备
non-skid device for vehicle tire 车辆轮胎防滑装置
non-skid finish 防滑面层
non-skid floor 防滑楼地面;防滑楼板;防滑地面;防滑地板
non-skid mat 防滑垫
non-skid material 防滑材料
non-skid paint 防滑油漆;防滑漆
non-skid pavement 抗滑路面;不滑路面
non-skid pile 抗滑桩
non-skid property 防滑性;防滑性能
non-skid quality 防滑特性;不滑特性
non-skid ribbed tile 防滑肋瓦;防滑螺纹砖;防滑螺纹瓦
non-skid surface 防滑面;不滑面;不滑路面
non-skid surface treatment 路面防滑处理
non-skid surfacing 防滑表面处理
non-skid thread 防滑线
non-skid tire 防滑轮胎
non-skid tire chain 防滑胎链
non-skid tread 防滑花纹
non-skid treatment 防滑处理
non-skid varnish 防滑脱油漆
non-skrink cement 抗缩水泥
non-slaking 不水化的;不水解的
non-slaking clay 非湿化性黏土
non-slaking loess 非湿化性黄土
non-slewing jib mobile crane 移动式旋转悬臂起重机
non-slewing road crane 城市道路开行的旋转起重机
non-sliding cutting 无滑动切割
non-sliding vane pump 不滑动叶片泵
non-slip 防滑梯级;不滑的;防滑(动)的
non-slip abrasives 防滑磨料
non-slip aggregate 防滑集料;防滑骨料
non-slip batten 防滑条
non-slip cleat 防滑木
non-slip concrete 防滑混凝土
non-slip deck covering 甲板防滑敷面
non-slip differential 无滑移的差速器;转动差速
non-slip drive 非滑动传动
non-slip emery insert 金刚砂防滑条
non-slip floor 防滑楼地面;防滑楼板;防滑地面;防滑地板;防滑桥面;防滑楼面
non-slip granolithic 防滑混凝土铺面
non-slip insert 防滑条
non-slip lath 防滑条
non-slip mat 防滑垫
non-slip nosing 防滑条;楼梯踏步防滑条
non-slip paint 防滑涂料
non-slip paving tile 防滑铺地砖;防滑地砖
non-slippery 不滑移的;防滑的
non-slippery paint 防滑移油漆
non-slippery rib(bed)tile 防滑移肋瓦

non-slipping 不滑移
non-slipping block 防滑块
non-slipping paint 防滑油漆
non-slipping preparation for belt 传动带防滑制剂
non-slipping spur 防滑块
non-slip plane 非滑移面
non-slip point 临界点;中性点
non-slip protection 防滑护面
non-slip ramp 防滑坡道
non-slip region 无滑移区
non-slip safety walk 防滑走道
non-slip strip 防滑条
non-slip surface 防滑面
non-slip terrazzo 防滑水磨石
non-slip tile 防滑地砖
non-slip tread 防滑楼梯;防滑梯级;防滑踏步
non-sloping 平坦的;无坡度的;无斜坡的
non-slump concrete 干硬性混凝土
non-slumping soil 非湿陷性土;非湿化性土
non-slurry pelletizing 干法造粒
non-soap grease 无皂润滑油;非皂基润滑脂
non-software support 非软件支持
non-soil volume 土孔隙体积
non-solidrotor gyroscope 非固体转子陀螺仪
non-soluble 不溶(解)的
non-soluble matter 不溶解物(质)
non-soluble substance 不溶性物质
non-solute 不溶质
non-solution 无解
non-solvent 非溶剂
non-solvent impregnated varnish 非溶性饱和漆
non-solvent induced phase separation 非溶剂致相分离
non-sorted 无分选的;未分选的;未分类的;不分类的
non-sorted polygon 非分选多边形土
non-space-sensitive 对空间不敏感的
non-sparkability 无火花性
non-sparking 不产生火星的;防发生火花的;不燃的
non-sparking floor 不发火地面
non-sparking tool 无火花工具;不产生火花的工具
non-special 非特殊的
non-specific 非特异性的
non-specific action 非特异作用
non-specific adsorption 非特性吸附
non-specifically adsorbed ions 非特性吸附离子
non-specifically adsorbed species 非特性吸附物种
non-specification 非规格化
non-specification material 非规范材料
non-specific resistance 非特殊性抵抗力
non-specific toxin 非毒性毒素
non-specific volume request 非专用卷申请;非指定卷请求
non-spectral colo(u)r 谱外色;非光谱色
non-specular reflectance 非镜面反射
non-specular surface 非镜面
non-speculative 非投机性的
non-spheric(al) 非球形的
non-spheric(al)bubble 非球形气泡;非球形空泡
non-spheric(al)lens 非球面透镜
non-spheric(al)nucleus 非球形核
non-spheric(al)particle 非球形颗粒
non-spheric(al)surface 非球面
non-spheric(al)symmetry of the source 震源非球形对称
non-spheric(al)wavefront 非球面波前
non-spiking behavio(u)r 非尖峰性能
non-spiking output 非峰值输出信号
non-spillway dam 非溢流坝
non-spin cable 不自转钢缆
non-spin hoisting cable 不旋转提升钢绳
non-spinning 不旋转的
non-spinning hoisting rope 不旋转提升绳
non-spinning rope 无股钢丝绳
non-spinning wire rope 交互捻钢丝绳;普通捻钢丝绳;不自转钢丝绳
non-split anode 不分瓣阳极
non-splitted level 不分裂能级
non-splitted sampling 不分流试样
non-spontaneous generation 非本能性发生
non-spontaneous reaction 非天然性反应
non-sprayable 不可喷雾的;不可喷涂的
non-spreading ridge 非扩张海岭
non-spring 非弹性
non-spring loaded 非弹簧加力的;非弹簧支承的;

非弹簧加载的
non-sprocketted 无齿卷盘的
non-square fiber length distribution 不等长纤维分布
non-square matrix 非方阵;非方形矩阵
non-stabilized rate 不稳定产量
non-stabilized rectifier 非稳压整流器
non-stabilized track 未稳定的轨道
non-stable condition 不稳定条件
non-stable propagation 失稳扩展
non-stable star 不稳定星
non-stable strategy 不稳定策略
non-stack company 非盈利公司
non-staff tachometer 无标尺视距仪
non-staging drive 隔行驱动;非阶梯驱动器
non-stagnant basin 非滞水海盆
non-staining 不染色;不弄脏;不着色的;不污染性;不污染的;不生锈;不可染的
non-staining anti-oxidant 无污染抗氧剂
non-staining cement 含氧化铁少的水泥;不着色的水泥;不沾染水泥;白色硅酸盐水泥;白(色)水泥
non-staining mortar 无灰砂浆;不污染砂浆
non-staining specimen 不染色标本
non-standard 非标准的;不合标准
non-standard analysis 非标准分析
non-standard axial cutting tool 非标准轴向刀具
non-standard bearing 非标准型轴承
non-standard carbide cutting tool 非标准合金刀具
non-standard chemical machinery 非标准化工设备
non-standard clay roof(ing)tile 非标准黏土屋面瓦
non-standard condition 非标准状况;非标准条件
non-standard cutting tool 非标准刀具
non-standard elevator 非标准电梯
non-standard equipment 非标准设备
non-standard equipment of chemical engineering 非标准化工设备
non-standard explosion prevention device 非标准防爆器
non-standard file 非标准文件
non-standard geologic profile 非标准地质剖面
non-standard hours 非标准工时
non-standardization 非标准化
non-standardized technique 非标准化技术
non-standard label 非标准标号
non-standard label handling 非标准标号处理
non-standard lift 非标准电梯
non-standard material 非标准材料
non-standard member 异型构件
non-standard method of settlement 非标准清算法
non-standard model 非标准模型
non-standard motor 非标准型电机
non-standard plug ga(u)ge 非标准塞规
non-standard pressure parts 非标准受压件
non-standard probability computation 非标准概率计算
non-standard product 非标准品
non-standard production 非标准产品
non-standard profile 非标准剖面;非标准截面;非标准轮廓;非标准外形
non-standard propagation 反常波传播;反差传播
non-standard section 非标准地段;非标准剖面;非标准截面;非标准断面
non-standard series 非标准系列
non-standard sheet of geologic(al)map 不规整地质图图幅
non-standard size 非标准尺寸
non-standard spoilage 非正常破损
non-standard state system 非标准态系统
non-standard structural element 非标准构件
non-standard thread 非标准螺纹
non-standard unit 非标准单位;非标准件
non-standard virgin material 非标准原始物料
non-staple 非主要的
non-staple food preparation room 副食加工间
non-staple food processing factory 副食加工厂
non-staple food(stuff) 副食品
non-state economy 非国有经济
non-state industrial enterprise 非国有工业企业
non-static 无静电干扰的;不引起无线电干扰的
non-static model 非静态宇宙模型
non-stationarity 非平稳性;非定常性
non-stationary 非平稳
non-stationary bilinear system 非平稳双直线系统
non-stationary dredge(r) 非定泊式挖泥船

non-stationary envelop(ed) function 非平稳包络函数
non-stationary flow 非稳定流;非恒定流;非定常流;不稳定流
non-stationary frequency 非固定频率
non-stationary front 不稳定的波浪前锋;非固定的工作前线
non-stationary iterative method 不定常迭代法
non-stationary load 非恒定负荷
non-stationary method 非平稳法;不定常法
non-stationary model 非稳定模型
non-stationary motion 非定常运动;非常定运动
non-stationary noise 非稳态噪声;不稳定噪声
non-stationary process 非平稳过程;非固定法
non-stationary random function 非平稳随机函数
non-stationary random model 非平稳随机模型
non-stationary random process 非平稳随机过程
non-stationary self-focusing 非稳态自聚焦
non-stationary series 非平稳数列
non-stationary spectrum 非平稳谱
non-stationary state 非定态
non-stationary stochastic process 非平稳随机过程
non-stationary system 非平稳系统
non-stationary thermal conductivity 不稳定热传导
non-stationary time series 非平稳时间序列
non-stationary trend 非平稳趋势
non-stationary two-dimensional flow 不稳定平面流
non-stationary wave 非平稳波
non-stationary way of operation 移位作业方式
non-stationary wind 不稳定风
non-station tunnel 公路隧道;无车站隧道
non-statistical 非统计性的
non-statistical behavio(u)r 非统计行为
non-statistical error 非统计误差
non-statutory audit engagement 非法定审计约定
non-steady 不稳定
non-steady aerodynamics 非定常空气动力学
non-steady bed load 不稳定底质
non-steady current 非稳定流;非恒定流;不稳定流;变量流
non-steady deformation process 不稳定变形过程
non-steady dredge-cut 不稳定挖槽
non-steady flow 非稳定流;非恒定流;不稳定流
non-steady form 不稳定水
non-steady motion 非定常运动
non-steady state 不稳定状态;非稳定态
non-steady state characteristics 不稳定状态特性
non-steady state laser operation 激光器不稳定工作状态
non-steady state membrane potential 非稳态势
non-steady state simulation 非稳态模拟
non-steady supercritical flow 不稳定超临界流
non-steaming economizer 不沸腾式省煤器
non-steerable wheel 不可操纵轮
non-steer drive axle 不转向的驱动轴
non-stellar astronomical object 非恒星天体
non-stereoscopic 非立体的
non-sterile raw wastewater feed 未消除原废水供水
non-stick button 自复位按钮;自复式按钮
non-stick button control 自复式按钮控制
non-sticking 不黏附的
non-sticking package paper coating 防黏包装纸用涂料
non-stickness 不粘连;不黏结;无黏性
non-stick push button 自复式按钮
non-stick system of control 自动控制系统
non-stick working 自动操作
non-sticky 不黏的
non-sticky and slightly plastic 不黏滞微可塑
non-sticky/non-plastic 不黏滞不可塑
non-sticky/plastic 不黏滞可塑
non-sticky soil 非黏性土
non-sticky/very plastic 不黏滞极可塑
non-stochastic assumption 非随机假定
non-stochastic effect 非随机性效应
non-stochastic explanatory variable 非随机解释变量
non-stochastic forecasting model 非随机预测模型
non-stochastic method 非随机方法
non-stochastic model 非随机模型
non-stochastic variable 非随机变量
non-stock corporation 无股票公司

non-stocked 无林地
non-stoichiometric 非化学计量的
non-stoichiometric ceramics 非化学计量陶瓷
non-stoichiometric composition 非化学计量组成
non-stoichiometric compound 非计量化合物;非化学计量化合物
non-stoichiometric crystal 非化学计量晶体
non-stoichiometry 非计量性;非化学计量性;偏离化学计量
non-stop 不间断的
non-stop character 直达性
non-stop chuck 不停产卡盘
non-stop concreting 混凝土连续浇筑(法);混凝土连续浇注(法)
non-stop crossing of opposite trains 不停车交会
non-stop dumping 行进中卸载;不停车卸载
non-stop flight 直达飞行
non-stop meet 不停车会车【铁】
non-stop operation 连续操作
non-stop overtaking 不停车越行
non-stop passing 不停车越行
non-stop run 连续工作;不停站运行;不间断工作
non-stop surveying 连续测量
non-stop switch 直达开关;不停开关
non-stop train 直达列车;通过列车
non-stop voyage 直达航程
non-storage 非储[贮]藏的;非积储式
non-storage area (库房中的)非堆货区
non-storage calorifier 非蓄热性热风炉;非蓄热性加热器;快速加热装置
non-storage device 非存储设备
non-storage display 非存储式显示装置
non-storage gas water heater 非蓄热式煤气热水器
non-storage pick-up device 无存储摄像装置
non-storage routing 无调蓄演算;非调蓄演算;非槽蓄演算
non-storage space 非存储空间
non-storage water heater 非蓄热式热水器;流动型热水炉;快速热水器;快速热水炉
non-straight factor 非直线系数
non-straight line correlation 非直线相关
non-straightness 不直度
non-stranded rope 无股钢丝绳
non-stratified 非成层的
non-stratified crown 不分层型树冠
non-stratified rock 非成层岩(石)
non-stratified sample 非分层样本
non-stressed concrete 不受力混凝土;无应力混凝土
non-stress meter 无应力计
non-stress strain meter 无应力应变计
non-stretched 非扩大的;非展宽的;非延伸的
non-striated fiber 无横纹纤维
non-strictly competitive two-person game 不严格竞争二人对策
non-strip(ping) additive 防剥落外加物;防剥落添加剂;(沥青的)防剥落剂
non-strip(ping) agent (沥青的)防剥落剂
non-structural 非结构的;不用于结构上的
non-structural adhesive 非结构用黏合剂
non-structural bonding adhesive 非结构性胶结剂;非结构性胶黏剂;非结构性结构剂
non-structural component 非结构构件;非结构部件
non-structural connection by tack welding 用定位焊接的非结构性连接
non-structural damage 非结构性破坏
non-structural decision 非结构化决策
non-structural design 非结构设计
non-structural effect 非结构效应
non-structural glued lumber product 非结构胶黏木制件
non-structural lightweight concrete 非结构用轻质混凝土
non-structural measure 非工程措施
non-structural measure of flood control 非工程性防洪措施
non-structural member 非受力的构造杆件;非结构(构)件;非受力的结构杆件;非结构杆件
non-structural metalwork 非承重金属结构
non-structural regression relation ship 非结构性回归关系
non-structural sealant 非结构密封膏
non-structural top screed 非结构性的顶面灰浆准条(定灰浆厚度的施工工具);表面覆盖层;铺面(混凝土)

non-structural unemployment 非结构性失业
non-structured operating system 非结构操作系统
non-subject to the restrictions 不在此限
non-submerged bank 不受淹河岸
non-submerged filter 非淹没滤池
non-submerged groin 非淹没式丁坝
non-submerged thin-plate weir 非淹没滤薄壁堰
non-subscripted variable 无下标变量
non-substitution of invoice transfer 不替换发票的转让
non-subterranean termite 非地下白蚁
non-sudden environmental risk 无骤发环境风险
non-sulfide flo(a)tation 非硫化矿浮选
non-sulfurized carbon steel 非硫化处理的碳钢
non-sulphate sulphur 非硫酸盐硫
non-supercharged engine 非增压式发动机
non-superimposable mirror image 不重叠镜像;不能重叠的镜像
non-supervisor mode 非管理状态
non-supported membrane 无支撑膜
non-supporting 不负担的;不支撑的;无支承的
non-supporting diaphragm wall 自立式地下连续墙
non-supporting retaining structure 自立式挡土结构
non-supressed ion chromatography 非抑制型离子色谱法
non-surface active 非表面活性的
non-surface-active agent 非表面活性剂
non-surfaced 未做面层的;未铺路面的
non-surfactant 非表面活性剂
non-surging spring 防振气门弹簧
non-suspended weight 非悬挂重量
non-sustainable 不能持久的
non-sustaining slope 逆流坡度;非持续坡度;反坡度
non-swapable 不可交换的
non-swelling 不溶胀的;不膨胀的;非膨胀性的
non-swelling lime-alkali gel 无膨胀钙-碱凝胶
non-switched connection 非交换连接
non-switched data link 非交换数据链路
non-switched line 非转接线;非交换线路
non-switched network 无交换网络;非交换网络
non-switched point-to-point line 非交换点到点线路
non-symbiotic 非共生的
non-symbiotic nitrogen fixation 自生固氮
non-symmetric(al) aberration 非对称像差
non-symmetric(al) adjustment 不对称调整
non-symmetric(al) aquifer 非对称性水体
non-symmetric(al) arch(ed) dam 非对称拱形坝;不对称拱坝
non-symmetric(al) cycle 非对称循环
non-symmetric(al) design 不对称花纹
non-symmetric(al) distribution 非对称分布
non-symmetric(al) line 不对称线路;不对称线
non-symmetric(al) loading 不对称加载;不对称荷载;非对称荷载
non-symmetric(al) network 不对称四端网络
non-symmetry powder diagram 非对称粉末图
non-symmetric(al) relation 非对称关系
non-symmetric(al) section 不对称断面
non-symmetric(al) semi-controlled bridge 不对称半控制电桥
non-symmetric(al) transistor 非对称晶体管
non-synchronized defibrillator 非同步型去颤器
non-synchronizing 非同步的
non-synchronous 非同步的;不同步的
non-synchronous computer 非同步计算机
non-synchronous data transmission channel 异步数据传输通道
non-synchronous digital hierarchy network element 非同步数字传输网络单元
non-synchronous initiation 不同步起动
non-synchronous motor 异步电动机;非同步电动机
non-synchronous network 非同步网络
non-synchronous starting 非同步起动
non-synchronous superposed arcuate structure 非同期叠加弧形构造
non-synchronous surveying 非同步测量法
non-synchronous timer 异步定时器
non-synchronous vibrator 异步振子
non-synchrotron radiation source 非同步加速辐射源
non-synergistic effect 非协同效应
non-syphon 防虹吸存水弯

non-systematic(al) code 非系统码
non-systematic(al) error 偶然误差;非系统性误差
non-systematic(al) joint 非系统节理
non-systematic(al) part 非系统性部分
non-systemic(al) code 非系统码
non-system personnel 非系统人员
non-tackiness 非黏结性;无黏结性
non-tacky 非黏性的
non-tacky conveyor belt 防黏运输带
non-tacky toner 不发黏墨粉
non-tangible 不记名的
non-tapered key 非锥形键
non-target biological effect 非目标生物效应
non-target organism 非靶生物
non-tariff 非关税
non-tariff barrier 非关税壁
non-tariff-barrieraction desk 非关税障碍服务台
non-tariff company 非收费标准协定公司
non-tariff restriction 非关税限制
non-task 非任务
non-tax 非征税
non-taxable exchanges of property 免税财产交换
non-taxable-goods for special use 特殊用途免税物品
non-taxable income 非税收益;非课税收益;非课税收入
non-taxable interest income 免税利息收入
non-taxable receipts 非课税收入
non-taxable securities 免税债券
non-taxable transference 免税转让
non-tax incentives 非税收性鼓励
non-tax liability 非税性负担额
non-tax objective 非税目的
non-tax payment 非税款负担;非征税支出;非税支付;非税性负担的支付金额
non-tax receipts 非征税收入;非税收入
non-tax revenue 非税收入;非税金收入
non-technical 非技术性的
non-technical approach to a technical problem 技术问题的非技术途径
non-technological depreciation 非工艺性损耗
non-tectonic 非构造的
non-tectonic fault 非构造断层
non-tectonic joint 非构造节理
non-tectonic motion 非构造运动
non-tectonite 非构造岩
non-telescopic form 非套推式支架;非伸缩式模板;非穿行式钢模板;不能伸长的模板
non-telescopic rotating mast crane 无伸缩的旋转型桅式起重机
non-temporary data set 非临时数据集;永久数据组
non-tension analysis 无拉力分析
non-tensioned 不加拉力的;不加张力的
non-terminal 非终止
non-terminal class 非终结符类
non-terminal node 非终结符节点;非终结节点
non-terminal olefin 非末端烯烃
non-terminal position 非终结符位置
non-terminal symbol 非终止符(号);非终结符(号)
non-terminal vocabulary 非终结符号表
non-terminal word 非终结字
non-terminating decimal 无尽小数
non-terminating interruptable state 非终结中断状态
non-thermal 非热能的
non-thermal background 非热背景
non-thermal bremsstrahlung 非热韧致辐射
non-thermal decimetric emission 非热分米波发射
non-thermal ground 非热异常地面
non-thermal process 非热过程
non-thermal radiation 非热射频辐射;非热辐射
non-thermal radio source 非热射电源
non-thermal source 非热源
non-thermal spectrum 非热谱
non-thermal spring 非温泉
non-thermal velocity 非热速度
non-thermoartesian system 非热自流系统
non-thermodynamic(al) equilibrium 非热动平衡
non-thermoplastic filament yarn 非热塑性长丝
non-thermoplastic blending mixer 非恒温配料搅拌器;非恒温式掺和搅拌器
non-thermostatic mixer valve 非恒温混合阀
non-thermostatic mixing valve 非恒温搅拌阀;非恒温混合阀
non-thermostatic shower mixer 非恒温喷淋混合器;非恒温淋浴器
non-thread kelly 无细扣方钻杆
non-threshold logic 非阈值逻辑
non-thrust quaywall 无侧推力岸壁
non-tidal 不感潮的;无潮汐的;非潮汐的
non-tidal alluvial river 无潮汐冲积河流
non-tidal alluvial stream 无潮汐冲积河流
non-tidal basin 有闸港池;湿坞
non-tidal compartment 无潮汐区;无潮区;非感潮区
non-tidal current 非潮汐(海)流;非潮流
non-tidal dock 不感潮港池
non-tidal drift 非潮汐(海)流
non-tidal river 无潮河流;非潮汐河流;不感潮河段
non-tidal section 无潮(河)段
non-tidal waters 无潮水域
non-tide point 无潮点
non-tied loan 无条件贷款;不加限制的贷款;不附带条件的贷款
non-tied rate 不附带条件的费率
non-tillage system 免耕法
non-till cultural treatment 免耕法
non-tilting 非倾侧式的
non-tilting concrete mixer 非倾侧式混凝土拌和机;非倾侧式混凝土搅拌机
non-tilting drum (搅拌机的)倾斜鼓筒
non-tilting drum concrete mixer 非倾侧式筒型混凝土拌和机
non-tilting drum mixer 鼓形自落式搅拌机;非倾侧式鼓筒式搅拌机;直卸料式拌和机
non-tilting mixer 非倾侧式拌和机;直卸式拌和机;非倾侧式拌和机;不倾斜搅拌机
non-tilting-negative-plane rectifier 底片平面无倾斜纠正仪
non-tilt mixer 非颠倒式拌和机;不倾斜搅拌机
non-tilt seismograph 非倾斜地震仪
non-time delay 无时间延迟;瞬息作用
non-time delay fuse 非延时熔断器
non tire chain 防滑车胎链
non-topographic(al) 非地形测量的
non-topographic(al) photogrammetry 非地形摄影测量
non-toppling block 不倒滑车
non-touch dryer 无接触干燥机
non-touching loop rule 不接触回路法则
non-toxic 无毒(性)的
non-toxic effect level 无毒性作用剂量水平
non-toxicity 无毒性
non-toxic paint 无毒性涂料;无毒漆
non-toxic pigment 无毒颜料
non-toxic plasticizer 无毒增塑剂
non-toxic pollutant 无毒污染物
non-toxic salt 无毒盐类
non-toxic tar 无毒焦油(沥青)
non-toxic toy lacquer 无毒玩具喷漆
non-tractor drill 非拖拉机式钻机
non-trade company 非商业公司
non-traded commodities 非贸易商品
non-traded goods 非贸易商品;非贸易财货
non-trade-off Phillips curve 非交替的菲利普斯曲线
non-trading body 非商业团体
non-traditional exporter 非传统出口商
non-traditional exports 非传统出口商品
non-traditional machining 特种加工
non-traffic deterioration 非交通性损毁
non-transferable letter of credit 不可转让信用证
non-transferable utility 非转移效用
non-transfered arc 非过渡电弧
non-transfer expenditures 不能转移的支出
non-transient water supply 非过境给水
non-transition metal 非过渡金属
non-transitive 非传递的
non-transitive dependency 非传递相关性
non-transitive relation 非可递关系
non-translational 非平移的
non-transparency 不透明性;不透明度;不透光性能
non-transparent 非透明的;不透明的;不透光的
non-transparent mode 非透明模式;不透明方式
non-transparent quartz combustion tube 不透明石英燃烧管
non-transparent quartz tube 不透明石英管
non-transportable radionuclide compound 不可转移性放射性核素化合物
non-transportable resources 不可运输的资源
non-transportation function 非交通性的功能
non-transportation revenue 非运输收入
non-transportation use 非运输用途
non-tratified form 不分层树型
non-traumatic hemostatic forceps 无损伤止血钳
non-traumatic round-bodied needle 无损伤性圆体针
non-traversable gallery 非人行廊道
non-treatable refuse 不能处理的垃圾
non-treatment 非沥青处治(路面)
non-treatment plant 非处理工厂
non-trivial condition 非无效条件;非平凡条件
non-trivial function 非无效函数;非平凡函数
non-trivial grammar 非无效文法
non-trivial partition 非普通划分
non-trivial solution 非无效解;非平凡解;非零解;非常规解
non-tronite 绿脱石;囊脱石
non-tropic(al) monsoon 非热带季风
non-tropic(al) sprue 非热带性口炎性腹泻
non-trussed purlin(e) roof 无桁架的檩条屋顶
non-tunable 非调谐的;不能调换的;不可调谐的;不可调
non-turbulent 非湍流的;非扰动的
non-turbulent diffusion 非湍动扩散
non-turbulent flow 非湍流
non-turnable container 一次性包装;非回收集装箱
non-twistable 不可扭转的
non-twisted 非扭曲的;无扭曲的;非扭转的
non-typical 非定型的
non tyre chain 防滑车胎链
no number 无号码
non-umeric application 非数据应用
non-under command light 故障指示灯;失控信号灯
non-uniform 非均匀;不均匀(的)
non-uniform air gap 空气隙不均匀
non-uniform approach 非均匀逼近
non-uniform beam 非均匀激光束;非等截面梁;不均匀光束;变截面梁
non-uniform bearing structure 非均匀布承重结构;非均匀承重结构
non-uniform capacity 不均匀电容
non-uniform charging 装料不均匀
non-uniform circuit 不均匀电路;不均匀传输线
non-uniform combustion 不均匀燃烧
non-uniform compact source 非均匀致密源
non-uniform continuous beam 变截面连续梁
non-uniform corrosion 不均匀腐蚀
non-uniform deformation 非均匀变形
non-uniform dielectric 非均匀电介质
non-uniform distribution 分布不均;非一致分布;非均匀分布;不均匀分布
non-uniform distribution load 不均匀分布荷载
non-uniform electric(al) field 不均匀电场
non-uniform electric(al) magnetic field 不均匀电磁场
non-uniform electromagnetic field 不均匀电磁场
non-uniform encoding 非线性编码;非均匀编码
non-uniform face 不均匀面
non-uniform field 不均匀场
non-uniform film thickness 膜厚度不均
non-uniform flow 紊流;非匀流;非等速流;不均匀流;不等(速)流;变速流
non-uniform foundation 非均质地基;不均匀地基
non-uniform frequency response 不均匀频率响应
non-uniform friction 非均匀摩擦
non-uniform function 非等值函数;非单值函数
non-uniform grading 不均匀级配
non-uniform hardening 不均匀淬硬;不均匀淬火
non-uniform inversion 不均匀反转
non-uniformity 非均质;非一致性;非均匀度;不均一性
non-uniformity coefficient 不均匀系数
non-uniformity of powder characteristics 粉末性质不均匀
non-uniform laser beam 不均匀激光束
non-uniform lift 不均匀分布的升力
non-uniform load 非均布荷载
non-uniform load bearing structure 非均布承载结构
non-uniformly convergence 非均匀收敛
non-uniformly convergent 收敛不一致(的);非一致收敛;不均匀收敛(的)

non-uniformly distributed stress 非均布应力
non-uniformly distributive coupler 非均匀分布耦合器
non-uniformly graded aggregate 级配不均匀骨料;不均匀级配集料
non-uniform magnetic field 不均匀磁场
non-uniform magnetization 不均匀磁化
non-uniform material 不均匀物质
non-uniform medium 非均匀介质
non-uniform mixing 不均匀搅拌
non-uniform motion 不均匀运动;不等速运动;变速运动
non-uniform movement 不均匀移动;不均匀沉陷;不等速运动
non-uniform pile 变截面桩
non-uniform pressure 不均匀压强;不均匀压力;不等压力
non-uniform probability distribution 非一致概率分布
non-uniform product 不均匀产品
non-uniform quantization 非均匀量化
non-uniform radiance 不均匀辐射强度;不均匀辐射
non-uniform rock pressure 围岩偏压
non-uniform rotation 不均匀转动
non-uniform rotor blade 变速叶轮叶片
non-uniform sampling 非均匀取样
non-uniform sand 不均质砂;不均匀砂
non-uniform scale 非均匀标尺;不等分标尺
non-uniform scale division 不均匀刻度;不等分标尺刻度
non-uniform sediment 非均质泥沙
non-uniform settlement 不均匀沉降
non-uniform shaft 不均匀轴;不均匀轴;变截面轴
non-uniform shell 不匀壳
non-uniform shock wave 不均匀激波
non-uniform shrinkage 不均匀收缩
non-uniform signal quantization 不均匀信号质量
non-uniform soil 非均匀土;不均质土;不均匀土(壤)
non-uniform solid surface 不均匀固体表面
non-uniform source aggregate 不均匀光源
non-uniform source of light 不均匀光源
non-uniform speed 非均匀速率
non-uniform stream 非均匀流
non-uniform stress 非均布应力;不均匀应力
non-uniform stress field 不均匀应力场
non-uniform structural member 非均匀结构杆件;非均匀结构构件
non-uniform structure 非均匀结构
non-uniform supporting structure 非均匀支撑结构
non-uniform surface 不均匀表面
non-uniform surface load 非均匀表面荷载
non-uniform texture 不均匀结构
non-uniform thermal conductivity 非均匀热导率
non-uniform tint 镀色泽不一致;镀层色泽不一致
non-uniform transmission 不均匀传输
non-uniform transmission line 不均匀传输线
non-uniform velocity 不等速度;不等流速
non-union 骨不连接;非工会的;不愈合
non-union shop 自由雇佣企业(指雇佣职工不必限于工会会员)
non-uniplanar bending 异面弯曲
non-unique model 非唯一模型
non-uniqueness 非唯一性;不唯一
non-unity 非同一式的;非同式
nonuple 九倍的;九个一组的
non-uplift pile 压力桩;受压桩
non-urban air 非城市空气
non-urban land 非都市土地
non-urban pollution sources 非城市污染源
non-urban source 非城市(污染)源
non-use 停用
non-use value 非使用价值
non-utility 不用;无用
non-utilization 未利用
non-utilized cargo payload 亏载
non-vacuum electron-beam welding 大气压电子束焊接
non-valent 无价的
non-valve filter 无阀滤池
non-vanishing matrix 非零矩阵;不消失矩阵
non-vanishing matrix element 不消失矩阵元
non-vanishing vector 非零向量
non-variable head torch 无调节喷头的喷灯

non-vectored 非向量的
non-vectored interrupt 非向量中断
non-vector interrupt 非中断向量;非向量中断
non-vehicle 未稀释漆料
non-ventilated 不通风的
non-ventilated cable trough 不通风电缆槽
non-ventilated flat roof 不通风平顶
non-ventilated motor 无风扇电机;自冷电动机
non-verbal communication 非口头信息交流
non-vertebrate fossil 无脊椎动物化石
non-vessel operating common carrier 非船主航商
non-vessel operation carrier 无船承运人
non-vessel owner 非船主
non-vessel owning common carrier 非船东普通承运人
non-viable 不能生活的
non-vibrated fresh concrete 未经振捣的新浇混凝土
non-vibrating coil 无振线圈
non-vibration parts 非振动零部件;无振动零部件
non-vibrator ignition system 非振动式发火制
non-violent death 非暴力死
non-violent fault creep 非剧烈断层蠕变
non-viscous 无黏性的;不黏的
non-viscous alluvium 无黏性冲积层
non-viscous flow 无黏性流;非黏滞流;非黏性流
non-viscous fluid 无黏性流体;非黏滞流体;非黏性流体
non-viscous liquid 理想液体;无黏性液体;非黏滞流体
non-viscous neutral 非黏性中性油
non-visible property 无形财产
non-vital circuit 非安全电路
non-vital circuit relay 非安全电路继电器
non-vitreous 非玻化的(吸水率大于百分之十)
non-vitreous ceramic tile 非玻璃质瓷瓦;非透明陶瓷瓦
non-vitreous grains 非玻化颗粒
non-vitreous tile 无釉面瓷砖;无玻璃质瓷砖
non-vitrified 非玻化的(吸水率大于百分之十)
non-vitrified grains 非玻化颗粒
non-voided concrete beam 实心砼梁;实心混凝土梁
non-void subset 非空子集
non-volatile 无挥发性的;非易失性的;非挥发的;不挥发(的)
non-volatile acid 非挥发性酸;不挥发性酸
non-volatile component 非挥发组成;非挥发成分;不挥发组成;不挥发成分
non-volatile contaminant 非挥发性污染物
non-volatile content 不挥发放物含量;不挥发成分
non-volatile diluent 不挥发稀释剂
non-volatile element 非挥发元素
non-vehicle fuel 不挥发性燃料;不挥发燃料
non-volatile hydrocarbon 不挥发性烃
non-volatile hydrocarbons 非挥发性烃类
non-volatile lubricant 不挥发润滑剂
non-volatile mass storage 非易失性海量存储器
non-volatile matter 加热涂料的残余;非挥发性物质;非挥发物;不挥发物(质)
non-volatile memory 永久储器;固定存储器;非易失性储存器
non-volatile oil 非挥发油;不挥发油
non-volatile oils 不挥发油类
non-volatile organic carbon 非挥发性有机碳
non-volatile organics 非挥发性有机物
non-volatile organohalogenated compound 非挥发性有机卤代化合物
non-volatile poisons 非挥发性毒物
non-volatile portion 不挥发部分
non-volatile random access memory 非易失性存储器
non-volatile recording 长时间记录
non-volatile residue 焦渣;不挥发残留物
non-volatile semi-conductor memory 不挥发性半导体存储器
non-volatile solid 非挥发性固体
non-volatile substance 不挥发性物质
non-volatile suspended solid 非挥发性悬浮固体
non-volatile swelling agent 不挥发的溶胀剂
non-volatile total organic carbon 非挥发性总有机碳
non-volatile vehicle 不挥发载色剂;不挥发性展色剂;不挥发性调漆料
non-volatilisation 非挥发

non-volatility 不挥发性
non-volcanic exogenic fragment 非火山质外生碎屑
non-volcanic geothermal area 非火山型地热区
non-volcanic ridge 非火山脊岭
non-voluntary license 非自愿许可
non-wage labor cost 非工资人工成本
non-walking-way gutter 屋檐滴水槽
non-washable 不能冲洗的;不能洗的
non-washed 不冲洗的
non-washing plate and frame filter press 非洗涤式板框压滤机;不可洗式板框压滤机
non-waste technique 无废技术
non-waste technology 无废工艺
non-water-carriage toilet facility 无排水的大便器
non-water level decline of recharge later 补给层水位不下降
non-water release from weakly permeable layer 弱透水层水量不释放
non-watertight 不隔水的;不防水的;不防渗的;透水的;非水密的;漏水的
non-watertight bulkhead 非水密隔舱;非水密舱壁
non-watertight door 非水密门
non-watertight floor 非水密肋板
non-water tunnel 无水隧洞
non-waxy crude 不含蜡原油;不含蜡石油
non-weak acid-base group 非弱酸碱基
non-wear 无磨耗
non-weathering 不风化的
non-weight-carrying 非承重的;非载重的
non-weight-carrying floor block 非承重楼板
non-weight-carrying floor brick 非承重地坪砖;非承重地面砖;非承重楼面砖
non-weight-carrying panel 非承重墙板;非承重间壁
non-weight-carrying partition (wall) 非承重隔(墙);非承重隔断
non-weight code 无权码
non-weighted code 非加权码
non-weighted number system 非加权数系统
non-weldable 不可焊接的
non-weldable steel 不可焊接的钢
non-welding 不可焊的;不焊合的;非焊接的
non-welding character 不可焊接性;不焊合性
non-wet room 干燥房间;非潮湿房间
non-wettable 不(可)湿润的;不(能)浸润的
non-wetted 不润湿的
non-wetted diamond 未镶嵌牢的金刚石
non-wetting 不湿润的
non-wetting metal 不湿润金属;不浸润金属
non-wetting phase 非湿相
non-white noise 非白噪声
non-white spectrum 非白色光谱
non-wood based panel 非木质人造板
non-working current 不工作电流
non-working days 不能作业天数
non-working platform 崖道;非工作平台
non-working side 非工作侧
non-working slope 非工作帮
non-working slope surface 非工作帮坡面
non-working stage 舞台的一部分但不起舞台作用;非工作阶段
non-working time 不能作业时间
non-work trip 生活出行
non-woven 无纺织物
non-woven asbestos cloth 无纺石棉布
non-woven carpet 非织造地毯
non-woven cloth 无纺布
non-woven fabric 非(织造)织物;非织织布;无纺织物;无纺(土工)布;非纺织纤维布
non-woven fabric filter separation activated sludge reactor 非纤维织物滤池分离活性污泥反应器
non-woven fiber 非织造纤维
non-woven floor covering 非织造地毯
non-woven geotextile 无纺(型)土工织物
non-woven glass fabric 无纺玻璃布
non-woven glass fiber tissue 非织造玻璃纤维(薄)毡
non-woven glass mat 非织物玻璃棉
non-woven mat 非织毡
non-woven material 无纺材料
non-woven media 无纺布滤材
non-woven scrim 无纺玻璃纤维网格布;非织造网格布

non-yellowing 不泛黄的;不泛黄性;不发黄的
non-yellowing polyurethane resin 不泛黄聚氨酯树脂
non-yielding 非可缩性;不屈服的;刚性的
non-yielding arch 非可缩性拱
non-yielding prop 刚性支柱;刚性支承
non-yielding retaining wall 刚性挡土墙;不屈服挡土墙;不发生位移的挡土墙;不动挡土墙;不变形挡土墙
nonyl acetate 乙酸壬酯
nonyl alcohol 壬醇
nonyl carbinol 癸醇
non-zero 非零
non-zero baseline 非零基线
non-zero baseline concentration 非零基线浓度
non-zero coefficient 非零系数
non-zero digit 非零位
non-zero electric(al) potential 非零电势
non-zero element 非零元素
non-zero expectation 非零期望
non-zero integer constant 非零整常数
non-zero load 非零负载;非零负荷
non-zero minimum-B field configuration 非零最小B场位形
non-zero pinch-off voltage 非零夹断电压
non-zero probability 非零概率
non-zero queue 非零排队
non-zero restriction 非零约束
non-zero solution 非零解
non-zero sum 非零和
non-zero-sum game 非零和博弈
non-zero-sum situation 非零和情况
non-zero-sum twoperson game 非零和二人对策
non-zero temperature 非零度温度
non-zero term 非零项
non-zero test function 非零检测函数
non-zero transverse index 非零横向指数
non-zero value 非零值
non-zinc epoxy primer 无锌环氧底漆
non-zonal term 非带项
non zone 禁止超车区
no-observed-adverse-effect level 无觉察到的有害效应水平;无毒性反应的最高浓度
no observed effect concentration 未检出浓度
noodle producing line 面条生产线
noodle style 傻瓜式(建筑);面条式(建筑)
no-off time 接通和间歇时间
nook 隐蔽角落;偏僻地带;偏僻地方;隅;隐蔽处;转角处;工作面的外露角;角落;凹角处
nook-shaft 凹角柱身
noon 中午;正午
noon constant 正午常数
noonday 中午
nooning 午休;午餐
noon interval 正午间隔
noonkambahite 硅钾钡钛矿
noon-mark 正午标;午标
noon report 正午船位报告
noon sight 正午瞄准;太阳中天观测
noon tide 午潮
no-operation 停止操作指令
no-operation bit 空操作位;无操作位
no-operation code 空操作码;非操作码
no-operation instruction 无操作指令
no-operation instruction blanking instruction 空白指令
no orders 无汇票
noose 活结环;套索;绞索;活结
noosphere 人类圈
no overloading 不准超载
no-overtaking zone 禁止超车区
no-pack 无衬垫纸型纸
no-padding tower 无填料塔
no paging 未标页数
nopal 仙人掌
no paragraph indention 不缩排
no-parallax viewfinder 视差修正瞄准器
no-parallel 无视差
no-parking 禁止停车;不准停车
no-parking marker 禁止停车标志;禁止停车标牌
no-parking restriction 不准停车的限制
no par value 非平价;非票面价值
no-passing line 不准超车线
no-passing zone 禁止超车区;禁超车区

no peddler 禁止摆摊
no perfect multicollinearity 不完全多重共线性
no pollution control 无污染控制
no pressure difference 无压差
no-pressure relay valve 无压继动阀
no-productive time 不作业时间;不生产时间;停产时间
no-productive time day 停工日;停机日
no pull 零位张力;无张力;冲力
no puncture tire 不漏气轮胎
no puncture tyre 不漏气轮胎
no radio 无线电讯号消失
noragathenic acid 原贝壳杉酸
noraghe (史前撒丁岛独有的)石砌的小屋与圆塔围成建筑体系
noral (浇铸铝合金铸件用的)铝浆;防腐蚀厚铝漆
no-raster 无光栅
norbergite 块硅镁石
norbornene 降冰片烯
norcoleucine 己氨酸
nordenskioldine 硼锡钙石;硼钙锡矿
Nordhausen sulphuric acid 发烟硫酸
nordite 硅钠锶镧石
nordmarkite 英碱正长岩
nordstrandite 诺三水铝石
nordstromite 辉硒铋铜铅矿
no reasonable alternatives 无合理的替代方案(废物处置)
no record 无记录
no reinforcement needed 无须补强
Norelac resin 劳拿锦纶树脂
no relation 没有关系
no-repair limit 免修极限尺寸
no reply 无回音;无答复
no-reply call 无应答呼叫
no requirement 无要求
no-reserves range 无备份油量的航程
noresidue at harvest time 无残留
no response 不响应
norethindrone acetate 醋炔诺酮
norethisterone enanthate 庚酸炔诺酮
no return point 无返回点
Norfolk-Island pine 诺福克南洋杉
Norfolk latch 棘轮锁;诺福克插销;压开锁;诺福汀锁
Norgaard gravimeter 诺伽重力仪
Nori 阿克灵顿工程砖(英国)
noria 戽水车;戽车;多斗挖土机;多斗挖泥机
noria basin 戽水车供水池;戽水车汲水池;供水池;库水车用水池
Nories' nautical tables 诺利斯航海表
norilskite 铂铁镍齐
no risk 不负风险
no risk after discharge 卸船后不负风险
no risk after shipment 装船后不负责任
no risk till overrail 过栏杆前不负责任
norite 苏长岩
noritegabbro 苏长辉长岩
nor-leucine 正亮氨酸
norlutate 醋炔诺酮
norm 限额;规范;定额;标准矿物成分;标准
normability 可模性
Normagal 铝镁耐火材料
normal 直角的;正位;正交的;簧片的静止交点;当量的;垂直的;常态;额定;标准;标称的
normal abandonment 正常报废
normal abundance 正常丰度
normal acceleration 正交加速度;正常加速度;垂直加速度;法向加速度;标准加速度
normal acceleration of gravity 正常重力加速度;标准重力加速度
normal accelerograph 法向过荷自记仪;法向过荷自记器
normal accelerometer 法向加速表
normal acid 正酸
normal acting relay 正常动作继电器
normal action 正常活动的
normal activity 正常生产
normal activity cost account 正常生产成本会计
normal address 标准地址
normal aeration 正常通气性;正常掺气量
normal age-class 模式令级分配
normal agglutinin 正常凝集素
normal aggregate 普通集料;普通骨料
normal aiming error 正常瞄准误差

normal air 标准空气;标准大气
normal-air cabin 增压舱
normal algebra 正规代数
normal algebraic variety 正规的代数簇
normal algorithm 正规算法
normal alkane 正烷烃;正构烷烃
normal alkane/isoalkane 正构烷烃/异构烷烃
normal alloy 正常合金
normal ambient temperature 标准周围温度
normal Amsterdam level 阿姆斯特丹零点
normal analysis 常量分析
normal and half-reverse lock system 定位和半反位锁闭制
normal angle 正常角(面);常角;法向角;法线角
normal angle camera 常角摄影机
normal angle lens 常用镜头;常角透镜
normal angle photograph 常角像片
normal angle projector 常角投影器
normal annual precipitation 正常年降水;多年平均降水(量)
normal annual runoff 正常年径流;多年平均径流(量)
normal anomaly 正异常
normal anticlinorium 正常复背斜
normal application 常用制动
normal approximation 正态近似;正态逼近
normal area 标准地区
normal astrograph 标准天体照相仪;标准体摄影仪
norma lateralis 侧面观
normal atmosphere 正常大气压;常压;标准大气
normal atmospheric recording 标准大气记录
normal atmospheric environment 标准大气环境
normal atmospheric pressure 正常大气压(力);标准大气压(力)
normal atmospheric temperature 常温
normal attribute 正规属性
normal axis 垂直轴;法线轴
normal axonometric projection 正轴侧投影
normal azimuth picture 正常方位图像
normal background 正常背景
normal bacteria 正常细菌
normal balance 正常差额
normal band 基带;额定带
normal barometer 标准气压计;标准气压表
normal barometric pressure 正常气压
normal base flow 正常基流;正常地下潜流;多年平均基流;常年基流
normal base line 正常基线
normal base thickness 基圆柱法面齿厚
normal beach 正常海滩
normal beam method 法向柱法
normal beam technique 纵波法(探伤)
normal bed 正常河床
normal bend 直角弯头;法向弯管
normal bending test 弯曲试验
normal bend test 正弯管试验;法线弯管检验;法向弯管检验
normal benzine 标准汽油
normal binary 普通二进制;常规二进制;标准二进制
normal bivariable 正态双变量
normal block 正规分程序;正规(试验)块体;普通砖块
normal boiling point 标准沸点
normal bolt 标准螺栓
normal bond 正规固定;正规黏合;正规砌合;普通(砖石)砌合
normal Bouguer gradient 正常布格梯度
normal brake application 正常制动
normal brake stop 正常制动停车
normal braking 正常制动
normal breaking 正常破损;正常破坏;正常断裂;正常破裂
normal brick 普通砖;标准砖;标砖
normal bridging 正常桥接
normal brightness 正射亮度;正射(法向)亮度
normal brood 正常产卵
normal bud 定芽
normal budget 正常预算
normal buffer 正常缓冲区
normal burden rate 正常负荷率
normal burette 标准滴定管
normal burning 正常煅烧
normal butane 正丁烷
normal butanol 正丁醇

normal-butyl 正丁基
normal cable 正规电缆
normal calomel electrode 当量甘汞电极;标准甘汞电极
normal capacity 额定功率;正常生产能力;正常容量;额定容量;标准容量
normal carbon chain 直碳链
normal carrier system 正规载波系统
normal carrying capacity of vehicle 车辆标准载重能力
normal cartographic(al) grid 标准地图格网
normal case photography 正直摄影
normal cast 正规浇铸;正规浇筑;普通(混凝土)预制件
normal cast block 标准浇块;正规浇块;普通预制块
normal cast concrete 正常浇筑混凝土;普通预浇混凝土
normal cast-iron drain(age) pipe 正规的铸铁排水管;标准铸件排水管
normal cast-iron waste pipe 标准铸件废水管
normal cast product 正规铸造件;正规浇筑产品
normal cast tile 普通预制(空心)砖瓦
normal category 正规范畴
normal cathode drop 正常阴极电压降
normal cell 标准电池
normal cellulose 正常纤维素
normal cement 正常水泥;普通水泥;标准水泥
normal cetane 正构十六烷
normal chain 直链;正(规)链
normal charge 标准药包;正常装药;标准装药
normal charging current 正常充电电流
normal charging period 正常充电时间
normal chart 标准图
normal circular pitch 法向周节
normal circular thickness 法向弧线厚度
normal circulation 正循环
normal-circulation rotary drilling 正循环旋转钻进
normal clear 正常进行
normal clearance 正常间隙
normal-clear system 正常开放制;正常开放系统
normal climate 正常气候
normal climb 正常上升
normal clinker 正常熟料
normal closed gate 常闭关闭门
normal closed signal 正常关闭信号
normal close valve 常关阀;常闭阀(门)
normal closing hours 通常的关闭时间
normal closure 正规闭包
normal code 正电码
normal coefficient 正规系数
normal coldest three-month period 正常最冷三个月时段
normal college 师范学院
normal colo(u)r photograph 标准彩色相片
normal colo(u)r sight 正常色视力;正常色觉
normal colo(u)r vision 正常彩色视觉
normal combustion 混合物的完全燃烧
normal combustion velocity 正常燃烧速度
normal command channel 正常指挥系统
normal compaction facies 正常压实相
normal compaction zone 正常压实带
normal compact operator 正规紧算子
normal compass 标准罗盘;标准罗经
normal competition 正常竞争
normal compiler writer 标准编译程序编制者
normal complex 法丛
normal component 正交分力;垂直分力;法向分力;正常成分;正态分量;垂直分量;法向分量;正常组分;标准部件
normal component of force 法向分力
normal component of stress 应力的法向分量
normal component stress 垂直分应力
normal compressive stress 正常压应力;正压应力;法向压应力
normal concentration 规定浓度;克当量浓度;当量浓度;标准浓度
normal concrete 普通混凝土
normal concrete block 普通混凝土砌块
normal concrete build element 普通混凝土建筑构件
normal concrete floor 普通混凝土楼板
normal concrete hollow block 普通混凝土空心块
normal concrete lintel 普通混凝土过梁
normal concrete wall 普通混凝土墙

normal condition 正常状态;正常条件;正常情况;基准状态;常规条件;标准情况
normal conditional locking 定位条件锁闭
normal conductive magnetic levitation 常导磁悬
normal cone 法锥;法线圆锥;标准锥面
normal congestion 自然阻塞
normal congruence 法汇
normal conic(al) equidistant projection 正轴等距离圆锥投影
normal conjunctive form 标准的逻辑乘法形式
normal consistence 标准稠度
normal consistency 正常和易性;正常稠度;标准稠度
normal consistency for cement paste 水泥净浆标准稠度
normal consolidated soil 正常压密土
normal consolidation 正常固结;标准固结
normal consolidation line 正常固结线
normal consolidation soil 正常压密土
normal construction contingency 正常施工临时费用
normal contact 正常接触;定位(接)点
normal contact Riemannian manifold 正规切触黎曼流形
normal continued fraction 正规连分数
normal continuous rating 额定连续出力
normal continuous sea-service rating 正常持续海上运动功率
normal contour 标准等高线
normal contrast 正反差
normal contrast grade 标准反差等级
normal control 正常控制
normal control field 标准控制字段
normal control wire 定位控制线
normal cooling 正常冷却;常规冷却
normal coordinate 简正坐标(数);正(则)坐标;正规坐标;法(向)坐标;标准坐标
normal coordinate system 基准坐标系
normal correction 正常校正;正常改正
normal correlation 正态相关;常态相关
normal correlation function 正态相关函数
normal correlation surface 正态相关(曲)面
normal corrosion 一般腐蚀
normal cost 正常成本
normal cost accounting method 普通的成本核算方法
normal cost pricing 标准成本定价
normal cost standard 正常成本标准
normal country 平坦地区;微起伏地区
normal covalency 正常共价
normal covering 正规覆盖
normal critical slope 正常临界坡度;正常临界比降
normal crossing 普通交叉
normal crossing shoal 正常过渡段海滩
normal crossover 正常渡线
normal crosssection 正剖面;垂直断面;正断面;正常剖面;横截面;法向断面
normal crown 正常路拱
normal crown section 正常路拱路段;无超高路段
normal cube 正立方体;标准立方体;标准混凝土立方试块
normal curb 直面缘石;正常路缘
normal curie point 正常居里点
normal current 正常电流
normal curvature 法曲率
normal curve 正态曲线;正交曲线;常态曲线;标准曲线
normal curve fitting 正态曲线拟合
normal curve of distribution 常态分配曲线
normal curve of error 正态误差曲线
normal curve of frequency 常态次数曲线
normal cut X 切割
normalcy 正常状态
normal cycle 常态旋回
normal cycle time 标准循环时间
normal cyclic operation of coal mines 煤矿正规循环作业
normal cylindric(al) conformal projection 正轴等角圆柱投影
normal cylindric(al) equidistant projection 正轴等距离圆柱投影
normal danger signal 正常关闭信号
normal danger system 正常危险制
normal data flow 常规数据流

normal data interchange 标准数据交换
normal data line 正常数据行
normal deflection 直角偏转
normal deformation 法向形变
normal demand 正常需求
normal density 正态密度;正常密度;标准密度
normal density function 正态密度函数;常态密度函数
normal depletion curve 正常退水曲线;正常亏损曲线;正常亏水曲线
normal depreciation 正常折旧
normal depth 正常水深;正常深度;正规深度
normal depth line 正常水深线
normal depth of flow 正常水深
normal derivative 法向导数
normal design 正常设计
normal deviate 正态离差;正态偏差;常态离差
normal deviate medium 正态中位绝对差
normal deviation 正态偏差;常态离差
normal device 电位探测装置
normal diameter 名义直径;标称直径
normal diametral pitch 法向径节
normal dimension 公称尺寸;标称尺寸
normal dip 正常倾斜;区域倾斜
normal-dip-slip fault 正向滑动断层
normal direct current restoration 正常直流分量恢复
normal directed-tree 正规有向树
normal direct factor 正规直因子
normal direction 正常方向;垂直方向;法线方向
normal direction flow 正向流
normal direction of rotation 正常转向
normal discharge 正常流量;正常放电;最大效率流量
normal discharge curve 正常流量曲线
normal disconnected mode 正常断开方式
normal disjunctive form 标准的逻辑加法形式
normal dispersion 正常色散;正常散布;正常频散;正常弥散;正常扩散;正常相关
normal displacement 正常位移;正常排水量;倾向滑距;法向位移
normal distribution 法向分布;正态分布;正常分布;高斯分布;常态分布
normal distribution curve 正态分布曲线
normal distribution diagram 正态分布图
normal distribution function 正态分布函数
normal distribution rate 正常分配率
normal divide 正常分水岭;正常分界线
normal division algebra 正规可除代数
normal divisor 正规子群
normal documents 正常单据
normal dog 定位锁簧
normal domain 正规数域
normal domestic sewage 一般家庭污水(量);正常生活污水;正常浓度家庭污水
normal double wrench 标准双头扳手
normal dropout voltage 正常释放电压
normal duty 正常负荷
normal duty conveyer 中型运输带
normal-duty duplex rake classifier 正常负荷双联扒式分级机
normal duty oil 正常负荷机油
normal dynamic(al) height 正常动高;正常动力高
normal dynamic(al) pressure of distribution system 配水系统的法向压力
normal earning 正常收益
normal earth 正常地球
normal earthquake 正常地震;浅震;浅源地震
normal effect 正常效应
normal elastic modulus 正弹性模量
normal electric(al) field 正常电场
normal electrode 标准电极
normal electrode potential 标准电极势;标准电极势
normal elemental time 正常的基本时间
normal elevation 正常水位
normal ellipsoid 正常椭球;标准椭球
normal ellipsoidal head 标准椭圆形封头
normal elliptic integral 正规椭圆积分
normal encoding 正常编码
normal end 标准端
normal endomorphism 正规自同态
normal energy level 正常能级
normal energy state 正常能态
normal engine speed 正常机速

normal entire function 正规整函数
normal entry 正常输入;正常入口
normal entry/exit 正常进/出
normal envelope 正常态范围
normal environment 正常环境
normal environmental lapse rate 正常环境直减率
normal environment condition 正常环境条件
normal epoch 正向期;正磁极期;正常期
normal equation 正则方程;正态方程;正规方程;法线方程;法方程式;标准方程(式)
normal equation of correlates 联系数法方程式
normal equivalent deviate 正态等差
normal erosion 自然冲刷;正常侵蚀;正常冲蚀;流水侵蚀;常态侵蚀
normal erosion action 正常侵蚀作用
normal erosion cycle 自然冲刷周期;正常侵蚀周期;正常侵蚀旋回;正常冲刷周期
normal error curve 正态误差分布曲线
normal error integral 正则误差积分
normal error law 正态误差规律;正态误差法则;正态误差定律
normal error of satellite orbit 卫星轨道法向误差
normal error recovery 正规错误校正
normal estimator 正规估计量
normal evapo(u)ration 正常蒸发;多年平均蒸发(量)
normal excitation 正常激磁
normal execution mode 正常执行态
normal execution sequence 正常执行顺序
normal exposure 正常曝光
normal extension field 正规扩域
normal eye 正常眼力【测】
normal failure period 正常失效期
normal fall 正常落差
normal fall method 正常落差法
normal fall rating 正常落差水位-流量关系曲线
normal family 正规族
normal family of functions 正规函数族
normal fanshaped fold 正常扇状褶皱
normal fault 正常断层
normal fault scarp 正断崖
normal fault trap 正断层圈闭
normal fence diagrams 正栅拦图
normal field 正向场;正常场
normal field capacity 正常原位持水量
normal field reduction 正常场校正
normal file 顺排档
normal filter 标准滤光片
normal findings 正常所见
normal finishing hydrated lime 普通饰面用熟料石灰
normal fixed quantity of materials stock 材料的储备定量
normal flame 正常火焰
normal flange 正常轮缘
normal flight 正常飞行
normal flight-line 正常飞行路线
normal flood 正常注水;标准注水
normal flora 正常植物区系;正常微生物区系
normal flow 正向流动;正常(径)流量;常规流
normal flow direction 正常流动方向
normal flow pattern 正常流态
normal flow rate 正常流速
normal fluid 正常流体;标准液体
normal flux array 正常磁通分布
normal flux magnet 常规磁体
normal flying 正常飞行
normal flying angle 正常飞行角
normal-focus earthquake 正态震源地震;正常震源地震
normal-focus mechanism 正断层震源机制
normal-focus shock 正常震源地震
normal fold 正(常)褶皱;对称褶皱
normal food chain 正常食物链
normal force 正向力;正交力;法向力
normal force between particles 土粒间的法向力
normal force characteristic 法向力特性
normal force coefficient 法向力系数
normal force distribution 法向力分布
normal forest 正规余树;正规林;模式林;标准林
normal forging sequence 正常锻造程序
normal form 正规形式;正常形式;规格化形式;范式;标准形式
normal format 正规形式;正规格式;标准格式

normal formation pressure 正常地层压力
normal form factor 标准形数
normal form for matrix 矩阵标准形式
normal form message 正常格式通信
normal form of a straight-line 直线的法线式
normal form of the characteristic 特征的正规形式
normal form-quotient 标准形率
normal formula 正规公式
normal fracture 正裂
normal frame 正规标架
normal free air anomaly 正常空间异常
normal freezing 正常凝固
normal freezing distribution 正常凝固分布
normal freezing equation 正常凝固方程
normal frequency 正态频率;正常频率
normal frequency combustion 正常频率燃烧
normal frequency curve 正态频率曲线;正态分布曲线;高斯频率曲线
normal frequency distribution 正态频率分布;常态频率分配;常态分布;常态次数分布
normal fuel system 主燃料系统
normal full load 正常满载;正常满(负)载
normal full-pool elevation 正常满库高度;正常满库水位
normal function 正(态)函数;正(常)函数
normal fundamental region 正规基本区域
normal gain 正常增益
normal gas capacity 标称气量
normal gas rate 额定用气量;公称用气量
normal Gaussian distribution function 正态高斯分布函数
normal general utility 正常利用率
normal geodesic line 正规测地线
normal geographic(al)coordinate 正常地理坐标
normal geopotential 正常地球位
normal geopotential number 正常重力位差数
normal girth 标准干围
normal glass 标准玻璃
normal glow 正常辉光
normal glow discharge 正常辉光放电
normal good hearing 正常听力
normal goods 正常商品
normal gradation 正常层次
normal graded bedding structure 正粒序层理构造
normal graded structure 正粒序构造
normal graded volcanic mud ball structure 正粒序火山泥球构造
normal gradient 正常梯度;法线梯度
normal grading 正递变
normal granite 石英二长岩
normal gravity 正常重力;标准重力
normal gravity field 正常重力场
normal gravity formula 正常重力公式
normal gravity line 正常重力线
normal gravity on the earth spheropotential 正常地球位面上的正常重力
normal gravity value 正常重力值
normal grazing 正常放牧
normal grid 直角坐标网;标准格网
normal gross takeoff weight 正常起飞总重量
normal gross weight 正常总重
normal ground joint 标准磨口接头
normal groundwater depletion 正常地下水亏耗
normal groundwater depletion curve 地下水正常退水曲线;地下水正常亏水曲线
normal growing stock 模式蓄积
normal growth 正常生长
normal habit 晶体惯态;正常习惯
normal habitat 正常(生长)环境
normal hardening cement 正常硬化的水泥
normal haul 正常土方运输(费)
normal hauling position 正常牵引拖曳情况
normal head 正常水头
normal hearing 正常听力
normal heating degree days 标准采暖度日
normal heavy concrete 普通重混凝土
normal height 正常高;法高
normal height of burst 正常炸高
normal helix 正交螺旋线;正螺旋线
normal hexane 正己烷
normal high tide level 正常高水位;正常高潮位
normal high water(level) 正常高水位;正常高潮位
normal histogram 常态直方图
normal hollow block 正常空心砌块;普通空心块

normal hollow tile 正规空心瓦
normal homologue 正同系物
normal horse power 正常功率;标称马力;额定马力
normal hour 标准工时;正常工时
normal humidity 正常湿度
normal hydrated lime 正常水化石灰;单水化石灰
normal hydrocarbon 直链烃
normal hydrogen electrode 标准氢电极
normal hydrologic(al) value 水文正常值
normal hydrostatic(al) pressure 正常静水压力
normal hyperbolic equation 模双曲方程
normal identifier 正常标识符
normal illumination 正常照明;正常光照;垂直照度
normal illumination method 直照法
normal image 正像
normal impact 正向冲击
normal impact effect 正常弹着效应
normal impact material 标准抗冲击材料
normal impedance 正入射阻抗;常态阻抗;标准(化)阻抗;标称阻抗
normal inching turning 正转点动
normal incidence 正入射;法向入射
normal incidence absorption coefficient 正入射吸收系数
normal incidence pyrheliometer 正入射日射强度表
normal incidence reflectivity 正入射反射率
normal incidence sound absorption 法向入声吸收系数
normal incidence sound absorption coefficient 垂直入射吸声系数
normal incidence transmission loss 垂直入射隔声量
normal incident absorption coefficient 正入射吸声系数
normal incised meander 正常深切曲流
normal inclusion 正常包裹体
normal income rate 正常盈利率;正常收益率
normal increment 模式生长量
normal indication 正常指示
normal indication lamp 定位表示灯
normal indication relay 定位表示继电器
normal indication wire 定位表示线
normal indicator diagram 正常示功图
normal induction 正常磁感应强度;正常(磁)感应
normal induction strength 正常磁感应强度
normal inertial force 法向惯性力
normal inflation 正常充气压力
normal inhibition 正常抑制
normal input cause 正常输入原因
normal input keying 正常输入键控
normal inspection 正常检查;常规检查
normal integral 正常积分;正常积分
normal intensity 标准强度
normal intensity of light 法向发光强度
normal interest rate 一般利率
normal interlocking 正常联锁;定位联锁
normal internal force 法向内力
normal interval 正常间隔
normal inverse matrix 正常逆(矩)阵
normal investigation of hydrometric(al) station 水文测站常规调查
normal investment practice 一般投资惯例
normalised band level difference 规定频带强级差
normalised characteristic loop 标准法特性回线
normalised condition 正规条件
normalised matrix 正规化矩阵
normalised tempering 正常回火
normal isopleth 平均等值线
normal issue 自动补给
normality 正态(性);正规性;正常整体标准;规度;规定浓度;当量浓度;常态
normality condition 正交条件;规范条件
normality law 正交定律
normality rate of aids maintenance 航标维护正常率
normality rate of aids-to-navigation 航标正常率
normality test 正态性检验;正态检验
normalizable 可规范化的
normalizable function 可规范化函数
normalizable kernel 可规范化核
normalization 正火;正规化;规格化;规范化;规度化;归一化;标准化
normalization condition 正规化条件;规范化条件;归一化条件;标准化条件

normalization constant 归一化常数
normalization constraint 归一化约束
normalization criterion 归一化标准
normalization factor 正规因子；正规化因子；规格化因子；归一化因数；归一化系数；标准化因子
normalization method 归一化法；归化法
normalization microroutine 规格化微程序
normalization of law 法律的规范性
normalization of scores 分数正态化
normalization of wave function 波函数归一化
normalization operation 规格化操作
normalization routine 正规化例行程序；规格化程序
normalization theorem 正规化定理
normalization transformation 规格化变换；正态变换
normalization unit 归一化单位
normalize 正态化；正火；正常化；规格化；规范化；归一化；常化
normalized 正火的；规格化的；规范化(的)
normalized admittance 归一化导纳；标准化导纳
normalized adsorption spectrum 标准化吸收谱
normalized and tempered steel 正火及回火钢
normalized areal productivity 规格化面生产量
normalized band level difference 标准化频带极差
normalized bit weight 标准化钻压
normalized central moment 规格化中心矩
normalized chain complex 正规化链复形
normalized characteristic loop 标准化特性回线
normalized chemical ion balance technique 标准化学离子平衡技术
normalized cochain 正规上链
normalized cofactor 正规余因子
normalized condition 正火状态；正规条件；规范化条件
normalized constraint 正规化约束
normalized coordinates 规格化坐标；正规化坐标；归一化坐标；标准化坐标
normalized co-spectrum 标准化同谱线
normalized cost function 规范价值函数
normalized coupling coefficient 归一化耦合系数
normalized covariance 标准化协方差
normalized crossed product 正规化的交叉乘积
normalized current 归一化电流
normalized curve 正常曲线；标准化曲线
normalized detectivity 归一化探测率
normalized device coordinates 规格化设备坐标
normalized distance 正常距
normalized distribution 正规化分布；归一化分布
normalized eigenfunction 正规化特征函数
normalized energy 归一化能量
normalized equation 正规化的方程
normalized factor 归一化因子；标准化因子
normalized filter 标准化滤波器
normalized floating point 规格化浮点
normalized footstep sound transmission level 正规化的脚步声传播声级
normalized force 标准化作用力
normalized form 规格形式；规格化形式；规范化形式；标准型；标准形式；标准公式
normalized frequency 归一化频率
normalized function 规范化函数
normalized geodesic 正规化测地线
normalized impact sound level 规定冲击声级；标准撞击声级
normalized impedance 归一化阻抗；标准化阻抗
normalized inverse derivative 正规反导数
normalized irradiance 归化辐照度
normalized level difference 正规化(声)级差
normalized matrix 正规化矩阵
normalized mode 正则化模态；正规模式；标准化振型；正规模型
normalized noise isolation class 归一化噪声隔绝类
normalized noise reduction 归一化噪声降低
normalized number 规格化数
normalized obsorption 正常吸收(能力)
normalized ordinate 标准化纵坐标
normalized orthogonal functions 归一化正交函数
normalized output 标称输出(功率)
normalized over-pore water pressure curve 归一化超孔压曲线
normalized parameter 归一化参量
normalized penetration rate method 标准化钻速法
normalized power pattern 归一化功率方向图
normalized price 定价；标准价格

normalized Q(value) 归一化Q值
normalized ratio 归一化比值
normalized reactance 标准化电抗
normalized resistance 标准化电阻
normalized response 规格化响应；标准化响应
normalized response spectrum 标准化反应谱；标准反应谱
normalized rotary speed 标准化转速
normalized signal 标准化信号
normalized speed time 标准化钻时【岩】
normalized steel 正火钢
normalized structure 正火组织；规范化结构；常化组织
normalized susceptance 归一化电纳
normalized system of inequalities 规范不等式组
normalized Taylor coefficient 规范的泰勒系数
normalized throughput 规范化吞吐量
normalized transfer function 归一化传递函数
normalized unit 标准单位
normalized value 标准值
normalized vector 正规向量
normalized voltage 归一化电压
normalized zone 正火区
normalizer 规整器；标准部件
normalize tempering 正火回火处理
normalizing 正火(处理)；规格化；正常化；归一；标准化
normalizing annealing 常化退火
normalizing device 校准仪
normalizing factor 归一化因数
normalizing furnace 正火炉；常化炉
normalizing line 正火作业线；常化作业线
normalizing parameter 标准化参数
normalizing pass 退火焊道
normalizing-quenching-tempering 正火淬火回火
normalizing selection 正常化选择
normalizing temperature 正火温度
normalizing treatment 正火处理
normalizing white 基准白
normal jam 自然阻塞
normal jet 标准气动量规
normal key 定位钥匙
normal kurtosis 正规峰态
normal lamp 定位表示灯
normal landing 正常着陆
normal latitude 地理纬度；标准纬度
normal lattice 正规格
normal law 正态法则；正态律；正态分布律
normal law error 正则误差
normal law of errors 常态误差定律；正态误差定律；正常误差定律；平均误差律
normal lead content 铁的正常含量
normal lead silico chromate 正硅铬酸铅
normal length 正常长度；标准长度
normal level 正常水平；正常能级
normal lift 一次浇筑的正常厚度(混凝土)；正常提升高度；正常上升高度；正常起吊高度
normal lighting 正常照明
normal lighting system 正常照明系统
normal line 垂直线；法线；标准线
normal linear scale 正态线性比(例)尺
normal line method 垂线法
normal liquid 正常液体
normal load 标称负载；正常载荷；正常荷载；正常负载；正常负荷；垂直载荷；垂直负载；质量负担；法向荷载；额定装载量；额定负载
normal load condition 正常荷载条件
normal load draught 正常吃水
normal loading 垂直荷载；标准荷载
normal loading capacity 车辆标准载重能力
normal locking 定位锁闭
normal lock magnet 定位锁闭电磁铁
normal logarithmic distribution 对数正态分布
normal logarithmic probability series 正态对数机率级数
normal loss 自然损耗；正常损耗；交易损失；途中自然减重
normal low water 正常低(潮)水位
normal low water level 正常低(潮)水位
normally applied brake 常闭式制动器
normally aspirated 正常吸气的；正常通风的
normally burning lime 正烧石灰
normally clear signal 正常开放信号
normally closed 常断开的；常闭合的；常闭的

normally closed auxiliary contact 常闭辅助触点
normally closed contact 动断触点；常闭接点；常闭触点
normally closed contactor 常闭接触器
normally closed interlock 休止辅助触头；常闭联锁装置
normally closed push button switch 常闭按钮开关
normally closed valve 常闭阀(门)
normally consolidated 正常固结的
normally consolidated clay 正常固结黏土
normally consolidated soft soil 正常固结软土
normally consolidated soil 正常压密土；正常固结土
normally consolidated soil deposit 正常固结沉积土
normally de-energized relay 正常失励继电器
normally de-energized stick relay 正常失励吸持继电器
normally distributed observations 正态分布的观测值
normally engaged gear 常啮合齿轮
normally engaged gear lock ring 常啮合齿轮锁环
normally functioning 功能正常的
normally hydrated cement 正常水化水泥
normally hydrated lime 正常水化石灰
normally insulated container 正常隔热容器；标准保温集装箱
normally loaded soil 正常固结土
normally off 常闭状态
normally on 常导通状态
normally open 常开状态
normally open circuit 正常断开电路
normally open contact 动合触点；常开接点；常开触头；常开触点；常开启接点
normally open contactor 常开接触器
normally open interlock 工作辅助触头；常开联锁装置
normally open push button switch 常开按钮开关
normally open rate 常开选通电路
normally open released brake 常开式制动器
normally open valve 常开阀
normally without charges 一般情况下免费
normal magnetic field 正常地磁场；正常磁场
normal magnetic flux density 正常磁密度
normal magnetic induction 正常磁感应
normal magnetization 正向磁化
normal magnetization curve 正常磁化曲线；标准磁化曲线
normal magnetogram 正常磁照图
normal magnetograph 正常地磁记录仪
normal magnification 正常放大(率)
normal magnitude 正常星等
normal man 标准人
normal map 标准图
normal map grid 标准地图格网
normal mapping 正规映射
normal margin of start-stop apparatus 启闭设备的标准改正
normal marine water 正常海水
normal market 顺价
normal maximum level 正常高水位；正常高潮位
normal meander 正常河曲
normal mean thrust 法向平均推力
normal measure 正规测度
normal mechanisms 正断层机制
normal message handling 正常信息处理
normal metamorphism 区域变质作用
normal meter 标准米
normal method of carrying out offshore diving 进行近海潜水的常规方法
normal microseismic noise 正常微震噪声
normal mixture 正常混合物；标准混合物
normal mode 正态模式；正规方式；正常状况；正常振荡模；正常方式；简正振型；简正模式；简正方式；标准模式
normal mode analysis 简正模式分析；简正模分析法；简正波分析法；标准振型分析
normal mode approach 标准振型法
normal mode equation 标准振型方程
normal mode field 简正模场
normal mode flow table 标准模式流程表
normal mode helix 简正模螺旋线
normal model 正态模型
normal mode method 正则振动型法；简正方法
normal mode noise 常态噪声

normal mode of vibration 正常振型;固有振型;简正振动模式;简正振动方式;标准振型
normal mode propagation 简正波传播;振荡波传播
normal mode surface wave 正态面波
normal mode vibration 简正振动
normal mode wave 简正波
normal mode(wave)theory 简正波理论
normal module 法向模型;法向模数;法面模数
normal modulus 中定伸强力;法向模型;法向模数
normal modulus of stress 法向应力模量
normal moisture 正常湿气;正常潮气;正常湿度;标准湿度
normal moisture capacity 正常湿度;正常含水量;正常持水量
normal moisture content 正常含水量
normal mortar 普通砂浆
normal mo(u)lding condition 标准成型条件
normal moveout 正常时差
normal moveout correction 动校正
normal multivariate analysis 正态多元分析
normal negative 标准底片
normal neighbo(u)rhood 正规邻域
normal network cause 正常网络条件
normal neutral plane 几何中性面
normal number 正规数
normal nut 标准螺母
normal nylon 正规耐纶
normal objective 常用物镜
normal offset 正支距
normal of meridian 子午圈的法线
normal on-load operation 正常带载操作
normal ooid 正常鲕
normal open 常通;常开
normal open circuit 正常开回路
normal opened contact 常开接点
normal opening 正常孔隙;正常开度;标准宽度;标准开度
normal open valve 常开阀
normal operating amperage 正常运转电流值
normal operating condition 正常工作状态
normal operating contact 定位动作接点
normal operating current 正常运转电流
normal operating expenses 正常工作开支
normal operating level 正常操作水平;正常运行库水位
normal operating limit 正常运转范围
normal operating loss 正常运行损失
normal operating period 正常工作时间
normal operating pressure 正常工作压力
normal operating range 正常运转范围
normal operating time 正常条件下的使用期限
normal operation 正常运转;正常运行;正常运算;正常营运;常规操作
normal operation condition 正常运行条件;正常工作情况
normal operation curve 正常运用曲线
normal operation loss 正常运行损失
normal operation mode 正常运行方式
normal operator 正规算子
normal opsonin 不耐热调理素
normal orbit 正常轨道
normal order 良序
normal order of crystallization 正常的结晶顺序
normal ore 普通矿石
normal orientation 正常取向;正常定向;标准方位
normal orifice 标准板孔
normal orthogonal system 标准正交系
normal-orthometric height 正常正高
normal osmosis 法向渗透
normal output 正常输出功率;正常输出;正常出力;正常产量;额定输出功率;标准产量
normal overhead 正常公里费用
normal overhead rate 正常间接费率
normal overload 常规超载
normal oxidizing solution 标准氧化溶液
normal packet function 正常波包函数
normal paper 中性像纸
normal parameter 正常参数
normal path 正常电导
normal pearlite 标准珠光体
normal peneplain 正常准(侵蚀)平原
normal period 标准周期
normal permeability 正常磁导率;标准磁导率;标准磁导率

normal phase 正常相
normal phase method 正相法
normal phase sequence 正常相序
normal phenomenon 正常现象
normal phosphate 正磷酸盐
normal photoelectric(al)effect 正常光电效应
normal photography 垂直摄影
normal pH value 正常pH值
normal picture 正常图像
normal pigging 正常清管工作
normal pin 止动销;垂直制动销;垂直销
normal pitch 垂直节距;常用屋面坡度;法向节距;标准螺(纹)距;标准距;标准间距
normal pitch auger 标准螺距螺旋
normal place 标准位置
normal plane 垂直面;法面;标准面
normal plane curvature 法面曲率
normal plate 标准板
normal-plate anemometer 垂直叶板风速计
normal plotting accuracy 正常绘图精度
normal plumb line 正常垂线
normal plywood 标准木质胶合板
normal point 正位点;标准点
normal point control relay 道岔定位控制继电器
normal point load 法向集中荷载
normal polar contact 极性定位接点
normal polarity 正接;正极性;标准极性
normal polarity armature 正极性衔铁
normal polarization 正常极化;常规极化
normal polygon 正规多角形
normal polytropic compression 正常多变压缩
normal pond level 正常蓄水位;水库正常(蓄)水位
normal pool level 正常蓄水位;正常供水位;全供水位;水库正常(蓄)水位
normal population 正态总体;正规总体
normal porcelain 普通瓷料
normal pore size 标称孔径
normal Portland cement 普通水泥;普通硅酸盐水泥;普通波特兰水泥
normal position 正常位置
normal position indicator 正常位置表示器
normal position of telescope 正镜测量;正镜【测】;望远镜标准位置
normal position of the cone 圆锥形正轴位置
normal position repeating relay 定位复示继电器
normal potential 正常电位;标准电位
normal potential function 正常位函数
normal potential of gravitation 引力正常位
normal potential of gravity 重力正常位
normal power 基本电源
normal power supply 正常供电
normal precast ware 正规预制品
normal precipitation 正常降水(量)
normal preparedness 正常准备状态
normal pressure 正压(力);正常压强;正常压力;垂直压力;常压;法向压力;标准压强;标准压力
normal pressure acid-base method 常压酸碱法
normal pressure and temperature 常压常温;标准压力和温度;正常压力和温度
normal pressure angle 正常压力角
normal pressure burner 常压燃烧器
normal pressure distribution 法向压力分布
normal pressure flange 常压法兰;常压凸缘
normal pressure head 正常压力水头
normal pressure rapid dyeing machine 常压快速染色机
normal pressure reactor 常压反应器
normal pressure surface (地下水的)常压面
normal pressure synthesis 常压合成
normal pressure tank car for chemical industry 常压化工专用槽车
normal price 正常价格
normal price of labo(u)r 劳动的标准价格
normal priority 一般优先权;标准优先权
normal probability 正态概率
normal probability curve 正态概率曲线;高斯曲线;常态或然率曲线;常态概率曲线;变量概率分布曲线
normal probability density 正态概率密度
normal probability distribution 正态概率分布
normal probability function 概率正态分布函数
normal probability paper 正态概率纸;正态概率表
normal probe 直探头
normal procedure 正常办理手续;常规方法

normal procedure call 正常过程调用
normal procedure deflection 正向试验法的弯沉;正常试验法的弯沉;常规试验的挠度
normal proceed 正常进行
normal process 正态过程;常规工艺
normal process(ing)section 常规处理剖面
normal production probability 常规生产式概率
normal production program(me) 正常生产程序
normal production rate 正常生产率
normal products 正品
normal profile 法向剖面;正常剖面;正常轮廓;标准剖面
normal profile of soil 正常土壤剖面
normal profit 正常利润
normal projection 正轴投影
normal projection method 法线投影法
normal propagation 正常传播;正比传播
normal pulse 平脉
normal quantities 正态量
normal radiation 正常辐射
normal radius of curvature 正规曲率半径;法曲率半径
normal random digit 正规随机数字
normal random function 正态随机函数
normal random number 正态随机数;正规随机数
normal random process 正态随机过程;正规随机过程
normal random variable 正态随机变数;正规随机变数
normal range 正常范围
normal rapping 法向振打
normal rate 规定利率;普通运价
normal rated heat load 标准额定热负荷
normal rated load 标准额定负荷
normal rated power 正常额定功率;额定功率;标准额定功率
normal rated thrust 正常推力;额定推力
normal rate of return 正常利润率
normal rating 额定出力;标准规格;标准定额
normal ray 正常光线
normal reach 正常运行距离;正常范围;正常(河)段
normal reaction 正交反力;正常反应;垂直反力;法向反作用力
normal reactive force 正交反力
normal reading distance 明视距离
normal real form 正规实形式
normal recession curve 正常退水曲线;正常亏损曲线;正常亏水曲线;标准退水曲线
normal reciprocal geothermal gradient 正常互易地温梯度
normal recoil length 正常后座距离
normal reducing solution 标准还原溶液
normal reduction potential 标准还原电势
normal reflection 正反射;垂直反射;法向反射
normal refraction 正常折射;折光正常
normal region 标准地区
normal reinforced concrete 正常加筋混凝土
normal relaxation behavio(u)r 正常松弛性能
normal release 正常释放;正常解锁
normal release condition 正常释放条件
normal relief angle 法向后角
normal representation 正规表示
normal requirement 正常需要
normal reserves 正常储备
normal reservoir elevation 水库正常(蓄)水位
normal reservoir level 正常蓄水位
normal residual 正规剩余的
normal resistance 正常电阻
normal resistivity device 梯度电极系
normal response 法向反作用力;正规响应;正常响应;常规响应
normal response mode 正常应答方式;标准响应方式
normal retirement 正常报废
normal return 正常报酬
normal return of investment 正常投资回收
normal revenue and expenditures 正常收支
normal reverberation time 正常混响时间
normal reversed time scale 正反时代表
normal reverse transfer 正反转换
normal revolution 正常转数
normal ring 正规环
normal ripple 正常波痕
normal rock pressure 岩体法向压力;岩石正压力

normal roll pan tile 标准波形瓦;常规波形瓦
normal roll pressure 轧辊正压力
normal root 定根
normal round 普通循环
normal route 正常进路
normal running 正常运转;正常运行
normal running condition 正常运行条件;正常工作条件
normal running fit 转动配合
normal running load 正常运行负荷
normal running speed 正常运转速度;正常运行速度
normal running temperature 正常运转温度;正常操作温度
normal runoff 正常径流量;正常径流
normal safety freeboard 正常安全超高
normal saline 生理盐水
normal saline solution 当量盐液
normal sample 正规样品;正规样本
normal sampling inspection 正常抽样检验;正常抽验
normal sand 标准砂
normal saturation curve 正常饱和曲线
normal scale 正态比例;标准比例尺
normal schedule 正常进度表
normal school 师范学校
normal scour 正常冲刷
normal scour cycle 正常冲刷周期
normal screw 标准螺纹
normal-seated cabin 正常坐姿驾驶舱
normal seawater 标准海水
normal section 横截面;正截面;正截口;正断面;法向截面;法截面;法剖面
normal section azimuth 法向截面方位角;法截线方位角
normal section line 法截线
normal sector 正扇形域
normal sedimentation 正常沉积
normal seed stand 模式母树枝;标准采种标分
normal segregation 正偏析;正常偏析;标准偏析
normal semifinite weight 正规半有限权
normal sensibility 正常灵敏度;标准灵敏度;标准感度
normal-separation fault 正向错位断层
normal sequence 正常顺序;正常层序
normal sequence of operation 正常动作程序
normal sequence space 正规序列空间
normal series 正规列
normal series of age gradations 模式令级序列
normal series without repetition 无重复正规列
normal service 正常服务
normal service condition 标准使用条件
normal service rating 正常使用功率;常用服务功率
normal set 正常凝固的
normal-setting 正常凝结
normal-setting cement 正常凝结的水泥
normal settlement 限额结算
normal settling cement 正常凝结的水泥
normal setup 正常配置
normal shape brick 普型砖;标准砖
normal sheaf 平行射向
normal shear structural system 直扭构造体系【地】
normal shift 正常移距;正变位
normal shoal 正常浅滩
normal shock 正激波
normal shock compression 正激波压缩
normal shock diffusion 正激波扩散
normal shock wave 正交冲击波;正激波
normal shrinkage 正常损耗
normal shutdown 正常停机;正常停工
normal side clearance angle 旁锋正留隙角
normal side rake angle 旁锋正刀面角
normal side relief angle 旁锋正后让角
normal sigmoid curve 正态 S 型曲线
normal silver sulfate 硫酸银
normal simple algebra 正规单代数
normal simultaneous stage relationship curve 正常同时水位关系曲线
normal single head wrench 标准单头扳手
normal sinking 普通凿井法
normal sitting position 正常架设姿态
normal size 标准大小;标准尺寸;正常尺寸
normal-slip fault 正常断层;正(滑)断层

normal slope 法线斜率
normal slump 标准坍落度
normal soil 正常土
normal solution 正(常)解;当量溶液;标准溶液
normal sonde 电位电极系
normal space 正规空间;标准间隔
normal span 正常跨距;正常跨度
normal specific speed Francis turbine 正常比速混流式水轮机
normal spectral measure space 正规谱测度空间
normal spectrum 匀排光谱;正态谱;正常谱;正常光谱
normal speed 正常速度
normal-speed screen 正常余辉荧光屏;正常衰减屏
normal spheroid 正常椭球
normal spheropotential 正常等位面
normal spheropotential number 正常地球位数
normal spheropotential surface 正常椭球等位面
normal spin 正常旋转
normal spinel 正尖晶石
normal spoilage 正常破损
normal spring tide 正常大潮
normal square 矩规
normal stage 正常水位
normal-stage punch(ing) 正规穿孔;卡片偶数行穿孔
normal stand 标准林分
normal standard cost 正常标准成本
normal standard quantity of material stock 材料的经济储备定量
normal star 正常恒星
normal state 正常状态;正常状况;常态;标准状态
normal static magnification 标称静态放大(倍数)
normal stationary process 正态平稳过程
normal stationary sequence 正态平稳序列
normal steel 正常结构钢;正常钢
normal steel casting 普通铸钢
normal stereogram 正立体图
normal stereographic(al) projection 正球面投影
normal stiffness coefficient 法向刚度系数
normal stochastic process 正态随机过程;正规随机过程
normal stock 正常库存量
normal stock method 正常存量法
normal storage 正常蓄水量
normal storage capacity 正常储量;正常库容
normal storage water level 正常蓄水位
normal straight blades snips 直刃剪
normal straight sequence 正常层序
normal strain 正应变;法向应变;标准应变
normal stratification 正常层序
normal stratigraphic sequence 正常层序
normal stream flow 正常水流;正常流量;正常川流
normal stress 正应力;正胁强;垂直应力;法向应力;标准应力
normal stress component 法向应力分量
normal stress effect 法向应力效应
normal structural steel 正常结构钢
normal structure 标准组织;标准结构
normal subgroup 正规子群
normal substitution 正常取代
normal subsystem 正规子系统
normal succession 正常层序
normal supercharging 正常增压
normal superphosphate 普通过磷酸钙
normal superposition 正常叠加;正常层序
normal support 正常支承;标准支承
normal surface 法面;法线速度面;正态曲面;正交面;垂直面
normal switch 定位道岔
normal switch position 道岔定位
normal synchronous speed 正常同步速度
normal system 正规系
normal tail water level 正常尾水位;正常尾水位
normal tangent conic(al) equivalent projection 正轴等积圆锥投影
normal tangent cylindric(al) equivalent projection 正轴等积圆柱投影
normal tape 普通扁
normal tellurigram 正常地电图
normal temperature 正常温度;常温;标准温度
normal temperature and pressure 常温常压;标准温度和压力

normal temperature cure 常温固化
normal temperature curing 常温养护(混凝土)
normal temperature distribution 正常温度分布
normal temperature lapse rate 正常温度递减率
normal temperature pressure 标准温度(下)压力
normal temperature washing agent 常温清洗剂
normal tensile stress 法向拉伸应力;单纯抗张应力;正常张应力;正常拉伸应力;法向应力
normal tension 正常张力;法向拉力;标准拉力
normal termination 正常终结;正常结束
normal ternary 标准三进位计算制
normal test 标准试验
normal test pressure 正常试验压力
normal thermometer 标准温度计;标准温度表
normal thickness 正常厚度
normal thickness of blade 叶片法向厚度
normal thickness on constant chord 定法向齿厚
normal thorium sulfate 硫酸钍
normal thread angle 垂直螺纹齿形角
normal thread form 标准螺纹形状
normal threshold of audibility 正常听觉;正常感觉(阈限)
normal threshold of feeling 正常感觉(阈限)
normal threshold of hearing 正常听力阈值
normal throw 正落差
normal thrust 垂直推力;法向推力;标准推力
normal tidal beach 经常感潮滩地
normal tideless beach 正常不感潮海滩
normal tile 普通瓦
normal timber life 木材正常使用年限
normal time 正常时间;纯劳动时间;标准时(间)
normal time response 正常时间响应;正常时间特性
normal timing 正常定时
normal to 垂直于
normal to a curve 曲线的法线;曲面法线
normal to an emergent wave surface 出射波面法线
normal to a surface 曲面的法线;表面的法线
normal to hyper-surface 超曲面的法线
normal tolerance class 普通公差等级
normal tolerance distribution 正态耐性分布
normal tooth 正常齿
normal tooth thickness 法向齿厚
normal topological group 正规拓扑群
normal topological space 正规拓扑空间
normal top-water level 正常洪水位;正常高水位;正常高潮位
normal torque 正常转矩
normal to screen 垂直于屏幕
normal to surface 曲面法线
normal to the stratification 层理的法向(线)【地】;垂直于层理方向的
normal to the surface 垂直于(法向)片面的
normal to track level 垂直于轨面
normaltoxin 标准毒素
normal trace 正规迹
normal traffic 正常通报
normal traffic growth 正常交通增长;运量的正常增长;交通量正常增长
normal traffic increment 正常增长交通量
normal traffic position 正常通报位置
normal train 正规列车
normal trajectory 理想弹道;标准弹道
normal transformation 正态变换;正规算子;正规变换
normal travel time curve 正常走时曲线;正常时距曲线;正常传播时间曲线
normal tree 正规树;规范树
normal trichromatic vision 正常三色视觉
normal trichromatism 正常三色视觉
normal trilinear coordinates 正规三线坐标;法三线坐标
normal twill 正规斜纹
normal type 正常型;正常体重
normal type brick 普型砖
normal type wagon 标准型货车
normal undercarriage 普通底架
normal unthreading 正常退带
normal upstate 正常工作状态
normal uranium 天然铀
normal utility 正常利用率
normal vacuum 标准真空
normal valency 正价;正常价【化】
normal valley 正常谷

normal valuation 正规赋值
normal value 正常(价)值;标准值
normal value of rock 岩石正常值
normal valve 定位阀
normal variable 正常变量
normal variation 正规变分;正规变差
normal variety 正规簇
normal vector 法向量;标准矢量
normal vector of a curve 曲面法线向量
normal vector of a surface 曲面的法线矢量
normal velocity 正常速度;法向速度;额定速度
normal velocity component 法向速度分量
normal velocity distribution 正常流速分布;法向速度分布
normal ventilation 常规通风
normal vertical deflection 正向垂直偏转
normal vibration 正常振动;简正振动
normal video 标称视频
normal viscosity 正常黏度
normal vision 正常视觉
normal voltage 正常电压;额定电压;标准电压
normal-voltage rating 正常电压下的额定值
normal volume 正常体积;标准体积
normal wage 正常工资
normal water 正常水;普通水
normal water capacity 正常持水量
normal water consumption 标准耗水量
normal water flow 稳定涌水
normal water level 正常水位;常水位;标准水位
normal water level elevation 正常水位高程
normal water level ga(u)ge 常水位计
normal water stage 常水位
normal water storage elevation 正常蓄水高度;正常蓄水高程
normal water yield of mine 矿坑正常涌水量
normal wave 正激波
normal wear 正常损耗;正常磨损;正常磨耗
normal wear and tear 正常磨损;正常磨耗
normal weight 正常重量;当量重量
normal weight aggregate 普通重量集料;普通重量骨料
normal weight concrete 常重混凝土;常规重量混凝土;普通混凝土
normal weld(ing) 正常焊接;常规焊接
normal well pattern 标准井网
normal width 标准宽度
normal work(ing) 正常运行;正常工作
normal working condition 正常工作条件
normal working day 正常工作日
normal working frequency 正常工作频率
normal working hours 正常作业时间;正常工作小时数
normal working load 正常工作荷载
normal working point 正常工作点
normal working pressure 正常工作压力
normal working state 正常工作情况
normal working temperature 正常工作温度
normal working time 正常工作时间
normal working voltage 正常工作电压
normal wrench fault 正平移断层
normal year 中水年;正常年(度);平水年;平均年
normal yield 正常年景;模式收获量;法正获量
normal yield table 模式收获表;标准收获表
normal Zeeman effect 正常塞曼效应
normal zero 标准零点
normal zoning 正向分带;顺向分带
Norman arch 诺曼式拱
norman(-bolt) 龙门柱销;缆柱销;舵端防松销
Norman brick 诺曼砖(规格为2.5412英寸)
Norman crypt 诺曼式地穴
Norman Gothic style 诺曼哥特风格
norman pin 龙门柱销;缆柱销;舵端防松销
Norman roofing tile 诺曼屋面瓦;诺曼式屋顶瓦
Norman slab 诺曼玻璃块
Norman style 诺曼建筑风格;诺曼式;诺曼风格
Norman(style)architecture 诺曼式建筑
Norman vault 诺曼拱顶;诺曼连拱式圆屋顶
Norman window 诺曼式窗(户)
normarkite 英碱正长岩
normative 标准的
normative analytic model 标准分析模型
normative annex 标准的附录
normative approach 规范化处理
normative documents 标准文件

normative element 标准成分
normative forecast 标准预测
normative forecasting technique 规范预测技术
normative mineral 标准矿物
normative model 规范模型;标准模型
normative reference 参考标准
normative requirement 标准的要求
normative scale 标准尺度
normative tax structure 标准税收结构
normative usage 一般惯例;常规;标准用法
normatron 模型计算机;典型计算机
norm basic data 定额基础数据
norm cost 定额成本
normed current assets 定额流动资产
normed current funds 定额流动资金
normed liability 定额负债;定额负载
normed linear space 赋范线性空间
normed requisition 定额领料单
normed space 赋范空间
normed vector space 赋范线性空间
normergia 反应力正常
normergic 反应力正常的
normergic reaction 常态反应
normergy 反应力正常
norm expenses 标准费用
norm following behavior 标准仿效行为
norm for detailed estimates 预算定额
norm for estimating labo(u)r requirements 劳动定额
norm for estimating material requirements 材料定额
norm for materials consumed 材料消耗定额
norm for preliminary estimates 概算定额
norm-fund other than commodities 非商品定额资金
norm-hour 定额小时
norm in kind 实物定额
norm in money 货币定额
norm management 定额管理
norm management system 定额管理制度
normobaric 正常气压的
norm of advanced average 先进平均定额
norm of a matrix 矩阵(的)范数
norm of consumption 消耗定额
norm of material consumption 物资消耗定额;材料消耗定额;实物消耗定额
norm of output 产量定额
norm of planned consumption 计划消耗定额
norm of reaction 反应量;反应规范
norm of rough estimates for a structure 概算定额
norm of technological economy 技术经济定额
norm of the civil law 民法规范
norm of vector 向量模;向量范数
norm of working hour 工时定额
normogram 正常图
normolineal 正线的
normomorph 形态正常
normotension 压力正常
normothermia 正常体温;温度正常
normotopic 位置正常的
norm price 指标价格;额定价格
norm reducing method 减模法
norms for approximate estimate 概算定额
norms for preliminary estimate 概算定额
normvitrinite 正常镜质体
no road 非通行道路;不通行
norol 汽车坡路停车防滑机构;防滑装置
no roll 汽车停车防滑机构
no-roll landing 无滑跑着陆
norsethite 钡白云石
Norske Geotechnical Institute 挪威土工研究所
north 北的
North African architecture 北非建筑
north aisle 走走廊(教堂);北侧通道
North America 北美洲
north America-Eurasia separation 北美大陆与欧亚大陆分离
North American anti-cyclone 北美高压
North American architecture 北美洲建筑
north American bivalve province 北美双壳类地理区
north American brachiopod realm 北美腕足动物地理区系
north American carbonates belt margin and intraclastic belt 北美碳酸岩带邻缘和内碎屑带
north American coral realm 北美珊瑚地理区系
North American high 北美高压
North American Lake Management Society 北美湖泊管理协会
North American Midcontinent-Tunguskan region 北美中大陆—通古斯区
north American nautiloid region 北美鹦鹉螺地理大区
North American Nebula 北美洲星云
North American plate 北美板块
north American platform 北美地台
north American realm 北美区系
north America paleoclimate epoch 北美古气候期
north America paleocontinent 北美古陆
Northampton sands 北安普敦砂岩
north-and-south line 坐标纵线
north arrow 指北针;指北箭头
North Atlantic bottom water 北大西洋底层水
North Atlantic current 北大西洋海流
North Atlantic Deepwater Oil Terminal 北大西洋深水油港
North Atlantic drift 北大西洋漂流
North Atlantic flow 北大西洋暖流
North Atlantic Ocean 北大西洋
North Atlantic Ocean Station 北大西洋海洋观测站
north Atlantic realm 北大西洋区系
North Atlantic stream 北大西洋海流
North Atlantic track 北大西洋航线
North Atlantic Treaty Organization 北大西洋公约组织
north Atlantic trilobite realm 北大西洋三叶虫地理区系
North-Baikal polarity superchron 北贝加尔极性超时
North-Baikal polarity superchronzone 北贝加尔极性超时间带
North-Baikal polarity superzone 北贝加尔极性超带
north bound 向北航行
north bound node 升交点
north by east 北偏东
north by west 北偏西
North Cape Current 北角海流
north cardinal mark 北方位标
North Carolina bearing method 美国北卡罗来纳承载板法
North Carolina bluestone 美国北卡罗来纳青石
North Carolina method 美国北卡罗来纳法(设计柔性路面厚度的一种方法)
north Cathaysian floral region 华夏北方植物地理大区
North-Caucasus orogeny episode 北高加索构造作用幕
north celestial pole 天球北极
North China arid subtropical to tropical zone 华北干旱亚热带-热带
North China coral province 华北珊瑚地理区
North China faunal province 华北动物地理区
North China fold system 华北褶皱系
North China geosyncline system 华北地槽系
North China great regression 华北大海退
North China non-marine ostacode province 华北非海相介形类地理区
North China region 华北大区
North China sea basin 华北海盆
North China semiarid subtropical to tropical subzone 华北半干旱亚热带-热带亚带
North China-Southeast Asia Plate 华北-东南亚板块
North China transgression 华北海浸
North China upland 华北高地
North Dakota cone 美国北达科他州圆锥体(用于试验土基承载量)
North Dakota cone method 北达科他州圆锥试验法(测定土基承载力)
North Dakota cone test 北达科他州圆锥试验
North Dakota design method 美国北达科他州柔性路面设计法
northeast[NE]东北
northeast by east 东北偏东
northeast by north 东北偏北
northeast drift current 东北漂流
northeaster 东北风;东北大风;东北信风
northeastern 东北的

northeast monsoon 东北季风
northeast monsoon current 东北季风暖流;东北季风海流
northeast pacific ocean 东北太平洋
northeast storm 东北风暴
northeast trade drift 东北信风吹流
northeast trade(wind) 北半球信风;东北信风
northeast trending fault 东北走向断层
northeastward 向东北;在东北;东北方向;朝东北的
north ecliptic pole 北黄极
north elevation 北立面(图)
north equatorial current 北赤道海流
north equatorial warm current 北赤道暖流
norther 强北风
northerly 北风
northerly limit 北图廓
northerly turning error 北转误差
Northern Alxa tectonic segment 北阿拉善构造段
northern anthracnose 北方炭疽病
northern celadon 北方青瓷(以别于越窑青瓷而得名)
northern China humid warm temperate to subtropical zone 北部潮湿暖湿—亚热带
northern foot 北方尺(1尺=13.12英寸)
northern frigid zone 北寒带
Northern hemi-sphere 北半球
Northern Himalayan geosyncline 北喜马拉雅地槽
northern hole 北方臭氧层空洞
northern ice cap 北极冰盖
northern Korean sea 北朝鲜海
northern latitude 北纬;北向
northern lights 北极光
northern limit 北界线
northern marginal deep fracture zone of the Sino-Korean platform 中朝地台北侧深断裂系
northern middle latitude 北半球中纬度
northern Pacific bivalve subprovince 东北太平洋双壳类地理亚区
Northern pacific ocean basin block 北太平洋海盆巨地块
Northern pine 云杉松木;北方松
northern platform depression 北部台坳
northern polarity 北磁极性
northern polar region 北极区
northern reference star 北天参考星
northern slope 北向坡
Northern Song official ware 北宋官窑
northern stratosphere 北方平流层
northern subpolar latitude 近北极纬度
northern suburbs 北郊
northern Vietnam-Guangxi shallow sea 北越—广西浅海
northern Vietnam sea 北越海
northern Vietnam sea basin 北越海盆
northern white cedar 北美香柏
northern white pine 北美乔松;北方白松
north European bivalve province 北欧双壳类地理区
north European paleoclimate epoch 北欧古气候期
north-facing facade 面北的立面(建筑物);朝北立面(建筑物);建筑物向北的正面
north-facing wall 朝北的墙;北墙
north-facing window 朝北的窗;北窗
north-finding instrument 寻北器;指北仪
northflowing river 北流河
northflowing stream 北流河
north foehn 北方焚风
north frigid zone 北极带
north galactic pole 北银极
North Gandise piedmont trough system 冈底斯北山前坳陷系
north geographic(al)pole 地理北极
north geomagnetic pole 地磁北极;北极
northgoing river 北流河
northgoing stream 北流河
north Hebrides trench 北赫步里底海沟
north heliographic(al)pole 日面北极
north hemi-spheric-Australian floral realm 北半球—大利亚植物地理区系
Northill anchor 诺思希尔式锚
north Indian ocean bivalve subprovince 北印度洋双壳类地理亚区
northing 向北驶;指北偏差;北向纬度差;北进
North Italian Quatrocento 北意大利14世纪时期

north Kalimantan basin 北加里曼丹盆地
north latitude 北纬;北向
north light 北窗;朝北窗;北极光
north-light barrel-vault shell roof 北面采光的筒形薄壳拱顶
north-light cupola 北面采光圆屋顶
north-light cylindric(al)shell 北面采光圆筒形薄板
north-light glazing 北窗采光玻璃
north-light gutter 北面采光天沟
north lighting 北面采光技术
north-light roof 北面采光的锯齿形屋顶;朝北采光屋面;北面采光屋盖;锯齿形屋顶;北向采光(锯齿形)屋顶
north-light roof glazing 表面采光屋顶装配玻璃;表面采光屋顶天窗
north-light roof roof-light 表面采光屋顶的天窗
north-light saucer dome 北面采光碟形窗
north-light truss 北向采光的桁架;北向采光的屋架;锯齿式屋架
north magnetic pole 磁北极;北极;北磁极
north magnetic pole position 北磁极位置
north northeast[NNE] 北北东;北东北
north northwest[NNW] 北北西;北西北
north orientation 北向定位;北向方位;北向
North Pacific basin 北太平洋海盆
North Pacific bivalve region 北太平洋双壳类地理大区
North Pacific current 北太平洋海流;北半球极海流
North Pacific drift 北太平洋飘流
North Pacific experiment 北太平洋观测计划
North Pacific ocean 北太平洋
north pal(a)eo-Mediterranean region 古地中海北部大区
north point 北点
North Polar Circle 北极圈
north polar distance 北极距
north polar sepuence 北极星序
north polar spur 北银极支
north pole 磁针指北端;北(天)极
North Pole ice 北极冰
north porch 北门廊(教堂);北侧门廊
north reference pulse 指北参考脉冲
North River bluestone 北河青石(产于纽约,主要用于筑路)
north roof building 北屋房屋;北屋建筑
North Sea 北海
North Sea basin 北海盆地
North Sea gas 北海天然气
north-seeking end 红端(磁铁指北极)
north-seeking force factor 指北力系数
north-seeking gyro 指北陀螺【航海】
north-seeking instrument 指北仪
north sky light 北向白昼光
north slope 北坡
north-south asymmetry 南北不对称性
north-south component 南北分量
north-south diwa region 南北地洼区
north-south middle point 坐标纵线中点
north-south negotiation 南北谈判
north-south structural zone 南北构造带
north-south tectonic system 南北构造系
north-south trade 南北贸易
north-stabilized indicator 指北针
North Star 北极星
north subtropical zone 北亚热带
north temperate bivalve realm 北温带双壳类地理区系
north temperate zone 北温带
North Temple pagoda 北寺塔(苏州)
north Tethyan region 特提斯北侧大区
north Tethyan trough 北特提斯海槽
north transept (建筑物的)北侧翼部
north tropic 北回归线
north tropic zone 北热带
northupite 氯碳钠镁石
north-upward presentation 北向朝上表示法
north upwarping region 北部隆起区
north Vietnam-Zhujiang transgression 北越—珠江海浸
northward 向北;朝北的
northward component 北向分量
northwest 西北
northwest and north China semiarid subtropical zone 西北-华北半干旱亚热带

northwest by north 西北偏北
northwest by west 西北偏西
northwest China 西北区
northwest corner method 西北角法
northwester 西北大风;西北风
northwesterly wind 西北风
northwestern 西北的
northwest lighting 西北光斜照法;西北光线法
northwest monsoon 西北季风
northwest pacific ocean 西北太平洋
northwest shipworm 西北蛀船虫
northwest wind 西北风
north wind 北风
North Zhili Sea 北直隶海
norton floor 刚铝石地面;刚铝石地板
Norton pulpstone 人造磨石
Norton's equivalent 诺顿等效
Norton's gear 诺顿齿轮
Norton's theorem 诺顿定理
Norton's theory 诺顿理论
Norton's tube well 诺顿管(深)井
no-run takeoff 无滑跑起飞
noruron 草完隆
Norusto 诺拉斯托(一种防锈漆)
Norway 挪威
Norway pine 挪威红松
Norway spruce 挪威云杉
Norwegian architecture 挪威建筑
Norwegian basin 挪威海盆
Norwegian clay 挪威黏土
Norwegian cut 挪威式掏槽
Norwegian Maritime Directorate 挪威海事管理局
Norwegian quartz 挪威白石英
Norwegian Sea 挪威海
Norwegian stave church 挪威式板条教堂
Norwich Crag 诺威奇砂质泥灰岩
Norwood blender 屋面铺细粒机
Norwood hopper 屋面铺细粒机上料斗
no sale 非销售
no sale final 非销售终结
nosazontology 病因学
no-screen film 无屏胶片;无屏蔽膜
no screw balance 无平衡螺钉摆轮
nose 岬;机头;翘头;山嘴;墩端;导板头;船首;池窑作业部;背斜鼻;鼻(子);鼻锥凸头;鼻端;浊锋
nose air intake 前进气口
no-seam 无缝
no seam back stitches 不踏回针
nosean(e) 勤方石
nose angle 尖端夹角;前端角;头部锥角;刀尖角;鼻角;包角(蜗壳)
nosean phonolite 勤方石响岩
nosean trachyte 勤方粗面岩
nose balance 轴向补偿
nose bar 压尺
nose batten 头部支肋
nose bit 曲柄钻钻头;突缘匙形钻;手摇扁钻
nose bushing 吸料模的钢套
nose cabin 前舱
nose cap 鼻锥帽
nose circle (轴的)圆端;轴圆端
nose cock 水管旋塞;水管小龙头;煤气旋塞;煤气小龙头;弯嘴旋塞;弯嘴龙头
nose compartment 前舱
nose cone 前锥体;头锥;鼻锥
nosed 突缘的
nose dive 暴落
nose down 前倾
nose end 管口端;孔端
nose ender 迎面强风
nose fillet 前缘嵌条
nosegay 小花束
nose gear nacelle 前轮舱
nose girder 导梁;鼻梁
nose hangar 机头棚
nose height 鼻高
nose in 倾斜
nose key 小木楔;钩头楔;暗楔
nose length 鼻长
noseless 无喷嘴的
nose line 梯段坡度线(各级踏步凸缘或前缘的连线)
noselite 勤方石
nose mudguard 车首挡泥板

nose nacelle 前舱
nose of chisel 凿尖
nose of cone（钻头的）牙轮锥顶；凸轮尖
nose of diaphragm wall 地下连续墙鼻
nose of fold 褶皱鼻
nose of frog 辙叉实际尖端；辙叉实际尖点
nose of groin 交叉拱前缘；丁坝端部；丁坝前端
nose of groyne 交叉拱前缘；丁坝端部；交叉拱尖端
nose of pier 桥墩棱体；桥墩尖端
nose of(the) punch 冲模头
nose of tool 刀片头；刀尖
nose of wing 翼前缘
nose ogive 头部卵形部分；拱形头部；蛋形头部
nosepiece 测头管壳；管口；接线头；显微镜镜鼻；显微镜管嘴；救火水龙头；物镜筒；端部喷嘴
nose pile 定位桩；导柱；标桩；放汽管口
nose pipe 放汽管
nose plate 锁盖板；船首护头板；分线盘
nose radiator 车头散热器
nose radius（交通岛的）端点半径；刀尖半径
nose rail 心轨；辙轨
nose rib 机头肋；前缘肋
nosering block 窑口圈砖
nose ring segment 扇形窑口护板
nose section 机头段；机头部分；前缘剖面
nose sill（钻塔前边横向的）基台木
nose spike 头部减震针
nose spinner 机头整流罩
nose steadying line 防摇索
nose suspended motor 鼻式悬挂电机
nose suspension 抱轴式悬挂
nose-suspension motor 前部悬挂电机
nose taper 端点斜坡
nose tip 鼻锥帽
nose to nose 面对面；顶对顶
nose-to-nose locking 对句锁闭
nose-to-tail 首尾相接
nose-to-tail drag curve 纵向阻力曲线；机身面积曲线
nose treatment（交通岛的）端点处理
nose wedge 暗楔
nose wedging 暗楔
nose wheel 前轮；车台鼻轮
nose wrapper plate 舵前缘板
no-shadow point 无银点
no-shear failure 剪切无破坏
no-shrink condition 无收缩状态
no-shrink condition of soil 土的无收缩状态
no-side-draft position 均衡牵引状态；非偏牵引位置
no signal 无信号
no signaled movement 非信号指挥的运行
no significants counts 不显著的
nosing 摇曳；楼梯踏步小突沿；机头整流罩；机车头部；突缘(饰)；凸边；踏步前缘；（机车摇曳引起的）侧向荷载；半圆线角；半圆线规
nosing bead 突缘圆饰
no single hypothesis 非单一假设
nosing line 梯段坡度线（各级踏步凸缘或前缘的连线）；突缘饰线；（楼梯的）踏口线
nosing method 拖拉架桥法
nosing motion of vehicle 车辆前后颠簸运动
nosing of tread 踏步突缘
nosing piece 楼梯踏步小突沿；（水道、管道等的）管嘴；突缘木条
nosing plane 凸缘刨；法兰刨
nosing strip 凸缘条；凸缘条线脚；法兰条
nosite 黝方石
no-skid pavement 抗滑动铺面
no-skid road 粗糙路面；不滑路
no-skid surface 抗滑动铺面；不滑（路）面
no-sky line 天空视见线
no-slip angle 临界角
no-slip point 中性点
no-slotted turbogenerator 无槽汽轮发电机
no-slump concrete 低坍落度混凝土；干（硬）性混凝土；无坍落度混凝土；不坍落混凝土
no smoking 禁止吸烟（标志）；不准吸烟
nosochthonography 疾病地理学
nosogeography 疾病地理学
nosometry 发病率计算法
no-spark 无火花
no-spark zone 无火花区
no-spillrule curve 防弃水线
no-spin lockup 自锁差速器

no spongy body 无海绵体
no standard 无规定
no standing 不准站立
no-steam consumption 空负荷汽耗
no-step metallization 无台阶的金属化
no-stick coating 不黏面涂层
nostopathy 怀乡病
no stopping 不准停车
no-strength temperature test 无强力温度试验
no-stress meter 无应力表
nostro 往账（本国银行在国外银行持有的外币账户）
nostro account 银行来往账户；我方账户
nostro deposit 我方存款
No such address 地址不详
no support 无支撑
no-support excavation 无支撑开挖
no surplus stock 无剩余贷款
no swell 无涌；道格拉斯零级浪
no-swell condition 无湿胀状态
no-swell condition of soil 土的无湿胀状态
nosykombite 斜霞正长岩
notable 显著的
not absolutely 不绝对
not acclimatized 水土不服
not accustomed to the climate 水土不服
no tag cell 无标记元
notal 背的
not applicable 不适用；不可适用
notarial acknowledgement 公证人认定
notarialacts 公证手续
notarial deed 公证证书；公证书
notarial protest certificate 拒绝证书
notarization 公证证书；公证
notarization of construction contract 施工合同的公证
notarize 公证
notarized draft 经公证人签章证明的汇票
notary 公证员；公证行；公证人
notary office 公证处
notary public 公证员；公证人
notary public office 公证处
notation 记法法；记号；记法；计数法；位置计数法；符号；表示法；标志法
notational convention 符号约定；标记法则
notational expression 记号表达式
notation constant 记法法常数；符号常数
notation for field node 场的节点记号
notation for list 列表的记号
notation index 符号索引
notation system 记数制；记数系统
notation variable 记法法变量
not at risk 不在保险范围内
not at the expense of production 不误生产
not beveled 无坡口
not-blown up 不饱满
not-busy interrupt 空间中断
notch 选择器标记；刻痕；开槽；接榫扣；入孔；缺口；山峡；槽口；编码孔；边槽；凹口；凹槽
notch action 刻痕作用
notch acuity 缺口锐度；切口锐度
notch adjuster 标记调整器
notch adjustment 刻痕调整；标记调整
notch amplifier 标记信号放大器；标度信号放大器
notch and tooth joint 齿槽连接
notch angle（作冲击试验用的）刻痕角
notch antenna 切口天线
not charged 免费的；不收费的
notch at arch springing 隧道起拱处缺口
notchback 客货两用汽车
notch-bar cooling bed 齿条式冷床
notch-bar hot bed 齿条式热床
notch-bar impact test 切口棒冲击试验；开槽试件冲击试验
notch bar test 凹槽试件冲击试验；切口棒试验
notch bending test 切口挠曲试验；凹口弯曲试验
notch block 开口滑车；凹槽滑车
notch board（测量水流量的）凹口板；带槽口楼梯斜梁；楼梯搁板；梯口板；梯级搁板（木楼梯的斜帮、承载踏步）；凹槽侧板；梯板；V形凹槽；凹板
notch breaking test 缺口断裂试验
notch break test for butt welded joint 对接接头缺口断裂试验

notch brittleness 刻痕脆性；缺口脆性；切口脆性；冲击切口脆性
notch brittleness test 缺口脆性试验
notch coefficient 缺口有效应力集中系数
notch connection 凹槽连接
notch contraction 切口收缩
notch curve 阶段曲线
notch cutting 切槽；切口口
notch dam 缺口量水坝；缺口坝；凹口坝
notch depth 切口深度
notch diplexer 天线共用器（锐截式）
notch drop 缺口跌水；凹口跌水
notch ductile steel 耐冲击钢
notch ductility 缺口延性；切口韧性
notch ductility test 缺口韧性试验
notched 开槽的；切口的；切槽的
notched and cogged joint 槽口榫接
notched and welded wide plate test 缺口焊接宽板试验
notched area 缺口面积；切口面积
notched bar 开槽钢筋；刻槽试杆；刻槽杆件；缺口试棒；切口试棒；切口杆件；凹口试杆
notched bar impact strength 冲击韧性值
notched bar pull test 缺口拉力试验
notched bar strength 切口拉力强度；齿形杆强度；齿口拉力强度
notched bar tensile test 缺口抗拉试验
notched bar test 刻痕试杆冲击试验；切口试样冲击试验；切口弯曲试验
notched bar value 缺口冲击值
notched beam 开槽梁；阶形梁；刻凹槽梁
notched belt 齿形皮带
notched brick 舌槽砖；开槽砖；刻凹槽砖
notched chisel 开槽錾子；开槽凿；带槽凿子
notched colter 缺口圆犁刀
notched column 有槽柱；阶形柱
notched disc 周缘凹口盘；缺口靶片
notched effect 切口效应
notched flange 有槽凸缘；有槽突缘；带槽凸缘；有槽法兰；带槽法兰
notched flat headed atomizer 缺口平头喷雾器
notched furnace 锯条底炉
notch-edged share 缺口铧；齿形铧
notched impact specimen 缺口冲击试样
notched impact strength 切口冲击强度
notched ingot 带缺口的铸锭
notched joint 开槽接合；插榫接合；凹口接头
notched mo(u)lding 凹口线脚；齿形线脚
notched plate 周缘凹口盘
notched sclerotinite 绿沟菌类体
notched scraper 齿形刮刀
notched semaphore blode 鱼尼形臂板
notched serrated sickle 锯齿式镰刀
notched sill 刻凹槽窗台板；刻有凹槽的门槛；刻凹槽门槛；齿形槛（水工消能设备）；齿坎；坎槛
notched specimen 刻槽试件；缺口试件；切口试样
notched speed sensing rail 刻槽测速钢轨
notched spike 倒刺肘形钉
notched spreader 开槽口的铺胶器
notched stage 有槽的平台
notched talcon 齿榫
notched tappet rot 刻缺锁簧杆
notched teeth syphilitic teeth 锯齿形牙
notched tensile specimen 缺口拉伸试样；切口拉伸试样
notched tensile strength 缺口抗拉强度
notched track diagram 刻槽轨道表示盘
notched undulated preheater 齿形波纹板预热器
notched V-belt 齿形三角带
notched weir 缺口堰；切口（量水）堰；狭缝堰
notch effect 应力集中效应；裂口效应；刻痕效应；刻槽影响；开槽影响；缺口应力集中效应；缺口效应；槽口影响
notch effect at high temperature 高温切口效果
notch embrittlement 缺口脆性；冲击切口脆性
notcher 钢材切割机；断屑器；刻痕器；开槽机
notches（控制器的）换级触点
notch expansion 标志扩展
notch factor 刻槽系数
notch fall 缺口落差；缺口降落
notch fatigue strength 切口疲劳强度
notch fatigue test 切口疲劳试验
notch filter 陷频滤波器
notch filter frequency range 陷波器陷除频率范围

notch flexure test 凹口抗弯试验;刻槽弯曲试验
notch frequency 陷波频率
notch ga(u)ge 凹槽标准规;凹口标准规;缺口式流量计
notch generator 标志信号发生器
notch geometry 缺口形状
notch grinder 磨槽机
notch groove 刻槽;凹槽
notch gun 泥炮
notch impact resistance 抗切口冲击性
notch impact strength 缺口冲击强度
notch impact strength value 切口冲击强度值
notch impact teat 缺口冲击试验
notch impact toughness 楔口冲击韧性;切口冲击黏度
notching 切口
notching 单榫接;槽口连接;下凹;做凹口;刻凹槽;开槽接合;开槽;木枕刻痕;多动式(继电器);程序开关
notching curve 下凹曲线;阶梯形曲线;切口曲线;下凹垂线
notching dies 开凹口模
notching filter 陷波滤波器;阶式滤波器
notching joint 槽式接合;槽口缝;凹槽节
notching machine 齿形刀片裁断机;开槽机
notching of splice bars 鱼尾板上道钉槽
notching press 切槽压力机;切凹口压力机;冲缺口压力机;冲槽机
notching punch 凹口冲头
notching ratio 换线比;切口比率
notching relay 加速继电器;分级继电器
notching spade 切口铲
notching time relay 多级式限时继电器
notching up 操纵杆加速操作
notching wire 齿痕钢丝
notch joist 开槽格栅;带槽口小梁;凹形桁条
notch ornament 凹口线脚;齿形线脚;凹口装饰
notch pier 缺口墩
notch pin 缺口销;凹口销
notch planting 铲隙栽植
notch plate 缺口板;切口(量水)堰板;槽口板;凹形板;凹板;V形槽口溢流闸板
notch pulse 标记脉冲
notch ratio 凹口比
notch sensitive 缺口敏感的
notch sensitiveness 切口敏感性
notch sensitivity 切口敏感度;缺口敏感性;切口灵敏度;凹痕感度
notch sensitivity factor 缺口敏感性系数
notch sensitivity ratio 切口敏感比
notch sensitivity test 缺口敏感性试验;缺口脆性试验;切口敏感性试验;切口脆性试验
notch shock test 切口震动试验;切口冲击试验;切口振动试验
notch spillway dam 切口溢流坝;凹口溢流坝
notch stanchion 梯柱
notch strength 切口强度;缺口强度
notch stress 切口应力;缺口应力
notch tensile strength 刻槽抗拉强度;切口拉伸强度
notch tensile test 切口拉伸试验
notch tension test 缺口拉伸试验
notch test 切口试验
notch toughness 刻痕韧性;冲击韧性;冲击韧度
notch toughness test 切口韧性试验
notch-type snapping roll 槽口式摘穗辊
notch value 冲击强度值
notch wedge impact 楔击缺口冲击试验
notch wheel 棘轮
not conform to 不符合
not conform to the rules 不合规定
not coupling coefficient of loading 装药不耦合系数
not dangerous 无危险的
not detectable 未检出(的);未测出的
not detected 未检测出的
note 票据;附注
note and bill discounted 贴现票据及汇票
not easily weathered 不易风化的
note at sight 见票付时票据
notebook 票据簿;记录簿
note broker 有价证券经纪人(股票等);票据等有价证券的经纪人
note collective 联合照会
noted 著名的

note dishonored 已拒付的票据
note example 示例
note form 手簿格式
note for payment of tonnage dues 吨位税通知书
note frequency 声频
note issurance facility 票据发行便利
note keeper 记录员
notekeeping 记录
not element 反相元件
not elsewhere specified 在别处未加说明;未加说明的;不另详述
note management audit 票据管理审计
note notice 票据到期通知书
not enough 不够的
no-tension rigidity 无张力硬度;无张力刚度
no-tension stiffness 无张力劲性;无张力劲度
note of authorization 授权声明;授权批准书;授权通知书
note of dishono(u)r 退票通知
note of hand 借条;期票;本票
note of marine protest 海损声明书;海事声明书
note of protest 海事声明;提出海事声明
note of surveying and mapping 测绘注记
note order payable to bearer 凭票即付持票人
note oscillator 音频振荡器
notepaper 便条纸
note payable 应付票据
note pulse 记号脉冲
not equal to 不等于
note receivable 应收票据
note receivable dishonored 已拒付的应收票据
note receivable protested 已拒付的应收票据
note register 票据登记簿
note renewals 转期票据
not erodible 无侵蚀性
notes and bills receivable discounted 已贴现应收票据
notes appended 附注
notes on discount 贴现票据
notes on drawings 图注
notes receivable ascollateral 以应收票据担保
notes receivable discounted 应收票据贴现
notes receivable from stockholder 应收股东据
notes receivable protested 已被拒付的应收票据
no test 无密押
notes to financial statements 财务状况说明书
notes to mariners 航行公告
notes to statistics 统计文字说明
note to bearer 凭票付款票据
not examined 未检查,未检验出的
not executable attribute 不可执行属性
not-for-profit organization 非营利组织;非盈利机构
not for sale 非卖品
not free 无间隙(机械零件之间)
not fully developed 没完全发育
not-go ga(u)ge 不过端量规
not good liquor 次液
not good quality 质量不好
not-go side 不过端
Notgrove freestone 诺特格罗夫易切岩
not habitable 不适宜居住的;不习惯的
not heat-treated steel wire 非热处理的钢丝
nothing available 无货供应
nothing between 无来往
nothingness explosive risk 无爆炸危险
no thoroughfare 禁止通行;禁止通过;不通行
no thoroughfare for vehicles 禁止车辆通行
no through traffic 无联运运输;无直达交通
no-throw 无捻丝线
no-throw wire rope 无捻多股钢丝绳
notice 警告牌;通告
noticeable wear 明显磨损
notice board 通知牌;警告牌;通告牌;布告牌;布告栏;布告板
notice by post 发函通知
notice day 通知期
notice deposits 通知存款
notice dishonour 停止付款通知
notice for accident indemnity 赔偿通知书
notice money 通知存款资金
notice of abandonment 弃船声明;委付通知书(海上保险)
notice of a call 催收通知
notice of action 执行公告

notice of arrival 到港通知
notice of arrival of goods 到货通知(书);催领通知
notice of award 得标通知书;中标通知
notice of award of the contract 签订合同通知书
notice of cancellation 注销通知;退保通知
notice of cessation 建筑工程停工通知
notice of change 更改通知单
notice of change employment 人事动态通知
notice of claim clause 索赔通知条款
notice of claims 索赔通知(书)
notice of clearance 结关通知;出港通知(书)
notice of damage 损害通知书
notice of default 违约通知
notice of defects 缺陷通知
notice of delay 延迟通知
notice of delivery paid 已付款收货通知单
notice of dishono(u)r 拒付通知;票据拒付
notice of due date 通知到期日
notice of exception 例外情况通知
notice of freight collection 运费托收通知
notice of hearing 审理通知书
notice of import 进口申请
notice of loss 损失通知书;损失通知单
notice of passenger number on train 乘车人数通知单
notice of payment 通知付款;付款通知书;付款通知单
notice of readiness 卸货准备完成通知书;卸货通知;准备装卸货通知;准备就绪通知书;就绪通知
notice of rescission 作废通告
notice of shipment 装船通知
notice of shipment goods 发货通知
notice of tax assessment 征税通知
notice of tax payment 纳税通知书
notice of tests 检验通知书
notice of the number of passengers on train 乘车人数通知单
notice of unsafe buildings 危险房屋公告
notice of vessel's arrival 船舶抵港通知;船舶到港通知
notice period 通知期限
notice plate 标记牌
notices by telex 电传通知
notice to airman 航空人员须知
notice to bidders 招标通知书;投标商须知;投标人须知
notice to commence 开工通知
notice to contractor 致承包商的通知
notice to employer and engineer 致雇主和工程师的通知
notice to mariners 航行通告;航海通告;航海公告;航道通告;航海公报;海员须知
notice to navigators 航行通告;航海通告;航道通告
notice to proceed 施工通知
notice to quit 迁出通知;租户迁出通知;停止出租通知
notification 通知(书);通告;报告书
notification charges 通知费用
notification clause 通知文句
notification of approval 批准通知
notification of award 中标通知书;评标(结果)通知书;授标通知
notification of award of the contract 签订合同通知书
notification of depreciation method 选用折旧法声明
notified party 被通知方;被通知方
notifier 通知者;通知人
notify address 通知地址
notify clause 通知条款
notifying bank 通知行;通知银行
notify lock 封闭通知
no-till agriculture 免耕农业
not imported empty 非空箱进口
not included 未包括在内
not in contact 未接触;不接触
not in contract 不包括在合同内;未订合同;不在合同中
not in detail 不详
not in stock 无库存;无存货;不在储[贮]存中
notion 概念
notional 概念上的
notional demand 名义需求
notional shipper 名义托运人

notion list 概念表
notion of style 风格概念;风格观念
notion-store 杂货店
not keeping contract 不履行合同
not known 不详
not known loss 尚未知道的损失
not long after 不久以后
not mined channel 未布雷航道
not more than 至多
not mud pumping 不冒浆
not negotiable 不可转让的;不可流通的;不得转让的
not negotiable check 不流通的支票
not normal 失常的
not noticeable earthquake 无感地震
not obvious 不明显
Notogaean 南界的
not ok liquor 次液
not on-duty casualty 路外人员伤亡
not on risk 不在保险范围内
No tooting 禁止鸣笛
not opencast for farmland cap 农田覆盖不宜露采
not operation 变反操作
notopleural suture 背侧缝
notorious possession 逆向占有;公开占有
no to scale 超出量程
not otherwise enumerated 未另列明;未另列名;未另计数
not otherwise provided 未另列名;别无规定
not otherwise provided for 未列项目
not otherwise provided/stated 未另列出
not otherwise rated 未定费率
not otherwise specified 未另说明
no touch 无接触;不接触
no touch relay 无触点继电器
no trace 无线索
no transshipment permitted 不准转运
no traps allowed 禁止张网
not ready 非准备好
Notre Dame de Paris 巴黎圣母院(法国)
not reentrant 不可重入的
not releasable to foreign nations 不可向国外发表
not reported 未报
not requiring additional reinforcement 不需要附加补强
not responsible 不承担责任
not responsible for anything lost 失物自理
not responsible for decay 腐烂不负责
not restricted 不受限制的
not reusable 不能重新使用的;不能再用的
not-reversible noting 不可逆的
not roundness 颗粒不圆度
not solidified 未凝固的
not soundproof 不吸声的;不隔声的
not specially provided for 未另有特别列出
not specified 未作说明;未规定的
not subjected to bending 不受弯曲
not subjected to pressure 非受压
not subject to cassation appeal 不得上诉
not subject to the limits 不在此限
not test 未试验
not to be covered 不应覆盖;不包括的
not to be laid flat 不可平放
not to be stowed below other cargo 怕压;不能堆压
not to be taken as an evidence 不足为证
not to be-tipped 勿倾倒
not to insure clause 无受益条款
not to order clause 不可转让条款
not to scale 不按比例
not to scale for diameter or length 没按直径或长度比例绘制
not trustworthy 无信用
not under command 失去控制
not up to standard 不合标准的
no turn 不准转弯
no-turn shuttle hauling 无调头的往返运输
not-water-carriage toilet facility 干厕;非水冲厕
no-twist finished stand 无扭精轧机座
no-twist finishing block 线材无扭精轧机组
notwithstanding the verdict 否定裁决
not work 失灵
noughts complement 零补数;基数补数
nought state 零状态
noumeite 硅镁镍矿

noumenon 实体
nouniformity 不一致性
nourishment 海滩重建;海滨重建
nourishment area 补给区
nourishment basin 补给盆地;补给流域
nourishment condition (冰川的)补给条件
nourishment source 补给水源
no-U-turn marking 禁止车辆调头标志
no-U-turn sign 禁止车辆调头标志
novacekite 镁种铀云母
novacite 天然微晶均密石英粉
novaculite 致密石英岩;微晶石英质燧石岩;天然微晶均密石英粉
novakite 砷铜银矿
Novalux 诺瓦勒克司(玻璃透镜)
no-valve filter 无阀滤池
novation 债务更新;债务的让度
novel biological hybrid system 新型生物混合系统
novel design 新颖设计
novel drilling method 新钻进方法
novel slewing-bridge type shiploader 新型旋转桥式装船机
novel tube 标准九脚小型管
novelty 下垂披叠板;新饰品;新产品;新颖性(指专利产品发明);新颖
novelty finish 美术涂装
novelty flooring 新颖拼花地板;新颖花纹地板
novelty of an invention 发明的新颖性
novelty pattern 新颖花样
novelty saw 刨削锯
novelty siding 新型木墙板;德国式檐板;下垂披叠板;企口壁板;企口板壁
novel wet digestion procedure 新湿式浸煮法
novenary 九进制的
novendenary 十九进制的
novia 埃及水车
novice 初学者;新手
novice designer 见习设计师
novice operator 见习人员;见习技术员;新的操作人员
novice traveller 初次旅游者
noviciate 新手
novit 诺维尔炸药
Novite 诺维特(硬质合金)
novochlorhydrate 水合氯醛
Novoid 诺瓦硅基粉末
Novokanstant 标准电阻合金
novolac 热塑性酚醛树脂;酚醛清漆
novolac epoxy 酚醛环氧树脂
novolac resin 合成酚醛树脂;线型酚醛树脂;热塑性酚醛树脂;酚醛清漆树脂
novolak 可溶酚醛清漆;酚醛清漆
novolak resin 线型酚醛树脂
no-voltage 零电压;无电压
no-voltage alarm 无电压报警器
no-voltage circuit-breaker 无电压自动断路器
no-voltage coil 无压线圈
no-voltage condition 无电压状态
not-voltage contact 零压接触
no-voltage cut-out 无压开关
no-voltage electromagnetic relay 无电压电磁式继电器
no-voltage protection 无压保护;失压保护
no-voltage protection test 无电压保护试验
no-voltage relay 零电压继电器;无电压继电器;失压继电器
no-voltage release 无压释放;失压脱扣器
now account 可转让提款指令账户
nowackiite 硫砷锌铜矿
nowadays mine surveying 矿山现状调查
nowadays mining depth 现已开采深度
nowcast 即时天气预报
no-wear 无损
no weathering exposure 无风化的曝晒;无风化露头
nowhere dense set 无处稠密集
no-wind heading 无风舷向【航海】
no-wind position 无风点;空中位置
no-wind position indicator 无风位置指示器
now-quiescent area of earthquake 目前地震平静区
now-toughened 已经韧化的
NO_x content 氮氧化物含量
noxious acid 有毒酸
noxious aquatic vegetation 有毒水生植被
noxious colo(u)ring matter 有害色素

noxious constituent 有害成分
noxious emissions 有害排放物
noxious fume 有害烟雾;有毒烟雾;有毒烟气
noxious gas 有害气体;毒气
noxious gas after explosion 爆后气体
noxious industry 有害工业
noxious liquid 有毒液体
noxiousness 有毒害性
noxious space 汽缸余隙
noxious stimulus 有害刺激(物)
noxious wastewater 有害废水;有毒废水
NO_x meter 氮氧化物测定仪
noxtane 五氯酚钠-钠-硼砂合剂(水溶性木材防腐剂)
noy 纳
no-year appropriation 无限期拨款
nozzle 铸口;割嘴;漏嘴;喷嘴;喷管;筒口;水管嘴;侧出口管;底注包孔口
nozzle adjusting screw 喷管调节螺钉
nozzle aerator 喷嘴通气器;喷嘴曝气器
nozzle air bleed 喷嘴排出空气
nozzle and socket list 接管表
nozzle angle 喷嘴角;喷嘴扩散角
nozzle area 喷嘴面积;喷管通路面积
nozzle area coefficient 喷嘴面积系数
nozzle area control system 喷嘴面积控制系统
nozzle area ratio 喷嘴面积比
nozzle atomization 喷嘴雾化
nozzle atomizer 喷嘴喷雾器
nozzle atomizing 喷嘴雾化
nozzle-baffle mechanism 喷嘴挡板机构
nozzle bell 承口
nozzle bitumen emulsifying machine 喷嘴式沥青乳化机
nozzle blade 喷嘴叶片;喷管隔片
nozzle blade cascade 喷嘴叶栅;导向叶栅
nozzle block 喷嘴组;喷管管夹
nozzle blower 喷嘴鼓风机
nozzle body 喷嘴壳体;喷嘴管体
nozzle bore 漏孔;喷嘴孔
nozzle boss 安装喷嘴用突出部
nozzle box 喷嘴箱;喷嘴室
nozzle box assembly 喷嘴箱部套
nozzle brick 注口砖;水口砖;烧嘴砖
nozzle bucket 喷嘴叶片
nozzle burner 燃油喷嘴
nozzle bushing (钻头的)喷嘴;套嘴
nozzle button (沸腾炉的)风帽
nozzle cage 喷嘴盒
nozzle cap 喷口盖
nozzle cap nut 喷油嘴紧帽
nozzle carrier 喷嘴座
nozzle cascade 喷嘴环
nozzle chamber 喷嘴箱;喷嘴内腔
nozzle chest 喷嘴箱
nozzle clamp 喷嘴夹
nozzle clearance 喷嘴间隙
nozzle clogging 喷嘴堵塞
nozzle closure 喷口隔板;喷口盖
nozzle cluster 喷口组;多喷嘴喷头
nozzle coefficient 管嘴系数
nozzle configuration 喷口形状
nozzle contour 喷管外形
nozzle-contour correction 喷口形状修正
nozzle-contraction-area ratio 喷嘴面积收缩率
nozzle control governing 喷嘴节流调速;喷管调节
nozzle control governor gear 喷嘴调节装置
nozzle control valve 喷嘴(控制)阀;喷嘴调节阀
nozzle cooler 丝根冷却器
nozzle cooling 喷管冷却
nozzle core 喷嘴插芯
nozzle coupling 喷口接合
nozzle cross-section 喷嘴截面
nozzle curvature 喷管曲率
nozzle cut-out 喷嘴阀开闭调速
nozzle cut-out governing 喷管调节
nozzle diaphragm 喷嘴装置;喷嘴环
nozzle die unit 强迫润滑拉模组合装置
nozzle dimensions for level ga(u)ge 液面计接管尺寸
nozzle-disc centrifuge 锥形盘式离心机;倒锥转盘式离心脱水机
nozzle discharge coefficient 喷嘴流量系数
nozzle divergence 喷口扩张

nozzle divergence angle 喷嘴扩张角
nozzle divergence loss factor 喷管扩张损失因数
nozzle division plate 喷嘴叶片
nozzle drier 热风喷射烘燥机
nozzle drop 喷嘴压降
nozzle edge 喷嘴切口;喷管切口
nozzle efficiency 喷嘴效率
nozzle-end 喷嘴根部;喷管端部
nozzle-end ignition 喷管端点火
nozzle entrance section 喷管入口段
nozzle entry 喷管入口
nozzle-exhaust plane 喷管出口面
nozzle exit 喷嘴;喷口截面;喷管出口
nozzle exit area 喷口面积
nozzle exit momentum 喷管出口动量
nozzle exit plane 喷管出口面
nozzle exit pressure 喷嘴压力;喷管压力
nozzle exit section 喷管出口截面
nozzle expansion ratio 喷嘴面积膨胀率;喷管膨胀比
nozzle filler block 喷管胀圈
nozzle filter 喷嘴过滤器
nozzle fittings 喷嘴配件
nozzle flame 喷嘴火焰;喷管火焰
nozzle flange 喷嘴法兰;喷嘴凸缘;接管法兰;喷嘴突缘
nozzle flange gasket(sur)face 接管法兰密封面;接管凸缘密封面
nozzle flap 喷嘴挡板系统
nozzle flapper 喷嘴挡板
nozzle-flapper mechanism 喷嘴挡板机构
nozzle flow coefficient 喷嘴流量系数
nozzle flow meter 喷嘴(式)流量计;喷管(式)流量计
nozzle form 喷嘴形状
nozzle for mounting of thermowell 温度计保护套管接管
nozzle for pressure ga(u)ge 压力计接管
nozzle for rapid filter 快滤池喷嘴
nozzle for thermometer of exhaust gas 排气温度计接管
nozzle friction coefficient 喷咀阻力系数
nozzle gate 水口闸板
nozzle governed 喷管调节
nozzle governing 喷嘴调节
nozzle group control 喷嘴组调节
nozzle grouping 喷嘴组
nozzle group valve 喷嘴阀
nozzle guide vanes 喷嘴导向叶片
nozzle head 喷嘴头
nozzle headbox 喷嘴式网前箱
nozzle heat transfer rate 喷管传热率
nozzle holder 喷嘴座
nozzle hole 喷嘴孔
nozzle hole diameter 喷嘴孔径
nozzle inclination 喷油嘴倾角
nozzle injection pressure 喷嘴喷射压力
nozzle inlet 喷嘴进口
nozzle inlet condition 喷管进口的气体参数
nozzle inlet Mach number 喷管进口气流马赫数
nozzle inlet temperature 喷管进口温度
nozzle inside diameter 喷嘴内径
nozzle interior contour 喷管内壁形状
nozzle in tube 管内喷嘴
nozzle jet 喷嘴射流
nozzle jet drier 喷嘴热风干燥器
nozzle jet separation 喷嘴喷气分离
nozzle liner 喷嘴内衬;喷嘴衬套;喷管衬垫
nozzle line strainer 喷嘴管路滤器
nozzle loss 喷嘴(水头)损失;喷管(水头)损失
nozzle loss factor 喷嘴损失系数
nozzleman 喷水工;喷射混凝土工;喷砂工;喷枪工;喷枪操作工;水力冲击机工人
nozzle manifold 多喷头插座
nozzle meter 管式流速计
nozzle metering characteristics 喷嘴调节特性
nozzle mixer 喷嘴混合器
nozzle-mix gas burner 喷射式混合气体燃烧器
nozzle mixing burner 引射式燃烧器
nozzle neck 接管颈;喷嘴颈
nozzle needle 喷嘴阀针;喷油针
nozzle noise 喷嘴噪声
nozzle of air supply 进气喷嘴
nozzle opening 喷嘴孔口;孔比;接管开孔;喷油孔;喷管口
nozzle opening pressure 喷嘴开启压力
nozzle opening ratio 接管开孔比
nozzle opening reinforcement 接管开孔补强
nozzle orientation 接管方位
nozzle orifice 喷嘴口;喷嘴的锐孔;喷口
nozzle orifice disc 喷嘴垫圈;喷嘴的喷孔圆盘;喷口调节盘
nozzle oscillating lever 喷摆动杆
nozzle outlet 管式排气口;喷口;(钻头)水眼
nozzle outlet air supply 喷口送风
nozzle outlet gas angle 喷嘴出气角
nozzle passage 喷嘴通道
nozzle passage friction 喷嘴通道摩擦
nozzle pipe 注射管;喷嘴管;喷射管
nozzle pitch 喷嘴节距
nozzle plate 喷嘴板;漏板;流量喷孔板;喷嘴(盖)板;喷口盖板
nozzle pocket 注口座
nozzle pore 漏孔
nozzle pressure 喷嘴压力;喷射压力;(钻头)水眼压力
nozzle pressure ratio 喷管压力比
nozzle process 喷嘴法
nozzle propeller 射流推进器
nozzle register 喷口控制器
nozzle regulation 喷嘴调节
nozzle regulator 喷嘴调节阀
nozzle reinforcement 接管补强
nozzle resonance 喷嘴共振
nozzle ring 喷管环;环状喷嘴;喷嘴环;喷气环;导叶栅
nozzle-rudder 导流管舵
nozzles and fittings 接管和配件
nozzles and other connections 接管及其他接头
nozzle screen 喷嘴滤油网;喷嘴挡板
nozzle seating block 喷管砖
nozzle section 喷嘴剖面
nozzle segment 喷嘴组;喷嘴弧段
nozzle separation 喷嘴气流离体
nozzle shape 喷管形状
nozzle shield 丝根冷却器
nozzle shroud 喷管罩
nozzle-sinker device 受油接头
nozzle size 喷嘴直径
nozzle sludge trap 喷嘴式岩粉捕集器(空气钻进时)
nozzle solenoid 喷管电磁阀
nozzle spray angle 喷嘴喷射角度
nozzle spray direction 喷嘴喷射方向
nozzles to shell 到壳体上接管
nozzle strainer 喷嘴滤网
nozzle-stream angle 喷嘴气流角
nozzle stub 短管
nozzle table 接管表
nozzle tester 喷射器性能试验装置
nozzle testing stand 喷嘴试验台;喷嘴试验器
nozzle throat 喷嘴喉部;喷管喉口;喷管喉部
nozzle throat area 喷嘴横截面;喷管喉部面积
nozzle throat pressure 喷管临界截面压力
nozzle thrust coefficient 喷管推力系数
nozzle tip 喷管头子;油嘴;喷嘴尖;喷油头;喷头
nozzle tube 喷嘴管;喷管
nozzle tube washer 嘴管垫圈
nozzle tube yoke 喷管架
nozzle type purifier 喷嘴式净化器
nozzle type slice 吸浆堰板
nozzle type spray dryer 喷嘴雾化型喷雾干燥器;喷嘴式喷雾干燥器
nozzle type turbine 喷嘴式水轮机;水斗式水轮机
nozzle type water wheel 喷嘴式水轮
nozzle valve 喷嘴阀
nozzle valve operating panel 喷嘴阀操作盘
nozzle valve rod 喷嘴阀塞杆;喷嘴阀杆
nozzle vane 喷嘴隔片;排气叶片
nozzle velocity 喷速;喷嘴速度;喷管出口速度
nozzle wake resonance 喷嘴尾流共振
nozzle wall 漏嘴壁;喷嘴壁
nozzle with a parallel shroud 圆筒形喷嘴
nozzle with conic(al)seat 锥形座喷孔
nozzle with spear 带小针喷嘴;针状喷嘴
nozzle wrench 喷嘴扳手
nozzling 锤头;打尖
n percent chanceflood n 百分数概率洪水
N-phenyl benzoic acid amide 苯酰替苯胺
npl-nipple 螺纹接管
N-precess 硅铁粉发热自硬砂法
n-propyl acetate 乙酸正丙酯;醋酸正丙酯
n-propylbenzene 正丙苯
n-propyl nitrate 硝酸丙酯
NPT-pipe thread 布立格螺纹
N-region 剩余电子区域
NR for decay 腐烂不负责
NRM wind scale 北洛基山脉风级
n-stage decision problem n 步决策问题
nsutite 六方锰矿
n-tetradecanol 正十四烷醇
nth order n 阶
nth polynomial correction n 次多项式校正
nth power n 次幂;n 次方
nth root n 次方根
N-truss 竖斜杆桁架;林维尔桁架;N 形桁架
N-type semi-conductor N 形半导体
nuance 细微差别;微细差别
Nuanetsi reversed polarity chron 努阿内齐反向极性时
Nuanetsi reversed polarity chronzone 努阿内齐反向极性时间带
Nuanetsi reversed polarity zone 努阿内齐反向极性带
nub 小瘤;小块
nu-bar 活动闸门;活栅门
nubbin cut-off machine 割头机
Nubian Sandstone 努比亚砂岩
nubilose 喷雾干燥器
Nuchar 纽恰尔牌活性炭
nuciform 胡桃状的
nuckonic steering 放射性敏感元件驾驶装置
nucleaction 核晶作用
nuclear absorption 核吸收
nuclear accelerator 核加速器
nuclear accident 核装置事故;核事故
nuclear activation analysis 核活化分析
nuclear activity 核活动
nuclear age determination 核年代测定(法);放射性年龄测定;放射性测定年代
nuclear agent 核化剂
nuclear-aid coal gasification 利用核能的煤气化
nuclear air burst 大气层核爆炸
nuclear air cleaning 核子大气净化
nuclear amplitude 核振幅
nuclear analysis of reactor core 反应堆芯核分析
nuclear angular momentum 核角动量
nuclear area 陆核;核区
nuclear association 核联合
nuclear attack submarine 核攻击潜艇
nuclear backscatter(ing) density test 放射性逆散射砂密度试验
nuclear bag fiber 核袋纤维
nuclear basin 陆核盆地
nuclear battery 核电池
nuclear belt scale 核子皮带秤
nuclear binding energy 核结合能
nuclear blast 核爆炸
nuclear blast combined injury 核爆炸复合伤
nuclear blast injury 核爆炸冲击伤
nuclear body 核体
nuclear boiler 核蒸汽发生器;核锅炉
nuclear bombardment 核轰击
nuclear burning 核燃烧
nuclear burst 核爆炸
nuclear cap 核冠
nuclear capture 核俘获
nuclear cargo ship safety certificate 核能货船安全证书
nuclear central generation station 核中心发电站
nuclear chain fiber 核链纤维
nuclear chain reaction 核链式反应
nuclear charge distribution 核电荷分布
nuclear chemistry 原子核化学;核化学
nuclear cloud 核云;放射云
nuclear collision 核碰撞
nuclear colony 核群
nuclear column 砥柱【地】
nuclear comment log 核水泥测井
nuclear compound 含环化合物
nuclear container ship 核集装箱船
nuclear converter 核转换器

nuclear counter 核辐射计数器
nuclear critical mass 核燃料的临界量
nuclear cycle 核周期;核循环
nuclear damage 原子核损伤;核损伤
nuclear data depositary 核数据库
nuclear debris 核碎片
nuclear decay 核衰变
nuclear decay equation 核衰变方程式
nuclear decay mode 核衰变模式
nuclear degeneration 核变性
nuclear delay 控制棒延迟
nuclear demagnetization 核去磁
nuclear density 核密度;核测密实度
nuclear density ga(u)ge 核子密度计
nuclear density meter 核密度仪
nuclear densometer 核子密度计;同位素密度仪
nuclear design of a reactor 反应堆物理设计
nuclear detection array 核检测台阵;核爆炸检测台阵
nuclear detector 核探测器
nuclear deterrent 核威慑力量
nuclear detonation 核爆炸
nuclear device 核子仪;核装置
nuclear direct sensing device 放射性直接探测器
nuclear direct sensitivity device 放射性直接探测器
nuclear disintegration 核蜕变
nuclear disintegration constant 核蜕变常数
nuclear disintegration energy 核蜕变能
nuclear disc 核盘;核板
nuclear dispersion 核色散
nuclear division 核分裂
nuclear double resonance method 核间双共振法
nuclear drilling 热核钻进
nuclear dust 核尘
nuclear dynamics 核动力学
nuclear ecology 核生态学
nuclear electric al power generation 原子能发电;核发电
nuclear electrical propulsion 核电推进装置
nuclear electric al quadrupole 核电四极矩
nuclear emission 核发射
nuclear emulsion 核乳胶;核乳剂
nuclear emulsion detector 核乳胶探测器
nuclear emulsion film 核乳胶片
nuclear energy 原子能;核能
nuclear energy absorption 核能吸收
nuclear energy fusion 核能熔化
nuclear energy level 核能级
nuclear energy metal 核能金属
nuclear energy plant 核能装置;核能发电站
nuclear energy power generation 核能发电
Nuclear Energy Research Bureau 核能研究局(美国)
nuclear energy resources 核能资源
nuclear energy source 核能源
nuclear energy structure 核能建筑物
nuclear energy surface 核能面
nuclear energy use 核能利用
nuclear engine 核发动机
nuclear engineering 核技术;核工程
Nuclear Engineering Society 核子工程学会
nuclear envelop 核膜
nuclear equation 核反应方程
nuclear equipment 核子仪
nuclear excavation 核爆炸挖土法
nuclear excitation curve 核激发曲线
nuclear explosion 核爆炸
nuclear explosion engineering 核爆炸工程
nuclear explosion for groundwater 核爆炸产生的地下水
nuclear explosion pollution 核爆炸污染
nuclear explosion seismology 核爆炸地震学
nuclear explosion stimulation 核爆炸刺激
nuclear explosive 核爆炸物
nuclear facility 核设施
nuclear fall-out 核散落物
nuclear fallout contamination 核微粒沾染
nuclear family 小家庭;核心家庭
nuclear field 核场
nuclear fission 核分裂
nuclear fission calorimeter 核分裂发热量测量卡计
nuclear fission product 核裂产物
nuclear fissure nuclear fission 核裂变
nuclear fluore scence 核共振荧光

nuclear fluorescence analysis 核荧光分析
nuclear force 核作用力;核力
nuclear force function 核力函数
nuclear force parameter 核力参数
nuclear fragment 核碎片
nuclear-free zone 无核区
nuclear fuel 核燃料
nuclear fuel after-treatment 核燃料后处理
nuclear fuel ceramics 核燃料陶瓷
nuclear fuel cycle 核燃料循环;反应堆燃料循环
nuclear fueled plant 核电站
nuclear fuel element 核燃料元素
nuclear fuel fabrication plant 核燃料制造厂
nuclear fuel form 核燃料状态
nuclear fuel industry 核燃料工业
nuclear fuelled-boiler 核锅炉
nuclear fuel magnetohydrodynamic(al) power generation 核燃料磁流体发电
nuclear fuel pebble 核燃料芯块
nuclear fuel pellet 核燃料芯块
nuclear fuel plate 板状燃料元件
nuclear fuel processing 核燃料处理
nuclear fuel reprocessing 核燃料再处理;核燃料后处理
nuclear fuel reprocessing analysis 核燃料后处理分析
nuclear fuel reprocessing plant 核燃料后处理工厂
nuclear fuel status and forecast 核燃料现况及预测
nuclear fusion 核熔融反应;核聚度;核聚变;热核反应
nuclear fusion physics 核聚变物理学
nuclear fusion reaction 核聚变反应
nuclear fusion reactor 核聚变反应堆
nuclear gamma-ray resonance adsorption 核伽马射线共振吸收
nuclear gas turbine 核能燃气透平
nuclear ga(u)ge 核子探头;核子计数器;核子测定仪
nuclear generated power 核电力工业
nuclear generator 核动力发电机
nuclear geochemistry 核地球化学
nuclear geology 核(子)地质学
nuclear geometric(al) cross-section 核几何截面
nuclear geophysical prospecting method 核物探法
nuclear geophysics 核地球物理学
nuclear graphite 核石墨
nuclear grid 原子能电站网
nuclear ground state 核基态
nuclear gyromagnetic ratio 核回转磁比
nuclear gyroscope 核回旋器
nuclear harmonic vibration 核谐振
nuclear hazard 核伤害
nuclear heat 核热
nuclear heating 原子能采暖;核能供暖
nuclear heating system 核供热系统
nuclear heat pumping 核热泵浦
nuclear inclusion body 核包涵体
nuclear induction 核感应
nuclear industry 核工业
nuclear industry waste(water) 核工业废水
nuclear injury 原子核损伤;核损伤
nuclear instability 核不稳定性
nuclear installation 核设施
Nuclear Institute 核研究所
nuclear instrument 核仪器
nuclear instrumentation 核子仪表设备;核检测仪表
nuclear interaction 核相互作用
nuclear isomer 核同质异能素
nuclear leakage 核泄漏
nuclear level controller 核液面控制器
nuclear level density 核能级密度
nuclear level indicator 核子料位指示器
nuclear level spacing 核能级间距
nuclear log cure plot 核测井曲线图
nuclear logging instrument 核测井仪
nuclear luminescent lamp 核发光灯
nuclear machine 核子仪
nuclear magnetic distance 核磁距
nuclear magnetic hydrodynamic(al) plant 核磁流体电站
nuclear magnetic logger 核磁测井仪
nuclear magnetic logging 核磁记录;核磁测井
nuclear magnetic moment 核磁矩

nuclear magnetic relaxation flowmeter 核磁豫弛流量计
nuclear magnetic resonance 核磁共振
nuclear magnetic resonance gyro 核磁共振陀螺
nuclear magnetic resonance spectrogram 核磁共振谱图
nuclear magnetic resonance spectrometer 核磁共振波谱仪
nuclear magnetic resonance spectroscopy 核磁共振波谱法
nuclear magnetic resonance spectrum 核磁共振谱
nuclear magnetic resonance thermometry 核磁共振测温法
nuclear magnetisation 核磁化
nuclear magnetism 核磁学;核磁性
nuclear magnetism log curve 核磁测井曲线
nuclear magnetism logging 核磁测井
nuclear magnetometer 核(子)磁力仪;核子磁力计;核磁强计
nuclear magneton 核磁子
Nuclear Material Management Regulations of the People's Republic of China 中华人民共和国核材料管理条例
nuclear matter 核物质
nuclear measurement (试验土特性的)核子放射量测;核子法量测
nuclear measuring technique 核测试技术
nuclear mechanical moment 核机械矩
nuclear membrane 核膜
nuclear metallurgy 核冶金学
nuclear meteorology 核气象学
nuclear method 核方法
nuclear microanalysis 核微分析
nuclear mode 核模(型)
nuclear moisture and density dynamic meter 核子动态湿度密度仪
nuclear moment 核矩
nuclear oceanography 核子海洋学
nuclear oil well logging 核油井记录
nuclear paramagnetic resonance 核顺磁共振
nuclear parameters 核参数
nuclear passenger ship safety certificate 核能客船安全证书
nuclear photoeffect 核光效应
nuclear photoelectric(al) effect 核光电效应
nuclear photon method 核光子法
nuclear physics 原子核物理学;核物理(学)
nuclear pile 核反应堆
nuclear pile wastes 核反应堆废物
nuclear plant 核工厂
nuclear plate 核板
nuclear point 核心点
nuclear poison 核毒物
nuclear polarization 核极化
nuclear pollutant 核污染物
nuclear pollution 核污染
nuclear pool 旋涡
nuclear pore 圆形核孔;核孔
nuclear potential 核位势;核力
nuclear potential energy 核位能
nuclear potential scattering 核的势散射
nuclear power 核动力
nuclear power economics 核动力经济学
nuclear powered 核动力的
nuclear powered aircraft carrier 核动力航空母舰
nuclear powered attack submarine 核动力鱼雷攻击潜艇
nuclear powered gasturbine 核动力燃气轮机
nuclear powered guided missile cruiser 导弹核巡洋舰
nuclear powered icebreaker 核动力破冰船
nuclear powered light source 核动力光源
nuclear powered propulsion 核动力推进器
nuclear power(ed) reactor 核动力反应堆
nuclear powered reciprocating engine 核动力往复式发动机
nuclear powered submarine 核潜艇;核动力潜(水)艇
nuclear powered submarine tanker 核动力潜水油轮
nuclear power generator 核电发电机
nuclear power industry 核动力工业
nuclear power joint corporation 核电合营公司
nuclear power plant 核动力装置;原子能发电厂;

原子能电站;核(发)电厂;核电站
nuclear power source 核能源
nuclear power station 核热电站;核(能)电站
nuclear power technology 核动力工艺学
nuclear power unit 核动力装置
nuclear precession magnetometer 核子旋进磁力仪
nuclear process 核过程
nuclear proliferation 核扩散
nuclear propelled ship 核动力船
nuclear propeller 核动力推进器
nuclear propulsion 核动力
nuclear proton 核质子
nuclear radiation 原子核辐射;核子辐射;核辐射
nuclear radiation burn 核武器热辐射烧伤
nuclear radiation chemistry 核放射化学
nuclear radiation detector 核辐射探测器
nuclear radiation effect 核辐射效应
nuclear radiation environment 核辐射环境
nuclear radiation injury 核辐射损伤
nuclear radiation level measuring instrument 核辐射物位测量仪表
nuclear radiation physics 核辐射物理学
nuclear radiation protection 核辐射防护
nuclear radiation safety 核辐射安全
nuclear radiation spectroscopy 核辐射谱学
nuclear radiation test 核辐射试验
nuclear radius 核半径
nuclear ralaxation 核弛豫
nuclear ramjet engine 核冲压式空气喷气发动机
nuclear raw material 核原料
nuclear raw-material technology 核原料工艺学
nuclear reaction 原子核反应;核反应
nuclear reaction analysis 核反应分析
nuclear reaction at high energy 高能核反应
nuclear reaction energy 核反应能
nuclear reaction excitation curve 核反应激发曲线
nuclear reaction kinetics 核反应动力学
nuclear reaction model 核反应模型
nuclear reaction particle analyzer 核反应粒子分析器
nuclear reaction products 核反应产物
nuclear reaction spectroscopy 核反应能谱学
nuclear reactor 原子核反应堆;核反应堆
nuclear reactor accident 核反应堆事故
nuclear reactor ceramics 核反应堆陶瓷
nuclear reactor containment 核反应堆防护外壳
nuclear reactor energy 核反应能
nuclear reactor engineering 核反应堆工程学
nuclear reactor facility 原子能反应堆设备
nuclear reactor instrumentation 核反应堆测量仪表
nuclear reactor kinetics 核反应堆动力学
nuclear reactor metallurgy 核反应堆冶金学
nuclear reactor physics 核反应堆物理学
nuclear reactor simulator 核子反应堆模拟装置
nuclear reactor structure 原子能反应堆结构
nuclear reactor technology 核反应堆技术
nuclear reactor wastes 核反应堆废物
Nuclear Regulatory Commission 核管制委员会(美国)
nuclear release 放射性产物排放
nuclear repulsion 核排斥
Nuclear Research Associates 核子研究联合公司(美国)
nuclear research center 核研究中心
Nuclear Research Council 核研究理事会(美国)
nuclear research reactor 核子研究反应堆
nuclear resonance 核共振
nuclear resonance fluorescence 核共振荧光
nuclear resonance magnetometer 核子共振磁力仪
nuclear resonance spectroscopy 核磁共振谱法
nuclear risk 核危险性
nuclear safety 核安全性
nuclear safety analysis 核安全分析
nuclear safety criterion 核安全标准
nuclear sap 核液
nuclear scattering 核散射
nuclear scattering cross-section 核散射截面
nuclear science 核科学
nuclear screening constant 核屏蔽常数
nuclear sediment density meter 核子泥沙比重计
nuclear sediment ga(u)ge 核子测沙仪
nuclear separation 核间距
nuclear services cooling water system 核装置冷却水系统

nuclear shell structure 核壳层结构
nuclear shield 核防护层
nuclear shielding 核屏蔽
nuclear shielding concrete 核屏蔽混凝土
nuclear shift 核移动
nuclear ship 原子(能)船;核动力舰船;核动力船
nuclear ship certificate 核动力船安全证明书
nuclear ship regulation 核动力船安全规则
nuclear shockwave 核冲击波
nuclear slurry density ga(u)ge 核子料浆密度计
nuclear soil moisture densometer 同位素含水量仪
nuclear soil moisture meter 核子土壤湿度计;同位素土壤含水量探测仪;同位素探测仪
nuclear soil test 核子土壤试验
nuclear solution 核溶解
nuclear source 核子源
nuclear spallation 散裂
nuclear spectrum 核能谱;核辐射谱
nuclear stability 核稳定性
nuclear star 中央星
nuclear statistical equilibrium 核统计平衡
nuclear steam generator 核蒸汽发生器
nuclear steam station 核蒸汽站
nuclear stimulated geothermal development 核激发法地热开发
nuclear stimulated geothermal power plant 核激发地热电站
nuclear stimulated geothermal well 核激发地热井
nuclear structure 原子核构造;核结构
nuclear structure physics 核结构物理学
nuclear structurology 核结构学
nuclear superheating 核过热
nuclear supersymmetry 核超对称性
nuclear surface burst 地面核爆炸
nuclear surface ship 水面核舰艇
nuclear surface tension 核表面张力
nuclear systems reliability engineering 核系统可靠性工程
nuclear technique 核子技术
nuclear test 核试验;放射性试验
nuclear test reactor 核试验反应堆
nuclear test site 核试验场
nuclear test suspension 核试验悬浮物
nuclear thermal 热核的
nuclear thermodynamics 核热力学
nuclear time scale 核反应时标
nuclear tipped 带有核推进器的
nuclear track 核径迹
nuclear track emulsion 核径迹乳胶
nuclear transition 核跃迁
nuclear transmission density test of sand 放射性直接传递砂密度试验
nuclear triode detector 核三极探测器
nuclear turbine 核能涡轮机;核能透平
nuclear turbo-jet 核能涡轮机喷气发动机;核能透平喷气发动机
nuclear turbojet power plant 核能涡轮喷气动力装置
nuclear underground burst 地下核爆炸
nuclear volcanic mud ball structure 有核火山泥球构造
nuclear waste 原子能工业废物;原子能工业废料;核废物;核废料
nuclear waste disposal 核废物处理;核废料处理
Nuclear Waste Policy Act of United States 美国核废料政策法令
nuclear weapon 原子武器;核武器
nuclear weapon technology 核武器工艺学
nuclear weapon test 核武器试验
nuclear winter 核冬天
nucleary 成核
nuclear yield 核武器威力
nuclear Zeeman effect 核塞曼效应
nucleate 形(成晶)核
nucleate boiling 核(态)沸腾;泡核沸腾
nucleated glass 微晶玻璃
nucleated habitat 居民集居点
nucleated settlement 核心聚落
nucleated twin 成核双晶
nucleat engineering 放射化学工程
nucleating agent 成核剂;成核催化剂
nucleating center 成核中心
nucleating glass 成核玻璃
nucleating point 成核点

nucleating power 形成晶核能力
nucleation 晶核形成;集结成核;集结;核子作用;人工造雨法;人工降雨法;成核(作用)
nucleation agent 成核(催化)剂
nucleation-and-growth transformation 成核成长转变
nucleation density 成核密度
nucleation energy 成核能
nucleation field 成核场
nucleation frequency 成核频率
nucleation instrument 核子成像仪器
nucleation phenomenon 成核现象
nucleation process 成核过程
nucleation rate 成核速率
nucleation ratio 成核比
nucleation site 形核位置;成核位置
nucleation stage 核化剂阶段;成核阶段
nucleation theory 成核理论
nucleation time 成核时间
nucleator 核;成核剂
nucleatr decay mode 核衰变模式
nucleic abundance 核素丰度
nuclei diffusion 核扩散
nuclei of condensation 凝结核
nuclei of crystallization 晶核;结晶核
nuclei of origin 起核
nuclei origin 起核
nucleocidin 核杀菌素
nucleocosmochronology 核子宇宙年代学;核纪年法
nucleogenesis 原子核的起源
nucleoid 拟核
nucleolinus 核点
nucleolus 核仁
nucleomicrosome 核微粒
nucleon 核子
nucleonic configuration 核子位形
nucleonic distance 核子距离
nucleonic instrument 核试验仪表
nucleonics 核子学;原子核工程学
nucleonics ga(u)ge 测厚规
nucleonic structure factor 核子结构因数
nucleon instrument 核子仪器
nucleon-nucleon scattering 核子相互散射
nucleon number 质量数
nucleon recoil 核子反冲
nucleophilic 亲核的
nucleoplasmic index 核质指数
nucleostratigraphy 核子地层学
nucleopore filter 核孔滤膜
nucleus 核;[复]nulei
nucleus band 核带
nucleus counter 计核器;核子计数器;核辐射计数管
nucleus crew 骨干船员
nucleus dorsalis 背核
nucleus dorsomedialis 背内侧核
nucleus flaw 钢轨核伤
nucleus formation 晶核生成作用
nucleus globosus 球状核
nucleus growth 晶核生长
nucleus interstitialis 间位核
nucleus lentis 晶状体核
nucleus of atom 原子核
nucleus of condensation 凝结核
nucleus of continent 大陆核
nucleus of crystal 晶核
nucleus of high-pressure 高压核心
nucleus of raphe 缝际核
nucleus olivaris 橄榄核
nucleus reaction control system 核反应控制系统
nuclide 核素
nudation 裸地
nude cargo 裸装货;裸(散)货
nude contract 无保证合同
nude ga(u)ge 裸规
nude ion ga(u)ge 裸露电离计
nude land 不毛(之)地
nude well 裸井
nude wire 裸线
nudger 自动推杆器
nudging (模制砖时)轻轻加压;用锤加工花岗石表面
nuee ardente 炽热云;炽热火山云
nuffieldite 硫铋铜铅矿
nugatory 没有价值的

nugget 矿块;块金【地】;焊点熔核;熔核;天然块金;天然贵金属块
nugget constant 块金常数
nugget constant of non-conditional simulation 非条件模拟块金常数
nugget effect 块金效应
nuggeting 寻找天然金块
nuggetizer 破碎机
nugget of conditional simulation 条件模拟块金常数
nugget size 焊点尺寸;熔核直径
nugget thickness 天然金块厚度
nu-gild 装饰用黄铜;饰用黄铜
nu-gold 饰用黄铜
nuisance 有害物(质);噪扰
nuisance analysis 公害分析
nuisance effect 公害效应
nuisance element 环境有害因素
nuisance free 无公害
nuisance-free environment 无妨碍物的环境
nuisance ground 垃圾场
nuisanceless 无公害的
nuisanceless closed loop technique 无公害闭路式工艺;无公害闭环式工艺
nuisanceless technology 无公害工艺
nuisance organism 公害组织;公害机体
nuisance parameter 多余参数;多余参量
nuisance tax 小额消费品税(美国);消费者所付小额税款;烦扰税
nuisance threshold 损害阈值
nuisance value 阻扰价值;超付价值
nujol mull 石蜡软布
Nukiyama point 拔出点
nukundamite 诺硫铁铜矿
null 零值;零位;静点;无感点
nulla bona 无可扣押之物
null adjustment 零位调整
nullaginite 努碳镍石
nullah 峡谷;水路
nullah bridge 沟壑桥;峡谷桥
null allocation 无效分配
null amplifier 指零仪表用放大器
null and void 无效(作废);依法无效;作废
null angle 零位偏角
null axis 零位轴
null balance 零位平衡
null-balanced receiver 零位平衡接收机
null built-in function 零内函数
null cargo 裸货
null carrier method 载波零点法
null character 空字符
null character string 空的字符行;空的字符串【计】
null circle 点圆;点图
null circuit 零电路
null class 零类
null cone 零锥
null-current circuit 零电流电路
null-current measurement 零电流测量
null curve 零曲线
null cycle 空转周期;空周期;不工作周期
null data area 空数据区
null data set 空数据组
null delimiter 空定义符
null depth 零位深度
null detection 零值指示;零值测试
null detector 消尽指示器;零值检波器;零点检测器
null direction 零方位
null drift 零点漂移
null electrode 零位电极
null element 零元(素)
null ellipse 零椭圆;点椭圆
null ephemeris table 零值星历表
null event 零事件;空事件
null field 零场;空信息组;空白字段;空白域
null file 空白文件
null fill-in 零值补偿;零插补
null flux magnet 零磁场磁体
null gate 零门;空门
null geodesic 零短程线;零测地线
null graph 空图;零图
null-gravity state 失重状态;无重状态
null-g simulator 失重模拟器
null hypothesis 虚无假设;虚假设;零假设;解消假

设;检验假设;无效假设
nullification 作废;取消
nullified bond 失效证券
nullifier 自动零位器
nullify 作废;使等于零
null indicating double-beam system 零示双光束系统
null-indicating oscilloscope 零点指示示波器
null indicator 零值指示器;指零器;零示指示器;零位指示器;零示器;空指示符;示零器
nulling 用橡木雕刻扇形的装饰术;线脚中的扇形饰;指爱
nulling coefficient 零消系数
nulling device 零示装置
nulling operation 零位调整
null instrument 平衡点测定器;零点指示器
nullity 零维数;零度数;零度;无效
null junction 零位连接
null lens 零透镜
null line 零线;空行
null link 空连接
null locator value 空定址值;空定位值;空定位符变量值
null matrix 零矩阵
null measurement of transconductance 补偿法互导测定
null meter 指零表
null method 消除法;指零法;零位法;衡消法;调零法;补偿法
null method of measurement 零位测量法;零差测量法
null modem 零调制解调器
null object 零对象
null offset 零点漂移
null operator 零算子
null pick-up 零位传感器
null point 哑点;零点
null pointer indication 无效指针标志
null pointer value 空指针值
null position 零位置;零位
null potentiometer 零值电位计
null process 空进程
null product 零积
null reading 指零读数;衡消读数
null reading cross pointer 指示零位交叉指针
null reading instrument 零位读数仪表
null record 空记录
null recurrent 零循环
null reply 空应答
null representation 空白表示法
null sequence 零序(列)
null sequence reactance 零序电抗
null set 零测度集;空集(合);零集
null setting 调零装置
null sharpness 消音锐度;消声锐度;指零锐度;静音灵敏度
null shift 零漂
null-shift method 零移动法
null space 零空间;零化空间
null spaced neutron log 零源距中子测井
null space of matrix 矩阵(的)零空间
null spacing neutron-neutron log curve 零源距中子-中子测井曲线
null spectrophotometer 零值分光光度计
null state 零状态
null statement 空语句
null string 空行;空串
null suppression 消去零;零抑制
null surface 零曲面
null symbol 空符号
null tensor 零张量
null term 空项
null torque 零位力矩
null transformation 零变换
null type bridge circuit 指零电桥电路;平衡电桥电路
null-type impedance meter 零式阻抗计
null-type infrared spectrophotometer 零位红外分光光度计
null-type yaw probe 对零式三孔方向测针
null value 空值
null vector 零向量
nulon 尼龙;聚酰胺纤维
Nuloy 纽络(硬质合金)

Nultrax 直线运动感应式传感器
number 号(码);数字;数目;数量;数;标号;编号
numberable 可数的
number abstract 编号清单
number account 不列户名
number address code 数字地址码;数地址码
number analysis 数值分析
number and name of fault 断层编号和名称
number and name of fold 褶皱编号和名称
number and name of landslide 滑坡名称及编号
number and percentage 数量和百分比
number approximation 数值近似(法)
number arriving in one unit of time 单位时间内的到达数
number attribute 数字属性;数属性
number average 数均
number average degree of polymerization 数量平均聚合度
number average molecular weight 数量平均分子量;数均分子量
number axis 数轴
number base 内部二进计数制;数基
number bus 数字总线;数码总线
number calculation 数值计算
number calculation process 数值计算过程
number class modulus N 对模 N 的同余类
number comparing device 数值比较装置
number computation 数值计算
number concentration 计数浓度
number control oscillator 数控振荡器
number conversion 数转换
number converter 计数制变换器;数转换器
number cruncher 数值计算研究机
number crunching 咬嚼数字
number cutter 套数齿轮刀具
number density 数密度;数量密度
number designation 号码;数字标号;赋值数
number dialing 拨号
number differentiation 数值微分(法)
numbered 达到限定值的
numbered access 计数访问
numbered button 编号按钮
numbered card 已编号卡
numbered overlay 编号迭片
numbered rules 号码排列的规定
numbered tick 坐标网延伸短线
numberer 手拨号码机
number estimate 数值估计
number field 数域
number-flux relation 计数—流量关系
number for books system 图书编号法
number format 数字格式;数格式
number for matched loading 配装号
number generator 数码发生器;数据键盘;数据发生器;数发生器
number grain method 数粒法
number group 号码组
number group connector 号码组接线器;号码群接线器
numberical density 数字的密度
numberical fault 数值误差
numberical reading 读数
numberical scale 数字的尺度
number identifier 用户号码识别装置
number index 代号索引
number information 查号台查号
numbering 计数;号数测定;位次编排;打号;编号
numbering aperture 计数口径
numbering area 编号区
numbering beads 号码牌
numbering cycle 循环编号
numbering device 编号器
numbering die 印记冲模;数字冲模
numbering directory 号码簿;号码表
numbering hammer 号锤
numbering in anti-clockwise direction 逆时针方向编号
numbering in clockwise direction 顺时针方向编号
numbering in daily series 逐日编列流水号数
numbering in result list 成果表中编号
numbering machine 计数机;号码机;编号印字机;编号机
numbering machine ink 号码机油墨
numbering of annex 附录编号

numbering of clause 条款编号
numbering of division and subdivision 单元与分单元编号
numbering of documents 文件编号
numbering of elements 单元体编号
numbering of equation 公式编号
numbering of field book 手簿编号
numbering of figure 图表编号
numbering of foot-note 脚注编号
numbering of marks 航标编号
numbering of name of dangerous goods 危险货物品名编号
numbering of nodes 节点编号
numbering of note 注释编号
numbering of note to table and figure 图表注释编号
numbering of part 部分编号
numbering of points 道岔编号
numbering of sections 书帖编号
numbering of subclause 分条款编号
numbering of subcommittees 分委员会编号
numbering of table 列表编号
numbering of technical committees 技术委员会编号
numbering of tracks 股道编号
numbering of wheels 车轮编号
numbering plan 编号计划；编号方案
numbering plan area 编号计划区
numbering plan area code 编号区代码
numbering schedule 编号一览表
numbering scheme 编码制
numbering splicing 编线焊接
numbering stamp 编号印字器
numbering system 编号系统
number integration 数值积分
number light 号码灯
number line 实数直线
number marking 号码标志
number method 数值计算法
number of acceptable sample 合格样品数
number of accessible states 温和状态数
number of acres grown per farm 每个农场种植的英亩数
number of actual word 实在字数
number of adit 平峒数
number of adjacent chains in a net 网重数
number of allocated passenger train 车底数
number of altitude 高度标记；高程注记
number of ampere turn of magnetic circuit 磁路安匝数
number of ampere-turns 安匝数；安培匝数
number of analysis and experiment samples 分析试验样品(个)数
number of angle domain 角域个数
number of animals for sale 出栏率
number of annotated photographs 调绘相片数量
number of anode plates 每槽电极片数
number of arrival 乘客到达人数
number of artificial recharge engineering 回灌工程数量
number of axle passages 轮轴通过次数
number of axles 轴号；轴数
number of axles braked 制动轴数
number of back-pumping 回扬次数
number of balance area 勾衡区编号
number of berths 泊位数
number of big diameter borehole 大孔径钻孔数
number of blank 空白单元数
number of blows 贯入击数；打击次数(桩)；冲击数；锤击数；锤击次数
number of blows in pile driving 锤击次数
number of blows per minute 每分钟击数
number of bolt hole 螺栓孔数量
number of bolt holes in flange 法兰螺栓孔数；凸缘螺栓孔数
number of boreholes 钻孔数
number of borings 钻孔数
number of both longitudinal and cross feeds 纵横进刀量种数
number of building cost 建筑造价指数
number of building stor(e)ys 建筑层数
number of carriages in a train stock 车辆编成辆数
number of cars per train 每列车厢数
number of casts 出铁次数

number of cells containing geologic(al) characteristic X 存在地质特征 X 单元数
number of cells containing mineral deposit Y 存在矿床 Y 单元数
number of cells containing XY 存在 XY 单元数
number of chamber 室数
number of change of polarity 极性转换次数
number of channels 信道数；频道数；通道数；波道数
number of channels measured 测道数
number of chargesable words 计费字数
number of chart 海图编号
number of checking hole of grouting 检查孔数
number of chemical factor 化学因子数
number of clear days 晴天天数
number of coach-units per train 每列车厢数
number of coats 涂层道数
number of coils 弹簧圈数
number of columns 行列数
number of conditions 条件数；(大地测量中的)多余观测数
number of consumption to arrange in order 消耗量所居位数
number of control photographs 控制相片数量
number of copies 图像份数
number of critical revolution 临界转数
number of cross feeds 横进刀量种数
number of cycles 循环次数；旋回数目
number of cycles per second 每秒周期数
number of cylinders 汽缸数
number of data-pairs 数据对数目
number of days of snow cover 积雪天数
number of days of snowfall 降雪天数
number of days' sales in receivables 应收账款平均收现日数
number of death traces 死道数
number of defects 缺陷数
number of deformation parameters test of rock mass 岩体变形试验点数
number of degrees of freedom 自由度数
number of delivered empty cars 排空数
number of depature passengers 发送乘客人数
number of depth point 深度点数
number of derricks 吊杆数【计】
number of determined flow cross-sections 测流断面数
number of diameter changing 变径次数
number of diaphragm 光圈号数
number of diesel cylinder 主机缸数
number of difference grid point 差分格点数
number of dimension 维数
number of direction 方向个数
number of divided segment 管片衬砌环分块数目
number of divisional map 图幅代号
number of divisions 分度数；剖分数目
number of dock 车位号
number of dot loading test 点荷载试验数
number of drawdown sequence 水位落程的顺序号
number of dynamic(al) penetration test 动触探点数
number of dynamic load of serviceable wagon 运用车动载量
number of earthquake 地震次数
number of effective coils 工作圈数
number of effective slip system 有效滑移系数目
number of elastic modulus test 弹模试验数
number of electric(al) prospecting section 电测剖面条数
number of elements 只数
number of employment 就业人数
number of empty cars delivered 交出空车数
number of empty cars received 接入空车数
number of end point of the section 剖面终止号
number of equivalent binary digits 等价二进制数位位数
number of exploratory trench 探槽数
number of export to arrange in order 出口量所居位数
number of extra lengths 多字长级别数
number of extra shorths 短字长级别数
number of fault segments 断层分段数
number of fault zone 断层带编号
number of feeds 进给量级数
number of field shear test 野外大剪试验点数
number of filter net 滤网号

number of filter test 渗水点数
number of film 胶片数
number of fire 火灾次数
number of fish recaptured 重捕鱼数
number of floor levels 楼层数；层数【建】
number of fold 褶皱编号
number of gears 齿轮数
number of generations 出现次数
number of geologic(al) points 地质点数
number of geomorgraphy point 地貌点数
number of ground stress determination 地应力测点数
number of grouting(bore)holes 灌浆孔数；注浆孔数目
number of hatches in operation 开工舱口数
number of heads of screws 螺丝头数
number of holes for holding cutting tool 刀具孔数
number of homogeneous subarea 均匀区段编号
number of hooks per unit 每单元弯钩数
number of hours worked 工作时数
number of house properties 房产编码；房产号数
number of hydrogeologic(al) point 水文地质点数
number of hydrogeologic(al) unit 水文地质单元编号
number of image 映射次数
number of impact structure 撞击构造编号
number of import to arrange in order 进口量所居位数
number of inbound and outbound car handled at station 车站办理车数
number of independent component 独立组分数
number of independent slip system 独立滑移系数目
number of individuals 个人数量
number of inflection point 拐点数
number of inland vessels available 内河可用船舶数
number of input traces 输入卷数
number of in situ soil samples 原状土样个数
number of interference fringes 干涉条纹数
number of investigated points 调查点数
number of investigation 调查个数
number of ionization 电离数
number of irrigation 灌水次数
number of items 项数
number of iteration 迭代次数
number of jobs available 就业范围
number of junction points 节点数
number of known deposits 已知矿床号
number of lamina 层片数
number of land drilling 土钻孔数
number of lanes 车道数量
number of lateral pressure test 旁压试验数
number of layers 涂料层数
number of leads 螺纹线数
number of leaves 叶片数
number of leveling course 水准路线数
number of liability accident for luggage 行李责任事故件数
number of liability accident for parcel 包裹责任事故件数
number of lines 线路数
number of links per bucket (链斗式多斗挖掘机的)每斗链节数
number of load cycles 装载循环次数
number of loaded cars delivered 交出重车数
number of loaded cars received 接运重车数
number of loading cycle 加载循环数
number of loading test 荷载试验数
number of load repetitions 荷载重复作用次数
number of lockages 过闸次数
number of locomotive round trips made in the district per day 机车供应台次
number of locomotives required by goods trains 货运机车需要台数
number of longitudinal feeds 纵向进给量种数
number of longitudinal well rows 纵向井排数目
number of long-term observation points 长期观测点数
number of machine-days in good condition 完好台日数
number of machine drilling 机钻孔数
number of mapping photographs 测图相片数量

number of mass transfer units 传质单元数
number of measurements 测回数;应变标志测量数目
number of metric threads cut 公制螺纹头数
number of module threads cut 模数螺纹头数
number of moisture measuring probes 湿度测量探状数目
number of mooring legs 锚腿数
number of motors 电机数量
number of (motor) vehicles 汽车数
number of moving point 滑动点数
number of nodes 节点数;单元体个数
number of non-nodes well 非节点井数
number of nudes well 节点井数
number of observation 观测次数
number of observation point for test 试验观测点号
number of observation shaft well 观测竖井编号
number of observation tunnel 观测平硐编号
number of observation well for test 试验观察孔号
number of observed points 观测点数
number of observed quantities 观测数
number of occurrence 反复次数
number of offset recognized 辨认错动次数
number of orebodies 矿体数
number of outgoing circuits 出线回路数
number of output to arrange in order 开采量所居位数;产量所居位数
number of over-proofed sample 超差样品数
number of packages 件数;货物件数
number of parameter division 参数分区数
number of particle 粒粒数
number of partitions 分区数
number of passenger casualties 旅客伤亡人数
number of passengers arrived 旅客到达人数
number of passengers carried 运送旅客人数
number of passengers carried per passenger carriage 客车载客人数
number of passengers carried per passenger train 旅客列车载客人数
number of passengers dispatched 旅客发送人数;发送旅客人数
number of passengers originated 旅客发送人数;发送旅客人数
number of passengers transported 旅客运送人数
number of passenger train set required 列车车底需要数
number of passenger train stock needed 旅客列车车底需要数
number of passes 涂漆道数;程数
number of photos in the sheet 图幅内相片数
number of physical factor 物理因子数
number of pictures per second 每秒帧数
number of pieces 件数
number of piles 桩数量
number of plants per unit area 单位面积株数
number of plates 板片数量;塔板数
number of plies 层数
number of pole-pairs 极对数
number of poles 极数
number of precipitation days 降水天数
number of predicted deposits 预测矿床号
number of profiles 剖面条数
number of pumping sections 取水段数
number of pumping strata 取水层数
number of punchings 冲数
number of quantitative analysis 定量分析值
number of quantum state 量子状态数
number of radial pipe 辐射管根数
number of rainy days 降雨日数
number of reagent dropping point 投源点数
number of regime observation line 动态观测线条数
number of regime observation point 动态观测点编号
number of rejects 废品数;废品率
number of repaired locomotive 检修机车修竣台数
number of repaired wagon 修竣车数
number of replication 重复数;重复次数
number of reserve well 备用水井数
number of responsible accident of freight traffic 货运事故件数
number of reversals 颠倒次数
number of revolution 转数

number of revolution per minute 每分钟转数
number of revolution per unit of time 单位时间转数
number of revolutions 旋转次数
number of rock-bolt 锚杆数
number of rock sample 岩样数
number of roller passes 碾压遍数
number of rolling coverage 碾压遍数
number of route 路线编号
number of route profile 路线剖面图号
number of runoff zone 径流带条数
number of runs 游程个数
number of saddle suppport 鞍座数目
number of sample for umpire analysis 仲裁样品数
number of samples 样本数目;标本数;样品数;样本组数
number of samples for external examination 外检样品数
number of samples for internal examination 内检样品数
number of satellites 卫星数
number of scan(ning) lines 扫描行数
number of seats 座位数
number of section 截面号数
number of segments 管片衬砌环分块数目
number of seismic survey point 地震测点数
number of serviceable carriages needed 运用客车需要数
number of serviceable cars held kept 运用车保有量
number of serviceable cars turnround 运用车工作量
number of sets 机组台数;套数
number of sets of sample 样品组数;抽样组数;采样组数
number of shallow wells 浅井数
number of shorts in array 组合激发炮数
number of shot-holes 炮眼数
number of sieve tray holes 筛板孔数
number of simultaneous observation 水位统测井点数目
number of slip system 滑移系数目
number of slots 箱位数;槽数
number of soil sample 土样数
number of sonic logging test 声波测点数
number of spatial distribution 空间分布数
number of specimens 标本块数
number of spring 泉水编号
number of spring mouth 泉口个数
number of stages 倍增级数
number of standard penetration 贯入试验锤击数
number of standard penetration test 标贯数
number of standard wagons(at a length of kilometres each) 列车计长
number of standup 站立人数
number of starting point of the section 剖面起始号
number of starts 起动次数;头数
number of static cone penetration test 静触探点数
number of station 测站编号
number of statistic(al) point 节理统计点数;节理统计点编号
number of steps of speeds 变速级数
number of stories 楼层数;层数【建】
number of strips in the sheet 图幅内航线数
number of strokes 行程次数;冲击数
number of strokes per minute 每分钟行程数
number of substations 分站数目
number of super-helical turns 超螺旋圈数
number of survey line 测线编号
number of survey point 测点号
number of swabs 抽子个数
number of swirls of water flow 水流环量
number of tanks 油柜数
number of tape 纸带编号
number of taper 锥度级数
number of teeth 齿数
number of test-pieces 试件个数
number of test points 试验点数
number of test segments 试验段数
number of test strata 试验层数
number of test wells 试验孔号;试验钻孔数

number of thaw-collapse pit 融陷坑数量
number of the ground stress measuring hole 地应力测量孔编号
number of the observation line for test 试验观测线编号
number of the ore reserve to arrange in order 拥有资源所居位数
number of theoretic(al) plates 理论板数;理论塔板数
number of threads 螺纹扣数
number of threads per unit length 单位长度螺纹扣数
number of times 次数
number of tons of freight carried 运送货物吨数
number of tons of freight originated 发送货物吨数
number of train operation accident 行车事故件数
number of transfer node 换乘节点数
number of transfer units 传热单元数
number of transversal well rows 横向井排数目
number of trial pits 试坑数
number of trips 航行次数
number of tube passes 管程数
number of turnover of current funds 流动资金周转次数
number of turns 转(弯)数;匝数
number of turns per centimeter 每厘米螺纹圈数
number of turns per inch 每英寸螺纹圈数
number of turns per unit length 单位长度螺纹圈数
number of units in sample 样本单位数
number of unknown 未知数
number of vane shear test 十字板剪切试验点数;十字板剪力试验点数
number of ventilating times per unit 通风换气次数
number of vertical shaft 竖井数
number of vertical stack 纵向叠加次数
number of vibrations 振动数
number of volcanic cones 火山锥数目
number of wagon loaded 装车数
number of wagons in train 列车编成
number of wagons loaded 装车数
number of wagons unloaded 卸车数
number of water pressure test hole 压水试验孔数
number of water pumping test hole 抽水试验孔数
number of water sample 水样数
number of water way 水槽数目(钻头)
number of weighted function curve 加权函数曲线数
number of well operation 开动水井数
number of well pipe 井管根数
number of well rows 水井排数
number of Whitworth's thread cut 惠氏螺纹头数
number of winding heads 缠绕头数
number of windings 匝数
number of work cycles 工作循环数;工作周期数
number of working days per year 每年工作天数;每年工作日
number of zero up-crossing 跨零波数
number one 第一号
number operator 数算符
number piece 工件数目
number plane 实数平面
number plate 号数板;号盘;号牌;车辆牌照;编号牌
number-plate lamp of car 车牌灯
number-plate light 牌照灯
number plate support 牌照板支架
number range 数值范围
number recorder 号码记录器
number record printer 数字记录打印机
number repetition 号码重拨
number representation 记数(制);数字表示法;数表示法
number representation system 记数系统;数制
number required 需用台数;要求数目
number revolving stamp 数字回转打印机
number roll 数字滚筒
number scale 数记;数标
number sender 发号器
number sequence 数序(列)
number sieve 数筛
number sizes 标号尺寸
numbers of bit used 用钻头数

numbers of blows 击数
numbers of compact 击实数
numbers of diesel engine 柴油机数
numbers of generator 发电机数
numbers of hoist gear 绞车挡数
numbers of injures and deaths 伤亡事故人数
numbers of layers 层数【地】
numbers of liner 缸数
numbers of mud pumps 泥浆泵数量
numbers of nozzle 喷嘴数量
numbers of pump valve 凡尔数
numbers of shift 轮班次数
numbers of stratification 层数【地】
numbers of time-velocity couple 时间-速度对数
number stamp 数字印模
number stamp set 数码压模
number stay 牌照板支架
number storage 数字存储器
number switch 号控机;数字开关
number symbol 数字标记
number system 记数制;记数系统;数系
number system base 数基
number system for steel 钢号
number tag 号签
number tape 数字带
number telling method 数字显字法;数字显示法
number template 数字模片
number the material 号料
number-theoretical method 数论法
number-theoretic function 数论函数;数论泛函
number theory 数论
number the probe of moisture measurement 湿度测量探头编号
number to be printed 印刷量
number token 数记号
number transfer bus 数字传送干线;数值传输线
number triple 三重数组
number two ranking variety 排列的第二号品种
number type 数据类型
number unobtainable tone 空号音
number window 计数窗
number with one digit 一位数字;个位数
numerable 数得清
numeral 码子;数字(的);船材尺码编号;数码
numeral aspect 数字示像
numeral dial lock 号码锁
numeral indicator 数字表示器
numeralization 数字化;编码化
numeral light 数字表示灯
numeral order 编号次序
numeral pennant 数字旗
numeral signal aspect 数字信号显示
numeral system 记数制;命数系统
numeral type dial lock 数字拨号锁
numeral wheel 数轮
numerate 数
numerated 分数的
numeration 记数;命数法
numeration of azimuth circle 方位度盘计算法
numeration system 记数系统;命数系统;数(字)制
numeration table 数字表
numerator 示号器;分子;分数
numeric(al) 数值的
numeric(al) ability tester 计数能力测定仪
numeric(al) abundance 多余度
numeric(al) accuracy 数值精度
numeric(al) address 数字地址
numeric(al) aerial triangulation 解析法空中三角测量
numeric(al) algebra 数值代数
numeric(al) alphabetic 数字字母符
numeric(al) analog(ue) method 数值模拟方法
numeric(al) analogy 数字模拟
numeric(al) analysis 数值分析
numeric(al) analytic compiler 数字解析测图仪
numeric(al) aperture 数值口径;数值孔径;数码孔
numeric(al) approximation 近似数值;平面近似;数值近似(法);数值逼近
numeric(al) area 数字区域
numeric(al) argument indicator 数字变元指示符
numeric(al) assembling machine 数控装配机
numeric(al) bit 数字位
numeric(al) bit data 数字位数据
numeric(al) calculation 数值计算

numeric(al) capacity 数字容量
numeric(al) catalog 码号目录
numeric(al) ceiling 最高限额
numeric(al) change 数值变化
numeric(al) character 数字电码组合;数字
numeric(al) character data 数字字符数据;数值字符数据
numeric(al) characteristic 数值特性
numeric(al) characteristic curve 数值特征曲线
numeric(al) characteristics of direction(al) data 方向数据的数字特征
numeric(al) character set 数字字符组;数字字符集
numeric(al) character subset 数字字符子集
numeric(al) check 数值校验
numeric(al) class 数字类;数值等级
numeric(al) class test 数字类测试
numeric(al) code 数字码;数字代码;数字代码;数(值)码
numeric(al) coded character set 数字编码的字符集
numeric(al) code input 数字代码输入
numeric(al) coding 数字编码;数值编码
numeric(al) coefficient 数字系数
numeric(al) colo(u)r difference 数值色差
numeric(al) comparison 数值比较;数字比较
numeric(al) computation 数值计算
numeric(al) computation of eigenvalues 特征值的数值计算
numeric(al) constant 常量;数字常数;数值常数
numeric(al) control 数字控制;数字计算控制;数值控制;数控
numeric(al) control center 数控中心
numeric(al) control computer 数字控制计算机;数控计算机
numeric(al) control cutting 数控切削;数控切割
numeric(al) control cutting machine 数控切割机
numeric(al) control deep hole drilling machine 数控深孔钻床
numeric(al) control device 数字控制装置;数字控制设备;数值控制装置;数值控制设备;数控设备
numeric(al) control drawing machine 数控绘图机
numeric(al) control filament winder 数字控制绕线机;数控绕线机
numeric(al) control language processor 数值控制语言处理机
numeric(al) controlled turret punch press 数字控制回转头压力机
numeric(al) control machine 数字控制机床;数值控制机;数控机床
numeric(al) control milling machine 数控铣床
numeric(al) control orthophotoscope 数控正射投影纠正仪
numeric(al) control press 数控冲床
numeric(al) control robot 数控机械手;数控机器人
numeric(al) control system 数字控制系统;数控系统
numeric(al) control tape 数控带
numeric(al) conversion code 数字转换代码;数字变换码;数值转换代码;数值变位码
numeric(al) coordinates 数字坐标
numeric(al) coprocessor 数字协处理器
numeric(al) correction 数字校正;数字改正
numeric(al) damping factor 数值衰减因子
numeric(al) data 数字资料;数字数据;数据
numeric(al) data base 数字数据库
numeric(al) data code 数值数据代码;数据码
numeric(al) data item 数字数据项
numeric(al) decrement 衰减率
numeric(al) designation 数字番号
numeric(al) designation of optic(al) glass 光学玻璃标号
numeric(al) determinant 数字行列式
numeric(al) differential 数字微分
numeric(al) differentiation 数字微分;数值微分(法)
numeric(al) digit 数位
numeric(al) display 数字指示;数字显示
numeric(al) display device 数字显示装置;数字显示器件
numeric(al) display unit 数字显示装置
numeric(al) distance 数值距离
numeric(al) distortion correction 数字法畸变校正
numeric(al) eccentricity 偏心率;数值偏心(率)
numeric(al) edited character 数字编辑字符

numeric(al) edited item 数字编辑项
numeric(al) equation 数字方程;数学方程
numeric(al) equivalents 数值当量
numeric(al) error 数字误差;数值误差
numeric(al) estimate 数值估计
numeric(al) estimate of probability 概率的数值估计
numeric(al) evaluation 近似计算
numeric(al) example 数字例题;数值例;数样本;数例
numeric(al) experiment 数值实验
numeric(al) experimentation 数值试验
numeric(al) expression 数值表示法;数值表达式;数式
numeric(al) factor 数字因数;数值因子
numeric(al) fault 数值误差;数值错误
numeric(al) field 数字域;数字栏
numeric(al) field data 数字域数据;数域数据
numeric(al) field descriptor 数域描述符
numeric(al) field form 数域形式
numeric(al) flow visualization 数值流观测
numeric(al) forecast(ing) 数值预报(法)
numeric(al) frequency 数字频率
numeric(al) function 数值函数;数函项
numeric(al) geomechanics 数值岩土力学
numeric(al) graph 数值图解
numeric(al)-graphic(al) method 数值图解法
numeric(al) index 数标
numeric(al) index of the vector group 矢量组的数标
numeric(al) indication 数字指示
numeric(al) indicator 数字显示器
numeric(al) indicator tube 数字显示管
numeric(al) information 数字信息
numeric(al) information processing 数字信息处理
numeric(al) information processing system 数字信息处理系统
numeric(al) instability 数字不稳定性;数值不稳定性
numeric(al) integrating 数值积分法
numeric(al) integration 数值积分(法)
numeric(al) integration error 数值积分误差
numeric(al) interpolation 数字内插;数值内插法
numeric(al) invariants 不变数
numeric(al) inversion 数字转换
numeric(al) item 数字项
numeric(al) keyboard 数字键盘
numeric(al) keypunch 数字键控孔机;数字键(控)穿孔
numeric(al) line 数字注记线
numeric(al) list of charts 海图图号索引
numeric(al) list of commodities 货物号码清单
numeric(al) literal 数字字面文字;数值文字
numerically 数字上
numerically coded instruction 数字编码指令
numerically controlled 数控的
numerically controlled benders 数字控制折弯机
numerically controlled boring machine 数字控制镗床
numerically controlled comparator 数控坐标量测仪
numerically controlled drilling machine 数字控制钻床
numerically controlled engine lathe 数字控制普通车床
numerically controlled filament-winding machine 数控绕线机
numerically controlled grinding machine 数字控制磨床
numerically controlled layout machine 数字控制划线机
numerically controlled machine 数控机床;数字控制机
numerically controlled machine tool 数字控制机床
numerically controlled measuring machine 数字控制测量机
numerically controlled milling machine 数字控制铣床
numerically controlled punching machine 数字控制冲床
numerically controlled shears 数字控制剪床
numerically controlled table 数字控制工作台
numerically controlled turret lathe 数字控制转塔车床
numerically equal 在数值上相等

numerically equivalent 数值等价的
numeric(al) magnitude 计算震级
numeric(al) map 数字地图
numeric(al) mapping 数字化测图
numeric(al) measure 数量
numeric(al) method 数值计算法;数值方法;数值法
numeric(al) method analysis 数值法分析
numeric(al) method of groundwater resources evaluation 地下水资源评价数值法
numeric(al) methods of integration 数值积分法
numeric(al) mode 数值模式
numeric(al) model 数字模型;数值模型
numeric(al) modeling center 数值模拟中心
numeric(al) modelling 数字模拟;数值模拟
numeric(al) move 数字移动
numeric(al) mutation 数突变
numeric(al) name 数字代号
numeric(al) notation 数字符号
numeric(al) number 数值
numeric(al) operand 数字操作数
numeric(al) operator 数字算子
numeric(al) order 表数次序;数字顺序;数字次序;数目顺序
numeric(al) orientation 数字定向
numeric(al) paneling 数值分段法
numeric(al) part 数字部分;数值部分
numeric(al) patent index 专利号索引
numeric(al) perforator 数字凿孔机
numeric(al) photogrammetry 数字摄影测量学
numeric(al) picture 数字图像
numeric(al) pictured variable 数值图像变量
numeric(al) picture format item 数值图像格式项
numeric(al) picture specification 数值图像说明
numeric(al) plotter 数控绘图机
numeric(al) point 数值点
numeric(al) portion 数字部分
numeric(al) positional control 数字位置控制
numeric(al) precision 数值精度
numeric(al) prediction 数值预报
numeric(al) printer 数字打印机
numeric(al) problem 数值问题
numeric(al) procedure 计算方案;数字计算法;数值(处理)方法
numeric(al) process 数字处理
numeric(al) process control system 数字过程控制系统
numeric(al) processing 数值处理
numeric(al) proportion 数值比例
numeric(al) protection for overhead contact lines 数字式架空牵引网保护装置
numeric(al) punch 数字穿孔(机)
numeric(al) quadrature 数值求面积;数值求(面)积(法);数值积分
numeric(al) range 数值范围
numeric(al) rating system 实数制
numeric(al) reading 读数
numeric(al) read-out 数字读出;数值读出
numeric(al) recognition system 数字识别系统
numeric(al) recording device 数字记录器件
numeric(al) rectifying enlarger 数字纠正放大机
numeric(al) reference 数值基准
numeric(al) relation 数值关系
numeric(al) relationship 数值关系
numeric(al) representation 数字表示法;数值法;数值表示;数的表示
numeric(al) restitution 数字测图
numeric(al) result 计算结果
numeric(al) roll 数字卷筒
numeric(al) scale 数字刻度;数字比例尺;数值刻度
numeric(al) search 数值搜索
numeric(al) selection 选号
numeric(al) selector 数字选择器
numeric(al) series 数值级数
numeric(al) shift 数字换挡
numeric(al) signal 数字信号;数值信号
numeric(al) simulation 数字模拟;数值模拟
numeric(al) solution 近似解;数值解
numeric(al) solution of algebraic(al) equation 代数方程的数值解法
numeric(al) solution of integral equation 积分方程的数值解法
numeric(al) stability 数字稳定性;数值稳定性;数值的稳定性
numeric(al) statement 数目表;统计(报)表;统计;数学统计或说明
numeric(al) step by step method 数值的逐步逼近法
numeric(al) stereocompilation 数字立体测图
numeric(al) stereo triangulation 解析法立体三角测量
numeric(al) study 数值研究
numeric(al) switch 号控机;数字开关
numeric(al) symbol 数字符号;数字标记;数符号
numeric(al) system 数系
numeric(al) table 数值表;数表
numeric(al) tabular 数值表
numeric(al) tabulator 数字制表机
numeric(al) tacheometry 解析法视距测量
numeric(al) tape 数字带
numeric(al) taxonomy 数字分类学;数值分类学
numeric(al) technique 数值技术
numeric(al) tensor 数值张量
numeric(al) term 数值项
numeric(al) time 给定序列时间
numeric(al) treatment 数值处理
numeric(al) trigonometry 数值三角学
numeric(al) unit 数值单位
numeric(al) value 数字值;数值
numeric(al) value of a real number 实数的绝对值
numeric(al) variable 数值变量
numeric(al) weather facility 数字气象站
numeric(al) weather forecasting 数值天气预报;天气数值预报
numeric(al) weather prediction 数字天气预报;数值天气预报
numeric(al) word 数目字
numeric-pictured arithmatic item 数形算术项
numerics in contours 等高线注记
numer mite 数字控制钻床
numerology 数学学
numeroscope 数字记录器;示数器
numerous 大批量的
numerous minute echoes 密集低小微波型
numerous small and low echoes 密集低小微波型
numerous ways 多方面
nummulitic limestone 货币虫灰岩
nummulitic sandstone 货币虫砂岩
nummulitic sea transgression 货币虫海海浸
nunatak 冰原岛峰;冰原石山
nun buoy 锥形浮标;菱形浮标;纺锤形浮筒;纺锤形浮标
nuncital 氨基甲酸叔己脂
nunc pro tunc 追溯既往
nuncupative will 口头遗嘱
Nunivak event 努尼瓦克事件
Nunivak normal polarity subchron 努尼瓦克正向极性亚时【地】
Nunivak normal polarity subchronzone 努尼瓦克正向极性亚时间带
Nunivak normal polarity subzone 努尼瓦克正向极性亚带
nunnery 女修道院;尼姑庵;庵
nunnery church 尼姑庵;女修道院
nun's choir 修女唱诗班
nuolaite 钇杂铌矿
nuplex 核动力综合企业
nurag(h)e 截头圆锥形石建筑;(史前撒丁岛独有的)石砌的小屋与圆塔围成建筑体系
nuraghi (史前撒丁岛独有的)圆顶石屋;圆锥形石建筑
nural 努拉铝合金
Nuralite 努拉岩(一种石棉沥青)
Nuremberg gold 铜铝金装饰用合金(纽伦堡合金);铜铝金装饰合金
Nuristan massif 努里斯坦地块
nurling tool 刻边工具
Nuroz 聚合木松香(商品名)
nurse and health keeper's visiting bag 保健箱
nurse balloon 补助气球
nurse call system 护士呼叫系统
nurse crop 保护作物
nurse graft 靠接
nurs(e)ling 苗木
nurse room 护士室;哺乳室
nursery 育婴室;育苗场;养鱼池;养蚕场;种苗场;苗圃;苗床;孵卵器;保育院;保育室
nursery area 繁殖区域
nursery bed 苗圃;苗床
nursery day school 日托托儿所
nursery drill 苗圃播种机
nursery forest bed 林木苗圃
nursery for trees 树木苗圃
nursery garden 苗圃;育苗区;幼儿园
nursery-grown plant 苗木
nurseryman 园主;园丁;苗圃工作者
nursery planter 苗圃播种机
nursery reproducing 繁殖苗圃
nursery room 哺乳室;托儿室
nursery school 幼儿园;苗圃;托儿所
nursery separator 试验区用脱粒机
nursery sock growing 育苗
nursery sod 繁殖场草皮;苗圃草皮
nursery stock 苗木;定植苗;出圃苗
nurse school 幼儿园
nurse's hostel 保育员住房;护士住房
nurse's office 护士办公室
nurse's school 护士学校
nurse's station 护士室;护士办公室;护士站
nurse's teaching block 护士教学区
nurse tree 保育树;保护树
nursing facility 养育设施;保育设施
nursing home 托儿所;小型疗养院;私人医院;私立疗养所;私立病院
nursing room 哺乳室
nursing the bit 维护钻头
nursing tube 供给管
nursing unit 保育单位;护理单位;看护单位;护理单元
nurture 抚育
Nuseal 努希尔(一种沥青防水剂)
Nusselt number 罗素数
Nust 诺斯特漆(一种金属防锈油漆)
nut 小煤块;螺母;螺帽;坚果
nut anchorage 螺帽固定;螺纹锚固;螺帽锚固;螺帽锚碇
nut and bolt 螺母及螺栓
nut-and-sleeve flare fitting 螺母套管式扩口管接头;三件式扩口管接头
nutant character 突变体性状
nut arbo(u)r 螺纹刀杆
nutating antenna 盘旋馈入天线
nutating-disc flowmeter 盘式流量计
nutating-disk meter 摆盘式流量计;圆盘式旋转流量计;摆盘式水表
nutating feed 盘旋馈电
nutation 下俯;章印;章动
nutational constant 章动常数
nutational ellipse 章动椭圆
nutation angle 盘旋角
nutation drive 章动传动
nutation drive units 章动传动装置
nutation in longitude 黄经章动
nutation in obliquity 交角障动;倾角章动
nutation in right ascension 赤经章动
nutation of inclination 倾角章动
nutation of longitude 黄经章动
nutation period 章动周期
nutation shaft 章动轴
nutation wheel 章动轮
nutator 章动器;盘旋馈入装置
nut bolt 螺钉;带帽螺栓;带螺的螺栓;带螺帽螺栓
nut buoy 胡桃形浮标
nut cap 死螺母;封紧帽;盖帽;螺帽盖
nut-castellating machine 冠形螺母开槽机
nut check 螺母锁紧
nut clamp 螺母夹
nut coal 小块煤
nut coke 焦丁
nut collar 螺栓垫圈
nut cover 螺帽母
nut cracker 螺母破碎机;铡碎坚果的粉碎机;角锥钻头
nut driving machine 拧螺母机
nut eyebolt 带螺帽环首螺栓;吊环螺栓
nut facing machine 螺母端部加工机
nut fastening machine 拧螺母机
nut for cross slide 横滑板螺母
nut for excess pressure head check 超压位防松螺母
nut forging machine 螺母锻造机
nut for maximum pressure head check 最大压位防松螺母

nut former 螺母锻压机
nut for plain cylinder cap 简单汽缸罩螺母
nut key 螺帽扳手
nut lathe 螺母车床
nut lock 螺母锁紧动螺帽；螺母锁紧；螺母保险；螺帽锁；螺帽固定器
nut lock bolt 螺母锁紧螺栓
nut locking 螺母锁紧
nut locking device 螺母锁紧装置
nut-lock washer 螺母锁紧垫圈；螺帽止松垫圈；止松垫圈；锁紧垫圈
nut machine 螺母机
nut making machine 螺母制作机
nut mandrel 螺母心轴
nut mill 脱壳磨机
nut mordant 螺帽的酸洗剂
nut of anchor 锚挡尾球
nut oil 胡桃油
nut piercing 螺母冲孔
nut planking machine 自动高速螺母制造机
nut-pliers 螺帽钳
nut press 螺母压机
nut retainer 螺母护圈
nutrient cycle 营养物循环
nutrient fluid 培养液
nutrient leaching 浸出养分
nutrient medium 培养基
nutrient-poor habitat 贫营养生境
nutrient removal plant 除营养物装置
nutrient-rich lake water 富营养湖水
nutrient-rich sediment 富营养物沉淀
nutrient-rich water body 富营养水体
nutrient salt 盐类营养物
nutrients analyzer for seawater 海水营养盐分析器
nutrient solution 营养液
nutrient specification 营养规格
nutrient status 营养状况
nutritional ingredient 营养成分
nutritional status 营养状况
nutrition canteen 保健食堂
nutritionist 营养学家
nutrition period 营养期
nutrition pot 营养钵
nutritive chamber 营养室
nutritive cube 营养土块
nutritive layer 营养层
nutritive medium 营养液
nutritive proportion 营养比例
nutritive ratio 营养比
nutriture 营养状况
nut-runner 自动螺帽扳手；螺帽紧松器；螺帽扳手；上螺母器
nut-running tool 螺帽松紧工具
nuts and washers 螺母和垫圈
nutsch 吸滤器
nutsch filter 吸滤器；吸尘器
nut screw 螺母
nut seat 螺母支承面
nut-setter 螺母固定器；机动扳钳
nut shaping machine 螺母加工机；螺母成型机
nutshell tire 坚果外壳籀圈

nut socket 地脚螺母套；预埋螺母管
nuts of an anchor 有杆锚孔上下的杆箍
nut spinner 快动螺母扳手；套筒扳手柄
nut switch 螺帽形开关
nut tap 螺母制作；螺母丝锥；螺母丝攻
nut tapper 螺母攻丝机
nut torque 螺母扭矩
nut-tree 胡桃树；坚果树
nut type 螺帽式
nutty slack 结核煤渣
nutty structure 核状结构
nut-wrench 螺帽扳手；螺母扳手
nu value 光色散值
nuvistor 超小型抗震管；超小型抗震电子管
Nux Pseudocerasi 樱桃核
N-value N 值
n-valued 多值的
n-valued logic 多值逻辑
N-W tectonic system 北西构造系
NW-trending structure 北西向构造带
nyctinastic movement 感运动
nyctipelagic plankton 夜浮游生物
nyctometer 暗视计
nyctoplankton 夜浮游生物
n-year dependable yield n 年内可靠供水量
n-year flood n 年一遇洪水
n-year low water n 年一遇枯水
n-year peak discharge n 年一遇洪峰流量
n-year peak flow n 年最大洪水
nyererite 泥碳钠钙石
Nykrom 高强度低镍铬合金钢
Nylander's test 铋试验
Nylatron 石墨填充酰胺纤维（商品名）
nylon 聚酰胺纤维；锦纶；尼龙；人造纤维
nylon-11 coating 尼龙-11 粉末涂料
nylon bag 尼龙布袋
nylon-base insulator 尼龙底绝缘板
nylon bearing 尼龙轴承
nylon belt 尼龙夹层皮带
nylon brush 尼龙刷
nylon bush 尼龙衬套
nylon carpet 尼龙地毯
nylon cleaning 尼龙清理刷
nylon cloth 尼龙布
nylon-coated metal 尼龙覆面钢
nylon coating 尼龙涂层
nylon cord 尼龙索；尼龙绳
nylon cord tyre 尼龙线织轮胎
nylon dam 尼龙坝
nylon epoxy 尼龙环氧
nylon-epoxy adhesives 尼龙环氧黏合剂
nylon fabric 尼龙织品；尼龙织物；尼龙结构
nylon fiber 尼龙纤维
nylon filled 装有尼龙纤维的
nylon film 尼龙薄膜
nylon filter anti-scour system 尼龙滤层防冲系统
nylon-fuel cell 尼龙燃料箱
nylon gasket 尼龙填圈；尼龙填料
nylon gear 尼龙齿轮
nylon grill(e) 尼龙格栅
nylon line 尼龙缆

nylon marline 尼龙细绳
nylon monofilament 尼龙单丝
nylon mo(u)lded gear 尼龙铸造齿轮
nylon mo(u)lding powder 尼龙铸粉
nylon net 尼龙网
nylon net dam 尼龙网坝
nylon paint 聚酰胺树脂涂料；尼龙涂料
nylon phenolic resin 尼龙酚醛树脂
nylon phenolics 尼龙酚醛塑料
nylon pin coupling 尼龙柱销联轴器
nylon plastics 尼龙塑料
nylon plug 尼龙栓
nylon rail pads 尼龙轨下胶垫
nylon-reinforced belt 尼龙加固带
nylon resin 锦纶树脂；尼龙树脂
nylon-roller pump 尼龙滚子泵
nylon rope 尼龙绳
nylon screen 尼龙筛网
nylon sieve 尼龙筛
nylon sling 尼龙链钩；尼龙吊索；尼龙吊具；尼龙吊带
nylon staple 尼龙钉
nylon thread 锦纶丝线
nylon tube 尼龙管
nylon waste 尼龙废料
nylon-weft yarn 尼龙纬线
nylon wire 尼龙线；尼龙丝；尼龙绳
nylon woven fabric 尼龙织物
nylon yarn 尼龙线；尼龙丝；尼龙纱
nymphaeum （古罗马）休息场所
Nyquist noise 热噪声
Nyquist's contour 乃奎斯特围线
Nyquist's criterion 乃奎斯特准则；乃奎斯特判据
Nyquist's demodulator 乃奎斯特解调器
Nyquist's diagram 乃奎斯特图
Nyquist's filter 乃奎斯特滤波器
Nyquist's flank 乃奎斯特截面
Nyquist's frequency 乃奎斯特频率
Nyquist's instability 乃奎斯特不稳定性
Nyquist's interval 乃奎斯特间隔
Nyquist's interval bandwidth 乃奎斯特前沿带宽
Nyquist's limit 乃奎斯特极限
Nyquist's locus 乃奎斯特轨迹
Nyquist's noise 乃奎斯特噪声
Nyquist's plot 乃奎斯特图
Nyquist's rate 乃奎斯特速度；乃奎斯特率
Nyquist's sampling interval 乃奎斯特取样间隔
Nyquist's sampling rate 乃奎斯特取样频率
Nyquist's sampling theorem 乃奎斯特取样定理；乃奎斯特采样定理
Nyquist's stability criterion 乃奎斯特稳定度准则；乃奎斯特稳定度判据
Nyquist's stability theorem 乃奎斯特稳定性定理
Nyquist's theorem 乃奎斯特定理
Nyquist's theorem of noise 乃奎斯特噪声定理
Nyquist's theory 乃奎斯特理论
nyronic acid 芥酮酸
nysconitrine 硝酸甘油
nytril fibre 奈特里尔纤维
Nytron 碳氢（化合物）硫酸钠清洁剂；碳氢化合（硫酸钠清洁剂）

O

oak 橡树;橡木;栎树;栎木
oak bark 栎树皮
oakbark tannin 栎树皮丹宁
oak block 橡木垫块
oaken 栎木制的
oakermanite 镁黄长石
oak extract 栎木浸液
oak fence 橡木栅栏
oak floor 橡木地板;栎木地板
oak floor cover(ing) 栎木楼面覆盖面材料;栎木地板铺面材料
oak frame 基台木
Oakley-Fulthorpe test 奥克利—富尔索谱试验
oakmoss oil 橡苔油;栎藓油
oakmoss resin 栎藓树脂
oak panel 栎木镶板
oak parquet(ry) 栎木拼花地板;栎木拼花地板
oak picket 栎木栅栏桩;橡木篱桩
oak plywood 栎木胶合板
oak post 栎木桩;橡木桩
oak shingle 栎木(外)墙面板;栎木屋顶板;橡木屋顶板;栎木瓦
oak sill 橡木窗台板;栎木窗台板
oak slat fence 栎木板条栅栏;橡木条篱笆;橡木条栏栅
oak sleeper 橡木枕
oak spar 栎木杠子
oak stake fence 栎木桩柱栅栏;橡木栅柱
oak strip floor cover(ing) 栎木条板地板铺面层
oak threshold 栎木门槛;橡木槛
oak trim panel 栎木边框板
oakum 油麻丝;敛缝油麻絮;麻絮(丝);麻刀;填(缝麻)絮
oakum gasket joints for pipes 管道接缝—麻丝垫圈接缝
oak varnish 栎棕色凡立水;橡木清漆
o-aminoazotoluene 邻氨基偶氮甲苯
o-aminobenzoic acid 氨茴酸
o-anisidine 邻茴香胺
oar 橹;桨状物;桨(船)
oarage 桨设备
oar blade 桨叶
oar block 桨叉
oar handle 桨柄
oar hole 桨门
oarlock 桨门;桨叉
oarlock block 桨插垫木
oar loom 桨柄
oar palm 桨的扁平部分
oar propelled lifeboat 划桨救生艇
oar propeller 机械桨
oarsman 桨手;划桨人
oarsmanship 划艇术;划船术
oar's number 桨位
oar value 票面价值
oasis 绿洲;键岛;火星暗斑;沃洲;水草田;沙漠绿洲
oasis farming 绿洲农业
oast 烘麦芽窑;烘房;烘炉
oasthouse 烘麦芽窑;烘房;烘炉
oath of arrival 到港誓言
oath of entry 货物进口誓言
oatmeal 燕麦粉(一种止水材料)
oatmeal paper 麦片纸
oat-propelled lifeboat 划桨救生艇
Obach cell 奥巴赫电池
Obach process 奥巴赫法
obcordate 倒心形的
obducted sheet 仰冲岩席
obduction 仰冲作用
obduction orogen belt 仰冲造山带
obduction zone 仰冲带
obeche 奥瑟契树;非洲轻木
obelion 顶孔矢状缝交点
obelisk 方尖石柱;尖柱;方尖形;方尖塔;方尖碑
obelize 加剑号
obelus 剑号
o-benzyl-p-chlorophenol 邻苄基对氯酚
Obermayer's reagent 奥伯迈耶试剂

Ober's test 奥伯氏试验
obervation ward 观察病室
oberwind 奥伯风
obesity 多脂
obex 闩
obiter 附带
obiter dictum 附言;附带意见
object 客体;目的;目标;物体;物件;对象;反对
object abeam 物标正横
object angle 物角
object base 目标程序库
object beach 物光束
object beam 物体光束
object blindness 物体盲
object brightness coefficient 景物亮度系数
object carrier 载物体;载物玻璃(显微镜)
object-centered representation 对象中心表示法
object character 景物特征
object classification 对象分类法
object code 结果代码;目标(代)码
object code compatibility 目标(代)码兼容性
object code field 目标码符号组
object-code header 特征码索引
object code listing 目标代码表
object code module 结果代码模块;目标代码模块
object code program(me) 目标代码程序
object colo(u)r 物体色
object command 结果指令
object computer 执行(用)计算机;目标(程序)计算机
object configuration 目标程序配置;目标程序结构
object contrast 景物反差
object coordinates 地物坐标
object cost 目标成本
object deck 目的卡片组;目标叠
object deck for a procedure segment 过程段的目标卡片叠
object defining aperture 物体限制孔径
object definition 目标定义;对象定义
object depth 物深
object design 目标设计
object detection 目标检测
object distance 物距
object fidelity 目标逼真度
object file 目标文件
object focal point 物焦点;物方焦点
object focus 物方焦点
object function 原函数;目标函数
object glass 接物镜;物镜
object glass collar 物镜环
object height 物体高度;地物高度
object identification 目标识别
object illumination 目标照度
object-image coincidence method 物像重合法
object-image function 目标像函数
object-image transform 物像变换
objecting measurement 客观量度
object instance 目标示例
objection 拒绝;反对
objectionable 不许可的
objectionable character 名声不佳
objectionable coating 故障形结皮
objectionable constituent 有害成分
objectionable feature 缺点
objectionable flicker 有害闪烁
objectionable (flow of) current 不合适电流
objectionable impurity 有害杂质
objectionable literature 违禁文献
objectionable name 有争议地名
objectionable odo(u)r 令人不愉快的气味
objectionable oil oxidation products 油氧化后生成的有害物
objectionable oxidation state 不适宜的氧化态
objective 经营目标;目的;目标
objective analysis 客观分析
objective and fair 客观公正
objective-and-task method 目标任务法
objective angle of image field 像场角

objective angular field 物镜视角范围;物镜视场
objective aperture 物镜孔径
objective appraisal 客观的评价
objective astigmatic corrector 物镜像散校正器
objective attribute 客观属性
objective balance 保留余额
objective braking 目的制动【铁】
objective cap 物镜盖
objective carrier 物镜架
objective cause 客观原因
objective changer 物镜转换盘
objective collar 物镜环
objective condition 客观条件
objective data 客观数据
objective distance 目标距离;物距
objective end 接物端
objective environment 客观环境
objective evidence 客观证据;客观凭证
objective exchange value 客观交换价值
objective fact 客观事实
objective factor 客观因素
objective failure probability 客观破坏概率
objective fidelity criterion 客观保真度标准
objective field of view 物镜视野;物镜视场
objective forecast(ing) 客观预报
objective function 目标函数
objective function equation 目标函数方程
objective glass aperture 物镜孔径
objective glass collar 物镜环
objective gloss 正反射光泽;镜面光泽
objective grating 物镜光栅;物端光栅
objective holder 物镜座;物镜框
objective indicator 客观指标
objective judgment 客观判断
objective law 客观规律
objective lens 物镜透镜;物镜
objective lens adjustment 物镜调整
objective lens magnification 物镜放大率
objective lens of photography 摄影物镜
objectively correct preference 客观地修正偏好
objective magnification 物镜放大
objective management 目标管理
objective map 进攻态势图
objective mount 物镜框
objective noise meter 客观声级计;客观噪声计;绝对噪声测定表
objective of environment(al) policy 环境政策目标
objective of pollution control 污染控制目标
objective photometer 客观光度计
objective planning 目标计划
objective prism 物镜棱镜;物端棱镜
objective prism spectrum 物端棱镜光谱
objective probability 客观概率
objective probability distribution 客观概率分布;客观(性)概率分配;客观事物的概率分布
objective program(me) 目标程序
objective program(me) instruction 结果程序指令;目标程序指令
objective program(me) library 目标程序库
objective range 目标距离
objective ranking 仪器评定
objective rationality 客观的合理性
objective reading 客观读数(法)
objective reality 客观现实
objective sample 客观样品
objective set 目标体系
objective shutter 物镜快门
objective sign 客观征象
objective slide 物镜筒
objective standard 客观标准
objective statement 用途费用报表
objective symptom 客观症状
objective system of environment(al) decision 环境决策目标体系
objective table 载物台
objective tax 客体税
objective test 客观测验
objective things 客观事物

objective tube 物镜筒
objective use value 客观使用价值
objective utility 客观效用
objective value 客观价值
objective value in use 客观使用价值
objective value theory 客观价值学说
objective variable 客观变数;客观变量
objective view 配景
objective with variable focal length 可变焦距物镜;变焦距物镜
objective zone 目标区
objectivism 客观性
objectivity 客观性
objectivity principle 客观性原则
objectivization 客观化;实体化
object language 目标语言;对象语言
object lens 物镜
object lesson 实际教训;具体例子;经验教训;实物教学
object library 目标程序库
object line 可见轮廓线;外围线;地形线;外形线;轮廓线
object list 结果表;对象表
object locating system 目标定位系统
object machine 目标计算机
object map 目标映像;目标图
object marking 固定物标示;立面标记
object mathematical model 对象数字模型
object metamerism 物体条件配色
object micrometer 物镜测微计
object module 结果模块;目的程序模块;目标模块
object module card 目标模块卡片
object module data set 目标模块数据集
object module library 目标(模块)程序库
object of assessment 课征对象
object of civil legal relationship 民事法律关系的客体
object of civil right 民事权利客体
object of entry 记入项目标;记入目的
object of expenditures 支出用途
object of expenditures budget 按支出用途编制的预算
object of invention 发明项目;发明的项目
object of observation 观察对象
object of private right 民事权利客体;民事权利的客体
object of regulation of economic law 经济法调整对象
object of taxation 征税对象;课税品;课税对象;纳税对象;赋税客体
object-oriented 面向目标的;面向对象的
object-oriented approach 面向目标的方法
object-oriented construction 面向对象的结构
object oriented programming 目标定向程序设计;面向对象程序设计
object pack 目标穿孔卡片叠;目标程序穿孔卡片叠
object phase 目标阶段;目标程序执行阶段
object plane 物(平)面
object plane scanner 物平面扫描仪
object planning 目标规划
object plate 检镜片
object point 物点
object program(me) 目标程序
object program(me) address 目标程序地址
object program(me) code generation 目标程序码的生成
object program(me) development 目标程序开发;目标程序编制
object program(me) language 目标程序语言
object program(me) library 目的程序库;目标程序库
object program(me) list 目标程序表
object program(me) module 目标程序模块
object program(me) optimization 目的程序优化;目标程序优化
object program(me) preparation 目标程序准备
object program(me) size 目标程序长度
object record 结果记录
object reflection 地物反射
object relationship 对象关系
object response 对象特性
object routine 结果程序;目的程序;目标程序
object run 目标操作
object selection 物体选择

object set 目标集(合)
object slide 物镜筒;物镜滑筒
object snap 目标捕捉
object space 物空间;物方物界;物方
object space coordinate system 物方空间坐标系
object spectrum characteristic 地物波谱特性
object staff 准尺【测】;花杆【测】;测量准尺;标杆
object stage 载物台架;载物台
object table 目标表
object tape 目标带
object tape format 目标带格式
object teaching 直观教学;实物教学
object time 目的程序时间;目标程序执行时间;目标程序时间
object vane 目标照准器
object variable 目标变量
object wave 物体光波;物体波
object weight of cell 单位对象权
oblate 扁球形的;扁圆(形的);扁球状的;扁平状;扁的
oblate ellipsoid 扁型椭球体;扁椭圆(体);扁椭球;扁球形;扁球体;扁球面
oblate ellipsoid head contour 扁椭球封头曲面
oblate ellipsoid of rotation 旋转扁椭球
oblateness 扁球形;扁心率;扁平;扁率
oblateness of the earth 地心扁率;地球扁率
oblate ring 一种铜制扁圆形挠性活塞环
oblate sphere 扁球形;扁球面
oblate spheroid 扁球形;扁球体;扁球(面)
oblate spheroidal 扁圆球形的
oblate spheroidal coordinates 扁球面坐标
obligate 专性的;负有责任
obligate aerobes 专性需氧微生物;专性需氧菌;专性好氧菌;专性好氧微生物
obligate aerobic 专性需氧的
obligate aerobic bacteria 专性好气细菌
obligate anaerobe 专性厌氧细菌;专性厌氧微生物
obligate anaerobic 专性厌氧;专性嫌气细菌
obligate anaerobic bacteria 专性厌气细菌;专性厌气细菌;专性嫌气细菌
obligate autotroph 专性自养生物
obligate autotrophy 专性自养
obligate chemoautotroph 专性化能自养生物;化能自养生物
obligated balance 保留余额
obligated parking lot 规定附设停车场
obligate fund to 拨款给……
obligate halophyte 专性盐生植物
obligate heterotroph 专性异养生物
obligate parasite 专性寄生物
obligate photoautotroph 专性光能自养生物
obligate saprophyte 专性腐生物
obligate symbiont 专性共生物
obligate symbiosis 专性共生
obligate thermophilic bacteria 专性嗜热菌
obligate thermophilic organism 专气高温微生物
obligation 债务;证书;契约;义务;责任
obligation assumed 负担的义务
obligation authority 专款授权;使用专款授权
obligation incurred 承担债务;承付款项;已发生债务;承担款项
obligation of appropriation 拨款的规定用途
obligation of client 业主义务
obligation of compensation for losses 赔偿损失的义务
obligation of giving claim notice 通知损失的义务
obligation of organizations to provide housing 住房分配
obligation outstanding 待支保留数
obligation principle 责任原则
obligation sale 销售义务
obligations under capital leases 资本租赁债务
obligation to compensation 赔偿的义务
obligation to pay tax 纳税义务
obligative 专性的
obligative heterotrophic bacteria 专性异养细菌
obligator anaerobic bacteria 偏性嫌氧性细菌
obligatory 强制性的
obligatory annual contribution 每年度必须交纳的会费
obligatory arbitration 强制仲裁
obligatory expenditures 必支费
obligatory parasite 专性寄生物
obligatory point 约束点

obligatory presentation copy 纳本
obligatory right 债权
obligatory rule 强制性规则
obligatory term 约束性条款;义务年限;强制性条款
oblige 债权人
obliged vessel 义务船;救援船;让路船;避让船
obligee 债主;债权人;权利人
obliger 负有义务的人
obligor 欠债者;债务人;负债人
oblique 斜的;倾斜的;偏斜的
oblique aberration 斜像差
oblique aerial photograph 倾斜空中照片
oblique aerial photography 倾斜航空摄影
oblique air photograph 倾斜摄影航片;倾斜航空照片
oblique air photograph strip 倾斜航空照片带
oblique air survey 倾斜航摄测量
oblique angle 斜角;偏斜角
oblique-angled 斜角的
oblique-angled axis 斜角轴
oblique-angled bond 斜角砌合
oblique-angled cocking 斜角翘起;斜角竖起
oblique-angled cracking 斜角裂缝;斜角裂开
oblique-angled end cocking 斜角端部翘起
oblique-angled end cogging 斜角端部榫接
oblique-angled halved joint with butt ends 粗端部的斜角对搭接
oblique-angled joint 斜接;斜角接头
oblique-angled junction 斜角汇合;斜角连接
oblique-angled masonry pattern 斜角圬工模样;斜角圬工形式
oblique-angled pipe junction 斜角连接管子
oblique-angled projection 斜角投影
oblique-angled rib 斜肋
oblique-angled saddle junction piece 斜角咬口连接件;斜角鞍形连接件
oblique-angled scarf (joint) 斜接榫;斜嵌接
oblique-angled scarf (joint) with wedge 楔形斜嵌接
oblique-angled strength 斜角强度
oblique-angled triangle 斜三角形
oblique angle earth pushing process 斜角推土法
obliqueangle intersection 钝交叉;钝角交叉(口)
oblique angle of flow 流向偏角
oblique arc 斜弧
oblique arch 斜交拱;斜拱
oblique arc measurement 斜弧度测量
oblique ascension 斜交升交点赤经
oblique asteroid 斜星形线
oblique astigmation 斜像散
oblique astigmatism 斜像散
oblique axis 斜轴(线);斜交轴;斜角坐标系
oblique axis mount 斜轴座架
oblique axonometric projection 斜轴测投影
oblique axonometry 斜轴测射影法;轴测斜投影
oblique bearing pile 斜向受力桩
oblique bedding 斜层理;斜层
oblique bed process 斜推法
oblique bending 斜向受弯;偏心弯曲;不对称弯曲
oblique berthing 斜向驶靠;斜向靠泊;斜驶靠泊
oblique bond 人字形黏合;钝角黏合;斜角黏合;斜纹砌块
oblique boom 斜挡栅
oblique bridge 斜桥;斜交桥
oblique brush 斜柄刷
oblique butt joint 斜平头接合;斜对接
oblique butt weld 斜对焊
oblique camera 倾斜照相机
oblique care 斜轴
oblique chart 斜轴图
oblique circuit 交叉回路
oblique circular cone 斜圆锥(体)
oblique cleat 斜割角
oblique cleft grafting 斜劈接
oblique coast 斜向海岸
oblique coastline 斜向海岸线
oblique cocking 斜向翘起的
oblique collision 斜向碰撞
oblique column 斜轴柱
oblique compartment 倾斜车厢
oblique compression 斜压缩
oblique compression joint 斜接头
oblique cone 斜锥;斜圆锥(体)
oblique convergence 斜收敛角

oblique coordinates	斜(角)坐标;斜交坐标
oblique coordinate system	斜(角)坐标系
oblique cord	斜索
oblique cordon	斜生单干形
oblique course	斜向
oblique coverage	有倾斜摄影资料地区
oblique cracking	斜裂缝
oblique crank	斜曲柄
oblique crank web	斜曲柄
oblique cross	斜口四通
oblique crossing	斜形交叉;斜交叉道(路);斜交(叉)
oblique cut	斜切
oblique cut grafting	片搭接
oblique cutting	斜刃切削;斜插;斜切
oblique cutting edge	斜切削刃
oblique cylinder	斜柱;斜置汽缸;斜圆柱体;斜(面)圆筒
oblique cylindric(al) orthomorphic chart	斜轴圆柱正形投影地图;斜圆柱正形投影图
oblique cylindric(al) orthomorphic projection	斜圆柱正形投影;斜切渐长海图
oblique dark ground illumination	倾斜暗场照明
oblique derivative	斜微商
oblique dislocation	斜向错动
oblique displacement	斜位移
oblique distance	斜距
oblique dovetail	斜鸠尾
oblique drawing	斜视图
oblique eccentric load(ing)	斜偏心荷载
oblique embedding	倾斜嵌入
oblique equation	倾斜时差
oblique equator	斜轴投影赤道;斜切投影赤道;倾斜赤道
oblique error	倾斜误差
oblique evapo(u)ration	斜向蒸发;斜向蒸镀
oblique expansion wave	斜向扩散波
oblique extension	斜向拉伸
oblique facial cleft	面斜裂
oblique fault	斜断层
oblique fillet	斜填角;斜角焊缝
oblique fillet weld	斜角焊接;斜交角焊缝;斜贴(角)焊缝;斜贴(角)焊接
oblique fillet welding	斜角焊接缝
oblique fire	斜射
oblique flow	斜向水流;斜流
oblique fold	斜褶皱
oblique force	斜(向)力
oblique fracture	斜断
oblique girdle	斜环带
oblique gnomonic projection	斜切心射投影
oblique grain	斜纹理;斜木纹
oblique graticule	斜轴投影经纬网
oblique grooving iron	槽刨包铁
oblique gutter	斜向排水沟
oblique halved joint with butt ends	粗端斜对搭接
oblique halving	斜嵌;斜对接
oblique height finder	斜高测量仪;倾斜测高仪
oblique helicoid	斜螺旋面
oblique hill shading	斜照晕
oblique hobbing	斜向滚齿
oblique hole	斜钻孔;斜炮眼;斜炮孔;斜炮洞
oblique hydraulic jump	斜水跃
oblique hydromagnetic shock wave	斜磁流体冲击波
oblique illumination	斜照法;斜射照明;倾斜照明
oblique illumination method	斜照法
oblique incidence	斜投射;斜入射;倾斜入射
oblique incidence coating	斜向涂敷
oblique incidence method	斜射法
oblique incldence reflectivity	斜入射反射率
oblique incidence sounding	斜向探测;斜投射探测;变频脉冲斜向探测
oblique incidence transmission	斜入射传输;斜入射传播
oblique incident absorption coefficient	斜入射吸收系数
oblique inflow	上升斜流
oblique intersection	斜交交叉口
oblique ionogram	斜向探测电离图
oblique joint	斜折合接;斜(向)对接;斜接头;斜接合
oblique key	斜键
oblique lamination	斜交纹理
oblique landform	倾斜地形
oblique lattice	斜点阵
oblique letter	斜体字
oblique light	斜照光
oblique lighting	斜侧照明
oblique line	斜线(道);倾斜线
oblique line overlap	直线式迭接的倾斜航空照片
oblique load(ing)	斜荷载;偏压
obliquely approaching target	迂回接近的目标
obliquely layered medium	斜层介质
obliquely slotted motor	斜槽式电动机
obliquely slotted pole	斜槽磁极
obliquely transverse dredging method	斜向横挖法【疏】
obliquely zygomorphic	斜向对称的
oblique magnetization	斜交磁化;斜向磁化
oblique march	斜行进
oblique masonry bond	斜纹砌块
oblique Mercator	斜轴墨卡托
oblique Mercator chart	斜圆柱正形投影图
oblique Mercator projection	斜轴墨卡托投影;斜圆柱正形投影;斜切渐长海图;斜坡墨卡托投影
oblique Mercator projection chart	斜轴墨卡托投影地图
oblique meridian	斜轴投影纬线;斜轴投影经线
oblique name	斜体字注记
obliqueness	倾度;歪斜
oblique nosing	斜的突缘
oblique notch	斜槽口
oblique notching	开斜槽
oblique nozzle	斜交接管
oblique observation	倾斜观察
oblique offset	斜支距
oblique orbit	斜轨道
oblique palmette	斜生多干形
oblique parallel	斜轴投影纬线;斜纬线
oblique parallelepiped	斜六面体
oblique pencil	斜锥;斜光锥
oblique perspective	斜透视
oblique photogrammetry	倾斜摄影测量
oblique photograph	倾摄照片;倾斜照片
oblique photography	倾斜摄影;倾斜航空照相
oblique pianoforte	斜式钢琴
oblique pile	斜桩
oblique pile driver	斜桩打桩机
oblique pillow-block bearing	斜支座(止推)轴承
oblique plane	斜面;任意平面;斜画面
oblique plane triangle	斜平面三角形
oblique plotter	倾斜测图仪
oblique plotting instrument	倾斜测图仪
oblique plotting machine	倾斜绘图机;倾斜测图仪
oblique plummerblock bearing	斜座止推轴承
oblique polarization	斜极化
oblique pole	斜轴投影极点
oblique position	斜位;倾斜位置
oblique prism	斜棱柱;斜角棱镜
oblique profile	斜剖面
oblique progradational mode	斜交前积模式
oblique projection	斜(轴)投影;斜投影
oblique projection chart	斜轴投影地图
oblique projection graticule	斜轴投影格网
oblique projection latitude	斜轴投影纬度
oblique projection longitude	斜轴投影经度
oblique projection pole	斜投影极
oblique projector	倾斜投影器
oblique propagation	倾斜传播
oblique pull	斜拔;斜拖;斜拉
oblique pyramid	斜棱锥(体)
oblique radio transmission	无线电波倾斜传播
oblique rake	倾斜式搂草机
oblique ray	斜射线
oblique reel-side delivery rake	倾斜轮式侧向搂草机
oblique reflected lighting	斜反射照明法
oblique reflection	斜反射
oblique rhumb line	斜轴投影无变形线;倾斜投影恒向线
oblique rib	斜肋
oblique rolling	斜轧
oblique rotation factor solution	斜旋转因子解
oblique saddle junction piece	鞋形连接件
oblique scale	倾斜(比例)尺
oblique scan	倾斜扫描
oblique scarf	斜角嵌接
oblique scarfing	斜嵌;斜铺毛地板
oblique scarf joint	斜嵌接(头)
oblique scarf with wedge	楔形斜嵌;楔形斜接缝;楔式斜嵌(接)
oblique sea	斜浪
oblique section	斜截面;斜断面;倾截面
oblique section view	斜剖视图
oblique segmentation	斜裂
oblique seismic reflection configuration	斜交地震反射结构
oblique serif	斜衬线
oblique setting	错箱
oblique shear(ing)	斜向剪切(作用)
oblique shock	斜震震;斜激波
oblique shock front	斜交激震波前;斜交激震波面
oblique shock wave	斜冲击波;斜激波
oblique side grafting	斜侧接
oblique sketch master	倾斜草稿底图
oblique-slip fault	斜滑断层;倾斜滑移断层;斜向滑动断层
oblique slot	斜槽
oblique sphere	斜天球;斜球体;斜交球
oblique spheric(al) triangle	斜球三角形;斜角球面三角形;任意球面三角形
oblique step	倾斜梯级
oblique stereographic(al) projection	斜球面投影
oblique strength	斜向强度
oblique stress	斜(向)应力
oblique stroke	斜线(道)
oblique strophoid	斜环索线
oblique subduction	斜插俯冲
oblique superposed folds	斜跨叠加褶皱
oblique system	斜晶系;斜角系
oblique tabled scarf	斜叠嵌接
oblique tabled scarf with key	斜交镶嵌接;斜叠键嵌接
oblique take-off	斜行航迹起飞
oblique tangent cylindric(al) conformal projection	等角斜轴切圆柱投影
oblique tenon	斜榫;倾榫
oblique T joint	斜接 T 形接头
oblique to grain	斜纹
oblique tooth	斜齿
oblique tooth gear	斜齿轮
oblique towing test	斜拖试验【船】
oblique trajectories	斜交轨迹
oblique translation	斜向直移
oblique triangle	任意三角形
oblique valve	倾斜活门
oblique vessel	让路船
oblique view	斜视图
oblique viewing	倾斜观察
oblique visibility	斜能见度;倾斜能见度;倾斜可见距离
oblique visual range	倾斜可见距离
oblique wave	斜(向)波
oblique weft	斜纬
oblique weir	斜堰
oblique wide-angle multiplex projector	广角倾斜投影测图仪
oblique wind	斜风
oblique window	斜窗
oblique windshield	倾斜挡风玻璃
oblique wing wall	斜翼墙
obliquity	斜度;倾斜(度);倾度
obliquity effect	倾斜效应
obliquity factor	倾斜因子;倾斜因素;倾斜因数
obliquity of action	作用斜度
obliquity of armo(u)r plate	装甲板倾斜度
obliquity of ecliptic	黄赤交角
obliquity of the ecliptic	黄道斜度
obliterate	清除
obliterated corner	湮没角点
obliterated data	擦除数据
obliterating oscillator	消音振荡器
obliterating power	消除功能;消声功能;消声本领;湮没力;涂抹力
obliteration	清除;磨损;涂除;消失作用;灭迹
obliteration of lakes	湖泊消失
obliterator	涂抹器
oblong	矩形;椭圆形;长方形(的)
oblong aperture	椭圆孔隙;椭圆孔口;扁长孔隙;扁长孔口
oblong dome	椭圆形穹顶;长椭圆形穹顶
oblong exit slit	长方形出射狭缝
oblong-hole screen	长孔筛;椭圆孔筛

oblong mesh 长方形筛孔;长方形网眼;长方孔网
oblong-mesh reinforcement 长方形网格钢筋
oblong ovate 长椭圆形
oblong pin anchor shackle 扁销卸扣
oblong-punched plate 长方眼钢板;长方孔板
oblong punching 冲长方(筛)孔
oblong shape 长方形
oblong shaped 长方形的
oblong size 横向像幅
oblong slice slate pencil 长方片滑石笔
oblong slot 长圆孔
oblong spike 长方形的穗
oblong-type machine 正流槽型长方形浮选机
obnoxious growth 有害生长
obnoxious species 有害种
oboe 双簧管;阿波耶系统
obovoid 倒卵形的
oboyerite 水碲氢铅石
obpyramidal 倒金字塔形的
obpyriform 倒梨形的
o-bromoaniline 邻溴代苯胺
obround design 长圆形设计
obround vessel 长圆形容器
obscission 脱落现象
obscission zone 脱落区
obscuration 阴暗;遮光效果;模糊
obscuration band 装饰边
obscuration of smoke 烟雾的不透光性
obscure 模糊(的);不明显的
obscure camera 暗箱
obscured glass 压花玻璃;闷光玻璃;毛(面)玻璃;乳色玻璃;不透明玻璃
obscure displayed 显示隐瞒
obscured light 遮蔽灯
obscured number 无名数
obscured sector 遮蔽扇形
obscured variable 屏蔽变星
obscure glass 防视玻璃
obscure ray 模糊射线;不可见光线;暗线
obscure wire glass 暗嵌丝网玻璃
obscuring 酸蚀
obscuring device 挡视线装置;防视装置
obscuring window 不透明玻璃;闷光玻璃窗
obscurity board operating 暗屏运行
obscurometer 模糊仪(测水泥细度)
obsequent fault-line scarp 逆断线崖
obsequent flow 逆向流动
obsequent landform 逆地貌
obsequent river 逆向河(流);反向河
obsequent slope 逆向坡
obsequent stream 逆向河(流);逆向(小)河
obsequent valley 逆向谷
observability 可观察性;可观测性
observability condition 可观测性条件
observability index 可观测性指标
observable 可探测;可观察量;可察觉的
observable condition 可见条件
observable controls 可观察控制
observable error term 可观测的误差项
observable quantity 可观测的量
observable temperature rise 可测温升
observable variable 可观测变量
observance of good seamanship 优良船艺惯例
observant party 守约方;守约的一方
observation 注视;观测;观察;监视;测天
observation accuracy 观测精度
observation accuracy of base station network 基点网观测精度
observation accuracy of gravimeter 重力仪观测精度
observation accuracy of ordinary station 普通点观测精度
observation adjustment 观测平差
observation aeroplane 观测机
observational astronomy 观测天文学
observational borehole 观测钻孔
observational buoy system 观测用浮标系统
observational catalogue 观测星表
observational check 外部检查;外观检查
observational data 观察数据;观测资料;观测数据;观测资料数据;实测数据
observational day 观测日
observational dispersion 观测弥散度
observational equation 观测方程

observational error 观测误差;观察误差
observational grid 观测网格
observational material 观测资料
observational measurement 定性测量
observational need 观测需要
observational network 观测网
observational place 观测站
observational point 观测站
observational post 观测站
observational procedure 观测步骤
observational record(ing) 观测记录
observational selection 观测选择
observational station 观测站
observational synthesis 综合观测结果;观测的综合工作
observation and calculation book 观测及计算手簿
observation and foecast of movement of groundwater and salt 地下水盐分输移观测与预测
observation angle 观测角
observation aperture 观察孔径
observation apparatus 观测仪器
observation at fixed hours 定时观测
observation at interval 异时观测
observation balloon 观测气球
observation base 观测基线
observation board 观测板;监察台
observation boat 观察船;观测船;测量船
observation book 观测手簿
observation borehole 观测孔
observation by incident light 用入射光观测
observation by reflected light 用反射光观测
observation by transmitted light 用透视光观测
observation car 观察车;游览车厢
observation check 外形检查
observation circuit 监听电路;监视电路;监测电路
observation coach 瞭望车
observation component 观测分量
observation condition 观测条件
observation data 实测数据
observation deck 观测台;观测甲板
observation depth 观测深度
observation desk 观测台;试验台
observation device 观测装置;观测设备
observation differential 观测误差
observation distance 通视距离
observation door 窥视孔;看火门;观察孔;观察门
observation earth system 观测地球系统
observation efficiency 观测效率
observation elevator 观光电梯
observation equation 观察方程;误差方程(式)
observation eyepiece 观察目镜
observation field 露天观测场地
observation floor 观测楼
observation forms 记录表格
observation frequency 观测次数;观测频率;观测频度
observation gallery 观察平台;观察眺台;观测廊道
observation gondola 景观缆车
observation grid 观察点网络
observation group 考察团
observation hole 观察孔;看火孔;观测钻孔
observation hole of dynamic(al) behavio(u)r 动态观测孔
observation index 观测指数
observation in groups 分组观测
observation instrument 观测仪器
observation item 观察项目;观测项目
observation items on testing field 试验场观测项目
observation line 观察线;观测线
observation line for regime 动态观测线
observation line number 观测线号
observation line number for test 试验观测线条数
observation line of pumping test 抽水试验观测线
observation measurement 观察;观测
observation method by two instrument heights 两次仪器高法
observation method of subfabric and dislocation 亚构造和位错观测方法
observation mine 视发水雷
observation-minus-analysis 观测资料不足分析
observation mirror 观察镜
observation monument 观测点;观测标志;观测标石;测站标石
observation net for regime 动态观测网

observation network 站网;测站网
observation of amplitude 出没方位幅的观测
observation of bank caving 岸坡崩塌观测
observation of bank ruin of reservoir 库岸再造观测
observation of depth ga(u)ge 观察深度表
observation of flow pattern 流态观测
observation of foundation stress 地基应力观测
observation of movement of mud-rock flow 泥石流动态观测
observation of movement of wind-drift sand 风沙流动观测
observation of navigation obstruction 航行障碍物观测
observation of reversing point of the plummet 垂球回动点观测
observation of saturation line in soil dam 土坝浸润线观测
observation of sea 海洋观测
observation of seashore sediment transport 海岸输沙观测
observation of settlements 沉陷观测;小住宅区的观测
observation of site 现场视察
observation of slope stability 边坡稳定性监测;边坡稳定性观测
observation of stage 水位观测
observation of synchronous tide stage 同步潮位观测
observation of the work 视察工作;现场视察;现场检验;工程检查;工程监视
observation of water pressure 孔隙水压力观测
observation panel 观察窗;门扇亮子
observation parallax 观察视差;观测视差
observation period of recovered water level 恢复水位观测时间
observation pillar 观测墩
observation pipe 观察管;观测管
observation place 观察位置;观测地点
observation plate 观测板
observation platform 观察台;观测台;观测平台;观测架
observation point 观察点;观测所;观测点;测点
observation point for regime 动态观测点
observation point number 观察点号;观测点号
observation point number for test 试验观测点数
observation point of artificial outcrop 人工露头观测点
observation point of attitude element 产状要素观测点
observation point of cleavage 劈理观测点
observation point of drill hole 钻孔位观测点
observation point of geomorphology 地貌观测点
observation point of hydrogeology 水文地质观测点
observation point of joint 节理观测点
observation point of karst shaft 岩溶竖井观测点
observation point of lineation 线理观测点
observation point of mine 矿坑调查点;矿井观测点
observation point of natural outcrop 天然露头观测点
observation point of Quaternary 第四纪地质观测点
observation point of schistosity 片理观测点
observation point of spring 泉水调查点
observation point of spring water 泉水观测点
observation point of subterranean river outlet 暗河出口观测点
observation point of surface water 地表水观测点
observation point of waste water 污水观测点
observation point of well 水井调查点
observation port 观察孔;观测孔;看火孔
observation post 观测哨;观测墩;观测点
observation procedure 观测方法;观测程序
observation profile line 观测剖面线
observation railcar 轨道检测车;瞭望轨道车
observation range 通视距离
observation reading 实测读数
observation record for test 试验观测记录
observation reduction 实验结果处理
observation report 观测报告
observation result 观测结果
observation room 观测室
observation route 观测线

observation route of geologic(al) mapping 地质填图观察路线
observation satellite 观察卫星;观测卫星
observation seaplane 海上观测飞机
observation seismogram 观测地震图
observation series 观测系列
observation set 测回【测】
observation ship 勘探船
observation spot 观测点;观测场
observation station 观察位置;观察站;气象台;观测站
observation station for regime 动态观测站
observation station of rock stratum movement 岩层移动观测站
observation stor(e)y 观测楼层
observation submersible 观察型潜水器
observation system 观测系统;监测系统
observation system simulation experiment 观测系统模拟试验
observation table 观察台
observation tank 观察柜;观察舱
observation target 测量觇标
observation telescope 观察望远镜;观测望远镜
observation tower 观测塔;瞭望塔
observation tube 观察管
observation tube fixing screw 抽筒固定螺旋
observation value 观测值
observation variable 观测变量
observation ward 观察室
observation watch 观测表
observation well 验潮井;观察孔;观测井;观测井
observation well design 观测井设计
observation well for soil moisture regime 土壤湿度动态观测井
observation well number for test 试验观测孔数
observation well of groundwater 地下水观测井
observation window 观察窗;观测窗
observation zone 观察区;观测区
observatory 瞭望台;观象台;观测站;观测台;观测所;气象台;天文台
observatory dome 天文台圆顶
observatory post 观测所
observatory-remote 遥测台
observe 遵守;观察
observed altitude 观测高度
observed altitude of lower limb (天体)下边观测高度
observed altitude of upper limb (天体)上边观测高度
observed anomaly 观测异常
observed azimuth 观测方位角
observed bearing 观测方位
observed boat 观测船
observed borehole 观测孔
observed clock time 观测钟面时
observed criterion 观测标准
observed cumulative failure probability 累积失效概率的观测值
observed current 实测流(量)
observed data 观测资料;观测数据;实验资料;实测资料;实测数据
observed depreciation 房地产使用价值的贬值;勘定折旧
observed depth 观测深度;实测深度
observed differential 观测误差
observed direction 已观测方向;观测方向;实测方向
observed discharge 实测流(量)
observed discharge pollutant loading 实测排放污染物负荷
observed distribution 观测分布
observed failure rate 观测故障率;失效率的观测值
observed fall 实测落差
observed flood 实测洪水
observed flood discharge 实测洪水流量
observed flow 实测流(量)
observed flow velocity 实测流速
observed frequency distribution 经验频率分布;实测频率分布
observed gravity value 观测重力值
observed instrument 观测仪器
observed items 观测项目;实测项目
observed latitude 观测纬度
observed life table 估定耐用年限表
observed line 观测线

observed longitude 观测经度
observed mean repair time 平均修复时间的观测值
observed mean value 观测均值
observed number 观察数
observed object 已观测目标;观测目标
observed parameter 观测参数
observed point 观测点;视准点
observed position 观测船位;实测位置
observed price 通行价格
observed profile 观测剖面(图);实测(纵)剖面;实测纵断面
observed quantity 观测量
observed ratio 观测比
observed ray 观测方向
observed reading 观察读数;观测值;观测读数;测量值
observed result 观察结果
observed seismology 观测地震学
observed series 实测系列
observed stage 观测水位;实测水位
observed standard 观测标准
observed storm pattern 实测暴雨类型
observed strength 实测强度
observed temperature 观测温度;实测温度
observed tidal height 实测潮位
observed tide height 实测潮位
observed topocentric quantity 站心观测量
observed trend in coefficients 系数的观察趋势
observed value 已观测值;观察值;观测值;实测值;测量值
observed water depth 实测水深
observed water-level 实测水位
observed wave height 实测波高
observed wind 实测风
observed window 观测窗
observed zenith distance 观测顶距
observe laws 遵守法律
observer 观察者;观测者;观测员;监场员;测者
observer bias 观察者偏差
observer member 观察员;国际标准组织的观察员
observer meridian 测站子午圈
observer's apparatus 观察用装置
observerscope 二人用内窥镜
observer screen 观察屏
observer's eye 观测透镜
observer's handbook 观察员手册;观测手册
observer's meridian 子午圈;观测者子午线;测者子午圈;测者子午圈
observe the proprieties 遵守礼节
observe the superficial venule 辨络脉
observing apparatus 观测仪表
observing cage 观测笼
observing curtain 观测幕
observing eyepiece 观测目镜
observing frequency 观测频率;观测频度
observing line 视准线
observing outline 观测纲要
observing pattern 观察模式
observing point 观测点
observing prism 观测棱镜
observing program(me) 观测程序
observing scaffold 观测架
observing ship station 观测船站
observing system 观察系统
observing tent 观测幕
observing theodolite 观测经纬仪
observing time 观察时间
observing tube 观测筒
observing variance 观测方差
obsidian 黑曜岩
obsidian dating 黑曜岩水化年代测定法
obsidian hydration dating 黑曜岩水化年代测定法
obsidian hydration dating method 黑曜岩水化测年法
obsidianite 熔融石;似曜石
obsidian porphyry 黑曜斑岩
obsidian ware 深色炻器
obsolescence 过时;陈旧;报废
obsolescence allowance 折旧
obsolescence and depreciation 陈旧与折旧
obsolescence charges 设备废弃费用
obsolescence clause 更新条款
obsolescence cost 废弃成本

obsolescence free 长时适用
obsolescence of assets 资产陈旧
obsolescent 需废弃的;逐渐要废弃的
obsolescent housing 过时住房
obsolescent property 陈旧资产
obsolescent road 废弃的道路
obsolescent road salvage 废弃路旧料
obsolete 过时的;退化的;废退的;废弃的
obsolete assets 陈旧资产
obsolete chart 旧海图
obsolete edition 旧版
obsolete equipment 过时设备;陈旧设备
obsolete grid 旧格网
obsolete housing 陈旧住房
obsolete information 已废信息;过时情报
obsolete map 过时地图;旧图
obsolete material 过时材料;报废材料
obsoleteness 陈旧
obsolete property 陈旧资产
obsolete road 废弃的道路
obsolete securities 已关门企业的股票
obsolete sheet 旧图
obsolete stock 陈旧存货
obsolete techniques 陈旧技术
obsolete wharf 陈旧码头
obsorbing complex of soil 土壤吸收性复合体
obstacle 障碍(物)
obstacle approximation 障碍模拟
obstacle avoidance 排除故障
obstacle avoidance laser radar 激光防撞雷达
obstacle avoidance laser system 激光防撞系统
obstacle avoidance sensor 防撞传感器
obstacle belt 障碍地带
obstacle clearance 障碍间隙
obstacle clearance box 障碍间隙框
obstacle clearance limit 超越障碍极限
obstacle construction 障碍设置
obstacle detection 障碍物探测;障碍探测
obstacle detouring 越障测量
obstacle diffraction 障碍绕射
obstacle diffraction loss 障碍绕射损耗
obstacle flag 障碍物标示旗
obstacle flow 障碍流
obstacle free zone 无障碍物地带
obstacle gain 障碍增益
obstacle-height to wheel-diameter ratio 障碍高度与车轮直径比
obstacle indicator 障碍指示器;故障指示器
obstacle light 障碍灯
obstacle limitation surface 净空限制面
obstacle marking 障碍标志;障碍标记
obstacle of vibration 振动危害
obstacle overcome ability 通过阻碍能力
obstacle performance 越障性能
obstacle plan 障碍设置计划
obstacle profile 障碍轮廓尺寸
obstacle quasi-static crossing 准静态越障
obstacle scour marking 障碍痕
obstacle shadow 障碍影
obstacle to drainage 排水障碍
obstacle to navigation 碍航物;碍航
obstacle to traffic 交通障碍物
obstacle to visibility 视界障碍
obstacle zone 障碍地带
obstetrical and gynecological hospital 妇产科医院
obstetrical department 产科
obstetrical room 助产室
obstetrical ward 产科病房
obstetric chair 产科椅;妇科用椅
obstetric delivery bed 产床
obstetric scoop 产勺
obstetric table 产床
obstinacy 顽固
obstruct 阻挡;阻碍;妨碍
obstructed 阻塞的
obstructed cavity 阻塞性空洞
obstructed channel 淤塞河道;淤塞河槽;阻塞河道;阻塞河槽;渠道堵塞
obstructed crest of spillway 闸检式区洪道堰顶;设闸溢洪道顶;设闸的溢洪道
obstructed spillway 闸控溢洪道;设闸溢洪道
obstructed valley 阻挡交通的山谷
obstructing sediment dam 拦渣坝
obstructing works 拦截工程

obstruction 阻塞;阻碍物;遮断;障碍物;障碍;挡水构筑物
obstruction approach marking 接近障碍物告示;接近障碍物标示
obstruction beacon 障碍物信标;障碍物标桩
obstruction buoy 障碍物浮标;障碍浮筒;沉航浮标;碍航物浮标
obstruction chart 障碍图
obstruction clearance 碍航物与船只之间的距离;安全通过净空;障碍物间隙
obstruction clearance criterion 净空条件
obstruction factor 阻塞因子
obstruction fading 障碍衰落
obstruction free flow area 有效流动截面;有效过水(断面)面积
obstruction ga(u)ge limit 障碍物限界
obstruction-guard 护栏;排障器
obstruction lake 不通航的湖;壅塞湖
obstruction lamp 障碍物标志灯
obstruction light 碍航标志灯;障碍物灯标;障碍物标志灯;障碍灯(标);航行阻障警告灯;碍航标灯
obstruction lighting 碍航物警告灯
obstruction mark(er) 碍航物标志;障碍物标示
obstruction marking 障碍物标志;障碍物标示
obstruction muffler 阻碍消声器
obstruction of track 线路阻塞;线路折断
obstruction of vision 视力的阻碍
obstruction period of navigation 碍航时间
obstruction period of traffic 阻车时间
obstruction restriction surface 限制障碍物的地面
obstruction signal 故障信号
obstruction spacing 障碍物间距
obstruction theory 障碍理论
obstruction to a homotopy 同伦的障碍
obstruction to navigation 阻航;航行障碍物;碍航
obstruction to traffic 交通障碍物
obstruction to vision 能见度变坏;视程障碍
obstruction warning 障碍报警
obstructive action 障碍作用
obstructive halogenation 阻抑卤化
obstructive scenery 障景;抑景
obstruct the traffic 阻碍交通
obstruct the view 阻碍视线
obsuring window 不透明玻璃窗
obtainable accuracy 可达准确度;可达精度;可达到的精度
obtain a high price 卖好价钱
obtain carbon from the air 从空气摄取碳
obtained mean 实得平均值;实得平均数
obtain employment 就业
obtain raw material locally 就地取材
obtain satisfaction of a claim 获得保证金退款
obtical tunnel filter 光学隧道滤光片
obturage 封严
obturating cup 密封碗
obturation 充засу
obturator 填充体;障塞器;紧塞具;密封装置;气密装置;填充体;充塞物;封闭器;闭塞器;闭孔器
obturator branch 闭孔支
obturator canal 闭膜管
obturator foramen 闭孔
obturator membrane 闭孔膜
obturator pad 紧塞垫
obturator piston ring 活塞密封环
obturator pouch 闭孔囊
obturator ring 紧塞圈;活塞环;闭塞环
obturator spindle 紧塞轴
obturator spindle spring 紧塞轴弹簧
obtuse 钝头的;钝角的
obtuse angle 钝角
obtuse angle arch 两内心挑尖拱;钝角拱
obtuse angle bevel gear 钝角伞齿轮
obtuse-angle bevel gearing 钝角锥齿轮传动
obtuse angle blade 钝角叶片
obtuse angle crossing 钝角辙叉
obtuse angle triangle 钝角三角形
obtuse angle wedge 钝角楔
obtuse arch 钝中心角拱;钝拱
obtuse bisectrix 钝角等分线
obtuse buoy 截锥形浮标
obtuse frog 钝角辙叉
obtuse quoin 钝形对角
obtuse quoin of (masonry) wall 墙的钝形转角石块
obtuse squint 钝角转角隅石
obtuse triangle 钝角三角形
obversion 转弯
obviate 排除(障碍、危险)
obvious fault 明显故障
obvious mark 明显标志
obvious target 明显物标
obvolvent 抱合
ocala 欧卡拉石灰石
occasional 临时(的);偶然的
occasional cause 偶然原因
occasional chair 备用椅
occasional extreme value 偶然极端值
occasional fault 偶然故障
occasional fixed time calls 临时定时通话
occasional flooding 临时洪水淹没
occasional fog signal 临时雾号
occasional income 偶然收入;非经常收入
occasionalism 偶因论
occasional light 临时光;临时灯标;偶用灯;随用灯
occasional line 不定期线
occasional lubrication 不定时润滑
occasional mo(u)lding pit 通气造型坑
occasional noise pulse 偶然噪声脉冲
occasional occurrence 非经常发生
occasional occurrence of earthquake 非经常性地震
occasional overdraft 临时透支
occasional regeneration 偶然再生
occasional service 临时业务;不定期航线
occasional storm 稀有暴雨;偶遇暴雨;偶尔暴雨;罕见暴雨
occasional survey 临时检验;临时检查
occasional survey of cargo gear 起货设备临时检验
occasional table 备用茶几;备用(小)桌
occidental topaz 黄水晶
occlude 封闭;遮断;夹杂
occluded air 封闭空气
occluded cyclone 锢囚气旋
occluded depression 锢囚低压
occluded foreign matter 夹附杂质
occluded foreign substance 夹杂物
occluded front 锢囚锋
occluded gas 吸留气体;夹入气体;包藏气
occluded oil 吸着油
occluded water 吸收水;吸留水
occluder 遮光板;光线限制器
occluding frame 咬合架
occluding paper 咬合纸
occluding relation 咬合关系
occlusal adjustment 咬合调整
occlusal balance 咬合平衡
occlusal correction 咬合调整
occlusal margin 咬合缘
occlusal surface 咬合面
occlusion 吸留物;咬合;掩蔽现象;夹杂物;停汽;闭锁;闭塞;包藏
occlusion compound 吸留化合物;包藏化合物
occlusion water 封存水;封闭水
occultation 掩源;遮掩;遮蔽
occultation angle 掩食角
occultation band 掩带
occultation curve 掩食曲线
occultation of satellites 卫星被掩
occultation time 掩食时刻
occulter 遮蔽体
occulting disk 掩蔽区(日冕仪);遮掩面
occulting light 隐显灯;明灭相间灯;明暗相间灯;明暗光;明暗灯(标);顿光;等明暗光
occulting quick flashing light 快闪明暗光;明相间闪光灯光;明暗急闪光;长联急闪光
occult light 歇光
occult mineral 隐蔽矿物
occult peak 隐蔽峰
occupable space 可占用空间
occupancy 占用
occupancy agreement 房屋租赁契约
occupancy change 调换房屋类型
occupancy classification or occupancy 房屋分类及其所起的作用
occupancy content 居住在一房屋内的最多人数
occupancy control 占有率控制
occupancy expenses 占用费用;被占用费用
occupancy factor 占有因数;人均住房面积数
occupancy indication 占线表示
occupancy indicator "有人使用"显示牌;有人无人牌
occupancy limitations 居住单元人数限制
occupancy load 占用荷载;使用荷载;使用负载
occupancy permit 使用执照;占用许可证;居住证;使用许可税
occupancy per person in dwelling 居住区密度
occupancy quota of circulating capital 流动资金占用额
occupancy rate 住房占用率;占有率;占用率;居住率;房屋占用率;房屋居住率;住房占有率
occupancy rate of circulating capital 流动资金占用率
occupancy rate of fixed capital 固定资金占用率
occupancy rate of storage area 库场利用率
occupancy sprinkler system 居住自动喷洒灭火器;住处自动喷洒灭火系统
occupancy standards 居住标准
occupancy study 车位占有调查
occupant 居住者;占有人;实际占有人
occupant load 高峰使用荷载;居住荷载
occupant mortgagor 住户抵押人
occupant restraint system 乘车安全束护系统
occupating coefficient 使用率
occupating crossing 专用铁路平道
occupation 职业;占有;占用;居住
occupational 职业的
occupational accident 职业性意外事故
occupational allergic reaction 职业性变态反应
occupational area 职业领域
occupational asthma 职业性哮喘
occupational cancer 职业(性)癌
occupational cancer risk 职业(性)致癌危险性
occupational carcinogen 职业性致癌物
occupational cataract 职业白内障
occupational chemical erosion 职业性化学侵蚀
occupational contact dermatitis 职业性接触皮炎
occupational contraindication 职业禁忌症
occupational deafness 职业性耳聋
occupational dermatitis 职业性皮炎;工业(性)皮肤病
occupational dermatosis 职业性皮肤病
occupational dermatosis of mines 煤矿职业性皮肤病
occupational disease 职业病
occupational dust 职业性尘末;工业尘末
occupational dysaudia 职业性听力障碍
occupational eczema 职业性湿疹
occupational environment 职业环境
occupational exposure 职业性照射;职业性暴露;职业(性)接触
occupational factor 利用系数
occupational freight warrant 占领区货运保单
occupational hazards 职业危害
occupational health 职业卫生;职业健康;职业保健
occupational hearing loss 职业性听力损失
occupational history 职业史;履历
occupational hygiene 劳动卫生
occupational index 就业指数
occupational infectious disease 职业性传染病
occupational lead poisoning 职业性铅中毒
occupational licensing 发放职业许可证
occupational mobility 职业流动性
occupational myopia 职业性近视
occupational negligence 职业性过失
occupational neurosis 职业性神经官能症;职业神经症
occupational noise 职业噪声
occupational noise exposure 职业性噪声接触
occupational noise source 职业性噪声源
occupational ophthalmopathy 职业性眼病
occupational pattern 就业结构
occupational permit 开业许可证
occupational poisoning 职业中毒;职业性中毒
occupational pollution control planning 行业污染控制规划
occupational psychiatry 职业精神病学
occupational psychology 职业心理学
occupational radiation 职业性照射
occupational radiation dose 职业辐射剂量
occupational radiation hazard 职业性辐射危害
occupational radiation protection 职业性辐射防护
occupational radiation worker 职业性辐射工作人员

occupational Reynaud's phenomenon 职业性雷诺现象
occupational risk 职业性危害
occupational safety 职业安全
Occupational Safety and Health Act 职业安全与卫生法;职业安全与卫生条例
Occupational Safety and Health Act of American 美国职业安全与卫生条例
Occupational Safety and Health Administration 美国职业安全与卫生管理局;美国劳动部职业安全与卫生局
Occupational Safety and Health Administration of Department of Labor 劳动部职业安全与卫生管理局(美国)
occupational satellite 职业卫星
occupational shift 职业转换;职业变化
occupational specialization 职业专门化
occupational stigmata 职业特征
occupational stress 职业性紧张
occupational structure 职业结构
occupational tax 职业税;开业许可税
occupational therapy 职业治疗;职业疗法;工作疗法
occupational toxicology 职业毒理学
occupational transformation 职业转换
occupation coefficient 占空系数;使用率
occupation disease clinics 职业病门诊
occupation disease prevention and treatment center 职业病防治所
occupation number 占有数;填充数
occupation of funds 资金占用
occupation of road 道路占用
occupation period 工作周期;使用时间;占有时间;占用时间
occupation permit 入伙纸
occupation principle 占领原则
occupation probability 占有几率;占有概率
occupation ratio 占有比
occupation road 专用道路
occupation standard 行业标准
occupation tax 开业税
occupation theory 占领学说
occupiable 可占有的
occupiable room 可暂住的房间
occupied amount of labo(u)r 劳动占用量
occupied area 占有面积;占用面积;在使用的面积
occupied band 占有能带
occupied block 占用区段
occupied factor 存在因子
occupied fallow 半休闲地
occupied frequency bandwidth 占有频带宽度
occupied funds 资金占用
occupied level 占有能级;满能级
occupied orbital 占有轨道
occupied point 设站点
occupied population 就业人口
occupied rate of current capital 流动资金占用率
occupied room 占用室
occupied section 占用区间
occupied state 已占态;占有态;填充状态
occupied station 施测站点;设站点
occupied territory 占领地
occupied track 占用线;停车股道
occupier 住户;用户
occupy 占有;占用;从事
occupy ones fractions rate per cent 占本级产率百分数
occupy total sample per cent 占全样百分数
occupy total sample rate per cent 占全样产率百分数
occur 发生
occurred region of tectonic movement 构造运动发生的地区
occurred stratigraphic(al) age 产出的地层时代
occurred structure layer 所出现的构造层
occurred tectonic stage 所发生的构造阶段
occurrence 具体值;事件;当前值;出现;产状【地】
occurrence classify of hydrothermal eruption 水热喷发的产状分类
occurrence classify of spring 温泉的产状分类
occurrence condition of groundwater 地下水出现条件
occurrence mode of orebody 矿体产状
occurrence number 出现次数
occurrence of contact plane 接触面产状

occurrence of geothermal water 地下热水赋存状态
occurrence of hits 命中率
occurrence of igneous rock igneous facies 火成岩的产状岩相
occurrence of igneous rocks 火成岩产状
occurrence of intrusive body 侵入体产状
occurrence of migmatite 混合岩产状
occurrence of minerals 矿物产状
occurrence of nodule 结核产状
occurrence of oil field water 油田水产状
occurrence of pore water 孔隙水产状
occurrence probability 出现几率
occurrence rate 出现率;发生率
occurrence rate of liability accident for luggage 行李责任事故发生率
occurrence rate of liability accident for parcel 包裹责任事故发生率
occurrences of geothermal fluid 地热流体赋存状态
occurrence time 发生时间
occurrent number 发生次数
occurs 重现
occurs number 重现号
ocean 大洋
ocean acoustic(al) topography 海洋超声测图法
ocean air (mass) 海洋气团
ocean air traffic control 海上空中交通管制
ocean and rail 海陆联运
ocean anti-cyclone 海洋反气旋;洋面反气旋
ocean area reconnaissance satellite 海洋侦察卫星
oceanarium 海洋水族管;海水水族馆;大型水族馆
ocean-atmosphere corrosion 海洋大气腐蚀
ocean-atmosphere exchange 海洋大气间的物质能量交换
ocean-atmosphere heat exchange 海洋大气热交换
ocean-atmosphere interaction 海洋大气相互影响
ocean-atmosphere interface 海洋大气界面
ocean-atmosphere relation 海洋大气关系
ocean-atmospheric (inter)change 海面气交换
oceanaut 海洋潜水员
ocean bank 大洋浅滩
ocean barge 海驳;出海驳船
ocean barging 海上驳运
ocean basin 海洋盆地;海盆;大洋盆地
ocean basin block 海盆巨地块
ocean basin floor 大洋盆地
ocean bathymetric(al) chart 世界大洋水深图;大洋水深图
ocean beach 海滩
ocean bill of lading 海运提(货)单;海洋提(货)单
ocean boarding boat 海洋交通船
ocean boarding ship 海洋交通船
ocean boarding vessel 远洋交通船;海洋交通船
ocean-borne 海运的
ocean bottom 洋底
ocean bottom earthquake 海底地震
ocean bottom explosion 海底爆炸
ocean bottom feature 洋底地形
ocean bottom hydrophone 深海海底水听器装置
ocean bottom hydrophone instrument 海底水听仪器
ocean bottom information 海底资料
ocean bottom microseism 海底微震
ocean bottom multiple 海底多次波
ocean bottom profiler 海底地层剖面仪
ocean bottom reference cable 水底参考电缆
ocean bottom scanning sonar 洋底扫描声呐;海底扫描声呐
ocean bottom seismograph 海洋地震仪;海底地震仪
ocean bottom seismometry 洋底测震术
ocean bottom station 海底观测站;海底观察站;海底测站
ocean bottom topographic meter 海底地貌仪
ocean canyon seachannel 海底峡谷;海底峡沟
ocean car ferry boat 海洋车辆渡船
ocean cargo 远洋货物
ocean carriage 远洋运输;海运
ocean carrier 远洋船(舶);海运承运人
ocean charges 海运运费
ocean chart 海洋图
ocean chemistry 海洋化学
ocean circulation 海洋环流;大洋环流
ocean clean on board bill of lading 已装船海洋清洁提单

ocean climate 海洋气候;大洋气候
ocean colloid chemistry 海洋胶体化学
ocean colo(u)r and temperature sensor 海水颜色与温度传感器
ocean commerce 海外贸易
ocean culture system 海洋养殖系统
ocean current 洋流;海(洋)流
ocean current chart 海流图
ocean current denudation 洋流剥蚀作用
ocean current determination 海流测定
ocean current dynamics 海流动力学
ocean current energy 海流能
ocean current generation 海流发电
ocean current meter 海流计
ocean current monitor 洋流检测仪
ocean current observation 海流观测
ocean data acquisition station 海洋数据采集站
ocean data acquisition system 海洋资料收集系统;海洋数据采集系统
ocean data buoy 海洋资料浮标;海洋数据观测浮标
ocean deep 海渊
ocean delta 大洋三角洲
ocean deposit 远洋沉积(物)
ocean depth 海洋深度;大洋深度
ocean development 海洋开发
ocean development technique 海洋开发技术
ocean discharge criterion 污水排海标准
ocean disposal 远洋废物处理;海洋配置;海洋排放;海洋(废物)处置
ocean disposal of wastewater 污水海洋处置
ocean dredging 海上挖泥
ocean drilling project 海洋钻探计划
ocean dumping 海洋投弃;海洋倾弃;海洋倾废
ocean dumping area 海洋倾倒区
ocean dumping control 海洋倾弃控制
ocean dumping site 海上抛泥区
ocean dumping surveillance system 海洋倾废监视系统;海上废物倾倒监测系统
ocean dynamics 海洋动力学
Oceanean pottery 大洋洲陶器
ocean earthquake 海洋地震
ocean economics and technology office 海洋经济和技术处
ocean eddy 海洋湍流
oceaneering 海洋工程学;海洋工程
ocean energy 海洋能
ocean engineering 海洋工程(学);海岸工程(学)
ocean engineering survey 海洋工程测量
ocean environment 海洋环境
ocean environment sensing equipment 海洋环境传感器
ocean expedition 海洋考察
ocean exploitation 海洋开发
Oceanexpo 海洋博览会
ocean fishery 海洋渔业
ocean fishery research vessel 海洋水产调查船
ocean fishing 海洋捕鱼
ocean-floor 洋底
ocean-floor basalt 洋底玄武岩
ocean-floor deposit 海底矿床
ocean-floor drilling 海底钻探;海底钻井
ocean-floor metamorphism 洋底变质作用;海底变质作用
ocean-floor mining 海底采矿
ocean-floor spread(ing) 海底扩张;洋底扩张
ocean-floor spreading hypothesis 海洋底扩大假说;海底扩张假说
ocean-floor structure 海洋底构造
ocean-floor work site 海底工作区
ocean food chain 海洋食物链
ocean food resources 海洋食物资源
ocean freight 远洋运输;远洋货运;海运运费;海运费
ocean freightage 远洋运价
ocean freight rate 海运运费率;海洋运费率
ocean front 海滨;滨海地带;沿海地带
ocean geography 海洋地理学
ocean geotechnics 海洋土工学
ocean geotechnique 海洋土工学
ocean geotectonics 海洋大地构造学
ocean-going 远洋航行的;远洋的;远航的;出海的
ocean-going area 远洋航区;远洋舱区
ocean going commerce 海外贸易
ocean-going dry cargo freighter 海洋干货轮
ocean-going fleet 远洋船队

ocean-going freighter 远洋货轮
ocean-going goods 远洋货物
ocean-going icebreaker 海上破冰船
ocean-going liner 远洋客轮
ocean-going platform 远洋航船上甲板；海上钻探平台；海上钻井平台
ocean-going pusher-barge combination 远洋顶推驳船队；出海顶推船队
ocean-going shipping 远洋航运
ocean-going tanker 远洋油轮
ocean-going trade 远洋贸易
ocean-going tug 远洋拖轮
ocean-going vessel 海船；远洋船；远洋轮(船)；远洋航轮；远洋船舶
ocean-gray 淡银灰色
ocean green 海洋绿
ocean-grey 淡银灰色
ocean greyhound 远洋特快班轮；快速远洋船
ocean harbo(u)r 海洋港口
ocean heat flow radiometer 海洋热流辐射计
ocean heat transport experiment 海洋热量输送实验
ocean hydrodynamics 海洋流体动力学
Oceania 大洋洲
Oceanian 大洋洲的
oceanic 海洋的
oceanic absorption 海洋吸收
oceanic abyss 海的深渊；深海区
oceanic acoustics 海洋声学
oceanic anoxic event 海洋缺氧事件；海水缺氧事件
oceanic anti-cyclone 副热带高压
oceanic bank 大洋滩
oceanic basalt 海洋玄武岩；大洋(型)玄武岩
oceanic basin 洋盆
oceanic bottom 大洋底
oceanic cartography 海洋制图学
oceanic chart 大洋图
oceanic circulation 大洋环流
oceanic climate 海洋性气候；海洋气候
oceanic coastal states 大洋沿海国
oceanic community 海洋群集
oceanic condition 海洋状况；海洋情况；海况
oceanic crust 海洋地壳；洋底壳；海底地壳；大洋地壳
oceanic crust dynamic project 洋壳动力学计划
oceanic current 海流；大洋流
oceanic cycle 海洋循环
oceanic deep 海渊
oceanic delta 海洋三角洲
oceanic deposit 海洋沉积(物)；大洋沉积
oceanic depth 海渊
oceanic distribution 海洋分布
oceanic economics 海洋经济学
oceanic environmental remote sensing 海洋环境遥感
oceanic era 原始海洋时代
oceanic fauna 海洋动物群落；海洋动物区(系)
oceanic floor 海底
oceanic-floor fracture zone 洋底破裂带
oceanic flora 海洋植物区系
oceanic flow 海流
oceanic forecast 海况预报；海况预报
oceanic fracture 大洋断裂带
oceanic fracture zone 海洋破裂带
oceanic front 海洋峰【气】
oceanic general circulation mode 海洋环流模型；海洋环流模式
oceanic geomorphology 海洋地貌学
oceanic geosyncline 大洋地槽
oceanic glacier 海洋性冰川
oceanic heat flow 海洋热流
oceanic hemisphere 海洋半球；水半球
oceanic high 海洋高压；副热带高压
oceanic industry 海洋业
oceanic invertebrate 海洋无脊椎动物
oceanic island 海洋(中弧)岛
oceanic-island basalt 海岛式玄武岩
Oceanic Islands 大洋洲
oceanicity 海洋性；海洋度
oceanic lake 海洋湖
oceanic layer 大洋层
oceanic lithoshpere 海洋岩石层；大洋岩石圈
oceanic margin 大洋边缘
oceanic mean heat flow 海洋平均热流
oceanic mean mantle flow 海洋平均地幔热流

oceanic meteorology 海洋气象学
oceanic mining 海洋矿业；海洋采矿
oceanic moderate climate 海洋性温和气候
oceanic name 海洋名称
oceanic noise 海洋噪声；海鸣
oceanic observation 海洋观测
oceanic observation satellite 海洋观测卫星
oceanic physics 海洋物理(学)
oceanic plankton 海洋浮游生物
oceanic plate 海洋板块；大洋板块
oceanic plateau 海底高原；大洋台地
oceanic plotting sheet 远洋作业图
oceanic pollutant 海洋污染物
oceanic pollution 海洋污染
oceanic prediction 海洋预报；海况预测；海况预报
oceanic province 海洋区(域)；洋区；大洋区
oceanic province coral reef 大洋区珊瑚礁
oceanic ridge 海底脊；洋脊；大洋(中)脊
oceanic ridge belt 洋脊地震带
oceanic ridge seismotectonic zone 大洋海岭地震构造带
oceanic riffs 大洋中隆
oceanic rift 大洋裂谷
oceanic rift system 大洋裂谷系
oceanic rise 海底隆起；大洋隆起
oceanics 海洋工程学
oceanic science 海洋科学
oceanic sea 海洋
oceanic sediment 海洋沉积(物)
oceanic slab 海洋板块
oceanic sounding 海洋测深
oceanic sounding chart 海洋水深图
oceanic stratigraphy 海洋地层学
oceanic stratosphere 海洋同温层；海洋平流层；大洋平流层；大洋冷水圈
oceanic surface 海面
oceanic swell 大洋涌浪
oceanic temperature waters 海洋温水域
oceanic tholeiite 大洋拉斑玄武岩
oceanic tide 海洋潮(汐)；海潮；大洋潮汐
oceanic transform fault 大洋转换断层
oceanic trench 深海槽；大洋沟
oceanic troposphere 海洋对流层；大洋对流层
oceanic turbulence 大洋湍流
oceanic volcano 海底火山；洋底火山
oceanic water 海洋水
oceanic waters 海洋；大洋水域
oceanic wave 大洋波浪；大洋波动
oceanic zone 大洋带
ocean incineration of toxic waste 有害废物海洋焚化
ocean inlet 入海口
ocean instrument calibration center 海洋仪器校准中心
ocean investigation ship 海洋调查船
ocean island 大洋岛
oceanite 大洋岩
oceanity 海洋性；海洋度
oceanium 海洋群落
oceanization 洋化作用；大洋化作用
ocean lane 远洋航路；海上航线；海上航路；航线【航海】
ocean life 海洋生物
ocean life by stages 海洋生命分期
ocean liner 远洋客轮班轮；远洋轮；定期远洋船
ocean magnetic anomaly 大洋磁异常
ocean management 海洋管理
ocean mapping 绘制海洋图
ocean marginal drilling project 大洋边缘钻探计划
ocean marine insurance 远洋水险；海运保险
ocean meteorological and oceanographic(al) monitoring system 海洋气象监控系统
ocean meteorological chart 海洋气象图
ocean meteorological observation 海洋气象观测
ocean meteorological station 海洋气象站；海洋气象观测站
ocean meteorology 海洋气象学
ocean mining 海洋采矿
ocean mixed layer experiment 海洋混合层实验
ocean monitoring 海洋监测
ocean navigation 远洋航行
ocean navigation station 海洋导航站
ocean noise meter 海洋噪声计
ocean observation apparatus 海洋观测装置

ocean observation tower 海洋观测塔
ocean observation vessel 海洋观测船；海洋定点观测船
oceanodromous 远洋回游的
oceanodromous migration 海洋洄游
oceanogenic 海洋成因的
oceanogenic sedimentation 海洋沉积作用
oceanographer 海洋学者
oceanographer deep 海洋学者海渊
oceanographic(al) 海洋学的
oceanographic(al) atlas 海洋地图集
oceanographic(al) blank chart 空白海图；海洋空白图
oceanographic(al) buoy 海洋观测浮标；海上浮标
oceanographic(al) cartography 海洋制图学
oceanographic(al) cast 采水测温
oceanographic(al) chart 海洋图
oceanographic(al) chemistry 海洋化学
oceanographic(al) condition 海洋水文条件；海洋(水文)情况
oceanographic(al) current meter 海洋流速仪；海流计
oceanographic(al) data 海洋学资料；海洋学数据
oceanographic(al) data interrogating station 海洋数据通信询问站
oceanographic(al) depth recorder 海洋水深记录仪
oceanographic(al) design criterion 海洋设计标准
oceanographic(al) diagram 海洋水文要素图(表)；海洋水文图
oceanographic(al) element charts 海洋要素图
oceanographic(al) element diagram 海洋要素图表
oceanographic(al) engineering 海洋工程
oceanographic(al) equipment 海洋调查设备
oceanographic(al) exploration 海洋调查；海洋查勘
oceanographic(al) facility 海洋调查设备
oceanographic(al) factor 海洋因素
oceanographic(al)-geologic(al) condition 海洋水文地质情况
oceanographic(al) hydrological data 海洋(学)水文资料；海洋水文数据
oceanographic(al) hydrological element 海洋水文要素
oceanographic(al) hydrological forecast 海洋学水文预报；海洋水文预报
oceanographic(al) hydrological observation 海洋学水文观测；海洋水文观测
oceanographic(al) hydrology 海洋水文学
oceanographic(al) information center 海洋情报中心
oceanographic(al) instrumentation 海洋仪器
oceanographic(al) level(l)ing 海洋水准测量
oceanographic(al) map 海洋图
oceanographic(al) medicine 海洋医学
oceanographic(al) model 海洋学模式
oceanographic(al) observation 海洋观测
oceanographic(al) observation of the Pacific 太平洋海洋观测
oceanographic(al) physics 海洋物理(学)
oceanographic(al) publication 海洋出版物
oceanographic(al) recording station 自记海洋水文站
oceanographic(al) research 海洋研究
oceanographic(al) research buoy 海洋研究浮标
oceanographic(al) research equipment 海洋学研究设备
oceanographic(al) research instrument 海洋学研究仪器
oceanographic(al) (research) ship 海洋调研船；海洋调查船；海洋水文调查船；海洋研究船
oceanographic(al) research vessel 海洋科学研究船
oceanographic(al) satellite 海洋研究卫星；海洋卫星
oceanographic(al) sedimentation 海洋沉积作用
oceanographic(al) station 海洋水文站；海洋观测站
oceanographic(al) station location 海洋考察站定位
oceanographic(al) submersible 海洋科学用深潜器
oceanographic(al) survey 海洋考察
oceanographic(al) survey(ing) 海洋测量；海洋调查

oceanographic(al) vessel 海洋考察船
oceanographic data 海洋数据
oceanographic dredge(r) 海洋地质采样器
oceanography 海洋学;海洋地理学
oceanography from space 空间海洋学
oceanology 海洋学;海洋水文学;海洋开发技术
ocean ooze 海洋淤泥
oceanophysics 海洋物理(学)
ocean optics 海洋光学
ocean organic chemistry 海洋有机化学
ocean organism 海洋有机物
ocean-oriented 与海洋有关的
oceanosphere 海洋图
ocean outfall 入港河口;污水排海口
ocean outfall dispersion 排海口污水扩散
ocean palace 远洋豪华客轮
ocean passages for the world 世界大洋航路
ocean physicochemistry 海洋物理化学
ocean platform 海洋台地;海洋平台;海上平台
ocean pollutant 海洋污染物
ocean pollution monitor 海洋污染监测器
ocean pollution monitoring 海洋污染监测
ocean pollution prevention 海洋防污法规
ocean pollution surveying 海洋污染调查
ocean port 海洋港口;海港
ocean power generation 海洋发电
ocean range vessel 远洋船;跟踪船
ocean reef 大洋暗礁
ocean region code 洋区码
ocean research vessel 海洋调研船;海洋调查船
ocean resource development 海洋资源开发
ocean resources 海洋资源
Ocean Resources Conservation Association 海洋资源保护协会
ocean ridge 洋脊;海岭;海脊;大洋脊
ocean robot 海洋自动记录仪
ocean route 远洋航路;海洋航线;海洋航路
ocean routing 远洋航线
ocean salvage vessel 远洋救助船
ocean sampling 海水采样
ocean sampling system 海水采样系统
oceansat 海洋卫星
ocean satellite 海洋卫星
ocean service 远洋航线
ocean ship 远洋船舶
ocean shipping 海运
ocean shipping agency 远洋运输代理公司
ocean shipping procedure 海运手续
ocean shipping supply corporation 外轮供应公司
ocean shipping trade 海洋运输业
ocean soil 海洋土
ocean scunding chart 大洋水深图
ocean space 海洋空间
oceansphere 海洋圈
ocean station 海洋观测站;海洋定点观测站
ocean station vessel 海洋水文气象船;海洋气象观测船;海洋观测船;海洋定点观测船
ocean structure 海洋结构物
ocean surface 海面
ocean surface chemistry 海洋表面化学
ocean surface current 表层洋流;表层海流
ocean surface temperature 海洋表层温度
ocean surface wave 海洋表面波
ocean surveillance 海上监视
ocean surveillance satellite 海洋监视卫星
ocean surveillance system 海洋监视系统
ocean survey(ing) 海洋测量;海洋调查;大洋调查
ocean survey ship 海洋测量船
ocean swell 海洋涌浪
ocean tanker 油轮
ocean temperature 海洋温度;海水温度
ocean temperature differential power 海水温差发电
ocean terminal 海洋转运油库;海洋基地;海港
ocean thermal energy conversion 海洋热能转换
ocean thermology 海洋热学
ocean tidal power 海洋潮汐能;海潮能
ocean tide 海潮
ocean tow 远洋拖航
ocean towing 远洋拖带
ocean trade 远洋运输;海运业
ocean traffic 海洋交通
ocean tramp 远洋不定期货船;不定期远洋货轮;远洋不定期货轮

ocean transport 远洋运输;海运
ocean transportation 远洋运输;海洋运输
ocean transportation insurance 海洋运输保险
ocean transport department 海运部
ocean transport statistics 海运统计
ocean trench 海沟
ocean trough 海槽
ocean tug auxiliary 海上拖运辅助船
ocean tug-boat 远洋拖轮
oceanus procellarum 风暴洋
ocean vessel 远洋船舶
oceanward 向海的
ocean warning 海洋水温升高
ocean waste disposal 海洋倾弃;海洋废物处置
ocean waste disposal control 海洋倾弃控制
ocean waste treatment 海洋废物处理
ocean water 远海海水;大洋水
ocean watering 海洋航线;海洋航路
ocean-water isotope composition 海水同位素成分
ocean wave 远洋波;海洋波;海浪
ocean wave electric(al) generator 海洋波浪发电机
ocean wave energy 海洋波浪能
ocean wave forecasting 海浪预报
ocean wave spectrometer 海洋波浪光谱仪
ocean wave spectrum 海浪谱
ocean waybill 海运运单;海运提(货)单;海上货运单
ocean weather data 海上气象数据
ocean weather observation ship 海洋气象观测船
ocean weather ship 海洋天气船;海洋(水文)气象船;海洋气象观测船;海上气象观测船
ocean weather station 海洋天气站;海洋气象站
ocean weather station observation 定点观测
ocean weather vessel 海洋水文气象船
ocellar lobes 单眼叶
ocellar texture 眼斑结构【地】
ocellar triangle 单眼三角区
oce steel 无渗碳体钢
ocher 赭石;赭色;黄褐色;土黄(颜料)
o-chloroaniline 邻氯苯胺
o-chloro-benzoic acid 邻氯苯甲酸
o-chlorophenol 邻氯苯酚
Ochoan 奥霍统【地】
Ochoan (stage) 奥霍阶
ochre = ocher
ochre brown 赭褐(色)的
ochre mutation 赭石突变
ochreous 赭色的;赭土的
ochre pollution 赭石污染
ochre puree 赭石泥
ocntrol-metering point 控制记录点
ocotea oil 樟油;奥寇梯木油
ocpan 锡基白合金
Ocrate concrete 奥克拉特混凝土;四氟化硅混凝土;赭土混凝土
Ocrate process (of treatment) 奥克拉特处理法
ocrating 混凝土表面四氟化硅处理;四氟化硅处理混凝土表面
o-cresol 邻甲酚
o-cresolphthalein 邻甲酚酞
o-cresolsulfon-phthalein 邻甲酚磺酞
OC soil 超固结土
octabin 八角形箱
octabromide test 八溴化物试验
octabromo diphenyl 八溴联苯
octachlorinated dibenzofurnan 八氯二苯并呋喃
octachlorobiphenyl 八氯联苯
octachlorocyclopentene 八氯环戊烯
octachlorodipropylether 八氯二丙醚
octachloropropane 八氯丙烷
octachlorostyrene 八氯苯乙烯
octacontane 八十烷
octacosane 二十八烷
octad 八价的
octadecanamide 硬脂酰胺
octadecanoic acid 十八酸
octadic 八进制的
octaethylcy clotetrasilazane 八乙基环四硅氨烷;八乙基环丁硅氨烷
octafluorocyclobutane 八氟环丁烷
octafluoropropane 八氟丙烷
octagon 八角形;八边形
octagonal 八角形的
octagonal aisle 八角形甬道;八角形会堂侧廊;八角耳堂

octagonal bar 八角棒材
octagonal bar steel 八角型钢;八角条钢
octagonal block 八边形地区;八边形街区
octagonal building 八角形建筑物;八角楼
octagonal chamber 八角形房间;八边形房间
octagonal chapter-house 八角形僧侣会堂;八角形教士会堂;八角形教师会堂
octagonal chimney 八角形烟囱
octagonal coil 去磁线圈
octagonal column 八角柱
octagonal contour 八边形等值线
octagonal cupola 八角形化铁炉;八角形冲天炉;八角形圆顶;八边形穹顶
octagonal dome 八角穹顶
octagonal donjon 八角形城堡;八边形中世纪城堡主楼
octagonal dungeon 八角形地牢;(八边形城堡上的)防御坚塔;八角形城堡
octagonal element 八边形元件
octagonal figure 八角形;八边形
octagonal footing 八角形底脚;八角形基础;八边形基座
octagonal ground plan 八边形底层平面图;八边形地面图
octagonal hammer 八方锤
octagonal ingot 八角钢锭
octagonal keep 八角看守所
octagonal lantern 八角形灯笼;八角形天窗;八角灯
octagonal mosaic tile 八边形马赛克(瓷)砖
octagonal nut 八角形螺母
octagonal opening 八边形空隙
octagonal pavilion 八角亭
octagonal pierhead 八角形堤头
octagonal pile 八角形桩;八边形桩
octagonal pillar tungsten carbide bit 八角柱状硬合金钻头
octagonal prism 八角棱镜
octagonal pyramid 八边形金字塔
octagonal sealing pile 八角形封闭桩
octagonal shaft 八角钻杆
octagonal site 八面体晶位
octagonal spire 八角形尖塔;八角尖塔
octagonal steel 八角钢材;八角钢
octagonal steel bar 八角钢条
octagonal tile 八边砖片
octagonal tower 八角塔
octagonal turret 八角塔楼
octagonal vault 八角拱顶
octagonal window 八角窗
octagonal wire 八边线材
octagon bar 八角型钢;八角棒材
octagon block 八角形块体
octagon column 八面柱
octagon-head(ed) screw 八角形头的螺钉
octagon ingot area 八角形钢锭浇注台
octagon lantern 八角形天窗
octagon screw 八角形螺钉
octagon steel 八角钢
octahedral 八面(体)的
octahedral angle 八面体的角
octahedral borax 八面硼砂
octahedral cleavage 八面体解理
octahedral coordination 八面体配位
octahedral copper ore 赤铜矿
octahedral crystal 八面(体)晶体
octahedral deviated stress 八面体偏应力
octahedral field 四面体场;八面体场
octahedral hybrid orbital 八面体杂化轨道
octahedral interstice 八面体空隙;八边形空隙
octahedral iron ore 磁铁矿
octahedral layer 八面体层
octahedral layer charge 八面体层电荷
octahedral linear strain 八面体线性应变
octahedral normal stress 八面体正应力;八面体法向应力
octahedral plane 八面体(平)面
octahedral shear(ing) strain 八面体剪(切)应变
octahedral shear(ing) stress 八面体剪(切)应力
octahedral sheet 八面体晶片
octahedral site 八面体座
octahedral site preference energy 八面体择位能;八面体位置优先能
octahedral stone 八面体金刚石
octahedral strain 八面体应变

octahedral stress 八面体应力
octahedral stress and strain 八面体应力和应变
octahedral stress criterion 八面体应力理论
octahedral stress path 八面体应力路径
octahedral stress plot 八面体应力图
octahedral structure 八面(体)结构
octahedral void 八面体空隙
octahedritc 八角形二氧化钛晶体
octahedrite 锐钛矿;八面(陨)石;八面体陨体;八面结构铁镍合金
octahedron 八面体;正八面体;[复]octahedra
octahedron group 八面体群
octal 八面的;八进制的;八脚;八边的
octal addition 八进制加法
octal address 八进制地址
octal base 八脚管座;八角管底
octal code 八进制码
octal counter 八进位计数器
octal debugger 八进制调试程序
octal debugging 八进制调试
octal debugging technic 八进制调试技术
octal decoder 八进位译码器
octal digit 八进制数(字);八进位数字
octal display format 八进制的显示格式
octal format 八进制格式
octal indication 八进制指示
octal loading program(me) 八进制装入程序
octal memory 八进制存储器
octal multiplication 八进制乘法
octal notation 八进(制)记数法
octal number 八进(制)数;八进位数字
octal number complement 八进制数补码
octal number system 八进(位数)制
octal numeral 八进制数
octalox 狄氏剂
octal sequence 八进制序列
octal socket 八脚管座;八角管座
octal system 八进制
octamethylcy-clotetrasiloxane 八甲基环四硅氧烷
octamethylene glycol 辛二醇
octamonic amplifier 倍频放大器;八倍频放大器
octane 辛烷
octane corrector 辛烷值校正器
octane curve 辛烷曲线
octane diacid 辛二酸
octane dicarboxylic acid 辛烷二甲酸
octanedioyl 辛二酰
octane enhancer 辛烷值提高剂
octane fade 辛烷值下降
octane improvement 改善辛烷值
octane laboratory method 辛烷值实验室测定法
octane level 辛烷值
octane number 辛烷值
octane number determination 辛烷值测定
octane number improver 辛烷值增进剂
octane number nomogram 辛烷值列线图
octane number requirement 辛烷值的要求
octane number response 辛烷值感应
octane number test engine 辛烷值试验机
octane performance 辛烷值特性
octane rating 辛烷额定值
octane rating index 辛烷值指数
octane ratio 辛烷值
octane requirement 对燃料辛烷值的要求
octane research method 辛烷值研究法
octane scale 辛烷值标度
octane selector 辛烷值选择器
octane unit 辛烷单位
octane upgrading 辛烷值提级
octane value 辛烷值
octangular 八角形的;八边形的
octanol 辛醇
octanone 辛酮
octant 八分圆;八分仪;八分区
octant anechoic chamber 卦限消声室
octant anechoic room 卦限消声室
octant angle 八分角
octant arc 八分弧
octant-homogenizing system 八分均化法
octaoxacyclotetracosane 八氧杂环二十四烷
octaphenyluranocene 八苯基环辛四烯铀
octaploidy 八倍性
octarius 量磅;品脱(容量名,1品脱=1/8加仑)
octastyle 八柱式建筑;八柱式(的)

octastyle portico 八柱式门廊
octastyle temple 八柱式圣殿;八柱式庙宇
octastylos 八柱式建筑物
octavalent 八价的
octave 倍频程;八度
octave analyzer 倍频分析器
octave-band 倍频(程)带
octave-band analyzer 倍(频)带分析器;倍频程分析器
octave-band filter 倍频(带)滤波器
octave-band level 倍频带极
octave-band oscillator 倍频带振荡器
octave-band pressure level 倍频带压级;倍频程带声压级
octave-band sound-pressure level 倍频带声压级
octave-bandwidth 倍频程带宽
octave coverage 倍频范围
octave filter 倍频滤波器
octave frequency band 倍频带;倍频程带程
octavo 八开
octel 含四乙铅抗爆剂
octet 八位字节;八位位组;八位二进制数字;八位二进制数的位组;八角体
octet alignment 按八进制校准
octet data 八位组数据
octet rule 八角定则
octet theory 八隅学说
octet timing signal 八进制定时信号
octobolite 辉石
octocalcium phosphate 磷酸八钙
octocosoic acid 二十八烷酸
octode 八极管
octodecimo 十八开本
octodenary 十八进制的
octofoil 八叶饰;八叶形;八瓣形
octofoil figure 八瓣形
octohydroxy-stearic acid 八羟基硬脂酸
Octoid gear 奥克托齿轮
octonary 八进制的;八倍的
octonary number system 八进制
octonary signaling 八进制信号发送
octopartite vault 八区式穹窿;八节头筒拱;八瓣穹隆
octoploid 八倍体
octopolar system 八极体系
octopus card 一卡通
Octopus system 章鱼系统【航海】
octroi 入市税
octual sequence 八进制系列
octuple 八重;八维;八倍(的)
octuple space 八维空间
octuplex telegraphy 八路电报
octupole 八极
octyl 辛烷基
octylacetic acid 癸酸
octylbenzene 辛基苯
octyl benzoate 苯甲酸辛酯
octylene 辛烯
octylene glycol titanate 钛酸辛二醇酯
octyl phenol 辛基酚
ocular 目镜
ocular accommodation 目镜调节
ocular blind 目镜光阑
ocular box 目镜盒
ocular circle 出射光瞳
ocular estimate 目测(法);外观评价;草地目测法
ocular-estimate by plot method 样方目测地
ocular estimate method 目估法
ocular estimation 目测;目估
ocular estimation method 目测法
ocular glass 目镜
ocular head 目镜头
ocular lens 接目镜;目镜目测(法);目估;目镜
ocular measurement 目测
ocular micrometer 接目镜测微计;接目测微尺;目镜千分尺;目镜测微器;目镜测微计;目测测微计;测微目镜
ocular piece 目镜
ocular prism 目镜棱镜
ocular spectroscope 目视分光镜;目镜分光镜
ocular surface 视角面
ocular system 视觉器官
ocular thread 目镜蛛丝;目镜螺纹
oculocardiac reflex 阿施内氏反射
oculogravic illusion 眼重力错觉【测】;重力异常幻视

oculo-gyral illusion 眼旋(转)错觉【测】
oculus 圆孔,小圆窗;眼形物;眼洞窗;牛眼窗
ocurrence of coal seam 煤层赋存
Oda metal 奥达铜镍耐蚀合金
odd 零星;零头;奇数(的);多数的;单数(的);额外(的)
odd address 奇地址
odd and even parity function 奇偶校验函数
odd and even price policy 均等价格政策
odd byte 奇偶数组
odd channel 奇数通路
odd-come-short 零头;零碎物(件);碎屑
odd component 奇支;奇分支;奇分量
odd-controlled gate 奇数控制门
odd cycle 奇数周;奇回路
odd days 单日
odd elements 奇数元素
odd-even 奇偶
odd-even check 奇偶校验;奇偶检验
odd-even counter 奇偶计数器;奇偶计数管;二元计数管
odd-even flip flop 奇偶双稳态电路
odd-even interleaving 奇偶交叉存取;奇偶交叉
odd-even logic 奇偶逻辑
odd-even predominance of ling-chain n-alkanes distribution 长链正烷烃分布奇偶优势
odd farm implements 不完全的工具
odd function 奇函数
odd harmonic 奇次谐音;奇次谐波
odd-integer 奇整数
oddity discrimination 单数辨别
odd job 零碎工作;零活;零工;临时工;打散工;零星工作
odd jobber 临时工
odd-job man 零工;勤杂工
odd-leg calipers 单向卡钳;单脚规;半内径规
odd level 奇数级;奇数层
odd-line interlacing 奇行扫描;奇数隔行
odd location 奇数存储单元
odd loop 奇数环;奇偶环
odd lot 零星货物;零批;不成套的东西
odd lot broker 零沽经纪人
oddly pinnate 奇数羽状的
odd man 临时工
odd mass number 奇质量数
oddment 库存物;陶管连接配件;零碎物
odd mode 奇数模
odd money 零钱
oddness 奇异
odd number 奇数;单数
odd-numbered days 单日
odd-numbered pass 奇数轧制道次
odd-odd 奇奇
odd-odd nucleus 奇-奇核
odd-parity 奇字称(性);奇同位
odd parity check 奇数奇偶检验
odd partition 奇分拆
odd parts 多余零件;残留物;废零件
odd permutation 奇置换;奇排列
odd pinnate 奇数羽状复叶;奇数羽状的
odd-pinnate leaf 奇数羽状复叶
odd-pinnately compound leaf 奇数羽状复叶
odd pitch 奇数螺距
odd-pitch roof 高跨比为单数的屋顶;怪坡屋顶;不规则坡度屋顶
odd-pitch screw 非标准螺纹
odd price 零售价格
odd quantum number 奇量子数
odds 优势;差别;不等式;比数
odds and ends 零星东西;残余
odd-shaped 异形的;畸形的
odd-shaped plate blank 畸形板坯
odd sheets 不合格纸
oddside 假箱;砂型假箱
oddside board 单面模板;模板
oddside pattern 向外凸的半个型板;单面模板
odds ratio 优势比;机会比例;让步比;胜负概率比;得失比率;比值比
odd state 奇态
odd substitution 奇代换
odds waste materials 零星废料
odd symbol 奇符号
odd symmetry 奇对称
odd term 奇字称性态

odd test 抽试
odd thread 奇螺纹
odd winding 奇数线组
odd work 额外工作
ode(i)on 音乐演奏厅;(古希腊或罗马的)小型剧院
Oden balance 比重天平;奥登天平
odenite 钛云母
odeom 希腊-罗马式戏院
Oder-Berger colloid mill 奥德尔-别格胶体磨
Odessey protractor 奥特赛量角器
odeum 小型音乐厅;演奏厅;(古希腊、罗马的)奏乐堂
odevity 奇偶性;奇偶数
o-dichlorobenzene 邻二氯苯
o-diethylbenzene 邻二乙基苯
o-diisopropylbenzene 邻二异丙苯
o-dimethylbenzene 邻二甲苯
odinite 拉辉煌岩
odiomorph 自形晶
odograph 自记计程仪;自动计程仪;自动航程记录仪;自动测距仪;里程记录器;里程计;里程表;计积仪;计步器;航向自记仪;航线记录仪;航程记录仪;航程记录器
odometer 自动计程仪;轮转计;路码表;里程仪;里程碑;计步器;速度表;车距计;测距计;测程轮
odometry 测程法
odontoceramotechny 瓷牙制作术
odontograph 画齿规;齿面描记器
odontograph method 画齿法
odontoid 齿形的
odontolite 齿绿松石
odontometer 齿轮公法线测量仪
odo(u)r 气味;臭味
odo(u)r air dilution ratio 恶臭稀释倍数
odo(u)r and taste removal 除味和除臭
odo(u)r and taste 臭味
odo(u)rant 加臭剂;赋香剂;恶臭物质;香料;有味物质;增味剂;添味剂
odo(u)r at low-temperature 低温臭气
odo(u)r barrier 臭气隔绝;气味绝缘;气味屏障
odo(u)r causing chemicals 发生气味的化学品
odo(u)r concentration 气味浓度;臭气浓度
odo(u)r control 除臭;气味控制;臭气控制
odo(u)r control additive 臭味拟制剂;解臭剂
odo(u)r destruction 恶臭去除
odo(u)r discharging chimney 排臭气烟囱
odo(u)r effect 臭气影响
odo(u)r effect on people 恶臭对人的影响
odo(u)r emission 气味散发;臭气发散
odo(u)r evaluation 恶臭评价
odo(u)r free 无臭;无气味;无臭气
odo(u)r froom diesel exhaust 柴油机排气的臭气
odo(u)r generation 气味物质
odo(u)riferous 有气味的
odo(u)rimeter 气味测量计
odo(u)rimetry 气味测量法;气味测定法;臭味测定(法);测臭法
odo(u)r index 气味指数;臭数
odo(u)r inhibitor 气味抑制剂
odo(u)r intensity 气味强度;臭味强度
odo(u)r intensity index 臭味强度指数
odo(u)r intensity method 气味强度法
odo(u)riphore 发臭团
odo(u)rization 加臭
odo(u)rizer 加臭器
odo(u)rless 不臭的;无气味的
odo(u)rless mineral spirit 无臭石油溶剂
odo(u)rlessness 无气味
odo(u)rless paint 无气味涂料;无臭油漆;无臭(味)涂料
odo(u)rless solvent 无臭溶剂
odo(u)r measurement 恶臭测定
odo(u)r measuring device 臭气测定装置
odo(u)r nuisance 臭气污染;气味干扰;气味公害;臭味公害;恶臭公害
odo(u)r of oil 石油的气味
odo(u)r operator 恶臭实验操作人员
odo(u)rous 有气味的;臭的
odo(u)rous cargo 气味货
odo(u)rous compound 恶臭化合物;有气味的化合物
odo(u)rous material 有味物质;恶臭物质
odo(u)rousness 恶臭浓度;气味浓度
odo(u)rous substance 恶臭物质

odo(u)r panel evaluation 气味的感官评定
odo(u)r pollutant 恶臭污染物
odo(u)r pollution 臭气污染;恶臭污染
odo(u)r-producing component 气味组分
odo(u)r release 排臭
odo(u)r removal 除臭(味)
odo(u)r removercoating 除臭涂料
odo(u)r sampling bag 恶臭采样袋
odo(u)r sampling method 恶臭采样法
odo(u)r source 臭气源;恶臭发生源
odo(u)r strength 气味强度;恶臭强度
odo(u)r suppression 气味抑制
odo(u)r synergism and counteraction 恶臭的协同与相抑制作用
odo(u)r test 气味试验;臭味试验
odo(u)r test bag 气味试验袋
odo(u)r threshold 气味限度;臭阈值;臭气阈
odo(u)r threshold concentration 臭气阈值浓度
odo(u)r threshold value 臭味阈值;恶臭阈值
odo(u)r-tight 气密的;气味密封
odo(u)r-tight cover 防臭盖
odo(u)r treatment 臭气治理
odo(u)r treatment by acid and alkali spray 酸碱喷洒除臭
odo(u)r treatment by fire 燃烧脱臭治理;燃烧除臭处理
odo(u)r treatment by ion(ic) exchange 离子交换除臭处理
odo(u)r treatment method 臭气治理方法
odo(u)r unit 臭气单位
O-D survey 起讫点调查
Odum 奥杜(一种西非硬木)
oecus (古罗马住宅中的)公寓、房间或大厅;(古罗马)大房间
oedema 水肿;浮肿
oedometer 里程计;里程表;里程仪;计距器;计积仪;压缩仪;固结仪;渗压仪;车距仪
oedometer curve 压缩曲线;固结曲线
oedometer settlement 压缩沉降;固结沉降
oedometer test 固结试验
oedometric modulus 压缩模量;固结模量;侧限压缩模量
oedometric test 渗压试验;固结试验;压缩试验
oedotriaxial test method 固结law三轴仪联合试验法
Oehman's apparatus 奥曼测斜仪(用照相法记录罗盘和倾斜器读数)
Oehman's survey instrument 奥曼测量仪器
oeil de boeuf 卵形窗;圆(形)窗;牛眼窗;椭圆窗
oeil-de-boeuf window 圆窗
oeillet 孔眼;箭孔
oelic acid 精制油酸
oellacherite 钡白云母
OEM coating 在线涂装涂料
oenanthal 庚醛
oenometer 酒精定量计
oeolotropic 各向异性的
Oerlikon gearing 奥立康制外摆线锥齿轮啮合
oermissible concentration limit of pollutant 污染物质极限容许浓度
oerstedmeter 磁场强度计;奥斯特计
oersted(oe) 奥斯特(磁场强度单位)
oertz rudder 合叶流线型舵
oestradiol benzoate 苯甲酸雌二醇
Oetling freezing method 奥梯林冻结凿井法
of bad repute 名誉不佳的
of-bearer 取砖器
Ofe 奥费(西非加纳产硬木)
of even date 同一日期的
off air 消失
off-airways 偏离航路
offal 废物;次品;碎屑;废品;废料;废弃物
offal boat 垃圾船
offal-timber 残材;废木
offal wood 次木料
off and fair 拆下敲直
off and on 时断时续;断断续续地
off-angle 斜的
off-angle drilling 斜孔钻进法;钻斜孔法
off an object 远离物标;在物标方向
off-axis 离轴
off-axis aberration 轴外像差;离轴像差
off-axis angle 偏轴角
off-axis chromatic aberration 轴外色差;偏轴色差
off-axis collimator 偏轴平行光管

off-axis cross-polarization 轴外交叉极化
off-axis distribution 离轴分布
off-axis effect 离轴效应
off-axis ellipsoidal mirror 离轴椭球镜
off-axis elliptical mirror 离轴椭圆镜
off-axis English mounting 偏轴英国式装置
off-axis error 离轴误差
off-axis feedback 离轴反馈
off-axis fluorescence 离轴荧光
off-axis Fresnel holography 离轴菲涅耳全息术
off-axis hologram 离轴全息图
off-axis holography 离轴全息术;离轴全息摄影
off-axis laser resonator 离轴激光共振腔
off-axis method 离轴法
off-axis mode 轴外模;离轴模式
off-axis parabolic mirror 离轴抛物面反射镜
off-axis plane wave 离轴平面波
off-axis point 轴外点
off-axis reference wave 离轴参考波
off-axis reflecting collimator 离轴反射式准直仪
off-axis space bandwidth product 离轴空间带宽乘积
off-axis spheric(al) mirror 离轴球面反射镜
off-axis sputtering 偏轴溅射
off-axis zone plate lens 离轴波带透镜
off-balance 倾倒;失去平衡;不平衡的
off-balance bit 偏心钻(头)
off-balance-sheet 资产负债表外
off-balance-sheet finance 不列入资产负债表的融资
off-balance sheet financing 资产负债表外投资;资产负债表外的资金融通;资产负债表项目外筹资
off-band observation 离带观测
off-base storage 基地外库存
off beam axis scattering 离光束轴散射
off-beam discrimination 偏轴鉴别
off bear 移开;取走;除去
off-bearing conveyer 侧向输送器
off-blast 停风
off-blast period 停风期
off-boarding platform 非候车站台(美国)
off-board market 交易所外市场
off-bottom 近离孔底;全部溶化
off-bottom spacing 离孔底距离
off bound ramp 驶出坡道;出境坡道
off-budget 预算外
off-budget year 非编制预算年度
off-button 脱位旋钮
off cam 脱离协联(的);非协联(的)
off cam runaway speed 脱离协联时的飞逸转速
off-capacity 容量不足
off cathode current 熄火阴极电流
off-center 脱离中心;离心;中心错位;偏心;中心不正;偏离中心;远离的
off-center application of force 偏心施力
off-center arrangement 偏心配置
off-center arrestment 偏心拦阻减速
off-center burner 偏心燃烧器;偏心喷燃管
off-center circuit 偏心电路
off-centered 偏离中心的;偏置的;不平衡的
off-centered coil 偏心线圈
off-centered dipole 偏心振子;偏心偶极子
off centered display 偏心显示
off-centered plan position indicator 偏心式平面位置显示器
off-centered radial sweep 偏心径向扫描
off centering 中心偏移;偏心
off-centering control tube 扫描中心调整管
off-centering field 偏心场
off-centering winding 偏心绕组
off-centering yoke 偏心线圈
off-center load(ing) 偏心荷载;偏心负载;偏心荷重;偏心加载
off-center operation 偏心运用;不平衡运行
off-center plan(e) display 偏心平面显示
off-center plane position indicat 偏心平面位置显示器;雷达偏心平面显示器
off-center portion of channel 航道偏离中心部分
off-center position 偏心位置
off-center rivet 偏心铆钉
off-center waterway 偏离中心水眼
offcentring 车辆曲线通过偏移
off channel 河槽外的;离岸航道
off-channel reservoir 河外水库
off-channel spreading ground 管渠外弥散场

off-chip marker 片外标志
off-clip 脱边
off-colo(u)r 非标准颜色;不正常色
off colo(u)red 不正常颜色的;不标准颜色的;颜色不正的;变色的
off-colo(u)r gasoline 变色汽油
off-colo(u)r gem 次级宝石
off-colo(u)r industry 炭黑工业
off-colo(u)r material 变色材料
off-contact 触点断开;开路触点
off-control data 失控数据
off-count mesh (支数不规则的)钢丝网;长孔筛
off coupon 不实行配给
off-course 偏出航向;偏离航线;偏航
off-course absorption 偏离吸收
off-course alarm 偏航警报器
off-course correction 偏航修正
off-course indication 偏航指示;偏离航向指示
off-course signal 偏航信号
off-course track 偏航航迹
off-critical amount 偏离临界的量
off-cut 截切不合格的;截片;下脚料
off-cut pipe 裁下的管段;下脚管;切余管
off-cycle 中止循环;停止运转周期;非周期的
off-cycle defrosting 中止循环法除霜(冷藏库);停止运转周期融霜
off-day 非工作日;休假日;预防修理日
off delay 延迟断开
off-depot storage 站外储存
off-design 与设计不符;偏离设计条件的;非设计工况;非设计的;非计算的;不经济的
off-design behavio(u)r 非设计特性;非设计工作规范
off-design condition 偏离设计条件;非设计条件;超出设计条件
off-design displacement 非设计状态排水量【船】
off-design performance 非设计性能
off-design point 非设计点
off-diagonal 非对角(线)的;对角线外的;不在对角线上的
off-diagonal elements 非对角线元素
off-dimension 尺寸不合格
off-duty 不当班的;备用的
off-duty factor 频宽比
off-duty runway 备用跑道
off-effect 中止效应;光止效应;撤光效应
off-emergency 紧急断开;紧急开关
Offenbuch faience 奥芬巴哈锡釉陶器
offence of bribery 行贿受贿罪
off-end driven ring frame 车尾传动细纱机
offender 事故原因
offending 不精细的
off-energy 损失能量
offense against sine condition 违反正弦条件
offensive cargo 违禁货物
offensive fill(ing) of ground 垃圾填方;不卫生的填土
offensive industry 有害工业;有毒工业
offensive odo(u)r 令人不愉快的气味;恶臭
offensive odo(u)r control 臭气防止
offensive odo(u)r control law 恶臭管制法
offensive trade 厌恶性行业
offensive tunnel 攻击性隧道
offer 要约;开盘;卖价;卖方索价;喊价;提供;出售;出价;发盘;发价;报盘;报价
offer a high price for 出高价买
offer and accept by post 用信件报盘及接受
offer and counter-offer 发盘和还盘
offer a price 出价
offer by description 凭规格报盘;附说明书的发盘
offer by tender 通过投标发盘;用投标报价的;通过投标报价
offer curve 提供曲线
offered capacity 提供容量
offered market 卖价市场
offered price 要价;卖主的开价;卖方报价
offeree 接受报价者;接盘人;买主;受要约人;受盘人;被报价人
offerer 出价人;发盘人;发价人;报盘人;报价人
offer financial assistance 捐助
offer for sale 供销价;提供出售
offer good until withdrawn by seller 卖方撤回有效报价;卖方撤回前有效的报盘
offer guidance by supplying information 信息引导

offer in a bid 报价
offering 出售物;插入的
offering chapel 捐献财物的教堂
offering consultation service 提供咨询服务
offering distributor 插入式分配器
offering for security 提供担保
offering information on manufacture 提出制造资料
offering list 供货单;报价单;价目表
offering of bonds 出售债券
offering or accepting bribes 行贿受贿
offering price 报出价格
offering sheet 报价单;价目表
offering table 捐献清单;捐献台
offering telex 提供电传
offering wanted 索盘
offer item 发盘项目;平价项目;提供出售项目
offer loans 发放贷款
offer on same terms and conditions 按同样条件报盘
offeror 要约人;出价人;发盘人
offer price 卖价;卖方报价
offer rate 卖价汇率;卖价;卖方开价;卖方报价
offer sample 发盘附样品
offers good until withdrawn 撤回有效报价
offer sheet 出价单;发价单
offer signal 预告信号
offer subject to buyer's inspection or approval 卖方看货后定发盘
offer subject to change 价格随时可变发盘
offer subject to sample approval 看样本后决定发盘;看样本后决定出价
offer subject to seller's confirmation 卖方确认有效报价;卖方确认后有效发盘
offer to buy 认购
offertory box (做礼拜的)捐款箱;(教堂里的)柱形捐献箱
offertory window (教堂圣堂右侧的)小矮窗
offer to sell 报盘;销货要约
offer under seal 密封发盘;签字蜡封式的要约
offer validity 报价期限
offer with engagement 实盘
offer without engagement 虚盘;不受约束出价
offer with string attached 附带条件的报盘
off-factory facility 厂外设施
off-flavo(u)r 异味;臭气
off-flow of dredge pump 泥泵脱流
off-focus detection technique 离焦探测技术
off formal 离位
off-formwork concrete 不用模板的混凝土;脱模混凝土
off-formwork concreting 无模板浇筑混凝土;无模板浇注混凝土
off frozen 解冻
off-gas 气态废物;尾气;抽气;废气
off-gas cleaning equipment 废气净化设备
off-gas line 废气管道
off-gas pump 抽气泵;排气泵
off-gas scrubber 废气洗涤器
off-ga(u)ge 等外;非标准的;不均匀厚度;不合格的;不按规格的
off-ga(u)ge goods 超尺码货物;超尺寸货物;超限货物
off-ga(u)ge material 等外轧材
off-ga(u)ge plate 等外板;不合格板
off-ga(u)ge strip 不合格的带材
off grade 低级的;等外品;等外的;不合格产品;不合格;等外
off-grade alumina 不合格氧化铝;不合格的氧化铝
off-grade glass 不合格玻璃
off-grade heat 不合格熔炼炉次
off-grade iron 等外铁;等外生铁;不合格铁
off-grade metal 等外金属;不合格金属
off-grade product 等外(产)品
off-grade size 等外品尺寸
off-grade sponge 不合格海绵金属;不合格的海绵金属
off-grain 脱бест(金刚石)
off-ground 停止接地;接地中断
off-grounded 未接地的
off-ground hauler 空中运输设备
offhand 无人管理的;未经批准的;自动的
off-hand grinding 手工磨光;手工研磨;手持工件磨光

off-hand process 自由成型
off-hand working 无模成型法
off-hap 平移断层;分错距
off harbo(u)r 港外;离港
off-heat 废品;熔炼废品
off-highway 越野的
off-highway end dump truck 越野(型)后卸汽车
off-highway hauler 越野运输车;越野牵引车
off-highway hauler for quarries 采石工人乘坐的越野运输车
off-highway lorry 越野载重汽车
off-highway manoeuvrability 越野机动性;非公路(行驶)机动性
off-highway truck 禁止通行的载重汽车;越野载重(汽)车;越野卡车;公路外行驶车辆
off-highway tire 越野轮胎
off-highway vehicle 越野汽车
off-highway wheel crane 越野轮式起重机
off-hire 停租
off-hire certificate 停租证书
off-hire clause 租期中止条款;停租条款
off-hire survey 退租检验
off-hook 摘机;离钩
off-hook service 离钩业务
off hour 业余时间
office 机关;办事处;办公室
office accommodations 办公用具
office allowances 办公费
office and residence of the president 总统府
office area 办公区
office-at-home 居家办公室
office automation 办公自动化;办公室自动化
office automation of environment(al) management 环境管理办公自动化
office automation system 办公室自动化系统
office block 办公楼街区;办公区;办公楼;办公大楼
office building 办公建筑;写字楼;办公楼;办公大楼
office building for rent 出租办公楼
office calculation 室内计算
Office Cherifien des Phosphates (摩洛哥)磷酸盐公司
office chief 办公厅主任;办公室主任
office class 局等级
office clip 回形针
office code 控制电码
office-compiled 内业编制的
office computation 室内计算
office control 内业检核
office copy 文件附本;存根;公文正本
office cubicles 办公小间
office dealing with outward goods traffic 发送货物办公室
office display terminal 办公用显示终端
office dumbwaiter 办公楼送公文升降机;公文升降机
office duplicator 办公用复印机
office duty hours 办公时间
office enclosed circuit TV 办公室闭路电视
office engineer 内业工程师;内勤工程师
office entrance 办公楼进厅;办公楼入口处
office equipment 内业装备;办公设备
office equipment repairs and maintenance expenses 营业设备维修保养费
office expenses 事务费用;办公费
office expense account 办公费用开账户
office expense book 办公费账簿
office floor 办公楼层
office for incoming and outgoing mail 收发室
office furniture 办公器具;办公家具
office furniture and fixtures 办公器具及装置
office hours 营业时间;办公时间;上班时间
office hut 临时办公室;临时木板办公房
office-identification 内业判读
office illumination 办公室照明
office information system 办公信息系统;办公室信息系统
office landscape 办公室空间景观布置;景观式办公室布置
office landscape screen 办公室屏风;办公室隔扇
office layout 办公室室内设计
office layout application 办公室布局应用程序
office lighting 办公室照明
office machine 事务用计算机;事务计算机
office management 事务管理

office manager 行政主管;业务经理
office-master drawing 正式原图
office occupancy 办公用房
office of civil air defence 人防办公室
office of commercial attache 商务专员
office of commercial representatives 商务代表处
office of departure 发运地办事处
office of entry 入境办事处
office of executive management use 行政管理办公用品
office of exit 离境办事处
Office of Hazardous Materials Transportation 危险品运输部
Office of Interstate Land Sales Registration 州际土地出售注册办公室
Office of Management and Budget 行政管理和预算局(美国)
office of nature reserves 自然保护区管理机构
office of origin 原寄局
Office of Overseas Scientific and Technical Research 海洋科学技术研究办公室
Office of Pesticide Program(me) of USEPA 美国环境保护局农药规划处
office of research 研究室
Office of Research Experiment 欧铁联盟研究所
office of technical assessment 技术评价办公室
Office of the charged' affaires 代办处
office of the president 总统府
Office of Thrift Supervision [美]节俭监督办公室
Office of Trade Ombudsman 贸易申诉局
office operation 室内作业;室内加工
office overhead 办公管理费
office park 办公大楼园区
office partition 办公室隔墙
office perforator 办公用打孔器
office portion 办公楼侧壁
office practice 业务实习
office premium 总保费
office price 正式价格
office printing 复印
office printing machine 办公用印刷机
office procedure 管理方法;办公程序
office procedure manual 办公手册
office quieting 办公环境安静措施
officer 职员;干事;工作人员;官员;公职人员;公务员
officer car 公务车
office record 内业记录
office revolution 办公室革命
officer in charge 主管官员;值班驾驶员;负责人
officer of the day 值日员
officer of the deck 舱面值班员
officer of the watch 值班驾驶员
officer on watch 值班驾驶员
officers and crews 全体船员
office salaries 职员薪金
office skyscraper 办公用摩天大楼;摩天办公大楼
office space 办公区;办公空间
office staff 办公室职员
office stor(e)y 写字间楼层;办公室楼层;办公楼层
office submission 公开投标
office supervisor 办公室负责人
office supplies 办公用品;办公用品
office telephone 公务电话机
office terminal 管理终端;办公室终端
office test 验收
office time 办公时间
office tower 办公大厦
office trailer 办公用挂车
office type partition 办公室式隔断
office typewriter 办公室打字机
off-ice wind 下冰风
office wing 办公室侧翼;办公室侧厅
office work 内业工作;内业;室内作业;室内工作
official 职员;正式的;官方的;法定的;法定
official acceptance 正式验收
official account 正式决算
official agent 正式代理人
official appraiser 官方鉴定人
official appreciation 法定升值
official approval 正式批准
official assignee 正式代理人;法定受让人;不定交让人
official authority 官方
official authorization 正式核准

official bill of lading 正式提单
official call 公务通话
official capital 国家资本;官方资本;法定资本
official classification 正式分类
official competences 法定权限
official compound 法定化合物
official contract 正式契约;正式合同
official copy 正式文本
official correspondence 业务函件
official counter quotation 国家牌价
official devaluation 法定贬值
official development assistance 政府开发援助;官方发展援助
official discount rate 法定折扣率;法定贴现率
official documents 公文;公函
official exchange rate 官方汇率;外汇官价;法定汇率
official exchange rate system 法定汇率制度
official fees for registering documents 登记费
official fixed price 法定固定价格
official gazette 政府公报;公报
official holiday 法定假日;法定假期
official hours 规定时间
official information 官方信息
official inventory date 正式盘存日期
official invoice 正式发票
official letter 公文;公函
official licence to do business issued by the local authority 开业合法证书
official liquidator 清理人;法定财产清算人
official log book 船志
official lower limit of fluctuation 法定汇率波动下限
officially guaranteed private export credit 官方担保的私人出口信贷
official magnolia 厚朴(植物)
official mansion 官邸
official map 市政公务地图;市政现状图
official market 官方市场
official market quotation 法定牌价;官方牌价
official master drawing 正式原图
official materials testing institute 法定材料试验研究所
official notary 官方公证人
official note 公文
official number 船舶登记编号
official parity 法定平价
official par of exchange 法定汇兑平价
official plan 正式计划;官方计划
official price 法定价格;法定固定价格
official quotation 法定证券价格;法定汇价
official rate 官价;法定利率;法定汇价
official rate of exchange 法定汇率
official receipt 正式收据
official receiver 清理受理员;破产事务官
official record 正式记录
official referee 正式仲裁人;官方仲裁人;官方公证人
official register 官方注册
official report 正式报告
official reserves 法定储备金
official residence 官宅;官邸
official return 正式报告;法定利润率
official sample 法定样品;法定试样
official seal 公章;封印
official sea trial 正式试航
official selling 法定出售商品
official service documents 正式业务文件
official signature 正式签署
official speed 规定速度
official speed limit 规定速度极限
official speed trial 正式试航
official standard 正式标准;法定标准
official statement 正式声明
official submission 公开投标;公开投票
official tariff 官方规定运费率
official telegram 公务电报
official telegraph dictionary 标准电码本
official telephone 公务电话
official test 正式试验;法定试验
official text 正式文本
official time of receipt 正式收报时间
official trial 正式试航
official upper and lower limit of fluctuation 波动的法定上下限

official upper limit of fluctuation 法定汇率波动上限
official visa 公务签证
official visit 正式访问
official ware 官窑;关窑(器)
officiate 主持;担任;行使职务;执行公务;司仪;裁判
officina 药房
offing 引水区域界线外附近海面;近岸安全距离;海面;外视海面;从岸能望见的远方海面;远景
offing wave 海面波
off-interval 关闭间隔
off-iron 铸铁废品;等外生铁
off-job accident 岗位外事故
off-key 不正常的
off lake rate 出栏率
off land 离岸
offlap 海退超覆;退覆
offlapping splitting 退覆式分岔
offlap sequence 退覆层序
offlet 路边引水沟;出水口;放水管
off-level 错层;层外;非规定层上
off-limit 界限限度外;界限范围外;止步;禁止通行;超出范围的
off-limit file 隔离文件
off-line 离线;机外;脱线;脱机;不在铁路沿线的
off-line analysis 离线分析
off-line calciner 离线分解炉
off-line coating 离线涂装;离线涂覆
off-line computer 离线计算机;脱机计算机
off-line control 离线控制;脱机控制;间接控制
off-line cutting machine 离线切割机
off-line data processing 离线数据处理;脱线数据处理;脱机数据处理
off-line data reduction 独立数据处理
off-line digitizing 脱机数字化
off-line drawing 脱机绘图
off-line editing 脱机编辑
off-line equalization 离线调节;离流程线调节
off-line equipment 离线设备;脱机设备
off-line fault detection 脱机故障检测
off-line installation 线外装置
off-line mode 脱机方式
off-line operation 离线操作;脱机作业;脱机操作
off-line output 间接输出
off-line plan generation 产生脱机方案
off-line plot 脱机绘图
off-line plotter 线外绘图仪
off-line process 脱机处理
off-line regulation 离线调节
off-line running monitor 跑偏监视器
off-line station 线外车站
off-line system 离线系统;脱机系统
off-line test(ing) 离线试验;离线测试;脱机测试;间接试验
off-line working 离线工作;脱机工作
off-load 卸载(的);卸货(的);非荷载的;下船(货物);卸下;卸料;减载;无荷载
off-load deaeration 无荷载除气
off-loader 卸料机;卸载机;卸材机
off-loading 减荷;卸下
off-loading point 卸载点
off-loading ramp 用装料台卸载
off-load period 卸荷期;非荷载期;卸载期;无荷载期;无负荷期
off-load regulation 卸荷调节
off-load tap 无载抽头
off-location alarm 偏位警报
off-loom take-up motion 脱机卷取动作
off lying 外围
off-lying danger 外围障碍物
off-lying island 远离海岸的岛;外围岛(屿)
off-lying sea 远海
off-melt 废品;熔炼成分不合格
off-mike 远离扩音器
off-net call 脱网呼叫
off-normal 越界;离位;偏位
off-normal contact 离位触点;离位触点
off-normal functioning 不正常情况
off-normal loading 不正常加荷
off-normal lower 下限越界
off-normal position 非正常位置;不正常位置
off-normal upper 上限越界
off odo(u)r 异味;气味不对;脱臭
off oil 次等油

off-on control 开关控制
off-on wave generator 键控制波浪发生器；键控信号发生器
off-order mode 非正常模式
off-path 反常路径
off-peak 正常的；非高峰的；非峰值以外的；非峰荷的；非繁忙；额定
off-peak commuting 非高峰时间通勤
off-peak demand 峰荷外的需求
off-peak electric(al) energy 非峰值电能
off-peak electric(al) supply 非高峰期供电
off-peak hot water heater 夜间储[贮]水式热水器
off-peak hours 轻负荷时；非高峰工作小时；非交通拥挤时刻；非高峰时间；非高峰期；非高峰荷时间；非繁忙时间
off-peak load 峰外负载；非高峰荷载；非高峰负荷；非峰值负载
off-peak period 非高峰时间；非高峰期；非峰荷期
off-peak power 非高峰功率
off-peak pumping 非高峰抽水
off-peak running 非高峰运转
off-peak service 非高峰供气
off-peak system 非高峰系统
off-peak tariff 非高峰价目表；非高峰收费表；非高峰期价格
off-peak time 非高峰时间
off-peak working and travel(l)ing 非高峰期工作和旅游
off period 开断时期；截止周期；间歇期间；淡季
off-plumb 不垂直
off-port 港外；断路位置
off-port flame 喷口外部火焰
off-position 关闭状态；释放位置；断路位置；断开位置；不工作状态；关闭位置；开路位置
off-position interlock 断路位置联锁
off-premises extension 备用电话分机
off-print 抽印本；单行本
off production 停产
off-punch 偏穿孔；不定位穿孔
off quality 品质不合格
off-railway line 不在铁路沿线的
off-ramp 驶出坡道；匝道驶出段
off-rating 超出额定值；非标准条件下；不正常状态；不正常条件
off-record loans 未登记借款
off resistance 断开电阻
off-resonance 离开共振
offretite 钾沸石
off-river impoundment 河外蓄水
off-river storage 河外蓄水；非河上水库
off-road equipment 越野设备
off-road equipment tyre[tire] 越野设备用轮胎
off-road ground tyre[tire] 越野车辆用轮胎
off-road haulage 越野运输
off-road mobility 越野机动性
off-road motorcycle 越野摩托车
off-road motor transport equipment 越野汽车运输装备
off-road operation 越野作业
off-road parking 路外停车
off-road traction 越野牵引
off-road train 越野车队
off-road transporter 越野运输工具
off-road vehicle 越野车；驶离路外车辆；路外行车；越野车辆
off-round pass 立椭圆孔型
off-running time 收车时间
offsaw 锯后
off-saw size 锯出尺寸
off-scale reading 读数
off-scheduled repair 计划外的修理
off-scourings 烂东西；垃圾；渣滓；废料
offscum 废渣
off-sea 离海的
off-sea fishery 外海渔业
off-season 淡季；非生产季节；闲季
off-sea wind 向岸风；离海风；海风
offset 型值(表)；阴阳榫接缝；乙字头；乙字管；终端连接板；制胶版；支距；绿偏移；曲线垂距；区距；逆转；偏置；偏移；偏心；位置偏移；透印蹭脏；调整偏差；水平断距距离；水平错距；失调；剩余偏差；抵销；冲抵(会计用语)；错移；错位；冲销；冲减；倾斜；残留误差；残留变形；放水管；不重合；补偿值；拔磙

offset account 抵销账户；冲销账户；备抵账户
offset adjustment 炮检距调整
offset against accounts 账款抵消
offset alignment 偏移对准
offset angle 偏置角；偏斜角；偏向角；偏角；补偿角
offset angle of fly jib (起重机的)副臂安装角
offset anomaly 分支异常
offset arc 偏移弧
offset arch 突拱
offset axis 偏移轴；位移轴
offset beam 偏移射束
offset behavio(u)r 偏移特性
offset bend 柱中竖直钢筋微弯；平移弯筋；纠偏弯筋；迂回管(弯头)；迂回管
offset binary code 补偿二进制码
offset blanket 橡皮滚筒
offset block 拐角混凝土砌块
offset box wrench 偏置套筒扳手
offset cam 偏置凸轮；偏心凸轮
offset canal 支渠
offset carrier 偏置载频；偏离载波
offset carrier frequency 偏移载波频率
off-set Cassegrain antenna 偏置卡塞格伦天线
offset charge 分散装药
offset chopping bit 冲击式偏心钻头
offset coefficient 偏移系数
offset colo(u)r 着色不均
offset component 偏差分量
offset conduit nipple 偏置管道上的喷嘴；偏置管道上的乳头
offset core 错动的岩芯
offset correction 偏差校正
offset course 阶宽墙基；突层(砌体)
offset crank 偏心曲柄
offset crank mechanism 偏心曲柄机构
offset crankshaft 偏置式曲轴；偏心曲柄轴
offset credit 抵销信贷
offset current 失调电流；补偿电流
offset cylinder 偏置汽缸
offset data 补偿数据
offset-delay relationship 绿时差—延误关系式
offset dial 偏移度盘；补偿度盘
offset digging 梯式挖沟机偏离机心作业
offset direction 偏移方向；偏离航向
offset disc harrow 偏置式圆盘耙；拖式圆盘路耙
offset distance 支距；偏距；差距
offset ditch 偏位明渠；偏位明沟
offset drilling 内插钻井；补充钻进；边界钻孔
offset duplicating 胶印复制
offset electrode 偏心式焊条
offset error 偏移误差
offset eyepiece 偏移目镜；补偿目镜
off-set facility 辅助设施；非生产设施
offset feature of ground object 地物错动特征
offset finish 错口
offset footing 宽阶底脚；偏心基础；踏步式基础；单面地脚；大方脚
offset foundation 偏心基础；踏步式基础
offset frame 偏置框架
offset frequency 偏移频率；偏频
offset gate 偏置栅极；补偿栅
offset gravure coater 补偿槽辊涂布机
offset ground zero 偏离爆心投影点
offset grouser shoe 偏置的履带板
offset guiding 偏置导星
offset guiding device 偏置导星装置
offset hand lever 曲手把
offset height 拔磙高度
offset hexagon open-end wrench 偏置六角开开口扳手
offset hexagon screwdriver 偏置六角形螺钉起子
offset hinge 长翼合页；长翼铰链；长脚铰链；长脚合页
offset hinges with ball-bearing 角尺轴承铰链；滚珠轴承偏置铰链
offset hologram 离轴全息照片；离轴全息图
offset house 错层建筑
offset hydraulic excavator 拐臂液压挖掘机
offset impedance relay 偏置阻抗继电器
offset impression 胶印
offset ink 胶印油墨
offset interchange 错位互通立交
offset intersection 错位(式)交叉
offset item 补偿项目；补偿项(目)

offset joint 偏置连接；错缝
offset lens 补偿透镜
offset limit 屈服极限
offset line 支线；支距线
offset link 奇数链接头链节；偏置链节
offset link with movable lug 带活舌偏置链节
offset lithography 胶版印刷
offset-lith-printing 平版胶印
offset local time 补偿地方时
offset machine 胶印机
offset method 永久变形(应力)测定法；平行位移测条件屈服强度方法；直角坐标法放点；支距法；偏装法；偏置法；偏距法；偏差调整法；残余变形(测定)法；补偿法
offset-mho relay 偏置姆欧继电器
offset modulus 条件模量
offset-mounted 错开装配的
offset mounting 偏置悬挂
offset nose cone 偏置头锥
offset nozzle 偏置喷管；非对称喷管
offset of edges of plates at joints 在连接处板边的偏差
offset of lead curve 支距
offset of plate edges at joints 连接处板边的偏移
offset of strata 断错地层
offset of tooth trace 齿线的偏移量
offset of U-bend 弯管的弯出距
offset of veins 断错岩脉
offset of wheel 车辆偏心
offset optic(al) square 偏置光学直角头
offset paper 胶版纸
offset parameter 补偿参数
offset parting surface 带凹凸分型面
offset pipe 乙字管；偏置管；制偏管；偏心管；迂回管
offset piping 偏心管
offset pivot 偏置轴销；铁门偏置铰链
offset plate 胶印版
offset plotting 支距标绘法
offset plow 偏置悬挂犁
offset position 偏移位置
offset post card 高级胶印卡纸
offset press 胶版印刷机；胶印机
offset printer 胶印机
offset printing 胶版印刷；胶印；胶片印刷；平板印刷
offset printing ink 胶版油墨
offset printing paper 胶版印刷纸
offset printing process 胶印法
offset printing room 胶印室
offset procedure 屈服过程
offset proofing press 胶印打样机
offset pull (与运输方向偏移的)偏牵引
offset punt 瓶底歪斜(玻璃制品缺陷)
offset qualification 偏移限定
offset ratchet screwdriver 偏置棘轮的螺钉起子
offset reserves 冲账准备
offset ridge 断错海岭
offset river 断移河
offset rod 支距杆；支距尺
offset roller 胶印辊筒
offset roll of calender 压延机旁滚筒；压延机侧辊
offset rotogravure coater 胶辊转印式涂布机
offset rudder 偏置舵
offsets 船体线型值表；船体尺度平衡表
offset scale 支距尺；偏置尺
offset screwdriver 偏头螺丝刀；曲柄起子；偏置旋凿；偏置螺丝刀；弯头螺丝刀
offset seam 模缝线偏移(玻璃制品缺陷)
offset selection 绿偏移选择【道】
offset sheets 船体型值表
offset shotpoint 偏置爆炸点
offset signal 偏置信号
offset signal method 偏置信号偏置法
offset socket wrench 弯头套筒扳手
offsets policy 补偿政策
offset spring bracket 支出弹簧托架
offset stacker 分选卡机
offset stacking 偏码(集装箱)
offset staff 支距尺；支距杆；偏距尺
offset stake 支距桩；外移桩【测】
offset stock 胶印材料
offset stream 断层水系
offset surface 偏移面
offset swather 侧置式割晒机
offset switch 偏移开关

offset table 型值表
offset tangent modulus 补偿正切模量
offset tappet 偏置挺杆
offset the amount used for imported technology 弥补进口技术需要
offset the difference 抵消差额
offsetting 支距法测量;支距(测)法;偏置法;偏心距;偏位;错移;透印;传印;蹭脏;蹭背
offsetting assets and liabilities 资产负债对冲
offsetting combustion in shaft kiln 立窑内偏火
offsetting deduction 相抵减项;抵冲扣除项目
offsetting distance 偏心距离
offsetting entry 冲抵分录
offsetting error 冲销误差;冲销差错
offsetting plow attachment 犁的侧置安装架
offsetting poor alignment 错位
offsetting receipts 补偿收入
offsetting spring 复位弹簧
offset to a debt 对借款的抵消
offset tool 偏刀;鹅颈刀
offset toolholder 偏置刀架
offset to the loss 对损失的补偿
offset track 偏离轨道;偏离磁道
offset tumbling mixer 偏心式倾翻出料搅拌机
offset-tune 失调的
offset type engine 偏置式发动机
offset uncertainty and change 消除不确定性与变化
offset unit 偏置单元;偏置器件;偏置构件;偏置部件
off-set utilities 辅助工程
offset variable 位移变量
offset voltage 失调电压;补偿电压
off-set wave 偏移波
offset well 排水井;对比井;补偿井;边界井
offset wind 离岸风
offset wrench 斜口扳手
offset yield strength 条件屈服强度;残余变形屈服强度;补偿屈服强度
off-shade 非标准色(调);败色
offshoot 岩枝;支脉;主干分支;泻水台;潜水面;旁系;飞檐;地下水面;承雨线脚
offshoot program(me) 小方案;分方案
offshore 在远海沿岸海面;向海的;远滨;离开海岸;离岸的;近海(的);近岸(的);海上;深外
offshore acoustic(al) satellite integrated system 近海声呐卫星组合装置;近海声呐卫星集成装置
offshore administrative boundary 领海界
offshore airport 海上机场
offshore anchor 开锚(外挡);外挡锚
offshore anchorage 海面锚泊;海面锚地
offshore anchor-ground 近岸锚地
offshore area 向海区;近海水域;近海区(域);近岸海域;海界区
offshore artificial island 海上人工岛
offshore auxiliary vessel 海上钻探辅助船;海上钻井辅助船
offshore bank 近海开发银行;滨外沙洲;滨外浅滩
offshore banking 境外金融
offshore banking center 境外金融中心
offshore banking transaction 离岸银行业务
offshore banking unit 离岸银行单位
offshore bar 沿岸沙坝;近海沙坝;海岸砂坝;海岸沙洲;滨外沙洲;海岸坝;沿岸沙坝;岸外坝
offshore bar pool trend 岸外沙坝油气藏趋向
offshore barrier 沿滩沙洲;沿滩沙坝;沿滩沙堤;沿滩沙坝;沿滩沙坝;滨外沙坝;滨外沙障;滨外沙堤;滨外沙坝;滨外海堤;滨外沙堤
offshore barrier-lagoon system 滨外堤—泻湖体系
offshore beach 滨外(沙)滩;滨外沙堤
offshore beach terrace 滨外滩阶地
offshore belt 滨外带
offshore bench 近岸海蚀台;海岸台;滨外海蚀台;滨海浪蚀台
offshore berth 离岸泊位;近海泊位
offshore boring 近海钻探;近海钻井;海床钻探;海上钻探;海上钻井;滨外钻孔;远岸钻探;海底钻探;海底钻井
offshore boring island 海上钻探人工岛
offshore boring platform 海上钻探平台
offshore breakwater 离岸防波堤;外海防波堤;岛式防波堤;岛堤
offshore breeze 离岸风
offshore captive insurance company 离岸控股保险公司

offshore center 离岸业务金融中心
offshore coal field 海底煤田
offshore combinate 海上联合作业基地
offshore completion 海上完井
offshore construction 近海建筑;海上建筑物;滨海结构物
offshore corrosion 海上腐蚀
offshore cost 国外价格;海外价格
offshore current 沿岸流;离岸流;近岸流
offshore deep-water berth 离岸深水泊位
offshore deep-water terminal 离岸深水码头
offshore deposit 远滨沉积;近滨沉积(物);滨外沉积
offshore development 近海开发
offshore discharge 近海卸货
offshore disposal 海洋处理
offshore diving 近海潜水
offshore dock 离岸式浮坞;岸式浮坞;升船浮坞;单坞墙式浮(船)坞;岸边浮坞
offshore dollar 境外美元
offshore dollar market 境外美元市场
offshore dredging 近海疏浚
offshore drill 海洋钻机
offshore drilling 海上钻探;近海钻探;近海钻井;海洋钻探;海洋钻井;海底钻探;海底钻井;滨外钻探;远岸钻探
offshore drilling and mining 近海钻探和采矿
offshore drilling assembly 海上钻探装置;海上钻井装置
offshore drilling barge 海上钻探船;海上钻井船
offshore drilling engineering 海上钻探工程;海上钻井工程
offshore drilling equipment 海上钻探设备;海上钻井设备
offshore drilling island 海上钻探人工岛;海上钻探塔(架);海上钻探平台;海上钻探岛;海上钻井塔(架);海上钻井平台;海上钻井岛;海面钻井岛
offshore drilling operation 海上钻探作业;海上钻探操作;海上钻井作业;海上钻井操作
offshore drilling platform 海上钻探平台;近海钻探平台;近海钻井平台;海上钻探台;海上钻井台
offshore drilling rig 海上钻探钻机;海上钻探装置;海上钻井装置
offshore drilling ship 海上钻探船;海上钻井船
offshore drilling structure 海上钻探构筑物;海上钻井结构;海上钻井结构;海面钻井结构
offshore drilling system 海上钻探系统;海上钻井系统
offshore drilling technology 海上钻探技术;海上钻探工艺;海上钻井技术;海上钻井工艺
offshore drilling tender 海上钻探补给船;海上钻井船;海上钻井补给船
offshore drilling unit 海上钻探装置;海上钻井装置
offshore drilling well 海上钻井
offshore drilling work 海上钻探工作
offshore drydock 浮坞
offshore dumping 海洋倾弃;海洋倾废
offshore dumping area 离岸排放垃圾区
offshore earthquake 近海地震;海洋地震;海上地震
offshore engineering 近海工程
offshore environment 近海环境
offshore exploitation 海上开发
offshore exploration 近海勘探;海上勘探;外海勘探;水上探验;水上勘探;水上勘察
offshore facies 近岸相;滨外相;滨海相
offshore features 海上水文特征
offshore field 海上油田;滨外油田
offshore financial center 离岸业务金融中心;境外金融中心;离岸金融中心
offshore fisheries 近海深水捕鱼
offshore fishery 近海渔业
offshore fishing 远洋捕捞作业;近海捕捞;深海捕鱼
offshore fishing grounds 海洋渔场
offshore fixed steel platform 近海固定平台
offshore floating dock 靠岸(式)浮(船)坞;海上浮(船)坞;离岸浮船坞
offshore floating drilling 海上浮式钻井
offshore floating platform 近海浮式平台
offshore floating terminal 海上浮式卸船作业站
offshore flow 近岸流
offshore fund 国外资金;海外投资;离岸基金;境外资金
offshore gas 近海天然气;海上天然气;海上采气;大陆架天然气
offshore geophysical survey 海上地球物理测量

offshore gravity meter 海洋重力仪
offshore hydrological feature 海岸水文学特征;海上水文要素
offshore industry 海洋工业
offshore installation 海上装置;近岸设施;海洋设备;海上钻探装置;海上钻井装置
offshore insurance company 海外保险公司
offshore investment center 境外投资中心
offshore island 沿海岛屿
offshore island-type berth 离岸岛式码头;离岸岛式泊位
offshore island-type breakwater 离岸岛式防波堤
offshore island-type terminal 离岸岛式码头
offshore jack-up drilling vessel 自升式钻探船;自升式钻井船
offshore line 近岸航线
offshore loading 海上系泊装油
offshore location 海上定位
offshore marine structure 离岸水工建筑物;海上水工结构(物);海上水工建筑物
offshore maritime structure 离岸水工建筑物
offshore mining 海底开采;海底采矿;水下开采
offshore mobile drilling rig 移动式近海钻探平台;移动式近海钻井平台
offshore mooring 近海系泊;海上停泊;海上锚泊;外海锚泊;离岸锚碇
offshore mooring device 海中停泊设备;海中碇泊设备
offshore mooring island structure 海上系泊岛式建筑物
offshore movement 离岸运动
offshore mutual fund 国外共同投资基金
offshore navigation 近海航行
offshore nuclear power plant 海岸核电站
offshore nuclear power station 近海核电站
offshore observation installation 海上测站
offshore observation station 海上观测站;海上测站
offshore office 海外办事处
offshore oil 大陆架石油;滨外石油
offshore oil delivery 海上油管输送;海上管路输油;海底油管输送
offshore oil-drilling 海洋石油勘探;海洋钻井
offshore oil-drilling platform 海上钻探油井工作平台;海上钻井油井工作平台
offshore oil exploitation 海上石油开采
offshore oil exploitation projects 海洋石油开发企业
offshore oil exploration insurance 近海石油开发保险
offshore oil field 海底油田;海上油田;近海油田
offshore oil industry 海上石油开采业
offshore oil reserves 海上石油储[贮]藏量
offshore oil resources 近海石油资源
offshore oil rig 海上石油钻机
offshore oil spillage 近海溢油;海岸溢油
Offshore Oil Spill Pollution Fund 海上油井溢油污染基金
offshore oil spills 近海油溢;岸外油溢
offshore oil terminal 油港
offshore oil well 滨外油井
offshore operation 海上作业
offshore patrol vessel 近海巡逻船
offshore petroleum resources 海洋石油资源;海上石油资源
offshore pile 近海桩
offshore piled platform 近海桩基础
offshore pipeline 海洋管线;近海管道;海上管线;海上管道;海底输油管;海底管道;水下管线;水下管路
offshore placer 滨海砂矿
offshore placer survey 滨外矿产调查
offshore platform 近海石油平台;离岸石油平台;近海平台;海上作业平台;海上钻探平台;海上钻台;海上钻井平台;海上平台
offshore pollution 海岸污染
offshore pollution liability agreement 海岸污染责任协定
offshore position 离岸位置
offshore power system 海上电力系统
offshore procurement 国外订购;国外采购
offshore producing 海上开采
offshore production 境外生产;海上开采
offshore production system 海上采油系统
offshore production well 海上油田

offshore prospecting 近海勘探;滨外勘探;滨海探测
offshore purchase 国外订购;国外采购;海外采购
offshore quay 离岸式码头
offshore radar platform 沿海雷达平台;海上雷达站
offshore radio relay station 近海无线电中继站
offshore rate 境外利率
offshore reef 离岸礁
offshore reinsurance 境外再保险
offshore ridge 滨外脊
offshore rig 海洋钻井设备;海洋钻机;海上钻探设备;海上钻探平台;海上钻井设备;海上钻井平台;海上钻机;浅海钻井设备
offshore route 近洋航线
offshore safety 海上安全
offshore sand bar deposit 滨外沙坝沉积
offshore sea wall 外海(防潮)堤;近海防波堤
offshore sediment 近海沉积(物);滨外沉积(物);滨海沉积(物)
offshore sediment transport 离岸输沙
offshore seism 海洋地震
offshore seismic exploration 海上地震勘测
offshore services 海上服务
offshore sewage outfall 离岸入海的污水排水口
offshore shell branch [美]境外空壳分行
offshore shoal 滨外浅滩
offshore shooting 海上地震勘测
offshore site investigation study 近海现场勘察研究
offshore slope 滨外坡
offshore soil mechanics 离岸土力学;近海土力学;滨海土力学
offshore sourcing 海外设厂
offshore spawning ground 近海产卵场
offshore spectral density function 外海波谱密度函数
offshore station 海上观测站;海上测站
offshore storage tank 海上储油柜
offshore strip 近海带;滨外带
offshore structure 离岸结构;近海建筑物;近海结构(物);近海建筑物;海洋钻架;海洋建筑物;海上建筑物
offshore supply vessel 近海供应船
offshore support ship 近海补给船
offshore support vessel 近海补给船
offshore surveillance 海上监视
offshore surveillance system 海上监视系统
offshore survey 近海测量
offshore tanker berth 离岸油轮泊位
offshore tanker terminal 近海油轮码头;近海油轮终端;近海油港;海上中转油罐
offshore technology 近海技术;海洋技术
offshore terminal 近海式码头;近海转运码头;海上中转站;岛式码头
offshore terrace 近海陆地;滨外陆地;滨后陆地
offshore track 近海航路
offshore trade transactions 境外贸易往来
offshore trough 近岸海槽;滨外海槽
offshore tug supply ship 近海拖带供应船
offshore unloading 海上卸油
offshore water movement 离岸水流运动
offshore waters 近海水域;海岸区(域);海岸水域;滨水水域;滨海水域
offshore wave 近海波(浪)
offshore wave condition 外海波况
offshore wave data 外海波浪资料
offshore wave recorder 海上自记测波仪;海上自记测波计;波浪计
offshore weather 海洋天气
offshore well 海上油井
offshore well drilling platform 近海油井钻井平台;滨外油井钻井平台;近海钻探平台;近海钻井平台
offshore wellhead assembly 海洋钻探井口装置
offshore wharf 离岸式码头
offshore wind 向海风;陆风(从陆地吹向海洋的风);离海风;离岸风;海风;吹开风
offshore wind power system 海上风力发电系统
offshore work platform 海上作业平台
offshore works 近海工程;海上工厂
offshore zone 离岸区;近海海区;近海地带;近海带;大陆架海区;滨外带
off-shutter concreting 无模板浇注混凝土
off-shuttering concrete 脱模混凝土;无模板混凝土;混凝土表面处理

off-side 远侧;对侧
off-side of bearing 轴承的出口端
offside overtaking 里档超车(不合规章的超车)
off side piping 外管
off-side reflection 侧面反射
off-site 装置外;工地外;厂区外
off-site benefit 间接效益
off-site construction 基地外建设
off-site control 装置外控制
off-site cost 外围设施费用
off-site engineer 场外工程师;非现场工程师
off-site fabrication 非现场装配;现场外预制;场外预制
off-site facility 工地外设施;厂外设施;装置外设施
off-site hazardous waste facility 非现场有害废物设施
off-site impacts of soil erosion 土壤侵蚀对其他地方的影响
off-site improvement 产权范围外的场地改善
off-site management cost 场外管理费
off site production 现场外的生产检查
off-site screening and storage 离现场筛分和储[贮]存
off-site special purpose rail 对外铁路专线
off-site surveillance 场外监视;非现场监视
off-site traffic and transport 对外交通运输
off-site work 场外工程
off-size 等外品;不合规格的;尺寸不合格;尺寸不符合要求;非规定大小;非规定尺寸;不合尺寸的
off slip 解除初保单
off-smelting 熔炼成分不合格
off-sorts 等外品
off sounding 深海区;在一百拓等深线外航行
off-specification 不合格的;不合格
off-specification goods 不合规格产品
off-specification material 等外品;不合规格的物料
off-specification requirements 超出规范要求
off-speed signal 速率偏差信号
offspring 后代;次级粒子;二次粒子
off-spur tunnel 岔洞;分岔隧道
off-square 经纬缩率差;斜方(轧件缺陷);不成正方形
off-stage 幕后(的);离开屏幕(的);离开舞台(的);不在舞台上(的)
off-standard 非标准(的);不合标准的
off-standard product 非标准品
off-standard size 非标准尺寸
off-standard voltage 非标准电压;不合标准电压
off-state 截止态;断开状态
off-state current 异常电流;关态电流;闭态电流
off-state time 断开时间
off status 关状态
off-stoichiometric combustion 不按化学计量燃烧
off-stream 停运
off-stream impoundment 河外蓄水
off-stream (pipe)line 停用管线;备用管线;旁流管线
off-stream storage 不使用仓库;备用仓库;河外蓄水;非河上水库;非河道水库
off-stream unit 停用设备;停工装置;备用设备
off-street 不靠街面的
off-street bicycle parking space 路外自行车停车场;街道外自行车停车场
off-street entrance 路外入口;街道外进口
off-street parking 路外停车;街道外停车;街道停车;街后停车
off-street parking area 街外停车场;路外停车场
off-street parking lot 街外停车场;路外停车场
off-street parking space 街外停车场;路外停车场
off structure well 构造边界井
off-sulphur iron 去硫铸铁
off sum 废渣
off-system unit 制外单位
off-take 承购;泄水口;泄水处;取出;排污管;排水沟渠;排水沟口;排气口;排出;出口(指水、气、汽、油等);出风道;分支;分接
off-take drift 排水平峒
off-take main 排水干管
off-take pipe 排水管(道)
off-take piping 排水管(道)
off-take point 排水点
off-take structure 分水建筑物
offtaking 引取水量
off taste 异味

off technical requirements 超出规范要求
off-test 试验结束
off-test product 未经检验的产品;不合格产品
off the beam 正横方向
off-the-book loan 账外借款
off-the-book property 账外财产;账外资产
off the book quotation 黑市价(格)
off the coast 在远离沿岸海面;离开海岸
off the course 偏离航向
off the gold standard 放弃金本位
off-the-job accident 岗位外事故
off-the-job training 脱产培训
off the land 在离岸方向;离开陆地
off the payroll 被解雇
off the peg 有定型产品的;现成的
off the port 靠近港
off-the-rack 现成的
off the record 非正式的
off-the-road test 越野性试验
off-the-road tire 越野轮胎;工程机械轮胎
off-the-shelf 现成产品;标准产品;现货供应;现成的;流行的;即用产品;成品的;畅销货;畅销的;非指定设计的
off-the-shelf camera 现用摄影机
off-the-shelf equipment 准军用设备;定型设备
off-the-shelf gear 现成设备;准军用设备
off-the-shelf hardware 现成设备;定型硬件
off-the-shelf integrated circuit 现成集成电路
off-the-shelf item 现有项目;现用项目;准军用产品;实用项目
off-the-shelf part 现成零件
off-the-shelf price 货架交货价
off-the-shelf solution 现成解决办法
off-the-shelf system 现用系统
off-the-shelf transponders 成批生产的运输机
off-the-shelf transports 成批生产的运输机
off-the-site work 场外工作
off the stock 下水了的(船舶);已完成的;已发送的
off-time 关机时间;非操作时间;空闲时间;停止时间;停机时间;断开时间;非正常状态;非定时的;不正常时间;不工作时间;不当班时间
off-time report 非定时报告
off-time wages 非工作时间工资
offtrack 偏离磁迹;出轨
off-track controller 跑偏控制器
off-track error 偏离轨道误差
off-track freight station 市区货运站
off-track housing 轨外房屋
off tracking 跑偏
off-track machine 地面机械【铁】
off-track safety test 脱轨安全综合试验
off transistor 不导通晶体管
off-trigger tube 截止触发管
off-tube 闭锁管
off-type 劣形;非典型
off-type variety 非典型品种;变劣品种
offward 向海方向
off-white 近白色;近于纯白的;米色;灰白色(的)
off working state 非工作状态
of glacial-marine sediment fraction by stages 浮冰碎屑分期
of great draught 吃水深的
of no consequence 无关紧要
of super-ior grade 超等
of the same colo(u)r 一色
oft-repeated 多次重复的
of unknown etiology 病原不明的
ogalloy 含油轴承
ogas awara high 小笠原(副热带)高压
ogdohedron 八面体
ogdohedry 八面体性;八半面像
ogee S形线脚;S形饰;双弯曲线;葱形饰;反曲线饰;反弧形(即溢流面曲线)
ogee arch 内外四心桃尖拱;葱头形拱;双弯形屋顶;S形拱
ogee bit 双弯曲嚼子
ogee bracket foot 双曲线(家具)腿;S形(家具)腿
ogee curve S形曲线;双弯曲线
ogee dam 渥奇溢流坝;反弧形溢流坝
ogee diversion dam 反弧形引水坝
ogee doorway 葱形门道
ogee downstream slab 下游向S形板
ogee gable 葱形山头;葱形山墙;葱头形山墙
ogee gutter 葱形檐槽;船底沟槽

ogee joint 管子插承接合
ogee mo(u)lding S 形装饰线脚;波状花边;枭混线【建】;双曲线条;葱形线脚
ogee plane (iron) 双弯曲刨刀;双弯曲刨铁;凹槽刨;浅脚刨
ogee roof 双弯形屋顶;S 形屋顶;渥奇式屋顶;葱形屋顶;葱头形屋顶
ogee roof gutter 葱形屋檐槽;S 形屋顶雨水沟
ogee spillway 双曲形溢洪道;双弯形溢流堰;渥奇式滚水坝;实用溢洪道;葱形溢洪道;反弧形溢流道;S 形溢洪道
ogee washer S 形边垫片
ogee weir S 形低坝;S 形堰;实用溢洪堰;反弧形溢流堰
ogee window 葱头形窗
ogival 卵形的;尖拱的;尖顶式的;尖顶部;蛋形
ogival arch 尖拱;葱形拱;尖顶拱;内外四心桃尖拱;葱头形拱
ogival dome 尖顶穹隆;葱形圆顶;葱头形穹顶
ogival plain tile 尖拱平砖
ogival-shape body 卵形体
ogival shell 尖瓶形壳体
ogive 累计频率曲线【水文】;葱形穹顶;拱形体;卵形拱;法国挑尖拱;头部尖拱;尖顶部;尖冰拱纹;葱形穹顶拱;葱形穹顶;分布曲线
ogive curve 拱形曲线
ogive distribution 累积频率分布
oglesby blue granite 细粒黑云花岗岩
Ogmensteel (钢铁工程中的)混凝土外壳加固方法
O. Graf visco(si)meter 格氏黏度计
ohba-gusu oil 樟树油
Ohio cofferdam 俄亥俄州式双层式围堰
Ohlson galactic pole 奥尔森银极
ohm 欧姆;欧(电阻单位)
ohmad 欧马德(旧的电阻单位)
ohmage 欧姆阻抗;欧姆值;欧姆电阻
ohmal 铜镍锰合金
ohm ammeter 欧姆安培计;欧安计
Ohmax 高欧姆铁铬钼电阻合金
ohmer 直读式电阻表;电阻表
ohm ga(u)ge 电阻计;电阻表;欧姆计;欧姆表
ohmic 欧姆的
ohmic base resistance 基极欧姆电阻
ohmic collector resistance 欧姆集电极电阻
ohmic contact 欧姆接触
ohmic electrode 欧姆电极
ohmic heating 欧姆加热
ohmic leakage 漏电阻
ohmic loss 欧姆损失;欧姆损耗;铜耗;电阻损失;电阻损耗
ohmic polarization 欧姆极化
ohmic resistance 直流电阻;欧姆电阻
ohmic-resistance loss 欧姆电阻损耗
ohmic resistance test 欧姆电阻试验
ohmic resistor 线性电阻器;欧姆电阻器
ohmic value 欧姆值
ohmilite 水硅钛锶石
ohm-m/cm 欧姆—米/厘米
ohmmeter 欧姆计;欧姆表;电阻表
ohm range 欧姆量程
ohm relay 欧姆继电器;电阻继电器
ohm's law 欧姆定律;电阻定律
ohm's law wheel 欧姆定律轮
ohm type reactance relay 欧姆型电抗继电器
ohm unit 电抗元件
o-hydroxy acetophenone 邻羟基苯乙酮肟
oikocryst 主晶
oil 油剂;含油气煤盆地;涂油
oil abrasion 油磨
oil absorbency 吸油性能
oil absorbent adhesive 吸油性黏合剂
oil absorber 油吸收剂
oil-absorbing 吸油的
oil-absorbing coating 蓄油表层
oil-absorbing polymer 吸油性聚合物
oil absorption 吸油能力;吸油量;吸油的;吸油作用;油(的)吸收
oil absorption material 吸油材料
oil absorption process 吸油方法
oil absorption test 吸油试验;吸油性能试验;吸油试验
oil absorption tower 油吸收塔
oil absorption (value) 吸油量;吸油值
oil absorption velocity of felt 原纸吸油速度
oil absorption volume 吸油量

oil absorptiveness 吸油性能
oil abundance zone 石油富集带
oil accident 漏油事故
oil accumulation 石油积聚;油藏;油床;石油聚集
oil accumulator system 蓄油器系统
oil acid free plastic gloves 耐油酸手套
oil acne 油脂性粉刺
oil addition 添加润滑油;加油
oil additive 油的添加剂
oil adsorption content 吸油量
oil adsorption material 吸油材料
oil aeration 油发泡
oil ag(e)ing 润滑油的氧化
oil air pump 油泵;油气泵
oil alkyd 石油醇酸
oil alkyd paint 石油醇酸油漆;油醇酸树脂漆
oil alkyd resin 石油聚酯树脂;石油醇酸树脂
oil ammoniac 氨草油
oil analysis 原油分析;油分析
oil and bulk carrier 运油和散(装)货船
oil and gas 油和煤气
oil and gas accumulation of series trap 系列圈闭的油气聚集
oil and gas accumulation of single trap 单一圈闭的油气聚集
oil and gas-bearing formation 含油气建造
oil and gas bearing trap 含油气圈闭
oil and gas cutting 油气侵
oil and gas drilling 石油天然气钻探
oil and gas field 油气田
oil and gas field exploitation 油田和气田开发
oil and gas inflow 油气侵
oil and gas lease 石油天然气勘探开采租约;石油和天然气勘探开采租约
oil and gas pipeline 输油输气管道
oil and gas pool 油气藏
oil and gas production 产油气量
oil and gas prospect 含油气远景
oil and gas province 油气省
oil and gas region 含油气区
oil and gas reserves 油气储量
oil and gas resources 石油及天然气资源
oil and gas separator 油气分离器
oil and gas telemetry 石油和天然气遥测
oil and grease 油脂;油类
oil and grease distributor 润滑油和润滑脂分配器
oil and grease removal 油类与油脂去除
oil and grease resistance 耐油脂性(能)
oil-and-moisture trap 油水分离器
oil and oil correlation 油-油对比
oil and show correlation 油-油苗对比
oil and smoke cleaner 净油烟机
oil and soap insecticide 油皂杀虫剂
oil and source rock correlation 油母岩对比
oil-and-water quenching tank 油水淬火槽
oil-and-water separating equipment 油水分离装置
oil-and-water separator 油水分离器
oil-and-water trap 油水收集器;油水阱;油水分离器
oil anti-foaming agent 润滑油的抗沫剂
oil antioxidant 油脂抗氧化剂
oil appliance 石油用品;石油使用;燃油器
oil application 浇油
oil area 含油面积;含油地带
oil arm 集油臂
oil asphalt 油沥青;石油(地)沥青
oil atomization 油雾化
oil atomizer 雾化喷油器;油喷雾器;喷油器;喷油管
oil baffle 挡油圈;挡油板
oil baffle disc 挡油盘
oil baffle washer 挡油垫圈
oil bag 油袋;镇油油袋
oil ballast 油渣
oil ball bearing theory 油膜滚珠轴承理论
oil bank 聚油带
oil bank break through 含油带向井突穿
oil barge 油船;油驳;供油驳
oil barrel 油桶
oil barrel pump 油桶打油泵
oil barrier 油栏
oil base 油品的基;油基;石油基地
oil based binder 油类黏结剂
oil-based building mastic 油基建筑用玛琋脂;油基建筑用胶黏剂

oil-based ca(u)lking compound 油基嵌缝料
oil-based coating 油基漆
oil based core 油基泥浆钻取的岩芯
oil-based ink 油基印墨
oil-based mastic 油基脂膜;油基胶黏剂
oil-based mastic joint sealer 油基玛琋脂接缝密封垫
oil-based medium 油基介质
oil-based mud 石油基泥浆;油基泥浆;石油基钻泥
oil-based paint 油基(油)漆;油底漆
oil-based paint coating 油基油漆涂层;油基油漆涂膜
oil-based primer 油性底漆
oil-based putty 油基油灰
oil-based scumble glaze 油基薄涂法使用的瓷釉
oil-based stain 油基色斑;油基污点
oil-based varnish 油基清漆
oil-based vehicle 油基溶剂
oil basin 油箱;油池;储[贮]油器;含油盆地
oil bath 油浴(器);油浸浴;油浸池;油浸槽;储[贮]油器;油池;油槽
oil bath air breathing 油浴式滤清器通气管
oil bath air cleaner 油浴式空气滤清器
oil bath clutch 油浴式离合器
oil bath filter 油浴过滤器
oil bath gear 油浴齿轮传动
oil bath gearbox 油浴式齿轮箱
oil bath lubrication 油浴润滑;油槽润滑
oil bath oiling 油浴法
oil bath resistance furnace 油浴电阻炉
oil bath technique 油浴技术
oil bath type air cleaner 油浴式空气滤清器;油浴式空气净化器
oil bath type bearing 油浴润滑式轴承
oil bath type oil circuit breaker 油槽式充油断路器
oil-bearing 含油(的)
oil-bearing bedil coat 油层
oil-bearing boundary 含油边界
oil-bearing crop 油料作物
oil-bearing depth 油层深度
oil-bearing emulsifying wastewater 含油乳化废水
oil-bearing emulsifying wastewater treatment 含油乳化废水处理
oil-bearing formation 含油层
oil-bearing formation casing 油层套管
oil-bearing inter margin 含油内边界
oil-bearing medium 含油介质
oil-bearing outer margin 含油外边界
oil-bearing reservoir 储油层
oil-bearing sandstone 含油砂岩
oil-bearing seeds 油籽
oil-bearing shale 油页岩;含油页岩
oil bearing stratum 含油矿层;含油岩层
oil bearing structure 储油构造
oil bearing trap 含油圈闭
oil bearing tree 木本油料树
oil-bearing wastewater 含油废水
oil-bearing wastewater treatment 含油废水处理
oil-bearing wastewater treatment method 含油废水处理方法
oil-bearing water bed 含油水层
oil-bearing wax 含油蜡
oil bed 油层
oil belt 油带;含油地带;含油带
oil belt duck 重型防水帆布
oilberg 超级油轮
oil berth 油码头;油轮泊位
oil bilge collecting tank 油污水汇集柜;舢集油柜
oil binder 油质黏合剂
oil black 石油炭黑
oil blackeite 蓝化钢
oil blackening 蓝化
oil-black plant 石油炭黑厂
oil blast 油灭弧
oil-blast explosion chamber 喷油灭弧室
oil-blast switch 油灭弧开关
oil bleaching 油脱色;油漂白
oil bloom 油花;油的荧光
oil blow-up type impulse breaker 喷油熄弧式脉冲断路器
oil blue 油蓝
oil body 油体;油的体度;润滑油的黏度
oil bodying 油的聚合
oil boiler house 燃油锅炉房
oil-bond distemper 油料色粉涂饰

oil-bonded core 油砂芯
oil-bonded sand 油砂
oil boom 拦油栅；拦油浮栅；拦油臂架；集油臂架；围油栏
oil boom tubing 油栅管
oil booster pump 油压油泵；增压油泵
oil borehole 油井
oil borne abrasion 油中杂质引起的磨损
oil-borne preservative 油溶性防腐剂；含油防腐剂；油性防腐剂
oil-borne preservative solution 含油防腐剂溶液
oil-borne stains 油媒污渍
oil-bound 含油的
oil-bound distemper 油性色粉涂饰；油性墙粉；油料色粉涂饰；油粉；含油水浆涂料
oil-bound film 油膜；润滑油膜
oil-bound water paint 油性水彩颜料；油粉；含干性油的水性涂料
oil bowl 油杯
oil box 油箱；油盒
oil box type spindle rail 油箱龙筋
oil brake 油制动器；油压制动器；油压制动
oil brand 油品的牌号
oil-break circuit-breaker 油断路器
oil breaker 油开关
oil-break fuse 油熔断器
oil-breaking 油裂解
oil-break switch 油开关；油断路开关
oil breather 润滑油呼吸管；润滑油通气管；油枕
oil brightening 上油增艳处理
oil brush 油刷
oil bucket pump 移动式润滑油加注器
oil buffer 油压缓冲器；油缓冲器；液压缓冲器
oil buffer stroke 油缓冲器行程
oil-bulk-ore carrier 油—散货—矿石三用船；石油散货矿石三用船
oil bumper 油减振器；油缓冲器
oil bunker 油库；燃油舱
oil bunkering 装燃油
oil-bunkering dock 船用燃油码头
oil-bunkering facility 船用燃油设备
oil-bunkering port 加油港
oil-bunkering wharf 船用燃油码头
oil-buoyancy flo(a)tation 油浮选
oil burner 油灯；油燃烧器；油焰器；油喷嘴；燃油炉；汽灯；烧油船
oil-burning 燃油的；烧油的
oil-burning appliance 燃油设备；燃油工具
oil-burning assembly 燃油装置；燃油系统
oil-burning boiler 燃油锅炉
oil-burning boiler house 燃油锅炉房
oil-burning burner 烧油的燃烧器
oil-burning cooker 烧油烹饪器
oil-burning engine 使用比汽油重的燃料的发动机
oil-burning equipment 燃油设备
oil-burning fire 烧油的炉火
oil-burning heating 燃油采暖；燃油供暖
oil-burning heating system 燃油供热系统；燃油供暖系统
oil-burning locomotive 燃油机车
oil-burning stove 燃油炉
oil-burning warm air heater 燃油热风加热器
oil bush 充油套管
oil byproducts 油类副产品
oil cake 沥青混凝土路面(俚语)；油饼；豆饼
oil-cake breaker 油饼碎裂机
oil calorimeter 油量热计
oil can 小油桶；油罐；加油车；加油罐；运油车
oil canal 油道
oil-can auto-lock ratchet 油壶的自锁棘轮
oil-can lid 油壶盖
oil-can lubrication 油壶润滑
oil-canning 膨胀下陷；(油桶；油罐式变动；(金属屋面材料)皱缩；卷曲；白铁罐
oil-can ratchet autolock 油壶的自动盖头
oil capacitor 油浸电容器
oil capacity 油容量；容油量
oil car 油罐车；油槽车
oil carbon 油碳
oil-carbon sludge 油—炭质残渣；炭质残渣
oil cargo tank 货油舱
oil carrier 油轮；油船
oil carrying ship 油船
oil-carrying submarine 输油潜水艇

oil case 油箱
oil casing depth 油层套管深
oil-catalyst slurry 催化剂油浆
oil catch 油挡
oil catcher 油滴接斗；集油器；受油器；盛油器
oil-catch ring 挡油圈
oil ca(u)lking 油性堵(塞)缝
oil cavity 油室；润滑油沟；润滑孔
oil cellar 地下油库
oil cellar wedge 油盒楔
oil cell bottom 油箱底座；油箱沉积物
oil centrifuge 油离心机
oil chamber 油室；润滑油室；储油器
oil change 润滑油更换
oil change period 换油周期
oil changer 换油器(机油)
oil channel 油路；油槽
oil channel plug 油道塞
oil charge 充油
oil charge valve 充油阀
oil check valve 油止回阀
oil chuck 油压卡盘；液压卡盘
oil circuit 油路
oil circuit breaker 油开关；油浸断路器；油断路器
oil circulating lubrication 油循环润滑(法)
oil circulating pipe 循环油管
oil circulating pump 循环油泵
oil circulating reactor 油循环反应器
oil circulating system 油循环系统
oil circulation 循环油路；油循环
oil circulation cooling 循环油冷；油循环冷却
oil circulation ga(u)ge 润滑油路压力计
oil circulation lubricating system 循环润滑油系统；油循环润滑系统
oil circulation lubrication 循环润滑
oil circulation process 油循环过程
oil circulation system 油循环系统
oil circulation valve 循环油阀
oil cistern 油杯
oil clarifier 油料澄清器；净油器；净油池；润滑油澄清器
oil classification 油品分类
oil-clay mixture 油与白土的混合物
oil cleaner 油净化器；油过滤清器；滤油器
oil cleaner screen 滤油网
oil cleaning device 油液净化装置
oil cleaning system 油净化系统
oil cleaning tank 油净化箱
oil clearance 油隙
oil clot 油块
oil cloth 油布
oil cloth dryer 油布干燥器
oil clutch 油压离合器
oil clutch drive 液压离合器传动
oil clutch drive housing 液压离合器传动箱
oil coal 沥青煤；土沥青
oil coalescer 油聚结器
oil coat 油膜；油层；油布雨衣
oil-coated 涂油的
oil-coated stone 沥青砾石；油石
oil cock 油旋塞
oil coil 油旋管
oil coke 石油焦炭；油焦
oil cold test 油冷却试验
oil collecting pipe 集油管
oil collecting system 集油系统
oil collecting tank 集油罐
oil collecting tray 集油盘
oil collection vessel 浮油回收船
oil collector 集油器；油收集器；储油箱
oil colo(u)r 油质颜料；油性包料；油溶性颜料；油溶性染料；油画颜料
oil column 加油柱；油桶；油罐式变动；加油塔
oil combustion 油燃烧；石油燃烧
Oil Companies Institute for Marine Pollution Compensation Ltd. 石油公司海上油污赔偿协会
Oil Companies International Marine Forum 石油公司国际航运论坛
oil compass 油液罗经
oil compatibility 油的相容性；油的混合性；油的兼溶性
oil compensating relief valve 滑油平衡安全阀
oil composition 油的组成
oil composition analysis equipment 原油组分分析设备
oil compression cable 油浸电缆；高压充油电缆
oil concentration 油分浓度；乳化
oil concentration analyzer 油水浓度计；油分浓度计
oil concession 石油开采权
oil condense 油浸电容器
oil condition 油的情况
oil conduit 油管
oil connecting pipe 油连接管
oil connection 油管接头
oil conservation 石油资源保护
oil conservator 油枕；存油器
oil consumption 油耗量；油耗；耗油量；润滑油消耗量
oil consumption country 石油消费国
oil consumption ratio 耗油率
oil contactor 油接触器
oil container 油箱；储[贮]油器
oil containment boom 拦油臂架
oil-contaminated surface water 油污染地表水
oil-contaminated water 含油污水
oil contamination 油污；油污染
oil content 油分(指油的含量)；含油量
oil content meter 油分浓度计
oil content monitoring arrangement 油分监测装置
oil control 润滑油调节
oil controlled valve 油动阀；液动阀
oil control ring 护油圈；润滑油控制环
oil conversion process 油的转化过程
oil conveying hose 输油软管
oil conveying pipe 输油管
oil conveying pipeline 输油管道
oil conveying piping system 输油管系(统)
oil-cooled 油冷的；油冷却的
oil-cooled clutch 油冷式离合器
oil-cooled stator winding 油冷定子绕组
oil-cooled system 油冷法
oil-cooled transformer 油冷(却)变压器
oil cooled turbogenerator 油冷汽轮发电机
oil-cooled valve 油冷真空管
oil cooler 油冷却器；冷油器
oil cooler access cover 通机油散热器的舱口盖
oil cooler core 冷油器芯子
oil cooler flap 冷油器通风片
oil cooler jacket 冷油器套
oil cooler shutter 油冷却器百叶窗
oil cooler unit 滑油冷却装置
oil cooling 油冷却的；油冷(却)
oil cooling method 油冷却法
oil cooling pipe 油冷却管
oil cooling radiation 滑油冷却器
oil cooling radiator 冷油散热器
oil cooling system 滑油冷却系统
oil core 油芯；油砂芯
oil core tube 油芯管
oil corona 油电晕
oil corrosion 油品腐蚀；油腐蚀
oil corrosion test 油腐蚀试验
oil country tubular goods 油田用管材设备；石油工业用管材
oil cracker 油裂化器
oil crane 油鹤；加油塔；加油臂
oil crisis 石油危机
oil crop 油料作物
oil crust 油垢
oil cup 油杯
oil cup cap 油杯盖
oil cup cover screw 油杯盖螺钉
oil cup cover spring 油杯盖簧
oil current breaker 断油器
oil curtain 油幕
oil cut-off valve 油截止阀；断油阀
oil cut-off valve bonnet 断油阀帽
oil cut-off valve handle 断油阀柄
oil cut-off valve seat 断油阀座
oil cut-off valve stem 断油阀杆
oil cut-out valve 给油总阀
oil cut-out valve spindle 给油停闭阀杆
oil cylinder 油压缸；油缸
oil dag 石墨润滑剂；石墨润滑油；胶体石墨；黑铅油
oildag 胶体石墨；石墨悬浮油剂；石墨膏
oil damper 油阻尼器；油减震器
oil damping 油阻尼；油减震
oil darkening 油品变黑

oil dashpot	油减震器;油缓冲器;油缓冲罐;减振油缸;缓冲油壶
oil dash pot relay	注油壶继电器
oil decolorization	油品的脱色
oil decomposition	油分解;油的分解
oil deficit	石油赤字
oil deflector	导油圈;挡油圈;挡油板
oil deflector ring	挡油环
oil deformation valve	油膜光阀
oil degradation	油液降解
oil-degrading bacteria	油降解菌
oil degumming	油品脱除胶质
oil dehydrating	油品脱水
oil dehydrating plant	油脱水装置;油品脱水装置
oil dehydration plant	油的脱水装置
oil dehydrator	原油脱水器;油料脱水器
oil delivery	油品输送;石油输送
oil delivery pipe	输油管
oil delivery pump	滑油输送泵
oil delivery truck	运送油品的车辆
oil delustering	油消光
oil demand	石油需要量
oil demulsibility	油品的破乳化性
oil density	原油的密度
oil denuding	油的脱吸
oil depletion	石油矿藏耗减
oil depletion allowance	石油矿藏耗减优惠
oil deposit	油渣;石油矿床;油料沉积物
oil depot	油站;油库区;油库
oil-depth ga(u)ge	油位测量计
oil derivative	石油衍生物
oil derrick	油井钻探设备;油井塔架;油井架
oil derrick legs	油井架腿
oil desulfurization	原油脱硫
oil detergent	清洗机油的洗涤剂
oil deterioration	油品变质
oil development	石油开发
oil development equipment	石油开发装置
oil dewaxing	油脱蜡
oil diffusion pump	油气扩散真空泵;油扩散泵
oil diluent gasoline	润滑油稀释用汽油
oil dilution	油料稀释;油的稀释
oil dilution system	滑油稀释系统
oil dilution tank	油的稀释槽
oil dilution test	润滑油稀释量测定
oil dilution valve	滑油稀释阀
oil dipper	油勺;油杓;抽油器
oil dipping	油乳浊液浸渍处理
oil dipstick	油尺
oil dirt	油垢
oil disc	油盘;存油盘
oil-disc brake	液压盘式制动器;油盘制动器
oil discharge	卸油;泄油
oil discharge outlet	放油口
oil discharge pipe	卸油管;泄油管;放油管
oil discharge plug	放油塞
oil discharge valve	泄油阀;放油阀
oil discharging efficiency	卸油效率
oil discharging rate	卸油效率
oil disclo(u)ration	油污斑渍
oil dispenser	油分配器
oil dispersant	油分散剂
oil dispersant mixture	油分散剂混合液
oil-dispersing property	油分散性
oil disposal hydrant	排油栓;油龙头
oil-dissolving solvent	油溶性溶剂
oil distillate	石油馏分
oil distillate fraction	油的蒸馏物组分;焦油分馏
oil distributing box	润滑油分配箱
oil distributor	油量分配器;喷油机;配油器;受油器;分油器(轴向)
oil distributor pipe	油分配器的管道
oil dock	油码头;油船坞
oildom	石油产区
oil dope	油品添加剂;润滑油涂料
oil drag	油的摩擦阻力
oil drag loss	油膜摩擦损失
oil drain(age)	放油;油类流出口;排油;排油管
oil drainage cone	排油锥形筒;放油锥形筒
oil drain canal	放油道
oil drain cock	卸油阀;排油管龙头;放油旋塞;放油开关
oil drain drum	排油罐
oil drainer	泄油塞;放油塞
oil drain hole	放油孔
oil drain out	排油口
oil drain plug	放油旋塞;放油塞
oil drain pump	排油泵
oil drain tank	油污柜;聚油箱;聚油槽
oil drain valve	卸油阀;泄油阀;放油阀
oildraulic strut	油液支柱
oil drawing gun	抽油枪
oil dregs	油料的沉淀
oil dressing	浇沥青
oil drier	油干料;油催干剂
oil drilling	石油钻探
oil drilling equipment	石油钻井装置投资
oil drilling island	石油钻探平台;石油钻井人工岛
oil drilling platform	石油钻探平台投资;石油钻井平台投资;石油钻井平台
oil drilling waste	石油钻井废液
oil drip	滴油器
oil drip feed	滴油润滑
oil drip pan	油滴盘
oil drippings	油滴;滴油;油滴污垢
oil drive	油传动
oil-driven back thread	回油线
oil-driven pump	油动泵
oil drop	油点;油滴
oil droplet	油腺;油滴
oil dropping apparatus	给油箱
oil drum	油桶
oil drum stand	油桶架
oil drying	油干燥
oil dual thermometer	油用温度计
oil duct	油路;油沟
oil dumped pressure ga(u)ge	充油压力表
oil duties	石油进口税
oil economizer	省油器
oiled	上了油的
oiled and edge sealed	(混凝土胶合板模板的)涂油封边
oiled application maintenance	浇沥青养护;浇油养护
oiled birds	油污染海鸟
oiled can	油浸罐头
oiled canvas	油帆布
oiled clay base	油土基层
oiled cloth	绝缘油布
oiled dusts	加过油的粉剂
oiled earth	油土;浇油土;泼油土(用铺路油类或沥青处理过的土)
oiled earth construction	油土筑路法
oiled earth road	油土路
oiled earth surface	油土面层;油土路面
oiled hiding power	油性遮盖力
oiled leather	涂油革
oiled linen	油布
oiled ore	油化矿石
oiled paper	油纸;油绝缘纸
oiled paper strip	油浸纸条
oiled pick	油纬
oiled-plate method	油板
oiled sand	油砂
oiled sandy-clay	拌油砂土
oiled silk	油绸
oiled wood	油木
oil effect on aquatic life	石油对水生生物的影响
oil ejector	油喷射器;注油器;喷油器
oil ejector pump	油喷射泵
oil-electric(al)	内燃电动的
oil-electric(al) bus	内燃电力车;内燃电动车
oil-electric(al) car	内燃电力汽车
oil-electric(al) drive	内燃机发电驱动
oil-electric(al) engine	内燃发电机;柴油发电机;柴油发动机
oil eliminator	油分离器;除油器;分油器;分油机
oil embargo	石油禁运
oil-embolus	油栓子
oil emptying	油倒空
oil emulsion	乳化油液;油乳浊液;油状乳液;胶;油乳化液;油品乳化液
oil emulsion adjuvant	油乳佐剂;油乳化佐剂
oil emulsion cement	油乳水泥
oil emulsion composition	油乳胶组成
oil emulsion drilling mud	混油泥浆;乳化原油钻井液
oil emulsion mud	油乳胶浆;油乳化泥浆;乳化原油泥浆
oil enamel	油性瓷漆
oil-enclosed power	油封齿轮箱
oil engine	油机;内燃机;燃油机;柴油机;柴油发动机
oil-engine driven	内燃机驱动的
oil engine driven device	内燃机驱动装置
oil engine driven plant	内燃机驱动设备
oil entry profile	出油剖面
oil equipment	石油设备
oiler	油船;自动涂油器;注油器;加油器;加油工;加油船;机舱加油;润滑工;涂油机
oiler cap	油杯盖
oiler submarine	潜水油轮
oilery	油库
oilex	发热硅铁
oil exit pipe	排油管
oil-expansion vessel	扩油器;储[贮]油柜
oil expeller	榨油机
oil exploitation	石油开采
oil exploration	石油勘探
oil exploration data base	石油勘探数据库
oil extend	充油丁苯橡胶
oil-extended rubber	油充橡胶;充油橡胶
oil-extended styrene buadiene rubber	充油丁苯橡胶
oil-extender	油性填充剂;石油软化剂
oil extraction	用油萃取
oil extraction wastewater	采油废水
oil extractor	油提取器;油料抽提机;抽油机
oil face	油面
oil facility	石油信贷(为石油购买者提供的信贷);石油贷款
oil factor	吸油值
oil failure	油污
oil-fast	防油的
oil fastness	耐油性;抗油性
oil faucet	注油龙头
oil feed	供油;给油
oil feeder	油壶;注油器;给油器;加油器;调油器
oil feeding device	润滑油供给装置
oil feeding hole	添油孔
oil feeding reservoirs	供油箱
oil feeding socket	给油套节
oil feeding system	给油系统
oil feeding tank	供油箱
oil feed injector	油料注入器;进油注射器
oil feed pump	给油泵
oil felt pad	油毡垫
oil felt pad asphalt	油毡沥青
oil fence	油栅;油栏;拦油浮栅;围油栏
oil field	油田
oil field automation	油田作业自动化
oil field bit	油井钻头
oil field body	石油野外钻探车车身
oil field brine	油田盐水;油田水;油田卤水
oil field casing	油井套管
oil field cementing	油井灌浆
oil field emulsion	矿场乳化油;矿场石油乳化液;油田乳化油
oil field equipment	油田设备
oil field gas	油田气
oil field model	油田模型;电解模型
oil field oil-bearing wastewater	油田含油废水
oil field pump	油田泵
oil field rotary	油井转盘
oil field rotary bit	油井旋转钻头;石油井用旋转钻头
oil field separator	油气分离器
oil field tank	油田储[贮]罐
oil field waste(water)	油田废水
oil field waste(water) treatment	油田废水处理
oil field water	油田水
oil fighting trailer	防漏油拖车
oil-filled	充油的
oil-filled air-blast cooled	充油鼓风冷却的
oil-filled bushing	油浸套管;充油套管;充油衬套
oil-filled cable	浸油电缆;油浸电缆;充油(式)电缆
oil-filled cable oil	充油电缆油
oil-filled condenser	油浸电容器
oil-filled condenser type bushing	油浸电容器式套管
oil-filled ignition coil	油浸式点火线圈
oil filled naturally cooled	充油自然冷却(的)
oil-filled radiator	取暖油炉
oil-filled system	充油系统

oil-filled transformer 油浸变压器;油浸式变压器
oil-filled working space 注油工作空间
oil filler 油性填孔剂;油性填缝剂;油料加入器;注油器;加油口
oil filler cap 加油管;加油口;加油器盖
oil filler hole 注油孔
oil filler pipe 油料加入器管;注油管;加油管
oil filler plug 注油塞;加油塞
oil filler point 注油孔
oil filler screen 加油器滤清器;加油器滤网
oil filling 注油
oil filling opening 注油口
oil filling pipe 注油管;加油管
oil filling plug 注油塞
oil filling port 加油口
oil film 油膜
oil film breakdown 油膜破坏
oil film conductivity 润滑油油膜导电性
oil film extent 油膜范围
oil film filter 油膜过滤器
oil film light modulator projection 油膜光阀投影
oil film loss 油膜损耗
oil film of gasholder 储油罐油膜
oil film on coastal waters 沿海原油浮油油膜
oil film rigidity 油膜刚度
oil film rolls 采用油膜轴承的轧辊
oil film strength 油膜强度
oil film test 润滑油油膜强度试验
oil film thickness 油膜厚度
oil film vibration 油膜振动
oil film viscosity 油膜黏度
oil film wedge 油楔
oil filter 油过滤器;油性过滤器;油料滤清器;滤油器;机油滤清器
oil filter body 滤油器体
oil filter bracket 滤油器托架
oil filter cartridge 滤油器芯子
oil filter cartridge spring 滤油器芯子弹簧
oil filter cartridge support 滤油器芯子支器
oil filter casing 滤油器壳
oil filter clamp 滤油器夹
oil filter cover spring 滤油器盖弹簧
oil filter drain plug 滤油器放油塞
oil filter element 滤油器元件
oil filter gasket 滤油器垫密片
oil filtering unit 滤油设备
oil filter paper 油滤纸
oil filter plug 加油塞;注油塞
oil filter press 压力滤油机
oil filter screen 滤油(器滤)网
oil filter shell 滤油器壳
oil filter strainer 滤油器滤网
oil finder 石油探测器
oil finding 石油普查
oil finish 上油处理
oil finished terne plate 涂油白钢皮
oil finish of furniture 家具涂油处理
oil-fired 烧油的;燃油的
oil-fired air-furnace 油反射炉;燃油反射炉;烧油反射炉
oil-fired boiler 燃油锅炉;烧油锅炉
oil-fired burner 油燃烧器
oil-fired central heating 燃油集中供暖法
oil-fired drier 烧油干燥机
oil-fired fire 燃油火焰
oil-fired flame 石油火焰
oil-fired furnace 烧油炉子;油窑;油炉;烧油(窑)炉
oil-fired gas turbine 使用液体燃料的燃气轮机
oil-fired heater 燃油采暖器
oil-fired heating system 燃油供热系统
oil-fired kiln 烧油窑炉
oil-fired power plant 烧油(发)电机
oil-fired rotary kiln 烧油回转窑
oil-fired station 烧油电厂
oil-fired superheater 燃油过热器
oil-fired thermal plant 烧油电厂;燃油热电厂
oil-fired unit heater 燃油单元采暖器;燃油单元采暖器
oil-fired warm air heater 燃油热风加热器
oil fires 石油燃料造成的火灾;石油火焰
oil firing 烧油(的)
oil firing burner 油燃烧器
oil firing system 烧油系统
oil fish 多脂鱼
oil-fixative pigment 油定色颜料

oil flame 油焰
oil flash 油的闪蒸
oil flinger ring 隔油环
oil float 油浮
oil flo(a)tation 油浮选;全油浮选
oil float support 油浮筒
oil flooded bearing 油膜轴承
oil flow 油流
oil flow coefficient 油流系数
oil flow detection relay 油流检定继电器
oil flow in bearing 在轴承内的润滑油流动
oil flowmeter 油量计
oil flow problem 石油流动问题
oil foaming 油生成泡沫
oil fog 油雾
oil fog generator 油雾发生器
oil fogger 油雾化器
oil fogging 油雾化
oil fog lubrication 油雾润滑
oil foot(ing) 油脚
oil forge 烧油锻炉
oil for injection 注射用油
oil-forming shale 油母页岩
oil form sludge 污泥含油
oil for watch 手表油
oil foulant 油污垢
oil fraction 油的馏分
oil-free 缺油的
oil-free alkyd resin 无油醇酸树脂
oil-free compressor 无油压缩机
oil-free diaphragm pump 无油膜片泵
oil-free labyrinth compressor 迷宫式无油润滑压缩机;无油润滑迷宫式压缩机
oil-free lubrication 无油润滑
oil-free marine air compressor 船用无油空气压缩机
oil-free petrolatum 无油蜡膏
oil-free polyester 无油聚酯
oil-free ultra-high vacuum coating equipment 无油超高真空镀膜设备
oil-free ultra-high vacuum exhaust station 无油超高真空排气台
oil-free wax 无油蜡
oil fuel 油燃料;油类燃料;燃料(石)油;石油燃料
oil fuel bunker 燃油柜
oil fuel bunkering 装船用燃油
oil fuel bunkering port 加船用燃油港口;装船用燃油港口
oil fuel burner 燃油燃烧器
oil fuel cost 油料费;油料成本
oil fueled engines 燃油动力机具
oil fuel feed pump 给燃油泵
oil fuel injecting pump 燃油喷射泵
oil fuel injector needle-valve 喷油器针阀
oil fuel pump 油燃料泵
oil fuel settling tank 燃油沉淀柜
oil fuel ship 烧油船
oil fuel storage tank 燃油储[贮]罐;燃油储[贮]存柜
oil fuel supercharge pump 燃油增压泵
oil fuel tank 燃油舱;油料罐
oil fuel transfer pump 燃油燃烧泵;燃油料输送泵
oil fuel unit pump 燃油泵组
oil funnel 油漏斗
oil furnace 油炉
oil furnace black 油料炭黑
oil-fuse cutout 油熔丝断路器;油浸熔燃断路器
oil gag 油压压紧装置
oil gallery 油沟;(油泵的)回油孔
oil garlic 烯丙基硫
oil gas 油制气;油气;石油气;石油破坏蒸馏得到的烃气
oil-gas accumulation and pool 油气聚集和油气藏
oil-gas accumulation and trend zone 油气聚集带和趋向带
oil-gas and source rock correlation 油气源对比
oil-gas basin 油气盆地;含油气盆地
oil-gas bearing condition of trap 圈闭含油气性
oil-gas-bearing formation 含油气建造
oil-gas bed 油气同层
oil-gas boundary 气油界面
oil-gas contact 油气接触(面)
oil-gas exploration 油气勘探

oil-gas field accumulation and other unit 油气田油气聚集带及其他油气聚集单元
oil-gas field complex 组合油气田
oil-gas fields group of dome 穹隆油气田群
oil-gas field water 油气田水
oil-gas from petroleum 石油破坏蒸馏得到的烃气
oil-gas generator 油气发生器
oil gasification 油气化;油的气化
oil gasifier 油气化炉
oil-gas interface 油气界面
oil-gas migration 油气运移
oil-gas process 石油破坏蒸馏制造烃气的过程;油气化法
oil-gas seal transformer 充油封闭式变压器
oil-gas shielded type transformer 充油封闭式变压器
oil-gas shield type transformer 充油屏蔽式变压器
oil-gas show and solid bitumen 油气显示和固体沥青
oil-gas tar 油气焦油;油气柏油
oil-gas tar pitch 油气硬沥青;油气焦油脂;油气焦油的沥青
oil-gas transfer pump 油气混输泵
oil-gas-water mixed bed 油气水混层
oil-gas zone 含油气带
oil gathering 石油收集
oil gathering system 集油系统
oil ga(u)ge 油位计;油面计;量油计;油规;油表;油标(表);润滑油量计
oil ga(u)ge glass 油位玻璃管;玻璃油位表
oil ga(u)ge pipe 油压表油管;油面指示管
oil ga(u)ge stick 量油杆
oil ga(u)ging tape 油标钢皮卷尺
oil gear 油压传动装置;油齿轮;润滑齿轮;甩油齿轮
oil gear compressor 油齿轮压缩机
oil gear motor 径向回转柱塞液压马达
oil gear pump 回转活塞泵
oil genesis 石油生成;石油起源
oil geology 石油地质
oil gland 油封填料盖
oil gloss 油光泽
oil gloss paint 油性光漆;油光涂料;油光调合漆
oil grade 油的黏度号;油的牌号
oil grating 油篦子
oil gravity tank 重力油柜
oil/grease wastewater 油/油脂废水
oil grinding 油研磨
oil grinding machine 磨油机
oil groove 油路;油道;油槽;滑油槽;润滑油槽
oil groove attachment 切制油沟装置
oil groove milling cutter 油槽铣刀
oil guard 挡油圈;挡油器;防油器
oil gun 喷漆枪;注油器;油枪
oil hammer 油锤作用
oil handling 油的处理
oil handling receptacle 油的处理接受器
oil handling wharf 石油码头
oil hand pump 手压油泵
oil harbo(u)r 油港
oil hardened 油淬火的
oil-hardened steel 油淬硬化钢
oil hardening 油中淬火;油(淬)硬化;油淬火
oil-hardening steel 油淬钢
oil hazard 油料事故
oil header 燃油加热器;吸油器;油料分布器;润滑油分配器
oil head tank 重力油柜
oil-heated steamer 燃油加热蒸煮器
oil heater 油加热器;加油加热管;热油器;燃油加热器
oil heating 油热法
oil heating chimney 燃油供热烟囱
oil heating installation 油热设备
oil heating system 燃油供暖系统
oil heating unit 油加热装置
oil herbicide 石油除草剂
oil herder 水上浮油吸聚剂
oil hold 油舱
oil-holder 油舱;盛油容器;油杯;润滑器
oil hole 油位表;油孔;注油孔
oil-hole cover 注油孔盖
oil-hole drill 油孔钻(头)
oil-hole location marks 油孔配置标记
oil-hole plug 油孔塞
oil-hole screw 油孔螺钉

oil horizon 含油层
oil hose 输油软管
oil hose davit (油轮上)货油软管吊柱
oil hose inner diameter 油管内径
oil house 油库
oil hulk 石油驳船
oil hydraulic 液压的;油压的
oil hydraulic braking cylinder 液压制动缸
oil hydraulic circuit 油压回路
oil hydraulic cylinder 油压缸
oil hydraulic dismounting tool for conic(al) fit 锥度配合液压拆卸工具
oil hydraulic motor 油压马达;液压马达
oil hydraulic operated 液压操作的;液压操作的
oil hydraulic operating table 油泵式手术台
oil hydraulic power system 液压动力系统;油压动力系统
oil hydraulic press 油压机
oil hydraulic pump 油压泵
oil hydraulic system 油压系统
oil-hydrogen mixture 油与氢气的混合物
oil hydrometer 油品比重计;油比重计
oil hydrosol 油水溶胶
oil identification 油的鉴定
oil ignitor 油点火器
oil-immersed air-blast cooling 油浸强迫风冷
oil-immersed arc control device 油浸式电弧控制设备
oil-immersed breaker 油浸式断路器
oil-immersed circuit breaker 油浸断路器
oil-immersed current transformer 油浸式电流互感器
oil-immersed forced-air cool 油浸风冷式
oil-immersed multiplate disk brake 油浴式多片盘式制动器
oil-immersed natural cooling 油浸自然冷却
oil-immersed power transformer 油浸式电力变压器
oil-immersed reactor 油冷电抗器;充油电抗器
oil-immersed regulating transformer 油浸式调整变压器
oil-immersed rheostat 油浸变阻器
oil immersed self cool 油浸自然冷却(式)
oil-immersed solenoid 油浴电磁线圈
oil-immersed starter 油浸起动装置;油浸起动器;油浸式起动器
oil-immersed switch 油浸开关
oil-immersed torque motor 湿式力矩马达
oil-immersed transformer 油浸变压器;浸油变压器
oil-immersed type bearing 油浸式轴承
oil-immersed water-cool 油浸水冷式
oil-immersed water cooling 油浸水冷却
oil-immersion bath 油浸池
oil immersion cooling 油浸致冷
oil-immersion crushing method 油浸压碎法
oil-immersion diffractometer 油浸衍射仪
oil immersion lens 油浸镜头
oil immersion (method) 油浸(法)
oil immersion microscope 油浸显微镜
oil immersion microscope objective 油浸显微镜物镜
oil immersion objective 油浸物镜
oil-immersion reflectivity 油浸反射率
oil-immersion test 油浸试验
oil-impervious composition 不透油涂料
oil-imports quota 石油进口限额
oil-impregnated 油浸渍过的;浸过油
oil-impregnated bearing 含油轴承
oil-impregnated biolithite 油浸生物灰岩
oil-impregnated cable 油浸电缆
oil-impregnated clay 油浸泥岩
oil-impregnated condenser 油浸电容器
oil-impregnated conglomerate 油浸砾岩
oil-impregnated dolomite 油浸白云岩
oil-impregnated limestone 含油石灰岩;油浸石灰岩;油浸灰岩
oil-impregnated metal 含油轴承合金
oil-impregnated paper 油浸渍的纸
oil-impregnated rock 石油浸染岩石
oil-impregnated sandstone 含油砂岩;油浸砂岩
oil-impregnated wood 油浸渍的木材;油浸过的木材
oil impregnation 油浸
oil-in 油入口;进油口
oil index 油量指示器

oil indication 油标志
oil indicator 油腻指示器;油指示剂;示油器
oil-induced deficit 石油赤字
oil-induced symptom 油症
oil industry 石油工业
oil industry trends 石油工业发展趋势
oil industry wastewater 石油工业废水
oil-in emulsion 水中油型乳(化)液
oiliness 油性;油腻性;油感;含油性;润滑性
oiliness additive 油性添加剂;润滑性添加剂
oiliness agent 油性剂;油性添加剂
oiliness carrier 油性载体
oiliness compound 润滑油增效剂
oiliness degree 油性度
oiliness film 油性膜
oiliness improver 油性添加剂
oiliness index 油性指标
oiliness of lubricant 润滑剂的油性
oiliness test 油性试验
oiliness tester 油性试验器
oiling 油润滑;注油;加油(法);润滑;上油
oiling action 润滑作用
oiling chart 润滑系统图;润滑图;润滑点说明图
oiling device 注油装置
oiling groove 油沟;注油槽
oiling gun 注油枪
oiling hole plug 加油孔塞;加润滑油孔塞
oiling machine 注油机
oiling period 喷油周期
oiling point 润滑点
oiling ring 注油环;加油环
oiling rolls 涂油辊
oiling station 涂油装置
oiling subgrade 浇油路基
oiling system 注油装置;加油系统;滑油系统;润滑设备;充油系统
oiling tube 注油管
oil injection 油的喷射;注油;喷油
oil injection cooling 油喷射冷却
oil injection header 注油总管
oil injection nozzle 喷油嘴
oil injection pump 射油泵
oil injection system 射油系统
oil injector 油头;注油器
oil inlet 油入口
oil inlet pipe 进油管
oil in mass 油块
oil in place 油藏储[贮]量;地质储量
oil in reserves 管路或油罐内的储存油;储存油
oil insecticide 油类杀虫剂
oil inserter 插入注油器
oil-insoluble sludge 不溶于油类的残渣
oil inspection kit 成套检查油料仪器
oil inspector 油检查员
oil insulation 油绝缘
oil insulator 充油绝缘套管
oil interceptor 油收集器;油类捕集器;截油器;集油器
oil interest 石油业
oil interrupter switch 断油开关
oil in water 水中油;水包油
oil-in-water dispersion 油在水中弥散;水中油扩散
oil-in-water emulsion 水包油乳化液;油水相乳化液(油为内相,水为连续外相);水包油型乳剂;水包油型乳化液
oil in water solution 石油水溶液
oil-in-water type 水中油型;水包油型
oil-in-water type coolant 乳化冷却液
oil isoperms 油藏等渗透率图
oilite 吸油合金;含油轴承合金;石墨青铜轴承合金
Oilite alloy 奥利特轴承合金;奥利特合金
oilite bearing hinge 含油轴承合金铰链
oilite bronze 多孔青铜
oilite bushing 含油轴套
oil jack 油压千斤顶;油壶;液压起重器;抽油装置
oil jacket 油套;润滑油套
oil jet 喷油射流
oil jet hopper feeder 喷油料斗送料器
oil jet lubrication 喷油润滑
oil jet path 喷油通路
oil jet pump 射油泵
oil jetty 油码头;石油码头
oil keeper 油泵;油承
oil laboratory 油脂实验室

oil lagging 油夹套
oil lamp 油灯
oill-bulk-ore carrier 石油/散货/矿石三用船
oil lead 油道
oil leak accident 漏油事件;跑油事故
oil leak(age) 漏油;润滑油泄漏
oil leak detector 漏油检测器
oil leaking 漏油
oil-leg 油腿
oil length 油度;油长;含油率;含油度
oil length of varnish 清漆的油度;油漆的含油率
oil lens 透镜体油藏
oilless 无油的;不需加油的
oilless compressor 无油压缩机
oilless bearing 自润滑轴承;自动润滑轴承;免油轴承;无油轴承;石墨润滑轴承;不加油轴承
oilless bushing 自润滑轴瓦;自动润滑轴瓦;石墨润滑轴瓦
oilless circuit breaker 无油断路器
oilless hessian 无油过滤麻布
oilless pump 无油泵
oillet(te) 小枪眼;小观察孔;小堡眼;漏洞;换气孔;视孔;油孔;视眼;眼圈;(油罐的、油箱的)孔眼
oil level 油位;油面高度;油面
oil level cock 油面计开关
oil level control 油面控制机构
oil level dipstick 油面测杆
oil leveler 油标(表)
oil level ga(u)ge 油位计;油面指示器
oil level ga(u)ge float 量油尺浮子
oil level ga(u)ge glass 油位计玻璃;油位表玻璃;玻璃油位表
oil level indicator 油位指示器;油面指示器
oil level(l)er 油位表
oil level mark 油面标志
oil level pipe 油面管
oil level plug 油面测量计栓
oil level ragulator 油位调节器
oil level rod 油位测量杆
oil level sight ga(u)ge 直读式油位计
oil level tell-tale 油面信号灯
oill-feed line 输油管线
oil life 油品的寿命
oil lift 油压支承
oil lift bearing 油顶起轴承
oil light 煤油灯
oil lighter 油点火器
oil-limiter 限油器
oil line 油管路;输油管线;油道
oil-line extension 油管接长
oil-line plugging 油管堵塞
oil-line pump 输油管泵;输油泵
oil-line scavenge 油管清扫
oil-line steaming 油管蒸汽吹扫
oill-in-water emulsion 水中油型乳液;水包油型乳液;水包油乳化液
oill-in-water emulsion mud 水包油乳化泥浆
oill-in-water type emulsifier 水包油型乳化剂
oil liquor 凝析油
oil-loading arm 装油臂;输油臂(油管头)
oil-loading/discharging by gravity flow 自流装/卸油
oil-loading facility 装油设备;油品装卸设备
oil-loading rack 装油栈桥;石油灌装栈桥;装卸石油产品的栈桥
oil-loading rate 装油效率
oil-loading terminal 石油装船码头;装油站
oil loam 油腻子
oil lock 油封设备
oil log 油井记录;采油记录
oil loss 油品损耗;石油损耗
oil losses control 石油损失控制
oill-slurry-ore carrier 石油矿浆矿石三用船
oill-spill 水面浮油
oil lubricant 油类润滑剂;油基润滑剂
oil-lubricated bearing 油润滑轴承
oil lubricating piping 油润滑管路
oil lubricating pump 润滑油泵
oil lubricating system 稀油润滑系统
oil lubricating tissue 油滑组织
oil lubrication 油润滑
oil lubricator 油润滑器;油壶;油杯;注油器;加油器
oil macadam 用胶结材料处理的碎石路;灌油碎石;沥青砾石

oil macadam pavement 灌油碎石铺面
oil macadam road 灌油碎石路
oil main line 主油管
oilman 石油业者
oil manifold (液压系统的)分油系统;油歧管
oil manifold valve 油歧管阀
oil manometer 油压力计;油压计;机油压力表
oil mark 油渍;油斑
oil marly clay 油浸泥灰质泥岩
oil mass 油量
oil mass viscosity 油的体积黏度
oil mastic 油质胶结料;油质玛琋脂
oil mat 沥青面层;油垫
oil material contamination 油类污染
oil mat road 柏油路
oil meal 油渣饼碎片
oil measurement 油计量
oil measurer 油量计
oil-measuring ga(u)ge 油位指示器
oil measuring tape 量油尺
oil-medium 油介质的
oil meter 石油计量器;油量计;油量表
oil-metering pump 润滑油计量泵
oil-metering system 油计量系统
oil migration 石油运移;石油移动
oil mileage 油的行驶里程
oil mill 制油厂;榨油机;榨油厂
oil mineral 油矿
oil minimum breaker 少油开关;少油断路器
oil miscible concentrate 浓油剂;乳油
oil mist 油雾
oil mist detection 油雾探测;油雾检测
oil mist detector 油雾探测器;油雾检测器
oil mist filter 油雾滤清器
oil mist lubrication 油雾润滑
oil mist monitoring 油雾监测
oil mist system 油雾系统;油雾润滑系统
oil mist trapping 油雾捕集
oil mix 油拌式面层
oil mixed type surface 油拌混合料;沥青混合料面层
oil mixture 油混合物;含油混合物
oil mobility 油类流动性
oil-modified alkyd 油改性醇酸树脂
oil-modified alkyd resin 油改性醇酸树脂;油改性聚酯树脂
oil-modified maleic alkyd 油改性顺(丁烯二酸)酐醇酸树脂
oil-modified polyurethane varnish 油改性聚氨脂清漆
oil-modified resin 油改性树脂
oil-modified urethane 油改性聚氨脂
oil moistened air filter 油浸式空气滤清器
oil money 石油收入
oil mordant 石油媒染剂
oil motor 油马达;液压发动机;液压电(动)机;燃油马达
oil mud 油泥浆;油基泥浆
oil-mull technique method 油磨法
oil-naphtha solution 油的石脑油溶液
oil neutralizing agent 油中和剂
oil number 吸油量;润滑油牌号
oil of absinthe 苦艾油
oilofapple 戊酸戊酯
oil of badian 八角茴香油
oil of bananas 香蕉水
oil of bay 桂油
oil of bayberry 月桂果油
oil of brazil nuts 巴西果仁油
oil of cedar wood 柏木油
oil of cloves 丁香油
oil of dogfish 鲨油
oil of eucalyptus 桉叶油
oil of flaxseed 亚麻仁油
oil offtake 采油量
oil of lavender 薰衣草油
oil of maize 玉蜀黍油
oil of mirbane 密斑油
oil of mustard 芥子油
oil of paraffin 石蜡油
oil of pears 乙酸戊酯
oil of peppermint 薄荷油
oil of ricinus 蓖麻油
oil of sassafras 黄樟油

oil of star anise 八角茴香油
oil of turpentine 松节油
oil of vitriol 浓硫酸
oil of zedoary turmeric 术术油
oil-operated reverse gear 油动换向装置
oil-operated steel support 液压金属支架
oil-operated transmission 油压传动
oil-operated voltage regulator 油压操作电压调整器
oil orange 油橙
oil ore carrier 石油矿石运输船
oil orifice 油孔
oil origin 石油起源
oil-ostatic(al) cable 油压电缆;充油钢管电缆
oil-out 出油口
oil outlet 排油;出油口
oil outlet fitting 放油接头
oil outlet pipe 出油管
oil outlet temperature 油的出口温度
oil-out line 油流出管道
oil output 出油量
oil oven 燃油炉
oil-overflow 充满油的
oil overflow valve 油溢流阀
oil oxidation stability 油的氧化稳定性
oil package 石油产品包装容器
oil packaging 油品包装;油的包装
oil packing 油填料;防油包装
oil packing paper 包装油纸
oil pad 油密封垫;润滑填料
oil-pad casing (轴箱盖)浸油垫板
oil-pad lubrication 油毡润滑;油环润滑法
oil paint 油质涂料;油性涂料;油性色漆;油性漆;油涂料;油漆;油膜;油画颜料
oil paint coat 油漆涂层
oil painter 油工
oil painting 油画
oil painting on glass 玻璃油画
oil palm 油棕
oil pan 油盘;油底盘;集油盘
oil pan armor 油底盘装甲
oil pan clearance 油盘间隙
oil pan depth ga(u)ge 油盘深度计
oil pan drain cock 油盘放油旋塞
oil pan drain plug 放油塞
oil pan filler block 油盘盖封座
oil pan filler cap 油盘加油口盖
oil pan gasket 油盘衬垫
oil pan heater 油底壳预热器
oil pan packing 油盘密封垫
oil paper 油(浸)纸;蜡纸
oil passage 润滑油通路
oil passage to rocker shaft 摇臂轴油路
oil passage way 油路
oil paste 油彩;油膏
oil patch 油田区域;油斑
oil pavement 重质残油油地
oil penetration 渗油性能
oil penetration test 渗油性能试验
oil performance 油的工作特性
oil performance in engine 油品在发动机内工作特性
oil permeability 油渗透率
oil permit 客轮载运汽油特许证;石油装载许可证
oil pesticide 石油类农药
oil phased mo(u)ld cream 油态脱模剂
oil phase-out zone 油相消失带
oil pier 石油码头
oil pike 油笔
oil pipe 油管
oil pipe access trestle 油管栈桥
oil pipe control system 油管控制系统;油压系统
oil pipe depth 油管深
oil pipe fitting 油管配件
oil pipeline 油管线;输油管线;输油管道;石油管线;石油管道
oil pipeline pump 管线输油泵
oil pipe-type cooler 管式冷油器
oil pipe way 油管栈桥
oil piping 润滑油管系;油管道
oil piping installation 油管安装
oil piping layout 油管的布置;油导管
oil pistol 油枪
oil pit 油孔;油槽
oil pitch 柏油;硬质石油沥青

oil pitching 用沥青油修补
oil plant 油料作物;炼油厂
oil plant and forest pests 油料作物和森林害虫
oil plasher 泼油器;泼油机
oil-pneumatic pressure cell 油气动式压力盒
oil pocket 油包;储油箱
oil pointer 油面指示器;油标(表)
oil pole 油极
oil polish 上光油
oil polisher 油脂澄清离心机
oil polishing 油磨光
oil pollutant 油类污染物
oil polluted 油污染的
oil polluted seashore 油污染海滨;石油污染的海岸
oil polluted waters 油污染水域;油污染水体;石油污染的水域
oil polluting 油污的
oil pollution 油(类)污染;石油污染
Oil Pollution Act 石油污染法令;防止油污法令
oil pollution control 油污染控制;石油污染控制;石油污染防治
oil pollution damage 油污染损害;石油污染损害
oil pollution detection 油污染检测;石油污染检测
oil pollution emergency 石油污染事故;石油污染紧急情况
oil pollution emergency plan 油污染应急计划
oil pollution environmental control 油污染环境治理
oil pollution experiment 油污染试验
oil pollution indicator 油污染指示种
oil pollution of the ground 地面油污染
oil pollution of the sea 海洋油污染
oil pollution of waters 水域油污染
oil pollution remote-sensing system 石油污染遥感系统
oil pollution residue 油污染残渣;石油污染残留(物)
oil pollution surveillance system 油污染监视系统;石油污染监视系统
oil pond 储(贮)油池
oil pool 油床;油藏;油池
oil pool waters 有油层的海域;储油海域
oil-poor circuit-breaker 贫油断路器
oil port 油港;供油港(口);加油口;加油港
oil posted price 石油标价
oil post price 石油牌价
oil potential 油势
oil potential gradient 油势梯度
oil pot relief valve 油罐安全阀
oil power 石油动力
oil-prepared paint 油性调合漆
oil preservative 油性防腐剂
oil press 油压机;液压机;榨油机;榨油厂
oil pressing mill 榨油厂
oil-pressure 油压(力)
oil-pressure adjusting screw 油压调节螺钉;油压的调节螺丝
oil-pressure adjusting valve 油压调节阀
oil-pressure adjustment 油压调节
oil-pressure atomizing burner 油压雾化喷射器
oil-pressure bearing 油压轴承
oil-pressure brake 油压闸;油压制动器
oil-pressure burner 油压烧嘴;油压燃烧器
oil-pressure check valve 油压止回阀
oil-pressure control 油压控制
oil-pressure control relay 油压控制继电器
oil-pressure control system 油压控制系统
oil-pressure control unit 油压调整装置
oil-pressure cut-out 油压切断器;油压开关
oil-pressure cutter 油压断缆器
oil-pressure cylinder 压力油缸
oil-pressure damper 油压减振器
oil-pressure ga(u)ge 油压表
oil-pressure ga(u)ge connection 油压表连接管
oil-pressure ga(u)ge feed pipe 油压表进油管
oil-pressure ga(u)ge pipe 油压表管路
oil-pressure indicator 油压指示器
oil-pressure indicator lamp 油压指示灯
oil-pressure installation 油压装置
oil-pressure lubrication 压油润滑
oil-pressure lubrication pump 压力润滑油泵
oil-pressure operated 油压操作的
oil-pressure pipe 油压管
oil-pressure pump 油压泵

oil-pressure reducing valve 油压减压阀
oil-pressure regulator 油压调节器
oil-pressure relay 油压继电器
oil-pressure release valve 油压泄放阀;油压放松阀
oil-pressure relief valve 润滑系减压阀;油压安全阀
oil-pressure relief valve body 油压安全阀体
oil-pressure relief valve cap 油压安全阀盖
oil-pressure relief valve nut 油压安全阀螺母
oil-pressure relief valve plug 油压安全阀塞
oil-pressure relief valve plug gasket 油压安全阀密封垫
oil-pressure relief valve spring 油压安全阀簧
oil-pressure retention 油压保持性
oil-pressure selector 油压选择器
oil-pressure shut-off switch 油压遮断器;油压开关
oil-pressure stabilizer 油压稳定器
oil-pressure supply system 供油系统;油压装置;油压设备
oil-pressure switch 油压开关
oil-pressure switch screw 油压开关螺丝
oil-pressure system 油压系统
oil-pressure test for strength 油压强度试验
oil-pressure test for tightness 油密试验
oil-pressure tube 油压管
oil-pressure turbine governor 涡轮机油压调速器
oil-pressure type atomizer 油压型雾化器
oil-pressure unit 油压装置;油压设备
oil-pressure unit type car retarder 油压式缓行器
oil-pressure valve 油压阀
oil-pressure warning signal 油压警告信号
oil-pressure warning unit 油压报警器
oil price simulation model 石油价格模拟模型
oil primer 油质底漆;油性底漆;油质底层涂料;油性打底漆
oil primer coat 油性底漆涂层
oil primer surfacer 油性底漆和二道漆两用漆
oil print 油斑
oil process 油法
oil processing 石油加工
oil processing industry 炼油工业
oil processing unit 石油加工装置
oil process patenting 旧工艺专利;旧方法专利
oil producing area 产油区
oil-producing country 产油国
oil-producing profile 出油剖面
oil product 石油制品;石油产品
oil product carrier 成品油轮
oil production 产油量;石油生产;石油开采;采油(量)
oil production measuring 测原油产量
oil production platform 采油平台
oil production rate 原油产率
oil productivity 产油率
oil product quay 成品油码头
oil-product ship 成品油轮
oil product terminal 成品油码头
oil-proof 耐油的;不渗油的;防油渗井;防油的;不透油(的);不漏油
oil-proof asbestos rubber joint sheet 耐油石棉橡胶接合垫片
oil-proof concrete 耐油混凝土
oil-proof enamel 耐油搪瓷
oil-proofing 不透油的;抗油;耐油
oil-proof mortar 耐油砂浆
oil-proofness 耐油
oil-proof rendering 防油粉刷;防油打底;防油涂抹;防油底涂
oil-proof telecommunication cable 防油通信电缆
oil-proof test 耐油试验
oil property 油的性质
oil prospecting 石油勘探
oil prospector 石油勘探手
oil-protecting floor 耐油地面
oil protection wall 挡油墙
oil puddle 油团
oil pump 油泵
oil pump body 油泵体
oil pump body cover gasket 油泵盖密封垫
oil pump body gasket 油泵体衬垫
oil pump bracket 油泵支架
oil pump bracket bolt 油泵支架螺栓
oil pump bypass valve 油泵旁通阀
oil pump capacity 出油量
oil pump casing 油泵壳

oil pump cleaner 油泵滤(清)器
oil pump coupling 油泵连接
oil pump cover 油泵盖
oil pump cylinder 油泵筒
oil pump drive 油泵驱动
oil pump drive gear 油泵主动齿轮
oil pump drive gear sleeve 油泵主动齿轮套
oil pump driven gear 油泵从动齿轮
oil pump driven wheel spindle 油泵从动齿轮轴
oil pump drive shaft 油泵主动轴
oil pump drive shaft gear 油泵主动轴齿轮
oil pump drive shaft support 油泵主动轴支架
oil pump drive spindle 油泵传动轴
oil pump driving wheel 油泵主动齿轮
oil pump eccentric wheel 油泵偏心轮
oil pumper 油泵工作员
oil pump expansion plug 油泵膨胀塞
oil pump feed 油泵供油
oil pump float screen 油泵浮筒滤网
oil pump for prestressed steel bar drawer 钢筋预应力张拉油泵
oil pump gasket 油泵密封垫
oil pump gear 油泵齿轮
oil pump house 油泵房
oil pump housing 油泵外壳
oil pump idler gear 油泵空转齿轮
oil pump idler gear shaft 油泵空转齿轮轴
oil pumping 油品泵送;抽油
oil pumping can 泵式油壶
oil pumping system 油品泵送系统
oil pumping unit 抽油机
oil pump inlet valve 油泵进油阀
oil pump motor 油泵电动机
oil pump outlet line 油泵出油管
oil pump outlet tube 油泵出油管
oil pump outlet valve 油泵出油阀
oil pump pinion 油泵主动齿轮
oil pump plunger 油泵柱塞
oil pump pressure relief valve ball 油泵压力安全球阀
oil pump purifier 油泵净油器
oil pump regulating valve 油泵调节阀
oil pump relief valve 油泵安全阀
oil pump relief valve spring 油泵安全阀弹簧
oil pump retaining spring 油泵扣紧簧
oil pump rod 油泵拉杆
oil pump room 油泵室
oil pump safety valve 油泵保险阀
oil pump screen 油泵滤网;油泵过滤器
oil pump screen cover 油泵滤网盖
oil pump screen retainer 油泵滤网护圈
oil pump shaft 油泵轴
oil pump shaft gear 油泵轴齿轮
oil pump shaft guide 油泵轴导管
oil pump shield 油泵护片
oil pump spindle 油泵心轴
oil pump spindle bearing housing 油泵心轴轴承壳
oil pump spindle thrust bearing 油泵心轴止推轴承
oil pump spring 油泵簧
oil pump strainer 油泵滤(清)器
oil pump suction pipe gasket 油泵吸油管密封片
oil pump suction tube 油泵吸油管
oil pump testing device 油泵试验装置
oil pump washer 油泵垫圈
oil puncture 油击穿
oil puncture test 油击穿试验
oil purification 油品净化;油净化处理;油净化
oil purification plant 油的净化装置
oil purification system 润滑油净化系统
oil purifier 油分水机;滤油器;净油器;净油池
oil putty 油性腻子;油腻子;油基腻子;油灰
oil quality ga(u)ge 油量计
oil quality indicator 油量计
oil quantity 油量
oil quantity ga(u)ge 油量计
oil quantity indicator 油量指示器
oil quarters 老住处;老宿舍;老房租
oil-quench 油淬火
oil-quenched cut-out 油浸熔丝保险器
oil quenching 油中淬火;油强化;油急冷(处理);油淬
oil-quenching bath 淬火油池
oil-quenching steel 油淬火钢
oil-quenching tank 淬火油槽

oil radiator 油液散热器;油散热器;油辐射器
oil rationing 石油配给
oil-reactive phenolic resin 油变性酚醛树脂
oil-reactive resin 油反应性树脂
oil reactivity 油变性;油反应度;油活化性;油再生性;油反应性
oil receiver 油盘;油槽;集油器;受油器;储油器;油接受器
oil receptacle 油槽
oil recirculating system 润滑油再循环系统
oil reclaimer 油类精制器;净油器;净油池;废油再生装置
oil reclaiming 废油的再生
oil reclamation 油回收;废油再生
oil reconditioner 油液净化装置
oil record book 油类记录簿
oil recovery 油再生;油回收;油的回收;采油
oil recovery barge 残油收集船舶;残油回收驳
oil recovery system 油回收系统;润滑油回收系统
oil recovery vessel 水面浮油回收船;浮油回收船
oil rectifier 油蒸馏塔;炼油塔;石油蒸馏器
oil red 油溶红;油红
oil reference price 石油参考价格
oil refiner 净油器;净油池
oil refiner pack 油类精制的过滤器
oil refinery 石油炼制厂
oil refinery effluent 炼油废水;炼油厂废水
oil refinery (plant) 石油加工厂;炼油厂
oil refinery pump 石油蒸馏泵
oil refinery waste 炼油厂废物
oil refinery wastewater 炼油厂废水
oil refining cauldron 炼油锅
oil refining factory 石油炼油厂
oil refining industry 炼油工业
oil refining still 石油炼制蒸馏釜
oil refuse 油渣;废油
oil refuse incinerator 油垃圾焚烧炉;燃油垃圾焚化炉;废油燃烧炉
oil regeneration plant 石油再生厂
oil region 含油区
oil regulator 油量调节器
oil reining 炼油
oil relative permeability 油相对渗透率
oil release valve 减压阀;放油阀
oil relief valve 油压安全阀;减压阀;润滑系减压阀;放油阀;放油安全阀
oil removal 隔油池;去油;除油
oil removal treatment 除油处理
oil remover 集油器
oil removing 脱油
oil requirement 油的需要量;油的要求;油的规格
oil rerun unit 润滑油再蒸馏装置
oil reserves 油田储藏量;含油量;石油藏量;石油储(藏)量
oil reservoir 油箱;储罐;油储[贮];油腔;油床;储油器;储油库;储油池;储油层
oil reservoir cap 储油器盖
oil reservoir coverscrew 储油器盖止动螺钉
oil reservoir engineering 储油工程
oil reservoir pressure 储油器内压力
oil reservoir rocks 储油岩层
oil reservoir screen 储油器滤网
oil reservoir thickness 油层厚度
oil reservoir water 油层水;构造水
oil residence index 油迹滞留指数
oil residue 蒸馏石油的残渣;油渣;滑油残渣
oil residue recuperation 残油回收
oil resinite 油树脂体
oil/resin ratio 油和树脂比
oil resistance 抗油性;耐油性
oil resistance polyvinyl chloride 抗油性聚氯乙烯
oil-resistant 耐油的;抗油的
oil-resistant coating 耐油涂层
oil-resistant rubber 耐油橡胶
oil-resisting 防油的;耐油的;不透油的
oil-resisting test 耐油试验
oil resistivity 耐油性
oil resources 石油资源
oil-retainance 保油性
oil retainer (thrower) 挡油环;护油圈;润滑油保持环
oil-retaining bearing 含油轴承
oil-retaining disk 挡油盘
oil-retaining property 润滑油保持性能

oil-retaining ring 油环	oil seal with lip facing outward 外向油封	oil-soluble dye stuff 油溶性染料
oil retention of caulk 油膏保油性	oil search 石油普查	oil-soluble inhibitor 油溶性抗氧化剂
oil return check valve 回油止回阀	oil seasoning 热油干材料	oil-soluble phenotic resin 油溶性酚醛树脂
oil return hole (油泵的)回油孔	oil seat gasket 耐油垫圈	oil-soluble resin 油溶性树脂；油溶树脂
oil return line 回油管	oil sediment 油泥；油中沉淀物	oil-soluble surfactant 油溶性表面活化剂
oil return passage 油回流管道	oil seed 油料籽	oil-solution 油剂
oil return pipe 油回流管；回油管	oilseed crop 油料作物	oil-solvent 油溶剂
oil return valve 回油阀	oil seepage 油源；油渗出；油苗	oil-solvent blend 油与溶剂的混合物
oil-rich 富油的	oil separate chamber 油分离槽	oil-solvent preservative 油溶性防腐剂
oil rich variety 含油高的品种	oil separating 油的分离	oil source rock 油源岩；生物油岩
oil rig 油井钻探设备；抽油装置	oil-separating barge 隔油驳船；滤油驳船；分油驳	oil sources 油源
oil rights 采油权	oil-separating tank 隔油池；油水分离槽	oil space 油槽
oil rig supply vessel 油井架供应船	oil separation tank 隔油池	oil specification 油品规格；油料技术规格
oil ring 油环；抛油环	oil separator 油水分离器；油气分离器；分油器；隔油池；润滑油分离器；除油器；分油器；分油机	oil spill 溢出的油；漏油；漏出的油；浮选油；漂油
oil-ring bearing 油环(式)轴承；油环润滑轴承		oil spillage 溢油量；漏油量
oil-ring guide 油环导槽	oil separator for steam 蒸气的油分离器；蒸气的脱油器	oil-spillage cleaning agent 溢油处理剂
oil-ring lubrication 油环润滑		oil-spill chemicals 溢油处理剂
oil-ring plugging 油环卡住	oil settling tank 油沉淀柜	oil-spill clean-up method 溢油清除法
oil-ring retainer 油环挡圈	oil shale 油页岩	oil-spill collect vessel 漂油回收船；污油回收船
oil-ring self-lubrication 油环自动润滑	oil shale-bearing formation 含油页岩建造	oil-spill control 漏油控制
oil-ring sticking 油环黏结	oil shale cement 油页岩胶凝材料；油页岩水泥	oil-spill detection 油溢测量
oil riser 油提升管	oil shale distillation 油页岩干馏	oil-spill dispersant 溢油分散剂
oil rock 油岩；含油岩层	oil-shale eduction process 油页岩炼制过程	oil-spill fates model 溢油灾难模型
oil roller 油滚柱	oil-shale fuel 页岩油	oil spilling 溢油
oil roll machine 辊式涂油机	oil-shale group 油页岩群	oil spill on the sea 海上漏油
oil rotary pump 油旋转泵	oil-shale kerogen 页岩油母	oil-spill prevention, control and countermeasure 溢油预防、控制河防范措施
oil royalties 石油产地使用费	oil-shale resources 油页岩资源	
oils 油类；轻质石油产品	oil-shale retort 油页岩干馏釜	oil-spill project 防溢油污染工程
oil-salt basin 含油—盐盆地	oil-shale retorting 油页岩的干馏	oil-spill remover 油溢消除
oil salvage 废油再生	oil-shale retorting plant 油页岩干馏厂	oil-spill risk analysis 溢油风险分析
oil sample 油样	oil-shale slag 油页岩矿渣	oil spills on sea surface 海面溢油
oil sampling bottle 原油样瓶	oil-shale tar 油页岩焦油	oil-spill warning system 溢油警报系统
oil sampling cock 取油样旋塞	oil sheet 薄油层	oil spit hole 喷油(小)孔；承油孔
oil sand 油砂；含油砂层	oil shell 镇浪油弹	oil splitting plant 油脂水解车间
oil sand processing plant 油砂加工厂	oil shield 挡油罩；护油罩；防油罩	oil spoon 油匙
oil saponification value 油的皂化值	oil ship 油轮	oil spot 油点；油斑
oil-saturated hemp rope 油绳；浸油饱和的麻绳	oil shipment 油品装运；油料装运；石油装载	oil-spot glaze 油滴釉
oil-saturated sand 饱含油砂	oil shock 石油危机	oil spray 油类喷雾剂；油的喷淋；喷油(雾)
oil-saturated sandstone 石油饱和的砂岩	oil shock absorber 油压减震器；油减震器；液压缓冲器；油压阻尼器	oil sprayer 油雾喷射器；油喷淋器；喷油器
oil saturation 含油饱和率；含油饱和度		oil spray hole 喷油孔
oil save-all 防止溅油装备	oil shortage 石油的不足	oil spraying 油的喷淋
oil saver 油回收器	oil show 油显示；漏油显示；油源；油苗	oil spray lubrication 喷油润滑
oil save tank 集油罐	oil shrinkage factor 石油的收缩因子	oil spring 油泉
oil saving bearing 省油轴承	oil shrinkage loss 石油蒸发损失	oil spurts 油喷射
oil scale 油标(表)	oil sight-feed ga(u)ge 外观注油计	oil stability 油品安定性
oil scavenge pump 吸油泵	oil sight ga(u)ge 外视油量计	oil stack 燃油炉的烟囱
oil scavenger 回油器	oil sight glass 油位观察孔；油位观察玻璃；油面视镜	oil stain 油渍；油着色剂；油性色剂(用于木料着色)；油性染色剂；油性(着)色料；油性(着)色剂；油污迹；油污；油迹；油斑
oil scavenger pump 回油泵		
oil scavenger pump shaft 回油泵轴	oil-silk 防水布；油绸	
oil scavenging pump 排油泵	oil sinking agent 沉油剂	oil stain corrosion 油斑腐蚀
oil scavenging pump oil feed-back pump 回油泵	oil siphon 芯吸润滑器；油虹吸管	oil-stained rock 油渍岩石
oil scoop 油收集器；油勺；油类收集器	oil site 润滑点；润滑部位	oil-staining 油着色
oil scraper 刮油器；刮油刀	oil skimmer 油撇取器；污油回收船；撇油器	oil starter 油起动器
oil scraper piston ring 刮油活塞环	oil skimmer barge 撇油船	oil starvation 缺油
oil scraper ring 刮油圈；刮油环；括油环	oil skimmer vessel 撇油船	oil statistics 石油统计
oil scraping ring 刮油环	oil skimming and fire boat 油污收集消防船	oil-steam mixture 石油与水蒸气的混合物
oil screen 滤油(器滤)网	oil skimming barge 撇油驳船	oil steering clutch 液压转向离合器；液压转向离合器
oil-screen clogging 油筛堵塞	oil skimming tank 撇油罐；污油回收槽	oil still 油蒸馏器
oil scrubber 油洗涤器	oil-skin 防水油布；油布；油布雨衣	oil-still tube supports 石油蒸馏炉的炉管支架
oil scumble glaze 油基薄涂釉面	oil skinning 石油的皮囊装运	oil stock 油储藏
oil scupper 排油槽	oil skip 油匙	oil stock loss 油的储藏损失
oil seal 油封圈；油封	oil skipper 油勺	oil stock loss control 油储藏损失的控制
oil seal collar 油封挡圈	oil sleeve 油套	oilstone 油石；油砥石；细磨刀石
oil seal(ed) pump 油封泵	oil slick 油污；油膜；油花；油斑；水面浮油；浮油	oil stone dust 油石粉
oil-sealed stuffer 油封填料	oil slick boom 水面油栅	oil stone powder 油石粉
oil seal felt 油封毡	oil slick cleaning vessel 浮油清扫船	oilstone slip 磨刀小油石；弧口凿磨石；油磨石；油石
oil seal felt retainer 油封毡护圈	oil slick portion 油滑地段	oil stop 防堵油漏材料
oil seal gasket 油封垫	oil slick separator 油花分离器	oil stopper 油性填料；油性腻子
oil seal groove 油封槽	oil slinger 抛油环；甩油环	oil storage 油料储存；油的储藏；石油储[贮]存；石油储存；储油
oil seal guard 油封护挡	oil slinger gasket 抛油圈垫	
oil seal housing 油封填料箱	oilslip 油石；砂轮；磨刀石	oil storage barge 油罐驳；油驳；储油驳
oil-sealing arrangement 油封装置	oil slot 油槽	oil storage capacity 润滑油存储能力；储油量
oil sealing gland 油封填料盖	oil sludge 油泥；含油污泥	oil storage cellar 储油的地下油库；储油的地下室；地下储油池
oil seal leather 油封革	oil/slurry/ore ship 石油/水泥浆/矿砂三用船	
oil seal lip 油封件的唇部	oil smoke 油烟	oil storage facility 油罐；储[贮]油设备；储运设备
oil seal remover 油封拆卸工具	oil socket 油窝	oil storage reservoir 储油库
oil seal replacement 油封更换	oil-softener 油类软化剂	oil storage ship 储油船
oil seal retainer 油封护圈	oil soil 古土壤；油污	oil storage tank 燃料油箱；储油箱；储油罐；储油槽；储[贮]油槽
oil seal ring 油封圈	oil-solid system 油粒综合体	
oil seal rubbing speed 油封摩擦速度	oil solubility 油溶解度；油溶性；油可溶性	oil store 储油仓库；油库；储[贮]油库
oil seal spring 油封弹簧	oil-soluble 可溶于油的	oil storing 油料储存
oil-seal torque 油封阻力矩	oil-soluble concentrate 油剂	oil stove 油炉
oil seal tube 油封管	oil-soluble detergents 油溶性清净分散剂	oil strainer 油料过滤器；滤油网；滤油器
oil seal washer 油封圈	oil-soluble dyes 油溶染料	oil strain test 滤油试验

oil streak 油污条痕
oil stream 油流
oil string 油线;油绳;油管柱;油层套管
oil stripping 洗油脱苯;油纹
oil struck at 在……(深度)遇到油
oil structure 储油构造
oil substitute 油的代用品
oil sucking machine 抽油机
oil suction 吸油口
oil suction bell 油泵钟形罩
oil suction pipe 吸油管
oil suction pipe adapter 吸油管连接器
oil suction pump 吸油泵
oil suction tube 吸油管
oil sump 油腔;油底盘;油底壳;油池;油沉淀池;储[贮]油槽;集油槽;机油盘;储油槽
oil sump capacity 油底壳容量
oil sump strainer 油沉淀池过滤器
oil sump tank 油底壳
oil supercharging pump 增压油泵
oil supply 供油
oil supply bar 主供油管;操作油管
oil supply head 受油器
oil supplying 油的供应
oil-supplying equipment 供油设备
oil-supplying facility 供油设施
oil-supplying plant 供油设备
oil supply line 供油管;操作油管
oil supply pier 供油码头
oil supply pump 供油泵
oil supply quay 供油码头
oil supply rate 供油率
oil supply system 供油系统
oil supply tank 供油槽
oil supply tanker 供油船
oil surface 油面
oil surfacer 油整面涂料;油质整面涂料;油性中层漆;油性二道漆;油性二道浆
oil surface road 油面路;浇油路
oil surface temperature 油的表面温度
oil surface tension 油的表面张力
oil surplus 石油盈余
oil-susceptible 油易染的;油敏感的
oil swell 油溶胀
oil switch 油开关;油浸开关;油断路器
oil switch trip 油开关脱扣
oil syphon 虹吸润滑器
oil syphon lubricator 虹吸注油器
oil syringe 注油枪;润滑油枪
oil system 液压系统
oil tagging system 油示踪法
oil tank 油槽;油箱;油柜;油罐;油舱;储[贮]油罐;储[贮]油罐;储油罐;储油池
oil tank age 油库
oil tank area 油罐区面积
oil tank base support 油箱底座
oil tank breather 油箱通气管;油箱呼吸器
oil tank car 运油车;油箱车;油槽车
oil tank coil heater 油罐加热盘管
oil tank cooler 油冷却器
oil tanker 运油船;运油船;油轮;油罐车;油船;油槽;石油运输船
oil tanker accident incident 油船失事事件
oil tanker jetty 油船码头;油轮码头
oil tanker washing 油槽车冲洗
oil tank hatch 油舱口
oil tank heater 油槽加热器
oil tank hopper 油箱入口接受过滤器
oil tank ladder 油舱梯
oil tank paint 油舱漆
oil tank pump 浸没式油泵
oil tank radiator 油箱散热器
oil tank truck 油槽车
oil tank vehicle 油品槽车
oil tank vent 油箱通气口
oil tank wagon 油箱车
oil tank washing plant 油罐车洗刷所【铁】
oil tannage 鞣革
oil tanning 油鞣
oil tappet 液压挺杆
oil tar 柏油
oil tar lump 焦油块
oil tar pitch 焦油沥青
oil tar sludge 焦油油泥

oil tea 油茶
oil temper 油回火
oil temperature 油温度
oil temperature ga(u)ge 油温计;油温表
oil temperature protecture 油温保护
oil temperature regulator 油温调节器
oil-tempered hardboard 油热处理油质(纤维)板
oil-tempered wire 浸油回火钢丝
oil tempering 油回火;油钢化;浸油回头
oil temper wire 油回火钢丝
oil tender 供油船
oil terminal 油码头;油港;石油码头
oil tester 验油机;试油器
oil-testing equipment 油料分析试验设备
oil-testing machine 油料试验机
oil thermometer 油温计;油温表
oil thickening 油料稠化
oil thief 取油样器
oil thinning 油的冲淡
oil thixotropic behavio(u)r 油的触变行为
oil through 通油孔
oil throw 喷油
oil thrower 溅油圈;溅油环;抛油圈;抛油环;甩油装置
oil throwing 油的喷淋
oil throw ring 抛油圈
oil-tight 油密(的);不漏油的;油封;不透水;不漏油
oil-tight bulkhead 油密舱壁;不渗油舱壁
oil-tight cover 油密盖
oil-tight drip tray 油密滴油盘
oil-tight floor 油密肋板
oil-tight hatch 油密舱口
oil-tight hatchcover 油(密)舱盖
oil-tight joint 油密接头
oil-tightness 不透油性;油密性
oil-tight riveting 油密铆接;油密铆钉
oil-tight test 油密试验;油密封试验
oil-tight transverse bulkhead 油密横舱壁
oil-tight work 油密工作
oil tintometer 油品色调计
oil-to-coke replacement ratio 油焦置换比
oil-to-oil booster 全液压增压器
oil-to-oil intensifier 全液压增压器
oil-to-water heat exchanger 水冷热交换器
oil trace 油迹
oil transfer equipment 输油设备
oil transfer pump 润滑油输送泵;输油泵
oil transfer pump house 输油泵房
oil-transferring 石油输送;油输送
oil transfer system 输油系统
oil transport 运油业务;石油运输
oil transportation 输油
oil transport system 输油管系(统)
oil transshipment port 石油转运港
oil trap 油收集器;油圈闭;油捕;集油槽;圈闭;捕油器
oil trap with sloping plank 斜板隔油沉淀池
oil tray 油盘
oil-treated fabric 涂油织物
oil-treated surface 沥青处治面层;沥青处治路面;铺路油处治面层;铺路油处治路面
oil treater 油净化处理器
oil-treating board 油浸板
oil treatment (土路的)沥青处理;油处理;(土路的)铺路油处治
oil-treatment of coal 煤的油处理
oil-trip valve 油压断流阀
oil trough 油池;油槽;润滑油槽
oil-trough circuit-breaker 油槽断路器
oil truck 油槽车
oil true colo(u)r 油的本色
oil tube 油管
oil tube bracket 油管托架
oil tube drill 油管钻头
oil tube(-type) cooler 管式冷油器
oil tubing pressure 油管压力
oil type 油型;油品
oil-type air cleaner 油浴空气过滤器
oil-type analysis 油型分析
oil-type buffer 油压减震器
oil-type preservative 油型防腐剂
oil-type rust preventive 油型防锈剂
oil-type scumble glaze 油类薄涂法使用的瓷釉
oil-type servo-motor 油压式伺服电动机

oil-type stain 油类污斑
oil under pressure 压力油;受压油液
oil unit train 石油专用列车
oil-using units 用油设备
oil vacuole 油泡;含油液泡
oil vacuum pump 油真空泵;滑油真空泵;润滑油真空泵
oil valve 油阀
oil vane servomotor 叶片式油液伺服机构
oil vaporizer 油类蒸发器
oil vapo(u)r 油气
oil-vapo(u)r diffusion pump 油蒸汽扩散泵
oil-vapo(u)r pump 油蒸汽泵
oil-vapo(u)r space 油汽空间
oil varnish 油基清漆;油质清漆;油性清漆;清(油)漆
oil-varnished cambric 油漆布
oil varnish length-gallons 油漆长加仑
oil vehicle 油基媒液
oil vehicle paint 媒液漆料
oil vent 排油
oil vermilion 油朱红
oil vessel 油容器
oil viscosity 油的黏度;滑油黏度;石油黏度
oil volatility 油的挥发性
oil volume 油容量
oil volume adjustment 油量调整
oil volume restrictor 油量限流器;油量节流阀
oil vulnerability index 油污染致命性指数
oil wagon 装油车
oil wake 油迹
oil warm air heater 燃油暖气供热装置
oil warm-up torch 重油点火嘴
oil wash 油洗
oil washing 洗油
oil-washing air cleaner 用油洗涤的空气滤清器
oil-washing apparatus 油洗装置
oil waste 含油废(弃)物
oil waste liquor 含油废液
oil water 含油水
oil-water bed 油水同层
oil-water boundary 油水接触面;油与水的界面;油水界面;油水边界
oil-water consumption 油水耗量
oil-water contact 油水面
oil-water emulsion 乳化液
oil-water gas 石油水煤气
oil-water gas tar 石油水煤气焦油
oil-water heater 油热水器;燃油热水器
oil-water interface 油与水之间的交界面;油水界面;切水(油水分界面)
oil-water interface detector 油水界面探测仪;油水界面测定器
oil water mixture 油水混合物
oil-water ratio 油水比
oil-water separation 油水分离
oil-water separation agent 油水分离剂
oil-water separation system 油水分离系统
oil-water separator 油水分离器
oil-water sludge 油水沉渣
oil-water surface 油水分界面;油水面;油水接触面
oil-water two-phase flow model 油水两相渗流模型
oil-way 油通孔;润滑油沟;油路;润滑油槽
oil wear 油耗
oil wearing quality 油的防磨损性质
oil wedge 油楔
oil wedge action 油楔作用
oil weir 防油堤
oil well 油井;油槽;集油阱;石油井;油眼
oil-well bit 油井钻头
oil-well blasting 油井爆破
oil-well blowing 油井喷油;油井井喷
oil-well casing 油井套管
oil-well cement 油井用水泥;油井水泥
oil-well cementating 油井注水泥
oil-well cementation 油井注水泥
oil-well cement classification 油井水泥分类
oil-well cement clinker property 油井水泥熟料性质
oil-well cement experiment instrument 油井水泥实验仪器
oil-well cement of high density 高密度油井水泥;高比重油井水泥
oil-well cement of low-density 低密度油井水泥
oil-well cement physical property 油井水泥物理性质

oil-well derrick 油井井架
oil-well drilling 油井钻进
oil-well drilling machinery 凿油井机
oil-well drive 油藏驱动机制
oil-well grouting 油井灌浆
oil-well grouting cement 油井灌浆水泥
oil-well packing 油井封隔;油井堵塞
oil-well pipe 油井管
oil-well platform concrete 油井平台混凝土
oil-well plugging back 复灌注老油井
oil-well plunger pump 油井柱塞泵
oil-well pressure 油井压力
oil-well proppant 油井支撑物
oil-well pump 抽油泵
oil-well screen 油井滤者;油井过滤筛;油井衬管
oil-well shooting 油井射孔
oil-well spray shield 井喷护罩
oil-well swab 从油井汲取液体的活塞
oil-well temperature 油井温度
oil-well tubing 油管
oil-well wastewater 油井废水
oil wet core 油浸岩芯
oil wet phase 油相
oil wettability 油润湿性能
oil-wetted air cleaner 油浴式空气滤清器
oil-wetted sharpening stone 油润湿的磨石
oil wharf 油码头;石油码头
oil wheeling 转鼓加油
oil whip 油膜振荡;润滑油起沫
oil whipping 油膜振荡
oil whirl 油膜旋窝;旋转油膜;油膜涡动
oil white 油白;混合白颜料
oil whiting test 渗透探伤
oil wick 油绳;油芯;润滑油芯
oil wick adapter 油绳接头
oil wick lubrication 油绳润滑
oil window 油窗
oil wiper 刮油器
oil withdrawl 采油量
oily 油质;含油
oily discharge 油质废水排放;含油废水
oil-yielding shrubs 油料林木
oil-yielding stratum 出油层
oily layer 油层
oily liquid 油状液体
oily-looking dim amber 油光暗淡的琥珀
oily lubricant 油性润滑剂
oily luster 油脂光泽
oily moist 油雾;油薄雾
oily moisture 油潮气
oily pollutant 油质污染物;油性污染物;含油污染物
oily putty 油质腻子
oily residue 含油残物
oily seed 含油种子
oily sewage 油质污水;油性污水
oily sewerage 含油污水
oily sheet 油层
oily sludge 油质污泥
oily waste 油状废水;含油废(弃)物
oily waste liquor 油质废液
oily waster 含油废物;含油废品
oily wastewater 油质污水;油性污水;含油废水
oily water 含油水;含油的水
oily wax 含油蜡
oil zone 油带;含油地带;含油带
ointment 软膏
ointment box 软膏盒
ointment for gas protection 防毒气软膏
ointment jar 油膏缸;软膏罐
ointment mill 软膏磨
ointment of zinc oxide 氧化锌软膏
ointment roller 油膏碾磨器
ointment slab 软膏板
oisanite 透绿帘石
oissipation function 瑞利耗散函数
oiticica oil 奥蒂似;巴西果油
ojosa 蜂窝构造
ojuelaite 纤砷铁锌石
okaite 兰仿黄长岩
Okan 奥坎可可棕色硬木(尼日利亚产)
okanoganite 氟硼硅钇钠石
o'kaycrete 日本冈谷快速硬化水泥
okenite 硅灰石;水硅钙石
Oker 铸造改良黄铜

Okhotsk Sea 鄂霍茨克海
Okhotsk sea air mass 鄂霍茨克海气团
Okinawa trough 冲绳海槽
Okinawa Trough extensional fracture zone 冲绳海槽张裂带
okonite 奥科耐特
okoume 加蓬桃花心木;加蓬木
OK sheet 合格印样
oktastylos (古希腊神庙的)八柱式
OKwen 奥克卫英棕色硬木(尼日利亚产)
olafite 钠长石
olation 羟桥聚合作用
olation polymer 羟配聚合物
Olbers method 奥伯斯方法
old account 旧账;未了账目
old account period 旧的会计年度
old age 老年期
old-age benefit 养老金
old-age care 养老事业
old-age home 养老院
old-age pension 养老金
old Alps 古阿尔卑斯山
old Altay mountain 古阿尔泰山
old apartment 老套间;老公寓;旧公寓
old Appalachian mountain 古阿巴拉契亚山脉
old arable soil 熟地
old area 老区域;老空地
old balance 上期结余
oldbury stone 老坟砂岩
Old Caledonian orogeny 老加里东造山作用幕
old castle 古堡
old chain 旧链
old channel 故河道;古河道;古河床;古河槽;旧河道
Old Cimmerian orogeny 老细末里运动
old civilization 古代文明
old clay layer 老黏土层
old cliff 古海蚀崖
old coast 古海洋;古海岸
old concrete 陈的混凝土;老的混凝土;旧混凝土
old cost 原始成本
old Dutch process 老荷兰式方法;老荷兰式工艺
old dwelling 老年人住宅
old dwelling unit 老住房单元;老住房单位;老居住单元;老居住单位;老年人住宅
old English bond 英国老式砌砖法
older Dryas stade 中得利亚斯冰阶
older peat 老泥炭
Older's formula 奥尔特公式
Older Stone Age 旧石器时代
oldest Dryas stade 早得利亚斯冰阶
old fashioned 过时的;老式的;旧式的
old-fashioned building 老式建筑
old-fashioned outmoded 陈旧
old-field 弃耕地
old flat 老公寓套房;老公寓楼层
old folks' home 养老院;敬老院
old form 古老形式
old fustic 黄桑
old gold 暗淡金黄色
old Greek mo(u)lding 古希腊装饰线脚;古希腊饰
old growth 原始林;处女林;成熟原始森林;成熟林
Oldham coupling 奥尔德姆联轴节;十字滑块联轴器
oldhamite 陨硫钙石
Oldham-Wheat lamp 奥尔德姆-惠特灯
old hand 老钻工;老手;老工人;熟练工(人);熟手
Oldhaven beds 奥德赫温层
old horse 炉底结块
old hummock 多年冰丘
old ice 多年冰;陈冰
old industrial bases 老工业基地
old iron 废铁
old karst 老岩溶;老年岩溶;老年喀斯特
old lady 老妇人(俗称英格兰银行)
old land 古平原;古陆;久耕地;熟地
old landmass 古陆块
old land slide 古滑坡
old line 既有线
old living unit 老的生活单元
old loess 老黄土
old man 增加钻孔杠杆作用的工具;手持式轻便凿孔机支架
old map 古地图
old maude 大管扳手
old meander crescent 新月形废河弯

old metal 废金属
old model 旧型号
old mountain 老年山
old name 曾用名
old paper 废纸
old paper stock 废纸纸浆
old part 报废件
old people's apartment (unit) 老(年)人公寓
old people's community center 老(年)人社交中心
old people's dwelling 老(年)人住房
old people's home 老(年)人家庭;老(年)人住处;老年人之家
old people's housing 老人住房
old person's flat 老人公寓楼层;老人公寓套房
old plasticity chart 旧塑性图
old price 旧价
Old Red continent 老红砂岩大陆
Old Red earth 老红壤
Old Red land 老红大陆
Old Red sandstone 老红砂岩
old river 老年河
old river bed 古河床
old river course 故河道
old Roman tile 筒瓦;古罗马瓦;古罗马砖
old roof membrane 旧屋面膜;旧屋面防水层
old rose 深玫瑰红色
old ruins 遗迹;古迹
old sand 旧砂
old share 旧股份
Oldshye-Rush-ton column 奥尔德休-拉什顿塔
old silver 旧银器色
old sinking fund 旧偿债基金
old snow 老年雪;旧雪;陈雪
old soil 老土
old stage 老年期
old-stage landform 老年期地貌
old stock 陈货
Old Stone Age 旧石器时代
old stream 老年河
old style 旧历;老式的
old surplus 原有盈余;旧公积
old terms 旧条款;旧例
Old Tertiary[E] 早第三纪【地】
oldtimer 老钻工
old topography 老年地形
old town 古城
old-town grey 旧城衰退
old type 旧式
old urban area 旧城区
Olduvai event 奥杜威事件
Olduvai normal polarity subchron 奥杜威正向极性亚时
Olduvai normal polarity subchronzone 奥杜威正向极性亚时间带
Olduvai normal polarity subzone 奥杜威正向极性亚带
old valley 老年谷
old value 旧值
old volcanic rock 古火山岩
old wall 古城墙
old-well deeper drilling 油井延深钻进
old wives' summer 秋老虎
old woman's tooth 连接工的工具;槽底刨;沟刨(婆婆牙)
old wood 旧木材
old workings 老工作区
old-world 旧时代的
old world brachiopod realm 老世界腕足动物地理区系
old world realm 老世界区系
olea europea 橄榄树
oleaginous 油质的;含油
oleaginous base 油脂性基质
oleaginous material 含油物质
oleaginousness 含油量
oleander 夹竹桃
oleate 油酸制剂;油酸酯;油酸盐
oleatum 油酸制剂
olecranon 鹰嘴
olefiant gas 成油气
olefination 烯化作用
olefin(e) 烯烃;链
olefin(e) 烯属烃;链烯烃
olefin(e) acid 烯酸

olefin(e) alcohol 烯醇
olefin(e)-based 基于烯烃的
olefin(e) aldehyde 烯醛
olefin(e) cement 烯烃含量
olefin(e) complex 烯烃配位化合物
olefin(e) copolymer 烯烃共聚物
olefin(e) compound 烯烃化合物
olefin(e) conversion process 烯烃转化法
olefin(e) fiber 烯烃纤维
olefin(e) plastics 烯烃塑料
olefin(e) polymer oil 烯烃叠合成润滑油
olefin(e) polymer oil from Fisher-Tropsch gasoline 费托法烯烃聚合油
olefin(e) sulfonate 烯属磺酸酯
olefinic acid 烯酸
olefinic alcohol 烯醇
olefinic bond 烯键
olefinic carbohydrate 烯属碳水化合物
olefinic carbon 烯碳
olefinic content 烯烃含量
olefinic fuel 烯烃燃料
olefinic hydrocarbon 烯烃
olefinic link 烯键
olefinic polymerization 烯属聚合作用;烯烃聚合
olefinic terpene 链式萜烯
olefin polymer oil 烯烃聚合油;聚烯烃油
olefin resin 烯烃树脂
oleic acid 油酸
oleic acid series 油酸同系物
oleic alcohol 油醇
oleic collector 油酸捕收剂
oleic series 丙烯酸系列
olein 油酸酯;油精;甘油三油酸酯;三油酸甘油酯;三油精
oleinic acid 油酸
Olenekian 奥伦尼克阶
oleo-alkyd resin 含油醇酸树脂
oleo-bi 油酸铋
oleo-buffer 油压缓冲器;液压减震器
oleocasein paint 掺油酪蛋白颜料
oleocreosote 杂酚油油酸酯;油酸木溜油
oleo-cushion 油压缓冲器;液压减震器
oleocyst 油囊体
oleocystidium 油囊体
oleo-damper 油压缓冲器;液压减震器;液压避震器;液压减震器;油压减震器
oleodipalmitin 一油二棕榈脂
oleodistearin 一油二硬脂
oleodynamic(al) press 油压机
oleo fork 油压缓冲叉
oleo-gear 油缓冲器;油压减震器;液压减震器
oleogel 油凝胶
oleogranuloma 石蜡瘤
oleograph 仿油画石版画;石印油画
oleography 平版复印油画术
oleo-gum-resin 油胶树脂
oleo-infusion 油浸剂
oleol 油浸剂
oleo leg 油压减震柱
oleoma 石蜡瘤
oleomargarine 珠脂;人造黄油
oleometer 油比重计;油量计;验油计;量油计
oleo oil 人造黄油
oleopalmitate 油棕榈酸盐
Oleoparaphene 一六五矿油
oleophilic 亲脂的;亲油的
oleophilic coating 亲油性保护层
oleophilic group 亲油基
oleophilic resin 亲油树脂
oleophobic 疏油的
oleophobic colloid 疏油胶体
oleophobic effect 疏油作用
oleophobic finisher 疏油整理剂
oleophylic 亲油的
oleophylic colloid 亲油胶体
oleophylic fertilizer 亲油肥料
oleophylic gellant 亲油胶凝剂
oleophylic graphite 亲油石墨
oleoplast 造油体
oleo-pneumatic brake 油传动刹车
oleo-pneumatic compensator 油压气动补偿器
oleo-pneumatic shock absorber 油气减震器
oleo-pneumatic strut 油液空气减震支柱
oleo-pneumatic suspension 油气悬架

oleo-pneumatic suspension system 油压气动悬挂装置
oleoptene 玫瑰蜡
oleorefractometer 油折射计
oleoresin 松脂;天然含油树脂;油树脂;含油松脂;含油树脂
oleoresin aspidium 绵马油树脂
oleoresin from conifers 针叶树上的含油树脂
oleo-resinous 油基树脂的;含油树脂的
oleo-resinous adhesive 油树脂黏合剂
oleo-resinous coating 油性树脂涂料
oleo-resinous mix(ture) 含油树脂的混合剂
oleo-resinous paint 油性树脂漆;油性树脂色漆;油质树脂漆
oleo-resinous varnish 油性树脂清漆;油质树脂清漆;油基树脂清漆
oleoresins 干性油和天然树脂的掺配物
oleoresin varnish 油树脂清漆
oleo rubber shock (absorber) strut 油液橡皮减震柱
oleosol 油溶胶;润滑脂
oleosome 油质体
oleo spring shock (absorber) strut 油液弹簧减震柱
oleostasis property 油滞性
oleostearate 油硬脂酸盐
oleostearin(e) 油硬脂
oleostrut 油液减震柱;油缓冲支柱;油液空气减震柱;油压减振柱
oleo undercarriage 油减震式起落架;油减振器底盘
olesome 油滴颗粒
oleum calcis 潮化氯化钙液
oleum eucalypti 桉油
oleum gossypii seminis 棉油
Oleum Ligni Santali 白檀油
oleum ricini 蓖麻油
oleum succini 琥珀油
oleum volatile 挥发油
oleyl 油烯基
oleyl alcohol 油醇
oleylalcohol disulfate 油醇二硫酸盐
oleyl carboxylate 油烯基羧酸盐
oleyl sulfate 油烯基硫酸酯
olfact 气味单位
olfactometer 气味测定器
olfactometry 气味测定法;臭味测定(法);测臭法;恶臭测定法
olfactronics 气味测定学
olgite 磷钠锶石
Oliensis spot test 奥廉斯斑点试验
Oliensis test 奥廉斯试验
olifiant gas 成油气
oligarchic network 少数控制网(络)
oligarchy of finance capital 财政资本寡头
Oligocene epoch 渐新世
Oligocene period 渐新纪
Oligocene regression 渐新世海退
Oligocene series 渐新统
oligoclase 更长石;奥长石
oligoclase albite 奥钠长石
oligoclase basalt 更长玄武岩
oligoclase granite 更长花岗岩
oligoclase trachyte 更长粗面岩
oligoclasite 更长石岩;奥长岩;奥长石
oligodynamic 微量活动的
oligodynamic action 微动作用
oligodynamic(al) disinfection 微动力清毒
oligohaline 寡盐生物;少盐水
oligohaline water 低盐分水;低度盐水
oligohalobic 寡盐的
oligohydria 水缺乏
oligoklas 奥长石
oligomer 齐分子量聚合物;低聚物;低聚体
oligomictic 单岩碎屑的
oligomictic breccia 单成分角砾岩
oligomictic conglomerate 单成分砾岩
oligomictic lake 少循环湖;单一水层湖
oligomictic sediment 单岩碎屑沉积物;陆海沉积
oligomitic facies 海陆过渡相
oligomitic sediment 陆海沉积(物)
oligonite 锰菱铁矿
oligophotic zone 微明区
oligopnea 呼吸迟缓
oligopoly 卖主垄断;卖方寡占;卖方供应垄断;求过于供的市场情况;少数买主垄断
oligopoly price 卖主垄断价格;卖主控制价格
oligopsony 少数买主垄断
oligopsony price 买主垄断价格
oligosaprobic 寡污水腐的;贫污水性的
oligosaprobic region 低污染区
oligosaprobic river 低污染河流
oligosaprobic stream 低污染河流
oligosaprobic waters 轻污染水域;微污染水体;低污染水域;低污染水体;低腐生水域
oligosaprobic zone 寡污水腐生带;寡腐生物带;低污水带;低污染带
oligosaprobiont 微腐生物
oligosite 奥长岩
oligotrophication 贫营养化
oligotrophic bog 贫营养酸沼
oligotrophic brown soil 贫脊棕色土
oligotrophic condition 贫营养条件
oligotrophic habitat 贫营养生境
oligotrophic lake 不够供应足够营养的湖泊;寡营养湖泊;缺营养湖泊;缺养分湖;贫滋育湖泊;贫营养湖;富氧湖(泊)
oligotrophic lake type waters 贫营养湖型水域;贫营养湖型水体
oligotrophic mire 贫营养泥沼
oligotrophic soil 贫脊土
oligotrophic swamp 寡养沼泽;贫营养林沼
oligotrophic water 贫营养水
oligotrophic water body 贫营养水域;贫营养水体
oligotrophic zone 寡污养带
olistoglyph 层间滑痕
olistolith 滑动岩体
olistostrome 重力滑动堆积;泥砾岩层;滑塌堆积;滑塌沉积;滑积岩;滑积层;滑混堆积;滑动堆积
olistostromic melange 泥砾混杂岩
olivaceous 橄榄状的;橄榄色的
olivary 橄榄形的
olive 油橄榄
olive acanthus 橄榄叶形饰
olive black 黑橄榄
olive branch 橄榄枝
olive brown 橄榄棕色;黄褐色的
olive building 橄榄形大楼
olive butt 橄榄形铰链
olive carving 橄榄木雕刻品
olive colo(u)r 橄榄色
olive drab 草绿色;草黄色;橄榄黄褐色
olive green 橄榄绿(色)
olive green clear 淡橄榄绿
olive green deep 深橄榄绿
olive green paint 草绿漆
olive green pigment 橄榄绿颜料
olive hinge 橄榄形铰链
olive hole 橄榄形孔
olive infused oil 橄榄泡制油
oliveiraite 水钛锆石
olive knuckle butt 橄榄形肘状铰链
olive knuckle butt hinge 橄榄形接头可拆对接铰链
olive knuckle hinge 橄榄形肘状铰链
olivenite 橄榄铜矿
olive nut 橄榄核
olive oil 橄榄油
olive oil-mill effluent 橄榄油厂污水
olive oil refining 橄榄油精炼
olive orchard 橄榄园
olive products 橄榄制品
oliver 脚踏铁锤
Oliver filter 奥利弗型(真空)过滤机;真空圆筒滤器
oliver plum yew 篦子三尖杉
olive-shaped vase 橄榄瓶
olive-stone carving 橄榄
olivet 人造珍珠
olive tree 橄榄树
olivette 人造珍珠;橄榄园(林);强烈泛光灯;强力泛光灯
Olive vacuum filter 奥利弗真空过滤器
olive-yellow 橄榄黄色
olivine 橄榄石
olivine alkali gabbro 橄榄碱性辉长岩
olivine anorthodite-rich rock 富橄榄石钙长石岩
olivine-apatite rock 橄榄磷灰岩
olivine augite-phyric basalt 橄辉斑玄武岩
olivine basalt 橄榄玄武岩

olivine basaltic glass 橄榄玄武玻璃
olivine brick 橄榄石块料；橄榄石砖
olivine clinopyroxenite 橄榄单斜辉石岩
olivine diabase 橄榄辉绿岩
olivine dolerite 橄榄粒玄岩
olivine gabbro 橄榄辉长岩
olivine gabbro norite 橄榄辉长苏长岩
olivine hornblende pyroxenite 橄榄普通角闪辉石岩
olivine hornblendite 橄榄角闪石岩
olivine kimberlite 橄榄石金伯利岩
olivine leucitite 橄榄白榴岩
olivine nephelinite 橄榄霞岩
olivine-nodule 橄榄石岩球；橄榄石结核
olivine norite 橄榄苏长岩
olivine orthopyroxenite 橄榄斜方辉石岩
olivine pyroxene hornblendite 橄榄辉石角闪石岩
olivine pyroxenite 橄榄辉石岩
olivine pyroxenolite 橄榄辉石岩
olivine refractory material 橄榄石耐火材料
olivine refractory (product) 橄榄石耐火砖
olivine-rich hypersthene gabbro 富橄榄石紫苏辉长岩
olivine rock 橄榄岩；纯橄榄岩
olivine teschenite 橄榄沸绿岩
olivine theralite 橄榄霞斜岩
olivine tholeiite 橄榄拉斑玄岩
olivine type macrocrystalline glaze 橄榄石型巨晶釉
olivine websterite 橄榄二辉石
olivinfels 橄榄岩；橄榄石
olivinite 橄榄岩；角闪辉石橄榄岩
olivinization 橄榄石化
olivospinal tract 三角束
olla 陶制壶；砂锅
ollacherite 钡白云母
ollenite 绿帘金红角闪片岩
ollite 滑石
ollite limestone 鲕状灰岩
Olmec architecture 奥尔梅克建筑
olmsteadite 磷铌铁钾石
olo 外间距
olobby transfer 粒状过渡
olock encoding 分组编码
Olon 奥隆乳油硬木（产于黄金海岸）
olsacherite 硒铅矾
Olsen ductility test 奥尔逊延展性试验
Olsen memory 线选开关存储器；奥尔逊存储器
Olsen tester 奥尔逊试验机
olshanskyite 羟硼锶石
Olsmoloy 铜锌镍合金
olten slag 熔渣
Olympic bronze 奥林匹克青铜
Olympic bronze alloy 奥林匹克铜绿；奥林匹克青铜合金
Olympic green 铜绿；奥林匹克铜绿
Olympieion 奥林帕斯神庙
olympite 磷钠石
Olympus Mons 奥林匹斯山
Omalon carpet foundation 奥马伦地毯（软）垫
O-man 架空式机械手；大型万能机械手
Omazene 灭菌铜
ombree finish 深浅交映的油漆加工
ombrogram 雨量图
ombrograph 自记录雨量器
ombrology 测雨学
ombrometer 雨量器；雨量计；微雨计
ombrophile 好雨植物；适雨植物；喜雨植物
ombrophilous plant 喜雨植物；好雨植物；适雨植物
ombrophobe 避雨植物；嫌雨植物
ombrophore 碳酸文淋浴器
ombrophyte 适雨植物
ombroscope 报雨器
Omega 奥米伽（导航系统）
Omega fixing 奥米伽定位
Omega grate plate 带熟料兜槽的篦板
Omega hyperbola family 奥米伽双曲线族
Omega hyperbolic grid system 奥米伽双曲线网格坐标系
Omega lane identification 奥米伽巷识别
Omega meson 奥米伽介子
Omega navigation system 奥米伽导航系统
Omega positioning and locating equipment 奥米伽定位设备

Omega receiver 奥米伽接收机
Omega signal format 奥米伽信号格式
Omega system 奥米伽系统
Omega transmitting station 奥米伽发射台
omegatron 高频质谱仪；高频质谱计；高频回旋质谱仪；奥米伽器
omegatron ga(u)ge 回旋真空规
omegatron mass spectrometer 回旋质谱仪
omegatron tube 回旋计管
omeiite 峨眉矿
omen 预兆；前兆
ominidirect light 环照灯
omirimbi 旱谷
omission 遗漏；岩层缺失；缺位；缺失（地层）；省略
omission excepted 遗漏不在此限
omission factor 遗漏系数；省略因数
omission rate 漏查率
omission solid solution 缺位固溶体
omitted error 省略误差
omitted-size (type) grain concrete 间断级配的混凝土
omn dieb 每日
omnia 咬封盖
Omnia concrete floor 奥尼阿地板（一种可预制或现浇的轻质钢筋混凝土地板）
omni aerial 全向天线
Omnia floor 奥尼阿楼板
Omnia grid 奥尼阿网格
omniantenna 全向天线
Omnia rib floor 奥尼阿带肋楼板
Omnia roof 奥尼阿屋顶
Omnia trimmer concrete plank 奥尼阿托梁混凝土板
Omnia wide slab 奥尼阿宽的厚板
omni beacon 全向指向标
omnibearing 全向；全方位
omnibearing beacon 全方位信标
omnibearing converter 全向转换器；全向方位变换器
omnibearing distance facility 全向测距导航设备；全向方位距离导航设备
omnibearing distance nagivational system 全方位距离导航系统
omnibearing distance navigation 全向导航系统；极坐标（系）导航；全方位距离导航
omnibearing distance navigation system 全向方位距离导航系统
omnibearing distance station 全方位距离电台
omnibearing indicator 全向方位指示器；全方位指示器；方位角指示器
omnibearing line 全方位线
omnibearing selector 无线电定向标选择器；全向选择器；全方位选择器
omnibus 总括的；公共汽车
omnibus bar 母线；汇流条；汇流排；配电板汇流条
omnibus bill of lading 总提单；混合提单
omnibus book 文集
omnibus budget reconciliation act 统括预算调整法
omnibus calculator 多用计算装置
omnibus clause 总括条款；综合条款；统括条款
omnibus credit 总括贷款
omnibus law 多项定律
omnibus rod 母线
omnibus speaker circuit 公共通话电路
omnibus train 慢车；混合列车
omnibus volume 文集
omnicompetent 全权的
omnidirectional 全向的；无定向的
omnidirectional antenna 全向天线；全向辐射天线
omnidirectional beacon 全向式无线电指向标；全向导航台；全向信标
omnidirectional collector 全向收集器
omnidirectional effect 全向效应
omnidirectional flux 全向流量
omnidirectional horn loudspeaker 全向号筒式扬声器
omnidirectional intensity 总强度
omnidirectional light 全向灯光；全方向普照灯
omnidirectional lighting 全向灯光设备
omnidirectional maneuver 任意方向机动
omnidirectional measurement 总强度测量
omnidirectional microphone 全向麦克风；全向传声器
omnidirectional optic(al) scanner 全向光学扫描器

omnidirectional phased array 全向相控阵
omnidirectional pressure 全向压力
omnidirectional radar prediction 全向式雷达预测
omnidirectional radiator 全方位辐射器
omnidirectional radio beacon 全向无线电指向标；全向无线电信标；全方向无线电导航标；全向信标
omnidirectional radiometer 全向辐射仪
omnidirectional radio range 全向无线电距离
omnidirectional range 全向导航
omnidirectional range set 全向测距仪
omnidirectional range station 全向导航台；全向测向电台
omnidirectional ranging 全向测距
omnidirectional response 全向响应
omnidirectional rotating radio beacon 全向旋转无线电信标
omnidirectional scanner 全向扫掠设备
omnidirectional seismic trigger 无定向地震触发器
omnidirectional signal 全向信号
omnidirectional solar cell 全向太阳电池
omnidirectional sound source 全向声源
omnidirectional system 全向制
omnidirectional transducer 全向换能器
omnidirective antenna 全向天线
omnidistance 全程
omnifarious 各种各样的
omnifiltration system 总滤系统
omniforce 全向力
omnigraph 绘图器；缩图器
omniguide antenna 全向辐射槽缝式天线
omnilateral pressure 全侧向压力
omnimate 简化自动生成设备
omnimatic 简化的自动生产设备
omnimeter 全向经纬仪
omni mill 全铣
omnipotence 无限权威
omnipotent 全能的
omnirange 全向式无线电信标；全向导航台；全方向；全程
omnirange antenna 全向无线电信标天线
omnirange course indicator 全向无线电导航航道指示器；全方位航向指示器
omnirange digital radar 全向数字雷达
omniranger 全向信标
omnirange radio set 全向信标无线电设备
omnirange resolver 全向无线电信标分解器
omnirange servo 全向无线电信标伺服系统
omniseal 全密封
omniselector 航向选择器；导航器
omnitron 全能加速器
omnium-gatherum 混合物气；混合物剂
omnivisual range radio 全向视程无线电设备
omnivore 杂食性动物（总称）
omphacite 绿辉石
omphacite-eclogite 绿辉石榴辉岩
omphalos 盾中央的浮雕饰；中心点；[复]omphali
omtimeter 高精度光学比较仪
on a 24 hour basis 二十四小时地
on account 作为分期付款；记账交易；记账；赊账
on-account payment 赊付；暂付款
on a field basis 根据现场资料
on a flat rate 按定价
on again off again 时断时续
on agreement 按合同
on air 送风期；水泵空吸状态（不进水）
on-air light 播发信号灯
on a large-scale 大规模地
on a lee shore 船在下风岸
on allocation 待配给
on an average 平均
on and off 时断时续
on-and-off controller 双位式控制器；双位开关控制器
on-and-off switch 通断开关
on-and-off the site 进退场
on-and-off-type boiler control 开闭锅炉控制阀
on an even keel 艉尾吃水相等
on approval 试行推销
on a small scale 小规模地
on a turn-key basis 采取交钥匙的形式
on-axis 同轴的
on-axis chopper blade 同轴调制盘叶片
on-axis gain 轴向增益

on-axis sputtering 同轴溅射
on balance volume 净额成交量
on-base illuminator 显微镜座上装置式显微镜灯
on beam ends 船九十度倾侧
on-beam polarization 波束轴向极化
on behalf of 代表……
on berth 在停泊中;在泊位上
on-berth wind 吹拢风
on board 已装船;在船上;船上
on-board bill of lading 已装船提单;装船提单
on-board carriers freight 第二船运费
on-board checkout 船上检查
on-board checkout and management system 船上检查和数据管理系统
on-board checkout equipment 船上检查设备
on-board communication station 船上通信电台
on-board control 机止控制;随车控制
on-board converter 车载变流器
on-board data automatic reporting system 船载数据自动报告系统;船上数据自动报告系统
on-board data processor 船用数据处理机
on-board digital computer control 船用数字计算机控制
on-board endorsement 已装船背书
on-board endorsement bill of lading 已装船背书提单
on-board notation 已装船批注;装船记载
on-board optic(al) system 机载光学系统
on-board power supply 车载电力供应
on-board power supply network 车载供电网
on-board rail lubricator 车载钢轨涂油器
on-board ramp way 斜坡道
on-board ship 船载
on-board station 船上岗位表
on bound ramp 入境坡道
on bound traffic 入境交通
on budget 在预算内
on business 因公
on call account 透支账户;透支户;随时支取账户
on-call circuit 应需线路
on cam operating condition 协联工况
on cam relationship (导叶开度与桨叶角度的)协联关系;组合关系
on-carriage 续运
on-carrier 陆地承运人;相继运送人
once 一次
once again 再次
once-and-for-all change 一次性变动
once-and-for-all effect 一次性效应
once command 一次指令
once establishment control network 一次布网【测】
once-fired ware 一次烧成器皿;一次烧成的制品
once firing 一次烧成
once-in-a-lifetime exposure 终身一次照射量
once-in-ten-year flood 十年一遇洪水
once more 再次
on centers 中心间距(离);中心间距
once-reflected beam 一次反射束
once-reflected pulse 一次反射脉冲
once-run distillate 一次馏液
once-run kerosene 直馏灯油
once-run oil 页岩原油蒸馏时得到的最后馏分
once setting in drying-firing 一次码烧工艺
once-through 直通
once-through boiler 直流锅炉;直通锅炉;单流式锅炉
once-through circulation 直流循环
once-through conversion 单程转化
once-through coolant system 一次流过冷却系统
once-through cooling 直流式冷却;一次流过冷却;单程冷却;非循环冷却
once-through cooling system 一次性冷却系统;单程冷却系统
once-through cooling water system 直流冷却水系统
once-through corrector plate 单程改正透镜
once-through cracking 单积裂化;非循环裂化
once-through design 一次设计
once-through flowsheet 一次循环流程图
once-through freshwater 单向流动淡水
once-through fuel cycle 一次燃料循环
once-through material balance 一次通过物料平衡
once-through method 一次法

once-through operation 一次运算;一次操作;单循环;单程操作;非循环操作
once-through pipe still 一次蒸发管式炉
once-through process 一次通过过程;非循环过程
once-through refuelling 直通换料
once-through saline 单向流动咸水
once-through seal 一次流过密封
once-through system 一次通过非循环系统
once-through type steam generator 一次通过型蒸汽发生器
once-through uranium-fuel cycle 一次铀燃料循环
once-through water system 直流水系统
on chip memory 单片存储器
on chip oscillator 单片振荡器
on clay 拌土
on close 收市执行
oncoid 似核形石
oncoite 蠕绿泥石
oncolite 藻灰结核;核形石
oncolithes 球状叠层石
oncolitic dolomite 核形石白云岩
oncolitic limestone 核形石灰岩
oncologic hospital 肿瘤医院
oncomelania snail 钉螺
on-coming flow 来水
on-coming sea 迎浪
oncoming traffic 迎面车流;对向交通
oncoming traffic stream 对向交通车流
on company service 由公司免费运输
on-condition 接通条件
on contract terms 以承包方式
on cost 间接费(用);杂费
on course 在航线上
on-course beam 航线无线电波束
on-course curvature 航向曲率
on-course detector 航向对准指示器
on-course selector 航向选择器
on credit 信用交易;赊欠(交易);赊购费
on credit account 信用赊账
on current basis 按现值
on day-to-day basis 按日的
on deck 在甲板上
on deck bill of lading 甲板货提单;舱面提单
on deck bill of letter 甲板货提单
on deck cargo 舱面货(物)
on deck cell guide system 甲板上格导系统
on deck clause 舱面条款
on deck delivery 舱面交货
on deck operation 舱面作业
on deck risk 舱面险
on deck shipment 舱面装货
on deck stowage 甲板装载
on default 未履行责任的
on demand 要求时;根据买方通知发货;即付
on demand bond 凭要求付款的保证书
on-demand system 立即响应系统;应求系统;即时(回答)系统;请求系统;请求服务系统
on-detector system 直联检测器系统
onde yarn 波形纱线
ondine flow routing 联机流量演算
ondoad tap changing transformer 有载调压变压器
ondograph 高频示波器;电容式波形记录器;(电容式)波形记录器
ondometer 频率计;测波仪;测波器;波形测量器;波长计
ondoscope 辉光管振动指示器;示波器
ondrejite 白钠硅钙镁矿
ondulated 波动的
ondulateur 时号自记仪
ondulation 波浪式振荡
on duty 值班
on-duty casualty 路内人员伤亡
on-duty director 值班主任
on-duty room for water crane operator 水鹤值班房
on-duty time 值班时间;上班时间
one able to do light manual labour only 半劳动力
one-above-the-other layout 纵向连接装置
one action 单动作的
one-address 地址的;单地址;一地址
one-address code 一地址码
one-address computer 单地址计算机
one-address instruction 一地址指令
one-address system 单地址式

one-ahead addressing 先行寻址;重复寻址;重复定址;超一寻址法;超前寻址
one and a haft circuit breaker connection 倍半结线
one-and-half brickwall [37cm brickwall] 一砖半厚墙
one-and-two pipe combined heating system 单双管式采暖系统
one-and-two pipe layout 单双管式布置
one and zero code 一与零编码
one-arm anchor 犁锚;单爪锚;单臂锚
one-armed anvil 单臂铁砧
one-armed Johnny 小手摇泵
one arm swinging front loader 单臂旋转式前置装载机
one aspect 一端
on easy terms 以分期付款方式
one-at-a-time operation 时分操作
one-atmosphere diving system 常压潜水系统
one-atmosphere manned underwater structure 常压载人水下结构
one-atmosphere vehicle 常压潜水器
one axis extend and another compressed type 一轴伸长一轴缩短型应变椭圆
one-axis laser gyroscope 单轴激光陀螺仪
one-axle scraper 单轴铲运机
one-axle steering 单轴转向
one balcony type auditorium 单眺台式观众厅;单楼座观众厅
one bar chamber 标准大气舱
one-base method 单基线法
one-bath method 单浴法;一浴法
one-bay 单跨的
one-bay beam 单跨梁
one-bay girder 单跨大梁
one-bay slab 单跨板
one-bay two-hinged frame 单跨双铰链构架;单跨双铰框架
one-beaked anvil 单啄铁砧
one-bed 单铺;单床铺
one-bed dwelling unit 单人居住单元
one-bed flat 单人公寓
one-bed guest room 单床客房;单铺客房
one bedroomed dwelling unit 单间卧室住房单元
one bedroomed flat 单间卧室套房;单间卧室楼层
one bedroomed living unit 单间卧室居住单元
one bedroom unit 一卧室单元
one-bed yard 有一个台座的(预应力混凝土)工场
one-belt reversing countershaft 可逆转单带轮副轴
one berth cabin 单人舱间
one best-way 一种最好的方式
one bit 一位
one-bit machine 一位计算机
one-bit processor 一位处理机
one body approximation 单体近似法
one-body welding machine 同体式焊机
one-body welding set 同体式焊机;单机体电焊机组
one-bolt fastening 单螺栓固定
one-brick (masonry) wall 一砖墙
one-brick wall 一砖墙;二十四墙;单砖墙
one bridge over one river 一河一桥
one-brush bottle washer 单刷洗瓶器
one-bucket excavator 单斗挖土机
one bundle of piece goods 一捆匹头货
one cabin boat 单级客轮
one-can urethane coating 一罐装聚氨酯涂料;单组分聚氨酯涂料
one carbon unit 一个碳单位
one card microcomputer 单板微型机
one cell one board form 一鼓一板式(护舷)
one cell switching 单元选通;翻转
one-centered arch 圆拱;圆弧拱;单心拱
one chance pursuit game 一次追击机会对策
one-channel record 单道记录
one chip 单片
one-chip computer 单片计算机
one circuit set 单调谐电路接收机
one coat 单层;一道涂层;一道漆
one-coat brush finish 单层粉刷
one-coat enamel 一次搪瓷
one-coat paint 一道成活漆;底面合一漆
one-coat plaster(ed) ceiling 单层抹灰的天棚;单层抹灰的顶棚;单层抹灰的天花板;表层涂有灰泥的天花板

one-coat plastering 一次抹面
one-coat work 单层粉刷
one coil 一圈;单线圈
one coil transformer 单线圈变压器
one-colo(u)r marking machine 单色打标机
one column adder 一位加法器
one-column matrix 单列矩阵
one-column radiator 单柱散热器;单柱散热片;单柱暖气片
one-column still 单柱蒸馏塔
one-compartment mill 单仓磨
one-compartment septic tank 单室腐化池
one-compartment ship 一舱制船;一舱进水不沉船
one-component 单成分
one-component adhesive 单成分黏结剂;单成分黏合剂
one-component cement 单组分水泥;单成分水泥
one-component coating 单组分涂层
one-component etching primer 单组分反应性蚀刻底漆;单组分反应性磷化底漆
one-component material 单成分材料
one component paint 单组分漆
one component phase diagram 一元相图
one-component plastics 多组分塑料
one-component pretreatment primer 单组分预处理底漆
one-component primer 单组分底漆;单一成分底漆
one-component samples analysis 单组分样品分析
one-component sealant 单组分密封膏
one-component system 一元系统;单组分体系;单组分系统
one-component wash primer 单组分磷化底漆;单组分浸蚀底漆;单组分洗涤底漆
one consignment of goods 一批货(物)
one-course 单方向;单程;单层的
one-course concrete pavement 单层(水泥)混凝土路面;单层混凝土铺面
one-course construction 单层筑ցա;单层结构
one-course method 单层法
one-course pavement 一次性浇灌面层;单层路面
one course system 单作制
one-course trench 单层路槽
one-cover coat enamel 单搪
one crop economy 单一作物经济
one cycle freezing test 单循环抗冻试验
one-cycle multivibrator 单周多谐振荡器;单程多谐振荡器
one-cycle process 单循环过程
one-cylinder 单缸的;单缸
one data delay time 单一数据延迟时间
one-day biochemical oxygen demand 一日生化需氧量
one-day cement 超早强水泥
one-deck classifier 单板分级器
one decker 单甲板船
one-degree of freedom 一个自由度;次自由度;单自由度
one degree of freedom suspension 单自由度悬挂
one-degree system 单自由度体系
one derrick boom cargo handling 单吊杆装卸方式
one-development method 单一开拓法
one-digit adder 半加(法)器
one-digit delay 一位数延迟
one-digit number 一位数
one-digit random number 一位随机表
one-digit subtracter 半减(法)器
one-digit time 数字周期
one dimension 一维
one-dimensional 一维的;单向的;单维的
one-dimensional analysis 一元分析法;单维分析法
one-dimensional aperture 一维孔径
one-dimensional aperture synthesis 一维孔径综合
one-dimensional array 一维阵列;一维数组
one-dimensional Brownian motion 一维布朗运动
one-dimensional code 一维代码
one-dimensional compression 单维压缩
one-dimensional compression method 单向压缩法
one-dimensional consolidation 一向固结;一维固结;单向渗透固结;单向固结;单维固结
one-dimensional continuity equation 一维连续性方程
one-dimensional correlation 一维相关
one-dimensional crystal 一维晶体

one-dimensional cycle 一维循环
one-dimensional deflection 一维偏转
one-dimensional deflection-modulated display 一维偏转调制显示器
one-dimensional disorder 单向无序;一维无序
one-dimensional domain wall 一维畴壁
one-dimensional drainage 单面排水
one-dimensional element 一维元素;一维单元
one-dimensional element representation 一维单元表示法
one-dimensional estuary water quality model 一维河口水质模型
one-dimensional extended orebody 单向延伸矿体
one-dimensional failure 一维衰坏
one dimensional ferroelectrics 一维铁电体
one-dimensional filtering 一维滤波
one-dimensional flow 一元流;单向流(动);一维流(动);单向水流
one-dimensional Fourier transformation 一维傅立叶变换
one-dimensional fundamental form 一维几何形状;一维基本形
one-dimensional grating 一维光栅
one-dimensional heat transfer 一维传热;单向传热
one-dimensional hologram 一维全息图
one-dimensional laser 一维激光器
one-dimensional lattice 一维点阵
one-dimensional linear drift 一维一次漂移
one-dimensional magnification 一维放大
one-dimensional manifold 一维流形
one-dimensional manipulation 一维计算
one-dimensional mathematical model 一维数学模型
one-dimensional model 一维模型
one-dimensional neighborhood 一维邻域
one-dimensional operation 一维运用
one-dimensional optimization 一维最优化
one-dimensional paper chromatography 单向纸色谱法
one-dimensional path 一维通路
one-dimensional plasma 一维等离子体
one-dimensional point 一维点
one-dimensional probability distribution 一维概率分布
one-dimensional problem 一维问题
one-dimensional quadratic drift 一维二次漂移
one-dimensional quasi-nonlinear wave propagation 一维准非线性波传播
one-dimensional quasi-nonlinear wave propagation theory 一维准非线性波传播理论;一维非线性波传播理论
one-dimensional representation 一维表示法
one-dimensional search 一维搜索
one-dimensional shear propagation 一维剪切波传播
one-dimensional simplex 一维单形
one-dimensional source 一维源
one-dimensional space 一维空间
one-dimensional space-bandwidth product 一维空间带宽乘积
one-dimensional spheric(al) wave 一维球面波
one-dimensional state of stress 单量应力状态
one-dimensional statistic 一维统计量
one-dimensional stress 一元应力;一维应力
one-dimensional stretch 一向延伸
one-dimensional torsion group 一维挠群
one-dimensional transfer function 一维传递函数
one-dimensional transmission 单向传输
one-dimensional trend analysis 一维趋势分析
one-dimensional two-phase flow 一维二相流
one-dimensional wave 一维波
one-dimensional wave spectrum 一元波谱;一维波谱
one-dimensional weighted moving average 一维移动加权平均
one-directional excitation 单向激振
one-directional load(ing) 单向加荷;单向荷载
one direction excavation 单口掘进
one-direction pressing 单向压制
one-direction self-aligning ball thrust bearing 单列径向止推滚珠轴承
one-direction thrust ball bearing 单向止推滚珠轴承
one-direction thrust bearing 单向止推轴承

one division 一分度;单刻度;单分度
one-drum hoist 单卷筒提升机;单卷筒绞车
one-eared pottery vase 单耳陶罐
one-effect evapo(u)ration 单效蒸发
one-eighth bend 弧形弯管
one eighth blending system 八分之一均化系统
one-eighty 翻转一百八十度
one-electron bond 单电子键
one electron gun 单电子枪
one-electron reduction 单电子还原
one end 一端
one-end lift 一端举升器
one-energy-storage network 单储能网络
one-factor difference 单因子差异
one-factor variance analysis 单因素方差分析
one-family detached dwelling 一家独用住宅
one-family dwelling 单户住房
one-family house 一户的住宅
one-family unit 单户住宅;单户单元
on-effect 给光效应
one-figure number 单值数
one-flag signal 单旗信号
one-flank gear rolling tester 齿轮单面啮合检查仪;单面啮合齿轮检查仪
one flight 单跑【建】
one-floor finger system 辐射形平房系统
one-floor house 单层房屋;平房
one-floor warehouse building 单层仓库
one-flue boiler 单燃烧管锅炉;单烟道锅炉
one fluid theory 单流体理论
one fluke anchor 犁锚;单爪锚;单臂锚
one fluke mooring anchor 单爪锚
one-for-one 一对一
one-for-one exchange algorithm 一对一交换算法
one-for-one translation 一对一翻译
one-fourth pitch 四分之一高跨比
one-frequency selection system 单频选择制
one ga(u)ge reading per day 每日一次水尺读数
one generator 单发生器
one-grid valve 单栅电子管
one group 一组
one-group model 单群模型
one-group theory 单群理论
one-group treatment of reactors 反应堆单群处理
one half-cycle 半次全振动
one-half method 二分之一法
one-half period 半周期
one-half period rectification 半波整流
one-half stor(e)y 半层楼
one-hand block 一手可拿的砌块
one handed saw 单人锯
one-hand pot 单手把壶;单手把筒;单手把盆;单手把罐;单手把坩埚
one hand quick grip wrench 单手快速夹紧卡钳
one-hand tile 一手可拿的砖瓦
one head automatic arc welding machine 单弧自动电焊机
one-high straddle carrier 单层跨运车
one-hinged arch 单铰拱
one-hinged arch bridge 单铰拱桥
one-hinged frame 单铰框架
one-hole wedge 单孔楔子
one-hour cement 速凝水泥;喷射水泥
one-hour door 一小时(防火)门
one-hour rainfall 小时雨量
one-hour rating 一小时定额;小时功率;小时定额;一小时功率;一小时额定出力
one-hour value 一小时值;小时值
one-hour wall 一小时(防火)墙
one-household structure 一户结构
one-hundred kilogram 一百公斤;一百千克
one-hundred percent 十足准备计划;百分之百
one-hundred percent [100%] viscose staple fibre fabric 纯黏胶短纤维织物
one-hundred percent financing 百分之百为贷款资金
one-hundred percent inspection 全数检验;全部检验
one-hundred percent location 商业区中的最佳位置
one-hundred percent modulation 百分之百调制
one-hundred percent statement 百分率表
one-hundred precent premium plan 全部溢价计划
one-inch map 一英寸地图
one-inch plank 一英寸木板

one-inch scale 一英寸比例尺
one in ten sampling 十取抽样
one-jet 单射流;单喷嘴的
one-knot paint brush 单节油漆刷
one lane 单线(行驶)的;单车道
one-lane bridge 单车道桥(梁)
one-lane highway 单车道公路
one-lane operation 单车道操作
one-lane road 单车道道路
one-lane traffic 单车道单向交通
one-lane tunnel 单车道隧道
one-layer 一层的;单层(的)
one-layer compressibility 单层压密量
one-layer course 单层一皮(砖墙)
one-layered monolithic masonry wall 单层整体圬工墙
one-layer operation 单层施工(法)
one-layer roofing 单层卷材;单层屋面(材料)
one-layer theory (柔性路面的)单层体理论
one-layer winding 单层绕组
one leader with two stay 双柱导架
one-leaf 单页的;单扇的
one-leaf door 单扇门
one-leaf shutter door 单扇百叶门;单扇卷帘门
one-leaf sliding shutter door 单扇百叶扯门;单扇滑动卷帘门;单扇推拉式卷帘门
one-leaf wall 单片墙
one legged crane 单脚起重机
one-level 一级代码
one-level address 一级地址
one-level channel 一阶通路
one-level code 一级代码;绝对地址码
one-level formula 一级公式
one-level garage 单层车库
one level interrupt 一级中断
one-level linkage register 一级连接寄存器
one-level method 一级方法
one-level storage 一级代码;一级存储器;单级存储器
one-level store 一级存储器
one-level store technique 一级存储技术
one-level subroutine 一级子程序;一级子例行程序
one-lever control 单杆控制
one-lick 铲刨法;挨个儿轮法
one-light steel window 独块玻璃;单块钢窗
one-line adapter 单线适配器
one line delay 单线延迟
one-line diagram 主接线图;单线图;单线电路图
one-line graphic 单线图
one-line operation 单线操作
one-line scanning 单线扫描
one-lip hand ladle 单嘴手杓
one-literal rule 单基本式规则
one machine 单机
Onemack 奥涅马克剑式无梭织机
one-man atmosphere submersible 单人常压潜水器
one-man atmospheric diving suit 单人常压潜水服
one-man atmospheric submersible 单人常压潜水器
one-man borer 单人镗床
one-man business 一人公司
one-man chain saw 单人链锯机
one-man chamber 单人舱
one-man completely enclosed diving dress 单人全密式潜水服
one-man completely enclosed diving suit 单人全密式潜水服
one-man control 一人控制
one-man cross-cut saw 单人截锯;单人横割锯
one-man diving submersible 单人潜水器
one-man drill 单人钻机
one-man drilling 单人钻进;单人打眼
one-man drilling machine 单人手扶钻机;单人手扶凿岩机
one-man flask 手抬砂箱
one-man grader 单人平路机;单人平地机
one-man machine 单人操纵机
one-man one vote 一人一票
one-man operation 一人操作(的作业);一人能完成的作业;单人操作;单人操纵
one-man paver 单人操作摊铺机;单人操作铺路机
one-man picture 一人景像
one-man saw 单人锯

one-man scraper 一人手推刮土机
one-man stone 中块石
one-man submarine 单人潜水器
one-man survival chamber 单人救生舱
one-man vibrator 一人操作的振动器
onemeter 组合式毫伏安计
one meter depth temperature 一米深温度
one-mile telescope 一英里射电望远镜
one million 一百万
one-minute managing 一分钟管理
one mistake made in treatment 一逆
one motor travel(l)ing crane 单电动机吊车;单马达移动吊车;单马达吊车
on end 竖着
on end laying 竖砌
one-node network 单节点网络
one number time 数周期
one-of-a-kind system 单机系统
one-off design 特种建筑物设计
one-off houses 单间住房
one-off pattern 一次模
one of the most abundant substances 自然界中最丰富的物质之一
one of the prime reasons 主要原因之一
one of valuable property of water 水的宝贵特征之一
one oiler 加油长
one-one 一对一
one-one mapping 一一映射
one-one relationship 一对一关系
one-on-two 一比二坡度;一比二;二分之一坡
one-order adsorption kinetic equation 一阶吸附动力学方程
one out often code 十中取码
one-over-one address 一加一地址
one pack 一包;一堆;一捆;一组
one package 一罐装
one-package formulation 单组分配方
one-package heat cured urethane coating 单组分热固化聚氨酯涂料
one-package moist-cured urethane coating 单组分湿气固化聚氨酯涂料
one-package non-reactive lacquer urethane coating 单组分非反应漆型聚氨酯涂料
one-package prereacted urethane coating 单组分预反应型聚氨酯涂料
one-package thermal retardation cure silicon rubber 单包装保温固化硅橡胶
one-package wash primer 一罐装磷化底漆;单组分磷化底漆
one-pack (bonding) adhesive 单成分黏合剂
one-pack cement 纯水泥
one-pack coating 单涂层
one-pack system 单组分
one panel 单门板
one-panel door 平板门
one-parameter 单参数
one-parameter family of curves 单参数曲线族
one-parameter method 单参数法
one-parametric 单参数的
one-parametric loading 单参数荷载
one-part adhesive 单组分黏合剂;单组分胶黏剂
one-part formulation 单组分配方
one-particle exchange 单粒子交换
one-part line 单股绳;单股缆索;单股绳索
one-part sealant 单组分密封材料;单组分密封膏
one-pass 一次走刀
one-pass algorithm 一遍算法
one-pass clarifier 单程澄清池
one-pass fillet electrode 单道平角钎焊条
one-pass machine 联合筑路机
one-pass operation 一次完成操作;一次通过操作;一次扫视操作;一遍操作
one-pass roller 单道次轧制
one-pass scheme 一遍方案
one-pass soil mixer 一次完成作业的土壤路拌机;一次完成作业的土壤搅拌机
one-pass type 单程式
one-pass weld 单道焊;单层焊缝
one-pass welding 单道焊接
one-pen recorder 单笔记录器
one-period-ahead forecast 提前一期的预测
one-person game tree 单人博弈树
one-person household 一人住户

one-phase 单相的
one-phase design 一阶段设计
one-phase electric(al)-arc furnace 单相电弧电炉
one-phase reaction 单相反应
one-phase relay 单相继电器
one-phonon adsorption 单声子吸收
one-phonon process 单声子过程
one-piece 不可分割的;整块;整体(式的);单件
one-piece annular combuster 单件式环状燃烧器
one-piece auger 整体螺旋
one-piece blade 整片叶
one-piece casting 整体浇铸件;整块浇铸件;整体铸件;整铸件
one-piece construction 整体结构
one-piece core box 整体芯盒
one-piece cutter head 整体式切割头
one-piece die 整体金属型
one-piece floating dock 整体式浮(船)坞
one-piece flywheel 整体飞轮
one-piece-forged 整锻
one-piece guard rail 整体护(轮)轨
one-piece hatch cover 整体式舱口盖
one-piece housing 整体式壳体
one-piece mo(u)ld 整体铸型;单件模
one-piece panel 整块板;单镶板
one-piece pattern 整体模(具)
one-piece roller 单块轧制
one-piece runner 整体式转轮
one-piece sawed timber 整块锯材
one-piece section 整管套
one-piece set 一构件支护
one-piece stator 整体定子
one-piece water closet 整体式坐便器
one-piece wheel 整轧车轮
one-pier elbow draft tube 单墩肘形尾水管
one-pier elbow draught tube 单墩肘形尾水管
one-pin(ned) arch(ed) girder 单铰拱梁
one-pinned frame 单铰框架
one-pinned parabolic arched girder 单轴抛物形拱梁;单铰抛物形拱梁
one-pip area 单脉冲区
one-pipe 单管式
one-pipe boiler 单管锅炉
one-pipe circuit system 单管环行系统
one-pipe drop heating system 单管上行下给供暖系统
one-pipe drop system 单管上行下给系统
one-pipe forced heating 单管受迫循环供热;单管强制循环供热;强制式单管供暖
one-pipe forced system 单管强制循环系统
one-pipe gas meter 单管煤气表
one-pipe gravity circulation system 单管重力循环系统
one-pipe gravity heating 重力式单管供暖
one-pipe gravity system 单管重力循环系统
one-pipe gravity (type) heating 单管重力循环供热
one-pipe heater 单管加热器
one-pipe hot water heating system 单管热水供暖系统
one-pipe jet pump 单管喷射泵
one-pipe loop circuit heating system 水平单管采暖系统
one-pipe main ring system 单管主环系统
one-pipe plumbing 单管管道工程;单管排气
one-pipe plumbing system 单管卫生管道系统
one-pipe ring main system 单管主环系统
one-pipe ring system 单管环路系统
one-pipe steam heat-supply network 单管制蒸汽热网
one-pipe system 单管供热系统;单管(排污)系统
one-pipe up-feed system 单管下行上给系统
one-piping gravity circulation system 单管重力循环系统
one-piping system 单管环行系统
one-place operation 一元运算
one plunger compressed air-hydraulic lift 单活塞气压液力升降机
one plunger compressed air-hydraulic lifter 单活塞气压液力升降机
one-plus-one address 一加一地址
one-ply yarn 单股纱
one point 单点
one-point boundary condition 单点边界条件
one-point continuous method 一点连续法;单点

连续法
one-point distribution 点分布;单点分布
one-point gate 一点控制极
one-point landing (飞机)单点落地
one-point method (测定土的最大干密度的)一点法
one-point method of stream ga(u)ging 单点测流法;点测流法
one-point mooring 单点系泊
one-point mooring method 单点系泊法
one-point mooring system 单点系泊法
one-point perspective 一点透视
one-point pickup 单点拾音器
one-point press 单点压力机
one-point sensing 单点检测
one-point wavemeter 定点波长计;单点波长计
one-pole knife switch 单极闸刀开关
one-pole method 单极法
one-pole switch 单极开关
one-port network 单口网络;二端网络
one-port parametric amplifiers 单端输出参量放大器
one-position adder 全加器
one pot coating 单罐装涂料
one pot system 单罐装系统
one pour of concrete 混凝土一次浇注;混凝土一次浇灌
one-power unit excavator 单发动机式挖掘机
one price 一口价
one-probe method 单探针法
one-product line 单产品管线
one-product pipeline 单产品管线
one-projector method 单投影器定向法
one-projector method relative orientation 连续相对定向
one pulse time 单脉冲时间
one push start 一次起动
one pylon 独塔
one-quadrant multiplier 单象限乘法器
one-quarter bend 四分之一弯头;直角弯头;九十度转弯
one-quarter brick bond 四分之一砖砌合
one-quarter clay brick 四分之一砖
one-quarter pipe bend 四分之一管子弯头
one-quarter wave skirt 四分之一波长跨接线;四分之一波长变量器
one-rail double-car retarder 单轨条双车缓行器
one-rail single-car retarder 单轨条单车缓行器
one-range winding 单平面绕组
one replication 一次反复
one-room apartment 单间房公寓;一户住宅
one-room camera 暗室制版照相机
one-room dwelling 一室户住宅
one-room(ed) flat 一户住宅;单间公寓;一室户;一室住宅;单层公寓
one-room house 一室户住宅
one-room size floor slab 大型楼板
one-room system 一室体系;每户以一大室作灵活分间的住宅体系
one-rope suspension grab(bing) 单绳悬挂抓斗
one-round edge 单圆边
one round trip 一个往返运行
onerous contract 有偿契约
one rouse appropriation of fixed assets 固定资产有偿调拨
onerous loan 有偿贷款
one-row brick-on-edge arch 单排砖缘拱
one-row hatch 一列舱口
one-row matrix 单行矩阵
one-row shaker-digger 单行抖动式挖掘机
one-sack batch 一袋水泥拌料;用一袋水泥的混凝土拌料;一袋水泥混凝土拌料
one-sample problem 单样本问题
one sample shear test 单试体剪切试验;单点(法)剪力试验
one-sample test 单样本检验;单样本测试
one-sash(ed) window 单扇窗;单框格窗
one's complement 一的补码;反码的补码(二进制)
one's complementer 二进制求反器
one-screen 双驼峰砂
one-seater 单座汽车
one-second theodolite 一秒经纬仪
one-second transit 一秒经纬仪
one-sector growth model 一部门增长模型
one seeded fruit 单籽果

oneself 本身
one set 一副
one-setting 一次调整
ones fractions occupy 50 to 0.5 millimeter fractions rate 本级占50~0.5mm级产率
one-shaft orientation method 单井定向法
one-shaped concrete base 单锥混凝土基底
one-shell lithiumbromide absorption 单筒型溴化锂吸收式制冷机
one-shift operation 单班作业;单班操作
one shipment 一次装运
one-short game 一次有效的对策
one-shot 一次通过(编程序);一次启动;单稳态的;单触发;单脉线路;单镜头拍摄;单冲;冲息
one-shot battery 次性电池;一次电池
one-shot camera 单像摄影机;单镜头摄像机;分光摄影机
one-shot chemical grouting 单液化学灌浆
one-shot circuit 单触发电路;冲息触发电路
one-shot colo(u)r camera 单像彩色照相机
one-shot device 一次有效装置
one-shot exploder 单发起爆器
one-shot forming 一次冶成;一次形成
one-shot gun 单击铆钉枪
one-shot hammer 单击铆枪
one-shot job 一步作业
one-shot lining 一次喷涂衬里
one-shot lubricating system 一次性润滑系统;集中润滑系统
one-shot lubrication 一次性润滑系统
one-shot method 一次完成法;一次(化学)灌浆法
one-shot mo(u)lding 一步发泡成型
one-shot multiplier 串并行乘法器
one-shot multivibrator 单稳线路;单冲多谐振荡器;冲息多谐振荡器
one-shot operation 一次起动操作;单步操作
one-shot output 单稳输出
one shot request 单发请求
one-shot riveting hammer 单击铆枪;单击铆锤
one-shot switch 单触发开关;冲息开关
one-shot timer 单次计时器
one-shot trigger circuit 单程触发电路
one-shot type 单层式
one-shot urethane coating 一步法氨基甲酸酯(泡沫)涂层
one-side argument 片面之词
one-side butt welding 单侧平焊;单侧对接焊;单面对焊
one-side coated art printing paper 单面涂层美术印刷纸
one-side coated duplex board 单面涂层双层纸板
one-side coated paper 单面涂层纸
one-side colo(u)red cover paper 单面色薄皮纸
one-side colo(u)red paper 单面色纸
one-side connection 单面接头
one-sided 偏于一方;单方面的;单侧的;单侧
one-sided acceptance sampling test 一方验收取样试验
one-sided alternative 单侧替换
one-sided automatic welding 单边自动焊接
one-sided bottom switchyard 单面井底车场
one-sided connection 单面接头(不对称的);单面连接
one-sided contract 单面契约
one-sided cordon 单侧单干形
one-sided difference 单侧差分
one-sided dressing 单面修整
one-sided filter 单边滤波器
one-sided least-square inverse filter 单边最小平方反滤波器
one-sided limit 单侧极限
one-sided nozzle 单侧喷管
one-sided search 单侧搜索
one-sided spot welding 单面点焊
one-sided spraying 单面喷雾
one-sided step junction 单边突变结;单边阶跃结
one-sided sweep 单翼平铲
one-sided swing parting 单面甩车调车场
one-sided system 单侧系统
one-sided test 单尾检验(法);单侧检验
one-side dump box car 单侧倾卸车
one-sided view 片面的观点
one-side filter 单侧压力
one-side formwork 单面模板

one-side glazed paper 单面上光纸
one-side load 单侧荷载
one-side panel exposure 一面板暴露
one-side polishing 单面抛光
one-side selection 单方选择
one-side water pressure 单侧水压力
one side weld(ing) 单侧焊(接);单面焊(接)
one-side wet back booth 单侧外置水幕喷漆橱
one-signal method 单信号(方)法
one-site labo(u)r 工地劳动力
one-sixteenths bend 十六分之一弯头;二十二点五度弯头
one-size 均一尺寸;同粒度
one-size(d) aggregate 均匀颗粒集料;均匀颗粒骨料;同尺寸骨料;单粒级集料;单粒级骨料;等大集料;等大骨料
one-size material 均尺寸材料
one-size mixture 同粒度混合料
one-sole course 单层
one's own making parts cost 自制零件费
one-span 单跨距;单跨度
one-span beam 单跨梁
one-span frame 单跨框架;单跨构架
one-span girder 单跨大梁
one span heavy load 一孔重载
one span light load 一孔轻载
one-span two-hinged frame 单跨双铰框架;单跨双铰构架
one-spool engine 单路式发动机;单转子发动机
one's pores will be fine in texture 腠理以密
one-spot tuning 同轴调谐;单钮调谐;单点调谐
one spray coat application 一次喷涂应用;一次喷涂施用
one square meter 一平方米
one-stack plumbing 单竖管卫生设备;单烟囱铅直测量
one-stage 单级的
one stage alkyd process 一步醇酸合成法
one-stage amplifier 单级放大器
one-stage casting 一段铸塑
one-stage communication 一段通信
one-stage compressor 单级压缩机
one-stage configuration 单级配置
one-stage converter 一段转化炉
one-stage crushing 一段破碎;单段破碎
one-stage design 一阶段设计
one stage digestion 一级消化
one-stage flo(a)tation 一段浮选
one-stage grinding 单段研磨;单段磨矿
one-stage joint 单级接головой
one-stage pump 单级泵
one stage resin 一步法树脂
one-stage rocket 单级火箭
one-stage sampling 一阶段抽样
one-stage solenoid-operated valve 单级电磁控制阀
one stage suspension preheater 一级悬浮预热器
one-stage/three-yard marshalling station 单向横列式编组站[铁]
one-stage-three-yard unidirectional combined marshalling station 单向一级三场混合式编组站
one start screw 单头螺纹
one station junction terminal 一站枢纽
one-step 单步;一步
one step autotrophic nitrogen removal 单级自氧脱氮
one-step camera 一步成像照相机
one-step cross beam 单步梁
one-step crystallization 一步结晶法
one-step derivation 单步派生
one-step design 一阶段设计
one-step elution 一步洗脱法
one-step firing process 一次烧成
one-step job 一步作业
one-step method 单步法;单步方法
one-step operation 一步作业;单步操作
one-step photography 一步摄影;一步成像摄影(术);瞬时摄影
one-step predicative filter 单步预测滤波器
one-step pressure regulator 单级压力调节器
one-step purifier 一步净化器
one-step refining of crude oil 原油一次加工
one-step synthesis 一步合成
one-stone grading 单粒径级配

one-stone grading aggregate 等粒径集料；等粒径骨料；单粒级集料；单粒级骨料
one-stop batch(ing) plant 一次配料工厂
one-stop shopping 百货采购处
one-stop station 各种技术服务站
one-stor(e)y brick structure 单层砖结构
one-stor(e)y building 单层建筑(物)；单层房屋
one-stor(e)y finger system 指型平房体系；放射形平房体系；指型平房系统；放射形平房系统
one-stor(e)y frame 单层框架
one-stor(e)y furnace 单层炉
one-stor(e)y house 单层房屋；平房
one-stor(e)y warehouse 平房仓库；单层货栈
one-storied factory 单层厂房
one-story furnace 单层炉
one-strand continuous casting machine 单线式连铸机
one-strand roller 单线轧制
one-strand rope 单股钢丝绳；单股钢(丝)索
one stream preheater 单系列预热器
one-stroke 一次行程；单行程
one swing 半次全振动
one-swing method 单投影器定向法
one-swing method relative orientation 连续像对相对定向
one-swing shipbuilding berth 单翼船台
one-table machine 单工作台成型机
one-tailed test 单尼检验(法)；单侧检验
one-tape Turing machine 单带图灵机
one-tenth maximum wave 十分之一最大波高；十分之一大波
one-tenth octave 十分之一倍频程
one-tenth second theodolite 十分之一秒经纬仪
one-terminal-pair network 单口网络；二端网络
one-third bond 三分之一砌合
one-third maximum wave 三分之一最大波高
one-third octave band 三分之一频程
one-third of the area 三分之一的面积
one-third pitch 三分之一高跨比
one-third rule 三分之一法则
one-third track headspan 三分之一股道软横跨
one-third two-thirds rule 三分之一和三分之二原则
one thousand hour clause 一千小时条款
one/three track headspan 三分之一股道软横跨
one-through 直流
one-through circulation 直流循环
one-through cooling 直流冷却
one-through cooling system 直流冷却系统
one-through cooling water 直流冷却水
one-through cooling water system 直流冷却水系统
one-through design 一次设计
one-through operation 直流操作
one-through process 一次通过过程
one-through system 直流系统
one-throw crankshaft 单拐曲轴
one-tier wall 单层墙
one-time adjudication system 一次性裁决制度
one-time consumption 一次消费
one-time developer 一次显影液
one-time fuse 一次性保险丝
one-time operation 一次有效利用
one-time pad 一次装填
one-time premium 一次性奖金
one-time survey 一次性调查
one-to-many 一对多
one-to-many route control systems 一对一线路控制系统
one-to-one 一对一
one-to-one assembler 直接翻译程序
one-to-one copy 等大复制
one-to-one correspondence 一一对应的关系；一一对应；精确套合；对应
one-to-one function 一一对应函数
one-to-one mapping 一一映射；一一对应；一对一映射
one-to-one onto mapping 一对一映满的映射
one-to-one route control system 一对多线路控制系统
one-to-one slope 坡度；比一坡度
one-to-one transformation 一一对应
one-to-partial select ratio 一与半选比
one-to-two 一比二
one touch gelation 触发胶化

one-touch key 一触键；单触键
one touch switcher 单触自动转换开关；单触转换开关
one-to-zero ratio 一与零(输出)比
one transistor cell 单一晶体管器件
one-trip oil container 一次使用的油桶
one-tube receiver 单管接收机
one-tube system 单管系统
one-tubing 单管系统
one-turn head 单回路磁头
one-turn stair(case) 单转楼梯；单转(向)楼梯
one-twentieth of the sum 总和的二十分之一
one-type ripper 单齿耙路机；单齿松土机
one-unit (power) plant 单机组电站
one-unit relational panel survey 一单位关系固定样本调查
one-value(d) function 单值函数
one valve 单级阀
one-valve receiver 单管接收机；单阀接收机
one-variable function 一元函数；单元函数
one-variable statistic(al) model 单变量统计模型
one-velocity reactor 单能中子反应堆
on even keel 平浮；平吃水
one vertical cable plane 单竖索面
one-wall sheet-piling cofferdam 单排板桩围堰
one-wattmeter method 单瓦特计法
one wavelength 一个波长
one-way 单行(线)；单向(性)；单路；单程
one-way accelerator 单向加速器
one-way air valve 单向空气阀
one-way alternate current locomotive 单向交流机车
one-way alternate traffic 单向交替交通
one-way analysis of variance 单因素方差分析
one-way annual (carrying) capacity 单向年通过能力；单向流(动)
one-way anti-thrust pier 单向推力墩
one-way arch 单向拱
one-way automatic leveling device 单向自动平层装置
one-way board 单向控制板
one-way bottle non-returnable 一次瓶
one-way bus 单向总线
one-way cable duct 单向电缆通道
one-way carrying capacity 单向通行能
one-way channel 单行航道；单向信道；单向纹道；单向通道；单向波道；单线航道
one-way chromatogram 单向色谱
one-way circuit 单向电路
one-way circulation 单向循环
one-way clamp 单向钳位
one-way classification 一向分组；一向分类；单因素分组
one-way classification yard 单向编组场【铁】
one-way clutch 单向联轴节；单向离合器
one-way clutch mechanism 单向离合机构
one-way cock 止回旋塞；单向考克
one-way communication 单向通信
one-way conducting device 单向导通装置
one-way connection 单向连接
one-way container 一次性集装箱；单程集装箱
one-way continuous slab 单向配筋连续板；单向连续板
one-way crown 一面排水路拱；单向路拱
one-way cylinder 单作用油缸
one-way damp circuit 单向钳压电路
one-way dimension 单向尺寸
one way disc 垂直圆盘犁
one-way disc harrow 灭茬圆盘耙
one-way distance measuring 单向测距
one-way distribution 单向分布
one-way drainage 单面排水
one-way drawing 单向牵伸
one-way element 单向元件
one-way encryption 单相密码
one-way entrance 单行入口
one-way equality 单向相等性
one-way fashion 单向式
one-way feed 单向馈电；单路馈电
one-way feeding 单边供电
one-way finite automata 单向有穷自动机
one-way fired furnace 单向火焰炉；单室炉
one-way fired pit 单向均热炉
one-way fired soaking pit 单向焰均热炉

one-way flat slab system 单向无梁板系统；单向钢筋平板系统
one-way gradient 单坡
one-way heat pipe 单向导热管
one-way input 单向输入
one-way input tape 单向输入带
one-way interaction 单向交互作用
one-way joist construction 单向龙骨构造；单向次梁结构
one-way lane 单向车道
one-way laser 单向激光器
one-way layer-built total-sun method 单向分层总合法
one-way layout 一元配置；一法配置
one-way line 单向线路
one-way linkage 单向链接
one-way lock 单向船闸
one-way lockage 单向过闸
one-way lubricating system 单线润滑系统
one-way maximum peak hour passenger capacity 高峰小时单向最大输送能力
one-way maximum peak hour passenger volume 高峰小时单向最大客流量
one-way membrane 单向膜
one-way merge 单向合并
one-way mine 单翼开采
one-way miniature 单向小型的
one-way mining 单翼开采
one-way mirror 单向镜(从亮室看是镜，从暗室看透明)
one-way motion 单向运动
one-way movement 单向运动
one-way navigation(al) channel 单行航道；单向航道
one-way non-deterministic stack 单向不确定栈
one-way observation 单向观测
one-way only operation 单向工作；单路操作
one-way orientation 单向取向
one-way pack 单向装填
one-way package 一次性包装
one-way pair 单向双车道；双条单向道
one-way passenger capacity 单向客运能力
one-way path 单行道
one-way pavement 单向车行道
one-way permutation pit sinking 单排井孔沉井
one-way phase switcher 单向相位转换开关；移相器
one-way plough 单向圆盘犁；单向犁
one-way plow 单向犁
one-way radio 单向无线电通信
one-way ram 单作用油缸；单程油缸
one-way ramp 单向匝道；单向坡道
one-way ranging 单向测距
one-way reinforced 单向配筋的；单向钢筋的；单向布筋的；单面加筋(的)
one-way reinforced plate 单向配筋板
one-way reinforced slab 单向配筋板
one-way reinforcement 单向钢筋；单向布筋
one-way reinforcement system 单向配筋系统
one-way remote monitoring system 单路遥控监测系统
one-way repeater 单向转发器；单向中继器；增音器
one-way restricted zone 单向通行限制区间；单向通行区
one-way revolving spanner 单向活动扳手
one-way road 单行道；单向道；单向交通道路
one-way rolling mill 非可逆式轧机
one-way satellite system 单路卫星系统
one-way section 单向行驶线段
one-way service pipe system 单向供水系统；单向供水
one-way shear 单边剪切
one-way sign 单向通行标志
one-way signal 单组信号
one-way slab 单向平板；单向配筋板；单向钢筋混凝土板；单向板
one-way slab system 单向钢筋混凝土板系统；单向板系统
one-way slope 单坡坡度；单向横坡；单坡面；单面坡(度)
one-way snow plough 单行雪犁
one-way stack 单向栈

one-way stack automaton 单向栈自动机
one-way street 单行道;单向(交通)街道
one-way system 单向配筋系统;单向系统
one-way throttle valve 单向节流阀
one-way thrust ball bearing 单向止推滚珠轴承
one-way thrust bearing 单向止推轴承
one-way time aligned 单程时间对齐
one-way traffic 单向运输(量);单向行驶;单向行车;单向通信(业务);单向通话;单向联络;单向交通
one-way traffic lane 单向行驶航道;单向通行道
one-way traffic turn 单向转弯
one-way transit 单向过闸
one-way transmission 单向传送;单向传输;单向传递
one-way transmission line 单向传输线路;单路输电线
one-way transparent glass 单向透视玻璃
one-way trunk 单向中继线
one-way valve 单向阀;止回阀;逆止阀
one-way valve piston 单向阀活塞
one-way valve sleeve 单向阀阀套
one-way valve spring 单向阀弹簧
one-way volume 单向交通量
one-way water supply station 单向给水站
one week 一周
one-wheel assembly 单轮分段装配
one-wheel barrow 单轮手推车
one-wheel hoe 单轮手推锄
one-wheel landing gear 单轮起落架
one-wheel plough 独轮犁
one-wheel plow 单轮犁
one-wheel roller 单轮压路机
one who pollutes should be responsible for treating 谁污染谁治理
one-window one-piece panel 单窗墙板;单片玻璃门肚板
one-window panel 单扇窗镶板;单扇窗墙板
one-wing block 单翼楼
one-winged building 单翼住房;单翼建筑
one wire system 单线制
one-withe wall 烟囱单层隔墙;单烟道隔墙
one with the excellent qualifications 具有优越条件者
on exchange basis 在交换的基础上
one-year authority 限当年有效的(预算或拨款)授权
one-year budget 单年度预算
one-year growth twig 一年枝
one-year ice 年冰
one-year renewable term 一年一续的保险
on-farm facility 农田设施
on-farm performance 现场生产性能测定
on file 归档
on flat clay 在平坦的黏壤
on-floor placing machine 地面混凝土砌块成型机
onflow 流入;进气;支流;供气
on-flow pigging 顺流清管
on-flow pigging operation 顺流清管
on-ga(u)ge 合格的;标准的
on-ga(u)ge plate 合格板(材);标准板;合格板
on-glaze colo(u)r 釉上色料;釉上彩色料;釉上彩
on-going evaluation 进行中评价
on-going-geothermal field 正在运行的地热田
on-going operation 正在进行的工作
on-going project 进行中项目;正在进行中的项目
on-going results 现在成果
on-going training 岗上培训
on-going wave section 上行波剖面
on good authority 有确实可靠的根据
on-grade (混凝土)地面板;等内的
on grain 按软方向(金刚石在钻头上的定向)
onground swimming pool 地面以上的游泳池
ongueka oil 衣散油
on hand 库存;手边
on-highway truck 普通卡车
on-highway vehicle 公路上的车辆;公路上的汽车
onhire 供出租用
on-hook 挂机
on-ice wind 上冰风
on-impedance 开态阻抗;接通阻抗
on improved bounds 估界的改进
on-interval 接通间隔
onion 管根;板根

onion architecture 葱头式建筑
onion dome 葱形圆顶;葱头形穹顶;(教堂等的)洋葱形圆顶
onion green glaze 葱绿釉
onion helm 葱形顶盖
onions alloy 铅锡铋易融合金
onion skin paper 葱皮纸
onion tower 葱形塔楼
onion weathering 葱头状风化
onisciform 海蛆形
o-nitroaniline 邻硝基苯胺
o-nitroanisole 邻硝基苯甲醚
o-nitrophenol 邻硝基苯酚
o-nitrophenyl-fluorone 邻硝基苯基荧光酮
o-nitrotoluene 邻甲苯胺
on-job 在工地的
on-job doctorate 在职博士生
on-job forming 现场立模
on-job lab 工地实验室;工地试验室
on-job laboratory 现场实验室;工地实验室;工地试验室
on-job test 现场试验
on-job training 在职训练;在职培训
on-job trial 工地试验;就地试验
onlap 入侵超覆;上覆;超覆
onlap bedding mode 上超层理模式
onlap sequence 超覆层序
on-level 在规定层上
on level trim 平吃水装载
on light friable soil 在轻质松散的土壤上
on-limits area 开放地
on-line 联线;线内;在线;联线;联机;并入电网
on-line adapter 联线衔接器;联线适配器;联机接线器
on-line aerophotogrammetric triangulation 联机空中三角测量
on-line aerotriangulation 联机空中三角测量
on-line alarm 联机报警
on-line analog(ue) input 联机模拟量输入
on-line analog(ue) output 联机模拟(量)输出;在线模拟量输出
on-line analysis 线上分析;流线分析;联机分析
on-line analyzer 材料含量监控器
on-line application 联机应用
on-line automatic monitoring system 在线自动监测系统
on-line banking system 联机银行业务系统
on-line batch processing 联机批量处理;联机成批处理
on-line batch processing system 联机成批处理系统
on-line capacity 联机容量
on-line chemical cleaning 在线化学清洗
on-line cipher 联机密码
on-line cleaning 在线清洗
on-line clock 作业钟
on-line closed loop 联机闭环
on-line coating 直接涂漆;在线涂装;维修漆
on-line colo(u)rimetric analysis 流线比色分析
on-line colo(u)rimetry analysis method 流线比色分析法
on-line command language 联机命题语言
on-line communication 联机通信
on-line computation 在线计算;联线计算;联机计算
on-line computer 在线计算机;联机计算机
on-line computer system 联机计算机系统;联机计算机设备
on-line control 直接控制;在线控制;联线控制;联机控制
on-line control system 联机控制系统
on-line counting 在线计数
on-line data base system 联机数据库系统
on-line data collection 联机数据收集
on-line data handling system 联机数据处理系统
on-line data process(ing) 联机数据处理;在线数据处理;联线数据处理
on-line data reduction 联机数据整理;联机数据简化
on-line debug(ging) 联机调试;在线调试;联机程序调试;联机(程序)调查
on-line debugging technique 联机调试技术
on-line definite subject retrieval 联机定题检索
on-line delayed time system 联机延时系统
on-line diagnostics 联线诊断;联机诊断
on-line digital image restoration 联机数字图像复原
on-line digital input 联机数字输入
on-line digital output 联机数字输出
on-line digitising system 联机数字化系统
on-line digitizing 联机数字化
on-line discrete data acquisition system 联机离散数据获取系统
on-line distillation 在线蒸馏
on-line editing 联机编辑(法)
on-line equalization 在线调节
on-line equipment 直接装置;联机装置;在线装置;在线设备;联机设备
on-line fault detection 联机故障检测
on-line file 联机文件
on-line flow routing 联机洪水演算
on-line forecasting 联机预报
on-line graphic(al) editing system 图形联机编辑系统
on-line groundwater quality monitoring station 联机地下水水质监测站
on-line handling 联机数据处理
on-line help 联机求助
on-line image 联机图像
on-line implicit calculation 联机隐式计算
on-line input 联机输入
on-line inspection device 管道带气检测器
on-line integrated materials management system 线上综合物料管理系统
on-line interaction 联机交互
on-line interactive system 联机交互系统
on-line interrogation 联机咨询
on-line job 联机作业
on-line light 联机指示灯
on-line maintenance 带电维修;在线维修;不停机检修;不停产检修
on-line meter 在线仪表
on-line microwave digestion 在线微波消解
on-line mode 联机方式
on-line model 在线操作模型
on-line mode operation 联机作业法
on-line modification 在线修改
on-line monitoring 在线检测
on-line monitoring and control 在线监控
on-line monitoring technology 在线检测技术
on-line negotiated terms 按议定条件
on-line operation 线上作业;直接工作;在线操作;联用操作;联线操作;联机运算;联机操作;机内操作;并网运行
on-line optimization 连线最佳化
on-line output 直接输出;联机输出
on-line peripheral test 联机外围设备测试
on-line pigging 带气清扫
on-line price surveillance 价格监视线
on-line processing 线上作业;联线处理;联机处理;联机操作;就地加工
on-line processing optimization 联机处理优化
on-line processor 联机处理机
on-line program(me) debugging 联机程序调试
on-line query 联机询问(站)
on-line question and answer system 联机问答系统
on-line real-time control 联机实时控制
on-line real-time detection 在线实时检测
on-line real-time operation 联机实时操作
on-line real-time processing 联机实时处理
on-line real-time system 联机实时系统
on-line replaceable unit 在线可换单元
on-line reporting point 联网报告点
on-line respiration meter 联机呼吸计
on-line respirometer 联机沉淀器
on-line retrieval 联机检索
on-line retrieval of information over a network 网络信息联机检索
on-line scanning of images 影像联机扫描
on-line searching 联机检索
on-line self-check function 在线自检功能
on-line sensing cell 在线传感元件
on-line service 在线服务;联机服务
on-line solid phase extraction 在线固相萃取
on-line stand-alone support program(me) 联机独立后援程序
on-line storage unit 联机存储器
on-line subroutine 在线子程序;插入式子程序
on-line system 在线系统;联机系统

on-line systematization 联机系统化
on-line teller system 联机出纳系统
on-line terminal 联机终端装置;联机终端;连线端末机
on-line terminal test 联机终端测试
on-line test(ing) 直接试验;在线试验;联机试验;联机检验;在线测试;联机检验;联机测试
on-line test program(me) 联机检验程序
on-line test routine 联机检验程序;联机检查程序
on-line traffic model 联机交通模型
on-line Turing machine 联机图灵机
on-line ultrasound-assited membrane bioreactor 在线超声膜生物反应器
on-line unit 联机装置;联机设备
on-line voltammetric wastewater analyzer 联机伏安污水分析仪
on-line water monitoring 在线水检测
on-line water quality monitoring instrument 联机水质监测仪
on-line working 联机工作
on-link mode 链接方式
onload 装载;在荷载下;带负载的;带负荷
on-loading 装载
on-load regulation 带负载调节;带负荷调节;有载调压;带荷调节
on-load regulator 带负荷调节器
on-load switch 负荷开关
on-load tap-changer of transformer 变压器有载调压开关
on-load tap changing 带载调节接点
on-load tap changing transformer 负载时分接头转换变压器
on-load voltage 加载电压
on-load voltage ratio adjuster 加载电压调整装置
on-load washing 负荷时清洗
only-child family 独生子女户
only-coat fiber 全皮纤维
on merits 按实情
on mortgage (以房屋)作抵押
on-net call 网内呼叫
Onnian 昂尼阶
on-off 离合;开关;起停;通断
on-off action 开关作用;开闭作用;开闭动作
on-off circuit 开关电路
on-off control 开关控制;继电器式控制;起停控制;通断控制;双位置控制;双位控制;双位调节;二值控制
on-off control action 双位(置)控制动作;开关控制动作;通断控制动作
on-off controller 两位控制器;开关型控制器;开关控制器;双位控制器
on-off controller system 开关控制器系统
on-off control register 开关控制寄存器
on-off control switch 双位开关
on-off control system 断续控制系统;开停控制系统
on-off cycle controller 开关周期控制器
on-off cycling 停开循环
on/off damper 启停风阀
on-off element 开关元件
on-off ga(u)ge 开关测量器;双关测量计
on-off information 通断型信息
on-off keying 振幅键控;开关键控
on-off measurement 离合测量
on-off mode 开关方式
on-off operation 开关操作;接通-断开操作
on-off passengers 上下车旅客
on-off push button 合分按钮
on/off radio control 无线电遥控开关
on-off regulator 振荡式调节器;开合调节器
on-off service 开停系统
on-off servo 继电器随动系统
on-off servo device 开关伺服装置
on-off servo mechanism 开关伺服系统;开关伺服机构
on-off state 断续状态
on-off steering 手柄操舵
on-off switch 通断开关;双控开关
on-off system 开关系统
on-off test 接通-断开试验
on-off thermostat 自动调温开关;开关调温器
on-off (time) ratio 开关时间比
on-off-type signals 开关型信号
on-off valve 换向阀;离合阀;开关阀;转换阀;双位阀
ONOGO type time signals 国际式时号

onomasiology 名称学
onomasticon 专门词汇
onomastics 名称学
onomatology 命名学
Onondaga limestone 奥嫩达加灰岩
onoratoite 氯氧锑矿
on order 已订购;订购货物
on-order quantity 订购量
ono-rising-stem valve 暗杆阀门
ono shot pulse 单触发脉冲
onozote 加填料的硫化橡胶
on passage 在运输中;在运输途中
on-peak 峰值的
on-peak power 峰荷功率
on period 工作期间;接通时期
on position 制动状态;工作状态;接通位置;接近位置;合闸状态;插入位置;闭合状态;接入位置;通电状态;闭合位置
on-premises sign 竖于基地上的标志
on-premise stand by equipment 应急备用设备
on presentation 提示时
on probation 试用
on rail 铁路交货(价)
on ramp 驶入匝道;驶入坡道
on record 有史以来的;公开发表的;留有记录的;记录在案的
on request 根据要求;根据买方通知发货;见票交货;即寄;函索;承索
on-request system 应求系统
on resistance 开态电阻;接通电阻
on resistance temperature coefficient 开态电阻温度系数
on river storage 河上水库
on route 在运输中
on rust paint 带锈涂料
Onsager equation 昂沙格方程
on-sale 出售;代销
on sale or return 销售或退货
on sales 特价出售
on sample 作样品
on-scene 实地的;当场;发生事情的地点
on-scene commander 现场指挥;海难现场指挥员
on-scene communication 现场通信;海难现场通信
on schedule 准时;正点;按照进度表;按预定时刻表;按期;按计划
on-schedule rate 正点率
on-screen menu 屏幕菜单
on-service training 在职训练;在职培训
on-service training program(me) 在职培训计划
onset 起始;初动
onset and retreat 进退
onset of cracking 开裂;裂缝发生
onset potential 起始电势
onsetter 井底把钩工
onset time 开始时间;初动时间
on shade 近似颜色
on share 分摊盈亏
on shipboard 在船上
on shipper's risk 由发货人承担风险
onshore 近岸的;向岸的;向岸;在岸上
onshore current 向岸水流运动;向岸流
onshore facility 岸上设备
onshore financial market 在岸金融市场
onshore gale 向岸大风
onshore marine terminal facility 港口陆上设施;港口陆上设备
onshore marine terminal of harbo(u)r 港口陆上设施;港口陆上设备
onshore mooring equipment 岸上系泊设备
onshore/offshore sediment movement 向/离岸泥沙运动
onshore-offshore transport 横向输沙
onshore oil field 海滨油田
onshore oil reserves 内海石油蕴藏量
onshore pile 岸土桩
onshore wind 向岸风;向岸大风;海风;吹拢风
onshore zone 破冰区
on short time 开工时间不足
on sick leave 休病假
on-site 现场的;在现场;就地;当地;在工地上;场区内;厂区内
on-site assembly 现场组装
on-site benefit 直接效益
on-site casting yard 就地浇注场地;就地浇筑场地

on-site compaction 就地压实
on-site concrete 现浇混凝土
on-site condition 现场条件
on-site connection 现场连接;现场装接
on-site construction 就地施工;就地建造;现场施工
on-site course of aids-to-navigation 航标现场课程
on-site digital processing system 现场数字处理系统
on-site dispatching 现场调度
on-site disposal 就地处置;就地处理;就地出渣
on-site electrolytic chlorination 就地电解氯化
on-site environmental sampler 就地环境采样器
on-site examination 现场考察;现场调查;就地诊验
on-site generation 就地生产
on-site handling 现场装卸;现场操纵;现场加工;就地处理
on-site improvement 产权范围内的场地改善
on-site incineration 现场焚毁;就地焚化
on-site incinerator 就地焚化炉
on-site inspection 现场检验;现场检查;就地检验
on-site installation 现场安装;原位安装;在位安装
on-site installation work 现场安装工作
on-site maintenance 现场维修
on-site management 现场管理
on-site management of wastewater 就地污水管理
on-site measurement 测量场测量
on-site mixer 就地拌和机
on-site mixing 现场拌和;现场搅拌
on-site monitoring 现场监测
on-site office 现场办公室
on-site paving mixer 工地路面拌和机
on-site pile casting yard 现场浇桩场
on-site plant 现场工厂;陆用装置
on-site prefabricating plant 现场预制装配工厂
on-site prefabrication 工地装配;工地预装配
on-site processing 在位处理
on-site repair 现场维修
on-site reuse system 就地再用系统
on-site runoff 当地径流
on-site runoff use 当地径流利用
on-site sewage disposal system 就地污水处理系统
on-site sewage treatment system 就地污水处理系统
on-site staking-out 现场定线
on-site support 在位支承
on-site tender 现场投标
on-site training 现场培训
on-site transfer of technology 现场技术传授
on-site vehicle 当地车辆
on-site verification 现场核查
on-site wastewater disposal system 就地污水处理系统
on-site welding 现场焊接
on-site work 现场工作
on situ 在现场
on size error 长度错误
on soil of humid area 在潮湿地区的土壤上
on soundings 在一百拓等深线内航行
onspeed 达到给定速度
on-spot check 现场检查
on-state 开态;通电状态;接通的
on state current 开态电流
on status 开状态
on-stream 在生产中的;在流程中;开工;在运转期;在工作工程中;投产运转中的
on-stream analysis 连续分析
on-stream analyzer[analyser] 流程分析器
on-stream chromatograph 流程色谱仪;流程气相色谱仪
on-stream chromatography 流程色谱;流程气相色谱法
on-stream cleaning 不通车清垢
on-stream descaling 不通车清垢
on-stream inspection 运转中检查;不停工检查;不停车检查
on-stream instrument 流程监控器
on-stream maintenance 运转中维修;开工期间维修;不停工维修
on-stream mechanical cleaning method 不通车清洗方法
on-stream method 运转中清洗
on-stream operation 顺流清管
on-stream period 运转期;连续开工期限
on-stream pigging 顺流清管

on-stream pressure 操作压力
on-stream storage 河道上的水库;河上水库
on-stream time 连续开工期限;开工时间
on-street parking 沿街停车场;街面停车
on-street parking area 路内停车场
on-street parking space 路内停车场
on strike 沿走向
onstringence 色散增数
onstruction overhead 施工管理费
on structure well 石油构造井
on-target detector 目标照准指示器
Ontarian 安大略阶
ontariolite 中柱石
on term 定期付
on-test 合格;试验开始
on-the-air monitor 主监察器
on the balance 两讫;两抵
on the barrel head 付现
on the beam 正横方向;正横;航向正确
on the beam well 单独抽油井
on the berth 在船位
on-the-bottom gravimeter 海底重力仪
on-the-bottom sea gravimeter 海中重力仪
on the bow 在首舷方向;在船首舷处
on the drawing board 设计中的
on the ebb 正在退潮
on the field basis 根据现场资料
on-the-fly modification 动态修改
on-the-fly reassignment 动态再分配
on-the-go adjustment 运行中调节;不停机调试;不停车调整
on the ground 监视人;监督员(俚语)
on-the-ground non-travel(l)ing centering 地面式非移动式脚手架
on-the-ground survey 地面测量
on the higher grazing 在较高的放牧场上
on the hook 钢筋整理工
on the house 免费的
on-the-job 现场架模
on-the-job accident 现场事故;岗位事故
on-the-job assembly 现场组装;现场装配
on-the-job calibration 作业中校正
on-the-job condition 现场操作条件
on-the-job dependability 现场使用可靠性
on-the-job forming 现场立模;现场支模;现场模架
on-the-job professional training 专业人员在职培训;在职专业训练
on-the-job service unit 现场服务车
on-the-job test 业务训练;技术训练
on-the-job training 业务训练;在职训练;在职培训;岗位培训;岗上培训;技术训练
on-the-job travel speed 现场行驶速度
on-the-line 用现款;在工作中
on-the-line printer 高速行线打印机
on the low hills 在较低的山底
on the move dumping 连续卸载;不停车卸载
on-the-open order 开盘订单
on the payroll 被雇佣
on the port bows 在左舷船道
on the port quarter 在左舷船侧后半部
on the pressure side 高压侧
on the pretext of 假托
on the quarter 在尾舷方向;在船尾
on the rack 按同心圆摆放金刚石
on-the-road mixer 路上拌和机;就地拌和机
on-the-run 自重流下装车,在运转中
on the same land 在同块地上
on the shelf 废弃的
on-the-site experiment 现场实验
on-the-site inquiry 现地调查
on-the-site labo(u)r 工地劳动力
on-the-spot 当场的;现场的
on-the-spot administration 现场管理
on-the-spot appraisal 现场鉴定
on-the-spot audit 就地审计
on-the-spot collection 现场收购;就地收购
on-the-spot control 就地监督
on-the-spot determination 现场测定
on-the-spot disposal 就地处理
on-the-spot experiment 现场实验
on-the-spot inquiry 实地调查
on-the-spot inspection 实地考察
on-the-spot investigation 现场勘察;现场调查;实地考察

on-the-spot meeting 现场会(议)
on-the-spot processing 就地加工
on-the-spot sample analysis 炉前分析
on-the-spot training 就地培训
on-the-spot transfer of surplus labor force 就地转移劳动力
on-the-spot turning 就地转弯
on the stocks 筹备中
on the strike 沿走向
on the suction side 进口侧
on the track 在轨道上
on the verge of bankruptcy 濒临破产的边缘
on the way 发射准备完毕
on the way to 在进行中;在途中;将要完成
on the weather beam 在上风正横方向
on the whole 大体上
on tick 赊欠(交易)
on time 正点;接通时间;按时;按期;工作时间;接通持续时间
on time and within budget 准时按值
on-time teller terminal 财务情况报告机
ontoanalysis 个体分析
ontological analysis 本体论分析
onto mapping 映满的映射
on-track equipment 轨道走行机械
on-track rail lubricator 地面钢轨涂油器
on trial 试用中;试验中
on truck 装上卡车;卡车上交货
on trust 赊账
on turn-key terms 以交钥匙的形式
on-unit 接通单元;待执行的操作
onus of proof 举证责任
on-vehicle material 车轴携带器材
onvex function 凸函数
onward lending 转借
onward-lending aid 转借援助
onward transmission 转交
on watch 值班
on weight of solution 按流体重量
on welded vessels 在焊接容器上
onychite 雪花石膏
onyx 缟状燧石;缟状石灰华;缟玛瑙;黑宝石;平行条带状;石华
onyx marble 黑大理石;细纹大理石;条纹大理岩;条纹大理石;石华大理石
Oocystis lacustris 湖生卵囊藻
ooid 鲕石;鲕粒
ooidal 鲕石的;鲕粒的
ooide 鲕石;鲕粒
ookinete 动合子
oolite 鲕状岩;鲕石;鲕粒
oolite-bearing 含鲕粒的
oolite-bearing micrite 含鲕粒微晶灰岩
oolite limestone reservoir 鲕灰岩储集层
oolite-pelitic texture 鲕状泥质结构
oolite pellet 鲕球粒
oolith 鱼卵状;鲕石
oolitic 鲕状
oolitic aluminous rock 鲕粒铝质岩
oolitic-bearing micritic limestone 含鲕粒微晶灰岩
oolitic-bearing pelitomorphic aluminous rock 含鲕粒泥状铝质岩
oolitic chert 鲕状燧石
oolitic facies 鲕石相;鲕粒岩相
oolitic formation 鲕状层
oolitic lime 鲕状石灰
oolitic limestone 鲕状石灰岩
oolitic-micritic limestone 鲕粒微晶灰岩
oolitic nodule 鲕状结核
oolitic ore 鲕状矿石
oolitic pelitomorphic aluminous rock 鲕状泥状铝质岩
oolitic pelmicrite 鲕状团粒微晶灰岩;鲕状球粒微晶灰岩
oolitic phosphoraite 鲕状磷块岩
oolitic projection 鲕状突起
oolitic sand 鱼卵状砂;鱼卵石砂
oolitic shoal 鲕粒浅滩
oolitic siliceous rock 鲕状硅质岩
oolitic structure 鲕状构造
oolitic texture 卵状结构;鲕状结构
oomicrite 微晶鲕粒灰岩;鲕粒泥晶灰岩
oomicrudite 鲕粒微晶砾屑灰岩
oomicsparite 鲕粒微晶亮晶灰岩

oomo(u)ld 鲕模
Ooms-lttner kiln 乌姆斯—伊特纳窑(烧瓷釉黏土管的环窑)
oonopsis sop 向日葵
oopellet 鲕球粒
Oort constant 奥尔特常数
Oort limit of density 奥尔特密度极限
Oort's cloud 奥尔特云
Oort-Spitzer mechanism 奥尔特—施皮策机制
oospararenite 鲕粒亮晶砂屑灰岩
oosparite 亮晶鲕粒灰岩;鲕粒亮晶灰岩
oosparmicrite 鲕粒亮晶泥晶灰岩
oosparrudite 鲕粒亮晶砾屑灰岩
oosterboschite 硒铜钯矿
Oosterhoff Group 奥斯特霍夫星群【地】
ooze 海泥;海底软泥;软泥
ooze bottom 软泥底
ooze film 软泥薄层
ooze muck 软泥
ooze sampler 淤泥采样器;软泥采样器
ooze sucker 吸泥器
oozuanolite 蛋粪石
oozy 渗出的;泥浆的;软泥的
opa (古典庙宇中屋顶梁的)空穴
opacified 乳白的
opacified glass 乳白玻璃;乳浊玻璃
opacified glaze 乳浊釉
opacified glaze tile 乳浊釉瓦
opacified silica-aerogel 不透明的硅土气凝胶
opacifier 遮光剂;乳浊剂;不透明剂
opacifying agent 乳浊剂;不透明剂
opacifying effect 乳浊效应
opacifying power 乳浊能力
opacimeter 浊度仪;黑度计;透光率计;不透明度计;不透明度测试仪;暗度计
opacimetry 暗度测量
opacitas 不透明区
opacite 不透明物
opacity 阻光率;阻光度;浊斑;混浊;浑浊度;乳浊(性);不透性;不透明性;不透明区;不透明度;暗度
opacity coefficient 不透明系数;不透明度
opacity density table 光学密度表
opacity factor 不透光系数;遮光系数
opacity hiding power (油漆等的)遮盖力
opacity meter 乳浊度计;不透明性指标计;不透明度计
opacity monitor 浊度监测器;浊度监测计
opacity paper 不透明纸
opacity test 乳浊度测定
opacity to sound 隔声性
opaco 背阳面;背阳坡
opagueness 不透明度
opaion (希腊建筑的)花格平顶凹板;(古希腊或罗马的)屋顶出烟口
opal 蛋白石;乳色的;乳白玻璃
opal agate 蛋白玛瑙
opal bulb 磨砂灯泡;乳白灯泡
opal cement 蛋白石胶结物
opalescence 乳色;乳光;乳白色;乳白化;乳白光
opalescent 乳色的;乳白的;乳光的
opalescent glass 乳白色玻璃;乳白玻璃
opalescent glass fiber 乳白玻璃纤维
opalescent glaze 乳白釉面;乳浊釉
opalescent lacquer 珠光漆
opalesque 乳白的
opalesque glass 乳白玻璃
opal glass 乳白(色)玻璃;蛋白石乳色玻璃;玻璃瓷
opal glaze 乳白釉
opaline 乳光的;发乳白色的;玻璃白
opaline chert 乳白黑硅石;乳白燧石
opaline glass 乳白玻璃;乳光玻璃
opaline glaze 乳白釉
opaline green 蛋白石绿(浅灰绿色)
opaline quartz 乳白石英
opaline silica 蛋白硅石
opalite 不纯蛋白石;蛋壳釉面材料;蛋白岩
opalization 乳白化;蛋白石化
opalize 使成乳色
opalized wood 硅化木
opalizer 遮光剂;乳浊剂;乳浊剂;不透明剂
opal jasper 蛋白碧玉
opal lamp 乳白色灯泡;乳白色灯;乳白灯泡
opal lamp bulb 乳白灯泡;乳浊灯泡
opal oil 乳白油

opal wax 乳白蜡
OPA process 二异辛基磷酸萃取法
opaque 修改液;不透明(的)
opaque alcohol based ink 不透明乙醇基油墨
opaque appearance 不透明状
opaque area 不透明区
opaque atmosphere 不透明大气
opaque attritus 不透明杂质煤;不透明碎屑
opaque bar 不透明光栅刻线
opaque body 不透明体
opaque brightener 不透明颜色提浓物
opaque ceramic glazed tile 乳浊的陶瓷釉面砖;不透明陶瓷釉面砖
opaque cloud cover 不透明云遮系数
opaque coat 无光泽涂层;不透明涂层
opaque coating 阴像刻图膜;不透明涂层;不透明膜
opaque colony 不透明菌落
opaque colo(u)r 暗色;不透明色(彩)
opaque colo(u)r concentrate 不透明色母料
opaque coprolite 不透明粪粒体
opaque copy 阴片
opaque cover 不透遮条;不透明盖板
opaque crystal 墨晶;不透明晶体
opaque dark area of fluid 液性暗区
opaque detector 不透明探测器
opaque enamel 乳白搪瓷;乳浊搪瓷
opaque finish 不透明涂饰剂
opaque fluid 不透明流体
opaque fused silica 不透明熔融氧化硅
opaque glass 不透明玻璃
opaque glaze 不透明釉(瓦);乳浊釉
opaque glaze coat(ing) 不透明釉层
opaque glazing 施以不透明的釉
opaque ice 不透明冰
opaque illuminator 不透(明)照明器
opaque ink 墨汁
opaque layer 浑浊层;不透明层
opaque margin 暗窄带
opaque mask 不透明屏(幕)
opaque material 不透明材料;不透明物
opaque medium 不透射线媒质;不透明介质
opaque mineral 不透明矿物
opaque mirror 不透明反射镜
opaqueness 不透明性
opaque original 不透明原稿
opaque paint 遮光涂料;不透明油漆;不透明颜料;不透明涂料;不透明漆
opaque paper 不透明纸
opaque particle 不透明微粒
opaque photo cathode 不透明光电阴极
opaque pigment 不透明颜料
opaque plastics 不透明塑料
opaque plate 不透明板
opaque plate glass 不透明板玻璃
opaque point 不透明点
opaque porcelain glaze 乳浊瓷釉
opaque portion 不透明部分
opaque projector 投射式放映机;反射式放映机;反射幻灯机;反光投影器;不透明投影器
opaque quartz glass muffle 不透明石英炉胆
opaque reflecting layer 不透明反射层
opaques 不透明像
opaque screen 不透明屏(幕)
opaque substrate 不透明底基
opaque telop 图片放映机
opaque to infrared 不透红外线的
opaque water-based ink 不透明水基油墨
opaque watercolor 不透明水彩
opaque water-colo(u)r painting 水粉画
opaquing fluid 修改液
Opax 奥帕克斯
open 开设;开启;敞口的;敞开的;打开
openable burning appliance 敞式燃烧器
openable fire 敞式炉
openable furnace 敞式炉
open a block 打开分程序
open access 敞口式出入口
open account 银行开户;开型会计数据;记账交易;未清账;未决算的账目;未结算账目;未结平账户
open account agreement 清算账户协定
open a circuit 断开电路
open a credit 开立信用证
open a credit account 开预存记账户

open addressing 开寻址列;开型寻址;开式寻址
open aerial 露天天线;开端天线;室外天线
open aestivation 张开花被卷迭式
open a file 打开一个文件
open a hawse 清链
open a hole 清理钻孔
open air 露天(的);开启式;户外(的)
open air assembly unit 露天的集合单元
open air basilica 露天式罗马大会堂;露天大会堂
open air bath 露天浴;室外浴场;露天浴场
open air cinema 室外电影院;露天电影院
open air clinker storage area 露天堆放敦料场;露天堆放煤渣场;露天堆放熔渣场;露天炉渣堆积处
open air combinational apparatus 敞式组合电器
open air combined electric(al) apparatus 敞开式组合电器
open air concert hall 露天音乐厅
open air constitutional electric(al) apparatus 敞开式组合电器
open air corridor 露天走廊;露天回廊
open air dance floor 露天舞池
open air depot 露天堆栈
open air drying 自然空气干燥;室外干燥
open air exhibition space 露天展览区间
open air exposure 露天曝置;露天曝晒
open air facility 室外设施
open air grandstands and bleachers 露天看台
open air hydrant 室外消防栓;室外消防龙头
open air installation 室外设备;室外装置
open air intake 露天取水口
open air life 户外生活
open air market 露天市场
open air microphone 振速传声器
open air model 露天模型
open air museum 露天博物馆
open air operation 露天作业区
open air parking garage 露天停车库
open air parking of houses 公共汽车露天停车场
open air piping 露天管线
open air plant 露天车间;露天工厂
open air pool 露天游泳池;露天水池
open air repository 露天堆栈
open air restaurant 露天餐馆
open air space 室外余地;室外空地
open air spark 露天光源
open air sports facility 露天运动设施
open air stage 露天舞台
open air station 露天式电站
open air stockpile 露天料堆;露天储存
open air storage 露天存车场【铁】;露天场;露天储(贮)藏场;露天储存;露天仓库
open air storage space 露天堆场;货场
open air swimming pool 露天游泳池;室外游泳池
open air temperature 外界温度;室外温度
open air terrace 外门廊;外阳台
open air theater 露天剧场
open air track 室外跑道;天然标记
open air vestibule 开敞式前室
open air weathering 露天风化;露天(气候)老化
open aisle coach 开放式通道客车
open a mine 开矿
open ammunition space 露天弹药库
open amortisseur 开路阻尼绕组
open an account 开户
open anchorage 开敞锚(泊)地;敞锚地
open and close movement 开合运动
open and close valve 开闭阀
open and notorious 公开的
open and shut action 开关动作
open and shut control 开关控制;开关调节
open and shut controller 自动控制器
open and shut valve 双效开关;双位开关;双阀开关
open and slip mode 张开及滑移型
open-angle 斜张角;(桥墩间)张开角
open annealing 敞开退火
open antenna 露天天线;户外天线
open approach 开敞式引道
open approach drainage 开敞式引道排水
open apron system 敞开式停机坪系统
open arc 开环形山弧;明弧
open arc furnace 明弧炉
open arch 明拱
open architecture form 敞开式建筑形式
open arch spandrel 空拱肩

open arc lamp 明弧灯;室外弧光灯;敞式弧光灯
open arc welding 明弧焊(接)
open area 空旷地;有效筛孔面积;广场;露天堆场;空旷地;筛面有效筛分面积;地面区段
open area ratio 开口面积比;开孔面积比
open a settlement account 开立结算账户
open-assembly time (黏接的)晾置时间;敞开装配时间;暴露时间
open association 稀疏群丛
open auger 开式螺旋;敞式螺旋
open a well 打井
open back bottom drive press 后开式底传动压力机
open back inclinable mechanical press 后开式可倾机械压床
open back inclinable press 后开式可倾斜压力机;前后送料可倾式冲床
open back power press 后开式压力机
open back press with fixed bed 后开式固定台压力机
open back transfer press 后开式多工位压力机
open back two point power press 后开式双点压力机
open backwater 开式回水
open balcony 露天阳台
open ball 开球【数】
open-band twine 开式锭带纱线
open band yarn 反手纱
open bank 开户银行
open barbecue 露天烤炉
open barge 敞口驳(船);敞舱驳(船);敞仓货船
open basin 开敞(式)港池;无闸港池;外港池;敞开式港池;敞开式船坞;泊船池
open basket weave packing 开口编篮码法
open bay 开阔海湾;开阔海港;外湾;敞湾
open beading 波纹板冲压法
open beam construction 露梁构造;露梁结构
open bearing 开式轴承;开启轴承
open bed 开式床身
open bed chromatography 开床色谱法
open bell 开式潜水钟
open belt 开式传动皮带;开口皮带
open belt conveyer 开敞式皮带运输机
open belt drive 开口皮带传动
open belting 开口皮带传动装置
open bench method 敞床植法
open bending frame 开口弯曲框架
open berth 露天船台;开放锚位;开敞锚(泊)地;开敞泊位
open bevel knee 钝角肘板
open bid(ding) 开标;公开招标;开盘报价;一般竞争投标;开标开启承包书
open bill of lading [open B/L] 未填收货人的提单;不记名提单
open bin 敞式料仓
open binder 松级配结层;有隙黏结层;开式结合层
open binder course 开式结合层;松级配结层
open block square 空心矩形尺
open blockwork 开孔块体
open boarding 稀铺屋面板;开口铺板;开缝铺板法
open boat 无甲板船;敞舱船
open body 敞式车身
open bond 开押信号;开口契约
open book-account 计算账面借款
open book credit 往来账户赊欠;往来赊欠账户
open-bottomed bell 开式潜水钟;敞底井;底开式井
open-bottomed concrete spreader 底开式混凝土摊铺机
open bottom floating platform 开底式活动平台
open bottom mo(u)ld 底开式钢锭模
open bottom scoop 开底式铲斗
open-bottom vessel 单底船
open box section U形横截面
open breathing apparatus 开放式潜水呼吸器
open bridge 空式桥;敞式桥
open bridge floor 空式桥面;无渣桥面;无柞桥面
open bridge space 露天桥楼场所
open brine system 开式盐水系统
open brine tank 开式盐水箱
open bubble 开口气泡;砂眼
open bucket well 提水井
open building pit 露天施工坑
open burning 露天燃烧;露天焚烧
open burning coal 非结焦性煤

open burning dump 露天燃烧垃圾堆
open business 开业
open butt 开口对接
open butt gas pressure welding 开式加压气焊
open butt joint 留缝对接;开口对接缝;明对接
open butt weld 开口对(接)焊
open by brief-cable 简电开
open cabin 敞开式座舱
open cage (深井泵的)开口阀罩
open caisson 开口沉箱;沉井
open caisson and box caisson 沉井与沉箱
open caisson foundation 开口沉箱基础;沉井基础
open caisson foundation structure 开口沉箱结构
open caisson method 开口沉箱法;沉井法
open caisson with cross walls 有交叉间壁的开口沉箱;格墙沉井
open canal 明渠;明槽;(无闸的)通海运河
open car 敞篷小汽车;敞篷(汽)车
open carbon arc light 开式碳精弧光灯
open cargo 自由货物;未列名货物
open cargo lighter 无盖驳船
open carrier chain 杆条式链
open carrying structure 开式承力结构
open case inspection 开箱检查
opencast 空心注件;露天矿
opencast coal mining 煤矿露天开采
opencast crude powder-type anfo explosive 露天粗粉状铵油炸药
opencast equipment 露天矿设备
opencast explosive 露天矿用炸药
opencast fine powder-type anfo explosive 露天细粉状铵油炸药
opencast getting 露天矿采掘工作
opencasting-in 开放型浇注
opencast-iron chock 铸铁导缆钩
opencast mine map 露天矿图
opencast mining 露天开采作业;露天开采
opencast mining locomotive 露天矿用机车
opencast of mountain slope 山坡露天采矿
opencast pit building 露采场建设
opencast survey 露天矿测量
opencast work(ing) 露天开采;露天开采作业
open cathode 开顶阴极
open cavity 明缩孔
open cell 联孔;开路电池;(泡沫塑料)开孔;开放式单元;敞开式电解槽
open cell cellular material 开孔式多孔材料
open cell cellular plastics 开孔泡沫塑料
open cell foam 开口泡沫;开孔泡沫
open cell foamed 开孔发泡的;开孔泡沫塑料
open cell foamed plastics 开孔泡沫塑料
open cell material 开孔材料
open-cell method 空细胞(防腐)法
open cell process 开孔式处理(木材防腐处理)
open cell sealant backing 开孔密封膏衬材料
open center 中心开口
open center control 中心开口控制
open-centered type-P indicator 延时式平面位置指示器
open center hydraulic system 开心式液压系统
open centerline shrinkage 外中心缩尺
open center valve 开中心阀
open centrifugal pump 开式离心泵
open chain 开环形山链
open chain hydrocarbon(s) 开链烃(类)
open chaining 开链
open chain organic compound 开链有机化合物
open chamber needle 菌形阀
open chamber needle valve 蕈形阀
open channel 阳沟;开路波道;开阔(的)航道;开放水道;明水道;明渠;明槽
open channel acoustic flowmeter 明渠声学流量计
open channel constriction 明渠障碍;明渠收缩段
open channel dam section 明渠坝段
open channel development 明渠开发
open channel diversion 明渠导流
open channel drainage 明沟排水;明渠排水
open channel flow 明渠水流;明渠流;明渠道流
open channel hydraulics 明渠水力学
open channel hydrodynamics 明渠流体动力学
open channel meter 开式槽流量计;明渠流量计
open channel routing 明渠演算;河槽洪水演进
open channel sewer flow 明渠污水流
open channel velocity distribution 明渠流速分布

open charter 开口租船合同;开放佣船契约;货物未定租船合同;货-港未定的租船合同;未注明货种及卸货港的租船合同
open check 普通支票
open chock 开式导缆器;开口导缆钳;导缆钩
open circuit 开式液压回路;开式循环;开式电路;开式(分级)流程;开路;开流;断路
open circuit admittance 开路导纳
open circuit arc 开路电弧
open circuit cell 开路电池
open circuit characteristic 空载线路特性曲线;开路特性(曲线)
open circuit column 开路破碎循环
open-circuit comminution 开式(工艺)破碎(破碎筛分后大颗粒不再送回原破碎机再次破碎)
open circuit contact 断路接点
open circuit control 开路控制
open circuit cooling 开式冷却
open circuit crushing 开路破碎;开回路粉碎
open circuit current 开路电流
open circuit diving gear 开式回路潜水装置
open circuited delay line 开路延迟线
open circuited line 开路线
open circuit excitation 空载励磁;开路励磁
open circuit grinding 开式粉碎;开路粉碎;开路磨矿;开路粉磨
open circuit grinding system 开路粉碎系统
open circuit grouting 无回路设备的灌浆系统;开路灌浆;开环式泵送灰浆
open circuit impedance 开路阻抗
open circuit impedance parameter 开路阻抗参数
open circuit indicator 开路指示器
open circuit jack 开路插座;开路插孔
open circuit line 断开线路
open circuit losses 空载损耗
open circuit method of grinding 粉磨开流法
open circuit mill 开路磨机
open circuit operation 断开电路操作
open circuit parameter 开路参数
open circuit potential 断路电势
open circuit radiation 开路辐射
open circuit radiator 开路散热器
open circuit reduction 一次通过的粉碎;单程粉碎
open circuit rod mill 开式棒磨机
open circuit signaling 开路信号发送
open circuit size reduction 开流破碎
open circuit system 开路系统;开流系统;正常开路制
open circuit tap changer 空载分接开关
open circuit transition 开路换接过程
open circuit transition method 开路过渡法
open circuit tunnel 开式风洞
open circuit underwater breathing apparatus 开式潜水呼吸器
open circuit voltage 空载电压;开路电压
open circuit winding 断路绕组
open circulation 开式循环
open circulation cooling system 敞式循环冷却方式
open circulatory system 开放循环系
open city 开放城市;不设防城市
open classroom 开放式教室
open clay 多孔砂质黏土
open close 断通
open closed shelter deck ship 可开闭式遮蔽甲板船
open/closed ventilation system 开/闭式通风系统
open/close test of flash-point of liquids 燃料开杯/闭杯闪点测定
open coast 开阔海岸;开敞海岸;无碍航物的海岸;敞露海岸
open coastal economic areas 沿海经济开放区
open coastline 敞露海岸线
open coat 疏涂(饰)层;疏上胶层
open cockpit airplane 敞舱飞机
open code 开型码;开型代码;打开码
open coil 开路线圈
open coil annealing 松卷退火
open coil annealing furnace 松卷退火炉
open coil armature 开路线圈电枢;开路电枢
open coil armature winding 开圈式电枢绕组
open collector 集电极开路输出门
open-collector bus driver 集电极开路总线驱动器
open-collector gate 集电极开路门
open-collector structure 集电极开路结构
open collet 弹簧套筒夹头

open column 空心柱;开柱
open combustion chamber 开式燃烧室
open commitments 开放权益
open community 稀疏群落;开阔群落
open competition 公开竞争;开放竞争
open computation shop 开放计算站
open condition 开放条件
open conduit 露明管道;明铺管道;明流管道;明渠;明管(渠)道;明管;明沟;敞开的管道
open-conduit drop 露明管道落差;明渠落差
open conference 开放同盟
open construction 蒙皮结构;骨架;透空结构;敞开式构造
open contact 开放触点;断开接点
open contact tone 触点打开音响指示
open container 开顶集装箱
open contour 未封闭等高线
open contract 开放契约;条件不明的契约;开口合同
open conveyer 敞式输送机
open cooling 开式冷却
open cooling system 开式冷却系统
open core 开口铁芯
open core type 空心式;开口铁芯式
open cornice 露椽檐口;敞檐
open corridor 敞廊
open cost work 计时工资的工作;计日工资的工作
open country 旷野;空旷地区;开阔地
open court 公开法庭;露天庭院
open court of a quadrangle 四合院中敞开的庭院;四角形公开法庭
open courtyard 开敞式院落
open cover(age) 预约保险(单)
open covering 开覆盖
open crate package 亮格箱
open credit 信用往来;信用贷款;一般信用证;开信用证;普通往来款项;无担保(信用)贷款;分类信用证
open crib 空心木垛;无底木笼;叠木笼
open crossed check 无记名划线支票
open crossing 无人看守道口
open cross section 开口截面
open cross-section of checker 格子体流通断面
open cup 开杯;无盖杯
open cup flash point 开口杯闪点
open cup flash test 开口闪点试验;开杯闪点试验
open cup flash tester 开口闪点试验器;开杯闪点试验器
open curve 开曲线;明曲线
open cut 露天开挖;开槽施工;明挖;大开挖(施工);露天矿;明堑
open cut drain 明渠
open cut drainage 路堑排水;明挖排水;明沟排水;明渠排水
open cut exploration 露天勘探
open cut foundation 明挖基础;浅平基(础)
open cut glory-hole mining 露天矿漏斗开采法
open cut method 露天开采法;明挖法施工;明挖法
open cut mine 露天掘矿
open cut mining 露天开采(法);露天采矿
open cut quarrying of clay 露天挖掘黏土
open cut tailrace 明挖尾水渠
open cutter 开式绞刀
open cut tester 开杯试验器
open cutting 明堑;明开挖
open cut tunnel 露天坑道;明挖隧洞;明挖隧道;明洞
open cut tunnel(l)ing 明挖隧道工程
open cut volume 明挖方量
open cycle 直流循环;开(式)循环;开路循环;开环
open cycle control 开环控制
open cycle controller 开口电路调器;开环控制器
open cycle control system 开口式调节系统;开环控制系统
open cycle engine 开式循环发动机
open cycle gas turbine 开式循环燃气涡轮机;开式循环燃气透平
open cycle mechanism 开路循环机理
open cycle reactor system 开式循环反应堆系统
open cycle system 开环系统
open dam 活动坝
open database connectivity 开放数据库链接性
open database control 开放性数据库接口
open debt 未清债务;未清欠账
open deck (铁路桥梁的)桥台板;露天甲板;开敞

甲板
open-deck car park 露天多层停车场
open-deck garage 露天汽车修理厂；露天车库；无侧墙式车库
open-deck girder 空敞桥面大梁
open-deck parking garage 露天上承式停车库
open defect 开口缺陷；明显缺陷；外观缺陷；露明缺陷
open defects in wood 木材外露疵病
open delivery boot 开式输车开沟器
open delta 开口三角形(连接)
open delta connection 开式三角形接法；开口三角形接法；V 形接法
open depot 露天货场
open design 开放型设计
open development 公开发展
open die 开式模(具)；开口锻模；拼合式螺栓锻模；敞开式锻模
open die forging 开式模铸造；开式模锻造
open digester 敞式汽消化池
open discontinuity 外露缺陷
open distribution reservoir 露天式配水池
open district 房屋稀少区；空旷地段；未建成区
open ditch 阳沟；明渠；明沟
open ditch drainage 明渠排水；明沟排水
open diversion channel 导流明渠
open diving bell 开式潜水钟
open diving dress 开式潜水服
open dock 露天修船场；露天船坞；感潮港池；开敞(式)港池；无闸港池
open domain 开域
open door 开放门户；门户开放
open door policy 开放政策；门户开放政策
open dovetail 明鸠尾榫
open dovetailing 明鸠尾榫接合
open dowmwarped extracontinental basin 开阔型陆外下陷盆地
open drain 明路漏极；开口式排水管；开口式地下排水管；排水明沟；敞式地下排水管
open drainage 明沟排水；开口式排水管；开放引流法；开放导液法
open drain buffer 开路漏极缓冲器
open drained joint (混凝土护墙之间的)明排水接头
open drain system 明沟系统
open dredge caisson method 敞口挖泥沉箱法
open dredging 明挖；敞式疏浚；敞开疏浚
open dredging process 大开挖；明挖法
open drift ice 稀疏流冰群；稀疏浮冰群
open drive sampler 薄壁压入式取土器
open drying bed 敞开式干燥床
open dual economics 开放性双重经济
open dug foundation 明挖基础
open dug reservoir 露天水池
open dump 露天废物堆场
open dump body 敞倾车身；敞口翻斗车身
open dumping 露天垃圾场；露天堆积
open eaves 露明屋檐；露檐口
open economics 开发经济学
open economy 开放(式)经济
opened 断开的
opened contract 开口合同
opened recirculating cooling water system 敞开式循环冷却水系统
opened track circuit 开路式轨道电路
opened void 开口孔隙；开口空隙
opened washer 开口垫圈
open electrolytic circuit 开路电解系统
open-end 可扩展；开口端
open-end agreement 开口协定
open-end antenna 开端天线
open-end auger 悬臂式螺旋
open-end basket 小物品打捞篮
open-end block 顶端开张的砌块；H 形空心砌块；敞口混凝土浇筑块；顶端开口的砌块
open-end bond 开口契约
open-end bucket 活底吊桶；活动底铲斗；开口吊斗；活动铲斗
open-end caisson 沉井
open-end class 开口组
open-end classification 开口组分类
open-end concrete block 敞口混凝土浇筑块
open-end contract 不定额契约；数量待定的契约；数量待定的合同；开口合同
open-ended 可扩充的；开路型；开口端；开放；无限

额的；非限定的；不定额的
open-ended architecture 可扩充结构
open-ended bucket 活底斗
open-ended characteristic 可扩充特性
open-ended circulation 开底循环
open-ended coil 开口线圈
open-ended design 可扩展设计；可扩充设计
open-ended drain pipe 开式泄水管
open-ended grants 不限用途的财政补贴
open-ended hangar 敞开式机库；敞口木棚
open-endedness 为应付意外事件可能性而作准备的状态
open-ended pile 开口桩
open-ended pipe grouting 敞口管灌浆
open-ended program(me) 可扩充的程序
open-ended project 敞口项目
open-ended shed 敞开的工棚；敞开的货棚；敞露的住宅；敞露的茅屋
open-ended spanner 开口扳手
open-ended structure 可扩充结构
open-ended system 可扩充装置；可扩充系统；开端型系统
open-ended tubing squeeze method 敞口油管挤注法
open-ended well 开底井
open-end fund 公开投资基金；开放型基金
open-end hole 通孔
open-end investment company 公开投资公司；可赎股投资公司
open-end investment trust 开放信托投资
open-end lease 开口租约
open-end licence 开放式进口许可证
open-end line 终端开路线；终端开断线路
open-end loan 未确定数额的贷款
open-end method 敞开进路式矿柱回采法
open-end mortgage 不固定期限的抵押；开放抵押
open-end mutual fund 开放型共同基金
open-end of tubing 套管的开端
open-end pillar method 开端式煤柱回采法
open-end pipe pile 敞口管桩；开口管桩
open-end press 偏心压力机
open-end roll 悬臂辊
open-end shed 敞口木棚
open-end skip 开口料斗；开口翻斗
open-end slugging wrench 开口的冲制扳手
open-end socket 开口套绳头套环
open-end span 开空桥跨
open-end spanner 开口扳钳；开口扳手
open-ends project 敞口项目
open-end stage 开敞尽端式舞台
open-end test 开练试验
open-end traverse 不闭合导线【测】；支导线【测】；开导线
open-end trust 开口信托
open-end trust deed mortgage 敞开信托契据抵押；可加信托契据抵押
open-end well 无过滤的管井；开口井
open-end wrench 开口扳手
opener 直弯机；(信用证的)开证(申请)人；开启机构；开瓶扳头；瓶塞开启工具；扳直机
opener boot 开沟器
opener draw rod 开沟器拉导杆
opener shank 开沟器体
open estuary 开敞河口湾；未封冻河口；不冻河口
open excavation 露天开挖；明挖；明堑；明开挖
open excavation and backfill 明挖回填
open expansion tank 自由膨胀水箱；敞开式露天水箱
open face 自由面；临空面；坡面；无支护工作面
open face blasting 空心爆破作业；多面临空爆破；明挖爆破
open face crown 露面冠；开面冠
open faced vacuum frame 敞式抽气晒版机
open face gold crown 露面金冠
open fairlead 开式导缆器
open fastener 敞开型拉链
open fault 开口断层
open fence 露天栅栏；围墙
open fender hitching 碰垫桩活
open fiduciary loan 信用借款；信用贷款
open field irrigation 露地灌溉(法)；旷野灌溉(法)
open field test 空场测验
open file 开文件
open filter 敞露式滤池

open fire 露天焚烧；明火
open firing 明火加热
open fishladder 开敞式鱼梯
open fissure 开口裂缝
open flame burner 明焰烧嘴
open flame furnace 火焰直接加热炉；有焰炉；直接加热炉；明焰炉；敞式焰炉；敞开式火焰炉
open flame kiln 明焰窑
open flame method 明火试验法
open flame oil furnace 敞式烧油焰炉
open flash point 开口闪点
open flood 敞喷
open floodway 露天分洪道
open floor 露明格栅楼板；组合肋板；构架肋板；露格栅楼板；空心肋板；开敞肋板；明桥面；明露格栅楼板；敞肋楼板
open floor duct 敞开的楼面管道
open flow 畅流流量；敞喷
open flow capacity 畅喷能力；敞喷能力
open flow potential 畅喷产量；敞喷产油能力；敞喷产量(指石油)
open flow system 开管流动系统
open flow test 畅喷试验；敞喷试验
open flow well 畅喷气井
open flued appliance 开启式有烟道燃具
open flume 明槽
open flume setting (涡轮机叶片的)明槽装配
open flume turbine 开(敞)式水轮机；明槽式水轮机；敞式水轮机
open flume water turbine 开式水轮机
open fold 宽展型褶皱；开阔褶皱；敞开褶皱；敞开褶曲
open folding 敞开褶皱
open-foot pile 开口桩
open for business 开始营业
open forging 自由锻造
open forging steel 商用锻钢坯
open form 敞开式；开型
open formation 稀疏群系
open formwork 开式模壳；开式模板
open for signature 开放供签署
open forward position risk 开放远期部位风险
open fracture 张断裂；开口裂缝
open fracture density 开裂缝密度
open frame 空框架；空腹架构
open frame girder 空腹式大梁；空框桁梁；空腹桁架梁；空腹大梁
open-frame tool carrier 双梁通用机架
open frame transformer 开启型变压器
open freight storage 露天货场；露天堆货场
open freight yard 露天堆货场
open-front derrick (前边敞开的)轻便井架
open-fronted 前开式的
open-fronted tomb 前面敞开式坟墓
open-front press 开式压力机
open-front store 正面敞开的商店
open fuel injector 开式喷油器
open fuel nozzle 开式燃料喷嘴
open furnace 开式炉膛；敞炉；敞开炉
open furrow 明垄沟
open fuse 裸导火索；开敞型保险器；明线保险器；明保险丝；外露保险丝；开口型保险丝
open fuse cutout 明熔丝断路器
open fuselage 敞开机身
open fusing 敞炉
open game 开型对策；开始对策
open gap 开式放电器；开启间隙
open gap-cooled rotor 开启气隙冷却转子
open garden theater 园林露天剧场
open gas flame dryer 敞式煤气火焰烘干装置
open gate 通门
open gear 开式齿轮
open gearing 开式齿轮传动装置
open gear lubricant 开式齿轮润滑油
open gear pair 开式齿轮副
open gears 敞式齿轮装置
open general license 普通输入许可证
open general-purpose motor 敞开式通用电动机
open geothermal system 开放型地热系统
open gill set frame 单针制条机
open gill spreader 单针延展机
open goods truck 敞车
open goods wagon 敞车
open-governor oil pressure system (涡轮机的)

油压开式调速器系统
open gradation 公开分层次;公开分等级;开级配
open graded 间断级配的(常指混凝土中骨料);松级配的;开级配的
open graded aggregate 间断级配骨料;非连续级配集料;多孔粗晶集料;开(式)级配集料;开(式)级配骨料;间断级配集料;天然级配集料;粗集料;粗骨料
open graded asphalt(ic) concrete 粗集料沥青混凝土;粗骨料沥青混凝土;开式沥青混凝土
open graded ballast 天然级配道砟
open graded bituminous concrete 开式级配沥青混凝土;粗集料沥青混凝土;粗骨料沥青混凝土
open graded friction course 开级配抗滑层
open graded mineral aggregate 间断级配的矿物集料;间断级配的矿物骨料
open graded mix 开式级配混合料;松级配混合料
open graded road mix surface 开式级配路拌法修筑的路面;开式级配路拌路面
open grain 无年轮限制的木材;木材软纹;大孔隙性;大孔隙率;粗纹理;粗疏纹理
open grained 粗粒的;粗木纹的;纹理松的;大孔隙性的;疏木纹的
open-grained iron 粗晶生铁
open-grained steel 粗粒钢;粗晶粒钢
open-grained timber 粗纹理木材;疏纹木材
open grain pig 粗晶生铁
open-grain structure 开式颗粒结构;大孔结构;粗晶组织;多孔式结构
open grate area 炉排有效面积;箅子空隙面积
open grate surface 炉排有效表面
open grid 悬栅
open grid deck 敞式格构桥面;敞开式格构桥面
open groin 开口丁坝
open groove 开口轧槽
open ground 无保护层地面;裸露土层;露地;空地;开阔地;天然地面
open group 开口组
open groyne 开口丁坝
open gunwale 无靠手木条的舷缘;敞露舷缘
open half line 开半直线
open hand knot 环结
open harbo(u)r 自由港;开放港(口);开敞式港湾;无掩护港口;不冻港;开口港
open hash definition 开散列定义
open hash method 开散列法
open hash table 开散列表
open hash technique 开散列技术
open hatch 开舱
open hatch vessel 敞舱口船
open hawse 敞口式锚链孔
open head drum 大口桶
open-hearth 开式炉;平炉
open-hearth block 平炉换向器
open-hearth furnace 开炉;平炉
open-hearth furnace slag 开炉渣
open-hearth furnace steel 平炉钢
open-hearth furnace steel-making 平炉炼钢
open-hearth plant 平炉车间
open-hearth port 平炉喷口
open-hearth process 平炉冶炼法;平炉(炼钢)法
open-hearth rail 平炉钢轨
open-hearth rolling furnace 可倾式平炉
open-hearth slag 平炉钢渣
open-hearth slag oxidation 平炉渣氧化作用
open-hearth steel 西门子—马丁钢;平炉钢
open heart mo(u)lding (线脚中的)心形饰
open-heated steel 过热钢
open heater 给水加热器;开(放)式加热器;敞口热水器;敞口加热器
open height 开口高度
open herding 满天星;分散放牧
open here 从此开箱
open hole 裸孔;裸井;无井管孔;安装孔
open hole caliper 孔径仪;井径仪;井径规
open hole completion 裸眼完井;裸眼成井
open hole drilling 无套管钻探
open hole packer 裸眼封隔器
open hole pressure 开井压力
open hole shooting 裸眼射孔;开孔爆破;无堵塞爆破
open hole tensile strength 开孔抗张强度
open-hole-test 安装孔试验
open hole well 裸眼井

open hood 外部集气罩
openhook link chain 钩头链
open hooves 广蹄
open hopper 底开泥舱;敞口式泥舱
open hopper wagon 敞开式倾卸车
open house 公开证券交易所;经验交流
open housing 租售无障碍的住房(政策);住房租售的开放(政策)
open housing law 住房业主中禁止歧视法
open hydraulic circuit 开式回路
open hydraulic fill 水力填充;引水填充
open hydraulic structure 透空式水工建筑物
open ice 疏冰
openig up 林冠疏开
open impeller 显露叶轮;开敞式转轮;敞式叶轮
open impeller pump 开式叶轮泵
open impeller sludge pump 开式叶轮淤泥泵
open impeller type dredge pump 开式叶轮泥泵
open indent 开口订货单;开放委托购买;非指定订单;非特定订单;不指定的订购单
open industrial structure 敞开式工业结构;开敞式厂房
open inequality 开不等式
opening 掀板;张开的;辊缝;孔口;孔洞;孔;开拓(采矿);开孔【岩】;缺林地;通风横巷;跳间;洞室;洞口;窗口;窗孔;打开;邻近焊缝开孔
opening accounts 初期账户
opening across width of jaws 颚式破碎机进料口宽度
opening amperage 开路电流强度
opening an account 开立账户;开户头
opening and shutting installation 开闭装置
opening a new coil 开绳法
opening angle 开度
opening area 开口面积;开孔面积
opening a terminal 打开终端
opening balance 期初余额
opening bank 开证(银)行;发证银行
opening between rolls 辊距;辊缝
opening bit 扩孔锥;扩孔器;开眼钻头;开孔钻头;矫形钻头(修管子用)
opening box 分梳装置
opening bracket 开括号
opening bridge 开合桥
opening bulb turbine 开启型灯泡式水轮机
opening capacity 虎钳最大开度
opening casing 孔口牌
opening clause 开始条款
opening contact 断路触点
opening corner plater 开口角复板
opening corporation book 开立公司账簿
opening cut 扩大导洞
opening damper 通风闸
opening date 开盘日期;开设日期
opening die 自动开合板牙
opening diffracting 衍射孔
opening dimension 洞孔尺寸;开启尺寸
opening door 双扇门中先开启的门扇
opening entry 开始记录;开始分录
opening exercise 开幕式
opening exhibition 展览会开幕
opening for beam 梁上开洞
opening for drainage 排泄孔
opening for lighting 采光口
opening for spark plug 火花塞孔
opening for taking out cullet 掏碎玻璃孔
opening ga(u)ge 开度计
opening gear 开启装置
opening height 开口高度
opening in (门、窗的)内开
opening in flat head 在平板封头上的开孔;平盖开孔
opening in head 封头开孔
opening in ice 冰间水面
opening inventory 期初存货
opening in weather portions of the ship 船舶露天部分开孔
opening iron 修缝凿
opening joint 节理(张)开度;清理灰缝
opening leaf (折叠门的)铰接门扉;开合门扉
opening light 可开启窗;开启窗;开敞式窗洞
opening limit control 开度(限位)控制
opening limiter 开度限制装置;开度控制设备
opening limiting device 开度限制装置;开度控制

设备
opening location 巷道位置
opening material 减黏剂
opening mode crack 张开型裂纹
opening movement 开展活动
opening of a letter of credit 开出信用证
opening of an account 开立户头
opening of bids 公开开标;公开招标;公开标价;开启承包书
opening of book 开户
opening of business finance 企业财务公开
opening of crack 裂隙开张
opening of damper 闸板开度
opening of discontinuity 结构面张开程度
opening of gangway 通道口
opening of hole 孔的大小;孔径;开孔口
opening of jaws 卡盘卡爪张开量
opening of refuse channel 垃圾收集口
opening of ring 环的裂开
opening of tender 开标
opening of the cylinder 展开圆筒
opening of the outlet part 出射部分的孔径
opening of the pores 毛孔开放
opening of the telescope 望远镜视场角
opening of tuyere 风口
opening of ventilation adit 风道洞口
opening of wound 创口
opening out (门、窗的)对开
opening plug 打开塞
opening pontoon bridge 开合浮桥
opening pressure 开启压力
opening pressure of safety valve 安全阀开启压力
opening press with two inclinable poles 开式双柱可倾压力机
opening price 开盘价(格)
opening profit 营业利润
opening quotation 开盘价(格);开盘
opening rate 开盘汇率;开盘价(格)
opening rate, closing rate 开盘汇率与收盘汇率
opening ratio 刀盘开口率
opening ratio of closed shield 闭胸式盾构的开口比
opening reading 起始读数
opening reinforcement 开孔补强
opening reinforcement of external pressure vessel 外压容器开孔补强
opening revenue 经营收入
opening roller 分梳辊
opening roof 开口车顶
opening-roof wagon 车顶开口棚车
opening sale 敞开式销售
opening session 会议开幕
openings for connections to brazed vessels 用于钎焊容器的连接开孔
openings for drainage 用于排放的开孔
openings for pipe connections 用于管子连接的开孔
opening share 开沟器;开沟铲
opening side light 枢轴舷窗
opening signal 打开信号
opening size 洞口尺寸;门框尺寸;开口尺寸
opening slot 开割槽
opening snap 开放拍击音;开瓣锐声
opening speed 开启速度;回程速度;断开速度
openings screened 装有纱窗的洞口
openings subjected to rapid pressure 经受压力突然波动的开孔
opening station 自由导线端点【测】
openings through or near welded joints 通过或靠近焊缝处的开孔
openings through welded joints 通过焊接连接处的开孔
opening stock 期初存货
opening stroke 开程
opening system 开启系统
opening temperature 大气温度;室外温度
opening the books 开账;开立账簿
opening time 开启时间;开孔时间【岩】;曝光时间;断开时间;断电时间
opening-to-application ratio 就业职位求供比率
opening to traffic 开放运输;开放交通;对交通开放
opening trading 开盘
opening transaction 开盘交易;开盘价(格)
opening up 疏通;打通;采准;开采

opening-up engineering 采准工程
opening-up nitches and gutters 疏通沟渠
opening velocity 断开速度
opening washer 扩孔垫圈
opening width 原幅;平幅
opening width of door 门开启宽度
opening window 活动窗扇;孔道窗
open inhalation 开放吸入法
open inlet 开敞进口;不冻河口
open inner court 露天内庭
open inquiry of the offer 公开询价
open installation 露天装置
open insurance 开口保险
open intake 开敞式取水建筑物
open integration formula 开型积分公式
open interconnection 敞开相连接
open interest 开放权益;未平仓合约;未结清权益
open interpolated curve 开内插曲线
open interval 开区间
open iron work 铁花饰
open jawed spanner 开口爪板手
open jaw wrench 开口爪式扳手
open jeep 敞篷吉普车
open jet 自由射流
open jet tunnel 开式风洞
open jetty 高桩码头;高架桥;明桩栈桥;明桩码头;透空式栈桥码头;透空式码头;穿通码头
open jet wind tunnel 开口式风洞
open job shop 开放式车间
open joint 有间隙接头;张开节理;露缝接头;留隙焊头;离缝接合;空腔排水缝;开口接合;开口缝;开缝接头;胶合间隙;明缝接头;明缝;对接接头;带间隙接头;凹缝
open joint drainage 开缝式(盲沟)排水;开缝式沟渠;开缝式管道排水;开缝排水沟
open-jointed 接头不嵌缝底板
open-jointed gap 接头间隙
open-jointed pipe 明接管;不嵌缝接口管;不嵌缝的接口管;开口对接管
open joints for pipes 管道开口接缝
open joint tile drainage pipe 不嵌缝接头的陶土排水管;不接口陶土排水管
open joist 露明格栅;明露格栅;明格栅
open kier boiling 不加压煮炼
open kiln 侧墙固定式窑炉
open kitchen 开敞式厨房
open knot 开裂节疤
open knot in wood 木材中的开裂节疤
open ladder 露明梯井
open lagging 间隔挡土板
open lake 出流湖
open land 空地
open land area 有待开放的地段
open land exception 空地建设的例外办法
open landscaped office space 开敞布置办公室
open landscape layout 大空间布局
open lap winding 开式绕组;开式叠绕组
open launder 开式流槽
open layer 排水土层
open layout 开敞式布置(图);自由式布置
open lay up 敞模铺放
open lead 开拓水道;冰间航道
open leaf 开合门廊;(折叠门的)铰链
open lens aperture 发散透镜孔径
open letter of credit 开放信用证;无条款信用书
open levee 开口堤;缺堤
open license 开放许可证
open lifeboat 开敞式救生艇
open light 明火灯;开灯;可开可关的窗户
open lighter 敞舱驳(船)
open line 旷道;开通路线;地面铁路线;车站交通线;明线
open-line city 带形城市
open link 开口链;锚链末端链环;无档链环
open link chain 开口链;无档链
open link fuse 明熔丝片
open lock 开通闸
open loop 开循环;开型环路;开(口)回路;开环(回路);开(放)式回路
open loop bandwidth 开环带宽
open loop characteristic 开回路特性
open loop control 开式控制;开环控制;无反馈控制
open loop control circuit 开环控制线路
open loop control system 开路控制系统;开环控制系统
open loop describing function 开环描述函数
open loop domestic water heater 民用开环太阳能热水器
open loop frequency response 开环频率响应
open loop gain 开环增益
open loop numberial control system 开环数控系统
open loop-optimal feedback 开环最优反馈
open loop phase response 开环相位响应
open loop policy 开环方案;开环策略
open loop process control 开环过程控制
open loop response 开环响应
open loop system 开环系统;开环控制系统
open loop transfer function 开环传递函数
open loop voltage gain 开环电压增益
open machine 开启型电机;敞开式机器
open macro 开式宏程序
open manometer 开管压力计
open map 开映射
open mapping theorem 开映射定理
open marine algal facies 开阔海藻质相
open marine amorphous facies 开阔海无定形相
open marine platform facies 广海台地相
open market 一般市场;自由市场;公开市场;开放(性)市场
open market operations 公开市场操作
open market paper 公开市场票据
open market policy 市场开放政策
open market rate 公开市场汇率
open market rent 市面租金
open memory location 开放式存储单元
open mesh reef 开口网状礁
open mesh steel floor 网状钢面板
open method 开型法
open mine 露天矿
open mining 露天开采;露天开采作业
open mix 开级配混合料;粗集料混合料;粗骨料混合料
open mixer 敞口混合器
open mixolimnion 开敞式湖水混成层
open mixture 开级配混合料
open mode 打开状态
open model 开模型;开放模型
open mole 开口突堤;开口防波堤
open mooring 八字锚泊
open mooring buoy 敞海系船浮筒
open moraine 开放型冰丘
open mortgage 敞口抵押;未列限额的抵押
open mortise 开口榫;榫槽;狭槽榫;榫沟;明榫;键槽;滑动榫槽
open mortise and tenon joint 明榫槽接合
open motor 开敞型电动机;敞开式电动机
open mo(u)ld 明浇铸型;敞口铸型
open mo(u)lding 敞成型技术
open mouth 开口
open mouth tongs 开口钳
open navigation canal 开敞式运河
open neritic marine deposit 开阔浅海沉积
open neritic marine facies 开阔浅海相
openness 开放性
open network control 开式网络控制;开路网络控制
open newel 敞开的悬挑楼梯;螺旋梯梯井;露明旋梯中柱
open newel garage 敞开式停车库;露天停车库
open newel lot 露天停车场地
open newel stair(case) 露明中柱旋转楼梯;露井楼梯;无中柱螺旋形楼梯;露明井梯;无中柱螺旋梯
open node 开型节点
open numbering 开始编号;开放编号
open numbering system 开放编号制
open object 开放目标
open ocean 公海;开阔大洋;外海
open-ocean mooring platform 公海系泊平台
open ocean operation 公海作业
open of bids 开标
open off cut 开切眼【矿】
open of refuse channel 垃圾收集口;垃圾收集器
open order 开放式订货;无条件订购;未完成的订货;略式订货单
open oscillator 敞式振子
open outside court type 外院式平面布置
open oven 明炉

open pack 疏冰群
open pack ice 稀疏冰;稀疏流冰群;稀疏浮冰群
open pan drying 敞口盘式干燥
open pan mixer 开锅式拌和机;敞口盘式搅拌机;敞口盘式混合器
open parking deck 露天停车台
open parking ground 露天停车场;露天车场
open parking structure 汽车停车棚;开敞式停车结构;开敞式存车库
open pass 开口孔型
open pasture land 开阔牧场
open pattern 开式
open pattern return bend 开敞式回转弯头
open peristyle court 周围列柱的议会;周围列柱的朝廷;周围列柱的法院;周围列柱的庭院
open phase 开路相
open-phase protection 断相保护
open-phase relay 开相继电器;断相继电器
open pier 桩基码头;高桩码头;露天码头;开口防波堤斗式墩;开口防波堤;突堤码头;透空式突堤码头
open pier construction 空架突堤结构
open pier on pile 高桩码头
open pier type 空架突堤式
open pile 露天堆
open-piled type 透空桩基式;桩基式
open piled type quay 透空桩基式码头
open piling 透空桩排;透空桩基
open pipe 开口缩孔;开管;明缩管
open pipe piezometer 开口测压计;竖管式测压计
open piping 明配套
open pit 露天开采坑;露天采矿场;露天采掘场;采场
open pit bench 露天矿台阶
open pit bottom 露天矿底盘
open pit development 露天矿开拓
open pit haulage 露天矿运输
open pit incineration 露天坑式焚烧
open pit incinerator 露天坑式焚化炉
open pit limestone quarry 露天石灰石矿
open pit mine 露天采矿;露天采掘矿
open pit mine life 露天矿服务年限
open pit mine production exploration engineering 露天生产勘探工程
open pit mining 露天采矿;开矿;露天开采
open pit mining limit 露天矿开采境界
open pit mo(u)ld 敞开式模具
open pit sill pillar 露天矿砥柱
open pit slope 露天矿边坡
open plain 开阔平原
open plan 大开间平面;开敞式平面;未决定的设计方案;统间式平面
open-plan educational building (有外走廊的)教育建筑
open planning 自由式平面布置;开敞式布置;统间式设计
open-planning millwork 特殊规格门窗型材的制造及供应工厂
open plan office 统间式办公室
open plan school 开敞平面的学校
open plan system 开敞式平面布置方法
open plant 开放工厂
open plan workstation 敞开工作站
open plate wheel 开板轮
open platform 露天站台;露天平台;开阔台地
open platform facies 开阔台地相
open plug 开路插头
open plumbing 露明的铅管系统;敞开式卫生工程;敞开式管道系统
open policy 预约保险单;预定保险;开口保险单;开放政策;全险保(险)单;无定额保险单;船名未定保险单;不定值保险单
open pollination 自然传粉
open polyline 未闭合折线
open-pool reactor 水池型堆
open population 开放性群体;开发人口
open pore coating 打底油;本色涂漆;白坯涂漆;白荏涂漆
open-pored 开孔
open pore finishing 白坯涂装;白荏涂漆
open porosity 开口气孔率;开口空隙率;表面多孔性
open port 自由港;开放港(口);开敞式港湾;开敞港(口);通商自由港;通商口岸;通商港(口)

open position 晶格空胞;空头地位;开放部位;开断状态;未轧平头寸;未结清期货合同;断路位置
open pot 开口坩埚;敞口锅
open prebaked anode type cell 敞开式预焙阳极电解槽
open pregrinding circuit 开路预粉磨流程
open pressure 开井压力;敞喷压力
open price 公开标价
open price agreement 统一报价协定
open pricing 公开定价
open printing 开启打印
open production 开放性生产;建筑构件大规模成套生产
open propeller efficiency 螺旋桨敞水率
open pumping 开敞式泵送
open quay 透空式码头
open question 悬而未决的问题;未解决的问题;待研究问题;存在的问题
open radiator 开路辐射器
open rafter 露明椽;明椽
open rail(ing) 栏杆;露明栏杆
open railway crossing 无护栅铁路交叉道;不设护栏的铁路交叉口
open ramp 敞开引道
open range 自由放牧地
open rate 自由运费率;规定费率
open rate of exchange 开盘汇率
open reactor 敞口反应器
open recirculating cooling system 敞开式循环冷却系统
open recirculating system 敞开式循环系统
open refrigerator 开式冷冻机
open region 开区域
open relay 开启式继电器;敞形式继电器
open relieving platform 桩基减载平台
open repo 开放买回协议
open reserves 公开储备金
open reservoir 露天水库;露天水池;敞开油储[贮];不封闭油气层;不封闭储集层
open resonator 开式谐振腔;开端式谐振腔
open return bend U 形管
open return system 开式回水系统
open reverse circulating valve 打开反循环阀
open ring reef 开口环状礁;开放环礁
open riser 空竖标;无踢板
open riser stair(case) 露空踏步楼梯;无踢板楼梯;露明踏步楼梯;露明楼梯;漏空踏步楼梯
open riser step 漏空踏步
open river 开阔河道;开放河道;未封冻河道;通航河道;畅流河道;不冻河(流)
open river discharge 畅流河道流量
open river navigation 畅流河道航运
open river period 通航期
open river reach 宽阔河段
open river stage 河道通航期;河道流量航期
open road 无遮蔽锚地;开放锚(泊)地
open road capacity 无阻碍道路通行能力
open road ditch 明路沟;马路明沟
open road drain 道路排水明沟
open roadstead 开敞锚(泊)地;开放泊地;无遮蔽锚地;无屏障锚地
open rock 多孔裂岩石
open rod hydrodynamic(al) press 开杆式水压机
open rod press 柱式油压机
open roller chock 滚筒导缆钩
open roof 露木屋顶;无顶棚屋顶;开敞式屋顶;露明屋顶;露椽屋顶;露天屋面;无天花板屋顶
open root 有间隙根部
open roughing pass 开口粗轧孔型
open route 地面铁路线
open routine 开型(例行)程序
open routing 直接插入程序
open running 开路运行
open runway 已扫雪的跑道;开放的跑道
open saloon 开放式客厅式车厢
open sand 开(式)级配砂;松砂;粗砂;松散砂;多孔性砂土;多孔砂
open sand casting 明浇铸造;明浇铸件;敞开式砂型铸造;敞浇铸件
open sand filter 开敞式砂滤池;敞式砂滤器;敞露式砂滤池;敞开式砂滤池
open sand mo(u)ld 明浇砂型
open sand mo(u)lding 开箱造型;敞箱造型
open sandwich-type panel 夹心饰板;夹心镶板;夹心间壁

open sash 开扇
open scourer 平幅煮练机
open screening area 有效筛孔面积
open sea 开阔海(面);开敞海面;外海;大海;敞海
open sea aeroplane 海上飞机
open sea and ocean 公海;开阔海洋
open sea berth 外海泊位;敞海泊位
open sea canal 无闸通海运河;(无闸的)通海运河
open sea deposit 广海沉积
open sealed tender 开标
open-seam tube 明缝管;开缝管
open sea navigation 外海航行
open sea pier 外海突堤码头
open sea platform 敞海系船平台
open sea pollution 公海污染
open seas 公海
open sea shelf 广海陆架
open sea shelf facies 广海陆架相
open season 渔期;开放季节
open sea terminal 开敞式码头;外海码头;外海泊位;敞开式码头
open sea wave 敞海波浪
open section 空心断面;(桥梁的)开断面;地面铁路线
open sedan 活顶轿车
open segment 开型图段
open set 开集;石垛节间(未填充部分)
open setting 无匣钵装窑法;敞开装窑法
open sewer 污水明沟
open shackle 开式卡环
open shaft 露天竖井;敞口通风井;小天井;开敞式竖井;敞口竖井
open sheathing 疏散支柱;疏散背板;断续式挖方支撑
open shed 敞棚;露天货棚;敞式圈;敞开式棚架;敞壁畜棚
open sheeting 间隔式挡板;架空式支撑;开式挡板;间隔式撑板;断续式挖方支撑;敞口挡板
open shelf 开架书架
open shelf reading room 开架阅览室
open shelf system 开架式
open shell 开口壳
open shell and tube condenser 立式壳管式冷凝器;开式壳管冷凝器
open shelter decker 开式遮蔽(防浪)甲板船
open shelter-deck ship 开式遮蔽(防浪)甲板船
open shelving 露明搁板;露明隔板;开式搁板
open shield 开式挡板;开放式盾构;敞开式盾构
open shield method 敞胸盾构法
open ship 预约初保凭证;预约初保单;敞舱船
open ship system 开式船制
open shop 自由雇佣企业;开放工厂;开站(运行)【计】;开运行;开式计算站;开放式机房
open shore 开放海岸
open shortest path first 开放的最短路径优先路由协议
open shrinkage 明缩孔
open shuttering 敞开式模壳
open side container 侧壁全开式集装箱;侧壁敞开式集装箱
open-sided block 边侧敞开的大建筑物;边侧敞开的街区
open-sided building 边侧敞开的大建筑物;侧敞开式大楼
open-sided hydraulic planer 单臂液压刨床
open-sided milling machine 单臂铣床
open-sided multi-stor(e)y garage 侧敞开式多层车库;边侧敞开的多层修车厂;边侧敞开的多层汽车库
open sided planer 单柱刨床
open-side hydraulic planer 单臂液压刨床
open-side planer 单柱刨床;单臂(龙门)刨床
open-side planer milling machine 单臂龙门(刨)铣床
open-side planing machine 单臂(龙门)刨床
open-side press 单臂压力机
open-side setting 排料口张开时的宽度
open-side tool block 单臂刀架
open-side type 单柱式;侧敞开式
open-side vertical shears 开式剪切机
open sight 瞄准器;缺口表尺
open sign 开门营业招牌

open slating 花铺石板瓦;疏铺石板瓦;疏铺(石)板
open slot 开口槽;开缝,切缝;铁芯开槽
open slot bubble 开口气泡
open sluiceway 明渠泄水道
open soaper 平洗机
open socket 开式插口;开口锚环
open soffit 敞开的拱腹内面;敞开的露天下部
open solar absorption refrigerating machine 敞开式太阳能吸收制冷机
open sound 开阔海峡;开阔海湾
open space 无建筑物区域;开放空间;绿地;旷地;空地;开敞处所;敞露部位
open space deposit 岩石间沉积物
open space index 空地指数
open space land program(me) 旷地开放纲领
open space of screen 筛子有效面积
open space ratio 敞地比;空地率
open space zoning 绿地分区
open-spandrel 拱上敞开的空间;敞开的拱肩;空腹式拱;开腹拱桥
open-spandrel arch 空腹拱;敞肩拱
open-spandrel arch bridge 空腹拱桥;敞肩拱桥
open-spandrel girder 空腹大梁
open-spandrel pier 空腹拱墩
open-spandrel rib 空腹肋
open-spandrel stone arch bridge 空腹石拱桥
open-spandrel truss 空腹桁架
open spanner 开口扳手
open spillway 自流式溢洪道;开敞式溢流堰;敞式溢洪道;明溢洪道;敞式溢洪道
open spindle bearing 开式锭子轴承
open spiral auger 开口式螺旋管钻
open-spiral worm conveyer 敞开式螺旋输送机
open spongy body 敞开的海绵体
open spread 间隔排列
open spur dike 开口丁坝
open square butt joint 平头开口对接焊
open square strand passes 开口方形直通系统
open stack 露天青储[贮]堆;开架式书库
open stacking 稀垛
open stacking ground 露天堆场
open stacking method 稀垛法
open stacking yard 露天堆场
open stack management 开架管理
open stage 敞开式舞台;露天舞台;开敞式舞台
open staging 露天码头;敞开的脚手架;露天舞台
open stair(case) 开敞楼梯;敞开(式)楼梯;露明楼梯;露天梯
open stairway 露明楼梯;明楼梯
open stair well 露明梯井;露明楼井
open stalk 空腹
open stall 敞式熔烧炉;敞式焙烧炉
open standing test 静置试验
open standpipe piezometer 敞口竖管式测压计
open starter 开式起动器
open steam 直接水蒸气
open steel 沸腾钢;不完全脱氧钢
open steel flooring 露钢梁楼板;漏空钢板楼面
open stem 槽针针杆
open stern frame 开式尾柱;敞式尾架
open stock 非成套的零售商品
open stone mix 无砂大孔混合料
open stone mixture 未灌浆的碎石混合料
open stope 空场工作面
open stope method 空场采矿法
open stope mining 空场法开采
open stope mining method 空场采矿法
open stopping 空场采矿法
open storage 露天库场;露天堆场;露天储放
open storage area 露天堆场区
open storage dump 露天储库
open storage ground 露天堆场
open storage pile 露天货堆
open storage track 堆场铁路线
open storage yard 露天堆场
open stor(e)y 无顶棚楼板;棚露明楼板
open storm drainage 地面敞开式雨水排除
open stream 开阔河道;畅流河道;不冻河流
open stream discharge 畅流河道流量
open string 露明楼梯斜梁;露梯基;露明楼梯小梁;开行;开串
open stringer 木楼梯露明斜梁
open string stair(case) 露明梁楼梯
open structure 开式结构;透水结构;透空式结构;

透空结构;疏松结构;敞开构造;敞形结构
open style 开放式(仓储);开敞式
open subprogram(me) 直接插入子程序;开型子程序
open subroutine 直接插入子程序;开子程序;开型子程序
open superstructure 敞开式上层建筑
open support 栅篱
open surface 开式面层;开式级配面层;多孔面层;多孔表面;粗糙(表)面;敞开面
open surface condenser 开敞表面电凝器
open surface cooler 开面式冷却器;水流式冷却器
open surface cooling 开式冷却
open surface method 明צ式
open surface of sea 开阔海面
open surge ignition system 断路电浪式点火系统
open switch cabinet 开式开关柜
open switch point 不密贴尖轨【铁】
open switch yard 露天开关站
open symbol 开符号
open system 装配式房屋;开系;开式生态系统;开式;开路系统;开放体系;开放系统;通用建筑体系;敞开系统;非封闭系统
open system architecture 开放型系统结构;开放体系结构;开放式系统结构
open system interconnection 开放型系统互连;开放系统互连
open system interconnection reference model 开放系统互连参考模型;开放互连系统参考模型
open system management 开放系统管理
open system of hot water supply 开式热水供应系统;敞开式热水供应系统
open system of recirculated cooling water 开式循环冷却系统;敞开式循环冷却水系统
open system precasting 露天预制系统
open systems interconnection 开放式系统互连
open system test (土体冻胀试验等的)开式试验法
open tabular column 开口管柱
open tabular pile 开口管桩
open tank filter 敞滤槽
open tank method 开口槽法;敞槽法
open tank treatment (木材防腐的)敞罐浸油处理法;敞槽法
open tee joint 露缝T形接头;留缝丁字接头;有缝T形接头;开口形接头
open telegram 通电
open tender 公开招标;一般竞争投标;招标采购;公开投标;开标
open tendering 公开招标
open term 开项
open terrace 露天台地;露天阶地
open territory 空旷地
open test pit 露天探坑;露天试(验)坑;探坑
open texture 开式级配组织;松组织;多孔隙组织
open-textured 不致密的;松散的;开式级配的
open-textured bituminous macadam 开式级配沥青碎石路
open-textured grading 疏松级配
open-textured mix 开式级配混合料;疏松混合料;不密实的混合料
open-textured surface 多孔面层;开式面层
open-textured tar macadam 开式级配焦油碎石路
open the bid 公开招标
open the books 开立新账
open the budget 提出预算案;提出预算
open throat (辙叉的)有害空间;敞义有害空间
open throat shears 开式剪切机
open throughout 开通的(洞)
open tidal harbo(u)r 有潮港(口)
open tidal port 有潮港(口)
open-timbered 外露式木结构;敞式木结构
open-timbered gable 露木山墙
open-timbered roof 结构外露的木屋顶;无顶棚屋面;露明格栅板;敞式楼板;露木屋顶;露木椽屋顶
open-timbered structure 敞式木结构;外露式木结构
open-timber floor 露明格栅楼板;露木楼板
open-timber house 露木房屋
open timbering 架空式支撑;断续式挖方支撑
open-timber roof 露梁屋顶
open time (黏胶的)晾置时间
open-time of primer 打底涂料的凉置时间
open T joint 露缝T形接头;开口形接头;有缝T形接头
open-to-buy 补购额;许购定额

open-to-buy estimate 补进预计额;许购定额的估计
open-to-buy report 许购定额报告
open-top 开式灶面板;顶部开口
open-top arc furnace 顶部敞口电弧炉
open-top bin 敞开式料斗
open-top car 敞车;敞舱舢板
open-top chamber 开顶实验槽
open-top container 开盖集装箱;开顶集装箱;敞口盛料箱;敞顶集装箱
open-top culvert 开顶涵洞;透顶涵洞
open-top drum 开顶桶
open-top duplex lantern 开顶双灯
open-top (feeder) 明冒口
open-top fumigation 开顶式熏气
open-top ladle 开口式铁水罐;敞式铁水罐;敞口铁水罐
open-top mill 顶开口式轧机
open-top mixer 敞顶式槽形搅拌机;敞顶圆槽搅拌机;敞口式搅拌机;敞顶搅拌机;敞顶拌和机
open-top mold 敞口式钢模;敞口钢锭模
open topped kiln 敞顶窑
open-topped vessel 无盖容器
open-top railway car 敞车;顶部敞开的铁路货车
open-top riser 明冒口
open-top roll housing 开口式机架
open-top semi-trailer 敞顶式半拖车
open-top side-dump concrete hauling unit 顶开侧卸混凝土运输设备
open-top single lantern 开顶单灯
open-top trailer 开顶式拖车
open-to-public area 游览区
open to the outside world in an allround way 全方位开放
open to traffic 开放交通
open tracery 花格漏窗;网眼花格窗
open track 露天轨道;开放线;通行线(指铁路);车站交通线
open trade 未完结交易
open-train billet mill 横列式钢坯轧机
open-train mill 横列式轧机
open transaction 明买明卖
open transition 开路瞬变过程
open transport network 开放式交通线网;开(放)式传输网(络)
open traverse 开导线
open treaty 开放条约
open tree formation 开阔树林
open trellis bond 格子连接;格构连接;格构式砌合
open trellis masonry wall 格子坯土墙
open trellis wall 格构式墙
open trench 开槽施工;明渠;明沟
open trench system 明渠排水法
open trench tunnel 明挖隧洞;明挖隧道
open trough 开放式槽;敞口槽
open trough drainage 明沟疏干法
open trough type conveyer 敞式槽形输送机
open trough urinal 露天尿槽;露明小便槽
open truck 敞车
open truss steel joist 露桁楼板格栅
open tube 开(口)管
open tube diesel hammer 开管式柴油锤
open tube manometer 开口气压计
open tube process 开管法
open tube sampler 开口式管状取土器
open tubular capillary packed column 空心毛细管填充柱
open tubular column 开管柱
open tunnel 明硐
open-type 开式的;开启式;开口式;敞(开)式
open-type air slide 敞式空气斜槽
open-type airslide conveyer 敞开式空气输送槽
open-type asphaltic concrete 多孔沥青混凝土
open-type berthing structure 透空桩靠船建筑物
open-type breakwater 透空式防波堤
open-type compressor 开启式压缩机
open-type cooling tower 开式冷却塔
open-type double action crank press 开式双动曲柄压力机
open-type electric(al) machine 敞开式电机
open-type electric(al) motor 敞开式电动机
open-type equipment 非防爆设备
open-type fuse 敞式熔断器
open-type heat exchanger 敞开式热交换器
open-type hotplate 开式灶面板

open-type hot-water heat-supply network 开式热水热网
open-type impeller 开式叶轮;开敞式叶轮
open-type inclinable press 开式可倾压力机
open-type induction motor 敞开式感应电动机
open-type ladle 开式铁水罐
open-type machine 开敞型电机;开敞式电机
open-type manlift 敞开式载人提升机
open-type manual shield machine 敞开式手挖盾构
open-type mechanical shield machine 敞开式机械盾构
open-type mixture 不密实混合料;开式级配混合料
open-type nozzle 开式喷嘴
open-type paper machine 开式造纸机
open-type pier on piles 高桩码头
open-type propeller shaft 开式万向传动轴;开式螺旋桨轴
open-type quay 开敞式码头
open-type ship 开式船
open-type switchgear 开式开关柜
open-type system 开式系统;敞开系统
open-type thrust bearing 敞开式推力轴承
open-type toilet cubicle 开敞式厕所隔间
open-type track circuit 开敞式轨道电路;开路式轨道电路
open-type triaxial test 开式三轴试验
open-type truss 开式桁架;露式桁架;开露式桁架
open-type truss bridge 开式桁架桥
open-type urinal enclosure 开敞式小便隔间
open-type wharf 透空式顺岸桥式码头;透空式(横式栈桥)码头;开式码头;顺岸栈桥式码头
open-type wharf on concrete cylinder 透空式混凝土管柱码头
open union 开放式工会
open up 开发;开采;发展;涌现出;开辟;打通;开拓(采矿)
open up a market 开辟市场
open up by combined methods 综合开拓法
open up new sources of revenue 扩大财源
open valley 明屋谷;露明天沟;屋向天沟;屋面天沟
open valve 明装阀门
open vat 敞口瓮
open vehicle 敞篷车
open ventilation system 开式通风系统
open vessel 开式船;无甲板船
open-view lift 升降时可透视景物的电梯
open void 开型孔隙
open void formation 开孔构造
open wagon 敞车;无棚货车;无盖车;敞篷货车;敞篷车
open washer 开口垫圈
open water 开阔海面;开旷水面;无敞水面;无屏蔽水域;无冰水面;未封冻水面;地表水
open water-cooling tower 敞式水冷塔;敞开式冷却塔;敞式凉水塔
open water rating 明流-水位流量关系
open water return 开式回水
open waters 开阔水域;开敞水域;无障碍水域
open water surface 开阔水面
open water tank 开式水箱
open water test 敞水试验
open waterway 开发通航河道
open waveguide 开波导
open weather 暖和晴朗天气
open weave cloth 网格布
open web 空腹体;空腹的
open-web elevator 条杆带式升运器
open-web girder 格构大梁;空腹梁;空腹大梁;格构梁
open-web girder bridge 空腹梁桥
open-web joist 桁架式小梁;空腹轻钢龙骨;空腹格栅
open-web steel 空腹钢材;明腹板钢材
open-web steel joist 空腹钢梁;空腹轻钢梁;空腹钢桁架;空腹钢格栅;空腹轻钢龙骨;空腹钢龙骨
open-web stud 空腹板墙筋
open-web truss 空腹桁架
open-web truss bridge 空腹式桁架桥
open weir 开口防波堤;开敞式溢流堰;开敞式堰;非闸控堰
open weld 留间隙焊缝;空隙焊
open well 露明梯井;开敞(式)竖井;普通水井;大口(径)井;敞口竖井
open well calculation method 大井计算法

open well foundation 井筒基础；沉井基础
open well stair(case) 无中心柱螺旋形楼梯；露明梯井楼梯
open wharf 高桩码头；透空式顺岸码头；透空式码头
open-wharf mooring island 系泊岛式码头
open width scouring machine 开幅洗涤机
open width suction drier 平幅真空吸水干燥机
open width washer 开幅洗涤机
open winding 开式绕组；开路绕组
open window 开口；开窗口；开敞窗
open-window mask 撕膜蒙片
open-window process 撕膜法
open-window unit 赛宾(吸音单位，相当于一平方英尺的全吸音面)；敞窗单位
open winter 无雪冬季
open wire 架空(明)线；明线
open wire carrier communication 明线载波通信
open wire carrier communication system 明线载波通信系统
open wire carrier line 明线载波线路
open wire carrier system 明线载波系统
open wire circuit 明线线路；明线电路
open wire fuse 开启型熔断器
open wire line 明线(线路)
open wire loop 明线回路
open wire parameter 明线参数
open wire transmission line 明线传输线
open wire transposition 明线交叉
open wiring 明线布置；明线布线；明线；布明线
open wood land 开阔林地；疏林地；开阔树林
openwork 露天作业；露天开挖；透空结构；未设防工事；透孔装饰；镂空细工；有网状小孔的；露天开采作业；露天开采；透雕细工
openwork carving 透雕
openworked balustrade 格子栏杆
openwork gabie 网眼山墙；透孔山墙；漏空山墙
openwork gablet 网眼花山头；透空花山头；透空小山墙
openwork gravel 架空砾石层
openworking 透孔织物；透孔制品；露天开挖；露天作业；开放工作
openworking area 露天作业地段
openwork iron gate 花铁大门；透花铁门；铁栅栏门
openwork jetty 透空桩突堤
openwork panel 外通镶板
openwork rosette 镂空玫瑰花形饰；网状插座；漏空圆花窗；透空圆花窗；透雕圆花饰
openwork tracery 网眼窗格；网眼窗花；漏空花窗格；透花格窗
openwork vase 镂空花瓶；通(花)花瓶
openwork wall 漏墙
open yard 露天畜栏
opepe 黄木树(金棕色的，产于尼日利亚)
operability 可操作性；操作度
operable 可运转的；可营运地
operable partition 可开启的隔墙；活动隔墙；活扇式隔墙；活动式隔断
operable transom 可开启的门楣；可开启的门顶窗；活动门亮子
operable wall 活扇式隔墙；活扇式隔断；可开启的墙；活动隔墙
operable window 可开启的窗；活动窗
opera glass 观剧望远镜
opera house 歌剧院
operameter 计算轮转仪器；运转计；动速计
operand 运算数；操作数；被运算数
operand access 操作数存取
operand address 操作数地址
operand address field 操作数地址字段
operand addressing operation 操作数寻址操作
operand bit 操作数位
operand bus 操作数总线
operand call syllable 操作数调用字节
operand delay 操作数延迟时间；操作数延迟
operand designation 操作数指定
operand effective address 运算数有效地址
operand entry 操作数入口；操作数进入
operand error 操作数误差
operand fetch cycle 取操作数周期
operand fetch instruction 操作数取出指令
operand fetch module 取操作数功能块
operand field 操作数字段；操作数信息组；操作数区；操作数段
operand format 操作对象格式

operand length 运算数长度
operand page 运算数页面
operand part 操作数部分
operand-precision register 操作数精度寄存器
operand queue 操作数队列
operand specifier 操作数说明符
operand specifier type 操作数说明符类型
operand stack 操作数栈
operand stack register 操作数栈寄存器
operand store 操作数存储器
operand subfield 操作数子字段
operand sublist 操作数子表
operand table 操作数表
operand type 操作数类型
operand value 操作数值
operand word 操作数字
operand word-length 操作数字长
operant behavio(u)r 可控行为
operant conditioning 操作性条件反射
operant level 操作水平
operant response 操作反应
operate 运行；运算；开动；经营；操纵
operate according to a same formula 同一配方运转
operate at an overload 超负荷运转
operate at full throttle 在大油门下工作
operate by remote control 遥控运行
operate class instruction 操作类指令
operated 操作的
operated by alternating current 交流操作
operated control system 操作控制系统
operated digit 被操作数位
operated line 运营铁路线
operated service and repair 维护和修理
operated stack register 操作叠式寄存器
operated track 运营轨道
operate in series 并接运转
operate memory 操作存储器
operate miss 操作失误
operate mode 工作方式
operate power 操作功率
operate storage 操作存储器
operate time 闭合时间
operate under capacity 开工不足
operate voltage 工作电压
operating 运转；运营；工作；操作
operating accident 操作事故
operating accident rate 运行事故率
operating account 营业账(户)
operating account payable 应付运营款
operating account receivable 应收运营款
operating activities 经营活动
operating administration 经办的主管部门
operating advantage 操作上的优点
operating agency 经营机构
operating agreement 运行协定；经营合同
operating and maintenance cost 运行和维护成本
operating and maintenance supplies 修缮用物料；修缮用品
operating angle 工作角度
operating apparatus 运作装置；操作器械
operating area 工作面积
operating arm 摇杆；平衡杆；操作杆；操纵杆；操纵臂
operating arm shaft 操纵臂轴
operating assets 营业资产
operating assets turnover 营业资产周转率
operating autonomy 经营的独立性
operating bank 开业银行
operating bar 操纵杆
operating basis earthquake 运转依据地震；运行依据地震；容许运行地震
operating bellows 控制膜盒；操纵膜盒
operating bench of water crane 水鹤操纵台
operating bias 工作偏压
operating block 运算程序块
operating board 工作台；控制仪表板；控制台；操作台；操作盘；操纵台
operating bolt load 操作螺栓载荷
operating booster assembly 操纵助力器总成
operating bridge 工作桥；操作桥
operating bucket (水泥混凝土摊铺机的)摊铺斗
operating budget 营业预算；业务预算；经营预算
operating bust 操作错
operating button 操作按钮；操纵按钮

operating cab 控制室；调度台；操纵室
operating cable 工作电缆
operating cab operator cabin 驾驶室
operating capability 管理能力
operating capacity 工作容量
operating capital 营业资本
operating capital turnover ratio 营业资本周转率
operating cash balance 营业现金余额
operating ceiling 实际升限；实际上升限度
operating center distance 实际中心距离；实际中心距
operating chain 操作铁链；操纵链
operating chamber 工作腔；工作室；手术室
operating characteristic 运行性能；运行特性；运算特征；工作特征；工作特性；经营特点；(焊条的)使用性；使用特性(曲线)；操作特性
operating characteristic curve 运行特性曲线；作业特征曲线；工作特征曲线；经营特性曲线；接收概率曲线；操作特性曲线
operating characteristic function 运算特征函数
operating characteristics of an electrode 焊条使用性
operating characteristics of electric(al) apparatus 电器特性
operating characteristics of the test of mean 均值检验的运算特性
operating charges 经营收费；操作消耗量；运行费(用)
operating chart 运行图表；作业图；操作图(表)
operating circuit 工作电路；操作电路
operating code 操作码
operating code field 操作码区
operating code mode 操作码方式
operating code register 操作码寄存器
operating coefficient 营业系数；工作系数
operating coefficient of rig 机台利用系数【岩】
operating coil 操作线圈
operating comfort 工作舒适；操作舒适
operating command 操作命令
operating company 自营公司
operating condition 运转状态；运营条件；运行状况；运行条件；营运条件；营业条件；工作状态；使用条件；操作条件
operating console 控制台；操作台；操作控制台；操纵台
operating consumption 操作消耗量
operating contact 工作接点
operating control 操作控制；操纵装置
operating control point 工作基点；闸楼【铁】
operating cost 运营费(用)；运转成本；运营费(用)；运营成本；运行费(用)；营运成本；营业开支；营业费(用)；营业成本；业务成本；管理费；经营费用；使用费(用)；生产费(用)；操作费用；保养费(用)
operating cost model 经营成本模型
operating cost of railway handling 铁路装卸作业成本
operating cost of shunting 调车作业成本
operating cost of statistics of railway transport 铁路运输成本运营统计
operating cost rate based upon net sales 按销售净额计算的营业费用率
operating cost statement 营运成本表
operating crank 起动手柄；操作曲柄
operating crew 工班；工组；操作组人员；工作班人员；作业班组
operating current 吸动电流；工作电流；动作电流；操作电流
operating curve 操作曲线
operating cycle 现金循环；运转周期；营业周期；营业循环；工作周期；工作循环；经营周期；操作循环
operating cycle period 经营期间
operating cylinder 工作油缸
operating data 运行数据；生产资料；生产数据；生产记录；操作数据
operating day 工作日
operating decision 运行决策；营业决策；业务决策；作业决策；经营决策
operating deck 工作桥；操作平台
operating deficit 营业亏损；营业亏损
operating delay 运算延迟；操作延误(时间)；操作延迟(时间)
operating delay time 操作延迟时间
operating department 运营处；车务部

operating derrick 开采井架
operating desk 工作平台;操作平台
operating device 工作装置;控制设备;操作装置;操作机构;操纵装置
operating differential subsidy 营运差额补贴(美国);经营差异津贴
operating display 操作显示
operating distance 作用距离;工作距离
operating distance of track 线路运营长度【铁】
operating duty 运行方式;工况;操作制度;操作任务;操作规程
operating duty test 运行工作方式试验;运行工况试验;运行方式试验
operating earning 营业收益
operating earning rate 营业资本收益率;经营资本利益率
operating economy 经营经济;操作经济
operating effectiveness 经营效能
operating efficiency 运行效率;营运效率;作业效率;工作效率;充满系数;操作效率
operating efficiency of a pump 泵的充满系数
operating element 操作元件;操作要素
operating engineer 运行工程师;施工工程师;操作技术(人)员;操作机师;操作工程师
operating envelope 作业范围
operating envelope of compressor 压缩机特性曲线
operating environment 操作环境
operating equipment 操作装置;操作设备
operating error 操作误差
operating exchange capacity 工作交换容量
operating expenditures 运行作业费
operating expenses 运营费(用);运行费(用);运输费(用);营运费(用);营运成本;营业支出;营业开支;业务开支;管理费用;经营支出;经营管理费用;经营费(用)
operating expense budget 营业费用预算;经营费用预算
operating expense ratio 经营费率
operating experience 运行经验;操作经验
operating factor 运转系数;运转率;运行系数;运算因子;工作系数
operating failure 工作失误;操作失误
operating flexibility of tray 塔板操作弹性
operating floor 工作台;控制层;操作层
operating fluid 工作液体;工作液
operating force 操纵力
operating fork 操纵叉
operating frequency 运转频率;运行频率;工作频率;操作频率
operating front aperture 前工作孔径
operating function 运算函数;操作功能
operating fund 流动资金;经营资金;营业资金;业务基金
operating gallery 操作走廊;操作廊道;操作廊
operating gang 操作班组;工作队;工作班
operating gangway 操作廊道
operating gear 闸门开闭装置;操作装置;操纵机构
operating gearing 经营杠杆率
operating grip 操作手柄
operating guide 使用指南
operating handle 操作手柄;操作柄;操纵杆;操纵把
operating head 运行水头;工作扬程;工作水头;工作寿命
operating head of filter 滤池工作水头
operating holding company 营业控股公司;营业股权公司;经营性控股公司
operating hours 工作小时;运转小时;操作小时
operating house (活动桥的)开关房;工作室;操纵室
operating hydrology 运算水文学
operating illumination 工作照明
operating incident 操作故障
operating income 营运收入;营业收益;营业收入;经营收入
operating income and expenditures 营业收支
operating information 业务资料;操作资料
operating information system 业务信息系统
operating instruction 使用说明书;运行规程;业务规章;操作指南;操作指令;操作须知;运行指令;作业细则
operating instruction manual 使用指示书;使用说明(书);操作说明书
operating instrument 控制仪表
operating item counter 操作次数计数器

operating key 电报电键;工作电键
operating kilometrage 运营公里里程
operating kilometre 营运里程
operating knob 操作按钮
operating labo(u)r 工作工人
operating lease 营运租赁;作业租约;经营性租赁;使用租赁;使用权出租
operating ledger 营业收支分类账;营业股权总账
operating level 运行水位;工作电平
operating level earthquake 运行级地震;容许运行地震
operating lever 操作杠杆;操纵杆
operating leverage 经营杠杆(作用);营业杠杆率;经营资本的财务杠杆作用
operating lever connecting yoke 操作杆连接架
operating lever lock 操作杆闩
operating license 操作许可证;营业执照
operating life 运行寿命;工作寿命;使用寿命;使用期限
operating life test 工作寿命试验
operating limit 工作极限
operating limitation 操作极限
operating line 作业线;工作线(路);操作线;操作管路
operating line of cargo-handling 装卸作业线;货物装卸作业线
operating linkage 工作连锁;操作联动装置
operating liquid 工作液体
operating load 运转负荷;运行荷载;运行负载;运行负荷;工作荷载;工作负载;工作负荷
operating locomotive 运用机车
operating log 工作记录
operating loss 运用损失;运营亏损;营业损失;经营损失
operating luffing 工作变幅(起重机)
operating luminance test 工作照度试验
operating machinery 操作机械
operating machinery with chain and sprocket 链条式启闭机
operating machinery with crank arm and connecting rod 曲柄连杆式启闭机械
operating machinery with curved arm 弯臂式启闭机
operating machinery with disk rotor 圆盘式启闭机
operating machinery with winch and connecting rod 绞车推杆式启闭机械
operating maintenance 运行维修;运行维护;运行检修;工作维修;例行保养;日常维修
operating maintenance procedure 操作维修步骤
operating management 经营管理
operating management of stadium and gymnasium 体育场馆经营型管理
operating manual 使用指示书;作业手册;使用指南;操作说明书;工作手册;操作规程
operating margin 运行裕量;营业毛利;营业边际
operating mass 工作负载
operating mechanism 工作机构;策动机构;操作机构;操纵机构
operating medium 工质
operating mistake 操作错误
operating mode 运行状态;运行模式;运行方式;运算方式;工况;操作方式
operating motor 执行电动机;操作电动机
operating net assets 经营性净资产
operating noise factor 工作噪声因数
operating nut 工作螺母;操作螺母
operating objective 经营目标
operating of discontinuity 不连续控制动作
operating oil line 操作油管
operating order 操作顺序;操作指令
operating outlay 业务费用
operating overhead expenses 非操作费用
operating overload 工作超荷
operating panel 操作开关盘;操作配电盘;控制板;操纵指示板
operating part 操作部件;控制部件;操纵部件
operating party 工作班
operating pedal 操作踏板;操纵踏板;控制踏板
operating performance 运行性能;营业实绩;使用性能;操作手续
operating performance income statement 营业实绩收益表
operating performancer 工作性能
operating performance ratio 营业实绩比率

operating period 运行期;运行(持续)时间;经营期;操作周期
operating personnel 运行人员;营业人员;维护人员;生产人员;操作人员
operating phase 运行阶段;操作期
operating (pipe)line 运行管线;主管路
operating piston 控制活塞
operating pitch radius 工作节圆半径
operating plan 营业计划;生产计划;操作方案
operating platform 控制台;工作台;控制台;操作平台
operating point 运行点;工作点;操作点;操作地点
operating policy 运行政策;调度原则
operating position 工作岗位;吸合位置;工作位置;操作位置
operating post 工作岗位
operating power 运营功率;工作功率
operating power supply 工作电源
operating practice 工作方式;操作实践
operating pressure 运行压力;工作压力;实际压力;操作压力
operating pressure angle 实际压力角
operating pressure cycle 操作压力循环
operating pressure head 工作压力
operating principle 操作原理;工作原理
operating procedure 运用程序;运算程序;工作方法;调度程序;操作过程;操作程序;操作步骤
operating procedure control 操作程序控制
operating procedure of inbound train 到达解体列车技术作业过程
operating procedure of outbound train 始发编组列车技术作业过程
operating procedure of taking-out placing-in of local car 本站作业车的取送技术作业过程
operating procedure of transit train without resorting 无调中转列车技术作业过程
operating procedure of transit train with resorting 有调中转列车技术作业过程
operating process 操作过程
operating-process chart 操作过程图表
operating profit 运营利润;营运利润;营业利润;经营利润
operating program(me) 操作程序;运算程序;营运计划;作业计划;工作计划;经营计划方案
operating program(me) aids-to-navigation 航标作业计划
operating program(me) tape 操作程序带
operating provisions 企业规章;操作规程;运行规定;操作规定
operating pump 操作泵
operating radius 作用半径;工作半径
operating railway 运营铁路
operating range 运转范围;运载范围;有效距离;作用距离;工作距离;工作间隔;工作范围
operating rate 运行速率;运算速率;工作速率;工作速度;开工率;设备生产能力
operating rate of filter 滤池滤率
operating ratio 运行率;营业实绩比率;经营比率
operating ratio based upon revenues 按收入总额计算的营业费用率
operating receipt 营业收入
operating recommendation 操作建议
operating record 运行记录;操作记录
operating regulations 运行规章条例
operating reliability 运行可靠性;工作可靠性;操作可靠性
operating repair 运行期检修;日常修理;日常维修;日常维护检修
operating report 运行报告
operating requirement 运行要求;使用要求
operating reserve capacity 运行备用容量
operating reserves 开工准备
operating result 营业结果;运行结果
operating revenue 营业收入
operating rights of railway bureau 铁路局经营权
operating ring 工作环;控制环
operating risk 经营风险
operating rock arm 操纵摇臂
operating rod 操纵杆
operating rod bearing 操纵杆座
operating rod for driving a switch 动作连接杆
operating rod lever 操作杆臂
operating rod spring 复进簧
operating room 控制室;作业间;作业室;工作室;

手术室;操纵室;报房
operating rope 牵引索;起重索
operating rule 运行规程;营运章程;作业规则;操作规则;运行规则;操作规程
operating rule curve 运行曲线;调度曲线;(水库)操作规程曲线
operating rule curve of reservoir 水库运作曲线;水库运用曲线;水库调度曲线
operating safety 安全操作
operating schedule 操作程序;调度程序;操作顺序;运用程序;运行时间表;运行计划;运行表;作业计划;工作时间表;调度程序
operating scheme 作业时间表;操作图;操作计划;操作程序;运行规划
operating sensitivity 工作灵敏度
operating sequence 运行顺序;操作顺序
operating sequence diagram 操作顺序图
operating shaft 运转轴;操纵轴
operating shaft arm 运转轴臂
operating shock 工作振动
operating signal lamp 运行信号灯
operating situation 运转情况;运行情况
operating software development 操作软件开发
operating solenoid 操纵电磁铁
operating space 工作空间;工作范围;操作区;操作空间;操作空白
operating specification 使用规程;使用说明(书);操作说明书;操作规程
operating speed 运营速度;运行速度;运行车速;运算速度;工作速度
operating speed range 使用速度范围;运转速度范围
operating spider 操作十字架
operating spring 操作簧
operating staff 作业人员;操作人员
operating standard 运行规程;操作标准
operating state 操作状态
operating statement 营业损益表;经营报表;营运报告;营业表;营业报告;经营情况表;损益表
operating state value 工作状态值
operating station 工作站;操作站;营业站
operating statistics 经营统计
operating status 工况;操作状态
operating steam 运行蒸汽
operating storage 操作储备;运行储备;活库容;运行储[贮]存容;有效库容;工作库容
operating storage allocation 运行储存分配
operating stress 工作应力;使用应力
operating stroke 工作行程
operating strut 操纵杆
operating supervisor 操作控制器
operating supplies 业务供应品
operating surplus 营业盈余
operating switch 工作开关;操作开关
operating symbol 操作符(号)
operating system 运行系统;运算系统;作业系统;执行系统;操作系统
operating system algorithm 操作系统算法
operating system characteristic 操作系统特性
operating system component 操作系统的组成部分;操作系统成分
operating system configuration 操作系统情况;操作系统配置;操作系统结构;操作系统方式
operating system dynamics 控制系统动力学
operating system family 操作系统系列
operating system file 操作系统文件
operating system generation 操作系统生成
operating system kernel 操作系统核心
operating system keyboard command 操作系统键盘命令
operating system management 操作系统管理
operating system master 操作系统总体
operating system monitor 操作系统监控程序;操作系统监督程序
operating system nucleus 操作系统内核
operating system overhead 操作系统开销
operating system program(me) 操作系统程序
operating system software 操作系统软件
operating system supervisor 操作系统管理程序
operating system text editor 操作系统文本编辑系统
operating system time-sharing 操作系统分时
operating table 手术台;操作台
operating tank 操作罐

operating team 工作班
operating technical statement 生产技术说明
operating temperature 工作温度;操作温度
operating temperature range 工作温度范围
operating tension 工作电压
operating test 运行试验;操作试验
operating theater 手术(示教)室
operating time 运转时间;运营时间;运算时间;作业时间;使用寿命;动作时间;操作时间
operating time measurement method 操作时间测定法
operating time of dredge(r) 挖泥船运转时间
operating time of protection 保护动作时间
operating time ratio 运转时间比
operating track 厂内轨道
operating transducer 操纵传感器
operating tube plug 操作管塞
operating tube shield 操作管套
operating unbalance 工作不平衡
operating unit 运行机构;业务单位;管理部门;经营机构;调节机构;操作装置;操作单位
operating unit sealing load 工作密封比压
operating valve 控制阀;操作阀;操纵阀
operating valve body 操作阀体
operating valve body stud 操作阀体柱螺栓
operating valve cap 操作阀盖
operating valve gasket 操作阀垫圈
operating valve lever 操作阀杆
operating valve spring 操作阀簧
operating valve stem jaw 操作阀柄爪
operating variable 操作变量
operating velocity 操作速度
operating voltage 运行电压;工作电压;使用电压;操作电压
operating voltage indicator 工作电压指示器
operating waste 运用损失
operating water 操作中用水
operating water depth 工作深度【水文】
operating water level 工作水位;运行水位;操作水位
operating water nozzle 水喷嘴
operating wavelength 工作波长
operating weight 运转重量;运行重量;工作重量;使用重量
operating width 作业面宽度
operating workload 运营工作量
operating yard 操作场地
operatinng availability 可能利用率
operation 运转;运营;运行;运算;作业;开工;操作
operation according to regulation 按章操作
operation-address register 运算地址寄存器
operation against regulations 违章操作
operation against rules 违章操作
operation agreement 运行协定
operational 工作的
operational acceptance test 实用验收试验
operational accounting and analysis system 经营会计分析制
operational accuracy 有效精度
operational action 业务活动
operational activity 业务活动
operational address instruction 操作地址指令;操作地址指令;功能地址指令
operational address register 操作地址寄存器
operational advantage 操作利益
operational aid 运行器材
operational altitude 有效高度;工作高度
operational amplifier 运算放大器
operational amplifier block 运算放大器部件
operational analysis 运行分析;运算分析;运筹学;操作分析;运算微积
operational and control station 操作与控制站
operational and equipment manual 操作与设备手册
operational and maintenance 操作和维修
operational and maintenance manual 操作与保养手册;操作和维修手册
operational announcing system 操作通知系统
operational approach 业务方法
operational area 工作面;作业面积;运作地区;工作地区;运行区
operational areas of airports 机场使用范围
operational aspect 运行情况
operational assistance 业务协助
operational attribute 操作属性

operational audit 业务审计
operational auditing 系统审计
operational base 现用基地;工作基地
operational behavio(u)r 运转状况
operational blank 操作空格
operational board 操作仪表板;操作盘
operational brake 工作制动器
operational budgeting 业务预算编制
operational buffer 操作缓冲区
operational building 操纵间
operational buoys for ocean environment 海洋环境作业浮标
operational calculus 运算微积分;运算微积;算子演算
operational calculus method 运算微积法
operational call sign 现用呼号
operational capability 运算能力
operational capacity 运转能力;营业量;工作能力;工作量;经营能力;操作能力
operational capital 营运资本
operational card 操作卡(片)
operational ceiling 实用上升限度
operational center 运行中心
operational change 操作变化
operational change over 作业调换;作业转换
operational character 运算符;操作字符;操作符;控制符(号)
operational characteristic 工艺特性;运行特性
operational chart 操作示意图
operational check 运行控制;运行检查;功能检验
operational checkout 作业检查
operational circuit 运算电路
operational code 操作码
operational code table 操作码表
operational command 操作指令;运算指令
operational commutator 操作分配器
operational computer 程序计算机
operational concept 运营方案
operational condition 运行条件;营运条件;作业条件
operational conditioning 运行调节
operational contract 运行合同
operational control 运行控制;业务管理;操作控制器;操作控制;业务控制;经营控制
operational control center of COMSAT 通信卫星操作控制中心
operational control key 操作控制键
operational control panel 操作控制板
operational control switch 操作控制开关
operational control system 操作控制系统
operational control unit 操作控制装置;操作控制设备;操作控制单元
operational control valve 操作控制阀
operational corrosion control 腐蚀控制管理
operational cost 运营成本;经营成本
operational cost control 作业成本控制;经营成本管理
operational counter 操作计数器
operational coverage diagram 探测范围图
operational crew 工作班;操作人员
operational criterion 运算准则
operational curve 运行曲线
operational cycle 操作周期;作业周期
operational data 运算数据;营运数据;操作数据;运行资料;工作数据
operational decision making 业务决策
operational decoder 操作译码器;操作解码器
operational defect 运行故障
operational delay 运转迟滞;交通延误
operational dependability 运转安全度
operational design analysis 运算设计与分析
operational design and analysis 操作设计与分析
operational desk 操作台;操纵台
operational development 产品改进性研制;运行改进
operational differential amplifier 运算差分放大器
operational difficulties 运用困难
operational difficulty 使用上的困难
operational digital part 操作数码部分
operational digital technique 数字运算技术
operational distribution panel 作业配电板
operational earning 经营收入
operational econometric model 可供使用经济计量模型;可操作经济计量模型

operational efficiency 运行效率
operational environment 操作环境
operational environment satellite 环境实用卫星
operational ephemeris 工作星历
operational equation 运算方程式
operational equipment 运行设备;运行备件
operational evaluation 运行评价;使用评价
operational exception 操作异常;操作事故;操作例外
operational expenditures 营运支出;业务费用;经营费
operational experience 运行经验
operational expression 操作表达式
operational factor 操作因素;运行参数;作用参数
operational failure 运转故障
operational feasibility 实施可能性
operational field 操作字段;操作位;作业场所
operational flexibility 运行机动性;操作适应性;操作灵活性;运行灵活性
operational flowchart 操作流程图;操作程序图;运算流程图
operational flow sheet 操作顺序图
operational forecasting 运行预测;作业预报
operational form 运算形式
operational formula 运算公式
operational free board 运行超高
operational frequency 运用频率
operational function 营运职能
operational game 经营计划
operational gaming 运行方法;运算对策;操作规则
operational grounding 操作接地
operational ground support equipment 现用的地面支援设备
operational guidance 运行管理
operational guide computer 操作指导计算机
operational hazard report 运转事故报告
operational height 工作高度;可达高度;射击高度;额定高度
operational hydrology 运转水文学;操作水文学;作业水文学
operational ice criterion 通行冰况标准;通航冰况标准
operational impedance 运算阻抗
operational improvement 操作改善
operational incident 操作事故
operational indicator 业务指标;操作指示器
operational information system 操作信息系统
operational inspection 操作检查
operational inspection log 操作检查记录
operational instructions 操作规程
operational integrater 运算积分器
operational key 操作键;工作键;电极电键
operational keyboard 操作键盘
operational knocks 操作杂音
operational label 操作标记
operational level 工作量规
operational lever 操作手把
operational lever bolt 操作杆螺栓
operational life 使用年限;服役寿命
operational life of bridge 桥梁使用年限
operational limit 使用公差
operational line 操作线
operational load 操作荷载
operational log 运行记录;操作记录;钻孔记录
operationally technical assistance 操作技术指导
operational maintenance 日常保养
operational management 业务管理
operational manager 操作管理员
operational manual 操作指南;操作说明书;操作手册
operational mathmatics 运算数学
operational mechanism 运行机构
operational mechanism of economy 经济运行机制
operational message 操作信息
operational meteorological satellite 业务气象卫星
operational meteorological satellite system 现用气象卫星系统
operational method 运算方法;操作方法
operational mile 运行哩油量
operational mode 操作方式;运转模
operational mode selection button 操作方式选择按钮
operational monitoring 操作监测;运行监测
operational multiplier 运算乘法器

operational navigation chart 作战领航图
operational note 操作记录
operational notice 操作标志
operational number 操作号码
operational of vehicle 车辆运转
operational order 操作指令;操作次序;运算指令
operational order of evaluation 求值的运算顺序
operational panel 操作板
operational parameter 运行参数;作用参数;工作参数
operational part 操作码部分;操作部件;操作部分
operational pavement system 路面设计运筹体系
operational payload 运行有效载重
operational performance 操作性能;运转性能;运算性能;使用特性
operational period 运行周期;操作周期;作业周期;工作周期
operational phase 操作阶段
operational pipeline 操作流水线
operational plan(ning) 操作计划;业务计划;作业性规划;运行计划;作业计划;经营计划;经营措施
operational planning system 操作计划系统;作业计划制度
operational policy 经营政策;营运方针;经营方针
operational port 工作基地
operational power supply 操作电源
operational precision 运算精度
operational problem 运算问题;操作问题
operational procedure 作业方法;操作手续;操作程序;操作步骤;实施程序
operational process 运行过程;运算工序
operational process chart 操作工序图表;操作程序图
operational program(me) 作业计划
operational project 业务项目;施行中的项目
operational proof cycle 运转保证周期
operational radiation instrumentation equipment 现用放射性仪器设备
operational range 航空基地管辖范围;活动距离;活动范围;工作范围
operational rate 业务汇率
operational readiness 操作度;作业待命
operational readiness test 运转准备试验
operational record 运行记录
operational recorder 操作记录器
operational reduction factor 操作原因缩减系数
operational register 操作码寄存器;工作寄存器
operational regulations 操作规程
operational relay 操作继电器;工作继电器
operational reliability 运转可靠性;操作可靠性;运行可靠性;工作可靠性
operational report 经营报告
operational requirement 作战要求;使用要求;操作要求
operational research 运用研究;运筹学;工作研究
operational reserves 业务准备;业务储备金
operational resistance 工作运行承受力
operational resource inventory system 现用资源调查系统
operational room 操作室
operational rule 操作规则;操作规则;运算律
operational safety 运行安全;操作安全;使用可靠
operational safety equipment 作业安全设备
operational schedule 操作程序表;操作程序
operational scheme 运作流程;经营计划;操作方案;操作程序
operational semantics 操作语义
operational sequence 运行序列;运算程序;工作顺序;工序;操作工序;操作程序
operational sequence analysis 工作顺序分析;操作顺序分析
operational sequence control 操作序列控制;操作工序控制;操作程序控制
operational sequence diagram 操作程序图
operational sheet 运行流程;操作说明书;操作卡(片);操作单
operational shock 运行振动
operational signal 操作信号
operational site 运行位置;运行场所;工作场所
operational site staff 现场运行人员
operational situation 运转情况;运行情况
operational slant range 实际斜距离
operational specifications 操作规程
operational speed 运转速度;运算速度

operational speed of a computer 计算机的运算速度
operational sphericity 运算球度
operational stability 操作稳定性
operational stack register 操作栈寄存器
operational staff 工作职员
operational stage 运行阶段
operational standard 操作规范;操作标准
operational stand-by program(me) 备用操作程序
operational steps 操作步骤
operational stock 运行库存
operational stop(ping) point 运营停车点
operational storage 运行储备
operational store 运行库存;操作存储器
operational stowage clause 甲板装载选择权条例
operational stripping ratio 生产剥采比
operational stroke 工作行程
operational suitability 操作适用性
operational suitability test 运转适应性试验;适用性试验
operational supervisor 操作主管人;操作管理程序
operational supply 给水管理
operational support equipment 实用支援装备
operational switch 操作开关
operational symbol 运算符(号);操作符号;操作符(号)
operational system 运行系统;操作系统
operational system function 操作系统功能
operational system matrix 运算系统矩阵
operational systems model 运算体制模式
operational table 操作表
operational tank 操作罐
operational task 操作任务
operational taxonomic unit 运算分类单位
operational technician 操作技术(人)员
operational technique 操作技术
operational technology 运行技术
operational temperature 操作温度
operational test 运转试验;操作试验;适用性试验;实用试验
operational testing equipment 运行试验设备
operational test set 操作测试设备;工作状态测试设备;不停机测试设备
operational theory school 经营管理理论学派
operational throughput 运输量
operational time 操作时间
operational ton 操作吨(指装卸);操作量(指装卸)
operational tools 运算工具
operational transfer vector 操作传送向量
operational trial 业务测试【计】
operational trouble 运转故障;运行故障;运行事故;营运故障
operational under surcharged conditions 超负荷运转;超负荷运行
operational unit 运算元件;操作部件
operational use 业务使用
operational use time 有效使用时间;有效时间
operational version 使用方案
operational voltage 工作电压
operational waste 运用期弃水;运行期弃水;使用期弃水
operational water level 工作水位
operational water supply 给水管理
operational weight of equipment 设备操作重量
operational word 操作字
operational work 操作
operational zone 可使用面积;作战区域
operation analysis 作业分析;经营分析
operation and maintenance 使用与维护
operation and maintenance cost 运转和维护费用;运行维护费
operation and maintenance department 作业与维修部门
operation and maintenance manuals 使用和维修手册
operation and maintenance of heating network 热网运行管理
operation and management 经营管理
operation area 运行区
operation assistant 业务助理
operation at full load 满负荷运行
operation at part load 部分负荷运行
operation board 控制盘;操作盘;控制仪表
operation building 操作间

operation charges of arriving 到达作业费
operation charges of start off 始发作业费
operation check 操作校验
operation circuit 工作电路;运算电路;控制线路
operation cluster 仪表盘;仪表板
operation code 运算码;命令码
operation command 工作指令
operation control center equipment 运营控制中心设备
operation control center 运营控制中心;运行控制中心;运行管理中心
operation control key 运算控制键
operation controller 运算控制器
operation control system 操作控制系统;运算控制系统
operation control unit 运算控制设备
operation cost 营运费(用);营运成本;业务费用;作业成本
operation cost system 工作成本制度;经营成本制度
operation curve diagram 运行曲线图
operation cycle 运转周期;作业循环;工作循环
operation cycle time 作业循环时间
operation decision-making 经营决策
operation declaration 运算说明
operation decoder 运算译码器
operation demand 运转需求量
operation device 控制设备
operation diagram 运行曲线图
operation efficiency 工作效率
operation evaluation department 业务评议局
operation expenses 运营费(用);营运费(用);营运成本;事业费
operation facility 营运设施
operation field 指令字段
operation flow 业务流程
operation flow chart 经营流程图
operation forecasting and object programming 经营预测与目标规划
operation function 业务职能
operation funds 营运资金
operation gaming 经营对策
operation income 营运收入
operation in full capacity 达产期
operation in storage-yard 货场作业
operation instruction 操作指令;运行操作规范;生产指导
operation in tandem 相继操作;连贯操作
operation in the pit 井下作业
operation in warehouse 库内作业
operation irregularity 工作事故
operation job card 工件卡;派工卡;施工卡
operation kilometrage 运行里程
operation line of absorption process 吸收操作线
operation load(ing) 运行荷载
operation location 经营位置
operation management 运营管理;作业管理
operation manager 业务经理
operation map 作战图
operation material 工作物质
operation method expressed in graphs 图上作业法
operation mileage 运行里程
operation mode 运营模式;运营方式;运算方式;工作方式
operation mode selection button 运行方式选择按钮;运算方式选择按钮
operation mode selector 操作方式选择器
operation modes in autopilot 自动舵操作方式
operation navigation chart 战术航空图
operation notice 运转标志
operation number 作业编号
operation of aeration tank 曝气池运转
operation of appropriation 拨款程序
operation of a project 项目进行情况
operation of business 经营企业
operation of capital 资金的运用
operation office for train receiving-departure 运转室【铁】
operation of flo(a)tation 浮选作业
operation of fund 资金运作
operation of law 法律的应用
operation of numeric(al) method 数值法演题步骤
operation of precision tool 精密仪器的操作;精密仪器操作

operation of relay 继电器动作
operation of rolling 轧制过程
operation of treaty 条约实施情况
operation on borrowings 负债经营
operation on open sea 公海作业
operation on sea 海上作业
operation opening 作业窗;工作孔
operation organization 运营组织
operation panel 控制仪表
operation parameter 使用参数
operation part 指令操作码;指令部分
operation period 运营周期
operation plan for a single ship 单船作业计划
operation planning office 运营策划室
operation planning system 经营计划系统;业务计划体系
operation plan of aids-to-navigation 航标作业计划
operation plan of railway construction enterprise 铁路施工企业的经营计划
operation pressure 动水压线
operation procedure 作业程序
operation process chart 作业流程图;工作过程表
operation profit 营业利润
operation rack 作业线
operation rate 设备作业率;设备运转率
operation ratio 运行率
operation record 运转记录;工作记录;工作程序表;加工记录
operation recorder 工作记录器
operation regulation 运行调节
operation requirement 运转要求;使用要求
operation research 运筹学;作业研究
operation research model 运筹学模型
operation research techniques 运筹学方法
operation revenue 运输收入
operation right of transport within an administrative district 管内运输经营权
operation room 操作室;控制室;手术室
operations 业务营运
operations and checkout 操作与校正
operations area 操作台
operations building 车间;厂房;运行大楼
operations card 工序卡
operations center 运行中心
operation schedule 运行时间表;运行步骤;作业进度表;作业计划表;作业程序;工作进度(表)
operation scheme 经营组织;经营计划;运转计划;营运计划
operation sequence control 运算顺序控制;运算程序控制
operation sequence plan 工序计划
operations flowchart 操作程序图
operation sheet 工艺规程;施工说明书;使用说明书;说明书
operation shock 工作振动
operation sign 行车标志
operations improvement 操作改进
operations per minute 每分钟操作次数
operations record 工序卡
operations research 运筹学
operations sheet 工序卡
operation standard 作业标准
operation statement 损益
operation step 控制步骤
operations tower 指挥台
operation switch 操作开关;工作开关
operation symbol 运算符号
operation system 运算系统;业务系统;控制系统
operation test(ing) 运行试验;运转测试
operation time 运转时间;运行时间;运算时间;作业时间
operation time-limit system 限时工作方式
operation time speed 运算速度
operation token 运算标志
operation transform pair 运算变换对
operation twist 经营方法改变;经营办法
operation type 作业形式;作业分类
operation under negative pressure 负压操作
operation under positive pressure 正压操作
operation ventilation 运营通风
operation waste 运转弃水;运转废水
operation work indices of passenger station 客运站行车工作指标
operation work plan of passenger station 客运站行车工作计划
operative 运算的;工作的;动作的;操作的
operative cable credit 可凭使用信用证
operative condition 运转状态;工作状态;工作情况
operative constraint 有效约束
operative decision 业务决策
operative glide direction 作用滑移方向
operative glide plane 作用滑移面
operative installations 使用装置
operative plan 作业计划
operative property 运营的房地产
operative room 手术室
operative route 运行线路
operative sheet 工作图板
operative technology 实用技术
operative temperature 作用温度
operative weldability 工艺可焊性;操作工艺性
operator 运算符;作业员;装配工;执行人;执行机构;工作人员;经纪人;驾驶员;机务员;算子;算符;测图员;操作者;操作员;操作人员;操作工;操作符;操纵者
operator A 操作员甲
operator aid 操作者的辅助工具;操作者的助手
operator algebras 算子代数
operator-applied occurrence 算符应用性出现
operator-assisted call 操作员协调的呼叫
operator authorization record 操作员特殊记录;操作员确认记录
operator cab 司机室
operator class 算符分类
operator command 操作指令;操作员命令
operator command language 操作员命令语言
operator communication 操作员通信
operator communication control facility 操作员通信控制设备
operator communication manager 操作员通信管理程序
operator communication message 操作员通信信息
operator console 操作员控制台
operator console facility 操作员控制台设备;操作员控制台程序
operator control 操作员控制
operator control address vector table 操作员控制地址向量表
operator control command 操作员控制命令
operator control console 操作员控制台
operator control element 操作员控制单元
operator control function 操作员控制功能
operator control language 操作员控制语言
operator control panel 操作员控制台面板;操作员控制面板
operator control station 操作员控制站
operator control system 操作员控制系统
operator control table 操作员控制表
operator-defined occurrence 算符定义性出现
operator delay 操作者延误
operator digital part 操作数码部分
operator display 售票员显示
operator distance dialing 长途话员拨号
operator domain 算子域
operator dominance 算子优势
operator endomorphism 算子自同态
operator equation 算子方程
operator equation of state 状态算符方程
operator error 操作者误差;操作员错误
operator exception code 运算异常代码
operator external interrupt processor 操作员外中断处理程序
operator fatigue 操作者疲劳
operator function 算子函数
operator grammar 运算符文法;算符文法
operator group 算子群
operator guidance 操作控制
operator guidance code 操作员引导码;操作员导向码
operator guidance indicator 操作员引导指示灯
operator guide 操作指南
operator-held control box 遥控板
operator hierarchy 算术运算符优先权
operator homomorphism 算子同态
operator identifier 操作员识别码;操作员标识符
operator in charge 值班人员;领班人员
operator indicator 操作台指示灯

operator information area 操作员信息区
operator interaction 操作员与计算机的交互作用；操作员交互作用
operator intercept 操作员截取
operator interface control block 操作员接口控制块
operator interface equipment 操作员接口设备
operator interruption code 运算中断码
operator intervention 操作员干预
operator intervention section 操作员干预部分
operator isomorphism 算子同构
operatorless 无操作者；自动操作
operator license 司机执照
operator logical paging 操作员逻辑调页
operator logic paging 操作员逻辑分页
operator matrix 算子矩阵
operator message 操作者信息；操作员信息；操作信息
operator monitor 操作员监视
operator-monitor communication 操作员-监控程序间通信
operator notation 算子符号
operator on duty 值班人员
operator over-sight 操作员监督
operator panel 操作员屏
operator part 操作码部分；操作符部分
operator platform 操作者控制台；操作者平台
operator potential 算子位势
operator precedence 算符优先（数）
operator precedence case 算符优先情况
operator precedence grammar 算符优先文法
operator precedence matrix 算符优先矩阵
operator precedence method 算符优先法
operator precedence parser 算符优先分析程序
operator precedence system 算符优先方式
operator priority order matrix 算符优先顺序矩阵
operator pseudomonotone 算子伪单调的
operator register 算符寄存器；操作数寄存器
operator response field 操作员应答字段；操作员响应字段
operator ring 算子环
operator routine 运算程序
operator's access code 操作员存取码
operator's brilliance control 使用者亮度旋钮
operator's cab 值机室；驾驶间
operator's cabin 驾驶室（塔吊、起重机）；驾驶间（塔吊、起重机）；操作室
operator schema 操作模式
operator's circuit 座席电路
operator's compartment 驾驶室
operator's computer 调度员集中计算机
operator's console 操作员控制台；操作控制台；操作（人员）控制台
operator seat 司机座
operator service 接线员服务；操作员业务；操作员服务
operator set-up time 操作准备时间
operator's handbook 使用说明（书）；操作手册
operator's license 司机执照；驾驶人执照
operator's panel 机控制板
operator's platform 操作平台
operator's position 工作位置
operator's pulpit 操纵台
operator's station 操作员控制台
operator stack 运算栈；操作栈；操作符栈
operator station 操作员站
operator station task 操作员站任务
operator's terminal 接线员终端
operator swivel seat 操作人员的旋转座位
operator table 算符表；操作符表
operator theory 算子理论
operator utilization 工时利用率
operator-valued distribution 算子广义函数
operator-valued function 算子函数
operator verification table 算符校验表
operator vision 操作人员的视野
operator wave function 算子波函数
operatory 工作室
opereating specification 操作范围
operon 操纵子
operon network 操纵子网
operon theory 操纵子学说
oper-type bearing 非密封轴承
o-phenanthroline spectrophotometry 邻菲咯啉分光光度法

o-phenylethylenediamide 邻苯乙二胺
o-phenyl phenol 邻苯基苯酚
ophicalcite 蛇纹大理岩；蛇纹方解石；蛇纹白云石
ophidian 似蛇的
ophimottling 辉绿斑状结构
ophiolite 蛇纹岩；蛇绿岩
ophiolite complex 蛇绿岩套
ophiolite emplacement 蛇绿岩侵位
ophiolite suite 蛇绿岩套
ophiolitic eclogite 蛇绿榴辉岩
ophiolitic melange 蛇绿混杂体
ophite 纤闪辉绿岩
ophitic 辉绿状的；辉绿岩结构的
ophitic texture 辉绿结构
ophthalmic chair 检眼椅
ophthalmic lens 镜片
ophthalmic room 眼科室
ophthalmoleukoscope 旋光色觉镜；色觉检测计
ophthalmology 眼科
ophthalmophacometer 晶状体屈光计；晶状体测光计
ophthalmophasmatoscopy 分光眼镜检查
Opie para dox 奥皮氏奇异现象
opinion book 信用调查录；意见书
opinion polls 民意测验
opinion rating 评价等级
opisometer 量图器；量距仪（地图等）；计图器；曲线计；曲线长度仪（图上量长度用）
opisthaptor 后粘器
opisthodomos 后室；（在古庙神坛尾部的）内门廊；后厅（古典寺庙后进的小室）
opium 寄生群落；阿片
opium poisoning 阿片中毒
opiums poisoning 阿片类中毒
opli on market 期货买卖
opoka 蛋白土
Oppel zone 奥佩尔带
Oppenauer oxidation 奥本瑙尔氧化反应
Oppenheimer-Phillips reaction 奥本海默-菲利普斯反应
Oppenheimer-Volkoff limit 奥本海默-沃尔科夫极限
Oppermann wicket (dam) 奥本曼活动撑板坝；奥本曼旋转桁架木板坝
Opperman source 奥本曼光源
oppidan 城市居民；走读生；城市居住
opponent 竞争对手；竞争厂家；对手
opponent-colo(u)rs theory 对立色理论
opponent firm 竞争同行
opponent vessel 对方船
opportune moment 适当的时机
opportunist 机会致病菌
opportunistic cost 机会成本
opportunistic species 机会物种
opportunity 机遇；机会；时机
opportunity cost 机会费用；机会成本；替代成本
opportunity cost method 机会成本方法
opportunity cost of value 机会成本值
opportunity evapo(u)ration 可能蒸发（量）
opportunity for advancement 晋升机会
opportunity for other contractors 为其他承包商提供机会
opportunity industry 机会产业
opportunity map of liquefaction 液化机会图
opportunity model 机会模型
opportunity study 机会研究；投资机会研究
opportunity task 计划外任务
opportunity time 地面水的可能渗灌时间
opportunity value 机会价值
opposed 对面的
opposed alignment 对向定线
opposed anvil 对面顶
opposed anvil apparatus 对面顶装置
opposed anvil type 对顶砧式
opposed-balanced reciprocating compressor 对称平衡往复式压缩机
opposed-blade damper 对开叶片式气闸；反向板条缓冲器；对置叶片风门
opposed complex nappe 反倾复合推复体
opposed compressor 对置式压缩机
opposed current 逆流
opposed cylinder 对置汽缸
opposed-cylinder (type) engine 汽缸对置（式）发动机；对置汽缸（式）发动机
opposed-engine 对置式发动机

opposed firing 双冲燃烧
opposed flight auger 对送式螺旋
opposed jet element 对冲型射流元件
opposed left turn 对向左转
opposed left turn equivalent 对向左转当量
opposed movement 对向运行
opposed multi-blade damper 对开式多叶阀
opposed piston 双动活塞；对置活塞
opposed-piston diesel 对置活塞柴油机
opposed-piston diesel engine 双动活塞柴油机
opposed-piston engine 对置活塞发动机
opposed-piston pump 对置活塞泵
opposed piston type engine 对置活塞式发动机
opposed-plunger pump 对置柱塞泵
opposed ram steering gear 对置活塞式电动液压舵机
opposed shutter 双遮光器快门；反向快门
opposed turn 对向转弯
opposed-voltage protective system 起动保护系统；对消电压保护系统
opposing coil 反作用线圈；反接线圈
opposing current 逆流；对向车流
opposing face 对置面
opposing flow 反向流
opposing lane 反向车道
opposing movement 本地运行
opposing phase 对向相位
opposing reaction 相反作用；对抗反应
opposing spit 逆流沙嘴
opposing spring 反作用弹簧；反抗弹簧
opposing torque 反抗转矩
opposing traffic 对向交通
opposing value 相反价值；抵消价值
opposing vehicle 对向车辆
opposing water mass 振荡的水体；交锋水体
opposing wave 反向波
opposing wind 迎浪风；逆风
opposing winding 反作用绕组
opposite 相反的；对面的；对立的
opposite acting force 反向作用力
opposite angle distance 对角距
opposite angles 对角；对顶角；对等角
opposite angles ventilation 对角式通风（法）
opposite arc 反向弧
opposite bank 对岸距离；对岸
opposite blade damper 反向叶片式风门；反向叶片式挡板
opposite change 对换
opposite circular polarization 反向圆偏振
opposite colo(u)r 相反色
opposite compression 反向压力
opposite course 相反航向
opposite crank 对置曲柄
opposite direction 相对方向；敌对方向；反（方）向
opposite electricity 异性电
opposite faces 对拉面
opposite field 反向场
opposite folden plate hybrid anaerobic reactor 厌氧异波折复合式反应器
opposite folding 相反折弯；对折
opposite fold limbs 反向褶皱翼
opposite force 对向力；反向力
opposite gradient 反坡线；反坡（降）
opposite hand 对面
opposite in sign 符号相反的
opposite joint 相对式接头【铁】；相对式对接【铁】；正对接合；对接；对缝
opposite leaves 对生叶
oppositely directed photon 反向传播光子；反方向传播光子
oppositely directed travel(l)ing wave 反向行波
oppositely equal ranges 反向等点列
opposite movement 对向运行
opposite number 对应物
opposite orientation 反定向
opposite parity 反宇称性；反奇偶性
opposite phyllotaxis 对生叶序
opposite pitting 对列纹孔式
opposite planting 对植
opposite polarity 异极性；反极性
opposite-polarity noise 异极性噪声
opposite pole 异极性（晶体）；不同名极
opposite potential method 反电位法
opposite pressure 反向压力

opposite projectivity 反向射影
opposite route 对向进路
opposite route with overlapped section 对向重叠进路
opposite scenery 对景
opposite sense 反指向
opposite sequence 逆序列
opposite side 相对侧;另一侧;对面;对方;对侧;对边
opposite sign 异号;反号
opposite spin 反向自旋;反方向旋转
opposite tide 逆潮(流)
opposite transition 反向跃迁
opposite-type solute 相反溶质
opposite vertex 对顶
opposition 对生;对立;抵抗;反接
opposition duplex 反双工
opposition method 反接法
opposition of moon 月冲
opposition test 背靠背试验
opposition to a risky policy 反冒进
opposition to rash advance 反冒进
oppositi-petalous 对瓣的
oppress 重压
oppressed 被压的
oppression weather 沉闷天气
oppressive weather 闷热天气
oppsing field 反向场
oppsite pressure 反压力
opsearch 运筹学
opsonic action 调理作用
opsonic index 调理指数
opsonization 调理作用
optic 镜片
optic(al) 旋光;光学的
optic(al) aberration 光学像差;像差
optic(al) absorbing property 光吸收特性
optic(al) absorption 光学吸收;光的吸收(作用)
optic(al) absorption band 光吸收带
optic(al) absorption edge 光学吸收限;光吸收限
optic(al) access coupler 光学取数耦合器
optic(al) access network 光接入网
optic(al) accessory 光学附件
optic(al) achromatism 可见光消色差性
optic(al) acquisition 光学探测;光学捕获
optic(al) acquisition equipment 激光捕获装置
optic(al) active substance 光活性物质
optic(al) activity 光学活性;旋光性;旋光度
optic(al) activity of oil 石油的旋光性
optic(al) adaptive technique 光自适应技术;光学自适应技术
optic(al) admittance 光学导纳
optic(al) aid 光学辅助工具
optic(al) air mass 光学气团;光学空气质量;大气光学质量
optic(al) alignment 光学准直;光学装校;光学调准
optic(al) alignment equipment 光学校准设备
optic(al) alignment unit 光学校准设备
optic(al) allocation for pollutant load discharged 污染负荷排放优化分配
optic(al) altimeter 光学高度计;光学测高计;光测高度计
optic(al) amicron 光学次微(胶)粒
optic(al) ammeter 光学电流计;光学安培计
optic(al) amplification 光放大
optic(al) amplifier 光学放大器;光放大器;光电放大器
optic(al) amplifier bandwidth 光放大器带宽
optic(al) amplitude modulator 光振幅调制器;光调幅器
optic(al) analog 光学模拟
optic(al) analogue computer 光学模拟计算机
optic(al) analog(ue) device 光学模拟装置
optic(al) analog(ue) memory 光模拟存储器
optic(al) analog(ue) transform technique 光学模拟变换技术
optic(al) analysis 光学分析法;光学分析;光谱分析
optic(al) analytical instrument 光学分析仪器
optic(al) analyzer 光分析器;光分析器;光学检偏镜;光学检偏振器
optic(al) analyzer spectroanalyser 光谱分析仪
optic(al) and electro-optic(al) image processing equipment 光学及电子光学图像处理设备
optic(al) anemometer 光学风速计
optic(al) angle 光轴角;光角

optic(al) angle ga(u)ge 光学测角仪
optic(al) angular motion sensor 光学角运动传感器
optic(al) anisotropy 光学异向性;光学的各向异性
optic(al) anomaly 光学(性)反常;光性异常;光反常
optic(al) antenna 光学天线;光波段天线
optic(al) antenna angle 光学天线张角
optic(al) antenna aperture 光学天线孔径
optic(al) antenna gain 光学天线增益
optic(al) antipode 光对映体;旋光对映体
optic(al) apparatus 光学装置;光学仪器
optic(al) appearance 光学外观
optic(al) approach 用光学仪器接近
optic(al) arm 光学臂
optic(al) arrangement 光学装置
optic(al) art 光效应艺术
optic(al) artware 光学工艺品
optic(al) aspheric(al) surface 光学非球面
optic(al) astrometry 光学天体测量学
optic(al) astronomy 光学天文学
optic(al) astrophysics 光学天体物理学
optic(al) asymmetry 光学不对称
optic(al) attenuator 光学衰减器
optic(al) automatic ranging 光学自动测距
optic(al) automatic ranging finder 光学自动测距仪;光学自动测距计
optic(al) automatic titrimeter 光学自动滴定仪
optic(al) automatic tracer 光学自动跟踪器
optic(al) automatic tracking 光学自动跟踪
optic(al) avalanche laser 光学雪崩式激光器
optic(al) axial angle 光轴角
optic(al) axial figure 光轴图形
optic(al) axial interference figure 锥光干涉图
optic(al) axial plane 光轴平面
optic(al) axis 光轴;视轴
optic(al) axis of crystal 晶体光轴
optic(al) axis of lens 物镜主光轴
optic(al) axis of symmetry 光学对称轴
optic(al) baffle 光学隔板
optic(al) balance 光学天平;光学补偿
optic(al) band 光波段
optic(al) band gap 光禁带
optic(al) bandpass filter 光学带通滤光片;光带通滤光器
optic(al) bar 光学导杆
optic(al) barmonic generation 光学谐波振荡
optic(al) barrel 光学镜筒
optic(al) beacon 光指向标
optic(al) beam-direction control 光束方向控制
optic(al) beam expander 光束扩束器;光束扩展器
optic(al) beam flying 光束扫描
optic(al) beam riding 光波束引导
optic(al) beam scanner 光束扫描器
optic(al) beam splitter 光束分光镜
optic(al) behavior 光学特性;光的特性
optic(al) bench 光具座
optic(al) bevel protractor 光学斜度规;光学量角器
optic(al) binary 光学双星
optic(al) biosensor 光生物传感器
optic(al) bistability 光学双稳态
optic(al) black 光学黑体
optic(al) blacking 光学黑色涂料
optic(al) bleach 光漂白剂
optic(al) bleaching agent 荧光增白剂;荧光漂白剂;光漂白剂
optic(al) block 光学系统
optic(al) boroscope 光学缺陷探测仪
optic(al) branch 光频支
optic(al) branching device 光分路器
optic(al) branching filter 光分路滤波器
optic(al) bridge circuit 光桥路
optic(al) brightener 荧光增白剂;光学增亮剂;光学光亮剂
optic(al) brightness 光学亮度
optic(al) Brinell hardness tester 光学布氏硬度计
optic(al) bundle 光纤束
optic(al) cable 光缆;光导纤维电缆;光导电缆
optic(al) cable assembly 光缆组件
optic(al) cable connector 光缆连接器
optic(al) cable driver 光缆激励器
optic(al) cable sheath 光缆护层
optic(al) calculating machine 光计算机
optic(al) calculation 光学计算

optic(al) calibration 光学仪器检定;光学校准
optic(al) caliper 光学校准器;光学轮尺
optic(al) cardan link 光控万向关节;光学万向节
optic(al) card reader 光学读卡器
optic(al) carrier 光学载波
optic(al) cavity 光谐振腔
optic(al) cavity diode 光腔二极管
optic(al) cavity laser interferometer 光谐振激光干涉仪
optic(al) cavity mirror sensor 光腔镜式传感器
optic(al) cavity mode 光学共振腔模;光学共振腔模式
optic(al) cell 光学玻璃皿
optic(al) cement 光学胶合剂;光学胶;光胶
optic(al) centering and edging 光学定心磨边
optic(al) centering device 光学对中器
optic(al) center of lens 镜片光心
optic(al) center 光学中心;光心;视觉中枢
optic(al) centre of registering frame 像框光轴点
optic(al) ceramics 光学陶瓷
optic(al) change-over assembly 可更换光学组件
optic(al) channel 光信号通道;光信道;光通道;光波道
optic(al) channel-dropping filter 光通道衰减滤光器
optic(al) character 光学特性
optic(al) characteristics 光学特性
optic(al) character reader 光字符阅读机;光学字符读出器;光阅读机;光符号阅读器;光标阅读机
optic(al) character recognition 光学字符识别;光学阅读机;光学符号读出;光符(号)识别;光标识别
optic(al) character recognition common language 光学符号识别通用语言
optic(al) character recognition ink 光学文字识读油墨;红外(线)光油墨
optic(al) character recognition machine 光学字符辨识机;光学符号识别机;光符号识别机
optic(al) character recognition reader 光学字符阅读器
optic(al) character recognition system 光学字符识别系统;光符号识别系统
optic(al) character recognition teleprinter 光学字符识别电传打字机
optic(al) character scanner 光学字符扫描器
optic(al) chassis 光学装置底盘
optic(al) check 光学检验
optic(al) chiasma 视交叉
optic(al) chopper 光斩波器;光断续器
optic(al) cine duplicating machine 光学印片机
optic(al) circle 光学度盘
optic(al) circular table 光学圆分度台
optic(al) circulator 光环行器
optic(al) clinometer 光学倾斜仪;光学测斜仪
optic(al) clouration 光学着色
optic(al) coating 光学涂层;光学镀膜;光学镀层
optic(al) code reader 光学代码阅读器
optic(al) code reading wand 光码读数杆
optic(al) coherent radar 相干光雷达
optic(al) coincidence index 光重合比孔索引
optic(al) collector 聚光器;集光器
optic(al) collimator 光准准仪
optic(al) colo(u)r compositor 光学彩色合成仪
optic(al) colo(u)r filter 光学滤色玻璃
optic(al) colo(u)rimeter 光学色度计
optic(al) colo(u)rimetry 目视比色法
optic(al) combiner 光组合器
optic(al) communication 光学通信;光通信
optic(al) communication channel 光通信信道
optic(al) communication fiber 光通信纤维
optic(al) communication link 光通信线路
optic(al) communication receiver 光通信接收机
optic(al) communication system 光通信系统
optic(al) communication technique 光通信技术
optic(al) commutation circuit 光开关电路
optic(al) comparator 光学坐标量测仪;光学比色计;光学比较仪
optic(al) compensating system 光学补偿系统
optic(al) compensation 光学补偿
optic(al) compensation camera 光学补偿摄影机
optic(al) compensation plate 光学补偿板
optic(al) compensator 光学补偿器
optic(al) component 光学子星;光学元件;光学部件

optic(al) computer 光学计算机;光计算机
optic(al) computing 光学计算;光计算
optic(al) condenser 光学聚光器
optic(al) condition 光学条件
optic(al) condition of rectification 纠正光学条件
optic(al) conduction 光传导
optic(al) conductivity 光电导率
optic(al) conductor 光导体
optic(al) conductor loss 光导体损耗
optic(al) connector 光连接器
optic(al) constant 光学恒量;光学常数;光学常量
optic(al) contact 光学接触;光接触
optic(al) continuous wave reflectometer 光连续波反射计
optic(al) continuum 光连续区
optic(al) contour grinder 光学曲线磨床;光学轮廓磨床
optic(al) contrast 光学衬比
optic(al) contrast-enhanced image 光学反差增强图像
optic(al) contrast enhancement 光学反差增强
optic(al) contraster 光学增对器
optic(al) contrast modulation 光学反差调整
optic(al) contrast ratio 光对比度
optic(al) convergence[convergency] 光学聚焦
optic(al) converter 光学变换器
optic(al) convolution 光学卷积
optic(al) coordinate measuring apparatus 光学坐标量测仪
optic(al) corner reflector 光学直角反射器
optic(al) correction 光学修正;光学校正;视觉改正
optic(al) correlation 光学相关
optic(al) correlation method 光学相关法
optic(al) correlation technique 光学相关技术
optic(al) correlator 光学相关器;光相关器
optic(al) count 光学计数
optic(al) count difference 光计数差
optic(al) countermeasures 光学对抗
optic(al) counterpart 光学对应体
optic(al) counter-rotating wedges 反向旋转光楔
optic(al) coupler 光耦合器;光隔离器
optic(al) coupling 光耦合
optic(al) cross-section 光学截面;光截面
optic(al) crosstalk 光学串扰;串光
optic(al) crown 光学冕玻璃;光学玻璃
optic(al) crystal 光学水晶;光学晶体
optic(al) cup 视杯
optic(al) current meter 光学流速仪
optic(al) curve grinding machine 光学曲线磨床
optic(al) cycle 光学周期
optic(al) damage 光学损伤;光损伤
optic(al) damage threshold 光学损伤阈;光损伤阈值
optic(al) data 光学资料;光学数据
optic(al) data bus 光数据总线
optic(al) data corrector 光学数据校正器
optic(al) data digitizer 光学数据数字化器
optic(al) data handling 光学数据处理
optic(al) data handling device 光学数据处理机
optic(al) data link 光数据线路
optic(al) data processing 光学数据处理
optic(al) data processing device 光学数据处理机
optic(al) data processing system 光学数据处理系统
optic(al) data processor 光学数据处理机
optic(al) data transmission channel 光数据传输信道;光数据传输通道
optic(al) datatransmission channel 光学数据传输信道
optic(al) decoy 光学假目标;光假目标
optic(al) decrepitation method 光学爆裂法
optic(al) defect 光学缺陷;光缺陷
optic(al) defects of crystals 晶体光学缺陷
optic(al) deflection 光偏移;光偏转
optic(al) deflection apparatus 光学弯沉仪;光学变位仪
optic(al) deflection beam 光束偏转
optic(al) deflectometer 光学偏向计;光偏转仪;光学弯度计;光学弯沉仪
optic(al) deflector 光偏转器
optic(al) deformation 光学形变
optic(al) delay circuit 光延迟电路;光学延迟回路
optic(al) delay line 光延迟线
optic(al) demultiplexer 光解多路复用器

optic(al) densitometer 光学密度计;光密度计
optic(al) densitometric method 光密度法
optic(al) densitometry 光密度测量术
optic(al) density 光亮度;视觉密度;光学浓度;光(学)密度
optic(al) density colo(u)r 光密度色度
optic(al) density measurement 光密度测量
optic(al) density of smoke 烟的光学密度
optic(al) density slicing 光学密度分割
optic(al) density slicing image 光学密度分割图像
optic(al) depression 旋光性降低
optic(al) depth 光学深度;光深(度)
optic(al) design 光学设计
optic(al) designation 光学符号;光学标号
optic(al) designer 光学设计师
optic(al) design optimization 光学设计最佳化
optic(al) design procedure 光学设计程序
optic(al) detected magnetic resonance 光探测顺磁共振
optic(al) detection 光学探测
optic(al) detection magnetic resonance spectrometer 光探测磁共振波谱仪
optic(al) detector 光学探测器;光检测器;光辐射探测器
optic(al) detector technology 光探测技术
optic(al) deviation 光偏移
optic(al) device 光学装置;光学器件
optic(al) diameter 光学直径;光测粒径
optic(al) dichroism 光学二向色性
optic(al) dielectric(al) constant 光频介电常数;光介电常数
optic(al) differential rectification 光学微分纠正
optic(al) diffraction 光学衍射
optic(al) diffraction image 光衍射图像
optic(al) diffraction pattern 光学衍射图
optic(al) diffractometer 光衍射计
optic(al) digital audio player 光学数字音频唱机
optic(al) digital computer 光学数字计算机;光数字计算机
optic(al) digital tachometer 光学数字转速表
optic(al) dilatometer 光学膨胀计
optic(al) dimming 光学模糊
optic(al) diode 光二极管
optic(al) directional coupler 光学定向耦合器;光定向耦合器
optic(al) directional filter 定向滤光器
optic(al) direction and ranging 光学定向和测距
optic(al) direction finder 光学定向器
optic(al) direction finding 光学测向
optic(al) direct vision finder 光学直视寻像器
optic(al) disc 光学刻度盘;光学视面;光盘;视盘
optic(al) disk cartridge 光盘盒
optic(al) disk control card 光盘控制卡
optic(al) disk recorder 光盘录像机
optic(al) disk system 光盘系统
optic(al) dispersion attenuation 光色散衰减
optic(al) displacement 光偏移
optic(al) displacement factor 光学像移因数
optic(al) displacement sensor 光学位移传感器;光学式位移传感器
optic(al) display keyboard 光显示键盘
optic(al) display plotter 光显示绘图仪
optic(al) display system 光学显示系统
optic(al) display terminal 光显示终端设备;光显示终端
optic(al) distance 光学距离;光程;可见距离
optic(al) distance measurement 光学测距法;光学测距
optic(al) distance meter 光学测距仪
optic(al) distortion 光学应变;光学扭曲;光(学)畸变
optic(al) distortion analyzer 光畸变分析仪
optic(al) distortion correction 光学畸变校正
optic(al) distribution network 光配线网
optic(al) disturbance 光扰(动)
optic(al) dividing head 光学分度头
optic(al) dividing head with projection readout 光学投影读数分度头
optic(al) dividing table 光学分度台
optic(al) document reader 光文件阅读机
optic(al) Doppler effect 光学多普勒效应
optic(al) Doppler radar 多普勒光雷达
optic(al) dosimeter 光学辐射剂量计
optic(al) double 视双星

optic(al) double star 光学双星
optic(al) doublet 光学双线
optic(al) doublet star 光学双星
optic(al) driver 光盘驱动器
optic(al) dust instrument 光学测尘仪
optic(al) dynameter 光学倍率计
optic(al) echo 光学回波
optic(al) edge-enhanced image 光学边缘增强图像
optic(al) edge enhancement 光学边缘增强
optic(al) effect 光学效应
optic(al) efficiency 光效率
optic(al) elastic axis 光弹性轴
optic(al) electric(al) axial angle encoder 光电轴角编码器
optic(al) electric(al) converter 光电转换器
optic(al) electron 光学电子
optic(al) electronic integrated circuit 光电集成电路
optic(al) electronic reproducer 光声头
optic(al) electronics 光电子学
optic(al) electronic technology 光电子技术
optic(al) electronic transducer 光电换能器
optic(al) electron microscope 光学电子显微镜
optic(al) electron paramagnetic resonance Gauss meter 光学电子顺磁共振高斯计
optic(al) element 光学元件;光学零件
optic(al) emission line 光学发射线
optic(al) emission spectrograph 发光摄谱仪
optic(al) emission spectrography 发射光谱学;发光摄谱学
optic(al) emission spectrometry 光发射光谱法
optic(al) enantiomorph 旋光对映体
optic(al) encoder 光学同步发射机;光学编码器;光编码器
optic(al) endfinish 光学端面光洁度
optic(al) endoscope 光内窥镜
optic(al) energy 光能
optic(al) energy density 光能密度
optic(al) engineering 光学工程
optic(al) enhancement 光学增强
optic(al) enhancement interpretation 光学增强解释
optic(al) enlargement 光学放大
optic(al) equipment 光学设备
optic(al) equivalence 光学等效
optic(al) error 光学误差
optic(al) etching and electron bombardment 光刻电子轰击法
optic(al) evaluation device 光学检验装置;光学检验设备
optic(al) evaluation equipment 光学检验装置;光学检验设备
optic(al) evaluation facility 光学检验设备
optic(al) evaluation plant 光学检验装置
optic(al) exaltation 旋光性增强
optic(al) examination 光学检验
optic(al) excitation 光激发
optic(al) excited laser 光激励型激光器;光激发激光器
optic(al) exciting pulse 光激发脉冲
optic(al) exitation rate 光激励率
optic(al) experiment 光学实验
optic(al) exposure 照明度
optic(al) exposure head 光学绘图头;光曝光头
optic(al) exposure meter 光学曝光表
optic(al) extensometer 光学伸长仪
optic(al) eye level finder 光学直视水平瞄准器
optic(al) fabrication 光学加工
optic(al) false colo(u)r composite 光学法假色合成
optic(al) Farday rotation isolator 光学法拉第旋转隔离器
optic(al) fault locator 光故障定位器
optic(al) feedback 光反馈
optic(al) feedback image intensifying system 光学反馈图像增强系统
optic(al) feeler 光学接触器
optic(al) fiber 光学纤维;光纤;光导纤维
optic(al) fiber acoustic(al) sounding 光导纤维水声探测
optic(al) fiber bundle 光纤束
optic(al) fiber cable 光学纤维电缆;光缆
optic(al) fiber cable for hybrid communication

复合通信光缆
optic(al) fiber cable for installation in duct 管道敷设式光缆
optic(al) fiber cable for installation in duct or tunnel 管道光缆
optic(al) fiber cable with composite sheath 复合护层光缆
optic(al) fiber camera 光导纤维摄像机
optic(al) fiber cavity 纤维光腔
optic(al) (fiber) communication 光纤通信
optic(al) fiber composite connector 复合式光纤连接器
optic(al) fiber cord 光纤软线;光缆软绳
optic(al) fiber coupling 光纤耦合
optic(al) fiber dispersion 光纤色散
optic(al) fiber distributed data interface standard 光纤分布数据接口标准
optic(al) fiber faceplate 光纤面板
optic(al) fiber interferometer 光导纤维干涉仪
optic(al) fiber plate 光导纤维板
optic(al) fiber preform 光纤预成形
optic(al) fiber probe 光纤探头
optic(al) fiber resonator 光学纤维谐振腔
optic(al) fiber ribbon cable 带状光缆
optic(al) fiber ring 光纤环路
optic(al) fiber sensing 光纤传感
optic(al) fiber sensor 光纤传感器
optic(al) fiber strength proof 光纤强度筛选
optic(al) fiber temperature sensor 光纤温度传感器
optic(al) fiber tube 光导纤维管
optic(al) fiber waveguide 光纤波导
optic(al) fibre abruption tester 光纤断裂测试器
optic(al) fibre acoustic(al) sensor 光纤声压传感器
optic(al) fibre active connector 光纤有源连接器
optic(al) fibre amperemeter 光纤电流计
optic(al) fibre amplifier 光纤放大器
optic(al) fibre amplifier multiplexer 光纤放大器多路复用器
optic(al) fibre biosensor 光纤声压传感器
optic(al) fibre bundle 光纤束
optic(al) fibre cable stranding machine 光纤成缆机
optic(al) fibre cable with plastic jacket 塑料护套光缆
optic(al) fibre chemical sensor 光纤化学传感器
optic(al) fibre circuit 光纤线路
optic(al) fibre coating 光导纤维涂料;光纤涂层;光纤被覆
optic(al) fibre communication repeater 光纤通信中继器
optic(al) fibre compression sensor 光纤压力传感器
optic(al) fibre concentrator 光纤集中器
optic(al) fibre connector 光纤连接器
optic(al) fibre cord 光缆软绳
optic(al) fibre coupler 光纤耦合器
optic(al) fibre current sensor 光纤电流传感器
optic(al) fibre delay line 光纤时延线
optic(al) fibre demultiplexer 光纤解多路复用器
optic(al) fibre dispenser 光纤松放器
optic(al) fibre dispersion tester 光纤色散测试仪
optic(al) fibre displacement sensor 光纤位移传感器
optic(al) fibre distribution box 光纤分线箱
optic(al) fibre electric(al) field strength sensor 光纤电场强度传感器
optic(al) fibre flowmeter 光纤流量计
optic(al) fibre for direct burial 直埋电缆
optic(al) fibre fusion splicer 光纤熔接器;光纤熔接机
optic(al) fibre guidance 光纤制导
optic(al) fibre gyroscope 光纤陀螺仪
optic(al) fibre hazard 光纤危害
optic(al) fibre helical microbend sensor 光纤螺旋微弯传感器
optic(al) fibre hydrophone 光纤水听器
optic(al) fibre image sensor 光纤图像传感器
optic(al) fibre interferometric sensor 光纤干涉传感器
optic(al) fibre ion sensor 光纤离子传感器
optic(al) fibre jacket 光纤套层
optic(al) fibre junction 光纤接续

optic(al) fibre line-to-circle converter 光纤圆直变换器
optic(al) fibre link 光纤线路
optic(al) fibre magnetic field strength sensor 光纤磁场强度传感器
optic(al) fibre magnetostrictive sensor 光纤磁致伸缩传感器
optic(al) fibre merit figure 光纤品质因数
optic(al) fibre monitor 光纤监控器
optic(al) fibre multiplexer 光纤多路复用器
optic(al) fibre pick-off coupler 光纤发送耦合器;光纤传感耦合器
optic(al) fibre preform 光纤预制件
optic(al) fibre pressure sensor 光纤压力传感器
optic(al) fibre proof-tester 光纤检验机
optic(al) fibre pulse compression 光纤脉冲压缩
optic(al) fibre radiation damage 光纤辐射损害
optic(al) fibre radiation sensor 光纤射线传感器
optic(al) fibre resonator 光纤谐振器;光纤共振腔
optic(al) fibre rewinder 光纤复绕机
optic(al) fibre ribbon 光纤带
optic(al) fibre ringer 光纤信号器;光纤振铃器
optic(al) fibre rotator 光纤旋转器
optic(al) fibre sensor 光纤传感器
optic(al) fibre sound pressure sensor 光纤声压传感器
optic(al) fibre source 光纤光源
optic(al) fibre strander 光纤绞线器
optic(al) fibre system 光纤系统
optic(al) fibre temperature sensor 光纤温度传感器
optic(al) fibre tensile strength 光纤抗拉强度
optic(al) fibre thermometer on thermal colo(u)r effect 光纤热色效应温度计
optic(al) fibre transfer function 光纤传递函数
optic(al) fibre transmission 光纤传输
optic(al) fibre transmission facility 光纤传输设备
optic(al) fibre trap 光纤阱
optic(al) fibre ultrasonic sensor 光纤超声传感器
optic(al) fibre velocity sensor 光纤速度传感器
optic(al) fibre vibration sensor 光纤振动传感器
optic(al) fibre video trunk 光纤视频干线
optic(al) fibre voltage sensor 光纤电压传感器
optic(al) fibre wavelength division multiplexer 光纤波分复用器
optic(al) fiducial marks 光学光标
optic(al) field 光场
optic(al) field effect transistor 光场效应晶体管
optic(al) figuring 光学修整
optic(al) film-fibre coupler 光薄膜纤维耦合器
optic(al) filming 光学镀膜
optic(al) filter 光学滤光镜;滤色镜;滤光器
optic(al) filter box 滤光片组件
optic(al) filter gas analyzer [analyser] 光学滤波式气体分析器
optic(al) filter predetection 检波前光学滤光片
optic(al) finder 光学寻像器;光学取景器;取景器
optic(al) finishing 光学精加工
optic(al) firecontrol instrument 光学射击控制仪器
optic(al) fire director 光学射击指挥仪
optic(al) fixed reticule sight 光学固定环瞄准具
optic(al) flame detector 感应火灾探测器;光感器
optic(al) flat 光学平行平板玻璃;光学平面;光学平度;光学平板;光学测平仪;平晶
optic(al) flat ga(u)ge 光学平面规
optic(al) flat glass 光学平玻璃
optic(al) flat glass filter 光学玻璃滤色镜
optic(al) flatness 光学平直度;光学平度;光学光滑度
optic(al) flatness ga(u)ge 光学平直仪
optic(al) flint glass 火石光学玻璃
optic(al) fluid-flow measurement 光学流体流动测量法
optic(al) fluorite 光学萤石
optic(al) flux 光通量
optic(al) foam 光学泡沫
optic(al) focus 光焦点
optic(al) focusing 光聚焦
optic(al) focusing device 光学聚焦装置
optic(al) focus switch 光学聚焦转换开关;光聚焦开关
optic(al) fog 光学灰雾
optic(al) font 光学字体;光学识别字体;光识别字体;光符识别用字体

optic(al) foot mark plummet 光学底点对点器
optic(al) force 矫形力
optic(al) foresight 光学瞄准器
optic(al) Fourier analysis method 光学傅立叶分析法
optic(al) Fourier analysis of lineament 线性体光学傅立叶分析
optic(al) Fourier transform 光学傅立叶变换
optic(al) free induction decay 光学无感生衰减
optic(al) frequency 光频(率);光频段
optic(al) frequency branch 光频支路
optic(al) frequency converter 光频变换器
optic(al) frequency division multiplex 光频分复用
optic(al) frequency domain reflectometer 光频域反射计
optic(al) frequency mixing 光混频
optic(al) frequency response 光频响应
optic(al) frequency shift 光频移
optic(al) frequency translation 光频变换
optic(al) fuse 光引信
optic(al) fused silica 光学石英玻璃
optic(al) gain 光学增益
optic(al) gain coefficient 光学增益系数
optic(al) galaxy 光学星系
optic(al) gate 光闸
optic(al) ga(u)ge 光学压力计;光学测量仪器
optic(al) generation 光波振荡
optic(al) generator 光振荡器;光发生器
optic(al) glass 光学玻璃
optic(al) glass blank 光学玻璃毛坯
optic(al) glass fiber 光学玻璃纤维
optic(al) glass material 光学玻璃材料
optic(al) glass pressing process 光学玻璃压形法
optic(al) glass sheet 光学玻璃板
optic(al) goniometer 光学测向器;光学测角计
optic(al) graphical rectification 光学图解纠正
optic(al) grating 光栅
optic(al) grating measuring system 光栅测量系统
optic(al) grid lines 光学光栅刻线
optic(al) grinding 光学研磨
optic(al) groove 交叉沟
optic(al) guidance 最佳导引;光制导;光学制导;光学导航;视线引导;视光诱导
optic(al) guidance system 光学制导系统
optic(al) guiding system 光制导系统
optic(al) gyroscope 光陀螺仪
optic(al) harmonic 光学谐波;光谐波
optic(al) harness 光缆捆束
optic(al) harness assembly 光缆捆束组件
optic(al) haze 光学轻雾;光霾;大型闪烁
optic(al) head 光度头
optic(al) height finder 光学测高仪
optic(al) heterodyne 光学外差
optic(al) heterodyne detection 光频外差探测
optic(al) heterodyne detection system 光外差探测系统
optic(al) heterodyne radar 光频外差雷达
optic(al) heterodyne receiver 光外差接收机;光频外差接收器
optic(al) heterodyne repeater 光外差中继器
optic(al) heterodyne spectroscopy 光频外差光谱学
optic(al) heterodyning 光外差作用
optic(al) heterogeneity 光学均匀性;光学不均匀
optic(al) hologram 光学全息图
optic(al) hologrammetry 光学全息摄影测量
optic(al) holographic(al) recording material 光全息记录材料
optic(al) holographic(al) interferometry 光学全息干涉测量技术
optic(al) holography 光学全息术;光学全息摄影
optic(al) homer 光学自动跟踪头;光学寻的头
optic(al) homing 光学自动跟踪;光学自动导航
optic(al) homing device 光学跟踪装置
optic(al) homing head 光学自动导引头
optic(al) homing head axis 光学导引头瞄准轴
optic(al) homodyne 光学零差
optic(al) homodyne receiver 光学零拍接收机;光零拍接收机
optic(al) homogeneity 光学均匀性
optic(al) horizon 真视地平线
optic(al) horizon sensor 光学地平仪
optic(al) hygrometer 光学湿度表
optic(al) identification 光学证认

optic(al) identification instrument 光学鉴定仪器
optic(al) illusion 视错觉;错视
optic(al) image 光学影像;光学像
optic(al) image combinator 光学合像仪
optic(al) image converter 光学图像转换器
optic(al) image feedback 光学图像反馈
optic(al) image formation 光学图像形成;光学图像;光学构像
optic(al) image processing 光学图像处理
optic(al) image processing method 光学图像处理方法
optic(al) image processing product 光学图像处理产品
optic(al) image processor 光学图像处理器
optic(al) image rejection mixer 光图像抑制混频器
optic(al) imagery 光学图像
optic(al) image stabilized telescope 光学稳像望远镜
optic(al) image transmission device 光学图像传输装置
optic(al) image unit 光学图像输入器
optic(al) imaging device 光学成像装置
optic(al) imaging system 光学成像系统
optic(al) immersed detector 光浸没式探测器
optic(al) immersion 光学浸没
optic(al) immersion gain 光学浸没增益
optic(al) impedance discontinuity 光阻抗突变
optic(al) impression 影像
optic(al) inactivity 不旋光性;光学钝性;非旋光性
optic(al) index 光学指数;光学指标
optic(al) indicator 光(学)指示器;光学指示器;光学示功计;光线指示器;测微显微镜
optic(al) indicatrix 光折射椭球;光学特性曲线
optic(al) indicatrix place determination method 光率体位置测定法
optic(al) indicatrix symmetric(al) axis 光率体对称轴
optic(al) inertial guidance and navigation system 光学惯性制导和导航系统
optic(al) inertial sextant 光学惯性六分仪
optic(al) information 光学信息
optic(al) information handling 光学信息处理
optic(al) information processing 光学信息处理
optic(al) information processing technique 光信息处理技术
optic(al) information storage 光学信息存储;光学信息储存;光信息存储器
optic(al) information transfer 光学信息传递
optic(al) inhomogeneity 光学不均匀
optic(al) injection 光学注入
optic(al) injector 光注入器
optic(al) input 光输入;可见输入
optic(al) instrument 光学仪器;光学仪;光学器械
optic(al) instrument and meter 光学仪表
optic(al) instrument and meter shop 光学仪表车间
optic(al) instrumentation 光学仪器设备
optic(al) instrument design 光学仪器设计
optic(al) instrument making 光学仪器制造
optic(al) instrument mo(u)ld and fog auto-detector 光学仪器霉雾自动检测仪
optic(al) instrument positioning 光学仪器定位
optic(al) insulation 光学绝缘
optic(al) integrated circuit 光集成电路;集成电路
optic(al) integration 光学集成
optic(al) integrator 光学积分器
optic(al) intensity 光谱强度
optic(al) interaction 光学相互作用
optic(al) interconnection 光学互接
optic(al) interface adapter 光接口适配器
optic(al) interface device 光学接口设备
optic(al) interference 光学干涉;光干涉
optic(al) interference coating 光学干涉膜
optic(al) interference filter 光(学)干涉滤光片
optic(al) interference measurement 光学干涉测量法
optic(al) interference method 光波干涉法
optic(al) interference type gas analyzer 光干涉式气体分析器
optic(al) infernece coating 光波干涉膜
optic(al) interferometer 光学干涉仪;光学干涉仪
optic(al) interferometry 光学干涉量度法
optic(al) interpolation 光学插补
optic(al) interpolation system 光学内插系统

optic(al) interval 光学区间;光学间距
optic(al) invariant 光学不变量
optic(al) inversion 光学转向;偏振转向
optic(al) inversion system 光学转像系统
optic(al) ionization energy 光电离能
optic(al) isolation 光学隔离
optic(al) isolator 光学隔离器;光频隔离器;光隔离器
optic(al) isomer 光学异构体;旋光异构体
optic(al) isomerism 光学异构;旋光异构
optic(al) isotropy 光学各向同性
optic(al) jig boring machine 光学坐标镗床
optic(al) Kerr effect 光学克尔效应
optic(al) laboratory 光学实验室
optic(al) lamp 光学灯
optic(al) landing system 光学着陆系统
optic(al) lane 光路
optic(al) lantern 幻灯机
optic(al) laser 激光器
optic(al) laser radar 激光雷达
optic(al) laser ranging system 激光测距系统
optic(al) laying 光学瞄准
optic(al) laying equipment 光学瞄准器
optic(al) length 光程
optic(al) length scale 光学刻尺
optic(al) lens 光学透镜;光学镜片;透镜
optic(al) lens face 光学透镜面
optic(al) lens polisher 光学透镜磨光器
optic(al) lens system 光学透镜系统
optic(al) lens waveguide 光学透镜波导
optic(al) lever 光(学)杠杆;光学比长仪
optic(al) levitation 光学悬浮
optic(al) liberation 光学天平动
optic(al) light filter 杂散光滤光片;光学滤光器
optic(al) line of sight 光学视线;光学瞄准线
optic(al) line-scan 光学行扫描
optic(al) linescan device 光行扫描器
optic(al) line terminal 光线路终端
optic(al) link 光线路;光链路
optic(al) link module 光连接模件
optic(al) liquid 光学浸液
optic(al) local oscillator 光学本机振荡器
optic(al) lock-on 光学自动跟踪;光学锁定
optic(al) lofting 光学放样
optic(al) log 光学计程仪;光学测程器
optic(al) logic operations 光学逻辑功能
optic(al) look-on 光学跟踪
optic(al) loss 光损失
optic(al) loss coefficient 光学损耗系数
optic(al) loss test set 光损耗测试仪
optic(al) low-pass filter 光学低通滤波器;低通滤光器
optically active 旋光的;光学活性的;起偏振作用
optically active absorption band 光学活性吸收谱带
optically active exchanger 旋光离子交换剂
optically active form 旋光体
optically active gas 旋光性气体
optically active material 光学活性材料
optically active polymer 旋光性聚合物
optically active resin 光学活性树脂
optically active substance 旋光物
optically cascaded pulse 光学级联脉冲
optically coated 光学镀膜的
optically contacted etalon 光学接触标准具;光胶合标准具
optically coupled circuit 光耦合电路
optically coupled device 光耦合装置
optically coupled integrated circuit 光耦合集成电路
optically coupled isolator 光耦合隔离器;光隔离器
optically coupled laser 光耦合激光器
optically coupled pulse amplifier 光耦合脉冲放大器
optically coupled system 光耦合系统
optically denser medium 光密介质
optically derived transform 光导变换
optically effective atmosphere 光学有效大气
optically excited laser 光激发激光器
optically excited metastable level 光受激亚稳态能级
optically flat filter 光学平面滤光片;光学扁平滤光器;光平面滤光片
optically flat glass 光学平面玻璃
optically flat surface 光学平面

optically flat window 光学平面窗
optically focused beam 光学聚焦光束
optically focused solar beam 光学聚焦太阳光束
optically graduated circle 光学度盘
optically identified source 光学证认源
optically inactive 非旋光的;不旋光的
optically inactive substance 不旋光物
optically levitated particle 光学悬浮粒子
optically matched crystal 光匹配晶体
optically negative 光学性负的
optically non-linear crystal 光学非线性晶体
optically perfect reflector 理想反射器;理想反光镜
optically plane glass 光学平面玻璃
optically positive 光学性正的
optically powered sensor 光供电源传感器
optically preamplified receiver 光前置放大接收机
optically pumped laser 光抽运激光器
optically pumped liquid laser 光泵液体激光器
optically pumped magnetometer 光泵磁力仪
optically pumped solid laser 光抽运固体激光器
optically pure 光学纯的
optically read instrument 光学读数仪器
optically stimulated crystal 光激发晶体
optically thicker medium 光密介质
optically thick medium 光厚介质
optically thin medium 光薄介质
optically thinner medium 光疏介质
optically thin plasma 光学薄等离子体
optically tracked 光学跟踪的
optically transparent medium 光学透明介质
optically visible object 光学可见天体
optic(al) machinery 光学机械
optic(al) magnetic probe 光磁探针
optic(al) magnification 光学放大
optic(al) manufacture 光学加工
optic(al) mapping 光学映像
optic(al) mark 光学测标
optic(al) mark card reader 光符号卡读出器
optic(al) marking 光标;视觉信号
optic(al) mark reader 光字符阅读机;光学标志阅读器;光标阅读机
optic(al) maser 光学脉塞;光量子放大器;光激射器;激光器
optic(al) maser oscillator 激光振荡器
optic(al) mask 光学掩模
optic(al) matched filter 光学匹配滤波器
optic(al) matched-filter imagecorrelator 光学匹配滤波器相关器
optic(al) matched spatial filtering 光学匹配滤波
optic(al) materials 光学材料
optic(al) means 光学装置;光学方法
optic(al) measurement 光学测量
optic(al) measurement of distance 光学测距(法)
optic(al) measuring device 光学测量装置
optic(al) measuring head 光学测头
optic(al) measuring instrument 光学测量仪器
optic(al) measuring machine 光学测量机
optic(al) measuring system 光学量测系统;光学测量系统
optic(al) measuring technique 光学测量技术
optic(al) mechanical comparator 光学机械式比较仪
optic(al) mechanical compilation 光学机械编图
optic(al) mechanical deformation 光学机械变形
optic(al) mechanical electrical 光机电的
optic(al) mechanical plotter 光学机械测图仪
optic(al) mechanical projection 光学机械投影
optic(al) mechanical rectification 光学机械纠正
optic(al) mechanical scanner 光学机械扫描仪
optic(al) mechanical scanning thermal imager 光机扫描热像仪
optic(al) mechanical system 光栅系统
optic(al) medium 光学媒质;光学介质
optic(al) memory 光学存储器;光存储器;光储存器
optic(al) memory circuit 光存储电路
optic(al) memory crystal 光存储晶体
optic(al) memory element 光学存储元件;光存储元件
optic(al) memory glass semiconductor 光存储玻璃半导体
optic(al) memory material 光存储材料
optic(al) memory matrice 光学存储元件矩阵
optic(al) memory matrix 光学存储矩阵

optic(al) memory system 光学存储系统
optic(al) meteor 光学气象;大气光学现象
optic(al) method 吸光光度法;优化方法;光学法测湿润角
optic(al) method of analysis 光学分析法
optic(al) metrological instrument 光学计量仪器
optic(al) metrology 光学计量
optic(al) microdensitometer 光学显微密度计
optic(al) micrometer 光学测微器;光学测微计;光测微计
optic(al) micrometer theodolite 光学测微器经纬仪;光学测微经纬仪
optic(al) micrometry 光学测微法
optic(al) microscope 光学反射镜
optic(al) microscopy 光学显微术;光学显微镜(检查)法
optic(al) microwave spectroscope 光学微波分光镜
optic(al) mineralogy 光性矿物学
optic(al) mirror 光学反射镜
optic(al) mirror slip 光学镜片
optic(al) mixer 光学混频器;光混合器
optic(al) mixer unit 光混频器
optic(al) mixing 光混频
optic(al) mixing box 光混合盒
optic(al) mixing rod 光混合柱;光混合棒;混光棒
optic(al) mode 光学模;光频振动模式;光模
optic(al) model 光学模型;光模像;混浊晶球模型
optic(al) model of nuclear reaction 核反应光学模型
optic(al) mode of vibration 光学振动模
optic(al) mode scattering 光学模散射
optic(al) mode stripper 光模消除器
optic(al) modulation 光学调制;光学调节;光调制
optic(al) modulation transfer function 光学调制传递函数
optic(al) modulator 光学调制器;光调制器
optic(al) module 光学组件
optic(al) moisture analyzer 光学水分分析器
optic(al) moment 光矩
optic(al) monochromator 光学单色仪
optic(al) mosaic 光学镶嵌
optic(al) multichannel analysis 光学多通道分析
optic(al) multichannel analyzer 光学多通道分析器
optic(al) multimode dispersion 光多模色散
optic(al) multiplexer 光复用器;光多路复用器
optic(al) multiplexing 光多路传输
optic(al) multiplexing camera 光学多倍投影仪
optic(al) multiplexing transmission 光学多路传输
optic(al) multiplication 光学倍增
optic(al) multiport coupler 光学多端口耦合器
optic(al) multispectral canning 光学多谱段扫描
optic(al) multispectral scanner 光学多谱段扫描仪
optic(al) multisurface prism 光学多面棱镜
optic(al) navigation system 光学导航系统
optic(al) navigator 光学导航装置
optic(al) negative crystal 负光性晶体
optic(al) network unit 光纤网络设备
optic(al) noise 光学噪声
optic(al) noise autocorrelation 光学噪声自相关
optic(al) noise level 光学噪声级
optic(al) noise map 光学噪声图
optic(al) non-uniformity 光学非凹性;光学不均匀
optic(al) normal 光轴法线
optic(al) normal line 光学法线
optic(al) null double-beam farinfrared spectrometer 光学零位双光束远红外光谱仪
optic(al) null method 光学衡消法;光平衡法
optic(al) null principle 光学零位原理
optic(al) null system 光学零位系统
optic(al) nutation 光学章动
optic(al) object 光学天体
optic(al) observation 光学观察;光学观测
optic(al) oceanography 光学海洋学
optic(al) orbital rendezvous apparatus 轨道会合用光学装置
optic(al) orbiting device 光学轨道飞行器
optic(al) orientation 光学定向;光性方位
optic(al) orientation device 光学取向器
optic(al) orthophoto printer 光学正射相片晒印机
optic(al) orthophotoscope 光学正射投影纠正仪
optic(al) oscillator 光激射振荡器
optic(al) ozone sonde 光学臭氧探空仪;光臭氧探针

optic(al) pair 视双星
optic(al) panel 光具板
optic(al) pantograph 光学缩放仪
optic(al) parallax 光学视差;视差
optic(al) parallel 光学平行计;光学平晶
optic(al) parallelism 光学平行性
optic(al) parallel rays 光学平行光束
optic(al) parameter 光学参量;光参数
optic(al) parameter amplification 光参量放大
optic(al) parameter amplifier 光参量放大器
optic(al) parameter oscillator 光参量放大振荡器
optic(al) parametric amplification 光参量放大
optic(al) parametric noise 光参量噪声
optic(al) parametric oscillator 光参量振荡器;光参量振荡器
optic(al) particle decoy 光粒子假目标
optic(al) parts 光学零件;光学部件
optic(al) pass 光通路
optic(al) patch panel 光接线板
optic(al) path 光线路径;光程
optic(al) path computer 光程计算机
optic(al) path difference 光程差
optic(al) path distortion 光路畸变
optic(al) path fluctuation 光程起伏
optic(al) path length 光学距离;光程(长)
optic(al) path-length error 光程误差
optic(al) path mismatch 光程失配
optic(al) pattern 光(学)图样;光学图案;反光图案
optic(al) pattern recognition 光学图形识别
optic(al) pendulum wire 光学垂线
optic(al) penetration 光穿透
optic(al) penetrator 光穿透器
optic(al) performance 光学性能
optic(al) performance detector 光学性能检测仪
optic(al) phase 光学相位
optic(al) phased array 光频整相阵列
optic(al) phase deviation 光相应偏差;光相位偏差
optic(al) phase modulator 光相位调制器;光调幅器
optic(al) phase shifter 光移相器;光学移相器
optic(al) phenomenon 光学现象
optic(al) phonon 光学声子;光频声子
optic(al) photodetector 光检测器
optic(al) photographic(al) analogy recorder 光学照相模拟记录器
optic(al) photon 可见光子
optic(al) physics 光学物理
optic(al) pick-off 光学传感器
optic(al) pick-up 光学拾波器;光学摄像管
optic(al) picture 光学图
optic(al) picture center 像主点;图像光心;像片主点
optic(al) pipe 光导管
optic(al) placode 视基板
optic(al) plane 光轴面;光学平面
optic(al) plastic lens 光学塑料透镜
optic(al) plastics 光学塑料
optic(al) plate 视基板
optic(al) plumb 光测垂准器;光学垂准器;光测悬线;光测铅垂线;光测垂线;光测垂球
optic(al) plumbing 光学垂准
optic(al) plumbing instrument 光学对点器;光学垂准器
optic(al) plumb line 光学垂准线
optic(al) plummet 光学对中器;光学对点器;光学垂准器;光测悬锤;光测铅垂线
optic(al) plummet centering 光学垂线对中
optic(al) plummet centring 光学垂线对中
optic(al) polarization 光偏振
optic(al) polishing 光学抛光
optic(al) position encoder 光学位置编码器;光位置编码器
optic(al) power 光强(度);光功率
optic(al) power budget 光功率分配
optic(al) power composite cable 光纤电力线组合缆
optic(al) power density 光功率密度
optic(al) power divider 光功率分路器
optic(al) power efficiency 光功率效率
optic(al) power equalizer 光功率均衡器
optic(al) power meter 光功率计;光功率计
optic(al) power output 光功率输出
optic(al) power spectrum analysis 光能谱分析
optic(al) precision component 光学精密零件

optic(al) printer 光学印刷机;光学印片机
optic(al) printing 光学印片
optic(al) print process 光学晒印法
optic(al) prism 棱镜
optic(al) probe 光学探头
optic(al) probing 光探测
optic(al) processing 光学处理
optic(al) processing circuit 光学处理电路;光处理电路
optic(al) processing system 光学处理系统;光处理系统
optic(al) processor 光学处理机
optic(al) profile type 光学式
optic(al) projected display 光学投影显示器
optic(al) projecting colo(u)r enhancement 光学投影彩色增强
optic(al) projection 光学投影
optic(al) projection comparator 光学投影比较仪
optic(al) projection instrument 光学投影仪
optic(al) projection magnification 光学投影放大
optic(al) projection pantograph 光学投影缩放仪
optic(al) projection reading device 光学投影读数装置
optic(al) projection system 光学投影系统;投影系统
optic(al) projector 光学投影仪;光学投影机;投影系统
optic(al) propagation 光传播
optic(al) propagator 光学传播体
optic(al) property 光学性质;光学性能;光性能
optic(al) property of minerals 矿物光学性质;矿物光性
optic(al) protective coating 光学保护涂层
optic(al) protractor 光学量角器;光学测角器
optic(al) protractor level 光学测角水准器
optic(al) proximity detector 光学接近探测器
optic(al) pseudocolo(u)r composite image 光学假彩色合成图像
optic(al) pseudo-colo(u)r composition 假彩色合成光学法
optic(al) pseudo-colo(u)r processor 光学假彩色处理仪
optic(al) pulsar 光学脉冲星
optic(al) pulsation 光脉动
optic(al) pulse 光学脉冲
optic(al) pulse code modulation 光脉冲编码调制
optic(al) pulse compression 光学脉冲压缩
optic(al) pulse counter 光脉冲计数器
optic(al) pulsecounting system 光脉冲计数系统
optic(al) pulsed ranging 光学脉冲测距
optic(al) pulse generator 光脉冲发生器
optic(al) pulse height 光脉冲高度
optic(al) pulse reflectometer 光脉冲反射计
optic(al) pump 光泵
optic(al) pumping 光抽运
optic(al) pumping magnetometer 光泵磁力仪
optic(al) pumping radiation 光抽运辐射
optic(al) pumping system 光泵浦系统
optic(al) pump system 光抽运系统
optic(al) purity 光学纯度
optic(al) pyrometer 光学高温计;光学测高温计;光电高温计;光测高温计;光测高温计
optic(al) pyrometry 光学高温测量术;光学测高温
optic(al) quadrant 光学象限仪
optic(al) quality 光学质量;光学性能
optic(al) quality homogeneous glass 光学均匀的玻璃
optic(al) quantum generator 光量子发生器
optic(al) quartz 光学石英
optic(al) quenching 光猝灭
optic(al) radar 光学雷达;光学定向测距仪
optic(al) radar altimeter 光雷达测高计
optic(al) radar detection technique 光雷达探测技术
optic(al) radar transmitter 光雷达发射机
optic(al) radial table 光学转台
optic(al) radiation 光学辐射;光辐射;视辐线
optic(al) radiation comparator 光学辐射比较器
optic(al) radiation pyrometer 光学测高温计
optic(al) radome 光学天线罩
optic(al) radome absorption 光学天线罩的吸收
optic(al) range 直视距离;光视距;光射距;视线距离
optic(al) range finder 光学测距仪
optic(al) range finding technique 光学测距技术

optic(al) range gating 光学距离选通
optic(al) ranging 光学测距
optic(al) ranging system 光学测距系统
optic(al) rarity 光束
optic(al) reader 光阅读器;光学阅读器;光输入机;光读出器
optic(al) reading 光测读数
optic(al) reading equipment 光学阅读器
optic(al) reading head 光学读数头
optic(al) reading system 光学读数系统
optic(al) reading theodolite 光学读数经纬仪
optic(al) readout 光学读出装置;光读出
optic(al) readout device 光学读出装置
optic(al) readout system 光学读出系统
optic(al) receiver 光学检出器;光接收器;光接收机
optic(al) receiver aperture 光接收器孔径
optic(al) receiver assembly 光学接收装置
optic(al) receiving set 光接收器
optic(al) recombination coefficient 光复合系数
optic(al) reconstruction 光学再现
optic(al) record(ing) 光学记录(法)
optic(al) recording system 光学记录系统
optic(al) rectification 光学校正;光学纠正
optic(al) reduction 光学缩小转绘
optic(al) reference line 光学基准线;光学参考线
optic(al) reference plane 光学基准面;光学参考面
optic(al) refinements 视差矫正
optic(al) reflective sensor 光反射传感器
optic(al) reflectometer 光学反射计;光反射计
optic(al) reflector 反光镜
optic(al) reflectror 光反射器
optic(al) reflex view finder 反光取景镜
optic(al) refraction 光反射
optic(al) regenerative repeater 光再生中继器
optic(al) register 光学配准
optic(al) registering 光学配准
optic(al) registration 光学配准;光学记录;光对准
optic(al) relay 光继电器
optic(al) relief model 光学立体模型
optic(al) remote sensing 光学遥感
optic(al) remote sensing technology 光学遥感技术
optic(al) repeater 光中继器
optic(al) reproducer 光学还音器
optic(al) resin 光学树脂
optic(al) resin gum 光学树脂胶
optic(al) resolution 光学分辨能力;光学分辨率
optic(al) resonance 光学共振;光谐振
optic(al) resonator 光学谐振腔;光学共振腔;光学共振器;光谐振器
optic(al) resonator entropy 光学共振器熵
optic(al) retardation plate 光学延迟板
optic(al) return loss 光回波损耗
optic(al) righting reflex 视翻正反射
optic(al) ring adapter 光环适配器
optic(al) ring resonator 光环谐振器
optic(al) rod coupler 光学棒状耦合器
optic(al) rod multiplexer-filter 光学棒状复用滤波器
optic(al) rosette diagram 光学玫瑰图
optic(al) rotary dispersion 光旋色散
optic(al) rotary power 光旋能力
optic(al) rotary table 光学回转工作台;光学分度台
optic(al) rotating stage 光学转台
optic(al) rotation 偏振光偏振面转动;旋光性;旋光度
optic(al) rotator 光学旋转器;光旋转器
optic(al) rotatory dispersion 旋光色散;旋光分散;旋光色散谱
optic(al) rudiment 视原基
optic(al) sag 光折角
optic(al) sampling 光学取样
optic(al) satellite 光学卫星
optic(al) scale 光学度盘;光学秤;光学测微尺
optic(al) scale interval 光学测微刻度值
optic(al) scale theodolite 光学刻度(尺)经纬仪
optic(al) scanner 光学析像器;光学扫描仪;光(学)扫描器;飞点扫描器
optic(al) scanner bar code 光扫描器用条线代码
optic(al) scanning 光扫描
optic(al) scanning device 光学扫描装置
optic(al) scanning system 光学扫描系统
optic(al) scanning turbidimetry 光扫描浊度测定法;光扫描比浊法

optic(al) scattering 光学散射
optic(al) sccond-harmonic generation 光倍频
optic(al) scrambler 光学扰频器
optic(al) search 光学搜索;光搜索
optic(al) second harmonic 光学二次谐波
optic(al) second harmonic generation 光学二次谐波振荡
optic(al) section 光学切面;光学截面;光切面
optic(al) seeker 光学自动导引头
optic(al) seismograph 光学式地震计
optic(al) selection rule 光学选择定则
optic(al) sensibilization 光学增感(作用);光敏化作用
optic(al) sensing head 光学传感头
optic(al) sensitive 感光的
optic(al) sensitivity 光敏度
optic(al) sensitization 光学增感
optic(al) sensitizer 光学增感剂
optic(al) sensor 光(学)传感器;光量传感器
optic(al) sensor system 光学传感系统
optic(al) separator 光学选矿机
optic(al) shaft encoder 光轴编码器
optic(al) shift 光学偏移
optic(al) shutter 光闸;光学快门
optic(al) sight 光学瞄准;瞄准望远镜
optic(al) sighting device 光学照准装置;光学瞄准具
optic(al) sign 光性符号
optic(al) signal 光信号;视光标志
optic(al) signal carrier 光信载波;光信号载波
optic(al) signal detection 光信号探测
optic(al) signal distortion 光信号畸变
optic(al) signal entropy 光信号熵
optic(al) signal generator 光学信号发生器
optic(al) signal modulation 光信号调制
optic(al) signal modulator 光信号调制器
optic(al) signal processing 光信号处理
optic(al) signal processor 光学信号处理机
optic(al) silica glass 光学石英玻璃
optic(al) simulation 光学模拟
optic(al) slicing 光学切片
optic(al) slide rule 光学计算尺
optic(al) slits 光狭缝
optic(al) slope corrector 光学坡度改正器
optic(al) smoke detector 光敏烟雾探测器;光烟感器
Optic(al) Society of America 美国光学会
optic(al) solar simulator 光学太阳模拟器
optic(al) sorter 光电选矿机
optic(al) sorting 光学选矿;光电选矿
optic(al) sound head 光声头
optic(al) sound propagation 光声传播
optic(al) sound recorder 光录音机;摄影录音机
optic(al) sound recording 光学录声
optic(al) sound reproduction 光声播放
optic(al) sound rushes 光学声样片
optic(al) sound technique 光声技术
optic(al) source 光源
optic(al) spacedivision multiplexing 光空分复用
optic(al) spatial modulator 光空间调制器
optic(al) spatial phase modulator 光学的空间相位调制器
optic(al) specification 光学鉴定
optic(al) spectral region 光学光谱范围
optic(al) spectrograph 光学摄谱仪
optic(al) spectrography 光学摄谱学
optic(al) spectrometer 光学分光镜;光谱仪
optic(al) spectroscope 光学分光镜
optic(al) spectroscopy 光学光谱学;光谱学;分光法
optic(al) spectrum 光学光谱;光谱;光功率谱
optic(al) spectrum analyzer 光谱分析器;光谱分析仪
optic(al) spheric(al) surface 光学球面
optic(al) spherometer 光学球径仪
optic(al) splice 光学接头
optic(al) splitter 分光束镜;分光器
optic(al) splitting system 分光系统
optic(al) square 直角转光器;直角(旋光)器;光学直角器;光学角尺
optic(al) stabilizer 光学稳定器
optic(al) stalk 视柄
optic(al) standard 光学标准
optic(al) Stark effect 光学斯塔克效应
optic(al) stereoscopic model 光学立体模型

optic(al) stimulation 光的刺激作用
optic(al) stop 光阑
optic(al) storage 光电存储器;光存储器;光存储
optic(al) storage capacity 光存储容量
optic(al) strain ga(u)ge 光学应变计
optic(al) strain meter 光学应变仪;光学应变计
optic(al) strength 光学强度
optic(al) strobe 光学频闪观测仪
optic(al) stroke 光学选通
optic(al) subcarrier communication 光副载波通信
optic(al) subsystem 光学分系统
optic(al) subtraction 光学相减
optic(al) sun 光学太阳
optic(al) super-heterodyne receiver 超外差(式)光接收机
optic(al) superposition 光叠加
optic(al) superposition device 光学叠像装置
optic(al) support 光学座
optic(al) surface 光学表面
optic(al) surface curvature 光学表面曲率
optic(al) surface inspection ga(u)ge 光学元件表面疵病检查仪;光学表面检查仪
optic(al) surveillance system 光学监视系统
optic(al) switch 光学开关
optic(al) symmetry 光学对称
optic(al) synchro system 光学同步系统
optic(al) system 光学系统
optic(al) system design 光学系统设计
optic(al) table 光学台;光学平台
optic(al) tach(e)ometer 光学视距仪;光学转速计
optic(al) taper 光纤锥
optic(al) tape reader 光电纸带输入机;光电带输入机;光带阅读器
optic(al) tapoff 光分接;光分出
optic(al) target 光学板;光学靶
optic(al) target coordinator 光目标坐标方位仪;光标方位仪
optic(al) target sounding 光学目标探测
optic(al) technique satellite 光学技术卫星
optic(al) technology 光学技术
optic(al) telecommunication cable 长距离通信光缆
optic(al) telegraph 光电报
optic(al) telemeter 光学测距仪
optic(al) telemetry 光遥测术;光学远距离测量法;光学测距法
optic(al) telescope 光学望远镜
optic(al) temperature measurement 光学的温度测量
optic(al) terminus 光学终端头
optic(al) test 用光学仪器检查;光学试验;光检查
optic(al) test bed 光学试验台
optic(al) test chart 光学测试卡
optic(al) tester 光测试仪
optic(al) testing instrument 光学测试仪器
optic(al) testing with laser 激光光学检验
optic(al) test wedge 光学检验劈
optic(al) theodolite 光学经纬仪
optic(al) theorem 光学定理
optic(al) thickness 光学厚度
optic(al) thickness ga(u)ge 光学测厚仪
optic(al) thickness moninitor 光学厚度监控器
optic(al) thickness of thin-film 薄膜光学厚度
optic(al) thickness tolerances 光学厚度公差
optic(al) thin film 光薄膜;光学薄膜
optic(al) thin film monitor 光学薄膜监测仪
optic(al) thread tool microscope 光学螺纹工具显微镜
optic(al) time division multiplexing 光时分复用
optic(al) time domain reflectometer 光学时域反射计;光时域反射计
optic(al) time domain refractor 光时域折射器
optic(al) to electric(al) transducer 光电转换器
optic(al) tolerance 光学容限;光学公差
optic(al) tooling 光学工装
optic(al) tooling bar 光学工具轴
optic(al) tooling level 光学水准仪;光学水平仪
optic(al) tooling scale 光学工具尺
optic(al) tool measuring machine 光学工具检测仪
optic(al) tracer arm 光学描迹笔
optic(al) tracker 光学跟踪装置
optic(al) tracker system 光跟踪系统
optic(al) track guidance 光学跟踪制导

optic(al) tracking 光学跟踪;光跟踪
optic(al) tracking control unit 光学跟踪控制器
optic(al) tracking satellite 光学跟踪卫星
optic(al) tracking system 光学跟踪系统
optic(al) tracking theodolite 光学跟踪经纬仪
optic(al) track photocell 光声带用光电管
optic(al) tract lesion 视束损害
optic(al) train 光具组
optic(al) transceiver 光收发机
optic(al) transfer characteristic 光学传递特性
optic(al) transfer function 光学传递函数
optic(al) transfer function instrument 光学传递函数测定仪
optic(al) transformation 光学变换
optic(al) transimpedance 光转移阻抗;光互阻抗
optic(al) transistor 光学晶体管;光敏晶体管
optic(al) transit 光学经纬仪
optic(al) transition 光跃迁
optic(al) transition rate 光跃迁率
optic(al) transit square 光学经纬仪直角器
optic(al) transmission 光(的)透射
optic(al) transmission coefficient 光透射系数;光传输系数
optic(al) transmission line 光透射线;光传输线
optic(al) transmission system 光传输系统
optic(al) transmissometer 光学大气透射仪
optic(al) transmitter 光发送机
optic(al) transmitter and receiver 光收发机;光端机
optic(al) transmitter aperture 光发射系统孔径
optic(al) transmitting set 光发射器
optic(al) trapping 光学捕获
optic(al) triangulation 光学三角测量
optic(al) tube length 光管长度
optic(al) tubes 光学镜筒
optic(al) tunneling 光的隧道效应
optic(al) turbulence 光学湍流
optic(al) turn table 光学回转台
optic(al) type extensometer 光学伸长计
optic(al) type font 光字体;反光式字体
optic(al) uniaxial crystal 一轴晶
optic(al) uniformity 光学均匀性
optic(al) unit 光学装置;信号灯器
optic(al) variable 光学变星
optic(al) vernier 光学游标
optic(al) video disk 光视盘;光录像盘
optic(al) video disk system 光盘电视系统
optic(al) viewfinder 光学寻像器;光学取景器;光学瞄准器;光学录像机
optic(al) viewing device 光学观察装置
optic(al) visual binary 光学目视双星
optic(al) voice communicator 光话音连接器
optic(al) voidness 光学纯
optic(al) wand 光棒
optic(al) water masses 光学水团
optic(al) wave 光波
optic(al) waveform meter 光波波形分析仪
optic(al) waveguide 光导纤维管;光学波导;光波导管;光波导
optic(al) waveguide application 光波导应用
optic(al) waveguide connector 光波导连接器
optic(al) waveguide coupler 光波导耦合器
optic(al) waveguide detector 光学波导探测器;光波导探测器
optic(al) waveguide fiber strengthening 光波导纤维的增强
optic(al) waveguide fiber 光纤;光波导纤维
optic(al) waveguide isolator 光波导隔离器
optic(al) waveguide lens 光波导透镜
optic(al) waveguide prism 光波导棱镜
optic(al) waveguide reflector 光波导反射镜
optic(al) waveguide splice 光波导接头
optic(al) waveguide switch 光波导开关
optic(al) waveguide transmission 光学波导传输
optic(al) wavelength 光波长
optic(al) wavelength meter 光波长计
optic(al) wedge 光楔;光劈
optic(al) whitener 荧光增白剂
optic(al) window 光学窗口;光学窗
optic(al) window interferogram 光学窗干涉图
optic(al) working 光学加工
optic(al) workshop 光学车间
optic(al) yardstick 光学码尺
optic(al) zeroing 光学调零

optic(al) zone 视带
opticator 光学系统部分;光学扭簧测微仪
optic-encoded system 光学编码系统
optician 光学仪器商;光学家;光学技师
opticla mass memory 光学大容量存储器
optico-acoustic(al) phenomenon 光声效应
opticochiasmatic 视交叉的
optics 光学
optics and optic(al) instrument 光学元件和仪器
optics flux 光通量
optics Kerr cell 光学克尔盒
optics laboratory 光学试验室
optics of liquids 液体光学
optics of metals 金属光学
optics of super-conductor 超导体光学
optics of the sea 海洋光学
optics of thin-film 薄膜光学
optics technology 光学工艺学
optics transducer 光学传感器
optidress 光学修正
optidress projector scope 光学修正投影显示器
optima 最佳的
optimal 最优的;最适的;最理想的;最佳的
optimal acceleration regime 最佳加速状态
optimal activity level 最佳作业水准
optimal-adaptive 最适应的
optimal adjustment 最优调整
optimal allocation 最优配置
optimal allocation for pollution load discharged 排污负荷优化配置
optimal approximate 最佳逼近
optimal approximate value 最近似值
optimal approximation 最优近似;最佳逼近
optimal assembly 最优装配
optimal basic feasible solution 最优基本可行解
optimal basic solution 最优基本解
optimal bias 最佳偏压
optimal binary search tree 最佳二叉查找树;最佳对半检索树
optimal block design 最优区组设计;最优块本设计
optimal bunching 最佳聚束
optimal capital structure 最适当的资金结构
optimal car flow 最佳车流量
optimal carrying capacity 最适承载容量;最适承载能力
optimal cash balance 最优现金结存额;最适量现金余额
optimal characteristics 最佳特性
optimal choice 最优选择;最佳选择
optimal circuit 最佳电路
optimal compaction 最优压实度
optimal condition 最优条件
optimal consistency 最优稠度
optimal constant control 最佳常数控制
optimal control 最优控制;最佳控制
optimal control algorithm 最佳控制计算法
optimal control equation 最优控制方程
optimal control law 最佳控制律
optimal control problem 最优控制问题
optimal control system 最优控制系统;最佳控制系统
optimal control theory 优化控制理论方法;最优控制理论;最佳防治理论
optimal coordination 最佳协调
optimal coupling 最佳耦合
optimal criterion 优化准则
optimal crooked principle 最佳弯曲原理
optimal damping 最佳阻尼
optimal decision 最优决策;最佳判定;最佳方案
optimal decision rule 最佳决定律
optimal desensitization 最优钝化
optimal design 最优设计;最优方案;最佳设计(法);最佳方案
optimal design of control network 控制网优化设计
optimal deterministic schedule 最优确定调度
optimal distribution 最佳分布
optimal dose 最适量
optimal estimate 最佳估算
optimal estimation 最优估计
optimal estimation algorithm 最佳估计算法
optimal estimator 最佳估计量
optimal export tax 最适当出口税
optimal factor employment 要素最优使用
optimal feasible solution 最优可行解

optimal field-anode temperature 最宜场阳极温度
optimal filter(ing) 最优滤波
optimal finish time 最优完成时间
optimal fitting 最佳拟合
optimal fitting curve 最佳曲线
optimal fitting plane 最佳拟合面
optimal flow 最佳车流量;分布最佳的车流
optimal focal depth 最佳震源深度
optimal frequency 最优频率
optimal growing condition 最优生长条件;最佳生长条件
optimal growth condition 适宜生长条件
optimal information value 最佳信息值
optimal intensity 最适强度
optimal inventory 最优在库量;最适量存货;最佳库存量
optimal investment 最适宜的投资
optimality 最优性
optimality analysis 最优化分析
optimality condition 最优性条件;最优条件;最佳条件
optimality criterion 最优性标准;最佳性准则;最佳标准
optimality equation 最优性方程;最佳性方程
optimality level 最优值
optimality principle 最优性原理
optimality standard 最优性标准
optimalize 使最佳化
optimalizing 最优的
optimalizing control 自寻最优点控制;自动寻求最优控制;极值调节
optimalizing input drive 最佳状态的输入给定器
optimal land utilization 节约用地
optimal length 最适长度
optimal level of emission 最优排放水平
optimal life 最适寿命
optimal linear predictor 最优线性预报值
optimal load 最适负荷;最佳荷载
optimally coded program(me) 最佳编码程序
optimally controlled 最佳控制的
optimal mark-up rate 最优加价率
optimal model 最优模型
optimal moisture for carbonation 最优碳化含水量
optimal parameter 优化参数;最佳参数
optimal parameter selection 最佳参数选择
optimal path 最优路径;最佳路径
optimal performance 最优性能
optimal plant size 最佳工厂规模
optimal pointwise control 最佳点态控制
optimal policies 最适宜的政策
optimal policy 最优政策;最优策略;最佳策略
optimal prediction 最优推算
optimal price policy 最适当价格政策
optimal process 最优过程
optimal production vector 最适当生产向量
optimal proportions 最适比例
optimal proportion zone 最适比例带
optimal ramification 最佳分级制
optimal randomized replacement 最佳随机化代换
optimal rate of mark 最适当加价率
optimal ratio 最适率
optimal ray 最优射线
optimal reactor shutdown 最佳停堆
optimal reverberation time 最佳混响时间
optimal routing program(me) 最佳推算程序
optimal salinity 最适盐度
optimal scheduling 最优调度
optimal scheduling strategy 最优调度策略
optimal searching 最佳检索法
optimal searching algorithm 最优检索算法
optimal segmentation method 最优分割法
optimal selection diversity 最佳选择分集
optimal set 最优集合
optimal size of business 最优营业规模
optimal smoother 最优平滑器;最佳平滑滤波器
optimal solution 最优解;最佳解
optimal solution graph 最佳解图
optimal sorting 优分
optimal stochastic control 最佳随机控制
optimal stop 任选停止
optimal strategy 最优策略;最佳策略
optimal stress 最佳应力
optimal structural design 最优结构设计
optimal structure 最优结构

optimal structure of imports and exports 最优化进出口商品结构
optimal switch surface 最佳开关面
optimal system 最优系统
optimal temperature 最宜温度;最适温度
optimal trajectory 最佳轨道
optimal value 最优值;最佳值
optimal zoom climb 最佳轨道上升
optiman 最佳人选
optimat 优化规范
optimatic 光电式高温计
optimatic pyrometer 光电式高温计
optimeter 光学计;光学比较仪;光学比长仪;光度计;光电比色计;投影比较仪
optimeter tube 光学比较仪光管
optiminimeter 光学测微计
optimiser 优化程序
optimistic estimate 乐观估计
optimistic fuzzy automaton 乐观的模糊自动机
optimistic-pessimistic forward pruning 乐观悲观正向修剪
optimistic time 乐观时间
optimizable loop 可优化循环
optimization 优选法;优化;最优化;最佳条件选择;最佳化;求最佳参数
optimizational function 优化功能
optimization algorithm 最优化算法;最佳化算法
optimization approach 最优化方法
optimization attribute 最佳化属性
optimization calculation 优化计算;最优化计算
optimization characteristics 最优化特性
optimization control 优化控制
optimization criterion 最优化原则
optimization design 最优化设计
optimization design of sewer pipe system 排水管理系统最优化设计;污水管道系统优化设计
optimization design of unit operation in water treatment 水处理单元过程作业优化设计
optimization design of unit process of water treatment 水处理单元过程(最)优化设计
optimization design of wastewater treatment 废水处理最优化设计
optimization design of wastewater treatment system 污水处理系统优化设计
optimization design of water treatment system 给水处理系统最优化设计;水处理系统优化设计
optimization in design of unit processes in wastewater treatment 污水处理单元过程最优化设计
optimization in design of wastewater treatment system 废水处理系统最优化设计
optimization method 最优化方法
optimization methodology 最佳化方法论
optimization mode 优化模式
optimization model 优化模型;最优化模型;最优化模式
optimization of braking force distribution 制动力最佳分布
optimization of branch instruction 转移指令优化
optimization of environment(al) system 环境系统最优化
optimization of heat exchanger 换热器的最优化
optimization of instruction 教学最优化
optimization of radiation protection 辐射防护最优化;放射防护最优化
optimization of scalar instruction 标量指令优化
optimization of separation 分离优化
optimization of system design 系统设计的最佳化
optimization of systems with random parameters 随机参数系统最优化
optimization of vertical curves 竖曲线最优设计
optimization of water management system 优化水管理系统
optimization pass 优化遍数
optimization performance 最优化性能
optimization phase 最佳化阶段
optimization possibility 优化可能性
optimization problem 优化问题
optimization procedure 最优化方法;最优化处理过程;最优法处理过程
optimization suitability 优化适合性
optimization system 最佳化系统
optimization technique 优选技术;最优化技术;最佳化技术
optimization theorem 最佳化定理

optimization theory 最优化理论
optimize 优化;最佳化
optimized 最佳化的
optimized attenuation 最佳衰减
optimized crane productivity 起重机最优生产率
optimized design of pile foundation stiffness to reduce differential settlement 变刚度调平设计
optimized design of structure 结构优化设计
optimized dispatch 优化稠度
optimized drilling 优化钻进;最佳化钻探
optimized drilling program(me) 最优化钻探程序;最优化钻井程序
optimized drilling service 最佳化钻探服务
optimized method 优选法
optimized quadruple 优化四元组
optimized system(s) design 系统设计最优化
optimized vehicle 最佳车
optimizer 优化器;最佳控制器;最佳化器
optimizer control 最佳工况调节装置
optimizer system 最佳喷油定时自调系统
optimize the allocation of resources 优化资源配置
optimize the structure of expenditures 优化支出结构
optimizing 最优化;最佳化
optimizing capacity 优化容量
optimizing characteristics 优化特性
optimizing computer 优化计算机
optimizing control 最优控制;最佳控制
optimizing control action 优化控制动作
optimizing controller 最佳控制器
optimizing control mode 最优化控制方式
optimizing control system 最优控制系统;最优控制系统
optimizing control theorty 最优控制理论
optimizing decision making 最佳决策
optimizing level 最佳电平
optimizing method 最优化法
optimizing parameter value 优选参数值
optimizing phase 优化阶段
optimizing planning for equalization treatment of river outfall 河流排污口均匀处理的最优化规划
optimizing planning of regional waswater treatment for river water quality 河流水质区域污水处理优化规划
optimizing planning of river water quality 河流水质模拟优化设计
optimizing regulation 优化调度
optimizing sort performance 分类效率最佳化
optimizing translator 优化翻译程序
optimizing treatment planning of pollutant quantity at river outfall 河流排污口最优化处理规划;河流排污口排污量优化处理设计
optimum 最有利的;最优化;最优的;最适用性;最适条件;最佳的;最好的
optimum aerial target sensor 最佳空中目标选定传感器
optimum allocation 最佳配置;最佳分配;最佳配置
optimum allocation of resources 资源最适分派;资源最佳分配
optimum allocation of sample 样本的最优分配
optimum allocation of social resources 社会资源最佳配置
optimum alternative 最优方案;最佳方案
optimum angle 最佳角度
optimum angle of attack 最佳攻角
optimum angle of incidence 最佳倾角
optimum approach path 最佳进场航线
optimum array current 最佳天线阵电流
optimum artificial dispersion 最佳人工散布
optimum asphalt content 最佳沥青含量;最佳地沥青含量
optimum balance 最适度的平衡
optimum ball charge 最适当的装球量;最佳装球量
optimum band 最佳波段
optimum bare equipment cost of shift conversion 变换转化最优净设备费用
optimum beam 最束
optimum bend radii 最佳河湾半径
optimum bend radius 最佳弯曲半径;河湾最佳半径
optimum bias 最佳偏置
optimum bit hydraulic horse power 最优钻头水马力;最优缺头水功率
optimum bit weight 最优钻压
optimum break 精选粒度

optimum bunching 最佳聚束
optimum burning altitude 最佳照明高度
optimum burst correcting code 最佳突发错纠正码
optimum capacity 最优能力;最佳容量;最佳能力
optimum carrier frequency 最佳载频
optimum catch 最适渔获量
optimum channel dimension 最佳航道尺度
optimum charge 最佳装药
optimum city size 城市最佳规模
optimum climate 最适气候;最佳气候
optimum climb 最佳状态上升
optimum climb path 最佳上升航线
optimum code 最优编码;最优(代)码;最佳码;最佳代码;最佳编码
optimum coding 最佳编码;排程序合理化
optimum column efficiency 最佳柱效(能)
optimum column length 最佳柱长
optimum column temperature 最佳柱温
optimum combination 最优组合
optimum comfort angle 最舒适角度
optimum compaction 最佳致密作用;最佳压实
optimum concentration 最适浓度
optimum condition 最适条件;最佳状态;最佳条件;最佳工况;最适宜情况
optimum configuration 最佳外形
optimum consumption 最佳耗量
optimum consumption use 最佳耗用量;最佳耗水量
optimum consumptive use 最佳消耗量;最佳耗用量;最佳耗水量
optimum contact angle 最佳接触角
optimum control 最佳控制;最佳控制
optimum controller 最佳控制器
optimum control theory 最佳控制理论
optimum cost 最佳费用;最佳成本
optimum coupling 最佳耦合
optimum crash program(me) 最佳赶工计划
optimum criterion 最佳准则
optimum cure 最适硫化;最适处置;最佳处治;正硫化
optimum currency areas 效益最大的货币区
optimum curve 最优曲线;最适曲线
optimum-curve for operation 最佳工作曲线
optimum cutting condition 最佳切削条件
optimum cutting speed 最佳切削速度
optimum damping 最佳阻尼
optimum damping parameter 最佳阻尼参数
optimum data 最佳数据
optimum decision rule 最佳判定规则
optimum decision system 最佳决断系统;最佳决策系统
optimum density 最优密度;最适密度;最佳密度
optimum density range 最佳密度范围
optimum depth 最佳水深;最佳覆盖厚度
optimum depth of cut 最佳挖掘深度
optimum design 优化设计;最优设计;最佳方案;最佳设计(法)
optimum design of sample 样本的最优设计
optimum detecting filter 最佳检测滤波器
optimum development 最优开发;最佳开发
optimum development plan 最优开发计划;最佳开发方案
optimum diaphragm 最佳光阑
optimum distance 最佳距离
optimum distribution 最佳分布
optimum division 最优分割
optimum dosage 最优投量
optimum drawdown 最佳泄降深度;最佳泄降量;最佳泄降;最佳泄降;最佳降深(抽水)
optimum drift angle 最佳漂移角
optimum drying condition 最适干燥条件
optimum duration 最佳工期
optimum dynamic(al) characteristic 最优动力特性
optimum efficiency 最优效率;最佳效率
optimum environment 最佳环境
optimum environmental quality 最佳环境质量
optimum error detection code 最佳误差检测码
optimum estimate 最优估计
optimum estimation 最佳估计
optimum estimation value 最佳估算值
optimum evasive maneuver 最佳规避机动
optimum expectation 最优期望值;最佳期望值

optimum extraction 最佳提取
optimum extrusion angle of die 最佳挤压凹模角
optimum fatigue strength 最佳疲劳强度
optimum feed location 最宜进料位置;最佳进料点
optimum filter 最佳滤波器
optimum filtering 最佳滤波
optimum fine aggregate percentage 最佳含砂率
optimum fit 最佳拟合
optimum fitting method 最佳符合法
optimum flow 最佳流量
optimum flow rate 最优排量;最优泵量;最佳流速
optimum focus 最佳焦点
optimum focusing 最佳聚焦
optimum formation 合理队形
optimum formula 最佳公式
optimum frequency 最佳频率
optimum gain 最佳增益
optimum-gain frequency 最佳增益频率
optimum gas parameter 最佳燃烧参数
optimum gas velocity 最佳气体流速
optimum general feasible solution 最优基本可行解
optimum gliding angle 最佳滑翔角
optimum gradation 最优级配;最佳级配
optimum gradient method 最优斜量法;最优梯度法
optimum grind 最适当的最终研磨;最佳磨矿粒度
optimum grind size 最适当的最终研磨粒度
optimum ground elevation 最佳地面高程
optimum growing condition 最优生长条件;最佳生长条件
optimum growth temperature 最适生长温度
optimum guide length 最佳波导管长度
optimum heading 最佳航向
optimum height 最佳高度
optimum height of burst 最佳爆炸高度
optimum height of cut 最佳挖掘高度
optimum hoisting speed 最佳提升速度
optimum humidity 最适湿度;最佳湿度
optimum hydraulic cross-section 最佳水力横断面;最佳水力断面;水力最优断面
optimum hydraulic section 最佳水力断面
optimum illumination 最佳照度
optimum image coder 最佳图像编码器
optimum impedance 最佳阻抗
optimum import 适时进口
optimum intake contour 最佳进气口外形
optimum intercept course 最佳截击航向
optimum interception course 最佳截击航向
optimum interpolation 最优内插;最佳插值
optimum interval interpolation 最优(选)点插值
optimum inventory 最佳库存货;适当存货
optimum lattice 最佳栅格;最佳晶格
optimum length 最佳长度
optimum leverage 最适资本结构
optimum liberation point 最佳解离点
optimum light intensity range 最适光强度较差
optimum linear filtering 最佳线性滤波;最大线性过滤
optimum linear measurement 最佳线性测量
optimum linear prediction 最优线性预测值;最佳线性预测法;最佳线性预测值;最适线性预测法;最佳线性预测
optimum linear system 最优线性系统;最佳线性系统
optimum load factor 最优负荷因素
optimum load(ing) 最佳负载
optimum load resistance 最佳负载电阻
optimum lot size 最佳生产或订购的一次批量;最佳批量
optimum match(ing) 最佳匹配;最佳配合
optimum merging pattern 最佳归并模式
optimum milling rate 最佳磨矿速度
optimum mix 最佳配合比;最佳配合比
optimum mix design 最佳配合比设计
optimum mixing strength 最佳混合气强度
optimum mixture 最佳混合物
optimum mode of operation 最佳工作方式
optimum moisture 最适水分;最佳水分
optimum moisture content 最优含水量;最适含水量;最佳湿度;最佳含水量
optimum moisture level 最优湿度值
optimum moisture percentage 最适含水率
optimum moisture value 最优湿度值
optimum network 最优网络

optimum nose curve 头部最佳形状
optimum nozzle 最佳喷嘴
optimum number 最佳数
optimum of an economy 经济的最优化
optimum oil content 最佳含油量
optimum operating frequency 最佳工作频率
optimum operating point 最优操作点
optimum operating slope 最佳工作边坡
optimum operating voltage 最佳工作电压
optimum orbit 最佳轨道
optimum order point 最佳订货点;最佳订购点
optimum order quantity 经济采购量
optimum organization of labo(u)rs 劳动优化组合
optimum output 最佳输出
optimum packing 最优堆积
optimum packing of granular soil 粒状土最优堆积方式
optimum parameter 优化参数;最优参数;最佳参数
optimum particle packing 最优颗粒堆积
optimum penetration rate 最优钻速
optimum penetration speed 最佳切入速度
optimum performance 最佳性能;最佳特性
optimum performance curve 最佳特性曲线
optimum performance parameter 最佳性能参数
optimum pH 最适 pH 值;最佳 pH 值
optimum phase 最佳相位
optimum pH range 最佳 pH 值上下限
optimum picture quality 最佳图像质量
optimum plant size 最适当的工厂规模
optimum plot size 最适面积
optimum point 最优点
optimum point of coagulation 最优凝聚点
optimum polynomial 最佳多项式
optimum population 最适种群;最适人数;最佳人口;最合适人口数
optimum position 最佳位置
optimum potential difference 最佳电位差
optimum practicable technology currently available 最佳实用可行技术
optimum practical gas velocity 最实际气体流速
optimum prediction 最佳预测
optimum pre-ozonation dose 最佳预臭氧化剂量
optimum pressure 最适压
optimum procedure 最优化程序
optimum process performance 最佳工艺特性
optimum Proctor density 最佳普氏密度
optimum profile 最优纵断面线;最佳纵断面线
optimum program(me) 最佳程序
optimum programming 最优规划(方案);最优程序设计;最佳规划法;最佳程序设计
optimum proportion 最佳掺和量比
optimum proportioning 最优配合比;最佳配合比;最佳比例
optimum purchase 最优采购量
optimum pure predictor 最佳纯预测器
optimum quantity order 最佳订购单
optimum quantity service 最佳服务
optimum range 最佳幅度
optimum rate of interest 最优惠利率;最适利率
optimum ratio 优越(化)比例;最适比率
optimum ratio between the cations 阳离子间最适比率
optimum reactor design 最优反应器设计
optimum receiver 最佳接收机
optimum reception 最佳接收法
optimum recording current 最佳记录电流
optimum reflectivity 最佳反射率
optimum reflux 最适宜回流
optimum reflux ratio 最佳回流比率
optimum reinforcement 最佳配筋
optimum relaxation factor 最优松弛因子
optimum relaxation parameter 最优松弛参数
optimum relay servomechanism 最佳继电伺服机构;最佳继电器随动系统
optimum reorder quantity 最佳订购量
optimum resolution 最佳分辨率
optimum resolution bandwidth 最佳分辨带宽
optimum resolving power 最佳分辨力
optimum resonance 最佳谐振;最佳共振
optimum response 最佳反应
optimum reverberation 最佳交混回响;最佳混响
optimum reverberation time 最佳混响时间
optimum rigidity distribution 最佳刚度分布
optimum rotary speed 最优转速

optimum routing 最佳航线选定;最佳航线拟定
optimum rule 最佳判决定则
optimum running density 最佳操作密度
optimum sampling density 最佳采样密度
optimum satellite configuration 最优卫星配置
optimum scale 最适当规模
optimum scale integration 最佳规模集成
optimum scheme 最优方案;最佳方案
optimum seeking method 优选法
optimum seismic design 最优抗震设计
optimum sensitivity 最佳灵敏度
optimum separation factor 最佳分离系数
optimum setting 最优调整;最佳调整
optimum shape 最佳形状
optimum ship routing 最佳(船舶)航线
optimum short-delay time interval 最优微差时间间隔
optimum signal 最佳信号
optimum signal to noise ratio 最佳信号噪声比;最佳信噪比
optimum size 最佳粒度;最佳尺寸
optimum size of power station 电站的最优尺寸
optimum soil moisture 土壤最佳含水量;土壤最适含水量;土壤最佳湿度;土壤最佳含水量
optimum soil process 最优土法
optimum solution 最优解;最佳解法
optimum speed 临界速率;临界速度;最佳速率;最佳速度
optimum speed of mill 磨机最佳工作转速
optimum splitting of azimuth 最佳的方位间距
optimum stability 最佳稳定度
optimum stack 最佳叠加
optimum stacking 最佳聚积
optimum standard gravity 最佳标准重力
optimum start-up 最佳启动
optimum strategy 最优策略
optimum strength 最优强度
optimum strength distribution 最优强度分布
optimum stress 最佳应力
optimum stripping ratio 最佳剥采比;最佳剥比
optimum structural design 最优结构设计
optimum structure 最优结构;最佳结构;结构优化
optimum successive overrelaxation 最优逐次超松弛
optimum sustainable yield 最适持久产量
optimum switching function 最佳开关函数
optimum switching line 最佳开关线路
optimum system control 最佳系统控制
optimum table-speed selector 最优转速选择器
optimum tariff 最适关税
optimum temperature 最适温度;最佳温度
optimum temperature difference 最适宜温差;最适温差
optimum temperature gradation 最佳温度梯度
optimum temperature of solubilization 最佳加溶温度
optimum temperature profile 最适温度分布图
optimum temperature sequence 最优温度序列
optimum thickness 最佳厚度
optimum time of burst 最佳爆炸时刻
optimum tine angle 最佳的齿倾角
optimum tooth wearing height 牙齿最优磨损量
optimum track of ship routing 船舶最佳航线
optimum track ship-routing 最佳航线(选定);最佳航线拟定
optimum traffic frequency 最佳通信频率;最佳工作频率
optimum traffic light pattern 最佳选交通信号灯相模式
optimum transfer 最佳过渡
optimum transmission frequency 最佳传输频率
optimum transmitted current 最佳传输电流
optimum tuning point 最佳调谐点
optimum twist 最适捻象
optimum useful life 最佳使用年限
optimum utilization 最适度的利用
optimum value 最优值;最佳值
optimum value of standard deviation 标准差的最优值
optimum value of variance 方差的最优值
optimum velocity 最适宜流速;最佳速度;最佳流速;临界速度
optimum viewing distance 最佳视距
optimum voltage 最佳电压

optimum water content 最优含水量;最适含水量;最佳含水量
optimum watering 最适水分
optimum waveform 最佳波形
optimum wavelet analysis 优化子波处理
optimum weighted stack 最佳加权叠加
optimum wide band vertical stack 最佳宽带垂直叠加
optimum width-depth ratio 最佳宽深比
optimum workability 最佳和易性;最佳工作能力
optimum working efficiency 最佳工作效率
optimum working frequency 最佳工作频率
optimum working temperature 最佳工作温度
optimum work time 最优工作时间
optimum yield 最适产量;适宜产量;适当的产量
optimum zero-lag filter 最佳无滞后滤波器
option 选择项目;选择(权);选件;可选项;任选项;备选样机
option agreement 选择权协议
optional 可选择(的);任意(选择)的;可选的
optional additional charges 选港附加费
optional annex 任选附则
optional attachment 任选附件
optional attribute 任选属性
optional bill of lading 选择港提单;选港提单;任选港提单
optional blank 任选空白
optional block skip 跳过任选程序段
optional board 任选板
optional bond 可提前偿还的债券
optional calculation 选择性计算
optional capital 选择性资本
optional cargo 选择卸港货物;选(择)港货物;任意卸港货物;任选港货物
optional charges 选择卸货港(所增运)费;任选港附加费
optional clause 任选条款;任意选择条款;任选条款
optional consumption 选择性消费;随意消费
optional course 选修课
optional currencies loan 选择货币贷款
optional delivery 选择(卸货)港交货;任意卸货港交货;任选港交货
optional destination 任选目的港
optional destination and option fee 选择卸货港和附加费
optional dividend 任选股利发付方法
optional equipment 选购的附加设备;专用设备;供选购的设备;可选设备;附加设备;备选设备
optional exploration 择优勘探
optional extended area service 市郊选择通信业务
optional extras 任选附件
optional feature 任选特点;任选机构;任选功能;选定的特性
optional fee 任选港附加费
optional flange 任意法兰;任意凸缘
optional forward exchange contract 选择远期外汇契约
optional function 选择机能;任选功能
optional function board 任选功能板
optional gear ratio 选定传动比;可变齿轮速比
optional half instruction 任选半指令;条件停机指令;选择停机指令
optional halt 临时停车站
optional information 随意选择资料
optional instruction set 选择指令组
optional interface 任选接口
optional interrupt 任选中断
optional item 可选择项目
optional label 任选标号
optional legal tender 可选择法偿
optional local peripheral 任选局部外围设备
optionally-signed constant 任选记号常数
optional member 任选成员;任选成分
optional modes of settlement 人寿保险赔偿方式选择
optional multilevel interruption 任选多级中断
optional parallel instrument bus 任选并行仪器总线
optional parts 任选零件;任选件;可选择件
optional passive reflector pods 选择性被动反射器吊舱
optional pilotage 自由引水
optional port 选择港;任选港;任选目的港
optional port charges 选择卸货港费
optional position 任意位置

optional priority 任选优先
optional rate 选择收费制
optional requirement 任选要求
optional resident routine 随意驻留例行程序
optional right 选择权
optional right for cargo 货物选择权
optional route 方案进路
optional sampling 任意抽样
optional service 可选择受理业务
optional shipment 任选港的装船
optional stop 任停;条件停机
optional stop instruction 任选停止指令;条件停机指令;随意停机指令;选择停机指令
optional stop order 任意停机指令
optional stowage 选港装载
optional stowage clause 任意装载条款
optional string 任选串
optional suppression 随意消除(法)
optional surcharge 选择卸货港费
optional test 选定的试验;选定(项目的)试验
optional tires 备选轮胎
optional transmission gear 任意传动装置;任意传动齿轮
optional units 选择供应的附件
optional word 任选字;辅助字
option board 选择板;选配电路板
option bond 可提前偿还的债券
option buyer 选择权买方
option charges 选港费
option code 选择码
option contract 选择权合同
option dealing 选择权交易;期权交易
option delivery 选港交货
option delta 选择权变化率
option exchange 选购权交易所
option exchange contract 选择权外汇契约
option fee 附加费
option for additional space 可增加租赁面积的选择权
option for loading/discharging port 装卸货物港口的优化
option for shipment of general cargo 件货装运优化
option forward contract 期货的远期合同
option function 选择功能
option holder 选择持有人
option interface 选择接口
option list 功能选择表
option market 选购权市场;期权买卖市场
option money 权利金;定价金
option of building field 选择建筑基地
option of device 设备选择;设备选型
option of mine base 矿山基地选择
option on debt instruments 债务凭证期权
optionor 提供选择权的人
option panel 选择板;选片
option period 选择权有效期限
option program(me) 选择程序
option route 迂回进路
options clearing corporation 选择权清算公司
option seller 选择权卖方
options for optimization 可供优选的方案
options on foreign currency futures 外币期货期权
options on futures contract 期货合约的期权交易
options paper 决策资料
option standard 选择标准
option switch 选择开关
option table 选择表
option to purchase contract 优先购买合同
option to purchase leased property 变租用为购置的选择权
option trading 选购权买卖
option warrant 选购权证
option writer 选择权卖方
optiphone 特种信号灯
optitherm radiometer 光热辐射计
opto-acoustic 光声
opto-acoustic(al) absorption cell 光声吸收盒
optoacoustic(al) detection 光声检测
optoacoustic(al) detection method 光声检测法
optoacoustic(al) detector 光声探测器
optoacoustic(al) effect 光声效应
optoacoustic(al) modulation 光声调制
optoacoustic(al) modulator 光声调制器

optoacoustic(al) processor 光声处理机
optoacoustic(al) spectroscopy 光声光谱学
optoacoustic(al) transducer 光声变换器
optocoupler 光耦合器;光隔离器
optodeflector 光偏转器
optoelectric(al) scanning 光电扫描
optoelectric(al) shift register 光电移位寄存器
optoelectronic 光电子的
optoelectronic amplifier 光电子放大器
opto-electronic antenna 光电天线
optoelectronic cell 光电池
optoelectronic circuit 光电子电路
optoelectronic cold cathode 光电子冷阴极
optoelectronic converter 光电子转换器
optoelectronic coupler 光电耦合器
optoelectronic data processor 光电子数据处理机
optoelectronic data storage 光电数据存储
optoelectronic detector 光电检测器
optoelectronic device 光电设备
optoelectronic digital logic 光电数字逻辑
optoelectronic digital logic system 光电数字逻辑系统
optoelectronic diode 光电二极管
optoelectronic directional coupler 光电定向耦合器
optoelectronic director head 光电跟踪头
optoelectronic image processing 光电子图像处理
optoelectronic integrated circuit 光电子集成电路
optoelectronic isolator 光电子隔离器
optoelectronic matrix memory 光电矩阵存储器
optoelectronic memory 光电(子)存储器;光电储存器
optoelectronic modulator 光电子调制器
optoelectronic multiplex switch 光电转换开关
optoelectronic noise rejector 光电噪声抑制器
optoelectronic panel 光电子板
optoelectronic plate 光电子板
optoelectronic power generator 光电功率发生器
optoelectronic processor 光电处理机
optoelectronic pulse amplifier 光电脉冲放大器
optoelectronic receiver 光电接收机
optoelectronics 光学电子学;光电子学
optoelectronic scanner 光电子扫描器
optoelectronic scanning 光电扫描
optoelectronic shift register 光电移位寄存器
optoelectronic shutter 光电子学光闸
optoelectronic storage 光电子存储器
optoelectronic switch 光电子开关
optoelectronic system 光电(子学)系统
optoelectronic transistor 光电晶体管
optoelectronic transmitter 光电子发射机
optoeletronic isolator 光隔离器
optogalvanic effect 光电效应
optogalvanic signals 光电流信号
optogalvanic spectrum 光电流谱
opto-gyral illusion 眼旋【转】错觉【测】
opto-hybrid integrated circuit 光混合集成电路
opto-hybrid receiver 光混合接收机
opto-hybrid transmitter 光混合发射机
opto-ionics 光离子学
optoisolator 光隔离器
optokinetic response 视动反应
optomagnetic 光磁(的)
optomechanical line scan 光学机械行扫描
optomechanical line scan recorder 光学机械行扫描记录器
optometer 视力计
optometry 视力测定法
optominimeter 光学指针测微计;光学千分比较仪;光电测微计
opto-miniter 光电测微计
optomotor response 视动反应
optophotography 光学摄影
optoscale 光标度
opto-sensor 光传感器
optothermal 光热
optothermal detector 光热探测器
opto-thermodynamics 光热力学
optotransistor 光晶体管
optron 光导发光元件
optronic 光导发光的
optronic relay 光电子继电器
optronics 光电子学;光导发光学
opulent renaissance 丰富的文艺复兴时期风格
opus alexandrinum 大理石块铺面

opus antiquum 粗毛石砌体；有混凝土填芯的块石工；古罗马粗毛石砌法
opus emplectum 空斗石墙砌体；空斗石墙
opus figuratum 成型的作品
opus incertum 有混凝土填芯的块石工程；夹混凝土块毛石砌体；混凝土芯墙毛石砌体；混凝芯的块石工
opus interrasile 装饰雕刻法（一种将背景平雕或将花纹平雕的方法）
opus isodomum （古罗马、希腊整齐砌筑的）规格砌块；块石端砌
opus isodomum of granite blocks 整块花岗石块端砌；花岗石块端砌
opus latericium （罗马砖砌墙或贴砖的）混凝土墙；嵌砖混凝土墙
opus listatum 砖石筑墙；砖石砌墙
opus lithostratum （古希腊和罗马任何的）镶面石层；装饰性铺砌
opus mixtum 砖面砌圬工墙
opus musivum （用彩色玻璃或搪瓷的）罗马马赛克；墙面镶嵌装饰
opus pseudoisodomum 古罗马方石成行砌墙
opus quadratum 块石顺垒砌法；方石筑墙（古罗马建筑）
opus reticulatum 罗马墙上小角锥石网状镶面；方石网眼筑墙
opus rusticum 凿有节疤的粗面石头；粗面石工
opus sectile 砖花地面；马赛克砖；嵌小方块马赛克
opus signinum 有几何图形的罗马式马塞克；古罗马引水渠内灰泥或水泥抹面
opus spicatum 古罗马人字形砌砌；人字形嵌砖细工
opus tectorium 古罗马一种内墙抹面
opus tesselatum 古罗马大理石马赛克装饰铺面；嵌云石块细工
opus testaceum 古罗马砌体用碎瓷瓦贴面
opus vermiculatum 按图案雕成虫迹形马赛克装饰
oqaque coat 不透明层
ora 奥拉风
oracle sanctuary 传神言者庇护所；传神言者圣地
oracular temple 占卜神使庙宇
oraganic mineral ultrafiltration membrane 有机矿质超滤膜
oragenic soil 山地土壤
oral agreement 口头协议
oral audit report 口头审计报告
oral contract 口头合约；口头协议；口头合同
oral deed 口头契约
oral evidence 口头证据
oral form 口头形式；口头方式
oral instruction 口头指示
oral lethal dose 口服致死剂量
oral notice 口头通知
oral notification 口头通知
oral pleading 口头答辩
oral purchasing 口头采购
oral report 口头报告
oral representation 口头声明
oral testimony 口头证明
orange 橙色染料；橙色的
orangeade 橙汁
orange alizarin(e) lake 橙黄茜素色淀
orange and lemon cream 橙柠奶酪
orange base 橙碱
orange-brown 橙棕色；橙褐色
Orangeburg 奥林奇伯格图案
orange cadmium 镉橙
orange chestnut 橙色
orange chrome yellow 橙铬黄
orange clear 浅橙黄色
orange colo(u)r 橙色；橙黄色
orange-colo(u)red 橙色的
orange filter 橘黄色滤光镜；橙色滤光镜
orange free state 橙游离状态
orange indicating light 橙色指示灯
orange juice 橙汁
orange lac 虫漆片；橙色紫胶
orange lake 橙色色淀
orange lead 铅橙；橙色铅；橙铅
orange light 淡橙黄色
orange mineral 铅橙；无机橙红
orange oil 橙油
orange oil separator 橙油分离机
orange oxide 三氧化铀；橙色氧化物
orange pale 淡白橙色

orange-peel clamshell 梅花抓斗
orange peel 瓣形抓斗；粗状表面；橘皮纹饰；橘皮；橙皮；油漆皱皮
orange-peel bucket 荷花抓斗；多爪式抓斗；多瓣抓斗；瓣形抓斗；瓣形戽斗；莲花抓斗；橘瓣（式）抓斗；梅花抓斗
orange-peel dredge(r) 多瓣抓斗挖泥机；多瓣抓斗挖泥船；橘瓣抓斗式戽斗挖土机
orange-peel effect 橘皮纹
orange-peel excavator 瓣形抓斗挖土机；橘瓣式抓斗挖土机；三瓣式戽斗挖土机；三瓣或四瓣抓斗挖掘机；多瓣抓斗式挖土机
orange-peel finish 橘纹漆
orange-peel grab 橘瓣(式)抓斗；橘瓣形抓斗；荷花抓斗；多爪抓岩机；多瓣抓岩机；多瓣抓斗；瓣式抓斗
orange-peel grab for ore 多瓣矿石抓斗
orange peeling 橘皮纹饰；瓣形戽斗；油漆皱皮
orange-peel oil 橙皮油
orange-peel sampler 荷花抓斗式采样器；多瓣抓斗采样器
orange-peel shape dam 双曲拱形坝；双曲坝
orange pekoe 橙黄白毫
orange pigment 橙色颜料
orange red 橙红(色)
orange-red sienna 橙红色的浓黄土
orangery 养橙温室；柑橘园
orange sapphire 橙刚玉
orange-seed oil 橙子油
orange shellac 紫胶漆片橙（液）；橘黄紫胶漆；橙紫胶；橙虫胶
orange spectrometer 橙色分光计
orange-tan 橙褐色
orange toner 橙(色)调色剂
orange yellow 橘黄(色)；橙黄色(的)；橙黄
orangic index 有机物指数
orangite 橙黄石
oranium bronze 铝青铜
ora serata retinae 锯齿状缘
oratory 小礼拜堂；祈祷室
Oratory of Gallerus 伽勒里斯的祈祷室
orb 朴素的圆浮雕饰；盲拱；球形物
Orbal oxidation ditch 奥贝尔氧化沟
orbed 球状的
orbicular 回转椭圆形的；球状的；球形构造的
orbicular gabbro 球状辉长岩
orbicular structure 球状构造；球形结构
orbicule 微粒体；同心球粒
orbit 轨道运行
orbit adjusting subsystem 轨道调整子系统；轨道调整分系统
orbit adjusting system 轨道调整系统
orbital 轨域；轨函数；轨道的；环城公路
orbital acceleration 轨道加速度
orbital altitude 轨道高度
orbital angular momentum 轨道角动量
orbital angular momentum vector 轨道角动量矢量；轨道动量矩矢量
orbital arrangement 轨道排列
orbital basin 圆形池
orbital coordinate system 轨道坐标系
orbital correlation diagram 轨道相关图
orbital curve 轨道曲线
orbital data 轨道数据
orbital decay 轨道衰减
orbital degeneracy 轨道简并度
orbital determination 轨道计算
orbital direction 轨道方向
orbital dynamics 轨道动力学
orbital effect 轨道效应
orbital electron 轨道电子
orbital electron capture 轨道电子捕获
orbital elements 轨道元(素)；轨道要素；轨道根数
orbital ellipse 轨道椭圆
orbital engine 轨道式发动机
orbital equation 轨道方程
orbital hybridization 轨道杂化
orbital inclination (angle) 轨道倾角
orbital latitude 轨道纬度
orbital longitude 轨道经度
orbital magnetic moment 轨道磁矩
orbital maneuvering system 轨道机动系统
orbital method 轨道法
orbital method of satellite geodesy 卫星大地测量轨道法
orbital moment 轨道角动量
orbital motion 轨道运行；轨道运动
orbital-motion mixer 行星式搅拌机；行星式混合机
orbital movement 轨道运动
orbital node 轨道节点；轨道交点
orbital orientation 轨道定向
orbital pad sander 轻轨型垫式打磨机
orbital path 轨道
orbital period 轨道周期；轨道运动周期
orbital phase 轨道相位
orbital phase conservation 轨道位相守恒
orbital pipe welder 管子环缝自动焊机
orbital plane 轨道平面；轨道面
orbital precession 轨道进动
orbital prediction 轨道预测；轨道预报
orbital process 圆形法
orbital quantum number 轨道量子数
orbital radio telescope 轨道射电望远镜
orbital radius 轨道半径
orbital rendezvous 轨道会合
orbital running frequency 轨道运行频率
orbital sander 轨道磨光机；天体轨道式磨砂机
orbital space 轨道空间
orbital space station 轨道空间站
orbital stability 轨道稳定性
orbital symmetry 轨道对称性；轨道对称
orbital symmetry rule 轨道对称定则
orbital theory 轨函数说
orbital tilt 轨道倾斜
orbital timing 轨道定时
orbital tracking 轨道跟踪
orbital tracking data 轨道跟踪数据
orbital trajectory 轨道
orbital velocity 轨道运动速度；轨道速度；卫星绕主星运行的平均速度
orbital velocity of satellite 卫星轨道速度
orbital welding 轨迹焊
orbital workshop 轨道工作室
orbit circumference 轨道周长
orbit control 轨道控制
orbit correction 轨道改正
orbit determination 轨道计算
orbit elements of satellite 卫星轨道参数
orbiter 轨道飞行器
orbit imaginary 轨道影像
orbit improvement 轨道改进
orbiting astronomic(al) observatory 天体观测卫星；轨道天文观测站
orbiting collision 轨道碰撞
orbiting geophysical observatory 轨道地球物理观测站
orbiting geophysical observatory of artificial earth satellite 人造地球卫星地球物理观测站
orbiting reflection 轨道反射镜
orbiting satellite 轨道运行卫星
orbiting solar observatory 轨道太阳观测台
orbiting solar telescope 轨道太阳望远镜
orbiting telescope 轨道望远镜
orbit maneuver motor 变轨发动机
orbit motor 摆线转子马达
orbit navigator 轨道导航仪
orbit-orbit resonance 轨轨共振
orbitotomy 开眶术
orbit parameter 轨道参数
orbit parameters of satellite 卫星轨道参数
orbit photography 轨道摄影
orbit point 圆圈飞行中心
orbit pump 转子泵
orbit resonance 轨道共振
orbitron 轨旋管
orbit shift coils 轨道偏移线圈
orbit space 轨道空间
orbit television system 轨道电视系统
orbit transfer 轨道过渡
orbit-transfer problem 轨道转移问题
orbit trimming 轨道修正
Orcadian series 奥尔卡德统
orcelite 六方砷镍矿；褐砷镍矿
orchard 果园
orchard boom 果园用喷杆
orchard crop 果园作物
orchard cultivator 果园用中耕机
orchard grader 果园用分选机

orchard heater 果园加热器
orchard house 果树温室
orcharding 果树栽培
orchard irrigation 果园灌溉
orchard mist sprayer 果园用喷雾器
orchard plough 果园犁
orchard soil 菜园土;果园土
orchard tool 果园工具
orchard tractor 果园用拖拉机
orchard tree 果树
orchella 热带藻
orchestra 正厅前排贵宾席;乐池;正厅(指剧场正对舞台部分);剧场正厅
orchestra box 乐池
orchestra circle 正厅后座
orchestra floor 管弦乐队演奏楼层;管弦乐队演奏台
orchestra level 池座层
orchestra pit 乐池;音乐池
orchestra rail 乐池栏杆
orchestra shell 设于舞台上的反射音响的大型壳体;音响反射板;演奏席壳盖;(室内舞台或露天剧场的)壳形音质反射体;壳形演奏台;壳形演奏台
orchid 红门兰【植】
orchil 苔色素
orchis 淡紫色
orcin 地衣酚
order 序列;秩序;指令;购货单;命令;汇兑;汇票;顺序;顺位;数量级;定购;订单;等级;次序;法令
order action card 有序作卡
order adjoint 有序伴随
order adjoint linear form 有序伴随线性形式
order adjoint net 有序伴随网
order adjoint norm 有序伴随范数
order adjoint random measure 有序伴随随机测度
order adjoint semi-group 有序伴随半群
order adjoint subset 有序伴随子集
order and receiving register 订货与收货登记簿
order and reply telegraph 往复传令车钟
order at the best 以最好价格订购
order bill 定料单;订料单
order bill of lading 指定人提单;提示提单
order bill of materials 材料订单
order bit 命令位
order blank 配货单;订单;订货单
order block 指令群
order-book 指示簿;订货簿
order-bounded 序有界的
order-bounded linear form 有序有界线性形式
order button 指令按钮;命令按钮;呼叫按钮;传号电键
order by merging 归并排序
order cancelled 已注销订单
order card 指令卡
order check 指定人支票(俗称抬头人支票);记名支票;抬头支票
order cheque 指定人支票(俗称抬头人支票);台头支票;定名支票
order code 指令码;指令地址;操作码
order-complete set 有序完备集
order contractive operator 有序收缩算子
order control 订单控制
order convergent filter 有序收敛滤子
order convergent sequence 有序收敛序列
order cycle 订货周期
order-dependence 与次序有关
order diffraction 衍射阶
order-disorder distribution 有序无序分布
order-disorder transition 序转变
order distribution 顺序分配
order drawing 订货图样;订货图纸;外注图
order dual 有序对偶
ordered alloy 有序合金
ordered data 有序型数据
ordered domain 有序畴
ordered field 有序域
ordered-form 订货单
ordered issued to shutdown an enterprise 责令关闭
ordered magnetism 序磁性
ordered pairs 序对;订购零件
ordered ring 有序环
ordered series 有序级数
ordered set 有序集(合);顺序集
ordered solid solution 有序固溶体

ordered state 有序(状)态;有次序状态
ordered structure 有序结构
ordered subset 有序子集
ordered triple 三重序
ordered water 定向水
order effect 试样提出的顺序效果
order element 指令元素;指令元件
order encode 命令编码
order encoder 命令编码器
order enforced to suspend operation 责令停业
order entry application 订单输入应用
orderer 开证申请人(信用证);订货者;订货人;发布命令者
order executed 已执行的订单
order expired 已满期的订单
order for goods 订货单;订单
order for information 情报收集指示
order form 购货单;订货单;订单
order format 指令格式;指令安排形式
order for material 订料单
order-function 次序函数
order-getting cost 外购产品成本;厂外销售成本
order goods 订货单;订货
ordering 有序化;排列次序;定序;次序关系;成序;分类
ordering bias 排序偏差;顺序准确性检验;顺序拉开检验
ordering by merging 归并排序;合并排序
ordering contract 订货合同
ordering cost 订货费用;订货成本
ordering effect 建序效应;成序效应
ordering energy 成序能
ordering information 订货资料;订购须知
ordering in launching 投入顺序
ordering instruction 订货须知
ordering of materials 购料单
ordering of raw materials 原材料订购
ordering of vessel 航行通知
ordering opportunity 订货时机
ordering point 订购点
ordering point system 订货点方式
ordering vector 次序向量;次序矢量
order instrument 指示票据
order isomorphic 次序同构的
order key 传号电键
order length 订货长度
order line 指令线;记录线
order line equipment 联络电话设备
order list 命令表;订货单
orderly 勤杂工;通信员;清洁工;有秩序;整齐的;顺序的
orderly bin 废纸箱(路旁)
orderly effect of symmetry 对称的有序效应
orderly improvement 定向改良
orderly increase 有序增长
orderly marketing agreement 正常销售协定
orderly process of community change 群落类型转变顺序过程
orderly shutdown 正常停车
order mode 命令态
order modification 次序改变
order notify 订货通知
order number 指令编号;订单号码;订单号(数);次数
order number of tests 试验次数
order of accuracy 正确度;精度级;精度等级
order of a curve 曲线的阶
order of a determinant 行列式的阶
order of a differential equation 微分方程的阶
order of a differential operator 微分算子的阶
order of a group 群的阶
order of a matrix 矩阵阶
order of an equation 方程式阶
order of an infinitesimal 无穷小的阶
order of a permutation 排列的阶
order of a polynomial 多项式的阶
order of a tensor 张量的阶
order of commencement of work 开工令
order of complexity 复杂度
order of compound 化合物次序
order of connection 连接顺序;接线顺序
order of consignee 收货人指示
order of convergence 收敛的阶
order of crystallization 结晶顺序;结晶次序

order of degeneracy 退化次数
order of determinants 行列式阶
order of diffraction 衍射级;绕射级
order of discharge 解除破产命令;解除令
order of discontinuity 不连续阶
order of distribution of a cash 分配现金顺序;现金分配次序
order of evaluation 评价等级
order of feed 喂料顺序
order of feeding 加料次序
order of flattening 扁平度
order of fringes 等色条纹级数
order of goods 商品等级
order of harmonic 谐音阶;谐波次数
order of interference 干涉级
order of interference colo(u)rs 干涉色级序
order of light(house) 灯塔等级
order of liquidity 流动性订货
order of magnitude 震级大小;绝对阶;绝对值的阶;绝对值大小;恒星等;订单额度;大小等级型
order of magnitudes 数量级
order of matrix 阵阶
order of merge 归并顺序
order of merit 标准订单
order of payment 付款凭单
order of perturbation 摄动阶
order of phase transition 相变级数;相变的级
order of quality 优劣顺序
order of reaction 反应级(数)
order of reference to arbitration 发交仲裁令
order of reflection 反射级;反射次数
order of rivers 河流分级;河流等级
order of sceneries 序景
order of signs 符号序列
order of spectrum 光谱级
order of streams 支流级;河流分级;河流等级
order of stretching 张拉次序
order of suppression 停止令
order of the day 议事日程;风气
order of the investigation 勘测步骤
order of the layer 层序号
order of triangulation 三角测量等级
order of tributaries 支流等级
order of units 位数
order of work 工作次序
order of zeros 零点相重数
order one Bessel function of the 1st kind 第一类一阶贝塞尔函数
order one Bessel function of the 2nd kind 第二类一阶贝塞尔函数
order one imaginary coordinate Bessel function of the 1st kind 第一类一阶虚宗贝塞尔函数
order one imaginary coordinate Bessel function of the 2nd kind 第二类一阶虚宗贝塞尔函数
order on hand 现有订货
order paper 指示票据
order parameter 有序参数
order parity error 指令奇偶误差
order picking 拣货
order picking S/R machine 拣选型有轨巷道堆垛机
order-point 订货点
order-point control 库存高低点定量控制制度
order-point control system 订货点控制系统
order port 待命港
order preserving mapping 保序映射
order preserving transformation 保序变换
order processing 订货单处理
order production 按订单生产
order program(me) 命令程序
order quantity 订货量
order quantity analysis 订购量分析
order quantity report 订购量报表
order register 指令寄存器;指令记录器
order relation 次序关系
orders 订货
order sequence 命令序列
order sheet 订货单
order size 订货量
order slip 订货单;订货通知单
orders of architecture 建筑风格;古典建筑的格子柱式;建筑柱型;建筑柱式
orders of fold 褶皱级别
order sorter 分级器

order source list 指令源一览表
orders per second 每秒钟的指令数目；每秒指令数
order statistic 顺序统计量
order structure 指令结构；指令安排
order structure type 有序结构的种类
order tape 指令带
order ticket charges 订票费
order time 指令时间
order to helmsman 舵令
order tone 蜂音信号
order to pay 命令付款
order to proceed 开工通知
order to show cause 说明理由令
order to suspend 暂时停工命令
order transition 级跃迁
order under reference 参考订单
order wire 联络线；记录线；命令线；通信线；信号线
order-wire circuit 传号电路
order-wire distributor 指令线分配器；联络线分配器；通知线分配器；传号线分配器
order with customer's brand 来牌订货
order-writing 写出指令
order zero Bessel function of the 1st kind 第一类零阶贝塞尔函数
order zero Bessel function of the 2nd kind 第二类零阶贝塞尔函数
order zero imaginary coordinate Bessel function of the 2nd kind 第二类零阶虚宗贝塞尔函数
ordinanry splice grafting 普通合接
ordinal 按序的
ordinal flood 洪水序列
ordinal inspection 顺序检查
ordinalism 序数论
ordinal number 序数；序号
ordinal power 序数幂
ordinal product 序数积
ordinal ranking of probability 概率的顺序分级
ordinal relation 顺序关系
ordinal scale 顺序分级；序数尺度
ordinal schemes 洪水序列图
ordinal sum 序数和
ordinal utility 序数效用
ordinance 规格；条例；法令
ordinance bench mark 法定水准(标)点
Ordinance Datum 英国国家高程基准面；规定基准(面)；法定基准面；法定基准
ordinance load 规定载荷；规定荷载
ordinance on industrial safety and hygiene 工业安全及卫生条例
ordinance regulating carriage of goods by sea 海运法令
ordinance survey datum 规定基准(面)；法定基准
ordinarily hook 普通钩
ordinary accident 一般事故
ordinary accident insurance 普通事故保险
ordinary anchor 整体锚
ordinary annuity 正常收(支)款；普通年金
ordinary area of new tectogenesis 新构造运动一般地区
ordinary backfill 普通回填
ordinary beacon 寻常标【航海】
ordinary bedding 普通基床
ordinary benchmark 普通水准标石
ordinary bid 普通投标
ordinary binary 普通二进制
ordinary bond 单价键
ordinary brace 普通手摇钻；普通曲柄钻
ordinary brake 常用制动
ordinary breakage 通常破损
ordinary brick 普通黏土砖
ordinary budget 普通预算
ordinary builder's cement 普通建筑水泥
ordinary call 普通呼叫；普通电话
ordinary capital resource 普通资本来源
ordinary carbon steel 普通碳钢
ordinary carbon structural steel 普通碳素结构钢
ordinary casing 普通门窗框
ordinary cement 普通水泥
ordinary ceramics 普通陶瓷
ordinary charges 普通费用
ordinary chattering 正常反跳
ordinary check 普通支票
ordinary chondrite 普通球粒陨石
ordinary cinder aggregate 普通炉渣集料；普通炉渣骨料；普通矿渣集料；普通矿渣骨料
ordinary clay 普通黏土
ordinary clay brick 普通黏土砖
ordinary closer 约四分之一块砖
ordinary coating 普通感光膜
ordinary coherence function 常相干函数
ordinary commodities 普通矿产
ordinary composition series of groups 常合成群列
ordinary compound interest method 普通复利法
ordinary concrete 普通混凝土
ordinary concrete tower 混凝土寻常标
ordinary construction 一般建筑；一般构造；砖木结构；普通结构；普通建筑；半防火建筑
ordinary corporation 普通法人
ordinary course of trade 正常贸易途径
ordinary course of transit 通常运输线；通常的运输途径
ordinary credit 普通信用证
ordinary creditor 普通债权人
ordinary crossover 普通渡线
ordinary cut 普通掘进掏槽法
ordinary dense concrete 普通密实混凝土
ordinary depreciation 正常折旧；普通折旧
ordinary detonating cord 普通导爆索
ordinary differential 常微分
ordinary differential equation 常微分方程
ordinary disbursement 通常费用
ordinary drawing rights 一般提款权
ordinary everyday expenses 日常费用
ordinary expenditures 经常费用(用)
ordinary fault rift 正断层裂谷；正断层地堑
ordinary financial institution 普通金融机构
ordinary flash 通常闪光
ordinary flat wagon 普通平车
ordinary flood ing 常规漫灌
ordinary flow 常遇流量
ordinary fluorite 普通萤石
ordinary form 普通式样；普通形式
ordinary formwork 普通样板；普通模壳；固定模板
ordinary frame 普通肋骨
ordinary gear train 定轴齿轮系
ordinary glass 普通玻璃
ordinary glazing 普通玻璃窗；普通玻璃
ordinary glazing quality 普通玻璃质量
ordinary goods 普通货(物)
ordinary goods train 普通货物列车
ordinary grade 普通等级
ordinary gravel 普通砾石
ordinary-hazard contents 普通公害内容；一般易燃室内用品；普通火灾隐患
ordinary helicoid 常螺旋面
ordinary highway 一般公路
ordinary hinged hatch cover 普通绞链式舱盖
ordinary hour 平峰期
ordinary household 普通住户
ordinary ice 常见冰
ordinary income 正常收入；普通所得；通常收入
ordinary index 寻常折射率
ordinary index of refraction 寻常折射率
ordinary interest 普通利息
ordinary interest rate 自然利率
ordinary in toxicity 普通毒物
ordinary jamb lining 普通门窗侧板
ordinary Kriging method 普通克立格法
ordinary laid rope 普通大缆
ordinary lay 异向捻；普通扭捻；普通交捻法；普通反捻法；普通搓捻法(钢丝绳)
ordinary lay cable 普通捻钢索；普通捻钢丝绳
ordinary lay rope 普通捻钢索；普通捻钢丝绳
ordinary lead 正常铅
ordinary leakage 正常漏损；通常漏损
ordinary level(l)ing 普通水准测量
ordinary level(l)ing rod 普通水准尺
ordinary life 常规寿命
ordinary light 常光(线)
ordinary light ray 寻常光线
ordinary lime mortar 普通灰浆
ordinary lining process (from bottom to top) 先墙后拱衬砌法
ordinary link 单价键
ordinary load 常遇荷载
ordinary loan 一般贷款
ordinary logarithm 常用对数
ordinary loss 通常损失；正常损失
ordinary low-alloy steel 普通低合金钢
ordinary low-water level 普通低水位
ordinary luminaire 普通照明灯具
ordinary maintenance 日常维修
ordinary masonry 毛石砌体；普通圬工；普通砌体
ordinary minimum flood 常遇最小洪流
ordinary minimum flow 常遇最小流量
ordinary moor 一字锚泊；倒退锚一字锚
ordinary mortar 普通砂浆
ordinary neap tide 一般小潮
ordinary negligence 一般过失
ordinary non-metallic commodities 普通非金属矿产
ordinary oil-well cement 普通油井水泥
ordinary open space 一般绿地
ordinary paint 普通油漆
ordinary partner 普通股东
ordinary partnership 普通公司
ordinary pebble 小卵石；普通卵石
ordinary pitch 普通屋面坡度；常用屋面坡度
ordinary point of a curve 曲线的寻常点
ordinary polyconic(al) projection 普通多圆锥投影
ordinary Portland cement 普通水泥；普通硅酸盐水泥；普通波特兰水泥
ordinary practice 普通习惯；通常习惯
ordinary practice of seaman 海员通常做法
ordinary pressure 常压
ordinary private calls 普通私务电话
ordinary profit 经常利益
ordinary profit and loss 经常损益
ordinary purchasing 普通采购
ordinary pyroclastic rock clan 正常火山碎屑岩类
ordinary quality 普通等级
ordinary quality brick 普通砖
ordinary rate of profit 普通利润率
ordinary ray 寻常射线；寻常光线；普通射线；常光(线)
ordinary receipts 经常收入
ordinary record 普通记录
ordinary refractory concrete 普通耐热混凝土
ordinary reinforced concrete 普通钢筋混凝土
ordinary repair 正常维修
ordinary revenue 经常收入
ordinary ring 普通环衬
ordinary river irrigation scheme 普通河流灌溉计划
ordinary road 一般道路
ordinary rock drill 普通凿岩机
ordinary rubble masonry 普通毛石砌体；普通毛石圬工；粗型墙
ordinary rudder 不平衡舵
ordinary safety explosive 常规安全炸药
ordinary sailor 普通海员；二级水手
ordinary sea(s)man 普通海员；二级水手
ordinary segment 标准块管片
ordinary shaft kiln 普通立窑
ordinary shareholder 普通股东
ordinary sheathed explosive 常规安全炸药
ordinary shortage 通常短缺量
ordinary shuttering 一般百叶窗(材料)；普通模板；普通模壳
ordinary sidelight 圆形舷窗
ordinary sight distance 普通视距
ordinary slow passenger train 普通旅客列车
ordinary spiral steel tube benchmark 螺旋钢管普通水准标石
ordinary spoilage 正常破损
ordinary spring tide 普通大潮；一般大潮；正常大潮
ordinary steel 普通钢
ordinary steel tower 钢寻常标
ordinary steel tube benchmark 钢管普通水准标石
ordinary stiffness 普通劲度
ordinary storm 普通暴雨
ordinary street 普通街道
ordinary structural concrete 普通结构混凝土
ordinary structural steel 结构用普通钢
ordinary structure 普通结构
ordinary subroutine 常用子程序
ordinary superphosphate 普通过磷酸钙
ordinary symbol 寻常符号；普通符号；常规字符；常规符号
ordinary tax 经常税
ordinary telegram 平电
ordinary telegraph code 明码

ordinary temperature 常温
ordinary temperature depth 常温深度
ordinary tender 普通投标
ordinary theodolite 普通经纬仪
ordinary tidal level 平均潮位;平均潮面
ordinary tower 寻常标
ordinary toxicity test 普通毒物测定
ordinary transfer 普通转印法
ordinary transit 普通经纬仪
ordinary truck 普通卡车
ordinary twin type 普通对绞多心型
ordinary type 常用字体
ordinary tyre 普通轮胎
ordinary water 轻水;普通水;常水
ordinary water level 中水位河床;中水位;正常水位;平水位;常水位
ordinary water level ga(u)ge 常水位计
ordinary wave 寻常波;寻波;正常波;常波
ordinary-wave component 寻波光波分量;正常波分量
ordinary wear-and-tear 自然损耗
ordinary working hours 正规工作时间;正常工作时间;通常工作时间
ordinary year 常年
ordinate 纵坐标;纵标;支距;切线纵距;竖坐标;竖标距
ordinate and abscissa figure of crest 极值点纵横坐标图
ordinate axis 纵轴
ordinate datum 法定基准面
ordinate dimensioning 纵横尺寸标注
ordinate error of traverse 导线纵坐标误差
ordinate of a pipeline 管线走始
ordinate of principal point 像主点纵坐标
ordinates 船体分站(船舶线型图上的竖站)
ordinates of the bivariate normal distribution 二元正态分布纵坐标
ordination 规则;规格;排列
ordination number 原子序数;原子序
ordination of pump 水泵规格
ordinatograph 光学垂线坐标仪;光目标坐标方位仪;光标方位仪
ordinatometer 纵测量器
ordnance 军需品;军械;兵器
Ordnance benchmark 英国陆军测量局水准点
ordnance datum 规定基准;军用基面
ordnance datum Newlyn 英国纽林标准基准面;纽林标准零点(英国)
ordnance factory 兵工厂
Ordnance map 地形图(军用地图英国陆军测绘局)
ordnance storage 军械库
ordnance store 军械库;军械材料
ordnance supplies 军械补给品
Ordnance Survey 英国测绘局;陆军测绘局(英国);英国陆地测量部;陆军地形测量
Ordnance Survey Datum 陆军测绘局标准零点(英国);英国陆地测量部地图
ordnance waste 军用废物
ordnance works 兵工厂
ordonezite 褐锑锌矿
ordonnance 安排;配置;布局(建筑物的)
ordorant 有气味的
ordosite 鄂尔多斯岩
Ordovician limestone 奥陶纪石灰岩
Ordovician (period) 奥陶纪
Ordovician-Silurian glacial stage 奥陶-志留纪冰期
Ordovician system 奥陶系
Ordovician transgression 奥陶纪海浸
ordure car 粪车
ordurous car 粪车
ore 矿石内的脉石;矿(石)
ore assay 矿石分析
ore assay mill 矿石试验磨碎机
ore barrier pillar 矿山岩层;矿石栅栏支柱
ore-bearing coefficient 含矿系数
ore-bearing rate 含矿率
ore-bearing ratio 含矿率
ore-bearing structure 含矿构造
ore-bearing vienlet 含矿细脉
ore-bearing zone 含矿地带;含矿带
ore bed 矿层
ore beneficiation 选矿
ore bin 矿石储[贮]仓;矿石仓;储渣仓
ore block 矿块;矿段;块段

ore blocked out 划分矿体
oreblock method 块段法
ore blocks map 矿块图
orebody 矿体;矿床;矿石
orebody boundary 矿体边界
ore-body grade 矿体品位
orebody head 矿头
orebody occurrence 矿体产状
ore-body of small scale and dispersion 矿体规模小而分散
orebody tail 矿尾
ore body thickness 矿体厚度
ore boulder 矿石漂砾;大块矿石
ore boundary 矿体边界
ore box 矿仓
ore breaker 碎矿机
ore breaking 破碎矿石
ore-breaking plant 碎矿车间
ore bridge 矿石装卸桥
ore bucket 矿斗
ore/bulk/oil carrier 矿石—散货—石油三用船;油散矿兼用船
ore bunch 矿囊;矿巢
ore bunker 矿石斗;矿料斗;矿石储[贮]仓
ore burdening 矿石配料
ore cargo 矿石货
ore carrier 运矿船;矿石船;矿砂船
ore-carrying railroad 矿石运输铁路
ore-carrying railway 矿山铁路
ore-carrying vessel 运矿石船
ore chamber 矿囊
ore channel 矿石道路
ore charge 矿石试样剂量;矿石料
ore chimney 朱庄矿体;管状矿脉;矿筒
ore chute 矿筒;放矿溜口;放矿溜井;放矿溜道
ore cluster 矿群
ore column 矿柱
ore compositive use unsettled 矿产综合利用问题未解决
ore concentrate 精选矿;精矿砂
ore-concentrating structure 聚矿构造
ore-conduit structure 导矿构造
ore-containing structure 储矿构造
ore-containing structure of fault 断层容矿构造
ore-containing structure of fold 褶皱容矿构造
ore-containing structure of fracture 裂隙容矿构造
ore-containing structure of intrusion 岩体容矿构造
ore-containing structure of volcano 火山容矿构造
ore content 矿石品位
ore control 矿控
ore core length 矿心长度
ore core number 矿心编号
ore crusher 矿石自磨机;矿石破碎机;碎矿机
ore-cum-oil carrier 油/矿两用船
ore dampness 矿石湿度
ore deposit 矿岩;矿斗;矿床
ore deposit exploration strengthen 增强矿床勘探工作
ore deposition research strengthen 着重成矿地质研究
ore deposit mining program(me) 矿床开采程序
ore deposits of difficult drainage 难于疏干的矿床
ore deposits of easy drainage 易于疏干的矿床
ore deposits of fine intrusive veins 细脉浸染矿床
ore deposits of non-special drainage 不需要专门疏干的矿床
ore deposits of magmatic eruption 岩浆喷溢矿床
ore deposits of penta deposit formation 五元素建造镍矿床
ore developed 已开拓矿体
ore dilution 矿石贫化
ore-distributing structure 布矿构造
ore district 矿区
ore drawing 下放矿石
ore dressing 选矿
ore-dressing engineer 选矿工程师
ore-dressing flow 选矿流程
ore-dressing plant 选矿厂;选矿装置
ore-dressing practice 选矿作业
ore-dressing product 选矿产品
ore-dressing recovery percentage 选矿回收率
ore-dressing scheme 选矿流程
ore-dressing tailings 选矿尾矿

ore-dressing wastewater 选矿废水
ore-dump blockness 矿堆块度
ore-dump sampling 矿石堆采样
ore expectant 预测矿藏量
ore extraction 矿石回收率;矿石采出率
ore-extraction process 矿石提取法
ore field 矿田
ore-field anomaly 矿田异常
orefield structure 矿田构造
ore-finding mineralogy 找矿矿物学
ore flo(a)tation 矿石浮选
ore flo(a)tation promoter 矿石浮选促进剂
ore forecasting map based on remote sensing 遥感矿产预测图
ore formation 矿石形成;矿石建造;床;矿层;矿藏
ore-forming control of circular feature 环形体控矿
ore-forming control of circular intersection 环形体交叉控矿
ore-forming control of circular-linear zone 环形体-线性体带控矿
ore-forming control of linear-circular intersection 线性体-环形体交切控矿
ore-forming control of monolineament of lineament 线性体单交叉控矿
ore-forming control of monolinearment 单线性体控矿
ore-forming control of multilineament of lineament 线性体多交叉控矿
ore-forming control of multilinearment 多线性体控矿
ore-forming element 造矿元素;成矿元素
ore-forming fluid 成矿溶液;成矿流体
ore-forming process of contact metasomatism 接触交代成矿作用
ore-forming simulating experiment 成矿模拟实验
ore furnace 熔矿炉
oregano 奥里根诺
ore getting 矿山开采
oregonite 砷铁镍矿
Oregon larch 大冷杉;俄勒冈落叶松
Oregon maple 大叶枫;枫树
Oregon pine 花旗松;俄勒冈黄杉;美国松
ore grab unloading system 矿石抓斗装卸装置
ore grade 矿石品位分级;矿石品位;矿石品级;矿石晶位
ore grade andalusite 红柱石矿石
oregrade dumortierite 蓝线石矿石
oregrade kyanite 蓝晶石矿石
oregrade sillimanite 夕线石矿石
ore grain 矿粒
ore grinder 矿石研磨机
ore guide 矿化标志
ore handling 矿石处理;矿石加工;矿物加工;矿处理;矿物装卸;矿装卸
ore-handling bridge 高脚吊车
ore-handling capacity 处理矿石的能力
ore-handling plant 矿山加工厂;矿山处理厂;矿石装卸设备
ore harbo(u)r 矿石专用港;矿物港
ore hearth 矿石熔烧炉
ore hoist 矿石提升机
ore hold 矿砂舱
ore hole 有矿物钻孔
ore horizon 矿层
ore-hunting information by cell 单元找矿信息量
ore-hunting information by index state 标志状态找矿信息量
oreide bronze 高铜青铜
ore identification 矿石鉴定
ore industry sewage 矿山工业废水;采矿工业污水
ore industry waste 矿山工业废水;矿山工业废物
oreing 高碳钢矿石脱碳法
ore in sight 可靠矿石储量;可靠储量
ore intersection 矿体交错;矿石穿通点
oreithalion tropicum 热带高山植物
ore-lead age 矿铅年龄
ore-loading crane 矿石装卸吊车
ore-loading plant 矿石转载机
ore-loading port 装矿港
ore lode 矿脉
ore losses 矿石损失
ore losses ratio 矿石损失率
ore lost in this year 本年度损失矿量

ore low grade or ore of tall harmful constituent 矿石品位低或有害组分高
orema causis 木材露天慢腐
ore magma 矿浆;含矿岩浆;含矿岩
ore management 矿石质量管理
ore mass 矿石质量;砂矿
ore microscope 矿相显微镜
ore microscopy 矿相学
ore mill 磨矿机;磨矿石厂
ore mine 矿井
ore mined in this year 本年度开采矿量
ore mineral 金属矿物;金属矿石
ore mining 矿山开采;采矿
ore nature 矿石性质
ore nest 矿巢
ore of sedimentation 砂矿
ore of special type 特殊类型矿石
ore/oil carrier 油/矿两用船;矿砂/石油船
ore open pit mining 矿石的露天开采
ore order or ore sort 矿石品级
ore particle 矿粒
ore pass 矿道;溜矿井;放矿溜井;放矿溜道
ore pellet 球团矿砂;矿球
ore pellet carrier 矿砂球团船
ore petrology 矿石岩石学
ore picker 拣选工;手选工
ore pier (突堤式的)矿石码头
ore pile 矿石堆
ore pillar 矿柱
ore pipe 管状矿脉;矿筒
ore pocket 矿囊;矿袋
ore preparation screen 选矿筛子
ore processing 矿石处理
ore processing technologic(al) condition 矿石加工技术条件
ore puddling 铁矿搅炼
ore pulp 矿泥;矿浆
or equal 被承认等同
ore quality assessment 矿产质量评议
ore rank 矿石牌号
ore rate in exploration engineering 勘探工程见矿率
ore ratio 含矿率
ore receiving port 矿石收货港
ore recovery ratio 采矿回收率
ore reduction 矿石破碎
ore-reserve analysis 矿石储量分析
ore reserves 矿产储量;矿藏量;矿藏
ore reserves depleted 储量耗竭
ore rib 矿柱;矿壁
ore rock 含矿岩
ore room 矿房
ore run 矿层走向
ore sampling 矿石取样(法);矿石采样
ore scraper 扒矿机
ore-search 寻矿;找矿;探矿
ore segment 矿段
ore separation plant 选矿厂
ore separator 矿石分离器;拣矿器
ore sewage 矿山废水
ore ship 矿石船
ore shoot 矿柱;富矿体
ore-sintering machine 矿石烧结机
ore-sintering plant 烧结厂
ore sketch 矿石素描图
ore skip 提矿箕斗
ore slag 矿渣
ore sluice 矿砂溜槽
ore slurry 矿浆;矿粉浆
ore slurry pipeline 矿浆输送管道
ore smelting works 矿石熔炼厂
ore solution 成矿溶液
ore source-bed 矿源层
ore specimen 矿石标本
ore stockwork 网状矿脉
ore storage bunker 矿石料仓;储矿仓
ore storage yard 堆矿场;矿石储存库
ore structure 矿石构造
ore-surge bin 缓冲矿仓
ore tailings 尾砂;尾矿
ore tails 尾矿
ore terminal 矿体终止点;矿石转运站;矿石码头
ore texture 矿石结构
ore transfer car 运矿车

ore trend tunnel 沿脉坑道;沿脉
ore type 矿石类型
ore unloader 卸矿机
ore vein 矿脉
ore wagon 矿石车
ore washer 洗矿水
ore wet grinding mill 矿石湿法粉碎机
ore worked-out section 采空区
ore yard 矿场;堆矿场
ore zone 含矿带
Orford process 奥福德分层炼镍法
organ 机件;机关;机构
organburg 粗袋布
organ chamber 风琴室
organ coefficient 藏器系数
organdie = organdy
organdy 蝉翼纱;玻璃纱
organ gallery 风琴席
organge G 橙黄 G
organic 有机的
organic accelerator 有机促进剂
organic accumulation 有机蓄积;有机物堆积
organic acid 有机酸
organic acidity 有机酸性
organic acid soluble phosphorus 有机酸溶性磷
organic acid utilization test 有机酸利用试验
organic acid wastewater 有机酸废水
organic addition agent 有机添加剂
organic additive 有机添加剂
organic adhesive 有机黏接剂;有机胶黏剂
organic adhesive method 有机黏合法
organic adsorbent 有机吸附剂
organic aggregate 有机集料;有机骨料
organic aggregate concrete 有机集料混凝土;有机骨料混凝土
organic agricultural chemicals 有机农用化学物
organic agriculture 有机建筑理论
organic analogy of city 城市有机体论
organic analysis 有机分析
organic analytic(al) reagent 有机分析试剂
organic analytic(al) standard 有机分析标准
organic anion 有机阴离子
organic anticorrosion agent 有机系防腐剂
organic-aqueous interface 有机相水界面
organic architecture 有机建筑
organic arsine 有机胂
organic bactericide 有机杀菌剂
organic base 有机碱
organic bath 有机电解质
organic bentonite 有机膨润土
organic bentonite ore 有机质膨润土矿石
organic binder 有机黏结剂;有机结合料
organic binder bond 有机黏结剂
organic binding agent 有机黏结剂
organic bits 有机碎屑
organic body 有机体
organic bond 有机结合剂
organic bonded refractory 有机物结合耐火材料
organic-borane 有机硼烷
organic bottom mud 有机质底泥
organic bottom sludge 有机底泥
organic bromine 有机溴
organic bromine compound 有机溴化合物
organic building 有机建筑物
organic carbon 有机碳
organic carbon content 有机碳含量
organic carbon method 有机碳法
organic carcinogen 有机致癌物
organic catalysis 催化作用
organic cation 有机阳离子
organic cationic compound 有机阳离子化合物
organic cationic stabilizer 有机阳离子稳定剂
organic cement 有机胶结剂
organic chemical biodegradability 有机化学可生物降解性
organic chemical pollutant 有机化学污染物
organic chemical reagent 有机化学试剂
organic chemicals 有机化学物;有机化学药品
organic chemicals fiber 有机化学纤维
organic chemist 有机化学家
organic chemistry 有机化学
organic chemistry laboratory 有机化学试验室
organic chemistry of coal 煤有机化学
organic chloramines 有机氯胺

organic chloride 有机氯
organic chlorine compound 有机氯化合物
organic chlorine oil 有机含氯润滑油
organic clay 有机(质)黏土
organic clay of high plasticity 高塑性有机质土
organic clay of low plasticity 低塑性有机质黏土
organic coagulant 有机絮凝剂;有机混凝剂
organic coagulant aid 有机助凝剂
organic coating 有机油毡;有机涂料;有机涂层
organic coating agent 有机包膜剂
organic colloid 有机胶体
organic colloidal dispersant 有机胶体分散剂
organic colloidal iron 有机胶态铁
organic colloidal matter 有机胶体物质
organic colloid in soil 土壤有机胶体
organic colo(u)ring matter 有机色料
organic combustion 有机燃烧
organic completeness 有机完整性
organic complex 有机络合物
organic complexing agent 有机络合剂
organic component 有机物组分;有机物成分
organic components in seawater 海水有机质成分
organic composition 有机组分;有机组成;有机成分;生物成分
organic composition of capital 资本有机构成
organic compound 有机化合物
organic compound removal 有机化合物去除
organic compounds poisoning 有机化合物中毒
organic constant 有机常数
organic constituent 有机组合;有机组分;有机物组分;有机成分
organic constituent and dissolved gas 有机组分和溶解气
organic contaminant 有机污染物
organic contaminant removal 除有机污染物
organic contamination 有机污染;有机污染
organic content 有机质含量;有机物含量
organic content in aggregate 骨料有机质含量
organic content of soils 土壤的有机质含量
organic content of the sediment 底泥有机质含量
organic-continuous system 有机相连续系统
organic coolant 有机冷却剂
organic-cooled reactor 有机物冷却堆
organic corrosion inhibitor 有机阻蚀剂;有机防腐剂
organic counter tube 有机计数管
organic cross-linked bentonite 有机交联膨润土
organic crystal 有机晶体
organic cyanide 有机氰化物
organic cyanide wastewater 有机氰化物废水
organic cycle 有机循环
organic debris 有机岩屑;有机尾矿;有机碎屑
organic decay 有机腐朽
organic decentralization 有机疏散
organic degradation 有机降解
organic demand 有机需要量
organic deodorizer 有机脱臭剂
organic deposit 有机沉积矿床;有机沉积(物);生物沉积(物)
organic deposition coast 有机沉积海岸
organic derivatives 有机衍生物
organic detritus 有机岩屑;有机残渣
organic dielectric(al) functional material 有机介电功能材料
organic differential thermal analysis 有机差热分析
organic disease 器质性疾病
organic dissolved component 有机溶解组分
organic disulfide 有机二硫化物
organic dry weight 有机物干重
organic dust 有机尘埃;有机性粉尘;有机灰尘;有机粉尘
organic dye 有机染料
organic dye solution laser 有机染料溶液激光器
organic dyestuff 有机染料
organic dye-tracer 有机染料示踪剂
organic effect 有机作用
organic efficiency 回收效率
organic effluent 有机流出物;有机废水
organic electrochemistry 有机电化学
organic electrode process 有机电极过程
organic electrolyte 有机电解液
organic electrolyte cell 有机电解液电池
organic element 有机元素
organic element analysis 有机元素分析

organic element compound 元素有机化合物
organic energy 生物能
organic environment 有机环境
organic ester 有机酯
organic evolution 有机界进化;生物进化
organic extraction 有机萃取
organic facies 有机相
organic farming 生物耕作
organic felt 有机油毡;原纸
organic fertilizer 有机肥(料)
organic fiber 有机光学纤维;有机纤维
organic filler 有机填充物;有机填料
organic film condenser 有机薄膜电容器
organic fitness 有机界的合理性
organic flocculation 有机絮凝(作用)
organic flouro-compound 有机氟化合物
organic flucculant 有机絮凝剂
organic fluid 有机溶液;有机流体
organic fluorescent 有机荧光体
organic fluorescent paint 有机荧光漆;有机硅涂料
organic fluorine insecticide 有机氟杀虫剂
organic fluorine pesticide 有机氟农药
organic foam plastic 有机泡沫塑料
organic form 有机形态
organic form of nature 大自然的有机形态
organic forms of soil nitrogen 土壤氮的有机类型
organic fouling deposition 有机污着底泥
organic fraction 有机馏分;有机部分
organic fragment 有机碎屑
organic framework texture 生物骨架结构
organic free tube 无有机物加速管
organic friction material 有机耐磨材料
organic fungicide 有机杀真菌剂
organic gas 有机气体
organic genetic gas 有机成因气
organic geochemistry 有机地球化学
organic glass 有机玻璃
organic grease 有机润滑脂
organic group 有机基
organic growth substance 有机生长物质
organic halide 有机卤化物
organic halogen compound 有机卤(素)化(学)物;含卤有机化合物
organic hazardous waste 有机有害废物
organic herbicide 有机除草剂
organic horizon 有机质层
organic hydrochemistry 有机水化学
organic hydroperoxide 有机过氧化氢
organic hygrometer 有机物湿度计
organic impurity 有机杂质
organic impurity test 有机杂质试验;有机杂质检验
organic index 有机物指数
organic industrial wastewater 有机工业废水
organic inhibitor 有机抑制剂;有机缓蚀剂
organic insecticide 有机杀虫剂
organic insulation 有机绝缘
organic intermediate 有机中间体
organic iodine 有机碘
organic iodine compound 有机碘化合物
organic irrigation 有机冲洗
organicism 整体主义
organic lacquer solvent 有机油漆溶剂
organic limestone 有机石灰岩
organic linear space 有机组合直线空间
organic lining 有机材料摩擦衬片
organic liquid 有机液体
organic liquid coolant 有机液体冷却剂
organic liquid gel 有机液体胶体
organic liquid laser 有机液体激光器
organic loading 有机负荷
organic loading rate 有机负荷率
organic lubricant 有机润滑剂
organic luminophor 有机发光体
organic macromolecule heavy metal trapping flocculant 有机高分子重金属捕集絮凝剂
organic manure 有机粪肥
organic mass spectrometry 有机质谱分析;有机质谱法
organic mass spectroscopy 有机质谱分析;有机质谱法
organic mastic adhesive 有机脂黏合剂;有机脂胶黏合剂
organic mat 有机覆盖物
organic material 有机材质;有机材料

organic materials like plant/wood moss 有机物如植物(树木)苔藓等
organic matrix 有机基体
organic matter 有机质;有机物质
organic matter and constant 有机质和有机常数
organic matter basis 有机基
organic matter bearing inclusion 含有机质包裹体
organic matter biodegradation 有机物生物降解
organic matter by coastal current transportation 沿岸流搬运的有机质
organic matter by lake current transportation 湖流搬运的有机质
organic matter by ocean current transportation 洋流搬运的有机质
organic matter by river transportation 河流搬运的有机质
organic matter by storm wave transportation 暴风浪搬运的有机质
organic matter by wave transportation 波浪搬运的有机质
organic matter content 有机质含量
organic matter degradation 有机物降解(作用)
organic matter in sediment 沉积物中有机质
organic matter of a podzol 灰化土有机质
organic matter of expulsive oil and gas 排出油气中的有机质
organic matter of source rock 烃源岩中有机质
organic matter removal 有机物去除
organic matter separation efficiency 有机物分离效率
organic matter test 有机质试验
organic matter test for fine aggregate 细骨料含量试验
organic matter used as fertilizer 作为肥料使用的有机质
organic medium 有机媒质;有机介质
organic membrane 有机膜
organic mercurials 有机汞制剂;有机汞化合物(木材防腐剂)
organic mercury 有机汞
organic mercury determination 有机汞测定
organic mercury pesticide 有机汞杀虫剂
organic mercury poisoning 有机汞中毒
organic metal 有机金属
organic metal compound 有机金属化合物
organic micro-constituent 有机显微组分
organic microcontaminant 有机微污染物
organic micropollutant 有机微污染物
organic mineral aggregate 有机矿质团粒
organic minerals 有机矿物
organic mineral soil 有机矿质土(壤)
organic moderated and cooled reactor 有机慢化冷却堆
organic moderated reactor 有机冷却型原子反应堆
organic moderator 有机慢化剂
organic modified bentonite 有机改性膨润
organic molecule 有机分子
organic nitro-compound 有机硝基化合物
organic nitrogen 有机氮
organic nitrogen compound 有机氮化合物
organic nitrogen insecticide 有机氮杀虫剂
organic nitrogenous wastewater 含有机氮废水
organic nitrogen pesticide 有机氮杀虫剂
organic nitrogen source 有机氮源
organic non-linear optic(al) material 有机非线性光学材料
organic nutrition 有机营养
organic of aquatic biotic body 水生生物机体的有机物
organic of biolysis 生物残骸分解的有机物
organic oil 有机油类
organic oil-vapour pump 有机油扩散泵
organic origin 有机来源;有机成因
organic origin theory 有机成因说
organic over-oxide articles 有机过氧化物
organic oxidation 有机氧化
organic particle 有机颗粒
organic particulate matter 有机颗粒物
organic particulate phosphorus 有机颗粒磷
organic peracid 有机过酸
organic peroxide 有机过氧化物
organic pesticide 有机农药;有机合成农药
organic phase 有机相
organic phenol type additive 有机酚型添加剂

organic phosphate 有机磷酸盐
organic phosphite 有机亚磷酸酯
organic phosphor 有机磷光体
organic phosphorous compound 有机磷化合物
organic phosphorous insecticide 有机磷杀虫剂
organic phosphorus 有机磷
organic phosphorus pesticide determination 有机磷农药检测
organic pigment 有机颜料
organic planar waveguide 有机平面波导
organic plastics 有机塑料
organic poison 有机毒物
organic pollutant 有机污染物
organic pollutant accumulation 有机污染物累积
organic pollutant analysis 有机污染物分析
organic pollutant criterion 有机污染物标准
organic pollutant degradation 有机污染物降解
organic pollutant load 有机污染物负荷
organic pollutant removal 除有机污染物
organic pollution 有机质污染;有机(物)污染
organic pollution criterion 有机污染标准
organic pollution factor 有机污染因子
organic pollution index 有机污染指数
organic pollution of water body 水体有机污染
organic polymer 有机聚合物
organic polymer coagulant-aid 有机高分子助凝剂
organic polymeric flocculant 有机高分子絮凝剂
organic polysulfide 有机多硫化物
organic precipitation 有机沉积(物)
organic priority pollutant compound 优先有机污染物排放
organic profile 有机物外形
organic promoter 有机助催化剂
organic quantitative analysis 有机定量分析
organic quenched counter tube 有机淬灭计数管
organic radical 有机基
organic reactant 有机反应物
organic reaction 有机反应
organic reaction kinetics 有机反应动力学
organic reaction mechanism 有机反应机理
organic reactive extender 有机反应型增充剂
organic reagent 有机试剂
organic reconstruction 有机重组
organic reef 有机礁;生物礁
organic reef deposit 生物礁沉积
organic reef environment 生物礁环境
organic reef facies 生物礁相
organic reef trap 有机礁圈闭
organic refuse 有机废料;有机垃圾
organic reinforcing agent 有机补强剂
organic ring compound 有机环状化合物
organic rock 有机岩;生物岩
organic salt 有机盐
organic scintillating solution 有机闪烁溶液
organic scintillation detector 有机闪烁探测器
organic scintillator 有机闪烁体
organics concentration 有机物浓度
organic-seal double-glazed unit 有机物封填双层玻璃窗
organic sediment 有机沉积(物)
organic sedimentation effect 有机物沉淀效应;有机沉淀效应
organic selection 有机选择
organic semiconductor 有机半导体
organic sewage 有机污水
organic shield 有机屏蔽层
organic silicon 有机硅
organic silicon compound 有机硅化合物
organic silicone oil 有机硅树脂油
organic silicon foamed plastics 有机硅泡沫塑料
organic silicon furan resin 有机硅呋喃树脂
organic silicon insulating varnish 有机硅绝缘漆
organic silicon plastics 有机硅塑料
organic siliconresin 有机硅树脂
organic silicon rubber 有机硅橡胶
organic siloxane polymer 有机硅氧烷聚合物
organic silt 有机淤泥;有机粉土;有机质粉土
organic silt of low plasticity 低塑性有机质粉土
organics in water analyzer 水生有机物分析仪
organic slime 有机淤泥;有机矿泥;有机残渣
organic sludge 有机污泥
organic soil 有机(质)土(壤)
organic soil containing coarse grains 含粗粒有机质土

organic soil containing lightly coarse grains 微含粗粒有机质土
organic soil horizons 土壤有机质层
organic solid 有机固体
organic solute 有机溶质
organic solution 有机溶液
organic solvent 有机溶媒;有机溶剂
organic solvent degreasing 有机溶剂脱脂(法)
organic solvent poisoning 有机溶剂中毒
organic solvent preservative 有机溶液防腐剂;有机溶剂防腐剂
organic-solvent type preservative 有机溶剂型防腐剂
organic source material 有机生油物质
organic space 有机空间
organic species 有机物种
organic sphere 有机界;生物界
organics phosphorus pesticide wastewater 有机磷农药废水
organic stabilized chlorine dioxide 有机稳定性二氧化氯
organic stabilizer 有机稳定剂
organic substance 有机物(质);有机剂;有机基质
organic sulfate 有机硫酸酯;有机硫酸盐
organic sulfide 有机硫化物
organic sulfur 有机硫
organic sulfur compound 有机硫磺化合物
organic sulfur fungicide 有机硫杀真菌剂
organic superconducting antenna 有机超导天线
organic superconducting resonator 有机超导共振器
organic superconducting waveguide 有机超导波导
organic surface 有机面
organic surface active agent 有机表面活性剂
organic surface loading 表面有机负荷
organic surface loading of biologic(al) filter 生物滤池表面有机负荷
organic suspended sediment 有机悬沙
organic symbol 有机符号
organic synthesis 有机合成
organic synthesis cationic flocculant 有机合成阳离子型絮凝剂
organic synthesis plant 有机合成装置
organic synthetic(al) chemistry 有机合成化学
organic synthetic(al) pigment 有机合成颜料
organic terrain 沼泽地
organic theory 有机理论
organic theory of architecture 建筑学的有机理论
organic thermal control coating 有机耐热涂层;温控有机涂层
organic thinner 有机稀释剂
organic titanate 有机钛酸酯
organic toner 有机调色剂
organic toxicant 有机毒剂
organic toxic index 有机毒物指数
organic toxic substance 有机有毒物质
organic transport 制式运输工具
organic-type friction material 有机摩擦材料
organic ultrasonic probe 有机超声探头
organic vapo(u)r 有机物蒸气
organic vapo(u)r analysis 有机物汽化分析
organic vapo(u)r counter tube 有机汽体计数管
organic varnish 有机漆
organic waste 有机废料;有机废(弃)物
organic waste liquor 有机废液
organic waste(water) 有机废水;有机污水
organic wastewater containing refractory compound 含难生物降解化合物有机废水
organic wastewater containing sulfate 含硫酸盐有机废水
organic wastewater treatment 有机废水处理
organic wastewater with medium-low concentration 中低浓度有机废水
organic water pollutant 有机水污染物
organic weathering 生物风化
organic world 有机界
organic yellow 有机黄(颜料)
organic zeolite 有机沸石
organic zinc-rich primer 有机富锌底漆
organidin preparation 奥尔尔尼丁
organism 有机体;生物体;生物
organis matter of soil 土壤有机质
organism degradation bacteria 有机物降解菌

organism level effect model 生物体效应模型
organism multiphase photo catalysis degradation reaction 有机物多相光催化降解反应
organism of water transport 水运运输组织
organism pollution 生物体污染
organism pond process 生物塘法
organism quantity pyramid 生物量金字塔
organism resistance to pollution 生物抗污染性
organisms disembarkation 生物上陆事件
organism-sediment index 有机物-底泥指数
organisol 有机溶胶
organization 组织(机构);机构;团体;编制
organizational affiliation 人事关系
organizational arrangement for construction 施工组织设计
organizational change 组织革新
organizational communication 集体传播
organizational coordination 组织协调
organizational cost 行政费(用)
organizational expenses 组建费;开办费
organizational framework 组织机构
organizational integrity 机构完整性
organizational process flow chart 组织程序流程图
organizational statement 组织语句
organizational structure 组织机构
organization and methods 组织和方法
organization and planning of construction 建筑施工组织设计
organization and planning of production 生产组织计划
organization audited 被审单位
organization by location 区域型组织
organization chart 组织图表;组织机构图表
organization charts for internal control 内部控制组织图
organization cost 开办费;开办成本;筹备费
organization cost amortization 开办费摊销
organization cost distribution 开办费分配
organization definition 结构定义
organization development 组织发展
organization for approving 批准单位
organization for compilation 编制单位
organization for economic cooperation 经济合作组织
Organization for Economic Cooperation and Development 经济合作与发展组织
Organization for the Protection of the Maritime Environment 海洋环境保护组织
Organization for Trade Cooperation 贸易合作组织
Organization for Tropic(al) Studies 热带研究组织
organization for working out 制定单位
organization in subject departments 按学科划分
organization maintenance personnel 在编保养人员;建制维修人员
organization method put forward by Taylor 泰勒组织理论
Organization of African Unity 非洲统一组织
organization of car flow 车流组织
organization of cargo handling 装卸作业组织
organization of civil aviation transport 民航运输组织
organization of construction 建设单位;施工组织
organization of internal audit 内部审计组织
organization of labo(u)r 劳动组织
organization of mapping 制图单位
organization of passenger flow 客流组织
Organization of Petroleum Exporting Countries 石油输出国组织
organization of rail transport 铁路运输组织
organization of railway passenger traffic 铁路旅客运输组织
organization of station operation 车站工作组织
organization of surveying 施测单位
Organization of the Arab Petroleum Exporting Countries 阿拉伯石油输出国组织
organization of traffic control 调度部门
organization of train operation 铁路行车组织
organization of transportation passing through the dam 过坝运输组织
organization of works 工作安排;工程组织(机构)
organization of worker's position 工作岗位责任制
organization-owned housing 单位住房
organization-raised funds 自筹资金
organization scheme 组织计划;组织方案

organization status change report 组织情况变化报告
organization structure of railway statistics 铁路统计组织结构
organization tax 开办税;开办登记费
organization wide 全机构的
organize 组织
organized air exhaust 有组织排气
organized air supply 有组织进风
organized array 已组织的数组;已编排阵列
organized natural ventilation 有组织自然风
organize into groups 编组
organizer 组织者;发起人
organizing 组织
organizing committee 组织委员会
organizing production according to scheduling charts and tables 按指示图表组织生产
organ loft 机关顶楼;风琴席
organoacetoxysilane 有机乙酰氧
organo-additive 有机添加剂
organoalkoxy silane 烷氧基硅烷
organoaroxy silane 烷氧基硅烷;烃基烷氧基硅烷
organoarsenic chemical 有机砷化学物
organoarsenic pesticide 有机砷农药
organobentonite 有机膨润土
organoboration 有机硼化
organoboron compound 有机硼化合物
organocalie 选择性顶替
organochlorine 有机氯
organochlorine compound 有机氯化合物
organochlorine contamination 有机氯污染
organochlorine insecticide 有机氯杀虫剂
organochlorine insecticide poisoning 有机氯杀虫剂中毒
organochlorine pesticide 有机氯杀虫剂;有机氯农药
organochlorine pesticide poisoning 有机氯农药中毒
organochlorine pesticide pollution 有机氯农药污染
organochlorine pesticide residue 有机氯农药残留
organochlorine residue 有机氯残渣
organo-chlorix compound 有机氯化合物
organo-chromium finish 有机铬处理剂
organo-chromium reagent 有机铬试剂
organ of Corti 螺旋器
organofluorine poisoning 有机氟中毒
organofluorosilane 有机氟硅烷;烃基氟硅烷
organofluotie residue 有机氟残留
organo-free effluent 无有机额出流物;不含有机物的流出物
organogel 有机凝胶
organogen 有机物元素
organogenic 有机成因的
organogenic precipitation 有机沉积物
organogenous sediment 有机生成沉积物;有机沉积(物)
organo-halogen 有机卤素
organo-iron compound 有机铁化合物
organo-lead compound 有机铅化合物
organoleptic method 感官法
organolite 离子交换树脂;有机岩;有机碱交换料
organolithium compound 有机锂化合物
organo-magnesium compound 有机镁化合物
organo-magnesium halide 有机镁卤化合物
organo-manganese compound 有机锰化合物
organomercurial herbicide 有机汞除草剂
organomercurial transformation 有机汞转化(作用)
organo-mercuric compounds poisoning 有机汞化合物中毒
organomercurous fungicide 有机汞杀菌剂;有机汞防霉剂
organomercury compound 有机汞化合物(木材防腐剂)
organomercury germicide poisoning 有机汞杀菌剂中毒
organomercury pesticide 有机汞农药
organomercury pesticides poisoning 有机汞农药中毒
organometal 有机金属化合物
organometallic anti-knock 有机金属(化合物)抗爆剂
organometallic catalyst 有机金属催化剂
organometallic chemistry 有机金属化学

organometallic compound 有机金属化合物
organometallic fungicide 有机金属杀菌剂
organometallic lead antiknock compound 有机铅抗爆剂
organometallic monomer 有机金属单体
organometallic polymer 有机金属聚合物
organo-metallics 金属有机物(质)
organometallic stabilizer 有机金属稳定剂
organometallic substitution 有机金属取代作用
organometallic synthesis 有机金属合成
organometalloidal compound 有机准金属化合物
organon 研究法
organonitrogen 有机氮
organonitrogen compound 有机氮化合物
organonitrogen insecticide 有机氮杀虫剂
organonitrogen pesticide 有机氮杀虫药;有机氮农药
organo non-metallic compound 有机非金属化合物
organo-peroxide 有机过氧化物
organophilic bentonite 疏水膨润土
organophobic 疏有机性的
organo-phobicity 有机疏水性
organophosphate 有机磷酸盐
organophosphate resistant strain 抗有机磷害虫
organophosphite 有机亚磷酸酯
organophosphor 有机磷
organophosphorus 有机磷
organophosphorus agent 有机磷制剂
organophosphorus chemicals 有机磷化学物
organophosphorus compound 有机磷化合物;有机含磷化合物
organophosphorus determination 有机磷农药检测
organophosphorus ester 有机磷酸酯
organophosphorus herbicide 有机磷除草剂
organophosphorus insecticide 有机磷杀虫剂
organophosphorus insecticide poisoning 有机磷杀虫剂中毒
organophosphorus intoxication 有机磷农药中毒
organophosphorus pesticide wastewater 有机磷农药废水
organophosphorus poison 有机磷毒物
organophosphorus poisoning 有机磷中毒
organophosphorus residue 有机磷残渣;有机磷残留
organo-poly-siloxane 有基聚硅烷
organoradiomercurial 有机放射性汞制剂
organosedimentary 生物沉积的
organosilan(e) 有机硅烷
organosilicate 有机硅酸盐
organosilicate coating 有机硅酸盐涂层
organo-silicic oil 有机硅润滑剂
organosilicon 有机硅
organosilicon compound 有机硅化合物
organo-silicone 有机硅氧烷;有机硅
organo-silicone compound 有机硅化合物
organo-silicone rubber 有机硅橡胶
organo-silicone polymer 有机硅聚合物
organo-silicone wastewater 有机硅废水
organosiloxane 有机硅氧烷
organosol 有机溶胶
organosol coating 有机溶剂涂布
organosol lignin 有机溶剂木素
organosomatic index 有机体指数
organo-sulfate 有机硫酸酯
organosulfur 有机硫
organosulfur compound 有机硫化合物
organosulfur emissions 有机硫排放物
organosulfur germicide poisoning 有机硫杀菌剂中毒
organosulfur pesticide 有机硫农药
organotin 有机锡
organotin-carboxylate 有机锡羧酸盐
organotin compound 有机锡化合物
organotin compounds contamination 有机锡化合物污染
organotin-fungicide 有机锡杀菌剂
organotin germicide 有机锡杀菌剂
organotin germicide poisoning 有机锡杀菌剂中毒
organotin poisoning 有机锡中毒
organotin stabilizer 有机锡稳定剂
organo-transition metal compound 有机过渡金属化合物
organo-uranium complex 有机铀络合物
organouranium compound 有机铀化合物

organ-shaped facade 风琴形态的外观
organs of public security 公安机关
organs of supervision and management of water pollution prevention and control 水污染防治监督管理机关;水污染防治监督管理机构
organ timbering 排柱
organum axiale 轴器
organum chordotonale 弦响器
organum spirale 螺旋器
organ whistle 风琴音汽笛
orgatron 簧片电风琴
orginal hole 主孔;主井眼
orginal line 基准线
Orgueil meteorite 奥盖尔陨星
orickite 水碱黄铜矿
oricycle 极限圆
oriel 突窗;凸窗;凸肚窗
oriel bay window 凸肚窗
oriel window 凸出壁外的窗;凸肚窗
orient 排列方向;东亚;定向
orientability 可定向性
orientability of a sound signal 音响信号定向能力
orientable 可定向的
orientable elbow 可定向弯管
orientable manifold 可定向总管
orientable surface 可定向的曲面
orientable telescopic chute 可定向伸缩式溜槽
oriental alabaster 条纹大理岩;条带状大理石
Oriental Amethyst 东方紫水晶;紫刚玉
Oriental aquamarine 东方水蓝宝石
oriental arbour-vitae 侧柏
oriental architecture 东方(式)建筑
oriental army worm 黏虫
oriental cherry 樱花
Oriental chrysolite 东方贵橄榄石
oriental civet cat 大灵猫
oriental cupola 东方圆顶;东方拱顶
oriental dome 东方式圆顶
oriental emerald 东方祖母绿;绿刚玉
oriental focusing helium detector 定向聚焦测氦仪
oriental granite (一种产于美国明尼达州的黑色和灰色的)粉红色花岗岩
orientalism 东方民族特征
oriental linaloe 东方沉香油
oriental metal 沥青罩面防水钢板
oriental oak 栓皮栎
oriental pearl 东方珍珠
oriental plane-tree 法国梧桐
oriental porcelain 东方瓷器
oriental region 东洋区
oriental rug 东方地毯;(中东和远东的)手工小地毯;东方毛毯
oriental sweetgum 安息香香脂
oriental sweetgum resin 安息香香脂树脂
oriental topaz 东方黄宝石;黄宝石
oriental white oak 木槲树;槲栎
Oriental zoo-geographic(al) region 东洋动物地理区
orientate 定方位
orientated 祭坛东向的教堂
orientated deposition 定向沉积
orientated growth 定向生长
orientated microstructure glass-ceramics 定向微观结构微晶玻璃
orientated polymer 取向聚合物
orientated solidification 定向凝固
orientating accommodative apparatus 朝向调节器
orientating coring 定向取芯
orientating function 定位作用
orientating reflex 朝向反射
orientation 校正方向;取向;排列方向;定向;朝向(东方);方针;方向;方位
orientation adsorption 定向吸附
orientational force 取向力
orientational invariance 定向不变性
orientational measurement 定向测量
orientational mode 定向方式
orientational movement 定向运动
orientational point 定向点
orientational texture 定向织构
orientation analysis 定向分析;方位分析
orientation angle 定向角;定位角
orientation auxiliary equipment 辅助定向设备
orientation auxiliary plant 辅助定向设备

orientation behavio(u)r 定向行为
orientation boat 导向船
orientation by backsight 后视定向
orientation compass 定向罗盘
orientation connection survey 连接测量;定向连接测量
orientation contrast 取向衬度
orientation coring 定向取芯
orientation correction 定向改正;方位校正
orientation cycle 定向循环
orientation data 方位元素
orientation device 定向仪
orientation diagram 定向图;定位图;方位图
orientation disorder 取向无序
orientation effect 取向效应;定向效应
orientation element 定向元素;方位元素
orientation equipment 定向设备
orientation error 定向误差
orientation factor 取向因素
orientation figure 定向数字
orientation flange assembly 定向翼缘装置
orientation for place 位置定向
orientation identification 方位判定
orientation index 晶向指数
orientation inset 方位插图
orientation law 取向律
orientation lever 定向杆
orientation line 定位线;标定线
orientation mark 取向标记;参考点
orientation matrix 取向矩阵;定向矩阵
orientation mode 定向(方)式
orientation movement 取向运动;朝向运动
orientation nozzle 定向喷管
orientation of building 房屋朝向;房屋定向;房屋方位
orientation of cracks 地裂缝方向
orientation of earth fracture 裂缝方向
orientation of entrance 进口定位;进口定向
orientation of finite longitudinal strain 无有限伸缩线方位
orientation of force component 分力取向
orientation of groin 丁坝取向
orientation of intermediate strain axis 中间应变轴方位
orientation of intersecting direction 方向交会法测定
orientation of lineation 线理的方位【地】
orientation of maximum compressive stress 最大压应力轴方位
orientation of maximum strain axis 最大应变轴方位
orientation of mesh lines 网格线定向
orientation of minimum strain axis 最小应变轴方位
orientation of nozzles and other attachments 接管与其他附件方位
orientation of observation tunnel 观测平硐方法
orientation of organic glass 有机玻璃定向
orientation of plane-table 平板仪定向
orientation of precipitate 析出粒子的取向
orientation of profile 剖面方位
orientation of reference ellipsoid 参考椭球定位
orientation of regional stress 区域应力方位
orientation of related features 关连形体间的方向
orientation of route 路线方位
orientation of smeared crystal 滑抹晶体方位
orientation of space 空间定向
orientation of stress axes 应力轴方位
orientation of striations 擦痕方位
orientation of tectonic movement 构造运动定向性
orientation parameter 定向参数
orientation polarizability 取向极化率
orientation polarization 取向极化;定向极化
orientation projection 定向投影
orientation range 取向范围
orientation reference data 定向基准数据
orientation relationship 取向关系
orientation rule 取向法则
orientation star 定向恒星
orientation statistics 定向统计学
orientation survey 试点测量
orientation system 定向系统
orientation test 初级试验
orientation theory of thixotropy 触变性取向论

orientation threshold 方向辨别阈
orientation time 定向时间
orientation triangle 定位三角
orientation uniformity 取向均一性
orientation unknown 定向未知数
orientation value 标定值
orientation variables 定向变量
orientation while drilling 随钻定向(法)
orientator 定向器
oriented 排列定向;定向的
oriented adsorption 定向表面吸附;单分子层定向表面吸附
oriented bicrystals 定向双联晶
oriented blasting 定向爆破
oriented circuit 定向回路
oriented coagulation 定向絮凝
oriented communication network 有向通信网
oriented core 定向中心;定向岩芯
oriented core barrel 定向岩芯管;钻孔定向仪
oriented coupling 定向接头
oriented crystallization 定向结晶
oriented device 定向器
oriented diamond 定向金刚石
oriented diamond bit 定向金刚石钻头
oriented direction 有向距离
oriented fabric 颗粒定向组构
oriented film 定向膜;定向薄膜
oriented gamma sampling 定向伽马取样
oriented gamma spectrum sampling 定向伽马能谱取样
oriented graph 定向图
oriented-impregnated bit 定向孕镶金刚石钻头
oriented matrix 定向矩阵
oriented meiotic division 定向减数分裂
oriented nuclei 取向核
oriented photograph 定向像片
oriented picture 定向像片
oriented pipe 定向钻杆
oriented polymer loose tube fiber 定向聚合物松套光纤
oriented polyvinyl chloride pressure pipe 定向聚氯乙烯压力管
oriented rod 定向钻杆;定向杆
oriented sample 定向试样
oriented sampling compensation value 定向取样补偿值
oriented section 定向切片
oriented specimen 定向标本
oriented steel 各向异性钢片
oriented strand board 定向刨花板;定向纤维板
oriented survey 定向测量
oriented tape 定向带
oriented tree 有向树;定向树
oriented ventilation control 定向通风控制
oriented water 定向水
orienting 确定方向
orienting core barrel 定向取芯筒
orienting drill pipe 钻杆柱定向
orienting hole survey 定向钻孔方位测量
orienting knob 方位的柄
orienting line 方向线;标定线
orienting member 定向楔装置
orienting thrust motor 定向推力发动机
orienting well survey 定向孔测量
orientite 锰柱石
orient yellow 天然硫化镉
orifice 注流孔;针孔;管口;漏嘴;漏孔;量孔;孔口;孔(洞);节流孔;锐口;锐孔;出料孔
orifice amplifier 节流放大器
orifice area 节流孔口面积
orifice box 孔口水箱;孔口量水箱;静水池
orifice buffle 锐孔挡板
orifice cap 孔口帽
orifice check valve 小孔止回阀;小孔单向阀;量孔节流单向阀
orifice choke 孔板式节流器
orifice coefficient 流出系数;孔口系数;测量孔板系数
orifice control valve 锐孔调节阀
orificed anode 带孔阳极
orifice diameter of grouting hole 注浆孔直径
orifice equation 孔口方程;节流方程
orifice feeding 孔板送风
orifice feed tank 孔口投药箱;孔口投料箱

orifice flow 孔流;孔口流;孔口出水;孔口出流
orifice flowmeter 孔口流量计;孔板流量计;锐孔流速计
orifice gas 喷嘴气体
orifice gas flow meter 孔口流量计
orifice gas scrubber 锐孔气体洗涤器
orifice ga(u)ge 孔口流量计;孔板测流规
orifice hood 孔口罩
orifice intake 进口
orifice measuring device 孔口测流设备
orifice meter 孔口流量计;管孔流量计;管孔流差计;量水孔;孔板流量计;锐孔流速计;测流孔
orifice meter 孔口流量计
orifice-metering coefficient 孔口流量系数
orifice method 小孔法
orifice mixer 孔流混合器
orifice of pipe 管口
orifice of spring 泉眼
orifice outflow 孔口出流
orifice piece 管嘴
orifice plate 光阑;隔板;孔口测流板;孔板;节流孔板;节流板;锐孔板;垫圈;挡板
orifice plate flowmeter 孔板流量计
orifice plate unit 孔板组
orifice plug 节流塞
orifice proportioning feeder 孔板式定量注入装置
orifice restrictor 孔口节流器
orifice rheometer 管口式流动度测定仪
orifice ring 料碗;孔环
orifice size 节流面积
orifice sleeve 节流套
orifice spillway 孔口式溢洪道
orifice tap 喷管阀门;小孔塞子
orifice tube test 混凝土混合物的流动性测试
orifice (type) visco(si)meter 管孔黏度计;锐孔型黏度计;孔口黏度计;细孔黏度计;短孔黏度计
orifice union 节流组件
orifice valve 小孔阀
orifice well tester 孔板流量计;孔板测试器
orifice with full contraction 完全收缩孔(口)
orifice with suppressed contraction 不完全收缩口;不完全收缩孔
orificial 管口的
orificing factor 限流因子;配流因子;通量阻隔系数
orificium 管口
origin 原点;原产地;起源;起端;起点;始点;出发点;成因;发送源
origin address field 初始地址段
origin address field prime 本地始端地址字段
original 原著;原型的;原件的;正本;底片;初始的;本来的
original accelerogram 原始加速度记录图
original adsorbent 天然吸附剂
original argument 初始变元
original articles 原作
original artwork 原图
original asphalt 原生(地)沥青;天然石油沥青;天然地沥青
original author 原著者
original average reservoir pressure 原始油层平均压力
original base 起飞机场
original beach profile 海滩原地形(剖面)
original bed 原河床
original bedding 原有垫褥;原有基层;原生层理
original bench mark 水准基点;原始水准点;水准原点
original bid 标书;第一次标书
original bills of lading 有效提单;提单正本
original binding 原装
original block 原色版
original boards 原封皮
original breed 原种
original budget 原来预算
original building 原有房屋
original bunch 原始束
original capital 初始资本;初始成本
original capital cost 初始投资
original certificate 原始证件;货源证书;货物来源证
original channel 老航道
original character 初始字符
original chart 原图;原版海图
original circuit 原电路
original classification 原始分类

original cleavage 原生劈理
original cloth 原封皮
original code of sample 样品原始编号
original cohesion 原始凝聚力;天然凝聚力;初始凝聚力
original composition 原始组分
original concentration 原浓度;初始浓度
original condition 原始状态;原始条件
original condition equation 原始条件方程
original consequent lake 原生顺向湖
original consistency 最初稠度;初始稠度
original contract 原有合同;原合同
original contract construction 总承包施工
original contractor 原承包商;原承包人;总承包者;总承包商;总承包人
original coordinates 起始坐标
original copy 原稿;正本;原垫底
original copy for illustrative matter 图像原稿
original copy for type composition 文字原稿
original cost 原值;原始费用;原始成本;原价;原始成本
original cost existing machine 现有机器的原价值
original cost method 原始成本法
original cost of fixed assets 固定资产原值;固定资产原始成本
original cost to date 现值
original cost value 开始成本价值;开砂造型
original covers 原封面
original crest 起始坡顶线
original cross-sectional area 原始截面积
original crystallization ages 原始结晶年龄
original data 原始资料;原始数据
original data collection 原始资料收集
original data source 原始数据来源
original deep thermal water 未被冷水掺合的深部热水
original definition 初始定义
original deformation 起始变形
original density 原始密度;初始密度
original density range 原密度范围
original description 原始描述
original design 原设计;原设计;原方案
original detail 原有的零件
original dilation point 始膨点
original dip 原始倾斜;原生倾斜;天然倾斜
original directive 原始命令
original division plan 成因区划
original documents 原始文件;原始单据;正本单据;原始文献
original draught 原始吃水
original drawing 原图
original edition 原版
original edition in microfilm 缩微胶卷原件
original elevation 原始高程
original entry 原始记录
original environment 原生环境
original equation 原始方程
original equilibrium constant 固有平衡常数
original equipment 初始设备
original equipment cost 设备购置费
original equipment manufacture coating 在线涂装涂料;母体设备制造用涂料
original equipment manufacturer 初始设备制造厂家
original equipment manufacturer coating 工业产品涂料
original equity (购置房地产的)头款;(购置房地产的)首期现款
original error 原始误差
original evidence 原始凭证
original evidence of railway transport statistics 运输统计原始凭证
original expiration date 原定到期时间
original feed 原给矿
original field of temperature 初始温度场
original film 原片
original film density 原片密度
original forest 原始林;原生林;处女林
original form 原型;原底
original format 原幅
original formation pressure 原始地层压力
original form survey 现场测量
original founder 原创办人
original gas cap 原生气顶气
original ga(u)ge length 原隔距长度

original grammar 初始文法
original grating 原型光栅;原始光栅
original grid 原形光栅;原始光栅
original gross rate 原始总收益率
original ground 原始地面;原地面;天然地面
original ground level 原(始)地面标高;天然地面高程;原(始)地面高程
original ground surface 原始地表;原始地面
original harbo(u)r 老港(口)
original heading 起始航向
original height of bank slope 岸坡原始高度
original horizontality 原始水平
original horizontal stratification 原水平层理
original host rock 原生围岩
original image 原像
original impulse 起始脉冲
original inclination 起始倾角
original influence 原始影响;自然影响
original information 原始资料;原始信息
original input data 初始输入数据
original input stream 初始输入流
original inspection 初检验
original interstice 原有空隙;原始缝隙;原生空隙;原生间隙
original investment 原有投资;原投资额
originality 新颖;独创性;创造力;本原
originality innovation 创新
original landform 原始地形
original leakage 原有漏损
original letter of credit 信用证正本
original level ground surface 原始水平地面
original levels of reference 原始参考标高
original line 原始基准线;原线
original line of sight 原瞄准线
original literal material 原始文字材料
original location 真实外景
original loess 原状黄土
original lot 原始一批;原始地区
original maceral 原生显微组分
original machine 原型自动机床
original machine unit 原始机械装置
original magma 原岩浆;原始岩浆;母岩浆
original manufacturer 原制造厂
original manuscript 原稿
original map 原图;原始图件;编绘底图
original map editing 原图编辑
original marking 原始标记
original master 原始主盘
original material 原始材料
original materials for description 著录的原始资料
original matrix 原矩阵
original maturity 初定偿还期
original mineral 原生矿物
original model 原始模型
original model of stereoscopic map 立体地图原模型
original mother plate 原始母版
original mo(u)ld 原型
original negative 原底片
original net rate 原始净费率
original node 原结点
original nucleus 母核
original number 原来号数
original object 原物
original observation equation 原始观测方程
original of contract 合同正本
original offset 原点偏移距
original or authentic clause 正本条款
original orbit 原始轨道;起始轨道
original or extended maturity 原定或延长偿还期
original package 原包装
original paper 原论文
original paper for diazo reproduction 重氮晒图原纸
original pattern 原稿
original payee 原收款单位;原收款人
original phreatic level 初始潜水位
original picture 原图
original picture negative 原像负片
original plane landslide 原生面滑坡
original plan(ning) 原规划;原计划
original plot 实测原图
original plotting point 作图点
original point 原始基准点;原点

original pore 原生孔隙
original pore pressure 原始孔隙压力
original port 老港(口);始发港;出发港
original position 原始位置;原位;初始位置
original positive 原正片
original premium 原保险费;起保费
original premium rate 原定费率
original problem 初始问题
original procedure 原始过程
original production 原件
original program(me) 原始程序
original pulse 原始脉冲
original purchase money 原购入价
original reconnaissance 初勘;初次踏勘;初次勘察;初次勘测;初次勘测;草测
original record(ing) 原始记录
original river 顺向河;原有河道;原生河(流)
original river bed 原河床
original rock 原生岩(石);基岩;母岩
original rock stress 原岩应力
original rock temperature 原岩温度
original root system 原有的根系
original sample 原样
original sample weight 原样重量;样品原重
original scale 原比(例)尺
original scene 原始现场
original scheme 原始方案
original seabed depth 原海底深度
original seed 原种种子
original seed farm 原种圃
original sequence number 始发流水号
original shape 原样
original signal 原始信号
original size 原始尺寸;原来尺寸
original slope-angle of reservoir bank 库岸原始坡角度
original soil 原始土壤;原生土(壤);原地土(壤);未垦地
original soil rammed 原土夯实
original soil rammed foundation 原土夯实地基
original soil sample 原状土样
original soil sampling 取原状土样
original soil saw soil 原土
original solution 原溶液
original species 原始种
original specification 原始技术条件
original spectrum 原始谱
original stage 发生期
original standard 原始标准
original standard sea-water 原始标准海水
original state 原来状态;初始状态
original stiffness 初始刚度
original stiffness matrix 原始刚度矩阵
original stock 原始混合物
original store 原储[贮]藏量
original stream 原生河(流)
original strength 原来强度;固有强度
original stress 原应力;初应力
original structure 原结构
original subgrade 原状路基;原土路基
original surface 起始表面;初始表面
original survey 初测
original symbol 原始符号
original table 原始表
original tape 原始带
original temperature 初始温度;原有温度;原始温度
original temperature field 原始温度场
original text 原文
original thickness of soil stratum 原始土层原度
original time 初始时间;发震时间
original time of earthquake 发震时刻
original trace 原迹线
original treatment 初次处理;初始处理
original unit 原件
original value 原值;原始值;原始(价)值;原价;起始值;初始值
original value of fixed assets 固定资产原始价值
original variable 起始变形
original vegetation 原生植被
original velocity 起始速度
original velocity recording 原始速度记录
original version 原始方案
original void ratio 原始孔隙比;初始孔隙比
original voucher 原始凭证

original water-cement ratio 原始水灰比
original water content 原始含水量
original water line 原始水线
original wave 原始波;原浅形;基波
original wave group 原始波群
original weight 原重
original work 创作;原作;原著
original yield 最初产量
original zone 原生带
origin and destination model 起讫点模型
origin and destination survey 交通起讫调查
origin and distribution of nitrogen 氮的起源和分布
originate 首创
originate loan 议定贷款
originating 发源;发信
originating agency 发行机构
originating basin 起点流域;发端流域
originating house 策划投资商号
originating material 自产原料
originating mode 始发方式
originating party 开证(银)行
originating passenger train 始发旅客列车
originating point 起点
originating register 发send记录器
originating station 始发站
originating toll center 长途电话中心局
originating traffic 始发交通
originating train 始发列车
origination 起源
origination fee 开办费;创办费
origination railway and destination railway 发送路和到达路
originator 抵押提出者;原始承办人;创办人
originator detection pattern 发端检测码
originator of name of tectonic movement 构造运动名称的创始人
origin center 发源中心
origin classify of spring 温泉的成因分类
origin cohesion 原始内聚力;初始黏结力
origin/destination 起点/终点
origin-destination data 起讫点资料
origin-destination model 起讫点模型
origin-destination study 起终点调查;起讫点调查;OD调查
origin-destination survey 起终点调查;起讫点调查;OD调查
origin-destination survey of place of employment 工作地点起讫调查
origin-destination table 起讫点交通表
origin distortion 度盘零位误差
origin map of groundwater pollution 地下水污染成因类型
origin of agricultural system 农业系统的起源
origin of aquifer 含水层成因
origin of a word 词源
origin of bar 沙洲的起源
origin of control network 控制网原点
origin of coordinates 坐标原点
origin of country 产地国
origin of curve 曲线起点
origin of deviation 离差的原点
origin of earthquake 地震成因
origin of environment(al) law 环境法渊源
origin of force 力原点;力的作用点
origin offset 原点偏移距
origin of geothermal fluid 地热流体成因
origin of groundwater 地下水起源;地下水成因
origin of heat source 热源成因
origin of heights 高程原点
origin of inclusion 包裹体成因理论
origin of infection 传染源
origin of longitude 经度起算点
origin of oil 石油成因
origin of petroleum 石油成因
origin of rifling 膛线起点
origin of sinter 泉华的成因
origin of species 物种起源
origin of vector 矢量原点
origin of water pollution 水污染成因
origin of wave 波源
origins and patterns of global climate 全球气候的起源和模型
origin structure of mountain 山脉成因类型

origin structure type of river valley 河谷成因类型
origin system 始端系统
origin time of earthquake 发震时间
origin type 成因类型
origin types of groundwater 地下水成因类型
origin zone 起点区;出发区
orignal value of fixed assets 固定资产原值
orillon 古典门窗过梁上挑出的线脚
orimet 管口式流动度测定仪
O-ring 胶皮垫圈;O 形环;密封圈
O-ring joint O 形环接合;圆环形接头
O-ring seal 圆环密封;密封胶圈
Oriskanizn stage 奥里斯坎尼阶【地】
Oriskany stage 奥里斯坎尼阶【地】
Orizaba limestone 奥里扎巴灰岩
orizanol 阿魏酸脂
orle 柱顶圆角装饰;平椽
orleans 奥尔良呢
orlet 预制混凝土房屋
orlite 水硅铅铀矿
orlo 立柱楞条装饰;柱顶圆角装饰;平行凹槽间光滑表面;柱基底座;扁平底座
orloff 奥尔洛夫马
orlon 丙烯酸共聚物;奥纶
orlop 最下甲板;最下层;锚链舱底甲板
orlop beam 最下甲板梁
orlop beam plate 最下甲板横梁板
orlop deck 底甲板;下层甲板
orlop stringer angle bar 最下船侧纵通角
Ormasby filter 奥姆斯拜滤波器
ormolu 锌锡铜合金;锌青铜;铜锌锡合金;仿金铜箔;奥姆拉铜锌锡合金
ormolu furniture 镀金家具
ormolu varnish 镀金漆;仿金漆;金属颜料清漆
or more clause 或更多条款
ornament 装饰物;装饰品
ornamental 装饰的
ornamental acoustic(al) gypsum waffle slab 装饰性吸声华夫干状石膏板;装饰性音响石膏镶板
ornamental aggregate 装饰骨料
ornamental alumin(i)um 装饰铝材;装饰铝
ornamental arcade (不开洞的)装饰拱廊
ornamental arch 装饰券;装饰拱
ornamental architecture 盛饰建筑;华丽建筑
ornamental archivolt 装饰性拱缘;拱门饰
ornamental archway 坊表
ornamental area 装饰面积;装饰区
ornamental art 装饰艺术;装潢艺术
ornamental artificial stone 装饰性人工石;人工点缀石
ornamental balustrade panel 装饰栏板
ornamental band 装饰性条纹;装饰性带条;花边嵌条
ornamental barrel vault 装饰性筒形拱顶;弯形拱
ornamental board 装饰板
ornamental bond 装饰性连接;装饰性砌合;装潢砌筑
ornamental bracket 装饰性牛腿;装饰性托架;装饰墙基
ornamental brass alloy 装饰铜合金
ornamental brick 釉面砖;装饰砖;饰砖
ornamental brickwork 装饰性砖工;图案砌筑
ornamental bronze 装饰用青铜制品;装饰用铜
ornamental candlestick 装饰烛台
ornamental capacity 装饰容量
ornamental cast block 装饰性浇制块;装饰用预制块
ornamental cast stone 装饰性浇制石
ornamental ceiling 天花板装饰
ornamental ceiling board 装饰性天花板块;装饰性天花板材
ornamental ceiling tile 天花板用装饰面砖;装饰用天花板面砖
ornamental coat(ing) 装饰涂层;装饰面层
ornamental cold water paint 装饰用水溶性涂料;装饰用水性漆
ornamental column 装饰柱
ornamental concrete 装饰混凝土
ornamental concrete finishing 混凝土饰面
ornamental design 装饰图案设计
ornamental door 装饰门;站饰
ornamental effect 装饰效果
ornamental element 装饰构件
ornamental embossment 浮花装饰;浮雕细工;浮雕花样
ornamental enamel 装饰品搪瓷
ornamental engraving 镂花
ornamental feature 装饰优点;装饰特色
ornamental felted fabric 装饰用毡制织物
ornamental finish 装饰终饰;装饰精修;花样修饰
ornamental fittings 装饰用品;装饰零件;金属装饰品
ornamental fixture 装饰固定装置
ornamental floor finish 装饰性地板终饰;装饰性地板抛光
ornamental floor(ing) 装潢地板
ornamental flower 观赏花卉
ornamental flower bed 观赏花圃
ornamental foil 装饰用金属箔;装饰箔
ornamental foliage plant 观叶植物
ornamental forest 风景林
ornamental form 装饰格式;装饰形式;装潢式样
ornamental gable 装饰性三角墙;装饰性山墙;装饰山墙
ornamental garden 观赏植物园
ornamental gilding 装饰性镀金
ornamental glass 装饰(用)玻璃
ornamental glass brick 装饰玻璃砖
ornamental grill(e) 装饰格栅;铁花格
ornamental hardware 装饰五金;五金装饰品
ornamental heavy ceramics 装饰大型陶瓷;装饰结构陶瓷
ornamental hung ceiling 装饰吊顶棚
ornamental iron 装饰铁器;装饰用铁件
ornamental ironwork 装饰铁件
ornamental ironwork technique 装饰铁件工艺
ornamentalism 装饰艺术
ornamentalist 设计家;装饰家
ornamentalize 装饰
ornamental jamb 装饰的门框边框
ornamental joint 装饰缝
ornamental laminate(d) board 装饰层压板;装饰胶合板
ornamental lantern 装饰灯
ornamental light fitting 装饰灯具
ornamental lighting 装饰照明
ornamental link 装饰连接;装饰铰链
ornamental lock 装饰锁
ornamental luminaire (fixture) 装饰灯具
ornamental marble 装饰大理石
ornamental margin 装饰边缘
ornamental masonry 装饰体
ornamental masonry bond 装饰砖石砌法
ornamental metal 装饰金属
ornamental model(l)ed coat 装饰压花粉刷
ornamental model(l)ed stuccowork 装饰压花拉毛粉刷
ornamental motif 装饰主题;装饰花纹
ornamental motif taken from nature 取自然界的装饰主题
ornamental mo(u)ld(ing) 艺术造型;装饰造型
ornamental nail 装饰钉;饰钉
ornamental niche 装饰壁龛
ornamental paint 装饰油漆;装饰涂料
ornamental painting 装饰画;装饰性绘画
ornamental panel 装饰(护)壁板
ornamental partition 装饰隔断
ornamental patent stone 装饰优质石材
ornamental pattern 装饰图案
ornamental pattern brickwork 装饰性花色砖工
ornamental pavilion 装饰亭
ornamental paving 装饰铺面
ornamental perforated window 漏窗
ornamental perforation 装饰孔眼
ornamental pillar 装饰华表
ornamental plant 园艺植物;观赏植物
ornamental plantation 观赏林;风景林
ornamental plaster 华丽粉饰
ornamental plastic board 装饰塑料板
ornamental plastic sheet 装饰塑料(薄)板
ornamental plastic slab 装饰用塑料板
ornamental plate 装饰板;饰板
ornamental pool 装饰水池
ornamental porcelain 陈设瓷
ornamental portal 装饰门头
ornamental power 装饰特征
ornamental precast concrete product 装饰预制混凝土制品
ornamental precast concrete tile 装饰预制混凝土面砖
ornamental property 装饰特性
ornamental quality 装饰质量
ornamental railing 装饰扶手
ornamental reconstructed stone 装饰人造石
ornamental relief 装饰花纹;装饰浮雕
ornamental rib 装饰肋
ornamental ridge covering 装饰性屋脊
ornamental rowlock paving 装饰侧砌铺面
ornamental screen 装饰屏
ornamental sculpture 装饰雕塑
ornamental shaft 装饰井
ornamental sheet 装饰板
ornamental shrub 观赏灌木
ornamental sintered glass 装饰烧结玻璃
ornamental slating 花岗石板瓦
ornamental steel 装饰钢材
ornamental steel bar 饰花钢条(围墙用)
ornamental structural ceramics 装饰结构陶瓷
ornamental structure 装饰结构
ornamental style 装饰风格
ornamental surface 装饰表面;饰面
ornamental suspended ceiling 装饰吊顶棚
ornamental tablet 装饰牌
ornamental tie 柱箍;装饰性拉杆
ornamental tile 装饰面砖;装饰瓷砖
ornamental touch 装饰格调;装饰风格
ornamental tower 装饰塔
ornamental town gateway 装饰城门
ornamental tree 观赏树木
ornamental trim 装饰镶边;装饰贴脸板;装饰压缝条
ornamental tunnel vault 装饰筒形拱顶;装饰拱顶
ornamental turret 装饰塔楼
ornamental unit 装饰建筑单元;装饰构件
ornamental vault 装饰拱顶
ornamental vessel 庭园雕刻装饰
ornamental wagon vault 装饰拱顶
ornamental wall 装饰墙
ornamental wall bracket 装饰墙牛腿;装饰墙托架
ornamental ware 装饰器皿;陈设器皿
ornamental water(-carried) paint 装饰乳化涂料
ornamental well 装饰井
ornamental window 装饰窗
ornamental wire(d) glass 装饰嵌丝玻璃
ornamental wood mo(u)lding 装饰用木线脚
ornamental work 装饰工作;装饰工程
ornamental work of iron 铁花
ornamental wrought iron work 装饰用锻铁件
ornamentation 装饰品;砖石装饰性砌法;装饰(术);纹饰
ornamentation on wall 墙面装饰
ornamented 经装饰的
ornamented architecture 盛饰建筑
ornamented tile 装饰瓦
ornamented work 装饰细工
ornament rafter end 装饰椽头
ornate 雕琢过的;华丽的;高度装饰的
ornate form 华丽格式
ornithology 鸟禽学
oro 耐热油漆
orobiome 山地生物群系
Orodo 奥罗多红棕色硬木(尼日利亚产)
orogene 造山带
orogenesis 造山作用
orogenetic 造山的
orogenetic process 造山过程
orogenic 造山的
orogenic area 造山区
orogenic belt 造山带
orogenic cycle 造山循环;造山旋回
orogenic event 造山事件
orogenic metamorphism 造山变质作用
orogenic movement 造山运动
orogenic period 造山周期;造山期
orogenic phase 造山期;造山幕
orogenic province 造山省
orogenic region 造山区
orogenic sedimentation 造山沉积
orogenic sequence 山地土系列
orogenic stage 造山阶段
orogenic unconformity 造山不整合
orogenic zone 造山带
orogenic zone of Andes type 安第斯型造山带
orogenic zone of Caucasus type 高加索型造山带

orogenic zone of Cordilleran type 科迪勒拉型造山带
orogenic zone of Himalayas type 喜马拉雅型造山带
orogensis 造山运动
orogeny 造山作用;造山运动;造山成因
orogeosyncline 造山地槽
orogram 山形图
orograph 山形仪;地形仪
orographic(al) 地形的;山岳形态的
orographic(al) barrier 地形障碍
orographic(al) character 地形特性;地形特征;山势特性
orographic(al) characteristic line 地性线;地貌特征线;地貌结构线
orographic(al) cloud 地形云
orographic(al) colo(u)ring 分层设色法
orographic(al) condition 山岳形态条件;地形条件
orographic(al) convection 地形对流
orographic(al) deformation of front 锋面的地形性变形;锋的地形性变形
orographic(al) disturbance 地形扰动
orographic(al) divide 山脉分界线;地形分水线;地形分水岭;地形分水界;地形分界线;分水岭;地面分水岭;地面分水界
orographic(al) element 山地形态要素
orographic(al) factor 山地形态因素
orographic(al) fault 山形断层
orographic(al) influence 山脉影响;地形影响
orographic(al) lifting 气流地形举升;山岳抬升;山形隆起;地形抬升;地形上升
orographic(al) map 地势图
orographic(al) method 山形法
orographic(al) model 地形模式;山岳模式;山水地形样图;地形模型
orographic(al) occlusion 地形锢囚【气】
orographic(al) precipitation 地形性降水;地形降水
orographic(al) probable maximum precipitation 山区可能最大降水
orographic(al) rain(fall) 地形雨;地形性降雨;山岳雨;地形降雨
orographic(al) region 山地地区
orographic(al) storm 地形暴雨;山岳暴雨
orographic(al) thunder shower 地形雷雨
orographic(al) tinting 地貌分层设色表
orographic(al) uplift 山岳抬升;山形隆起;地形抬升
orography 山志学;山脉志;山岳形态学;山文学;地势表现法
orohydrographic(al) model 山水立体模型;高山水文地理模型
orohydrography 高山水文地理学;山脉水文学;山区水文(地理)学;山地水文学;地形水文学
orohydrology 山区水文学
Oroide 金色铜;铜锌锡合金
orology 山理学
orometer 高山气压计;气压高程仪;山岳气压计;岳高度计;山地气压表
orometry 山的高度测量
oronym 山岳名称
oropesa sweep 单船软扫【测】
orophyte 山岳植物
orophytia 亚高山植物群落
oropion 山脂土
oroseite 伊丁石
orotectonic line 山岳构造线
Orotron 奥罗管
Orowan's relation 奥罗万关系式
Oroya fever 奥罗亚热
orphan 孤立数据
orphanage 孤儿院
orphan asylum 孤儿院
orphan bank 地面废料场
orpheite 硫磷铅铝矿
orpiment 三硫化二砷;雌黄含量;雌黄
orpine 紫花景天
orrery 太阳系仪
orrotatomg smog 刺激性烟雾
Orsat analysis 奥萨特分析(法)
Orsat analyzer 奥萨特分析仪
Orsat apparatus 奥萨特分析器;奥萨特气体分析仪
Orsat flue gas analyzer 奥萨特烟气分析仪
Orsat gas analyzer 奥萨特气体分析仪
Orsat gas apparatus 奥萨特气体分析器
orseille 苔色素

orsellinic acid 苔色酸
orterde 硬土层;硬盘土
Orthatest 奥托比较仪
orthesis 矫正法
orthicon 正摄像管;低速电子束摄像管
orthicon image tube 正析像管
orthiconoscope 直像管
orthite 褐帘石
ortho-acetic acid 原乙酸
ortho acid 原酸;邻位酸
orthoalkaligneiss 碱性正片麻岩
ortho-aluminate 原铝酸盐
ortho-aluminic acid 原铝酸
ortho-aminobenzenesulfonic acid 邻氨基苯磺酸
ortho-aminobenzoic acid 邻氨基苯甲酸
ortho-aminotoluene 邻甲苯胺
ortho amphibolite 正角闪岩
ortho-amyl phenol 邻戊基苯酚
ortho-anisidine 邻氨基苯甲醚
orthoantigorite 正叶蛇纹石
ortho-antimonite 原亚锑酸盐
ortho-antimonous acid 原亚锑酸
orthoarenite 正砂屑岩
ortho-arsenate 正砷酸盐
ortho-arsenic acid 原砷酸;砷酸
ortho-arsenite 原亚砷酸盐
ortho-arsenous acid 原亚砷酸
ortho-axis 正交轴(线);正轴
orthobaric density 正压密度;规压密度;本压密度
orthobaric volume 规压容积;标准容积;本压容积
ortho-borate 原硼酸盐
ortho-boric acid 原硼酸;硼酸
orthobrannerite 斜方钛铀矿
orthocartograph 正射投影测图仪
orthcenter[orthocentre] 垂心;正中心;垂心
orthocentric surface 垂心曲面
orthochamosite 正鲕绿泥石
orthochem 原生化学沉积;正化组分;正化颗粒
orthochemical constituent 正化组分
orthochemical rock 正常化学岩
ortho-chloroaniline 原氯苯胺
orthochlorophenol 原氯苯酚
ortho-chlorotoluene 邻氯甲苯
orthochromatic 正色的;正染色的
orthochromatic coating 正色感光膜
orthochromatic emulsion 正色性乳剂;正色乳剂
orthochromatic film 正色纸;正色胶片;正色胶卷;正色感光片
orthochromatic filter 正色滤光器
orthochromatic paper 正色像纸
orthochromatic photographic(al) material 正色片
orthochromatic plate 正色片;正色底片
orthochromatism 正色性
orthochrome 原色母;原母色;正色
ortho-chromic acid 原铬酸
orthochromophil 中性染剂染色性的
orthochronological fossil 正年代化石
orthochronology 标准地质年代学
orthochronous 正确时间
ortho-chrysotile 正纤维蛇纹石
orthoclase 正长石
orthoclase basalt 正长玄武岩
orthoclase diorite 正长闪长岩
orthoclase feldspar 正长石
orthoclase ieucite basalt 正长白榴玄武岩
orthoclase nepheline basalt 正长霞石玄武岩
orthoclase porphyry 正长斑岩
orthoclase syenite 正长石正长岩
orthoclase trachybasalt 正长粗面玄武岩
orthoclasization 正长石化
orthoclastic 有彼此成直角之裂缝的;正解理
orthocline 直倾型
orthocobaltite 斜方辉砷钴矿
orthocode 垂向码
orthocomplement 正交补
orthocomplementation 正交补
orthocomplemented 正交补的
ortho compound 邻位化合物
orthoconglomerate 正砾岩
orthocorrection 垂向校正
orthocorrection digit 垂向校正位
orthocrasia 正常反应性
ortho-cresol 邻甲酚
orthocumulate texture 正堆积结构

orthocycle 正交圆
orthod 灰土
orthodactylous 直指的
orthodiagonal 正轴
orthodiagonal crystal 正轴晶体
orthodiagram 正影描记图
orthodiagraph 正影描记器;正透影器;正摄像仪
orthodiagraphy 正影描记术
orthodichlorobenzene 邻二氯苯
orthodinitrocresol 邻二硝基甲酚
orthodome 正轴坡面
orthodontic appliance 正畸器械
orthodontic materials 正畸材料
orthodox 旧式
orthodox approach 正统方法;传统方法
orthodox approach in econometric research 经济计量研究的传统方法
orthodox beam 正规束
orthodox casting 普通模铸
orthodox construction method 传统构筑法;传统建筑法
orthodox explosive 普通炸药;通用炸药
orthodox material 正统材料;正常材料
orthodox method 经典方法;正统方法
orthodox practice 固有方法;传统惯例
orthodox reinforced concrete 传统钢筋混凝土
orthodox scanning 正规扫描
orthodrome 大圆(弧);大圆圈线;大圆航线
orthodromic 顺向性
orthodromic bearing 大圆方位
orthodromic conduction 顺向传导
orthodromic illumination 正向照明
orthodromic line 大圆弧线
orthodromy 大圆航海术;大圆航法
ortho-effect 邻位效应
orthoenstatite 斜方顽辉石
orthoericssonite 斜方钡锰闪叶石
orthoester 原酸酯
orthoevolution 直向进化
orthoferrite 正铁氧体
orthoferrosilite 正铁辉石
orthoflow 正流式
orthoflow fluid cracking unit 正流流体裂化装置
orthoflow process 正流式过程
orthoforming 正流式
orthoforming process 正流式重整过程
orthoforming unit 正流重整装置
orthogamy 正常配偶
orthogenesis 直向演化
orthogenic 直向发生的
orthogeosyncline 正地槽
orthogeotropism 直向地性
orthogneiss 正片麻岩;火成片麻岩
orthogon 直角三角形;正方形;矩形;长方形
orthogonal 相互垂直的;直角的;直交的;正交的
orthogonal anisotropic plate 正交各向异性板
orthogonal anisotropy 正交各向异性
orthogonal antennas 正交天线
orthogonal array 正交装置;正交阵列;正交排列阵
orthogonal atomic orbital 正交原子轨道
orthogonal axes 笛卡尔坐标轴;直角坐标轴;正交轴(线)
orthogonal axis 正交轴(线)
orthogonal basis 正交基
orthogonal bordering 正交加边
orthogonal Cartesian coordinates 正交直线坐标
orthogonal characteristic function 正交特征函数
orthogonal circle 正交圆
orthogonal code 正交码
orthogonal comparisons 正交比较
orthogonal complement 正交余;正交补
orthogonal complement space 正交补空间
orthogonal complete reducibility 正交完全可约性
orthogonal component 正交分量;正交分力
orthogonal cone 正交锥面
orthogonal coordinates 直角坐标;正交坐标
orthogonal coordinate system 正交坐标系
orthogonal corresponding parametric system 正交对应参数系统
orthogonal curve 正交曲线
orthogonal curve system 正交曲线系统
orthogonal curvilinear coordinates 正曲线坐标
orthogonal curvilinear coordinate system 正交曲线坐标系

orthogonal cutting 正交切削;垂直切割;垂直割法
orthogonal decomposition 正交分解
orthogonal decomposition of vector 向量正交分解(法)
orthogonal deflection 正交偏转
orthogonal descriptor 正交描述符
orthogonal design 正交设计
orthogonal design method 正交设计法
orthogonal detection 正交检波
orthogonal dimension 正交维数
orthogonal direction 正交方向
orthogonal direct sum 正交直和
orthogonal elements 正交元素
orthogonal expansion 正交展开;正交函数展开
orthogonal experiment 正交试验
orthogonal experimental design 正交试验设计
orthogonal factorization method 正交因子法
orthogonal families 正交族
orthogonal family of function 正交函数族
orthogonal field 正交场
orthogonal field tube 正交场加速管
orthogonal filter 正交滤波器
orthogonal flat 正交平坦形
orthogonal for finite sum 选点正交
orthogonal frame 正交标架
orthogonal frame bundle 正交标架丛
orthogonal frequency division multiplexing 正交频分复用
orthogonal function 正交函数
orthogonal function expansion 正交函数展开(式)
orthogonal function system 正交函数系
orthogonal geometry 正交几何学;正交几何
orthogonal group 正交群
orthogonal groupoids 正交广群
orthogonal group pair 正交群对
orthogonal-guidance 正交导槽
orthogonal idempotent elements 正交幂等元
orthogonal impervious boundary 直交隔水边界
orthogonal increment 正交增量性
orthogonal intersection 正交
orthogonal involution 正交对合
orthogonality 相互垂直;直交性;正交(性)
orthogonality condition 正交条件
orthogonality criterion 正交(判别)准则
orthogonality of character 特征标的正交性
orthogonality of modes 模的正交性
orthogonality of trigonometric(al) function 三角函数的正交性
orthogonality property 正交性
orthogonality property of normal modes 正则振型的正交性
orthogonality property of the mode function
orthogonality rejection 正交关系
orthogonality relation(ship) 正交关系
orthogonality theorem 正交定理
orthogonalizable 可正交化
orthogonalizable code 可正交码
orthogonalization 相互垂直;直交化;正交化
orthogonalized condition equation 正交条件方程
orthogonalized plane wave 正交平面波
orthogonalizing process 正交化步骤
orthogonal joint 正交接合
orthogonal kets 正交刃
orthogonal Latin cubes 正交拉丁立方
orthogonal Latin rectangles 正交拉丁长方形
orthogonal Latin squares 正交拉丁方
orthogonal lattice 直角点阵
orthogonal law 正交律
orthogonal layout 正交表
orthogonal level transducer 液面式垂直度传感器
orthogonal linear transformation 正交线性变换
orthogonal lines 正交直线组
orthogonally anisotrophic 正交各向异性的
orthogonally associated 正交相伴的
orthogonal mapping 正交映射
orthogonal mate 正交配偶
orthogonal matrix 正交矩阵
orthogonal mean square fit 正交均方拟合
orthogonal measure 正交测度
orthogonal modal vector 正交模态矢量
orthogonal mode 正交模态;正交模式
orthogonal motion 正交运动
orthogonal nodal plane 正交节平面
orthogonal nozzle 正交接管

orthogonal observation 正视观测
orthogonal operations 正交运算
orthogonal operator 正交算子
orthogonal orthomorphisms 正交的正交射
orthogonal parametric system 正交参数系统
orthogonal parity check 正交奇偶校验
orthogonal parity check sum 正交奇偶校验和
orthogonal perspective 正透视图;正射透视
orthogonal pervious boundary 直交透水边界
orthogonal pervious-impervious boundary 直交透水隔水边界
orthogonal pinch 正交箍缩
orthogonal plane frame 正交平面框架
orthogonal point 正交点
orthogonal polarization 正交偏振;交叉极化
orthogonal polynomial 正交多项式【数】
orthogonal polynomial expansion 正交多项式展开
orthogonal polynomial method 正交多项式方法
orthogonal process 正交过程
orthogonal projection 正投影;正态投影;正射影;正交投影;正交射影(法)
orthogonal projection cartograph 正射投影测图仪
orthogonal projection map 正射投影地图
orthogonal projection point 正交投影点
orthogonal projector 正射投影仪
orthogonal quantities 正交元素
orthogonal quasi-groups 正交拟群
orthogonal ray 波向线
orthogonal reference axis 正交参考轴
orthogonal reflectivity 正交反射率
orthogonal regression 正交回归
orthogonal regression analysis 正交回归分析
orthogonal regression line 正交回归线
orthogonal reinforcement 正交配筋
orthogonal relation 正交关系
orthogonal representation 正交表示
orthogonal reticulate 直角网状的
orthogonal rheometer 正交流变仪
orthogonal rib 正交肋
orthogonal scanning 正交扫描
orthogonal section 正交剖面;正剖面
orthogonal series 正交级数
orthogonal set 正交集
orthogonal set of principal axis 正交主轴组
orthogonal signal 正交信号
orthogonal spur dike 正挑丁坝;正交丁坝
orthogonal square 正交方格
orthogonal stochastic process 正交随机过程
orthogonal straight-lines 正交直线
orthogonal strain 正交应变
orthogonal subspaces 正交子空间
orthogonal substitution 正交代换
orthogonal system 正交系(统)
orthogonal system of curves 正交曲线系;正交的曲线系
orthogonal system of hypersurfaces 正交的超曲面系;正交超曲面系
orthogonal table 正交表
orthogonal tensor 正交张量
orthogonal test 正交试验
orthogonal thickness 正交厚度
orthogonal thickness at given point 某点正交厚度
orthogonal trajectory 正交轨(迹)线
orthogonal transformation 正交变换
orthogonal transformation group 正交变换群
orthogonal translation 正交位移
orthogonal valley 正交走向谷【地】
orthogonal vector 正交向量;正交矢量
orthogonal view 正交视图
orthogonal viewing 正交观察
orthogonal viscoelastic body 正交黏弹体
orthogonal zone 直方带
orthograde 直体步行的
orthogranite 花花岗岩;正长花岗岩
orthograph 正投影图;正视图;正断面图
orthographic(al) 正射的;直角的;直线的;正交射影的;正交的
orthographic(al) chart 正射投影海图
orthographic(al) coordinates 正射坐标
orthographic(al) design interpretation 正投影图样说明
orthographic(al) drawing 正投影图
orthographic(al) photograph 正射像片
orthographic(al) plan 平面图

orthographic(al) position 正射位置
orthographic(al) projection 正投影;正态投影;正视图投影;正射投影;平行投影
orthographic(al) projection map 正射投影地图
orthographic(al) project map 立体投影图
orthographic(al) rectification 正射纠正
orthographic(al) relief method 浮雕式地貌立体表示法
orthographic(al) representation 正视表示法
orthographic(al) synthesis 正交合成
orthography 正投影法;正射影;正射法;正交投影;正交射影(法);面图投影;剖面
orthography of geographic(al) names 地名正字法
orthography property (of maps) 正形性
orthoguarinite 正片橱石
orthoheliotropism 直向阳性
orthohelium 正氦
orthohexagonal 正六边形的;正六边形
orthohexagonal axis 六角正交轴(线)
orthohydrous maceral 正常氢量显微组分
orthohydrous vitrinite 正常氢量镜质组
ortho-isomeride 邻位同分异构体
orthojector circuit-breaker 高压微量喷油断路器;正射断路器
orthojoaquinite 斜方硅钠钡钛石
orthokinesis 正动态
orthokinetic 同向移动
orthokinetic coagulation 同向凝结(作用)
orthokinetic flocculation 同向絮凝(作用)
ortholovenite 斜方钠钙锆石
orthomagmatic 正岩浆的
orthomagmatic (mineral) deposit 正岩浆矿床
orthomagmatic stage 正岩浆期
orthomapper 正射投影测图仪
orthomatrix 正杂基
ortho metamorphism 正变质作用
orthometamorphite 正变质岩
Orthometar lens 奥索曼泰透镜
ortho-methylnitrobenzene 邻硝基甲苯
orthometric 正交的
orthometric correction 正高校正;正高改正;竖高改正
orthometric drawing 正视画法
orthometric elevation 正高
orthometric error 正高误差
orthometric growth 直线增长
orthometric height 正高
orthometric system 正高系统
orthomic feldspar 正拟态长石
orthomicrite 正泥晶灰岩
orthomicrosparite 正泥晶亮晶灰岩
orthomigmatite 正混合岩
ortho mixture 不同纤维交织物
orthomode 正交方式
orthomode cavity 正模腔
orthomode coupler 正模耦合器
orthomode transducer 直接式收发转换器;正交模式转换器
orthomorphic 正形的;等角的
orthomorphic chart 正形图;正形投影海图;正形海图
orthomorphic mapping 正形投影制图
orthomorphic projection 正形投影
orthomorphism 正射;等角性
orthomorphy 正形
orthomylomite 正糜棱岩
orthonegative 正射负片
orthoneutrophil 中性染剂正染性的
Orthonic 奥索尼克磁性材料
Orthonik 镍铁磁芯材料
ortho-nitraniline 邻硝基苯胺
orthonitric acid 原硝酸
ortho-nitroaniline 邻硝基苯胺
orthonitroaniline orange 邻硝基苯胺橙
ortho-nitroanisole 邻硝基苯甲醚
ortho-nitrophenol 邻硝基酚
ortho-nitrotoluene 邻硝基甲苯
orthonormal 正交规格化的;正规化;标准化的
orthonormal basis 标准正交基(底)
orthonormal coordinates 标准正交坐标
orthonormal function 标准正交函数系;标准函数
orthonormal function expansion 规范正交函数

展开式;标准正交函数展开式
orthonormality 正规化;规格化正交性;标准正交性
orthonormalization 正交化;正规正交化;正规化;标准正交化
orthonormal matrix 标准正交矩阵
orthonormal mode 正交模态
orthonormal model 标准正交模型
orthonormal observation grid 标准正交观测格网
orthonormal polynomial 标准正交多项式
orthonormal portal 正洞门
orthonormal set 标准正交集
orthonormal statistical model 标准正交统计模型
orthonormal system 规格化正交系;标准正交系
orthonormal system of functions 规格化正交函数系
orthonormal tetrad 四维标准正交标架
orthonormal vector 正交单位向量
orthonormal vectors 正交单位向量组;标准正交向量组
ortho orientation 邻位取向;邻位定向
orthop(a)edic hospital 矫形医院
orthopan 全色的
ortho-panchromatic 正全色的;全色的
orthopan film 全色胶卷;全色胶片
ortho-para orientation 邻对位定向
orthopedic force 矫形力
orthopedics 整形外科
orthopedium 矫形医院
orthophoria 平视
orthophoric 正位的
orthophosphate 正磷酸盐
orthophosphoric acid 正磷酸
orthophot 正射投影装置
orthophoto 正射像片;正像片
orthophotobase 正射像片底图
orthophotograph 正射像片
orthophotography 正射投影法
orthophoto-image 正像片影像
orthophoto instrument 正射投影仪
orthophotomap 正射投影地图;正交像片地图;正射影像地图;正射投影像片组合图
orthophotomat 正射投影仪
orthophotomosaic 正射投影像片镶嵌图;正射像片镶嵌图
orthophotoplan 正射像片图
orthophoto plotter 正射投影绘图仪
orthophoto-projector 正射投影仪
orthophotoquad 正射影像仪
orthophoto-restitution 正射像片成像图
orthophoto scale 正像比例尺
orthophoto-scanner 正射像片扫描器
orthophotoscope 正射投影仪;正射投影纠正仪
orthophoto stereo-mate 立体正像片
orthophoto-system 正射像片系统
orthophoto-technique 正射投影技术
orthophoto-workshop 正射像片讨论会
orthophyre 正长斑岩
orthophyric 正斑状的
orthophyric texture 正斑结构
orthopictomap 正像形地图
orthopinacoid 正交轴面
orthopinakiolite 斜方硼镁锰矿
orthoplatform 正地台
orthopole 正极;垂极
ortho-position 正位;邻位
ortho-position moment 正位力矩
orthoprism 正棱锥(体)
ortho-projection equipment 正射投影仪
orthoprojector 正射投影仪
orthoptic(al) circle 切距圆
orthoptic(al) curve 切距曲线
orthoptoscope 正视镜
orthopyramid 正轴锥;正棱锥(体)
orthopyroxene 斜方辉石类
orthopyroxene granite 斜方辉石花岗岩
orthopyroxenite 斜方辉石岩
orthoquartzite 正石英岩
orthoquartzitic 正石英的
orthoquartzitic conglomerate 正石英岩质砾岩;正石英砾岩
orthoradial 直辐射的
ortho-relict texture 正残余结构
orthorhombic 斜方晶的;正交(晶)的
orthorhombic(al) and monoclinic pyroxene amphibolite 二辉角闪岩
orthorhombic(al) and monoclinic pyroxene granulite 二辉麻粒岩
orthorhombic(al) fabric 斜方组构
orthorhombic(al) lattice 正交晶格
orthorhombic(al) packing 斜方晶系排列(金属的)
orthorhombic(al) pyroxene 斜方辉石
orthorhombic(al) structure 斜方晶结构
orthorhombic(al) symmetry 斜方对称
orthorhombic(al) system 斜方晶系
orthorhombic system 正交晶系
ortho-rock 正变质岩;火成变质岩
orthoroentgenograph 正影描记器
orthoroentgenography 正影描记术
orthoron 波导放大器
orthoscanner 垂向扫描器
orthoschist 正片岩
orthoscope 正射投影仪;水检眼镜;缝隙纠正仪
orthoscopic 直线式;无畸变的
orthoscopic effect 正视立体效果
orthoscopic eyepiece 无畸变目镜
orthoscopic image 无畸变图像
orthoscopicity 保真显示性;保真显示器
orthoscopic observation 平行光观察
orthoscopic ocular 无畸变目镜
orthoscopic optic(al) system 无畸变光具组
orthoscopic rectifier 正射投影纠正仪
orthoscopic stereomodel 无畸变立体模型
orthoscopic stereopair 无畸变立体像对
orthoscopic stereo viewing 正视立体观察
orthoscopic system 无畸变系统
orthoscopic view 正视观察
orthoscopy 无畸变
orthose 正长石
orthoselection 直向选择;定向选择
orthosilicate 原硅酸盐;正硅酸盐
ortho-silieic acid 原硅酸
orthosis 整直法;矫正法
orthosite 粗粒正长岩
orthoskiagraphy 正影描记术
ortho-spectrum 正谱
orthostat 竖立的石块;丁边砌法;古典神庙下部的护墙石板或古建筑的石肋脚
ortho-state 正态
orthostatic 直立的
orthostatic position 直立位置
orthostatic reaction 直立反应
orthostatic tolerance 直立耐性
orthostereoscope 立体X线די影器;双筒立体显微镜
orthostereoscopy 正射投影立体观测
orthostichies 垂直纵列线
orthostichous 直列的
orthostichy 直列线;垂直纵列线
orthostratigraphy 正地层学
orthostyle 列柱式门廊;直线形列柱式(建筑);列柱式
ortho-sulfanilamide 邻磺胺
ortho-sulfate 原硫酸盐
ortho-sulfuric acid 原硫酸
ortho-symmetric 正交对称的
orthotaenite 正镍纹石
orthotectic 正溶的
orthotectic stage 正岩浆期
orthoterion 牵引器;牵伸器
ortho-test 正式试验;正规试验(法)
Orthotester 奥氏测微计;杠杆式比较仪
orthotic 整直的
orthotolidine 邻联甲苯胺
orthotolidine arsenite test 邻联甲苯胺亚砷酸盐试验
orthotolidine-oxalic acid 邻联甲苯胺草酸
orthotolidine reagent 邻联甲苯胺试剂
orthotolidine test 邻联甲苯胺试验
ortho-toluene sulfonamide 邻甲苯磺酰胺
ortho-toluic acid 邻甲苯甲酸
ortho-toluidine 邻甲苯胺
ortho-toluylic acid 邻甲苯甲酸
orthotomic circle 三圆的直交圆
orthotomic system 面正交系统
orthotonic type 正力型
orthotopic 正位的;常位的
ortho-torbernite 正铜铀云母
orthotronic error control 正交错误控制
orthotropic 正交各向异性的;正交转动对称的
orthotropic(al) deck bridge with closed steel ribs 闭合式钢肋正交类性板面桥梁
orthotropic bridge 正交桥
orthotropic curved plate 正交异性曲板
orthotropic deck 正交各向异性桥面
orthotropic deck bridge design 正交异性上承桥设计
orthotropic deck bridge with open ribs 开式钢肋正交异性板面桥梁
orthotropic deck structure 正交各向异性桥面结构
orthotropic material 正交各向异性材料
orthotropic particle 正交各向异性颗粒
orthotropic plate 正交异性板;正交各向异性板
orthotropic plate method 正交各向异性板法
orthotropic sector plate 正交各向异性扇形板
orthotropic shell 正交各向异性壳
orthotropic skew(ed) plate 正交各向异性非对称板
orthotropic slab 正交异性板
orthotropic solid 正交各向异性固体;正交各向异性体
orthotropic steel deck bridge 正交异性板面桥
orthotropism 直向地心性
orthotropy 异面异性;正交各向异性
orthotype 直模标本
orthovanadate 原钒酸盐
ortho-vanadic acid 原钒酸
orthox 正常氧化土
Orth's solution 奥尔特氏溶液
orthstat 竖立的石块;(古建筑的)石勒板
Ortman coupling 奥特曼联轴器
Orton cone 测温锥;标准测温熔锥;奥尔顿测温锥;奥尔顿耐火锥
orvillite 水锆石
Orvus 奥沃斯(一种加气剂)
oryctocoenose 生物化石群
oryzanolum 阿魏酸脂
Osaka Port 大坂港(日本)
Osaka tube 反射速调管
osannite 钠闪石
osar 蛇形丘
osar delta 蛇丘状三角洲
osarizawaite 羟铝铜铅矾
osarsite 硫砷锇矿
osazone reaction 费歇尔成脎反应
Osborne beds 奥斯博恩层
Osbornite 奥氏博尼特
osbornite 陨氮钛石
oscar 潜艇
osciducer 振荡传感器
oscillant share 振动犁铧
oscillated spray line 摆动式喷雾管
oscillating 振荡的
oscillating acceleration 振动加速度
oscillating adaptive control 振动适应控制
oscillating adsorption method 振荡吸附方法
oscillating agitator 摇摆式拌和机;振动式搅拌机;振动搅拌器;振荡搅拌器;振荡搅拌机;往复回转式搅拌器;摆动式搅拌器
oscillating air shaker 断续气流式抖动器
oscillating and eccentric sifter 摆动偏心筛分器
oscillating arc 振荡电弧
oscillating arm 摆臂
oscillating armature motor 振动电枢式电动机
oscillating arm mechanism 摆动臂机构
oscillating axle 摇动桥;摆动轴
oscillating bar 摇动筛条
oscillating-bar hay loader 摆动式装干草机
oscillating-base fan 摆动吹风机
oscillating batch charger 振动加料机
oscillating beacon 振荡信号;闪烁指向标;闪烁信标;闪光信标
oscillating beam 冲击梁
oscillating bearing 关节轴承
oscillating big gun 脉动喷射器管嘴
oscillating bit 偏心钻;振荡钻头
oscillating blade microtome 振动式超薄切片机
oscillating block slider crank mechanism 摆杆滑块曲柄机构
oscillating body 振动体
oscillating bond 振动键
oscillating brush 摆动路刷;摆动电刷
oscillating bubble 振动气泡
oscillating cam 摇动凸轮;摆动凸轮
oscillating cam gear 摆动凸轮盘

oscillating center 振荡中心
oscillating characteristic 振荡特性
oscillating charge 振荡电荷
oscillating chute 振动式斜槽
oscillating circuit 振荡电路
oscillating circular saw 摆动圆锯机；摆动式圆锯
oscillating coil 振荡线圈
oscillating colo(u)r sequency 振荡彩色顺序
oscillating component 振动部分
oscillating condition 振荡条件
oscillating contact 振动触点；振荡触点
oscillating contour 振荡电路
oscillating controller 振荡控制器
oscillating control servomechanism 振荡控制的随动系统
oscillating conveyer 振荡输送机；摇动式输送机；摇动式输送器；摇运机；振动式输送机
oscillating crank gear 摆动曲柄装置
oscillating crank lever 摆动曲柄杆
oscillating crystal 振荡晶体
oscillating crystal method 回摆晶体法
oscillating current 振荡电流
oscillating cylinder 摆动汽缸
oscillating cylinder viscometer 振荡圆筒式黏度计；振荡筒式黏度计
oscillating cylindric(al) valve 柱形摆动阀
oscillating deck 摆动板
oscillating deliver 振动溜槽
oscillating detector 振荡检测器
oscillating device mechanism 重锤摆动装置的机构
oscillating digger 摆动式挖掘机
oscillating discharge 振荡放电
oscillating discharge source 振荡放电离子源
oscillating disk method 振动盘法
oscillating disk viscometer 振动盘黏度计；振荡圆盘式黏度计
oscillating displacement pump 摆动式排量泵
oscillating distributor 振动分配器
oscillating divergent 振动发散的
oscillating double bond 摆动双键
oscillating drive 摆动式传动装置
oscillating drop 振荡滴
oscillating electric(al) field 振荡电场
oscillating electromotive force 振荡电动势
oscillating electron moment 振荡电子矩
oscillating engine 摆缸式发动机
oscillating fan 摇头风扇；摆式电扇；摆动电风扇
oscillating feeder 振动式投料器；振动进料器；振动给料器；振动加料器；往复式给料器；往复式给料机；摆动进料器
oscillating field 振荡场
oscillating finishing beam 摆动式抹平梁
oscillating flashboard 倾覆式闸板；舌瓣闸门；翻板闸门；舌瓣闸板；倾倒闸门
oscillating flow 振动流；脉动流；波动流
oscillating flow rate 波动流量
oscillating follower 摆动随件
oscillating foundation 摇摆支座
oscillating function 游移函数；振动函数；振荡函数；摆动函数
oscillating gate 摇摆式卸载闸门
oscillating gear 扇形摆动齿轮
oscillating granulator 摇摆式制粒机；振动碎粒机；振荡制粒机
oscillating grinding 微量纵磨削
oscillating groove grinding machine 摇摆式沟道磨床
oscillating guide 振动器
oscillating hay loader 摆动式装干草机
oscillating hitch 摆动关节连接装置
oscillating hydraulic jump 振动水跃
oscillating hysteresis 振动滞后
oscillating impinge plate 摆式冲击板
oscillating impulse 振荡脉冲
oscillating insulation 隔振
oscillating insulator 隔振器
oscillating jet 摆动射流
oscillating jet flowmeter 康达效应流量计
oscillating jet technique 振荡射流法(测动表面张力)
oscillating joint 振动连接
oscillating journal bearing 摆动轴颈轴承
oscillating jump 摆动水跃
oscillating lapping machine 摆动式精研机

oscillating launder 振动式流槽
oscillating link 振动键
oscillating linkage 振动键
oscillating liquid continuous extraction tower 往复回转式液体连续萃取塔
oscillating load 振动荷载；振动负载；交变荷载；交变负载；摆动荷载
oscillating loop 振荡波腹
oscillating machine for testing of lubricants 润滑剂振动试验机
oscillating magnet 振动磁体
oscillating magnetic field 振荡磁场
oscillating mast 摇腿；摆式支架
oscillating membrane 振荡薄膜
oscillating meter 振动式仪表
oscillating mill 振动研磨机
oscillating mirror 往复偏转反射镜
oscillating mirror scan system 振动反射镜扫描系统
oscillating mixer 振动搅拌机
oscillating mixers agitator 往复回转混合搅拌器
oscillating mode 振动型；振动模式；振荡
oscillating model of the second kind 第二类振动宇宙模型
oscillating motion 振荡运动；摆动运动
oscillating motor 摇动马达
oscillating movement 振荡运动；摆动运动
oscillating neutral 振荡中性(点)
oscillating part 振荡部分
oscillating period 振荡周期
oscillating phase 振荡相
oscillating pickup system 振动传感系
oscillating-pipe sprinkler 摆动管(式)喷灌机
oscillating piston 振荡活塞
oscillating piston compressor 旋转活塞压气机
oscillating piston liquid meter 环形活塞流量计
oscillating plane 振动面；摆动面
oscillating plate 振荡板
oscillating plate column 振动板式柱
oscillating plough 摆动犁
oscillating pneumatic roller 摆动式气胎路碾；摆式轮胎压路机；摆动式气胎压路机
oscillating point 振动点；振荡点
oscillating potential difference 振荡电位差
oscillating pulse 振荡脉冲
oscillating pump 摆动泵
oscillating quantity 振荡量；摆动量
oscillating rake 振动耙
oscillating range 振荡范围
oscillating reactions 振荡反应
oscillating region 振动范围；振荡区域
oscillating regulator 振荡式调节器
oscillating riddle 往复筛；摆动筛
oscillating rod end block 往复杆头块
oscillating roller 振动式镇压器；摆动式辊
oscillating roller twister 往复罗拉加捻装置
oscillating rolling mill 振动式轧机；摆动式轧机
oscillating sampler 振动式取样器；摆动式取样器
oscillating sander 振动式磨光机；往复式砂带磨床；摆式砂带磨床
oscillating satellite 振荡卫星
oscillating saw 往复锯；摆(式)锯；摆动锯
oscillating scraper 振动式刮土板
oscillating screen 振荡式筛网；振动筛(筛去泥浆中固体颗粒用)
oscillating screening centrifuge 立式振动离心脱水机
oscillating screw 摆动螺钉
oscillating separator 振动分选机
oscillating sequence 振荡级数
oscillating series 振荡级数
oscillating shaft 振动轴
oscillating sieve 摇摆筛
oscillating sieving machine 振动筛粉机；振动筛分机
oscillating sort 振动分类法；振荡分类；交替分类
oscillating spindle lapping machine 摆动精研机
oscillating sprinkler 摆动式喷灌机
oscillating steering axle 振荡转向轴
oscillating strainer 摆动式粗滤机；摆动式除渣机
oscillating straw rake 抖动式逐稿器
oscillating subsoiler 振动式土铲
oscillating system 振动系统
oscillating thermal precipitator 振动热沉淀器

oscillating thermionic valve 振荡热离子真空管
oscillating thinner 摆动式间苗机
oscillating time 振动时间
oscillating track 摆动航迹
oscillating transformer 振荡变压器
oscillating traverse motion 往复摆动运动
oscillating treatment controller 摇摆调节器
oscillating trough 振荡槽
oscillating tube 振荡管
oscillating turret 摇摆炮塔
oscillating type 往复回转式
oscillating unit 振荡单元
oscillating unsteady flow 波动不稳定流
oscillating valve 摇动阀
oscillating vector 振荡矢量
oscillating voltage 振荡电压
oscillating washer 摆动垫圈
oscillating water mass 振荡水质；振荡水体
oscillating wave 振荡波；往复波
oscillating weight 振荡砣；摆轮锤
oscillating weight bolt 自动锤栓
oscillating weight bolt spring 自动锤栓簧
oscillating weight bridge 自动锤夹板
oscillating weight cam 自动锤凸轮
oscillating weight guiding tube 自动锤导套
oscillating weight jewels distance piece 自动锤带钻隔片
oscillating weight lower bridge 自动锤下夹板
oscillating weight pinion 自动锤齿轴
oscillating weight ring 自动锤衬套
oscillating weight ring spring 自动锤衬套簧
oscillating weight tension washer 重锤张力垫圈
oscillation 振荡(作用)；来回振动；颤振；摆动
oscillation absorber 缓冲器；减震器
oscillation amplitude 振荡幅度；波幅；摆幅
oscillation antidamping 振荡反阻尼
oscillation antinode 振荡波腹
oscillation bearing 摆动轴承
oscillation burning 振荡燃烧
oscillation camera 回摆照相机
oscillation center 振荡中心；摆动中心
oscillation chute 摇运机；摇动溜槽
oscillation circuit 振荡回路；振荡电路
oscillation constant 振动常数；振荡常数
oscillation counter 振荡计数器
oscillation cross ripple mark 摆动交错波痕
oscillation damping 消振；振动阻尼
oscillation damping material 消振材料
oscillation direction indicator 摆动指示器
oscillation due to discharge 放电振荡
oscillation effect 振荡效应
oscillation energy 振动能量
oscillation flanking transmission 振动侧向传递
oscillation fork 振动叉
oscillation form 振荡形式
oscillation-free motion 无振运动
oscillation frequency 自由振荡频率；振动频率；振荡频率；脉动频率；摆动频率
oscillation function 振荡函数
oscillation hysteresis 振动滞后
oscillation impedance 振动阻抗
oscillation impulse 振荡脉冲
oscillation indicator 振荡指示器
oscillation insulation 隔振
oscillation intensity 振动强度；振荡强度
oscillation in the pitch mode 纵向摇摆(车辆)
oscillation in the roll mode 横向摇摆
oscillation isolator 隔振器
oscillation-less 无振动
oscillation level 振荡电平
oscillation limit 振荡限度；振荡界限；振荡极限；振荡范围；脉动界限；脉动极限；脉动范围；波动界限；波动极限；波动范围；摆动界限；摆动极限；摆动范围
oscillation loop 振荡波腹
oscillation loss 振荡损失
oscillation lubricator 振动润滑器
oscillation method 振荡法；回摆法
oscillation observation 摆动观测
oscillation of a population 种群波动
oscillation of dislocation 位错的振动
oscillation of gas 气体摆动
oscillation of sound pressure 声压振动
oscillation period 振动周期；振荡周期；脉动周期；

摆动周期
oscillation phase 振荡相位
oscillation phenomenon 振动现象
oscillation photograph 回摆图
oscillation photograph indexing 回摆图指标化
oscillation photograph symmetry 回摆图的对称
oscillation photography 振动照相法
oscillation pickup 拾振器;振动传感器;振荡传感器;脉动传感器;测振器
oscillation pile-up 振荡堆垒
oscillation pin 摆卡
oscillation power tester 振荡功率测试仪
oscillation quantum number 振动量子数
oscillation ripple 振动波痕;摆动波痕
oscillation ripple mark 振荡波痕;摆动波痕
oscillation screen 振荡筛
oscillations in the horizontal scan 水平扫描振荡
oscillation source 振源
oscillation space 振颤空间
oscillation stroke 摆动行程
oscillation suppression 振荡抑制
oscillation suppressor 振荡抑制器
oscillation system 振荡系统
oscillation tank circuit 振荡槽路
oscillation test 振动试验
oscillation theory 振动理论;颤动说
oscillation transformer 振荡变压器
oscillation transmission 振动传递
oscillation trough 振荡槽
oscillation tube 振荡管
oscillation volume 振颤容积
oscillation wave 振荡波;摆动波
oscillator 振子;振动子;振动式输送机;振动器;振荡器;继续器;摆动仪
oscillator actuating voltage 振荡器动作电压
oscillator amplifier 振荡放大器
oscillator amplifier unit 振荡放大器
oscillator box 振荡器柜
oscillator circuit 振荡线路
oscillator circuit at very-high frequency 甚高频振荡电路
oscillator coil 振荡线圈
oscillator crystal 振荡器晶体
oscillator density 振子
oscillator divider 振荡分频器
oscillator doubler 振荡倍频器
oscillator drift 振荡频率漂移
oscillator filter 振荡器滤波器
oscillator frequency 振荡器频率
oscillator frequency tolerance 振荡器频率容限
oscillator grid 振荡栅极;振荡栅
oscillator group 振子群
oscillator house 振荡器柜
Oscillatoria ogardhii sp. 阿氏颤藻;阿氏颤蓝细菌
oscillator input 振荡输入
oscillator intensity 振幅强度
oscillator load 周期性荷载;反复荷载;振荡器负载
oscillator loading 发生器负载
oscillator-mixer-first detector 振荡混频第一检波器
oscillator operating region 振荡器工作区域
oscillator out 振荡器输出
oscillator output 振荡器输出
oscillator padding 本机振荡器统调
oscillator plate 振荡片
oscillator power 振荡器功率
oscillator priming 振荡器触发
oscillator quartz 振荡器石英
oscillator section 振荡器部分
oscillator signal 振荡信号
oscillator stage 振荡级
oscillator strength 振子强度
oscillator supply 振荡器电源
oscillator sync 振荡同步
oscillator timer 振荡计时器
oscillator tracking 振荡器统调
oscillator tube 振荡管
oscillator voltage 振荡器电压
oscillator wavemeter 振荡器式波长计
oscillator with ceramic filter 陶瓷滤波器振荡器
oscillator with transformer feedback 变压器回授振荡器;变压器反馈振荡器
oscillatory 振荡的
oscillatory acceleration 震动加速度;振动加速度
oscillatory axial movement of rotary kiln 回转窑纵向窜动
oscillatory circuit 振荡电路
oscillatory combustion 振动燃烧
oscillatory control surface 振动控制面
oscillatory convection 振荡对流
oscillatory current 振动流;振荡电流
oscillatory damped 振荡性衰减
oscillatory degree of freedom 振荡自由度
oscillatory differential equation 振动微分方程
oscillatory discharge 振荡放电;振动放电
oscillatory displacement 振荡性位移
oscillatory excitation 振荡激发
oscillatory extinction 波动消光
oscillatory formation 振荡型式
oscillatory-frequency multiplier 振荡频率倍增器
oscillatory impulse 振荡脉冲
oscillatory initial value 振荡的初始值
oscillatory instability 振荡不稳定性
oscillatory laser output 激光振动输出;激光振荡输出
oscillatory load 震动荷载;振动荷载;交变荷载;交变负载;摆动负载
oscillatory magneto-absorption 振荡磁吸收
oscillatory magneto-resistance 振荡磁阻
oscillatory mixer 振动搅拌器
oscillatory motion 往复运动(海洋风生波);振动;振荡运动;脉动;波动;摆动
oscillatory motor 摆动式液压马达
oscillatory movement 振荡运动
oscillatory movement stage 震荡运动阶段
oscillatory normal stress 振荡正应力
oscillatory occurrence 振荡现象
oscillatory period 振荡周期
oscillatory potential 振荡电势
oscillatory power 振荡功率
oscillatory pressure 脉动压力
oscillatory process 振荡过程
oscillatory pulse 振荡脉冲
oscillatory regime 振荡状态
oscillatory region 振荡区域
oscillatory response 振动性响应;振荡干扰
oscillatory ripple mark 摆动波痕
oscillatory series 振荡级数
oscillatory shear 振动剪切;振荡剪切
oscillatory shear flow 振荡剪切流
oscillatory spin 振动螺旋
oscillatory stress 振荡应力
oscillatory surge 振荡性电涌;振荡(性)冲击
oscillatory system 振动系统;振荡系统
oscillatory torque 振动力矩
oscillatory twinning 振荡孪生
oscillatory type 振荡式
oscillatory viscoelastometer 振荡黏弹计
oscillatory wave 振幅波;振动波;振荡波;往复波
oscillector 振荡选择器;振荡频率选择器
oscillight 显像管;电视接收管
oscillion 振荡器管;三级振荡管
oscillistor 半导体振荡器
oscillogram 振荡图;示波图;波形图
oscillogram analyzer 波形图分析仪
oscillogram trace reader 示波图读出器
oscillogram tube 示波管
oscillograph 录波器;示波仪;示波器
oscillograph camera 示波器照相机
oscillographic display 示波显示
oscillographic integrator 示波积分器
oscillographic method 示波法
oscillographic polarography 示波极谱(法)
oscillograph photography 示波器照像(术)
oscillograph printer 示波器打印机
oscillograph record(ing) 示波图记录
oscillograph recording system 示波器记录装置
oscillograph record reader device 示波记录读码器
oscillograph trace 示波图;示波器迹线
oscillograph tracing 示波带上的记录曲线
oscillograph vibrator 示波器振子
oscillography 示波
oscillometer 脉动仪;示波仪;示波器;示波计;摆动仪
oscillometric 示波的
oscillometric titration 振量滴定;高频滴定(法);示波滴定
oscillometry 振量法;高频指示;示波测量术
oscillopolarography 示波极谱(法)
oscilloprobe 示波器探头;示波器测试头
oscillorecord camera 示波图记录照相机;示波图摄影机;示波器用摄影机
oscilloscope 录波器;示波仪;示波器
oscilloscope camera 示波器照相机;示波器(用)摄影机
oscilloscope coupling 示波器探针
oscilloscope-detector 示波器探测器
oscilloscope display test 示波器显示检测
oscilloscope pattern 示波图
oscilloscope photograph 示波器照相
oscilloscope probe capacitance 示波器探头电容
oscilloscope screen 示波器荧光屏
oscilloscope test 示波器试验
oscilloscope trace 示波图;示波器描迹;示波器迹线
oscilloscope tube 录波管
oscilloscopic chromatography 示波色谱法
oscilloscopic comparison 示波比较法
oscilloscopic polarography 示波极谱(法)
oscilloscopy 示波术
oscillosynchroscope 同步示波仪;同步示波器
oscillotron 阴极射线示波管;示波管;示波管
osculate 密切
osculating circle 密切圆
osculating coordinates 密切坐标
osculating cubic curve 密切三次曲线
osculating element 密切轨道根数;吻切根数
osculating ellipse 吻切椭圆
osculating epoch 吻切历元
osculating interpolation 密切插值法
osculating orbit 密切轨道;吻切轨道
osculating orbital elements 密切轨道根数
osculating orbital ellipse 密切轨道椭圆
osculating orbit element 密切轨道元素
osculating parabola 密切抛物线
osculating plane 密切(平)面;吻切平面
osculating point 密切点;吻切点
osculating quadric 密切二次曲面
osculating sphere 密切球面
osculation 密切
osculation point 密接点
osculatory 接触的;密切的;绝缘层;密贴层
osculum 细孔;海绵排水孔
Oseen-Goldstein correction 奥森-戈乐茨改正
Oseen's approximation 奥森近似式
Oseen's force 奥森力
osegood-Haskins test 奥哈二氏试验
o-semicarbazone 邻缩氨基脲
osglim lamp 氖氦辉光灯
osier 柳条;柳枝
osier bed 柳林;柳园
osier mattress 柳枝排
osier twig 垂柳
osier willow 框柳;青刚柳
Osirian column 奥希利斯柱(古埃及神像)
Osiris pillar 奥赛列斯(古埃及的主神之一)柱
oskar 蛇形丘
Oslo Convention 奥斯劳条约
Oslo pile shoe 奥斯劳桩靴
Oslo's phenomenon 奥斯劳氏现象
Oslo-type crystallizer 奥斯劳型结晶器
osmanthus edges 桂花边
osmate 锇酸盐
Osmayal 欧斯马铝锰合金
osmelite 针钠钙石
osmic 锇的
osmic acid 锇酸
osmification 锇处理
osmiridisulite 富硫铱锇矿
osmiridium 铱锇合金;等轴锇铱矿
osmiridium alloy 锇铱合金
osmium filament 锇丝
osmium fluoride 八氟化锇
osmium lamp 锇丝灯
osmium ore 锇矿
osmium pen alloy 锇铑铂笔尖合金
osmium tetroxide 四氧化锇
osmograph 渗透压记录仪
osmolar 渗透的
osmol(e) 渗透压摩尔;渗透压克分子;渗摩
osmolity 渗透度
osmometer 渗压计;渗压表;渗透压计;渗透计
osmometry 渗压测定(法);渗透压力测定法;渗透压测定法
osmondite 奥氏体变态体

osmophilic 耐高渗透压的;趋渗的
osmoplastic method 渗透塑剂法
osmoreceptor 渗透压感受器
osmoregulation 渗压调节;渗透压调节
osmoregulator 透射调节器
osmoregulatory 调节渗透的
osmoregulatory mechanism 渗透压调节机制
osmoregulatory migration 渗透调节回游
osmosalts 防腐盐;渗渍盐剂;渗透盐剂(一种木材防腐剂)
osmoscope 渗透仪;渗透试验器
Osmosekaolin 电泳精选黏土
osmose process 渗透法
osmoses diffusion process 渗透扩散法(木材防腐处理)
Osmoseton 电泳精选黏土
osmosis 渗析;渗透作用;渗透(性)
osmosis and osmotic pressure 渗透和渗透压
osmosis fouling 渗透污垢
osmosis preservation 渗透性防腐
osmosis treatment 渗透处理
osmos tube 渗透管
osmotaxis 趋渗性
osmotic 渗透的
osmotic balance 渗透天平;渗透平衡
osmotic bleaching 渗透漂白
osmotic buffering 渗压缓冲
osmotic cell 渗透池
osmotic coefficient 渗析系数;渗透系数
osmotic concentration 渗透浓度
osmotic effect 渗透作用;渗透影响;渗透效应
osmotic energy 渗析能
osmotic equation 渗透方程
osmotic equilibrium 渗透平衡
osmotic equivalent 渗透当量
osmotic exchange 渗透交换
osmotic flow 渗透流
osmotic force 渗析力;渗透力
osmotic fragility 渗透脆性
osmotic injector 渗透压注射器
osmotic membrane 渗透膜;渗透隔膜;防渗薄膜
osmotic method 渗透压力法;渗透方法
osmotic movement 渗透运动
osmotic permeability 渗透性
osmotic phenomenon 渗析现象;渗透现象
osmotic potential 渗透势防渗;渗透势
osmotic pressure 浓差压;渗压;渗析压力;渗透压强;渗透压力;渗透压
osmotic pressured difference 渗压差
osmotic pressure effect 渗透压效应
osmotic pressure gradient 渗(透)压梯度;渗透压(力)梯度
osmotic pressure head 渗析压头
osmotic pressure potential 渗析压势
osmotic regulation 渗透压调节
osmotic relation 渗透关系
osmotic repulsive pressure 渗透排斥力
osmotic resistance 渗透阻力
osmotic separation 渗透分离
osmotic shock 渗透性冲击;渗透的猝度;渗透冲击
osmotic suction 渗透吸力
osmotic treatment 渗透处理
osmoticum 渗压剂
osmotic value 渗透值
osmotic water 渗透水
osmotic water transport 渗透水迁移
osmotic work 渗透功
osmotropism 向渗性
osnaburg 粗平纹棉布;低支纱棉布
osnode 自密切点
oso echinate 刺牡蛎
osophone 奥索风助听器;奥索风
oso strea 牡蛎(属);蠔
Osos wind 奥索斯风
OSPA system 钢筋束锚固式预应力混凝土后张法
osplatarsenite 砷铂锇矿
Osprey method 奥斯泼雷锻造
osram 灯泡钨丝;锇钨灯丝合金
osram lamp 钨丝灯(泡)
ossature 房屋或其局部的骨架
ossein 骨胶原
osseine 生胶质;骨胶
osseous lamella 骨板
ossuary 纳骨处

ostensible agency 虚假代理;名义代理
ostensible agent 名义代理人
ostensible authority 名义代理权
ostensible partner 名义上的股东;名义合伙人
ostensory 圣餐盒
osteocalla 多孔石灰华
osteolite 土磷灰石
osteolith 骨磷灰石;土磷灰石
osteomalacia 骨软化病
osteopathia striata 纹状骨
osteopoiki losis 脆弱性骨硬化
osteoprosis 骨质疏松症
oster 管子丝扣扳钳;管子扳牙
Osterberg sampler 奥斯特柏格式取样器
oster chaser 管螺纹梳形板牙
osteria 路边客栈
Osterloh's calculation table 奥斯托罗计算表
Osterstrome liquid phase Process 奥斯托斯脱罗莫液相过程
ostiary (天主教的)主教;看门人
ostiole 小入口;小门
ostiolum 小门;小孔
ostium 门口;复[ostia]
ostium primum 原中隔孔
ostium secundum 第二中隔孔
ostracode faunal province 介形虫动物地理区
ostracode limestone 介形虫灰岩
ostracum 介壳
Ostrea 牡蛎属
ostreotoxismus 牡蛎中毒
ostria 奥斯特风
Ostrya carpinifolia 欧洲铁木
Ostwald 奥斯特瓦尔德
Ostwald colo(u)r system 奥斯特瓦尔德色系
Ostwald double cone 奥斯特瓦尔德颜色圆锥体
Ostwald-Folin visco(si)meter 奥福二氏黏度计
Ostwald ripening 奥斯特瓦尔德熟化
Ostwald's adsorption isotherm 奥斯特瓦尔德吸附等温线
Ostwald's colo(u)rimeter 奥斯特瓦尔德比色计
Ostwald's colo(u)r solid 奥斯特瓦尔德色图
Ostwald's colo(u)r system 奥斯特瓦尔德色系(列)
Ostwald's curve 奥斯特瓦尔德曲线
Ostwald's decade rheostat 奥斯特瓦尔德十进变阻器
Ostwald's diagram 奥斯特瓦尔德图
Ostwald's dilution law 奥斯特瓦尔德稀释定律
Ostwald's equation 奥斯特瓦尔德电离方程式
Ostwald's glass Capillary Viscometer 奥斯特瓦尔德玻璃毛细管黏度计
Ostwald's gravimeter 奥斯特瓦尔德比重计
Ostwald's indicator theory 奥斯特瓦尔德指示剂理论
Ostwald's law 奥斯特瓦尔德定律
Ostwald's process 奥斯特瓦尔德方法
Ostwald's purity 奥斯特瓦尔德纯度
Ostwald's ripening 奥斯特瓦尔德催熟;奥斯特瓦尔德成熟
Ostwald's theory of indicator 奥斯特瓦尔德指示剂理论
Ostwald's universal-photometer 奥斯特瓦尔德万能光度计
Ostwald's U-tube 奥斯特瓦尔德 U 形毛细管黏度计
Ostwald's visco(si)meter 奥斯特瓦尔德(U 形毛细管)黏度计;奥氏黏度计
Ostwald-type visco(si)meter 奥斯特瓦尔德式毛细管黏度计
osumilite 大隅石
Osward pipette 奥斯华德氏吸管
otaphone 助听器
otavite 菱镉矿
otentiometer recorder 分相器式记录器
o-terphenyl 邻羟基苯
other accident 其他事故
other accounts payable audit 其他应付款项审计
other accounts receivable audit 其他应收款项审计
other assets 其他资产
other auxiliary steamers 辅助船舶投资
other budget 其他预算
other building 其他建设
other business income 其他业务收入
other charges 营业外费用;其他费用
other classification of oil field water 油田水其他分类

other classification of trop 圈闭其他分类
other climate index 其他气候标志
other coefficient of equipment 设备其他参数
other construction costs 工程建设其他费用
other cost 公杂费;其他费用
other crystal 其他晶体
other deduction 其他扣款
other deposits of cave 洞穴其他堆积物
other direct expenses 其他直接费
other-domain resource 非本域资源
other end of double working turnout 双动道岔乙端【铁】
other engineering property of rock and soil 岩土的其他工程性质
other equipment 其他设备
other expense 其他支出;其他费用
other expense budget 其他支出预算
other factors 其他因素
other features 其他特征
other financial report 其他财务报表
other forms 其他形态
other income 其他收益;其他收入
other income budget 其他收入预算
other income of passenger transport 客运其他收入
other indirect expenses 其他间接费
other indirect indicators 其他间接标志
other industrial equipment depreciation expenses 其他生产设备折旧费
other insurances permitted 可另有保险
other interest rate 其他利率
other interpreted ore guides 解译其他找矿标志
other investment 其他投资
other investment audit 其他投资审计
other labo(u)r expense 其他人员工资
other liabilities 其他负债
other manufacturing overhead 其他制造费用
other marks 其他标志
other matters concerned 其他有关事项
other mesial colo(u)r 其他中间色
other methods 其他方法
other minor element 其他微量元素
other miscellaneous receipts 其他杂项收入
other money capital audit 其他货币资金审计
other nations or districts 其他国家或地区
other parameter of estimated reserves 估算储量的其他参数
other party 对方
other people's money 别人的资金
other petrologic(al) conception 其他岩石学概念
other petrologic(al) mineral composition 岩石的其他物质组份
other pollution 其他污染
other property of minerals 矿物其他性质
other real estate 其他不动产
other related parameter 其他有关参数
other reserve term 其他储量术语
other revenue 其他收益;其他收入
other sector expenses 其他部门支出
other sedimentary rocks 其他沉积岩
other sedimentary structure 其他沉积构造
other sedimentary texture 其他沉积结构
other sensors 其他遥感器
other shape 其他形状
other side 对方
other tax 其他税
other tax rate 其他税率
other texture 其他影纹
other-than-earth satellite 非地球卫星
other than ring type 非环形的
other thematic maps 其他专题图
other type gravimeter 其他类型重力仪
other values 其他产值
other yields 其他收益
Othmer process 奥斯默法
otitic barotrauma 中耳气压伤
Otocatalytic degradation 奥多催化降解
otolaryngological department 耳鼻喉科
o-tolidine 邻联甲苯胺
o-toluamide 邻甲苯甲酸
o-toluidine 邻甲基苯胺;邻甲苯胺
otophone 助听器;奥多风
Ottawa sand 渥太华砂

Ott current meter 奥特流速仪
Ottelia acuminate Dandy 海菜花
ottemannite 斜方硫锡矿
otter board 网板
otter trawl 网板拖网
otter trawler 拖网船
Otto bottom current meter with two wheels 奥托双叶轮式海底流速仪
Otto cycle (内燃机的) 四冲程；等定容循环；奥托循环
Otto (cycle) engine 四冲程循环内燃机；四冲动机；等容循环发动机；奥托循环发动机
Otto-Lardillon method 奥托-拉狄隆方法
ottoman 矮凳；横棱纹织物
Ottoman architecture 土耳其式建筑；奥托曼建筑（土耳其）
Ottonian architecture 奥托尼昂建筑（德国）
Ottonian church 奥托尼昂教堂（德国）
Ottonian ecclesiastical architecture 奥托尼昂宗教建筑（德国）
Ottonian Renaissance 奥托尼昂文艺复兴式（德国）
Otto type fuel 奥托型发动机燃料
ottrelite 硬绿泥石
ottrelite-slate 硬绿泥石板岩
Ott's sign 奥特氏征
otwayite 羟碳镍石
O-type dowel 环形暗销；圆形齿环
Ouachita marine trough 沃希托海槽
ouachitite 黑云母方沸碱煌岩
ouari 瓦利风
ouatite 锰土
oubliette 秘密地牢；土牢；地牢（仅在牢顶有出口）
Ouchterlony technique 奥脱洛尼氏技术
Oudia technique 奥登氏技术
Oudin current 奥丁电流
Oudin resonator 奥丁谐振器
Oudin test 奥丁试验
oued 干谷；枯水河；旱谷
Ou kiln 瓯窑
oulet hole 出口孔
ounce 英两；盎；盎司
ounce-foot 盎司英尺
ounce-inch 盎司英寸
ounce metal 高铜黄铜；铜币合金；盎司铸造铜合金
ouncer transformer 小型变压器；袖珍变压器
ounce-weight 盎司数
ounderground section 地下段
ounding machine 捣碎机
oundy mo(u)lding 浪花饰线脚；波形线条；波形线脚
our ability to service debt 财政偿还能力
ourayite 硫铋铅银矿
ourdoor closet 室外厕所
ouricury oil 小冠椰子油
ouricury wax 小冠巴西棕蜡
our letter 敝函
ouropardo wax 高密度特硬蜡
oursinite 硅钴铀矿
our time 散处时间
outage 运转中断；预留容积；余积；油量差；供电中断；减耗量；排油孔；排液孔；排出量；停电；事故率；灯标失明；储运损耗(量)
outage ga(u)ge 排量表(安装在油罐上)
outage method 罐口量油法
outage of running 运行中断
outage rate 停机率；停电率
outage time 停工时间；事故性停工时间
out and back haul 往返运输
out and home 往复航程
out and home voyage 出航和回航
out-and-in bond 一顺一丁砌砖法；凹凸砌合；凹突砌合
out and in search 来回搜索
out-and-out 版面超出切刀线
out and removal of coating 撕膜
out at sea 在航海途中；出海
outback 内陆(的)；边远地区
outbalance 胜过；优于失去平衡；重于
outband 外镶边石；外侧带；外圈框石；外挂的
out-band signalling 带外信令；带外信号方式
out-batter pile driving 斜桩敷设；仰打(桩)
outbid 出高价；超过
outbid each other 互相抬价
outbliette (一种有凝气门入口的) 地下室

outboard 舷外的；舷外；在舷外；在外舷；机外；外侧；船外
outboard ball bearing 外置球轴承
outboard bearing 外伸轴承；外置轴承
outboard boom 舷外斜桁；伸出舷外的栏杆
outboard coil 舷外消磁绕组
outboard door 外侧舱门
outboard engine 外侧挂机；舷外发动机；外装发动机
outboard float 外侧浮筒
outboard mechanical seal 外置机械密封
outboard motor 艇外马达；舷外推进器；舷外挂机；舷外发动机；挂式推进器；挂桨；挂机(艇)；外装马达；外装电动机；外置马达；操舟机
outboard profile 船舶侧视图
outboard propeller 外侧螺旋桨
outboard recorder 外部记录程序
outboard screw 外侧螺旋桨；船外推进器
outboard shaft 外侧轴
outboard shot 舷外链节；固定无杆锚的短链(在锚链孔)；接锚短链
outboard stay 外牵索
outboard tank 舷侧水柜
outboard turning 外转式
outboard turning screws 外旋式双推进器
outboard type 外向式
outboard universal joint 半轴外侧万向节
outboard work 舷外作业
outbond 墙面顺砖砌合；横叠水的；外砌的
out-bond of wall 墙面顺砖砌合
out-bond wall 外砌墙
outbond wall board 外挂墙板
outbound 开往外国的；出港的；出港(机场)
outbound bearing 外伸支承
outbound cargo 外运货(物)；出口货(物)；出境货(物)
outbound course orientation 外飞航线定向法
outbound craft 挂桨艇
outbound engine lead track 机车出库线
outbound freight 外运货(物)；出口(船)货
outbound freight house 发送仓
outbound lane 出境车道
outbound parcel traffic 发送包裹运输
outbound platform 出发站台；发送站台
out-bound track 出发线【铁】
outbound traffic 外向交通；对外交道；出境交通
outbound train 出发列车
outbound train made up 自编出发列车
out branch 出支路
outbreak 断裂；溃决；突然蔓延；冲决；爆燃
outbreak area 爆发地区
outbreak control 暴发流行病控制
outbreaking of fire 发生火灾
outbreak in investigation 爆发调查
outbreak investagation 暴发调查
outbreak of disease 疾病蔓延；疾病流行；疾病发作
outbreak of fire 起火；火灾的爆发
outbreak period 爆发期
outbreak survey 暴发调查
out-budget expenses 预算外支出
outbuild 建造过多；超建
outbuilding 副屋；外屋；外围建筑；附属建筑(物)；附属工程；附属房(屋)；附属房
outburst 溃决；气喷；喷发；瓦斯突出；突发；冲破；冲决；爆炸；爆发
outburst bank 浪蚀海岸；海堤中段；海岸迎冲段；冲击岸堤
outburst of a population 种群爆发
outburst of gas 井下瓦斯严重
outburst passenger flow 突发性客流
out cable (预应力混凝土的) 外露张拉钢索
out chamber 接待室
outclass 超过船级检验年限的船；远远优于；胜过；远远超过
out code 关闭指令【计】
outcome 结果；产物；产量
outcome function 出现函数；成果函数
outcome of test 试验结果
outcome-oriented method 结果指向法
outcome variable 结果变量
outcoming geothermal fluid 喷出的地热流体
outcoming power 出射功率；出现功率
outcoming signal 输出信号
out-commuter 出城通勤交通
outconnection 流线改连接

outconnector 改接符；流线外接符；流线改接符；外接符
outcrop altitude of rock 基岩出露高程
outcrop area of aquifer 含水层出露面积
outcrop line 露头线
outcrop map 露头图
outcrop of coal bed 煤层露头
outcrop of fault 断层露头
outcrop of ground water 地下水露头
outcrop of water 水流出逸口；水流出露口
outcrop on surface 地面露头
outcropped halo 出露晕
outcrop(ping) (地层的) 露头；地表露头；出露地表；露出地面的岩层；岩层露头
outcropping rock 露头岩石
outcrop sampling 露头采样
outcrops of conglomerate 砾岩露头
outcrop spring 露泉
outcrop stripping 露头剥离
outcrop surveying 露头测量
outcrop water 露头渗出水；渗出水
outcurve 向外拐弯
outcut 切口
outdated 过时的；过期的；老式的
outdated building 老式建筑
outdated data 陈旧资料
outdated line map 过时线划图
out-dated specification 过时规范；旧规范
out-degree 出度；引出次数
outdiffusion 向外扩散；外扩散
outdiffusion effect 外扩散效应
outdistance 追越
outdo 高过
outdoor 户外(的)；室外的；表面的
outdoor activities 户外活动
outdoor advertising along highway 沿公路的露天广告(牌)
outdoor advertising sign 室外广告标志
outdoor advertising structure 室外广告建筑
outdoor aerial 室外天线
outdoor ag(e)ing characteristics 耐室外气候老化性
outdoor air 室外空气
outdoor air design condition 室外空气计算参数
outdoor-air intake 新鲜空气吸入口；室外空气吸入口
outdoor air temperature 室外空气温度
outdoor air temperature sensing device 室外空气温度传感装置
outdoor antenna 室外天线
outdoor architecture 室外建筑
outdoor arena 露天活动场所；户外活动场所
outdoor arrester 室外避雷器
outdoor assembly 露天会场
outdoor assembly shelter 露天会场的风雨棚；室外集合的风雨棚
outdoor basement stair(case) 地下室的户外楼梯；地下室的室外楼梯
outdoor bath 室外浴池
outdoor bathing place 浴场
outdoor bed 露地苗床
outdoor blending bed 露天预均化堆场；露天混合料床
outdoor boiler 露天锅炉；室外锅炉
outdoor camera 门外摄影
outdoor cinema 露天电影院
outdoor clinker storage area 露天熔渣堆场
outdoor closet 室外厕所
outdoor coating 室外涂料
outdoor commercial advertising device 室外商业公告装置；露天商业广告装置
outdoor condition 室外条件
outdoor corridor 室外走道；室外阳台
outdoor critical air temperature for heating 采暖室外临界温度
outdoor critical illuminance 室外临界照度
outdoor current transformer 户外电流互感器；户外电流变流器
outdoor design dry-bulb temperature for summer air-conditioning 夏季空调室外计算干球温度
outdoor design hourly temperature for summer air-conditioning 夏季空调室外计算逐时温度
outdoor design mean daily temperature 室外计算日平均温度
outdoor design relative humidity for summer

ventilation 夏季通风室外计算相对温度
outdoor design relative humidity for winter air conditioning 冬季空气调节室外相对湿度
outdoor design temperature 室外设计温度
outdoor design temperature for calculated envelope in winter 冬季围护结构室外计算温度
outdoor design temperature for summer ventilation 夏季通风室外计算温度
outdoor design temperature for winter ventilation 冬季通风室外计算温度
outdoor design wet-bulb temperature for summer air-conditioning 夏季空调室外计算湿球温度
outdoor dining terrace 露天用餐凉台
outdoor distributing box 室外配电箱;室外配电盒
outdoor distribution equipment 户外配电装置
outdoor durability 室外耐久性
outdoor emulsion paint 室外用乳胶漆
outdoor environment 户外环境;室外环境
outdoor equipment 露天设备;户外设备;室外设备
outdoor exposure 室外暴露
outdoor exposure test 室外曝晒试验;室外曝露试验;大气暴晒试验;风化试验
outdoor facility 露天设施;室外设备
outdoor fire hydrant 室外消火栓
outdoor furniture 室外家具;庭园家具
outdoor garden 庭园
outdoor glass of double-glazing units 外层窗玻璃
outdoor grafting 户外嫁接
outdoor handle 门外手柄
outdoor heating 室外供暖;室外供热
outdoor high voltage transformer 户外高压变压器
outdoor home decoration 室外住宅装饰
outdoor humidity 室外湿度
outdoor humidity sensor 室外湿度探测器
outdoor hydrant 室外消防栓
outdoor illumination 室外照明
outdoor-indoor transmission 室外-室内传声损失
outdoor infrared beam detector 户外式红外光束检测器
outdoor installation 室外设备;露天设备
outdoor insulation 室外绝缘
outdoor insulator 户外绝缘子
outdoor lampholder 室外灯座;室外灯柱
outdoor lamp-socket 室外灯插座
outdoor leaner's pool 室外教学游泳池
outdoor lighting 户外照明;室外照明
outdoor living area 室外起居平台
outdoor living patio 室外起居平台
outdoor location 外景;露天设备;室外安装;室外装置;室外设备;室外安装地点
outdoor meteorological parameter 室外气象参数
outdoor milking bail 室外挤奶台
outdoor mosaic finish 室外马赛克面层
outdoor motor 露天电动机;户外型电动机;室外型电动机
outdoor museum 露天博物馆
outdoor music stand 露天音乐台
outdoor opening 通向室外的门窗口;采光口
outdoor operation 户外作业
outdoor paint 室外用油漆;外用(油)漆
outdoor painting 室外油画;室外刷漆
outdoor pane 外层窗玻璃
outdoor pedestal type post insulator 户外针式支柱绝缘子
outdoor phase modifier 室外调相机
outdoor photography 室外摄影
outdoor pick-up 室外摄像
outdoor picture control point 像片野外控制点
outdoor pile 室外堆积
outdoor pipeline 室外管线
outdoor pipe trench 室外管沟
outdoor piping 室外管线
outdoor plant 露天电站;露天车间;露地植物;户外式电站;室外装置
outdoor planting 室外栽培;室外栽植
outdoor plasticinsulated flexible copper wire 户外用铜芯塑料软线
outdoor platform 室外平台
outdoor pool 露天游泳池;室外水池
outdoor power station 露天(发)电站;露天发电厂;户外式电站
outdoor prepiping 室外管道预留接头
outdoor public telephone cable 室外市话电缆
outdoor rail 门轨;外门围栏

outdoor rainwater system 雨水外排水系统
outdoor recreational resources 户外游息资源;户外活动设备
outdoor room 户外房间(露台、帐篷等)
outdoor school 露天学校
outdoor shooting areas 室外打靶场
outdoor shot 室外摄影
outdoor sign 室外标志
outdoor single pole isolating switch 户外单极隔离开关
outdoor sitting space 户外起居室;室外起居空间
outdoor sound pressure level 露天声压级
outdoor space 室外空地;户外空间;室外空间
outdoor sport 野外活动
outdoor sports facility 室外运动设施
outdoor stage 露天舞台
outdoor stair(case) 室外楼梯
outdoor steps 室外台阶
outdoor stockpile 露天料堆
outdoor storage 露天堆(货)场;露天储存;室外储存
outdoor storage area 露天堆放区;露天堆场
outdoor storage yard 露天堆场
outdoor stress crack life 室外应力龟裂寿命
outdoor substation 露天变电所;室外配电变电所;室外变电站
outdoor support porcelain insulator 户外支柱电瓷瓶
outdoor swimming pool 露天游泳池;室外游泳池
outdoor switchgear 户外开关设备;室外交换装置;室外分配装置
outdoor switching station 露天开关站;户外式开关站
outdoor switch yard 露天开关站;户外式开关站
outdoor teaching pool 露天教学游泳池
outdoor temperature 室外温度
outdoor temperature sensing device 室外温度传感设备
outdoor temperature sensor 室外温度探测器
outdoor tennis court 室外网球场
outdoor termination end 户外终端头
outdoor termination module 室外出线端子
outdoor terrace 露天平台
outdoor test field 露天试验场
outdoor theater 露天剧场
outdoor toilet 室外厕所
outdoor track 室外跑道
outdoor transfer 户外变压器
outdoor transformer 室外变压器;露天变压器;户外变压器
outdoor transformer station 露天变压器站;户外变电所
outdoor trench 室外管沟
outdoor tuning coil 室外调谐线圈
outdoor type 户外型
outdoor type bushing 室外线用套管
outdoor type generator 室外发电机
outdoor type switch gear 室外开关装置
outdoor type transformer 户外式变压器
outdoor use telephone 户外用电话
outdoor varnish 外用清漆;室外用清漆
outdoor weathering 露天风化作用
outdoor wintering 露天越冬
outdoor wiring 户外布线
outdoor work 户外作业;室外工作
outdoor working 露天作业
outdoor working basset 露天作业
out double peak 外双峰
outdraw 外拉伸
outdrill 超钻
outdrive set 舷外推进机
out edge of vane 叶片出水边
out-emission 向外排放
out end 发射端;排水口;出料口;卸载端;机尾
out-end table 出料台
outer 外面的;外层的;外侧的;外部的
outer addendum 大端齿顶高
outer air defense 外围防空
outer aisle 边跨;侧廊;外侧廊
outer Antarican geosyncline 外南极洲地槽
outer antisubmarine screen 外层防潜幕
outer-arc ridge 外弧脊【地】
outer atmosphere 高空大气(层);外层大气;上层大气
outer atmosphereric temperature 外大气层温度

outer axis 外轴
outer Baical mountains 外贝加尔山地
outer band 吊杆顶圈
outer bank 河曲凹岸;外堤;外岸
outer banquette 迎水面马道;迎水面护堤;外戗道;外护堤
outer bar 口外沙堤;口外沙坝;外滩;外沙坝;外拦江砂
outer bark 外树皮;外皮
outer barrel (双层岩芯管的)外管
outer basin 外港池
outer bay 外跨
outer bay biotope 外湾生境
outer beach 外海滩
outer bearing 外轴承
outer bearing cover 外轴承盖
outer belt 外环【道】
outer belt rail 窗台外镶条
outer bending moment 外弯矩
outer berm(e) 迎水面马道;迎水面护堤;外戗道;外护堤
outer berthing area 外停泊区
outer blind 外百叶窗
outer boarding station 港外检疫泊地
outer border 外图廊;外画廊
outer bottom 外底;外层底板
outer boundary line of ore body 矿体外边界线
outer brake 外制动器
outer brake hub 外刹车毂;外制动毂
outer breakwater 外海防波堤
outer brick 外皮砖
outer buoy 港外浮标
outer Caecilian gulf 外贝加尔海湾
outer capillary water 外毛管水
outer casing 外壳罩
outer casing housing 外罩
outer Cathaysian marine trough 外华夏古陆海槽
outer Caucasus mountains 外高加索山地
outer chlorinated rubber paint 室外氯化橡胶漆
outer chordal thickness 大端弦齿厚
outer circumferential highway 外环路
outer city 外城;市郊;罗城;郊区
outer closure dike 迎水侧截流戗堤(围堰)
outer coat 外表面涂层
outer coating 外部涂层;外套;外敷层;(涂料的)外层
outer coat of paint 外层油漆;外层涂料
outer collar 凸辊环;边辊环
outer column 后柱;外柱
outer conformal radius 外保角半径
outer connection 外转车道
outer connection ramp 外接式匝道;外接式坡道
outer container 外容器
outer continental shelf 外(大)陆架;大陆棚外半部;大陆架外缘
Outer Continental Shelf Lands Act 外大陆土地使用条例
outer control limits 外管理界限
outer core 外核
outer core barrel length 外岩芯筒长度
outer core barrel size 外岩芯筒尺寸
outer corner 外角
outer corona 外日冕
outer court 外庭院;外院
outer cover 外壳
outer covering of tree trunk 树干的表皮
outer cross wall 外部横墙
outer cutting bit 外管钻头(双层岩心钻的)
outer cycloid 外摆线
outer cylinder 外置刹车缸;外端汽缸
outer dead-center 外死点
outer dead point 外死点
outer deck sash 外顶栅窗
outer defense 外围防御;室外防御工事
outer diameter 外直径;外径
outer diameter fitting 外径配合
outer diameter kickers 外径掏槽刀;外径边刃(指钻头)
outer dike 外排水道(外渠的)
outer dimension 外形尺寸
outer ditch 外部沟槽;室外沟槽
outer dock 外港池;泊船池
outer dome 外层穹隆;室外穹隆
outer door 外门;外层门

outer drainage 外排水
outer-eaves corbel bracket 外檐斗拱
outer edge 外缘;外边界
outer elevator 外侧升降舵
outer end of table 工作台外端
outer enveloping profile 外包络面
outer face 外面
outerface of spindle 主轴外表面
outer fan deposit 外扇沉积
outer fiber 外层纤维
outer field generator 外层旋转发电机
outer fire box 外火箱
outer fish plate 外(侧)鱼尾板
outer fissure 外沟
outer fix 目的地定位
outer flame 外层焰
outer flank 外侧
outer flat keel 外平板龙骨
outer floor slab 外部楼板
outer force 外力
outer fort 外堡垒;外线堡;外炮台
outer gallery 外廊;外部走廊
outer garden 外苑
outer gate 底阀;外大门;外闸门;外庙门;外城门
outer gate barrel 灯泡座;外环门;导水机构外环(贯流式水轮机的)
outer gear 外齿轮
outer gimbal axis of gyro 陀螺仪外框架轴
outer gimbal ring 外平衡环
outer gimbal suspension of gyro 陀螺的悬挂外框架
outer gland sealing 外密封
outer gloss paint 外层有光泽漆;室外有光泽漆
outer guide 外导轨
outer handrail 外部扶手;室外扶手
outer harbo(u)r 外港
outer harbo(u)r area 外港区
outer harbo(u)r line 外港界线
outer head (复式闸门的)外闸首
outer head cover 外顶盖
outer hearth 壁炉地面的外部
outerhigh 外陆缘高地
outer hoop 外环路
outer housing 外套
outer hull 外艇体
outer hydration sphere 外水合层
outer inspection 外观检验;外观检查
outer insulation 外部隔音层;外部保温层;外部绝缘层
outer jacket 外保温层;外保护层;外夹层;外套
outer keel 外平板龙骨;外龙骨
outer Lagrangian point 外拉格朗日点
outer landing ship areas 外侧登陆舰区
outer lane 超车车道;外侧车道;边缘车道
outer layer 外层
outer lead bonding 外部引线接合
outer leaf 外壳;外层
outer limit 宽打期限;外缘;外限
outerline 外线;轮廓
outer lining 外隔板;外贴脸;外衬;安全衬
outer loading 外部荷载
outer locator 远距导航台;外部探测器
outer lock light 过渡舱照明灯
outerlock of recompression chamber (加压舱的)前舱
outer lock viewport 过渡舱观察窗
outer loop 外匝道;(立体交叉的)外匝道;补充电路
outer lot line court 沿外界线的天井
outer main 外路干线
outer mantle 外地幔;地幔上部
outer mapping radius 外映射半径
outer marble 外层大理石
outer margin 外缘
outer mark 远距指点标
outer marker 指点符号;外指点标
outer marker beacon 远距指点标
outer-marker crossing height 通过远距信标台的高度
outer masonry wall 外部砖(石)墙
outer masonry wall column 外部砖(石)墩柱;外部砖(石)墩墙
outer masonry wall facing 外部砖(石)墙饰面
outer masonry wall lining 外部砖(石)墙衬砌
outer mast 外门架【机】

outer material 外覆盖层建筑材料
outer measure 外测度
outer molecular layer 外分子层
outer mosaic finish 外部马赛克饰面
outermost 最外面;最外层的
outermost fiber 最外层纤维
outermost layer 最外层
outermost line 外图廓线
outermost orbit 最外轨道
outermost reflector 最外部反射层;表面反射层
outermost shell 最外层
outermost surface 最外层表面
outer mo(u)ld vibration method 外模振动法
outer multiplication 外乘(法)
outer nave aisle 外侧廊
outer neritic zone 外浅海区
outer normal 外法线
outer nozzle 外喷嘴
outer oil paint 外层油漆
outer oil seal 外轴承油封
outer orientation 外方位;外部定向;外部定位
outer pack 外包
outer paint 外层涂料;外用漆;室外用油漆
outer paint coat 外表油漆层;外表涂料层
outer painting work 外层涂刷工作
outer-pane 外层玻璃
outer panel 外部墙板
outer peak 外峰
outer photoeffect 外光电效应
outer pipe (双层岩芯管的)外管
outer pit aperture 纹孔外口
outerplanar 外平面的
outerplane graph 外平面图
outer planet 带外行星
outer planetary gear 外行星齿轮
outer plank 外列板
outer plating 外列板
outer plexiform layer 外网状层
outer point 外点
outer port 外港
outer post 承舵柱
outer precast block 外部预制块
outer precast brick 外部预制块
outer prestress 外缘预应力;外侧预应力
outer primer 外部底漆
outer product 外积;向量积
outer pulpit 外部控制台
outer pump housing 外泵壳
outer rabbet line 外嵌接线;下嵌接线;槽口外边线
outer race 外座圈
outer-race ball track 外环滚道
outer radiation belt 外辐射带
outer radius 外半径
outer rail 外轨
outer rail super-elevation 外轨超高
outer rancering 外座圈
outer rear cover 外后盖
outer reef-arc 外礁弧
outer region of boundary layer 边界层外区
outer rereal 窗外帮
outer reticle circle 外光环
outer reveal 窗樘筒子板;门樘筒子板;(窗框的)外侧壁
outer ring 外环;外围
outer ring and roll assembly 滚轮滚针轴承
outer ring road 外环路
outer ring spacer 外隔圈
outer ring with rib 带挡边外圈
outer ring with single raceway shoulder 分离型外圈
outer road(stead) 港外锚地
outer room 外间
outer screen (多层排列筛分装置的)上筛层;电缆外编织层
outer sealing 外部嵌缝
outer sedimentary arc 外沉积弧
outer separation 外分隔(路段);高速公路旁加宽道路;干铁附设便道
outer separator 外侧分隔带;外分隔带
outer separator casing 选粉机外壳
outer separator cone 选粉机外锥
outer set screw 外固定螺钉
outer shaft 外部竖井;外轴;外侧轴
outer sheath 外护套

outer sheet 外层板材
outer shelf deposit 外大陆架沉积
outer shell 外套管;外壳
outer shoe 外托板;外支块;外滑脚
outer shore 滨外;外海
outer shoreline 外滨线
outer side 外档
outer-side bracket arm 单材拱
outer side of a vertical surfaces 外旋
outer sill 外坎板
outer skin 外表面;外壳
outer skin of curtain walling 幕墙外皮
outer slatted blind 外百叶窗
outer sleeve 外套筒;外套管
outer slide dwell 外滑块停歇时间
outer slope 礁外坡;迎水坡;外坡
outer solid bowl 实体外转鼓
outer space 宇宙空间;外层空间;外部空间;太空
outer space communication 外层空间通信
outer sphere 外逸层;随动球
outer sphere complex 外层络合物
outer spindle 外轴
outer spyhole 外窥孔
outerstack 外砌物
outer stairrial 外楼梯扶手
outer stay 外撑条
outer stay cone 座环外圈(贯流式水轮机)
outer stone 钻头外侧刃金刚石
outer strake 外列板
outer string 楼梯外侧小梁;楼梯外侧斜梁;外侧楼梯帮
outer stringer 外侧楼梯斜梁
outer string of stair(case) 楼梯外侧小梁
outer stripping material 外剥离物
outer structural system 外墙构造;外墙结构系统
outer stud 外墙筋
outer sublittoral 外次大陆架(水深五十~二百米)
outer sublittoral zone 下浅海地带
outer suburban district 远郊区
outer suburbs 远郊
outer support 后立柱;外支架
outer surface 木材离心面(木表)
outer surfacing waterproofing layer 外贴式防水层
outer-sync 不协调;不同步
outer-tank attachment 箱外紧固件
outer term 外项
outer thimble 外套筒
outer tidal delta 外潮三角洲
outer tier 外层
outer tower 外架
outer transport area 外运输区
outer tube 外套管;外胎;(双层岩芯管的)外管
outer tunnel 外隧道
outer tympan 外压纸格
outer tire 外胎
outer undercoat(er) 外部底漆
outer urban region 远郊区
outer varnish(ing) 室外清漆;外罩清漆
outer veneer 外饰面板
outer vinyl paint 外层乙烯树脂漆
outer volcanic ridge 外火山脊岭
outer vortex 外螺旋线
outer wall 外墙;外壁
outer wall block 外墙砌块
outer wall component 外墙构件
outer wall construction 外墙砌筑;外墙结构
outer wall finish 外墙饰面
outer wall lining 外墙衬砌
outer wall slab 外墙板
outer wall (sur)facing 外墙面层
outer wall tile 外墙面砖
outer wall unit 外墙单元
outer ward 外层守卫区
outer water pressure 外水压力
outer waterway 外列板水沟
outer wheel bearing 外轮轴承
outer wheel hub 外轮毂
outer wheel path 外轮迹线;外档车道
outer window 外窗
outer window blind 外层百叶
outer window check 外窗槽口
outer window frame 外窗框(防暴风雨用)
outer window rabbet 外窗槽口
outer window sill 外窗台

outer wingman 外僚机
outer work 外部工作;外部工程
outerwork function 外功函数
outer wrap 外缠绕层
outerwrap jacket 外包套
outer wrapping 外缠绕层
outer withe 外部烟囱隔板
outer wythe 外部烟囱隔板
outer zone 外区;外带
outer zone of southwest Japan 西南日本外带
outface 倾向坡
outfall 河道出口;渠口;排泄口;排污口;排水口;出水口;出口管道
outfall arm 河口支汊
outfall channel 排水渠
outfall cleaning device 出口清洗装置
outfall concentration 污水排出口浓度
outfall construction 出口管施工
outfall ditch 排水渠(道);排水沟
outfall drain 集水沟;拦水沟
outfall fan 沟口冲积扇;河口扇形地;河口冲积扇;冲积锥;冲积扇
outfall head work 排水建筑
outfall loss 跌水损失;出口损失
outfall of river 河道出口
outfall pipe 出口管
outfall riser hydraulics 立管排污口水力学
outfall sewer 排污总管;下水道;污水排泄总管;污水干管;出水总管;出水污水道
outfall structure 排水结构;出口建筑物
outfall temperature 出口温度
outfall to rainwater head 雨水斗弯头
outfall work 出口工程;排水工程;泄水工程
outfan 输出端
out-feed channel 出料槽
out file 输出文件
outfire 灭火
outfit 舾装品;舾装;装备;配备品;配备;成套装置;成套装备;成套用具;成套设备;成套备用工具;备用工具
outfit afloat 水上舾装
outfit allowance 设备折让
outfit car 宿营车
outfit insurance 配备品保险
outfit of equipment 装备总体
outfitter 旅行野营用具店;服装用品店;舾装工;装配工
outfitting 装配;配备;舾装设备;舾装工程;舣装
outfitting pier 舾装码头;舣装码头;待装码头
outfitting shop 舾装车间
outflow 流出物;流出量;流出(口);外流;出流(量)
outflow arm 河口支汊
outflow boundary 出流边界
outflow boundary condition 出流边界条件
outflow channel 出水槽;泄水渠(道);放水渠道;放水渠
outflow concentration 出流浓度
outflow device 出流装置
outflow from lake 湖泊出流量
outflow from reach 河段出流量
outflow from reservoir 水库放水量;水库出流量
outflow from the cross-section(al) of aquifer 含水层断面流出
outflow head 流出水头
outflow heater 外流加热器
outflow hydrograph 出流过程线
outflowing 泄出
outflowing river 出流河
outflowing sediment 流出泥沙;出沙量
outflowing stream 出流河
outflowing tee 出流三通
outflowing T-piece 出流三通
outflow installation of dry dock 干坞抽水设备
outflow jet 排出流束
outflow lake 外流湖;外来流湖
outflow margin 出流余量
outflow of surface water 地表水的流出量
out flow pattern 实际驶离车流图式;面状出水点
outflow point 出流点
outflow rate 出流率;出流量;出流量
outflow storage 蓄泄(水库)
outflow storage factor curve 蓄泄因数曲线(水库)
outflow tract 流出道

outflow tube 出水管
outflow variable 出流变化
outflow velocity 流出速度;排出流速度;出流流速
outflow volume 流出量;流出总量;排放量;出水总量;出流总量;出流
outflow vortex 漩涡出流;泄水涡流;出流漩涡
outflow weir 出流堰
outflux sediment 流出泥沙
outfold 倒转褶皱
outfoot 抢先
out force 外力
out freight 去程运费
outgage 斜捻沟(木甲板接缝的斜坡口);出口(指水、气、汽、油等)
outgas 脱气;去气;释气
outgassing 漏气;放气;除气(作用)
outgassing coefficient 排气系数
outgassing index 气泡指数
outgassing rate 释气率;除气率
outgate 溢流口;输出门(电路的);出口(指水、气、汽、油等)
out-ga(u)ge goods 超限货物
outgo 支出项目;支出
outgoing 流出;启程;出射;出港
outgoing access 出网访问
outgoing air 排出空气;排出的空气
outgoing air station 国际航空站
outgoing barrel 发送室
outgoing cable 引出电缆;出局电缆
outgoing call 去话呼叫;出局市话
outgoing circuit 输出电路
outgoing compartment 发送室
outgoing control 输出控制
outgoing dock 发图室
outgoing ebb 退潮流
outgoing feeder 输出馈路;对外馈电线
outgoing feeder cubicle 馈出线开关
outgoing feeder transformer 出线输电变压器
outgoing group 引出组
outgoing group selector 出局选组器
outgoing highway 对外公路
outgoing invoice 销售发票
outgoing jack 输出塞孔
outgoing junction 出口继线
outgoing laser beam 出射激光束
outgoing line 引出线;出站管线;出线;发送管线;发车线【铁】
outgoing loading platform 地图图库装卸车台
outgoing long-wave radiation 向外长波辐射
outgoing manifold 发送汇管
outgoing material 取样
outgoing mirror 输出镜;射出镜;出端反射镜
outgoing neutron 出射中子
outgoing particle 出射粒子
outgoing pipeline 发送管线
outgoing point 发送点
outgoing probability 出射几率
outgoing quality 出厂质量
outgoing radiation 射出辐射
outgoing route 输出路径;发车进路
outgoing run time 输出运行时间
outgoing rural line 区内出线
outgoings 开支;开销
outgoing siding 发车线【铁】
outgoing signal 输出信号;发打信号;发出信号
outgoing station 发送站;发车站
outgoing steamer 离港船
outgoing tide 落潮;退潮
outgoing track 发车线【铁】
outgoing traffic 离境交通;出区交通
outgoing train 始发列车;出发列车
outgoing trunk 单向中继线;出中继线;出中继线;出局中继线;出局干线
outgoing trunk circuit 出中继器
outgoing trunk switch 出中继键
outgoing valve 发送阀
outgoing water 冒出的水;涌出的水
outgoing wave 输出波
outgrowth 生成物;产物;侧淀积
outgush 涌流;涌出(水、油);喷出;排出
outhaul cable 开启桥升吊索;升吊索
outhouse 户外厕所;屋外厕所;外屋;附属小屋;附属建筑物;附属房(屋);附属房
outiet stationary jam 出纸口滞留卡纸

outing flannel 软绒布
out in reckoning 估计错误
outinsulation 外墙热绝缘;外绝缘
out island 外围岛屿
outlawed debt 已失时效的债务;失时效的债务
outlay 支付;支出;开支;经费;外置;费用;表面移植物
outlay accounts 支出账户
outlay cost 实际支出成本;消费投资成本;支出成本;实支成本;付现成本
outlay curve 支出曲线;实际成本曲线;费用曲线
outlay for advertisement 广告费
outlay for debt payment 偿还债务支出
outlay for liquidation 清理支出
outlay for loan payment 偿还债务支出
outlay tax 支出课税
outleakage 漏出量
outlet 引线口;引出口;流出口;排油孔;排水口;输出端;电源插座;出线口;出线;出路;出口截面;出口(指水、气、汽、油等);放水口
outlet air speed 出口风速
outlet air velocity 出口风速
outlet angle 出口角(度)
outlet aperture 出口孔径;出口孔
outlet assembly 出布装置
outlet bellmouth 钟形出水口;喇叭形出(水)口
outlet blade angle 叶片出口角(螺旋桨)
outlet boundary 出口边界
outlet boundary condition 出口边界条件
outlet box 接线盒;线头匣;引出箱;接线匣;接合线盒;出线匣;出线盒【电】
outlet branch 放水支管
outlet bucket 虹吸道出口;出口下缘;出口消能反弧段;出口消力床;出口反弧段
outlet capacity 泄水能力;泄水孔容量
outlet case 引出箱;排出箱
outlet chamber 输出腔;出水井;出流室;出流井;放水室
outlet chamber of inverted siphon 倒虹管出水井
outlet channel 出水渠;泄水渠(道);排水沟;出水槽;出流渠;出口渠(道)
outlet cock 放水旋塞;放出旋塞
outlet collar 卸料口颈圈
outlet compartment 排出口小室
outlet conduit 泄水道;出水管道;出口管道;泄水管;排水沟
outlet cone 出料锥体;出料锥斗
outlet connection 出口连接管;出口接头
outlet control 出口控制
outlet control gate 出口控制闸门
outlet control structure 泄水(控制)建筑(物);湖出口建筑物;河(出)口(控制)建筑物
outlet-control water heater 出口控制式热水器;出口控制热水器
outlet culvert 阴沟出口
outlet damper 出口挡板
outlet data 出口数据
outlet delay jam 出纸口延迟卡纸
outlet device 排水装置;出水装置;出口装置
outlet diameter 排出口直径
outlet dimension 出口尺寸
outlet discharge 泄水孔流量;排泄口流量;出口流量
outlet discharge flange 出口法兰;出口凸缘
outlet draft 出口抽力
outlet drain 泄水沟;排水道
outlet drain line 出口排水线
outlet duct 排出道
outlet duct for tailings 粗粉返回管
outlet element (水库的)泄水机构
outlet elevation 出口标高
outlet end 卸料端;输出端
outlet engineering 出口工程【给】
outlet entrance 泄水孔;泄水渠进口;泄水孔进口;泄水管进口;泄水道进口;泄流孔
outlet flange diameter 出口法兰直径;出口凸缘直径
outlet flow condition 出口流态
outlet flow rate 出口流量
outlet flue 出口烟囱
outlet for investment 投资场所
outlet for water in a sprinkler system 喷灌系统中的出水口
outlet from reservoir 水库排泄口
outlet gate 泄水闸门;排水闸(门);放水闸(门)
outlet ga(u)ge 出口测站

outlet glacier 从陆冰流出的流冰
outlet guide 出口导板
outlet head 卸料端磨头;出口扬程(水泵)
outlet header 排水干管;出口总管
outlet head loss 出口水头损失
outlet hole 排泄孔;排放孔;放水孔
outlet hose 出口软管
outlet hose nozzle 水龙接嘴;水龙带接嘴
outlet inlet main road 城市出入主干道
outlet line 输出线(路);出口管线
outlet line filter 输出线路滤器
outlet loss 出口损失;出口水头损失
outlet masonry 出口砌物
outlet monitoring 排污口监测
outlet nipple 出口螺纹接套
outlet non-return valve 出口止回阀
outlet nozzle 出口喷嘴
outlet of branched channel 汊道出口
outlet of discharge pipeline 出泥管出口
outlet of drainage ditch 排水沟出口
outlet of lake 湖泊出口
outlet of sewer 下水道出水口;污水管出口;污水出口
outlet of spring 泉眼;泉水出口
outlet of tile drain 排水瓦管出口
outlet of ventilating system 通风系统出口
outlet opening 排出孔;出口孔
outlet operating temperature 出口操作温度
outlet orifice 出口孔
outlet parameters 出口参数
outlet pigtail 出口引管
outlet pipe 流出管;放出管;泄水管(道);排出管;出水管;出口管
outlet point 排出孔
outlet pollution 出口污染
outlet port (小炉的)喷烟口;泄水口;排烟出口;出水口;出口(指水、气、汽、油等);废气排出口
outlet portal 出口门
outlet pressure 出口压力
outlet regulating gate 泄水控制闸板;泄水控制闸门
outlet regulation 河口整治
outlet retaining wall 出口挡土墙
outlet road 出路;出城道路
outlet roller 出口辊
outlet sale 市场销售
outlets capacity 出线容量
outlet seam 毛出
outlet section 出口段;出流断面;出口断面
outlet selector 输出选择器
outlet sewer 污水管出口;污水出口;出水管
outlet side 排出端;出口端;出口侧
outlet side of pump 泵的排出端
outlet size 出口尺寸
outlet sleeve 出口套管
outlet sluice 泄水闸;泄水道;排水闸(门);出口排水闸;冲沙闸(门);放水闸(门)
outlets of grit chamber 沉沙池出口
outlets of thermal water 热水出口
outlet sound absorber 出口消声器
outlet structure 出水构造;排水构筑物;出口建筑物
outlet structure for closed drain 地下管道出口建筑物;下水道溢出口建筑物
outlet submerged culvert 压力式涵洞
outlet tail piece 出气尾管
outlet temperature 出口温度
outlet terminal 引出端;输入接线端;输出接线端;输出端子;出线端
outlet testing block 出线选试器
outlet through the curb 路缘排水口
outlet tile 出口瓦管
outlet tower 排水塔
outlet transition 泄水孔渐变段;泄水孔过渡段;出口渐变段
outlet trough 排水槽;出流水槽
outlet trunnion 卸料端轴颈;卸料端空心轴(颈)
outlet tube 出水管
outlet tubing 出水管
outlet tunnel 泄水隧道
outlet union nipple 出口螺纹接管
outlet union nut 出口联结螺母
outlet valve 排泄阀;放出阀;泄水阀(门);压力水出口阀;压力阀;排气阀;出流阀;出口阀
outlet velocity 出口速度;出口流速
outlet-vent 出口通气孔

outlet ventilator 排气扇;气窗;排气窗;通风器出口
outlet voltage 出线端电压
outlet waste 废水排水口
outlet water 排出水;出流水;废水
outlet weir 出水堰
outlet works 泄水建筑(物);泄水构筑物;泄水工程;河口工程;排水设施;排水出口工程;出口建筑物;出口工程【给】
outlet zone 卸料区;出口区
outlier 逸出值;老围层;界外值;外露层;蚀余柱;分离物;非正常值
outlier error 超限误差
outlier of the first kind 第一类离群值
outlier of the second kind 第二类离群值
outlier of the third kind 第三类离群值
outline 外形(线);要点;概要;概略;轮廓;廓;画轮廓;画草图;提纲;素描;大纲
outline area of mice sheet 云母片轮廓面积
outline assembly drawing 外形装配图
outline chart 轮廓图
outline covering the inspection jobs 检验提纲
outline dimensional drawing 外形尺寸图
outline dimensions 轮廓尺寸
outline draft 线划原图
outline drawing 轮廓图;略图;草图;外形图
outline driving 环行圈植法
outline edition 填充图
outline evaluation of technical economics 概略技术经济评价
outline image 外形轮廓像
outline inboard profile 纵剖面布置图
outline input 轮廓线输入;外线输入
outline letters 空心字
outline light 轮廓光;轮廓照明灯
outline lighting 泛光照明;轮廓照明;节日彩灯;外廊采灯
outline map 轮廓线图;轮廓略图;略图
outline map for filling 填充地图
outline map of regional structure 区域构造纲要图
outline map of structure 构造纲要图
outline of arch 拱曲线
outline of drainage basin 流域轮廓;流域简图;流域概述;流域边界
outline of one's book on soils 一本土壤方面的书的提纲
outline of plate tectonic 板块构造总论
outline of process 生产过程简图
outline of scanned area 扫描区域目标轮廓;扫描面积的轮廓
outline of site investigation 勘察工作纲要
outline of station 车站轮廓
outline of tooth 齿外形
outline of vault 拱顶曲线
outline overall and installing dimensions 外形及安装尺寸
outline plan 轮廓规划图;平面示意图;大略规划;粗略规划
outline profile 侧影图;侧视轮廓图
outline program(me) 规划纲要;概略工程计划;轮廓规划;提纲;大纲
outline sketch 轮廓草图;略图
outline specifications 简要材料设备一览表;简要标准规范(设计、施工等);简要施工说明;规定提纲
outline study 概略研究
outlining 描绘轮廓
outlining terrazzo patterns 勾画轮廓的水磨石图案
outloading 卸货;将卸货物;出货
outloading conveyer 出货输送机
outlook 视野;预计;展望;景色
outlooker (支持探出山墙房顶的)撑杆;山墙挑檐;出山撑杆
outlot (抵押没包括进去的)小块土地
outlying air base 海外航空基地
outlying area 郊区;城市外围区;边远地区
outlying area homes mortgage insurance 郊区房屋抵押保险
outlying business district 外围商业发展区;市郊商业区
outlying danger 离岸障碍物;伸出的危险物
outlying district 边远地区
outlying island 外围岛屿
outlying points 远距道岔
outlying sands 港外沙洲;离岸沙洲
outlying service point 对外服务点

outlying siding 区间岔线
outlying switch lock 远距道岔闭锁器
outlying zone 外围地区
outmigration 迁出
out-milling 对向铣切
out-moded equipment 陈旧设备;过时设备
outmoded notion 陈旧观点
outmost 最外层的
out movement 对向运动
out of action 停运的
out of adjustment 未校正的
out of alignment 出线的;不成一直线;偏离
out-of-balance 失去平衡(的);不平衡的
out-of-balance current 不平衡电流
out-of-balance load 不对称负荷;不对称负载
out-of-balance potential 失衡电压;位能失衡
out-of-balance weight 失衡载重;偏离重心位置的载重
out of band 不同频道信号传输;不同道信号传输
out of band noise monitor 带外噪声监测器
out-of-bounds 禁止入内的;越限;界限之外;边境之外
out-of-bounds area 禁地
out of cash 无现款
out of center 离中心;偏心;偏离中心
out of center of wheel 车辆偏心
out-of-commission 停用;不起作用;后备的
out-of-commission runway 废弃的跑道
out of contior 不能控制的
out of control 在管理上不加以控制的;无法操纵(的);失去控制的;失控;非控制;无法控制
out-of-core 反应堆外
out-of-course 途中运缓
out of court agreement 不受法庭干预的协议
out-of-date 落后;陈旧;过时的;过期的
out-of-date check 失效支票
out-of-date film 过期胶卷
out-of-date map 过时地图
out-of-dateness 过时;陈旧
out of door 露天的;户外(的);野外
out-of-door drying 露天晒干
out of drawing 图误;错绘
out-of-face maintenance 全面维修
out-of-face raise 全面起道
out-of-face surfacing 全面起道捣固
out-of-fashion 放弃的
out-off-band 带外
out-off-ga(u)ge load 超限装载
out-of-flatness 不平整
out-of-focus 离焦;不聚焦;焦点失调
out-of-focus appearance 离焦状态
out-of-focus image 离焦像;离焦图像;散焦图像
out-off of supply 停止供电
out-of-frame 帧失调;帧失步
out-of-frame alignment time 帧失步调整时间
out-of-frame second 帧失步秒
out of funds 缺乏资金
out of ga(u)ge 超限;不能计量
out of ga(u)ge cargo 超尺码货物;超尺寸货物
out-of-ga(u)ge cars and limited train 超限及限速列车
out-of-ga(u)ge freight 超限货物
out-of-gear 脱开的;脱离齿轮;不工作的;脱开的;切断的
out-of-gear mesh 脱离开的
out-of-gear worm 脱落蜗杆
out of hold 在舱外
out of keeping with 不相合
out of keeping with a country's tradition 不合国情的
out-of-kilter algorithm 不按次序演算法
out-of-level 不水平;不平坦
out-of-line 离开定线;超行;不在一直线上;不相符
out-of-line coding 线外编码;非常规排列程序
out of one's line 外行
out-of-operation 失去操纵的;不运转的;停运的
out-of-order 动乱;发生故障;不正常的;紊乱;无顺序;无次序的;出故障的
out-of-order execution 乱序执行
out-of-oriented vibration 面外振动
out-of-out 外廓尺寸;包外尺寸
out-of-phase 脱相;失相;不同相;相位差;异相(位);离相;不同相位
out-of-phase component 异相成分

out-of-phase cross correlation 异相交叉相关
out-of-phase current 异相电流；不同相位电流；不同相电流
out-of-phase difference effect 相位差影响
out-of-phase differential effect 差动影响
out-of-phase energy spectrum 异相能谱
out-of-phase signal 异向信号；异相信号
out-of-phase vibration 异相振动
out-of-phase voltage 异相电压
out-of-pile rig 反应堆外台架
out of plane 平面外
out-of-plan freight traffic 计划外运输
out of plumb 不垂直度；非铅直(的)；不精确(的)；不垂直；歪斜
out-of-pocket 实际花费；亏损；缺乏资金；现金支付
out-of-pocket cost 实际成本；现金成本；现付费用；实际现金支出；实际费用；实付费用；付现成本
out-of-pocket cost or expenses 实际现金支出
out-of-pocket expenses 现付费用；现金支付实际花费；现金支付开支；实际支出；零星杂项费用
out-of-pocket manufacturing cost 付现制造成本
out-of-position 偏移(的)；偏离预定位置；位置不当(的)；未正确就位(的)；不在正确位置的
out-of-position 不在适当位置
out-of-position welding 焊位不当
out of print 绝版
out-of-print map 绝版地图
out-of-range 出界
out-of-range number 超位数
out of range value 超量值；超范围值
out of red 有盈余
out-of-repair 破损；失修(的)
out of rig 舾装不全的
out-of-round 不圆(的)；变径差
out-of-roundness 椭圆度；失去圆度；不圆性；不圆度
out-of-round of roller 辊不圆
out-of-schedule train 临时列车
out-of-season fruiting 季节外结果
out-of-sequence services 不符顺序的运营；常规外的服务；非正常服务(业)
out of service 暂停服务；停用；不能使用的；不能工作的；停止运行；报废的
out-of-service jack 不用的塞孔；不工作塞孔
out-of-service time 修理时间；失效时间；非服务时间
out of shape 形状不规则的；走样的；失去正常形状的；变形的
out of sight 视界(以)外
out-of-size 尺寸不合格的；不合规定尺寸的；非正常大小的
out-of-specification piece 不符合技术规格样品
out-of-square 倾斜的；广场外；歪斜
out of square for wire 网歪斜
out-of-square mesh 网眼不正
out of station transfer 站外换乘
out-of-step 不同步
out-of-step operation 失步运行；失步操作
out-of-step switching 失步开合
out of stock 已售完；缺货；无存货
out of stock rate 缺货率
out of straight 不直
out-of-supply 停止供电
out-of-the country job 国外职业工作；国外职业活动
out of the money 较现值不利；未到价合约
out-of-the-way district 偏僻地区
out-of-tolerance 超差
out-of-tolerance parts 超差零件
out of town 城外的；出城
out-of-towner 郊区居民
out of trim 失衡
out-of-true 不精确；不正确；出差错；有毛病
out of true of wheel 车轮不精确
out of truth 有毛病；不准确度
out of use 无用的；不能使用的；报废
out-of-wind 没有扭曲；平正面；不翘扭平面
out-of-work 停止不用的；不工作的；不能工作的；失业；失效的
out-opening 向外开的门窗；外窗；外门
out-opening door 外开门；向外回旋门
out-opening window 外开窗
out operater 出算符
out of work 失效的
out opting 拒绝接受特别提款权分配额的权力
out orbit 外层轨道
out-ot-phase current 反相电流

out-out business 外围业务
outparcel (抵押没包括进去的)小块土地
outpatient clinic 门诊所
outpatient department 门诊部
out perform 运转能力；工作性能优于；超额完成
out per man 每人工时产量
out phase 相位不重合；相位(位)；反相；不同相
out-phase component 异相分量
out-phase system 异相制
out-phasing modulation 反相调制；异相调制
out phasing modulation system 移相调制方式；反相调制方式
outpocketing 外包缝合法
out poise 重于；多于
outporch 外阳台
outport 小港；外阳台；外港；输出港；出航港；出发港
outport bank deposit 外埠同业存款
outport surcharge 偏僻港附加费；小港运费附加费
outpost 前哨基地；前哨地区；分支机构
outpost well 探边井
outpouring 流出；喷溢
out-primary 初级绕组线头
outpulsing register 输出脉冲记录器
output 流输入输出；计算输出；排出量；输出；生产量；出量；产出；搏出量
output acoustic(al) power 输出声功率
output admittance 输出端导纳；输出导纳
output amplifier 输出放大器
output amplifier impedance 输出放大器阻抗
output amplifier stage 输出放大级
output amplitude 输出振幅；输出信号振幅
output amplitude level 输出幅度电平
output and finance plan 生产财务计划
output angle 输出角
output architecture 输出设备结构
output area 输出区
output array 输出阵列
output at full gate opening 导叶全开时的出力
output attenuatoin 输出衰减
output attribute 输出属性
output axis 输出轴
output axle nose 主动桥的主动轴颈
output back-off 输出补偿
output bars 输出条
output beam 输出束；输出光束
output beam coupling 输出光束耦合
output beam energy 输出光束能量
output beam power 输出光束功率
output block 输出信息组；输出信息块；输出区；输出分程序；输出存储区；输出部件
output borrow 输出借位
output break 输出中断
output budgeting 编制产出预算
output buffer 输出缓冲器
output buffer storage 输出缓冲存储器
output bundle 输出束
output bus 输出总线
output bus check 输出总线校验
output busd river 输出总线驱动器
output cable 输出电缆
output capability 输出能力
output capacity 输出能力；输出量；生产能力；出产量；产量
output carry 输出进位
output category 输出分类
output cavity 输出空腔谐振器
output channel 输出信道；输出通道
output characteristic 输出特性
output characteristic curve of hydraulic torque converter 液力变矩器输出特性曲线
output chip 输出片
output choke 输出端扼流圈
output choke coil 输出端扼流圈
output circuit 输出电路
output circuit impedance 输出电路阻抗
output class 输出种类
output coefficient 利用系数；产出系数；比力矩系数
output command 输出指令
output conductance 输出电导
output connection 输出连接；输出接线
output constant 输出常数
output contactor 输出端接触器
output control 生产管制
output controllability 输出可控性

output controller 输出调节器
output control lever 输出操作杆
output cost 可变费用
output cost ratio 生产成本比率
output coupling 输出耦合
output coupling device 输出耦合装置
output coupling factor 输出耦合系数
output current rating 输出电流额定值
output data 输出数据
output data strobe 输出数据选通
output data structure 输出数据结构
output decoupling zero 输出解耦零点
output delay 输出延迟
output demodulator 输出解调器
output derivative feedback 输出微分反馈
output detector 输出探测器
output device 输出装置；输出设备；出件装置
output diaphragm 输出光阑
output digit 输出数字；输出数位
output digital data 输出数字数据
output disable 禁止输出；输出中止
output display area 输出显示区
output display unit 输出显示设备；输出显示器
output distributed processing 输出分布处理
output distribution of service 服务的产出分配
output disturbance 输出干扰
output-doubling plan 翻番计划
output drift 输出信号变化
output driver 输出激励器
output due to initial condition 由于起始条件的输出
output due to input 由于输入作用的输出；零状态输出
output edit 输出编辑
output efficiency 效率系数；输出效率
output electrode 输出电极
output element 输出元件；输出环节
output enable 允许输出；可输出的
output end 输出端；输出电路
output end mirror 输出端面镜；输出端反射镜
output energy 输出能量
output energy filtering 输出能量滤波
output energy of pulse 脉冲输出能量
output energy stability 输出能量稳定度
output ensemble 输出集
output equation 输出方程
output equipment 输出装置；输出设备
output error 输出误差；输出错误
output error rate 输出的差错率
output estimation 产量估计
output executive 输出执行
output face 输出面
output factor 输出因素；输出因数；输出系数；输出率；出力因素；出力因数
output feedback 输出反馈
output file 输出文件
output film 输出胶片
output film format 输出胶片幅面
output filter 输出滤波器
output flange 输出轴凸缘；输出法兰盘
output flow 输出流量
output flow range 输出流量范围
output flyback 输出回描；输出端回程
output force 输出力
output form 输出形式；输出格式
output format 输出格式；信息输出公式
output format specification 输出格式说明
output formatter 输出格式程序
output frequency of standard oscillator 基准振荡器输出频率
output from decision-assisting model 决策支援模型的输出
output function 输出操作
output gap 输出隙
output gear 驱动齿轮；输出齿轮
output governor 输出功率调节器；输出调节器
output handler 输出处理机
output header 输出标题
output hopper 输出漏斗
output horse power 输出功率；输出马力
output hum level 输出交流声电平
output hunting loss 输出的搜索损失
output identification 输出标记
output impedance 输出阻抗

output impulse 输出脉冲
output impulse frequency of photoelectric(al) tachometer 光电测速计输出脉冲频率
output in bed measure 下方产量
output increment 输出增量
output index 产量指数
output indicating voltage 输出端指示电压
output indicator 输出指示器;产量计
output information 信息输出;输出信息
output inhibitor 输出禁止
output in metal removal 出屑量
output-input ratio 输出功率与输入功率比;产出投入比率
output in stock removal 出屑量
output instruction 输出指令
output intensity 输出强度
output interelectrode capacitance 输出端极间电容
output interface 输出接口
output interface adapter 输出接口器;输出端接口器
output interleaving control 输出交叉控制
output interrupt register 输出中断寄存器
output-investment ratio 产出投资比;产量-投资比
output jack 输出塞孔
output job stream 输出作业流
output lead 引导装入程序;引出线;输出引线;输出端
output level 输出级;输出电平
output level instability 输出电平不稳定性
output light flux 输出光通量
output limit 输出极限
output limited 受输出限制的
output-limiter 输出限制器
output line 输出线(路)
output linear group 输出线性部分
output link 输出网络节
output list 输出表
output listing format 输出列表格式
output load(ing) 输出负载
output loading factor 输出负载因数;输出负载率
output loss 出力损失
output machine variable 输出计算机变量
output magazine 接卡箱
output matrix 输出矩阵
output measure 输出测度
output mechamism 输出机构
output medium 数据输出装置
output meter 输出计;输出测量表;输出表
output-meter adapter 输出计附加器;输出表接合器
output method of depreciation 产量折旧法
output mirror transmission 输出镜透射率
output module 输出模块
output module valve 输出模块阀
output monitor 输出监视器
output monitor interrupt 输出监督程序中断
output multiplexer 输出多路调制器
output network 输出网络;输出网路
output node 输出节点
output noise 输出噪声
output norm 定量定额
output of cargo-handling machinery 装卸机械作业量
output of coal under unified central planning 统配煤产量
output of column 塔的生产能力
output of commodities 商品产出
output of consumer goods 消费品产量
output of current period 本期产量
output of electricity 发电量
output of hearth area 单位炉床面积产量
output of livestock product 畜产品产量
output of pick-up 拾震器的输出
output of plant 工厂生产量;工厂产量
output of power station 电站出力
output of the object program(me) 目标程序的输出
output of turbine 汽轮机功率;涡轮机功率;汽轮机出力;涡轮机出力
output on loading 装载生产率
output option 输出任选
output order 输出指令
output-oriented budgeting 侧重产出的预算编制
output pad 输出衰减器;输出端衰减器
output panel 输出控制台
output parameter 输出参数
output parameter address 输出参数地址
output part 输出部分
output passenger flow 输出客流
output per actual man-day for loading/discharging 装卸实际工作工日产量
output per day 每日产量;日产量
output per hectare 每公顷产量
output per hour 小时作业产量;小时生产率
output peripheral control 输出外围控制
output peripheral equipment 外部输出设备
output per litre 升功率
output per machine-hours 装卸机械台时产量
output per man-hour 工时效率;工时产量;人时产量
output per man shift 每人每班劳动生产率;每人每班产量
output per manshift 工班劳动生产率
output per minute 每分钟输出;每分输出量
output per resharpening 每次磨锐钻头的进尺率;每次磨锐钎头的进尺率
output photon 输出光子
output polarity 输出端极性
output polarization 输出偏振
output port 泄水口;输出端;出油口
output power 输出功率
output power at the wheel rim 轮周输出功率
output power density 输出功率密度
output power meter 输出功率计
output power of a motor vehicle 动力车的输出功率
output power of visual transmitter 图像发射机输出功率
output power spectrum 输出功率谱
output power stability 输出功率稳定度
output pressure 输出压力;出口压力
output printer 输出记录器;输出打印机
output process 输出进程;输出过程
output processor 输出处理机
output program(me) 输出程序
output pulsation 输出波动
output pulse 输出脉冲
output pulse amplitude 输出脉冲幅度
output pulse frequency 输出脉冲频率
output pulse width 输出脉冲宽度
output punch 输出穿孔机
output puncher 输出穿孔机
output quantity 输出量
output queue 输出排队;输出队列
output quota 生产定额;产品定额;产量定额
output quota of handling machinery 装卸机械生产定额
output radiation 输出辐射
output range 输出范围
output rate 生产率
output rate of dredge(r) 挖泥船生产率
output rate per hour 每小时输出率
output rating 输出额定值;输出率;额定输出功率
output ratio 输出率;产出比
output reactance 输出电抗
output record 输出记录
output recorder 输出记录器
output reference 输出基准
output reflection coefficient 输出反射系数
output register 输出寄存器
output regulation 输出调节
output regulator 输出调节器
output regulator problem 输出调节器问题
output relay panel 输出继电器控制板
output report 输出报告
output resistance 输出电阻
output resolution 输出分解
output resonant circuit 输出谐振电路
output resonator 输出共振器
output revolutions 输出转数
output ripple 输出脉动;输出干扰声
output routine 输出例行程序;输出程序
output scale 输出比例尺
output screen 输出屏
output sequence number 输出顺序号
output servo-mechanism 输出伺服机构
output shaft 输出轴
output signal 输出信号
output slit 输出狭缝
output sort 输出分拣
output speed 输出速率;输出速度
output spike 输出尖峰
output stability 输出稳定度
output stacker 输出收卡机;输出接卡箱
output stage 末级
output standard 标准时间
output string 输出字符串
output subroutine 输出子程序
output sun gear 中心齿轮输出;中心齿轮驱动;太阳齿轮输出;太阳齿轮驱动
output system 输出系统
output table 图形显示台;输出台
output tape 输出纸带
output tape punch 自动纸带穿孔机;输出纸带穿孔机
output target 生产指标
output terminal 输出端
output test 功率试验
output time series 输出时间系列
output token 输出标记
output torque 输出转矩;输出扭矩;输出力矩
output traffic control 输出信息量控制;输出流量控制
output transducer 输出变送器
output transfer 输出传送
output transformer 输出变压器;输出变量器
output transformerless 无输出变压器;无变压输出器
output translator 输出变换器
output-truncated condensation 被截输出缩合
output trunk 输出中继线
output tube 输出管
output turbine 输出功率透平
output unit 输出装置;输出设备;输出机;输出单元;输出部件;产量单位
output value 产值
output value of system 系统输出值
output variable 输出量;输出变量
output variation 输出变化
output vector 输出矢量
output voltage 输出电压
output volume 排量
output wave 输出波
output waveguide 输出波导
output wavelength 输出波长
output well 输出井
output winding 输出线圈;输出绕组
output window 输出窗
output work 输出功
output work queue 输出排队
output writer 输出程序
output writing program(me) 输出记录程序
out ramp 出口坡道;外坡道
outrange 超量程;超出作用距离(范围)
outreach 伸出长度;极限伸距;起重机臂伸出长度;起重机臂;外伸距(起重机);伸距(起重机臂的);吊杆跨距
outreach of crane 起重机臂伸出长度
outrigged 伸出舷外的
outrigger 舷外支架;舷外斜木;舷外浮材;支腿;稳定支撑;外伸支架;突起物;挑梁;伸出梁;船外铁架;承力外伸支架
outrigger base 支腿底板;梁外撑架底座;悬臂梁底板;平衡板
outrigger beam 悬臂梁;挑梁;挑出的梁
outrigger hydraulic cylinder 支腿油压缸
outrigger jack 撑脚千斤顶;支腿千斤顶;支腿支撑器(汽车式起重机等);支腿螺杆;支撑式千斤顶;支撑起重器
outrigger pad 支腿垫板
outrigger plate 平衡板
outrigger platform 悬臂型平台
outrigger pressure 支腿压力
outrigger scaffold 外挂脚手架;悬臂脚手架;挑出(式)脚手架
outrigger shaft 延伸轴
outrigger shore 悬臂支柱;外挑支撑;主结构凸出物的支柱
outrigger stabilizer 支撑稳定物
outrigger wall set flangepole 墙上的悬臂旗杆
outrigger wheel 挑出轮
outright costs 全部费用
outright cost of an undertaking 企业的全部费用
outright expense 全部费用
outright merger 公开合并

outright rate 期汇净价
outright sale 卖断
outright sell 卖断交易
out-rudder 外侧舵
outrun 超过;追过
outrun report 卸货报告
outrush 高速流口;高速流出;出口压头
outsail 超越他船
outscriber 输出记录机
outset 向外流;离岸流
outset angle 偏斜角
outsharn 外矽卡岩
outshoot 凸出
outshore 远距离海岸;海上
outshore longitudinal girder 外岸纵主梁(台式码头)
outshot 突出部分;伸出部分;废品;凸出部分;侧屋
outside 向海上;在外面;在外侧;外面的;外部的;外部;室外
outside air 新风;室外空气
outside-air intake 室外空气入口;室外空气进口
outside air intake duct 室外空气进风道
outside air temperature 指示气温;室外空气温度
outside-and-inside cal(l)ipers 内外卡钳
outside antenna 室外天线
outside appurtenances 外部配件
outside architrave 外侧门头线;外部框缘
outside atmospheric pressure 室外大气压力
outside auditing 外部审计
outside auxiliary lifting pontoons 傍台升浮筒
outside awning blind 室外遮阳篷
outside awning blind with expanding arms 带伸臂的室外遮阳板
outside axle-box 轴颈箱
outside band noise monitor 带外噪声监测器
outside bank slope 河岸外坡
outside barrel (双层岩芯管的)外管
outside base 外墙底板
outside blade 外切刀齿
outside blind 外百叶
outside board 舷外
outside brick 外皮砖
outside bridge wall 外桥砖
outside bridge wall block 外桥砌块
outside broadcast 室外广播
outside broker 一般经纪人;界外经纪人
outside cable 室外电缆
outside cal(l)iper ga(u)ge 外卡规
outside cal(l)ipers 外径测径器;外卡尺;外卡钳;外卡规;外卡
outside capital 借入资本
outside carrier 外部运输设备
outside casing 门窗外衬饰;外侧贴面;外侧框缘;外压条;外门窗框
outside casing liner 外壳衬板
outside casing plate 外壳板
outside cellar steps 户外地下室阶梯
outside chaser 外螺纹梳刀
outside chlorinated rubber paint 外氯氯化橡胶漆
outside circle 齿顶圆
outside clearance 外间隙(钻具与孔壁之间的间隙)
outside clearance ratio 外距比;外间隙比(取样器)
outside clinch 外转结;外活套结
outside coat 罩面层
outside coating 外涂层
outside coat of paint 罩面漆
outside column 外柱
outside contracting approach 采用外界承包方法
outside core 外形芯
outside corner 外墙角;阳角
outside corner mo(u)lding 外角线脚;外角边条;外侧拐角线脚
outside-corner trowel 阳角抹子
outside corridor 外走廊
outside-coupled pipe 接籀连接的管子
outside crawler 外侧履带
outside cutter (钻头的)外侧切削具
outside cylinder 外汽缸
outside decorator 厂外装饰家
outside-deviation 外面偏差
outside diameter 外直径;外径
outside diameter character 外径代号
outside diameter circle 顶圆直径

outside diameter controlled polyethylene pipe 控制外径的聚乙烯管
outside diameter fitting 外径配合
outside diameter ga(u)ge 外径标准尺寸
outside diameter of bailer 提捞筒外径
outside diameter of bubble cap 泡罩外径
outside diameter of collars 钻铤外径
outside diameter of drill pipe 钻杆外径
outside diameter of gear 齿轮外径
outside diameter of joint 接头外径
outside diameter of penetrometric pod 触探杆外径
outside diameter of swab 抽子外径
outside diameter of tray 塔板外径
outside diameter runout 外圆跳动量
outside diameter tolerance 外径公差
outside dimension 外尺寸
outside director 外界董事(非股东董事)
outside door 外门
outside door frame 外门框
outside drawing 外观图;外形图
outside drop of sewer manhole 外落式下水道人孔;外落式下水道检查井
outside drop type of sewer manhole 外落式下水道检查井
outside-drum filter 外滤式过滤器;外滤式过滤机;外鼓式过滤器;外鼓式过滤机
outside easing 外门窗框
outside effect 外部效果
outside embankment toe 外堤坝趾;外堤岸趾
outside end of arbour 柄轴外端
outside estimate 最高估计;最高的估计
outside extension 外部电话分机
outside face 外部立面
outside facing 外侧贴面;窗外框
outside fillet (weld) 外角焊缝
outside film coefficient 外表面放热系数
outside finance 外部资金
outside financing 外来投资
outside finish 外整修;外侧装饰;表面修整;屋外整修;外粉刷;外部装修;出面修整
outside finish of building 建筑外部装修
outside fire protection 外防火保护
outside firm 外单位
outside fix 旁心三分角线定位法;外定位
outside flange 外法兰;外凸缘
outside flashing light 房间外警告闪光
outside floor slab 室外楼板
outside flow 外部水流
outside-flush drill-pipe 外平钻杆
outside flush(ing) water 室外冲洗水
outside foundation line 基础墙外皮线;基础(墙外)边线;外侧基础线
outside frame 外门框
outside gallery 外走廊
outside ga(u)ge 外径规
outside girder (桥的)最外侧大梁
outside glazing 外装玻璃法;外装窗玻璃;外镶玻璃
outside gloss paint 外涂有光漆
outside gouge 外弧口凿;(锐口在凸面的)半圆凿
outside groove 边槽
outside gutter 外排水管
outside handrail 室外扶手;室外(门)把手
outside heating device 外部加热设备
outside help 外援
outside humidity 室外湿度
outside hydrant 室外给水栓;室外消防栓
outside illumination 室外照明
outside in 从外侧向内侧
outside-inside mating diameter 内外配合直径
outside insulating of building 建筑物外侧保温
outside insulation 外隔声层;外保温层;外部绝缘
outside-journal axle 外轴颈
outside kitchen 室外厨房
outside lap 进气余面;外余面
outside layer 外层
outside lead circle of pile 板桩外导向圈
outside leader 外水落管;外落水管
outside lead ga(u)ge 外螺纹导程仪
outside lever 罐外控制杠杆(压缩空气)
outside lighting 室外照明;室外照明
outside line organization 跨专业组织
outside lining 外层镶衬;外侧贴面;外气套;外隔

板;外衬(砌);外侧贴面;外部镶衬(框架);窗框;框架外饰板;框架外衬;框架外饰板;框架外部构件
outside link 外拉杆
outside lip surface 外唇面
outside loading 外荷载
outside loan 信托贷款;信托放款
outside locking device 外锁闭
outside loop 外循环;倒飞筋斗
outside marble 大理石贴面
outside market 界外市场
outside masonry (wall) 外层砖石墙
outside masonry wall column 外层砖石墙柱
outside masonry wall facing 外层砖石墙面
outside masonry wall lining 外层砖石墙衬砌
outside measurement 外侧尺寸;外包尺寸
outside micrometer 外径千分尺;外径千分表;外径测微规
outside-mixing burner 扩散式喷燃器
outside monkey baord (钻塔的)外伸二层台
outside network 外管网;外部网;外部管道系统
outside nuclei 外加晶种
outside of a fouling point 警冲标外方【铁】
outside of a fouling post 警冲标外方【铁】
outside of railway loading and unloading charges 路外装卸费
outside order expenses 外部定货费
outside packing 外包装
outside paint coat 外层涂料;外层油漆
outside painting work 外层涂刷油漆工
outside panel 外墙板
outside pipe (双层岩芯管的)外管
outside pitch line length 齿顶高度
outside plank 曲面板
outside plate 外板
outside port 过驳港
outside post type container 外柱式集装箱
outside power 外部动力;外功率;外力
outside power supply 外部动力供应
outside precast brick 外墙预制块
outside pressure ga(u)ge 舱外压力表
outside price 场外价格
outside primary 初级线圈外端
outside primer 室外底漆
outside prism 路基外侧的弃土堆
outside processing materials 委托加工材料
outside protection 外部保护
outside pulpit 室外控制台
outside quality control 外部质量控制
outsider 门外汉;非会员
outside race 外座圈
outside radius 外半径;齿顶圆半径
outside rail 外轨;室外栏杆;室外扶手
outside relative humidity 室外相对湿度
outside reveal 外门窗口的侧墙面
outside rolling hitch 外转结
outside room 外侧房间
outside run 室外布线
outside scaffold(ing) 搭外脚手架
outside screw and yoke valve 外螺杆和手轮阀
outside screw non-bonnet 无压盖外螺纹
outside seal 外部密封
outside sealing 外部嵌缝
outside sealing of well 封井
outside shaft 外部竖井
outside sheet 外层薄板
outside shell 外壳
outside shell member 外壳构件
outside shell of firebox 火箱外壳
outside shutter 外面百叶窗;外开百叶窗
outside single-point thread tool 单刃外螺纹车刀
outside skin 外壳;外皮
outside skin type container 外板式集装箱
outside slatted unit 外百叶
outside slope 迎水坡;外坡
outside source 外部信号源
outside specialist 外来专家
outside spinning 外旋压
outside spline 外侧水槽
outside spline ga(u)ge 测花键外径的环规
outside sprinkler 外部喷水灭火器
outside square 外方角
outside staging 舷外作业板架
outside stairrail 室外楼梯扶手

outside stair(case) 室外楼梯
outside stairway 室外楼梯
outside standpipe 室外(给水)立管
outside stile 门窗外边梃;门窗边梃
outside stitching machine 缝外线机
outside stone (钻头的)外侧刃金刚石
outside storm system 外排水系统
outside strake 外列板
outside string 楼梯外栏栅;外梯梁;外侧楼梯帮
outside stringer 外梯梁(不与墙连接的楼梯斜梁)
outside studding plate 外墙竖龙骨顶板;外侧间柱木板
outside supplier 外购补给品
outside surface 木材离心面(木表)
outside system 外墙构造
outside tap 室外龙头;丝锥;打捞锥;打捞器
outside temperature 外界温度;室外温度
outside temperature sensor 室外温度探测器
outside the planned market 超出国家计划市场之外
outside-the-station equipment 站外设备
outside tier 外壳
outside tile 外部面砖;外部瓷砖
outside torque 外力偶
outside torsion(al) moment 外扭矩
outside transmitter 室外发射机
outside trim 窗外框;室外装饰;窗外框
outside tube (双层岩芯管的)外管
outside tube diameter 外管直径
outside tud 外墙筋
outside tunnel controlling survey 硐外控制测量
outside tunnel work 硐外作业
outside twist(ing) moment 外扭矩
outside undercoat(er) 外部用的底层涂料
outside upset pipe 外加厚管子
outside use 外部使用
outside vapo(u)r deposition process 外气相沉积法
outside vapo(u)r phase oxidation process 管外气相氧化法
outside varnishing 外罩清漆
outside veneer 外饰面板
outside venture 营业外经营;业外营业;附属业务
outside venture capital 外部投入资本
outside vibrator 外模振荡器
outside view 外观图;外形图
outside view drawing 外观图
outside vinyl paint 外部乙烯树脂漆
outside walk 室外通道
outside wall 外墙;(钻头的)外侧面;外壁
outside wall block 外墙砌块
outside wall construction 外墙构造
outside wall facing 外墙面层
outside wall finish 外墙饰面
outside wall lining 外墙衬砌
outside wall member 外墙构件
outside wall panel 外墙板
outside wall surface 外墙面
outside wall tile 外墙面砖
outside wall unit 外墙单元;外墙构件
outside wear 外径磨损
outside weld(ing) 外焊;缝
outside widening 外侧加宽
outside window 外窗
outside window frame 外窗樘
outside window sill 外窗台
outside wind pressure 室外风压;洞外风压力
outside wire 火线
outside wiring 室外布线;室外架线
outside with 外层单砖墙
outside work 户外工作;室外工作;外活
outside wythe 外层单砖墙
outsize 特大
outsize determination 非标准尺寸测定
outskirt 城市外围;关厢;近郊;市郊
outsole 基底;外底
out-spent 废的
outspread intersection 扩散式交叉口
out square knee 钝角肘板
outstanding account 应收账目;悬账;欠账;未销账款;未清账目;未清账款;未清余额;未结算账目;未付账款
outstanding account payable 应付未付账款
outstanding advance for estimated tax 未预付估计税金

outstanding amount 欠款
outstanding amount of debt 债务总额
outstanding balance 未偿债额;未用余额;未清余额;待结款项未清余额
outstanding balance of account 余额报表
outstanding balance of borrowed money 未清借款余额;未偿借款余额
outstanding borrowing 未偿还借款
outstanding capital 未偿还资本
outstanding capital stock 净发行股票;净发行股本
outstanding check 未兑现支票
outstanding cheque 未兑现支票
outstanding claim 未决赔款;未解决的赔款
outstanding contract 尚待完成的合同
outstanding contribution 未清摊款
outstanding cycle of subgrade soil 地基土的卓越周期
outstanding debt 未偿债务
outstanding design 优秀设计
outstanding drawing 未还提款
outstanding dues 未付的应付款
outstanding expenses 应付未付费用;未付费用
outstanding external debt 未偿外债
outstanding feature 显著特征;特色
outstanding file 出口押汇未了卷
outstanding flange 突出的翼缘;突缘
outstanding guarantee credit 未偿担保信贷
outstanding item 未偿贷款
outstanding leg 凸出的立柱;伸出的支架;突出肢
outstanding leg of angle 角钢的伸出肢
outstanding leg of angle bar 角钢突出肢
outstanding loan 未偿贷款
outstanding loan portfolio 未偿贷款总额
outstanding loss 未决赔款;未解决的赔款;保险的未决赔款
outstanding loss reserves 未付清赔款准备金
outstanding obligation 债务余额;未偿债务
outstanding of deposits 存款余额
outstanding on loan 已发放的贷款
outstanding order 未交货订单
outstanding point 方向标;明显地点;明视地点;明视地点
outstanding principal 未偿还本金
outstanding principal of investment trust fund 投资信托基金未偿本金
outstanding problem 未解决的问题
outstanding rainfall 特大降雨量;特大暴雨量
outstanding recommendation against class 有关船级社遗留问题
outstandings 未清(算)账目;未结算账目;未偿贷款;未付账
outstanding securities 已发行证券
outstanding shares 已发行股票
outstanding stock 已发行股票;未偿股份
outstanding work 未完成的工作
outstation 野外靶场;分站;分局;支局;外围站
outstation transmission unit 站外传输单元
outstep boring 油田边缘的油井;边缘镗孔
outstep drilling 甩开钻井
outstretch 拉长;扩张
outstrip 超过
outsulation 外绝缘;外墙热绝缘
out-swinging 外旋转的
out-swinging casement 外开扇;外开窗扇;外开窗扉
out-swinging casement window 外形竖铰链窗
out-swinging side hung casement 边轴外旋转门式窗
out-swing ventilator 外开通风窗
out switch 输出开关
out time 处置时间
out-to-out 全长(度);总宽度;总长度;满外尺寸;外到外尺寸;外廓尺寸;包外尺寸;最大尺寸
out-to-out dimension 外皮到外皮尺寸;轮廓尺寸;总尺寸;满外尺寸;外包尺寸
out-to-out distance 总尺寸;轮廓尺寸;满外尺寸;外沿间距离;外包尺寸
out track build-up 外轨抬高
outturn quantity 卸出数量
outturn report 货物溢短残损单;卸货报告
outturn sample 卸货样品
outturn weight 卸货重量
outvalue 价值超过;比有价值
outward 向外;外面的
outward and home freight 往复运货

outward appearance 外观;外表
outward bill 出口汇票
outward-bound 开往外地;开往外埠;向外开出
outward bounder 外航船;出国航行船
outward-bound road 向外开出的路;离港的道路;离站的道路
outward-bound ship 开往国外的船;外航船;出国船
outward bulge 船外腹
outward bulging 向外鼓;船外腹
outward call 发出催缴通知
outward cargo 外运货(物);出口物资;出口船货;出港货物
outward charges 出港费
outward clearance certificate 开航结关证书
outward collection 出口托收
outward current 外向电流
outward declaration 轮船出口报告
outward documentary bill 出口跟单汇票;出口押汇
outward flow 外向水流;外流
outward flow turbine 外流式水轮机;外流涡轮;外流式涡轮机
outward freight 销货运费;外运运费;外运货(物);出口物资;出口船货
outward goods 出口船货
outward inspection 外表检查
outward-looking development policy 向外发展政策
outwardly 外表上
outwardly opened casement 外开扇;外开窗扇;外开窗扉
outwardly opened door 外开门
outward manifest 出口舱单
outward material 外运材料
outward opening 伸出的凉廊;外开式;外翻门
outward opening door 外开门
outward opening window 外开窗
outward passenger 出境旅客;出发港旅客
outward pilotage 出港引水费
outward port charges 出口港口手续费;出港手续费
outward processing 外运加工
outward reinsurance 分出再保险
outward rotation 外旋
outwards 超过
outward sailing 出航
outward seepage 外渗
outward service 对外业务
outward-swinging side hung (window) casement 外开窗
outward-swinging window 外开窗
outward thrust 向外推力;外向推力;外推力
outward turning screws 外旋式双推进器
outward voyage 外航;出航;出国航行
outware clearance 出口许可证
outwash 刷净;冲刷;冰水沉积
outwash apron 冰水沉积扇;冰水沉积平原
outwash coast 冰水沉积海岸
outwash cone 冰水沉积锥
outwash delta 冰水沉积三角洲
outwash deposit 冰水平原沉积;冰水沉积(物)
outwash drift 冰水沉积
outwash fan 洪积扇;冰水扇形地;冰水沉积扇
outwash gravel 冰水冲积砾石
outwash plain 冰水沉积平原;冰川沉积平原;冰水冲平原
outwash seepage 出渗
outweigh 重于;多于;超重
outweight 重量超过
outwell 流水
outwindow 伸出的凉廊
outwork 野外工作;露天作业;户外工作;厂外工作;城堡;防御工事
outworn 过时的;破损
outworn building 陈旧建筑
outworn wharf 陈旧码头
ouvala 干宽谷;灰岩谷
ouvrell'oeil 危险可凝点
Ou ware 瓯窑
ova (雕刻在线脚上的)卵形装饰
ovaide leaf 卵圆形叶
oval 鸭蛋圆;卵形线;卵形;椭圆形;椭圆的
oval arch 椭圆形拱;椭圆拱
oval-body section 卵形体截面
oval bow machine 椭圆形连铸机
oval brad 椭圆角钉

oval brad head nail 椭圆平头钉
oval calotte 椭圆曲面
oval cam 卵形凸轮；椭圆形凸轮
oval chuck 椭圆夹头
oval column 椭圆形柱
oval copy 椭圆形样板
oval countersunk fillister head machine screw 半沉头机器螺钉
oval countersunk fillister head wood screw 半沉头木螺钉
oval countersunk head rivet 椭圆埋头铆钉；半沉头铆钉
oval countersunk screw 椭圆埋头螺钉
oval court 椭圆庭院；椭圆形庭院
oval (cross-) section 椭圆(形)截面；蛋形截面
oval defect 卵形缺陷
oval door knob 椭圆门执手
oval-eccentric gearing 椭圆—偏心齿轮传动
ovalene 卵苯
oval fillister head screw 球面圆柱头螺钉
oval flange 椭圆凸缘；椭圆法兰
oval flaw 钢轨核伤
oval flowmeter 椭圆流量计
oval foramen 卵圆孔；博塔洛氏孔
oval gear 椭圆齿轮
oval gear flowmeter 椭圆齿轮流量计
oval grid 椭圆形栅极
oval groove 椭圆槽
oval groove seal contact face 椭圆槽密封面
oval head brass rivet 椭圆头铜铆钉
oval head nut 半埋头螺母
oval head rivet 椭圆头铆钉
oval head screw 椭圆头螺丝；椭圆螺钉；半埋头螺旋
oval head split rivet 椭圆头开口铆钉
oval hexagon head bolt 卵形六角螺栓
oval in outline 外形椭圆
ovalisation 成椭圆形
ovality 卵形度；椭圆度
ovalization (deformation) 椭圆化；成椭圆形
ovalizing balance 椭圆摆轮
oval kiln 椭圆形窑
oval knob door furniture 椭圆形执手门的家具
oval knot 椭圆节；椭圆节
oval ladder step 两端椭圆的绳梯板
oval lavatory 椭圆形洗面器
oval lost head nail 椭圆地板钉；椭圆形埋头钉
oval luthern 椭圆形老虎窗
ovally grooved seal face 椭圆槽密封面
oval muffler 椭圆形消声器
ovalness 椭圆度
oval of Cassini 双级方程曲线
oval of zero velocity 零速度卵形
ovaloid 卵形面；似卵形；封头处曲面
oval orifice 椭圆孔式
oval pass 椭圆轧槽
oval pattern 鱼盘型
oval pin shackle 扁销卸扣
oval piston 椭圆形活塞；椭圆活塞
oval plate 椭圆形板
oval reel 椭圆绕线轴
oval repeater 椭圆孔型用围盘
oval ring 椭圆垫
oval rivet 椭圆头铆钉
oval scale 椭圆尺
oval sectional pin shackle 具有椭圆形截面插销得卸扣
oval-section steel 椭圆形断面钢筋
oval-shaped 椭圆形的
oval-shaped column 椭圆形立柱
oval-shaped roughing pass 椭圆粗轧孔型
oval-shaped sewer 卵形下水道
oval-shaped sprocket 椭圆形星轮
oval slab 椭圆形板
oval slotted countersunk head unloosing screw 半沉头不脱出螺钉
ovals of Cassini 卡西尼卵形线
oval-square passes 椭圆-方孔型系统
oval strand 椭圆形股绳
oval strand wire rope 椭圆股钢丝绳
oval tubing 卵形管
oval-turning device 车削椭圆装置
oval varnishing brush 椭圆形油漆刷
oval window 卵圆窗；卵形窗

oval wire 椭圆形钢筋；椭圆(形断面)钢丝
oval-wire brad 椭圆钢丝曲头钉
ovate leaf 卵圆形叶
ovate-oblong 长卵形的
ovb 落水
oven 炉(子)；烤炉；加热炉；烘箱；烘干箱；恒温箱
oven ag(e)ing 炉内老化；加热老化
oven battery 炉组
oven block 炉组
oven bottom 炉底；炉床
ovencerite 水磷铝钙镁石
oven chamber 炉室
oven cleaner 烘炉洗净剂
oven coke 炉焦
oven colo(u)rs (glaze) 炉彩釉
oven cure 烘炉硫化
oven-dried 烘干(的)
oven-dried aggregate 烘干集料；烘干骨料
oven-dried box 烘干箱
oven-dried sample 烘干标本；烘干样；烘干试样
oven-dried soil 烘干土
oven-dried weight 烘干重(量)
oven-dried wood 烘干材；炉烘木材
oven-dry 烘干的
oven-dry aggregate 烘干骨料；烘干集料
oven-dry density 烘干容重度；烘干密度；全干密度；炉干密度
oven drying 烘炉干燥；烘干
oven drying method 烘干法
oven drying of soil 土的烘干
oven drying test 烘干试验；烘干试验法
oven-dry soil 烘干土
oven-dry timber 窑干木料；炉烘木材；烘干木材
oven-dry weight 烘干重(量)
oven-dry weight aggregate 集料烘干重量
oven-dry wood 窑干木料；烘干木材
oven flue 炉灶烟道
oven fork 炉叉
oven gas 焦炉气
oven glass 烹饪用玻璃；炊具玻璃
oven glassware 玻璃烘具
oven inner liner 加热炉内衬
oven Jun glaze 炉钧釉
oven loss test 炉热损失试验；炉加热损失试验；加热损失试验；烘干损耗试验
oven port 炉门
oven process 炉法；烘箱灭菌法
ovenproof glass 防火玻璃；耐热玻璃
oven soldering 炉焊(接)；炉中软钎焊
oven-sole block 炉底砌块
oven-sole brick 炉底砖
oven stability test 耐热稳定试验
oven stack 排气烟囱
ovenstone 耐火石
oven tar 焦炭炉柏油；焦炭炉焦油
oven test 炉热试验；耐热试验；焙烧试验
oven-top ware 耐热玻璃炊具
oven-to-table ware 耐热餐具
oven-type boiler 烘箱式锅炉
oven-type furnace 箱式炉；烘箱式锅炉
oven wall 炉墙；暖墙(设采暖烟道的墙壁)；暖墙；火墙
oven wall liner 炉墙衬里
ovenware 耐烤玻璃器皿；烹调器皿；厨房用器皿
oven weathering test 恒温箱风化试验
over absorbed burden 多摊间接费
over absorbed cost 多摊成本
over absorbed expenditures 多摊费用
over-accelerator 促进剂过量
over accuracy 总精确度
overaction 作用过度
overact mining 开采过甚
over-adaptation 超适应
overaeration 过量曝气；过度曝气
overage 超过额；超额的负载；超额；超出；附加物品
over-aged 过度时效的
over-aged machine 超期服役机械
overaged product 过剩产品
overag(e)ing 过时效
overage rent income 分成租金收入
over-age vessel 超龄船(舶)
overaging 过度时效
overagitation 过度搅拌
over-aking rule 超车规则

overall 总的；全体；全面的；全部的
overall absorption coefficient 吸收总系数
overall accuracy 总准确度；总精度；总和精度；全部准确度
overall adjustment 整体平差【测】
overall amplification 总放大率
overall analysis 全面分析
overall analysis of economic benefits 经济效益综合分析
overall application 全面施用
overall appraisal of a project 项目综合评价；项目的总评价
overall approximate estimate 总概算
overall approximate estimate cost 总概算造价
overall approximation formula 整体逼近公式
overall architecture 总的体系结构
overall arrangement 统筹安排；布局
overall arrangement of exploratory engineering 勘探工程总体布置
overall arrange system 总体部署方案
overall attenuation 总衰减；总减少；净衰减
overall attenuation level 总衰减电平
overall attenuation measurement 总衰减测量
overall average 总平均(值)
overall balance 总平衡；总差额；综合平衡；全面余额；全部差额
overall balance and unified arrangement 综合平衡
overall balance of urban economy 城市经济综合平衡
overall balance on liquidity basis 清偿基础总差额
overall balancing 总合平衡；综合平衡
overall bonding 整体黏结
overall borrowing cost 借款总费用
overall breadth 总宽(度)
overall brightness-transfer characteristic 总亮度传输特性
overall budget 综合预算；全面预算
overall budgetary estimate of construction project 建设项目总概算
overall budgetary operation 全面预算活动
overall budget level 预算总(金)额
overall business forecast 全盘商业情况预测
overall capitalization rate 总资本化率
overall carbon content 总碳量
overall cell reaction 总电解反应
overall central information problem 全面中央信息问题
overall characteristic 总特性
overall charge 总负责；总开支；全部费用
overall check 总检；全面检查
overall coefficient 综合系数；总系数；全系数
overall coefficient of heat 传热总系数
overall coefficient of heat transfer 总传热系数；综合传热系数
overall collapse 全部破坏；全部崩溃
overall collection efficiency 总收集效率；总除尘效率
overall combined feed ratio 总综合喂料比；总综合进料比
over-all commitment 承诺款项总额
overall composition 总成分
overall computing speed 总计算速度
overall computing velocity 总计算速度
overall consideration 总报酬；全盘考虑；全面考虑
overall construction plan 施工总进度计划
overall construction site plan 施工总平面图；施工总布置图
overall construction time 全部施工时间
overall consumption 总消耗
overall contract 总承包
overall contrast 总反差
overall core recovery 总岩芯回收率
overall cost 总费用；总成本；全部成本
overall cost control 全面费用控制；全面成本管理
overall cost of capital 综合资本成本
overall cost rate 综合费用指标
overall coverage ratio 综合偿债保障率
overall credit 总信贷
overall cross-section 总截面
overall depth 总高度
overall depth of section 断面总深度
overall design 总体设计；总体计划
overall design management 设计总体组
overall design of pavement 【路面总体设计】

overall development 综合开发;全面发展
overall diameter 总直径;机座外径;全径;外包直径
overall diameter of electrode 焊条总直径
overall differential protection 整体差动保护装置
overall dilution ratio 总贫化率
overall dimension 通体尺寸;外界尺寸;极限尺寸;全尺寸;最大尺寸;总尺寸;外形尺寸;外廓尺寸;外包尺寸;轮廓尺寸
overall discing 全面耙松
overall drawing 总图
overall drilling speed 经济钻速
overall drilling time 总钻眼时间
overall driving safety 全面行车安全
overall (dust) collection efficiency 总收尘效率;总集尘效率
overall economic effect 全面经济效果
overall economic potential 整体经济潜力
overall effective tax rate 综合有效税率
overall efficiency 总效率;总有效利用系数;整机效率;遍计效率
overall efficiency of separation 除尘效率
overall elasticity modulus 总的弹模量;综合弹性模量
overall elongation 总延伸(量)
overall energy 总能量
overall erection time 总安装时间
overall error 总误差;总体误差
overall estimated cost 预算总造价
overall estimation of capital construction 基本建设概算
overall evaluation 综合评价
overall excavation 满堂开挖
overall expansion ratio 总膨胀比
overall external dimension 外侧最大尺寸;外部尺寸
overall financial plan 综合财务计划;全面财务计划
overall financing for insurance 统筹保险基金
overall firing performance 齐爆性能
overall flaw detection sensitivity 相对探伤灵敏度
overall floorage 总建筑使用面积;总建筑面积;建筑总面积
overall floor area 总建筑面积
overall floor depth 楼板总厚度
overall f-number 总光圈大小
overall form 总外形
overall formation constant 总形成常数
overall frequency response 总频率响应;整机频率响应
overall gain 总增益
overall gradient 总比降
overall head 总水头
overall head allowance 总水头容许值
overall heat balance 总热平衡
overall heat consumption 总热耗;综合热耗
overall heating 总热值
overall heat transfer 传热系数
overall heat transfer coefficient 总传热系数
overall heat transmission 总传热
overall heat transmission coefficient 总传热系数
overall height 总高(度);整机全高;建筑高度
overall height of crane 起重机全高
overall height of installation 安装总高度
overall housing 全套的住房;总的住房;包括一切的住房
overall limitation 全面限制
overall impression method of selection 一般印象人员选择法
overall indicator 综合指示器
overall inspection 全面检验;全面检查
overall installing capacity 总装机容量
overall intermodulation 总互调
overall investigation 综合调查;全面调查
overall investigation method 全面勘察法
overall kiln setting density 总装窑密度
overall labo(u)r 全部劳动消耗
overall layout 总体布置
overall layout of construction 施工现场总体布置图
overall layout of port 港口总体布置
overall length 船长;总长(度);整机全长[机];全长
overall length of bridge 桥梁全长
overall length of broach 拉刀全长
overall length of chassis 底盘全长
overall length of crawler 履带全长
overall length of pile 桩的总长度
overall length of tunnel 隧道全长

overall level 总电平
overall lift 总升力
overall limitation period 一般时效期
overall limits 全部限额
overall line attenuation 线路净衰减
overall loading 满载;满负载
overall loan 总贷款
overall location loss 全程定位衰耗【铁】
overall location operation loss 全程工作衰耗【铁】
overall logistic(al) demand 总的物资需求量;总补给量;后勤物资需求总量
overall look 全貌;全景
overall loss 总损失;总损耗
overall losses ratio 总损失率
overall loss measurement 总损耗测量
overall magnetic field 总体磁场
overall mass coefficient 总ành质系数
overall material balance 总物料平衡;综合物料平衡
overall mean velocity 断面总平均流速;断面平均速度;断面平均流速;断面平均风速
overall measurement 总测量;外尺寸
overall merit 总评价;综合评价
overall model 总体模型;整体模型
overall model test 整体模型试验
overall noise 总噪声
overall noise level 总噪声级
overall noise temperature 总噪声温度
overall number 总数
overall objective 整体目标
overall operating speed 总操作速度
overall operational cost 总运行费
overall or comprehensive budget 全面或综合的预算
overall output 总产量
overall output per manshift 每班产量
over-allowance 尺寸上偏差
overall parallax 全视差
overall parallax angle 全视差角
overall parameter measuring method for pavement roughness 路面粗糙度综合参数测定法
overall pattern 总体图式
overall pavement design 路面总体设计
overall payment agreement 总支出协议;总支出协定
overall percent utilization 总利用率;总利用率
overall performane 总指标
overall permeability 总渗透率
overall perspective 全貌;全景
overall picture of continued pollutant loading 连续污染物负荷总图
overall plan 总体设计;总体方案;总进度(表);总体规划;总体计划;综合规划;全面规划
overall planning 综合计划;整体规划;全面规划;统筹规划;总体规划;总体计划
overall planning and allround consideration 统筹兼顾
overall planning and proper arrangement 统筹兼顾适当安排
overall planning management 全面计划管理
overall planning of works 工程总体策划
overall planning scheme 总体设计方案
overall planning stage of city 城市总体规划阶段
overall plant efficiency 车间总效率;工厂总效率;企业总效率;设备总效率;电站总效率
overall plate efficiency 总塔板效率
overall population distribution 总体人口分布
overall population growth rate 人口总增长率
overall population mean 总体人口均值
overall position of exchange 总的外汇头寸
overall power consumption 综合电耗
overall pressure 总压力
overall pressure drop 总压(力)降
overall pressure ratio 总压比
overall price index 综合物价指数
overall production 总产量
overall professional responsibility 全面专业责任
overall program(me) 总进度
overall project 总体方案;总计划;综合计划;综合方案;全面计划方案;全面计划;全面规划方案;全面方案
overall project management 项目全面管理
overall property 综合性能;全部性能
overall propulsive efficiency 总推进效率

overall pulse 总脉冲;全脉冲
overall quantity 总体数量
overall raising of social insurance funds 统筹社会保险基金
overall rate 总速率
overall rate of return 综合收益率;全面回报率
overall ratio 总传动比
overall rationing system 包干制
overall reaction 总反应
overall reaction rate 总反应速率
overall reconnaissance 普查
overall recovery 总回收率
overall reduction 总去除率;总降低作用;总除去率【给】
overall reduction ratio 总减速比
overall refrigerating effect 总冷却效应
overall regulation of funds 统一调剂资金
overall reinforcement area 钢筋总面积
overall reinforcing area 钢筋总面积
overall reliability 总可靠性;总可靠度
overall remuneration 全部薪酬
overall resolution 总分辨率
overall resolution efficiency 总分离效能指标;总分离效能
overall response 总响应
overall rigidity 总刚度
overall risk 综合风险;综合险
overall rolling width 压实宽度;总滚压宽度
overalls 工作服;工装裤
overall sampling 全面抽样
overall screen surface area 总筛分面积
overall sediment model 全沙模型
overall sediment model test 全沙模型试验
overall selection 总选择性
overall site layout 施工现场总平面布置图
overall size 总尺寸;轮廓尺寸;外形尺寸;外包尺寸
overall slope 总坡度;总比降
overall solution 一揽子解决办法;总体解法;共同解法;全面解决办法
overall speed 总速率;区间速度;区间车速
overall spread 全距
overall stability 总稳定性;整体稳定性;整体稳定
overall stability constant 稳定积;总稳定常数
overall stability of reinforced earth 加筋土总稳定性
overall statistical survey 全面调查
overall steel area 钢筋总面积
overall stiffness 总劲度;总刚度
overall strain 总应变
overall strategy 全面策略;全面战略
overall strength 总体强度;总强度;整体强度
overall stress 总应力
overall stripping ratio 平均剥采比
overall structure 总体结构;整体结构
overall system 整个系统;完整系统
overall system adjustment 综合调整
overall system gain 总系统增益
overall target 综合指示器
overall test 总体测试;整机试验
overall theoretic(al) power 理论总功率
overall thermal efficiency 总热效率;综合热效率
overall thermal resistance 总热阻
overall thickness 总厚度
overall thickness of coal bed 煤层总厚度
overall thickness of coal seam 煤层总厚度
overall thickness variation 总厚度变化
overall time delay 总延迟时间
overall time limit 一般时效期
overall timing method 全部计时法
overall transfer characteristic 总传输特性
overall transfer function 总传递函数
overall transmission loss 总透过损失
overall transportation 综合运输
overall transportation complex 综合运输系统;综合运输体系
overall travel speed 区间车速
overall travel time 全行程时间;区间车行时间
overall uniform sampling fraction 统一抽样比
overall unit protection 单元总体保护装置
overall utilization of petroleum 石油综合利用
overall velocity 总速度
overall ventilation efficiency 总通风效率
overall vessel 超龄船(舶)
overall view 全图;全貌;全景

overall viewer 全貌窥视窗
overall volume of construction 总建造体积;结构总体积
overall watering 进行全面灌水;全面浇水
overall weight 总重(量)
overall weldability 综合可焊性;使用焊接性
overall width 最大宽度;总宽(度);整机全宽;全宽(度);外形尺寸宽度
overall width of bridge 桥梁全宽
overall width of chassis 底盘全宽
overall width of crane 起重机全幅
overall width of road 道路总宽度
overall yield 总收率;总回收率;总产率
overamplification 过量放大;过分放大
over-and-above the assigned profit norm 超额利润分成
over-and-over addition 逐次相加;重复相加
over-and-short account 溢缺额账户;超额和短缺账户
over-and-under absorbed overhead 制造间接费分摊余缺
over-and-under chain gear 内外径啮合链传动装置
over-and-under controller 自动控制器
over-and-under culvert 上下分流廊道(船闸中部)
over-and-under scale 有正负刻度的秤
over-and-under type 升降式
overannealed 过度退火的
overanneal(ing) 过度退火(的)
over-applied expenses 多分配费用
over-applied factory overhead 多摊派间接制造费;多分配的生产费用
over-applied manufacturing over-head 超支制造间接费用
over-applied over-head 超额摊派间接费用
overarch 在上面做成拱形;上架拱圈;架设拱圈;覆以拱
over arm 悬臂;横臂;悬梁;横杆
overarm brace 横梁支架
over arm support 支架;撑杆
overbake 过烧;烘烤过度
overbaking 过度焙烧;焙烧过度
overbalance 超重;超值;超(出)平衡;过平衡;价值超过;失去平衡;失均衡;不平衡
overbalanced super elevation 过超高
overbalance of exports 出超;顺差
overbalance of imports 逆差;贸易入超
overband separator 越带分离器
overbank 洪水河槽;河漫滩;倾斜过度;滩地;大坡度转弯;泛滥
overbank area 中水河槽以上断;漫滩区;河滩区域;泛滥区(域)
overbank contraction 河滩压缩;河滩缩窄
overbank deposit 越岸沉积;漫滩泥沙;漫滩沉积(物)
overbank flood plain 漫滩水流;河漫滩;滩地
overbank flow 溢岸流;漫滩流(量)
overbanking 翻摆
overbank plain 漫滩平原
overbank sediment 漫滩沉积(物)
overbank splay 漫滩冲积扇
overbank storage 漫滩蓄水量;滩地槽蓄量
overbank turbidite 漫滩浊积岩
overbar 上划线
overbased calcium sulfonate 高碱性磺酸钙
over-based detergent 高碱性清洁剂
overbasic 高碱性的
overbate 过度软化
over beam 过梁;悬梁
overbend(ing) 过度弯曲(的)
overbiased amplifier 过偏压放大器
over bid 提高标价;出价过高
overblasting 喷砂过度
overblowing 过吹
overblown steel 过吹钢
overboard 向船外;落水
overboard discharge 向船外排水
overboard discharge line 舷外排出管
overboard discharge valve 舷外排水阀
overboard fall 吊货杆牵索
over-board outreach of dipper dredge(r) 铲斗挖泥船舷外伸距
overboard steam drain 蒸汽排出
overborrowing 借入过多
overbottom pressure 余压

over-bought 超储
over-bought position 超买部位
over brace 横杆支架
overbreak 岩石松动;龟裂;(隧洞的)超挖;超控;超爆;过度断裂;(地层结构上的)塌方
overbreakage 超挖度;超挖范围;过度破碎;过度崩落;挑顶
over-break control 超挖控制;超爆破控制
overbridge 人行天桥;上跨桥;高架桥;跨线桥;旱桥;天桥
overbridge magnetic separator 过桥式磁选机;过桥式磁铁分离器;过桥式磁力分离机
overbrimming 满溢
overbrowsing 过度啃牧
overbudgeting 过细的预算
overbuild 建筑在顶上;建筑在上面;重建;建造过多;建筑过多;超额建设;建屋过多
overbuilding freeway 高架高速干道;越过建筑物的高速公路
over-build section 超额建断面
overbuilt 建造过多的
overbunching 过聚束
overburden 冲积土(壤);残积层;积土层;泥浆中杂质;排土桥;上覆荷载;浮盖层;覆盖层表土
overburden aerial ropeway 露天开挖系统空运钢丝索道
overburden amount 剥离量
overburden bit 表土层(钻进用)钻头
overburden blasting 覆盖层爆破;剥离爆破
overburden conveying bridge 运输桥(露天开挖系统);栈桥(露天开挖系统)
overburden depth 覆盖深度
overburden drag scraper 剥离铲运机
overburden drilling 覆盖层钻进;表土层钻进
overburden earth pressure 上覆土(重)压力
overburden equipment 剥离设备;剥土设备
overburden excavator 剥离型挖掘机;剥土电铲;表土挖土机
overburden funicular railway 空中缆车(露天开挖系统);登山铁路(露天开挖系统)
overburdening 超荷载
overburden layer 表土;剥离层;上覆层;地面积土;覆盖层
overburden load 土荷载;覆盖土荷载;覆盖荷载
overburden locomotive 开挖面机车
overburden map 覆盖层地质图
overburden material 剥离物
overburden mining 剥离工作
overburden on top slab 顶板覆土
overburden power drag scraper (machine) 剥离用动力刮铲机
overburden pressure 超载压力;超载;自重压力;过载压力;积土压力;上覆岩层压力;覆盖土压力;覆盖层自重压力
overburden pressure gradient 上覆压力梯度
overburden ratio 剥离比;剥采比
overburden ratio of deep approximate boundary limits 深部似境界剥采比
overburden ratio of drill hole 钻孔剥采比
overburden ratio of mine 矿山采剥比(黑色冶金)
overburden ratio of open pit 露天矿剥采比
overburden removal 剥除表土;剥离工作
overburden rock 覆盖岩层;覆盖岩
overburden slope 剥离层边坡;覆盖层边坡;剥离层坡度
overburden soil 剥离层土壤;覆盖层土壤;剥离土;覆盖土
overburden stream 超负荷河
overburden stress 超载应力
overburden strip 剥离;剥开
overburden stripping 覆盖层剥离;覆盖层剥除;剥去表土
overburden thickness 覆土厚度;覆盖层厚度
overburden tip and disposal installation 剥离层排土场装置;剥离层排土场设备
overburden transport 废石料运输
overburden wagon 剥离物运送货车
overburn 焙烧过度
overburned brick 过烧砖;过火砖
overburned clinker 过烧熟料
overburning 过烧;烧损;烧毁;烧坏;烧过火;焙烧过度
overburnt 过烧的
overburnt brick 过火砖

overburnt lime 过多燃烧的石灰;过烧石灰;过火石灰
over-buy 超买
over canal bridge 运河桥
overcanopy 用帐篷遮盖
over cap 顶盖
overcapacity 超负荷;超承载能力;过量;能力过剩;后备生产率;生产能力过剩;设备过剩
overcapacity of pump 泵的超负荷运行;泵的超负荷
overcapitalization 资本过多;投资过多;资本过剩;投资过正
overcapitalize 投资过度
over capture 滥捕
over-car antenna 通信车天线;车上天线
over carburization 渗碳过量
overcarburizing 过渗碳;过度渗碳
over-carriage 错装港货
over carrier 余载户;超量运送人
overcarry 多运
overcarrying 结转下期
overcarrying cost 结转下期费用
overcast 支撑架空管道拱形支架;架空支架;架空管道支架;多云;低沉;风桥
overcast day 阴日
overcasting 溢流浇注
over-casting staff 测量杆;测量标杆
over-cast loader 铲斗后卸式装载机
overcast sky 阴天
over-ceiling 超过上限
overcenter clutch 不沿发动机轴向布置的联轴器
overcenter engagement 偏心啮合
overcenter mechanism 偏心自锁机构
overcenter type clutch 杠杆压紧式离合器;偏心自锁常开式离合器
overcentric clutch 不沿发动机轴向布置的联轴器
overcentre type clutch 偏心自锁常开式离合器;杠杆压紧式离合器
overcharge 充电过度;超载;超收;超额装载;超额需要;装载过多;过重;过载;过量充电;过负荷;过充电;加料过多;计费过多;收费过多;上覆荷载;额外收费
overcharge and account 溢开账款
overcharge claim 索回多收运费的要求
overcharge load 过载负荷
overcharge price 索要虚价
overcharge protector 过充电保护器
overcharge sheet 多收费的单据
overcharging 过充电
overcheck 套格子花纹
overchlorinated 过氯化的
overchlorination 过氯化(作用)
overchute 溢流斜槽;跨渠槽;渡槽
overcircuit breaker 过载断路器
overclassification 分等过高
overclayey soil 过黏土(壤)
overcloak 覆盖搭接部分的挡水板;搭盖护墙板;搭接边缘;覆盖挡水板
over coalcutter 上部截煤机;上部截槽截煤机
over coarse-grained soil 巨粒土
overcoat 涂饰;涂层;大衣;外套;涂油漆;多涂一层油漆;保护膜;涂刷
overcoatability 再涂性;面漆配套性
overcoat block 盖砖;挡块砖
overcoating 超敷;罩光涂层;面漆涂层;外敷层;涂油漆;涂饰;饰面;保护涂料;保护涂层;涂刷
overcoating resin 罩面用树脂
overcoil 挑框游丝;摆轮游丝的末圈
overcolo(u)r 着色过浓
overcome friction 克服摩阻力
overcoming grade 可越过的坡度
overcommutation 超前换向;过度整流
overcompacted 过分压密的;过度压实的
overcompacted clay 超压实黏土
overcompacting 过度压实的
overcompaction 过分压密;过度压实;过度压密;过度夯实
overcompensate 过度补偿;补偿过度
overcompensated basin 过补偿盆地
overcompensated meter 过补偿电度表
overcompensated optic(al) fiber 过补偿光纤
overcompensation 过补偿;代偿过度;超量补偿
overcompound 过配合;过复绕;过复励
overcompound dynamo 过复激电机
overcompounded generator 超复激发电机

overcompound excitation 超复励;超复激励;过复绕激励
overcompound generator 超复绕发电机;过复激式发电机;高复激发电机
overcompression 过压缩
over-compression cloud chamber 超压云室
overconditioning 过度调湿处理
overconservative 过分保守的
over-consolidated 过度固结的
over-consolidated clay 超压密黏土;超固结黏土;过度固结黏土
over-consolidated soil 超压密土;超固结土;先期固结土
over-consolidated soil deposit 超固结土层;超固结土沉积物
over-consolidation 超压密;超固结作用;超固结;过度固结
over-consolidation clay 超固结黏土
over-consolidation effect 超固结效应
over-consolidation ratio 超固结比;超压密比
over construction site plan 施工总平面图
over-control 超调现象
overconvergence 过度收敛
over cooking of oil 油热炼过度
overcooled 过冷却的
overcooling 过冷却;过冷(的)
overcooling austenite 过冷奥氏体
overcooling extent 过冷却度
overcooling fluid 过冷液体
overcore 应力解除钻进;套钻;掏心钻进
overcoring 套钻;套取岩芯
over-coring method 套孔法
overcorrect 过校正
overcorrection 再调整;过分校正;过度修正;过度改正
overcorrection coefficient 过度校正系数
overcorrection factor 过度校正系数
over-cost 超出成本
overcount 计数过度
overcoupled circuit 强耦合电路
overcoupled transformer 过耦合变压器
overcoupling 过临界耦合
overcover 遮盖物
overcrack 过度裂化
over-critical 超临界的
over-critical electric(al) field 超临界电场
overcropping 过度种植;耕种过度
overcrossing 跨线桥;上跨交叉;跨越
overcrowded 过度拥挤
overcrowded dwelling 过度拥挤的住宅
overcrowded household 居住拥挤户
over-crowding 居住密度过高;过密;过分拥挤;工业过分集中;太拥挤
over-crowding city 人口过密集城市;人口稠密城市
over-crowding population 稠密人口
overcrowding sleeping 过密就寝(每人三平方米以下)
overcrowned 路拱过大的
overcrushing 过度破碎
overcultivation 过度种植;耕种过度
overcure 过度硬化;过度熟化;过度硫化;固化过度
overcuring 过熟化
overcurrent 过载电流;过流;过量电流;过电流(的)
overcurrent breaker 过载断路器
overcurrent circuit breaker 过电流断路器
overcurrent factor 过电流系数
overcurrent protection 过载电流保护装置;过载保护;过流保护;过电流保护;电流过载保护
overcurrent protective breaker 过电流保护开关;过电流保护闸
overcurrent protective device 过载保护装置【电】;过流保护装置
overcurrent protective unit 过电流保护装置
overcurrent protector 过流保护器
overcurrent relay 过载继电器;过流继电器;过电流继电器
overcurrent relay system 过电流继电保护装置;过电流继电保护系统;过电流继电保护方式
overcurrent release 过电流释放
overcurrent time protection 过流时限保护
overcurrent time relay 过电流时间继电器
overcurrent trip 过电流脱扣器
overcut 超径切削;割断;挑顶;上部截槽
over-cut fiber 超长纤维
over cutter 超挖工具
overcutting 超挖量;超挖;过度切削;过度刻划
overcutting jib 上截盘
overdam 堤坝淹没;堤坝漫顶;坝淹没;坝漫顶;过度壅水
over-dammed 壅水过高
overdamming 过压密实;堤坝淹没;过度壅塞;过度壅水
overdamped 过阻尼的
overdamped condition 过湿度条件;超过阻尼条件;过湿状态
overdamped galvanometer 大阻尼电流计
overdamped galvanometer seismograph 过阻尼电流计地震仪;大阻尼电流计地震仪
overdamped gravity meter 过阻尼型重力仪
overdamped response 过阻尼响应
overdamped system 超阻尼系统;超阻尼体系
over-damp(ing) 超阻尼;过阻尼;过度阻尼
over-damping system 超阻尼系统;超阻尼体系
overdecorated 过分装饰的
over-deep boring rod 超深井钻杆
overdeepening 过量下蚀;过度下蚀;深切
over-delicate 超灵敏的
over-dense plasma 超密等离子体
overdensity of developing well 开采井密度过大
over depreciated 折旧过多
over-depreciation 超提折旧;超额提取折旧
over-depth 附加水深;航行附加吃水;超深
over-depth allowance 允许超深
over-depth of boring 超钻深度
over-depth of dredging 挖泥超深
over-depth of water (航道等的)富余水深
overdesign 超裕度设计;超要求设计;过于安全的设计;保险(系数大的)设计;安全系数设计
overdesigned 过分保险设计的;过分保守设计的;设计裕量过大的;设计余量过大的;超过设计的
over-determined 超定(的);过定的数;过定的
over-determined equation 超定方程
over-determined linear equations 超静定线性方程组;超定的线性方程组
over-determined problem 超定问题
overdeveloped barrier 过分发展的势垒
overdeveloped city 过大城市
overdeveloping 显影像过度
overdevelop(ment) 发展过热;显影过度;超量冲洗;显影像过度;过量抽水;过度显影;过度开发;过度发展;过度发育;过度发达
over-digging 超载多抓
over-dimension cargo 超尺码货物;超尺寸货物
over-dimension container 超规格集装箱
over-dimensioned 超尺寸的
over-dimensioning 尺寸过大;超过规定尺寸;超尺寸的
overdischarge 过量放电;过放电
overdisperse 不平均分布;过度散布
over-dispersion 超扩散
overdistension 膨胀过度
over-dominance 超显性;过显性
over-dominance hypothesis 超显性说;超显性假说
overdoor 门头饰(板);门头花饰;门头镶板;门头饰画;门口上方的装饰墙面
overdosage 过度剂量
overdose 过量用药;过量;过度用药;过度剂量
overdosing 过度加剂量
overdraft (地下水的)过度抽汲;透支;超级吃水;超量取水;超过许可吃水;超采;轧件上弯;过度通风;过度抽取;航行附加吃水;透支;上部通风(装置);富裕水深
overdraft account 透支账户;透支户
overdraft by outport correspondents 外埠同业透支
overdraft for export bills 出口单据押汇透支
overdraft from the national bank 向银行透支
overdraft interest 透支利息
overdraft kiln 过度通风的窑;上部通风的立窑
overdraft of groundwater 地下水过量开采
overdraft on current account secured 往来抵押透支
overdrafts on current account unsecured 往来透支
overdraft system 透支制
overdrainage 过度排水
overdrape 厚窗帘
overdraught 超量取水;过度通风;上压;上部通风(装置);超采;富裕水深
overdraw(ing) 张拉过度;透支;超支
overdrawing account 透支户
overdrawing negotiation 超额押汇
overdredging depth 挖泥富裕深度
overdress 薄外衣;过度装饰
overdried 过干的
over-drilling 超钻;超限钻进
over-drive 过驱动;过激(励);加速移动;激励过度;增速传动;超速转动;超速行驶;超速器;超速传动
over-drive capability 超过激励能力
over-drive capacitor 超速电容器
over-drive clutch 超转离合器;超速离合器
over-drive clutch pawl return spring 超速离合器爪回动弹簧
over-drive factor 过激系数
over-drive gear 高速挡齿轮;增速器;增速齿轮;超速传动齿轮
over-drive housing adapter plate 超速器壳接板
over-drive housing gasket 超速器壳垫密片
over-drive idler gear 超速空转齿轮
over-driven 超速传动的
over-driven amplifier 超载放大器;过激励放大器
over-driven pile 锤击过度的桩;超打桩;过深桩;打过头的桩
over-drive pinion cage 超速小齿轮罩
over-drive protection 过激保护
over-drive shaft 超速传动轴
over-drive shaft bearing 超速传动轴轴承
over-drive shift valve 高速挡换挡阀
over-drive shift valve accumulator 高速挡换挡阀蓄能器
over-drive speed 高速挡速度;超越速度
over-drive stationary gear 超速固定齿轮
over-drive suppression 超速抑制
over-drive system 超速传动系统;增速传动装置
over-driving 超深(桩);超打
over-driving of wood pile 超深木桩;木桩的过夯
over-driving pile 超打桩
overdry 过干;过分干燥
overdue 过期未付;过期的;误期
overdue account 逾期未付账款;逾期账款;过期账款
overdue account receivable 应收逾期账款
overdue arrival 逾期到达
overdue bill 过期汇票
overdue check 过期支票
overdue debt 过期债务
overdue debt bank loan 过期未还银行借款
overdue fine 过期付款;滞纳金
overdue interest 逾期利息
overdue list 误期表
overdue maintenance 过期的维修
overdue payment 逾期付款;过期应付款
overdue penalty for breach of faith 逾期违约金
overdue premium 逾期保费
overdue tax payment 滞纳税款
overdue vessel 误期船只
over-dwelling 居住密度过高;过密拥挤(住房)
over-dye 重染色
overeating 过度饱食
overedger 包缝机
overedging machine 包缝机
over-emptying 过泄(闸门)
over-end unwinding 管顶退解
over-entry certificate 超付进口税证书;退回超额进口税证书
over-equipment 装备过度;设备过剩
overestimate 过高估计;过度估价;估计过高
overestimation 过高估计;过度估价
overestimation of investment in kind 实物投资估价过高
overetching 过腐蚀
overexaggerate 过分夸大
over-excavate 超挖
over-excavation 超挖(量)
overexcitation 过载;过励(磁);过激(励);过度激发;过调制
overexcitation limiter 过励(磁)限制器
overexcitation protection 过励磁保护(装置)
overexcite 过励(磁)
overexcited 过励磁的
overexcited machine 过励(磁)电机
overexertion 用力过度

overexpanded nozzle 过膨胀喷嘴
overexpansion 过膨胀;过度膨胀
over-expenditures 超支
overexploitation 过量开采;过度开发
overexposed hologram 过曝光全息图
over-exposure 感光过度;超量曝光;过曝(露);过度曝光;过度感光;曝光过量;曝光过度
over-exposure portion 曝光过度部分
over-exposure region 曝光过度区;过曝光区
overfall 运河储水处;溢水道;溢流沟;溢流道;溢流;溢出口;扰流;突升海水深度;挡水堰
overfall channel 溢流水道
overfall crest 溢洪堰顶;溢洪道顶(部)
overfall dam 溢流堰;溢水坝;溢流坝
overfall dike 溢流堤;溢流丁坝;过水丁坝
overfall gap 溢流口
overfalling water 溢流水
overfall land 溢流地
overfall pipe 溢流管
overfalls 急流碎浪;急流浪花;逆流碎波
overfall section 溢流段截面
overfall spillway 溢流堰;溢流式溢洪道;溢洪道;坝顶溢洪道
overfall spring 溢流泉
overfall type fish pass 溢流式鱼类通道
overfall valve 溢流阀
overfall weir 溢流堰
overfall with air admission 掺气溢流堰;进气溢汽堰(水流下有空气通路);补气溢流堰
overfatigue 过于劳累;过劳;过度疲劳;疲劳过度
overfault 逆断层;上冲断层
over-feed 超越进给;喂料过多
over-feed burning 上给料式燃烧
over-feed combustion 上给料式燃烧
over-feed-firing 上饲式燃烧
over-feeding 过量进料;过量加料;过量供给
over-feeding technique 超喂进纱技术
over-feed pin stenter 超喂针板拉幅机
over-feed rate 超喂率
over-feed roller 超喂辊
over-feed tenter 超喂拉幅机
overfelt 压制毛毡
overferilization 过度施肥
overfill 超填;超罐;溢出;过满;过量回填;过充满;灌注;满出
over-filling 超罐;过度充盈
overfinishing 过度修整
overfire 过度燃烧;烧毁
overfire air 上部引入的燃烧空气;二次空气;二次风
overfire air fan 火impl风扇;二次风机
overfiren air port 二次风口
overfiring 过度燃烧;过烧;烧损
overfishing 过分捕捞;过度捕捞;捕捞过度
overflocculation 过度絮凝
overflow 存储上溢;充满;超过业务量;超出额;溢流口;溢土(量);溢水口;溢出;涨满;漫流;漫出;设备超载;上溢;泛溢
overflow alarm 超位报警器(水箱水位);溢满信号器;溢流警报
overflow alarm bit 溢出报警位
overflow area 溢出区
overflow baffle 溢流挡板
overflow ball mill 溢流式球磨机
overflow basin 溢流池
overflow bellmouth 钟形溢流口;喇叭形溢流口
overflow bin 溢流仓
overflow bit 溢出位
overflow block 溢出块
overflow box outlet 溢流液出口
overflow bridge 溢水桥;漫水桥
overflow bucket 溢出桶
overflow bunker 溢流仓
overflow buttress 溢流扶壁;溢流扶垛
overflow call 全忙呼叫
overflow call meter 溢流呼次数计
overflow capacity 溢流能力;溢流道排水量
overflow cargo 超溢货物
overflow cascade washers 冲淋装置
overflow chamber 溢流井;溢流闸室;溢流堰闸室;溢流水箱;溢流室
overflow channel 泄洪渠;溢水道;溢流水道;溢流渠(道);溢流河道;溢洪道;泛滥河道
overflow check 溢出检查
overflow check indicator 溢流指示器;溢出检查指示器

overflow chute 溢水沟;回料溜子;粗粉溜子;溢水陡槽
overflow cock 溢流管嘴;溢流阀;溢流栓
overflow cofferdam 过水围堰
overflow components of delay 溢流延误
overflow components of stop rate 溢流停车率
overflow condition 溢出条件
overflow condition test 溢出条件测试
overflow conduit 溢流管
overflow connection 溢流停止;溢出连接
overflow contact 溢出接点;全忙接点
overflow control device 溢流控制装置;溢流控制设备
overflow crest 溢流峰值;最低溢流量;溢流堰顶;溢洪道顶(部)
overflow critical depth 漫流临界水深
overflow crown 溢流拱顶
overflow dam 滚水坝;溢流堰;溢流坝;过水坝;滚水坝
overflow depth 溢流水深;溢流深度
overflow detection system 溢流检测系统
overflow detector 溢出检测器
overflow dike 溢流堤;溢流丁坝;过水丁坝;过水堤
overflow discharge 泄洪量;溢流流量;溢流量
overflow discharge opening 溢流卸料孔
overflow drain 溢出漏;溢漏
overflow earth and rock cofferdam 过水土石围堰
overflow earth dike 溢流土堤
overflow earth-rock fill(ed) dam 过水土石坝;溢流土石坝
overflow edge 溢流堰顶缘;溢流口
overflowed land 潮滩地;淹没地;泛滥地区
overflow electric(al) power 溢流损失电力;溢流等效电力
overflow entry 溢出项;溢出登记项
overflow error 溢出错误;溢出误差
overflow file area 备存文件区
overflow filter 溢流滤池
overflow fixture 溢流器具
overflow flag 溢出标记
overflow flood spillway 洪水溢流道;溢洪道
overflow from surface 地面溢出
overflow gallery pipe 溢流收集管
overflow gate 溢流闸门;溢流式闸门;溢流门;气门
overflow governing 溢流量调节
overflow gravity dam 溢流重力坝
overflow groin 淹顶丁坝;溢流丁坝;淹没丁坝
overflow gutter 溢流排水口
overflow handling 溢出处理
overflow hash 溢出散列
overflow head 溢流水头
overflow height 溢流高度
overflow hole 溢流孔
overflow indication 溢出指示
overflow indicator 溢满指示器;溢出指示器;满溢指示器
overflow(ing) 溢流;漫溢;满溢;过流
overflowing frequency 溢流次数
overflowing nappe 溢流水舌
overflowing rate 溢流比率
overflowing sheet 溢流水层
overflowing sheet of water 溢流形成的片状水流
overflowing spring 溢流泉;溢出泉
overflowing time 溢流时间
overflowing water 溢流水
overflow installation 溢水设备;溢流设备;过流设备
overflow jig 溢流跳汰机
overflow ladle 溢流桶
overflow land 溢流地;漫水地;漫流滩地;河滩地;泛滥区(域);泛滥地
overflow launder 溢流槽
overflow level 溢流水位;漫槽水位;泛滥水位
overflow line 溢流管线;溢流线
overflow lip 溢流嘴
overflow loss 溢流损失
overflow main line 溢流总管
overflow meadow 泛滥草甸
overflow mechanism 溢流装置
overflow meter 溢呼表;全忙计数器
overflow mo(u)ld 溢流式铸模
overflow nappe profile 溢流水舌截面
overflow nipples 溢流短管
overflow of computer 计算机溢出

overflow of record 记录溢出
overflow oil 溢油
overflow operation 溢出操作
overflow orifice 溢流孔
overflow outlet 溢流出口;满溢出口
overflow panel 溢流面板
overflow passage 溢水通路;溢流通路;溢流通路;溢流道
overflow pavement 过水路面
overflow pipe 溢水管;溢流管;排水管(道);调节管;放流管
overflow pipes with adjustable rings 带调整环的溢流管
overflow piping 溢流管
overflow plate 溢流板
overflow pool 联合溢出域
overflow port 溢流口;溢流孔;溢出口
overflow position 溢出位
overflow-prevention 漫溢预防
overflow probability 溢出概率
overflow process 溢流法;溢出过程
overflow pulse 溢流脉冲;溢出脉冲
overflow queue 溢流排队
overflow rate 溢流率;过流率;表面负荷;溢流速度
overflow rating 溢流率
overflow record 溢出记录
overflow register 溢出寄存器
overflow reservoir 溢流水库;溢流式水库
overflow rim 溢流边缘
overflow river 漫槽河流;洪水河;泛滥河流;泛滥河
overflow river channel 泛滥河道
overflow rock-fill dam 堆石溢流坝;堆石滚水坝
overflow rule 超量规则;最少运费确定规则
overflow screw conveyer 溢流绞刀输送机
overflow seating 加座;备用座
overflow section 溢流截面;溢流断面;溢流段
overflow sequential access method 溢出的顺序存取方法
overflow sewer 溢流排污管
overflow silo 溢流仓
overflow siphon 溢流式虹吸管;溢流虹吸管
overflow slab 溢流面板;溢流板
overflow sluice 溢流下料槽
overflow spillway 溢流堰;溢流排洪道;溢洪道;洪水溢洪道;洪水溢流道;顶部溢流道
overflow spillway dam 溢流坝
overflow spillway section 溢流道截面;溢洪道截面
overflow stage 洪水期
overflow stand 溢流挡墙
overflow standpipe 溢流竖管;竖井溢洪道
overflow station 溢流式水电站
overflow status 溢出状态
overflow status set 溢出状态组
overflow strainer 溢水滤筛
overflow stream 溢流河道;漫槽河流
overflow structure 溢水结构;溢水构筑物;溢流建筑物
overflow system 溢流系统
overflow table 溢出表
overflow tank 溢水箱;溢流塔;溢流柜;溢流罐;溢流槽;溢流(高架)水箱
overflow tap 溢流栓;溢流龙头
overflow tower 溢流塔
overflow trap 溢流阱;溢出自陷;溢出中断;溢出捕俘
overflow trough 溢水槽;溢流斜槽;溢流水槽;溢流槽;排水槽
overflow tube 溢水管;溢流管
overflow tubing 溢流管
overflow tunnel 溢流隧洞;溢流隧道
overflow type 溢出类型
overflow type power house 溢流式厂房
overflow type power plant 溢流式水电站;溢流式电站
overflow type spillway 溢流式溢水口;溢流式溢洪道
overflow valley 泛滥河谷
overflow valve 溢流阀
overflow vessel 溢流容器
overflow wall 溢流墙
overflow water 溢流水
overflow water heater 溢流水加热器
overflow weir 溢流堰;滚水坝;明流堰;河岸溢洪道
overflow well 溢水井;溢流井
overflow with chaining 使用链的溢出

overflow with internal chaining 使用内链的溢出
overflume 越渠槽渡;渡槽
overflush 过分洗井
over-flux 超通量;超额通量
overfocus 远焦点;过焦点
overfold 折页过头;倒转褶皱
over-freight 超载货物;超运费;超过租船合同货量的运费;载货过多
over-frequency 超频率;超过频率;超过额定频率
over-frequency protection 超频保护装置
over-frequency relay 超频继电器
overfrothing 浮选过多起泡
overfueling 燃料供给过量
over-fulfil 超额完成;超额生产
over-fulfil a production target 超产
over-fulfil a quota 突破定额
over-fulfilment bonus 超额完成计划奖金
over-fulfil the quota 超过定额
over-full demand 超充分需求
overgain 加深木纹
overgassing 过度析出气体
over-gate capacity 超开度容量(水轮机)
overga(u)ge 超过规定尺寸;飞边;正偏差轧制;正偏差;轮距超出范围;等外尺寸的;等外
overga(u)ge notice 轮轨间距出错通知
overgear 超速档;加速齿轮(系);加速变速装置
overgild 镀金(于)
overglass 玻璃灯罩
overglaze 釉面(的)
overglaze colo(u)r 搪瓷颜料彩釉;釉面颜色;釉面颜料;重釉色彩
overglaze decoration 二次上釉装饰的瓷罩
overglazing 厚釉;上釉过厚
overgound plant 地面电厂
overgrain 加深木纹的再油漆
overgrainer 过度析皂器;过度粒化器;深漆木纹的工具;木纹漆刷;纹理刷;粉饰漆
overgraining 过度粒化;刷水色(上清漆前的工序)
overgrate air 二次风
over-graze 超载放牧
overgrazing 过度放牧
overgrind(ing) 研磨过度;过度粒化;过细粉磨;研磨过细;过碎;过度碎磨;过度碾磨;过度磨碎;过度粉碎
overgrinding brush 深漆木纹刷;过度粒化饰纹刷
overground 过度粉碎的;研磨过度的;过磨的;过度研磨的;过度粉碎;地上
overground cable 架空电缆;架空电线
overground installation 地上装置;地面敷设
overground power station 架空动力站
overground route 陆路
overground section 地面铁路线
overgrow 蔓生
over-grown bed 丛草河床
over-grown farmland 草荒
over-grown lake 丛草湖泊;杂草丛生湖;多草湖
overgrowth 过度生长;漫生;附生;浮生
over-growth cement texture 次生加大胶洁物结构
overhand 从内脚手架砌外墙;支撑的;伸出水面的;手举过肩的;平式缝接;从内墙砌砖
overhand knot 反手结
overhand mining 上向台阶式开采
overhand stoping 上向梯段回采法
overhand wharf tie 反手系扎
overhand work 里脚手砌法;内脚手砌墙法;在内脚手架上砌砖作业
overhang 撑出;悬挂;悬垂物;悬垂;张臂式;美元或存货的过剩部分;前倾边;外伸;突出;调伸长度;挑出屋顶;挑出楼层;伸出;伸臂;电枢端部接线;倒悬
overhang bracket 翅托
overhang crane 悬挂起重机;悬臂起重机;高架起重机;吊装式起重机
overhang door 吊门
overhang echo 悬垂回波
overhang for sun-shading 遮阳挑檐
overhang grinding wheel 外伸砂轮
overhangine-type steam hammer 单柱汽锤
overhanging 前探的;在轴端的;突出的;悬伸的;出挑(建);悬出;凸出伸臂
overhanging action 悬索作用
overhanging angle (type) step 悬挑的扇形踏步
overhanging arm 外伸臂
overhanging arm of bascule bridge 仰开桥桥翼

overhanging bank 悬岸;峭壁;陡岸
overhanging beam 伸出梁;悬索梁;悬臂平板梁;悬臂梁;外伸梁;挑梁;伸出臂梁;伸臂梁
over-hanging bow 飞剪型船首;飞剪式船首
overhanging cliff 悬岩;悬臂式峭壁
overhanging dam 悬臂平板坝
overhanging deck 悬伸甲板
overhanging eaves 挑檐;飞檐
overhanging footway 悬臂式走道;悬臂式人行道
overhanging gable roof 悬山
overhanging girder 悬臂桁架梁
overhanging hammer 单臂锤
overhanging in length 调伸长度
overhanging length 悬挑长度;自由长度;梁的伸长;伸出长度
overhanging pendant switch 外伸悬垂式按钮
overhanging pile driver 悬臂(式)打桩机
overhanging pipe 悬臂管;悬空架管
overhanging press 悬臂冲床
overhanging ripple 偏菱形波痕
overhanging rock 危岩
overhanging roof 挑水屋面;悬挑屋顶;外伸屋顶;挑出屋顶;大屋顶
overhanging scaffold 悬空脚手架
overhanging section of ventiduct 风道悬臂段
overhanging siphon spillway 悬臂式虹吸管溢洪道
overhanging slope 陡坡;倒坡
overhanging span 侧伸跨度
overhanging spring 外伸弹簧
overhanging stair(case) 悬挑式楼梯
over-hanging stern 悬伸船尾
overhanging stor(e)y 挑出楼层
overhanging support 悬锤支座
overhanging truss 悬臂式桁架
overhanging wall 悬墙
overhang lap 主筋悬伸搭接(焊接网)
overhang roll 悬辊
overhang roll type mill 悬臂辊式轧机
overhang sash 悬挑窗扇
overhang sash window 上悬外撑窗
overhang wheel 外伸轮
overhardening 过度硬化;过硬的
overhaul 彻底检修;彻底检查;超运;拆修;拆卸修理;拆开检查;检配;修理;追越他船;冷修;检修;分解修理;大检修
overhaul and repair 检修与修理;检查与修理;大修和修理
overhaul and un-wheeling repair shed 大修架修库
overhaul and upgrading of track 线路大修
overhaul a rope 检查绳索
overhaul a tackle 清理绞辘
overhaul belt conveyer 长距离胶带运输机;长距离胶带输送机
overhaul charges 超运费;超距运费;超程运费;修理费(用);大修理费
overhaul distance 超远距;加运程
overhaul fees 大修费(用)
overhauling 修配;卸修
overhauling check 大修检查
overhauling weight 清理吊货绞辘重铁
overhaul inspection 拆检;大修检查
overhaul instruction 大修指令
overhaul life 大修周期;大修期限;大修理周期
overhaul manual 检修手册
overhaul of track 线路大修
overhaul period 工作寿命
overhaul pipeline 大修工艺过程
overhaul repair funds 大修基金
overhaul schedule 大修计划表
overhaul shop 机修车间
overhaul stand 检修台;大修台
overhaul yardage 超运土方(量);超距土方;加运土方量
Overhauser effect 奥因好塞典效应
overhead 营业开支;仰面;总的;高架的;开销;经常(管理)费(用);架空;企业一般管理费;天花板;塔顶馏出物;塔顶;顶盖;垂向净空;前留分;拔头油;管理费
overhead access 开销接入
overhead account 管理费用账户
overhead air cooler 塔顶空气冷却器
overhead allocation 管理费分配;间接费用分配;间接费分配
overhead analysis 间接生产费分析

overhead and ears in debt 负债累累
overhead arm locking lever 跨臂锁紧手柄
overhead balance (窗扇的)顶部平衡器
overhead beam discplough 高架圆盘犁
overhead belt conveyer 架空式输送带;高架胶带输送机
overhead bin 高架仓;吊挂箱
overhead bit 附加位;辅助位;辅助比特;服务位
overhead bucket 上置式秤斗斗;上置式盛料桶
overhead bucket dredge(r) 高架链斗挖泥船
overhead budget variance 间接费用预算差异
overhead cabin 高架仓;高架小屋
overhead cable 架空线;高架线;飞线;架空电线
overhead cable line 架空电缆线路
overhead cable way 高空索道;架空索道
overhead cam 上凸轮;顶凸轮
overhead camera 吊式复照仪
overhead camshaft 上凸轮轴;悬置式凸轮轴;架空凸轮轴;顶置凸轮轴
overhead carrier 架空输送设备;吊运式输送斗
overhead catenary system 接触网
overhead chain conveyer 架空链索输送机
overhead charges 通常开支;总开销;总管理机构开支;杂项开支;杂项费用;经常费(用);普通用费;管理费
overhead clearance 竖向净空;(跨线桥的)顶上净空;跨线桥净空;净空高(度);净空
overhead clearance for navigation 通航净空
overhead coal bunker 高位煤斗
overhead coil 顶盘管
overhead communication line 架空通信线路
overhead compass 倒置罗经;倒挂罗经
overhead component 间接费用
overhead concealed closer 架空暗闭锁器;顶部暗藏式闭门器
overhead condenser 塔顶产物冷凝器
overhead conductor 架空线;高架电缆;空空接触轨;架空电线;架空导线
overhead construction 架空线路建设
overhead contact 架空接触线
overhead contact line 架空接触线;张力补偿架空线;架空接触网;架空导线
overhead contact line/catenary 接触悬挂【电】
overhead contact line equipment 接触网
overhead contact line mast 架空接触线支柱;架空接触线桅杆
overhead contact line with catenary 链型悬挂
overhead-contact system 加空接触网系统
overhead convection(al) type pipe still 顶部对流式管式炉
overhead convection type 向上对流式
overhead convection type of pipe still 向上对流式管式炉
overhead conveyer 高架输送带;高架运送机;高架运输机;悬挂式输送机;架空输送设备;架空式输送机;高架输送器;架空式输送机
overhead-conveying system 吊挂运送系统
overhead cooler 高架冷却器;冷却水包
overhead copying camera 吊式制版相机
overhead cost 行政管理费用;一般费用;总开支;直接成本(包括材料工时在内);杂项开支;杂项费用;管理费;开销费用;间接生产成本;间接成本
overhead counter shaft 架空副轴
overhead counterweight 衡重式仰开桥
overhead counterweight bascule bridge 衡重式开合桥
overhead crane 行车;高架起重机;桥式起重机;桥式吊(车);天车
overhead crane girder 桥式起重机大梁
overhead crane winch 架空起重绞车
overhead crossbar 跨头横杆
overhead crossing 线岔【电】;高架交(道);立体交叉;跨越;空中交叉;架空交叉;架空敷设;上跨立体交叉;上跨交叉
overhead crossing highway 高架立体交叉公路
overhead damping 人工降雨
overhead device for closing 关门器
overhead distribution 间接成本分配;架空配线;架空配电;上分式;上分配
overhead distribution sheet 间接费分配表
overhead ditch 天沟
overhead door 吊门;上升卷门;升降门;上翻门;翻门
overhead door closer 关门器
overhead door holder 翻门定门器

overhead door of garage 车库翻门
overhead door of the roll-up type 卷门
overhead door of the swing-up type 上旋门
overhead door operator 翻门开关
overhead door track 翻门轨道
over head drilling 上向孔钻进；顶板凿岩
overhead driving gear 天轴传动
overhead drum 蒸馏釜顶冷凝器
overhead easement 高空使用权
overhead eccentric jaw crusher 高架偏心颚式破碎机
overhead eccentric type jaw crusher 架空偏心颚式破碎机
overhead efficiency standard 间接费用效率标准
overhead efficiency variance 间接费用效率差异
overhead electric cable 架空电缆
overhead entrance 架空引入
overhead expenditures 经常性开支
overhead expenses 杂项费用；管理费用；管理费；经常费(用)
overhead explosion-proof crane 防爆型桥式起重机
overhead feed hopper 悬挂给料斗
overhead fillet weld 仰焊角焊缝
overhead fillet welding 仰角焊
overhead filling rack 高位装油栈桥
overhead fire tube boiler 架空式火管锅炉
overhead floor bridge 跨线天桥
overhead frame plow 高架犁
overhead frog 架空(电线)辙叉
overhead gantry 高架龙门架
overhead garage door 上翻式汽车库门
overhead gaurd 车顶炉罩
overhead glazing 高窗
overhead grabbing crane 高架抓斗吊车
overhead gravity pipeline 架空重力管路
overhead ground wire 架空地线；避雷线
overhead ground wire support 架空地线支撑
overhead guard 顶罩
overhead heater 高架加热器
overhead heating system 架空供热系统
overhead highway 高架公路
overhead highway bridge 高架公路桥
overhead hood 高悬罩
overhead illumination 顶部照明
overheading crossing 立体交叉路(口)
overhead ingot turning and weighing machine 架空式钢锭回转秤量机
overhead inversion 高空逆温
overhead irrigating machine 人工降雨设备
overhead irrigation 人工降雨灌溉；喷水灌溉；喷灌
overhead isolation crane 绝缘型桥式起重机
overhead isolator 杆架隔离开关
overhead joint 顶部接缝
overhead lane control signal 顶上车道管理信号
overhead laying 架空敷设
overhead lifting gear 悬挂式起重装置
overhead light 顶灯
overhead lighting 顶部照明；顶部采光
overhead lighting batten 灯吊架
overhead line 架空线(路)；架空明线；架空管道
overhead line crossing 线叉叉夹
overhead line insulator 悬式线路绝缘子
overhead line knuckle 架空线钩лих
overhead line mast 架空线立柱；架空线桅杆
overhead line observation coach 接触网检查车
overhead line pylon 架空电线塔架
overhead lines insulator 悬式棒形绝缘子
overhead live cable 高架带电电缆
overhead loader 上倾式装载机；上顶装渣机；高架装载机；高架卸料机；翻斗式装载机
overhead loader with armo(u)red chain conveyer 装有链式输送机的上倾式装载机
overhead loading 高架装车
overhead loading rack 上装料点
overhead main 架空干管；上分式干管
overhead manipulator 架式机械手；大型万能机械手
overhead mark 顶点
overhead method 内务操作方式
overhead monorail 高架单轨铁路；高架单轨道；单梁起重机
overhead monorail crane 高架单轨起重机
overhead multiplier 摊费乘数(劳务的定价法)
overhead network 架空线路

over-head of material 材料费用
overhead oil loading rack 装油栈桥
overhead outlet 架空引出
overhead pan conveyer 高架盘式输送机
overhead passing 架空走道；上跨通道；上跨交叉
overhead pipe 高架管道；架空管道
overhead pipe grid 架空管网
overhead pipeline 架空管线
overhead pipeline clearance 架空管道净空
overhead piping grid 架空管网
overhead plate support 顶部滤板支撑
overhead pore 架空孔隙
overhead position 仰焊位置
overhead position weld(ing) 仰焊(接)；顶焊
overhead post office line 架空通信线路
overhead power cable 架空电缆
overhead power line 架空电力线(路)
overhead power transmission line 架空电力线(路)
overhead projector 高架投影仪；高架投影机；投影幻灯机
overhead protection for scaffolding 脚手架上面保护装置
overhead rail 吊轨；高架铁道；架空轨道
overhead rail collector 架空轨道集电器
overhead railway 高架铁路；高架铁道
overhead railway bridge 高架铁路桥
overhead rall milker 单轨吊车式挤奶装置
overhead rate 管理费率；间接费分摊率；间接费率
overhead reapportionment 一般管理费用再分派
overhead reservoir 压头水箱；高位水库；高架储[贮]水池；高架式储液器；顶置油箱
overhead revolving slab charging machine 架空式板坯回转装料机
overhead road 高架道路
overhead roadway 高架车道
overhead ropeway 架空索道
overhead saw 悬式锯；吊截锯
overhead scaffold 架空脚手架
overhead scale frame 吊挂式秤架
overhead screed 大混凝土顶部找平层
overhead service crane 检修用桥式起重机
overhead service entry 高架供电引入线
overhead shovel 翻斗式装载机；过顶铲；后卸式装载机
overhead shower 上翻式淋浴器
overhead sign 高架标志；架空式标志；门式标志
overhead slewing crane 圆轨旋转式桥吊
overhead sonar 回音测冰仪；冰下声呐
overhead spending variance 间接费耗用差异
overhead spray 顶棚喷水器
overhead sprinkler 顶喷式喷灌机
overhead sprinkling irrigation system 架空喷灌系统
overhead sprinkling system 高架喷灌系统
overhead stacking crane 堆垛用桥式起重机
overhead standard 间接费用标准
overhead static cable 架空避雷线
overhead stopping 向上顶蚀作用；向上回采矿石法
overhead storage 高架水箱；高水塔
overhead storage water tank 高架蓄水箱
overhead structure 上空结构；顶部结构；高架结构；架空结构；顶部构造；上部结构
overhead support 高架式支座
overhead suspension 悬吊；悬空支架；架空悬置
overhead system 高架系统；架空线路系统；架空电网
overhead system of heating 高架供暖系统
overhead take-off 机顶篷布；架空线出线(端)
overhead tank 压力罐；压力槽；高位槽；高架水箱；高架水塔；人工降雨罐
overhead tax 一般性的税
overhead telephone line 架空电话线
overhead time 整理操作时间；开销时间；多余时间；辅助操作时间；额外时间
over-head tnuh scale 车秤
overhead tower 架空电线塔架
overhead track line 架空轨道
overhead track way 架空轨道
overhead traffic 开销通信量
overhead tramway 架空电车道
over-head transformer 高架变压器
overhead transmission gear 架空传动装置

overhead transmission line 高架输电线路；架空输电线路
overhead transmission line cable 架空传输电缆
overhead transportation 高架运输；架空运输
overhead transportation structure 高架运输结构
overhead transporter 单塔式架空缆车
overhead travel(l)er 桥式起重机
overhead travel(l)ing crane 桥式吊车；移动式高架起重机；高架移动式起重机；架空移动式起重机；桥式起重机；天车
overhead travel(l)ing crane with hoist 电动葫芦桥式起重机
overhead travel(l)ing drilling machine 悬动钻床
overhead travel(l)ing grab crane 高桥移动式抓斗起重机；桥式抓斗机(车)
overhead trolley 高空吊运车；高架小车；架空吊车；架空电车；单轨起重机；单轨吊车
overhead trolley conveyer 高空吊运输送机；悬挂式运输机；悬挂式输送机；吊链输送机；吊篮运输机
overhead trolley line 架空电车线
overhead trolley system 架空电车系统
overhead truck scale 架空车秤
overhead-type garage door 车库翻门；上跨式车库门；卷帘门；吊门
overhead-underground network 架空—埋地混合网络
overhead value analysis 间接费用价值分析
overhead valve 顶置气门；上置阀门；顶置式气门；顶阀
overhead valve engine 顶置气门发动机
overhead variance 间接制造成本差异；间接生产费差异
overhead view 顶视图；俯视图；俯视
overhead waiting hall 高架候车厅
overhead walkway 行人天桥
overhead warning line 拦河警告线
overhead water-storage tank 高位蓄水池
overhead water tank 高架水箱
overhead weld(ing) 仰焊(接)
overhead wire 架空索；架空电线；架空线
overhead wire carrier equipment 明线载波机
overhead wire locomotive 桥式电线机车
overhead wiring 架空布线
overhead work 架空作业
overhead working truck 高空作业车
overheap 堆积过多
overheat 使过热；过热
overheat alarm 过热警报器
overheat allowance 加热限度；(涂膜的)加热容许性
overheat awitch 过热开关
overheat blowing 过热送风
overheat cutoff device 过热切断装置
overheated 过热的
overheated economy 过热的经济
overheated steel 过热钢
overheated structure 过热组织；(焊接部分的)过热结构
overheated zone 过热区
overheater 过热器
overheat for economic development 经济发展的过热症
overheating 过分加热；过热
overheating fault 过热故障
overheating layer 过热层
overheating protection 过热保护
overheat protection 过热保护
overheat protective relay 过热保护继电器
overheat protector 过热保护器
overheat spot 过热点
overheat steam 过热蒸
overheavy firing 强化燃烧
over-height cargo 超高货
over-height detector 超载探测仪；超高检测器
over-height mark 超高标志
over-height of groundwater level 地下水位过高
over-height vehicle detection 超高车辆检测
over-height vehicle detection system 超高汽车检测设施
over hoist alarm device 起升高度超越报警器
over hoisting 过高吊装；过高升起；过高举起
overhours 加班时间
overhouse aerial 屋顶天线
overhoused 房子太大的
overhung 悬垂的；悬臂式的；外伸的

overhung crank 外伸曲柄
overhung door 悬挂式门；悬垂门；吊门
overhung governor 起重调节器
overhung height 悬吊高度
overhung latrine 外悬式厕所
overhung load 悬臂荷载
overhung pinion 外伸小齿轮
overhung turbine 悬臂式透平
overhung-type centrifugal compressor 外悬式离心式压缩机
overhung-type motor 悬装式电动机
overhung wall winch 挂墙绞车
overhydration 水合过度
overhydrocracking 深度加氢裂化
overich mixture 过富拌和物
overimprovement 过度开放
over-inflated condition 充气状态
over-insulation 超绝缘
over-insurance 保额过高的保险；超值保险；超额保险
overintensive agriculture 过度密集农业
overinterest 过期利息
over-interrogation gate 查询限制器
overinvestment 投资过度；过度投资；投资过剩
over-invoicing 高报货价
overirradiation 过度辐照；超量灌溉
overirrigation 过量灌溉；过冷灌溉；灌水过多；滥灌
overissue of bonus 滥发奖金
overite 水磷铝钙石
overjacking 补充张拉(钢筋)
overlacquer 罩光喷漆
overlade 过负荷；装载过多；过载；超载
over-laden 超负荷的；超载的
overlaid 贴面
overlaid plywood 贴面胶合板
overlaid seam 搭缝
overlaid veneer 贴面薄板
overland 超卸(货物)；陆上的；陆路的；多卸；地面的；地表；飞越陆地
overland belt 地面长距离皮带运输机
overland bill of lading 水陆联运提单
overland cable 陆地电缆
overland canal 陆上运河
overland carrier 陆运工具
overland common point 陆运共通地点；内陆共同点；内陆公共点；冰陆联运转换点
overland common point rate 陆运共通地点优惠运费率
overlanded 短缺及损坏报告
overlanded and short-landed report 短溢报告
overlanded cargo 溢卸货
overland flow 地表流；地面泛流；漫滩流(量)；漫流；漫地水流；坡面漫流；地面漫流；地面径流；地表漫流；地表径流(量)；表面径流(量)
overland flow hydrograph 漫流过程线；地面漫流过程线；地表径流过程线
overland flow-land treatment system 地表漫流土地处理系统
overland flow method 漫流法
overland flow system 漫流系统；地表漫流系统
overland-flow treatment 地表径流处理系统
overland freight 陆运货运；陆路运费
overland freight haulage (港口货物的)陆上疏运
overland propagation 地面传播
overland route 陆运路线；陆路；旱路
overland runoff 地面径流；地表漫流；地表径流(量)
overland runoff coefficient 地面径流系数；地表径流系数
overland section 筑堤蓄水段
overland service 陆路运费
overland trade 陆路贸易
overland transit 陆上运输线；陆上转运；陆上集散
overland transit empty 外地集装箱回空
overland transit full 外地装箱回送
overland transportation 陆运；陆地运输；内陆运输
overlap 折叠；交叠；交错；互搭；焊瘤；漆膜的搭接复盖；铺层；喷刷路搭接；叠覆；搭接部分；搭接；重叠选择；重叠摄影；重叠操作；重叠；超期还船；超覆；覆蔽；飞边
overlap access 重叠访问
overlap action 重叠作用；重叠动作
overlap adjustknob 搭接调整旋钮
overlap and wastage of bars 钢筋的搭接及损耗

overlap angle 重合角
overlap blade 重叠叶片
overlap capacitance 叠加电容
overlap changeover 交叉转换
overlap channel 重叠通道
overlap coefficient 重叠系数；覆盖系数
overlap control 重叠控制
overlap control button 重叠控制按钮
overlap cut 重叠切削
overlap days 超出天数
overlap edge 搭接边
overlap factor 重叠因子
overlap fault 逆掩断层；冲断层；超覆断层
overlap fetch architecture 重叠取指结构
overlap fetching 重叠取指
overlap fold 超覆褶皱
overlap image 重叠像
overlap indicator 重叠指示器
overlap integral 重叠积分
overlap joint 搭接；搭接接头；重叠接合；重叠层错
overlap kernel 重叠核
overlap length 搭接长度；搭接长度；重合弧长度
overlap loop 重叠循环
overlap method 重叠法
overlap mode 重叠方式；并联方式
overlap movement 重叠运行
overlap of spectral lines 谱线重叠
overlap operand 重叠操作数
overlap orthogonal spread 重叠正交排列
overlap pairs 重叠像对
overlap panel 重叠面板
overlapped 重叠的
overlapped access 重叠访问
overlapped account 重复账户
overlapped channel 重叠通道
overlapped data channel 重叠数据通道
overlapped execution 重叠执行
overlapped joint weld 搭头焊缝
overlapped joint welding 搭头焊缝
overlapped line 套轨线路
overlapped memories 重叠存储器
overlapped memory access 重叠存储器存取
overlapped operation 重叠操作
overlapped operation(al) buffer 重叠操作缓冲器
overlapped siding 搭叠鱼鳞板
overlapped span of control 重叠用户定义域控制；重叠的控制范围
overlapped valve 正遮盖阀；正余面阀；正叠量阀
overlapping 覆盖；交叉；交错；重叠(操作)；复盖
overlapping aerial inside 重叠向内
overlapping aerial outside 重叠向外
overlapping angle 重叠角
overlapping arcs 重叠弧
overlapping area 重叠面积；重叠区；覆盖重叠区
overlapping astragal 门扇门挡条；遮盖半圆饰；门扇盖缝条；互搭圈线
overlapping attribute 交叠属性
overlapping band 交叠能带
overlapping boards 搭接鱼鳞板；鱼鳞板；搭接板
overlapping booster 侧置助推器
overlapping business operations 业务交叉
overlapping code 重叠密码
overlapping colo(u)rs 重叠色彩
overlapping community 重叠群落
overlapping construction schedule 交叉施工进度表
overlapping contrast 重叠对比；重叠对
overlapping corner-beam 搭接角梁
overlapping covers 卷边封面
overlapping curve 磨削交叉花纹；网纹
overlapping data grouping 重叠数据组
overlapping debt 重复债务
overlapping degree 重叠度
overlapping device 重叠装置
overlapping distortion 重叠畸形
overlapping drawing 重叠图
overlapping fold 搭接褶皱
overlapping gate 重叠栅
overlapping generation 重叠世代
overlapping grid 重叠坐标格
overlapping grid template 坐标格网模片
overlapping hologram record 重叠全息图记录
overlapping image 重叠图像
overlapping interface 重叠面

overlapping inversion 重叠倒位
overlapping investment 重复投资
overlapping joint 搭接连接；搭接缝
overlapping leaves 叠生叶
overlapping length 搭接长度
overlapping levels 重叠能级
overlapping line 重叠线段；套线【铁】
overlapping magnetic anomaly 叠加磁异常
overlapping mean 相继平均值
overlapping multiprocessor 重叠式多重处理机
overlapping of grating spectra 光栅光谱重叠
overlapping of reinforcement bars 钢筋搭接
overlapping of sounds 声音复叠；声音叠复
overlapping of strand 重叠纹
overlapping orbitals 重叠轨道
overlapping or interlacing of lines 弓背连接线
overlapping part 搭接部分
overlapping peak 重叠峰
overlapping phase 重叠相位；交叠相位
overlapping photographs 重叠像片
overlapping pictures 重叠画面
overlapping pile 叠置桩
overlapping placement 错缝(混凝土)浇筑块；错缝浇筑
overlapping plank sheeting 搭接厚板壁；搭接木望板
overlapping pluvial fan 上叠式洪积扇
overlapping pulse 重叠脉冲
overlapping region 重叠区
overlapping regulator 重叠调整器
overlapping rod 重叠控制棒组
overlapping roller conveyer 双层滚轴输送机
overlapping run 搭头焊道
overlapping seam 搭接接缝
overlapping segment 重叠段
overlapping sensitivity 重叠灵敏度
overlapping set 交叠集
overlapping spectrum 重叠光谱
overlapping splitting 超覆式分岔
overlapping strake 搭接外列板
overlapping strip 重叠摄影带
overlapping surge 重叠涌浪
overlapping suture 重叠缝合
overlapping symbol 复合符号
overlapping thickness of effective circles 浆孔有效圈交叠厚度
overlapping tile 搭接瓦
overlapping time 重叠时间
overlapping toes 重叠趾
overlapping tooth action 重合轮齿啮合
overlapping virtual image 重叠虚像
overlapping zone of sweep(ning) trains 扫海趟重叠带
overlap processing 重叠处理
overlap protection block section 保护区段
overlap ratio 重合度；重叠比
overlap regulator 重叠调整器
overlap section 锚段关节；重叠区段
overlap segment 重叠程序段
overlap signalling 交叠信号方式
overlap span 锚段关节；分段间隙
overlap spread 重叠排列
overlap test 重叠试验
overlap track circuit 叠加轨道电路
overlap unconformity 超覆不整合
overlap-welded steel pipe 搭头焊接钢管
overlap welding 搭接焊；互搭焊接
overlap zone 重叠带
over-large 尺寸过大
over-lashing 超强系梁【船】
overlaxation factor 过弛豫因子
overlay 罩面；重复占位；加重物；熔敷层；涂层；涂；贴面；套图塑料膜；叠层；道路覆盖层；覆盖图；覆盖(板)；复面；反复使用；表面修整用席
overlayable segment 可覆盖段
overlay approach of environmental impact assessment (环境影响评价的)图形叠置法
overlay area 覆盖区(域)；重叠区
overlay contact 重叠接点
overlay controller 重叠控制器；重叠控制程序
overlay control program(me) 覆盖控制程序
overlay declaration 覆盖说明
overlay defining 重叠定义
overlay directory 覆盖目录

overlay drafting 叠加画面;重叠绘图法
overlayed plywood 贴面胶合板
overlay face 复面
overlay file 覆盖文件
overlay flooring 镶木地板;硬木拼花地板;终饰地板
overlay glass 层叠玻璃;套料玻璃
overlaying 套色(玻璃);套料(玻璃);重复占位;堆焊;镀
overlaying deposit 表沉积层
overlaying of information 信息覆盖
overlaying slope 仰坡
overlaying surfaces 重叠面
overlaying welding 堆焊;补焊
overlay life 罩面使用年限
overlay mat 表面薄毡
overlay metallization 敷镀金属
overlay method 叠置片法;叠加法
overlay module 覆盖模块
overlay of pavement 罩面【道】
overlay paper 养护覆盖纸(混凝土);覆盖纸(混凝土);白垩纸
overlay path 覆盖路径;重叠通路;重叠路径
overlay percentage 覆盖百分率
overlay program(me) 覆盖程序;重叠程序
overlay region 重叠区
overlays 重复占位段;重复占位程序段
overlay segment 重叠程序段;程序重叠段
overlay segment area 覆盖段区
overlay sheet 覆面片材;贴面片材;贴面板;表面毡
overlay strategy 覆盖策略
overlay structure 重叠结构
overlay supervisor 重叠管理程序;覆盖管理程序
overlay supervisor routine 覆盖管理程序的例行程序
overlay technique 叠铺法
overlay transistor 层叠型晶体管
overlay tree 重叠树形图;重叠树;树形重叠;覆盖树(形图)
overlay tree structure 覆盖树形结构
over-ledge flow 滑梁水
over-ledge-flow-induced rapids 滑梁水险滩
over-lending 超贷
over-length 总长(度);过长;后备长度;超长
over-length fiber 超长纤维
over length of trailer 拖车总长度
overlevel 过电平
over level tripping 越级跳闸
over-lift 超采原油
over-limed 加灰过量的
over-limed cement 石灰过多水泥;石灰含量过多的水泥;多石灰水泥
over-limit rate 超标率
overline 重行;重叠行;跨线(的);跨路的;加班时间;眉题;上划线
overline bridge 跨线桥;天桥
overline bridge for ice transportation 输冰桥
overline bridge for passenger 旅客天桥
over-line rate 超标率
overload 超重;超负载;超负荷;装载超重;过负荷;过度负荷;超载;过载;负载超过
overload alarm 超负荷警报装置;过载警报装置
overload alarm control 过载警报控制
overload alarm mechanism 超载信号报警器
overload and damage list 溢短残损单
overload capability 超负荷能力;过载能力
overload capacity 过载额定量;超载容量;超载能力;超负荷容量;超负荷能力;过载容量;过载能力
overload characteristic 超载特性
overload circuit 过载电路
overload circuit breaker 过载断路器;超载断路器
overload clutch 过载离合器;安全离合器
overload condition 过载状态
overload coupling 超载安全离合器;安全联轴器;安全联轴节
overload crack 超载裂缝;过载裂缝
overload current 过载电流
overload cutout 超负荷断路器;过载保护;过载开关
overload detector 过载检测器
overload device 过载装置
overloaded 装载过多
overloaded cargo 超载货物
overloaded river 多沙河流;多泥沙河流
overloaded stream 多沙河流;多泥沙河流
overload efficiency 过载效率

overloader (装载斗作业时超过车顶的)斗式装载机;高吊回转式装载机;翻转式装载机;翻斗式装载机
overload factor 过载系数
overload firing 超负荷燃烧;过负荷燃烧
overload forward current 过载正向电流
overload friction clutch 超载摩擦离合器;防超载摩擦离合器
overload indicator 超载指示器;超负荷指示器;过载指示器
overloading 超额装载;过量装药;超载;负载;负载超过
overloading blasting 过量装药爆破
overloading characteristic 过载特性
overloading operation 超负荷运转
overloading penalty for breach of faith 超载违约金
overload level 超载级;超载级;过载电平
overload light 超荷标志灯;过载信号灯
overload limit switch 超载限制开关
overload margin 过载定额
overload meter 超载测量计
overload number of semi-cushioned seats on passenger train 硬座超成定员
overload of cutter motor 绞刀马达超载
overload of swing winch 横移绞车超负荷
overload operation 超载运行;超负荷运转;超负荷运行
overload output 超负荷输出
overload point 超载点;过载点
overload protection 超载防护;超负荷保护;过载保护;过荷保护;过负载保护;防止超载
overload protection device 超负荷安全装置;过载保护装置
overload protection load 超负荷保护装置
overload protective device 防过载装置
overload protective relay 过载保护继电器
overload protective system 过载保护系统
overload protective unit 过载保护装置
overload protector 超负荷保险装置;超负荷保护装置;超负荷安全装置;过载保护装置
overload provision 超载规定
overload rate 超员率
overload rating 额定载
overload recovery time 过载恢复时间
overload relay 超载继电器;超负荷继电器;过载继电器
overload release 过载释放;过载继电器;过载断路器;安全器
overload release hitch 超载松脱联结器
overload relief valve 超载泄放阀;超载安全阀;超负荷安全阀;过载溢流阀;过载减压阀;过载安全阀
overload relieve valve 过载溢流阀;过载减压阀;过载安全阀
overload repeater relay 过载中间继电器
overload river 超载河
overload running test 过载试验
overload safety device 过载安全装置
overload shearing clutch 剪销式安全离合器
overload simulator 过载模拟程序
overload spring 过载弹簧
overload stop 过载限制器
overload strength test 超载强度试验
overload switch 过载开关;过载断路器
overload test 超载试验;超负荷试验;过载试验;荷重试验;破坏性试验
overload time relay 过载限时继电器
overload trial 过载试验
overload trip 过载跳闸
overload trip device 超载脱开装置;过载跳闸装置
overload value 超负荷值
overload valve 超载阀;过载阀
overload warning device 超负荷警报装置;过载警报装置
overload wear 过载磨损
over-loan 超额贷款
over loan situation 借款过多状况
overlock machine 锁缝机;包缝机
over logging 计量偏多
over-long 过长
overlook 瞭望;监视;监督;俯视
overlook and oil sand 无视油砂
overlubricate 过量润滑
overluminous star 光度过大星

overlying 上覆(的)
overlying atmosphere 上覆大气
overlying base course 上覆垫层
overlying bed 上覆地层;上覆层;覆盖层
overlying deposit 表沉积层;表层沉积物;上覆沉积物
overlying earth 地表土
overlying formation 上覆地层
overlying ground 地面覆盖
overlying impermeable strata 上覆非渗透地层
overlying impervious bed 上覆不可透水层
overlying layer 覆盖层
overlying measures 超覆岩层
overlying mortgage 次级抵押
overlying of ore body 矿体上界面
overlying pervious bed 上覆可透水层
overlying roadway 上层路
overlying rock 上覆岩层;覆盖岩层;覆盖岩
overlying seam 覆盖接缝;覆盖矿层;上覆层
overlying sediment 上覆沉积物
overlying soil 上覆土层;上覆土
overlying stratum 上覆(地)层;覆盖(土)层;盖层;覆层
overlying wall rock 上部围岩
overlying water 上覆水
over maintenance 维修过剩
overman 工头;工长;总管(指管理人员);监工;监督者
overman bit 管状钻头
overmantel 壁炉墙额饰;壁炉架额饰
overmany 过多的
overmastication 过度研磨
overmatuer source rock 过成熟源岩
overmature dry gas 过成熟干气
overmature stage 过成熟阶段
overmaturity 过熟
over-maximal 超最大值的
overmeasure 裕量;余量;高估;留量
overmilling 过度研磨
overmix 过度混合;拌和过度;过度拌和
overmixing 过度搅拌;搅拌过度的;拌和过度的
overmoded pipe 多模光导管
overmodulation 过调制
overmodulation capacity 过调制容量
overmodulation indicator 过调制指示器;过调指示器
over-mortared concrete mix 灰浆过量的混凝土混合料
overmull 过混(混砂时间过多)
overneutralizing 过补偿
overnight at call 日拆资金
overnight bag 短途旅行提包
overnight case 短途旅行提包
overnight concrete 快硬混凝土
overnight finishing 隔夜抹光
overnight loan 即刻贷款
overnight millionaire 暴发户
overnight parking 整夜停车;通宵停车
overnight pond 小水池(灌溉)
overnight service 通宵检修;通宵服务
over-norm 超定额
overnourished profile 供砂过量的剖面
overnutrition 过度营养
over on bill 附加运费
overornamented 过度装饰(的);过分装饰的
over-over communication 交替通信
over-over mode 交替方式
overoxidized heat 过氧化熔炼
over-packaging 超强包装
over-packing 超紧密堆积作用
over-paid 超额支付
overpaint 油漆面层;多刷油漆;全面涂漆于
overpass 溢流挡板;立体交叉桥;越过;超路;立交桥;跨线桥;旱桥;上跨路;上跨立体交叉
overpass bridge 立体交叉桥;跨线桥;上跨桥
overpass bridge under the hump 峰下跨线桥
overpass for pedestrians 人行立交;人行天桥;步行天桥
overpass grade separation 上跨铁路立体交叉
overpass ramp 上跨交叉的斜坡道;立体交叉的斜坡道;上跨交叉匝道
overpay 多给;多付;多缴款
overpayment 超付;超额偿付;逾期付款;多付(给)
overpeaked compensating amplifier 过峰补偿放

大器
over-period allocation 跨期摊配
overpickling (板材等的)过酸洗;过度酸洗;酸洗过度
over-pigmentation 颜料过度淀积(作用);颜料过度淀积
over-pin 测量用滚柱
over-pin gear measurement 齿轮跨针测量
overpoling 还原过度
over-populated zone 人口过剩区
overpopulation 个体数过剩;种群过密;过布居;人口过剩;人口过多
over-porous paint film 多孔漆膜
overpotential 超电压;超电位;超电势;过电压
overpotential resistance 超电压电阻
overpotential test 过压试验
overpour gate 上溢式闸门
overpour head gate 上溢引水闸门
overpouring water 溢流水;上溢水
overpower 超功率;过载;过功率;过负荷;配用功率过大(的驱动装置)
overpower protection 过载保护
overpower relay 过功率继电器;过负荷继电器
overpower scram 超功率时快速停堆
overpower valve 超力阀
overprediction 预报过高
overpressure 超压(力);压力增长;压力上升;过压;过剩压力;剩余压力
overpressure cooling 过压冷却
overpressured hydrothermal system 超压水热系统
overpressured zone 超高压带
overpressure limit for vessel used 容器的超压极限
overpressure method 过压试验法
overpressure protection 超压保护
overpressure pumping 憋泵
overpressure release valve 超压安全阀;超压放松阀
overpressure resistant 耐压的
overpressure test(ing) 超压试验
overpressure turbine 超压水轮机;反击式涡轮机;反击式透平;反击式水轮机;反击式汽轮机
overpressure valve 过压阀
overpressurization 产生剩余压力;过量增压
over-prestressed 超预应力的;预应力超限的;预应力过张的
over-prestressed beam 超预应力梁
overpriced 价格过高的
overprint 烙印;加印;描印;添印;套印;套加印;叠印;叠加的
overprint information 套印资料
overprinting 重印;印在空白区的标记;叠加
overprinting coating 印铁罩光涂层
overprinting data 套印资料
overprinting fold 叠加褶皱
overprinting lacquer 印铁罩光漆
overprinting varnish 印铁清漆;纸张罩光清漆;罩光油
over-privileged 超水准(指经济收入)
overproduce 超额生产;生产过剩
overproduction 生产过剩
overproduction of commodities 商品过剩
overproduction theory 生产过剩理论
over-program(me) 重复编程
over-proof 超过额定的;超差;超标准(的)
over-proofed sample 超差样品
overprotection 过度防护
overpulverization 过度粉碎
overpumped well 过度抽水井
overpumping 超量取水;过量抽水;过度抽水;上射式水泵
overpumping of well 井的过量抽水
overpunch 上部打孔;三行区打孔;附加穿孔;额外打孔
overpunched numeric 附加穿孔的数字
overpunched numeric string 附加穿孔的数字串
overpunching 上部穿孔;附加穿孔
over-purlin(e) insulation 檩上保温层
over-purlin(e) lining 冷淬火;过度淬火
overquenching 冷淬火;过度淬火
over-quota 超定额
over-quota loan 超定额借款
overradiation 辐射过度
over-rake 锚泊中浪从船首打上
overramming 过度夯实;过度捣实
over-range 超出额定范围;超量程;超出正常界限;超出额定界限;超出额定范围
overrange protection 过量程保护
over-ranging 超出额定界限;超出额定范围
overrate 超过定额值;超定额;逾限率;过高评价;过高定额;估价过高;估量过大
overrating 定额过高
over-rating voltage 超出额定值的电压
overreach 延长动作;超越;延长动作时间
over-reclaim and cultivate wasteland 过渡垦植
over reclamation on steppe 滥垦草原
over-recovery 超量恢复
overreduced steel 过还原钢
overreduction 过度还原;还原过度
overrefine 过度精制
overrefined 过度细致的
over-registration 重复登记
overregulate 过调节
over-reinforced 钢筋过多的;配筋过多的;超配钢筋的;超量配筋的;超筋
over-reinforced beam 超筋混凝土梁;超筋梁
over-reinforced concrete 配筋过多的混凝土;超配筋(的)混凝土;超量配筋混凝土
over-reinforced design 超筋设计
over-reinforced section 超配筋断面
over-relaxation 超松弛(法);过于松弛;过度松弛
over-relaxation factor 超松弛因数
overresponse 反应过度
over-restoration 超量恢复
overrich 过浓的;过富的
overrich mixture 超富灰拌和料;过富混合物;多水泥拌和物
over-rich oiled surface 多油路面(沥青过多);沥青过多的路面
over ridden ticket 过站车票
override 超控;超驰;超乘
override billing 超驰清单
override control 超驰控制;过调节控制
override facility 人控功能
override interrupt 最优先中断【计】
over ride key 对消钮
override limit 重叠限额
override pressure 调压差值
overrider 转分手续费;保险杆(汽车)
override selector valve 超驰选择阀
override switch 超越控制开关
override system 备用系统
overriding 掩覆;上驮;叠置
overriding brake 超控刹车;超控制动
overriding commission 追加佣金;代办佣金
overriding control 超越控制;超驰控制
overriding credit 背对背信用
overriding factor 首要因素
overriding of an emergency brake application 使紧急制动无效
overriding plate 上驮板块;上冲板块
overriding process control 超驰过程控制
overriding royalty 追加租金(按营业额追加的租金)
overriding terrestrial control 地面超越控制
overriding-(type) clutch 单向离合器
overrigged 过多的装备
overrigid 过刚的;多余杆;具有多余杆件的(结构);太死板的
overripe 过熟;太熟
overripe snow 过软雪
overripe wood 过老木材;过熟材
overrised vessel 高干舷船
overriver 过河的;跨河的
overriver level(l)ing 越河水准测量;跨河水准测量;过河水准测量
overriver structure 跨河建筑物
overroad 跨路的
overroad stay 跨路拉线
overroasting 过烧;焙烧过度
overroll 过度碾压
overrolling 过度碾压;滚动;碾压过度的;翻转
overroof 顶盖;覆以屋顶
overrun 超越限度;超越;超限运转;超限运动;超限;超过正常范围;超过范围;超过额定界限;超额材积;超出;超程;场端安全区;溢流;过度运行;漫出
overrun a signal 冒进信号
overrun brake 超越刹车阀;超限制动器;超速控制器;过载控制阀
overrun clutch 超越离合器
overrun controlled brake 刹车自控闸
overrun coupler 超越联轴器
overrun coupling 超越联轴节;超速联轴节
overrun critical depth 漫溢临界水深
overrun error 超越误差
overruning brake 超速自行控制器
overrun lamp 照相散光灯
overrunning 超曳现象
overrunning clutch 超限运转离合器;超速离合器
overrunning coupler 超越离合器
overrunning coutch 超越离合器
overrunning governor 限速器
overrunning of signal 冒进信号
overrunning third rail 顶接触网第三轨
overrunning torque 超速运转扭矩
overrunning weed 杂草丛生
overrun on cost 费用超支
overrun safety arrestment 防止冲出跑道的安全拦阻装置
overs 筛渣;多印;超印;放放纸
oversail 伸出;突出
oversailing 连续突腰线;连续突腰层(墙上突出之砖层或石层)
oversailing brick course 挑砖层
oversailing bricks 挑砖
oversailing conoid 凸出的锥曲面
oversailing course 束带层;腰线;连续突腰线;连续突腰层(墙上突出之砖层或石层);突出砖层;挑砖层;挑出的砖(石)层
oversailing floor 挑出地板
over-sampling 采样过密
oversanded 多砂的;含砂过多的
oversanded cement 含砂过多的水泥;多砂水泥
oversanded mix(ture) 多砂混合物;多砂混合料;多砂拌和料;多砂拌和物;多沙拌和物
oversanding concrete 多砂混凝土
oversandy soil 过砂土壤
oversaturated 过饱和的;硅石的
oversaturated cycle 超饱和周期
oversaturated intersection 超饱和交叉口
oversaturated rock 过饱和岩
oversaturated steam 过饱和蒸汽
oversaturated vapo(u)r 过饱和蒸气
oversaturation 过饱和现象;过饱和
oversaturation soil 超饱和土
overscan 过扫
overscanning 过扫描
oversea 国外;海外;往海外的
oversea bridge 海上架桥
oversea broadcast 海外广播
oversea Chinese city 华侨城
oversea communication 海外通信
oversea container limited information sheet 海外集装箱公司须知
Oversea Container Line 海外集装箱运输公司
Oversea Container Ltd. 海外集装箱运输公司
oversea contract engineering risks insurance 海外承包工程险
oversea cooperation program(me) 海外合作项目
oversea investment 海外投资
over seal 顶封
oversea line 国际航线;国际航空公司
oversea market 国外市场;国际市场;海外市场
oversea project 海外工程项目
oversea purchase 国外订购
overseas allowance 海外津贴
overseas and domestic telecommunication service 国内外电信业务
overseas branch 海外办事处;国外分行
overseas Chinese venture 侨资企业
oversea's container 海上运输集装箱
overseas debt 外债
overseas discharge and replacement depot 海外卸货及代换站
overseas services 海外业务
overseas-funded enterprise 外资企业
overseas-funded firm 外资商行
oversea shipper 国外托运人;国外货主
oversea shipping trade 远洋运输业
overseas investment credits 海外投资信贷
overseas legal reserves 海外法定准备金
oversea small hydropower 国外小水电
overseasoning 过多干燥;过多陈化

Overseas Private Investment Corporation 海外私人投资公司
overseas training 海外培训
oversea(s) transmission 洲际传输;海外运输
oversea telegram 海外电报
oversea trade 海外贸易;对外贸易
oversea traffic 海外交通
oversea works 海外工程
oversea works contract 海外工程承包
overseeding 过量播撒;过度催化;过度播云(人工降雨)
overseer 管理人;工头;监视程序;监工;监督人(员)
over-sell 超卖
oversensitive 过灵敏的
oversensitive control 过灵敏控制
overset 翻转
oversew 缝合;对缝
overshadow 覆盖
oversheet 叠置片
overshelves 搁架
over shipped tow packages 溢装两件
overshoot 超射(度);超调(量);超出规定;镶边;溢出;逸出;过量;过调节;过冲;尖头信号;脉冲跳增
overshoot a landing field 落地过高(飞机)
overshoot clipper 冲限制器
overshoot current 过冲电流
overshoot distortion 过冲失真
overshooting 落地过高(飞机)
overshoot landing 落地过高(飞机)
overshoot pulse 过冲脉冲
overshoot ratio 尖头信号幅相对值
over/short landed cargo report 溢/短卸货报告单
overshot 突出部分(建筑物);超射;上射(式)的;上击(式)的;打捞筒【岩】;打捞器
overshot assembly 钻具打捞器;(绳索取芯的)打捞器
overshot duct 凹印墨斗
overshot guide 打捞筒导向器
overshot head (绳索取芯的)打捞头;(绳索取芯的)打捞矛
overshot loader 上倾式装载机
overshot rotary feeder 捞筒式转鼓给料机
overshot spillway 坝顶泄槽(溢洪道)
overshot stacker 上抛卸式堆草机
overshot tank filling 自顶部注入油罐内
overshot tubing seal divider 打捞筒油管密封分压器
overshot water wheel 上射式水轮;上击式水轮
overshot wheel 上射式水轮机
overshot with bowl 带引锥的打捞筒
overside 从船边
overside delivery 船边交货;转驳运输;直接过驳;过驳交货;外挡过驳;输入港水上船边交货
overside delivery clause 船边交货条款;过驳交货条款;吊钩下交货条款
overside discharging 过驳卸货
overside discharging and loading 直接换装;外挡装卸
overside goods 过驳货
overside lifting 侧方起吊
overside loading 过驳装货
overside port 船侧装卸口
overside transfer 外挡过驳
oversight 看漏;管理;观察;监视;监督;失察
oversight function of a parent company 母公司监督作用
oversimplification 过分简化
oversintered 过烧的
oversite concrete 找平混凝土;满堂混凝土垫层;满堂红混凝土垫层;垫层混凝土;地基混凝土板层
oversite excavation 地基地挖方
over-size 尺寸过大;超径;超过尺寸;超大件;超差;过大(的);加大;大载重量的;安全系数过大的
over-size aggregate 超径集料;超径骨料;超尺寸集料;超尺寸骨料;过大的集料;过大的骨料
over-size brake 过大的刹车;过大的制动器
over-size brick 大型砖;大尺寸砖
over-size chute 粗粉溜子
over-size clinker 过大的熟料
over-size clutch 过大的离合器
over-size cobbles 超径鹅卵石(10毫米);过大的鹅卵石
over-size core 超径岩芯(用薄壁钻头取得的)

over-size dirtproof bearing 超尺寸防尘轴承
over-size discharge 粗料出口
over-size distribution 累积粒度分布
over-sized materials 过大的材料;粗粒;超径材料;特大材料;不合格材料
over-sized particle 筛上料
over-sized product 加码;筛余物
over-size factor (筛子的)超大颗粒系数;(颗粒的)超大颗系数;筛上产品率
over-size grains in cement 水泥粗粒
over-size gravel 超径砾石
over-size hole 超径孔
over-size in the undersize 筛下中的过大者
over-size material 粗粉;过大的材料;筛上料
over-size mold 超尺寸的模子
over-size of hole 超径【岩】
over-size package 过大包装
over-size particles 粗粉
over-size percentage 筛余百分率
over-size piece (木头等的)大块
over-size piston 加大尺寸活塞
over-size piston ring 加大活塞环
over-size pneumatic tire 大型充气轮胎
over-size powder 筛上粉末
over-size product 过大产品;筛上物
over-size return 粗粉回磨;回粉
over-size return belt 回料胶带
over-size return conveyer 回料输送机
over-size rod 导向钻杆
over-size section 过大剖面
over-size stone 过大石
over-size vehicle 超型载重汽车;超型车辆;超大型载重汽车;大载重量汽车;大型汽车;大型车辆
over-size weld 超尺寸焊缝
overslabbing 混凝土板铺覆(道路)
overslaugh 沙滩;妨碍
overslipping 岩层倾覆;滑过;地层倒转
overslung car frame 上挂式轿厢框架
oversnow machine 雪橇
oversnubbing 过度缓冲
oversoaking 过度渗入
over-sold 卖空;超卖;空头
over-sold position 超卖持有额;空头
over-sowing 草地补播
overspan bridge 上跨桥
overspanning operation 跨挡作业
over sparse city 人口过稀城市
overspecialization 超特化;专用化程序过高
over-speed 超转速;超速运行;超速;过速
over-speed alarm 超速警告
over-speed braking electro-magnetic valve 超速制动电磁阀
over-speed cableway 超速索道
over-speed chopper 超速限制器
over-speed condition 超速状况
over-speed control 超速控制;超速控制
over-speed detection device 超速监测设备
over-speed detection unit 超速监测装置
over-speed drive 超速传动
over-speed governor 超速调节器;限速器
over-speeding 超速转动
over-speed limit device 限速装置;限速器
over-speed limiter 超速限制器
over-speed lowering 超速下降
over-speed power 超速功率
over-speed preventer 超速限制器
over-speed protection 超速保护;过速保护
over-speed protection (device) 超速防护(装置);超速保护(装置)
over-speed relay 超速继电器
over-speed safety gear 超速制动器
over-speed shut-off 超速停车装置
over-speed test 超速试验
over-speed trip mechanism 超速跳闸机构
over-speed trip pin 超速脱扣销
over-speed valve 主汽阀
over-spend 超支
overspill 人口外溢;人口过剩;溢出物;外溢
overspill in industrial centers 工业中心人口过剩
overspin 过自旋
oversplash 飞溅的水(防波堤顶部);过度溅泼;溅浪;飞溅
overspray 喷逸;喷溅;飞漆
overspray fastness 不喷逸性;不飞漆性

overspread 蔓延;遮盖;漫布;覆盖
over-spun wire 缠弦
over-square engine 超宽发动机;超方形发动机;超方发动机
over-stability 超稳定性;超安定性;过稳定性;过度稳定性
overstabilization 过自旋
over-stable 超稳定(的)
overstain 过染;过度染色
overstand 机上余尺
over-standard 超定额
overstandard of bacterial virus content 细菌病毒含量超标
overstandard of toxic metallic element 毒性金属元素超标
overstandard rate of pollutant 污染物超标率
overstatement 浮计
oversteepening 削陡作用;削峭作用;过陡作用;逆转作用
oversteer(ing) 过度转向;过度操纵
over-step 超越;超覆;超出;越权;逾越;横越
over-step one's authority 超越权限
over-step thrust propagation 上叠式逆冲扩展
overstitching machine 包缝机
overstock 库存过剩;存货过多;储备过多;超储;供应过剩;供过于求;进货过多;积压
over-stocked 过密的
over-stocked materials 超储物资
overstocking 过度放牧;过大放牧(度)
over-stooked pasture 载畜过多的草场
over-stored materials 超储材料
overstor(e)y 上层木;上层;顶层(指楼房);天窗;高侧窗
overstor(e)y window 高侧窗
overstowage 货物倒装;不良装载(没有按到港次序装载)
over-stowed 堆垛过高
over-strain 超限应变;残余应变;用力过度;过载;过于劳累;过度应变;极限应变;疲劳过度;瞬态失真
over-strain ag(e)ing 过应变时效;冷作时效
over-strained 超限应变的;过度变形的
over-strained state 过度变形状态
over-straining 过大应变
over-strength 超强度;人员超编制
over-strength material 超强度材料
over-stress 加超限应力;过载;超载;超应力;过度应力;过度紧张;使受力过大;超限应力;过应力
over-stressed 超应力的;应力超限的;过压的;过张的;超限应力的
over-stressed area 超应力区
over-stressed rock mass 应力超限的岩体;过应力岩体
over-stressing 超负载;应力超限;超限应力;过应力
over-stress testing 过应力测试
overstretch 过拉伸;过度伸长
overstretched 过大伸长
overstretching 超张拉(预应力筋)
overstroke 扳过头
overstuffed 多油的
overstuffed furniture 加厚软垫家具
oversubmergence 覆盖
over-subscription 超额认购;认购超额
over-super-elevation 过超高
oversupply 过度供应;供应过多;供应过度;供过于求
over-swing 超出规定;逸出;过冲;偏转过度
overswing limiting 过幅限制
over-synchronous 超同步的
over-synchronous braking 再生制动(作用)
overtailing course 砖石挑出层
over-take 超越他船;追越;赶上
overtaken vessel 被追越船
overtaking 越行;超车
overtaking collision 追撞
overtaking lane 超越车道;超车(车)道
overtaking light 尾灯
overtaking meter 接近速度指示器
overtaking of trains 列车越行
overtaking prohibited sign 禁止超车标志
overtaking rule 越车规则;超车规则
overtaking sight distance 超车视距
overtaking situation 追越局面
overtaking station 越行站
overtaking the other 追越他船

overtaking vessel 追越船
overtaking visibility distance 超越能见度距离;超车能见度距离
overtaking wave 追浪;尾浪
overtamping 振捣过度;捣固过度
over-tax 超额征税;课税过重;起额征税;收税过重;负担过度
over-temperature 过热温度
over-temperature control 超温控制
over-temperature cutout 过热保险器
over-temperature plant 超温植物
over-temperature relay 过热继电器
over-temperature signal 过热信号
over-temperature trip device 超温限制器
overtempering 过度回火
overtension 电压过高;超限应力;超应力;超电压;过压;过电压
over-tensioning 超张拉
overtension relay 过电压继电器
overtest 过量试验
over-the-counter trading 现货交易
over the ground 经过陆运;对地
over the horizon 地平以上;超地平
over-the-horizon communication 超视距通信
over-the-horizon detection system 超视距探测系统
over-the-horizon optic(al) communication 超视距光学通信
over-the-horizon radar 超远程电离层雷达;超视距雷达;超地平线雷达
over-the-horizon system 视距外通信方式
over-the-load carrier 下悬式搬运车
over-the-road common carrier 长途汽车运输公司
over-the-road truck 长途货车
over the sea 经过海道
over the side 在舷外
over-the-telephone market 通过电话交易的市场
over-the-top action 翻转
over-the-top tedder 草条翻转式翻草机
over-the-top vapour 塔顶排出的蒸汽
over-thin 更细;超薄
overthrow 废除;推翻;倾覆;门柱贴活装饰物
overthrow distortion 过冲失真
overthrust anticline 掩冲断层背斜;逆掩断层背斜;上冲断层背斜
overthrust block 逆掩断块;逆掩断层推覆体;上冲断块
overthrust distance 逆掩断矩
overthrust fault 逆掩断层;掩冲断层;上冲断层;掩冲大断层;逆掩大断层
overthrust fold 逆掩褶皱
overthrust mountain 逆掩断层山
overthrust nappe 逆掩推覆体;逆掩断层推覆体
overthrust plane 逆掩断层面
overthrust sheet 逆掩岩席;逆掩层
overthrust slice 掩冲岩片;上冲推覆体;上冲片体
overthrust trap 逆掩断层圈闭
overtide 倍潮
overtighten 拉得过紧;拧得过紧
over-tile 脊瓦;盖瓦;槽瓦
over-tile and under-tile roofing 覆瓦-底瓦屋面
overtime 加工时间;超限时间;超时;加班时间;加班;额外工作时间
overtime allowance 加班津贴
overtime and night differential 加班费和夜勤津贴
overtime arrangement 加班成本费用;加班办法
overtime bonus 额外工时奖金
overtime check 过时检查
overtime compensation 加班工资
overtime duties 加班工作
overtime hours (长途电话中的)超过的时间;加点;加班;加班时数
overtime job 加班加点
overtime or oddshift bonus 超时或临时加班奖金
overtime pay(ment) 加班工资;加班费
overtime premium 超时补加费;加班津贴;加班奖金
overtime premium earnings 加班奖金收入
over-time ticket 过时车票
overtime work 加班加点;加班工作;加班
over-tipped face 岩层侧转面;岩层倾覆面
overtone 谐波;盖底色;墨色;涂色
overtone band 泛音谱带;泛频谱带
overtone crystal 谐波压电晶体;谐波晶体
overtone frequency 谐波频率;泛音频率

overtone method 泛音法
overtone oscillator 泛音振荡器
overtones mode 谐波振型
overtonging 丝扣拧得过紧
overtop 超出;过高
overtop flue 上烟道
overtop-gear 超速挡装置
overtopped 受压木
overtopped cofferdam 过水围堰;漫顶围堰
overtopped dam 溢流坝;溢水坝;溢流堰;漫顶坝
overtopped earth and rock cofferdam 过水土石围堰
overtopped stem 被压杆
overtopping 放倒(塔杆);越浪;越顶(洪水);溢顶;淹过;漫溢;漫顶
overtopping discharge 漫顶(流)量
overtopping of highway 淹过路面;洪水淹过路面
overtopping of spillway 溢洪道溢流
overtopping rate 漫顶率
overtopping stage 溢水期;漫顶水位
overtopping water 越顶水量
overtopping (water) stage 漫顶水位
overtopping wave 越顶波浪
overtrades 高空信风
overtrading 超过运用资金的购买;业务发展过快;交易过量
overtraining 过度训练;培训过度
over-trap fault 超覆断层
over-travel 超行程;超冲程;超程;再调整;过调;提升过紧;大移动
over-travel-limit switch 限位开关;行程开关;终端开关
over travel(l)ing crane 桥式吊(车)
overtree sprinkler 树冠喷灌机
over-t retrieval plot 成功检索曲线
over-trowelling 抹光过度
overturn 倾覆;湖水对流;倾倒;水体翻转
overturned 倒转的
overturned anticline 倒转背斜
overturned bed 倒转岩层;倒层;倒覆地层;倒地层
overturned coal seam 倒转煤层
overturned fold 倒转褶皱
overturned limb 倒转翼
overturned succession 倒转层序
overturning 挡土墙倾覆;倾覆
overturning effect 倾覆作用
overturning factor of safety 倾覆稳定系数
overturning fold 倒转褶皱
overturning force 倾覆力
overturning load 倾翻荷载
overturning moment 倾斜力矩;倾覆力矩
overturning moment at base 基底的倾倒力矩
overturning of rail 钢轨倒
overturning of retaining wall 挡土墙倾覆
overturning of vessel 船只倾覆
overturning of water body 水体翻转
overturning skip 翻斗
overturning speed 颠覆速度
overturning stability 倾覆稳定(性)
overturning stability piers and abutments 墩台倾覆稳定
overturning stability test 倾覆稳定性试验
overturning test 车辆倾翻试验
overturning thrust 倾覆推力
overturn of lakes 湖水翻转
overturn period 翻转期
overturn protection 防后翻装置
overturn the thread 拧断螺纹
over-type worm gear 蜗轮蜗杆传动
overuse 过度使用;过度利用
overuse equipment 拼设备
overvaluation 高估(价值);计价过高
over-valued idea 超价值观念
over-velocity test 超速试验
overventilation 过度换气;换气过度
overvibrate 过度振捣;过度振动
overvibration 振荡过头;超振动作用;超振捣;震捣过度;振动过度(混凝土);过度振捣;过度振动作用;过度振动作用;过度振动;过度振捣
overvibration of concrete 混凝土振动过度
overview 概述
overview schedule 宏观进度计划
overvoltage 超额电压;超电压;最高电压;过压;过电压

overvoltage alarm 过压报警(器)
overvoltage condition 过压状态
overvoltage crowbar 过压保护
overvoltage diverter 过电压分流器
overvoltage electrode 超压电极
overvoltage fuse 过压熔断器
overvoltage multiple 过电压倍数
overvoltage of no-load line switching 空载线路切合过电压
overvoltage protection 过压保护;过电压保护
overvoltage protective device 过压保护装置;防过电压装置
overvoltage protective system 过压保护系统
overvoltage protector 过电压保护器
overvoltage region 过压范围
overvoltage relay 过压继电器;过电压继电器
overvoltage release 过压断路器
overvoltage suppressor 超压抑制器
overvoltage test(ing) 超压试验
overvoltage threshold 过压阈值
overvulcanization 过硫作用;过度硫化
overwash 越浪;越堤冲岸浪;漫顶;冰水沉积
overwash plain 冰水洪积平原;冰水沉积平原
overwash pool 浪成堤后池
overwash zone 冲溅区;溅浪区
overwater 飞越水面
overwater boring 水上钻探
overwater construction 水上施工
overwater construction machinery 水上施工机械
overwater drilling 水上钻探
overwater exploration 水上勘探;水上勘察
overwater flight 水上飞行
overwater investigation 水上勘探;水上勘察
overwater line 吃水线以上
overwater refraction 水面折射
overwater reserves 水上储量
overwater search 水上探测;水上搜索
overwater wind speed 水面风速
overwater work 水上作业
overweight 超重;超限重量;过重
overweight cargo 超重货(物)
overweight coating 过量涂布涂层
overweight payload 超重有效荷载
overweight times 超重次数
overweight ton 超重吨数
overweight train 超重列车
overweight vehicle 超载车辆;超重型车;超重车辆;大载重量汽车
overweld(ing) 过焊;过渡焊接
overwet land 沼泽地;过湿地
overwet mix 过湿混合料;过湿拌和(料)
overwhelming 占优势的;不可抵抗的
overwhelming superiority 绝对优势
over-width 超宽
over-width cargo 超宽货
overwind 过度卷绕;上卷式
overwind clearance 过卷距离
overwinder 过卷防止器
overwinding 卷绕过紧;满过头;(为吊重物而在滚筒上的)绳索
overwinding apparatus 过卷防止器
overwind prevention 防止提升过卷
over wings diameter 翼展
overwintering 越冬;越年;过冬
overwintering grass 越年生牧草;越冬牧草
overwintering plant 越冬植物;多年生植物
overwood 上木;上层木
overwork 过劳;工作时间过长;工作过度;使用过度
over-wound cake 满过头
overwrite 改写;重写;冲掉改写;冲掉
overwrite error 重写误差
overwrite noise 重写噪声
overwriting 超记录
overwriting error 重写错误
overy cutting 采伐过度
overyear 隔年
overyearing 越年;越冬;过冬
overyear regulation 多年调节
over years 多年
overyear storage 多年调节库容;多年蓄水;多年平均库容
overyear storage capacity 多年平均库容量
over-yielding mix 超制成量的混合料
over-zone feeding 越区供电

ove tails 尾矿粉
ovhl [overhaul] 检修;大修
ovification 卵形成
oviform 卵形的;卵形
oviposition 产卵;产蛋
ovoid body 卵形体;卵形物;卵形
ovoid briquette 煤球
ovoid mo(u)lding 卵形饰
ovolo 凸圆线脚;圆突形线脚装饰;凸出四分之一圆（饰）
ovolo mo(u)lding 馒形线脚;四分之一周缘饰;凸圆形装饰线条
ovonic device 玻璃开关
ovonic electroluminescence 双向场致发光
ovonic memory switch 双向记忆开关
ovonics 交流控制半导体元件;双向开关半导体器件
ovonic switch 双向半导体开关
ovonic threshold switch 双向阈值开关
ovostab 苯甲酸雌二醇酯
Ovshinsky effect 奥维辛斯基效应;奥氏效应
Ovshinsky glass semi-conductor 奥氏玻璃半导体;奥维辛斯基玻璃半导体
Ovshinsky semi-conductor 奥氏半导体
ovulite 卵化石
ovum 卵形饰;卵形装饰;卵石饰;[复]ova
owe 欠债;负债
owelty of exchange 交换土地的补差款
owelty of partition 租户之间的补差款
Owen fracture zone 欧文破裂带
Owen process 桉油浮选法
Owen's jet dust counter 欧文粉尘计数器
Owen's organ 欧文器
owing 未付的
OWL 常水位
owler 走私者
owling 走私;晚上捕鱼
owl-tail ridge ornament 鸱尾【建】
own-accord crushing 自行压碎;自行挤碎
own capital 自有资本
own code 专用码;固有码
own coding 专用编码;扩充工作码
own core 自带芯
own demand 自用电量
owned capital 自有资金
owned equipment 自有施工机械
owner 拥有者;业主;物主;所有者;所有人;船东;发包人
owner-architect agreement 业主与建造师契约;委托人与建造师合同;业主一建筑师协议书
owner-builder 自建业主
owner code 箱主代号(集装箱)
owner-contractor agreement 业主与承包人合同;业主与承包人契约;业主一承包人协议书
owner-controlled firms 业主控制的企业
owner-controlled insurance 业主自承保险
owner-driver cabby 车主出租汽车
owner indicator 自显指示剂
owner interview survey 车主访问调查
owner-occupant 屋主住户;自住业主
owner-occupation with public subsidy 自建公助
owner-occupied 业主自用的
owner-occupied house 自用住房
owner-occupied housing 自住的房屋;房主自用住宅
owner-occupier 私房自住者
owner of cargo 货主
owner of freight 货主
owner of goods 货主;货物所有人
owner of record 注册业主
owner of the project 项目的所有者
owner-peasant 自耕农
owner process 业主建设进程
owner record 主记录;文件编写人记录
owner's capital 业主资本
owner's duty 业主义务
owner's engineer 甲方代表;业主工程师
owner's equity accounting 所有者权益核算
owner's equity audit 所有者权益审计
owner's equity of state 国家所有者权益
ownership 物主身份;所有制;所有权;产权;占有权
ownership advantages 所有权优势
ownership and use rights of land 土地所有权和使用权

ownership capacity 自置生产能力
ownership capital 自有资本
ownership equity 所有权益
ownership hy the whole people 全民所有制
ownership interest 所有者权益
ownership of land by the state 土地国有制
ownership of property 物权;财产所有权
ownership of trade mark 商标权
ownership rule 所有权法则
owner's inspector 业主代理人;业主监工员;甲方检查员
owner's liability insurance 业主义务保险;业主责任保险;所有者责任保险
owner's load and count 货主自装自理
owner's obligation 业主义务
owners pro hac vice 法定船东代表
owner's representative 业主代表
owner's rights 业主权利
owner's risk 甲方风险;业主风险;自负风险;货主自负风险;危险由货主自负
owner's risk clause 货主负担风险条款
owner's risk deterioration 货主自负腐烂变质险
owner's risk of breakage 货主自负破损险;货主负责破损险
owner's risk of damage 货主自负海损险;货主负担损害危险
owner's risk of fire 货主自负火险
owner's risk of freezing 货主自负冻坏险
owner's risk of leakage 货主自负渗漏险
owner's risk of shafing 货主自负擦损险
owner's risk of shifting 货物移动由货主负责
owner's risk of wetting 受湿由货主负责
owner's technical personnel 业主的技术人员
owner's wagon 货主货车
owner's wharf 货主码头
owner-user berth 业主专用舶位;货主码头
owner-user ship 业主专用船
owner-user terminal 业主专用码头
owner-user vessel 业主专用船
owner will carry mortgage 卖方愿意收回抵押;业主愿意收回抵押
own fund 自有资金
own jitter 自抖
own load 自重
own plane 固有平面
own profit and loss responsibility 独立核算责任
own quantity 固有量
own right-of-way 特别优先通行权
own risk 自我风险;自负风险;自保
own ship 本船
own type 固有类型
own-weight 自重
own-weight moment 自重力矩
ow release pesticide system 缓释型农药
owyheeite 脆硫锑银铅矿;脆硫锑银矿
oxacid 含氧酸
oxacyclobutane 氧杂环丁烷
oxalacetate 草醋酸酯;草醋酸盐
oxalacetic acid 草乙酸;草醋酸
oxalaldehyde 草醛
oxalamide 草酰胺
oxalate 乙二酸盐;草酸盐;草酸
oxalate calculus 草酸盐结石
oxalate extract 草酸盐提出取
oxalate method 草酸盐法
oxalate precipitation method 草酸沉淀法
oxalate treatment 草酸盐处理
oxaldehydic acid 乙醛酸
oxaldiureide dioxime 草酸二脲二肟
oxalene 草烯
oxalic acid 乙二酸;草酸
oxalic acid ozonation 草酸臭氧化
oxalic acid treatment 草酸处理
oxalic ester 草酸酯
oxalic methylene blue stain 草酸美蓝染色法
oxalimide 草酰亚胺
oxalism 草酸中毒
oxalite 草酸铁矿
Oxally 奥克萨力;包层钢
oxaloacetamide 草乙酰胺;草酸乙酰胺
oxaloacetic acid 草酰乙酸
oxalomolybdic acid 草钼酸
oxalopropionamide 草酰丙酰胺
oxalosuccinic acid 草酰琥珀酸;草丁二酸

oxalyl 草酰
oxalyl chloride 乙二酰氯;草酰氯
oxalylurea 草酰脲
oxamethane 草氨酸乙酯
oxamic acid 草氨酸
oxamic acid hydrazide 氨基草酰肼
oxamic hydrazide 草氨酰肼
oxamide 草酰胺;草酸铵
oxamidic acid 草氨酸
oxamido- 草酰氨基
oxammite 草酸铵石
oxamoyl 氨基草酰
oxamyl 草氨酰
oxane 恶烷
oxanilic acid 苯胺羰酸
oxanilide 草酰替苯胺
oxanthranol 蒽二酚
oxanthrol 蒽酚酮
oxanthrone 蒽酚酮
oxazole 恶唑
oxazolidine 恶唑烷
oxazolidone 恶唑烷酮
ox ball 旗杆顶块;桅顶球
oxbow 马蹄形弯道;U形牛颈弯
oxbow effect 辊弯曲效应
oxbow lake 弓形湖;牛轭湖;废河道湖
oxbow loop 弓形河湾
oxbowshaoed depression 牛轭洼地
oxbowshoed depression 马蹄形洼地
oxbow swamp 弓形沼泽;牛轭湖沼泽
oxbowu 形河曲;牛轭(湖)形弯道
oxdiazole 恶二唑
oxer 牛栏
oxetone 螺二氧己烷
ox-eye 卵形孔;圆形孔;卵形窗;牛眼窗;卵形老虎窗
oxeye mo(u)lding 凹形断面线脚
oxeye (window) 牛眼窗
Oxford-blue 暗蓝色;紫蓝色
Oxford clay 牛津黏土
Oxfordian 牛津阶【地】
Oxford inflator 牛津吹张器
Oxford unit 牛津单位
oxgall 混合胆汁
oxhead 牛头饰;牛头尊
oxhide glue 牛皮胶
oxiacalcite 草酸方解石
oxibendazole 奥苯达唑
oxic-settling-anaerobic process 好氧一沉淀一厌氧工艺
oxidability 可氧化性
oxidable 可氧化的
oxidant 氧化物;氧化剂
oxidant concentration 氧化剂浓度
oxidant model 氧化剂模型
oxidant pump 氧化剂泵
oxidant recorder 氧化剂记录器
oxidant smog 氧化剂烟雾
oxidant swirler 离心式氧化剂喷嘴
oxidapatite 氧磷灰石
oxidate 氧化组沉积;氧化物沉积
oxidation 氧化作(用)
oxidation accelerator 氧化促进剂
oxidation agent 氧化剂
oxidational losses 氧化损失
oxidation and reduction in natural water bodies 天然水体氧化还原作用
oxidation brick 氧化砖
oxidation catalysis 氧化催化
oxidation channel treatment 氧化沟处理
oxidation characteristic 氧化特性
oxidation-coagulation process 氧化-混凝工艺
oxidation colo(u)r 氧化染料
oxidation column 淋水塔;氧化塔
oxidation couple flocculation technology 氧化偶合絮凝技术
oxidation degree 氧化阶
oxidation deodorizing 氧化脱臭
oxidation desulfuration tower 氧化脱硫塔
oxidation desulfurization 氧化脱硫
oxidation desulphation tower 氧化脱硫塔
oxidation ditch 氧化沟;氧化槽
oxidation-ditch extended aeration process 氧化延时曝气法
oxidation-ditch performance 氧化运行特性

oxidation ditch process 氧化工艺
oxidation drying 氧化干燥
oxidation effect 氧化效应
oxidation equivalent 氧化当量
oxidation film 氧化膜
oxidation flame 氧化焰
oxidation-flocculation process 氧化絮凝工艺
oxidation index 氧化率
oxidation-inhibited grease 氧化抑制润滑脂;抗氧化润滑脂
oxidation inhibitor 氧化抑制剂;抗氧化剂
oxidation kinetics 氧化动力学
oxidation lagoon 氧化湖
oxidation layer 氧化层
oxidation loss 氧化损耗
oxidation mask 氧化掩模
oxidation material 氧化材料
oxidation number 氧化值;氧化数
oxidation of coal 煤氧化作用;煤氧化
oxidation of odo(u)r by ozone 臭气臭氧氧化法
oxidation oven 氧化炉
oxidation period 氧化期
oxidation pit 氧化针孔
oxidation polymer 氧化聚合
oxidation polymerization 氧化聚合作用
oxidation pond 氧化塘;氧化池
oxidation pond effluent 氧化塘出流水;氧化塘出水
oxidation pond process 氧化塘工艺;氧化塘法
oxidation potential 氧化势;氧化电位
oxidation pretreatment 氧化预处理
oxidation prevention 氧化预防
oxidation process 氧化过程;氧化法
oxidation process of wastewater 废水氧化处理
oxidation product 氧化产物
oxidation promoter 氧化促进剂
oxidation rate 氧化速率;氧化率
oxidation reaction 氧化反应
oxidation reduction 氧化还原作用;氧化还原
oxidation-reduction analysis 氧化还原分析
oxidation reduction catalyst 氧化还原催化剂
oxidation-reduction cell 氧化还原电池
oxidation-reduction discontinuity 氧化还原电势突变
oxidation-reduction electrode 氧化还原电极
oxidation-reduction equilibrium 氧化还原平衡
oxidation reduction half-cell reaction 氧化还原半电池反应
oxidation-reduction indicator 氧化还原指示剂
oxidation-reduction in soil 土壤氧化还原(作用)
oxidation-reduction method 氧化还原法
oxidation-reduction pair 氧化还原电对
oxidation-reduction potential 氧化还原电位;氧化还原电势
oxidation-reduction potential of seawater 海水氧化还原电位
oxidation-reduction process 氧化还原过程
oxidation-reduction reaction 氧化还原反应
oxidation-reduction system 氧化还原系统
oxidation-reduction titration 氧化还原滴定(法)
oxidation resistance 抗氧化性;抗氧化能力;抗氧化;耐氧化能力
oxidation resistant 抗氧化剂;抗氧化的;不生锈的
oxidation resistant carbon-carbon composite 耐氧化碳—碳复合材料
oxidation resistant coating 抗氧化涂层
oxidation-resistant steel 抗氧化钢
oxidation retarder 氧化抑制剂
oxidation scale 铁鳞
oxidation sewage plant 氧化污水处理厂
oxidation sewage process 氧化污水处理法
oxidation stability test 氧化稳定性试验
oxidation stabilometer 氧化平衡仪
oxidation stain 氧化污染;氧化色斑;氧化变色
oxidation starch 氧化淀粉
oxidation state 氧化状态;氧化态
oxidation sterilizing 氧化消毒
oxidation susceptibility 氧化性能
oxidation tank 氧化槽
oxidation technology 氧化工艺
oxidation temperature 氧化温度
oxidation test 氧化试验
oxidation test at high temperature 高温氧化试验
oxidation tint 氧化色
oxidation tower 氧化塔

oxidation treatment 氧化处理
oxidation treatment of wastewater 废水氧化处理法
oxidation tube 氧化管
oxidation-typed bactericide 氧化型杀菌剂
oxidation value 氧化值
oxidation zone 氧化作用带;氧化带;氧化层
oxidative assimilation 氧化同化作用
oxidative attack 氧化侵蚀
oxidative decolo(u)rization 氧化脱色
oxidative decomposition 氧化分解
oxidative degradation 氧化降解(作用)
oxidative phosphorylation 氧化磷酸化(作用)
oxidative scission 氧化裂断
oxidative system 氧化系统
oxidative systems of various groups of organisms 不同类生物氧化系统
oxide 氧化层;氧化皮;氧化物
oxide accretion 氧化物结块
oxide addition 氧化物添加剂
oxide adsorbent 氧化物吸附剂
oxide agent 氧化剂
oxide agents and organic over-oxide articles 氧化剂和有机过氧化物
oxide asphalt 氧化(地)沥青
oxide base cermet 氧化物基金属陶瓷
oxide cathode 氧化物阴极
oxide-ceramic laser 氧化物陶瓷激光器
oxide-ceramics 氧化物陶瓷;氧化陶瓷
oxide-ceramic shell mo(u)ld 氧化物陶瓷型壳
oxide-coated 涂氧化物的
oxide-coated cathode 氧化物阴极;氧化物涂覆阴极
oxide coated filament 敷氧化物灯丝
oxide coating 氧化物涂料;氧化性涂料;氧化物覆膜法;氧化镀层;氧化的涂层
oxide colo(u)r 氧化物颜料;氧化物色料
oxide compound 氧化物
oxide core 氧化物磁芯;氧化铁铁芯
oxide deposit attack 氧化物沉积侵蚀
oxide dispersion strengthening 氧化物弥散强化
oxide-dissociation chemical interference 氧化物—离解化学干扰
oxided lead ore 氧化铅矿石
oxided Pb-Zn ore 氧化铅锌矿石
oxide etch 氧化腐蚀
oxide etching 氧化物刻蚀
oxide-facies iron formation 氧化物相铁建造【地】
oxide film 氧化膜
oxide-film arrester 氧化膜避雷器
oxide-film treatment 氧化膜处理
oxide fuel reactor 氧化物燃料堆
oxide fume 氧化物烟气
oxide glass 氧化物玻璃
oxide-induced stress 氧化引起的应力
oxide ion 氧化离子
oxide isolation 氧化物隔离
oxide layer 氧化层
oxide magnet 氧化物磁铁;氧化物磁石
oxide magnetic compact 氧化物磁石
oxide matrix 氧化物基体
oxide membrane 氧化膜
oxide mineral 氧化物矿物;氧化无机物
oxide nuclear fuel 氧化物核燃料
oxide of alumina 刚玉;矾土
oxide of alumin(i)um 矾土;氧化铝
oxide of chromium 氧化铬
oxide of iron 氧化铁
oxide of nitrogen 氮的氧化物;氧化氮
oxide ore 氧化物矿石
oxide paint 氧化物涂料;氧化物油漆;氧化金属涂料
oxide particle 氧化物颗粒
oxide passivation 氧化物纯化法
oxide phosphor 氧化物磷光体
oxide platelet 片状氧化物
oxide precipitate 氧化沉淀物
oxide resistor 氧化物电阻
oxide semiconductor 氧化物半导体
oxide skin 氧化物层;氧化皮
oxides of hydrogen 氢的氧化物
oxide solid 氧化固体
oxide spot 氧化物斑点
oxide structure 氧化物结构
oxide surface 氧化表面
oxidetic 酸结合力的

oxide tool 氧化物(陶瓷)刀具
oxide window 氧化层窗孔
oxidic film fire coat 氧化膜
oxidiferous 含氧化物的
oxidimetry 氧化还原滴定法;氧化测定法;氧化(还原)滴定(法)
oxidizability 氧化性;氧化能力;可氧化性
oxidizable 可氧化的
oxidizable compound 可氧化化合物
oxidizable material 易氧化材料
oxidizable nitrogen 可氧化氮
oxidizable salt 可氧化盐
oxidization 氧化
oxidize catalytic filter media 氧化催化过滤介质
oxidized 氧化的;生锈的;被氧化的
oxidized asphalt 吹制沥青;氧化(地)沥青
oxidized asphaltic bitumen 氧化地沥青
oxidized bitumen 氧化(地)沥青
oxidized bronze 氧化青铜
oxidized bronze finish 氧化青铜饰面
oxidized cellulose 氧化纤维素
oxidized coal 氧化煤
oxidized coal-tar 氧化煤焦油
oxidized copper ore 氧化铜矿石
oxidized linseed oil 氧化亚麻仁油
oxidized lubricant 氧化润滑剂
oxidized magnesium test (of lime) (石灰的)氧化镁试验
oxidized metal 氧化金属
oxidized microcrystalline wax 氧化微晶蜡
oxidized Mo ore 氧化钼矿石
oxidized nickel ore 氧化镍矿石
oxidized oil 氧化油
oxidized ore 氧化矿石
oxidized ore zone 氧化矿带
oxidized paraffin 氧化石蜡
oxidized resin 氧化树脂
oxidized rosin 氧化松香
oxidized rubber 氧化橡胶
oxidized Sb ore 氧化锑矿石
oxidized sediment 氧化沉积物
oxidized shale 页岩残渣;氧化页岩
oxidized sludge 氧化污泥
oxidized sludge process 氧化污泥法
oxidized species 氧化物物种
oxidized spot 热迹
oxidized starch 氧化淀粉
oxidized still 氧化炉
oxidized streak 氧化斑痕
oxidized turpine oil 氧化松节油
oxidized wastewater 氧化废水
oxidized zinc ore 氧化锌矿石
oxidized zone 氧化带
oxidized zone of sulfide deposit 硫化矿床氧化带
oxidize nozzle 氧化剂喷嘴
oxidize pump impeller 氧化剂泵叶轮
oxidizer 氧化剂
oxidizer compartment 氧化剂舱
oxidizer distributor 氧化剂分配器
oxidizer pump 氧化剂泵叶轮
oxidizing 吹氧(法);吹气(法);氧化的
oxidizing ability 氧化能力
oxidizing ability of the atmosphere 大气的氧化能力
oxidizing acid 氧化性酸
oxidizing action 氧化作用
oxidizing agent 氧化剂
oxidizing air 助燃空气
oxidizing anti-corrosion agent 氧化防腐剂
oxidizing atmosphere 氧化气氛
oxidizing barrier 氧化障
oxidizing capacity of the atmosphere 大气的氧化能力
oxidizing catalytic reactor 氧化催化反应器
oxidizing chamber 氧化室
oxidizing condition 氧化条件
oxidizing electrode 氧化电极
oxidizing environment 氧化环境
oxidizing firing 氧化烤色
oxidizing flame 氧化焰;氧化性火焰
oxidizing force 氧化力
oxidizing furnace 氧化炉
oxidizing installation 氧化装置
oxidizing material 氧化物

oxidizing period 氧化期
oxidizing phthalic alkyd resin 氧化型邻苯二甲酸醇酸树脂
oxidizing potential 氧化电位
oxidizing power 氧化能力;氧化力;氧化本领
oxidizing process 氧化工艺
oxidizing property 氧化性质
oxidizing reaction 氧化反应
oxidizing recirculation channel 氧化再循环沟槽
oxidizing/reducing power 氧化还原能力
oxidizing roast 氧化焙烧
oxidizing roasting 氧化煅烧
oxidizing solid 固体助燃剂
oxidizing solution 氧化溶液
oxidizing stability 氧化稳定性
oxidizing substance 氧化物物质
oxidizing temperature 氧化温度
oxidizing type alkyd resin 氧化型醇酸树脂
oxidizing zone 氧化层;氧化带
oxidogenic substrate 氧化系底物
oxido-indicator 氧化还原指示剂
oxido-reduction 氧化还原
oxidosome 氧化体
oxid-reduction equivalent 氧化还原当量
Oxigest 延时曝气法或接触消化法的活性污泥处理装置
oxime 肟
oximide 草酰亚胺
oxinane 氧杂环辛烷
oxiniacic acid 氧烟酸
oxirane 氧杂环丙烷
oxirane ester resin 环氧乙烷酯
oxirine 氧杂环丙烯
oxiron 过氧化聚丁二烯
Oxiser 延时曝气法的活性污泥处理装置
oxisol 氧化土
oxitol 苯基溶纤剂
oxivor 水合氯氧化铜
oxo acid 含氧酸
oxo alcohol 氧化法合成醇类
oxocane 氧杂环辛烷
oxocin 氧杂环辛三烯
oxocinol 氧杂环辛三烯醇
oxo-compound 氧基化合物;含氧化合物
oxodic 泥炭沼泽群落的
oxolan 草脲胺
oxolation polymer 氧络聚合物
oxole 氧杂环戊二烯
oxonane 氧杂环壬烷
oxonin 氧杂环壬四烯
oxoniu mion 水合氢离子
oxo-process 氧化合成法;氧化法
oxosilane 氧硅烷;硅氧烷
oxozone 双氧气
oxscull 牛头饰
ox-skull 牛头骨饰
oxstall 牛舍
oxter piece 支撑杆;垒石墙用的木柱;垂杆
oxter plate 船尾板
oxyacetone 乙酰甲醇;氧丙酮
oxyacetylene 氧乙炔;氧炔;氧乙炔;氧乙炔气
oxyacetylene blowpipe 氧炔焊吹管;氧气乙炔吹管;氧炔喷焊器;氧炔吹管;氧炔喷焊机;氧化乙炔焊枪
oxyacetylene burner 氧气乙炔燃烧器
oxyacetylene cutter 氧乙炔切割器;氧炔切割机;氧炔断割器;氧化乙炔切割器;氧化乙炔割断器
oxyacetylene cutting 氧炔切割;氧炔切割;氧炔切割;氧炔切割断;气割(法);气割;氧炔气切割
oxyacetylene cutting equipment 氧气乙炔切割设备
oxyacetylene cutting torch 气割枪;氧炔割枪
oxyacetylene flame 氧乙炔焰;氧炔焰;氧炔焰;氧化乙炔焰
oxyacetylene flame spraying 氧炔火焰喷涂
oxyacetylene hose 氧乙炔软管
oxyacetylene powder gun 氧乙炔粉体枪
oxyacetylene process 氧乙炔切割程序;氧炔焦化材面防腐法
oxyacetylene rod gun 氧乙炔棒枪
oxyacetylene scarfing 氧乙炔火焰清理
oxyacetylene scarfing machine 氧乙炔火焰清理机
oxyacetylene torch 氧乙炔焊炬;氧气乙炔焊;氧炔喷焊器;氧炔焊炬;氧炔火焰;氧炔焊枪;氧化炔焊枪

oxyacetylene treatment 氧乙炔处理
oxyacetylene welder 氧乙炔焊接器;氧乙炔焊机;气焊机
oxyacetylene welding 氧乙炔焊接;氧气乙炔焊;氧化乙炔焰焊接;气焊;氧乙炔焊接气焊(法);氧(化炔)焊;气焊氧乙炔焰焊接
oxyacetylene welding of mild steel 低碳钢氧乙炔气焊
oxyacetylene welding outfit 氧乙炔气焊机;氧炔(气)焊机;气焊机
oxy-acid 羟基酸;含氧酸
oxy-apatite 氧磷灰石
oxy-arc cutting 氧气弧切;电弧氧气切割;氧焰切割;氧气电弧切割;氧弧切割
oxy-arc welding 氧弧焊
oxy-austenite 含氧奥氏体;氧化奥氏体
oxybasiophitic 酸辉绿结构的
oxybenzene detector 苯酚检测器
oxybiontic 需氧的
oxybiosis 需氧生活
oxybiotic bacteria 有氧生物菌
oxybiotic organism 需氧性生物
oxybiotite 氧黑云母;高铁黑云母
oxyboosted blast 富氧鼓风
oxycarbide 氧碳化物
oxy-carbonitride 氧碳氮化物
oxycatalyst 氧化催化剂
oxycel 氧化纤维素
oxycellulose 氧化纤维素
oxychloric cement 镁氯氧水泥
oxychloride 氧氯化物;氯氧化物
oxychloride (bonded) wheel 氧化镁结合剂砂轮
oxychloride cement 索勒尔胶结料;菱苦土水泥;氯氧(化物)水泥
oxychloride terrazzo 含氯氧化物的水磨石
oxychlorination 氧氯化
oxychlorination method 氯氧化法
oxychlorium cement 氯镁水泥
oxychlorosene 氧氯苯磺酸
oxychromatic 嗜酸染色的
oxychromatin 嗜酸染色质
oxycoal gas 灯用煤气;氧煤气
oxycoal gas flame 氧煤气焰
oxycompound 含氧化物
oxy-cytting 氧乙炔切割
oxyde nitreux 一氧化二氮
oxydetic 酸结合力的
oxydipropionitrile 氧二丙腈
oxydizing flame 氧化焰
oxydol 双氧气
oxydrolysis 氧化水解(反应)
oxydum nitrosum 一氧化二氮
oxyferrite 氧化铁素体
oxy-fired furnace 纯氧助燃窑炉
oxyfluoride 氟氧化物
oxy-flux cutting 氧熔剂电弧切割
oxy-fuel burner 富氧燃烧器;富氧喷枪
oxy-fuel combustion 富氧燃烧
oxy-fuel fired furnace 全氧助燃窑
oxy-gas flame 氧煤气焰
oxygen-18 drift 氧—18漂移
oxygen absorbed 吸氧量;耗氧量
oxygen absorbent 吸氧剂
oxygen absorption 氧吸收
oxygen absorption barotrauma 氧吸收性中耳气压损伤
oxygen acceptor 受氧体
oxygen-acetylene cutting 氧乙炔切割;氧气乙炔切割;氧化切割;气割下料
oxygen-acetylene cutting equipment 氧乙炔切割设备
oxygen-acetylene welding 气焊;氧炔焊;氧乙炔焊
oxygen acid 含氧酸
oxygen-activated sludge 纯氧活性污泥
oxygen-activated sludge process 纯氧活性污泥法
oxygen-activated sludge system 氧活化污泥系统;纯氧活性污泥系统
oxygen air supply 供氧气量
oxygen analyzer 氧气分析仪;氧气浓度分析器;含氧量分析仪
oxygen anion generator 负氧离子发生器
oxygenant 氧化剂

oxygen apparatus 氧气设备
oxygenate 充氧气;充氧
oxygenated 充了氧的
oxygenated actived sludge plant 充氧活性污泥装置
oxygenated actived sludge process 充氧活性污泥法
oxygenated aqueous solution 充氧水溶液
oxygenated asphalt 氧化(地)沥青
oxygenated oil 氧化油
oxygenated organics 氧化有机物;含氧有机物
oxygenated solution 含氧溶液
oxygenated solvent 含氧溶剂
oxygenated water 含氧水;充氧水
oxygenation 充氧作用;充氧
oxygenation capacity 充氧能力;充氧量
oxygenation efficiency 充氧效率
oxygenation in river 河流充氧
oxygenation rate 充氧速率
oxygenation system 充氧系统
oxygen atom 氧原子
oxygenator 充氧器
oxygen balance 氧平衡
oxygen balance evaluation 氧平衡评价
oxygen balance of explosive 炸药氧平衡
oxygen-bearing copper 含氧铜
oxygen bleaching 氧漂白
oxygen blower 鼓氧机;鼓氧器;吹氧机;吹氧器
oxygen blowing 吹氧
oxygen-blowing converter 吹氧转炉
oxygen-blown converter 吹氧转炉
oxygen blown steel 吹氧钢
oxygen bomb 氧气瓶
oxygen bomb ag(e)ing test 氧弹老化试验
oxygen bomb calorimeter 氧弹量热器
oxygen bomb test 氧弹试验
oxygen bottle 氧气筒;氧气瓶;氧化气瓶
oxygen-breathing apparatus 供氧设备;输氧呼吸器;氧气呼吸器;氧气设备
oxygen buffer 氧缓冲
oxygen canister 氧气罐
oxygen capacity 氧容量;充氧量
oxygen-carbon atomic ratio 氧碳原子比
oxygen carrier 载氧体;含氧物质
oxygen carrying capacity 携氧能力
oxygen cell 氧气电池;充气电池
oxygen-charging station 充氧站
oxygen cleaned valve 洁净的氧阀
oxygen coefficient 含氧系数
oxygen combining power 氧结合力
oxygen compound 氧化合物
oxygen compound of sulphur 硫氧化物
oxygen concentration 氧浓度;含氧浓度
oxygen concentration in atmosphere 大气中氧浓度
oxygen condtion 好氧条件
oxygen consuming capacity 耗氧能力
oxygen consuming coefficient 耗氧系数
oxygen consuming content 耗氧量
oxygen consuming matter 耗氧物质
oxygen consuming quantity 耗氧量
oxygen consuming rate 耗氧速率
oxygen consumption 氧消耗;耗氧量;氧的消化;耗氧(量)
oxygen consumption ga(u)ge 耗氧计
oxygen consumption rate 耗氧速率
oxygen consumption rate curve 耗氧速率曲线
oxygen consumption type of corrosion 耗氧型腐蚀
oxygen consumptive organics 耗氧有机物
oxygen consumptive pollutant 耗氧污染物
oxygen-containing gas 含氧气体
oxygen content 氧含量;含氧量
oxygen content in clear bottle 白瓶含氧量
oxygen control 氧气设备操纵机构
oxygen controller recorder 氧调节记录仪
oxygen converter 氧化转炉
oxygen converter steel 顶吹氧转炉钢
oxygen convertible phthalic resin 氧转化型邻苯二甲酸醇酸树脂
oxygen convertible resin 氧转化型树脂
oxygen corrosion 氧腐蚀
oxygen-cut 氧气切割
oxygen cutter 氧气切割机;气割炬

oxygen cutting 氧气切割
oxygen cutting machine with magnetic wheel 磁轮氧气切割机
oxygen cutting tip 氧切割喷嘴
oxygen cutting valve 氧切割阀
oxygen cycle 氧循环
oxygen cylinder 氧气筒;氧气瓶;氧化气瓶;储氧钢筒
oxygen cylinder truck 氧气筒搬运车
oxygen debt 氧债
oxygen decarbonizing 氧除碳法
oxygen decompression 吸氧减压(法)
oxygen deficiency 氧亏;缺氧量;缺氧
oxygen deficiency indicator 氧亏指示器
oxygen deficient 氧的欠缺;缺氧的
oxygen deficient condition 缺氧条件;缺氧情况
oxygen-deficient environment 缺氧环境
oxygen deficient indicator 缺氧指示器
oxygen deficit 氧亏;氧不足;亏氧(量);缺氧(量);欠氧
oxygen deficit in subterranean atmosphere 地下空气缺氧
oxygen deficit ratio 氧亏率
oxygen demand 需氧量;氧需要;氧气需求
oxygen demand index 需氧量指数
oxygen demanding carbonaceous material 需氧含碳物质
oxygen-demanding constituent 需氧组分
oxygen-demanding material 需氧物质
oxygen-demanding pollution 需氧污染;需氧处理
oxygen-demanding waste 需氧废物
oxygen-demanding waste(water) 需氧废水
oxygen depletion 氧消耗;氧亏耗;氧耗竭;耗氧量;缺氧
oxygen depletion safety shut off device 防止不完全燃烧装置
oxygen diffusion rate 氧扩散率
oxygen difluoride 二氟化氧
oxygen dissociation curve 氧解离曲线
oxygen dissolution rate 溶氧率
oxygen dissolution reaction 溶氧反应
oxygen distribution 氧分布
oxygen economy 节约氧气
oxygen effect 氧效应
oxygen electrode method 氧电极法
oxygen eliminating agent 除氧剂
oxygen-enriched air 富氧空气
oxygen-enriched combustion 增氧燃烧
oxygen-enrichment 富氧
oxygen environment 氧气环境
oxygen equilibrium 氧平衡
oxygen equilibrium curve 氧平衡曲线
oxygen equilibrium evaluation 氧平衡评价
oxygen equilibrium model 氧平衡模式
oxygen equilibrium parameter assessment 氧平衡参数评价
oxygen equipment 氧气设备;供氧设备
oxygenerator 制氧机
oxygen exhauster 排氧装置
oxygen explosive 液氧炸药
oxygen filling station 灌氧站
oxygen flame-grooving equipment 氧焰开槽设备
oxygen flask method 氧瓶燃烧法
oxygen flow indicator 氧气示流器
oxygen flow meter 氧气流量计
oxygen free 无氧的;不含氧
oxygen-free copper 无氧铜
oxygen-free environment 无氧环境
oxygen-free flux 无氧焊剂
oxygen-free high conductivity copper 无氧高导电性铜
oxygen-free zone 无氧区
oxygen from the air 从空气得到氧
oxygen fugacity 氧逸散度
oxygen furnace steel 吹氧炼钢
oxygen gas 氧气
oxygen gas boosting 吹氧
oxygen ga(u)ge 氧气压力表
oxygen generating plant 制氧车间
oxygen gouging 火焰清铲
oxygen hose 氧气(软)管;供氧软管
oxygen humidifier 氧气湿化计
oxygen-hydrogen 氢氧焊枪
oxygen-hydrogen blowpipe 氢氧焊枪

oxygen-hydrogen weld 氢氧焊
oxygenic photosynthesis 生氧光合作用
oxygen index 需氧指数;氧指数
oxygen inhalation apparatus 氧气吸入器
oxygen inhalator 氧气吸入器
oxygen inhaler 吸氧器
oxygen inhibition 氧阻聚(作用)
oxygen inlet 氧气入口
oxygen installation 氧气设备;氧气厂;氧化厂
oxygen intake 吸入氧气;氧吸收;摄氧量
oxygen intoxication 氧中毒
oxygen ion mobility 氧离子迁移率
oxygen ions 氧离子
oxygen isotope 氧同位素
oxygen isotope epoch 氧同位素阶段
oxygen isotope geothermometer 氧同位素地质温度计;氧同位素地热温标
oxygen isotope thermometer 氧同位素温标
oxygen isotopic geothermometer 氧同位素地质温度计;氧同位素地热温标
oxygenize 充氧
oxygenized air 富氧空气
oxygen jet 氧气喷嘴
oxygen jet steel 吹氧炼钢法
oxygen jetting 吹氧
oxygen lack 氧不足;亏氧(量);缺氧
oxygen lance 氧焰切割器;氧气枪;氧气切割枪;氧气切割器;氧气喷枪;氧气割炬
oxygen lance concrete drilling 用氧气烧枪在混凝土上打孔
oxygen lance cutting 氧矛切割
oxygen lancing 氧焰切割
oxygen limitation safety device 氧气限量安全装置
oxygen limiter 氧气限量安全装置
oxygen line 氧气供应管道
oxygen log 氧测井
oxygen log curve 氧测井曲线
oxygen loss coefficient 耗氧系数
oxygen machining 氧气切割
oxygen-making plant 制氧车间
oxygen mask 氧气面罩;氧气面具;供氧面具
oxygen measurement set 氧气测量仪
oxygen measuring burette 氧气量管
oxygen meter 氧气测量计;测氧计
oxygen microelectrode 氧微电极
oxygen minimum layer 最小含氧层
oxygen monitor 氧气监视器
oxygen monitor controller 氧气监护控制器
oxygenolysis 氧化分解
oxygenotropism 向氧性
oxygen partial pressure 氧分压
oxygen plant 制氧车间
oxygen point 氧点
oxygen poisoning 氧中毒;氧气中毒
oxygen poor layer 贫氧层
oxygen pressure indicator 氧气压力表
oxygen pressure regulator key 氧气压力调节器钥匙
oxygen probe 氧探针;氧探测器
oxygen-producing capacity 产氧能力
oxygen-producing equipment with electrolysis 电解水制氧设备
oxygen-producing plant 制氧车间;制氧车厂
oxygen production potential 产氧势
oxygen production rate 产氧率
oxygen propellant 富氧推进剂
oxygen protection apparatus 氧气防护器
oxygen purity 氧的纯度
oxygen radical anion 氧游离基阴离子
oxygen rate 含氧率
oxygen ratio 氧比;酸度系数
oxygen receiver 氧气瓶
oxygen recorder 电磁测氧仪
oxygen regime 含氧情况
oxygen removal by combustion 燃烧除氧
oxygen requirement 需氧量;氧需要量
oxygen reservoir 储氧囊
oxygen-rich gas 富氧气体
oxygen-rich layer 富氧层
oxygen sag 减氧曲线
oxygen-sag curve 氧下垂曲线;溶解氧下跌曲线;氧垂曲线
oxygen saturation 氧饱和(度)
oxygen saturation capacity 氧气饱和量;饱和量

oxygen saturation value 氧饱和值
oxygen scavenger 氧气清除剂;去氧剂;除氧剂
oxygen sensor 氧传感器
oxygen sequence 氧序
oxygen source 氧源
oxygen starvation 电介质腐蚀;缺氧腐蚀
oxygen station 氧气站
oxygen status 氧气状况
oxygen status of soil solution 土壤溶液内氧化状况
oxygen steel 吹氧钢
oxygen steel furnace 吹氧炼钢炉
oxygen steel making process 氧气炼钢法
oxygen-steel process 氧化炼钢法
oxygen supply 氧源;供氧
oxygen supply control 供氧调节装置
oxygen supply ga(u)ge 储氧计
oxygen supply volume 供氧量
oxygen survey 氧气法
oxygen system 供氧系统
oxygen tank 氧气筒;储氧箱;储氧箱
oxygen tension 氧张力;氧气张力
oxygen tent 氧气帐
oxygen tolerance 氧容限;耐氧量
oxygen tolerance test 氧敏感试验
oxygen top blowing 顶部吹氧法炼钢
oxygen torch 氧气炬
oxygen toxicity 氧气中毒;氧毒性
oxygen transfer efficiency 传氧速率;氧转移效率;氧输送效率
oxygen transfer exponent 氧转移指数
oxygen transfer rate 氧转移率
oxygen-transfer trailer 氧气运输拖车
oxygen transporting system 氧气运输系统
oxygen unit 氧气装置
oxygen uptake 吸氧;摄氧量
oxygen uptake rate 吸氧率;氧吸收率;摄氧速率
oxygen utilization 氧利用;耗氧量
oxygen utilization coefficient 氧利用系数
oxygen utilization rate 耗氧速率
oxygen valve 氧气活门;氧气阀
oxygen weld(ing) 氧气焊接;氧焊接
oxygon 锐角三角形
oxygonal 锐角三角形的
oxyhalide 卤氧化物
oxyhalides hydroxyhalides 氧卤化物和氢氧卤化物
oxyhalogenide 卤氧化物
oxhdroxide 氧氢氧化物
oxy-helium dive 氦氧潜水
oxy-helium diving 氦氧潜水
oxy-helium-nitrogen diving 氦—氮—氧潜水
oxy-helium-nitrogen saturation diving 氦—氮—氧饱和潜水
oxyhelium saturation diving 氦氧饱和潜水
oxyhexaster 锐六角星
oxyhornblende 玄武岩角闪石;玄闪石
oxyhydrogen 氢氧(的);氢氧爆炸气
oxyhydrogen blowpipe 氢氧吹管;氢氧焊枪
oxyhydrogen cutting 氢氧切割
oxyhydrogen flame 氢氧焰
oxyhydrogen flame welding 氢氧焰焊接
oxyhydrogen gas 氢氧爆炸气
oxyhydrogen saturation dive 氢氧饱和潜水
oxyhydrogen saturation diving 氢氧饱和潜水
oxyhydrogen torch 氢氧焊枪
oxyhydrogen voltameter 氢氧电量计
oxyhydrogen weld 氢氧焊
oxyhydrogen welding 氢氧焰焊接;氢氧焊接
oxyhydroxide 氢氧化合物
oxykerchenite 氧纤磷铁矿
o-xylene 邻二甲苯
oxylene (使木材防火的)化学浸透法
oxyliquid 液氧炸药
oxyliquid method 液氧法
oxylophyte 酸土植物;酸性植物;酸性土指示植物
oxyluminescence 氧化发光
oxy-lung 封闭循环式氧呼吸器
oxymagnite 氧磁铁矿
oxymeter 量氧计
oxyn 氧化干性油
oxynaphthoic acid 羟萘甲酸
oxynatural 氧气天然气
oxynatural gas burner 氧气天然气烧嘴
oxynatural gas cutting 氧天然气切割
oxynitrate 含氧硝酸盐

oxynitride 氮氧化合物
oxynitrocarburizing 氧碳氮共渗
oxyopia 视力锐敏
oxypathy 氧中毒
oxyphenisatin 酚丁
oxyphile tectono-geochemical field 亲氧构造地球化学场
oxyphilic plant 喜酸植物
oxyphilous plant 避酸植物
oxyphobe 嫌酸植物;避酸植物
oxyphosphate of copper cement 磷氧铜黏固粉
oxyphoto bacteria 生氧细菌
oxypolygelatin 氧化聚明胶
oxy-propane cutter 氧丙烷切割器
oxy-propane cutting 氧丙烷切割;氯丙烷切割
oxyresinite 氧化树脂体
oxyrygmia 暖酸
oxysalt 含氧盐
oxysemifusinite 氧化半丝质体
oxysphere 氧圈
oxy stop 气氧控制器
oxysulfide sulfoxide 硫氧化物
oxytalan 耐酸纤维
oxytalanolysis 耐酸纤维溶解
oxytank 曝气池
oxythermal coating 氧热涂层
oxythermal repair of refractory 耐火材料的氧热修补
oxytol acetate 醋酸溶纤剂
oxytolerant 耐氧的
oxy welding 乙炔焊接
oyamalite 稀土锆石
oyanalite 大山石
Oyashima disturbance 大八州变动
Oyashima disturbance group 大八州地壳变动群
Oyashio current 寒流;亲潮(日语)
oylet (采光的、通气的)枪眼;孔眼;小孔,窥视孔
oyster 牡蛎;透镜形零件
oyster bed 牡蛎养殖场;牡蛎场
oyster boat 采牡蛎船
oyster cloister 牡蛎养殖架
oyster dredge(r) 牡蛎采集船;牡蛎耙网
oyster farm 牡蛎养殖场
oyster farming 牡蛎养殖
oyster fork 牡蛎叉
oyster gray 牡蛎灰;浅灰色
oystering 木纹理罩面
oyster land 牡蛎养殖场
oyster lime 蛎壳灰
oyster opener 剥牡蛎器具
oyster raft 牡蛎筏
oyster rake 牡蛎耙
oyster reef 牡蛎礁;蚝礁
oyster reef coast 牡蛎礁海岸
oyster rock 蚝壳礁
oyster sand 牡蛎砂
oyster shell 蛎壳;牡蛎;贝壳;蚝壳;牡蛎壳
oyster shell concrete 蚝壳混凝土;牡蛎壳混凝土
oyster shell lime 蚝壳石灰;牡蛎壳石灰
oyster shell scale 牡蛎蚧;牡蛎介壳虫
oyster shell whiting 贝壳粉
oyster tongs 牡蛎铁
oyster white 灰乳色
ozalid 蓝图
ozalid paper 氨熏晒图纸
ozalid print 氨熏晒图
ozalid process 重氮盐印相法
ozamin 苯并红紫
oz av 常衡盎司
oz cast iron 铈硅钙球墨铸铁
oz-ft 盎司英尺
oz-in 盎司英寸
ozobenzene 臭氧苯
ozocerite 石蜡;地蜡
ozokerite = ozokerite
ozokerite deposit 地蜡矿床
ozokerite pitch 硬石蜡
ozonagram 臭氧图(电化学探空仪得出的臭氧分布)
ozonation 臭氧化
ozonation biological activated carbon 臭氧氧化生物活性炭
ozonation-biological activated carbon process 臭氧氧化生物活性炭工艺
ozonation by-products 臭氧氧化副产物
ozonation process 臭氧氧化法
ozonation process in wastewater treatment 废水处理的臭氧氧化法
ozonation sludge minimization technology 臭氧氧化污泥最小化技术
ozonation treatment of wastewater 废水臭氧氧化处理注
ozonator 臭氧发生器
ozone 臭氧
ozone absorption 臭氧吸收
ozone-activated carbon biofilm 臭氧活性炭生物膜
ozone-activated carbon method 臭氧活性炭法
ozone-activated carbon-photocatalysis process 臭氧活性炭—光催化工艺
ozone ag(e)ing 臭氧老化
ozone attack 臭氧侵蚀
ozone balance 臭氧平衡
ozone behavio(u)r 臭氧特性
ozone-biofilm filtration 臭氧—生物膜过滤
ozone biological activated carbon process 臭氧氧化生物活性炭工艺
ozone box 臭氧老化试验箱
ozone budget 臭氧收支
ozone-carbon reaction 臭氧—炭反应
ozone change 臭氧变化
ozone cloud 臭氧云
ozone-coagulation process 臭氧混凝土工艺;臭氧组合工艺
ozone concentration 臭氧浓度
ozone-containing air 含臭氧的空气
ozone content 臭氧含量
ozone control 臭氧控制
ozone crack 臭氧龟裂
ozone cycle 臭氧循环
ozone-damaging emission 损害臭氧的排放物
ozone data 臭氧资料
ozone decline 臭氧减少
ozone decomposition 臭氧分解
ozone decrease 臭氧减少
ozone degradation 臭氧降解(作用)
ozone depleter 消耗臭氧物质
ozone-depleting potential 臭氧消耗潜能值
ozone depletion 臭氧耗竭
ozone-depletion potential 臭氧消耗潜能值
ozone-destroying potential 臭氧消耗潜能值
ozone destruction 臭氧破坏
ozone deviation 臭氧离差
ozone dilution effect 臭氧稀释效应
ozone diminution 臭氧减少;臭氧消耗
ozone disinfection 臭氧消毒法;臭氧消毒
ozone disinfection by-products 臭氧消毒副产物
ozone dosage 臭氧用量
ozone drop 臭氧消耗;臭氧减少
ozone enhanced biologically active carbon 臭氧强化生物活性炭
ozone episode 臭氧异常事件
ozone equilibrium 臭氧平衡
ozone field 臭氧场
ozone-forming species 产生臭氧物质
ozone generator 臭氧发生器
ozone hole 臭氧洞;臭氧层空洞
ozone increase 臭氧增加
ozone index 臭氧指数
ozone-indigo disulphonate spectrophotometry 臭氧—二磺酸靛蓝分光光度计
ozone-induced cracking 臭氧龟裂
ozone-inorganic adsorbent 臭氧—无机吸附剂
ozone instrument 臭氧机
ozone intoxication 臭氧中毒
ozone lamp 臭氧灯
ozone layer 臭氧层
ozone layer degradation 臭氧层退化;臭氧层的衰减
ozone layer depleting potential 臭氧层消耗潜能量
ozone layer depletion 臭氧层耗竭
ozone layer destroy 臭氧层破坏
ozone layer loss 臭氧层损耗
ozone loss 臭氧损耗
ozone machine 臭氧发生器;臭氧机
ozone map 臭氧分布图
ozone measuring device 臭氧测量装置
ozone meauring station 臭氧测量站
ozone meter 臭氧计
ozone minimum 臭氧浓度极小值
ozone model 臭氧模型
ozone modification 人为影响臭氧
ozone modifying substance 臭氧改性物质
ozone monitor 臭氧监测仪
ozone monitoring 臭氧监测
ozone monitoring station 臭氧监测站
ozone observation 臭氧观测
ozone observational record 臭氧观测记录
ozone observing station 臭氧观测站
ozone oxidation method 臭氧氧化法
ozone oxidation process 臭氧氧化法
ozone paper 臭氧纸
ozone partial pressure 臭氧分压
ozone poisoning 臭氧中毒
ozone pollution 臭氧污染
ozone poor air 臭氧浓度低的空气;缺少臭氧的空气
ozone precursor 臭氧化前体
ozone preoxidation 臭氧预氧化
ozone producer 臭氧制造者;臭氧产生者
ozone profile 臭氧的高度分布曲线
ozone proof 耐臭氧性
ozone reaction column 臭氧反应塔
ozone reactor 臭氧反应器
ozone record 臭氧记录
ozone-reducing substance 减少臭氧物质
ozone reduction 减少臭氧
ozone regime 臭氧规律
ozone research manager 臭氧研究管理人
ozone residual 剩余臭氧
ozone resistance 抗臭氧性;耐臭氧性
ozone rich air 臭氧浓度高的空气
ozone scavenger 臭氧清除剂
ozone science 臭氧科学
ozone shield 臭氧屏蔽
ozone's horizontal distribution 臭氧的地平方向分布
ozone sink 臭氧吸收汇
ozone-slow sand filtration biofilm 臭氧—慢速砂滤池生物膜
ozonesonde 臭氧探空仪
ozone sounding 臭氧探测
ozone sounding station 臭氧探测站
ozone spectrophotometer 臭氧分光光度计
ozone spectroscopy 臭氧光谱学
ozone station 臭氧站
ozone steady state 臭氧平衡;臭氧的稳定状态
ozone's vertical distribution 臭氧的高度分布;臭氧的垂直方向分布
ozone synthesis 臭氧合成法
ozone tailgas treatment 臭氧尾气处理
ozone technology 臭氧技术
ozone thickness 臭氧层厚度
ozone thinning 臭氧层逐渐变薄
ozone toxicity 臭氧毒性
ozone transport 臭氧输送
ozone treatment 臭氧处理
ozone treatment of water 水的臭氧处理
ozone trend 臭氧趋势
ozone unit 臭氧单位
ozone value 臭氧值
ozone ventilating plant 臭氧通气设备
ozone ventilation plant 臭氧通气设备
ozone-water contact reactor 臭氧—水接触反应装置;臭氧—水接触反应器
ozonidate 臭氧剂
ozonidation 臭氧化;臭氧化作用
ozonidation method 臭氧氧化法
ozonide 臭氧化物
ozonization 臭氧消毒;臭氧化作用;臭氧化;臭氧处理
ozonization degradation 臭氧化降解
ozonization method 臭氧化法
ozonization plant 臭氧处理装置;臭氧消毒装置
ozonize 用臭氧处理[给];臭氧化;臭氧处理
ozonized air 臭氧化空气
ozonizer 臭氧消毒机;臭氧化器;臭氧发生器
ozonlysis 臭氧分解
ozonlysis reaction 臭氧分解反应
ozonogram 臭氧分布图
ozonolysis 臭氧裂解反应
ozonometer 臭氧计
ozonometry 臭氧测术;臭氧测定法
ozonoscope 臭氧检验器;臭氧测量器
ozonosphere 臭氧圈;臭氧层

P

pabstite 硅锡钡石;锡钡钛石
pace 运动速度;最多车速段;梯台;步距;步幅;步调
pace-board 踏步板
pace car 开路车
pace clock 步时钟脉冲
Paceco portainer 帕斯柯岸壁集装箱装卸桥
pace counter 记步器
pace lap 试车圈
pace length 步距;步长
pacemaker 领跑者
pace method 步法
pace of technologic(al) innovation 科学技术进步的速度
pace out 步测出(一段距离)
pace pulse 步脉冲
pacer 定速装置;步测者
pacer clock 定速时钟
pacesetter 标兵
pace tally 步测计数器
pace voltage 跨步电压
pache 色彩图例;颜色标准;工程量平面及图解
pachimeter 测量机;弹性切力极限测定计;测重计
pachnolite 霜晶石
pachometer 测厚计
Pachuca agitator 帕秋卡搅拌槽
Pachuca tank 帕秋卡槽
pachymeter 厚度计;测厚计
Pacific-Africa crustal-wave system 太平洋—非洲地壳波系
Pacific American Steamship Association 太平洋美国汽船协会
Pacific American Tankship Association 太平洋美国油轮协会
Pacific anticyclone 北太平洋高压;太平洋高气压;太平洋反气旋
Pacific arctic front 太平洋北极风
Pacific area communication system 太平洋地区通信系统
Pacific Area Travel Association 太平洋地区旅游协会
Pacific Basin Economic Co-operation Committee 太平洋地区经济合作委员会
Pacific Basin Economic Council 太平洋地区经济理事会
Pacific coastal trough 滨太平洋海槽
Pacific coast cypress 太平洋岸柏树
Pacific Coast Gas Association 太平洋沿岸天然气协会
Pacific coast spruce 太平洋岸云杉
Pacific Coast Standard Time 太平洋沿岸标准时
Pacific Coast West Bound Conference 太平洋西海航运联盟
Pacific coast yellow cedar 太平洋岸黄雪松
Pacific converter 派西菲克型直接成条机
Pacific coral-volcano chain 太平洋火山—珊瑚岛链
Pacific crustal cupola 太平洋壳状
Pacific daylight time 太平洋夏季时间
Pacific Economic Cooperation Conference 太平洋经济合作会议
Pacific faunal region 太平洋动物区
Pacific Gas, Electric(al) Co. 太平洋煤气电力公司(美国)
Pacific geosyncline 太平洋地槽
Pacific high 北太平洋高压;太平洋高压
Pacific international line 太平洋国际航线
Pacific iron 吊杆坐转轴;鹅颈弯;鹅颈头
Pacific Mail Steamship Company 太平洋邮船公司
Pacific meteorological network 太平洋气象观测网
Pacific North West Container Line 太平洋西北集装箱班轮公司
Pacific Ocean 太平洋
Pacific Ocean area 太平洋区
Pacific Ocean current 太平洋海流
Pacific Ocean route instructions 太平洋航路指南
Pacific orogeny 太平洋运动
Pacific plate 太平洋板块
Pacific polar front 太平洋北极风
Pacific red cedar 太平岸太平红松

Pacific region 太平洋区
Pacific ridge seismotectonic zone 太平洋海岭地震构造带
Pacific Rim 太平洋带状地区
Pacific ring of fire 太平洋活火山带
Pacific salmon 太平洋鲑
Pacific satellite 太平洋区域通信卫星
Pacific silver fir 太平洋岸银杉
Pacific Standard Time 太平洋标准时
Pacific suite 太平洋(岩)套
Pacific temperate faunal region 太平洋温带动物区
Pacific Time 太平洋时间
Pacific-type continental margin 太平洋型大陆边缘
Pacific-type orogeny 太平洋型造山运动
Pacific wind system 太平洋风系
Pacific yew 太平洋岸紫杉
pacing 定步;步测
pacing factor 决定因素;决定性因素
pacing group 定步组
pacing group size 定步组大小
pacing item 决定因素
pacing response 定步响应
pacing technology 起决定作用的技术
pack 装填;流冰群;填塞;打包;穿孔卡片叠;包装法;包装;包裹法;组装;填密;剥蚀
package 小件;一揽子;装箱;壳套;合同包;程序包;标准部件;包装;包件;包(裹);组装
package air conditioner 整体式空气调节器
package arrangement 一揽子方案;一揽子安排
package assembly 组装结构
package barter 一揽子易货
package base 外壳基座
package basis 件装规则;计件标准;包装计算标准;包装规则;包装法
package bid 组合投标;一揽子要价
package board 插件板
package boiler 快装式锅炉;可移式锅炉
package car 零担货车;包裹车
package card 组合卡片
package cargo 件货
package circuit 封装电路
package contract 一揽子(承包)合同
package conveyer 件货输送机
package count 封装数
package counter 包装计数器
packaged 成套的;成套
packaged air conditioner 组合式空调器;整装空调机;冷风箱;冷风机;成套空调机组
packaged and bulk 包装的和散装的
packaged bearing 盒装轴承
packaged boiler 组装锅炉;整装锅炉;快装锅炉
packaged building 成套建筑物
packaged building program(me) 成套建筑计划
packaged circuit 封装阻容
packaged concrete 干包装混凝土;干配料混凝土(按水比加水即可使用);袋装搅拌料
packaged deal 整套工程;整批交易;成批交易;承包人整套工程;一揽子交易;一揽子承包;总价交易;总承包
packaged deal project 整套工程项目
packaged design 紧凑设计;成套设计
packaged dry combined materials 袋装干混合料
packaged dry mortar 袋装砂浆干料
package deal contract 一揽子交易合同
package dealer 设计负责人;承包负责人;总揽承包人;总负责人;总承人
package deal sewage-treatment plant 整体交易污水处理装置;成套污水处理装置
package debug aid system 程序包调试辅助系统
package density 存储密度;包装密度
packaged equipment 小型移动式装置
package design 包装设计
packaged goods regulations 包装商品规则
packaged grout 袋装灌浆干料
packaged heating boiler 快装采暖锅炉
packaged incinerator 封装式焚化炉
packaged industrial treatment plant 快装工业废水处理厂

packaged in place pile 就地灌注桩
package dissipative resistance 管壳耗散电阻
packaged lumber 包装好的木材;成组木材
packaged magnetron 封装磁控管
packaged modularised feed unit 模块化组装式给水装置
packaged mortar 袋装砂浆干料
packaged paint 原装漆
packaged plant 小型装置;组装设备;可移动装置;密封装置
packaged power plant 小型移动式电站
packaged production 小批(量)生产
packaged program(me) 组合程序
packaged reducer 小型减速机
packaged rig 成套钻探设备
packaged sewage treatment plant 快装污水处理厂
packaged sound attenuator for air-conditioning system 用于空调系统的组装式消声器
packaged stacking 成捆堆垛法
packaged substation 整装式变电所;成套式变电所
packaged system 成套系统
packaged technology 成套技术
packaged timber 成组木材
packaged tour 包费旅行
packaged transistor 密封式晶体管;封装晶体管
packaged treatment plant 快装处理厂
packaged unit 小型装置;组装式机组;整装机组;整体装置;可移动装置
packaged vacuum unit 小型真空设备
packaged water chiller 成套水冷却器
packaged water system 快装给水厂
package file management program(me) 程序包文件管理程序
package for rolling stocks 车辆设备包
package freight 零担货(物);包装货(物)
package goods 件货;包装零售的货物
package handling 包装物件搬运;件货装卸
package in damaged condition 包装破损
package inspection and verification 封装检验
package insurance 整批保险
package investment 一揽子投资
package job contract 整套承包合同
package kitchen 小厨房;紧凑式厨房;组装式厨房
package lead 组件引线
package limitation of liability 承运人赔偿责任限制
package loan 一揽子贷款
package mill 打包用带钢轧机;打包带钢轧机
package mortar 干配料砂浆
package mortgage 总体抵押;包括家用设备在内的住宅贷款
package number 包装件号
package of communication system 通信系统包
package of environmental control system 环控系统包
package of equipment 成套设备
package of escalator and elevator 电梯扶梯包
package of fire alarming system and building automatic surveillance system 防灾报警及设备监控系统包
package of flexible turnout 可挠型道岔设备包
package of joint turnout 关节型道岔设备包
package of platform screen door 屏蔽门包
package of power supply equipment 供电设备系统包
package of service program(me) 服务程序包
package of signal(l)ing equipment 信号设备包
package of veneer 镶面板饰作;镶面板包装
package or units covered by the contract 合同规定的一批或各单位
package plan 一揽子计划
package plant 打包工场
package plant export 整厂输出
package price 一揽子价格;组合价格;混合价格
package printing 包装印刷
package program(me) 一揽子计划
package project 一揽子项目
package proposal 一揽子建议
packager 打包机;包装机

package reactor 快装式反应堆
package saving 综合储蓄
package sealing with laser 激光封装
package sewage-treatment plant 装配式污水处理厂;成套污水处理设备;小型拼装污水厂
package shell 组装外壳
package solution 一揽子解决办法
package stability 储[贮]存稳定性;储藏耐久性;包装稳定性
package store 不供坐饮的小酒店
package technique 组装技术
package technology 包装工艺
package terminal 封装引出线
package test 封装测试
package tour 旅行社包办的旅游;包费旅行
package transfer 成套转让
package transport 成建运输
package treatment plant 小型处理装置;移动式污水处理装置;成套处理厂
package trim 商品门窗细木工;商品门窗细工;预制门窗贴脸
package trust deed (mortgage) 全部不动产的信托抵押借据
package tunnel kiln 装配式隧道窑
package turbine 快装式涡轮机
package type air cooler 便携式空气冷却器
package type construction 预制装配构造;轻便式构造;预制装配式结构;装配式施工;轻型装配机构
package-type sprinkler system 组件式喷水灭火系统
package-type steam generator 快装式锅炉
package unit 成套设备;成套机组;小型装置单元;整套装置;独立装置;成套装置
package weight 成套机重量
packaging 集装;打包;封装;包装(法)
packaging capacity of waste 废物包装容量
packaging conveyer line 包装流水作业线
packaging cost 包装成本
packaging density 组装密度;封装密度
packaging facility 包装设备
packaging industry 包装工业
packaging ink 包装印刷用油墨
packaging machine 打包机;封装机
packaging material 包装材料
packaging paper 包装纸
packaging plant 包装设备;包装工厂
packaging quality 包装质量
packaging standard 包装标准
packaging system 包装系统
packaging unit 包装设备;包装工厂
packaging waste 打包废物
pack alloy 压铸铝合金
pack and bale 叠架捆货法
pack animal 驮畜
pack annealing 装箱退火(板材);叠式退火;成捆退火
packback diamond drill (可背运的)轻便金刚石钻机
packbasket 背篓
pack-basket shop 背篓商店
pack builder 煤矿井下充填工
pack calorizing 固体铝化(处理)
pack carburizing 固体渗碳;固体渗碳;包围渗碳法
pack chromizing 固体渗铬
pack cloth 打包粗麻布;包装用布
pack compression 致密充填;填料收缩
pack duck 打包帆布(黄麻制);包装用布
packed 填充的;成组的
packed absorber 填充式吸收器
packed absorption 填充式吸收塔
packed absorption column 填充式吸收塔
packed absorption tower 填充式吸收塔
packed air handling unit 柜式空调机
packed array 组装数组;紧缩数组;合并数组
packed attribute 压缩属性
packed bed 填充床;填充层;填料床
packed-bed dedusting scrubber 填料床洗涤除尘器
packed bed denitrification 填充床脱氮作用
packed bed filter 滤清器;填充层过滤器
packed-bed process 填充床法
packed-bed reactor 填料床反应器
packed bed scrubber 填充床洗涤器
packed boat 小型油船
packed bottom 填实座

packed brick 包装砖
packed broken rock soling 填充碎石作为铺砌基础
packed bulk density 包装密度
packed capillary column 填充毛细管柱
packed cargo 件货;包装货物
packed chord (型钢的)弦杆组合构件;组合弦杆
packed column 填充塔;填充柱;填充塔
packed column for absorption 吸收填充塔
packed data 集的数据
packed deal 一揽子交易;一揽子承包
packed decimal 压缩(式)二进制(数)
packed decimal number format 压缩十进制数格式
packed decimal system 组合式十进制
packed density 填充密度;搞实容重
packed distillation column 填充(蒸馏)塔
packed drain 袋装砂井
packed extraction column 填充式萃取塔
packed extraction tower 填充式萃取塔
packed extractor 填充提取塔;填充萃取塔
packed factor 填充因素
packed format 合并形式
packed freight 成件包裹货物
packed goods 件货
packed heat insulation 填充热绝缘;填充保温
packed height 填充高度
packed hole assembly 组合钻具(由稳定器、钻铤、扩器组成);满眼钻具组合;满眼钻具
packed hole drilling 满眼钻进
packed hole drilling technology 满眼钻进工艺
packed hole technique 用接近井径钻铤钻进技术(一种防井斜技术措施)
packed ice 密集冰
packed injection 填塞注浆
packed-in-situ pile 钻孔灌注柱
packed joint 包垫接头;填实缝;填塞(接)缝
packed layer 填充层
packed parcel 小件包装商品
packed rail beam 轨束梁
packed reaction column 填充反应柱
packed reaction tower 填充反应塔
packed refuse 打包垃圾
packed sand drain 袋装砂井;袋砂井
packed slip joint 填函连接
packed snow avalanche 固结雪崩
packed snow runway 压实的雪跑道
packed soil 夯实土(壤);素土夯实;搞实土(壤)
packed solid 充实固体
packed space 填充容积
packed spray tower 填充喷淋塔;填充淋塔
packed stone 挤紧石块;挤紧的石块;垛紧的石块
packed stone revetment 填石护坡;填石护岸
packed tower 填料塔;填充塔
packed tower aeration 填料塔曝气;填料曝气塔
packed tower filter 填料塔式滤池;填料塔式过滤器
packed volume 填充容积;堆砌体积
packed weight 包的毛重;带包装重量;包装重量
packer 压土机;灌浆器;煤矿井下充填工;填塞圈;堵水塞;堵塞器;搞锅;打包机;采油封隔器;包装压紧机;包装机;包装工
packer bin 包装机料斗
packer feed bin 包装机喂料斗
packer grouting 栓塞灌浆(法);分段灌浆法
packer guarantee clause 卖方保证条款
packer-head pipe-making machine 打包机头式混凝土制管机
packer-head process 混凝土管垂直浇制法;打包法;立式旋转挤压制管工艺
packer hopper 包装机料斗
packer permeability test 压水试验
packer(-plug) 封隔器
packer ring 密封圈
packer rubber 封隔器橡胶碗
packer rubber grab 橡胶碗打捞器
packer's bay 包装室;包装间
packer seat 封隔器座
packer surge bin 包装机缓冲料斗
packer system 填塞注浆
packer test 压水试验;钻孔压水试验
packer with expanding shoe 带膨胀鞋的封隔器
packery 罐头厂
packet 小件行李;信息包;小捆;消息包;一捆;传输组;方块体;包裹;包
packet adapter 分组适配器
packet address recognition 包地址识别

packet assembler and disassembler 包装和解装程序
packet assembler disassembler 打包和解包设备
packet assembly 包装装配
packet assembly/disassembly 分组装/拆
packet bit 包信息位
packet boat 邮政船;客货班轮
packet control 包控制
packet data network 分组数据网络
packet day 邮件截止日;邮船开航日
packet delay 包延迟
packeted oil 填充油
packet error detection 包错误检测
packet file 粗方锉
packet folio 大四开图纸
packet format 分组格式
packet framing 包成帧
packet gateway 包连接器
packet handler 分组处理器;包处理器
packet interleaving 包交替
packetize 分组化;分成包;包化
packetized voice network 分组化话音网
packet length 包长度
packet length selection 分组长度选择;包长度选择
packet level error control 包级错误控制
packet level logical interface 包级逻辑接口
packet martensite 束状马氏体
packet mode 分组方式;包方式
packet mode operation 包方式操作
packet mode terminal 分组式终端;包方式终端
packet multiplexer 分组多路复用器
packet multiplexing technique 包多路转换技术
packet network 包交换网
packet processing 包处理
packet processing flow 分组处理流程
packet radio network 分组无线网
packet repeater 分组中继器
packet retransmission 包重传输
packet retransmission interval 包重发间隔
packet sequencing 分组排列;包排序;包定序
packet ship 客货班轮
packet switch 信息包交换;包交换
packet-switched data network 包交换数据网络
packet switched data transmission service 包交换数据传输服务
packet switching 分组接转;分组交换;包交换
packet switching center 分组交换中心;包交换中心
packet switching data network 分组交换数据网络
packet switching exchange 包交换
packet switching network 分组交换网络;分封式交换网络
packet switching processor 分组交换处理机
packet switching protocol 包交换协议
packet switching service 包交换业务
packet switching unit 包交换部件
packet system charge 包交换方式负荷
packet technology 包交换技术
packet terminal 包交换终端
packet transfer procedure 包传输程序
packet transmission 群发送;包传输
packet transmission controller 分组传输控制器
packet type switch 组合式开关
pack film 密封胶卷
packfong 锌白铜
pack for shipping 包装以备发货
pack-hardening 装箱渗碳硬化;装箱渗碳;固体渗碳硬化
pack-hardening furnace 装箱渗碳炉
pack heating furnace 叠板加热炉
pack hole 充填孔
pack horse 驮马
pack-house 肉类果品加工包装厂;堆栈;仓库;包装车间;包装成品工厂;库房
pack ice 流冰群;冰块;密集冰块;大块浮冰(群)
pack ice zone 浮冰区域
pack in bags 袋装
pack in dozen 一打装
packing 填料;填装;填实;填塞料;填密;填集;填充(包装);楔子;装色谱柱;淋水填料;效集;加填充物;集合体;货物装箱;盘根;图像压缩;填集;垫渣;垫料;搞固;衬垫;背衬;包装物;包扎;剥蚀
packing and commodity protection 包装与商品保护
packing and labelling regulation 商品包装和标

签规定
packing and marks clause 包装标志条款
packing and presentation 包装装潢
packing and wrapping expenses 包装费
packing arrangement 存储排列
packing article audit 包装物审计
packing assembly 密封装置
packing bearing 垫层支座
packing bin 包装机喂料仓
packing block 填塞块;衬层
packing board 包装纸板
packing bolt 密封螺栓;填密螺栓
packing box 装料箱;密封盒;货箱;货物箱;填料箱;填料函;打包箱;包装箱;包装盒
packing box compensator 填料箱补偿器
packing bush 填料压盖衬套
packing cargo 包装货(物)
packing carrier 皮带盘垫套
packing case 转运货物的箱子;密封盒;货箱;填密盒;填料箱;打包箱;包装箱;包装盒
packing chamber 填料盒;填料函
packing charges 装箱费;包装费
packing chest 包装箱
packing chromatography 填充色谱(法)
packing clamp 压紧箍;封严箍
packing clause 包装条款
packing cloth 打包麻布
packing coefficient 级配系数;密实系数
packing coefficient of checkwork 格子体填充系数
packing collar 垫圈
packing column 填料塔
packing compact 装模
packing company 包装公司
packing compound 填孔剂
packing compressor 填料压缩器
packing concrete 捣实混凝土
packing concrete in forms 模内捣实混凝土;模壳捣实混凝土;捣实模内混凝土
packing cone 紧塞锥
packing container 包装箱
packing container test method 包装货物试验方法
packing cord 填料绳
packing cost 包装成本
packing counter 包扎柜台
packing course 嵌片层;衬垫层;填实层;填层;垫层
packing degree 填实度
packing density 充填度;装填密度;敛集密度;夯实密度;群集密度;排列密度;填充密度;堆积密度;捣实密度;存储密度;充填密度
packing department 包装车间
packing design 包装设计
packing dimensions 装箱尺寸
packing disposal 填埋处置;填埋处理
packing dryer 填充物干燥器
packing edge 密封齿缘
packing expander 垫料胀圈
packing extra 包装费(用)另计
packing factor 记录因子;填料因子;填充因子;填充系数;堆积系数;存储因子;填充因数
packing felt 毡衬;毡垫
packing filler 填充料;充填料
packing filling 填充物
packing filter 填料
packing firm 包装公司
packing flange 止水法兰;止水凸缘;密封涨圈;密封凸缘
packing fluid 封填液;封隔液
packing for cartridge cover 滤筒顶盖垫圈
packing force 压紧力
packing for discharge cover 卸渣门垫圈
packing for foot valve 底阀座垫片
packing fraction 敛集率;紧束分数;填充系数;填充比
packing fraction loss 敛集率损耗
packing gas chromatography 填充气相色谱法
packing gasket 密封衬垫
packing gland 密封压盖;密封套;密封衬垫;填函盖;填料压盖;填料函(压盖);填料盖;填料函料
packing gland lock spring 填充块压盖弹簧
packing gland nut 填函压盖螺母;填料函压盖螺母
packing grease 密封润滑脂;密封黄油
packing groove 盘根槽;填料槽
packing gun 驮载炮
packing hardening 填料硬化

packing head 填料头;封隔盖
packing height 填料高度;填充高度
packing holder 密封圈座
packing hook 填料钩
packing house 罐头厂;肉食包装厂;屠宰加工厂;食品加工厂;食品罐头厂;副食品加工场;包装车间;肉类冷藏所;肉食加工厂
packing house byproduct 食品加工副产物
packing house pitch 仓库沥青
packing house wastes 肉类加工废物
packing house waste(water) 肉类加工厂废水;食品加工厂废水
packing index 填集指数;堆积指数
packing industry 包装工业
packing inspection 包装检验
packing joint 密封连接
packing knife 填料割刀
packing layer 填料层;填充层
packing leakage 密封漏泄;汽封漏汽
packing leather 皮碗
packingless 无密封的
packingless pump 无密封的泵
packing letter of credit[L/C] 打包信用证
packing list 装箱清单;装箱明细表;装箱单;包装清单;包装明细表
packing list in lain paper 无卖方名称的包装清单
packing loan 打包贷款
packing lubrication 填充润滑
packing machine 打包机;包装压紧机;包装机
packing machinery 包装机械
packing maker 密封器;密封接合器
packing manner of sample 样品包装方式
packing mark 包装标志
packing material 密封填料;密封材料;填实料;填塞料;填料物;填充物;填充材料;充填料;衬料;包装材料;包装用料
packing measurement 装箱尺寸;包装尺寸
packing medium 填料介质;填充介质;包装填料
packing method 存储方法;填料方法;填法
packing module 包装模数
packing module dimensions 包装模数尺寸
packing needle 捆针;打包针
packing non-uniforming factor 填物不均匀因数
packing note 包装记录单
packing nut 压紧螺母;紧塞螺母;密封螺母;填密螺母;衬垫螺母
packing nut wrench 填密螺母扳手
packing of a chromatography column 填充色谱柱
packing of aggregate 集料填实性
packing of column 柱填充
packing of data 数据群集
packing off 密封
packing of orders 指令组合
packing of pipe joints 管接头填塞;管缝填塞;管接头密封;堵塞管缝
packing of sleepers 轨枕衬垫
packing of stuffing box 填料箱密封填料
packing of tube 管子填料
packing paper 垫(料)纸;包装纸;包皮纸
packing paper ink 包装纸用油墨
packing pawl 装填器
packing pick 填集料镐
packing piece 垫块;填塞件;衬垫(物);填塞块;填密片;衬片
packing plant 肉食加工厂;屠宰加工厂;食品加工厂;食品罐头厂;副食品加工厂;包装车间
packing plate 密封垫圈;密封垫;板;垫板;包装板
packing press 包装机;填料压机;打包机;包装压紧机
packing procedure 填充手续
packing proximity 填集趋近度
packing radius 堆积半径
packing ratio 装填率;填充比
packing restrainer 填料压板
packing retainer 填密函盖
packing retainer spring 填密函盖弹簧
packing ring 轴封环;胀圈;密封圈;密封环;活塞环;填密圈;填料环;填料环;填料隔圈;垫圈
packing rock 填充用块石;堆砌块石
packing routine 压缩程序
packing rubber 橡胶垫圈;填料橡胶
packing scale 包装秤
packing screw 填料螺旋起子;衬垫螺旋
packing seal 填料密封

packing service 封装服务;包装业
packing shed 打包间;包装室;包装间
packing sheet 填装片;包装纸
packing shift 打包班组;包装班组;手工包装作业线
packing size 填充物尺寸
packing sleeve 圆环轴套
packing slip 装箱单
packing space 填密空间;填充空间
packing specification 货物内容说明书;包装单;包装规格
packing spool 填隙环
packing spring 密封圈弹簧;填密簧;填料弹簧
packing stick 包装填充用板条
packing strip 轴瓦垫片;填密片
packing structure 填充物结构
packing stuffing 集装箱装箱作业
packing supplies 包装用品
packing support 填料支承架;充填物支架;填充物支撑
packing surface 密封面
packing table 包装台
packing theory 紧凑理论
packing tightening 充填物的紧密化
packing timber 包装用材
packing tool 捣实工具
packing tower 填料塔
packing treatment 填埋处理
packing unit 包装工厂
packing up 包装
packing-up and assembly 收工集合
packing vessel 内装填料的容器
packing void 装填空隙度
packing void fraction 填料空隙度
packing volume 填充容积;包装体积
packing washer 压紧垫圈;密封环;密封垫圈;垫圈
packing water seal 水密封口
packing worm 填料螺旋起子;填料钩
pack in sacks 袋装
packless 无填料的;无填充的;无密封的;无衬垫的;无包装的;未包装的
packless diaphragm valve 无填塞膜片阀
packless joint 无填料接合
packless matrix 不垫背纸型
packless pump 无填料泵
packless seal 无包扎密封
packless valve 无套阀;无填料阀
pack line 包装线;包装作业线
pack man 小贩
pack marks 堆垛痕迹;堆垛痕迹
pack mill 叠板轧机
pack of bricks 砖砌体
pack of clay brick 黏土砖的包装
pack off 解雇
pack on all sails 张满全风帆
packplane 货舱能更换的飞机
pack powder calorizing 固体铝化(处理)
pack pressure dynamometer 充填体测压计;充填体测力计;充填体内压力测定器
pack radio section 军用无线电设备
pack road 驮马道
pack-rolled 叠轧的
pack-rolled sheet 叠轧薄板
pack-rolling 叠板轧制
pack rope 打包绳
packsand 密实砂;细砂岩;细粒砂;易破碎的细粒砂岩
pack set 结块;压实凝结;结拱
pack-sintering 装箱烧结
pack stability 储[贮]存稳定性
packstone 泥粒(状)灰岩;泥粒(石)灰岩
pack test 包装测验
pack thread 打包线;包装绳
pack tilting device 叠板翻转机
pack twine 包扎麻绳
pack type dialyzer 平板型透析器
pack unit 箱装部件;部件;步话机;背包电台
pack up 收拾;打包;包装垫高;包装垫块;收工
pack wall 废石垛墙
packway 土路;马道
packwork 打包工作;填梢工作;填塞作业;填塞工程
PA container 轨道滑行集装箱
pact 协定;合同;公约;契约;条约
pacteron 铁碳磷母合金

pad 制动块;固定衰减器;梁垫;活动房屋区的小路;缓冲垫;缓冲层;起落场地;台垫;垫片;垫;床垫;充填;衬垫;插入;覆板
padauk 紫檀硬木(产于非洲、缅甸、印度和安达曼群岛)
padauk furniture 红木家具
pad bearing 垫块支座;垫层支座;带油垫轴承;衬垫轴承
pad control 衰减器控制;衰耗控制;衰耗
paddauk 花梨木
padded armchair 沙发椅
padded bill 虚报的账单
padded bit 槽顶钻头
padded cell 装有衬垫的病房;软垫病室
padded door 装有衬垫的门;衬垫门;衬垫板
padded room 软垫病室
padded shoulder 垫肩
padder 轧辊;浸扎车;浸染机;微调电容器;涂漆用橡胶片;垫整电容
padder extractor 轧水机
padding 虚报账目;虚报开支账目;橡胶片涂漆;统调;填料;填积层(焊接);填充物;填充物;填充数;填充;垫料;大石块
padding bit 填充位
padding character 填充字符【计】
padding cloth 衬布
padding data 装填数据
padding machine 压染机;打底机
padding mangle 轧染机;浸染机;初染机
padding material 填密材料
padding operation 填充操作
padding steel reinforcement 垫筋
paddle 小铲;两头桨;搅料桨;桨;明轮桨;划桨;平桨;太阳电池中片;短桨
paddle aeration system 桨板(式)曝气系统
paddle aeration tank 叶轮式曝气池
paddle aerator 叶轮式曝气池;桨式曝气器;桨板(式)曝气器
paddle agitator 桨式搅拌机;叶片搅拌器;桨式搅动器
paddle and anvil 拍打成型
paddle badminton 板羽球
paddle bar 闸门
paddle beam 明轮梁;明轮推进器承梁;明轮架
paddle bitumen emulsifying machine 叶桨搅拌式沥青乳化机
paddle board 桨板
paddle board of water wheel 水轮叶板
paddle boat 明轮船
paddle box 明轮推进器罩;明轮壳
paddle conveyer 刮板式输送机
paddle dampening device 叶轮式减振装置
paddled conveyor chain 刮板输送链
paddle door 闸门;漏斗门;放水门
paddle engine 明轮发动机
paddle feeder 桨叶式给料机;桨叶给料器
paddle gate 闸门
paddle gummer 叶片式除粉机
paddle hole 闸门输水口;进水出水孔
paddle jig 桨叶式跳汰机
paddle level switch 扳钮开关
paddle loader 桨式装载机
paddle mechanism 桨板搅拌机制
paddle mill 叶片式洗矿机
paddle mixer 叶片(式)搅拌机;叶片式拌和机;桨叶搅拌机;桨式混合机;桨轮搅拌机;桨轮式拌和机
paddle mixer shaft 叶轮搅拌机轴
paddle passenger steamer 明轮客轮
paddle pump 叶片泵;桨式泵
paddler 明轮船
paddle reaction chamber 叶轮式反应室;轮叶式反应室
paddle reaming bit 翼状扩孔钻头
paddle roller 叶片辊
paddle screw conveyer 叶片式螺旋输送器
paddle shaft 叶轮轴;桨轴;明轮轴
paddle star (强制式搅拌机的)星形桨架
paddle steamer 明轮船
paddle stirrer 桨(叶)式搅拌机;叶板搅拌器;桨板搅拌器
paddle-type agitator 叶片搅拌器;桨式搅动器;桨式搅拌器
paddle-type bunker discharge carriage 叶轮式料斗卸料车

paddle-type coiler 叶片式卷取机
paddle-type elevator 刮板升运器
paddle-type mixer 转臂式混砂机;桨式搅拌混合机;桨叶式搅拌器;桨式拌和机
paddle-type nodulizer 叶片式成球机
paddle-type solar array 桨叶式方阵
paddle-type stirrer 桨叶式搅拌器
paddle valve 闸板阀
paddle vessel 明轮船
paddle wheel 叶轮;桨轮;明桨;径向直叶风扇轮
paddle wheel aeration 桨轮曝气(污水处理池用)
paddle wheel aerator 叶轮曝气器;叶轮式搅拌机;叶轮式混合器
paddle wheeler 装有明轮的船;明轮船
paddle wheel fan 叶轮风扇;叶轮式通风机;离心式鼓风机
paddle wheel house 明轮罩
paddle wheel propeller 明轮推进器
paddle wheel scraper 叶轮式铲运机
paddle wheel sensor 叶轮传感器
paddle wheel steamer 叶轮式汽轮机;明轮汽船;明轮船
paddle worm conveyer 叶轮式螺旋输送机
paddling 划动
paddling-machine 搅麻机
paddling pool 儿童浅水池;搅拌池
paddling process 搅拌法
paddock 临时堆集(矿石、石料);畜牧围场;围起的土地;小牧场;草地中围地;驯马围场;围场
paddock grazing 划片轮牧场
paddock grazing method 划片轮牧法
paddock holder 挂锁支架
paddock sheave 铲斗的滑车轮
paddress 调换数据地址;程序转移地址
paddy bit 张刃式钻头
pad dy(e)ing 轧染
paddy field 水田;稻田
paddy field boat 水田机耕船;水田船
paddy field harrow 水田耙
paddy field pesticide 水稻田用农药
paddy fields with dense hydrographic(al) system 水网稻田地
paddy-pounder 碾米机
paddy rice 稻谷
paddy's hurricane 无风而下微雨
paddy soil 水田土
pad eye 眼板;垫板孔眼;扣板
pad eye for mooring 系缆眼板
pad footing 垫块状基础;衬垫基础
pad foundation 块形基础;独立基础;垫式基础
pad hook 板柄钩
pad ink 打印色
pad insertion method of packing 垫板起道法
pad keeper 垫限位器
padlock 挂锁;扣锁;荷包锁;铲斗滑轮
pad lubrication 填料润滑
pad lubricator 填料润滑装置;垫式润滑器
pad material 垫料
Padnos method 帕德矛斯法
pad of equalizer bar 平衡梁鞍座垫
pad pore of filter 反滤层孔隙
pad printing 凹版移印
pad-rolling 滚垫法
pad sander 衬垫砂磨机
pad saw 鸡尾(手)锯;手锯;斜形狭圆锯;小圆锯;弓锯;嵌锯
pad shoe 千斤顶后座
padstone 梁端垫块;垫石;承梁垫石
pad support 垫块支承网;垫托
pad support stone 支承垫石
pad the log book 在钻探报表上记进尺
pad-transfer printing 墨辊转移印刷
pad value 衰减值
pad welding 垫块焊接
paediatric hospital 儿童医院
paediatrics 小儿科;儿科
paediatric ward 儿科病房
Paeonia peregina 欧洲牡丹
Paeonia suffruticosa 牡丹
paeudocotunnite 氯铅钾石
pagan arena 竞技场(古罗马)
pagan basilica 非基督教的长方形教堂;教堂(罗马)
pagan dom 异教徒居住区

pagan temple 非基督教神殿
page 有平行铁杆的横梁;短而薄的楔子;小木楔;编辑页
pageable dynamic(al) area 可调页动态区
pageable nucleus 可调页核心程序
pageable partition 可调页分区
pageable region 可调页区
page addressing 页面寻址
pageant 华饰;露天表演
pageant arena 露天表演场
page boundary 页面边界;页界
page change 更换页面
page copy 页面复制;页拷贝;纸页式副本
page cord 捆版线
paged addressing 按页寻址
page description language 页面描述语言
page down 往下翻页
paged segmentation 页式分段
page entry 续页入口;页面入口
page exit 页面出口
page facsimile recorder 纸页传真记录器
page fault 页面出错
page feed-out 纸页送出
page frame 页框
page ga(u)ge 版面量规
page grid 页面格网
page image 静止图像
page-in 进页面
page layout 页面格式
page mode 页式
page number of report 报表页面数
page-out 出页面
page printer 页式打印机
page proof 拼好版的打样;单页清样
pager 寻呼机
page recording 页式记录
page register 页寄存器
page search 分页检索
page station 呼唤人员站
page swap file 页交换文件
page swapping 页面交换
page table look-up 查页表
page template 页面模板
page transmission 按页发送
paginal translation 对译
paginate 标页数;标页号
pagination 标注页码
paging 页面调度;广播传呼;调度页面;分页【计】
paging channel 寻呼信道
paging receiver 寻呼接收机;播叫接收机
paging system 呼唤系统;分页系统
paging telephone set 寻呼电话【船】
paging view terminal 页式显示终端
paging zone 寻呼区
pagium 冰积土演变
pagoda 佛塔;宝塔
pagoda for Buddhist relics 舍利(子)塔
pagoda shaped jar 塔形罐
pagoda stone 叶蜡石;宝塔石
pagoda-tree 塔状树木(如槐、榕树等);榕树
pagodite 叶蜡石;寿山石;冻石
pagon 冰内生物
pagoscope 测霜仪
pago stick 退滩装置
paha 小冰川脊;冰积丘
pahoehoe flow 结壳熔岩流
pahoehoe lava 绳状熔岩
pahoehoe lava structure 绳状熔岩构造
paid 已支付的;收讫;付讫
paid absence allowance 假期补贴
paid annual vacation 薪资照付的年度休假
paid area 付款区;付费区
paid case book 现金支出账簿;现金支出账
paid cash 支付现金
paid check 已付支票
paid dearly for 高价购买
paid debt 还清的贷款
paid dividends 已付股息;已付股利
paid fire department 职业消防局
paid firefighter 职业消防战斗员
paid home 送货到户费已付
paid in 已缴;认缴;实缴
paid in advance in quarterly instalments 分季预缴
paid-in capital 已认股本;缴入股本;实缴股本;已

缴资本;投入资本;实收资本
paid-in capital increase 实收资本增加
paid-in capital in excess of par value 超面值缴入资本
paid-in capital in excess of stated value 超设定值投入资本;超定值缴入资本
paid-in investment 实投资本
paid-in portion 实缴部分
paid-in slip 存款单
paid-in stock 已收股本
paid labo(u)r 有酬劳动
paid leave 薪资照付的假期
paid man 职业消防战斗员
paid off 付清
paid on presentation 交单付款
paid other expenses 已付其他费用
paid public holidays 薪资照付的例行假日
paid rent 已付租金
paid salaries 已付薪金
paid service indication 纳费业务标识
paid sick leave 薪资照付的病假
paid-up 付讫
paid-up capital 缴清股本;已缴资本;已缴股本;已付资本;已付股本;实收资本;实收股本
paid-up cheque 付讫支票
paid-up insurance 已付款的保险
paid-up insurance premium 已清缴保费;已缴清保费;已缴付的保费
paid-up license fee 已缴许可证费
paid-up loan 已还清的贷款;还清的贷款
paid-up policy 保费付足的保险单
paid-up share 已缴股份;缴足股本
paid-up stock 已缴股本;实收股本
paid-up transport revenue 已缴运输进款
paid use of state-owned resources 国有资源有偿使用
paid vacation 照付薪资的休假;带薪休假
paid voucher file 已付凭单档案
paid vouchers 已付款凭证
paid waste book 废物支出簿
pail 小漆桶;桶;提桶;水桶;吊桶
pail closet 桶厕;便桶式厕所;便桶间
pail for used dressings 废物桶
paillasse 草褥;圬工基座
pail latrine 旱厕;便桶式厕所
paillette 装饰中的小块彩色金属箔取得闪烁效果
paillon 测厚金属箔
pailoo 牌楼(中国)
pail privy 便桶
painfully loud sound 疼痛的强声;声的痛阈
painite 硅硼钙铝石;铝硼锆钙石
pain roll crusher 光辊破碎机
paint 油漆;漆;涂色;涂抹;涂料
paintability 可涂刷性;可涂色性;可涂漆性;涂覆性能
paintable 可涂刷的
paintable finish 可涂色性加工;可涂漆性加工
paint agitator 油漆搅拌器
paint air gun 油漆喷枪
paint-and-batten process 涂油漆法
paint and varnish naphtha 油漆溶剂油;漆用石脑油
paint and varnish remover 脱漆剂
paint atomizer 油漆喷雾器
paint baking 漆层烘烤
paint baking oven 烤漆炉;烤漆烘箱;烘漆炉
paint base 油漆基料;涂料载色体;油漆打底;漆基;底漆
paint binder 漆黏合剂
paint blower 压缩空气喷涂器;喷枪;喷漆器
paint box 色彩调制箱;颜料盒;绘画箱
paint bridge 桥架(副剧舞台上绘制布景用的)
paint brush 油漆刷;漆刷子;涂料刷;画笔;漆刷
paint brush bristle 涂料刷毛
paint brush ferrule 涂料刷箍
paint brush handle 涂料刷柄
paint brush ring 画笔套圈
paint bucket 油漆提桶;漆罐
paint burning 燃烧除漆;加热清除油漆
paint can 漆罐
paint catapult 涂料弹涂机
paint checking 油漆裂缝
paint chemist 油漆化学家;涂料化学家
paint chipping chisel 刮漆刀;铲漆凿
paint circulating system 油漆循环系统
paint clay 色调;油漆黏土

paint-coat 油漆面层;涂层;漆层
paint coat cushion 涂油垫层;油漆垫层
paint-coated 着色的
paint-coated plate 涂层片
paint coat failure 漆膜损坏
paint coating 涂漆;涂漆层
paint conditioner 调漆器
paint conditioner tank 调漆罐
paint consistency 涂料稠度;油漆稠度
paintcrete 牢固性油漆(用于混凝土或石料表面)
paint defect 油漆疵病
paint degradation 油漆变质
paint-destroying agency 油漆破坏剂;涂料破坏剂
paint display 油漆涂绘手法
paint distributing roller 涂料分配辊
paint-drainage hole 多余油漆流出孔
paint dryer 油漆干燥剂;油漆催干剂;涂料催干剂
painted 染色的;涂漆的
painted china 彩色瓷器
painted corridor 彩绘回廊
painted decoration 油漆装饰;彩绘装饰
painted decorative finish 油漆装饰面层
painted earthen and lacquer animal 彩陶漆动物
painted earthen animal 彩陶动物
painted egg 彩画蛋
painted eggshell 彩蛋
painted enamel 涂刷的瓷漆;着色瓷釉;绘画珐琅;画珐琅
painted enrichment 涂刷的增饰;油漆饰面
painted film 漆膜
painted glass 印花玻璃;彩绘玻璃;彩色玻璃;彩玻璃
painted lacquer 彩漆
painted motif 油漆的主题
painted ornamental finish 涂刷的装饰面层;油漆饰面
painted plastic piece 涂膜塑料片
painted pleasure boat 画舫
painted porch 彩画游廊
painted pottery 彩陶
painted road stripe 涂漆标线;油漆标线
painted sculpture 彩塑
painted sheet metal 油漆的金属板
painted shell 彩蚌
painted silk fan 彩绘绢扇
painted steel 涂漆钢板
painted surface 涂漆面
painted tile mural tablet 瓷砖壁画
painter 油漆工(具);画家;漆工;艇首缆;不洁雾
painter and decorator 油漆装饰工
painter line 系艇索
painter's brush 油漆匠用刷子;画刷
painter's ca(u)lk 漆工油膏
painter's glue 油漆匠用胶水
painter's knife 漆工刀
painter's labo(u)rer 油漆小工
painter's mitt 涂漆手套
painter's naphtha 白节油;调漆油
painter's overalls 漆工工作服
painter's putty 油漆填料;油漆腻子;漆用腻子;漆刷油灰;漆工油灰
painter's shop 油漆车间
painter's spattle 油漆刮刀
painter's tool 油漆工具;漆工工具
painter's torch 漆刷喷灯;漆工喷灯;漆工火炬;油漆工喷灯
painter's work 油漆工程
painter trade 油漆行业
paint failure 涂料失效
paint filler 油漆填料;油漆底层;腻子
paint film 漆膜;涂膜;涂漆膜;涂料膜
paint film failure 涂膜破坏
paint film polishing media 漆膜研磨材料
paint filter 颜料过滤器;油漆过滤器
paint finish 油漆罩面
paint for building construction purposes 建筑用油漆;建筑用涂料
paint for container 集装箱涂料
paint for galvanized iron 镀锌钢板涂料
paint for metal 金属用油漆;金属用涂料
paint formulation 油漆配方;涂料配方
paint for plastics 塑料用涂料
paint for preservation of structure 结构防护油漆;结构防护涂料

paint for road marking 路线用漆
paint frame 布景架;涂料膜
paint from nature 写生画
paint grade 宜涂漆木制品
paint-grade talc 涂料级滑石
paint grade titanium dioxide 涂料用钛白;涂料用二氧化钛
paint grinder(mill) 涂料研磨机
paint guide 涂枪(喷枪)导管
paint guide cover 涂料(喷枪)导管盖
paint gun 喷漆枪;油漆喷枪
paint harling 漆石涂装法;甩石涂装法;甩漆布;甩漆饰;护壁粗铸件;干黏石面
paint-holding property 保漆性
paint hose 油漆软管;涂料软管;输漆软管
paint-in 油漆抗议
paint industry 油漆工业;涂料工业
painting 油漆工作;油画;绘画;涂装;涂油漆;涂刷;涂漆;图画;刷漆;上色;上漆
Painting and Decorating Contractors of America 美国涂装与装饰承包人协会
painting area 涂漆面积
painting brush 漆刷;画笔
painting colo(u)r 油漆的颜色
painting compressor 压气喷漆枪
painting contractor 油漆承包人
painting damp walls 湿墙涂漆
painting defect 油漆缺点
painting device 涂刷设备
painting environment 涂装环境
painting gun 喷漆枪
painting gun nozzle 涂料喷枪喷嘴
painting in free-sketch style 写意画
painting in oil 油画
painting in watercolo(u)r 水彩画
painting knife 调色刀
painting machine 油漆机;涂漆机;彩印机
painting machine for road marking 路面标记喷涂车
painting machinery 油漆机械
painting material 涂料
painting nozzle 喷漆嘴
painting of pipe 管子涂料
painting of tube 管子涂料
painting on glass 玻璃的彩饰
painting out 涂版
painting position 油漆台位
painting practice 油漆操作;涂刷操作
painting procedure 涂料施工程序
painting process wastewater 油漆过程废水
painting punt 舷侧工作平底小船
paint ingredients 涂料成分;油漆成分
painting room 漆房;画室
painting schedule 涂装规范
painting scheme 油漆计划;涂刷计划
painting screen 画屏风
painting silk 画绢
paintings in booklet 画册
painting spray(ing) 喷漆
painting squeezing action 涂料受挤压作用
painting system 油漆次序;涂刷次序
painting technology 涂装工艺;涂漆工艺
painting utensils 绘画用具
painting waste 喷漆废水
painting work 涂刷工作;油漆工作
painting workshop 喷漆车间
paint kettle 小油漆桶;油漆桶
paint line 水线带;标色水线【船】;涂漆标线
paint liquid 涂料调和剂
paint locker 油漆间
paint loft 阁楼;布景桥架
paint manufacturer 油漆制造厂;涂料制造厂
paint manufacturing industry 涂料制造工业;油漆制造工业
paint mark 涂漆标志
paint marking 涂漆标线
paint membrane 涂料膜;油漆膜
paint mill 磨涂料机;涂料碾盘;涂料碾磨机
paint mist 漆雾
paint mitt 涂漆手套
paint mixer 涂料混合器;涂料混合机;调漆机
paint mixing 调漆
paint mixing room 涂料混合室
paint naphtha 油漆溶剂油;调漆油

paint nozzle 涂料喷嘴
paint oil 涂料干性油；油漆溶剂油；清油；涂料油；调漆油
paint on plaster 抹灰面油漆
paint-on technique 涂抹技术
paint on timber 木材面油漆
paint out 用油漆涂抹；用油漆盖死
paint oven 烤漆烘箱；烘漆炉
paint pail 油漆桶
paint particle 油漆颗粒；涂料颗粒
paint paste 油漆膏；厚漆
paint patch panel 油漆比较试验板；涂料比较试验板
paint peeling 油漆剥落
paint pigment 油漆颜料
paint plant 油漆车间；涂漆场
paint pot 油漆罐；油漆桶；漆桶
paint power 漆粉
paint prepared for use 配制供使用的油漆；配制供使用的涂料
paint pressure tank 涂料加压罐
paint primer 髹漆底涂；涂漆底；涂底漆；底层油漆
paint process wastewater 油漆过程废水
paint production plant 涂料生产设备
paint pump 涂料泵；油漆泵
paint remover 去漆工具；去涂料剂；去漆水；去漆剂；清洗油漆剂；清除涂料剂；脱漆剂；涂料清除剂；除漆剂
paint remover in paste form 浆状去漆剂
Paint Research Institute of American 美国涂料研究学会
paint residue 油漆沉淀物；涂料沉淀物
paint resin 漆用树脂
paint resist temperature 油漆及涂料等的耐热温度
paint rock 铁赭土；铁质岩；铁赭石
paint roller 油漆辊；油漆滚子；油漆滚筒；滚筒漆刷；滚漆筒；漆滚筒；涂滚辊；涂覆滚子
paint roller brush 滚筒漆刷
paint roller mill 涂料磨机
pain trolley 手摇缆车
paint room 储漆室
paint scheme 油漆配色
paint scraper 涂料刮刀；油漆刮刀；刮漆机；刮平刀；漆铲
paint scrubber 洗漆罐；洗漆器；洗漆剂；洗漆工
paint shaker 油漆搅拌器
paint shop 油漆作；油漆作坊；油漆行；油漆间；油漆车间
paint skin 漆皮；油漆面层
paint sludge 油漆污泥；涂料污泥
paint soil 铁赭土
paint solvent 油漆溶剂
paint spatula 拌漆角刀
paint spillage 油漆溢出；油漆溅出
paint spray 油漆喷枪；喷漆橱；喷漆器
paint spray booth 喷漆室；喷漆间；喷漆房
paint sprayer 涂料喷射机；喷漆机；漆喷枪；喷涂机；喷漆机
paint sprayer-bulb 喷漆球
paint spray gun nozzle 涂料喷枪喷嘴
paint spraying 喷油漆；喷涂料
paint spraying apparatus 喷漆装置；喷漆设备
paint spraying device for bridge maintenance 桥梁维修喷漆设备
paint spray(ing) gun 喷漆枪；喷漆器
paint spraying machine 喷漆机
paint spraying outfit 喷漆工具
paint spraying pistol 涂料喷枪；喷漆枪
paint spraying process 油漆喷涂法
paint spray(ing) shop 喷漆车间
paint spraying system 喷漆设备；喷漆法
paint spray mask 喷漆用面具
paint spray nozzle 涂料喷嘴
paint spray pump 油漆喷涂泵；涂料喷涂泵
paint spray room 喷漆室；涂料喷涂室
paint stage 油漆用脚手架或吊架
paint staining 油漆方法
paint storage 油漆库
paint store 涂料仓库；油漆仓库；油漆间
paint strainer 滤漆筛
paint streaming 颜料流线
paint striker 长油漆刷
paint strip 脱漆剂
paint stripper 脱漆剂；除漆剂
paint stripper in paste form 浆状除漆剂

paint stripper in powder form 粉状除漆剂
paint system 油漆做法；涂装体系；涂料(配套)系统
paint tank 油漆罐；涂料罐
paint technician 油漆技师；涂料技师
paint technology 油漆工艺；涂料工艺
paint temperature 喷漆温度
paint test 着色试验法(观测水流形状)
Paint Testing Manual of ASTM 美国材料试验学会涂料试验手册
paint the door green 把门漆成绿色
paint thermal decomposition oxidation 涂刷热分解氧化法
paint thinner 稀料；油漆稀释剂；漆料稀释剂；涂料稀释剂
paint thixotropy 涂料触变性
paint vehicle 油漆媒液；漆料
paint wastewater 油漆废水
paintwork 油画；油漆作业；油漆工作；油漆工程；油漆工
paintwork shop 涂料车间；油漆车间
pair 一双；双腔；对偶；电线对；成对
pair annihilation 对湮没
pair assembly 双机组
pair bond 配对结合
pair cable 双绞电缆；双股电缆
pair case 双重外壳
pair-cast cylinder 汽缸成对铸造式缸体
pair conversion 成对转换
pair delay 成对延迟
paired 配对的；成对的
paired arches 对拱
paired-bond dissociation 双键离解
paired booster 双联助推器
paired brush 成对电刷
paired buret 配对滴定管
paired burette 配对滴定管
paired cable 双芯电缆；双芯电缆；双线电缆；双绞电缆；对绞电缆
paired columns 双柱；双排柱；对柱；成对柱
paired comparison 对偶比较；成对对比；成对比较
paired comparison system 成对比较法
paired cross 成对交配
paired data 配对资料；对比资料
paired design 配对设计
paired difference test 配对差检验
paired disparity code 成对不均等性代码
paired echo 双回波
paired electrons 成对电子
paired experiment 配对试验
paired eyepiece 双目镜
paired I-band mitochondria 成对明带线粒体
paired ion chromatography 离子对色谱法；离子对层析
paired lattice 成对晶格；成对点阵
paired-line cross 双十字线
paired-line patterns 双线图案
paired-line target 双线标板
paired metamorphic belt 双变质带【地】
paired multiplication 双联乘法器
paired multiplier 双乘法器
paired observation 对比观察
paired pace-maker 成对起搏器
paired pilasters 双壁柱；对半柱；成对壁柱
paired pile 成对桩
paired plot 配对小区
paired pulses 成对脉冲
paired sampling 配对取样；配对抽样
paired towers 双塔式建筑
paired t-test 对偶 t 检验
paired variables 成对变量
paired windows 对窗
pair emission 对发射
pair gain system 线对增容系统
pair glass 双层中空玻璃；双层玻璃
pairing 配对；双合；对偶；成双；成对
pairing function 配对函数
pairing method 对比法
pairing model 对模型
pairing the rods 水准标尺配对
pair landing 双机着陆
paira magenta 偶合品红
pair mast girder 龙门桅横桁
pair masts 龙门桅(杆)；双柱桅
pair oar boat 双桨舟；双桨艇

pair observations 双观测
pair of aerials 成双天线；复合天线
pair of anti-parallel 反并联臂对
pair of bellows 手用吹风器
pair of columns 成对柱；双柱
pair of compasses 圆规
pair of control parameter 控制参数对
pair of curves 曲线对
pair of diminutive towers 成对小塔
pair of elements 元素对
pair off 逐对分开；成对分置
pair of faults 层错对
pair of gateways 双门道；大门口成对建筑物
pair of gears 齿轮副
pair of glasses 双层玻璃
pair of half scoops 抓斗；抓泥斗
pair of magnets 一对磁石；一对磁铁
pair of minarets 一对回教尖塔；(伊斯兰教寺院建筑的)成对叫拜楼
pair of module 模块对
pair of nippers 钳子
pair of pages 双页
pair of pictures 像对
pair of piers 两个码头；两个墩子
pair of piles 一对电堆；一对桥墩；一对柱；成对桩
pair of plates 硬片像对
pair of points 偶点
pair of pylons 两个(机场)定向塔；大门口成对塔形建筑物
pair of rafters 成对椽子
pair of ribs 两个肋
pair of rollers 滚轮对
pair of semi-detached houses 有公共隔墙的两间毗连房屋
pair of small towers 成对小塔
pair of stations 台对
pair of steps 绳梯；梯子；双折梯
pair of stereoscopic picture 立体像对
pair of tracks 双轨【铁】
pair of turrets 一对角楼；一对塔楼；成对角楼
pair of wheels 轮对
pair of windings 绕组对
pair parking 车尾相对的停车方式
pair-point 偶点
pair potential 对势
pair rolling 钢板的成对合轧
pair-selected ternary 成对选三进制
pairs of observation 双观测
pairs of trains 列车对数
pairs of water molecule 水分子偶
pair spectrometer 电子偶分光计
pair split 线对错接
pair stations 台对
pair test 两点检验
pair-to-pair capacity 线对间电容
pair transistor 双晶体管
pair trawler 双拖网渔轮；双拖网渔船；对拖网渔船
pair tube 配对管；对偶管
pairwise correlation 两两相关
pairwise disjoint set 两两不相交
pairwise independence 两两独立
pairwise independent events 两两独立事件
pairwise orthogonal 两两正交
pairwise uncorrelated 两两不相关
paisanite 钠闪微岗岩
paisley 佩斯利涡旋纹花呢(苏格兰制)
Pajari apparatus 帕亚里钻孔测斜仪
paktong 白铜
palace 宫殿；皇家花园；王宫
palace architecture 宫殿建筑
Palace at Ctesiphon 泰西封宫殿(伊拉克)
Palace at Ferusz-Abad 费鲁兹—阿西德宫(位于伊朗波斯利斯，建于公元 250 年)
palace car 豪华客车
palace complex 宫殿集合体
palace construction 宫殿施工；宫殿建筑
palace de Fontainebleau 枫丹白露宫(法国)
palace de Luxembourg 卢森堡宫(法国)
Palace Garden 故宫御花园
palace hotel 豪华旅馆
palace lantern 宫灯
palace of culture 文化宫
Palace of Cyrus the Great at Pasargadae 帕萨尔加德的赛勒斯大帝宫(伊朗)

Palace of Delights 欢喜宫
Palace of Diocletian at Split 斯普利特的第渥克利欣宫(南斯拉夫)
Palace of King Minos at Knossos 克诺索斯米诺王宫(希腊)
Palace of Nations 国家宫;民族宫
Palace of Persepolis 佩斯波莱斯宫(波斯古建筑)
Palace of Sargon 萨尔贡王宫(亚西利亚古建筑);萨尔贡王宫
Palace of Sarvistan 萨维斯坦宫(波斯古建筑)
Palace of Shapur 1 at Bishapur 比沙布尔的沙布尔一世宫殿(波斯古建筑)
Palace of States 国家宫
Palace of the Arts 艺术宫
Palace of the Dawn 曙光宫
palace of the emperor 皇宫
Palace of the Emperors at Rome 罗马皇宫
Palace of the Popes at Avignon 阿维尼翁山上的神父宫(法国哥特式建筑之一)
Palace of Whitehall 白宫官邸
palace terrace 宫中阳台;宫中平台
palace wing 宫殿侧翼
palacheite 赤铁矾
Palaearctic region 古北区
pal(a)eo-Africa 古非洲大陆
pal(a)eoamphibole 古角闪石
pal(a)eoanchropology 古人类学
pal(a)eo-Antarctica 古南极大陆
pal(a)eoanthropologist 古人类学家
pal(a)eo-Asia 古亚细亚洲
Pal(a)eoathaysian 古华夏系【地】
pal(a)eo-Australia land 古澳大利亚大陆
pal(a)eobasin 古盆地;古海盆
pal(a)eobathymetric map 古等深线图
pal(a)eobiochemistry 古生物化学
pal(a)eobioecology map 古生态图
pal(a)eo-calcite 古方解石
Pal(a)eocene epoch 古新世
Pal(a)eocene series 古新统
pal(a)eochannel 古河道;古河槽
pal(a)eocirculation 古环流
pal(a)eoclay 老黏土
pal(a)eoclimate 古气候
pal(a)eoclimate epoch 古气候期
pal(a)eoclimate zone 古气候带
pal(a)eoclimatic sequence 古气候序列
pal(a)eoclimatic zonation 古气候分带
pal(a)eoclimatologic(al) map 古气候图
pal(a)eoclimatologic(al) unit 古气候单元
pal(a)eoclimatologic-pal(a)eohydrologic(al) map 古气候—古水文图
pal(a)eoclimatologist 古气候学家
pal(a)eoclimtology 古气候学
pal(a)eocolatitude 古余纬度
pal(a)eo-continent topography 古地形
pal(a)eocool-temperate zone 古寒温带
pal(a)eocrustal stress field 古地应力场
pal(a)eocrust of weathering 古风化壳
pal(a)eocrystic ice 古老冰;古结晶冰
pal(a)eocurrent 古水流;古流;古海流
pal(a)eocurrent direction 古流向
pal(a)eocurrent map 古流线图
pal(a)eodelta 古三角洲
pal(a)eodepth 古深度
pal(a)eodirection of geomagnetic field 古地磁场方向
pal(a)eodose 古剂量
pal(a)eodynamic(al) map 古动力图
pal(a)eoearthquake 古地震
pal(a)eoecological event 古生态事件
pal(a)eoecological reconstruction 古生态的再现
pal(a)eoecology 古生态学
pal(a)eoeffusive 古喷出岩
pal(a)eoenvironment 古环境
pal(a)eoequator 古赤道
pal(a)eoequator position 古赤道位置
Pal(a)eogene 古第三纪
Palaeogene system 下第三系
pal(a)eogeographic(al) 古地理的
pal(a)eogeographic(al) control 古地理控制
pal(a)eogeographic(al) event 古地理事件
pal(a)eogeographic(al) feature 古地理特征
pal(a)eogeographic(al) map 古地理图
pal(a)eogeographic(al) method 古地理法
pal(a)eogeographic(al) pole 古地极
pal(a)eogeographic(al) stage 古地理时期
pal(a)eogeographic(al) unit 古地理单元
pal(a)eogeography 古地理(学)
pal(a)eogeologic(al) 古地质的
pal(a)eogeologic(al) map 古地质图
pal(a)eogeology 古地质学
pal(a)eogeomorphologic(al) map 古地貌图
pal(a)eogeothermal gradient 古地温梯度
pal(a)eoglacial map 古冰川图
pal(a)eo-Greenland 格陵兰古陆
pal(a)eo-ground temperature condition 古地温条件
pal(a)eohuman fossil 古人类化石
pal(a)eohumidity 古湿
pal(a)eo-hydrochemical condition 古水化学成分
pal(a)eo-hydrodynamic(al) condition 古水动力条件
pal(a)eo-hydrogeochemical map 古水文地球化学图
pal(a)eo-hydrogeologic(al) map 古水文地质图
pal(a)eohydrology 古水文学
pal(a)eoichnology 古遗迹学
Pal(a)eoid 上古
Pal(a)eoid tectonic system 上古构造体系
pal(a)eointensity of geomagnetic field 古地磁场强度
pal(a)eo-island 古岛
pal(a)eokarst 古岩溶;古喀斯特
pal(a)eokarstic surface structure 古岩溶面构造
pal(a)eokinetic system 古运动区系统
pal(a)eolandscape picture 古景观图
pal(a)eolatitude 古纬度
pal(a)eolatitude zonation 古纬度分带
pal(a)eolength of time units 古计时单位长度
pal(a)eolimnology 古湖沼学
pal(a)eolithologic map 古岩性图
pal(a)eo-longitude position 古经向位置
pal(a)eo-lowland 古低地
pal(a)eolysocline 古溶跃面
pal(a)eomagnetic 古地磁的
pal(a)eomagnetic belt 古地磁条带
pal(a)eomagnetic chronological scale 古地磁年代表
pal(a)eomagnetic data processing 古地磁数据处理
pal(a)eomagnetic dating 古地磁年龄测定
pal(a)eomagnetic dating method 古地磁测年法
pal(a)eomagnetic field 古地磁场
pal(a)eomagnetic mean direction 古地磁平均方向
pal(a)eomagnetic measurement 古地磁测量
pal(a)eomagnetic north pole 古地磁北极
pal(a)eomagnetic north pole position 古地磁北极位置
pal(a)eomagnetic pole 古地磁极;古磁极
pal(a)eomagnetic pole position 古地磁极位置
pal(a)eomagnetics 古地磁学
pal(a)eomagnetic south pole 古地磁南极
pal(a)eomagnetic south pole position 古地磁南极位置
pal(a)eomagnetic stratigraphic(al) unit 古地磁地层学单位
pal(a)eomagnetic stratigraphy 古地磁地层学
pal(a)eomagnetic time scale 古地磁年表
pal(a)eomagnetism 古地磁(学);古磁学
pal(a)eomagnetism analysis 古地磁分析
pal(a)eontologist 古生物学家
pal(a)eonucleus 古陆核
pal(a)eo-oceanic and pal(a)eoclimatic event 古海洋与古气候事件
pal(a)eo-oceanic current 古洋流
pal(a)eo-oceanographic(al) environment parameter 古海洋环境参数
pal(a)eo-oceanographic(al) reconstruction 古海洋再造
pal(a)eo-oceanography and pal(a)eo-climatology 古海洋学与古气候学
pal(a)eooceanology 古海洋学
pal(a)eo-ocean topography 古海洋地形
pal(a)eo-orientation 古方位
pal(a)eo-Pacific ocean 古太平洋
pal(a)eo-Pacific-Wusuli gulf 古太平洋—乌苏里海湾
pal(a)eosubpolar zone 古副极地
pal(a)eo-peneplain 古准平原
Pal(a)eophyte 古生代植物
pal(a)eophytogeographic(al) map 古植物地理图
pal(a)eoplacer 古砂矿
pal(a)eoplain 古平原
pal(a)eo-plateau 古高原
pal(a)eoplatform 古地台
pal(a)eorecipitational map 古降水量图
pal(a)eosalinity 古盐度
pal(a)eoseismic cliff 古地震陡崖
pal(a)eoseismic collapse 古地震崩塌
pal(a)eoseismic data 古地震资料
pal(a)eoseismic ditch 古地震沟
pal(a)eoseismic fault 古地震断裂
pal(a)eoseismic fracture 古地震裂缝
pal(a)eoseismicity 古地震活动性
pal(a)eoseismic landslip 古地震滑坡
pal(a)eoseismic liquefaction deformation 古地震液化形变
pal(a)eoseismic remainder deformation 古地震剩余形变
pal(a)eoseismic sand vein 古地震砂脉
pal(a)eoseismic wedge 古地震楔
pal(a)eoseismogeology 古地震地质
pal(a)eoseismology 古地震学
pal(a)eosere 古生代演替系列
pal(a)eoshoreline 古海岸线
pal(a)eoslope 古斜坡;古坡向
pal(a)eosoil 古土壤
pal(a)eosome 古城体;古成体
pal(a)eospecies 古物种;古生物种
pal(a)eostructure 古构造
pal(a)eostructure map 古构造图
palaeostructure of coal-accumulation 聚煤古构造
pal(a)eosubtropical zone 古亚热带
pal(a)eo-swamp 古沼泽
pal(a)eotectonic 古构造的
pal(a)eotectonic map 古构造图;古地质构造图
pal(a)eotectonics 古构造;古地质构造学
pal(a)eotectonic stress 古构造应力
pal(a)eotectonic stress field 古构造应力场
pal(a)eotectonic stress field map 古构造应力场图
pal(a)eotectonic system 古构造体系
pal(a)eotemperature 古温度
pal(a)eotemperature map 古温度图
pal(a)eoteperature determination 古温度判定
Pal(a)eotethys 古特提斯
pal(a)eothermocline 古温跃面
pal(a)eothermometry 古温度测定
pal(a)eo-tide 古潮汐
pal(a)eotologic taxon 古生物分类
pal(a)eotopographic(al) 古地形的
pal(a)eotopographic(al) map 古地形图
pal(a)eotopographic(al) unit 古地形单元
pal(a)eotopography 古地形学;古地形
pal(a)eotropic(al) 古热带的
pal(a)eotropic(al) region 旧热带植物区;古热带区
pal(a)eotropic(al) zone 古热带
pal(a)eoturbidity current 古浊流
pal(a)eoturbulence 古水深度
pal(a)eo-upland 古高地
pal(a)eovolcanic 古火山的
pal(a)eovolcanic pal(a)eontology 古火山学
pal(a)eovolcanic rock 古火山岩
pal(a)eovolcano 古火山
pal(a)eowarm-temperate zone 古暖温带
pal(a)eo-water-energy 古水能量
pal(a)eo-water temperature 古水温
pal(a)eoweathered crust and pal(a)eosoil analysis 古风化壳古土壤分析
pal(a)eowind 古风
pal(a)eowind map 古风图
Pal(a)eozoic 古生代的
Pal(a)eozoic Alps 古生代阿尔卑斯山
Pal(a)eozoic basin 古生代盆地
Pal(a)eozoic era 古生代
Pal(a)eozoic erathem 古生界
Pal(a)eozoic group 古生界
Pal(a)eozoic layer 古生代地层
Pal(a)eozoic ocean 古生代海洋
palaestra 体育场;(古希腊的)练习角力及各种竞技的公共场所;健身房
palaetypal rock 古相岩

palafitte 古代湖上桩承棚屋;古代湖上桩承茅舍;古代湖上桩承住房;瑞士山中倾斜屋顶的农舍;瑞士山中倾斜屋顶的木屋
palagonite 橙玄玻璃
palagonite tuff 玄玻凝灰岩
Palais de Vesailles 凡尔赛宫(法国)
Palais du Louvre 罗浮宫
palaite 红锰磷石
palankeen 轿
palanquin 轿子
Palaoinonestes sinensis 中华小长臂虾
palarization multiplexing 偏振复用
Pal-Asia 古亚洲大陆
Pal-Asiatic system 古亚洲断裂体系
Pal-Asiatic tectonic domain 古亚洲构造域
palasite 橄榄石铁陨石
palasome 主岩;主矿物;基体
palatial 富丽堂皇的;宫殿(似)的
palatial architecture 宫殿式建筑
palatial furnishings 宫室内陈列品
palatial hall 富丽堂皇的厅堂
palatial house 皇宫般的房屋;皇宫
palatial style 宫殿式;宫殿风格
palatium 宫殿(尤指古罗马之皇宫)
palazzo 总督宫(意大利威尼斯);庄严的大府邸
palba wax 帕尔巴蜡
pale 尖板条;浅色;淡;苍白色;范围
palea 托苞;扁刚毛
palearctic 古北区的
paleate 被鳞片的
pale blue 鸭蛋青;淡蓝色;粉青
pale-boiled oil 浅色聚合油;浅色熟油;聚合亚麻子油;浅色清油
pale brick 红砖;未烧透砖
pale colo(u)r 浅色;淡色
pale crepe 苍皱橡胶
pale drying oil 淡干油
pale fence 桩篱式围篱;竖管栅栏;竖板栅栏;板篱
pale fencing 桩排;栅栏;围栅;板篱
pale green 苍绿;浅绿
pale green bottle 浅绿瓶
pale green glass 淡绿色玻璃
pale green glaze 浅绿釉
pale grey-blue 浅灰青
pale layer 灰白层
paleochannel 埋藏河道
paleo-clay 老黏土
paleo-clay layer 老黏土层
paleo-cohesive soil 老黏性土
paleocurrent system of basin 盆地古流体系
Paleogene period 早第三纪【地】
paleogeographic(al) type of coal formation 聚煤古地理类型
pale oil 浅色润滑油;苍色油
paleo-India 印度古陆
paleo-Indo-China 印支古陆
paleolith 旧石器
Paleolithic Age 旧石器时代
paleo-loess 老黄土
pale paint 浅色漆;淡漆
pale pink 苍桃红色
pale pinkish grey 藕色
pale purple 青莲色;淡紫色
pale purple flower 淡紫色花
pale red 淡红;淡红色;苍红(色)
palermoite 柱磷锶锂矿
pale rosin 浅色松香
pale shade 弱色
palestra 体育场;希腊或罗马体育训练馆;[复]palestrae
pale straw yellow 浅草黄的
pale tone 浅色调
palette 溶残席;坯板;调色盘;调色板
palette board 调色板
palette for painter 绘画用调色板
palette knife 刮铲;调色刀;调漆刀;调墨刀
pale yellow 苍黄色;浅黄色
pale yellow filter 淡黄色滤光镜
pale yellow glaze 浅黄釉
Pali 巴利红褐色硬木(印度产)
palichnology 古遗迹学
palid 铅基轴承合金;巴里特合金
palification 用桩加固土壤;用加固地基;桩加固地基);打桩

paligorskite 坡缕石
paligorskitization 坡缕石化
palimpsest sediment 变质沉积物;变余沉积物
palimpsest structure 变质岩构造;变余构造;变异构造
palimpsest texture 残留结构;变余结构
paling 做围篱;打桩;垂直外模板;篱笆桩;围篱;栅板;栅栏圆柱;栅栏尖板条;尖板条;木栅;木栅栏;板篱
palingenesis 再生作用
palingenesis way 再生作用方式
palingenetic drainage 再生水系
palingenetic ore-forming theory 再生成矿说
palingenetic regeneration 重演再生
palingenetic river 再生河
palingenetic stratabound deposit 再生层控矿床
palingenetic stream 再生河
paling fence 桩篱式围篱;竖管栅栏;竖板栅栏
palinspastic map 再造图;位置复原图;复原图;复原地图
palisade 栅栏;木栅栏;围篱;板篱;岩壁
palisade Altay 阿尔泰峭壁
palisade block 竖砌池壁砖
palisade check dam 木栅拦沙坝;木栅谷坊
palisade fence 栅栏篱笆
palisade layer 栅栏层
Palisade's disturbance 帕利塞德变动
Palisade's sill 帕利塞德岩床
palisade zone 栅状带
palisander 红木
palisander wood 红木
palisandre 红木
palisandre wood 红木
Palitzsch's buffer solution 巴氏硼酸盐缓冲液
Palium 铝系轴承合金
pall 制动绞盘;遮盖物;棘爪
Palladian 帕拉第奥式(建筑)
palladiana 水磨石面层
Palladian architecture 帕拉第奥式建筑
Palladian classicism 帕拉第奥古典式建筑
Palladianism 帕拉第奥学派;帕拉第奥建筑主义;帕拉第奥建筑形式
Palladian motif 帕拉第奥式建筑特色;帕拉第奥式建筑处理手法
Palladian motif window 帕拉第奥式建筑特色窗
Palladian motive 帕拉第奥式建筑;帕拉第奥式风格
Palladian revival 帕拉第奥式建筑的复兴
Palladian terrazzo 帕拉第奥式水磨石
Palladian window 帕拉第奥式窗
palladic 钯的;钯制的
palladinite 钯华
palladious iodide 碘化亚钯
Palladium 智慧女神雅典娜的神像
palladium black 钯黑
palladium catalyst 钯催化剂
palladium chloride 氯化钯;二氯化钯
palladium copper 钯铜
palladium difluoride 二氟化钯
palladium diiodide 二碘化钯
palladium dydroxide 氢氧化钯
palladium gold 钯金
palladium gold paste 钯金浆料
palladium iodide 碘化钯
palladium-iron catalytic reduction method 钯—铁催化还原法
palladium ore 钯矿
palladium silver 钯银
palladium-silver method 钯银法
palladoarsenide 斜砷钯矿
palladobismutharsenide 铋砷钯矿
palladseite 硒钯矿
pallasite 橄榄陨铁
pallesthesiometer 水下扬声器
pallet 制模架;集装架;棘爪;抹子;货盘;货板;平板架;托盘;托板;调节瓣;提水板;随行托板;随行夹具;随行工作台;垛材盘;吊盘;底模板;锤垫;草垫
pallet belt conveyer 板架式输送带
pallet board 货盘;货架;承砖坯板;坯板
pallet box 托盘箱
pallet brick 楔尖坯板;嵌条砖;带槽砖
pallet cargo 托盘货物
pallet carrier 集装箱运输车;集货架运输车

pallet changer 集装架交换装置;随行托板交换装置;随行夹具交换装置
pallet cleaner 集装箱(底板)清洗机
pallet container 托盘集装箱
pallet conveyance ship 托盘船
pallet conveyance vessel 托盘船
pallet conveyer 板架式运输机;集装箱输送机;平板输送机;板式运送机;板架式输送机
pallet conveyer mo(u)ld machine 滑板输送式造型机
pallet deck 货盘甲板
pallet drier 托板式干燥器
pallet elevator 运载托盘升降机
pallet feed 夹具送料
pallet feeder 板式送料器
pallet for door-to-door transportation 联运托盘;门到门运输用托盘(集装箱)
pallet for freight container 集装箱用托盘
pallet fork 托盘式叉齿
pallet for through transportation 联运托盘
pallet frame 托盘架
pallet grab 集装箱抓具;集装箱抓具
pallet handling 托板化运送;集装箱运输;货盘货物装卸
palletisation 货盘化;托盘码垛
palletised cargo 用货盘装运货物
palletised load 托盘化货物;用集装箱运的货物;用货盘装运
palletiser 托盘组织装机
palletization 集装箱化;装运托盘化;夹具化;码垛堆放;货盘化;货板化;托盘码垛;托盘化(运输)
palletize 夹板装载;码垛堆积;托盘化
palletized bag 托板装载的袋装水泥
palletized cargo 架装货物;货板成组化;托盘化货物;托盘成组货物;使用货盘的货物
palletized load 用货盘装运
palletized mo(u)lding compound 粒状模塑料
palletized resin 粒状树脂
palletized sack 托板装载的袋装水泥
palletized storage 托盘化储[贮]藏
palletized transfer line 随行夹具式组合机床自动线
palletized transfer machine 棘爪式连续自动工作机床
palletized transport 托盘化运输
palletized truck 架装货车
palletized unit load 托盘化组合装运
palletizer 造粒机;码堆机;托盘组织装机;托盘码垛机;堆列铺设机;叠包机
palletizing 制粒;造粒;夹板装载;码堆积;装运
palletizing machine 堆垛机;叠板机
pallet jewel 棘爪宝石;叉瓦钻
palletless package brick transporting 无托盘包装的砖块运输
pallet lifter 集装箱起重设备
pallet load 托盘货
pallet loader 抬板机
pallet loading 集装箱装运
pallet master file 随行夹具主文件
pallet-molding 砂模制砖法
pallet-moving walk 自动人行道
pallet packing 托盘包装
pallet pool 托盘联营组织
pallet return 集装箱回收;货盘回运
palletron 高压电子谐振器
pallet ship 托盘货船;托盘船;包装货船;托盘运输船
pallet shipment 托盘装运
pallet ship system 托盘船装卸方式
pallet shrink package 托盘热收缩包装
pallet shuttle system 随行夹具梭动系统
pallet sling 托盘吊索
pallet stock 货板用木
pallet stone 叉瓦钻
pallet stowage 托盘装载波
pallet system 货盘系统;托盘式集装运
pallet traffic 托盘运输
pallet trailer 托盘拖车
pallet transport 托盘运输
pallet transverse girder 集装箱横(大)梁
pallet truck 货架装货车;集装箱货车;码垛车;托盘搬运车;叉式自动装卸车
pallet-type moving wamp 板片式活动人行道;踏板式电动走道
pallial groove 套膜沟
palliasse 草荐;草垫;坑工基座

palliative 防腐剂;镇静剂;减轻剂;减尘剂
palliative dust laying 减轻积尘措施;防尘措施
palliative measures 减少积尘措施
pallograph 船舶振动记录仪;船舶振动记录器
Pall ring 陶瓷填充圈;鲍尔环
palm 锚爪;棕榈;掌状物
palmae bark 棕树皮
palm and needle 帆缆针线包
palm and needle whipping 绳头合扎
palmately lobed 掌状圆裂的
palmately parted 掌状深裂的
palmate-pinnate 掌状羽状的
palm bark 棕皮
palm branch 连茎棕榈叶
palm butter 棕榈油
palm capital (古埃及的)棕榈形柱头饰
palm coir mat 地棕垫
palm colo(u)r 棕色的
palm column 棕榈叶式柱
palmella 不定群体
palm end 扁头
Palmer and Barber formula panel 帕尔默－巴伯公式板
Palmer granite 帕尔默石(一种产于美国缅因州带浅红色的花岗岩)
palmeter 帕耳计
palmette 棕叶状花饰;棕叶(花)饰【建】
palmette training 多干形整枝
palmetto fiber 沙巴尔纤维
palmetum 棕榈园
Palm function 帕姆函数
palm grease 棕榈油;贿赂
palm grip hand knob 星形手钮
palm house 热带植物温室;温室(栽培棕榈等用)
palmic acid 棕榈酸;十六(烷)酸
palmierite 钾钠铅矾
palmiform 棕榈树叶状
palming off 仿冒
palmitamide 棕榈酸酰胺
palmitate 十六烷酸;棕榈酸酯
palmitic acid 棕榈酸;软质酸;软脂酸
palmitinic acid 棕榈酸;软脂酸;十六(烷)酸
palmitoleic acid 棕榈油酸;抹香鲸烯酸
palm kernel 棕仁
palm kernel oil 椰子仁油;棕榈仁油;棕榈核油
palm-leaf openwork screen 芭蕉罩
palm leaf ornament 棕榈叶装饰;棕叶装饰
palm-like lobe 指形叶
palm oil-mill effluent 棕榈油厂废水
palm pile 掌式桩
palm plate 棕板
palm pulp oil 棕榈油
palm rope 棕索;棕绳
palm shaft column 棕榈树杆般的柱
palm stay 掌状撑条;掌舵型撑;掌式撑条
palmula 爪垫
palm vault(ing) 棕榈叶形拱顶;扇形穹顶
palm wax 棕榈蜡
palm whipping 绳头缠扎
palmy days 全盛时期
Palomar Sky Survey 帕洛玛天图
palosapis 菲律宾龙脑香
palouser 帕罗塞尘暴;尘暴
palpable 可触的
palpable coordinate 显坐标
palsabog 泥炭丘沼泽
palstage 古地理时期;古地理阶段
palstance 角速度
palta 鳞片
Palu 帕卢紫褐色硬木(印度产)
paludal 沼泽的
paludification 沼泽化
palus 沼泽;河边低地;[复]pali
palus epidemiarum 疫沼
palus putredinis 腐沼
palus somnii 梦沼
palustric acid 长叶松酸
palygorskite 绿坡缕石;坡缕石
palynology 抱粉学
palynostratigraphy 孢粉地层学
pamatol 木材防腐剂
pament 麦芽作坊地面砖
p-Aminobenzene sulfonic acid 氨基苯磺酸
pamite 浮油松香

pampa (尤指阿根廷的)南美大草原;潘帕草原
pampakopetra fibre 高级石棉纤维
pampas 大草原
pampero 帕姆佩罗风
pamphlet 小册子
pampiniform 蔓状的
pampiniform plexus 蔓状丛
pampre 葡萄饰【建】
pan 重矿物淘选盘;锅;卷涂机储[贮]漆盘;海盆;浅箱;平锅;盆状凹地;土盘;淘盘;摄全景;底盘;凹池
panache 彩色花纹效应;穹顶中两支承券间的三角形部分;羽毛饰【建】
panadapter 景象接收器;扫调附加器
panagglutinable 全可凝集的
panagglutination 全凝集
panagglutinin 全凝集素;泛凝集素
panalarm 报警设备;(有灯光和振铃的)报警系统
pan algebraic curve 泛代数曲线
panalyzer 全景分析仪
Panama Canal 巴拿马运河
Panama Canal tolls 巴拿马运河航行费
Panama Canal tonnage 巴拿马运河吨位
Panama Canal zone 巴拿马运河区
panama passage 巴拿马地峡
panama rubber 美洲橡胶
Panamax 巴拿马型(船)
Pan-American 泛美的
Pan-American cell 泛美型浮选机
Pan-American Center for Health Engineering and Environmental Science 泛美卫生工程与环境科学中心
Pan-American Health Organization 泛美卫生组织
Pan-American jig 泛美跳汰机
Pan-American pulsator jig 泛美自动阀脉动跳汰机
Pan American World Airways 泛美航空公司
pan and roll roofing tile 意大利式屋面盖瓦;平头筒形屋面瓦;板瓦和筒瓦
pan and roll tile 意大利式瓦
pan-and-tilt function 俯仰功能
pan arrest 盘托;托盘
panasqueiraite 羟氟磷钙镁石
Panathenaic frieze 古希腊式柱间雕带(神庙饰带)
Panathenaic way 像古希腊盛大节日般的状况(祭神庆典)
panautomorphic granular 全自形粒状
panadapter = panadaptor
panavision 宽银幕电影
pan bench 蒸发盘组
pan bow 弯弓
pan breaker 坚土破碎机;盘式粉碎机;松底土器;硬土层破碎机
pan breeze 焦砟;煤渣;炉渣
pancake 沥青路面油饼;(灌浇沥青不均匀而形成的)油饼;油包;平螺旋状;盘形混凝土块;扁平形的
pancake auger 扁平螺旋钻
pancake coil 盘形线圈;盘香管;饼形线圈;扁平线圈
pancake collapse 平倒塌
pancake engine 水平对置式发动机;扁平发动机
pancake forging 扁平锻造
pancake helix 扁平螺旋线圈
pancake ice 荷叶冰块;饼状冰块;饼状冰;莲叶冰;荷叶冰;冰饼
pancake landing 平坠降落(飞机)
pancake light 饼状灯
pancake motor 短轴型电动机;扁平型电动机
pancake reactor 扁平型反应堆;扁平反应堆
pancake synchro 扁平型同步机
pancake winding 盘形绕组;扁平(型)绕组
pan-calcimed gypsum 锅(盘)中燃烧的石膏
pan car 铸模车
pan ceiling 镶板天花板;镶板平顶;镶板天棚
Pancelet graphical construction 庞塞莱特图解法(土压力)
Pancelet wheel 庞塞莱特水轮机
panchayat forest 村有林
panchlimation 泛顶极群落
panchpat 邦奇帕染色丝绸
panchromate 全整色
panchromatic 全色的
panchromatic coating 全色感光膜
panchromatic emulsion 全色性乳剂
panchromatic film 全色软片;全色胶片;全色薄膜
panchromatic infrared film 全色红外片

panchromatic minus blue photography 全色减蓝摄影
panchromatic negative 全色负片
panchromatic photograph 全色像片
panchromatic photographic(al) material 全色片
panchromatic plate 全色感光板;全色干片;全色底片
panchromatic rendition 全色重显
panchromatic response 全色响应
panchromatic safety roll film 卷筒胶卷
panchromatic television camera 全色电视摄像机
panchromatic vision filter 全色滤光镜
panchromatism 全色性
panchromatograph 多用色谱仪;多能色谱仪
panchrome stain 全可染剂
pancking 失速平坠着陆
panclimax 泛演替顶极;泛顶极群落
pan closet 盆式便器;盘式便器
pancock coil 旋管
pan coefficient 蒸发皿系数
pan construction 肋板结构;密肋梁式楼板结构;盘形结构
pan control 全景调整
pan conveyer 槽式输送机;平板输送机;槽式输送器;吊盘输送机
pancratic 可变焦距的
pancratic eyepiece 可调节目镜
pancratic lens 可调节物镜;可调焦透镜;可调光焦透镜
pancratic(lens) system 可调透镜组
pancratic magnification 可调放大率
pancratic telescope 变焦望远镜
pan crusher 碾盘式破碎机;盘式粉碎机
panda 熊猫
panda car 警察巡逻车
panda crossing 路面涂成"熊猫"斑形的人行横道
pandaite 水钡钽烧绿石
pandal 公共集会用的临时棚舍
pandemic 广泛流行的
pandemic disease 大范围流行病
pandermite 白硼钙石
pan design 镶板设计;墙板设计
pandicon 直列指示管
pandocheum 综合医院
pandown 镜头下移;摄影机垂直摄全景
pan-dried 盘里炒干的
Pandrol fastenings 潘得罗扣件
Pandrol flexible fastenings 潘得罗弹簧扣件
pan dryer 盘式烘干器;盘式烘干机;盘式干燥器
pane 一片;一方;门窗上的方格玻璃;窗玻璃片
pane hammer 顶边锤;石工锤;斧锤;尖锤
panel 小组调查会;(仪表面板;仪表板;栅格;格构分格;间;间架;名册;嵌板;盘区;贴板;步架;表盘;板条;板料;镶板;门心板
panel absorbent 薄板吸声体;薄板式吸声体
panel absorber 薄板吸声器;板状吸声装置;板式吸音器
panel air system 板式通风系统
panel area (相邻测深垂线间的)节间面积
panel arrangement 镶板安排;镶板布置;格板安排;格板布置
panel assembly 仪表板总成;板件总成
panel base 镶嵌图片
panel beam 格子梁
panel beating 钣金加工
panel bed 面板座
panel board 合成板;高压板;镶嵌板;镶板;仪表板;表盘板;帕尔默－巴伯公式板
panel board box 线路箱;配电盘线路箱
panel board cabinet 线路箱
panel body 带棚车身
panel bolt 面板螺栓
panel boundary 盘区边界
panel box 线路箱;配电箱;配电盘
panel brick 预制砖墙;预制砖板;焦炉用长形硅砖;护墙砖板;护墙板砖;大板砖
panel building shop of ship 船体小分段车间
panel ceiling 镶板平顶;板材吊顶;镶板天花板;镶板顶板
panel ceiling heating 平顶辐射板采暖;格板平顶采暖
panel center of pressure 板壁压力中心
panel clamp 样板夹
panel clip 板夹

panel code 信号布板码
panel coil 螺旋板热交换器;散热盘管
panel coker 成漆板焦化器
panel coking test 成漆板焦化试验
panel connection 顶板连接
panel construction 镶板施工;墙板施工;预制板材施工(法);壁板式构造
panel construction method 镶板施工方法;墙板施工方法
panel construction system 镶板建筑系统;墙板施工系统
panel container 盘式集装箱
panel cooler 壁板式制冷装置;平板冷却器;板式冷却器
panel cooling 嵌入式冷却系统;辐射散热
panel covering 镶板(细工)
panel curtain wall (非承重的)外饰面挂墙板;(非承重的)外饰面墙板;板式幕墙
panel curvature 屏曲率
panel-cut 盘刻;板纹
panel cut out drawing 面板开口图
panel decoration 装饰板
panel design 嵌花花纹
panel discussion 专家小组讨论;讨论会(学术方面)
panel display 面板显示;平板显示
panel divider 面板接continue件
panel door 镶板门;格子式镶板门;格板门;拼花板门;壁挂式采暖炉
paneled ceiling 格子式平顶;格子平顶;格子顶棚;嵌板平顶
paneled form 小块拼合模板
paneled framing 镶板框
paneled in oak 铺有橡木镶板;铺有柞木镶板
paneled lining 门框周围衬砌;门框周围衬里
paneled partition 镶板隔断
panel end lap 板端搭接
panel erection 墙板架设;墙板安装
panelescent lamp 薄板式荧光灯
panel facade 墙板立面;墙板外貌;镶板立面;镶板外貌
panel face 板面
panel fence 格板围篱;镶板围篱
panel fire 壁挂式采暖炉
panel for economic assessment 经济评估小组
panel for environmental assessment 环境评估小组
panel form 组合模板;格形模板;小块拼合模板
panel formwork 格板模板工程;镶板模板工程
panel-for-panel curtain wall 幕墙墙板
panel for scientific assessment 科学评估小组
panel for technical assessment 技术评估小组
panel frame 格板架;墙板架
panel gate 分段垛间闸门;分段垛间闸坝
panel ga(u)ge 壁板划线规
panel girder 花格梁;格子梁;方格梁;花梁
panel glass 板玻璃
panel grade 板材等级
panel grid 采暖管的预制嵌板;格板格栅
panel heater 壁挂式采暖器;板式散热器;板式取暖器;嵌入式供暖器;平板加热器;辐射板供暖器;壁柱式采暖器
panel heating 嵌入式供暖;平板加热器加热;辐射板供暖;隐蔽式供暖;壁板式采暖
panel heating and cooling 嵌入式供热和供冷
panel heating system 供热片供暖系统;辐射式供暖系统
panel heating unit 辐射采暖板组件
paneling 镶板细工;木板饰面;装饰板;镶补板;盘区开采;门心板
paneling door 镶板门
paneling lumber 镶板材
panel insert 板门镶嵌件
panel instrument 面板仪表
panel interchangeability 镶板的可互换性;墙板的可互换性
panel interrupt mask on 面板中断屏蔽
panelized construction 板式结构;板式构件
panelized house 预制装配房屋
panelized housing 大板式住房
panelized roof system 嵌装式屋面系统
panelized-tube-wall section 水冷壁管屏
panel jack 面板插口
panel joint 板格结点;格接
panel lamp 仪表板小灯;仪表盘灯

panel landing 盘区车场
panelled ceiling 嵌板平顶;格子式平顶
panelled door 肚板门;框档门;镶板门
panelled framing 墙板构架;嵌板框架;镶板门框架
panel length 节间长度
panel light 仪表灯;仪表板照明灯
panel lighting 区格式照明
panel(l)ing 镶板细工;装门心板;镶面板;嵌板(细工);分段法
panel lining 镶板配套装修
panel load 节间荷载;节点荷载
panel masonry wall 格板圬工墙;墙板圬工墙
panel material 顶板材料
panel meter 面板式仪表;嵌镶式仪表
panel meter/alarm unit 板式测量报警装置
panel methods 小组测试法
panel mo(u)ld 镶板模;间壁模;镶模板;墙板模
panel mo(u)lding 镶板线条饰
panel-mounted 装在面板上的
panel mounting 面板安装
panel mount variable optic(al) attenuator 面板安装可变光衰减器
panel node 节点
panel number 面板编号
panel of arbitrators 仲裁小组
panel of experts 专家小组
panel of pavement 路面板
panel of vibrated brickwork 振动砖砌护墙板
panel on earth resources and environment 地球资源和环境小组
panel on meteorology of the stratospheric(al) and mesosphere 平流层和中层气象学小组
panel on weather and climate 天气和气候小组
panel-painting 薄板油画;镶板油漆
panel partition 板式隔断
panel patch 修补用板片
panel pattern 镶板花式模型
panel pier 节格式桥墩;节间桥墩
panel pin 镶板钉;板销
panel planer 刨板机
panel planer and thicknesser 单面木工压刨床
panel plating method 镀板法
panel point (桁架的)节点;桁架节点;人工标志(点);区格点
panel point of a truss(ed girder) 桁架大梁的节点
panel product 木材板制品(胶合板、木屑板等);木板制品
panel radiator 嵌入式散热器;辐射板放热器;板式散热器
panel reinforcement 补强圈补强
panel roof 格板屋顶
panel room 控制间
panel rough grinding machine 管屏粗磨机
panel saw 板条锯;板锯;带锯
panel screen 板面屏幕
panel service 面板服务
panel shell 格板薄壳;镶板薄壳
panel shuttering 格板式闸门;模板
panel siding 镶板壁板;护墙板
panel soffit 镶板底面;格板底面
panel sound absorber 板振荡吸声装置
panel spacing 板间间隔;板间伸缩缝
panel spalling 板面散裂
panel spalling test 抗热震性试验;嵌板散裂试验;平板法剥落试验
panel stiffness 板刚度
panel strap 垫片;衬板;紧固夹板;嵌条
panel strip 盖缝条;板条;压缝条;嵌条
panel structural system 镶板结构系统;墙板结构系统
panel study 抽样调查;访问调查
panel switch 面板开关;平装开关
panel system 大板(建筑)体系;分组安装制
panel-to-panel connector 墙板联结件
panel tracery 哥特垂直式花格窗
panel tracking digitizer 板式跟踪数字化仪
panel truck 小型载货篷车
panel turnover machine 翻板机
panel type air cleaner 排板式空气过滤器
panel type air filter 平板型空气过滤器
panel type ammeter 面板型电流表
panel type automatic switch board 面板型自动交换机
panel type construction 框架式结构;框格式结

构;框格式建筑;预制墙板式结构
panel type heating system 辐射采暖系统
panel type house 预制墙板(式)房屋;预制板式房屋
panel type instrument 屏型仪表;盘形仪表
panel type resonator 格板式共振器
panel type structure 框格式结构;框格式建筑
panel type switchboard 面板式开关板
panel type voltmeter 盘用电压表
panel van 厢式货物运输车
panel wall 隔墙;框格墙;节间墙;幕墙;大板墙;板式墙
panel wall block 板式墙街坊;板式墙住房
panel wall building 板式墙建筑;幕墙建筑
panel wall of vibrated brickwork 振动砖墙板;振动砖墙壁板
panel wall system 墙板体系
panel wall unit 预制墙板
panel washer 洗平板机;平板状垫片;平板状垫圈
panel washing machine 平式洗涤机
panel with window opening 有窗孔的墙板
panel work 构架工程;镶格工作;构架作业;嵌板工作
panel zone 节间区
pane of a hammer 锤的顶边
pane of a roof 屋顶天窗玻璃
pane of glass 玻璃板;门窗玻璃块;玻璃片
pane separation 窗玻璃片隔条
panethite 陨磷碱锰镁石;磷镁钠石
pan evapo(u)ration 盘式蒸发
pan evapo(u)rimeter 盘式蒸发仪
pan factor 蒸发器系数;蒸发皿系数
pan feeder 盘式给料器;盘式给料机;盘式给矿机;板式喂料机;盘式给料机
pan filter 盘式过滤器;盘式过滤机
pan floor 格板地板
pan floor panel 格板地板用的镶板
panformation 硬土层;泛演替顶极;泛群系
pan form stair(case) 盘模楼梯
pan formwork 盘式模板
pan fraction 筛余百分率;筛盘分级;筛分
pan furnace 罐炉
pan furrow 犁沟底
Pangaea breakup 泛大陆解体
Pangea 泛古陆
pangeosyncline 泛地槽
pangeosyncline stage 泛地槽阶段
pan grid 盘式格栅
pan grinder 盘式碾矿机;轮碾机;盘碾;盘式磨矿机;盘磨
pan grinding 碾磨
pan handle 狭长投影;狭长地带;混凝土基础(跳)梁;锅柄
pan head (铆钉或螺栓的)锅形头;锥形头;截锥头;皿形头;盘形头;弓架
pan head bolt 锅头螺栓;盘头螺栓;截锥头螺栓
pan head hole 盘形穴
pan head rivet 盘形头铆钉;锅头铆钉;平顶铆钉;平头铆钉;盘头铆钉
pan head rivet with tapered neck 半埋式铆钉
pan head screw 皿形螺钉;锅头螺钉;盘头螺钉;大柱头螺钉
pan humidifier 盘式加湿器
panhydrometer 通用液体比重计;通用比重计
panhygrous 全湿的
panic bar 门栓
panic bolt 紧急出口栓;紧急保险螺栓;太平门栓
panic button 应急按钮;紧急保险开关;紧急按钮
panic buying of goods 抢购物资
panic devices on doors 安全门推开装置;紧急装置
panic dump 应急转储;紧急转储【计】
panic dump routine 应急转储程序
pan ice 浮冰;饼状冰
panic equipment 应急设备
panic exit device 太平门栓
panic exit mechanism 太平门机械装置
panic grass 黍
panic hardware 紧急出口栓;太平门栓
panicle 圆锥花序
panicled thyrsoid cyme 密伞圆锥花序
panic-proof lock 安全门门锁
panic purchase 抢购风潮
panic relief installation 紧急脱机装置
panic selling 怕地段衰落而引起的竞卖
panic stop 紧急保险止动器;紧急制动
panidiomorphic 全自形的;全同形的

panidiomorphic granular 全自形粒状
panidiomorphic granular texture 全自形粒状结构
panimatric photo 综合法绘图
pan irrigation 根圈灌溉
Panisaj 巴尼瑟灰褐色硬木(印度产)
pan jig 盘式跳汰机
pan joist floor 盘式格栅地板
panking 铺板
pan left 镜头左转
pan lehr 小车式退火窑
panlite 聚碳酸酯树脂
pan mill 碾碎机;磨石;碾盘式碾磨机;碾盘式磨机;碾磨盘
pan mixer 锅式拌和机;立轴式搅拌机;盘式搅拌机;盘式拌和机;盘式混合器
panne velvet 平绒
pannier 突出部分建筑;壁柱顶托座;牛腿;托臂(模板);篮状饰;柳条篓子;蒌框;石笼
pannier condenser 侧向式冷凝器;侧向布置式凝汽器;背篮式凝汽器
pannikin 小盘;金属小杯
panning 全景拍摄;淘洗重矿物工艺;淘盘选;淘盘洗选法;随动摄影
panning angle 摇头角度
panning head 摇头
panning plate 淘沙盘
panning study 规划研究
pannus 碎片云
pan of balance 天平盘
pan of distilled water 装蒸馏水的盆
panoplay 摄全景动作
panopticon 圆形监狱;中心辐射式全景建筑;望远显微两用镜;珍品展览室
panoram 全景装置;全景镜;全景图
panorama 周视图;全景装置;全景(图)
panorama camera 全景摄影机
panoramagram 全景图;全影立体照片
panorama lift 观光电梯
panorama theatre 全景电影院
panoramic 全景的
panoramic adapter 全景适配器;扫调附加器
panoramic aerial camera 全景航摄仪
panoramic aerial photography 全景航空摄影
panoramic amplifier 扫调放大器
panoramic attenuator 全景衰减器
panoramic camera 全景摄影机;全景照相机
panoramic cinema 全景电影院
panoramic comparator 扫调比较器
panoramic comparison 扫描比较;全景比较;扫调比较
panoramic detector program(me) laboratory 全景探测计划实验室
panoramic distortion 全景畸变
panoramic drawing 全景透视图
panoramic effect 全景效应
panoramic emulsion 全景乳剂
panoramic glass window 全景玻璃窗
panoramic infrared photo system 红外(线)全景照相系统
panoramic lens 全景镜头
panoramic monitor 全景监视器;扫调监视器
panoramic parallax stereogram 全景视差立体图
panoramic perspective 全景透视
panoramic photograph 全景照片;全景像片
panoramic photographic(al) distortion 全景摄影畸变
panoramic-photograph map 全景照片图
panoramic photography 全景摄影;全景摄像
panoramic picture 全景像片;全景图像
panoramic profile map 全景廓线图
panoramic projector 透视投影器
panoramic radar 环视雷达;全景雷达
panoramic receiver 全景扫调接收机;全景接收机;扫频接收机;扫调接收机
panoramic screen 全景宽银幕;扫调屏
panoramic sight 周视瞄准镜
panoramic sketch 全景图
panoramic technique 全影技术;全景技术
panoramic videomapping 全景显示
panoramic view 全景图;全景
panoramic wallpaper 画景墙纸
panoramic window 全景窗
pan ore feeder 盘式给矿机
pan panel floor 格式地板

pan pattern 格子模型
pan pelletizer 盘式制粒机
panplain 大泛滥平原;泛准平原
panplane 大泛滥平原;联接泛滥平原
pan plate 格子板
panplatform 泛地台
panplatform stage 泛地台阶段
pan process 油再生法;直接蒸气(再生)法
panradiometer 辐射热仪;全波段辐射计
pan-range 扫调范围
panrest 锅架
pan rivet head 锅形铆钉;皿形铆
pan roll 压碎机
pan-roofing tile 板瓦
pan scale 锅垢;盘秤
pansclerosis 全硬化
pan scraper 铲斗
Panseri alloy 盘斯里铝硅合金
pan shaker 簸动运输机
pan-shaped base 凸缘底板
pan-shot 全景摄影;全景摄像;全景拍摄
pan shovel 勺斗式机铲
pan slab 矩形格栅板
pan soffit 格子底面
pan soil 硬土;坚土
panspermatism 病原普遍存在说
panspermia 病原普遍存在说
pan steps 盘模踏步
pan support 锅架
pant 整流罩
pantachromatic 全消色差的
pantagraph engraving machine 缩放仪刻模铣床
Pantal alloy 潘塔尔铝合金;耐蚀铝基合金
pan tank 倾入槽
pantechincon 家具仓库;大型仓库
pantechincon van 家具搬运车
pantelegraph 传真电报
pantellerite 碱流岩
panthalassa 泛古洋;泛大洋
panthalassic type 广分布海洋生物类模型
Pantheon 罗马万神殿(建于120~124年);万神庙;伟人祠;大教堂
Pantheon at Rome 罗马万神殿
pantheon dome 万神殿式圆屋顶
pan tile 槽形屋面瓦;波形瓦;平波S形瓦;平瓦
pantiled roof 瓦屋顶;波形瓦屋面;波形瓦屋顶;波形瓦屋板;波瓦屋顶
pantile lath 波形瓦挂瓦条;槽形瓦挂瓦条
pantile plate 波形瓦片;波形瓦板
pantiling 波形瓦盖的屋顶;以波形瓦盖屋顶
panting 挠振现象;拍击
panting action 挠振作用;脉动作用
panting arrangement 防挠振结构;抗拍结构
panting beam 加强梁;防挠梁;补强梁;抗拍梁;抗击击梁
panting deck 防挠振甲板
panting frame 加强框梁;加强框架;防挠振肋骨;抗折肋骨
panting strains 折击变形
panting stress 拍击应力
panting stringer 防挠纵梁;补加纵桁;抗拍纵材
pantobase 全能起落
pantobase aircraft 任意场地起落飞机
pantochromic 两色或更多色的
pantochromism 多色(现象)
pantodrill 自动钻床;全能钻床
pantograph 线条仪;异型加工机;缩放仪;缩放器;受电弓;电杆架;导电弓(架);放大尺;比例绘图仪;比例绘图器;比例画图仪;比例画图器;比例放图仪
pantograph adherence pressure 受电弓接触压力
pantograph arrangement 受电弓配置
pantograph base 伸缩继电器座;缩放仪台
pantograph bounce 受电弓离线
pantograph coach 受电弓车
pantograph collector 架式集电器
pantograph contact pressure 受电弓接触压力
pantograph control 受电弓控制
pantograph control mechanism 受电弓控制机构
pantograph copying grinder 缩放仪仿形磨床
pantograph cylinder 受电弓汽缸;受电弓控制机构
pantograph damper 受电弓阻尼器
pantograph dropping device 受电弓降弓装置
pantograph gate 伸缩式拉门;缩放仪式拉门

pantograph head 受电弓弓头
pantographic 使用缩图器的
pantographic(al) reduction 使用缩图器绘成的缩图
pantograph jack 菱形架式伸缩千斤顶
pantograph link 缩图器连杆
pantograph operating signal 集电弓运行信号
pantograph pan 受电弓滑板
pantograph ratio 缩放比
pantograph sway 受电弓偏侧量
pantograph trolley 伸缩集电器
pantograph well 受电弓承槽
pantograph workshop 受电弓检修间
pantography 图形缩放
pantograver 缩放式刻图仪
pantometer 经纬测角仪;万能测角仪;万测仪
pantomography 全体层照相术;曲面层摄影
pantomorphism 结晶全对称(现象);全形性;全对称性;全对称现象
pantopaque 碘苯酯;碘苯十一酸乙酯
pantoplankton 泛浮游生物
pantoscope 广角照相机;广角透镜
pantoscopic camera 全景摄影机;全景照相机
pantoscopic lens 大角度透镜
pantostrata 连续层
pantosurface aircraft 任意场地起落飞机
pantothenate 泛酸盐
pantothenic acid 泛酸;本多生酸
pantropic 泛嗜的
pantropical plant 泛热带植物;分布在全热带的植物
pantry 配膳室;厨房与餐厅间的服务室;食品室;餐具室;备餐室
pantry man 餐厅管理员
pantry room 配膳间
pantry sink 备餐间洗涤槽;备餐间污水池
pantry window 备餐室窗;备餐间送货窗
pants 桩锤帽;钢板桩锤垫板
pants press 热模精压
pan-type air outlet 盘式送风口
pan-type bed-load sampler 盘式推移质采样器;盘式底砂采样器
pan-type bed-material sampler 盘式河底质采样器
pan-type car 盘式卸载车
pan-type concrete mixer 盘式混凝土搅拌机
pan-type humidifier 盘式增湿器;盘式加湿器
pan-type mixer 盘式混料机;盘式混合机
pan-type pelletizer 盘式造粒机
pan-type sampler 盘式河砂采样器;盘式采样器
pan-type stair(case) 盘式楼梯
pan-type tread 盘式踏板
pan-type vehicle 盘式卸载车
pan-up 镜头上移
pan vibrator 表面振捣器;振动盘;平板振捣器
pan visco(si)meter 盘式黏度计
pan vulcanization 罐中硫化法
panzer 铠装运輸机
panzer actimometer 湿差电感光计
pao-ferro 巴西铁树
paoh-tah 佛塔
paolovite 斜方锡钯矿
pap 奶头状物;乳胶体;檐沟出水孔
papagayo 帕帕加约风
papagoite 硅铝铜钙石
papal cross 天主教的十字架
papal tree 菩提树
Papanicolaou's vaginal smear technique 巴氏染色法
paper 论文;票据
paper adhesive 纸张黏结剂
paper air cleaner 纸质空气滤清器
paper aircraft 设计图上的飞机
paper air filter 纸滤气器;纸空气过滤器
paper analog(ue) recording 纸卷模拟记录
paper analysis 纸上分析
paper and board manufacture 纸和纸板制造
paper-and-cotton insulated cable 纸棉绝缘电缆
paper-and-ink-compatibility test 纸张与油墨适(应)性试验
paper-and-pencil error 书写错误
paper and pulp industry 造纸纸浆工业;造纸和纸浆工业
paper and pulp industry wastewater 造纸和纸浆工业废水
paper and pulp wastewater 造纸和纸浆废水

paper applying machine 铺纸机
paper back 普及本；平装书；平装本
paperback book rack 平装书陈列架
paper-backed 用纸作底层的
paper-backed alumin(i)um 纸镀铝
paper-backed lath 有纸垫的板条；油毡面板条
paper-backed metal lath 纸背金属条板
paper backing 纸背
paper bag 纸袋
paper bag machine 纸袋机
paper bag package 纸袋包装
paper base 纸基
paper-based laminate 纸质层压板；碎纸塑料；层压板；胶纸板
paper birch 纸皮桦
paper blanket 纸质表面层；纸毯
paperboard 纸板
paperboard container 纸板箱
paperboard gasket 纸质垫片
paperboard grade(stock) 纸板用
paperboard industry 纸板工业
paperboard mill 纸板厂
paperboard mill wastewater 纸板厂废水
paperboard products 纸板制品
paperboard property 纸板性能
paperboard test 纸板性能测定
paper bolts 造纸木材
paper-bound edition 平装本
paper-bound volume 平装本
paper box 纸箱
paper box package 纸箱包装
paper bridge 纸桥
paper bushing 纸套管
paper cable 油纸皮电缆；纸绝缘电缆
paper calculation 纸上计算
paper capacity 一次印张数
paper card reading mechanism 纹板链传动机构
paper carrying roll 引纸辊
paper cart (混凝土道路工程的)展纸车
paper carton 包装纸板盒
paper cassette 存纸盒
paper ceiling 糊纸顶棚；纸糊顶棚；纸糊平顶
paper certifying its status as a legal person 法人资格凭证
paper characteristic 纸质
paper chromatoelectrophoresis 纸色层电泳法
paper chromatogram 纸色谱图
paper chromatographic(al) method 纸色谱分析法
paper chromatographic-nephelometric method 纸色谱浊度分析法
paper chromatographic scanner 纸色谱扫描器
paper chromatography 纸上色谱法；纸上色层分析法；纸上色层分离法；纸色谱(法)；纸层析(法)
paper chromatography spectrophotofluorometry 荧光分光光度纸层析；纸色谱法—荧光分光光度法
paper chute 纸质斜槽
paper clasp 办公用纸夹子
paper clay 造纸用陶土；薄层黏土
paper clip 回形针
paper cloth bag 纸塑复合袋
paper coal 薄层煤；页状煤
paper coating 纸张涂料；纸张涂层；纸面涂布；裱纸层
paper colo(u)r 纸张颜色；纸张颜料；染纸颜料
paper colo(u)ring 纸张染色
paper condenser 纸介质电容器
paper conditioning 纸张处理
paper cone 宝塔纸管
paper constant 纸常数
paper consumption 纸张消耗量
paper conversion 纸张加工
paper converter 纸制品加工机
paper copy 平装本
paper core 纸芯管
paper-core(d) cable 纸芯电缆
paper-core enamelled type cable 纸包漆皮绝缘电缆
paper-core plywood 纸芯胶合板
paper-core screened cable 纸绝缘屏蔽电缆
paper-covered veneer 覆纸的表层；覆盖纸的镶面
paper-covered wire 纸绝缘线
paper cover mo(u)lding 纸盖线脚；纸盖模型

paper credit 纸币信用
paper cure 纸板养护
paper curing 纸板养护
paper curing machine 纸张整饰机
paper currency 纸币
paper-cut 剪纸
paper-cut silhouette 剪影
paper cutter 切纸器；切纸机；裁纸器
paper cutting machine 切纸机
paper defect 纸病
paper deflector 导纸板
paper delivery 供纸量；出纸
paper destructor 废纸焚烧炉
paper dish 纸盘
paper-disk chromatography 圆纸色谱
paper disposal shaft 纸张处理井筒
paper disposer 纸处理器
paper distortion 纸张变形
paper drain 纸板排水；排水纸板
paper drain preloading method 纸板排水预压法
paper drinking cup 纸杯
paper drum 卷纸筒
papered wall 糊纸的墙
paper electrophoresis 纸上电泳(法)；纸电泳法；纸电泳
paper electrophoresis apparatus 纸上电泳仪
paper electrophoresis method 纸电泳法
paper electrophoretic separation 纸(上)电泳分离(法)
paper-element filter 纸芯滤清器
paper exit sensor 出纸口传感器
paper extraction 抽纸
paper-faced 纸面的
paper-faced gypsum board 纸面石膏板
paper face overlay 塑料贴面
paper factory 造纸厂
paper fastener 纸张扣钉
paper feed 输纸
paper-feeding mechanism 输带机构
paper feed key 给纸键
paper felt 纸毡；薄毡
paper felt duck 造纸帆布
paper fiber 造纸纤维(指适宜造纸的纤维)
paper fiber sludge 纸纤维污泥
paper filler 纸充填物
paper filter 纸过滤器
paper filter element 纸滤芯
paper finger 纸夹
paper finishing 纸张装饰
paper flaw 纸病
paper foil 纸基金箔
paper folding machine 折页机
paper for covering books 书皮纸
paper for Hausler type roof(ing) 铺设豪斯勒式屋面用纸
paper form 厚纸板盒；硬纸模板；纸模
paper for mica tape 云母带纸
paper formulation 配浆比率
paper gasket 纸垫片；纸衬
paper glass 纸杯
paper-glosser 纸面加光机
paper going velocity 走纸速度
paper gold 纸黄金
paper grade 像纸反差等级
paper guide 输纸机；导纸机构；导纸板
paper guide flap 导热器
paper hanger 裱糊工(人)
paperhanger's brush 糊纸刷
paperhanger's hammer 糊墙纸用锤
paperhanger's lath 糊纸工人用的板条
paperhanger's scissors 糊墙纸用剪刀
paperhanging 糊纸；壁纸；墙纸；贴壁纸；贴墙纸；裱糊墙纸
paperhanging adhesive 墙纸黏结剂
paperhanging paste 糊墙纸用浆糊
paper-holder 压纸尺；纸卷；卫生纸匣；卫生纸盒；手纸匣；手纸盒
paper hygroscope 纸张湿度计
paper incinerator 纸张焚化炉
paper industry 造纸工业
paper industry sewage 造纸工业污水
papering 用纸裱糊；糊纸；墙纸；裱糊
papering (interleaving) machine 铺纸机
papering work 糊纸工作

paper ink relation 油墨纸张关系
paper in reel 纸卷
paper insulated 纸绝缘的
paper-insulated cable 纸质绝缘电缆；纸绝缘电缆
paper-insulated enamelled wire 纸质绝缘漆包线
paper-insulated enamel wire 纸绝缘漆包线
paper-insulated lead-covered cable 纸质绝缘铅包电缆
paper-insulated lead-sheathed cable 纸绝缘铅包电缆；铠装电缆
paper insulated underground cable 纸绝缘地下电缆
paper insulation 纸绝缘
paper interleaving machine 垫纸机
paper jam 卡纸；堵纸
paper kite 风筝
paper knife 切纸刀；裁纸刀
paper knitting 纸线编织
paper laminate 叠合纸板
paper lamp-shade 纸灯罩
paper-latcher 卷装保温材料工人
paper layer 纸层
paper-lead cable 铅包纸绝缘电缆
paperless item processing system 无凭单项目处理系统
paper liner 纸衬片
paper location 纸(带)上定线；图上选址；图上定线
paper location of line 纸上定线(法)
paper loss 账面损失
paper machine 抄纸机；造纸机
paper machine room 造纸车间
paper machine screen 造纸机前筛浆机
paper maker 造纸工
paper-makers' alum 造纸明矾
paper-making 纸的生产；造纸业；造纸
paper-making canvas 造纸帆布
paper-making machinery 造纸机
paper-making raw material 造纸原料
paper-making waste 造纸废物；造纸废料
paper-making wastewater 造纸废水
paper-making wastewater treatment 造纸厂废水处理
paper manufacture 纸张生产；造纸
paper manufacturing machinery 造纸机械
paper matrix 纸型
paper maunfacturing industry 造纸工业
paper meal 纸粉；纸屑
paper method 纸上作业法；图上作业法；室内作业法
paper mill 造纸厂
paper mill(foam) oil 造纸厂泡沫油
paper mill waste 纸浆废液；造纸厂废水
paper model 纸模型
paper money 纸币
paper money inflation 纸币通货膨胀
paper money standard 纸币本位
paper money system 纸币制度
paper mo(u)ld 硬纸模板；纸型；纸模；纸板
paper mulching 纸覆盖
paper negative 纸质底片
paper out 出纸
paper-out sensor 出纸检测器
paper overlaid 糊纸的
paper overlaid plywood 华丽板
paper overlay 木纹纸；贴面纸
paper overlay board 纸质贴面板
paper pallet 纸托盘
paper partition chromatography 纸分配色谱(法)
paper peat 薄层泥炭；薄层泥煤；页状泥炭
paperphenol 酚醛纸
paper-pile 纸堆
paper piler 堆纸器
paper print 像片；纸质蓝图
paper processor 冲纸机
paper profit 账面利益；账面利润
paper pulp 纸浆
paper pulp fibered plaster 纸筋灰
paper pulp filter 纸浆过滤器
paper pulp lime plaster finishing 纸筋灰面罩
paper pulp mill 纸浆厂
paper-punch 纸张穿孔器
paper recovery 纸回收
paper recycling 纸的循环
paper reel 垫纸卷筒

paper refuse 废纸;纸垃圾
paper resist 露花纸版;镂花纸板
paper roll 纸辊;纸卷
paper roller 滚纸筒;记录纸卷筒;记录纸滚筒
paper sack 纸袋
paper salvage 废纸利用
paper scrap 废纸;纸屑
paper sculpture 纸制装饰
paper separator 纸质分隔器
paper shale 纸状页岩;薄页岩;薄层页岩
paper shears 剪纸机
paper sheathing 纸衬
paper sheathing board 厚纸质衬板
paper sheet 纸片
paper sheeting 纸板;建筑用纸板
papershelled almond 薄壳扁桃
paper shredder 切废纸机
paper shrinkage 纸张收缩
paper side guide 挡纸侧板
paper size 纸张开数;纸张规格;纸张尺寸
paper sleeve 纸套管
paper slew 超行距走纸
paper sliding-door 糊纸推拉门
paper sliding-screen 纸隔扇
paper smoothness tester with vacuum pump 带真空泵的纸张平滑度测定仪
paper snow fence 纸制的雪栅栏
paper space 图纸空间
paper spar 薄片方解石
paper speck 纸斑
paper speed 纸速
paper stain 纸印
paper stencil printing 纸版印花
paper stock 纸料;造纸材料
paper strength 纸张强度
paper stretch 纸张伸长率
paper strip 条形纸
paper strip chromatography 条形色谱法
paper strip method 纸条法;纸片法
paper strip mixed lime mortar 纸筋灰
paper strip mixed mortar plaster 纸筋灰粉刷
paper strip plaster 纸筋石灰
paper stuff 纸浆
paper substrate 印花纸坯
paper surface efficiency 纸面效率
paper tape 纸带
paper tape coder 纸带编码器
paper tape decoder 纸带译码器
paper-tape equipment 纸带机
paper tape input 纸带输入
paper tape punch(er) 纸带凿孔机;纸带穿孔机
paper tape reader 纸带输入机;读带机;光电输入机
paper tape recording 纸带记录
paper tape reperforator 纸条复凿孔机
paper tape sprocket 纸带中导孔
paper tapestry 纸挂毯
paper tape transcriber 纸带转录机;纸带读出机
paper technology 造纸工艺学
paper template 纸模片
paper test 纸张性能测定
paper tester 纸张性能测定仪
paper textile 纸制织品;纸织品
paper thermochromatography 纸上热色谱法
paper thickness 纸张厚度
paper-thimble dust sampler 纸套粉尘取样器
paper throw 超行距走纸
paper throw character 跑纸符号
paper to bearer 不计名票据
paper towel dispenser 纸巾分配器
paper transfer 移花;贴花纸
paper trimmer 裁纸刀
paper tube 纸套;纸筒;纸管
paper-used 用纸
paper valve bag 自封口纸袋
paper wad 纸填
paper wadding 纸填料
paper wall 纸糊墙
paper warehouse 纸库
paper washer 纸垫
paper waste(water) 造纸废水
paper web 卷纸筒
paper weight 纸张重量;纸压铁;图镇;镇尺
paper winder 卷纸机;垫纸卷取机
paper window 纸糊窗

paper with a neutral image tone 中性色调像片
paper with silkworm eggs 蚕纸
paper with warm image tone 暖色调像片
paper wood 纸材;造纸木材
paper work 日常文书工作;文件单据工作;书面(资料)工作
paper worker 造纸工
papery 胶质纸板
paper' yarn 纸线
paper-yarn fabric 纸线织物
paper yellow 纸黄(色)
papier-mache 纸浆(模);纸模纸板;纸型;混凝纸(用于制造盒、盆、盘的纸质可塑材料)
papier-mache box 混凝纸板做的盒子
papier-mache facade 虚饰门面
papilionaceous corolla 蝶形花冠
papillary wave 脉首波
pappe 纸板
pappus 冠毛
Pappus' theorem 巴普斯定理
papreg 层压纸板;胶压纸板;塑化纸板
papyriform 埃及柱头(纸莎草式样)
papyrograph 复写纸
papyroline paper 裱在布上的纸
papyrotype 锌版复制
papyrus 纸莎草;纸草
papyrus-bud capital 纸莎草芽状柱头;纸莎草芽状柱帽
papyrus capital 纸莎草饰柱头;纸莎草形柱头
papyrus column 纸莎草饰柱;纸莎草形柱
papyrus half-column 纸莎草形半柱
par 面值;平水【建】;票面额;同等;等量;平价
para-academic 半学术性
para-acetaldehyde 三聚乙醛
para acid 仲酸;对位酸
para-alumohydrocalcite 副水碳铝钙石
para-aminoacet anilide 对氨基乙酰苯胺
para-aminoacetophenone 对氨基苯乙酮
para-aminoazobenzene hydrochloride 对氨基偶氮苯盐酸盐
para-aminobenzoic acid 对氨基苯甲酸
para-aminobenzoic acid sodium salt 对氨基苯甲酸钠
para-aminobenzoyl hydrazine 对氨基苯甲酰肼
para-amino benzyl cellulose 对氨基苄基纤维素
para-aminodiphenyl 对联苯基胺
para-aminomethyl benzoic acid 对氨甲基苯甲酸
para-aminophenol 对氨基苯酚
para-aminophenol hydrochloride 对氨基苯酚盐酸盐
para-aminophenol sulfate 对氨基苯酚硫酸盐
para-aminophenol wastewater 对氨基苯酚废水
para-aminophenyl ethyl ether 对氨基苯乙醚
para-aminophenylmethylether 对氨基苯甲醚
para-amino-salicylate 对氨基水杨酸
para-amino sulfuric acid 对氨基硫酸
para-aminotenzaldehyde 对氨基苯甲醛
para-aminotert-utylbenzene 对氨基特丁基苯
para-aminotoluene 对甲苯胺
para amphibolite 副角闪岩
paraanisic acid 对茴香酸
paraanisidine 对茴香胺
para-arsanilic acid 对氨基苯胂酸
paraballoon 充气天线
parabanic acid 仲酸脲;草酰脲
parabasal apparatus 副基器
parabasal body 动质体;副基体
parabasalt 普通玄武岩
parabedded 似层状
parabema 希腊教堂至圣所旁的房间
parabenzoquinone 对苯醌
parabenzoquinone dioxime 对苯醌二肟
parabiont 共生生物
para-biphenylamine 对联苯基胺
parabituminous 副烟煤
parablastesis 副变质岩
parablepsia 错视
parablgm decision analysis 示范决策分析
parabola 抛物线;抛物面反射器
parabola control potentiometer 抛物线控制电位器
parabola cupola 抛物线穹形屋顶
parabola dome 抛物线穹形屋顶
parabola equation 抛物线方程式

parabola frustum 抛物线截角锥体
parabola-growth 抛物线生长
parabola mass spectrum 抛物线质谱
parabola mass-spectrograph 抛物线质谱仪
parabola mass spectrography 抛物线质谱法
parabola mass-spectrometer 抛物线质谱仪
parabola of fit 拟合抛物线
parabola of higher order 高阶抛物线
parabola optimization of single factor 单因子抛物线优选法
parabola (roof) truss 抛物线屋面桁架
parabola vault 抛物线拱顶
parabolic(al) 抛物线的;抛物的
parabolic(al) aerial 抛物面天线
parabolic(al) antenna 抛物面形天线;抛物天线;抛物面天线
parabolic(al) approximation 抛物线逼近法
parabolic(al) arc 抛物线弧
parabolic(al) arch 抛物线拱
parabolic(al) arch bridge 抛物线拱桥
parabolic(al) arch dam 抛物线拱坝
parabolic(al) arched girder 抛物线起拱大梁
parabolic(al) arch section 抛物线拱断面
parabolic(al) arch support 抛物线拱支架
parabolic(al) asymptote 渐近抛物线;抛物线渐近线
parabolic(al) baseline correction 抛物线基线校正
parabolic(al) beam 抛物线梁
parabolic(al) bearing pressure distribution 轴承压力的抛物线形分布
parabolic(al) body 抛物线体
parabolic(al) cable 抛物线形钢丝束
parabolic(al) catenary 抛物线悬链线
parabolic(al) characteristic 抛物线特性
parabolic(al) chord truss 抛物线(形折弦)桁架
parabolic(al) collineation 抛物线性直射
parabolic(al) conchoid 抛物蚌线
parabolic(al) condenser 抛物线面聚光器
parabolic(al) conoid 抛物线圆拱
parabolic(al) conoid shell 抛物线圆拱壳体;抛物线圆穹壳体
parabolic(al) coordinates 抛物线坐标
parabolic(al) cross-section 抛物线形(横)断面
parabolic(al) crown 抛物线路拱
parabolic(al) cupola 抛物线圆屋顶
parabolic(al) current 抛物线波电流
parabolic(al) curve 抛物线(形)曲线;抛物曲线
parabolic(al) cusp 抛物尖点
parabolic(al) cylinder 抛物线体;抛物(线形)柱面;抛物线筒柱
parabolic(al) cylinder antenna 抛物柱面天线
parabolic(al) cylinder coordinates 抛物柱面坐标
parabolic(al) cylinder function 抛物柱面函数
parabolic(al) cylindrical equation 抛物柱面形方程
parabolic(al) cylindrical surface 抛物柱面
parabolic(al) delay distortion 抛物线形延迟失真
parabolic(al) detection 抛物线检波
parabolic(al) dish 抛物柱面;抛物面
parabolic(al) distribution 抛物线分布;抛物分布
parabolic(al) distribution load 抛物线形分布荷载
parabolic(al) dome 抛物线穹窿
parabolic(al) dune 新月形沙丘;抛物线沙丘
parabolic(al) edge 抛物线形边缘
parabolic(al) equation 抛物形方程;抛物线方程
parabolic(al) flight 抛物线飞行
parabolic(al) flow characteristic 抛物线流量特性
parabolic(al) flow-concentration model 抛物线形流量—密度模型
parabolic(al) folium 抛物叶形线【数】
parabolic(al) form 抛物线形式;抛物线模板
parabolic(al) formula 抛物线公式
parabolic(al) frame 抛物线构架;抛物线刚构
parabolic(al) function 抛物线函数
parabolic(al) geometry 抛物几何
parabolic(al) girder 抛物线形大梁;抛物线梁;抛物线大梁
parabolic(al) governor 抛物线调节器
parabolic(al) grade 抛物线形坡度
parabolic(al) haunched girder 抛物线加腋梁
parabolic(al) haunched slab 抛物线加腋板
parabolic(al) headlamp 抛物线形大灯
parabolic(al) homology 抛物性透射
parabolic(al) hopper 抛物线形料斗
parabolic(al) horn 抛物线形喇叭筒;抛物线形喇叭
parabolic(al) hyperboloid shell 双曲抛物面壳

体;双曲抛物面薄壳
parabolic(al) hypercylinder 抛物超柱
parabolic(al) increase allocation method 抛物线递增分配方式
parabolic(al) index fiber 梯度折射率光纤
parabolic(al) interpolation 抛物线内插法;抛物插值法
parabolic(al) interpolator 抛物线插补器
parabolic(al) kettle feed bin 抛物线形喂料仓
parabolic(al) law 抛物线定律
parabolic(al) layer 抛物层
parabolic(al) lens 抛物线透镜
parabolic(al) load 抛物线形荷载;抛物线形分布荷载
parabolic(al) loading 抛物线加载;抛物线加荷
parabolic(al) measuring weir 抛物线形堰
parabolic(al) method 抛物线法
parabolic(al) metric geometry 抛物度量几何
parabolic(al) metric group 抛物度量群
parabolic(al) microphone 抛物面反射器式传声器;抛物面反射传声器;抛物面传声器
parabolic(al) mirror 抛物柱面镜;抛物线形镜子;抛物面镜;抛物面反射器;抛物面反射镜
parabolic(al) mirror control 抛物柱面镜控制
parabolic(al) model 抛物线模型
parabolic(al) nose 抛物线形头部
parabolic(al) nozzle 抛物线形喷嘴
parabolic(al) operator 抛物形算子
parabolic(al) orbit 抛物线轨道
parabolic(al) partial differential equation 抛物形偏微分方程
parabolic(al) path 抛物线轨迹
parabolic(al) plane 抛物平面
parabolic(al) plug 抛物线形插销
parabolic(al) point 抛物点
parabolic(al) point of a surface 曲面的抛物点
parabolic(al) profile 抛物线纵断面
parabolic(al) projectivity 抛物性射影
parabolic(al) projector 抛物面聚光灯
parabolic(al) quadratic surface 抛物形二次曲面
parabolic(al) quadric hypersurface 抛物形二次超曲面
parabolic(al) radiator 抛物面辐射器;抛物镜面天线
parabolic(al) radio telescope 抛物面射电望远镜
parabolic(al) rank 抛物形秩
parabolic(al) reentry 以抛物线速度再入
parabolic(al) reflector 抛物柱面反射镜;抛物线形反射镜;抛物面镜;抛物面反射器;抛物面反射镜;抛物面反光罩
parabolic(al) regression 抛物回归
parabolic(al) relation 抛物线关系
parabolic(al) rib 抛物线式拱肋;抛物线拱肋
parabolic(al) Riemann surface 抛物线黎曼曲面
parabolic(al) roof truss 抛物线屋顶桁架
parabolic(al) scanning 抛物面扫描
parabolic(al) section 抛物线断面
parabolic(al) sector 抛物扇形域
parabolic(al) segment 抛物线段
parabolic(al) shading 抛物线形黑斑
parabolic(al) shape 抛物线形
parabolic(al) shell 抛物线薄壳
parabolic(al) solution 抛物线解法
parabolic(al) space 抛物空间
parabolic(al) spiral 抛物螺(旋)线
parabolic(al) spiral curve 抛物线形螺旋曲线
parabolic(al) spring 抛物线形弹簧
parabolic(al) steel eccentricity concrete beam 混凝土梁中钢筋的抛物线偏心距
parabolic(al) sub-algebra 抛物子代数
parabolic(al) sub-group 抛物子群
parabolic(al) tendon 抛物线形预应力筋
parabolic(al) torus antenna 抛物环面天线
parabolic(al) trajectory 抛物线轨道
parabolic(al) transformation 抛物(形)变换
parabolic(al) transition curve 立方抛物(线)形缓和曲线
parabolic(al) truss 抛物线形桁架;抛物线桁架
parabolic(al) type 抛物(线)形
parabolic(al) vault 抛物线拱顶;抛物线穹顶
parabolic(al) velocity 抛物线速度
parabolic(al) velocity distribution 抛物线速度分布
parabolic(al) velocity reentry 以抛物线速度再入
parabolic(al) vertical curve 抛物线形竖曲线
parabolic(al) voltage 抛物波电压

parabolic(al) wave 抛物线形波
parabolic(al) waveform 抛物线波形
parabolic(al) weir 抛物线堰;抛物线形堰
parabolic(al) wire 抛物线形钢丝束
parabolic-reflector microphone 抛物面反射镜式传声器
paraboloid 抛物线;抛物面反射器;抛物面;椭圆抛物面
paraboloidal 抛物线体;抛物面的
paraboloidal antenna 旋转抛物面反射器
paraboloidal coordinates 抛物面坐标
paraboloidal mirror 抛物面镜
paraboloidal reflector 旋转抛物面反射器;抛物面反射镜;抛物面反光罩
paraboloid(al) roof 抛物面形屋顶
paraboloidal shell 抛物面薄壳
paraboloidal shell of rotational symmetry 对称回转抛物面薄壳
paraboloidal shock 抛物线形激波
paraboloidal surface 抛物柱体曲面;抛物柱面曲面
paraboloidal umbrella roof of folded shell 抛物面形伞状的折板薄壳屋顶
paraboloidal umbrella shell roof 抛物面形伞状薄壳
paraboloid calotte 抛物曲面拱顶
paraboloid condenser 抛物面聚光镜;抛物面聚光器
paraboloid dish (antenna) 抛物面天线
paraboloid lens 抛物面透镜
paraboloid method 抛物面法
paraboloid of revolution 旋转抛物面;回转抛物面
paraboloid reflector 抛物面反射器
paraboloid revolution 抛物线形旋转面
paraboloid roof 抛物面屋顶;抛物面屋顶
paraboloid shell 抛物面形壳;抛物面壳(体)
paraboloid umbrella roof 抛物面伞形屋顶
parabromoacetanilide 对溴乙酰苯胺
parabromoacetophenone 对溴苯乙酮
para-bromoaniline 对溴苯胺
parabromoaniline sulfate 对溴苯胺硫酸盐
para-bromoanisole 对溴苯甲醚
parabromobenzoic acid 对溴苯甲酸
parabromobenzoyl chloride 对溴苯甲酰氯
parabromomandelic acid 对溴苦杏仁酸
para-bromop-henacyl bromide 对溴苯酰甲基溴
para-bromophenylhydrazine 对溴苯肼
parabromotoluene 对溴甲苯
parabscan 抛物面扫描器
parabundle 伞投勉资
paracardo 亚轴节
paracarmine 副洋红;副卡红
paracarmme 副胭脂红
paracelsian 副钡长石
paracentesis 穿刺
paracentric 超出中心的
paraceratubae 侧蜡管
paracetaldehyde 三聚乙醛
paracetamol 对乙酰氨基酚
paracetanol 醋氨酚
parachlor nitraniline red 对氯邻硝基苯胺红
para-chloroacetophenone 对氯苯乙酮
parachloroaniline 对氯苯胺
para-chloroben-zenesulfonyl chloride 对氯苯磺酰氯
para-chlorobenzoic acid 对氯苯甲酸
para-chlorobenzoyl chloride 对氯苯甲酰氯
para-chlorobenzyl chloride 对氯苯基氯
para-chlorobromobenzene 对氯溴苯
para-chloromandelic acid 对氯苦杏仁酸;对氯扁桃酸
para-chloromercuribenzene sufonate 对氯汞苯磺酸
para-chloromercuribenzoic acid 对氯汞苯甲酸
para-chloronitrobenzene 对氯硝基苯
para-chloro-penyl methyl sillicone 对氯苯基甲基硅油
para-chlorophenol 对氯酚;对氯苯酚
para-chlorophenoxyacetic acid 对氯苯氧基乙酸;对氯苯氧基醋酸
para-chlorophenoxyacetic ester 对氯苯氧乙酸乙酯
parachlor-orthonitraniline 对氯邻硝基苯胺红
para-chlorotoluene 对氯甲苯
parachlor red 对氯邻硝基苯胺红
parachor 克分子等张比容;摩尔等张体积;等张比容
parachute 降落伞;提升器断绳保险;断绳防坠器

parachute beacon 降落伞式指向标
parachute cam 断绳防坠器抓爪
parachute carrier lifeboat 降落伞救生艇
parachute container 降落伞套
parachute distress signal 降落伞遇难信号
parachute flare 降落伞火号
parachute gear 断绳保险装置
parachute pack 降落伞包
parachute response 降落伞反应
parachute set 降落伞无线电台
parachute tower 跳伞塔
parachute vault 降落伞式拱顶;伞形圆形屋顶
parachute weather buoy 空投气象浮动站;海洋气象自动观测浮标
Paracide 对二氯苯(熏蒸杀虫剂)
paracite 方硼石
paraclase 断层
paraclimax 亚群系
paracoagulation test 副凝固试验
paracompact space 仿紧空间
para-compound 对位化合物
paracon 聚酯类橡胶质
paraconformity 准整合;假整合;似整合
paraconglomerate 副砾岩
paracoquimbite 紫铁矾;副针绿矾
paracostibite 副硫锑钴矿
paracoumarone-indene resin 香豆酮—茚树脂;聚库玛隆树脂
paracraton 准克拉通【地】
paracresol 对甲酚
paracril 丁腈橡胶
paracrystal 类晶体;次晶
paracrystalline state 次晶态
paracrystallinity 次晶度
paracrystals 不全晶体
para-curve 抛物线
paracyanic acid 雷酸
paracyanobenzoic acid 对氰基苯甲酸
paracycle of relative change of sea level 海平面相对变化准周期;海面相对变化准旋回
paracycle of relative variation of sea level 海平面相对变化准周期
paracyclone dust cleaner 对流旋风除尘器
paracyclone dust collector 串式旋流集尘器
paracyclophane 对环芳烷
paracyl reflector 抛物面反光罩
paracytic type 平轴式气孔
paracytic type of stomata 平裂形气孔
paradamite 副羟砷锌石
parade 公共散步场所
parade ground 检阅场;阅兵场;练兵场;表演场
paradelta 准三角洲
para-derivative 对位衍生物
paradesmose 副连丝
paradesmus 副连丝
paradibromobenzene 对二溴苯
paradichlorobenzene 对二氯苯(熏蒸杀虫剂)
paradichlorobenzol 对二氯苯
paradiethylaminobenzaldehyde 对二乙氨基苯甲醛
paradigm 范例
paradiiodobenzene 对二碘苯
paradimethoxybenzene 对二甲氧基苯
paradime thylaminobenzaldehyde 对二甲氨基苯甲醛
paradinitrobenzene 对二硝基苯
paradise 修道院花园或公墓;教堂天井;教堂前面的庭院;乐园;天堂
paradise of children 儿童乐园
paradoorasite 副砷锑矿
parados 掩体后土墙;(希腊剧院的)乐队入口;背墙(防弹片杀伤而在堑壕或掩体后构筑的土垛)
paradox 反论
paradox gate 附环闸门;附环移滚阀
paradoxical 奇异的
paradoxical embolus 反常栓子
paradoxical phase 反常相
paradoxical reaction 反常反应
paradoxic metastasis 逆行转移
paradoxite 肉色正长石
paradox of value 价值矛盾
paradrop 空投
para-dye 偶合染料
para-ecology 埋藏学
paraedrite 金红石

paraequilibrium 平衡感觉障碍
para-ester wastewater 对位酯废水
paraethoxyacetophenone 对乙氧基苯乙酮
paraethoxybenzoic acid 对乙氧基苯甲酸
paraethoxychry-soidine 对乙氧基柯衣定;对乙氧基黄吡精
paraethyl phenol 对乙基苯酚
parafacies 成岩亚相;成岩次相;副相;分相
parafe 草签
parafeed coupling 旁馈耦合
paraffinaceous 石蜡的
paraffinaceous petroleum 含石蜡石油
paraffin base 石蜡基;石蜡底子
paraffin(e) 链烷(属)烃;馏出蜡;石蜡烃煤油
paraffin(e) acid 石蜡酸
paraffin(e) aldehyde 石蜡醛
paraffin(e)-asphalt petroleum 石蜡沥青基石油;石蜡沥青混合基原油
paraffin(e) base 石蜡底子;石蜡基
paraffin(e)-base asphalt 石蜡基沥青
paraffin(e)-base concrete curing compound 石蜡基混凝土养护合成物;石蜡基混凝土养护剂
paraffin(e)-base crude oil 石蜡基原油
paraffin(e) base lubricating oil 石蜡基润滑油
paraffin(e)-base oil 石蜡基石油
paraffin(e) base petroleum 石蜡基石油
paraffin(e) bath 蜡浴;石蜡浴
paraffin(e) bath apparatus 蜡浴器
paraffin(e) block 石蜡块
paraffin(e) bowl 蜡碗
paraffin(e) chlorination unit 石蜡氯化装置
paraffin(e) coal 含石蜡烟煤;石蜡煤
paraffin(e) coating 石蜡涂层;石蜡封盖
paraffin(e) content 含蜡量;石蜡含量
paraffin(e) content test 含蜡量试验
paraffin(e) crude oil 石蜡基原油
paraffin(e)-cut section 石蜡切片
paraffin(e) cutter 切蜡器
paraffin(e) degreasing 石蜡脱脂
paraffin(e) deposit 石蜡矿
paraffin(e) dirt 石蜡垢
paraffin(e) distillate 石蜡馏分
paraffined paper 石蜡纸
paraffin(e) duck 石蜡防水帆布
paraffin(e) embedding 石蜡打底
paraffin(e) emulsion 石油乳液;石蜡乳剂
paraffin(e) emulsion sizer 石蜡乳液浸润剂
paraffin(e) filter 石蜡滤波器
paraffin(e) flux 石蜡渣液
paraffin(e) grade wax 商品石蜡
paraffin(e) hydrocarbon 链烷烃;石蜡系碳氢化合物
paraffin(e) immersion method 石蜡浸渍法
paraffin(e)-impregnated wood 石蜡浸注材
paraffin(e) intermediate crude 石蜡中间基石油
paraffin(e) jelly 石蜡油膏;凡士林
paraffin(e) knife 刮蜡器
paraffin(e) lubrication 石蜡润滑
paraffin(e) mass 石蜡块
paraffin(e) melter 熔蜡器
paraffin(e) method 石蜡切片法
paraffin(e) mixed concrete 掺蜡混凝土
paraffin(e) molle 软石蜡
paraffin(e) motor 石蜡发动机
paraffin(e) naphtha 石蜡基石油精
paraffin(e) naphthenic oil 石蜡—环烷型石油
paraffin(e) oil 煤油;石蜡油;液状石蜡;液体石蜡
paraffin(e) pan 蜡锅
paraffin(e) paper 石蜡纸
paraffin(e) press 石蜡压滤机
paraffin(e) pressed distillate 石蜡馏分
paraffin(e) radiation shielding wall 石蜡防射线屏蔽墙
paraffin(e) refined wax 纯石蜡
paraffin(e) scale 未精制石蜡;粗石蜡;鳞状蜡
paraffin(e) scraper 石蜡刮削器
paraffin(e) section 石蜡切片
paraffin(e) series 石蜡系
paraffin(e) shale 石蜡页岩
paraffin(e) slack wax 粗石蜡;含油石蜡
paraffin(e) soap 石蜡皂
paraffin(e) solution 石蜡溶液;石蜡乳液
paraffin(e) spreading unit 摊蜡器
paraffin(e) stain 石蜡染色剂

paraffin(e) sweating 石蜡发汗
paraffin(e) test 石蜡浸透探伤
paraffin(e) waterproofing 石蜡防水
paraffin(e) wax 硬石蜡;蜡;粗石蜡
paraffin(e) wax and oil emulsion 石蜡和油乳液
paraffin(e) wax melting point test 石蜡熔点试验
paraffin(e) wax oxidation 石蜡氧化
paraffin(e) wax quality test 石蜡质量试验
paraffin(e) wax white 白石蜡
paraffin(e) wire 浸蜡线;石蜡绝缘线
paraffin(e) xylol 石蜡二甲苯溶液
paraffinic 石蜡族的
paraffinic acid 石蜡族酸
paraffinic-base crude 石蜡基原油
paraffinic-base(crude) petroleum 石蜡基石油
paraffinic crude 石蜡基原油
paraffinic gases 石油尾气
paraffinic hydrocarbon 石蜡系烃
paraffinicity 石蜡含量;链烷烃含量
paraffinicity curing compound 石蜡含量
paraffinic oil 石蜡油;石蜡型石油
paraffinic solvent 烷烃溶剂
paraffinoma 石蜡瘤
paraffins 烷烃烃;石蜡
paraffinum durum 硬石蜡;固体石蜡
paraffinum liquidum 矿脂;石蜡油
Paraflex wire rope 帕拉弗雷克斯钢丝绳
paraflocculus 旁绒球
paraflow 抗凝剂
para-fluoroaniline 对氟苯胺
parafluorobenzoic acid 对氟苯甲酸
para-fluorophenylacetic acid 对氟苯乙酸
paraflutizide 对氟噻嗪
paraform 多聚甲醛
paraformal-dehyde 多聚甲醛;仲甲醛
parafuchsin 偶合品红;副品红
parafunction 机能异常
paragearksutite 副钙铝氟石
paragenesis 共生
paragenetic 前成的
paragenetic association of minerals 矿物共生组合
paragenetic diagram 共生次序图
paragenetic light 前灯
paragenetic relation 共生关系
paragenetic sequence 共生次序
paragenetic twin 共生双晶
paragenic ore 共生矿
paragentic mineral 共生矿物
parageosyncline 准地槽;副地槽
paragglutination 副凝集
paragglutination reaction 副凝集反应
paragglutinogen 副凝集原
paraggulation phenomenon 副凝集现象
paraglomerate 类砾岩
paragneiss 水成片麻岩;副片麻岩
paragntic mineralogy 共生矿物学
paragon 圆形大珍珠;纯粹钻石(100 克拉以上的);典范
paragonite 钠云母
paragonite schist 钠云母片岩
paragraph ship 分舱货船
paraguanajuatite 副硒铋矿
Paraguay 巴拉圭
Paraguay tea 巴拉圭茶
Paragutta 巴拉格塔合成树脂;合成树胶;假橡胶
parahelium 仲氦
parahilgardite 副氯羟硼钙石;副氯硼钙石
parahopeite 副磷锌矿
parahydroxyacet-ophenone 对羟基苯乙酮
parahydroxyaniline acetaldoxime 对羟基苯胺乙醛肟
parahydroxyazobenzene-parasulfonic acid 对羟基偶氮苯对磺酸
parahydroxyben-zaldehyde 对羟基苯甲醛
para-hydroxybenzene 对羟基苯
para-hydroxybenzoic acid 对羟基苯甲酸
parahydroxydiphenyl 苯羟二苯
parahydroxydi-phenylamine 对羟基二苯胺
para-hydroxy phenyl ethyl ketone 对羟基乙酮
para-hydroxypropiophenone 对羟基苯丙酮
paraiodoaniline 对碘苯胺
paraiodoanisole 对碘苯甲醚
parajamesoite 副脆硫锑铅矿
parakeldyshite 副硅钠锆石

parakhinite 副碲铅铜石
Paral 帕拉尔黄褐色硬木(产于缅甸、印度)
paralagoon 边缘泻湖
paralaurionite 副羟氯铅矿
paraldehyde 仲乙醛;三聚乙醛;副醛
paraldol 仲醛醇
paraleucaniline 副白苯胺
paraliageosyncline 滨海地槽;陆缘地槽;大陆边缘地槽
paralic 海陆交互的
paralic basin 近海盆地
paralic coal-bearing formation 近海含煤建造
paralic coal deposit 近海煤层【地】;近海煤系
paralic environment 近海环境
paralic sedimentation 近海沉积作用
paralic swamp 沿海沼泽;近海沼泽;海滨林沼
paralic zone 沿岸地带;近海地带
paralimnion 湖底近岸带
Paralla 帕拉拉号【船】
parallactic 视差的
parallactic angle 星位角;位角;视差角;时差角
parallactic base 视差基线
parallactic correction 视差修正
parallactic data 视差数据
parallactic displacement 视差位移
parallactic ellipse 视差椭圆
parallactic equation 月角差
parallactic error 视差误差
parallactic grid 视差格网
parallactic inequality 月角差;视差不等
parallactic instrument 视差量测仪
parallactic libration 视差天平动
parallactic measurement 视差测量
parallactic microscope 视差显微镜
parallactic motion 视差运动;视差移动;视差动
parallactic mounting 赤道(式)装置
parallactic movement 视差移动
parallactic net 视差网
parallactic orbit 视差轨道
parallactic photograph 视差像片
parallactic polygon 视差导线
parallactic polygonometry 视差等线测量
parallactic refraction 折射视差【测】;视差折光差
parallactic shift 视差移动;视差位移;视差偏移
parallactic slide 视差滑尺
parallactic table 视差表
parallactic triangle 天文三角形;视差三角形
parallactic wedge 视差楔
parallactoscopy 视差镜术
parallageosyncline 海滨地槽
Parallax 倾斜线;视差
parallax adjustment 视差调节
parallax age of tide 视差潮龄
parallax and refraction 视差与折光差;视差与蒙气差
parallax arm 视差尺
parallax bar 视差量测杆;视差杆;视差尺
parallax clearance 消除视差
parallax colo(u)r tube 视差式彩色荫罩管
parallax compensation 视差补偿
parallax compensator 视差补偿器
parallax computation 视差计算
parallax computer 视差校正计算机
parallax converter 视差变换器
parallax correcting finder 视差校正取景器
parallax correction 视差校正;视差改正量;视差改正
parallax correction finder 校准视差指示器
parallax correction setter 偏差校正器
parallax determination 视差测定
parallax difference 视差较【测】;视差差异
parallax displacement 视差位移
parallax effect 视差效应;视差现象
parallax equation 视差方程
parallax error 视差误差;视差
parallax factor 视差因子
parallax-free 无视差
parallax-free model 无扭曲模型;无扭曲立体模型
parallax-free motion 无视差运动
parallax heighting 视差测高程
parallax in altitude 高度视差;视差改正量
parallax inequality 月视差不等;视差不等
parallax instrument 视差量测仪
parallax measurement 视差测量
parallax microscope 视差显微镜

parallax of cross-hairs 十字丝视差;交叉瞄准线视差
parallax offset mechanism 视差校正器;视差补偿机构
parallax of reticule 十字线视差;十字丝视差
parallax of wires 十字丝视差
parallaxometer 视差计
parallax panoramagram 视差周视体视图片
parallax photogrammetry 视差照相测量术
parallax range transmitter 视差校正发射机
parallax reading error 读数视差
parallax refractometer 视差折射计
parallax rod 视差杆
parallax second 秒差距
parallax slide 视差滑尺;上下视差滑尺
parallax star 视差星
parallax table 视差表
parallax wedge 视差楔;视差光楔
paralled laminate 平行层积
paralled surface grinding machine 双端面磨床
parallel 平行;纬线;等维度圈;赤纬;并行考虑;并行;并列的;并联的
parallel absorbent baffle 平行吸声板
parallel access 并行访问;并行存取
parallel accumulator 并行累加器
parallel action 并行作用
parallel active tracking program(me) 并行主动跟踪程序;并行有效跟踪程序
parallel activity 平行活动
parallel adder 并行加法器
parallel addition 并行加法
parallel adit 平行坑道;平行导坑
parallel advance 平行推进
parallel aerial photography 并列航空摄影
parallel-affine 平行仿射
parallel algirithm 并行算法
parallel alignment 平行度调准
parallel alley 平行小路
parallel amplifier 并联放大器
parallel anchor mooring 并联锚系碇
parallel and pipeline processing 并行和流水线处理
parallel angle 平行角
parallel antenna 平行天线
parallel application 平行安装
parallel approach 平移驶靠【船】;平行驶靠
parallel arc 平行圆弧;平行弧
parallel arc furnace 并联电弧炉
parallel arc measurement 平行圈弧度测量;纬线弧线测量
parallel arc weld 并列弧焊
parallel arithmetic unit 平行运算器
parallel arithmetic processor 并行算术处理器
parallel arithmetic unit 并行运算器;并行算术运算器;并行四则运算器
parallel-arm suspension 平行臂式独立悬架
parallel arrangement 并联电路;并联配置;并联连接;平行排列;并联装置;并联排列
parallel arrangement of (bore) holes 钻孔平行布置
parallel arrangement of transmission lines 输电线路平行布置
parallel arrangements of moving stair(case) 自动楼梯平行布置
parallel arrangement type junction terminal 并联式枢纽
parallel array processor 并行阵列处理机
parallel assignment algorithm 并行分配算法
parallel asynchronous computer 并行异步计算机
parallel attack 平行线法
parallel automatic exposure system 并联自动曝光系统
parallel-averted photograph 等偏摄影像片
parallel-averted photography 等偏摄影
parallel avertence 同偏
parallel axes 平行轴
parallel-axes gears 平行轴齿轮
parallel-axes shaving 平行轴剃齿法
parallel axiom 平行公理
parallel axis 平行轴
parallel(-axis) extinction 平行消光
parallel-axis geara 平行齿轮系
parallel-axis theorem 平行轴定理;平行移轴定理
parallel baffle muffler 平行隔板式消声器
parallel band 平行带
parallel banks method 平行边坡法

parallel bars 平行杆
parallel barter 平行易货
parallel basin 平行状流域
parallel beach 平行海滩
parallel beam 平行束;平行射线束;平行梁;平行光束
parallel bearing 滑动轴承;普通轴承
parallel bedding 平行层理
parallel bench vice 平行台用虎钳
parallel berth 顺岸码头
parallel binary accumulator 并行二进制累加器
parallel binary computer 并行二进制计算机
parallel bisection 并联中割
parallel block 平行台;平行垫铁;并行块;并行分程序
parallel boarder 平行边线
parallel body 平行舯体;平行船中体;船中平行体
parallel boom truss 平行弦杆桁架
parallel border 平行边线
parallel boresighting 平行校靶
parallel branch 并联支路
parallel buffer 并行缓冲器
parallel build 平行成型
parallel build package 平行成型卷装
parallel bumps 平行布置的人造凸起障碍物
parallel burning 带状烧除
parallel bus 并联总线
parallel by character 并行字符
parallel cabin 平行壁座舱
parallel canal 平行渠道
parallel capacitor 并联电容器
parallel capacitor circuit 并联电容电路
parallel capping 平行顶盖
parallel carrier 平行托架;平行夹头
parallel carry 并行进位
parallel cascade action 并行级联作用;并联串级动作
parallel cellular chain 并行单元链
parallel cellular organization 并行单元组织
parallel chain of command 平行的指挥系统
parallel channel 并行通道;并联通路
parallel channel printer 并联通路打印机
parallel-chap cut 平行龟裂掏槽
parallel cheek pliers 平行面夹钳
parallel cheese 平行筒子
parallel chord method 平行弦法;并行弦法
parallel chords 平行弦
parallel-chord truss 梯形桁架;平行弦桁架
parallel chorisis 平行辟分
parallel circle 线圈;平行圈;纬圈
parallel circle of declination 纬度圈
parallel circuit 关联电路;平行线路;平行流程;分电路;并联线路;并联环路;并联电路
parallel circuit design 并联回路设计;并联电路设计
parallel circuit firing 并联电路起爆
parallel circuit power transmission system 并联电路输电方式
parallel circular plate chamber 圆形平行板电离室
parallel circular saw 多圆盘锯
parallel clamp 平行夹头;平行垫铁
parallel classes 平行类
parallel clause 平行条款
parallel cleavages 平行劈理
parallel cline 平行渐变群
parallel cluster 平列发动机簇
parallel collimator 平行光管法
parallel column 并联柱
parallel combination 并联组合
parallel communication interface 并行通信接口
parallel comparator 并行比较器
parallel compensating condenser 并联补偿电容器
parallel compensation 反馈补偿
parallel computation 并行计算
parallel computation environment 并行计算环境
parallel computation technique 并行计算法
parallel computer 并行计算机;并行操作计算机;并联计算机
parallel condenser compensation device 并联电容补偿装置
parallel connected meter 并联水表
parallel connected thermocouple 并联热电偶(装置)
parallel connected type reactor 并联式电抗器
parallel connection 并行连接;并联连接;并联接头;并联
parallel connection method 并联接法

parallel connection of the motors 电动机并联法
parallel connecting by-pass tunnel 平行坑道;平行导坑
parallel construction 平行结构
parallel contracting 平行承发包
parallel control 平行控制;并行控制;并联控制
parallel control structure 并行控制结构
parallel convey 并行传送
parallel coordinates 平行坐标
parallel coping 平行压顶;平行盖顶
parallel coupling 并联耦合
parallel course 平行航向;等维航向
parallel course computer 航线计算机;并行航线计算机
parallel crank 平行曲柄
parallel crank four bar linkage 平行四杆回转链系
parallel crank mechanism 平行曲柄机构
parallel crossover 平行渡线【铁】
parallel curb parking 平行路边停车
parallel current 并流;并联电流
parallel current dryer 并流干燥器;气流平行干燥机
parallel current jet condenser 并流注水凝汽器
parallel current system 顺气流系统
parallel current tunnel dryer 并流式隧道干燥器
parallel curvature 平行曲面
parallel curve 平面曲线
parallel curved rib arch bridge 平行曲线肋拱桥
parallel curve rib arch 平行曲线肋拱
parallel cut 平行挖掘;平行截割;平行切割;直眼掏槽;平行切割;平行截割
parallel-cut blasting 平行空炮孔掏槽爆破
parallel dam 顺坝
parallel data 并行数据
parallel data adapter 并行数据适配器
parallel database 并行数据库
parallel data controller 并行数据控制器
parallel data processing 并行数据处理
parallel deactivation 平行失活
parallel deal 平行交易
parallel decimal accumulator 并行十进位累加器
parallel decimal adder 并行十进制加法器
parallel decomposition 平行分解
parallel design 平行设计
parallel determination 平行测定
parallel digital computer 并行数字计算机
parallel dike 导流堤;平行堤;顺河堤;顺坝
parallel direction 平行方向
parallel displacement 平行位移;平行变位
parallel displacement fault 平移断层
parallel distance 纬距
parallel distortion 平行四边形畸变
parallel distribution in the large 整体平行分布
parallel ditch system 平行沟(渠)系统
parallel docking 平行对接
parallel double pick insertion 平行双纬引送
parallel drainage 平行(排)水系(统);并行水系
parallel drainage pattern 平行状水系
parallel drainage system 平行状水系
parallel drifting bands 平行漂移带
parallel drift method 平行的导坑方法
parallel drum 平行转鼓
parallel drum type capstan 平行卷筒式牵引轮
parallel drum winding 圆筒形绞筒提升
parallel ed 平行的
parallel edge 平行边
paralleled path 平行通路
parallelehedra 平行面体
parallel electric(al) fields 平行电场
parallel entry 平行布置进风巷道;并行输入;并行入口;并联入口
parallelepiped 平行六面体【数】;平行六边形
parallelepipedal cut stone 切成平行六面体的石材
parallelepiped and maximum likelihood classification program(me) 平行六面体和极大似然分类程序
parallelepiped classification 平行六面体分类;平行多面体分类
parallelepiped(on) of forces 力的平行六面体
parallel escalators 平行式自动扶梯
parallel evolution 平行演化
parallel excitation 并激
parallel executable statement 并行可执行语句
parallel execution 并行执行
parallel experience 类似经验

parallel exposure 平行接近距离
parallel extended route 并行扩充路径
parallel extinction 平行吸光
parallel fabric 直纹布
parallel-faced type of ga(u)ge 平行面规
parallel faults 平行断层;平列断层
parallel-fed motor 并联供电电机
parallel feed 平行进给(法);平行进给;平行进刀;侧向馈送;并流送料法;并联馈电
parallel feedback 平行反馈;并联反馈
parallel feed back integrator 并联反馈积分器
parallel feedback operational amplifier 并联反馈运算放大器
parallel feeder 并联馈路
parallel feeding 混合进料;并行送料;并行进料
parallel fibrous texture 平行片状结构
parallel field 平行场
parallel fillet weld 填角焊分缝;侧角焊分缝
parallel filling-in process 并行填充过程
parallel filter 并联滤波器
parallel financing 平行融资
parallel fire suppression 修线扑火
parallel firing 并联点火
parallel fissure 平行缝隙
parallel flange beam 平行翼缘梁
parallel flanged girder 平行缘梁
parallel flange truss 平行弦桁架
parallel flo(a)tation circuit 平行浮选回路
parallel flow 平行流(动);同向流;并行流;并流
parallel flow basin 平行流水池;并流池
parallel flow condenser 顺流冷凝器
parallel flow dryer 平流干燥器;并流干燥器;平行流干燥器;顺流烘干机
parallel flow drying 顺流烘干;并流干燥
parallel flow eddy 平行流涡流
parallel flow engine 轴流式燃气轮机
parallel flow evapo(u)rator 并流蒸发器;平行流蒸发器
parallel flow gas turbine 分流式燃气轮机
parallel flow heat exchanger 并流式热交换器;并流式换热器;并流换热器
parallel flow heating furnace 顺流式连续加热炉
parallel flow nozzle 平行流喷嘴
parallel flow rotary dryer 并流旋转干燥器
parallel flow superheater 并流式过热器;顺流过热器
parallel flow suspension preheater kiln 顺流热交换悬浮预热器窑
parallel flow turbine 轴流式涡轮(机);轴流式透平;轴流式水轮机;并流式涡轮机;并流汽轮机
parallel flow type spray dryer 并流式喷雾干燥器
parallel fold 平行褶皱;同心褶皱;同心褶曲
parallel folding 平行折页
parallel forces 平行力
parallel four wire feeder 平行四线式馈线
parallel full adder 并行全加器
parallel gap reflow soldering 平行间隙回流焊接
parallel gap weldig 平行间隙焊接
parallel gas feed 双路供气
parallel gate 并联门
parallel ga(u)ge 平行规
parallel-gear drive 平行齿轮传动
parallel gears 平行齿轮
parallel generator 并联发电机
parallel generator theorem 并联发电机定理
parallel girder 平行梁
parallel girder bridge 平行梁桥
parallel gliding 平行滑翔;平行滑移
parallel grading method 平行级配法
parallel grain 平行纹理
parallel grating 平行光栅
parallel grinding wheel 平行砂轮
parallel-groove clamp 接触线固定线夹;并沟线夹;平行槽线夹
parallel group 平行链晶
parallel growth 平行连生;平行交互生长;并生
parallel guidance mechanism 平行导杆装置
parallel guide 平行导轨
parallel gutter 箱形雨水槽;平行雨水槽;平行天沟;并行雨水槽
parallel half adder 并行半加器
parallel half-subtracter 并行半减器
parallel hand tap 等直径丝锥

parallel heading 平行进风平巷;平行导坑
parallel heave 平行平错
parallel hedra 平行面体
parallel helical gears 平行螺旋齿轮
parallel hob 等径滚铣刀
parallel hobbing cutter 圆柱形齿轮滚刀
parallel hogback flood plain 平行鬃岗式河漫滩
parallel holding ground 顺截河埂储[贮]木场
parallel hole 平行钻孔;平行炮孔
parallel hole blasting 平行孔掏槽爆破
parallel hole connection 炮孔并联
parallel hole cut 平行炮孔掏槽
parallel hole drilling 平行孔钻进
parallel hole method of sublevel stoping 平行炮孔分段回采
parallel homeomorphy 平行同形
parallel hook-up 并联接线
parallel hydraulic circuit 并联液压回路
parallel impedance 并联阻抗
parallel impervious boundary 平行隔水边界
parallel import 平行进口
parallel in 并行输入;并网;并联输入
parallel index line 平行指向线
parallel index technique 平行标线法
parallel inductance loading 并联电感负载
parallel induction 平行诱导
parallel information processor 平行信息处理机
paralleling 并列
paralleling earth pushing process 并列推土法
paralleling switch 并联开关
parallel input 并行输入
parallel input data 并行输入数据
parallel input mode 并行输入方式
parallel input-output card 并行输入一输出插件板
parallel in serial out 并行输入串行输出
parallel insertion 并行插入
parallel installed filters 并联过滤器
parallel instruction control unit 并行指令控制器
parallel instruction queue 并行指令排队器
parallel intake entry 平行进风巷
parallel interface 并行接口
parallel interface element 并行接口元件
parallel intergrowth 平行交互生长;平行互生
parallel inverter switch 并联变换开关
parallelism 类似;平行(性);平行度;对应;并行性
parallelism detection 平行检测
parallelism of multiprocessing 多重处理的并行性
parallelism of optic(al) axis 光轴平行度
parallelism tolerance 平行度公差
parallelization 平行化
parallelization tester 平行度试验仪
parallel jaw 平行钳口;平口虎钳
parallel-jawed clamp 平行咬夹
parallel-jaw pliers 平虎口钳
parallel-jaw vice 平口台钳;平口虎钳
parallel jetties 平行导流堤
parallel justification 并行齐行
parallel key 平行键;平键
parallel knot 平行纽结
parallel ladder track 平行梯线
parallel laid web 平行铺设纤维网
parallel laid wire rope 平行捻钢丝绳
parallel lamellar structure 平行纹理构造
parallel laminate 顺纹层压板;平行迭层板;顺纹层材
parallel-laminated 平行层压的
parallel-laminated veneer 平行木纹胶合板
parallel lamination 同向纹理层压
parallel lamp 平行光源
parallel latitude sailing 等维航迹计算法
parallel lattice cupola 平行格构的圆顶
parallel lattice dome 平行格构的穹隆;平行格构穹顶
parallel lattice girder 平行格构大梁
parallel law 平行四边形定律
parallel lay 顺捻;平行捻
parallel lay rope 等捻绳索;平行捻绳
parallel lift 平行四边形悬挂装置;平行四边形起落机构
parallel lift linkage 平行四边形悬挂装置
parallel light 平行光
parallel light beam 平行光束
parallel light density 平行光密度
parallel line assay 平行线测定
parallel line interface 并行线路接口

parallel line oscillator 平行线振荡器
parallel line resolution 平行线分辨率;平行线分辨力
parallel lines 并行线路;并联线路;平行(直)线
parallel link 并行链接
parallel linkage 平行连接;平行连杆机构
parallel link hitch 平行四连杆悬挂装置
parallel link lift 平行四杆悬挂装置
parallel link suspension 平行杠杆式独立悬架
parallel loading 并行加载
parallel loan 平行贷款
parallel locks 复线船闸
parallel logic 并行逻辑
parallel logic circuit 并行逻辑电路
parallell-to-grain compressive strength 顺纹抗压强度
parallelly travel(l)ing cable crane 平移缆索起重机;平行移动式缆机
parallel machine 并行作用式计算机;并行机;并联机
parallel machine file 平行机用锉
parallel machine operation 机组并联运行
parallel machine processing 并行机处理
parallel main storage 并行主存储器
parallel market 平行市场
parallel memory 并行存储器
parallel memory access 并行存储器存取
parallel merging 并行合并
parallel method 平行线法
parallel method orientation 平行方向法
parallel middle body 平行中体
parallel mill 平行铣刀
parallel misalignment 平行性偏差;不平行性
parallel mixing circuit 并联混合电路
parallel mobile mode 平行移动方式
parallel mode 平行水系模式;并行式;并行方式
parallel model 并行模型
parallel mode symbol 并行处理记号
parallel modulation 并联混合调制
parallel motion 平行运动
parallel motion device 平行移动装置
parallel motion linkage 平行四边形悬挂装置
parallel motion mechanism 直尺平行运动机构;平行运动机构
parallel motion protractor 全能绘图仪;全能绘图机;平移角尺;平行运动分度规;平行尺;分度规
parallel mount 平行安装
parallel movement 平行运动
parallel multi-blade damper 平行多叶阀
parallel multiplier 并行乘法器
parallel nozzle 平行喷嘴
parallel of altitude 地平纬圈;地平经圈;等高圈
parallel of celestial latitude 黄纬平行圈
parallel of declination 赤纬圈;赤纬平行圈
parallel of latitude 黄纬圈;黄经圈;纬线;纬度线;纬度圈;等维圈
parallelogeotropism 直向地性
parallelogram 平行四边形
parallelogram distortion 平行四边形畸变
parallelogram eccentric mechanism 平行四边形偏心机构
parallelogram identity 平行四边形恒等式
parallelogram law 平行四边形法则
parallelogram lever 平行四边形框架
parallelogram linkage 平行四边形悬挂装置
parallelogram linkage steering 四边形联杆转向机构
parallelogram mechanism 平行四边形机构
parallelogram of forces 力的平行四边形
parallelogram of velocities 速度的平行四边形
parallelogram plate 平行四边形的板
parallelogram ripper 平行四边形裂土器;平行四边式松土机
parallelogram slab 平行四边形(楼)板
parallelogram stand 平行四边试验台
parallelogram steering linkage 平行四边形转向杆系
parallelogram type 平行四边形型
parallelogram type tipper 平行四边形松土器
parallelogram yard 平行四边形车场
parallelohedron 平行多面体
parallelometer 平行仪;平行四边形检测仪
parallel operation 平行作业;平行操作;并行操作;并列运行;并联运行;并联供电;并联操作
parallelopipedon 平行六面体
parallel oscillation 平行振荡;并联振荡

parallel oscillatory circuit 并联振荡电路
paralleloscope 平行镜
parallelosterism 同晶型组与化学成分的关系
parallelotope 超平行体
parallel packet 平行束
parallel padding 并联统调
parallel pantograph 平行四边形缩放仪
parallel parking 纵列式停车;平行停靠;平行停车;顺列式停车
parallel passage reactor 并行通道反应器
parallel pattern 平行型
parallel perspective 平行透视
parallel pervious and impervious boundary 平行透水隔水边界
parallel pervious boundary 平行透水边界
parallel phase resonance 并联相位共振
parallel photography 平行摄影
parallel photometric(al) analysis 平行光度分析
parallel pin 平行销;等直径销
parallel pipe 平行双管线;平行管首道;平行管;并流管道
parallel pipelined processor 并行流水线式处理机
parallel pipelines 平行管线
parallel plane grinding machine 平面磨床
parallel planer 平行刨床
parallel planes 平行平面
parallel planing machine 平行刨床
parallel plastometer 平行盘式可塑仪
parallel plate 平行玻璃板;平行板;测量仪底盘
parallel plate chamber 平行板电离室
parallel plate condenser 平行板电容器
parallel plate counter 平行极板计数管;平行板电极计数器
parallel plate electrode 平行板电极
parallel plate electrostatic precipitator 板式电除尘器
parallel plate interceptor 平行板拦截器
parallel plate interferometer 平行板干涉仪
parallel plate laser 平行平板腔激光器
parallel plate line 平行板线
parallel plate loading 平行板加载
parallel plate micrometer 平行玻璃板测微器
parallel plate model 平行板模型
parallel plate oil interceptor 平行板式油水分离器;平行板隔油池
parallel plate oil separator 平行板式油分离器
parallel plate oscillator 平行板振荡器
parallel plate plastometer 平行板塑性计;平行板式塑性仪
parallel plate precipitator 板式除尘器
parallel plate viscometer 平行板黏度计
parallel plate waveguide 平行板波导
parallel pointing 平行射向
parallel polarization 平行极化
parallel polarized 平行偏光
parallel pore model 平行孔模型
parallel port 并行口;并行接口
parallel pricing 平行定价法
parallel procedure structure 并行过程结构
parallel process 平行工序
parallel processing 并行加工;并行处理
parallel processing efforts of P-and S-wave 纵横波并行处理
parallel processing system 并行处理系统
parallel processing system evaluation board 并行处理系统评价电路板
parallel processor 并行处理机
parallel processor control unit 并行处理机控制器
parallel processor operating system 并行处理机操作系统
parallel processor software 并行处理机软件
parallel processor system hardware 并行处理机系统硬件
parallel processor system software 并行处理机系统软件
parallel profile method 平行剖面法
parallel program(me) 并行程序
parallel program(me) model 并行程序模型
parallel programming 并行程序设计;并线编程
parallel programming language 并行程序设计语言
parallel projection 平行投影;平行射影
parallel publication 相重出版物
parallel pump 卧式往复泵

parallel pyrolysis 并联热解
parallel quay 顺岸码头
parallel radio tap 并联无线电窃听
parallel ramp 平行坡道;平行匝道
parallel rate of exchange 平行汇率
parallel rays 平行光线
parallel reactance 并联电抗
parallel reaction 平行反应
parallel reactor 并列反应器
parallel read 并行读出
parallel reamer 平行铰刀
parallel-rectangular-bar reticle 平行矩形带调制盘
parallel-rectangular frame 平行矩形框架;平行的矩形框架
parallel-rectangular reticle blade 平行矩形条调制盘叶片
parallel rectifier 并联整流器
parallel redundancy 并行冗余
parallel register 并行寄存器
parallel regulator 平行调节器;分流调节器;并联调节器
parallel relay adder 并行继电器加法器
parallel relay network 继电器并联网络
parallel reliability 并联可靠性
parallel representation 并行表示
parallel requisition 并联条件
parallel research memory 并行探索存储器
parallel reservoirs 并联水库
parallel resistance 并联电阻
parallel resonance 并联谐振;并联共振
parallel resonance filter 并联谐振滤波器
parallel resonance frequency 并联谐振频率;并联共振频率
parallel resonance impedance 并联谐振阻抗
parallel resonant circuit 并联谐振电路;并联共振电路
parallel resonant interstage 并联谐振耦合
parallel rewriting 并行重写
parallel ring 平行环
parallel ring accumulator 并行环累加器
parallel ring register 并行环形寄存器
parallel ring type register 平行环式寄存器
parallel ripple mark 平行波痕
parallel rivet 平行铆钉
parallel road 平行路;平行滩列
parallel rod 平行连杆;平行杆
parallel rod array 平行棒阵列
parallel-rod oscillator 平行棒振荡器
parallel-rod screen 平行杆筛;平行棒条筛
parallel-rod tuning 平行杆调谐
parallel roller 平行滚柱
parallel roller bearing 平行滚柱轴承
parallel rolling 等断面轧制
parallel roll profile 平辊型
parallel route 平行进路【铁】
parallel row joint 并排(铆钉)接头
parallel ruler 平行规;平行尺;平行线尺【测】;一字尺
parallel run controller 平行控制器
parallel running 平行运转;平行操作;并行运转;并行运行;并联运转;并联运行
parallel runway 平行跑道
parallel-runway approach 平行跑道进场着陆
parallels 平行线;垫铁
parallel sailing 同纬度航法;等维计算法
parallel scale 纬线尺度
parallel scann(ing) 并行扫描;平行扫描
parallel scheduling 并行调度
parallel search 平行搜索;并行检索;并行查找
parallel search memory 并行检索存储器;并行查找存储器
parallel search storage 并行搜索存储器;并行检索存储器;内容定址存储器
parallel search type memory 并行检索型存储器
parallel section 平行剖面;平行截面;平行截口;平行断面
parallel section across river 横跨河流的平行断面
parallel section method 平行断面法
parallel separation 平行隔距(离)
parallel sequential computer architecture 并行时序计算机体系结构
parallel sequential moving method 平行顺次移动方式
parallel serial 并串行
parallel serial computer 并串行计算机

parallel serial conversion 并串行转换;并串行变换
parallel serial mode 并串行方式
parallel serial operation 并串行操作
parallel serial register 并串行寄存器
parallel serial switch 并串联转换开关
parallel series 混联;多路串联;并串联
parallel series blasting circuit 并串联爆破线路
parallel series connection 并串联接线法
parallel series converter 并串联变换器
parallel series switch 并串联转换开关
parallel series wiring 并串联接线
parallel server 并行服务器
parallel service channel 并联服务通道
parallel servos 并联伺服机构
parallel session 并行会话;并行对话
parallel shaft (speed) reducer 平行轴减速器
parallel shank end-mill with coarse teeth 粗齿直柄立铣刀
parallel sheaf pattern 平行束状
parallel shears 平行剪
parallel shift 平行移动
parallel shift register 平行移位寄存器
parallel shot 一致性检查
parallel-sided combustion chamber 圆筒形燃烧室
parallel-sided precast foundation pile 边平行的预制基础桩
parallel siding 等厚边壁板;方边壁板
parallel simulation 并行模拟
parallel sinkage 平行下沉
parallel slide 平行滑尺
parallel slide valve 平行闸阀
parallel slit 平行缝隙
parallel-slit interferometer 平行缝干涉仪
parallel slot 平行槽;槽边平行槽
parallel sphere 平行(天)球
parallel spin 平行自旋
parallel spiral 平行螺旋
parallel spires 平行螺旋线
parallel spoked reticle 平行辐条状调制盘
parallel spring 平行簧片
parallel stair(case) 两跑楼梯;平行楼梯
parallel standard 平行本位;复本位制
parallel standard system 平行本位制
parallel stay cable 平行拉索
parallel-stays 平行拉线;平行性拉线
parallel steer 平行转向
parallel stock guide 平行式导料装置;平行导料装置
parallel stockpile 平行料堆
parallel storage 平行存储器;并行存储器
parallel straight frame 平行直梁式车架
parallel straight lines 平行直线
parallel strand lumber 单板条层压材
parallel-strand stand stay cable 平行钢绞线拉索
parallel strip 平行带;平行板条
parallel-strip line 带状传输线;平行条状线
parallel structure 平行构造
parallel subspaces 平行子空间
parallel summing network 并联加法网络
parallel surface 平行(曲)面
parallel switching relaying system 并联开关继电系统
parallel symbol 并行符号
parallel system 平行系统;平行式布置;平行布置系统;排水系统布置—平行分区式系统;并行系统;并行方式;并联系统;并联式
parallel system of forces 力的平行系统
parallel table 对照图表;对照表
parallel tangent method 平行切线法;平行切线方法
parallel tangents 平行切线
parallel tap 并联抽头
parallel task spawning 并行任务派生
parallel tensor field 平行张量场
parallel terminal 并行终端设备;并行终端
parallel test(ing) 平行试验;平行检验
parallel texturing 双根平行变形
parallel thread 平行螺纹;普通螺纹
parallel throw 平行落差
parallel to cross track level 平行于轨面
parallel to grain 顺纹
parallel to-serial converter 并串联变换器;并串联转换器
parallel to the axis 与轴平行的;平行于轴线
parallel to the axis of tilt 像横线
parallel to the grain 平行于纹理

parallel trace of lines 平行迹线；线的平行踪迹
parallel track 等维航线
parallel track circuit 并联式轨道电路【铁】
parallel tracking 平行循迹；平行随纹机构
parallel training 平行训练法
parallel training wall 顺岸导墙；顺岸导流堤；导流堤；平行导流堤
parallel train working diagram 平行运行图
parallel transfer 并行传送
parallel transformation 平行变换
parallel transgression 平行海进
parallel translation 平移；平行移动
parallel transmission 同时传输；并行传送；并行传输；并联输送
parallel transmission of information 信息并行传输
parallel-travel(l)ing cableway 平行移动式缆道
parallel trench 平行堑壕
parallel triple filament method 平行三灯丝法
parallel truss 并联桁架；平行桁架
parallel-tube amplifier 电子管并联放大器
parallel tuned circuit 并联调谐电路
parallel twin 平行双晶
parallel twin-screw extruder 平行双螺杆挤出机
parallel type 纬线型
parallel-type acceleration lane 平行型加速车道
parallel-type inverter 并联式变流器；变联式变流器
parallel-type worm 柱形蜗杆
parallel unconformity 假整合；平行不整合
parallel variation 平行变异
parallel vector 平行向量
parallel vector operation 并行向量运算
parallel ventilation branch 平行风路
parallel vice 平行虎钳；平口虎钳
parallel watersheds method 并行流域法
parallel waves 平行波
parallel wharf 顺岸码头
parallel wind(ing) 平行卷绕；复绕；并绕；并联绕组
parallel wing 拱肩墙；平行翼状物
parallel wing wall 平行翼状墙
parallel wire 平行双线
parallel wire cable 平行线电缆；平行钢丝索；平行钢丝绳
parallel wire line 平行双线有线电路
parallel wire resonator 平行线谐振器
parallel wire strand 平行钢丝束
parallel wire system 平行线系统
parallel wire unit 平行钢丝(束)张拉设备(预加应力用)；平行钢丝束
parallel wiring 并联接线
parallel with axis of alluvial-pluvial fan 平行冲洪积扇轴
parallel with river 平行河流
parallel with river valley 平行河谷
parallel word recognizer 并行字识别器
parallel working 平行作业
parallel works 导流设施；导流工程
parallel wound bobbin 平行卷绕筒子
parallel wound coil 并绕线圈
paralleogram of forces 力的平行四边形
parallergin 副变应原
paraloc 参数器振荡电路
paralogy 形似性
paralstonite 三方钡解石
paralyser = paralyzer
paralysis 停顿
paralysis of upper extremities 瘫池
paralysis period 不灵敏周期
paralyzer 并行分析程序；阻滞剂
param 氰胍
paramagnet 顺磁(性)物(质)；顺磁体
paramagnetic 顺磁的
paramagnetic absorption 顺磁吸收
paramagnetic alloy 顺磁合金
paramagnetically doped 顺磁法掺杂的
paramagnetic analytical method 顺磁分析法
paramagnetic anisotropy 顺磁各向异性
paramagnetic body 顺磁体
paramagnetic broadening 顺磁展宽
paramagnetic center 顺磁中心
paramagnetic complex 顺磁性配合物
paramagnetic compound 顺磁化合物
paramagnetic contribution 顺磁贡献【物】
paramagnetic crystal 顺磁晶体
paramagnetic Curie point 顺磁居里点
paramagnetic Curie temperature 顺磁居里温度
paramagnetic current 顺磁电流
paramagnetic effect 顺磁效应
paramagnetic Faraday's effect 顺磁法拉第效应；法拉第顺磁效应
paramagnetic Faraday rotation 顺磁法拉第旋转
paramagnetic impurity 顺磁杂质
paramagnetic iron 顺磁铁
paramagnetic linewidth 顺磁线宽
paramagnetic material 顺磁性物质；顺磁材料
paramagnetic moment 顺磁矩
paramagnetic ore 顺磁矿石
paramagnetic oxygen analyzer 顺磁性氧分析仪
paramagnetic particle 顺磁粒料
paramagnetic property of oxygen 氧的顺磁性
paramagnetic relaxation 顺磁松弛；顺磁弛豫
paramagnetic resonance 顺磁谐振；顺磁共振；电子顺磁共振
paramagnetic resonance absorption 顺磁谐振吸收
paramagnetic resonance method 顺磁共振法
paramagnetic resonance spectrometer 顺磁共振谱仪
paramagnetic resonance spectroscopy 顺磁共振波谱法
paramagnetic resonance spectrum 顺磁共振(波)谱
paramagnetic resonance spectrum parameter 顺磁共振谱参数
paramagnetic salt 顺磁盐
paramagnetic scattering 顺磁散射
paramagnetic shielding 顺磁屏蔽
paramagnetic shift 顺磁位移
paramagnetic shift reagent 顺磁性位移试剂
paramagnetic solid state maser 顺磁性固体微波激射器
paramagnetic spectrum 顺磁光谱
paramagnetic substance 顺磁性体；顺磁(物)质
paramagnetic susceptibility 顺磁(性)磁化率
paramagnetic system 顺磁系统
paramagnetic Zeeman effect 顺磁塞曼效应
paramagnetism 顺磁性
paramagnetism mineral 顺磁性矿物
paramarginal resources 准边界资源
para-massive structure 准块状构造
parambulator 计程车
paramedic 辅助医务人员
paramedical personnel 辅助医务人员
paramelaconite 副黑铜矿
parament 大厅装饰品；大厅家具
paramertic modulator 参量调制器
parametamorphism 负变质作用
parameteorology 参数气象学
parameter 计算指标；残材清理；参数；参量；参变数；参变量
parameter adjustment 参数平差
parameter adjustment control 参数装定装置；参数调整装置
parameter adjustment with constraint 附条件参数平差
parameter amplifier 参数放大器；参量放大器
parameter association 参数联系
parameter attribute 参数属性
parameter attribute list 参数属性表；参数特征表
parameter block 参数块
parameter bound 参数界限
parameter bound and length 参数界和长度
parameter card 参数卡(片)
parameter computer 参数计算机
parameter control cost 参数控制费用
parameter correlation 参数相关
parameter curve 参数曲线
parameter-definition 参数定义
parameter delimiter 参数分界符；参数定义符；参数定界符
parameter descriptor 参数描述符
parameter descriptor list 参数描述符表；参量描述表
parameter detection 参数检测
parameter diagnosis 参数诊断
parameter earthquake 参数地震
parameter equation 参数方程
parameter estimation 参数估值；参数估计；参量估计
parameter estimation of water quality model 水质模型参数估值
parameter field 参数场
parameter format 参考格式
parameter-free test 无参数检验；非参数检验
parameter identification 参数识别
parameter influence coefficient 参数影响系数；参量影响系数
parameter instruction 参数指令
parameterization 参数化法；参数化
parameterization of radiation 辐射的参数化
parameterize 参数化
parameter language 参数语言
parameter length 参数长度
parameter list 参数表
parameter matrix 参数矩阵
parameter measurement 参数测量
parameter mode 参数模式；参数联系模式
parameter mode display 参量式显示
parameter model 参数模型
parameter of a population 总体参数
parameter of bit rate 钻头尺寸参数
parameter of borehole construction 管井结构参数
parameter of computational resource potential 计算资源潜量的参数
parameter of consistency 一致性参数
parameter of curve 曲线参数
parameter of distribution 分布参数
parameter of drilling and blasting 钻爆参数；凿岩爆破参数
parameter of drilling tool 钻具参数
parameter of electric(al) machine 电机参数
parameter of estimated reserves 估算储量参数
parameter of explosion crater 爆破漏斗参数
parameter of Flinn's K-value 弗林参数 K 值
parameter of gradient survey 梯度测量参数
parameter of heating medium 供热介质参数
parameter of isotope determination 同位素测量参数
parameter of local ventilation 局部通风参数
parameter of magnetic field 磁场参数
parameter of marine depositional environment 海洋沉积环境参数
parameter of material 物料参数
parameter of moisture transportation in aeration zone 包气带水分运移参数
parameter of orthogonal thickness 正交厚度参数
parameter of potential survey 电位测量参数
parameter of quantitative evaluation of exploration type 勘探类型定量评价参数
parameter of regularity 正则性参数
parameter of saturation 饱和参数
parameter of seismic ray 地震射线参数
parameter of shear strength 抗剪强度参数
parameter of specular gloss 镜面光泽参数
parameter of strength 强度参数
parameter of substitutional layer 代替层参数
parameter of substitution layer for A-type curve A 型曲线代替层参数
parameter of substitution layer for K-type curve K 型曲线代替层参数
parameter of substitution layer for Q-type curve Q 型曲线代替层参数
parameter of the influence of intermediate principal stress 中间主应力影响系数
parameter of topographic(al) correction of combined profiles 联合剖面地改参数【地】
parameter of track laying 轨道结构参数
parameter of trap measurement 度量圈参数
parameter of water quality model 水质模型参数
parameter optimization 参数最优化；参数优选；参数优化；参数寻优
parameter oscillation 参数振荡
parameter pack(et) 参数包
parameter passing 参数传送
parameter perturbation technique 参数扰动法
parameter plane 参数平面；参变平面
parameter plane method 参数平面方法
parameter position 参数位置
parameter potentiometer 参数电位计
parameter-ratio 标轴率
parameter related to digital calculation 数值计算有关参数
parameter research 参量研究
parameter resonant over-voltage 参数谐振过电压

parameter selection menu 参数选择表
parameter sensitivity 参数敏感度
parameter set 参数集
parameter setting instruction 参数置放指令
parameters for reserve calculation 储量计算参数
parameter shift 参数偏移
parameters of coal accumulation basin 聚煤盆地参数
parameters of cost schedules 成本表列参数
parameters of finite segment length 微元段长度参数
parameters of magnetic body shapes 磁性体形状参数
parameters of pollution 污染参数
parameters of rock property 岩石物性参数
parameters of soil sampler 取土器的基本参数
parameters of tide 潮汐参数
parameters of water quality assessment 水质评价参数
parameters optimized drilling 参数优化钻进
parameter space 参数空间
parameter statement 参数语句
parameter statistical inference 参数统计推论
parameter storage 参数存储器
parameter study 参数研究
parameter substitution 参数代换
parameter switch conversion 参数转换
parameter table 参数表
parameter tag 参数标记
parameter test(ing) 参数测试;参数调试
parameter tracking 参数跟踪
parameter transformation 参数变换
parameter transition 参数转移
parameter value 参数值
parameter variance 参数方差
parameter variation 参数变值法;参数变换;参数变化
parameter vector 参数向量
parameter voltage stabilizator 参数稳压器
parameter voltage stabilizer 参数稳压器
parameter weighting 参数加权
parameter well 参数井
parameter word 参数字
paramethylaminophenol 对甲氨基酚
paramethylaminophenol sulfate 对甲氨基酚硫酸盐
paramethylnitrobenzene 对硝基甲苯
paramethyl red 对甲基红;副甲基红
parametral plane (晶体的)标轴平面
parametral ratio 晶标轴率
parametric(al) 参量的;参数的
parametric(al) acoustic(al) array 声参量阵
parametric(al) adjustment 参数平差
parametric(al) amplification 参数放大;参量放大
parametric(al) amplifier 参数放大器;参量放大器
parametric(al) analysis 参数分析;参量分析
parametric(al) and non-parametric(al) classification 参数分类法和非参数分类法
parametric(al) array 参量阵
parametric(al) assumption 参数假设
parametric(al) borehole 参数井
parametric(al) boundary condition 参数边界条件
parametric(al) classification 参量分类
parametric(al) component 参数元素
parametric(al) conversion 参量转换
parametric(al) converter 参量变频器
parametric(al) correlation 参数相关
parametric(al) curve 参数曲线
parametric(al) curve method 参量图法
parametric(al) damping 参数阻尼;参量阻尼;参考阻尼
parametric(al) delimiter 参数分界符;参数定义符
parametric(al) dependence assumption 参数相关性假设
parametric(al) derivative 参数导数
parametric(al) design 参数设计
parametric(al) detection 参数检测
parametric(al) determination 参数测定
parametric(al) device 参量器件;参考器件
parametric(al) differentiation 参数微分法
parametric(al) diode 参数(放大)二极管;参量(放大)二极管
parametric(al) down-conversion 参量下转换
parametric(al) downconverter 参量下变频器
parametric(al) electronic computer 参量(元件)电子计算机
parametric(al) element 参量元件
parametric(al) emission 参量发射
parametric(al) equation 变数方程;参数方程;参量方程
parametric(al) excitation 参数激励;参数激发;参量激励;参量激发;参考激励
parametrical face 参数面
parametric(al) fault 参数故障
parametric(al) fluorescence 参量荧光
parametric(al) frequency converter 参量变频器;参量频率变换器
parametric(al) frequency divider 参量分频器
parametric(al) frequency multiplication 参量频率倍增;参量倍频
parametric(al) function 参数函数
parametric(al) gain 参量增益
parametric(al) generator 参量发生器
parametric(al) geometric design 参数化几何设计
parametric(al) hydrology 参数水文学
parametric(al) hypothesis 参数假设
parametric(al) image conversion 参量图像变换
parametric(al) image converter 参量图像转换器
parametric(al) inference 参数推断
parametric(al) infrared image converter 参量红外变像管
parametric(al) instability 参变不稳定性
parametric(al) interaction 参量互作用;参变性作用
parametric(al) inversion 参数反演
parametric(al) latitude 参数纬度;参量纬度
parametric(al) latitude method 改化纬度法
parametric(al) linear planning 参数线性规划
parametric(al) linear programming 参数线性规划;参量线性规划;参考线性规划
parametric(al) longitude 参量经度
parametrically induced resonance 参变诱导共振
parametric(al) method 参数法
parametric(al) method of frequency stabilization 参量稳频法;参考稳频法
parametric(al) method seasonal adjustment 季节调整参数
parametric(al) mixer 参量混频器;参考混频器
parametric(al) mixing 参量混频
parametric(al) mode 总体参数模
parametric(al) model 参数模型;参量模型
parametric(al) modulator 参考调制器
parametric(al) multiplier 参量倍增器
parametric(al) network 参数格网
parametric(al) nonlinearity 参数非线性
parametric(al) optics 参变光学
parametric(al) optimized value 参数最佳值
parametric(al) oscillation 参量振荡
parametric(al) oscillator 参数振荡器;参量振荡器;参考振荡器
parametric(al) oscillator threshold condition 光参量振荡阈值条件
parametric(al) phase-locked oscillator 参变器
parametric(al) photodiode 参量(放大)光电二极管
parametric(al) plane 参数平面;参变平面
parametric(al) point 参数点
parametric(al) potentiometer 参数电位计
parametric(al) preamplifier 参量前置放大器
parametric(al) procedure 参数过程
parametric(al) process 参变过程
parametric(al) product planning 参数产品计划
parametric(al) programming 参数规划(法);参数程序设计
parametric(al) pumping 参数泵
parametric(al) pump separation 参数泵分离
parametric(al) receiver 参考接收器
parametric(al) regression coefficient 参数回归系数
parametric(al) relationship 参数间关系
parametric(al) representation 参数表示
parametric(al) resonance 参量谐振;参量共振
parametric(al) resonance heating 参量共振加热
parametric(al) seismometer 参量地震计
parametric(al) selection 参数选择;参量筛选
parametric(al) singular point 参数奇点;流动奇点
parametric(al) solution 参数解
parametric(al) sonar 参量声呐
parametric(al) sounding 参数测深
parametric(al) space 参数空间
parametric(al) stabilization channel 恒参信道
parametric(al) statistic(al) analysis 参量统计分析
parametric(al) statistical inference 系数统计推论
parametric(al) statistics 参数统计学;参数统计;参量统计学
parametric(al) subharmonic oscillator 参量次谐波振荡器
parametric(al) subroutine 参数子程序
parametric(al) synthesis 参数综合
parametric(al) system 参量系统
parametric(al) test 参数检验
parametric(al) up-conversion 参量上转换
parametric(al) up-converter 参量上转换器;参量上变频器;参数上变频器
parametric(al) user 参数用户
parametric(al) variable 参数变量
parametric(al) variation 参数变值法;参数变化;参量变化
parametric(al) variation channel 变参信道;参数信道
parametrics 参数学
parametrization 参量化
parametron 参数器;参数激励子;参变元件;参变管;变感元件;变参元件;变参管
parametron computer 变参计算机
parametron core 参数器铁芯
parametron logical circuit 参数器逻辑电路;变感元件逻辑电路
parametron shift register 变参移位寄存器;变参数移位寄存器
para-midophenol 对氨基酚
paramisan sodium 氨基水杨酸钠
paramitome 透明质
paramnesia 记忆错误
paramo 美洲热带高山草地
paramontroseite 副黑钒矿
paramorph 同质异形(晶)体;同质异晶(体);像像
paramorphism 全变质作用;同质异形性;同质异晶性;同质异晶现象;同质假象
paramos 高寒带
paramount 首要的;卓越的;防火石膏墙板;最高的
paramount clause 首要条款
paramount factor 首要因素
paramp 参量放大器
paramudras 超结核
paranaphthalene 蒽
Parana pine 柏拉那松木;巴西杉
paranatrolite 副钠沸石
paranbutlerite 副铁矾
paranema 副丝
para-nitraniline red 对硝基苯胺红;对位红
paranitroacetanilide 对硝基乙酰苯胺
paranitroaniline 对硝基苯胺
paranitroanisole 对硝基苯甲醚
paranitrobenzaldehyde 对硝基苯甲醛
paranitrobenzenesulfonic acid sodium 对硝基苯磺酸钠
paranitrobenzoic acid 对硝基苯甲酸
paranitrobenzoyl chloride 对硝基苯甲酰氯
para-nitrochloro-benzene 对硝基氯(化)苯
paranitrophenetole 对硝基苯乙醚
paranitrophenol 对硝基酚
paranitrophenol sodium salt 对硝基酚钠
paranitrophenylacetic acid 对硝基苯乙酸
paranitrophenylhydrazine 对硝基苯肼
paranitrotoluene 对硝基甲苯
paranotal expansion 侧背扩张
paranotum 侧背板
Paranox 巴拉诺克斯(一种润滑油多效添加剂)
paranthelion 远幻日
paranthracite 高变质无烟煤
paranza trawl 对拖网(意大利沿海)
paraoctyl phenol 对辛基苯酚
para-ortho conversion 邻位变换
paraos 小型近海航船(印度尼西亚一带)
paraoxone 对硝基磷酯
parapatry 侧地性
parapet 人行道;护栏;栏杆;防浪(矮)墙;界墙;屋顶以上的隔火墙;胸墙;矮护墙;女儿墙【建】;短墙
parapet brick 砌筑防浪墙的黏土砖;砌筑女儿墙的黏土砖;砌砖墙
parapet building component 女儿墙建筑构件
parapet clay brick 女儿墙黏土砖
parapet displacement 女儿墙位移
parapeted terrace 有栏杆的阳台

parapet facing 护墙饰面;女儿墙饰面;护墙饰
parapet grill(e) 女儿墙格栅;格栅
parapet gutter 箱形天槽;女儿墙排水沟;箱形槽;匣形天沟;匣形水槽;女儿墙天沟;外边内天沟
parapet lining 女儿墙衬砌;女儿墙涂底
parapet masonry(wall) 砖砌檐墙;圬工女儿墙
parapet of window 窗槛;窗台
parapet panel 护墙墙板;女儿墙节间;女儿墙墙板
parapet piece 压檐木
parapet railing 胸墙
parapet signs 胸墙标志
parapet skirting 护墙墙裙;女儿墙踢脚板;女儿墙脚泛水;女儿墙壁脚板;女儿墙勒脚
parapet slab 护墙板
parapet stone 拦墙石
parapet wall 护墙;胸墙;压檐墙;女儿墙【建】;防浪(胸)墙
parapet wall block 女儿墙砌块
parapet wall height 女儿墙高度
parapet water proofing 女儿墙防水
paraphase amplifier 倒极放大器;倒相放大器;分析放大器
paraphase coupling 倒相耦合
para-phenetidine 对氨基苯甲醚
para-phenylene diamine 对苯二胺;对苯二胺
para-phenylene diamine dihydrochloride 对苯二胺盐酸盐
para-phenyl phenol 对苯基苯酚
paraphernalia 随身用具
para-phthalonitrile 对苯二甲腈
paraphthaloyl chloride 对苯二甲酰氯
para-phthalphenol 对苯二甲酚
paraphysial 侧丝的
paraphysis 隔丝;侧丝;侧棒
parapierrotite 斜硫锑铊矿
paraplasm 副质
para-plastic rubberized asphalt sealing strip 橡胶树脂沥青密封条
paraplatform 准地台【地】
paraplex 增塑用聚酯
para-position 对位
paraprofessional 专业人员助手
parapropanol 对丙胺酚
para-propionylphenol 对羟基苯丙酮
paraquartzite 副石英岩
paraquinonechlorimide 对苯醌氯亚胺
pararabin 副阿拉伯胶素
pararammelsbergite 副斜方砷镍矿
para red 偶合红;黄光颜料;褐红颜料;橙红色涂料;对位红
para-red developing range 毛巾红显色机
parareducine 副还原碱
para-relict texture 负残余结构
pararenite 副砂屑岩
para-ripple 对称波痕;超波痕
pararock 副变质作用;水成变质岩;副变质岩
pararocks 水成变质岩类
pararosaniline 对玫瑰红;副蔷薇苯胺
pararosaniline acetate 副蔷薇苯胺乙酸盐
pararosaniline base 副品红碱
pararosaniline chloride 副品红
pararosaniline dye 副品红染料
pararosaniline hydrochloride 盐酸副品红;副蔷薇苯胺盐酸盐
Para rubber 帕拉(橡)胶
para-rubber tree 橡皮树
parasagittal 旁矢状面的
parascenium 后台两翼建筑物;(古希腊剧场台向前伸出的)两翼凸台
paraschacherite 斜方汞银矿
paraschist 水成片岩;副片岩
paraschoepite 副柱铀矿
paraschozite 副磷钙锌矿
paraselene 近幻月;幻月
paraselenic circle 幻月环
parasheet 多角形降落伞
parashenion (古希腊剧场后台向前伸出的)两翼凸台
parasilicate 副硅酸盐
parasite 寄生阻力;寄生植物;寄生(生)物;寄生虫;寄生振荡;寄生效应;寄生现象
parasite index 寄生虫指数
parasite plant 寄生植物
parasite pumping 不规则泵送

parasitic(al) absorption 寄生俘获
parasitic(al) amplitude modulation 寄生调幅
parasitic(al) animal 寄生动物
parasitic(al) antenna 寄生元件;无源天线
parasitic(al) autooscillation 寄生自振荡
parasitic(al) bacteria 寄生细菌
parasitic(al) capacitance 寄生电容
parasitic(al) capture 寄生俘获
parasitic(al) chain 寄生链
parasitic(al) circuit 寄生电路
parasitic(al) component 寄生成分
parasitic(al) cone 寄生火山锥
parasitic(al) crater 寄生火山口
parasitic(al) current 寄生电流
parasitic(al) disease 寄生虫病
parasitic(al) drag 废阻力
parasitic(al) element 寄生元件;寄生单元;二次辐射体
parasitic(al) error 寄生误差;粗误差
parasitic(al) feedback 寄生反馈
parasitic(al) fold 寄生褶皱;从属褶皱;附属褶皱
parasitic(al) force 二次力
parasitic(al) fungi 寄生真菌
parasitic(al) host 寄生宿主
parasitic(al) image 寄生像
parasitic(al) inductance 寄生电感
parasitic(al) infestation 次生灾害
parasitic(al) insect 寄生性害虫
parasitic(al) lasing 寄生激光作用
parasitic(al) light 杂光
parasitic(al) loss 附加损失
parasitically excited antenna 寄生振子天线
parasitic(al) mode of life 寄生生活方式
parasitic(al) mode suppressor 寄生模抑制器
parasitic(al) moment 次弯矩
parasitic(al) noise 寄生噪声
parasitic(al) nutrition 寄生营养
parasitic(al) oscillation 寄生振荡
parasitic(al) parameter 寄生参数
parasitic(al) plant 寄生植物
parasitic(al) potential 寄生潜力
parasitic(al) reflector 寄生元件
parasitic(al) resonance 寄生共振
parasitic(al) root 寄生根
parasitic(al) scattering 寄生散射
parasitic(al) signal 寄生信号
parasitic(al) site 寄生部位
parasitic(al) stopper 寄生振荡抑制器;寄生效应限制器
parasitic(al) stress 次应力
parasitic(al) suppression 寄生抑制
parasitic(al) suppressor 寄生振荡抑制器;寄生振荡抑制;寄生抑制器
parasitic(al) vibrator 无源振子
parasitic(al) volcano 寄生火山;侧火山
parasitic(al) wave 寄生波
parasitic(al) worm 寄生虫
parasitic(al) zoonosis 动物寄生传染病
parasiticide 杀寄生物的;杀寄生虫剂
parasitism 寄生现象;寄生关系;寄生
parasitization 寄生作用
parasitoid 拟寄生
parasitoidism 虫寄养性
parasitosis 寄生虫病
paraskenion (古希腊剧场后台向前伸出的)两翼凸台;舞台两侧耳房(古希腊)
parasol 阳伞
para-spectrum 仲谱
paraspurrite 副灰硅钙石
parastas 壁角柱;(在壁端柱终止的)墙端;底座似的墙
parastas ray 近轴光线
parastas region 近轴范围
parastat 防静电器
para state 仲态
para-state enterprise 半国营企业
parastochastic function 超随机函数
parastratigraphy 副地层学
parastratotype 副层型
para-sulfamoylbenzoic acid 对胺磺酰苯甲酸
para-sulfamoylbenzoic acid sodium salt 对胺磺酰苯甲酸钠
parasymbiosis 假共生
parasymplesite 副砷铁石

parasynapesis 平行配合
para-synchronous surveying 准同步测量法
parasyndesis 平行配合;平行联合
parasystole 并行收缩;并行收缩
paratacamite 三方氯铜矿;副氯铜矿
paratactic(al) lines 平行线
paratectonic crystallization 构造期结晶
paratellurite 副碲矿
paratenic host 转续宿主
parater-butylphenol 对特丁基苯酚
paratergite 侧背片
paraterphenyl 对三联苯
para-tert-butylcatechol 对叔丁基邻苯二酚
para-tert-butyphenol 对叔丁基苯酚
para-tertiary butyl phenol 对叔丁基苯酚
para-tertiary butyl phenol phenolic resin 对叔丁基苯酚醛树脂
Paratethys 拟特提斯(古地中海边缘区);类特提斯【地】
parathiocresol 对硫代甲酚
parathion 对硫磷
para-tillite 准冰碛岩
para-toluene sulfonamide 对甲苯磺酰胺
para-toluene sulfonate 对甲苯磺酸盐;对甲苯磺酸
para-toluenesulfonic acid 对甲苯磺酸
para-toluenesulfonic acid ammonium salt 对甲苯磺酸铵
para-toluenesulfonic acid sodium salt 对甲苯磺酸钠
paratoluensulfonyl chloride 对甲苯磺酰氯
paratoluic acid 对甲基甲酸
para-toluic acid para-toluylic acid 对甲基甲酸
para-toluidine 对甲苯胺
paratoluidine hydrochloride 对甲苯胺盐酸盐
paratoluidine sulfate 对甲苯胺硫酸盐
paratolunitrile 对甲苯甲腈
Paraton(e) 巴拉东(一种黏度添加剂)
para-toner 对位红色原
paratope 互补位
paratory 教堂中进行准备工作的地方
paratransit 辅助交通
paratransit vehicle 辅助公共汽车
paratriptica 防衰剂
paratrophy 寄生营养;偏寄生营养
paratype 副型;副模式;副模标本
paratyphoid fever 副伤寒
paraumbite 副水硅锆钾石
parautochthonous granite 准原地花岗岩
paravane 防水雷器
paravauxite 副磷铁铅矿;副蓝磷铝铁矿
paravent 挡风屏障;挡风屏帘
paravesical pouch 闭孔囊
parawedge louver 楔形遮光栅照明窗
parawollastonite 副硅灰石
parawydartite 副羧钙铀矿
para-xenol 苯基苯酚
paraxial 近轴
paraxial approximation 近轴近似法
paraxial beam 近轴光束;旁轴光束;旁轴电子束
paraxial constant 近轴常数;旁轴常数
paraxial electron beam 近轴电子束
paraxial focal plane 近轴聚焦面
paraxial focus 近轴焦点;旁轴焦点
paraxial image 近轴像
paraxial imagery 近轴成像
paraxial marginal ray 近轴边缘光
paraxial mode 近轴模
paraxial monochromatic light 近轴单色光
paraxial optics 近轴光学
paraxial ray 近轴光线;旁射光线;旁轴光线
paraxial ray equation 傍轴射线方程
paraxial ray tracing 近轴光速跟踪
paraxial region 近轴区;近轴范围;傍轴区
para-xylene 对二甲苯
parazoon 动物寄生虫
par bond 平价债券
parbuckle 套拉绳;上下索
parbuckle screw 松紧螺钉
parcel 小宗货物;小包;一块;批件;盘丝;部分;包裹;包
parcel automatic sorting machine 包裹自动分拣机
parcel bag 包裹袋
parcel-bulk shipment 整船散货
parcel car 包裹车

parcel cargo 一票货
parcel check room 包裹寄存室
parcel collection 小搬小运
parcel corner 分界点
parcel-gilt 局部镀金的
parceling-out 土地分配
parcel land 小块土地;局部用地;部分土地
parcellary plan 地籍图
parcel limits 地界
parcelling 划分并分配;打包;镶板
parcel list 零担货清单;包件清单;包裹单
parcel lot 一票货;零担货(物)
parcel office 包裹室;包裹房
parcel of land 一片土地;一块土地
parcel plan 分部计划
parcel plating 局部电镀;部分电镀
parcel post 邮包邮务处;邮寄包裹;包裹邮件;包裹邮递
parcel postage 包裹邮费
parcel posting counter 包裹收寄窗口
parcel posting machine 包裹收寄机
parcel post insured 保价邮政包裹
parcel post receipt 包裹收据
parcel post risks 邮保风险
parcel rate 小包运价;包裹运费率;包裹收费率
parcel receipt 包裹收据
parcel revenue 包裹运输收入;包裹收入
parcel room 包裹室
parcel's characteristic file 地块特征文件
parcel's file 地块文件
parcel shelf 包裹架
parcel shipment 包裹运输
parcels plan 部分计划
parcel stamp 包裹戳记
parcels train statisitics 行包专列统计
parcel survey 地块测量
parcel tanker 零担散液船;多隔舱油轮
parcel ticket 包裹票
parcel tying machine 包裹捆扎机
parcel van 厢式载货汽车;铁路包裹车;件货货车
parch 干透
parch crack 拉裂;干裂;切边裂纹
parched field 晒干的田地
parch heat 炎热
parchment 衬垫沥青纸毡;羊皮纸
parchment bean 带壳咖啡豆
parchment corn 带壳玉米
parchment imitation 仿羊皮纸
parchmentized fibre 硬化纸板
parchment paper 硫酸纸;假羊皮纸;充羊皮纸
parclose (中世纪教堂中分隔空间的)隔断;围屏隔离幕;屏障【建】;围绕观众台的矮护墙
parclose screen 屏幕
par collection 可按票据面值收取的制度;可按面值收取的制度
Parece Vela basin 帕里西维拉海盆
pare down expenses 节省开支
pareja 小型对拖网
pareja trawl 小型对拖网
parekklesion 小教堂
parenchyma 软细胞组织
parenchyma of wood 木材的薄壁组织
parenchymotelinite 薄壁结构镜质体
parent 本源
parent aircraft 母机;运载飞机
parental home 儿童教养院;儿童教育院
parental magma 母岩浆
parental school 儿童教育院
parent area 本源面积
parent chain 母体链
parent channel 主干渠
parent city 母城(城市规划用语);原有城市
parent committee 直属上级委员会
parent company 总公司;母公司
parent compound 母体化合物
parent compressional wave 母压缩波
parent crystal 原始晶体
parent directory 母体目录
parent distribution 母体分布
parent element 母体元素
parent environment 母体环境
parent exchange 母电话局
parent firm 总公司;母公司
parent fish 亲鱼

parent focusing 母聚焦
parent fraction 母体部分
parent glass 原玻璃;基础玻璃
parent glaze 基础釉
parenthesis-free notation 无括号表示法
parenthesis nesting 括号重数;括号的嵌套
parenthesized expression 括号表达式
parent hydrocarbon 前体碳氢化合物
parent-investor 母公司投资人
parent ion 母体离子;母离子
parent ion stability 母离子稳定性
parent isotope 原生同位素
parent lattice 基质晶格;母点阵;母材晶格
parent map 原图纸;原图;基本图;底图
parent material 基体材料;母体材料;母材料;亲本材料;原材料;母质;母料
parent material horizon 母质层
parent material lomdform association 母质地形组合
parent matrix 黏合剂;母体
parent member 原物件
parent metal 基本金属;基体金属;基焊料;母体金属;母材;母板;底层金属;底材
parent metal test specimen 母材试件
parent molecule 母体分子
parent moment 母体矩
parent name 源名
parent nitramine 母体硝胺
parent nucleus 母核
parent patent 本源专利
parent peak 母体峰;母峰
parent plate 母板;原板
parent pollutant 原污染物
parent population 母体;母集团;亲鱼群体
parent radioisotope 母(体)放射性同位素
parent rock 基岩;母岩
parent's bath 父母的浴室
parent segment 母段
parent ship 母型船;母舰;母船
parent soil 原状土
parent soil material 土壤母质
parent statement 母公司报表
parent station 主台
parent statistical unit 母体统计单位
parent stock 母体原料
parent subsidiary relationship 母子公司关系
parent substance 母质
parent tube 基管
parent vehicle (集装箱变换系统的)基础车
parergon 附属装饰;辅助装饰
Pareto Criterion 帕累托标准
Pareto diagram 帕累托排列图
Pareto improvement 帕累托改进
Pareto index 帕累托指数
Pareto minimum 帕累托极小
Pareto multiplier 帕累托乘子
Pareto nonoptimality 帕累托非最优状态
Pareto optimal boundary 帕累托优边界
Pareto rationality 帕累托合理性
paretta 干黏石面;乱石饰面;表面有突出疙瘩的粗陋浇注坯工;一种干黏土石
par example 举例
par exchange rate 官方平价汇率;平价汇率
parfacies 亚相
parfocal 等焦距
parfocal distance 齐焦距离
parfocality 齐焦性
parfocality adjustment 齐焦调整
parfocality adjustment sleeve 齐焦调节套筒
parfocalization 齐焦
parfocal objective 齐焦物镜
pargasite 韭闪石
parge 灰泥;石膏;砂浆涂层;涂灰浆;粗涂灰泥
parge board 山墙封檐板
parge coat 灰泥抹面层;砂浆涂层处理
parget 灰泥;粗纸筋;粗涂灰泥
parget coat 粗涂灰泥
parg(et)ing 搪灰泥(在烟道或炉膛中涂抹灰泥);做线条;抹灰(光面);抹砂浆层;抹粗灰浆;涂灰泥;粗灰泥层;粗涂灰浆;粗涂;粗灰浆层;古式外部灰泥作业;打底的墙面砂浆层
pargetry 拉毛凸粉刷;平表面抹灰;装饰抹灰;凸纹粉刷
parget-work 灰泥饰面抹灰;外部抹灰作业;凸纹粉刷

parhelic circle 幻日环
parhelion 幻日
parhelium 仲氦
parian 仿大理石
Parian cement 仿云石水泥;巴利安白粉
parianite 帕里安奈沥青
Parian marble 巴利安灰大理石
Parian plaster 巴利安灰浆;巴利安粉刷;巴利安石膏
Parian porcelain 巴利安瓷
Parian ware 白色瓷器;巴利安瓷器
paring 削下来的皮;凿削;切片;切边
paring chisel 削凿(刀);凿削;薄凿;扁凿
paring gouge 削面凹凿;圆凿(子);弧口凿;弧口薄凿
paring hammer 弯曲用锤
paring knife 削皮刀
paring machine 剥皮机
paring stone 建筑石
pari passu 同步增长
pari passu provision 同步条款
peripheral zone 周围带
Paris black 巴黎黑;巴黎骨炭
Paris black pigment 巴黎黑颜料
Paris blue 天蓝(色);巴黎蓝
Paris blue pigment 巴黎蓝颜料
Paris cement 巴黎水泥
Paris Convention 巴黎公约
Paris green 碱性甲基绿;翠绿;翡翠绿;巴黎绿
parish 教区
parish center 教区中心
parish church 教区教堂
parish hall 教堂
parish house 教区大厦
parish level 教区水平;教区标准
parish room 教堂
parish territory 教区
Parisian alloy 巴黎合金
Parisian cement 巴黎水泥
parisite 氟碳钙铈矿;氟菱钙锌矿
parison 雏形;料泡;玻璃半成品;型坯
parison maker 型坯工
Parison mo(u)ld 初型模
parison swell 型坯膨胀
Paris plaster 熟石膏;巴黎石膏
Paris red 巴黎红
par issue 平价发行
Paris violet 巴黎紫
Paris white 亮粉(碳酸钙);碳酸钙粉;巴黎白(粉)
Paris yellow 巴黎黄
Paris yellow pigment 巴黎黄颜料
par item 可按面值收取的票据
parities according to season 季节差价
parity 相等;光洁度;均势;奇偶性;同位;同等;等价;平价
parity bit 奇偶校验位;奇偶位;奇偶检验位
parity character 奇偶符号
parity check 一致奇偶校验;奇偶检验;奇偶(性)检验
parity check bit 奇偶校验位
parity check code 奇偶校验码;奇偶监督码
parity check digit 奇偶检验位;一致校验数字
parity check equation 奇偶校验方程;一致校验方程
parity checker 校验器;奇偶校验器;奇偶检验器
parity checking capability 奇偶检验能力
parity check interrupt 奇偶检验中断
parity check symbol 奇偶检验符号
parity check system 奇偶校验系统;奇偶校验方式;奇偶检查制
parity check unit 奇偶检查装置
parity clause 平价条款;平等条款
parity code 奇偶校验码;奇偶检验码
parity coefficient 平价系数
parity control 奇偶控制
parity count character 奇偶计数符
parity detection 奇偶检验
parity digit 奇偶校验位;奇偶位;奇偶数字
parity disable 非奇偶性
parity duty 平衡关税
parity error 奇数误差;奇偶性错误;奇偶检验误差;奇偶错误
parity error flag 奇偶错误标记
parity flag 奇偶标记
parity generation 奇偶发生
parity generator 奇偶生成器;奇偶发生器
parity grid 平价栅;平价网络

parity income 同酬;等值收入
parity index 平价指数;对等指数
parity indicator 奇偶显示器
parity interrupt 奇偶性中断;奇偶校验中断
parity logic 奇偶逻辑
parity matrix 一致校验矩阵;奇偶校验矩阵
parity of authority and responsibility 权责平衡
parity of gold and silver 金银比价
parity of price 比价
parity of representation 相等平价
parity of silver 银平价;白银平价;白银比价
parity of treatment 平等待遇
parity price 平价(价格);平衡价格;对等价格
parity price policy 平价政策
parity program(me) 平价方案
parity rate 平价指数
parity rate of commodity 商品比价
parity rate of manufactured goods 工业品比价
parity ratio 平价比率;比价率
parity return 等值报酬
parity sheet of foreign exchange 外汇平价表
parity state 奇偶校验状态
parity track 奇偶校验道;奇偶孔道
parity transformation 奇偶变换
parity unit 折实单位;折实单价
parity value 等价
park 园林;公园;园林景区;停驻;停车场
park administration 公园管理
park and ride 存车换乘
park architecture 园林建筑艺术;公园建筑学;园林建筑(学)
park avenue 花园大道
park bench 公园长凳
park block 停车区;多层停车库
park boundary 公园园界
park bridge 公园桥梁
park cemetery 墓园
parked vehicle 停驻车辆;停放的汽车
park encroachment 蚕食公园用地
parker 停放的车辆;停驻车辆
parkering 磷化处理
parkerise (加接触剂的)磷化金属防锈处理;磷酸盐保护膜(防锈)处理;磷酸盐被膜(防锈)处理
parkerising 磷化
parkerite 硫铋镍矿
parkerized 磷酸盐膜防锈处理的;磷化处理的
parkerizing 磷酸盐(被膜防锈)处理;磷化处理
parkerizing process 磷酸盐防锈处理;磷化(过程)
Parker-Kalon 木螺钉
Parker-Kalon screw 薄板螺钉
Parker's alloy 帕克尔镍铬黄铜
Parker's cement 帕克尔水泥;天然水泥
Parker's process 帕克尔法
Parker's truss 帕克尔桁架(上弦呈多边形);曲弦桁架
park forest 森林公园
park highway 公园公路;公园大路;林阴公路
parking 公园;停车处;停车场
parking access 停车场进出口
parking accumulation 停车累计量;停车场存车累计数
parking accumulation study 存车累计调查
parking allocation 停车布置
parking and riding system 存车换乘系统
parking apron 停车场;停机坪;停机场;停车道
parking area 停车区;停车场;停车场地;停车场
parking area sign 停车场标志
parking area with charges 收费停车场
parking arrangement 停车排列方式;停车布置
parking at right angle 垂直停车场
parking attraction volume 停车吸引量
parking ban 禁止停车
parking ban road 禁止停车的道路
parking ban sign 禁止停车标志
parking bay 停车带;港湾式停靠站;港湾式停车处;停车位;停车加宽处;停车港
parking block 停车区
parking brake 汽车手刹车;停车制动器;停车闸;存车制动
parking brake cam 停车闸凸轮
parking brake shoe 停车闸瓦
parking brake verification test 制动系统试验
parking building 车库建筑;多层停车场
parking capacity 停车容量

parking census 停车调查
parking compound 停车院子;停车场地
parking configuration 停车布置;停车排列方式
parking control area 停车控制区;停车管制区
parking court 停车小院(子)
parking deck 停车楼面;停车坪;停车平台
parking demand 停车需要量;停车需求量;停车实数
parking disc 停车标签
parking duration 停车延续时间;停车时间;停车持续时间
parking elevator 立体停车场提升装置
parking facility 停车装置;停车设施
parking fee 停车费
parking fine 停车罚款
parking floor 停车层
parking forbidden 禁止停车
parking garage 停车库;室内停车场;多层停车库;车库
parking grade 停车道坡度
parking ground 停车场
parking index 停车指数;车辆存放指数
parking interview 停车场调查
parking interviews method 停车场询问法
parking inventory 停车位清单
parking inventory study 编制停车场清册调查
parking lamp 停车灯
parking lane 泊车线;驻车道;停车道
parking lay-by 路旁停车场;路侧停车场;路边停车场
parking light 停放汽车信号灯;停车灯
parking line 停车线
parking load 停车延续时间;停车负荷
parking location 停车位
parking lock 停车区
parking lot 集装箱堆场;停车区;停车地段;停车地点;停车场;多层停车场
parking lot lane 停车段车道
parking maintenance day 停车场检修日
parking meter 汽车停放收费计;汽车停放计时器;停靠表;停车收费计;停车计时器;停车计费表;船舶靠泊速度指示器
parking meter zone 计时停车区
parking need 停车位需求(量)
parking orbit 驻留轨道
parking ordinance 停车规则
parking origin-destination survey 停放车辆起讫点调查
parking period 停车时间
parking place 停车处;停车场
parking point 停车场地;停车地段;停车地点;停车处
parking port 存车港湾
parking ramp 停车斜面;停机坪;停机线
parking ratio (商业建筑的)停车比率
parking regulation 停车规则
parking restriction 停车限制
parking roof 停车屋顶
parking set 停车位
parking shed 停车棚
parking sign 停车标志
parking signal 停车信号
parking sign street 停车标志街
parking space 停车空地;停车间距;停车位;停车场地;停车场;停放场地【铁】;车位
parking space limit 停车净空限界
parking space limit marking 停车限界标志
parking square 停车广场
parking stall 停放场所(汽车);停车位;车位
parking stall depth 车位进深
parking stall dimension 车位尺寸
parking stall width 车位宽度
parking standard 停车定额;停车标准
parking station 停车站;公交停靠站
parking stor(e)y 停车楼层
parking street 停车街道
parking strip 停车地带;停车带
parking structure 停车场建筑;停车场机械设施;停车楼
parking study 停车调查
parking supply 停车供应(量)
parking survey 停车调查
parking tender 停泊管理员
parking terrace 停车平台
parking tier (多层停车场的)一层;停车层
parking time 停车时间
parking time limit 停车时限

parking tower 停车塔楼
parking turnout 停车处的让车岔道
parking turnout and rest area 路侧避车道和休息区
parking turnover 停车场周转数
parking turnover rate 停车周转率;停车场周转率
parking vehicle 停驻车辆
parking voucher 停车收据;停车凭证
parking way 林阴道
Parkinson's vice 巴氏虎钳
park land 公共绿化;疏树草原;疏树草地;疏林草原;疏林草地;稀树草原;热带疏林;温带疏林
parklike 公园般的
park nursery 公园苗圃
park path 公园小路
park railing 公园铁栅栏;公路铁栅墙
park road 林阴路;公园(道)路
park sightseeing bus 游览车
park spring 淡褐色砂石(一种产于英国约克郡的石材)
park strip 带形公园
park system 园林系统
parkway 花园路;风景区道路;园道;公园路;公园大道;林阴路;风景区干道;风景公路
Parlanti casting process 帕兰蒂铝模铸造法
parlatory 寺院中可接待客人的房间
parliament 议会
parliamentary company 公益事业
parliament block 国会区段
parliament building 议会大厦;议会大楼;国会大厦
parliament butt 长翼铰链
parliament heel 缓慢倾侧
parliament hinge H形铰链;H形合页;长翼铰链;长脚铰链;长脚合页
parliament house 议会大厦;议会大楼;国会大厦
Parlin cup 巴林杯
Parlin cup viscosity 巴林杯黏度
parlir milker 固定式挤奶室
parlodion 火棉胶片
parlor 客厅;接待室;会客厅;会客室
parlour = parlor
parlo(u)r car 花车;豪华铁路客车;豪华客车
parlo(u)r house 妓院
parma 中绒
Parmalee wrench 双层岩芯管拆卸专用钳
parmatol 木材防腐剂
parna 风成黏土
parnauite 砷铜矾
parnitene 反苯环丙胺
parochial 教区的
parochial church 教区教堂
parochial house 教区住房
parochial school 教区学校
parochor 等张体积
parodos (古戏剧院观众厅两旁的)侧廊;(古戏剧院观众厅两旁的)侧门
par of exchange 汇兑平价;外汇汇率
parol agreement 口头协议
parol contract 口头协定;口头协议
parol evidence 口头证据
parol evidence rule 口头证据法规
paroline 液体石油膏
parol promise 口头允诺
paronite 石棉橡胶板
par or rate of foreign exchange 外汇行市
par or stated value 面值或设定价值
paroxypropion 对羟苯丙酮
paroxypropione 对羟基苯丙酮
paroxypropiophenone 对羟基苯丙酮
paroxysmal eruption 突发性喷发
parpend 系石;穿墙石块
parpend stone 系石;穿墙(丁头)石
parpent 系石;穿墙石块
parpoint (厚度逐渐减小的)成层粗石圬工
parquet 镶木;镶花地板;席纹地面;正厅(指剧场正对舞台部分);剧场正厅;嵌木地板;拼花木板;拼花地板
parquet block 镶嵌木板条块;镶嵌木块;拼花木块;拼花地板块
parquet border 拼花地板镶边;拼花边缘
parquet circle 剧院或音乐厅的正厅后排;楼下后厅;戏院正厅后排;正厅后座
parquet floor 镶木地板;席纹地板;拼花地板
parquet flooring 拼花地板铺设;拼花地板;硬木拼花地板

parquet floor layer 镶木地板工
parquet mosaic 拼花马赛克
parquet plant 镶木厂
parquet polishing 镶木地板打光
parquet polishing machine 镶木地板打光机
parquetry 木地板花式;地板镶花模式;镶木地板,镶木工作;镶嵌细工;镶嵌饰;镶木细工;镶木法;镶花细工;拼花木作
parquetry block 镶木地板细工的木块;镶木地板的板块
parquetry composite 镶木细工的拼镶;镶木地板的拼镶
parquetry floor cover(ing) 铺设拼花地板;铺设镶木地板
parquetry flooring in basket weave pattern 席纹地板
parquetry layer 镶嵌层
parquetry panel 嵌花镶板
parquetry sealing 镶木地板的密封
parquetry staves 镶木细工桶板
parquetry timber 镶木地板用木材
parquetry wood 镶木地板细工木料
parquet sander 拼花地板磨光机
parquet sanding machine 拼花地板砂光机
parquet sealing 镶嵌层
parquet solid hardwood floor 同质硬木拼花地板
parquet strip 拼花(小)板条;拼花地板条
parquet strip flooring 镶木条形地板;条形地板;小半条地板;拼花地板条
parquet timber 镶嵌木料
parquet varnish 镶木地板清漆
parquet wood 镶木地板
parquet work 镶嵌木工;镶木细工;镶木工作;拼花木作;镶木厂
Parr calorimeter 帕尔量热器
par regime 平价制度
parrel 壁炉及其装饰;绳环;帆桁固定器
parrel truck 滑珠环
parrot blue 鹦鹉蓝
parrot green 鹦鹉绿;砷铜绿
Parr turbidimeter 帕尔浊度计
Parry arcs 彩晕轮
parscope 气象雷达示波器
parsec 秒差距
parse list 剖析表
parser 分析算法;分析程序
parser construction 分析程序结构
parser generator 分析程序的生成程序
parse tree 分析树
parsettensite 红硅锰矿
Parseval's theorem 巴塞伐尔定理
Parshall (measuring) flume 巴歇尔量水槽;巴歇尔测流量装置;巴氏计量槽
parsing algorithm 分析算法
parsing a simple-variable declaration 简单变量说明的分析
parsing a structure declaration 结构说明的分析
parsing davit 滑轮吊架;吊锚杆
parsing phase 分析阶段
parsing process 分析过程
parsing program(me) 分析程序
parsing sentence 分析句子
parsing structure reference 结构引用的分析
parsing technique 分析技术
parsing time 分析时间
parsonage 牧师住宅
parsonage building 牧师俸地;牧师住宅
Parson's brass 锡基锑铅铜合金;派森黄铜
Parsons Duncan process 帕森斯—邓肯铸锭法
parsonsite 三斜磷铅铀矿
Parson's number 帕森数
Parson-stage steam turbine 帕森汽轮机级
parstelin 反苯环丙胺
part 局部;片段;部分
part a line 拉断绳索
partan algorithm 平行切线算法
part and parcel (of) 主要部分
part assembler 部件装配工
part assembling drawing 部件装配图
part assembly 部件装配
part assembly drawing 部件装配图;部分装配图
part by part 详图;逐项
part by volume 体积分量;体积分;体积比例;按体积计的份数

part by weight 重量份;重量比例;质量比例
part cargo 未满载
part charge 局部装药;分段装药;部分装药
part circle sprinkler 扇形喷灌机
part-colo(u)red painting 彩画
part correlation 部分相关
part-cut-part-fill section 半挖半填路段;半挖半填(式)截面;半挖半填(式)断面;半填半挖(横)断面;半填半堑
part-cut-part-fill subgrade 半挖半填式路基;半填半挖式路基
part design flexibility 部件设计的灵活性
part dislocation 部件错位
part drawing 零件图
part drawing number[part dwg No.] 零件图号
parted casing 断裂套管柱
parted pattern 分块模样
part element drill 散装钻机
part elevation 局部立面
parterre 一块平地;剧场池子;正厅后座;剧院楼下正厅(后)座;花坛
parterre box 剧院楼下正厅后座包厢
Partex penetrant process 帕太克斯浸油探伤法
part face blast 工作面局部爆破;半工作面爆破
part fill in and ram 部分加入并加夯;部分浇灌和捣实(混凝土)
part fuse 部分熔совет
partheite 帕水硅铝钙石
Parthenon 巴台农神殿(公元前438年建于希腊雅典)
Parthenon pediment 巴台农神殿的山墙檐饰
Parthian architecture 帕提亚建筑;安息建筑
parti 设计总方案;建筑设计草图;建筑设计的基本总方案
partial absorption 部分吸收
partial acceptance 局部认付
partial account settlement 部分结账单
partial adjustment 局部调整;部分平差
partial adjustment hypothesis 局部调整假设
partial adjustment model 局部调整模型
partial admission 局部进汽;局部进气;局部供给
partial-admission turbine 局部进气式透平;局部进气式涡轮机;部分进气式涡轮机;非整周进水式水轮机
partial aeration 部分曝气
partial agglutinogen 副凝集原
partial agonist 部分激动剂
partial air conditioning 局部空调;部分空调
partial air oxidation 部分空气氧化
partial air pressure 限制空气压力
partial alkalinity 偏碱性
partial allocation method 部分分配法
partial allotment 部分分配
partial analysis 局部分析;局部分析;偏分析;部分分析
partial anatexis way 部分深熔作用方式
partial annealing 不完全退火
partial approach 局部方法
partial area 部分面积
partial area contribution 局部产流
partial area runoff 局部产流;部分面积径流
partial arithmetic 部分运算
partial aromatic solvent 部分芳烃溶剂
partial arrangement drawing 零件布置图
partial assembly drawing 装配分图;零件装配图
partial audit 部分审计
partial autocorrelation 偏自相关
partial automatic 半自动的
partial automatic checker 半自动检验
partial automation 部分自动化;半自动化
partial automatization 部分自动化
partial auxiliary view 局部辅助投影图
partial average 部分海损
partial award 部分裁决
partial awning-deck vessel 局部天幕甲板船
partial balance 部分结算
partial basement 部分地下室;局部地下室
partial bearing 半轴承
partial bed load movement 局部底砂运动;局部底质运动
partial belief 部分置信
partial bid 局部要价;局部投标;投标承包部分项目;部分投标
partial bleaching 部分退色法
partial blind shield 半挤压盾构

partial board 厚芯合板
partial boring machine 自由断面掘进机
partial breaking of waves 局部波浪破裂;局部破浪
partial bulkhead 局部隔堵;局部舱壁;半堵壁;半舱壁
partial calcination 部分烧结;部分煅烧
partial carbonization 局部碳化
partial cargo 部分货物
partial carry 部分进位;保留进位
partial cash advanced 预付部分现款
partial cash transaction 部分现金交易
partial casing 部分门窗框;部分外壳
partial cementation 局部胶结;部分灌浆
partial census 局部调查
partial charter 部分租船
partial charter party 部分租约
partial circulation operation 小循环运行
partial closure 局部关闭(闸门)
partial clover leaf 部分苜蓿叶式;不完全的四叶式(指交叉口)
partial cloverleaf interchange 部分苜蓿叶式立体交叉
partial coagulation 部分凝固
partial coherence 局部相干性;部分相干性
partial coherence factor 部分相干因子
partial coincident picture 部分重叠图像
partial collapse 局部破坏;局部崩塌;部分倒塌
partial colo(u)r blindness 不全色盲
partial combustion 部分燃烧
partial common control system 部分共同控制式
partial complex 不完整综合体
partial compression test 部分压缩试验
partial computability 部分可计算性
partial concrete casing 部分混凝土模壳
partial concrete encasement 部分混凝土饰面;部分混凝土外壳
partial concrete haunching 部分混凝土镶边;部分混凝土加腋;部分混凝土起拱
partial concrete sheath coat 部分混凝土外皮
partial condensation 部分冷凝
partial condensator 部分冷凝器
partial condenser 分凝器;部分冷凝器
partial conductor 次导体
partial conjugation 局部接合
partial container freighter 部分集装箱货船
partial contents 部分内容
partial continuity 部分连续性
partial contract 分合同;分包制
partial contraction 不完全收缩
partial contraction joint 局部收缩缝;局部伸缩缝
partial control 局部控制
partial control of access 局部出入口限制;部分出入限制
partial cooperation 部分合作
partial correctness 部分正确性
partial correlation 偏相关;部分相关
partial correlation analysis 偏相关分析
partial correlation coefficient 净相关系数;偏相关系数
partial correlation ratio 偏相关比
partial cost 部分成本
partial cover plate 局部盖板
partial cracking 部分裂化
partial cross-section 部分截面;部分断面;分截面
partial crushing 部分破碎
partial cultivation 局部垦殖;局部耕作
partial cut-off 半截防渗墙;局部截水墙;局部齿墙;部分截水墙;不到防渗基岩的截水墙
partial cutting 部分采伐
partial damage 部分损坏
partial decomposition 部分分解
partial decoupling 部分去偶
partial deformation theory 局部变形理论
partial degeneracy 部分简并性
partial delivery 零批交货;分期交货;分批交货;部分交付
partial denominator 偏分母
partial density 部分密度
partial dependency 部分依赖
partial derivation 部分派生
partial derivative 偏微商;偏导数
partial desiccation 半脱水;半干燥
partial desulfurization 部分脱硫法
partial detonation 部分爆炸

partial devaluation 部分贬值
partial difference 偏增量;偏差(分)
partial difference equation 偏差分方程
partial differential 偏微分
partial differential equation 偏微分方程
partial differential equation language 偏微分方程语言
partial differentiate 偏微分
partial differentiation 偏微分;偏导法
partial discharge 局部放电;节间流量;(两测流垂线间的)部分流量;部分放电
partial discharge test 局部放电试验
partial disconnection 半断接
partial dislocation 局部位错;偏位错;部分位错;部分脱错
partial dislocation of crystal 晶体局部位错
partial dispersion 部分色散;部分频散
partial dispersion factor 部分色散因子
partial dissociation 部分离解
partial distillation 局部蒸馏;部分蒸馏
partial diversion 部分分水
partial divide 部分分水岭
partial dominance 部分显性
partial double bottom 局部双层底
partial double tax system 准两税制
partial drainage 部分排放;局部排放
partial drainage spring 部分排泄型泉
partial drawing 部分图;局部图
partial drive pulse 半选驱动脉冲
partial drop of pressure 局部压(力)降;局部电压降
partial drought 小旱
partial dump 局部转储;部分转储
partial duration series 部分历时系列
partial duration series method 部分历时序列法;基准水位(法)
partial duty port 降能旁通槽
partial earth 部分接地;不完全接地
partial eclipse 偏蚀;偏食(日、月)
partial eclipse of the Sun 日偏食
partial elasticity 不完全弹性
partial elimination 部分消失法
partial enclosure 局部密封罩
partial enlarged detail 局部放大图
partial enlarged drawing 局部放大图
partial enlarged view 局部放大地图
partial enumeration method 部分列举法
partial environmental anomaly 局部性环境异常
partial equilibrium 局部均衡;部分均衡
partial equilibrium analysis 局部均衡分析
partial equilibrium economics 局部均衡经济学
partial equilibrium type 局部均衡型
partial erase 部分删除
partial erosion 局部冲刷
partial error 部分误差
partial ester 偏酯
partial evacuation 局部疏散
partial eviction 部分迁动
partial excavation 局部开挖;部分挖掘
partial excavation method 分部开挖(施工)法
partial extension 部分扩建
partial extraction method 偏提取方法
partial face machine 自由断面隧道掘进机
partial factor 分项系数;局部系数
partial factor(s) method 分项系数法
partial failure 局部破坏;局部故障;部分失效;部分破坏
partial firing 部分点火
partial fixed 局部约束的;局部固定的
partial fixing 部分装配;部分嵌图;部分嵌固;部分固定
partial floor 半肋板
partial flow filter 支管过滤器;分流过滤器
partial force 分力
partial fraction 部分分数;部分分式
partial-fraction expansion 部分分数展开
partial full-duplex 部分双工
partial function 部分函数
partial fusion 部分熔融
partial general view 装配分图;零件装配图;分装配图
partial grade separation 部分立体交叉
partial granulation 部分成粒(不完全成粒)
partial graph 部分图
partial group 部分群

partial grouting 部分灌浆
partial habitat 局部生境;部分环境区
partial half-life 部分半衰期
partial haunching 部分镶边;部分起拱
partial heating 局部采暖
partial heat of dilution 稀释微分热
partial heat of solution 溶解微分热
partial hip (四坡屋顶戗脊的)部分角梁;半斜屋脊
partial homogenization temperature 部分均一温度
partial hydrolysis 部分水解
partial ignorance 部分无知
partial image 分图像
partial image rectification 局部图像纠正
partial immersion bath 局部浸浴
partial impregnation 局部浸渍
partial impression tray 部分印模托盘
partial income statement 部分损益计算书
partial increment 偏增量;偏差
partial increment deviation 偏增量变差
partial inequality coefficient 偏不等系数
partial information maximum likelihood estimator 部分信息极大似然估计量
partial innovation 部分革新
partial insurance 部分保险
partial integration 部分积分
partial interception ratio 部分截流率
partial interest 局部利益
partial interpenetrating polymer network 部分互穿聚合网络
partial interpretation 部分解释
partial inversion 局部回返;部分反转
partial investment 部分投资
partial ionization 部分离子化;部分电离
partialization vacuum box 部分真空盒
partial jet hole bottom reverse circulation core barrel 喷射式孔底反循环取芯钻具
partial journal bearing 半围轴承
partial least squares 偏最小二乘
partial light bath 局部光浴
partial lining 部分衬砌
partial liquefaction 部分液化
partial load 局部负载;局部荷载;分载;部分加载;部分荷载
partial load coefficient 荷载分项系数
partial load factor 部分荷载系数
partial loading 部分加荷
partial load operation 低负荷运行;部分负荷运行
partial load run 部分负荷运行
partial loss 部分损失;部分漏失;部分海损
partial luminous flux 部分光通量
partial lunar eclipse 月偏蚀
partially access route 半封闭线路
partially adhered 部分黏结的
partially aerated combustion 部分预混(式)燃烧;本生燃烧
partially air dried 局部晾干的
partially amortized mortgage 部分摊还抵押;部分分期偿还、其余到期偿还的抵押贷款
partially automatic 部分自动的
partially automatic control 部分自动控制
partially automatic rectifier 半自动刻图纠正仪
partially biological treatment 不完全生物处理
partially bonded 部分黏结的
partially bonded concrete overlay 部分结合式混凝土加厚层
partially closed hood 半密闭罩
partially closed resonator 部分闭合谐振腔
partially closed structure (路面的)部分封闭式结构
partially coherent light 部分相干光
partially compacted concrete 部分捣实混凝土
partially completed tree 部分完全树
partially computable 部分可计算的
partially consistent observation 部分一致观测值
partially constructed syntax tree 部分构造语法树
partially crushed gravel 部分破碎砾石
partially dextrinised starch 部分糊精化淀粉
partially disclosed principal 半公开当事人
partially distributed uniform load 局部均布荷载
partially downed weir 半淹式堰坝
partially drained loading 部分排水加荷
partially dry steam 部分干燥蒸汽
partially elastic 部分弹性的
partially enclosed apparatus 半封闭型电器
partially evacuated gas 部分抽空气

partially exempt obligation 免除部分责任;部分免除责任
partially exposed basement 半地下室
partially filled aperture 部分连续孔径
partially fixed 部分嵌固的
partially fixed(-end) beam 部分固端梁;部分嵌固梁
partially fixed(-end) column 部分固端柱
partially fixed joint 半嵌固连接;部分固定的节点
partially fixed point 半固定点
partially fixed support 半固定支点
partially grouted resin bolt 部分树脂锚固锚杆
partially halogenated alkanes 部分卤化烃烷
partially halogenated chloride fluoride compound 部分卤化氯氟碳化合物
partially halogenated hydrocarbon 部分卤化碳氢化合物
partially hanging rudder 半悬挂舵
partially hydroluzed polyacrylamide 不完全水解聚丙烯酰胺
partially improved road 简易公路
partially intergrown knot 部分混生节
partially inverted file 部分倒排文件
partially ionized plasma 部分电离等离子体
partially linked block design 部分相联区组设计
partially miscible organic solvent 不完全溶混有机溶剂
partially mixed cell 部分混合室
partially mixed concrete 缩拌混凝土
partially occupied band 不满带
partially ordered set 半序集(合)
partially ordered space 半序空间
partially ordered system 半序系
partially ordered task 半序任务
partially owned subsidiary 部分占有的子公司
partially participating stock 有限参加股
partially penetrated well 不完全井;不完整井
partially penetrating well 非完整井;非完全井;不完整井;不完善井;不完全渗水井
partially plane polarized 部分平面偏振
partially polarized light 部分偏振光
partially polarized radiation 部分偏振辐射
partially prestressed 部分预加应力的
partially prestressed column 部分预应力柱
partially prestressed concrete bridge 部分预应力混凝土桥梁
partially prestressed concrete girder 部分预应力混凝土梁
partially processed 局部处理的
partially qualified name 部分限定名
partially reflected reactor 部分反射层反应堆
partially reflecting layer 部分反射层
partially reflecting mirror 部分反射镜
partially reflecting surface 局部反射面
partially reinforced 局部配筋的
partially reinforced concrete masonry 局部配筋混凝土砌体(圬工)
partially reinforced masonry wall 局部加筋砖墙砌体
partially restrained beam 半约束梁
partially saturated soil 部分饱和土
partially secured liabilities 部分担保负债
partially selected cell 半选单元
partially selected core 半选选心
partially self-checking circuit 部分自检查电路
partially separated system 局部分流排水系统;(排水系统的)部分分流制;部分分流系统;部分分离系统
partially silvered mirror 半镀银反射镜
partially silvered mirror splitter 局部镀银镜式分离器
partially sloping quay 半斜坡式码头
partially sloping wharf 半斜坡式码头
partially slotted floor 部分漏缝地面
partially solidified metal 半固态金属
partially specified finite-state 部分说明的有限状态
partially specified finite-state machine 局部说明的有限状态机
partially spent 部分报废
partially sprinklered space 部分有洒水的地区
partially stabilized zirconia 部分稳定氧化锆
partially stretched 部分拉伸的
partially submerged orifice 半潜孔口
partially suppressed contraction 局部不完全收缩

partially tempered glass 局部强化玻璃;局部钢化玻璃;区域钢化玻璃
partially tensioned 部分受拉的
partially transmitting mirror 部分透射镜
partially transparent 半透明
partially treated indsutrial waste 部分处理的工业废料
partially underhung rudder 半悬挂舵
partially unemployed 不完全失业;半失业者
partially vacuumed cavity 部分真空的空洞
partially vented 部分通气的
partially weathered zone 部分风化带
partial mast 歉年
partial match retrieval 部分匹配检索
partial melt(ing) 部分融化
partial miscibility 部分溶混性;部分混溶性;部分互溶;不完全混溶性
partial mixing 部分混合
partial model 局部模型;不完全的模型
partial molal free energy 偏摩尔自由能
partial molal volume 偏摩尔体积
partial monopoly 部分独占
partial multicollinearity 部分多重共线性
partial multiple correlation coefficient 偏多重相关系数
partial new products 部分新产品
partial nitrification efficiency 短程硝化效率
partial node 不全节;次波节
partial noise exposure index 部分噪声暴露指数
partial number concentration 计数浓度
partial objectives for 分目标
partial obsolescence 部分陈旧
partial occupancy 部分使用(建筑未完工);部分占用
partial opening 局部开启
partial operation 局部操作
partial order 部分(有)序;半序
partial ordering 次序关系;部分定序;半定序
partial ordering relation 半定序关系
partial oxidation 部分氧化
partial oxidation cracking 部分氧化裂化
partial oxidation fermentation 部分氧化发酵
partial oxidation reaction 不完全氧化反应
partial package boiler 部分组装锅炉;半快装锅炉
partial parasite 局部寄生;半寄生植物
partial paved shoulder 部分铺砌式路肩
partial payment 分期付款;部分支付
partial penetrating and interfering wells formula 非完整干扰井公式
partial penetrating well coefficient 非完整井系数
partial penetrating well formula of unsteady phreatic flow 潜水非稳定流不完整井公式
partial penetration 部分穿透
partial pivot 部分主元素;部分主元
partial plan of standard cost accounting 标准成本会计部分计划
partial pluton 分部深成岩体
partial pneumatic shield 局部气压盾构
partial polarization 部分偏振
partial pondage 局部蓄水量;局部调节容量
partial population curve 人口密度曲线
partial potential temperature 分位温
partial precalcining 部分预分解
partial prefabrication 部分预测;半装配
partial pressure 局部应力;局部压力;气体分压;分压强;分压(力);部分压力
partial pressure analysis 分压强分析
partial pressure analyzer 分压强分析器;分压分析器
partial pressure culvert 半压力式涵洞
partial pressure difference 分压差
partial pressure evapo(u)ration(process) 分压蒸发法
partial pressure ga(u)ge 分压量规;分压计
partial pressure gradient 分压梯度
partial pressure mass spectrometer 分压强质谱计
partial pressure measure ga(u)ge 分压强测量规
partial pressure of carbon dioxide 二氧化碳分压
partial pressure of oxygen 氧分压;氧的局部压力
partial pressure of sulfur 硫分压
partial pressure of water vapo(u)r 水蒸气分压力
partial pressure sensor 分压力传感器
partial pressure vacuum ga(u)ge 分压真空计;分压强真空规
partial prestress(ing) 部分张拉;部分预(加)应力

partial prestressing design 部分预应力设计
partial product 部分乘积
partial productivity indices 局部生产率指数
partial-product word 部分积字
partial projection drawing 局部投影图
partial prospected area 部分勘探区
partial purification 部分净化
partial quotient 偏商
partial reactive force 部分反作用力
partial read pulse 半读脉冲;半选读出脉冲
partial recognition 部分识别;不完全识别
partial reconnaissance 局部勘测
partial record(ing) station 记录不完整测站;观测项目不全测站
partial recursive function 局部递归函数;部分递归函数
partial recursive predicate 部分递归谓词
partial reduction 部分还原
partial reflection 部分反射
partial reflectivity 部分反射率
partial reflux 部分回流
partial reflux operation 部分回流操作
partial regression coefficient 净回归系数;偏回归系数
partial regression equation 偏回归方程
partial regulation 局部整治;不完全调节
partial reinforcement 局部配筋
partial release 部分释放;部分松弛(预应力筋)
partial release clause 部分取消抵押条款
partial relief 局部缓和;部分卸荷;部分减低水区
partial renewal 局部更新
partial repair 局部修理
partial repair method 分部修理法
partial replacement plank floor 木底板部分更换
partial resonance 部分谐振
partial response 部分响应
partial response code 部分响应编码
partial restraint 局部约束;部分约束
partial retirement of shipping documents 分批赎单
partial retraction 部分退缩
partial-reversal processing 半逆向显影
partial reverse circulation 局部反循环
partial revision 部分修测【测】
partial roasting 部分焙烧
partial roundabout 半环式交叉;半环式广场
partial safety factor 分项安全系数;分部安全系数;部分安全系数
partial safety factor for action 作用分项系数
partial safety factor for resistance 抗力分项系数
partial safety factor for strength of material 材料强度分项系数
partial safety factors of pile 桩的分部安全系数
partial sales 分批销售
partial saponification number 部分皂化值
partial saturation 局部饱和;部分饱和;不完全饱和;不安全饱和
partial scour(ing) 局部冲刷
partial screw earth auger 短螺旋成孔机
partial scroll case 非完整圆式蜗壳;不完整蜗壳;不完全蜗壳
partial sea damage 局部海损
partial secondary trunking 部分二次中继法
partial section 局部视图;部分断面
partial section core hole 分段取心钻孔
partial section lining 分断面衬砌
partial section tunnel boring machine 部分断面隧道掘进机
partial section tunneller 分断面隧道掘进机
partial section view 局部剖视图;局部剖面图
partial segmentation 部分分裂;不全分裂
partial select input pulse 半选输入脉冲
partial select output 半选输出
partial sequence 部分序列
partial serial correlation coefficient 偏序列相关系数
partial series 部分系列
partial sewage treatment 部分污水处理
partial sewerage system 局部排水系统
partial shear crack 部分剪切裂缝
partial shear deformation stage 局部剪切变形阶段
partial shipment 分批装运;分批装船;分批交货;部分装运;部分装货
partial shoulder 部分铺砌式路肩

partial side wear of rail 钢轨边缘偏磨
partial simulation 局部模拟;实体模拟;实体仿真
partial slash disposal 清理部分剩余物
partial smoke detection system 部分感烟探测系统
partial softening 部分软化
partial solar eclipse 日偏食
partial solid solution 部分固溶体
partial solubility 部分溶解度
partial solution 部分溶液;部分解
partial solution of decision model 决策模型的部分解
partial solvability 部分可解性
partial specific volume 微分比容
partial spent 局部消耗
partial splice 局部拼接;部分连接
partial sprinkler system 局部喷水灭火系统;局部安装的喷水灭火系统
partial stack 部分叠加
partial state trading 部分国有贸易
partial sterility 部分不实
partial sterilization 局部灭菌;部分灭菌
partial storage 部分储存;局部库容;部分蓄水
partial storage of ore 部分留矿
partial structural perturbation 部分结构扰动
partial submergence 部分下沉;局部淹没;部分淹没;部分浸水
partial subsidence 局部下沉;局部沉淀
partial substitution 部分替换;部分替代
partial sum 部分和(数)
partial summation 部分求和
partial super-elevation 部分超高
partial superstructure 局部(船)上层建筑;局部船楼;岛式上层建筑
partial survey 非全面调查
partial suspension of solids 固体部分悬浮
partial swing 局部回转;局部摆动
partial switching 部分翻转
partial symmetry 不全对称
partial synthesis 部分合成
partial system 局部系统
partial system failure 部分系统故障
partial system test 个别部分试验
partial taper 局部渐缩;部分줄缩变段
partial tempering 部分强化
partial tensioning 局部拉伸
partial thawing 局部解冻;部分融化
partial thermoremamet magnetization 部分热剩余磁化强度
partial tide 分潮
partial track 分径迹
partial transform 偏变换式;部分变换式
partial transverse ventilation 部分横向通风
partial treatment 局部处理
partial tree 部分树
partial turbine 分流水轮机
partial umbel 分伞形花序
partial use 部分投产
partial vacuum 局部真空;部分真空;不完全真空;半真空
partial vacuum box 部分真空盒
partial valence 余价
partial vapor pressure 水蒸气分压力
partial variance 偏方差
partial variation 部分变分
partial vibration 部分振动
partial view 局部图;局部(视)图;部分视图
partial vitrification 局部玻璃化
partial volume 分体积;部分体积;部分容积
partial vulcanization 部分硫化
partial water drive 部分水压驱动
partial water meter 分流水表
partial wave 部分波
partial-wave method 分波法
partial wintering 部分越冬
partial wiring diagram 部分接线图
partial word 部分字
partial write operation 部分写操作
partical partition 不完全隔墙
participant 参与方;参与发行证券集团的银行或证券行
participate in tender 参加投标
participating 参与盈余分配
participating agency 参加机构
participating bond 参与分红债券;参加债券;分享

participating capital stock 参与分红股
participating carrier 同盟船公司
participating certificate 参加证书
participating dividend 参与分红股利;分享股息
participating house 参加商号
participating loan 共同贷款人
participating member 正式成员
participating policy 分享保单
participating preferred share 参与分红优先股
participating preferred stock 参与分红优先股;参加优先股;分红优先股
participating stock 分享股
participation 参与放款;分享
participation certificate in loan 贷款参加证书
participation certificates 参与证券(一种抵押证券);参与证单
participation coefficient 振型参与系数
participation factor 振型参与系数
participation fee 参与费
participation in the profit 参与分红
participation loan 共同贷款;共同参与贷款;参与放款
participation mortgage 参与抵押
participation oil 参与油;参股油
participation rate 参加率
participation sale certificate 参与销售证券
participation system 参与制
participative management 职工参与管理(法);参与管理
participative model of organization 参与组织模型
participator 参与者
participatory audits 分头进行的审计
participatory group 参与团体
particle 质点;粒子;颗粒;微粒;碎粒
particle acceleration 质点加速度;粒子加速
particle accelerator 粒子加速器
particle-against-particle crushing 颗粒相互破碎作用;颗粒相互破碎法
particle aggregate 颗粒聚集
particle aggregation 颗粒的聚集
particle analyzer[analyser] 粒子分析器
particle angularity 颗粒棱角性
particle association 颗粒组合
particle astronomy 高能粒子天文学
particle attraction theory of plasticity 可塑性颗粒吸引学说
particle attrition 颗粒磨损;颗粒磨耗;颗粒摩擦
particle beam 粒子束流
particle beam mass spectrometry 离子光束质谱法
particle beam weapon 粒子束武器
particle bed 颗粒料层
particle binding 颗粒结合
particle board 刨花板;碎木胶合板;颗粒板;木屑纤维板;木屑板;碎屑胶合板;碎木板;碎粒板;芯板材
particle-board core 刨花板芯板
particle board core plywood 刨花板芯胶合板;碎料板芯胶合板
particle board core stock 刨花板芯材;刨花板芯料
particle board furniture 刨花板家具
particle board panel stock 刨花板镶板材;刨花板板料
particle board shelving 刨花板搁板
particle board solid core door 碎料胶合板实心门
particle board underlayment 刨花板垫层;刨花板垫板
particle board with a veneer overlay 单板贴面刨花板
particle boundary 粒径界限
particle catalyst oxidation 粒子催化氧化
particle characterization reactor 颗粒表征反应器
particle charging 粒子荷电;尘粒荷电;尘粒带电
particle chemical oxygen demand 颗粒化需氧量
particle cloud 粒子云
particle coagulation 颗粒凝聚
particle coalescence 颗粒聚结
particle collector 颗粒收集器;集尘装置
particle collision 颗粒碰撞
particle concentration 粒子浓度;颗粒浓度;粉尘浓度
particle-conservation equation 粒子数守恒方程
particle count 粒子计数
particle counter 颗粒计数器

particle counting device 颗粒计数器;粉尘计数器
particle counting penetrometer 颗粒计数透度计
particle cross linking 颗粒交换
particle data 粒径数据
particle density 颗粒密度
particle destabilization 颗粒扰动
particle detection 粒子探测
particle detector 粒子探测器;粒子检测器;颗粒检测器
particle diameter 粒子直径;粒径;颗粒直径
particle diameter of sediment 泥砂粒径
particle diffusion coefficient 颗粒扩散系数
particle dispersion 微粒分散胶体
particle dispersoid 颗粒分散
particle displacement 质点位移
particle distribution 粒子分布;颗粒分布
particle distribution function 粒子分布函数
particle distribution limit 粒子分布极限
particle dropout 尘粒甩落
particle dynamics 质点动力学
particle electrode 粒子电极
particle emission 粒子发射
particle encounter 粒子碰撞
particle energy 质点能量;粒子能量
particle environment monitor 颗粒环境监测器
particle fall 沉降物;颗粒沉降;散落物
particle fall velocity 颗粒沉(降)速(度)
particle-filled composite 颗粒填充型复合材料
particle fineness 粒子细度;颗粒细度
particle flux 粒子通量;粒子流
particle flux density 粒子流(量)密度
particle free sediment 颗粒自由沉降
particle free sedimentation 颗粒自由沉淀
particle grading sampler 粒子分级采样器
particle group 粒群
particle hardness 颗粒硬度
particle having a finish 有包覆膜的(颜料)粒子
particle-hole 粒子空穴
particle horizon 粒子视界
particle image velocimetry 粒子图像测速法
particle in cell method 格网中的质点法;格网质点法
particle index 颗粒指数
particle index test 颗粒指数试验
particle integration rate 粒子集结率
particle interceptor trap 颗粒物拦截阱
particle interface 颗粒界面
particle interference 颗粒迁移率;颗粒流动性
particle interference in concrete mixes 混凝土搅拌时的颗粒间干扰行为
particle isolation method 颗粒分离法
particle limit 颗粒限度
particle material to be screened 筛分出来的颗粒物料
particle mean size 颗粒平均大小
particle measurement 粒度测定
particle mechanics 质点力学
particle microstructure 颗粒显微结构
particle mix(ture) 颗粒混合物
particle mobility 粒子迁移率;颗粒迁移率;颗粒流动性
particle motion 质点运动
particle motion filter 质点运动滤波器
particle movement 颗粒错动
particle moving velocity 颗粒流动速度
particle number density 平均密度
particle of suspended matter 悬浮物质的颗粒
particle optics 粒子光学
particle orientation 颗粒取向
particle-oriented paper 定向颗粒纸
particle packing 颗粒堆积
particle pair production 粒子偶生成
particle panel materials 刨花板料
particle-particle interaction 粒子-粒子相互作用
particle path 质点运动轨迹;粒子迹;颗粒轨迹;水质点轨迹
particle physics 粒子物理学
particle population 粒子数
particle porosity 颗粒孔隙率;微粒孔隙率
particle precipitation 颗粒沉淀
particle property 粒子性质
particle radiation 粒子辐射
particle radius 颗粒半径
particle-reinforced intermetallic matrix compos-

ite 颗粒增强金属间基复合材料
particle-reinforced metal matrix composite 颗粒增强金属基复合材料
particle-reinforced titanium matrix composite 颗粒增强金属钛基复合材料
particle removal 去除颗粒
particle-removal mechanics 颗粒去除机理
particle repulsion theory of plasticity 可塑性颗粒排斥学说
particles 碎木料
particle scattering 粒子散射
particle scattering factor 粒子散射因数
particle scattering function 粒子散射函数
particle sediment 颗粒沉降
particle segregation 颗粒离析
particle selection by air flo(a)tation 气浮法的颗粒选择
particle separator 颗粒分离器
particle setting velocity 颗粒沉(降)速(度)
particle settling 颗粒沉降
particle settling velocity 粒子沉降速度
particle shape 粒形;颗粒形状;粉粒形状
particle shape factor 颗粒形状系数
particle shape index 颗粒形状指数
particle shape of aggregate 集料粒状
particle shape test 颗粒形状试验
particles in suspension 悬浮质点
particle size 砂子粒径;微粒尺寸;粒子大小;粒径大小;粒径;粒度;颗粒大小;颗粒尺寸
particle-size accumulation curve 粒度累计曲线;粒径累积曲线
particle-size analysis 筛析;颗粒(级配)分析;粒度分析;泥砂粒分析
particle-size analysis of fluvial sediment 河流泥砂颗粒级配分析
particle-size analyzer 微粒分析器;粒度分析仪;粒度分析器
particle-size average 粒度平均值;平均粒度
particle-size bracket 粒级;颗粒分级
particle-size category 颗粒分类
particle-size class 粒级
particle-size classification 粒径分类(法);按粒度分类
particle-size classifier 粒度分级器
particle-size control 粒度控制
particle-size curve 粒度分布曲线;颗粒组成曲线
particle-size determination 粒径测定;颗粒尺寸测定;粒度测定
particle-size distribution 颗粒级配;质点大小分布;粒子大小分布;粒径分布;颗粒分布;颗粒大小分布;颗粒尺寸分布
particle-size distribution chart 粒径分布曲线(图);颗粒级配曲线
particle-size distribution curve 粒径分布曲线;颗粒级配曲线;颗粒大小分布曲线
particle-size distribution determination 粒径分布测定
particle-size distribution diagram 颗粒大小分布图
particle-size distribution mode 粒度分布方式
particle-size distributor 颗粒分选器
particle-size effect 颗粒大小效应;颗粒影响
particle-size efficiency 局部过滤效率
particle-size factor 粒度系数
particle-size fraction 粒度级
particle-size fraction of sample 样品粒度
particle-size gradation analysis 粒径级配分析
particle-size grading 粒度分级
particle-size limit 粒径极限;颗粒尺寸极限
particle-size minotor 粒径检测器
particle-size of dust 灰尘粒径
particle-size of fume 烟气粒径;烟气粒度
particle-size of sludge 污泥颗粒粒径
particle-size range 粒度范围
particle-size ratio 颗粒比
particle-size reduction 粒度减小
particle-size scale 粒径比例
particle-size segregation 颗粒离析
particle-size study 粒级分析;颗粒粒度研究
particle sizing 颗粒分级;粒度分布;粒子筛选;颗粒筛分
particle sliding 颗粒滑移
particles of aggregates 骨料颗粒
particle spacing 颗粒间隙
particles per cubic(al) centimeter 每立方厘米的

粒子数
particle stated contaminant 颗粒态污染物
particle strength 颗粒强度
particle structure 颗粒结构
particle surface 颗粒表面
particle surface area 颗粒表面积
particle suttling velocity 颗粒沉降速度
particle swarm 颗粒群
particle terminal fall velocity 颗粒终降速度
particle texture 集料颗粒表面质地；颗粒质地
particle-thickness technique 质点厚度测量技术
particle track 粒子径迹
particle-track method 粒子径迹法
particle velocity 质点速度；颗粒速度；水质点速度
particle wear 颗粒磨损；颗粒磨耗；颗粒摩擦
parti-colo(u)r 杂色
parti-colo(u)red 杂斑的
parti-colo(u)red painting 彩画(古建筑)
particular 详细数据；详细说明；详细资料；特色；细节；特殊比例尺
particular average 特别海损；单独海损
particular case 特殊情况
particular charges 特别费用
particular commodity 特定商品
particular commodity rate 特定货物运费率
particular data 单要素资料
particular deduction 特殊扣除项目；特殊扣除
particular drilling condition 特殊钻井情况
particular equivalent form 特殊的等价形式
particular form 特殊方式
particular function 特定函数
particular integral 特解；特别积分
particular international law 特殊国际法
particular investigation stage 详细调查阶段
particularistic Gothic 特殊哥特式建筑风格；哥特式建筑风格
particularity 特质；特殊性；细目
particularity annual output 分项年产量；分项基建投资
particularity annual value 分项年产值
particularize 特殊化；详述
particular kind of commodity 特殊商品
particular leg 特定航程
particular lien 特定留置权；特别留置权
particularly compact 特别致密
particular operation 特殊操作
particular package 特殊包装
particular paper 验货单
particular pollutant 特性污染物
particular pollutant concentration 特性污染物浓度
particular-postlude 具体尾部
particular processing section 特殊处理剖面
particular proposition 特定命题；特称命题
particular reaction 局部反应
particular resonance 局部谐振；局部共振
particular risk 特定危险性
particulars 详细资料；详细数据；事项
particular service requirement 特殊使用条件
particular sheet 明细表
particulars of amendment 修改事项
particulars of cargo 货物明细单
particulars of insurance 保险细目
particular solution 特解
particular sphere of producion 特殊生产领域
particular strain 特殊的品种
particular surroundings 特殊环境
particular tare 实际皮重
particular territory 特殊厂区
particular use-value embodied in services 服务中的特殊使用价值
particular value 特值
particular water supply 局部供水
particulary 特殊性
particulate 粒子；颗粒状物；微粒子；散粒
particulate asbestos material 颗粒石棉材料
particulate carbon 颗粒碳
particulate characteristic 颗粒特征
particulate chemical monitor 颗粒化需氧量
particulate cloud 粒子云
particulate composite 颗粒复合材料；粉末复合材料
particulate control 悬浮微粒控制；尘埃控制
particulate emission 颗粒散发物；颗粒排放量；微粒排放
particulate filler 颗粒填料

particulate filler composite 粒状填料复合材料；粉末复合材料
particulate filter 颗粒物过滤器；微粒过滤器
particulate fluidization 散式流态化
particulate form 颗粒态
particulate grouting 微粒灌浆
particulate inorganic material 颗粒无机物
particulate inorganic phosphorus 颗粒无机磷
particulate Kjeldahl nitrogen 颗粒克耶定氮量
particulate loading 大气中的尘烟荷载；微粒负荷
particulately fluidized bed 散式流化床
particulate mass analyzer 颗粒物分析器
particulate mass concentration 颗粒物质浓度
particulate materials 粉料
particulate matter 颗粒物质；颗粒物；微粒子材质
particulate mechanics system 粒体力系
particulate metal concentration 颗粒金属浓度
particulate nitrogen 颗粒氮
particulate organic carbon 颗粒有机碳
particulate organic material 颗粒有机物
particulate organic matter 有机粉尘
particulate organic nitrogen 颗粒有机氮
particulate phase 颗粒相；微粒相；微粒相；散式相
particulate phase-inversion polyvinylidene fluoride membrane 颗粒相—逆增聚偏氟乙烯膜
particulate phosphorus 颗粒磷
particulate pollutant 粒子污染物；粒状污染物；颗粒污染物
particulate pollution 粒状物污染
particulate precipitator 微粒沉淀器
particulate radionuclide 微粒放射性核素
particulate removal 微粒的去除
particulates 大气尘粒
particulate scattering 颗粒散射；微粒散射
particulate separation equipment 固体颗粒物分离装置
particulate silica 微粒硅土
particulate size 颗粒物粒度
particulate species concentration 颗粒性物种浓度
particulate stated contaminant 颗粒态污染物
particulate substance 颗粒物质
particulate system 粒体系
particulate technology 颗粒工程
particulate trap 微粒收集器
particulate waste 粒子状废物；颗粒状废物
parties liable on a bill of exchange 负汇票上的债务的人
parties to a bill 票据当事人
parties to a contract 签约人
parties to the project 工程的有关当事人
part in bending 受弯部分
parting 掀板；裂理；裂层；剖截；挖割；脱膜；砂箱分界线；道路岔口；岔口；分开；分金
parting agent 隔离剂；模型润滑剂；脱膜剂；脱膜隔离涂料；分型剂；分离剂
parting bead 隔条；隔扇条(窗框)；隔窗线条；推拉窗扇中梃；双悬窗扇中梃
parting beat 隔片
parting clay 夹层黏土
parting compound 脱模剂；隔离剂；离型粉；阶梯形分型面；分型剂
parting dust 分型粉
parting face 分型面；分离面
parting fence 分隔围栏
parting flange 分离法兰；分离凸缘
parting gate 分型面内浇口
parting handling 拼合处理
parting knife 切割刀
parting lath 薄隔板
parting line 缝缝线；分型线；分模线
parting lineation 裂线理；剥离线理
parting-line gating 分型面上浇注系统
parting-line gating system 水平浇注系统(铸造)
parting-line print 水平芯头；分型面芯头
parting line zone 分型区
partingl turnover 道岔【铁】
parting material 分离材料
parting medium 脱模剂；分型剂
parting-off grinder 切断研磨机；切断模具
parting-off tool 造型修刀；切断刀；切断车刀
parting of the pass 孔型锁口
parting plane 裂开面；界面；际面；分型面；分离面；分割面
parting plane lineation 分离面线理

parting planing tool 切断刨刀
parting powder 分离粉；隔离粉；脱模剂；离型粉；分型粉
parting pulley 拼合轮；拼合滑轮；拼合滑车
parting rail 中横档
parting sand 夹层砂；夹层砂土；砂隔层；分型砂；分离砂
parting shears 剪槽铁机
parting slip 垂直活动窗扇重量分隔片；双悬窗扇中梃；分离平衡室；分车带；中导板；隔片；吊锤箱中间隔板；吊窗隔条
parting slotting tool 切断插刀
parting stop 双悬窗分隔木条；隔条；防断保险绑绳
parting strap 防断保险绑绳
parting strip 双悬窗扇中梃；推拉窗扇中梃；分隔条；分车带
parting strip of bridge deck 桥面分车带
parting surface 分模面
parting tool 剖刀，切凿；开槽工具；开裂工具；截刀；切割刀具；切割刀；切刀；三角锉(刀)；三角凿；刀具
parting wall 分隔墙
parting wheel 切割用砂轮
part in tension 受拉部分；部分受拉
part interchangeability 部分的互换性
partition 分隔物；中间壁；瓜分；隔离物；隔开；间墙；区分；分区；分片；分块；分隔(舱壁)；分割；分拆；分布；间壁
partition allocation method 分区法
partition balancing 分区平衡；分段平衡
partition beam 间隔梁；间壁梁
partition block 空心砌块间壁；隔墙砖；隔墙砌块；隔墙块(体)；空心砌块隔断墙；空心块间壁
partition blockwork 隔墙块砌筑；隔墙砌筑工作；隔墙堵塞工作
partition board 隔扇；隔墙板；隔板；分隔板；分隔间壁
partition board mixer 隔板式混合槽
partition brick 隔墙砖
partition building component 隔墙构件
partition bulkhead 间壁
partition cap 隔墙上槛；隔墙墙帽
partition capacitance 部分电容
partition capacity 分配容量
partition chromatography 分溶层析法；分配色谱法；分配色层法；分配层析法
partition coefficient 划分系数；配分系数；分配系数；分离系数
partition column 分配柱；分离柱
partition combination 分区组合
partition communication region 分区通信区
partition compaction 分段压实量
partition component 隔墙构件
partition control descriptor 分区控制说明符；分区控制描述符
partition control table 分区控制表
partition coping brick 隔墙顶砖
partition coverings 间壁覆盖；填实空间
partition curve 分布曲线；分配曲线
partition density 分配密度；分配比重；分离密度
partition device 分隔装置
partition door 格扇
partitioned 分段的
partitioned access method 分区存取(方)法；分片存取法
partitioned-area reservoir data 分区储集层数据
partitioned data base 分区数据库
partitioned data organization 分区数据组织；分区数据结构
partitioned data set 分区数据集；分片数据集；分段数据集；部分数据集
partitioned date 分区数据
partitioned emulation programming 分区仿真程序设计；部分模拟程序设计
partitioned file 分区文件
partitioned file access 分区文件存取
partitioned hash function 分区杂凑函数
partitioned logic 分块逻辑
partitioned matrix 配分矩阵；多块矩阵
partitioned mode 分区方式；分块方式
partitioned operation 划分运算
partitioned organization 分区组织；分区结构；分块结构
partitioned radiator 分段式放热器

partitioned regenerator 分隔式蓄热室
partitioned segmentation 分割式段落(调度);分割段落式
partitioned sequential file 分区式顺序文件
partitioned space 分隔空间
partitioned tank furnace 分隔式池窑
partition effect 分配效应;分配效果
partition end cap 隔墙端帽;隔墙端部装头;隔墙端部装饰
partition equilibrium 分配平衡
partition factor 隔热系数
partition-free 无隔墙的
partition function 划分函数;配分函数;分配函数;分拆函数
partition head 隔墙顶木条;隔墙顶杆;隔墙上槛
partition height 隔墙高度
partition infilling 隔墙内填充物
partitioning 架设隔断;架设隔墙;格板材料;划分;打隔断;分块
partitioning adjustment 分区平差
partitioning glazing 装饰隔墙玻璃;玻璃隔墙;玻璃隔断
partitioning method of estimation 分割估计法
partitioning model 分配模型
partitioning of a matrix 矩阵分块
partitioning of matrices 矩阵分块
partitioning of space 空间分隔
partitioning system 隔墙体系
partition insulator 绝缘隔板;绝缘垫;绝缘导管
partition isotherm 分配等温线
partition law 分配定律;分配定理
partition line 双晶缝合线;缝合线
partition liquid 分配液(法)
partition load 分区装入;部分装入;部分加载
partition map 划分图
partition member 隔墙构件
partition method 隔离方法;分离法;分类法
partition noise 电流分配噪声;分配噪声
partition non-rated 不耐火隔断;矮隔墙
partition number 分区数
partition of an integer 一整数的分解
partition of a set 集合的分划
partition of elements 元素配分
partition of energy 能量分布
partition of land 土地划分
partition of load 负载分布;负荷分配
partition of native style 板壁
partition of Ni between mineral pair as a function of temperature 矿物对间镍配分与温度函数关系图
partition of outcome variable 结果变量划分
partition of saline soil 盐渍土的分区
partition of unity 单位分解
partition panel 隔墙板;隔断板
partition plate 隔墙上槛;隔墙板;隔仓板;隔板;喷嘴叶片;金属板隔墙;隔墙顶部垫板
partition post 隔断立柱
partition pot 隔墙花座
partition profile 隔墙纵断面图
partition programming 分块规划;分块的规划
partition queue area 分区排队区域
partition queue element 分区队列元(素);部分队列单元
partition ratio 分配比(率);配分比
partition redefinition 分区重新定义
partition ring 分隔垫圈
partition running 分区运行;分段运行
partition save area 分区保存区
partition screen 隔断
partition section 工地剖面;隔断截面
partition-segmented memory manager 分段式存储器管理程序
partition size 分配粒度;分级粒度
partition slab 隔墙板
partition slotter 隔挡机
partition specification table 分区说明表
partition standard label 分区标准标号
partition stud 隔墙立柱
partition studding board 骨架式隔墙覆盖板;骨架式隔墙板;骨架式隔墙覆面板
partition system 隔断系统;隔墙系统
partition test 分析检验;分类检验;分割检验;分拆检验
partition theorem 分割定理

partition tile 隔墙用空心砖;内隔墙用的空心砖;非承重隔墙(空心)砖;不承重隔墙(空心)砖;黏土空心砖;隔墙砖;隔墙用瓷砖
partition tree 划分树
partition trim 隔墙修饰
partition user 区分用户区域
partition value 分割值
partition wall 隔膜;隔墙;间壁;分隔墙
partition wall beam 隔墙托梁;隔墙承托梁;承托隔墙梁
partition(wall) block 隔墙砌块
partition(wall) brick 隔墙砖
partition(wall) panel 隔墙镶板;隔墙墙板
partition(wall) penetration 隔墙渗透;隔墙穿透
partition(wall) shape 隔墙形状
partition(wall) tile 隔墙砖
partition(wall) unit 隔墙构件;隔墙单元
partition window 间隔窗
partiversal 半锥背斜
part-length rod 局部控制棒
part list 零件清单
part load 低负荷;部分荷载;部分负荷
part load behavio(u)r 低负荷性能
part load efficiency 低负荷效率
part load organization station 零担货物组织站
part load performance 部分负荷性能
part-load transfer increase revenue 零担中转给收入
part-load transshipment station 零担货物中转站
partly 部分地
partly adjustable weir 半调节堰
partly alternating stress 部分交替应力
partly buried support 半埋式支座
partly closed slot 半闭口槽
partly clouded 少云
partly cloudy 少云
partly completed stock 库存半成品
partly computerized 半计算机化
partly cover torn 部分包装破损
partly-developed negative 部分显影底片
partly dissolved fossil 部分溶解化石
partly finished goods 半成品
partly incomplete aquifuge 隔水层局部缺失
partly mechanized shield 半机械化盾构
partly miscible liquid 部分混溶液
partly mixed estuary 部分混渗型河口
partly mounted 半悬挂式的
partly open mo(u)ld 半敞开式模具
partly ordered set 半序集(合)
partly paid share 部分已缴款股份
partly penetrating well 不完整井
partly pulsating stress 部分脉动应力
partly rational expectation 部分理性预期
partly renew 部分换新
partly self-supported 部分自给
partly solid jetty 半实体突堤
partly soluble 部分可溶的
partly submerged intake structure 半淹没式取水口建筑物
partly submerged orifice 局部淹没孔口;部分淹没孔口
partly underhung rudder 半悬式舵
partly unemployed 半失业
partly variable cost 部分变动成本
partly with suspended ceiling 局部有吊顶
part method 分解学习法
part name 零件名称
partner 股东;合作者;伙伴;合伙人;合股人;补板
partner in charge 联营体主办人
partner's equity of a joint venture 投资人权益
partnership 无限公司;合伙营业;合伙;合股;全体合伙人
partnership agreement 合伙契约
partnership assets 合伙人资产
partnership at will 自愿合伙
partnership business 合营企业
partnership capital 合伙资本
partnership income 合伙经营收入
partnership or individual 集体或个体
partner's investment interests 投资人权益
partner's loan 合伙人贷款
partner's personally solvent 合伙人个人偿债能力
partner's salary 合伙人薪金
partner star 伴星

part number[P/N] 零件(编)号;件号;部件号
part of deposit 区段(指矿段)
part of detail in enlarged scale 局部放大详图
part of elevation with enlarged scale 立面局部放大
part of rarefaction 疏部
partography 纸分配色谱(法)
part operation 操作码部分
part owner 股东;船的共有人
par-tox 窗框防腐剂
part packed 部分包装
part per billion 十亿分之一(10^{-9})
part per million 百万分之一(10^{-6})
part per trillion 亿万分之一(10^{-12});万亿分之一
part per volume 体积百分数
part plan 局部平面
part pooling 部分集合经营
part program(me) 部件加工程序;部分加工程序
part receipt 部分收据
part reducing atmosphere 部分还原性介质
part requirement card 零件规格卡片
partridge feature spot 灰被釉
partridgeite 方铁锰矿
partridge-wood 红色硬木
parts 元件;组装零件;零件;机件;部件
parts and accessories 零部件
parts and accessories for imported equipment 配套工程
parts and auxiliary equipment 配套件
parts and components 配套件;散件
parts and components manual 零件与元件手册
parts and components required for maintenance 维修备品和零配件
parts and equipment cost 零配件费
parts book 零件册(簿)
parts by weights 重量配合比;重量份数;按重量计的份数;按重量
parts caddy 零件搬运车
parts catalog(ue) 零件目录(表);备件目录
parts common 通用零件
parts counter 零件计数器
parts depot stock 备件库储备
parts depreciation 零件磨损
parts distribution 零件分配
parts drawing 零件图
parts dry(weight) 干重
part section 局部剖面
part-sectioned view 局部剖视图;局部剖面图
part-section excavation 分部开挖
parts failure rate 零部件故障率
parts fitting 零件的配合
part sheet 部分图幅
part shipment 分批装运
part shipment credit 分批装运信用证
parts inventory 零件清单
parts kit 整套维修零件
parts list 零件目录(表);零件明细表;零件单;备件明细表
partsman 备件员
parts manufacturer 零件制造者
parts memo 零件备忘录
parts number 零件号码;零件编号;部件号码
parts of a block 滑车部件
parts of a building 建筑配件
parts of drilling tool 钻具部件
parts of line 缆绳的分段
parts of toothed harrow 钉齿耙部件
parts peculiar 专用零件
parts per billion 十亿分率;十亿分之几
parts per hundred million 亿分率;一亿分之几
parts per million 百万分之几
parts per thousand 千分之一;千分之几
parts per trillion 一万亿分之几;万亿分之几;百亿分之几
parts rack 零件架
parts release notification 配套认可书
parts renewal 零件修复
parts requirement list 零部件规格一览表
parts requisition and order request 零件请购与订货申请
parts selection matching 零件选配
parts service 零件供应站;零件服务站
parts storage bin 零件储仓
part stack segment number 部分叠加段数

parts tag 零件标签
parts transportation truck 部件搬运车
part-subassemble standard 零部件标准
part-swing shovel 局部回转挖土机
part-tenant farmer 半自耕农
part the casing 孔内拆开套管柱
part throttle 局部节流;部分开启节流
part-time 非全日的
part-time application 短时使用;部分时间使用
part-time employees 非整班雇员
part-time employment 非全日制就业;非全日性工作
part-time farming 季节性耕作
part-time idleness 半失业
part-time job 零星工作
part-time job system 非全时工作制
part-timer 兼职者;零工;非全日工
part-time shift 非全日班次
part-time staff 兼职人员;非全日(制)工作人员
part-time vocational education 非全日职业教育
part-time work 打工
part-time worker 零工;兼职工;兼职者
part to be cut out 被切割件
part wall 公用隔墙
part winding start 部分绕组启动
part-winding starter 分线圈启动器
part winding starting 部分绕组启动法
part-winding start motor 部分线圈启动电动机
party 一方;工作班;班组
party A 甲方;契约甲方
party adjudged bankrupt 被法院宣布为破产的当事人
party arch 界拱;公共拱;共用拱门;公用拱门
party B 乙方;契约乙方
party C 丙方;契约丙方
party chief 队长(外业)
party colo(u)red 杂色的
party colo(u)rs 表面涂刷各种颜色
party concerned 有关方;有关当事人;当事方
party corbel 共用牛腿;共用托臂;防火隔墙(悬挑的)
party driveway 进入毗邻两地的路
party fence 隔墙;通墙;通防篱;共用(围)篱;共有隔墙;共用栅篱;共用(隔)墙;共用隔篱
party giving the mandate 委托人
party in a dispute 争议当事人
party in a lawsuit 诉讼当事人
party in breach (default) 违约的合同当事人
party issuing contract 发包人
party line 宅基界线;共用线(路);共线;合用线
party line bus 合用总线
party line telephone 共线电话
party masonry wall 户内墙;共用圬工墙
party member 班组成员
party on line work 路线测量队
party parapet 共用矮隔墙
party passengers travel(l)ing on different trains 分乘
party report 班报表
party tent 集会帐篷
party to a contract 合同有关方(面);合同当事人
party wall 共有隔墙;共用(隔)墙;界墙;分户墙
party wire 共用(电话)线;公用电(话)线
partzite 水锑铜矿
parure 一套宝石饰物
parvafacies 同时相
par value 票面价值;面值;市场价值;平价
par-value capital stock 面值股票
par-value method 面值法;票面值法
par value of bond 债券票面价值
par value of exchange 汇兑平价
par value of foreign exchange 外汇平价
par value share 面值股票
par value stock 面值股票
par value system 平价制度
parvial field 分场
parvis 寺院前庭;寺院前院;教堂门廊(房间)
parvis turret 教堂前庭塔楼
parwelite 硅砷锑锰矿
Pasadenan orogeny 帕萨迪运动
Pascal 帕(斯卡)(压力单位:牛顿/平方米)
Pascalian fluid 帕斯卡流体;非黏性流体
Pascal's distribution 帕斯卡分布
Pascal/second 帕斯卡/秒

Pascal's lag distribution 帕斯卡滞后分布
Pascal's lag model 帕斯卡滞后模型
Pascal's law 帕斯卡定律
Pascal's law of fluid pressure 帕斯卡液压定律
Pascal's liquid 帕斯卡流体
Pascal's principle 帕斯卡原理
Pascal's theorem 帕斯卡定理
Pascal's theory 帕斯卡原理
Pascal's triangle 帕斯卡三角
Paschal table 复活节表
Paschen's law 帕邢定律
pascoite 橙钒钙石
pascuum 牧场
pas-de-souris 城堡护城河到城门口的阶梯
pash back wheel 退圈轮
Pasquill-Gifford diffusion parameter 帕斯奎尔—吉福德扩散参数
Pasquil's practical diffusion formula 帕奎尔实用扩散公式;帕奎尔经验扩散公式
Pasquill-Turner stability category 帕斯奎尔—特纳尔稳定度分类法
pass 狭路;哑口[道];轧制道次;轧道;轧槽;孔型;通过;沙洲洞;道次(轧制);滨外礁洞;遍数;隘口
passable bill 通用票据
passable conduit 能通行管沟
passage 水路;航路;航道;通行权;走廊;孔道;通路;通过;通道
passage aisle 小通道
passage appliance 通道设施
passage area 流道断面;交通面积;通路面积
passage bed 过渡岩层;过渡层
passage boat 载客渡轮;载客渡船;交通船;定线交通船
passage clear 通行无阻
passage cock 通路旋塞
passage detector 通过型检测器
passage fare 旅费;通行费
passage flow disturbance 水流扰动传播
passage for freight handling 运货通道
passage for heat-transfer medium 传热介质通道
passage form 中间类型
passage for product 成品流道
passage grave (由巨石长道引向用土盖起的)墓室
passage money 旅费;通行费
passage of air 空气通道
passage of apical layer collapse 顶板冒落通道
passage of aquifer 含水层通道
passage of artificial water filling 人工充水通道
passage of a screen 滤网;过筛;筛孔
passage of borehole 钻孔通道
passage of bottom layer break through 底板突破通道
passage of chip 切屑排出通路
passage of crowd 人流
passage of current 电流通路
passage of earthquake fracture 地震裂隙通道
passage of fault 断层通道
passage of harbo(u)r 港口通道;港口航道
passage of heat 导热;传热通道
passage of heavy fracture 强裂隙通道
passage of karst collapse 岩溶塌陷通道
passage of karst collapse column 岩溶陷落柱通道
passage of karst fissure 岩溶裂隙通道
passage of karst louver 岩溶天窗通道
passage of karst pipe 岩溶管道通道
passage of natural water filling 天然充水通道
passage of time 时间推移
passage of title 所有权的转移
passage of waste water infiltration 污水入渗通道
passage of water 水的通道
passage of water inundation 充水通道
passage order 航行调度
passage prohibited 禁止通行
passage rock bed 过渡岩层
passage section 流道截面
passage set 直通插销
passage signal 通航号志;通行信号
passage space 建筑物内部的通道(走廊、楼梯等)
passage survey 航道调查
passage ticket 船票
passage ventilation duct 通过式通风管
passageway 走廊;通行道;通道;过道;楼道;通路(走廊)
passageway transfer 通道换乘

passage width 车厢通道宽度
passage with poison protection 防毒通道
pass alert safety system 通道安全报警系统
pass a line 传递绳索
pass a messenger 传递引绳
passameter 杠杆(式)卡规;外径(指示)规;外径精测仪
passanger duty 通过税
pass angularity 孔型的顶角
pass a parameter 传送参数
pass a plot program(me) 议决小区试验方案
passavant 免税通行证;免税通告片
passband 通频带;通带
passband filter 通带滤光片;通带滤波器
passband limiting switch 带宽限制开关
passband sharing 通带共用制
pass band utilization factor 通频带利用系数
pass bandwidth 通带宽度
pass book 顾客赊账簿;银行存折;存折
passbook loan 以存款作抵押的贷款
pass box 周转箱
pass-by 旁通;绕道通过
pass-by circuit 旁路
pass-by track 会车线
pass by value 按值传送
pass-check 通行证(车辆);入场券
pass contour 孔型外形
pass convexity 孔型凸度
pass-course 旁近路线;通行路线
pass door 通行门;穿堂门
passed data set 交接数据集;传送数据组
passed dividend 空过股息;空过股利
passementerie 金银饰带
passenger 旅客(设施);旅客货物和列车轮渡;乘客
passenger access 旅客通道
passenger accidental injury 旅客意外伤害
passenger accommodation 载客车辆;客运设备;客舱
passenger accumulated rate 旅客积累率
passenger agent 旅行社;旅客服务处
passenger aircraft 航线客机
passenger and cargo ship 客货轮;客货船
passenger and freight marketing 客货营销
passenger and freight statistical modernization and data sharing 客货统计现代化与信息共享
passenger and luggage handling level 顾客与行李接送平台
passenger apron 客机坪
passenger automobile 客车;载客汽车
passenger berth 旅客铺位;卧铺(位)
passenger block 旅客区段;客运大厦
passenger board on or alight from train 旅客乘降
passenger boat 客轮;客船;定期客轮
passenger bridge (机场)旅客登(下)机桥;旅客(天)桥;旅客人行道
passenger building 旅客住房;站房
passenger bus 大客车
passenger cabin 旅客舱;客舱
passenger capacity 客运能力;客运量
passenger capacity area 载客面积
passenger car 小汽车;轿车;客车
passenger car equivalent 客车换算值
passenger cargo liner 客货班轮
passenger cargo vessel 客货轮;客货船
passenger car inspection depot 旅客列车检修所
passenger car kilometer 客车公里
passenger carrier 客船
passenger carrying capacity 载客量;载客定额;客位数
passenger carrying line 高速客运线路
passenger car servicing point 客车整备所
passenger car technical servicing depot 客车技术整备所(库列检)[铁]
passenger car track 客车停留线
passenger car unit 小客车交通量单位(用于计算道路通行能力);小客车单位
passenger casualty 旅客伤亡
passenger casualty accident rate 旅客伤亡事故发生率
passenger casualty accident record 旅客伤亡事故记录
passenger casualty accident responsibility 旅客伤害事故责任
passenger certificate 乘客额定证书

passenger character(istic) 出行者特征
passenger chartered 包车乘客
passenger clerk 客运管理员
passenger coach 客运车辆;客车
passenger coaches turn-around siding 转向线【铁】
passenger compartment installation 客室安装
passenger concourse 旅客大厅;旅客通道
passenger conveyer 载客输送机;自动人行道
passenger conveyer belt 载客运输皮带
passenger counter 旅客接待柜台;乘客计数器
passenger deck 旅客甲板
passenger depot 客运站
passenger displacement 旅客位移
passenger displacement ship 排水型客轮
passenger display 乘客显示器
passenger duty 客运税
passenger elevator 载人电梯;载客电梯;客用电梯;客梯;乘客电梯
passenger elevator car 载客电梯厢
passenger elevator door 载客电梯门
passenger elevator entrance 旅客电梯入口
passenger equipment 客运设备
passenger escape 旅客紧急撤离
passenger evacuation 乘客疏散
passenger extra 临时旅客列车;临时客车
passenger facility 旅客方便设施;旅客设施;旅客设备;客运设备
passenger fares 客运费
passenger ferry 旅客摆渡;载客渡轮;载客渡船;旅客渡轮;旅客渡船;客运轮渡;客运船;旅渡;人渡
passenger flow 客流潮;客流
passenger flow collector-distribution point 乘客集散点
passenger flow diagram 客流图
passenger flow diagram for each direction 客流方向图
passenger flow direction 旅客流向;客流流向
passenger flow forecast 客流预测
passenger flow form of oblique line 客流斜线表
passenger flow investigation 客流调查
passenger flow paths 旅客流线
passenger flow route 旅客路线
passenger flow statistics 客流统计
passenger flow volume 旅客流量;客流量
passenger foot-bridge 天桥;人行跨线桥;旅客人行桥;旅客天桥;人行天桥
passenger gangway 旅客上下船用的引桥;旅客上下船用的舷梯;旅客上下船用的跳板
passenger guide system 旅客向导系统
passenger handling 客运服务
passenger-handling building 接待旅客的建筑物
passenger-handling point 接待旅客站;接待旅客点
passenger harbo(u)r 旅客港;客运港
passenger high platform 旅客高站台
passenger information display system 乘客信息显示系统
passenger information indication sign 旅客导向牌
passenger information indication system 旅客导向盘
passenger interface module 乘客接口模块
passenger intermediate platform 旅客中间站台
passenger-kilometer 旅客公里;人公里;旅客周转量
passenger landing 客运码头
passenger liability insurance 旅客责任险
passenger lift 载人电梯;载客电梯;客用电梯;客梯;乘客电梯
passenger lift door 乘客电梯门;载客电梯门
passenger lift shaft 客运电梯井;乘客电梯井
passenger liner 远洋客轮;邮轮;客运班轮;客运班机;客班轮;定期客轮;定期客船
passenger-liner fleet 定期客轮队
passenger link load 区间客流负荷
passenger list 乘客单
passenger loading apron 旅客装载跳板
passenger locomotive 客运机车
passenger luggage hold 旅客行李舱
passenger main platform 旅客基本站台
passenger mentality 乘客心理
passenger-mileage 乘客周转量;客运周转量
passenger miles 旅客周转量
passenger movement 客流
passenger on board 船上乘客
passenger oneself responsibility 旅客自身责任

passenger operation 乘客操纵
passenger overpass (bridge) 旅客跨线桥;人行天桥
passenger-place kilometers cost 客位公里成本
passenger plane 客机
passenger platform 旅客站台;客运站台
passenger pontoon 旅客浮码头
passenger pool 客运联营
passenger port 客运码头;客运港
passenger quay 客运码头
passenger railway station 客运车站【铁】
passenger railway stock 铁路客车
passenger ramp 旅客斜坡道;旅客坡道;旅行斜坡道
passenger receipts 客运收入
passenger refuge adit 避难通道
passenger refuge way 人行避难通道
passenger room 旅客室;旅客舱;客房
passenger rope 旅客缆索道
passengers' accommodation 旅客舱室
passengers and freight service 客运和货运业务
passengers carried 运送旅客人数
passenger seat 乘客座椅;旅客座位
passengers' entrance 旅客入口
passenger service 旅客服务处;客运业务
passenger service equipment 客运服务设备
passenger service system 乘客服务系统
passenger service vehicle 载客汽车
passenger shelter 旅客雨棚;候车室;乘客雨棚
passenger ship 客轮;客船
passenger shipping company 旅客航运公司
passenger ship safety certificate 客船安全证书
passenger ship with deluxe cabin 带豪华舱客轮
passenger side window 旅客边窗
passengers' lounge 旅客休息厅
passengers' overpass 旅客天桥
passenger space 载客舱位;客位
passenger special line 客运专线【铁】
passengers per annum 每年乘客量
passengers' pull gear 防险闸装置
passenger station 客运站;客运车站
passenger station building 旅客站舍;旅客站房
passenger station complex 客运站综合体
passenger station of international through traffic 旅客联运站
passenger steamer 客轮
passenger stock 客运车辆
passenger stopping point 旅客乘降所
passenger submarine 潜水客轮
passenger subway 行人地道;旅客地道;人行地道
passenger sundry expenses 客运杂费
passenger survey 客流调查
passenger tender 旅客交通艇
passenger terminal 候车室;客运枢纽;客运码头;客运终点站
passenger terminal building 客运枢纽站主楼;终点站候车室;终点站房;客运终点站
passenger ticket 客票
passenger ticket revenue 旅客票价收入;客票收入
passenger traffic 客流;客运交通;客运
passenger traffic accident 客运事故
passenger traffic communication system in station 车站客运通信系统
passenger traffic density 客运密度
passenger traffic indics 旅客运输指标
passenger traffic line 客运线路
passenger traffic only line 客运专线【铁】
passenger traffic railway 客运铁路
passenger traffic revenue 客运收入
passenger traffic statistics 旅客运输统计;客运统计
passenger traffic unit cost 客运单位成本
passenger traffic volume 客运量
passenger train 客车;旅客列车;客运列车
passenger train and freight train yard longitudinal arrangement 客货纵列式
passenger train arriving at destination station 终到旅客列车
passenger train attendant 列车乘务员
passenger train beyond bureau driving increased revenue 客车跨局开行加给收入
passenger train car 客运车厢
passenger train control 客运调度
passenger train crew 旅客列车乘务组
passenger train crews originated 旅客列车乘务工作组

passenger train depreciation expenses 客车折旧费
passenger train dispatching 客运调度
passenger train ferry 旅客列车轮渡
passenger train formation 旅客列车编组
passenger train-hour 旅客列车小时
passenger train-kilometer 旅客列车公里
passenger train locomotive 客运机车
passenger train move detaied diagram 旅客列车运行详图
passenger train operation equipment 客运列车运转设备
passenger train operator 旅客列车驾驶员;驱动器
passenger train passing track 旅客列车会让线
passenger train path 旅客列车运行线
passenger train prior than goods train 先客后货
passenger train revenue 旅客列车收入
passenger train running plan 旅客列车运行方案
passenger train siding 旅客列车到发线
passenger train stock 旅客列车车底
passenger train stock recomposition 车底改编
passenger train stock servicing 车底整备
passenger train stock servicing point 客车整备所
passenger train stock servicing shed 客车整备库(棚)
passenger train stock servicing yard 旅客列车整备场
passenger train stock storage track 客车车底停留线
passenger train stock working program(me) 客车车底周转图
passenger train that arrives in the morning when departing in the evening 夕发朝至旅客列车
passenger train timetable 旅客列车时刻表
passenger train track 旅客列车到发线
passenger train track stock cleaning yard 客车洗涤场
passenger train travel(l)ing speed 旅客列车行进速度
passenger train washing plant 客车洗刷所【铁】
passenger train working graph 旅客列车运行方案图
passenger transference 旅客换乘
passenger transit speed 旅客运送速度
passenger transport 旅客运输;客运
passenger transport department 客运局
passenger transport expense on per kilometer 单元客运支出
passenger transport expenses 客运支出
passenger transport plan 旅客运输计划
passenger transport turnover 旅客运输周转量
passenger travel 旅客周转量
passenger tunnel 行人通道;旅客地道
passenger turnover 旅客周转量;客运周转量
passenger vehicle 小客车;客车;载客车辆
passenger vessel 客轮;客船
passenger volume of peak hours 高峰客流量
passenger waiting room 候船室;候车室
passenger walkway 旅客走道
passenger way-bill 客运价格表
passenger wharf 客运码头
passenger working train 旅客列车开行方案
passenger yard coach yard 客车整被场
passenger zone 客运区;乘客上下车区域;乘客地带
pass entry 入账
passer 成品检查员
pass filter 过滤器;滤波器
pass for genuine 冒充真货
pass from contractor to employer 从承包人转移给雇主
pass gate 筏道闸门
pass highway 直通公路
passholder 持证乘客
pass horse 渡槽支架
passify 钝化
passimeter (车站的)自动售票器;(车站的)自动售票机;旋转式栅门;杠杆式内径指示计;计步器;内径指示规;内径精测仪;步数计;步程计
passing 传递
passing area 超车地带
passing arguments to entry name 变元转向入口名字
passing bay 让车道(道路加宽部分);错车道
passing beam 近光

passing bridge 通过桥梁
passing capillary water 过路毛细水
passing clear 安全通过
passing clear distance 安全通过距离
passing cross-traffic 通过横向交通
passing current limit 极限通过电流
passing distance 超车距离
passing hollow 月牙槽
passing key 共用电锁
passing lane 公路人车车道;超车(车)道
passing lifting slings under bottom 攻穿千斤顶
passing load 瞬时荷载;路过荷载
passing loop 越行线;会让线;让车回线
passing machine 筛滤机
passing maneuver 超车动作
passing minimum sight distance 最短超车视距;安全超车视距
passing off 仿冒
passing of (the) risk 风险转移
passing of title 商品主权的转移
passing place 经过地点;会船区;会船加宽段(运河);会车段;让船道;让车道(道路加宽部分);错车道;避车道;铁路岔道
passing point 越行站;相会点(美国)
passing punch 预冲孔冲头;顶孔冲头
passing round 绕越
passings 搭接盖片;搭接长度
passing-screen size 过筛粒径;过筛粒度
passing service facility 通行服务设施
passing shower 小阵雨;过境阵雨
passing siding 越行线;会让线;会车线;让车侧线
passing sight distance 超车视距
passing signal 通行信号
passing situation 两船通过情况
passing spring 传递簧
passing station 越行站;会越站【铁】;会让站
passing strake 连通列板
passing tally 点数理货
passing through a lock 过闸
passing through curve 通过曲线
passing through method 路线穿越法;横穿越法
passing through method mainly be used tracing method be used as supplementary 穿越法为主,追索法为辅
passing through the main line 通过正线
passing through the siding 通过侧线
passing tonnage 通过吨位;通过吨数
passing track 副车道;越车道;超车道;错车道;越行线;会让线;让车线;全让线
passing tracking 让车道(道路加宽部分)
passing-type furnace 贯通式炉
passing unregistered 内部泄漏量
passing vehicle 越行车辆;超越车辆
passionflower 大丽花
passire soil former 惰性成土因素
passivant 钝化剂
passivated oil 耐用油;耐氧化油
passivated transformer oil 变压器耐用油
passivating 钝化
passivating action 钝化(作用)
passivating agent 钝化剂
passivating coat 钝化涂膜;钝化涂层
passivating film 钝化膜
passivating process 钝化工艺
passivating treatment 钝化处理
passivation 抗腐蚀作用;钝化处理;钝化(作用)
passivation coat(ing) 钝化膜层;钝化膜
passivation effect 钝化效应
passivation glass 钝化玻璃
passivation glass coating 钝化玻璃涂膜
passivation layer 钝化层
passivation of steel reinforcement 钢筋钝化
passivation potential 钝化电势
passivation reagent 钝化剂
passivator 抗腐蚀物;钝化剂
passive 钝态的
passive absorption 被动吸收
passive absorptive rate 被动吸收值
passive action 被动作用
passive activation analysis 无源活化分析
passive-active cell 钝化活化电池
passive-active data simulation 无源有源数据模拟
passive-active detection and location 无源有源探测定位

passive adaption 消极适应
passive adsorption of organisms 生物被动吸收
passive agglutination test 被动凝集试验
passive antenna 无源天线
passive anti-roll system 被动式抗横摇系统
passive arbitrage 被动套利
passive armor 被甲
passive Arthus reaction 被动阿瑟斯反应
passive assets 固定无形资产
passive attention 被动注意
passive automatic night time tracking system 被动式夜间自动跟踪系统
passive blockage 被动禁止兑换
passive bond 无息债券;消极债券;无息公债
passive capture 被动袭夺
passive catch 不活动挡(阀门分配机构)
passive cation 被动阳离子
passive chemical ionization mass spectrometry 无源化学电离质谱法
passive circuit 无源电路
passive closed loop 无源闭合环路
passive collection 无源收集
passive commerce 依赖外国船的进出口贸易;无船的运输业;被动贸易
passive component 无源元件;无源设备;被动部件
passive condition 钝态
passive constraint 消极约束
passive content 非活动的储料量
passive continental margin 被动大陆边缘
passive debt 无息债务;无息债券
passive decoy 被动假目标
passive defect 原因不明的故障
passive detection 无源探测;无源检波;被动探测
passive detector 无源探测器;辐射指示器
passive device 无源器件;被动元件
passive diffusion 被动扩散
passive dispersal 被动散布
passive display 被动显示;被动式显示
passive display system 无源显示系统
passive earth pressure 土抗力;被动土压力
passive earthquake 被动地震
passive earth resistance 被动土抗力;无源大地电阻;土阻力
passive earth thrust 被动土推力
passive edge 被动边
passive electric(al) circuit 无源电路
passive electrode 接地电极;收集电极
passive element 寄生元件;无源元件;被动元素
passive equipment 被动装置
passive excercise 被动操练
passive exercise 被动运动
passive face of flexible wall 柔性墙的被动面
passive failure 被动破坏
passive fare table 待用票价表
passive fault 死断层;不活动断层
passive fault block 不活动断块
passive fibre optics element 被动纤维光学元件
passive field shielding 被动场屏蔽
passive filter 无源滤波器
passive fire defence 消极防火;被动防火
passive fire protection system 被动消防系统
passive fire safety equipment 被动消防安全设备
passive fixed information transponder 无源固定信息应答器
passive flow 无源流动
passive flow adjustment 被动流量调节
passive flow zone 被动流动带
passive fold 被动褶皱
passive front 不活跃锋;被动锋【气】
passive gamma-ray analysis 无源伽马射线分析
passive graphics 被动型制图学;被动图形学
passive gravitational mass 被动引力质量
passive guidance 被动引导
passive hardness 耐磨硬度;纯态硬度
passive highway grade crossing protection 被动式平交道口防护
passive homing 被动追踪;被动式寻的
passive homing guidance 被动寻的制导
passive hot water system 被动式热水系统
passive imaging system 被动成像系统
passive improvement trade 被动加工贸易
passive indication device 无源指示装置
passive infection 被动侵染;被动感染
passive infrared 被动红外

passive infrared decoy 被动红外假目标
passive infrared equipment 被动红外装置
passive infrared homing head 被动式红外自动导向头
passive infrared range finder 被动(式)红外测距仪
passive injection 被动贯入
passive instrument 被动式仪器
passive insulation 消极隔振
passive intercept tracking system 被动拦截跟踪系统
passive interval 被动时间间隔
passive intervention 消极干预
passive isolation 消极隔振
passive laser 无源激光器
passive laser beam communications satellite 被动激光通信卫星
passive lateral earth pressure 横向被动土压力
passive learning 被动学习
passive localization 无源测距
passive measurement 被动测量
passive memory 被动式存储
passive metal 惰性金属;钝化金属
passive microwave 无源微波;被动微波
passive microwave radio meter 无源微波辐射计
passive microwave remote sensing image 无源微波遥感图像;被动微波遥感图像
passive mode 被动方式
passive mode-locking 被动锁模
passive monocular viewer 被动式单目观察仪
passive movement 被动运动
passive network 无源网络
passive network station 无源网络站
passive neutron analysis 无源中子分析
passive night-vision device 被动式夜视仪;被动式夜视装置
passive ocean margin 被动大洋边缘
passive optical component 被动光学元件
passive optical filter 被动滤光片
passive optic(al) network 无源光网络
passive optical satellite surveillance 被动式光学卫星监视
passive optical seeker technique 被动光学寻的器技术
passive orienting action 被动方位作用
passive penetration 被动贯入
passive permafrost 原永冻层
passive pile 被动桩
passive pollutant 潜在性污染物;惰态污染物
passive position 被动体位
passive pressure 被动压力;被动土压力
passive probe 无源探测器
passive processing trade 被动加工贸易
passive Q-switch 被动 Q 开关
passive radio meter 无源辐射计
passive Rankine failure zone 被动朗肯破坏区
passive Rankine pressure 朗肯被动压力;被动朗肯土压力
passive Rankine state 朗肯被动土压力状态
passive Rankine zone 朗肯被动区
passive reabsorption 被动重吸收;被动回吸收
passive recording device 被动式记录设备
passive reflector 无源反射器
passive regenerator 被动换热器
passive reinforcement 非受力钢筋
passive relay station 无源中继站
passive remote sensing 自然源遥感;被动式遥感
passive resistance 土抗力;无源电阻;被动抗性;被动抗力;被动地压
passive retrodirective element 被动后向反射器
passive rupture 被动破坏;被动断裂
passive satellite 无源(人造地球)卫星;被动卫星
passive satellite electrooptic detector 被动卫星光电探测器
passive secretion 被动分泌
passive sensitization 被动敏化
passive sensor 被动式传感器;被动传感器
passive sign post 无源信号标杆法
passive smoke control system 被动式烟气控制系统
passive soil pressure 土抗力;被动土压力
passive soil resistance 被动土压力
passive solar energy heating system 被动式太阳能采暖系统
passive solar energy system 被动式太阳能系统

passive solar house 被动式太阳房
passive sonar 噪声定位声呐;无源声呐;被动(式)声呐
passive sounding 无源探测
passive source method 被动源法
passive state 钝态;钝化状态;被动状态;被动局面
passive state of plastic equilibrium 被动状态的塑性平衡;被动塑性平衡状态
passive station 从站;被动站
passive surface 钝化(表)面
passive surface of sliding 被动滑动面
passive switch 被动(式)开关
passive system 无源系统;被动装置;被动系统;被动式
passive thrust of earth 土的被动压力;被动土压力;被动土推力
passive torpedo 被动式声导鱼雷
passive tracker 被动式跟踪装置
passive trade 消极贸易;被动贸易
passive trade balance 贸易逆差
passive traffic 被动运输
passive transducer 无源传感器
passive transfer 被动转移;被动传输
passive transference 被动转运
passive transfer test 被动转移试验
passive transport 被动转运;被动运输;被动输送
passive trust 消极信托
passive type 被动类型
passive unit pressure of soil 单位被动土压力
passive variable information transponder 无源可变信息应答器
passivity 钝性;钝态;钝化
passivity limit 被动限
pass judgment on 评定
passkey 撞锁钥匙;总钥匙;碰锁钥匙;万能钥匙
passlamp 弱光灯;近光灯
pass lifting slings under a wreck 在沉船下穿引千斤索
pass line 轧制线;受装索
pass master 符合要求
pass of satellite 卫星通过
passometer 计步器;内径指示规;步数计;步程计;步测计
pass on 传达
pass on the port side 从左舷通过
pass on the starboard side 从右舷通过
pass-out steam 旁通蒸汽
pass-out steam engine 旁路汽轮机
pass-out turbine 抽气式汽轮机
pass-over mill 递回式轧机;二辊周期式薄板轧机
passover offset 迂回管
pass partition 分程隔板
pass-partout 万能钥匙
pass point 连接点;控制点;加密点【测】;传递点
passpoint templet 连接点模片
passport 护照;通航护照
pass road 狭路;通道
pass schedule 孔型系统;孔型安排
pass sequence 焊道顺序;焊道程序
pass set-up 孔型调整
pass sorting 通过分类法
pass symbol 山隘符号【地】
pass test 通过测试;通过试验;通过检验;试验合格;检验合格;测试通过
pass the requirements 符合规格
pass-through 隔墙上的洞口;借涨价将原料成本的增加转嫁给消费者;让渡;送菜窗口;传递窗;厨房与餐厅间递菜用的传递口
pass-through accounts 入账
pass-through facility 贯通装置
pass-through opening 通道口
pass-through partition connector 穿通式连接条
pass-through security 转换抵押
pass-through window 传递窗口
pass trestle 渡槽支架
pass unimpeded 畅通
pass valve 直通阀
pass waterway 通过水道
pass without examination 免验放行
password 口令
pass work on to next shift 交班
pass year 上一年
past and clear 让清
pasta pectini 果胶糊

pastas 泥膏剂;敞廊;细长内廊(古希腊)
past due 过期
paste 净浆浆体水泥;浆(糊);基墨;糊状物;糊(剂);厚漆;软膏;漆膏;涂胶;砂浆
paste aggregate bond 浆体集料黏结;水泥浆集料结合力;水泥浆骨料结合力
paste aggregate interface 浆体集料界面
pasteboard 多层纸板;硬纸板;厚纸板;纸板;胶纸板;胶合纸板
paste brush 浆糊刷;裱糊刷
paste carburizing 糊状渗碳
paste cathode 涂糊阴极
paste coating 涂刮;色浆涂布
paste colo(u)r 浆色
paste colo(u)r concentrate 色浆
paste content 水泥浆的含量
paste content of concrete 混凝土中水泥含量
paste cutting compound 糊状润滑剂
pasted 浆糊的;胶的;膏的
pasted-end sack 粘贴纸袋
paste diffusion method 糊剂扩散法(木材防腐处理)
paste dipping 糊状树脂浸渍
pasted plate 糊制蓄电池极板;涂浆型极板
paste dryer 灰浆干燥剂;漆干剂;燥漆;膏状干料;膏状催干剂;灰浆干燥器
paste dye 浆状染料
paste expansion 水泥浆体膨胀
paste explosive 糊状炸药
paste extruder 糊膏挤压机
paste filler 腻子;膏状填料;膏状填孔剂;膏状体质颜料;糊膏灌装机;嵌缝膏
paste flux 浆状助熔剂
paste for hanging 裱贴浆糊
paste forming property (of lime) (石灰的)成浆性
paste form of dyes 浆状染料
paste for paper hanging 墙纸粘贴剂;墙纸浆糊
paste for wall paper hanging 墙纸粘贴剂
paste gold 金膏
paste in Japan 日本漆浆
paste in oil 油性色浆
paste-in-oil basic lead carbonate 油糊碱性碳酸铅
paste-in-oil white lead 油糊白铅
paste jewel 人造宝石
pastel 菘蓝染料;淡而柔和的色彩;彩色粉画;彩色笔;彩粉画;粉笔画
pastel book 彩色画簿
pastel chalk 彩色粉笔
pastel colo(u)r 中间色;粉彩色;淡色
pastel-colo(u)red concrete 彩色混凝土
pastel drawing 蜡笔画
pastel fixative 彩色粉画固着剂
paste-like 浆糊状
paste lining gravity dam 浆砌石重力坝
pastellist 粉画家
pastel painting 彩色油漆;彩色图画
pastel paper 粉彩纸
pastel pencil 彩色画笔
pastel shade 彩色阴影;清淡优美的色调;浅色相;轻淡色调
pastel tone 柔和的色调
paste lubricant 糊状润滑剂
paste material 糊状材料
paste matrix 浆糊基体;已硬化(的)水泥浆体;水泥石
paste metal interface 水泥浆金属界面
paste mill 磨浆机
paste mixer 调浆机;调糊机
paste mixing equipment 拌浆机
paste mo(u)ld 衬碳模
paste mo(u)lding 糊塑法
paste mo(u)ld machine 衬壁吹泡机
paste mo(u)ld ware 衬碳模成型的制品
paste-on-paste 泥糊堆雕;堆花
paste paint (in oil) 厚漆;漆膏;糊状油漆
paste pigment 浆状颜料
paste-plate accumulator 涂浆式极板铅蓄电池
paste polish 糊状清漆;虫胶清漆
paste polymer 浆状聚合物;糊状聚合物
paste-pot 浆糊桶
paste PVC resin 糊状聚氯乙烯脂
paster 涂胶纸;涂胶机;贴笺纸
paste red lead 红铅油膏
paste residue 浆糊余渣

paste resin 膏状树脂;糊状树脂;软树脂;树脂浆
paste roller 磨浆机
paste shrinkage 水泥浆收缩
paste soap 膏状皂
paste solder 糊状钎焊料
paste spot 浆糊斑点
paste spread 涂胶
paste table 浆糊桌
paste-type paint remover 糊状去漆剂
Pasteur-Chamberland filter 巴斯德—张伯兰滤器
Pasteur effect 巴斯德效应
Pasteur filter 陶瓷过滤器;巴斯德滤器
Pasteur flask 巴斯德烧瓶
pasteurization 巴氏消毒法;巴斯灭菌(消毒)法;低热消毒法;低热灭菌;巴斯德消毒法;巴斯德灭菌法;巴氏杀菌
pasteurize 巴氏杀菌
pasteurized soil 消毒土壤
Pasteurizer 巴斯德消毒器;巴氏灭菌器
pasteurizing varnish 低温杀菌清漆
Pasteur's method 巴斯德法
Pasteur's salt solution 巴斯德盐溶液
pasteur's theory 巴斯德学说
paste volume 水泥浆含量(混凝土)
paste volume of concrete 混凝土含浆量
paste white lead 糊状白铅
paste wood filler 木材腻子;木材填缝膏
pastfixation development 定影后显影
past frontal fog 锋后雾
pastiche 庸俗装饰;不合适装饰
pasties 乳饰
pastil(le) 软锭剂
pasting 粘贴;胶合;贴板;裱涂;裱糊
pasting and applying to wall 刷浆并铺贴在墙上
pasting brush 裱糊刷
pasting machine 抹灰机;裱糊机
pasting of core 砂芯胶合
pasting of starch 淀粉的糊化
pasting table 裱糊桌
pastometer 巴氏消毒温计
pastomiter 巴氏消定温计
pastophorium 古教堂中至圣所两侧的房间
pastophory 教堂圣atre器室;教堂圣器室
pastoral area 牧区
pastoral farming 畜牧业
pastoral industry 牧区畜牧业
past pointing test 过指试验
past precipitation 过去降水量;前期降水(量)
pastry 面色铺
pastry oven 多层点火熔炉
past service life 已超过服务年限
pasturage 畜牧业;牧场;放牧
pasture 牧草;放牧地
pasture animal husbandry 草原畜牧业
pasture ground 牧场
pasture husbandry 放牧业
pasture land 草场;牧场
pasture management 牧地管理;牧场管理
pasture of sea 海洋牧场
pasture plant 牧草
pasture pollution 草场污染
pasture productivity 草原生产能力
pasture renovation 牧场更新
pasture ripper 草地松土器
pasture spelling 草地休闲
pasture unit 牧地单位
pasture with foragegrass 矮草型放牧
pasturing 放牧
pasturing area 牧区
past weather 过去天气;过去的天气
pasty 膏状的;浆糊状的;糊状
pasty lubricant 糊状润滑
pasty mass 糊状物质
pasty stock 浆料;糊状料
pat 水泥饼;试饼;扁块
patagium 飞膜
patana 山地草场;山坡草地
patand 柱底梁;压条;地梁;地槛;底木;柱脚
pataphysics 超然科学
Patapsco formation 帕塔普斯岩层
patch 修补搪料;小片物;小片水面;小块浮冰;小块地;流冰原;片滩;塞片;插入码;浮冰区;补丁(片);斑纹;修补
patch area 插入区

patch bay 配电盘;接线架;配线盘;插线架;插头安装板
patch block 贴砖
patch board 接线架;接线板;接线盒;接插板;转插板;配电盘;转接插件;配线板;插头板
patch board verifier 接线板检验器
patch bolt 埋头螺栓;打补钉用螺栓;补件螺栓
patch burning 局部用火
patch canker 斑溃疡
patch clamp 修管夹箍
patch code 插入码
patch compound 修补剂
patch concrete 修补混凝土
patch cord 连接电缆;插入线;软线(配电盘的);接插线
patch diagram 方框图
patch emulsion 修补乳胶
patcher 修饰工;修补工
patches of alluvium 小块冲积地;冲积地段
patch ice 块冰
patchiness 斑点状分布性
patching 修炉;炉衬修补;临时性接线;贴补(集装箱等);搪衬;砂型修补;封补破洞;补坑;修补
patching a damaged vessel against flooding 堵漏【船】
patching approach 补丁方法
patching blanket 封舱毯子
patching block 衬砖
patch(ing) board 转接插件;转接板;接线盘;接线板
patching cement 补缀粘胶
patching composition 修补用组合物
patching compound 修补混合物;修补配合料
patching concrete 修补混凝土
patching curve 插入线;插入曲线
patching emulsion 修补乳胶;修补乳化液
patching end panels 端板贴补
patching flat panel roof 顶板贴补(集装箱)
patching gun 修炉衬喷枪
patching hole 修补洞;补孔
patching machine 修补机;喷浆机
patching material 修补材料;修补料
patching mix(ture) 修补配合料;修补混合料;修炉浆料
patching mortar 修补用砂浆;填补砂浆
patching operation 修补工作;修补工程
patching panel 接线盘;编排板
patching plank 封舱木板
patching plant 修补工厂;修补设备
patching plaster 修补灰浆
patching ratio 补坑率
patching side panels 侧板贴补(集装箱)
patching tap 修理用丝锥
patching technique 修补技术
patching unit 修补元件;修补单元
patching up 垫版
patching work 补缀器;修补工作
patch kettle 修补用小锅炉
patchlike pattern 斑状模式
patchline base 补充基线
patch load 加载
patch method 分块法
patch mortar 修补砂浆
patch of crizzle 表面微裂纹
patch of visible sky 小块可见的天空
patch panel 接线板;接插板;配线盘;插头板;配电盘
patch planting 孔状植树;片状栽植
patch plaster 修补石膏;修补灰浆
patch plate 修补板料;接合板
patch-plug 接插头;转接插头
patch pot 补坑用(沥青)桶
patch reef 礁坪;补丁礁
patch reef trap 补丁礁圈闭
patch repair 路面坑槽修补
patch routine 插入程序
patch snow cover 斑点状积雪
patch test 斑贴试验;斑试验;斑片试验
patch thermocouple 接触热电偶
patch tile 盖砖;挡块砖
patch-typed auto-thermal thermophilic aerobic digestion process 批式自热高温好氧消化工艺
patch weld 补孔焊缝;补焊
patch work 修补工作;修补工程;零星修补工作;混杂物;拼凑物;修补作业;拼补织物;补花
patchwork print 补缀型印花样

patchwork repair 零星修补
patchwork specialist 修补工作专家
patchwork survey 小地区填补测量
patchy 不调和的;有斑纹的;满是补丁的
patchy adhesion 片状粘连
patchy continuity 可燃物的杂凑连续性
patchy corrosion 成片锈蚀
patchy runway 已修整的跑道
patchy shadow 絮状阴影
pate 浆状泥料
pate dure 高温瓷
pate frit 熔块瓷
paten 薄金属圆盘
patency 开放;出现期
patent 专利品;专利特许证;专利;批准专利
patentability 专利性
patent abridgement 专利说明书节录本
patent abstract 专利文摘
patent abstracts bibliography 专利文摘目录
patent abstract series 专利文摘丛刊
patent agent 专利代理人
patent agreement 专利协议
patent anchor 无杆锚
patent and trademark notices 专利和商标公告
patent and trademark office 专利局
patent applicant 专利申请人
patent applicant service 专利申请人服务
patent application 专利申请(书)
patent application service 专利申请服务
patent application specification 专利申请说明书
patent article 专利品
patent attorney 专利律师;专利代理人
patent axe 专利斧
patent Axminster carpet 阿克斯明斯特专利地毯
patent-back carpet 特制底地毯;专利背衬地毯
patent binder clip 扎筋夹
patent block 通用滑车
patent blue 专利蓝
patent boards 专利墙板;专利板材;专利建筑板材;特制建筑板材
patent bulletin 专利公报
patent bush hammer 石面修琢锤;薄刃石锤;薄刷斧
patent certificate 专利证(书)
patent chain hoist 神仙葫芦
patent claim 专利申请范围;专利权限
patent claw 薄爪凿
patent cloth 蜡布
patent coated paper 特制光纸;特别光纸
patent collection 专利专藏
patent concordance 专利对照
patent concordance index 专利对照索引
patent cooperation treaty 专利权合作条约
patent defect 表面缺陷;明显缺陷
patent documents 专利文献;专利文件
patent documentation 专利文献
patent door 定型门
patent drawing 专利图样;专有图样;专有图案
patent drier in paste 燥漆
patent dryer 燥漆;漆头;干漆
patent duct forceps 未闭导管钳
patented claim 专利申请权限;专利申请
patented floor 专利地板
patented invention 专利发明
patented steel wire 铅淬火钢丝
patented wire 铅淬火钢丝;铅淬钢丝
patentee 专利权所有人;专利权持有者;专利持有人;专利持有者;专利持有人;取得专利权者
patent expenses 专利费用
patent family service 相关专利优先权服务
patent fee 专利费
patent file 专利资料档
patent floor 专利地板
patent flour 上等面粉
patent for invention 发明专利权
patent fuel 专利燃料
patent-fuel plant 专利燃料厂
patent glass 平板玻璃
patent glazing 干镶玻璃窗(法);无油灰缝镶装玻璃;专利装玻璃配件;无油灰镶装玻璃法;专用装玻璃法
patent glazing bar 上等玻璃条;专利的无油灰窗玻璃隔条;专利的玻璃装配隔条
patent glazing roof 专利的无油灰窗玻璃屋顶;专利玻璃装配的屋顶

patent granted 批准的专利
patent green 专利绿
patent hammer 面石修饰锤;石面修琢锤;薄刷斧
patent-hammered 面石修饰的
patent holder 专利权持有者;专利权持有人;专利持有者;专利持有人
patent index 专利索引
patent information 专利情报
patent information activities 专利情报活动
patent information service 专利情报服务
patent infringement 专利侵权;侵犯专利权;侵犯专利
patenting furnace (进行韧化处理的)加热炉;铅淬火炉
patentizing 钢丝韧化处理
patentizing furnace 铅淬火炉
patent journal 专利杂志
patent knotting 补疤液
patent law 专利权法;专利法
patent leather 黑漆皮;漆皮;漆革
patent library 专利图书馆
patent license 专利(权)许可证;专利特许使用权
patent license agreement 专利许可证协定
patent licensing 专利许可
patent literature 专利文献
patent log 拖曳式计程仪(船尾舵);拖曳式航速计
patent marking 专利权的标示;专利标记
patent material 专利材料
patent notices 专利通告事项
patent number concordance 专利号对照索引
patent of addition 增补专利
patent office 专利局
patent of invention 发明专利
patentor 专利权授与人;专利权的授与者
patent pending 专利有效期
patent period 出现期
patent pick 专利鹤嘴锄;专利手镐
patent plaster 特制熟石膏;特制粉饰;干硬性砂浆;干硬性灰浆;特制灰泥
patent plate 专利板料;专利平板玻璃
patent pool 共享专利企业;专利共享;专利共享企业
patent-pool agreement 专利共享协定
patent pooling 专利共同使用制
patent protection 专利保护
patent protective 特许护板;保护板
patent protector 专利护板;特许止漏器
patent report 专利报告
patent retrieval 专利检索
patent right 专有权;专利权;特许权
patent rolls 专利登记簿
patent roof glazing 无油灰屋顶玻璃
patent roofing glazier 特种屋面玻璃工
patent room 专利室
patent royalty 专利权使用费
patent runs out 专利期限已满
patents act 专利法
patents allocation index 专利号码分类索引
patents coverage 专利收录范围
patent slip 缆车滑道;升船滑道;船台滑道;船排
patent sounding machine 手摇测深仪
patent specification file 专利资料档
patent specifications 专利说明书
patent stone 人造石
patent stone floor covering 铸石地面饰面
patent stone shop 人造石工场
patent stone skin 人造石表皮(层)
patent stone stair(case) 人造石楼梯
patent stone tile 人造石砖瓦;铸石版面砖
patent stone tile floor cover(ing) 人造石地面砖
patent stone waterproofer 专利的石材防水剂
patent stone work 精巧的石材工作;人造石工作
patent system 专利制
patent technology 专利技术
patent yellow 专利黄
patera 插座(接线盒);圆盘(花)饰;[复]paterae
paterage 圆花饰
pateraite 黑钼钴矿
Patera process 帕特拉法
pateriform 浅碟形的
paternity test 试验
paternoite 水硼镁石
paternoster 链斗式升降机;串珠状线脚;念珠;串盘式输送机(一般用于运送小物件)
paternoster bucket elevator 多斗提升机

paternoster conduit 串珠状通道
paternoster elevator 链斗式提升机;链斗式提升机;链斗式升料机;斗槽式提升机
paternosteric migmatite 串珠状混合岩
paternoster lakes 联珠湖;串珠湖
paternoster lift 链斗式提升机;链斗式升降机
Paternoster pump 佩特诺斯特泵;链斗式提水机;链斗式疏浚机;链斗式泵
paternoster structure 串珠状构造
pate-sur-pate 泥浆堆花浮雕
pate tendre 低温瓷(器)
path 小路;小道;线圈分支;轨线;路径;途径;通路;道路
path analysis 路径分析
pathbreaking 探索性的
path-by 旁通管
path clearance 路径空隙
path coefficient 路径系数;径联系数;道路系数
path combination 流程组合
path component 道路连通区
path compression 通路压缩
path control 路由控制;路径控制;通路控制
path control method 轨迹控制法
path curve 轨迹线;轨线
path delay 路径延迟
path diagram 路径直方图
path difference (直达声与反射声的)行程差;路径差;程差
path equation 道路方程
pathergy 过敏反应性
path expression 路径表达式
pathfinder 领航机;航迹指示器;维修指南;探索者;导引装置
pathfinder element 指引元素;探途元素;导向元素
pathfinder index 探途指标
path-finding 路由寻找;路由选择;领航
pathfinding apparatus 引导装置
path-gain factor 通路增益系数;道路增益系数
path generation method 通路形成法
path-goal theory of leadership 途径—目标领导理论
path gravel 砾石路;走路石子
path information unit 路由信息单元;路径信息单元
path information unit programming 路由信息单元程序设计
path in woodland 林间小道
path layer network 通道层网络
path length 轨迹长度;路径长度;波程长度;通路长度
path-length antenna 行程天线
path length correction 波程长度改正
path-length of binary tree 二叉树的通路长度
pathless switch 无通路开关
path line 轨迹线;流线线;径迹;迹线
path loss 路径损耗
path matrix 路径矩阵
path model 路径模型
pathname 路径名
path of action 啮合轨迹
path of a fluid element 流体元迹线
path of chain 链路
path of contact 啮合线(齿轮传动)
path of electrode 运条途径
path of energy transformation 能量转换途经
path of filtration 渗径
path of flow line 流线轨迹
path of haul 运料路线
path of infiltration 渗透途径
path of insertion 就位道
path of integration 积分路径
path of jet 射流路径;射流航迹;射流轨迹
path of light 光斑点
path of loading 加载路径
path of particle 质点轨迹
path of percolation 渗透线;渗透途径;渗透路径;渗滤路径;渗流线路;渗流途径;渗流路径;渗径
path of plume 烟羽路径
path of ray 光路
path of seepage 渗透途径;渗透路线;渗径
path of seismic waves 地震波路径;地震波途径
path of sliding 滑动路径
path of storm center 暴雨中心路径
path of stressing force 应力的行径

path of stretching force 拉伸力的行径
path of tensioning force 拉伸力的行径
path of transmission 传播途径
path of travel 通路;通道;走道
path of typhoon 台风路径
pathogen 病原物;病原
pathogen contaminantion of water body 水体病原体污染
pathogen-contaminated sewage 含病原菌污水
pathogen contamination 病菌污染
pathogenesis 致病原因
pathogenetic bacteria 病原细菌
pathogenetic fungi 病原菌
pathogenetic organism 病原有机体
pathogenic bacteria 致病菌
pathogenic cocci 病原性球菌
pathogenic environmental factor 致病环境因素
pathogenic factors 发病因素
pathogenic fungus 病原霉菌
pathogenicity 致病原因;致病性;病原性;病态
pathogenic microorganism 致病微生物;病原微生物
pathogenic occupational factor 致病职业因素
pathogenic organism 致病生物体;病原生物
pathogenic pollution sources 病原体污染源
pathogenic solid waste 含病菌病毒固体废物
pathogen pollution of water body 水体病原体污染
pathogens 致病菌;病原体
pathologic(al) environmental hydrogeology 病理环境水文地质学
path on departure 离港誓言
path overhead 通道开销
path plotting 路径测绘
path problem 路程问题
path recorder 航迹记录器【船】
path roller 筑路路碾
path search 路线检索
path sensitization 路径敏化;通路敏化
path sensitizing 通路敏化法
path sensitizing test generation 路径敏化测试产生法
paths of gamma photos 伽马光子路径
paths of neutrons 中子路径
path surround 小径围绕;周围小道
path-switched ring 通道倒换环
path termination point 通道终结点
path tile 铺小路的砖
path tracing method 通路追迹法
path tracking 轨道跟踪
pathway 小路;轨迹;径路;航迹【航空】;通道;施工便桥;弹道;便道;小道
pathway of a pollutant 污染物的路径
pathway of heat loss 散热途径
pathway of pollutants 污染物途径
pathways of pollutant discharge 污染物排放通道
path width 航迹宽度【航空】
pathwise 顺向的;顺的;顺道的
patience 椅突板
patient care area 病人护理区
patient division 病人分隔;病区
patient emergency measuring device 病人危急测量设备
patient quarters 病房
patient's bathroom 病人浴室
patient's garden 病人花园
patient's room 病人房间
patient unit 病人单人(居住)单元
patient vicinity 可以接近病号的区域
patin 柱脚;柱底座
patina 古翠;锈色;年久而产生的柔和色彩;木器等表面的光泽;荒漠结皮;铜锈;铜绿(锈);沙漠盐壳;风化变色
patina green 铜锈绿
patination 生绿锈
patinous chloride 氯化亚铂
patio 内院;庭院;天井
patio awning 庭院遮篷
patio block 凉台砌块;庭院砌块;便道砌块
patio door 庭院门;滑动式外门;庭院式窗
patio door fitting 庭院门零件;庭院门装配
patio door furniture 庭院门配件
patio door hardware item 庭院门上小五金件
patio home 带庭院住宅
patio house 内院式住宅

patio housing 天井式住宅
patio masonry wall 庭院圬工墙
patio process 混汞法
patio umbrella 庭院遮阳伞
patland 垫底横木(早期英国)
patobionts 林地动物
patocoles 林地常居动物
pat of cement 水泥试饼;水泥饼
pat of cement-water paste 水泥浆体试饼
patois 土语
patoxene 林地偶居动物
patriarchal cross 两横臂十字架;主教十字架
patrimonial sea 承袭海
patrimony 世袭财产
patrinene 败酱烯
patrix 阳模;上模
patrix and matrix 阳模和阴模;模具
patrol 巡线;巡护员
patrol area 巡逻区
patrol camera 巡逻用照相机
patrol car 巡逻车
patrol cleaning 巡回清扫;清洁巡逻
patrol desk 通信值班台
patrol grader 巡路平地机;养路(用)平地机;平路机
patrol hydrofoil missile ship 导弹巡逻水翼艇
patrol inspection 巡回检查
patrol kettle 移动式沥青釜(养路用)
patrolling 巡逻
patrolling section 巡回区
patrol maintenance 巡回养护;巡回养路【道】
patrolman 巡逻者;巡逻员;巡逻工;巡逻员;巡逻工;外勤人员
patrol-rescue boat 巡逻救助艇
patrol route 巡逻路线
patrol sloop 巡逻炮舰;护卫舰
patrol station 巡逻站
patrol survey 高速巡线
patrol system 巡回(养路)制
patrol telephone 巡线电话
patrol time 巡守时间
patrol torpedo boat 鱼雷快艇
patrol vessel 巡逻艇
patron 主顾
patronite 绿硫钒石;绿硫钒矿
patronite lode deposit 绿硫钒脉状矿床
patron operated ticket checker 验票机
pat stain test (鉴定沥青混凝土含油量的)饼块染迹试验
pattee 中间小两头大
patten 柱基;柱底座;墙脚;平板;柱脚
patten of grind mill 研磨机挡板
patter 木抹子;镘板
pattern 形式;样式;帧面;刮板;格局;款式;模子;模型;模式;花样;喷束直径;图样;图形;图像;图案;定型航线;方式
pattern adaptive writing 模式自适应写入
pattern allowance 模型余量;模型留量;模型公差
pattern analysis 模式分析
pattern and drawing 模型及图样
pattern arrangement 井网布置
pattern assembly 模组(熔模铸造用)
pattern board 模型底板
pattern bolting 系统锚固
pattern bond 花纹砌合;砌筑形式(砖墙或圬工);砌花墙
pattern brick masonry work 有装饰性的砖的砌砌(工);拼成图案的砖工
pattern bush 空心直浇口棒
pattern cam 模凸轮
pattern card 纹板
pattern chain 纹板链
pattern checking 模型检验
pattern class 图形类别;模式类别
pattern classification 模式分类
pattern classification of seepage movement 渗流运动形态分类
pattern classifier 模式分类器
pattern coating 木模涂层;木模漆
pattern colo(u)r 帧面色;木模涂色;图像色;图案色
pattern cope 上模
pattern correction 图形改正
pattern crack(ing) 网状裂缝;龟裂;规则裂缝;网状裂纹
pattern cutter 成型刀

pattern description 图形描述；模式描述
pattern description language 模式描述语言
pattern design 模型设计；花纹设计；图案设计
pattern development 按样板加工下料(钣金工)；放样
pattern die 型模；压型熔模
pattern discrimination 模式辨认
pattern displacement 光栅移位；光栅位移；光束位移；图像变位
pattern display 铺路块料；图形显示器
pattern distortion 影色畸变；光栅失真；图像失真；图像畸变
pattern distortion allowance 模型变形公差
pattern dowel 模型合销
pattern draft 起模斜度；拔模斜度
pattern draw(ing) 起模；下模；脱模
pattern drawing mechanism 起模机构
pattern draw travel 起模行程
pattern drilling 定样钻眼(法)
pattern drive 按一定井网驱动
pattern dwarf hedge 图案矮篱
patterned brick 画像石
patterned carpet 花纹地毯
patterned diffuser 图案散光罩
patterned finish 花纹饰面
patterned floor cover(ing) tile 花式地板面砖；拼成图案的地面覆盖材料
patterned floor(ing) (finish) tile 拼花地板砖
pattern edged plate 饰边盆景；花边玻璃
patterned glass 压花玻璃；花纹玻璃；图案玻璃
patterned grain 图案形纹理
patterned ground 花纹表层；网纹土；图案土；多边形土
patterned hole 图形孔眼
patterned leaf 花样叶
patterned lumber 型材
patterned sheet 压纹板；压花板
patterned veneer 装饰镶板；拼花镶面板
patterned wallpaper 花样壁纸
patterned wire(d) glass 嵌丝玻璃；花式嵌丝网玻璃
pattern efficiency of boreholes 钻孔布置效率
pattern embossing 压花
pattern enumeration 模型列举；模式列举
patterner 制模工人；木模工
pattern eucalypt 桉树分布类型
pattern evaluation 模型评定
pattern facility 图形设备
pattern filling 图案填充
pattern flood 面积注水
pattern flooding 按一定井网注水
pattern floor cover tile 花式地面砖
pattern for diffusion 扩散模式
pattern format 标准图形格式
pattern forming 模型加工
pattern for reference 参考样本
pattern frequency 标尺频率
pattern function 模型函数；模式函数；方向图函数
pattern generagor 测视图案信号发生器
pattern generation 模型形成；模式生成
pattern generator 图形发生器；图案发生器
pattern glass 拼花玻璃；装饰花纹玻璃
pattern-glazed 拼花玻璃装配
pattern glue 木模胶
pattern grading and cutting machine 放样剪样机
pattern grouping 一型多模铸造法
pattern half 半分模型
pattern handling statement 模式处理语句
pattern harmonization 散布面调整
pattern heating 局部屏蔽加热
pattern identification 模式辨认
pattern information 模式信息
pattern information processing 模式信息处理
pattern information processing system 模式信息处理系统
patterning 图形；图案结构；形成花纹；制作布线图；图像重叠；图案装饰；图案形成
pattern intensity 类型强度
pattern language 标准图形语言
pattern layout 钻孔布置方案；模型设计(图)；花纹设计；炮眼布置方案；图案设计
pattern lumber 制模板材
pattern maker 制模工人；模型工；翻砂工(人)；制模工
pattern-maker's contraction 制模收缩

pattern-maker's hammer 模型工用锤
pattern-maker's lathe 木工车床；模型工车床
pattern-maker's rule 模型工用尺；缩尺；制模尺；模型工缩尺
pattern-maker's saw 狭条锯；制模锯
pattern maker's shop 制模车间
pattern-maker's shrinkage 模型收缩公差
pattern-making 制模工作
pattern-making material 模型材料
pattern master 主模(型)；母模
pattern match 镶板拼装法；组装法；装配法
pattern matching 模型匹配；模式匹配；图形匹配
pattern matching operation 模式核对运算
pattern matching statement 模式核对语句
pattern metal 制模合金；模型金属
pattern mill 模型铣床
pattern miller 模型铣床
pattern milling machine 模型铣床
pattern mo(u)ld 压制图案模型；母模；图案塑膜
pattern move 模式移动
pattern movement mechanism 模板移动机构
pattern multiplication 方向图相乘
pattern name 图形名称
pattern noise 图像噪声
pattern number 模型数
pattern of active faults 活动断层类型
pattern of a round 炮眼排列方式
pattern of blasting 爆破模式
pattern of coming pressure 来压方式
pattern of consumption 消费方式
pattern of crustal movement 地壳运动形式
pattern of deformation 变形的模式；变形图
pattern of dip isogons 等倾斜形式
pattern of enterprise 企业结构
pattern of equipotentials 等位线图
pattern of events 现象特征
pattern of exposure 接触模式；曝露模式
pattern of flow 流型；流态
pattern of import and export merchandise 进出口商品结构
pattern of interest rate 利率模式
pattern of investment 投资结构
pattern of juxtaposition 并列式
pattern of light 光的图像
pattern of light and shade 光与阴影的模式；光影图形
pattern of load 荷载形式
pattern of loading 加荷的模式；荷载图(示)
pattern of metallogeny 成矿方式
pattern of mineral deposit dewatering 矿床疏干方式
pattern of need 需要构成
pattern of organization 组织类型
pattern of ownership 所有制形式；所有制结构
pattern of pollution 污染的模式
pattern of preferred orientation 优选方位形式
pattern of price 价格构成
pattern of production 生产结构
pattern of relay 接力方式
pattern of river network 河网形式；河网类型
pattern of salvage value 残值类型
pattern of scour 冲刷形态
pattern of seawater intrusion 海水入侵方式
pattern of sewage system 下水道系统形式
pattern of spacing 布井方式
pattern of starting production 投产方式
pattern of symbol 符号模式
pattern of temperature 温度的分布
pattern of tide 潮型
pattern of trade 贸易格局
pattern of variability 变化模式
pattern of vortex 涡线形式
pattern of vortex line 涡线涡线图谱
pattern of wells 钻孔布置；井孔分布
pattern of world trade 世界贸易模式
pattern operand 模式操作数
pattern operation editing sequence 模式操作编辑序列
pattern operator 模式操作字(符)
pattern paint 美术漆
pattern paper 服装图样纸
pattern path 花纹路
pattern paving 花样铺面
pattern plan 模型设计

pattern plaster 制模石膏
pattern plate 模板
pattern plate bolster 模板框
pattern plate insert 型板镶块
pattern plate mo(u)lding 模板造模
pattern primitive 模式基元
pattern printout 图形输出
pattern recognition 模式识别；模式辨认；图识别；图像识别
pattern recognition processing 模式识别处理
pattern recognition program(me) 模式识别程序
pattern recognition system 模式识别系统
pattern recognization 图形识别
pattern recognizer 模式识别器；模式识别机
pattern recongnition method 模式识别法
pattern reference point 图形参考点
pattern reinforcing rib 模型加强筋
pattern repeat cycle 模式重复循环；图像重复循环
pattern representation 图形表示法
pattern resin 制模树脂
pattern room 制模车间
pattern screw 起模螺钉
pattern scribing 组装划线法；装配划线法
pattern search 模型搜索；模式搜索
pattern search method 模型搜索法；模式搜索法
pattern selection 模型选择
pattern selector 测试图选择器
pattern-sensitive fault 特殊数据组合故障；特定模式故障
pattern shooting 组合爆炸；组合爆破；图样爆破
pattern shop 制模车间；模型工场；模型车间
pattern shrinkage 模型收缩公差
pattern shuttle 模板梭动机构
pattern size 图形大小
pattern smoothing 方向图平滑
patterns of exchange rate 汇率类型
patterns of foreign trade 外贸类型
patterns of seismic model analysis 地震模拟分析模式
pattern solid angle 方向图立体角
pattern spray 脱模液
pattern staining 暗斑；结构痕迹显现；粉末电沉积图案
pattern stock 木模毛坯料
pattern storage 模型库
pattern store 木模仓库；模型库
pattern string 模式串
pattern stripping 漏模；脱模
pattern stripping device 漏模装置
pattern swelling 模型膨胀
pattern taper 取模斜度；拔模斜度
pattern theory 模式理论
pattern tracer 描花纹机
pattern varnish 模型上漆
pattern velocity 图案速度
pattern veneer 装饰镶板
pattern wax 模蜡
pattern weaver 试样织工
pattern well spacing 按标准井网布置钻孔
pattern wheel 提花轮
pattern winding 模型绕组；标准绕组
pattern wire glass 嵌丝玻璃
pattern wood 模型用木材
Patterson projection 帕特孙投影(法)
Patterson synthesis 帕特孙综合法
pat test 试饼试验；试块试验；试饼法；饼形试件试验；饼块试验
Pattinson's white lead 帕廷森铅白
patucidisperse system 少量分散系统
paucilithionite 钾锂云母
paucity 贫乏；少许；少量；微量
paul 制轮爪；制动爪；卡子；倒齿；掣爪；防倒轮掣子
Paul-Bunnell test 保尔—邦内尔试验
Pauli exclusion principle 不相容原理
Pauli girder 鱼腹形支架；鱼腹形梁；泡利梁
paulin 篷布；舱盖布；防雨帆布；防水帆布
paulingite 方碱沸石；鲍林沸石
Pauling rule 鲍林定则
Pauling's exclusion principle 鲍林不相容原理
Pauling's rules 鲍林规律；鲍林规则
Pauli's anomalous moment term 泡利反常矩项
Pauli's equation 泡利方程
Pauli's exclusion principle 泡利不相容原理
Pauli's matrices 泡利矩阵

Pauli's rearrangement theorem 泡利重排定理
Pauli's spin space 泡利自旋空间
Pauli's spin susceptibility 泡利自旋磁化率
Pauli-type electrodialyzer 泡利式电渗析器
paulmooerite 勃砷铅石
Paulownia 桐树
Paulownia imperials 泡桐
Paul trap 保罗捕获
paumelle hinge 鲍物勒铰链;活动式马鞍合页(又称 H 形合页);可折铰链;单结合点门铰链
paunch 防摩席
paunch mat 垫席
pauperization 贫困化
pauper labor 贫困劳工
pause button 暂停按钮
pause control 停顿控制
pause days 休产日
pause input/output 暂停输入/输出
pause instruction 暂停指令
pause line 停留线
pause mode 暂停工作状态
pause period 停歇时间;停机时间
pause sequence 中止序列
pause status 暂停状态
pause switch 间歇开关
pavage 铺设;铺筑;铺路税
pave 铺路
paved apron 铺面停机坪;铺墙裙板;有铺面的码头前沿;铺面码头前沿;有路面的码头前沿地带
paved area 铺筑过的场地
paved bed 铺石花坛
paved channel 铺砌渠道;铺面渠道
paved concrete track 整体灌注式轨道;混凝土联合铺道机
paved crossing 铺砌式道口;道口铺面
paved ditch 铺砌沟;铺砌街沟
paved drain 铺砌的泄水沟
paved feedlot 硬地面肥育地;水泥地面肥育场
paved floor 块料地面;铺砌桥面;铺砌地板
paved flooring 铺砌地面
paved footway 铺面人行道
paved ford 铺砌过水路面
paved garden 铺地园
paved gutter 铺砌式道沟;铺砌马路边沟;铺砌街沟;铺砌沟;铺砌道路边沟
paved inlet 铺砌进水口
paved intersection 铺砌式交叉
paved invert 有衬砌的倒拱底板;密铺沥青材料的仰拱;铺砌仰拱;衬砌底拱
paved-invert corrugated pipe 底拱沥青铺砌的波纹管
paved "leak-off" 铺砌的泄水沟
paved lot 水泥地面肥育圈
paved median 中间铺砌部分
paved play area 铺筑的游戏场地
paved prism 行道灯上的棱形玻璃镜
paved road 铺面路;有铺面的路
paved runway 混凝土铺面跑道;铺砌的跑道;铺面跑道
paved shoulder 铺砌式路肩;铺砌的路肩
paved shoulder marking 路肩铺装标示
paved walk 铺筑的走道
paved with cobblestone 卵石铺砌
paved with tiles 瓷砖铺面
pave light 从穹形玻璃屋顶进入的光线
Paveloveskee formula 巴甫洛夫斯基公式
Paveloveskee function 巴甫洛夫斯基函数
pavement 路面;人行道;铺砌层;铺面;铺盖;便道
pavement base 路面基层;铺面基层
pavement base plate 路面底板(接缝处)
pavement bed 路基
pavement behaviou(u)r 路面作用;路面状态;路面状况
pavement border 铺装路面界限
pavement breaker 路面破碎机;掘路机
pavement brick 铺面砖;路面砖
pavement cement 路面水泥
pavement classification number 道面分级号码(机场)
pavement component 路面组成部分
pavement concrete 路面混凝土;铺面混凝土
pavement construction 路面施工
pavement construction survey 路面施工测量
pavement core drilling machine 道路岩芯钻机

pavement cross-section 路面断面
pavement deflection 路面变位;路面弯沉;路面挠度
pavement depression 路面沉陷
pavement design 路面设计
pavement design method of the Federal Aviation Administration 美国联邦航空局道面设计法
pavement distress 路面损坏;路面破坏;路面恶化;路面病害
pavement durability 路面耐久性
pavement edge curb 路面缘石;路边缘石;平式缘石
pavement edge effective superelevation 路边有效超高值;路边实际超高值
pavement edge line 路面边线;车道外侧线
pavement element 路面组成单元
pavement ends sign 路面铺装终止标志
pavement entrance 路面入口
pavement evaluation 路面评价;路面评定
pavement evenness 路面平整度
pavement failure 路面破坏
pavement flag 路面石板
pavement foundation 路面基础
pavement geometry 路面几何特性
pavement glass 铺面玻璃
pavement grade 路面坡度
pavement grip 路面防滑;抗滑性
pavement improvement 路面改建
pavement in rows 横排式铺砌;条砌
pavement invert 涂沥青的波状铁管
pavement layer 路面(结构)层
pavement laying 路面铺设(材料)
pavement life 路面寿命
pavement-life study 路面使用寿命研究
pavement light 沿人行道地下室采光窗;路面灯光;路面采光窗;路面采光;地下室顶光;地窖顶窗;窗井顶部采光窗
pavement-light glass block 路面采光砖
pavement line 有路面的路线
pavement maintenance 路面维修
pavement management 路面管理
pavement management system 路面管理体系
pavement marker 路面标钉
pavement marking 路面标线;路面标记;交通标志
pavement marking machine 路面划线机;路面标线机
pavement material 路面材料
pavement milling machine 路面铣削机;路面铣切机;刨路机
pavement mixture 路面混合料
pavement of asphalt 沥青路面;石油沥青路面
pavement of clinker 缸砖路面
pavement of cobble stones 大卵石路面
pavement of concrete block 混凝土(方)块路面
pavement of flagstones 石板铺面
pavement of highway 公路路面
pavement of riprap 乱石铺面;乱石路面;乱石护面
pavement of road 道路路面
pavement of stone blocks 块石路面;石块路面
pavement of the wharf slope 斜坡码头护岸
pavement overlay 路面板(人行道)
pavement paint striper 路面油漆划线机
pavement patching materials (for concrete pavement)(水泥混凝土路面的)路面补强材料
pavement performance 路面性能;路面使用能力
pavement performance survey 路面性能调查研究
pavement pouring 浇灌路面
pavement prism 路面采光砖;(地下室采光用)折光物体;人行道地下室采光棱镜;采光玻璃块
pavement proper 路面主体
pavement pumping 路面(板)抽吸作用;路面(板)唧泥作用;路面喷浆;路面抽水作用
pavement quality 路面质量
pavement quality concrete 路面品位混凝土(PQ 混凝土)
pavement ram 铺路夯
pavement recapping 路面翻修
pavement reflectivity 路面反射性
pavement rehabilitation 路面修复;路面维修
pavement removal 路面剥除
pavement restoration 路面修复
pavement retreading 路面复拌
pavement rideability 路面通行能力;路面通行性;路面行车质量
pavement roller 压路机
pavement roughness 路面粗糙度;道面粗糙度

(机场)
pavement sample 路面取样
pavement saw 路面切缝锯
pavement scarifier machine 耙路板;耙路机;翻路器
pavement sealer 路面封缝料
pavement serviceability 路面耐用性;路面服务能力
pavement service valve 井内的关断阀;(设在路边的)用户管阀
pavement settlement point 路面沉陷测试点
pavement sinkage 路面沉陷
pavement skid-resistance test 路面抗滑性试验
pavement slab 路面平板;路面板;铺面板;铺路板(材)
pavement slab connector 路面板接头
pavement slab pumping 路面(板)翻浆;路面(板)唧泥(作用)
pavement smoothness 路面光滑性;路面光滑度;路面平整度
pavement spalling 路面碎裂;路面剥落
pavement spreader 路面沥青喷洒车
pavement stability 路面稳定性
pavement stake 路面桩
pavement stiffness 路面刚度
pavement strengthening 路面加固;路面补强
pavement strip 路面标线
pavement stripping 路面车道划线
pavement structural evaluation 路面结构评价
pavement structure 基础铺砌层;路面结构;护面结构;铺面结构
pavement structure design 路面结构设计
pavement structure layer 路面结构层
pavement subgrade blanket 路面底层
pavement-subgrade interface 路面—路基交界面
pavement surface 路面层
pavement system 路面系统
pavement thickness 路面厚度
pavement trafficability 路面通行性;路面通行能力
pavement type island 路面式安全岛;平台安全岛;平式安全岛
pavement undersealing 路面基层处治
pavement vibrator 路面振动器
pavement wear 路面磨损;路面磨耗
pavement width 路面宽度
pavement width transition 路面宽度变化过渡区间
pavement wood block 木块铺面;木块路面
pavemill 路面铣削机;路面铣切机;刨路机
pave prism 行道灯上的菱形玻璃
paver 路面砖;铺砌工(人);铺面材料;铺路机;无釉铺地砖
paver beam 混凝土摊铺机的梁
paver boom 桁梁;摊铺机桁架(混凝土)
paver bucket 摊铺机卸料斗
paver-finisher (路面的)摊铺整修机
paver-finisher without hopper 无摊铺斗的路面整铺机
paver frame 摊铺机的机架
paver lath 普通机床
paver-mixer 铺摊拌和机
paver's hammer 铺路工的整面锤
paver's level 铺砌工用的水平仪
paver tile 铺路面砖
pavestone 铺路石
pave with tiles 瓷砖铺面
pavier 铺料机
pavilion 阁;穹形物;亭子;底侧翻光面
pavilion bridge 亭桥
pavilion distillator 蒸馏水器
pavilion-like 帐幕般的;亭子般的
pavilion-like wood pagoda 楼阁式木塔
pavilion of the Fragrance of Buddha 颐和园佛香阁(北京)
pavilion on terrace 榭
pavilion roof 亭式屋顶;四坡屋顶;攒尖式屋顶;攒尖顶;棱锥形(方攒尖)屋顶
pavilion school 帐幕式学校
pavilion system 成组分隔式(医院)
pavilion type 分隔式
pavilion-type classroom 帐幕式教室
pavilion-type structure 帐幕式结构
pavillion roof 攒尖顶【建】
pavimentum (用面砖、大理石、石头和硬质砖等材料与水泥铺筑而成的)古典式人行道;夯实混凝土路面(古罗马)
paving 铺砌工程;铺砌;铺面;铺路面

paving aggregate 铺路集料;铺路骨料
paving asphalt 铺路沥青;铺路用地沥青
paving asphalt cement 铺路用沥青水泥;铺地沥青水泥
paving bank 有护面的河岸;砌面河岸
paving beetle 打夯机;夯具;铺路夯具
paving block 铺块;路面条石;铺面块体;铺路块料;铺地块材;铺砌块
paving block of rubble breakwater 堆石堤护面层
paving breaker 路面切割机;路面破碎机;路面破坏机
paving breaker drill 破碎路面钻;筑路用凿岩机;路面钻孔机;掘路钻
paving brick 路面砖;铺面砖;铺路砖;铺地(面)砖;地面砖
paving brick clay 铺地砖土
paving brick on edge 侧砖铺砌;侧砖铺面
paving brick on headers 丁砖铺地
paving cement 铺路水泥
paving chipper 路面破碎机;路面切割机
paving clinker 铺地熔渣
paving clinker brick 铺路熟料砖
paving cobble 铺路石
paving company 筑路公司
paving concrete 铺路混凝土
paving contractor 铺路承包人
paving course 铺砌层
paving equipment 路面机械
paving expansion joint 铺面伸胀缝;铺路伸缩缝
paving finisher 整面机;铺面修整机
paving flag 铺石板路;铺方块石;铺面块体;铺路石板;铺路薄片石
paving glass 铺路玻璃
paving in echelon 梯形铺砌
paving in lozenge form 菱形铺砌
paving in rows 横列铺砌
paving in setts 石块铺砌;小方石铺砌;石块铺面
paving in stone blocks 小方石铺砌;块石铺砌的路面
paving layer 铺装层
paving machine 铺砌机;铺路机
paving machinery 路面机械
paving material 铺面材料;铺路材料
paving mixer 混凝土摊铺拌和机;铺路拌和机;摊铺(料)拌和机
paving mixture 铺面混合料;铺路混合料
paving notch 铺砌的谷道
paving of dike face 坝面铺砌
paving of engineering bricks 工程砖铺砌
paving of natural setts 天然小方石铺路
paving pattern 铺砌图案
paving plant 铺筑设备;铺装路面设备;铺地植物
paving plant train 铺路设备列车
paving project 铺路计划;铺路规划
paving rammer 铺砌夯具;铺路夯锤
paving repair 路面修补
paving rock 护面块石
paving sand 铺路砂
paving sett 铺路小方石;铺路石块;铺面石板
paving sett making machine 生产铺路小方石的机器
paving sett tester 铺路小方石测试仪
paving slab 铺砌条石;铺面板材;铺路板(材)
paving sloping way 实体斜坡道
paving spread 摊铺;混凝土摊铺
paving spreader 混凝土摊铺机;摊铺机
paving stone 铺面石板;铺石;铺砌石;铺路石;铺路薄片石
paving stone block 砌路用石块
paving system 铺面系统
paving tamper 铺路夯
paving tile 镶面瓷砖;路面砖;铺面砖;铺面(面)砖;铺地(面)砖;地面砖
paving train 铺路机组;联合铺路机;铺路机动列车;铺路机械列车;拖带式联合铺面机组
paving unit 铺路块
paving vibrator 铺路振动器
paving with flags 石板铺砌;石板铺面
paving with mortar 砂浆铺砌
paving with pebbles 卵石铺砌;用卵石铺砌;卵石铺面
paving with setts 小方石铺路
paving with tiles 花砖铺面
paving wood 铺面木块

paving wood block 铺面木块
pavio(u)r 路面砖;铺砌工(人);铺面工人;铺砌砖;铺路机;铺路工;铺地砌块;无釉铺地砖;铺路材料
pavio(u)r brick 路面砖
pavio(u)r labo(u)r 铺路工
pavio(u)r's hammer 铺石锤;铺砌锤
pavio(u)r's level 铺路机用水准仪
pavit 冷敷沥青工艺
pavonazzo 彩条大理石
pavonite 块硫铋银矿
pawl 制轮爪;制动爪;棘爪;棘皮;卡子;(与棘轮相配合的)推爪;倒齿;掣子;掣爪;掣手
pawl and ratchet mechanism 棘爪和棘轮机构
pawl bed 掣手底盘
pawl casing 爪罩;棘爪罩
pawl feed(ing) 前进棘爪
pawl limb 掣手底盘
pawl operating spring 棘爪操作弹簧
pawl pin 爪销
pawl plate 制动爪板
pawl rack 止动器架;掣轮器齿板
pawl rim 掣手底盘
pawl ring 掣手底盘
pawl spring 爪簧;制动簧片
pawl type cooling bed 爪式冷床
pawl washer 爪式垫圈;带鼻防松垫圈
pawn 有顶盖的长廊;有顶盖的通道
pawnbroker 典当商
pawnshop 当铺
Pawsey stub 波西短线
paw wheel 棘轮
Paxboard 帕斯板(一种绝缘板)
Paxvboard and paxfelt 热绝缘板和毡
paxite 斜方砷铜矿
Paxton gutter (用来收集暖房或温室内外潮气的)帕斯顿沟槽
pay 薪水;薪金;薪俸;报酬
pay a bill 付账
payable 应付的
payable account 应付账款
payable after date 发票日后定期付款
payable area 可采区
payable at a definite time 定期付款
payable at a fixed time 定期付款
payable at destination 到达港支付
payable at fixed period after date 出票后定期付款
payable at fixed period after sight 见票后定期付款
payable a week after sight 见票后一星期付款
payable balance 可付余额
payable by instalment 分期支付
payable ground 可采矿床
payable in account 赊账付款
payable in exchange 可用汇票支付
payable on completion of discharging 卸完时支付
payable on receipt 收到后付款
payable on the installment 分期付款
payable period 应付期间
payable quantity 按量付款
payoble rest date 带薪休息日
payable to bearer 付给来人
payable to holder 付执票人
payable ton 计费吨
pay according to one's post 职务工资
pay according to the arranged price 照价付款
pay A corporation 付甲公司
pay after arrival 后付
pay after tax 扣除税款后支付额
pay agent 保险理赔代理人
pay a more effective role 起到更有效的作用
pay an indemnity 给予赔偿
pay a price 付出代价
pay a ransom 遣散费
pay a retaining of guaranteed wage 保留工资
pay a rope 放松绳索
payasat 缅甸塔
pay a seam 填塞缝隙
pay as may be paid 依约照付
pay as may be paid thereon 照原保险单赔偿
pay-as-you-earn 所得税预扣法;依照入息支付;预扣所得税;收支相等
pay-as-you-enter 上船后即付;结关后即付
pay-as-you-go 分期付款;费用发生拨款制;按程收费
pay-as-you-go account 每张支票登账需加付费用

的支票往来户
pay-as-you-go-plan 分期付款计划;按程收费计划
pay-as-you-see 投币式电视;收费电视;见货后即付款
pay-as-you-use 费用支用拨款制
pay-as-you-view 投币式电视;收费电视
pay attention to 注意
payawut 缅甸佛堂
payback 归还欠款;投资回收率;收回
payback loans within a set time 限期偿还贷款
payback method 偿付方法;回收期法;还本期评估法
payback of added investment 附加投资偿还年限
payback period 偿付期限;偿还期;偿还年限;收回期
payback period by investment difference between lines 差额投资偿还期
payback period method 回收年限法
payback period of investment 投资回收期;投资偿还年限
payback period of loan 借款偿还期;贷款偿还期
pay bed 产油层
pay before shipment 装船前付现;装船前付款
pay bill 工资单
pay by instalments 分批付款
pay by the piece 按件计付工资
pay card 工资卡片
pay cash 现金付款
pay cash in transit summarily 汇缴途中款
pay check 工资支票;薪水支票
pay chief administrative expenses 上缴上级管理费
pay compensation 退赔
pay date of bond interest 债券付息日
pay day 支付日;交割日;发薪日;发(放)工资日
pay differential 工资级差
pay-dirt 控方;控制土方;丰矿土;发横财;富矿砂;可采矿石;含矿泥砂
pay down 先支付部分贷款(分期付款购货);用现金支付;即时支付;预付款
pay duty 纳税
pay earnest money 付定钱
payee 验收人;抬头人;受款人;收款人;付款人
payee's account 受款人账户
pay envelope 工资袋
payer 付款人;支付者
payered pavement 成层路面
payer for honour 荣誉付款人
pay fine 缴罚款
pay first rule 先付规则;互保协会再赔
pay for 负担费用;补偿
pay for a loss 赔偿损失
pay for another 代付
pay forfeit money 缴罚款
pay for holidays 休假工资
pay for illness leave 病假工资
pay formation 有价值开采的矿层;填土石方法
pay for the hire of 付……租费
pay for the loss of cash or goods entrusted to one 赔账
pay for the removal of houses 房屋迁移费
pay freeze 工资冻结
pay full value 付足价值
pay grade 薪金等级
pay gravel 可采砂金;含矿砾
pay heavy price 付高价
pay horizon 产油层
pay in 解款入银行;解款;把款项解入
pay in advance 预付款
pay in book 付款簿
pay in cash 现金交付;现金支付
pay in cross check 转账支票支付
pay in full 一次付清;全额付清;付清
paying agent 付款银行
paying bank 付款行;付款银行
paying by check 支票支付
paying certificate 付款凭证
paying in advance 预付货款
paying-in slip 缴款通知单
paying list 付款清单
paying load 营运运输能力
paying-off 放线;开卷
paying out 出水口;付出;付出;支付
paying out machine 放线机
paying passengers per day 每天购票乘客人数
paying quotation 应付外汇汇价
paying remuneration according to standard out-

put 标准产量计酬
paying shell 填缝勺
paying thickness 值得开采的(矿层)厚度
paying unit 支付单位
paying well 富油井;经济上有利的井;高产油井;含量丰富的井
pay-in kind 支付实物;实物缴纳
pay-in slip 进账单;缴款通知单
pay into 解款入银行;解款
pay-in warrant 缴款书
pay item 付款项目
pay land taxes 付土地税
pay later 延期付款
pay library 收费图书馆
pay limit 最低可采极限
pay line 最小开挖线;设计开挖线;计价线
pay list 薪水账;工资名册;工资单;工资表
payload 载货重量;有用负荷;有效载荷;有效负荷;有效负荷;营运重量;营运运输能力;载重;工资负担;净载重量;收费载重;负荷量;付费重量
payload and apogee performance 有效载重与远地点的关系
payload bay 有效载重舱
payload build-up 仪表装配
payload capability 有效负载能力
payload capacity 有效负载(容)量;有效载能力;有效承载力;有效载重量;最大有效载荷;净载重能力
payload capsule 有效荷载舱
payload carrier 有效荷载载体
payload communications system 有效负载通信系统
payload deployment and retrieval system 有效荷载展开和恢复系统
payloader 运输装载机
payload fraction 有效载重量与车重之比
payload instrument 作为有效载荷的仪器
payload mass 有效负载质量
payload-mass ratio 有效负载质量比
payload of freight container 集装箱载重
payload of train 列车牵引净重
payload range 业载航程
payload ratio 收费载重率;有效载荷比
payload-release point 投物点
payload rope 承载索
payload sensor 有效荷载传感器
payload space 载货容量;载货容积
payload-to-vehicle weight ratio 有效荷载与车辆总重比;有效载重量与车船自重比
payload volume 有效负载体积
payload weight 有效荷载重量
payload weight distribution 有效载重量在车桥上的分配
payload-weight ratio 载重自重比
paymaster 工资结算员;发款员
payment 付款方法;支付的款项;付款(额);付款(额);付出;拨付
payment according to labo(u)r contributions 按劳动分红
payment according to work 按劳付酬
payment adjustment 支付调整
payment after delivery 交货后付款
payment after due date 到期后偿付
payment after inspection 验货后付款
payment after inspection clause 验付条款
payment after termination 终止后的付款
payment against delivery 交货时付款
payment against documents 凭单(证)付款
payment against documentray letter of credit 凭单付款信用证
payment against documents credit 凭单据付款信用证
payment against documents through collection 凭单托收付款
payment against draft credit 凭汇票付款信用证;凭汇票付款的信用证
payment against goods shipped on consignment 寄售付款
payment against presentation of shipping documents 凭运单付款
payment against provision 暂定金额支付
payment agreement 支付协议;支付协定
payment allocated-out 有偿调出
payment at maturity 到期日付款

payment at regular fixed time 定期交款
payment at sight 见票即付;即期付款
payment balance 支付差额
payment based on land shares 土地报酬
payment bond 付款担保
payment by bill 汇票付款
payment by cash 现款支付
payment by check 支票付款
payment by collection 托收付款方式
payment by draft 汇票支付;汇票付款
payment by hour 计日工资
payment by instalments 分期付款;分期摊付
payment by measurement 按方计费;采用计量支付
payment by results 最后付款;按效果付酬;按劳付酬
payment by the hour 计时工资
payment by transfer 拨付
payment cap 最大还款额
payment certain 肯定支付
payment condition 付款条件
payment contractor 支付合同
payment documents 支付凭证;支出单据
payment domiciled 指定地付款
payment draft credit 凭汇票付款的信用证
payment estimate 支付预算
payment for claims 索赔金支付
payment for honour 名义支付
payment for insurance 保险费支出
payment for labo(u)r 劳动报酬
payment for the use of state funds 资金占用费
payment forward 预付货款
payment guarantee 付款保证
payment in advance 预付款;预支工资;预付(款项);提前支付
payment in arrear 未付的尾数;拖欠款
payment in balance 收支不平衡
payment in cash 现款支付;现金付款;现金支付
payment incurred for out-of-duty casualty 路外人员伤亡垫款
payment incurred for uniform 制服垫款
payment in due course 依期付款;届时付款;期满兑付
payment in event of suspension 暂时停工情况下的支付
payment in foreign currencies 外币支付
payment in full 一次付清;全付;全部付讫;付讫
payment in installments 分期付款
payment in kind 实物支付
payment in product 交付产物
payment instruction in foreign exchange 外汇票证
payment instrument 支付凭证
payment in term 定期支付
payment line 计价线;最小开挖线;支付方式
payment of advance 预付款支付
payment of claims 理赔金
payment of compensation 补偿费
payment of contract 合同付款
payment of customs 关税的支付
payment of duties 交税
payment of duty 付税
payment of fees 缴纳费用;费用的缴纳
payment of honour 荣誉付款
payment of interest 付息
payment of oversea accounts 海外账款的支付
payment of retention money 保留金的支付
payment of shares 交付股金
payment of technology transfer 技术贸易支付方式
payment of the current premium 本期保险金的支付
payment on account 赊销;分期付款;分期偿还
payment on a receipt 凭收据付款
payment on arrival of shipping documents 见货运单据付款
payment on delivery 交货付款
payment on demand 要求付款;见票即付
payment on invoice 凭发票付款
payment on land shares 土地分红
payment on receipt credit 凭收条付款信用证
payment on shipment 交货付款
payment on statement 凭账单付款
payment on termination 终止时的支付
payment on terms 定期付款
payment order 付款委托书;付款通知(书);付款

通知单;付款命令
payment position 收支状况
payment prior to delivery 交货前付现;交货前付款
payment quantities 付款量
payment ratio 股息发放率
payment received 收讫
payment respite 延期付款
payments and receipts 线下收支
payments balance with surplus 收支顺差
payment schedule 付款日程表
payment side 付方
payments in arrears 拖欠分期款项
payments inbalance 收支不平衡
payments on long-term debt 长期债务还本付息
payments problems 收支困难
payments to contractor 对承包人的支付
payments to employer 对雇主的支付
payment term 付款期(限);支付条款;支付条件
payment to consultant 对咨询人的支付
payment to consulting engineer 对咨询工程师的支付
payment transaction application layer 支付交易应用层
payment under reserves 保留付款
payment upon arrival, cost, insurance, freight and war-risks payment upon arrival 到岸价格加战争险货到付款价;到岸价格加战险货到付款价
payment voucher 支出凭单;付款凭证
payment withheld 拒付款
Payne's process 潘恩木材防火法
pay no commission 不付佣金
pay-off 解聘;还清;清偿;船头偏离风向;偿还;偿债;偿清;付清;分配盈利;续料;送料;盈利;支付;报酬
payoff a score 付账
payoff escrow 清偿契据
payoff expected 支付期望的
payoff foreign debt ahead of schedule 提前全部付清外债
payoff function 支付函数;报偿函数
payoff gross investment 投资回收总额
payoff matrix 支付阵;支付矩阵;报偿矩阵
payoff method 支付方法
payoff method for capital investment 资本投资回收年数法;投资回收年数法
payoff one's creditors 偿清债务
payoff period 清偿期限(债务的);投资回收期;补偿期
payoff period method 返本期法
payoff reel 展卷机;开卷机
pay off the mortgage 归还抵押借款
pay on a piecework basis 按件计酬
pay on application 请求时付款
pay on arrival 到付
pay on delivery 交货时付款;货到付款
pay on due date 依期付款
pay one's dues 还账
pay one's score 付账
pay on receipt 收到货后付款
pay on return 返回后退付;返回后付款
pay on the spot 现付
pay order 支款凭证;支付凭证;支出凭证
pay-ore 富矿(石)
pay-out eye 导出眼
payout period 还付时期;投资返本年限法
payout period determination 确定偿还期
pay-out ratio 盈利率
pay-out the line 放线
payout time 投资回收期;投资返本年限法
pay packet 工资袋
pay pause 薪资冻结;薪金冻结;工资冻结
pay period 工资支付期;发工资期
pay profit 利润解缴
pay quantity 支付定额;结账工程量;付款工程量
pay rate per hour 小时工资率
pay ready money 付现钱;付现金
pay relationship 工资关系
pay rent 付租金
pay reparations 赔款
pay rock 含矿岩(浆)
payroll 在职人员名单;发薪簿;发资单;薪工;工资(单);工资表;工资单;发放工资额
payroll account 薪工账户;工薪账户
payroll accounting 薪工会计;工资核算

payroll analysis 薪工分析;工薪分析
payroll audit 薪工审计;工薪审计
payroll check 薪工支票;工资支票
payroll clearing account 薪工结算账户
payroll clerk 薪工结算员
payroll computer 计算机编制并保持工资记录
payroll control 薪工管理
payroll cost 薪工成本;工薪成本
payroll deduction plan 发放薪工时的扣款计划
payroll department 薪工管理部门
payroll department expenses 薪工管理部门费用
payroll distribution 薪工分配
payroll distribution book 薪工分配簿
payroll distribution journal 工资分配日记账
pay roller 受薪者;受津贴者;依靠工资为生者
payroll forms 工资表格
payroll fund 薪工周转金
payroll inventory control 工资清单核对
payroll journal 工薪日记账
payroll payment 发薪
payroll records 工资核算
payroll register 薪工登记簿;工薪表
payroll-related basis of allocation 按工薪分摊费用
payroll sheet 薪工清单
payroll slip 薪工通知单
payroll system 薪工制度;工资管理系统
payroll tax 薪工税;工资税;工薪税;就业税
paysage 风景画;乡村景色
pay sand 含矿砂;含石油砂;油砂层;生产层;产油层;富含油砂层
pay scale 薪资标准;工资标准;薪工标准
pay schedule 工资表
pay school 私立学校
pay-section 有利断面(隧道)
pay senior administrative agency expenses 上缴上级支出
pay-sheet 工资表;发薪簿;工资单
pay slip 支付传票
pay stain test 试饼染迹试验(鉴定沥青混凝土含油量)
pay standard 供给标准
pay station 公用电话间;公用自动收费电话亭
pay station board 公用电话台
pay station set 公用电话机
pay streak 产油层;富线;富矿线;富矿脉
pay sundry fees on late 迟交运杂费
pay system of using state-owned land 国有土地有偿使用制度
pay tax 纳税;完税
pay telephone 自动收费公用电话
pay-television 收费电视
pay terms 合同条款
pay the interest accrued 支付所生利息
pay the reckoning 付账;付清账
pay thickness 富矿层
pay through the nose 付高价;被敲竹杠
pay to bearer 付持票人
pay toilet 收费厕所
pay to order 凭指示付给
pay to self 向签票人本人支付
pay to the order of A bank 付给甲银行
pay tribute 交纳贡势
pay up 全部清偿;全部付清;按时付清;付清
pay up a debt 还债
pay warrant 支付命令
pay zone 产油层
Pb-Ag ore 铅银矿石
Pb-bearing copper ore 铅铜矿石
pbenicate 用酚消毒
p-benzoquinone 对苯醌
p-benzoquionone dioxime 对苯醌二肟
PBJ cable 树脂沥青黄麻皮电缆
Pb-Pb isochron 铅—铅等时线
Pb-U dating 铅铀年代测定法
Pb-Zn ore 铅锌矿石
P-code 精码
PC[prestressed concrete]预应力混凝土
pcrmanent lubrication 持久润滑
pct 每百分
pcu 小客车(交通量)单位
Pd-Fe bimetal 钯铁双金属
P-diagram P 偶图
p-dimethy 对二甲胺基偶氮苯
PD project 参与设计项目

PD technique 参与设计技术
pdythene film 聚氯乙烯膜
pea 锚爪尖距
pea cannery wastewater 豌豆罐头工厂废水
peace footing 平时编制
peaceful use of seabed 海床和平利用
peace officer 治安员
peach-black 桃花芯黑
peach blossom 桃红色
peach blow 桃色釉;(中国瓷器上的)桃红色釉
peach karnel oil 杏仁油
peach(red)桃红色
peach tan 桃棕色
peach tree 桃树
pea coal 粒煤
pea coat 水手短外套
peacock 孔雀
peacock blue 孔雀蓝
peacock coal 彩斑块煤;发闪光煤
peacock copper 孔雀铜矿
peacock green 孔雀绿
peacock green glaze 孔雀绿釉
peacock ore 孔雀铜矿;斑铜矿
peacock's eye 孔雀眼(木材缺陷);孔雀眼花纹
peacock's-eye veneer 孔雀眼饰面板
peacock stone 孔雀石
pea coke 焦末;焦屑;豆粒大小的焦炭
peafowl 孔雀
pea gravel 绿石砂;小卵石;小砾(石);小豆石;细砾石;细砾;豆状石砾;豆石;豆砾石
pea gravel concrete 细石混凝土;豆石混凝土
pea gravel grout 绿豆砂灰浆;绿豆石灰浆;豆砾石灰浆;小豆石灰浆;豆砾石压力灌浆
pea-green 淡绿色的;青豆色的;豆青;豆绿色;豌豆绿色;淡绿色
pea grit 小豆石;绿豆砂;豆状砂岩;豆石
peairon ore 豆铁矿
pea jacket 水手短外套
peak 最高点统计;最高产量;高峰;尖峰;尖点;突出部分;顶峰;船首狭窄部分;齿顶尖;锋顶;峰端
peak absorption 峰值吸收
peak absorption method 峰值吸收法
peak acceleration 最大加速度;峰值加速度
peak amount 最大量;高峰量
peak amplification 峰值放大
peak amplitude 峰值振幅
peak analysis 峰值分析
peak-and-hold 峰值保持
peak angle 峰角
peak annual output 最高年产量
peak arch 尖拱;尖形拱
peak area 峰面积
peak area of peak 峰面积比
peak asymmetry 峰不对称性
peak attenuation 峰值衰减
peak automatic gain control 峰值自动增益控制
peak band 顶端铁箍
peak base 峰基线;峰底
peak brake power 最大制动功率
peak broadening 峰加宽
peak broading effect 展宽效应
peak bulkhead 尖舱隔堵
peak capacity 高峰容量;尖峰容量
peak cathode current 阴极电流最大值;峰值阴级电流
peak-charging effect 峰值充电效应
peak chlorine loading 氯的最大浓度
peak chopper 波峰斩波器
peak claimed 峰需量
peak clipper 峰值削波器;峰值限制器;峰值限幅器
peak clipping 削峰
peak-clipping circuit 削峰电路
peak coefficient 峰值系数
peak concentration 最大浓度;高峰浓度;峰收缩
peak concentration of tracer 示踪剂峰值浓度
peak concentration time 最大浓度时间;峰值浓度时间
peak consumption 高峰用量;高峰热耗;高峰耗水量;高峰负荷
peak consumptive use 日耗水高峰
peak contact 齿尖接触
peak cringle 帆后顶角索耳
peak critical void ratio 最大临界孔隙比
peak cross-section 最大截面

peak crystal temperature 结晶峰值温度
peak current 高峰电流;峰值电流;峰电流
peak current rating 额定峰值电流
peak curve 峰值曲线;放热峰
peak day-load curve 高峰日负荷曲线
peak day sendout 最大日供气量
peak daytime demand 白昼高峰需求
peak demand 最高要求;最大需要量;最大需求量;高峰需(水)量;高峰时间;高峰耗量;高峰负荷;尖锋需(电)量
peak demand capacity 峰荷容量
peak demand heating plant 高峰供热厂
peak demand time 需水高峰时间
peak density period 高峰时间;高峰负荷期间;交通高峰期
peak design capacity 高峰设计容量
peak detection 峰值检测;峰值检波;峰位检测
peak detector 峰值检波器
peak-detector circuit 峰值检波电路
peak diameter 最大直径
peak diamond performance 金刚石最大效能
peak die load 模具最大负荷
peak difference current 峰值差动电流
peak discharge 最大流量;高峰流量;洪峰流量;顶峰流量
peak displacement 峰值位移
peak distortion 最大失真;峰畸变
peak dose 峰值剂量
peak-earthquake-shock acceleration 峰值地震动加速度
peaked amplifier 建峰放大器
peakedness 峰度
peaked noise 脉冲噪声
peaked roof 尖屋顶
peaked trace 尖部扫迹
peaked wave 尖峰波
peak effect 峰值效应
peak efficiency 最高效率;高峰效率
peak electric(al)energy 电能高峰
peak energy 高峰能量;峰值能量
peak energy measurement 峰能量测量
peak envelope detection 峰值包络检波
peak envelope power 峰值包络功率;包络线峰值功率
peaker 加重高频设备;脉冲整形器
peak error 峰值误差
peak factor 振幅因数;振幅系数;峰值因数;峰荷因数
peak field 峰值场
peak fire season 火险高峰季节
peak firing temperature 最高烧成温度;最高燃烧温度;峰值燃烧温度
peak flame temperature 最高火焰温度
peak flood 洪峰流量;洪峰
peak-flood discharge 洪峰排放
peak-flood intervals 洪峰间歇
peak flow 最高流速;最大流量;高峰流量;洪峰流速;洪峰流量;峰值流量;尖峰流量
peak flow frequency 高峰流量频率
peak flowmeter 峰值流量计
peak flow period 高峰流量期间
peak flow rate 最高流速
peak frame 尖舱肋骨
peak generating capacity 高峰发电(容)量
peak gust 最大阵风;峰值阵风
peak half width 半峰宽(度)
peak handling 峰值储备
peak hardness 最大硬度;峰值硬度
peak height 峰值高度;峰高
peak-holder 峰值保持器
peak-holding 峰值保持;极值保持
peak hold switch 峰值保持开关
peak hour 高峰时期;高峰时间;高峰时刻;高峰小时;高峰负荷时间;高峰时刻;峰荷时间(指供水、供电等)
peak hour boarding rate 高峰小时上车率
peak hour call attempt 高峰小时呼叫尝试
peak hour clearway 高峰小时的高速公路
peak hour factor 高峰小时系数;高峰交通时间系数
peak hour flow 高峰小时交通
peak hour line 高峰线路
peak hourly volume 高峰小时交通量
peak hour passenger 高峰小时旅客
peak hour passenger movements 高峰小时旅客吞吐量(候机楼)

peak hour ratio 高峰小时比率
peak hour speed 高峰小时时速
peak hour traffic 高峰小时交通;高峰小时车流
peak hour urban clearway 高峰时禁止停车道路
peak hour value 峰值;高峰时间数值
peak identification 峰鉴别
peak impulse voltage 脉冲峰值电压
peak indicator 峰值指示器
peaking 引入窄脉冲;引入尖脉冲;高频提升;脉冲修尖;峰化
peaking capacity 额定最大容量
peaking circuit 信号校正电路;校正电路
peaking coil 校正线圈;建峰线圈
peaking control 高频补偿控制
peaking effect 建峰效果
peaking factor 高峰系数
peaking load 高峰负荷;最大负荷;高峰荷载
peaking load allowance 高峰允许负荷
peaking network 高频建峰网络;高频补偿网络
peaking plant 调峰电厂
peaking pulse 削尖脉冲
peaking service 高峰负荷运行;高峰负荷运动
peaking strip 线绕式脉冲传感器
peaking transformer 峰值变压器
peaking unit 尖峰机组
peaking value 最高值;最大值;峰值
peaking variation factor 高峰变化系数
peaking wet-weather flow 高峰雨天流量
peak intensity 峰值强度;峰度强度
peak inverse anode voltage 最大反向阳极电压
peak inverse current 反峰值电流
peak inverse voltage 反峰值电压
peak joint 桁架顶端接头;桁架顶接头;屋脊节点;桁架顶中央接头
peak joint of truss 桁架顶中央接头
peak keep-adjuster 峰值保持器
peak lag 洪峰滞时;洪峰峰后
peak let-through current 最高通过瞬间电流
peak level 峰值级;峰值电平;峰顶面;峰顶高程;分水岭高程
peak light intensity 峰值光强度
peak limiter 峰值限制器;峰位限制器
peak limiter circuit 峰值限制电路
peak limiting 峰值限制
peak line 桅顶至斜桁间短索;峰线;峰顶线;波峰线
peak load 高峰负荷;高峰荷载;最高载重;最高荷载;最大载重;最大负荷;最大负荷;尖峰荷载;尖峰负荷;峰荷(载);峰负载
peak load allowance 高峰允许负荷
peak load calorifier 尖峰加热器
peak load condition 峰值负载状态
peak load controller 峰荷控制器
peak load gas 调峰气
peak load generator 峰荷发电机
peak load hydroelectric(al) plant 峰荷水电站
peak load hydroelectric(al) project 尖峰负荷水力发电规划
peak load hydropower plant 峰荷水电站
peak load(ing) 峰值负载;峰值负重;峰值负荷
peak load operation 最高负荷操作
peak load period 高峰负荷时间;峰荷期间;高峰负荷周期
peak load plant 峰荷电厂
peak load power 峰荷电力
peak load power plant 高峰负荷发电厂
peak load power station 高峰负荷发电所;高峰荷发电站
peak load pressure 最大荷载压强
peak load pricing 高峰负荷定价
peak load station 峰值负载发电站;峰负载发电厂;尖峰负荷电站
peak load time 尖峰负荷时期;尖峰负荷时间
peak load unit 调峰机组
peak magnetizing force 最大磁化力
peak margin 峰值容限
peak match(ing) 峰值匹配;峰值匹配
peak maximum 峰最大值;峰顶点
peak metal temperature (卷材涂装)最高板温
peak moment 最高力矩
peak monthly price 最高月份价格
peak negative anode voltage 阳极反向峰值电压
peak odo(u)r 顶香剂
peak of a curve 曲线高值;曲线的峰值
peak of back voltage 反峰电压

peak of dry gas generation 干气生成峰
peak of flight 迁入高峰期
peak of flow 洪峰
peak of loading 加负荷的峰值
peak of negative pressure 负压力峰值
peak of oil generation 石油生成峰
peak of stripping 剥离高峰
peak of wet gas generation 湿气生成峰
peakology 峰论
peak output 最高产量;最大输出功率;最大流出量;尖峰出力
peak output of pulse 脉冲峰值功率
peak overlap 峰重叠
peak overpressure 最大剩余压力;最大超压
peak overshoot(-swing) 尖峰超越量
peak particle velocity 峰值质点速度
peak pass amplifier 高峰信号放大器
peak-peak 最高峰(值)
peak-peak detector 峰—峰检波器
peak-peak load 高峰最大荷载
peak pendant 桅顶至斜桁间短索
peak performance 最大生产率
peak period 高峰期(间);高峰时期
peak plane 峰顶面
peak plateau 峰坪
peak plate current 板极电流峰值
peak point 峰值点
peak point voltage 峰点电压
peak pore pressure ratio 峰值孔压比
peak position 峰位
peak potential 峰值电位
peak power 最大出力;高峰功率;尖峰出力;峰值荷载;峰值功率;峰值负重;峰值负荷;最大功率
peak power control 峰值功率控制
peak power handling 峰值功率储备
peak power limitation 最大功率限制
peak power meter 峰值功率计
peak power output 最大输出功率;峰值功率输出
peak pressure 最高压力;最大压力;峰压
peak pressure indicator 最高压力指示器
peak production 最高产量;高峰产量
peak pulse amplitude 脉冲峰值幅度;峰值脉冲幅度;最大脉冲振幅
peak pulse power 最大脉冲功率
peak pulse voltage 脉冲峰值电压;脉冲峰值
peak pumping speed 最大抽吸速度
peak quantity 高峰数量
peak radiant intensity 峰值辐射强度
peak rate of flood discharge 洪峰最高流速
peak rate of flow 最大流量
peak rate of runoff 最大径流量;径流峰值
peak ratio 高峰小时流量比;峰比值;峰比率
peakreach 峰顶河段
peakreach canal 峰顶河段
peak-reading device 峰值读数装置
peak-reading diode voltmeter 峰值二极管电压表
peak-reading voltmeter 峰值电压表
peak recognition 峰值识别
peak regulation 调峰;峰值调整;峰荷调节
peak reservoir 峰值储备
peak resolution 峰分辨率
peak response 最大灵敏度;峰值响应;峰值反应
peak responsibility 峰值负载能力
peak reverse voltage 最大反向电压;峰值反向电压
peak risk 巨额危险(保险)
peak runoff 峰值径流量;最大径流量;最大径流;径流峰值
peak scours 泥炭冲洗
peak searching 峰值检索
peak season 旺季
peak-seeking 寻峰
peak segment 峰段
peak selling period 旺季
peak sensitivity 最大灵敏度
peak separation 高峰间隔;脉冲间隔;峰值间距;峰顶间距;波峰间隔
peak shape 峰形
peak shaving 高峰调节
peak-shaving gas 调峰气
peak shear dilation angle 峰值剪胀角
peak shearing resistance 最大抗剪强度;最大剪切阻力
peak shear strength 最大剪(切)强度;峰值抗剪强度;峰值剪切强度

peak shift 峰漂移度;峰移位
peak shifting effect 移峰效应
peak shock pressure 最大冲击压力
peak sideband power 峰值边带功率
peak sign 建筑商标记;端墙承包人标记;材料商标记
peak signal 最强信号;峰值信号
peak signal level 最大信号电平
peak size 最大尺寸
peak sound pressure 峰值声压;最大声压
peak sound pressure level 最高声压级
peak span 桁跨索
peak spectral detectivity 峰值光谱探测率
peak spectral power 峰值光谱功率
peak spectral response 峰值光谱响应
peak splitting 峰的分裂
peak spreading 峰扩散
peak stability number 最大稳定参数
peak stage 最高水位;峰顶水位;峰顶阶段;波峰阶段
peak strain 峰值应变
peak strength 最大强度;强度峰值;峰值强度
peak stress 应力峰(值);最大应力;峰值应力
peak stress intensity 峰值应力强度
peak strip 绕线式脉冲传感器
peak stripping ratio 高峰剥采比
peak suction 最大真空度;最大负压;最大低压区;低压峰值
peak supply 高峰供应
peak symmetry 峰对称(性)
peak tailing 峰拖尾
peak tank 纵倾平稳水舱;尖舱
peak temperature 最高温度;峰值温度
peak temperature rise 最高温升
peak tidal current velocity 最大潮流速度;最大潮流流速
peak time 高峰时间;高峰期(间);成峰时段
peak time in rainy season 雨季峰期时间
peak-to-average ratio 最大值与平均值之比;峰值均值比
peak to background ratio 峰比
peak-to-mean ratio 峰值均值比;最大值和平均值之比
peak-to-peak 由极大到极小;正负峰间;振荡总振幅;峰至峰
peak-to-peak amplitude 正负峰间幅值;正负峰幅值;正反峰间隔值;峰间幅值;峰—峰振幅;峰—峰幅度;倍幅;双振幅
peak-to-peak current 峰间电流
peak-to-peak force 峰间力
peak-to-peak ratio value 峰比
peak-to-peak ripple voltage 波纹峰间电压;波纹电压全幅值
peak-to-peak separation 峰间距
peak to peak square wave 峰—峰矩形波
peak-to-peak value 正负峰间幅值;正负峰间的倍幅值;全幅值;双幅值;峰—峰值
peak-to-peak voltage 正负峰间电压;峰间电压
peak torque 最大转矩
peak-to-trough amplitude 全振幅;峰谷振幅;单振幅
peak-to-trough ratio 峰谷比
peak-to-valley height 峰谷高度
peak-to-valley ratio 峰谷比;峰值谷值比
peak traction 最大牵引力
peak traffic 高峰运输;高峰交通(量)
peak traffic flow 高峰交通流;高峰车流
peak traffic period 高峰交通期
peak transmittance 峰值透过率
peak twenty-four-hour precipitation 二十四小时高峰降水量
peak unit flow 最大单位流量;单位过程线峰值
peak use 高峰用量;高峰耗水量
peak use rate 最大利用率;最大耗水率
peak-utilization curve 峰荷利用曲线
peak-valley-load difference 峰谷负荷差
peak value 高峰值
peak value of anomaly 异常峰值
peak vector envelope 峰值向量包线
peak velocity 峰值速度
peak vibration monitor 尖峰振动探测计
peak voltage 最大电压;峰值电压;峰压
peak voltmeter 峰值伏特计;峰值电压表
peak volume 高峰交通(量)
peak volume change 最大体积变化;高峰交通量变化

peak wave height 最大波高
peak wavelength 峰值波长
peak wavelength in turbulent spectrum 湍流谱极大尺度
peak wheel-load 轮载峰值
peak white 白色电平峰值
peak white detector 白色峰值电平检测器
peak white limiter 白色信号峰值限制器;白色信号峰值电平限制器
peak white limiting circuit 白色信号峰值限制电路
peak white raster 白色电平光栅
peak width 区域宽度;色谱峰区域宽度;峰宽(度)
peak width at half-height 半宽(度);半高峰宽;区域半宽度
peak with heading 伸头
peak withstand current 峰值耐受电流
peak working pressure 最大工作压力
peak workload period 工作最忙时期
peaky curve 尖顶曲线;有峰曲线;峰形曲线
peaky curve of regime 多峰动态曲线
peak year 最高纪录年份
peak year intensity of construction 高峰年施工强度
peak yield 最高产量
peaky sea 汹涌海面
peal 敲响
peal to arbitration 诉诸仲裁人
peamafy 坡莫高导磁率合金
pean 锤头;斧头;公螺纹
pean hammer 剁石锤;尖头锤;斧锤
pean-hammered finish 斧锤琢面
peanut 花生(状)
peanut fiber 花生纤维
peanut gallery 剧院的顶层楼座;剧场的上层楼厅
peanut nodule 花生状结核
peanut oil 花生油
peanut tube 小型管
peapod dinghy 双头小艇
pear 梨属【植】;梨形物;梨树
pearceite 硫砷铜银矿;砷硫银矿
pear drop 梨花形饰
pear-ended spur dike 梨形丁坝
peariform slot 梨形槽
pearl 珠状物;珍珠色;珍珠
pearl ash 珍珠灰;粗碳酸钾
pearl beading 串珠饰【建】
pearl black 珠状炭黑
pearl boat 采珠艇;采珍珠艇
pearl braid 波状花线编带
pearl composition 粒状组织
pearl culture 珍珠养殖
pearled 有珍珠色彩的;用珍珠装饰的
pearlescent coating 珠光涂料
pearlescent effect 珠光效应
pearlescent lacquer 珠光漆
pearlescent pigment 珠光颜料
pearl escing agent 珠光剂
pearl essence 鱼鳞粉;珍珠素;珠光粉
pearl fabrics 双反面针织物
pearl farm 珍珠养殖场
pearl filler 碳酸钙填料
pearl finish 珠光漆
pearl fisher 采珠业者
pearl fishery 采珠场;采珍珠场;采珠业
pearl fishing 采珠业
pearl glaze 珍珠釉
pearl glossy 珠光
pearl glue 颗粒胶
pearl grain 珍珠克拉(珍珠的重量单位)
pearl gray 珠蓝色;珠灰色;珠光灰
pearl hardening 硫酸钙块
pearling agent 珠光剂
pearl ink 珠光油墨
pearlite 珠光体;珍珠岩
pearlite absorbent ceiling 珍珠岩吸声平顶
pearlite acoustic(al) ceiling 珍珠岩吸声平顶
pearlite aggregate 珍珠岩骨料
pearlite asphalt concrete 珍珠岩沥青混凝土
pearlite binder 珍珠岩黏结剂
pearlite colony 珠光体团
pearlite concrete 珍珠岩混凝土
pearlite concrete backup wall 珍珠岩混凝土衬墙
pearlite concrete roof 珍珠岩混凝土屋顶
pearlite concrete roof slab 珍珠岩混凝土屋顶板;珍珠岩混凝土屋面板
pearlite critical cooling rate 珠光体临界冷却速度
pearlite expansion plant 膨胀珍珠岩厂
pearlite gravel 珍珠岩砾石
pearlite-gypsum ceiling board 珍珠岩石膏吊顶板
pearlite insulating concrete 珍珠岩隔音混凝土;珍珠岩隔热混凝土
pearlite insulating material 珍珠岩绝热材料
pearlite insulation 珍珠岩绝缘
pearlite iron 珠光体可锻铸铁
pearlite lightweight concrete 珍珠岩轻质混凝土
pearlite loose fill 珍珠岩松散填充料;珍珠岩松填
pearlite loose fill insulation 珍珠岩填隔热;珍珠岩松散填充绝热材料
pearlite malleable cast-iron 珠光体可锻铸铁
pearlite ore 珍珠岩矿
pearlite plaster 珍珠岩砂浆;珍珠岩泥;珍珠岩灰浆
pearlite popping 珍珠岩爆裂
pearlite porphyry 珍珠斑岩
pearlite sound absorbent ceiling 珍珠岩吸声平顶
pearlite steel 珠光体钢
pearlite wall 珍珠岩墙
pearlitic cast-iron 珠光体铸铁
pearlitic cementite 珠光体渗碳体
pearlitic grain 珠光体晶粒
pearlitic heat resistant steel electrode 珠光体耐热钢焊条
pearlitic iron 珠光体生铁;珠光体可锻铸铁
pearlitic steel 珠光体钢
pearlitic structure 珠状构造
pearlitic texture 珠状结构;珠光体结构
pearl lacquer 珠光漆
pearl lamp 珍珠灯
pearl lugger 采珍珠船
pearl luster 珠光光泽
pearl magnetic anomaly 串珠状磁异常
pearl mica 珠光云母
pearl mill 珠磨(机);瓷球球磨机
pearl millet 御谷
Pearl Modque at Agra 阿拉格(印度)的珍珠清真寺
pearl moss 鹿角莱;鹿角菜胶
pearl mo(u)lding 珠式线脚;链珠饰
pearl plant 漆草
pearl polymerization 悬浮聚合(法);成珠聚合(法)
pearl powder 珍珠粉
pearl rack 波纹畦编
Pearl River 珠江
Pearl River datum 珠江基面
pearl sinter 硅华
pearl spar 白云石
pearlstone 珍珠岩
pearl white 锌钡白;鱼鳞粉;珍珠白
pearly 珠状的
pearly luster 珍珠光泽
pearlyte 珠光体
pear oil 梨油;醋酸戊酯
pear peel glaze 梨皮纹釉
pear-push 悬吊式按钮
pear-shaped curves of zero velocity 梨形零速度线
pear-shaped cutter 梨形铣刀
pear-shaped dam 梨形坝
pear-shaped dike 梨形堤
pear-shaped dome 梨形穹隆
pear-shaped heart 梨形心
pear-shaped mixing drum 梨形拌和筒
pear-shaped mo(u)lding 梨形线脚
pear-shaped profile 梨形纵断面;梨形轮廓
pear-shaped shaft 梨形轴;梨形柱身
pear shaped syringes 梨形洗涤器
pear shaped thimble 牛眼环;绳端心环
Pearson and Spearman correlation coefficient 皮尔逊—斯皮尔曼相关系数
Pearson-Neuman-James method 皮尔逊—纽曼—詹姆斯波浪推算法
Pearson's air elutriator 皮尔逊空气颗粒分级器
Pearson's coefficient 皮尔逊系数
Pearson's coefficient of meansquare contingency 皮尔逊均方列联系数
Pearson's coefficient of skewness 皮尔逊偏态系数
Pearson's correlation coefficient 皮尔逊相关系数
Pearson's criterion 皮尔逊准则
Pearson's curve 皮尔逊曲线(水文统计用)
Pearson's distribution 皮尔逊分布
Pearson's distribution curve 皮尔逊分布曲线
Pearson's formula 皮尔逊公式(计算斜面风压)
Pearson's frequency distribution 皮尔逊频率分布
Pearson's goodness-of-fit criterion 皮尔逊拟合优度准则
Pearson's linked relative method 皮尔逊环比法
Pearson's product-moment correlation coefficiency 皮尔逊积矩相关系数
Pearson's Type I distribution 皮尔逊一型分布
Pearson's Type III distribution 皮尔逊三型分布
Pearson's Type V distribution 皮尔逊四型分布
pear-switch 悬吊开关;梨形拉线开关
pear tree 梨树
pear wood 梨木
peasant plant protection 农业植保员
peasant population 农业人口
peasant pottery 农民用陶器
peasants voluntary labo(u)r 民工建勤
Pease's electric(al) tester 皮斯电火花闪点测定器
peasey hut 夯土建筑物;土墙;土墙茅舍
pea shingle 绿豆砂;豆粒砾石
pea souper 黄色浓雾
pea soup fog 浓雾;黄色浓雾
pea stone 豆石;豆砾石
pea stone concrete 豆石混凝土
peat 泥炭状的;泥炭土;泥炭堆积物;泥炭;泥煤
peat adsorption 泥炭吸附
peat attribute 泥炭属性
peat auger 泥炭钻
peat ball 泥炭球
peat bank 泥煤堆
peat bed 泥炭沼(泽);泥炭田;泥炭地;泥炭层;酸沼
peat blasting 泥炭爆破法;泥煤爆破法;开采泥炭爆破法;开采泥煤爆破法
peat bog 泥炭沼(泽);泥炭田;酸沼;泥沼
peat breccia 泥炭角砾
peat brick 泥炭砖;泥煤砖
peat brown 土褐色
peat building board 泥炭建筑板材
peat burning power station 燃煤泥发电站;燃泥煤发电厂
peat charcoal 泥炭;泥煤焦炭
peat cinder 泥煤炉渣;泥炭炉渣
peat clay 泥炭质黏土
peat clinker 泥炭熔块;泥炭熟料
peat coal 泥煤
peat coke 泥煤焦炭
peat concrete 泥炭混凝土
peat content 泥炭含量
peat cube 泥炭
peat cutter 泥炭切割刀
peat cutting machine 泥炭切取机
peat deposit 褐煤
peat diagenesis 泥炭成岩作用
peat dust 泥炭末;泥炭粉;泥煤尘
peatery 泥炭产地
peat excavator 泥煤挖掘机
peat fertilizer 泥炭肥料
peat fiber 泥炭纤维
peat fiber cord 泥炭纤维绳索
peat filler 泥炭填充料
peat formation 泥炭的形成
peat foundation 泥炭地基
peat fulvic acid 泥炭黄腐酸
peat gas 泥煤气体
peat-gley soil 泥炭—格列土;泥炭潜育土
peat hag 泥煤地
peatification 泥炭化作用
peatification stage 泥炭化阶段
peat insulating board 泥炭隔热板
peat island 泥炭岛
peatland 泥炭田;泥炭地
peat layer 泥煤层
peat litter 泥炭碎屑
peat-litter privy 泥炭碎屑简易厕所
peatman 泥炭矿工
peat meal 泥炭粉
peat moor 泥炭沼(泽);酸沼
peat moss 泥炭沼(泽);泥炭苔(藓)
peat mo(u)ld 泥炭土
peat paraffin 泥煤石蜡
peat pitch 泥炭质沥青;泥炭硬沥青
peat plant block 泥炭腐殖质育苗钵

peat podzol 泥炭灰壤
peat powder 泥炭粉
peat press 泥炭压榨机
peat profile 泥炭剖面
peat-reek 泥炭烟
peat-sapropel 泥炭质腐泥
peat scour 泥炭腹泻
peat seeding pellet 泥炭压制播种饼
peat seeding starter 泥炭压制播种饼
peat soil 泥沼(质)土;泥炭状土壤;泥炭土
peat spade 泥炭铲
peat stratum 泥炭层
peat structure 泥炭结构
peat tar 泥炭沥青;泥炭焦油;泥煤焦油;泥煤柏油
peat-tar acid 泥炭焦油酸
peat-tar pitch 泥炭硬沥青;泥炭焦油脂;泥煤焦油沥青;泥炭油脂;泥煤沥青
peat trap 水封弯头
peat wadding 泥炭填絮
peat wax 泥炭蜡
peaty 泥炭似的;泥炭的;含泥炭的
peaty clay 泥炭质黏土;泥炭黏土
peaty earth 泥炭土
peaty fertilizer 泥肥
peaty fibrous coal 纤维炭;丝状炭
peaty ground 泥煤矿区;泥炭矿区
peaty meadow 泥炭草甸
peaty moor 泥炭沼泽
peatype sampler 盘式河砂取样器
peaty pitch coal 泥炭沥青硬煤;弹性泥炭
peaty soil 泥沼(质)土;泥炭状土壤;泥炭土
peaty spoil 泥炭地
Peaucellier linkage 波赛利连杆机构
peau de sole 双面横棱缎
peavey 压脚子;钩棍;尖头搬钩;长撬棍
pebble 小石子;小卵石;细砾;严重桔皮;中砾;卵石状锈蚀;卵砾石;砾;透明水晶;鹅卵石
pebble accretion 卵石堆积
pebble analysis 卵石分析
pebble apron 卵石散水
pebble ball 卵石球
pebble ballast 卵石道砟
pebble bar 卵石滩
pebble beach 卵石滩;卵石海滩
pebble bed 卵石层
pebble-bed clarifier 卵石床澄清池;卵石床沉淀池;卵石滤床澄清池
pebble-bed gas cooled reactor 煤球炉式气冷反应堆;球形燃料气冷反应堆
pebble-bed reactor 球形燃料反应堆
pebble-bed shield 砾石层屏蔽
pebble bond 卵石黏合;砾石黏合
pebble breccia 中砾角砾岩
pebble clarifier 卵石澄清池
pebble coal 卵石煤;球状煤
pebble concrete 砾石混凝土;卵石混凝土
pebbled 多卵石的
pebble dash 干撒豆石饰面(外墙);干粘石;干粘卵石饰面;小卵石涂抹;干粘卵石;灰泥卵石涂层
pebble dashing 墙面加撒豆石工;抛小石粗面加工;抛小石粗面
pebble-dash plaster 卵石抹面;卵石抹灰;细砾泼撒抹面
pebble debris grade 卵石—碎石粒组
pebbled porphyritic texture 卵斑结构
pebbled sandstone 卵石砂岩
pebble filling 卵石填层
pebble flint 鹅卵形燧石
pebble fuel element 卵石形燃料元件;煤球形燃料元件
pebble gravel 卵石
pebble ground 卵石层
pebble matrix filter 砾石填质滤池
pebble mill 卵石球磨机;砾磨机
pebble packing 卵石填料
pebble pavement 卵石铺面;卵石路面
pebble paving 卵石铺面;卵石路面
pebble plastering 卵石粉刷
pebble powder 粒状火药
pebble rapids 卵石急滩
pebble riverbed 卵石河床
pebble rock 砾岩
pebble roundstone 圆卵石;卵石
pebble shoal 卵石浅滩

pebble stone 圆石;中砾;卵石;中砾岩
pebble stove 石球式热风炉
pebble-strewn deflation pavement 风蚀卵石盖层
pebble texture 中砾结构
pebble tube mill 卵石管磨机;砾石管磨
pebble wall(ing) 卵石墙
pebble ware 斑纹陶器
pebbling 铺卵石(路面);磨成小球;橙皮状表面
pebbly 多卵石的;多石子的
pebbly bottom 卵石底
pebbly braided river 砾石质网状河
pebbly sand 含砾砂
pebbly sandstone 含中砾砂岩
pebbly structure 卵石构造
Pebidian 贝比迪亚岩系
pecan 山核桃木
peck 啄孔(木材瑕疵);配克(粒状物容量单位,合9.09公斤或2加仑);斗(古建筑);凹穴
pecked finish 点凿琢面
pecked line 点虚线
pecker 尖头器具;接续器;鹤嘴锄;穿孔针
pecker block 支承
peckerhead 电视机上的接线端(俚语)
pecker neck (架塔用的)螺母扳手
pecker punch 侧壁取土器
pecking 啄斩石面
pecking motor 步进电动机
pecking order 权势等级
peckings 临时性铺砌用的劣质砖;欠火劣形砖
pecky 有啄痕的;腐霉的(木材)
pecky cyp(ress) 虫蛀柏树
pecky timber 有蛀孔木材;有霉斑木材
Peclet number 佩克莱数
pecoraite 镍纤蛇纹石
Pectacrete 彭克混凝土(一种憎水水泥)
pectin 黏胶质;果胶;黏胶质
pectinate 篦齿状的
pectinate line 梳状线;齿状线
pectinate mode 栉状水系模式
pectin content 果胶含量
pectin substance 果胶物质
pectization 胶化
pectolite 针钠钙石
pectoral girdle 肩带
peculiar 特有财产
peculiar and non-standard items 特殊的和非标准的项目
peculiar motion 本动
peculiar motion component 本动分量
peculiar part 特殊零件
peculiar proper motion 本动
peculiar setup 特殊配置
peculiar spectrum 特殊光谱
peculiar variable 特殊变星
peculiar velocity 本动速度
pecuniary condition 财政(或)经济状况
pecuniary difficulty 财政困难
pecuniary economics 财政经济学
pecuniary embarrassment 财政困难
pecuniary exchange 以货币购物
pecuniary offence 罚款
pecuniary penalty 罚金
pecuniary resources 财力
ped 小土粒;团块;土壤自然结构体
pedal 脚蹬的;踏板(卫生洁具配件);垂足
pedal accelerator 脚踏加速器;脚踏板加速器;加速踏板
pedal adjusting collar 踏板调整环
pedal adjusting link 踏板调整连杆
pedal adjustment 踏板调整
pedal adjustment machine 踏板调整机
pedal arm 踏板臂
pedal bale breaker 天平式松包机
pedal base 踏板座
pedal bracket 踏板托架
pedal brake 脚踏制动;脚踏闸;脚刹车;踏板制动
pedal bushing 踏板轴衬
pedal clearance 踏板间隙
pedal control 踏板控制
pedal crank 踏板曲柄
pedal cross shaft 踏板横轴
pedal cross shaft level 踏板横轴杆
pedal curve 垂直线;垂足曲线
pedal cycle 自行车;脚踏车

pedal cycle track 脚踏车路线;自行车路线
pedal cyclist traffic 骑自行车者交通量
pedal disc 脚盘
pedal-dynamo 脚踏发电机
pedal evener 天平杆给棉调节器
pedal feed motion 天平杆给棉装置
pedalfer 淋余土;铁铝土
pedal gear 踏板齿轮
pedal generator 脚踏发电机
pedal guard 踏板安全装置
pedal lever 踏板式杠杆;踏杆;踏板式扭杆
pedal line 垂足线
pedal pin 踏板销
pedal pull back spring 踏板拉后弹簧
pedal pull rod 踏脚拉杆;踏板拉杆
pedal push rod 踏板推杆
pedal push rod guide 踏板推杆导承
pedal push rod knob 踏板杆头
pedal rail 天平杆横轨
pedal regulating motion 天平调整装置
pedal retaining collar 踏板护圈
pedal return spring 踏板回动弹簧
pedal-rod 踏板拉杆;踏板杆
pedal-rod bracket 踏板杆托架
pedal-rod grommet 踏板杆护环
pedal-rod return spring 踏板杆回动弹簧
pedal roller 天平罗拉
pedal rotary thresher 脚踏脱粒机
pedals 方向舵脚蹬
pedal shaft 踏板轴
pedal spindle 脚踏轴
pedal spindle nut 脚踏轴螺母
pedal steer(ing) 蹬踏板转向
pedal stop 踏板停止器
pedal surface 垂足面
pedal thresher 踏踏脱粒机
pedal triangle 垂足三角形
pedal valve 脚踏阀;踏板阀
pedal wheel 脚踏轮
pedate leaf 鸟足状叶
pedately 鸟足状的
pedately cleft 鸟足状半裂的
pedately veined 鸟足状脉的
pedatifid 鸟足状裂的
pedatilobed 鸟足状浅裂的
pedatinerved 鸟足状脉的
pedatipartite 鸟足状深裂的
pedatisect 鸟足状全裂的
peddler 行商
peddler car 沿途零担车
pedesis 布朗运动
pedestal 消隐电平;细颈柱;熄弧脉冲电平;柱脚;轴承台;支座;支架;扩底的;基座;基架;托架;台座;底盘
pedestal abutment 墩座
pedestal ashtray 台座式烟灰缸
pedestal base 轴架底座
pedestal bearing 轴承架;支承轴承;高轴承;架座
pedestal bearing insulation 轴承架绝缘
pedestal bearing type 托架轴承式
pedestal binder 轴架夹板;车驾托板
pedestal body 轴架体;轴架
pedestal bolt 轴架螺栓
pedestal boulder 柱顶石;柱顶砾
pedestal brace 轴架拉条
pedestal cap 轴架盖;轴承座盖
pedestal carriage 支托座
pedestal closet (搁在架子上的)抽水马桶;坐式(抽水)马桶
pedestal concrete pile 扩底混凝土桩
pedestal control 基座控制;基底电平调整
pedestal controller 中央控制台;中央操纵台
pedestal control tube 消隐脉冲电平调整管
pedestal cross brace 轴架横拉条
pedestal drinking fountain 台座式饮水器
pedestal fair leader 立式导缆钳
pedestal foot(ing) 柱脚;座墩底脚;扩座柱脚;阶形基础;基座底脚;护底柱脚;台座基脚;台座基础;台座底脚;活地板
pedestal for block instrument 闭塞器座
pedestal foundation 阶形基础;台座基础
pedestal frieze 台座雕带;柱加雕带;(建筑上的)基座腰线
pedestal generator 基准电压发生器

pedestal generator circuit 台座发生器电路
pedestal grinder 立柱磨床
pedestal height 消隐脉冲高度
pedestal jaw 轴箱切口
pedestal lavatory 落地式脸盆；立式洗面器；台座式盥洗盆；台地盥洗盆
pedestal level 消隐脉冲电平；消隐电平；熄灭脉冲电平；基准电平；区分电平；封闭电平
pedestal mo(u)lding 台座线脚；台踢线；底座线脚
pedestal optic(al) comparator 柱架式光学比测仪
pedestal pan 坐式便盆；坐便器
pedestal pile 扩脚桩；扩底钻孔灌注桩；扩底桩；护底桩；球形扩脚桩；大头桩；爆破桩
pedestal-processing amplifier 基准脉冲处理放大器
pedestal pulse 基座脉冲；台座脉冲
pedestal ring 座圈
pedestal rock 柱顶石；基岩；蘑菇石
pedestal sight 托架瞄准具
pedestal sign 可动底盘标志杆；柱座标志；可移动带盘标志杆
pedestal signal 柱座信号；柱座(式)信号(灯)；柱式信号灯
pedestal sit 台阶
pedestal socket 吊艇柱座
pedestal stone 柱石；石柱
pedestal supported panel flooring 架空板材地面（活动地板的常规做法）
pedestal swing hydraulic hammer 座式旋转液压锤
pedestal table 独腿桌；支柱桌
pedestal tank 支座水箱
pedestal technique 无坩埚技术；台基技术
pedestal tie bar 支座系杆
pedestal tie bar bolt 轴架系杆螺栓
pedestal tie bolt 轴架系紧螺栓
pedestal type 台座式
pedestal-type bearing 托架轴承
pedestal-type centrifugal compressor 支承式离心压缩机
pedestal-type drinking fountain 台座式饮水器
pedestal-type impeller agitator 台座式叶轮搅拌器
pedestal-type secondary clock 座式子钟
pedestal-type squeezer 座架式压铆机
pedestal urinal 立式小便器；台座式小便池
pedestal washbasin 台式洗面器；立式洗面器；落地式脸盆
pedestrain work 步行工作
pedestral rock 菌状石
pedestrian accidents 车辆撞伤行人事故
pedestrian actuated signal 行人揿钮信号；行人调控信号
pedestrian actuated signal sign 行人揿钮信号标志
pedestrian area 步行区
pedestrian barrier 行人护栏
pedestrian barrier panel 行人阻挡栏板；行人分隔带
pedestrian bridge 行人天桥；行人桥；人行桥
pedestrian city center 可以步行的市中心
pedestrian clearance interval 行人过街清尾时间
pedestrian congestion 行人拥挤；行人密集
pedestrian control 行人道交通管制；行人(交通)管制；人行道交通控制；人行道交通控制器
pedestrian control device 行人通行控制器
pedestrian control fence 行人管制栅栏
pedestrian-controlled cold emulsion sprayer 手推冷乳剂喷洒器
pedestrian-controlled vibrating roller 手推移动的振动压路机
pedestrian conveyer 行人传送带
pedestrian count 行人(观测)计数
pedestrian crossing 行人过路线；人行横道；人行过道；步行交叉
pedestrian crossing facility 人行过路设施
pedestrian crossing line 人行横道线
pedestrian crossing sign 人行过街标志
pedestrian crosswalk 人行过街道；人行横道
pedestrian deck 人行道路面(板)；立体人行道；人行平台；步行平台
pedestrian detector 行人检测器；行人探测器；行人感知器
pedestrian distribution system 客流运送系统
pedestrian enclave 步行区
pedestrian facility 人行交通设施
pedestrian ferry passenger 行人渡口旅客
pedestrian flow 行人流；行人交通；人流
pedestrian fork lift truck 人行手推叉车

pedestrian group 行人群
pedestrian guardrail 行人护栏；人行道栏杆
pedestrian island 行人安全岛；人行岛；安全岛
pedestrianize 只供行人使用；禁止车辆通行
pedestrian lane 人行小路；小巷(道)
pedestrian light control 行人色灯(信号)控制
pedestrian load(ing) 人群荷载
pedestrian loading island 乘客候车站台
pedestrian mall 林阴步道；人行林阴路；步行街（道）
pedestrian malls 徒步购物区
pedestrian metal guardrail 行人金属防护栏
pedestrian module 步行信号调制
pedestrian motor vehicle traffic accident 行人车祸
pedestrian network 人行道路网
pedestrian operated traffic signal 行人扳动的交通信号
pedestrian over-bridge 人行天桥
pedestrian overcrossing 人行天桥
pedestrian overcrossing screen 人行天桥护网；人行桥护网
pedestrian overpass 人行天桥
pedestrian pass 人行过道；人行通道
pedestrian passageway 人行道
pedestrian passenger 人行横道线；人行横道
pedestrian path 步行道；步道；人行道
pedestrian period 人行时期
pedestrian phase 行人信号相
pedestrian precinct 行人专用区；步行区
pedestrian protection 行人保护
pedestrian psychology 行人心理
pedestrian push-button control 行人按钮控制
pedestrian rail 行人栏杆
pedestrian ramp 人行道斜坡；人行坡道
pedestrian recall 行人请求绿灯
pedestrian(refuse) island 行人安全岛
pedestrian road 行人专用道
pedestrian road-crossing tunnel 地下过街人行道
pedestrian safety 人行交通安全
pedestrian segregation 行人与车辆分流；人车分流
pedestrian separation fence 行人隔离栏；行人隔栏
pedestrian signal 行人(交通)信号
pedestrian signal display 行人信号显示
pedestrian skyway 人行天桥；行人天桥
pedestrian space 散步区；行人占地面积
pedestrian stair(case) 行人楼梯；人行楼梯间
pedestrian street 行人街道；步行街
pedestrian subway 地下(人行)道；人行隧道；人行地道；旅客地道
pedestrian subway interchange 地下人行道立体交叉
pedestrian survey 行人调查
pedestrian traffic 行人通行；人行交通；人行道交通
pedestrian traffic door 人行交通门道；行人上下车门道；人行交通入口
pedestrian traffic volume 行人交通量
pedestrian traffic way 人行交通道路；步行街(道)
pedestrian traffic zone 行人上下车地带
pedestrian tube 人行隧洞；人行隧道
pedestrian tunnel 人行地道；人行隧道；地下通道；地下道
pedestrian underpass 人行地道；高架桥下人行道；人行地道；地下人行道；步行隧道
pedestrian vehicle accident 行人车祸
pedestrian vehicular conflict 行人车辆冲突点
pedestrian volume 行人交通量
pedestrian walk 人行道
pedestrian way 道路用地外的行人专用道；人行道
pedestrian zone 行人地带
pedestrian zone marker 行人地区标(志)线；行人地带标(志)线
pede window 教堂十字平面底部的窗；十字架底的窗
pedgee 接受抵押人
pedicab 三轮车
pedicel 梗节；花柄
pedicellaria 叉棘
pedicle 花梗
pediculicide 灭虱药
pedigree farm 纯种牧场
pediment 人字墙；山墙；三角形楣饰；大门门头饰（常由肘托、檐口、三角饰或其他楣饰组成）；麓原；山头【建】；山前侵蚀平原；山前平原；山麓缓斜平原

pediment apex 门窗过梁上三角形檐饰的顶尖；人字脊
pediment arch 三角拱饰；人字拱；弧形拱；山形拱；三角形拱饰；三角或弧形拱饰；三角拱
pedimentation 麓原形成作用；山麓夷平作用
pedimented 有山墙的；人字形的
pedimented door frame 有檐饰的门框
pedimented window 有檐饰的窗
pediment foot 山墙基础
pediment glacier 山麓冰川；山岳冰川
pedimeter 计步器；步程计；步测计
pedimetry 步测法
pedion 单面晶；单面
pedion crystal 单面晶体
pediplain 联合山麓侵蚀面；山麓侵蚀平原
pediplanation 山麓夷平作用；山麓侵蚀作用
pediplane 山麓侵蚀面平原
pedocal 钙层土；钙土
pedoclimax 土壤顶极
pedode 硅质空心团块
pedodynamometer 脚力测定器
pedogenesis 土壤发生；成土作用
pedogenetic relation 土壤发生
pedogenic 成土的
pedogenic process 土壤发生过程；成土作用
pedogeochemical anomaly 土壤(圈)地球化学异常
pedogeochemistry 土壤地球化学
pedogeography 土壤地理学
pedography 土壤描述学
pedohydrology 土壤水文学
pedologic(al) 土壤的
pedologic(al) chart 土壤图
pedologic(al) classification 土壤学分类
pedologic(al) classification system 土壤学分类系统
pedologic(al) feature 土壤形成物
pedologic(al) map 牧业图；土壤(分布)图
pedologic(al) profile 土壤纵断面
pedologic system of soil classification 土壤学的土分类法
pedologic(al) unit 土壤单位
pedologist 土壤学家
pedology 基础土壤学；土壤学
pedometer 里程表；记步仪；计步器；计步测距器；步计数；步程计；步测计
pedometry 步测法
Pedomicrobium manganicum 锰土微菌
pedon 单个土体
pedo relict 土壤残留
pedosphere 土壤圈；土界；表土层
pedotheque 土壤样品
pedotubule 土壤管状物
pedrail 履带
pee 锚爪尖部
Peedee Belemnite standard PDB 标准
peek 取数；抽看
peek-a-boo 同位穿孔
peek-a-boo card 同位穿孔卡(片)
peek-a-boo check 相同位穿孔检查
peek-a-boo system 同位穿孔(检索)系统
peel 小型堡寨（中世纪英国）；堡塔；旋切；悬臂料装杆；装料杆；揭片；桨掌面
peelable coating 可剥性涂料；可剥性涂层
peelable paint 可剥(性)漆
peelable type 可剥型
peel adhesion 剥离黏着力
peel back 脱壳
peel coat film 撕膜片
peel coat material 撕膜材料
peeled 去皮的
peeled billed-wood 剥皮圆材
peeled veneer 剥落的镶面板
peeled yarn 剥皮纱
peeler 旋切用原木；去皮工；坯料去皮车床；坯料剥皮机；剥皮机
peeler block 旋切木段
peeler core 层板原木的木芯
peeler log 旋切用原木；去皮圆木；旋切原木；胶合板材
peeler-regeneration system 激生再生系统
peeling 铸件表皮；落砂清理；鳞剥；去皮；起皮；喷丸；脱釉；脱皮；剥釉；表层剥落；剥皮
peeling action 剥皮作用
peeling failure 剥离破坏

peeling knife 去皮刀;剥皮刀
peeling layer method 剥层法
peeling machine 剥皮机
peeling off 剥离;剥开;揭皮;掉皮;剥片;清铲;剥落
peeling strength 撕裂强度;剥离强度
peeling test 剥离试验
peeling tester 剥离试验仪
peel off 剥落
peel off area 剥离区
peel off pack 剥离型面膜
peel ply 剥离层
peel strength 撕裂强度
peel test specimen 剥离试样
peel tower 堡塔
peen 公螺纹;锤头;锤尖;锤顶;锤的尖头
peened surface 喷丸处理表面;锤击强化表面
peener 喷丸装置;喷丸机
peen hammer 剁石锤;尖头锤;尖锤;夯砂锤;斧锤
peen-hammered face of stone 斧垛石面
peen-hammered finish 斧锤琢面;尖锤琢面
peen hardening 喷丸硬化;弹射硬化
peening 用锤尖加工;敛紧(钻头上的金刚石);喷珠硬化;喷丸硬化法;喷丸硬化;喷丸清理;喷射加工硬化法;锤击
peening effect 喷丸强化效应;喷砂强化效应
peening machine 喷丸机;喷砂机;锤击机
peening tool 清棱凿;锤打工具
peening wear 冲击磨损
peen plating 扩散渗镀法;金属粉末扩散渗镀法
peen rammer 夯砂锤;扁头夯砂锤
peep at 窥视
peeper 定位放大器(镶嵌用)
peep hole 猫眼(俗称);展望孔;观察孔;观测孔;窥视孔;窥孔;观测镜;看火孔
peephole block 看火孔砖
peephole mask 窥孔掩蔽;同位孔隐字
peephole optimization 窥孔优化技术;窥孔优化
pee pot 安全帽(俚语)
peep-sight 取景;照准器
peep-sight alidade 测斜照准仪
peep-sight compass 觇孔照准罗盘
peep window 窥视窗
peer goal-setting 同级制订目标
peer layer 同等层
peer layer communication 同等层通信
peerless pile 可抽出外壳的混凝土桩;去模混凝土桩
peer rating 相等评价法
peer review 同行审查;同行评定
peer-to-peer 同畴;对等式;点对点
Peet's valve 皮梯式闸门截止阀
peg 小木桩;木栓;木橛;木钉;栓钉;钉住;钉;标记
peg adjustment 桩正法;两点校正法【测】;木桩校正法【测】
peg adjustment method 桩正法
pegamoid 防水布;人造革
pegamtitic elements 伟晶岩元素
peg-and-cup dowel 模型定缝销钉及套
peg and socket 栓钉式
peganite 磷矾土
Peganum harmala 骆驼蓬
pegboard 有销孔的硬度质纤维板
peg dowel 栓销
peg down 约束;用木钉钉牢
peg-foot roller 羊足压路机
peg gate 下浇口;直水口
pegged 设桩的
pegged exchange rate system 固定汇率制度
pegged leggers 木钉裤
pegged mattress 用木桩定位的沉排
pegged out 放样;定界;用木桩标线
pegged point 桩点
pegged point mapping 桩点法测图
pegged price 限定价格;控制价格
pegged rate 固定汇率
pegged station 桩标点
peggies (10英寸×10英寸的)石板瓦;(长宽不一的)石板瓦
pegging 销子连接;限定汇率;筛孔卡粒
pegging out 标桩;立桩;打桩【测】;桩定;定线;定界;打桩定线;放样;放线;标桩定位
pegging out a curve from the chord 弦线法标定曲线
pegging out a curve from the tangent 切线法标定曲线

pegging out of railway line 铁路线打桩(放线)
pegging out the line 路线之标概
pegging rammer 型砂捣锤
pegging the exchange 固定汇价
peg graft 插接
peggy timber 有蛀孔木材;有毒斑木材
peg ladder 木钉梯
peg-leg 木支柱
peg-leg multiple reflection wave 微屈多次反射波
pegmatite 结晶花岗岩;黑花岗岩;伟晶岩
pegmatite group 伟晶岩类
pegmatitic granite 伟晶花岗岩
pegmatitic gypsum ore 伟晶石膏矿石
pegmatitic mineral 伟晶矿物
pegmatitic ore deposit 伟晶岩矿床
pegmatitic orefield structure 伟晶岩矿田构造
pegmatitic ore-forming process 伟晶岩成矿作用
pegmatitic phase 伟晶相
pegmatitic stage 伟晶岩阶段
pegmatitic structure 文象构造;伟晶构造
pegmatitization 伟晶作用;伟晶岩化(作用)
pegmatoid 伟晶岩状;伟晶岩相;似伟晶岩的
pegmatolite 正长石
pegmatophyre 花斑岩;文象斑岩
pegmatophyric 花斑状的;文象斑状
peg method 两点校正法;木桩校正法;标桩校正法
peg model 标桩模型
peg mold 移动模板(线脚抹灰用)
peg of crossing centerline 中心桩【测】
peg out 用木桩标线;放样;定线;定界
peg-peg multiple reflection 微屈多次反射
peg plan 纹钉图
peg planting 戳孔造林
peg point 放样点;定界点;桩位;桩点
peg-raker saw 双切齿锯
peg-shaped tooth 钉状牙
pegs of wall centre line 轴线控制桩【测】
peg stake 木橛;木标桩
peg stay 套栓式窗撑;套栓式风撑;窗扇支撑
peg switch 记次转换开关;记钉开关;标记转换开关;栓转电闸
peg table 木钉;木桩
peg test 木桩校正法【测】
peg to 放样;定线;定界
peg tooth 三角齿
peg-tooth drum 钉齿式脱粒滚筒
peg tooth saw 三角齿锯
peg-tooth saw blade 三角齿锯片
pegtop paving 平铺砌石;小石块铺砌;螺形砌石路面;小石块砌筑陀螺形路面;小石块铺砌;小方石铺面
peg to windward 尽量逆风航行
Pegui 勃固黄褐色硬木(巴西产)
peg welding 栓柱焊接
peg wire 螺母防松铁丝
pehrmanite 铁塔菲石
peimidine 贝米丁
peiminane 贝母烷
peiminone 贝母酮
pein 锤尖;锤顶;公螺纹;锤的尖头
Peine girder 派尼大梁
Peine section 派尼截面;宽翼缘断面
Peine (sheet) pile 派尼钢板桩
peisleyite 裴斯莱石
pejorative 恶化的
pek 油漆;涂料
Peking coordinate system 北京坐标系
Peking flowering crab 海棠【植】
pekoite 皮硫铋铜铅矿
pel 图元
pelabuan 半开敞锚地
pelagic 浮游的
pelagic abyssal sediment 远深海沉积物
pelagic(al) fish concentration 中上层鱼群
pelagic(al) fishes 中上层鱼类
pelagic(al) fish stocks 中上层鱼类群体
pelagic(al) region 中上层区
pelagic(al) resources 中上层资源
pelagic(al) zone 中上层区
pelagic animalcule 海洋微生物
pelagic clay 远洋性黏土
pelagic community 远洋生物群
pelagic deposit 远洋沉积(物);深水沉积物;深海沉积(物)

pelagic division 远洋区
pelagic environment 远洋环境;浮游环境
pelagic facies 远海相
pelagic fauna 远洋动物区系;水层动物区系;深海动物群;深海动物区系
pelagic fish 浮游鱼
pelagic fishery 远洋渔业
pelagic fishes 大洋性鱼类
pelagic food chain 远洋食物链;水层食物链
pelagic island 大洋岛
pelagic larva 海面幼虫;浮游幼体
pelagic limestone 远洋灰岩
pelagic mode of life 水层生活;浮游生活
pelagic ooze 远洋软泥
pelagic organism 远洋生物;水层生物;浮游生物
pelagic period 浮游期
pelagic phase 浮游生活阶段
pelagic pollution 远洋污染
pelagic realm 远洋域;远洋区;深海区
pelagic region 远洋区域;大洋区;浮游区
pelagic resources 远洋资源
pelagic sediment 远洋沉积(物);深水沉积物
pelagic sedimentation 深水沉积作用
pelagic species 远洋种;上层种
pelagic survey 远海测量
pelagic suspension 远洋悬浮
pelagic tide 大洋潮汐
pelagic trawl 深海拖网
pelagic zone 水层带;大洋区
pelagite 海底锰块
pelagopholus 栖海面的
pelagos 水层生物
pelagosite 潮间带盐壳
Pelasgian construction (古代小亚细亚的)巨石结构
pele 剥离;去皮;堡塔
pelean cloud 热云团云;炽热火山云
peleeite 斑苏玄武岩
Pele's hair 火山毛
Pele's tears 火山泪
Pele-tower 皮尔塔(英国北部和苏格兰的一种避难用的小塔或房子)
pelgaium 海面群落
pelhamite 蛭石
pelican hook 铰链钩;滑钩;速脱钩
pelican slip 滑钩
Peligot blue 佩利果特蓝
Peliocan hook 佩肯活钩
pelite 黏土岩;铝质岩;泥质岩;变质泥岩
pelite-gneiss 泥质片麻岩;泥片麻岩
pelitic 泥质的
pelitic anhydrite-gypsum ore 泥质硬石膏—石膏矿石
pelitic cement 泥质胶结
pelitic content of road metal 道砟含泥量
pelitic gneiss 泥质片麻岩
pelitic gypsum 泥质石膏
pelitic gypsum-anhydrite ore 泥质石膏—硬石膏矿石
pelitic hornfels 泥质角页岩
pelitic limestone 泥质灰岩
pelitic sandstone 泥质砂岩
pelitic siltstone 泥质粉砂岩
pelitic strain bands 泥质应变带
pelitic texture 泥质结构
pelitic tuff 泥质凝灰岩
pelitite ore 泥石硬石膏矿石
pelitomorphic aluminous rock 泥状铝质岩
pella 斗篷
pellagra 糙皮病
pellagrosarium 糙皮病疗养院
pellant 去垢的
pellet 小丸;小球;小粒;小弹丸;圆片;渔网浮标;钙团粒;料球;粒素;晶片;球粒;切片;切晶片;片状器件;片剂;团粒;弹丸;(激光或电子束的)靶丸
pelletal-bearing micritic limestone 含球粒微晶灰岩
pelletal-micritic limestone 球粒微晶灰岩
pellet aluminous rock 球粒铝质岩
pellet-bearing pelitomorphic aluminous rock 含球粒泥状铝质岩
pellet bit 霰弹作用
pellet bit design 霰弹钻头设计
pellet bit development 霰弹钻头研制
pellet bit jet pump 霰弹钻头喷射泵
pellet bit texture 霰弹钻头结构

pellet blender 丸粒掺和机
pellet bonder 球式结合器;球式接合器
pellet cloud 霰弹聚集
pellet compression 靶丸压缩
pellet cooler 颗粒冷却器
pellet corona 靶丸电晕
pellet counting circuit 霰弹计数回路
pellet density 颗粒密度
pellet dryer 料粒干燥器;丸状干燥器
pelleted catalyst 粒状催化剂
pelleted limestone 球粒灰岩
pelleted mineral wool 矿棉球
pelleted pitch 球状硬沥青
pelleter 制片机;制粒机
pellet extrusion modeling 颗粒金属挤压成型法
pellet floc 粒么絮体;粒状絮体
pellet formation 球形结构;小球形成
pellet head 造粒机头
Pelletier's green 佩尔蒂绿;永固绿
pellet impact bit 霰弹冲击钻头
pellet impact drilling 霰弹冲击钻进
pelleting 压丸;造粒;加药丸
pelleting machine 造粒机;成粒机
pelleting property 成片性能
pellet interference 霰弹碰撞
pelletisation 粒化
pelletizating 球团
pelletization 细矿团化;制丸;制粒;造粒;粒化;球团化;球粒化;成球
pelletize 制粒;粒化;成球
pelletized 球状的;丸状的
pelletized anthracite 无烟煤煤砖
pelletized concentrate 粒化母料
pelletized feed 入窑料球
pelletized granule 颗粒剂
pelletized material 球状物料
pelletized pigment 颗粒化颜料
pelletized raw meal 成球生料
pelletized slag 粒状矿渣
pelletized type aggregate 球团化集料;球团化骨料
pelletizer 霰弹制备设备;压球机;压片机;制丸设备;制粒机;造粒机;球形造粒;成球机
pelletizing 造粒(作用);成球
pelletizing capacity 成球能力
pelletizing drum 成球筒
pelletizing plant 成球设备
pellet-lime mud 球粒灰泥
pellet mill 制片机
pellet module 片状微型组件
pellet mo(u)lding 链珠形花边;小球饰;链珠(形)线脚;丸子饰
pelletoid 似团粒;似球粒
pelletoidal phosphoraite 球粒状磷块岩
pellet ore 球团矿
pellet ornament 圆形浮雕装饰品
pellet part 球状部件
pellet pelitomorphic aluminous rock 球粒泥状铝质岩
pellet phosphoraite 球粒磷块岩
pellet power 丸状炸药;霰弹功率
pellet rate 霰弹补给速度
pellet rate recorder 霰弹喷射速度记录器
pellet receiver 颗粒收集器
pellet softening 丸状软化
pellet system 霰弹法
pellet texture 球粒结构
pellet-to-diameter ratio 霰弹直径比
pellet-to-nozzle ratio 霰弹—喷口比
pellicle 照相胶片;菌膜;胶片;薄膜;表膜;半透膜
pellicle mirror 薄膜镜;半透镜
pellicle reflector 薄膜反射器
pellicular adsorbent 薄膜吸附剂
pellicular electronics 薄膜电子学
pellicular formation 薄膜形成
pellicular front 渗透面;附着水面;薄膜水带;薄膜面;薄膜峰
pellicular ion-exchanger 薄膜型离子交换剂
pellicular micro beads 薄壳型微珠载体
pellicular moisture 附着水;薄膜水(分)
pellicular packing 薄壳型填充剂
pellicular resin 薄膜型树脂;薄膜树脂
pellicular salt 盐壳;薄膜盐壳
pellicular stage 薄膜阶段
pellicular support 薄壳载体

pellicular water 附着水;薄膜水(分);薄层水
pellicular zone 薄膜水层;薄膜带
pell-mell 混乱的
pell-mell block 乱块;乱块;抛筑的块体
pell-mell construction 杂块砌的建筑;乱块建筑;混凝土块建筑;抛筑构造
pell-mell placed 抛筑的
pell-mell-placed concrete block 抛筑的混凝土块体;方块抛填
pell-mell placing 乱抛;抛筑
pell-mell structure 杂糅构造
pellodite 冰泥砾岩;冰川泥岩
pell off 剥落
pelloid layer 防卷曲层
Pellon 佩纶(无纺织物)
pellonxite 生石灰
pellucidity 明晰性;透明性;透明度
pellumina 氧化铝膜
pellyite 硅铁钙钡石
Pelmatozoa 有柄部
pelmet board 门窗帘;窗帘板;门窗帘板
pelmet(box) 门窗帘匣;窗帘箱;窗帘盒
pelmet lighting 窗帘匣顶泛光照明
pelmicrite 泥质微晶灰岩;球粒微晶灰岩;球粒泥晶灰岩;微晶球粒灰岩
pelmicsparite 球粒微亮灰岩
pelochthium 泥滩群落
pelogloea 浮游生物淤泥
peloid 球状粒
pelophyte 沼泽植物
pelorus 哑罗经;罗经刻度盘
pelorus stand 哑罗经座
pelota 回力球
pelouse 草甸
pelsparite 亮晶球粒灰岩;泥晶球粒灰岩;球粒亮晶灰岩
pelsparrudite 球粒亮晶砾屑灰岩
pelt 迅猛前进;毛皮
Peltier effect 帕尔帖效应
pelting rain 倾盆大雨
pelting rain humidity 倾盆大雨的湿度
Pelton's turbine 培尔顿水轮机;水斗式水轮机
Pelton (water) wheel 水斗式水轮机;冲斗式水轮;培尔顿(水斗式)水轮机
Pelton wheel case 培尔顿水轮外壳
peltry 皮货;毛皮;皮囊
pelvimeter for external measurement 外测量器
pelvimeter for outlet measurement 出口测量器
pelyte 泥质岩
Pemberton level(l)ing rod 彭佰顿水准尺
Pembroke table 彭布罗克折面桌
pen 描笔;牲畜栏;牲口圈
penal interest 违约利息
penalized feasible decom-position method 罚可行分解法
penal sum 赔偿费;罚款总额;罚金总额
penalties for environmental damage 对造成环境危害的罚款
penalty 罚款;罚金
penalty-and-bonus clause 赏罚条款;惩罚奖励条款
penalty audit 罚款审计
penalty cargo 高费率货物
penalty clause 赔偿条款;违约罚金条款;惩罚条款;罚款条款;罚款罚金
penalty cost 违约成本;惩罚价值;罚支支出
penalty dockage 延期使用码头费;使用码头延期罚金
penalty exchange rate 惩罚汇率
penalty factor 惩罚因子;罚因子
penalty for breach of faith 违约金
penalty for delay 延误罚款;误期罚款
penalty for late shipment clause 延迟装运罚款条款
penalty for non-performance of contract 违约罚款
penalty for violation of regulation 违章处罚
penalty function 罚函数;补偿函数
penalty function algorithms 罚函数算法
penalty function method 罚函数法
penalty method 惩罚法;补偿法
penalty of breach of contract 违约罚款
penalty on delayed delivery 延滞交货罚金;延迟交货罚金
penalty period 违反规定的停车时间;停车逾时
penalty provision 罚则
penalty rate 惩罚率

penalty run 补做试验
penalty sum 违约罚款
penalty tax 税款滞纳罚金
penalty term 处罚条款
penalty test 追加试验;追补试验
pen-and-ink drafting 清绘着墨
pen-and-ink drawing 钢笔画
pen-and-ink posting 用钢笔过账
pen-and-ink recorder 自动记录器;笔墨记录器
pen-and-ink test 纸张吸墨试验
Penang 槟城(马来西亚)
pen arm 自记器笔杆;笔尖杆;笔杆(记录仪表)
pen automatic recorder 笔式自动记录仪
pen bow 小圆架
pen-breeze 炉渣
pen carriage 笔架
pencatite 水镁石大理岩;水滑结晶灰岩;水滑大理石
pencel 小三角旗
pen centering 描笔中心调整
pencil 光锥(体);毛撮;射束
pencil beam 锥形光束;尖向束;集束性光束;锐向性射束;铅笔状射束;笔形波束;铅笔形波
pencil-beam antenna 笔形射束天线;锐锥形射束天线;锐方向性波束天线;铅笔束天线;笔形波束天线
pencil carriage 绘图笔架
pencil-case 铅笔盒
pencil cedar 铅笔松
pencil chuck 绘图笔夹头
pencil cleavage 铅笔劈理;笔状劈理【地】
pencil coaxial cable 细芯铜轴电缆
pencil compasses 圆规
pencil core 通气芯;大气冒口芯;笔状泥芯
pencil core bit 中心小水口全面金刚石钻头
pencil diamond 玻璃刀
pencil draft 铅笔原图
pencil drawing 铅笔画
pencil drawn 铅笔绘的
penciled edge 半圆边缘圆弧棱边
pencil edging 圆边加工
pencil eraser 铅笔橡皮
pencil follower 笔式跟踪数字化器
pencil-follower digitizing table 感应笔跟踪数字化器
pencil gate 铅笔形浇口
pencil glide 铅笔式滑动
pencil gneiss 石笔片麻石
pencil-grade talc 铅笔级滑石
pencil guide 铅笔销子导轨
pencil hardness 铅笔硬度
pencil hardness test 铅笔硬度试验
pencil lacquer 铅笔漆
pencil lead holder 笔头夹板
pencil light beam 锥形光束
pencil-like cursor 笔式跟踪头
pencil(l)ing 铅笔线花饰;铅笔线花样;砖灰缝涂白;涂料勾缝(砂浆缝加涂饰);画白线;墙缝白漆线条;彩色勾缝
pencil mixer 笔形混频管
pencil of curves 曲线束
pencil of light 光束
pencil of lines 线束
pencil of matrices 矩阵束
pencil of parallel rays 平行光束
pencil of planes 面束
pencil of rays 光束;光流
pencil of X rays X 射线束
pencil ore 笔铁矿
pencil-outlined wash drawing 铅笔淡彩
pencil rocket 超小型火箭;小型火箭
pencil rod 细的光面圆钢筋;盘条;吊筋(悬挂式顶棚);细钢筋;细棒材
pencil round 细磨圆角
pencil-rounded 细磨圆角(的)
pencil scratching tester 铅笔硬度试验机
pencil shape beam radar 锐方向性射束式雷达
pencil sharpener 转笔刀
pencil sketch 铅笔画;铅笔草图
pencil slit 铅笔窄板条
pencil stock 铅笔坯料
pencil stone 石笔石;滑石
pencil structure 笔状构造
pencil tester 铅笔硬度计
pencil tube 超小型管;笔形管

pencil tube amplifier 超小型管放大器；笔形管放大器
pencil-tube oscillator 笔形管振荡器；超小型管振荡器
pencil-type probe 笔形测针
pencil-type thermocouple probe 笔形热偶测针
pencil-work 铅笔图画
pen-code 光笔编码
pen connect 笔塞孔
pen container 栏式集装箱；家畜集装箱；牲畜货柜；牲畜集装箱
pencraft 笔法
pend 悬挂；悬垂吊着
pendage 地层倾斜
pendant 悬饰；悬架式按钮台；悬垂的；下垂物；下垂体；尖旗；短吊绳；洞顶下垂体；顶棚悬挂体；顶棚下垂体；吊饰；吊架；吊灯架
pendant boss 悬饰浮雕
pendant box 悬吊按钮站
pendant chain 悬挂锚链；吊链
pendant cliff 悬崖
pendant cloud 管状云
pendant collet 条轴套夹；四开环
pendant continuous row fluorescent fixture 吊装荧光灯带
pendant control 控制板
pendant control box 悬吊开关盒
pendant control switch 悬挂式开关；悬挂式按钮
pendant cord 悬吊弦；悬吊绳；吊绳
pendant fitting 悬吊件
pendant fixture 悬吊装置
pendant formation 悬垂构造
pendant group 侧基
pendant halyard 桅顶旗索
pendant hole 壳端孔
pendant keystone 下垂的拱顶石
pendant lamp 挂灯；吊灯
pendant lampholder 悬挂的灯座
pendant lamp-socket 悬挂的灯座
pendant light 吊灯
pendant light fitting 吊灯配件；悬挂式灯具；悬吊灯具
pendant line 锚头缆
pendant luminaire 悬挂式照明器；吊灯
pendant luminaire fixture 悬挂式照明设备
pendant mark 测线标志
pendant mounting channel 悬挂式安装管路
pendant plow 步犁
pendant point 悬挂点
pendant post 悬柱；墙架立柱；壁架柱；悬饰柱
pendant push 悬吊开关；悬吊按钮；垂悬式按钮
pendant reheater 悬吊式中间再热器
pendant signal 下垂式信号装置；悬挂式信号机；吊灯信号
pendant sprinkler 悬挂式喷水器；下垂洒水喷头
pendant superheater 悬挂式过热器；立式过热器；屏式过热器
pendant switch 悬架式电闸；悬挂开关；拉线开关；垂悬式开关；悬吊开关
pendant tackle 短索绞辘
pendant terrace 连岛砂坝
pendant-type luminaire 悬挂式照明器
pendant-type monorail scale 单轨吊秤
pendant vault 悬挂式穹顶；圆顶穹隆
pendant walkway 悬挂式走道
pendant wire 悬索；测绳
pendant yard-tackle 桁端绞辘
pen defect 露明缺陷
pendeloque 枝形吊灯的梨形玻璃垂饰；犁形玻璃垂饰；犁形宝石
pendence 坡度；斜面
pendency 悬垂；垂下
pendent 未定的；悬吊的；垂挂的；滴形装饰；垂挂装饰；悬的；尖旗；三角旗；短索；短挂绳；吊饰
pendent cloud 漏斗云
pendent drill 下垂式钻机；垂悬式钻机
pendent drop method 悬滴法
pendent forms 悬挂式模板
pendentive 穹隅（圆屋顶过渡到支柱之间的渐变曲面）；帆拱
pendentive bracketing 穹角形托架；隅折
pendentive cradling 穹顶弧形肋
pendentive dome 三角穹顶；方墙四角支承拱上的穹顶

pendent pipe 悬吊管
pendent post （中世纪英国屋架的）下伸小柱；吊柱；墙架立柱；壁架柱
pendent rock 悬吊岩
pendent signal 悬垂式信号装置；悬垂信号灯
pendent terrace 连岛砂坝
pendent tube 悬吊管
pendent-type air hoist 悬挂式气动吊车
pendice 屋顶房间；平屋顶上的小屋
pendicle 垂饰
pendiculated 小墩支撑的
pendicule 小墩（充当支座）
pending 悬而未决的
pending claim 未决诉讼
pending interrupt condition 待处理中断条件
pending patent 申请中的专利
pen drawing 钢笔画
pen drawing oscilloscope 笔绘示波器
pendular dynamometer 摆锤式功率计；摆动式测力计
pendular inclination sensor 摆锤式倾斜传感器
pendular movement 钟摆运动
pendular movement of mediastinum 纵隔摆动
pendular oscillation 摆动
pendular ring 摆动环
pendular rotatory test 摆动旋转试验
pendular stage 环膜水阶段；触点水阶段；摆动水位
pendular water 悬着水；环膜水；触点水
pendulate 振摇
pendulation 摆动
pendulosity 摆性
pendulous 悬垂的
pendulous abdomen 悬垂腹
pendulous accelerator 悬垂式加速计
pendulous accelerometer 悬垂式加速计；摆式加速表
pendulous device 摆动装置
pendulous gyro accelerometer 摆式陀螺加速表
pendulous gyro integrating accelerator 摆式陀螺积分加速表
pendulous gyroscope 悬锤陀螺仪；摆式陀螺仪
pendulous heart 悬垂心
pendulous multideck screen 多层悬筛
pendulous pick-up 悬挂式传感器
pendulous plant 垂枝植物
pendulous type trigger 摆式触发器
pendulous vibration 悬摆振动
pendulum 振动体；摆动锤
pendulum accelerometer 摆式加速度计
pendulum-action spreader 摆动式撒布机
pendulum alidade 摆式照准仪
pendulum anemometer 摆式风速表
pendulum apparatus 摆仪
pendulum assembly 钟摆钻具
pendulum astrolabe 垂摆等高仪
pendulum balance 摆动天平
pendulum bar 摆杆
pendulum basis 摆仪基准
pendulum bearing 摆轴支座；摇摆式支座；摆柱支座；摆式轴承；摆动轴承；摆动支座
pendulum bearing for bridge 桥梁摆轴支座
pendulum bob 摆锤
pendulum camera 垂摆摄影机
pendulum car 摆式车辆
pendulum circular saw 摆动圆锯；摆动式圆锯
pendulum clinker cooler 悬挂摆动箅床式熟料篦冷机
pendulum clinometer 单摆横倾指示器；垂针式斜度仪；垂针式测斜仪；摆式斜度仪；摆式测斜仪；摆式测坡仪
pendulum clock 摆钟
pendulum concrete vibrator 摆动式混凝土振动器；摆式混凝土振捣器
pendulum cone 摆锥
pendulum control 摆式调节
pendulum conveyer 摆式运送机；摆式输送机
pendulum counterbalance 摆式配重
pendulum crane 摆式吊车
pendulum crusher 摆式破碎机
pendulum damper 摆式配重；摆式减振器
pendulum damping gear 飞摆阻尼机构
pendulum damping test 摆杆阻尼试验
pendulum day 摆日
pendulum decelerometer 摆式减速计

pendulum determination 摆仪测量
pendulum deviation 重力摆偏差
pendulum drill-assembly 摆式钻具（防止孔斜）
pendulum dynamometer 动摆测力器；摆动式测力计；摆动测力器；摆锤式功率计；摆锤式测功器
pendulum effect 摆动效应
pendulum feeder 摆式喂料器；摆式喂料机；摆式给料器；摆式给料机
pendulum flexure 摆仪架弯曲；摆的挠曲
pendulum floater 摆式浮标
pendulum force 钟摆力
pendulum governor 摆式调速器；摆动调节器；摆调节器
pendulum gradient indicator 垂针式斜度仪；垂针式测斜仪；摆式测斜仪；摆式测坡仪
pendulum gravimeter 摆式重力仪
pendulum gravity measurement 摆仪重力测量
pendulum gravity survey 摆仪重力测量
pendulum grip gear 摆式安全器
pendulum guide 摆式滑块
pendulum hammer 摆锤打桩机
pendulum hardness 摆杆硬度；摆锤硬度；摆测硬度
pendulum hardness tester 摆杆硬度试验机；摆杆硬度计；摆式硬度试验机
pendulum hydrometer 摆式流速仪
pendulum impact machine 摆锤冲击机
pendulum impact method 摆锤冲击法
pendulum impact strength 摆锤（法）冲击强度
pendulum impact test 摆锤（式）冲击试验；摆锤（法）冲击试验
pendulum impact tester 摆式冲击试验机；摆动冲击试验机；摆锤（式）冲击试验机
pendulum impact testing machine 摆式冲击试验机；摆锤（式）冲击试验机
pendulum inclinometer 单摆倾斜仪；摆式倾斜仪
pendulum instrument 摆仪
pendulum jigger 摆锤打光机
pendulum leader 摆式导管
pendulum leaf 摆动门扉
pendulum length 摆长
pendulum level 自动安平水准仪；垂针式水平仪；摆水准
pendulum-like movement 摆样运动
pendulum line 垂线
pendulum machine 摆锤打桩机；摆锤冲击机
pendulum magnetometer 摆式磁强计
pendulum manometer 摆式压力计
pendulum mass 摆质量
pendulum meter 摆式振动积算表
pendulum method 摆法
pendulum mill 摆式轧机
pendulum model 钟摆式
pendulum motion 摆动运动
pendulum motor 飞摆电动机
pendulum multiplier 拉伸冲程加长器；摆动式倍增器
pendulum observation 重力摆观测
pendulum oiler 摆锤润滑器
pendulum period 摆周期
pendulum pile driver 摆式打桩机
pendulum pipe hanger 摆式悬吊管子
pendulum plastometer 摆式塑性计
pendulum plate 摆动板
pendulum pliers 摆动式钳
pendulum plumb bob 摆式垂球
pendulum point 摆仪重力点
pendulum press 摆式压力机；摆锤压榨机
pendulum principle 摆原理
pendulum pulverizer 摆锤粉磨机
pendulum pump 摆泵
pendulum rectifier 振动式整流器
pendulum relay 振动子继电器；摆动子继电器
pendulum resistant balance 抗摆秤
pendulum rhythm 钟摆状节律；钟摆律
pendulum-rocker hardness 摆杆硬度
pendulum rod 摆杆；摆锤杆
pendulum roller 摇摆转子；底革打光机；摆锤辊子
pendulum rolling machine 摆压光机
pendulum rotary compressor 摆式回转压缩机
pendulum saw 吊截锯；摆锯
pendulum scale 摆秤
pendulum sclerometer 摆测硬度计
pendulum sclerometer hardness 摆式硬度计测定的硬度

pendulum seismograph 摆(动)式地震仪
pendulum sensor 摆式传感器
pendulum sextant 锤摆摆水准六分仪;摆式六分仪
pendulum shaft 测锤竖井;摆轴
pendulum shutter 摆动闸门
pendulum side bearing 摆块式旁承
pendulum slip 摆动分隔片
pendulum spray 摆式喷雾器
pendulum spring 悬挂弹簧;飞摆弹簧;摆簧;摆弹簧
pendulum stabilizer 悬摆稳定器
pendulum stanchion 摇摆式支座;摆轴(支)柱
pendulum station 摆仪重力点
pendulum strength tester 摆锤式强力试验机
pendulum support 摆仪座;摆式支撑
pendulum survey 摆仪测量
pendulum suspension 摆式悬挂
pendulum table 摆动升降台
pendulum tester 摆式(试验)仪;摆锤式试验仪
pendulum tiltmeter 摆式倾斜仪
pendulum trigger 摆式触发器
pendulum-type coder 摆式发码器
pendulum-type distributor 摆动式撒布机
pendulum-type dynamometer 摆式测力计
pendulum-type friction machine 摆锤式摩擦试验机
pendulum-type impact machine 摆式冲击机
pendulum-type mechanical rake 摆式耙路机
pendulum-type portable skid resistance tester 轻便摆式抗滑测验仪
pendulum-type raker 摆式耙路机
pendulum-type ring roll pulverizer 摆式磨粉机
pendulum-type sampler 摆式取样机
pendulum-type tachometer 摆式转数计
pendulum-type time relay 摆式时间继电器
pendulum-type wave generator 摆式造波器
pendulum value 摆仪重力值
pendulum vector 摆向量;摆矢
pendulum vibration 摆动;摆的振动
pendulum visco(si)meter 摆锤(式)黏度计
pendulum wire 垂线;摆线
penecontemporaneous 准同生的;准同期的;沉积后到固结前的
penecontemporaneous deformation 同生变形
penecontemporaneous deformation structure 准同生变形构造;准同期变形构造
penecontemporaneous faulting 准同期断裂
penecontemporaneous fold 准同期褶皱
peneplain 准平原;基准面平原;侵蚀高原
peneplain deposit 准平原沉积
peneplanation 准平原作用
peneplane 准平原
pen equation 笔头差
pen error 笔尖误差
peneseismic country 准震区
peneseismic region 准震区
penetrability 贯穿性;贯穿能力;贯穿本领;可穿透性;渗透性;渗透力;穿透性
penetrable 可渗透的;能贯穿
penetralia 宫中密室;内部房间;最深处(尤指房屋的内院、高宇的内殿)
penetrameter 透度计;透光计
penetrameter sensitivity 透度计灵敏度
penetrance 贯穿性;可穿透性;透射;通过;穿透
penetrant 贯入料;渗透剂;穿透物;穿透剂;穿透的
penetrant inspection 着色检查;浸透检查;浸渗检验;渗透检验
penetrant method (超声波探伤的)透过法
penetrant test 渗透检验
penetrate 透视;通过;穿透;穿入
penetrated asphalt ballast 充填式沥青道床
penetrated crack 穿透裂纹
penetrated force 贯入力
penetrated mortise and tenon 长榫
penetrated mortise hole 穿横板的榫眼
penetrated record 贯入记录
penetrated set 贯入度
penetrated test 贯入度试验
penetrating 贯入;贯入度;渗透
penetrating ability 贯穿能力
penetrating agent 渗透剂
penetrating aid 防水剂
penetrating asphalt 贯入用沥青;渗入沥青
penetrating capacity 穿透能力
penetrating component 穿透成分

penetrating concentration 透过浓度
penetrating cone 贯入锥;触探锥;触探
penetrating depth 贯穿深度
penetrating face (钻头的)克取面
penetrating finish 浸漆处理;渗透性清漆;渗透清漆涂层
penetrating floor(ing) sealer 渗透地板密封剂(渗入木材的漆)
penetrating fluid 浸注液;渗透液
penetrating force 贯入力;穿透力
penetrating item 贯穿件
penetrating ligature 贯穿结扎
penetrating luminous flux 透射光通量
penetrating oil 贯入用沥青;渗透油
penetrating oil pot 油浸锅
penetrating power 渗透力;贯穿(能)力;贯穿本领;穿透能力;穿透率;穿透本领
penetrating pricing 最满意定价
penetrating quality 贯入度;贯入性
penetrating radiation 贯穿辐射;穿透(性)辐射
penetrating sealer 渗透性密封材料;渗透保护层
penetrating sealing 渗透密封
penetrating soft soil 穿过软土
penetrating stain 渗透着色(剂);渗透染色剂
penetrating strategy 渗透策略;穿透策略
penetrating tie 穿插枋
penetrating time 纯钻进时间
penetrating type sealant 混凝土防冻用的透入式封面料
penetrating varnish 渗透清漆
penetrating vibrator 插入式振动器;插入式振捣器
penetrating wave 贯穿波
penetrating well 贯穿井;渗水井;穿透井
penetration 针入度;贯入度;贯穿;浸透;焊透;熔透;熔深(指焊接);侵彻度;透射深度;透入(性);透过;穿透深度;穿透率;穿透力;穿透(度);插入浸透
penetration ability 贯穿能力
penetration advance 钻孔进度
penetration asphalt 渗透用沥青
penetration bead 穿透焊珠;穿透熔焊;穿透小珠;熔深焊道
penetration binder 灌入结合料
penetration bituminous macadam surface 贯入式沥青碎石面层;贯入式沥青碎石路面
penetration breccia 贯入角砾岩
penetration capacity 穿透能力
penetration coat 灌浇层;贯入层
penetration coefficient 贯入系数;贯穿系数;穿透系数
penetration coefficient of sheet pile 板桩贯入系数
penetration complex 透入络合物
penetration concrete 灌入混凝土;贯入混凝土
penetration cone 贯入锥
penetration construction 贯入法(施工);用贯入法筑路
penetration control 穿透深度控制
penetration convection 穿透对流
penetration course 用贯入法的结构层;灌浇层;灌浆层;贯入层;浇灌层
penetration cross-section 贯入截面
penetration curve 渗透曲线
penetration dampness 透湿度
penetration depth 压痕深度;贯入深度;透入深度;渗透深度;射入深度;穿透深度;插土深度
penetration dish 贯入碟;针入度保温蝶;沥青针入度保温碟
penetration distance 贯入深度;渗透距离
penetration drilling of well completion 造井机具
penetration drying type ink 渗透干燥型油墨
penetration dy(e)ing 渗透染色
penetration effect 穿透效应
penetration factor 穿透因子
penetration feed 延伸速度(钻进时);给进速度(钻进时)
penetration force 贯入力
penetration funnel 贯穿漏斗
penetration fusion welding 深熔焊接
penetration ga(u)ge 穿透深度计
penetration gneiss 贯入片麻岩
penetration-grade asphalt 黏稠沥青;膏体(石油)沥青
penetration-grade bitumen 膏体(石油)沥青;黏稠沥青(英国)

penetration hardness 压痕硬度
penetration index 针入度指数;贯入(度)指数;渗透指数
penetration inspection 渗透探伤
penetration layer 贯入层
penetration limestone 贯入石灰岩
penetration limit 贯入限度;贯入极限
penetration load 贯入荷载
penetration-load curve 贯入度—荷载曲线
penetration log 钻速记录
penetration macadam 灌入法碎石路;灌沥青碎石面层;灌沥青碎石路面;灌浆碎石路;贯入式碎石路
penetration macadam with coated chips 上拌下贯式(沥青)路面
penetration material 穿透物质
penetration metamorphism 贯入变质(作用)
penetration method 灌入法;灌浆(方)法;贯入法【岩】
penetration needle 贯入针;土壤密实度测定针
penetration number 透入值;(沥青的)针入度(测定)值;渗透值
penetration of bitumen 沥青针入度
penetration of dampness 透湿度;水分渗透作用;水分渗透
penetration of echo sounder 回声测深仪的穿透能力
penetration of electric(al) field 电场渗透
penetration of fracture 裂缝深度
penetration of frost 冰冻深度
penetration of hardness 淬硬深度
penetration of joints 节理穿层性
penetration of pile 桩贯入度;桩的贯入量;桩的贯入度;沉桩
penetration of radiation 照射透入
penetration of rain 雨水渗透
penetration of spaces 空间渗透
penetration of splinters 弹片穿透深度
penetration of the tool 进刀
penetration of two parallel barrel vaults 两个平行筒状拱顶的贯穿
penetration of water 渗水
penetration oil 贯入沥青
penetration pavement 贯入法(修筑)路面
penetration peg 穿透钉
penetration per blow 每锤击贯入度;每击贯入度;每锤贯入量;每冲击一次套管的延深(用锤夯入)
penetration per revolution 每转进尺量【岩】
penetration per rig-month 钻探效率
penetration piece 贯穿件
penetration pile 贯入桩
penetration piston 贯入柱塞(CBR试验用)
penetration pit 贯穿坑
penetration pricing 最低价格定价法;最低定价法
penetration priming 渗透性底油;渗透性底漆
penetration probability 穿透概率
penetration probe (测硬化混凝土的)贯入阻力探针
penetration quantity 贯入量
penetration range 夜间能见距;夜视距离
penetration rate 钻探效率;钻入速度;钻进速度;凿孔速度;贯入量;机械钻速;焊透率;穿透速度;穿透率;穿进速率
penetration rate meter 钻速表
penetration rate monitor 钻机钻速监控器
penetration rate probe (测硬化混凝土的)贯入阻力探针
penetration ratio 针入度比;熔合比
penetration record 贯入记录
penetration resistance 压入阻力;压入抗力;贯入阻力;贯入阻抗;抗穿透力;抗渗性;抗穿穿强度;抗穿透性;耐穿透性;动贯入阻力;(钻头钻入层时的)吃入阻力
penetration resistance curve 贯入阻力曲线
penetration resistance of cohesive soil 黏性土的贯入阻力
penetration resistance test 针入度试验(沥青的);贯入度试验
penetration resistance value 贯入阻力值
penetration riveting 贯入铆接
penetration rock 贯入岩石
penetration rod drilling 贯入杆钻孔
penetration sleeve 穿通套筒;穿通套管
penetration slip surface 贯入滑动面
penetration sounding 贯入触探;触探
penetration speed 钻进速度;钻机穿进速度;给进

速度(钻进时)
penetration strain 贯穿应变
penetration strength 贯入强度;贯入力;抗穿透强度
penetration structure 穿透构造
penetration surface course 灌沥青面层;贯入式面层;表面渗入层;表层渗入层
penetration surface treatment 贯入式表面处治;贯入式表面处理;表面渗入处理
penetration technique 贯入技术;贯入试验法
penetration test 锥入度测定;(沥青的)针入试验;贯入试验;贯入度试验;透入度试验;透度试验;刺度试验;穿透试验;触探试验
penetration test after loss-on-heating 加热损失后针入度试验
penetration test apparatus 长杆贯入仪
penetration testing (沥青的)针入度试验;贯入(度)试验
penetration testing needle 贯入试针
penetration texture 穿透结构
penetration theory 渗透理论;穿透理论
penetration through brickwork 砖墙渗漏
penetration through wall 墙体渗透性
penetration time 钻进时间;渗透时间
penetration tip 贯入(顶)端
penetration treatment 贯入处治(道路);灌沥青处理;灌浆处理
penetration twin 贯穿孪晶;穿插双晶;穿插孪晶
penetration type cement-bound macadam 贯入式水泥结碎石
penetration value 贯入值
penetration varnish 渗透清漆
penetration viscometer 压入法黏度计
penetration water source 渗透型水源地
penetration weld 熔透焊缝
penetration work 灌浇工作;浇灌工作
penetrative 能贯穿
penetrative cell 穿透单元
penetrative convection 贯穿对流;透射对流;穿透对流
penetrative element 透入性要素
penetrative fabric 透入性组构
penetrative lineation 透入性线理
penetrative radiation 贯穿辐射
penetrative rock 贯入岩(层)
penetrative shear 透入剪切
penetrator 圆锥(体);针入度仪;针入度计;贯入器;贯入计;穿头
penetrator ram 推顶杆
penetrex 防水无色液体
penetrometer 针入硬度计;针入度仪;针入度计;针穿硬度计;贯入仪;贯入器;贯入度计;贯密计;透光计;透度计;穿透仪;穿透计;触探计;触探器;稠密计;稠度仪;稠度计
penetrometer number (沥青的)针入度(测定)值;稠度计测定值
penetrometer of bitumen 沥青针入度仪
penetrometer test 触探试验
penetron 重电子(即介子);伽马射线穿透仪
penetron tube 射线穿透管
penetro-visco(si)meter 贯入式黏度计;针入式黏度计
Pen-Farrel 彭—法心轴机器护罩
penfieldite 六方氯铅矿
penfold 牲畜围栏;牛栏;一磅装(盒子)
pen galvanometer 笔式电流计
penguin 滑走教练机
pen holder 钢笔杆;笔杆
penicilium 毛丛
penicillate 毛笔形的;画笔状的
penicillate maxilla 毛下颚
penicilliary 毛笔形的
penicilliform 毛笔形的
penicillin formulation effluent 含青霉素成分的废水
penikisite 磷铝镁钡石
penikkavaarite 暗钠闪辉长斑岩
peninsula 半岛
peninsula-base kitchen cabinet 半岛状厨房橱柜
peninsular garden 半岛园
peninsular pier 宽突堤码头
peninsular platform 半岛式月台
peninsular quay 半岛式码头
penitential cell 牢房;教养所
penitentiary (寺院中的)忏悔室;教养院;教养所;监狱

penitent ice 残余冰
penitent snow 残余雪
pen knife 削笔刀;小刀;单开小刀
penkvilksite 水短柱石
penlight 钢笔式手电筒
Penman's method (计算蒸发能力的)彭曼法
pen metal 含锡黄铜(铜85%,锌13%,锡2%)
pen motor 记录笔驱动电机;记录笔电动机
pennant 悬饰;小三角旗;奖旗;尖旗;三角旗;短挂绳;风三角
pennant diagram 尖旗图(连续梁图解法)
pennant grit 粗砂粒;三角形砂粒
pennantite 锰绿泥石
pennant line 缆绳
pennant rod 小三角旗的旗杆
pennant staff 三角旗杆;船首旗杆
pennant wire 缆绳
pen nib 钢笔尖
pennine 叶绿泥石
Pennine lime 彭奈恩山石灰
Pennine Range 彭奈恩山脉
penning 用锤尖锤击;块石铺砌的路面;块石垫层;石块铺砌
penning gate 直升式(矩形)泄水闸门
Penning ga(u)ge 彭宁冷阴极电离真空计
Penning pump 彭宁放电泵
penninite 叶绿泥石
Pennsylvania bluestone 宾夕法尼亚青石
Pennsylvania drilling 宾夕法尼亚钻进法
Pennsylvania drilling method 宾夕法尼亚钻进法
Pennsylvania Era 宾夕法尼亚纪【地】
Pennsylvania farmhouse colonial architecture 宾夕法尼亚农舍殖民地建筑(美国)
Pennsylvanian system of drilling 钢绳冲击钻进(方)法
Pennsylvania slate 灰黑色板岩;宾夕法尼亚板岩(美国)
Pennsylvania System 宾夕法尼亚系【地】
Pennsylvania truss 折弦(再)分析架;宾夕法尼亚式桁架
pen number 笔号
penny 便士;分(钉子的长度单位)
penny arcade 花费极少的游乐场所
Pennycock 彭尼科克屋面装配玻璃
pennydog 杂工工头
penny gaff 低票价剧场;低级音乐厅
penny-in-the-slot 自动售货机
penny post 邮政
penny-shaped crack 铜板形裂缝
penny stone 平扁小(块)石;扁平小石
pennyweight 便士重
Penokean 安尼米基系
pen oscillograph 描笔式示波器;笔式示波器
penpit 大不列颠的史前六居
pen plotter 笔式绘图机;墨水笔绘图机;笔式绘图仪
pen point 笔尖
pen recorder 印码机油墨记录器;描笔式记录器;笔式自动记录器;笔式记录器;笔尖记录器
pen recording polarograph 笔式极谱仪;笔式极谱仪
pen record oscillograph 笔录示波器
penroseite 硒铜镍矿
Penrose process 彭罗斯过程
pen ruling 直线笔划线
Penryn 彭林淡灰色花岗岩(英国康沃尔产)
pen scribe 笔尖划线法
pension 恤养金;养老金;退休金;膳宿公寓
pensionable age 退休年龄
pension allowance 养老金
pension basis 退休金计算基础
pension contributions 养老金缴款
pensioner 养老金领取者
pension fund 退休金专款;恤养基金;养老金基金
pension fund deductions 纳税时的养老金扣除额
pension fund reserves 恤养基金准备;养老金基金准备
pension plan 养老金计划;退休金计划
pension plan accounting 养老金计划会计
pension pool 养老金联营
pension rights 养老金领取权
pension schemes 养老金计划
pension trust 养老金信托
Pensky-Martens closed (flash) tester 宾斯克—

马丁密闭式闪点试验器;宾斯克—马丁密闭式闪点测定仪
Pensky-Marten's flame point 宾斯克—马丁法闪点
Pensky-Martens flash-point 宾斯克—马丁密闭式闪火点
Pensky-Martens flash-point test 宾斯克—马丁密闭式闪点试验
Pensky-Martens flash point tester 宾斯克—马丁密闭式闪点试验器;宾斯克—马丁测定仪
Pensky-Martens closed tester 宾斯克—马丁密闭式测定仪
pen speed 记录笔速度
pen-stabling 散养牛棚
penstock 引水道;压力(输)水管;压力钢管;给水栓;救火龙头;水闸;水门
penstock bracing 压力管道支撑
penstock buried 埋入式压力水管;埋入式压力管道;埋入式压力钢管;埋管
penstock bypass chamber 压力管道旁通管室
penstock course 压力管道管节;压力水管管节
penstock dam 水闸坝;闸门坝
penstock dam gallery 水闸坝瞭望塔;水闸坝坑道
penstock drain 压力管道排水;闸门排水
penstock drain header 压力管道排水总管
penstock footing 压力水管管座
penstock gate 压力管道闸门
penstock hydraulics 压力管道水力学
penstock intake 压力管道进水口
penstock manifold 水闸坝支管;水闸坝歧管;压力管道支管;压力管道分叉管
penstock pier 压力管道支墩
penstock pipe 压力管道;水闸管道
penstock reducer 压力管道调压井
penstock reflection time 压力管道反射时间
penstock section 压力管道管段;压力管道断面
penstock shaft 压力井;闸门轴;压力管井
penstock tangent 压力管道直线段
penstock valve 压力钢管阀(门)
pen system 光笔系统
pent 坡屋;庇檐;单斜顶棚
penta 五氯苯酚
pentabasic 五元碱性的
pentabasic acid 五元酸
pentabasic alcohol 五元醇
pentaborane 五硼烷
pentabromization 五溴化反应
pentabromochlorocyclohexane 五溴一氯环己烷
pentabromodiphenyl oxide 五溴二苯醚
pentabromoethyl benzene 五溴乙基苯
pentabromoniline 五溴苯胺
pentabromophenol 五溴苯酚
pentabromotoluene 五溴甲苯
pentacarboxylic acid 五羧酸
pentacene 并五苯
pentachlorodiphenyl 五氯联苯
pentachloroethane 五氯乙烷
pentachloronitrobenzene 五氯硝基苯
pentachlorophenate poisoning 五氯酚中毒
pentachlorophenol 五氯苯酚七十四(一种防虫剂);五氯苯酚
pentachlorophenol wastewater 含五氯苯酚废水
pentachlorothiophenol 五氯硫酚
pentacite 季戊四醇醇酸树脂;季戊醇酸树脂
pentacle 五角星饰;五角星
pentacle of Solomon 所罗门的五角星饰
pentacontane 五十烷
pentacyclic 五轮列的
pentacyclictriterpane 五环三萜烷
pentacyclicus 五轮列的
pentad 五天(期);五个一组;五个接触点
pentadecagon 十五边形
pentadecane 十五烷
pentadecanoic acid 十五酸
pentadecanol 十五烷醇
pentadecanone 十五烷酮
pentadecanoyl 十五酰
pentadecendioic acid 十五烯二酸
pentadecene diacid 十五烯二酸
pentadecylenic acid 十五烯酸
pentadecylic acid 十五酸
pentadiene 戊二烯
pentaerythrite 季戊四醇
pentaerythrite tetranitrate 季戊四醇四硝酸酯
pentaerythritol 四甲醇甲烷;季戊四醇

pentaerythritol abietate 季戊四醇松香酯
pentaerythritol ester 季戊四醇酯
pentaerythritol fatty ester 季戊四醇脂肪酸酯
pentaerythritol maleic resin 马来酸季戊四醇树脂
pentaerythritol phosphite 季戊四醇亚磷酸酯
pentaerythritol phthalic resin 邻苯二甲酸季戊四醇树脂
pentaery thritol tallate 妥尔油酸季戊四醇酯
pentaerythritol tetran-haptanoate 庚酸季戊四醇酯
pentaerythritol tetranitrate 季戊四醇四硝酸酯
pentaerythritol tetrastearate 季戊四醇四硬脂酸酯
pentaethrite tetra-acetate 季戊四醇四乙酸酯
pentaethylene hexamine 五乙烯六胺；五亚乙基六胺
pentafluorobenzoic acid 五氟苯甲酸
pentagon 五连形；五角形；五边形
pentagonal 五角形的；五方的；五边形的
pentagonal bastion 五角形堡垒
pentagonal bipyramid 五角双锥
pentagonal dodecahedron 五角十二面体
pentagonal ground plan 五角形地平面图
pentagonal indenter 五边形压痕器
pentagonal number 五角数
pentagonal prism 五角棱柱
pentagonal screen 五角形防护幕
pentagonal section 五角形截面；五边形截面
pentagonal tile 五角形砖
pentagonal tower 五角形塔楼
pentagonite 五角石
pentagonum 五边形
pentagram 五角星形；五角星
pentagrid of converter 变频器用五极管
pentahapto 五合
pentahedron 五面体
pentahydrite 镁黄铁矿；五水泻盐
pentahydroborite 五水硼钙石
pentahydrocalcite 五水碳钙石
pentahydroxyhexoic acid 半乳糖酸
pentaiodide 五碘化物
pentaiodoethane 五碘乙烷
pentalane 并环戊烷
pentalene 并环戊二烯
pentalobal cross-section 五叶形截面
pentalpha 五角星形
pentamer 五聚物
pentamethide 有机金属化合物
pentamethine 二碳花青
pentamethine cyanme 五甲炔花青
pentamethoxyl red 五甲氧基红
pentamethylene 环戊烷
pentamethylenediamine 戊二胺
pentamethylene glycol 亚甲基二醇；戊二醇
pentamethylene tetramine 亚甲基四胺
pentamethyl pararosaniline 五甲基副品红
pentamethyl violet 五甲紫；巴黎紫
pentamirror 五面镜
pentamul 乳化剂
pentandioic acid 戊二酸
pentane 正戊烷；戊烷
pentane dicarboxylic acid 庚二酸；戊二羧酸
pentanedioic acid 戊二酸
pentanediol 戊二醇
pentanedione 乙酰基丙酮
pentane lamp 戊烷灯
pentangular 五角的
pentanizer 戊烷馏除塔
pentanizing column 戊烷馏除塔
pentanizing tower 戊烷馏除塔
pentanol 戊醇
pentan-thiol 戊硫醇
pentaoxacyclooctadecane 五氧杂环十八烷
pentaphenylethane 五苯基乙烷
pentaprism 五棱镜；五角棱镜
pentapropyl 五丙基
pentaptych 五幅一联画
pentasol 工业戊醇
pentastyle 五柱式
pentastyle building 五柱式建筑
pentastyle temple 五柱式庙宇
pentastylos 五柱式建筑
pentee 电梯机器房；附加房屋；披屋；阁楼
Pentelic marble 彭代里孔山的大理石（希腊）
pentenoic acid 戊烯酸
pentenol 戊烯醇

penthemeron 五天(期)
penthouse 屋顶小楼；单斜顶棚；遮棚；阁楼；披屋；屋檐；屋顶机房；屋顶公寓；屋顶房间；风雨棚
penthouse apartment 屋顶公寓
penthouse apartment unit 房间公寓单元；阁楼公寓单元
penthouse combustion chamber 上层燃烧室
penthouse flat 披屋套间
penthouse living unit 披屋居住单元
penthouse machine room 阁楼机房
penthouse roof 披屋顶；单坡屋顶
penthrite 烈性炸药；季戊炸药；季戊四醇四硝酸酯
pentice 披屋；坡屋；平屋顶上的小屋；屋顶房间
pentile 波形瓦
pentlandite 硫镍铁矿；镍黄铁矿
pentode 五极管
penton 聚二氯季戊醚；片通（一种氯化聚醚塑料）
penton-rubber 氯化聚醚塑橡胶
pentosan 多缩戊糖
pen touch 笔痕
pentoxide 五氧化物；五氯化物
pentration energy loss 贯穿能量损失
pent roof 庇屋顶；单坡屋顶；单斜屋顶
pentrough 水槽
pent-up 被拦住
pent-up demand 被抑制的需求
pent-up water 壅高水；封存水；封闭水
pentyl 季戊炸药
pentyl amine 戊胺
pentylidyne 次戊基
pentyne 戊炔
pen type 笔型
pen-type barn 散放牛舍
pen-type dosimeter 笔型剂量计
pen-type pulse tester 脉冲测试笔
penultimate certificate 工程预决算书
penumatic double-pen recorder 气动双针记录器
penumbra 半影；半阴影；画面浓淡相交处
penumbra cone 半影锥
penumbral eclipse 半影食
penumbral lunar eclipse 半影月食
penumbra theory 半影原理
penumbra region 半影区
pen-writing oscilloscope 笔录示波器
pen writing recording instrument 笔式记录器
peon 日工；雇农
peonage 雇佣日工
peony 牡丹(红)
peony brocade 牡丹绸
peony design 牡丹纹
peony red 牡丹（红）色；浅紫红色
people flow rate 人流速率
people mover 运客工具
people mover system 运客装置系统
people move transit system 小型快递交通系统
people outside the authorized personnel quota 编外人员
people power 人员力量
people's air defence 人防
people's air defense subway 人防连通道
People's Heroes Monument 人民英雄纪念碑
people's livelihood 民生
people throughout 行人通过量
people-to-people trading activity 民间贸易
Peorian 皮奥里间冰期
peperino 碎晶凝灰岩
peplomer 膜粒
pepper-and-salt 椒盐色的
pepper and salt fracture 局部珠光体化断口
pepperbox turret 胡椒盒式塔楼（圆平面尖顶或圆顶）
pepper-corn rent 象征性租金
peppermint camphor 薄荷醇
peppermint essence 薄荷精
peppermint oil 薄荷油
peppermint oil dementholized 薄荷素油
peppermint soap 薄荷皂
peppermint test 薄荷油试验
pepper sand 胡椒子砂
pepper sludge 微粒悬浮酸渣
pepper tree 漆椒树
peppery 涂层上凸起的小块或颗粒
pepple coating 斑纹漆
peptide 缩氨酸

peptisation = peptization
peptizate 胶溶体
peptization 解胶作用；解胶；胶溶(作用)；分散
peptizator 胶乳剂；胶溶剂
peptizer 胶溶剂；凝固剂；塑解剂
peptizing 胶束吸附；胶溶作用
peptizing agent 胶溶剂；胶化剂；塑解剂
peptizing power 胶溶能力
peptizing property 胶溶性
pepyrus column 有纸莎草茎图案的圆柱
peracetic acid 过乙酸；过氧乙酸；过醋酸
peracid 过酸；高酸
peracidity 过酸性
peralcohol 过氧化合物
peralkaline 过碱性
peraluminous 过铝质
perambulation 巡视；踏勘；查勘
perambulator 间距规；手推车；转轮测距仪；路程计；测器程；测程轮；测程车；步程仪
per ampoule 每安瓿
per animal 每匹
per annum 每年；按年计
per annum rate 年率
perbasic 高碱性的
perbenzoic acid 过苯甲酸
per biennium 每两年(期)
perbituminous coal 高氢烟煤
per blow 每次冲击
per blow rate 每次冲击率
perborate 过硼酸盐
perborate of soda 过硼酸钠
perboric acid 过硼酸
perbromic acid 过高溴酸
perbromo-acetone 六溴丙酮；全溴丙酮
perbromo-ethane 六溴乙烷；全溴乙烷
perbromo-ether 全溴乙醚
perbromo-ethylene 全溴乙烯；四溴代乙烯
perbunan 丁氰橡胶；别布橡胶
Perca fluviatilis 河鲈
per capita 每人；每年平均值；人均；按人均人数分配；按人均人口分配；按人计算
per capita calculation 每人计算
per capita consumption 按人头平均计算的消费；人均消耗量；按人口平均计算的消费；每人消耗量
per capita consumption daily 每人每日消耗量；每人每日耗(水)量
per capita consumption expenditures 人均消费支出
per capita demand 人均需求量
per capita demand method 人均需求法
per capita disposable real income 人均可支配实际收入
per capita dose 每人剂量
per capita energy consumption 人均能源消费量
per capita expenditures 人均支出
per capita gross domestic product 人均国内生产总值
per capita gross national product 人均国民生产总值
per capita gross product 人均总产量
per capita household electricity consumption 人均用电量
per capita income 每人所得；人均收入；人均国民收入
per capita index 人均指标
per capita living space 人均居住面积
per capita national income 人均国民收入
per capita output 人均产量
per capita per day 每人每日
per capita product 人均产值
per capita quantity index 人均物量指数
per capita rate 人均税率
per capita real output 人均实际产量
per capita retained profit 人均留利
per capitat level 居民人均税额
per caput dose equivalent 每人平均剂量当量
percarbonate 过碳酸盐；过碳酸钾
percarbonic acid 过酸酯
perceived colo(u)r 知觉色
perceived delay 感觉延误
perceived model 感觉模型；视模型
perceived noise decibel 感觉噪声分贝；可感噪声分贝；可闻噪声分贝
perceived noise level 感觉噪声级；可感噪声级

perceived noise level in decibels 感觉到的噪声分贝
perceived noisiness 噪度
perceived pseudoscopic model 感觉的假立体模型
perceived stereomodel 感觉的立体模型
per cent 每百分;百分之几;百分率;百分比
percent absorption 吸光百分率
percent abundance 百分率中相对分布量
percentage 延伸率;百分数;百分率;百分比
percentage absolute humidity 绝对湿度百分率
percentage absorptivity 百分吸收系数
percentage addition 百分数附加
percentage afforestation 造林百分比
percentage agreement 按总造价百分比计费的合约;按百分比收费协议;按造价百分比付款合约
percentage air void 含气率
percentage allocation 按百分比分摊
percentage allowance over-standard time 超过标准时间的宽度百分率
percentage analysis method 百分率分析法
percentage annual import 进口量占百分比
percentage area method 百分面积法
percentage articulation 传声百分清晰度
percentage balance sheet 百分率资产负债表
percentage bar chart 构成柱状图;百分条图
percentage bar diagram 百分条图
percentage bar graph 百分比柱形图表
percentage basis 百分率基准;百分比基准
percentage beam modulation 束调制度
percentage berth occupancy 泊位利用率
percentage between screens 中间率(相邻筛号之间的成分百分率)
percentage bridge 百分法电桥
percentage bulking (砂的)湿胀率
percentage by volume 体积百分比;容积浓度百分率;体积百分数;体积百分数;按体积计百分率
percentage by weight 重量百分比(率);按重量计百分数
percentage cash flow 现金流量率;资产股息率
percentage chart 百分率图
percentage circular chart 百分圆图
percentage comparison 百分率比较
percentage composition 组成百分比;结构百分率;成分百分比;百分组成;百分含量
percentage concentration 浓度百分率;百分浓度;百分比浓度
percentage conductivity 百分电导率
percentage consolidation 固结率
percentage content value 百分比含量值
percentage contract 按总价百分比收费合同;提成合同;按百分比收费合同
percentage core recovery 岩芯获得率;岩芯回采率
percentage correction 百分率校正
percentage cover 覆盖百分率
percentage curve 分布曲线
percentage depletion 消耗百分比;百分率折耗;百分法折耗
percentage depletion allowance 百分率折耗减免
percentage deviation 百分偏差
percentage differential protection 比率差动保护装置
percentage differential relay 比率差动继电器;百分差动继电器
percentage distortion 畸变系数
percentage distribution 相对分布;分配百分比
percentage dot map 百分点地图
percentage elongation 伸长百分率;百分(比)伸长率
percentage error 相对误差;误差百分率;百分(率)误差;百分比误差
percentage extraction 回采率;萃取(百分)率
percentage fee 按总价百分比计费;按造价百分率付款;百分数补偿费用;百分比费用
percentage fibre extraction 剥麻率
percentage fines 细粒含量百分率
percentage fraction weight 筛分粒径组分重量百分比;筛分粒径组分重量百分率;筛分颗粒组分重量百分比;筛分颗粒组分重量百分率
percentage frequency 百分频率
percentage humidity 饱和度;湿度百分数;湿度百分;百分(数)湿度
percentage inclination 倾斜率;倾斜百分率;坡度率;坡度百分率
percentage increase 增长率;增长的百分数;百分数增加;增产百分率
percentage in weight 重量百分数
percentage line drop 电路电压降百分数
percentage liquid phase 液相百分数
percentage loading of mill 磨机填充率;球磨的填充系数
percentage loss 损失率;损耗比;百分率损耗
percentage map 百分率图
percentage method 百分率法;百分法
percentage modulation 调制度;调制百分比
percentage of absentees to total staff in service 缺勤率
percentage of accumulated sieve residues 累计筛余百分数;累计筛余百分率
percentage of accuracy 准确度百分数;精度百分数
percentage of bed-expansion 膨胀率【给】;床膨胀率
percentage of bright sunshine 日照率
percentage of broken grain 谷粒破碎率
percentage of carbon 碳的百分数
percentage of class A goods 正品率
percentage of clean-up 清舱量比重
percentage of coal thickness 含煤系数
percentage of complement 配套百分率
percentage of completion 完工百分率
percentage-of-completion basis 完工百分率制
percentage-of-completion inventory valuation 完工百分率制的存货估计
percentage-of-completion method 完工百分率法;完工程度毛利计算法
percentage of consolidation 压缩百分率;固结百分数;固结百分率;固结百分比
percentage of content 含量百分比
percentage of core recovery 岩芯取样率;岩芯采收率;岩芯采取率
percentage of coverage 覆盖率
percentage of damage 破碎率;损失的百分(比)率
percentage of damaged goods 残损率
percentage of damaged house 房屋损害率
percentage of defective goods 次品百分率
percentage of direct transshipment 直取比重;直接换装比
percentage of divided seam recovery 分层采取率
percentage of double tracks to total route kilometres 铁路双线率;铁路复线率
percentage of double track total route kilometers 双线占线路总里程的百分率
percentage of drag reduction 减阻率
percentage of dust 粉尘的百分率
percentage of earbearing tiller 成穗率
percentage of electric(al) sections to total route kilometers 电化率
percentage of elongation 伸展率
percentage of employees living with dependents 职工带眷比
percentage of empty container 空箱比重(集装箱)
percentage of fines 细粉的百分率;细粉率;精粉率
percentage of fine sand 细砂百分含量
percentage of forest cover 森林覆盖率;森林覆被率
percentage of germination 发芽率
percentage of glass 玻璃质比率
percentage of grade 坡度百分率
percentage of gravel 砾石百分含量
percentage of green area 绿地百分数;绿地百分比
percentage of higher-category reserves 高级储量百分比
percentage of hollow filament 空心率
percentage of humidity 湿度百分数;湿度百分率
percentage of hydrocarbons detected 检测含烃百分比
percentage of idle time to utilization time 闲置时间和利用时间百分比
percentage of inclination 倾斜率
percentage of land-covered forests 森林覆盖率
percentage of liveweight growth 增重率
percentage of loading of mill 磨机研磨体填充系数
percentage of locomotive under repair 机车检修率
percentage of log correction 计程仪改正率
percentage of loss 损耗率
percentage of machine-days on good condition 机械台日完好率
percentage-of-mean-annual method 多年平均值百分率法
percentage of mineral dilution 矿物贫化率
percentage of moisture 含水率;含水百分数;湿度百分数;湿度百分率
percentage of moisture content 含湿百分率
percentage of non-productive stay 非生产性停泊所占比重
percentage of open area 开孔百分率
percentage of opening 筛孔面积百分率
percentage of oxygen saturation 氧饱和率;氧饱和百分率
percentage of particle breakage 颗粒破碎率
percentage of pass 总产品率
percentage of passenger seats utilization 客座利用率
percentage of passenger seats utilization per car 客车客座利用率
percentage of passenger seats utilization per train 列车客座利用率
percentage of passing 过筛百分数;过筛百分率
percentage of passing weight 过筛重量通过率
percentage of perfectness of handling machinery 装卸机械完好率
percentage of pole embrace 极弧系数
percentage of politics restraint 政治干预概率
percentage of porosity 孔体积百分数
percentage of product 品级率
percentage of products sent back for repair 返修率
percentage of profit 利益率
percentage of punctaulity of trains despatched to total trains 列车出发正点率
percentage of purity 纯度
percentage of reinforcement 钢筋百分比(率);含钢筋率;配筋率
percentage of reject 拒收百分率
percentage of root mean square amplitude 均方根振幅百分比
percentage of runoff 径流率
percentage of sales analysis 销售百分率分析;销售百分比分析
percentage of sales approach 销售百分比法
percentage of sales forecasting method 销售额百分比预测法
percentage of sales method 销售百分比法
percentage of sand 含砂率;砂率;砂的百分含量
percentage of saturated(dissolved) oxygen 溶解氧饱和百分数;溶解氧饱和百分率
percentage of saturated water content 饱和含水率
percentage of saturation 饱和率
percentage of sediment 泥砂含量;泥砂百分率;含泥量
percentage of separate sieve residue 分计筛余百分数
percentage of shale content 含砼率
percentage of short-circuit impedance 短路阻抗电压百分比
percentage of shot hole(depth in blasting) 炮眼利用率
percentage of simultaneous drainage 同时排水百分率
percentage of simultaneous water demand 同时给水百分率
percentage of sludge moisture content 污泥含水量百分数
percentage of soil moisture content 土壤含水率
percentage of storage 蓄水率;蓄水度;库存率;存储率
percentage of success 成活率
percentage of surplus suites 建余比【建】
percentage of swell and shrinkage 胀缩率
percentage of the budgeted figure 完成预算的百分比
percentage of thresh 脱粒率
percentage of trucks 货车率
percentage of unitization 成组比重
percentage of unloaded car kilometers 空车走行率
percentage of urban greenery coverage 城市绿化覆盖率
percentage of utility loss 效用损失率
percentage of utilization of fixed assets 固定资产利用率
percentage of vegetation 植被覆盖百分率;植被百分数

percentage of voids 孔隙百分率；孔隙率；空隙率；空隙百分率
percentage of wagon loading capacity utilized 货车载重利用率
percentage of wear 磨耗(百分)率
percentage of wear and tear 折旧率
percentage of weight of salt to ice 冰内加盐百分数
percentage of workable coal thickness 可采含煤系数
percentage of worker's attending work 工人出勤率
percentage open area of screen 筛的有效面积百分率；筛孔面积百分率
percentage overlap 重叠百分比
percentage over-size curve 粗料百分率曲线
percentage paid off 付讫百分率；付清百分率
percentage passing 通过率；通过百分率
percentage passing screen 过筛百分数；过筛百分率
percentage pie chart 百分比圆形图
percentage point 百分点(百分率中相当于1%的单位)
percentage precision 百分精密度
percentage purity 纯度百分率
percentage reactance 百分电抗
percentage recovery 采收率
percentage recovery of coal core 煤芯采取率
percentage recovery of core 岩芯采取率
percentage recovery of full hole 全孔采取率
percentage recovery of ore core 矿心采取率
percentage recovery of roundtrip 回次采取率
percentage reduction 去除百分数
percentage reduction in area 面积缩小率；断面收缩率
percentage reduction of area 断面收缩率；截面缩小百分率；面积收缩率
percentage reduction of diameter 直径缩减率
percentage refusal density test 百分比抗沉密实度试验
percentage registration 百分比读数
percentage reinforcement 配筋率
percentage relative humidity 相对湿度百分率；百分相对湿度
percentage removal 去除百分率
percentage rent 按承租人收入的百分比计租；按百分比计租金
percentage reserves 备用率
percentage residue 残余百分率
percentage restraint differential protection device 比率约制式差动保护装置
percentage retained 筛余率
percentage saturation 饱和百分数；饱和百分率
percentages distribution 百分比分配
percentage sieving 过筛百分数；过筛百分率
percentage similarity 相似率
percentage size pick-up 上浆率
percentage speed variation 速度变化率
percentage spread 宽展率
percentage standard deviation 百分均方偏差；百分标准偏差
percentage standard error 百分标准误差
percentage statement 百分率表(比较表)
percentage subsidence 沉降率
percentage support 顶板面积支护率
percentage syllable articulation 音节明晰度
percentage synchronization 同步百分比
percentage tare 皮重率；百分比皮重
percentage tax 比例税
percentage test(ing) 挑选试验；抽查；百分率试验
percentage transmission for sample 样品的百分透光率
percentage transmittance 百分透射率；百分透过率
percentage undersize curve 筛下细料百分率曲线
percentage utilization 利用率
percentage variation 百分比变化
percentage voids 空隙百分率；孔隙百分数
percentage weight of grading fraction 粒径分级的重量百分比
percentagewise 用百分数方式；用百分率方式；按百分计算
percent articulation 百分清晰度
percent break 断路百分数
percent by volume 容积百分比；体积百分比；体积百分数
percent by weight 重量百分比；重量百分数
percent by weight in volume 重/容百分比
percent by weight in weight 重/重百分比
percent coarser 大于某一粒径的颗粒百分率；超径百分数
percent compaction 压实系数；压实百分数；压密百分比
percent concentration 浓度百分率；百分浓度
percent consolidation 固结度；固结百分数；固结(百分)率；固结百分比
percent conversion 转化率
percent correction 百分校正
per cent death loss 死亡百分率
percent defective 不合品率；不合格率
percent error 百分误差
percent finer 小于某粒径的颗粒百分率
percent fines 细粉率；砂率；细粒含量；含细粉料百分率；粉尘百分率
percent frequency effect 频散率；视频散率
percent harmonic 谐波百分比
percent ignition loss 烧失量百分率
percentile 粒度百分位数；百分之一；百分位(数)；百分比下降点；百分位数值
percentile curve 百分位数曲线；按百等分分布曲线
percentile method 百分位数法
percentile sound level 百分声级
percentile speed 百等分数速限
percent infiltration 漏风百分数
percent modulation 调制百分比
percent moisture 含水率
percent moisture content 百分含水率
percent of accuracy 准确度
percent of actual wagon-group loadings to those planned 成组装车计划完成百分率
percent of actual wagon loadings carried by through trains organized at loading points to those planned 装车地直达运输计划完成百分率
percent of auxiliary to total locomotive 机车辅助走行率
percent of baffle cut 折流板切口分数
percent of call lost 呼损率
percent of circulation expenses 流通费用率
percent of computed spillway flood-elevation curve 计算溢洪量百分数—(水库)高程关系曲线
percent of cutting volume 岩屑体积百分比
percent of empty of loaded wagon kilometers 空车走行率
percent of enlargement/reduction 缩放百分率
percent of excess air 过剩空气率
percent of ink transfer 传墨率
percent of net income to stockholders 纯益与股东权益百分比
percent of open area 开孔率
percent of pass 合格率
percent of passed sample 样品合格率
percent of punctuality of passenger trains running on time to total passenger trains 旅客列车运行正点率
percent of return air 回风百分比
percent of sanctioned alternations in wagon loadings to total planned 批准变更的装车数占总装车数百分率
percent of shear cut 切裁率
percent of sodium 含钠百分率
percent of swell and shrinkage 胀缩率
percent of unctuality of goods trains received on time to total goods trains 货物列车正点率
percent of unpassed sample 超差率
percent of wagon loading of irrational transport to total wagon loadings accomplished 不合理运输占总装车数百分率
percent of wear 磨损百分率
percent of wearing-out tooth 掉落牙齿百分比
percent open area 孔眼面积百分数
percent pack 装箱率
percent passing 过筛百分数；过筛百分率；合格率
percent per annum 年百分率
percent pigment volume 颜料体积百分比
percent primary air 一次风百分率
percent proportion content of soluble salt 易溶盐含量百分比
percent ratio error 变比误差率；变比误差百分数
percent reactance drop 百分电抗压降
percent recovery 总馏出率
percent reduction 减缩率；还原率；缩减率
percent retained 筛余率；筛留百分数；筛留百分率
percent ripple 波纹度；波纹百分比
percent sampling inspection 百分比抽样检查
percent sand 含砂率
percent saturation 饱和率；饱和度；饱和百分比(率)
percent saturation of market 市场饱和度
percent shot 渣球百分比含量
percent similarity of community 共性相似率
percent size (筛分后的)大小区分率
percent slope 倾斜百分率
percent sodium 含钠百分率
percent swell and shrinkage 胀缩百分率
percent symbol 百分比符号
percent-test portaprobe 百分检测携带式探测计
percent thickness recovery 厚度回复百分数
percent transmission 透光百分率
percent transmittance 透光百分率
per centum 每百
percentum dosage 用百分比配料
percent utilization 利用率
percent utilization of capacity 用百分数表示的生产能力利用系数
percent voids 孔隙百分数
percent volume to volume 容积百分比
percent weight in volume 重量体积百分比
percent wet-out 润湿率
perception 感觉；理解力；洞察力
perception of brightness 亮度感
perception of depth 景深感(觉)
perception of relief 立体感
perception of solidity 实体感
perception-reaction time 感觉反应时间
perception threshold 感应界限
perception time 感觉时间
perceptron 视感控器
perceptual effect 感受效果
perceptual knowledge 感性认识
perceptual stage 感性阶段
perceptual threshold 感觉阈
perch 杆；栖息；浮筒顶标
perch basin 栖息盆地
perch beacon 标杆
perch bolt 汽车弹簧钢板螺杆
perched aquifer 栖留含水层；上层含水量；表层水层；滞水含水层
perched basin 坡栖盆地
perched block 坡栖岩块
perched boulder 栖岩坡；坡栖岩块
perched ground water 上层滞水；静滞地下水；静止地下水；滞水；栖留水；栖留地下水
perched ground water-table 静滞地下水面；静止地下水位
perched karst 悬挂岩溶
perched lake 悬水面湖；滞水湖
perched permanent ground water 常静滞地下水；永久滞水
perched river 栖留河
perched spring 上层滞水泉；滞水泉；栖留泉
perched stream 滞水河；滞流河；栖留河
perched subsurface stream 栖留地下河；地下栖留河
perched temporary ground water 暂时悬着地下水；暂时滞水
perched water 潜水；栖留水；表层流；上层滞水；静止地下水
perched water table 滞水水位；滞水面；静止地下水位；栖留潜水面；栖留地下水位；栖留地下水面；上层水位；上层滞水面；地下水位；水位；上悬潜水面
perch groundwater 滞水
perching 铲软
perching knife 铲刀
perchlorate 过氯酸盐；高氯酸盐
perchlorate explosive 高氯酸盐炸药
perchlorethane 全氯乙烷
perchlorether 全氯乙醚
perchlorethylene 四氯乙烯(一种除油渍剂、干洗剂)；全氯乙烯
perchloric 高氯的
perchloric acid 高氯酸
perchloric acid bath 高氯酸钾电解液

perchloric acid ether 高氯酸乙酯
perchloric aicd anhydride 七氧化二氯
perchloride 高氯化物
perchlorinated 全氯化的
perchlorinated hydrocarbon 全氯碳化物
perchlorinated polyvinyl chloride 全氯代聚氯乙烯
perchlorination 全氯化(作用)
perchlormethyl mercaptan 全氯甲硫醇
perchloro-butadiene 全氯丁二烯
perchlorocarbon 全氯碳化物
perchloroethane 六氯乙烷
perchloro-ethylene 全氯乙烯;四氯乙烯
perchloro-ethylene lacquer 过氯乙烯清漆
perchloro-ethylene resin 过氯乙烯树脂
perchloro-hydrocarbon 全氯代烃
perchloromethane 过氯甲烷;全氯甲烷
perchloronaphthalene 全氯萘
perchloroparaffin 全氯化石蜡
perchloropropane 八氯丙烷
perchloro-vinyl 高氯乙烯
perchlorovinyl primer 过氯乙烯底漆
perchloryl fluoride 高氯酰氟;过氯酰氟
perchromate 过铬酸盐
perchromic acid 过铬酸
percivalite 钠铝辉石
percle 高级密织薄纱
perclene 四氯乙烯
perclose (中世纪教堂中分隔空间的)隔断;(围绕观众台的)矮护墙;围栏
Perco copper sweetening 培柯铜盐脱硫
percolate 渗滤液;渗出液
percolate corrosion 渗滤溶蚀
percolate-flake runoff mixed corrosion 渗滤—片状径流溶蚀
percolating 渗透;渗滤
percolating bed 渗层;渗透过滤床;渗滤层;渗床
percolating filter 生物滤池;渗透滤器;渗透滤层;渗滤过滤器;渗滤器;渗床;滴漏滤池
percolating hose 渗透水带
percolating volume 渗透水量
percolating water 渗滤水;渗透水;渗漏水;渗滤量
percolation 渗滤作用;渗透;渗滤;渗漏;渗过
percolation apparatus 渗滤仪;渗滤仪;测渗仪
percolation area 入渗面积
percolation basin 渗水池
percolation bed 渗透层;渗滤床;渗滤层;渗床
percolation capacity 渗透能力;渗漏能力
percolation coefficient 渗透系数
percolation control 渗透控制
percolation deasphalting method 渗滤脱沥青法
percolation extractor 渗滤式浸出器;渗滤器
percolation filter 重力过滤器;渗滤池
percolation flow 渗入流量
percolation friction 渗透摩阻力;渗滤摩阻力;渗滤摩擦
percolation ga(u)ge 渗透计;渗漏计;测渗仪;测渗计
percolation gradient 渗透梯度;渗透坡降
percolation head 渗透水头
percolation layer 渗滤层
percolation leaching 渗滤沥滤选矿法
percolation line 渗流线;渗透线;浸润线;饱和线
percolation loss 渗漏损失
percolation method 渗透法
percolation path 渗透途径;渗漏线;渗流路线;渗流路径
percolation pit 渗滤坑;渗坑;渗井
percolation pond 渗滤池
percolation pressure 渗透压力;渗透压;渗滤压力;渗流压力
percolation prevention 渗滤预防
percolation process 渗滤法
percolation rate 渗滤速率;渗透速度;渗滤率;渗漏率;渗漏量
percolation ratio 渗滤率;渗漏率
percolation ration 渗滤量
percolation resistance 渗透阻力
percolation tank 渗滤槽
percolation test 渗透试验;渗滤试验;渗漏试验;地面渗透试验
percolation treatment 渗滤处理
percolation-type sludge bed 渗滤式污泥床
percolation velocity 渗滤速度
percolation water 渗透水;渗滤水;渗漏量

percolation well 渗水井;渗滤井;渗井
percolation zone 渗透带;渗漏带
percolator 滤池;过滤器;咖啡壶;渗滤器;渗滤池;渗流器
Percolex 帕柯勒斯(一种专利防水剂)
percoline 渗流线路
percolised hose 衬里渗透水带(具有特殊的衬里和覆盖层)
per compass 根据罗经
per contra 反之
percrystalline 过晶质
percrystallization 透析结晶(作用)
percurrent 到顶的
percussion 撞击;叩打;击发;碰撞;冲击
percussion action 撞击作用;碰撞作用;冲击作用
percussion and grabbing method 冲击抓法
percussion arrangement 击发装置;冲击机构
percussion bit 撞击钻头;冲击(式)钻头
percussion bit design 冲击钻头设计;冲击钻头结构
percussion (bore) hole 冲击钻孔
percussion borer 冲击式钻机;冲击式钎子
percussion boring 撞钻;冲击钻探;冲击钻孔;冲击钻进;冲击式钻探;冲击钢索钻
percussion boring apparatus 撞钻装置
percussion boring machine 冲击钻机
percussion boring method 撞钻法;冲击钻法
percussion boring rig 冲击钻机
percussion cap 炸药帽;冲击起爆管;雷管;碰炸帽
percussion cone 冲击锥印痕;冲击裂纹
percussion coring 冲击式取芯
percussion crusher 反击式破碎机
percussion cutting machine 冲击式切割机
percussion device 撞击装置
percussion digging method 冲击式挖掘法
percussion drill 冲击钻(机);撞击钻孔;冲击式钻机;冲击式凿岩机
percussion drill bit 冲击式钻头
percussion drill hammer 冲击钻锤
percussion drilling 冲击钻进;冲击钻探;冲击钻孔;冲击钻井;冲击式凿岩;冲击钻
percussion drilling machine 风动凿岩机;冲击钻孔机
percussion drilling method 冲击钻法
percussion drilling point 冲击钻尖端
percussion drilling sampling 冲击钻探采样
percussion drill (rig) 冲击钻机
percussion drill rod 冲击式钻杆
percussion feeder 冲击型进料器
percussion figure 撞击图像
percussion frequency 撞击频率
percussion fuse 击发雷管;冲击雷管;冲击保险丝
percussion fuze 触发引信
percussion grinder 撞碎机
percussion hammer 冲击式凿岩机;冲击锤
percussion hammer firing mechanism 击针锤击发装置
percussion hand boring 手摇冲击钻孔;人力冲击钻进;手摇冲击钻
percussion hand drilling 人力冲击钻进
percussion height 冲击高度
percussion jumper 冲击式钎子
percussion mark 撞痕【地】;击痕;碰痕
percussion mechanism 击发机构
percussion oscillation drilling 冲击振动凿岩
percussion piston 撞击活塞
percussion powder 炸药;冲击起爆炸药
percussion power press 撞压机
percussion press 撞压机
percussion primer 撞击起爆药包
percussion resonance 叩响
percussion reverse circulation drill 冲击反循环钻机
percussion riveting 冲击式铆接
percussion riveting machine 撞击铆钉机;冲击式铆钉机;风动铆钉机
percussion rock drill 冲击岩芯钻;风动凿岩机
percussion rotary drill 冲击回转钻机
percussion-rotation drilling 冲击回转钻进
percussion scar 击痕;碰痕
percussion screw press 直接电动螺旋压力机;直接电动螺旋锤
percussion sidewall sampling 冲击式井壁取芯
percussion size reduction 反击破碎
percussion stem 冲击式钻杆

percussion system of drilling 冲击钻进法
percussion table 振动台;碰撞式摇床
percussion test (bore) hole 冲击试验孔
percussion tool 冲击工具;冲击钻进钻具;冲击钻进工具
percussion tool weight 冲击式钻具重量
percussion-type boring machine 冲击式钻机
percussion-type pile engine 冲击式打桩机
percussion-type preliminary mill 冲击式粗磨机
percussion-type sidewall coring 冲击式孔壁取芯器
percussion-type tie tamper 冲击式砸道机;冲击式打夯机
percussion-type track tamper 冲击式轨道捣固机
percussion-type vibratory pile driver 振动冲击沉桩机
percussion wave 冲击波;脉首波
percussion weld 冲击焊缝;冲击焊接
percussion welder 储能焊机;冲击焊机;冲击电焊机
percussion welding 锻接;撞击焊接;储[贮]能焊;冲击焊(法);冲击(电阻)焊接
percussion welding machine 冲击焊机
percussive action 破碎作用;冲击作用
percussive air machine 风钻;风动凿岩机
percussive bit 冲击钻头
percussive boring 冲击钻探;冲击法;冲击钻进
percussive cap 撞击帽
percussive chisel 冲击钻头
percussive coring bit 冲击钻进取芯钻头
percussive drill 冲击式钻机
percussive drilling 冲击钻孔;冲击钻探;冲击钻进;冲击式凿岩
percussive energy 撞击能
percussive force 撞击力;冲力;冲击力
percussive machine drilling 机械冲击钻进(法)
percussive method of riveting 撞击铆接法
percussive open well machine 冲击式打井机
percussive piston tool 有冲击式活塞的工具
percussive pneumatic method of riveting 风动冲击铆接头
percussive pneumatic tool 风动冲击工具;冲击式风动工具
percussive pressure of surroundings 冲击围岩压力
percussive rock drill 冲击式凿岩机
percussive rod boring 冲击杆钻探
percussive-rotary drilling 冲击式旋转工具;冲击旋转钻井;冲击回转钻进
percussive size reduction 反击破碎
percussive sound 震音;击发声
percussive test 冲击试验
percussive tool 冲击机具
percussive welding 撞击焊接
percussopunctator 梅花针
percylite 氯铜铅矿
per day 每天;每日
perdicarbonate 过碳酸氢盐
per diem 每天;每日;日率制;按日计
per diem allowance 每日津贴
per diem expenses 每日费用
per diem fee 计日工资
per diem rate 日息率;按日计利息
perdistillation 透析蒸馏
perdurability 延续时间
perduren 硫化橡胶;硫化橡胶
pereletok 双季冻土;残冻层
perennation 多年生
perennial 终年的;多年生植物;多年生的;常年的;长年的
perennial artesian water 持续自流水;常年自流水
perennial base flow 常年基流
perennial branch 多年生枝
perennial broadleaf weed 多年生阔叶杂草
perennial canal 常年渠道;常流河槽;常年河槽
perennial channel 常年水道;常年河槽
perennial cover 常年积雪
perennial crop 多年生作物
perennial crop irrigation 多年生作物灌溉
perennial drainage 常流水
perennial fitting 完全配合
perennial flow 常年流(量);常年水流;常年径流
perennial fluctuation of water-level 水位的多年变幅
perennial forest underground water area 多年冻结潜水区

perennial frost 永久冻土;多年冻土
perennial frost soil 永冻土
perennial grass 多年生牧草;多年生草
perennial groundwater regime 地下水多年动态
perennial ice 多年冰
perennial interrupted river 常年间断河流
perennial interrupted streamflow 常年间断水流
perennial irrigation 常年灌溉
perennial lake 常年湖
perennially frozen lake 常年封冻湖
perennially frozen soil 多年冻土
perennially navigable waterway 常年通航航道
perennial maximum air temperature 多年最高气温值
perennial maximum runoff 多年最大径流量
perennial minimum air temperature 多年最低气温值
perennial minimum runoff 多年最小径流量
perennial over-draft (地下水的)常年过量抽取;常年过度抽取;常年超采
perennial period(ic) spring 常年周期泉
perennial plant 多年生植物;宿根植物
perennial pond 常年池沼
perennial recharge 常年补给
perennial river 常年河流;常年河;常年河;长流河
perennial runoff 常年径流
perennial snowfield 常年雪原
perennial spring 常年泉(水);长年泉
perennial stream 常年河流;常年河;常流河
perennial streamflow 恒流河川;常年水流;常年流量
perennial water course 常年航线;多年河;多年水道;常年含水层;长期不变水道
perennial weed 多年生杂草
perennial yield 常年出水量;常年产水量
perester 过酸酯
peretaite 锑钙矾
perfect absorber 完全吸收体
perfectape merging unit 穿孔纸带归并机
perfect automated cataloguing technique 全自动编目技术
perfect automatic 全自动的
perfect automatic batcher 全自动(混凝土)搅拌机
perfect automatic batcher plant 全自动拌和厂
perfect balance 完全平衡
perfect binding 胶粘装订;无线装订
perfect black body 理想黑体;绝对黑体
perfect code 理想码
perfect combustion 完全燃烧;安全燃烧
perfect combustion catalyzer 助燃添加剂
perfect comparator 理想比较仪
perfect competition 完全竞争
perfect competition market 完全竞争市场
perfect condition 理想状态
perfect correlation 完全相关
perfect crystal 完整晶体;完美晶体
perfect cube 完全立方
perfect detector 理想探测器
perfect diamagnetism 完全抗磁性
perfect dielectric 理想电介质
perfect differential 全微分;完整微分
perfect diffuser 全漫射体;全漫射面;全扩散体
perfect diffusion 理想的散射;完全散射;全漫射;全扩散
perfect discharge 完全出料法
perfect dislocation 全位错;完全位错
perfect dust collection 完全除尘
perfect eaquility in distribution 分布完全均等
perfect elastic body 理想弹性体
perfect elasticity 理想弹性体;完全弹性
perfect elastic material 理想弹性材料;完全弹性材料
perfect elastoplastic material 理想弹塑性材料
perfect filling 全充填
perfect fit 完全拟合
perfect fitting 完全配合
perfect flower 雌雄芯同花;两性花
perfect fluid 理想流体;理想流动;完全流体;非黏性流体
perfect focusing mass spectrometer 完全聚焦质谱仪;完全聚焦质谱仪
perfect form 美好的形式;完好的模板
perfect frame(work) 完整构架;理想构架;静定构架;全栓架
perfect fuzzy variable 完全模糊变量

perfect gas 理想气体;理想流体
perfect gas constant 理想气体常数;理想气体常量
perfect gaseous mixture 理想气体混合物
perfect gas equation 理想气体状态方程式
perfect-gas law 理想气体定律
perfect graph 完全图
perfect hydraulic jump 完全水跃
perfect inequality in distribution 分布完全不均等
perfecting press 双面印刷机
perfecting title 使产权文书完备
perfection of crystals 晶体完整程度
perfections 木完
perfect isomorphism 完全类质同象
perfect jump 理想的交替跨接
perfect lens 理想透镜
perfect light source 理想光源
perfect liquid 理想液体;完全液体
perfectly aligned seat 对准座
perfectly diffuse radiator 全漫辐射体
perfectly diffuse reflector 理想漫反射器;全漫反射体
perfectly diffusing plane 全扩散面;完全扩散面
perfectly diffusing reflection 均匀扩散反射
perfectly diffusing sphere 全漫射球
perfectly diffusing surface 完全扩散面
perfectly elastic 完全弹性的
perfectly elastic plastic system 完全弹塑性体系
perfectly elastic solid 完全弹性固体
perfectly elasto-plastic body 完全弹塑性体
perfectly elasto-plastic system 完全弹塑性体系
perfectly inelastic collision 完全非弹性碰撞
perfectly killed steel 全脱氧钢
perfectly mobile component 全活动性组分
perfectly plastic 完全塑性的
perfectly plastic mechanism 完善塑性机理
perfectly plastic theory 完善的塑性理论
perfectly representative sample 具有完全代表性样本
perfectly round shape 浑圆形
perfectly water-proofing plywood 去防水胶合板;完全耐水胶合板
perfect mixing 理想混合;全混;完全混合;充分混合
perfect monopoly 完全垄断;完全独占
perfect moon tide 完全太阴潮
perfectness 完美
perfect number 完全数
perfect obligation 完全债务
perfect oligopoly 完全垄断
perfect one's plan 完善计划
perfect optic(al) system 理想光学系统
perfect order 完全有序
perfect overfall 完全溢流堰
perfect overflow 完全溢流堰
perfect pendulum 数学摆
perfect plastic 完全塑性的
perfect plasticity 完全塑性
perfect plastic yield 完全塑性屈服
perfect plate 理想板
perfect press 双面印刷机
perfect prestress 理想预应力;完善的预应力
perfect prestressing 完善的预加应力;完全预应力
perfect prognostic 理想预报
perfect radiation 完全辐射
perfect radiator 完全辐射体
perfect reflecting diffuser 理想漫反射面
perfect reflection 全反射
perfect register 对花准确
perfect restraint 完全受约束;全约束
perfect safety 确保安全
perfect score 满分
perfect set 完备集
perfect soil 理想土体
perfect solution 理想溶液;完美溶液
perfect square 整方;完全平方
perfect state 理想状态
perfect stretching 完全受拉伸
perfect synchronism 完全同步
perfect tensioning 完全受拉
perfect the construction of urban infratruture 完善城市基础设施建设
perfect thread 精密螺纹;全螺纹
perfect transfer press 全连续自动压力机
perfect transmitting body 完全透明体
perfect vacuum 高度真空;理想真空;绝对真空;完全真空
perfect vacuum pump 高真空泵
perfect weir 完全堰
perfemic 过铁镁质
perferrate 过铁酸盐;高铁酸盐
perflation 换气;通风
perflecto-comparator 反射比较仪
perflectometer 全反射测量计;反射显微镜;反射头
per floor 每层【建】
perfluorination 全氟化(作用)
perfluoro alkoxyalkane copolymer plastic lined metal pipe 全氟烷氧基烷烃共聚物塑料衬黑色金属管
perfluoroalkoxy polymer 全氟烷氧基聚合
perfluoroalkoxy resin 全氟烷氧基树脂
perfluoroalkylation 全氟烷基化(作用)
perfluoroallene 全氟丙二烯
perfluorocarbon 全氟烃;全氟碳化物
perfluorocarbon tracer 全氟化碳示踪剂
perfluoro-compound 过氟化物;全氟化物
perfluorocyclicether 全氟环醚
perfluorocyclobutane 全氟环丁烷;八氟环丁烷
perfluorodimethyl cyclobutane 全氟二甲基环丁烷
perfluoroethane 六氟乙烷;全氟乙烷
perfluoroether 全氟乙醚
perfluoroethylene 全氟乙烯
perfluoroethylene-propylene copolymer plastic lined metal pipe 全氟乙烯—丙烯共聚物塑料衬黑色金属管
perfluoro-hydrocarbon 全氟代烃
perfluorokerosense 全氟煤油
perfluorokerosine 全氟煤油
perfluoroorganometallic compound 全氟有机金属化合物
perfluoropropane 八氟丙烷
perfluoropropylene vinylidene fluoride copolymer 全氟丙烯二氟乙烯共聚物
perfluorosulfonic acid membrane 全氟硫酸膜
perfluorpospbutylene 全氟异丁烯
perfoliate 穿叶的
perfoliate-leaf 贯穿叶
perfoot 极点
per-foot-bit cost 每英尺钻头成本
per-foot-of hole drilled 每英尺钻孔
perforate 凿孔;打孔;穿孔;冲孔
perforate bowl 离心机筛孔转鼓;带孔转鼓
perforated 多孔的;穿孔的
perforated acoustical tile 多孔隔音板
perforated aeration disc 穿孔曝气盘
perforated air duct 多孔风管
perforated alumin(i)um 穿孔铝板
perforated alumin(i)um strip 穿孔铝片
perforated arch brick 多孔砖拱
perforated area of screen plate 筛板筛孔面积
perforated asbestos cement 多孔石棉胶泥
perforated asbestos-cement pipe 多孔石棉水泥管
perforated backing 穿孔的底衬;多孔垫层
perforated baffle 多孔挡板
perforated baffle plate 多孔消力板;多孔导流板
perforated bars 带孔导杆
perforated base 多孔基底
perforated baseboard 多孔板
perforated baseplate 钻孔底板
perforated basket centrifuge 过滤式离心分离机
perforated block 多孔砌砖;多孔块体
perforated board 多孔板;穿孔板
perforated bottom 多孔底板;多孔底;穿孔底板
perforated bowl 带孔转鼓
perforated (bowl) cylinder 多孔转鼓
perforated box caisson breakwater 开孔箱形波堤
perforated breakwater 多孔防波堤
perforated brick 孔板砖;空心砖;多孔砖;多孔空心砖
perforated brick masonry(work) 多孔砖圬工
perforated brick with grip-slot 带有抓槽的空心砖
perforated caission breakwater 带孔沉箱防波堤
perforated caisson 多孔混凝土沉箱
perforated caisson breakwater 开孔沉箱防波堤
perforated calcium silicate brick 多孔硅酸钙砖;多孔灰砂硅酸盐砖
perforated cap 穿孔盖板
perforated casing 有射孔套管;有孔眼的套管;有孔井管;钻孔套管;滤管(井的);多孔井壁管;带眼的套管

perforated casing well 多孔井壁管井
perforated ceiling 多孔天花板；多孔平顶；穿孔板平顶；穿孔板顶棚
perforated ceiling air supply 多孔天花板送风
perforated ceiling board 多孔板平顶板；穿孔吊顶板
perforated ceiling tile 多孔平顶板
perforated cellular brick 多孔空心砖
perforated channel 多孔水槽
perforated chaplet 箱式泥芯撑
perforated clay brick 多孔黏土砖
perforated clay engineering brick 多孔的工程用黏土砖
perforated clay pipe 多孔瓦管
perforated concrete 多孔混凝土；大孔混凝土
perforated concrete block 多孔混凝土砌块
perforated concrete breakwater 多孔混凝土防波堤
perforated concrete brick 多孔混凝土砖
perforated concrete caisson 多孔混凝土沉箱
perforated concrete caisson breakwater 多孔混凝土沉箱防波堤
perforated concrete crib 多孔混凝土石笼
perforated concrete pipe 多孔混凝土管；大孔混凝土管
perforated concrete tube 多孔混凝土管
perforated cone 锥筒筛
perforated corrugated alumin(i)um 多孔波形铝
perforated corrugated metal pipe 多孔波形金属管
perforated cover plate 多孔盖板
perforated cylinder 筛筒；圆筒筛；筛孔筒；多孔滚筒
perforated deck(ing) 穿孔的桥面
perforated deflecting cap 穿孔转向器
perforated design 穿孔图案
perforated diaphragm 多孔隔膜
perforated drain pipe 排水花管；多孔排水管；穿孔排水管
perforated drum 多孔转鼓
perforated electrode 多孔电极
perforated facing 多孔罩面板；多孔面层
perforated feed pipe 有孔给水管
perforated fiberboard 穿孔纤维板
perforated fibre hardboard 多孔硬质纤维板
perforated fibrous plaster sheet 多孔纤维粉刷板
perforated film 有孔胶片；电影胶片；打孔胶片；穿孔胶卷
perforated floor panel 多孔地板镶板
perforated flushing pipe 多孔冲洗管
perforated glass 多孔玻璃；泡沫玻璃
perforated glass block 泡沫玻璃砖块；多孔玻璃砖块
perforated glass membrane 多孔玻璃膜
perforated grids 格孔栅；多孔栅
perforated grill 多孔网格；穿孔栅
perforated gypsum board 多孔石膏底板；多孔石膏板；穿孔石膏板
perforated gypsum ceiling board 穿孔石膏吊顶板
perforated gypsum lath 穿孔石膏条板
perforated gypsum panel 穿孔石膏板
perforated gypsum plasterboard 多孔石膏灰泥板；多孔石膏纤维板；穿孔石膏板；多孔石膏粉刷板
perforated hardboard 穿孔硬质纤维板；多孔硬质纤维板
perforated head 有孔封头
perforated hexagonal block 六角形多孔块体
perforated ladle 漏勺
perforated lath(ing) 穿孔的板条
perforated lime-sand brick 多孔灰砂砖
perforated liner 穿孔的衬板；多孔的衬套；穿眼的衬管
perforated litho master 穿孔头靠模
perforated marble slab 多孔大理石板
perforated mat 孔眼薄毡
perforated mat sheet(ing) 多孔毡毯护墙板
perforated metal 穿孔金属板
perforated metal pan 多孔金属盘
perforated metal pipe 多孔金属管
perforated metal screen 大孔筛；圆孔筛板；多孔金属筛；穿孔金属筛
perforated metal strip 穿孔的金属条
perforated method 穴施法
perforated mirror 穿孔镜
perforated mixer 多孔混合器；波纹式混合器
perforated page size 穿页页面大小
perforated pan aerator 穿孔盘曝气器
perforated panel 多孔板；多孔镶板；穿孔板
perforated panel air outlet 多孔板送风口

perforated panel ceiling 穿孔板平顶；穿孔板顶棚
perforated panel sound-absorbing construction 穿孔板吸声结构
perforated paper 穿孔的纸张
perforated paper tape 穿孔纸带
perforated pattern 穿孔图案
perforated pipe 有孔管；花管（进水管有滤网的一段）；多孔管；穿孔管
perforated pipe completion method 贯眼完成法
perforated pipe distributor 多孔管式分配器；多孔管分布器；穿孔管布水器
perforated pipe sprinkler 多孔管喷水灭火装置
perforated piping 多孔管
perforated piping distributor 多孔管布水器
perforated piping underdrain 多孔管式地下排水管
perforated plasterboard 穿孔的石膏板；穿孔的灰泥板
perforated plate 多孔板；穿孔板；冲孔板；蜂窝板
perforated-plate column （在孔眼上包网状衬垫的）穿眼管柱；多层孔板蒸馏塔
perforated plate distillation column 孔板蒸馏塔
perforated-plate distributor 多孔板分布器
perforated plate extraction tower 筛板萃取塔
perforated-plate extractor 多孔板萃取塔；多孔板萃取器
perforated plate for cooling 冷却段多孔板
perforated plate for drying 干燥多孔板
perforated plate for test sieves 试验用筛穿孔板
perforated plate sieve 板孔筛
perforated plate tower 孔板塔；筛板塔
perforated plate type atomizer 多孔板式喷雾器
perforated plate with oval perforations 带有椭圆形穿孔的穿孔板
perforated plywood 多孔胶合板；多孔板；穿孔胶合板
perforated polyurethane mesh 冲孔聚氨酯筛板
perforated propeller 多孔式螺旋桨
perforated protector 多孔保护装置
perforated ring 喷雾环；多孔环
perforated roller 有孔滚筒；多孔滚筒
perforated roofing base sheeting 多孔屋面底板
perforated roofing deck(ing) 多孔屋面面板
perforated roofing substrate 穿孔的屋面基层；穿孔的屋面下层
perforated screen 多孔的围屏；多孔板筛；冲孔筛
perforated sheet(ing) 多孔护墙板；穿孔板
perforated sheet lath 金属拉网（抹灰用）；板条网眼钢皮；网眼钢皮
perforated sheet metal lath(ing) 网眼钢皮；穿孔金属薄板条
perforated sheet of gypsum plasterboard 多孔石膏粉刷板
perforated signature 透字签名
perforated skimmer 漏勺；多孔撇渣器
perforated slab 多孔板
perforated slag 多孔炉渣；多孔熔渣
perforated slip 凿孔纸带
perforated sprinkler 多孔管喷灌装置
perforated steam spray 喷汽器
perforated steel plate screen 多孔钢板筛
perforated steel sleeve 带孔管鞋（注水泥用）
perforated steel tube 多孔钢管
perforated stone 透水石；多孔石
perforated strainer 多孔滤管
perforated substance 穿质
perforated substrate 穿孔的地下（基）层
perforated substructure 穿孔的下层结构
perforated tail pine 筛管
perforated tape 凿孔纸带；穿孔带
perforated tape code 穿孔纸带代码；穿孔带码
perforated tape reader 穿孔纸带读入装置；穿孔带阅读器；穿孔带读出器
perforated tile 瓦管；多孔瓦；多孔砖
perforated tracery 漏花窗格；多孔窗花格
perforated tray 多孔板塔盘
perforated tube 多孔管
perforated tubing 多孔管
perforated underdrain pipe 穿孔管地下排水管
perforated under layer 通风次层；通风下层；通风底层；多孔次层；多孔下层；多孔底层
perforated visible soffit 穿孔的可见底面
perforated wall 有孔墙；花墙；漏墙；花格墙（指用各种砌块砌成的）；穿孔墙；有漏孔的墙
perforated wall centrifuge 过滤式离心机

perforated washer 多孔水洗器；多孔水槽清洗器
perforated well 多孔管井
perforated well casing 多孔钻井管
perforated well point 穿孔管井点
perforating 打孔；穿孔性的
perforating action 成孔作用
perforating branches 穿支
perforating completion 射孔完井
perforating device 打孔装置
perforating die 穿孔冲模；冲孔模
perforating equipment 射孔装置；穿眼设备；打孔设备；穿孔设备
perforating fiber 贯穿纤维
perforating grill 穿孔格箅
perforating gun 射孔枪
perforating machine 打孔机；穿孔机
perforating press 穿孔压力机；冲孔机
perforating rate 穿孔速率
perforating roller 穿孔辊
perforating speed 凿孔速率
perforating tool 穿孔工具
perforating wound 穿孔创
perforation 孔眼；射孔；打眼；穿孔眼(线)；穿孔；冲孔；齿孔
perforation of screen 屏幕的开孔
perforation pattern 打洞的模式；打洞的式样
perforation plate 麻黄型穿孔板
perforation rate 穿孔速率
perforation rim 穿孔缘
perforation strip 穿孔带
perforator 剪票钳；螺旋钻；打洞机；钻孔机；凿孔器；凿孔机；打孔器；穿孔器；穿孔机
perforator unit 打眼设备；穿孔设备
perform 预成型；履约；履行；实行；施行
performability 可执行性；可运行性
perform a contract 履行契约
performance 性能；效能；运行；作业；功能；履约；经营成果；实绩；操作性能
performance accounting 业绩评价会计；经营会计；经营成绩管理会计
performance adjustment 性能调试
performance analysis 性能分析；工效分析
performance and cost evaluation 性能与成本估价
performance and fairness 效率与公平
performance audit 业绩审计
performance bond 性能保证书；履约保证书；履约保证金；履约保证金；履函；完工保证书；行动债券
performance budgeting 职能预算
performance calculation 功率计算；特性计算
performance characteristic 性能特征；运转特性；运行性能；运行特性；工作特征；工作特性；动作特性；操作特性；表现特征
performance characteristic curve 运行特性曲线
performance chart 性能图；性能曲线；工作特性图；操作性能图
performance checking 性能检查
performance code 功能法规；功能准则
performance coefficient 性能系数
performance company 演出公司
performance computation 特性计算
performance correlation 性能相互关系
performance-cost-dependability 性能、可靠性与价格
performance cost ratio 性能价格比
performance criterion 性能准则；性能标准；功能准则
performance curve 性能曲线；运行曲线；工作特征曲线；特性曲线
performance data 性能数据；运行数据；功能数据；特性数据
performance density 表观密度
performance design 性能设计
performance diagram 运行图；运行曲线图
performance estimation 性能估计
performance evaluation 性能评价；性能估价；业绩评价；经营成绩评价；绩业评价；操作评定
performance examination 性能检验
performance factor 性能因数；运行系数；工况因数
performance figure 性能指数；性能指标；性能数字；质量指标；工作指数
performance function 性能函数；功能函数
performance guarantee 履行合同保证人；性能保证；质保金；履约担保；履约保函
performance guarantee money 履约保证金

performance guarantee test 运行保证考核
performance index 性能指数;性能指标;业绩指数;作业完成指数,质量指标;工作指数;特性指数;生产性能指数;成绩指标
performance index of internal combustion engine 内燃机性能指标
performance indicator 效益指标
performance level 性能指数;性能指标
performance-linked pay 按劳付酬
performance load 运行荷载;运行负荷
performance loading relationship 性能荷载关系
performance loss 特性损失
performance manual 性能手册
performance margin 工作安全系数
performance measure 性能指数;性能指标
performance measurement 性能测定
performance monitor 性能监视器
performance monitoring 性能监察
performance number 性能数;功率值;特征数;特性数
performance of a bit 钻头工作性能(总进尺,单位进尺成本)
performance of analog(ue)-to-digital convertor 模数转换器性能
performance of a well 油井生产率
performance of blasting material 爆破材料性能
performance of boiler 锅炉性能
performance of combination vehicle 连接车的运行性能
performance of contract 履行合同
performance of exploration tunneling equipment 坑探设备性能
performance of exploration tunneling instrument 坑探仪表性能
performance of propellant 推进剂性能
performance of public vehicles 公用车辆性能;公共交通的运行;公共交通的服务
performance of the adsorbent 吸附剂性能
performance parameter 性能参数
performance pedigree 性能系谱
performance period 运行周期;实现周期
performance per liter 升功率
performance plot 特性图
performance-plus 性能更好的
performance prediction 性能预测
performance-proved product 使用合格的制品
performance provision 性能规定
performance qualification test 性能合格试验;技能合格考试;操作合格考试
performance range 性能范围
performance rating 考绩标准;成绩评定
performance rating scale 考绩等级尺度
performance record 运行图
performance reduction 性能换算;特性归化
performance requirement 性能要求
performance risk 履行合同风险
performance sampling 作业取样
performance security 履约保证;履约担保;履约保证金
performances of test piece for welding 焊接试板性能
performance specification 产品性能说明书;性能说明书;性能规格;规范;设计说明书;设计任务说明书
performance standard 性能标准;作业标准;功能标准;生产定额
performance standard zoning 功能标准区划
performance table 性能表
performance test 性能试验;性能考核;运转测试;运行试验;运行测试;特性试验;使用试验;使用测试;生产性能测定;生产考核;操作测验;作业测验
performance testing 运行试验;功能试验
performance testing station 成绩测验站
performance test stand 性能试验台
performance trait 生产性能的性状;生产性能特征
performance trial 性能试验;产量鉴定试验
performance under saturated condition 饱和条件下的性状
performance value 品度值
performance variable 操作变量
performance variance 实绩差异
performed foam 成品泡沫(材料)

performed joint filler 嵌缝料
performer 制锭机;执行器
performeter 工作监视器
performic acid 过甲酸
perform one's duty 履行职责
Perform tray 泊尔佛姆塔板;网孔塔板
perfo-rockbolt 多孔岩栓
perfume 香水;芳香剂
perfume box 香水盒
perfumed ink 香味油墨
perfume fixative 香料固定剂
perfume oil 芳香油
perfume plant 香料植物
perfumery 化妆品制造厂;香水制造厂;香料厂
perfuming detergent 加香洗涤剂
perfusate 灌注液
perfusion 灌注(法);灌流
perfusion cannula 灌注套管
perfusion catheter 灌流导管
perfusion culture 灌注培养
perfusion fixation 灌流固定
perfusion flow 灌流流量
perfusion scanning 灌注扫描
perfusion tube 输液管
pergament 假羊皮纸
Pergamum frieze 佩伽马雕带(小亚细亚的遗迹)
pergamyn paper 耐油纸
pergelisol 永久冻土;永冻土
pergelisol table 永冻土面
perget 烟囱内抹面;石膏花饰
pergola 藤架;藤顶凉亭;藤荫小径;蔓藤棚架
pergola pillar 花架柱
per gross ton 每一毛吨
per gyrocompass 根据陀螺罗经
per gyrocompass course 根据陀螺罗经航向
perhafnate 高铪酸盐
perhalide 全卤化物
perhalogenated 全卤化的
perhalogenated alkane 全卤化烷烃
perhalogenated fluorocarbon 全卤化碳氢化合物
perhamite 磷硅铝钙石
per hatch per day 每日每舱口
per head 按人计算
per head per day 每人每天
per hectare 每公顷
per hour 每小时
per hour change coefficient 每小时变化系数
per hour factor 每小时耗油量
per hour per direction 每小时单方向
per hour wage 每小时工资;小时工资
perhumid climate 过湿气候
per hundred output included wage 每百元产值工资含量
perhydride 过氢化物
perhydroanthracene 全氢化蒽
perhydrogenated rosin 全氢化松香
perhydrol 强双氧水
perhydrous coal 高含氢煤
perhydrous vitrinite 高氢镜质组;高氢镜质体
perhydrous vitrite 高氢微镜煤
perhydroxyl radical 过羟基
periaktos 剧场换幕机具(古希腊)
periapsis 近拱点
periapsisial aisle 近拱点过道;近拱点侧廊
periareon 近火点
periastron 近星点
periastron advance 近星点进动
periastron effect 近星点效应
periblain 外皮质煤
peribolos 圣地围墙;教堂围墙;庙宇围墙;寺院围墙;圣殿围墙;有围墙寺院;有围墙内院
peri-bone pad 抱骨垫
pericarp 果皮
pericenter 近中心点;近心点;近星点
pericentric inversion 臂间倒位
periclase 方镁石
periclase brick 方镁石砖
periclase marble 方镁石大理岩
periclase porcelain 方镁石瓷
periclasite 方镁石
periclinal 穹状的;弯状的
periclinal bedding 围斜层理
periclinal division 平周分裂
periclinal structure 穹状构造;穹顶构造;围斜构造

periclinal wall 平缝壁
pericline 穹顶;肖钠长石;周缘;盆状向斜;围斜构造
pericline law 肖钠长石律
pericline twin law 肖钠长石双晶律
pericontinental area 陆缘区
pericontinental sea 陆缘海
pericycloid 周摆线
pericyclonic ring 气旋周环
pericynthion 近月点
periderm 软木
peridosphere 橄榄岩石圈
peridot 贵橄石;橄榄石;电气石
peridotite 橄榄岩;橄榄石
peridotite deposit 橄榄岩矿床
peridotite for chemical firtilizer 化肥用橄榄岩
peridotite group 橄榄岩类
peridotite shell 橄榄岩壳(层);上地幔
peridotitic glass 橄榄玻璃
peridrome 古圆柱式寺院中绕圣堂墙;(古希腊建筑的)围廊
peridromos 环绕围柱式建筑外窄通道
perifocus 近主焦点;近心点;近焦点
periform 梨状的
perigalacticum 近银心点,[复]perigalactica
perigean spring tide 近地点大潮
perigean tidal current 近地点潮流;近地点潮流
perigean tide 近月潮;近地点潮
perigean tide range 近月潮差;近地点潮差
perigee 近地点
perigee kick 近地点喷火
perigee of satellite orbit 卫星轨道近地点
perigee-to-perigee period 近地点周期
periglacial 冰缘的;冰缘的;冰川边缘
periglacial climate 冰缘气候
periglacial geomorphology 冰缘地貌学
periglacial landform 冰缘地貌
periglacial zone 冰缘地带
perigon 360度角;周角【数】
perihelical motion 近日点移动
perihelic conjunction 近日点合
perihelic opposition 近日点冲
perihelion 近日点
perihelion distance 近日距;近日点距离
perihelion effect 近日点效应
perihelion motion 近日点移动
perihelion precession 近日点进动
perijove 近木木点
perikinetic aggregation 碰撞凝结
perikinetic coagulation 异向凝结(作用)
perikinetic conjunction 异向连接
perikinetic flocculation 碰撞凝结
perikinetic motion 异向运动
perikon detector 双晶体检波器
peril insured against 承保的风险
perilla oil 紫苏子油;荏油
peril point 危险点
perils of land transit 陆上运输危险
perils of sea 海上风险;海难
perilune 近月点
peril work 危险工作
perimagmatic 岩浆缘的
perimartian 近火点
perimercurian 近水点
perimeter 范围界线;周边长度;周界;周长;周边;视野仪表;视野计
perimeter area ratio 周界面积比
perimeter beam 圈梁;周边梁
perimeter blasting 圆周爆破;周边爆破;光面爆破;轮廓爆破
perimeter bonding technique 周边砖石砌合技术
perimeter bracket 周边突伸;周边挑出
perimeter car park 外围停车场;周边停车场
perimeter column 周边柱
perimeter connection 周边连接
perimeter control cable 环形控制电缆
perimeter diffuser 边沿散热器
perimeter ditch 围护沟;围沟
perimeter drainage system 周围排水系统(建筑物);周边排水系统(建筑物)
perimeter drilling 周边钻探
perimeter effect 周界影响;周界效应;周边界影响
perimeter flashing 周边泛水
perimeter frame 周边框架
perimeter grouting 围蔽灌浆;周围压力灌浆;周

灌浆
perimeter heating 周边供暖;周边采暖(建筑物)
perimeter heating system 周边取暖系统;周边供暖系统
perimeter insulation 包边绝缘材料
perimeter isolation 分离周边
perimeter lighting wiring 环形照明线路
perimeter masonry wall 周边圬工墙
perimeter masonry wall column 周边圬工墙柱
perimeter masonry wall facing 周边圬工墙贴面
perimeter masonry wall lining 周边圬工墙衬砌
perimeter of a circle 圆的周长
perimeter of airway 风巷周长
perimeter of bar 钢筋周长
perimeter of big aperture well 大口井边长
perimeter of irrigation 灌溉周界
perimeter of well group 井群周边长度
perimeter prestressed concrete column 周边预加应力的混凝土柱
perimeter road 周边道路
perimeter seal 周边封口
perimeter shear 周切力;周界剪;周剪力;环向剪力
perimeter structural system 周边结构系统
perimeter structure 围护结构
perimeter system 周边系统
perimeter tensioned concrete column 周边受拉伸的混凝土柱
perimeter trench 围护沟;周护沟(防止基础受冲刷)
perimeter wall 周边墙;围护墙
perimeter wall(building) component 周边墙构件
perimeter wall construction 周边墙施工
perimeter wall facing 周边墙饰面
perimeter wall lining 周边墙衬砌
perimeter warm air heating system 周边热风供暖系统;热气供暖系统
perimeter zone 周边地带;周边区
perimetrial joint 周界连接;环形连接
perimetric joint 周边接缝
perimetric pattern 周边式;外延长度
perimetry 周边视野检查法;视野检查;视野测量法
perimorph mineral 包被矿物
Perinea-carboniferous system 石炭二叠系
per injection 每次注射
perinuclear body 核环体
period 用期;周期;时期;时间段;分期
period allowed for conveyance 运行期限
period-amplitude relation 周期变幅关系
period analysis 期间分析
period-and-level start-up control 按周期和基准起动控制
period architecture 当代建筑
periodate 高碘酸盐;过碘酸盐
periodate lignin 高碘酸盐木素
periodate oxidation 高碘酸盐氧化
periodate oxidized cellulose 高碘酸盐氧化纤维素
period at maximum temperature 最高温度期
period average 长期平均
period between inspections 检测期间隔
period between overhauls 大修期间隔
period change 周期变化
period charges 期间费用
period cleaning serving inspection 定期清洁维修
period control 周期调节
period cost 期间成本;当期成本
period cut 定期采伐量
period-density relation 周期密度关系
period depletion 递减期
period discontinuity 周期突变
period down-wave 周期下降波
period-eccentricity relation 周期偏心率关系
period expense 期间费用
period expenses audit 期间费用审计
period for recovery of investment 投资回收期
period for shipment 装船期
period-frequency characteristic 周期频率特性
period frequency of cycle 周期数
period furniture 古式家具
periodic(al) 周期的;间歇的;周期式;定期的;周期性的
periodic(al) accounting 定期会计核算;期间会计核算
periodic(al) acid 过碘酸;高碘酸
periodic(al) activated sludge system 周期性活性污泥系统
periodic(al) adjustment 定期调整
periodic(al) allocation of cost 费用按期分摊;期间成本分配
periodic(al) alternate load 周期性交变荷载
periodic(al) anaerobic baffled reactor 周期性折流板厌氧反应器
periodic(al) analysis 定期分析
periodic(al) annealing furnace 周期式退火炉
periodic(al) annealing lehr 间歇式退火窑
periodic(al) antenna 周期性天线
periodic(al) attack 周期性发作
periodic(al) attenuation 周期衰减
periodic(al) audit 期间审核
periodic(al) bill 定期汇票
periodic(al) block 定期作业分区;轮伐更新区
periodic(al) blowdown 定期排污
periodic(al) boiler survey 船舶锅炉定期检验
periodic(al) boundary condition 周期边界条件
periodic(al) budget 定期预算
periodic(al) budgeting 定期预算的编制
periodic(al) calculation of unit expenditures 定期单位支出计算
periodic(al) calibration 定期校准
periodic(al) calibration test 定期校准试验;定期标定试验
periodic(al) cavity 周期性空蚀
periodic(al) chain 周期链
periodic(al) change 周期(性)变化
periodic(al) change characteristics of sand-wave 沙波周期性变化特性
periodic(al) chart 周期表
periodic(al) check 定期检查
periodic(al) circuit 周期性电流回路
periodic(al) classification 周期分类(法)
periodic(al) cleaning 定期清理
periodic(al) coming pressure 周期来压
periodic(al) comonent 周期分量
periodic(al) control 周期控制
periodic(al) copolymer 嵌段共聚物
periodic(al) coupe 轮伐区;分期伐区
periodic(al) crystallization 间歇结晶;反复结晶(作用)
periodic(al) current 周期性流(水);周期性海流;周期(水)流;周期电流;振荡电流;间歇流;潮流
periodic(al) cutting-area 分期采伐面积
periodic(al) damping 欠阻尼;周期阻尼
periodic(al) decimal 循环小数
periodic(al) deformation of riverbed 河床周期性变形
periodic(al) deformation of stream bed 河床周期性变形
periodic(al) deposit 定期存款
periodic(al) dipole fluctuation 周期性偶极起伏
periodic(al) discharge 周期性放电;定期排污
periodic(al) discharge monitoring report 定期排污监测报告
periodic(al) distribution curve 周期分布曲线
periodic(al) docking 周期性进坞
periodic(al) docking survey 定期坞检
periodic(al) drainage 周期排水;周期排污;定期排污
periodic(al) drainage monitoring report 定期排污监测报告
periodic(al) drift 定期漂移
periodic(al) dumping 周期性转储
periodic(al) dumping time-sharing 周期性转储分时
periodic(al) duration 周期历时
periodic(al) duty 周期性工作方法;循环工作;周期运行(方式);周期性工作;周期性负载;周期工作方式;周期负载
periodic(al) economic crises 周期性经济危机
periodic(al) elongation 周期延长
periodic(al) equation 周期方程
periodic(al) error 周期性误差;周期误差
periodic(al) examination 定期检测
periodic(al) excitation 周期激振
periodic(al) family 周期族
periodic(al) fever 周期热
periodic(al) field focusing 周期场聚焦
periodic(al) flood 周期性洪水;周期泛溢
periodic(al) flow 周期性流动;周期流
periodic(al) fluctuation 周期性波动
periodic(al) flushing 定时冲洗
periodic(al) frequency modulation 周期频率调制
periodic(al) frozen ground 季节冻土
periodic(al) function 周期函数
periodic(al) furnace 周期式作业炉;间歇式炉;分批作业炉
periodic(al) gas lift 周期性气体升液器
periodic(al) grazing 周期(性)放牧
periodic(al) group 周期类
periodic(al) health examination 定期健康检查
periodic(al) heat transfer 周期性传热
periodic(al) incident wave 周期性入射波
periodic(al) increment 周期生长量;定期生长量
periodic(al) index room 期刊目录室
periodic(al) index space 期刊目录室
periodic(al) inequality 周期差
periodic(al) input/ouput 周期性输入/输出
periodic(al) inspection 小修(理);周期性检查;周期检查;定期检验;定期检查
periodic(al) inspection and repair 定期检修
periodic(al) inspection report 定期检查报告
periodic(al) instability 周期性不稳定性
periodic(al) interrupt 周期性中断
periodic(al) inventory 定期盘点;定期盘存
periodic(al) investigation 定期调查
periodic(al) key 周期索引
periodic(al) kiln 周期作业窑;周期性窑;周期加热窑;间歇窑;分批加热炉
periodic(al) law 元素周期律;周期律
periodic(al) line 链路;周期线
periodic(al) load 周期(性)荷载;周期性负载
periodic(al) load line inspection 载重线定期检验
periodic(al) load line renewal 载重线合格证定期更新
periodic(al) lubrication 定期润滑
periodically 周期地
periodically flooded area 周期性洪泛区
periodically lubricated axle-box 定期润滑的轴箱
periodically lubricated bearing axle-box 定期润滑的轴箱轴承
periodically operated switch 周期动作开关
periodically running 周期作业
periodically waterlogged 周期性积水的
periodic(al) maintenance 定期养护;定期维修;定期维护;定期检修
periodic(al) maintenance of container 集装箱定期维修保养
periodic(al) maintenance operation 定期维修作业
periodic(al) mean annual increment 定期平均年生长量
periodic(al) mean annual intermediate yield 定期平均年疏伐量
periodic(al) measurement 定期测量;定期测定
periodic(al) meeting 定期会议
periodic(al) migration 定期迁移
periodic(al) model 周期模型
periodic(al) monitoring 定期监测
periodic(al) motion 周期运动
periodic(al) movement 周期运动;周期性活动
periodic(al) observation 定期观察;定期观测
periodic(al) one-way sign 定时单向通行标志
periodic(al) operation 定期运行;周期性运转;定期运行
periodic(al) orbit 周期轨道
periodic(al) ordering 控制存货的定期进货法
periodic(al) ordering system 定期订货系统
periodic(al) oscillation 周期摆动;周期振动;周期(性)振荡;周期性波动
periodic(al) outbreak 周期性暴发
periodic(al) overhaul 定期修理;定期大修
periodic(al) pattern 周期方向图
periodic(al) payment 定期付款
periodic(al) payment plans 定期付款计划
periodic(al) payment with present value of one yuan 一元现值定期付款;每元现值每期付款数
periodic(al) peak load 周期性峰荷
periodic(al) performance report 定期工作报告
periodic(al) perturbation 周期摄动
periodic(al) phenomenon 周期现象
periodic(al) plankton 周期性浮游生物
periodic(al) point 周期点
periodic(al) pollution 周期性污染
periodic(al) potential 周期电势
periodic(al) precipitation 间歇沉淀

periodic(al) process 周期性过程
periodic(al) propeller shaft survey 尾轴定期检验
periodic(al) property 周期性质;周期性
periodic(al) publications 期刊
periodic(al) pulse train 周期性脉冲群;周期脉冲列
periodic(al) quantity 周期量
periodic(al) random process 周期随机过程
periodic(al) rating 周期性负载额定强度
periodic(al) reaction to temperature 温周期
periodic(al) reading room 期刊阅览室
periodic(al) recurrence 周期性流行
periodic(al) recurrent 循环的
periodic(al) recurrent event 周期循环事件
periodic(al) repair 定修;定期修(理);定期检修
periodic(al) repair of locomotives 机车定期检修工作
periodic(al) report 定期报告
periodic(al) resonance 周期谐振;固有谐振
periodic(al) resurvey 周期性复查
periodic(al) resynchronization 循环同步法;周期同步法
periodic(al) return 周期性循环
periodic(al) reversal 周期性换向;周期换向
periodic(al) reverse 周期性换向
periodic(al) reverse current (电镀的)周期反向电流
periodic(al) review 定期审查;定期盘点;定期复查
periodic(al) ringing 周期振铃
periodic(al) rolling 周期断面轧制
periodic(al) roof pressure 周期顶板压力
periodic(al) room 杂志室;期刊阅览室
periodic(al) sampling 周期性取样;周期抽样;定期采样
periodic(al) scanning 定期扫描
periodic(al) section 周期断面型钢
periodic(al) seizure 周期性发作
periodic(al) selection 周期性选择作用
periodic(al) sequence 循环序列;周期序列
periodic(al) settling 周期性下沉
periodic(al) silting and scouring 周期性冲淤
periodic(al) solution 周期解
periodic(al) spiral 周期螺旋
periodic(al) spontaneous pressure 周期自然冒落
periodic(al) spring 周期(性)泉;间歇泉;虹吸泉
periodic(al) stack 期刊书架
periodic(al) stand depletion 定期林分疏伐量
periodic(al) state 周期状态
periodic(al) statement 定期报表
periodic(al) stochastic process 周期性随机过程
periodic(al) stresses 周期性应力
periodic(al) structure 周期性结构
periodic(al) suburban ticket 市郊定期客票
periodic(al) surface 周期曲面
periodic(al) survey 周期性检验;定期检验;定期检查
periodic(al) system 周期系(统)
periodic(al) system of elements 元素周期系
periodic(al) table 周期表
periodic(al) table of elements 元素周期表
periodic(al) tank furnace 间歇作业窑
periodic(al) temperature variation 周期性温度变化
periodic(al) tenancy 定期租赁
periodic(al) tendency 周期性趋向
periodic(al) term 周期项
periodic(al) test 周期试验;定期试验
periodic(al) testing 周期性检测
periodic(al) ticket 定期票
periodic(al) time 周期时间;周期
periodic(al) undulation 周期性起伏
periodic(al) unsteady flow 周期性不稳定(水)流
periodic(al) up-wave 周期上升波
periodic(al) variable 周期变量
periodic(al) variation 周期性变化;周期性变动;周期变化
periodic(al) variation of shoal 浅滩周期性变化
periodic(al) verification 周期性检定
periodic(al) vibration 周期振动
periodic(al) visit to the site 定期下工地
periodic(al) volcano 间歇火山;周期性火山
periodic(al) vortex 周期漩涡;周期涡流
periodic(al) water flow 周期性水流
periodic(al) wave 周期波
periodic(al) wind 周期风

periodic(al) work 周期性工作
periodic(al) working 分期开采
periodic(al) yield 定期收获量
periodicautomatic-reclosing equipment 周期性自动重合闸装置
periodicity 周期(性);周波;间歇性;间发性
periodicity component 周期性成分
periodicity concept of accounting 会计的时间观念
periodicity condition 周期条件
periodicity fruiting 周期性结果
periodicity of a pseudo-random number sequence 伪随机数序列的周期性
periodicity of community群落的周期性
periodicity of dosing 投配周期
periodicity of earthquake 地震周期性
periodicity of function 函数周期性
periodicity of tectonic movement 构造运动周期性
periodide 高碘化物
period in advance 提前期
period indicator 周期指示器
period inspection work card 周期检查工作卡
period in the train(working) diagram 运行图周期
periodism 周期性;周期型
periodization 周期化
period-luminosity-colo(u)r relation 周光色关系
period-luminosity curve 周光曲线
period-luminosity relation 周期关系;周光关系
period map 历史分期图
period-mean density relation 周期平均密度关系
period meter 周期计;周期测量仪;回声探测器
period number 期号
periodo-carbon 全碘化碳
periodo-ethane 全碘乙烷
periodo-ether 全碘乙烯
period of acceleration 加速周期
period of adjustment 调整时期;调整期间
period of advance notice 预先通知期限
period of aeration 曝气周期
period of amortization 折旧期;摊还期
period of analysis 分析期限
period of architecture 建筑时代
period of availability 有效时期
period of a variable star 变星周期
period of back-pumping 回扬周期
period of balanced production 均衡生产年限
period of circulating decimal 小数循环节
period of circulation 流通期
period of complete rotation 满转周期
period of complete turnover of locomotive 机车全周转时间
period of concentration 集流时间;集流期;汇流期
period of construction 工程期限;建设期;施工期限
period of contact 接触时间;闭合期间
period of contraction 紧缩时期
period of cost recovery 成本回收期间
period of crop-rotation 轮作周期
period of curing 养护期
period of custody and control 保管期
period of depreciation 折旧期;耐用年限;贬值年限
period of design 设计期限;设计年限;设计进度(表)
period of dipping and heaving 上下浮动周期
period of disease 发病期
period of dormancy 休眠期
period of drainage 排水周期
period of drought 干旱周期;干旱(时)期
period of ebb and flood 涨落潮周期
period of ebb and flow 涨落期
period of element 元素周期
period of elongation 伸长期
period of employment 雇佣期
period of eustatism 海平面变化周期
period of expansion 发展期
period of flaming 火焰期
period of flood and ebb 涨落潮周期
period of flushing 涨潮时间;涨期间
period of folding 褶皱期
period of forecast 预测的时间间隔
period of forecasting 预报时效
period of free storage 免费储存期限
period of full draw-down 全消落期(水库);全落时期
period of glaciation 冰期
period of grace 优惠期

period of groundwater highly mineralized 地下水高矿化期
period of groundwater low mineralized 地下水低矿化期
period of guarantee 保证期限
period of half change 半变期
period of half decay 半衰期
period of hardening 硬化期;凝固期
period of heating 加热期
period of high water level 高水位期
period of ice-cover 封冻期;封冰期
period of ice drifting 流冰期;冰封期
period of immaturity 非生产期
period of incubation 潜伏期
period of initial 初凝时间
period of initial setting 初凝期
period of inspection and repair 检修周期
period of lease 租借期限
period of light variation 光变周期
period of limitation 期限
period of line 扣押期
period of load development 负荷增长期
period of loading 正在期;装料期;加荷期;负重期
period of low flow 枯水期
period of low-water level 低水位期
period of maintenance 维修期
period of maximum temperature 最高温度期
period of meander 曲流期;蜿蜒期
period of melting 熔化期
period of minimum consumption 最小负荷(交通)
period of modern deposition 新泥砂沉积期;现代沉积期
period of motion 运动周期
period of movement 活动时期
period of natural vibration 自振周期;固有振动周期
period of navigation obstruction 碍航时间
period of neotectonic movement 新构造运动时期
period of non-linear system 非线性系统周期
period of optic(al) cycle 光学循环周期
period of orogenesis 造山期
period of oscillation 振荡周期;摆动周期
period of pendulation 摆动周期
period of pitching 纵摇周期
period of precession 岁差周期
period of precession of equinox 岁差周期
period of production 生产期限
period of pulsation 脉动周期
period of qualifying service 合格服务期限
period of quality guarantee 质量保证期
period of reappearance 重现期
period of recharge 回灌周期
period of record 记录周期;观测时期;记录时期;记录年限
period of recovery 恢复时期
period of regime change cycle 动态变化周期的长度
period of repeating decimal 小数循环节
period of reservoir with full water 满库(水)时间
period of responsibility 责任期
period of retardation 滞洪期;迟延时期
period of retention 停蓄期;滞留期;固位期;停留期
period of retention extraction reserves 回采矿量保有期限
period of retention geologic(al) reserves 地质储量保有期限
period of retention ore reserves 矿产储量保有期限
period of retention ore reserves for production 生产矿量保有期限
period of retention prepared reserves 采准矿量保有期限
period of retention reserves for development 开拓矿量保有期限
period of return investment 投资回收周期
period of revolution 公转周期
period of rhythmic light 灯光周期
period of rise 涨洪期;上升期
period of rising tide 涨潮期间
period of roll(ing) 横摇周期
period of rotation 自转周期;转动周期
period of sampling 取样周期;抽样周期;采样周期
period of satellite 卫星周期
period of satellite orbit 卫星轨道周期
period of saturation 饱和期

period of schooling 学习期限
period of seiche 静震周期;假潮周期;湖面波动周期
period of service 使用寿命;服务期
period of settlement 决算期
period of simple harmonic motion 简谐运动周期
period of slack sales 销售淡季
period of slow wave structure 慢波(线)周期
period of snow accumulation 积雪期
period of spawning 产卵期
period of spring discharge decreasing 泉流量下降期
period of spring discharge increasing 泉流量增加期
period of stagnation 滞水期;停滞时段
period of storage 储[贮]存时间
period of supply 供应期
period of swell 涌浪周期
period of tectonic movement 构造运动时期
period of test(ing) 试验周期
period of thaw 解冻期;融雪期;融化期
period of tillage 耕作季节
period of time 时间周期
period of transformation 自转周期;转变期;成型期;变换周期
period of transition 过渡时期;转变时期;转换时期
period of transmission 发送周期
period of treatment 处理期间
period of uncovering 潮退期
period of undulation 起伏周期;波动周期;摆动周期
period of use 利用期;使用期
period of validity 有效期;预报时效
period of vegetation 植物生长期
period of vibration 周期振动;振动周期
period of warming 加湿时期
period of warranty 担保期;保证期;保险期
period of water-level decline 水位下降期
period of water-level raising 水位上升期
period of water source operation 水源地投产年限
period of waves 波周(期);波的周期;波动周期
period of weakening 路基软化时期;软化时期
period of without rainfall 旱期
periodogram 周期图
periodogram analysis 周期图分析
periodograph 周期图仪
periodometer 调和分析仪
period oscillation 周期摆动
period oscillator 周期振荡器
period parallelogram 周期格子
period per second 周/秒
period planning 期间计划
period prevalence 期间患病率
period project 期间计划
period range 周期区段
period ratio 周期比
period-redius relation 周径关系
period regulation 周期性调节;周期调节
period regulator 周期调节器
period resolution 周期分辨率
period section steel 周期断面钢材
period shut-down 周期保护停推
period spectrum 周期波谱
periods per second 赫
period stipulated for delivery 约定交货期
period style 时代式样;时代风格
period when hunting prohibited 禁猎期
period without rainfall 枯水期;无雨期;枯季
perioval precipitation rate 环沉率
peripedinent 外缘山麓侵蚀面
peripherad 向周围;向外周
peripheral 周围性的;周围的;围柱式的;外围
peripheral air defense 外围防空
peripheral air ducting 周边空气管
peripheral apparatus 外围设备
peripheral aquifer 外围含水层
peripheral area 周围地区
peripheral bar 周边拉杆
peripheral basin 周缘盆地
peripheral beam 圈梁
peripheral block 周边砌块
peripheral blowing 腰风;侧鼓风
peripheral buffer 外围缓冲器;外部设备缓冲器
peripheral bus 外围总线
peripheral cam 圆周凸轮
peripheral car park 外围停车场;市缘停车场

peripheral cell 围室
peripheral chain system 周边链
peripheral circuit 周边电路;外围电路
peripheral clearance 周围间隙;周边间隙
peripheral coating of shaft kiln 立窑的粘边
peripheral collision model 边界碰撞模型
peripheral column 檐柱
peripheral column supported purlin(e) 正心桁
peripheral compartment 外周室
peripheral component interconnect bus 外设部件互联总线
peripheral component of velocity 圆周分速度
peripheral compressor 周边式压缩机
peripheral control 外围设备控制;外围控制
peripheral control line 外围控制线路
peripheral control program(me) 外围设备控制程序;外部设备控制程序
peripheral controller 外围设备控制器
peripheral control unit 外部控制器
peripheral coordinator 外围设备调整机
peripheral core 圆心
peripheral current 周边流
peripheral data bus 外围数据总线
peripheral data register 外围数据寄存器
peripheral decoding 外围设备译码
peripheral depression 边缘洼地
peripheral device 外围设备
peripheral dike 周边围堤
peripheral discharge 周边卸料;周边出料;边缘卸料
peripheral discharge mill 周缘出料磨;边缘卸料磨机
peripheral discharge opening 周边卸料孔
peripheral discharge screen 周边卸料筛
peripheral disk 外部盘
peripheral divide 周缘分界线
peripheral drive 边缘传动(装置)
peripheral envelope 周包膜
peripheral equipment 外转设备;外围设备;外部设备;附加设备;辅助设备
peripheral fault 周缘断层;周围断层;环周断层;边缘断层
peripheral feed 周边投料;边缘喂料
peripheral feed settling tank 周边进水沉淀
peripheral field 外围视野
peripheral flooding 边外注水
peripheral flow 周边流
peripheral force 圆周力
peripheral function translator 外围功能转换器;外部函数翻译程序
peripheral geosyncline 周缘地槽
peripheral granule 周颗粒
peripheral graphic(al) device 外围绘图设备
peripheral grinding 圆周磨削
peripheral halo 空心晕
peripheral hoist 圆周升降机
peripheral hole 周边炮眼;周边炮孔;圈定炮眼
peripheral hot stamping press 外围热冲压压力
peripheral inhibition 周围抑制
peripheral initiation 周围起爆
peripheral instrumentation 外围设备
peripheral interface 外围设备接口;外围接口
peripheral interface adapter 外围接口转接器;外部接口适配器
peripheral interface channel 外部设备接口通道
peripheral interface module 外围接口组件
peripherality 周边性
peripheral jet 圆周喷射;切向喷射
peripheral jet craft 周边喷射气垫艇
peripheral jet hovercraft 周边射流气垫船
peripheral joint 沿周缝;周边缝
peripheral layer 周围层
peripheral length 圆周长度;周边长度
peripheral length of circular features 环形体周长
peripheral loading 周边荷载
peripheral locality 外围地点
peripheral lowland 边缘低地
peripheral-marking machine 周边标志机
peripheral memory 外存储器
peripheral milling 圆周铣削;周缘铣
peripheral moraine 周碛;围碛
peripheral mount support 周边支撑
peripheral of ring structure 环状构造的边缘破裂带
peripheral operation 外围操作;外部操作
peripheral order buffer 外部指令缓冲器

peripheral output device 外围输出设备
peripheral part 边缘部分
peripheral passage 循环通路
peripheral peocessing unit 外围处理机
peripheral plane 周边平面
peripheral pressure 切线压力
peripheral processor 外围处理机
peripheral pump 漩涡泵;叶轮泵;周边泵
peripheral radar 外围雷达
peripheral reaction 圆周反应
peripheral region 周围区
peripheral resistance 外周阻力
peripheral rod 周围杆
peripheral routine 外围程序
peripheral-slot discharge 周边缝排料
peripheral speed 圆周速度;周速(度);周边速度;边周速度;边缘速率
peripheral stress 圆周应力;周向应力;孔周应力
peripheral subsystem channel 外围子系统通道
peripheral support computer 外围支援计算机
peripheral synclinal belt 边缘向斜带
peripheral system 周围系统
peripheral taxiway 环形滑行道
peripheral teeth 周边铣齿
peripheral tie beam 腰箍;周边联系梁
peripheral transfer 外围设备间传输;外围传送;外部传送
peripheral tube 外围管
peripheral (turbine) pump 涡流泵
peripheral type 边缘型
peripheral unit 外围设备;外部设备;附属设施;附属设备
peripheral velocity 圆周速率;圆周速度;周边速度;轮缘速度
peripheral vision 周围视觉;间接视;视觉边线;边界视力
peripheral weir 圆周堰;周边堰
peripheric drift 周壁导坑
peripheric equipment 外部设备
peripheric receptor 终末感受器
peripheric unit 外部设备
periphery 圆柱表面;周线;周围;周边;外周;外围;四周
periphery beam 圈梁;外墙梁;地脚梁
periphery cam 平凸轮;盘形凸轮
periphery fillet welding 贴角周边焊
periphery grouting 周边灌浆
periphery of a town 城镇周边界线;城镇外围地区
periphery seal 周围封闭
periphery wall 外围墙;外墙
periphyton 周丛藻类;周丛生物;固着生物;水中悬垂生物
periplutonic metamorphism 边缘深成变质作用
peripneustic 侧气门式的
periposeidon 近海王点
peripteral 周围有气流的;围柱式的;围柱式殿;四周界柱式(建筑);列柱式的
peripteral building 围柱殿式建筑;围柱式建筑
peripteral colunnade 围柱殿的柱廊
peripteral hexastyle temple with fourteen columns on the flanks 两侧翼有14根柱的周柱六角式庙宇
peripteral octastyle temple with seventeen columns on the flanks 两侧翼有17根柱的周柱八角式庙宇
peripteral temple 围柱式神庙;围柱殿式庙宇
periptere 柱围房屋
peripteros 围柱式建筑物;单排围柱殿;周柱式建筑物;围柱式建筑
periptery 围柱式房屋;四周界柱式(建筑);柱围房屋;围柱式建筑
perisarc 围鞘
periscian region 日影环绕区
periscope 熔炉观测镜;潜望镜
periscope angle 潜望镜视角
periscope automatic stellite navigator 潜望镜自动卫星导航仪
periscope azimuth circle 潜望镜方位圈
periscope bending compensation 潜望镜弯曲补偿
periscope binocular 潜望镜式双筒望远镜
periscope center 潜望镜部位
periscope condition 潜望状态
periscope depth 潜望深度;潜望镜深度
periscope fairing 潜望镜导流罩

periscope feather 潜望镜浪花
periscope guard 潜望镜护板
periscope height 潜望高度
periscope hole 潜望镜孔
periscope housing 潜望镜罩
periscope image stabilization 潜望镜图像稳定
periscope optic(al) system 潜望镜光学系统
periscope radar 潜望雷达
periscope rising distance 潜望镜升距
periscope service speed 潜望镜使用航速
periscope stand 潜望镜平台
periscope tracking 潜望镜跟踪
periscope wake 潜望镜航迹
periscopic 凹面凸面透镜的
periscopic antenna 潜望镜天线
periscopic depth 潜望最大深度
periscopic drift angle 潜望式偏航测角器
periscopic drift-angle sight 潜望式偏航角观测器
periscopic drift sight 潜望镜式偏流计
periscopic inspection of drill holes 潜望镜检查钻孔;钻孔下潜望检视;钻孔潜望镜检查
periscopic lens 潜望透镜;凹凸透镜
periscopic range-finder 潜望测距仪
periscopic sextant 潜望(镜式)六分仪
periscopic sight 潜望镜;潜望式测微器;潜望瞄准镜
periscopic television 潜望电视
periscopic vision 潜望镜观察
periscopic wind ga(u)ge 潜望式风速计
periscopic wind ga(u)ge sight 潜望式风速观测器
periselen 近月点
perish 碎散;毁灭;灭亡;腐烂
perishability 易朽性;易腐烂性
perishable cargo 鲜货;易朽烂货物;易腐货物
perishable cargo ship 鲜货船
perishable commodities 易腐烂商品
perishable consumer goods 一次性消费产品
perishable freight 易腐(烂)货物;易冻货物
perishable goods 易腐物品;易腐货物;易冻货物
perishable inventory 易腐物储存
perishable items 易腐食品
perishable material 易朽材料;易腐材料
perishableness 易朽性
perishables 易腐物质;易腐烂物质;易腐货物
perishable supplies 易腐补给品
perishable tool 消耗性工具
perished steel 过渗碳钢;脆弱钢
perisphere 周层;球形建筑物;外围层;势力范围
perispheric 正圆球形的
peristalith 环绕墓穴的竖立石;石碑圈
peristaltic 蠕动的
peristaltic chargescoupled device 蠕动电荷耦合器件
peristaltic pump 蠕动泵
peristaltic rushes 蠕动波
peristaltic wave 蠕动波
peristele 石柱
peristerite 蓝彩钠长石;晕长石
peristerium 第二祭坛天盖;内祭坛天盖
peristylar 列柱廊
peristyle 周柱式;列柱庭院;列柱式;列柱(廊);边廊式
peristyle building 周柱式建筑
peristyle ceiling 周柱式天花板;周柱式天顶;周柱式天棚
peristyle column 列柱中庭的柱;周柱式中的柱
peristyle court 周柱式庭院
peristyle garden 廊柱园
peristyle house 周柱式房屋
peristylium 周柱廊中庭(古罗马);列柱廊中庭(古罗马);有柱廊庭院;列柱庭院;[复]perisylia
perisutural basin 近缝合带盆地
perite 氯氧铋铅矿
peritectic crystal 包晶
peritectic granulation 包晶微细化
peritectic point 转熔点;包晶反应点;包晶点
peritectic process 转熔过程
peritectic reaction 转熔作用;转熔反应
peritectics 转熔
peritectic spheroidizing 包晶球化
peritectic structure 包晶结构
peritectic temperature 转熔温度
peritectic texture 包橄结构
peritectoid 转熔体;包析体(的);包晶体;包晶(的)
peritectoid reaction 包析反应;包晶反应

peritery 四周有柱的建筑
peritidal area 潮缘区
peritidal complex 潮缘复合体
peritidal rock 潮缘带岩石
peritral 四周有柱的房屋
peritreme 孔缘
perittogamy 随机质配
peritubular membrane 管周膜
periuranian 近天王点
peri-urban 城市近郊
peri-urban area 市区周围地带;城市周围地区
peri-urban population 城市周围人口
peri-urban road 城市周围道路;城市近郊道路;近郊环路
perk 快速动作
per kilogram 每公斤
Perking brass 铸造用锡青铜;波京黄铜
perking switch 快动开关
Perkin's jig 柏琴夹具
Perkin's mauve 苯胺紫
Perkin's synthesis 柏琴合成
perknite 辉闪岩类;辉闪岩
per like day 非连续计算的滞期日
perlimonite 褐铁矿
Perlin viscosity cup 珀林黏度杯
perlit 波利特铸铁
perlite 珠光体;高强度珠光体铸铁
perlite aggregate 珍珠岩集料;珍珠岩骨料
perlite concrete 珍珠岩混凝土
perlite core 膨胀珍珠岩芯板
perlite deposit 珍珠岩矿床
perlite plaster 珍珠岩灰泥;珍珠岩灰浆
perlitic structure 珍珠构造
perlitic texture 珍珠结构
perloffite 磷铁锰钡石
perlon-1 卡普隆(商业名)
perlucidus 漏隙云
perm 泊姆(渗透速度单位)
perma 层压塑料
permacem 防水水泥漆(商品名)
permaclad 碳素钢板上覆盖不锈钢板的合成层板
permactron 双注电子波管
permafrost 永久冻土;永冻土;多年冻土;冻结土;常冻
permafrost area 永冻地区;冻土区
permafrost bottom 多年冻土下限
permafrost degradation 永冻降解;永冻层融化(过程)
permafrost drilling 永冻土钻孔;冻土层钻进
permafrost ecosystem 永久冻土生态系统;冻土生态系统
permafrost frozen layer 永冻层;恒冻层
permafrost horizon 永冻层
permafrostology 冻土学
permafrost region 冻土区
permafrost soil 永久冻土;永冻土
permafrost table 永冻土面;永冻(土)层深度线;永冻层层面线;多年冻土面
permafrost type 多年冻结型
permafrost zone 永冻带;多年冻土带;冻土带
permafrozen ground 永冻土
permag 清洁金属用粉
permagnag 测磁计
Permali 泊马利(一种浸树脂胶合木板)
permalloy 高导磁率镍铁合金;透磁合金;坡莫(镍铁)合金
permalloy compensation 坡莫合金补偿
permalloy strip 坡莫合金传感器
permanence 持久性;耐久性;永久性;安定度
permanence condition 不变条件
permanence of continents 大陆永恒(说)
permanence of distribution 分布的不变性
permanence of lognormality 对数正态不变性
permanence of normality 正态不变性
permanence theory 永恒理论
permanency 永久性
permanency leakage 永久性渗漏
permanent 固定数据文件;固定标准地
permanent absorbing phase strip 永久性吸收相位条带
permanent account 实账户(即资产账户负债账户及资本账户)
permanent action 永久作用
permanent agreement 永久性协定
permanent agriculture 不休闲耕作

permanent amictic lake 永久封冻湖
permanent anode 永久阳极
permanent appointment 正式任命
permanent appropriation 永久拨款
permanent array 固定台阵
permanent assets 固定资产
permanent aurora 夜气辉
permanent awning 常设天幕
permanent axis 恒轴(线)
permanent-backed resinshell process 树脂砂覆砂造型
permanent back-water region 常年回水区
permanent ballast 固定压载(物);恒压载
permanent bank protection 永久性护岸
permanent bar 永外磁棒;永久磁棒
permanent basis 平时编制
permanent bearing 永久支座
permanent bed 固定河床;稳定河床
permanent bench mark 永久水准(基)点;固定水准点
permanent blanket of unfused batch 永久性未熔生料毯
permanent blasting wire 固定引爆线
permanent blue 永久蓝
permanent blue pigment 永久蓝颜料
permanent blue toner 永久蓝色原;永久蓝调色剂
permanent bracing 永久撑杆;固定拉杆;永久支撑
permanent bridge 永久性桥(梁)
permanent building 永久性建筑(物)
permanent bunker 常备煤舱;常备燃料储[贮]舱
permanent buoyant tank 永久浮力舱
permanent caisson 永久性沉箱
permanent-capacitor motor 永久电容器式电动机
permanent capital 永久资本
permanent casting 硬模浇铸
permanent cast light(weight) concrete form 浇制轻质混凝土固定模板
permanent cave 永久屏蔽室
permanent change in dimensions on heating 重烧尺寸变化
permanent change of station 永久性调动
permanent character 永久特性
permanent chute 永久漏口
permanent circulation 永久环流
permanent closed contact 常闭触点
permanent commission 常务委员会
permanent committee 常务委员会;常设委员会
Permanent Committee for the South Pacific 南太平洋常设委员会
permanent community 永久群落
permanent compaction 永久性压实;永久性压密
permanent-completion packer 永久完井封隔器
permanent component 固有部分
permanent compression 永久压力
permanent compression set 永久压缩变形;压缩永久变形
permanent connection 永久连接;固定连接;不可卸连接
permanent construction 永久性结构;永久性建筑(物)
permanent contraction 永久性收缩;残余收缩;残存收缩
permanent control 永久性控制;稳定控制
permanent cortex 永久皮质
permanent counting station 永久性观测站;常设计数站
permanent coupling 永久性连接;刚性联轴节
Permanent Court of Arbitration 常设仲裁法庭
permanent curl 永久卷曲
permanent current 持久流;恒定流;恒流;恒定电流;稳定海流;定常海流;持恒电流;常流
permanent current capital 固定性周转资本
permanent dam 永久坝;固定坝
permanent damage 永久性破坏
permanent data file 永久性数据文件
permanent data set 永久性数据集
permanent defence 永久性防御工程;永久防御工事
permanent deflection 永久弯曲;永久翘曲;永久挠曲;永久挠度;永久变位;剩余变位;残余弯沉
permanent deformation 余面应变;余留应变;永久应变;永久性变形;永久形变;永久变形;残余形变;残余变形
permanent deviation 永久偏移
permanent dipole 永久偶极子

permanent dipole orientation 永久偶极定向
permanent disability 永久伤残
permanent display 连续显示
permanent disposal 永久处置
permanent distortion 残留变形
permanent documents 永久文件
permanent drift 永久偏移
permanent drillable bridge plug 可钻的永久桥塞
permanent drillable packer 可钻永久式封隔器
permanent drilling island 永久钻井岛；永久式钻井岛
permanent drilling well 永久钻井
permanent drop 固态转速率；固态转差率
permanent dunnage 固定垫舱物
permanent dye 持久染料；不变色染料
permanent dynamic(al) speaker 永磁电动式扬声器；永磁动圈式扬声器
permanent earth 永久地网
permanent echo 固定目标回波；不变回波
permanent efflux 持久流向
permanent elasticity 永久弹性
permanent electric(al) dipole moment of molecule 分子固有电偶极矩
permanent elongation 永久伸长；持久伸长；残余伸长
permanent embossed finish 耐久性拷花整理
permanent energy 恒定能量
permanent error 永久(性)误差；永久错误；固定误差；固定错误
permanent eruption 持续性喷发
permanent expanding 永膨胀
permanent expansion 永久膨胀；持久膨胀；残余膨胀；非可逆膨胀
permanent experiment 长期试验
permanent extension 永久延伸；永久拉伸
permanent failure 永久失效
permanent fast violet 永久坚牢紫
permanent fault 永久(性)故障；固定故障；常发故障
permanent feedback 固态反馈
permanent fender 固定舷木护栏；固定护舷材
permanent field 恒定场
permanent file 永久文件
permanent filling 永久性充填
permanent financing 永久筹资
permanent finish 永久性整理
permanent flame 持久火焰
permanent flightlocks 永久性梯级船闸
permanent flooded tank 永久淹水舱
permanent flow 恒定流；稳定流；常年水流流量；常定流
permanent flow valve 恒流阀
permanent form 永久性模板；定型式样；固定模板；持久性形式
permanent form(work) 永久性模板
permanent fortification 永久性防御工事
permanent foundation 永久性基础
permanent frozen earth 永久冻土区
permanent garbage can 固定式垃圾箱
permanent garden 永久花坛
permanent gas 永气体；永久气体
permanent geodetic beacon 永久性测量觇标；永久大地觇标
permanent glow signal 永久信号；不灭信号
permanent grassland 永久性草地
permanent green 永久绿
permanent green pigment 永久绿颜料
permanent green toner 永久绿色原；永久绿调色剂
permanent groundwater level 固定地下水位面；固定地下水位；常年地下水位
permanent guest 长住旅客
permanent hack 固定的晒砖场
permanent hardness 永久硬度
permanent hardness of water 水的永久硬度
permanent hard water 永久硬水
permanent haulage line 固定运输路线
permanent head tank 恒水头水池
permanent hearing defect 永久性听觉损伤
permanent hearing loss 永久性听觉丧失
permanent heating 固定加热
permanent heterozygote 永久杂合体
permanent hoist 永久提升机
permanent hotbed 固定温床
permanent housing 永久性住宅
permanent humidity 恒定湿度

permanent hydrometric station 永久水文站；基本水文站
permanent ice foot 永久冰壁；多年贴岸冰
permanent impairment 终身丧失工作能力的损伤
permanent income 永久收益；长期收入
permanent ink 永固油墨；不褪色油墨；不变墨水
Permanent International Association of Navigation Congress 国际航运会议常设委员会
Permanent International Association of navigation Congress 国际航海会议常设委员会
Permanent International Association of Road Congress 国际道路会议常设委员会
permanent investment 永久投资；长期投资
permanent investor 长期投资人；长期投资者
permanent irrigation 永久性灌溉
permanent joint 永久接合；固定接头
permanent label 永久性标志；永久性标签
permanent lake 常年湖
permanent level filter 恒水头滤池
permanent lighting 常设照明
permanent limitation 永久性限制
permanent linear change after heating 重烧后永久线变化
permanent liner 永久衬里
permanent liner change on reheating 重烧永久线变化率
permanent line stake 永久线路桩
permanent lining 永久衬砌；二次衬砌；固定衬套；固定衬砌；固定衬板
permanent lining material 永久性衬砌材料
permanent lining support 永久性衬砌支撑
permanent load(ing) 永久性加荷；永久载重；永久(性)荷载；固定荷载；恒载；死荷载；长期负载；长期负荷
permanent loan 长期贷款；永久贷款
permanent longitudinal magnetism 水平纵向永久磁性
permanent loop junctor 固定环路连接器
permanent loss 永久损耗
permanent lubrication 持续润滑
permanent luminescent pigment 恒效发光颜料
permanent lustred finish 耐久性上光整理
permanently absorbed water 永久吸收水；永久吸附水
permanently attached equipment 永久性附属装置
permanently elastic 永久地弹性
permanently extinct lake 永久干涸湖
permanently flowing water course 常年流水层
permanently frozen ground 永冻土
permanently frozen soil 永久冻土；永冻土
permanently locked envelope 永久锁定信件
permanently lubricated axle-box 持续润滑的轴箱
permanently lubricated bearing axle-box 持续润滑的轴箱轴承
permanently moored 固定停泊的
permanently nonhardening 永久地不结硬
permanently plastic 永久地塑性
permanently resident volume 永久(性)常驻卷
permanently ventilated lobby 永久通风的大厅
permanent-magnet 永久磁铁；永磁体
permanent-magnet alloy 永磁合金
permanent-magnet alternator 永磁发电机
permanent-magnet ammeter 永磁式安培计
permanent-magnet chuck 永久磁性吸盘
permanent-magnet-erasing head 永磁抹音头
permanent-magnet field generator 永磁发电机
permanent-magnet focusing 永磁聚焦
permanent-magnet generator 永磁发电机
permanent-magnetic ferrite 硬磁铁氧体
permanent-magnetic field 永磁场
permanent-magnetic lens 永磁透镜
permanent-magnetic microscope 永磁显微镜
permanent-magnetic property 永磁性质
permanent-magnet instrument 永磁式仪表
permanent-magnetism 永磁(性)；永久磁性
permanent-magnetization 永久磁化
permanent-magnet machine 永磁电机
permanent-magnet material 永久磁性材料；永磁材料
permanent-magnet meter 永磁式仪表
permanent-magnet moving coil instrument 永磁动圈式仪表
permanent-magnet moving-iron instrument 永磁动铁式仪表；极化电磁系仪表

permanent-magnet off-centering 永磁偏心
permanent-magnet pole 永磁极
permanent-magnet steel 永磁钢；永久磁钢
permanent-magnet stepper motor 永磁步进电动机
permanent mailing address 永久通信地址
permanent management arrangement of INTELSAT organization 国际电信卫星组织永久管理安排
permanent map record 地图图形记录
permanent mark 永久标志
permanent marker 永久性标志；永久性标记
permanent meadow 永久性草原；永久草甸
permanent member 永久性成分
permanent memory 永久性存储器；永久存储器；固定存储器
permanent merchant 永久商人；常驻商
permanent mobile home space 永久性的活动房屋位置
permanent modification 永续变异
permanent moisture 永久潮湿
permanent monument 永久性木桩；永久线路桩
permanent mooring 固定系泊；长期系泊
permanent mooring buoy 常设锚泊浮筒
permanent mooring system 永久性锚泊系统
permanent mortgage 永久抵押(抵押十年以上)
permanent mo(u)ld 永久性铸模
permanent-mo(u)ld casting 硬模铸造
permanent-mo(u)lding method 熔模法
permanent mounting 永久封固
permanent multi-use house 永久性多种用途猪舍
permanent navigation facility 永久通航设施
permanent navigation lock 永久船闸
permanent navigation structure 永久通航建筑物
permanent network juncture 永久性网络结点
permanent non-slip protection 永久防滑保护
permanent notice 永久性通告
permanent nursery 永久苗圃；固定苗圃
permanent object 永久性对象；永久目标
permanent occupancy 长期租用房间
permanent of a square matrix 方阵的永久性
permanent openings 永久巷道
permanent open space 永久空地
permanent orange 永久橙；黄光颜料橙
permanent order 固定顺序
permanent output 长期输出功率
permanent parasite 终生寄生物
permanent partial disability 永久性局部伤残
permanent partition (wall) 永久隔墙
permanent pasture 永久草地
permanent patch 永久性修补
permanent peacock blue 永久孔雀蓝
permanent piling 永久性的桩材(防护)
permanent plankton 长期性浮游生物
permanent plant 永久设备；固定设备
permanent planting 定植
permanent pleating 耐久褶裥加工
permanent poison 永久性毒物
permanent pool 永久库容
permanent population 常住人口
permanent post 永久性兵营
permanent precipitation 永久沉淀
permanent preparation 持久标本
permanent press 耐久压烫
permanent-press resin 永压树脂
permanent prestress 永久预应力；永存预应力
permanent prestressing 永久预加应力
permanent production platform 固定生产平台
permanent project 永久性工程
permanent property 永久性物资；永久性财产
permanent pump 主抽水泵
permanent radio station 固定无线电台
permanent ramp 固定坑线
permanent red 永久红
permanent red pigment 永久红颜料
permanent red tone 永久红色原；永久红调色剂
permanent regulation structure 永久性整治建筑物
permanent relationship 永恒的关系
permanent repair 永久修理；永久性修理；经常维修；大修
permanent reserves 静储量
permanent resident 常住居民
permanent river 永久河；常流河
permanent river bed 永久河床
permanent riverermanent stream 常年河流

permanent roofing 永久性屋面
permanent sample plot 永久标准地
permanent sash 固定窗扇;固定窗框
permanent saving 长期储蓄
permanent seat 固定基座
permanent segment 永久段;固定程序段
permanent segment area 永久分段区域
permanent set 永久凝固;永久变形(率);终凝;固定伸长;固定沉淀;残余形变;残余变形
permanent set stress 永久残余应力;永久变余应力
permanent settlement 永久沉陷
permanent shiplock 永久性船闸
permanent shore 永久支柱;永久支撑;永久顶撑;永久撑木
permanent shrinkage 永久收缩
permanent shuttering 永久模壳;永留模板;永久性模板
permanent sign 永久性标志;永久性记号
permanent signal 不灭信号灯
permanent sluice 永久性泄水孔
permanent snow-cover 永久积雪
permanent snow-field 常年积雪区;永久雪原;常年积雪区
permanent snow-limit 永久雪线
permanent snow-line 永久雪线
permanent sound 留置探子
permanent spare system 固定备用方式
permanent specimen 持久标本
permanent speed 稳定速度;正常转速;恒定转速
permanent speed restriction section 恒定限速区段
permanent-split capacitor motor 固定分相电容器式电动机
permanent sprinkler system 固定式喷灌系统
permanent stability 永久稳定性;耐久性
permanent staff 在编人员
permanent standard point 永久水准(基)点
permanent state 固定状态;持久状态
permanent station 永久测站
permanent station network 固定台网
permanent storage 永久储[贮]存量;永久性库容;永久性存储器;永久存储器;固定存储器
permanent storage area 固定存储区
permanent stores 永久性保存;常备物料
permanent strain 残余变形;永久应变;永久变形
permanent stream 永久河(流);常流河
permanent stress 固定应力;永久应力
permanent stress annealing 应力退火
permanent stretch(ing) 永久伸长;永久拉伸
permanent structure 永久性建筑(物);永久性结构
permanent subsidy 经常补助费
permanent supplementary artificial lighting 常设人工辅助照明
permanent supply ditch 永久性供水沟渠
permanent support 永久支护;固定支护;固定支撑
permanent supporting system 内部砌壁(隧道);永久性支护系统
permanent sustain 永久支撑
permanent symbol 永久符号
permanent symbol table 永久符号表;固定符号表
permanent tack 永久性回黏
permanent tangible property 耐久性有形资产
Permanent Technical Committee 常设技术委员会
permanent testing 长期试验
permanent thermocline 恒定斜温层;海洋主温跃层
permanent three-phase short circuit current 稳态三相短路电流
permanent threshold shift 永久性听阈位移
permanent tidal station 长期验潮站
permanent time signal 连续时号
permanent tooth 恒齿
permanent trestle 永久栈道
permanent twist 永久扭转
permanent type detector 永久型检验器
permanent undercure 永久性欠固化
permanent varied flow 稳定变速水流
permanent vegetation 永久性植被
permanent vegetation cover 永久植被
permanent ventilation 恒定通风
permanent violet 永久紫
permanent waste storage 废物的永久性储藏
permanent water 永久水;稳定水源
permanent water flow 长期性流水
permanent water ga(u)ge 永久性水尺
permanent water sealing 永久性止水

permanent water supply system 固定供水竖管系统
permanent water table 永久潜水面;永久地下水位;永久地下水面
permanent wave 永久波;驻波
permanent way 轨线;线路(上部建筑物);永久性道路;轨道线路;路面;路基
permanent way and structures 线路和建筑物
permanent way construction machinery 线路施工机械(指铺轨、铺渣等机械)
permanent way district 工务(分)段
permanent way gang 线路工区
Permanent Way Institute 英国养路学会
permanent way maintenance 工务
permanent way maintenance department 工务处
permanent way program(me) 线路工作程序
permanent way roller 铺钢轨用辊子
permanent way work 线路工作
permanent way workshop 工务工厂
permanent weight 永久重量
permanent well completion 常用井;长期使用的井
permanent whipstock 固定造斜器
permanent white 永久白;钡白;永固白;硫酸钡
permanent wilting 永久凋萎
permanent wilting coefficient 永久凋萎系数
permanent wilting percentage 永久凋萎含水率
permanent wind 恒定风
permanent working enclosure 永久工作舱
permanent works 永久性工程;永备工事
permanent yellow 永久黄
permanent yellow pigment 永久黄颜料
permanent zone 永久载重区
permanent zoophankton 长期性浮游生物
permanganate 高锰酸盐
permanganate-chlorine treatment 高锰酸盐氯处理
permanganate demand 高锰酸盐需要量
permanganate index 高锰酸盐指数
permanganate method 高锰酸盐法;高锰酸盐滴定法
permanganate number 高锰酸盐值
permanganate permutite 锰砂
permanganate value 高锰酸盐值
permanganic acid 高锰酸
permanganimetric method 高锰酸钾法;高锰酸钾滴定法
Permanite 泊玛奈特(一种防潮层)
per man per shift 每班每人
Permant 泊曼特铁镍合金
Permason 泊玛森(一种木材防腐剂)
perma-split capacity motor 固定分相的单相电容式电动机
permasyn motor 永磁同步电动机
permatron 磁控管;磁场控制管
permavent 永久通风的天窗
Permax 泊马克司镍铁合金
permeability 渗透率;浸水率;通气性;渗透性;导磁系数;磁导系数;磁导率;穿透率
permeability alloy 坡莫合金;透磁合金
permeability and acid resisting binder 抗渗性耐酸胶结料
permeability apparatus 透气率测定仪;透气法比表面积测定仪;渗透计
permeability barrier 透性障;渗透阻挡层
permeability-block method 渗透率区域法
permeability bridge 磁导率电桥
permeability cell 透气性试样室;透气室;透气管;渗透仪试样盒
permeability coefficient 渗透系数;渗滤系数
permeability coefficient of dispersion test method 弥散试验法渗透系数
permeability constant 渗透常数
permeability curve 渗透曲线;磁导率曲线
permeability determination 渗透性测定
permeability determination of soil 土的渗透性测定
permeability distribution 渗透分布
permeability equation 渗透方程式;渗透(性)方程
permeability factor 渗透因素;渗透系数
permeability for gas 透气性
permeability for liquids 液体渗透性
permeability heat 导热性
permeability indicator 磁导系数指示器
permeability limit 渗透限度
permeability measurement 透气性测量
permeability measurement of rock 岩石渗透性

质测定
permeability meter 透气性测定仪;渗透计;磁导率计
permeability method 透气性试验法;渗透性试验法
permeability modulation 磁导率调制
permeability number 穿透数
permeability of aquifer 蓄水层渗透能力;含水层渗透能力
permeability of ballast 道砟的渗透性能
permeability of concrete 混凝土渗透性
permeability of free space 自由空间导磁率
permeability of ground surface 地面透水性
permeability of heat 导热性;透热性
permeability of heat conducting 导热性;导热度
permeability of permanent magnet 永磁磁导率
permeability of reservoir 储集层渗透率
permeability of rock to air 岩石的空气渗透系数
permeability of soil 土壤的可渗透性
permeability of sphere 球体导磁率
permeability of strata 地层渗透性
permeability of strata in supply area 供给区岩层渗透率
permeability pressure difference 渗压差
permeability pressure gradient 渗压梯度
permeability profile 渗透率剖面
permeability rate 渗透率
permeability rating of air 透气率
permeability reducing admixture 抗渗剂;减渗剂;防渗(外加)剂
permeability reducing agent 防渗外加剂;减渗剂;防渗剂
permeability requirement 透水性要求
permeability resisting expansive cement 抗渗性膨胀水泥;抗渗透性无收缩水泥
permeability resisting non-shrinkage cement 抗渗性无收缩水泥;抗渗透性膨胀水泥
permeability-saturation 渗透率与饱和率
permeability seal 不渗透遮挡层
permeability separatory 渗透分离设备
permeability separatory apparatus 渗透式分离设备
permeability surface area 表面积渗透率
permeability survey 渗透性测量
permeability tensor 渗透张量
permeability test 耐久渗透试验;渗透性试验;渗透试验
permeability tester 渗透性试验机
permeability testing cup 渗透性试验杯
permeability testing machine 渗透仪;渗透性试验机
permeability to air 透气性
permeability to diffusion 扩散渗透性
permeability to gas 渗气率;透气性
permeability to heat 导热性;热的可透度;透热性;导热度
permeability to water 透水系数;透水性
permeability to water vapo(u)r 蒸汽渗透性;透水汽性
permeability-tuned inductor 导磁率调谐电感线圈;磁性调谐可变电感线圈
permeability tuning 磁性调谐;磁调谐;磁导系数调谐
permeable 可渗透的
permeable base 渗透基层;渗透基底;透水层;透水基底
permeable bed 透水岩床;透水基床;透水层;渗水基岩
permeable blanket 透水铺盖层
permeable breakwater 透水式防波堤;透水防波堤;透空式防波堤;透水防波堤
permeable channel 渗水通道
permeable coefficient 渗透系数
permeable confining bed 可渗透隔水层;可渗透封闭层;透水隔层;透水封闭层
permeable constant 透水常数
permeable course 透水层
permeable crib dike 透水木笼堤;透水格笼堤;渗水格笼堤
permeable dam 透水坝
permeable dike 透水堤;透水坝;透水式导流堤;透水木笼堤;透水格笼堤透空堤;渗水丁坝
permeable embankment 透水路堤
permeable fault 导水断裂;导水断层
permeable foundation 透水层
permeable groin 格栅丁坝;透水丁坝;渗水丁坝
permeable ground 透水围岩;透水土层;透水土

(壤);透水层;渗水土层
permeable groyne 透水防波堤;格栅丁坝;透水折流坝;透水丁坝
permeable joint 透水缝;渗透接缝;渗透缝;渗水接缝;渗水缝
permeable layer 可透层;透水层
permeable material 透水材料;渗透性材料;渗水材料
permeable medium 可渗透介质;磁导介质
permeable membrane 透性膜;透膜;渗透膜
permeable overburden 透水覆盖层
permeable passage 透水通道
permeable pile dike 透水桩坝;透水桩堤
permeable plastics 可透塑料
permeable plate 透水板
permeable reactive barrier 可透活性垒
permeable reservoir 渗透性热储
permeable revetment 透水式护岸
permeable riverbed 透水河床
permeable rock 可透性石;可透透岩(层);透水岩(石);透水性岩石;透水性岩层;渗透岩石;渗水岩石
permeable rock bed 渗水岩层
permeable rocks 透水岩类;渗水岩类
permeable rod 透磁棒
permeable shell 透水壳体;透水壳层;透水坝壳
permeable slab 透水板
permeable soil 可透性土壤;透水性土壤;透水土壤;渗透性土(壤);渗水土
permeable soil subgrade 渗水土路基
permeable spur 透水丁坝;渗水丁坝
permeable stake 透水栏栅
permeable stratum 透水岩层;透水(地)层;渗透层
permeable structure 透水结构;透水建筑物;透水构造;导水构造
permeable subsoil 透水底土(层)
permeable test 渗透试验
permeable to steam 透蒸汽的
permeable to water 透水的
permeable training structure 透水导治建筑物
permeable wave absorber 透水消波设备;透水消波器
permeable works 透水建筑物;渗水建筑物
permeable zone 渗透区;渗透带
permeably-lined furnace 渗透排烟炉;渗透衬里炉
permeameter (apparatus) 渗透仪;渗透性试验仪;渗透计;磁导仪
permeameter cell 渗透仪试样盒
permeametry 渗透测粒法
permeance 透过;渗入;磁导
permeance factor 磁导系数
permeants 穿行者
permeaselectivity 选择渗透性
permeate 渗透物
permeate agent 渗透剂
permeate flux 渗透通量
permeater 透气性测定仪
permeate water 渗透水
permeating cold air 渗透冷空气量
permeating deformation test 渗透变形试验
permeating medium 浸渗介质;渗透介质
permeating the plant 渗进作物内部
permeation 透过;渗透作用;渗透;渗入;渗气
permeation chromatography 渗透色谱法
permeation coefficient 透水系数
permeation consolidation type of soil 土的渗透固结类型
permeation constant 渗透常数
permeation flaw detection 渗透探伤法
permeation fluid mechanics 渗流力学
permeation grouting 渗透灌浆
permeation limit 渗透极限
permeation rate 渗透速率;渗透率
permeation separation 渗透分离
permeation time lag 渗透时间滞后
permeation tube 渗透管
permeator 渗透器;渗透计
Permendur 波门杜尔铁钴合金
Permendur alloy 波门杜尔合金
Permenorm 波曼诺(铁镍)合金;铁镍合金
Permeometer 织物透气性测试仪
permetallurgical cement 熔渣水泥
per meter 每延米
Permian basin 二叠纪盆地

Permian clay 二叠纪黏土
Permian limestone 二叠纪石灰岩
Permian period 二叠纪
Permian system 二叠系
permill(age) 千分率;千分比;千分率比
per mille 千分之一;每千
permineralization 完全矿化
permingeatite 硒锑铜矿
per minute 每分钟
Perminvar 波尼瓦尔恒导磁率合金;波尼瓦尔(铁镍钴)合金
permissble creep strain 容许蠕变应变
permissible 许用的;许可的;容许的
permissible ammonium nitrate explosive 安全硝铵炸药
permissible approach speed 允许接近速度
permissible ballast pressure 道床容许压力
permissible blasting device 安全起爆装置;安全爆破装置
permissible blasting material 安全爆破器材
permissible blasting unit 安全起爆装置;安全起爆器
permissible building area 建筑(面积和高度)限定区
permissible canal velocity 渠道容许流速
permissible carbon dioxide content 二氧化碳容许含量
permissible cartridge 安全药筒;安全药卷
permissible clearance 允许间隙
permissible coalcutter 防爆截煤机
permissible concentration limit 极限容许浓度;容许浓度限度
permissible concentration of carbon monoxide 一氧化碳容许浓度
permissible concentration of radioactivity 放射性允许浓度
permissible contaminant limit 允许污染极限
permissible contamination 可容许污染;容许污染
permissible control 容许调节
permissible criterion 许可标准;容许基准;容许标准
permissible current 允许电流;容许电流
permissible cutting speed 容许切削速度
permissible deficiency of super-elevation 容许欠超高
permissible deflection 容许挠度;容许偏差
permissible deformation value of foundation 地基容许变形值
permissible design 容许设计
permissible deviation 允许偏差;允许离差;容许偏差;容许差度
permissible deviation of constructional elements 建筑构件的容许偏差
permissible dimensional error 容许尺寸误差
permissible dip value of foundation 地基容许倾斜值
permissible discharge 容许泄水量;可容许排污;容许排放;容许流量
permissible discharged water of industry 工业允许排水量
permissible dose 容许量;容许剂量
permissible draft 容许吃水
permissible draft head 容许吸出水头
permissible drawdown 工作深度;容许泄降深度;容许消落深度;容许降(落)水位;容许降落深度
permissible dustiness 允许含尘量
permissible dynamite 安全炸药
permissible electric(al) equipment 防爆电器设备
permissible equipment 防爆设备
permissible error 允许误差;允许故障率;公差;可容许误差;容许误差
permissible explosive 核准炸药;合格炸药;安全炸药
permissible explosives for coal mines 煤矿炸药
permissible exposure 容许照射
permissible exposure limit 容许曝露极限;容许辐照限度;容许曝露极限
permissible feed 容许供给
permissible flame lamp 火焰安全灯
permissible flexibility 许可挠性;许可挠度;容许挠性
permissible for types of steels 容许的钢材类型
permissible gross combination weight 容许汽车连接总重量
permissible height 容许高度

permissible hydraulic gradient 允许水力梯度
permissible lateral deviation 容许离线偏差
permissible length 许用长度
permissible level 允许数值;容许含量;容许标准
permissible level of human exposure 人体容许曝露水平;人体容许曝露程度
permissible light 安全灯
permissible limit 允许极限;容许限量;容许限(度);容许极限
permissible limit of harmful substance 有害物质容许限值
permissible limit of pollution 容许污染限度;容许污染极限
permissible limit of vibration 容许振动极限
permissible load 许用荷载;允许负荷;容许载重;容许荷载;容许负载;容许负荷;安全荷载
permissible loss 容许损耗
permissible machine 防爆矿山机械
permissible manufacturing deviation 容许加工偏差
permissible maximum speed 最高容许速度;容许最高速度
permissible mean maximum pressure 容许平均最大压力
permissible mine equipment 许用矿山设备
permissible mine locomotive 防爆式矿用机车
permissible mining modulus method 允许开采模数法
permissible motor 密闭(型)电动机;防爆电(动)机;安全电动机
permissible noise level 容许噪声级;容许噪声标准
permissible orbit 容许轨道
permissible out of roundness 允许不圆度
permissible overload 容许过载
permissible payment period 准许偿付期间
permissible pipe velocity 管道允许流速
permissible point 许可的水准
permissible point-load 容许点荷载;容许点负荷;容许集中荷载
permissible precision 容许精度
permissible pressure 容许压强;容许压力
permissible pressure difference 容许压(力)差
permissible probability of failure 容许破坏概率
permissible range 容许范围
permissible rates of water entry 容许进水率
permissible response rate 容许适应率
permissible revolutions 容许转数;许可转数
permissible rope length of drum 卷筒容绳量
permissible sample size 允许试样量;允许试验量
permissible settlement 容许沉降(量)
permissible settlement difference of foundation 地基容许沉降差
permissible settlement of foundation 地基容许沉降量
permissible shapes 许用形式
permissible shear 容许剪力
permissible soil bearing capacity 容许土(壤)承载力
permissible soil bearing pressure 容许土(壤)承载力
permissible soil pressure 容许土压力;容许承载力
permissible solution 容许解(法)
permissible speed 允许速度;容许速率;容许速度
permissible speed of a line 线路容许速度
permissible storage battery locomotive 防爆蓄电池机车;防爆式蓄电池电机车
permissible stress 许用应力;允许应力;准许应力;容许应力
permissible stress design 容许应力设计
permissible stress method 容许应力法
permissible sulfur 允许硫含量;容许硫含量
permissible temperature 许可温度;容许温度
permissible tolerance 容许现差;容许耐力;容许含量;容许公差;容许变量
permissible type 允许形式
permissible unsharpness 容许不清晰度
permissible value 准许价值;容许值
permissible variation 容许变量;容许变化
permissible vehicle dimensions 容许车辆尺寸
permissible velocity 许可流速;允许速度;容许速度;容许流速;安全流速
permissible velocity of flow 允许流速
permissible vibration 允许振动
permissible voltage fluctuation 允许电压升降

permissible water cycle of industry 工业允许水循环利用率
permissible wear 容许磨损;允许磨损
permissible working load 容许使用荷载;容许工作荷载
permissible working stress 容许工作应力
permissible yield 许可产量;容许渔获量;容许开采量
permission 许可;允许;准许;容许
permission approach speed 容许接近速度
permission aspect 准许显示
permission for duty free importation 免税进口许可证
permission indication 准许指示;准许显示
permission to deal 参加交易的许可
permission to land 允许着陆
permission to take off 允许起飞
permissive 许可的
permissive action 允许作用
permissive action link device 允许行动联络装置
permissive aspect 许可显示
permissive attachment 容许路签附加装置
permissive automatic block system 容许自动闭塞制;容许自动闭塞系统
permissive block 允许区截;容许闭塞制;容许闭塞
permissive card 容许通行证
permissive control 许可调节
permissive coupling speed 容许连接速度;容许挂接速度
permissive host 许可性寄主;许可寄主
permissive light 容许表示灯
permissive mode 允许模式
permissive output level 容许输出水平
permissive period of transport 容许运输期限
permissive provision 非约束性条款
permissive release 允许解除
permissive run 容许运行
permissive signal 容许信号
permissive speed in district 区段允许速度
permissive speed of train passing through station 通过车站允许速度
permissive speed on track 线路允许速度
permissive staff 容许路签
permissive stop 容许停车
permissive temperature 许可温度
permissive timing system 容许定时制
permissive token block instrument 容许凭证闭塞机
permissive-wage adjustment 容许调整工资制
permissive waste 容许耗损;允许废品;通常损耗
permit 许可证;允许(证);准许(证);执照;容许;通行证
permit building 建造许可证;核准建筑
permit deformation 容许变形
permite alumin(i)um alloy 耐蚀铝硅合金
permited mixing area 容许混合区
permit for pollutant discharge 污染物排放许可证
permit holder 持证人
permit occupancy 准许用地;使用许可证
permit of compliance 合质许可
permit of waste dumping 废弃物倾倒许可证
permits 许可证费(用)
permits for cutting forest and trees 林木采伐许可证
permit system 许可证制度
permit system for discharging pollutants 排污许可证制度;污染物排放许可证制度
permit system of pollutant discharged 排污许可证制度
permit(tance) 许可
permitted band 导带
permitted dynamite 安全黄色炸药
permitted explosive 容许爆炸;容许爆破;许用炸药;安全炸药
permitted exposure time 允许暴露时间
permitted heat loss 允许热损失
permitted leak 允许渗透量
permitted line 容许谱线
permitted load 容许荷载
permitted payload 车辆容许载货质量
permitted speed 限速
permitted stress 容许应力
permitted structure 符合所在地区全部法令的结构物

permitted term 认可术语
permitted transition 容许跃迁
permitted use 许可使用
permitted vacuum height of intake 允许吸上真空高度
permittee 持证人
permittivity 介电系数;介电常数;电容率
permittivity of free space 自由空间电容率
permittivity of vacuum 真空电容率
permit to discharge 卸货准单;排放许可证
permit to fishery and trade 国际渔业贸易许可证
permit to land 起货证
permit to mechanical work 机械许可工作证
permit to open hatch 开舱单
permit to shipping 装船准单
permit to unlade 卸货准单
permit to unload 卸货准单
permit zoning 区划许可证
permium on bonds 债券收回溢价
permivar 镍铁钴高导磁率合金;镍铬钴高导磁率合金
Permo-Carboniferous 石炭二叠过渡期
Permo-Carboniferous period 石炭二叠纪
Permoglaze 珀莫釉(一种透明防水的墙壁涂料);(透气防水的)墙壁涂料
Permolite 珀莫利特(一种木材防腐剂)
permolybdate 过钼酸盐
permolybdic acid 过钼酸;高钼酸
permometer 透气性测定仪;透气度测定仪
permonosulfuric acid 过一硫酸
permonosulphuric acid 过一硫酸
per month 每月
permselective 选择性渗透的
permselective membrane 选择性渗透膜
permselectivity 选择通透性
permutability 转置性;置换性;可置换性
permutable 可排列的
permutate 重排列
permutation 重新配置
permutation and combination 排列与组合
permutation code 置换码;互换码
permutation decoding 置换解码
permutation group 置换群
permutation index 置换索引
permutation lock 转换锁;互换锁
permutation matrix 置换矩阵;互换矩阵;排列矩阵
permutation modulation 置换调制;变换调制
permutation symbol 排列符号
permutation table 置换表;排列表
permutation without repetition 不重复排列
permutation with repetition 重复排列
permutator 转换开关;机械换流器
permute 滤砂软化
permuted code 置换码
permuted-title index 循环置换标题索引
permutite 滤水砂;滤砂;软水砂;人造沸石
permutite A 阴离子交换膜
permutite base exhange 滤砂离子交换
permutite filter 沸石滤池
permutite pozzolana 人造火山灰(质材料)
permutite softener 沸石软化器
permutoid reaction 交换体沉淀反应
perna 全氯萘
pernambuco 棘云实红木
pernam cotton 伯南棉
pernetti 支托架
pernicious 有毒的
pernicious effluent 有害废水
pernicious gas 有害气体;有毒气体
pernicious habits 恶习
pernicious influence 恶劣影响
perniciousness 危害性
pernicious vomiting 恶阻
pernio 冻疮
perniosis 冻伤病
pernitric acid 过硝酸
pernitroso-camphor 二亚硝基樟脑
Pernmet 波梅特铜钴镍合金
perobrazhenskite 斜方水硼镁石
peroganism 超个体
peroid of transition 过渡阶段
perolene 热交换有机液体
peronate 被厚毡毛覆盖的
perovskite 钙钛矿色;钙钛矿
perovskite ore 钙钛矿矿石

perovskite structure 钙钛矿结构
perovskite-titanomagnetite deposit 钙钛矿钛磁铁矿矿床
perovskite type structure 钙钛矿型结构
peroxamine 过氧化胺
peroxidation 过氧化作用;过氧化反应
peroxide 过氧化物
peroxide bleach 过氧化物漂白剂
peroxide bleaching 过氧化物漂白
peroxide bleaching liquor 过氧化物漂白液
peroxide bond 过氧链
peroxide bridge 过氧桥
peroxide-catalyst 过氧化物催化剂
peroxide-coagulation process 过氧化物混凝工艺
peroxide cure 过氧化物硫化
peroxide decomposer 过氧化物分解剂
peroxide effect 过氧化物效应
peroxide number 过氧化值
peroxide of hydrogen 过氧化氢
peroxide of iron 过氧化铁
peroxide value 过氧化值
peroxide vulcanization 过氧化物硫化
peroxide vulcanizing agent 过氧化物硫化剂
peroxidization 过氧化作用
peroxidizing property 过氧化性
peroxyacetic acid 过氧乙酸
peroxyacetyl nitrate 过氧乙酰硝酸酯;过氧酰基硝酸酯
peroxybenzoic acid 过苯甲酸
peroxy benzoyl nitrate 过氧苯酰硝酸酯
peroxybutyryl nitrate 过氧丁酰硝酸酯
peroxychromates 过铬酸盐
peroxydol 过硼酸钠
peroxyformic acid 过氧甲酸;过甲酸
peroxyl radical 过羟基
peroxy-monosulfate 过一硫酸盐
peroxy nitrate 过硝酸盐
peroxy nitric acid 过氧酰硝酸;过硝酸
peroxy-phosphoric acid 过磷酸
peroxypropiony nitrate 过氧丙酰硝酸酯
peroxy radical 过氧游离基
peroxy salt 过酸盐
per package limitation of liability 按包计算的责任范围
per pass 每通过一次
perpective three-dimensional display 体视显示
perpen(d)穿墙石(块);拉结石;贯砖;贯石(贯穿墙壁而露其两端的长石);贯墙石
perpendicular 正交的;垂直的;垂直的;垂距
perpendicular anisotropy 正交各向异性
perpendicular arch 垂直式拱(圈)
perpendicular architecture 垂直哥特式建筑;哥特式建筑(公元12~16世纪时盛行于西欧)
perpendicular axis theorem 垂直轴定理
perpendicular bisector 垂直平分线;中垂线;垂直等分线
perpendicular cathedral 垂直哥特式大会堂;垂直哥特式大教堂;哥特式大教堂
perpendicular component 正交分量;垂直分量
perpendicular construction 垂直式建筑
perpendicular cut 垂直切割
perpendicular cut of jack rafter 小椽子的垂直切割
perpendicular directional line 垂直方向线
perpendicular displacement 正交位移;竖向位移
perpendicular distance 垂直距离
perpendicular distribution of population 人口垂直分布
perpendicular drain(age) system 垂直式排水系统
perpendicular drive 垂直驱动
perpendicular drop method 垂线法
perpendicular error 垂直差
perpendicular Gothic 垂直哥特式
perpendicular heave 直交平错
perpendicular heterogeneous slab 正交各向异性板
perpendicular incidence 直射
perpendicular interpenetration 垂直穿插
perpendicularity 正交;垂直状态;垂直性;垂直度;垂直
perpendicular joint 正交接合;垂直接缝;正交连接;竖缝
perpendicular layout 垂直式布置
perpendicular layout of sewers 垂直式下水道布置;垂直式沟渠布置
perpendicular line 正交线;垂直线;垂线

perpendicular linear polarization 垂直线偏振
perpendicular magnetic recording 垂直磁记录
perpendicular magnetization 垂直磁化
perpendicular oceano-graphic(al) platform 直立式海洋考察平台
perpendicular offset 交距;垂直支距;垂直截距;垂直支距
perpendicular parking 横列式停车
perpendicular period 垂直式时代
perpendicular plane 垂直面
perpendicular radiation 垂直放射
perpendicular recording 垂直记录;垂直磁记录;垂向记录
perpendicular scan 垂直扫描
perpendicular section 垂直截面
perpendicular separation 垂直隔段;直交隔离
perpendicular shaft 竖式导井;垂直导井
perpendicular shaft rotary angle 立轴回转角
perpendicular shaft rotary speed 立轴预转速
perpendicular slip 直交滑距;垂直滑距
perpendicular spur dike 正挑丁坝
perpendicular style 英国晚期哥特式建筑;哥特式建筑风格;垂直式建筑风格
perpendicular system 排水系统布置—垂直式系统
perpendicular throw 直交落差;垂直落差
perpendicular to axis of alluvial-pluvial fan 垂直冲洪积扇轴
perpendicular to bed surface 垂直层面的
perpendicular to grain 截纹
perpendicular to river 垂直河流
perpendicular to river valley 垂直河谷
perpendicular to strike of abundant ground water zone 垂直富水带走向
perpendicular to the strike of aquifer 垂直含水层走向
perpendicular tower 垂直(哥特式)塔楼
perpendicular tracery 垂直式窗花格;垂直窗格
perpendicular well 竖井
perpend or transversal joint 直缝或竖缝
perpends 竖缝
perpend stone 控石;贯石(贯穿墙壁而露其两端的长石);贯墙石;穿墙石(块)
perpend wall 全部用丁头砖石砌的墙;单石墙;薄墙
perpetual annuity 永续年金
perpetual budget 永续预算
perpetual calendar 永久日历;万年历
perpetual day 极地永昼
perpetual frost climate 永冻气候
perpetual inventory 永续盘存
perpetual inventory account 永续盘存账户
perpetual inventory sheet 材料连续结存单
perpetual inventory system 永续盘存制
perpetual lamp 永亮灯;长命灯
perpetual lease 永租权;永久租权;永久租借权
perpetual load 永久载重;永久荷载
perpetual loan 永久贷款;无期贷款
perpetually frozen soil 永久冻土;永久冻土壤;多年冻土
perpetually renewable lease 永久可重新签订的租约
perpetual machine of the first kind 第一类永动机
perpetual motion 永恒运动
perpetual motion machine 永动机
perpetual motion machine of the first kind 第一类永动机
perpetual motion machine of the second kind 第二类永动机
perpetual motion machine of third kind 第三类永动机
perpetual night 极地长夜
perpetual resource 可再生能源
perpetual shadow 全年阴影
perpetual snow 永久积雪;万年积雪
perpetual snow measurement 积雪测量
perpetuities 永续年金
perpeyn 系石;拉结石;穿墙(丁头)石
perpeyn wall 墙墩;墙壁支柱
perphosphoric lime 过磷酸石灰
perplanar 超平面
per power of attorney 委托书
per procuration 凭委托书
perps 竖缝
perquisite 特殊待遇;附加福利;非经常性额外收入;额外所得

perrierite 珀硅钛铈铁矿
perron 露天梯级(建筑物入口);露天台阶;室外台阶;室外楼梯;石台座
perron landing 露天平台登台处
perron step 室外台阶踏板
Perrotine 波若丁印花机
perrycot 硅酸盐石料;波特兰石
perryite 陨硅铁镍石;硅磷镍矿
persalic 过硅铝质
persalt 过酸盐
per sample 根据样品;每种试样
per second 每秒钟
per second each second 每秒每秒
perseveration 持续动作
per share 每股
per share earning ratio 每股收益率
per share earnings 每股盈余
Pershbecker furnace 佩希伯克尔(转式)炉
per shift 每班
per-shift drilling depth 班进尺【岩】
per ship 每船
Persian 男像柱
Persian architecture 波斯建筑
Persian blinds 波斯百叶窗;百叶窗
Persian blue 波斯青(色料)
Persian carpet 波斯地毯
Persian column 波斯人形柱;波斯柱
Persian corbelled arch 波斯突拱
Persian faience 波斯陶器
Persian Gulf 波斯湾
Persian Gulf basin 波斯湾盆地
Persian gulf red 铁红
Persian lilac 花叶丁香
Persian porcelain 波斯瓷器
Persian red 波斯红(色料)
Persian rug 波斯地毯
Persian screw 螺旋排水机
Persian style 波斯式建筑
Persian three colo(u)red ware 波斯三彩
Persian wheel 波斯式水车;波斯水轮;波斯式戽斗
persic oil 杏仁油
persiennes 可调叶板的外百叶窗;百叶帘
persilicic 过酸质;硅石的
persimmon juice 柿油;柿液;柿漆;柿红
persimmon juice work 柿漆饰面
persimmon shape wash basin 翻口面盆
persimon red 柿红
persistant oscillation 非衰减摆动;等幅摆动
persistatron 冷持管
persisted state 持久状态
persistence 持续(性);持久度;长期留存
persistence characteristic 余辉特性;持久特性;残留特性
persistence eliminator 余辉消除器
persistence of energy 能量守恒;能量不灭
persistence of pesticides 农药耐久性;农药持久性;农药的长期留存
persistence of pollutant 污染物的持久性
persistence of river 河流持续性
persistence of vision 余像现象;视觉驻留;视觉暂留
persistence tendency 持续性趋势;持续性倾向;持续型趋势
persistency 持久度
persistency of pollutant 污染物持久性
persistent 持续的
persistent-cause forced outage 持久性强迫断电
persistent characteristic 持久特性
persistent chemicals 持久性化学物质
persistent current 持续电流
persistent current memory cell 持续电流存储单元
persistent fire 耗时长的火灾
persistent gas 持久性毒气
persistent insecticide 长效杀虫药
persistent internal polarization 持久内偏振
persistent irradiation 持久辐照
persistent line 住留谱线;光谱最后线
persistent material 持久性物质;长效原料
persistent organic pollutant 难降解有机污染物
persistent organochlorine pesticide 持久性有机氯农药;长效有机氯农药
persistent oscillation 连续振荡;等同摆动;等幅振荡;等幅摆动;持续振荡;非衰减摆动
persistent pesticide 持久性农药;长效农药;残留性农药

persistent pollutant 持续性污染物;持久性污染物
persistent pollution 持久性污染
persistent radiation 持久辐射
persistent retention of moisture 积水
persistent spectral hole burning 持续光谱烧孔
persistent state 回归状态;回归;持续状态
persistent substance 持久性物质
persistent toxicant 持久性毒物;持久性毒剂
persistent toxic pesticide 持久毒性农药
persistent toxic substance 持久性毒性物质
persistent tremor 持久性震颤
persistent violator 经常违章者
persistent war gas 持效毒气
persistent wave 连续波;等幅波
persister 冷持管
persisting type 迁延型
persistron 持久显示器
persitent organic pollutant 持久性有机污染物
person accountable for an estate duty 负责缴纳遗产税者
person acquiring the property 财产受让人
personal accident 人身意外伤害
personal accident insurance 人身意外伤害保险;人身意外保险
personal accounts 人名账户
personal administration 人事管理
personal air sampler 工作人员空气取样器;个人空气取样器
personal alarm device 个人报警装
personal alarm locator 个人报警定位器
personal alert safety system 个人安全报警系统
personal allowance 个人容限
personal and instrumental equation 人仪差
personal and instrumental error 人仪差
personal autobiography 履历(表)
personal automation 个人自动化
personal bias 人差【测】
personal breathing apparatus safety line 个人呼吸器安全绳
personal buoyancy aids standard 个人助浮标准
personal carrier 人员紧急运载工具
personal channel 人员流动渠道
personal check 私人支票
personal circuit 专用线路
personal code 个人代码
personal committed failure 人为故障
personal communication 个人通信
personal communication network 个人通信网
personal communication service 个人通信业务
personal communication system 个人通信系统
personal communicator 个人通信器
personal computer 个人计算机;个人电脑
personal computer front end 小型计算机调谐器
personal computer roundness instrument 个人计算机圆度仪
personal computer television 电脑电视
personal computing 个人计算
personal computing technology 专用计算技术
personal connection 人事关系
personal correction 人差改正;测者误差改正量
personal cost 人事费(用)
personal damage 人身事故
personal data terminal 专用数据终端
personal decision analysis 个人决策分析
personal decision making 个人决策
personal development 人的能力发展
personal difference 人差【测】
personal digital assistant 个人数字助理
personal dose monitor 个人剂量计
personal dosimeter 个人剂量计;个人照射量计
personal dosimetry 个人剂量学
personal effects 私有财产;动产;个人财产;私人物品;私人所有物
personal equation 观测者误差;个人在观察上的误差;个人观察误差;人差【测】
personal equipment 个人装备
personal error 观测者误差;个人误差;人为误差;人差【测】;测者误差;操作人员误差
personal estate 私有财产;动产
personal expenditures 人员费用
personal expenses 人员费用
personal factor 人为因素;人的因素
personal feature 个体特征
personal flo(a)tation device 人员漂浮设备

personal guarantee 人的担保
personal handheld computer 个人便携式计算机
personal history 个人史
personal hygiene 个人卫生
personal identification code 标识用户代码
personal identification data 个人识别数据
personal identification number[PIN] 个人身份号码;个人识别号码
personal income tax 个人所得税
personal information terminal 个人信息终端
personal injury 个人伤害;人身伤亡
personal injury accident 人身受伤事故
personal injury by accident 人身事故
personal injury insurance 个人损害保险
personal insurance 人身保险
personal interview 面谈(调查)
personal interview method 当面询问法
personalistic decision theory 个性决策理论
personality 个性
personality card socket 专用插件插座
personality module 个性模块
personal ledger 人名分类
personal liability insurance 个人责任保险
personal living unit system 个人居住单元体系
personal locator beacon 个人定位器信标;人员定位(信)标
personal management 人事管理
personal management analysis 人事管理分析
personal module 专用组件
personal monitor 个人检测器;个人监测仪
personal parallax 人视差
personal probability 主观概率;人定概率
personal property 个人财产;人的财产;动产
personal property broker 动产经纪人
personal property loan 动产贷款
personal prophylaxis 个人预防
personal proprietor 独资业主
personal protection 个人防护
personal protection equipment 个人安全保护装置;个人安全保护设备
personal radio telephone 个人无线电话机
personal rapid traffic 个体快速交通
personal rapid transit 快速列车;快速客运
personal rapid transit system 小型快递交通系统;小型高速客运系统
personal records 人事档案
personal records system 人事档案系统
personal residence 个人居所
personal responsibility 人事责任
personal safety 人身安全
personal safety equipment 个人防护设备
personal safety supplies 人身安全防护用品
personal sampler 个人采样器
personal selling 销货员推销;人员推销
personal service 个人劳务;人事费(用)
personal service establishment 生活服务设施
personal signal(l)ing system 私人通信系统
personal space 个人空间
personal subsidies 个人津贴
personal sub-system 人控部分
personal tax 个人纳税
personal tax rate 个人税率
personal tax relief 个人税收减免
personalty 动产
personal-type airport 私人(用)飞机场
personal use 私用
personal water system 自供水系统
personate 假面状的
personator 冒名者
person challenged 被异议人
person concerned 当事人
person for approving 批准者
person for compilation 技术负责人;编制者
personification 人格化
person in charge 主管
person in charge of fire 防火负责人
person in charge of owner's project 工程甲方负责人
person in charge of the project 工程主持人
person insuring 投保人
person machine interface 个人机器接口
personnel 人员
personnel accommodation portion 职工住宿部分;职工膳食供应区

personnel administration 人事管理
personnel administrator 人事主管人
personnel aisle 人员通道
personnel alarm monitor 个人报警监测器
personnel appliance 人员输送车
personnel appointment and removal 人事聘免
personnel assignment 人员调配
personnel automated data system 小型自动数据系统
personnel barrier 防护栏(杆)
personnel block 职工区段;办公大楼
personnel building 职工用房
personnel canteen 职工小卖部;职工食堂
personnel card index 人事卡片索引
personnel carrier 人员携带者;人员输送车
personnel-carrying hoist 载人升降机;乘人电梯
personnel changing room 职工更衣室
personnel chart 人事组织图表
personnel decision 人事决策
personnel decontaminated waste 人员去污废物
personnel department 人事部门
personnel dining room 职工食堂
personnel director 人事主管人
personnel distress device 个人求救装置
personnel distress locator 个人求救定位器
personnel dosimeter 个人剂量计
personnel entrance 职工入口
personnel equipment data 大型设备资料
personnel evacuation 人员疏散
personnel exchange service center 人才交流服务中心
personnel expenditures 人事费(用)
personnel forecast 人事预测
personnel function 人事功能
personnel house 职工住宅
personnel information 人事资料
personnel job hoist 工地载人升降机
personnel loading chart 人员负荷图;人力负荷图
personnel location system 人员测位系统
personnel lock 职工衣帽箱(带锁的);人闸
personnel management 人员管理;人员编制;人事管理
personnel management information system 人事管理信息系统
personnel manager 人事主管人
personnel mobility 人员流动
personnel mobility law 人才流动法规
personnel monitoring 人体辐射剂量检查
personnel not directly engaged in production 非直接生产人员
personnel occurrence report 人员事故报告
personnel office 人事科
personnel placement 人员安排
personnel policy 人事政策
personnel profile 人事登记资料
personnel property 动产
personnel protection 人身保护
personnel records and reports 人事记录与报表
personnel recruitment 人事招募
personnel recruitment program(me) 人工招募计划
personnel-related basis of allocation 按人数分摊费用
personnel replacement cost 人员更换成本
personnel requisition 人员申请
personnel safety guard 人员安全防护
personnel scheduling 人员配备计划
personnel section 人事科
personnel selection 人员选择
personnel selection consultant 人员选择咨询师
personnel service 人员服务
personnel specification 人员规范
personnel supervisor 人事管理员
personnel system 人事制度
personnel toilet 职工盥洗室;职工厕所
personnel traffic 职工交通(量)
personnel training 人员训练;人员培训
personnel transfer capsule 人员载运舱
person on probation 试用人员
persons eligible for tax exemption 税收减免的纳税人
person-time 人次
person-time of exposure 暴露人时
person-to-person call 定人呼叫

person-to-person communication 个人通信
person to person phone call 叫人电话
person trip 行人行程
person-year 人年(劳动量单位,一个人在一年内完成的工作量)
person-year of exposure 暴露人年
person-year of observation 观察人年
persorption 吸混(作用);多孔性吸附
Persoz hardness 珀苏兹硬度
perspective 配景;远景;透视图;透视画法;透视的;透视
perspective air photograph 倾斜摄影航片
perspective air survey 倾斜航展测量
perspective anomaly 透视近点角
perspective axis 透视轴
perspective azimuthal projection 透视方位投影
perspective cell 远景单元
perspective center 灭点;视点;透视点;透视中心
perspective chart 透视图;透视投影地图
perspective collineation 透视变换
perspective construction 透视的建筑(图)
perspective correspondence 透视对应
perspective distortion 透视畸变
perspective drawing 透视图
perspective drawing instrument 透视绘图仪
perspective evaluating method by analysis minerogenetic geologic(al) conditions based on remote sensing 遥感成矿地质条件分析远景评价方法
perspective evaluating method by comprehensively analysing remote sensing-geonomy date 遥感地学资料综合分析远景评价方法
perspective figure 透视图形
perspective geometry 画法几何;透视几何(学);投影几何(学)
perspective grating 透视光栅
perspective grid 透视投影格网;透视格网
perspective grid method 透视格网法
perspective line 透景线
perspective mapping 透视映射
perspective membrane 选择渗透膜
perspective model 透视模型
perspective net 透视网
perspective normal cylindric(al) projection 正轴透视圆柱投影
perspective oblique cylindric(al) projection 斜轴透视圆柱投影
perspective of the ground 地形原图
perspective picture 透视画
perspective plan 透视图
perspective plane 投影图;透水平面;透视(平)面
perspective point 透视点
perspective position 透视位置
perspective projection 立体投影;几何投影;无穷远点射投影;透视投影;点射投影
perspective ray 透视线
perspective region resources estimation 远景区资源量估计
perspective rendering 透视渲染
perspective representation 透视法;透视表示法
perspective scale 远近比例
perspective sketch 透视草图
perspectives on water 水资源的展望
perspective stereo-symbol 透视立体符号
perspective surface 透视面
perspective symbol 透视符号
perspective threedimensional display 立体透视显示(器)
perspective transformation 透视变换
perspective view(ing) 透视;立体投影;透视图
perspective viewing transformation 透视图变换
perspective volume symbol 透视立体符号
perspectivity 选择透过性;明晰度;透视性
perspectograph 透视制图;透视纠正仪;透视绘画器;透视画仪器
perspectometer 透视画仪器
perspex 有机玻璃瓦楞天窗;透明塑料;有机玻璃;透明塑胶;塑胶玻璃;防风玻璃
perspex spatula 有机玻璃调刀
perspiration 出汗
perspiration resistance 耐汗性
per square inch 每平方英寸
perssonel authorization 人员定额
perstalith 环绕墓穴的竖立石

per standard compass 根据标准罗经
per stor(e)y 每层【建】
perstylar 周柱廊式的;列柱廊式的
perstylos 周柱廊式建筑;列柱廊式建筑
persuader 诱导器
persuader roll 摇杆辊子
persuasive 有说服力的
persulfate 过硫酸盐;过磷酸盐;过二硫酸盐
persulfate method 高硫酸盐法
persulfide 过硫化物
persulfuric acid 过硫酸
persulphate 过硫酸盐
persulphide 过硫化物
persulphuric acid 过硫酸
perted loudspeaker 敞膜式扬声器
per ten thousand converted ton-kilometer employed fixed assets cost 每万换算吨公里占用固定资产原价
per ten thousand current funds finished converted ton-kilometre 每万流动资金完成的换算吨公里
per ten thousand current funds finished transport revenue 每万流动资金完成的运输收入
per ten thousand ton-kilometre included wage 每万吨公里工资含量
per ten thousand transport industry output employed fixed assets cost 每万元工业产值占用固定资产原价
per ten thousand transport revenue employed fixed assets cost 每万元运输收入占用固定资产原价
perthiocarbonic acid 过硫碳酸
perthiocyanic acid 过硫代氰酸
perthite 条纹长石
perthite-bearing nepeline syenite 含条纹长石霞石正长岩
perthite syenite 条纹长岩正长岩
perthitic texture 条纹结构【地】;条纹长石结构
perthosite 淡钠二长岩
per thousand 每千
pertica 中世纪教堂圣坛节日悬挂圣物的梁
per time unit 每时间单位
pertinax 熔结纳克斯胶;胶纸板;酚醛塑料
pertinency factor 有关因数
pertinent data 相应的资料;有关数据
pertinent detail 有关的细节
pertinent literature 切题文献;对口资料
pertinent parameter 有关参数
pertinent passage 有关章节
pertmuational isomer 置换异构体
per ton of dry matter produced 生产每吨干物质
per tour 每一回次
pertrified wood 硅化木
pertungstate 过钨酸盐
perturation method 微扰法
perturbation 扰动;微扰动波;微扰;扰
perturbation analysis 摄动分析
perturbation calculus 小扰动法计算
perturbation coefficient 扰动系数;摄动系数
perturbation equation 干扰方程;扰动方程;微扰方程;摄动方程
perturbation factor 摄动因素
perturbation method 扰动法;微扰方法;摄动法;摄动法
perturbation of earth's gravitational field 地球引力场摄动
perturbation of velocity 速度变动
perturbation of vertical 垂线扰动
perturbation orbit 摄动轨道
perturbation quantity 微扰量
perturbation sample 微扰样品
perturbations of daily schedule 每日故障计算表
perturbation symbol 摄动符号
perturbation term 摄动项
perturbation theorem 摄动定理
perturbation theory 扰动理论;微扰理论;摄动理论
perturbation wave 扰动波
perturbative force 摄动力
perturbative function 摄动函数
perturbator 扰动器
perturbed distribution 扰动分布
perturbed edges picture 边缘紊乱图像
perturbed elements 受摄根数
perturbed equation of motion 受摄运动方程

perturbed field 畸变场
perturbed Keplerian notion 受摄开普勒运动
perturbed motion 受摄运动;受扰运动
perturbed periodic(al) potential field 受扰周期势场
perturbed problem 扰动问题;摄动问题
perturbed surface 激励面;受扰面
perturbed term 摄动项
perturbed triplet 扰动型三合镜
perturbed water surface 小扰动水面
perturbing body 摄动体
perturbing force 扰动力;摄动力
perturbing potential 扰动位
perturbing term 摄动项
Peru architecture 秘鲁建筑
Peru basin 秘鲁海盆
Peru-Chile ocean basin block 秘鲁—智利海盆巨地块
Peru-Chile trench 秘鲁—智利海沟
Peru current 秘鲁海流
per unit 每一设备的;每一单元;每一单位;每单位量;每单位;标么值
per-unit area 每一单位面积;每单位面积
per-unit calculation 单位值计算
per-unit catchment area 单位集水面积
per-unit flow 流量标么值
per-unit gain 增益标么值
per-unit gate position 导流开度相对值;导流开度标么值
per-unit length 单位长度
per-unit quantity 单位量
per-unit synchronous reactance 单机同步电抗
per-unit system 相对值系统;单位法;标么制;单位系统;单位量制度
per-unit value 按单位值
per-unit volume 每一单位体积;每单位体积;每单位容量
per-unit weight 每一单位重量;每单位重量
Peru subduction zone 秘鲁俯冲带
Peru-type subduction zone 秘鲁型俯冲带
Peruvian architecture 秘鲁建筑
peruviol 橙花油醇
pervade 弥漫
pervaporation 渗透蒸发
pervaporization 膜蒸发;渗透蒸发
pervapo(u)ration 全蒸发(过程)
pervapo(u)ration membrane 全蒸发膜
perveance 导流系数
perverse incentive 反常动力
perverse migration 反常移动
perversion 侧向弯曲
perverted image 反像
pervial 能透过的;可透水的
pervibration 内部振捣
pervibrator 内部振捣器;插入式振捣器;插入式混凝土振捣器
pervious 可透过的;可渗透的;透水的;穿透的
pervious backfill material 透水性回填材料
pervious bed 反滤层;透水基床;透水底层;透水层
pervious blanket 排水层;排水垫层;透水铺盖层;透水铺盖;透水护坦;透水垫层
pervious boundary of apical layer 顶板透水边界
pervious boundary of bottom layer 底板透水边界
pervious boundary of fan-shaped intersection 扇形相交的透水边界
pervious breakwater 透水防波堤
pervious cesspool 透水污水池;污水渗井
pervious concrete 透水混凝土
pervious constant 透水常数
pervious course 透水层;渗透层
pervious crib 透水木笼坝;透水笼堤
pervious crib dike 透水木笼坝;透水格笼坝
pervious cushion 透水垫层
pervious dike 透水堤(坝)
pervious drainage layer 多孔排水层
pervious fence 透水栏栅
pervious foundation 透水地基
pervious groin 透水丁坝
pervious ground 透水围岩;透水土层;透水土(壤)
pervious material 透水材料;渗水材料
perviousness 可透性;透水性;透过性;渗透性
perviousness of concrete 混凝土的渗透性
perviousness test 透水度试验

perviousness to air 透气性
perviousness to diffusion 扩散渗透性
pervious notice 预先通知
pervious overburden 透水覆盖层
pervious pile dike 排桩透水丁坝;排桩透水堤
pervious plate 透水板
pervious revetment 透水式护岸;透水护岸
pervious rock 透水岩(石);透水性岩石;透水性岩层
pervious rocks 透水岩类
pervious sand gravel 透水砂砾(石)层
pervious seam 漏缝
pervious shell 透水壳体;透水壳层;透水坝壳
pervious slab 透水板
pervious soil 透水性土壤;透水土(壤)
pervious spur 透水丁坝
pervious stratum 透水岩层;透水地层;透水层
pervious structure 透水结构;透水建筑物
pervious subsoil 透水下层土;透水基层;透水底土(层);渗透性地基土
pervious surface 透水地面
pervious to air 空气渗透;透气
pervious to moisture 湿气渗透;透湿
pervious to water 渗水;水渗透;透水层
pervious training structure 透水导治建筑物
pervious wall 透水墙
pervious works 透水建筑物
pervious zone 透水区;透水地带;透水带
per week 每周;每星期
per-word error probability 每字误差概率
per working hatch per day 每日每适宜舱口
per yard 每码
perylene sodium sulfonate 对苯二甲酸二甲酸磺酸钠
peryphyton 水生附着生物
pessimism fuzzy automaton 悲观的模糊自动机
pessimistic design 悲观设计法
pessimistic estimation 悲观估计
pessimistic performance time 保守估定成效时间
pessimistic time 悲观时间;保守估定的时间
pessimistic time estimate 悲观的时值估计
pest 有害生物
pest and disease control 除害灭病
pest control 瘟疫控制;害物防治;害虫防治;虫害防治;防治虫害;病虫害防治
pest control chemical 害物防治药剂;害虫防治药剂
pest control equipmemt 植保机具
pest control measure 害物防治措施;害虫防治措施
pest control method 病虫害防治法
pest hole 疫区
pest house 隔离医院;传染病医院
pesticidal chemical 农药药剂
pesticide 杀害物药剂;杀虫药;杀虫剂
pesticide abuse 滥用杀虫剂;滥用农药
pesticide-added fertilizer 农药肥料
pesticide adjuvant 农药辅助剂
pesticide analysis 农药分析
pesticide analytical model 农药分解模型
pesticide chemistry 农药化学
pesticide combination 农药混合物
pesticide concentration 农药浓度
pesticide-containing wastewater 含农药废水
pesticide contamination 农药污染
pesticide contamination of groundwater 地下水农药污染
pesticide contamination of soil 土壤农药污染
pesticide contamination potential 农药污染势;潜在农药污染
pesticide control 农药控制
pesticide degradation 农药降解
pesticide detection 农药检测
pesticide drift 农药飘失
pesticide effect 农药影响
pesticide effectiveness 农药药效
pesticide environmental chamber 农药环境模拟室
pesticide factory 农药厂
pesticide formulation 农药剂;农药剂型
pesticide metabolism 农药代谢
pesticide monitoring 农药监测
pesticide pathway and fate 农药行径和归宿
pesticide phytotoxicity 农药药害
pesticide poisoning 农药中毒
pesticide pollution 农药污染;农药公害
pesticide processing plant 农药厂
pesticide production wastewater 农药生产废水

pesticide registration 农药注册
pesticide residue 农药残余;农药残留物;农药残留量;农药残留
pesticide residue analysis 农药残留量分析
pesticide residue hazard 农药残余危害;农药残留危害性
pesticide residue tolerance 农药允许残留量;农药残留允许量;农药残留容许量;农药残留容限量
pesticide runoff 农药流失
pesticide runoff simulator 农药径流模拟器
pesticides 农药
pesticide safety precaution scheme 农药安全预防方案
pesticide tolerance 农药容许量;农药耐量
pesticide toxicity 农药毒性
pesticide toxicology 农药毒理学
pesticide transport 农药输移
pesticide transport and runoff model 农药输移和径流模型
pesticide use emulsifier 农药用乳化剂
pesticide waste 农药废料
pesticide wastewater 农药废水
pesticide wastewater treatment 农药废水处理
pesticide year 农药年度
pesticon 高灵敏度摄像管
pestilence 瘟疫;时延;传染病
pestilential pathogen 瘴气
pesting 研磨
pestis fulminans 暴发性鼠疫
pestis major 暴发性鼠疫
pestle 研磨;研棒;捣锤;杵
pestle mill 捣锤
pestle stamp 捣杵
pestling 用杵捣;研磨;捣碎
pest management 病虫害控制
pest management system 虫害管治制度
pestoration of unsafe building 危险房屋的修复
petal 花瓣;翼瓣
petal catcher 膜瓣收集器
petaling 淡褐色硬木(印度产);花瓣形
petalite 叶状锂长石;透锂长石
petal-like 花瓣状的
petaloid 花瓣状的
petaloid volcanic mud ball structure 花瓣状火山泥球构造
petarasite 氯硅锆钠石
petasus 阔边帽(古希腊);赫尔墨斯有翼帽
pet cock 小旋塞;小活塞;小型旋塞(阀);排气阀;小龙头;减压开关;扭塞;排泄开关;手压开关;排气栓
Peterhead 彼得红色黑云母花岗石装饰
peterlineum 焦油护木剂;蒽油防腐剂
peter man 渔夫;荷兰渔船
Peterman test 彼得曼氏试验
peter-net 岸边网
peterologic(al) cavity 岩石的空洞
peterologic(al) enclave 岩石的包体
peter out 消耗掉;渐大至消失;逐渐枯竭
Petersburg standard 彼得斯堡木材体积标准(1标准计量单位=165立方英尺)
peters mill 盘辊磨
Peterson coil 彼得森线圈
Peterson current meter 彼得森海流计
Peterson grab 双半筒型海底取样抓斗
Peterson graph 彼得森图
petit average 港务杂费
Petite truss 交叉斜杆的普腊桁架
petit grain oil 橙叶油
petition 申请
petitioned work project 申请的工程
petitioner 请求人
petition for naturalization 入籍申请
petition in bankruptcy 破产申请
petition of right 权利申请书
petitory action 要求确立产权的诉讼
petit-point articles 纳纱制品
Petit's canal 小带间隙
petization 胶溶作用
petizator 胶溶剂
petoscope 运动目标电子探测器
petrail 木结构房屋的横梁
petralol 液体石油膏
petrean 硬化的;化石的;多岩的;岩质的
petrichor 潮土油

petrideserta 岩下荒漠群落
petri dish 皮特里盘;培养皿
petrification 化石[地];石化(作用)
petrification of sand 固砂;砂的固化
petrified forest 化石森林
petrified rose concretion 重晶石玫瑰花状
petrified soil 石化土
petrified wood 木化石;石化木
petrifoid 防水透明液
petrify 转化为石质;石化
petrifying liquid 硬化液;石化液;防潮液(喷墙用)
petrifying water 硬化水
petrissage 揉捏法
petrium 砾石群落
petroacetylene 石油乙炔
petro-agriculture 石油农业
petrobenzene 石油苯
Petrobia lateens 潜岩螨
petroblastesis way 扩散变晶作用方式
petrocarbon 石油碳
petrochemical complex 石油化学总公司;石油化工总公司;石油化工总厂;石油化工联合工厂;石油化工公司
petrochemical compound 石油化学化合物
petrochemical corporation 石油化学总公司;石油化工总公司;石油化工公司;石化公司
petrochemical derivatives 石油化工产品
petrochemical engineering 石油化工
petrochemical engineering (designing) institute 石油化工设计院
petrochemical extinguishing system 石油灭火系统
petrochemical industrial bureau 石油化学工业局
petrochemical industry 石油化学工业;石油化工工业;石油化工;石化工业
petrochemical industry corporation 石油化学工业总公司
petrochemical industry organic wastewater 石化有机废水
petrochemical industry waste(water) 石化工业废水
petrochemical intermediate 石油化学中间产品
petrochemical manufacture 石油化学产品的生产
petrochemical materials 石油化工原料
petrochemical odo(u)r 石油化学产品的气味
petrochemical origin 蜡石化学成因
petrochemical plant 石油化学工厂;石油化工装置;石油化工厂
petrochemical processing 石油化学加工
petrochemical processing plant 石油化学加工装置
petrochemical process unit 石油化工工艺装置
petrochemical products 石油化工产品
petrochemical reaction 石油化学反应
petrochemicals 石油化学制品;石油化学产品;石油化工制品;石油化工产品;石化产品
petrochemical storage 石油化学产品的储藏
petrochemical storage vessel 石油化学品的储藏容器;石油化学产品的储藏器
petrochemical waste 石油化学废弃物;石油化工废物
petrochemical waste disposal 石油化学废物处置;石油化学废料处置
petrochemical wastewater 石油化工废水;石化废水
petrochemical wastewater reuse 石化废水回用
petrochemical wastewater treatment 石化废水处理
petrochemical works 石油化工厂
petrochemistry 岩石化学;石油化学
petrochemistry data analysis 石油化学数据分析
petrochemistry data base 岩石化学数据库
petroclastic 碎屑的
petroclastic rock 碎屑岩
petrocole 石栖动物
petrocurrency 石油通货
petro-dollar 石油美元
petrofabric 岩组;岩石组构;组构
petrofabric analysis 岩相分析;岩石构造分析
petrofabric diagram 岩组图;组构图
petrofabrics 岩组学
petrofacies 岩相
Petroff equation 佩特罗夫摩擦系数公式
petrofilling station 加油站
Petro-forge 高速锤
petro-forge machine 压缩空气高速锤
petro-forging machine 佩特罗式高速锤
petrogas 液体丙烷;石油丙烷

petrogenesis 岩石成因学;岩石成因(论);成因岩石学
petrogenetic element 造岩元素
petrogenetic grid 成岩格子
petrogenetic type 岩石成因类型
petrogenic 造岩的
petrogenic classification 岩石成因分类
petrogenic element 造岩元素
petrogenic grid 岩石成因格网
petrogenic mineral 造岩矿物
petrogenic temperature determination method of melting inclusion 熔融包体成岩温度测定法
petrogeny 岩石发生学;岩石成因学
petroglyph 岩石雕刻
petroglyphy 岩石雕刻术
petrograd standard 软木材体积标准
petrograph 岩石雕刻
petrographer 岩相学家;岩石学者;岩石学家
petrographic(al) analysis 岩相分析;岩石分析
petrographic(al) characteristic 岩石特征;岩石特性
petrographic(al) classification 岩土分类;岩石分类
petrographic(al) classification analysis 岩石分类分析
petrographic(al) classification for engineering geology 岩石工程地质分类
petrographic(al) composition 岩石成分
petrographic(al) examination 岩石鉴定
petrographic(al) facies 岩相
petrographic(al) interpretation 岩相分析;岩相解释;岩相译释
petrographic(al) investigation 岩相研究
petrographic(al) microscope 岩相显微镜;岩石显微镜;偏光显微镜
petrographic(al) microscopy 偏光显微镜技术;岩相显微镜技术
petrographic(al) preparation 岩石试件制备
petrographic(al) province 岩区
petrographic(al) quantitative analysis 岩相定量分析
petrographic(al) series 岩系
petrography 岩石上刻像或文字;岩相学;岩类学;描述岩石学
petrol 汽油;石油
petrolage 石油处理法
petrol-air mixture 汽油空气混合气
petrol/alcohol mixture 汽醇混合物
petrol asphalt 石油沥青
petrolat 矿脂
petrolatum 轴承包装油;矿脂;软石脂;凡士林;不定型石蜡
petrolatum album 白矿脂
petrolatum gauze 凡士林纱布
petrolatum oil 矿脂
petrolax 液体矿脂
petrol balance 汽油秤
petrol barge 汽油驳船;石油驳船
petrol barrel 石油桶
petrol bomb 凝固汽油弹
petrol bowser 汽油加油车
petrol burner 汽油喷灯
petrol can 汽油桶;汽油壶
petrol carrier 油船
petrol cock 汽油旋塞;汽油阀(门)
petrol coke 石油焦
petrol compacting and finishing screed 内燃式夯样机
petrol concrete vibrator for mass work 大体积混凝土内部振捣器
petrol consumption 汽油消耗
petrol container 汽油罐
petrol delivery 汽油的供应
petrol depot 汽油库
petrol-depth ga(u)ge 汽油深度计;汽油表
petrol dewage 汽油废水
petrol dripping 汽油油滴;漏油;滴油
petrol-driven 汽油发动机驱动的
petrol-driven roller 汽油发动机驱动的路碾;汽油发动机驱动的压路机
petrol-driven truck 汽油发动机驱动的载重汽车
petrol drum 汽油桶
petrol dump 汽油库
petrol duties 石油税
petrol-electric(al) bus 汽油机电机公共汽车
petrol-electric(al) car 汽油电车

petrol-electric(al) drive 汽油机—电机驱动
petrol-electric(al) generating set 内燃机驱动的发电机
petrolene 沥青脂;可溶性质;软沥青质;软沥青;软地沥青;石油烯
petrol engine 汽油引擎;汽油机;汽油发动机
petrol engine boat 汽油机船
petrol engine exhaust emission 汽油发动机废气污染物
petroleous 含石油的
petroleum 液体天然沥青;石油(产品);石油磺酸酸铝泡沫剂
petroleum accumulation 石油储备
petroleum acids 石油酸
petroleum additive 石油添加剂
petroleum analysis 石油分析
petroleum anchorage 油船锚地
petroleum and natural gas origin 石油及天然气成因
petroleum aromatics 石油芳烃
petroleum asphalt 石油(地)沥青
petroleum base 油基;石油碱
petroleum base additive 石油基添加剂
petroleum base rust preventive 石油基防锈剂
petroleum benzene 焦苯;石油挥发油;石油苯
petroleum benzin 石油醚
petroleum benzine 汽油原油
petroleum berth 油码头;油轮泊位
petroleum bitumen 石油沥青
petroleum black 石油炭黑
petroleum bloom 石油荧光
petroleum briquet 石油砖
petroleum bulk station 配油站
petroleum bulk storage buildings 储[贮]散装(石)油建筑物
petroleum butter 石油膏;凡士林
petroleum chemical plant 石油化工厂
petroleum chemicals 石油化学制品;石油化学产品;石油化工制品
petroleum chemistry 石油化学
petroleum chromatography 石油色谱
petroleum coal 固体石油
petroleum coke 石油焦炭
petroleum composition 石油组成
petroleum conservation 石油资源节约
petroleum contaminant 石油污染物
petroleum contamination 石油污染
petroleum cracking 石油裂解;石油裂化
petroleum cracking process 石油裂化过程
petroleum crude 原油
petroleum crude oil 石油原油
petroleum cut 石油馏分
petroleum degrading microorganism 石油分解微生物
petroleum deposit 石油矿床
petroleum derivative 石油衍生物
petroleum-derived hydrocarbon 石油衍生烃
petroleum-derived products 石油衍生物
petroleum distillate 石油蒸馏液;石油馏出物
petroleum distillation 石油蒸馏
petroleum distillation residue 石油蒸馏残余物
petroleum drilling 石油钻探;石油钻井
petroleum drying oil 石油系合成干性油
petroleum emulsion 石油乳液;石油乳剂
petroleum engine 汽油引擎;汽油机
petroleum engineering 石油钻采工程;石油工程
petroleum ether 石油醚
petroleum ether insoluble 石油醚不溶液;不溶于石油醚
petroleum ether soluble 能溶于石油醚
petroleum-ether soluble matter 石油醚提出物
petroleum exhaust 石油废气
petroleum expansion 石油加热膨胀
petroleum extraction 石油提炼;石油开采
petroleum facility 石油设施
petroleum feeder 石油喷射器
petroleum fermentation 石油发酵
petroleum flux 石油稀释剂;石油软化剂
petroleum formation 石油生成
petroleum fraction 石油馏分;石油分馏物
petroleum fuel 石油燃料
petroleum fuel oil 石油燃料油
petroleum gas 石油气体;石油气;石油挥发气体
petroleum gas compressor 石油气压缩机

petroleum gas oil 石油气体油
petroleum genesis 石油形成
petroleum geochemical analysis and isotopic geochemistry 石油地球化学分析及同位素地球化学
petroleum geochemical classification of kerogen 干酪根石油地球化学分类
petroleum geochemistry 石油地球化学
petroleum geologic(al) hole 石油地质钻孔
petroleum geology 石油地质学
petroleum grease 石油润滑脂;石油润滑剂
petroleum hydrocarbon 石油烃
petroleum hydrocarbon compound 石油烃类化合物
petroleum hydrocarbon oil 石油烃油
petroleum hydrocarbon resin 石油烃树脂
petroleum industry 石油工业
petroleum industry wastewater 石油工业废水
petroleum installations 石油设施
petroleum institute 石油学院
petroleum intelligence 石油情报
petroleum intermediate product 石油中间产品
petroleum intersectional service 石油区间勤务处
petroleum jelly 矿脂;石油膏;石油冻;凡士林
petroleum jelly white 白凡士林
petroleum jetty 石油装卸码头;石油码头
petroleum law 石油法
petroleum legislation 石油法案
petroleum leve 轻石油
petroleum liquids 石油产品
petroleum locomotive 石油机车
petroleum lubricant 石油润滑剂
petroleum lubricating grease 石油润滑脂
petroleum metering instruments 石油计量仪器
petroleum microbiology 石油微生物学
petroleum microorganism 石油微生物
petroleum migration 石油扩散;石油迁移;石油运移
petroleum motor oil 石油车用润滑油
petroleum naphtha 溶剂汽油;轻汽油;石脑油;粗汽油
petroleum non-hydrocarbon compound 石油非烃化合物
petroleum odour 石油气味
petroleum oil 石油油料;石油系油类
petroleum oil coke 石油焦
petroleum-oil column 石油蒸馏塔
petroleum ointment 矿脂
petroleum operations 石油作业
petroleum origin 石油起源
petroleum origin theory 石油成因说
petroleum paraffin 石油石蜡
petroleum pier 石油码头
petroleum pipeline 原油管线;输油管;石油管线;石油管道
petroleum pipeline survey 输油管道测量
petroleum pipeline system oil pipeline system 石油管路系统
petroleum pitch 石油硬沥青;石油沥青;石油类焦油脂;石油类柏油脂
petroleum pitch coke 石油沥青焦炭
petroleum pollutant 石油污染物
petroleum pollution 石油污染
petroleum pool 储油层
petroleum port 石油港口;油港
petroleum posted price 石油计税标价;石油标价
petroleum processing 石油加工
petroleum processing flow 石油加工过程流程
petroleum processing waste 石油加工废(弃)物
petroleum producer 石油企业家
petroleum product 石油制品;石油产品
petroleum production 采油
petroleum production by underground combustion 地下燃烧法采油
petroleum production country 石油生产国
petroleum production engineering 采油工程
petroleum prospecting 石油勘探
petroleum prospecting by microbiology 微生物学石油勘探
petroleum province 含油区;石油省;石油产区
petroleum pump 原油泵;重油泵;石油泵
petroleum recovery 石油再生;石油回收
petroleum refinery 炼油厂;石油炼厂
petroleum refinery waste 石油加工废料;石油加工厂废物

petroleum refinery wastewater 石油加工厂废水
petroleum refining 石油提炼;石油炼制
petroleum refining capacity 石油炼制能力
petroleum refining furnace 石油炼制炉
petroleum refining industry 炼油工业;石油炼制工业
petroleum refining industry sewage 石油炼制工业污水
petroleum refining wastewater 炼油废水;石油炼制废水
petroleum reforming 石油重整
petroleum refrigerant 石油制冷剂
petroleum region 石油区
petroleum reserve 石油储(藏)量
petroleum reservoir 油储[贮];储油层
petroleum residual oil 石油渣油
petroleum residue 炼制残油;石油脚;石油残渣
petroleum resin 石油树脂
petroleum resources 石油资源
petroleum resources conservation 石油资源保护
petroleum resources survey 石油资源调查
petroleum revenue tax 石油收入税
petroleum science 石油科学
petroleum seismic prospecting 石油地震勘探
petroleum shale 油页岩
petroleum ship 油船;石油运输船
petroleum situation 石油动态
petroleum soap 石油皂
petroleum solid 石油固体
petroleum solvent 石油溶剂
petroleum specialties 特殊石油产品;特种石油产品
petroleum spirit 轻质烃混合物;汽油;石油溶剂油;石油精
petroleum still 石油蒸馏釜
petroleum storage depot 油库
petroleum substitute 石油代用品
petroleum sulfonate 磺酸盐;石油磺酸盐;石油磺酸油
petroleum sulfur analyzer 石油含硫量分析器
petroleum supplement 石油代用品
petroleum surveying ship 石油勘探船
petroleum sweetening 石油脱硫
petroleum tailings 蜡渣;石油蒸馏残余物
petroleum tank 油罐;油槽
petroleum tank wagon 铁路油槽车
petroleum tar 石油沥青;石油焦渣;石油残渣油;石油柏油
petroleum terminal 石油码头
petroleum tester 石油测定器
petroleum testing method 石油试验法
petroleum thinner 石油稀释剂
petroleum times 石油时代
petroleum transportation 石油运输
petroleum trap 石油圈闭
petroleum vapo(u)r 石油蒸气
petroleum voltage test equipment 原油高压物性试验设备
petroleum waste 石油废物
petroleum wastewater 石油废水
petroleum wax 石油蜡;石蜡
petroleum well drilling wire rope 石油钻井用钢丝绳
petroleum wharf 石油码头
petrol filler lid 汽油加注口盖
petrol filling station 汽油加油站;加油站
petrol filter 汽油滤(清)器;汽油过滤器
petrol fume 汽油蒸气
petrol ga(u)ge 汽油表
petrol-ga(u)ge dial 汽油表刻度盘
petrol generating set 汽油生产设备;汽油生产装置
petrol generator 汽油发电机
petrol hand hammer rock drill 汽油机驱动手扶冲击式钻机
petrol hoist 汽油驱动的提升机
petrol hose 汽油胶管
petrolic ether 石油醚
petrolic machinery plant 石油机械厂
petroliferous 含石油的
petroliferous bed 含油岩层;含油层
petroliferous sandstone 含油砂岩
petroliferous shale 含油页岩
petroliferous strata 含油岩层
petrolift 燃料泵
petrol immersion vibrator 汽油驱动插入式振捣

器;汽油驱动插入式振动器
petrol inceptor 防石油存水湾
petrol-intercepting trap 隔汽油具
petrol internal vibrator 汽油驱动的内部振动器
petrolization 石油洒播法;石油处理
petrolize 柏油铺路
petrol-level ga(u)ge 汽油表
petrol level indicator 汽油液面指示器
petrol motor 汽油机;汽油发动机
petrol motor car 汽油机汽车;汽油机车
petrol motor roller 汽油路碾;汽油压路机
petrol needle vibrator 汽油驱动杆状振动器
petrologen 油母页质
petrologic(al) classification 岩石学分类
petrologic(al) colo(u)r 岩石颜色
petrologic(al) fusain 岩化丝炭
petrologic(al) indication 岩石学标志
petrologic(al) indices 岩石学指数(系数/参数/公式比例)
petrologic(al) mineral composition 岩石的矿物成分
petrologic(al) phase equilibrium 岩石相平衡
petrologic(al) tone 岩石的色调
petrologic province 沉积岩区
petrologist 岩石学者;岩石学家
petrology 岩石学;岩理学
petrology of coal 煤岩学
petrology of metamorphic rocks 变质岩岩石学
petrology of ocean-floor 海底岩石学
petrol ointment 石油软膏
petrol operated pile driving hammer 汽油机驱动操作打桩锤
petrolo-shale 含油页岩
petrol pervibrator 汽油机驱动的内部振动器
petrol pipe 汽油管
petrol poker vibrator 汽油机驱动内部振捣器;汽油机驱动插入式振捣器;汽油机驱动内部振动器;汽油机驱动插入式振动器
petrol-powered lift truck 汽油机驱动的自动装卸车
petrol-powered winch 汽油机驱动的绞车
petrol-pressure ga(u)ge 汽油压力计
petrol-proof 防汽油的
petrol pump 汽油机驱动泵;汽油泵
petrol pump case 汽油泵壳
petrol resistance 耐汽油性
petrol roller 汽油压路机;汽油路碾
petrol sediment bulb 汽油沉降器
petrol separator 汽油分离器
petrol sewage 汽油废水
petrol spirit 汽油
petrol spud vibrator for concrete 汽油机驱动的插入式混凝土振动器
petrol starting engine 汽油启动机
petrol station 加油站;汽油站;汽油加油站
petrol storage 汽油库
petrol storage stability 汽油储存量的稳定性
petrol strainer 汽油滤(清)器
petrol tank 汽油箱;汽油桶
petrol tank car 轻油罐车
petrol tanker 汽油油槽车
petrol tractor 汽油机拖拉机
petrol trap 隔汽油具;汽油滤(清)器;汽油分离器;石油放油弯管;防石油存水湾
petrol vapo(u)r 汽油气
petrometallogeny 岩石成矿作用
petromictic conglomerate 岩块砾岩
petromodel 岩石模式
petromorphology 岩石形态学;构造岩形学
petronate 一种石油磺酸盐
petronol 液体石油脂;液体石蜡
petro-occipital fissure 岩枕裂
petrophysics 岩石物理学
petrophytes 石生植物
petropine 组合式无缝地板
petropols 石油树脂
petropons 石油干性油
petroresins 石油树脂
petrosa 岩部
petrosal foramen 岩孔
Petrosan 贝洛生(一种加气剂商品名)
petrosapol 石油软膏
petroseite 硒铅铜镍矿
petroselinic acid 岩芹酸
petrosio 液体矿脂

petrosphere 地壳;岩石圈
Petrostroma 石质海绵属
petrotectonic assemblage 岩石构造组合
petrotectonics 岩石构造学;构造岩组学;构造岩石学
petrous 石质的
petrous apex 岩部尖
petrous branch 岩支
petrous coast 石质岸
petrous portion 岩部
petrous pyramid 锥部
petrovicite 硒铋铅汞铜矿
petrox 油酸铵皂化的石蜡油
petroxolin 油酸铵皂化的石蜡油;液体石蜡
petrumite 仿石油漆
petscheckite 铌铁铀矿
pet shop 宠物商店
petterdite 砷铅矿
Petter engine 彼得发动机
Petter motor oil test 皮特发动机润滑油试验
Petterson-Nansen water bottle 南森瓶
Petthan 佩丹黄红色硬木(缅甸产)
petticoat 裙状物;桅套;桅裙
petticoat insulator 裙状绝缘子;裙式绝缘子
petticoat pipe 圆锥形漏斗管;烟罩
petticoat pipe base 烟罩座
petticoat value 裙形阀
Pettit truss 普蒂再分式桁架;再分普蒂桁架
pet toy 动物玩具
petty average 港务杂费
petty cabotage 沿岸小型航运
petty cash 零用钱;小额现金;零用现金;备用金
petty cash current deposit 小额活期存款
petty cash deposit 小额现金存款
petty cash fund 零用现金基金;定额备用金
petty cash journal 零用现金日记账
petty cash payments 小额现金支付
petty commodity economy 小商品经济
petty deposits 小额存款
petty expenditures 杂费
petty imprest 小额预付款
petty insurance 小额保险
petty lessor 小土地出租者
petty loan 小额贷款
petty loss and profit 小损益
petty pilferage 小窃
petty street garden 小游园
petty sum 小额
petty sum wasting property 小额耗用财产
petunia 深紫红色;喇叭花
petun(t)se 瓷泥;白墩子
pet valve 小型旋塞(阀);小阀;试验阀;小型阀门
petzite 碲金银矿
Petzval condition 佩兹伐条件
Petzval curvature 佩兹伐曲率
Petzval lens 佩兹伐透镜
Petzval sum 佩兹伐曲率
Petzval surface 佩兹伐曲面
peucine 树脂;沥青
peucinous 树脂性的;沥青性的
peuroseite 硒铅铜镍矿
PE value of the environment 环境的 PE 值
pew 教堂闭合板凳;教堂靠背长凳;教堂中的一排座位(靠背固定的长凳)
pew chair 添加折叠式椅子(在教堂长凳子旁)
pew hinge 长翼铰链(有蛋形关节饰的)
pewter 锡蜡;铅锡锑合金
pewter solder 锡焊料
pexitropy 冷却结晶作用
pez 地沥青
Pfaffian differential equation 波发夫微分方程
Pfalzian orogeny 法尔茨造山运动【地】
pferidekraft 马力米制单位
Pfisher rotor weigh feeder 费斯特转子称重喂料机
p-fluoro aniline 对氟苯胺
Pfund cryptometer 芬德遮盖力计
Pfund film ga(u)ge 芬德涂膜厚度计
Pfund hardness number 芬德硬度(值)
Pfund hardness tester 芬德硬度计
Pfund indenter 芬德压头
Pfund series 芬德系【地】
pgt per gross ton 每毛吨
phacellite 钾霞石
phacoidal structure 透镜形构造;扁豆状结构
phacoids 扁豆体

phacolite 岩脊;扁菱沸石
phacolith 岩脊;岩鞍;扁豆状
phacometer 透镜折射率计
pH adjustment pH 值调整
phaeism 暗型
phaenerophyta herbaceae 草本高位芽植物
phaeodium 暗块
Phaeophyceae 褐藻纲
phaeton 游览车;敞篷汽车;敞篷车;二马四轮便马车
phaiozem 灰色森林草原土
phaiozem and black soil area 灰色森林土及黑土区
phakometer 透镜组聚光能力测定计;数测定计
phakoscope 数测定计
P-hammer P 型回弹仪
phaneric 显晶质的
phaneric texture 显晶结构
phanerite 显晶岩
phaneritic 显晶岩的
phanerocrystal 斑晶
phanerocrystalline 显晶质的
phanerocrystalline rock 显晶岩
phanerocrystalline texture 显晶质结构;显晶结构
phanerogam 显花植物
phanerogamic wood 显花植物式木材
phaneromer 显晶质的
phaneromere 显粒岩
phaneroplasm 明体
phaneropore 显露气孔
phanerosis 显现
phanerous 显明的
Phanerozoic geochronologic(al) scale 显生宙地质年代表
Phanerozoic time 显生代【地】
Phanerozoic Time-Scale 显生代地质年代表
phanic 显明的
phanodorn 非罗多
phantasm 幻影;幻象
phantasmatology 幻象学
phantastic architecture 奇异的建筑
phantastran 固态延迟管
phantastron 延迟管;幻象延迟管
phantastron divider 幻象电路分频器
phantom 影痕;决断层;模型;幻影;幻象;人造模型;缺失地层;陀螺罗经随动部分;仿真
phantom aerial 假天线
phantomatic fringe pattern 虚条纹图
phantom balance 幻象电路平衡
phantom bottom 虚假河底;虚假海底;假海底;伪底
phantom circuit 幻象线路;幻象电路
phantom circuit balance 幻象电路平衡
phantom circuit repeating coil 幻象电路转电线圈
phantom connection 模拟电路接线;幻象电路接线
phantom crystal 先成晶体;阴影晶体;幻晶体
phantom drawing 幻想景色;幻画
phantom dumping 非法轻弃
phantomed cable 幻象线路电缆
phantom element 随动部分;从动构件
phantom freight 在售价上加的假设运费
phantom gear 随动齿轮
phantom group 幻象群
phantom horizon 假定平行线
phantom indication 虚假指示
phantoming 架成幻路
phantom line 点画线;虚线;假想线;幻线
phantom load 虚负载;假想荷载;假设负荷
phantom loading method 虚负载法
phantom material 体模材料
phantom part 随动部分
phantom powering 幻路供电
phantom repeating coil 幻路转电线圈
phantom ring 随动环
phantom signal 雷达幻影;幻象信号
phantom stem 随动部分锚颈;船首中锚
phantom target 假目标;幻目标
phantom transposition 幻路交叉
phantom view 幻想景色;幻影图;幻景;部分剖视图
phaoplankton 透光层浮游生物
pharaonite 镁钾钙霞石
pharmaceutical analysis 药物分析
pharmaceutical chemistry 制药化学
pharmaceutical factory 药厂;制药厂
pharmaceutical glass 药用玻璃
pharmaceutical industry 制药工业
pharmaceutical plant 制药厂

pharmaceutical source 药物污染源
pharmaceutical technology 制药工艺(学)
pharmaceutical waste 制药废料
pharmaceutical wastewater 制药废水
pharmaceutical wastewater containing antibiotics 含抗生素制药废水
pharmacolite 毒石
pharmacosiderite 毒铁石;毒铁矿
pharmacy 药房;药店
pharology 灯标学
pharos 航标灯;灯塔
pharosage 辐射照度
pharos of Alexandria 亚历山大半岛灯塔(古代七大奇观之一)
phase 周相;可执行程序段;局面;阶级;阶段;物相;位相;灯相;分散相
phaseable marks 定相标记
phase adapter 相位转换附加器;换相器
phase address system 相位地址系统
phase adjusted stack section 相位调整叠加剖面
phase adjustment 相位调准;相位调整;定相调准;定相调整
phase advance 相位超前
phase advanced condenser 进相电容器
phase advancer 相位超前补偿器;进相机;超前补偿器
phase after-phase start-up 逐相起动;按相起动
phase age 月相大不等潮龄;潮龄
phase ambiguity 相位多值性
phase amplitude discriminator 相幅鉴别器
phase analysis 相分析;阶段分析;物相分析
phase and chronology of fault 断层期次和时代
phase angle 相移角;相位角;相角;位相角
phase angle difference 相角差
phase angle error 相角误差
phase angle indicator 相角指示器
phase angle of a current transformer 电流互感器的相角差
phase angle of a voltage transformer 电压互感器的相角差
phase angle of the escapement 擒纵轮相位角
phase angle voltmeter 相角电压表
phase area 相面积
phase array scanning 相控阵扫描
phase average 相平均
phase back 推迟
phase bank 相阻
phase behavio(u)r 相状态;相特性
phase behavio(u)r of greases 润滑脂的相特性
phase belt 相带
phase bit 定相位
phase bleed 相流失
phase boundary 相界
phase boundary crosslinking 相界交联
phase boundary potential 相界电势;界面电位
phase caving method 阶段崩落法
phase change 相的变化;相(化)
phase-change coefficient 相变系数
phase change failure protection 换相(故障)保护(装置)
phase change material 相变材料
phase change of mineral 矿物相变
phase change on heating 加热相变化
phase changer 相移装置;移相器;换相器;变相器
phase change switch 换相开关
phase change theory 相变理论
phase change time 换相时间
phase change zone 相变区
phase changing plate 变相板
phase changing range 变态间隔
phase changing superplastic forming 相变超塑成型
phase characteristic 相位特征;相位特性
phase check 相位检查
phase coding 相位编码
phase coefficient 相位系数;位相系数
phase coincidence 同相
phase comparator 相位比较器
phase comparison 相位比较
phase comparison localizer 比相定位器
phase comparison protection 相位比较式保护;相差保护
phase comparison protection unit 相差保护装置
phase comparison relay 相差继电器

phase compensating device 相位补偿装置
phase compensator 相位补偿器
phase composition 相组成;混合相
phase compression 相移压缩
phase conductor 相导体
phase constant 相位常数;相位常量;周相常数
phase constituent 相成分
phase contact 同相性接触
phase contact area 相界面积
phase contact bond 同相接触联结
phase contacting area 相界面;相分界面
phase contour 等相角线
phase contrast 相衬;相位反差;相差衬托;相差
phase contrast cinematography 相差电影摄术式
phase contrast condenser 相差聚光镜
phase contrast electronic microscope 相差电子显微镜
phase contrast microscope 相差显微镜;相衬显微镜;位相显微镜
phase contrast microscopy 相衬显微术;相差显微术;相差显微镜检查
phase contrast rule 相衬定则
phase control 相位控制;色调节
phase control device 相位控制装置
phase controller 相位控制器;相位控制机
phase control of grid voltage 栅极电压相位控制
phase conversion 相转变;换相
phase converter 相位变换器;换相器;换相变流机;变相机
phase coordinates 相坐标
phase correction 相位校正
phase correlation 相位对比
phase crossover 相位交点
phase current 相电流
phase curve 相位曲线;位相曲线
phase cycle 相位周期
phased antenna 相控天线
phased application 分阶段间隔操作
phased array 相阵;相控天线阵
phased-array 3-D radar 相控阵三坐标雷达
phased-array antenna 相控阵天线
phased array jammer 相控阵干扰机
phased array of dipoles 同相偶极天线阵
phased array radar 相控阵列雷达
phased array sector scanner 相控阵式扇形扫描显像仪
phased array space tracking radar 相控阵空间跟踪雷达
phased class 相关复合类
phased development 阶段发展
phased development plan 分期开发计划
phase decomposition 相分解
phase delay 相位滞后;周相延迟;相延迟
phase delay measurement 相位延迟测定
phase demodulation 鉴相
phase depending on rectification 相敏整流;相敏鉴流
phase detection 相位检测
phase detector 相位检测器;相位比较器;鉴相器;检相器
phase determination 相位确定;测相
phased evacuation 分段疏散
phase development 阶段发育
phase deviation 相位偏移;相偏移;相对频偏;移相变位
phased evolution 阶段演进
phase diagram 信号相运行图;相转变图;相位图;相图
phase diagram for a binary solid solution 二元固体溶液相图
phase diagram of equilibrium 平衡相图
phase difference 相位差;相差
phase difference between traces 道间相位差
phase difference indicator 相差指示器
phase difference measurement 相位差测量
phase difference method 相位差法
phase digestion 分期消化
phased implementation 阶段性实现
phase dipole 相偶极子
phase discrimination 相位鉴别
phase discriminator 鉴相器
phase disengagement 相分离
phase disentrainment 相夹带剔除
phased isolation ditch 定期隔离沟;分期隔离沟法

phased isolation ditch technology 分期隔离沟法技术
phase displacement 相移;相位移动;相位移;相位移;相位偏移;相位差
phase distortion 相位失真;相位畸变
phase distortionless 位相无畸变
phase distribution 相分布
phase distribution of field 场相位分布
phased laser array 激光相控阵
phasedless target 无相差靶标
phased operation 分阶段操作
phase down 逐步减缩
phased program(me) 逐步实施计划
phased program(me) of works 施工进度表
phased project planning 阶段项目设计;分阶段工程设计;分阶段方案设计
phased project planning guideline 阶段项目规划指南
phased reduction 分期减少
phase drift 相位移动;相位漂移;相位偏移
phased stack 调相叠加
phased treatment 分期处理
phase duration 相延续时间
phase encoding 相位编码
phase encoding technique 相位编码技术
phase enrichment coefficient 相富集系数
phase equalization 相位均衡
phase equation 相方程式
phase equilibrium 相平衡
phase equilibrium computer 相平衡计算机
phase equilibrium condition 相变平衡条件
phase equilibrium diagram 相平衡图
phase equilibrium isotope effect 相平衡同位素效应
phase equilibrium line 相平衡曲线
phase equipment 信号显示装置
phase error 相位误差
phase error of satellite 卫星相位差
phase error of sighting object 照准目标相位差
phase factor 相因数;相位因数
phase failure protection 相故障保护
phase failure relay 断相继电器
phase fault 相间短路
phase feeding 分饲
phase filter 相滤波器
phase filtering 相位滤波
phase filtering method 相过滤方法
phase fluctuation 相位波动
phase focus(s)ing 调相聚焦
phase frequency characteristic 相频特性
phase frequency response 相频响应
phase Fresnel's lens 菲涅尔相位透镜
phase front 相前;波阵面
phase function 相函数
phase generator 相移发生器
phase grating 相栅
phase group 相组
phase history 相位变化过程
phase hologram 相位全息图
phase holography 相位全息术
phase hunter 寻相器
phase in 逐步投产;逐步使用;逐步采用;分阶段引进;按程序逐步改进
phase indicator balancing control 相位指示器平衡控制
phase inequality 相位月潮不等;相位不等
phase inequality of height 潮高月相不等
phase inequality of time 潮时月相不等
phase information 相位信息
phase in-lock 相锁
phase instrument for IP 相位激电仪
phase insulation 相间绝缘
phase integral 相积分
phase into 逐步投入
phase inversion 相转化法;相位倒转;转相;换相;位相倒置;倒相
phase inversion system 反相方式
phase inversion temperature 相变型温度;转相温度;倒相温度
phase inversion temperature of emulsion 乳状液相转变温度
phase inverter 倒相器;倒相电路;反相器;反相放大器

phase in verter circuit 反相电路
phase jump 相位跃变;相位突变;位相突变
phase keystone 相位梯形
phase lag 相位滞后;相位落后;位相滞后;迟角(潮汐);迟潮时间;潮夕迟角
phase lag angle 相位滞后角
phase lag of tide 潮汐位相滞后
phase lead 相位提前;相位超前;超前角
phase leading 相位超前
phase-lead network 超前网络
phaseless aperture synthesis 无相位孔径综合
phase library 阶段库;段库;程序段库
phase line 相线
phase linear amplifier 线性相位特性放大器
phase linear receiver 线性相位特性接收机
phase load 相负荷
phase localizer 相位式定位器;比相定位标
phase lock 同相;锁相
phase-locked control 相位同步控制
phase-locked demodulator 锁相解调器
phase-locked detection 锁相
phase-locked detector 同相检波器
phase-locked Dopper tracking system 同相多普勒跟踪系统
phase-locked loop 锁相回路;锁相环(路)
phase-locked loop detector 锁相环路检波器
phase-locked oscillator 锁振荡器;锁相振荡器
phase-locked subharmonic oscillator 参数器
phase-locked technique 锁相技术
phase logic 相逻辑;状态逻辑;阶段逻辑
phase map of oil and gas 天然气和石油相图
phase margin 相位余量;相位容量;相补角
phase margin of control system 控制系统的相补角
phase match 相匹配
phase matched for parametric(al) processes 参量过程相位匹配
phase measuring 相位量测
phase measuring equipment 测相设备
phase measuring system 相位量测系统
phase meter 相位计;相位差仪;相位表;相差计
phase method 相位法
phase microscope 相衬显微镜;位相显微镜
phase microscopy 相位显微镜技术
phase modifier 整相器;调相器;调相机
phase-modulated beam 调相光束
phase-modulated signal 调相信号
phase modulation 相位调制;相位调节;调相
phase modulation carrier telegraph 调相式载波电报
phase modulation constant 调相常数
phase modulation detector 调相检波器
phase modulation receiver 调相接收机
phase modulation recording 调相记录
phase modulation system 调相制
phase modulation unit 相位调节装置
phase modulation wave 调相波
phase modulator 调相器
phase module 相位模块;相调制
phase monitor 相位指示器
phase movement matrix 相位运行表
phase name 阶段名
phase-null 零相位
phase null detector 零相位检测器
phase-null meter 零相位计
phase number 相数
phase object 相物体
phase objective 相物镜;相差接物镜
phase of equality 均等相
phase of exploration 勘探阶段;勘察阶段;勘测阶段
phase of periodic(al) quantity 周期相量
phase of prosperity 繁荣阶段
phase of secondary magnetic field 二次磁场相位
phase of sediment transportation 泥砂移动状态
phase of the cycle 循环阶段
phase of the moon 月相
phase of tide 潮相位
phase of total magnetic field 总磁场相位
phase of transport of sediment 泥砂移动状态
phase of use 使用阶段
phase of waves 波相
phase-only hologram 纯相位全息图
phase order 相序
phase oscillation frequency 相振荡频率

phase out 逐步撤出;逐渐停止
phase-out of ozone-depleting substances 逐渐停止消耗臭氧物质
phase-out production 停产
phase-out regime 逐步不用规定
phase overload protection 单相过负荷保护
phase oxidation 相氧化
phase partition 相间隔板;相分配法
phase permeability 相渗透率
phase plane 相位平面;相平面
phase plane analysis 相平面分析
phase plane diagram 相平面图
phase plane method 相平面法
phase plate 相移片
phase point 相点
phase position 相位
phase pre-equalization 相位预均衡
phase program(me) 阶段性计划
phase pulse 相位脉冲
phase quadrature 转像相差;转像差;方照
phaser 相移器;移相器
phase ratio 相比率;相比
phase reaction 相反应
phase-reference alternator 基准相位同步发电机;参考相位同步发电机
phase region 相域
phase regulator 相位调节器;调相机
phase relationship 相位关系;相态关系
phase relay 相继电器
phase resistance 相阻;相电阻
phase resolution 相位分辨率
phase resolver 分相器;分析器
phase resonance 相位共振;相共振
phase response 相响应;相位反应
phase retardation 相延迟
phase reversal 相位倒转;倒相;反相
phase-reversal coding 反向编码
phase-reversal modulation 倒相调制
phase-reversal protection 倒相保护
phase-reversal relay 反相继电器
phase-reversal transformer 倒相变压器
phase-reversing connection 反相连接
phase revolution 相位周值
phase rotation 相序
phase-rotation relay 反相继电器
phase rule 相律;物相定律
phase rule equilibrium 相律平衡
phase sample 相位采样
phase sampling 相位采样
phase segregation 相偏析
phase segregator 分相绝缘器
phase selection 相位选择
phase selector 选相器;相位选择钮
phase sensitive 相位灵敏的;相敏
phase sensitive amplifier 相敏放大器
phase sensitive circuit 相敏电路
phase sensitive demodulator 相敏解调器
phase sensitive detector 相敏整流器
phase sensitive element 相敏元件
phase sensitive modulator 相敏调制器
phase sensitive parametric(al) seismometer 相敏参量地震计
phase sensitive rectifier 相敏整流器;相敏检波器;鉴相器
phase sensitive trigger 相敏触发器
phase sensitivity 相位灵敏度
phase separated interpenetrating polymer network 相分离互穿聚合物网络
phase separation 相分离;析相作用;分相
phase separation glaze 分相釉
phase separation process 相分离工艺
phase separation spinning 相分离纺丝
phase separation tank 分相槽
phase separation temperature 析相温度
phase separator 分离器;分析器
phase sequence 信号相位顺序;相序
phase-sequence generation diagram 相序形成图
phase-sequence relay 对称分量继电器
phase-sequence reversal protection 反相保护
phase servo system 相位伺服系统
phase setting 相位调整
phase setting circuit 移相电路
phase shift 相转变;相移;移相;周相移动
phase shift characteristic 相移特征曲线;相移特性

phase shift circuit 相移电路;移相电路
phase shift coder 相移编码器
phase shift coding 相移编码
phase shift decoder 相移译码器
phase(-shift) delay 相位延迟
phase shift discriminator 相移鉴频器
phase shift distortion 相移失真
phase shifter 移相式振荡器;移相器
phase shift filter 移相滤波器
phase shift frequency curve 相频特性曲线
phase-shifting autotransformer 移相用自耦变压器
phase-shifting circuit 移相电路
phase-shifting control 移相控制
phase-shifting device 移相器
phase-shifting modulation 移相调制
phase-shifting transformer 移相变压器;移相变换器
phase shift keying 相移键控;相移键控
phase shift magnetometer 相移磁强计
phase shift microphone 相移传声器
phase shift migration 相位移偏移
phase shift modulator 相移调制器
phase shift mutation 相移突变
phase shift network 相移网络
phase shift omnidirectional radio range 相移全向无线电指向标
phase shift oscillator 移相式振荡器;相移振荡器;线振荡器
phase shift trigger 移相触发器
phase site 相座
phase solubility 相液解度;相溶解度
phase solubility analysis 相溶解度分析;相溶度分析
phase space 相空间
phase space cell 相空间元件
phase space factor 相空间因子
phase space integral 相空间积分
phase space optics 相空间光学
phase spectrum 相位谱;相谱;位相谱
phase spectrum curve 相位谱曲线
phase speed 相位速度;相速(度)
phase speed method 相速度法
phase split 相位时间比
phase-split enclosed bus 分析封闭母线
phase-split network 分析网络
phase splitter 分相器;分相电路;分析器
phase splitting 分相
phase splitting amplifier 分相放大器;分析放大器
phase splitting device 分析装置
phase splitting stage 分相离级
phase stability 相稳定性;相位稳定性;自动稳相原理
phase-stable acceleration 稳相加速
phase-stable accelerator 稳相加速器
phase-stable angle 稳定相角
phase-stable area 稳定相区
phase state 相态
phase state of migration 运移相态
phase state of migration and solubility of oil and gas 运移相态和油气溶解度
phase step 阶跃
phase structure 相结构
phase-swept interferometer 扫相干涉仪
phase switcher 移相器;移相开关;换相器
phase synchroniazed oscillator 同相振荡器
phase synchronism 相位同步
phase terminal 相引线;相端子
phase thickness of thin-films 薄膜的相位厚度
phase time modulation 相位时间调制
phase titration 相滴定;分相滴定
phase-to-earth 对地
phase-to-phase 相位间的;相间
phase-to-phase quick-break overcurrent protection 相间速断过流保护
phase-to-phase voltage 相间电压;相电压
phase tracker 相位跟踪器
phase transfer function 相位传递函数
phase transfer method 相转移法
phase transfer model 相转移模型
phase transformation 相转变;相位转换;相变;物相变化
phase transformation genesis of earthquake 相变成因
phase transformation range 相变范围
phase transformation toughened ceramic 相变增韧陶瓷
phase transformation toughened zirconia 相变

增韧氧化锆
phase transformer 相位变换器;变相器
phase transition 相转换;相转变;相变
phase transition energy 相变能量
phase transition temperature 相变温度
phase transition under high-pressure 高压相变
phase trapezoid 相位梯形
phase-tuned tube 调相管
phase unbalance 相不平衡
phase-under-voltage relay 低相电压继电器
phase variable 相变量
phase variation 相转变;相位变化;相变异;位相变易
phase velocity 相位速度;相速(度)
phase velocity method 相速度法
phase velocity of wave 波相速度;波的相速(度)
phase voltage 相电压;对中性线电压
phase voltage against earth 对地相电压
phase voltage regulator 相电压调整器
phase volume 相体积
phase volume ratio 相体积比
phase volume theory 相体积理论
phase wave 相波
phase winding 相绕组
phase wound motor 相绕式电动机
phase wound rotor 线绕式转子
phasic change 阶段变化
phasic development 阶段发育
phasic difference 位相差
phasic heterogeneity 阶段异质性
phasic state 位相状态
phasic theory 阶段性理论
phasing 定相;分段处理法
phasing adjustment 定相准准;定相调整
phasing arrangement 相位布置
phasing capacitor 定相电容器
phasing commutator 定相互器;定相变换器
phasing design 分相设计
phasing device 定相器
phasing diagram 分相图
phasing lamp 定相灯
phasing line 定相线
phasing of industrial investment 分阶段安排工业投资
phasing out of obsolete machinery 淘汰旧机器
phasing plug 调相插塞
phasing ring 定相环
phasing sequence 定相序列
phasing signal 整相信号;图像位置调整信号;定相信号
phasing stub 调相短截线
phasing switch 调相开关
phasing synchro 调相同步机;定相自动同步机;变相同步机
phasing technique 调相技术
phasing transformer 相移变压器;移相变压器
phasitron 调相管
phasmajector 单像管
phasograph 测量相位畸变的电桥
phasometer 相位仪;相位计;相位表;功率因数计;功率因数表
phasor 相量;矢量;复数量
phasor argument 相量幅角
phasor current 相量电流
phasor diagram 相量图;矢量图
phasor difference 矢量差
phasor equation 相量方程式
phasor function 相量函数;矢量函数
phasor power 相量功率
phasor power factor 相量功率因数;基波功率因数
phasor product 相量积
phasotron 回旋加速器
phasotropy 氨基氢振动异构
phastograph 塑性计
phathalic acid 酞酸
phaunouxite 芳水砷钙石
pheasantry 养鸡场;鸡舍
phellem 木栓;栓皮栎
phellem layer 木栓层
phelloderm 绿皮层;栓内层
phellogen 木栓形成层
phenacemide 苯乙酰脲
phenaceturic acid 苯乙酰甘氨酸
phenacite 似晶石;硅铍石
phenacyl 苯甲酰甲基

phenacyl alcohol 苯甲酰甲醇
phenacyl bromide 苯甲酰甲基溴
phenacyl chloride 苯甲酰甲基氯
phenacyl ester 苯乙酮酯
phenacyl halide 苯甲酰甲基卤
phenacylidene 苯甲酰亚甲基
phenadoxone 苯吗庚酮
phenaglycodol 氨苯甲基了烷二醇
phenakite 硅铍石
phenakite ore 似晶石铍矿石
phenalgin 氨苯
phenanthrene 菲
phenanthrene ring 菲环
phenanthrenol 菲酚
phenanthrophenazine 二苯吩嗪
phenarsazine chloride 吩吡嗪化氯
phenate 石炭酸盐;酚盐;苯酚盐
phenazine 吩嗪
phenazine dye 吩嗪染料;吖嗪染料
phenazinone 吩嗪酮
phenazinyl 吩嗪基
phenazopyridine 苯基偶氮吡啶二胺
phenbutamide 苯磺丁脲
phencyclidine 苯环利定
phenelzine 苯乙肼
phenethium 酚乙铵
phenethyl 苯乙基
phenethyl acetate 醋酸苯乙酯
phenethyl alcohol 苯乙醇
phenethylamine 苯乙胺
phenethyldiguanide 苯乙双胍;苯乙基二胍
phenetidine 氨基苯乙醚
phenetole 苯乙醚
phenexion 氯烯磷
phenformin 苯乙双胍;苯乙基二胍
phengite 含硅白云母;多硅白云石;多硅白云母
phenicarbazide 苯胺脲
phenindamine 苯茚胺
phenindione 苯茚酮;苯茚二酮
pheniprazine 苯异丙肼
phenmetrazine 苯甲吗啉
phenmetrazine hydrochloride 吩美嗪盐酸盐
phenobarbitone 苯巴比通
phenobutiodil 碘芬布酸
phenocopy 拟表型;表型模拟;表现型模拟
phenocritical period 表型临界期
phenocritical phase 表型临界期
phenocrystal 斑晶
phenocrystalline 斑晶质
phenocrystal mineral 斑晶矿物
phenocryst-free cryptocrystalline texture 无斑隐晶质结构
phenodiazine 二氮
phenoecological spectrum 物候生态谱系
phenogram 物候图;表型图
phenogroup 表现型群
phenol 石炭酸;酚;苯酚
phenol acclimated activated sludge 经酚类化合物驯化的活性污泥
phenol acetaldehyde resin 苯酚乙醛树脂
phenol aldehyde laminated cloth board 酚醛布板
phenolate 酚盐;苯酚盐
phenolated iodine solution 酚制碘溶液
phenolate group 酚盐基
phenolate process 酚盐净化法
phenol benzoate 苯甲酸苯酯
phenol biodegradation 酚生物降解
phenol bismuth 苯氧二羟铋
phenol-blocked isocyanate 酚类封端异氰酸酯
phenol camphor 酚樟脑
phenol cement 苯酚水泥
phenol coefficient 酚系数
phenol coefficient method 酚系数法
phenol compound 酚类化合物
phenol-containing 含酚的
phenol-containing coking wastewater 含酚焦化污水
phenol-containing wastewater 含酚污水
phenol content 酚含量
phenol degradation 酚降解
phenol degrading bacteria 酚降解菌
phenol-derivative herbicides 酚类除草剂
phenol derivatives 苯酚衍生物
phenol diiodide 二碘酚

phenol disulfonic acid 苯酚二磺酸
phenol disulfonic acid method 酚二磺酸分光光度法
phenol disulfonic acid spectrophotometry 酚二磺酸法
phenol effluent 含酚污水
phenol ether 苯酚醚
phenol extraction 苯酚萃取
phenol fabric washer 苯酚纤维垫圈
phenol fiber 酚纤维
phenol-fiber cam 酚纤维凸轮
phenol formaldehyde 酚醛树脂
phenol formaldehyde glue 酚醛树脂胶
phenol formaldehyde plastics 酚醛塑料
phenol formaldehyde resin 苯酚甲醛树脂
phenol formaldehyde resin modified 改性酚醛树脂
phenol furfuraldehyde plastics 苯酚糠醛塑料
phenol furfural resin 苯酚糠醛树脂;苯酚糠醛树脂
phenol-furfurol mo(u)lding compound 苯酚糠醛压塑粉
phenol glue 酚醛胶
phenol glycerin 酚甘油
phenol hardener 酚醛固化剂
phenolic acid 酚酸;酚醛酸
phenolic acid resistance paint 酚醛耐酸漆
phenolic adhesive 酚醛黏结剂;酚醛(树脂)黏合剂;酚醛树脂黏结剂
phenolic adhesive paint 酚醛胶合漆
phenolic alcohol 酚醇
phenolic aldehyde 酚醛
phenolic aldehyde acid-proof varnish 酚醛耐酸漆
phenolic aldehyde insulation varnish 酚醛绝缘漆
phenolic alkyd varnish 酚醛醇酸清漆
phenolic aniline resin 酚苯胺树脂
phenolic anti-oxidant 酚类抗氧剂
phenolic baking insulating paint 酚醛烘干绝缘漆
phenolic bonding medium 酚醛黏结介质
phenolic cement(ing agent) 酚醛胶凝剂;酚醛树脂水泥
phenolic cloth laminate 酚醛布板
phenolic compound 酚类化合物;酚化合物
phenolic constituent 含酚组分
phenolic effluent 含酚废水
phenolic epoxy resin 酚醛环氧树脂
phenolic ester 酚酯
phenolic ether 酚醚
phenolic fiber 酚纤维;酚醛纤维
phenolic fire-retardant paint 酚醛防火漆
phenolic floor paint 酚醛地板漆
phenolic foam 酚醛泡沫
phenolic formaldehyde resin 酚醛树脂
phenolic formaldehyde resin bonded laminate sliding track 胶木(片层压制)滑道
phenolic glue 酚醛胶
phenolic group 酚基
phenolic hydroxyl group 酚式羟基
phenolic insulating paint 酚醛绝缘漆
phenolic laminate 酚醛胶合板
phenolic laminated board 酚醛层压板
phenolic mastic 酚醛胶泥
phenolic mastic pointing 酚醛胶泥勾缝
phenolic medium 酚醛介质
phenolic mo(u)lding compound 酚醛模压混合剂
phenolic mo(u)lding material 酚醛模压材料
phenolic mo(u)lding powder 电木粉;酚醛模压粉
phenolic novolac 可溶可熔酚醛树脂
phenolic paper 酚醛纸
phenolic plastics 酚醛塑料
phenolic pollutant 酚污染物
phenolic pollution 酚污染
phenolic putty 酚醛腻子
phenolic ready mixed paint 酚醛调和漆
phenolic resin 酚醛树脂;防水水胶
phenolic resin adhesive 酚醛树脂黏剂;酚醛树脂胶合剂;酚醛树脂胶;酚醛树脂黏结剂
phenolic resin baking varnish 酚醛树脂烘干清漆
phenolic resin bonded 酚醛树脂黏结的
phenolic resin bonded plywood 酚醛树脂胶合板
phenolic resin cement 酚醛树脂黏结剂;酚醛树脂胶黏剂;酚醛树脂胶泥;酚醛树脂合剂;酚醛树脂胶;酚醛树脂水泥
phenolic resin coating 酚醛树脂涂料
phenolic resin enamel 酚醛树脂瓷漆

phenolic resin fibre 酚醛树脂纤维
phenolic resin foam 酚醛树脂泡沫(塑料)
phenolic resin glue 酚醛树脂胶
phenolic resin paint 酚醛树脂漆
phenolic resin plastics 酚醛树脂塑料
phenolic resin production wastewater 酚醛树脂生产废水
phenolic resin varnish 酚醛树脂凡立水;酚醛树脂清漆
phenolic resin vehicle 酚醛树脂载色剂;酚醛树脂展色料;酚醛树脂媒液
phenolics 酚醛塑料;酚醛树脂
phenolic ship cabin paint 酚醛船舱漆
phenolic-silicone resin 酚醛—缩硅酮树脂
phenolic synthetic(al) resin cement 酚醛合成树脂水泥;酚醛合成树脂胶
phenolic synthetic(al) resin putty 酚醛合成树脂油灰;酚醛合成树脂腻子
phenolic tubing 酚醛管
phenolic varnish 酚醛清漆
phenolic vehicle 酚醛溶剂
phenolic waste (liquor) 含酚废液
phenolic wastewater 含酚污水;含酚废水
phenolic water drum 苯酚水槽
phenolization 酚处理
phenol lagoon pollution 氧化塘酚污染;含酚水塘污染
phenol mass concentration 苯酚质量浓度
phenol-novolac epoxy resin 可溶可熔酚醛环氧树脂
phenological calendar 物候历
phenological chart 物候图
phenological dates 物候期
phenological isolation 物候的隔离
phenological observation 物候观测
phenological phase 物候期
phenological spectrum 物候谱(系)
phenological succession 物候演替
phenology 物候学;物候现象;物候
phenol oil 酚油;酚润滑油;苯酚润滑油
phenol ointment 酚软膏
phenol photodegradation 酚光降解
phenol photooxidation 酚光氧化
phenolphthalein 酚酞
phenolphthalein alkalinity 酚酞碱度
phenolphthalein indicator 酚酞指示剂
phenolphthalein sodium 酚酞钠
phenolphthalein test 酚酞试验
phenolphthalein test paper 酚酞试纸
phenolphthalexone 酚酞络合剂
phenolplast 酚醛塑料
phenol plastics 酚醛塑料
phenol poisoning 石炭酸中毒;酚中毒
phenol pollutant 酚污染物
phenol pollution 酚污染
phenol reagent 酚试剂
phenol reagent method 酚试剂法
phenol red 酚红;苯酚磺酞
phenol removal process 除酚工艺;除酚法
phenol-resin glue 酚醛树脂胶
phenol resinifying solution 酚醛树脂塑化液
phenol resin laminate 酚醛树脂叠层
phenol resinous glue 酚脂胶黏剂
phenol resins 酚树脂
phenol resin sand 酚醛树脂砂
phenol resion laminate 酚醛树脂层板
phenol resorcinol 苯酚—间苯二酚树脂胶黏剂
phenols 酚类
phenol safranine 酚藏红花
phenol storage 苯酚储[贮]罐
phenol sulfatase 苯硫酸酶
phenolsulfonic caid 酚磺酸
phenol-sulfonic acid resin 苯酚磺酸树脂
phenol sulfonphthalein 酚磺酞;酚红;苯酚磺酞
phenolsulfonphthalein standards 酚红标准液
phenolsulfonphthalein test 酚碘酞试验
phenol sulfuric acid 苯酚硫酸酯
phenolsulphonphthalein injection 酚磺钛针
phenol tricarboxylic acid 苯酚三羧酸
phenol value 酚值
phenol value of activated carbon 活性炭的吸酚值
phenol waste(liquor) 含酚废液
phenol waste(water) 含酚废水
phenol wastewater disposal 含酚污水处理

phenol wastewater treatment 含酚污水处理;含酚废水处理
phenol water disposal 含酚水处置;酚水处理
phenol water extraction 酚水提取法
phenol water treatment 含酚水处理
phenolysis 酚解
phenomenonl difference 形式上的差别
phenomenonl sea 汹涛;九级浪
phenomenon of coagulation 凝结现象
phenomenological analysis 现象分析
phenomenology 现象学
phenomenon 现象;天象
phenomenon of bank suction 岸吸现象
phenomenon of diffraction 衍射现象
phenomenon of diffusion 扩散现象
phenomenon of fluorescence 荧光现象
phenomenon of interference 干涉现象
phenomenon of rupture 断裂现象
phenomic lag 表型迟延现象
phenonaphthazine 苯并吩嗪
phenones 苯某酮
phenonium ion 苯氧离子
phenoperidine 苯哌利定
phenophase 物候期
phenophenanthrazine 苯并菲嗪
pheno-plast 酚醛塑料
pheno-plastics 酚醛塑料;酚基塑料
phenoquinone 苯酚合苯醌
phenospectrum 物候谱(系)
phenostal 草酸二苯酯
phenothalin 酚酞
phenothiazine 硫代二苯胺;吩噻嗪
phenothiazine type anti-oxidant 吩噻嗪型抗氧化剂
phenothiazinyl 吩噻嗪基
phenothiazone 吩噻嗪酮
phenothioxin 吩噻恶
phenotype 显型;表型;表现型
phenotypical plasticity 表型可塑性
phenotypical ratio 表现型比值
phenotypical reversion 表型反转
phenotypic classification 表型分类
phenotypic correlation 表型相关
phenotypic lag 表型迟延现象
phenotypic mixing 表型混合
phenotypic number 表型值
phenotypic ratio 表型比值
phenotypic score 表现型等级
phenotypic selection 表型选择
phenotypic variance 表型方差
phenotypic variance recording 表型方差记录
phenoweld 改性酚醛树脂黏合剂
phenoxazine 吩恶嗪
phenoxetol 苯氧基乙醇
phenoxide 酚盐
phenoxide titanium 苯氧基钛
phenoxthine 吩噻恶
phenoxy acetaldehyde 苯氧基乙醛
phenoxy-ace-tamide 苯氧基乙酰胺
phenoxy acetic acid 苯氧基乙酸
phenoxy acetone 苯氧基丙酮
phenoxybenzamine 苯氧苄胺
phenoxybenzamine hydrochloride 苯氧苄胺盐酸盐
phenoxy benzoic acid 苯氧基苯甲酸
phenoxy-butyronitrile 苯氧基丁腈
phenoxyethanol 苯氧乙醇
phenoxyl 苯氧基
phenoxy propandiol 苯氧基丙二醇
phenoxypropane diol 苯氧基丙二醇
phenoxypropyl penicillin potassium 苯丙西林
phenoxy resin 苯氧树脂;苯基树脂
phenpiazine 对二氮萘
phenprobamate 苯丙氨酯
phenprocoumon 苯丙香豆醇
phenprofone 苯丙砜
phenprofonum 苯丙砜片
phenpromethamine 二甲基苯乙胺;苯丙甲胺
phentermine 苯丁胺
phenthiazone 吩噻嗪酮
phenthiol 苯硫酚
phentriazine 苯并三嗪
phenyl 苯基
phenylacetaldehyde 苯乙醛
phenyl acetaldehyde dimethyl acetal 苯乙二甲缩醛

phenyl acetamide 苯乙酰胺
phenyl acetate 醋乙酸苯酯
phenyl acetic acid 苯乙酸;苯基乙酸;苯醋酸
phenyl-acetone 苯基丙酮
phenylacetyl 苯乙酰
phenylacetyl chloride 苯乙酰氯
phenylacetylene 苯乙炔
phenylacetylglutamine 苯乙酰谷氨酰胺
phenylacetylglycine 苯乙酰甘氨酸
phenylacrylic acid 苯乙丙烯酸
phenylalaninase 苯基丙氨酸酶
phenylalanine 苯基丙氨酸;苯丙氨酸
phenylalanine mustard 苯基丙氨酸氮芥;苯丙氨酸氮芥
phenyl aldehyde 苯甲醛
phenyl-allylene 苯丙炔
phenylamide pesticide 苯胺杀虫药
phenylamine 苯胺
phenylaminocadmium dilactate 乳胺镉
phenylaminopropionic acid 苯基丙氨酸
phenylaniline 苯基苯胺
phenylanilineurea 苯基苯胺脲
phenylanthracene 苯基蒽
phenyl-arsenimide 苯砷亚胺
phenyl-arsenite 苯亚胂酸盐
phenyl-arsenoxide 苯亚胂氧化物;苯基氧化砷
phenyl-arsine oxide 苯亚胂氧化物;苯胂化氧
phenyl-arsine sulfide 苯胂硫
phenylate 苯醚
phenylating agent 苯基化剂
phenylation 苯基化作用
phenylazo- 苯偶氮基
phenylazobenzene 偶氮苯
phenylbenzene 联二苯
phenylbenzoic acid 苯基苯甲酸
phenylbenzyldimeth lammonium chloride 苯基苄基二甲基氯化铵
phenylborine 苯基甲硼烷
phenylboronate 苯基硼酸盐
phenylbutazone 苯基丁氮酮;苯丁唑酮;苯丁唑啉
phenyl carbamate 氨基甲酸苯酯
phenylcarbamoyl 苯氨羰基
phenylcarbinol 苯甲醇
phenyl cellosolve 乙二醇苯醚;苯基溶纤剂
phenylchinaldine 苯基甲基喹啉
phenylchloride 苯基氯
phenyl-chloroform 苯基氯仿
phenyl chlorosilane 苯基氯硅烷;苯基硅烷
phenyl compound 苯基化合物
phenyl concentration 苯酚浓度
phenylcyanamide 苯氨腈
phenyl cyanide 苄腈;苯基氰
phenylcyclohexane 苯基环己烷
phenylcyclohexanol 苯基环己醇
phenyldicarbinol 苯二甲醇
phenyldichlorarsine 二氯化苯胂
phenyldichloroarsine 苯基二氯胂
phenyldichlorophosphine 苯基二氯磷
phenyl didecyl phosphite 亚磷酸苯基二癸酯
phenyl-difluoride 二氟化苯
phenyl diglycol carbonate 苯基二甘醇碳酸酯
phenyl-dihydroxyarsine 苯亚胂酸
phenyl-diiodide 二碘苯
phenyldimethanic acid 苯二甲酸
phenylene 苯撑
phenylene-diacetic acid 苯二乙酸
phenylenediamine 苯二胺
phenylene-diamine oxidase 苯二胺氧化酶
phenylene-diarsonic acid 苯二胂酸
phenylene diazo 苯双偶氮基
phenylene diazosulfide 苯双偶氮硫
phenylene diisocyanate 亚苯基二异氰酸酯
phenylenedimethylene 苯二甲基
phenylene silicone rubber 苯撑硅橡胶
phenylethane 乙苯
phenylethanol 苯基乙醇
phenylethanol amine 苯乙醇胺
phenylethyl acetate 醋酸苯乙酯
phenylethyl acetourea 苯乙基乙酰脲
phenylethyl amine 苯乙胺
phenylethyl butyrate 丁酸苯乙酯
phenylethyl carbinol 苯乙丙醇
phenylethylene 苯乙烷;苯亚乙烯;苯亚乙基
phenylethylhydrazine 苯乙肼

phenyl fluoride 苯基氟
phenylfluorone 苯基荧光酮
phenylfluorone spectrophotometry 苯芴酮分光光度法
phenylformic acid 苯甲酸；安息香酸
phenylglycine-o-carboxylic acid 苯基甘氨酸邻羧酸
phenylglycocoll 苯基甘氨酸
phenylglyoxal 苯甲酰甲醛
phenylglyoxylic acid 苯酰甲酸
phenyl group 苯基
phenyl-halide 苯基卤
phenylhydrazine 苯肼
phenylhydrazine levulinic 苯肼乙酰丙酸
phenylhydrazine-p-sulfonic acid 苯肼对磺酸
phenylhydrazine sulfate 苯肼硫酸盐
phenylhydrazine urea 苯肼脲
phenyl hydrazones 苯腙类
phenyl hydrazoquinoline 苯肼基喹啉
phenyl-hydrogen sulfate 硫酸苯氢酯
phenylhydroxylamine 苯基羟胺
phenylic acid 石炭酸；酚
phenyl isocyanate 异氰酸苯酯
phenyl isothiocyanate 苯基芥子油
phenyl ketone 二苯甲酮
phenyllactazam 苯基内酰联胺
phenyl-lactic acid 苯乳酸
phenyl-magnesiumhalide 苯基卤化镁
phenylmalonate 苯丙二酸
phenylmercuric acetate 乙酸苯汞；醋酸苯汞
phenylmercuric acetate poisoning 醋酸苯汞中毒
phenylmercuric acrylate 丙烯酸苯汞
phenylmercuric bromide 苯基溴化汞
phenylmercuric chloride 苯基氯化汞
phenylmercuric hydroxide 氢氧化苯汞
phenylmercuric iodide 苯基碘化汞
phenylmercuric nitrate 硝酸苯汞
phenylmercuric oleate 油酸苯汞
phenylmercuric propionate 丙酸苯汞
phenylmercuric salicylate 水杨酸苯汞
phenylmercuric salt 苯基汞化卤
phenyl mercuriethanolammonium acetate 醋酸苯汞乙醇铵
phenylmercury borate 硼酸苯汞
phenylmercury chloride 氯化苯汞
phenylmercury succinate 琥珀酸苯汞
phenyl-methane 甲苯
phenyl methyl acetate 乙酸苄酯
phenyl methyl carbinol 苯基甲基醇
phenyl methyl glycidyl ether 苯基甲基缩水甘油醚
phenyl methyl ketone 苯基甲酮
phenyl methyl silicone 苯甲基硅酮
phenyl methyl silicone oil 苯基甲基硅油
phenyl monochloride 一氯（代）苯
phenyl mustard oil 苯基芥子油
phenyl neopentyl phosphite 亚磷酸苯新戊酯
phenyl-nitramine 苯硝胺
phenyl nitromethane 苯基硝基甲烷
phenylnitrone 苯基硝酸灵
phenylo boric acid 苯硼酸
phenylog 联苯物
phenylol 苯酚类
phenyl salicylic acid 苯基水杨酸
phenyloxamic acid 苯胺羰酸
phenyl-paraffin alcohols 苯基脂醇
phenyl pentane 苯基戊烷
phenyl phenol 苯基苯酚
phenyl phenol phenolic resin 苯基苯酚酚醛树脂
phenyl phosphate 磷酸苯酯
phenyl phosphine 苯膦
phenylphosphonic acid 苯膦酸
phenyl phosphorodiamidate 苯基磷二酰胺
phenyl polychloride 多氯代苯
phenyl polyhalide 多卤代苯
phenyl potassium 苯基钾
phenylpropanol 苯丙醇
phenylpropanolamine 苯丙醇胺
phenyl propionate 丙酸苯酯
phenylpropionic acid 苯基丙酸
phenylpropyl 苯丙基
phenylpropyl alcohol 苯丙醇；苯丙醇
phenylpropyl aldehyde 苯丙醛；苯丙醛
phenylpropyl ketone 苯丙基甲酮
phenylpyrazolone 苯吡唑酮
phenylpyridine 苯基吡啶

phenyl pyruvic acid 苯（基）丙酮酸
phenylquinoline 苯基喹啉
phenyl salicylate 水杨酸苯酯；萨罗
phenyl salicylic acid 苯基水杨酸
phenyl silane resin 苯基硅烷树脂；苯硅烷树脂
phenyl silicone 苯基硅酮
phenylsiloxane 苯基硅氧烷
phenylsuccinic acid 苯基丁二酸
phenyl-sulfamic acid 苯氨基磺酸
phenylsulfamoyl 苯氨基磺酰
phenylsulfinyl 苯亚磺酰
phenyl sulfone 二苯砜
phenylsulfonyl 苯磺酰
phenylthioalcohol 苯硫酚
phenylthiocarbamide 苯基硫脲
phenylthiocarbonimide 苯基芥子油
phenylthiohydantoic acid 苯异硫脲基乙酸
phenylthioisocyanate 苯异硫氰酸
phenylthiosemicarbazide 苯氨基硫脲
phenylthiourea 苯硫脲
phenyl toluene sulphonate 甲苯磺酸苯酯
phenyl trichlorosilane 苯基三氯硅烷
phenyl tri-ethoxysilane 苯基三乙氧基硅烷
phenylurea 苯基脲
phenyl urethane 苯氨基甲酸乙酯
phenyl vinyl ether 苯基乙烯基醚
phenyl-xanthogenic acid 苯基黄原酸
phenyl-β-naphthylamine 苯基贝塔萘胺
phenyt benzoate 苯甲酸苯酯
phenythilone 苯西酮
phenytoin 二苯乙内酰脲
pheophoebin 脱镁叶绿酸
pheophorbide 脱镁叶绿素
phermitocorundum 熔剂刚玉
phermitospinel 熔剂尖晶石
pheromone 信息素
phethenylate 苯噻妥英
pneumatic fibre separator 风动纤维分离机
phial 管形瓶；管瓶
phiale 古希腊船；喷水池（教堂前）
phialiform 碟形
phi grade scale 中粒级标度；φ粒度分级标准
Philadelphia level(l)ing rod 费拉德尔菲亚水准尺
Philadelphia rod 费拉德尔菲亚水准标尺
Philadelphia Stock Exchange 费城证券交易所
philanthropic hospital 慈善医院
philanthropic institution 慈善机构
philbrick 防水溶液
philharmonic hall 音乐厅
Philibert roof truss 菲利波特屋架
Philibert truss 菲利波特桁架
philippeion 腓力比圆殿（位于希腊奥林比亚）
philippeum = philippeion
Philippine basin 菲律宾海盆
Philippine cedar 菲律宾雪松
Philippine ebony 菲律宾乌木
Philippine geosyncline 菲律宾地槽
Philippine mahogany 菲律宾红木；菲律宾红柳安
Philippine ocean basin block 菲律宾海盆巨地块
Philippine Sea 菲律宾海
Philippines plate 菲律宾板块
Philippine teak 菲律宾柚木
Philippine trench 菲律宾海沟
Philippine-type structure 菲律宾入字形构造
Philips band 菲利浦谱带
philipsbornite 菲利浦博石
Philips curve 菲利浦曲线
Philips driver 十字螺丝刀
Philips ga(u)ge 菲利浦真空计
Philips hot-air engine 菲利浦热空气发动机
Philips process 菲利浦法
Philips screw 带十字槽头的螺钉；菲利浦螺钉
Philips screwdriver 菲利浦式旋凿
Philips stabilization function 菲利浦稳定函数
Philips sulfur dioxide monitor 菲利浦二氧化硫监测仪
philipstadite 淡铁镁闪石
Philleo factor 菲里系数
Phillips head 十字槽头螺钉；十字槽螺母
Phillips head screw 十字槽螺钉头
phillipsite 斑铜矿；钙十字沸石
Phillips recess 螺丝头十字凹槽
Phillips recessed-head screw 十字槽埋头螺钉
Phillips screw 十字槽头螺钉；十字槽头螺钉

Phillips screwdriver 十字头螺丝刀
philosophy 基本定律
philosophy of architectural design 建筑设计原理
philosophy of art 艺术哲学
philosophy of ecology 生态学哲学
philosophy of environment(al) sciences 环境科学哲学
philosophy of measurement 测量方法
Philplug adaptor 菲尔插头
philsim 菲尔西母合金
pH-indicator 酸碱指示剂
phi rotation φ角
phi scale 中标度
phlebite 脉成岩类
phlegma 冷凝液
phlobaphene 红粉
phlobaphinite 鞣质体
phloem fibre 韧皮纤维
phlogistication 除氧（作用）；除氧
phlogiston 燃素
phlogopite 金礦石；金云母
phlogopite beforsite 金云母镁云碳酸岩
phlogopite deposit 金云母矿床
phlogopite leucitite 金云母白榴岩
phlogopite marble 金云母大理岩
phlogopite olivine sharn 金云母橄榄石矽卡岩
phlogopite rauhaugite 金云母白云（石）碳酸岩
phlogopite serpentinite 金云母蛇纹岩
phlogopite sharn 金云母矽卡岩
phlorite 红块高岭土
phloroglucin(e) 间苯三酚
phloroglucinol 间苯三酚
phlorol 邻乙基（苯）酚
phloxine 焰红染料
phloxine lake 弗洛克辛色淀
phluometry 几何辐射度学
pH meter 氢离子浓度计；氢离子计；酸碱计；pH值测定仪；pH计；酸度计
phobic solvent 疏质子溶剂
Phoebus 太阳神（希腊神话）
phoelectron 光电子
phoenicite 红铬铅矿
phoenicochroite 红铬铅矿
phoenix 凤凰饰
Phoenix alloy 菲尼克斯合金
phoenix caisson 凤凰式沉箱
phoenix column 弓形槽钢组合圆柱
phoenix tree 梧桐
pH of maximum adsorption 最大吸附pH值
phogtmagnetoelectric 光磁电的
phokoscope 透镜组聚光能力测定计
pholad(idae) 蛀石（海）虫；穿石贝
pholas 甲壳饰
pholerite 大岭石
pholidolite 硬金云母
pholidosis 鳞序
Pholin's alloy 菲林合金
pholographic(al) deconvolution 全息去褶积
phon 响度单位；方（声音响度单位）
phonatome 鸣谱
phonautograph 声波振动记录仪；声波记振仪
phone booth 电话间；公用电话亭
phone duct 电话线管道
phone exchange 电话交换（台）
phone installation 电话装置；电话设施
phone jack 电话插口
phone line 电话线
phonemeter 通话记录器；测声仪；测声计
phone operator 电话接线员
phonepatch 电话插接线
phone point 电话安装点；电话插座；电话插塞
phone set 电话机
phone socket 电话插座；电话接线盒
phonetic typewriter 口授打字机
phonetic value 音值
phonetic writing 音标
phone traffic 电话传达
phonevision 电话电视
phoney 假货
phoney loans 空头债
phonic motor 蜂音电动机
phonics 声学
phonite 霞石
phonmeter 声强度计；测声器

Phono-bronze 铜锡系合金；福诺青铜
phonochemical reaction 声化反应
phonochemistry 声化学
phonodeik 声波显示仪；声波显示器
phonoelectric(al) bronze 高导电率锡镉青铜
phonogram 标音字母
phonograph 声波自动记录仪
phonograph cutter 刻纹头
phonographic(al) recorder 录音机
phonolite 响岩
phonolite group 响岩类
phonolitic texture 响岩结构
phonolitic trachyte 响岩质粗面岩
phonolitic trachyte porphyry 响岩质粗面岩斑岩
phonological observation 物候观察
phonological season 物候季节
phonological simulation model 物候仿真模型
phonolysis 声解作用
phonometer 音强度计；响度计；声响度计；声强计；测音仪；测音计；测声器
phonometry 声强测量法
phonomotor 声发动机
phonon 声子
phonon coupled level 声子耦合能级
phonon emission 声子发射
phonon excitation 声子激发
phonon model 声子模型
phonon parametric(al) amplifier 声子参量放大器
phonon reflecting grating 声子反射光栅
phonon spectrometer 声子谱仪
phonon spectrum 声子谱
phonon thermal conductivity 声子热导率
phonooptic(al) ceramics 声光陶瓷
phonophone 振子频率调节器
phonophonesis 超声波透入
phonophore 报话合用机
phonophote 音波发光机；声波发光机
phonophotogram 声波照相记录
phonophotography 声波照相术；声波照相法
phonopore 报话合用机
phonoprojectoscope 声音投映示波器
phonoreception 声感
phonorecord 唱片
phonoscope 验声器；测声器；验电器；声波自动记录仪(记录振动波形)
phono-selectoscope 声频选择器(滤去低频声波)
phonosensitive 感声的
phonosynthesis 声合成
phono-telemeter 声测距仪
phonotelemetry 无线电声学遥测术；声遥测(技)术；声学遥测(技)术
phonotropism 向声性
phonovision 电视电话
phonozenograph 水声定位器；声测角计；声波定位器；声波测向器
phon scale 方的标度
phony mine 假地雷
phopshonic acid 膦酸
phoral 铝磷合金
phorate 甲拌磷
phorbin 脱镁叶绿母环类
Phoredox process 福列德克斯工艺
phoresis 电泳现象
phoresy 携播；传运
phorogenesis 滑移作用
phorographic record 照相记录法
phorone 二异亚丙基丙酮
phoronomics 纯运动学
phoronomy 纯运动学
phos-copper braze alloy 磷铜合金硬焊料
phosfolan 硫环磷
phosgene 光气；碳酰氯
phosgene poisoning 光气中毒
phosgenismus 光气中毒
phosgenite 角铅矿
phoshofibrite 纤磷石
phosinaite 磷硅铈钠石
Phosnic bronze 福斯利克含磷镍青铜；磷镍青铜
phospham 磷胺
phosphamidon 磷胺
phosphaminase 氨基磷酸酶
phosphate 磷酸盐矿石；磷酸盐
phosphate acid ion 磷酸根
phosphate based solution 磷酸盐溶液

phosphate-bearing sandstone 含磷砂岩
phosphate binder 磷酸盐黏合剂；磷酸盐胶结料
phosphate-bonded brick 磷酸盐结合砖
phosphate-bonded refractory castables 磷酸盐耐火浇注料
phosphate bronze bar 磷青铜棒
phosphate bronze casting 磷青铜铸件
phosphate bronze tube 磷青铜管
phosphate buffer 磷酸盐缓冲液
phosphatebuffer buffer system 磷酸盐缓冲系统
phosphate cellulose 磷酸纤维素
phosphate cement 磷酸盐水泥
phosphate chalk 磷灰岩(土)；磷钙石
phosphate chromate treatment 磷酸铬酸(盐表面)处理(法)
phosphate chrome 磷铬黄
phosphate coagulation 磷酸盐凝聚(作用)
phosphate coating 磷酸盐处理；磷化处理；磷酸盐涂层；磷化膜；磷化护层(法)；磷化层
phosphate compounds 磷酸盐化合物
phosphate-containing detergent 含磷酸盐洗涤剂
phosphate-containing detergent apparatus 含磷酸盐洗涤器
phosphate crown glass 磷酸盐冕(牌)玻璃
phosphate deficiency index 缺磷指数
phosphate desulfurization 磷酸盐脱硫
phosphate ester 磷酸酯；磷酸盐酯
phosphate etch primer 磷酸盐反应性底漆；磷酸盐蚀刻底涂料
phosphate fertilizer 磷酸盐肥料；磷肥
phosphate fertilizer plant 磷肥厂
phosphate fiber 磷酸盐纤维
phosphate-free detergent apparatus 不含磷酸盐洗涤器
phosphate glass 磷酸盐玻璃
phosphate group 磷酸基
phosphate industry waste 磷酸盐工业废料
phosphate investment 磷酸盐包埋料
phosphate ion 磷酸盐离子；磷酸根离子
phosphate ions removal 除磷酸根离子
phosphate melt 磷酸盐熔体；磷酸盐熔化物
phosphate method 磷酸盐法
phosphate of lime 磷酸钙
phosphate opacified glaze 磷酸盐乳白釉
phosphate optic(al) glass 磷酸盐光学玻璃
phosphate phosphorus 磷酸磷
phosphate pickling agent 磷化处理液
phosphate plasticizer 磷酸酯类增塑剂
phosphate porcelain 磷酸盐瓷
phosphate process 磷酸盐法
phosphate proteciton coating 磷酸盐保护膜
phosphate radical 磷酸根
phosphate rock 磷盐岩；磷酸岩；磷酸岩；磷钙土
phosphate sediments 磷酸盐沉积物
phosphate softening 硫酸盐软化法
phosphate treatment 磷酸盐处理；磷化处理
phosphate wastewater 磷酸盐废水
phosphatic cement 磷酸盐胶结物
phosphatic concretion 磷质结核
phosphatic deposit 磷质沉积
phosphatic manure 磷肥
phosphatic nodule 磷质结核；磷酸盐结核
phosphatic rocks 磷质岩
phosphatic sediment 磷质沉积
phosphatide 磷脂
phosphating 磷酸盐化；磷酸盐处理；磷酸膜防(腐)蚀处理；磷酸敷层；磷酸处理(用热磷酸保护金属表面)
phosphating agent 磷酸盐化剂；磷化剂
phosphating coat 磷化(处理)膜
phosphating process 磷酸盐处理法
phosphating treatment 磷酸盐表面处理
phosphatization 磷酸盐化
phosphatized 磷化处理的
phosphatizing 渗磷
phosphatoptosis 磷酸盐沉积
phosphenyl 苯膦基
phosphide 磷化物
phosphide eutectic 磷共晶
phosphine 磷化氢；碱性染革黄棕
phosphine poisoning 磷化氢中毒
phosphinic acid 次磷酸
phosphinidyne 次磷基
phosphite 亚磷酸盐

phosphite antioxidant 亚磷酸酯类抗氧剂
phosphoantimonomolybdic heteropolyacid 磷锑钼杂多酸
phosphoarsenic heteropoyacid 磷砷杂多酸
phosphoferrite 磷铁锰矿；水磷铁石
phosphoglycerate 磷酸甘油酯
phosphogypsum 磷石膏
phospholipid 磷脂
phospholipids 磷脂类
phospholipids concentration 磷脂浓度
phospho-lipin 磷脂
phosphomolybdenum blue 磷钼蓝
phosphomolybdenum blue colo(u)rimetry 磷钼蓝比色法
phosphomolybdenum blue spectrophotometry 磷钼蓝分光光度法
phosphomolybdic acid 磷钼酸
phosphomolybdic heteropolyacid 磷钼杂多酸
phosphomolybdic pigment 磷钼颜料
phosphonate 膦酸盐
phosphonic acid 氨甲基膦酸
phosphonic salt 膦酸盐
phosphonitrilic resin 磷腈树脂
phosphonoacetone 二氧磷基丙酮
phosphono propionic acid 膦酰基丙酸
phosphophyllite 磷叶石
phosphor 荧光物质；荧光剂；磷光质；磷光物质；磷光体；黄磷；无机发光材料
phosphoraite 磷块岩
phosphoramidic acid 氨基磷酸
phosphorated 含磷的
phosphorated material 含磷材料
phosphor bronze 磷青铜
phosphor bronze wire cloth 磷铜丝布
phosphor burning 荧光粉烧伤
phosphor coated screen 荧光屏；磷光屏
phosphor-coated tube 荧光管；涂磷管
phosphor colo(u)r response 荧光粉彩色响应
phosphor copper alloy 磷铜合金
phosphor-decay time 磷光余辉时间；磷光衰减时间
phosphor dot 荧光点；磷光点
phosphor dot array 荧光点阵；磷光点阵；嵌镶荧光屏
phosphor dot faceplate 磷光点平面
phosphor dot plate 磷嵌镶屏
phosphor dots 磷光体点阵
phosphor dot trio 三色磷光点组
phosphorescence 荧光现象；磷叶石；磷光(现象)；磷肥
phosphorescence afterglow 荧光余辉；磷光余辉
phosphorescence analysis 磷光分析
phosphorescence band 磷光带
phosphoarescence effect 磷光效应
phosphorescence intensity 磷光强度
phosphorescence of dyes 染料的磷光
phosphorescence quenching 磷光猝灭
phosphorescence spectrum 磷光谱；磷光光谱
phosphorescence variation detector 磷光变化探测器
phosphorescent 磷光质；磷光性；发磷光的
phosphorescent coating 夜光涂料；磷光涂料
phosphorescent decay 磷光余辉
phosphorescent detector 磷光辐射探测器
phosphorescent glow 磷光现象；磷光
phosphorescent light 磷光
phosphorescent material 发磷光物质
phosphorescent paint 磷光涂料；磷光漆
phosphorescent pigment 夜光颜料；夜光涂料；磷光颜料
phosphorescent screen 荧光屏；磷光屏
phosphorescent sign 发磷光的标志
phosphorescent substance 磷光体
phosphorescope 磷光镜
phosphoretic steel 磷钢
phosphoria formation 含磷组
phosphoric 磷的
phosphoric acid 磷酸
phosphoric acid-based flux 磷酸基助焊剂；磷酸基助熔剂
phosphoric acid by furnace process 热法磷酸
phosphoric acid by wet process 湿法磷酸
phosphoric acid-diatomite catalyst 磷酸硅藻土催化剂
phosphoric acid ester pesticide 磷酸酯类农药

phosphoric acid method 磷酸法
phosphoric acid pickling method 磷酸浸渍(除锈)法
phosphoric amine-containing wastewater 含磷胺废水
phosphoric bronze 磷酸铜
phosphoric crown glass 磷酸冕玻璃
phosphoric flint glass 磷酸铅玻璃;磷酸火石玻璃
phosphoric manure 磷肥
phosphoric(pig) iron 高磷生铁
phosphorimetric analysis 磷光分析法
phosphorimetry 磷光光度法;磷光分析法;磷光测量(法)
phosphorise = phosphorize
phosphoriser 钟罩
phosphorism 磷中毒;慢性磷中毒
phosphorite 磷灰岩;磷钙土
phosphorite deposit 磷块岩矿床
phosphorite facies 磷质岩相
phosphorite ore 磷块岩矿石
phosphorite rock 磷灰岩
phosphorite terminal 磷矿码头
phosphorization 增磷;磷化作用
phosphorized copper 磷铜
phosphorize 磷化
phosphor laser 磷光体激光器
phosphor material 制备磷光体的材料
phosphoro-amidate 氨基磷酸酯;氨基磷酸盐
phosphorodithioate 二硫代磷酸酯
phosphorodithioic acid 二硫代磷酸
phosphorogen 磷光增强剂
phosphorograph 磷光像;磷光画
phosphorography 磷光照相术;磷光画法
phosphorolysis 磷酸分解(作用)
phosphorometer 磷光计
phosphorophotography 磷光照相术;磷光摄影学
phosphoroscope 磷光计;磷光测定器
phosphorosilicate glass 磷硅玻璃
phosphorous 含磷的
phosphorous anhydride 三氧化二磷
phosphor(ous) bronze 磷青铜
phosphorous compound 磷化合物
phosphor(ous) copper 磷铜
phosphorous hydride 磷化氢
phosphorous model of lake and reservoir 湖泊水库磷模型
phosphorous pentoxide 五氧化二磷
phosphorous recovery 磷回收
phosphorous reduction 磷降低
phosphorous removal 脱磷(法);除磷酸盐
phosphorous removal efficiency 除磷效率
phosphor pattern 荧光图案;磷光图样
phosphorpenia 磷质减少
phosphor persistence 荧光体余辉;荧光粉余辉
phosphor plate 磷光板
phosphor powder 荧光粉;磷光粉
phosphor powder saturation 磷光粉饱和
phosphor printing 涂覆磷光体印刷法
phosphor removal 脱磷
phosphorroessierite 水重磷镁石;磷氢镁石
phosphor saturation 荧光粉饱和度
phosphor-solder 磷光焊剂
phosphor tin 磷锡合金
phosphorus adsorption 磷吸附
phosphorus alloy 磷合金
phosphorus and nitrogen uptake 磷和氮吸收
phosphorus anhydride 三氧化二磷
phosphorus banding 磷光带
phosphorus-bearing formation 含磷建造
phosphorus-bearing slag 含磷炉渣
phosphorus-bearing waste 含磷废料
phosphorus-bearing wastewater 含磷废水
phosphorus bromonitride 二溴氮化磷
phosphorus chloride 氯化磷
phosphorus-containing formation 含磷建造
phosphorus-containing inorganic ion exchanger 含磷无机离子交换剂
phosphorus-containing sandstone 含磷砂岩
phosphorus-containing slag 含磷炉渣
phosphorus-containing waste 含磷废料
phosphorus-containing wastewater 含磷废水
phosphorus content graduation 磷含量分级
phosphorus copper 磷铜
phosphorus cycle 磷循环

phosphorus deprivation test 磷剥夺试验
phosphorus detector 磷检测器
phosphorus dichloride 二氯化磷
phosphorus donor complex 磷供体化合物
phosphorus-doped optic(al) fiber 掺磷光纤
phosphorus doped silica 掺磷二氧化硅
phosphorus enrichment ratio 磷浓缩率
phosphorus fixation 磷固定
phosphorus form 磷形态
phosphorus glass 磷玻璃
phosphorus gypsum residue 磷石膏废渣
phosphorus-hydro-polymaleic anhydride water treatment reagent 含磷水解聚马来酸酐水处理剂
phosphorus index 磷指数
phosphorus nitride 氮化磷
phosphorus-nitrogen ratio 磷氮比
phosphorus ore 磷矿石;磷矿
phosphorus-oxychloride 磷酰氯
phosphorus oxyfluoride 三氟氧化磷
phosphorus pentabromide 五溴化磷
phosphorus poisoning 磷中毒
phosphorus pollution 磷污染
phosphorus powder 磷粉
phosphorus release 磷释放;释磷
phosphorus removal 脱磷
phosphorus retention coefficient 磷滞留系数
phosphorus segregation 磷偏析
phosphorus sesquisul phute 倍半硫化四铁
phosphorus source 磷源
phosphorus sulfide 硫化磷
phosphorus sulfofluoride 三氟硫化磷
phosphorus thiochloride 硫氯化磷
phosphorus trifluoride 三氟化磷
phosphorus trifulfide 三硫化二磷
phosphorus trioxide 三氧化二磷
phosphorus yellow 黄磷
phosphorylamide 磷酰胺
phosphorylation 磷酸化
phosphoryl trianilide 磷酰三苯胺
phosphosiderite 斜红磷铁矿
phosphosilicate 磷硅酸盐
phosphotungstic acid 磷钨酸
phosphotungstic green 磷钨绿
phosphotungstic pigment 磷钨颜料
phospho-tungstomolybdic lake 磷钨钼酸色淀
phosphuranylite 福磷钙铀矿
phos-silver braze alloy 磷银合金硬焊料
phosteam process 蒸汽磷化(处理)法
phost line 寄生管
phot 辐透【物】;厘米烛光(照度单位)
photallochromy 光照变色性
photelectricity 光电学
photetch 光刻技术
photetching 光蚀
photic 光的;透日光;受光
photicity 光化率
photic layer 透光层
photicon 移像光电摄像管;光电摄像管;高灵敏度摄像管
photic region 光射海区;透光海区;透光层
photics 光学
photic stimulator 光刺激器
photic zone 透光层
photino 光微子
photino β-decay 超光子贝塔衰变
photinos 超光子
photion 充气光电二极管
photistor 光敏三极管;光敏晶体管
photo 像片;拍照
photoabsorption 光致吸附;光吸收
photoabsorption band 光吸收带
photoabsorption coefficient 光吸收系数
photoabsorption cross-section 光吸收截面
photoabsorption system 光吸收系统
photo-achromatism 照相消色差
photoacoustic(al) deflection technique 光声偏折技术
photoacoustic(al) detection technique 光声检测技术
photoacoustic(al) microscope 光声显微镜
photoacoustic(al) signals 光声信号
photoacoustic(al) spectroscopy 光声光谱学
photoacoustic(al) technique 光声技术
photoactinic 光化射线的;光化的

photo actinic action 光化作用
photoactivate 光敏化
photoactivated oxidation 光激氧化(作用)
photoactivation 光敏作用;光活化(作用)
photoactivation analysis 光活化分析
photoactivation reaction 光活化反应
photoactivator 光活化剂
photoactive 光敏的;光敏化;光活
photoactive catalysis 光化催化剂
photoactive polymer 光活性聚合物
photoactive substance 光敏物质;光活物质;感光物质
photoactivity 光活性
photoactivity detector 光敏化测试器;光活性测定仪
photoactor 光电变换器材;光电变换件
photoaddition 光化加成反应
photoaffinity label(l)ing 光亲和记记
photo aircraft 航摄飞机;航空摄影飞机
photoalidade 像片视准量角仪;光电照准仪;摄影照准仪
photoalidade compilation 像片量角仪编图
photoalidade method 像片量角仪编图法
photoallergic reaction 光变态反应
photoallergy 光变应性
photo altitude 摄影高度
photoammeter 光电流表;光电安培计
photoamplifier 光放大器;光电放大器
photoamplifier circuit 光电放大电路
photo analysis 像片分析;光电分析
photoangulator 摄影量角仪
photoanode 光电阳极
photo-assisted degradation 光辅助降解
photoaudio generator 光电式音频信号发生器
photoautotroph 光能自养生物
photoautotrophic 光合自养的
photoautotrophic bacteria 光能自养菌
photoautotrophic microorganism 光营养微生物
photobacteria 发光细菌;发光菌
photo base(line) 像片基线
photo-base orientation 像片基线方向
photobeat 光频差拍;光拍
photobehavior 感光特性
photobinocular 光电双筒望远镜
photobiology 光生物学
photobleaching 光致漂泊;光致褪色;光褪色;光漂泊
photobleach material 光致退色材料
photo calibration 光度校准
photo calibration mirror 测光校准镜
photocapacitance 光致变容;光电容
photocard board 厚像纸
photo carriage 像片盘
photocarrier 光载流子
photo carrier plate 像片盘
photocartograph 摄影制图仪;摄影测图仪(器)
photocartography 摄影制图
photocatalysed 光催化的
photocatalysis 光催化(作用)
photocatalyst 光催化剂
photocatalyst fixation 光催化剂稳定
photocatalyst modification 光催化剂改性
photocatalytic activity 光催化活性
photocatalytic decolo(u)ration 光催化脱色
photocatalytic degradation 光催化降解
photocatalytic degradation activity 光催化降解活性
photocatalytic efficiency 光催化效率
photocatalytic oxidation 光催化氧化
photocatalytic oxidation method 光催化氧化法
photocatalytic oxidation treatment 光催化氧化处理
photocatalytic process 光催化工艺
photocatalytic property 光催化性能
photocatalytic rate 光催化速率
photocatalytic reactor 光催化反应器
photocatalytic thin-film cascade reactor 薄膜阶梯式光催化反应器
photocatalytic treatment 光催化处理
photocatalyzed destruction 光催化蜕变
photocatalyzer 光催化剂
photocathode 光阴极;光电阴极
photocathode effective diameter 光阴极有效直径
photocathode efficiency 光阴极效率
photocathode luminous sensitivity 光阴极光灵敏度

photocathode-phosphor image intensifier 光阴极磷光体像增强器
photocathode radiant sensitivity 光阴极辐射灵敏度
photocauterization 放射烙术
photocautery 放射烙术
photocell 光电元件；光电管；光电池；光灯管；电光池
photocell amplifier 光电管放大器
photocell-controlled automatic door 光电控制自动门
photocell control plastic flow limit instrument 光电管控制塑液限流仪
photocell illuminometer 光电管照度计
photocell indicator 光电管指示器；光电管显示器
photocell light check 光电校验
photocell matrix 光电矩阵
photocell pick-off 光电管传感器；光电池
photocell pick-up 光电管受光器
photocell pyrometer 光电池高温计；光电池高温计
photocell relay 光电管继电器；光电池继电器
photocell sensitivity 光电管灵敏度；光电池灵敏度
photocell telemetering commutator 光电管遥测换向器
photocell type length measurement system 光电测长系统
photocell type reset equipment 光电式转换装置
photocell whiteness meter 光电池式白色度计
photocenter 光心；像片中心
photo centering 像片归心
photocentric orbit 光心轨道
photoceram 感光玻璃
photoceramic process 陶瓷照相法
photoceramics 陶瓷印像法；陶瓷照相法
photocharting 像片编图；摄影制图
photochemical 光化学的
photochemical absorption 光化吸收（作用）
photochemical absorption law 光化吸收定律
photochemical action 光化作用
photochemical activation 光化学激活；光化活化
photochemical activity 光化（学）活性；光化活化
photochemical addition 光化加成
photochemical aerosol 光化学气溶胶
photochemical bioclimatology 光化学生物气候学
photochemical blooming 光化热弥散
photochemical catalysis 光化学催化（作用）；光化催化
photochemical catalyst 光催化作用；光催化剂
photochemical cell 光化电池
photochemical chlorination 光化学氯化
photochemical combustion 光化学燃烧
photochemical decomposition 光化分解
photochemical degradation 光化学降解；光化降解
photochemical dissociation 光化离解
photochemical effect 光化学效应；光化效应
photochemical efficiency 光化效率
photochemical engraving 光化刻图
photochemical equivalent 光化当量
photochemical etching 光化腐蚀
photochemical excitation 光化激发
photochemical filter 光化滤光镜
photochemical fixation 光能化学固定；光化学固定
photochemical fog 光化学烟雾
photochemical induction 光化诱导；光化感应
photochemical irrdiation 光化学辐照
photochemical isotope separation 光化同位素分离
photochemical kinetics 光化动力学
photochemically induced oxidation 光化学诱导氧化
photochemically reactive organic material 光化学反应性有机物
photochemical nitrification 光化学硝化作用
photochemical oxidant 光化学氧化剂
photochemical oxidant index 光化学氧化剂指数
photochemical oxidation 光化学氧化
photochemical oxidation method 光化学氧化法
photochemical ozonization 光化臭氧化
photochemical phenomenon 光化现象
photochemical pollutant 光化学污染物
photochemical process 光化学处理；光化学过程；光化学工艺
photochemical processing 光化学加工
photochemical Q-switch 光化学 Q 开关
photochemical reaction 光化作用；光化学反应；光化反应
photochemical reactive solvent 光化学反应性溶剂

photochemical reactivity 光化反应性
photochemical reduction 光化学还原法
photochemical sensitization 光化敏化
photochemical separation 光化分离
photochemical smog 光化学烟雾
photochemical smoke 光化学烟雾
photochemical transformation 光化学转化
photochemical treatment 光化学加工
photochemiluminescence 光化学发光
photochemistry 光化学
photochemistry of ozone 臭氧的光化学
photochlorination 感光氯化
photochopper 光斩波器；光线断路器
photochromatic ink 光致变色油墨
photochrome 彩色照相；彩色照片；彩色像片
photochromic 光色玻璃；光敏材料；光致变色的；光色敏的；光色
photochromic compound 光色化合物
photochromic crystal 光色晶体
photochromic effect 光色效应
photochromic fibre 光致变色纤维
photochromic film 彩色胶片
photochromic filter 光致变色滤光器
photochromic glass 光致变色玻璃；光色玻璃；彩色照相玻璃
photochromic glass fiber 光致变色玻璃纤维
photochromic glaze 光致变色釉；变色釉
photochromic lens blank 光致变色镜片毛坯
photochromic material 光致变色材料；光色材料；变色材料
photochromics 光色体；光色材料
photochromic spectacle glass 光致变色眼镜玻璃
photochromism 光致变色性；光致变色现象；光致变色；光色性；光色度
photochromo lithography 彩色照相版平印术
photochromy 光色性；彩色像影；彩色照相术
photochronograph 照相记时仪；照片记时仪；记录摄影机；活动物体照相机；动体照相机
photochronography 摄影记时仪
photoclino-dipmeter 摄影测斜仪
photoclinometer 摄影测斜仪
photocoagulation 光致凝结；光凝固术；光凝固；光焊接
photocoagulator 光凝结器；光凝固器；光焊接机
photocolorimetric method 比色法
photocolo(u)r 光赋色
photocolo(u)ration 光呈色反应
photocolo(u)rimeter 光比色计；光斑比色计
photocolo(u)rimetric determination 光比色测定
photocolo(u)rimetry 光比色法；光斑比色法
photocolo(u)r material 光赋色材料
photocolo(u)r paper 彩色照相纸
photo combustion 光燃烧
photocommunication 光通信
photocomposer 照相排字机
photocomposition 光电排版
photocon 光导元件；光导器件
photoconductance 光电导率；光电导
photoconducting cell 内光电效应光电管
photoconduction 光致变色法；光电导
photoconduction effect 光电导效应
photoconduction of glass semi-conductor 玻璃半导体光电导
photoconduction sensitivity 光电导灵敏度
photoconductive camera tube 光导式摄像管
photoconductive cell 光电导管；光导电池
photoconductive ceramics 光电导陶瓷；光导陶瓷
photo conductive compensator 光电导补偿器
photoconductive current 光电导电流
photoconductive detector 光电导检测器；光电导探测器
photoconductive device 光电导器件
photoconductive disintegration 光致蜕变
photoconductive effect 光电导效应；光导效应
photoconductive element 光敏元件；光电导元件
photoconductive film 光敏薄膜；光电导膜
photoconductive gain factor 光电导增益因子；光电导增益系数
photoconductive image transducer 光电导图像变换器
photoconductive infrared detector 光电导红外线探测器
photoconductive lag 光电导弛豫
photoconductive layer 光电导层

photoconductive material 光电导材料
photoconductive meter 光电导计
photoconductive method 光电导法
photoconductive mixing 光电导混频
photoconductive optic(al) sensor 光导式光传感器
photoconductive pick-up tube 光导式摄像管
photoconductive polymer 光导聚合物
photoconductive property 光电导性能；光电导特性；光导特性
photoconductive response 光电导响应
photoconductive sensor 光导式传感器
photoconductive target 光导靶
photoconductive thermoplastic 光导热塑体
photoconductive time constant 光电导时间常数
photoconductive tube 光电导管
photoconductivity 光电导性；光电导率
photoconductivity switching 光电导开关
photoconductography 光电导刻蚀术
photoconductor 光敏电阻；光敏导体；光电导体；光导体
photoconductor array 光电导体阵列
photoconductor array axis 光电导体阵列轴
photoconductor drum 光导鼓
photoconductor gap 光电导体间隙；光导体间隙
photoconductor lag 光电导体惰性
photoconductor surface 光电导体表面
photoconductor-thermoplastic film 光电导热塑料膜
photoconstriction coefficient 光致收缩系数；光颈缩系数
photoconstriction effect 光致收缩效应；光颈缩效应
photo-contour map 等高线影像地图
photo-contour process 摄影绘制等高线法
photo-control 像片控制
photo-control base 控制底图
photocontrol diagram 像片联测略图
photocontrol point 像片控制点
photo-control survey 像片控制测量
photoconverter 光转换器
photo coordinates 像片坐标
photo coordinate system 像片坐标系
photo copier 晒像机
photocopy 影印副本；影印；照相复制本；照相版；摄影拷贝；照相复制
photocopying paper 像纸
photocopying process 照相翻拍法
photocopy machine 照相复印机
photo corner 像片角
photo correlation 光相关性
photo correlation spectroscopy 光交互作用分光法
photocount distribution 光电元分布
photo counter 光计数器
photocoupled gate 光耦合门电路
photocoupler 光隔离器；光电耦合器
photocreep 光蠕变
photo crosslinking 光致交联
photocurable coating 光固化涂料
photocured paint 光固化涂料
photocuring 光固化
photocurrent 光电流
photocurrent carrier 光电载流子
photocurrent generator 光电发生器
photocurrent noise 光电流噪声
photocurrent signal 光电流信号
photod 光电二极管
photodarkening effect 光暗化效应
photo darkroom 暗室
photodarlington 光敏达林顿放大器
photo Darlington detector 光电复合晶体管检测器
photodechlorination 感光去氯
photodecolo(u)ration 光消色反应
photodecolo(u)rization 光褪色
photodecomposition 光解作用；光致分解；光分解（作用）
photo defectoscope 光电探伤仪
photodegradable polymer 光降解聚合物
photodegradation 光致降解；光致降级；光降解（作用）
photodegradation efficiency 光降解效率
photodegradation mechanism 光解机理；光致简并机理
photodegradative 光降解的
photodensitometer 光密度计
photodensitometric quantitation 光密度定量

photodensitometry 光密度分析法；光密度测量法
photodepolarization 光去极化
photodepolymerization 光解聚(作用)
photodeposition 光分解沉积
photodetachment 光致分离；光电分离
photodetection 照相检测法；光电探测；光电检测；摄影检测
photodetector 光探测器；光检测器；光电探测器；光电检测器
photodetector array 光检测器矩阵；光电探测器阵列
photodetector head assembly 光电探测头装置；光电探测头
photodetector pick-off 光探测器
photodetector responsivity 光电检波器响应度
photodetector signal-to-noise ratio 光电检波器信噪比
photodeveloping 光显影
photodevelopment 光显影
photodevice 光电探测器
photodichroic 光致二向色
photodichroic matched spatial filter 光致二向色匹配空间滤波器
photodichroic material 光致二向色材料
photodichroism 光致二向色性；光二色性
photodichromic memory 光致二向色存储器
photodielectric(al) effect 光致介电效应
photodiffusion effect 丹伯效应
photodigital memory 光数字存储器
photodigitizer 像片数字化；光电数字化器
photodigitizing system 图像数字仪
photodigitizing unit 光电数字化器
photodimer 光二聚物
photodimerization 光二聚
photodiode 光控二极管；光电二极管
photodiode array 光电二极管阵列；光导二极管阵
photodiode circuit 光电二极管电路
photodiode coupler 光电二极管耦合管
photodiode dark current 光电二极管暗电流
photodiode detector 光电二极管检测器
photodiode displayer 光电二极管显示器
photodiode sensor 光电二极管传感器
photodisintegration 光致蜕变；光致分裂
photodissociation 光致离解；光离解(作用)；光解作用
photodissociation laser 光致离解激光器
photodosimetry 照相剂量学
photodraft 照相复制图
photodrawing 光电制图
photo duodiode 光电双二极管
photo duplicate 照相复制品；照相复制(本)
photo-duplicating 照相复制
photo-duplication 照相复制法
photodynamic(al) action 光动力学作用
photodynamic(al) effect 光动力学效应
photodynamics 光动力学
photo editor 像片编辑员
photoeffect 光(电)效应
photoeffect detector 光效应检测器；光电效应检波器
photoefficiency of photocathode 光电阴极的光电效率
photoelastic 光测弹性的
photoelastic analysis 光弹性分析
photoelastic analyzer 光弹性分析仪
photoelastic apparatus 光弹性仪；光弹性设备
photoelastic bakelite 光弹性酚醛塑料
photoelastic bench 光测弹性台
photoelastic biaxial ga(u)ge 光弹性双向应变计
photoelastic body 光弹性体
photoelastic coating 光弹性涂层；光弹性镀层
photoelastic coating method 光弹涂敷法；面层光弹试验法
photoelastic coefficient 光弹系数
photoelastic coefficient elasticoptic(al) coefficient 光弹性系数
photoelastic constant 光弹性常数；光弹性常量；光弹常量
photoelastic effect 光弹作用；光弹性效应；光弹效应；压光效应
photoelastic equipment 光弹性仪
photoelastic experiment 光弹试验
photoelastic freezing method 光弹性冻结法
photoelastic fringe pattern 光弹性条纹图
photoelasticimeter 光测弹性仪

photoelastic investigation 光弹性研究
photoelasticity 光弹性学；光弹性(力学)；光测弹性学；光测弹性
photoelasticity technique 光弹技术
photoelasticity test 光弹性试验
photoelastic load indicator 光弹性测力计
photoelastic material 光弹(性)材料
photoelastic meter 光弹性仪
photoelastic method 光弹性法；光弹法；光测弹性(试验)法
photoelastic method of stress determination 光弹性应力法
photoelastic model 光弹性模型；光弹模型
photoelastic model method 光弹模型法
photoelastic mode-locking 光弹性锁模
photoelastic model test 光弹性模拟试验
photoelastic phenomenon 光弹性现象
photoelastic photograph 光测弹性照相
photoelastic pressure sensor 光弹性压力传感器
photoelastic reflection method 反射光弹法
photo elastics 光测弹性
photoelastic sensibility 光弹性灵敏度
photoelastic sensor 光弹性传感器
photoelastic strain analysis 光弹应变分析
photoelastic strain ga(u)ge 光弹性应变仪；光弹性应变计
photoelastic strainmeter 光弹性应变仪；光弹性应变计
photoelastic stress 光测弹性应力
photoelastic stress analysis 光弹性应力分析
photoelastic stress analysis method 光弹性应力分析法
photoelastic stress determination 光弹应力测定(法)
photoelastic stress ga(u)ge 光弹性应力计
photoelastic stressmeter 光弹性应力计
photoelastic stress method of stress determination 光弹性确定应力法
photoelastic study 光弹性应力分析
photoelastic system 光弹性系统
photoelastic technique 光弹性技术
photoelastic tensor 光弹性张量
photoelastic test 光弹性试验；光弹试验
photoelastimetric method 光弹性(测定)法
photoelastimetry 光弹性测定术
photoelastometry 光测弹性术
photoelectrcataly oxidatio 光电催化氧化
photoelectrcatalysis 光电催化
photoelectrcatalytic degradation 光电催化降解
photoelectrcatalytic reactor 光电催化反应器
photoelectret 光驻极体
photoelectric(al) 光电效应的；光电的
photoelectric(al) absolute magnitude 绝对光电星等
photoelectric(al) absorption 光电吸收
photoelectric(al) absorption analysis 光电吸收分析法
photoelectric(al) absorption coefficient 光电吸收系数
photoelectric(al) absorption index 光电吸收系数
photoelectric(al) activity 光电激活度
photoelectric(al) alarm 光电报警器
photoelectric(al) controller 光电控制器
photoelectric(al) amplifier 光电放大器
photoelectric(al) analysis 光电分析(法)
photoelectric(al) analytical balance 光电分析天平
photoelectric(al) astrolabe 光电等高仪
photoelectric(al) astrometry 光电天体测量
photoelectric(al) astrophotometry 光电天体光度学
photoelectric(al) attenuation coefficient 光电衰减系数
photoelectric(al) auto-collimation 光电自准直
photoelectric(al) auto-collimator 光电自准直仪；光电自准直平行光管；光电自动照准仪
photoelectric(al) brightness 光电亮度
photoelectric(al) cartridge 光电式拾音头
photoelectric(al) cathode 光电阴极
photoelectric(al) cell 光电池；光灯管
photoelectric(al) cell amplifier 光电管放大器；光电池放大器
photoelectric(al) cell-operated relay 光电电眼；光电信号器；光电池继电器；光电控制装置
photoelectric(al) cell unit 光电池组

photoelectric(al) chopper 光电斩波器
photoelectric(al) circle reading 度盘光电记录
photoelectric(al) circle recording 度盘读数记录
photoelectric(al) circuit 光电(元件)电路
photoelectric(al) coded disk 光电编码盘
photoelectric(al) colo(u)r analyser 光电析色器
photoelectric(al) colo(u)r comparator 比色仪
photoelectric(al) colo(u)rimeter 光电色度计；光电彩色计；光电比色计
photoelectric(al) colo(u)rimetric analysis 光电比色分析
photoelectric(al) colo(u)rimetry 光电色度学；光电比色(分析)法
photoelectric(al) colo(u)ring 光电比色法
photoelectric(al) colo(u)r meter 光电比色计
photoelectric(al) colo(u)r method 光电颜色测定法；光电比色测定法
photoelectric(al) colo(u)r pyrometer 光电比色高温计
photoelectric(al) comparator 光电比较仪；光电比测器
photoelectric(al) constant 光电(效应)常数
photoelectric(al) control 光电控制
photoelectric(al) control equipment 光电控制设备
photoelectric(al) controller 光电控制器
photoelectric(al) control unit 光电控制装置
photoelectric(al) conversion 光电转换
photoelectric(al) converter 光电转换器
photoelectric(al) counter 光电式计数器
photoelectric(al) crystal 光电晶体
photoelectric(al) current 光电流
photoelectric(al) cut-off register control 光电切断点控制
photoelectric(al) densimeter 光电密度计
photoelectric(al) densitometer 光电(式)显像密度计；光电(式)密度计
photoelectric(al) detection 光电探测；光电检测
photoelectric(al) detector 光电(管式)探测器；光电(式)检测器；光电感察器；光束式探测器
photoelectric(al) device 光电元件；光电装置；光电器件
photoelectric(al) direct reading spectrometer 光电直读光谱仪
photoelectric(al) door closer 光电自动关门器
photoelectric(al) door opener 光电开门器
photoelectric(al) double-slit interferometer 光电双狭缝干涉仪
photoelectric(al) dust detector 光电式烟尘检测器；光电粉尘检测器
photoelectric(al) dust meter 光电量尘计；光电测尘器
photoelectric(al) effect 光电效应
photoelectric(al) effect attenuation coefficient 光电效应衰减系数
photoelectric(al) efficiency 光电效率
photoelectric(al) electron multiplier tube 光电电子倍增管
photoelectric(al) element 光电元件
photoelectric(al) emission 光电子发射；光电放射；光电发射
photoelectric(al) emissivity 光电发射率
photoelectric(al) emittance optic(al) sensor 光电子发射式光传感器
photoelectric(al) emitter 光电发射体
photoelectric(al) encoder 光电编码器
photoelectric(al) equation 光电效应方程
photoelectric(al) examination of dust sampler 光电测尘仪
photoelectric(al) exposure meter 光电曝光表
photoelectric(al) eye 光电眼；光电信号器；光电控制装置；光电池继电器
photoelectric(al) fatigue 光电疲劳；光电管老化
photoelectric(al) fire detector 光电消防探测器
photoelectric(al) flame detector 光电式火焰探测器
photoelectric(al) flame-failure detector 光电火陷故障探测器
photoelectric(al) flame failure device 光电式火焰切断器
photoelectric(al) fluorimeter 光电荧光计
photoelectric(al) follow-up 光电随动系统；光电跟踪系统装置
photoelectric(al) follow-up system 光电跟踪系统

photoelectric(al) Fourier transformation 光电傅立叶变换
photoelectric(al) function generator 光电(式)函数发生器
photoelectric(al) galvanometer 光电电流计
photoelectric(al) gas analyzer 光电气体分析仪
photoelectric(al) glossmeter 光电光泽计
photoelectric(al) grader 光电式分级机
photoelectric(al) guider 光电导星镜
photoelectric(al) high-speed recording pyrometer 光电高速记录高温计
photoelectric(al) hygrometer 光电湿度表
photoelectric(al) igniter 光电点火器
photoelectric(al) illuminometer 光电式照度计
photoelectric(al) image converter 光电图像变换管;光电变像管
photoelectric(al) image tube 光电像管;光电成像管
photoelectric(al) indicator 光电指示器
photoelectric(al) input machine 光电输入机
photoelectric(al) inspection 光电检查
photoelectric(al) instrument 光电仪器
photoelectric(al) integrator 光电积分器
photoelectric(al) interaction 光电相互作用
photoelectric(al) intrusion detector 光电侵入检测器;光电防盗报警器
photoelectric(al) ionization 光致电离;光电电离
photoelectric(al) level transducer 光电式液位传感器
photoelectric(al) light curve 光电光变曲线
photoelectric(al) lighting control 光电照明控制
photoelectric(al) light wave comparator 光电光波比较仪
photoelectric(al) line setting instrument 光电定线仪
photoelectric(al) liquid-level indicator 光电液位指示器
photoelectric(al) luxmeter 光电照度计;光电勒克司计
photo electrically controlled cutting machine 光电控制切割机
photoelectric(al) magnitude 光电星等
photoelectric(al) material 光电材料
photoelectric(al) measuring device 光电测量装置
photoelectric(al) measuring microscope 光电测量显微镜
photoelectric(al) membrane manometer 光电膜测压计
photoelectric(al) meter 光电表;电磁测光表
photoelectric(al) method 光电法
photoelectric(al) method for measurement of reflectivity 反射光电测定法
photoelectric(al) micrometer 光电式测微计
photoelectric(al) micrometer microscope 光电测微显微镜
photoelectric(al) microphotometer 光电显微光度计;光电测微光度计
photoelectric(al) microscope 光电显微镜
photoelectric(al) miniature propeller current meter 光电微型旋桨流速仪
photoelectric(al) mixer 光电混频器
photoelectric(al) mixing experiment 光电混频试验
photoelectric(al) multiplier 光电倍增器;光电倍增管
photoelectric(al) multipurpose water-level meter 光电式多功能水位计
photoelectric(al) nigrometer 光电黑度计
photoelectric(al) nucleus counter 光电核计数器
photoelectric(al) numeral tube 光电数码管
photoelectric(al) obscuration-type smoke detector 减光型光电感烟探测器
photoelectric(al) observation 光电观测
photoelectric(al) paper tape reader 光电纸带输入机
photoelectric(al) particle counter 光电粒子计数器
photoelectric(al) particle size meter 光电颗粒分析仪
photoelectric(al) performance tester 光电性能检测仪
photoelectric(al) phenomenon 光电现象
photoelectric(al) photometer 光电光度计
photoelectric(al) photometer method 光电光度法
photoelectric(al) photometry 光电光度测量(法);光电测光
photoelectric(al) pick counter 光电织物密度分析器
photoelectric(al) pick-off 光电探测器
photoelectric(al) pick-up 光电拾音器
photoelectric(al) plethysmograph 光电式体积描记器;光电容积描记器
photoelectric(al) plethysmography 光电容积脉波检查
photoelectric(al) plethysmometer pick-up 光电容积描记图测量换能器
photoelectric(al) polarimeter 光电偏振计
photoelectric(al) polarimetry 光电偏振测量
photoelectric(al) position photosetter 光电定位排字机
photoelectric(al) potentiometer 光电电势计
photoelectric(al) power converter 光电功率变换器
photoelectric(al) precision encoder 光电精密编码器
photoelectric(al) property 光电性能
photoelectric(al) proportionality law 光电正比定律
photoelectric(al) pyrometer 光电高温计
photoelectric(al) radial velocity 光电视向速度
photoelectric(al) radial velocity spectrometer 光电视向速度仪
photoelectric(al) radiationdetector 光电辐射探测器
photoelectric(al) raw material commodities 光电原料矿产
photoelectric(al) reader 光电阅读器;光电读数器;光电读出器
photoelectric(al) reading 光电读出
photoelectric(al) receiver 光电接收器
photoelectric(al) recorder 光电记录器
photoelectric(al) record(ing) 光电记录
photoelectric(al) recording instrument 光电记录仪;光电记录器
photoelectric(al) red threshold 光电红色阈
photoelectric(al) reflectometer 光电反射计
photoelectric(al) register control 光电定位控制
photoelectric(al) register controller 光电定位控制器
photoelectric(al) regulator 光电调节器
photoelectric(al) relay 光电继电器
photoelectric(al) relay meter 光电式继电器电表
photoelectric(al) relay rack 光电继电器架
photoelectric(al) response 光电响应
photoelectric(al) safety device 光电式安全装置
photoelectric(al) scanner 光电探伤仪;光电扫描器
photoelectric(al) scanning 光电扫描
photoelectric(al) scanning machine 光电扫描机
photoelectric(al) scanning moire grid 光电扫描云纹栅
photoelectric(al) scanning monochromator 光电扫描单色仪
photoelectric(al) scleroscope 光电硬度计
photoelectric(al) sediment analyzer 光电泥砂颗粒分析仪
photoelectric(al) seismometer 光电地震仪;光电地震计
photoelectric(al) sender 光电发送器
photoelectric(al) sensing equipment 光电感应设备
photoelectric(al) sensitivity 光电灵敏度
photoelectric(al) sensor 光电传感器
photoelectric(al) separation 光电选矿
photoelectric(al) separation method 光电选矿法
photoelectric(al) separator 光电选矿机
photoelectric(al) sextant 光电六分仪
photoelectric(al) siltmeter 光电沉淀计;光电泥淤仪;光电泥砂量测器
photoelectric(al) sludge-level indicator 光电污泥面指示仪
photoelectric(al) smoke-density control 光电烟密度控制
photoelectric(al) smoke-density controller 光电烟密度控制器
photoelectric(al) smoke detector 光电式感烟探测器
photoelectric(al) solar magnetograph 太阳光电磁象仪(法);光电测光
photoelectric(al) solid-state device 光电固体器件
photoelectric(al) sorter 光电分类器
photoelectric(al) spectrocolo(u)rimeter 光电光谱比色计;光电分光测色计
photoelectric(al) spectrometer 光电频谱仪;光电分光计
photoelectric(al) spectrophotometer 光电(式)分光光度计
photoelectric(al) spectrophotometry 光电分光光度测量
photoelectric(al) spectropolarimeter 光电旋分光光度计
photoelectric(al) speedometer 光电速度仪
photoelectric(al) sphygmograph 光电脉搏描记器;光电脉压计
photoelectric(al) star photometer 恒星光电光度计
photoelectric(al) steelscope 光电析钢仪
photoelectric(al) storage 光电存储器
photoelectric(al) surface 光电面
photoelectric(al) switch 光电开关
photoelectric(al) tach(e)ometer 光电转速计
photoelectric(al) tape reader 光电纸带阅读器;光电输入机;光电读带器;光电带读出器;穿孔带光电输入机
photoelectric(al) technology 光电技术
photoelectric(al) temperature 光电温度
photoelectric(al) template colo(u)rimeter 光电模板比色计
photoelectric(al) threshold 光电阈
photoelectric(al) threshold contrast 光电阈衬比
photoelectric(al) timer 自动光电定时器;光电定时器
photoelectric(al) titration 光电滴定
photoelectric(al) tracing flame cutting machine 光电跟踪气割机
photoelectric(al) tracker 光电跟踪器
photoelectric(al) tracking 光电跟踪
photoelectric(al) tracking system 光电跟踪系统
photoelectric(al) transducer 光电变换器
photoelectric(al) transfer fibre 光导纤维
photoelectric(al) transformation efficiency 光电转换效率
photoelectric(al) transformer 光电变换器
photoelectric(al) transit instrument 光电中星仪
photoelectric(al) translating system 光电读数系统
photoelectric(al) transmissometer 光电透射计
photoelectric(al) triode 光电三极管
photoelectric(al) tristimulus colo(u)rimeter 光电三色激励比色计
photoelectric(al) tube 光电管
photoelectric(al) tube amplifier 光电管放大器
photoelectric(al) turbidimeter 光电浊度仪;光电浊度计
photoelectric(al) turbidimetry 光电比浊法
photoelectric(al) type ammeter 光电式电流计
photoelectric(al) type smoke detector 光电感烟探测器
photoelectric(al) viewing tube 光电显像管
photoelectric(al) visco(si)meter 光电黏度计
photoelectric(al) weft-finder 光电探纬器
photoelectric(al) whirl measuring apparatus 光电转速测量仪
photoelectric(al) width ga(u)ge 光电宽度计
photoelectric(al) work function 光电功函数
photoelectric(al) yarn regularity tester 光电式纱线均匀度检验仪
photoelectric(al) yield 光电子产额
photoelectricity 光电学;光电
photoelectricity phenomenon 光电现象
photoelectric scattering smoke detector 散射型光电感烟探测器
photoelectrochemical 光电化学的
photoelectrochemical effect 光电化学效应
photoelectrochemistry 光电化学
photoelectrolytic cell 电解光电池
photoelectromagnetic 光电磁
photoelectromagnetic cell 光电磁光电管
photoelectromagnetic detector 光电磁探测器;光电磁检测器
photoelectromagnetic effect 光电磁效应;光磁电效应
photoelectromagnetic photodetector 光电磁光检测器;光电磁光电检测器
photoelectrometer 光电计;光电比色计

photoelectrometric agent 光电比色剂
photoelectron 光电子
photoelectron analytic(al) spectrometer 光电子分析光谱仪
photoelectron beam 光电子束
photoelectron counter 光电子计数器
photoelectron counting 光电子计数
photoelectron device 光电子发射装置
photoelectron discharge 光电子放电
photoelectron escape time 光电子逸出时间
photoelectron-Fenton oxidation method 光电芬顿氧化法
photoelectronic 光电子的
photoelectronic amplification 光电子放大
photoelectronic device 光电子设备;光电子器件
photoelectronic diode 光电子二极管
photoelectronic emission 光电子发射
photoelectronic liberation 光电子逸出
photoelectronic pulse amplifier 光电子脉冲放大器
photoelectronics 光电子学
photoelectronic spectrum 光电子谱
photoelectronic transducer 光电子换能器;光电子变换器
photoelectronic travel(l)ing waveguide 光电子行波管
photoelectron image 光电子像;光电子图像
photoelectron limited resolving power 光电子限制的分辨本领
photoelectron-luminescence 光控场致发光
photoelectron microscopy 光电子显微术
photoelectron-multiplier-tube 光电子倍增管
photoelectron noise 光电子噪声
photoelectron spectrograph 光电子能谱仪
photoelectron spectroscope 光电子能谱仪
photoelectron spectroscopy 光电子能谱(法);光电子能(谱)法;光电子光谱学;光电分光法
photoelectron spectrum 光电子能谱(法)
photo-electron-stabilized-photicon 移像光电稳定摄像管;光电稳定摄像管
photoelectron statistics 光电子统计
photoelectron stream 光电子流
photoelectrophoretic camera 光电泳照相机
photoelectrostatic image 光学静电像
photoelectro-type 光电铸版
photoelement 光元素;光电管;光电池
photoelement photoelectric(al) cell 光电管
photoelement with external photoelectric(al) effect 外光电效应光电管
photoelimination 光致消除
photoemission 光电子发射;光电放射;光电发射
photoemission microscope 光电发射显微镜
photoemission pick-off 光发射传感器
photoemissive camera tube 光电发射摄像管
photoemissive cathode 光电发射阴极
photoemissive cell 光电放射管;光电发射元件;光电发射管
photoemissive detector 光发射探测器;光电发射探测器;光电发射检测器
photoemissive device 光电发射器件
photoemissive element 光电放射元件;光电发射元件;光电池
photoemissive layer 光电发射层
photoemissive material 光电发射材料
photoemissive photodetector 光电发射光电检波器
photoemissive relay 光电放射(式)继电器
photoemissive silicon detection array 光电发射硅探测阵列
photoemissive tube 光电发射管
photoemissive tube photometer 光电发射管光度计;光电发射管光电检波器
photoemissivity 光发射能力;光电发射能力
photoemitter 光发射体;光电发射体
photo emulsion 感光乳剂;照相胶
photoemulsion layer 摄影乳剂层
photoengraving 影印版;照相制版;照相凸版印刷;照相蚀刻法;照相感光制版;光刻
photoengraving pattern 光刻图案
photo enlargement 像片放大
photo enlarger 光放大器
photo enlarging paper 光放大纸
photoequilibrium 光致平衡
photoetched evapo(u)ration mask 光刻蒸发掩膜
photoetched slow wave structure 光刻慢波结构
photo etch(ing) 光刻;影印版;照相版;光蚀刻

photoetching 光蚀刻法;光刻法;光刻
photoetching machine 光刻机
photoetching method 光蚀刻法;光刻法
photoetching technique 光刻技术
photoetch method 光刻蚀法
photoetch pattern 光刻图案
photoetch-resist resin 光致抗蚀剂树脂
photoevent 光子事件
photoexcenter 像片偏心
photoexcitation 光激
photoexcited nucleus 光激核
photoexciting motion 光激运动
photoexitation 光激励
photoexposure 摄影曝光
photo exposure enlarging meter 像片曝光放大表
photo exposure machine 曝光机
photo-exposure positive 曝光正片
photoextinction 比浊分析法
photoextinction method 消光浊度法;浊度分析
photoextrusion reaction 光致突变反应
photofabrication 光制成型;光刻成型法;光(电)加工;照相化学腐蚀制造法
photofabrication photolithography 光刻法
photofacsimile 光电传真
photo-Fenton degradation 光芬顿降解
photo-Fenton oxidation 光芬顿氧化
photo-Fenton reaction 光芬顿反应
photoferro-electrics 光铁电体
photofibre and light cable jointer 光纤光缆连接器
photofibre coupler 光纤耦合器
photo-field emission 场致光电发射
photofigure analyzer 光度图形分析仪
photofission 光致裂变;光致分裂
photofixation 光固定;定位
photoflash 闪光灯;照相闪光灯;闪光照片;闪光灯照片
photoflash battery 闪光灯电池
photoflash bomb 照相炸弹;照相闪光弹
photoflash composition 摄影闪光混合物
photoflash lamp 照相用闪光灯;摄影闪光灯
photoflash light 摄影闪光灯
photo flight 航摄飞行
photoflood 照相泛光灯;超压强光灯;溢光灯;摄影泛光灯
photoflood bulb 超压强烈溢光灯泡
photoflood emulsion 人工光源乳剂
photoflood lamp 溢光灯;照相散光灯;摄影(泛光)灯;超压强烈溢光灯
photoflood lamp bulb 超压强烈溢光灯泡
photofluerescence phenomenon 光激荧光现象
photofluorography 荧光照相;荧光屏图像摄影;荧光屏摄影术
photofluorometer 荧光计
photofluoroscope 荧光屏
photoformer 光电函数发生器;光电管振荡器;光电波形发生器
photogalvanography 光电摄影
photogalvanometer 光电流计
photogalvanometric registration 光电流记录
photogelatin 摄影明胶
photogen 页岩煤油
photogene 余像;余影
photo generator 光电信号发生器
photogenic 光原性的
photogenic bacteria 发光(细)菌
photogenic colony 发光菌落
photogenous 光原性的
photogeologic(al) map 航摄地质图
photogeology 影像地质学;航摄地质学;摄影地质学;地质摄影
photogeology method 摄影地质法
photo geometry 像片几何学
photogeomorphology 航摄地貌学
photoglow tube 充气光电管
photogoniometer 像片量角仪;光电测角计;摄影测角仪
photogoniometer for pairs of photographs 双像量角仪
photogoniometric method 像片量角法;光测角法
photogram 传真电报;测量照片;物影照片
photogrammeter 摄影测量员
photogrammetric accuracy 摄影测量精度
photogrammetric aerial triangulation 像片空中三角测量

photogrammetrically numerical restitution 航测数字测图
photogrammetric apparatus 摄影测量装置;摄影测量仪(器)
photogrammetric automatization 摄影测量自动化
photogrammetric base 摄影测量基线
photogrammetric bridging 摄影测量控制加密
photogrammetric(al) camera 航空摄影机
photogrammetric(al) camera lens 航摄仪镜头
photogrammetric computor 摄影测量计算机
photogrammetric condition equation 摄影测量条件方程
photogrammetric control 摄影测量控制
photogrammetric control point 加密点【测】
photogrammetric coordinates 摄影测量坐标
photogrammetric coordinate system 摄影测量坐标系(统);地面摄影测量坐标系
photogrammetric coordinatograph 摄影测量坐标仪
photogrammetric distortion 摄影测量畸变
photogrammetric engineer 摄影测量工程师;测量摄影师
photogrammetric flight 摄影测量飞行;航摄飞行
photogrammetric(al) flight calculator 航摄飞行计算尺
photogrammetric(al) flight sortie 航摄飞行架次
photogrammetric instrument 摄影测量仪(器)
photogrammetric interpolation 摄影测量内插
photogrammetric intervalometer 空中摄影定时器
photogrammetric map 摄影测量地图;摄影测绘图
photogrammetric measurement 摄影测量
photogrammetric method 照相测量方法
photogrammetric network 摄影测量网
photogrammetric objective 摄影测量物镜
photogrammetric parallax 摄影测量视差
photogrammetric platting 摄影测绘图
photogrammetric plotter 摄影测量绘图仪;摄影测量绘图器
photogrammetric procedure 摄影测量法
photogrammetric process 摄影测量过程
photogrammetric pyramid 摄影测量锥形法
photogrammetric recording 摄影测量记录
photogrammetric refraction 照相测量大气折射
photogrammetric sketch 摄影测量草图
photogrammetric stereo-camera 测量立体摄影机
photogrammetric surveying apparatus 摄影测量设备
photogrammetric technology 摄影测量技术
photogrammetric triangulation 像片三角测量;摄影三角测量
photogrammetric unit 摄影天顶筒;摄影测图仪(器)
photogrammetrist 摄影测量员;摄影测量学(专)家
photogrammetry 照相测量学;照相测量术;摄影测量学;摄影测量术;摄影测量法;摄影测量;摄影测绘学
photogrammetry and remote sensing 摄影测量与遥感
photogrammetry by intersection 交会摄影测量学
photograph 像片;照片;图片;摄影测图仪(器);摄影
photograph album 影集
photograph annotation 像片调绘
photograph axis 像片坐标轴
photograph base 像片基线
photograph camera 摄影机
photograph center 像片中心
photograph collection 照相档
photograph data memory 照相数据存储器
photograph decentration 像片离心
photograph developer 显影剂;显像剂;显色剂
photographer 摄影师
photographer studio 照相馆
photographic(al) 照相的;摄影资料地区
photographic(al) absorption coefficient 照相吸收系统
photographic(al) aircraft 航空摄影飞机
photographic(al) airplane 摄影飞机
photographic(al) albedo 照相反照率;摄影反照率
photographic(al) altitude 照相高度
photographic(al) amplification 照相放大
photographic(al) analysis 像片分析
photographic(al) apparatus 照相器材
photographic(al) astrometry 照相天体测量学;摄影天体测量学

photographic(al) astronomy 照相天文学;摄影天文学
photographic(al) astrophotometry 照相天体光度学;摄影天体光度学
photographic(al) astrospectroscopy 照相天体光谱学;摄影天体光谱学
photographic(al) aviation 航摄飞行
photographic(al) aviation calculator 航摄飞行计算尺
photographic(al) barograph 摄影气压仪;摄影记录气压计
photographic(al) base(line) 摄影基线
photographic(al) borehole survey(ing) 钻孔摄影测量;钻孔摄影测绘
photographic(al) brightness 照相亮度
photographic(al) bromide picture 溴化纸像片
photographic(al) bromide print 溴化纸像片
photographic(al) bundle of rays 摄影光速
photographic(al) camera 航空摄影机;照相机
photographic(al) characteristic curve 感光度特性曲线
photographic(al) chemical 照相药品
photographic(al) chemistry 照相化学;摄影化学
photographic(al) circle reading 度盘读数照明记录
photographic(al) coating 感光膜
photographic(al) coefficient 摄影系数
photographic(al) collage 剪辑照片
photographic(al) compartment 照相设备舱
photographic(al) control 照片控制点定位法
photographic(al) copy 照相复印;复印片
photographic(al) copying 照相复制
photographic(al) coverage 有航摄资料地区;照片拍摄范围;摄影资料地区
photographic(al) coverage regulator 像片重叠调节器
photographic(al) dark room 照相暗室
photographic(al) data 摄影资料
photographic(al) data memory 照相数据存储器
photographic(al) data process 摄影数据处理
photographic(al) data recording 数据摄影记录
photographic(al) data reduction 像片数据压缩
photographic(al) datum plane 像片基准面;摄影基准面
photographic(al) day 航摄日
photographic(al) dead land 航摄死区
photographic(al) decoration 照相装饰法
photographic(al) densitometer 照相密度计
photographic(al) density 片片黑度;照相密度;照相底片密度;摄影密度
photographic(al) density control 影像密度控制;照相密度控制
photographic(al) detail 影像细部
photographic(al) detection 摄影探测
photographic(al) detector 照相探测器
photographic(al) developer 显影剂;显像剂;显影液
photographic(al) digitizing system 摄影测量数字化系统
photographic(al) distance 像片上的距离
photographic(al) documents 照相文献
photographic(al) dosimeter 摄影剂量仪
photographic(al) dosimetry 照相剂量测定法
photographic(al) effect 照相效应
photographic(al) effective area 像片有效面积
photographic(al) emulsion 感光乳剂;照相乳胶液;照相乳胶;照相乳剂;摄影乳剂
photographic(al) emulsion detector 照相乳剂检测器
photographic(al) emulsion layer 摄影乳剂层
photographic(al) emulsion technique 照相乳胶技术
photographic(al) engineering 摄影工程
photographic(al) enhancement 图像增强
photographic(al) enlargement 照相放大
photographic(al) enlargement section 照相放大组
photographic(al) ephemeris 照相星历表
photographic(al) equipment 照相器材
photographic(al) equipment plant 摄影器材厂
photographic(al) extension 像片联测
photographic(al) facsimile 照相复制
photographic(al) field 景物
photographic(al) field camera 地面摄影机
photographic(al) field lens 摄影物镜

photographic(al) film 照相胶片;照相胶卷;照相底片;感光胶片;摄影软片;摄影胶片
photographic(al) film badge 照相胶片佩章
photographic(al) film dosimeter 照相胶片剂量计
photographic(al) filter 照相滤器
photographic(al) fixing 定影
photographic(al) flight 航摄飞行;摄影飞行
photographic(al) flight calculator 航摄飞行计算尺
photographic(al) flight-line 照相航线
photographic(al) flight planning 航摄飞行计划
photographic(al) flying height 摄影航高
photographic(al) fog 灰雾;底片雾
photographic(al) formula 摄影公式
photographic(al) glass 照相底板用玻璃
photographic(al) grain 照相晶粒
photographic(al) graininess 照相乳剂粒度
photographic(al) granularity 摄影颗粒性
photographic(al) gun 照相枪
photographic(al) hill shading 直照晕渲【测】
photographic(al) image 像片;摄影影像;影像剖视仪
photographic(al) image base 像片基线
photographic(al) imagery 摄影成像
photographic(al) inclinometer 摄影测斜仪
photographic(al) industry wastewater 照相工业废水
photographic(al) infrared 摄影红外
photographic(al) instrument 摄影仪器
photographic(al) integration 照相累积;摄影积分
photographic(al) intelligence 照相情报
photographic(al) intelligence data 照相情报资料
photographic(al) intensifier 照相强化剂
photographic(al) interpretation 像片判读
photographic(al) interpretation data 照相判读资料
photographic(al) interpretation report 照相判读报告
photographic(al) interpreter 像片判读员
photographic(al) laboratory 照相暗室
photographic(al) laboratory technician 摄影室技术员
photographic(al) latent image 照相潜像
photographic(al) layer 感光层;乳剂层
photographic(al) layout drawing 摄影打样图
photographic(al) lens 照相镜头;航摄仪镜头
photographic(al) library 像片库
photographic(al) light curve 照相光变曲线
photographically catalyzed nucleation 光致晶核形成;见光度
photographically recorded reading 摄影记录读数
photographically reproduced reprint 照相复制重印本
photographic(al) magnetograph 照相磁像仪
photographic(al) magnitude 照相星等
photographic(al) map 影像地图;摄影图;摄影制图;摄影地图
photographic(al) mapping 像片测图;像法测图;摄影测图
photographic(al) mask 照相掩模
photographic(al) masking 光掩蔽;摄影遮光
photographic(al) material 照相材料
photographic(al) material deformation 摄影材料变形
photographic(al) measurement 照相测量
photographic(al) memory 照相存储器
photographic(al) method 摄影方法
photographic(al) micrography 显微摄影
photographic(al) mission 航摄资料分布示意图
photographic(al) nadir 像片天底点
photographic(al) negative 底片;照相负片;照相底片
photographic(al) noise 照相噪声
photographic(al) normal point 像片标准点
photographic(al) nucleation 光致успенdcastrophotos;见光度
photographic(al) objective 摄影物镜
photographic(al) observation 摄影观测
photographic(al) optical system 摄影光学系统
photographic(al) output device 照相输出设备
photographic(al) panorama 照相全景;全景像片
photographic(al) paper 像纸;照相纸;印相纸
photographic(al) paper recording 照相纸记录
photographic(al) parameter 显影参数
photographic(al) perpendicular 摄影主光轴
photographic(al) photometer 照相光度计;摄影光度计

photographic(al) photometry 照相光度学;照相光度术;照相光度测量;照相测光;光电光度学;摄影光度学
photographic(al) plan 像(片)平面图
photographic(al) plane 摄影平面
photographic(al) plate 照相底片;照相底版;摄影干版;摄影底片;摄影底版
photographic(al) plate detection 照相底片探测
photographic(al) plate number 底片号
photographic(al) plate reading 照相底片读数
photographic(al) plot 摄影测量原图
photographic(al) plotter 照相绘图机
photographic(al) plotting 摄影测量测图
photographic(al) plotting head 光绘曝光头
photographic(al) pod 摄影机吊舱
photographic(al) point 摄影测量加密点
photographic(al) positive 正片;照相正片
photographic(al) print 照相印刷;影印
photographic(al) printing 照相印花;照片印刷
photographic(al) printing nuit 照相印刷设备
photographic(al) procedure 摄影方法
photographic(al) process 照相法
photographic(al) processing 摄影处理
photographic(al) processing equipment 照相处理设备
photographic(al) pyramid 像片锥形法
photographic(al) quality 照片质量;摄影质量
photographic(al) radiant 照相辐射点
photographic(al) radiant point 摄影辐射点
photographic(al) reading 照相判读
photographic(al) reconnaissance 摄影勘察;摄影勘测
photographic(al) record 照片记录法
photographic(al) recorder 屏幕录像机;照相录器;胶片记录器;摄影记录器
photographic(al) recording 照相记录;光录声;摄影录音;摄影记录
photographic(al) recording acceleration 照相记录加速度仪
photographic(al) recording instrument 照相记录仪
photographic(al) recording polarograph 照相式极谱仪;摄影极谱仪
photographic(al) record system 摄影记录系统
photographic(al) rectification 像片纠正;摄影测量纠正
photographic(al) reducer 照相减薄剂
photographic(al) reduction 照相缩小
photographic(al) registration 摄影记录
photographic(al) remote-sensing system 摄影遥感系统
photographic(al) report 照相报告
photographic(al) reprint 照相复印
photographic(al) reproduction 摄影复制
photographic(al) resection 摄影测量交会
photographic(al) resolution 像片分辨率;照相分辨率
photographic(al) resolving power 摄影分辨率
photographic(al) rocket 摄影火箭
photographic(al) rotation technique 照相旋转技术
photographic(al) scale 像片比例尺;照片比例(尺);摄影比例尺
photographic(al) science 摄影科学
photographic(al) screen 滤色屏
photographic(al) sensing system 摄影遥感系统
photographic(al) sensitivity 显影敏感性;感光度;照相灵敏度
photographic(al) sensitizer 照相敏化剂
photographic(al) sensitometer 感光度测定仪
photographic(al) sensory unit 照相记录装置
photographic(al) silver recovery 照相废银回收
photographic(al) sketch 像片略图
photographic(al) solar magnetograph 太阳照相磁像仪
photographic(al) sortie 照相架次;航摄飞行架次
photographic(al) sound recoder 光录声机;光学录音机;摄影录音机
photographic(al) sound recording 摄影录音
photographic(al) sound recording head 光学录音头
photographic(al) spectrophotometry 照相分光光度法;照相分光光度测量

photographic(al) spectroscope 照相分光镜
photographic(al) spectrum 照相感光范围
photographic(al) speed 照相速度;摄影速率;摄影速度
photographic(al) spool 胶片卷轴
photographic(al) star catalogue 照相星表
photographic(al) station 摄影站
photographic(al) stencil 照相制版
photographic(al) stereopair 立体像对
photographic(al) storage 光图像存储器;照相存储器;数据摄影记录装置;摄影存储器
photographic(al) store 照相存储
photographic(al) strip 摄影航线
photographic(al) studio 照相室;摄影室
photographic(al) study of combustion 燃烧过程的摄影研究
photographic(al) sunshine recorder 摄影日照计
photographic(al) survey 摄影测量学;摄影测量
photographic(al) surveying 摄影勘察;摄影勘测;摄影地形测量;摄影测量术;摄影测量
photographic(al) surveying camera 测量摄影机
photographic(al) suspension 摄影机悬挂装置
photographic(al) tape 摄影胶带
photographic(al) technical squadron 照相技术中队
photographic(al) technique 摄影技术
photographic(al) technology 摄影工艺
photographic(al) telescope 照相望远镜
photographic(al) template 摄影模板
photographic(al) theodolite 摄影读数经纬仪
photographic(al) thermal neutron image detector 热中子成像照相探测器
photographic(al) thermometer 摄影温度计
photographic(al) tilt 摄影倾斜角
photographic(al) track 照相迹速
photographic(al) tracking 照相跟踪;摄影跟踪
photographic(al) tracking station 摄影经纬仪跟踪站
photographic(al) trailer 活动暗室
photographic(al) transit 摄影读数经纬仪
photographic(al) transparency 摄影正片
photographic(al) triangulation 照相三角测量
photographic(al) typesetter 摄像排字机
photographic(al) velocity 摄影速度
photographic(al) vignetting 渐晕照相法
photographic(al) waste 照相废液
photographic(al) weaving 照相机织造
photographic(al) window 可摄影窗口
photographic(al) zenith telescope 摄影天顶望远镜
photographic(al) zenith tube 照相天顶筒;摄影天顶望远镜
photograph index map 标有控制点的像片索引图
photographing 饰墙后底层图迹在面层上透视;隐现凹凸
photographing base(line) 摄影基线
photograph library 照片收藏室
photograph meridian 像片主纵线
photograph montage 摄影拼接;摄影剪辑;像片镶嵌
photograph number 航摄像片编号
photograph orientation angle 摄影取向角
photograph overlap 像片重叠
photograph parallel 像水平线;像横线
photograph perpendicular 像片垂线;摄影机主光轴
photograph plumb point 像底点
photographs of progress 进度照片
photograph transmission system 传真传输系统
photography 照相术;摄影学;摄影术;摄影测量学
photography by artificial light 灯光摄影
photography organization 摄影单位
photography room 照相室
photography time 摄影时间
photography with quartz lens 石英透镜摄影
photograve 摄影凹版
photogravure 影印凹版;照相制版凹版法;照相凹版印刷;照相凹版
photogravure ink 凹版用油墨
photogravure press 凹版印刷机
photogravure printing 照相凹版印刷
photogravure proofing machine 凹印打样机
photo-grey glasses 变色眼镜
photogrid 坐标变形试验
photogun 光电子枪

photogyration 光回转
photohalide 感光性卤化物
photoHall effect 光霍耳效应
photohalogenation 光卤化作用;光卤化
photohardening 光硬化
photohead 光电传感头
photohead plotter 光电传感头绘图机
photoheliogram 太阳照片;太阳全色照片
photoheliograph 太阳照相仪;太阳全色照相仪
photohobia 不耐光
photohole 光穴
photo-holoelasticity 全息光弹(方)法
photohyalography 照相蚀刻术;光照蚀刻术
photoimpact 光碰撞;光控脉冲;光电脉冲;光冲量
photoinactivation 补体光灭活作用;光灭活作用;光钝化作用
photo index 像片索引图;相片索引图
photo indexing 像片检索
photoinduced 光致;光感生的
photoinduced carrier 光生载流子
photoinduced chemistry 光感应化学
photoinduced Hall effect 光生霍尔效应
photoinduced strain 光致应变
photoinduction 光诱导;光感应
photoinitiation 光引发(作用)
photoinitiator 光致引发剂
photoinjected charge 光注入电荷
photoinjection 光注入
photoinjection electro-photography 光致注入电子摄影
photoinstrumentation 照相仪
photointerpretation 像片判读;照片解释;照片判读
photo interpretation key 像片判读标志
photo interpretative technique 像片判读技术
photointerpreter 相片判读者;图像判读者;照片判读员;光电判读仪
photointerpretive program(me) 光电翻译程序
photointerrupter 光电遮断器;光斩波器
photoion 光离子
photoionization 光致电离;光离子化;光化电离
photoionization cross-section 光子电离截面;光致电离截面
photoionization detector 光致电离检测器;光离子化检测器;光电离检测器
photoionization efficiency 光致电离效率
photoionization ion source 光致电离离子源
photoionization radiation 光致电离辐射
photoionization source 光致电离源
photoionization transition 光电离跃迁
photoisland grid 镶嵌器
photoisolated 光电隔离的
photoisolator 光隔离器
photoisolator trigger 光隔离触发器
photoisomer 光致同分异构体
photoisomeric change 光化学异构化;感光异构变化
photoisomerism 感光特性异构
photoisomerization 光致异构化;光化学异构化
photojournalism 摄影报道
photojunction battery 光电结电池
photojunction cell 光结电池
photokey 航摄面积示意图;航摄地区略图
photokinesis 光动现象;光动态
photokinetic 趋光的
photoklystron 光电速调管
photokymograph 光转筒记录器;记录照相机
photolability 不耐光
photo laboratory 摄影实验室
photolayer trace 光敏层中径迹
photolens chromatism 摄影透镜色差
photo-level 发光电平表
photo-line production 影像轮廓线法
photolithographic 光刻法的
photo-lithographic(al) method 摄影刻印术
photolithography 影印法;照相平印术;照相平版印刷术;光刻蚀法;光刻
photo-litho plate 照相平版
photolocking 光锁定
photolofting 光学放样
photolog 摄影记录
photologic synthesizer 光逻辑合成器
photology 光学;成像学
photolometer 光电比色计
photoluminescence 光致发光;光照发光;光激发光

photoluminescence detector 光致发光探测器
photoluminescence efficiency curve 光致发光效率曲线
photoluminescence powder 光致发光粉
photoluminescence thermometer 光致发光温度计
photoluminescent 光激荧光的
photoluminescent detector 光致发光探测器
photoluminescent efficiency curve 光致发光效率曲线
photoluminescent exit marking system 荧光安全出口标志系统
photoluminescent powder 光致发光粉
photoluminescent thermometer 光致发光温度计
photolysis 光解(作用);光分解(作用)
photolysis gas chromatography 光解气相色谱
photolyte 光解质
photolytic 光解的
photolytic cycle 光解循环
photolytic silver 光解银
photolyzable film 光解薄膜
photomacrograph 宏观照相;宏观照片;宏观相片;粗型照片
photomacrographic magnification 宏观摄影放大;低倍摄影放大
photomacrography 宏观摄影术
photomagnetic 光磁的;光磁
photomagnetic disintegration 光磁蜕变;光磁分裂
photomagnetic effect 光磁效应
photomagnetic memory 光磁存储器
photomagnetism 光磁性
photomagnetoelectric(al) effect 光磁电效应
photomagnetoelectric(al) theory 光电磁理论
photomania 光性躁狂
photomap 空中摄影地图;像片平面图;相片地图;影像地图;照相地图;航测地图;摄影制图
photomapper 光制图仪
photomapping 像片测图
photomapping equipment 摄影测绘设备
photo marking equipment 光学划线机
photomask 遮光模;光掩模
photo masking 光掩蔽
photomaterial 摄影材料
photomaton 自动摄影机
photomechanical 照相制版法
photomechanical art publication 照相制版美术品
photomechanical copy 翻版
photomechanical copying 照相制版复制
photomechanical copying equipment 复制机
photomechanical effect 光机械效应;光机效应
photomechanical method 光学机械法
photomechanical paper 照相制版纸
photomechanical plant 胶印厂
photomechanical plate 照相制版印刷版
photomechanical plate making 照相制版
photomechanical print 照相制版印刷品
photomechanical printing 影印
photomechanical process 照相制版工艺;照相制版
photomechanical reduction 照相缩小
photomechanical treatment 摄影处理
photomechanics 照相制版印刷术;照相制版工艺
photomemory 光存储;光电存储器
photomeson 光介子
photometallic etching 光金属蚀刻
photometeor 大气光学现象
photometer 光度计曝光表;光度计;曝光表;散发仪;测光仪;分光计
photometer axis 测光轴
photometer bench 光力测度台;光度计座;光度计架;测光导轨
photometer constant 光度计常数
photometer eyepiece 光度计目镜
photometer field 光度计场
photometer head 光度头;光度计头;光电头
photometer head with equality of brightness prism 等亮度棱镜的光度头
photometer head with equality of contrast prism 等衬度棱镜的光度头
photometering 光度测定
photometer screen 光度计屏;曝光表玻璃罩
photometer test plate 光度计试验板
photometric(al) 计光的
photometric(al) absolute magnitude 绝对测光星等
photometric(al) analysis 光度分析

photometric(al) analyzer 光度分析器
photometric(al) aperture 光度孔径
photometric(al) bench 光度台
photometric(al) brightness 光照度;光度学亮度
photometric(al) brightness scale 光度亮度划分;光度亮度范围
photometric(al) calibration 光度校正
photometric(al) colo(u)r contrast 光度的色衬比
photometric(al) computer 测光计算机
photometric(al) copying 光学复制
photometric(al) cube 测光用立方体;光度计棱镜
photometric(al) curve 测光曲线
photometric(al) data 测光数据
photometric(al) determination 光度测定
photometric(al) error 光测误差
photometric(al) gas analyzer 光谱式气体分析器
photometric(al) head 光度计头
photometric(al) integrating sphere 光度积分球
photometric(al) integrator 光度积分器
photometric(al) intensity 光强度;光强
photometric(al) measurement 光度测量;光电测定法
photometric(al) method 光度(计)法;光度测量法;光度测(定)法
photometric(al) paper 光度测定纸
photometric(al) paradox 光度佯谬
photometric(al) parallax 光度视差
photometric(al) period 光度周期
photometric(al) phase 光度相;光度位相
photometric(al) pyrometer 光测高温计;光度高温仪
photometric(al) quality 光度质量
photometric(al) quantity 光度值;光度量
photometric(al) radiation equivalent 光度等效值
photometric(al) readout unit 测光读数仪
photometric(al) receiver 光度计
photometric(al) relationship 光度关系
photometric(al) requirement 光度要求
photometric(al) scale 光度标
photometric(al) scanning 光度扫描
photometric(al) sorter 光电分选机
photometric(al) standard 测光用标准光源;光度标准(器)
photometric(al) term 光度学术语
photometric(al) test plate 测光试验板;光度试验板
photometric(al) titration 光度滴定
photometric(al) titrimetry 光度滴定法
photometric(al) transfer function 光度传递函数
photometric(al) unit 光度单位;光度学单位
photometric(al) wedge 测光楔;光度测量楔
photometrology 摄影测量术
photometry 光度学;光度滴定法;光度测定法;光测学;计光术;测光学;测光术;测光法
photomicrocopy 缩微照相复印本
photo microdensitometer 影像微密度计
photomicrograph 显微照片;微观照片
photomicrographic(al) camera 显微照相机
photomicrographic analysis 照相显微分析
photomicrographic apparatus 显微照相设备
photomicrography 显微照相术;显微摄影(术);缩微照相术;粗型照相术
photomicrometrology 显微照相摄影测量术
photomicroscope 显微照相机;显微摄影机;照相显微镜;照相显微镜;摄影显微镜
photomicroscopy 显微照相术
photomigration technique 向光移动技术
photomilling 光铣法
photomixer 光混频器
photomixer diode 光混频二极管
photomixing 光混频
photomixing technique 光混频技术
photomodified airplane 改装的航空摄影飞机
photomodulator 光调制器
photomontage 片镶嵌图;集成照片;照相拼版;合成照片
photomorphism 光形态建成;光形态发生
photomorphogenesis 光形态建成;光形态发生
photomorphosis 光形态建成;光形态发生
photomosaic 像片镶嵌(图);照相镶嵌;照片拼接;感光特性镶嵌幕
photo mosaic index 像片镶嵌索引图
photomotive force 光动势
photomountant paste 像片镶嵌胶
photomultifier with low-noise 低噪声光电倍增器

photomultiplier 光电倍增器
photomultiplier cell 光电倍增器;光电倍增晶体管;光电倍增管
photomultiplier counter 光电倍增计数器
photomultiplier detector 光电倍增(管)探测器;光电倍增管检测器
photomultiplier light pen 光电倍增管光笔
photomultiplier monitor 光电倍增管监控器
photomultiplier noise pulse 光电倍增管噪声脉冲
photomultiplier tube 光电倍增管
photomultiplier tube microphotometer 光电倍增管光度计
photomultiplier with high quantum efficiency 高效率光电倍增管
photomultiplier with wide spectral response 宽光谱光电倍增管
photomultisensor 光电多路传感器
photomural 大幅照片;(装饰性的)大幅像片;照相壁画
photon 光子;光电子
photon absorption 光子吸收
photon activated switch 光子起动开关
photon activation analysis 光子活化分析
photon amplification 光子放大
photon beam 光子束
photon bombardment 光子轰击
photon bunching 光子聚束
photon camera 光子摄像机
photon capture 光子俘获
photon component 光子分量
photon converter 光子转换本
photon correlation 光子相关
photon correlation spectrometer 光子相关光谱仪
photon correlation spectroscopy 光子关联能谱法
photon counter 光子计数器
photon counting 光子计数
photon counting detector 光子计数检测器
photon counting mode 光子计数模式
photon counting radiometer 光子计数辐射计
photon counting spectrophotometer 光子计数分光度计
photon counting statistics 光子计数统计学
photon-coupled array 光子耦合阵列
photon-coupled isolator 光子耦合隔离器
photon-coupled isolator device 光子耦合隔离器
photon-coupled pair 光子耦合对
photon coupling 光子耦合
photon crosslink 光子交联
photon curve 光子曲线
photon degeneracy 光子简并度
photon density 光子密度
photon detector 光子探测器;光子检波器
photon drag detector 光子牵引探测器
photon drag effect 光子牵引效应
photon drag effective 光子牵引效应
photon drag photodetector 光子牵引光探测器
photon echo 光子回波
photon effect 光子效应
photonegative 负光电导性的
photonegative effect 负光电效应
photon emission 光子发射
photon emission noise 光子发射噪声
photon emission spectrum 光子发射谱
photon energy 光子能量
photon engine 光子发动机
photonephelometer 光电浊度仪;光浊度计
photoneutrino 光生中微子
photoneutrino effect 光致中微子效应
photoneutrino process 光生中微子过程
photon excited atom 光子激发原子
photonexcited electron 光子激发电子
photon exitance 光子出度
photon field 光子场
photon fluctuation 光子起伏;光子波动
photon flux 光子通量
photon flux density 光子通量密度
photon gain 光子增益
photon gas 光子气(体)
photon-generated carrier 光生载流子
photonic vision 明视觉
photon-induced action 光子感生作用
photon-induced fluctuation 光子感应波动
photon intensity 光子强度

photon irradiance 光子辐射度
photonitrosation 光亚硝(基)化
photon liberation 光子释放
photon lifetime 光子寿命
photon limited sensitivity 光子限灵敏度
photon limited signal 光子限信号
photon log 光子测井
photon meter 光子计
photon momentum 光子动量
photon monochromator 光子单色仪
photon noise 光子噪声
photon noise limit 光子噪声极限
photonoise 光噪声
photon-phonon collision 光子声子碰撞
photon-photon interaction 光子光子相互作用
photon population 光子布居
photon quenching 光子猝灭
photon radiation 光子辐射
photon recoil effect 光
photon sail 太阳帆
photon scattering 光子散射过程
photon sphere 光子层
photon statistics 光子统计学
photon stream 光子流
photon structure 光子结构
photon theory 光子理论
photon thermal conductivity 光子热导率
photon trapping 光子俘获
photonuclear absorption 光核吸收
photonuclear activation analysis 光致核活化分析
photonuclear reaction 光核反应
photonuclear reaction threshold 光核反应阈
photonucleation 光致晶核形成;见光度
photo number of geologic(al) outcrop 地质露头照片数
photon velocity 光子速率
photon yield 光子产额
photo-offset 照相胶印法
photo-offset copy 影印本;影印
photo-offset print 摄影胶版印刷
photooptic(al) coding 光学照相编码
photooptic(al) memory 光存储器
photooptic(al) receiving unit 光接收器
photooptic(al) tracking data 光学摄影跟踪数据
photooptics 摄影光学
photoorganotrophy 光有机营养
photoorthotropic elasticity 光正交各向异性弹性
photooscillogram 光波形图
photo-oxidant 光(化学)氧化剂
photooxidant stabilizer 防光老化剂;光氧化稳定剂
photooxidation 光致氧化(作用);光氧化(作用);感光特性氧化作用
photooxidation decolo(u)rization 光氧化脱色
photooxidation reaction 光氧化反应
photooxidation treatment 光氧化处理
photo-oxidation treatment of wastewater 废水光氧化处理法
photooxidative decolo(u)rization 光氧化脱色
photooxidative degradation 光氧化降解
photooxydation actinometer 光致氧化感光计
photooxygenation 光致生氧
photoparametric(al) diode 光参量二极管
photoparesthesia 光觉异常
photopatch test 光斑片试验
photopathy 避光性
photopeak 光峰;光电峰
photopen recorder 光电光笔记录器
photoperceptivity 感光能力
photoperiod 光周期;光照期
photoperiodic(al) 光照期的
photoperiodic(al) aftereffect 光周期余效
photoperiodic(al) induction 光周期诱导
photoperiodic(al) inhibition 光周期抑制
photoperiodic(al) perception 光周期感应
photoperiodic(al) reaction 光周期反应
photoperiodic(al) regime 光周期状况
photoperiodic(al) response 光周期反应
photoperiodic(al) response variety 感光性品种
photoperiodic(al) sensitivity 光周期的敏感性
photoperiodicity 光(照)周期性
photoperiodism 光周期性
photoperiod sensitivity 光照敏感性
photoperspectograph 摄影透视仪
photophase 光相应

photophilic 嗜光的;喜光的;适光的
photophilous 喜光的;适光的
photophil phase 光感期
photophobic 避光的
photophobic photolabile 不耐光的
photopholic 嫌光的
photophone 光音器;光音机;光线电话机;光电话机;光电传声器
photophoresis 光致运动;光泳现象
photophoshorescene phenomenon 光激荧光现象
photophosphorescence 光致磷光
photophosphorylation 光致磷酸化作用;光磷酸化(作用)
photopia 光适应性;光适应
photopic condition 光适应条件
photopic curve 适光曲线
photopic range 适光范围
photopic response 白昼视觉响应
photopic vision 昼光视觉;可见光视觉;明视(觉);明光视觉;日光视觉;白昼视觉
photopiezo-electric(al) effect 光压电效应
photopigment 光色素;光敏色素;感光色素
photoplan 像平面图;像片图;像片平面图;像片平面
photoplane 航摄飞机;摄影平面
photo plane coordinate system 像平面坐标系
photo-planimetric(al) method 航测综合法
photo-planimetric(al) method for field point layout 综合法全野外布点
photo-planimetric(al) method of monophotogrammetry 综合法单张像片测图
photo-planimetric(al) mapping 综合法测图
photo-planimetric(al) method of photogrammetry 综合法像片测图
photo-planimetric(al) method of photomap mapping 综合法像片测图
photo-planimetric(al) method of single photograph mapping 综合法单张像片测图
photoplast 简易像片量测仪
photoplastic effect 光塑性效应;光范性效应
photoplastic film 光塑胶片
photoplasticity 光塑性力学;光塑性
photoplasticity method 光塑性法
photoplastic material 光塑材料
photoplastic measurement 光塑性测量
photoplastic model test 光塑性模拟试验
photoplastic recording 光塑记录;感光塑料记录
photoplastics 光敏塑料
photoplate 照相底片;乳胶片
photoplethysmograph 光体积描记器
photoplotter 像片测图仪
photoplotting apparatus 摄影测图仪(器)
photoplotting instrument 摄影地形图
photopography 照相地形图
photo-point control 像片点联测
photopolarimeter 光偏振计
photopolarimetry 照相偏振测量
photopolarography 光极谱法
photo polygon 像片导线
photopolygoniometry 像片导线测量
photo polygonometric traverse 像片导线
photopolymer 光致聚合物;光聚合物;感光聚合物
photopolymer coating 感光性树脂涂层
photopolymer holography 光聚合物全息术
photopolymerisable 光致聚合的;光聚合的
photopolymerization 光致聚合(作用);光聚作用;光聚合(作用);光化聚合(作用)
photopolymerizer 光聚合剂
photopolymer plate 感光聚合物印版
photopositive 正趋光性;正光电导性;光正性的
photo pot 光穴
photo potential 光势;光生电位
photopotentiometer 光电电位器
photopredissociation 光致预离解
photopret 像片判读仪
photo-principal distance 像主距
photoprint 影印
photoprinter 晒印员;影印机;光印机;晒像机
photoprinter-rectifier 纠正晒像机
photoprocess 光学处理
photoprocessing 光学加工;摄影处理
photoprocessing waste solution 光学处理废液
photoproduced meson 光生介子
photoproduct 光化产品;摄影成果

photoproduction 光致作用;光生
photoproduction cross-section 光致作用截面;光生有效截面
photo profile in well 井下摄影剖面
photo projector 像片投影仪
photoproof 像片校正
photoprotection 光照保护法;光损伤防护;防光
photoptometer 光敏计;光觉计
photoptometry 辨光测验法
photoradiochromatography 光放射色谱法
photoradiogram 无线电传真电报;无线电传真图片
photoradiometer 光辐射计
photoreaction 光致反应;光(化)反应
photoreactivation 光照逆转作用;光照逆转;光活化作用;光复活作用
photoreactive chlorophyll 光反应性叶绿素
photoreactivity 光反应性
photo-reactor 光反应器
photoreader 光电阅读器;光电输入机;光电读出器
photoreading 图像判读;光读数
photoreceiver 光接收器;光接收机
photo-reception 像片接收;光感受
photoreceptor 光受器;光敏接收器件;光感受器;感光(受)器
photorecombination 光致复合;光复合
photorecombination laser 光复合激光器
photorecon 摄影侦察
photorecon equipment 摄影侦察设备
photoreconnaissance 航空摄影侦察;摄影侦察;摄影勘察;摄影勘测
photoreconnaissance pilot 摄影勘测员
photoreconnaissance satellite 摄影侦察卫星
photorecorder 照相记录器;摄影记录器
photorecovery 光致复活作用;光照活化作用
photo rectification 像片纠正
photorectifier 光电检波器;光电检波管;光电二极管
photoredox reaction 光氧化还原反应
photoreducer 像片缩小仪;缩小仪
photoreduction 像片缩小;照相缩版;光致还原;光还原作用
photoreflective polarization coupler 光反射偏振耦合器
photorefraction 光折射
photorefractive coupler 光折射耦合器
photorefractive effect 光折射效应;光折变效应
photorefractive intermodal exchanger 光折射模内交换器
photorelay 光控继电器
photorelief map 摄影地形图
photorepair mechanism 光修复机制
photorepeater 影印机;照相复印机
photoreportage 摄影报道
photo-reproduction means 照相复制设备
photoresist 光阻材料;光致抗蚀剂;光敏抗蚀剂;光刻胶;感光性树脂;感光耐蚀膜;感光保护膜
photoresistance 光敏电阻
photoresistance cell 光敏电阻元件
photoresistance compensator 光敏电阻补偿器
photoresistance relay 光敏电阻继电器
photoresist-containing wastewater 含光致抗蚀剂废水
photoresist flow 光致抗蚀通道
photoresistive cell 光电导管
photoresistor 光敏电阻器;光敏电阻;光电导管
photoresistor-cell relay 光敏元件继电器
photoresist process 光致抗蚀工艺
photorespiration 光呼吸作用
photoresponse 光响应
photoreversal 光致复活作用;光照逆转(作用);光照活化作用
photo-revised map 像片修测图
photoroentgenography 荧光射线摄影
photorotary encoder 光致旋转编码器
photo scale 像片比例尺
photo scale figure 像片比例尺数字
photoscan 光扫描图
photoscanner 摄影扫描器
photoscanning 光扫描
photoscanning system 光扫描系统
photoscan system 像片扫描系统
photoscattering particle-size analyzer 光散射粒径分析仪
photoscintillater 光闪烁器
photoscope 透视镜荧光屏;透视镜

photoscreen method 光屏法
photo-second 辐透秒(曝光单位)
photosedimentation 消光法(粒径分析方法之一);测光沉淀法
photosedimentation analysis 照相沉淀物分析
photo sedimentation apparatus 沉淀摄影器
photosedimentometer 光透式粒度测定仪
photosensibility 感光度
photosensing marker 光电读出标记;光电检测标志
photosensing material 光敏材料
photosensitiser 感光剂
photosensitive 光致变色的;光敏的
photosensitive adhesive 光敏胶黏
photosensitive area 光敏区;光敏面积
photosensitive cathode 光敏阴极
photosensitive cell 光敏电池;光电管
photosensitive coating 光敏镀层;感光层
photosensitive device 光敏装置
photosensitive diode 光敏二极管
photosensitive emulsion 光敏乳胶;感光乳剂
photosensitive fiber optic(al) sensor 光敏光纤传感器
photosensitive film 光敏膜;光敏薄膜
photosensitive glass 光敏玻璃
photosensitive glass-ceramics 光敏微晶陶瓷
photosensitive layer 光敏层
photosensitive material 感光材料
photosensitive material additive 感光材料助剂
photosensitive materials 光敏材料
photosensitive medium 光敏媒质
photosensitive mosaic 光敏镶嵌屏
photosensitive paper 像纸
photosensitive plane 像片感光面
photosensitive polymer 光敏聚合物
photosensitive polymerization 光敏聚合
photosensitive portion 感光部分
photosensitive resin 光敏树脂;感光树脂
photosensitive resist 光敏抗蚀剂;光刻胶
photosensitive resistance 光敏电阻
photosensitive sealant 光敏封闭剂
photosensitive sealer 光敏封闭剂
photosensitive semiconductor 光敏半导体
photosensitive semiconductor sensitizer 光敏半导体增敏剂
photosensitive surface 光敏(表)面;感光面
photosensitive switch 光敏开关
photosensitive tube 光敏管
photosensitivity 光敏(感)性;光感性;感光灵敏度
photosensitivity test 光敏感试验
photosensitization 光致敏作用;光敏(化)作用;光感作用;感光作用
photosensitized initiation 光敏引发(作用)
photosensitized oxidation 光敏化氧化作用
photosensitized reaction 光敏反应
photosensitizer 照相增感剂;光敏剂;光敏化剂;感光剂
photosensitizing coating 光敏涂料;光敏涂层
photo sensitometer 感光仪
photosensor 光敏器件;光传感器;光敏元件;光电探测器;光电传感器
photosenstizer 光敏材料
photoseparating uranium isotopes 光分离铀同位素
photoseparation 光分离
photosetting 照相排版;光固化
photosignal 光信号
photosignal channel 光信号通路
photo-skinned model 航摄像片贴面模型
photosonochemical degradation 光声化学降解
photosource 光源
photo space auxiliary coordinate system 像空间辅助坐标系
photo space coordinate system 像空间坐标系(统)
photosphere 光球(层)
photospheric eruption 光球爆发
photospheric facula 光球光斑;光斑
photospheric model 光球模型
photospheric network 光球网络
photospheric radiation 光球辐射
photospheric spectrum 光球光谱
photospheric surface 光球表面
photospheric telescope 光球望远镜
photospheric temperature 光球温度

photospot 聚光；摄影聚光灯
photostability 耐光性
photostabilization 光稳定化
photostable 见光安定的；耐光的；不感光的
photostage 光照阶段
photostat 复印机；影印法；直接影印件；直接影印机
photostat copy 晒印图；影印本
photostat print 照相复印
photostat printing 复印机印制
photostereograph 立体测量仪；立体测量仪
photostimulation 光激发作用；光刺激
photostimulator 光刺激器
photostratigraphy 摄影地层学
photostress method 光敏涂层应力分析法
photo strip 摄影航线
photostudio 摄影棚；摄影场；照相馆；照相室
photosurface 光敏面
photo-surface model 航摄像片贴台模型
photosurface spectral response 光敏面光谱响应
photoswitch 光敏开关；光控开关；光控继电器
photosynchronous pick-up 光电同步拾波器
photosynthesis 光作用；光能合成；光合作用
photosynthetic(al) ability 光合能力
photosynthetic(al) acting 光合作用
photosynthetic(al) action 光合作用
photosynthetic(al) activity 光合活性
photosynthetic(al) algae 光合藻类
photosynthetic(al) apparatus 光合作用仪
photosynthetic(al) area 光合面积
photosynthetic(al) aulfide oxidation 光合硫化物氧化作用
photosynthetic(al) autotroph 光合自养生物
photosynthetic(al) bacteria 光合细菌
photosynthetic(al) conversion efficiency 光合转化效率
photosynthetic(al) cycle 光合循环
photosynthetic(al) efficiency 光合效率
photosynthetic(al) layer 光合层；透光带
photosynthetic(al) number 光合比值
photosynthetic(al) organism 光合生物
photosynthetic(al) oxygen 光合氧
photosynthetic(al) oxygenation 光合供氧
photosynthetic(al) phosphorylation 光合磷酸化作用
photosynthetic(al) pigment 光合颜料；光合色素
photosynthetic(al) process 光合过程
photosynthetic(al) productivity 光合生产率
photosynthetic(al) quotient 光合商
photosynthetic(al) ratio 光合商
photosynthetic(al) reaction 光合反应
photosynthetic(al) system 光合系统
photosystem 光合系统；光合体系
photosystem activities 光合系统活性
photo tach(e)ometer 光电转速计
photo-tacheometry 摄影视距测量法
phototactic movement 趋光运动
phototactic reaction 趋光反应
phototactic rhythm 趋光节律
phototape camera 光电带录像机
phototape reader 光(电)带读出器；光电穿孔带读出器
phototatic 趋光的
phototched mask 光刻掩模
phototelegram 传真电报
phototelegraph 传真电报机
phototelegraphic(al) receiver 传真接收机
phototelegraphy 传真电报术
phototelephone 传真电话；光电话机；光电电话；传像电话
phototelephony 光线电话
phototelescope 照相望远镜
phototelescopic technique 摄影望远镜装备
photo-theodolite 摄影读数经纬仪；摄影经纬仪；照相经纬仪
photo-theodolite camera 照相经纬仪
photo-theodolite objective 摄影经纬仪物镜
photo-theodolite survey 摄影经纬仪测量
photothermal 光热
photothermal cabinet 光热箱
photothermionic 光热离子的
photothermionic image converter 光热离子变像管
photothermoelasticity 光热弹性；光热弹性
photothermoelastic method 光热弹性法
photothermography 光热照相术

photothermomagnetic 光热磁性的
photothermometry 光测温学
photothermy 光热作用
photothyristor 光闸流管
phototimer 照相计时器；光亮计时器；光电定时器；曝光计；摄影计时器
photo tonality 摄影的明暗部分；摄影色调
phototonus 光紧张
phototopography 摄影地形测量；照相地形测量；摄影测量学
phototoxicity 光毒性
phototoxis 光线损害
phototransformation 光转化
photo transformer 像片纠正仪
phototransistor 光电晶体管；光敏晶体三极管；光敏晶体管
phototransistor circuit 光电晶体管电路
phototransistor detector 光电晶体管检测器
phototransit 摄影读数经纬仪；照相经纬仪
phototranslating system 光电转换系统
photo transmission 图片发射
phototriangulation 像片三角测量；摄影三角测量
photo triangulation instrument 像片三角测仪
phototriangulator 摄影三角测量仪
phototriode 光电三极管
phototroller 光控继电器
phototronic photocell 光电子光电池
phototroph 光养型；光能利用菌
phototrophic 光养性的
phototrophic bacteria 光能利用菌
phototrophy 光营养
phototropic 向光性的；向光的
phototropic body 光变色体
phototropic glass fiber 光敏玻璃纤维
phototropic material 光色材料
phototropic movement 向光运动
phototropism 向光性；光色互变(现象)；光变性；趋光性
phototropy 光致变色现象；光色随入射波长的变化；光色互变(现象)
phototube 光电管；光灯管
phototube cathode 光电管阴极
phototube relay 光电管继电器
phototube-type fractional collector 光电管型分次收集器
phototvulcanization 光硫化
phototype 照相版术；摄影制版；摄影凸版
phototype machine 照相制版机
phototypesetter 照相排字机
phototype travel(l)ing wave tube 光电型行波管
phototypography 照相印刷术
photounit 光电元件
photovalve 光电元件；光电管
photovariator 光敏电阻
photovaristor 光敏变阻器；光电变阻器
photoviscoelasticity 光黏弹性
photovisual achromatism 光化视觉消色差性
photovisual magnitude 仿视星等
photovoltage 光电压
photovoltaic 光生伏打的
photovoltaic cell 阻挡层光电池；光生伏打电池；光伏管；光电池
photovoltaic converter 光伏变换器；光电能量变换器；光电变换器
photovoltaic detection 光生伏打探测
photovoltaic detector 光生伏打探测器；光电探测器
photovoltaic device 光生伏打器件
photovoltaic effect 光生伏打效应；光伏效应；光电势效应；光电伏打效应
photovoltaic glass 光致电压玻璃；光电玻璃
photovoltaic infrared detector 光伏红外探测器；光电红外探测器
photovoltaic junction 光生伏打结
photovoltaic meter 光生伏打曝光表；光生伏打计；光伏计
photovoltaic mode 光电形式
photovoltaic optic(al) sensor 光伏式光传感器
photovoltaic photodetector 光生伏打光电检测器；光伏式光检测器
photovoltaic pick-up tube 光生伏打摄像管
photovoltaic process 光生伏打过程
photovoltaic radiation detector 光电流辐射探测器
photovoltaic sensor 光伏式传感器
photovoltaic technology 光电技术

photovoltaic turbidimeter 光电浊度仪；光电浊度计；光电测砂仪
photox cell 氧化亚铜光电池
photoxide 光氧化物
photoxylography 照相木版印刷术
photoyield 光产额
photozincograph 照相锌版印刷品
photpolymer 光聚合物
photronic cell 光电池
photronreflectometer 光电反射计
phoxim 锌硫磷；腈肟磷
phragmites peat 芦苇泥炭
phragmobasidium 多隔担子
phragmocone 闭锥
phragmoid 成膜的
phragmoplast 成膜体
phragmosome 成膜粒
Phragmosporae 多隔孢子类
phrase 词组
phrasing 分节法
phreatic 准火山的；潜水的；地下的
phreatic aquifer 潜水含水层；地下含水层；地下含水层；无压地下水含水层
phreatic cycle 潜水周期(指地下水)；潜水位变化周期；地下水(涨落)循环；地下水位变化周期
phreatic decline 潜水下降；地下水位降低
phreatic deep buried zone 潜水深埋带
phreatic discharge 潜水排出；地下涌水流量；地下水涌流量；地下水流量；地下水出流量
phreatic divide 潜水水界；潜水分水岭；潜水分界线；潜水分界面；地下水分线；地下水分水岭；地下水分界
phreatic eruption 蒸汽喷发
phreatic evapo(u)ration 潜水蒸发
phreatic fluctuation 潜水面升降；潜水动态；潜水波动；地下水位涨落；地下水位升降变化；地下水面波动
phreatic gas 准火山瓦斯
phreatic groundwater 潜水(指地下水)；无压地下水
phreatic high 最高地下水位；高地下水位；潜水高水位；潜水动态曲线峰点；地下水上含水层
phreatic hydrothermal eruption 潜水水热喷发
phreatic level 潜水水位
phreatic line 潜水线(指地下水)；渗流线；地下水线；地下水位线；地下水流线；地下井水线
phreatic low 最低地下水位；潜水动态曲线谷点；地下水下降含水层；地下水下部含水层；地下水凹面；低地下水位
phreatic overflowing zone 潜水溢出带
phreatic permeability 渗水性
phreatic rise 潜水上升(指地下水)；地下水位上升；地下水上升
phreatic sinking zone 潜水下沉带
phreatic surface 地下水水位；潜水面(指地下水)；地下水面
phreatic water 自由水；潜水(指地下水)；无压地下水；地下水
phreatic water discharge 潜水出流；地下水涌出；地下水流；地下水出流
phreatic water flow 地下水流
phreatic water in pluvial fan 洪积扇潜水
phreatic water in sand dune 沙丘潜水
phreatic water in valley 河谷潜水
phreatic water level 潜水位；潜水水位；地下水位；地下水水面
phreatic water of continental salinization 大陆盐化潜水
phreatic water regime in irrigation area 灌区潜水动态
phreatic water regime on bank 岸边潜水动态
phreatic water regime on divide 分水岭潜水动态
phreatic water surface 地下水水位
phreatic water table 无压地下水位；地下水位
phreatic water well 潜水井
phreatic water zone of continental salinization 大陆盐化潜水带
phreatic wave 潜水波
phreatic zone 潜水带；地下水层；饱和层
Phreatoicidae 槽穴科
phreatophyte 潜水植物；潜水湿生植物；地下水湿生植物
phreatophytes 深根植物
pH-recorder 氢离子浓度记录仪
phroparaffine 重质蜡

phrygana 常绿短灌木丛
Phrygian marble 弗利吉亚大理石
pH-stat method 酸碱法
pH test paper 氢离子指数试纸
phthalaldehyde 苯二醛
phthalaldehydic acid 苯醛酸
phthalate 邻苯二甲酸盐;酞酸盐
phthalate anhydride 邻苯二甲酸酐
phthalate ester 邻苯二甲酸酯
phthalate resin paint 邻苯二甲酸盐树脂涂料
phthalate resin varnish 邻苯二甲酸盐树脂清漆
phthalic acid 酞酸;邻苯二甲酸;苯二甲酸
phthalic acid resin 邻苯二甲酸树脂
phthalic anhydrate 邻苯二甲酸酐
phthalic anhydride 邻苯二甲酸酐
phthalic resin 邻苯二甲酸树脂
phthalic resin enamel 邻苯二甲酸树脂瓷漆;醇酸瓷漆
phthalide 苯酞
phthalimide 邻苯二甲酰亚胺
phthalocyanine 酞菁素;酞菁
phthalocyanine blue 酞菁蓝
phthalocyanine dye 酞菁染料
phthalocyanine dyestuff 苯二甲酸染料;酞化青染料
phthalocyanine green 酞化青绿;酞菁绿
phthalocyanine pigment 酞菁颜料;酞菁系颜料
phthalonitrile 邻苯二甲腈
phthalophenone 二苯代酚酞
phthalphenol 苯二甲酚
phthalylhydrazine 邻苯二甲酰肼
phthalylsulfacetamide 邻苯二甲酰磺乙酰胺
phthanite 致密硅页岩
phtoremediation 植物修复
phtovolatilization 植物挥发法
phugoid 长周期振动
phugoid curve 起伏运动曲线;长周期振动曲线
phugoid mode of metion 长周期运动
phugoid motion 起伏运动
phugoid oscillation 浮沉振荡
phunki 成熟紫胶
phuralumite 柱磷铝铀矿
phurcalite 束磷钙铀矿
pH value 酸碱度;pH 值;氢离子浓度负对数值
pH-value computer 土壤酸度探测器
pH value determination 酸碱值测定
pH value meter of mud 泥浆 pH 值计
pH-value of oil field water 油田水酸碱度
pH value of water 水的酸碱值
phycobiont 共生藻
phycochrome 淡水藻色素
phycocolloid 藻胶
phycological biochemistry 海藻生物化学
phycological ecology 海藻生态学
phycological systematics 海藻分类学
phycology 藻类学
phycomycete 藻菌(植物)
phycophyta 藻类植物
phycoplast 藻膜体
Phygon 二氯醌(杀菌剂)
phyiscal state of the environment 环境的自然状况
phyletic classification 系统分类
phyletic extinction 线系绝灭
phyletic speciation 线系物种形成
phyllarenite 叶砂屑岩
phylliform 叶形
phyllite 鳞片状矿物;千枚岩
phyllite aggregate 千枚岩集料;千枚岩骨料
phyllite metasiltstone 千枚岩变质粉砂岩
phyllite metavolcanitic rock 千枚状变质火山岩
phyllite mylonite 千枚糜棱岩
phyllite quartzite 千枚石英岩
phyllites 化石叶
phyllite slate 千枚板岩
phyllitic granite 千枚岩状花岗岩
phyllitic slate 千枚状板岩
phyllitic structure 千枚岩状构造;千枚状构造
phyllochlorite 叶蠕绿泥石
phylloclade 叶状枝
phyllocladene 扁枝烯
phyllocrystalline rock 结晶千枚岩
phyllode 叶状柄;瓣步带
phylloid cladode 叶状枝
phyllomorphic stage 自生成矿变化阶段;晚期成岩作用阶段

phyllonite 千枚糜棱岩
phyllosilicate 页硅酸盐;层状硅酸盐
phyllosilicate minerals 层状构造硅酸盐矿物
phyllotelinite 叶结构镜质体
phyllotheca 隔壁内墙
Phyllotreta cupreata 铜色黄条跳甲
phyllotungstit 叶铁钨华
phyllovitrinite 叶镜质体
phyllovitrite 微叶镜煤
phylocoenogenesis 群落系统发育
phyloephebic 最优势期的
phylogenesis classification 系统分类
phylogenetic 系统发生的
phylogenetic chart 系统发生图
phylogenetic classification 系谱分类
phylogenetic lines 种族发育谱系
phylogenetics 系统发生学
phylogeny 系统发育;系统发生
phylozone 系统发生带
physaliatoxin 水母毒素
physaliform 气泡样的
physalite 硅酸铝矿
physallization 泡沫形成;成泡作用
physeteric acid 抹香鲸酸
physetoleic acid 抹香鲸烯酸
physical 物质的;实体的
physical absorption 物理吸附(作用)
physical absorption and deposition of pollutant 污染物的物理吸附与物理沉淀
physical acoustics 物理声学
physical address 物理地址;实际地址
physical address block 实际地址区
physical age 实际年龄
physical ageing 物理老化
physical agent 物理因素
physical alteration 自然改变
physical amplification factor 物理放大因数
physical analog(ue) 物理模拟;实物模拟
physical analog(ue) method 物理模拟方法
physical analysis 物理分析
physical analysis of sewage 污水物理分析
physical and biotic environment 物质和生物环境
physical and chemical environment 理化环境
physical and chemical examination 理化检查
physical and chemical examination of sewage 污水水质物理化学试验
physical and chemical identification 理化鉴定
physical and chemical indices 物理化学指标
physical and chemical inspection 理化检验
physical and chemical property 理化性能
physical and chemical treatment 理化处理
physical and mechanical property 物理机械性能
physical and physico-chemical analysis 物理及物理化学分析
physical antioxidant 物理性防老剂;物理防老剂
physical appearance 物理性质;物理外观
physical aptitude examination 体格适应能力检查
physical argument 实变量
physical arrangement 书型排架法
physical aspect 物理性质;物理性能
physical aspirating 物理除尘
physical assay 物理分析
physical assets 有形资产;实物资产
physical assets under specific fund 专用基金项下的实物资产
physical astronomy 物理天文学
physical atmosphere 物理大气压
physical attack 物理侵蚀
physical attributes of terrain 地面自然特征
physical base address 实际基地址
physical behavio(u)r 物理性质;物理属性;物理性状;物理行为
physical blowing 物理发泡
physical blowing agent 物理发泡剂
physical boundary 自然边界线;物理边界
physical budget 实物预算;实际预算
physical budget deficit 实际预算赤字
physical capital 物质资本;实物资本
physical capital goods 实物资产;实物资本财物
physical catalyst 物理催化剂
physical cause 实质因素
physical change 物理变化
physical channel 物理通道;实际通道
physical characteristic 自然特征;物理性质;物理特性;体型特征

physical characteristics of wastewater 废水物理特性
physical chart 地势图
physical check 物理检验
physical checkup 体格检查
physical-chemical degradation 物化降解
physical-chemical denitrification 物化脱氮处理
physical-chemical examination 物化检验
physical-chemical parameter 物化参数
physical-chemical pretreatment 物化预处理
physical-chemical process 物理化学法;物化法
physical-chemical property 物化性质
physical-chemical purification 物化净化(法)
physical-chemical sewage treatment 物化污水处理
physical-chemical system 物化系统
physical-chemical treatment 物化处理
physical chemistry 物理化学
physical chemistry adsorption 物理化学吸附作用
physical chemistry of high polymers 高分子物理化学
physical chemistry of high polymer systems 高聚物系统物理化学
physical chemistry of scatter systems 分散体系物理化学
physical circuit 实体电路
physical circumstance 具体情况;实际情况
physical classification 物理分类
physical clay 物理性黏粒
physical cleaning method 物理清洗法
physical climate 物理气候;地理气候
physical climate system 物理气候系统
physical climatology 物理气候学
physical colo(u)r 天然色
physical commodity 实体商品;实物商品
physical compatibility 物理相容性
physical completion date 确实完成日期;实际完成日期
physical composition of investment 投资实物构成
physical concept 物理概念
physical conception 物理概念
physical condition 外界条件;自然条件;物理状态;物理条件
physical condition survey 环境条件调查
physical conformation 体型结构
physical constant 物理常数
physical constituents of wastewater 废水的物理组分
physical constitution 健康素质
physical constraint 自然条件限制
physical construction 机械结构
physical contact 直接接触;体接触
physical contact connector 物理接触连接器
physical control 物理控制;物理防治(法)
physical corrosion 物理腐蚀
physical corundum 物理刚玉
physical count 实地计数
physical criterion 物理标准
physical damage 物理性损坏;物理性破坏;实质性损害
physical damage only 只负责自然损失
physical data 自然环境资料;物理数据
physical database design 物理数据库设计
physical data independence 物理数据独立性;数据实际结构独立性
physical decolo(u)rization 物理脱色
physical dedusting 物理除尘
physical degradation 物理性剥蚀;物理降解
physical delivery 实际交货
physical demand 实际需求
physical depreciation 物理衰减;实物折旧;实际折旧
physical description notes 外形描述附注
physical design (机械的)结构设计;物理设计
physical design decision 物理设计决定
physical destructive effect 直接破坏效果
physical desulfurization 物理脱硫
physical detail 地形详图
physical deterioration 物理腐蚀
physical determination 物理测定
physical development 物理显影
physical device 物理设备;实际设备
physical device number 物理设备号
physical device table 实际设备表

physical diazo process 金属重氮法
physical dimension 结构尺寸;几何尺寸;天然尺寸;实际尺寸
physical disintegration 物理性崩解;物理蜕变
physical distribution 物资调运;实体分配
physical distribution center 货物储运中心
physical distribution management 实体分配管理
physical double star 真双星;物理双星
physical driving force 物理驱动力
physical drought 物理性干旱;物理干旱
physical dryness 物理干燥
physical ecology 物理生态学
physical education 体育
physical education statistics 体育统计
physical effect 物理效应
physical effort 体力
physical electronics 物理电子学
physical element 物理根数
physical end 实际结束
physical energy 物理能;体力
physical entanglement 物理缠结
physical entity 物理实体
physical environment 自然环境;物质环境;物理环境
physical environment deterioration 自然环境恶化
physical equation 物理公式
physical equilibrium 物理平衡
physical equipment 物理设备
physical error 物理误差
physical evapo(u)ration 自然蒸发
physical evidence 物证证据
physical examination 物理性检验;物理检验;体格检查
physical examination of water 水的物理检测
physical exercise 体育活动
physical exertion 体力劳动
physical experimental equipment 物理试验设备
physical exploration 物理勘探(法)
physical explosion 物理爆炸
physical facility 物质设施
physical factor 自然因素;物理因子;物理因素
physical fault 物理故障;实际故障
physical feature 自然特征;天然地形;物理性质
physical field change 物理场变化
physical file format 物理文件格式
physical file name 物理文件名字;实际文件名称
physical file structure 物理文件结构
physical fire 物理火灾
physical fire model 物理火灾模型
physical fitness test 身体健康检查
physical foamer 物理发泡剂
physical foaming 物理发泡
physical foaming agent 物理发泡剂
physical form 自然形状;外形;外观;天然形态
physical format 物理格式
physical fragmentation of rock 岩石物理破碎
physical function 实际职能
physical gas analyzer 物理式气体分析器
physical geochemistry 物理地球化学
physical geodesy 物理大地测量学
physical geographic(al) climate 自然地理气候
physical geographic(al) planning 自然地理规划
physical geographic(al) regionalization 自然地理区划
physical geographic(al) survey of watershed 流域自然地理调查
physical geographic(al) zoning 自然地理区划
physical geography 自然地理学
physical geologic(al) environment 自然地质环境
physical geologic(al) phenomenon 自然地质现象;物理地质现象
physical geology 物理地质学;地貌学;地貌地质学
physical geomorphology 物理地貌学
physical half-life 物理的衰期;物理半衰期
physical heat of slag 熔渣物理热
physical heterogeneity 物理不均匀性
physical holding of stock 实物库存
physical horizon 物质地平;物理地平
physical hydrographic factor 物理水文因素
physical hygrometer 干湿球温度表;物理湿度计
physical incompatibility 物理不相容性
physical index 物理指标
physical injury 物理性损伤;物理伤害
physical input-output 实物投入产出;实际输入输出

physical inspection 体格检查
physical interaction 物理相互作用
physical interface 物理接口
physical interference 物理干扰
physical inventory 实际盘存;实际库存;实地盘存
physical inventory book 实地盘存簿
physical inventory control 实地盘存控制法
physical inventory method 实地盘存法
physical inventory system 实地盘存制
physical isomer 物理异构物
physical isomerism 物理异性;物理异构;物理性异构
physical labo(u)r 体力劳动
physical landscape 自然景色;自然景观;天然风景
physical law 物理定律
physical law generator 物理定则产生器
physical layer 物理层
physical layout 外形布置
physical length 实际长度
physical level of programming 程序设计的物理层
physical libration 物理天平动
physical life 物理寿命;实用年限;实际寿命;物理特性保持年限;实用寿命;实际使用寿命;实际使用年限;实际年限
physical life-time of building 房屋使用期限
physical limit 物理限制
physical location 物理单元;实际位置
physical location of documents 文献的收藏地点
physical logging 物理探查记录;物理勘探(法)
physically adsorbed species 物理吸附物种
physically damaged amount 机械破坏量
physically disabled person 残疾人员
physically handling 搬运
physical magnification factor 物理放大因数
physical make-up 装帧
physical map 自然地图;自然地理图集
physical market 现货市场;实物市场
physical measure 实物计量
physical measurement 物理测量;物理测定
physical-mechanical property of rock 岩石物理力学性质
physical-mechanical property of soil 土的物理力学性质
physical mechanics 物理力学
physical mechanism 物理机理
physical medium 实际媒体
physical memory 物理内存;实体存储器
physical memory location 实际存储单元
physical memory space 实际存储空间
physical metallurgical 物理冶金的
physical metallurgy 物理冶金(学)
physical meteorology 物理气象学
physical method 物理法
physical mineralogy 物理矿物学
physical mix(ture) 天然混合物;物理状态混合物
physical model 直观模型;物理模型;实物模型;实体模型
physical modelling 物理模拟
physical model of polarized unit 极化单元的物理模型
physical modification 物理变性
physical motor pool 汽车集中调度场
physical name 实体名字
physical network arrangement 物理网络结构
physical network configuration 实际网络配置
physical network resource 物理网络资源
physical node 物理节点
physical non-linearity 物理非线性
physical numeric(al) address 实际数字地址
physical objective 具体目标
physical obstacle 天然障碍
physical obstruction 外界障碍
physical oceanographic(al) atlas 海洋自然地理图集
physical oceanographic(al) survey 自然海洋调查
physical oceanography 海洋物理(学);物理海洋学
physical operations 物理操作
physical optics 物理光学
physical organic chemistry 理论有机化学;物理有机化学
physical orientation 物理定向
physical parameter 物理参数
physical parameter of natural gas 天然气主要物理性质参数

physical parameter of oil field water 油田水物理参数
physical pendulum 物理摆;复摆;振摆仪
physical phase analysis 物相分析
physical phenomenon 物理现象
physical photometer 客观光度计;物理光度计
physical plan 实物计划
physical plane 实体平面;实际平面
physical planning 体型规划;物资调运计划;物资调运规划;物资调用计划;土地整治计划;实体规划;实际规划
physical plant 建设项目中的动产和不动产
physical plant configuration 装置的物理构型
physical plastic limit 物理塑限
physical poison 物理毒剂
physical pollutant 物理(性)污染物
physical pollution 物理污染
physical pollution index 物理污染指数
physical pollution of water body 水体物理污染
physical pool 汽车集中调度场
physical precipitation 机械力除尘器
physical presentation statement 外形说明
physical prevention and treatment 物理防治(法)
physical price 实物价格
physical problem 实际问题
physical process 物理过程
physical profile serial 体格分类等级
physical profile serial code 体格分类等级代码
physical property 有形财产;物质性质;物理性质;物理性(能);物理特征;物理特性
physical property of coal 煤的物理性质
physical property of liquid 液体物理性质
physical property of minerals 矿物物理性质
physical property of minerals and rocks 岩矿物理性质
physical property of natural gas 天然气物理性质
physical property of oil field water 油田水物理性质
physical property of petroleum 石油物理性质
physical property of sediments 沉积物物理性质
physical property of soil 土的物理性质
physical property of solid bitumen 固体沥青的物理性质
physical property of spring 泉水物理性质
physical property of water 水的物理性质;水的物理特性
physical property parameter of oil/gas and water 油气水物性参数
physical property parameter of reservoir 油层物性参数
physical property test 物理性能试验
physical prospecting 物理勘探(法)
physical protection 人体防护
physical pseudomorphy 物理性假同晶(现象)
physical purification 物理净化(作用)
physical purification of water body 水体物理净化
physical quality of water 水的物理检验
physical quantity 物理量;实物量
physical radiation effect 物理辐射效应
physical realizability 物理上可实现性
physical record 物理记录;实际记录
physical record length 实际记录长度
physical recovery 完整回收
physical reference surface 自然参考面
physical refining process 物理加工过程
physical relief 自然起伏;自然地形
physical relocation 实际搬迁
physical remediation 物理修复
physical resin loss 树脂机械损耗
physical resistance 物理抗性
physical resonance 物理共振能
physical resources 物质资源;物力资源
physical ripening 物理成熟
physicals 实际货物
physical sampling 物理取样
physical sand 物理砂粒
physical science 自然科学;物理科学
physical sea-level 自然海平面
physical security 生理保证
physical sedimentation 自然沉淀
physical sedimentation tank 自然沉淀池
physical seismology 地理地震学
physical self-purification 物理自净化作用

physical separated 实物分隔
physical separation 物理隔离;物理分选;物理分离
physical sequential access 物理顺序存取;实际顺序存取
physical service 实际服务
physical simulation 物理模拟;实体模拟;实体仿真
physical simulation system 物理模拟系统
physical size 机械结构尺寸;物理尺度;外形尺寸;天然尺寸;实际尺度
physical smuggling 实物走私
physical softener 物理软化剂
physical solvent 物理溶剂
physical sorting equipment 物理分选装置
physical speed 实际速度
physical stability 物理稳定性
physical standard of living 实际生活标准
physical state 物态
physical stratification pattern 物理层理形式
physical stratigraphy 物理地层学
physical strength 体力
physical stress 物理性应力
physical structure 物理结构;物质结构
physical surface 自然表面
physical surface of the earth 地球自然表面
physical surface point 地面点
physical swap-out 实际换出【计】
physical system 因果系统;物理系统;实体系统
physical system time 实际系统时间
physical test(ing) 物理试验;物理检验法
physical testing sample 物理检验试样
physical theory 物理理论
physical thermogenesis 物理生热作用
physical tolerance 物理耐性
physical ton of handling 装卸自然吨
physical tracer 物理示踪剂
physical tracing 物理示踪
physical training 体育锻炼
physical transformation 物理转化
physical transient area 实际暂驻存储区
physical transportation 机械搬运
physical transport capacity 实际运输能力
physical transport of organism 生物的物理搬运;生物的机械搬运
physical treatment 物理处理(法)
physical treatment unit operation 物理处理单元操作
physical unit 物理装置;实物单位;实际部件
physical unit of goods 货物有形计算单位;货物的实物计算单位
physical unit operation 物理单元操作
physical unit process 物理单元处理法
physical upgrading 物理选矿;物理浓集
physical value 重置价值
physical vapour deposition 物理气相沉积
physical variance 实际差异
physical verification 实际验证
physical visual binary 物理目视双星
physical volume 实物体积;实物数量;装订卷数
physical waterproofing 物理防水;自然防水(性)
physical wear and tear 实际磨损
physical weather analysis 物理天气分析
physical weathering 机械风化(作用);物理风化(作用)
physical world 物质世界
physical yield limit 自然产水量极限
physician 医生;医师
physicist 物理学家
physico-chemical 物理化学的;物化的;理化的
physico-chemical analysis 物理化学分析;物化分析
physico-chemical barrier 物理化学屏障
physico-chemical bound water 物理化学结合水
physico-chemical characteristic 物理化学特性
physico-chemical condition of water body 水体物理化学条件
physico-chemical constant 物理化学常数
physico-chemical degradation 物化降解
physico-chemical denitrification 物化脱氮处理
physico-chemical environment 理化环境
physico-chemical environment index 理化环境指标
physico-chemical examination 理化检查;物化检验
physico-chemical geology 物理化学地质学
physico-chemical identification 理化鉴定
physico-chemical index 理化指标
physico-chemical indices 物理化学指标
physico-chemical inspection 理化检验
physico-chemical method 物理化学方法
physico-chemical method of sewage treatment 污水物化处理法
physico-chemical parameter 物理化学参数;物化参数
physico-chemical pretreatment 物化预处理
physico-chemical process 物理化学过程
physico-chemical property 理化性能;物理化学性质;物化性质
physico-chemical purification 物理化学净化;物化净化(法)
physico-chemical sewage treatment 污水物理化学处理;污水物化处理;物理化学污水处理;物化污水处理
physico-chemical system 物化系统
physico-chemical treatment 理化处理;物理化学处理(法);物化处理
physico-chemical treatment of wastewater 污水物化处理
physico-chemical unit process 物理化学单元过程
physico-chemico-biological treatment of municipal sewage 城市污水物化生物处理
physico-chemistry 物理化学
physiographic(al) analysis 地文分析
physiographic(al) economical condition of ore district 矿区自然地理经济条件
physiographic(al) unit 地文单位
physico-mathematical model 物理数学模型
physico-mechanical characteristic 物理力学特性
physico-mechanical property 物理力学性质
physico-statistic(al) prediction 物理统计预报
physics block 医疗区段
physics laboratory 物理试验室;物理实验室
physics of cement 水泥物理学
physics of materials 材料物理学
physics of super-high energies 超高能物理学
physics relate to building(construction) 建筑物理
physics-structure coalification 物理—结构煤化作用
physio-chemical process 物化法
physio-chemical treatment of night soil 粪便物理化学处理
physio-chemical treatment of wastewater 废水物理化学处理
physiogeographic(al) 自然地理的
physiogeographic(al) map 自然地理图集
physio-geologic(al) function 物理地质作用
physiohomic(al) homogeneity 外貌均匀度
physiognomy 自然规律学;外貌
physiognomy of community 群落外貌
physiographic(al) 自然地理的;地文的
physiographic(al) balance 地文平衡
physiographic(al) circular features 地文环形体
physiographic(al) climax 地文演替顶极
physiographic(al) condition 自然地理条件;地文条件
physiographic(al) cycle 自然地理循环;地文循环
physiographic(al) demarcation 自然区划
physiographic(al) ecology 自然地理生态学;地文生态学
physiographic(al) environment 自然地理环境;地文环境
physiographic(al) factor 自然地理因素;地文因素
physiographic(al) formation 自然地理群系
physiographic(al) geology 地文地质学
physiographic(al) map 地文图
physiographic(al) parameter 自然地理参数;地文参数
physiographic(al) province 地文省;地文区;地理区(域)
physiographic(al) region 地文区域;地理区(域)
physiographic(al) regionalization 自然区划
physiographic(al) succession 自然地理演替
physiographic(al) symbol 地貌形态写景符号
physiographic(al) unit 地文单位
physiography 自然地理学;地文学
physiography ecology 自然地理生态学
physiologic(al) acidity 生理酸性
physiologic(al) acoustics 生理声学
physiologic(al) alkalinity 生理碱性
physiologic(al) available water 生理有效水
physiologic(al) climatology 生理气候学
physiologic(al) ecology of aquatic animal 水生动物生理生态学
physiologic(al) effect of noise 噪声生理效应
physiologic(al) effect of radiation 辐射的生理作用
physiologic(al) parallax 生理视差
physiologic(al) psychology 生理心理学
physiologic(al) security 生理保证
physiology 生理学
physiology of labo(u)r 劳动生理学
physiolysis 自然分解
physio-mechanical property of rock 岩石物理力学性质
physique 体格
physisorption 物理吸着;物理吸附(作用)
phytal zone 沿岸浅水带
phytanate 植烷酸
phytane 植烷
phytane/n-18 alkane ratio 植烷/正十八烷
phytanic acid 植烷酸
phyteral 植物煤素质
phytic acid 植酸
phytoaeron 空中微生物群落
phytobenthon 水底植物
phytobiochemistry 植物生物化学
phytochemistry 植物化学
phytochrome 光敏色素
phytocide 除莠剂;除草剂
phytoclimate 植物气候
phytoclimatic district 植物气候地区
phytoclimatic zone 植物气候带
phytoclimatology 植物小气候学;植物气候学
phytocoenology 植物群落学
phytocoenose complex 群落复合体
phytocoenosium 植物群落
phytocoenosium type 植物群落型;群落类型
phytocoenostics 植物群落学
phytocommunity 植物群落
phytodegradation 植物降解
Phytodiniales 植甲藻类
phytoecology 植物生态学
phytoedaphon 土壤微生植物(群落)
phytoextraction 植物萃取法
phytogeneous rock 植物岩
phytogenetic soil 植成土
phytogenic 植物成因的
phytogenic argillaceous texture 植物泥质结构
phytogenic rock 植物岩
phytogenic soil 植成土
phytogeographer 植物地理学家
phytogeographical region 植物地理区
phytogeography 植物地理(学)
phytograph 群落结构图
phytograph index 植物图解指数
phyto-group 植物群类
phyto-indicator 指示植物
phytokarst 植物岩溶
phytol 植物醇
phytolite = phytolith
phytolith 植物岩;植物化石
phytology 植物学
phytometer 植物蒸腾计;蒸腾计
phytometry 植物测法
phytomicroorganism 植物微生物
phytopal(a)eontology 古植物学
phytopaleontology 植物化石学
phytopathological inspection 植物病害检验
phytopathy 植物病
phytopedology 植物土壤学
phytophenology 植物物候学
phytophily 适草性
phytophoric rock 植游岩
phytoplankton 植物性浮游生物;浮游植物
phytoplankton abundance 浮游植物丰度
phytoplankton biomass 浮游植物生物量
phytoplankton bloom 浮游植物增殖;浮游植物大量繁殖
phytoplankton bottle 浮游植物采样瓶
phytoplankton community 浮游植物群落
phytoplankton equivalent 浮游植物当量
phytoplankton outburst 浮游植物勃发
phytoplankton population 浮游植物种群;浮游植物群
phytoplankton production 浮游植物生产量
phytopneumonoconiosis 植物尘肺
phytoprecipitin 植物沉淀素

phyto-sanitary 植物检疫
phytosanitary certificate 消毒证明文件
phyto sanitary control 植物卫生管理
phytoscopic 拟植物
phytosociology 植物群落学
phytostabilization 植物固定法
phytosterol 植物甾醇类
phytotaxonomy 植物分类学
phytotoxic air pollutant 植物毒性空气污染物
phytotoxin 毒植物素
phytotransformation 植物降解
phytotrichobezoar 植物毛粪石
phytotron(e) 人工气候室;植物人工气候室
phytoxanthin 叶黄素
phytoxylin 植木胶
pial tree 菩提树
pia mater 软膜
pianlinite 偏岭石
piano block 钢琴型砌块
piano hinge 琴式铰链;钢琴铰;长铰链;连续合页;长合页
piano-key push 琴键式按钮
piano lacquer 钢琴漆
pianola roll 穿孔纸卷
piano line 表面细条道;表面细条纹
piano nobile 有正式接待室和饭厅的楼层;有客厅楼层;主要楼层
piano wire 高强钢丝;钢琴(钢)丝
piano wire concrete 高强钢丝混凝土;钢弦混凝土
piano-wire screen 钢丝筛(网)
piassava 棕榈叶纤维
piaster board nail 灰板钉
piaton wrench 活塞扳手
piauzite 板沥青
piazza 拱廊;步廊;广场;内天井
pibal 高空测风报告;测风气球观测;测风气球
Picard iteration method 比卡迭代法
picaroon 海盗船
picaroon hole 卡钩头伤孔
piccolo 短笛
picea 云杉属;针枞
Picea abies 挪威云杉
picein wax 真空封蜡
Piche atmometer 湿纸蒸发表
Piche evapo(u)rimeter 毕谢蒸发器
picite 土磷铁矿
pick 检选;尖镐;集束辊;挑拣;拾取;采集;采;十字镐
pick-a-back 背负式运输;机载
pick-a-back assembly 重叠装配
pick-a-back conveyer 可伸缩运输机
pick accumulator 储[贮]纬器
pick and click test (for soil cement) 凿、敲试验(简易测定水泥硬度)
pick and dip 新英格兰砌砖法;东方砌砖法;挤浆法;泥瓦工砌砖技巧
Pickard core barrel 毕氏岩心管
pickaroom 扳钩
pick axe 鹤嘴斧;十字镐;十字斧;丁字斧;镐(头);鹤嘴镐
pickaxe handle 十字斧柄
pick blow 钎子冲击
pick-blowing nozzle 喷纬嘴
pick box 截齿座
pick breaker 针碎机;锤式破碎机;齿式破碎机
pick count 纬密度
pick counter 织物分析镜
pick density 纬密
pick device 采集装置
picked device state 拾取设备状态
pick-dressed ashlar 细琢方石;斩石
picked dressing 凿平修饰;粗琢石面(饰)
picked out 点饰(用颜色小斑点衬托浮雕图形)
picked port 选定港;被选择港口
pickel 冰杖
picker 检出器;拣选工;起模针;十字镐
picker cylinder 采摘筒;纺锭滚筒
picker feed 切向进给
pickeringite 镁明矾;镁铝矾
picker knives 拾取刀
picker level 木桩高
picker test 风镐试验
picker work 用风镐开采
picker zone 采棉区
picket 用栅围住;用桩围住;短测杆;尖桩条;尖桩木窄
立板;尖桩;尖木桩;尖顶桩
picketage 定标
picket boat 雷达警哨舰
picketed lead 拦鱼导栅
picket fence 木桩栅栏;尖桩围栅;尖桩篱栅;木栅(栏);磁篱
picket-fence retic(u)le 栅状调制盘
picket-fence sludge rake 栅栏式污泥耙
picket-fence stirrer 栅栏式搅拌器
picket-fence type sludge thickener 栅栏式污泥浓缩器
picket gate 拦鱼栅;导水栅门(水工试验用)
picket line 纠察线;前哨线
picket-point method 桩点法
picket ship 巡逻艇;巡逻船
Pickett's formula 比克特公式
picket type snow fence 尖桩式雪栅
pick flowers 攀折花木
pick glass 密度镜
pick hammer 鹤嘴锤;十字锤;风镐
pick handle 镐柄
pick-hoe 镐锄
pick identifier 拾取标识符
picking 咬起;垮落前落石;拣选;提货
picking belt 分选带(垃圾);筛分带
picking belt conveyer 拣选输送带;拣选胶带输送机
picking claim 挑剔索赔
picking conveyer 拣选输送机;手选胶带输送机;挑选输送器
picking disk 旋转挑选台
picking drum 梳筒滚筒;梳刷滚筒
picking machine 采摘机
picking motion 投梭装置
picking of first arrival times 初至拾取
picking platform 采集台
picking rate 采摘速度;采集速度
picking stick 打梭棒
picking surface 采摘面
picking table 选筛台;排棉台;选矿床;摇床;手选台
picking time 采收期
picking-up 咬底;油漆咬底;涂层边缘接合;撬出;起毛;灰浆层划痕;带出(指轮胎路面材料);打底
picking-up coat 拉毛粉刷层;拉毛粉刷
picking up of paint 蘸漆量
picking up of sediment 携带泥砂
picking wood 选出木材
picking zone 采棉区
pickle 稀酸液;浸渍用盐水;酸洗液;冰斧
pickle brittleness 浸蚀脆性;氢脆
pickled pine 酸处理松木
pickled plate 酸洗钢板
pickled sheet steel 酸洗薄钢板
pickled wire 酸洗过的金属线
pickle-free 非酸洗;非酸蚀
pickle house scratch 酸洗痕
pickle jar 泡菜坛
pickle lag 酸浸时滞性试验
pickle line processor 精除鳞机
pickle liquor 酸洗液;废酸浸液
pickle patch 酸洗斑点
pickler 酸洗装置;酸洗液;酸洗设备
pickle solution 酸洗液
pickle stain 酸洗污斑;酸洗色斑
pickle wastewater 酸浸废水
pickling 浸酸;木材及木船;酸洗
pickling acid 酸洗(用)酸;酸浸酸
pickling acid waste 废酸洗液
pickling agent 酸洗剂;酸浸剂
pickling bath 盐腌盆;酸洗槽;酸洗池
pickling bay 酸洗工段
pickling blister 酸洗泡
pickling brittleness 酸洗脆性
pickling cell 酸洗池
pickling department 酸洗车间
pickling inhibitor 酸洗抑制剂
pickling line 酸洗作业线;酸洗机组
pickling liquor 酸液
pickling machine 酸洗机
pickling of metal 金属酸洗法
pickling oil 防锈油
pickling plant 酸洗车间
pickling process 酸洗法;酸浸法
pickling rinse water 酸洗冲洗水
pickling tank 酸浸槽
pickling test 酸洗检查
pickling tub 酸洗池
pickling unit 酸洗装置
pickling vat 酸洗桶;酸洗槽;酸洗缸
pickling waste acid liquor 废酸浸液
pickling waste liquor 酸洗废液
pickling waste(water) 酸性废水;酸洗(浸)废水;酸浸废水;酸化废水
pickling zone 酸洗工段
picklock 撬锁工具
pick-mattock 鹤嘴锄;钢镐
pick mining machine 冲击式截煤机
pick mode 拾取设备的操作方法
pick-off 截止;拾取;自动脱模装置;机电传感器
pick-off coupling 切向耦合;拾取耦合
pick-off electrode 灭火电极
pick-off gear 选速装置;可换齿轮
pick-off gearbox 速度变速箱
pick-off roll 引离辊
pick-off signal 截止信号
pick-off solenoid 分离电磁线圈
pick-off unit 取件装置
pick out 挑选;剔除
pick photograph 控制片
pick point 镐头
pick pole 镐柄
pick punch 冲孔器
pick setting 安装截齿
pick test 取样试验;抽样试验;抽样检验;抽验
pick tester 拉毛试验仪
pick-up 债券买卖提高收益率;给料量;受感器;始动值;拾波器;拾音器;拾波;传波器;(电极的)黏污;邻近干扰;拾起;读出传感器;待励单元;吃刀
pick-up amplifier 选择器放大器
pick-up and delivery 接送零担物业务
pick-up and delivery service 接送货物服务
pick-up and drop off point 小港站
pick-up and drop point 起吊点
pick-up and drop train 摘挂列车
pick-up antenna 接收天线;拾取信号天线
pick-up arm 拾振器臂;拾取器
pick-up arrangement 起吊方式
pick-up arrangement of piles 桩的起吊方式
pick-up baler 干草拣拾压捆机;拾拾压捆机
pick-up belt 拔取皮带
pick-up bridge 堆场装卸桥
pick-up brush 集流刷;集电刷
pick-up bus 沿途接送的汽车
pick-up calibration 传感器校准
pick-up camera 摄像机
pick-up camper 小汽车旅行住房
pick-up capturer 吸棉嘴
pick-up carrier (城市污水处理场的)截水沟;采掘运载
pick-up cartridge 拾声器心座;拾声(器)头
pick-up chopper blower 风扇式检拾切碎机
pick-up coach 轻便客车
pick-up coil 电动势感应线圈;传感线圈;拾音线圈;耦合线圈;拾振器线圈;拾波线圈;测向线圈
pick-up crane 轻便起重机
pick-up current 接触电流;起动电流;始动电流;触动电流
pick-up cylinder 滚筒式拾捡器
pick-up cylinder chopper 滚筒式拾捡切碎机
pick-up device 集束卷绕装置;受丝卷绕装置;拾取装置;摄像装置;拾波器
pick-up electrode 信号极
pick-up factor 接收效应;拾音系数;拾波因数
pick-up fire fighter 临时消防战斗员
pick-up fitting 对接装配
pick-up frame 轻便起吊架;轻型起重机架
pick-up gear 吸取装置;钩取装置;钩起装置
pick up goods 提货
pick-up grab 小物品打捞器(钻孔内);小物件打捞器(钻孔内)
pick-up hole 压力测量孔;测量孔
pick-up hook 脚扣
pick-up lens 摄像镜头
pick-up line 捡拾信号线
pick-up load 启动负荷
pick-up loader 拾拾装载机
pick-up loader and cleaner for olives 橄榄捡拾装载清理机
pick-up loop 耦合圈;拾波线圈;拾波环

pick-up magnet 捡拾磁体
pick-up mechanism 捡拾机构
pick-up method 捡拾方法
pick-up of engine 发动机(的)加速性能
pick-up of grains 木纹显露
pick-up of paint 漆刷蘸漆量
pick-up of valve seat 阀座翘陷
pick-up piston 传感器活塞
pick up point 起吊点;着力点;桩吊点
pick-up points for precast concrete piles 预制混凝土桩吊装点
pick-up potentiometer 测量传感器电位计
pick-up press 捡拾压捆机
pick-up probe 接收探测头;接收探测器
pick-up procedure 整理过程
pick-up pump 真空泵;抽出泵
pick-up ring 自动联合器环
pick-up roll 引辊;取料辊;取胶辊;上料辊
pick-up roller 捡拾辊
pick ups 打捞工具
pick-up sage 刷平流淌(油漆);刷薄(油漆过厚)
pick-up sensitivity 摄像灵敏度
pick-up separator 捡拾筛选机
pick-up service 上门收货服务
pick-up slot 销槽
pick-up spectral characteristic 摄像光谱特性
pick-up speed 加快速度
pick-up suction 抽真空
pick-up the casing 托住套管
pick-up the heat 热的回收利用
pick-up the lab 付账
pick-up the suction 抽真空
pick-up time 吸合时间;工作时间;启动时间;拾取时间;吸起时间
pick-up tongs 拾件钳
pick-up truck 小吨位运货汽车;轻型载货汽车;轻型货车;小型运货汽车;轻运货车
pick-up tube 摄像管
pick-up tube preamp 摄像管预放器
pick-up tube with electron zoom 变倍率摄像管
pick-up unit 传感器;捡拾装置;拾波器
pick-up value 吸起值
pick-up velocity 扫描速度;起动流速;拾取速度;起动转矩
pick-up voltage 起始电压;始动电压;拾取电压
pick-up well 加速井
pick-up winding 拾取线圈;拾起绕组
pick-up window 拾取窗口
pick with handle 镐
piclear unit 图像清除器
picling bath 酸洗槽
Picloram 毒莠定(一种除草剂)
picnic area 野餐地
picnic baskets 野餐篮
picnic flask 野餐热水瓶
picnic site 野餐营地;野餐场所;郊游野餐营地
picnometer 比重瓶;液体比重计;比重管
picnostyle 密柱式(古希腊、古罗马神庙的,柱间为一点五米柱径的形式)
pico 微微(10^{-12});皮可
picoampere 微微安(培)
picocurie 微微居里
picocurie liter 微微居里/升
picofarad 皮法拉;微微法(拉)
pico-gram 微微克
picolinamide 吡啶酰胺
picologic 微微逻辑电路
picometer 微微米
picoperine 吡哌乙胺
picophytoplankton 超微型浮游植物
picoplankton 超微型浮游生物
picosecond 微微秒
picotite 铬尖晶石
picotpaulite 辉铁铊矿
picotrace analysis 皮痕分析
picot yarn 毛圈花线
pico-watt 微微瓦
picranisic acid 苦味酸
picratol 比克拉托
picric acid 苦味酸;苦酸味
picrite 橄苦岩;苦橄岩
picrite basalt 苦橄玄武岩
picrite dolerite 苦橄粒玄岩
picrite porphyrite 苦橄玢岩

picritic glass 苦橄玻璃
picritic picrite 苦橄斑岩
picritic teschenite 苦橄沸绿岩
picrochromite 铬镁尖晶石
picrocrichtonite 镁钛铁矿
picroepidote 镁绿帘石
picrofluite 氟镁石
picroknebelite 镁锰橄榄石
picrolite 纤维状蛇纹岩;硬蛇纹石;叶蛇纹石
picrolite asbestos 硬蛇纹石石棉
picrolonate 苦酮酸盐
picronigrosin 苦味酸苯胺黑
picropharmacolite 镁毒石
picrotephroite 镁锰橄榄石
pictandcime 镁方沸石
Pictest 靠表;杠杆式千分表
pictite 木屑石
pictoamesite 镁绿泥石
pictogram 象形图;图解
pictograph 古代石壁画;象形文字;图像统计图表;统计图表
pictographic(al) map 解释图
pictoline 等密度线
pictomap 等密度线影像地图
pictorial 绘画的;报表设计
pictorial chart 象形图
pictorial composition 图画的构成;图像合成
pictorial computer 帧型计算机;图解式计算机
pictorial computer with courser 图示航行计算机
pictorial data representation 图形数据表示
pictorial detail 图像细节
pictorial diagram 直观图;图画示意图;示意图;实物电路图
pictorial directory 区域平面示意图
pictorial display 图像显示
pictorial display definition mode 图像显示定义方式
pictorial edition 有插图的出版物
pictorial holography 图像全息术
pictorial information 图像信息
pictorial infrared photography 红外(线)成像照相;图像红外线摄影术
pictorial map 写景图;图示地图
pictorial marking for handling of non dangerous goods 装卸指标标志
pictorial markings for handling of packages 包装储运图示标志
pictorial model 图像模型;图画模型
pictorialness 图示性
pictorial pattern recognition 画像图形识别;图案识别
pictorial position indicator 图像位置显示;图示位置显示器
pictorial projection 图像投影
pictorial representation 图像表示;直观图
pictorial section 图像分排
pictorial statement 图形决算表;图形报表
pictorial surface 图像面
pictorial symbol 形象符号
pictorial symbolization 图形标记
pictorial unit chart 象形单位图
pictorial view 示图
picts' house 地下石室
picture 影像(图);帧面;图像
picture altitude 图像高度
picture altitude control 图像高度控制
picture amplifier 图像放大器
picture amplitude 图像幅度
picture analysis 图像分析
picture analyzer 图像分析器
picture and blanking signal 图像和消隐信号
picture and waveform monitor 图像和波形监视器
picture angle 视角
picture angular field 摄影视场
picture aperture 图像孔径
picture archive 照片资料馆
picture area 像幅;帧面积;图像面积
picture-area shift 图像移动
picture aspect ratio 帧面宽高比;图像宽高比
picture automata 图片自动机
picture background 图像背景
picture bandwidth compression 图像带宽压缩
picture black 图像黑部分
picture blanking 图像消隐

picture blurring 图像模糊
picture book 图画书
picture border 图像边缘
picture bounce 图像跳动
picture breakdown 图像干扰
picture break-up 图像撕裂
picture brightness 图像亮度
picture carrier 图像载波;图像信号载波
picture carrier plate 图像片盘
picture carrier trap 图像载波陷波器
picture carrier turn table 承片转盘
picture center 图像中心
picture centring 图像中心;图像定心
picture centring adjustment mechanisms 图像定心调整机构
picture channel 图像信道;图像通道
picture characteristics 图像特性
picture charge 图像电荷
picture charge pattern 图像电位分布图
picture-chasing 图像跟踪
picture coding 图像编码
picture coil 图像调整线圈
picture collection 图画专藏
picture colo(u)r quality 图像彩色质量
picture communication 图像通信
picture composition 图像合成
picture compression 图像压缩
picture contact point 片片连接点
picture content 图像分量
picture contour 图像轮廓
picture contrast 构像反差;图像黑白对比;图像对比度;色调对比
picture control 图像调整
picture control coil 图像控制线圈
picture control desk 图像控制台
picture control point 像片控制点
picture crosstalk 图像串扰
picture cup apart 图像分割
picture current 图像电流
picture cut-off voltage 图像截止电压主
picture data 图像数据
picture database 图像数据库
picture decoding 图像解码
picture defect 图像缺陷
picture definition 图像清晰度
picture deflection 图像位置调整帧偏转
picture degradation 图像变劣
picture demodulator 图像信号检波器
picture depth 图像深度
picture description 图像描述
picture description grammar 图像描述文法
picture description instruction 图形描述指令
picture description language 图像描述语言
picture detail 图像细节
picture detail factor 图像细节因数
picture detail region 图像细节区
picture diagonal 屏面对角线;图像对角线
picture diagram 图示意图
picture dictionary 图解词典
picture dimension 图像尺寸
picture displacement 图像位移
picture display 图像显示
picture display definition mode 图像显示定义方式
picture distortion 图形畸变;图像畸变
picture disturbance 图形干扰;图像跳动
picture dot 像点;像素
picture-dot interlacing 像点交错;跳点扫描
picture drome 电影院
picture drop-out 图像失落;图像消失
picture edge 图像边缘
picture editor 像片编辑员
picture element 像质;像元;像素;像点;画素
picture element replenishment method 像素充实法
picture element signal 像素信号
picture enhancement 图像增强
picture facsimile apparatus 图像传真机
picture fault 图像缺陷
picture feature 图像特性
picture-field 像场
picture-field warning 像场缩小
picture file 图形文件;画面文件;图画资料档
picture flyback 帧逆程;帧回描
picture flyback time 帧逆程时间

picture format 像幅;帧型;帧格式
picture format item 图像格式项
picture forming glass 镶嵌图案玻璃
picture frame 像帧;像框;显像帧面;镜框
picture frame filter 图框滤波
picture-frame form 钢木定形模板;胶合板模(浇混凝土用);胶合板钢框架模板
picture-frame stage 景空式戏台;景框式戏台;景架式戏台;景架式戏台;镜框式戏台;框式舞台
picture framing 厚边
picture framing glass 景框式玻璃;景架式玻璃;镜框玻璃
picture frequency 像频
picture-frequency adjustment 帧频调整
picture frequency setting 帧频调整
picture gallery 绘画展览室;画廊
picture gate 曝光口;摄影机片门
picture geometry 图像几何形状
picture glass 镜框玻璃;镶嵌图案的玻璃;画框玻璃
picture grain 图像颗粒结构
picture halftone transition 图像半色调跃迁
picture hall 电影院
picture height 祯面高
picture height control point 像片高程控制点
picture hook 挂画钩
picture horizon 像片地平线
picture horizontal control point 像片水平控制点;像片平面控制点
picture house 电影院
picture hum 图像交流干扰;图像哼扰
picture identification 图像识别
picture illumination circuit 像片照明电路
picture image 照相图;图像映像
picture impairment 图像缺陷
picture in digital form 码式图像
picture information 图像信息
picture input transformer 图像输入变压器
picture instability 图像不稳定性
picture in straw 麦秸画
picture intelligence 图像信号
picture interference 图像干扰
picture inversion 图像黑白反转;图像反转
picture jacket 有图封
picture jitter 图像跳动
picture lag 图像惰性
picture language 图像语言
picture level 图像电平
picture level control 图像电平控制
picture line 图像扫描线
picture lock 图像同步;锁像
picture locking technique 图像锁定技术
picture made of sorghum stalk 高粱秆画
picture matching monitor 图像匹配监视器
picture material 图像素材;图像内容
picture measurement 像片量测
picture measuring 像片量刚
picture measuring apparatus 像片量测仪
picture memory 图像存储器
picture modulation 信号调制(深度)百分率;图像调制
picture modulation percentage 图像调制制度
picture modulation swing 图像调制摆幅
picture monitor 信号监视器;图像监视器
picture monitoring receiver 图像监视接收机
picture mo(u)lding 挂镜线(条);画凡饰;墙上方挂镜线条
picture noise 显像噪声
picture number 航摄像片编号
picture of large image scale 大面积图像
picture orbiting facility 图像旋转设备
picture orientation 图像旋转
picture original 图像原样
picture outline distortion 图像外形失真
picture output 图像信号输出
picture output signal 图像输出信号
picture output transformer 图像输出变压器
picture overlap 像片重叠
picture palace 电影院
picture parts 图形部分
picture pattern 图像图案
picture-percent 图像百分率
picture period 帧周期
picture-phone 电视电话;图像电话;模拟电视电话;伴像电话

picture pick-up system 摄像系统
picture plan 像片平面图
picture plane 成像面;图面;投影面;画面;像面;显像面;帧面
picture plane coordinate system 像平面坐标系
picture point 像点
picture point control 像素控制
picture point selection 像片选点
picture point time 像素传送时间;像素显示时间
picture position 图像位置
picture processing 图像处理
picture processing implementation 图片处理技术实现
picture property 图像特性
picture quality 像片质量;图像质量
picture quantizer 图像数字转换器
picture rail 挂镜线(条);画镜线
picture raster 图像光栅
picture rate 摄影速度
picture ratio 图像纵横比;图像宽高比
picture receiver 图像信号接收机;电视接收机
picture recorder 图像记录程序;图像记录器;图像印录器
picture recording 录像
picture registration 图像配准;图像重合
picture repetition frequency 帧频
picture replenishment 图像补充
picture reproducer 图像重显装置;图像再现设备
picture reproduction 图片复制
picture resolution 像片分辨率;图像分解力
picture retention 图像残留
picture rich in detail 清晰图像
picture roll 图像滚动
picture rotate control 图像旋转控制
picture scale 像片比例尺;图像比例尺
picture scanner 图像扫描器
picture scanning 图像扫描
picture scanning frequency 帧频
picture screen 显像屏;图像屏幕
picture search 图像搜索
picture seeking 图像搜索
picture segment 图形片段
picture sensor 图像信号接收器;图像传感器
picture separation 影像分离
picture-shading signal 图像黑斑补偿信号
picture shape 图像形状
picture sharpness 图像锐度;图像清晰度
picture shift 图像偏移
picture show 画展
picture showing 图像显示
picture siganl store 图像信号存储
picture signal 图像信号
picture signal amplifier 图像信号放大器
picture signal amplitude 图像信号振幅
picture signal band 图像信号频带
picture signal distribution amplifier 图像信号分配放大器
picture signal gain 图像信号增益
picture signal generator 图像信号发生器
picture signal monitor 图像信号监视器
picture signal polarity 图像信号极性
picture signal property 图像信号性质
picture simultaneous system 帧同时制
picture size 像幅;图像尺寸
picture slip 图像滑动
picture smear 图像失真
picture source 图像源
picture space 图像空间
picture space coordinate system 像空间坐标系(统)
picture specification 图像说明
picture specification character 图像说明字符
picture specification table 图像区分表
picturesque romantic style 富有浪漫色彩的建筑风格
picturesque stretch of road 风景路段
picture start 图像起始
picture steadiness 图像稳定性
picture sticking 图像保留
picture storage tube 图像存储管
picture strip 图像带;扫描行
picture structure 图像结构
picture superimposition 图像重合;图像配准
picture sweep unit 图像扫描

picture switching 图像切换
picture switching error 图像切换误差
picture synchronization 帧同步;场同步
picture synchronization control 帧同步控制
picture synchronization signal 图像同步信号
picture synchronization transmission system 图像同步传输系统
picture synchronizing signal 图像同步信号
picture synthesis 图像合成
picture taking wavelength 摄影波长
picture tearing 图像撕裂
picture telegraphy 传真电报
picture theatre 电影院
picture tile 有图瓷砖
picture timebase 帧扫描
picture tonal quality 图像色调质量
picture tone 图像色调
picture trace 像片迹线
picture track 画道
picture transfer converter 图像标准转换设备
picture transmission 图像传输
picture transmission period 图像传送
picture transmission system 图像传输系统
picture transmitter 图像发射机;电视发射机
picture transmitter device 图像发送装置
picture transmitter power 图像发射机功能
picture traverse 图像移动;图像宽度
picture tube 显像管
picture tube anode 显像管阳极
picture tube hood 显像管罩
picture unit 图像单元
picture varnish 图画清漆
picture video detector 图像视频检波器
picture white 图像最白部;图像白色
picture width 帧面宽度;图像宽度
picture width control 帧宽度调节
picture window 画廊;观景窗;观花窗;外景窗;陈设图画的窗;威尼斯式窗;眺望窗;借景窗;陈画窗
picture wire 挂图像用金属线
picture with high definition in the corners 四角清晰度高的图像
picture without contrast 平淡图像
picture-writing 象形文字
picul 担(中国和东南亚国家的一种质量单位,1担 $= 5 \times 10^4$ 克);百市斤(1市斤=500克)
picul stick 扁担
pie 盘形绕组
piebald 花斑状的
piebaldism 斑驳状态
piece 一张;一片;一件;切片
piece angle 接头用角钢
piece basis 计件标准;包装计算标准
piece by piece 一件一件地
piece by price system 按件计价制
piece-cargo 件货
pieced beam 拼合梁
pieced carving 透花
pieced column 拼合柱
pieced parapet 镂空女儿墙
pieced timber 胶合木材;组合木材;拼接木材;修补过的木材;拼接木材;拼合木料
pieced wall 漏墙;(用各种砌块砌成的)花格墙
piece-dyed fabric 匹染织物
piece dy(e)ing 匹染;整块染色(地毯)
piece goods 散件货;布匹;件货;匹头货
piece list 零件表;零件明细表
piece mark 装配构件记号;零件标号
piecemeal 断片;分块;片段
piecemeal building 零星建筑物
piecemeal determination 逐段确定法
piecemeal development 零星建筑;零星开发
piecemeal replacement of major units 主要设备的逐渐替换
piecemeal salt dome 底辟盐丘
piece mo(u)lding 分段组合模制
piece mo(u)ld process 分部件模铸的混凝土构件;组装模具法
piece number 零件编号;件号
piece of anode[anodic] plate 阳极片
piece of baggage 行旅包
piece of water 一片积水
piece-per-hour rate 每小时定额件数
piece price 计件工资单价
piece production cost 单件生产费用

piece rack 零件架
piece rate 计件工价;计件工资率;计件工资单价;单件生产时间
piece rate pay 计件工资
piece rate principle 按件计酬原则
piece rate system 计件工资制;分率系统
piece rate wage 计件工资
piece-root grafting 根接
pieces 毛束
pieces damaged from extra work 额外工作的损坏工件
pieces-damaged stock 库存损坏工件
pieces number 零件号码
pieces-spoiled stock 库存损坏工件
piece wage rate 计件工资制
piece wages 计件工资
piecewise 分段地
piecewise adjustment 分段调整
piecewise analysis 分段分析
piecewise analytic(al) function 分段解析函数
piecewise approximation 逐段逼近;分段近似
piecewise approximation method 逐段逼近法
piecewise bilinear function 分段双线性函数
piecewise constant function 分段常值函数
piecewise continuity 分段连续性
piecewise continuous differentiable function 分段连续微分函数
piecewise continuous function 分段连续函数
piecewise design procedure 分段设计法
piecewise Fourier spectrum 分段傅立叶频谱
piecewise function 分段函数
piecewise interferometric generation 分段干涉量度振荡
piecewise iteration 逐段迭代
piecewise linear 逐段线性
piecewise linear approximation 折线法;分段折线近似法
piecewise linear discrimination function 分段线性判别函数
piecewise linear function 分段线性函数
piecewise linear function generator 分段线性函数发生器
piecewise linear interpolation 分段线性插值
piecewise linear iteration 逐段线性迭代;分段线性迭代
piecewise linearization 分段线性化
piecewise linear machine 分段线性机
piecewise linear mapping 分段线性映射
piecewise linear model 分段线性模型
piecewise linear regression 逐段线性回归;分段线性回归
piecewise linear regression model 分段线性回归模型
piecewise linear stretch 分段线性扩展
piecewise linear system 分段线性系统
piecewise linear topology 组合拓扑
piecewise polynomial 分段多项式
piecewise polynomial approximation 分段多项式近似
piecewise polynomial interpolation 分段多项式插值
piecewise regression 分段回归
piecewise regression model 分段回归模型
piecewise regular function 分段正则函数
piecewise smooth 分段光滑
piecewise smooth curve 分段光滑曲线
piecewise smooth function 分段光滑函数
piece wood 木片
piecework 论件计酬工作;件工;计件工作;单件生产
piecework cost 单件成本
pieceworker 计件工(人)
piecework job 计件工作
piecework premium wage 计件奖励工资
piecework price 单件产品价格
piecework programming 计件程序设计
piecework rate 计件工资率;单件生产率
piecework system 计件制
piecework wage on a quota plus overfulfillment basis 定额超产计件工资
piecework wage plus basic salary 底薪加计件工资
piecework wage system 计件工资制
pie chart 圆形图;圆形统计图;圆图;圆盘图
piecing 接头线(抹灰);接头纹(抹灰);拼接木材
picture geometry fault 图像几何畸变

pied 杂色的;斑驳的
pied a terre 临时休息处;备用房屋
pie diagram 圆形图;饼分图
piedimont 山墙
piedmont plain 山前(侵蚀)平原
piedmont 形成于山麓的;山前地带;山麓
piedmont alluvial deposit 山麓冲积物;山麓冲积层
piedmont alluvial plain 山麓冲积平原
piedmont angle 山麓角
piedmont belt 山麓地带
piedmont bench 山麓梯级
piedmont benchland 山麓阶地
piedmont bulb 山麓冰川舌瓣;冰川底部冰的扩展
piedmont denudation plain 山麓剥蚀平原
piedmont denudation surface 山麓剥蚀面
piedmont deposit 山麓堆积
piedmont depression 山前洼地;山前拗陷
piedmont environment 山麓环境
piedmont facies 山麓相
piedmont flat 山麓梯级
piedmont glacier 山麓冲积平原;山麓冰川
piedmont gravels 山麓砾石
piedmont ice 山麓冰;山脚冰
piedmont interstream flat 山前侵蚀平原;山前平原
piedmontite 红帘石
piedmontite-schist 红帘石片岩
piedmont plain 山麓平原
piedmont scarp 山麓断(层)崖
piedmont slope 山麓坡地;山前坡地
piedmont stairway 山麓梯级
piedmont steps 山麓梯级;山前梯地
piedmont stream 山区河流
piedmont treppe 山麓梯级
piedouche 小台座
piegeonhole parking structure 鸽舍式停车结构
pie graph 饼图形;圆形图;饼状图
Pieler's lamp 皮勒尔型沼气检验灯
pien(d) 铲口;突角;尖脊;棱角;尖棱
pien(d) check 棱角槽口;棱角扣榫;踏步板槽口;石级间搭接扣
pien(d) joint 搭接扣接缝
pien(d) rafter (四坡屋顶的)面坡椽
piezoid 石英片
pie plate 铺平木
pier 支架砖(窑窟槽钢支架用);闸墩;码头;墙垛;墙墩;突码头;突堤;凸式码头;墩;窗间墙;窗间壁
pier abutment 桥墩;墩式桥台;岸墩
pierage 码头税;码头费
pier and arch system 墩拱系统
pier and panel system 柱墩划板组
pier and panel wall system 柱墩幕墙系统;壁式框架结构系统
pier apron 码头前沿
pier arcade 码头拱廊
pier arcading 码头上的连拱饰
pier arch 墩拱;墩顶拱
pier-arch system 墩拱系统
pier base 桥梁墩座;桥墩基座;桥墩基础;墩座
pier base plate 码头底板
pier basilica 码头上长方形交易所
pier bath 单间浴室
pier bent 墩式排架
pier between two windows 窗间墩;窗间砖石墩
pier block 砌墙块材;实心块材
pier body 桥墩体;墩体;墩身
pier bond 墩台砌筑;码头砌筑
pier bonding 墩台砌合法
pier brace retaining wall 支墩式挡土墙
pier break-water 桥墩分水尖
pier bridge 途栈桥;码头栈桥
pier buttress 支撑扶壁的墩
pier cantoné 哥特式复合墩
pier cap 桥墩顶座;坐垫;墩帽;墩顶
pier capping beam 墩盖梁
pier cavity 墩空穴;墩空腔
pierce 刺穿;穿入
Pierce circuit 振荡器电路
pierced 开洞的;有孔的;穿通的;贯穿的;墩帽
pierced arcade 串联的拱廊
pierced billet 荒管;穿轧
pierced bond 穿孔砌墙法
pierced brick 穿孔砖
pierced buttress 穿孔式扶壁

pierced carving 雕空
pierced concrete block screen wall 穿孔的混凝土砌块护墙
pierced floor finish 穿孔的地板修饰
pierced louver (安装在门板上的)百叶窗
pierced marble slab 穿孔的大理石板
pierced masonry wall 穿孔的圬工墙
pierced metal tile 穿孔的金属瓦片
pierced panel 穿孔板
pierced panel wall 有联镶板墙
pierced plank 穿孔板
pierced shear wall 并联剪力墙;有孔抗剪墙;开洞剪力墙
pierced steel plank(ing) 穿孔钢板
pierced wall 有花洞的墙;透空墙
pierced work 漏空的花饰;漏空的花边
Pierce instability 皮尔斯不稳定性
Pierce lenses 皮尔斯透镜组
piercement diapir 刺穿底辟【地】
piercement dome 盐丘
piercement salt dome 刺穿盐丘
piercement structure 底辟构造;刺穿构造
piercement-type salt dome 刺穿型盐丘
piercer 冲床;锥孔器;锥子;钻孔者;钻孔器
piercer roll 穿孔机轧辊
Pierce-type extractor 皮尔斯型引出器
piercing 贯通;熔化穿孔;冲孔
piercing a tunnel 隧道掘进
piercing by electric(al) arc 电弧穿孔
piercing die 冲孔模
piercing drill 热力钻孔;穿心钻;穿孔钻
piercing efficiency 热力钻进效率
piercing error (隧道的)贯通误差
piercing fold 盐丘;挤入褶皱
piercing jet 火焰喷射
piercing machine 钻孔机;冲孔压力机
piercing mandrel 穿轧芯棒;穿孔芯棒;冲头
piercing mill 穿轧机;穿孔机
piercing of a tunnel 隧道掘进
piercing plug 金属锥芯
piercing point 刺穿点
piercing pole 尖形杆
piercing press 冲孔机
piercing rate 热力钻进速度
piercing saw 弓锯;钢丝锯;硬细石(镶嵌细工用)
piercing saw blade 钢丝锯条
piercing staff 穿孔(探)杆
piercing test 贯穿试验
pier column 墩柱;支柱
pier concrete 桥墩混凝土
pier contraction coefficient 闸墩收缩系数
pier coping 桥墩帽石
pier core 柱墩心
pier crane 码头起重机
pier crew 陆上班;码头带缆工人
pier dam 丁坝
pier drainage 桥墩排水设施;桥墩排水
pier drilling 浅海边钻井;桩墩钻井;栈桥钻探;水上钻井
pier dues 码头税;码头费
piered buttress 墩式扶壁
piered louver (装在门板上的)百叶窗
piered plank 穿孔板
pier fender 码头防撞设施;码头防撞设备;桥墩护栏;码头护木
pier finger 码头突堤
pier-finger type 廊道指形式(客机坪)
pier footing 墩式基础;桥墩基础;桥墩底座
pier footing form 桥墩基础模箱
pier form(work) 墩台模板
pier for steel aqueduct 钢管支墩
pier foundation 桥墩基础;墩式基础;墩基础
pier foundation by box caisson 沉箱墩台基础
pier foundation(structure) 墩式基础
pier glass 窗间镜;窗间墙镜;穿衣镜
pier guard 码头防护(桩)
pier head 堤头;码头;桥墩墩头;明墩墩头;闸墩头(部);突堤堤头;墩帽;墩顶;防波堤头;突堤头;突堤前端;突码头前端
pierhead light 港口口门灯标
pierhead line 港口建筑线;码头边线;透空桩水工建筑物界限;突堤建筑线;透空式码头建筑物建筑界限
pierhead mast and light 堤头桅杆

pierhead pontoon 码头趸船;码头驳船;浮码头
pierhead sea wall 堤坝前防浪墙
pierhead trestle 码头栈桥
pierhead wharf 突堤码头前缘泊位
pier height 墩高
pier hole 墩孔
pier hose 墩头
pier impost 码头拱墩
pierite 白云石
pier load 墩负载
pier loss 桥墩损失
pierman 码头装卸工(人);码头工人;码头带缆工人
pier mark 墩标
pier master 码头长
pier moment 墩上力矩
pier nose 闸墩首部墩尖;闸墩前端;墩尖墩尖;桥墩鼻尖
pier nose cutwater 分水尖(桥墩)
pier nosing form 墩尖模板
pier of rigid frame construction 框架墩
pier orientation 桥墩方位
pier pocket 墩穴
pierr dacrydium 陆均松
pierre perdue 混凝土抛石基;毛石基础;抛石基床;抛石(基);水冲抛石
pierrotage 框架间用的石块砂浆填充料(美国南部殖民地时期);石块填充料
pierrotite 硫锑铊矿
piers and slips 墩式修船台
pier-satellite type 廊道卫星式(客机坪)
pier section 墩截面
pier shaft 墩身
pier-shaft jump formwork 桥墩跳升滑模(板)
pier-shaft lift formwork 桥墩跳升滑模(板)
pier shape 磁悬浮导轨梁形状
pier shed 前方货场;前方仓库
pier side 码头前沿
Pierson-Moskowitz wave spectrum 皮尔逊—莫斯科维奇波谱
pier spacing 桥墩间距
pier structure of high level platform supported on pile 高桩码头结构
pier strut 墩柱
pier stud 墩柱
pier system 突堤式码头系统
pier table 桥墩顶台
pier template 墩顶板;桩柱顶板;墩帽
pier-to-door 码头到户
pier-to-house 码头到门;码头到户
pier-to-inland depot 码头到内地仓库
pier to pier 码头到码头
pier to pier service 港际集装箱运输业务;码头到码头运输业务
pier tray 桥墩台托盘
pier type 桥墩形式
pier-type footing 墩式基础
pier-type foundation 墩式基础
pier-type hydroelectric(al) station 支墩式水电站
pier underpinning 墩式托换
pier with approach bridge 引桥式码头
pier with approach trestle 引桥式码头
pier with concave side 前沿下凹的码头
piesoelectric(al) stabilizer 压电稳频器
piestic interval 地下水位等深线间距;地下水等高线间距;等压线间距;等压线高差
piestic water 压力水
piestic wave 压力波
pietra dura 硬细石(镶嵌细工用);佛罗伦萨马赛克饰面;钢丝锯
pietraverdite 阿尔卑斯浅绿凝灰岩
pietrisikite 高温地蜡
pie wag(g)on 押犯警车;流动小吃车
pie winding 盘式绕组;饼式绕组
piexe handling time 工件装卸时间
piezallochromy 压致变色
pieze 千帕(斯卡)(压力单位);毕西
piezo 压力
piezo-balance type dust monitor 压电天平式粉尘计
piezocalibrator 压电石英校准器
piezocaloric coefficient 压热系数
piezocaloric effect 压热效应
piezocell 压电电池
piezoceramic noisecancelling microphone 压电陶瓷抗噪声传声器
piezochemistry 压力化学;高压化学
piezochrom(at)ism 受压变色
piezochromism 受压变色
piezocone 压锥水;孔压触探仪;可测孔隙水压力测头
piezocone test 孔压静力触探试验
piezocontrol 压电整频
piezocoupler 压电耦合器
piezocrystal 压晶;压电水晶;压电晶体
piezocrystallization 压结晶作用;加压结晶
piezocrystallzation way 压结晶方式
piezodialysis 压力渗析法;加压渗析
piezodialysis membrane 压渗膜;加压渗透膜
piezoeffect 压电效应
piezoelectric 压电
piezoelectric(al) accelerometer 压电式加速度计;压电加速表
piezoelectric(al) activity 压电活性
piezoelectric(al) angular rate sensor 压电角速度传感器
piezoelectric(al) apparatus 压电仪
piezoelectric(al) axis 压电轴
piezoelectric(al) buzzer 压电蜂鸣器
piezoelectric(al) calibrator 压电校准器
piezoelectric(al) cartridge 压电拾音头
piezoelectric(al) cell 压电式压力盒;压电式传感器
piezoelectric(al) ceramic discriminator 压电陶瓷鉴频器
piezoelectric(al) ceramic gassparking plug 压电陶瓷气体火花塞
piezoelectric(al) ceramic material 压电陶瓷材料
piezoelectric(al) ceramic receiver 压电陶瓷受话器
piezoelectric(al) ceramic ring 压电陶瓷环
piezoelectric(al) ceramics 压电陶瓷
piezoelectric(al) ceramics microphone 压电陶瓷传声器
piezoelectric(al) charging 压电充电
piezoelectric(al) coefficient 压电系数
piezoelectric(al) compliance 压电顺度
piezoelectric(al) condenser strain ga(u)ge 压容应变计
piezoelectric(al) constant 压电系数;压电常数
piezoelectric(al) control 压电效应控制;压电控制
piezoelectric(al) coupler 压电耦合器
piezoelectric(al) coupling 压电耦合
piezoelectric(al) coupling coefficient 压电耦合系数
piezoelectric(al) coupling constant 压电耦合常数
piezoelectric(al) crystal 压电晶片;压电晶体
piezoelectric(al) crystal filter 压电晶体滤波器
piezoelectric(al) crystal unit 压电晶体
piezoelectric(al) cutter 压电刻纹头
piezoelectric(al) cutterhead 压电刻纹头
piezoelectric(al) deformation constant 压电变形常数
piezoelectric(al) detector 压电式探测器;压电检测器
piezoelectric(al) device 压电器件
piezoelectric(al) direct effect 正压电效应
piezoelectric(al) driver 压电推进器
piezoelectric(al) effect 压电效应
piezoelectric(al) effect strain ga(u)ge 压电效应变仪
piezoelectric(al) equation 压电方程
piezoelectric(al) filter 压电滤波器
piezoelectric(al) flowmeter 压电流量计
piezoelectric(al) frequency 压电频率;压电晶体频率
piezoelectric(al) ga(u)ge 压电式测量器;压电计;压电压力计
piezoelectric(al) generator 压电发电机
piezoelectric(al) gyroscope 压电陀螺仪
piezoelectric(al) hysteresis 压电滞后
piezoelectric(al) ignition 压电点火
piezoelectric(al) ignition system 压电晶体点火
piezoelectric(al) indicator 压电指示器
piezoelectric(al) light modulator 压电光调制器
piezoelectric(al) loudspeaker 压电式扬声器
piezoelectric(al) material 压电材料
piezoelectric(al) matrix 压电矩阵
piezoelectric(al) measuring system 压电量测系统
piezoelectric(al) microphone 压电式微音器;压电式传声器;压电传声器;晶体传声器
piezoelectric(al) modulus 压电系数;压电模数;压电模量
piezoelectric(al) monitor 压电监视器
piezoelectric(al) oscillation 压电振荡
piezoelectric(al) oscillator 压电振荡器;压电式振荡器;晶体振荡器
piezoelectric(al) oscillograph 压电式示波器;压电示波器;晶体示波器
piezoelectric(al) phonometer 压电声强计
piezoelectric(al) pick-up 压电拾音器;晶体拾音器;石英拾音器
piezoelectric(al) polymer transducer 压电聚合物换能器
piezoelectric(al) pressure ga(u)ge 压电(式)压力计
piezoelectric(al) pressure meter 压电压力计
piezoelectric(al) probe 压电探针
piezoelectric(al) property 压电特性
piezoelectric(al) pump 压电泵
piezoelectric(al) pyrometer 压电高温计
piezoelectric(al) quartz 压电水晶;压电石英
piezoelectric(al) quartz airborne dust sampler 压电石英法飘尘取样器;压电石英法飘尘监测仪
piezoelectric(al) quartz crystal 压电石英晶体
piezoelectric(al) quartz crystal resonator 压电石英晶体共振器
piezoelectric(al) quartz deposit 压电水晶矿床;压电石英矿床
piezoelectric(al) quartz detector 压电石英检测器
piezoelectric(al) quartz filter 压电石英滤波器
piezoelectric(al) quartz mass sensor 压电石英质量敏感器
piezoelectric(al) quartz plate 压电石英片
piezoelectric(al) receiver 压电式接收器
piezoelectric(al) relay 压电式继电器
piezoelectric(al) resonator 压电谐振器;压电式谐振器;压电共振器
piezoelectric(al) seismometer 压电式地震仪;压电式地震计;压电地震检波器
piezoelectric(al) sender 压电变换器
piezoelectric(al) source 压电源
piezoelectric(al) stabilizer 压电石英稳频器
piezoelectric(al) stack 压电堆
piezoelectric(al) stiffness constant 压电劲度常数
piezoelectric(al) strain ga(u)ge 压电应变仪;压电效应变仪
piezoelectric(al) strain sonstant 压电胁变常数
piezoelectric(al) stress 压电胁强
piezoelectric(al) stress coefficient 压电胁强常数
piezoelectric(al) stress constant 压电胁强常数
piezoelectric(al) susceptibility 压电灵敏度;压极化率
piezoelectric(al) system 压电系统
piezoelectric(al) tensor 压电张量
piezoelectric(al) transducer 压电转换器;压电换能器;压电传感器
piezoelectric(al) transformer 压电变压器
piezoelectric(al) transition 压电跃变
piezoelectric(al) transmission 压电传递
piezoelectric(al) traverse effect 横向压电效应
piezoelectric(al) type vibration ga(u)ge 压电式振动计
piezoelectric(al) vibration 压电振动
piezoelectric(al) vibration generator 压电振动器
piezoelectric(al) vibration machine 压电振动机
piezoelectric(al) vibrator 压电振子;压电振动器;石英振动片
piezoelectric(al) voltage 压电电压
piezoelectric(al) voltage constant 压电电压常数
piezoelectric(al) weft-detector system 压电式探纬装置
piezoelectricity 压电学;压电性;压电现象;压电
piezoelectrics 压电学;压电体
piezoelectron 压电电子
piezoga(u)ge 压力计
piezograph 水压线图;水压线圈
piezograph engine 水压机
piezograph lift 水压提升机
piezograph valve 水压阀
piezoid 压电石英
piezoisobath 加压等深线
piezojunction transducer 压电结型换能器
piezolighter 压电点火器
piezoluminescence 压电发光
piezomagnetic 压磁性

piezomagnetic coefficient 压磁系数
piezomagnetic effect 压磁性;压磁效应
piezomagnetic material 压磁材料
piezomagnetic method 压磁法
piezomagnetic moment 压磁矩
piezomagnetic sensitivity constant 压磁灵敏度常数
piezomagnetic stress ga(u)ge 压磁应力计
piezomagnetism 压磁效应
piezomagnetization 压磁化
piezometamorphism 压力变质(作用)
piezometer 压力计;压力表;压电计;流压计;压强计;孔隙水压力仪;孔隙水压力计;水压(力)计;测压计;测压管
piezometer head 测压管水头
piezometer opening 测压孔
piezometer orifice 压力仪;测压孔
piezometer ring 流压环;均压环;环形压力计;环形流压计
piezometer surface 测压管水面
piezometer tip 孔隙水压测头
piezometer tube 孔压管;测压管
piezometer type 测压计型
piezometer type thermometer 压力式温度计
piezometric 量压的
piezometric(al) transducer 压电换能器
piezometric conductivity 导压系数
piezometric contour 等压线
piezometric elevation 测压管高程
piezometric head 测压水头;测压管水头
piezometric head line 测压水头线;测压管水头线
piezometric height 测压高度
piezometric level 液压梯度线;压力计平面;水压面;测压水位;测压管水位
piezometric line 测压管水位线;自由水面(坡)线;水压坡降线;水力坡降线
piezometric method 测压法
piezometric nest 测压孔组
piezometric pipe 测压管
piezometric potential 承压水头;测压位能;测压水头
piezometric pressure 测压管压力
piezometric regime 渗压状态
piezometric rise 测压管水头上升
piezometric rise test 测压管水头上升试验
piezometric slope 测压坡度
piezometric stage 测压管水位
piezometric surface 水压面;测压面;测压管液面;测压管水面
piezometric tube 测压管
piezometric well 测压井
piezometry 压力测定;流体压力测量法;加压法
piezooptic(al) 压光的
piezooptic(al) coefficient 压光系数
piezooptic(al) effect 压光效应
piezooptic(al) property 压光性质
piezooscillator 压电振荡器
piezophony 压电送受话器;晶体送受话器
piezoprobe 压探头
piezoquartz 压电石英;压电晶体
piezoquartz manometer 加压水晶压力计
piezo receiver 压电受话器
piezoremanent magnetization 压剩余磁化
piezoresistance 压敏电阻;压力电阻效应;压电电阻
piezoresistance coefficient 压电电阻系数
piezoresistance effect 压电电阻效应
piezoresistance measurement 压电电阻测量
piezoresistive accelerometer 压电电阻加速度计
piezoresistive material 压敏电阻材料
piezoresistive pick-up 压电电阻拾声器;压敏电阻传感器
piezoresistive pressure transducer 压阻压力变送器
piezoresistivity 压电电阻率
piezoresistor 压敏电阻;压敏电阻;压电电阻器
piezoresonator 压电谐振器
piezosensor 力敏元件
piezothermoluminescence 压力热致发光
piezotropic fluid 压性流体
piezotropy 压性
pig 铸块;管道除垢器;生铁块;生铁;扫线球;锭
pig alert 清管器位置信号器
pig and cattle slurry 猪牛粪水
pig and ore process 生铁矿石法
pig bed 铸床;生铁槽
pig breaker 铸锭破碎机;铁块破碎机;碎块机;生铁打碎机

pig-casting machine 生铁浇铸机;铸锭机;生铁锭机
pig copper 生铜;粗铜锭
pigenetic concretion 次生结核
pigeoholed wall 蜂窝墙
pigeon hole 剧院看台的顶排座位;鸽栅式文件分类架;小房(间);绞盘杆承孔
pigeonhole corner 凹凸不齐的墙角
pigeon-holed arch 多孔小拱
pigeon-hole(d) bond 鸽舍砌筑(法)
pigeon-holed foundation wall 地垄墙
pigeon-holed masonry wall 有鸽舍的墙
pigeon-holed wall 花格墙;多孔墙
pigeon-house 鸽舍
pigeonite 易变辉石
pigeonry 鸽棚
pigeons 噪扰
pigeon-toed 内八字脚
pig farm 养猪场;猪场
pig fitted with wire brushes 清管刷
piggery 养猪场;猪场
ppiggery waste(water) 猪圈废水;猪场废水;养猪场废水
piggin 汲水桶;长柄勺
pigging material 清管器材料
pigging (of pipe) 管道清扫
pigging operation 清管
pigging process 清管法
pigging system 清管系统
piggyback 附加有效荷载的;驮背运输;骑背钉;背负运输;工作面链式输送机;机载
piggyback anchor 背负锚
piggyback board 级联式插件板
piggyback car of trailers 背负式平车
piggyback collection (废物的)人工收集
piggyback container 集装箱
piggyback control 分段控制;级联控制
piggyback distribution 分段运销
piggyback formwork 驮背式模架
piggybacking 背负式生产
piggyback operation 驮背式集装运输;子母车运输
piggyback plan 驮背式运输方式
piggyback service 挂式驮运业务;铁路平车运输
piggyback system 汽车驮运系统;铁路驮背系统;背负式运输法
piggyback tandem mortgage 双人贷款
piggyback terminal 背负式运输枢纽
piggyback thrust propagation 背驮式逆冲扩展
piggyback traffic 驮背式联运交通;集装箱公路运输
piggyback transport 开上开下运输
piggyback traveller 背负式起重行车
piggyback truck 装在铁路平车上运输的载重汽车
piggyback twistor 分段磁扭线;背负双绞扭存储器
piggyback twistor memory 分段磁扭线存储器
piggy packer 装卸搬运工;夹运装卸机;铁路驮背式装卸车
pig house 猪舍
pig iron 铸铁;生铁;毛铁;坯铁
pig-iron alloy 合金生铁
pig-iron barrow 运生铁手车
pig-iron for steel making purposes 炼钢生铁
pig-iron ladle 铁水罐
pig-iron mixer 生铁搅拌器
pig launcher 清管器发送器
pig lead 铸铅;铅锭;粗铅锭
pig line 钓丝
pig lug 金属把(手)
pig machine 铸铁机
pigment 色素;色料
pigmentability 颜料着色性;颜料捏合性;可着色性
pigment absorption 颜料的吸收;颜料调和
pigment agglomerate 颜料结块
pigment(ary) dyestuff 有机颜料
pigmentary size 色素胶水
pigmentation 颜料捏合;颜料淀积;着色(作用);加颜料;色素形成(作用)
pigmentation of cement 水泥着色
pigment binder ratio 颜料漆料比;调色比;颜基比
pigment binder system 颜料—漆料系统
pigment binding capacity 颜料染色能力
pigment blend 颜料溶合;颜料混合
pigment bordeaux 颜料枣红
pigment brown 颜料棕
pigment carrier 颜料载体
pigment carrier medium 颜料载色剂

pigment chrome yellow 颜料铬黄
pigment colo(u)r 颜料色素;颜料
pigment colo(u)ration 涂料着色;色剂呈色
pigment compatibility 颜料的相容性
pigment concentrate 浓加工颜料
pigment concentration 颜料的浓度
pigment content 颜料含量
pigment dispersant 颜料分散剂
pigment dispersant demand 颜料分散剂需量
pigment dispersion 颜料的散开
pigment double coating 颜料双层包膜
pigment drawout 颜料刮涂检验法
pigment dulling 颜料发暗
pigment dyeing 涂料染色
pigment dye(stuff) 颜料化染料
pigmented 加颜料的;着色的
pigmented cement 彩色水泥;加颜料的水泥
pigmented coating 色漆
pigmented compound 填料混炼胶
pigmented compound curing 加色剂薄膜养护(混凝土路面)
pigmented concrete 着色混凝土
pigmented finish 着色装饰;着色涂料;着色饰面
pigmented flatting varnish 着色无光漆
pigmented floor(ing) varnish 着色地板清漆
pigmented garden tile 着色花园砖瓦
pigmented glass 有色玻璃;颜色玻璃;着色玻璃
pigmented mastic asphalt 着色沥青玛瑞脂
pigmented mortar 着色砂浆
pigmented paint 加颜料的油漆
pigmented paper 色纸
pigmented resinous powder 有色树脂粉末
pigmented rubbing varnish 加颜料的揩擦清漆
pigment extender 颜料增量剂;体质颜料
pigment extending capability 使颜料增量的能力
pigment exuding 颜料渗色
pigment fast violet 颜料坚牢紫
pigment fibre yarn 无光化纤纱
pigment figure 色素图案;色纹
pigment fineness 颜料的细度
pigment finish 涂颜料
pigment flo(a)tation 浮色
pigment flushing process 颜料冲洗方法
pigment foil 颜色箔
pigment for colo(u)ring cement 水泥着色颜料
pigment for colo(u)ring concrete 混凝土着色颜料
pigment formation 颜色配方
pigment for protective coats on facades 建筑立面保护层用颜料
pigment grade 颜料等级
pigment grain 颜料粒子
pigment granule 色素粒
pigment green 颜料绿
pigment grind 漆浆
pigment industry 颜料工业
pigment industry wastewater 涂料工业废水
pigment industry wastewater treatment 涂料工业废水处理
pigmenting property 着色性能
pigment in natural fats 天然油脂色素
pigment mineral 颜料矿物
pigment mineral commodities 矿物颜料矿产
pigment miscibility 颜料的可混性
pigment oil stain 油性色浆
pigmentolysis 解除色素(作用)
pigment orange 颜料橙
pigment packing factor 颜料吸油系数
pigment padding 悬浮体轧染(法)
pigment pad dyeing 悬浮体轧染(法)
pigment particle globularity 颜料颗粒的球度
pigment paste 颜料浆;颜料糊;颜料膏;漆浆;色浆
pigment percentage 颜料份
pigment photogenic property 颜料发光能力
pigment pollution 颜料污染
pigment powder 颜料粉
pigment precursor 色素母体
pigment preparation 预分散颜料;颜料制备物;加工颜料
pigment printing 涂料印花
pigment printing paste 涂料色浆
pigment purple 颜料红紫
pigment rayon 无光人造丝
pigment red 颜料红

pigment resin emulsion colo(u)r 涂料乳胶色浆
pigment resin printing 涂料印花
pigment rubine 颜料玉红
pigment scarlet 大红粉
pigment settlement 颜料沉积
pigment stain 颜料污点
pigment test of welds 焊缝着色试验
pigment transfer 涂料转移
pigmentum 涂剂
pigment volume 颜料体积
pigment volume concentration 颜料占涂料总体积的百分比;颜料体积浓度;颜料容积浓度
pigment volume ratio 颜料体积比;颜料容积比;颜料比容
pigment waste 颜料厂废物
pigment wastewater 颜料厂废水
pigment weight concentration 颜料重量浓度
pigment weight percent 颜料重量比
pigment wettability 颜料润湿性
pigment-wetting agent 颜料润湿剂
pigment wiping filler stain 木器用颜料填孔着色剂
pigment yellow 颜料黄
pig metal 金属铸块;金属块;金属锭
pig mo(u)ld 锭模
pig mo(u)lding machine 铸锭机
pig on pork 本家汇票
Pigou effect 皮古效应
pigpen 塔顶台板栏杆;猪圈
pig receiver station 生料接收站
pig signal 清管信号
pig signaller 清管信号发生器;清管器位置信号器
pig skin 返粗
pigsty(e) 木墩;框架;木垛;猪圈
pigsty stacking 井字形堆垛法
pigsty timbering 空心垛式支架;木垛支护
pigtail 软导线;引出端;螺旋管段;卷尾支管;绞合;柔韧铜辫;盘管;尾缆;绳尖;不整齐的绳索现象
pigtail cable 尾纤
pigtailed photodiode 带尾纤的光电二极管
pigtail splice 绞线接头;捻合;平行扭接
pig tin 锡块;锡锭
pig train 成串清管器
pig trap launch 清管器发送
pig trap receive 清管器接收
pig unit 养猪场
pig up 生铁增碳
pig wrack 角叉(藻)胶
pike 收税栅;大路
pike lifter 草垛提升机
pike man 用镐的矿工;税道关卡看守人
pike noise 尖噪声
pike peak 尖峰;山嘴
pike pole 杆钩;杆叉;撬杆;桶杆;吊杆;长搭钩;扳钩
pike staff 小杆钩
pil 涨潮时形成的小(海)湾
pila (有考古价值和趣味的)灰泥;(支撑顶梁的)柱顶方块材;(意大利教堂的)圣水池
Pila-alginite 皮拉藻类体
pilaster 柱墩;(桥台前墙的)扶壁;墙壁墩子;附壁柱;壁柱;半露柱;挨墙柱
pilaster base 壁柱基础
pilaster block 壁柱块;半露柱块;柱墩砌块;柱墩砌块
pilaster capital 壁柱头;壁柱帽
pilaster capping 壁柱帽顶装饰;壁柱头;壁柱帽
pilastered 加壁柱的
pilastered facade 加有壁柱的(建筑)立面
pilaster facade 壁柱正面;壁柱外观
pilaster face 壁柱正面
pilaster in pairs 成对壁柱
pilaster mass 短壁柱;壁柱墩子
pilaster side 壁柱侧面
pilaster strip 壁柱头;无帽壁柱;长壁柱;壁柱
pilastrade 一列壁柱;一排半露方柱;一系列壁柱;一排壁柱
pilastre 壁柱;半露柱
Pilbara nucleus 皮尔巴拉陆核
Pilbara old land 皮尔巴拉古陆
pilchard oil 沙丁鱼油
Pilcher's stop rot 皮尔彻木材防腐剂
pile 坐滩;桩木;桩;库存品;堆码;堆;(木头等的)大块;反应堆
pile abutment 桩式桥台
pile acting as a column 柱桩
pile action 桩的支承作用;桩承作用

pile adhesion 桩的表面摩擦力
pile anchor 拉力桩;桩锚
pile anchoring 桩锚碇
pile and cylinder jetty 桩筒式码头
pile and fascine 桩篱
pile and nap lifting machine 起绒机
pile and pile foundation 桩和桩基础
pile and plank retaining wall 柱板式挡土墙;插棒式挡土墙;插板式挡土墙
pile and sheet-pile 主板桩结构
pile and sheet-pile driving work 桩和板桩打桩工作
pile and stone dike 木桩堤坝;木桩堆石堤
pile and waling groin 木桩横撑丁坝;排桩丁坝
pile and waling groynes 桩固横撑丁坝;透水丁坝;桩及横撑丁坝
pile and waling permeable groin 木桩横撑透水丁坝
pile band 桩头铁箍;桩箍
pile barge 驳船桩
pile beacon 桩形立标;柱形立标【航海】
pile-bearing 桩支承
pile bearing capacity 桩的承载能力;桩的承载量
pile bending moment 桩的弯曲力矩
pile bent 排桩;桩的横向构架;桩排架;桩桥台
pile bent abutment 桩式桥台;桩式木桥台
pile bent bridge 桩排架桥
pile bent pier 桩排架桥墩;排架桩墩
pile bent sill 下盖梁
pile block 垫桩;顶桩;桩顶;桩垫
pile body 桩体;桩身
pile bolt 桩螺栓
pile boom 码头护木
pile boring rig 桩钻孔机
pile breakwater 桩式防波堤
pile bridge 桩式桥;桩桥;桩承桥;桩桥
pile building 岸边房屋;水上房屋;水上房屋
pile built in place 就地灌浇桩;就地灌注桩;钻孔灌注桩
pile bulb 桩底圆趾
pile bulkhead 桩式驳岸;桩构堵壁
pile butt 桩的粗端(一般指原木端)
pile butt elevation 桩端高度;桩端标高
pile by percussion drill method 冲孔成桩法
pile cage 桩钢筋笼;桩的钢筋骨架
pile cage spacer 桩钢筋骨架保护层垫环
pile cam 毛圈三角
pile capacity 单桩承载力;桩的承载能力;桩的承载量;桩的承载力
pile capacity formula 桩的承载力公式;桩承载力公式
pile cap beam 桩顶联系梁
pile cap coefficient 承台效应系数
pile cap girder 桩顶大梁
pile cap(ping) 桩帽
pile capping beam 桩帽梁;桩承台梁
pile carpet 绒头地毯
pile casing (tube) 桩的套管;桩套;桩管;护筒
pile casting 浇制桩
pile centre line 桩中心线
pile charring 土法炭化
pile-clump dike 桩簇坝;簇桩坝;桩簇丁坝
pile-clump lashing 桩簇捆绑
pile-clump spur dike 桩簇丁坝;桩簇坝
pile cluster 桩组;桩群;群桩
pile coating 植绒
pile cofferdam 桩(式)围堰;排桩围堰;板桩围堰
pile collar 桩箍
pile columns 群桩柱;群桩
pile completely in the ground 基桩
pile compressive stress 桩的压应力
pile concrete 制桩混凝土
pile connected with melted sulphuric paste 硫磺胶泥接桩法
pile core 桩芯
pile corrosion 桩的腐蚀
pile cotton blanket 棉绒毯
pile cover 桩帽
pile cross-section 桩的截面面积
pile crown 桩头;桩帽
pile-crown ring 桩顶箍
pile crush 绒头抗压性能
pile crushing 绒头压陷(地毯)
pile curtain 桩排
pile cushion 桩垫

pile cut-away 截桩
pile cutoff 桩截段;桩段
pile-cylinder dolphin 桩筒系船柱
pile-cylinder pier 桩筒式突堤;管柱码头
piled 桩基的
piled anchor 桩式锚
piled bridge 桩基桥
piled bulk produce 散堆货物
pile(d) dike 桩(式)堤;桩挡土墙;桩围堰;桩堤;木桩坝;排桩丁坝
piled dolphin 桩基靠船墩;靠船簇桩
pile delivery 纸堆式收纸
pile density 毛圈密度
pile depth 桩深
piled fendering 垫桩;护桩;防撞桩;桩式护舷
piled footing 桩承底脚;桩基(础)
piled foundation 桩基(础);有桩基础
pile diameter 桩径;桩的直径
pile diaphragm 桩隔墙;桩隔板
piled jetty 桩式突码头;桩式防波堤;桩式栈桥码头;桩式突堤;桩基突堤码头;桩基码头;桩基防波堤;栈桥式码头
pile dolphin 系船桩;桩基靠船墩;柱形灯桩【航海】;灯桩(上设灯标)
piled pier 桩基栈桥;桩基码头;桩式桥墩;桩基突堤码头
piled quay 桩承驳岸;桩承码头
pile drawer 拔桩器;拔桩机
pile drawing machine 拔桩机
pile drill 桩钻孔机
pile drilling machine 桩钻孔机
pile driven platform 桩基平台
pile driver 打桩工;桩架;打桩机
pile driver barge 打桩船
pile driver crane 打桩起重机
pile driver hammer 打桩锤;桩锤
pile driver hose 打桩机软管
pile driver lead 打桩机导(向)柱
pile driver tower 打桩机架;打捆机架
pile driving 打桩
pile driving abutment 打桩台
pile driving analyzer 打桩分析仪
pile driving barge 打桩船
pile driving boat 打桩船
pile driving by vibration 振动打桩;振动沉柱法
pile driving by water-jet 射水打桩
pile driving crane 打桩起重机
pile driving curve 打桩曲线
pile driving damage 打桩损害
pile driving depth 打桩深度
pile driving diagram 打桩曲线
pile driving efficiency 打桩效率
pile driving engine 打桩机
pile driving equipment 打桩设备
pile driving formula 打桩公式;动力公式;锤击桩公式
pile driving frame 打桩架
pile driving hammer 桩锤;打桩锤
pile driving helmet 桩箍;桩帽;打桩帽
pile driving machine 打桩机
pile driving material 打桩材料
pile driving method 打桩(方)法
pile driving operation 打桩作业
pile driving over water 水上沉桩
pile driving plant 打桩机;打桩设备
pile driving platform 打桩平台
pile driving pontoon 打桩船
pile driving reaction 打桩反力
pile driving record 打桩记录
pile driving resistance 打桩阻力
pile driving rig 打桩机具;打桩设备;打桩机
pile driving stress 打桩应力
pile driving test 打桩试验
pile driving time 打桩时间
pile driving tool 打桩工具
pile driving tower 打桩架
pile driving vessel 打桩船
piled rock 石堆
pile drum 桩的鼓状部位
piled sand 堆状积砂
piled stone hill 掇山
piled trestle 桩基栈桥;桩基突堤码头
pile dwelling 水上屋;岸边住房(半水半陆,打桩支承)

pile dynamics 桩的动力学
pile eccentricity 打桩的偏心度;桩的倾斜度;桩的偏心距;桩偏心距
pile engine 打桩器;打桩机
pile envelope 反应堆外壳
pile equipment 打桩设备
pile extension 接桩
pile extraction[extracting] 拔桩
pile extraction resistance 拔桩阻力
pile extractor 拔桩器;拔桩机
pile fabricating yard 制桩场(地)
pile fender 桩柱碰垫;码头碰垫桩;护桩;防撞桩;防冲桩
pile fendering 防撞桩
pile ferrule 桩箍
pile fission 反应堆分裂
pile fit 适宜采用桩基的
pile flutter 桩的颤动
pile follower 送桩(机)
pile foot 桩脚
pile footing 桩承底脚;桩基(础);低桩承台
pile force ga(u)ge 桩用振动钢丝测力计;桩力计
pile for driving 要打的桩
pile foreman 打桩组长;打桩领班
pile fork 桩定位筋
pile formula 桩柱公式;桩载公式;桩承载力公式;打桩公式
pile foundation 桩基础
pile foundation grillage 桩基格排
pile foundation stiffness 桩基刚度
pile foundation structure 桩基结构
pile foundation work 桩基工程
pile frame 打桩(机)架;桩架;打桩架
pile frame-bent abutment 框架式桥台
pile freeze 桩的凝固
pile friction 桩表面摩擦;桩摩擦力
pile grating 格式桩基承台
pile grillage 桩格排;桩群;桩基承台
pile grillage foundation 桩格排基础
pile groin 排桩丁坝
pile group 桩组;桩群;群桩
pile groyne 排桩丁坝
pile guide 导桩
pile-guiding frame 导桩架
pile gun 堆测试枪
pile hammer 打桩机;桩锤;打桩锤
pile hammer blow 桩锤击
pile hammer drum 桩锤卷筒
pile hammer guide 桩锤导架
pile hammer head 桩锤头
pile hammer rail 桩锤围栏
pile hammer rope 桩锤缆索
pile hammer tower 打桩塔
pile hammer weight 桩锤重;桩锤重量
pile harrow 桩式耙
pile head 桩头
pile head elevation 桩端标高;桩端高度
pile head protection 桩头保护
pile heave 桩的隆起;桩的回升;桩回升
pile height 桩长;绒高(从衬垫至地毯面高度);桩高;反应堆高度
pile helmet 桩顶帽;桩帽;桩盔
pile holder 桩夹
pile hole 桩孔
pile hoop 桩箍
pile hurdle dike 透水桩坝;透水桩堤;透水排桩丁坝;透水排桩堤
pile hydrant 桩式消防栓
pile integrity test 桩的完整性试验;桩的非破损性试验
pile jacketing 桩的套护
pile jacking 压桩;桩压
pile jacking-in method 静压沉桩法
pile jetti 射水沉桩法
pile jetting 小冲沉桩;水冲法沉桩;水冲沉桩(法);射水沉桩法;射水沉桩
pile keeper 桩夹
pile layout 桩位布置;桩的布置
pile leader 导桩架
pile lifter 垛板升降台
pile lighthouse 桩柱灯塔
pile lining 桩壁
pile load 桩荷载
pile load capacity 桩承重力

pile loading 桩的加荷
pile load(ing) test 桩载试验;桩基荷载试验;桩的静载试验;桩(的)承载试验
pile location 桩位
pile loop 毛圈
pile mandrel 桩芯
pile manufacture 制桩;桩的制作
pile material 制桩材料;反应堆材料
pile monkey 桩锤;打桩锤
pile moorings 系船桩组
pile mo(u)ld 桩模
pile neutron 反应堆中子
pile No. 桩号
pile of bags 袋垛
pile off-gas system 反应堆除气系统
pile of freshly placed concrete 新浇混凝土桩
pile of rubble 碎岩堆;碎砖堆;碎石堆
pile of slide prevention 抗滑桩
pile-on property 染深性能
pile oscillator 堆振荡器;反应堆振荡器
pile overdrive 超打桩;过深打桩
pile pad 桩垫
pile partly in the ground 部分埋在地下的桩
pile pedestal 桩底脚;打桩台座
pile penetration 桩的贯入量;桩的贯入度;桩的贯入
pile period 反应堆周期
pile pier 排架墩座;桩支码头;桩基栈桥;桩基码头;桩墩
pile pitching 斜桩缆索;斜桩
pile-placing method 桩的灌浇法;打桩(方)法;置桩(方)法
pile plank 板桩;木板桩
pile-planking 打板桩墙;桩上铺板
pile-plank retaining wall 柱板式挡土墙
pile plant 打桩机
pile platform 桩台
pile platform type lock 桩台式闸
pile platform type lock wall 桩台式闸墙
pile point 毛圈针片;桩尖
pile point bearing 桩尖支承
pile point bearing capacity 桩尖承载力
pile point resistance 桩尖阻力
pile poison removal 反应堆毒物排除
pile position 桩位
pile positioning 桩的定位
pile press machine 压桩机
pile protection 桩套
pile puller 拔桩器;拔桩机
pile pulling 拔桩
pile pulling machine 拔桩机
pile pulling test 拔桩试验
pile quay 桩式码头;桩基码头
piler 堆垛装置;集草器;堆垛机;堆垛工
pile raft 桩筏基(础)
pile-raft foundation 桩筏基础
pile rail 毛圈导轨
pile rake 桩的倾斜度;桩的倾斜
piler bed 堆垛机
pile reaction 桩的反力
pile redriving test 桩复打试验;桩重新打入试验
pile relaxation 堆松弛
pile retaining pin 毛圈护针
pile retapping test 桩复打试验
pile ring 桩顶箍;桩头铁箍;桩环;桩箍
pile row 排桩;板桩
piles battressed in same direction 单向斜桩
piles by percussion 冲孔桩
pile screwing 把桩旋入土中
pile-screwing capstan 螺旋桩式绞盘;螺旋桩式绞盘
pile section 桩段
pile setting 吊桩就位;整绒(地毯整理);竖绒(使绒毛竖起)
piles formed by drilling 钻孔桩
piles formed by grabbing 抓孔桩
pile shaft 桩身
pile shaft frictional forces 桩身摩阻力
pile shape 桩形
pile shearer 剪绒头机
pile sheathing 围堰板桩;打板桩
pile shoe 桩靴;板桩靴
pile-sinking 沉桩;打桩
pile sinking in water 水中沉桩
pile sinking with the water jet 水射沉桩;水力冲射沉桩

piles in pair 成对桩
pile size 桩的尺寸;桩的大小
pile skin 桩表皮
pile-soil system 桩土体系
pile spacer bar 桩定位筋
pile spacing 桩距;桩间距
pile splice 桩的拼接(物);桩接头;接桩
pile splicing 接桩;桩的拼接
pile staging (storage yard) 桩的堆放(场)
pile start-up 反应堆启动
pile static loading test 静力压桩试验
pile stem 桩身
pile stockade 桩围栏;桩栅(栏)
pile stoppage point 桩止点(桩的最后打入深度)
pile stress 桩应力
pile structure 桩基结构;桩基建筑物
pile structure offshore platform 桩基式近海平台
pile support 板桩支架
pile-supported 桩支承的
pile-supported building 干闸建筑
pile-supported continuous footing 桩承连续基础;桩基连续基础
pile-supported continuous foundation 桩承连续基础
pile-supported footing 桩承基脚;桩承基础
pile-supported foundation 桩基连续基础
pile-supported platform 桩承台
pile-supported quaywall 桩承台(式)岸壁
pile-supported raft (foundation) 桩承筏阀基础
pile-supported structure 桩支承结构
pile-supported wharf wall 桩承台式码头岸壁
pile tally 码垛
pile tear resistance 毛圈撕裂强力
pile technology 核反应堆工艺学
pile tensile stress 桩张应力;桩拉伸应力
pile test 桩荷载试验;桩承载试验
pile test load 试桩荷载
pile tip 桩脚;桩尖;桩端
pile tip elevation 桩顶标高
pile tip resistance 桩尖阻力
pile toe 桩底(即桩尖);桩脚
pile toe resistance 桩端阻力
pile to jacket link 桩与支撑桁架相联结
pile tolerance 桩位公差
pile top 桩顶;桩盖顶
pile top displacement 桩顶位移
pile torque shear test 桩的扭剪试验
pile transfer function 反应堆传递函数
pile trestle 桩栈架;桩构栈道;桩构排架;桩承栈桥;排桩栈桥
pile-type abutment 桩式桥台
pile type dolphin 桩式靠船墩
pile-type pier 桩式桥墩
pile ultimate (bearing) capacity 桩的极限承载力
pileum 鸟冠
pile unload 桩的卸载
pile-up 装满;增水【水文】;堆积叠起;积存;堆挤;堆积;堆存;扳键
pile-up effect 脉冲堆积效应;堆垒反应
pile-up of dislocation 位错堆积
pile-up process 堆积过程
pile-up pulse 累积脉冲
pile-up valve 集成阀
pile variable 反应堆变量
pile vibrosinking 振动沉桩
pile vibrosinking method 振动沉桩法
pile wall 板桩岸壁
pile water jet 射水沉桩
pile weave 起绒织物;割绒织物;绒头织物;叠叠组织
pile weir 桩堰
pile welding 堆焊
pile wharf 桩式码头;桩基码头
pile winding 分层叠绕线圈
pile with bulk shaped base 扩脚桩(爆炸成型的);蒜头桩
pile with horizontal timber 主桩套板结构
pile with wings 带翼桩
pile wood 桩木;导木
pile work 桩工;围桩工程
pileworm 蛀木虫
pile yard 桩堆放场;制桩场(地)
pile yarn 绒头纱线
pilfer 偷窃
Pilger mill 皮尔格式轧管机

Pilger seamless-tube mill 皮尔格式无线钢管轧机
Pilger tube-reducing process 皮尔格缩管法
pilgrimage altar 朝圣者旅途中的祭坛
pilgrimage chapel 朝圣者旅途中的小教堂
pilgrimage house 朝圣者旅途中的住房
pilgrim rolling process 周期式轧管法
Pilgrims ship 朝圣船
Pilgrims way 朝圣路
pili annulati 环状发
pilier cantone 哥特式复合墩
pilification 打加固板桩
piliform 毛样的
piling 桩工；桩材；排桩；刷涂堆漆；堆墨；打桩工程；打桩
piling adhesive 打桩胶
piling and burning 堆烧；堆聚烧除
piling and layout of stones 叠石
piling and stacking machine 堆垛机
piling bar 钢桩；钢柱
piling barge 打桩船
piling baseline 沉桩基线
piling bay 堆垛跨
piling beam 钢板桩
piling bin 堆放架
piling box 垛板箱
piling by vibration 振动沉桩法
piling cap 桩帽
piling car 堆垛车
piling cluster 桩组
piling contractor 打桩承包者；打桩承包人；打桩承包公司；打桩承包工
piling cooling bed 堆垛冷却台
piling crane 叠板堆垛吊车
piling crown 桩帽
piling density pench grade 分级堆积密度
piling deviation 沉桩偏位
piling footing 桩基(础)
piling foundation 桩基(础)；打桩的基础
piling frame 打桩架
piling guide 桩锤导架
piling height 堆积高度
piling home 把桩打到止点
piling job 打桩工
piling layer 桩基层
piling lead 桩锤导架
piling machine 堆垛机；堆布机
piling method 打桩(方)法
piling noise 打桩噪声
piling of rockeries 叠山
piling of the stator cores 定子铁芯堆叠
piling permit 打桩执照
piling pipe 桩管；直管
piling plan 桩位平面图；桩位布置图；打桩布置图
piling plant 打桩设备；打桩机
piling reaction 打桩反力
piling resistance 打桩阻力
piling rig 打桩机
piling rise 板桩上升
piling sculpture 堆雕
piling stress 打桩应力
piling-supported 桩基的
piling survey(ing) 沉桩测量
piling test 打桩试验
piling tube 直管
piling up 堆起
piling-up batch by batch 分批堆垛
piling wall 桩垣；桩墙
piling winch 打桩绞车
piling works 打桩工程
Pilkington twin process 双面研磨法
pill 小球；终端片
pilla 石柱；砖柱
pillar 柱；主柱；中竖框(门窗)；矿柱；基柱；基墩；煤柱；台柱(子)；石柱；驳柱(船)；标柱，标石；支柱
pillar-and-panel work 柱式采矿
pillar-and-stall method 房柱式采矿法
pillar bearing 柱支座；墩支座
pillar blast 煤柱爆破
pillar-bolt 柱形螺栓；螺撑；伸出的双头螺栓；撑架螺栓
pillar-box 邮筒；信筒
pillar brace 立柱牵索
pillar bracket bearing 柱式轴承架
pillar bulkhead 支承舱壁
pillar buoy 柱形浮标；杆状浮标

pillar burst 矿柱压碎
pillar crane 转柱式起重机；转柱起重机；柱式起重机；立柱式旋臂起重机；塔式起重机
pillar die set 导柱式模架
pillar drain (土坝内的)柱状排水
pillar drilling machine 柱形钻床；柱式钻床；柱架钻床
pillar drive 地层支护平巷
pillared hall 立柱式会堂；立柱式大厅
pillared rectorite adsorption 层柱累托石吸附
pillaret 小柱
pillar extraction 矿柱回采；煤柱回采
pillar extraction drilling 煤柱上打眼
pillar file 柱锉
pillar form(work) 柱模板
pillar for side panel and door 侧板和门柱
pillar fountain 柱式饮水池
pillar harrow 桩式耙；柱水耙；柱式耙；杆齿耙
pillar hydrant 柱式给水栓；桩式消防栓；柱式消火栓；柱式消防栓
pillaring 回收煤(矿)柱；柱系列；烟柱；高炉冷料柱；冷料柱；回收矿柱法
pillar insulator 柱形绝缘子
pillar jib crane 转柱悬臂起重机；转柱挺杆起重机
pillar knees 支柱肘板；支柱肘板
pillar ladder 支柱梯【船】；支柱梯【船】；舱内支柱兼升降梯柱
pillar line 煤柱线；煤柱放顶线
pillar man 煤矿井下充填工；砌垛工
pillar method 柱式采矿法
pillar method of mining 柱式采矿法
pillar method of stoping 柱式采矿法
pillar method of working 柱式采矿法
pillar mining 柱式开采法；煤柱回收
pillar of victory 凯旋柱
pillar piscina (教堂中的)圣水喷的墩柱
pillar plate 底盘
pillar platform 柱形平台
pillar post 信筒
pillar press 柱式压(力)机
pillar-protected surface 矿柱保护范围
pillar quay 柱式码头；墩柱式码头
pillar-recovery operation 回收支柱工作
pillar reinforcement 柱筋
pillar reservoir 直立储筒
pillar-row type under-ground diaphragm wall 柱列式地下连续墙
pillar shaper 柱式牛头刨床；柱式成型机
pillar shaping machine 柱架牛头刨床
pillar split 矿柱巷道
pillar stand 柱座
pillar stiffener 支撑柱
pillar stone 柱状石纪念碑；隅石；柱石，奠基石
pillar structure 柱状构造
pillar support 柱支座；柱架；柱基(础)；矿柱支撑；墩支座；支柱
pillar switch 柱式开关
pillar system 柱式采矿法
pillar tap 立柱式水龙头；立柱式水嘴；柱式龙头；支柱式龙头
pillar type 柱状形
pillar type basin tap 立柱式洗脸盆水龙头
pillar type tap 立柱式水龙头
pillar valve 柱形阀
pillar wharf 墩柱式码头
pillar with clustered shafts 杆簇柱；集体墩柱
pill-box 小房屋；永备发射点；碉堡
pill-box antenna 抛物柱面天线；抛物盒天线
pill-box counter 球形盒式计数器
pillet 小丸
pill heat 首次熔炼
pilling 起球
pilling home 把桩打到止点
pill of alumin(i)um 铝珠
pillow 轴承；枕块；褥垫；托木
pillow block 缓冲支承件；轴座；轴枕；轴台；中间轴轴承；支撑轴承
pillow block bearing 座架；支承
pillow block bearing cup 轴台轴承杯
pillow block breather 轴台通气管
pillow block cap 立式轴承箱盖
pillow block cover 轴台盖
pillow block cover gasket 轴台盖衬
pillow block flange 轴台凸轮

pillow block for ball-bearing 滚珠轴承座
pillow capital 枕垫形柱头
pillowcase tank 折叠罐
pillow deformation 枕形变形
pillow distortion 枕形失真；枕形畸变
pillow distortion equalizing 枕形畸变校正
pillowed 向外弯出的；枕垫形的
pillow joint 球形接头；球形接合
pillow-jointing 球形连接
pillow lava 枕状熔岩
pillow lava flow 枕状熔岩流
pillow structure 枕状构造
pillow test 枕形气密试验；枕形抗裂试验
pillowwork 枕垫形凸块表面装饰处理
pill test 小球测定法
pill transformer 匹配短截线
pilolite 镁坡缕石
pilot 新产品试制；引水员；引示信号；引示波；引火器；引航(员)；指引；领水员；领航(员)；领港员；驾驶仪；驾驶【船】；排障器；导正销，导洞
pilot actuated regulator 间接作用调节器
pilotage 引航技术；引航(费)；领航，领港(费)
pilotage certificate 引航证
pilotage charges 引水费
pilotage chart 领航图
pilotage district 航标导航水域；引航区
pilotage dues 领港费
pilotage ground 引航员上下船地段
pilotage inward 领航入港费；入港领港费
pilotage outward 出港引航费
pilotage rate 引航价率；引航费(率)
pilotage tariff 领航费价目表
pilotage waters 航标导航水域；引航水域；引航区；领航水域
pilot air 先导气流
pilot air shutter 引火嘴调风板
pilot alarm 导频告警器
pilot alternator 辅助交流发电机
pilot amplifier 导频放大器
pilot anchorage 引水锚地
pilot-and-reamer bit 中部凸出的十字钻头
pilot angle 引导角；控制角；排障器角铁
pilot arc 维持电弧；导引电弧
pilot arc current 维弧电流
pilot area 试验区
pilot a ship into or out of a harbor 领港
pilot authority 引航管理机关；引航管理机构；引航管理当局
pilot automatic dead reckoning equipment 自动舵航迹积算仪
pilotaxitic texture 交织结构
pilot balloon 导向气球；导航气球；测风(气球)
pilot balloon observation 测风气球观测
pilot balloon theodolite 测风经纬仪
pilot balloon transit 测风经纬仪
pilot band filter 导频带通滤波器
pilot bar 排障器杆；排障杆
pilot base 排障底板
pilot beam 缓冲梁；导向射束
pilot bearing 导轴承
pilot bearing spacer 导轴承隔片
pilot bell 引示铃；监听铃；监视铃
pilot bit 超前钻头；导向钻头；领眼钻头；前导钻头；定向钻头
pilot blast-hole bit 导向孔爆破钻头
pilot block system 引示制；指示制
pilot boat 引航船；领航船
pilot bob 竖井垂锤
pilot boiler 试验锅炉
pilot borehole 导向钻孔
pilot boring 先行试钻；试钻；打导洞；超前钻探
pilot brace 排障器拉条
pilot bracket 指示灯插座
pilot bridge 驾驶室(又称驾驶台)【船】
pilot brush 选择器电刷；控制刷；测试刷
pilot burner 火种燃烧器；引燃喷嘴；点火燃烧器；导燃烧嘴；辅助烧嘴
pilot cable 引航电缆；引导电缆；导引电缆
pilot cable duct 引示电缆套管；导向电缆套管
pilot calculation 试算
pilot carrier frequency 导频频率
pilot car sign 先导车标志
pilot casing 超前套管
pilot casting 试制铸件；标准铸件

pilot cell 引示电池;指示用电池;领示电池;控制元件;驾驶室
pilot channel 子沟;溢洪缺口;引流槽;引河;子槽;领流通道;导频信道;导频通道;导流河槽
pilot chart 领航图;引水海图;引航图;航线指南图;航路资料图(美版);航空气象图
pilot circuit 控制线路;控制手柄;控制电路;导频电路
pilot clock 监视用钟
pilot cloth 海员厚绒呢
pilot column 中间试验柱;中间试验桩
pilot concrete lane 机场混凝土导向跑道
pilot concrete strip 机场混凝土导向带;机场混凝土导向跑道
pilot connection 控制连接
pilot contactor 辅助接触器
pilot contract 临时协定
pilot control 领示控制;导频控制
pilot-controlled type valve 液控式阀
pilot controller 领控控制器;导频控制器
pilot current 导频电流
pilot cut 引河;裁弯引槽
pilot cut-off 导流裁弯;裁弯导槽;引河
pilot cutter 引航快艇;引航船
pilot delay line 领示延续线;控制延迟线
pilot depot 试验站
pilot detecting level meter 重锤式料位计
pilot device 试生产装置
pilot differential relay 引线差动继电路
pilot director 航向指示器;导航仪
pilot drift 超前导坑;先进导坑;导洞掌子面;导洞;超前巷道
pilot drill 导向孔钻机;定心钻
pilot drill hole 超前钻孔;导向钻孔
pilot drive 导洞开挖;导洞掘进
pilot dues 引航费
piloted head 导燃喷头器
piloted ignition 引火着火
piloted multidiameter reamer 带导向柱异径铰刀
piloted ship 被引航的船
pilot engine 清路机车【铁】;向导机车;向岸机车;引导机车;清道机车;启动发动机;调车机车;辅助机车;辅助发动机
pilot exciter 导频激励器;副励磁机;副激磁机;副磁机;辅助励磁机
pilot experiment 中间试验
pilot fairway 引水航路
pilot fee 引航费
pilot filter 导频滤波器
pilot firm 企业负责人
pilot fit 导向配合
pilot flag 引航旗;招请引水员旗号;招请引航员旗号
pilot flame 引燃火焰;起动火舌;点火火焰
pilot flood 试注水
pilot flow 先导流量;操纵流量
pilot frequency 领示频率;导频
pilot frequency oscillator 导频振荡器
pilot frequency switching 导频倒换
pilot furnace 中间工厂试验炉;试验窑炉
pilot gas burner 导燃气烧嘴
pilot ga(u)ge 引导式塞规;引导塞规
pilot generator 测速发电机;辅助发电机;辅极发电机
pilot guide 操作指示器
pilot guide bit 导向钻头
pilot head 导燃喷燃器
pilot heading 指向导洞;超前掘进工作面;超前导洞
pilotherm 恒温器;双金属片恒温控制器
pilot hole 导眼;先导孔;试坑;定位孔;导向钻孔;导向孔;导洞;超钻
pilot-hole cover 地下坑道超前钻孔群
pilot house 驾驶室;操纵室;操舵室
pilot house bell 驾驶室报时钟
piloti (房屋的)架空底层用支柱;钢筋混凝土重型柱
piloti construction 鸡腿建筑
pilot ignition 引燃
pilot indicator 驾驶仪;导频指示器;导电平指示器;导航指示器
piloting 领航;引航技术
piloting cable 导航海底电缆
piloting district 引航区
piloting waters 引航区;引航水域
pilot injection 引燃喷射
pilot investigation 试点调查
pilot ion 比较离子

pilotis 高楼架空底层用支柱;桩基(水下或陆地群桩基础)
pilotis building 群桩基础建筑;鸡腿式建筑
pilot jack 招请引航员旗号
pilot kiln 中试窑;试验窑炉
pilot ladder 引航员梯
pilot lamp 信号灯;引示灯;引导灯(桩);指示灯;领航灯;监视灯;度盘灯;导向灯;表盘灯
pilot lane 航空跑道;引导跑道
pilot launch 领航艇
pilot left ship 引航员离船
pilotless 无人驾驶
pilot level 导频电平
pilot lever 控制手柄
pilot license 引航员执照
pilot light 信号灯光;引火;引导灯(桩);指示灯;领示灯;领航信号灯;领航灯;监视灯;度盘灯;导向灯
pilot-line production 初步小规模生产
pilot locomotive 辅助机车
pilot machine 领航仪
pilot management 引航管理
pilot material 试生产材料
pilot method 导洞法
pilot method of excavation 导洞开挖法
pilot mill 导向铣刀
pilot model 先导模型;引导模型;试验性模型;试选样品;试验模型
pilot model study 探索性模型研究;试验性模型研究
pilot model system 试验性模型系统
pilot motor 分马力电动机;控制电动机;伺服电动机;辅助电动机;辅导电动机
pilot nail 临时钉;定位钉;安装钉
pilot narrow-band filter 导频窄带滤波器
pilot nut 导枢帽;导枢螺母
pilot ocean data system 试验性海洋资料系统
pilot oil 调节油;操纵油
pilot oil pressure 控制油压
pilot oil regulating valve 控制油的调节阀
pilot operate 光导操纵
pilot-operated 远距离操纵的;随动操纵
pilot-operated check valve 液控单向阀
pilot-operated controller 自动控制器;辅助控制器
pilot-operated pressure control 先导式压力控制阀
pilot-operated type valve 液控式阀
pilot-operated valve 导阀;导阀;液压控制动作阀
pilot operation 引导操作
pilot oscillator 主控振荡器;导频振荡器
pilot piece 柱脚保护罩(防止车辆冲击)
pilot pile 中间试验桩;排障桩;导桩
pilot pin 准销;定位针;定位栓;导销
pilot pin operating limit switch 导销操纵的行程开关
pilot pipe 超前套管
pilot piston 导向活塞
pilot plant 中试厂;小规模试验工厂;小型试验性工厂;中试装置;中间试验厂;中间工厂;试验性设备;试验(性)工厂;试生产装置;试验工厂;半工业性装置
pilot plant installation 中间试验厂设备
pilot plant investigation 中试厂研究
pilot plant kiln 中间试验炉
pilot plant reactor 中间规模堆
pilot plant scale 试验厂规模
pilot plant scale production 试验性生产
pilot plant study 半生产性研究
pilot plant test 中间试验
pilot plant unit 小型试验设备
pilot plant work 中间工厂研究
pilot plough 除雪犁
pilot plow 除雪犁
pilot plunger 导阀柱塞
pilot pressure 向导压力;预示压力
pilot pressure pump motor 辅助压力泵起动机
pilot process 中试过程;试验性过程
pilot production 试制;试验性生产;试生产;生产试验
pilot production expenditures 试生产费用
pilot production line 试验性生产流水线;试生产线
pilot production unit 试验生产设备
pilot project 试验项目;试验计划;试点项目;试点工程
pilot protection 引线保护装置
pilot pulse 引导脉冲;导频脉冲
pilot pump 中试泵

pilot punch 定位打孔器;定距侧刀(冲切侧边缺口的凸模);导向冲头
pilot radiostation 引航无线电台
pilot raise 超前
pilot ram 先导压头
pilot reactor 试验性反应堆
pilot reamer 导向扩孔器;导径铰刀
pilot reamer bit 扩眼导向钻头
pilot reaming bit 中心导向式金刚石扩孔钻头
pilot receiver 导频接收器
pilot regulator 导频控制器;导频调节器
pilot relay 引示替续器;控制继电器;导频继电器
pilot relief valve 液控溢流阀;控制安全阀
pilot run 试运行
pilot-safety device 燃气炉安全(送气)装置
pilot safety valve 液控安全阀
pilot sample 试制样品;试选样品
pilot scale 小规模;中试规模;中间工厂规模
pilot-scale device 中间试验厂装置
pilot-scale equipment 中间试验厂装置
pilot-scale facility 半生产性装置;中间试验厂装置
pilot-scale nanofiltration membrane separation 中尺度滤膜分离
pilot-scale system 中试规模系统
pilot-scale test 小规模试验;扩大试验;中试试验;半工业性规模试验
pilot scheme 试验计划
pilot screw 导向螺钉
pilot's cruising ground 引航员巡航区
pilot seam 调整焊缝;试验焊缝
pilot selector switch 自动操舵开关
pilot sensor 引频感测器;导频传感器
pilot servomotor 引航伺服机;中间接力器;副接力器
pilot shaft 导井;超前井
pilot sheet 领航图
pilot shoulder reamer 阶梯状扩孔器;多级扩孔器
pilot signal 引频信号;引导信号;指示信号;信号;控制信号;控速信号;导频信号
pilot signal light 领航信号灯
pilot signal switch 领示信号开关
pilot sleeve 导向套筒;导向套管
pilot's lien 引航员留置权
pilot spark 指示火花;触发火花
pilot's post 火场向导岗位
pilot spotting telescope 检像望远镜
pilot's sailing direction 航行方位
pilot stage 样品试制阶段
pilot station 引航站
pilot's trace 航迹【航空】
pilot streamer 先导闪流;导频流光
pilot study 中试(研究);中间试验研究;扩大试验研究;探索研究;探索性研究;试点研究
pilot subcarrier 导频副载波
pilot-supervisor room 驾变监控舱
pilot surfacing lane 铺导航跑道路面
pilot survey 试验调查;试生产测量;试点调查;初步勘察
pilot switch 控制开关;辅助开关
pilot system 先导系统;引导系统;试验性系统;试验系统
pilot tap 导向丝锥;带导柱丝锥
pilot tape 引导带
pilot test 小规模试验;小型试验;先试验;样品试验;中试;中间试验;探索性试验;初步试验;半工业性试验
pilot testing 中间试验
pilot tone 导频音
pilot transfer 接送引航员
pilot trench 导沟;导水沟
pilot truss 脚手架的桁架
pilot tube 指示灯;超前套管;风速指示仪
pilot tunnel 导向隧道;导坑;导洞;超前导洞;前群导洞
pilot tunnel method 隧道导洞掘进法;导坑法
pilot type bit 塔式钻头;多级钻头;导向钻头
pilot type plug bit 导向锥形钻头
pilot unit 中间试验厂;试验性设备;试生产装置;导向单元
pilot valve 导阀;先导伺服阀;主导阀;控制阀;给油门;伺服控制阀;操纵阀;分油活门
pilot valve governing 操纵阀调节
pilot valve governor 导阀调节器
pilot vessel 引航船;领航船

pilot voltage 标准电压
pilot voltmeter 送电端电压表
pilot waste stabilization pond system 中型污水稳定塘系统；中等污水稳定池系统
pilot waters 引航水域；领航水域
pilot wave 领示波；领波；导频波
pilot wedge 导向楔
pilot well 试验井
pilot wheel 手轮；导轮；操纵轮；操舵轮
pilot wire 引示线；领示线；控制线；导引线；操作线
pilot wire protection 导引线保护(装置)
pilot wire regulator 领示线调节器
pilot wire trans mission regulator 领示线自动增益调节器
pilot workings 超前巷道
pilt zone 试验区
pilus 菌毛
pimco (固定建筑板材用的)钢夹
pimelate 庚二酸盐
pimelic acid 庚二酸
pimelic dinitrile 庚二腈
pimelite 脂镍皂石
pi mill 竖井磨煤机
pimp 电焊工助手(俚语)；竖井清洁工
pimpernel 琉璃繁缕属(草本植物)
pimple 鹤嘴锄；尖锄；疙瘩(陶瓷缺陷)；表面斑点
pimple mound 小残丘
pimple plain 残丘平原
pimpling 粗糙度
pin 销(钉)；细杆；轴针；轴销；栓销；大头针；插脚
pinacoid 轴面体；平行双面
pinacoidal 轴面的
pinacoidal cleavage 轴面解理
pinacoidal joint 轴面解理
pinacol 邻二叔醇类；四甲基乙二醇
pinacolone 频哪酮
pinacotheca 画廊；美术馆；绘画陈列馆
pinaculum 古希腊或罗马建筑中有脊的屋顶
pinakiolite 硼镁锰矿
pinakoid 轴面体
pinakothek 美术馆
pinal schist 绢云母片岩
pin and box 公母扣(接头)
pin-and-eye connection 螺栓联结；螺栓连接；铰链连接
pin-and-eye type reamer 牙轮式扩孔器
pin-and-plug assembly 引线插头装置
pinang 槟榔
pin arched girder 销接拱梁
pin arrangement 插脚排列
pin assignment 引线分配
pinax 古希腊或罗马剧院舞台后柱间装饰板
pin barrel 带针圆筒
pin-based column 铰端柱
pin bearing 滚珠轴承；固定铰支座；销轴承；滚柱轴承
pin-bearing strength 销承压强度
pin-bearing support 滚柱支承
pin beater mill 凸轮式冲击研磨机
pin bit 手摇钻头
pinboard 转接板；插销盘；插接控制盘；插接板
pinboard computer 插接式计算机
pinboard machine 针孔板计算机；插盘机；插接板计算机
pinboard matrix 插接式矩阵电路
pinboard programmer 插接式程序装置
pinboard programming 插接式程序设计；插接板程序设计
pin bolt 销钉
pin boss 销座；销壳
pin breaker 针碎机；齿碎机
pin-break shank 带安全销的支柱
pin-break shovel 带剪销安全器的锄铲
pin-break standard 剪销式炉柄
pin bush(ing) (活塞销的)销衬套；销套
pince-ciseaux 有刃镊
pin center 销中心线
pincer gun 杠杆式焊钳
pincerlike mode 钳状水系模式
pincers 钳状物；有齿镊；夹子；夹钳；钳子；铁钳；手钳；拔钉钳
pincers-like grab 钳式抓斗
pincer spot welding head 杠杆式点焊钳
pincer-type bale lifter 抓爪式草捆提升机

pincette 夹架
pinch 狭缩；箍缩；夹紧；镊子；捏；坍缩；收缩效应
pin chain 销接链
pinch-and-swell shape 串珠状
pinch and swell structure 膨缩构造
pinch area 压紧区域
pinch bar 折皱(带钢缺陷)；起钉器；爪杆；爪棍；尖头长杆；撬棍；撬杆；铁辊；尖撬棍
pinchbech 冒牌货
pinch beck 金色铜锌合金；假名饰物；金色黄铜
pinch bending 预压痕弯曲
pinch bolt 销；系紧螺栓
pinch clamp 弹性夹
pinchcock 活嘴夹；夹紧旋塞；弹簧夹
pinch cock clamp 弹簧节流夹
pinch column 收缩柱
pinch compression 等离子体箍缩
pinch current 收缩效应电流
pinch-discharge 箍缩放电
pinched coating 折皱涂层
pinched resistor 挤压电阻
pinched stem 靶茎
pinch effect 箍缩效应；夹紧效应；收缩效应
pinchers 钢条折叠缺陷；条钢折叠缺陷；条钢折叠；夹锭钳；(沟槽的)木板支撑；铁钳
pinch fit 公差配合
pinch frequency 夹断频率
pinching 受热开裂(耐火砖的)；手工捏练成型
pinching action 挤压作用
pinching of strata 地层尖灭；地层变薄
pinching-out 变薄；压出；冷拔
pinching screw 压紧螺钉；夹紧螺钉
pinching style 尖灭方式
pinch instability 箍缩不稳定性
pinchite 氯氧汞矿
pinch machine 箍缩机
pinch marks 挤压纹痕
pinch-off 夹断；交链断裂
pinch-off blades 夹紧板
pinch-off conditioning 夹止调节；夹断调节
pinch-off die 修边模
pinch-off effect 箍缩效应
pinch-off seal 压紧密封；夹断封接；铜管封接
pinch-off voltage 箍断电压；夹断电压
pinch out 压出；挤出；尖灭
pinch pass 开坯孔型；精冷轧；平整道次
pinch phenomenon 缩颈现象
pinch plane 紧压面；夹紧面
pinch plate 扣板
pinch plate rail fastening 扣板式扣件
pinch-preheated 收缩预热的
pinch resistor 收缩电阻器
pinch rod 测杆
pinch roll 压轮；领纸辊；拉辊；夹送辊
pinch roller 压紧轮；压带轮；紧带轮；夹送轮；摩擦导布辊
pinch roll unit 拉辊装置
pinch screw seal 螺旋压紧封闭
pinch spalling 机械剥落；铲剥
pinch-swell boudin 肿缩型石吞肠
pinch-swell mullion 肿缩式窗棂
pinch trim collar 夹切挡环
pinch wheel 紧带轮；夹送轮
pin clamp 插销口
pin clip 连接销的弹簧卡
pin compatible 管脚互换性；插脚兼容
pin configuration 管脚排列
pin-connected 栓接的；枢接的；铰接的
pin-connected construction 铰接结构
pin-connected frame 枢接框架；铰接框(架)；栓接框架；铰接框架
pin-connected joint 铰节点；螺栓联结；螺栓连接；铰链连接
pin-connected node 铰节点
pin-connected plywood girder bridge 钉板梁桥
pin-connected truss 枢接桁架；铰接桁架；栓接桁架
pin-connected truss bridge 枢接桁架桥
pin connection 销接头；销接合；销钉连接；榫头；销钉接合；引线连接；管脚连接；螺栓连接；销接结合
pin connection in bottom boom 桁架下弦的销钉连接
pin connection in top boom 桁架上弦的销钉连接
pin contact 针孔式触点

pin coupling 销连接；外螺纹半锁接箍
pin crusher 棒磨机
pincushion 针垫
pincushion correction circuit 枕形失真校正电路
pincushion corrector 枕形校正器
pincushion distortion 正畸变；枕形失真；枕形畸变
pincushion distortion equalizing 枕形畸变校正
pincushion field 枕形致偏磁场；枕形磁场
pincushion magnet 枕形失真调整磁铁
pin disc mill 滚柱盘磨机
pindone 杀鼠酮
pin dot matrix 针点矩阵
pin down 牵制；船首前倾
pin drift 柱销拆卸器
pin drill 针头钻；销钉钻；销孔钻；尖头钻
pine 灌木病；松树
pineapple 松果形装饰品；菠萝形装饰品；卵形鳞状尖顶饰
pine bark 松树皮
pine barren 松林泥炭地
pine board 松木板
pine cone 松木圆锥体
pined rafter 戗脊椽木
pine forest 松林
pine gum processing effluent 松脂价格废水
pine leaf oil 松叶油
pin electronics control 插销式电子控制
pin end 公扣端；铰接端；销接端
pin-ended 铰接端的
pin ended beam 固端梁
pin-ended column 铰端(支)柱；栓端柱
pin-ended portal frame 门形刚架；铰接式柱脚；铰接式柱脚门形刚架；铰接柱脚门式刚架
pin-ended strut 栓端支杆；销接支柱
pinene 蒎烯
pine needle oil 松叶油
pine nut 松子；松球
pine oil 松油；松树油；松叶油；松节油
pine pitch 松焦油沥青
pine plywood 松木胶合板；松木夹板
pine rams blowout preventer 管闸板防喷器
pine resin 松香；松脂
pine rocker bearing 铰支承
pinery 菠萝园；菠萝温室；松林
pine-seed oil 松子油
pine shingles 松木墙面板；松木屋顶板；松木瓦
pine sleeper 松木枕
pine tar 木焦油；松木焦油(沥青)；松焦油
pine tar pitch 松木焦油脂；松焦油沥青
pine tree 松树
pine-tree antenna 松树形天线
pine-tree array 松树形天线阵；松树式天线阵
pine-tree crystal 松枝晶；树枝状晶体；枝晶
pine-tree oil 松树油
pine-tree resin 松树脂
pine-tree structure 枝晶组织
pine-tree wall 松墙
pinetum 松林；松树园；[复]pineta
pine wall 松墙
pine-wood 松木
pine-wood flooring 松木地板
pine-wood oil 松木油
pine-wood tar 松木焦油(沥青)
pin expansion test 插头膨胀试验
pin exploration 钎探检验
piney 暗红色硬质松木(印度产)
piney tallow 松木硬脂
pin face wrench 叉形带销扳手
pin feather 毛锥
pin feed 针孔传输；导针送纸
pinfeed form 针孔传输形式
pinfeed platen 针孔馈送压纸卷轴
pin filler 栓钉垫(板)
pin fin 钉头
pinfold 牲口栏
pin for fixing gyroposition 陀螺定向销
pin for steady arm 定位器销钉
pin for suppressing metal 紧固销钉
pin frame 铰接框架；销接框架
pin function 插头功能
ping 声呐脉冲；中频声脉冲；来自声呐设备的脉冲信号；水声脉冲；声脉冲(信号)
pin ga(u)ge 销规
pin gear 针轮

pin gearing 针轮啮合
pin gear shaper 针齿轮刨齿机
pinger 音响信标;脉冲定深器;水声测位仪;声脉冲发射器;声波发射器
pinger location 声波发射定位
pinger proof echo-sounding system 声脉冲回声探测系统
pinging sonar 脉冲声呐
ping jockey 声呐兵
ping-listen-train method 脉冲监听序贯法
pingo 冰锥;冰核丘
pingo ice 地冰丘冰
ping-pong 往复式
pin groove 销子槽
pin guard 纹钉保护器
pinguite 绿脱石
pinhead blister 针孔
pinhead Morocco 摩洛哥针头玻璃
pinhead-size water droplet 针头水滴
pinhead tile 细粒的小砖
pin hinge 销钉铰链;销铰轴芯铰链;销铰;铰链;枢轴合页;抽心铰链;抽心合页
pin-hold lens 膜孔透镜
pin-hole (木材的) 蛀虫孔;销孔;销钉孔(眼);引线孔;螺丝孔;毛孔;小孔;针眼;针孔;针尖状气孔;大头针孔
pin-hole borer 凿船虫
pin-hole burner 针状火焰燃烧器
pin-hole camera 针孔照相机;针孔摄影机
pin-hole collimation 针孔准直
pin-hole defect in lumber 木材中小孔缺陷
pin-hole detector 微孔检测器
pin-hole grinder 销孔磨床
pin-hole imaging 针孔成像
pin-hole leak 小漏;针眼渗漏;针孔裂缝
pin-hole lens 针孔透镜;针孔透镜
pin-hole plotter 打孔机;打孔器
pin-hole porosity 针孔状疏松
pin-hole stowage 风车式装载
pin-hole test 针孔试验(分散性土)
pin-holing 小气孔;油漆麻点;针孔;起针孔(釉面);针眼
pin-index system 管脚符号系统
pining 钉扎
pining rod 收缩棒
pin in groove 锁销
pin insulator 针形绝缘子;针式绝缘子
pinion 波形瓦;小齿轮;减速驱动齿轮;游星齿轮;小牙轮;链齿
pinion-and-rack steering 小齿轮和齿条转向装置
pinion-and-rack steering device 齿轮齿条式转向装置
pinion-and-rack steering gear 齿轮齿条式转向传动机构
pinion barrel 小齿轮室
pinion bearing 小齿轮轴承
pinion bearing spacer 主动齿轮轴承隔圈
pinion cage 小齿轮架
pinion carrier 主动齿轮套
pinion cutter 小齿轮铣刀;插齿刀
pinion drive 小齿轮传动
pinion drive shaft 小齿轮传动轴
pinion file 带齿小锤;锐边小锉
pinion frame 行星齿轮架
pinion gear 小齿轮;游星齿轮
pinion gearing 小齿轮传动装置
pinion-gear shaper 小齿轮刨齿机
pinion grease 小齿轮润滑脂;传动齿轮用润滑脂
pinion guide 小齿轮导承
pinion head 齿轮齿顶
pinion housing 小齿轮外罩;齿轮机座
pinion leaf 小齿轮的一个齿叶
pinion mate 小齿轮对
pinion quill shaft 小齿轮套筒轴
pinion rack 齿臂板;啮合齿条;齿杆
pinion ratio 传动齿轮比
pinion ring 压装齿圈
pinion shaft 小齿轮轴
pinion shaft bearing 小齿轮轴承
pinion shaft outside bearing 小齿轮外侧轴承
pinion shimming 主动齿轮调整(用垫片)
pinion stand 小齿轮座;小齿轮架;小齿轮箱;齿轮机架
pinion stand gear 小齿轮传动

pinion steel 齿轮钢
pinion stop 齿轮停止器;齿轮定位器
pinion-type steering gear 齿轮式转向机构
pinion unit 齿轮传动装置;齿轮传动系
pinion wall 块垛墙;扶墙
pinion wheel 小齿轮
pinion wire 小齿轮线坯;齿轴丝
pinite 块云母
pin jack 针形插孔;管脚插孔;接触插孔
pin joint 销钉连接;销榫接法;销接头;销接;公接头;铰链接合;铰节;铰接;外螺纹半锁接头;枢接(合)
pin jointed 枢接合的;活节连接的;铰链接合的;铰接的
pin-jointed arched girder 铰接拱梁
pin-jointed bar 铰接杆件
pin-jointed frame 铰接框架
pin-jointed platform 铰接平台
pin-jointed purlin (e) 铰接体系;铰接檩条
pin jointed truss 销接桁架;铰接桁架
pin joint ship 铰接船
pink 小孔;饰缝;尖叉帆船;桃红色;石竹;淡红色;粉红(色)
pink aggregate 粉红石集料;粉红石骨料
pink blossom 粉花
pink colo(u)r 铬锡红;粉色;粉红色
pink-colo(u)red 桃红色
pin key 销键
pink fused alumina 铬刚玉
pink garden 石竹园
pink glass 粉红色玻璃
pink glaze 桃红釉
pink gold 金银铜镍装饰合金
pink group 粉红色类(载体)
pinkiness 粉红
pinking 发动机燃料爆震;震性
pinking roller 压花滚刀
pinking shear 锯齿剪切机
pink lauan 红柳安
pink luster 粉红光泽
pink noise 红色噪声
pink noise of internal combustion engine 内燃机发爆噪声
pink nose 粉红鼻
pink knot 小木节;小树节;针节
pink primer 淡红色底漆
pink priming R. M. paint 木料粉红打底漆
pink salt 锡盐;氯锡酸铵
pink slip 解雇通知书
pink stern 尖尾(船型);尖船尾
pink-stern ship 尖尾型船
pink support 粉红色载体
pink tea 午后茶会
pinkus metal 铜合金
pinlay 钉嵌体
pin-lift 顶杆
pin-lift arrangement 顶杆装置
pin lifting 顶杆起模
pin-lifting device 顶箱机构
pin-lift mo(u)lding machine 顶箱式造型机;顶杆式起模造型机
pin-lift pattern stripping arrangement 顶杆起模装置
pin-lift stripper 顶杆式起模机构
pin link 带轴链节
pin load 销负荷
pin lug coupling 销耳式接口
pin mark 针标;承烧痕
pin matrix board 插接矩阵板
pin metal 针用黄铜
pinnace 二桅小船
pinnacle 小尖塔;小尖顶;羽状水;高峰;尖礁;尖峰;尖顶;海塔;山顶;顶峰;攒尖饰
pinnacle canopy 尖帽罩;尖拱天盖
pinnacle(d) iceberg 峰形冰山;尖塔形冰山
pinnacle nut 六角槽顶螺母
pinnacle on the sea 海中岩峰
pinnacle reef 尖头礁;塔礁;宝塔礁;宝石礁
pinnacle rock 尖形礁石
pinnacle terminating in conic(al) form 圆锥体尖顶
pinnacle terminating in pyramidal form 金字塔形尖顶
pinnate drainage 羽状水系;羽状水;羽毛状水系

pinnate drainage pattern 羽状排水系统;羽状水系型
pinnate fibro-lamella 羽毛状梗子层
pinnate joint(ing) 羽状节理
pinnately compound leaf 羽状复叶
pinnate mode 羽状水系模式
pinnate tension joint 羽状张拉节理
pinnate vein 羽状脉
pinnatifid 羽状半裂的
pinned bond 轨端接头
pinned connection 铰节点;铰接合
pinned conveyer 针带式输送器
pinned end 铰接端;栓接端
pinned joint 铰节点
pinned purlin(e) 铰接檩条;铰接檩条
pinned system 蛇节(连接)方式;活节(连接)体系
pinner 小块石;(砌体中支承块石材用的)垫石
pinning 支撑;填隙石;棋盘格嵌石;花石墙壁;销钉连接;销住;打小桩
pinning-in 塞嵌碎石片;镶入;填塞
pinning machine 订商标机
pinning up 打楔
pinnoite 柱硼镁石;柱镁石
pin number 插脚数
pin nut 销螺母;带销螺母
pin of a hinge 铰链销钉
pin of universal joint 万向节销
pinoid pits 松型纹孔
pinoleum blind 外百叶窗;外遮阳;外卷帘;木条卷帘
pinolin 松脂醇
pinoline 松香烃
pinolite 松球菱镁片岩
pin packing 枢孔填板;枢孔填料;枢孔垫片
pin pallet 销钉式擒纵叉
pin pallet escapement 销钉式擒纵机构
pin pallet escapement balance 销钉式擒纵摆轮
pin pallet escape wheel 销钉式擒纵轮
pin pallet fork 销钉式擒纵叉
pin pallet lever escapement 销钉式擒纵机构
pin-photo diode 针形光电管
pin pile 锚碇桩
pin plate 销板;栓接板;枢板
pin pliers 销钳
pin plug 圆柱插头
pinpoint 准确定位;针尖;精确定点;极精确;目标定点;刺点
pinpoint accuracy 超高精度;高精确度;高度精确性
pinpoint arrester 针尖放电避雷器
pinpoint blasting 抛掷爆破
pinpoint charge 抛掷药包
pinpoint control 精确定位控制
pinpoint detail 微小细部
pinpointed focus 最佳焦点
pinpointed forceps 针尖镊
pinpoint flame 针焰点
pinpoint floc 针尖状絮凝体
pinpoint gate 针孔型浇注口
pinpoint photograph 刺点像片;目标像片
pinpoint photography 精确定点摄影
pinpoint program(me) 非常详细的程序表
pinpoint reference 精确定位
pinpoint target 点目标
pinpoint technique 精密技术
pinpoint the trouble 正确测定故障
pinpoint welding 点焊
pin-prick 刺点
pin pull 抽辊钮
pin punch 小直径凸模;尖冲头
pin punching register 针穿孔寄存器
pin punching system 针穿孔记录系统
pin push-off machine 顶箱式造型机
pin rack 销齿条
pin rail 衣帽架;带栓横梁;(舞台栓住布景绳用的)有栓横梁;带栓横板
pinrail 舞台大桥上的栏杆
pin rail rack 挽索插栓架
pin rammer 扁头砂舂;扁头夯砂锤
pin registration 销钉定位
pin removal 拔销器
pin retainer 挡料销保持器
pin riveting 双敲铆接法;双敲铆法
pin rocker bearing 铰接支座;摇杆轴承;铰支承
pin rod 销杆
pin roll 销轴;销辊

pin roller (测量齿厚或节圆直径用的)滚柱
pin-roller treatment 针辊处理
pins and line 砌砖墙用的铅直线及水平线
pin seal 销钉连接
pin sensing 探针式读出;探计式读出
pin shank 销钉杆
pin-shaped electrode 针形状电极;针形极;针形电极
pin slit 销子槽
pin slot 销子槽
pin-socket contact 针孔式触点
pinsonic quilting 超声波衲缝法
pin spanner 带销扳手;叉形扳手
pin splice 销钉拼接(板)
pin spot 细光束聚光灯
pin spot-light 细光束聚光灯
pin stop 挡料销
pin stripe 细条子
pin-stripe duck 细条子帆布
pin stripe effect 细条效应
pint 量磅
pint[pt]品脱(容量名,1 品脱 = 1/8 加仑)
pintadoite 钙钒华
pin tap 公锥
pintaux nozzle 轴针式喷嘴
pintel 链节销;枢栓
pin thread 外螺纹
pin timbering 锚杆支护法
pin tin 曲柄钻头
pintle 针栓;链节销;扣钉;配流轴;锁紧舵栓;枢栓;舵栓
pintle atomizer 针形喷雾器;针栓喷雾器
pintle bearing 舵针轴承;底枢轴承
pintle bushing 底枢轴瓦
pintle cap 底枢盖
pintle chain 斜栓节链;扣钉链;铰接链;短环链;扁节链;扁环节链
pintle hinge 扣钉铰链
pintle hitch 挂索驱动装置
pintle hole 扣钉孔
pintle hook 销栓挂钩;牵引钩;销钩【机】
pintle hook lock 钩锁卡
pintle nozzle 针式喷嘴
pintle plate 扣钉板
pintle score 舵钮空挡
pintle spring 扣钉弹簧
pin-to-box pipe 公母扣管子
pin-to-box sub 公母扣(接头)
pin-to-box tube 公母扣管子
pin together 用螺栓连接
pin tongs 带箍的电工钳;带箍钳;四瓣
pin tong space 接头切口
pin tooth 滚削齿;嵌齿
pin toothing 钝齿啮合
pin-to-pin coupling 双头公扣接头
pin truss 栓接构架;枢接构架
pint-size economy tractor 微型经济牵引车
pin tumbler 弹子锁销子;销钮锁式开关
pin tumbler lock 枢轴杠杆锁
pin tumbler mechanism 弹子锁机构
pin type attachment 销式连接
pin type carbide bit 针状硬合金钻头;针状碳化物钻头
pin type insulator for high-voltage line 高压线路针式绝缘子
pin type insulator for low-voltage line 低压线路针式绝缘子
pin type mill 钉式粉磨机
pin type safety joint 销式安全接头连接
pin type support insulator 针形支柱绝缘子
pin type tunnel kiln 插针式隧道窑
pin-up 钉在墙上的东西;可钉在墙上的
pin-up lamp 壁灯
Pinus Sinensis 油松
Pinus Sylvestris 欧洲赤松
pin valve 针阀
pin vise 针钳;针夹具
pin wheel 钝齿轮;销轮;针轮
pinwheel ground plan 针轮平面图;直升机地平面图
pinwheeling 就地旋回
pin with groove 带槽销
pin with round head 圆头销钉
pinworm hole 蛀眼
pin wrench 销钉扳手;带销扳手

pinyon pine 矮松
pioases 沃洲;水草田
pioneer 多孔无水砌块;首创;开拓工作
pioneer architecture 原始建筑
pioneer bore 隧道导洞
pioneer community 先锋群落
pioneer cut 开段沟
pioneer drift 超前导洞
pioneer drift method 导坑法
pioneer enterprise 先驱企业
pioneer hole 钻井导孔;隧道导洞
pioneer industry 先行性行业;先驱工业
pioneering 勘察;施工整地;垦荒;开拓工作
pioneering action 创举
pioneering architect 首创建造师;开拓型建造师
pioneering spirit 首创精神
pioneering stage 初创阶段;草创时期
pioneering undertaking 创举
pioneering work 开拓工作;创举;先行工程
pioneer invention 开创性发明
pioneer metal 耐蚀镍铬合金
pioneer patent 开创性专利
pioneer plant 先驱植物;先锋植物
pioneer population 先驱群体
pioneer product 先驱产品;新产品
pioneer road 临时便道;开荒道路;荒区道路;拓荒道路;施工便道
pioneer species 先驱种
pioneer stage 先锋阶段
pioneer tunnel 超前小平巷;超前导洞;测辟导洞;隧道导洞
pioneer tunnel method 超前导洞掘进法
pioneer well 勘探钻孔;拓荒井;(油田的)第一眼井;导井;初探井
pionnate 粘分子子团
pion π 介子
piotine 皂石
pioting fan 摇头风扇
pioury 印度黄
pip 尖脉信号;尖脉冲;带尖的支撑(窑具);反射点;报时信号
pipage 输送费;沿管蒸馏;管线输送的;管系;管道运输;管道系统;管道输送
pip amplifier 脉尖放大器;脉冲放大器
pip coding 脉尖编码
pip displacement 标记点位移
pipe 铸缩管;管子;管状物;管状矿脉;管;溶管;大酒桶;重皮;长口哨
pipe accessories 管道附件
pipe adapter 管接头
pipe aerator 管式曝气器
pipe alignment 管道定线;管子安装方位
pipe alley 管道;窄巷;管沟;管道沟
pipe anchor 管锚(墩)
pipe and cable locator 导管及电缆定位仪
pipe and duct snipping machine 管道与风道洗涤器
pipe and strainer underdrain 滤管地下水排水管渠
pipe and strainer underdrainage 滤管地下水排水管渠
pipe and tube fittings 管子配件
pipe and tube mill 轧管机
pipe angle 管子弯头;管子接头
pipe anvil vice 管子台虎钳
pipe aqueduct 管道渡槽;管道桥;导水管
pipe arch 管拱;钢管肋拱
pipe arrangement 管线布置;管系图;管系(布置);管路布置;管道布置;配管
pipe away 脱开管道;发出开船信号;用管道输出;汲出(用管子)
pipe badger 管道清理器
pipe band 管子夹箍;管夹带
pipe bank 管簇;管组
pipe barrel 圆柱体;筒状物;管筒
pipe base 管支座;管基
pipe basket 钻杆装运车架
pipe batten 舞台吊幕横管
pipe bearing 管道支承
pipe beater 管状逐搞器
pipe becoming stuck 卡钻
pipe bedding 管道垫层
pipe bell 管子承(接)口
pipe bend 时形管;管子弯头;管子弯曲;管肘;管形弯头;管弯头;管弯

pipe bender 弯管器;弯管机
pipe bending instrument 弯管器
pipe bending machine 弯管机
pipe bending mandrel 弯管芯棒;弯管芯子
pipe bending pliers 弯管钳子
pipe bending roll 弯管辊
pipe bending shoe 弯管胎具
pipe bending shop 弯管车间
pipe bending spring 弯管弹簧
pipe bending tool 弯管工具
pipe berth 管架床铺
pipe bit 管钻头;管状钻头;套管钻头
pipe blocking 管堵
pipe body internal pressure strength 管体抗内压强度
pipe body tensile strength 管体抗拉强度
pipe body thread joint strength 管体丝扣连接强度
pipe body yield strength 管体屈服强度
pipe bonding adhesive 管子胶黏剂
pipe-boom crane 管型臂起重机
pipe bore 管子内径
pipe bore side 管子内侧
pipe-borne noise 工作产生的噪声
pipe box 管套
pipe bracket 管子托架;管子固定件;管托架;管托;管钩
pipe bracket for lightweight block partition 轻质砌块隔墙上的管托
pipe branch 斜叉三通管;管接头;支管;管子分支;歧管
pipe break 管道破裂
pipe bridge 管桥;管道桥;管道吊架;架管桥
pipe bridge beam 管桥梁
pipe brush 管刷;清管刷
pipe buffer (焊管前的)管口打磨工
pipe buggy 托管材小车
pipe bundle 管束
pipe burner 棒状燃烧器;管状燃烧器
pipe burring reamer 去管口毛刺绞刀
pipe burst 管子裂口;管子破裂;管子爆裂
pipe bursting 管道炸裂
pipe bursting strength 管子破裂强度
pipe bushing 管衬套
pipe callipers 测管厚卡钳;测管厚卡尺
pipe canal 管沟
pipe cap 管盖;管帽;管端帽盖
pipe capacity 管道容量;管道通过能力;管道输送量
pipe car 管车
pipe carrier 管托;管架;导管支座
pipe casing 管围壁;管套;套管
pipe casting 管子铸件
pipe-casting in trenches 沟内铸管;管沟中浇灌管子
pipe cavity 缩孔;缩管
pipe centering apparatus 管子找中装置
pipe chamfering machine 钢管管道倒棱机
pipe chaplet 管子芯撑
pipe chase 卧管槽;管槽;管子槽
pipe chocking 管子塞头;管子堵头
pipe choking 管阻塞
pipe chuck 管子卡盘
pipe chute 斜管;管形放矿槽;溜管
pipe circulating system 管道环流系统
pipe circulation line 循环管道
pipe clamp 管夹(板);管卡(子);管箍
pipe clay 制陶白黏土;制陶管土;制管土;管土
pipe cleaner 通管器;洗管器
pipe cleaning 管道清洗
pipe cleaning agent 管道清洁剂
pipe cleaning device 清洗管道工具
pipe cleaning machine 清管机
pipe cleaning pig 清管器
pipe clip 管卡(箍);管座;管子卡(钳);管钳子;管夹
pipe clip connection 管夹接头
pipe clogging 管道阻塞
pipe close 管塞(子)
pipe closer 管子堵头;管塞头
pipe coater 涂管机
pipe coating 管子涂层;管子涂料;管面涂层
pipe coating process 管子涂层过程;管道涂敷法
pipe coil 旋管;线圈;盘旋管;盘管;盘点;盘肠管
pipe collar 管箍;管道泛水
pipe column 钢管混凝土柱;管子柱;管柱;管束
pipe compartment 管子格间
pipe compensator 导管调整器;导管补偿器

pipe concrete 混凝土管
pipe conduit 管路
pipe-connected 用管连接的
pipe connecting device 接管装置
pipe connecting flange 接管法兰;接管凸缘
pipe connection 管子连接(件);管子接头;管连接;管接
pipe connection clip 接管夹
pipe connector 管子连接件
pipe connection 管接头
pipe constant 钻管体积校正系数
pipe constriction 管道收缩
pipe-cooler 管道冷却器;管式冷却器
pipe cooling surface 管子冷却面
pipe cooling system 冷却管道系统
pipe core 管子型芯;管芯
pipe coupling 套管接头;管子偶接;管子连接;管(子)接头;联结节;偶联管
pipe cover 管壳
pipe covering 管套;管道覆盖物;管子覆盖;管外保护层;管顶覆盖层;管道保温层;管覆盖层
pipe covering mate 管道包扎毡
pipe covering tape 管子覆盖带
pipe cover test(ing) 管道隔热层性能试验
pipe cradle 管架
pipe cross 十字管;四通;十字管接头
pipe cross section 管子断面;管道断面
pipe crown 管顶;弯管顶
pipe culvert 管渠;圆管涵;管涵;管道排水涵洞;管道涵洞;管道;涵管
pipe curshing machine 管压平机
pipe cushion 管支墩;管垫
pipe cutter 管子切割器;管子割刀;割管机;截管器;截管机;切管机;切管机;切管刀
pipe cutter with cutting wheels 有切管轮的切管机
pipe cutting 管子切割
pipe cutting and threading 截管车丝
pipe cutting and threading machine 切管和车丝机;切管和车丝纹机
pipe cutting machine 管子车丝机;节管机;切管机;切管器
pipe cutting pliers 切管钳
pipe cutting wheel 割管轮
pipe cylinder 管柱
pipe cylinder dolphin 管柱系船墩
piped 管的
pipe debris rack 管制拦污栅
piped end 收缩头
pipe dented 管子瘪入;管子凹陷
pipe derrick 管式桅杆起重机
pipe design 管道设计
pipe detector 埋管位置探测器
pipe diameter 管径
pipe diaper 管子接头麻布
pipe die 管子丝口扳牙;管子丝扣扳钳;管模;车丝板牙
pipe ditch 管沟
piped log 管腐原木
piped medical gas supply system 管道医用气体供应系统
piped metal 缩孔金属
pipe dog 管子钳;管子搭头;管子搭钩;管钳;缩孔钳
pipe dolly 带辊架管子
pipe dope 管子润滑剂;管子涂料;管具涂料
pipe drain 管沟;管式排水;管道排水;排水管(道)
pipe drainage 管道排水;排水管(统);排水管道
pipe drainage system 管道排水系统
pipe drainage work 管道排水工程
pipe drain grade line 管底纵坡度线
pipe drawing 拔管;管线图纸;布管平面图;管拉拔
pipe drill 管钻;内螺纹车丝锥
pipe drive head 套管打入头
pipe drive shoe 套管打入鞋
pipe driving 下套管
piped services 管道公用设施
pipe duct 套管;管子通道;管渠;管沟;管道
piped water 管子给水
piped water supply 管道供水
pipe earthing 管道埋地;管道埋土
pipe elbow 管形弯头;弯头
pipe elbow meter 管肘流速计
pipe elevator 提管器
pipe elimination 除缩孔法
pipe enamel 地下煤气管沥青涂料;管壁涂料

pipe end 管端
pipe end aligner 管端对准器
pipe end chamfering machine 管端倒棱机
pipe end to handtight face length 管端至手旋紧面长度
pipe end washer 管端垫圈
pipe engineering 管道工程
pipe erector 管子工;管道工;管道安装工
pipe expander 胀管机;扩管机
pipe-expanding machine 扩口机
pipe expansion bend 管子补充弯头;管子膨胀弯头
pipe expansion joint 管子膨胀接头
pipe expansion test 管口扩张试验
pipe explosion 管子爆裂;管爆裂
pipe extraction 拔管
pipe extractor 拔管器
pipe extrusion 管子挤压(法)
pipe factor 钻管校正系数;钻管体积校正系数;水管因素
pipe fastener 管子卡钳;管子卡
pipe ferrule 管箍
pipe filter 管式过滤器
pipe finder 探管仪
pipe finger 钻杆立根指梁
pipe fish 海龙;尖嘴角
pipe fitter 管道修理工;管道安装工;装管工人;管子工;管道装配工;管道工人;水暖工
pipe fittings 接管零件;管道安装工作;管子安装;管头头;管道零件;管子配件;管件;管(道)配件;管道接头;管材配件;管材附件
pipe flange 管子凸缘;管子法兰(盘);管子凸缘(盘);管凸缘;管(道)法兰;管子凸缘
pipe flange joint 管子法兰接头;管子凸缘接头
pipe flashing 管道泛水
pipe floating the surface 用管子排除(混凝土)屋面积水
pipe flow 管流;管道水流
pipe forceps 管夹;管子钳;管钳
pipe forcing method 管道压入法
pipe-forcing system 推压管系统;推顶管系统;顶管机
pipe for concrete placing by gravity 自落式混凝土浇筑管
pipe for conducting air 通风管道
pipe for hydraulicking 水力冲挖土管
pipe fracture 管子损坏;管的损坏;管道破裂
pipe frame 架管框架;管架
pipe frame house 铁管架温室
pipe friction 管(道)摩擦;管壁摩擦;管道摩阻(力)
pipe friction coefficient 管道摩阻系数
pipe friction factor 管道摩阻系数
pipe friction formula 管道摩阻力公式
pipe friction loss 管道摩擦水头损失
pipe friction-loss coefficient 管道摩阻损失系数
pipe friction resistance 管道摩阻(力)
pipe funnel 管道漏斗
pipe furnace 管式炉
pipe gallery 管线廊道;管廊;敷管廊道
pipe gang 管道工程队
pipe gasket 管垫圈;管道接头垫圈;法兰垫圈;凸缘垫圈
pipe ga(u)ge 管厚规
pipe glue 管子胶黏剂
pipe go-devil 刮管器;清管器
pipe grab 抓管工具;管子抓取器;管材提升夹板
pipe grapple 管卡;管夹
pipe grid 管网;管状栅极;管栅式分布器
pipe grip 管子钳;管道夹握器;管钳
pipe gripper 管子夹钳;管扳手;管钳
pipe groove clearer 活塞槽清洁器
pipe groove machine 管道沟槽开挖机
pipe grout 管内灌浆
pipe guardrail 管式护栏
pipe guide 移动导管(装置);滑动卡圈;滑动导向轴承;导管装置
pipe handling machine 铺管机
pipe handrail 管子扶手;管子栏杆
pipe hand truck 手推管道运输车
pipe hanger 管子吊具;管子吊架;管道吊筋;管道吊架;挂管钩;吊管器;吊管架;吊管钩
pipe hauling trailer 运管拖车
pipe head 管头;管道进口端
pipe header 联箱箱;带连接管的配水干管

pipe heading 管道进口端
pipe heated hotbed 水管加热温室
pipe heater 管式炉
pipe heating tape 环绕管子的电热带
pipe hoister 吊管机
pipe holder 管式储气罐;管支架;管道托架
pipe hook 拧管子用旋转扳手;管子钩;管钩;吊管钩
pipehook for wall mounting 钉入墙内的管钩
pipehook hydraulics 管子扳手液压系统
pipehook installation 管子扳手装置
pipehook insulation 管子扳手绝缘
pipe hoop 管箍
pipe hydraulics 管流水力学
pipe incrustation 管子水垢;管子积垢;管子结壳
pipe in series 串联管路
pipe-insert 水管套座
pipe inside diameter 钻杆内经
pipe installation 埋管;管道铺设;下管;管子安装;管线铺设;管系安装;管道敷设;管道安装;导管装置;安装管子;安装管道
pipe insulated joint 导管绝缘接头
pipe insulating section 管道热绝缘截面
pipe insulation 管子绝缘;管绝缘;管道保温材料;管道保温;绝缘管壳;水管绝缘
pipe insulation grade loose fill 管道绝缘松填等级
pipe insulation grade section 管道绝缘等级断面
pipe irrigation 引水管灌溉
pipe jack 管子拉紧装置
pipe jacking 顶管法;管子推顶法;管子顶托;管道顶入;顶管施工;顶进法
pipe-jacking and thrust boring method 顶管法
pipe jacking method 顶管法
pipe-jacking survey 顶管测量
pipe-jacking system 顶管机;顶管顶推系统
pipe-jacking tunnel(l)ing machine 顶管掘进机
pipe jaw 接杆
pipe-jib crane 管型悬臂起重机
pipejoining position 接管方位
pipe joint 管接头;管接合;管接;结合杆;导管接头
pipe joint assembly 管道接头组合;管子接头总成;管道接头总装
pipe joint cement 管子接头防漏铅油
pipe joint clamp 管接头夹
pipe joint composition 管子连接用涂料;管接头处垫料
pipe joint compound 管道接头填塞料;管子接缝填料
pipe joint(ing) 管子接头
pipe jointing clip 接管夹;接管头
pipe joint(ing) compound 管道接头填料;管子胶合剂
pipe jointing ring 管子连接密封环;管子连接密封圈
pipe jointing tape 管子连接密封条;管子连接密封带
pipe joint nozzle 管接嘴
pipe joint piece 管子连接件
pipe joint seal(ing) compound 管道接头封口胶;管道接头密封胶
pipe joint set in cement mortar 管道接头用水泥砂浆嵌实
pipe jumbo 架管台车;布管台车
pipe junction 管道联结节;管子连接;管子接头;短管接头
pipe key 管式钥匙
pipe lagging 管道隔热套;管道绝缘套;管外保护层;管道保温
pipe lagging cloth 管道包扎布
pipe lagging loose fill 松填管道隔热套
pipe lamination 缩孔分层
pipe lateral with nozzle 带喷嘴支管
pipe lateral with orifices 带孔口支管
pipe lathe 管子加工车床;管车床
pipe launher 管状水槽
pipe layer 埋管机;吊管机;敷管机;布管工人;管道工;管道敷设机;管道安装工;铺管工;吊管机
pipe layer attachment 敷管机;管道安装工附件;吊管机
pipe laying 管道敷设;管道安装;埋管工作;铺设管道;布管;安装管道
pipe-laying barge 敷设管道船;管道敷设驳;布管驳船
pipe-laying crane 铺管机
pipe-laying device 敷设管子设备;敷设管子工具
pipe-laying drawing 铺管图
pipe-laying gear 铺管设备
pipe-laying machine 管道安装机;埋管机;敷管

机;铺管机
pipe-laying man 管道安装工
pipe-laying operation 铺管作业
pipe-laying plan 敷设管子平面图;敷设管子计划
pipe-laying plant 敷设管子施工设备
pipe-laying plow 敷管用犁
pipe-laying progress chart 铺管进度图表;敷设管道进度表
pipe-laying reel ship 卷筒铺管船
pipe-laying site 铺管现场
pipe-laying system 管道铺设系统
pipe-laying tractor 铺管牵引车;吊管拖拉机
pipe-laying trencher 开沟铺管机
pipe-laying vessel 铺管船
pipe-laying winch 敷设管道绞盘;敷设管道绞车
pipe layout 管系布置;管道布置
pipe leak 管子泄漏;管子漏接
pipe leakage 管道泄漏;管道渗漏;管道漏水
pipeless electric(al) drill 无管电钻
pipeless furnace 无管锅炉;无管式火炉;无管火炉
pipeless heating and ventilation 无管式供暖与通风;无管式采暖与通风;无管式采暖通风;无管供暖与通风设备
pipeless heating system 无管式供暖系统;无管式采暖系统
pipeless system of heating 无管式供暖系统;无管式采暖系统
pipeless triple valve 无管三通阀
pipe lifter 管子提升器
pipe lift truck 运管叉车
pipe-like conduit 管状通道
pipe-like passage 管状通道
pipe line 管路;管线
pipeline agitator 液体连续搅拌机
pipeline anchorage 管线固定
pipeline architecture 流水线结构
pipeline blind 管线上的插入板;管线上的流量孔板
pipeline blockage 排泥管堵塞
pipeline breathing valve 管线通气阀
pipeline bridge 管道桥;管线桥
pipeline burying machine 管线埋设机;埋管机
pipeline by-pass gate 管道旁通闸门
pipeline centring device 管道找中器
pipeline cleaner 管道清洁器;管道清理器;清管器
pipeline coating 管道涂料
pipeline compressor 增压压缩机;管道压缩机
pipeline concrete anchor 混凝土管道锚固;混凝土管座
pipeline concrete support 混凝土管道支座
pipeline connection 管道接头;管道接口
pipeline construction 管线敷设;管道施工
pipeline contractor 管道承包人;管道建设者
pipeline control 流水线控制
pipeline conveyer 管道输送
pipeline coordination 管网综合
pipeline corridor 管线廊道
pipelined computer system 流水线计算机系统
pipelined functional unit 流水线功能部件
pipelined microprogramming system 流水线微程序设计系统
pipeline dredge(r) 吸扬(式)挖泥船
pipeline drying pig 管道干燥清管器
pipelined system 流水线系统
pipeline enclosure 管道间
pipeline equipment 管道设备
pipeline excavator 挖沟机;管沟挖掘机
pipeline failure 管道破裂
pipeline filter 管线过滤器
pipeline flange 管道凸缘;管道法兰
pipeline float 排泥管浮筒
pipeline flow efficiency 管流效率;管道效率
pipeline flowmeter 管道流量计
pipeline gas 管道(煤)气
pipeline ga(u)ge 管道流速计
pipeline gradient 管线坡度
pipeline head 管线首站
pipeline heater 管道(式)加热器
pipeline identification 管线鉴别;管线标志
pipeline inspection pig 管道清理器;管道清洁器
pipeline-inspection vehicle 管道检查用潜水器
pipeline inspector 管线检验员
pipeline installer 管道支座;管道安装者
pipeline inventory 商品供应线上的库存量
pipeline jetty 管道突堤;管道堤

pipeline lagging 管道保温绝热层
pipeline laying 管道安装;管道布置;管道铺设
pipeline laying technique 管道铺设技术
pipeline layout 管路布置;管路布置
pipeline length 管段
pipeline maintenance 管道维护;管道养护
pipeline mark 管线标
pipeline materials 管道材料
pipeline mixer 管道混合器
pipeline net 管网;管道网
pipeline network 管道网;环状管网
pipeline noise 管道噪声
pipeline oil 洁净油
pipeline oil run 管道输油量
pipeline operation 流水线操作;流水线运算
pipeline operation report 管线运行报告
pipeline organization 流水线结构
pipeline parts 管道配件
pipeline pig 清管器
pipeline plugging pig 管道堵塞器;堵管器
pipeline positioning 管道定位
pipeline pressure 管线压力;管道压力
pipeline processor 流水线处理机;导管处理机
pipeline project 拟议中项目
pipeline protection 管道保护
pipeline pump 管道泵;管线泵
pipeline pumping station 管道泵站
pipe liner 管道(安装)工;管套;套管;井筒;铺管工
pipeline reactor 管道式反应器
pipeline reclamation 管道修理
pipeline refrigeration 管道式制冷法;管道制冷
pipeline rehabilitation 管道恢复能力
pipeline relocation 管线改移
pipe(line) resistance 管道阻力
pipeline restraint 管线制约
pipeline rider 管线警卫
pipeline river crossing 过河管线
pipeline room 管道间
pipeline run 管道输送量
pipeline run report 管线运行报告
pipeline scraper 刮管器
pipeline spans 管道跨距;管线跨距
pipeline spreading work 管道敷设工程
pipeline stiffener 管道支肋
pipeline stopcock 管道旋塞
pipeline storage 管道储存量
pipeline supervision by overlight 空中巡视管线
pipeline support 管架
pipeline surge 管道压力波动
pipeline survey 管道测量;管线测量
pipeline suspension bridge 管道悬索桥
pipeline system 管系;管路系统;管线系统;流水线系统;流水线式;流水线方式
pipeline system milking parlor 管道式榨奶挤奶厅
pipeline telemetering 管道遥测
pipeline to move solids 输送固体管线
pipeline transmission system 管线传输系统;管道输送系统
pipeline transport(ation) 管道输送;管道运输
pipeline transport of concrete 混凝土管道输送
pipeline trench(ing) 管线沟槽;管线地沟
pipeline tunnel 管线隧道
pipeline turbine 管线用燃气轮机
pipeline unit 流水线部件
pipeline valve 管道阀(门)
pipeline valve actuator 管道阀驱动器
pipeline vibration eliminator 管道消振器
pipeline walker 管道巡视员
pipeline walking 巡线
pipeline work 管道工程;敷设管道
pipeline wrapping 管道绝缘;管道绝缘
pipelining 管道涂层;管道内衬;管路输送;管道安装;管道敷设;管衬里;流水线作业;流水线数据处理;流水线技术;流水线化;流水线法;流水线操作;敷设管线;管套筒
pipelining erection 管道安装
pipelining machine 铺管机
pipelining practice 铺管业务
pipelining processing 流水线处理
pipelining segment 流水线段
pipe load 管道荷载
pipe location 管道定位;管道位置
pipe locator 管位探测器;管道定位器;埋管位置探测器;探管仪

pipeloop 循环管线
pipe loss 管道压损;管道热损;管道损失;管道水头损失
pipe machine 管螺纹切丝机
pipe main 干管
pipe-making machine 制管机
pipe-man 管道工(人);管子工;管工
pipe managing and working device 自动化管子加工机
pipe mandrel 管道标志;管子芯轴
pipe manhole 管道人孔;管道进入孔;管道检查井
pipe mast 圆筒桅杆;管状桅杆
pipe material 管料;管材
pipemidic acid 吡哌酸
pipe mill 制管厂;管材轧机;钢管轧机
pipe-mill rolling stand 钢管轧机机座
pipe mould 管道模制
pipe mo(u)lding 管子铸造
pipe-mover 移管机;管道移动器;喷灌装置管道移动器
pipe nail 瓦状泥芯棒;弧面泥芯棒;管夹钉
pipe net 管网
pipe net alignment 管网定线
pipe network 管网;管道系统
pipe network analysis 管网分析
pipe network analyzer 管网分析仪
pipe network drawing 管网图纸
pipe network grid system 环状管网
pipe network leakage 管网漏失量
pipe network parts 管网配件
pipe network system 管网系统
pipe newel 管状楼梯扶手中柱
pipe nipple 管子螺纹接管;管(子)接头;管螺纹接套;管接圈
pipe nipple both end plain 两头都是平头的管子接头
pipe nut 管子螺帽
pipenzolate bromide 吡哌喷酯
pipe offset 乙字(形连接)管;偏置管;迂回管
pipe of variable form 变形管
pipe on blocks 垫块支托的管道
pipe opening 管口
pipe orebody 筒状矿体
pipe ore sampler 管状矿采取器
pipe organ effect 管风琴效应
pipe orifice 管形孔口;管口;管孔
pipe orifice protection 管护口
pipe outside diameter 钻杆外径;管子外径
pipe packing 管子垫圈
pipe pedestal 管支墩
pipe pendant 管吊环
pipe pendant lamp 管吊灯
pipe penetration 管道贯入(度)
pipe penstock 引水管道
pipe picker 管状分离轮
pipe piece detail 管件图
pipe pier 钢管支撑;钢管墩(柱)
pipe pile 管桩;钢管桩;空心管桩
pipe-pile foundation 管桩基础
pipe pile hole 管桩孔
pipe piling foundation 管桩基础
pipe placing 搬移管道;敷设管道
pipe plan 管系图
pipe plant 制管厂
pipe plastics 塑料管子
pipe plate 管状薄板;制管钢板
pipe pliers 管子钳;管钳
pipe plug 管子塞头;管塞(子);管堵(头);丝堵
pipe pointer 磨尖机
pipe position indicator 管位指示器
pipe precipitator 管式沉淀器
pipe press 管式压力机
pipe pressure 管压力
pipe protection 管道保护
pipe protection against corrosion 管子防腐
pipe prover 管子液压试验器;管子检验器;管道水力检验仪;管道水力检验计;管道检验仪
pipe pulling and snubbing device 不压井起下油管装置
pipe pump 管道泵
pipe purging 管道清洗
pipe purlin(e) 管状桁条;管状檩条
pipe pusher 推管装置;(打管的)钻工;管道推进器
pipe pushing 顶管法;顶管

pipe-pushing pit 管道推进地坑
pipe pushing unit 顶管用千斤顶
pipe quenching 喷水冷却
piper 管道工
pipe rack 管子堆场;管支架;管架;立根垫;摆管架
pipe-rack dryer 管架干燥器;热管承放式干燥器
pipe racker (操作管子的)钻工
pipe racking stand 立根垫;摆管架
pipe racking system 立根摆管机构
pipe rail 管式护栏
pipe rams (防喷器的)闸板
pipe range 管道;管道网;管道范围;管长
piperazine 哌嗪
pipe reamer 管子铰刀;管口铰刀;管铰刀
pipe reducer 管道变径接头;大小头;渐缩管
pipe reducer of drill rod 钻杆异径接头(两头,一公一母)
pipe reducing bushing 缩小管径衬套
pipe reduction 缩管
pipe relined with cement mortar 用水泥砂浆重衬里管道
pipe repair clamp 修管箍
pipe repair stand 管子修理架
pipe resistance 管阻力;管路阻力
pipe resistance coefficient 管流阻力系数
pipe resonance 声管共振
pipe-resonator muffler 插管共振式消声器
pipe retainer 管具承座
pipe return bend 管道回转弯头
pipe revetment 管子保护层
piperidinium salt 哌啶盐
pipe rigging 管线支架;管道安装
pipe ring 管环;管箍;管道(吊)环
pipe riser 管子提升器
piperitenol 薄荷烯醇
piperitenone 薄荷烯酮
pipe rivet 管形铆钉
pipe roller 管式滚筒;滚管机
piperonyl cyclonene 增效环
pipe roof 管棚
pipe roof protection 管棚护顶
pipe run 管路装置;管线;管路;管段;管道走向;管道;导向管
pipe rust removal 管道除锈机
piperylene rubber 戊二烯橡胶
pipe saddle 管座;管道基座;鞍形管座;鞍形管夹
pipe sag 管子凹下
pipe sampler 管状取样器;取样管
pipe sampling 用管子取样;管道取样
pipes and tubes 各类管子
pipe saw 管锯
pipe scaffold(ing) 管子脚手架;安装管道脚手架;钢管脚手架
pipe scale 管垢
pipe scraper 刮管刀;刮管器;刮板式清管器;管子刮刀
pipe screw-cutting lathe 管子螺纹车床
pipe seal 管接头密封;管缝
pipe sealing 管子密封;封管
pipe sealing compound 封管化合物
pipe seam 缩孔起层
pipe section 管子断面;管壳
pipe section machine 制管机;管壳机
pipe segregation 缩孔偏析;缩管偏析
pipe separator 管道分隔器
pipe setback 立根架;立根垫
pipe setting 下管
pipe sewer 管沟;污水管
pipe sewer line 下水管道;污水管道
pipe shaft 管式工作坑;管道井;管制井筒;管形井筒;管身;管子竖井(隧道通气孔)
pipe-shaped cave 管道状溶洞
pipe-shaped rubber belt conveyer 管形胶带输送机
pipe-shed 管棚
pipe-shed support 管棚护顶;管棚
pipe sheet 管板
pipe shell 管壳
pipe shoe 管鞋
pipe-shoe bit 管靴钻头;管鞋钻头
pipe shop 管子商店;管子车间
pipe shroud 管套
pipe sinking 下管
pipe sinking by boiling 托盘下管
pipe sinking by floating 浮力下管【岩】

pipe sinking method 沉管法
pipe size 管子尺寸;口径;管道口径
pipe sizing 管道计算;管道尺寸选择
pipe sizing chart 管子尺寸选择表
pipe skeleton 管形骨架;管形框架
pipe sleeve 管套;管接箍;套管
pipe sleeve with gasket 管翼垫片;垫衬套管
pipe small-diameter tunnel 小管径隧道
pipe socket 管子承(接)口;漏斗口;承口
pipe socket wrench 管子套筒扳手
pipes of water supply and drainage 给排水管道
pipe soil 管道土壤
pipe soil sampler 管状取土器
pipe space 管距;管道间
pipe spacer 定距管
pipe spacing 管距
pipe spanner 管钳
pipe spanner wrench 管子钳
pipe spigot 管子插口;管子喇叭口;管子接口;管子承接口;管接口
pipe spinning machine 管形拉丝机
pipe spreader 轴套管
pipe standards 管(道)标准
pipe standard specification 管子标准规格
pipe steel 管钢
pipe steel reinforcement 钢管配筋;管式钢筋
pipestep-taper pile 逐级缩径钢壳与钢管组合桩
pipe still 管式炉;管式蒸馏器
pipe-still distillation 管式炉蒸馏
pipe-still heater 管式炉加热器;管式加热炉
pipe stock 管扳手
pipe stock and dies 管子绞板(包括板牙和板牙架)
pipestone 烟斗石
pipe stop(per) 管塞(子);管道堵头;龙头
pipe storage 管式储气罐
pipe storage of gas 管束储气
pipe storm drain 雨水管
pipe straightener 调直管子工具;直管器;管子矫直机;调直器;调管器
pipe straightener machine 直管机;钢管矫直机
pipe strap 管吊带;管子支吊管;管环;吊管带;管卡
pipe stress analysis 管子应力分析
pipe string 管柱;管道支线
pipe stroke of crane 吊管钩
pipe subdrain 暗沟管;地下排水管道
pipe subway 管式地下铁道
pipe summit 管顶
pipe supplier 管材供应商
pipe supply contract 管材供应合同
pipe support 管道支座;管座;管支架;管架;管道支(撑)架;管材支护
pipe support bracket 管托架
pipe support concrete 管架底座混凝土
pipe-supported pipeline 架空管道
pipe suspension 管子吊挂
pipe-swabbing machine 管材清理机
pipe swedge 管材整形器
pipe swing 管子软连接;管子柔性连接
pipe switch 管道转辙器
pipe swivel 管子的旋转接头
pipe syphon 虹吸管
pipe system 管网;配水管网;排水管系(统)
pipe system layout 管系布置
pipe tap 锥管螺纹;管子螺丝锥;管用丝锥;管螺纹丝锥;管壁孔;内螺纹车丝锥;分接管
pipe tap drill 管用钻孔攻丝复合丝锥;管螺纹丝锥;内螺丝锥
pipe tee T形管节;三通;丁字管(节)
pipe temperature sensor 水管温度探测器
pipe tester 管道试验器
pipe testing 管材试验
pipe-testing machine 管材试验机;试管机
pipe thawing machine 管道融冰机
pipe thermocouple 管形热电偶
pipe thermometer 管路温度计
pipe thimble 管套
pipe thread 管子螺纹;管螺纹;管端螺纹
pipe thread cutter 管螺纹刀
pipe thread cutting machine 管螺纹切割机;管纹机;管材车丝机
pipe thread die 管螺纹板牙
pipe threading 管口螺丝;管子车丝;管口车丝;管端螺纹
pipe-threading lathe 管子螺纹车床

pipe-threading machine 管子切丝机;管子刻丝机;管子滚丝机;管(子)车丝机;管螺纹车床
pipe-threading tool 割管螺纹刀具
pipe thread plug ga(u)ge 管螺纹塞规
pipe thread protector 管子护丝;管螺纹护套
pipe thread ring 管螺纹环
pipe thread tap 管丝锥;管螺纹公锥
pipe thrusting method 顶管法
pipe tobacco 斗烟叶
pipe tongs 接管器;管子钳;管钳
pipe tongs wrench 管钳扳手
pipe tool 管加工工具;修管工具;装管工具
pipe top 管顶
pipe-to-soil potential 管对地位势;管道土壤电势
pipe train 管子生产线
pipe transport 管道运输
pipe transportation 管道运输
pipe transport of concrete 混凝土管道输送
pipe trench 管道地沟;管沟;管道沟
pipe trestle 管道支架
pipe trolley (从钻塔中拖出管子的)小车
pipe-troughing distribution system 管槽结合式配水系统
pipe truck 运输管道的卡车
pipe truss 钢管桁架
pipet(te) 吸移管;吸液管;吸量管;吸管;移液吸(移)管;球管
pipet(te) analysis 吸管分析法
pipet(te) blower 吸管洗涤吹干橡皮球
pipet(te) box 吸管盒
pipet(te) connection 吸管接头
pipet(te) joint 吸管接头
pipet(te) method 吸管法;移液管法
pipet(te) rack 吸液管架
pipet(te) stand 吸管架;吸液管架
pipet(te) sterilizer 吸管消毒器
pipet(te) support 吸利管架;吸管架
pipet(te) system 定体积比分量系统
pipettor 吸移管管理器
pipe tunnel 管式隧道;管子隧道;管弄;管沟;管道隧洞
pipe turnbuckle 导管反正扣
pipe twist 修管器;管钳
pipe-type aerator 管式曝气器
pipe-type cable 管状电缆
pipe-type character 管子类型标记
pipe-type classifier 管式分粒器;管式分级机
pipe-type cooler 管式冷却器
pipe-type culvert 管式涵洞
pipe-type drain(ing) 管式排水
pipe-type dryer 管式干燥器;管式干燥机
pipe-type electric(al) precipitator 管式电收尘器
pipe-type electrostatic dust-precipitator 管式静电收尘器
pipe-type electrostatic precipitator 管式电除尘器
pipe-type feeder 管式加料机
pipe-type filter 管式过滤器
pipe-type heater 管式炉
pipe-type holder 管式储气罐
pipe-type outlet 管式出水口
pipe-type precipitator 管式电除尘器;管式除尘器;管式沉降器;管式沉淀器
pipe-type press 管式压力机
pipe-type radiator 管式散热器
pipe-type rail 管式栏杆
pipe-type roller 管式滚筒
pipe-type sample 管式取样品;管式采样
pipe-type sampler 管式取样器
pipe-type scaffold(ing) 管式脚手架
pipe-type sleeve 管式套筒
pipe-type still 管式蒸馏器;管式炉
pipe-type stove 管式热风炉
pipe-type subway 管式地(下)铁(道)
pipe-type tunnel 管式隧道
pipe-type furnace 管式炉
pipe-type type precipitation 管式沉淀器
pipe-type underdrain 管式暗沟
pipe-type valve tray 管式浮阀塔盘
pipe-type vibrator 管式振动器
pipe underdrain 管式暗沟
pipe union 管子联管节;管子接头;管箍;联管节
pipe used for shells 用作壳体的管子
pipe valve 管阀
pipe-ventilated motor 管道通风电动机

pipe vibrator 管式振动器
pipe vise 管子台虎钳;管子虎钳;管钳
pipe vise chain 管钳链条
pipe wall 管壁
pipe-wall ga(u)ge 管壁厚度测量仪
pipe-wall rail 靠墙管子扶手
pipe warps 管子翘曲
pipe warping 管子绝缘
pipe way 油管栈桥;管状地下通道
pipe weep 管子渗水;排水管(混凝土砌体背后排水用)
pipe welder 钢管焊接机;焊管机
pipe welding 管道焊接
pipe-welding machine 钢管焊接机;焊管机
pipe wiper (清除管外泥浆的)橡皮刮板;擦管器
pipe wiper rubber 橡皮刮尼板(清除管内泥浆用)
pipe with access eye 有检查孔的管道
pipe with a string on it 报废管子
pipe with base 有底座的管道
pipe with cover-plate 加盖管
pipework 管件;管道系统;管道工程;输送管线
pipework basement 管道工程基础
pipework cellar 管道工程地窖
pipework condensation 管冷凝
pipeworking device 管子加工机具
pipework system 管道系统
pipe worms 管子螺纹
pipe wrapping 管子包扎
pipe wrapping cloth 管道包扎布
pipe wrapping mate 管道包扎毡
pipe wrench 管钳(子);管子扳手;管扳手;美式管钳;大管钳
pipe yard 管材堆(置)场
pipe-zone backfill 管道区回填料
piping 中心喉管;管涌(现象);管线;管系统;管道输水;管材;接管;铺管;配管
piping and fittings 管子和配件;管子及配件
piping and instrumentation diagram 配管自控流程图
piping and instrument diagram 带控制点工艺流程图
piping arrangement 管线布置;管系图;管系布置;管路布置;管道布置
piping by heave 管涌隆起;隆起管涌
piping by subsurface backward erosion 隆眼管涌
piping course 管道敷设层
piping design 管道设计;配管设计
piping diagram 管线图;管系图;管路图;管道系统图;管道布置图;配管图
piping distribution system 管式配水系统
piping drawing 布管图;配管图;管线图;管路图;管道图
piping drop 管道(向下)立管
piping effect 管涌
piping element framing 管件构架
piping erection 管道安装
piping erection drawing 管道安装图
piping erosion 管涌侵蚀;管道侵蚀
piping filter 管道过滤器
piping fracture 管道断裂
piping gallery 管线廊道
piping hanger 管道挂钩;吊管钩
piping housing 管壳
piping in concrete 混凝土管道
piping installation 管道施工;管道安装
piping joint 管道接头;管道接口
piping layout 管系布置;管路图;管路布置;管道布置;输送管路图
piping line 管系
piping lining 管道衬砌;管道衬里;管道衬层
piping lining machine 铺管机
piping load 管道荷载
piping loss (of heat) 管道热损失
piping machine 安管机
piping-main scheme 母管制
piping manifold 管路歧管
piping network 管路网;管道网
piping of concrete 管送混凝土;管道输送混凝土;混凝土管道
piping of river 河流引水管道
piping oil 管道输油
piping packing 接管填料
piping parts 管道部件
piping plan 管线图;管道详图;管道图;管道平面图

piping porosity 管状孔隙
piping ratio 管涌比
piping riser 管道竖管
piping schema 喷水管布置
piping shaft 管道井
piping standard connected to the vessel 与容器连接的配管标准
piping system 管线系统;管路系统;管路网;管道系统;配水管网
piping trench 管沟
piping water 泉水;管道水;管涌水
piping work 铺管工作;铺管工程
pip integrator 脉尖积分器;脉冲积分器
pipkin 小瓦锅;小金属锅
Pipli 比布利灰褐色硬木(印度产)
pip matching 脉冲刻度校准;标记调节
pip-matching circuit 标记匹配电路
pipper 标线盘小孔
pipper image 标线盘小孔像
pip voltage 夹断电压
pipy 管状的
pique 灯芯布;凹凸(纹)织物
pique crepe 凹凸绉
piquia fat 白胡桃油
piracetam 吡乙酰胺
piracy 截夺;海上劫掠;侵犯他人财产;夺流;非法翻印;版权侵害行为
piracy of groundwater 地下水袭夺
piragua 双桅平底船;独木舟
pirated stream 袭夺河;被夺河(流)
pirate river 截夺河
Pira test bench 庇拉试验台
pirating 窃水
pirazolac 吡拉唑酸
pirce elasticity of demand 需求价格弹性
piridoxilate 吡多酯
pirigarnite 石榴二辉麻粒岩
piriklazite 斜长二辉麻粒岩
pirlindole 吡吲哚
pirn 卷绕
pirn winding 绕组
pirogliride 吡咯格列
piroheptine 吡咯庚汀
pirolazamide 吡拉酰胺
piromerite 软钾镁矾
Piroplasmea 梨质纲
pirprofen 吡咯洛
Pirquet's reaction 划痕试验
pirquiasite 皮硫锡锌银矿
pirquinozol 吡喹诺唑
pirssonite 钙水碱
Pisa 比萨(意大利城市)
pisanite 铜绿矾
pisay 夯土建筑;土坯砖;草筋泥
piscary (在他人渔区内的)捕鱼权;捕鱼场
piscatology 捕鱼学
piscator 捕鱼人
piscicide 鱼类灭绝
piscicultural 养鱼的
pisciculture 养鱼
piscina 洗圣器池;洗礼盆;鱼塘;养鱼池塘
piscine (古罗马教堂的)浴池;冷水浴池
piscivorous 食鱼的;食浮游生物的
pise 砌墙泥;捣实土;捣实土(壤)
pise construction 夯土结构;夯土建筑
pise de terre 捣实土建筑;干打垒建筑
Pi slab 双T板
pisolite 火山豆;豆状岩;豆状石灰岩;豆石;豆粒
pisolith 豆石
pisolith-prlitic texture 豆状泥质结构
pisolitic 豆状的
pisolitic aluminous rock 豆粒铝质岩
pisolitic-bearing pelitomorphic aluminous rock 含豆粒泥状铝质岩
pisolitic limestone 豆状灰岩
pisolitic ore 豆状矿石
pisolitic pelitomorphic aluminous rock 豆粒泥状铝质岩
pisolitic phosphorite 豆状磷块岩
pisolitic structure 豆状构造
pisolitic texture 豆状结构
pisolitic tuff 豆状凝灰岩
pissasphalt 软(地)沥青;天然(软)沥青
pissasphaltum 天然软沥青

piss coat 涂刷不当的油漆涂层(俚语);浅涂油漆涂层
pissoir 街道厕所(欧洲)
pistachio 浅黄绿色
pistacite 绿帘石
Pista coagulation process 皮斯塔混凝过程
Pista desilting trap 皮斯塔除砂井
piste 简易道路;便道;小路
pistol 金属喷镀器;枪式焊接器;喷敷枪
pistol brick 手枪形砖
pistolgraph 快速摄影机
pistol grip 喷枪手柄;枪柄;手枪式把手
pistol lance 手枪式喷枪
pistol range 枪式焊接区域;手枪射击场
pistol-type branch 手轮式水枪
pistol-type stapler 手枪式钉器;射钉枪
piston 阳模;活塞;汽缸活塞
piston accelerator 活塞加速器
piston accumulator 活塞式蓄压器;活塞蓄能器
piston acoustic source 活塞式声源
piston action of train 列车活塞作用
piston-actuated 活塞驱动的
piston-actuated valve 活塞驱动阀
piston air 活塞风
piston air drill 活塞式风动凿岩机
piston air valve 活塞式气阀;活塞式空气阀
piston and diaphragm pump 活塞与隔膜泵
piston attenuator 活塞式衰减器
piston barrel 活塞筒
piston blower 活塞压气机;活塞式鼓风机
piston blowing engine 活塞式鼓风机
piston body 活塞体
piston bolt 活塞螺栓
piston boss 活塞销座子;活塞销座;活塞销壳
piston bottom 活塞底
piston buffer change valve 活塞式缓变阀
piston bushing 活塞衬套
piston cap 活塞帽
piston chamfer 活塞倒棱
piston clearance 活塞余隙;活塞间隙
piston clearance ga(u)ge 活塞余隙规
piston coil spring 活塞式盘簧
piston compression ring 活塞压缩环;活塞环
piston compressor 活塞式压气机;活塞式压缩机
piston concrete pump 活塞式混凝土泵
piston connecting rod 活塞杆
piston cooling 活塞冷却
piston cooling jet 活塞冷却喷管
piston corer 活塞取样管;活塞取(岩)芯器
piston core sampler 活塞式取样器;活塞式取土(样)器
piston coring 活塞式岩芯取样
piston cross-head 活塞十字头
piston cross-head joint pin 活塞十字头连接销
piston cross-head pin 活塞十字头销
piston cross-head slide bar 活塞十字头摇杆
piston crown 活塞头;活塞帽;活塞顶
piston crown contours 活塞顶轮廓
piston cup 活塞皮碗
piston cup expander 活塞皮碗扩张器
piston curl 活塞涨圈;活塞圈;活塞环
piston-cylinder apparatus 活塞一汽缸装置
piston-cylinder balance 活塞缸式秤
piston-cylinder type 活塞缸式
piston-cylinder ultra high pressure and high temperature device 活塞缸式超高压高温装置
piston-cylinder ultra high pressure device 活塞缸式超高压装置
piston damper 活塞式减振器
piston deep-well pump 活塞式深井泵
piston deluge valve 活塞型雨淋阀
piston diameter at ring lands 活塞环仑的活塞直径
piston displacement 活塞位移;活塞排量;活塞冲程
piston displace ment volume 活塞位移容积
piston drill 活塞式凿岩机;活塞式冲击机;活塞穿孔机
piston drive 活塞传动
piston drive sampler 活塞取样器;活塞式取土(样)器;活塞驱动采样器
piston effect 活塞作用
piston effort 活塞作用力
piston engine 活塞引擎;活塞式发动机;活塞发动机
piston-engined 活塞发动机的
piston expander 活塞底缘扩张器
piston expansion engine 活塞式膨胀机

piston extruder 活塞式挤泥机
piston extrusion machine 活塞式挤压机
piston face 活塞表面
piston feeler set 活塞测隙规
piston flow 柱塞流;活塞流
piston fluid meter 活塞式流量计
piston follower 随动活塞
piston force 活塞力
piston ga(u)ge 活塞压力计;活塞式压力计;活塞量规
piston gland 活塞压盖
piston governor valve 活塞调节阀
piston governor valve cage 活塞调节阀座
piston governor valve retaining spring 活塞调节阀扣紧簧
piston governor valve strainer 活塞调节阀滤器
piston groove 活塞槽
piston groove cleaner 活塞槽清洁器
piston gudgeon pin 活塞销
piston guide 活塞导承
piston head 活塞头;活塞顶
piston head end 活塞头端
piston head plate 活塞顶板
piston heat set screw 活塞头止动螺钉
piston hydromotor 活塞液压马达
piston hydrophone 活塞式水听器
piston hydroreducer 活塞式液力减压器
piston inserter 装活塞器
piston intensifier 活塞式增压器
piston jig 活塞跳汰机;活塞式跳汰机
piston junk ring 活塞压环
piston junk ring bolt 活塞压环螺钉
piston knock(ing) 活塞撞击;活塞爆震;活塞爆击;敲缸
piston land 活塞环岸
piston leather packing 活塞皮垫圈
pistonless steam pump 蒸汽抽水机
piston-like action (钻头包泥后的)似活塞作用
piston load capacity 活塞负荷能力
piston machine 活塞式机械
piston meter 活塞式流速计;活塞式计量仪表;活塞流量计
piston models 活塞模型
piston motor 活塞(式)液压马达
piston nut 活塞螺母
piston only 活塞本体
piston operated valve (气动式液压传动的)活塞操纵阀
piston packing 活塞密封;活塞圈;活塞密封(圈)
piston packing ring 活塞密封环
piston path 活塞行程;活塞冲程
pistonphone 活塞式发音器;活塞发声仪;活塞发生器
piston pin 活塞销
piston pin bearing 活塞销轴承
piston pin boss 活塞销座
piston pin bushing 活塞销衬套
piston pin circlip 活塞销簧环
piston pin clamping spring 活塞销卡簧
piston pin compressor 活塞圈压缩器
piston pin drift 活塞销冲头
piston pin end 活塞销止动螺钉
piston pin end cap 活塞销盖
piston pin hole 活塞销孔
piston pin hole reamer 活塞销孔铰刀
piston pin knock 活塞销爆击
piston pin land 活塞圈两槽间的隔板
piston pin locking screw 活塞销锁紧螺钉
piston pin lock ring 活塞销锁环
piston pin retainer 活塞销护圈
piston pin retaining ring 活塞销锁环
piston pin set screw 活塞销锁紧螺钉
piston pin snap ring 活塞销锁圈
piston-pipe sampler 活塞式管状取土器
piston plug 柱塞
piston plunger pump 柱塞泵
piston positioner 活塞式定位器
piston power 活塞动力
piston pressure 活塞压力
piston pressure ga(u)ge 活塞压力表
piston pressurization 活塞压送
piston puller 活塞式拉晶机
piston pump 活塞式水泵;活塞泵
piston push rod 活塞推杆

piston push rod check nut 活塞推杆防松螺母
piston reciprocating mud pump 活塞往复式泥浆泵
piston reciprocating rock drill 活塞往复式凿岩机
piston refrigerator 活塞式制冷机
piston removal 卸活塞器
piston ring 压缩环;涨圈;活塞环
piston ring back 活塞环内表面
piston ring back clearance 活塞环内表面间隙(指与活塞底之间)
piston ring breakage 活塞环断裂
piston ring carrier 活塞环垫块
piston ring chamfer grinding machine 活塞环倒角磨床
piston ring clamp 活塞环夹
piston ring compressed gap 活塞环压缩隙
piston ring end gap 活塞环端间隙
piston ring ends 活塞环端
piston ring expander 活塞环撑涨器
piston ring extractor 活塞环拆卸器
piston ring face (与汽缸接触的)活塞环外表面
piston ring file 活塞环锉
piston ring fitting 活塞环的装配
piston ring flutter 活塞环振动
piston ring free gap 活塞环自由开口间隙
piston ring gap 活塞环间隙;活塞环隙
piston ring grinder 活塞环磨床
piston ring grinding machine 活塞环磨床
piston ring groove 活塞环槽
piston ring groove cleaner 活塞槽清洗器
piston ring inserter 活塞环压入器(装配时用)
piston ring joint 活塞圈卡
piston ring land 活塞环棱
piston ring lock 活塞环端接合
piston ring machine 活塞环端面磨床
piston ring material 活塞环材料
piston ring packing 活塞环密封
piston ring pin 活塞环销
piston ring pliers 活塞环张口钳;活塞环安装钳
piston ring remover 活塞环拆卸器
piston ring scuffing 活塞环磨损
piston ring seal 活塞环式密封(件)
piston ring side 活塞环端
piston ring side clearance 活塞环端隙
piston ring slot 活塞环槽
piston ring spreader 活塞环拆卸器
piston ring tension 活塞环张力
piston ring tongs 活塞环张口钳;活塞环
piston ring top land 活塞环槽脊;活塞端环槽脊
piston ring travel 活塞环行程
piston ring wall 活塞环壁(活塞环径向厚度)
piston ring width 活塞环宽度
piston ring with taper face 锥面活塞环
piston rod 活塞杆
piston rod boss 活塞杆毂
piston rod bottom nut 活塞杆底螺母
piston rod cap 活塞杆盖
piston rod cotter 活塞杆制销;活塞杆栓
piston rod cross head 活塞杆十字头
piston rod for air pump 气泵活塞杆
piston rod gland 活塞杆压盖
piston rod guide 活塞杆导向器;活塞杆导承
piston rod guide cap 活塞杆导板盖
piston rod nut 活塞杆螺母
piston rod oiler 活塞杆加油器
piston rod packing 活塞杆密封填料;活塞杆密封
piston rod pin 活塞杆销
piston-rod scraper ring 活塞杆胀圈;活塞杆刮油环
piston rod split pin 活塞杆开尾销
piston rod stuffing box 活塞杆填密函
piston rod stuffing box gland 活塞杆填料函压盖
piston rod thrust 活塞杆推力
piston rubber 活塞用橡胶
piston sampler 活塞式取样器;活塞式取土(样)器;活塞取样器;活塞取土器
piston seal 活塞密封(圈)
piston skirt 活塞裙;活塞侧缘
piston skirt clearance 活塞侧缘隙(活塞侧缘和汽缸之间的余隙)
piston skirt expander 活塞外缘涨器;活塞裙扩大器
piston slap 活塞松动;敲缸
piston slide valve 活塞式滑阀;活塞滑阀
piston speed 活塞运动速度
piston spring 活塞弹簧
piston stamp 活塞式捣矿机;活塞式碎矿机

piston stem 活塞杆
piston stem nut 活塞杆螺母
piston stop 活塞限位器
piston stop screw 活塞头止动螺钉
piston stop washer 活塞抵冲垫圈
piston striker 活塞冲击器
piston stroke 活塞行程;活塞冲程
piston supercharger 活塞(式)增压器
piston surface 活塞表面
piston swept volume 活塞排量;汽缸换气量;汽缸工作容积
piston Tee ring 活塞丁字环
piston test 活塞试验
piston thrust 活塞推力
piston tohead clearance 活塞底和汽缸头间隙
piston tool 活塞打钉器
piston top 活塞头
piston travel 活塞行程;活塞冲程
piston travel limiter 活塞行程限制器
piston-tuned circuit 活塞调谐电路
piston type 活塞式
piston-type actuator 活塞式传动装置
piston-type air compressor 活塞式空压机
piston-type air starter 活塞式气动起动器
piston-type applicator 活塞式挤泥器
piston-type area meter 活塞式流速计
piston-type bottom hole pressure recorder 活塞式井下压力计
piston-type concrete pump 活塞式混凝土(输送)泵
piston-type (core) sampler 活塞(式)取样器;活塞取土(样)器
piston-type damper 活塞式减振器
piston-type domestic water pump 家用水泵;活塞式家用水泵
piston-type drill 活塞式凿岩机
piston-type flowmeter 活塞式流量计;活塞式流量表
piston-type fuel pump 活塞式燃料泵
piston-type gasholder 干式储气罐;活塞式储气罐
piston-type holder 活塞式储气罐
piston-type internal combustion engine 活塞式内燃机
piston-type knapsack sprayer 活塞(泵)式背负喷雾机
piston-type metering pump 活塞式计量泵
piston-type mortar pump 柱塞式灰浆泵
piston-type paint sprayer 活塞式喷漆器
piston-type poppet valve 活塞式提升阀
piston-type pressure controller 活塞式压力调节器
piston-type pressure ga(u)ge 活塞式压力表
piston-type pressure ga(u)ge for load 负荷活塞式压力表
piston-type pressure regulator 活塞式压力调节器
piston-type pressure sensor 活塞式压力传感器
piston-type pulsator 活塞式脉动器
piston-type pump 活塞式(抽水)泵
piston-type rctary pump 活塞式回转泵
piston-type reducing 活塞式减压阀
piston-type regulating valve 活塞式调节阀
piston-type sensor 活塞式传感器
piston-type shock absorber 活塞式减振器
piston-type sliding valve 活塞式滑阀
piston-type soil sampler 活塞式取土(样)器
piston-type spray pump 活塞式喷雾泵
piston-type tunnel ventilation 活塞式隧道通风
piston-type variable area flowmeter 活塞式可变截面流量计
piston-type washbox 活塞式跳汰机
piston-type wave generator 活塞式生波机
piston valve 活塞式阀门;活塞阀
piston valve body 活塞阀体
piston valve bush 活塞阀衬套
piston valve cover 活塞阀盖
piston valve cylinder 活塞阀筒
piston valve distribution 活塞阀分配器
piston valve gear 活塞阀机构
piston valve liner 活塞阀衬
piston valve pump 活塞阀泵
piston valve spider 活塞阀十字辐
piston valve spindle 活塞阀杆
piston valve stem 活塞阀杆
piston visco(si)meter 活塞黏度计
piston wall 活塞壁
piston water meter 活塞式水表
piston with cast-iron ring 带铸铁环的活塞

piston with slipper 带滑块的活塞
piston work relax 活塞松动
pit 现期交易部(美国粮食交易所);乐池;样洞;正厅后座;矿坑;交易场;麻坑;溶洞;取土坑;纹孔;蚀坑;砂孔;地坑;底坑;凹坑;凹点;凹斑(蚀像)
pit analysis 铸坑样品分析
Pitanco 皮炭可(一种绝缘材料)
pit and quarry industry 坑口采石工业;露天采石工业
pit annealing 槽口退火
pit annulus 纹孔环
pit aperture 纹孔口
pit asphalt 天然沥青;坑地沥青;天然石油沥青;天然地沥青
pit auger 储[贮]料坑螺旋
pit ballast 采石场碎石
pit bank 井口平台
pit-bank level 坑口构筑物水平面;井口构筑物水平面
pit barring 孔壁支木;井壁支木
pit bin 坑槽式煤仓
pit bing 矿山堆场
pit blade 挖坑铲
pit board 井壁支护板;坑板;基坑支托板
pit border 缘孔纹;纹孔缘
pit bottom 露天采场底面;井筒工作面(凿井时)
pit broker 现场经纪
pit burner 烤窑用灯头;点火燃烧器
pit canal 纹孔道
pit car 矿车
pit car loader belt 矿车皮带运输机
pit car oil 黑色润滑油
pit cast cast-iron flange pipe 坑铸铸铁法兰管;坑铸铸铁凸缘管
pit casting 地坑铸造;地坑铸件
pit casting process 坑式铸管法
pit-cast pipe 坑铸管;沟内浇铸管道;沟内灌注管道
pit cavity 纹孔腔
pitch 硬质沥青;硬焦油脂;纵摇(船的);纵倾角;管中心距;螺栓间距;沥青;节距;间距;码头上堆货处;倾入;倾伏角;坡度;投掷;树醪;钉距;齿距;侧伏角;俯仰;刨刀夹角;柏油
pitch acceleration 俯仰(角)加速度
pitch accumulative error 周节累积误差
pitch adder 俯仰角加法器
pitch adjustment (电动刨路机的)纵向倾角调整;纵向倾角调节
pit chamber 纹孔室
pitch and roll buoy 测波浮筒;测波浮标;自动测波浮标
pitch angle 纵摇角;螺旋升角;螺旋角;螺距角;节锥半角;节面角;俯视角
pitch angle detector 俯仰角测定器
pitch angle distribution 投射角分布
pitch angle of orebody 矿体侧伏角
pitch angle redistribution 螺距角再分布
pitch arc 弧线节距;弧线
pitch arm 挖土铲斗定位杆;调节臂;调节杆;(可伸缩的)活动杆
pitch asphalt 混合沥青
pitch attitude 俯仰姿态
pitch attitude control 俯仰姿态控制
pitch axe 鹤嘴斧
pitch axis 俯仰轴;倾斜轴
pitch binder 沥青黏合剂
pitch-bitumen 硬沥青地沥青混合料
pitch-bitumen binder 硬沥青地沥青结合料
pitch black 沥青黑色;深黑色;漆黑
pitch blende 沥青晶质铀矿;沥青铀矿;沥青油矿
pitch block 楼梯斜度样板;硬沥青;节圆柱;节圆距
pitch board 坡度板;支承楼梯踏步三角板;楼梯踏步三角板;楼梯踏步三角定线板
pitch boiler 硬沥青熬煮锅
pitch-bonded basic brick 沥青结合碱性耐火砖;焦油砖
pitch-bonded brick 沥青结合耐火砖
pitch bonding 沥青结合的
pitch brace 调节杆系杆;挖土铲斗定位杆;倾斜轴
pitch cake 沥青硬块
pitch chain 节链;短环链
pitch chord ratio 节弦比
pitch circle 节圆;接圆;齿节圆
pitch circle diameter 节圆直径
pitch coal 烟煤;沥青煤;沥青褐煤;焦油煤
pitch coil 节距线圈

pitch coke 沥青焦炭
pitch cone 节锥;分圆锥;分度锥;分度齿轮
pitch cone angle 节锥角
pitch cone distance 节锥母线长度
pitch cone radius 锥距;节锥半径
pitch control 节距控制;节距调节
pitch control additive 色调控制剂
pitch control link 螺距联杆
pitch-control nozzle 俯仰控制喷管
pitch control oil pump 调螺距油泵
pitch control rod 螺距调节杆
pitch cooler 沥青冷却器
pitch correction 螺距修正
pitch curve 齿轮节线;分度曲线
pitch cylinder 节圆柱
pitch damper 纵向填平板;俯仰阻尼器
pitch-damper servo 俯仰减摆器
pitch-dark 漆黑
pitch diameter 中径;中间径;螺纹中径;节圆直径;节径;电极圈直径
pitch diameter of thread 螺纹中径
pitch-diameter ratio 螺距直径比
pitch dimension 踏步斜长;倾斜量;斜跑尺寸;梯段标杆
pitch earth 沥青土
pitched dressing 琢边石工;琢方石
pitched face 砌石护坡
pitched felt 油毡;沥青毡;柏油毡
pitched felt coat 沥青毡层
pitched foundation 铺砌基础
pitched island 砌石护坡分水岛
pitched paper 沥青纸;油纸;柏油纸
pitched roof 斜屋面;斜屋顶;坡屋顶
pitched roof area 坡屋顶面积
pitched roof truss 坡顶屋架
pitched slope 铺砌斜坡;砌石护坡;砌石边坡
pitched stone 斜琢边石
pitched stone work 铺砌石块工作
pitched truss 倾斜桁架;坡顶屋架
pitched turbine type agitator 斜叶涡轮式搅拌器
pitched work 块石护坡工程;砌石护坡
pitch emulsion 沥青乳胶
pitch-epoxy glue 沥青环氧胶
pitcher 厚垛圬工凿;铺路(方)石;护坡石料;瓶状叶;陶瓷碎坯;水瓶;水罐;大水罐
pitcher barrel sampler 通式取样
pitcher house 酒窖
pitcher mo(u)ld 素坯粉制成的模具
pitcher pump 罐形吸入;罐形泵;罐形吸水泵;浅井手摇泵
pitch error 螺距误差;节距误差;航向角误差;齿距误差
pitchers 碎坯;瓷渣;废瓷渣
pitcher tee 罐形三通
pitch face 槌加工面;斜凿面;凿面
pitch-faced 凿面的(石料);凿毛的面
pitch-faced masonry 凿面石建筑
pitch-faced stone 粗凿石;凿面石料
pitch factor 节距因数
pitch fiber 沥青浸渍纤维材料
pitch fiber pipe 沥青纤维管;沥青浸渍纤维管;沥青石棉管
pitch filler 硬沥青填料;硬焦油脂填(缝)料;脂填缝料;沥青填缝料
pitch filling 煤沥青填料
pitch flock 木髓灰
pitch follow-up system 俯仰随动系统
pitch-fork 干草叉;草叉;长柄叉
pitch ga(u)ge 螺纹样板;螺距规
pitch gear 径节齿轮
pitch glance 沥青光泽
pitch grit 硬沥青块
pitch grounded surface 沥青碎石路面
pitch grouted macadam 硬焦油(脂)碎石路
pitch grouted surface 灌沥青碎石路面;灌沥青碎石路面
pitch grouted surfacing 灌沥青碎石路面层;灌沥青碎石路面
pitch grout(ing) 灌沥青;硬沥青灌浆;硬焦油沥青灌浆
pitch gyroscope 俯仰陀螺仪
pitch helix 节距螺旋线
pitchhole 雕琢镶嵌凹坑;石面凹坑
pitch-impregnated basic brick 浸渍沥青碱性耐火砖
pitch-impregnated brick 沥青浸渍耐火砖

pitch impregnation 浸渍沥青
pitch indicator 螺距指示器;俯仰指示器
pitching 插桩;砌石护坡;块石护坡;海漫;前后颠簸;砌筑护坡;铺砌;船舶纵摇
pitching above water 水上砌石护面
pitching angle 俯仰角;纵摇角
pitching anti-cline 伏背斜【地】
pitching axis 俯仰轴
pitching block 护坡石;铺路石块
pitching borer 开眼钎头;开孔钻头
pitching chisel 锤凿;宽边凿;斧凿
pitching coil chain 节距精确的环链条
pitching error 纵摇误差
pitching fold 倾覆褶皱;倾伏褶皱;伏褶曲
pitching machine 发球机
pitching method 铺砌法
pitching model 俯仰模型
pitching moment 俯仰力矩
pitching motion 纵摇;纵向运动
pitching of pile 斜桩斜度;插桩
pitching of slope 护坡;斜坡砌筑;斜坡铺面
pitching period 纵摇周期
pitching piece 出梁;楼梯平台支承小横梁;裙梁;梯口梁
pitching pile 插桩
pitching slab 护面板
pitching stone 手铺砌石(块);铺砌片石
pitching syncline 倾覆向斜;伏向斜
pitching the threads 用沥青封丝扣(为防止漏水)
pitching tool 斧凿
pitch instrument 俯仰仪
pitch interval 音程
pitch knot 树脂节;树脂心节
pitch ladle 沥青勺
pitch lake 沥青湖
pitch lap 沥青抛光盘
pitch line 铆钉距线;中心线;节线;齿条节线;齿距线
pitch line speed 节圆圆周速度
pitch line velocity 节线速度
pitch macadam 沥青碎石路;沥青碎石
pitch mastic 硬沥青砂胶;沥青脂砂胶;沥青玛琋脂
pitch mastic flooring 硬沥青玛琋脂地面;沥青玛琋脂地面;硬沥青玛琋脂面层;沥青砂胶地面
pitch mop 沥青刷子
pitch motion 俯仰操纵机构
pitch nozzle 俯仰喷嘴
pitch number 齿距数
pitch of arch 拱矢高;拱矢高;拱高(度);拱的高跨比
pitch of a roof 屋顶高跨比;屋顶坡度
pitch of baffle 折流板间距
pitch of beat 拍频音调
pitch of bolts 螺栓间距
pitch of boom 起重机臂倾斜度;吊杆倾斜度;臂倾斜度;(起重机、挖掘机)臂的倾斜角
pitch of boreholes 钻孔间距
pitch of buckets 斗距
pitch of centers 顶尖高度
pitch of chain 链环距
pitch of corrugations 波纹高跨(比);波状高跨比
pitch of cutter teeth 铣刀齿节
pitch of drills 钻头轴距
pitch of fins 片距
pitch of hinge 枢纽侧伏角
pitch of holes 钻井斜度;钻孔间距;钻井倾角;孔间距
pitch of holes in pattern 排孔节距
pitch of nails 钉距
pitch of orebody 矿体侧伏方向
pitch of piles 桩的间距
pitch of pipe 管子斜度
pitch of propeller 螺旋桨螺距;推进器螺距
pitch of rise and fall 升降距
pitch of rivets 铆钉距;铆钉间距
pitch of roof 房顶高跨比;屋面高跨比;屋顶斜度;屋顶坡度;层面坡度
pitch of scan(ning) 扫描行宽
pitch of screws 螺距;螺纹距
pitch of sleepers 轨枕距
pitch of stair(case) 楼梯斜度;楼梯坡度
pitch of teeth 轮齿距;齿距
pitch of the aircraft 飞机俯仰
pitch of the corrugation 波纹节距
pitch of the rivets 铆钉距

pitch of threads 螺距；螺纹距
pitch of transposition 换位节距
pitch of tunnel support 隧道支护间距
pitch of turn 匝距
pitch of vibration 振动周期
pitch of waves 波长；波距
pitch of weld 焊缝距；焊缝中心距；焊缝接距
pitch of well 焊缝中距
pitch of worm 蜗杆节距
pitch(o)meter 螺距规；节距规
pitch-on metal 浸涂沥青的软钢板；涂沥青薄钢板
pitch ore 沥青铀矿；沥青铜矿
pitchover 程序转弯
pitch overboard 抛出船外
pitch paver 沥青摊铺机
pitch payer 沥青勺
pitch paying 浇注沥青
pitch peat 硬沥青泥炭；沥青状泥炭；沥青质泥炭；沥青泥炭
pitch percent 节距系数
pitch period 齿节周期
pitch phase detector 俯仰状态探测器
pitch pine 刚松；北美油松；含脂松木；多脂松；北美油松木
pitch pipe 斜管；调音管
pitch piping 斜管
pitch plane 节平面；节面
pitch plate 斜板
pitch play 齿隙
pitch pocket 沥青槽；油眼(木材缺陷)；松脂眼；松脂孔；树脂囊；树脂孔
pitch point 节点；齿轮(啮合)节点
pitch poisoning 沥青中毒
pitchpole 大颠簸
pitch polishing 沥青抛光模具
pitch-polymer damp course 树脂聚合物防潮层
pitch pot 沥青锅
pitch program(me) 俯仰角变化程序
pitch radius 节圆半径
pitch rate gyroscope 俯仰角速度陀螺
pitch ratio 螺距比；间距比(率)
pitch ray 木射线；树脂髓线
pitch resin 沥青树脂
pitch rod 调节杆
pitch-row 孔距
pitch scale 俯仰角度标
pitch seam 树脂缝；松脂裂纹；树脂痕
pitch sensing accelerograph 俯仰运动荷自记器
pitch sensing accelerometer 俯仰运动加速表
pitch spacing of (bore) holes 钻孔间距
pitch speed 节距速度
pitch spread 节距张度
pitch stabilizer 减纵摇装置
pitch stick 螺距调整杆
pitch stiffness 纵向安定性；纵稳定性过大
pitchstone 珠石；松脂岩(石)；松脂石
pitchstone porphyry 松脂斑岩
pitch streak 树脂斑；松脂条纹；松脂裂纹；树脂条纹；树脂痕
pitch surface 节面
pitch system 径节制
pitch tester 节距检验器
pitch thread 螺距；径节螺纹；齿节螺纹
pitch time 节拍时间
pitch trim compensator 音调调整补偿器
pitch tube 斜管；树脂管
pitch tubing 斜管
pitch valve 斜阀
pitch varnish 沥青凡立水；沥青清漆
pitch wheel 相互啮合的齿轮
pitchy 多树脂的；漆黑的；黏性的；用沥青覆盖的；涂有沥青的；沥青的
pitchy lumber 多树脂木材
pitchy road 沥青路；柏油路
pitchy timber 多脂树木材
pitchy wood 多树脂木材；多脂树木材
pit coal 烟煤；矿产煤；坑煤；煤炭
pit corrosion 点蚀；穴蚀；点状腐蚀；麻点腐蚀；点腐蚀
pit cover 坑盖；井盖
pit covering 井罩；坑罩
pit crane 翻坯吊车；翻板坯吊车
pit crater 锅状火山口
pit-culture 坑种

pit deepening 采场降深
pit-depth measuring tape 测井卷尺
pit digging 挖坑
pit door 拉引室门
pit drainage method 矿井排水方法
pit drainage system 矿井排水系统
pit engineering No. 坑探工程编号
pit equipment 矿山设备
pit excavation 基坑挖掘；基坑开挖
pitfall 陷阱；陷坑
pit fire 矿中火灾
pit flock 木髓斑
pit for escalator 自动扶梯坑
pit foundation 坑基；井式基础
pit foyer 正厅后排休息室
pit frame fire 井架火灾
pit furnace 坑式炉；坑炉；井式加热炉；地炉
pit gear 矿井设备；井筒设备
pit gravel 未过筛砾石；未筛选的砾石；未筛分砾石；山砾石
pith 木髓物；木髓；树木的髓心
pit hatch 矿井升降口
pith-ball electroscope 木髓球验电器
pit-head 矿井口；矿坑口
pit-head bath 坑口浴池；坑口浴场
pit-head committee 煤矿委员会(由煤矿主和矿工共同推派代表组成)
pit-head power plant 坑口电站
pit-head price 煤的矿山价格
pit-head price of coal 煤的框架口交货价格
pi-theorem π 定理
pith fleck 木髓条纹；髓状斑纹
pit-hill 井口地面；矿井口建筑
pith knot 空洞节；髓节
pit hole 坑孔；坑洞；石洞(储存气体用)；小凹坑
pith ray 髓线；木材线
pith shake 髓裂
pit immersion test 试坑浸水试验
pit incineration 坑式焚化
pit kiln 炼焦炉；炉坑
pit lathe 落地车床；地坑车床
pit lime 石灰坑
pit limit 露天开采境界
pit liner 机坑衬里；槽衬垫
pit loading test 试坑荷载试验
pit log 流速计程仪
pitman 矿工；锯木工；转向器臂；连接杆；机工；钳工；采石工(人)
pitman arm 转向壁；转向垂臂；连接杆臂；连杆臂
pitman bearing (宾夕法尼亚钢丝绳钻进的)连杆轴承
pitman box 连杆轴承
pitman coupling 连杆头；拉杆连接头
pitman-cut-ter bar alignment 连杆切割器对位
pitman eccentric bearing (颚式破碎机的)连杆偏心轴承
pitman head 摇杆头；连杆头
pitman-knife alignment 连杆切割器对位
pitman lead 连杆偏前量
pitman pin 连杆销轴
pitman shaft (轧碎机的)连杆轴；连接轴；连杆轴；曲柄轴
pitman shaft bushing 连杆衬套
pitman socket 连杆窝形夹板
pitman strap 连杆头托座；连杆颊板
pitman stub 连杆头
pitman-type cutter 连杆式切割器
pitman wheel 连杆偏心轮
pit membrane 纹孔膜
pit method 坑井(渗水)法
pit moisture 新出矿湿度
pit mo(u)lding 地上造型；地坑造型
pit of wood cell 木材细胞纹孔
Pitometer 皮托(流速)计；(测量流速的)皮托管压差计；皮托管测压器
Pitometer log 皮托管式计程仪；皮托式计程仪
Pitometer measurement 皮托管量测
pit orifice 纹孔口
pito-static tube 风速管
Pitot 空速管；皮托管
Pitot curve 总压曲线；全压曲线
Pitot ga(u)ge 皮托压差计；皮托流速测定管；皮托计
Pitot hole 总压孔
Pitot investigation 速度场测定

Pitot log 皮托管式计程仪；水压计程仪；液压计程仪
Pitot loss 皮托损失；总压损失；全压损失
Pitot mast 空速管柱
Pitotmeter 皮托压差计；皮托流速计；皮托管
Pitotmeter survey 皮托计测量
Pitot pressure 皮托管压力；水压计程仪的压力
Pitot pressure ga(u)ge 皮托压力计
Pitot sphere 皮托球体
Pitot-static difference 总压和静压差；全压静压差
Pitot-static rake 梳状动静压管
Pitot-static system 全压静压系统
Pitot static traverse 皮托管静压测定
Pitot static tube 皮托静压管；动静压同轴皮托管；皮托管静力管
Pitot-static tubing 皮托静压管沟
Pitot survey 皮托计调查
Pitot traverse 皮托管捡；梳状管
Pitot tube 皮托流速计；皮托流速测定管；皮托管
Pitot tube anemometer 皮托管式风速计
Pitot tube cover 皮托管壳
Pitot tube flowmeter 皮托管式流量计
Pitot tube measurement 皮托管测量
Pitot tube method of measuring flow 皮托管测流法
Pitot tube mouth 全静压管入口；皮托管(进水)口
Pitot tube traverse 皮托管行程
Pitot tube velocity survey 皮托管风速测量计
Pitot-Venturi folw element 皮托—文丘里管测流元件
Pitot-Venturi tube 皮托—文丘里管
pit pair 纹孔对
pit pattern 麻点图像
pit permeability method 试坑渗水法
pit planer 落地刨床；地坑刨床
pit planting 穴状栽植；穴植
pit practice 铸坑操作
pit privy 坑厕
pit prop 临时木支柱；坑道支柱；临时支柱(矿井)；坑木
pit pump 矿井泵
pit pump equipment 矿井排水设备
pit quarry 露天采矿坑；露天采石场
pit rail 矿用钢轨
pit recharge 坑槽补给
pit room 回采工作面数
pit rope 矿用钢丝绳
pit run 未筛分的土石料；毛料；未筛的；石场未分选料；原采出的
pit-run aggregate 未筛(分)骨料；未筛集料；统货集料
pit-run clay gravel 天然坑泥木
pit-run earth 坑取土料；未取选料
pit-run fines 坑石屑；未筛分石屑；采石场石屑
pit-run gravel 未过筛砾(石)；未筛选的砾石；未筛(分)砾石；坑取砾石；天然级配砾石；滩砾；采自料坑的砾石
pit-run material 未筛料；未筛选涂料；未筛选料；未筛选(的)材料；采自料坑的材料；坑取(材)料
pit-run ore 原矿石；未筛分矿石
pit-run sand 坑砂；坑砾石；未过筛砂；未筛(分)砂
pit sampler 坑探取样器；探坑取样器；槽探取样器；槽式底砂采样器
pit sampling 浅井采样；探坑取样
pit sand 坑砂
pit saw 直锯；龙锯；坑锯；双人用竖拉大锯；大锯
pitsaw file 半圆锉
pit sawing 双人锯(木)
pit scale 落地磅秤；矿山秤；地磅秤；地磅
pitshaft 井筒
pit shooting 坑中爆炸
pit silo 农业储[贮]仓；农业矿渣
pit slag 坑口矿渣
pit slope 露天采场边坡
pit source 料坑；(骨料的)坑矿料源
pit storage 窖藏
pit sweeper 井清理(输送)机
pittcoat 棬裙
pitted 带有纹孔的；有坑洼的；麻斑的
pitted corrosion 孔蚀；点蚀
pitted pebble 麻面卵石
pitted secondary wall 纹孔次生壁
pitted skin 麻面(混凝土)
pitted surface 有坑槽的路面；麻面(混凝土)
pitted vessel 纹孔导管
pitted wear 麻面磨损

pit teeming 坑铸法
pit test 铸坑样品试验;坑探
pit-through 蚀穿(气体)
pitticite 土砷铁矾
pittied split 用油灰填塞裂隙
pitting 锈斑;钻孔侵蚀(池窑耐火材料);孔蚀;麻点,纹孔式;点状腐蚀;凹陷;凹蚀;凹痕
pitting attack 麻点腐蚀;点蚀
pitting coefficient 汽蚀系数
pitting corrosion 麻点腐蚀;点蚀;点腐蚀;孔蚀;孔腐蚀;坑蚀;膜孔型腐蚀;麻点状腐蚀
pitting corrosion test 点腐蚀试验
pitting due to cavitation 点蚀;锈斑;麻洞;汽蚀坑;汽蚀剥蚀
pitting factor 孔蚀系数;孔蚀率;点蚀系数
pitting guarantee 汽蚀保证
pitting initial geologic(al) logging 坑探工程原始地质编录
pitting potential 局部腐蚀势;麻蚀电位
pitting resistance 麻点阻力;凹痕阻力;抗气蚀性
pitting sampling 坑探采样
pitting surface 麻面(混凝土)
pitting test 点蚀试验
pitting type 凹陷型
pitting volume 汽蚀失重;汽蚀坑体积
pittmentized paint 无颜料白漆
pit torus 纹孔塞
pi-T transformation 星形三角变换
pit transplanting 穴移植
Pittsburgh Plate Glass Company 匹兹堡平板玻璃公司(美国)
Pittsburgh plus pricing 匹兹堡加价法
Pittsburgh process 无槽垂直引上法
Pitt's patent 皮特专利(指门拉手与心轴的连接)
pit-type 立柱可移式龙门型
pit-type carburizing furnace 坑式渗碳炉
pit-type crucible furnace 坑式坩埚炉;坑式坩埚炉
pit-type electric(al) furnace 坑式电炉
pit-type gas carburizing electric(al) furnace 井式气体渗碳电炉
pit-type weighbridge 地坑式地磅
pituita 稠黏液
pituitary 稠黏液的
pituitous 稠黏液的
pit underpinning 矿井支柱;坑式托换
pit ventilator 矿井通风机
pit water 矿坑水;矿井水
pitwood 坑木;矿用木材
pitwork 井筒水泵设备
pit yard 矿场
pityrol 糠馏油
Pitzer equation 皮查方程式
Piura basin 皮乌拉盆地
piuri 印度黄
pivalate 特戊酸盐
pivot 转轴;轴销;轴榫;轴尖;中支枢;中枢;支枢;合叶心;枢纽;枢
pivotability 枢转性
pivot action of the spuds 定位柱的定心作用;定位桩的定心作用
pivotal 中枢的
pivotal axis 枢转轴
pivotal bearing 枢轴支座;铰接支座
pivotal cooler 可旋转冷却器
pivotal coordinate analysis 主坐标分析
pivotal element 枢轴元素
pivotal element equation 枢轴元素方程
pivotal fault 转动断层;旋转断层;枢轴断层;枢纽断层
pivotal harbo(u)r 枢纽港
pivotal interval 枢轴间隔
pivotal line 枢轴线
pivotally attached 铰接的
pivotal point 枢折点;交会点;枢轴点
pivotal port 枢纽港
pivotal rate 中心汇率
pivotal substation 枢纽变电所
pivotal value 枢轴值
pivot anchor bolt 支枢锚碇螺栓
pivot angle 摆动角
pivot arm 旋转臂
pivot axis 枢轴线;摆动轴
pivot beam lock 枢轴式犁辕锁定器
pivot-beam ripper 转梁铰接式松土机
pivot bearing 枢轴支承;轴尖支承;中心轴承;立式止推轴承;枢轴承;摆动支座
pivot bearing lubricant 钻石轴承润滑油
pivot bearing of bridge 桥梁枢轴支承
pivot belt 转向皮带
pivot bolster 中枢摇枕
pivot bolt 主销;轴枢螺钉;枢轴螺栓
pivot bracket 万向吊环;横销;轴头;转环
pivot bridge 平转桥;旋转桥;旋开桥;中间支点的开合桥;平旋桥;平开桥
pivot-bucket conveyor-elevator 枢轴吊斗提升输送机
pivot burnisher 枢轴研磨机
pivot center 旋转中心;转动件旋转轴;摆心
pivot clearance 轴肩间隙
pivot column 基准列
pivot cup 枢轴环
pivoted 旋转的;枢轴的
pivoted arm 转臂;枢杆
pivoted armature 枢轴衔铁
pivoted bearing 旋转轴承;自位轴承;自动调整轴承;立式推力轴承
pivoted bogie 转向架;转动车架
pivoted bolt 活节螺栓;转柄螺栓
pivoted brace 柱形支撑外钩
pivoted bucket carrier 翻斗式运料机;转斗式运输机
pivoted bucket conveyer 枢轴斗式输送器
pivoted carriage 回转刀架
pivoted casement 旋转窗;摇窗;悠哒窗(弧形转动窗)
pivoted conveyer 枢轴式皮带运输机;旋转式输送机
pivoted detent 装有枢轴的棘爪;枢轴式擒纵机构;枢轴掣子;枢制动器
pivoted detent escapement 枢轴式天文钟擒纵机构
pivoted dog 支杠锁簧
pivoted door 旋转门;转轴式门
pivoted end column 铰支柱;铰端(支)柱
pivoted engine 枢轴支承发动机
pivoted equalizer bar(履带底盘半刚性悬架的)平衡横梁
pivoted factor 枢轴因素
pivoted float 活动游标;活动浮子;球形浮标
pivoted gate 支枢式闸门;转动式闸门
pivot edge 支承刃
pivoted jib crane 旋转式悬臂吊车
pivoted lever 回转杆
pivoted loop 回转环
pivoted motor 悬吊翻转电动机
pivoted-pad guide bearing 支柱式导轴承;刚性支柱式导轴承;螺栓顶瓦导轴承
pivoted-pad thrust bearing 支柱式推力轴承
pivoted relay 支点继电器;枢轴继电器
pivoted sash 旋转窗扇;旋窗窗扇
pivoted scuttle 枢轴旋动舷窗
pivoted shoe 轴枢瓦
pivoted shoe thrust bearing 自动调整式推力轴承;支柱式推力轴承
pivoted shutter 转动式遮光窗
pivoted ventilation window 枢转式通风口
pivoted weir 活动堰;翻板堰
pivoted window 摇窗;转轴式窗;旋窗;旋转窗
pivot element 枢轴元素;主元
pivot file 加工轴颈用锉
pivot flank 旋回固定侧
pivot gate 旋转门;旋转式闸门
pivot hinge 轴头式铰链;轴铰链;尖轴铰链;门头合页;铰铰;吊轴合页
pivot holder 枢形夹座
pivot-hung window 中悬窗;翻窗;中旋窗
pivoting 绕枢旋转;尼浮【船】;船首突然右倾现象
pivoting beam 枢轴梁
pivoting bearing 心轴支承;枢轴承;立式推力轴承;中心支承
pivoting belt 回转胶带
pivoting colter 自动定位圆犁刀
pivoting cradle 转托支架
pivoting eccentric 偏心轮
pivoting fan 摆动吹风机
pivoting high 枢轴高压
pivoting mirror 旋转镜
pivoting moment 转动力矩
pivoting motion 旋转运动
pivoting motor 铰链吊挂式发动机
pivoting point 旋转点;支承点;铰链点;枢心;中心点;支点;旋转调头支点
pivoting side light 枢轴舷窗
pivoting slide rest 回转刀架
pivoting transom(e) 转动的楣窗
pivot joint 球形枢颈;车轴关节
pivot journal 旋转轴颈;轴颈;枢轴颈
pivot knuckle joint 旋转轴颈接合;旋转轴接合;枢轴铰链接合
pivot-leaf gate 铰转门;转动式闸门;活瓣式闸门
pivotless breaker 无触点臂的断电器
pivot light 摇窗
pivot location 枢轴位置
pivot loupe 尖头放大镜
pivot mechanism 转轴装置
pivot number 主数
pivot of tongue 尖轨枢轴【铁】
pivot on 支承在枢轴上
pivot operation 主元运算
pivot pad 枢轴垫
pivot pier 平旋桥中心支墩;旋转桥台;转轴支墩;开合桥跨墩
pivot pier protection 平旋桥中墩护桩;平旋桥中墩护石
pivot pin 转向主销;旋转轴;主销;中心承枢;枢销;枢轴
pivot pin cap 枢销盖
pivot pin lock pin 枢销锁销
pivot pitch 中心点间距
pivot plow 翻转犁
pivot point 旋转轴支点;支点;照准点
pivot point layout 交会测点;垂直定位法;测深垂线定位
pivot-point line 支线点;(测流断面的)固定起点线
pivot-point screw 枢轴螺旋
pivot polisher 枢轴抛光机
pivot port 中枢运港
pivot post 定位桩;定位柱
pivot post bearing 定位柱支架;定位柱支架
pivot post group 定位桩群;定位柱群
pivot reamer 浮动铰刀
pivot ring(底环下面的)支持环
pivot row 基准行
pivot screw 枢轴螺钉
pivot selection 主元选择
pivot shaft 铰销;枢轴
pivot slewing crane 枢轴起重机
pivot socket 枢承
pivots polishing mcahine 轴颈抛光机
pivot station 照准点
pivot-steer 支枢转向的
pivot steering 支枢转向;支枢转动
pivot stud 转向主销
pivot stud angle 转向主销倾角
pivot suspension 枢轴支承
pivot tank 中央储[贮]槽;中央储槽
pivot transformation 取主枢移
pivot transmitter 枢轴式发射机
pivot tripod 枢轴三脚架
pivot tube 空心铰销;空心合页心
pivot turn 原地转弯
pivot type ball bearing 冲压碗形向心推力球
pivot type davit 俯仰式吊艇柱
pivot window 旋转窗
pix 圣体容器;(奉献给罗马教堂主持人的)装饰匣
pixel 像元;像素;图元
pixel area 像素面积
pixel array 像素阵列;图元阵列
pixel array cache 像素缓冲区
pixel array dimensions 像素数组大小
pixel-by-pixel basis 以像素为基础
pixel loading 像素输入
pixel scale 像素比例尺
pixel scan 像素扫描
pixel size 像元大小
pixis 圣体容器;放圣餐的器皿
pixlock 图像锁定
pixlock playback mode 图像锁定重放方式
piypite 钾铜矾
pizotifen 苯噻啶
pkwy[parkway] 林阴大道;园林大道
placage 光墙的建筑处理;墙面装饰
placanticline 长垣
placard 铭牌;(有凸起图案的)装饰抹灰;广告牌;

招贴;名牌;布告;标语牌
place 座位;地点;场所;放置;部位;安置;安放
placeability (混凝土的)和易性;可灌注性;混凝土的可灌注性;浇筑性
placeability of concrete 可浇筑性;混凝土可灌(注)性;混凝土浇筑性;混凝土和易性;混凝土和易度
placeability of mix 拌和物浇筑性;混凝土拌和物的和易性;拌和物可灌性
placeable 可灌注的;可浇筑的
place advance 工作面掘进;工作面超前
place an order for goods 订货
place brick 未烧透的砖;欠火砖;生烧建筑砖;低质建筑砖;(未烧透的)等外砖;半烧砖
place clearing 清理场地
placed concrete 浇筑混凝土;浇灌混凝土
placed material 填料
placed riprap 抛石
placed rockfill 干砌石块;干砌块石;砌石;安放块石;填石
placed stone facing 干砌石块护面;砌石护面
place environment(al) monitoring sensor 区域性环境监测传感器
place for sightseeing 游览胜地
place historic(al) interest 古迹
place hour angle 地方时角
place identification sign 地名牌
place in layers 分层铺砌
place in location near assembly 放置在便于装配的地方
place in operation 开工;交付使用;投入运行;投入生产;投产
place in position at use point 放置在使用地点
place in service 投入运行;投入使用
place into the soil 埋入土壤
place mark 碎部点
placement 键接(合);方位;安置
placement algorithm 布置算法
placement cost (建筑物内部的)布置费
placement density 铺筑时的密实度;铺路时的密实度;填方密实度
placement drawing (建筑物内部的)布置图
placement in layers 分层填筑
placement lift 浇筑层高度;混凝土浇筑高度
placement memorandum 信息备忘录
placement moisture 回填含水率;填筑湿度;填筑含水率;铺筑含水率
placement moisture content 填料含水率;铺筑时含水率
placement number 位置号数
placement of aids-to-navigation 航标配布
placement of caisson 沉箱安放
placement of concrete 混凝土浇筑;浇筑混凝土
placement of lettering 注记字列
placement of marks 标志配布
placement of shotcrete 施喷(混凝土)
placement point 布置场所;装设点;浇筑点
placement policy 布局规则
placement program(me) 布局程序
placement property 浇灌性能
placement rule 布局规则
placement site 浇筑现场
placement treatment 定点处理
placement water content 施工时含水率
place mix 通路组合
place name 地名
place name bank 地名数据库
place name data base 地名数据库
place name gazetteer 地名录
place name index 地名索引
place name sign 地名牌
place name spelling 地名译音
place name standardization 地名标准化
place of abode 居住区
place of acceptance or receipt 接受或收货地点
place of amusement 娱乐场所
place of application of force 力的作用位置
place of arbitration 仲裁地;仲裁的地点
place of assembly 集会的场所;装配场;人员聚集场所
place of awaiting transportation 旅客等候处
place of business 商业地点;营业地点;营业所
place of delivery 交货地;交付地
place of despatching 发货地(点)

place of destination 到达处;目的地
place of detention 隔离所
place of dispatching 发货地(点)
place of effective management 实际管理机构所在地
place of embarkation 登机场所;乘船码头;装船码头;装货场;登船栈桥
place of entertainment 娱乐场所
place of examination 探伤地点
place of export 出口处
place of historic(al) interest 古迹
place of interest 名胜
place of jurisdiction 裁判管辖地
place of loading 装货地点;装货处
place of origin 产地;发源地
place of outdoor assembly 室外集会场所
place of outstanding architectural merit 建筑价值突出的广场;建筑优点突出的广场
place of payment 支付地点;付款地点
place of production 产地
place of public accommodation (供游乐、饮食等的)公共场所
place of public amusement 公共娱乐场所
place of public entertainment 公共娱乐场所
place of public resort 公共休养(场)所
place of receipt 接货地点
place of refuge 避难场所
place of residence 居住地点;居住场所
place of seal 盖章处;盖印处;封印处
place of seepage and leakage 渗漏水处
place of shipping 起运地点;启运地点
place of survey 检验地点
place of taking in charge 接受货物负责处理的地点
place of tax payment 纳税地点
place of worship 礼拜堂;教堂
place on 插入;套上;放上
place open to the public 开放场所
place or court 建筑群中间的空地
place order 订货
place out of service 停止工作
placer 浇注机;浇筑机;浇筑工(人);投资者;砂矿;放置器
placer accumulation 冲积矿床
placer concentration 砂矿床富集作用
placer deposit 砂矿床;砂矿
placer dredge(r) 采砂矿船;采矿矿船
place refuge 避难所
placer formation 砂矿床
placer gold 砂金
place risks 空间的风险
placer mine 砂金矿
placer mineral 砂积矿石;重砂矿物
placer-mining 淘金;砂矿淘采;砂(积)矿开采
placer prospecting drill 砂矿勘探钻机
placer sampling 砂矿采样
placer sand 漂砂;砂(积)矿砂
places 巡逻车
places of historic interest and scenic beauty 名胜古迹
place system 广场系统
place theory 部位学说
place to bury rock lump 溶蚀岩片埋放位置
place to place comparision 不同地区对比
place under repair 安排修理
place under separate control 分口管理
place value 分配值
placid out-flow gas 瓦斯平稳泄出
placid out-flow methane 瓦斯平稳泄出
placing 放置;安装;铺设;浇筑;出盘
placing and trowel(l)ing 随打随抹
placing an order 发出订单
placing broker 分保经纪人
placing bucket 浇灌料斗;混凝土浇灌斗;浇筑斗(混凝土)
placing buoys 设置浮筒
placing concrete 浇筑混凝土;浇灌混凝土
placing concrete against natural ground 地模混凝土浇筑
placing concrete lining 混凝土支护安装
placing condition 浇筑条件
placing crew 灌注队;混凝土灌注队;浇筑工组
placing density 装窑密度
placing depth 埋深
placing drawing 布筋图

placing filling 抛堆土石料
placing hard-surface 铺设硬质路面
placing head to head 头对头安放
placing in service 列入运用
placing into reserves 列入备用
placing method 浇筑方法
placing of cables 钢缆敷设
placing of concrete 浇筑混凝土;浇灌混凝土;混凝土灌注
placing of dry-mix shotcrete 干拌喷射混凝土的灌注
placing of material 材料堆放
placing of reinforcement 架设钢筋
placing of shotcrete 喷射混凝土灌注
placing of steel bars 配筋
placing of the facing 护面铺装
placing of tube 沉管就位
placing on production 投入生产;投入开采
placing operation (建筑物内部的)安装工作
placing period 装配周期;安装周期
placing plant 混凝土浇灌设备;混凝土灌注设备
placing point 摊铺现场(筑路工程);布置场所;安装现场
placing procedure 放置步骤
placing rate 铺管速度
placing sand 垫砂
placing sequence 布置程序
placing site 铺管现场;铺管工地
placing speed 浇筑速度
placing speed of concrete 混凝土浇灌速度
placing temperature (路面的)摊铺温度;浇筑温度;入仓温度
placing the fill(ing) 填充填料
placing the roofing 铺屋面;盖屋顶
placing train 混凝土浇筑列车
placing under water 水下浇灌混凝土;水下浇灌
placing velocity 浇灌速度
placing volume 工作容积
PLA Datum 伦敦基准面
plafond 彩绘天花板;柱顶飞檐顶棚;具有装饰的天花板;天花板上的彩画
plage 海滨胜地
plage area 谱斑面积
plage corridor 谱斑走廊
plage flare 谱斑区耀斑
plage radiation 谱斑辐射
plaggen 耕土
plagiaplite 斜长细晶岩
plagifoyaitarkite 斜白霞斑岩
plagiocephalia 斜头
plagiocitrite 斜橙黄石
plagioclase amphibolite 斜长角闪岩
plagioclase amphibolite facies group 斜长角闪岩相组
plagioclase-arenite 斜长石砂碎屑岩
plagioclase arkose 斜长石砂岩
plagioclase feldspar 斜长石
plagioclase gabbro 斜长辉长岩
plagioclase gneiss 斜长片麻岩
plagioclase granite 斜长花岗岩
plagioclase hornblende schist 斜长角闪片岩
plagioclase rock 斜长岩
plagioclasite 斜长岩
plagioclastic 斜解理
plagioclastic-arenite 斜长碎屑岩
plagioclimax 偏途演替顶级
plagiodacite 斜长英安岩
plagiodont 斜齿型
plagiogeotropism 斜向地性
plagiogranite 斜长花岗岩
plagiogranite porphyry 斜长花岗斑岩
plagiograph 斜缩图仪
plagiohedral 斜面型
plagiophototaxy 斜趋光性
plagiophototropism 斜向光性
plagiophyre 斜长斑岩
plagiosere 偏途演替系列
plagiotrachyte 斜长粗面岩
plagiotropism 斜向性
plague 瘟疫
plague of hail 雹灾
plague spot 疫区
plague voltage 破坏电压
plaid 方格织物;方格墙布;方格花纹

plain 净料；海平原；平滑；平海底；单纯的
plain acoustic(al) tile 光面天花板
plain antenna 直接耦合式天线；简单天线
plain antenna system 简易天线设备
plain arch 单拱；光面拱；弧形纹；平拱；粗拱
plain area 平原地区；平面面积
plain artificial stone 光面人造石；普通人造石
plain asbestos yarn 纯石棉纱
plain ashlar 琢方石；琢面石板；光面石块；饰面石板
plain asphalt 纯沥青；普通沥青；纯地沥青
plain background 清晰背景；平坦路基
plain ball 普通钢珠
plain ball bearing 普通滚珠轴承
plain bar (of reinforcement) 光面钢筋；光圆钢筋；无节钢筋；无纹钢筋；扁钢
plain-barrel(l)ed roll 平辊
plain bar type ga(u)ge 普通杆式量规；普通杆式规
plain basic steel 普通碱性钢
plain batten door 平板门
plain beam 板梁；实覆梁
plain beam bridge 板梁桥；实覆梁桥
plain bearing 滑动轴承；普通轴承
plain bearing axle-box 轴箱滑动轴承
plain bearing material 滑动轴承材料
plain bearing pillow bolck 普通轴承轴台
plain bed 无槽机座
plain bed lathe 普通床身式车床
plain bending 平面弯曲
plain bevel gear 普通伞齿轮
plain blank board 无空白模件板
plain block 平块
plain board 平板
plain boiler tube 普通锅炉管
plain bolt 普通螺栓
plain boltel 朴素凹圆饰
plain bond 无担保债券
plain bone screw 无槽接骨螺钉
plain braid 平编带
plain brake drum 平制动鼓
plain brake drum shaft 平制动鼓轴
plain-break breaker 直接熄弧断路器
plain brick paver 平面铺地砖
plain brick wall 清水砖墙
plain brick wall pointed with cement mortar 清水砖墙水泥勾缝
plain bumper 平面防撞装置；缓冲器；防撞减振器；振实机；振动台
plain bush 普通轴衬
plain bushing 普通轴瓦
plain butt joint 平头对接头；平筋平接缝；无筋平接缝；纯平接缝
plain butt weld 平边对接焊；对焊；平头对接焊缝；无坡口对接焊
plain can 素铁罐
plain carbon steel 碳素钢；普通碳钢；单纯碳素钢
plain carbon tool steel 碳素工具钢
plain cast iron 普通铸铁
plain cast iron roll 平面铸铁轧辊
plain cast iron share 普通铸铁犁铧
plain cast stone 光面铸石；简单铸石；光面人造石；简单人造石
plain cement 净水泥；清水泥；纯水泥
plain cement paste 纯水泥浆
plain cement roofing tile 清水泥屋面瓦
plain center-type grinding machine 普通中心式磨床
plain channel 平原航道
plain chisel 扁钻
plain chromium steel 无纹铬钢
plain clamp 普通夹具
plain clay roofing tile 平陶瓦
plain climate 平原气候
plain clinometer 钻杆下部装的测斜仪；普通倾斜仪
plain coagulation 自然沉淀
plain coast 平原海岸
plain code 明码
plain code telegram 明码电报
plain coil 光线圈
plain colo(u)r 素色
plain colo(u)r glaze 素三彩
plain colo(u)r tile 单色瓦
plain column 光面柱
plain combustion chamber 普通燃烧室
plain compression ring 普通压缩环

plain concrete 普通无筋混凝土；无筋混凝土；素混凝土
plain concrete bedding 素混凝土垫层
plain concrete block 素混凝土块
plain concrete column 素混凝土柱
plain concrete foundation pier 无筋混凝土墩台基础；素混凝土墩台基础
plain concrete pavement 素混凝土路面
plain concrete pier 素混凝土墩
plain concrete roofing tile 平混凝土屋面瓦
plain concrete single base 无筋混凝土单个基础；素混凝土单个基础
plain concrete structure 素混凝土结构
plain concrete surface 素混凝土面层
plain conductor 单金属导线；金属裸线；裸线；裸导线；普通导线
plain conduit 简单导管
plain connecting rod 普通连杆
plain coordinates 明码坐标
plain cord 白坯绳
plain core barrel 单层岩芯管
plain corner joint 平头角焊接
plain country 平原地区
plain coupler 普通套管接头
plain cover 素封面
plain crosshead guide 平型十字头导框
plain cube 简单立方体
plain cup wheel 杯形砂轮
plain cushion 平垫
plain-cut joint 平刮(灰)缝
plain-cut nippers 普通剪钳
plain cutter 平铣刀
plain cutter holder 普通刀架
plain cylindric(al) form 普通圆柱形
plain cylindric(al) furnace 普通锅炉
plain cylindric(al) plug ga(u)ge 普通圆柱塞规
plain dandy 无图案水印辊
plain deep drawing 简单深拉延
plain-denotation 简单标志
plain denudation 蚀原作用
plain detonator 普通雷管
plain disk 普通圆盘
plain dividing apparatus 普通分度装置
plain dividing head 普通分度头；水平分度头
plain door 光面门；平面门
plain dressing 光面修整；平面抹灰；朴素修饰【建】
plain drifting 转动漂动
plain drill 扁头凿；扁头凿子；扁头钎子
plain drill collar 扁头凿套环；扁头凿卡环
plain earth 素土
plained 澄清的
plain edge 光面边缘；光边；平边
plain edge tile 平砖
plain elbow 不带边弯头
plain embossed design 素色轧花
plain end 平管口；无坡口；(未套纹的)管材的平滑端
plain end joint 平头端焊接
plain-end pipe 平头管子；平端管材
plain end reducer 平端渐缩管
plain-ended screw 平端螺钉
plain-end steel pipe 平端钢管
plain-end tube 平端管
plain equal cross 不带边周径四通
plain equal side outlet elbow 不带边周径三向弯头
plain equal tee 不带边周径三通
plain extractor roll 普通脱水辊
plain face 平光表面；平面
plain-face corrugated flange 平面波纹法兰；平面波纹凸缘
plain feed 普通进给
plain fill 素填土
plain film 平片
plain fin 平直翅片
plain fishplate joint 平鱼尾板结合
plain fishtail bit 普通鱼尾钻头
plain fit 三级精度配合
plain fixed gap ga(u)ge 普通固定式间隙规
plain flange 平面法兰；平面凸缘；对接法兰；对接凸缘
plain flat 平腰线
plain flat strip mo(u)lding 压缝条；平整腰线
plain flood 平原洪流
plain floor cover(ing) tile 无花纹铺地瓷砖
plain floor tile 无花纹铺地瓷砖；普通铺地砖

plain-form 平面形状；平面形态
plain fracture 平裂
plain-frame core box 框架式芯盘
plain friction bearing 活动摩擦轴承
plain furnace 平火胆
plain fuse cut-out 敞露式熔丝保险器
plain galvanized steel sheet 镀锌铁皮
plain gap 空气隙放电器；简易放电器
plain gate 平面闸门；平板闸门
plain gear 正齿轮；平齿轮
plain girder 光面梁；宽腹梁；实腹梁；板梁
plain glass 无条纹玻璃；防护白玻璃
plain glass roof(ing) tile 平玻璃屋面瓦
plain glass spectacles 平光眼镜
plain glazed coat(ing) 简单釉面层
plain glazed finish 简单釉面装饰
plain glazing 单纯上釉；单装配玻璃
plain grade 普通级精度
plain grate 平炉排
plain gravity slide 纯重力滑动
plain greening 平原绿化
plain grinding 普通磨削；平磨
plain grinding machine 普通磨床；外圆磨床
plain grinding machine internal 普通内圆磨床
plain grinding wheel 普通砂轮
plain groove 哑槽；无音槽
plain ground 清晰背景
plain helical milling cutter 普通螺旋铣刀
plain hinge 单铰
plain hocton spheroid 球形罐
plain hollow mill 普通空心铣刀
plain hook 平钩
plain hopper bottom tank 简单锥形底槽
plain horn 普通喇叭管
plain hot rolled wire 热轧光面钢丝
plain hypo 普通定影液
plain indexing 普通分度
plain indicator 平面图显示器
plaining 提纯；澄清
plaining agent 提纯蚀原剂；澄清剂
plain insert 光滑轴瓦；平轴瓦
plain intersection 简易交叉；平面交叉
plain invoice 普通发票
plain iron copper alloy 纯铁铜合金
plain joint 平整接；平接
plain jolter 振实台；振实机
plain jolting machine 振实造型机
plain journal bearing 活动轴承；普通颈轴承
plain knee-and-column type drilling machine 普通升降台式钻床
plain knife 光刃刀片；平刃刀片
plain labyrinth packing 平齿型迷宫气封
plain-laid 普通捻向
plain laid rope 平捻绳
plain landform 平原地貌
plain language 普通语言；普通文字
plain lap 简易搭接
plain lap joint 普通搭接接头
plain lath 素灰石膏条板
plain lead-covered cable 光皮铅包电缆
plain lift 普通起落机构
plain line 平原线
plain link 平链节
plain linoleum 素色亚麻油地毡；单向油毡
plain linoleum cement 单色油地毡水泥
plain live axle 普通动轴
plain live rear axle 普通传动后轴
plain location 平原地区选线
plain loop 单股会让氹
plain loop for ceilings 天花板的平线圈；天棚的平线圈
plain macadam 清水碎石路(无结合料的碎石路)；纯碎石路；纯石子路
plain masonry 普通砌体；无筋砌体；素砖石砌体；纯圬工
plain meandering river 平原游荡性河流
plain meandering stream 平原游荡性河流
plain metal 普通金属
plain miller 平面铣床
plain miller machine 平面铣床
plain milling 平铣
plain milling cutter 平铣刀；平面铣刀
plain milling machine 普通铣床；平面铣床
plain mineral oil 普通矿物油；普通矿物润滑油

plain mitre joint 平斜接
plain module board 无孔插件板
plain mortar 素砂浆
plain motion mechanism 平面运动机构
plain mo(u)lding 窗线脚
plain net 平织网
plain nozzle 普通喷嘴
plain nut 普通螺母
plain of abrasion 磨蚀平原
plain of accumulation 堆积平原
plain of denudation 剥蚀平原
plain of erosion 侵蚀平原
plain of marine denudation 海蚀平原
plain of marine erosion 海蚀平原
plain oil 普通润滑油
plain overlap seam 简单搭接缝
plain panel 无装饰的墙面板；无装饰的镶板
plain paper 无花纹墙纸；单色墙纸
plain paper copier 普通纸复印机
plain patent stone 纯人造石
plain paving brick 平面铺地砖
plain pedestal 整体轴架
plain pendant 吊托架；素垂饰
plain pick-up transfer 普通传递装置
plain pile 光面桩
plain pilot 简单导向杆；普通导向钻头
plain pin 卡钳钉；平销钉
plain pipe 光面管
plain pipe-end 未攻丝管端
plain pipet(te) 无刻丝移液管；未刻移管
plain plate 光面钢板；平铁板
plain plating 平添纱；简单添纱
plain post 光面柱
plain post cap 简易柱帽
plain press 直接传动压力机；普通压榨
plain presser 平压片
plain pressure ring 平接压力环
plain quality 平纹密度
plain radial drilling machine 普通摇臂钻床
plain rail 普通横木；门窗横头；门窗横挡
plain reconstructed stone 素色熔合宝石；素色人造石
plain reflectography 普通反射复制法
plain region 平原区；平原地区
plain reinforcement (bar) 光面钢筋；无节钢筋；无筋钢筋
plain reinforcement rod 无节加强筋
plain reinforcing bar 光面钢筋；无节钢筋
plain resistance 欧姆电阻
plain ring ga(u)ge 普通环规
plain river 平原河流
plain roll 平面轧辊
plain roll crusher 平辊破碎机
plain rolled glass 园艺用玻璃；压延玻璃；平玻璃瓦
plain rolled horticultural glass 无花纹园艺用的压延玻璃
plain roller 光面辊
plain roller glass 平板玻璃瓦
plain roof 炼钢炉平顶
plain roofing tile 平屋面瓦
plain roof tile 屋顶平瓦
plain round bar 光面圆钢材；光面圆钢条；光面圆钢筋
plain round mo(u)lding 纯圆饰
plain round nut 普通圆螺母
plain round rod 光面圆棒
plain run board 平踏脚板
plain sail 正常帆
plain sailing 平安航行
plain sandblast 简易喷砂
plain saw bench 平锯台
plain-sawed 弦锯的；平锯的
plain-sawed clear oak 弦锯无节疤橡木材
plain-sawed lumber 弦锯材；平锯木材；弦切板
plain-sawed select oak 弦锯精选橡木材
plain-sawing 直锯(法)；平锯(法)；弦锯法
plain-sawn 平锯木
plain-sawn clear oak 弦锯无节疤橡木材
plain-sawn lumber 弦锯材
plain-sawn select oak 弦锯精选橡木材
plain-sawn timber 弦锯木(材)
plain scarf 普通嵌接；平面斜接
plain screw 普通螺钉；平钉
plain section 平刃动刀片；无螺纹部分

plain sedimentation 单纯沉淀作用；自然沉积；自然沉淀；净沉淀；简单沉降；普通沉淀
plain sedimentation tank 简单沉降槽
plain segment 无针弧面
plain sequence 正序
plain settlement (污水的)简单沉淀
plain settling tank 自然沉淀池
plain shade 一色；表色
plain shaft 普通轴
plain shaking screen 平面摆动筛
plain shank 普通开沟器体
plain shed 普通梭口
plain shedding tappet 平纹开口踏盘
plain sheet 白纸；平板；平铁皮；素铁
plain sheet iron 无楞铁皮
plain sheet of gupsum plaster board 石膏粉饰的光板
plain shoe 平端管头
plain sidemilling cutter 普通侧铣刀
plain slider 平导针片
plain slide rest 普通刀架
plainsman 平原居民；平民居民
plain snap ga(u)ge 卡规；普通外径规
plain soil 纯土；素土
plain soil cushion 素土垫层
plain solid brick masonry 无筋实心砖砌体
plain spark-gap 普通放电器
plain spheroid 无弧折的球形储罐
plain spindle 轻型轴
plain square cut off ends 端部切平齐
plain stage 普通载物台
plain stage of slope 边坡平台
plain steel 普通钢；碳素钢
plain steel bar 碳素钢条
plain steel rim 简单钢轮缘
plain steel wire 光圆钢丝
plain stem 无螺纹节
plain stirrup for ceilings 光面顶棚吊钩
plain straight wheel 普通盘形砂轮
plain strain 简单应变；单面应变
plain stream 平原河流
plain stress 简单应力
plain stringer 简单楼梯梁
plain stripper 带漏板的造型机
plain strip shingle 平板条屋顶木瓦；平板条屋顶板
plain suction dredge(r) 普通吸扬式挖泥船
plain surface grinder 平面磨床
plain surface grinding machine 平面磨床
plain surface machine 光滑面电机
plain tail 平直末端
plain tail pipe 平直尾喷管
plain tank 平滑油箱
plain tape 平纹带
plain taper attachment 普通锥形附件
plain taper sunk key 普通锥形埋头键
plain tappet 平纹踏盘
plain tee joint 平头丁字接头；平头丁字焊接
plain terrain 平原区
plaintext 明文
plain theodolite 普通经纬仪
plain thermit 铝热剂
plaintiff 原告
plain tile 平瓦；无楞瓦；板瓦
plain tile roof 平瓦屋顶
plain tile roof cladding 平瓦屋面覆盖；平瓦屋面层
plain time rate 无奖金计时工资
plain tint 普染
plain tire 无缘轮胎
plain tool rest 普通刀架
plain tooth cross-cut saw 通常齿横割锯
plain top glove 平口手套
plain total enclosure 普通全封闭式机壳
plain towed roller 光面路碾；拖式光面压路机
plain tract 下游河段；平原河段；冲积河段
plain transfer line 直接输送式组合机床自动线
plain transit 普通经纬仪
plain triangle 三角定规
plain tricolo(u)r 素三彩
plain tri-metal bearing 三金属轴瓦
plain triple valve 普通三通阀
plain trolley 小推车；手摇缆车
plain tube 光滑管
plain tube economizer 光管省煤器
plain turning 顶尖车工

plain turning slide 纵刀架
plain type 简易型；简易式；普通式
plain type breaker 简单型断路器
plain under-ridge tile 简单的脊瓦
plain value 简单值
plain valve head 平阀头
plain veneer 平滑单板
plain vise 简式老虎钳；普通虎钳；平口钳
plain view drawing 平面图
plain wall 清水墙
plain wall paper 白墙纸
plain washer 平垫圈；平垫
plain water 淡水
plain water extinguisher 清水灭火器
plain water-filled roller 平面充水镇压器
plain water flush 清水洗孔；清水洗井
plain waterway 平原航道
plain weave 平纹
plain weave carpet 平纹交织地毯
plain web 光面腹板
plain web beam 平腹工字梁
plain web(bed) arched girder 平腹拱形大梁
plain web(bed) three-hinge(d) arched girder 平腹三铰拱架
plain wheel 平面轮
plain whipping 绳头缠扎
plain white body 素胎；白坯
plain wire 无筋钢丝；光面钢丝
plain-wire-cut brick 直切砖；直切砖
plain wire strainer 铁丝拉紧器
plain wood 素木；漆器用材
plain wood of lacquerware 漆器材
plain work 石面斫平(工作)；平缝；石凿平；凿平的石面；石面
plain zed 光面 Z 形钢；光面 Z 形铁
plain zone 平原地区
plaisance 它旁游憩屋；(公寓附近的)凉亭或娱乐室
Plaisancian 普莱桑斯阶【地】
plait 折叠；卷起
plait-band 纽索饰；辫状带
plaiter 折布机
plaiting 折布
plaiting apparatus 落布装置；甩布架
plaiting channel deposit 辫状水道沉积
plait mill 卷料机
plait of onion 葱辫
plait point 褶点
plan 线路平面；进程表；设计图；设计；方案
planalto 高平原(绝对高度 200～600 米)
plan and section 平面图和剖面图
plan and specification 施工图纸及技术说明书
planar 平面的；二度的；二维的
planar air-gap generator 平面气隙发电机
planar array 平面阵
planar array pattern pumping well 抽水井的平面布置形式
planar array radar 平面阵列雷达
planar array radar antenna 平面阵列雷达天线
planar bearing structure 平面支承结构
planar ceramic tube 平陶瓷管
planar chromatography 平面色谱法
planar configuration 平面形状
planar construction 平面结构
planar control grid 平面控制栅极
planar cross-bedding 平面交错层理
planar cross bedding structure 板状交错层理构造
planar cross set 平面交错层组
planar defect 面缺陷
planar density 面密度
planar device 平面器件
planar diode 平面型二极管；平面二极管
planar discontinuous groups 平面不连续群
planar distribution map of testing borehole 试验钻孔平面布置图
plan area 平面面积；横断面(面)积；水平投影面积；直线面
planar electrode tube 平面电极管
planar electromechanical coupling factor 平面机电耦合系数
planar element 面状要素；平面元
planar epitaxial method 平面外延法
planar fabric 面状组构
planar factor 晶相因数
planar film 平面膜

planar flow 平面流(动)
planar flow structure 流面构造；片板状流动构造
planar fluence 平面注量
planar flux density 平面通量密度
planar force system 平面力系
planar fracture method 面状断裂法
planar frame 平面框架
planar frame element 平面框架构件；平面骨架构件；平面框架构件
planar frame structure 平面骨架结构；平面构架结构；平面框架结构
planar graph 平面图形；平面图
planar growth structure 晶面长大组织
planar helix winding 平面螺旋缠绕
planaria 涡虫
planar interface 平界面
planar inversion 平面反演
planar iris 平面可变光阑；平面光阑
planarity 可平面性
planarization 平坦化
planarization agent 表面流动控制剂
planar junction 平面连接器；平面结
planar load-bearing structure 平面承重结构
planar mask 显像管荫罩；荫罩板
planar method 剥层法
planar module 平面组装插件
planar molecule 平面分子
planar motor 平面电动机
planar multiplicity factor 晶面多重性因数
planar net density 面网密度
planar network 平面网络
planar orientation 沿面取向
planar outline of earth fracture 裂缝平面形状
planar package 平面组装
planar parallel system 平面平行系统
planar phased array 平面相控阵
planar point 平点
planar point of a surface 曲面的平点
planar prediction 平面预测
planar problem 平面问题
planar process 平面程序
planar-radial flow 平面径向流
plan arrangement 平面布置
plan arrangement of standard heavy trucks 标准车队的横向布置
planar reflector 平面反射器
planar reflector billboard array 平面反射器同相多振子天线
planar rotational system 平面转动系统
planar shear 平面剪切
planar shear deformation 平面剪切变形
planar slide 平面滑动(地铁)
planar solution 平面解
planar state of stress 平面应力状态
planar stator motor 平面定子电动机
planar stress problem 平面应力问题
planar structure 平面结构；延展面构造
planar supporting structure 平面支承结构
planar technique 平面技术
planar tectonic stress field 平面构造应力场
planar theory of elasticity 平面弹性理论
planar transistor 平面晶体管
planar two-dimensional frame system 二维平面刚架体系
planar warm anomalies 面状暖异常
planar water 吸附水
planar water flow test 顺面水流试验
planar wave system 平面波系
planar weight-carrying structure 平面承重结构
planar winding 纵向缠绕；平面缠绕
planar zigzag structure 平面锯齿结构
plan as a whole 统筹
planation 夷平作用
planation surface 夷平面；侵蚀面
plan boundary layer 平面边界层；二元边界层
planceer 天花板；拱圈底面；(檐或楼梯的)底板
planceer piece 楼梯底板；挑檐底板
planch (作木地板用的)木料；木地板；平板
plan checked and approved as its arriving 随到随批计划
plan checked and approved by focus 集中计划
plancheite 纤硅铜矿
planchet 小金属圆片；圆片；金属坯料
planchet casting 模板铸件

planchette 平板绘图器；平板仪
planching 铺设地板
plancier 拱底板；挑檐底板；拱圈底面；楼梯底面；底面；底板；檐底板；天花板
plancier level 下皮标高
plancier piece 挑檐底板；出檐天花板
Planck's constant 普郎克常数；蒲朗克常数
Planck's law 普郎克定律
plancon 木墩；八角(形)粗木料
plan concerning manpower and pay scales 劳动工资计划
plan contour figure of induced polarization gradient array 激电中梯平面等值线图
plan deposit 投标文件押金
plan detail 平面细节；平面细部
plan diagram of special geologic(al) body 特殊地质体平面图
plan dimension 平面图尺寸
plan dimension of standard heavy trucks 标准车队的平面布置
plan display 平面位置显示
plan-do-check-action cycle 计划—实施—检查—总结工作循环(处理循环)；计划—实施—检查—总结工作(处理)
plan-do-check-action management 计划—执行—检查—处理工作法
plan drawing 平面图
plane 平坦；平刨；平面；投影；磁芯屏蔽；飞机；法国梧桐；刨子；刨平；刨
plane abrasion 平磨
plane affine transformation 平面仿射变换
plane algebraic curve 平面代数曲线
plane angle 面角；平面角
plane angle of a dihedral angle 二面角的平面角
plane antenna 平天线
plane array 平面阵
plane asbestos cement slate 石棉水泥平板
plane ashlar 光面琢石；平(面)琢石
plane atmospheric wave 大气平面波
plane baffle 平面障板
plane bearing 平面轴承；平面支座；平面支承
plane bed 平坦基床；平坦河床；平面岩层
plane belt 平面传送带；平面传递带
plane bending 平面弯曲
plane bevel wheel 平斜轮
plane bit 刨刀
plane bolter 方眼筛
plane bottom hammer 平底锤
plane buoyant jet 平面浮射流
plane cam 凸轮
plane carrier 航空母舰
plane Cartesian coordinates 笛卡儿平面坐标
plane cement tile 平面花砖；平面水泥瓦
plane chart 平面海图
plane compound-slide system 水平双滑系统
plane concave antenna 平凹形天线
plane concave head-on photomultiplier 凹镜窗式光电倍增管
plane concrete 纯混凝土；无筋混凝土；素混凝土
plane concrete tile 平面混凝土瓦
plane conductor 平板导体
plane configuration 平面构型
plane conformal geometry 平面保形几何学
plane contact lay wire rope 面接触钢丝绳
plane control point 平面控制点
plane coordinate azimuth 平面坐标方位角；坐标方位角
plane coordinate intersection tables 图廓点直角坐标表
plane coordinates 平面坐标
plane coordinate system 平面坐标系(统)
plane coordinate system of mining area 矿区平面坐标系统
plane coordinate triangle 平面坐标三角形
plane crossing method 平面交会法
plane cross-section 水平截面
plane cross-section assumption 平截面假定
plane curve 平曲线；平面曲线
plane curve bridge 平面曲线桥
plane curve location 平面曲线测设
plane curve of autocorrelation function 自相关函数平面图
planed 刨光的
planed all round 全部刨光的

plane deformation 平面变形
plane deformation state 平面变形状态
plane design 平面设计
plane design of city urban road 城市道路平面设计
plane edge 刨平边
planed fault 平削断层
planed flare 刨削的斜展面
plane diagram 平面图
plane diffraction grating 平面衍射光栅
plane diffuser 平面扩压器；平面雾化器；平面吹气器；平面扩散器
plane distortion 平面畸变
plane distortion state 平面变形状态
plane distribution 面状分布
plane distribution shape of sand drain 砂井平面布置形状
planed joint 刨光接头
planed lumber 刨光材；刨光的木材
planed matchboard 刨光企口板；刨光拼合板
planed reuse 计划再用
planed signal 平面信号；二维信息
planed slab 机刨板材
planed timber 光面木(料)；刨平木料；刨光木料；刨光木材
planed-to-caliper hardboard 厚度整平硬质纤维板；极平硬质板
plane earth 平面地
plane earth attenuation 地平面损耗
plane earth factor 地平面因数
plane edge 刨边
plane elasticity 平面弹性力学
plane elastic system 平面弹性系统
plane electrode 平面电极
plane electromagnetic wave 平面电磁波
plane element 平面元
plane elliptic arc 平面椭圆弧
plane evolute 平面渐屈线
plane expanded metal 多孔式钢板网；平面钢板网
plane fault 顺层断层；层面断层；平面断层
planeferrite 普蓝铁矾
plane field 流形上的平面场；平面场
plane figure 平面图形；平面图
plane figure of anomaly interpretation 异常解释平面图
plane filtering 平面滤波
plane flow 平面流(动)；顺面流
plane foliation 分层面；成层面；层状面
plane force system 平面力系
plane formula 平面式
plane four bar mechanism 平面四杆机构
plane fracture 平面破裂；平面断裂；平裂
plane frame 平面框架；平面桁架
plane frame element 平面构架构件
plane frame structure 平面框架结构
plane frame work 平桁架；平面框架结构；平面构架
plane-free boundary 自由平面边界
plane front 平面波阵面
plane gasoline 航空汽油
plane gate 平板大门；平面闸门；平板闸门
plane geodesy 平面测量测量学
plane geometry 平面几何学；平面几何形状
plane girder 平面桁架；平面大梁
plane glass roof(ing) tile 平面玻璃屋面瓦
plane goniometer 平面测角计
plane grate 平面炉箅
plane grating 平行光栅；平面光栅
plane-grating spectrograph 平面光栅摄谱仪
plane grinder 平面研磨机；平面磨床
plane grinding 平面磨削
plane group 面网
plane hangar 飞机库
plane harmonic equation 平面简谐方程
plane harmonic motion 平面谐波运动
plane harmonic wave 平面谐波
plane heat-source method 平面热源测定法
plane hole survey 飞机洞库测量
plane hydromagnetic wave 平面磁流体波
plane included angle 平面夹角
plane inset of slope 斜井平车道调车场
plane interface 平面分界面
plane involute 平面渐伸线；平面渐开线
plane iron 刨铁；刨刃
plane isoline map of magnetic anomaly 磁异常平面等值线图

plane isometric line map 平面等值线图
plane joint 平面接缝
plane joint of weakness 弱面缝
plane kerosene 航空煤油
plane knife 刨刀
plane lamina 平面薄片
plane lapping 平面研磨
plane laser designator 机载激光指示器
plane law 面律
plane layer 平面层
plane layered medium 平面分层介质
plane layout 平面规划
plane linear perspective 平面直线透视
plane liner 平衬板
plane link 平面杆
plane map blasting 爆破平面图
plane mirror 平面镜;平面反光镜
plane motion 平面运动
plane movement 平面运动
plan end lumber 平端材
planeness 平整度;平面度
plane net 平面网
plane network 平面格网
plan environmental assessment 规划环境评价
plan environmental impact assessment 规划环境影响评价
plane of a structure 结构平面
plane of bars 杆件平面
plane of bending 弯曲面;弯曲平面;变曲面
plane of best definition 最清晰平面
plane of break 破裂面;断裂面;崩落面
plane of buoyancy 漂浮面;浮面
plane of cleavage 解理面;断裂面;裂开面;劈裂面
plane of collimation 视准面
plane of composition 组合面
plane of contact 接触面
plane of coordinates 坐标面
plane of couple 力偶面
plane of crystal 晶体表面
plane of datum 基准面
plane of deflation 风蚀面
plane of delineation 图像平面
plane of denudation 剥蚀面
plane of departure 发射平面
plane of direction 瞄准面
plane of disruption 破碎面;断裂面
plane of easy slip 易滑移面
plane of elements 构件平面
plane of equal settlement (埋管的)等沉陷面;(埋管的)等沉降面
plane of failure 破坏(平)面
plane of field lens 场镜面
plane of flattening 延展面
plane of flexure 挠曲面
plane of flo(a)tation 漂面;水线面
plane of flow 流动平面
plane of fold 褶皱面
plane of fracture 折断面;破裂面;断面
plane of gravity 重心面
plane of homology 透射平面
plane of illumination 光照面
plane of incidence 入射面
plane of inlet 入口平面
plane of isostatic compensation 均衡基面
plane of kinking 弯结面
plane of lateral corrosion 侧蚀面
plane of lattice 晶格面
plane of light incidence 光线入射面
plane of loading 加荷面
plane of maximum shearing stress 最大剪(切)应力面
plane of mirror symmetry 镜面对称面
plane of no-deformation 无伸缩剖面
plane of no distortion 无畸变面;不变歪平面
plane of normal section 法截面
plane of optic(al) axis 光轴面
plane of oscillation 振荡面;振动面;摆动面
plane of perspectivity 透视平面
plane of plummet oscillation 垂球摆动面
plane of polarization 极化平面;极化面;偏振面;偏振面
plane of principal axis 主轴面
plane of principal shearing stress 主剪应力(平)面
plane of projection 射影面积;投影面

plane of propagation 传播面
plane of radial shear 径向剪力面
plane of reference 基础面;参考平面;参考面
plane of reflection 反射面;反射面
plane of refraction 折射面
plane of registering frame 像框面
plane of regression 回归平面
plane of rotation 转动面
plane of rupture 折断面;破裂面;断面;断裂面
plane of saturation 饱和面
plane of schistosisty 片岩面;片理面
plane of sharpest focus 锐聚焦面
plane of shear 剪切面;剪力面
plane of shear fracture 剪裂面;扭裂面
plane of sight 视准面
plane of sliding 滑动面;坍滑面
plane of slip 滑动面
plane of stratification 层面;分层面
plane of symmetry 对称面
plane of synchronization 同步平面
plane of tensile fracture 张裂面
plane of the easel 承影面
plane of the meridian 子午线平面
plane of the sky 天球切面
plane of thrust 冲断面
plane of transposition 转置平面
plane of unconformity 不整合面
plane of uniform distortion 均匀变歪剖面
plane of uniform expansion or contraction 等伸缩剖面
plane of vapo(u)rization 汽化平面
plane of vibration 振动平面;振动面
plane of vision 视场面
plane of weakness 危险面;最小抵抗面;弱面;脆弱面;薄弱面
plane-of-weakness joint 假缝;槽缝;弱面缝
plane of yaw 偏航平面
plane orientation number 平面取向数
plane parallel 平行平面的;成平行面的
plane-parallel atmosphere 平面平行大气压
plane parallel glass plate 面平行玻璃板
plane parallel layer 平行平面层
plane parallel plate 平行玻璃板
plane parallelplate ionization chamber 平行板电离室
plane parallel to the profile plane 侧平面
plane perpendicular to projection plane 投影垂直面
plane photogrammetric network 平面摄影测量网
plane photogrammetry 平面摄影测量(学)
plane plastic deformation 平面塑性变形
plane plate 平面绘图仪;平板仪;平板
plane plate micrometer 平行玻璃板测微器
plane point set 平面点集
plane Poiseuille flow 平面泊肃叶流
plane polar coordinate 平面极坐标
plane polariscope 平面偏振光镜;平面偏光仪
plane polarization 线偏振;平面偏振
plane polarization light 平面偏振光
plane polarization linear polarization 平面极化
plane polarization wave 面偏振波
plane polarized 平面偏振的
plane polarized antenna 平面偏振天线
plane polarized beam 平面偏振的辐束线
plane polarized electromagnetic wave 平面偏振电磁波
plane polarized light 面偏振光;平面偏振光
plane polarized method 平振法
plane polarized wave 面偏振波;平面偏振波
plane polarizer 平面偏光镜
plane position indicator 环视扫描显示器;平面位置显示器;平面示位图
plane post 空运邮件
plane potential function 平面势函
plane print 平版印刷
plane problem 平面课题
plane-profile map 平纵断面图
plane progressive wave 平面行波;平面前进波
plane projection 平面投影
plane projection method 平面投影法
planer 粗加工面;刨刀痕面;地面整平机;路刮;路机;平路机;刨工;刨床(工)
plane radio 航空无线电台
plane radiography 平面X线照相术

planer and lathe tool grinding machine 刨刀和车刀磨床
planer center 刨床转度卡盘
planer cutting tool 刨床切削刀具
plane rectangular coordinates 经纬距;平面直角坐标
plane reflection 平面映射
plane reflector antenna 平面反射器天线
plane repeater indicator 平面位置辅助指示器
plane resection 平面后方交会(法)
planer fixture 刨削夹具
planer ga(u)ge 刨规
planer heads 刨床刀架;刨床架
plane right coordinates 平面直角坐标
planerite 土绿磷铝石
plane riverbed 平坦河床
plane(r) knife 刨刀
plane rockbed 平面岩层
plane rod element 平面杆单元
plane roll 滚轧面
planer parts 刨床部件
planer saw 刨锯
planer slide 刨床滑板
planer structure 平面结构
planer table 刨床工作台
planer tool 刨刀
planer-type grinder 刨式磨床
planer-type machine 多刀刨床
planer-type miller 刨式铣床
planer-type milling machine 刨式铣床
planer-type surface grinder 门式平面磨床
planer vise 刨床用虎钳
planer work 刨工工作
planer worm gear 平面包络蜗轮蜗杆传动
plane sailing 平面航迹计算法;平面航海法
planes at zero 水平舵
plane seal housing 平板密封套
plane search for an immobile hider 对不动隐藏者的平面搜索
plane section 平截面;剖面;水平断面
plane-serveying 平面测量法
plane set-hammer 方锤
plane shear failure 平面剪切破坏
plane sheet reflector 平板反射器
plane similarity transformation 平面相似变换
plane sketch 平面示意图
plane skylight 采光板
plane slanted bottom 平面斜底
plane slide 平面滑动;刨床滑板
plane sliding 平面滑动
plane sliding bearing 平板滑动支座
plane slip 平面滑动
plane source of earthquake 平面震源
plane state of strain 平面应变状态
plane state of stress 平面应力状态
plane stock 刨座;刨身;刨架
plane strain 平面应变;平面胁变
plane strain apparatus 平面应变仪
plane strain compression test 平面应变压缩试验
plane strain condition 平面应变条件
plane strain ellipsoid 平面应变型椭球体
plane strain extension test 平面应变拉伸试验
plane strain fracture 平面应变断裂
plane strain fracture toughness 平面应变断裂韧性;平面应变断裂韧度
plane strain instability 平面应变不稳定性
plane strain loading 平面应变加载
plane strain shear diagram 平面变形剪切图;砂的平面变形剪力图
plane strain state 平面应变状态
plane strain-stress singularity 平面应变应力奇点
plane stress 平面应力
plane stress analysis 平面应力分析
plane stress distribution 平面应力分布
plane stress fracture 平面应力断裂
plane stress state 平面应力状态
plane structure 平面结构
plane substrate 平面衬底
plane sundial 平面日晷
plane surface 平面;夷平面;平曲面
plane surface circumscribed by the water line 水线定界线的平面
plane surface of sliding 滑动平面
plane survey(ing) 平面测量

plane survey preliminary calculation 平面测量概算
plane symmetric(al) isomerism 平面对称异构现象;平面对称异构
plane symmetry 面对称
plane symmetry flow 平面对称流动
plane system 平面体系
plane systematic sampling 平面等距抽样
plane system of forces 平面力系
planet 行星
plane table 平板;大平板仪【测】;平板仪;平板绘图仪;平板测绘仪
plane-table alidade 大平板仪;平板照准仪
plane-table board 平板测图板
plane-table compass 平板仪罗盘
plane-table fixing 平板仪定位
plane-table for making 划线平台
plane-table level tube 平板仪水准器
plane-table map 平板仪测量图;实测原图
plane-table mapping 白板测图
plane-table method 平板仪测图法;平板仪测量法
plane-table operation 平板仪作业;平板(仪)测量
plane-table operator 平板仪作业员
plane-table paper 平板图纸
plane-table photogrammetry 平板(仪)摄影测量
plane-table plate 平板仪测图板
plane-table plotting 平面表格绘图
planetabler 平板仪作业员
plane-table sheet 测板图幅;平板图纸
plane-table station 平板仪测站
plane-table survey(ing) 平板(仪)测量
plane-table tachymetric survey 平板仪视距测量
plane-table tachymetry 平板(仪)视距法
plane-table topographical survey 平板仪地形测量
plane-table traverse 平板仪导线
plane table triangulation 平板三角测量;图解三角测量
plane-table triangulation 平板仪三角测量
plane-table tripod 平板仪三角架
plane-tabling 平板仪测图;平板仪测量;平板测量的;平板测绘
planet agitator 回绕式搅拌器;灰泥搅拌机
planetarium 天象仪;天文台;天文馆
planetarium type vibrator 行星式振动器
planetary 行星的
planetary aberration 行星光行差
planetary agitator 行星式搅拌器
planetary astronomy 行星天文
planetary atmosphere 行星大气
planetary axle 行星齿轮轴
planetary ball mill 行星式球磨机
planetary ball swaging 行星式钢球冲击缩管机
planetary beater 行星式搅拌器
planetary blade lift 带有行星减速器的刮刀举升机构
planetary boundary layer 行星边界层;地表边界层
planetary brake 行星式制动器
planetary cage 行星传动装置箱
planetary cam 行星凸轮
planetary camera 行星式摄影机
planetary capsule 行星研究仪器舱
planetary carrier 行星齿轮架
planetary circulation 行星环流;大气环流
planetary closer 筐篮式合缆机
planetary coiler 行星式圈条器
planetary compulsory mixer 行星式搅拌器;行星式拌和机
planetary configurations 行星相互位置图
planetary cooler 行星式冷却机;多筒冷却器;多筒冷却机
planetary drive 行星传动;行星齿轮传动
planetary drive axle 行星齿轮驱动轴
planetary drive set 行星齿轮驱动装置
planetary drive system 行星齿轮驱动系统
planetary ecosystem 行星生态系统
planetary electrons 行星式电子
planetary engineering 行星工程
planetary environment 行星环境
planetary final drive 行星齿轮终传动
planetary final reduction 行星齿轮终减速
planetary frame 行星框架
planetary gear 行星齿轮
planetary gear apparatus 行星齿轮装置
planetary gear carrier 行星齿轮架
planetary geared drum 行星齿轮传动滚筒
planetary geared hoist 行星齿轮式吊车;行星齿轮传动提升机
planetary gearing 行星齿轮传动
planetary gear mechanism 行星齿轮传动机构
planetary gear mixer 行星齿轮混合器
planetary gear ratio 行星齿轮传动比
planetary gear set 行星齿轮组
planetary gear train 行星(齿)轮系
planetary gear transmission 行星齿轮传动
planetary gear type inverter 行星齿轮型变速器
planetary gear type transmission 行星齿轮变速箱
planetary geology 行星地质学;星质学
planetary geometry 行星几何学
planetary gravitational field 行星重力场
planetary high frequency vibrator 行星式高频振动器
planetary hoist 神仙葫芦;差动绞辘
planetary hot mill 行星式热轧机
planetary hub gear 行星齿轮轮毂
planetary information 行星信息
planetary jet 行星急流
planetary landform 星体地貌
planetary landing 行星上登陆
planetary light deviation 行星光行差
planetary mapping 行星测图
planetary mill 行星式碾磨机;行星式轧机;行星式磨机;多筒磨机
planetary milling machine 行星式铣床
planetary mixer 行星式搅拌机;行星式搅拌器;行星式混合器
planetary mixing and kneading machine 行星式搅拌捏和机
planetary motion 行星运动
planetary orbit 行星轨道
planetary overrunning clutch 行星齿轮传动单向离合器
planetary paddle mixer 行星式桨状混合器
planetary-parallel gear 行星平行轴二级减速齿轮
planetary-parallel reduction gear 行星平行轴减速齿轮
planetary perturbation 行星摄动
planetary pin 行星齿轮轴
planetary pinion 行星小齿轮;行星传动齿轮
planetary power steering 行星齿轮转向装置
planetary precession 行星岁差
planetary precession of the equinoxes 行星岁差
planetary rear axle 行星齿轮后轴
planetary reducer 行星减速装置;行星减速齿轮装置;行星减速齿轮
planetary reduction 行星齿轮减速
planetary reduction final drive 行星齿轮最终传动
planetary reduction gear 行星减速器;行星减速齿轮
planetary rolling 行星轧机
planetary rolling mill 行星式轧机
planetary rotating paddles (搅拌机的)行星式旋转叶片
planetary rotation engine 行星式旋转发动机
planetary rotation machine 行星旋转机
planetary sampler 多筒采样器
planetary screw extruder 行星式螺杆挤压机
planetary seismology 行星地震学
planetary space 行星空间
planetary space navigational system 行星空间导航系统
planetary spindle 行星心轴
planetary steer(ing) 行星齿轮转向
planetary steering axle 行星齿轮转向轴
planetary steering device 行星齿轮转向机构
planetary steering gear 行星转向器
planetary stirrer 行星式搅拌器;行星式拌和机
planetary stirring machine 行星式搅拌器;行星式搅拌机
planetary strander 筐篮式捻股机
planetary stranding machine 行星式钢丝绳捻股机
planetary stream 行星流星雨
planetary surface analysis 行星表面分析
planetary system 行星系
planetary system of winds 行星风系
planetary tectono-geochemistry 宇宙构造地球化学
planetary temperature 行星温度
planetary thread miller 行星式螺纹铣床
planetary thread milling machine 行星式螺纹铣床
planetary timing gear 行星齿轮定时装置
planetary transmission 行星式传动装置;行星传动;行星变速器
planetary traversing mechanism 行星齿轮转向机构
planetary tube 行星式磨筒;行星式冷却筒
planetary tube coldrolling mill 行星式冷轧管机
planetary tuning gear 行星齿轮分配机构
planetary type 行星式
planetary type concrete mixer 行星式混凝土搅拌机
planetary type grinding machine 行星式研磨机
planetary type mixer 行星式混合机
planetary type winding machine 行星式缠绕机;立式缠绕机
planetary upperatmosphere 行星高层大气
planetary vorticity effect 行星的涡度效应
planetary wave 行星波;罗斯比波;长波
planetary wheel 行星轮
planetary wheel carrier 行星齿轮轴架
planetary wind system 行星风系
planet cage 行星齿轮机构壳体;行星传动箱;行星齿轮箱
planet carrier 行星齿轮支座
planet differential 行星齿轮
planet differential gear 行星差动齿轮
planet escape 脱离行星引力区
planetesimal hypothesis 行星尘假说
planet-fall 着陆
planet gear 行星齿轮系;行星齿轮
planet gear carrier 行星齿轮架
planet gearing 行星齿轮;行星式齿轮传动
planet gear speeder 行星增速器
planet gear speed reducer 行星齿轮减速机
plane tile 冠瓦;平(板)瓦
planet-landing 着陆
planet location diagram 行星位置图
planetocentric 行星中心的
planetocentric coordinates 行星中心坐标
planetocentric coordinate system 行星中心坐标系
planetographic 行星面的
planetographic coordinates 行星表面坐标
planetography 行星表面学
planetoid 小行星
plane tomography 平面断层照相法
plane to plane communication 飞机内通信
planet pinion 行星小齿轮;行星传动齿轮
planet pinion carrier 行星小齿轮架
planet pinion gear 行星小齿轮
planet power drive 行星齿轮动力驱动
planetrain 飞行列车(高速磁浮车)
plane transpiration 植物蒸腾
plane tree 法国梧桐;悬铃木
plane triangle 平面三角形
plane trigonometry 平面三角学
plane truss 平面桁架
plane truss works 平面桁架工程
planet sensor 行星位置测量仪
planet surface exploration 行星表面探测
planet tracker 行星跟踪仪
planet type closer 筐篮式捻股机
planet-type grinding mill 行星式粉碎机
planet type strander 筐篮式合绳机
plane untwining 平面疏解【铁】
plane upsetting 平面镦粗
plane valving surface 平面配流面;配流盘表面
plane vibration 平面振动
plane wall 平直墙
plane wave 面波;平面波;二维波
plane wave Born approximation 平面波玻恩近似
plane wave generator 平面波发生器
plane wave initiation 平面波起爆
plane wave interaction 平面波相互作用;平面波相互干扰
plane wave quantum 平面波量子
plane wave reciprocity calibration 平面波倒易校准
plane wave reflection coefficient 平面波反射系数
plane wave resolution 平面波的分解
plane wave scattering coefficient 平面波散射系数
plane wave signal 平面波信号
plane wave solution 平面波解
plane wave system 平面波系
plane wood 刨身
plan figure of contour 等值线平面图
plan figure of contour of self-potential method 自然电位等值线平面图
plan figure of electric(al) parameter 电性参数平面图

plan file 图夹
plan flood 计划洪水
planform 轮廓;外形;平面图
plan-form geometry 平面形状;平面形态
planform of river 河流的平面形态
plan for the development of science and technology 科技发展规划
plan for the formation of freight trains 铁路货物列车编组计划
plan for the purchase of export commodities 出口商品收购计划
plan geometry 平面几何(学)
planicom 解析测图仪
planification 整平术
planiform 平面的
planigraph 缩放图器
planigraphy 平面断层照相法;缩放图法;断面X射线照相术;层折X射线照相法
planimegraph 面积比例规;缩图器;比例规
planimeter 积分仪;木面积仪;面积仪;面积计;求面积仪;求积仪;平面仪;平面求积仪;测面仪
planimeter constant 求积常数
planimeter drum 求积仪侧轮
planimetering 平面测量
planimetric(al) 平面图的;平面的;平面测量的
planimetric(al) accuracy 平面控制精度
planimetric(al) adjustment 平面平差
planimetric(al) air survey 平面航空摄影测量
planimetric(al) arm 面积仪方向导杆;求积仪描迹臂;平面导杆
planimetric(al) base 面积仪平面控制基线;平面控制基线
planimetric(al) base map 平面底图
planimetric(al) compilation 地物测绘
planimetric(al) control point 平面控制点
planimetric(al) coordinates 平面坐标
planimetric(al) detail 平面碎部点
planimetric(al) details file 平面文件
planimetric(al) digitizing 平面位置数字化
planimetric(al) displacement 平面位移
planimetric(al) element 地物要素
planimetric(al) error 平面位置误差
planimetric(al) feature 地物;地面形态;地貌
planimetric(al) line 轮廓线;平面线
planimetric(al) map 平面地图;平面布置;无等高线地图
planimetric(al) map of condensate gas pool 凝析气藏平面图
planimetric(al) map of gas pool 气藏平面图
planimetric(al) map of oil and gas pool 油气藏平面图
planimetric(al) map of oil pool 油藏平面图
planimetric(al) mapping 平面测量图
planimetric(al) method 平面测量法
planimetric(al) motion 平面运动
planimetric(al) movement 平面运动
planimetric(al) plate 测图板
planimetric(al) point 平面控制点;地物点
planimetric(al) position 平面位置
planimetric(al) reconnaissance 平面位置勘测
planimetric(al) rectangular coordinates 平面直角坐标
planimetric(al) rectangular coordinate system 平面直角坐标系
planimetric(al) representation 平面位置表示法
planimetric(al) scale 平面比例(尺)
planimetric(al) sketch 平面略图
planimetric(al) survey 平面测量;平面图测量
planimetry 求积法;平面几何(学);面积法;平面测量法;测面法;平面测量学
planing 材料表面整平;弄平;创(削)
planing and jointing machine 刨平和接合两用机
planing and mo(u)lding machine 刨平和造型两用机
planing blade 刨刀
planing capacity 刨削容量
planing control sheet 设计控制表
planing cutting speed 刨削速度
planing jig 刨床夹紧装置
planing knife 刨版刀
planing length 刨程
planing machine 地板刨光机;刨床
planing machine platen 刨床工作台
planing machine table 刨床工作台

planing mill 刨削加工厂;刨削车间
planing mill products 木工厂制品;刨光木制品(如天花板、墙板、地板条等)
planing of stee plate 钢板刨边
planing operation 刨削加工;刨削操作
planing skip 漏刨
planing slide bearing 刨台导轨
planing speed (滑艇的)滑行速度
planing tool 刨刀
planing tool carriage 刨床刀架
planing vertical surface 垂直面刨削
planing work 刨工工作
planish 磨平;刨平;矸光;碾平
planished sheet 精轧薄板;平整薄板
planisher 打平锤;精轧机座;打平器
planish extra-bright 特亮抛光
planishing 精轧;打平;碾平
planishing hammer 矸光铁;展平锤;敲平锤;打平锤
planishing knife 手拉平口刨
planishing pass 平整道次
planishing roll 精轧辊
planishing stand 精轧机座
planisphere 平面球体图;球体投影图;平面球形图;天球平面图
planisphere map 半球图
planispheric astrolabe 寻星仪
planispiral 平旋的壳
planispiral shell 平旋壳
planizer 平面扫描头
plank 木板;厚木板;厚木;厚板;软质木板;条板;板
plankage 垫板租费
plank and beam framing 板梁式框架
plank base 木板基层
plank bed 木板床;板床
plank bridge 木板桥
plank buttress 板状干基;板状板根
plank carrier 木板搬运器
plank casing 厚木板(门窗的)边框
plank check dam 木板拦砂坝;木板谷坊
plank covering 目标覆盖
plank crossing 木板交叉口;木板交叉道
plank door 厚木板门
plank drag 木板路刮;木刮板;木板耢;木板刮器;木板刮
planker 拖板
plank facing 木面板
plank floor cover(ing) 毛地板覆盖层
plank floored bridge 木板桥
plank floor(ing) 木桥面;木地板;毛地板;木底板;木板假顶;木板地面
plank floor planer 地板刨平机
plank flume 木渡槽;木板水槽;木板渡槽;木板槽
plank foundation platform (放在澡堂、洗衣间地板上和门口的)木条踏脚;(放在澡堂、洗衣间地板上和门口的)木条垫
plank frame 木板框架;木门框;木框(门窗);板钉框架
plank house 厚木板房
plank hurdle works 木板栏栅
planking 用木板支撑;装镶木板工作;木壳板;铺板;地板;船壳板;板材;安装木板
planking and strutting 立柱钉板(临时木构);挖土支撑;方方支撑;板架支撑
planking and stuffing 板架支撑
planking shutter 挡土板
planking stuff 镶板材料
planking to prevent rock-fall 护坡插板
plank in prestressed clay 预应力黏土板
plank lagging 木背板
plank lath(ing) 木板条
plank nail 毛地板钉;木板钉
plank of a wagon side 敞车侧墙板
plank-on-edge floor 直接铺设楼面;实心木地板;密铺排木搁栅地板;立板楼盖
plank partition 厚板隔墙
plank pavement 木板路面
plank pile 木板桩;厚板桩
plank piling 打木板桩
plank platform 木板平台
plank road 木板路;厚板道
plank roadway 木板车道
plank roof 木屋顶;板屋顶;木条屋面
plank roof truss 木屋架
plank run(way) (用于摊铺混凝土的)木跳板

plank sheathing 木望板;木模(板);木护壁;木板护壁;厚板壁
plank sheer 游艇甲板流水沟;舷缘材;甲板边板;舷缘板;船缘方木
plank sheet pile 薄板桩
plank split 板裂缝
plankter 浮游生物有机体
plankton 浮游微生物
plankton association 浮游生物组合
plankton bloom 浮游生物的大量繁殖
plankton centrifuge 浮游生物离心机
plankton community 浮游生物群落
plankton counting chamber 浮游生物计数框
plankton-eating 食浮游生物的
plankton equivalent 浮游生物当量
plankton feeding fish 浮游生物食性鱼
plankton haul 浮游生物拖网采集
plankton hyponeuston 浮游性水面下漂浮生物
planktonic animal 浮游动物
planktonic arthropod 浮游节肢动物
planktonic crustacean 浮游甲壳类
planktonic rotifer 浮游轮虫
plankton inorganism 浮游生物
plankton microscope 浮游生物显微镜
plankton multiplier 浮游生物增殖因素
plankton net 浮游生物网
plankton organism 漂游生物;漂浮生物
plankton sampler 浮游生物采样器
plankton type community 浮游生物型群落
planktosphaeria gelatinosa 浮球藻
plank track 木板走廊;木板路
plank trim 木板框子;厚木板门窗框
plank truss 木屋架;木桁架;木桁架
plank truss bridge 木桁架桥
plank tubbing 木板井壁
plank-type grating 板式格栅
plank way 栈道
plank wood pile 厚木板桩
plan layout 平面布置
plan location 平面位置
plan management 计划管理
plan miller 平面铣床
plan milling 平面铣
plan milling machine 平面铣床
plan model 规划模型
plannar structure 成面构造
planned 计划中
planned adjustment 计划调节
planned allocation of commodities 计划调拨商品
planned annual profit 年度计划利润
planned balance 计划差额;事先平衡
planned budget 事先安排的预算
planned commodity economy 有计划商品经济;计划商品经济
planned cost 计划支出;计划成本
planned cost calculation 计划成本计算
planned cost of work completed 完成工作计划成本
planned cost variance of materials 材料计划成本差异
planned coverage 计划成因区
planned cutting 计划采伐
planned depth 计划水深;设计水深
planned development 计划开发;计划发展
planned dimensions of channel 计划航道尺度
planned distribution of passenger tickets 票额分配计划
planned district 计划地区
planned economy 计划经济
planned expenditures 计划支出
planned figure 计划数字
planned fleet 规划船队
planned freight traffic 计划内运输
planned investment 计划投资
planned limit 计划限额
planned loan 计划贷款
planned maintenance 计划检修;计划保养;计划养护;计划维修
planned management 计划管理
planned norm 计划定额
planned outage 计划停机
planned outlay 计划支出
planned output 计划生产量
planned overlay (程序的)计划覆盖

planned overlay structure 计划覆盖结构
planned parenthood 计划生育
planned payment 计划付款
planned performance 预期成效
planned performance rate of financial expenditures 财务支出计划完成率
planned performance rate of financial income 财务收入计划完成率
planned performance rate of investments 投资计划完成率
planned performance rate of labo(u)r wages 劳动工资计划完成率
planned performance rate of ore reserves 储量计划完成率
planned performance rate of project work 项目工作计划完成率
planned period 有计划的周期；计划时间；计划期间；计划期限
planned population 规划人口
planned preventive repair 预期检修
planned price 计划价格
planned production 计划生产
planned profit 计划利润
planned project 规划项目
planned purchase and supply 统购统销
planned quantity for water supply 计划供水量
planned reconnaissance 计划勘测
planned regulation 计划调节
planned residential development 规划住宅发展区
planned schedule 计划进度表；计划进度安排
planned section 设计断面
planned ship-type 规划船型
planned ship-width 规划船宽
planned slope 植被边坡
planned-stage construction 按计划步骤施工；按计划施工
planned statement of standardized current capital 定额流动资金计划表
planned subsidence 计划沉降
planned supply 计划供应
planned surface 平整地面
planned target 控制数字；计划指标
planned times of turnover of current capital 流动资金计划周转次数
planned timetable 计划时间表
planned timing 计划定时；计划调速
planned tooth 刨成齿
planned transport 计划运输
planned unit development 有规划的地段建设；有规划的地段发展；规划发展单元
planned unit price 计划单价
planned unit residential development 规划居住发展单元
planned upkeep 计划维修
planned urban development 规划的城区发展；市区发展规划
planned value 计划价值
planner 规划工作者；规划者；规划师；规划人员；计划员；计划人员
planning 平面布置；整平；计划；滑行；平面设计；编制计划；刨平；刨工
planning activity 计划行动
planning administration 规划机构
planning advance 前期付款；计划预付拨款
planning and analysis for production 生产计划与分析
planning and controlling department 计划调控部
planning and decision accounting 计划和决策会计
planning and design office 规划设计办公室
planning and design work 规划设计工作
planning and estimating 设计与估计
planning and preparation 筹划
planning and scheduling 计划调度
planning area 规划区
planning arrangement 计划安排
planning basis 规划的基础
planning bench 刨台
planning blade 刨刀；刮刀(平地机)
planning board 规划局；规划委员会；计划委员会
planning boat 滑行船
planning budget 计划预算
planning bureau 规划机构
planning by objective 按目标制订规划
planning card index 设计卡片索引

planning center line 规划中线
planning chart 规划海图；计划图；计划航线图；计划表
planning commission 规划委员会；计划委员会
planning competition 计划竞赛；规划竞赛
planning concept 规划构想
planning consent 设计获准
planning control 计划控制；计划管理
planning cost 计划成本
planning criteron 规划标准
planning cutter 刨刀
planning data 规划数据；计划数据
planning department 计划科
planning directive 规划指向；规划指令
planning documents 规划文件；计划文件
planning economy 计划经济
planning estimates 计划概算
planning exploration 规划性勘察
planning flexibility 计划适应性；计划灵活性；计划机动性
planning forecast 规划预测
planning for incidents 事故的应变计划
planning for one-way street 单向交通街道规则
planning for the handicapped 残疾人设计
planning function 规划功能；计划职能
planning goal 规划目标
planning graph 规划图
planning grid 房屋布置图；模数比尺方格；平面草图设计网络；规划网络；模数规格；平面设计网络；设计网络
planning group 规划分类；计划组；规划组
planning horizon 规划远景
planning horizontal surface 水平刨削面
planning horizon theorem 计划层次原理
planning idea 规划设想
planning index 计划指标
planning information 设计资料
planning investigation 规划调查
planning item 规划项目
planning law 规划法
planning legislation 计划法规；规划法规；规划法令
planning level 规划水平
planning machine 平整机；刨平机
planning machine operator 刨工
planning map 规划图；规划地图
planning method 规划方法
planning method for parts from other designs 借用件计划方法
planning method for standard parts 标准件计划方法
planning mill 木材加工场；刨木厂
planning mine scale 设计生产规模
planning mining depth 设计开采深度
planning nature 计划职能
planning net(work) 设计网络；计划网络
planning of blasting procedure 爆破设计
planning of building 房屋设计
planning of central district 中心区划区
planning of city centre 市中心规划
planning of comprehensive water pollution control 水污染综合控制规划；水污染综合防治规划
planning of distribution of productive forces 生产力布局规划
planning of dry dock 干坞(的)规划
planning office 规划局；规划办公室；规划机构
planning of grouting 灌浆计划
planning of highway location 路线规划
planning of parks and greens 公园及绿化规划
planning of port distribution 港口布局规划
planning of project finance 项目资金的规划；项目资金的制订
planning of residential area 居住区规划；住宅区规划
planning of soil and water conservation 水土保持规划
planning of statistic(al) experiments 统计试验方案
planning of trees and shrubs 树种规划
planning of water logging control 治理规划
planning operation 规划实施
planning party 规划组
planning permission 规划许可证；计划允许
planning perspective 规划远景

planning phase 规划阶段；设计阶段
planning principle 规划原则
planning production scale 设计生产规模
planning programming and budgeting system 计划程序和预算系统
planning-programming-budgeting-accounting system 规划—设计—预算—核算系统
planning-programming-budgeting system 编制预算计划系统；规划及设计—预算系统；规划编程预算系统；设计、计划和预算体系；计划、规划和预算综合编制法
planning proposal 规划建议
planning region 规划地区；规划区
planning reserves existence 设计储量失实
planning restriction 规划限制；计划限制
planning right of cost 财务支出计划权
planning section chief 计划部门主管
planning shipform 规划船型
planning space 规划区
planning stage 规划阶段
planning standard 规划标准
planning standard specification 规划标准(技术)规范
planning summary sheets 计划一览表
planning supervisor 计划主管人
planning survey 规划调查；规划测量；设计测量
planning team 规划组
planning telpherage standards 拟建架空索道规格
planning theory 规划理论
planning time horizon 计划工作时间长短
planning to anticipated requirements 规划预期要求
planning transport pipeline meterage 拟建输送管道规格
planning unit 规划单位
planning work 规划工作；计划工作；刨工
planning zone 规划区
plano-aspheric corrector 施密特校正镜
planoblast 浮游水母
plan oblique drawing 斜方投影图
planoclastic rock 等粒碎屑岩
plano-concave 平凹(的)
plano-concave lens 平凹透镜；平凹镜片
plano-concave plate 平凹板
plano-conformity 平行整合
planoconic 平锥形的
plano-convex 平凸的；平凸
plano-convex lens 平凸镜片；平凸透镜
plano-convex spotlight 平凸型聚光灯
plan of building structure 建筑结构平面图
plan of capital construction 基本建设计划
plan of column grid 柱网平面
plan of diaphragm wall 地下连续墙平面图
plan of distribution 分配计划
plan of economic and use zones 经济和经营区平面图
plan of empty wagons after unloading and transshipment 卸空车计划
plan of equipment arrangement 设备平面布置图
plan of erosion 侵蚀平原
plan of flow filaments 水流平面图
plan of foundation construction 建筑基础平面图
plan of ground movement 地面移动图
plan of hydrocarbon detection 油气检测平面图
plan of land utilization 土地规划
plan of layered seismic facies 分层地震相平面图
plan of loaded and empty wagons delivered and received at junction station 分界站货车出入计划
plan of major repair 大修计划
plan of management 施业计划
plan of material requisition 材料申请计划
plan of mining areas 矿区平面图
plan of necessities 各项需用计划
plan of number of empty serviceable wagons to be kept 空车保有量
plan of number of goods trains 货物列车数计划
plan of number of loaded wagons for delivery at junction station to be kept 移交重车保有量计划
plan of number of local wagons for unloading to be kept 管内工作车保有量计划
plan of number of wagons not for traffic use 非运用车计划

plan of number of wagons on hand 现在车计划
plan of observation points 观测点位置图
plan of operation 营运计划;业务规划;作业计划
plan of optimal groundwater mining 地下水最佳开采方案
plan of power transmission and telecommunication 输电及通信线路图
plan of production and finance 生产财务计划
plan of profit delivering 利润上缴计划
plan of provision of holes 预留孔平面图
plan of qualitative indices of utilization of wagons 货车运用质量指标计划
plan of quantitative indices of wagon utilization 货车运用数量指标计划
plan of redemption 分期偿还计划
plan of road and water system 道路水系平面图
plan of sandstone percentage 砂岩百分比平面图
plan of schedule 进度计划
plan of seismic facies 地震相平面图
plan of seismic T_0 isochrone 地震等 T_0 平面图
plan of sewage treatment plants 污水厂平面布置
plan of sewerage system 排水管道布置图;下水道平面图
plan of site 总平面图;总布置图
plan of steam and gas piping 动力管网图
plan of streets 街区设计图
plan of subsidence area 沉降区平面图
plan of supply 供应计划
plan of technical indices for freight traffic 运输工作技术计划
plan of technology progress 技术进步计划
plan of terminal yard 枢纽调车场平面图
plan of the accident site 事故发生地点平面图
plan of the mixing materials 拌和料配比设计
plan of the overburden 覆盖层平面图
plan of through trains and wagon-group loadings organized at loading points 直达列车和成组装车计划
plan of transportation system 交通运输图
plan of utilizing foreign capital 利用外资计划
plan of ventilation adit 通风洞平面
plan of wagons utilized for loading 使用车计划
plan of water and drainage piping 给排水管网图
plan of weave 意匠图
plan of wiring 线路图;线路图
plan of work in goods traffic 工作量计划
plan of works 计划(任务)书
planograph 用平版印
planographic(al) printing 平版印刷
planography 平版印刷术
planogrinder 龙门磨床
plano-guidance 平面导向台
plano-guidance desk 平面导向
planoid 超平面
planoid gear 偏轴面齿轮
planology 计划学
planometer 测绘仪;面积仪;平面求积仪;平面规;测平仪
planometric 平面图像投影
planometry 测平法
planomiller 刨式铣床
planomilling machine 刨式铣床
planomural 平壁
plano-parallel 平面平行的
planophyre 层斑岩
planophyric 层斑状
plan oriented 面向计划的
plan-oriented market economy 适应市场经济的计划
planosol 黏磐土;湿草原土
planosphere 平面球形图
plano-type grinder 门式平面磨床
plano-type surface grinder 刨式平面磨床
plan-parallel structure 平行构造
plan period 规划期限
plan pollution source 面污染源
plan position 平面位置
plan position accuracy 平面位置精度
plan position antenna 平面位置天线
plan position display 平面位置显示器
plan position fixing 平面位置固定
plan position indication 平面位置显示
plan position indication approach 平面位置指示进场

plan position indication method 平面位置显示法
plan position indication prediction 平面位置显示器预测
plan position indication sonar 平面位置指示声呐
plan position indicator 平面位置显示器
plan position indicator radar 平面位置指示雷达
plan position indicator repeater 平面位置显示器中继站;外接平面位置显示器
plan position indicator scan 平面位置显示扫描
plan position indicator scope 平面位置显示器;平面位置雷达显示器
plan position information 平面位置信息
plan position map 平面位置图
plan position measurement 平面位置测量
plan position radar indicator 平面位置雷达指示器
plan position repeater 平面位置遥示器
plan position set 平面位置显示器
plan problem 平面问题
plan-profile sheet 平面—剖面图;平面—侧面图
plan radius 平面半径
plan range 平面距离
plans 设计图
plan schedule 进度计划
plan section 平原段
plan serves suggestions 参考性计划
plans for making up losses 亏损弥补方案
plans for river basins 流域规划
plans for the development of high technology industries 高技术产业发展规划
plan shape 底层平面图模型
plan sheet 平面图
plansifter 平筛;平面筛
plan sink 平面吸收源
plan sketch 设计图;计划图;简图;草图;点位略图;平面(示意)图
plan symbol 面状地物符号
plant 植物;栽种;公用通信事业全部设施;工程设备;工厂;机械设备;设备;插置设备;插置
plant accessories 电厂附属设备
plant acid 植物酸
Plantagenet style 英国金雀花王朝式(建筑)
plantago mucilage 车前胶
Plantain 车前属【植】
plant air 工艺空气;工厂用压缩空气
plant airfield 工厂机场
plant alkaloid 植物碱
plant analysis room 车间分析室
plant and equipment budget 工厂设备预算
plant and equipment investment 工厂和设备投资
plant and transport manager 设备与运输主任;设备与运输经理
plant appraisal 厂房估价;厂场估价
plant appreciation 厂场资产升值
plant area 装置占地
plant area exchange 厂区电话站
plant area trunk exchange 厂区长途电话站
plant arrangement 车间布置
plant ash 草木灰
plant assets 厂场资产
plant assets appreciation 工厂资产增值
plant association 植物群丛
plantation 种植园;种植;植树造林;人工林
plantation density 定植密度
plantation farm 种植场
plantation house 种植园房屋
plantation of rubber 橡胶种植园
plantation planting 农场绿化;农场种植
plantation plowing 垦殖耕作
plantation pole 长度单位
plantation rubber 天然橡胶;种植园橡胶
plantation sods 种植园中草皮
plant availability 设备利用率
plant baller 营养土钵压制器
plant balling machine 营养土钵压制机
plant battery 厂用蓄电池
plant bed 植床;苗床
plant benthon 水底植物
plant binder 树枝扎捆装置
plant biochemistry 植物生物化学
plant bowl 蔬菜碟;植物皮;植物壳
plant box 苗木箱
Plant Breeding and Acclimatization Institute 植物育种与驯化研究所
plant-breeding station 植物育种站

plant building 厂房
plant bulk 装置整体;装置尺度
plant by-product 植物副产品
plant canopy 林冠
plant canopy community 种植地界
plant capacity 装置容量;装配容量;工厂生产能力;工厂生产量;工厂设备容量;工厂设备能力;工厂能量;设备容量;设备能力;设备功率;电站容量;电厂容量
plant capacity utilization 设备能力利用
plant carpet 植物地毯
plant cavitation factor 装置汽蚀系数
plant chemotaxonomy 植物生化分类学
plant chromatograph 工业色谱仪
plant climate 植物气候
plant-climate interaction 植物与气候的相互作用
plant climatology 植物气候学
plant colo(u)ring matters 植物色素
plant column 安装支柱;竖立柱子
plant community 植物群落
plant compartment 设备隔间
plant condensate 天然气加工厂凝析油
plant condition 生产条件
plant construction 工厂施工;工厂建筑
plant constructional material 设备结构材料
plant consumption 植物水消耗;植物耗水量;厂用电消耗
plant contractor 成套设备承包人
plant control 设备控制
plant cooled clinker 工厂冷却的熟料
plant cost 直接建厂费用(包括工程设计费和建筑施工费)
plant cost estimating 设备费预算
plant cover 植物覆盖;植被
plant culture 植物栽培
plant damage 植物损害
plant damage map 植物损害图
plant decoration 工厂装饰
plant dedicated to project 工程专用设备
plant density 植物密度
plant depot 器材堆场;承包人仓库;设备修配工场
plant depot repair 工厂仓库修理
plant depreciation 机械折旧
plant design 装置设计;工厂设计;车间设计
plant discharge 电站泄流量;电站发电流量
plant disease 植物病害
plant disease control principle 植物病害防治原则
plant diseases and insect pests 植物病虫害
plant disembarkation 植物上陆
plant distribution 植物分布
plant drainage 工厂排水
plant dryer 植物标本干燥器
plant dust 植物性粉尘
plant ecology 植物生态学
planted border 绿化路边
planted mo(u)lding 贴附线脚;贴饰;安装线脚【建】
planted roof 种植屋面
planted slope 有植被的边坡;绿化边坡
planted stop 镶嵌门挡;贴附式门挡
plant efficiency 装置效率;工厂效率
plant efficiency rate 施工机械使用效率
plant effluent 工厂污水;工厂废水
plant electrical consumption 厂用电(力)
plant emergency organization 工厂防险队
plant engineer 运行工程师;工厂工程师;设备工程师
plant engineering 工厂设备配套工程;工厂工程(学);设备运转技术;设备安装使用工程;设备安装工程
plant engineering department 工厂工艺科;工厂设备科
plant engineering division 工厂设备安装部门
plantenolic acid 车前烯醇酸
plant enrichment 设备施用与装配技术
plant environment 植物环境
plant equipment 固定设备;工业设备;厂场设备
plant equipment package 工厂成套设备
plant equipment renewal 厂商设备更新
planter 安装人;播种机;容纳埋入涵管的槽;栽值机;花盆;水下树墩
planter lever 播种装置接合杆
Plante-type plate 铅极板
plantex 描图布
plant executive 工厂负责人

plant expansion 工厂扩建
plant extract 植物浸出液;植物提取物
plant facility 工厂设施
plant factor 总设备利用率;工厂设备利用率;设备使用系数;设备利用系数;电站容量因数;发电利用率;发电厂利用率
plant failure 装置故障
plant feed conveyor system 工厂供料输送系统
plant feed unit 栽植装置
plant fiber 植物纤维
plant fiber filtering medium 植物纤维滤材
plant fire proection organization 工厂消防组织
plant food 植物营养;植物需要的氧分
plant for construction 施工设备;施工机具
plant forest 营造
plant form 工厂模型
plant formation 植物群系
plant for processing grain feeding 谷物饲料加工工厂
plant for smoke adsorption 吸收烟灰设备
plant generating capacity 电站发电容量
plant generating set 厂用发电机组
plant generator room 厂用发电机室
plant geography 植物地理(学)
plant growing 植物栽培
plant growing quite tall 长的高大的植株
plant hand 工厂操作人员
plant head works 厂头构筑物
plant health 植物检疫
plant heat rate 全厂热耗率
plant-hire rate 设备租用率
plant hole digger 种植穴挖掘机
plant hunter 植物标本采集者
plant indicator 指示植物
plant industry 设备安装工业
planting 铺基础垫层;打底;植树;绿化
planting-and-cultivation attachment 栽植与中耕附加装置
planting and micro-climate 种植与小气候
planting and noise suppression 种植与降低噪声
planting apparatus 栽植器
planting bar 栽植锹
planting bed 种植池
planting belt 植树带;栽植带
planting borer 栽植挖穴机;栽植挖坑机
planting box 植树箱;种植箱;基础底层涵管
planting brick 营养砖;栽植砖
planting bush 植灌木
planting climber 植爬蔓植物
planting composition 栽植配置
planting cord 栽植绳
planting depth 栽种深度
planting design 绿化设计
planting distance 植距;栽植距离
planting easement 路边植树权
planting flowers and plants 植花草
planting for roadside protection 路旁防护栽植
planting furrow 栽植沟
planting growing 植物栽培
planting hatchet 栽植锄
planting hoe 植树镐
planting hole 栽植穴
planting hole machine 栽植挖穴机
planting in residential area 居民区绿化
planting in rows 列植;成行列植
planting in single-row 单行栽植
planting in triangle 三角形植树
planting-line 植畦
planting machine 栽植机
planting material 栽培材料
planting mattock 栽植镐;栽植锄
planting of layers 压条法
planting of plants 工厂绿化
planting of slip 插木造林
planting out 造林;栽植;荒山造林
planting-peg 植棒
planting percent 成苗率
planting pit 定植穴
planting plan 植树计划
planting point 植树点;栽植点
planting scheme 种植计划
planting screen 绿篱
planting season 种植期
planting series 造林区划

planting site 栽植地
planting sleeves 植树套
planting sods 种植草皮
planting soil 种植土;果园土
planting standard 绿化标准
planting stick 栽植标桩
planting stock 栽植材料;定植苗
planting strip 植树带
planting system 栽植方式
planting tree 植树木
planting tree when giving birth to a baby 产婴植树
planting trench 栽植沟
planting tube 植树套
planting turf 种植草皮;植草皮
planting-wedge 植楔
planting with ball 带土栽植
planting work 种植工程
plant in rows 条植
plant investment 工厂投资
plant investment turnover 工厂投资周转率
plant irrigation with artificial light 配有人造光的植物灌溉
plant kingdom 植物界
plant label 植物名标牌;植物名牌
plant lag 调节对象滞后
plant layout 工艺布置;工厂设备布置;工厂规模;工厂布置;设备布置;设备布局;电站布置;车间布置
plant layout by products 以产品为对象的车间布局
plant layout department 工厂设计部门
plant layout drawings 工厂设计图纸
plant ledger 工厂资产分类账;厂场设备分类账
plant ledger card 工厂资产分类账卡
plantlet 小苗
plant life 电厂寿命;电厂使用年限
plantling 幼植物
plant load demand 工厂用电需求;电站负荷需求;电厂厂用电需求;厂用电负荷
plant load factor 设备负载因数;电站容量因数
plant location 电站位置;厂址;厂地
plant location model 工厂位置模型
plant location programming 厂地规划
plant loss 电站损失
plant machining shop-order 工厂机加工任务单
plant maintenance 工厂维持费
plant maintenance expenses 厂场维持费
plant maintenance work 机械设备的维修工作;机械设备的保养工作
plant management department 工厂管理处
plant material 植物物质;植物体
plant mix 工厂拌和;厂拌混合料;厂拌和物;工厂搅拌
plant-mix base 厂拌混合料底层
plant-mixed 工厂拌和的;厂拌的;工厂搅拌的
plant-mixed asphaltic macadam 厂拌(石油)沥青碎石
plant mixed bituminous concrete 厂拌沥青混凝土
plant-mixed bituminous macadam 厂拌沥青碎石(路面)
plant-mixed bituminous mix(ture) 厂拌沥青混合料
plant-mixed bituminous pavement 厂拌沥青路面材料
plant mixed bituminous surface 厂拌沥青路面
plant mixed bituminous surfacing 厂拌沥青路面
plant-mixed concrete 厂拌混凝土;工厂搅拌的混凝土
plant-mixed macadam 厂拌碎石(路面)
plant-mixed material 厂拌材料
plant-mixed mortar 厂拌砂浆
plant-mixed process 厂拌法
plant-mixed sandy-gravel concrete aggregate 厂拌砂砾混凝土集料;厂拌砂砾混凝土骨料
plant-mixed stabilization method 厂拌稳定土的方法
plant-mixed surface treatment 厂拌混合料表面处治;厂拌混合料表面处治
plant-mixed tar macadam 厂拌柏油碎石
plant mixed truck 厂拌混凝土运送车
plant mixer 工厂拌和机;工厂拌和机
plant mixing 厂拌;集中搅拌;工厂搅拌
plant-mix(ing) material 厂拌材料
plant-mix(ing) method 厂拌法;集中搅拌法;工厂搅拌法

plant-mix(ing) process 厂拌法
plant-mix(ing) pavement 厂拌混合料路面;厂拌材料铺面
plant-mix(ing) surfacing 工厂搅拌铺面材料
plant mixture 厂拌混合料
plant monitoring 设备检查
plant morphologic(al) and mutational changes 植物形态变化及突变
plant morphology 植物形态学
plant motif 植物花纹图案
plant mulching 植物覆盖
plant network 厂用电力网;厂内网络;厂内电力网
plant nursery heating 苗圃供热
plant nutrient 植物营养
plant nutrition 植物营养
plant nutritious substance 植物营养物
planton density 浮游生物密度
plant operation 装置的操作;设备的使用
plant organism 植物有机体
plant origin 植物来源
plant ornament(al) feature 工厂装饰特色;工厂装饰特征
plant ornament(al) finish 工厂装修终饰
plant out 移植
plant overhead 工厂管理费用
plant path effect 斜射线效应
plant pathogenic fungi 植物病原真菌
plant pathogenic microorganisms 植物病原微生物
plant pattern 工厂样板;工厂模型
plant peak capacity 工厂最高生产能力
plant percent 工厂苗木成活率
plant personnel required 装置定员
plant physiological ecology 植物生理生态学
plant pigment 植物色素
plant piping 工厂管路;工厂管道系统;铺设厂内管线
plant pit 树穴
plant plankton 浮游植物
plant planning 设备计划
plant poison 植物毒
plant population 植物群体
plant pot for road use 路用栽植盆器
plant power consumption 工厂电消耗;工厂电耗量
plant practice 工厂实习
plant process piping 工厂和工艺管道安装
plant product 植物产品
plant production 工厂生产
plant protection 植物保护;植保
plant protection service 植物保护服务;植保服务
plant protection unit 植保站
plant protector 植物保护器
plant purification 植物净化
plant quarantine 植物检疫
plan tract 平原段
plant register 设备一览表;设备明细表;固定资产登记表;设备记录;设备登记
plant remain 植物残余物
plant requirement 植物需水量;工厂需用量
plant residue 植物残体
plant resin 工厂树脂
plant resin-based concrete curing agent 工厂树脂基混凝土养护剂
plant resin-based concrete curing compound 天然树脂基混凝土养护剂
plant resin-based mastic 天然树脂基的砂胶;天然树脂基的玛琋脂
plant room 工作间;机房
plant rubber 天然橡胶
plant safety rules 工厂安全条例
plant sample 厂拌样品
plant sanatation 植物检疫
plant sandblast damage 植物沙爆危害
plant-scale 工厂规模
plant-scale equipment 工厂用设备;工厂规模设备;生产用设备
plant-scale operation 工业生产
plant-scaler 工业规模(的)
plant school 苗圃;苗木培育圃
plant science 植物科学
plant scroll 种植蔓生植物
plant service 工厂设施;厂用动(力);厂用电(力)
plant setter 栽苗机
plant setting 植物栽植
plant setting machine 栽苗机

plant setting mechanism 栽植装置
plant sewer 装置下水道
plant site 厂址
plant size 工厂规模
plant-size equipment 生产型设备
plant spare 工厂备件
plant standard 全厂定额
plant station 机械站;机械设备厂;机械设备站
plant substation 厂用变电所
plant succession 植物演替
plant supervision 生产监督;生产管理
plant supply 厂用电源
plant surface microglora 植物表面的微植物群落
plant taxonomy 植物分类学
plant temperature 植物温度
plant test 工厂(技术)试验;工业设备试验
plant thermal efficiency 全厂热效率
plant-to-row method 对比法
plant track 工厂专用线
plant transpiration 植物蒸发(作用)
plant trash 植物残体
plant trees 栽树
plant trial 工厂条件下试用
plant trough 种植槽;种植盆
plant tub 栽种花木的桶;栽种花木的盆;栽植桶
plant two-rows method 双行定植法
plant unit 植物单元
plant unit control 装置控制
plant upkeep work 车间保养工作;工厂保养工作
plant use 工厂应用;生产使用
plant use factor 厂用率
plant utilization budget 工厂设备利用情况预算表
plant utilization factor 装置利用系数
plant virus 植物病毒
plant wastewater 工厂废水
plant waterway 电站输水道
plant-wide bonus plan 全厂性奖励计划
plant-wide overhead rate 全厂间接费用率
plant wilt 植物萎蔫
plant without storage 径流式水电站;径流电站;无库容电站;无库容的水电站
plan type 规划类型;计划类型
planula larva 浮浪幼虫
planum 平面
planus 平的
plan utility 平面利用效果
plan view 俯视图;平面图;顶视图
plan view of road 路线平面图
plan with contour lines 有等高线的平面图
plaque 陶瓷画屏;饰板;蚀斑;垫饼;瓷花金属板;匾;斑块
plaque-a-jour enamel 透底珐琅
plaque assay 空斑试验
plaque assay method 蚀斑检定法
plaque diffuser 平板式扩散器
plaque forming unit 空斑形成单位;蚀斑形成单位
plaque-infested port 有疫情的港口
plaque rail 板轨;装饰品搁板
plaques 试验样板
plaquette 金属印模;小匾额;小饰板;饰板
plaque-type penetrameter 斑点式贯入计;斑点式贯入度仪
plasbestos 沥青乳化液防水剂
plaschair 塑料座(钢筋定位用)
plasclip 保护层塑料垫环;塑料夹(钢筋定位用)
plash 积水坑
plashing 编结篱笆;编篱
plaskon 脲醛热固树脂
plaslip 塑料夹(钢筋定位用);保护层塑料垫环
plasm(a) 等离子气(体);等离子体;等离子区;原生质;深绿玉髓
plasma accelerator 等离子体加热器;等离子加速器
plasma-activated 等离子激活
plasma activated chemical vapo(u)r deposition 等离子体激活化学气相沉积
plasma analyzer 等离子体分析仪
plasma arc 等离子弧
plasma arc coating 等离子电弧敷层
plasma arc cutter 等离子电弧切割机
plasma arc cutting 等离子弧切割;等离子焰切割;等离子电弧切割(法)
plasma arc cutting machine 等离子电弧切割机
plasma arc furnace 等离子炉;等离子电弧炉
plasma arc melting 等离子熔炼

plasma arc remelting 等离子弧重熔
plasma arc surfacing 等离子弧堆焊;等离子弧堆焊
plasma arc welder 等离子弧焊机
plasma arc weld(ing) 等离子弧焊(接);等离子电弧焊接
plasma argon arc welding machine 等离子氩弧焊接机
plasma astrophysics 等离子天体物理学
plasma balance 等离子区平衡
plasma blowing 等离子喷吹法
plasma body 等离子体
plasma boiler 等离子体加热器
plasma boundary region 等离子边界区域
plasma cathode 等离子体阴极
plasma chemical degradation 等离子体化学降解
plasma chemical vapo(u)r deposition 等离子体化学气相沉积;等离子化学气相沉积
plasma chemistry 等离子体化学
plasma chromatography 等离子体色谱法
plasma cloud 等离子体云
plasma collapse 等离子体破坏;等离子体崩溃
plasma column 等离子弧柱
plasma confinement 等离子体约束
plasma core 等离子体核心
plasma-coupled device 等离子体耦合器件
plasma degumming machine 等离子去胶机
plasma deposited ceramics coating 等离子体沉积(金属)陶瓷覆盖层
plasma deposited cermet coating 等离子体沉积金属陶瓷覆层
plasma desmearer 等离子体去污机
plasma diagnostic instrument 等离子体诊断仪
plasma diagnostic technique 等离子体诊断技术
plasma diffusion 等离子体扩散
plasma diode 等离子体二极管
plasma direct reading spectrometer 等离子体直读光谱仪
plasma discharge 等离子区放电
plasma display 等离子体显示器;等离子体显示
plasma display panel 等离子体显示板
plasma disturbance 等离子体扰动
plasma drift 等离子体漂移
plasma dynamics 等离子体动力学
plasma eater 等离子体吞食器
plasma echo 等离子体回波
plasma ejection 等离子体抛射
plasma emission spectrometry 离子发射光谱法
plasma enforced chemical vapo(u)r deposition 等离子体增强化学气相沉积
plasma engine 等离子体发动机
plasma engineering 等离子工程
plasma etching 等离子腐蚀
plasma filament 等离子体线柱
plasma flame 等离子热熔喷镀
plasma focus 等离子体焦点
plasma frequency 等离子体频率
plasma frequency meter 等离子体频率计
plasma furnace 等离子体加热炉
plasma gas 等离子体气体
plasma gas chromatography 等离子体气相色谱法
plasmagel 胶浆
plasma generated condensation aerosol 由等离子体产生的冷凝气溶胶
plasma generator 等离子体发生器;等离子发生器;等离子发电机
plasmagram 等离子体色谱图
plasmaguide 等离子体波导管;充等离子体波导管
plasma gun 等离子体枪;等离子枪;等离子喷枪
plasma heating 等离子体加热;等离子加热
plasma heat treatment 等离子热处理
plasma incineration technique 等离子体焚烧技术
plasma injector 等离子体注入机
plasma instability 等离子体不稳定性
plasma interferometer 等离子体干涉仪
plasma ion oscillation 等离子体离子振荡
plasma jet 等离子体射流;等离子体枪;等离子体喷注;等离子体喷焰;等离子体发动机;等离子喷射;等离子流
plasma jet coating 等离子喷镀层
plasma jet cutter 等离子切割机
plasma jet cutting 等离子切割
plasma jet excitation 等离子体喷注激发
plasma laser 等离子体激光器
plasmalemma 细胞膜

plasma light source 等离子体光源
plasma linac 等离子体直线加速器
plasma loading characteristics 等离子体负载特性
plasma magnetron 等离子体磁控管
plasma mantle 等离子体幔
plasma mass spectrometry 等离子体质谱法
plasma membrane 质膜
plasma metallizing 等离子体金属喷涂
plasma MIG welding 等离子熔化极气体保护焊
plasma nozzle 等离子体气体喷嘴
plasma orifice gas 等离子气(体)
plasma oscillation 等离子体振荡
plasma oscillation analyzer 等离子体振荡分析仪
plasma panel 等离子体显示屏面;等离子面板
plasmapause 等离子体层顶
plasma physics 等离子体物理学
plasma pinch 等离子体籀缩
plasma plating 等离子弧喷镀
plasma plug 等离子火花塞
plasma polymerization 等离子体聚合;等离子聚合(作用)
plasma pressure tensor 等离子体的压力张量
plasma printer 等离子体记录仪
plasma probe 等离子体探测器;等离子区探测器
plasma propulsion 等离子体推进
plasma-pump 等离子泵
plasma quantometer 等离子体光量计
plasma radiation 等离子体辐射
plasma reaction 等离子体反应
plasma reactor 等离子体反应堆
plasma recombination 等离子体复合
plasma rocket 等离子体火箭
plasmaron 等离子体极化子
plasma scope 等离子体显示器
plasma sheath 等离子体壳层;等离子层
plasma sheet 等离子体片;等离子片
plasma simulation 等离子体模拟
plasma sintering 等离子体烧结
plasma source 等离子源
plasma source ion implantation 等离子体源离子注入
plasma spectrometry 等离子光谱法
plasmasphere 等离子体层;等离子层
plasma spray coating 等离子喷涂
plasma sprayed coating 等离子喷镀层
plasma spraying 等离子喷涂;等离子弧喷涂
plasma spray zirconia 等离子喷涂氧化锆
plasma squeezing 等离子体挤压
plasma state 等离子态
plasma stock wave 等离子体冲击波
plasma stream 等离子流
plasma stripper 等离子去胶机
plasma surfacing 等离子堆焊
plasma tail 等离子体尾
plasma technology 等离子技术
plasma thermodynamic(al) equilibrium 等离子体的热力学平衡
plasma torch 等离子体焊炬;等离子焰炬;等离子体焰炬;等离子喷枪;等离子火焰;等灯离子炬
plasma treatment 等离子处理
plasmatron 等离子体加速器;等离子体管;等离子体发生器;等离子流发生器;等离子管;等离子电焊机
plasma turbulence 等离子体湍流
plasma turbulence acceleration 等离子体湍动加速
plasma turbulence reactor 等离子体湍动反应区
plasma-type waveguide 等离子体型波导
plasma wave 等离子体波
plasma weakly non-linear theory 等离子体弱非线性理论
plasma welding 等离子焊接;等离子焊
plasma welding torch 等离子弧焊枪;等离子焊炬
plasmid 质粒
plasmid chimera 质粒嵌合体
plasmid engineering 质粒工程
plasmodium 变形体
plasmogen 原生质
plasmoid 等离子状态;等离子状态;等离子粒团
plasmon 等离子体振子;等离子激元
plasmon decay neutrino process 等离子体激元衰变中微子过程
plasmon model 等离子体激元模型
plasome 微胶粒
Plasser-Theurer 普拉塞—陶依尔

plastaleke 沥青防水合成物
plastalloy 细晶低碳钢；铁合金粉末
plastelast 塑弹性物
plaster 胶泥；抹面灰浆；抹灰工人；抹；镘灰；灰泥；灰浆；热石膏；墙粉；熟石膏；灰饰工；粉饰
plaster aggregate 灰浆用集料；灰浆用骨料；灰浆集料
plaster arch 抹灰的门卷；不加装修的门洞
plaster area 抹灰面积
plaster bandage 石膏绷带
plaster bandage in function(al) position 功能位石膏
plaster base 抹灰基层；抹灰底层；粉刷打底
plaster baseboard 抹灰底层板
plaster-base finish tile 灰泥基底饰面砖；面砖粗糙背纹；粗糙背纹面砖
plaster-base nail 钉石膏板的平头钉
plaster bat 石膏垫板
plaster bead 灰泥护角；灰泥墙角护条；抹灰护角条；护角条
plaster bed 石膏床
plaster blasting 外部装药爆破(法)
plaster block 石膏块；石膏镜盘
plasterboard 灰胶纸柏板；糊墙纸板；粉饰板；纸面石膏板；灰浆板；石膏灰泥板；石膏(大)板
plasterboard ceiling 灰泥板顶棚
plasterboard drier 石膏板干燥机
plasterboard forming station 石膏板成型站
plasterboard for secondary processing 作中间产品的石膏板
plasterboard nail 石膏板钉；纸面石膏板钉
plasterboard panel partition (wall) 灰泥配板
plasterboard sheet 石膏灰泥板片
plasterboard sheet partition wall 灰胶纸柏板(薄板)隔墙；石膏灰泥(薄)板隔墙
plasterboard stud partition 石膏板木龙骨隔断
plasterboard wall 灰胶纸柏板墙；石膏纤维板墙
plaster bond 灰泥黏结剂；抹灰墙粉黏结剂
plaster bonding agent 灰浆黏合剂
Plaster brass 普拉斯特黄铜
plaster carving 石膏雕刻
plaster cast 管型石膏夹；石膏注件；石膏型；石膏模(型)
plaster casting 石膏型铸造；石膏浇注
plaster ceiling (panel) 灰泥吊顶；抹灰顶棚；抹灰天花板
plaster ceiling (slab) 灰泥吊顶；抹灰顶棚；抹灰天花板
plaster coat 灰浆层；抹灰层
plaster coat(ing) of truss 桁架涂料层
plaster concrete 石膏混凝土
plaster contractor 粉刷承包人；抹灰承包人
plaster corner 灰泥墙角
plaster cornice 灰泥檐口线脚；抹灰挑檐
plaster cove 凹圆线脚；灰泥凹圆面；凹圆灰线脚；抹灰凹圆角；灰泥抹灰线脚
plaster crack 石膏裂缝；灰泥裂缝
plaster cupola 灰泥穹顶；灰泥圆屋顶
plaster dab 灰泥小块；灰泥团；灰泥涂抹
plaster dot 灰泥斑点
plastered 抹灰泥的；涂有灰泥的
plastered both sides 双面抹灰泥；两侧抹灰泥
plastered brickwall 混水砖墙
plastered brickwork 抹灰砖砌体；混水墙
plastered ceiling 桁架顶棚；抹灰顶棚
plastered counter ceiling 灰泥悬吊平顶(起隔声绝热作用)
plastered ship 水泥结构船
plastered wall 抹灰墙；混水墙
plastered wood-lath ceiling 灰条平顶；板条抹灰顶棚
plasterer 泥水匠；泥水工；抹灰工；粉刷工
plasterer lath hammer 抹灰板条锤
plasterer scaffold(ing) 抹灰工用脚手架
plasterer's darby 抹灰工刮尺；抹灰用镘刀
plasterer's float 抹灰抹子；抹灰用抹子
plasterer's hair 抹灰麻刀
plasterer's hammer 抹灰用锤
plasterer's hatchet 粉刷工用刮刀
plasterer's hawk 抹灰用板；抹灰工灰板；灰板
plasterer's labo(u)rer 抹灰辅助工；杂工；彩色工
plasterer's larry 粉刷工用的拌浆铲；粉刷工用的拌浆锄
plasterer's lath hammer 抹灰用板条锤

plasterer's lath(ing) 抹灰板条
plasterer's putty 抹灰工细灰膏；细石灰膏；抹灰腻子；粉刷用外层灰
plasterer's tool 泥水匠工具；抹灰工具
plasterer's trowel 粉刷工用的瓦刀；抹灰镘刀
plaster externally 外墙面抹灰
plaster figure 石膏像
plaster fillet 楞角；石膏线脚
plaster finish 粉刷饰面
plaster floor 灰抹地板
plaster for capping 压顶灰浆(圆柱试件抗压试验时,手抹面找平用)
plaster for X-ray rooms (用硫酸钡为集料的)防X射线抹灰
plaster frieze (古典建筑柱石横梁与挑梁之间的)灰泥雕带
plastergon 木丝墙板
plaster ground 抹灰定厚木条；冲筋；抹灰准木；靠尺板；靠尺；抹灰找准木条；抹灰用靠尺
plaster guard 灰浆挡板；防灰浆罩
plaster head 抹灰护角条；墙角护条
plaster impression 石膏印模
plastering 涂灰泥；抹面工作；抹浆；抹灰工作；抹灰；灰泥面；刷浆；石膏制品；粉刷
plastering action (灰浆的)造壁作用
plastering agent 胶结剂；黏结剂
plastering defect 抹灰面疵病
plastering machine 灰浆喷射器；抹灰机
plastering material 抹灰用建筑材料；抹灰材料
plastering mortar 抹面砂浆
plastering on metal lath 钢丝网抹灰
plaster ingredient 灰浆配料
plastering sand 抹灰用砂；抹灰砂
plastering tool 抹灰工具
plastering trowel 抹灰抹子；抹灰镘刀；抹灰泥刀；抹灰刀
plaster joint 灰缝
plasterjointless floor(ing) 灰泥无缝地板
plaster key 板条间灰泥
plaster kiln 石膏窑
plaster lath(ing) 抹灰(用)板条；灰板条；石膏条板
plaster lime 石膏粉刷
plaster manufacturer 灰泥厂商；灰泥制作者
plaster match 石膏胎模
plaster material 塑性材料
plaster method 抹灰方法
plaster mixer 灰泥搅拌机；石膏搅拌机；石膏浆混合器
plaster mixing and pouring machine 石膏模联合成型机
plaster mix(ture) 灰浆混合料；灰泥混合料；灰浆涂料；抹灰拌和
plaster mixture based on lime 以石灰为基底的抹灰用料
plaster model 石膏模(型)
plaster mortar mixer 灰浆砂浆搅拌机
plaster mo(u)ld 石膏模(型)
plaster mo(u)ld casting 石膏模造法
plaster-mo(u)lded cornice 粉饰线条挑檐
plaster mo(u)lding 抹灰线脚；石膏模注型法
plaster mo(u)ld method 石膏模型法
plaster mo(u)ld vacuum casting 石膏模真空注浆法
plaster mo(u)ld vacuum casting line 石膏模型真空注浆生产线
plaster of Paris 建筑石膏；煅石膏；熟石膏；烧石灰；烧石膏；半水石膏
plaster of Paris bandage 石膏绷带
plaster of Paris bed 石膏床
plaster of Paris finish 熟石膏饰面
plaster of Paris mortar 熟石膏灰泥
plaster on expanded metal lath 钢板网抹灰
plaster on metal lath 金属眼板上抹灰
plaster on metal lath(ing) 钢丝网上抹灰；网眼钢皮上抹灰；金属眼板抹灰；金属拉网上抹灰
plaster on plasterboard 在石灰泥板上抹灰
plaster on reed lath(ing) 芦苇条抹灰
plaster on wood(en) lath(ing) 抹灰板条上抹灰；灰板条上抹灰；板条抹灰
plaster pad 石膏开裂监视标志；石膏开裂监视标志
plaster panel 石膏板
plaster pattern 石膏模(型)
plaster plant 灰泥厂
plaster primer 抹灰层底涂料；抹灰底漆；灰泥用底漆

plaster putty 抹灰用油灰；抹灰用油灰
plaster quarry 采石膏场
plaster reinforcement 灰浆加筋
plaster reinforcing mesh 抹灰增强网
plaster relief model 石膏地形模型
plaster rendering 灰浆涂抹；灰浆打底
plaster retarder 石膏缓凝剂
plaster ring 抹灰垫圈；灰泥环
plaster rock 天然石膏；生石膏
plaster room 石膏房
plaster sand 石膏医疗室
plaster scheme 粉刷方案；抹灰简图
plaster scratch coat 粗灰墁层
plaster screed 准条；抹灰冲筋
plaster set 假凝(现象)；石膏浆凝结
plaster's float 灰抹子；抹灰抹子
plaster's hair 抹灰麻刀
plaster's hawk 抹灰用板
plaster shears 石膏剪
plaster shooting 推土机推土；灰浆喷射；糊泥爆破法；(孤石的)覆土爆破
plaster shot 二次爆破炮眼
plaster skim coat 粉刷的表层
plaster slab 灰浆板；石膏板
plaster's lath(ing) 抹灰板条；灰泥板条
plaster slip 石膏浆
plaster slurry mixing station 石膏料浆站
plaster sprayed with oil paint 喷油浆粉刷
plaster spreader 摊石膏器
plaster's putty 筛滤灰膏
plaster staff 灰浆护角条；墙角护条
plaster stain 抹灰污斑
plaster statue 石膏像
plaster stone 石膏；生石膏
plaster stop 石膏压条；抹灰挡板
plaster store 熟石膏仓库
plaster's trowel 灰抹子；铁抹子
plaster stuff 熟石膏拌和物
plaster surface 粉刷面层
plaster system 抹灰次序；抹灰方式
plaster tablet 石膏板
plaster-throwing machine 抹灰喷射机
plaster trowel 抹泥刀
plaster wainscot cap 灰泥护墙板帽；护墙板抹灰压顶
plaster wall(ing) 灰浆砌墙；抹灰墙
plaster waste mo(u)ld 废石膏模型
plaster work 抹灰作业；墁农；抹灰工作；抹灰工程；粉刷工作
plastery 灰泥状的
plastic 可塑的；胶质物；塑造的；塑性的；塑料面砖；塑料；成型的；范性的
plastic addition 添加塑料；掺塑料
plastic additive 塑料添加剂
plastic adhesive 塑性黏结剂
plastic adjustment 塑性调整
plastic admix(ture) 塑料掺和剂
plastic after effect 塑性后效
plastic aggregate 塑性集料；塑性骨料；塑料集料；塑料骨料；塑性粒料
plastic airplane 塑料飞机
plastic alloy 塑料合金
plastically conceived 形象构思的；塑性想像的；塑性构思的
plastically deforming area 塑性变形区
plastically flexible roof 塑性弯曲顶板
plastic analysis 塑性分析
plastic and glass fiber pipes 塑料玻璃纤维管
plastic and gunny mixed spun bag 麻塑交织袋
plastic and polumer waste 塑料和聚合物废物
plastic and resin waste(water) 塑料和树脂废水
plastic anisotropy 塑性各向异性
plastic armour plate 塑料装甲板
plastic article 塑性制品
plastic arts 塑造术；造型艺术
plastic asphalt cement 塑性沥青膏；塑性沥青胶泥
plastic assault boat 塑胶攻击艇
plasticate 塑炼
plasticating rate 增塑速率
plastication 增塑；塑化(作用)
plasticator 压塑机；塑炼机
plastic awning 塑料天篷
plastic bag 塑料袋；塑胶袋

plastic bag silage 塑料袋青储[贮]
plastic bar 塑性杆件
plastic barrier material 塑料防潮材料;塑性隔绝材料
plastic base 塑料片基;塑料基板
plastic-based 塑性底座的
plastic-base mirror 塑性底座镜
plastic bathroom inlet gull(e)y 浴室内塑料进排水口
plastic bathroom outlet 浴室内塑料排水口
plastic bath(tub) 塑料浴盆
plastic beam bending 塑性梁弯曲
plastic bearing 塑料轴承
plastic bearing structure 塑性承重结构
plastic bed divider 花坛分界隔板
plastic behavio(u)r 塑性状态;可塑性;塑性特征;塑性行为
plastic bending 塑性弯曲
plastic bending moment 塑性弯矩
plastic bending strength 塑性抗弯强度
plastic binder 塑料黏结剂
plastic binding 塑料圈装订本
plastic biological shielding wall 塑料辐射隔离器
plastic board 塑胶板;塑胶板
plastic boat 塑料艇;塑料船
plastic-bodied car 塑料车身小客车
plastic body 塑料车身;塑胶体;塑料车厢;塑性体
plastic-bonded explosive 塑料黏结炸药
plastic-bonded wallpaper 涂塑墙纸
plastic-bond fireclay 塑性耐火黏土
plastic bonding 塑料连接
plastic bonding adhesive 塑性胶黏剂;塑性黏合剂
plastic bonding emulsion 塑性黏结乳剂
plastic bound 塑性边界;塑性界限
plastic box 塑胶箱
plastic bronze 黏塑性青铜;塑性青铜;塑性铅青铜
plastic buckling 塑性失稳;塑性压曲
plastic building component 塑性建筑构件
plastic building material 塑性建筑材料
plastic bumper 塑料缓冲器
plastic buoy 塑料浮标
plastic bushing 塑料套管
plastic cable 塑料皮电缆
plastic cable clip 塑料电缆夹
plastic cable end box 塑料电缆终端盒
plastic cage 塑料保持架
plastic calculation 塑性计算;塑性分析
plastic capacitor 塑料电容器
plastic capsulation 塑料封装
plastic capsule 塑料封壳
plastic case 塑料盒;塑料管壳
plastic casing 塑料壳体;塑料盒;塑料封装
plastic caulk 塑料油膏
plastic ceiling 塑料顶棚
plastic ceiling dome light 塑料穹隆采光
plastic ceiling light cupola 塑料采光圆屋顶
plastic ceiling panel 石膏天花板
plastic ceiling plaster 塑料顶棚粉饰
plastic ceiling saucer dome 塑料碟形穹隆
plastic cement 塑性黏接料;塑化水泥;塑性黏结料;塑性水泥;塑性黏结剂;塑性黏结剂;塑性胶;塑胶
plastic cement cold glaze 塑料水泥冷上釉;塑胶冷上釉
plastic cement grout 塑性水泥浆
plastic cementing agent 塑性黏结剂;塑性胶黏剂
plastic centroid (钢筋混凝土设计的)塑性矩心;塑性中心
plastic chamber 塑料室
plastic channel (section) 塑料槽
plastic character 造型特性
plastic chrome ore 塑性铬矿
plastic cistern 塑料蓄水池;塑料储[贮]水器
plastic clad composite steel sheet 塑料复合薄钢板
plastic cladding 塑料覆盖层
plastic clad fibre 塑料色层光纤
plastic clad optic(al) fiber 塑料包层光纤
plastic clad silica optic(al) fiber 塑料包层石英光纤
plastic clay 可塑黏土;塑性黏土;塑性土;不合格砖泥;重黏土
plastic clay grinding mill 塑性黏土碾磨机
plastic clay pan soil 塑性硬黏土;塑性黏磐土
plastic clay soil 塑性黏土

plastic clip 塑性夹
plastic cloth 塑料布
plastic coal 塑性煤
plastic coat 塑料涂层
plastic-coated 塑料涂层的
plastic-coated chain link fencing 塑胶覆盖围篱
plastic-coated cotton glove 涂塑料的棉手套
plastic-coated curtain rod 塑料涂层窗帘杆
plastic-coated fabric 涂塑布
plastic-coated fabric manufacturing aggregate 涂塑布机组
plastic-coated fiber 塑料涂层纤维
plastic-coated glass screen 涂塑玻璃纤维窗纱
plastic-coated hardboard 涂塑料硬纸板;塑料纤维板
plastic-coated hose 包塑软管
plastic-coated material 石膏粉面材料
plastic-coated metal hose 包塑金属软管
plastic-coated metals 塑料涂敷金属
plastic-coated sheet steel 塑性护面钢板
plastic-coated steel 塑料涂覆的钢材
plastic-coated strip 塑料覆面带材
plastic-coated tube 塑料被覆管
plastic-coated wallpaper 涂塑墙纸
plastic-coated wire rope 包塑料钢丝绳
plastic coating 塑料套;塑料膜;塑料敷层;塑料护面;塑料涂料;塑料涂层;塑料涂层;塑料涂层;塑性盖层;塑胶涂层
plastic coating technology for steel tube 钢管涂塑工艺
plastic coefficient 塑性系数
plastic cohesive soil 黏塑性土壤;塑性黏性土
plastic cold bonding agent 塑性冷黏合剂
plastic cold water paint 塑料水性漆;塑料水溶性涂料
plastic collapse 塑性破坏
plastic colo(u)r 造型配色
plastic colo(u)r-relief presentation 自然光照地貌立体表示法
plastic column 塑料立柱
plastic combustion method 塑料燃烧法
plastic commutator 塑料换向器
plastic component 塑性构件;塑料部件
plastic composition 复合塑料;塑料复合材料
plastic compound 塑性混合物
plastic compression 塑压
plastic concrete 塑性混凝土;大坍落度混凝土
plastic concrete cut-off wall 塑性混凝土防渗墙
plastic concrete reinforcement distance piece 混凝土配筋塑料隔距块
plastic conduit 塑料导线管;硬质塑料管;塑料导线管
plastic cone 塑料圆锥体
plastic consistency 可塑稠度;塑性结持度;塑性稠度
plastic consistency of concrete 混凝土塑性稠度
plastic constraint 塑性约束
plastic construction 塑性结构;塑料建筑
plastic construction material 塑性建筑材料
plastic container 塑性容器;塑料容器
plastic content 含塑量
plastic cord 塑料线
plastic core 塑料芯
plastic core binder 树脂型芯黏结剂;树脂砂芯黏结剂
plastic core-rubber-tires 胶轮
plastic corrugated board 波纹塑料板
plastic corrugated film 波形塑料膜
plastic corrugated tile 塑料波形瓦;塑料波纹瓦
plastic cover 塑料覆盖层;塑料涂层
plastic-covered copper wire 包塑铜线
plastic-covered distributing cable 包塑配线电缆
plastic-covered steel cord 包塑钢线
plastic covering 塑料覆盖层
plastic crack 收缩裂缝
plastic cracking 收缩裂缝;塑性开裂;塑性裂纹(指混凝土)
plastic cracking of concrete 混凝土塑性开裂
plastic cupola 塑料圆屋顶
plastic cushion (of tie bar) 塑料垫条
plastic decoration 塑料装饰
plastic decorative board 塑料饰面板
plastic deflection 塑性变位
plastic deflexion 塑性挠度
plastic deformation 塑性弯沉;柔性变形;塑性变形;塑变;范性形变

plastic deformation area 塑性变形区范围
plastic deformation metamorphic way 塑变变质方式
plastic deformation stress 塑性变形应力
plastic deformation theory 塑性形变理论
plastic deformation zone 塑性变形区域
plastic delivery pipeline 塑料排泥管线
plastic design 极限设计;塑性(理论)设计;按塑性理论设计
plastic design method 按塑性理论设计方法;塑性设计法
plastic-detritus 塑性岩屑
plastic diffuser 塑料散光罩
plastic dip coating 塑料浸涂层
plastic disperse system 塑性分散系
plastic dispersion 塑性扩散(现象)
plastic distemper paint 塑料调和漆
plastic divider strip 塑料分隔板条
plastic dome 塑料穹隆
plastic domed roof-light 塑料穹隆顶天窗
plastic door 塑料门
plastic door leaf 塑料门扇
plastic door pull handle 塑料门拉手柄
plastic dosimeter 塑料剂量计
plastic dowel 塑料合缝钉
plastic downbuckling 塑性下弯
plastic drain 塑料排水带;塑料板排水
plastic drainage pipe 塑料泄水管;塑料排水管
plastic drain board 塑料排水板
plastic drain layer 塑料排水管敷置机
plastic drawing material 绘图塑料
plastic dressing paint 表面处理塑料涂料;整修塑料油漆
plastic dual lateral block 塑料双层支管块
plastic earth pressure 塑性地压
plastic effect 造型效果;立体感;塑性效应;浮雕效应
plastic eggcrate ceiling 塑料格片顶棚
plastic element 塑料构件
plastic elongation 塑性延伸
plastic embedding 塑性包埋
plastic embossed model 塑料地形模型
plastic embossing machine 塑料压花机
plastic emulsion 塑料乳胶;塑性乳剂
plastic emulsion flooring 塑料乳胶地面
plastic emulsion paint 塑料乳胶漆
plastic encapsulated coil 塑料包封线圈
plastic end cap 塑料管帽
plastic energy 塑性能
plastic engraving 塑料片刻图
plastic equilibrium 塑性平衡
plastic eraser 擦字胶
plastic explosive 可塑炸药;塑性猛烈炸药
plastic extruder 塑料挤出机
plastic fabric 塑料织物
plastic facade 塑料正面
plastic-faced 塑料贴面
plastic-faced die 塑料面模具
plastic-faced hardboard 塑料贴面纤维板;塑料贴面硬纸板
plastic-faced plywood 塑料贴面胶合板;塑料面胶合板
plastic facing 塑料铺面;塑料饰面
plastic factory 塑料工厂;塑料厂
plastic factory wastes 塑料厂废物
plastic factory wastewater 塑料厂废水
plastic failure 塑性破坏
plastic failure criterion 塑性失效准则
plastic fan 塑料叶片;塑料风扇
plastic fat 软化脂
plastic fatigue 塑性疲劳
plastic fence 塑料栅栏;塑料篱笆
plastic fibre 塑料纤维
plastic fibre optics 塑料纤维光学
plastic figure in painting 立体感图像
plastic file 塑性锉
plastic filler (用于填塞木缝孔洞的)塑性填充物;塑性填料
plastic filler method 塑料间隙规测定法
plastic-filling instrument 成型充填器
plastic film 塑料膜;塑料胶片;塑料薄膜;薄膜塑料
plastic film capacity 塑料膜电容器
plastic film coating 塑料膜敷层
plastic film condenser 塑料膜电容器
plastic film covered wire 塑料薄膜包线

plastic film curing 塑料薄膜养护
plastic film for roofing 塑料薄膜屋顶
plastic film for wrapping 包装用塑料薄膜
plastic film mulching 塑料薄膜覆盖
plastic film package 塑料薄膜包装
plastic filter 塑料芯滤清器;塑料滤水器;塑料滤色镜;塑料滤波器;塑料过滤器
plastic filter cloth 塑料滤层布;塑料滤布
plastic filter medium 塑料滤料;塑料过滤器介质
plastic fines 塑性细颗粒
plastic finger plate 路标;塑料指板
plastic finish 塑料终饰;塑性修整;塑料面层
plastic fireclay 塑性耐火黏土
plastic fittings 塑料零配件;塑料配件
plastic fixture 塑料装置;塑料设备
plastic flake 塑料薄片
plastic flashing piece 塑料闪光块;塑料挡水板
plastic float 塑料浮子
plastic floor 塑料地板;塑胶铺面;塑胶楼面;塑料地面
plastic floor covering 塑料地板铺设
plastic flooring 塑胶地板;塑料地板铺设;塑料地面
plastic flow 范性流动;塑性变形;软流;塑料流动;塑料流变;塑性流;(晶粒的)塑性滑移;塑流;范性流变
plastic flow condition 塑流条件
plastic flow curve 塑性流动曲线
plastic flow deformation 塑性流动变形
plastic flow equation 塑性流动方程
plastic flow loss 徐变损失;塑流损失
plastic flow model 塑性流模型
plastic flow of concrete 混凝土塑性流(动)
plastic flow of steel 钢的塑流;钢的塑性流动
plastic flow range 塑性流幅
plastic fluid 宾汉姆流体;塑性液体;塑性液;塑性流体
plastic fluidity 塑性流度
plastic flume 塑料渡槽
plastic foam 多孔塑料;海绵状塑料;泡沫塑料;泡沫乳胶;塑性泡沫塑料;塑料泡沫;发泡塑料
plastic foam board 泡沫塑料板
plastic foam concrete 塑性泡沫混凝土
plastic foam cupola 泡沫塑料圆屋顶
plastic foam dome 泡沫塑料穹隆
plastic foam insulant 泡沫塑料绝缘
plastic foam insulation 泡沫塑料隔热材料;泡沫塑料隔声层;泡沫塑料隔热层
plastic foam lightweight concrete 泡沫塑料轻质混凝土
plastic foam plaster baseboard 泡沫塑料灰泥底板;泡沫塑料灰泥踢脚板
plastic foam seal(ing) 泡沫塑料焊接;泡沫塑料密封
plastic foam sheet 泡沫塑料薄板
plastic focusing fibre 塑料聚焦光纤
plastic foil 塑料薄膜
plastic foil base 塑料薄膜基底
plastic forceps 整形镊
plastic form 塑性造型
plastic formation 塑性岩层;塑性地层
plastic forming 可塑成型
plastic formwork 塑料模板
plastic fountain basin 塑料喷水池
plastic fracture 塑性破裂;塑性破断;塑性断口;塑性断裂
plastic-fracturing theory 塑性断裂理论
plastic frame 塑性框架;塑料框
plastic framed plaque 塑料边框板
plastic framed window 塑料框窗;塑料窗
plastic friction 塑性摩擦
plastic frozen soil 塑性冻土
plastic fuel tank 塑料油箱
plastic furniture 塑料家具
plastic fuselage 塑料机身
plastic gasket 塑料垫圈;塑料垫片;塑料衬垫
plastic gasket joint 塑料衬片接头
plastic gear 塑料齿轮
plastic-glass fiber reinforcement 塑料玻璃纤维加劲材
plastic glazing 有机玻璃;塑料玻璃
plastic glove 塑料手套
plastic glue 合成树脂胶;塑胶
plastic goods 塑料商品
plastic-grade talc 塑料滑石
plastic grass 塑料草绳

plastic grating 塑料光栅
plastic greenhouse 塑料温室
plastic grip 塑料夹子;电木夹子
plastic ground 塑性地面;塑性围岩;塑性地基
plastic ground pressure 塑性地压
plastic guard rail 塑料护栏;塑料护轨
plastic guide 塑料导向器
plastic gutter 塑料檐槽;塑料天沟
plastic halftone 网目片
plastic hammer 塑料手锤
plastic handrail 塑料扶手
plastic heat shield 塑料隔热板
plastic helmet 塑料工作帽
plastic hinge 塑性铰;塑性铰链;塑胶铰链
plastic hinge length 塑性铰长度
plastic hinge line 塑性铰线
plastic hinge method 塑性铰接法
plastic hinge position 塑性铰位置
plastic hinge zone 塑性铰区
plastic hollow floor filler 塑料空心地板填料
plastic hood 塑料套
plastic hook 整形钩
plastic hose 塑料水龙带;塑料软管
plastic house 塑料单宅房屋
plastic housing 塑料罩
plastic hysteresis 塑性滞后
plastic igniter cord 软点火线
plasticimeter 塑性计
plastic impervious membrane 塑料板防水层
plastic-impregnated 浸渍塑料
plastic impregnation 塑胶浸渍处理
plastic impregnation aggregate for insect screening 涂塑窗纱机组
plastic incinerator 塑料焚烧炉
plastic index 可塑指数;塑性指数
plastic index of bituminous mixture 沥青混凝土的塑性指数
plastic industry 塑料工业
plastic industry wastes 塑料工业废水
plasticine 造型材料;蜡泥塑料;塑泥;代用黏土
plastic inelasticity 塑性非弹性
plastic injection mo(u)lding machine 塑料注射模型孔
plastic insert fitting 塑料插入管件
plastic instability 塑性失稳;塑性不稳定性
plastic insulated cable 塑料绝缘电缆
plastic insulated wire 塑料绝缘线
plastic insulated wire cord 塑料绝缘电线
plastic insulating material 塑料绝缘材料
plastic insulating tape 塑料绝缘胶带
plastic insulating tube 塑料绝缘管
plastic insulation 塑料绝热材料;塑料绝缘材料;塑料绝缘体;塑料隔声材料
plastic insulation material 塑料绝缘材料
plastic insulator 塑性绝缘体
plastic interlock 板桩塑料连锁接头
plastic intrusion 塑性干扰
plasticisation 增塑;塑化
plasticise 增塑;塑化
plasticised 塑化的;可塑的
plasticised concrete 塑性混凝土;可塑混凝土
plasticised fabric 塑性化纺织物
plasticised mortar 塑化灰浆
plasticised mortar plasticized mortar 塑化砂浆
plasticiser 增韧剂;柔软剂
plasticising 增塑;塑化
plasticising action 增塑作用
plasticising agent 塑化剂;增塑剂
plasticising frost protection agent 增塑防霜剂;增塑防冻剂
plasticising of concrete 混凝土的塑化
plasticising oil 增塑油
plasticising resin 塑化树脂
plasticism 造型学
plasticity 增塑性;可塑性;塑性;受范性;受范力学;范性学
plasticity agent 增塑剂;塑化剂
plasticity and consistency of cohesive soil 黏性土的可塑性与稠度
plasticity and liquidity test 塑液阳联合试验
plasticity chart 塑性图;塑限图
plasticity classification of soil 土的塑性分类
plasticity coefficient 塑性系数
plasticity condition 塑性状态;塑性条件

plasticity grade of soil 土的可塑性分级
plasticity index 塑性指标;增塑指数;塑性指数;塑性值
plasticity index of rock 岩石塑性指数
plasticity index of soil 土的塑性指数
plasticity law 塑性定律
plasticity measurement 可塑性测定
plasticity modulus 塑性模量
plasticity needle 塑性试针
plasticity number 可塑值;可塑性指数;可塑度;塑性指数
plasticity of asphalt 沥青塑性
plasticity of concrete 混凝土塑性
plasticity of rock 岩石的塑性
plasticity of soil 土的可塑性
plasticity processing 塑性加工
plasticity range 塑性阶段;塑性范围
plasticity ratio 塑性比
plasticity-recovery number 塑性恢复值
plasticity-recovery number-Williams 威廉斯可塑性恢复值
plasticity retention index 塑性保持指数
plasticity retention percentage 塑性保持率
plasticity stability 塑性稳定性
plasticity test 可塑性试验
plasticity value 可塑值
plasticity water 塑性水
plasticization 增塑;塑化;塑炼
plasticized 塑化的;可塑的;增塑的
plasticized cement 塑化水泥
plasticized concrete 塑化混凝土;可塑混凝土
plasticized melt-spinning 塑化熔体纺丝
plasticized mortar 塑化灰浆
plasticized pitch 塑化(硬)沥青;增塑材料
plasticized polyvinyl chloride 软聚氯乙烯;塑化聚氯乙烯
plasticized polyvinyl chloride sheet 增塑聚氯乙烯片材
plasticized Portland cement 塑化硅酸盐水泥;塑化波特兰水泥
plasticizer 增塑剂;增韧剂;软化剂;柔韧剂;塑化剂;成型剂
plasticizer adhesive 增塑黏合剂
plasticizer migration 增塑剂迁移
plasticizer resistance 耐增塑剂性
plasticizer wastewater 增塑剂废水
plasticizing 可塑;增塑;塑化
plasticizing admixture 增塑混合物
plasticizing agent 增塑剂;增韧剂;塑化剂
plasticizing component 增塑组分
plasticizing effect 塑化效果
plasticizing efficiency 可塑效率
plasticizing frost protection agent 增塑防霜剂;增塑防冻剂
plasticizing of concrete 混凝土的塑化
plasticizing oil 增塑油
plasticizing rate 增塑速率
plasticizing resin 塑化树脂
plastic jacket 塑料罩
plastic-jacketed optic(al) fiber 套塑光纤
plastic jamming pressure 塑料片基抗划力
plastic joint filler 接缝塑料填料
plastic koalin 塑性高岭土
plastic lag 塑性延缓
plastic laminate 塑料叠合板;塑料层压板
plastic-laminated 塑料胶合的
plastic-laminated hardboard 塑料硬质高压板;塑料硬质纤维板
plastic-laminated heat shield 塑料层压避热层;塑料层压热防护层
plastic-laminated plywood 塑料层压胶合板
plastic-laminated surfaced hardboard 塑料层压纤维面板
plastic-laminating of steel 塑料钢层压板;塑料层压钢
plastic language 造型语言
plasticlast 塑性碎屑
plastic latch 塑料弹簧锁
plastic layer 可塑层;塑料覆盖薄膜铺放机
plastic layer maximum thickness Y graduation 交质层最大厚度 Y 分级
plastic layer maximum thickness Y value 胶质层最大厚度 Y 值
plastic layer volume curve 胶质体体积变化曲线

plastic-laying machine 塑料薄膜覆盖机
plastic lay-in panel 塑料嵌入式护墙板
plastic lead 塑性铅
plastic lens 塑料透镜
plastic lens system 塑料透镜系统
plastic letter 塑料铅字
plastic lifeboat 塑料救生艇
plastic liferaft 塑料救生筏
plastic light cupola 塑料采光圆屋顶
plastic light-diffusing ceiling 塑料漫(射)光顶棚
plastic light fitting 塑料采光装置
plastic light fixture 塑料采光设备
plastic limit (土的)塑性含水率;塑性限度;塑性界限;塑性极限;塑限
plastic limit design 塑性极限设计;塑限设计
plastic limit load 塑性极限荷载
plastic limit of soil 土壤塑限;土壤湿度塑限;土的塑限
plastic limit state (土的)塑限状态
plastic limit test 塑限试验
plastic-lined metal fitting 塑料衬金属管件
plastic-lined metal pipe 塑料衬金属管
plastic-lined steel pipe 塑料衬里钢管
plastic line marking machine 划塑料线机(道路)
plastic liner 塑料衬里
plastic lining 塑料镶边;塑料衬里;塑料衬
plastic lip(ping) 塑料凸缘;塑料翼缘;塑料镶边
plastic load 使荷载进入塑性阶段的荷载
plastic-load approach 塑性荷载法
plastic load bearing structure 塑性荷载的承重结构
plastic load-carrying structure 塑性荷载的承重结构
plastic loaded structure 塑性阶段的承重结构
plastic loss 蠕变;徐变损失;塑性损失
plastic luminaire (fixture) 塑料采光装置
plastic lyophilic sol 塑性亲液溶胶
plastic made tile 塑料瓦
plastic magnesia 塑性苦土;塑性氧化镁
plastic making 可塑法成型;可塑成型法;塑性成型法
plastic making material 塑性划线材料(道路)
plastic map 塑料地图
plastic mass 塑性物质;塑料
plastic match-plate 塑料双面模板
plastic material 可塑性物质;塑性物质;塑料材料;塑料填料;塑料
plastic matrix 塑料基质
plastic mechanics 塑性力学
plastic media 塑料介质
plastic media filter 塑料滤料滤池
plastic member 塑性杆件
plastic membrane 塑料涂层;塑料薄膜
plastic memory 塑性记忆
plastic metal 高锡含锑轴承合金;塑性合金
plastic-metal bearing alloys 塑料金属轴承合金
plastic-metal-ceramic composite 塑料金属陶瓷复合材料
plastic-metal powder product 塑料金属粉末制品
plastic-metal wire 成型金属丝
plastic metamorphism 塑性变质
plastic method 塑性分析法;塑性阶段设计法
plastic methods of structural analysis 结构分析的塑性方法
plastic microballs 塑料粉球
plastic mix 塑性混合
plastic mixture 塑性混凝土混合料
plastic mobility 塑性淌度
plastic model 塑料模型
plastic-moderated reactor 塑料慢化堆
plastic modified 由塑性发生的变化
plastic modified asphalt 塑性体改性沥青
plastic modified asphalt membrane 塑性体改性沥青油毡
plastic modified coal tar ca(u)lking compound 塑料油膏;塑料改性焦油油膏
plastic modulus 塑性模量
plastic mold package 塑料模压封装
plastic moment 塑性力矩;塑性弯矩;塑性阶段弯矩
plastic moment distribution method 塑性力矩分配法
plastic monoblock 塑料整体件
plastic monomer 塑料单体
plastic mortar 塑性灰浆;软练砂浆;软练;塑性砂浆;有塑料成分的灰泥;塑性胶砂;软练胶砂
plastic mortar cube 软练砂浆立方体
plastic mortar strength test 软练胶砂强度试验法

plastic mo(u)ld 塑料模型;塑料模板;塑料模
plastic mo(u)ld cooling system 塑料模制冷却系统
plastic mo(u)lded gear 塑料铸造齿轮
plastic mo(u)ld for concrete 混凝土塑料模型
plastic mo(u)lding 塑料装饰嵌线(条);可塑成型;塑料铸模
plastic mo(u)lding press 塑料制品成型机
plastic-mounted lamination 塑料片叠层地形模型
plastic mounting channel 塑料装槽;塑料装配线
plastic nature 塑性特征;塑性
plastic net 塑料网
plastic nozzle 塑料喷嘴
plastic numeral 塑料符号
plastic-nylon boom 塑料尼龙水栅
plasticon 聚苯乙烯薄膜
plastic optic(al) fibre 塑料光纤;塑料光学纤维
plastic optics 塑料光学
plasticorder 塑性变形曲线描记仪;塑度计;塑度计
plastic organic clay 塑料有机黏土
plasticostatics 塑性体静力学
plastic overlay 塑料覆面
plastic packaging 塑料封装
plastic packing 塑料填密
plastic packing ring 塑料填密环;塑料密封圈
plastic pail 塑料桶
plastic paint 皱纹漆;干酪素拉毛涂料;塑性涂料;塑料漆
plastic painting 塑性涂料
plastic panel 塑料镶板;塑料护墙板
plastic panel in a wall 外墙采光塑料板
plastic paper 塑料纸
plastic partition (wall) 塑料隔墙
plastic parts 塑料零件
plastic paste 塑料泥团;塑料糊
plastic pattern 塑料模型
plastic phase 塑化相
plastic phosphor 塑料磷光体
plastic pigment 塑料颜料
plastic pipe 塑料管子;塑料管
plastic pipe cap 塑料管帽
plastic pipe creep 塑料管蠕变
plastic pipe fitting 塑料管件
plastic pipe joint 塑料管接
plastic piping 塑料管
plastic piping system 塑料管道系统
plastic pit setter 塑料筒装置
plastic plaster 塑料砂浆
plastic plate 塑料片;塑料板
plastic plate drain preloading method 塑料板排水预压法
plastic plate mo(u)lding machine 塑料复制版浇铸机
plastic platform 塑性平台
plastic plating 真空塑料喷涂;塑料金属喷镀
plastic plow 塑料贴面犁
plastic polishing 塑料抛光
plastic pollution 塑料污染
plastic potential 塑性位能;塑性势
plastic potential surface 塑性势面
plastic potentiometer 塑料电位计
plastic preformed gasket 塑料预制垫片;塑料预制垫圈
plastic preformed gasket joint 塑料接头垫片;塑料接头垫圈
plastic pressing 塑压成型;湿压成型
plastic pressure-sensitive tape 塑料压敏带
plastic print 塑料蓝图
plastic printing 塑料印刷
plastic printing ink 塑料印刷用油墨
plastic process 塑性砖坯;塑性压制
plastic processing equipment 塑料加工设备
plastic product 塑料制品
plastic profile 塑料外形;塑料异型材
plastic propellant 塑料推进剂
plastic property 塑性特征;塑性
plastic protective coating 塑料保护涂层
plastic pull handle (门窗上用的)塑料拉手柄
plastic pump 塑料泵
plastic putty 塑料油灰;树脂系腻子
plastic pyramid 塑料四面体;塑料立锥体;塑料金字塔
plastic quench 塑料淬火剂淬火
plastic radiation shielding wall 塑料辐射防护措施

plastic rail 塑料门窗冒头;塑料门窗横挡;塑料轨道;塑料轨条;塑料围栏
plastic rainwater articles 塑料雨水附件
plastic rainwater goods 塑料雨水管件
plastic rainwater gutter 塑料雨水屋檐槽;塑料雨水槽
plastic rainwater outlet 塑料雨水斗
plastic range 可塑范围;塑性区;塑性范围
plastic range of stress 塑性范围内应力
plastic range test 塑性范围试验
plastic recoding medium 塑性记录媒质
plastic recoil 塑性回弹
plastic recovery 塑性恢复
plastic redistribution of moment 弯矩塑性重分布
plastic refractory 耐火浆料;塑性耐火材料
plastic refuse bag 塑料垃圾袋
plastic region 可塑范围;塑性范围
plastic reinforced primacord 塑料加固的导爆索
plastic reinforcement distance piece 塑性混凝土配筋间隔块
plastic reinforcement spacer 塑性混凝土配筋间隔物;塑性混凝土配筋定位件
plastic relief map 塑料立体图;塑料立体地图
plastic relief model 塑料立体模型
plastic remover 塑料薄膜揭除机
plastic repair 成型修复
plastic resilience 塑性回弹
plastic resin 塑料树脂
plastic resin-based jointless cover(ing) 塑性树脂基铺成的无缝地面
plastic resist 塑料抗蚀剂
plastic resistance 塑性阻力
plastic resonance 塑性共振
plastic riverbed lining 塑性河床护面
plastic roadline 塑料道路条纹;清晰的(道)路(划)线
plastic rock 塑性岩(石)
plastic rod 塑料棒
plastic roller shutter 塑料卷帘百叶窗
plastic roof(ing) foil 塑料屋顶薄膜
plastic roofing sheet 塑料屋面片材
plastic roof-light 塑料天窗
plastic roof panel 塑料屋顶板
plastic roof shutter 塑料屋顶雨水槽
plastic rose 塑料圆花窗
plastic rotation 塑性转动
plastic rupture 塑性破裂
plastics 塑料;塑料制品;电木
plastic sack 塑料袋
plastic sand 塑性砂
plastic sandwich 塑料夹层结构
plastic sandwich system 塑料夹层系统(铺砌法)
plastic saucer dome 塑料碟状穹隆;塑料浅穹隆(天窗采光用)
plastic scintillant 塑料闪烁体
plastic scintillation counter 塑料闪烁计数器
plastic scintillator 塑料闪烁体;塑料闪烁器
plastic scintillator detector 塑料闪烁体探测器
plastics-coated 塑料贴面的
plastics-coated waterproofing 塑料防水
plastics coating 塑料涂料;塑料盖层
plastics cover 塑罩
plastics covered handrail 塑料面扶手
plastic screw anchor 塑料螺钉锚座
plastic screw plug 塑料螺塞
plastic scribing 塑料片刻图
plastic seal 塑料密封
plastic sealant 塑料密封胶;塑料密封材料;塑料密封膏
plastic sealing compound 塑料密封剂;塑料密封混合剂;塑料封闭剂;塑料封闭混合剂
plastic sealing gasket 塑料密封垫片;塑料密封垫圈
plastic sealing material 塑料密封材料
plastic sealing ring 塑料密封环
plastic section 塑性断面;塑料型材;塑料部件
plastic section coefficient 塑性断面系数
plastic section modulus 塑性截面模量;塑性断面模量
plastic sentiment 造型感情
plastic set 塑性凝结;塑性沉降
plastic setting 塑性定型
plastic settlement crack 塑性沉淀裂缝;塑性沉陷裂缝
plastics for horticultural use 塑料在园艺上的应用

plastic shading 缔约程序;地貌晕线法;地貌晕翁法	plastic structural system 塑料结构系统	plastic waterproof membrane 塑料防水膜
plastic shape 塑料模型;塑料型材	plastic structure 塑性结构	plastic water stop(ping strip) 塑料止水条;塑料止水带
plastic shear 塑性剪切	plastic stuffing 塑料填म	
plastic shear modulus 塑性剪切模量	plastic substance 可塑性物质	plastic wave 塑性波
plastic sheath 塑料外壳;塑料护皮	plastic sulphur 弹性硫	plastic wax 塑性石蜡
plastic sheath cable 塑料外皮电缆;塑料皮电缆;塑料铠装电缆	plastic support 塑性支承	plastic weight-carrying structure 塑性承重结构
	plastic supporting structure 塑性承重结构	plastic welding 塑性焊接;塑态焊接;塑料焊接;塑焊
plastic sheathing 塑性套管	plastic surface 塑性路面	plastic white line composition 清晰的白线合成物(道路上)
plastic sheet 塑料片;塑料薄板;塑料板	plastic-surfaced 塑料贴面;塑料涂层	
plastic sheet flooring 塑料卷材地面	plastic-surfaced hardboard 塑料纤维面板	plastic white line markings 清晰的白线标志(道路上)
plastic sheeting 塑料薄膜	plastic surgery 整形外科;矫形外科	
plastic sheeting for roofing 塑料屋面薄板;塑料屋面薄片	plastic surgery hospital 整形外科医院;矫形外科医院	plastic window 塑料窗
	plastic suspension 塑料悬浮;塑料悬挂;塑性悬浮体	plastic window frame 塑料窗框
plastic shell 塑料薄壳	plastics wall finish 塑料墙面	plastic window gasket 塑料窗衬垫
plastic shielding wall 防辐射的塑料墙	plastics waterproofing 塑料防水	plastic window light 塑料玻璃窗
plastic shim material 塑性垫片材料	plastic swimming pool 塑料游泳池	plastic window section 塑料窗外形;塑料窗剖面(图)
plastic shrinkage 塑性收缩	plastic system 塑料系统	
plastic shrinkage crack(ing) 塑性缩裂;塑料收缩裂缝;塑性收缩裂缝	plastic tab 塑料薄片	plastic window yarn 塑料窗纱
	plastic tank 塑料罐	plastic wing 塑料机翼
plastic shrinkage cracks of concrete 混凝土塑性收缩缝	plastic tanker 塑料槽车	plastic wire 塑料线;塑料丝
	plastic tape 塑料带	plastic wood 塑化木(材);硝酸纤维素(木粉)浆;填缝浆膏;塑性填料;塑料木粉浆;塑料木粉膏
plastic shuttering 塑料模壳;塑料模板	plastic temperature range 塑性温度范围	
plastic siding 塑料板壁;塑料墙板	plastic tent 塑料帐篷	
plastic sign 造型语言	plastic testing machine 塑性试验机	plastic work done 塑性功
plastic silo 塑料青储[贮]器;塑料布青储[贮]器	plastic textile 塑料布	plastic working 塑性加工
plastic silt 塑性淤泥	plastic theorem 塑性定理	plastic woven bag 塑料编织袋
plastic silty clay 塑性粉质黏土	plastic theory 塑性理论	plastic yield 塑性拉伸;塑流点;塑变值
Plastics Institute of America 美国塑料学会	plastic theory of failure 塑性破坏理论	plastic yielding 塑性屈服
plastic skylight 塑料天窗	plastic theory of limit design 极限设计塑性理论;塑性极限设计理论	plastic yield-point 塑性屈服点;塑料流点;塑料的固性极限
plastic slate 塑性石板		
plastic slatted roller blind 塑料条板卷帘	plastic theory of reinforced concrete 钢筋混凝土塑性理论	plastic yield stress 塑性屈服应力
plastic sleeper bearing plate 塑料轨枕垫板		plastic yield test 塑流试验
plastic slip (晶粒的)塑性滑移;塑性滑动	plastic theory of reinforced concrete design 钢筋混凝土塑性理论设计	plastic yield value 塑性屈服值;塑流值
plastics machinery 塑料机械		plastic zone 塑性区;塑性范围;塑性带
plastics meeting face 塑料贴接面	plastic thermistor 塑料热敏电阻器	plastic zone method 塑性区法
plastic sock 塑料防护软管	plastic thermoluminescence detector 塑料热释光探测器	plastid 质体;成型粒
plastic soil 可塑土;塑性土		plastification 塑性化;塑化(作用)
plastic soil cement 塑性水泥土	plastic thickening 塑性增厚	plastifier 增塑剂;软化剂;塑化剂
plastic soil cement lining 塑性水泥土混合衬砌;塑性灰土护岸	plastic thinking 造型构思	plastify 增塑;塑炼
	plastic though theory 塑性盆地理论	plastifying 塑性化
plastic sol 塑性溶胶	plastic tie-plate 塑料轨枕垫板	plastifying agent 塑化剂
plastic solar still 塑料太阳蒸馏锅	plastic tile 塑料瓦(片);塑料(瓷)砖;塑料面砖	plastifying oil 塑性化油
plastic solid 柔性固体;塑性固体	plastic tile flooring 塑料块地面	plastiga(u)ge (测曲轴轴承游隙用的)塑料线间隙规
plastic solidification 塑料固化	plastic tined 镶塑料的	plastigel 增塑凝胶;塑性凝胶
plastic solution 塑性溶液;黏性溶液	plastic tip hammer 镶塑料头的小锤	plastigel mo(u)ld 塑性凝胶模
plastic spacer 塑料定位器(钢筋用)	plastic-tired 塑料轮胎的	plastigraph 塑度计
plastics pipe 塑料管;塑料管材	plastic torque modulus of section 弯矩塑性截面模量	plastilock 用合成橡胶改性的酚醛树脂黏合剂
plastic sponge underlay 塑料海绵底层		plastiment 塑性灌缝水泥浆
plastic squeezer 塑料挤压机	plastic torsion 塑性扭力	plastimeter 塑度计
plastics removal 排塑	plastic track 塑料跑道	plastimets 金属塑料复合材料
plastics replica 塑料复型	plastic track detector 塑料径迹探测器	plastipaste 增塑糊
plastics spray coating 塑料喷涂	plastic trim 塑料修饰	plastisol 增塑糊;增塑溶胶;塑性溶胶;塑溶胶;塑料溶胶;塑料分散体
plastic stability 塑性稳定性	plastic T-square 塑料丁字尺	
plastic stability theory 塑性稳定理论	plastic tube 塑性管;塑料管	plastisol ink 热熔油墨
plastic stabilizer 塑性稳定剂	plastic tube for incline measuring 塑料测斜管	plastispray 塑料粉末喷涂
plastic stage 应力应变曲线的塑性阶段;塑性状态;塑性阶段	plastic tubing 塑料管材;塑料管	plastocene 油泥
	plastic tweezer 塑料镊子	plasto-concrete 塑料混凝土
plastic stairrail 塑料楼梯栏杆;塑料楼梯扶手	plastic type explsoive 塑性炸药	plasto elastic 塑弹性的;塑性弹性的
	plastic-tyred 塑料轮胎的	plastoelastic body 弹塑性物体;弹塑性体
plastic state 可塑态;黏流态;塑性状态;塑性态;塑态	plastic up-and-over door fo the roll-up type 塑料卷升式升降门	plastoelastic composition 塑弹性物质;塑弹性材料;塑弹性合成物
plastic-steel combination 塑料钢组合		
plastic-steel laminate 塑料钢层板	plastic valve 塑料阀	plastoelastic deformation 塑弹性变形;塑弹形变
plastic stickness 塑性黏度	plastic veneer 塑料贴面板;塑料饰面板;塑料饰面	plasto-elasticity 塑弹性
plastic stiffness 塑性刚度	plastic veree 塑饰料面	plastoelastic system 弹塑性体系
plastic stiffness matrix 塑性刚度矩阵	plastic vibration 塑性振动	plastogel 塑性凝胶
plastic stopper 塑料阻止器;塑料门挡;塑料塞子	plastic viscosity 塑性黏度;塑流黏度;塑黏性	plastogel coating 塑性溶胶涂布
plastic strain 塑性应变;塑性胁变;塑性变形;塑料应变	plastic viscosity equation on tackmeter 黏性计的塑性黏度方程	plastoglage 合成油漆
		plastograph 塑性变形图描记器;塑性形变记录仪;塑性变形曲线描记仪;塑性变形记录仪;塑度计
plastic stratum 塑性岩层	plastic-viscous flow 塑性黏滞流(动);黏塑性流动	
plastic straw 塑料细管	plastic-vitreous fragment 塑性玻屑	
plastic strength 塑性强度	plastic vitreous surfacing 塑性水泥涂釉面	plasto-hydrodynamic(al) lubrication 塑性流体动压润滑
plastic stress 塑性应力	plastic volumetric(al) strain 塑性体积应变	
plastic stress distribution 塑性应力分布	plastic wall cover(ing) 塑料墙面涂料	plastoleum 地板蒙皮
plastic stress range 塑性应力阶段;塑性应力区域;塑性应力范围	plastic wall panel 塑料墙板	plastomer 塑性体;塑体;硬塑料;可塑物;塑性高分子物质;塑料塑性体;塑料
	plastic wall paper 塑料墙纸	
plastic stress redistribution 塑性应力重分布	plastic wall plaster 墙面塑料粉刷	plastomer modified asphalt 塑性体改性沥青
plastic stress-strain diagram 塑性应力一应变图	plastic ware 塑料器具	plastomer modified asphalt membrane 塑性体改性沥青毡
plastic stress-strain matrix 塑性应力一应变矩阵	plastic washer 塑料洗涤器;塑料垫圈	
plastic stress-strain relation 塑性应力一应变关系	plastic waste 塑料废物;废塑料	plastometer 可测侧向应力固结仪;柔性塑仪;塑性计;塑度计;范性计
plastic strip 清晰的条带;清晰的条纹;路上清晰的条带;路上清晰的条纹;塑料板条	plastic water bar 塑料止水条	
	plastic water paint 塑料水溶性颜料;塑料水粉涂料	plastometer constant 塑性常数;塑度计常数
plastic structural cladding 塑料外挂板;塑料围护结构	plastic water pipe 塑料水管	plastometer index 胶质层指数
	plastic water pipeline 塑料输水管道	plastometric set 塑性变形
plastic structural gasket joint 塑料联结垫片;塑料联结垫圈	plastic waterproofing sheet 塑料防水片材	plastometry 塑性测定法;塑料测定法
		plastoponic 塑料种植法

plasto-rubber 增塑
plastotype 塑型
plasto-visco-elastic body 塑—黏—弹性体
plasto-viscous flow 塑性—黏滞流动
plastumen 沥青石油合成物(用于修补屋顶)
plastyle lens 塑玻透镜
plat 张(厚钢板);地区图;地皮;地段(平面)图;地段
platan(e) 法国梧桐;悬铃木
platarsite 硫砷铂矿
platband 横箍带;道路花草边缘;横线条;平边;扁突线条
plate 印板;浅盘;平盘;片状体;片状;涂覆金属;电容器板;标度盘;版层;块板【地】;板材;板(极)
platea (古罗马的)宽通道或宽街道
plate accretion 板块增大
plate action 平板效应;平板作用
plate adapter 硬片转换盒
plate aerator 平板式曝气池
plate aerial 平板天线
plate air conduct 铁皮风管
plate air preheater 板式空气预热器
plate alignment 平板找正
plate amalgamation 混汞析金法;平板混汞法;板上混汞法
plate analysis 平板计算;平板分析
plate anchor 槛锚;锚板螺栓;地基螺栓;下槛锚固螺栓
plate anchoring system 板固定系统
plate and angle column 角钢和板组合柱;平板角铁柱
plate and angle separator 腹板角钢间分隔板
plate and channel column 钢板槽钢组合柱
plate and cone viscometer 板锥式黏度计
plate and frame press 板框(式)压滤机
plate and frame type filter press 板框(式)压滤机
plate and knot 门执手与盖板
plate and sheet piler 垛皮机,垛金属板机
plate and tube condenser 板管式冷凝器
plate and wire electrofilter 板式除尘器
plate anemometer 压板风速表;平板风速计
plate antenna 平板天线
plate apron conveyer 铁板输送机,链板式输送机
plate arch 平板拱
plate arch aqueduct 板拱渡槽
plate arched girder 平板拱梁
plate armo(u)r 装甲板;钢板装甲
plate asphalt 沥青板
plateau 坪值;高原(台地);海高原;平稳状态;平复时期;平台期;停滞时期;大盘;[复]plateaux
plateau age 坪年龄
plateau basalt 高原玄武岩
plateau characteristic 坪特性曲线;坪特性
plateau-climate 高原气候
plateau cure 正硫化
plateau curve 平坦曲线
plateau earthquake 高原地震
plateau equation 平稳方程
plateau eruption 高原喷发
plateau escarpment 高原坡地
plateau frigid zone 高原寒带
plateau glacier 高原冰川
plateau gravel 高原砾石
plateau karst 高原岩溶
plateau lake 高原湖沼;高原湖(泊)
plateau length 坪长
plateau mountain 高原山;山原
plateau of accumulation 堆积高原
plateau of curve 曲线的平稳段
plateau of denudation-accumulation 剥蚀堆积高原
plateau of erosion 侵蚀高原
plateau period 高产稳产期
plateau plain 高原平原
plateau potential 平台期电位
plateau problem 坪问题
plateau ring structure 环形高原构造
plateau section 高原地区
plateau slope 坪斜
plateau station 高原测站
plateau subfrigid zone 高原亚寒带
plateau subtropic(al) mountainous region 高原亚热带山地
plateau temperate zone 高原温带
plate automatic bender 板材自动弯曲机
plateau transformer 高原变压器

plateau tropical north-fringe mountainous region 高原热带北缘山区
plateau T wave 前拱形 T 波
plateau value 平稳值
plateband 墙柱上的凸出线条缘饰
plate battery 阳极电池组;板状电池;板极电池
plate beam 板梁
plate bearing 平板支座;板支承
plate bearing structure 板支承结构
plate bearing test 板垫法承载力试验;(土壤的)板承压试验;钢板承载试验;平板加载试验;平板荷载试验;平板承载试验;承载板试验;荷载试验;板承(载)试验
plate bearing value (土基的)板承值
plate behind armo(u)r 装甲背板
plate belt 平面带输送
plate-belt feeder 板式给矿机;板式喂料机;板式给料机
plate bender 弯板机;板料折弯机
plate bending and shearing machine 板料折弯剪切两用机
plate bending machine 弯板机;板料弯曲机
plate bending rolls 卷板机;弯板机;辊式弯板机;弯曲滚板机
plate bending stand 弯板台
plate beveller 刨板边机
plate bin 金属板料仓;金属板料斗
plate blank 板坯;刻图片
plate-blasting shop 钢板喷砂车间
plate block 挂钩砖;平板砖
plate bolt 锚板螺栓;大帽螺栓;板极钉
plate boundary 板块界限;板块结合带;板块界
plate boundary activity 板块边缘活动性
plate boundary earthquake 板块边界地震
plate brake 薄片制动器
plate brass 黄铜板
plate brushing machine 刷(金属)板机
plate bubble 照准部水准器
plate buffer 圆板缓冲器
plate by-pass capacitor 板极旁路电容器
plate cab(in) 金属板小室;金属板座舱
plate calender 平板砑光机
plate cam 平板凸轮
plate camera 硬片照相机;干板照相机
plate cap 阳极帽;盖板;板极引线;板极帽
plate capacitance 板极电容
plate carrier 硬片暗盒
plate-cavity tuning shaft 板极谐振腔调谐轴
plate ceiling 石膏板顶棚
plate center 底片中心
plate change 硬片递换;换底片
plate changing magazine 换暗盒
plate characteristic 板极特性
plate chimney 金属板烟囱
plate choke coil 板极扼流圈
plate-chopping machine 板式切割机
plate circuit 阳极回路
plate circuit breaker 板极断路器
plate circuit detector 板极检波器
plate circuit efficiency 板极电路效率
plate circumference 平板周边
plate clamps 直压板夹具;钢板夹钳
plate clarifier 斜板沉淀池
plate cleaner 平板选煤器;平板分选机
plate clearance 板极间距
plate clippings 金属板剪切;金属板裁剪
plate clock 圆盘式挂钟
plate closer 钢板压环
plate closet 壁橱;餐具室;餐具壁橱;盆碟室
plate clutch 圆盘离合器;闸片离合器;片式摩擦离合器;盘式离合器
plate coil (exchanger) 螺旋板热交换器;螺旋式换热器
plate coils 钢板套管
plate collision 板块碰撞
plate column 塔板式蒸馏塔;多层柱;板式蒸馏塔;板式塔
plate compactor 振动捣实器
plate concrete vibrator 平板式混凝土振动器
plate conductor 阳极导体
plate conjunction 像片联测
plate connection 板材连接
plate connector 组合插座
plate constant 底片常数

plate constant variance 底片常数方差
plate construction type 板结构类型
plate contact freezer 平板接触式冷冻器
plate container 金属板容器;金属板集装箱
plate control 阳极调整
plate convergence 板块会聚
plate conveyer 平板输送带;平板运输机;板式输送机
plate cooler 冷却板
plate coordinate 摄影坐标
plate copying 晒版
plate correction 层间改正
plate count 平皿计数;平板测数
plate count agar 平面培养计数琼脂
plate count technique 平皿计数技术
plate-coupled 板极耦合
plate-coupled multivibrator 板极耦合多谐振荡器
plate coupling 片式离合器
plate-covered step 金属片覆盖的梯阶
plate cover(ing) 平板覆盖
plate cowl 金属板外壳;金属罩
plate cramp 金属板夹钳
plate crane 铁板起重机;板材吊车
plate crusher 颚式轧碎机;颚式轧碎机
plate crystal 片状晶体;片状结晶
plate culture 平皿培养;平板培养
plate current 阳极电流
plate-current cut-off 板流截止
plate cut 底片开挖(基础开挖中的一种);橡子搁置口;板切口;垫板割切;橡子檐截口
plate cut-off current 板极截止电流
plate cutter 中厚板剪切机
plate cutting shop 钢板切割车间
plated adjustable turning chair 电镀活动靠背转椅
plated amberian 乳白衬里琥珀玻璃板
plated bar 熟铁条
plated beam 叠镀梁
plated box pile 焊接板箱形桩
plated cermet coating 电镀金属陶瓷层
plated chromium finish 电镀铬面
plated circuit 印刷电路;印刷板电路
plated construction 板结构
plated diamond tools 电镀金刚石工具
plate decoupling 板极去耦
plate deflection 平板挠度
plate depth 平板厚度
plate destruction 板块消亡;板块消减
plate detection 阳极检波;屏极检波;板极检波
plate-detection type vacuum tube voltmeter 板极检波真空管电压表
plate-detection voltmeter 板极检波伏特计
plate detector 阳极检波器;板极检波器
plated fabrics 添纱织物
plate different movement 板块差异运动
plate diffuser 板式扩散器
plate dimension 图幅;底片尺寸
plate discharger 板形放电器
plate displacement 平板位移
plate distortion 底片变形
plate distributor 分配器盘
plated mast 空心钢桅;铆接板桅
plated metal 沉积金属
plated mo(u)lding 贴附饰;安装上的线脚
plate door 隔舱壁上装的可移动式铁门
plate doubler 金属板并丝机;金属板折叠机
plate dowel 片销;片栓;片暗销;板销;板榫
plated parquet 板构镶木细工
plated printed circuit 印刷电路;平板式印刷电路
plate-driving force 板块驱动力
plated rudder 复板舵
plate drum 金属板滚筒;金属板卷筒
plate dryer 多层干燥器;多层干燥机
plate drying table 金属板干燥台
plated structure 板结构
plated-through hole 金属化通孔;金属化孔;穿透镀膜的小孔
plate duct 金属板管道
plate duplicate 复制版
plated wire 镀磁线
plated wire memory 镀线存储器;磁膜线存储器
plated wire storage 镀线存储器
plated with brass 镀以黄铜的;镀黄铜的
plate dynatron oscillator 板极负阻管振荡器
plate earthing 板接地
plate edge 板边

plate edge bevel(l)ing machine 斜边刨床
plate edge planer 板材刨边机
plate edge planing machine 板边刨床
plate edges cutting 板边切割
plate efficiency 塔板效率;板效率;板极效率
plate efficiency factor 塔板效率因子
plate electrode 电极板;板状电极;板极
plate electrophoresis 板电泳(现象)
plate electrostatic precipitator 板式除尘器
plate element 板块要素
plate engraving ink 雕刻凹版油墨
plate enlargement ring 涨圈簧
plate erector 钢板工
plate etching 玻璃板蚀刻
plate evapo(u)rator 金属板蒸发器
plate exchanger 板式换热器;薄片式热交换器;板形换热器
plate facing 平板饰面;平板衬砌
plate factor 底片因子
plate failure 平板破坏
plate feeder 圆盘给料机;板式喂料机;板式进料机;板式给料器;板式给料机
plate ferrule 管板密套
plate-filament capacity 板丝电容
plate filter 片式滤清器;板式过滤器;板式过滤机
plate filter press 板式压滤机
plate-fin cooler 散热片(式)冷却器
plate-fin exchanger 板翼热交换器
plate-fin heat exchanger 板翅式换热器
plate finish 平板加工
plate finishing machine 铅版整修机
plate flanging machine 板材弯边机
plate flattening machine 板材矫正机
plate floor 平板楼面;平板肋板;实肋板
plate flow sheet 板流程图
plate-folding machine 栅栏式板夹机
plate-follower 阳极跟随器
plate for fixing work 固定工件夹板
plate format 板式结构
plate-form ga(u)ge 成型样板
plate-form roof 平屋顶;平塔顶;截头屋顶
plate-form spring 板簧
plate form trailer 板式拖车
plate form(work) 金属板模型;金属板模板
plate for standing seams 竖缝板
plate for ultra-microminiaturization 超微粒干版
plate for walls 墙板
plate frame 像框;板肋
plate-frame type oil filter 板框式滤油器
plate freezer 平板式冻结机
plate funnel 金属板烟囱;金属板漏斗
plate fuse 片状熔断器
plate gate 船坞插板门
plate ga(u)ge 板卡尺;板材量具;板规;样板
plate girder 板制桁架;由钢板和角钢拼制的梁;钢板梁;巨梁;板桁;板大梁;铁板梁
plate girder bridge 板梁桥
plate girder chord 板梁弦杆
plate girder deck bridge 板梁上桥
plate girder depth 板梁高度
plate girder flange splice 板梁翼缘拼接;板梁翼缘接合
plate girder joist 工字板梁
plate girder span 板梁宽度
plate girder stalk 板梁腹板
plate girder web 板梁腹板
plate girder web splice 板梁翼缘拼接;板梁翼缘接合;板梁腹板拼接;板梁腹板接合
plate glass 镜子玻璃;厚玻璃板;厚板玻璃;平板玻璃;图案玻璃;玻璃板;板玻璃
plate glass door 平板玻璃门
plate glass furnace 平板玻璃池窑
plate glass panel 镜面玻璃壁板
plate glass shelf 玻璃搁板
plate glass tank 平板玻璃池窑
plate glass window 玻片观察孔
plate glazing 平板压光
plate glazing calender 平板砑光机;平板抛光机
plate grab 抓板器;钢板爪;吊板钩
plate graphitic heat-exchanger 板式石墨换热器
plate grid 第二栅极
plate-grid capacitance 板极栅极电容
plate-grid capacity 板栅电容
plate grinder 磨版机

plate grinding machine 磨版机
plate grip 板材夹具
plate gripping tongs 板材夹钳
plate groove exchanger 板槽式换热器
plate gutter 平板排水沟;平板檐槽
plate-handling crane 板材运输起重机
plate heat exchanger 膜片式热交换器;板式热交换器
plate height 塔板高度;板高(度)
plate hinge 板铰
plate hoist 厨房餐具升降器;厨房用升降机
plate holder 硬片暗匣;底片盒;承片框;板夹
plate hood 金属板盖;金属板罩
plate hopper 金属板漏斗;金属板料斗
plate hypothesis 板块假说
plate ice 荷叶冰;饼状冰;冰饼
plate impedance 电子管内阻;板极阻抗
plate inductor 屏路扼流圈;板极扼流圈
plate inspection 板材检查
plate inverse voltage 极板反电压
plate iron 中厚钢板;铁板
plate jams 板块挤压
plate jib 实体式起重机臂;板式挺杆
plate jig 平板式夹具;板式钻模
plate jiggering machine 平盘旋坯机
plate joint 平板接合
plate junction 板块结合带
plate juncture 板块结合带;板块边界
plate keel 平板龙骨
plate keying 板极键控法
plate layer 线路维修工;线路工区工人;铺路工【铁】;养路工
platelayer's ga(u)ge 轨距规
platelayer's troll(e)y 养路工手摇车;巡道车(铁路)
platelet 小晶片;片状物;片晶;小片状体
plate level 照准部水准仪;照准部水准器;上盘水准器;度盘水平
plate-level ga(u)ge 板式液面计
plate-leveling machine 板材校平机;钢板矫平机
plate level(l)er 中厚板矫直机
plate library 底片库
plate-like 似片状的;板状的
plate-like body 板状体
plate-like clay 板状黏土
plate-like crystal 板状晶体
plate-like particle 片状粒子;板状颗粒
plate limiter 板极限制器
plate line 支承椽子的一部分墙
plate line level bucket capacity 板线斗容【疏】
plate lining 平板镶衬
plate link chain 平滑板链;板链;板环链
plate linseed oil 板状亚麻籽油
plate liquid cooler 板式液体冷却器
plate load 平板荷载;阳极负荷
plate load bearing structure 平板承重结构
plate load(ing) test 荷载试验;承载板加压试验;平板荷载试验;平板加载试验
plate-loading test procedure 荷载试验方法
plate load test 板荷载试验;平板荷载试验;板材荷载试验
plate lock 木壳销;板销
plate lug 极板接线片
plate-machine 制皿机
plate magazine 底片暗盒
plate maker 刻版机
plate-making 制版
plate manometer 膜片压力计;膜片压力表;片式压力表
plate margin 板块边缘
plate mark 磨料伤痕;抛光盘刻痕
plate marking 板式标号;板材标号
plate marking machine 板材划线机
plate measurement 硬片量测
plate measuring machine 底片量度仪
plate metal work 钣金工
plate mica 云母板
plate micrometer 公法千分尺
plate mill 制板厂;制板机;轧板机;磨盘式磨粉机;板材轧机
plate mill roll 钢板轧辊
plate mill stand 钢板轧机机座
plate modulated 屏调制
plate modulation 阳极调制;板极调制;板极调幅
plate modulus 平板模数
plate moment 板的力矩

plate motion 板块运动
plate mo(u)lding 模板造型;模板造型;板造型
plate movement 板块运动
platen 压纸卷轴;压板;机床工作台;平板;台板;承纸卷轴
plate nadir 像片天底点
plate nadir point 天底点
plate nail 钩头道钉
platen airheater 板式空气预热器
platen cover 模板罩
plate negative 硬片
plate neutralization 板极中和
platen heater 电热板
platen heat exchanger 屏式热交换器
plate nip 印版接口
platen knob 滚筒旋钮
platen machine 平压印刷机
plate nodule 板状结核
platen press 印压机;平压印刷机;平板硫化机
platen-pressed chipboard 平板碎纸胶合板;平板粗纸板;平板刨花板
platen-pressed particle board 平压碎纸板;平压刨花板;平压颗粒板;平行碎料板
platen reheater 屏式中间再热器
platen release lever 滚筒释放杆
platen superheater 屏式过热器
plate nut 带铆接凸缘螺母
platen variable knob 滚筒可变旋钮
platen variable mechanism 滚筒可变机构
plate oblique thrust 板块斜冲运动
plate of a laminated spring 叠板片簧
plate of constant depth 等厚度板
plate of tripod 三角架顶
plate oil 凹版油墨用油;凹版墨料
plate oil-filter 板框滤油机
plate orifice 盘形孔板
plate orifice meter 板孔流量计
plate out 一次性起霜
plate overlap method 底片重叠法
plate packing 板片密封垫
plate panel 镶板
plate panel partition (wall) 镶板隔墙
plate parquet 板构镶木细工;有背架的镶木地板
plate partition (wall) 板分隔墙
plate pewter 锡锑合金板
plate piler 板堆垛机
plate plane 像面
plate plumb point 天底点
plate-polishing machine 板抛光机
plate potential 板极电位;板极电势
plate precipitator 平板除尘器
plate press 板式压滤机
plate printing ink 凹版油墨
plate problem 平板问题
plate processing 板材加工
plate processing shop 板材加工车间
plate protector 片状避雷器
plate-pulsed transmitter 板极脉冲调制发射机
plate-punching machine 板材冲床;冲孔机
plater 钢板工;冷作工;钣金工;电镀工人;壳板装配工;金属板工;喷镀工人;涂镀装置;涂层装置;镀层装置
plate rack 餐具架;盘碗架;存板架;餐具橱;盘架
plate radiator 片式散热器
plate rail (室内陈设瓷器的)线脚槽;平轨;装饰品格架;盘碗壁架;板轨
plate rainwater gutter 金属板雨水槽
Plater brass 布拉特黄铜
plate rectifier 板极检波器
plate regenerator 板式回热器
plate reservoir 金属板容器;金属板储[贮]器
plate resistance 板极电阻
plate-resistance bridge 测量电子管内阻的电桥;板阻电桥
plateresque 豪华装饰;仿银器(华丽)装饰
plateresque architectural ornamentation 富丽装饰【建】
plateresque architecture 仿银器(华丽)装饰建筑
plateresque ornamentation 华丽装饰
plateresque style 银匠式建筑风格(16世纪西班牙)
plate ribbed arch bridge with circular openings 圆筒片拱桥;圆洞片拱桥
plate rigidity 承载板试验刚度;板承刚度

plate roll 金属板碾;金属板滚筒;弯板机;卷板机
plate roller 滚板机
plate roller gate cylinder 金属板圆筒;金属板碾筒装砂石口
plate rolling 中厚钢板轧制
plate rolling mill 平板轧机
plate rolls 直板机
plate roof 板材屋顶
plate roofing 板材铺屋面
plate rotation 平板转动
plates 板材
plate saddle hanger 板式鞍形吊架
plates and bubble caps 塔板和泡罩
plate saw 平板圆锯;板锯
plate scale 底片比例(尺)
plate scrap 板材碎片
plate screen 孔板挡污栅;平板筛;平板格网;板孔筛
plate-screen modulation 板帘调制
plate screw 金属板螺桨;板式螺旋;底板螺旋
plate scrubber 板式洗涤塔
plate scullery 洗碗室;洗盘间;洗碟室
plate scum collector 金属板沉淀池浮渣收集器
plate setting 铺板
plate settler 斜板沉淀池
plate shackle 平钩环;扁卸扣
plate shaking mechanism 极板振打机构
plate shale 板状页岩;板页岩
plate-shaped drill 盘式凿岩机;碟形钻头凿岩机
plate shearing machine 剪板机
plate shears 剪板机;剪板器
plate-shear test 平板剪切试验
plate shelf 墙上装饰品支架
plate shop(mill) 制板厂;制板工厂
plate silo 金属板筒仓
plate size 板的尺寸
plate sketch 透写图
plate skin 金属板表皮
plate slab 钢板坯
plate slab mill 板坯轧机
plate slap 印版晃动
plates of unequal thickness 不等厚度的平板
plate span 平板跨度
plate spring 钢板弹簧;片弹簧;板簧;板弹簧
plate spring assembly 板簧总成
plate spring pressure ga(u)ge 板簧压力表
plate squaring shears 门式剪板机
plates specifications 板材标准
plate-stalk girder 腹板大梁
plate-stalk stiffness 腹板劲度
plate-stalk strength 腹板强度
plate-stalk structure 腹板结构
plate-stalk supported on three sides 三边支承的腹板
plate-stalk supporting structure 腹板承重结构
plate-stalk(sur)facing 腹板表面处理
plate-stalk testing machine 腹板试验机
plate-stalk theory 腹板理论
plate-stalk thickening 腹板加厚
plate-stalk translation 腹板平移
plate-stalk type of construction 腹板构造类型;腹板构造形式
plate-stalk wall 腹板墙
plate staple 骑马钉板
plate steel 钢板;钢板;厚钢板
plate steel case 钢板机壳
plate-steel liner 钢板衬砌;钢板衬垫
plate stem 组合船首材;平板船首柱
plate stiffener 加劲板
plate stiffness 板的劲度
plate stock 板材
plate stone 石板;板岩
plate straightener 整直滚板机;钢板矫直机
plate-straightening machine 整直滚板机;中厚矫直机
plate-straightening roll 板料矫平机;直板滚板机
plate stratigraphy 板块地层学
plate streak 平均画线法;平板画线法
plate-strenghtening machine 钢板矫直机
plate stretcher 金属薄板矫直机;中厚拉伸矫直机
plate strip 板条
plate structure 薄板结构;板式结构;板块构造
plate subduction 板块消亡;板块消减;板块俯冲
plate supply 阳极电源;板极电源
plate supply filter 板极电源滤波器

plate-supported belt feeder 有托板的胶带给料机
plate suspension roof(ing) 钢板悬吊屋顶;悬吊式板屋顶
plate-tamping machine (混凝土块生产中的)平板夯捣机
plate tank 金属板容器;硬片洗印槽;板极振荡回路
plate tectonic 板块构造的
plate tectonic cycle 板块构造轮廓
plate tectonic model 板块构造模型
plate tectonic model for metallogenesis 板块构造成矿模式
plate tectonics 板块构造
plate tectonic theory 板块构造学说
plate terminal 板极引出线;板极端子
plate test 板材试验
plate testing machine 板材试验机
plate theory 塔板理论;薄板理论;板块理论
plate thickness 板材厚度;平板厚度
plate tile 平板瓦
plate tin 镀锡钢板
plate-to-plate calculation 逐板计算
plate-to-plate impedance 板极间阻抗
plate-to-plate spacing 极板间隔
plate tower 层板蒸馏塔;板式蒸馏塔;板式塔
plate tracery 板型窗花格;平板窗花格;平板窗格;板制窗花格
plate train 平板列车
plate transformer 板极变压器
plate tuning oscillator 板极调谐振荡器
plate turn 平转弯
plate turnover 板材翻转机
plate-type 板式
plate-type absorber 隔栅式吸声装置
plate-type blanking die 板式落料模
plate-type changeover valve 板式转换阀
plate-type condenser 板式冷凝器
plate-type contraction joint 金属片伸缩缝
plate-type delivery valve 平板式增压阀
plate-type dome 板型圆顶
plate-type electrostatic dust-precipitator 板式静电收尘器
plate-type electrostatic precipitator 板式静电除尘器
plate-type electrostatic separator 板式静电分选机
plate-type evapo(u)rator 板式蒸发器
plate-type exchanger 平板式热变换器
plate-type feeder 板式给料机
plate-type heat condenser 板式冷凝器
plate-type heat exchanger 板式(热)交换器;板片式换热器
plate-type martensite 片状马氏体
plate-type precipitator 板状沉降机;板式除尘器
plate-type pressure filter 板式压滤机
plate-type pressurized oil filter 板式加压滤油机
plate-type regenerator 板式热交换器
plate-type thread rolling machine 板式滚丝机
plate-type tread 板式梯级
plate-type vibrator 平板式振动器;平板式振捣器
plate-type vibratory compactor 振动压实盘
plate valve 平板阀(门);片状阀;片阀;盘形阀;板式活门;板式阀;板阀
plate variometer 板极变感器
plate-vibrating compactor 平板式振动器;平板振动压实机;平板式振捣器;平板式振动压实机
plate vibration 台式振动
plate vibrator 平板振动器;平板振动机;平板振捣器;台式振动器
plate vibratory tamper 平板式振动夯压器
plate voltage 阳极电压;板极电压
plate-wagon 平板货车(铁路)
plate warmer 碗碟保温器;菜保温器
plate washer 洗碗器;洗碟器;平垫圈;板垫圈
plate wave 板波
plate-webbed arched girder with three hinges 三铰腹板拱梁
plate-web girder 腹板梁
plate weight-carrying structure 钢板承重结构
plate-welded spiral case 钢板焊接蜗壳
plate wheel 盘轮
plate whirler 拷版机;烘版机
plate width 板宽
plate winding 板极线圈
plate wire 镀线
plate work 钣金加工;钣金工工作;板材加工;板

加工工作;板构造物;板结构;带边梁板
plate-work and welding section 铆焊工段
plate working 板材处理
plate-working machine 板材加工机械;钢板精整机
platfond 顶棚
platform 承台;楼梯平台;月台;装卸台;站台;站平台;旅客站台;举重台;讲台;浅台;平台(式);平盘;台基;地台
platform(al) folded belt 台褶带
platform alignment 平台校准
platform-and-stake body 平台带栏条车体
platform announcement point 站台广播
platform assembly 装饰台总成;装卸台组件
platformate 铂重整产品
platform at portal 洞口平台
platform awning 站台雨棚
platform balance 台秤;磅秤
platform barrier 站台栅栏
platform based container 台架式集装箱
platform based container open sided with complete super structure 带有完整上部结构的敞侧台架式集装箱
platform based container with incomplete super structure and fixed corner post 带有不完整上部结构和固定角柱的台架式集装箱
platform based container with incomplete super structure and fixed ends 带有不完整上部结构和固定端结构的台架式集装箱
platform based container with incomplete super structure and folding ends 带有不完整上部结构和折叠端结构的台架式集装箱
platform bench scale 平台式台秤
platform body 平板车厢;平板车身
platform bollard 闸顶系船柱;闸底系船柱;平台系船柱
platform bridge 天桥
platform canopy 站台雨棚
platform car 站台手推车;平台车;平车;平板货车
platform carbonate rock association 台地碳酸盐岩组合
platform car kiln 台车式窑
platform classification 地台类型
platform compass 平台罗经
platform container 搁架式集装箱(只有柱架及底板);平台集装箱
platform conveyer 平台运搬器;板式运输机;板式输送机;平台输送机
platform crane 平台式起重机;台车起重机;月台起重机
platform crane car 平板吊车
platform deck 平台甲板
platform deep fracture 地台的深断裂[地]
platform dimensions 月台空间;平板货尺寸
platform-diwa reform type 地台—地洼改造型【地】
platform-diwa superimposition type 地台—地洼叠加型【地】
platform drier for bagged grain 袋装谷物平台干燥机
platform drill 台式钻机;台式凿岩机
platform drop 平台的下降
platform dryer 平台式干燥机
platformer 铂重整装置
platform erection 平台调准;平台安装
platform escapement 标准擒纵机构组件
platform evolution 地台的演化
platform facies 陆架相;台地相;地台相
platform fault 平台断层
platform field 台田
platform floor 车厢底板
platform for crane repair 吊车修理平台
platform form 站台形式
platform for remote sensing 遥感平台
platform frame 平台型厂架
platform framing 平台式构筑;平台框架
platform gantry 平台式龙门架;平台高架移动起重机
platform gate 站台栅门
platform grade 站台坡度
platform header 木楼梯平台丁头梁
platform height 站台高度
platform hoist 平台升降机;平台式起重机;平台起重机;平台卷扬机;挖掘吊盘绞车
platform ice foot 潮间(带)台状冰脚
platform in an underground station 地铁站台

platforming 铂重整
platforming process 铂重整过程
platforming reactions 铂重整反应
platform jubilee wagon 轻型平板车
platform ladder 平台梯架;平台式梯
platform length 站台长度
platform level 站台层
platform life truck 举高平台车
platform line 站台线
platform lorry 平板载货汽车;平台式载重汽车;平板车
platform loudspeaker system 站台扩音器系统
platform-marginal shoal facies 台地边缘浅滩相
platform-marginal organic reef facies 台地边缘生物礁相
platform-marginal slope deposit 台地边缘沉积
platform-marginal slope facies 台地边缘斜坡相
platform number 遥感平台编号
platform of grade crossing 道口平台
platform of hump crest 峰顶平台【铁】
platform of rigid construction 刚性平台
platform of slope 边坡平台
platform of sluice 船闸闸底
platform on piles 桩承台
platform on the crest of hump 峰顶平台【铁】
platform post 站台标
platform quay 平台式码头
platform quay wall 桩承台式(式)岸壁
platform railing 工作台栏杆;平台栏杆
platform reef 平台礁
platform region 地台区
platform road 站台线;站台边装卸线
platform roof 截头屋顶;月台棚;平屋顶;站台棚;平台层顶
platform roofing 月台顶棚;站台雨棚;站台防雨棚
platform scale 台秤;地中衡;地磅;磅秤
platform screen door 站台屏蔽门;屏蔽门
platform screen door control room 屏蔽门控制室
platform screen door system 屏蔽门系统
platform sea 地台海
platform semitrailer 平床式拖车;平板式半挂(拖)车;平板半挂车
platform servo 平台伺服机构
platform shed 站台雨棚
platform shelter 站台雨棚
platform side board 车厢边拦
platform sling 平板吊架;吊货盘
platform spring 平台弹簧
platform stage 地台阶段
platform stair(case) 平台式楼梯;油平台楼梯;层间双折楼梯;无井楼梯
platform structural layer 地台构造层
platform structure 站台结构
platform subway 站台地道;旅客站台地道
platform ticket 站台票
platform top arc bow 车厢帐篷弧形杆
platform track 站台线
platform trailer 平台拖车;平板拖车;平板挂车
platform truck 平台式起重车;平台车;平板式载重汽车
platform truck scale 汽车地磅
platform trunk 平台式运输车
platform tunnel (火车站的)进出站地道;月台隧道;月台地道;站台地道;旅客站台地下道
platform type 站台式;平台式
platform-type alumin(i)um-bearing formation 地台型含铝建造
platform-type coal-bearing formation 地台型含煤建造
platform type formation 地台型沉积建造
platform-type magmatic formation 地台型岩浆建造
platform-type manganese-bearing formation 地台型含锰建造
platform-type manure barrow 平台式厩肥小推车
platform-type metallogenic formation 地台型成矿建造
platform-type oil and gas-bearing formation 地台型含油气建造
platform-type oil shale-bearing formation 地台型含油页岩建造
platform-type phosphorus-bearing formation 地台型含磷建造
platform-type vibrator 振动台;板式振动器

platform-type working craft 平台型作业船
platform underpass 站台地道
platform vibrator 振动台;平台振动器;板式振动器;板式振捣器
platform wagon with sides 带栏板的平板拖车
platform wall 桩承台(式)岸壁
platform weighing machine 台式称量机;台秤
platform weighing scale 台秤
platform weight machine 台秤
platform wharf 平台式码头
platform wharf wall 桩承台式码头岸壁
platform-wicket 剪票口
platform width 站台宽度
platic instrument 成型质充填器
platina 铂金;粗铂
platinammines 铂氨络物
platinate 铂酸盐
platine 锌铜合金
plating 喷镀;外包金属板;镀金属;电解沉淀;电镀;船壳板
plating action 镀敷作用;电镀作用
plating balance 电镀槽自动断流装置
plating bath 电镀槽
plating department 电镀车间
plating efficiency 平板效率
plating industry 电镀工业
plating industry sewage 电镀工艺污水;电镀工业废水
plating iron 镀铁
plating line 镀覆机组作业线
plating liner 肋骨衬条
plating numerals 钢板编号
plating on plastics 塑料电镀
plating on steel 钢材涂层
plating-out 电解法分离
plating paper 印纹纸
plating rack 电镀支架
plating residue 电镀残渣
plating rinse wastewater 电镀清洗废水
plating room 电镀间
plating sequence 装版顺序
plating shop 电镀车间
plating solution 电镀液
plating system 钢板连接方式;钢板连接方法
plating tank 电镀槽
plating vat 电镀槽
plating waste 电镀废料
plating wastewater 电镀废水
platinian carrollite 含铂的硫铜钴矿
platiniancopper 含铂自然铜
platinian vsotskiter 含铂的硫钯矿
platinic 铂的
platinic acid 铂酸
platinic chloride 氯铂酸
platinic hydroxide 氢氧化铂
platinic sulfate 硫酸铂
platinic sulfide 二硫化铂
platiniferous 含铂
platiniferous sand 铂砂
platiniridium 铂铱矿;铂铱合金
platinite 硫硒铋铅矿;赛白金;代白金
platinite alloy 高镍铁合金
platinization 披铂;镀铂作用
platinize 镀铂
platinized asbestos 载铂石棉;披铂石棉;铂石棉
platinized carbon electrode 镀铂碳电极
platinized charcoal 披铂炭
platinized platinum 铂片
platinizing bath 镀铂浴
platino 金铂合金(含铂百分之十一);金铂合金
platinochloride 氯铂酸盐
platinocyanide 铂氰化物
platinode 铂极
platinogold 铂金页
platinoid 假铂;假白金;镍铜锌合金电阻丝;铜镍锌电阻合金;铂状的;铂铜
platinoid solder 铜镍锌焊料
platinoid wire 铜镍锌合金丝
platinoiridita 铱铂合金
platinoiridium 铱铂合金
platinotron 高原管;磁控放大管;铂管
platinotype 铂印相法
platinous bromide 溴化亚铂
platinous chloride 二氯化铂

platinous cyanide 氰化亚铂
platinous hydroxide 氢氧化亚铂
platinous iodide 碘化铂
platinous sodiumm chloride 氯亚铂酸钠
platinum 铂金钴度量;白金
platinum alloy 铂合金
platinum analyzer 铂含量分析仪
platinum anode 铂阳极
platinum asbestos 铂石棉
platinum-based alloy 铂基合金
platinum-based catalyst 铂基催化剂
platinum-bearing quartz vein 含铂石英脉
platinum bichloride 二氯化铂
platinum black 铂黑;白金黑粉
platinum blue 铂蓝
platinum boat 白金舟
Platinum bolometer 铂测辐射热计
platinum bromide paper 溴化白金纸
platinum bronze 铂青铜
platinum bronze alloys 铂青铜合金
platinum bushing 铂金衬套
platinum catalyst 铂催化剂
platinum cathode 铂阴极
platinum chain 白金链
platinum chloride 氯化铂
platinum-cobalt method 铂钴法
platinum-cobalt scale 铂金钴度量
platinum-cobalt standard solution 铂钴标准液
platinum compounds 铂化合物
platinum cone 铂锥体;铂金锥
platinum contacet process 铂触法
platinum contact 铂触点;铂接点
platinum crucible 铂坩埚;白金坩埚;铂金坩埚
platinum dichloride 二氯化铂
platinum dicyanide 二氰化铂
platinum difluoride 二氟化铂
platinum diiodide 碘化铂;二碘化铂
platinum dish 白金杯
platinum ear-pick 白金耳勺
platinum electrode 铂丝电极;铂电极;白金电极
platinum electroplating 电镀铂
platinum family element analysis 铂族元素分析
platinum family elements 铂族元素
platinum filament 铂丝
platinum foil 铂箔
platinum-free glass 无铂玻璃
platinum-gold-copper quartz vein 含铂金铜石英脉
platinum-group elements 铂族元素
platinum-group glass 铂族玻璃
platinum-group metal alloy 铂族金属合金
platinum-group metal chalcogenides 铂族金属硫属化物
platinum-group metal halides 铂族金属卤化物
platinum-group metal ore 铂族金属矿
platinum-group of metals 铂族金属
platinum iodide 碘化铂
platinum iridio 铂铱耐蚀耐热合金;铂铱合金
platinum-iridium catalyst 铂铱催化剂
platinum-iridium needle 铂铱针
platinum-iridum alloy 铂铱合金
platinum isotopes 铂同位素
platinum-lamp 白金丝的白热电灯
platinum loop 铂环;白金圈;白金环
platinum loss 铂耗量
platinum luster 铂光泽(彩)
platinum metal alloy 铂族合金
platinum metals 铂类金属
platinum method 白金法
platinum microelectrode 铂微电极
platinum mirror coating 铂镜面镀层
platinum Mohr 铂黑(金属粉末)
platinum needle 白金针
platinum-nickel alloy 铂镍合金
platinum ore 铂矿石
platinum paper 铂盐相纸
platinum photocell 铂光电元件
platinum placer 砂铂矿床
platinum-plated copper 镀铂铜
platinum plating 镀白金
platinum-platinum nickel couple 铂铂镍热电偶
platinum-platinum-nickel thermocouple 铂铂镍温差电偶
platinum-platinum-rhodium thermocouple 铂铂铑热电偶;铂铂铑温差电偶

platinum-platinum thermoelement 铂铂热敏元件
platinum potassium chloride 铂氯酸钾
platinum powder 铂粉
platinum process 白金法
platinum product 铂材
platinum pyrometer 铂高温计
platinum reforming 铂重整
platinum resist 铂光泽彩露花(陶瓷装饰法)
platinum resistance sensor 铂电阻敏感器
platinum resistance temperature needle 铂电阻测温探针
platinum resistance temperature sensor 铂电阻感温计
platinum resistance thermocouple 铂电阻热电偶
platinum resistance thermometer 铂阻温度计;铂电阻温度计
platinum resistor 铂电阻器
platinum-rhenium reforming 铂铼重整
platinum-rhodium gauze pad 铂铑丝网垫
platinum-rhodium heating element 铂铑发热体
platinum-rhodium thermocouple 铂铑热电偶
platinum ring 铂环
platinum sensor 铂传感器
platinum sheet 铂片
platinum-silica-alumina catalyst 铂硅氧化铝催化剂
platinum-silver 铂银合金
platinum-silver alloy 铂银合金
platinum sparkler 铂闪烁点
platinum sponge 铂绒;铂棉;铂海绵
platinam steel 白金钢
platinum still 铂蒸馏釜
Platinum strip bolometer 铂带测辐射热计
platinum-substitute crucible 代铂坩埚
platinum sulfate 硫酸铂
platinum tetrachloride 四氯化铂
platinum thermocouple 铂热电偶
platinum thermometer 铂丝温度计
platinum-tin catalyst 铂锡催化剂
platinum-tungsten alloys 铂钨合金
platinum wire 铂丝;白金丝
platinum wire electrode 铂丝电极
platinum wire liquid level sensor 铂丝液位传感器
platinum wire temperature sensor 铂丝温度敏感器
platinum wire thermometer 铂丝温度计;铂丝温度表
platinum-wound tube furnace 白金丝管式炉;白金丝管式锅炉
platinum yellow 铂黄
platnam 普拉特纳姆镍铜合金
platometer 面积仪;平面求积仪;测面仪
Platonic year 大年;柏拉图年(分点绕黄道一周约为25800年)
platonychia 扁平指甲
platoon 车队
platoon commander 小队长
platoon diffusion 车队散布
platoon diffusion theory 车队散布理论
platoon dispersion 车队散布
platoon dispersion theory 车队散布理论
platoon identification 优先进入线路的信号联动系统
platoon identification scheme 交通信号联动系统
platoon length 车队长度
platoon metering 车队限流
platoon movement 车辆成队行进
platoon speed 车队速度
platt 墙上横木
platted mo(u)lding 编织物状的装饰线脚;饰编织物状的线脚
platter 大浅盘;大碟
plattering 窑顶
platting 绘制地图;编织物
plattnerite 块黑铅矿
platy 板状的
Platybelondonidae 铲齿象亚科
platycelian 前凹后凸的
platycelous 前凹后凸的
platycephalic 扁头的
platycephalous 扁头的
platy flow structure 片板状流动构造;板状流动构造;板流构造
platy grain 片状颗粒
platy jointing 板状节理
platykurtic distribution 低峰态分布

platy kurtosis 平低阔峰;低峰态
platy limestone 板状灰岩
platy mineral 片状矿物
platy-monocrystal 片状单晶
platynite 硫硒铋铅矿
platyonychia 扁平指甲
platy rock 板岩石
platy soil structure 板状土壤结构
platysome 平板体
platystencephalia 扁长头
platystencephalism 扁长头
platy structure 板状构造
platy talc 扁平滑石
platy texture 板状结构
plauenite 钾正长石
Plauen limestone 普劳恩石灰石(德国)
plausibility check 逻辑检查
plav 芦苇沼泽;河滩
play 空隙;闪动
playa 干盐湖;干荒盆地;河口沙地
playa deposit 干盐湖沉积
play adjustment 隙缝调整
playa lake 雨季浅水湖;雨季湖
playback 读出;重演;重放;放音;放声;反演
playback accuracy 再现精度
playback amplifier 重放放大器;放音放大器
playback button 重放按钮
playback control 重放控制
playback control duty 重放控制占空因素
playback equalizer 重放均衡器
playback facility 播放设备
playback head 读头;重放磁头
playback mode 再现状态;重放状态
playback robot 学演机器人
playback system 回放系统
play button 工作按钮
player 局中人
playfairite 普硫锑铅矿
playfield park 运动公园
play film 铁水面花纹
playful use of space 空间娱乐
playground 运动场;游戏场;儿童游戏场
playground equipment item 运动场设备项目;运动场设备
play-guide 影剧院预售票处
playhouse 儿童游戏室;剧场
play in axle-box 轴箱隙隙
playing fast and loose with funds 乱支乱用
playing field 游戏场;运动场
play in the couplings 联轴节窜动量
play lot 儿童游戏场地;儿童活动园地
play movement 接头间隙
play of colo(u)r 色彩闪变;变色;变彩
play-off beam 抹迹射束;回扫射束
play of forces 力的起伏
play of light and shadow 光与影的运用
play of movement 间隙
play of valve 阀隙;阀门缝隙;阀的间隙
play out 尖灭
play pipe 游管
play-pulverulent structure 片状粉块结构
playroom 娱乐室;文娱室;文娱活动室;游戏室;儿童游戏室
play sculpture 游戏雕刻
play smash 破产
play street 游戏街道
play switch 工作开关
play the role of van-guard 发挥带头作用
play up steel production 突出钢铁生产
play yard 游戏场
plaza 广场;集市场园
plazas de toros 斗牛场
plaza system 广场系统
pleached 树枝编织的
pleached alley 交叉窄道
pleasance 游乐场;游乐园;庭园
please advise 请指导
please confirm 请确认
pleased 满意的
please exchange 请交换
please note 请注意
please pay 请付款
please reply at once 请即刻答复

please turn over 请阅后页;请看背面
pleasure boat 游艇;游船
pleasure boating 驾驶游艇
pleasure city 观光城市
pleasure craft 游艇
pleasure cruise boat 游览船
pleasure diving 娱乐潜水
pleasure dome 富丽堂皇的大厦
pleasure-garden 游览公园;游乐公园
pleasure ground 娱乐场
pleasurehouse 娱乐室
pleasure steamer 游览汽艇;游览船;观光船;旅游船
pleasure traffic 游览车;游览交通
pleasure trip 游览
pleat 打褶;褶叠层;褶层
pleated cushion 折垫
pleated filter 折叠过滤器
pleated prefilter 织物颗粒过滤器
pleated sheet structure 折叠片状结构
plectenchyme 密丝组织
plectenchyminite 假薄壁菌质体
plectostele 编织中柱
pledge 质押;受抵押人;抵押器;抵押品;抵押;否定保证
pledge assets 抵押资产;出质资产
pledged account loan 抵押账户贷款
pledged assets 抵押资产;已抵押资产;质押资产
pledged savings account mortgage 担保储蓄账户抵押
pledgee 质权人;抵押权人;接受抵押者;承押人
pledge holder 抵押品持有人
pledge of immovable 不动产抵押
pledge of movable 动产抵押
pledge of obligation 债权抵押
pledge of protect forest 护林公约
pledger 抵押者;出质人
pledge securities 抵押有价证券
pledget 捻缝麻絮;纱布
pledge the credit of 假借信用
pledging of accounts receivable 以应收账款为抵押的(信贷)
pledgor 抵押人
plegmatium 侧褶区
pleietrepie 多效的
pleiobar 高压区;高压等值线;高气压等值线
pleiochasium 多歧聚伞花序
pleiomerous 多基数的
pleiomorphism 多态性
pleion 正距中心【气】;正距平区【气】
pleionomer 均低聚物;同性低聚物
pleiotomy 多分枝式
pleiotropy 多效性
Pleistocene epoch 更新世【地】
Pleistocene glaciation 更新世冰川作用【地】
Pleistocene period 更新世时期【地】
Pleistocene series 更新统【地】
pleistoseismic area 高地震区
pleistoseismic zone 强震带
plemontite 红帘石
plenary capacitance 全电容
plenary session 全体会议
pleniglacial period 盛冰期
plenilune 满月时期;望月
plenipotentiary 有全权的
plenish 装备用具(房间);安装设备(房屋)
plenishing nail 钉木板的大钉;安装用钉
plentiful discharge 丰水(流)量
plentiful runoff 丰水径流(量)
plenum 进气增压;全体会议;送气;充实;充气增压
plenum air 进入空气
plenum air chamber 高气压室
plenum barrier (悬挂顶棚结构中隔墙顶上的)隔声屏障
plenum box 充气箱
plenum chamber 静压室;风室;增压室;高压气室;静压箱;送汽包;集气箱;稳流室;通风室;通风恒压室;送气室;充水室;充气室;充风式
plenum chambered craft 气室型地板效应艇
plenum duct 送气通道;送风管道
plenum-exhaust ventilation 抽压联合通风
plenum fan 送气风扇
plenum grid 送气格栅;送气框格;送气格子;送风格子
plenum heater 空气加热器

plenum heating 空气加热;热风采暖;送风加热
plenum heating system 热风采暖系统
plenum hopper 通风漏斗;送气斗
plenum in 进入空气
plenum method of ventilation 压力通风法
plenum process 打气送风法
plenum process of tunnel(l)ing 隧道压气掘进法
plenum space 空气恒压室
plenum system 空气调节系统;压力通风系统;供气方式;恒压送风系统;通风系统;送气系统
plenum system of air supply 空气进入系统
plenum type preform machine 淋喷式预成型设备
plenum velocity 静压箱风速
plenum ventilation 送气通风
plenum ventilation method 压力通风法
plenum ventilation system 送风式通风系统
plenum with monolith concrete floor and nozzles 充水室充水加整体混凝土顶板和喷嘴
plenum with monolith concrete floor and porcelain spheres 充水室充水加整体混凝土顶板和瓷球
plenum with porous plates 充水室加多孔板
plenum with precast concrete blocks nozzles 充水室加预浇混凝土块喷嘴
plen-wood system 楼板下送气系统
pleochroic halo 多色光晕;多色光环;五色晕;多色晕;多色性晕
pleochroic halo age 多色晕年龄
pleochroic halo dating 多色晕年代测定法
pleochroic halo method 多色晕法
pleochroic halo ring 多色环
pleochroic halo width 多色晕宽度
pleochroism 多向色性;多色性;复色现象
pleochromatic 变色的
pleocrystalline 全晶质
pleocrystalline texture 全晶质结构
pleodont 实齿
pleomorph 多形晶
pleomorphic 多形的;多晶的
pleomorphism 同质异形(现象);同质多形现象;同质多晶(现象);多型性;多形性;多态性
pleonaste 镁铁尖晶石
plerome 中柱原
plerotic water 地下水
plesiostratotype 近层型
plessite 合纹石
Plessy's green 普雷赛绿
plethora 树木节瘤
plethysmogram transducer 容积描记图换能器
plethysmograph 容积描记器;体积描记器;体积描绘仪
plethysmography 容积描记法;体积描记法
Pleuroceridae 侧角螺科
pleuroclase 氟磷镁石
pleuston megalo-plankton 大型浮游生物
Plewa vent(ing) duct 普勒瓦通风管道
plex format 丛格式
plexglass 安全玻璃
plex grammar 丛状文法
plexicoder 错综编码器
plexiform 折叠状花饰;网状花饰;丛状的
plexiform layers 丛状层
Plexiglass 普雷克斯玻璃;有机玻璃;聚异丁烯酸树脂;聚甲基丙烯酸甲酯树脂;耐热有机玻璃;树脂玻璃;丙烯有机玻璃
Plexiglass model 有机玻璃模型
Plexiglass trough 有机玻璃水槽
plex processing 丛处理
plex programming 丛的程序设计;丛程序设计
plex structure 丛结构
plexus 超细纤维丛
pleysteinite 十字山石
pleyston 水面浮游生物
plfield replaceable unit 插件
pliability 可弯性;可挠性;柔顺性
pliability test 沥青韧性试验;韧性试验;弯曲试验
pliable material 可弯曲材料
pliableness 柔韧性
pliable paraffin 石蜡敷糊
pliable support 让压支架
pliancy 可挠性;柔韧性
plicated layer 褶皱层
plicated structure 皱纹构造
plication 细褶皱;褶皱作用

plication structure 皱纹构造
Plident-platoon identification scheme 车队识别方案
Plident system 优先进入线路的信号联动系统
plied yarn 合股线;合股纱
pliers 老虎钳;夹钳;钳子;手钳
pliers for radio mechanics and electricians 成套钳
pliers spot welding head 钳式点焊头;点焊钳
plier wrench 钳子扳手
Plimber(ite) 普林柏木板(一种建筑用木板)
p-limit 比例极限
Plimsoll's disk 载重线标圈
Plimsoll's line 载重线标志;吃水深度标记;干舷标志
Plimsoll's mark 载重线标志;载重吃水线标志;载货吃水标志;载货吃水线;干舷标志;吃水深度标记
plint 操练椅
plinth 砖墙压顶;柱基(础);勒脚;接头座;底座;础石;操练椅
plinth block 基底石块;基底砌块;门贴脸座;贴脸墩;踢脚墩(堵头);底座木块;门磴座
plinth board 门基板
plinth brick 勒脚砖;基座砖
plinth course 勒脚层;墙基石;墙基层;底座层;砖勒脚顶层
plinth for anchor arm 锚臂连接底座
plinthite 杂赤铁土
plinthitic horizon 网纹层
plinth level 柱座标高
plinth masonry wall 基础圬工墙
plinth masonry work 勒脚场工作业
plinth stone 地伏;柱础;基石;底石
plinth tile 勒脚瓦
plinth wall 勒脚;台脚
plinth walling 勒脚砌体;起勒墙
pliobond 合成树脂结合剂
Pliocene epoch 上新世【地】
Pliocene period 上新世时期【地】
Pliocene series 上新统【地】
pliotherm period 较高温期
pliotron 功率电子管
plioweld 橡胶金属结合法
plis diapir 底辟褶皱
Plochere colo(u)r system 普洛彻莱颜色系统
plodder 螺旋挤压机
ploidy 倍性;倍数性
plombage 充填术
plombierite 泉石华;温泉淬石
Plondrel 灭菌磷
plot 小区;小块土地;样方;计算;基址;区分;图表;地块;地段;测绘板;分划;绘制
plot a course 标绘航线
plot a curve 标绘曲线
plot a grid 展绘坐标网
plot a map 绘制地图
plot area 小区面积;占地面积
plot arrangement 小区排列
plot assignment 土壤小区划分
plot a traverse 测设导线
plot command 绘图指令
plot contour 描绘等高线
plot data 绘图数据;数控绘图
plot distance 标绘距离
plot error variance 小区机误变量
plot experiment 小区试验
plot file 绘图文件
plot form 分畦形式
plot interval 绘图间隔
plot lead 石墨
plotless method 非标准地取样法
plot line 建筑红线;地界线;定线;地基界线
plot map 像片索引图
plot mark 标记
plot measurement 小区测验
plot medium 绘图介质
Plotnikow effect 射线纵散射效应
plot observation 小区观测
plot observer 测绘员
plot of joint frequency 节理频率图
plot of land 基址
plot of screen test 筛分试验图
plot of sieve test results 筛子试验结果图
plot(o)mat 自动绘图机;自动绘图仪
plot origin 绘图原点
plot overlay 绘图薄膜

plot pen 描图笔;绘图笔
plot plan 地盘图;小区设计;总图;基址图;区划图;位置图;地区图;地段(平面)图;坝区平面图
plot planning 总平面布置;小区规划;建筑用地规划
plot point 展点;标绘点;标定点
plot program(me) 计划程序
plot ratio 分割比;容积率
plot runoff 小区径流
plot sampling core 中心取样
plot seeder 苗圃与试验地播种机
plot sign 小区设计
plots of kinetic biochemical oxygen demand 动态生化需氧量图
plot spraying equipment 地面喷雾器
plot survey 内业处理测量成果
plot table 测绘板;标图板;标绘板
plottable error 展绘误差
plottage 一块地皮的面积
plottage increment 小块土地合并后增值;并地增值
plottage value 由合并土地所增加的价值
plotted 标绘的
plotted curve 曲线图
plotted function 函数表
plotted line 规划线
plotted point 标出点
plotted sounding 测深图
plotter 描绘器;绘图者;绘图员;绘图用具;绘图仪;绘图器;绘图机;绘迹器;测图员;标图仪;标绘器
plotter controller 绘图机控制器
plotter-coordinatograph ratio 测图仪坐标联动比
plotter-digitizer 测图仪数字化器
plotter driver tape 绘图机驱动
plotter hood 绘图器遮光罩
plotter interface 绘图器接口设备
plotter interpolation 绘图机插补
plotter orientation 绘图桌定向
plotter output 绘图输出
plotter position 标定位置
plotter step size 绘图机步距;绘图机步长
plotter tape 绘图纸带
plot test 小区试验
plot the control points 展绘控制点
plot the details 展绘碎部
plot the photograph 描绘像片
plotting 绘制;作图;制图;绘图;绘曲线;画曲线;填图;标图;标绘
plotting accuracy 测图精度
plotting a curve 绘制(一条)曲线
plotting anti-collision 绘图避碰
plotting apparatus 绘图器;测图仪
plotting basic software 绘图基本软件
plotting beam 绘图射束
plotting board 描绘板;绘图板;输出函数表;测绘装置;测绘板;标图板;曲线板
plotting camera 测图镜箱
plotting chart 空白定位图;航行作业图;海洋空白图;标绘海图
plotting curve 绘制曲线
plotting data 绘图数据
plotting device 曲线绘制仪;绘图装置;曲线绘制器
plotting equipment 绘图装置;绘图仪器;绘图设备;绘图机
plotting error 展绘误差;绘图误差
plotting file 绘图文件
plotting film 绘图薄膜
plotting grid 绘图格网
plotting head 绘图头;反射作图器;反射描绘器;反射图器(雷达附件)
plotting instrument 立体测像仪;标绘仪器;绘图仪;绘图器;测图仪;标图仪器
plotting interval 标绘间隔时间
plotting lens 测图镜头
plotting light 绘图照明灯
plotting machine 展点仪;绘图员;测图仪
plotting material 绘图材料
plotting method 地图编绘法;测图方法
plotting mode 绘图方式
plotting model 绘图模式
plotting objective 测图物镜
plotting of angle by chords 弦线绘角法
plotting of angle by tangents 切线绘角法
plotting of control points 控制点展绘
plotting of epicentre 震中标绘
plotting office 绘图室

plotting of iso-density 等密度曲线描绘
plotting of points 展点
plotting of seismic data 地震数据标绘
plotting of the results 试验结果绘制线图
plotting of triangulation point 三角点展绘
plotting paper 绘图纸;绘图用方格纸;绘图方格纸;特种绘图纸;方格(绘图)纸;标绘纸;比例纸
plotting pencil 绘图铅笔
plotting plate 标绘板
plotting program(me) 绘图程序
plotting rate 绘图率
plotting ratio 绘图比(例)
plotting room 航海室;标图室
plotting scale 绘图比例尺;制图比例尺;测图比例尺;标图尺
plotting scale paper 绘图方格纸
plotting seismic data 标绘地震数据
plotting sheet 空白海图;空白定位纸;图板;标绘图
plotting speed 绘图速度
plotting square 直角坐标展点尺
plotting system 绘图系统;测图系统
plotting table 描绘板;绘图桌;图形显示图;输出函数表
plotting table orientation 绘图桌定向
plotting tablet 标图板
plotting tape 绘图纸带
plotting technique 绘图方法;测图方法
plotting unit 描绘设备;绘图桌设备;绘图设备
plotting velocity 绘图速度
plotting writer 绘图记录器
Plough 北斗七星
plough 窄式刮刀;沟刨;路犁;犁形器具;排雪器;爬犁
plough ability 可耕性
plough back 利润再投资
plough back profit 再投资用的利润
plough bed 耕作底层
plough board 犁板(平路机铲刀)
plough bolt 皿头方颈螺栓;防松螺栓;皿头方形螺栓
plough carriage 轮架
plough coulter 犁刀
plough deflector 装煤偏导板
plough digger 犁式挖沟机
ploughed 沟刨的
ploughed-and-tongued joint 企口接合;榫槽连接;嵌榫拼接;嵌榫接合;榫连接
ploughed bead 沟槽凸圆线脚;沟槽凸饰;沟槽凸饰
ploughed bead mo(u)lding 沟槽圆珠饰
ploughed blade (采煤机的)割刀;(采煤机的)刨刀
ploughed depth 翻耕深度
ploughed fallow 翻耕
ploughed field 耕地
ploughed land 耕地
plough feeder 犁式给料器
plough groove (物体表面的)沟槽
plough handle 犁把手
plough harrow 犁耙
plough horizon 耕作层;土壤耕作层
ploughing disc harrow 耕作圆盘耙
ploughing in green 压青;翻青
ploughing machine 犁田机
ploughing of cultivated land 熟地翻耕
ploughing season 耕种期;耕种季节
plough land 可耕土
plough layer 耕作层;耕翻层;耕层
plough pan 犁底磐;犁底层
plough plane(r) 槽刨;沟刨
ploughshare 铧;犁头;犁铧;开沟器
ploughshare mixer 犁耙式混料器
ploughshare vault 六勒拱顶
plough steel 犁钢
plough steel wire 特号制钢丝
plough steel wire rope 特号钢丝绳
plough strip 开槽木板条;抽屉边板
plough the sea 破浪前进
plough-type ditcher 犁形挖沟机
plough-type raking machine 犁形耙刮机
plough wind 犁风
plow 路犁;犁形器具;平地机
plow anchor 单爪犁锚
plow and tongue joint 企口接合;槽榫接合
plow back 以利润再投资
plow back earnings 再投资利润
plow back profits 用于再投资的利润
plow base 犁体

plow beam 犁梁
plow bit 凿子
plow blade 刮板;犁片
plow body 犁体
plow body grinding machine 犁体曲面磨床
plow bottom 犁体
plow bottom for sod and clay 草皮黏土地犁体
plow carriage 轮架
plow carrier 集电器架
plow-depth control 犁深调节
plowed-back earnings 扣薪(作年终分红之用)
plowed bead 沟槽凸圆线脚
plowed soil 翻耕土壤
plow for reclamation work 开荒犁(装有近似螺旋型犁体)
plow groove 板边头开槽;沟槽
plow harrow 犁耙
plowing 犁入(塑性岩石破碎机理)
plowing-back 资本化
plowing direction 耕作方向
plowing in the hold 平舱
plowing position 耕作位置
plowing season 耕种期
plowing sweep 沟耕铲
plow-in optic(al) cable 直埋式光缆
plow packer 犁后碎石镇压器
plow pan 耕层底心土层;犁底层
plow plane 沟刨;槽刨
plow point 犁头铁
plow reverser 双向犁翻转机构
plow scraper 梨形卸料器
plows curvature 犁面曲度
plow share 犁头;犁铧;开沟器
Plowshare geothermal plant 普洛谢尔式地热电站
plowshare twist 犁铧的穹顶;犁铧的穹隆
plow sole 犁底层
plow standard 犁柱
plow steel 缆绳钢;锋钢;犁钢
plow steel rope 高碳钢索
plow steel wire 铅淬火高强度钢丝
plow steel wire rope 铅淬火高强度钢丝绳
plow stick 犁柄
plow strip 抽屉边板;开槽木条
plow tail 犁尾
plow the ocean 破浪前进
plow tripper 梨形卸料器
plow type feeder 犁式给料器
plow-type trenching machine 犁式挖沟机
plow width 犁幅度
plow wind 犁风
ploybasic 多代的
ploy component sandstone 复成分砂岩
ployhalogenocarbon 多卤碳化合物
plstoleum 楼面覆盖层
pluck 支撑痕(陶器上);辊印;缺釉痕;陶器上的支撑痕;釉花
plucked finish 削平石面;颗粒状粗面;粗磨面
plucker 抓包机;摘取装置;采集装置
plucking 陶器上的支撑痕;冰蚀作用;冰川拔削作用;拔蚀作用;拔蚀;釉花
plucking cycle 采摘周期
plucking shears 采摘剪
plucking time 采摘时间
plug 旋塞芯;芯杆;芯棒;岩颈;压槽子砖的砖;柱销;给水栓;救火龙头;锚链孔塞;丝堵;栓销;疏水塞子;塞子;塞住;塞;堵塞物;带顶头轧制的;船底塞;沉积栓;插销【电】;插头;反相制动
plug adapter 插头;插座式转接器;插座合器;插塞;塞子;插塞式转换器
plug adapter lamp holder 插口灯座
plug a hole 堵塞漏洞
plug-and-feather hole 插楔开石孔
plug-and-feather method 半圆铁棒夹楔劈石法
plug and feathers 裂石楔;插楔开石工具
plug and filler 填补片
plug-and-jack type connector 插头插孔式连接器
plug and play 即插即用
plug and socket 插塞和插孔;插头插座
plug and socket arrangement 插接口装置
plug and socket connection 插销式连接
plug and socket connector 插销接头;插头插座
plug and wedge 裂石楔
plug back 封闭
plug-back cementing 井底灌水泥浆

plug base 插座
plug bit 不取心钻头;不取心金刚石钻头
plugboard 插头板;转接板;接插板;接线盘;接线板;插线盘;插线板;插(件)板;控制板;配线板;穿孔板
plugboard chart 插接图;插接板控制图
plugboard controlled capstan lathe 插销盘控制式转塔车床
plug bolt 膨胀螺栓
plug bond 塞钉式钢轨接续线;插头接线
plug box 塞绳式交换机;插线盒;插头箱;插塞盒
plug braking 反相制动;反相序制动;反接制动
plug brick assembly 塞砖装配
plug bush 塞套;插套
plug cartridge 塞线塞
plug-center bit 扩孔钻;带芯杆钻头;插头中心钻
plug clogging 滤布堵塞
plug cluster 多插头端子板
plug cock 全油道塞;柱塞开关;龙头;旋塞开关;旋塞;转阀;塞阀;有栓旋塞
plug compatible 插头兼容性;插接兼容
plug-compatible mainframe 插接兼容机
plug-compatible manufacturer 兼容设备制造商;插件兼容制造厂家
plug-compatible memory 插接兼容存储器
plug-connected attenuator 插接式衰减器
plug connection 插接销;插头连接
plug connector 插塞式连接器;插塞接头
plug contact 插头
plug cord 塞绳;插头(软)线
plug core 塞子
plug cover fuse cutout 塞式熔断器
plug cut-out 塞式保险器;插塞式保险器
plug cutter 塞形钻孔器
plug diameter 柱塞直径
plug die 浮动塞棒
plug dome belonite 岩塞丘
plug drain valve 栓塞式排水阀
plug drawing 钢管定径拉拔法
plug drill 冲击凿岩机
plug-driving gun 孔塞驱进枪
plug effect 土塞作用
plug end fitting 终端接头
plug expander 斜轧扩径机
plug flow 活塞式流动;活塞流;平推流;推流;栓塞流(动);塞状流;塞式流动;塞流
plug-flow activated sludge process 推流式活性污泥法
plug-flow aeration 推流式曝气
plug-flow aeration pond 推流式曝气池
plug-flow conveying 栓流输送
plug-flow reactor 活塞流反应器;平推流反应器;平流反应堆;推流反应器
plug-flow system 活塞流系统
plug-flow type pneumatic conveyer 栓流式气力输送装置
plug for air pump 气泵旋塞
plug forming 模塞助压成型;模塞成型
plug for rotor cleaning 转鼓清洗螺塞
plug for seal 堵塞密封
plug for water pipe 水管塞
plug fuse 螺旋熔断器;熔线塞;熔丝塞(子);插头保险丝;插式熔断器;插塞式熔丝;插入熔断器;插式保险器
pluggability 可插性
pluggable unit 可插件
plug ga(u)ge 圆柱塞规;塞规;塞尺
plugged bit 带岩芯的钻头
plugged impression 嵌块
plugged lumber 已刮灰木材;抹过腻子的木料
plugged plywood 填充夹板
plugged program(me) 插入程序
plugged-program(me) computer 插入(式)程序计算机
plugged-program(me) machine 插入式程序计算机
plugged screen 堵塞的(过)滤器
plugged steel 加盖钢;封顶钢
plugged with brick 用砖塞孔
plugger 凿岩机;填塞物;锤击式钻岩机
plugger drill 风镐
plugger hole 向下炮眼
plugging 钻孔回填;止水;沸沸(玻璃液);加盖;填塞法;堵塞出铁口;钉木楔;带顶头轧制的;封孔;

封堵;封底(沉井);反相制动
plugging ability 密封能力;堵塞能力
plugging agent (油的)堵火剂;止水剂;堵塞剂;封堵剂
plugging back 堵封
plugging bar 打夯机;捣砂锤
plugging block 塞子砖
plugging chart 插接图;插接板控制图
plugging chisel 钢钎;星形长凿;嵌缝凿(刀);刀形凿
plugging compound 填充料
plugging degree 堵塞程度
plugging depth 下塞深度
plugging device 封堵装置
plugging effect 土塞效应
plugging inhibitor 防堵塞剂
plugging material 堵塞物;堵漏材料
plugging of catalytic bed 催化床堵塞
plugging relay 停车继电器;防逆转继电器
plugging site 下塞位置
plugging switch 逆流制动开关
plugging test 堵漏试验
plugging the gap 填补空隙
plugging thief zone 堵漏失层
plugging-up 闭塞;堵塞
plugging-up device 闭塞装置
plugging-up line 障碍试验线;闭塞线
plug go-devil 隔离塞
plug hatch 隔热舱盖;密封舱盖;密封盖
plug hole 浇筑孔;塞孔;船底放水孔;插塞头;插塞孔
plug-in 插座;插入式的
plug-in amplifier 插入式放大器
plug-in assembly 插入部件
plug-in board 插入式电路板;插卡;插件
plug-in bobbin 插入式线圈管
plug-in bond 插入接头
plug in box 插接箱
plug in cable 插入电缆
plug-in card 插入卡;插件板
plug-in change-oversystem 插入转换制
plug-in chassis 插入式机架
plug-in circuit 插入式电路
plug-in circuit card 插入式电路板;插件板
plug-in coil 插换线圈
plug-in-component 插入部件;插入式元件
plug-in connection 插入连接;插座连接
plug-in connector 插入式接头;插件
plug-in console 插式控制台
plug-in controller 插入式控制器
plug-in cord 插入式电缆;插入式塞绳
plug-in coupling 塞入式接头
plug-in coupling transformer 插入式耦合变压器
plug-in device 插塞装置
plug-in discharge tube 插入式放电管
plug-in duct 插入式配线导管
plug-in electronics 插入式电子设备
plug-in inductor 插入式线圈;插入式感应线圈;插换感应线圈
plug-in interface card 插入式接口插件
plug-in mixed duplexer 插入式混合天线收发换开关
plug-in mixer-duplexer 插入式混合天线开关
plug-in module 插入组件;插入式保险器;插件
plug-in mounting 插装;插入式固定;插入式安装
plug-in package 可替换部件;插入式保险器
plug-in point 插座;插头;插入接点;插接点
plug-in pump 插装式泵
plug-in relay 插入式继电器;插入式继电器
plug-in sensor 插入传感器
plug-in sheet 插件片
plug-in socket 插入式电插座;卡口插座
plug-in strip 闸刀开关;闸刀;刀形开关;插入片
plug-in switching block 插入式切换组件
plug-in system 插入式装置;插入(式连接)系统;插换制
plug-in tabulator 插入式制表机构
plug-in terminal 插入终端
plug-in termination 插入终端
plug-in transformer 插入式变压器
plug-in trouble 插接不良
plug-in type memory 插入式存储器
plug-in type supply circuit 插入式供电电路
plug-in tyre busbar 插接母线
plug-in unit 插件;插入元件;插入单元;插入部件

plug-in valve 插装式阀
plug in voltage tester 插入式电压表
plug-in weld 插头焊接
Plug-it 墙上插头(商品名)
plug jack 插孔
plug jet 气动塞规;气动测头
plug key 旋塞式开关;插塞式开关;插塞开关;插塞电键
plug lines 芯棒划痕
plug load 插塞荷载
plug machine screw tap 机用中丝锥
plug mandrel 定径心轴
plug mill process 自动轧管法
plugmold 配线踢脚板;配线护墙板
plug of clay 泥塞
plug of cloth 滤布孔堵塞
plug of felt 油毛毡填充料
plug of pipe 管子堵头
plug of piping 管子堵头
plug of tube 管道堵塞;管子堵头
plug of tubing 管子堵头
plug-on connection 插座连接
plug over-ride prevention 防止插入
plug packer 堵塞(封隔)器
plug pin 塞钉;插销;插钉
plug pipe tap 塞状管丝锥
plug pliers 塞钳
plug point 家电插头
plug program(me) patching 变程板
plug pump 柱塞泵
plug receptacle 插座
plug reef 插头礁
plug retainer 插头护圈
plug reversal 反相制动;反接制动
plug rod 柱塞杆
plug rolling 芯棒轧管法;自动轧管
plug roll process 芯棒轧管法
plug screw 塞子螺钉;堵头螺钉;螺塞
plug screw ga(u)ge 阴螺纹规
plug seat 插座;插孔
plug-selector 塞绳式选择机
plug shot 小炮眼
plug sleeve 塞套;插头接套
plug socket 塞孔;插座;插孔
plug spanner 塞扳子
plug stick 塞杆
plug stopping 反流制动
plug switch 插头开关;插塞开关;插接开关
plug switchboard 插塞式交换台
plug tap 插入接头;插入大龙头;旋塞式龙头;转接式龙头;二锥;二道螺丝攻
plug tapp 中丝锥
plug tenon 加塞榫;尖榫;插塞榫;塞榫;插榫
plug tester 火花塞试验装置
plug thread ga(u)ge 螺纹塞规
plug thumb screw 蝶形螺塞
plug tube-rolling mill 自动轧管机
plug-type 插入型;插入式的;插入式
plug-type area meter 插塞式面积流量计
plug-type fuse 插入式熔丝
plug-type nozzle 塞式喷管
plug-type pycnometer 塞形密度计;塞形比重计
plug-type rail bond 塞钉式钢轨连续线;插入式连接线
plug-type resistance bridge 插塞式电阻电桥
plug-type resistor 插塞式电阻器
plug-type spoiler 插式扰流片
plug-type valve 塞阀
plug up 塞住
plug valve 旋塞(式)阀(门);栓塞阀;塞阀
plug weld 电铆焊;铆焊;塞焊焊缝
plug weld(ing) 插头焊接;电铆焊;塞焊(接);铆焊;塞焊(焊)缝
plug wire 插线
plug with earthing contact 具有接地的电插座
plug wood 埋木法
plug wrench 火花塞专用扳手
plum 毛石;蛮石(大体积混凝土用);高额利润;梅红色;毛石块料(混凝土用);混凝土用毛石块料;大石子
plumangite 普鲁曼奇矿
plumasite 刚玉更长岩;刚玉奥长岩
plumb 线铊;管子安装;铅锤;吊线
plumbagine 石墨粉;石墨

plumbaginous 含石墨的;石墨的
plumbaginous schist 石墨片岩
plumbago 白花丹属;石墨;黑铅;碳陶质耐火材料;炭精
plumbago clay 石墨质耐火黏土
plumbago crucible 石墨坩埚
plumbago dressing 石墨涂料
plumbagolike pyrolusite 石墨状软锰矿
plumbago pot 石墨坩埚
plumbago process 黑铅法
plumbago refractory 石墨(质)耐火材料
plumb aligner 铅锤对准器
plumb and level 水平垂直尺
plumbate 高铅酸盐
plumb bevel 垂球测斜面
plumb bob 铅锤球;垂球;测锤
plumb bob cord 铅锤线
plumb-bob-level 垂球水准器
plumb bob method 悬锤测量法
plumb bob section 悬锤机构
plumb bob string 垂球线
plumb bob vertical 铅锤(垂)线
plumb bond 直缝砌法;全顺砖砌体;铅直式砌合
plumb bond pole (砖的)垂直接缝定位杆
plumb column 垂直柱
plumb cut (屋架椽子下部的)垂直切口
plumb datum 铅锤基准
plumbean 正铅
plumbed 铅垂的
plumbed manifolds 铅垂流形
plumbeous zirconate-titanate ceramics 锆钛酸铅陶瓷
plumber 管子工人;管子工;管工;铅管工;水暖工;白铁工
plumber black 管工用黑漆;管工黑漆;管工黑胶
plumber block 轴台;支承块
plumber block bolt 轴台螺栓
plumber block keep 中间轴承盖
plumbering 接管工程
plumber's black 管工黑油
plumber's black soil 管工黑油
plumber's dope 管工用油灰
plumber's friend 揣子【建】
plumber's furnace 喷灯
plumber's helper 揣子【建】
plumber's knife 管子工刀
plumber's mate 水暖工助手
plumber's rasp 管工粗锉
plumber's round iron 管道工用圆烙铁
plumber's scraper 三角刮刀
plumber's shop 白铁工车间;管子车间
plumber's smudge 管工焊接灰
plumber's soil 管工黑油;管工黑胶
plumber's solder 管工焊料;铅锡焊料
plumber's tool 管工工具;管子工的工具
plumber's union 管子连接套
plumber's works 管道工程;管子工场
plumber's tool 管工工具
plumber white 铜锌镍合金
plumbery 管子工场;管工车间;铅器;铅管工场;铅管工厂
plumbian atokite 含铅的锡铂钯矿
plumbic 高铅的
plumbic acid 高铅酸
plumbi carbonas 碳酸铅
plumbic chloride 氯化高铅;四氯化铅
plumbic ocher 高铅赭石;高铅赫石
plumbic sulfate 硫酸高铅
plumbiferous 产铅的;含铅的
plumbing 用铅锤检查垂直度;管件安装;管件;管道工程;给排水图纸;接管工程;铅管工程;铅锤测量;铅垂线校正;卫生管道工程;室内管工;室内管道工程;室内给排水工程;铅锤;测深;房屋管工;房管系统;波导设备;波导管
plumbing agent 膨胀剂
plumbing appliance 管道工程用具;管道工程设备
plumbing appurtenance 管道工程装置;管道工程建筑;管道工程附件;管道工程配件;管道工程附属设备
plumbing arm 移点器【测】;垂臂
plumbing bar 垂准杆;垂杆
plumbing base 两垂线间距
plumbing branch 管道工程分支
plumbing column survey 柱子吊装测量

plumbing contractor 管道工程承包者;管道工程承包人
plumbing discipline 给水排水专业
plumbing drain 污水管路
plumbing equipment 垂准设备
plumbing fittings 管件安装;卫生洁具配件;卫生管道配件;卫生设备;卫生洁具
plumbing fixture 水暖配管件;室内管道工程固定装置;卫生设备;卫生器具;卫生洁具
plumbing fork 移点器【测】;对心器;定心器
plumbing installation 管道装置;卫生管道装置;卫生管道安装;卫生工程管道铺设;卫生工程管道安装
plumbing instrument 垂准器
plumbing iron 管道工程使用的电加热焊铁
plumbing line 垂直线;铅直线;垂准线
plumbing method 垂准法
plumbing mirror 垂准镜
plumbing network 房屋管网;室内管网
plumbing noise 铅锤噪声;卫生设备噪声
plumbing pipe 卫生管道
plumbing piping 卫生工程管理;卫生工程管道
plumbing plant 垂准设备
plumbing pole 垂准杆
plumbing rod 对中杆
plumbing run 卫生管线
plumbing services 卫生管道设备;卫生管道装置
plumbing snake 柔性管缆(疏通排水管和排气管用);管道疏通器
plumbing staff 对点杆
plumbing survey 垂直度测量
plumbing system 污水系统;室内管道系统;污水粪便系统;卫生设备系统;卫生工程系统;室内管沟系统;室内管工系统
plumbing system appliance 用水设备;卫生设备
plumbing trap 管道存水弯;卫生管道存水弯
plumbing unit 冷热水装置;卫生设备组合;卫生器具;卫生间设备
plumbing wire 垂准线
plumbing wire bracket 垂球定中夹
plumbing work 室内布线;管道工程;卫生设备工程
plumbi nitras 硝酸铅
plumbi oleas 油酸铅
plumbious chromate 铬酸铅
plumbism 铅中毒
plumbite 亚铅酸盐;铅酸盐
plumbite of soda 铅酸钠
plumbite process 铅酸钠精制过程
plumbite treatment 铅酸盐处理
plumb joint 薄铁板搭接焊缝;锡焊接;锡焊搭接;灌铅搭头;灌铅搭缝;填铅接合(缝);垂直接缝
plumb level (有铅垂线的)水准仪;准直水平仪;铅垂水准器;水平仪
plumb line 铅垂线;垂直线;垂准线
plumb line deflection 线划位移;垂线偏差
plumb-line deviation 铅垂线偏差;垂线偏差
plumb line hook 垂球悬挂钩
plumb line position 垂线位置
plum-blossom needle 梅花针
plum-blossom pattern 梅花形
plum-blossom pyonex 梅花针
plumb monoxidum 一氧化铅
plumbness 垂直状态;垂直
plum bob 铅锤
plumbobetafite 铅贝塔石
plumbocuprite 砷铜铅矿
plumbodolomite 铅白云石
plumboferrite 铅铁矿
plumbogummite 水磷铝铅矿
plumbojarosite 铅铁矾
plumbomicrolite 铅细晶石
plumboniobite 铅铌铁石
plumbons acid 铅酸(二价铅)
plumbopalladinite 铅钯矿
plumbopyrochlore 铅烧绿石
plumboresinite 水磷铅铝矿
plumbotellurite 斜方碲铅石
plumbotsumite 羟硅铅石
plumbous bichromate 重铬酸铅
plumbous cyanide 氰化铅
plumbous oxide 一氧化铅;氧化铅
plumbous plumbate 三氧化二铅
plumbous subacetate 碱式乙酸铅
plumbous subcarbonate 次碳酸铅

plumbous sulfide 硫化铅
plumbous sulphate 硫酸铅
plumbous tungstate 钨酸铅
plumb perpendicular 用垂球测定垂直度;铅垂线校正
plumb pile 垂直桩
plumb pneumatic jig 转阀式风力跳汰机
plumb point 天底点;垂准点
plumb point triangulation 铅锤点三角测量;像底点三角测量
plumb position 铅垂位置;垂球位置
plumb post 垂直支柱;挖掘工程支柱;垂直柱
plumb rule 靠尺;垂线尺;垂规
plumb rule and bob 垂线尺和铅锤
plumbsol 银锡软焊料
plum colo(u)r 梅红色
plum design 梅形茶具
plume 烟羽;卷流;黑火头;喷流
plume boundary 烟缕边界
plume center line 烟缕中线
plume contamination 烟羽污染;烟缕污染
plume diffusion 烟羽扩散
plume height 烟羽高度;烟缕高度
plume laser detector 烟羽激光探测器
plume-like inflorescence 羽状花序
plume moisture 烟羽湿度;烟缕湿度
plume of bubble 气泡卷流
plume of contaminant 污染物流束;污染物流股;污染物烟缕
plume opacity 烟羽不透明度;烟缕不透明度
plume pattern 烟羽型
plume rise 烟羽抬升;烟羽上升;烟缕上升
plume rise by buoyancy force 热力抬升
plume rise by source momentum 动力抬升
plume rise formula 烟羽抬升公式
plume rise height 烟羽抬升高度
plume rise model 烟羽抬升模式;烟羽上升模型;烟缕上升模式
plume shape 烟羽形状
plume spread 烟缕伸展范围
plume structure 羽状构造
plume trap 排烟罩
plume travel time 烟缕漂移时间
plumety 薄细布
plum green 梅子青
plummer block 支撑轴承;推力轴承
plummer block bearing 架座
plummer block cover 轴台盖
plummet 铅垂线;铅垂球;垂线仪;垂球;测铅
plummet body 铅锤
plummet convergence 垂线收敛
plummet-distance 垂线间距离
plummet lamp 垂灯
plummet level 铅垂水准;定垂尺尺
plummet method 测锤法
plummet weight 铅锤重量
plummet wire 锤球线
plum needle 梅花针
plumose structure 羽状构造
plump 鼓起的
plump line 粗线
plump seed 饱满籽粒
plum rain 梅雨
plums 填料;圬工石
plum stone 毛石;蛮石
plum vase 梅瓶
plunalsite 硅铝铅石
plune 羽状
plunge 急降;盲目投资;倾入;前后颠簸;跳进;掉进;大坡度倾斜;负债
plunge angle 倾伏角
plunge angle of hinge 枢纽倾伏角
plunge angle of orebody 矿体倾伏角
plunge angle of striation 断层擦痕倾伏角
plunge area 倾伏区
plunge area of fold 褶皱构造倾伏地段
plunge bath 大浴池;浴池;池浴
plunge battery 浸液电池
plunge cut 全面进给法
plunge-cut milling 插进式铣削
plunge grinding 横磨;全面进磨法;成型磨削
plunge lake 瀑布湖
plunge line 波浪破碎线
plunge of fault striation 断层擦痕倾伏向

plunge of fold 褶皱倾伏(角)
plunge of orebody 矿体倾伏方向
plunge point (波浪的)破碎点;(波浪的)卷波点;下冲点;潜入点(入库泥砂);波浪破碎点
plunge pond 跌水潭
plunge pool 消力池;浴池;跌水潭;跌水坑;水池
plunger 阳模;压模冲头;撞针杆;钟罩;供料机冲头;模冲;活柱塞;活塞;皮碗泵;揣子【建】;冲杆;插棒式铁芯;插棒
plunger-actuated 柱塞作用的
plunger adaptor 冲头接头
plunger barrel 柱塞筒
plunger breaker 卷跃碎浪
plunger bushing 柱塞套;柱塞衬筒;泵塞套筒
plunger cam 冲头凸轮
plunger carrier 冲头支架
plunger case 柱塞缸
plunger centering knob 冲头定中心旋钮
plunger chip 冲头
plunger compartment 柱塞室
plunger connecting rod 柱塞连杆
plunger contact 插棒接点
plunger control spring 柱塞控制簧
plunger core machine 挤芯机
plunger coupling 滑动套万向接头
plunger crank 柱塞传动曲柄
plunger diaphragm pump 柱塞隔膜泵
plunger displacement 泵塞排量
plunger duster 柱塞式喷粉机
plunger electromagnet 吸入式电磁铁;插棒式电磁铁
plunger elevator 液压举升器;液压顶升机;液压电梯;柱塞升机
plunge-reversal mode 倾伏倒转性模式
plunger eye 柱塞套圈;柱塞扣眼
plunger feeder 柱塞给料机;活塞式给料机
plunger follower spring 柱塞导筒弹簧
plunger force 柱塞力
plunger forward time 柱塞推进时间
plunger fuel pump 圆柱式燃油泵
plunger gear 喷油泵柱塞齿轮
plunger guide 柱塞导承
plunger helix 柱塞螺旋槽
plunger hydraulic elevator 柱塞液压提升机
plunger jig 活塞跳汰机;活塞式跳汰机
plunger key 插棒式按键
plunger knob 柱塞旋钮
plunger latch 锁栓
plunger leather 柱塞皮封
plunger lever 柱塞杆
plunger lever pin 柱塞杆销
plunger lever shaft 柱塞杆轴
plunger lift 柱塞行程
plunger line switch 插棒式预选器;插棒式寻线机
plunger lock 转辙锁闭器;转辙器锁闭器
plunger lower helix 柱塞下螺边
plunger machine 柱塞铸模机
plunger magnet 插棒式磁铁
plunger mo(u)ld 柱塞模
plunger motor 柱塞液动机;柱塞式液压电动机
plunger mo(u)lding 活塞式压铸
plunger over-current relay 插棒式过电流继电器
plunger overtravel 柱塞超冲程
plunger pin 柱塞销;跳针
plunger piston 柱塞;状活塞
plunger pitman 柱塞连杆
plunger pole 柱塞杆
plunger press 柱塞式压捆机
plunger pressure 冲杆压力
plunger pump 圆柱式泵;往复泵;柱塞泵
plunger pump ram 柱塞泵撞头
plunger pusher 柱塞式推进机
plunger relay 吸入式继电器;活塞式继电器;插棒式继电器
plunger relief valve 柱塞式溢流阀
plunger retainer 阳模环
plunger return spring 柱塞回动弹簧
plunger setting 柱塞式铁芯调整
plunger shear knife 柱塞剪切刀
plunger size 柱塞尺寸
plunger spring 柱塞弹簧
plunger spring cup 柱塞弹簧皮碗
plunger stop 柱塞运动暂停机构

plunger stroke 柱塞行程
plunger switch 柱塞转换开关
plunger tappet 柱塞挺杆;柱塞杆
plunger tester 压力强度测定仪
plunger tip 冲头
plunger travel 柱塞行程
plunger type 柱塞式
plunger type brake 插棒式制动器;插棒式电磁铁制动器
plunger type briquetting machine 柱塞式团矿机
plunger type cell 插棒式电池
plunger type clay gun 柱塞式泥炮
plunger type control valve 柱塞式控制阀
plunger type cylinder 柱塞缸
plunger type die casting machine 活塞式压铸机
plunger type dust gun 活塞式喷粉枪
plunger type extruder 活塞式挤坯机;活塞式挤出机
plunger type fuel pump 柱塞油泵
plunger type hydraulic motor 柱塞液压马达
plunger type instrument 插棒铁芯式仪表
plunger type knapsack sprayer 背负式活塞喷雾器
plunger type line switch 插棒式寻线机
plunger type overcurrent relay 活塞式过电流继电器
plunger type overvoltage relay 活塞式过电压继电器
plunger type pickling machine 柱塞式酸洗机
plunger type portable extinguisher 柱塞式手提灭火器
plunger type pump 柱塞泵
plunger type relay 插棒式继电器
plunger type undervoltage relay 活塞式欠电压继电器
plunger type visco(si)meter 柱塞式黏度计
plunger type washbox 活塞式跳汰机
plunger type wave generator 冲击式造波机;冲击式生波机
plunger upper helix 柱塞上螺边
plunger valve 柱塞阀
plunger wear 柱塞磨损
plunging 压入法;内缘翻边;跳进的;波浪卷跃破碎
plunging angle 倾伏角
plunging anticline 倾覆背斜;倾伏背斜
plunging breaker 卷跃碎浪;卷涌的浪;卷破波;卷浪
plunging cliff 倾覆崖;倾伏崖;半淹海蚀崖
plunging coast 下沉海岸
plunging crown 倾伏褶额
plunging end 倾伏端
plunging end area of syncline 向斜的向心倾伏末端
plunging fire 俯射
plunging fold 倾覆褶皱;倾伏褶皱
plunging nappe for spillway 溢洪道下泄水舌
plunging of hinge 枢纽倾伏向
plunging shot 翻转照准
plunging siphon 滴管;吸量管
plunging syncline 倾覆向斜;倾伏向斜
plunging terrace 潜伏阶地
plunging trickling filter 瀑布式滴滤池
plunging water 瀑布式喷嘴暖气池
plunging wave 卷波;卷跃波浪;卷破浪
plun-jet 气动量复
plural cyclone 多级旋风分离器
plural executive 兼任执行人
plural gel 复合凝胶
pluralism 多元论;多元化社会;多元性
plurality 多元;复数性;复数
plurality correlation 复相关
plurality dies 通用螺纹切头
plurality of control 复杂控制
plural mo(u)ld operation 多模操作
plural population 复数人口
plural process 多次过程
plural service 不同源供电
plural water supply 复合给水
pluramelt 包层钢板
pluramelt process 复合金属法
pluridyscrinia 多种分泌障碍
pluriennial regulation 多年调节
plurilocular 复室的
plurivorous parasite 多主寄生物
Pluronic polyols 复合多元醇
plus 正型
plus and become symbol 插值符号

plus angle 仰角
plus caster 正后倾
plus count 加法计数
plus declination 正赤纬
plus deviation 正偏差
plus distance 补距
plus driver 十字(形)螺丝刀;十字螺丝起子
plus earth 阳极接地;正极接地
plus effect 正效果
plus electricity 正电
plus end 正端子
plus equal key 加号等号键
plus grade 上升坡度;正坡;升坡;上坡
plush 长毛绒织物
plush cam 毛圈三角
plush carpet 剪绒地毯
plush copper ore 毛赤铜矿
plus infinity 正无穷大
plus internal standard method 加内标法
plus involute 正渐开线
plus lap 正遮盖
plus line 正线
plus material 面料;筛面料
plus mesh 正筛孔;颗粒大于筛的;筛余料;筛上物;筛上料
plus mineral 正矿物
plus minus 加减符;正负;加减
plus-minus screw 调整(用)螺钉;调整螺杆
plus-minus switch 加减分接开关
plus-minus tolerance 正负公差
plus No. x material 大于某筛号的材料
plus or minus clause 允许有一定增减的条款
plus plate 正极
plus point 导线加桩点
plus pressure 正压(力)
plus pressure furnace 加压炉
plus pressure process 正压力方法
plus screw 十字槽头螺钉;带十字槽头的螺钉
plus seed stand 采种区
plus sheet 补充印数
plus side 加侧
plus sieve 筛上
plus sight 后视
plus sign 正号;加号
plus size of a screen 筛上颗粒尺寸
plus stand 正型林分
plus strand 正链
plus symbol 加符号
plus tapping 正抽头
plus thread 正螺纹
plus time shift 正时间移动
plus tolerance 正公差
plus variable structure 固定加可变结构
plus zero 正零
plus zone 正号区;加号区
pluteus 矮墙(罗马建筑女儿墙)
pluto 放射性检查计
plutocrat 富豪
plutology 地球内部学
pluton 深成岩体
plutonic 深成的
plutonic acid 钚酸
plutonic breccia 深成角砾石
plutonic earthquake 深震;深源震;深源地震;深成地震
plutonic facies 深成相
plutonic intrusion 深成侵入(体)
plutonic melting flow region 深熔流动区
plutonic metamorphism 深成变质(作用);深部区域变质(作用)
plutonic mineral 深成矿物
plutonic plug 塞紧的软木塞;岩栓
plutonic rock 深成岩;深层岩
plutonic water 深成水
plutonism 火成说;火成论;深成现象
plutonist 火成论者
plutonite 火成岩;深成岩
plutonium carbonitride 碳氮化钚
plutonium fire 钚火灾
plutonium fission 钚核裂变
plutonium pile 钚堆
plutonium producing reactor 钚生产反应堆
plutonium recycle test reactor 钚再循环试验反应堆
plutonium regenerating reactor 钚再生反应堆

plutonium sesquioxide 三氧化二钚
plutonyl 钚酰
plutor(e)深成岩体
Pluvex 普鲁维克斯(一种沥青防水层)
pluvial 多雨的
Pluvial Age 洪积时期;雨期
pluvial conglomerate 洪积砾岩
pluvial denudation 雨水冲蚀
pluvial erosion 洪水侵蚀
pluvial fan 洪积扇
pluvial index 雨量指数;降雨指数
pluvial lake 雨成湖;洪积湖
pluvial landform 洪积地貌
pluvial-moisture ratio 雨湿比
Pluvial period 洪积纪
pluvial period 雨期;雨季;多雨期
pluvial phase 洪水相
pluvial placer 洪积砂矿
pluvial region 多雨地区
pluvial soil 洪积土
pluvial terrace 洪积阶地
pluviifruticeta 常雨灌木群落
pluviilignosa 常雨木本群落
pluvio-fluvial denudation 雨水河水剥蚀作用
pluviogram 雨量曲线
pluviograph 雨量仪;雨量图;雨量计;自记录雨量器;自记录雨量计
pluviometer 雨量器;雨量计
pluviometer station 雨量测站
pluviometric coefficient 雨量系数
pluviometric gradient 雨量梯度
pluviometry 雨量学;雨量测定法;测雨法
pluvioscope 雨量器;雨量计
pluviothermic ratio 雨温比
pluvious insurance 雨水险;雨淋险
pluvious policy 雨淋保险单
pluvium 洪积层
plvmetal 包铝层板
ply 折叠;绞合;片层;往返行驶;单板层
ply adhesion 层间附着力
ply bamboo 多层竹板
ply bond strength 层间黏合强度
ply bracing panel 多层支撑胶合板
plycast 熔模壳型
plyclip H 形铝支架条(胶合板连接用)
ply-constructed belt 分层棉织胶带
Plydek 普莱德克(一种塑性地板材料)
plyers 拔管小车;老虎钳
Plyform 胶合板(混凝土)模板(商品名);制胶合板
ply glass 胶合玻璃;夹玻璃纤维的玻璃板;纤维夹层玻璃;胶合玻璃
plying cement 叠层黏合剂
plying limit 航区限制;航行区域
plying machine 并线机
plymax 镶铅装饰胶合板;镶铝装饰用胶合板;金属贴面胶合板
plymetal 胶合金属板;包层金属板;包层层板;金属贴面胶合板;金属(面)胶合板;夹金属胶合板;涂金属层板
plymold 模塑胶合板
plynya 冰湖的
ply plastics 弯形胶合板;层合塑料
ply-rating 层数(轮胎帘布的);轮胎层级;线网层率
ply rolling 叠轧
ply separation machine 层离机
ply steel 复合钢
Plysyl 普莱希尔(一种永久性胶合板组成的模板)
plywood 胶合板;粘板;三合板;层压木(板);层压板;层板
plywood adhesive 胶合板用黏结剂;胶合板用黏剂
plywood box beam 胶合板箱形梁
plywood case 三夹板箱
plywood caul 胶合板垫板
plywood ceiling 胶合板平顶;胶合板顶棚
plywood concrete forms 胶合板混凝土模板
plywood connecting plate 胶合板角撑板;胶合板连接板
plywood construction system 胶合板结构系统
plywood container 胶合板集装箱
plywood core 胶合板芯板
plywood corner plate 胶合板角板
plywood covering 层板蒙皮
plywood decking 胶合板铺面

plywood diagonal deck plating 木夹板或斜张甲板
plywood diaphragm 胶合隔墙;胶合板隔墙
plywood door 夹板门;胶合板门
plywood door panel 胶合板门镶板;胶合板嵌板
plywood exterior sheathing 胶合板外部衬板
plywood-faced door 胶合板门;夹板门
plywood facing 胶合板面;胶合板墙面
plywood factory 胶合板工厂;胶合板厂
plywood fish plate 胶合板鱼尾接合板
plywood floor 胶合板箱底
plywood flooring 胶合板地板
plywood flush door 胶合板光面门;胶合板平面门
plywood forms coating 胶合板模板涂层;胶合板模板涂料
plywood form(work) 胶合板模壳;胶合板模板
plywood formwork coating 胶合板模板涂料
plywood for structural use 结构用胶合板
plywood frame 胶合板肋骨
plywood girder 胶合板大梁
plywood gusset 胶合板节点板
plywood gusseted trusses 胶合板撑板架
plywood gusset plate 胶合板节点加固板;胶合板节点角撑板;胶合板节点联结板
plywood joint plate 胶合板角撑板;胶合板接合板;胶合板接缝板
plywood knee bracket plate 胶合板带膝形托架;胶合板带膝形托座
plywood-lined form(work) 胶合板衬里的模壳;胶合板衬里的模板
plywood metal sheet 金属面胶合板;夹层金属薄板
plywood mo(u)ld 胶合板模板
plywood panel 胶合板板格;胶合板板条
plywood panel board construction purposes 建筑用的胶合板镶板(门窗上的六格板)
plywood panel ceiling 胶合板顶棚
plywood panel for concrete form 胶合板模型板
plywood panel foundation 胶合板基础
plywood panel size 胶合板幅面
plywood parquet 胶合板块;胶合板镶嵌楼面;胶合板地板块
plywood-phenolic resin 酚醛树脂胶合板
plywood plant glue waste 胶合板厂废胶水
plywood plate 胶合板
plywood portal frame 胶合板门式构架;胶合板门式刚架
plywood roof decking 胶合板屋顶铺面
plywood roof sheathing 胶合板望板;胶合板承梁板;胶合板屋面板
plywood sawer 胶合板锯工
plywood shear wall 胶合板抗剪墙;胶合板抗剪墙
plywood sheathing 胶合板盖板;胶合板覆盖;胶合板望板;胶合板衬板;贴铺胶合板
plywood sheet 胶合板薄片
plywood shell 胶合木壳体
plywood shuttering 胶合板模壳;胶合板模板
plywood shuttering coating 胶合板模板百叶窗;胶合板模板涂层;胶合板模板涂料
plywood side planking 舷侧合板
plywood siding 胶合板墙板;胶合板壁板;胶合板壁
plywood soffit 胶合板挑檐底面
plywood square 胶合板地板块;胶合板镶块;胶合板块
plywood system 胶合板系统
plywood underlayment 胶合板垫层;胶合板衬底
plywood-veneered door 胶合板门;夹板门
plywood wall facing 胶合板墙面
plywood wall sheathing 胶合板墙面板;胶合板墙体夹衬板
plywood web 胶合板腹板
plywood with boilproof glue lines 具有耐高温胶缝的胶合板
plywood with metal core 金属夹芯胶合板
p-mercuribenzoate 对苯甲酸汞;安息香酸汞
pmh aval 长椭圆形
P-Mn ratio 磷锰比
p-m rotor 永磁转子
PMT 旁压试验
pneu-bin panel 充气墙
pneudraulic autopilot 气动液压自动驾驶仪
pneudyne 气动变向器
pneulift 气动升降机
pneumacator 气压测深器;气动发送器
pneumacator draft ga(u)ge 吃水计

pneuma-lock 气动锁紧;气动夹紧
pneumathode 气囊
pneumatic 气压的;气体力学的;气体动力学的;气动的
pneumatic accumulator 气力蓄能器
pneumatic actuator 气压传动装置;气动执行机制;气动执行机器;气动执行机构;气动传动
pneumatic adder-subtractor 气动加减器
pneumaticae 含气小房
pneumatic aeration 压气曝气;气压曝气;气动曝气
pneumatic agitator 气动搅拌器
pneumatic air lock valve 气力锁风阀
pneumatic air machine 风动机械
pneumatic alarm set station 气动报警给定器
pneumatic alloy spider 气动三叶片
pneumatically 靠压缩空气
pneumatically applied 气动施工
pneumatically applied asphalt emulsion mortar 气力喷乳化沥青砂浆
pneumatically applied concrete 喷射混凝土
pneumatically applied mortar 气压喷浆;喷射用砂浆;喷射砂浆
pneumatically assisted 压缩空气辅助的;气力辅助的
pneumatically boosted 压缩空气助推的;气力助推的
pneumatically controlled 压缩空气控制的;气力控制的
pneumatically controlled aggregate bin gates 压缩空气控制的骨料仓闸门;气力控制的骨料仓闸门
pneumatically controlled gate valve 气控阀门
pneumatically controlled loader 气动上料器
pneumatically controlled sample leak 气动控制进样漏孔
pneumatically driven 气动的;风动的
pneumatically operated 气动操作的;风动的
pneumatically operated batching plant 压缩空气操纵的配料设备;气力操纵的配料设备
pneumatically operated butterfly-valve 气动操作蝶阀
pneumatically operated cargo valve 气动货油阀
pneumatically operated concrete placer 混凝土压气浇筑设备
pneumatically operated dial feed 气动转盘送料
pneumatically operated direction valve 气动操纵换向阀
pneumatically operated grass seed drill 气动操纵式草籽播种机
pneumatically operated switch 气动开关
pneumatically placed concrete 喷射混凝土;喷射浇灌混凝土
pneumatically placed mortar 气压浇筑砂浆;气压浇注砂浆;气压浇灌砂浆
pneumatically released pilot valve 气动操纵导阀
pneumatically sensed 气压传感的
pneumatically tired buggy 气胎手推车;充气轮胎小车
pneumatic analog(ue) 气动模拟
pneumatic analog(ue) computer 气动模拟计算机
pneumatic analog(ue) simulator 通风模拟设备
pneumatic apparatus 风动工具
pneumatic applied concrete 喷射浇筑混凝土;喷射混凝土
pneumatic applied mortar 喷射水泥砂浆
pneumatic architecture 充气建筑
pneumatic arm 气动臂
pneumatic ash conveyer 风力输灰装置
pneumatic ash removal system 气力除灰系统
pneumatic atomization 气力弥雾
pneumatic atomizer 气压雾化器
pneumatic automatic 气力自动装置
pneumatic automatic control 气式自动控制
pneumatic autopilot 气动式自动驾驶仪
pneumatic auxiliary console 气动辅助支架
pneumatic bag 气囊
pneumatic ballast 风动分渣车
pneumatic ballast hopper wagon 风动卸渣车
pneumatic ballast machine 风动卸渣机
pneumatic barrier 压气式防波堤;空气帘;气压防波堤;气泡帘;气动隔板
pneumatic batch conveyer 间歇式气力输送机;分批式气力输送机
pneumatic bearing 气动轴承
pneumatic beetle 气夯

pneumatic bench sand rammer 风动台式砂箱捣实机
pneumatic(al) blending 空气搅拌
pneumatic boat 橡皮充气艇;充气快艇
pneumatic bone 含气骨
pneumatic brake 气制动(器);气闸;气压制动器;气力制动器;气动闸;风闸
pneumatic brake actuator 气动制动执行机构
pneumatic brake application 空气制动施加
pneumatic brake bell 气闸钟
pneumatic brake controller 气制动控制装置
pneumatic brake door 气闸门
pneumatic brake operation device 气压式制动操纵装置
pneumatic brake tank 气闸室
pneumatic brake with electric(al) control 电控制的气闸
pneumatic braking 压缩空气制动;气压制动
pneumatic breaker 气锤;气动破碎机
pneumatic breakwater 气幕防波堤;压气式防波堤;空气帘防波堤;空气防波堤;气压(式)防波堤;喷气式消波设备
pneumatic breakwater curtain 气泡防波帘
pneumatic bridge hygrometer 气动桥式湿度表
pneumatic bumper 气压减振器
pneumatic cactus grab 风动多爪抓岩机
pneumatic caisson 压气沉箱;气压沉箱
pneumatic caisson foundation 气压沉箱基础;沉箱基础
pneumatic caisson method 气压沉箱法
pneumatic caisson pier 气压沉箱桥墩
pneumatic caisson pile 压力灌浆管桩
pneumatic caisson quay wall 气压沉箱式岸壁
pneumatic calking 气压填缝;气压嵌缝;气压堵缝
pneumatic capper corker 气动塞瓶机
pneumatic capstan 气动主动轮
pneumatic carrier 气力输送管
pneumatic ca(u)lker 风动密缝凿
pneumatic ca(u)lking 气动堵缝;风动凿密
pneumatic ca(u)lking tool 风动捻缝工具
pneumatic cell 铬电池;空气搅拌池;气体探测管;气流电池;气动式电池;气拌池
pneumatic cell without pump body 浅型充气浮选槽;无泵体的充气式浮选槽
pneumatic cement conveyer 水泥压气输送机
pneumatic cement handling 压缩空气水泥运送;气力水泥运送;压缩空气水泥装卸;气力水泥装卸;水泥的气动运送;水泥的气力输送
pneumatic chaff dispenser 气动式干扰物投放器
pneumatic chain saw 风动链锯
pneumatic chalking hammer 铲平风锤
pneumatic chamber feeder 仓式气力输送机
pneumatic chipper 气凿;风铲
pneumatic chipping hammer 气动錾;风凿;风錾
pneumatic chisel 气凿;风动凿;风动镐
pneumatic chuck 气动卡盘;气动夹头
pneumatic chucking 气动卡紧
pneumatic chute 气动溜槽
pneumatic chute system 气动溜槽系统
pneumatic circuit 气动回路;气压系统;气压管路
pneumatic circular saw 风动圆片锯
pneumatic clamp 气动压板
pneumatic classification 风力分级
pneumatic classifier 气力式选粒机;气力分级器
pneumatic clay 气化黏土
pneumatic clay digger 气动黏土挖掘机
pneumatic cleaner 气动吸尘器;风力选矿机
pneumatic cleaning 气动净化;气力清洁;气流式清选;风力选矿;风力清洗;风力清除;风力精选
pneumatic cleaning system 气流除尘系统
pneumatic clearer 气流清洁器
pneumatic clock 气压钟
pneumatic cloth feeder 气压导布装置
pneumatic clutch 气动离合器
pneumatic clutch brake 气动离合式刹车
pneumatic coal hammer 落锤风镐
pneumatic coal picker 气动煤镐
pneumatic compacting and finishing screed 风动夯样板
pneumatic compactor 气夯;风动夯(具);风动捣实器
pneumatic compass 气动罗盘
pneumatic components 气动元件
pneumatic components for compressed air gen-

eration 产生压缩空气气动元件
pneumatic compression riveter 气动铆钉枪
pneumatic concentration 风选法
pneumatic concentration of asbestos 石棉风选
pneumatic concentrator 风选法;风力选矿机
pneumatic concrete breaker 气动混凝土破碎机;风动混凝土破碎机
pneumatic concrete coveying 气力混凝土输送
pneumatic concrete handling machine 气力混凝土运输机
pneumatic concrete placer 混凝土浇筑机;气压混凝土浇灌机;气动式混凝土浇灌机;气动混凝土浇筑机
pneumatic concrete placing 气力混凝土浇筑
pneumatic concrete placing machine 气力混凝土浇筑机
pneumatic control 空气控制;气压控制;气力操纵;气控;气动控制;气动调节
pneumatic control instrument 气控仪表
pneumatic controller 压缩空气控制器;气压式控制装置;气压式调节计;气动控制器;气动调节器
pneumatic control manifold 气动控制屏
pneumatic control of liquid level 液面的气动控制
pneumatic control relay 气压控制继电器
pneumatic control system 气压控制系统;气动控制系统
pneumatic control unit 气动控制单元
pneumatic control valve 气动控制阀;气动调节阀
pneumatic conveyer 空气输送装置;气压输送装置;气动输送机;气动运输;气动管道输送机;气动管道输送器;风动输送机;气压输送机;气流输送器;风动运输设备;气压传送机;气流输送机
pneumatic conveyer dryer 气流干燥器;气流输送干燥机
pneumatic conveyer system 气力输送系统;气力输送系统;气动管道装卸系统
pneumatic conveyer with pressure 压送式风力输送器
pneumatic conveyer with suction 吸送式风力输送器
pneumatic conveying 压缩空气输送;空气输送;气压力输送;气力输送;气动输送
pneumatic conveying device 气动吹送装置
pneumatic conveying dry 气流干燥
pneumatic conveying dryer 气流干燥机;气流干燥设备
pneumatic conveying equipment 气动输送设备
pneumatic conveying plant 气动吹送装置
pneumatic conveying system 风动输送系统
pneumatic core breaker 风动型芯破碎机
pneumatic core sampler 压气钻芯取样器
pneumatic counter 气动式(交通量)计数器
pneumatic counter balance 压缩空气平衡装置
pneumatic coupler control 车钩气动控制
pneumatic coupler system 车钩空气系统
pneumatic crane 气动起重机;气动吊车;风动起重机
pneumatic cranking 气力起动
pneumatic cushion 气垫;气压式缓冲器;气囊式缓冲器
pneumatic cushioning 气压减振
pneumatic cushion shock absorber 气垫减振器
pneumatic cut-off winder 气压自停卷布机
pneumatic cutter 气动切割嘴;风动剪
pneumatic cylinder 压缩空气瓶;气压缸
pneumatic cylinder cock 气筒旋塞
pneumatic cylinder under constant pressure 稳压风缸
pneumatic damping 气体阻尼
pneumatic deception device 充气伪装装置
pneumatic decking 气动除冰
pneumatic deduster 风动收尘器
pneumatic dedusting 压缩空气除尘;气压除尘
pneumatic de-icer 气动除冰设备
pneumatic deicing 气动除冰
pneumatic demolition pick hammer 气动拆毁镐(拆除房屋等用);风镐
pneumatic descaling apparatus 风动除鳞装置
pneumatic despatch 气力输送法;气力输送
pneumatic detector 气压式探测器;气动式检测器;气体检测器;气动探测器;气动测车器
pneumatic diaphragm control valve 气动薄膜调节阀
pneumatic diaphragm pump 风动隔膜泵
pneumatic die cushion 气力模垫

pneumatic differential pressure transducer 气动压差变送器
pneumatic differential transmitter 气动差动变送器
pneumatic digger 风锹;风动铲具;气动铲;风镐;风铲
pneumatic discharge 气动卸料
pneumatic discharge apparatus 气动卸料装置;气动送料装置;风送装置
pneumatic discharge channel 空气卸料槽
pneumatic discharged car of bulk cement 散装水泥气力卸料车厢
pneumatic discharger 气动卸料机
pneumatic discharge vehicle 气动装卸货车;气动卸载式载重汽车;气动卸货车
pneumatic dispatch 气力输送法;气力输送;气力传送;气力递送
pneumatic distributor 气动输送分配器;气动分配器
pneumatic dolly 气动捣棒;气动铆钉顶;气动铆钉托
pneumatic door opener 气动开门器
pneumatic draw-off device 气流牵拉装置
pneumatic dredge pump 气力泥泵
pneumatic dredge(r) 气动挖泥船
pneumatic drill 风镐;气钻;气动钻;气动排钟式播种机;风钻;风动钻车;风动钻
pneumatic drill cradle 风钻槽形托架
pneumatic drill (hammer) 风动凿岩机;气动凿岩机
pneumatic drilling 空气钻眼;气动钻进
pneumatic drilling machine 气动钻机;风动钻机
pneumatic drilling method 气动钻法
pneumatic drill leg 风动凿岩机腿架;气动凿岩机腿架
pneumatic drill sharpener 锻钎机
pneumatic drive 压缩空气驱动;气力驱动;气动驱动;气动传动
pneumatic-driven pavement breaker 风动式路面破碎机
pneumatic driver 风动打桩机
pneumatic dryer 气流式干燥器;气流式干燥机;气流干燥机;风力干燥机
pneumatic drying 气流干燥;通风干燥
pneumatic drying system 气动干燥装置
pneumatic dual tired wheel 双充气轮胎的轮子;双充气轮胎的车轮
pneumatic duct system 气力管道系统
pneumatic dust removal system 气动除尘装置
pneumatic earth hammer 气动夯土机;气动打夯机
pneumatic ejector 气压喷射泵;压气喷射泵;气压喷射器;气动喷射泵;风动喷射器
pneumatic elastic wheel 弹性气压车轮
pneumatic-electric(al) convertor 气电转换器
pneumatic-electric(al) lamp 压气电灯
pneumatic-electric(al) signal converter 气电信号转换器
pneumatic electronic fail-safe system 气动电子式损坏安全性系统
pneumatic element 气动元件
pneumatic elevating conveyer 气力提升机;气力升送机
pneumatic elevating gear 气力升降装置;气力提升装置
pneumatic elevator 压缩空气升降机;气压升降机;气吸机;气流式升运器;气力提升机;气动升降机;风动提升机
pneumatic elutriator 气流输送分级机
pneumatic emptying 气动卸料
pneumatic end loaded fender 充气端部加载护舷
pneumatic engine 压缩空气发动机
pneumatic engineering 气动工程;气动技术
pneumatic equipment 气动设备;气动元件
pneumatic equipment exhaust noise 气动设备的噪声
pneumatic excavation 压气施工法
pneumatic excavator 气动掘凿机;风动钻岩机;风动挖掘机;气动掘进机
pneumatic external vibrator 气动外部振动器
pneumatic fed 气压喷射(混凝土);压气喷射(混凝土)
pneumatic feed 气动喂料;风动给料
pneumatic feed column 气动推进凿岩机支架
pneumatic feeder 气力送料器;气力加料机;风动给料机
pneumatic feeding 风动送料;气动给料
pneumatic feedleg 气动(钻进)腿架;气腿子;气腿
pneumatic feed motor 气动进给马达;气动钻进马达

pneumatic fender unit 充气护舷部件
pneumatic fertilizer distributor 气动式撒肥机
pneumatic fiber-fineness indicator 气流式纤维细度仪
pneumatic fill assembly 气动进料设备
pneumatic (filled) fender 充气护舷;充气橡胶防撞装置;充气碰垫
pneumatic fill(ing) 风力充填
pneumatic filling roof control 风力充填法控制顶板
pneumatic filter 气压滤池
pneumatic finisher 气力修整器;气动修整器;风力修整机
pneumatic fire detection system 气动火警感察系统
pneumatic firedoor 风动炉门
pneumatic flipover bucket loader 气动摆斗式装载机;气动上翻式装载机
pneumatic float 气动抹子;气动泥刀
pneumatic flo(a)tation 泡沫浮选;充气浮选
pneumatic flo(a)tation cell 充气式浮选槽
pneumatic flo(a)tation method 充气浮选法
pneumatic flocculation 气动絮凝
pneumatic fluidizing conveyer 空气输送斜槽
pneumatic flush tank 气压冲洗水箱
pneumatic fly-ash removal apparatus 风动除尘装置
pneumatic forging hammer 压缩空气锤
pneumatic forming 气压成型
pneumatic foundation work 压气(沉箱)基础工程
pneumatic ga(u)ge 气动测量仪表
pneumatic ga(u)ging 气动测量
pneumatic gear rotary drilling machine 气动齿轮钻机
pneumatic governor 气压式调速器;气力式调速器;气力调节器
pneumatic grab 风动抓岩机
pneumatic grain conveyer 气流式谷物输送器
pneumatic grain elevator 吸粮机
pneumatic grain handling equipment 气力装卸谷物设备
pneumatic grain handling machine 吸粮机
pneumatic grain handling plant 气力装卸谷物设备
pneumatic grain handling unit 气力谷粒装卸装置;气力谷粒运送装置;气力粒料装卸装置;气力粒料运送装置;气力装卸谷物设备
pneumatic grain unloading system 谷物抽吸卸载系统;散粮气力卸货系统
pneumatic grass cutter 风动割草机
pneumatic grinder 气动研磨机;风动研磨机;风动手提砂轮机;风动砂轮机;风动砂轮
pneumatic grinding machine 风动研磨机;风动砂轮
pneumatic gripping 气动夹钳;气动夹紧器
pneumatic gun 气锤;气枪;喷射枪;风动铆钉枪
pneumatic hack saw 气动弓形钢锯
pneumatic hammer 空气锤;气锤;风镐;风动凿;风动锤
pneumatic hammer drill 风动凿岩机
pneumatic hammer pile driving 气锤打桩机;气锤打桩
pneumatic hammer's disease 气锤病
pneumatic hand hammer rock drill 气动手持凿岩机
pneumatic handling 气压卸货;气力装卸;气力输送;气力操作
pneumatic handling plant 气力装卸设备
pneumatic hand riveter 气动铆锤;铆钉枪
pneumatic hand-sprayer 手提气压喷雾器
pneumatic hay conveyer 气流式干草输送器
pneumatic high-pressure control 气动高压控制
pneumatic high-speed duster 气力式快速喷粉机
pneumatic hoist 气压提升机;压缩空气提升机;空气提升机;气压卷扬机;气动起重机;气动卷扬机;气动绞车;风动绞车
pneumatic holder-on 气动铆钉顶座
pneumatic holder-up 气动铆钉顶座
pneumatic hold-on 风动冲压顶把
pneumatic homogenization 空气搅拌;气力均化
pneumatic homogenizing 空气搅拌;气力均化
pneumatic hopper loader 气动加料斗
pneumatic hose 压风胶管;供气胶管;软风管;气路软管
pneumatic-hydraulic 气动液压的
pneumatic-hydraulic autopilot 气动液压自动驾

驶仪
pneumatic-hydraulic controller 气动液压控制器
pneumatic-hydraulic control system 气动液压控制系统
pneumatic-hydraulic pin puller 风动液压拔料机
pneumatic ice delivery system 气压输冰系统；气动输冰系统；风动输冰系统
pneumatic impact breaker 气动冲击破碎机
pneumatic impact hammer 气动冲击锤
pneumatic impact spanner 气动扳手
pneumatic impact wrench 风动扳手
pneumatic impingement nozzle 冲击式气动雾化喷嘴
pneumatic indicator 气动指示器
pneumatic injection 压气灌浆；风力注射不燃灭火；风动喷射
pneumatic installation 气压设备；气动设备安装；风动设备
pneumatic internal vibrator 气动内部振动器
pneumatic irregularity tester 气流式不匀率试验仪
pneumatic jack 管子气动起重器；气压千斤顶；气压顶升器；气力千斤顶；气力起重器；气动千斤顶
pneumatic jig 风筛；风力跳汰机
pneumatic jigging 风力跳汰；风力淘选
pneumatic key 气动电键
pneumatic knapsack 背负式气力喷雾器
pneumatic knockout 气动落砂机
pneumatic lapper 气动研磨器
pneumatic lathe dynamometer 车床气动测力计
pneumatic level control 气动液位控制
pneumatic level ga(u)ge 气压水位计
pneumatic lift 气动提升机；气动升降机
pneumatic lifting bag 气力提升袋
pneumatic lifting device 气力升送机；气力提升机
pneumatic lifting pump 气力提升泵
pneumatic line 压缩空气管路
pneumatic linear actuator 气动缸
pneumatic line for concrete distribution 混凝土的管道吹送（分布）
pneumatic line type heat detector 线型气动感温探测器
pneumatic linkage 气压联动装置；风压联动装置
pneumatic loader 气动装载机械；风动装药器
pneumatic loading 风动装药
pneumatic loading device 气控装料设备
pneumatic loading machine 气动装料机
pneumatic loading system 气压加载系统
pneumatic lock 压缩空气闸门；气动闸门
pneumatic logic elements 气动逻辑元件
pneumatic logic members 气动逻辑部件
pneumatic loudspeaker 气流扬声器；气流传声器
pneumatic low-pressure control 气动低压控制
pneumatic machinery 风动机械
pneumatic marking pen 风动标志笔
pneumatic measurement 气流测定
pneumatic method 压气施工法
pneumatic method of sinking 压气沉箱凿井法
pneumatic micrometer 气动量仪；气动测微仪
pneumatic mixer 气动混料机
pneumatic mixing 气动混合
pneumatic mixing and stirring 气动混合搅拌
pneumatic mortar 压力砂浆；喷射用砂浆；喷射用灰浆
pneumatic mortar applying 压力喷浆
pneumatic motor 气动马达；气动电动机；风动马达
pneumatic-mounted 安装在充气橡胶轮胎上的
pneumatic mud drilling 充气泥浆钻进
pneumatic multijet weaving machine 多喷嘴喷气织机
pneumatic nailing machine 气动铆钉机
pneumatic nebulizer 气动喷雾器
pneumatic nozzle 气动雾化喷嘴
pneumatic nozzle link 气压调节杆
pneumatic nut wrench 风力螺母扳手
pneumatic oil switch 气动油压开关
pneumatic operation 气力装卸；气力操作
pneumatic operational amplifier 气动运算放大器
pneumatic operator 气动操作机构
pneumatic oscilating bone drill 风动骨钻
pneumatic outlet 气体出口
pneumatic overhead loader 气动后卸式装载机
pneumatic overshot loader 气动摆式铲斗装载机
pneumatic paint brush 气压喷漆刷；气压喷漆器
pneumatic panel 配气板

pneumatic pantograph control 受电弓气动控制装置
pneumatic parameter 气压参数
pneumatic pavement breaker 风动路面破碎机；气动路面破碎机
pneumatic paving breaker 压气式路面破碎机；风动式路面破碎机
pneumatic paving rammer 气动铺路夯（实机）
pneumatic peener 气动喷砂机
pneumatic perforator 气压打孔机；气压凿孔机；气压穿孔机
pneumatic pick 气动挖土机；气动锤；气压凿岩机；气镐；气动挖掘机；松土机；风镐；风动镐
pneumatic picker 气动镐；风镐
pneumatic pick hammer 气动镐；风镐
pneumatic pick mining 风镐落煤法
pneumatic pier 气压沉箱桥墩
pneumatic piezometer 气压式孔隙水压力仪；气压式测压计；气动式测压计
pneumatic piezometer tip 气动式孔压测头
pneumatic pile 气压桩
pneumatic pile driver 气动打桩机；气动打桩机
pneumatic pile hammer 气动打桩锤；风动打桩锤
pneumatic piling 压入式打桩；气压打桩
pneumatic pillow 压气垫座
pneumatic pipe 风力输送管
pneumatic pipeline pig 气动清管器
pneumatic pipework/installation 气动管道工程/安装
pneumatic piping 气力输送管道
pneumatic piston 压风机活塞；气动活塞
pneumatic piston actuator 气压活塞传动装置
pneumatic placed concrete 气压浇筑混凝土
pneumatic placement 风动给料设备；风动填筑；气动浇筑
pneumatic placer 压气喷浆机；气力压送机；风动浇筑机
pneumatic placing 压气浇筑（混凝土）；气动浇筑；气压浇灌；喷射浇灌
pneumatic plant 压缩空气设备；压风设备
pneumatic plaster-throwing machine 气动灰浆喷射机
pneumatic pole tamper 气动塞药棒；气动捣固棒
pneumatic positioner valve 气动阀门定位器
pneumatic positioning relay 气动调位继电器
pneumatic power 气力；气动功率
pneumatic power amplifier 气动功率放大器
pneumatic power drive 气力传动
pneumatic power drive lift 气力传动提升机
pneumatic preparation 气动选；风力选
pneumatic press 气动压力机
pneumatic pressure 气压
pneumatic-pressure cell 气压浇搅拌池
pneumatic pressure device 压缩空气加压机构
pneumatic pressure pile 气力灌浆管桩
pneumatic pressure regulator 气动调压器
pneumatic pressure tank 气压罐；气压箱
pneumatic pressure transmitter 气动压力传感器
pneumatic pressurizing 气动加压
pneumatic printing 气动印刷
pneumatic printing frame 真空晒版机
pneumatic process 气压法（隧道施工）；气压沉箱法
pneumatic program(me) set station 气动程序给定器
pneumatic pruning shears 气动修枝剪
pneumatic pulldown 气动输带装置
pneumatic pulsator 气动式脉动器
pneumatic pump 空气泵；气压泵；气流泵；气力输送泵；气动输送泵；气动泵；抽气泵；仓式泵；风动泵
pneumatic pump dredge(r) 气动泵挖泥船
pneumatic pumping 压气抽水；气压气送
pneumatic puncher 风动冲击式截煤机
pneumatic punching machine 气动冲床；风动冲压机
pneumatic pusher 风动推杆
pneumatic pyrometer 气动高温计
pneumatic rabbit 气动传送器
pneumatic rabbit channel 气动速送器管道
pneumatic rabbit hole 风动传送装置孔道
pneumatic ram 气动夯实；气压夯锤；气力夯；气夯锤；气夯；风捣锤
pneumatic rammer 气动夯实机；气动锤；气锤；风动振捣机；风动锤；风冲子
pneumatic rate-of-rise tubing 气动差温式探测管

pneumatic recorder 气动记录器
pneumatic refractometer 气动折射计
pneumatic regulating valve 气动调节阀
pneumatic regulator 气动调节器
pneumatic relay 气压继电器；气动替续器；气动继电器
pneumatic remote control 压缩空气遥控；远距离气动控制
pneumatic remote control system 气动远距离控制系统
pneumatic remote indicator 气动遥示器；气动远距离指示器
pneumatic restriction 气流限制
pneumatic retarder 风动缓行器
pneumatic reverse gear 气力回动装置
pneumatic reversing valve 气动逆向阀；气动换向阀
pneumatic ribbon type receiver indicator 气动色带指示器
pneumatic rivet buster 风动铆钉铲
pneumatic riveter 铆钉枪；风动铆钉机；气动铆接机；气动铆钉枪；气动铆钉（钉）机；风铆机；风动铆钉枪；风动铆（接）机；气压铆钉器
pneumatic riveter machine 风动铆机；气压铆钉机
pneumatic rivet hammer 气压铆钉锤
pneumatic riveting 气压铆接；气压铆钉；气动铆接；风动铆接
pneumatic riveting appliance 气铆工具
pneumatic riveting gun 气压铆钉枪；气动铆钉枪
pneumatic riveting hammer 气压铆钉锤；气铆锤；气动铆钉锤；气动铆锤；风动铆（钉）锤
pneumatic riveting machine 风动铆机；气压铆钉机
pneumatic rivet machine 气压铆钉机
pneumatic robot 气动机器人
pneumatic rock drill 风动凿岩机；气动凿岩机
pneumatic rocker type shovel(loader) 气动摇臂（式）装载铲；气动摇臂（式）装载机
pneumatic rod extractor 气动钻杆提器
pneumatic rod puller 风动钻杆拔取器
pneumatic roller 气胎路辗；轮胎式压路机；气胎辊；气动式碾压机；气动路碾
pneumatic rolling 用橡胶辊滚压；用橡胶辊碾压
pneumatic rotameter 气动转子流量计
pneumatic rubber fender 充气橡胶碰垫；充气橡胶护舷；充气橡胶防撞装置
pneumatics 气体力学；气动装置；气动力学；气动技术
pneumatic sabre saw 马刀形气动锯；气动线锯
pneumatic sander 气力喷砂机；气动打磨机
pneumatic sand rammer 气动型砂捣器机
pneumatic sand system 气力送砂装置
pneumatic saw 气力锯
pneumatic scaling hammer 气动除锈气锤；除锈气锤
pneumatic scarifier 压气翻路机
pneumatic scavenging gear 气动吹扫装置；气动排气装置；气动吹除装置；气动换气装置
pneumatic screw conveyer 气力螺旋输送机
pneumatic screw press 气动螺旋压砖机
pneumatic scrubber 气动洗涤器
pneumatic seal 气力密封；充气水封
pneumatic sealing method 压气法（隧道施工）
pneumatic seat 气垫座椅
pneumatic sensor 气流传感器
pneumatic separation 气力分选；气动分离；风力分离
pneumatic separator 气力式分离机；气动分离器；风力清选机；风力分选机
pneumatic servo 气压伺服装置；气动伺服装置；气动舵机；阀门定位器
pneumatic servoelement 气动伺服元件
pneumatic servomechanism 气动伺服机构
pneumatic servooperation 气动伺服操作
pneumatic servosystem 风动伺服装置；气动伺服系统
pneumatic setting element 气动给定元件
pneumatic settlement cell 压气式沉降仪；气压式沉降仪；气压式沉降盒
pneumatic sett paving rammer 气动小方石铺路夯实机
pneumatic sewage ejector 气动式污水排水器
pneumatic sewage system 气压排水系统
pneumatic shaft sinking 压气沉箱凿井；气压舱沉井
pneumatic shaker 气力抖动器
pneumatic sheaf conveyer 气流式干束输送器

pneumatic shield driving 气压盾构推进;压缩空气盾构推进
pneumatic shield with compressed air 气压盾构
pneumatic shock absorber 气压吸振器;气动减振器;气力减振器;气动式减振器;气垫缓冲器;气垫避振器
pneumatic shock absorber strut 气压减振柱
pneumatic shot blasting machine 气动喷丸机;喷丸清理机
pneumatic shovel 风动装渣机
pneumatic shutter 气动快门
pneumatic shuttle 气压传送装置;气动速送容器;风箱
pneumatic simulator 气动仿真器
pneumatic sine wave generator 气动正弦波发生器
pneumatic single-pen recorder 气动单针记录器
pneumatic sinking 气压基础下沉工作;气压沉箱下沉
pneumatic size analysis 风力筛分分析
pneumatic slide 气动传送机
pneumatic slide valve 气动滑阀
pneumatic sliding door drive 推拉门压力驱动
pneumatics machine 风动机械
pneumatic soil rammer 气动夯土机
pneumatic sounder 气压测深仪
pneumatic spade (hammer) 风铲;气动铲;气锹;风动铲
pneumatic spader 气动铲土机
pneumatic splint 充气夹板
pneumatic sprayer 气力喷雾机;气泵喷雾器
pneumatic spraying machine 气力式喷雾机
pneumatic spray painting 气动喷漆;风动喷漆
pneumatic spring 空气弹簧;气压弹簧;带气垫的板簧
pneumatic stapler 气动装订;气动钉机
pneumatic starter 气动起动器;气动起动装置;缩空气起动机;气动起动器
pneumatic starting device 压缩空气起动装置
pneumatic starting valve 压缩空气起动阀
pneumatics technique 气动技术
pneumatic steel 喷气炼钢
pneumatic steel caisson 压气钢制沉箱;钢制气沉箱
pneumatic steel-making 转炉炼钢
pneumatic steering gear 气动式舵机;气动操舵装置
pneumatic stirrer 风动搅拌机
pneumatic stirring 压气搅拌(法)
pneumatic storage tank 气压水箱
pneumatic stowing 气动装载;气动充填;风力充填
pneumatic stowing machine 气动装载机;气动充填机
pneumatic stowing pipe 气动充填的管子
pneumatic stowing plant 气动装载设备;气动充填设备
pneumatic stowing tube 气动充填的管子
pneumatic stripper 气流抄针器
pneumatic structure 气承建筑;气结构;充气结构
pneumatic suction 气吸机
pneumatic suction cleaning system 吸风式清洁装置
pneumatic suction conveyer 风力吸送器
pneumatic suction conveying system 吸气式输送系统
pneumatic sump pump 风动水坑抽水泵;风动潜水泵
pneumatic surge chamber (机车的)压气式均衡风缸;气压调压室
pneumatic suspension 空气悬浮
pneumatic switch 气动开关
pneumatic switchboard 气动开关板
pneumatic system 气动系统
pneumatic system of sewerage 气压排(除)污水系统;气压排(除)污水法
pneumatic tacker 气动钉机
pneumatic take-down 气流牵拉
pneumatic tamper 气动打夯机;气动夯;气夯;风动打夯机
pneumatic tamping machine 气动夯实机
pneumatic tank 气压罐;压气瓶;储气箱
pneumatic tank switcher 气动储箱切换开关
pneumatic tank water closet 气压水箱冲洗池
pneumatic tape transport 气动带传送
pneumatic target 气动靶
pneumatic teletransmitter 气动远程发送器

pneumatic tension device 气流张力装置
pneumatic tester 气流仪
pneumatic test(ing) (管线的)空气查漏检验;压气试验;气压试验;气密试验;气动试验
pneumatic test stand 气压试验台
pneumatic thickener 气压增稠器;气压脱水机;气动稠化器
pneumatic (tie-)tamper 气压轨枕夯具;风动捣固机
pneumatic time delay relay 气流延时继电器
pneumatic timing system 气动正时系统
pneumatic tip(ping) wagon 气动翻斗车
pneumatic tire 气胎;充气轮胎
pneumatic-tired 充气轮胎的
pneumatic-tired buggy 气胎手推车
pneumatic-tired (bull) dozer 充气轮胎推土机
pneumatic-tired mobile crane 充气轮胎式流动起重机
pneumatic-tired roller 轮胎式压路机;气胎压路机;气胎碾;充气轮胎压路机
pneumatic-tired tractor 充气轮胎拖拉机;轮胎式拖拉机
pneumatic tire rim 气胎轮辋
pneumatic tissue 储气组织
pneumatic-to-current converter 气电转换器
pneumatic toggle link 气动肘杆
pneumatic toggle press 气动肘杆压力机
pneumatic tool 气压工具;气动工具;风动工具
pneumatic tool accessory 气动工具附件
pneumatic tool plant 气动工具厂
pneumatic traction 气力牵引
pneumatic tractor 气动牵引车
pneumatic train/door control 气动车门控制
pneumatic transducer 气动式传感器
pneumatic transducing element 气动变送元件
pneumatic transfer device 气力输送装置;气动移动装置
pneumatic transfer system 气动输送系统;气动传送系统
pneumatic transfer tube 气动传送管
pneumatic transmission 压气式传动;气动传动;压气传动;气压传动;气力输送;气动变送
pneumatic transmission and control 气动变送及控制
pneumatic transmission lag 气压输送滞后
pneumatic transmission system 气动传输系统
pneumatic transmitter 压气传送器;气动传感器;气动传杆器
pneumatic transmitting rotameter 气动传送旋转流量计
pneumatic transport 气压输送;气体输送;气力输送;气动输送;风动输送
pneumatic transportation 气动运输;风力输送
pneumatic transportation by pipeline 管道气压运输
pneumatic transportation groove 空气输送斜槽
pneumatic transport device 气力输送装置
pneumatic transport pipe 空气输送管
pneumatic transport placer 压气运送设备;气压输送浇筑机;喷送器
pneumatic transport pump 气力输送泵;气力运输泵
pneumatic transport system 气力输送系统
pneumatic trash system 风力输送废物装置
pneumatic traversing guide control 气动式横动导丝控制
pneumatic tri-pen recorder 气动三针记录器
pneumatic trough 集气槽
pneumatic trowel 气动抹子;气动泥刀
pneumatic tube 压缩空气输送管;压缩空气管道;气压运输管;气压(输送)管;气动输送管;气动(导)管;风力输送管
pneumatic tube conveyer 风力输送器;气流管式输送机;空气输送管
pneumatic tube detector 充气管式检测器
pneumatic tube fire alarm system 气压管式火灾报警系统
pneumatic tube installation 风管装置;气力输送管装置
pneumatic tube system 气动输送管系统
pneumatic tubing 压缩空气管道
pneumatic turning jack for wheel set 风动轮对转辘
pneumatic type 气化型
pneumatic type mortar pump 气动式灰浆泵
pneumatic type spray dryer 气流喷雾式干燥器

pneumatic tyre 气胎;充气轮胎
pneumatic-tyre bulldozer 气轮式推土机
pneumatic-tyred bucket (elevator) loader 充气轮胎斗式装载机
pneumatic-tyred crane 充气轮胎起重机
pneumatic-tyred drilling jumbo 充气轮胎钻车
pneumatic-tyred excavator 充气轮胎挖掘机;轮胎式挖掘机
pneumatic-tyred loader 充气轮胎装载机
pneumatic-tyred mobile crane 汽车式起重机;轮胎式活动起重机
pneumatic-tyred roller 轮胎压路机;气胎压路机;气胎(路)碾
pneumatic-tyred shovel 充气轮胎正铲
pneumatic-tyred shovel loader 充气轮胎挖掘装载机
pneumatic-tyred slewing crane 充气轮胎悬臂起重机
pneumatic-tyred trackway system 充气轮胎(行)车道系统
pneumatic-tyred tractor 充气轮胎拖拉机
pneumatic-tyred tractor-drawn roller 拖拉机牵引的充气轮胎压路机
pneumatic-tyred vehicle 充气轮胎车辆
pneumatic-tyred vibrated finishing screed 气动振动匀面板;气动振动整平板
pneumatic-tyred wagon drill 钻探车;汽车式钻车
pneumatic tyre expander 气张胎器
pneumatic tyre rim 气胎轮缘;气胎轮辋
pneumatic tyre (wheel) 充气轮胎
pneumatic unit 气动元件
pneumatic unit combination instrument 气动单元组合仪表
pneumatic unloader 气吸机
pneumatic valve 气压阀;气动阀
pneumatic valve positioner 气动阀位控制器
pneumatic vertical conveyer 气力提升机;气力垂直输送机
pneumatic vibrator 气动振捣器;压缩空气振动器;气压振动器;气动振动器;风动振动器;风动振捣器
pneumatic vibrator unit 气动振动装置
pneumatic vibratory knockout 气动落砂器
pneumatic waer measuring equipment (混凝土搅拌输送车的)气动量水装置
pneumatic water barrel 压气提水桶;抽气排水筒
pneumatic water ga(u)ge 压气水位计
pneumatic water supply 压气供水;气压给水
pneumatic water supply equipment 气压给水设备
pneumatic water supply installation 气压给水设备
pneumatic water supply system 压气供水系统;气压给水系统
pneumatic water supply tank 气压给水箱
pneumatic wave generator 气压生波器(波浪水工模型用)
pneumatic weighing system 气力称重系统
pneumatic weighting 气流加压
pneumatic wheel fender 充气轮胎碰垫;充气轮胎护舷
pneumatic wheel roller 气胎(路)碾
pneumatic winch 气动卷扬机;气动绞车;压气绞车
pneumatic with electric(al) control brake 电控制的气闸
pneumatic work 气压沉箱工程;压气沉箱掘进工作;气压工程
pneumatic wrench 气动扳手
pneumatogen apparatus 氧气发生器
pneumatogenic 气成的
pneumatogenic enclosure 气成包体
pneumatogenic mineral 气成矿物
pneumato-hydatogenetic 气液成的
pneumato-hydratogenetic 气水成的
pneumato-hydrothermal deposit 汽化热液矿床
pneumatoilitc solution 气成热液
pneumatolito-hydrothermal metasomatites 气成热液交代变质岩
pneumatology 气体力学
pneumatolysis 气化(作用)
pneumatolytic 气化的
pneumatolytic clay 气化黏土
pneumatolytic deposit 气化矿床;气成矿床
pneumatolytic differentiation 气化分异作用
pneumatolytic-hydrothermal deposit 汽化热液矿床
pneumatolytic hydrothermal metamorphism 气

化热液变质作用
pneumatolytic metamorphism 气化变质
pneumatolytic mineral 气化矿物;气成矿物
pneumatolytic ore 气化矿
pneumatolytic replacement 气化交代
pneumatolytic stage 气化期
pneumatolytic water 气态水
pneumatophore 救生氧气袋;出水通气根
pneumatophorus 救生氧气袋
pneumatosaccus 浮囊袋
pneumech silo 气动机械卸料库
pneumeractor 柜液面指示器;测量石油产品量的记录仪
pneumoconiosis 肺尘病;尘肺
pneumoelectric 气电
pneumo-electric(al) convertor 气电转换器
pneumo-electric(al) ga(u)ge 气电量规
pneumo-fathometer 空气探测仪
pneumography 充气造影术
pneumohydraulic 气动液压的
pneumohydraulic operation 风动水力操作
pneumokoniosis 尘肺;肺尘病
pneumo-mechanical flo(a)tation cell 空气机械浮选机
pneumonoultramicroscopicsiliccovolcanoconiosis 矽肺病
pneumo-oil switch 气动油压开关
pneumoroentgenogram 充气造影照片
pneumo-slurry transport 气压输泥
pneumotaxis 趋气性
pneumotomography 充气X线体层照相术
pneumotransport 气动输送
pneumozonal texture 气泡环带结构
pneuncatic plant 气动装置
pneutronic 气动电子的
pneu-vac drier 真空气流干燥器
pniladelpnus sp 山梅花
p-nitrophenol 对硝基苯酚
P-number 不溶解挥发性脂肪酸值
Pnyx 古雅典城内集会场所
PO[purchase order] 购货订单
poach 加水拌匀;侵入偷窃;漂洗;踩成泥浆
poacher 违禁捕鱼者
poacket welting machine 袋口踏边机
poaptor 纵向力操纵装置;操纵装置
poaque silica glass 不透明石英玻璃
PO Box 邮政信箱
Pocabontas coal 珀卡煤(一种烧铆钉用煤)
poche 建筑图涂黑部分(图例)
Pochet furnace 波歇炉
pocill 小杯
pocillum 小杯
pock 麻点
Pockel's cell 普克尔盒
Pockel's effect 普克尔效应
pocket 小块地区(死胡同);预留梁洞;钱袋;矿袋;空器;溶蚀坑;袋状海湾;储卡箱;储卡机;凹口;凹处
pocket accumulator 袖珍蓄电池
pocket air of plastics 塑料中的气泡
pocket alterter 袖珍报警器
pocket alterter charger 袖珍报警器充电器
pocket alterter transmitter 袖珍报警器发送器
pocket amalgamator 袋式混汞提金器
pocket ammeter 携带式安培计;便携式安培计
pocket anemometer 袖珍风速标;便携式风速仪;轻便风速计;轻便风速表
pocket aspiration psychrometer 轻便通风干湿仪;轻便通风干湿计
pocket atlas 袖珍地图集
pocket barker 小型鼓状剥皮机
pocket barograph 袖珍气压计
pocket beach 湾头小滩;湾状滩;袋状滩;袋形海滩
pocket bell 袖珍无线电传呼机
pocket bimetal thermometer 袖珍双金属温度计
pocket block 水口座砖
pocket book 袖珍本
pocket boom 袋状河硬
pocket bore 小钻
pocket borer 小钻;小型钻机
pocket builder 制栓机;布筒机
pocket butt 轻便铰;凹槽平缝铰链
pocket calculator 袖珍(式)计算器;袖珍计数机;微型计算机;便携(式)计算器

pocket camera 袖珍照相机
pocket card 推销员许可证;经纪人许可证
pocket chamber 袖珍式放射线测量仪;袖珍剂量计;袖珍电离室;携带式剂量计
pocket chisel 袖珍凿(子);门窗扁凿;凹槽錾子
pocket chronometer 袖珍天文表;天文怀表
pocket clay 高硅黏土;矿袋土;囊状黏土
pocket cod 袋形水域
pocket compass 袖珍罗盘
pocket computer 袖珍计算机;便携(式)计算机
pocket concrete penetrometer 袖珍式混凝土贯入仪
pocket conveyer 袋式输送器;袋式输送机
pocket corridor 袋形走道
pocket cylinder 窝眼筒
pocket dedusting 布袋除尘
pocket disk 窝眼圆盘
pocket divider with sheath 带套袖珍两脚规
pocket dosimeter 携带式放射线测量计
pocket drill 轻便小型钻机
pocket edition 袖珍版
pocket elevator 斗式提升机;袋式提升机
pocket feeder 转叶式供料器;回转定量加料器
pocket filling and sealing packager 袋充填封口包装机
pocket filter 便携式过滤器;扁袋集尘器
pocket flap 兜盖
pocket flashlight 闪光灯;袖珍灯;手电筒
pocket floor 闸底
pocket folder 栅栏式折页机
pocket grinder 压榨碎木机
pocket-handkerchief 手帕
pocket icebreaker 小型破冰船
pocket instrument 袖珍仪表;轻便仪表
pocket in wall 墙内座梁;墙内梁座(墙上安装梁的凹穴)
pocket-knife 小折刀
pocket lamp 袖珍灯;手电筒
pocket lamp bulb 手电筒灯泡
pocket leaf 袋状叶
pocket lens 手持放大镜
pocket level 携带式水平仪;袖珍水平仪;小型水平仪
pocket line 袋绳
pocket magnetometer 袖珍磁力仪
pocket magnifier 袖珍放大镜
pocket map 袖珍地图
pocket measuring stereoscope 袖珍量测立体镜
pocket memory 袖珍(式)存储器
pocket meter 袖珍式仪表;袖珍测试计;量袋器
pocket microscope 袖珍显微镜
pocket milling 凹坑铣削
pocket monitor 袖珍监测仪;小型监测器
pocket multimeter 袖珍万用电表
pocket of air 气穴;气坑
pocket of poverty 贫困区域;贫困地带
pocket penetrometer 袖珍(式)贯入器;袖珍(式)贯入仪;小型贯入仪
pocket piece 凹槽板;吊窗锤箱;吊窗锤匣板
pocket piston 小活塞;袖珍活塞
pocket plane 轻便刨
pocket planting 袋植
pocket plate 袋形极板
pocket plum 小型型芯座
pocket radio 袖珍式无线电设备
pocket receiver 袖珍接收机
pocket recorder 袖珍录音机
pocket rot 囊腐(木材缺陷);穴腐(木材内);袋状腐朽;袋腐;(木材的)腐孔
pocket rule 折尺;木折尺
pocketscope 轻便示波器;袖珍示波器
pocket setting 棚架装窑;箱形料架;格架装窑
pocket sextant 袖珍六分仪
pocket-shaped holder 斗形容器
pocket shear meter 小型十字板剪切仪;小型十字板仪
pocket-size 袖珍的
pocket-sized soil testing instrument 袖珍土壤试验仪
pocket-sized transit 袖珍经纬仪
pocket-size electronic flash unit 袖珍式电子闪光装置
pocket-size reel 袖珍带盘
pocket-size thickness ga(u)ge 小型测厚仪
pocket slide cal(l)pier 携带式(游标)卡尺
pocket sliding door 内插式滑门

pockets of soil 土囊
pockets of swamp material 湿土囊
pockets of water 水囊
pockets on the cover of the work 书卡袋
pocket spring 穴泉;小型涌泉
pocket stereoscope 袖珍立体镜
pocket stereo view 袖珍立体镜
pocket storage 洼地蓄水(量);洼地储积(量);凹地储[贮]水
pocket tape 钢皮卷尺;钢卷尺
pocket telephone 便携式电话机
pocket tester 小型测试器
pocket theodolite 袖珍经纬仪
pocket thermometer 袖珍温度计;袋式测温器
pocket tidal stream atlas 袖珍潮流图集
pocket transit 袖珍经纬仪
pocket transmitter 袖珍发报机(救生艇属具)
pocket tripod 小型三角架
pocket type filter 扁袋式集尘器
pocket type switch 组合开关
pocket valley 袋形谷
pocket ventilation roll 套管式毛毯热风干燥辊
pocket voltmeter 小型伏特计;袖珍伏特计;便携式伏特表
pocket watch 袖珍式表
pocket well 泄水阱
pocket wrench 小型扳手
pockety ore body 囊状矿体
pockety residuum 袋状风化壳
pocking 水泡
pock mark 麻点;斑点;麻坑
pock marked 麻点;痘痕
pock marking 麻点;油漆麻点
pockwood 愈疮木
pocky 有麻点的;有麻洞的;表面不平的
pocky cloud 悬球状云
Pocono sandstone 波科诺砂岩
poco oil 水薄荷油
pocosin 浅沼泽
p-octylphenyl salicylate 水杨酸对辛苯酯
pod 吊装预制浴室;(钻头的)纵槽;密集小群;浅洼地;吊舱;扁透镜矿体
POD analysis 波氏分析
pod auger 有纵槽的螺旋钻;纵槽钻头
Podbielniak analysis 波氏分析;波氏法精密分馏
Podbielniak extractor 波特贝尔尼克萃取器;波氏萃取器
pod bit 勺形钻头
pod cooling unit 舱内冷却装置
pod drag 短舱阻力
podetium 柱体
podger 可调节扳手
podiform 豆荚状;遍透镜状;扁透镜状
podiform fold 豆荚状褶皱
podium 墩座墙;裙房;管足;列柱墩座;墩座;[复]podiums 或 podia
podium block 墩座石块
podium temple 列柱式寺院建筑;列柱式神庙建筑;墩柱式寺院建筑;墩柱式神庙建筑
podo 波多软质木材(产于南半球)
podolite 碳酸磷灰石
podsol 灰壤;灰化土
podsolization 灰壤(化)作用
podsol (soil) 灰壤
pod test in green house 冬季盆栽试验
podunk 偏僻小城镇;不重要的偏僻小镇
podwer used in testing for acid 测酸性的一种药粉
podzol 灰化土(壤);灰壤
podzolic 灰化的
podzolic brown earth 灰化棕壤
podzolic horizon 灰化层
podzolic peat 灰化泥炭;灰化泥煤
podzolic soil 灰壤;灰化土壤;灰化土
podzolic soils 灰化土类
podzolic yellow earth 灰化黄壤
podzolization 灰壤(化)作用;灰化作用
podzolization of soils 土壤灰化作用
podzolized bog soil 灰化沼泽土
podzolized laterite 灰化红土;灰化红壤
podzolized meadow soil 灰化草甸土
podzolized red earth 灰化红土;灰化红壤;炭化红壤
podzolized yellow earth 灰化黄壤
podzol soil 灰壤
podzol type of soil formation 灰壤式成土作用

podzoluvisol 灰化土及生草灰化土
poecile (古雅典大会场的)拱廊
poeciloblast 变嵌晶
poetic architecture 富有诗意的建筑艺术
Poetsch freezing process 波特什冻结凿井法
Poetsch process 波特什盐水冻结凿井法
Poggendorff compensation method 波根多里弗补偿法
Poggendorff's first method 恒流直流电位计
Poggendoriff bichromate cell 波根多里弗重铬酸盐电池
pogonip 冻雾;冰雾
pogo stick (带弹簧的)水准尺;风动捣实器(俚语)
Pogson magnitude scale 波格森星等标度
Pogson ratio 波格森比值
Pogson scale 波格森标度
Pohle air lift pump 波勒空气提升泵
Pohlman method 波尔曼超声探像探伤法
Pohl's test 波尔氏试验
poid 形心曲线
poidometer 重量计;巨重快测计;加料计;称重计;称量计
poikile (古雅典大会场的)拱廊
poikilitic 嵌晶状
poikilitic cementation 嵌晶胶结物
poikilitic cement texture 嵌晶胶结物结构
poikilitic structure 嵌晶构造
poikilitic texture 嵌晶结构;包含碎屑结构;包含结构
poikiloblast 变嵌晶
poikiloblastic 变嵌晶状
poikilocrysal 嵌晶
poikilohydric plant 变水植物
poikilonymy 名称混乱
poikilophitic texture 辉绿嵌晶结构
poikilosmotic 变渗(透)压的
poikilosmotic animal 变渗性动物
poikilotherm 变温动物
poikilothermal animal 变温动物
poikilothermic 不定温的
poikilothermic animal 变温动物
poikilothermy 温度变化适应性;变温性
poikilotope 嵌含晶
poikilotopic fabric 嵌含的组构
poikoblastic texture 变嵌晶结构
poilite 石棉水泥(墙)板;石棉水泥瓦
Poincare elements 庞加莱根数
Poincare group 庞加莱群【地】
Poincare sheroid 庞加莱椭球体
Poincare sphere 庞加偏振球
Poincare franc 庞加利法郎
poine 草甸菌素
Poinsot ellipsoid 惯性椭球
point 小岬角;小岬;针尖;辙尖;尖端,尖、点,齿尖;磅音
point a brick wall 勾墙缝
point accuracy 点位精度
point active load 点动荷载
point adhesion method 点粘法
point analysis 点分析
point-and-cup suspension 尖杆杯座式悬挂
point-and-flat suspension 尖杆平座式悬挂
point-and-groove suspension 尖杆槽座式悬挂
point-and-slot support 尖杆缝槽式支架
point angle 锥尖角;顶角
point anomaly 点异常
point anomaly value 点异常值
point approximation 点近似法;单点近似法
point-area relationship 点面关系曲线
point at infinite 无穷远点
point bar 弓弧边坝;尖砂坝;曲流沙坝;凸岸沙嘴;凸岸边滩;沙嘴;点砂坝;边滩
point bar deposit 凸岸沙嘴沉积;曲流内侧尖形沙洲沉积;边滩沉积
point bar facies 边滩相
point bar sequence 曲流砂坝层序
point bar trap 点砂坝圈闭
point base 点位底图
point bearing 点支承;端支承
point-bearing capacity (桩的)允许承载力;桩端承载能力;桩端承载量;桩端(重能)力
point-bearing pile 尖端支承桩;支承桩;端支桩;点支承桩
point bit 锥形钻头
point-blank 近距离平射

point block 尖轨顶铁;点式住宅;轨尖块;塔式建筑群
point-block housing 塔式高层住房
point bonded bolt 前端黏结型锚杆
point brilliance 光点亮度;光点辉度;光源耀度
point building 塔式建筑
point-by-point calculation 逐点计算
point-by-point computation 逐点计算
point-by-point indicator 逐点示功器
point-by-point integration 逐点积分;近似积分
point-by-point interpolation 逐点内插
point-by-point mapping 逐点测图
point-by-point measurement 逐点测量
point-by-point method 逐点计算法;逐点测定法
point-by-point rectification 逐点纠正
point-by-point sailing 逐点航行
point-by-point scanning 逐点析像;逐点扫描
point-by-point survey 逐点测量
point-by-point test 逐步试验
point cathode 点状阴极
point charge 集中药包;点电荷
point circle 穿孔周期;齿顶圆
point clamp 道岔勾锁器
point concentration 点浓度
point-connectivity 点连通度
point contact 点接触
point-contact cement texture 点接触胶结物结构
point-contact lay wire rope 点接触钢丝绳
point-contact rectifier 点接触整流器
point-contact transistor 点接触晶体(管)
point-continued type speed control system 点连式调速系统
point convective model 点对流模型
point coordinates 点素坐标
point correspondence 点对应
point counter 尖端式计数管;点端式计数管
point counter tube 点端式计数管
point-counting analysis 标点计数分析
point-count method 点计数法
point covering number 点覆盖数
point cross point 交点
point cut-off 切断点
point cycle 点周期;穿孔周期
point damping factor 桩尖阻力系数
point data 定点资料;点据;单站资料
point defect 点缺陷
point deformation 桩尖变形
point density 点密度
point dependence domain 点依赖域
point-dependent segmentation 按点分割
point design 解决关键问题的设计;指定设计;符合规定要求的设计
point-designation grid 点标定格网
point desprit 点子花六角网眼纱
point detector 转辙检测器;联锁箱;道岔锁闭检查器
point-device 非常精确的
point diagram 点图
point diameter (螺纹的)前端直径
point dictionary 点位词典
point-diffraction interferometer 点衍射干涉仪
point discharge 尖端放电;点污染源排放;点流量
point distribution 调运点
point distribution analysis method 点分布分析法
point distribution function 点分布函数
point-down method 六角钢立轧法
point draft 山形穿法
point drip reaction 点滴反应
point-duty 值勤;站岗;交通指挥
point-duty policeman 值岗交通警
pointed 削尖;尖的
pointed arch 哥特式尖拱;哥特式拱;尖形拱;尖顶拱;二心内心桃花尖拱
pointed arched corbel-table 哥特式挑出面层;尖拱式挑出面层
pointed arch of brickwork 砖砌二心内心挑尖拱
pointed ashlar 尖琢石;粗凿麻面方石
pointed barrel vault 哥特式筒形拱顶;尖头式筒形拱顶
pointed bit 尖头钻
pointed body 尖头体
pointed body shape 尖头体形状
pointed bolt 尖头螺栓
pointed bottom pottery 锥底陶器

pointed chisel 尖凿
pointed compasses 指针罗盘
pointed cone 尖锥;尖头锥体
pointed contact 点接触
pointed corrosion 点蚀
pointed dome 尖圆屋顶;尖头圆屋顶;葱头形穹顶
pointed door opening 尖头(式)门洞;尖拱式门洞
pointed dressing 尖凿修整
pointed drill 尖钻;尖凿
pointed element 带尖构件;尖头构件
pointed end 尖端
pointed finish 点凿面
pointed funnel bit 漏斗形尖钻头
pointed hammer 尖头锤;尖锤
pointed horse shoe arch 尖头马蹄形拱
pointed in white cement 白水泥勾缝
pointed joint 勾缝
pointed joint of random rubble (成层砌筑的)乱石砌法
pointed journal 锥形轴颈
pointed peaky pulse 尖顶脉冲
pointed pediment 三角形檐饰
pointed peg 尖桩
pointed pile 尖头桩
pointed plain tile 尖头平瓦
pointed polishbrush 尖形抛光刷
pointed probe 尖探针
pointed punch 尖头穿孔机
pointed rivet head 尖锐圆锥头
pointed rod 尖头棒;尖头杆
pointed rope 尖头绳
pointed Saracenic arch 回教式尖拱;撒拉逊尖拱
pointed shell 尖顶壳体
pointed shell dome 葱头形薄壳顶
pointed shoe 水准尺尖端
pointed shovel 尖形挖掘铲
pointed soldering copper 尖焊铜
pointed stake 尖桩
pointed steel hammer 尖头钢锤
pointed stone arch bridge 尖顶石拱桥
pointed style 哥特式;尖拱式;尖头式
pointed tip 尖头电极
pointed tool 尖头工具
pointed tooth 尖齿
pointed transverse barrel vault 尖头横向筒形拱顶
pointed transverse tunnel vault 尖头横向筒形拱顶
pointed transverse wagon (head) vault 尖头横向斜拱顶
pointed tree 尖顶树
pointed trowel 勾缝镘刀;尖形泥刀;尖形抹子
pointed wire-rod 尖头钢丝棒
pointed work 点凿加工;琢凿加工
point effect 尖端效应
point efficiency 点效率
point elasticity 点弹性
point elasticity of demand 需求的点弹性
point electrode 尖端极;尖端电极;点焊电条
point element method 点元法
pointel(le) 菱形铺砌;斜铺方砖地面
point emanation source 点状射气源
point emission source 点排放源
point-ended pier 尖端形桥墩
pointer 指针;指示学
pointer arm 指示器
pointer counter 指针式计数器;刻度计数器
pointer counterbalance 指针平衡锤
pointer dial 指针式刻度盘;指针式计数盘
pointer frequency meter 指针式频率计
pointer galvanometer 指针式检流计;指针电流计
pointer ga(u)ge 指针表
pointer instrument 指针式仪表
pointer justification count 指针调整计数
point erosion source 点冲刷源
pointer reading 指针读数
pointer register 指针式计量器
Pointers 指极星
pointer's telescope 瞄准手用望远镜
pointer stop 指针挡销
pointer tip 指针尖
pointer to stack 堆栈的指针
pointer to symbols 符号指示字
pointer type 指针式;指针类型
pointer-type indicator 指针式指示器
point estimate 点估计

point estimation 点值估计；点估计
point exposure 点曝光
point file 触点锉刀
point filler 尖轨间隔铁
point filter 点筛选器
point focusing 光聚焦
point focusing monochromator 点聚焦单色器
point force 有心力；汇交力；集中力
point form 沙嘴形状
point for measuring temperature 测温点
point-foussed 聚焦于一点的
point frame 点框架
point frequency 点频率
point gamma 灰度系数的微分值
point ga(u)ge 量棒；轴尖式量规；针形水位计；水位测针；测针【测】
point ga(u)ge reading 测针读数
point gear 指针机构
point gearing 点啮合
point geometry 点素几何学
point grid 测点网
point group 晶族；点阵，点群；点集
point group analysis 点群分析
point group parameter precision determination 点阵参数精密测定
point hardening 局部淬硬；局部淬火
point harmonization 瞄准调整
point hologram 点全息图
point hours 标准点时
point house 点式住宅
point identification 像点辨认
pointillism 点彩画法
point image 像点；点像
point impact 点撞击
point independence number 点独立数
point indicator 辙尖指示器；辙尖指示器
pointing 制尖；指瞄准；指点；瞄准；嵌填灰缝；嵌填；勾缝
pointing accuracy 照准精度；目标指示精度；瞄准精度
pointing backwards 针尖反向
pointing by bisection 二等分照准
pointing by sandwiching 双丝照准
pointing calibration 指向校准
pointing chisel 点凿
pointing control 指向控制
pointing correction 指向改正
pointing device 瞄准机构
pointing error 指向误差；指示误差；照准误差；瞄准误差
pointing forwards 针尖同向
pointing in direction 水平瞄准
pointing instrumentation 瞄准仪
pointing joint 嵌缝；勾缝
pointing knife 制钉刀
pointing line 视准线
pointing machine 勾缝机；仿真雕刻机；镶嵌机；轧尖机、锻尖机
pointing masonry 勾缝砌体；勾缝坑工；清水墙；嵌灰坑工
pointing mortar 砌合砂浆；勾缝砂浆；坑工砂浆
pointing of holes 孔的布置（钻孔或炮眼）
pointing of tag end 锻头
pointing problem 瞄准问题
pointing rule 勾缝尺
pointing sill 活闸门槛
pointing stuff 勾缝料
pointing system 瞄准系统
pointing template 嵌填勾缝模板
pointing tool 勾缝工具；倒棱工具
pointing trowel 尖镘刀；勾缝馒刀；勾缝溜子；勾缝刀；溜子
pointing velocity 瞄准速度
pointing with cement mortar 水泥砂浆勾缝
pointing with putty 油灰勾缝
point initiation 点引发
point in normal position 道岔定位
point in reverse position 道岔反位
point insertion 插接点
point-integrated sampler 积点采样器；积点采水样器
point-integrating sampler 点积式取样器；点采集器
point-integrating sediment sampler 积点式泥砂采样器

point interpolation 插点法
point intersection 交会测点法
point in the sample space 样本空间的点
point involution 点列对合
point iterative method 点迭代法
point iterative process 点迭代过程
point jammer 针对型干扰发射机
point-junction 点结合
point junction transistor 点结合型晶体管
point Kriging 点克立格法
point lamp 点光源
point lattice 点晶格
point layout 布点方案
point layout for aerial triangulation blocks 区域网布点
point layout for aerial triangulation strips 航线网布点
pointless 钝的
pointless screw 平端螺钉
point lever 转辙柄；尖握柄
point light 点光
point light(ing) source 点光源
point light(ing) lamp 点光源
point lightning protector 尖端型避雷器；尖端放电避雷针
point live load 集中动荷载
point load at the crown 作用于拱顶的集中荷载
point load(ing) 点荷载；集中荷载；集中负载；点负荷
point load(ing) device 点荷载仪
point loading index 集中荷载指数
point load(ing) index of rock 岩石点荷载指标
point load(ing) method 点荷载法
point load(ing) strength 点荷载强度
point load(ing) tensile test 点荷载拉伸试验
point load(ing) test 点荷载试验
point load pile 端承桩
point load strength index 岩石点荷载强度指标
point load stress 集中荷载应力
point load system 集中荷载体系
point machine 道岔机
point marker 刺点器；点标记符
point marking device 刺点仪
point mass 点质量
point matching 点匹配
point matching method 对点纠正法；点匹配法
point matrix 点矩阵
point measurement 地点观测
point measuring method 点测定方法
point method 点算法
point micrometer 点测头千分尺
point mode 多点式；点式
point-mode display 点式显示
point model for traffic 交通点状模型
point mode of digitizer 数字化仪点方式
point mortar joint 勾灰缝
point mutation 点突变
point name 点名
point nose pliers 尖头钳
point number 点号
point-numbering keyboard 点号键盘
point number of common reflection area element 共反射面元的反射点数
point object 点状地物；点目标
point of action 作用点
point of admission 进气点
point of application 作用点；着力点；力的作用点；加药点；加力点；使用点；施力点
point of application of force 施力点
point of arrival 到达点
point of attachment 附着点
point of attack 作用点；浸蚀点；腐蚀点
point of attention 注视点
point of average response 成绩基准法
point of bearing 支点
point of beginning 初始点【测】
point of bisection 二等分点
point of break 折断点
point of breakthrough 贯通点
point of burst 爆裂点
point of certainty 待定点
point of change 变换点
point of change of gradient 变坡点；倾斜变换点；坡度转折点；坡度变换点

point of change slope 变坡点
point of clamping 夹紧插座；夹紧点
point of collection 挖进面；挖掘面；作业面
point of commencement of the railway 路线起点【铁】
point of commencement of the route 路线起点
point of compass 罗盘点
point of compound curvature 复曲线(起)点
point of compound curve 复曲线(起)点；复曲线连接点
point of compression 压缩点
point of concentration 集中点；集流点，汇流点
point of conflict 交通冲突点
point of confluence 汇流点；汇合点；合流点
point of congelation 胶凝点；冻凝点
point of connection 连接点
point of contact 接触点
point of contraflexure 转折点；拐点；反弯点；反挠点
point of crossing 交叉点
point of cultural interest 文物遗址
point of curvature 挂弯点；曲线起点；弯曲起点
point of curve 曲线起点；直圆点
point of curve change 变曲点
point of curve to curve 三心复曲线的半径变化点
point of cusp 尖点
point of delivery 交货地点
point of departure 起始点；起航点
point of derailment 车轮脱轨起点
point of destination 目的点；目的地；航程终点
point of detonation 爆震点
point of deviation 转向点；分歧点；偏离点
point of discharge 卸料点；排污点
point of discontinuity 突变点，不连续点
point of divergence 辐散点；分支点；分界点
point of egress 逸出点
point of elbow 肘端
point of emergence 露头点；后节点
point of emergence of dislocation 位错露头点
point-of-entry device 进水点装置
point-of-entry industry 进水点工业
point-of-entry system 进水点装置
point-of-entry water treatment device 进水点水处理装置
point of equilibrium 均衡点
point of evapo(u)ration 气化点；起始蒸发点
point of exertion 着力点
point of fabrication 钢结构制作中心
point of failure 破坏点；断裂点；破损点
point-of-failure restart 故障点再启动
point of fault 故障点
point of fixity 固定点；嵌固点
point of flammability 闪点
point of fluidity 流源；流限；流化点；(金属的)屈服点
point of fluidization 流化点
point of force application 力作用点；加力点；施力点
point of force concurrence 力交会点
point of fossils 化石点
point of free delivery 已付费交货地点
point of frequency accidents 多事故发生点
point of frog 辙叉实际尖端；辙叉实际尖点
point of geochemical prospecting 地球化学勘探工作点
point of geophysical prospecting 地球物理勘探工作点
point of gradient change 变坡点
point of gradient change on convex section 凸形变坡点
point of gross leakage 严重渗漏点
point of hock 小节端
point of holes 炮眼布置
point of hot water consumption 热水消耗量顶点
point of hydrology 水文工作点
point of ignition 着火点；燃烧点
point of impact 撞击点
point of impingement 碰撞点
point of incidence 入射点
point of incipient fluidization 流态化起点
point of inflection 转折点；拐点；弯曲点；反弯点；变曲点
point of inflexion 回归点；反曲点；变曲点
point of inflexion on a curve 曲线的拐点
point of inflow 入流点
point of initial leakage 初始渗漏点

point of interest 名胜
point of interruption 中断点
point of intersection 相交点;交点;交叉点
point of intersection of members 杆件的(相)交点
point of intersection of tangents 切线交点
point of invocation 调用点
point of irradiation 照射点
point of junction 连接点;交汇点;交点;交叉点;会合点;汇合点
point of land 地角
point of laying up 系泊船处
point of light 光点
point of load 荷载点;集中荷点;负荷点
point of load application 荷载作用点;荷载点
point of load collection 荷载集中点
point of loading 荷载作用点;装车地点
point of loading and unloading 装卸(地)点
point of lowest burning 燃烧最低点
point of lubrication 润滑点
point of magnetic modification 磁变态点
point of magnetic transformation 磁性转变点
point of maximum curvature 最大弯曲点;最大曲线率
point of maximum load 最大负载点;拉伸弹性极限;极限拉伸强度
point of measurement 度量点;量测点
point of mineralization 矿化点
point of minimum solubility 最小溶度点
point of mirror 控制点
point of neighbo(u)rhood 邻域指示器
point of no flow 零速点
point of no-return 不可返回点;不返回临界点(指航线临界点)
point of observation 观测点
point of onset of fluidization 流态化起点
point of origin 起点;始点;发货地(点)
point of osculation 密切点
point of perception 发觉危险的时间地点
point of perigee 近地点
point of pile 桩尖;桩端
point of placement 填筑点;填筑地点;布放点
point of pole 杆塔顶点
point of possible collision 可能碰撞点
point of potential collision 潜伏冲突点
point of pouring 浇筑温度
point of puncture 穿刺点
point of radial intersection 辐射交会点
point of recalescence 再辉点
point of reciprocity 互易点
point of reference 参考点;控制点;假定水准标点;基准点;水准点
point of reflection 反射点
point of reflection on the curve 曲线的反弯点
point of resurgence 恢复点
point of reverse curvature 反向曲线点;反曲线点
point of reverse curve 反向曲线的连接点;反曲线点
point of rigid support 固定点
point of rise (洪水的)起涨点;上升点
point of rotation 旋转点
point of rupture 挠曲点;断裂点
point of safety 安全地点
point of sale 销货点
point-of-sale terminal 售货点终端;销货点终端机;销售点终端机
point of satellite meridian passage 卫星中天点
point of saturation 饱和点
point of secant 割点;曲中点
point of self-oscillation 自振点
point of shipment 装运站;装船地点
point of shipment delivery 装运货物输送地点
point of shipping 装运地点
point of shoulder 肩端
point of sight 观测点;瞄准点;视点
point of slippage 打滑点
point of solidification 固化点;凝固点
point of split-off 分离点
point of springing 起拱点
point of spud 桩尖【疏】
point of square control network 方格网点
point of square grids 方格点;方格网点
point of stagnation 停滞点
point of sulphuration 硫化点
point of support 支撑点;桩计算用);支点;支承点
point of suspension 悬点

point of switch 辙叉尖(端);尖轨尖(端);尖轨轨尖
point of symmetry 对称点
point of tangency 接触点;曲线终点;切线点;切点
point of tongue 尖轨轨端
point of topography 地貌点
point of transition 转折点;转变点;过渡点;临界点
point of tyre-road 轮胎与路面接触点
point of unearth relics 出土文物点
point of use 应用地点
point-of-use device 用水点装置
point-of-use equipment industry 用水点装置工业
point-of-use industry 用水点工业
point-of-use system 用水点装置
point-of-use treatment 用水点处理
point-of-use treatment device 用水点处理装置
point of vertical curve 竖曲线起点
point of vertical intersection 坡度变换点
point of view 观点
point of vision 视点
point of water distribution 配水点
point of water inrush 涌水点;决口处
point of weld 焊接点;焊点
point of zero charge 零点电荷;零电荷点
point of zero distortion 等角点
point of zero flow 流量零点;断流点
point of zero moment 零弯矩点;零力矩点;弯矩零点
point of zero shear 零剪力点
point of zero voltage 电压零点
point operation 点运算
point orientation of diamond 金刚石角定向
point-outside-polygon 多边形外点
point particle 点粒子
point pinnacle reef pool group 点礁油气藏群
point plane precipitator 针板式电集尘器
point plot 展点图
point-plot mode 点记录方式
point plotter 点绘图仪
point-point angular correlation 点与点的角关联
point pole 点极
point pollution 点污染
point pollution source 点污染源
point position 单点定位
point position design 点位设计
point-positioning data base 点位数据库
point precipitation 点降水量;单站降水量;单站(降)雨量
point pressure 接触点压;点压力
point probe-dynamic(al) characteristic method 点探动态特性法
point probe-static characteristic method 点探静态特性法
point process 点过程
point processing 图像点处理
point projection microscope 点投影显微镜
point projection X-ray microscope 点投影式X射线显微镜
point projector electron microscope 点投影显微镜;点投影电子显微镜
point projector microscope 点投影显微镜
point quadrat method 样方测定法
point radiator 点状辐射体
point rail 尖轨;心轨;辙尖轨
point rainfall 点雨量;单站降水量;单站(降)雨量;单点雨量
point reconnaissance 定点勘测
point reef 点礁
point reflection 点反映
point relaxation 点松弛
point relaxation method 点松弛法
point relay 点继电器
point representation 点线表示法;点位表示法;点表示
point resistance 尖端阻力;顶端阻力;顶点阻力
point resistance force 总端阻力
point resistance frequency 点阻频率
point resistance of pile 桩尖阻力;桩端阻力
point resistance pressure 单位端阻力
point sample 点样本;点样;单点样
point sampler 点样采集器;点取样器;点采样器;单点取样器
point sampling 样点抽样(法);点抽样;点采样
point sampling method 点样抽样法;单点抽样法
point sampling theory 点抽样理论

points bound segment 点线部分
point scanner 点扫描仪
point scattering 点散射
point selection 选穴法;选点
point sensor 点传感器
point separator 辙尖隔离块
point separator of switch 尖轨间隔铁
point series 点系列
point set 点集
point setting 道岔转换
point set topology 点集拓扑学
point shop 尖轨车间
point sink 点汇
point slope form 点斜式
point slope method 点斜法
pointsman 扳道工(人);转辙工;扳道工【铁】;转辙工人
points of entry 入境地点
points of the back 背部穴
points of tie 连测对比点
point sound source 点声源
point source 散烟点;点源;点污染源;点光源;点辐射源
point source control 点源控制;点污染源控制
point source detoxification 点污染源去毒化
point source discharge 点源排放;点污染源排放
point source discharge of toxic 点污染源有毒物质排放
point source discharge permit 点污染源许可排放量
point source dislocation 点源位错
point source distribution 点源分布
point source distribution of earthquake generation 地震发生的点源分布
point source epidemic 点源流行
point source field 点源场
point source lamp 点光灯
point source light 点光源灯
point source load(ing) 点污染源负荷
point source map 点源图
point source model 点源模型
point source of air pollution 空气污染点源
point source of light 点光源
point source of pollution 点污染源
point source of sediment contamination 底泥污染点源
point source of water pollution 水污染点源
point source outfall 点污染源排污口
point source pollution 点源污染;点污染源污染
point source radiator 点源辐射体
point source theory of earthquake generation 地震发生的点源理论
point specifier 小数点区分符
point spectrum 离散谱;点谱
point spread function 点散播函数;点扩散函数;点分布函数
points scatter 点子分散
point-steel hammer 尖钢锤
point stream mode 点流记录方式
point stress state 一点应力状态
point stretcher 尖轨连(接)杆
point-supported 定点支承的
point surveying 点测量法
point-suspended 点形悬挂的;点式悬挂;单点悬挂的
point suspension 点形悬挂
point switch 尖轨转辙器;尖轨转换器
point symbol 点状符号;点符号;不依比例符号
point-symmetric(al) graph 点对称图
point symmetry 点对称
point symmetry group determination 点群测定
point system 工分制度;评分法;点系
point system on small quotas 小指标计分奖
point target 点状目标;点目标
point technique 瞄准技术
point template 尖轨样板;点样板
point tie 穿吊顺序
point tool 锤头;凿头
point toothing 点啮合
point-to-plane discharge 尖端平面间放电
point-to-plane technique 点面法
point-top mountain 尖顶山
point-to-point 逐点的;逐点;点对点;点到点;依航标航行

point-to-point circuit 点间通信线路
point-to-point communication 点到点通信
point-to-point connection 点到点连接;点到点连接
point-to-point control 逐点控制
point-to-point controlled robot 点位控制机器人;点控制的机器人
point-to-point control system 点到点控制系统
point-to-point correspondence 点对应
point-to-point data link 定点数据传输线
point-to-point dodging 逐点自动匀光
point-to-point line 点到点线路;两点间线路;闭合线
point-to-point link 点到点链路;点到点连接
point-to-point linkage 点对点通道
point-to-point measurement record 点测记录
point-to-point method 逐点测定法
point-to-point network 点到点网络
point-to-point operation 两点间操作
point-to-point protocol 点到点协议
point-to-point radio equipment 点对点无线电通信设备
point-to-point segment 点到点分段
point-to-point speed 直线运动速度;平移速度
point-to-point system 点到点方式
point-to-point transmission 两点间传输;两点间、两地点间传输;点到点传输
point-to-set mapping 点到集的映射
point tower block (单独耸立的)塔式高层建筑
point trace 点迹
point transfer 转刺(点)【测】
point transfer device 转点仪
point transfer matrix 点变换矩阵
point transposition 按点换位
point triad 三基色荧光点组
point type fire detector 点型感温探测器
point type heat fire detector 点型感温火灾探测器
point type speed control system 点式调速系统
point unit heat transfer coefficient 质点传热系数
point unit vortex 点涡
point value 工分值
point vanishing 消灭点(透视)
point variogram 点变差函数
point velocity 点速(度);点流速
point vortex 自由涡流,点涡旋
point water ga(u)ge 针形水尺
point water quality measurement 点水质测定
point welding 凸焊;点焊接;点焊
point welding machine 点焊机
point width 刀尖距;刀顶距
pointwise 逐点的;点
pointwise availability 逐点可应用度
pointwise connection 点态连通;点态连接
pointwise convergence 点态收敛
pointwise discontinuous function 点态不连续函数
pointwise infimum 点式下确界
pointwise Lagrangian operator 点态拉格朗日算子
point write mode 点写入模型
pointy 非常尖的
Poir(e) dam 波里坝;针坝
Poirriers' blue 普瓦里埃氏蓝
Poirriers' orange 普瓦里埃氏橙
poise 泊(黏度单位);平衡块;秤锤
poise beam 秤杆
poised riverbed 河床稳定河
poised state 平衡状态
poised stream 稳定河流
poise error 平衡误差;偏重
poiser 氧化还原反应缓冲剂
Poiseuille's equation 泊萧叶方程
Poiseuille's space 缓动层
poise weight 秤锤;砝码
poising action 平衡作用
poising agent 平衡剂
poising callipers 平衡卡规
poising tool 平衡工具;摆轮平衡工具
poison 毒药;毒物;反应堆残渣
poison atmosphere 含毒空气
poison brick 毒虫砖
poison computer 毒性分析计算机
poison control 毒品管理
poison control center 毒物控制中心
poison cupboard 毒药橱
poison fog 毒雾
poison gas 毒气
poisoning 中毒

poisoning by agricultural chemicals 农药中毒
poisoning of reactor 反应堆中毒
poison leakage 毒品泄漏
poison limit 毒物极限
poison material 有毒物质;有毒物
poison oak 有毒栎
poisonoak 太平洋漆树
poison of catalyst 催化毒物
poison of tracer 示踪剂毒性
poisonous 有毒的
poisonous articles 毒害品
poisonous articles and infective articles 毒害品和感染性物品
poisonous colo(u)r 有毒颜料
poisonous effect 毒效
poisonous fish 有毒鱼
poisonous gas 有毒气体
poisonous glaze 有毒釉
poisonous liquid 有毒液体
poisonous matter test 毒性检验
poisonous metal 有毒金属
poisonousness 毒性
poisonous plant 有毒植物
poisonous substance 有毒物质;毒物
poisonous waste 有毒废物;有毒废水
poisonous wastewater 有毒污水
poison range 毒物范围
poison tower 除毒塔
poisonwood 毒漆树
Poisson-Arago spot 泊松·阿喇戈光斑
Poisson binomial trials model 广义二项试验模型
Poisson-Boltzmann equation 泊松·波耳兹曼方程
Poisson distribution of earthquake occurrences 地震发生的泊松分布
Poisson process with stationary increments 具有平稳增量的泊松过程
Poisson's approximation 泊松逼近
Poisson's arrival 泊松到达
Poisson's binomial distribution 泊松二项分布
Poisson's boundary condition 泊松边界条件
Poisson's bracket 泊松括号
Poisson's bright spot 泊松亮斑
Poisson's coefficient 泊松系数
Poisson's constant 泊松常数
Poisson's density function 泊松密度函数
Poisson's diffraction 泊松衍射
Poisson's distribution 泊松分布
Poisson's distribution table 泊松分布表
Poisson's equation 泊松方程
Poisson's false-target model 泊松伪靶模型
Poisson's flow 泊松流
Poisson's formula 泊松公式
Poisson's function 泊松函数
Poisson's index of dispersion 泊松离差指数
Poisson's input 泊松输入
Poisson's integral formula 泊松积分公式
Poisson's kernel 泊松核
Poisson's law of large number 泊松大数定律
Poisson's model 泊松模型
Poisson's noise 泊松噪声
Poisson's number 泊松数;泊松值
Poisson's parenthesis 泊松括号
Poisson's probability table 泊松概率表
Poisson's process 泊松过程
Poisson's ratio 横向变形系数;泊松比
Poisson's ratio correction 泊松比修正
Poisson's ratio effect 泊松比效应
Poisson's ratio of rock 岩石的泊松比
Poisson's rectangular wave 泊松方波
Poisson's renewal pulse process 泊松更新脉冲过程
Poisson's sampling 泊松抽样
Poisson's solution 泊松解
Poisson's statistics 泊松统计
Poisson's stream 泊松流
Poisson's sum formula 泊松求和公式
Poisson's summation formula 泊松求和公式
Poisson's transform 泊松变换
Poisson's wave equation 泊松波动方程
poitevinite 泼水铁铜矾
poitrail 木结构的高强横梁
poke 捅;存数;拨火
poke hole 搅拌孔
poke peak 山顶

poker 火钳;拨火(铁)棒;搅拌杆
poker bar 搅拌杆;钎
poker burn 斑伤
poke rod 搅拌棒;捅料棍;拨动杆
poker picture 烙画
poker shaped case (混凝土振动器的)振动头
poker test 拨火棒试验
poker type pendulum concrete vibrator 插入式混凝土振动器
poker vibrator 插入式振荡器(混凝土);振动棒;插入式振捣器
poker vibrator with handle 有手柄的振动筛
poker with motor 装有电动机的振动棒(混凝土捣实用)
poker work 烙画;烫画
poket-through 公用事业穿墙洞
poket-through construction 公用事业穿墙洞结构
poke weld(ing) 手动挤焊;手推点焊;手动(焊钳)点焊;手点焊;钳点焊
pokey 监狱
poking 棒触;插棒;搅动(混凝土)
poking bar 搅拌棍
poking in the kiln 捅窑内物料
Pokko movement (disturbance) 六甲变动
pokrovskite 半水羟碳镁石
polacke 波拉克风
polaity formula 极性式
Polak machine 布拉格冷室压铸机
Poland flexible pavement design method 波兰柔性路面设计法
Polanyi adsorption potential 波兰尼吸附势
polar 极坐标的;极地的;地极的
polar action 极地效应
polar activation 极性活化作用;极性活化
polar additive 极性添加剂
polar adhesive 极性黏结剂
polar air 极地空气
polar air mass 极地气团
polar altitude 仰极高度;极高
polar anemia 极地贫血
polar angle 矢量角;子午角;极坐标角;极距角;极角
polar anti-cyclone 副极地高压;北极高压;极地反气旋
polar arc 极弧;磁极弧度
polar arm 极臂
polar attachment 寻极仪
polaraulic selector 脉冲极性选择器
polar automatic weather station 极地自气象观测站
polar axis 极轴
polar bar 极臂
polar bear 北极熊
polar bearing 极方位角
polar binding 极性键联;极性键合
polar body 极体
polar bond 极性键
polar candle-power diagram 按向分布图
polar cap 极帽;极盖;极地冰冠
polar cap absorption 极冠吸收;极盖吸收
polar cap ice 极地冰
polar caps 极冠
polar cell 极体;极地环流
polar change brightening 变极增白整理
polar chart 极区海图;极区(地)图
polar circle 极圆;极圈
polar circulation 极地环流
polar climate 极地气候
polar code 极性电码
polar colo(u)r 极性色料
polar comparator 极坐量测仪
polar compound 极性混合物;极性化合物
polar condition 极条件【测】
polar cone 极锥
polar continental air 极地大陆空气
polar continental air mass 寒带大陆性气团;极地大陆气团
polar control 联合操纵系统;极坐标法控制
polar control system 极控制系统
polar coordinate in the space 空间极坐标
polar coordinate method 极坐标法
polar coordinate method with electromagnetic distance measurement 电磁波测距仪极坐标法
polar coordinate navigation system 极坐标导航系统
polar coordinate paper 极坐标图纸;极坐标纸

polar coordinate position 极坐标定位
polar coordinate position-fixing system 极坐标定位系统
polar coordinate robot 极坐标式机器人
polar coordinates 极坐标
polar coordinate system 极坐标系统;极坐标系
polar coordinatograph 极坐标仪
polar covalence 极性共价
polar covalent bond 极性共价键
polar crane 回转式吊车
polar crystal 极性晶体;铁磁晶体
polar current 极地气流
polar curve 光分布曲线;极坐标曲线
polar cushion 极垫
polar cyclone 极涡;极地涡旋
polar cylindric(al) coordinate system 柱面极坐标系
polar day 极昼
polar decomposition 极坐标分解;极分解
polar desert soil 极地漠境土
polar diagram 极线图;极坐标图
polar diagram recorder 极坐标记录器
polar dial 极日规
polar diameter 极直径
polar displacement 极位移
polar dissolved organic compound 极性溶解有机化合物
polar distance 极距
polar distribution 极坐标分布;极角分布;极分布
polar duplex 极化双工
polar easterlies 极地东风带
polar-easterlies index 极地东风指数
polar echo 极化回波;偏振回波
polar ecosystem 极地生态系统
polar effect 极效应
polar electrojet 极电急流
Polar equal-area net 赖特网
polar equation 极坐标方程(式)
polar expansion 线性膨胀
polar experiment 极地试验
polar explosive 极地炸药
polar figure 极性图
polar filament 极丝
polar finder 寻北器
polar flattening 极向扁率
polar flattening of ellipsoid 椭球极扁率
polar force 极向力;极力
polar form 极坐标形式;极型
polar-frequency coded track circuit 极频电码轨道电路
polar front 极锋
polar front belt 极锋带
polar front jet 极锋急流
polar front stream 极锋急流
polar front theory 极锋学说
polar geographic(al) orbiter 极轨道地球物理卫星
polar glacier 极地冰川
polar gnomonic chart 极切心射投影海图
polar gnomonic graticule 极切心射投影标线
polar graph paper 极坐标纸
polar gravity 极重力
polar grid 极区格网
polar group 极性基(团)
polar guide 极性波导管
polar heliostat 极向定日镜
polar high 极地高压;副极地高压;北极高压
polar high belt 极地高压带
polar hole 极孔
polar ice 极地冰;极冰
polar ice cloud 极地冰晶云
polar ice sheet 极地冰原;极地冰盖
polarimeter 旋光仪;旋光计;极化计;极化表;偏振计;偏振器;偏振镜应力仪;偏光计
polarimetric analysis 旋光分析
polarimetric sensor 偏振式传感器
polarimetry 旋光分析;旋光测定法;偏振测定法;偏振测定;测偏振术
polarimicroscope 旋光显微镜
polar impulse 极性脉冲
polar impulse track circuit 极性脉冲轨道电路
polar index 极性指数
polar indication 极性指示
polar intersection 极交会
polar invasion 寒潮

Polaris 北极星
Polaris almanac 北极星历
polariscope 旋光镜;光测偏振仪;极化光镜;起偏振镜;偏振镜;偏振光镜;偏振光镜;偏振器;偏光计;偏光(应力)仪;分极镜
polariscope test 偏振测试
polariscope tube 偏振光镜管
polariscopic 旋光镜的;偏振镜的
polariscopy 旋光镜检查;偏振镜检查;偏振测定法
Polaris correction 北极星高度求纬度改正量
polarised light 偏振光
polariser 起偏振镜
polarising microscope 偏振光显微镜
polarising microscope in reflected light 反射光偏振显微镜
Polaris monitor 北极星监视仪
Polaris-toting crane 波拉瑞氏起重机
polaristrobometer 精密旋光计;精密偏振计;偏振计
polarite 斜方铅铋钯矿;铋铅钯矿
polarite filter 滤池
polariton 极化声子;电磁声子;电磁激子
polarity 极性现象;极性带;极性;偏光性;配极
polarity-changing code 直流转极电码
polarity charging 定极性充电
polarity check 极性检查
polarity check circuit 极性检查电路
polarity chron 极性时
polarity chronzone 极性时间带
polarity coincidence correlator 极性重合相关器
polarity control 极性控制
polarity correlator 极性相关器
polarity diversity 极化分集
polarity effect 极性影响;极性效应;极化效应
polarity epoch 极性时纪;极性期
polarity event 极性事件
polarity event and polarity epoch 极性事件与极性期
polarity excursion of geomagnetic field 地磁场极性偏移
polarity filtering 极化滤波
polarity finder 极性测定器
polarity hyperchron 极性巨时
polarity hyperchronzone 极性巨时间带
polarity hyperzone 极性巨带
polarity indicator 极性指示器
polarity in metasomatism 交代极性
polarity interval 极性间隔
polarity inverse 极性倒转
polarity inversion 极性转换;极性颠倒;极性变换
polarity inverter 极性变换器
polarity inverting amplifier 极性反转放大器
polarity lamp 极性检测灯
polarity mark 极性标记
polarity microchrconzone 极性微时间带
polarity microchron 极性微时
polarity microzone 极性微带
polarity of chain 分子链的极性
polarity of feedback link 反馈链的极性
polarity of feedback loop 反馈环的极性
polarity of subduction 俯冲极性
polarity of transformer 变压器绕线方向;变压器极性
polarity relay 极性继电器
polarity reversal 极性颠倒;极性倒转;磁极倒转反转;磁极倒转反向
polarity reverser 极性换向器
polarity-reversing switch 极性转换开关
polarity separation technique 极性分离技术
polarity splitter 极性分离器;分极器
polarity subchrconzone 极性亚时间带
polarity subchron 极性亚时
polarity subzone 极性亚带
polarity superchron 极性超时
polarity superchronzone 极性超时间带
polarity syperone 极性超带
polarity test 极性试验
polarity tester 极化指示器
polarity time scale 极性倒转年表
polarity transition 极性过渡
polarizability 极化性;极化强度;极化率;极化度
polarizability catastrophe 极化率突变
polarizable 可极化
polarization 光偏振;极化(作用);偏置;偏振化(作用);偏振;偏光;配置变换;电极化度

polarization analyzer 光偏振分析器;检偏振光镜;检偏镜
polarization angle 偏振角
polarization apparatus 旋光器;偏振器
polarization axis 极化轴
polarization band 极化谱带
polarization battery 极化电池组
polarization capacity 极化额
polarization cell 极化电池
polarization changer 极化变换器
polarization character 偏光性
polarization coefficient of rocks and minerals 岩矿石的极化系数
polarization colo(u)r 偏光色
polarization colo(u)rimeter 偏光比色计
polarization cross 偏振交叉
polarization current 极化电流
polarization curve 极化曲线;偏振曲线
polarization degeneracy 极化简并度
polarization direction 偏振化方向
polarization direction correlation 极化和方向的关联
polarization diversity 极化分集法;偏振分集
polarization effect 极化效应;偏振效应
polarization ellipse 极化椭圆;偏振椭圆
polarization error 极化误差
polarization fading 极化衰落
polarization figure 偏光图
polarization fluorometer 偏振荧光计
polarization holography 偏振全息摄影
polarization index 极化指数
polarization index test 极化率试验
polarization intensity 极化强度
polarization interference filter 干涉偏振滤光器;偏振干涉滤光器
polarization interference microscope 偏光干涉显微镜
polarization interferometer 极化作用干涉仪;偏振干涉仪
polarization isocline 偏振等倾线
polarization-magnifying lens 偏光放大镜
polarization-maintaining coupler 保偏耦合器
polarization-rnaintaining fiber coupler 保偏光纤耦合器
polarization maintaining optic(al) fiber 偏振保持光纤;保偏光纤
polarization-maintaining single-mode fiber 保偏单模光纤
polarization matching 极化匹配
polarization measurement 极化测定;偏振测量
polarization microscope 极化显微镜;偏振显微镜
polarization modulation 偏振片调节
polarization objective 偏光物镜
polarization of electromagnetic wave 电磁波的极化
polarization of electrostrictive material 电致伸缩材料的极化
polarization of laser output 激光输出偏振度
polarization of light 光偏振;光的偏振
polarization of polymer 聚合物极化
polarization of the sky light 天空光偏振
polarization parameter 偏振参量
polarization phenomenon 极化现象
polarization photoelasticity 偏光弹性学
polarization photometer 偏振光度计
polarization plane 偏振面;偏光面
polarization potential 极化势;极化电位;极化电势
polarization potential difference 极化电位差;电极电位差
polarization power 极化力
polarization-preserving coupler 保偏耦合器
polarization-preserving optical fiber 保偏光纤
polarization-preserving single-mode coupler 保偏单模耦合器
polarization-preserving single mode fiber 保偏单模光纤
polarization receiver 偏振接收机
polarization receiving factor 接收天线极化因数
polarization reversal 极化反转
polarization rotation 极化旋转;偏振旋转
polarization selectivity 偏振选择性
polarization spectrum 偏振谱
polarization state 偏振态
polarization switch 偏振开关;变极开关;变极点火

开关
polarization tensor 极化张量
polarization transfer 极化转移
polarize 使极化
polarized ammeter 极化安培计
polarized antenna 偏振天线
polarized area 极化区域
polarized attachment 极化电源插头
polarized beam 极化束;偏振波束
polarized bell 极化铃
polarized burst 偏振暴
polarized capacitor 极化电容器
polarized ceramics 极化陶瓷
polarized circuit 极化电路
polarized component 偏振分量;偏振成分
polarized degree 偏振度
polarized dipole magnetization 极化偶极子磁化
polarized direction 偏振方向
polarized electrode 极化电极
polarized fluorescence 极化荧光;偏振荧光
polarized glass 偏光玻璃
polarized glass screen 偏振光玻璃网板
polarized indicator electrode 极化指示电极
polarized induction 极化感应;极化磁感应
polarized-latching type relay 栓锁型极化继电器
polarized light 偏振光;偏光
polarized light analog 偏振光模拟
polarized light filter 偏光滤光器
polarized light microscope 偏振光显微镜
polarized light microscopy 偏振光显微术;偏光显微(镜)术
polarized magnet 永久磁铁;极化磁铁
polarized magnetic amplifier 极化磁放大器
polarized meter 极化计;极化表
polarized monostable type relay 单稳态型极化继电器
polarized neutrons 极化中子
polarized observation 偏振观察
polarized pattern viewer 偏振片检影器
polarized plate viewing system 偏振板观察系统
polarized plug 极性插头;极化插头;定位插头
polarized point 极化点
polarized radiation 极化辐射;偏振辐射
polarized receptacle 极化插座
polarized-reflected light 偏振反射光
polarized-reflected light microscope 偏反光两用显微镜
polarized relay 有极继电器;极化继电器
polarized return-to-zero recording 极性归零制记录;极化归零记录
polarized shear wave velocity 极化剪切波速
polarized signal 极化信号
polarized spectrum 偏振光谱
polarized spot 极化点
polarized state 极化状态
polarized state of light 光的偏振状态
polarized type 极化式
polarized-vane type ammeter 极化铁叶式安培计
polarized viewing 偏振观察
polarized wave 极化波;偏振波
polarizer 极化器;极化剂;下偏光镜;起偏振器;起偏器;偏振器;偏振镜;偏光镜;偏光片;偏振镜
polarizing angle 起偏角;偏振角;布儒斯特角
polarizing battery 极化蓄电池
polarizing coil 极化线圈
polarizing disc 偏振片
polarizing eyepiece 起偏振目镜;偏振目镜
polarizing film 偏振片
polarizing filter 极化滤波器;偏振滤光镜;偏振滤光片
polarizing glass 极化窗;偏光镜
polarizing interference 偏光干涉
polarizing microscope 偏振光显微镜;偏光显微镜
polarizing microscope identification 偏光显微镜鉴定
polarizing microscopy 偏光显微(镜)术
polarizing mirror 极化反射镜;偏光反射镜
polarizing monochromator 极化单色仪;起偏振单色器;偏振单色仪
polarizing operator 配极运算子
polarizing optic(al) pyrometer 偏振光测高温计
polarizing optics 偏振光零件
polarizing plate 起偏振片
polarizing prism 起偏振棱镜;偏振棱镜

polarizing pyrometer 偏振光光测高温计
polarizing reflector 偏光反射镜
polarizing resonance decoupling method 偏共振去耦法
polarizing screen 偏振光镜;偏光镜
polarizing spectacle 偏振镜;偏光镜
polarizing spectrophotometer 偏振分光光度计
polarizing target 极化目标
polarizing unit 极化部件
polarizing voltage 极化电压
polar keying 极性键控
polar latitude 极黄纬;极地纬度
polar light 极光;偏光
polar line 极线
polar linkage 极性键
polar liquid 极性液体
polar liquid phase 极性液相
polar longitude 极黄经
polar low 极涡;极地涡旋;极地低压
polar map 极区图;极区地图
polar marine air 极地海洋空气
polar maritime air 极地海洋空气
polar maritime airmass 极地海洋气团;极区海洋气团
polar meteorologist 极地气象学家
polar meteorology 极地气象学
polar method 极坐标法
polar migration 极移动
polar mobility 极性迁移率
polar modulation 极性调制;极化调制
polar modulus of section 极截面模数
polar molecule 有极分子;极性分子
polar moment 极力矩;极矩
polar moment of inertia 惯性极矩;极惯性矩
polar monomer 极性单体
polar motion 极移;地极移动
polar mount 极化固定架
polar mounting 极式装置
polar navigation 极性导航子;极区航行;极区导航
polar net 极坐标投影法;极网
polar night 极夜
polar normal 极法线
polar nucleus 极核
polar number 极性价数
polar oblateness 极向扁率
polar of polar coordinates 极坐标极点
polarogarphy 极谱分析(法)
polarogram 极谱图;极谱
polarograph 极谱仪;极谱分析仪;极化记录器
polarographic(al) analysis 极谱分析(法)
polarographic(al) maximum suppressor method 极谱最大抑制法
polarographic analyzer 极谱分析仪
polarographic circuit 极谱电路
polarographic determination 极谱测定
polarographic maximum 极谱极值
polarographic measurement 极谱测量
polarographic method 极谱法
polarographic oxygen analyzer 溶解氧极谱分析仪
polarographic scanner 极谱扫描器
polarographic study 极谱研究
polarographic titration 极谱滴定(法)
polarographic wave 极谱波
polarograph recorder 极谱记录器
polarography 极谱仪测定法;极谱学;极谱法
polaroid 人造偏振片;偏振片
polaroid adjustment knob 偏振片调置钮
polaroid-anomalies 极化子异常
Polaroid camera 波拉德照相机;一次成影照相机;即显胶片照相机
polaroid film 即显胶片;人造偏振片
polaroid filter 偏振滤光镜;偏振滤光片
polaroid foil 人造偏振箔
polaroid glass 偏光镜
polaroid glasses 偏光眼镜
polaroid polarizer 人造起偏振镜;偏振光镜
polaroid sector 扇形偏振片
polarometer 极谱计
polarometric titration 极谱滴定(法);极化滴定(法)
polaron 极化子;极化学;偏振子
polar optic(al) scattering 极性光学模式散射
polar orbit 两极轨道;极轨道
polar orbiting geographic(al) observatory 极轨道地球物理观测台
polar orbiting satellite 极轨道卫星;极地轨道卫星
polar orthographic(al) projection 极正射投影
polar outbreak 寒潮
polar ozone 极地臭氧
polar pantograph 极坐标缩放仪
polar performance 极坐标演算
polar period 极周期
polar pesticide 极性农药
polar phase 极性相
polar photoelastic test 偏光弹性试验
polar pigment 极性颜料
polar pitch 极距
polar planimeter 定极求积仪;定极面积仪
polar plate 极板
polar platform 极地空间平台
polar plot 极坐标图;速矢端迹
polar plotting method 极距法
polar plume 极羽
polar polymer 极性聚合物
polar positioning system 极坐标定位系统
polar projection 极投影
polar projection coordinate network 极投影坐标网
polar projection map 极投影图
polar prominence 极区日珥
polar pulse 极性脉冲
polar pulse track circuit 极性脉冲轨道电路
polar radiation 极辐射
polar radius 极半径;地球极半径
polar ray 极射线
polar reaction 极性反应
polar reciprocal surface 配极曲面
polar reciprocation 极反演变换
polar region 近极区域;极区;极地
polar response curve 极坐标响应曲线
polar ring 极环
polars 高偏振星
polar satellite 极轨道卫星
polar second moment 极惯性矩
polar second moment of area 截面极惯性矩
polar semi-axis 极半径
polar semi-diameter 极半径
polar sensitive 极向灵敏;极性灵敏
polar ship 极地船
polar solar wind 极区太阳风
polar solvent 极性溶剂
polar solvent liquid 极化溶剂液体
polar spheric(al) coordinates 球面极坐标;球极坐标
polar spheric(al) triangle 极球面三角形
polar spot 极斑
polar stimulation 极刺激
polar stratospheric cloud 极地平流层云
polar stratospheric cloud abundances 极地平流层云量
polar subset 极子集
polar support 极性载体
polar surface 极曲面
polar survey 极坐标测量法
polar sway 地极颤动
polar symmetry 极对称
polar symmetry axis 极性对称轴
polar system 极系;配极系
polar tangent plot 极性切线图
polar telescope 天极仪
polar temperature limination 次生温度分层
polar tide 极潮汐
polar tip 极尖
polar transmission 极性传输;极化传输
polar transport 极性运输
polar transposition 极性交叉
polar triangle 极三角形
polar trough 极地槽
polar variation 极变化
polar vector 极向量;极限量;极矢(量)
polar vibrational mode 极化振动模
polar view 极面观
polar vortex 极涡;极地涡旋
polar wabble 极移
polar wandering 极游动;地极迁移
polar-wandering chart 极移图
polar-wandering curve 极移曲线
polar wandering path 极游移路线
polar water 极地水

polar wind 极风
polar wind belt 极风带
polar winding 极向缠绕
polar wobble 地极摆动
polar year 极年
polar zone 极区；极地带；寒带
polaxis 极轴；极化轴
polcard 碳酸钙盐
Polcon Varge superfilter 纤维超滤机
polder 新坝地；迁地；海涂围垦；围圩；围垦地；低地；围海造陆；圩垸；吹填地；冲积畦地
polder dike 围堤；围垦堤
polder embankment 圩堤
polder land 堤围泽地
polder method 筑堤疏干围垦法
polder reclamation 筑堤疏干围垦法
polder region 筑堤围垦区
Poldi hardness tester 泊尔迪（携带）式硬度计
poldine metilsulfate 泊尔定甲硫酸盐
pole 杆塔；杆；极点；木杆；电线杆；电杆；磁极；棒
pole altitude 极高
pole-amplitude 极幅
pole-and-chain 撤桩机；调桩机；调桩机；撤桩机
pole and stake awning 杆桩遮篷
pole arc 极弧；磁极弧
pole-arm 横担（电杆）；线担（电杆）
pole armature 凸极电枢
pole assignment 极点配置
pole at infinity 无穷远极
pole axis 极轴
pole base 电杆底板
pole beacon 柱形立标【航海】；杆标
pole biased relay 偏极继电器
pole body 极身
pole bolt 极螺栓；磁极螺栓
pole boundary 磁极边界
pole brace 电杆的拉线；电杆支撑；电杆
pole bracket（cantilever）悬臂支架；支架；悬臂；磁极支架；撑架
pole cap 杆帽；杆顶帽
pole carrier 杆式货物挂车
pole cell insulation 极身绝缘
pole chain 六十六尺测链；量地测链；蛇链
pole-change 极变换
pole-change generator 变极发电机
pole-change motor 换极电动机；变极电动机
pole-change motor starter 马达变极启动器
pole changer 转向开关；转换器；换极器；换极开关
pole change type alternate current generator 换极式交流发电机
pole changing 换极
pole-changing control 变极调速
pole-changing motor 换极电动机；变极电动机；变极式感应电动机
pole-changing speed regulating 变极调速
pole changing switch 换极器；换极开关
pole-changing time 转极时间
pole-changing value 转极值
pole chute 纵木滑道
pole clearance 杆距；极隙；极间间隙
pole climber 铁镫；脚扣；上杆脚扣
pole coil 磁极线圈
pole construction 柱式结构；底层柱式结构
pole-control focus lamp 杆控聚光灯
pole core 电极铁心；磁极铁芯
polectron 聚乙烯咔唑树脂
pole damper 磁极阻尼器；磁极减振器；磁极阻尼器
pole derrick 墙装动臂式起重机；单桅杆起重机；小型滑轮式起重机
pole diagram 极点图
pole dislocation 杆位错
pole distance 杆距；极距；天体极距
pole dolly 辘车
pole dominance 极点为主性
pole drain 木杆排水沟
pole drill 钻杆冲击钻机
pole drywall blade 板墙刮刀
pole element 磁极件
pole embrace 极弧空间范围
pole erecting working car 立杆作业车
pole extension 极延伸部分；极端
pole face 磁极面；极面；极端
pole-face bevel 极面斜角
pole-face shaping 极面整形

pole figure 极像图
pole finder 极性试验器
pole finding paper 极谱纸
pole fittings 电杆附件
pole flattening 极扁率
pole float 长杆浮标；杆状浮标；浮标杆
pole flux 磁极通量
pole footing 桩柱底脚；杆基
pole form 磁极形状
pole-frame construction 木构架结构；木骨架建筑
pole gap 极隙；极间空隙
pole head 极靴
pole height 电杆高度
pole hook 钩篙；钩杆；杆钩
pole horn 极尖；后极尖；电极凸出部分
pole house construction 支架房屋构造
pole inside figure 内极位置
pole insulation 磁极绝缘
pole joint 磁极连接
pole ladder 木工梯；圆木梯；单杠梯
pole lathe 脚踏车床
pole leakage 磁极漏泄
pole line 架空电线；地上杆线
pole line hardware 架空线路金属附件
pole man 云梯操纵手
pole mark 杆示标志；杆标
pole mast 单杆桅
pole mattress 木沉排
pole mounted oil switch 电杆上油开关
pole mounted switch 杆上开关
pole-mounting lightning arrester 杆装式避雷器；杆式避雷针
Polenske number 波伦斯基值
Polenske value 不溶解挥发性脂肪酸值；波伦斯基值
pole of angular momentum 角动量极；动量矩极
pole of inaccessibility 冰极
pole of inertia 惯性极
pole of orbit 轨道极
pole of rotation 旋转极；自转极
pole of side equation 边方程极点
pole of the ecliptic 黄极
pole of the equator 赤极
pole-on object 极向天体
pole-on star 极向恒星
pole orbit 极轨道
pole outside figure 外极位置
pole pair 杆对；极对；磁极对
pole paper 极谱纸
pole piece 脊板；极靴；极芯；极片；栋木；磁极片；极部
pole piece spacing 极靴间隔；磁极间隔
pole pin 杆销
pole pitch 极距；磁极距
pole plant 电杆倾放设备
pole plate 下樑；檐樑；挑檐杆；磁极钢板；磁极板；椽木垫板；承椽板
pole plate system 挑檐桁；磁极板系统
pole plough 简易犁
pole point 极点
pole press 立柱压机
pole prod 电杆腐朽测试器
pole pruning saw 长杆修树枝锯
pole reagent paper 电极试纸
pole reflector 标杆反光镜
pole retriever 自动降杆器
pole reverser 换极开关
pole-reversing key 倒极性开关
pole ring 环首杆；端头带圈的杆；磁极环
pole road 木杆道
pole sander 杆式打磨器
poles and towers 杆塔
pole scaffold(ing) 撑杆脚手架；杆支脚手架；木杆脚手架
pole setting 立杆
pole shank 极芯
pole sheet 磁极叠片
pole shim 磁极和电枢之间的铁填片
pole shoe（电机的）磁极片；极靴
pole-shoe contact 极靴触点
pole-shoe leakage 极靴漏泄
pole-shoe spreader 极靴撑
pole shore 支柱；垂直支撑；单独立柱
pole shunting 磁极分路
pole shut-off 关闭杆
pole signal 杆式标志；杆标

poles of magnetic needle 磁针极
poles of the earth 地球的两极
poles of the heaven 天球两极
pole spacing 杆距；杆挡
pole span 杆距；杆挡；极距
pole spar 实心杆
pole spinning machine 电杆捆扎机
pole spraying 杆件喷涂；电杆喷涂
pole spring 杆簧
pole spring cap 杆簧盖
pole spur 极靴凸出部分
Pole Star 北极星；指极星
polestar recorder 北极星记录器
polester 聚酯
polester-faced acoustic(al) tile 聚酯薄膜吸音板
polester-glass fibre sheet 聚酯玻璃丝板
Pole's tile 普尔瓦
pole strength 磁极强度；极强（指磁极）
pole support 支杆
pole switch 杆装开关；杆上开关；架空安装开关；极柱式开关
pole tent 立柱帐篷
pole terminal 极靴
pole tide 极点潮
pole tie 圆木轨枕；杆材轨枕
pole tip 极靴头；极靴；极尖
pole tool 钻杆冲击钻进工具
pole-tool method 钻杆冲击钻进法；杆钻冲击钻进法；钻冲击钻进法
pole top transformer 杆顶变压器
pole trailer 运电杆拖车；长货挂车；长挂车；长材拖车
pole trailer system 架杆式拖车方式
pole tram-way 木杆轨道
pole tree 杆材树
pole type substation 杆塔式变电所；架杆式变电站
pole type switch 杆式开关
pole type transformer 杆式变压器；杆装变压器；架杆式变压器；电杆上变压器
pole-vault stand 撑杆跳高架
pole wag(g)on 杆连式拖车；木材车
pole wandering 极移
poleward migration 极向迁移
pole winding 磁极绕组
pole with cross arms 分线杆
pole-zenith-star triangle 定位三角形
pole-zero assignment 极一零点配置
pole-zero compensation circuit 极一零点补偿电路
pole-zero configuration 极一零结构
polhemusite 硫汞锌矿
polhoide 本体极迹
polhode cone 体锥面
polhody 极轨迹
polianite 黝锰矿
police 公安人员
police boat 水警艇
police box 岗亭
police building 公安大楼；公安大厦
police hostel 公安招待所；公安休养所；公安疗院；公安宿舍
police launch 水警艇
police law 管制法
police office 警察局
police post 派出所
police power 政府执法权
police protection 治安保护
police stand 岗亭；交通指挥台
police station 公安局；警察局
police substation 派出所
policies issued basis 保险单签发基础
policlinic 门诊部
policy 策略；方针；保单
policy analysis 政策分析
policy appraisal 政策评价
policy clause 保险条款；摆平投标
policy cost 政策性成本；方针性成本
policy coverage 政策范围
policy decision 决策；方针决策
policy depreciation 政策性折旧
policy evaluation 政策评估
policy fee 保险单签发手续费
policy forecasting 政策预测
policy for industry 工业政策
policy for internal observance 内部政策
policy holder 投保方；保险客户；保险单持有人；保

单持有人
Policy Holders Protection Board 保险履约保护委员会
policy improvement algorithm 策略改进算法
policy loan 保险单贷款
policy-maker 政策制定者
policy-making 政策制定
policy-making body 决策机构
policy-making in a macrocosmic sense 宏观决策
policy-making studies 决策学
policy manual 保险手册
policy mix 政策配合
policy objectives 政策目标
policy of air transportation insurance 空运保险单
policy of changing, transferring, renting and selling 改、转、租、买政策
policy of economic control 经济统制政策
policy of insurance 保险单
policy of one control 一控双达标政策
policy of opening 开放政策
policy of prevention in the first place and integrating prevention with control 预防为主、防治结合的政策
policy of product-use price differentials 不同用途差别定价政策
policy of seclusion 闭关自守政策
policy of tightening up environmental management 强化环境管理政策
policy of title insurance 地契保险的保险单
policy on resettlement 移民政策
policy planning 政策规划
policy prognosis 非确定型决策
policy proof of interest 信用保单;保险单权益证明
policy reserves 保险单责任准备金
policy restraint 政策干预
policy space 策略空间
policy stability 政局稳定
policy surveillance 政策监督
policy unstability 政局动荡
policy varial 政策性变数
polimerisado oil 巴西果油聚合油;巴西果油厚油
poling 插板;支撑;立杆;极性调整;插木还原;插板临时支撑软地层法
poling board 电缆沟的壁板;隧道洞顶防塌护板;堰板;支撑板;沟槽竖撑板;挡土板;垂直挡土板;撑板;插板;隧洞矢板
poling board method 撑板开挖法;撑板掘进法
poling board method of tunnel(l)ing 隧道插板掘进法;隧道撑板掘进法
poling plank 堰板
poling plate 掘进(刀)板;支撑板;插板;钢插板
poling selection 架线路选择方式
Polis 城邦;早期城市(古希腊的城市国家)
polish 研光;高纯度水处理;磨光剂;磨;泡立水;抛光剂;上光剂;打磨;擦亮;擦光
polishability 抛光性
polishable 可琢磨的;可磨光的
Polish accumulator technique 波兰累加器法
Polish architecture 波兰建筑
polish attack 抛光浸蚀
Polish bond 波兰式砌合
polish brush 抛光刷
polished 磨光的
polished aggregate 磨光集料;磨光骨料
polished alumin(i)um finish 抛光铝饰面
polished asbestos-cement board 磨光的石棉水泥板
polished-barrel rod 抛光圆棒
polished bolt 抛光螺栓
polished brass 黄铜抛光;抛光黄铜
polished brass glass cutter 抛光铜质玻璃割刀
polished coefficient 磨光系数
polished concrete pavement 磨光的混凝土路面
polished core bit 抛光的岩芯钻头
polished diamond 抛光的金刚石
polished die 抛光模
polished dressing 修饰打光;磨光
polished drum coating 抛光鼓涂布
polished edge arising 抛光边加工
polished effluent 深度污水处理后出水
polished face 磨光石面;磨光面
polished filter 最终净化滤池
polished finish 抛光处理;石面磨光;抛光饰面;磨光面;镜面光泽面
polished finish of stone 石面磨光;石面打光
polished foil 抛光金属箔
polished glass 磨光玻璃;抛光玻璃
polished gold 磨光金
polished part 抛光件
polished piston 抛光活塞
polished plate 磨光板;磨光玻璃(片);磨光玻璃;磨光平板玻璃
polished plate glass 磨光平板玻璃
polished pond 最终净化塘
polished press plate 磨光压板
polished rod head 抛光钻头
polished rod head counterweight 深井泵光杆头平衡重量
polished rough coating 抛光的粗底层
polished sample 磨光试样;抛光样品
polished section 光片;磨光片
polished silver 抛光银
polished silvered plate glass 磨光涂银玻璃
polished slab 抛光板
polished specimen 磨光标本;抛光试件
polished steel bar 抛光的钢条
polished stone 抛光的石面
polished stone coefficient 石料磨光系数
polished stone value 石料磨损值;石料磨光值
polished surface 精加工表面;磨光面;抛光面
polished thin section 光薄片;磨光薄片;磨光薄膜
polished value 石料磨光值
polished wire(d) glass 磨光夹丝玻璃
polished work 抛光表面;磨光表面;抛光石料;石面磨光
polish emulsion 乳液抛光剂;上光乳剂
polisher 抛光剂;擦光器;磨盘;抛光盘;抛光器;抛光机;抛光工;水处理终端过滤器;擦亮剂
polish fault 磨光断层
polish grind 碾磨打光;磨平打光
polishing 最终精制;磨光;抛光;上光;打光;擦亮
polishing agent 抛光剂;上光剂
polishing agent for monocrystalline silicon 单晶硅抛光剂
polishing apparatus 抛光设备
polishing barrel 抛光筒
polishing belt 抛光带
polishing block 抛光台
polishing brush 抛光刷;磨光刷;磨版刷
polishing characteristics 磨光性
polishing clamp 抛光夹具
polishing cloth 抛光用呢;抛光布
polishing coefficient 磨光系数
polishing composition 抛光剂
polishing compound 擦亮剂;抛光剂;抛光油;抛光(混合)剂;抛光膏
polishing cone 抛光锥
polishing cream 抛光膏;擦亮乳膏
polishing department 抛光工段;抛光车间
polishing disk 磨盘;抛光盘
polishing drum 抛光筒
polishing effect of traffic 交通(对路面的)磨光作用
polishing equipment 抛光设备
polishing felt 抛光毡;抛光用毡
polishing file 细锉;抛光锉
polishing filter 增泽过滤器;精滤器
polishing fluid 抛光液
polishing fluosilicate 抛光氟硅酸盐
polishing hammer 抛光锤
polishing hardness 抗磨硬度
polishing iron 铁磨光器
polishing lathe 磨光机;抛光车床
polishing line 磨光生产线;抛光生产线
polishing machine 角抛(磨)光机;磨损性;抛光机;抛光机
polishing marks 抛光磨痕
polishing material 光泽剂;磨光物;磨光剂;磨光材料;抛光材料;擦亮剂
polishing medium 抛光剂;抛光介质
polishing metal 抛光金属
polishing of diamonds 金刚石抛光(用细粒金刚石钻头钻进极坚硬细粒岩石时)
polishing of thermoplastic floor tile 热塑地板花砖抛光
polishing of wood 木材打光
polishing oil 擦光油
polishing operation 最终净化操作【给】
polishing paper 砂纸;抛光纸
polishing paper disc 抛光纸片;纸片抛光
polishing paste 抛光膏;擦车油膏
polishing plant 抛光工厂;抛光车间
polishing pond 精度处理塘;深度污水处理塘
polishing powder 研磨粉;抛光粉
polishing process 优质处理;深度处理
polishing roll 抛光(轧)辊
polishing rouge 抛光用铁丹粉;抛光用红丹粉;擦亮铁丹
polishing runner 抛光盘
polishing scratch 磨痕
polishing section 抛光车间
polishing slate 磨石片
polishing solution 抛光液
polishing stage 三级处理(指废水处理)
polishing stake 抛光研磨器
polishing stick 磨光棒
polishing stone 磨光石
polishing technique 磨光术
polishing test 磨光试验;抛光试验
polishing the tape 抛光布带
polishing tool 抛光工具
polishing treatment 最终净化处理;抛光处理
polishing-type wax 抛光蜡;擦车蜡
polishing unit 抛光设备
polishing varnish 抛光漆;亮光清漆;可抛光清漆;可打磨清漆;抛光漆;打光清漆;擦亮清漆
polishing wax 打光蜡;擦车蜡
polishing wheel 磨光轮;抛光砂轮;抛光轮
Polishki Registry Statkow 波兰船舶登记局
polish layer 打光层数;打光层次
Polish notation 波兰表示法
polish of diamond 金刚石磨光
polish paste 糊状抛磨料;抛磨膏
Polish Register 波兰船级社
polish resistant aggregate 抗磨光集料;抗磨光骨料;防滑集料;防滑骨料
polish rod (深井泵的)光杆;磨光的钻杆
polish rod clamp 夹紧器;深井泵拉杆钩环
polish susceptibility 磨光敏感性
polish-susceptible 易磨光的;容易抛光的
polish-susceptible aggregate 易磨光集料
polish the still wall 釜壁抛光
polish till dry 干抛光
political appraisal of a project 项目政治评价
political atlas 政区地图集
political center 政治中心
political environment 政治环境
political geography 政治地理
political lecturer 报告员
political map 行政区划图;政区地图
political risk 政治风险
political strike 政治罢工
political unit 行政区
politic tuff 细疑灰岩
politure 磨光;泡立水;抛光剂;抛光
poljie 岩溶盆地;喀斯特洼地;灰岩盆地;坡立谷;[复]poljia
polka dot 圆点花纹
polka-dot method 点光栅法
polka dot paint 圆点花纹多彩涂料
polka-dot raster 跳点光;点光栅
polkovicite 硫锗铁矿
poll 敲击火石;挂号通信;登记通信;锤头平面;探询;查询
Pollaczek's formula 波拉泽克公式
Pollaczek-Spitzer identity 波拉泽克—斯皮策恒等式
poll adze 带平头扁斧
Pollak's test 波拉克氏试验
Pollapas 坡拉帕斯(一种木材防腐剂)
pollard 截梢树;剪顶树;无顶树;头木;截梢树
pollarding 修剪树冠;截土
pollard system 头木作业法
poll code 注册码;登记码
polled face 显出火石的墙面
polled sample statistics 合并样本计量
pollen 花粉
pollen analysis 花粉分析
pollen analysis method 花粉分析法
pollen cells 药室
pollen chamber 储[贮]粉室
pollen peat 富花粉泥炭
pollen spectrum 花粉谱
pollen stratigraphy 花粉地层学

pollen wall architecture 粉壁构造
pollen wall stratification 粉壁分层
pollen wall structure 粉壁结构
pollination 传粉
polling 转态过程；终端设备定时询问；轮询；剪短；投票表决；探询；查询
polling discipline 轮询规定
polling interval 轮询间隔；查询间隔
polling list 轮询表；登记通信表；登记表；查询表；发送设备登记表
polling method 按序询问法
polling period 巡检周期
polling rate 巡检速度
polling routine 查询程序
polling set-up 轮询装置
polling table 登记表
polling technique 轮询技术
poll network 轮讯网
Pollock commutator 集锁式换向器；波洛克换向器
Pollock construction 集锁式结构
pollopas 塑料玻璃
pollster 民意测验者
polltacker 民意测验者
poll tax 人头税；人丁税
poll the experience 交流经验
pollucite 铯沸石
pollutant 沾污物；污染物
pollutant accumulation 污染物累积
pollutant analysis 污染物分析
pollutant analyzer 污染物分析仪；污染物分析器
pollutant assement 污染物评价
pollutant assessment index 污染评价指数；污染评价指标
pollutant band 污染带范围
pollutant burden 污染物负荷(量)
pollutant burden pattern 污染物负荷模式
pollutant characterization 污染物表征
pollutant chemical degradation 污染物的化学降解
pollutant chemistry 污染物化学
pollutant concentration 污染物浓度
pollutant concentration control 污染物浓度控制
pollutant concentration field 污染物浓度场
pollutant concentration level 污染物浓度水平
pollutant concentration limit 污染物浓度限度
pollutant concentration prediction 污染物浓度预报
pollutant constituent 污染物成分
pollutant content 污染物含量
pollutant control technology standard 污染物控制技术标准
pollutant degradation 污染物降解(作用)
pollutant discharge 污染物排放；污染物排出
pollutant discharge concentration 污染物排放浓度
pollutant discharge elimination system 污染物排放消除体系
pollutant discharge loading 污染物排放负荷
pollutant discharge quality 污染物排放质量
pollutant discharge standard 排污标准；污染物排放标准
pollutant dispersal 污染物散布
pollutant dispersion 污染物散布；污染物扩散；污染物传播；污染物弥散
pollutant disposal 污染物处置；污染物处理
pollutant distribution 污染物分布
pollutant dynamics 污染物动力学
pollutant effect 污染物的影响
pollutant effect beyond national boundary 超国境污染物影响
pollutant effect man and society 污染对人类和社会的影响
pollutant effect on climate 污染物对气候的影响
pollutant effect on fishes 污染物对鱼类的影响
pollutant effect on man and society 污染物对人和社会的影响
pollutant emission 污染排放；污染发散物
pollutant emission standard 污染物扩散标准
pollutant equilibrium 污染物(质)平衡
pollutant flux 污染物通量
pollutant generation quantity 污染物生产量；污染物产生量；污染产生量
pollutant identification 污染物识别；污染物鉴定；污染物鉴别
pollutant index 污染指数；污染物指数；污染物指标
pollutant in rain 雨中污染物
pollutant in snow 雪中污染物

pollutant interaction with environment 污染与环境相互作用；污染物与环境的相互作用
pollutant in water body 水体污染物
pollutant level 污染物浓度；污染水平
pollutant load 污染物浓度
pollutant loading 污染物负荷(量)
pollutant loading in equivalent standard 等标污染负荷
pollutant loading rate 污染负荷率
pollutant load per unit 污染物单位负荷；单位污染物负荷
pollutant mass 污染物质量；污染物团
pollutant monitoring 污染物监测
pollutant offset policy 污染物补偿政策；污染补偿政策
pollutant pathway 污染物行径；污染物路径；污染物的传播途径
pollutant persistency 污染物的持久性
pollutant quantity in pure base 污染物折纯碱量
pollutant recipient 污染物接受者
pollutant reduction 减少污染物
pollutant removal 去除污染物；除污染物
pollutant removal efficiency 污染物去除系数
pollutant removal rate 污染物去除率
pollutant runoff simulation model 污染径流模拟模型
pollutant-sediment interaction 污染物沉淀物相互作用
pollutant self-purification 污染物的自净作用
pollutants in combined sewage 合流污水中的污染物质
pollutants in rainfall 雨水中的污染物质
pollutants in wet weather flow 雨天流污水中的污染物质
pollutants mixing in water body 水体污染物混合
pollutant sources 污染源
pollutant source criterion 污染源基准
pollutant source identification 污染源识别；污染源鉴定；污染物来源鉴别
pollutant source region 污染源地
pollutant source type 污染源类型
pollutant standard index 污染物标准指数；污染物标准指标
pollutant start value 污染起始值
pollutant substance 污染物质
pollutant surveillance 污染物监视；污染物监测
pollutant susceptibility 污染物易感性；污染物感受性
pollutant target 污染物指标；污染物目标
pollutant target exposure 污染物目标暴露
pollutant transformation 污染物转化
pollutant transport 污染物输移
pollutant transport model 污染物输移模型
pollutant transport prediction 污染物输移预报
pollutant transport simulation model 污染物输移模拟模型
pollutant transport transformation 污染物输移转化
pollutant treatment 污染物质处理
pollutant unit area load 单位面积污染负荷
pollutation discharge coefficient 排污系数
pollutation effect on fishes 污染对鱼类的影响
polluted agricultural land 污染农田；被污染的农田
polluted air 污浊空气；污染空气
polluted area 污染区
polluted atmosphere 污染的大气
polluted belts along river bank 沿岸水域污染
polluted by sewage 污水污染；受污水污染
polluted environment 污染环境；污染的环境；被污染的环境
polluted estuary 污染河口；污染的河口
polluted fishes 污染鱼类；被污染的鱼类
polluted flow 污染水流
polluted ground 污染工地
polluted groundwater 污染的地下水
polluted groundwater zone 污染地下水带；污染地下水区
polluted industrial wastewater 污染工业废水；生产污水
polluted lake 污染湖泊
polluted load 污染负荷
polluted matter 污染物质；污染物
polluted overburden 被污染的覆盖层
polluted rain 污染雨水
polluted reach 污染河段
polluted river 污染河流；污染的河流；受污染河流

polluted river water 污染河水；污染的河水；被污染的河水
polluted river water treatment 污染河水处理
polluted sediment 污染底泥
polluted stream 污染河流；污染的河流
polluted stretch of river 河流污染蔓延
polluted substance 污染物质
polluted surface water 污染地表水；被污染的地表水
polluted water 污水；污染水；污染的水；受污染水
polluted water algae 污染水藻
polluted water recovery 污染水回收
polluted waters 污染水域；污染水体；污染的水域
polluted water sources 污染水源
polluted waters quality 污染水体水质
polluted waterway 污染水道；污染的水道；污染的排水沟
polluted zone from shore 岸边污染带
polluter 污染物质；污染者；污染环境者
polluter pays policy 污染者负担政策
polluter pays principle 污染者支付原则；污染者负担原则；污染者付款原则；污染者偿还原则
pollute the environment 污染环境
polluting effect 污染效应
polluting investigation of groundwater 地下水污染调查
polluting oil 污染油
polluting property 污染性质；污染特性
polluting runoff 污染径流
polluting strength 污染强度
pollution 沾污；污染
pollution abatement 消除污染；减轻污染；污染治理；污染抑制；污染消除
pollution abatement equipment 减轻污染设备；减除污染的设备；污染治理设备；防尘设备
pollution-absorbing plant 吸污植物
Pollution Abstracts 污染文摘(期刊)
pollution accretion 污染积成物；污染堆积
pollutional 污染的
pollutional abatement 污染消除
pollutional condition 污染状态；污染状况；污染条件
pollutional consequence 污染后果
pollutional contribution 污染后果
pollutional effect 污染效应
pollutional equivalent 污染当量
pollutional front location 污染锋面位置
pollutional index 污染指数
pollutional indicating organism 污染指示生物
pollutional indicator 污染指示器；污染指示剂
pollutional load 污染负荷；污染量
pollutional matter 污染物质
pollutional river 污染河流
pollutional stream 污染河流
pollution area 污染区
pollution assessment 污染评价
pollution assimilation 污染同化
pollution atlas 污染图册；污染地图集
pollution belt 污染带
pollution beyond national boundary 超国境污染
pollution biology 污染生物学
pollution by antimony 锑污染
pollution by arsenic 砷污染
pollution by asbestos 石棉污染
pollution by beryllium 铍污染
pollution by biodegradable waste 可生物降解废物污染
pollution by carbon oxides 碳氧化物污染
pollution by chemical fertilizer 化肥污染
pollution by chromium 铬污染
pollution by cobalt 钴污染
pollution by copper 铜污染
pollution by dredging 疏浚污染
pollution by farm chemicals 农药污染
pollution by fluorine 氟污染
pollution by gasoline additives 汽油添加剂污染
pollution by human activity 人类活动污染
pollution by iron 铁污染
pollution by lead 铅污染
pollution by livestock effluent 牲畜污水污染；牲畜排出物污染
pollution by manganese 锰污染
pollution by molybdenum 钼污染
pollution by nickel 镍污染
pollution by nitrogen oxides 氮氧化物污染
pollution by organochlorine pesticide 有机氯农

药污染
pollution by organonitrogen pesticide 有机氮农药污染
pollution by organophosphorus pesticide 有机磷农药污染
pollution by oxygen consumptive substance 耗氧物质污染
pollution by particulates 颗粒物污染
pollution by pesticides 农药污染
pollution by plant nutritious substance 植物营养物污染
pollution by plasticizer 增塑剂污染
pollution by polychlorinated biphenyls 多氯联苯污染
pollution by polycyclic(al) aromatic hydrocarbons 多环芳烃污染
pollution by salts 盐类污染
pollution by sediments 底泥污染;沉积物污染
pollution by sulfur 硫污染
pollution by sulfur oxides 硫氧化物污染
pollution by tin 锡污染
pollution by toxic chemicals 有毒化学物质污染
pollution by trace metal 微量金属污染
pollution by turbidity 混浊污染
pollution by vanadium 钒污染
pollution by zinc 锌污染
pollution capacity 污染容量;污染能力
pollution carpet 污染毡状层
pollution-carrying capacity 污染容纳量
pollution-carrying resource 有污染资源
pollution-causing industry 产生污染的工业
pollution characteristic 污染特征
pollution charges 排污收费;排污费
pollution chemistry 污染化学
pollution code 污染法规
pollution coefficient 污染系数
pollution concentration 污染浓度
pollution control 污染治理;污染控制;污染防治;防止污染
Pollution Control Act 污染控制法令
Pollution Control Agency 污染控制局
Pollution Control Board 污染控制局
pollution control bond 控制污染债券
pollution control equipment 污染控制净化装置
pollution control facility 污染控制装置;污染控制设施
pollution control function 污染控制作用;污染控制效能
pollution control guide 污染控制指南
pollution control index 污染控制指数;污染控制指标
pollution control instrumentation 污染控制仪器
pollution control legislation 污染控制立法
pollution controller 污染控制器
pollution control management 污染控制管理
pollution control measures 污染控制措施
pollution control of solid wastes 固体废物污染控制
pollution control plan 防污计划
pollution control planning 污染控制规划
pollution control planning of industrial area 工业区污染控制规划
pollution control plant 污染控制装置
pollution control policy 污染控制政策
pollution control regulation 环境保护法;污染控制法规;污染控制条例
pollution control ship 污染控制船;污染监控船
pollution control structure 污染控制建筑物
pollution control system 污染控制系统
pollution control technology 污染控制技术
pollution counter 污染计数器;污染计数计;污染计数管
pollution-creating source 引发污染源;污染产生源;创生污染源
pollution criterion 污染标准;污染基准
pollution cycle 污染周期;污染循环
pollution damage to marine environment 海洋环境污染损害
pollution danger 污染威胁
pollution degree 污染程度
pollution deposit 污染沉积(物)
pollution detection 污染检测
pollution determination 污染测定
pollution discharge 排污

pollution disease 污染疾病
pollution distribution 污染分布
pollution dose 污染剂量
pollution ecology 污染生态学
pollution ecology of waters 水域污染生态学
pollution economics 污染经济学
pollution effect 污染影响;污染的后果
pollution effect on man and society 污染对人类和社会的影响
pollution energy 污染型能源
pollution engineering 污染工程
pollution engineering technique 污染工程技术;防污工程技术
pollution episode 污染敏感时刻;特别容易受污染威胁的时候;污染事件
pollution episodic model 污染事件模型
pollution evaluation 污染评价
pollution experiment in model ecosystem 受控生态系统污染试验
pollution factor 污染系数
pollution feature 污染特征
pollution free 无污染(的)
pollution-free automobile 无污染汽车
pollution-free energy soruce 无污染能源;清洁能源
pollution-free engine 无污染发动机
pollution-free equipment 无污染装置
pollution-free food 无公害食品
pollution-free fuel 无污染燃料
pollution-free installation 无污染装置
pollution-free pesticide 无污染农药
pollution-free plastics 无污染塑料
pollution-free production 无污染生产
pollution-free pulping 无污染制浆
pollution-free technology 无污染工艺
pollution-free vegetable 无公害蔬菜
pollution from land-based sources 陆源污染;陆地来源的污染
pollution hazard 污染危害
pollution history 污染史;污染过程
pollution horror 污染恐怖
pollution-immune 不受污染影响的
pollution incident 污染事故
pollution incident statistics 污染事件统计
pollution index method 污染指数法
pollution indication ion 污染性离子
pollution indication organism 污染指示生物
pollution indicator 污染指示物
pollution indicator plant 环境污染指示植物
pollution-induced disease 污染引起的疾病
pollution in limited area 有限区污染;局部污染
pollution in rain and snow 雨雪中的污染物
pollution intensity 污染强度
pollution-intensive goods 强烈造成污染的货物;污染密集货物
pollution interaction 污染作用
pollution investigation 污染调查
pollution in wide areas 大面积污染;广域污染;广泛污染
pollution level 污染水平;污染级别;污染程度
pollution liability 污染责任;污染应负的责任
Pollution Liability Agreement Among Tanker Owners 油轮船东污染责任协议
pollution liability insurance 污染责任保险
pollution license 污染物排放许可证
pollution limit 污染限度;污染极限
pollution load (污水对河流等的)污染负荷
pollution load allocation model 污染负荷分配模型
pollution load coefficient 污染负荷系数
pollution load influence 污染负荷影响
pollution loading 污泥负荷量
pollution loading amount 污染负荷量
pollution loading index 污染负荷指数
pollution loading rate 污染负荷率
pollution load of non-point source 非点污染源污染负荷
pollution loss cost 污染损失费用
pollution management 污染管理
pollution measurement 污染测量;污染测定
pollution mechanism 污染机制;污染机理
pollution meteorology 污染气象学
pollution microbiology 污染微生物学
pollution mixing zone 污染混合带
pollution model 污染模式
pollution model of groundwater 地下水污染模型

pollution monitoring 污染监测
pollution monitoring and assessment 污染监测与评价
pollution monitoring network 污染监测网
pollution norm 污染标准
pollution nuisance 污染公害
pollution of bathing place 浴场污染;游泳池污染
pollution of beach 海滨污染
pollution of chromium 铬污染
pollution of coastal waters 沿岸水域污染;海岸水域污染
pollution of controlled waters 可控水体污染
pollution of deep well water 深井水污染
pollution of estuary 河口污染
pollution offset system 污染补偿制度
pollution of groundwater 地下水污染
pollution of heavy metal 重金属污染
pollution of herbicide 除草剂污染
pollution of lake 湖泊污染
pollution of natural water 天然水污染
pollution of ocean originating on land 陆源海洋污染
pollution of oil ship 油船污染
pollution of reservoir 水库污染
pollution of river 河流污染;河道污染
pollution of seawater 海水污染
pollution of stream 河流污染;河川污染
pollution of surface water 地表水污染
pollution of the environment 污染环境
pollution pathway 污染途径
pollution permit 污染许可证
pollution physiological effect 污染生理效应
pollution-plagued city 污染危害的城市
pollution plume 污染烟缕;污染的烟羽
pollution potential 污染势;污染潜势
pollution potential index 潜在污染指数
pollution prediction 污染预测;污染预报
pollution prediction of solid wastes 固体废物污染预测
pollution prevention 防污染;防污;污染防治
pollution prevention and control ordinance 污染防治条例
pollution prevention and control technology 污染防治技术
pollution prevention at the source 源头污染预防;水源防污
pollution prevention industry 污染防治工业
pollution prevention ordinance 污染防治条例;防污法令
pollution prevention system 防污系统;防污染系统
pollution prevention technique 污染防治技术;防污技术
pollution product charges 污染产品收费
pollution-prone 易于受到污染的
pollution proof insulator 防污绝缘子
pollution receptor 污染受体
pollution reducing 减少污染;减轻污染
pollution reduction 减少污染;减轻污染
pollution region 污染区
pollution-related 与污染有关的
pollution-related disease 与污染有关的疾病;污媒疾病
pollution-related victim 污染受害者
pollution relationships 污染关系
pollution release 污染物释放
pollution rent 污染权利转让
pollution research and monitoring area 污染调查监测区
pollution-resisting plant 抗污染植物
pollution-resisting species of trees 抗污染树种
pollution restriction 污染限制
pollution right 污染权
pollution risk 污染危险;污染风险
pollution risk analysis 污染风险分析
pollution risk frequency 污染危险频率
pollution scale 污染标度
pollution-sensitive typology 按对污染的敏感性分类型;污染敏感类型
pollution sign of water quality 水质污染征兆
pollution sink 污染汇集地
pollution source 污染源
pollution source assessment 污染源评价
pollution source contribution ratio 污染源贡献率
pollution source control 污染源控制

pollution source control strategy 污染源控制对策
pollution source criterion 污染源基准
pollution source identification 污染源识别
pollution source location 污染源位置
pollution source of water body 水体污染源
pollution source recognition 污染源识别
pollution source release history 污染源排放史
pollution sources assessment model of river 河流污染源评价模型
pollution source with continuous discharge 连续排放污染源
pollution source with instantaneous discharge 瞬时排放污染源
pollution source with interval discharge 间断排放污染源
pollution standard index 污染标准指数；污染标准指标
pollution study 污染研究
pollution surveillance 污染监视
pollution survey 污染调查
pollution survey report 污染调查报告
pollution tax 污染税
pollution tester 污染测试器
pollution threat 污染威胁
pollution tolerant organism 耐污生物；耐污染生物
pollution tolerant plant 耐污染植物
pollution tolerant species 耐污染物种
pollution toxicity 污染毒生；污染毒度
pollution type 污染类型
pollution under gusty condition 阵风污染
pollution warning system 污染警报系统；污染报警系统
pollution washoff coefficient 除污系数
pollution water 污染水
pollution waters 污染水域；污染水体
pollution yielding traffic condition 产生污染的交通条件
pollution zone 污染区；污染地带；污染带
pollution zone width 污染带宽度
pollutograph 污染图
Polmer-Bowlus flume 帕—鲍氏计量槽
polo court 马球场
polocyte 极体
polo ground 马球场
poloidal coil 极向场线圈
poloidal magnetic field 角向磁场
polonaise rug 丝织地毯
Polonceau roof truss 法国式屋架
Polonceau truss 法国式桁架
polonium ion 钋离子
Poltal 波尔塔尔复杂铝合金
poltechnic(al) 多种工艺的
poltechnic(al) college 工业大学
poltechnic(al) exhibition 工艺展览会
poltechnic(al) institute 工业大学
poltechnic(al) school 工艺学校
polular name 俗名
polup 珊瑚虫
poly 多聚
polya 坡立谷
polyacetal 聚缩醛
polyacetylen 聚乙炔
polyacidic base 多元酸碱；多酸价碱
polyacrylamide 聚丙烯酰胺
polyacrylamide clear agent 澄清剂
polyacrylamide flocculant 聚丙烯酰胺絮凝剂
polyacrylamide gel 聚丙烯酰胺凝胶
polyacrylamide gel electrophoresis 聚丙烯酰胺凝胶电泳
polyacrylamide hydrazine 聚丙烯酰胺肼
polyacrylate 聚丙烯酸盐
polyacrylate acrylic ester 聚丙烯酸酯
polyacrylate modified cement 聚丙烯酸酯改性水泥
polyacrylic acid 聚丙烯酸
polyacrylic acid-polyvinylodene fluoride blending membrane 聚丙烯酸聚偏氟乙烯共混膜
polyacrylic desalination 聚丙烯酸类脱盐
polyacrylic plastics 聚丙烯酸类塑料
polyacrylic resin 聚丙烯酸树脂；聚丙烯酸化物树脂
polyacrylic rubber 聚丙烯酸化物橡胶
polyacrylics 聚丙烯酸化物
polyacrylonitrile 聚丙烯腈
polyacrylonitrile copolmer 聚丙烯腈聚合物
polyacrylonitrile fiber 聚丙烯腈纤维；腈纶

polyacrylonitrile fiber wastewater 聚丙烯腈纤维废水
polyacrylonitrile ultrafiltration membrane 聚丙烯腈超滤膜
polyacrylonitrile wastewater 含聚丙烯腈废水
polyaddition 加成聚合
polyaddition reaction 加成聚合反应
polyaddition resin 加聚树脂
polyadelphite 锰铁榴石
Polya distribution 玻利亚分布律
polyaerylate resin 聚丙烯酸酯树脂
Polya frequency distribution of order two 波利亚二阶频率分布
polyagglutination 多凝集反应
polyalcohol 多元醇
poly aleinic anhydride 聚马来酸酐
polyalkane 聚链烷
polyalkoxide 聚烷氧化物
polyalkoxysilane 聚烷氧基硅烷
polyalkylated 多烷基化
polyalkylbenzene 多烷基苯
polyalkylene amide 聚亚烷基酰胺
polyalkylene glycol 聚亚烷基二醇
polyalkyl synthetic(al) lubricant 聚烷基合成润滑剂
polyallomer 异质同晶聚合物；异质同晶共聚物
polyallomer resin 异质同晶共聚树脂
polyallyl alcohol 聚烯丙醇
polyalumin(i)um 聚合硅酸铝
polyalumin(i)um calcium chloride 聚合氯化钙铝
polyalumin(i)um chloride 聚氯化铝；聚合氯化铝；多氯化铝
polyalumin(i)um chloride containing calcium 含钙聚合氯化铝
polyalumin(i)um chloride flocculant 聚氯化铝絮凝剂
polyalumin(i)um chloride sulfate[sulphate] 聚硫氯化铝
polyalumin(i)um coagulant 聚铝混凝剂
polyalumin(i)um ferric chloride 聚合氯化铝铁
polyalumin(i)um ferric phosphate 聚合磷酸铝铁
polyalumin(i)um ferric silicate 聚合硅酸铝铁
polyalumin(i)um ferric silicate sulphate 聚合硫酸铝硅铁
polyalumin(i)um ferric sulphate 聚合硫酸铝铁
polyalumin(i)um ferric sulphatochloride 聚合硫酸氯化铝铁
polyalumin(i)um organic modified sulfate[sulphate] 聚有机改性硫酸铝
polyalumin(i)um phosphate 聚合磷酸铝
polyalumin(i)um silicate chloride 聚合硅酸氯化铝
polyalumin(i)um silicate sulfate 聚硅酸硫酸铝
polyalumin(i)um sulfate 聚合硫酸铝
polyalumin(i)um sulfate modified bentonite 聚硫酸铝改性膨润土
polyalumin(i)um sulfate modified bentonite coagulant 聚硫酸铝改性膨润土混凝剂
polyamide 聚酰胺；尼胺
polyamide-epoxy 聚酰胺环氧树脂
polyamide fiber 聚酰胺纤维
polyamide fiber reinforced cement 尼龙纤维增强水泥
polyamide-imide 聚酰胺酰亚胺
polyamide-imide resin 聚酰胺酰亚胺树脂
polyamide plastics 聚酰胺塑料
polyamide resin 聚酰亚胺树脂胶；聚酰胺树脂
polyamide-thickened lubricant 聚酰胺稠化润滑油
polyamidine 聚脒
polyamine 聚胺；多胺
polyamine-methylene resin 聚亚甲胺树脂；聚甲烯树脂
polyaminobismaleimide resin 聚氨基双马来酰胺树脂
polyaminoresin 聚胺树脂
polyampholyte 高分子两性电解质；两性电解质
polyamphoteric electrolyte 聚两性电解质
polyaniline 聚苯胺
polyanion 聚阴离子
polyanionic surfactant 多阴离子表面活性剂
polyanion sensor 聚阴离子传感器
Polya process 波利亚过程
polyarch xylem 多元木质部
polyaromatic hydrocarbon 多芳烃
Polyart 高密度聚乙烯合成纸

polyarylate 多芳基化合物
polyarylester 聚芳酯
polyarylether 聚芳醚
polyaryloxysilane 聚芳氧基硅烷
polyarylsulphone 聚芳砜
Polya's distribution 波利亚分布
Polya's enumeration theorem 波利亚计数定理
Polya's theorem 波利亚定理
polyatomic acid 多元酸
polyatomic alcohol 多元醇
polyatomic phenol 多元酚
polyatomic ring 多节环
polyatron 多阳极计数放电管
polyautography 石印术
polyaxial stress 多轴应力
polyaxial test 多轴试验(三向部等压)
polyaxon 多轴骨针
polyazanaphthlene 多氮杂萘
polyazelaic polyanhydride 聚壬二酸酐
polyazo dye 多偶氮染料
polyazo pigment 多偶氮颜料
poly bag 玻璃纸袋
poly bag rubber 塑料袋橡胶
polybar trailer type row-crop sprayer 多杆拖车式中耕作物喷雾机
polybase 多碱
polybasic 多元的；多碱的
polybasic acid 多元酸；多碱价酸；多价酸
polybasic alcohol 多元醇
polybasic carboxylic acid 多元羧酸
polybasic ester 多元酸酯
polybasic ion 多元碱离子
polybasite 硫锑铜银矿
poly-benzamide 聚对苯酰胺
polybenzimidazole resin 聚苯并咪唑树脂
polybenzoimidazolone 聚苯并咪唑酮
polybenzothiazole 聚苯并噻唑；聚苯并咪唑
polybinary system 多二进制
polyblend 高聚物混体；高聚物共混体；聚合物共混；聚苯乙烯橡胶混合物；塑料橡胶混合物；复合高聚物
Polybor-chlorate 复合氯酸盐
polyboring 多用钻孔
polybromide 多溴化合物
polybrominated biphenyl 多溴(化)联苯
polybutadiene 聚丁乙烯；聚丁二烯
polybutadiene-acrylic acid-acrylonitrile 聚丁二烯丙烯酸/丙烯腈
polybutadiene-acrylic acid copolymer 聚丁二烯丙烯酸共聚物
polybutadiene acrylonitrile 聚丁二烯丙烯腈
polybutadiene composite propellant 聚丁二烯复合推进剂
polybutadiene rubber 聚丁二烯橡胶
polybutadiene styrene 聚丁二烯苯乙烯
polybutadiene styrene resin 聚丁二烯苯乙烯树脂
polybutene 聚丁烯
polybutene base 聚丁烯基
polybutene tape 聚丁烯胶带
polybutylene glycol adipate 聚己二酸丁二醇酯
polybutylene glycol succinate 聚丁二酸丁二醇酯
polybutylene pipe 聚丁烯管
polybutylene plastics 聚丁烯塑料
polybutylene resin 聚丁烯树脂
polybutylene tape 聚丁烯胶带
polybutylene terephthalate 聚对苯二甲酸丁二酯
polybutylene tube 聚丁烯管
polybutyl methacrylate emulsion 聚甲基丙烯酸丁酯乳液
polycalcium 多种钙
polycaprolactam 聚己内酰胺
polycaprolactone 聚己内酯
polycaprolactone glycol 聚己内酯二醇
polycaprolactone polyol 聚己内酯多元醇
polycarbafil 玻璃纤维增强聚碳酸酯
polycarbam(in)ate 聚氨基甲酸酯
polycarboimide 聚碳酰亚胺
polycarbonate 聚碳酸酯
polycarbonate membrane filter 聚碳酸酯薄膜滤片
polycarbonate plastics 聚碳酸酯塑料
polycarbonate resin 聚碳酸酯树脂
polycarbonate sheet 聚碳酸酯板
polycarboranesiloxane 聚碳硼烷硅氧烷
polycarboxylate cement 聚羧酸黏固粉

polycarboxylic acid 聚羧酸
polycarpea roberti 旋状白鼓丁
polycarpous 多心皮的
polycatalytic degradation 聚催化降解
polycationic acrylamide 阳离子型聚丙烯酰胺
poly cell approach 多单元法
Polycels dye 波利赛尔斯染料
polycentered arch 多心拱
polycentered arch bridge 多心拱桥
polycentric 多心的
polycentrism 多中心论
polycephalous root 多头根
polyceptor 多受体
polycerams 聚合陶瓷
poly-change speed motor 多速电动机
poly-channel amplifier 多路放大器
polychannel tunnel kiln 多孔窑
polychasium 多歧聚伞花序
polycheiria 多手畸形
polychlamydeous chimaera 复层嵌合体
polychloral 吡啶和三氯乙醛产物
polychloride 多氯化物
polychlorinated biphenyl 多氯联苯
polychlorinated biphenyl compound 多氯联苯化合物
polychlorinated biphenyl pollutant 多氯联苯污染物
polychlorinated biphenyl pollution 多氯联苯污染
polychlorinated biphenyls 聚氯联苯；多氯化联苯类；多氯代联苯
polychlorinated dibenzofuran 多氯氧
polychlorinated hydrocarbon 多氯烃
polychlorinated triphenyl 多氯三联苯
polychlorization 多氯化反应
polychlorobenzoic acid 多氯苯甲酸
polychlorobutadiene 聚氯丁二烯
polychlorodifluoroethylene 聚二氟氯乙烯
polychloroesters hard foam plastics 聚氯酯硬质泡沫塑料
polychloroether 聚氯醚
polychlorohydrocarbon 多氯化烃
polychloroparafin 多氯化石蜡
polychloropinene 聚氯蒎烯
polychloroprene 聚氯丁烯；聚氯丁二烯
polychloroprene foam 聚氯丁烯泡沫
polychloroprene rubber 氯丁橡胶
polychlorosulfate alumin(i)um 多氯硫酸铝
polychlorotrifluoroethylene 聚三氟氯乙烯
polychroism 多向色性；复色现象
polychromasia 多染色性
polychromatic 多色的
polychromatic finish 多色饰面；多色面漆；虹彩装饰；虹彩涂料；彩色装饰；彩饰表面；彩色光面
polychromatic glass 多色彩饰玻璃
polychromatic lidar 多色激光雷达
polychromatic light 多色灯；多色光；复色光
polychromatic optic(al) transfer function 复色光学传递函数
polychromatic opulence 丰富多彩
polychromatic radiation 多色频辐射；多色辐射
polychromatic radiator 多色辐射器
polychromatism 多色（现象）
polychromatophilia 多染色性
polychromator 多色仪
polychrome 多色(的)
polychrome brick 彩色砖
polychrome decoration 多色装饰；多色彩饰
polychrome decorative painting 彩画
polychrome figurecurve vase 五彩人物角瓜式瓶
polychrome film 彩色胶卷
polychrome finish 彩色装修
polychrome glazed-brick decoration 彩色釉面（瓷）砖装饰
polychrome methylene blue 多色亚甲基蓝
polychrome painting 新彩
polychrome pottery 彩饰陶器
polychrome sculpture 彩色雕塑
polychromy 多色性；彩色艺术；多色画法；多色；彩艺术；彩饰法
poly-clastic psammitic texture 复屑砂状结构
poly-clastic tuff 复屑凝灰岩
polyclimax 多元演替顶极；多元顶级；多级相
polyclimax theory 多顶极学说
polyclinal chimaera 多成分嵌合体
polyclinal fold 多斜褶皱
polyclinic 多科医院；综合医院；分科门诊所
poly-coated 涂上聚乙烯的
polycomplex 络聚剂
polycomponent 多组分
polycomponent conglomerate 复成分砾岩
polycomponent system 多组分系统；多组分体系；多元系(统)
polycompound 多组分化合物
polyconcave pore 曲凹孔
polycondensate 缩聚物
polycondensation 缩聚作用
polycondensation reaction 浓缩反应；浓缩作用；凝聚作用；凝聚反应
polycondensation resin 浓缩树脂
polyconic(al) chart 多锥投影图
polyconic(al) conforming projection 正形多圆锥投影
polyconic(al) projection 多锥(体)投影；多圆锥投影
polyconic(al) projection chart 多锥投影海图；多圆锥投影地图
polyconic(al) projection grid 多圆锥投影格网
polyconic(al) projection of USA 美国多圆锥投影
polyconic(al) projection with merdianal intervals on same parallel decrease away from central meridian by equal difference 等差分纬线多圆锥投影
polycore cable 多芯电缆
polycore flexible wire 多芯软电线
polycormic 多干的
polycotyledon 多子叶植物
polycoumarone resin 香豆酮茚树脂
polycrase 复稀金矿
polycrystal 多晶体；多晶
polycrystal diamond 聚晶金刚石
polycrystal diamond bit 聚晶金刚石钻头
polycrystal diamond compact bit 聚晶金刚石复合体钻
polycrystalline 多晶质的；多晶的
polycrystalline aggregate 多晶聚合体
polycrystalline area 多晶面积
polycrystalline ceramics 多晶陶瓷材料；多晶陶瓷
polycrystalline compact diamond 多晶体金刚石
polycrystalline copper 多晶铜
polycrystalline diamond 聚晶金刚石；多晶金刚石
polycrystalline diamond bit 聚晶金刚石钻头
polycrystalline diamond compact 金刚石复合片；烧结金刚石
polycrystalline diamond compact bit 金刚石复合片钻头；复合片钻头
polycrystalline diamond compound bit 多晶金刚石复合钻头
polycrystalline diamond wire drawing die 多晶金刚石拉丝模
polycrystalline fibre 多晶纤维
polycrystalline germanium 多晶锗
polycrystalline glacier ice 复晶(质)冰川冰
polycrystalline graphite 多晶石墨
polycrystalline growth 多晶生长
polycrystalline ingot 多晶锭块
polycrystalline irons 多晶形铁粉
polycrystalline laser 多晶激光器
polycrystalline manmade carbon fibre 多晶人造碳纤维
polycrystalline material 多晶体物质；多晶材料
polycrystalline oxide ceramics 多晶氧化陶瓷
polycrystalline polymer 多晶型聚合物；多晶形聚合物
polycrystalline rod 多晶棒
polycrystalline silicon 多晶硅
polycrystalline structure 多晶结构
polycrystalline substance 底部结构
polycrystalline substrate 多晶衬底
polycrystalline unit 多晶体
polycrystallinity 多晶性；多晶结晶度
polycrystallite 混杂晶体
polyculture 多种作物栽培；多种养殖
polyculture of fish and rice 稻田养鱼
poly curve 多变曲线
polycyanoalkylsiloxane 聚氰基烷基硅氧烷
polycycie geosyncline 多旋回地槽
polycycle 多循环；多旋回
polycycle engine 多冲程发动机
polycycle sampling 多循环取样
polycyclic(al) 多循环的；多旋回的；多间隔期的；多环的
polycyclic(al) aromatic compound 多环芳(香)烃化合物
polycyclic(al) aromatic hydrocarbon 多环芳(香)烃
polycyclic(al) aromatic hydrocarbons 多环芳(香)烃类
polycyclic(al) compound 多环化合物
polycyclic(al) cycloalkene 多环烯烃
polycyclic(al) hydrocarbon 多环烃
polycyclic(al) landform 多循环地形
polycyclic(al) metallogenesis 多旋回成矿作用
polycyclic(al) organic matter 多环有机物
polycyclic(al) orogenesis 多旋回造山运动；复相造山运动
polycyclic(al) sampling 多循环取样
polycyclic(al) stale 多环式中柱
polycyclic(al) tectogenesis theory 多旋回构造运动学说
polycyclization 多环化(作用)
polycycloalkane 多环烷烃
polycyclohexyl methacrylate 聚甲基丙烯酸环己酯
polycyclonaphthenic acid 多环烷酸
polycyclopentadiene 聚茂；聚环戊二烯
polycyclo-rubber 环化橡胶；多环橡胶
polycylinder 多柱面；多圆柱体
polycylindrical endovibrator 波长调节筒
polydactylia 多趾畸形；多指畸形
polydactylism 多趾畸形；多指畸形
polydactyly 多指趾畸形
polydiallyl phthalate 聚邻苯二甲酸二烯丙酯
polydichlorostyrene 聚二氯苯乙烯
polydiene 聚二烯
polydiene rubber 二烯橡胶
polydienes 聚二烯类
polydiethylene glycol adipate 聚二甘醇己二酸酯
polydiethylene siloxane 聚二甲基硅氧烷
poly-dimensional 多维的
polydirectional 多方向性；多方向的
polydirectional microphone 指向性可调传声器
polydirectional splitting 多向分岔
poly-disc 重直多圆盘犁
poly-disc tiller 多圆盘翻土机
polydisperse adsorption 多分散性粉尘
polydisperse aerosol 多分散气溶胶
polydispersed aerosol 多相分散气溶胶
polydispersed pollutant 多相分散的污染物；多分散污染物
polydisperse latex 多分散乳胶
polydisperse particle 多分散粒子
polydisperse system 杂散系统；多分散(体)系
polydispersion index 多分散指数
polydispersity 聚合度多分散性；多分散性
polydispersity index 多分散指数
polydispersoid 多分散
polydispersoid colloid 多分散胶体
polydomain 多畴
polydontia 多牙
polydrill 多用钻孔
polydymite 硫镍矿
polye 坡立谷
polyelectrolyte 高分子电解质；聚合物电解质；聚合电解质；聚电解质
polyelectrolyte complex 聚电解质络合物
polyelectrolyte enhanced ultrafiltration 聚电解质强化槽滤
polyene 聚烯；多烯(烃)；多聚烃
polyene primer 多烯底漆
polyene varnish 多烯清漆
polyenic acid 多烯酸
polyenic compound 多烯化合物
polyepichlorohydrin 聚表氯醇
polyepichlorohydrin rubber 氯醇橡胶
polyepoxide 多环氧化物
polyepoxysuccinic acid 聚环氧琥珀酸
polyester adhesive 聚酯黏合剂；聚酯胶黏剂
polyesteramide 聚酯酰胺；聚酰胺酯
polyesteramide material 聚酰胺酯物料
polyester and acetate fibre dyeing effluent 聚酯醋酸盐纤维染色废水
polyester base 聚酯胎体；聚酯片基
polyester bath 聚酯塑料浴盆

polyester board 聚酯板
polyester bodning agent 聚酯黏结剂
polyester carpet 涤纶地毯
polyester cast(ing) resin 聚酯釉质树脂；聚酯铸塑树脂
polyester coat(ing) 聚酯涂层
polyester compound 聚酯合成物；聚酯化合物
polyester concrete 聚酯混凝土
polyester concrete panel 聚酯混凝土护墙板；聚酯混凝土镶板
polyester cord 聚酯绳索
polyester corrugated sheet(ing) 聚酯波纹片
polyester drawing film 聚酯绘图薄膜
polyester drawing sheet 聚酯绘图片
polyester elastic filament oil 涤纶弹力丝油
polyester fabric 聚酯织物
polyester fiber 聚酯纤维；聚酯织物；聚酯硬纸板；聚酰纤维；涤纶
polyester filament oil 涤纶长丝油
polyester film 聚酯软片；聚酯胶片；聚酯薄膜
polyester finish 聚酯终饰
polyester fleece 聚酯毡
polyester floor cover(ing) 聚酯地板材料；聚酯楼面覆盖层
polyester floor(ing) 聚酯地板材料；聚酯楼面覆盖层
polyester form(work) 聚酯模板
polyester gate 聚酯门
polyester-glass fibre sheet 聚酯玻璃钢板；玻璃纤维增强聚酯板；聚酯胶结玻璃纤维板；聚酯胶结玻璃丝板
polyester glycol 聚酯二醇
polyesterification 聚酯化（作用）
polyester imide 聚酯酰亚胺
polyester isocyanate 聚酯异氰酸酯
polyester laminated film 聚酯层合薄膜
polyester mat 聚酯毡
polyester mortar 聚酯灰浆；聚酯砂浆
polyester paint 聚酯漆
polyester plasticizer 聚酯增塑剂
polyester plastics 聚酯塑料；聚酯类塑料
polyester plate 聚酯片
polyester polyol 聚酯型多元醇
polyester polyurethane 聚酯型聚氨酯
polyester powder coating 聚酯粉末涂料
polyester printing 聚酯薄膜图
polyester putty 聚酯油灰
polyester reinforced oxidized bitumen sheet 聚酯增强氧化沥青油毡
polyester resin 聚酯树脂
polyester resin adhesive 聚酯胶黏剂
polyester resin board 聚酯树脂板
polyester resin coated 聚酯树脂胶涂的
polyester resin coat(ing) 聚酯树脂涂料
polyester resin compound 聚酯树脂合成物；聚酯树脂化合物
polyester resin concrete 聚酯树脂混凝土
polyester resin corrugated light-admitting board 聚酯树脂透光波纹板
polyester resin corrugated sheet(ing) 波纹聚酯树脂薄片
polyester resin-faced 聚酯树脂饰面的
polyester rubber 聚酯橡胶
polyester sheet 聚酯薄片
polyester urethane 聚酯型氨基甲酸酯
polyester wastewater 聚酯废水
polyester window 聚酯窗
polyester wire enamel 聚酯漆包线漆
polyethene building sheet 聚乙烯建筑板材
polyethenoxy ether 聚氧乙烯醚
polyether 聚醚；多醚
polyether diisocyanate 聚醚二异氰酸酯
polyetherether keton 聚醚醚酮
polyether glycol 聚醚二醇
polyetherimide 聚醚酰亚胺
polyether polyol 聚醚型多元醇
polyether polyurethane 聚醚型聚氨酯
polyether rubber impression material 聚醚橡胶印模材料
polyether sulfone 聚醚砜
polyether surfactant 聚醚类表面活性剂
polyether wastewater 聚醚废水
polyethlenimine 聚氮丙啶
polyethoxylated alkyl phenol 聚乙氧烷基酚
polyethylene 聚乙烯

polyethylene agricultural film 聚乙烯农用薄膜
polyethylene bathroom gulley 聚乙烯浴室排水管；聚乙烯浴室水落管
polyethylene bathroom inlet 聚乙烯浴室进水口
polyethylene bathroom outlet 聚乙烯浴室出水口
polyethylene cable 聚乙烯绝缘电缆
polyethylene coated polyester fabric 聚乙烯涂层聚酯织物
polyethylene coated polyethylene fabric 聚乙烯涂层聚乙烯织物
polyethylene coated steel pipe 聚乙烯涂层钢管
polyethylenediamine cellulose 聚乙二胺纤维素
polyethylene dielectric 聚乙烯电介质
polyethylene diphthalate condenser 苯二甲酸聚乙烯介质电容器
polyethylene disulfide 聚二硫化乙烯
polyethylene drainage plate 聚乙烯排水板
polyethylene fabric 聚乙烯织物
polyethylene fiber 聚乙烯纤维
polyethylene film 聚乙烯薄膜
polyethylene glycol 聚乙二醇；聚氧乙烯；聚氧乙二醇
polyethylene glycol adipate 聚己二酸二乙酯
polyethylene glycol ether 聚乙二醇醚
polyethylene glycol maleate 聚马来酸乙二醇酯
polyethylene glycol monostearate 聚乙二醇单硬质脂酸酯
polyethylene glycol nonylphenyl ether 聚乙二醇单甲醚丙烯酸酯
polyethylene glycol stearate 聚乙二醇硬脂酸
polyethylene glycol succinate 聚丁二酸乙二醇酯
polyethylene high density 高密度聚乙烯
polyethylene hydroperoxide 聚乙烯化过氧氢
polyethyleneimine 聚乙烯亚胺
polyethylene isophthalate 聚间苯二甲酸乙二醇酯
polyethylene laminated paper 聚乙烯层合纸
polyethylene large diameter profile wall drain pipe 聚乙烯大直径异形壁排水管
polyethylene large diameter profile wall sewer pipe 聚乙烯大直径异形壁污水管
polyethylene-lined 聚乙烯衬里
polyethylene membrane 聚乙烯薄膜
polyethylene modified asphalt 聚乙烯改性沥青
polyethylene oil 聚乙烯润滑油
polyethylene oxide 聚环氧乙烷
polyethylene oxide admixture 聚烯氧化掺和剂
polyethylene oxide polymer 聚氧化乙烯聚合物
polyethylene pipe 聚乙烯管
polyethylene pipe fitting 聚乙烯管件
polyethylene plastics 聚乙烯塑料
polyethylene plastic waterproofing membrane 聚乙烯塑料防水层；聚乙烯塑料防水板
polyethylene polyamine 多乙烯多胺
polyethylene-polyvinyl acetate 聚乙烯聚乙酸乙烯酯
polyethylene protection film 聚乙烯保护膜
polyethylene protection sheeting 聚乙烯保护板材
polyethylene resin 聚乙烯树脂
polyethylene sheet(ing) dampproof(ing) course 聚乙烯薄膜防潮层
polyethylene squeeze type wash bottle 聚乙烯挤压型洗瓶
polyethylene tarpaulin 聚乙烯盖布
polyethylene terephthalate 聚对苯二甲酸乙二酯
polyethylene terephthalate wastewater 聚对苯二甲酸乙二酯废水
polyethylene terephthslate 聚乙烯对苯二甲酸酯
polyethylene thermoplastic high-pressure irrigation pipeline system 聚乙烯热性塑性塑料高压灌溉管道系统
polyethylene tube 聚乙烯管
polyethylene tubing 聚乙烯管道；聚乙烯管
polyethylene wax 聚乙烯蜡
polyethylhydrosilicone 聚乙基氢硅氧烷
polyethyl methacrylate 聚甲基丙烯酸乙基酯
polyethylmethacrylate point 聚甲基丙烯乙烯尖
polyethyl silicone 聚乙基硅氧烷
polyexponent 多变指数
polyferose 多糖铁
polyferric alumin(i)um sulfate 聚合硫酸铝铁
polyferric chloride 聚合氯化铁
polyferric chloride flocculant 聚氯化铁絮凝剂
polyferric chloride sulfate 聚硫氯化铁
polyferric phosphonate chloride 聚合磷氯化铁

polyferric phosphonate sulfate 聚合磷酸铁
polyferric silicate sulfate containing boron 含硼聚合硅硫酸铁
polyferric sulfate 聚合硫酸铁
polyferroalumin(i)um chloride 聚合氯化铝铁
Poly-Filter 波利·菲耳特丙纶机织土建布
poly-fluid theory 多流体理论
polyfluorhydrocarbon 多氟化烃
polyfluoride 多氟化物
polyfluorinated ether 多氟化醚
polyfluorobenzene 多氟化苯
polyfluoroprene 聚氟丁二烯；氟丁橡胶
polyfoam 泡沫塑料
polyfocal projection 多焦点投影
polyfoil 多叶饰
Polyfon 木质素磺盐酸钠
polyformaldehyde 聚甲醛
polyfunctional 多作用的
polyfunctional alcohol 多官能醇
polyfunctionality 多官能度
polyfurfurylidene acetone 聚亚糠基丙酮
polygamous 多偶性
polygen 多价的
polygene 多源的；多时代的；多成因；复成的
polygeneric dyes 多色染料
polygene sediment 多源沉积物
polygenetic 多源的；多时代的；多色的；多成因的；复成的
polygenetic compound antimony deposit 多因复成锑矿床
polygenetic compound copper deposit 铜的多因复成矿床
polygenetic compound deposit type 多因复成矿床类型
polygenetic compound gold deposit 金的多因复成矿床
polygenetic compound iron deposit 多因复成铁矿床
polygenetic compound lead zinc deposit 多因复成铅锌矿床
polygenetic compound metallogenesis 多因复成成矿作用
polygenetic compound molybdenum deposit 钼的多因复成矿床
polygenetic compound ore deposit 多因复成矿床
polygenetic compound tin deposit 锡的多因复成矿床
polygenetic compound uranium deposit 多因复成铀矿床
polygenetic compound wolfram deposit 多因复成钨矿床
polygenetic sediment 多源沉积物
polygenetic topography 复合地形；复成地形
polygenetic volcano 复成火山
polygene volcano 复成火山
polygenic deposit 多源沉积物
polygenic sediment 多源底泥；多源沉积物
polygenous 多成因的
polygenous two-splitting 多次二分岔
polygeosyncline 复地槽
polyglass 苯乙烯玻璃（塑料）；苯乙烯塑料
polyglass tire 合成纤维玻璃丝轮胎
polyglycerin 聚甘油
polyglycerol 聚甘油；聚丙三醇
polyglycerol ester 聚甘油酯
polyglyceryl phthalate 聚邻苯二甲酸甘油酯
polyglycol 聚乙二醇；聚二醇
polyglycol distearate 聚乙二醇二硬脂酸酯
polyglycol ether 聚乙二醇醚
polygon 龟裂形土纹；多角形；多边形
polygonal 多角形的；多边形的；多边的
polygonal angle 导线角
polygonal annular foundation 多边环形基础
polygonal arch 多角拱；多边形拱
polygonal area 龟裂区域
polygonal bar 多边形棒
polygonal bay window 多角窗；八角窗
polygonal block 多角形砌块；多角形建筑（物）
polygonal block method 多边形块段法
polygonal bond 多面体砌合
polygonal bowstring girder 折弦拱形（大）梁
polygonal bowstring truss 折弦拱形桁架；折线弓弦桁架
polygonal building 多角形建筑（物）

polygonal cavity block 多角形空心砌块
polygonal cavity tile 多角形空心砖
polygonal centralized church 多角形中央教堂
polygonal chart 多边图
polygonal choir 多角形唱诗班席位
polygonal church 多角形教堂
polygonal church with central space 有中央广场的多角形教堂
polygonal column 多面柱;多角柱
polygonal condition 多边形条件
polygonal connection 多边形接线
polygonal course 导线【测】
polygonal crystal 多面体晶体
polygonal cupola 多角形圆顶阁;多角形圆屋顶
polygonal curve 多角曲线
polygonal diminutive tower 多角形装饰性塔楼;多角形装饰性角楼
polygonal dome 多角形穹顶
polygonal dome with circular sector 多边形圆扇面穹顶
polygonal domical vault 多角方形平面拱顶
polygonal donjon 多角形城堡主塔
polygonal drum 多角形鼓状物;多角形圆筒
polygonal dungeon 多角形地牢;多角形土牢;多角形城堡主塔
polygonal earth 多边形土
polygonal element 多边形单元
polygonal equipment 直链形悬挂
polygonal exposed concrete column 多边形外露混凝土柱;多面形外露混凝土柱
polygonal fair-faced concrete column 多角平直面的混凝土柱;多边平直面的混凝土柱
polygonal figure 多面形的;多边形的;多角形的
polygonal folded plate roof 多边形折板屋顶
polygonal frame 多边形结构;多角形框架
polygonal function 折线函数
polygonal graph 多边图
polygonal ground 龟裂地面;网纹土;多边形土;多边形龟裂土;地面龟裂
polygonal ground plan 多角形地面图;多角形底层平面图
polygonal hall-choir 有唱诗班席位的多角形大厅;有唱诗班席位的多角形会堂
polygonal hall-quire 有唱诗班席位的多角形大厅;有唱诗班席位的多角形会堂
polygonal height traverse 多角高程导线
polygonal hipped-plate roof 多面形四坡板屋顶;多角形四坡板屋顶
polygonal hollow block 多角形空心砌块
polygonality 多边性
polygonal keep 多角形防守所;多角形堡垒;多角形城堡主塔
polygonal line 折线
polygonally closed traverse 多边形闭合导线
polygonal masonry 皮虎石砌体;多角石砌体;多角石砌圬工;多角砌体
polygonal masonry bond 多角形砌体砌合;多边形砌体砌合
polygonal measurement network 多边形测量网络
polygonal method 多角形法;多边形计算矿量法;导线法;导线测量法
polygonal network 多角形网络;导线网
polygonal network plan 导线网扩展图
polygonal nozzle 多角形喷嘴
polygonal orifice 多角形孔口
polygonal ornament 多角形装饰品
polygonal pier 多角支柱;多角墩
polygonal plan 多角形地面图;多角形底层平面图
polygonal plate 多边平板;多角平板
polygonal plug ga(u)ge 多边形孔的塞规
polygonal point 导线点
polygonal pot 多角形盆;多角形罐
polygonal prismatic shell roof 多角形折板屋顶;多面形棱柱薄壳屋顶;多边形棱柱薄壳屋顶
polygonal pulley 多角形滑车
polygonal quire 多角形唱诗班席位
polygonal random rubble masonry 蛛网形缝乱毛石砌体;毛石圬工;多角形毛石砌工
polygonal ring 多边线圈;多边形环;多角环
polygonal ring structure 多角环形构造
polygonal rod 多角形棒
polygonal roof 多边形(坡)屋顶;攒尖顶【建】
polygonal rubble 多边形毛石;乱毛石墙砌体;虎皮石(墙)

polygonal rubble facing 毛石贴面
polygonal rubble masonry 多角毛石砌体
polygonal screen 多边形筛
polygonal sett paving 多边形石块铺砌;多角形石块路面
polygonal shell roof 折板壳屋顶
polygonal slab 多角形平板
polygonal small tower 多角形小型角楼
polygonal soil 多角形土
polygonal spire 多角形尖塔
polygonal steel arch 多角形钢拱
polygonal structure 多边形结构
polygonal support 多边形支柱
polygonal termination 多角形终端设备;多角形终点站
polygonal tilted-slab roof 多面形倾斜屋顶
polygonal top-chord 折线形上弦杆;多边形上弦杆
polygonal traverse 多角导线;多边导线
polygonal traversing 多角测量
polygonal truss 折弦桁架;多边形桁架
polygonal turret 多角形角楼;多角形塔楼
polygonal vault 多角形拱顶
polygonal voltage 边电压
polygonal wall 多角形墙
polygonation 多角形测量法
polygon definition file 闭合曲线定义文件
polygon grinding machine 多角形磨床
polygon hob 多棱零件加工滚刀
polygonization 多角形化;多边形化;多边化
polygonization boundaries 多边化边界
polygonization crack 多边化裂纹
polygonized quartz 多边化石英
polygonized structure 多边化结构
polygon lathe 多边形仿形车床
polygon leg 导线边
polygon mat 多边形块
polygon mesh 多边形网络
polygon method 多边形法;多边形法
polygon milking parlor 多边形挤奶厅
polygon milling attachment 多边形铣削附件
polygon mirror 光学多面体
polygon misclosure 导线闭合差
polygon of forces 力多边形
polygon of railway network 多边形铁路网
polygon of velocities 速度多边形
polygonometric point 导线点
polygonometric method 导线测量法
polygonometric network 导线测量网
polygonometry 多角长法;导线测量
polygonous 多角的;多边形的
polygon search 多边形检索
polygon-section forward calculation method 多边形截面法
polygon structure 多边形结构
polygon support 多边形支架
polygon truss 线形桁架
poly grab 多瓣抓斗
polygradient magnetic separator 多梯度磁选机
polygram 多字组;多字母;多边图形
polygram casting 复制铸造
polygraph 多种波动描记器;多功能记录仪;复写器
polyhalide 多卤化物
polyhalide water 高盐分水
polyhaline 多盐性;多盐水
polyhalite 杂卤石
polyhalite rock 杂卤石岩
polyhalogenated 多卤化的
polyhalogenated hydrocarbon 多卤烃;多卤碳氢化合物
polyhalogenated pollution 多卤化污染
polyhalogenoalkylsiloxane 卤烷基聚硅氧烷
polyhalogenocarbon 多卤烃
polyhedra 多面体
polyhedral 多面体的;多面体;多面的
polyhedral angle 立体角;多面角
polyhedral body 多角体
polyhedral cone 多面锥体
polyhedral convex cone 多角凸锥
polyhedral edge detector 图像多边检测器
polyhedral game 多面体对策
polyhedral pore 多面体孔
polyhedral projection 多面投影;多面体投影
polyhedral region 多面角区域
polyhedral shell 多面体薄壳

polyhedral structure 多面体结构
polyhedron 可剖分空间;多面体;多角体
polyhedron face 多面体的面
polyhexamethylene adipamide 聚己二酰己二胺
polyhexaphenylformic acid 聚六苯甲酸
polyhindered phenol 多位阻酚
polyhydrate 多水水化物;多水合物
polyhydrazide 聚酰肼
polyhydrazone 聚腙
polyhydric acid 多元酸;多碱价酸
polyhydric alcohol 多元醇
polyhydric alcohol conjugated polyether wastewater 多元醇共轭聚醚废水
polyhydric alcohol conjugate polyether 多元醇组合聚醚
polyhydric phenol 多元酚
polyhydrone 多聚水
polyhydroxyalkanoate 聚羟基烷酸酯
polyhydroxy benzoic acid 聚苯乙酯
polyhydroxy ether 聚羟基醚
polyhydroxy(lated) compound 多羟基化合物
polyhydroxy monobasic acid 多羟基一元酸
polyhydroxy phenol 多羟基苯酚
polyhydroxyvalerate 聚羟基戊酸
polyimide 聚酰亚胺
polyindene resin 聚茚树脂
polyion 高分子型离子;聚离子;多价离子
poly-iron 多晶形铁粉
polyiron 树脂羧基铁粉
polyisobutylene 聚异丁烯
polyisobutylene film 聚异丁烯薄片;聚异丁烯薄膜
polyisobutylene-isoprene 聚异丁烯戊二烯
polyisobutylene rubber 聚异丁烯橡胶
polyisobutylene sheet(ing) 聚异丁烯膜;聚异丁烯薄片;聚异丁烯薄膜
polyisobutylene synthetic(al) rubber 聚异丁烯合成橡胶
polyisocyanate 多异氰酸酯
polyisocyanate foam 聚异三聚氰酸盐泡沫
polyisocyanurate 多异氰脲酸酯
polyisomerism 多同型(现象)
polyisoprene 聚异戊二烯
polyjet 多射流
polyjet valve 多射阀
polylaminate 多层的
polylaminated sealing process 多层封接法
polylaminate wiring technique 多层布线技术
polylateral 多边形的
poly-lens objective 多透镜物镜
polyline 折线
polyline colo(u)r index 折线颜色索引
polyline index 折线索引
polyliner 多孔衬套
polyline representation 折线表示法
polylith 多块
polylitharenite 复岩屑砂屑岩
Polyloom 波利洛姆烯烃纤维
polymagnetsium ferric chloride 聚合氯化镁铁
polymaleic acid 聚马来酸
polymarker 多重记号
polymarker colo(u)r index 多重记号颜色索引
polymarker facility 多重记号设备
polymarker representation 多重记号表示法
polymer 聚合物;聚合体;多聚体
polymer alloy 聚合物合金;掺混聚合物
polymer aluminate flocculant 高分子铝酸盐絮凝剂
polymer-augmented water flooding 聚合物强化水驱动
polymer backbone 聚合物主链
polymer based composite 高分子复合材料
polymerbitumen 聚合沥青
polymer bitumen sheeting 聚合物沥青油毡
polymer blend 共混聚合物;高分子共混物;掺混聚合物
polymer carbon 聚合物碳
polymer catalyst 高分子催化剂
polymer cement 聚合物水泥
polymer cement concrete 聚合物水泥混凝土
polymer cement mortar 聚合物水泥砂浆
polymer-cement ratio 聚合物水泥比
polymer chain 聚合物链
polymer chemistry 高聚物化学;高分子化学
polymer-clad silica fiber 聚合物包层石英纤维
polymer coagulant 高分子凝结剂

polymer coagulation 高分子混凝
polymer-coated pigment 聚合物包膜颜料
polymer-coated yarn 涂塑纱
polymer colloid 高分子胶体
polymer compexation 高分子络合
polymer concrete 聚合物混凝土
polymer content 聚合物含量
polymer crystal 高分子晶体
polymer damping material 高分子阻尼材料
polymer degradation 聚合物降解
polymer dispersion 聚合物乳液;聚合物分散体
polymer drilling fluid 聚合物钻孔液;聚合物钻井液
polymer drugs 高分子药物
polymer electrolyte 高分子电解质
polymer emulsion 聚合物乳胶;聚合物乳液
polymer ferrocement 聚合物钢丝网水泥
polymer ferro-cement with steel fiber 钢纤维增强聚合物钢丝网水泥
polymer fibre 锦纶
polymer film 高分子膜
polymer film detector of fire smoke 聚合物薄膜烟感火灾探测器
polymer film overlay system 高分子膜层系统
polymer film smoke fire detector 聚合物薄膜烟感火灾探测器
polymer filter 聚合物滤色片
polymer flocculant 聚合物絮凝剂
polymer fume fever 聚合烟尘雾热;聚合物烟热器
polymer gel 聚合物凝胶
polymer generative chemistry 高分子生成化学
polymer-homologous series 同系聚合物
polymer-homologue 同系聚合物
polymeric 聚合的
polymeric acceptor 聚合物(接)受体
polymeric additive 聚合(物)添加剂
polymeric adsorbent 吸附树脂
polymeric alumin(i)um ferric sulfate 聚合硫酸铝铁
polymeric alumin(i)um zinc chloride 聚合氯化铝锌
polymeric based material 聚合基材料
polymeric catalyst 高分子催化剂
polymeric chain 聚合链
polymeric coating 聚合物涂层
polymeric complex flocculant 高分子络合絮凝剂
polymeric compound 高分子化合物
polymeric dielectrics 聚合物电介质;高分子电介质
polymeric diphenyl methane diisocyanate 聚合型二苯甲烷二异氰酸酯
polymeric donor 聚合物给体
polymeric dye 高分子染料
polymeric ferric sulfate 聚合硫酸铁
polymeric flocculant 高分子吸附;高分子絮凝剂
polymeric high molecular material 高分子材料
polymeric homologue 同系聚合物
polymeric isocyanate 多异氰酸酯
polymeric material 聚合物材料;聚材料容
polymeric material vessel 有机合成材料容器
polymeric matrix 聚合物基material
polymeric membrane 高分子膜;聚合隔膜
polymeric membrane for separation 高分子分离膜
polymeric microfiltration membrane plasma 高分子微滤膜空腔
polymeric modifier 聚合物改性剂
polymeric paper 合成纸
polymeric phosphate-alumin(i)um chloride 聚磷氯化铝
polymeric pigment 聚合颜料
polymeric plasticizer 聚合物增塑剂;高分子增塑剂;高分子型增塑剂
polymeric reagent 聚合物试剂
polymeric separator 聚合分离器
polymeric structure 高分子结构
polymeric surface active agent 高分子表面活性剂
polymeric surfactant 聚合物表面活性剂
polymeric type epoxy plasticizer 高分子量环氧增塑剂
polymeric viscosity modifier 聚合物黏度调节剂
polymeride 聚合物
polymer impregnated concrete 聚合物浸渍混凝土;聚合物注入混凝土
polymer impregnated fibre concrete 聚合物纤维浸渍混凝土
polymer impregnated mortar 聚合物浸渍砂浆

polymer impregnated precast concrete 聚合物浸渍预制混凝土
polymerisable 可聚合的
polymerisate 聚合产物
polymerism 聚合现象
polymerizability 可聚合性
polymerizable 可聚合的
polymerizable monomer 可聚合单体
polymerizable oligomer 可聚合低聚物
polymerization 聚合作用;聚合现象;聚合化;聚合(反应)
polymerization accelerator 聚合加速剂
polymerization analysis 聚合分析
polymerization at normal temperature 常温聚合
polymerization autoclave 高压聚合釜;聚合高压釜
polymerization catalyst 聚合催化剂
polymerization chemicals 聚合剂
polymerization degree 聚合度
polymerization dye 聚合染料
polymerization exponent 聚合指数;聚合率
polymerization inhibitor 阻聚剂
polymerization initiator 聚合引发剂
polymerization intermediate 聚合中间体
polymerization mechanism 聚合历程;聚合机理
polymerization modifier 聚合改性剂
polymerization pipe 聚合管
polymerization plant 聚合装置
polymerization process 聚合过程
polymerization processor 聚合处理机
polymerization promotor 聚合促进剂
polymerization rate 聚合速度
polymerization regulator 聚合调节剂
polymerization resin 聚合树脂
polymerization retardation 聚合阻滞
polymerization retarder 阻聚剂;聚合阻滞剂;聚合抑制剂
polymerization stabilizer 聚合稳定剂
polymerization stopper 阻聚剂
polymerization technique 聚合技术
polymerization time 聚合时间
polymerization under hot-water 热水聚合
polymerization unit 聚合装置;聚合设备
polymerized 聚合的
polymerized dehydrated castor oil 聚合脱水蓖麻油
polymerized fatty acid 聚合脂肪酸
polymerized iron chloride 聚合氯化铁
polymerized oil 熟油;聚合油;集合油
polymerized resin 聚合树脂
polymerized rosin 聚合松香
polymerized substance 聚合物
polymerized tung oil 聚合桐油
polymerizer 高温焙烘机;聚合器;聚合剂;聚合釜
polymerizing agent 聚合剂
polymerizing power 聚合能力
polymer macromolecule 聚合物大分子
polymer-making autoclave 高压聚合釜
polymer matrix 聚合物母体;聚合物基体材料
polymer melt 聚合物熔体
polymer melting temperature 聚合物熔体温度;聚合物熔融温度
polymer modification 聚合物改性
polymer modified asphalt 聚合物改性沥青
polymer modified cement mortar 聚合物改性水泥砂浆
polymer modified concrete 聚合物改性混凝土
polymer modified glass fibre reinforced concrete 聚合物改性玻璃纤维增强混凝土
polymer modified mortar 聚合物改性砂浆
polymer modified oil 聚合物改性油
polymer molecular weight 聚合物分子量
polymer morphology 聚合物形态学
polymer mortar 聚合物灰浆;聚合物砂浆
polymer network 聚合物网络
polymer optic(al) fibre 聚合物光学纤维
polymer orientation 高分子定向(作用);聚合物取向
polymer phase 聚合相
polymer physics 高聚物物理学
polymer plant 聚合装置
polymer plastics 聚合物塑料
polymer Portland cement concrete 聚合物水泥混凝土
polymer pump 聚合泵
polymer quenching 聚合物淬火

polymer reaction 高聚物反应
polymer resin 聚合树脂
polymer rheology 聚合物流变学
polymer roofing 聚合物屋面
polymer science 高分子科学
polymer sealer 聚合物密封剂
polymer semiconductor 聚合物半导体
polymer series 聚合物系列
polymer slurry 聚合物泥浆
polymer solid electrolyte 高分子固体电解质
polymer solution 高分子溶液;聚合物溶液
polymer-solvent interaction 聚合物溶剂相互作用
polymer stabilization 聚合物稳定(土壤);聚合物加固(作用)
polymer stabilizer 聚合物稳定剂
polymer thin film 聚合薄膜
polymer waste 聚合物废料
polymer waterproof material 高分子防水材料
polymery 多出式
polymetal crust 多金属结壳
polymetallic 多金属的
polymetallic crust 多金属结壳
polymetallic deposit 多金属矿床
polymetallic mud 多金属泥
polymetallic nodule 多金属结核
polymetallic oxide-hydroxide 多金属氧化物氢氧化物
polymetallic sulfide 多金属硫化物
polymetal nodule 多金属结核
polymetamorphism 多相变质(作用)
polymeter 多功能测量仪;多能湿度计;多能湿度表;多功测计;复式物性计
polymethacrylanmide 聚甲基丙烯酰胺
polymethacrylate 聚甲基丙烯酸酯
polymethacrylate gel 聚甲基丙烯酸酯凝胶
polymethacrylic acid 聚甲基丙烯酸
polymethin dye 聚甲炔染料
polymethine 聚甲炔
polymethlene polyphenyl isocyanate 聚亚甲基聚苯异氰酸酯
polymethyl acrylate 聚丙烯酸甲酯
polymethyl acrylic acid 聚甲基丙烯酸
polymethylene 聚甲烯
polymethylene diisocyanate 多亚甲基二异氰酸酯
polymethylene glycol 聚甲二醇
polymethylene polyphenylene polyisocyanate 多亚甲基多亚苯基多异氰酸酯
polymethylene polyphenyl isocyanate 多芳基多异氰酸酯
polymethylhydrosiloxane 聚甲基氢硅氧烷
polymethyl methacrylate 有机玻璃;聚甲基丙烯酸甲酯
polymethyl methacrylate base injecting paste 甲凝(注浆材料)
polymethyl methacrylate base injection paste 聚甲基丙烯酸甲酯堵漏液
polymethyl methacrylate resin 聚甲基丙烯酸甲酯树脂
polymethylmethacrylic acid 聚甲基丙烯酸
polymethyl pentene 聚甲基戊烯
polymethylphenyl siloxane fluid 苯甲基硅油
polymethyl styrene 聚甲基苯乙烯
polymicrotome 多片切片机
polymictic agglomerate 复成分集块岩
polymictic breccia 复成分角砾岩
polymictic conglomerate 复成分砾岩
polymictic lake 常混流湖;常对流湖
polymictic rock 多杂质岩;复矿碎屑岩
polymictic tuff 复成分凝灰岩
polymictic volcanic breccia 复成分火山角砾岩
polymictic welded agglomerate 复成分熔结集块岩
polymictic welded breccia 复成分熔结角砾岩
polymictic welded tuff 复成分熔结凝灰岩
polymierization agent 聚合剂
polymignite 铌铈钇钙矿
polymine 聚乙烯亚胺
polymineralic 多矿物的
polymineralic clay 复矿物黏土
polymineralic claystone 复矿物黏土岩
polymineralic rock 多矿岩;多矿物岩;复矿岩
polymineral rock 多矿物岩
polymitosis 多次有丝分裂
polymnite 树林石
polymodal current rose 多向水流玫瑰图
polymodal distribution 多众数分布;多峰分布
polymodol population 多峰总体

polymolecular 多分子的
polymolecularity 高分子性;多分子性;多分散性
polymolecular layer 多分子层
polymorph 同质异象(变)体;同质多像;多形体;多形晶;多晶型物
polymorphic 多种组合形式的;多形体系;多态的;多晶型的
polymorphic change 多形性变化
polymorphic code 多形代码
polymorphic colony 多态群体;多态菌落
polymorphic form 多晶型物;多晶型
polymorphic inversion 多晶转变
polymorphic modification 同质多像变体;多形体
polymorphic programming language 多形程序设计语言
polymorphic substance 同质多晶型物;多晶型物
polymorphic system 多形系统
polymorphic transformation 多形性转变;多晶转变;多晶型转变
polymorphic transition 多晶型转变
polymorphic transition temperature 同质多像转变温度
polymorphism 同质多晶(现象);多型现象;多形性;多形现象;多形变态;多态性;多态现象;多色性;多晶型现象;多晶型;多晶现象
polymorphous 多形的;多晶型的
polymorphous layer 多形层
polymorphy 多晶型现象
polynary 多元的
polynary system 多元系统;多元系
polyneopentyl glycol adipate 聚新戊二醇己二酸酯
polynia 冰沼湖;冰湖的
polynite 多泥石
polynitrobenzene 聚硝基苯
polynomial 多项式
polynomial adjustment 多项式平差
polynomial and special function program(me) 多项式及特殊函数程序
polynomial approximation 多项式近似值(法);多项式近似法;多项式近似;多项式逼近
polynomial approximation method 多项式逼近法
polynomial arithmetic 多项式计算
polynomial checksum 多项式检查和
polynomial code 多项式代码
polynomial coefficient 多项式系数
polynomial complexity 多项式复杂性
polynomial complexity algorithm 多项式复杂性算法
polynomial computer 多项式计算机
polynomial continuous game 多项式连续对策
polynomial correction 多项式改正
polynomial counter 多项式计数器
polynomial curve 多项式曲线
polynomial curve fitting 多项式曲线拟合
polynomial deflation 多项式降价
polynomial degree 多项式的次数
polynomial distribution 多项式分布
polynomial equation 多项式方程
polynomial equation solver 代数方程解算装置
polynomial error model 多项式误差模型
polynomial expansion 多项式展开
polynomial expression 多项式
polynomial extension 多项式扩展
polynomial factorization 多项式因式分解
polynomial fitting 多项式拟合法
polynomial form 多项式形式
polynomial formulation 列多项式
polynomial function 多项式函数
polynomial hashing 多项式散列
polynomial in several elements 多元多项式
polynomial interpolating function 多项式插值函数
polynomial interpolation 多项式内插法;多项式插值法
polynomial-like function 拟多项式函数
polynomial method 多项式法
polynomial model equation 模型多项式方程
polynomial modeling 多项式模拟
polynomial regression 多项式回归
polynomial sort 多步分类
polynomial surface 多项式曲面
polynomial syntax 多项式文法
polynomial with non-negative coefficients 非负系数多项式
polynuclear 多环的

polynuclear aromatic hydrocarbon 多核芳香化合物
polynuclear compound 多环化合物
polynuclear hydrocarbon 多核化合物
polynuclear metal complex 多核金属化合物
polynuclear phenol-glycidyl ether derived resin 多环酚缩水甘油醚衍生树脂
polynucleated city 多核心城市;多中心城市
polynya 冰沼湖;冰穴;冰间湖
polyodontia 多牙
polyoestrous 多求偶期的
polyol blend 混合多元醇
polyolefin 聚烯烃
polyolefin acid 聚烯酸
polyolefine 聚烯烃树脂
polyolefin pipe 聚烯烃管
polyolefin plastics 聚烯烃塑料
polyolefin plastomer 聚烯烃塑性体
polyolefin resin 聚烯烃树脂;聚烯树脂
polyolefins 聚烯烃类
polyolphosphonate 聚多元醇膦酸酯
polyol resins 多羟基树脂(类)
polyorganosiloxane 聚有机硅氧烷
polyorganostannosiloxane 聚有机锡硅氧烷
polyorganotitanosiloxane 聚有机钛硅氧
polyoxybiontic 多酸性的
polyoxyethylene 聚氧乙烯
polyoxyethylene alcohol 聚氧化乙烯醇
polyoxyethylene alkyl amine 聚氧乙烷基胺;聚氧化亚乙基烷基胺
polyoxyethylene alkyl ether 聚氧乙烷基醚;聚氧化亚乙基烷基醚
polyoxyethylene glycol 聚乙二醇
polyoxyethylene lauryl ether 聚氧乙烯月桂基醚
polyoxyethylene methylene 聚氧甲烯
polyoxyethylene octyl phenyl ether 聚氧乙烯辛基苯醚
polyoxyethylene oxide 聚氧化乙烯
polyoxyethylene type surfactant 聚氧乙烯型表面活性剂
polyoxymethylene 聚甲醛
polyoxymethylene fiber 聚甲醛纤维
polyoxyphenylene coating 聚苯氧涂料
polyoxypropylene glycol 聚丙二醇
polyoxypropylene plastics 聚甲醛塑料
polyoxytetramethylene glycol 聚丁二醇
polyp 珊瑚虫
polypbasic 多相的
polypedon 土壤群体;多立体剖面
polyphagous animal 多食动物;多全性动物
polyphase 多相(的)
polyphase alternator 多相交流发电机;多相交流电机
polyphase asynchronous motor 多相异步电动机
polyphase circuit 多相电路
polyphase commutator machine 多相整流子电机
polyphase commutator motor 多相换向器电动机
polyphase compound commutator motor 多相复励换向器电动机
polyphase converter 多相变流机;多相换流机
polyphase coupling system 多相耦合方式
polyphase current 多相电流
polyphase current vibrator 多相电流振动器
polyphase deflecting coil 多相偏转线圈
polyphase dispersion 多相分散
polyphase equilibrium 多相平衡
polyphase flow 多相流(动)
polyphase freezer 双相冷却致冷机
polyphase generator 多相发电机
polyphase induction motor 多相感应电动机
polyphase induction regulator 多相电感调节器
polyphase inverter 多相变流器
polyphase kilo-watthour meter 多相电度表
polyphase liposomine 多相油酯质体
polyphase merging 多步合并
polyphase meter 多相仪表;多相计数器;多相电度表;多相测试仪表
polyphase microinstruction 多步微指令
polyphase microstructure 多相显微组织
polyphase motor 多相电动机
polyphase power 多相功率;多相电力
polyphase power factor meter 多相功率因数表
polyphase reaction 多相反应

polyphase rectifier 多相整流器
polyphase rotary convertor 多相旋转变流器;多相旋转变流机
polyphase series commutator motor 多相串励换向器电动机;多相串激换向器电动机
polyphase series motor 多相串激电动机
polyphase shunt commutator motor 多相并励换相器电动机;多相并激换向器电动机
polyphase shunt motor 多相并励电动机;多相并激电动机
polyphase sinusoid 多相正弦波
polyphase solid rotor reluctance machine 多相固态转子磁阻电机
polyphase sort(ing) 多相分类;多步排序
polyphase structure 多相组织
polyphase symmetric(al) system 多相对称系统
polyphase synchronous generator 多相同步发电机
polyphase system 多相制;多相系(统);多相(体)系
polyphase transformer 多相变压器
polyphase watthour meter 多相瓦特小时计
polyphase wattmeter 多相瓦特计
polyphase winding 多相绕组
polyphasic orogenic cycle 复相造山旋回
polyphenol 多酚
polyphenol pharmacy wastewater 多酚制药废水
polyphenyl 聚苯
polyphenyl chloride compound 聚氯苯基化合物
polyphenylene coating 聚苯醚涂
polyphenylene ethyl 聚乙基苯
polyphenylene methyl 聚甲基苯
polyphenylene oxide 聚亚苯基醚;聚苯醚;聚苯撑氧
polyphenylene resin 聚苯树脂
polyphenylene sulfide 聚亚苯基硫醚;聚苯硫醚
polyphenyl ether 聚苯醚
polyphenyl ether resin 聚苯醚树脂
polyphenylmethylsiloxane 聚苯甲基硅氧烷
polyphenylsiloxane 苯基聚硅氧烷
polyphosphate 聚磷酸盐;多磷酸盐
polyphosphate accumulating bacteria 聚磷酸盐积累菌
polyphosphate accumulating organism 聚磷酸盐生物;聚磷酸盐积累生物
polyphosphate alumin(i)um ferric sulfate 聚磷硫酸铝铁
polyphosphate bacteria 聚磷酸盐菌
polyphosphate ester 聚磷酸酯
polyphosphate exchanger 聚磷酸盐交换剂
polyphosphate ferric sulfate 聚合磷硫酸铁
polyphosphate granule 聚磷酸盐颗粒
polyphosphate sodium 聚磷酸钠
polyphosphoric acid 多磷酸;多聚磷酸
polyphyletic evolution 多系列演化
polyphyletic theory 多元论
polyphyletism 多元论
polyphyodont 多换性齿
polypide 虫体
polyplant 聚合装置
polyploid 多倍体;多倍的
polyploidy 多倍性
polypods 多脚混凝土预制块(护岸用)
polypole induction motor 多极感应电动机
poly-pollution zone indicating organism 多污指示带种
polypore filter 多孔过滤器
polypots 塑料钵
poly-p-phenylene 聚苯
poly process 多变过程
polyprocessor 多重信息处理机
polyprocessor reticulum 多处理机交叉
polyprocessor system 多处理机系(统)
polyproplylene-fibre carpet 丙纶地毯
polypropylene 聚丙烯
polypropylene adipate 聚己二酸丙二醇酯
polypropylene-asphalt membrane 聚丙烯沥青面层(战地机场临时用)
polypropylene coated polypropylene fabric 聚丙烯涂层聚丙烯织物
polypropylene corrugated pipe 聚丙烯波纹管
polypropylene fiber 聚丙烯纤维
polypropylene fiber concrete 聚丙烯纤维混凝土
polypropylene fiber reinforced concrete 聚丙烯纤维增强混凝土
polypropylene fibre reinforced cement 聚丙烯纤维增强水泥

polypropylene fibrillated film fibre 聚丙烯薄膜纤化纤维
polypropylene film 聚丙烯胶片；聚丙烯薄膜
polypropylene glycol 聚丙二醇
polypropylene glycol ether 聚丙二醇醚
polypropyleneimine 聚丙烯亚胺
polypropylene logic array 聚丙烯金属
polypropylene oxalate 聚丙烯草酸
polypropylene oxide 聚氧化丙烯；聚环氧丙烷
polypropylene pipe 聚丙烯管
polypropylene plastics 聚丙烯塑料
polypropylene-plywood laminate 聚丙烯—胶合板层合板
polypropylene reinforced concrete 聚丙烯丝混凝土
polypropylene rope for ship 船用聚丙烯绳
polypropylene sheeting 聚丙烯软片；聚丙烯薄膜
polypropylene water reservoir 聚丙烯水库
polyprotonic acid 多元酸
Polypus 水螅体
polypyrimidone quinazolone 聚嘧啶酮—喹唑啉酮
polypyromellitimide 聚均苯四甲酰亚胺
polypyrrole 聚吡咯
polyquinazoline dione 聚喹唑啉二酮
Polyrad 包立拉德
polyranger 多量程仪表
polyreaction 聚合反应
polyrod aerial 介质天线
polyrod antenna 介质天线
polysaccharide 多聚糖
polysalt 高分子盐
polysaprobe 污水和聚丙烯船用缆绳；污水腐生生物
polysaprobic 重污水的；多污水腐生的
polysaprobic river 多污染河流
polysaprobic system 多污水腐生系统
polysaprobic zone 高度污染带；多污生物带；多污带；多腐生带
polysemy 多义性
polysilcate ferric alumin(i)um chloride 聚合硅酸氯化铝铁
polysilic alumin(i)um chloride 聚合硅酸氯化铝
polysilic alumin(i)um ferric chloride 聚合硅酸氯化铝铁
polysilic alumin(i)um ferric sulfate 聚合硫酸铝硅铁
polysilic aluminum ferric containing boron 含硼聚合硅酸铝铁
polysilicate 一聚硅盐酸；聚硅酸盐
polysilicate ferric chloride 聚合硅酸氯化铁
polysilicate flocculant 聚硅酸絮凝剂
polysilicic acid 聚硅酸
polysilicic acid-polyferric chloride flocculant 聚硅酸聚氯化铁絮凝剂
polysilicic acid with megnetsium and ferric i-rons 聚合硅酸镁铁
polysilicon 多晶硅
polysilicone 有机硅聚合物
polysilicone rubber 聚硅橡胶
polysilicones 聚硅类
polysilicon ferric sulfate 聚硅酸铁
polysilicon ferric sulfate flocculant 聚硅硫酸铁絮凝剂
polysilicon gate 多晶硅栅极
polysiloxane 聚硅氧烷
polysiloxane alumin(i)um soap grease 聚硅醚铝皂润滑脂
polysiloxane rubber 硅橡胶；聚硅氧烷橡胶
polysilsesquioxane 聚倍半硅氧烷
polysiphonous 多管的
Polysius double stream counter flow preheater 双逆流预热器
polysize specimen 多粒径试样
polysleeve 多路的
polysography 重复造影术
polysomy 多体性
polyspast 复式滑车；复滑车；滑车组
polyspast set of pulley 滑车组
polyspeed 均匀调节速度；多速度的；多速的
polyspermous fruit 多子果
polystachous plant 多穗状花植物
polystage 多级的
polystage amplifier 多级放大器
polystation Doppler tracking system 多站多普勒跟踪系统
polystele 多体中柱

polystenohaline 狭多盐生物
polystep reaction 多步反应
polyster synthetic(al) lubricant 聚酯合成润滑剂
polystomatous 多口的
polystream 多氯苯
polystylar edifice 多柱式建筑
polystyle 多柱式建筑；多柱式(的)
polystyle hall 多柱厅
polystyrene 聚苯乙烯；苯乙烯树脂
polystyrene acrylonitrile 聚苯乙烯丙烯腈
polystyrene board 聚苯乙烯板
polystyrene butadiene 聚苯乙烯丁二烯
polystyrene ceiling board 聚苯乙烯天花板
polystyrene ceiling tile 聚苯乙烯天花板
polystyrene concrete 聚苯乙烯膨珠混凝土；聚苯乙烯混凝土
polystyrene copolymer 聚苯乙烯共聚物
polystyrene core panel 聚苯乙烯填心板
polystyrene diffusion broadening 聚苯乙烯扩散展宽
polystyrene film 聚苯乙烯软片；聚苯乙烯薄膜；聚苯乙烯膜
polystyrene film capacitor 聚苯乙烯薄膜电容器
polystyrene foam 聚苯乙烯泡沫；发泡聚苯乙烯
polystyrene foam board for thermal insulation purpose 隔热用泡沫聚苯乙烯板
polystyrene foam cladding panel 聚苯乙烯泡沫覆盖板
polystyrene foam concrete 聚苯乙烯泡沫混凝土
polystyrene foam tile 泡沫聚苯乙烯砖
polystyrene-impregnated concrete 聚苯乙烯注入混凝土
polystyrene insulant 聚苯乙烯绝热材料；聚苯乙烯绝缘材料
polystyrene insulator 聚苯乙烯绝缘体
polystyrene latex 聚苯乙烯胶乳
polystyrene paint 聚苯乙烯涂料
polystyrene plastics 聚苯乙烯塑料
polystyrene plywood laminate 聚苯乙烯胶合板
polystyrene profile 聚苯乙烯纵剖面；聚苯乙烯纵断面
polystyrene resin 聚苯乙烯树脂
polystyrene sheet 聚苯乙烯片；聚苯乙烯板
polystyrene tile 聚苯乙烯空心砖；聚苯乙烯瓦管；聚苯乙烯瓦片
polystyrene tile adhesive 聚苯乙烯瓦涂层胶黏剂
polystyrene trim 聚苯乙烯门窗装饰
polystyrene void former 聚苯乙烯空心模
polystyrene wall tile 聚苯乙烯墙面面砖；聚苯乙烯墙面贴砖
polysuccinimide 聚琥珀酰亚胺
polysulfide 聚硫化物；聚硫；多硫化物
polysulfide base 多硫化物基
polysulfide coating 聚硫化物涂层；聚硫涂料；多硫化合物敷层
polysulfide cooking 多硫化物蒸煮
polysulfide liquid polymer 聚硫液体聚合物；多硫化物液体聚合物
polysulfide polymer 多硫化物聚合物
polysulfide process 硫合法
polysulfide rubber 硫化橡胶；硫合橡胶；聚硫橡胶
polysulfide synthetic rubber 多硫化物合成橡胶
polysulfide treating 多硫化物处理
polysulfonamide 聚砜酰胺
polysulfone 聚砜
polysulfone resin 聚砜树脂
polysulfur nitride 聚氮化硫
polysulphide 多硫化物
polysulphide-based rubber 多硫化合物橡胶
polysulphide-based rubber building mastic 聚硫胶建筑码琋脂
polysulphide-based sealant 硫化橡胶密封胶
polysulphide liquid polymer based sealant 聚硫基密封剂
polysulphide sealant 聚硫嵌缝膏；聚硫密封胶；多硫化物填封料
polysulphide sealant extruder 聚硫化物密封挤出机
polysulphide synthetic(al) rubber 聚硫合成橡胶
polysuperficiality 多面性
polysuspensoid 聚悬胶(体)
polysynthetic(al) twin 聚片双晶；复合双晶

polysynthetic(al) twinning 多合孪生
polytechnic 工艺学校
polytechnic college 工业大学
polytechnic exhibition 工业展览馆
poly-technic school 中等专科学校
polytene chromosome 多线染色体
polyterephthalate 聚对苯二甲酸酯
polyterephthalate plastics 聚对苯二甲酸酯塑料
polyterpene 多萜(烯)
polytetrafluoroethylene 聚四氟乙烯
polytetrafluoro ethylene coated glass fabric 聚四氟乙烯涂层玻纤织物
polytetrafluoro ethylene external mo(u)ld-release agent 聚四氟乙烯外脱膜剂
polytetrafluoro ethylene fibre 聚四氟乙烯纤维
polytetrafluoroethylene film 聚四氟乙烯薄膜
polytetrafluoroethylene gasket 聚四氟乙烯垫圈
polytetrafluoro ethylene packing 聚四氟乙烯盘根
polytetrafluoro ethylene plastic-lined ferrous metal pipe 聚四氟乙烯塑料衬黑色金属管
polytetrahydrofuran 聚四氢呋喃
polytetramethylene glycol 聚丁二醇
polytetramethylene hexamethylene urethane 聚亚己基氨基甲酸四亚甲酯
polytetramethylene oxide 聚四氢呋喃
polytetramethylene terephthalate 聚对苯二甲酸丁二醇酯
Poly-Tex pretreatment equipment 波莱—特克斯式预处设备
polythalamous 多室的
polythene 聚乙烯和聚丙烯船用缆绳；聚乙烯
polythene bagged plant 塑料袋育苗
polythene bathroom gulley 聚乙烯浴室上落管
polythene bathroom inlet 聚乙烯浴室进水口
polythene bathroom outlet 聚乙烯浴室出水口
polythene building sheet 聚乙烯建筑板材
polythene film 聚乙烯薄膜
polythene pipe 聚乙烯管
polythene protection film 聚乙烯保护膜
polythene rope 乙纶绳
polythene sheet 聚乙烯薄膜
polythene underlay roll 聚乙烯薄膜卷铺垫衬
polythioamide 聚硫酰胺
polythioester 聚硫酯
polythioether 聚硫醚
polythiol 多硫醇
polythionate 连多硫酸盐
polythionate process 连多硫酸盐法
polythionic acid 连多硫酸
polythiourethane 聚硫代氨基甲酸酯
polythylene plastics 苯乙烯塑料
poly tomography 复式体层照相术
polytope 可剖分空间；多面体
polytrifluoro chloro ethylene 聚三氟氯乙烯
polytrifluoromonochlorethylene 聚三氟一氯乙烯
polytrifluorostyrene 聚三氟苯乙烯
polytrimethylene adipate 聚己二酸亚丙基
polytrimethylene sebacate 聚癸二酸亚丙基酯
polytron 多稳元件
polytrope 多方球；多层球；多变性；多变曲线
polytrophic 多滋的
polytropic(al) 多向的；多变的
polytropic(al) atmosphere 多元大气
polytropic(al) change 多元变化；多向变化
polytropic(al) compression 多向压缩；多变压缩
polytropic(al) compression curve 多变压缩曲线
polytropic(al) core 多方球心
polytropic(al) curve 多变循环曲线；多变的曲线
polytropic(al) cycle 多变循环
polytropic(al) efficiency 多变效率
polytropic(al) equilibrium 多方平衡
polytropic(al) expansion 多样膨胀
polytropic(al) exponent 多变指数
polytropic(al) gas sphere 多方气体球
polytropic(al) head 多变压头
polytropic(al) index 多方指数
polytropic(al) model 多方模型
polytropic(al) process 多方过程；多变过程
polytropic change 体积压缩变化
polytropic index 体积压缩指数
polytropism 同质多晶型；同质多晶(现象)；多变性
polytropy 多变现象
polytype 多种类型的；多型；多晶型物
poly-type bucket 多瓣抓斗

poly-type grab 多瓣抓斗
polytypism 多型性;多型现象
poly-unit 聚合装置;叠合装置
polyunsaturated acid 多不饱和酸
polyunsaturated fat 多不饱和脂肪
polyunsaturated fatty acid 多不饱和脂肪酸
polyunsaturated oil 多不饱和油
polyunsaturation reaction 多不饱和反应
polyurea(s) 聚脲(类)
polyurethan(e) 聚氨酯;聚氨基甲酸乙酯
polyurethan(e) adhesive 聚氨酯黏合剂
polyurethan(e) base 聚氨酯甲酸乙酯
polyurethan(e) bonding adhesive 聚氨酯甲酸乙酯胶黏剂
polyurethan(e) cement 聚氨酯甲酸乙酯油灰
polyurethan(e) clearcolle 聚氨酯甲酸乙酯覆盖清漆
polyurethan(e) coated polyester fabric 聚氨酯涂层聚酯织物
polyurethan(e) coated polypropylene fabric 聚氨酯涂层聚丙烯织物
polyurethan(e) coating 聚氨酯涂料;聚氨酯敷层;聚氨基甲酸乙酯涂料
polyurethan(e) curative 聚氨酯固化剂
polyurethan(e) foamed cellular 聚氨基甲酸酯泡沫塑料
polyurethan(e) foam unit 聚氨基甲酸酯泡沫塑料块
polyurethan(e) extruder 聚氨酯挤出装置
polyurethan(e) fiber 聚氨酯纤维;聚氨基甲酸酯纤维
polyurethan(e) film 聚氨酯薄膜
polyurethan(e) finish 聚氨酯罩面漆;聚氨酯面漆;聚氨基甲酸乙酯面漆
polyurethan(e) floor cover(ing) 聚氨酯甲酸乙酯地板材料;聚氨酯甲酸乙酯楼板覆盖层
polyurethan(e) floor(ing) sealant 聚氨酯甲酸乙酯地板密封胶
polyurethan(e) flooring sealing material 聚氨酯地板嵌缝材料
polyurethan(e) foam 聚氨酯泡沫(塑料);聚氨脂(类)泡沫;聚氨基甲酸酯泡沫
polyurethan(e) foam board 泡沫聚氨酯板;聚氨酯泡沫板
polyurethan(e) foam pig 聚氨酯泡沫清管器
polyurethan(e) foam plastics 聚氨酯泡沫塑料
polyurethan(e) foam scraper 聚氨酯泡沫清管器
polyurethan(e) foam sheet 泡沫聚氨酯板;聚氨酯泡沫板
polyurethan(e) foam strip 泡沫聚氨酯条;聚氨酯泡沫条
polyurethan(e) foam unit 聚氨酯泡沫装置
polyurethan(e) grouting 聚氨酯类浆材
polyurethan(e) hard foam 聚氨酯甲酸乙酯硬泡沫
polyurethan(e) insulating core 聚氨酯隔声芯体;聚氨酯隔热芯体;聚氨酯绝缘芯体
polyurethan(e) insulating material 聚氨酯隔声材料;聚氨酯隔热材料;聚氨酯绝缘材料
polyurethan(e) lacquer 聚氨酯喷漆;聚氨酯清漆
polyurethan(e) latex 聚氨酯胶乳
polyurethan(e) mass 聚氨酯块
polyurethan(e) material 聚氨酯材料
polyurethan(e) paint 聚氨酯涂料;聚氨酯树脂漆
polyurethan(e) plastics 聚氨酯塑料
polyurethan(e) poured-in-place thermal break material 聚氨酯现场浇筑隔热材料
polyurethan(e) putty 聚氨酯腻子
polyurethan(e) resin 聚氨酯树脂;聚氨基甲酸乙酯树脂
polyurethan(e) resin adhesive 聚氨酯树脂胶合;聚氨基甲酸酯胶黏剂
polyurethan(e) resin coating 聚氨基甲酸酯树脂涂料
polyurethan(e) resin paint 聚氨基甲酸乙酯树脂
polyurethan(e) rigid foam 聚氨酯硬泡沫板
polyurethan(e) rigid foam building board 聚氨酯硬泡沫板材
polyurethan(e) rigid foam laminated board 聚氨酯硬泡沫胶合板材
polyurethan(e) sealant 聚氨酯密封剂;聚氨酯嵌缝膏;聚氨酯嵌缝材料;聚氨酯密封膏
polyurethan(e) sealer 聚氨酯封接
polyurethan(e) strip 聚氨酯条
polyurethan(e) surface coating 聚氨酯饰面层

polyurethan(e) surface coating material 聚氨酯(表面)涂料
polyurethan(e) thermoplastic elastomer 聚氨酯热塑性弹性体
polyurethan(e) thermoplastic non-elastomer 聚氨酯热塑性非弹性体
polyurethan(e) two-part bonding adhesive 聚氨酯两面黏接
polyurethan(e) varnish 聚氨酯清漆
polyurethan(e) waterproofing paint 聚氨酯防水材料
polyvalence 多价;多方面相关性
polyvalent 多价的
polyvalent alcohol 多元醇
polyvalent metal cation 多价金属阳离子
polyvalent notation 多价表示法
polyvalent number 多价数
polyvariant system 多变系
polyversity 综合大学
polyvingl acetate emulsion 聚醋酸乙烯乳液
polyvino 塑性花砖地面
polyvinyl 聚乙烯化合物的;聚乙烯
polyvinyl acetal 聚乙烯醇缩醛;聚乙烯醇缩乙醛
polyvinyl acetal resin 聚乙烯醇缩醛树脂
polyvinyl acetals 聚乙烯醇缩醛类
polyvinyl acetate 聚醋酸乙烯纤维素(一种混凝土外加剂);聚乙酸乙烯酯;聚醋酸乙烯酯;聚醋酸乙烯
polyvinyl acetate adhesive 聚醋酯乙烯酯胶黏剂;聚醋酯乙烯黏结剂;聚醋酯乙烯黏合剂
polyvinyl acetate bonding agent 聚醋酯乙烯黏合剂
polyvinyl acetate cement 聚乙酸乙烯水泥;聚醋酸乙烯水泥;聚醋酸乙烯水泥
polyvinyl acetate concrete 聚醋酯乙烯酯混凝土
polyvinyl acetate emulsion 聚醋酯乙烯酯乳液
polyvinyl acetate emulsion adhesive 聚乙酸乙酯胶黏剂乳液
polyvinyl acetate emulsion-cement mortar 聚醋酸乙烯乳液水泥砂浆
polyvinyl acetate emulsion paint 聚醋酸乙烯酯乳化漆
polyvinyl acetate floor cover(ing) 聚醋酸乙烯酯地板面层
polyvinyl acetate glue 聚醋酸乙烯酯胶
polyvinyl acetate latex 聚乙酸乙烯酯乳胶
polyvinyl acetate mortar 聚醋酯乙烯酯砂浆
polyvinyl acetate paint 聚醋酸乙烯涂料
polyvinyl acetate resin adhesive 聚乙烯酯乳液胶;聚乙酸乙烯树脂黏合剂;聚醋酸乙烯树脂胶
polyvinyl acetate suspension 聚醋酸乙烯酯悬浮液
polyvinyl alcohol 聚乙烯醇
polyvinyl alcohol concentration 聚乙烯醇浓度
polyvinyl amine 聚乙烯胺
polyvinyl bonding medium 聚乙烯胶合剂
polyvinyl bromide 聚溴乙烯
polyvinyl butyral 聚乙烯缩丁醛;聚乙烯醇缩丁醛
polyvinyl butyral resin 聚乙烯醇缩丁醛树脂
polyvinyl carbazole resin 聚乙烯基咔唑树脂
polyvinyl chloride[PVC] 聚氯乙烯
polyvinyl chloride acetate 聚氯乙烯酯;聚氯乙烯醋酸酯
polyvinyl chloride alloy 聚氯乙烯合金
polyvinyl chloride asbestos sheet 聚氯乙烯石棉卷材
polyvinyl chloride base 聚氯乙烯基
polyvinyl chloride board 聚氯乙烯板
polyvinyl chloride building board 聚氯乙烯建筑板材
polyvinyl chloride cable 聚氯乙烯绝缘电缆;聚氯乙烯电缆
polyvinyl chloride carpet 聚氯乙烯地毯
polyvinyl chloride ca(u)lking compound 聚氯乙烯接缝材料
polyvinyl chloride ceiling 聚氯乙烯顶棚;聚氯乙烯塑料天花板
polyvinyl chloride ceiling profile 聚氯乙烯顶棚型材
polyvinyl chloride ceiling trim 聚氯乙烯顶棚镶边条
polyvinyl chloride cement 聚氯乙烯胶泥
polyvinyl chloride coal tar pitch roofing 聚氯乙烯煤焦油沥青油毡;聚氯乙烯煤沥青毡
polyvinyl chloride coal tar roofing felt 聚氯乙烯煤焦油油毡

polyvinyl chloride coated fabric 聚氯乙烯涂层织物
polyvinyl chloride coating 聚氯乙烯饰面层
polyvinyl chloride conduit 聚氯乙烯管
polyvinyl chloride construction section 聚氯乙烯结构型材
polyvinyl chloride corer 聚氯乙烯取样管
polyvinyl chloride corrugated board 聚氯乙烯波纹板
polyvinyl chloride corrugated sewer pipe 聚氯乙烯波纹污水管
polyvinyl chloride cover(ing) 聚氯乙烯面层
polyvinyl chloride door and window 聚氯乙烯门窗
polyvinyl chloride drain pipe 聚氯乙烯排水管
polyvinyl chloride duct 聚氯乙烯线槽
polyvinyl chloride fabric 聚氯乙烯织物
polyvinyl chloride film 聚氯乙烯软片;聚氯乙烯薄膜
polyvinyl chloride filter media 聚氯乙烯滤布
polyvinyl chloride floor finish 聚氯乙烯地板面层
polyvinyl chloride flooring tile 聚氯乙烯铺地砖
polyvinyl chloride floor sheet 聚氯乙烯地面卷材
polyvinyl chloride floor tile 聚氯乙烯地板块
polyvinyl chloride flower box 聚氯乙烯花盆
polyvinyl chloride flower trought 聚氯乙烯花槽
polyvinyl chloride foam 泡沫聚氯乙烯;聚氯乙烯泡沫
polyvinyl chloride gasket 聚氯乙烯垫带
polyvinyl chloride gasketed sewer fitting 聚氯乙烯带垫污水管件
polyvinyl chloride heat stabilizer 聚氯乙烯热稳定剂
polyvinyl chloride hollow section 聚氯乙烯空心型材
polyvinyl chloride hollow trim 聚氯乙烯空心镶边条
polyvinyl chloride hollow unit 聚氯乙烯空心构件
polyvinyl chloride insulated sheathed copper core power cable 聚氯乙烯绝缘及护套铜芯电力电缆
polyvinyl chloride insulating film 聚氯乙烯绝缘膜
polyvinyl chloride insulation grade film 聚氯乙烯绝缘膜
polyvinyl chloride large-diameter plastic gravity sewer pipe 聚氯乙烯大直径重力污水管
polyvinyl chloride organosol 聚氯乙烯有机溶胶
polyvinyl chloride panel 聚氯乙烯板
polyvinyl chloride paste 聚氯乙烯糊
polyvinyl chloride pipe 聚氯乙烯管;PVC 管
polyvinyl chloride plastic drain water pipe 聚氯乙烯塑料排水管
polyvinyl chloride plastic pipe 聚氯乙烯塑料管
polyvinyl chloride plastics 聚氯乙烯塑料
polyvinyl chloride plastics torch 聚氯乙烯塑料气炬
polyvinyl chloride plastisol 聚氯乙烯塑性溶胶
polyvinyl chloride pressure pipe 聚氯乙烯压力管
polyvinyl chloride pressure rated pipe 聚氯乙烯压力级管
polyvinyl chloride prime window 聚氯乙烯主窗
polyvinyl chloride profile 聚氯乙烯型材
polyvinyl chloride rainwater system 聚氯乙烯雨水系统
polyvinyl chloride resin 聚氯乙烯树脂
polyvinyl chloride ribbed gravity sewer pipe 聚氯乙烯带肋重力污水管
polyvinyl chloride roofing sheet 聚氯乙烯屋面卷材
polyvinyl chloride roof tile 聚氯乙烯瓦
polyvinyl chloride rubber 聚氯乙烯橡胶
polyvinyl chloride sealing sheeting 聚氯乙烯密封板
polyvinyl chloride section 聚氯乙烯型材
polyvinyl chloride shape 聚氯乙烯型材
polyvinyl chloride sheathed cable 聚氯乙烯铠装电缆
polyvinyl chloride sheet 聚氯乙烯片材;聚氯乙烯片;聚氯乙烯卷材
polyvinyl chloride sheeting 聚氯乙烯薄板
polyvinyl chloride sheet roofing 聚氯乙烯片材屋面
polyvinyl chloride-tar mixture 聚氯乙烯焦油混合物
polyvinyl chloride thermal insulation 聚氯乙烯隔热层
polyvinyl chloride tile 聚氯乙烯瓦
polyvinyl chloride trim 聚氯乙烯镶边条
polyvinyl chloride unit 聚氯乙烯构件

polyvinyl chloride-vinyl acetate 聚氯乙烯乙酸乙烯酯
polyvinyl chloride wall board 聚氯乙烯墙板
polyvinyl chloride wall sheet 聚氯乙烯墙板
polyvinyl chloride waterbar 聚氯乙烯止水条
polyvinyl chloride waterstop 聚氯乙烯止水条;聚氯乙烯止水片;聚氯乙烯止水带
polyvinyl chloride window shape 聚氯乙烯窗型材
polyvinyl chloride window unit 聚氯乙烯窗构件
polyvinyl dichloride 聚二氯乙烯
polyvinyl ether 聚乙烯基醚
polyvinyl ethyl ether 聚乙烯基乙基醚
polyvinyl fiber wastewater 维纶纤维废水
polyvinyl fluoride 聚氟乙烯;聚氟乙丙
polyvinyl formal 聚乙烯缩甲醛;聚乙烯醇缩甲醛
polyvinyl formate 聚甲酸乙烯酯
polyvinyl furan resin 聚乙烯呋喃树脂
polyvinyl glue 聚乙烯胶
polyvinylidene chloride 聚偏氯乙烯;聚偏二氯乙烯
polyvinylidene chloride fiber 聚偏二氯乙烯纤维
polyvinylidene fluoride 聚偏氟乙烯
polyvinylidene fluoride asymmetric(al) membrane 聚偏氟乙烯非对称膜
polyvinylidene fluoride hollow fibre membrane 聚偏氟乙烯中空纤维膜
polyvinylidene fluoride membrane bioreactor 聚偏氟乙烯膜生物反应器
polyvinylidene fluoride memebrane 聚偏氟乙烯膜
polyvinylidene fluoride microporous membrane 聚偏氟乙烯微孔膜
polyvinylidene fluoride plastic-lined ferrous metal pipe 聚偏氟乙烯塑料衬黑色金属管
polyvinylidene fluoride resin 聚偏氟乙烯树脂
polyvinylidene fluoride ultrafiltration membrane 聚偏氟乙烯超滤膜
polyvinylidene resin 偏二氯乙烯树脂
polyvinyl iodide 聚碘乙烯
polyvinyl isobutyl ether 聚乙烯基异丁基醚
polyvinyl methyl ether 聚乙烯基甲基醚
polyvinyl methyl ether-maleic anhydride copolymer 乙烯基甲醚顺丁二酸酐共聚物
polyvinyl plastics 聚乙烯塑料
polyvinyl pyrrolidone 聚乙烯(基)吡咯烷酮
polyvinyl resin 聚乙烯(基类)树脂
polyvinyl sulfuric acid 聚乙烯硫酸
polyvinyl waterbar 聚乙烯止水条
polyvinyl waterstop 聚乙烯止水条
polywater 聚合水;反常水
poly-wire tire 合成纤维钢丝轮胎
polyxylene 聚二甲苯;聚对二甲苯
polyyne 聚炔烃
polyzoic 多个员的
polzenite 黄煌岩
pomegranate 石榴树;石榴【植】
pomelo 柚(子);球形顶饰
Pomet 玻密特烧结纯铁
pomet 烧结纯铁
pommel 圆头;球端;柱案;手夯;圆形顶端饰物;球形顶饰;球饰;前鞍
pommelled horse 鞍马
pomology 果树学
pomp 华丽
Pompeian architecture 庞贝建筑
Pompeian red 赭石红;庞贝红;铁红
Pompey red 铁红
pompholygometer 测气泡计
pompier 救火梯
pompier belt 带钩安全带
pompier chain 挂钩梯链(在楼房的窗口挂数架挂钩梯)
pompier ladder 救火梯
pom-pom 小型架式风钻;自动铆(钉)枪;自动枪
pompon 绒球(用于装饰)
pompous 豪华的
ponceau 朱红;丽春红染料;酸性红色染料;深红
poncelet 百千克米
Poncelet graphical construction 庞斯列图解法(土压力)
Poncelet wheel 下射曲叶水轮
pond 水塘;水坑;池塘;池
pondage 蓄水量;储[贮]水量;调蓄量;调节蓄水量;调节泄水;调节容量;水库蓄水量;池塘蓄水量
pondage action 调蓄作用
pondage capacity 蓄水能力;调节库容

pondage correction 蓄水量改正
pondage factor 调节因数(水库);调节系数(水库)
pondage limit (水库、水池的)极限调节容量;(水库、水池的)极限蓄水量
pondage method (确定土渠渗透损失的)蓄水法
pondage reservoir 调节水库;蓄水池
pondage topographic(al) map 蓄水区域图
pondage type power plant 蓄水池式发电厂;调节池式发电站
pond area 淹没区
pond culture 池塘养殖;池塘养鱼
pond ecosystem 池塘生态系统
ponded area 淹没面积;淹没地区
ponded basin 滞水盆地;滞流盆地
ponded lake 储[贮]水池;围堤造地;水库
ponded plain 堵塞平原
ponded river 堵水成河;阻塞河
ponded stream 阻塞河
ponded turbidite 堵塞浊积岩
ponded water 围堤蓄水
pond environment 水库环境
ponder 估量
ponderability 重量可称性;可称量性
ponderable 可称量的
ponderable amount 可估计量
pondermotive force 有质动力
ponderosa pine 美国黄松
ponderous fluid 有重量流体
pond filling dam 水力冲填坝
pond habitat 池塘生境
ponding 坑洼;水坑;拦坝;筑塘;泡水;蓄水;围水养护(法)(混凝土);堵水;堵塞作用
ponding area 壅水区
ponding curing 围水养护(混凝土)
ponding in biologic(al) filter 生物滤池堵塞
ponding method of curing 围水养护法
ponding method of curing concrete 混凝土养池养护法;混凝土泡水养护法
ponding of water 积水
ponding test 浸水试验
pond life 池塘生物
pond management 蓄水池管理
pond pine 晚星松;北大西洋松
pond pisciculture 池塘养鱼法
pond scum 池面浮渣
pond silt 塘泥
pond sludge 塘泥
pond tailings 池塘尾渣
pond weed 角果藻
pondy 多池塘的
pongee 充茧绸;茧绸
pongee cocoon 蓖麻茧
pongo 峡谷
ponograph 疲劳描记器
ponor 落水洞;溶井
ponsol dye 滂梭染料
ponsol red 滂梭红
ponsol yellow 滂梭黄
pontachrome 滂铬
pontamine colo(u)r 滂胺染料
pontamine diazo scarlet 滂胺偶氮猩红
pontesfordian Series 庞特斯福德统【地】
Pontian 蓬蒂阶【地】
pontianak 滂阡树脂
pontianak gum 节路顿胶;滂阡树脂胶
Ponti geosyncline 蓬蒂地槽
pontil 吹玻璃用的铁棒
pontium 深海群落
pontlevis 吊桥
pontocaine 丁卡因
pontol 仲醇和叔醇混合物
pontonier 架设浮桥
pontoon 浮码头;起重机船;平底船;趸船;潜水钟;潜水箱;浮舟;浮趸舟;浮箱;浮筒
pontoon aboard floating dock 母子船坞
pontoon and trestle 趸船及栈桥
pontoon at anchorage area 锚地趸船
pontoon barge 平底驳船
pontoon-bowed barge 方头驳船
pontoon breakwater 平底型防波堤
pontoon bridge 空驳;浮桥;舟桥
pontoon bridge horizontal closure 浮桥平堵流
pontoon causeway 浮箱栈桥;浮筒栈桥;浮筒式堤道;浮桥码头

pontoon crane 平底船(式)起重机;水上起重机;船边起重机;浮筒起重机;浮吊;驳船起重机
pontoon deck 停船甲板;底囤甲板
pontoon derrick 水上起重机;浮式起重机;浮吊
pontoon dock 组合式浮船坞;浮船坞
pontoon dock-gate 浮箱式坞门
pontoon dry dock 浮船坞
pontoon erection 浮运架设(法)
pontoon fixing 趸船码头固定
pontoon for pipeline 排泥管浮筒
pontoon girder 箱形大梁
pontoon hatch cover 箱形舱口盖;盒形舱口盖
pontoon lifeboat 水密双层底救生艇;浮箱式救生艇
pontoon manoeuvring winch 移锚绞车
pontoon-mounted concreting plant 浮船式混凝土搅拌设备;装在浮驳上的混凝土搅拌设备;水上混凝土拌和厂
pontoon on rollers 滚动箱形门
pontoon pier (head) 趸船码头;浮筒码头;浮码头
pontoon pile driver 水上打桩机
pontoon pile driving plant 水上打桩设备;水上打桩机
pontoon piling plant 打桩船
pontoon raft 箱式救生筏;救生筏
pontoon roof 浮顶
pontoon seaplane 双浮筒式水上飞机
pontoon siting 浮码选址
pontoon storage tank 浮式储[贮]罐
pontoon swing bridge 开合浮桥;旋启浮桥;开合式浮桥;浮旋桥
pontoon tank 底囤水舱
pontoon-tank roof 浮动槽顶
pontoon type barge 方驳
pontoon type bow 方形船头;方形船首
pontoon type breakwater 浮箱式防波堤;浮箱式防波堤
pontoon type grab dredge(r) 趸船型抓斗挖泥船
pontoon type hatch cover 箱形舱口盖
pontoon type vessel 方形船舶
pontoon type wharf 趸船式码头;浮式码头
pontoon type work boat 平底型工作船
pontoon under carriage 浮筒式起落架
pontoon wharf 浮码头
Pontryagin's maximum principle 庞特里雅金极大原理
pony 辅助的
pony axle 空转轴
pony beam 矮梁
pony car 小型轿车;单轨小车
pony circuit 短程电路
pony engine 小型调车机车;小火车头
pony girder 阳台横梁;矮梁;矮大梁
pony girder bridge 半穿式桥(梁)
pony hand-saw 小带锯
pony ladle 浇勺
pony mixer 换罐式混合机
pony motor 小型电动机;辅助电动机
pony packer 小直径封隔器
pony plate beam 矮板梁
pony rod 短抽油杆
pony roll 卷轴;卷线管;盘卷
pony rounder 预轧机
pony-roughing mill 中间轧机
pony-roughing pass 中间道次
pony-roughing stand of rolls 预轧机座
pony set 小棚架;小桁架
pony sill 动力机的枢格式底架
pony-size 小尺寸的
pony stand 预精轧机座
pony support 小型支架
pony-tail hair style 马尾式
pony truck 小型转向架;小车;拖车
pony truss 小桁架;矮桁架
pony truss bridge 半穿式桁架桥;矮桁架桥
pony turbine 小型透平;小型涡轮机
pood 普特(俄罗斯的质量单位,1 普特=16.38 千克)
pool 闸间河段;塘;水塘;水区;水坑;池沼;深槽(航道);储集场;池塘;备用物资储集场
pool-and-riffle sequence 滩槽序列
pool area 分级机的工作面积
pool based variable-rate tending system 一揽子浮动贷款制度
pool block 界砖;玻璃液流料槽
pool boiling 大容量沸腾;池式沸腾;池沸腾

pool capacity 沉淀池容量
pool car 合用小汽车
pool cargo 分摊承运的货物
pool cathode 液体阴极;汞弧阴极;池阴极;电弧放电液态阴极
pool-cathode tube 汞弧整流管;汞弧阴极整流管
pool conveyer 环形输送机
pool cushion 消能齿;缓冲池
pool deck 游泳池防滑面板
pool diagram 油气藏图示
pool diagram and its parameter 油气藏图示及参数
pooled 合并的
pooled buffer 联营缓冲;池形缓冲区
pooled product 汇集产品
pooled sample 有意选择的事例
pooled standard deviation 合并标准差
pooled variance 合并方差
pool effort 通力合作
pool embankment in garden 园林驳岸
pool fire 油池水
pool fish pass 水池式鱼道
pool floor 池底板
pool furnace 床炉;反射炉
pool geyser 塘式间歇泉
pool habitat 池塘生境
pool hall 游泳馆;弹子房
pooling 联营;集资经营;积水
pooling agreement 联营协议;联营协定;联合经营协议;合伙经营协议
pooling economy 联营经济
pooling function 集储功能
pooling funds 集中资金
pooling information 集合资料
pooling of capital 资金的统筹;集中资本
pooling of cross-section(al) and time-series data 截面数据与时间序列数据的合并
pooling of interest method 联营法
pooling of land 土地入股
pool level 水库水位
pooll-riffle sequence 深槽浅滩序列
pool of available nodes 可利用节点池
pool of buffer 一组缓冲
pool of channel floor 河底深槽
pool of labo(u)r power 劳动力资源
pool of mortgages 抵押集合
pool operations 操纵集团
pool pallet 通用集装箱
pool parking 合乘汽车停车处
pool pressure at starting influx 油藏开始注水时压力
pool queue 池队列
pool reach 闸间河段
pool reactor 池式反应堆
pool reinsurance 集团分保
pool room (赌场的)收赌注处;弹子房
pool scheme 联营方案;分摊方案
pool spring 深潭泉(水)
pool stage 渠化状态
Pool's tile 普尔瓦
pools zone 积水区
pool table 弹子台
pool tablet 玻璃液流料槽
pool the experience 交流经验
pool train 联营列车
pool tube 汞弧整流管
pool-type fish pass 槽式过鱼道
pool van 合用有篷货运车
pool wall hole 隐藏洞穴
Poon 彭木(一种产于印度安达曼群岛的红褐色硬木)
poop 后船楼;浪打船尾;尖锐脉冲;艉楼;尾楼【船】;船尾;冲打
poop bulkhead 尾楼舱壁;船尾楼舱壁
poop cabin 船尾楼舱室;尾楼舱室
poop deck 尾楼甲板
poop deck awning stanchion 船尾楼甲板天幕支柱
poop deck spring 尾楼甲板倒缆(靠码头用)
poop down 尾淹;船尾被浪击坏
poop framing 艉楼构架
poop front bulkhead 艉楼前舱壁;船尾楼前端舱壁
poop house 尾室
pooping 大浪冲上船尾
poop ladder 船尾梯
poop maneuvering room 船尾操舵室
poop rail 船尾楼栏杆

poop sheer strake 尾楼舷侧缘列板
poop sheet 书面声明
poop shot 风化层爆破;废炮
poop staff 船尾旗杆
poop superstructure 艉楼
poor 劣质的
poor aggregate 劣质集料;劣质骨料;不良级配集料;不良级配骨料
poor area 贫困地区
poor bearing stratum 不良承载层;软弱承载层
poor bond 黏结不良
poor boy 爪式原始岩芯钻头
poor-boy core barrel 现场用短截套管做成的岩芯管
poor casing seat 不合格套管座
poor clay 瘠黏土;贫黏土;贫瘠黏土
poor combustion 不完全燃烧;不良燃烧
poor-compactibility 压塑性不良;低成型性
poor compression 不足压缩
poor concrete 劣质混凝土;贫灰混凝土;少水泥(灰)混凝土;少灰混凝土
poor conductor 不良导体
poor conductor of heat 不良导热体
poor consolidation 固结不良
poor contact 接触不良;不良接点
poor core recovery 低岩芯采取率
poor correlation 弱相关
poor country 穷国
poor cracking stock 不易裂化的原料
poor crop 歉收;收成不佳
poor design 不良设计;设计不良
poor diamond 劣质金刚石
poor distribution 分布不均
poor drainage 排水不良;不良排水
poor efficiency 低效率;低效;低利用率
poor fit 配合不良
poor fit of switch point 尖轨不密贴
poor flow quality 流动性不良
poor focus 不良聚焦
poor gas 贫燃料气
poor geometry 不利几何条件
poor grade 劣级配
poor-graded 级配不良的
poor-graded soil 不良配土
poor grain size distribution 不良级配
poor ground 软弱地基;贫瘠地
poor growth 生长缓慢
poor hand 生手
poor harvest 歉收
poor holding ground 抓力不良锚地
poorhouse 贫民住房;养育院;贫民救济院;贫民院
poor ice content frozen soil 少冰冻土
poor insulation 不良绝缘
poor insulator 不良绝缘子
poor land 薄土;薄地
poor lighting 不良照明
poor lime 劣质石灰;劣石灰;贫石灰
poorly-compacted backfill 压实差的回填土
poorly directive antenna 弱定向天线
poorly drained soil 滞水土
poorly-drained stream basin 排水不畅流域
poorly graded 分选差的;级配不良的
poorly graded aggregate 不良级配集料;不良级配骨料;劣质级配集料;劣级配骨料
poorly graded gravel 劣级配砾石;级配差的砾石
poorly graded sand 级配差的砂;劣砂;白垩木
poorly graded soil 级配不良的土;不良配土
poorly heated 加热不好的
poorly marked 标记不明的
poorly sorted 级配不良的;分选差的
poorly sorted soil 分选性差的土
poorly-ventilated 通风不良
poor maintenance of equipment 设备维护不良
poor management 经营粗放
poor mix 贫混合物;劣质混合物
poor-mix concrete 贫配合混凝土;贫灰混凝土
poor mixing 不良混合
poor mixing ratio RCD concrete 少灰碾压混凝土
poor mixing ratio rolled concrete 贫碾压混凝土
poor mixture 劣混合物;稀混合气;劣质混合物;劣混合料;贫混合料;水泥少的混合料
poor mountains and torrential rivers 穷山恶水
poor mud 劣质泥浆
poor nutritional state 营养不良

poor of water quantity 水量贫乏
poor oil 劣质油;低致原油
poor ore 低级矿(石);粗矿(石);贫矿(石)
poor organization resulting in holding up the work 窝工
poor performance 性能差
poor pine 小松
poor products 劣品
poor quality 质量差的;劣质的
poor quality clay 劣质黏土
poor quality concrete 劣质混凝土
poor quality groundwater 劣质地下水
poor quality of the ores 贫矿
poor quality water 劣质水
poor rate 救济或资助贫民税
poor raw water 劣质原水
poor regulation 不良调节
poor river water 劣质河水
poor rolling track 难行线
poor running track 难行线
poor sales 销路不佳
poor separation 不良分离
poor soil 贫瘠土;贫瘠地
poor solvent 劣溶剂;不良溶剂
poor source rock 差烃源岩
poor's pyx (圣餐礼拜中的)奉献箱
poor stable 稳定性差
poor stage 河流枯水期
poor starting 起动不良
poor stop 制动不良
poor stratum 不良地层;不良地质
poor subgrade 劣质路基;软弱地基;不良路基
poor subsoil 劣质底土;软弱地基;软弱底土;不良地基
poor surrounding 不良环境
poor ties 离散度大读数
poor value 低品位矿石
poor visibility 能见度不良;通视不良;不良能见度
poor-water zone 贫水溢出带
poor weld 有缺陷焊缝;劣质焊缝
poor wire connection 导线接触不良
poor workmanship 工艺低劣;工程质量低劣;手艺低劣;施工质量低劣
pop 上托;二次爆破眼
pop action 突开动作
pop an empty stack 堆栈变空
pop architecture 流行建筑
pop art 流行艺术;通俗艺术
pop-blasting 小炮眼爆破;二次爆破
popcorn concrete 米花状无细骨料混凝土;米花状混凝土;多孔无砂混凝土
popcorn concrete terrace(d) house 大孔混凝土的街坊房屋
popcorn noise 跳跃噪声
popcorn polymer 米花状聚合物;端聚合物
popcorn polymerization 增链聚合(作用);米花状聚合(作用)
pop culture 通俗文化
pop down 压栈
pop gate 雨淋式浇口
pop hole 二次爆破眼;爆破孔
popholing drill 浅眼凿岩机
popin 波平
pop-in load 突加负载;进法荷载
pop instruction 上托指令
poplar 杨树;白杨树;白杨木
poplar and willow 杨柳
poplin 府绸(用于做帐帘)
pop-off 溢流冒口;出气帽口;出气口;爆点
pop-off flask 可折式砂箱;铰链式砂箱
pop-off valve 安全阀;爆um阀
pop-outs 坑穴(路面);气泡;突然爆裂;(混凝土表面上的)坑穴
pop-out windscreen 推出式挡风玻璃
Popov adjustment 波波夫平差
Popov criterion 波波夫判据
Popov line 波波夫线
Popov's condition 波波夫条件
Popov's method 波波夫方法
Popov's stability criterion 波波夫稳定性判据
popparoni 变形醋酯丝
popped out 胀ành现象
popped pearlite 轻质珍珠岩
popper 圆形疤;沸腾床的床身;爆点

poppet 下水支架;下水架;临时支柱(船舶下水时的);提头架;随转尾架;舢板舷侧桨门板;垫架
poppet foot valve 提升式底阀
poppet holes 绞盘插棒孔
poppet relief valve 提动式溢流阀
poppet stock 床尼
poppet valve 蕈线阀;圆盘阀;正动提升阀;盘杆阀;提升阀;提动阀;碟形阀
poppet-valve distribution 提升阀分配器
poppet valve gear 坠阀装置;提升阀装置
popping 木料迸裂;抹灰爆裂;糊炮二次爆破;汽船声;石灰胀裂(建筑砖缺陷);钉子挤出;爆音;爆裂孔眼;爆裂;爆孔
popping of gas 气体释放
popping pressure 阵发压力;爆发性压力;突开压力;安全阀起座压力
popping rock 岩塌石块;岩崩石块;凸出的石块;爆破飞石
popping the whip 急甩拖驳
poppy 罂粟花饰;深红色;深红的;顶花饰;罂粟花;芙蓉红
poppy head 凳饰(教堂坐凳装饰);罂粟花饰;顶花饰
poppy oil 罂粟油
poppy (seed) oil 罂粟子油
Pop rivet 波普空心铆钉;热铆铆钉
pop safety valve 快泄安全阀;紧急安全阀;急泄安全阀
pop-shoot(ing) 二次爆破;突然射出;小炮眼爆破;爆破
pop shop 当铺
pop shot 二次爆破炮眼;二次爆破眼;修边炮孔;小炮孔
pop test 突击式测验
popular 流行的
popular astronomy 大众天文学
popular ceramics 大众陶瓷
popular edition 普及版
popularity 流行
popularization 推广
popularize environmental protection knowledge 普及环保知识
popularizing agricultural technique 农业技术推广
popular pressure range 普遍压力范围
popular resort and historic spot 名胜古迹
popular science readings 科普读物
popular sink unit 大众化的洗涤盆;经济型洗涤盆
popular symbolism 通俗象征主义
popular ware 热门货(品)
populate 粒子数增加
populated 粒子数增加的
populated area 居住区;居民区
populated country 居民地区
populated place 居民点
populated place name 居民点名称
populated range 居住区
populated zone 居住区
population 组;总体;种群;占有数;粒子数;聚居群;集居数;母体;人数;人口;群体;群丛;全体;数目
population accident rate 人口事故率
population analysis 种群分析;人口分析
population attributable risk 人群归因危险度
population average 总体均值
population balance method 总体平衡法;综合平衡法
population biology 群体生物学
population carrying capacity of land according to quality 制土分民之律
population census 人口统计数字;人口普查
population change 人口变动
population characteristic 人口特征
population characteristic value 人口特征值
population composition 人口组成;人口构成
population composition by age 人口年龄构成
population concentration 人口集中
population control 种群控制;人口控制
population covariance 总体协方差
population density 种群密度;占据密度;人口密度;群体密度;栖息密度;虫口密度
population dependent on agriculture 依靠农业为生的人口
population depression 种群减退
population development 人口发展
population difference 粒子数差

population dispersal 种群疏散;居民疏散;人口疏散;群体散布
population dispersion 种群扩散;群体离散
population distribution 总体分布;人口分布
population distribution of a city 城市人口分布
population dynamics 种群力学;种群动态;居群力学;人口动态;人口动力学;群体力学;群体动态
population ecology 人口生态学
population effect 粒子数效应
population effect model 种群效应模型
population-employment sequence 人口就业序列
population environmental policy 人口环境政策
population equilibrium 种群均衡;人口平衡
population equivalence 人口当量(工厂污水量换算为人口污水量)
population equivalent 人口当量
population equivalent of industrial waste (water) 工业污水人口当量;工业废水人口当量
population estimate 种群估算;人口估算;人口估计
population expansion 人口膨胀
population experiment 群体试验
population explosion 人口骤增;人口爆炸;群体暴增
population forecast 人口预测
population geography 人口地理学
population growth 种群增长;人口增长
population growth form migration 人口机械(社会)增长
population hang-up 粒子数悬布
population-health sequence 人口健康序列
population implosion 人口内向暴增;人口内挤
population improvement 群体改良
population-income sequence 人口收入序列
population index 种群指数
population inversion 粒子数反转
population investment 人口投资
population key 人口资料
population law 人口规律
population map 人口地图
population mean 总体平均值;总体平均数;总体均值;总体均数;总平均;群体平均数
population median 总体中位数
population migration 人口迁移
population mobility 人口流动
population moment 总体矩
population movement 人口迁移
population not in labor force 非劳动人口
population of city 城市人口
population of installed appliances 在用燃具总数
population of levels 能级数
population of parameters 参数组;参数群
population of rotational levels 旋转水平的总体
population of samples 抽样总体
population parameter 总体参数;人口参数
population parameter space 总体参数空间
population percentile 人口百分位数
population pessure 群体压力
population planning 人口规划
population prediction 人口预测
population profile 人口概况
population projection method 人口预测法
population proportion 总体比例
population pyramid 人口金字塔;人口分布统计图表(如性别、年龄等)
population quality 人口素质
population quantity 种群数量
population rate 总体率
population ratio 粒子数比
population register 人口登记册
population relocated 迁移人口;动迁人口;搬迁人口
population reproduction 人口再生产
population risetime 粒子数增长时间
population sampled 被抽样的总体
population standard deviation 总体标准(离)差
population standardize rate 总体标准化率
population statistics 人口统计
population structure 人口结构
population theory 人口理论;全域理论
population threshold 粒子数阈值
population trend 人口趋势
population urbanization 人口城市化
population variance 总体方差;总离散;母体方差
population waiting for employment 待业人口
population-weighted average concentration 密度加权平均浓度

population with linear trend 具有线性趋势的总体
populous 人口稠密的
populus 杨树
populus tremula 欧洲山杨
pop-up 上托;突然升起;弹起装置
pop-up dialogue box 弹出式对话框
pop-up head 弹升式喷洒头;草地喷洒头
pop-up indicator 机械指示器
pop-up sewer opening 提起式下水口
pop-up waste 提起式下水口
pop valve 安全阀;压力容器上的安全阀;紧急阀;突开式安全阀;突开阀
poratable brushcutter 轻便除草机
P/O ratio 磷氧比值
porcelain 瓷质;瓷料;瓷
porcelain accessory 瓷配件
porcelain alloy cutter 瓷合金刀具
porcelain antenna insulator 瓷天线绝缘子
porcelain-arm insulator 瓷横担绝缘子
porcelain ball 瓷球
porcelain ball for graining machine 磨板机用瓷球
porcelain ball mill 瓷(制)球磨机
porcelain basal body 瓷基体
porcelain base denture 瓷托托牙
porcelain bath 瓷浴盆
porcelain bead 瓷珠
porcelain beaker 磁烧杯
porcelain boat 瓷舟
porcelain bobbin insulator 电瓷瓶;瓷绕线管绝缘子
porcelain body 瓷胎
porcelain brick 瓷砖
porcelain bushing (shell) 瓷套管;陶瓷套管
porcelain capacitor 陶瓷电容器
porcelain cartridge fuse 瓷壳保险丝
porcelain carving 瓷雕
porcelain casserole 有柄瓷蒸发皿;瓷勺皿
porcelain cement 瓷器胶合剂;陶瓷粘结剂;陶瓷粘合剂;瓷质胶粉;瓷器结合料;瓷胶
porcelain cement spatula 瓷粉调刀
porcelain chamber 瓷室
porcelain clad circuit breaker 瓷罩式断流器;瓷瓶式断路器
porcelain clad current transformer 瓷罩式电流互感器;瓷瓶式电流互感器
porcelain clad type circuit breaker 瓷绝缘子式断流器
porcelain clad type current transformer 瓷罩式变流器
porcelain clay 高岭土;陶土;瓷土
porcelain clay deposit 瓷土矿床
porcelain cleat 瓷夹板
porcelain clip 瓷夹
porcelain colo(u)r 瓷用色料;瓷色料
porcelain combustion boat 瓷燃烧舟
porcelain combustion tube 瓷燃烧管
porcelain condenser 瓷电容器
porcelain connector 瓷壳接线盒
porcelain core 瓷芯
porcelain covered connector 瓷接线盒
porcelain cross arm 瓷横担
porcelain crucible 陶瓷坩埚;瓷坩埚
porcelain cup 瓷盘绝缘子;瓷杯
porcelain cut-out 瓷断流器
porcelain cyclone 瓷旋流器
porcelain cylinder 瓷圆柱体
porcelain dinner-ware 瓷器餐具
porcelain discharge pipe 瓷排水管
porcelain disc resistor 瓷盘电阻器
porcelain dish 瓷皿
porcelain duct 瓷管道
porcelain earth 高岭土;瓷土
porcelain elbow tube 瓷弯管
porcelain enamel 搪瓷釉;搪瓷珐琅;搪瓷
porcelain enamel finish 搪瓷面层;搪瓷涂层
porcelain enamel frit 搪瓷熔块
porcelain enameling 上搪瓷釉;上瓷釉
porcelain enamel(l)ed building panel 搪瓷建筑板材
porcelain enamel(l)ed standard dome lamp 万能型灯
porcelain enamel panel 搪瓷墙板
porcelain enamel sanitary wares 搪瓷卫生洁具
porcelain enamel ware 搪瓷器(皿)
porcelaineous glass 仿瓷玻璃

porcelain eye 瓷眼
porcelain facing 瓷面
porcelain filling 瓷充填
porcelain filter 陶瓷滤头;陶瓷滤波器;陶瓷过滤器
porcelain filtering funnel 瓷过滤漏斗
porcelain filtering stick 瓷过滤棒
porcelain finish 搪瓷饰面
porcelain fitting 瓷配件
porcelain fixture 瓷灯具
porcelain for display 陈设瓷
porcelain form 陶瓷管
porcelain fracture 瓷断面
porcelain funnel 瓷漏斗
porcelain fuse 瓷壳保险丝
porcelain fuse box 瓷保险丝盒
porcelain glaze 瓷釉
porcelain handle 瓷把手
porcelain head screw 瓷头螺钉
porcelain hook 瓷衣钩
porcelain housing 瓷外罩
porcelain imitation mug with cover 带盖仿瓷杯
porcelain inlay 瓷嵌体
porcelain insulating 瓷绝缘
porcelain insulator 瓷绝缘物;陶瓷绝缘子;电瓷瓶;瓷质绝缘子;瓷瓶;绝缘子;瓷绝缘体
porcelain insulator for interior wiring 室内配线用瓷绝缘子
porcelainite 莫来石
porcelainization 涂瓷
porcelain jacket crown 瓷套冠
porcelain jar 瓷罐
porcelain jasper 瓷样石
porcelain jewellery 瓷首饰
porcelain knob 瓷柱;瓷栓;瓷钮
porcelain knob wiring 瓷珠布线
porcelain lamp holder 瓷灯头
porcelain lamp shade 瓷灯罩
porcelain lead-in bushing 瓷进线套管
porcelain light fitting 瓷灯具
porcelain liner (水泵的)陶瓷缸套
porcelain lining 瓷衬
porcelain luminaire (fixture) 瓷灯具
porcelain manufacture waste gas and water 陶瓷厂废气和废水
porcelain mask 瓷面具
porcelain media 瓷磨球
porcelain mill 瓷磨;瓷球磨机
porcelain mortar 瓷砂浆;瓷研钵
porcelain nozzle 陶瓷喷嘴;瓷质喷丝头;瓷质纺丝头
porcelain oblong fuse box 长方瓷保险丝盒
porcelainous 陶瓷的
porcelainous glass 仿瓷玻璃
porcelainous phase 瓷相
porcelain paper 瓷面纸
porcelain paste 未烧的陶瓷坯
porcelain pearl 瓷制珍珠
porcelain petticoat 瓷裙
porcelain pillow 瓷枕
porcelain pipe 瓷管
porcelain piping 瓷管
porcelain plate 瓷托
porcelain plug 瓷插头
porcelain plug fuse 瓷插入式熔丝
porcelain plugger 瓷粉充填器
porcelain pot 瓷罐
porcelain process 制瓷过程;制瓷工艺
porcelain radiator 陶瓷散热器
porcelain receptacle 瓷插座
porcelain ring filter 瓷环过滤器
porcelain roller 瓷辊
porcelain roof tile 瓷屋顶瓦
porcelain sandstone 瓷砂石
porcelain screw holder 瓷质螺丝灯头
porcelain screw socket 瓷质螺丝灯座
porcelain shell for cable 电缆瓷套
porcelain shell for cable terminal box 电缆终端盒瓷套
porcelain shell for capacitor 电容器瓷套
porcelain shell for circuit breaker 断路器瓷套
porcelain shell for current mutual inductor 电流互感器瓷套
porcelain shell for voltage mutual inductor 电压互感器瓷套
porcelain slab 瓷粉调板

porcelain sleeve 瓷套管
porcelain spar 中柱石
porcelain spark plug 瓷火花塞
porcelain split knob 分裂瓷钮
porcelain spool insulator 瓷绝缘子线轴
porcelain stone 瓷石
porcelain structure 瓷砖结构
porcelain support insulator 瓷支持绝缘子
porcelain telephone insulator 电话用瓷瓶
porcelain tendre 软瓷
porcelain thimble 嵌环
porcelain through insulator 瓷套管绝缘子;瓷套管
porcelain tile 铺地瓷砖;锦砖;铺地砖;瓷砖
porcelain tower 瓷塔
porcelain tub 瓷浴盆
porcelain tube 瓷管;瓷管
porcelain tubing 瓷管
porcelain type oil circuit breaker 瓷管型油断路器
porcelain utensil 瓷仪器
porcelain wall bushing 穿墙瓷套管
porcelain ware 瓷器;卫生瓷
porcelain wares 成套瓷器类货
porcelain waste pipe 瓷排水管
porcelain water proof cord switch 瓷防水拉线开关
porcelain with care 小心瓷器
porcelain with cloisonne enamel 景泰蓝瓷
porcelain worker's pneumoconiosis 陶瓷工尘肺
porcelein triangle 瓷三角架
Porcella 普色拉(一种防火油漆)
porcellaneous 瓷样的
porcellanite 陶瓷岩;白陶土
porcellanous 瓷样的
porch 大门内停车处;入口处;门口;廊室;廊;门廊;脉冲边沿;骑楼;过道
porch chamber 船屋;门厅
porch column 门廊柱
porched 有门廊的
porch enamel 甲板油漆(船上)
porch lattice 门廊格栅;门廊格构
porch paint 门廊漆
porch rafter 门廊椽
porch rail 门廊栏杆
porcupine 松软的钢索头
porcupine boiler 豪猪式锅炉;放射形锅炉
porcupine die 多孔冲模
porcupine roller 针辊
pore 细孔;孔隙;孔;毛孔
pore air pressure 孔隙气压力;孔隙空气压力
pore and crack detection 砂眼裂纹检验
pore and water in rock 岩石的空隙和水
pore angle water 孔隙水(指黏土接触间的角隅水)
pore aquifer of volcanic rock 火山岩孔洞含水层
pore brine 孔隙卤水
pore cement(ing material) 孔隙胶结材料;孔隙黏结材料
pore chain 管隙链;导管链
pore characteristic 空隙特性;孔隙特性
pore compressibility 孔隙压缩系数
pore conduction 空隙电导
pore content 空隙量;孔隙量
pore-creating 造孔;成孔
pore cross-seciton 孔隙截面
pore diameter 空隙直径;孔隙直径;孔径;气孔直径
pore diffusion 细孔扩散;孔隙扩散
pore disc 多孔盘
pore distribution 空隙分布;孔隙分布;孔分布
pore entry radius 过水孔道
pore filler 孔隙填料
pore filling 孔隙填充;空隙填充;填孔
pore-fissure aquifer 孔隙裂隙含水层
pore-fissure water 孔隙裂隙水
pore flow rate 孔隙流速
pore fluid 孔隙流体
pore fluid motion 孔隙流体运动
pore fluid pressure 孔隙流体压力
pore fluid resistance 孔隙流体阻力
pore fluid velocity 孔隙中流体速度
pore form 孔隙形状;空隙形状
pore formation 空隙形成;孔隙形成
pore-forming 造孔;成孔
pore-fracture system 孔隙裂缝系
pore-free 无气孔的
pore-free mass 无孔隙体
pore-free material 无孔材料

pore fungus 多孔菌
pore geometry 孔隙几何形状;孔几何学;微孔几何形术
pore geometry factor 孔隙几何系数
pore head 孔隙水头
pore ice 孔隙冰
pore in liberation 单元体内孔隙
pore interconnection 孔间通道
pore liquid 孔隙液体
pore matrix 充填砂浆(骨料中的);孔隙黏结料
poremeter 孔隙率计
pore moisture 孔隙水分
pore multiple 复管孔
pore negative center 孔负中心质
pore negative water pressure 负孔隙水压力
pore neutral stress 孔隙中和应力
pore noise 空隙电导噪声
pore of aquifer 含水层孔隙
pore of liberation separate 单元体间孔隙
pore of oxide film 氧化膜细孔
pore phase 孔隙相
pore plate 孔板
pore pressure 中性压力;孔隙压力
pore pressure cell 孔隙压力计;孔隙压力盒
pore pressure coefficient 孔隙压力系数
pore pressure correction 孔压修正
pore pressure dissipation 孔隙压力消失;孔隙压力消散
pore pressure dissipation test 孔隙压力消散试验
pore pressure ga(u)ge 孔隙压力计
pore pressure measurement 孔隙压力测量
pore pressure meter 孔隙压力计
pore pressure parameter 孔隙压力参数
pore pressure problem 孔隙压力问题
pore pressure ratio 孔隙压力比
pore pressure theory 孔隙压力理论
pore pressure transducer 孔隙(水)压力传感器
pore ratio 空隙比;孔隙比
pore ratio-pressure curve 孔隙比—压力(关系)曲线
pores content 孔隙含量
pore seal(ing) 空隙填封;孔隙填封
pore sealing preparation 空隙填封周边;孔隙填封准备
pore shape 孔隙形状;空隙形状;孔隙形态
pores in set cement paste 水泥石孔隙
pore size 孔眼尺寸;孔径大小;气孔尺寸
pore size distribution 细孔分布;孔隙大小分布;孔隙大小分配;孔径分布;气孔分布;气孔尺寸分布
pore size distribution analysis 孔径分布分析
pore size distribution of porous glass 多孔玻璃孔径分布
pore size gradation 孔级配
pore size refinement 孔隙尺寸改善
pores of stone 石孔隙
pore-solid ratio 孔隙系数;孔隙固体比
pore solitary 单管孔
pore solution 孔隙溶液;间隙溶液
pore space 孔隙空间;孔隙容积;孔隙;孔隙容量;气孔间距
pore-space filling 孔隙充填
pore structure 微孔结构;气孔结构;孔隙结构;孔结构;微孔构造
pore structure parameter 孔隙结构参数
pore suction 孔隙吸力
pore system 孔隙系统
pore tension 孔隙张力
pore texture 微孔结构
pore throat iso-efficiency radius 喉道等效半径
pore throat ratio 孔喉比
pore tortuosity factor 孔隙沟路曲折系数
pore treatment 孔隙处理(砖面防潮方法);孔隙处理
pore type 孔隙类型;空隙类型
pore type in soil 土中孔隙类型
pore velocity 孔隙流速
pore volume 空隙体积;细孔容积;孔隙体积;孔隙容积;孔隙容积;孔隙量;孔体积;孔容;空隙量
pore volume of catalyst 催化剂孔隙度
pore volume of unsaturation by mercury 汞未饱和的孔隙体积
pore wall 孔隙壁
pore water 孔隙水;间隙水;气孔水
pore water content 孔隙水含量
pore water head 孔隙水(水)头

pore water pressure 中性压力;孔隙水压(力)
pore water pressure cell 孔隙压力测压盒;孔隙水压力仪;孔隙水压力盒
pore water pressure ga(u)ge 孔隙水压力计
pore water pressure parameter 孔隙水压力参数
pore water tension 孔隙水张力
porgram relative branch 程序相对转移
Porifera 多孔动物门
poriform 毛孔状
porigelinite 多孔腐植体
poriness 孔隙度;多孔性
Porlan 波尔纶聚氨基甲酸酯纤维
Porlezzina 波勒齐纳风
porocel 高度粉碎的高活性白
porochrom 白色硅藻土载体
porodine 胶状岩;胶状岩
porology 孔隙学;空隙学
poromeric 多孔聚合物的
poromeric material 透气性化纤织物
poromerics 多孔人造革
porometer 气孔计
poroplastic 多孔塑性的
poroplastic box 多孔塑料箱
poroplastic membrane 多孔塑料膜
poroplastics 多孔塑料
pororoca 河口涌潮;河口高潮
poroscope 测孔计
porosimeter 岩石孔隙仪;孔隙仪;孔隙(率)计;孔率(性)计;孔度计;空度计;气孔测定仪;测孔(隙)仪
porosimetry 孔隙度测定
porosin 波诺生
porosint 多孔材料
porosis 空洞形成
porosity 孔隙性;孔隙率;孔隙度;孔度;空隙率;密集气孔;气孔(率);松孔;疏松度;多孔性;多孔率
porosity apparatus 孔率仪;气孔测定仪
porosity characteristic 孔隙特性
porosity chart 气孔图
porosity curve 孔隙度曲线
porosity factor 孔隙度因数
porosity feet 孔隙度英尺
porosity gradient 孔隙变化梯度
porosity inspection 孔隙率检查
porosity log 孔隙比测井
porosity measurement 孔隙率测定
porosity of closed void 闭型孔隙率
porosity of coal 煤孔隙率;煤的孔隙率
porosity of filter 过滤器孔隙率;过滤器孔隙率
porosity of grains 颗粒孔隙率
porosity of kerosene method 煤油法孔隙度
porosity of opened voids 开型孔隙率
porosity of particles 颗粒孔隙率
porosity of reservoir 储集层孔隙度
porosity of rock and soil 岩土的孔隙性
porosity of screen tube framework 滤水器骨架孔隙率
porosity of sediment before compaction 沉积物压实前孔隙度;沉积物压实后孔隙度
porosity of soil 土壤孔隙性;土的孔隙率;土的孔隙度
porosity overlay 孔隙比重叠图
porosity percent 孔隙比率
porosity ratio 孔隙比率;孔隙比
porosity sample 孔隙性样品
porosity test 吸潮试验;孔隙性试验;孔隙度试验;气孔检验
porosity tester 孔率检验器;气孔度测验器
porosity trap 孔隙性圈闭;地层圈闭
porosity unit 孔隙度单位
porous 疏松的素烧瓷体;素烧瓷的;可透水的;多孔状;多孔隙的;多孔的(物料)
porous absorbent material 多孔吸收材料
porous absorbent system 多孔吸收体;多孔吸收材料
porous absorber 多孔吸收体;多孔吸收材料
porous absorption 多孔性吸附
porous acoustic(al) absorption material 多孔性吸音材料;多孔性吸声材料;多孔性声吸收材料
porous acoustic(al) material 多孔吸声材料
porous adsorption material 多孔性吸声材料;多孔性吸声材料
porous aggregate 多孔填料;多孔集料;多孔骨料
porous agricultural film 多孔农用薄膜

porous air diffuser 多孔扩散器;多孔空气扩散器
porous anionexchanger 多孔阴离子交换剂
porous anion exchange resin 多孔性阴离子交换树脂
porous aquifer 孔隙含水层;多孔含水层
porous article 多孔制品
porous backed strip 多孔衬背带材
porous backfill(ing) 多孔回填(物);透水填充体;渗透回填
porous barrier 多孔隔板
porous bearing 多孔轴承
porous bed 透水岩床;透水基床;多孔床
porous block 多孔(滤)块;多孔滤板
porous block conveyer 装有多孔滤板的(气力)输送机
porous body 多孔体;多孔坯体
porous bond 多孔黏结料
porous breakwater 透水防波堤
porous brick 微孔砖;多孔砖;多孔性砖
porous bronze 多孔青铜
porous bronze alloy 多孔青铜合金
porous bronze bearing 多孔青铜轴承
porous bronze parts 多孔青铜零件
porous building material 多孔建筑材料
porous bushing 多孔衬套
porous carbide 多孔碳化物
porous carbon 多孔碳材;多孔炭
porous carbon electrode 多孔碳电极
porous carrier 多孔载体
porous casting 多孔铸件
porous cast iron 多孔铸件
porous catalyst 多孔催化剂
porous cell 素烧瓶;素烧瓷筒
porous cellulose acetate 多孔性醋酸纤维
porous cementation 孔隙胶结
porous ceramic diffuser 多孔陶瓷扩散器
porous ceramic filter media 多孔陶质过滤介质
porous ceramic membrane 多孔陶瓷膜
porous ceramic plate 多孔陶瓷板
porous ceramics 多孔陶瓷
porous ceramic sensor of moisture 多孔陶瓷湿度传感元件
porous ceramic tubular filter 多孔陶质管式过滤器;多孔陶质管式过滤机
porous channel 多孔式管路
porous chrome oxide 多孔氧化铬
porous chromium electroplating 多孔性电镀铬
porous clay cup 多孔陶土管
porous clay pipe drain 多孔黏土排水管
porous coating 多孔敷层
porous coke 多孔焦炭
porous compact 多孔坯块
porous concrete 泡沫混凝土;多孔混凝土
porous concrete drain 多孔混凝土排水装置;多孔混凝土排水系统;多孔混凝土排水管;多孔混凝土排水沟
porous concrete framework 多孔混凝土骨架
porous concrete pipe 多孔混凝土管
porous construction(al) material 多孔建筑材料
porous cooling 蒸发式冷却;发散式冷却
porous cup 多孔杯
porous-cup method 多孔杯法
porous cup tensiometer 多孔杯张力仪
porous dam 透水坝
porous dehiscence 孔裂
porous diaphragm 多孔隔膜
porous diaphragm cell 多孔隔膜电解池
porous diffuser unit 多孔扩散设备
porous diffusion muffler 多孔消声器;多孔材料扩散消声器
porous diffusor 多孔扩散消声器
porous dike 透水堤(坝)
porous disc 多孔石;透水圆板;(试验土质的)透水板
porous earthware pipe 多孔陶管
porous elastic medium 多孔弹性介质
porous elastic solid 多孔弹性固体
porous electrode 多孔性电极;多孔电极
porous fiber 多孔纤维
porous fiber tube 多孔纤维管
porous fill 多孔隙填土;多孔排水垫层
porous filling paste 填孔浆
porous film 多孔涂膜
porous filter 多孔过滤器
porous filter cylinder 素烧滤筒;多孔滤筒

porous filter plate 多孔过滤板
porous filter tube 多孔滤管
porous fissure medium 裂隙孔隙介质
porous flow 渗流
porous foamed plastics 多孔塑料
porous-foil electrolytic capacitor 多孔箔片电解电容器
porous formation 多孔隙岩层
porous foundation 多孔性地基
porous-free 无孔的
porous glass 多孔性玻璃;多孔玻璃
porous glass ionic membrane 多孔玻璃离子膜
porous glass membrane 多孔玻璃膜
porous granule 多孔粒子
porous graphite 多孔石墨
porous graphite-containing bronze bearing 多孔石墨青铜轴承
porous ground 多孔地层;多孔地层
porous hollow fibre membrane 多孔中空纤维膜
porous ice 多孔冰晶
porous iron 多孔铁
porous iron-copper bearing 多孔铁铜轴承
porous iron-copper parts 多孔铁铜零件
porous iron-lead-graphite bearing 多孔铁铅石墨轴承
porous iron matrix 多孔铁基体
porous layer 多孔层
porous-layer bead 多孔层珠
porous-layer open column 多孔层空心柱
porous-layer open tubular column 多孔开口管柱;多孔层空心管形柱;多孔层开管柱
porous-layer support 多孔层载体
porous limestone 多孔石灰石;多孔灰岩;多孔石灰岩
porous mass 多孔物质
porous material 多孔性材料;多孔物质;多孔材料
porous matrix 多孔基体
porous media 孔隙介质;多孔介质
porous media filtration 多孔滤料过滤
porous media flow 多孔介质流
porous membrane 素瓷膜;多孔膜;多孔薄膜
porous metal 多孔金属
porous metal filter 多孔金属过滤器
porous metal parts 多孔金属零件
porous metal plate 多孔金属过滤板
porous microbeads support 全多孔微珠载体
porous mo(u)ld 多孔塑模;多孔模型
porous natural stone 多孔天然石
porous nature 多孔性
porousness 空隙率;孔隙度;多孔性
porous nickel 多孔镍
porous nickel cup 多孔镍引爆杯
porous open-grain structure 多孔构造;大颗粒结构
porous packing 多孔型填充剂
porous paint 透性漆(潮气等可以透过的油漆)
porous paper 多孔纸
porous particle 多孔颗粒
porous pavement 有孔路面;(道路的)透水路面
porous pellet 多孔粒状硝酸铵
porous pipe 排水管;渗水管;多孔管
porous piping 多孔管
porous pit sinking 多排孔沉井
porous plastics 泡沫塑料
porous plate 透水板;素烧瓷板;多孔板
porous plate filter bottom 多孔板滤池底板
porous plug 多孔砖;多孔塞
porous plug ladle 多孔塞浇包
porous plug process 多孔塞吹氩除气法
porous polyethylene 多孔聚乙烯
porous polymer 多孔聚合物
porous polymer beads 多孔聚合物球
porous polystyrene 多孔聚苯乙烯
porous polystyrene gel 多孔聚苯乙烯凝胶
porous porcelain evapo(u)rimeter 多孔瓷蒸发器
porous pot 多孔瓶
porous powder metallurgy material 多孔粉冶材料
porous product 多孔制品
porous reactor 多孔反应堆
porous refractory metal 多孔耐火金属
porous reservoir model 孔隙油层模型
porous rock 透水岩(石);多孔岩(石);多孔隙岩石
porous rock layer 多孔岩层
porous rocks 透水岩类
porous rubber 海绵橡胶

porous sand 多孔性砂土
porous septum 多孔隔膜
porous silica brick 多孔硅砖
porous silicon-carbide refractory 多孔碳化硅耐火材料
porous sintered metal 多孔烧结金属
porous sintered product 多孔烧结制品
porous sintered tungsten carbide 多孔烧结碳化钨
porous skin 多孔蒙皮
porous slab 透水板；多孔板
porous soil 多孔土壤；疏松土(壤)；多孔隙土(壤)；多孔(隙)土
porous solid 多孔体
porous solid bed 多孔固体床
porous sound-absorbing material 多孔吸声材料
porous spring 孔隙泉
porous stainless steel 多孔不锈钢
porous steel parts 多孔钢质零件
porous stone 滤水石；透水石；多孔石
porous stone disc 多孔(石)滤板(土的三轴压力试验用)
porous strip 多孔带材
porous structural clay tile 多孔空心砖
porous structure 透水结构；透水建筑物；疏松结构；多孔结构；多孔构造
porous surface 多孔表面
porous surface tube 表面多孔管
porous tile 素烧瓷板；素烧板
porous tip piezometer 多孔探头式测压管
porous tricalcium phosphate 多孔磷酸三钙
porous tube 多孔管
porous tubing 多孔管
porous tubular support 多孔支撑管
porous vessel 单穿孔导管
porous wall 细孔壁；多孔壁
porous-wall cooling 多孔壁冷却
porous-walled breakwater 多孔墙式防波堤
porous wear-resistant parts 多孔耐磨零件
porous weld 疏松焊缝；多孔焊缝
porous wheel 多孔(砂)轮
porous wood 有孔材
porphyre 斑岩
porphyrin 叶啉
porphyrite 斑岩；玢岩
porphyrite-type iron deposit 玢岩式铁矿床
porphyritic 斑状；斑岩的
porphyritic bitumen 斑状沥青
porphyritic crystal 斑晶
porphyritic fabric 斑岩组构
porphyritic filler 斑岩基质
porphyritic granite 斑状花岗岩
porphyritic-like texture 似斑状结构
porphyritic melphyre 黑色斑岩
porphyritic paving sett 搬运铺路小方石
porphyritic rock 斑岩
porphyritic sand 斑岩砂
porphyritic structure 斑状构造
porphyritic texture 斑状组织；斑状结构
porphyritic tuff 斑岩凝灰岩
porphyrization 粉碎(作用)
porphyrize 研细
porphyroblast 变斑晶
porphyroblastic 斑状变晶
porphyroblastic texture 斑状变晶结构
porphyroclasite 碎斑岩
porphyroclast 残碎斑晶
porphyroclastic 碎斑状
porphyroclastic structure 碎斑构造；残碎斑状结构
porphyrogranulitic texture 斑粒结构
porphyroid 残斑状；残斑变质岩；残斑变岩
porphyroid granite 花岗岩斑岩；似斑状花岗岩
porphyroid neomorphism 残斑新生变形作用
porphyropectic fabric 斑晶胶凝型组织
porphyrotopic fabric 斑晶组构
porphyry 彩色花岗岩；斑岩
porphyry asphaltic concrete 斑岩沥青混凝土
porphyry bitumen concrete 斑岩沥青混凝土
porphyry column 斑岩柱
porphyry copper 斑状铜矿；斑岩铜矿
porphyry copper deposit 斑岩铜矿床
porphyry gravel 斑岩砂砾
porphyry molybdenum deposit 斑岩钼矿床
porphyry slab 斑岩石板
porphyry-type copper and molybdenum-bearing formation 斑岩型含铜钼建造
porphyry-type gold deposit 斑岩型金矿床
porphyry-type tin deposit 斑岩型锡矿床
porplim 叶吩
porpoise 前后振动
porpoising 浮沉运动
Porro-Kopper principle 波罗—科佩原理
Porro prism 波罗棱镜
Porro prism erecting system 波罗正像系统
Porro system 波罗全内反射棱镜系统
Porro telescope 波罗望远镜
Porro tester 波罗板
Porsal cement 波色尔水泥
port 装货门；港口；港；口岸；孔；进出口；极对；燃烧口；枪眼；汽门；端口；船侧开口；舱门；仓门；左舷
porta (古罗马的)城门
portabelt 轻便带式运输机
Port Abidjan 阿比让港(科特迪瓦)
portability 可移植性；可携性；可携带性；轻便性；便携性
portability of program(me) 程序移植性
portable 轻便的；手提式；手提的；便携的
portable acetylene generator 便携式乙炔发生器
portable acidity meter 便携式酸度计
portable acoustic(al) tracking system 便携式声学跟踪系统
portable aggregate plant 移动式集料制备车间；移动式集料加工设备
portable agitator 移动式搅拌机；移动式拌和机
portable air compressor 移动式空(气)压(缩)机
portable air painter 便携式空气涂漆机
portable air pollution analyzer 便携式空气污染分析仪
portable air velocity meter 便携式气流速度计
portable alignment ga(u)ge 携带式前轮定位仪
portable alignment gyrocompass 便携式校准用陀螺罗经
portable altar 可移动的圣坛；可移动的祭坛
portable alumin(i)um alloy cabinet 便携式铝合金箱
portable ambient air analyzer 便携式环境空气分析仪
portable ammeter 便携式安培计
portable amplifier 移动式放大器
portable analytic(al) instrument 手提式分析仪器；轻便分析仪器
portable anchor 轻便小锚；轻便锚
portable anesthesia apparatus 携带式麻醉机
portable antenna 移动式天线
portable apparatus 手提式器械
portable appliance 便携式器械；便携式器具；手提式仪器具；手提(式)仪表；便携式仪表；便携式设备
portable arc welding machine 便携式弧焊机
portable area monitor 便携式地面辐射监测仪
portable asphalt plant 移动式沥青制备设备；移动式沥青工厂；移动式地沥青(混合料)拌和设备；移动式地沥青(混合料)拌和厂
portable aspirating equipment 便携式吸气整备
portable auger conveyer 轻便式螺旋输送器
portable automatic gas monitor 便携式自动瓦斯监测仪
portable automatic tide ga(u)ge 轻便式自记潮位计；轻便自记验潮仪；便携式自动水位计；便携式自动潮汐计
portable awning 轻便天幕
portable axial flow ventilator 便携式轴流通风机
portable bag closer 手提封包机
portable ballast 可移动压载物
portable bar 活动杆
portable barometer 轻便气压计
portable batcher plant 移动式分批调拌设备
portable batch plant 移动式称量装置
portable batch plant with gradation control 移动级配控制的称量装置
portable bath 轻便槽
portable battery pH-meter 携带式直流pH 计
portable beacon 移动标灯
portable beam 舱口活动梁
portable belt conveyer 移动式皮带输送机；移动式带式输送机；轻便式运送机；移动式胶布输送机；活动输送带；活动带式输送机；轻便胶带输送机
portable bilge block 活动式垫木；活动侧盘木(船坞垫船用)
portable bin 移动式料斗；移动式料仓；移动式储[贮]料箱
portable blower 轻便式扇风机；轻便鼓风机；手提鼓风机
portable boiler 移动式锅炉；可移动锅炉；轻便锅炉；便运锅炉
portable boom conveyer 移动式长臂输送机；轻便臂式输送机
portable borehole deflectometer 轻便钻孔挠度计；轻便钻孔变位计
portable borer 移动式钻(孔)机
portable breaker 移动式破碎机
portable bridge 轻便桥；轻便活动桥
portable building 活动房屋；拼装式房屋
portable cabana 移动式棚屋；轻便式棚屋
portable cable 轻便缆
portable cable drilling 轻便缆索钻探
portable calculator 便携式计算器
portable camera 便携式照相机
portable camera transmitter 便携式摄像发射机
portable canopy 便携式盖缝
portable cantilever floor crane 轻便臂式起重机
portable carbon monoxide monitor 便携式一氧化碳监测器
portable catwalk 轻便栈桥
portable ca(u)lking applicator 便携式嵌缝机
portable cement pump 移动式水泥泵
portable centrifuge 携带式离心机
portable charge 轻便装药
portable chemical cylinder 轻便化学罐
portable ciphering unit 便携式加密机
portable circular saw 手提式圆锯
portable clearing saw 轻便除草机
portable cold room 移动式冷间；移动式冷藏间
portable collection system 移动式收集系统
portable communication system 便携式通信系统
portable compactor 手提击实仪
portable compass 轻便罗盘
portable compass indicator 便携式罗盘指示器
portable compensated ground monitor 便携式补偿法地面监测仪
portable compiler 简便编译器；便携式程序编译器
portable composite analyzer 便携式综合分析仪
portable compression 便携式加压舱
portable compressor 移动式(空气)压缩机；轻便式(空气)压缩机；便携式(空气)压缩机
portable computer 手提式计算机；便携式计算机
portable console 轻便型控制仪表板
portable contaminant collector 手提式污染物采集器
portable conveyer 轻便搬运机；移动式运输机；移动式输送机；轻便送机；轻便输送机
portable cooling unit 移动式冷却装置；移动式供冷机组；轻便冷却器
portable cord 移动线
portable counter 轻便计数管；手提式计数器
portable cover furnace 罩式炉
portable crack ga(u)ge 手提裂缝计
portable crane 轻便式起重机；移动式起重机；轻便起重机
portable crusher 移动式破碎机
portable crushing and creening plant 移动式筛分破碎装置
portable crushing plant 移动式扎石设备；移动式破碎设备；移动式破碎机；移动式破碎厂；轻便扎石厂
portable cup anemometer 轻便转杯风速表
portable cutting tool 便携式切割工具
portable cylinder 分置式油缸
portable dam 便携式工具
portable dance floor 拼装式跳舞地面；便携式跳舞地面
portable data loader 便携式数据加载器
portable data medium 便携式数据记录媒体；便携式数据记录介质
portable data recorder 手提式资料记录器
portable deck 活动甲板
portable deflectometer 手持式形变仪；手持式挠度仪
portable derrick 轻便钻塔；轻便井架
portable derrick crane 轻便式转臂吊车；台灵架；移动式人字起重机；轻便转臂起重机；轻便转臂吊机
portable detector 携带式检测器；小型检测器；轻

便检测器
portable diesel generating set 移动式柴油发电机组
portable digital radiometer 便携式数字辐射计
portable digital recorder 手提数字记录器
portable digital voltammeter 便携式数字伏安计
portable dipmeter 携带式栅陷振荡器;携带式地下水(位)测定仪
portable director 有定向转向架的移动式消防炮
portable dissolved oxygen meter 便携式溶氧监测仪
portable diving air compressor 便携式潜水空气压缩机
portable downdraft submersible pump 便携下吸式潜水泵
portable drill 移动式钻床;轻便钻机;轻便钻
portable drilling and tapping machine 轻便钻孔攻丝机
portable drilling machine 可移动式凿岩机;便携式钻机
portable drilling rig 移动式钻探设备;轻便钻机
portable drop tank 可搬运可拆式水箱
portable dryer 移动式烘炉
portable duct 移动式抽风管
portable dust cleaner 手提式吸尘器
portable duster 轻便式喷粉器
portable dust sampler 便携式尘埃取样器
portable dust suctorl portable dust arrestor 轻便吸尘器
portable earthing lead 便携式接地引
portable eddy-current crack detector 便携式涡流探伤仪
portable electric(al) dental engine 轻便电动钻牙机
portable electric(al) drill 轻便(式)电钻;手提式电钻
portable electric(al) drilling machine 轻便电钻机
portable electric(al) grinder 移动式电磨机
portable electric(al) pump 手提式电动泵
portable electric(al) saw 轻便电锯;手提式电锯;移动式电动切砖机
portable electric(al) set 轻型发电机组
portable electric(al) tool 小型电动工具;轻便电动工具;轻便电力工具;手提式电动工具
portable electric(al) welding equipment 移动式电焊设备
portable electrostatic painting machine 便携式静电喷漆机
portable elevator 移动式升降机
portable emergency transceiver 便携式应急收发信机
portable engine 移动式发电机;轻便发动机
portable equipment 可携带设备;便携设备
portable explosion-proof digital level ga(u)ge 便携式防爆数字液位仪
portable explosion-proof gas detector 便携式防爆瓦斯探测器
portable extinguisher 轻便灭火器;便携式灭火器;手提式灭火器
portable extinguishing equipment 轻便灭火设备
portable factory 现场工厂;小型工厂
portable fan 移动式风机
portable fence 可移动围栏;活动栅栏
portable field spectrometer 携带式野外能谱仪;便携式场能谱仪
portable field-strength meter 便携式场强仪
portable film developer 手提胶片显影剂
portable fire extinguisher 便携式灭火器;手提(式)灭火器;手提(式)灭火机
portable fire-fighting appliance 轻便灭火器
portable fire pump 手抬消防泵
portable flame thrower 手提式火焰喷射器
portable floor crane 轻便落地吊车
portable flume 活动量水槽
portable fluorescence apparatus 携带式X荧光仪
portable fluorescent lamp 便携式荧光灯
portable foam applicator 手提式泡沫喷口
portable foam cannon 移动式泡沫炮
portable foam device 手提式泡沫发生器
portable foam generating device 手提式泡沫发生器
portable foam monitor 移动式泡沫炮
portable form 活动模板
portable frame bridge 移动式构架桥;装配式构架桥;装配式刚架桥;轻便构架桥

portable furnace 移动式炉
portable galvanometer 便携式测流计
portable garage 移动式汽车库
portable gas 瓶装气
portable gas monitoring instrument 便携式气体监测仪
portable gasoline measuring 移动式汽油计量器
portable generator 移动式发电机
portable generator set 轻便发电机组;便携式发电机组
portable grain bin 便携式谷仓
portable grain bulkhead 散装谷物活动隔舱壁
portable grandstand (预制构件的)装卸式大看台
portable grinder 轻便磨床;移动打磨机;手提式砂轮机;手提式磨光机
portable grinding 移动式打磨
portable grinding wheel 移动式磨(砂)轮
portable grizzly 移动式格筛
portable hand appliance 轻便手用器械;轻便手用器具
portable hand-held sound level meter 手提式声级计
portable hand punch 小型手动穿孔机;便携式手动穿孔器
portable hardness tester 携带式硬度计
portable hatch beam 舱口活动梁
portable heater 移动式取暖器;轻便加热器
portable heating appliance 轻便式采暖用具
portable helium detector 轻便测氦仪
portable hoist 移动式绞车
portable hopper 移动式漏斗;移动式料斗;活动料斗
portable hopper car 移动式漏斗车
portable hose reel 便携式胶管卷盘
portable house 活动房屋
portable hydrant 移动式消防栓
portable hydrant flowmeter 携带式消防栓流量计
portable hydraulic centre-shaft rolling machine 轻便液压中轴式卷驱动机
portable impact breaker 移动式冲击破碎机
portable impact compactor 手提(式)夯击压实机
portable impactor 移动式夯击压实机
portable inclinometer 手提测斜仪
portable indicator 便携式指示器
portable infrared gas analyzer 便携式红外气体分析仪
portable ink jet printer 便携式喷墨打印机
portable installation 移动式装置;移动式设备
portable instrument 携带式仪表;便携式仪表;便携式化仪器;手提式仪器;便携式仪表;便携式器具
portable ionization chamber 轻便电离室
portable iontophoreser 轻便型离子透入器
portable irradiator 手提式辐照器
portable irrdadiation facilitiy 移动式辐照装置
portable irrigation hydrant 轻便灌溉放水栓
portable jaw crusher 移动式颚式破碎机
portable kettle 移动式沥青炉
portable ladder 轻便梯
portable lamp 行灯;移动灯;手提航行灯号;手提灯
portable lamp bulb 行灯灯泡
portable lamp plug socket 行灯灯座
portable laser 便携式激光器
portable laser rangefinder 便携式激光测距机
portable lathe 轻便车床
portable level recorder 移动式水位计
portable life support system 轻便式生命保障设备
portable light 行灯;移动灯;手提航行灯号;手提灯
portable lighter 行灯;轻便灯
portable lighting 移动式照明;便携式照明设备
portable linear ratemeter 携带式线性计数率计
portable liquid level detector 便携式液位检测器
portable loader 移动式装载机;轻便式装载机;轻便装载机
portable lock 挂锁
portable Loran assisted device 便携式罗兰辅助装置
portable luminaire 可移动式照明装置;手提灯具
portable machine capability analyzer 便携式机器能力分析仪
portable machine gun 手提机枪
portable machines 轻便机械;轻便机器
portable machine tool 轻便工具机
portable magnetic flaw detector 手提式磁力探伤仪
portable magnetic observatory 轻便地磁台
portable magnetic recorder 便携式磁记录器

portable magnetic susceptibility meter 便携式磁化率测定器
portable magnetic thickness ga(u)ge 轻便磁性测厚仪
portable manhole cover 便携式人孔盖
portable map board 轻便图板
portable mast 轻便井架
portable material elevator 移动式物料提升机
portable matrix sign 移动式矩阵标志
portable mercury detector 轻便测汞仪
portable metal ladder 便携式金属梯
portable meter reading system 轻便式读表系统
portable microfilm reader 手提式缩微胶卷阅读器
portable microscope 携带式显微镜
portable mill 移动制材厂
portable mining substation 移动式变电所
portable mircowave television transmission system 便携式微波电视传输系统
portable mixer 移动式搅拌设备;移动式搅拌机;移动式拌和机
portable mixing plant 移动式搅拌设备
portable mobile receiver 便携式接收机
portable mobile scaffold 移动式脚手架
portable mobile station 步谈机
portable moisture monitor 便携式水分监测仪
portable monitor 轻便监测仪;便携式监护仪
portable mortar mixer 移动式灰浆搅拌机
portable mo(u)ld 手提模
portable mo(u)ld dryer 移动式烘炉
portable muller 移动式混砂机
portableness 轻便性
portable network 流动台网
portable nuclear reactor 移动式核子反应堆
portable oceanographic(al) recorder 轻便测波仪
portable oceanographic(al) survey system 轻便式海洋测流系统
portable oil jack 携带式油压千斤顶
portable optic(al) fibre cable 移动光缆
portable oscillograph 便携式示波器
portable outlet header 临时排水干管
portable oxygen set 便携式氧气设备
portable pacemaker 携带式起搏器
portable partition 可移动的隔墙
portable pendulum tester 手提式摆式仪
portable pension 移动继续年金
portable personal computer 便携式个人计算机
portable pH-eH meter 轻便酸碱度及电位计
portable phone 移动式电话
portable photo facsimile transmitter 便携式照片传真发送机
portable pillar 活动支柱
portable pit sampler 轻便坑取样器
portable planer 轻便刨床
portable plant 移动式设备;移动式处理厂
portable plate 活动板
portable point ga(u)ge 便携式测针
portable poisonous gas monitor 便携式有毒气体监测仪
portable polyurethane foam kit 便携式聚氨酯发泡工具
portable power 移动式液压校正器
portable power tool 便携式电动工具
portable precise acidimeter 便携式精密酸度计
portable program(me) 可移植程序
portable pulling machine 移动式提管机;移动式套管起拔机
portable pulse rate indicator 携带式脉率指示表
portable pump 移动泵;可移动泵;轻便泵;轻便(抽水)泵;手提泵
portable pump carriage 轻便泵车
portable pumping unit 移动式水泵机组;可移动泵设备
portable pump trailer 手抬泵拖车
portable pyrometer 携带式高温计;轻便高温计
portable radar apparatus 便携式雷达装置
portable radar performance tester 便携式雷达性能检测器
portable radiation instrument 手提式放射能指示器
portable radiation monitor 携带式辐射监测器
portable radiation survey meter 便携式辐射监测仪
portable radio 便携式收音机
portable radioactive density indicator 轻便放射性密度指示器
portable radioactive logging apparatus 携带式

放射性测井仪
portable radio apparatus 手提无线电设备
portable radio equipment 手提式无线电设备
portable radio set 便携式收音机
portable radio station 便携式电台
portable radio-telephone equipment 手提无线电话设备
portable rail bending machine 轻便弯轨机
portable railroad 工地轻便铁道
portable railway 工地轻便铁道；轻便铁路；轻便轨道
portable rainfall simulator 轻便式降雨模拟器
portable ram 外置式油缸
portable ramp 移动式坡道
portable reader 手提式阅读器
portable readout unit 手提测读装置
portable receiver 便携式接收机
portable recorder 便携式录音机
portable recorder receiver 便携式收录两用机
portable rectifier 移动式整流器
portable remote hydrotemp 轻便遥测水温计
portable re-railer 便携式复轨器
portable resistance welder 移动式接触焊机
portable retort 手提式曲颈甑
portable rig 轻便钻机；移动式凿岩机支架；轻便钻探设备；轻便钻架
portable riveter 轻便铆(接)机
portable riveting forge 轻便铆钉炉
portable riveting machine 轻便铆(接)机
portable road beacon 移动式信号灯(筑路时用)
portable rock drill 轻型凿岩机
portable Rockwell hardness tester 便携式洛氏硬度计
portable rod 轻便式水准尺；轻便式标尺
portable roller (装橡胶胎后轮的)小型压路机
portable roller conveyer 移动滚式运输机
portable rotation viscometer 便携式旋转黏度计
portable router 手提剖孔机
portable saber saw 手提式往复锯
portable salinity-temperature meter 轻便温盐测量仪；轻便温盐测量计
portable sampler 轻便取样器；轻便采样器
portable sand conditioning unit 移动式混合砂处理装置
portable saw 轻便锯机
portable saw mill 移动锯木机
portable scaffold 轻便脚手架
portable scale 手提秤
portable scaler 轻便定标器；便携式定标器
portable scraper 轻便铲斗
portable screen 活动屏幕；活动屏风；活动滤筛
portable screen belt loader 移动式胶带筛分装载机
portable screening machine 移动式筛分机
portable seismic recorder 便携式地震记录仪
portable seismograph 手提地震仪；轻便地震仪
portable seismographic (al) station 流动地震台；便携式地震台
portable seismometer 手提地震计
portable server 轻便服务车
portable set 便携式设备
portable shear apparatus 轻便剪切仪
portable shelter 便携式掩蔽体
portable shunt 移动式道岔
portable sign 移动式标志
portable signal 可移号志
portable signal enhancement seismograph 信号增强型轻便地震仪
portable signal light 手提信号灯
portable single-zone temperature controller 便携式单区温度控制器
portable skid-resistance test device 手提式抗滑试验仪
portable slewing crane 轻便旋臂起重机
portable slotting machine 轻便插床
portable software 可移植软件
portable soil sampler 轻便取土器
portable soil tester 便携式土壤分析测试仪
portable solids mixer 便携式固体混合机
portable sonar 轻便声呐
portable sound grader 便携式声级计
portable sounding test 轻便触探试验
portable sound slide projector 便携式有声幻灯机
portable spectrometer 手提式光谱仪
portable spectrophotometer 便携式分光光度计
portable spotlight 移动式聚光灯

portable spot welder 移动式点焊机；轻便点焊机
portable sprinkler system 移动式喷灌系统
portable stacker 移动式装载机
portable staff 活动水尺；轻便水尺；轻便水准尺；轻便式标尺
portable staging 活动台架
portable stanchion 活动支柱
portable standard meter 携带式标准仪表
portable starter 轻便起动机
portable station 流动站；便携台；便携电台
portable steam crane 轻便蒸汽起重机
portable-steam engine 锅驼机
portable structure 轻便建筑
portable substation 流动变电站
portable suction basin 便携式帆布水槽
portable suction fan 携带式吸尘器
portable system organization 简便式系统结构
portable tachoscope 手提转速计
portable tackboard 轻便性布告板；轻便型布告牌
portable tank 移动式水箱；移动式罐
portable telephone 携带式电话；手提式电话
portable telescopic crane 轻便伸臂起重机
portable television transmitter 便携式电视发射机
portable temporary dead light 活动舷窗盖
portable terminal 便携式终端设备
portable tester 便携式测试仪
portable testing set 手提测试仪；便携式测试仪器；手提式测试器
portable testing unit 便携式检验设备
portable test kit 携带式测试设备箱
portable test set 携带式试验装置
portable theodolite 轻便经纬仪；便携式经纬仪
portable thickness ga(u)ge 厚薄千分尺；手提厚度计
portable thresher 移动式脱粒机
portable tiltmeter 轻便倾斜仪；手提式倾斜仪
portable timekeeper 携带式钟表
portable toolbox 手提式工具箱；手提工具箱
portable tool 轻便工具
portable tower 活动觇标
portable track 轻便轨道；轻便道轨
portable track truck 轻便轨道车(窄轨)
portable traffic signal 移动式交通信号
portable trailer-mounted drill rig 轻便拖车式钻机；拖挂式活动钻机；移动式轻便钻机
portable transit 轻便经纬仪；便携式经纬仪
portable transmitter 移动式发射机
portable truck 轻便手推车
portable truck scale 便携式卡车秤
portable turbidimeter 携带式浊度计
portable type 便携型；便携式；携带式；移动式；轻便式
portable ultra-high-speed water spray system 手提式超高速喷水系统
portable ultrasonic diagnostic scanner 便携式超声诊断扫描仪
portable ultraviolet radiator 便携型紫外线辐射器
portable unit 轻便装置；便携装置；便携式装置
portable universal drilling machine 轻便万能钻床
portable vacuumatic cleaner 移动真空吸尘器
portable vacuum cleaner 移动式真空吸尘器；移动式真空除尘器
portable vapo(u)r monitoring instrument 便携式蒸汽监测仪
portable ventilating equipment 轻便式通风设备
portable video-camera 便携式摄像机
portable visco(si)meter 携带式黏度计
portable water purifier 轻便净水器；手提净水器
portable water quality monitoring system 轻便水质监测装置
portable water storage dam 便携式储[贮]水坝
portable water treatment equipment 移动式水处理设备
portable wattmeter 便携式瓦特计
portable weighing machine 可移动秤台
portable weir 活动量水堰
portable welder 便携式焊机；移动式焊接机
portable welding equipment 便携式焊接设备
portable winch 轻便绞车；轻便绞车
portable X-ray analyzer 携带式X射线分析仪
portable X-ray equipment 携带式X射线设备
portable X-ray fluorescence spectrometer 轻便X射线荧光分光计
portable yard crane 轻便吊车

port accommodation 港口设施；港口容纳量
portacid 移酸滴管
port activity 码头活动
port activity problem 码头活动问题
port additional 港口附加费
port address 通路地址
port addressing 接口寻址
Port Adelaide 阿德莱德港(澳大利亚)
port administration 港务管理；港口管理
portadrill method 吸力钻井法
portadrill reverse circulation method 吸力反循环钻井法
port aft 左舷后部
portage 货物；(两条水路之间的)连水陆路；搬运物；搬运费；货物搬运；搬运
Portage group 波特奇群
port agency 港务机构；港口代理处
port aide 左边
portail 教堂的主入口处；教堂入口
portainer 集装箱码头起重机；门框式起重机；码头集装箱起重机；波泰纳；岸边集装箱装卸桥
port air duct 左舷通风干路；左舷风道
portal 正门；教堂入口；门形架；门架；门户；门的；入口；洞口(如月亮形壶门等)；峒门
portal air curtain system 洞门空气幕系统
portal arch 门式拱
portal architecture 正门建筑
portal area 洞口；门管区
portal beam 硬横梁
portal bent 门形排架
portal brace 桥门斜撑
portal bracing 门撑杆；柱间桥门联结系；门架支撑(系)；门撑(系)；桥门支撑；桥门联结系；桥门撑架
portal bridge 刚架桥；门架桥
portal bridge crane 龙门起重机；门式起重机
portal bridge crane with fixed jib 带固定臂架的门式起重机
portal bridge gallows 龙门架
portal building 入口建筑；洞口建筑
portal cable crane 门式缆索起重机
portal circulation 门脉循环
portal clearance 隧道口净空；门座净空；入口净空；桥洞净空
portal clearance of tunnel 隧道口净空
portal column 大门柱；正门柱；门柱
portal crane 门式吊车；港口起重机；门座起重机；门式起重机；岸边集装箱装卸桥
portal crane bucket 门式起重机吊斗
Port Aleciras 阿尔赫西拉斯港(西班牙)
portal effect 门架作用；桥门架效应
portal entrance 隧道入口
Port Alexandria 亚历山大港(埃及)
portal excavation 洞口开挖
portal frame 龙门架；门式钢架；门式钢架；门式刚架；门式吊架；门式框架梁；门式框架；桥门架
portal frame beam 门式框架梁
portal frame block 门架建筑
portal frame bridge with inclined supports 有斜支座的刚架桥；有斜支撑的刚架桥
portal frame building 门架建筑；门式框架建筑
portal frame compound unit 门架组合构件
portal frame construction 门式框架结构
portal framed structure 门式框架结构；门架式结构；门架结构
portal framed tower 门形索塔
portal frame effect 桥门架效应
portal frame pier 门式墩
portal frame span 门式框架跨度
portal frame structure 门式框架结构；门式刚架结构
portal frame tower(cable stayed bridge) (斜拉桥的)门式框架塔
portal frame with fixed ends 固定底门式刚架
portal frame with lean-to 带披的门式框架
portal gantry 起重机门架
Port Algeriers 阿尔及尔港(阿尔及利亚)
portal girder 门架大梁
portal grab ship unloader 门座抓斗卸船机
portal jib crane 高架起重机；龙门吊；门座旋臂起重机；门式旋臂起重机；门式旋臂吊机；门架式起重机；门吊
portal leg 门架柱
portal lobule 门小叶；门管小叶
portal method 门架法；门架式解法(计算柱剪力分

布的）
portal of entry 侵入门户
portal of tunnel 隧道入口
portal openwork gablet 正门的镂空花纹山墙
portal pier 门式桥墩
portal reclaimer 门座取料机；门式取料机
portal rigid frame 门形刚架；门桄
portal robot 门式机器人
portal scraper 门式刮料机
portal (slewing) crane 门座式起重机；门座回转式起重机
portal span 门架跨距
portal steel tower 门形铁塔
portal structure 门式结构；硬横跨；洞门建筑工程
portal strut 门式刚架撑杆；门式刚架支撑；门式刚架撑杆；门架横撑
portal system 门脉系统
portal-to-exhaust shaft 吸出式通风方式
portal-to-portal ventilation system 全射通风方式
portal tower crane 门座塔式起重机
portal travel(l)ing crane 门式移动起重机
portal-type frame 龙门架
portal-type pylon 门式塔架
Port Amsterdam 阿姆斯特丹港（荷兰）
port anchor 左舷锚；左锚
Port Anchorage 安克雷奇港（美国）
port anchorage 港区锚碇；港口锚泊
port anchor winch 左锚绞车
port and harbo(u)r 港口和海港
port and harbo(u)r authority 港务局
port and harbo(u)r construction 港口建设
port and harbo(u)r depth 港口水深
port and harbo(u)r engineering 港口工程
port and harbo(u)r law 港湾法；港口法
port and harbo(u)r planning 港口规划
port and harbo(u)r works 港口工程
port and starboard 左边与右边
port approach 进港航道
portaprobe simul-tester 同时检测携带式探测针
port apron 舷窗护板
Port-a-punch 便携式穿孔机
port arch 喷出口拱顶
port area 港区；港口区
portasaw 轻便锯
Port Ashdod 阿什杜德港（巴勒斯坦）
portasilo 移动式水泥筒仓
port assets 港口资产
port association 端口的结合
portative force 起重力
Port Auckland 奥克兰港（新西兰）
port authority 港务局；港务机关；港口管理局；港务管理机构；港务当局
port backland 港口腹地（使用港口的内陆）
Port Balboa 巴尔博亚港（巴拿马）
Port Baltimore 巴尔的摩港（美国）
port bar 港口沙洲；港口砂坝；港口拦（门）砂（坝）；门闩
Port Barcelona 巴塞罗那港（西班牙）
Port Bares Salaam 达累斯萨拉姆港（坦桑尼亚）
Port Basel 巴塞尔港（瑞士）
port basin 港池
Port Basrah 巴士拉港（伊拉克）
port beam 左舷；左舷正横
Port Beirut 贝鲁特港（黎巴嫩）
Port Belem 贝伦港（巴西）
Port Belfast 贝尔法斯特港（英国）
Port Benghazi 班加西港（利比亚）
Port Bergen 卑尔根港（挪威）
port berth 港口泊位
Port Bilbao 毕尔巴港（西班牙）
port block 港区；港口作业区；喷口砌块；喷口块
Port Bordeaux 波尔多港（法国）
Port Boston 波士顿港（美国）
port bottom 炉底
port bound 在港困守
port boundary 港界（线）
port bow 左前方【船】；船首左舷方向
port bower 左舷主锚；左舷锚；左首锚
Port Bremen 不来梅港（德国）
Port Buenos Aires 布宜若斯艾利斯港（阿根廷）
port building 港口建筑物
port bulk plant 港口油库
Port Burnia 伯尼港（澳大利亚）
Port Calcutta 加尔各答港（印度）

port capacity 港口吞吐量；港口容量；港口（吞吐）能力
port capacity for loading and unloading cargo 港口通过能力
Port Cape Town 开普敦港（南非）
port cap rake 小炉脖碴；小炉脖拱
port captain 驻港船长
Port Castries 卡斯特里港（拉丁美洲岛国圣卢西亚）
Port Cebu 宿务港（菲律宾）
port chain cable 左锚链
port charges 港务费；港口使费；港口费（用）；入港费；出港手续费
Port Charleston 查尔斯敦港（美国）
port chart 港域海图；港湾图；港区图；港口图
port charter 港口租船合同
Port Chicago 芝加哥港（美国）
port circulating pump 港口循环泵
port city 港湾城市；港口城市
port clause 港口条款
port clearance 结关港口；出口结关
Port Cleveland 克利夫兰港（美国）
Port Cochia 科钦港（印度）
port cock 汽缸排水阀
Port Colombo 科伦坡港（斯里兰卡）
port colony 港区聚居地；港区侨居地
port commerce statistics 港口业务统计资料
port commission 港务委员会
port commissioner's office 港务局
port communication 港口通信
port communication office 港口通信科
port complex 港区
port congestion 港口拥塞；港口拥挤；港口积压；港口堵塞
port congestion surcharge 港口拥塞附加费；挤港附加费
port connection 油口接头
port construction 港口建筑；港口建设；港口工程施工
Port Copenhagen 哥本哈根港（丹麦）
Port Corinto 考林托港（尼加拉瓜）
port coverage 火焰覆盖面积
port craft 港口工作船
port crane 港口起重机；左方起重机
port crown rake 小炉脖碴；小炉脖拱
portcul(l)is 城堡吊闸；吊门；用吊门关闭；城堡吊门
port customs 港口习惯
Port Damietta 达米埃塔港（埃及）
Port Damman 达曼港（沙特阿拉伯）
port deep tank 左深舱（油，水）
port depot 港口站；港口油库
port design 港口设计
port design and construction 港口设计和施工
port design capacity 港口设计能力
Port Detroit 底特律港（美国）
port development 港口开发；港口发展
port differential 不同港口运费差别
port director 港务局长
port disbursement 港口使费
port dispatching 港口调度
port dispatching telephone 港口调度电话
port district 港区
port drainage 港口排水
port drainage system 港口排水系统
port dredging 港口疏浚
port dues 港税；港口费（用）；入港税；入港费；船舶港务费
Port Duisburg 杜伊斯兰堡港（德国）
Port Durban 德班港（南非）
port duties 入港费；入港税
porte cohere 车辆通道入口；车辆门（道）；停车门廊；车道门廊
ported 装有气门的
ported air ring 环形风口
port efficiency 港口装卸效率；港口效率
Port Eilat 埃拉特港（巴勒斯坦）
porte-lumiere 轻便回照器
port end 喷出口端墙
port engine 滑阀式发动机
port engineer 总轮机长；港口工程师
port engineering 港口工程学；港口工程；海港工程
port engineering and construction 港口设计和施工
port entrance 港口入口（处）；进港口门；进港航道
port entrance jetty 进港口门导堤

port entrance light 港口口门灯标；进港口门灯标
port entry and exit visas 进出港签证
port environment 港口环境
port environment condition investigation 港口环境条件调查
port equipment 港口装卸设备；海港设备
porter 搬行李工人（车站）；看门人；搬运人；搬运工（人）（指码头作业工人）；搬运车
porterage 运费；搬运业；搬运费（用）
porterage company 搬运公司
porter bar 换辊套筒
Porter governor 波特（摆式）调速器
porter's apartment(unit) 门卫公寓
porter's dwelling 门卫住房
porter's flat 传达室；门卫公寓
porter's house 门卫住所；酒店
porter's lodge 门房；传达室；门卫住房
porter's room 接送室；候工室；搬运工室；门卫室
porter system 门卫电话设备
porter unit 门卫居住单元
port establishment 港口朔望高潮间隙；潮候时差
Porteus process for heat treatment 波梯斯热处理法
port extension 港口扩建（工程）
port facility 码头设备；港口设施；海港设施
port ferry 港内交通
portfire 引火具；导火筒
port flap 排水翼盖；排水口翼盖；排水孔带铰链的盖
Port Flixstowe 费利克斯托港（美国）
port floating crane 港口浮式起重机
portfolio 有价证券；资产组合；公文包；文件袋；投资搭配
portfolio acquisition 全部购买
portfolio assumed 未满期责任
portfolio ceded 分出未满期责任
portfolio investment 有价证券投资；证券投资；间接投资
portfolio investor 证券投资人
portfolio premium 未满期保费
portfolio selection 投资搭配选择
portfolio theory 投资搭配理论
portfolio transfer 未满期业务转移
portfolio withdrawal 未满期业务转出
port forward 左舷前
port frame 装货口加强肋骨；门框
Port Fremantle 弗里曼特尔港
Port Gdynia 格丁尼亚港（波兰）
Port Genoa 热那亚港
Port Ghent 根特港（比利时）
Port Godank 格但斯克港（波兰）
Port Gothenburg 哥德堡港（瑞典）
Port Haifa 海法尔港（以色列）
Port Haikou 海口港
port hand 左舷
port-hand buoy 左侧浮标；左舷浮标
port handling 港口装卸
port handling machinery 港口装卸机械
port handling operating area 港口作业区
port hand mark 左岸标志
port health officer 港口检疫（人）员
Port Helsinki 赫尔辛基港（芬兰）
port hinterland 港口腹地（使用港口的内陆）
Port Hobart 霍巴特港（澳大利亚）
port hoisting crane 港口提升式起重机
porthole 舷窗；舷孔；观察孔；孔口；看火孔；墙孔；汽门（钻头的）水槽；装货口；观察口；舱口
porthole die 孔道模；多孔拉拔模
port(hole) glass 舷窗玻璃；观察口玻璃
porthole swivel plate 观察口转板
porthole taper slide 观察斜滑板
porthole taper slide guide 观察口斜滑板导轨
port holiday 港口假日
Port Honolulu 火奴鲁鲁港（美国）
Port Houston 休斯敦港（美国）
portico 门廊
portico of columns 柱廊
portico with two columns 双柱式门廊
porticus 门廊；柱廊
port identifier 端口识别器
portiere 门帘；门窗帘
portiere rod 门帘杆；门窗帘棍
portil 硅酸钠盐
port illumination 港口照明
port inboard 左舷内

port industry 港口工业
porting arrangement 进排气口排列;气口布置
porting ring 节流孔板
port installation 港口设施;海港设备
portion 一部分;份;部分
portion chlorinator 定量加氯机
portion gate 分配门
portioning device 计量器
portion of marketing 市场份额
portion of pipeline 管段
portion of profits retained by enterprise 企业利润留成
portion of state expenditures for production and construction 生产建设性支出
portion-wise addition 分批添加
port island 港岛;人工港;人工岛港
port jamb 小炉侧墙
Port Jeddah 吉达港(沙特阿拉伯)
Port Karachi 卡拉奇港
Port Kelang 克朗港(马来西亚)
Port Khor Fakhan 豪尔法坎港(阿拉伯联合酋长国)
port king post 左舷吊柱
Port Kingston 金斯敦港(牙买加)
Port Kopar 科佩尔港(南斯拉夫)
port labo(u)r productivity 港口劳动生产率
Port Laltakia 拉塔基亚港(叙利亚)
port land area 港口陆域
Portland bed 波特兰层
Portland blast furnace cement 波特兰矿渣水泥;矿渣硅酸盐水泥;波特兰炉渣水泥
Portland blast furnace slag cement 矿渣硅酸盐水泥;矿渣水泥
Portland cement 硅酸盐水泥;普通水泥;波特兰水泥
Portland cement artificial marble 硅酸盐水泥人造大理石
Portland cement asbestos 石棉硅酸盐水泥
Portland Cement Association 波特兰水泥协会
Portland Cement Association design method for concrete pavement 波特兰水泥协会混凝土路面设计法
Portland cement clinker 硅酸盐水泥熟料
Portland cement concrete 硅酸盐水泥混凝土;普通水泥混凝土;波特兰水泥混凝土
Portland cement exterior plaster 硅酸盐水泥外墙粉刷
Portland cement for dam 大坝用硅酸盐水泥;硅酸盐大坝水泥
Portland cement for highway 硅酸盐道路水泥
Portland cement glazed coat(ing) 上釉硅酸盐水泥面层
Portland cement grout 波特兰水泥浆;硅酸盐水泥浆
Portland cement grouting compound 硅酸盐水泥灌缝浆
Portland cement imitation marble 硅酸盐水泥人造大理石
Portland-cement-lime mortar 波特兰水泥石灰砂浆;硅酸盐水泥石灰砂浆
Portland-cement-lime plaster 波特兰水泥石灰混合料
Portland cement man-made marble 硅酸盐水泥人造大理石
Portland cement manufactured marble 硅酸盐水泥人造大理石
Portland cement mortar 硅酸盐水泥砂浆
Portland cement mortar stucco 硅酸盐水泥砂浆拉毛粉刷
Portland cement paint 清水泥涂料;硅酸盐水泥涂料
Portland cement panel 硅酸盐水泥板
Portland cement pavement 普通水泥路面
Portland cement plaster 硅酸盐水泥浆
Portland cement refractory castables 硅酸盐水泥耐火浇筑料
Portland cement refractory concrete 硅酸盐水泥耐火混凝土
Portland cement-sand grout 硅酸盐水泥砂灌缝
Portland cement slurry 硅酸盐水泥砂浆
Portland clinker 硅酸盐水泥熟料;硅酸盐水泥熟料
Portland clinker composition 硅酸盐水泥熟料组分
Portland fly-ash cement 粉煤灰硅酸盐水泥;波特兰粉煤灰水泥
Portlandian 波特兰阶【地】
Portlandian stage 波特兰层

portlandite 氢氧钙石;羟钙石;熟料水化形成的氢氧化钙
Portland limestone 波特兰石灰石
Portland Port 波特兰港(美国)
Portland-pozzolana cement 波特兰火山水泥;硅酸盐火山灰水泥;火山灰质硅酸盐水泥
Portland-pozzolana concrete 波特兰火山水泥混凝土
Portland-pozzolanic cement 硅酸盐火山灰水泥
Portland-slag cement 波特兰矿渣水泥;矿渣硅酸盐水泥;矿渣硅酸盐水泥
Portland stone 波特兰石料;波特兰石
Portland trass cement 硅酸盐火山灰水泥
Port La Spezia 拉斯佩齐亚港
portlast 船舷上缘
port lens 舷窗厚玻璃
portlet 小港;渔港
port level 端口级
Port Liangungang 连云港港
port lid 排水口盖;排水口翼盖
port light 港湾灯塔;港入口处或突码头端小灯塔
port light glasses 左舷灯玻璃
port lighting 港口照明
port line 港区界线
port lintel block 炉顶砖块
port list 左舷倾侧
Port Liverpool 利物浦港(英国)
port locker 左边储物舱
port log 停泊日记
Port Long Beach 长滩港(美国)
Portmadoc bed 波马多克层
Port Madras 马德拉斯港(印度)
port main boiler 左主锅炉
port maintenance engineering vehicle 港口维修工程车
Port Malmo 马尔默港(瑞典)
port management 港口管理
port management body 港口管理机构
port manager 港务主任;港务经理
port map 港湾图
port mark 卸货港标志;港口标志;目的港标志
Port Marseilles 马塞港(法国)
port master 港务长
Port Mawei 马尼港
port medical officer 港口医务人员;港口检疫(人)员
port model 港湾模型
port model test 港湾模型试验
Port Mombasa 蒙巴萨港(肯尼亚)
Port Montreal 蒙特利尔港(加拿大)
port mouth 港湾口门;喷火口
Port Nantes 南特港(法国)
Port Naples 那不勒斯港(意大利)
port navigation 港内航行
Port Oakland 澳克兰港(美国)
port of arrival 抵达港;到达港
port of barges 驳船港
port of call 沿途停靠港;中途港;挂靠港;寄泊港;停靠港(沿途);停泊港
port of clearance 结关港口;船只结关证
port of coaling 装煤港(口)
Port of Dalian 大连港
port of debarkation 卸货港口;卸货港;下船港;目的港
port of delivery 卸货港;交货港;交船港
port of departure 启运港;启航港;始发港;出发港
port of destination 驶向港;终程港;目的口岸;目的港;到达港
port of discharge 卸货港码头;卸货港口;卸货港
port of dispatch 发货港
port of distress 避难港;避风港
port of distribution 集散地
port of embarkation 装载港;启运港;上客港
port of entry 报关海港;进入港;进口港;外贸港(口);输入口岸;输入港;抵达港;到达港
port of entry price 进口港到岸价
port of exit 出口港
port of export 运出港
port of fetching 抵港
port office 港务处;港务局;港务办公室
port officer 港务管理人员
Port of Huangpu 黄埔港
port of landing 登陆港
port of loading 发运港;装港;装货港;载货港(口)

port of loading for imports 进口装货港口
Port of London Authority 伦敦港务局
Port of Nantong 南通港
Port of New Yoke 纽约港
Port of New Yoke Authority 纽约港务局
port of origin 出口港;输出港;始发港;出发港
Port of Qingdao 青岛港
porto franco 免税出入港
port of recruit 接应港;补给港(口)
port of refuge 避风港;安全港(口)
port of registry 注册港;船籍港
port of reshipment 转运港
port of sailing 起航港;启船港;出发港
Port of Shanghai 上海港
port of shelter 避风港
port of shipment 出发港;装货港;起运港;发货港
port of shipping 海运港;装船港;装运港
port of transit 中转港(口)
port of transshipment 转运港;转口港;中转港(口)
port of unloading 卸货港;起货港
portoise 船舷上缘
porto marble 黑金大理岩
port opening 喷火口
port operating district 港口作业区
port operating organization 港口营运机构
port operation service 港口通信业务
porto power 港液压救援设备
Port Orford cedar 美国花柏;奥福德港雪松
Porto Rico trench 波多黎各海沟
Port Oslo 奥斯陆港(挪威)
port outboard 左舷外
port outlet 出口(指水、气、汽、油等)
port packer 港口用集装箱装卸机
Port Panjiang 潘姜港(印度尼西亚)
port pedestal crane 港口台架式起重机
port plan 港区图;港口平面图
port planner 港口规划人员
port planning 港口规划
port plant 海港设备
port plate 配流盘
port police (force) 港口公安所;港口公安处;港口公安局
port pollution 港口污染
port poniter 通路指示字
port power supply 港口供电
port practices 港口业务手续
port promotion group 港口业务发展组
port promotion section 港口业务发展组
port quarter 左后方(船);港务局及附属建筑物所在区域;船尾左舷方向
port radar 港口雷达
port radar installation 港湾雷达设备
port radar system 港口雷达系统
port radio station 港口无线电台
port railroad 港内铁路;港口铁路(美国)
port railroad station 港口车站【铁】
port railway 港区铁路;港区铁道;港内铁路;港口铁路(英国)
port railway station 港口车站【铁】
portrait 肖像
portrait format 肖像像幅
portrait of angel 天使像
portraiture 肖像画法
port rates 港税;港口税率;港口费率
portray 描绘
portrayal method 描绘法
portrayal of terrain 地形形态描术
port reach 港内河段
port reeve 装卸监督员;港务管理人员;港口管理人员
port refuge 避难港
port regulation 港口管理条例;海港法规;港章
port repair ship 港口修理船
port risk policy 港口停泊险
port risks 港内安全险;口险;港内停泊险
port risks insurance 停航保险
port road 港区公路;港口道路
port roof 气窗拱;喷出口顶
port rubber 舷窗橡胶
port rule 港口管理条例
port sanitary statement 港内疫情通告
Port San Juan 圣胡安港(波多黎各)
Port Santos 桑托斯港(巴西)
port sash 小舷门;排水口翼盖

port's cargo throughput 港口货物吞吐量
port scene 港口面貌
port service 港务
port service signal 港口服务信号
port sharing unit 端口共享设备；端口共享部件
Port Shuwaikh 舒威赫港（科威特）
portside 左舷侧；左舷的；左舷
portside aids 左舷航标；左侧航标；左岸航标
portside light 左舷灯
portside of channel 航道左侧
portside opening 驳门
portside pontoon 左舷侧浮箱
portside wall 小炉侧墙
port sign 港口号志；港口标志
port signal 港口信号
port sill 舷门槛
port silting 港内淤积
port site 港址
port site selection 港址选择；港口选址
port siting 港址选择；港口选址
port's maximum monthly tonnage of cargo transferred 港口最大月吞吐量
Portsmouth ball valve 朴斯茅茨式球阀
port speed 港口装卸效率
port stacker 移动式装载机
port state control 港口国管理
port station 港口站；港口车站
port station in advance 港前车站
port's territory lighting 港区照明
port's throughput capacity 港口吞吐能力；港口吞吐量
port storage system 港口堆存系统
port structure 港口建筑物；港口构筑物
port surcharge 港口附加费
port survey(ing) 港湾测量
port's volume of freight traffic 港口货物吞吐量
port swing line 左转弯线；左转向线
Port Szczecin 什切青港（波兰）
port tack 向左抢风调向
Port Tacoma 塔科马港（美国）
port tank 左柜
port tariff 港口费率
port tax 港口税
port telephone office 港口电话所
port telephone switching center 港口电话中心
port terminal 港口码头
port throughput capacity 港口吞吐能力
port time 停港时间
port time of rail car 铁路车辆在港停留时间
port timing 进排气孔口开闭时刻图；汽门定时
Port Tokyo 东京港（日本）
port-to-port 港到港；码头到码头
port-to-port basis 港至港条件
port-to-port container service 港际集装箱运输业务
port-to-port contract 港至港合同
port-to-port protocol 端口到端口协议；港至港运输业务
port to port service 码头到码头运输业务
port-to-port traffic 港到港交通
port tower 港口瞭望塔
port traffic 港内交通
port trafficability 港口通过能力
port trafficability of loading/unloading siding 港区铁路装卸线通过能力
port traffic department 港口运输部门
port traffic facility 港内运输工具
port transport 港内运输
port transportation tools 港内运输工具
Port Trieste 的里雅斯特港（意大利）
port trust 港口托拉斯
port tug 港作拖轮
portuary area 港区
Portugal orange 葡萄牙橙
Portuguese architecture 葡萄牙建筑
Portuguese bowline 双套结
Portuguese carpet 葡萄牙地毯
port usage 港口习惯
port user 港口用户；港口使用者
Port Valence 瓦朗斯港（法国）
Port Valencia 巴伦西亚港（西班牙）
Port Vancouver 温哥华港（加拿大）
portveyor system 移箱输送机方式
port walker 检查工（平板玻璃）；看边工
port warden 装卸监督员；港务管理人员；港口管理人员
port warehouse 港口仓库
port water area 港口水域
port water depth 港口水深
port waters 港口水域
port water supply 港口给水
port water supply system 港口给水系统
Port Wellington 惠灵顿港（新西兰）
port wheeled crane 港口轮胎式起重机
Port Wismar 维斯马港（德国）
port with bars 沙洲港
port works 港口建筑物；海港工程
porzite 莫来石
posada 小客栈；小旅店
posiode 正温度系数热敏电阻
posistor 正温度系数热敏电阻
posit 安置；安排
positex 阳电荷乳胶
position 形势；状况；职位；阵地；立场；位置；地位；船位、航向和航速；层位；额度；部位
position 0 零点
position above water 水上定位
position adjustment 位置调整
position advantage 位置优势
positional 位置的
positional accuracy 点位精度
positional adjustment 按照位置调整
positional astronomy 方位天文学；球面天文学
positional camera photography 定位摄影
positional code 位置码
positional controller 位式控制器
positional cylinder 定位缸
positional error 位置误差；定位误差
positional error coefficient 位置误差系数
positional error constant 位置误差常数
positional error detector 位置误差探测器
positional format 位置格式
positional game 位置对策
positional indicator 示位器
positional information 位置信息
positionally weighted binary 按位加权二进制
positional notation 位置记数法；位置计数法；位置表示法
positional number 位置数；按位记数
positional number system 按位记数系统
positional operator 位置算子
positional parameter 位置参数；定位参数
positional punch 定位穿孔
positional representation 位置记数法；按位表示
positional representation system 按位记数制
positional response of carbon microphone 碳粒传声器倾角特性
positional servomechanism 位置伺服机构
positional stability 位置稳定性；定位稳定性
positional title 职称
positional tolerance 位置公差；位置容限
positional tolerance zone 位置度公差带
positional welding 定位焊（接）
position ambiguity 位置模糊度
position and azimuth determining system 定位和定向系统
position angle 旋轮安装角；星位角；位置角；方位角
position approximate 概位
position area sketch 阵地地域草图
position attribution 位置属性
position balance 位置平衡
position balance system 位置平衡制
position buoy 指示浮标；雾中拖标；船位浮标
position busy relay 占位继电器
position by bearing and distance 方位与距定位法
position by dead reckoning 积算船位；推算船位
position by mixed observations 综合定位法
position by observation 观测船位
position by radar 雷达定位法
position by sun sights 日光定位法
position by terrestrial observation 陆标定位法
position check 位置检核
position chorograph 位置测定器
position circle 位置圆；位置圈；定位圆；船位误差圆
position classification 职位分类
position clock 座席钟
position code 位置代码；定位码
position code converter 位置电码变换器
position code distance 位置码距离法
position coder 位置编码器
position code sequential 位置码序列法
position command signal 位置指令信号
position compensation 位置补偿
position cone 定位锥
position control 位置控制；位置调节器
position control by maneuver 操船控制船位
position control cam 位置调节凸轮
position controller 位置控制器
position control lever 位置调节手柄
position control lever assembly 位调节杆总成
position control loop 位置控制回路
position control servomechanism 位置跟踪系统
position control signal 位置控制信号
position control system 位置控制系统
position coordinates 点位坐标
position correction 位置校正
position data 位置数据
position defilade 遮蔽阵地
position detection 位置检查
position detector 箱位检测器；位置检测器
position determination 定位
position determining equipment 定位设备
position deviation 位置偏差
position dialing 座席拨号
position diamond 摆放金刚石
position dilution of precision 位置精度几何因子
positioned plug valve 定位旋塞阀
positioned weld 定位焊；定位焊接；暂焊；平位焊缝
position effect 位置效应
positioner 转动换位器；转动定位器；焊接变位机；位置控制器；胎具；定位装置；定位器；反馈放大器；变位机
positioner and orientation 位置控制和定向
position error 点位误差；船位误差
position error of sheet corner 图廓点位移值
position error of traverse 导线点位误差
position error probe 偏位传感器
position estimated by sounding 测深辨位
position failure 位置失效；位置故障
position feedback 位置反馈；定位反馈
position feedback bridge 位置反馈电桥
position feedback signal 位置反馈信号
position filar micrometer 位丝测微计
position finder 测位仪；测位器
position finder ball 自位滚珠
position finding 位置测定；探测；定位；船位测定；定位测定
position finding by intersection method 交会定位法
position finding device 测位仪；测位器
position finding instrument 测位仪；测位器
position fix 定位
position fixed ends 固定端位置；位置固定端
position fixing 定坐标；定位；测定位置
position fixing and navigation(al) system 定位与导航系统
position fixing by polar coordinate method 极坐标法水深定位
position fixing error 位置测定误差
position fixing system 定位系统
position for gradual application of brake 制动机的制动位
position for moderate application of brake 制动机适度制动位
position for reception of fault notices 收受故障通知的座席
position for repairing a rolling stock 修车台位
position for the maneuver 占据机动起始位置
position function 位置函数
position game 位置对策
position gyro 位置陀螺
position head 高程水头；落差；位置（水）头；位势头；位势水头
position identification 位置识别
position independence code 位置无关码
position indicating grid 位置移动指示图
position indicating mark 示位标
position indicating switch 位置指示开关
position indication 位置指示
position indicator 位置指示器；示位器；定位仪
position indicator coil 位置指示器线圈

position indicator pointer 位置指针
position influence 位置影响
positioning 就位;配置;调位;定位
positioning accuracy 定位精度
positioning action 定位作用
positioning anchor 定位锚
positioning and location system 位置测定系统
positioning a stranded vessel to prevent shifting 难船固位
positioning band 定位弹带
positioning bolt 定位螺栓
positioning buoy 定位指示浮标
positioning by bearing and distance 方位距离定位(法)
positioning by laser 激光定位
positioning by radio direction finder 无线电测向仪定位法
positioning by radio goniometer 无线电测向仪定位法
positioning by wireless compass 无线电测向仪定位法
positioning camera 定位摄影机
positioning control 居中调整;中心控制;定位控制;定位操纵
positioning controller 定位调整器
positioning corn 定位销
positioning device 定位装置
positioning disc 导向盘
positioning dowel 定位销
positioning element 定位件
positioning error 定位误差
positioning fixture 定位夹具
positioning format item 定位格式项
positioning for size 按大小排位
positioning indicator 定位指示器;状态指示器
positioning information 定位信息
positioning key 定位键
positioning line 定位线
positioning mark 定位标记
positioning method by photographing satellite orbit 卫星轨道摄影定位法
positioning numerical control 定位数控
positioning of beams 射线位置调整;梁的定位
positioning of cast-in-place pile 灌注桩定位
positioning of line 线路放样
positioning of piston operated valve 活塞控制操纵阀的开启位置;活塞控制操纵阀的开启程度
positioning of tendons 钢丝束定位(预应力)
positioning pointer 定位指针
positioning projector 定位投影器
positioning radar 雷达定位
positioning radar accuracy 雷达定位精度
positioning relay 定位继电器
positioning screw 定位螺钉
positioning servo 定位伺服机构
positioning signal 定位信号
positioning-sounding system 定向测深系统
positioning space 定位间距
positioning spigot 定位栓塞
positioning stop 定位档
positioning structure 定位建筑物
positioning system 定位系统;自动位置调节系统;测网系统
positioning table 定位工作台
positioning time 定位时间
position instrument by bearings 方位定位仪
position-invariance 移位不变性
position invariant operation 位置不变性运算
position-keeping 定位
position lamp 指示灯
position learning time 确定距离时间
position length 位置长度
position length for repairing car 修车台位长度
position light 状态指示灯;锚位灯;航行灯
position light signal 位置色灯信号
position limit 位置极限
position limitation 位置限制
position limit price 入市限价
position line 定位线;位置线;船位线;方位线
position line table 高度方位表
position line transferred 转移船位线
position link 座席线路
position link controller 座席线路监控员
position load distributor 呼叫分配器

position loop 位置回路
position map of survey line 测线位置图
position mark 定位符号
position measurement 位置测量
position measuring instrument 位置测量仪表
position memory 位置记忆
position message 位置报告
position micrometer 位置测微器
position misclosure 坐标闭合差
position modulated 调位
position modulation scan 位置调制扫描
position notation positional notation 按位记数法
position number 位置号码
position of accident 出事地点
position of adjacent traces 近道位置
position of administrative division 行政区位置
position of assembly 安装装置;安装位置
position of bearings 支承位置;轴承位置
position of biont starting plane 生物初始面位置
position of biont terminal plane 生物终结面位置
position of biont transform plane 生物转换面位置
position of boundary of rock 岩石界面位置
position of building material field 产地位置
position of cave-in 井塌深度
position of circle 度盘位置
position of covered tools 埋钻深度
position of culmination 高点位置
position of deltaic apex 三角洲顶点位置
position of depression 凹点位置
position of depth point 深度点位置
position of equilibrium 平衡位置
position of experimental part 试验段层位
position of far traces 远道位置
position of fault 故障地点
position of focal plane 焦面位置
position of ground fix station 地面定位台位置
position of ground subsidence 地面沉降位置
position of groundwater parting 地下分水岭位置
position of heavenly body 天体位置
position of homogenous subarea 均匀区段位置
position of interesting point 交点位置
position of leakage sector 渗漏段位置
position of load 荷载位置
position of lost circulation zone 漏层井深
position of magnetic anomaly 磁异常位置
position of magnetic base station 磁测基点位置
position of magnetic general base station 磁测总基点位置
position of membrane 膜位置
position of nannofossil 超微化石带位置
position of observation 测站
position of pal(a)eoshoreline 古海岸线位置
position of peril 危急情势
position of platform 平台位置
position of plunging crown 倾伏端位置
position of processed interval 处理井段位置
position of related features 位置公差
position of ship 船位
position of spectator 观测站;见解;现点;立足点
position of sporopollen zone 孢粉带位置
position of station 测站位置
position of strength 实力地位
position of subsidence center 沉降中心位置
position of tectonic region 构造区位置
position of telescope 望远镜位置
position of the contours 等高线位置
position of the first trace 初始道位置
position of vadose overflow 渗流溢出点位置
position of welding 焊位;焊接位置
position of well rows 井排方位
position operator 位置算子;位置算符
position order 点位等级
position paper 意见书;专题报告
position parameter 位置参数;位置参量
position pilot lamp 座席引示灯
position plotter 航迹绘算器
position plotting sheet 定位作业纸
position pseudoallelism 位置拟等位性
position pull 插图位置
position pulse 定位脉冲;定时脉冲
position recorder 位置记录器
position reference system 位置基准系
position register 座席计次器
position regulator 位置调整器

position relay 位置继电器
position report 位置报告;船位报告
position reporting procedure 位置报告程序
position representation 位置表象
position roll stabilization system 滚动自动稳定系统
position sampler 脉位取样器;脉位采样器
position selector valve 换向阀;位置选择阀;方向控制阀
position-sensing unit 敏感件
position sensitive 位置灵敏
position sensitive detector 位置敏感探测器;位置灵敏探测器
position sensitive Geiger counter 位置灵敏盖革计数器
position sensor 位置传感器
position sheet 空白海图;编制头寸表
position signal indicator 位置信号指示器
position signal(ling) 位置信号
position signalling apparatus 位置信号器
position specializing in reception 专用于收受的座席
position specializing in transmission 专用于传递的座席
position strength formula 位置强度公式
position surface 位置面
positions wanted 求职广告
position switch 行程开关
position telemeter 位置遥测计
position telemetering 位置遥测术
position timing 位差调整
position title 职务称呼
position to be tested 试压部位
position tracker 位置跟踪器;船位跟踪器;反射作图器;反射描绘器
position trade 期货交易
position trader 期货买卖者
position trading 期货交易
position transducer 位置转换器;位置传感器;位置变换器
position transmitter 位置变送器
position triangle 位置三角形
position type telemeter 位置式远测装置;位置式遥测计
position underwater 水下位置
position updating 位置修正
position vacant 招聘
position value 期货价值
position variable 位置变量
position variation 位置变化
position vector 位置向量;位置矢量
position voltage 位置电压
position voltage control 位置电压控制
position wanted 待聘
position water meter 水位计
position weld(ing) 定位焊接
positive 正的
positive absorption 正吸收;正吸附
positive acceleration 正加速度
positive acceleration phase 正加速相
positive acceleration turn 正过载转弯
positive acknowledgement 肯定应答
positive acknowledge signal 肯定信号
positive-acting clutch 刚性离合器
positive-action low-bowl 施压铲土的浅铲斗;强制铲土的浅铲斗
positive after-image 正余像;正后像
positive after-potential 正后电位
positive air pressure control 正风压控制
positive alignment 正对准
positive allowance 正公差
positive alternating with magnetic anomaly 正负相间磁异常
positive altitude 仰角
positive amplitude modulation 正极性幅度调制;正极性调幅
positive and negative variable 正负变量
positive angle 正角
positive angle of elevation 仰角
positive anomaly 正异常
positive appliance 填筑机械
positive arc (六分仪的)主弧;正弧
positive area 相对上升区;正相区;正相单元
positive area of influence 正影响面积

positive armature 正极性衔铁
positive artesian head 正涌泉水头;正承压水头
positive aspect 有利方面
positive attitude 积极态度
positive axis 正轴;正半轴
positive azeotrope 正共沸混合物
positive back rake tipper bit 正斜镶嵌头
positive balance 正平衡;正均衡
positive bar 正杆
positive barrier to seepage 直达基岩的隔水层;直达基岩的防渗墙;完全防渗墙
positive battery metering 增压计量
positive beam 正粒子束
positive belt 正电输送带
positive bending 正弯曲
positive bending moment 正弯矩
positive bias 正偏置;正偏压;正偏流
positive biasing battery 正偏压电池
positive bilinear form 正双线性形式
positive birefringence 正双折射
positive blanking pulse 正消隐脉冲
positive block 绝对闭锁
positive blower 旋转鼓风机;正压鼓风机;增压鼓风机;离心鼓风机
positive booster 增压机;升压器;升压机
positive boundary 正边界
positive braking 有效制动
positive brush 正电刷
positive buoyancy 有效浮力;正浮力;浮托力
positive bus-bar 正母线
positive camber 正上挠度;正曲率;外曲面;外弧
positive carrier 正载流子;正载波;正荷载流子
positive cash flow 正现金流量
positive catalysis 正催化作用
positive catalyst 正催化剂
positive catch 强制啮合机构;强制啮合
positive caving 强制崩落
positive cerium anomaly 正铈异常
positive chain complex 正链复形
positive character 正光性
positive characteristic distortion 正特性失真
positive charge 阳电荷;正电荷
positive charge carrier 正电荷载流子
positive charge group 正电荷基
positive chemography 正化学照相法
positive chemotaxis 正趋化性
positive chuck 刚性夹盘
positive circle throw gyratory movement 强制旋摆运动;强制旋回运动
positive circle throw screen 刚性圆程振动筛
positive circle throw type screen 强制旋摆式筛分机
positive circulation 强制循环
positive clamping 正向箝位
positive clipping 正向削波
positive clock 正同步脉冲
positive clock pulse 正的同步脉冲
positive clockwise 取顺时针方向为正的
positive closing 强制关闭
positive closure 正闭包
positive clutch 牙嵌离合器;刚性离合器;非摩擦离合器
positive coastline 正性海岸线;正性岸线
positive cochain complex 正上链复形
positive collar 凸辊环
positive collector grid 正极集电栅
positive colloid 正胶体
positive colo(u)r film 彩色正片
positive column 阳极区;正柱区
positive-column screw instability 螺旋不稳定性
positive complex 正复形
positive conductor 正导线
positive cone 正锥
positive confinement 绝对密封
positive confining bed 正封闭层;上隔水层
positive confirmation 肯定式函证
positive constant 正常数
positive contact 阳极接头;正电接头
positive contrast 正反差
positive control 正控制;完全控制
positive-control thyratron 正压控制闸流管
positive cooling 人工冷却;补助冷却
positive coordinates 正坐标
positive copy 正像本

positive corpusc(u)le 正粒子
positive correction 正校正
positive correlation 正相关
positive Cotton effect 正考顿效应
positive course pitch 正大距
positive crankcase ventilation 曲轴箱强制通风
positive crankcase ventilation system 正压曲轴箱通风系统
positive crankcase ventilator 曲轴箱强制通风装置
positive creep 正向波蠕变;正向波漂移;正向波形不稳;正向波蠕变;正向波漂移;正向波不稳
positive crystal 正晶体
positive current 正电流
positive curvature 正曲率
positive curvature sector 正弯曲扇块
positive cut-off 伸入基岩的截水墙;完全截水墙
positive cycle 正循环
positive damping 正阻尼
positive define system 正定系统
positive definite 正定的;正定
positive definite differential form of degree two 正定二次微分形式
positive definite form 正定型
positive definite function 正定函数
positive definite Hermitian transformation 正定埃尔米特变换
positive definite inner product 正定内积
positive definite integral form 正定积分形式
positive definite kernel 正定核
positive definite matrix 正定矩阵
positive definite meter 正顶度规
positive definiteness covariance 正定协方差
positive definite operator 正定算子
positive definite quadratic form 正定二次型
positive definite second variation 正定第二变分
positive definite sequence 正定序列
positive definite symmetric(al) kernel 正定对称核
positive definite symmetric(al) matrices 正定对称矩阵
positive definite transformation 正定变换
positive delivery 压力输送
positive delivery of oil 油料的压力输送
positive density 正密度
positive developer 正片显影液
positive deviation 正偏差
positive diageotropism 正横向地性
positive diazo process 阳图重氮法
positive die 阳模
positive dihedral angle 正二面角;反角
positive direction 正向;正方向
positive discharge 强制卸土;强制卸载;脉动流出;强制卸料
positive-discharge bucket elevator 脉动卸料斗式提升机;强制排料斗式提升机
positive dislocation 正位错
positive dispersion 正色散
positive displacement 正向位移;正压排浆装置;(湿法喷射混凝土的)混合料输送装置;正向排浆装置;强制排量
positive-displacement air compressor 正排量空气压缩机
positive-displacement compressor 容积式压缩机;正位移压缩机;正排量式压缩机;容积式压气机
positive-displacement engine 容积式发动机
positive-displacement expansion engine 容积式膨胀机
positive-displacement fan 容积式风机
positive-displacement(flow)meter 容积式流量计
positive-displacement gear-type pump 正排水齿轮泵;正排量齿轮泵
positive-displacement grout pump 排液灌浆泵
positive-displacement hydraulic drive 正排量液压传动装置
positive-displacement hydraulic motor 容积式液压马达
positive-displacement meter 正压流量计;排代流量计;正压移动泵;正位移液体计量器;正位移计;(容积式的)正排水流量计;正排流量计;(容积式的)正排量流量计;容积式煤气表
positive-displacement metering valve 容积式计量阀;步进式计量阀
positive-displacement motor 正排量电动机;容积式液压马达;容积式液压电动机
positive-displacement of mixture 混合料输送装置

positive-displacement pump 正位移泵;容积式泵;变容真空泵;正排水泵;正排气泵;容积泵
positive-displacement refrigerator 容积式制冷机
positive-displacement screw type compressor 正排量螺旋式压气机
positive-displacement supercharger 给定式增压器
positive-displacement type 正排量型
positive-displacement type fuel pump 正排量式燃料泵
positive-displacement type water meter 容积式水表
positive distortion 正畸变
positive distribution 正广义函数
positive divisor 正除数
positive dobby 积极式多臂机
positive double complex 正二重复形
positive double-lip seal 刚性双唇密封
positive draft 正压通风;正压气流;正向气流;加力通风;人工通风;强制通风
positive drag 正阻力;正曳力
positive drainage 正向排水
positive drill hole 见矿钻孔
positive drive 正动;正传动;正车驱动;强制驱动;强制传动
positive drive lift 强制驱动电梯
positive driven type supercharger 确动式增压器
positive driving 确定传动
positive echo 正像重影;正回波
positive economics 实证经济学
positive edge clock 正边沿触发时钟脉冲
positive edge of cross cut 剖线的正边
positive electricity 阳电;正电
positive electrode 阳极;阳电极;正极;正电极
positive electron 阳电子;正电子
positive element 相对上升区;正相单元
positive elongation 正延性;正延长
positive emission 阳离子发射
positive emulsion 正性乳剂
positive energy 正能量
positive energy state 正能态
positive energy wave 正能波
positive engagement 螺丝扳手套上;螺丝扳手接合
positive error 正误差
positive europium anomaly 正铕异常
positive even numbers 正偶数
positive excess 正电荷过剩
positive exponent 正阶
positive eyepiece 正像目镜;正目镜
positive factor 积极因素;积极因素
positive feed 机械喂料;机械进料;强制喂料;强制进料
positive feedback 正反馈
positive feedback amplifier 正反馈放大器
positive feeder 正馈(电)线;正极馈电线
positive film 正像胶片
positive flo(a)tation 正浮选
positive force 正向力
positive form 正形;正地形
positive forming 阳模成型
positive frequency 正频率
positive frequency modulation 正调频
positive friction 正摩擦
positive function 正函数
positive functional 正泛函
positive g 正过载
positive gear 直接传动装置
positive-geared feed 压力进给
positive gearing 直接啮合;直接传动
positive geometry programming 正几何规划
positive geothermal anomaly 正地热异常
positive geothermal gradient 正地热梯度
positive getropism 正向地性
positive ghost 正像重影
positive glass 玻璃正片
positive glow 阳极辉光;阳极光
positive-going 朝正向变化;正向(的)
positive-going computation 正演计算
positive-going edge 正沿;上升沿
positive-going input 正向输入
positive-going noise 正程噪声
positive-going pulse 正向脉冲
positive-going sawtooth voltage 正锯齿形电压
positive-going sawtooth wave 正向锯齿波
positive-going signal 正向信号

positive-going sync pulses 正向同步脉冲
positive-going transition 正向变换
positive-going trigger 正向触发器;正脉冲触发;正极性触发
positive goodwill 肯定的商誉;可靠商誉
positive governing 直接控制;直接调整;强迫调整
positive-g period 正过载时间
positive grid 正栅极
positive grid characteristic 正栅极特性
positive grid current 正栅电流
positive grid multivibrator 正栅多谐振荡器
positive grid oscillator 正栅振荡器
positive ground 正极搭铁
positive ground system 正极搭铁系统
positive group 阳根;正基
positive g tolerance 正重力加速度容限;正过载容限
positive guard band 正色纯保护环
positive head 正水头
positive helicity 阳螺旋
positive Hermitian operator 正埃尔米特算子
positive high relief 正高突起
positive hole 空穴
positive homogeneity 正齐性
positive-homogeneous operator 正齐次算子
positive honing 刚性加压珩磨
positive ignition 强制点火
positive ignition engine 强制点火发动机
positive image 阳像;正像
positive image reproduction 正像重现
positive impedance 正阻抗
positive impulse 正脉冲
positive incidence 正迎角;正攻角
positive income effect 积极所得效果
positive incoordination 正失调
positive indication 正的指示
positive induction 正诱导
positive infinitely variable driving gear 正无级变速传动装置
positive infinitely variable unit 正无级变速器
positive infinity 正无穷大
positive information 有用情报
positive input value 正输入值
positive integral 正整数
positive integral 正整数项
positive intensified image 增强正像
positive interaction 正相互作用
positive interest 加仓费
positive interesting point 正交点
positive interference 正干扰
positive ion 阳离子;正离子
positive ion beam 正离子束
positive ion cloud 阳离子云
positive ion current 阳离子电流
positive ion density 阳离子密度
positive ion-exchange resin 阳离子交换树脂
positive ion linear accelerator 正离子直线加速器
positive justification 正码速调整
positive knockout 直接式打料
positive lag 正滞后
positive landform 正地形
positive landing 正型落点
positive lap 正余面
positive latex 阳性胶乳;阳电荷乳胶
positive lead 正表笔
positive leap second 正闰秒
positive lens 正透镜;正态透镜
positive lift force 正升力
positive limiting 正值限制
positive line 正线
positive linear functional 正线性泛函
positive linear mapping 正线性映射
positive locking 强制联锁
positive-locking differential 强制联锁式差速器
positive logic 正逻辑
positive low relief 正低突起
positively actuated valve 正动阀
positively asymptotically stable 正向渐近稳定的
positively biased 正偏置的
positively charged ion 正电性离子
positively charged membrane filter 带正电荷膜滤器
positively charged ultrafiltration membrane 带正电荷超滤膜
positively driven blower 正传动鼓风机

positively homogeneous 正齐次的
positively inclined impervious bottom bed 正坡倾斜隔水底板
positively non-linear correlation 非线性正相关
positively oriented curve 正定向曲线
positively oriented trihedral 正指向的三面形的
positively stable 正向稳定的
positive magnetic anomaly area 正磁异常区
positive magnetic bay 正湾扰
positive magnetic moment 正磁矩
positive magnetostriction 正磁致伸缩
positive mapping 正映射
positive mark polarity 正的传号极性
positive mask pattern 正掩模图
positive matrices 正矩阵
positive matrix 正矩阵
positive maximum 正极大(值)
positive measure 正测度
positive meniscus 正液面
positive meniscus lens 正弯月透镜
positive mesh 正啮合
positive meter 容积式煤气表
positive method 正片法
positive microfilm 缩微胶卷正片
positive mid high relief 正中高突起
positive mineral 正矿物
positive mixer 快速混合机
positive mixing 强制式拌和(法)
positive modular function 正模函数
positive modulation 正极性调制;正调制
positive moment 正弯矩;正力矩
positive moment capacity 正弯矩能力;正弯矩容量
positive moment reinforcement 正弯矩钢筋
positive momentum error 正动量误差
positive motion 确动;强制运动
positive motion cam 确切凸轮;确动凸轮
positive motion cylindric(al) cam 正运动圆柱形偏心轮
positive motion disk cam 确动盘形凸轮
positive motion plate cam 确动平板凸轮
positive mo(u)ld 阳模;密闭式模具;全压式塑模;不溢式模具
positive mounting 正片镶嵌
positive movement 相对向上移动;正向移动;正相运动;上升运动
positive movement of sea level 海面上升
positive negative bipolar 正负双极
positive negative control 正负控制
positive negative controller 正负控制器
positive/negative direction 正/负方向
positive negative logic 正负逻辑
positive negative switch 正负象转换开关
positive negative three level action 正负三位作用
positive nonzero integer 正非零整数
positive normal 正法线
positive number 正数
positive opening 强制打开
positive operator 正算子
positive orbit closure 正轨道闭包
positive order 正指令
positive-order component 正序分量
positive ordinate 正纵坐标
positive ore 肯定储量;正矿石
positive orientation 正定向
positive orientation of spin 正自旋取向
positive orthant 正像限
positive orthogeotropism 正直向地性
positive output 正输出量;正输出
positive ovality 正椭圆度
positive overlap 正重叠
positive paper 阳极纸
positive part 正部分
positive pawl 正棘爪
positive peak 正峰值
positive pedestal voltage 正极性消隐脉冲电压
positive penetrability 正穿透性
positive period 正周期
positive period method 正周期法
positive phase reactance 正相电抗
positive phase-sequence relay 正序继电器
positive phase-sequence voltage relay 正相序电压继电器
positive photoconductivity 正光导性
positive photograph 正摄影;正片

positive photoresist 正性胶;正性光刻胶
positive phototropism 正向光性
positive picture 正像;正片
positive picture modulation 正极性图像调制
positive picture phase 正极性图像
positive picture polarity 正极性图像
positive picture signal 正极性图像信号
positive plane 正平面
positive plate 阳极板;正片;正极板;正极
positive plate cam 确动平板凸轮
positive plate making 阳像制版
positive polarity 正极性
positive polarity transmission 正极性传送
positive pole 阳极;正极;板极
positive potential 正势;正电位;正电势
positive predictive value 阳性预告值
positive pressure 正压(力)
positive pressure air system 正压空气系统
positive pressure breathing 加压呼吸
positive pressure breathing apparatus 正压式呼吸器具
positive pressure burning 正压煅烧
positive pressure cabin 增压座舱
positive pressure control 正压力控制
positive pressure exhauster 正压排气机
positive pressure fan 正压风扇
positive pressure gating system 封闭式浇注系统
positive pressure ventilation 正压通风
positive pressure wave 正向压力波
positive primary photoelectric(al) current 正原始光电流
positive prime 自然回水;离心吸入
positive print 正像复印品;白底蓝图
positive probability 实证概率
positive process 正像工艺
positive prospecting indication 成矿有利标志
positive pulse 正脉冲
positive pulse counting 正脉冲计数
positive pulse stuffing 正脉冲塞入
positive punch 正穿孔
positive punch stripper 正冲头退料器
positive quantity 正量
positive radiation 正辐射
positive radical 阳根
positive radius of communication 有效通信半径
positive rail 正轨迹
positive rake 正前角;前倾度
positive rake angle 正前角
positive ray 阳极射线;正电射线;正射线
positive-ray method 阳离子射线法
positive-ray parabolas 正射线抛物径迹
positive reaction 正反应;正反力
positive reactivity 正反应性
positive real function 正实函数
positive real matrix 正实矩阵
positive receiving 正信号接收;正像接收
positive reception 正信号接收
positive recurrent 正递归的
positive refraction 正折射
positive regeneration 正再生;正回授;正反馈
positive regime 正平衡【水文】
positive regular 正则的
positive regulation 正压稳定
positive reinforcement 受拉钢筋;正弯矩钢筋;正挠钢筋
positive relativity 正相关
positive release wave 正泄放波
positive relief 正向地貌;正突起;正地形
positive replica 复制阳模
positive reserves 动储量;真实储量(指地下水、矿藏等);可靠储量
positive resistance 正电阻
positive resist pattern 正抗蚀图
positive response 肯定应答;肯定响应
positive result 良好结果;成果
positive-return cam action forming die 斜楔作用确定回程成型模
positive reversing 强制换向
positive rhythm 正韵律
positive ridigity 正刚度
positive rolling moment 右滚力矩;正向转动力矩
positive root 正根
positive rotary pump 正旋转式泵;正向旋转泵
positive rotation 正旋转;正转

positive scribing 阳像刻图
positive seal 十足密封
positive sealing system 可靠止水系统
positive semidefinite form 正半定形式
positive semi-definite function 半正定函数
positive semi-definite matrix 正半定矩阵；半正定矩阵
positive sense 正指向
positive sense-class 正向类
positive separation 正幅度差；分色正片
positive sequence 正序的；正序
positive-sequence component 正序分量
positive-sequence coordinate 正序坐标
positive-sequence current 正序电流
positive sequence polyphase system 正序多相系统
positive-sequence power 正序功率
positive series 幂级数
positive sharp wave 正尖波
positive shear(ing) 正剪力
positive shift 正向移动
positive shoreline 正性海岸线；正性岸线
positive side 正面
positive sign 正号
positive signal 正信号
positive skewed distribution 正偏态分布
positive skewed lognormal distribution 正偏态对数正常分布
positive skew(ness) 正偏移；正偏态；正偏斜
positive slope 正斜率；正坡度
positive sol 阳电溶胶
positive space 正空间
positive speed gearbox 正速变速器；减速箱
positive spike 正尖峰信号；正峰信号
positive spin 正旋转
positive spinel 正尖晶石
positive spinel type 正尖晶石型
positive spiral striation 正螺纹
positive square root 正平方根
positive stability 正稳性
positive static liquid head 正静液压头
positive stay 正拉索
positive step 正步
positive stereo 正立体
positive stiffness 正刚度
positive stop 限位挡块；主动停止
positive strand 正链
positive strap 正极连接片
positive streaking 正拖影；正拖尾；图像正拖尾
positive stress 正应力
positive stripe 阳条子
positive stripper 固定卸料板
positive subdefinite 次正定
positive supercharger 正增压器
positive surface charge 正表面电荷
positive surge 正涌浪；水位上涌；潮升过高(由于气象原因)
positive system 正系统
positive tappet 正挺杆；双向挺杆
positive taxis 正趋性
positive temperature coefficient 正温度系数；正温度变化系数
positive temperature coefficient ceramics 正温度系数陶瓷
positive temperature coefficient of resistivity 电阻正温度系数
positive terminal 阳极端；正极接线柱；正极端子；正端子
positive test 正检验
positive thread 阳螺纹；切制螺纹
positive throw eccentric type screen 强制偏心摇晃筛分机
positive thrust 正推力
positive top front rake 正前角
positive topographic(al) unit 正地形单元
positive torsion 顺扭转
positive total-flow gas meter 煤气总流量表
positive traction differential 无滑差牵引差速器
positive train-ground indication loop 车地信息接收环路
positive train identification 有车标志
positive train indication 车地信息接收
positive transfer 正迁移
positive transition 正跃迁

positive transmission 正极性传输；正调制传送
positive triggering pulse 正触发脉冲
positive tropism 正向性
positive tube 正栅压闸流管
positive twist 正扭
positive type 正类型
positive type function 正型函数
positive unloading stripper 固定卸料板
positive value 正值
positive valve 正压阀
positive variation 正变分
positive vector 正向量
positive ventilating 压力通风；强制通风
positive ventilation 压力通风
positive video signal 正视频信号
positive voltage 正压【电】
positive water-hammer gradient 正水锤梯度
positive wave 正波
positive wire 正线；正馈线；正极引线
positive work 正功
positive-working photoresist 正性感光胶
positive worm take-up motion 蜗轮式积极卷取运动
positive writing 正片记录
positive yellow 正黄
positive zero 正零
positive/zero/negative justification 正/零/负码速调整
positive zone pressurization 分区正压通风
positive-α bolometer 正电阻温度系数测辐射计；金属测辐射热计
positivity 正性
positivizaton 正函数化
positizing 正化
positizing polynomial 正化多项式
positrino 正微子
positron 阳电子；正电子
positron-annihilation energy 正电子湮没能量
positron-annihilation radiation 正电子湮没辐射
positron beam 正电子束
positron camera 正电子照相机
positron decay 正电子衰变
positron-decay energy 正电子衰变能量
positron disintegration 正电子衰变
positronium 电子偶素
positron radiation 正电子辐射
positron radiator 正电子辐射体
positron reduction reaction 正电子还原反应
positron scintigraphy 正电
positron tomography 正电子断层照相法
posnjakite 一水兰铜矾
possession by entireties 不可分的所有权
possession by supraeconomic measures 超经济占有
possession of site 占用现场
possession of site and access thereto 现场及其通道占用权
possession through non-usage 非使用占有
possessive interval for construction 施工天窗【铁】
possessor 占有人；持有者
possessory lien 占有性的留置权；占有留置权；财产留置权
possessory limit 地产界
possibilities list 可能事件表
possibility 可能性
possibility function 可能性函数
possibility of a fuzzy event 模糊事件的可能性
possibility of inaction 静止的可能性
possibility of liquefaction 液化可能性
possibility of parking 停车可能性
possibility of reverter (财产继承中的)可收回权
possibility of trouble 故障率；事故率
possible accretion 可能的淤积
possible accuracy 可能准确度；可能精确度
possible adsorbate 容许吸附质
possible capacity 可能容量；可能通行能力
possible carcinogenicity 可能致癌性
possible cause 可能原因
possible coast erosion 可能的海岸冲刷
possible consumption 可能消费量
possible deviation 可能离差
possible effectiveness 可能效力
possible error 可能误差；可能错误

possible explosive risk 有爆炸危险
possible groundwater pollution 潜在地下水污染
possible illegle act 可能的非法行为
possible infiltration 可能渗入
possible liquefaction zone 可能液化区
possible loading rate 容许负荷率
possible loss 可能的损失
possible magnitude of earthquake 可能发震震级
possible maximum precipitation 可能最大降水(量)
possible maximum stage 可能最高水位
possible method of mine 矿山可能规模
possible minimum stage 可能最低水位
possible mining mode 可能开采方式
possible ore 推测矿量；可能矿量；远景储量；地质储量
possible output 可能输出；可能储(藏)量
possible pass capacity 可能通过能力
possible place of earthquake 可能发震地点
possible precipitation 可能降水量
possible precipitation of solid 容许固体沉积物
possible price 可能的价格
possible range 可能范围
possible reserve 可能库存；可能储(藏)量；地质储量；推定储量
possible route 方案进路
possible solution 可能有效解决办法
possible sunshine duration 可照时数
possible time of earthquake 可能发震时间
possible traffic capacity 可通过能力；可能通行能力；可能交通量
possibly bursting water 可能突水
possibly-intended program(me) 可能(的)预期程序
Possion approximation 泊松近似值
possolanic lime 水硬石灰
possum-trot plan 分隔式住宅平面图；穿堂式平面
post 支柱；邮政；桩材；柱(子)；职务；职位；岗位；支台；矿柱；木支柱；贸易站；所；端子；登记(柱)；采矿机械撑柱；标杆
post-absorption 后吸收
post-absorptive state 吸收后状态
post-accelerating anode 后加速阳极
post-acceleration 后效加速度；后加速；后段加速
post-acceleration oscilloscope tube 后加速示波管
post-acceleration tube 后加速管
post-accelerator 后加速器
post-accident 事故后
post-accident fire 事故后火灾
post-adjustment 服务地点差价调整数
post-adsorption 吸收完毕
post-aerated burner 光焰燃烧器；扩散式燃烧器
post-aeration 后曝气
postage 邮资
postage paid 邮费已付；邮费付讫
post agitator 后搅拌机
Postal 国际邮政电报电话联合会
postal car 邮车
postal card 明信片
postal course 函授课程
postal express service 急邮
postal insurance 简易保险
postal kiosk 邮亭
postal knocker 邮件门环
postal map 邮政地图
postal number 邮件编码号
postal order 邮政汇票
postal route 邮路
postal route map 邮路地图
postal service revenue 邮运收入
postal transport charge receivable 应收邮运运费
postamplifier 后置放大器
post-and-beam building 柱梁式房屋
post-and-beam construction 柱梁结构；骨架结构；梁柱结构
post-and-beam framing 梁柱式框架；柱梁构架；梁柱构架
post-and-beam structure 柱梁结构
post-and-beam system 柱梁系统
post-and-block fence 柱夹块式围墙；柱板围墙
post and girder construction 梁柱结构
post and girt 梁柱木结构；木框架房屋；立柱围梁(结构)
post-and-lintel 简支梁柱；过梁柱；连梁柱；边柱和

过梁
post-and-lintel construction 立柱横梁结构;骨架结构;梁柱结构;柱梁结构;抬梁式构架
post-and-paling 木栅栏;木栅围篱
post-and-pane 立柱嵌板结构体系;砖木房屋;砖木结构;半木结构
post-and-panel 立柱镶板式
post and panel form 镶板式模板
post-and-panel structure 砖木结构;立柱镶(板)式结构;板柱结构
post-and-pan house 砖木房屋
post and petrail 立柱小栏杆;砖木建筑;半木建筑
post and rail 立柱栏杆
post and tele-communications office 邮电所;邮电局
post and timber 柱子与支架
post and wire fence 钢丝栅栏;立柱铁丝网围栏
post-annealing 焊后退火;氧割后退火;后退火
post-anneal treatment 后期退火处理
post a ship 公告某船动态
post assembling 组装后
postassium acetate 醋酸钾
postassium hypochlorite 次氯酸钾
postattack ecology 核袭击后生态学
post audit 事后审计
post baking 后烘干
post bearing 墙托架轴承
post billing system 事后付款系统
post boat 邮务艇;邮务船
postboost 主动段以后;关机后的
post-boost guidance 被动段制导
postbox 信箱;邮箱;邮筒
post brace 立柱支撑
post bracket 交叉柱托架;角柱托架
post brake 柱式制动器
post-breaking wave height 破碎后波高
postbuckling 超临界屈曲
postbuckling behavio(u)r 超临界屈曲状态;滞后压屈现象;后期压屈特性
postbuckling ductility 开裂后延性
postbuckling strength 纵向压弯后强度;滞后压屈强度
post burn 伤完时间
postburst 爆后
post-calcination treatment 后煅烧处理
post-Caledonian platform 后加里东地台
post-calibration 测后检定
post cap 柱帽;木柱铸铁帽
post carrier 柱桩运輸架
post cedar 北美翠柏;香肖楠
post-censal estimates 普查后估计
post central office 邮政总局
post chlorination 后加氯化;后氯化
post-chlorination treatment 后氯化处理
post clay 不纯硅质耐火黏土
post-climax 超演替顶极;后顶极群落;超顶极;超电极
post-closing balance sheet 结账后资产负债表;结账后平衡表
post-closing entries 结账后分录
post-closing trial balance 结账后试算表
post-coagulation 后混凝
post code 邮政编码;记入代码
post-code building 公布法规后修建的房屋
post-colo(u)ring 后着色
post colo(u)r value 后色值
post column 支柱
post-column reactor 后柱反应器;柱后反应器
post-combustion 后燃;补充燃烧;补充燃料
post completion audit 完工后审计
post-completion service 竣工后服务;完工后附加工程;竣工建筑附加业务
post-conceptual painting 后期概念绘画
post condenser 后冷凝器
post-construction 工程建成以后的;施工后的
post-construction monitoring 竣工后监测
post contraction 后期收缩
post-cooling 后期冷却(法)
post-cooling crack 延迟裂纹
post-cooling method 后期冷却(法)
Post correspondence problem 波斯特对应问题
post-cracking 后发开裂;次生裂缝
post-cracking ductility 裂后延性
post-cracking strength 裂后强度

post-cracking stress 裂后应力
post crane 塔柱起重机;转柱(式)起重机
postcrash 撞击引起的火灾
postcrash fire 摔毁后起火
postcritical 超临界的;超过临界点
post-Cronak 锌版剩余药层去除法
post crown 桩冠
post-crystalline 晶后的
post-crystalline deformation 晶后变形
post-cure 后硫化;后固化;二次硬化;二次硫化
postcure bonding 二次硬化黏接
post cured insulation 二次固化绝缘
postcure finish 后热定形整理
postdam 鱼堤;后堤
Postdam sandstone 波茨坦砂岩
post dated 填迟日期的;填迟日期
post-dated bill 填迟日期票据
post-dated check 日期填迟支票;填迟日期支票
post-dated cheque 填迟日期支票
post-defecation 二次澄清
post deflection acceleration 倾斜后加速度
post-deformation axial-ratio 标志体变形后轴率
post-deformation length of markers 标志体变形后长度
post-deformation structure 沉积后构造
post delineator 柱式轮廓标
post density 支柱布置密实度
post-depositional 沉积作用后的;沉积后的
post-depositional fold 同沉积褶皱
post-depositional process 沉积期后作用
post-depositional structure 后沉积构造
post-design evaluation 设计后评价
post-detection 后检波
post-detection band width 检测后带宽
post-detection dedispersing technique 检波前消色散技术
post-detection integration 检定后积分
post-detector 检波器后的
post detritus depositional remanent magnetization 沉积后沉积碎屑剩余磁化强度
post-diluvial period 后洪积世
post-disaster damage survey 灾后损害调查
post-disaster phase 灾后阶段
post-disaster reconstruction 灾后重建
post-disaster relief 灾后救济
post-disaster syndrome 灾后综合征
post-distillation bitumen 馏余沥青
post-drawing 后拉伸;支架拆除
post-dredging survey 浚后测量
post drill 支架式钻机;支架式凿岩机;支架式风钻;架式风钻
post driver 打桩机;竖杆机
post duty 岗位责任制
post-earthquake 余震
post-earthquake damage survey 震后灾害调查
post-earthquake deformation 震后变形
post-earthquake fire 震后火灾
post-earthquake functionality 震后功能
post-earthquake investigation 震后调查
post-earthquake land use planning 震后土地利用规划
post-earthquake movement 震后活动
post-earthquake slip 震后滑动
post echo 后回音
posted 已过账的
posted capacity 公布容量
post edit(ing) 事后编辑
post-edit program(me) 算后编辑程序
posted loading 标明桥梁载重;标明载重(桥梁)
posted price 牌价;标明的价格;标价
posted rate 挂牌汇价
post-effect 后辐照效应
post-effect of drought 干旱遗留症
post-elastic behavio(u)r 弹性后效
Postel's projection 波斯特尔投影
post emergence 萌芽后
post-emergence application 出土后施用
post-emergency evaluation 紧急过后的评价
post-emphasis network 后加重网络
post employment activities 聘后作业
post-emulsification penetrant 乳化性渗透液
post entry 事后报关
post-enumeration check 普查后核查
post-enumeration test 普查后检查

post-equalization 后置补偿;后均衡复元
poster 招贴画;招贴
poster bed 柱床
poster board 广告板
poster cloth 广告招贴布
poster colo(u)r 招贴画颜料;广告色;广告画
post-erecting car 立杆作业车
posterior 后面的;后的
posterior asynclitism 利次曼氏倾斜
posterior basal segment 后基底段
posterior border of petrosal part 岩部后缘
posterior branch 后支
posterior distribution 后验分布;后分布
posterior error bound 后验误差界
posterior estimate 后验估值;后验估计
posterior gray column of spinal cord 后灰柱
posterior information 后验信息
posterior layer 后层
posterior part 后部
posterior photocuring 后光固化
posterior probability 验后概率;后验概率
posterior probability density function 后验概率密度函数
posterior probability distribution 后验概率分布
posterior probability law 事后概率法则
posterior root 后根
posterior segment 后段
posterior spiracle 后气孔
posterior surface 后面
posterior value 终值
posterior wall 后壁
poster map 宣传(地)图
postern 边门;后门;便门;暗道
postero-anterior oblique position 后前斜位
postero-anterior[P-A] position 后前位
posterosuperior 后上
posterotransverse diameter 后横径
poster paint 广告涂料
poster paper 招贴纸;广告纸
poster session 海报会
post evaluation 项目完成后审核;后评估
post-event data reduction 震后数据整理
post-event inspection 震后检查
post-expansion 后期膨胀
post export financing 出口后融通
post failure behavio(u)r 后期破坏特性
post-fault equilibrium 故障后平衡
post fence 柱式围篱;柱式护栏;护柱
post filling 后填法
post-filter 最后过滤器;后过滤器
post-filtration 后过滤
post-filtration tank 后过滤池
post-filtration unit 后过滤装置
post fire critique 火灾后的评论
post fire hydrant 柱式消火栓;柱式消防栓
post fire inspection 火灾后检查
post-firing check 起动后检查
postfix notation 反向波兰表示法
postflare 大头墩;大头柱
post flash-over fire 袭燃后火灾
post-flood 灌后
post footing 桩基脚;杆桩基脚;杆柱基脚
post-formation theory 后成说;渐成论
post-formed plywood 热压成型胶合板
post forming 后续成型;热压成型;二次成型;后成型
post-freshet period 汛后期
Post Functionism 后功能主义
post gasket 杆衬
post gate 转柱门
post gatherer 挑料工
post-geosyncline stage 后地槽阶段
post-ginning cleaner 轧后净棉机
post-glacial 冰期后的;冰川期后的
post-glacial age 冰后期
post-glacial climate 后冰川期气候
post-glacial deposit 冰后期沉积
post-glacial period 冰后期
post-glacial period crust movement 冰期后处壳运动
post-glacial system 冰后系
post-graduate education 研究生教育
post-graduate reseach institute 研究生院
post-grouting 后灌浆
post grouting for cast-in-situ pile 灌注桩后注浆

post hanger bearing 柱式悬轴承;柱式吊架轴承
post head 接线柱【电】
post-heat 后热
post-heat current 焊后加热电流
post-heating 焊后加热
post-heating current 后热电流
post-heat temperature 后热温度
post-heat treatment 后热处理;随后热处理
post-Hercynian platform 后海西地台
post holder 挑料工
posthole 小孔;柱坑;柱洞
posthole auger 挖洞器;匙形取样器;勺钻;柱孔(螺旋)钻;勺形钻
posthole borer 挖洞器;匙形取样器;匙形取土器
posthole digger 挖洞器;匙形取样器;匙形取土器;柱穴挖掘机;柱坑挖掘机;大型螺旋挖坑机
posthole rig 多次修理过的钻探设备
posthole well 浅坑;浅井
post homogenizing 后均化
post house 邮政所;驿馆
posthumous fold(ing) 复活褶皱;再次褶皱
posthumous structure 复活构造
post hydrant 邮筒式消防栓;柱式消火栓;柱式消防栓
post hydrolysis 后水解
postiche 庸俗装饰;不合适装饰;多余装饰
posticum 后楹廊;(古庙神坛附近的)内门廊;后门廊;后门
post-ignition fire risk model 起火后火灾危险模型
post-ignition hood 二次点火器
post-ignition model 起火后模型
post import financing 进口后融通
post-impounding activity 蓄水后活动性
Post-Impressionism 后印象派
post increment 算后增量
postindexing 后变址
post indicator valve 后指示阀
post-industrial era 超工业时代
post-industrial society 后工业化社会
post-industry society 后工业化社会
post(ing) 过账;记入;登录;传递
posting a new item 传递一个新条款
posting cycle 过账循环
posting field 记录区域
posting journal entries 日记分录过账
posting mounted 安装在架上的
posting mounted drill 架柱式凿岩机
posting of building permit (在工地上)张贴建筑许可证
posting plumb 定位铅锤
posting reference 过账依据
posting square 定位规
postinjection 补充喷射
post-installation 装配后的;安装后
post-installation audit 安装后效果审核
post-installation review 运行考核;安装后检测
post insulator 装脚绝缘子;柱状绝缘子;柱形绝缘子;柱式绝缘子;低压通信绝缘子;棒式绝缘子
post in the ledger 转入总账
post-intrusion structure 侵入后构造
postique 庸俗装饰;(工程完工后附加的)多余装饰;不合适装饰
post-irradiation effect 后效应;后辐照效应
postis (古腊梅建筑构造中支撑门楣的)门侧壁
post jamb 柱壁;门窗框边框柱
post keeper 线路所值班员
post larva 后期仔鱼;后期幼体
post legged 后踏
post lift 柱式举升机
post light 柱灯
post light support 柱式灯架
Post machine 波斯特机器
post-magmatic 岩浆后期的
post-magmatic mineral deposit 岩浆期后矿床
post-magmatic ore deposit 岩浆期后矿床;岩浆后期矿床
post-main sequence stage 主序后阶段
postmaximum spectrum 光极大后光谱
post median 正中线后的;下午;午后的
post-Mesozoic platform 后中生代地台
postmetamorphic 后变态的
postmineral faulting 成矿后的断层
postmix 后混合
Post Modern Architecture 后现代建筑

Post Modern Formalism 后现代形式主义
Post-Modernism 后现代主义;后期现代主义
postmortem 算后检查;事后检查
postmortem dump 算后转储;停机后输出【计】
postmortem method 计算后检查法;算后检查法
postmortem program(me) 算后检查程序
postmortem routine 算后检查程序;事后剖析程序
post mo(u)ld 柱的模板(混凝土)
post-mo(u)lding operation 脱膜后加工
post-mounted drill 架柱式凿岩机
post multiplication 后乘;自右乘
post-Newtonian approximation 后牛顿近似
post-Newtonian celestial mechanics 后牛顿天体力学
post-Newtonian effect 后牛顿效应
post-Newtonian metric 后牛顿度规
post-normalization 后规格化
post-Norman 后诺曼底
postnotum 后背板
post-occupancy survey 使用后调查
post-occupation evaluation 使用后评估
post office 邮政局;邮局
post office box 邮政信箱
post office bridge 箱式电桥
post office cable 电信电缆
post office department 邮政部门
post office line 电信线路
post office red 邮局红
post office telephone cable 邮局电话线
post office tower 电信塔
post-operational assistance 投产后的援助
postorder for binary tree 二叉树的后根次序
postorder with degrees 带次数的后根次序
post-oreforming fracture 成矿后断裂
post-oreforming structure 成矿后构造
post-ore halo 矿化后晕
postorogenic 造山期后的;造山期后
postorogenic coalification 造山期后煤化作用;造山后煤化
postorogenic phase 造山后期
postorogenic rift 造山期后的裂谷
post-ozonation 后臭氧化作用;臭氧作用后
post paid 已付邮资
post pallet 带有角柱制模板;立柱式托盘;立式托盘;四角加柱的托盘;四角加柱的货板
post-Panamax 超巴拿马型
post panel 告示板
post parcel scale 包裹秤
post-photocuring 后光固性
post pipe vise 柱式管子虎钳
post-plasticity 后增塑性
post-platform stage 后地台阶段
postpleurella 侧板后片
Post-Pliocene 上新世后;晚上新世【地】
post pole 柱杆;孤立柱;电线杆
postponable cost 延期成本;延缓成本;可延期费用;可延期成本
postpone 延期;延后;延迟;暂缓;顺延
postpone a payment 延期支付
postponed installation of tunnel support 先挖后支法
postponed interest 延期利息
postponed result 延期结果
postponement 延期;搁置
postponement of payment 延期支付
postponement of tax 税款延期缴纳
postponement of taxation 延期课税
postponement of the taxation 延期纳税
postponement of the works 工程展期;工程延期
postpone parking number 延停车数
postposition 后置
postprandial diarrhea 餐后腹泻
post-precipitation 继沉淀;后沉积(作用)
post-precipitation fouling 后沉淀污垢
post pressionism 后期印象画派
post-pressure grouting technology 压浆技术
post-processing 后加工;后部工艺;后加工处理;后处理
post-processing method 后置处理法
post-processing system 后处理系统
post-processor 后处理程序;后处理机;事后处理程序
post processor program(me) 后置信息处理程序
post-production activity 生产后作业

post-production handling 后产品处理
post production service 生产后的维修
post-project 工程建成以后的;后续工程
post project evaluation 项目后评价;项目后评估
post project evaluation and follow-up 项目后评价和后继行动
post puller 拔柱机
post-qualification 资格后审
post-qualification study 对投标人资格的事后审查
postradiation 放射后的
post-rank salary system 职务等级工资制
post-reaction 后反应;补充反应
post-reactor 补充反应器
post read station 记录读出站
post-renaissance architecture 文艺复兴后建筑
post-repair behavio(u)r 修复后性能
post responsibility 岗位责任
post responsibility rule 岗位责任制
post result 事后成效
post-revolving mechanism 立柱回转机构
post road 驿路
post rod 顶杆
Post-Sakawa epeirogeny 后佐川造陆运动【地】
post-sale service 产品售后服务
posts and timbers 方柱;方木
postscenium (剧院的)后台
post-scoring method 后录音法
postscript 后记;附加语
post-secondary treated influent 二级处理后的进水
post-secondary treated sewage 二级处理后的污水
post-secondary treated wastewater 二级处理后的污水
post-seismic deformation 震后变形
post-seismic slip 震后滑动
post-seismic strain 震后应变
post-seismic stress 震后应力
post selection 补充拨号
post-set time 退出时间
post setting 柱的设置;架支柱
post shock 余震
post shore 柱支撑;单独立柱;立柱;杆柱;杆撑
postshot 后喷(浆)
post-shrinkage 后收缩
post-shrinkage spacing 支柱间距;后收缩空隙
post sinter 再烧结
post-sintering 重烧结;二次烧结
posts of shade 遮阴栅的支柱
post-solidification structure 凝固后构造
post spacing 柱距
postsphygmic 脉波后的
postsphygmic interval 脉后间期
postspiracular 气孔后的
post-splitting 后裂爆破
post spring 杆簧
post-stack migration 叠后偏移
poststack varying velocity diffraction migration 变速绕射叠后偏移
poststatement event 财务报表编制后发生的事项
post stone 细粒砂岩
post stone foundation 用柱垫石的基础
post stress 后张应力;后加应力
post-stressed 后张的;后加应力的
post-stressed concrete 后张法预应力混凝土;后张法混凝土
post-stressed wire 后加应力钢丝
post-stressing 后加应力;后张的
post-stretching 后张法(预应力结构);后加拉伸;后张拉
post-supported floor 短柱支撑的底层地板
post-sync field-blanking interval 同步信号后消隐间隔;场同步信号后消隐间隔
postsynchronize 后同步
post-synthesis phase 合成后期
post-taxation profit 纳税后利润
post-tectonic 构造后的
post-tectonic crystallization 构造后结晶
post-tectonic filling vein 后构造充填脉
post-telecommunication building 邮电大楼
post-telegram office 邮电所
post telephone and telegraph administration 电话电报管理局
post tender 招标后
post-tension 后加拉力;后加张力
post-tensioned 后拉的;预应力后张;后张的

post-tensioned bar 后张拉钢筋
post-tensioned beam 后张梁
post-tensioned brickwork 后张砌砖体
post-tensioned cable 后张拉钢丝索;后张钢丝索
post-tensioned concrete 后张法预应力混凝土;后张混凝土;后张法混凝土
post-tensioned concrete pile 后张法预应力混凝土桩;后张混凝土桩
post-tensioned concrete segmental bridge 后张法预应力混凝土节段拼装桥
post-tensioned concrete slab 后张法混凝土板
post-tensioned construction 后张拉结构
post-tensioned member 后张预应力构件
post-tensioned method 后张法
post-tensioned prestress concrete 后张法预应力混凝土
post-tensioned prestress concrete segmental bridge 后张预应力混凝土节段拼装桥
post-tensioned prestressing 后张预(加)应力
post-tensioned prestress reinforced concrete 后张法预应力钢筋混凝土
post-tensioned reinforced concrete 后张预应力钢筋混凝土
post-tensioned slab 后加拉力板;后张法预应力混凝土板
post-tensioned steel 后张钢筋
post-tensioned system 后张法;后张拉体系;后张拉设备
post-tensioned tendon 后张筋棘;后张钢筋束
post-tensioned unit 后张锚头
post-tensioned wedge anchorage 后张法楔形锚具
post-tensioned wire 后张拉钢丝
post-tensioning 后加拉力的;后张拉;后张;后拉
post-tensioning anchorage 后张法锚固
post-tensioning cable 后张钢丝索
post-tensioning field log 预应力后张现场记录
post-tensioning method 后张法
post-tensioning procedure 后张预应力法
post-tensioning system 后张拉体系
post-tensioning type 后张法
post-tensioning wedge anchorage 后张法楔形锚具
post-tensioning wire 后张钢丝;后张拉钢丝
post-tension unit 后张锚头
post-Tertiary 第三纪后
post-top luminaire 柱顶灯具
post transport 邮政运输
post transportation 邮件转运
post-treatment 后处理
post treatment equipment 后处理设备
post trigger 后置触发器
post-trigger delay 触发延迟
post truss 柱式桁架
post-type 柱式
postulate 公设;公理【数】
postulated bed rock 假想基础
postulated condition 预定条件;假设条件
postulated point 假设点
postulate of molecular chaos 分子混沌性假设
postulation 假设;假定
post under ridge pole 屋脊梁下支柱
posture in walking 行
post vulcanization 后硫化作用;后硫化
post wage system 岗位工资制
postwar 战后的
postwar architecture 战后建筑
postwar block 战后建造的房子
postwar building 战后建筑
postwar house 战后建的房屋
postwar housing 战后住房
post-weld annealed 焊后退火
post-welded heat treatment 焊接后热处理
post weld heat treated 焊接后热处理
post-weld heat treatment of connections 连接的焊后热处理
post-welding 焊后
post-weld(ing) heat treatment 焊后热处理
post-weld interval 焊后间隙
post-weld treatment 焊后处理
post-works 工程建成以后的
post-write disturb 写后干扰
post-write disturb pulse 写后干扰脉冲;存入后干扰脉冲
post yellow 铅铬黄
post yield 继续屈服

post-yield ductility 屈服后延性
post-yielding behavio(u)r 屈服后性状;屈服后性能
post-yielding effect 屈服后效应
post-yielding stiffness 屈服后刚度
post yoke (柱端的)连接叉
pot 锅穴;罐;壶;电位计(俚语);釜;捕鱼笼
potability of water 水的可饮用性
potable effluent resource application 废水用作饮用水资源
potable groundwater 可饮用地下水
potable type construction 轻便式构造
potable water 饮用水;可饮用水
potable water aquifer 可饮用水含水层
potable water intake 饮用水取水口
potable water network 饮用水管网
potable waterpipework for buildings 建筑中的饮用水管道
potable water pressure pipe 饮用水压力管
potable water quality standards 生活饮用水水质标准
potable water recharge 可饮用水补给
potable water reservoir 饮用水(蓄水)池
potable water resources 可饮用水水资源
potable water supply 饮水供应;饮水供给
potable water supply area 饮用水供应范围
potable water supply pipe 饮用水供水管
potable water system 饮用水系统;饮水系统
potable water tank 饮用水水箱
Pota fibre 波塔纤维
pot ale 锅酒糟
potamic 河道的;河川的
potamic pollution 河流污染
potamium 河流群落
potamobenthos 河流底栖生物;河底生物
potamoclastic rock 河流碎屑岩
potamogenic and lacustrine benthon 河水湖泊底栖生物
potamogenic deposit 江口沉积;河口沉积物
potamogenic rock 河流沉积岩
potamological 河流学的
potamologic geography 河流地理学
potamologist 河流学者
potamology 河流学;河川学
potamometer 流速仪;河川浮游生物;水力计
potamon zone 低地河段
potamoplankton 河流浮游生物
pot annealing 箱内退火;密闭退火
pot arch 坩埚预热炉;加温炉
potarite 汞钯矿;钯汞膏
potary 旋切
potash 钾碱;碳酸钾;草碱
potash alum 纤钾明矾;钾明矾;明矾
potash-bearing sandstone and shale 含钾砂页岩
potash-bearing shale 含钾页岩
potash bentonite 钾盐斑脱土;变蒙脱石
potash blue 钾碱蓝
potash feldspar 钾长石
potash feldspar actinolite schist 钾长阳起片岩
potash feldspar chlorite schist 钾长绿泥片岩
potash feldspar granite 钾长花岗岩
potash feldspar graphic(al) granite 钾长花岗斑岩
potash feldspar rock 钾长石岩
potash feldspathization 钾长石化
potash fertilizer 钾肥
potash glass 钾玻璃
potash-lead crystal glass 钾铅结晶玻璃
potash-lead glass 钾铅玻璃
potash-lime fertilizer 钾钙肥
potash-lime glass 钾钙玻璃
potash lye 苛性钾;钾碱液
potash manure 钾肥
potash mica 白钾云石
potash mine 钾碱矿
potash mine wastewater 钾矿废水
potash mining 钾碱开采
potash prussiate 赤血盐
potash salt 钾碱盐;钾盐
potash salt deposit 钾盐矿床
potash soap 钾肥皂;钾皂
potash water 含钾泉水
potash water glass 钾水玻璃
potassamide 氨基钾
potassic aplite 钾质细晶岩
potassic deposits 钾的沉积

potassic feldspar-bearing theralite 含钾长石霞斜岩
potassic feldspar porphyry 钾质长石斑岩
potassic felsite 钾质霏细岩
potassic fertilizer 钾质肥料
potassic glass 钾玻璃
potassic keratophyre 钾质角斑岩
potassic magnesian fertilizer 钾镁肥
potassic nepeline syenite 钾质霞石正长岩
potassic nepheline microsyenite 钾质霞石微晶正长岩
potassic nitrate 钾硝;硝酸钾
potassic pegmatite 钾质伟晶岩
potassic quartz keratophyre 钾质石英角斑岩
potassic rhyolite 钾质流纹岩
potassic-rhyolite group 钾质流纹岩类
potassic salt 钾盐
potassinm permanganate 过锰酸钾
potassium acid carbonate 碳酸氢钾
potassium acid oxalate 草酸氢钾
potassium acid sulfite 亚硫酸氢钾
potassium-adsorption ratio 钾吸附比率
potassium alcoholate 醇钾
potassium alum 铝钾矾;钾明矾;钾镁矾;钾矾
potassium aluminate 铝酸钾
potassium-alumin(i)um fluoride 氟化钾铝
potassium-alumin(i)um silicate 硅酸铝钾
potassium-alumin(i)um sulfate 明矾;硫酸矾
potassium aluminotrisilicate orthoclase 铝硅酸钾
potassium amalgam 钾汞齐
potassium amide 氨基钾
potassium aminochromate 氨基铬酸钾
potassium-argon age dating 钾亚年代测定法
potassium-argon age determination 钾亚年代测定法
potassium-argon age method 钾氩测年法;钾亚年代测定法
potassium-argon age 钾氩年龄
potassium-argon dating 钾氩测年
potassium-argon method 钾氩法
potassium-argon method of age determination 钾氩法年代测定
potassium arsenate 砷酸钾
potassium arsenite 亚砷酸钾
potassium arsenite solution 亚砷酸钾溶液
potassium balance 钾平衡
potassium base agent 钾基灭火剂
potassium-based mud 钾基泥浆
potassium bentonite 钾盐斑脱土
potassium benzene-diazotate 苯重氮酸钾
potassium benzoate 苯甲酸钾
potassium benzodisulfonate 苯二磺酸钾
potassium benzohydroxamate 苯甲羧肟酸钾
potassium bicarbonate 碳酸氢钾
potassium bichromate poisoning 重铬酸钾中毒
potassium bifluoride 氟化氢钾
potassium binoxalate 草酸氢钾
potassium biphthalate 苯二甲酸氢钾
potassium bisulfite 亚硫酸氢钾;酸式亚硫酸钾
potassium bisulphate 硫酸氢钾
potassium borohydride 氢化硼钾;氢硼化钾
potassium borohydride solution 硼氢化钾溶液
potassium bromate 溴酸钾
potassium bromide 溴化钾
potassium-calcium age dating 钾钙年代测定法
potassium-calcium age method 钾钙年代测定法;钾钙测年法
potassium calcium manure 钾钙肥料
potassium-calcium method 钾钙年龄法;钾钙法
potassium carbonate 碳酸钾;草碱
potassium carbonate glass 钾玻璃
potassium carbonate-lime glass 钾碱结晶玻璃
potassium channel window width 钾道窗宽
potassium chlorate 氯酸钾
potassium chloride 亚氯酸钾;氯化钾
potassium chlorplatinate 铂氯酸钾
potassium chromate 铬酸钾
potassium chrome alum 硫酸铬钾;钾铬矾
potassium chromic oxalate 草酸铬钾
potassium chromic sulfur 铬钾矾
potassium citrate 柠檬酸钾
potassium cobalticyanide 钴氰化钾;氰高钴酸钾
potassium cobaltinitrite 钴亚硝酸钾
potassium content anomaly 钾含量异常
potassium crhmium sulfate 钾铬矾

potassium cyanate 氰酸钾
potassium cyanide 氰化钾
potassium cyanide poisoning 氰化钾中毒
potassium-dating 钾氩测年法
potassium dichromate 重铬酸钾
potassium dichromate digestive solution 重铬酸钾消解液
potassium dichromate filter 重铬酸钾滤光器
potassium dichromate method 重铬酸钾法
potassium dichromate oxygen demand 重铬酸钾耗氧量
potassium dichromate ultraviolet photometry 重铬酸钾紫外分光光度法
potassium dihydrogen phosphate 磷酸二氢钾
potassium dihydrogen phosphate modulator 磷酸二氢钾调节器
potassium dioxalate 草酸氢盐钾
potassium disilicate 焦硅酸钾
potassium disulfate 焦硫酸钾
potassium disulfide 二硫化二钾
potassium dithionate 连二硫酸钾
potassium element colo(u)red map 钾元素彩色图
potassium ethyldithio-carbonate 黄原酸钾
potassium ethylxanthogenate 黄原酸钾
potassium feldspar 钾长石
potassium ferrate 高铁酸钾
potassium ferrate composite aids 高铁酸钾复合剂
potassium ferrate water treatment agent 高铁酸钾水处理剂
potassium ferric oxalate 草酸铁钾
potassium ferric sulfate 铁钾矾
potassium ferricyanide 铁氰化钾;赤血盐
potassium ferricyanide solution 铁氰化钾溶液
potassium ferrocyamde 黄血盐;亚铁氰化钾
potassium fertilizer 钾肥
potassium filter 钾滤色镜
potassium fire 钾火灾
potassium fixation 钾的固定
potassium fluoborate 氟硼酸钾
potassium fluoride 氟化钾
potassium fluosilicate 氟硅酸钾
potassium glass 钾玻璃
potassium gold cyanide 氰化亚金钾
potassium humate 腐殖酸钾
potassium hydrate 氢氧化钾
potassium hydride 氢化钾
potassium hydro-fluoride 氢氟化钾
potassium hydrogen diiodate 碘酸氢钾
potassium hydrogen fluoride 氟化氢钾
potassium hydrogen oxalate 草酸氢钾
potassium hydrogen phthalate 苯二甲酸氢钾
potassium hydrogen sulfite 亚硫酸氢钾
potassium hydroxide 苛性钾;氢氧化钾
potassium hydroxide number 氢氧化钾值
potassium hyperchlorate 高氯酸钾
potassium hyper-oxide 超氧化钾
potassium hypobromite 次溴酸钾
potassium hypophosphite 次磷酸钾
potassium iodate 碘酸钾
potassium iodide 碘化钾
potassium iodide starch test paper 碘化钾淀粉试纸
potassium ion 钾离子
potassium liberation 钾的释放
potassium linoleate 亚油酸钾
potassium loading test 钾负荷试验
potassium magnesium manure 钾镁肥
potassium magnesiumsulphate 钾镁硫酸盐
potassium manganate 锰酸钾
potassium mercuric iodide 碘化钾汞
potassium metabisulfite 焦亚硫酸钾
potassium metaperiodate 高碘酸钾;偏高碘酸钾
potassium metasilicate 硅酸钾
potassium metasomatism 钾质交代作用
potassium mineral 钾矿物
potassium molybdate 钼酸钾
potassium niobate 铌酸钾
potassium niobate-tantalate 铌钽酸钾
potassium nitrogen fertilizer 钾氮肥
potassium osmate 锇酸钾
potassium oxalate 草酸钾
potassium oxalatoantimonate 草酸
potassium oxide 氧化钾
potassium oxide-lead crystal glass 钾铅结晶玻璃

potassium percarbonate 过碳酸钾
potassium perchlorate 过氯酸钾;高氯酸钾
potassium perchlorate explosive 高氯酸钾炸药
potassium perchromate 过铬酸钾
potassium periodate 高碘酸钾
potassium periodate oxidation photometry 高碘酸钾氧化光度法
potassium periodate spectrophotometry 高碘酸钾分光光度法
potassium permanganate 高锰酸钾
potassium permanganate consumption 高锰酸钾消耗量
potassium permanganate method 高锰酸钾法
potassium permanganate oxygen demand 高锰酸钾耗氧量
potassium permanganate poisoning 高锰酸钾中毒
potassium permanganate test 高锰酸钾试法
potassium peroxide 过氧化钾
potassium peroxy-chromate 过铬酸钾
potassium peroxydisulfate 过硫酸钾
potassium persulfate 过硫酸钾;过硫酸;高硫酸钾
potassium persulfate digestion 过硫酸钾消解
potassium persulfate-molybdenum blue method 过硫酸钾钼蓝法
potassium persulfate oxidation 过硫酸钾氧化
potassium persulfate oxidation method 过硫酸钾氧化法
potassium phenate 苯酚钾
potassium phenolsulfonate 苯酚磺酸钾
potassium phenylformate 安息香酸钾
potassium phosphate 磷酸钾
potassium photocell 钾光电池
potassium platinate 铂酸钾
potassium platinichloride 铂氯酸钾
potassium potential 钾位
potassium propionate 丙酸钾
potassium pump 钾泵
potassium pyroarsenate 焦砷酸钾
potassium pyrophosphate 焦磷酸钾
potassium pyrosufite 焦亚硫酸钾
potassium pyrosulfate 焦硫酸钾
potassium release 钾的释放
potassium rhodanate 硫氰酸钾
potassium salt 钾盐
potassium saltpeter 钾硝石
potassium selenate 硒酸钾
potassium selenide 硒化钾
potassium silicate 钾水玻璃;硅酸钾;水玻璃
potassium silicofluoride 氟硅酸钾
potassium-sodium alloy 钾钠合金
potassium-sodium niobate 铌酸钾钠
potassium sodium niobate piezoelectric(al) ceramics 铌酸钾钠压电陶瓷
potassium sorbate 山梨酸钾
potassium sorbate wastewater 山梨酸钾废水
potassium stannate 锡酸钾
potassium succinate 丁二酸钾
potassium sulfate 硫酸钾
potassium sulfide 硫化钾
potassium sulfite 亚硫酸钾
potassium sulfite process 亚硫酸钾法
potassium sulphite process 亚硫酸钾法
potassium super-oxide 超氧化钾
potassium tantalate 钽酸钾
potassium tantallum niobate 铌酸钽钾
potassium tantallum niobate grating 铌酸光栅
potassium terephthalate 对苯二甲酸钾
potassium tetrathionate 连四硫酸钾
potassium thiocyanate 硫氰化钾
potassium thiosulfate 硫代硫酸钾
potassium titanate 钛酸钾
potassium tolerance test 钾耐量试验
potassium tungstate 钨酸钾
potassium vanadate 钒酸钾
potassium wolframate 钨酸钾
potassium xanthogenate 黄原酸钾
potassium zinc chromate 铬酸锌钾
potassium zinc silicate 硅酸锌钾
potato masher 土豆泥机;捣碎器;捣烂机
potato processing waste 马铃薯加工废物
potato pulp 马铃薯渣
potato roll 卷成线球状的水带
potato starch 马铃薯淀粉
potato waste(water) 马铃薯废水;土豆废水

pot bank 陶瓷工厂(英国旧用语)
pot bearing 锅状支承;盆式支座
pot-bottom 锅底石
pot broaching machine 筒形拉刀拉床
pot car 筐式送锭车
pot carriage 坩埚车
potcher 漂洗槽;漂白机
pot clay 坩埚黏土;陶土
pot colo(u)r 坩埚颜色;坩埚熔制的色料
pot core 罐形磁芯;壶形铁芯
pot crusher 罐式压碎机;罐式破碎机
pot culture 盆栽
pot die forming 空心模成型
pot dissolver 罐式溶解器
poteclinometer 连续井斜仪
potency 效能;效价强度
potency assay 效能测定
potency of accelerator 催速剂效力
potency test 效能试验
potency unit 效能单位
potential 位的;电位
potential absorptance 势吸收率
potential accident cause 事故征候
potential acidity 潜在酸度
potential acquisition valuation method 潜在获益评价法
potential active fault 潜伏活动断裂
potential alkali reactivity 潜在碱反应性
potential alternative 可能选用方案
potential and current transformer 电压电流变压器
potential anomaly 电位异常
potential anomaly curve 电位异常曲线
potential anomaly curve of mise-a-la-masse method 充电法电位异常曲线
potential anti-node 电压波腹;电位波腹
potential area 可能含油地区;勘探面积;勘探地区
potential assessment of mineral resources 矿产资源潜力评价
potential at sea level 海平面位
potential attenuator 电势衰减器
potential barrier 位能障碍;位垒;势垒
potential barrier layer 垫垒层
potential borrower 潜在的借款方
potential boundary 电位极限
potential break 潜在断裂
potential capacity 潜在通行能力
potential capacity of land 土地潜力
potential capillarity 毛管潜能
potential carbon of kerogen in source rock 母岩干酪根中有成烃潜力的碳
potential carcinogen 潜在致癌物质
potential cement analysis 潜在水泥成分分析
potential climax 潜在顶极群落
potential clinker composition 潜在熟料矿物组成
potential coefficient 电位系数
potential coefficient of the earth 地球位系数
potential collision point 潜在碰撞点
potential competition 潜在竞争
potential compound composition 潜在的化合物组成
potential contaminant source location 潜在污染源位置
potential contamination of water 潜在水污染
potential contamination source 潜在污染源
potential core 速度核心区
potential customer 潜在顾客
potential dam site 可能坝址;潜在坝址
potential decrease 电位降
potential deformation energy 变形位能
potential demand 潜在需求
potential demand of the market 潜在市场需求
potential density 势密度
potential-determining ions 确定电位离子
potential device 电位器;变压装置
potential diagram 电位图
potential difference 位差;势差;电位差
potential difference induced by secondary field 二次场感应电压
potential difference meter 电位差计
potential differential 位势差
potential distribution 电位分布
potential disturbance 位扰动
potential divider 分压器
potential drop 势降;电位降;磁势降

potential drop of internal resistance 内阻压降
potential-drop-ratio method 电位降比法
potential dynamic(al) method 动电位法
potential earthquake zone 潜在地震区
potential ecological risk assessment 潜在生态风险评价
potential ecological risk index 潜在生态风险指数
potential electrode 测量电极;势电极;电位电极
potential electrode series 电位电极系
potential electrolyte 潜在电解质;势电解质
potential emission 潜在排放
potential energy 位势;位能;势能
potential energy contribution 势能分布
potential energy of deformation 变形位能;变形势能;变形潜伏能
potential energy of gravitation 引力位能
potential energy of strain 应变势能
potential energy of twist 扭转潜伏能
potential energy shortage 潜在能源不足;潜在的能源不足
potential entry 潜在的进入
potential environmental carcinogen 潜在环境致癌物质
potential environmental impact 潜在还原作用
potential equalization 均压
potential equalizer 电位均衡器
potential equalizing jumper 均压跳线
potential equalizing net 均压网
potential equalizing wire 均压线
potential evapo(u)ration 蒸发能力;可能蒸发(量);潜在蒸发
potential evapo(u)rimeter 土壤蒸发计
potential evapo(u)rtranspiration 可能蒸散(发)
potential failure surface 潜在破裂面;潜在破坏面
potential fall 电压降;电位降
potential fertility 潜在的肥力
potential field 位场;势场
potential fire hazard 潜在火险
potential fishery resources 潜在渔业资源
potential flow 潜流;位流;势流
potential flow analyzer 电解槽
potential force 潜在力;位力
potential free contact 无压接触;干式无电位接点
potential free contact output 无压接触输出
potential function 位函数;势函数
potential galvanometer 电位检流计;测量电位的电流计
potential gradient 位能梯度;势梯度;势陡坡;电位梯度;电势梯度
potential gross income 可能总收入
potential gross national product 可达国民生产总值;潜在国民生产总值
potential groundwater yield 地下水供水蕴藏量
potential gum 原在胶;潜伏胶
potential hazard 潜在危险;潜在危害;潜在事故
potential hazardous materials 潜在有害物质
potential head 潜水头;位置头;位置水头;位头;位势头;位势水头;势头;势水头
potential health hazard 潜在健康危害物
potential heat value 潜热值
potential hill 位垒;势垒
potential hole 势穴
potential hump 势峰
potential hydroelectric(al) resources 水力资源蕴藏量
potential hydroenergy 潜在水力能量;水力位势;水力位能
potential image 势像
potential impact 潜在影响
potential impulse 电压脉冲
potential income 潜在收入;潜在利益
potential indicator 带电指示器
potential infiltration rate 入侵度
potential instability 位势不稳定性;对流不稳定(性)
potential irrigation 地下水灌溉
potentiality 潜势;潜力
potential jump 电位突变
potential landslide 将要滑动滑坡
potential lender 潜在的贷款方
potential line 势位线;等水位线;等势线
potential logging 电位测井
potentially active fault 潜在活动断层
potentially available nitrade 潜在有效硝酸盐

potentially environmentally detrimental 可能有害于环境的
potentially harmful substance 可能有害物质;潜在有效物质
potentially ozone-depleting substance 可能消耗臭氧层物质
potentially ozone-destroying process 可能破坏臭氧的过程
potentially plastic hinge zone 潜在塑性铰区
potentially polluting acitivity 可能造成污染的活动
potentially reactivity 潜反应性
potentially toxic chemcials 潜在有毒化学品
potentially toxic volatile organic compound 潜在有毒挥发性有机化合物
potential management 地下水管理
potential market 潜在市场
potential measurement 电位测量
potential measurement in self-potential method 自然电场法电位测量
potential mediator 电位介体
potential mineral 潜矿物
potential mineral composition 潜在矿物组成
potential mineral composition analysis 潜在矿物组成分析
potential mobile phosphorus 潜在活性磷
potential modulus 电位系数
potential motion of a fluid 势流
potential natural vegetation 潜在自然植被
potentialness 潜力
potential nucleus 潜在晶核;潜势晶核;潜力晶核
potential of deformation 变形势
potential of disturbing masses 扰动位
potential of external force 外力势
potential of hydrogen 酸碱度
potential of inertial centrifugal force 惯性离心力位
potential of materiel surface 实体表面位
potential of negative pressure 负压势
potential of penetration pressure 渗压势
potential of production 生产潜力
potential of random masses 扰动位
potential of regeneration 再生电位
potential of solid body 固体位
potential of vector field 向量场位
potential ore 潜在矿量;潜在储量
potential output 可能的产出量;可能产量;潜在产量
potential overpopulaion 潜在过剩人口
potential path length 潜程长
potential pattern 电位充电图
potential-pH diagram 电位酸碱度图;电势酸碱度图
potential phosphorus loading 潜在磷负荷
potential phosphorus recovery 潜在磷回收
potential plateau 势坪
potential pollutant 可能污染物(质);在污染物;潜伏污染物
potential pollutant source 潜在污染物源
potential pollution 潜在污染
potential pollution effect 潜在污染效应
potential pollution index 潜在污染指数
potential pollution source 潜在污染源
potential power 潜在功率;水能蕴藏量;动力蕴藏量
potential pressure head 水压差;水位差
potential production 潜在生产量
potential profile 电势分布图
potential pulse 电压脉冲
potential purchasing power 潜在购买力
potential rate of evapo(u)ration 可能蒸发率
potential-ratio method 电位比探测法
potential reaction 潜在反应
potential rectifier 电位整流器
potential reduction of time 缩短时间的可能性
potential refractive lindex 位折射指数
potential regrowth of bacteria 潜在细菌再生长
potential regulator 电压调节器;电位调节器
potential relay 电压继电器
potential resource amount 潜在资源量
potential resources 潜在资源;潜在的资源
potential rise 电位升
potential risk 潜在危险;潜在风险
potential sales 潜在销售量
potential scattering 势散射
potential science 潜科学
potential shift 势移
potential shrinkage 潜在收缩
potential site 可能坝址

potential slide 潜在滑坡
potential slide area 潜在滑坡区
potential slipperness 潜在滑溜性
potential slip surface 潜在滑动面
potential social productivity 潜在的社会生产率
potential solution tree 期望解树
potential source 压源;位源;势源;电压源
potential source of contamination 潜在污染源
potential source of pollutant 潜在污染物源
potential source of water pollution 水污染潜在源
potential source rock 潜在源岩
potential sources 潜在来源
potential stabilizer 电位稳定器
potential stock 潜在股票
potential strength 潜在强度
potential stress 静电强度;潜在应力
potential stress energy 应力势能
potential subsidence 可能沉降;潜在沉降
potential supply 潜在供给
potential surface 位势面
potential surface analysis 潜力面分析
potential surface of sliding 潜在滑动面
potential survey in mise-a-la-masse method 充电法电位测量
potential swell 膨胀势
potential switch 电压换接开关
potential taxpayer 潜在付税人
potential temperature 温位;位温;势温度
potential temperature effect 位温效应
potential temperature gradient 位温梯度
potential test 钻孔潜在产量测定试验
potential tester 电压试验器
potential theory 位势理论;位论;势能理论;势论
potential thermal effect 潜在热效应
potential titrimeter 电位差滴定计
potential to ground 对地电位
potential traffic 潜在交通
potential traffic capacity 可能通行能力
potential traffic flow 潜在交通流量
potential traffic volume 潜在交通量
potential transform 泊松变换
potential transformer 仪表(用)变压器;压互;电压互感器;测量用变压器;(测量仪表用的)变压器
potential transformer phase angle 电压互感器的相角
potential transmittance 势透过率
potential transpiration 可能蒸腾;可能散发
potential trouble measure 潜在故障预防措施
potential trough 位阱;势能槽;势谷
potential unemployment 潜在失业
potential utility 潜在效用
potential value 潜在价值
potential value of mineral resources 潜在资源价值
potential velocity 位速;潜速度
potential vertical rise 竖胀潜量
potential volume change 体变潜量
potential vortex 自由涡流
potential vorticity 位势涡量
potential wake 势伴流;船尾伴流
potential water head 势头;潜水(水)头
potential water-power resources 水能资源;水力蕴藏量;水利资源蕴藏量;水力资源蕴藏量
potential wealth 潜在财富
potential well 位阱
potential winding 电压线圈
potential wire 电压监察控制导线
potential working population 潜在工作人口
potential yield 可能产水量;最大可能出水量;可能回收率;可能出水量
potentiating agent 增效剂
potentiation 增效作用;增强作用;增毒作用;势差现象
potentioanalysis 电位分析
potentiodynamic(al) analysis 电位动力分析(法)
potentiodynamic(al) method 动电位法
potentiometer 电位器;电位计;电势计;递变电阻电位计;分压器;分压计
potentiometer braking 电位器制动
potentiometer braking controller 电位计式制动器
potentiometer card 电位计用图表
potentiometer circuit 电位器电路
potentiometer controller 电位器控制器
potentiometer function generator 电位计函数发

生器
potentiometer group 电位计组
potentiometer lamp 电位计灯泡
potentiometer measurement 电位计测量法
potentiometer method 电位差计法
potentiometer movable arm 电位计可动臂
potentiometer oil 电位计油
potentiometer output 电位计输出
potentiometer pyrometer 高温电位差计；热电温度计；电温度计；电位高温计
potentiometer resistance 电位器电阻；电位计变阻器
potentiometer slider 电位器滑臂
potentiometer theodolite 电位经纬仪
potentiometer transit 电位经纬仪
potentiometer type calculator 电位计式计算器；电势计式计算器
potentiometer type controller 电位计式控制器
potentiometer type resistor 电位器式电阻器
potentiometer type rheostat 电位计式变阻器
potentiometric amplifier 电位放大器；电势放大器
potentiometric analysis 电位分析法；电位分析；电势分析
potentiometric analyzer 电位分析仪
potentiometric cell 电位滴定池
potentiometric controller 电位控制器
potentiometric differential titration 电势差示滴定(法)
potentiometric extensometer 电势测微计；电测伸缩仪
potentiometric method 电势滴定法
potentiometric model 电解模型
potentiometric model study 电位测量模型研究
potentiometric optic(al) pyrometer 电位计式光测高温计
potentiometric pH meter 电位计式酸碱计
potentiometric pick-up 电位计传感器
potentiometric recorder 电位记录仪；电位记录器
potentiometric stripping analysis 电位溶出分析
potentiometric surface 静压水面；等势面；等测压面；测压管液面；测压管水面
potentiometric titration 电位滴定法；电位滴定；电势滴定
potentiometric titration apparatus 电位滴定设备
potentiometric titrimeter 电位滴定仪；电位滴定计
potentiometric transducer 电位换能器
potentiometric wheel 电势计轮
potentiometry 电位分析法；电位滴定法；电位测定法；电势分析法；电势测定法
potentiostat 恒电位器；电势恒定器
potent ozone depletion substance 强力的臭氧消耗物质
potette 无底坩埚
pot experiment 盆栽试验
pot fireman 控制黏稠料加热的工人
pot floor 空心地面；空心楼面；壶形楼面(一种由空心黏土砖组成的肋构楼面)；空心砖地板；空心砖楼板
pot for caustic soda concentration 固碱锅
pot for walls 砌墙用空心砖
pot furnace 地坑炉；罐炉；坩埚炉
pot gas 烧硫炉气体
pot-grown plant 盆栽植物
pot head 终端套管；端套
pothead insulator 套管绝缘子
pot head tail 交接箱电缆引入口；配线盒进线孔
pot-hole 路面坑槽；路面凹(坑)；圆洞；窝穴；坑洞；锅穴(坑)；壶穴；地穴；沼穴；海底圆谷；潭；水洞；竖井；冰川壶穴；凹坑；凹处
pot-hole patcher 坑洞填补工；坑洞填补机
pot-hole patching 坑洞修补
pot holes 坑槽
potholing 在构筑物上挖洞；路面上坑洼的形成
pot house 小酒店
Potier diagram 保梯图
potier reactance 保梯电抗
Potier's coefficient of equivalence 保梯等效系数
Potin 铜锌锡合金；波廷铜锌锡合金
pot incubation test properly 合理盆栽试验
pot insulator 罐形绝缘子
pot joint 滑块式万向节
pot kiln 坩埚窑
potkinomatic granite 构造纹花岗岩
pot lead 石墨

pot lid 壶盖
pot life 储[贮]放时间；储[贮]存期；坩埚使用寿命；凝结时间；活化寿命；活化期；适用期；保存期限(液体物)
potline 电解槽系列
potline current 电解槽系列电流
potlog hole 脚手架跳板横木孔洞
pot making machine 空心砖制造机
potman 烧沥青工人；熟练沥青工
pot masonry wall 空心砖墙
pot masonry work 空心砖砌体
pot melting furnace 罐熔炉
pot melting method for optic(al) glass 光学玻璃坩埚熔炼法
pot metal 有色玻璃；坩埚制玻璃；铸铁；铜铅合金；锅铁；低级黄铜；锅罐铸铁
pot-metal glass 全色玻璃；彩色装饰玻璃；有色玻璃；彩色装饰玻璃
pot mill 罐磨机；球形磨；瓷瓶球磨机
pot mo(u)ld 空心砖模
pot neoprene bearing 盆式橡胶支座
Potomac bed 波托马克层
potometer 蒸腾计；透明度仪；透明度计；散发计
potomylonite 初糜棱岩
pot-opal glass 坩埚熔制乳白玻璃
Potosi silver 波托西银合金
pot plant 盆栽植物
pot planting 营养钵栽植
pot plunger 压料塞
pot preheating oven 坩埚预热炉
pot process 罐法
pot products 锅类产品
pot quenching 固体渗碳直接淬火
pot-rack 暖瓶架；壶架；瓶架
pot roasting 锅焙烧
pot roof 空心砖屋顶
potroom 电解车间
pot rubber bearing for bridge 桥梁盒式橡胶支座
pot ruby 红宝石玻璃
Potsdam absolute point of gravity 波茨坦绝对重力点
Potsdam gravimetric system 波茨坦重力制；波茨坦重力系
Potsdam sandstone 波茨坦砂岩
Potsdam standard gravity 波茨坦标准重力
Potsdam system 波茨坦系
pot setting 坩埚安装
pot shelving 锅架；盆架
pot sherd 陶瓷碎片
pot sink 罐式洗涤盆；洗碗盆
pot stability 活化稳定性
pot steel 坩埚钢
pot step 空心砖踏步
pot still 蒸馏罐；罐式蒸馏器
potstone 粗皂石；块滑石
potted capacitor 密封式电容器
potted coil 屏蔽线圈
potted component 密封元件
potted concrete surface 有坑洞的混凝土路面
potted flower bed 盆栽花坛
potted landscape 盆景；山水盆景
potted landscape garden 盆景园
potted line 密封线
potted plant 盆景；盆栽花草；盆栽植物
potted surface 有坑洞的路面
potter 陶工；陶瓷工人
Potter-Bucky grid 可动铅格
potter clay 陶土
potter flower pot with perforated carving 带浮雕的陶瓷花盆
Potter oscillator 泊脱振荡器；波特振荡器
Potter process 波特钢丝镀锌法
potter's clay 黏土；陶土；陶塑用黏土
potter's earth 陶塑用黏土
potter's horn 陶工修坯用牛角刀
potter's lead 粗粒方铅矿
potter's plaster 陶瓷工业用石膏粉
potter stone 生石膏
potter's wheel 拉坯轮车；陶土用转盘；陶轮；陶钧
potters wheel 陶工旋盘
potter's work 陶器
Potter-type flowmeter 波特型流量计

potter wastewater 陶瓷制造废水；陶厂废水
pottery 陶器制造术；陶瓷厂；制陶工艺；陶器；陶瓷器；陶
pottery and porcelain 陶瓷
pottery-body stain 陶瓷坯体着色剂
pottery bracket with support 带瓷瓷撑
pottery casting 陶器铸坯；陶瓷器铸坯法
pottery clay 陶土；陶塑用黏土
pottery clay ore 陶瓷黏土矿石
pottery figurine 陶俑
pottery glaze 陶器釉
pottery gui 陶规簋
pottery industry 陶瓷工业
pottery jar 瓦罐；陶罐；瓷坛
pottery kiln 坩埚窑；陶器窑；陶瓷窑
pottery lathe 陶器旋床
pottery making equipment 制陶设备
pottery mosaic 陶土马赛克；陶瓷地砖
pottery mo(u)lding 陶器模制
pottery pillow 陶枕
pottery pipe 陶管
pottery pot 陶壶
pottery stone 陶石；瓷石
pottery tissue 陶瓷包装纸
pottery topped brass screw 瓷头铜螺钉
pottery turning 陶器旋坯
pottery ware 陶器器皿；陶器
pottery waste 陶厂废水
pottery wastewater 陶瓷加工废水；陶瓷厂废水
Potthoff's test 波特霍夫检验
potting 制陶；浇铸封入；陶器制造；封装；封入铸塑
potting compound 铸封材料；封装材料；封装化合物
potting process 罐烧法
potting resin 铸封用树脂
potting syrup 浇注液
pottsco 液体熔渣；熔融矿渣
Pott's curvature 波特氏弯曲
Pottsville series 波茨维尔统
potty industry 陶瓷工业
pot-type boiler 家用热水器；罐式锅炉
pot-type burner 罐式燃烧器
pot-type evapo(u)rator 蒸馏釜式蒸发器
pot-type ingot buggy 翻锭车
pot-type oil switch 壶式油断路器
pot-type piston 普通活塞；筒式活塞
pot valve 罐阀
pot valve pump 罐阀泵
pot wagon 坩埚车
pot wall 空心砖墙
pot watering system 盆栽灌水系统
poubaite 硒碲铋铅矿
pouce 亚麻屑
pouch 小袋；小包；陷凹；捕鱼笼
pouch line 团绳
pouf 厚垫榻；厚蒲团；厚褥垫
poughite 碲铁矾
Poulsen-arc converter 振荡器
poulter mudjack 波式压浆泵
poultice 泥敷剂
poulticing 局部隆起；起泡
poultry and livestock farm waste 禽畜养殖废物
poultry and livestock manure 禽畜排泄物
poultry breeding 家禽饲养
poultry business 家禽业
poultry dressing farm 家禽屠宰场
poultry dung 家禽粪
poultry establishment 家畜场
poultry farm 养鸡场；家禽饲养场
poultry farming 家禽业
poultry hall 家禽棚
poultry house 家禽舍；动物房(医院、研究所等)
poultry house waste 家禽饲养废水
poultry manure 家禽粪肥
poultry netting 家禽金属丝笼
poultry-packing factory plant 家禽加工厂
poultry plant waste(water) 家禽饲养场废水
poultry producer 家禽生产者
poultry wagon 家禽车
poultry wastes 家禽废水
poultry yard 家禽饲养场
poumar yarn count system 波马纱支制
pounce 吸墨粉；印花粉；炭粉；撒粉
pounced 用压痕或打孔来装饰的
pounce-pot 撒粉罐

pounce wheel 印花粉齿轮
pouncing 印花粉印图
pouncing paper 研磨纸
pound 英镑;养鱼池;击碎;牲口栏;磅
poundage 磅数;按磅收费数
poundal 磅达
pound atom 磅原子
pound bloc 英镑集团
pound brush 普通平刷;磅刷
pound-calorie 磅卡(路里)
pound-centigrade unit 磅—摄氏单位
pounded 夯实过的
pounder 夯锤;捣具;杵
pound-foot 磅·英尺
pound force 磅力
pound-inch 磅·英寸
pounding 击拍运动;捣碎;捣实;打碎
pounding damage 撞击破坏;击拍损伤
pounding method of concrete curing 混凝土养护池养护法;混凝土浸水养护法
pounding method of curing concrete 水浸养护混凝土法
pounding mill 捣磨机
pounding of traffic 交通拥塞
pounding of valve seat 阀座碎击
pounding of vehicles 车辆拥塞;车辆颠簸
pounding out 润滑油挤出
pounding-rotary drilling 冲击回转钻进
pounding the bottom 航行中撞底
pounding up machine 捣平机
pound-lock 蓄水闸门;蓄水闸
pound-mol 磅·摩尔
pound per bulk 磅每桶
pound per cubic(al) foot 磅每立方英尺
pound per cubic(al) inch 磅每立方英寸
pound per foot 磅每英尺
pound per inch 磅每英寸
pound per square foot per hour 磅每平方英尺小时
pound per square foot 磅每平方英尺
pound per square inch 磅每平方英寸
pounds per barrel 每桶磅数
pounds per inch in thickness 每英寸厚磅数
pounds per linear inch 每英寸磅数
pounds per square inch 每平方英寸的磅数
pounds per square inch absolute pressure (psia) 磅每平方英寸绝对压力
pounds per square inch-ga(u)ge 磅每平方英寸表压(力)
pound sterling 英镑
pound weight 磅重
pounson 软黏土质土壤
pour 浇铸;浇筑块;浇灌;倾注
pourability 可浇注性
pourable 可灌注的;可浇注的;可灌入的
pourable sealer 可浇注密封剂
pour area 浇注面积
Pourbaix corrosion diagram 波尔贝克斯腐蚀图
pour by pour 分块浇注
pour chute 浇筑溜槽
pour coat 灌涂层
pour cold 低温浇注
pour concrete 浇筑混凝土
pour consistency 灌注稠度
pour-depressant addition 抗凝添加剂
poured aerated concrete 灌注加气混凝土
poured architectural concrete 浇注建筑混凝土
poured asphalt 灌筑(地)沥青;灌注(地)沥青;浇注(地)沥青;浇灌性地沥青;摊铺地沥青
poured concrete 灌浇混凝土;浇筑混凝土
poured concrete foundation wall 浇注混凝土基础墙
poured concrete lining 喷射混凝土衬砌
poured concrete structure 现浇混凝土结构
poured construction 现浇混凝土工程(施工)
poured-in-place 就地浇筑;现场浇注;现场浇灌;现场灌筑;就地浇筑
poured-in-place concrete 就地灌注混凝土;就地浇筑混凝土;现场浇筑;现场浇灌
poured-in-place concrete calcony 现浇混凝土阳台
poured-in-place concrete eaves unit 现浇混凝土檐口构件
poured-in-place concrete frame 现浇混凝土框架
poured-in-place concrete lining 就地浇筑混凝土衬垫
poured-in-place concrete pipe 就地浇筑混凝土管
poured-in-place concrete rib 就地浇筑混凝土肋
poured-in-place concrete slab 就地浇筑混凝土板;现场浇混凝土板
poured-in-place core 就地浇心墙
poured-in-place floor 就地浇筑楼板;就地浇筑楼面
poured-in-place joint 现浇板缝
poured-in-place mortar 现浇砂浆
poured-in-place pier 就地浇筑(桥)墩
poured-in-place pile 就地浇筑桩
poured-in-place(pile) cap 就地浇筑桩帽
poured-in-place pipeline 就地浇筑管线;就地浇筑管路
poured-in-place reinforced concrete 现浇钢筋混凝土
poured-in-place shell-less pile 就地浇筑无壳桩
poured-in-place tension pile 就地浇筑抗拔桩;就地浇筑拉力桩
poured-in-place uncased pile 无套管现浇桩;无壳现浇桩
poured joint 灌注缝;浇筑(接)缝;浇注缝
poured joint filler 灌注式填缝料
poured jointing 灌浆缝
poured joints for pipes 管道灌注接缝
poured lead joints for pipes 管道灌铅接缝
poured short 未浇满
poured weld 铸焊焊缝
pourer 铸锭工;注子;浇注工(人)
pour from the bottom 底注
pour hopper 浇注斗
pouring 灌注;浇注;浇灌
pouring aisle 铸锭跨
pouring basin 浇注槽;浇口杯
pouring basket 中间包
pouring bay 铸锭间;浇铸间
pouring boom 浇注混凝土的吊杆;浇注混凝土的臂杆
pouring box 中间罐;中间包;浇注箱;冒口保温箱
pouring bush 浇口杯
pouring can 灌油罐;灌缝器;手提式喷壶
pouring chute 浇注混凝土的溜槽;混凝土浇溜槽
pouring compound 浇注混合物
pouring compound for stone sett paving 小石块铺面的浇注混合剂
pouring consistency 浇灌稠度
pouring cup 漏斗形浇口;浇口杯
pouring defect 浇注缺陷
pouring dish 盘形外浇口
pouring equipment 浇注设备
pouring floor (铸造车间的)浇注工作面
pouring funnel 灌油漏斗
pouring gang 浇注组
pouring gate 直浇口
pouring head 浇注压头;浇口道
pouring house 铸造浇注场
pouring in 浇入
pouring in catilever work 悬臂梁(施工的)混凝土浇注
pouring in lifts 分层浇注(混凝土)
pouring in vertically clamped mo(u)lds 垂直浇注
pouring jacket 套箱
pouring joint 浇注缝;浇注缝
pouring kettle 灌油壶;浇注勺
pouring ladle 浇注勺;浇注斗;浇注罐;浇注桶;浇注包
pouring lip 浇注唇
pouring loop 浇注环
pouring machine 浇注机;倒浆机
pouring nozzle 浇注嘴
pouring of concrete 混凝土浇注;混凝土灌注
pouring on flat 水平浇注;平面浇注
pouring orifice 注油口
pouring output 浇注混凝土的产量;浇注混凝土的工作量
pouring pit 浇注坑
pouring pit brick 铸锭耐火砖;钢桶衬砖
pouring pit refractory 铸锭用耐火材料;浇坑耐火材料
pouring platform 浇注台
pouring point 浇注点;倾倒点
pouring position 浇注位置
pouring pot 铸钢桶;灌注壶;浇筑壶;浇筑罐
pouring procedure 浇注方法
pouring process 倾倒(方)法
pouring rain 倾盆大雨
pouring reel 加勒特式线材卷取机;侧出料式卷线机
pouring rope 浇口串;(灌铅接头的)石棉填缝绳
pouring runner 石棉填缝浇口
pouring sequence 浇注(混凝土)顺序
pouring slot 浇料口
pouring speed 浇注速度
pouring station 放流站
pouring temperature 浇注温度
pouring tilt 铸型倾斜浇注法
pouring-tip 浇包嘴
pouring tower 灌注塔;浇筑塔;浇灌塔
pouring tower of concrete 混凝土浇筑塔
pouring truck 浇注车
pouring tube 注油管
pouring-type granular insulation material 灌注类颗粒的绝缘材料
pouring weight 重块
pouring with longitudinal joint 有纵向浇注缝的浇筑;柱状浇筑
pouring wool 玻璃棉填塞;原玻璃棉
pour inhibitor 抗凝剂
pour in layers 分层灌注
pour-in-place concrete 就地浇注混凝土
pour monolithically 整块浇筑;整块浇注;整块浇灌
pour mo(u)lding 浇筑成型法;浇注成型(法)
pour-out 倒出;倾倒
pour out in shape of stream 呈股涌出
pour plane 流倾面
pour plate 倾注平皿;倾倒平板培养基
pour plate culture 注皿培养法
pour plate method 倾注培养法
pour plug 注塞
pour point 浇注点;液体的最低温度流动点;流动点;流点;倾点
pour point depressant 抗凝剂;降凝剂
pour point depressant additive 倾点下降剂
pour point depressor 倾点下降剂
pour point temperature 倾点温度
pour point test 倾点试验
pour reversion 倾点回升
pour stability 凝点安定性
pour steel 浇注钢水
pour temperature point 流动温度点
pour test 流点试验;倾点试验;倾倒试验
pouzacite 叶绿泥石
poverty 贫乏
poverty index 贫困指数
poverty line 贫困线
poverty trap 贫困陷阱
powder 火药;散剂;尘土;粉末;粉剂;粉
powder absorbent 粉状吸收剂
powder activity 粉末活性
powder-actuated fastening tool 弹射紧固工具;射钉枪
powder-actuated tool 由起爆火药驱动的工具
powder additive for mortar 砂浆添加粉
powder aerosol 粉末气雾剂
powder agglomerates 粉块
powder appearance 粉末外形
powder applicator trolley 撒粉器小车
powder applying machine 喷粉机
powder arc method 粉末电弧法
powder ash air-entrained concrete 粉煤灰加气混凝土
powder bag 炸药包
powder bag filling machine 粉剂装袋机
powder bed 粉末层
powder bed coating 流动床粉末涂装
powder bender 机动弯筋机
powder binder applicator 粉末黏结剂撒粉器
powder blast 粉药爆破
powder blast development 药室掘进
powder blasting explosive 粉状炸药
powder blending 粉末混合
powder blower 喷粉器;吹粉器
powder blowing-in technique 吹样技术
powder blow in technique 撒料法
powder blue 紫藤颜料粉;氧化钴;浅蓝色;吹青;粉末绀青
powder bond mat 粉末黏结毡
powder box 炸药箱;火药箱
powder bridging 粉末桥接

powder camera 粉末照相机
powder cart 弹药车
powder casein 粉状酪蛋白
powder cementation coating technique 粉末包渗
powder chamber 药室
powder characteristic 粉末表征;粉末特征
powder charge 炸药;火药装药;装药量
powder charging 装药
powder chest 火药箱
powder classification 粉末分级
powder cleaner 粉状清洁剂
powder cleaning agent 粉状清洁剂
powder clutch 粉末离合器
powder coal 煤粉
powder coating 粉末面层;粉末涂料;粉末涂敷;粉末敷设层
powder coating process 粉末涂装法
powder colo(u)r sorting method 粉末色度鉴定法;粉末色度分选法
powder compact 粉末坯块;粉坯(块);粉末压块
powder compacting 粉末压制
powder compound 粉状混合料
powder compressing machine 粉末压片机
powder concentrate 母粉料;粉状母料
powder core 粉末铁芯
powder cork 粉状软木
powder corundum 金刚砂粉
powder crystal analysis 粉晶分析
powder cutting 氧熔剂切割
powder cutting process 铁氧焰精整
powder density 粉体密度;粉末密度;粉密度;料密度
powder density determination 粉末密度测定
powder deposite 火药沉着
powder-detergent 粉状洗涤剂
powder development 粉末显影
powder diagram 粉末衍射图
powder diagram indexing 粉末图指标化
powder diamond 粉末状金刚石
powder diffraction camera 粉末衍射照相机
powder diffraction method 粉末衍射法
powder diffraction pattern 粉末衍射形式;粉末衍射花样
powder dip coating 粉末浸涂法
powder drift 药室;装药巷道
powder-dust sampling meter 粉尘取样仪
powder dye 粉状染料
powdered 粉末状;磨成粉状的
powdered activated carbon 粉状活性炭;粉末活性炭
powdered activated carbon activated sludge process 粉末活性炭—活性污泥法
powdered activated carbon adsorption 粉末活性炭吸附
powdered activated carbon biological treatment 粉末活性炭生物处理法
powdered activated carbon enhanced sequencing batch reactor 粉末活性炭强化序批间歇式反应器
powdered activated carbon treatment 粉末活性炭处理
powdered activated carbon treatment process 粉末活性炭处理工艺
powdered activated charcoal 粉末活性炭
powdered additive 粉末状添加剂
powdered additive for concrete 混凝土的粉末状添加剂
powdered adhesive 粉末状胶黏剂
powdered admix(ture) 粉末状掺和物
powdered alloy 合金粉末
powdered alloy steel 合金钢粉
powdered alumin(i)um 铝粉
powdered anthracite 无烟煤
powdered asbestos 粉末状石棉
powdered asphalt 沥青粉;粉状(地)沥青
powdered asphaltic bitumen 粉末状沥青
powdered basalt 粉末状玄武岩
powdered bearing 粉末冶金轴承
powdered bronze 青铜粉
powdered calcium carbide 电石粉;粉末状碳化钙
powdered caouthouc 粉末天然橡胶
powdered carbon 碳粉;粉末炭
powdered carbon activated sludge pro-cess 粉末炭活性污泥法;粉末炭活性钙污泥法

powdered carbon adsorption 粉末炭吸附
powdered carbon content 碳粉含量
powdered carborundum 金刚砂粉;粉状金刚砂
powdered catalyst 粉末催化剂
powdered clinker 粉化熟料
powdered coal 煤粉;煤
powdered coal burner 煤粉燃烧器;煤粉燃烧炉
powdered coal flame 煤粉火焰
powdered coal fuel 煤粉燃料
powdered colo(u)ring agent 粉末着色剂
powdered colo(u)ring matter 粉末状着色材料
powdered colo(u)ring substance 粉末状着色物
powdered compact 粉末坯块
powdered concrete hardener 粉末状混凝土凝固剂
powdered densifying admix(ture) 粉末状增浓掺和物;粉末状致密掺和物
powdered distemper 刷墙粉
powdered emery 细金刚砂粉;金刚砂粉
powdered extrusion 粉末挤压(成型)
powdered ferrite 粉状铁氧体;粉状碳素体
powdered fodder 粉状饲料
powdered fuel 粉状燃料;粉尘燃料
powdered fuel plant 制煤粉装置
powdered glass 玻璃粉
powdered glue 胶合粉;粉状胶(黏剂)
powdered glue for wall paper 壁纸专用胶粉
powdered gold 金粉;粉化金
powdered gold bronze 金色青铜粉
powdered granite 花岗石粉;粉末状花岗石
powdered granular 粉末颗粒活性炭
powdered gypsum 石膏粉
powdered hardener 粉末硬化剂
powdered hibiscus 木槿根粉
powdered ink 粉状油墨
powdered insulant 粉末绝缘体
powdered integral waterrepeller 合成抗水粉
powdered iron 铁粉
powdered-iron core 粉末铁芯
powdered kaolin 高岭土粉;瓷土粉
powdered laundry soap 皂粉
powdered lead 铅粉
powdered leather 皮革粉末
powdered lime 潮解石灰;石灰粉;粉石灰
powdered limestone 石灰石粉末
powdered magnesium 镁粉
powdered marble 大理石粉
powdered material 粉状材料;粉末材料;粉末材料
powdered metal 粉末冶金材料;粉末金属
powdered mica 云母粉
powdered mineral 粉状矿物;粉末无机物
powdered natural rock asphalt 粉末状天然成岩地沥青
powdered natural rock asphalt mastic 天然石粉状沥青胶泥
powdered natural rubber 粉末天然橡胶
powdered ore 粉状矿(石)
powdered pesticide 粉状农药
powdered product 粉末产物
powdered pumice(stone) 轻石粉;浮石粉
powdered quartz 石英粉
powdered raw rubber 生橡胶粉
powdered reaction 粉末反应
powdered refractory 耐火粉料
powdered remover 粉状去漆剂
powdered resin 粉末树脂
powdered retarder 粉状缓凝剂
powdered rock phosphate 磷矿粉
powdered rubber 橡胶粉
powdered sample 粉末试样
powdered sandstone 砂石粉
powdered silver 银粉
powdered soap 肥皂粉
powdered soap dispenser 肥皂粉配出器
powdered solder 粉状焊料
powdered solid fuel 粉状固体燃料
powdered solids 粉状固体粒子
powdered spar 晶石粉末
powdered stripper 粉状除漆剂
powdered sulfur 硫黄粉
powdered talc 滑石粉
powdered talc surfacing 滑石粉面层
powdered toner 墨粉
powdered trass 火山灰
powdered white pigment 粉状白颜料

powdered zeolite 粉末沸石
powdered zinc 锌粉
powder electrode position 粉末电沉积涂装
powder emery 金刚砂
powder emulsion 粉末乳液
powder enamel 干法搪瓷;粉状搪瓷釉
powder enamel cup with landscape design 粉彩山水杯
powder enamel vase with tripod design 粉彩钟鼎瓶
powder engineering 粉体工程
powder explosion suppressant 干粉抑制剂
powder extinguishing system 干粉灭火系统
powder extraction metallurgy 粉末制取冶金
powder factor 炸药(有效)系数;单位炸药消耗量
powder factory 火药制造厂
powder feeder 送粉器;粉末加料器
powder feeding speed 送粉速率
powder filter 粉末滤光片
powder fines 粉末的极细部分
powder fire branch 干粉枪
powder fire extinguisher 干粉灭火器
powder fire extinguishing grenade 粉末灭火弹
powder fire monitor 干粉炮
powder fire truck 干粉消防车
powder flag 易爆危险品旗;爆炸品信号旗
powder flowbility 粉末流动性
powder flow control 粉末流动性控制
powder flow measurement 粉料流动性测定
powder flowmeter 粉末流量计
powder forging 粉末锻造
powder-form agent 粉尘剂
powder forming press 粉末制品压力机
powder full density process 粉末全致密化工艺
powder funnel 漏粉斗
powder fuse 炸药引信;炸药雷管
powder gas 火药气体
powder grain density 火药粒密度
powder grain size 火药粒大小
powder gravity 粉末比重;粉末密度
powder grinding method 粉末研磨法
powder grinding test 粉末研磨试验
powder gun 粉末喷枪;喷粉机
powder handgun 干粉炮
powder hole 干井(指没有油气的钻井)
powder hopper 粉末加料斗
powder hose and handgun 干粉软膏和干粉枪
powder house 炸药库
powder impregnation 粉末浸涂
powdering 敷粉;磨成粉;点子花饰;磨粉;上粉;撒粉;分型粉
powdering agent 隔离剂;打粉剂
powdering inlet 粉末入口
powdering machine 碾碎机;喷粉机
powdering of paints 涂料粉化
powdering of polishes 虫胶漆膜粉化
powder in mo(u)ld coating 模内粉末涂覆
powder installation 干粉灭火装置
powder insulation 粉末绝缘材料;绝热粉
powderization 粉化
powder jetting 粉体喷搅法
powder lance 炸药撞杆;炸药撞针
powder lancing 炸药撞杆切断
powderless etching 无粉腐蚀
powderless etching machine 无粉腐蚀机
powder-like structure 粉末结构
powder lime 粉石灰;粉状石灰;石灰面
powder-loading density 装药密度
powder lot 粉料
powder lubricant 粉末润滑剂
powder magazine 炸药库;火药库;火药舱
powder magnet 压粉磁铁;粉末磁铁
powderman 放炮药人;引爆人;炸药管理员;炮工;爆破员
powder manufacturing apparatus 制粉设备
powder mass 粉末物质
powder measurements 粉末材料的测量
powder metal 金属粉末
powder metal bearing 粉末冶金轴承
powder metal bit 粉末冶金法制胎体的钻头
powder metallography 粉末金相学
powder metallurgic approach 粉末冶金法
powder metallurgicl gear 粉末冶金齿轮
powder metallurgic method 粉末冶金法

powder metallurgic porous material 粉末冶金多孔材料
powder metallurgic press 粉末冶金压机
powder metallurgic process 粉末冶金法
powder metallurgic product 粉冶制品;粉末冶金制品
powder metallurgic technique 粉末冶金技术
powder metallurgy 粉冶零件;粉末冶金(学)
powder metallurgy press 粉冶用压机
powder metallurgy superalloy 粉末冶金高温合金
powder metal matrix 粉末金属胎体
powder metal mo(u)lding press 粉末金属成型压力机
powder metal press 粉末制品压力机;粉末制品成型压力机;粉末金属压机
powder method 粉样法;粉末法
powder microstructure 粉末显微结构
powder mill 火药厂
powder mine 装药坑道;装药洞室
powder mixer 粉料混合器
powder mixing 粉末混合
powder mix(ture) 火药混合物;粉料混合物
powder-moisture test 火药水分试验
powder monkey 爆裂药;起爆器;爆破工;装炸药的爆破工人;装药工(人);炸药发放员;爆炸手
powder morphology 粉末形态
powder mortar 粉末砂浆
powder mo(u)lding 粉末模塑法;粉末模制;粉末成型;粉料模塑
powder nozzle 干粉喷嘴
powder of resin 松香粉
powder outlet 粉料出口
powder packaging machine 粉包装设备
powder packing 粉末装填
powder paint 粉末涂料
powder particle 粉末颗粒
powder pattern 粉末图样
powder pesticide 粉剂农药
powder photography 粉末照相法;粉末摄影
powder pigment 粉状颜料
powder pistol 粉末喷枪
powder plasticizer 塑化粉剂
powder plastics coating 粉末塑料涂层
powder plastics mo(u)lding 塑料粉末磨塑成型
powder pocket 炸药袋
powder porosity 粉末孔隙度
powder post 朽材
powder post beetle 粉蠹;粉虫(木材蛀虫)
powder premix 粉状预混料
powder premixing 粉末预混合;粉末预混
powder preparation 粉末制备;粉料制备
powder press mo(u)lding apparatus 粉末模压装置
powder pretreatment 粉末预处理
powder processing 粉末加工
powder property 粉末性能
powder recovery tray 粉末黏结剂回收盘
powder reduction 粉末还原
powder release station 干粉施放站
Powder River basin 波德河盆地
powder roll feed arrangement 粉末轧制加料装置
powder rolling 粉末轧制成型
powder rolling process 粉末轧制法
powder rolls 粉末轧机
powder room 火药库;火药室(女厕所的)四笼间;女宾休息室;女盥洗室;化妆室;女卫生间
powder rubber 粉状橡胶
powder sack 炸药袋
powder salt 粉盐
powder sample plate 粉状样品板
powder scarfing 氧熔气刨;喷铁粉火焰清理
powder sensibility 炸药敏感度
powder-set fasteners 用枪打入的紧固件;膨胀螺钉;用枪打入的锚固件
powder shop 粉末车间
powder sieve vibrator 粉末筛激振器;粉末筛振动器
powder silk 火药袋绸
powder sintering 粉末烧结
powder sintering mo(u)lding 粉末烧结造型
powder size 粉末粒度
powder snow 粉雪;粉状雪;粉末状雪
powder space 药腔
powder specimen 粉末标本
powder sprayer 喷粉器

powder spray gun 粉末喷枪
powder spraying 喷粉;撒粉
powder spraying coating 粉末喷涂
powder spray(ing) method 粉末喷涂法
powder stearic acid 粉状硬脂酸
powder stick 药卷
powder stock 粉末原料
powder strip 金属粉末轧制带材
powder technology 炸药工艺学;粉体技术;粉末工艺
powder test 粉末试验
powder train 火药导火线;火药串;导火索
powder-type coating 粉状涂料
powder-type compound 粉末型混合料
powder-type explosive 粉状炸药
powder type fire extinguisher 干式灭火器;粉末灭火器
powder vessel 火药舱
powder washing 氧熔气刨;铁粉助熔表面清理
powder waste 粉末状废物
powder weld process 粉末焊接法
powdery 粉状的;粉末状的
powdery analysis 粉末分析
powdery diffractometry 粉末衍射(术)
powdery explosive 粉状炸药
powdery form 易成粉末状;粉末形状
powdery material 粉状物料
powdery-state 粉状
powdery structure 粉末状结构
powdery substance 粉末状物质
powdery talc 粉状滑石
powdery-tin skimmings 粉状锡渣
powdex 粉末树脂过滤器
powdiron 多孔铁
powdiron bearing 多孔铁轴承
poweder method 粉末法
powellite 钼钨钙矿;钼钙矿
Powell method 鲍威尔方法
Powell-wood 鲍威尔木材防腐工艺
Power 鲍尔铅基轴承合金
power 力量;幂;权力;势力;动力;乘方
power-absorbing element 吸能元件
power-absorption device 加载测功器
power-absorption unit 动力制动装置
power accumulator 储能器
power-actuated 机械传动的;机动的;动力驱动的;电力驱动的
power-actuated setting device 射钉枪
power-actuated tool 机动工具
power administration 电业管理局
power akte machine 动力铁鞋安置机
power alarm annunciator 电动报警器
power amplification 功率增益;功率放大率
power amplification coefficient 功率放大系数
power amplification ratio 功率放大系数
power amplifier 功率放大器
power amplifier stage 功率放大级
power amplifier tube 功率放大管;功率电子管
power analyzer 功率分析仪
power and control cable 配电和控制电缆
power and free conveyer 功率和落体(结合的)输送装置
power and free system 功率和落体(结合的)输送系统
power and navigation drawdown limit level 发电航运消落极限水位
power angle 功率角;功角
power-angle curve 功角特性曲线;功角曲线
power arc 电弧
power arm (杠杆的)力臂;动力臂
power arm ditcher 动力臂式挖沟机
power-assisted control 利用动力的控制
power-assisted door 动力操纵门
power-assisted steering 助力转向;动力转向
power-assisted steering gear 带助力器的转向装置
power-assisted steering ram 操向装置助力油缸;操向助力油缸
power attenuator 功率衰减器
power at the drawbar 牵引功率
power at the power takeoff 动力输出轴功率
power auger 电钻;机械钻;机动钻
power auger boring 电动螺旋钻孔;电动螺钻
power axle 驱动轴;传动轮轴
power balance 动力平衡

power balance equation 功率平衡方程;动力平衡方程(式)
power balance model 动力平衡模型
power bandwidth 功率通带
power bandwidth ratio 功率带宽比
power bar bender 机动弯筋机;动力钢筋弯曲机
power barrow 机动自卸汽车
power bay 电源架
power bender 机动弯筋机
power benefit 发动效益
power bill 动力消耗清单
power blade 动力传动的刮刀;动力传动的刀片
power blader 机刨
power board 配电板;供电瓶;配电盘
power boat 机动船;汽艇;机动艇;摩托艇
power box 机力卸载车厢
power bracket 功率范围
power brake 机动闸;机动制动器;动力制动器;电力制动器
power-brake changeover switch 动力制动转换开关
power breeder 动力增值反应堆
power broadenign 功率展宽
power broom 机械刷;机扫;机动扫帚;动力清扫器;扫路机
power broom drag 机动刮路刷
power budget 功率分配
power buggy 机动装卸车;动力小车;动力牵引车;电动小车
power buggy of concrete 混凝土机动运送车
power burner 大功率燃烧器
power cabin 机房
power cable 动力电缆;电源电缆;电力电缆
power cable line 动力电缆
power cable with concentric 带同心中性导线的电力电缆
power cable with flat steel wire armo(u)ring 扁钢丝铠装电力电缆
power calculation 功率计算
power calorizing 固体渗铝
power canal 水电引水渠;发电引水渠
power canal development 渠道引水式电站
power capability 可能功率
power capacitor 电力电容器
power capacity 功率容量
power capacity of stream 河流发电容量
power capstan 动力绞盘;电动绞盘
power car 动力车(辆);电机车;发动机吊舱
power carrier telephone 电力载波电话机
power carrying capacity 功率荷载容量
power cart 动力车
power-cart duster 机动车式喷粉机
power ceramic capacitor 高功率瓷介电容器
power chain saw 动力链锯
power change switch 功率变换开关
power channel 动力廊道;发电引水渠
power characteristic 功率特性;幂特性
power charges 动力费用
power chip(ping)s spreader 机动石屑撒布机
power choke coil 电源滤波扼流圈
power choring 零星杂物机械化;小型机械化
power circuit 电源电路;电力网;主电路;动力电网;电力电路
power circuit breaker 电力开关
power cleaning 喷吹清理
power-closed car door 升降机轿厢动力关闭门
power-closed car gate 升降机轿厢动力关闭闸门
power-closed door 动力关门的门
power-closed gate 动力关门的闸门
power clutch 机力操纵离合器;机动离合器
power coal 动力用煤
power coastdown 功率下降
power coefficient 功率系数
power collecting device test bench 集电装置试验台
power-compacted charge 团块炉料
power company 电力公司
power compensation differential scanning calorimetry 功率补偿差示扫描量热法
power component 有功分量;有功部分;实数部分
power concrete saw 机动混凝土锯
power condenser 电力电容器
power conditioner 动力调节器
power conditioning unit 功率调节器
power conduction 动力线

power conduit 压力管道;动力管道
power connection 连接电源;接电源;电源接头
power consumer 用电户;电力用户
power-consuming 消耗动力的
power consumption 能量消耗;能量吸收;功率消耗;功率损耗;功耗;耗电量;动力消耗(量);电力消耗(量);电耗
power consumption of activated sludge 活性污泥法耗能量
power consumption of drainage 排水动力消耗
power consumption of grinding 粉磨电耗
power consumption of hoisting 提升动力消耗
power consumption of transport 运输动力消耗
power contactor 电力接触器
power contour 等功率线
power control 功率控制;功率调整;电源控制装置;动力操纵;动力控制;助力操纵机构
power-control box 动力控制箱
power-control drive shaft 动力控制的驱动轴
power-control housing 动力控制架;动力控制壳;动力控制箱
power-control hydraulic units 电力控制的液压机组
power-controlled 动力操纵的
power-controlled lowering 动力下降
power controller 功率控制器;功率调节器
power-control motor 动力控制电动机
power-control ratio 功率放大系数
power-control rod 功率控制棒;动力操纵杆
power-control servo 助力器;功率控制伺服机构
power-control shaft worm gear 动力控制蜗杆传动(装置)
power-control unit 动力控制装置;动力控制单元;电源控制器;电动控制单元
power-control valve 动力控制阀
power conversion 功率变换
power conversion efficiency 功率转换效率
power converter 整流器;功率变流器
power cord 电源线;电源软线
power cost 能源费用;电费;电力价格;动力开支;动力费用;动力成本
power cost of technologic(al) process 工艺过程动力费
power crane 动力起重机;动力吊车;电动起重机;电动吊车
power curber 动力铺侧石机
power current 工频电流;强电流
power current cable 电缆
power current overhead transmission line 高架输电线
power current transmission line 输电线
power curve 功率曲线;势曲线;出力曲线
power cut 切断电源;停电;动力切断;电源切断
power cut-off 结束工作
power cut-off relay 电源切断继电器
power cut-out switch 电源切断开关
power cutter 机械切割工具
power cycle 动力循环
power cylinder 动力油缸;动力缸
power dam 动力利用的拦水坝;坝内式电站厂房
power demand 功率需(求)量;能的需要量;动力需求;电力需求
power density 功率密度
power-density distribution 功率密度分布
power density of electrode surface 电极圆表面比功率
power density spectrum 功率密度谱
power despactcher 电调
power detection 功率检波;强信号检波
power detector 功率指示器;功率检波器;功率测试器;强信号检波器
power developed 已发功率;实发功率
power development 功率开发
power device 功率器件
power diagram 功率图
power differential relay 功率差动继电器
power digger 动力挖掘机
power diode 功率二极管
power discharge 发电流量
power disconnecting switch 电力隔离开关
power disconnect switch 切断电源开关;电源开关
power dispatch 电力调度
power dispatcher 电调员
power dispatching telephone 电力调度电话
power dissipation 功率损耗;功率耗损;功率耗散

功耗
power distribution 动力分配;功率分布;配电
power distribution box 配电箱;动力配电箱;电力配电箱
power distribution building 配电房
power distribution panel 动力配电盘;电力配电箱
power distribution reliability 供电可靠性
power distribution room 配电房
power distribution station 配电站
power distribution system 动力分配系统;配电系统
power distribution unit 配电装置;配电部件;动力分配装置
power distributor 机动喷洒车;动力喷油车
power divider 功率分配器
power dividing 功率分流
power-dividing septum 功率分隔板
power door 机动门(水密滑动舱壁门)
power down stand-by 电源故障备用设备
power drag line 机械铲
power-drag scraper 机械铲;机动拖铲
power-drag scraper machine 动力牵引铲运机
power drain 耗用功率
power drawdown level 发电消落水位
power-drawn 机动的
power-drawn blader 机动平地机
power-drawn grader 动力平土机;动力平路机;机动平路机;机动平地机
power drift blast 药室爆破
power drill 电动钻(床);电钻;机力钻床;动力钻床;动力钻(机);动力凿岩机
power drive 动力驱动;机械传动;电力驱动;电力传动
power drive chain 传动链条
power-driven 电动的;机械传动的;机动的;动力驱动;电力驱动的
power-driven blader 机动平路机;机动平地机
power-driven bolt rod stretching machine 电动螺杆张拉机
power-driven butterfly valve 电动蝶阀
power-driven capstan 动力绞盘
power-driven crane with single beam 电动单梁起重机
power-driven drainage irrigation station 电动排灌站
power-driven fork truck 电动叉车
power-driven grader 动力平地机
power-driven hammer 电动锤
power-driven hand-tool 手持机动工具
power-driven irrigation and drainage equipment 排灌动力机械
power-driven Loyang spoon 机动洛阳铲
power-driven membrane pump 电动隔膜泵
power-driven pressure test pump for pipe ways 电动试压泵
power-driven pump 动力驱动泵
power-driven reciprocating-arm distributor 动力驱动往复臂式配水器
power-driven reciprocating harrow 动力摆动耙
power-driven road roller 动力驱动压路机
power-driven rotary sweeper 动力驱动旋转式清扫机
power-driven rotating distributor 动力驱动旋臂式配水器
power-driven screwdriver 电动螺丝刀
power-driven screw-jack set 电动螺旋起重机组
power-driven shears 电动剪切机
power-driven single-wheeled roller 动力驱动单轮压路机
power-driven squeegee 动力驱动橡皮刮板
power-driven stud 动力旋进的双头螺栓
power-driven system 动力驱动系统
power-driven threshe 动力脱粒机
power-driven tool 电动工具
power-driven truck 动力驱动载重汽车
power-driven valve 电动调节阀
power driven vessel 机动船
power-driven wheel 动力驱动的行走轮
power drive pump 动力传动泵
power driver 功率激励器;机动打桩机
power driver circuit 功率驱动电路
power drive shaft 传动轴
power drum 卷料机;卷取机;动力卷筒
power drum winch 动力绞车;动力卷扬机
power dumbwaiter hoistway 小型电动运货升降

机井道
power dump 切断电源;电源切断
power dump rake 机力横向搂草机
power-duration curve 功率历时曲线
power duster 动撒粉器;动力撒粉器
power earth auger 电动钻井机;电动土钻;机动钻土机;动力土钻
power economy 动力经济
powered 机械带动的;机力操纵;机动的
powered balloon 动力气球
powered barge 机动驳船
powered bender 机动钢筋弯曲机
powered cable logging 动力索道集材;动力钢索集材
powered contactor 电传动接触器
powered control 带动力的控制
powered craft 机动船
powered flight 动力飞行
powered grab 电动抓岩机
powered gypsum 石膏粉
powered hand tool 手持机动工具
powered industrial truck 工业用机动车
powered jig saw 动力细锯
powered lowering 动力下降
powered phase 主动段
powered phase guidance 主动段制导
powered photocatalyst 供电光催化剂
powered platform 电动施工平台;电动工作平台
powered rod bender 机动钢筋弯曲机
powered steering 动力转向
powered support 机械化支架
powered truck 动车
powered underwater research vehicle 自航式水下研究潜水器
powered wheel loader 电动轮装载机
powered whistle 动力号笛
power efficiency 功率效率;出力效率
power electronics 功率电子学
power elevation 机械提升设备;机械提升;机动提升
power elevator 电动升降机
power encomonics 动能经济学
power end 水泵动力端;电力端
power energy 电能
power energy head 动能高
power engineer 电力工程师
power engineering 力能工程;动力学;动力工程;电力工程
power enrichment valve 增力加浓阀
power equalizer 功率均衡器
power equipment 电力设备;动力设备
power equipment depreciation apportion and overhaul charges 动力设备折旧摊销及大修费
power equipment preaging 动力设备预陈化
power exchange 功率交换
power excursion 功率失常激增
power expenditures 功率耗损
power expenses 电力费
power exponent 幂指数
power export 功率输出;动力输出;能量输出
power extraction 功率削减
power facility 动力设备
power factor 炸药消耗率;功率因素;功率因数;功率系数;马力系数
power-factor adjustment 功率因数调整装置
power-factor compensation 功率因数补偿;功率系数的补偿
power-factor compensation capacitor 功率因数补偿电容器
power-factor compensation device 功率因数补偿装置
power-factor compensator 功率因数补偿器
power-factor control 功率因数调节
power-factor improvement 功率因数改善
power-factor indicator 功率因数指示器;功率因数计;功率因数表
power-factor measurement 功率因数测量
power-factor meter 功率因数计;功率因数表
power-factor of load 负载功率因数
power-factor regulating relay 功率因数调节继电器
power-factor regulator 功率因数调整器
power-factor swing 功率因数动荡
power fail 电源故障;电源断电;掉电
power-fail recovery system 掉电恢复系统
power fail restart 电源故障重新起动

power fail safe 电源保安装置
power failure 供电事故;供电中断;供电中断;供电故障;停电;断电;电源故障;电力故障
power-failure interrupt program(me) 掉电源中断程序
power-failure protection 电源断开保护
power fan 动力风扇
power fault 电源故障
power feed 自动进料;机械进料;机动进给;机动供料;机动给料;动力进给;动力供应;电力馈电
power feed drifter 机械推进式凿岩机
power feeder 电力馈线
power feeder system 供电系统
power feeding system 电力馈送方式
power feed lever 机动进给杆
power feed motor of drilling machine 钻机自动推进装置电动机
power feed rotary table 机动转台
power feed section 供电段
power feed shell 自动推进滑座
power feed system 主油系
power fine grader 地基整平机;动力精细平路机
power fire extinguisher 干粉灭火器;干粉灭火机
power float 电动(混凝土)平整机;机械抹灰板;机械抹灰;机动抹光机;机动镘板;动力镘板;电动抹子
power flow 动力流线;功率通量;功率潮流;上网潮流;电力潮流
power-flow distribution 能源分布;能流分布
power fluctuation 功率起伏;功率波动;功率变化;功率摆动
power fluid 传动液
power flux 功率通量
power form (电热张拉的)支柱模板
powerforming 强化重整;动力重整
power formula 乘方公式
power frame 电源机架
power-free conveyer 推杆运输链
power frequency 工业频率;工频;市电频率;电源频率
power frequency discharge voltage 工频放电电压
power frequency sparking test 工频火花电压试验
power frequency sparkover voltage 工频跳火电压
power frequency test 功率频率试验
power frequency withstanding voltage 工频耐压水平
power from ocean tides 潮力
power fuel 动力燃料;发动机燃料
powerful 强有力;强功能的;大功率的
powerful architecture 强功能结构
powerful computer 高效率计算机
powerful explosive 烈性炸药
powerful grip 高强度连接的过盈
powerful instruction set 强功能指令系统
powerful light source 强光源
powerful machine 强力电机
powerful operating system 强功能操作系统
powerful oxidant 强氧化剂
powerful peripheral 强功能外围;强功能外部设备
powerful receiver 高效接收机;高灵接收机
power function 功效函数;功率函数;幂函数
power function correction 幂函数校正
power function model 幂函数模型
power fuse 高电流熔断器;电力熔断器;电力保险丝;保险丝
power fuse unit 电源熔断器
power gain 功率增益
power gain function 功率增益函数
power gain tester 功率增益测量仪
power gallery 动力廊道
power gas 功率增益;动力瓦斯;动力气体;动力煤气;发生炉煤气
power generating machine 动力机械
power generating plant 动力设备;动力厂
power generation 动力产生;电力生产;发电
power generation by rubbish 垃圾发电
power generation by waste heat 余热利用发电;余热发电
power generation dispatching 发电调度
power generation from sea current 海流发电
power generation load rate 发电负荷率
power generator 动力发生器;电力发电机
power governor 功率调节器
power grader 机动平路机;机动平地机;动力平土机;动力平路机

power graph 功率曲线图
power grid 电力网
power grid detection 强信号栅极检波
power grid detector 高压电力网探测器
power gritter 机动石屑撒布机;机动铺砂机
power grizzly 动力格筛
power group enumeration theorem 幂群计数定理
power gun 机力油枪
power hack saw 动力弓锯
power hack saw blade 电动弓锯条
power hack sawing machine 电动弓锯机
power hammer 机动锤;动力锤;电锤
power-handling capability 功率使用容量;动力使用能力;额定功率值
power hand saw 手持机动锯;手持动力锯
power hay press 动力干草压捆机
power head 动力头
power hoe 机械耙
power hoist 自动吊车;机动卷扬机;动力吊车;动力吊车
power house 发电厂(房);动力房;主厂房;动力室;动力间;动力车间;动力厂;发电站;发电厂动力车间;电站岩洞
power-house assembly 发电厂机组
power-house at dam toe 坝后式厂房
power-house bridge crane support beam 厂房桥吊承重梁
power-house building 发电厂建筑;发电厂房
power-house cavern 发电厂洞穴
power-house inside the dam 坝内式厂房
power-house overall 电力厂工人外套
power-house shop cap 电力厂工作帽
power-house structure excavation 发电厂建筑挖掘
power-house substructure 电站厂房下部结构;发电厂地下结构
power-house substructure block 电站地下部分
power-house superstructure 发电厂(动力间)上部结构;电站厂房上部结构;发电厂地上结构
power-house superstructure block 电站地上部分
power-house within the dam 坝内式厂房
power hydraulics 泵液压
power import 动力运用;动力输入
power increment 功率增长值
power index 幂指数
power indicator 功率指示器;功率指示计;示功器
power inductance 电力电感
power induction 电力感应
power industry 动力工业;电力(工)业
power information 功率信息
powering 动力估计
power input 功率输入;输入功率
power input shaft 动力输入轴
power input to a machine 电机的输入功率
power installation 动力装置;电力装置;发电装置
power insulator 电流绝缘体
power intake 动力(电)输入端
power intake station 动力(电)输入站
power interchange 功率互换
power interference 电源干扰
power interlocking 动力联锁
power interlocking device 电力连锁装置
power interruption 供电中断
power inverter 功率变换器
power jet 主射口;主量孔;动力喷嘴;动力喷口
power jet valve 增压喷口阀
power kerber 机动铺装缘石机
power kerosene 拖拉机燃油;烈性燃油;动力煤油
power klystron 功率速调管
power law 幂(定)律
power law distortion 指数律失真;幂律失真
power law fluid 幂律流(体)
power law index 幂律指数
power lawn mover 动力剪草机
power law profile 幂律廓线
power law relation 幂函数关系式
power law wind profile 幂指数风廓线;风廓线幂指数律
power lead 电力引入线
powerless 无效的;无权的
power level 功率级;功率电平;功率等级;幂级;权级;声能级;动力水平
power level diagram 功率电平图
power level gain 功率电平增益

power level indicator 功率指示器;功率电平指示器;电平指示器
power leveling 功率调整
power level saftey system 功率保护系统
power lifetime 连续操作时间
power lift 自动起落机构;机械提升;机械式起落机构;电动升降机;动力提升
power lift control 机力提升器调节
power-lifter 动力提升机构
power-lift lever 动力起落操纵杆
power-lift plough 动力升降犁
power lighting substation 动力照明变电所
power lighting system 动力照明系统
power lighting transformer feeder cubicle 动力照明变压器馈电柜
power limitation 功率限制
power line 输电线;供电线;火线;输电线路;动力线路;动力线;电源线;电线;电网;电力线(路)
power-line carrier 输电线载波;动力线路载波;电力线载波
power-line carrier communication 电力线载波通信
power-line connection 电源接线
power-line fittings 电力线配件
power-line interference 电力线干扰
power-line noise 电力线噪声
power-line terminal 电源接线柱
power-line tower 输电线塔架
power-line voltage 电源电压
power link 联网;动力系统的连接线路;动力联合
power linkage 动力悬挂装置
power liquid 动力液体
power load 功率负荷
power loader 机械装载机;机动装卸机;机动装料斗;动力装载机
power load flow 电力潮流;潮流
power loading 电负荷;功率荷载;动力负载;动力负荷;负载功率
power loom 动力织机
power loss 功率损失;功率损耗;动力损失
power loss of electric(al) current 电流功耗
power loss quantity 功耗量
power lug 电线接头
power machine 机动搅拌机;动力机
power machine-building industry 动力机器制造工业
power machinery 动力机械
power magnification 功率倍数
power magnifier 功率放大器
power main 输电干线;动力干线;电源线;输电线
power-making 产生动力
powerman 发电机专业人员
powerman magnification 功率放大
powerman margin 功率储备
power market area 电力市场范围
power mean 幂平均值
power-measuring circuit 功率测量电路
power-measuring device 功率测量设备;功率测定装置
power mechanics 动力学
power meter 功率计;功率表;电功率表
power method 幂法
power mismatch loss 输入失配损耗
power mixer 动力拌和机
power modulation 功率调制
power monitor 功率监视器
power monograph 功率图集
power mover 机动割草机
power nailer 机动打钉器;机动打钉机
power navvy 机动挖泥机;机动挖掘机;挖土机;挖掘机;电动挖掘机;机动挖土机
power network 供电网(络);电力网
power network frequency 电力网频率
power number 功率指数
power of adsorption 吸附力
power of agitator 搅拌功率
power of a locomotive 机车动力
power of amendment 修改权
power of amplification 放大率
power of appointment 委任权;财产指定权;财产处理权
power of a test 检验的势;把握度
power of attorney 委托书;授权(文)书;代理(委托)权
power of attorney endorsement 委托背书

power of battery 电池功率
power of cementation 黏结能力
power of chemical transportation 化学搬运力
power of cohesion 内聚力
power of continuum 闭联集的势
power of curing (散块沥青的)固化能力
power of decision 自主权
power of detantion 扣押权
power of dredge pump 泥泵功率
power of electromagnet 电磁铁功率
power-off 切断电流;停车
power-off condition 切断电源状态;发动机停车状态
power-off contact 安全接点(开闭器)
power-off protection 断电保护
power-off rapping 停电振打
power-off relay 停电用(电源转换)继电器;故障继电器;电源切断继电器
power-off speed 无动力速率
power-off time 停电时间
power of gravity 重力能;重力
power of gripping 握力
power of hardening 凝固力
power of initial setting 初凝力
power of lens 透镜焦强;透镜焦度
power of migration 运移动力
power of mirror 反射镜的光焦度
power of motor 电(动)机功率
power of mud pump 泥浆泵功率
power of number 幂数
power of procuration 委任状;代理权
power of pump station 泵站容量
power of refraction 折射率
power of regeneration 再生能力
power of sale 销售权
power of source 电源功率
power of stereoscopic observation 立体观察能力
power of straining 应变能(量);应变功率
power of telescope 望远镜放大倍数
power of ten 十的幂
power of test 检验效能;检定力;检定的功效;检测能力
power of the assembling unit 机组功率
power of traction 牵引力
power oil 动力油
power-on 送电;开车(发动机);通电
power-on clear 电源接通清除
power-on condition 发动机开车状态
power-on driving pulley side 驱动皮带轮侧功率
power-on light 电源显示器
power-on/off button 电源开关
power-on reset 工作复位;加电复位
power-on self test 上电自检
power-on shaft 轴功率
power-on speed 发动机工作时速率
power-on test 通电试验
power-on time 供电时间
power-opened self-closing door 动力开启的自闭门
power-operated 机械传动的;机械操纵的;动力操作的;机力操纵的;动力驱动的
power-operated baling press 电动打包机
power-operated bar-bender 动力操作的弯筋机
power-operated buggy 机动斗车
power-operated bush hammer 机动凿石锤
power-operated control 能源控制
power-operated control system 动力操纵系统
power-operated conveying machine 电动传送机
power-operated crane 动力操纵的起重机
power-operated door 动力操作的门;动力操纵门
power-operated drilling carriage 机动钻车
power-operated fire door 电力操纵防火门
power-operated grease gun 自动润滑脂枪;机动黄油枪;自动黄油枪
power-operated hoist 自动提升装置;机动提升装置
power-operated rolling shutter 电动卷百叶
power-operated shearing machine 电动剪断机
power-operated skip (搅拌机的)自动倾翻
power-operated stirring gear 自动搅拌机;机动搅拌装置
power-operated trenching machine 动力操作挖沟机
power-operated valve 机械驱动阀;伺服电动机操纵阀
power-operated vertical hydraulic press 机动立式液力压榨机
power operation 功率操作
power oscillation 功率振荡
power oscillator 功率振荡器
power oscillograph 功率示波器
power-ourput device 功率输出器件
power outage 停电事故;动力故障
power output 功率输出;输出功率
power output amplifier 功率输出放大器
power output of turbine 透平功率输出;涡轮机功率输出
power output shaft 动力输出轴
power output system 功率输出系统;功率输出方式
power output tube 功率电子管
power pack 动力站(指水、汽、压缩空气);电源组;电源部分
power package 整装电源机组;动力机组
power pack system 动力机组系统
power panel 供电瓶;配电盘;配电板;电源板
power panel board 电力配电盘;动力配电盘
power parameter 功率参量
power pattern 功率方向图
power peak 功率尖峰
power penalty 功率损失;功率代价
power per cylinder 单缸功率
power per pound 机器的单位重量功率;每磅功率
power per unit 单机容量
power pipe 动力管道
power pipe cutter 动力切管机
power pipe hammer 机动桩锤
power pipeline 动力管系
power piping 动力管道
power piston 动力活塞
power plane 电动刨床;电源层
power planning 电力规划
power plant 动力装置;动力设备;动力(发电)厂;动力车间;发电站;发电所;发电设备;发电厂
power plant accessories 电厂附属设备;发电辅助设备
power plant building 发电厂建筑
power plant capacity 发电厂(设备)容量;电厂容量
power plant cooling water 发电厂冷却水
power plant cost 动力车间成本;发电厂成本
power plant discharge 电站发电流量
power plant efficiency 电厂效率
power plant expenses 动力车间费用;发电厂费用
power plant for emergency 备用发电设备
power plant for mobile radio apparatus 移动式无线电机用电源
power plant intake structure 发电站取水口构筑物
power plant load factor 电厂负荷因数
power plant nozzle 发动机喷管
power plant operated by currents 海流发电厂
power plant operator 电厂运行员
power plant research 发电装置研究
power plant unit 动力厂装置
power plant waste 发电厂废物
power plant wastewater 发电厂废水;电厂废水
power plant weight 发动机装置重量
power plant with concentrated fall 集中落差式电站
power plug 动力插头
power plunger pump 电动柱塞泵
power point 墙边插座;电源插座;电力插座;功率点
power point for cleaners 吸尘器电源插座
power pole 电力柱;电杆
power policy 能源政策
power potential 动力势;电势
power press 机动压力机;动力压力机;电动印刷机
power press with fixed bed 固定台压力机
power producer 核动力反应堆;动力发生器
power-producing 供应动力的;生产动力的
power production 电力生产;产生功率;发电;动力生产
power product theorem 幂乘积定理
power project 动力工程;电力计划;电力工程;发电站项目
power proof press 电动打样机
power-propelled 机动的
power-propelled vibrating machine 动力推进振捣机
power protection 功率保护装置
power pump 机动泵;动力泵
power punch 冲击钎子
power-raised derrick 机械竖立式钻塔;电动自升式钻塔
power rammer 电动夯实器;电动打夯机;内燃式打夯机;动力夯;打夯机
power range 功率范围
power rate 单位功率;动力费用;电价;电费率
power rating 功率额定值;功率;额定功率
power ratio 功率比
power reactor 动力反应堆
power rearwheel combine drive 后轮动力联合驱动
power receiving equipment 受电设备
power receptacle 电力插座
power recorder 功率自记器;功率记录表
power recovery turbine 功率回收透平
power rectifier 整流电源;功率整流器;电力整流器
power reduction 功率减小
power reels 带钢卷取机
power regulation 功率调整
power regulator 功率调节器
power-sail ship 机帆船
power relation 指数关系
power relay 功率继电器
power removal 动力除去
power requirement 需用功率;需电量;电耗
power reserves 动力储备;储备容量
power reservoir 发电站水库
power reservse gear 动力回动装置
power resistance 功率电阻
power resistor 功率电阻器;电力电阻器
power resources 动力资源
power response 最大功率输出频率特性
power responsivity 功率反应度
power revenue 发电收益;发电收入
power reverse 动力反向
power rheostat 功率变阻器;电力变阻器
power rig 内燃机驱动的钻机;动力设备
power ringing 交流铃;机振铃
power river 水力发电河流
power rivet 强力铆钉
power riveter 电动铆(接)机
power riveting 机打铆钉;机力铆接
power roller 机动压路机;机动碾碾;机械滚筒
power roll forming 强力旋压;变薄旋压
power room 动力车间
power rotary finisher 电动磨光机;电动旋转式磨光机
power rules 幂规律
power running 电力转动
power rush 功率冲击
power sander 动力磨砂机;电动磨光机;电动打磨器
Powers apparatus 鲍威尔斯仪
power save circuit 节电电路
power saving fluorescent lamps 节能灯管
power saw 动力锯床
powersaw boss 锯工班班长
power sawing 电锯;动力锯
Powers-Brunauer model 鲍威尔斯·布鲁诺尔模型
power scale 测功率
power scheme 能源规划;动力方案;电站规划;发电方案
power scraper 动力铲运机;电耙
power scraper excavator 动力铲运挖掘机;动力铲运挖掘装置
power scraper machine 机动铲运机;铲运机
power scraper shovel 电动刮料铲
power screen 机械筛;机动筛
power screw 传力螺钉;传动螺杆
power scythe 路肩割草机
power scythe attachment 动力割刀附加装置
power section 电源部分
power selsyn 功率自整角机;功率自动同步机;动力自动同步机
power selsyn motor 动力自动同步机;电力自动同步机
power selsyn system 动力自动同步系统
power semiconductor 功率半导体器件
power sense circuit 电源检测电路
power sensitivity 功率灵敏度
power separating filter 电源分隔滤波器
power series 幂指数;幂级数
power series logarithm 幂级数的对数
power series solution 幂级数解法
power service 供电业务
power servo 动力伺服机构

power servomechanism 强力伺服机构;动力随动系统
power set 供电装置;幂集(合);发电机组;动力装置
power setting 功率调定;调功率;动力调整
power shaft 动力轴
power shafting 动力轴系
power shaker 机械振动器
power shears 动力剪切机
power shift 动力变速;动力换挡
power shifting 机械调挡;机械变速
power shift transmission 动力换挡变速箱
power shirting 电力插座护板
power shortage 缺电
power shovel 掘土机;机铲;挖土机;动力挖掘机;动力铲;电铲;单斗挖土机;单斗挖掘机;单斗掘土机
power signal 动力信号
power signal generator 功率信号发生器
power signal(l)ing installation 电力信号装置【铁】
power silage unloader 机力青储[贮]料卸载机
power site 电站站址;电厂址
power skate 动力式制动铁鞋
power slewing 电力转动
power slips 自动卡瓦;动力卡瓦
power socket 电源插座
power source 能源;功率源;晶体管弧焊;动力能源;电源
power source for aids 航标能源
power source for indication lamp 表示灯电源
power source for relay control 继电器控制电源
power source for switch control 道岔控制电源
power source for switch indication 道岔表示电源
power source for switch operation 道岔动作电源
power source optimization 电源优选
power source room for communication 通信电源间
power source room for signal 信号电源间
power-spectral analysis 功率谱分析
power-spectral analyzer 功率谱分析仪
power-spectral computer 功率频谱计算机
power-spectral density 功效谱密度;功谱密度;功率(频)谱密度
power-spectral density function 功率频谱密度函数
power spectrum 功率谱;功率频谱;能谱
power spectrum curve 功率谱曲线
power spectrum envelope method 功率谱包线法
power spectrum value 功率谱值
power spinning 强力旋压
power splitter 功率分配器;功分器
power splitting 功率分配
power splitting transmission 动力分支传动;动力分路传动
power spray 粉末喷镀
power sprayer 动力喷雾器
power spraying pump 动力喷射泵
power spraying with hand lance 带手持喷枪的动力喷洒(装置)
power squeezer 机力压榨机
Powers remolding test 鲍威尔斯重塑试验
Powers spacing factor 鲍威尔斯间距系数
power-stabilized unit 动力稳定器
power stack 功率叠加
power stage 功率级
power starvation 供给不足
powerstat 变压调压器
power station 发电厂;发电站;发电所;动力站(指水、汽、压缩空气);动力厂
power station at dam toe 坝后式水电站
power station attendant 电厂值班人员
power station at the toe of the dam 坝后式水电站
power station chamber 地下发电站
power station coaling plant 发电站煤设备
power station data acquisition equipment 电站数据采集设备
power station equipment 电站设备
power station in cascade 梯级水电站
power station in river channel 河床式水电站
power station insulator 电站电瓷绝缘子;电站电瓷
power station load curve 电站负荷曲线
power station planning 电站规划
power station process control equipment 电站过程控制设备
power station steam turbine 电站汽轮机

power steering 液压转向装置;动力转向装置;动力转向
power steering gear 动力操舵装置
power steering pump 动力转向泵
power steering unit 动力转向装置
power storage 电力储[贮]存;供发电的储[贮]水量;储能;发电蓄水量;发电库容
power stream 可资利用水力的河流;主射流
power stretcher 电动(地毯)伸展机
power string tubing 工作液输送管
power strip 配电盘;供电板;馈电条;电源板
power stroke 工作冲程;膨胀冲程;提升行程;动力冲程
power structure 权力结构
power substation 供电所
power-sum symmetric(al) function 幂和对称函数
power supply 能量供应;动力源;动力供应;电源;供电
power supply agreement 供电协议
power supply arm 供电臂
power supply box 电源箱
power supply bureau 供电局
power supply cabinet 动力柜
power supply cable 供电电缆;电力供电电缆;电力电缆
powersupply circuit 电源电路
power supply circuitry 发动机供油线路
power supply computer 供电用计算机
power supply contract 供电合约
power supply controller 电调
power supply cord 电源线
power supply equipment 电源设备;供电设备
power supply filter 电源滤波器
power supply fluctuation 电源波动
power supply flutter 电源电压脉冲
power supply for relay control 继电器控制电源
power supply for switch control 道岔控制电源
power supply for switch indication 道岔表示电源
power supply for switch operation 道岔动作电源
power supply hum 电源交流声
power supply interface 供电接口
power supply kilowattage 供电量
power supply kit 装配整流电源的整套零件;成套电源
power supply main 供电干线
power supply monitor 电源监控器
power supply network 供电网(络)
power supply of electric(al) traction system 电力牵引供电系统电源
power supply optimization 电源优选
power supply panel 电源屏
power supply plant 动力车间
power supply plug 电源插头
power supply reliability 电源可靠性
power supply ripple 电源脉动
power supply room 电源室
power supply socket 供电插座
power supply system 供电系统;供电方式;动力供应系统;电源系统;电力系统
power supply system on electrified railway depreciation expenses 电气化铁路供电系统折旧费
power supply terminal 供电端
power supply through cable 电缆供电
power supply through conducting-sliding rails 滑触线供电
power supply unit 供电机组;供电装置;供电设备;动力设备;电源装置;电源部件;电源设备
power supply voltage 供电电压
power supply wiring 电源接线
power supply workshop 供电车间
power surplus 功率富裕量
power swing 功率起伏;功率波动;功率变化;功率摆动
power-swing-block releasing relay 功率振荡闭锁复归继电器
power swing phenomenon 指针振摆现象
power switch 动力开关;电源开关;电力开关
power switchboard 电源配电板;电力开关盘
power switch circuit 功率转换电路
power switch group 主电路组合开关
power switching 功率切换
power switching board 电源配电板
power switching distribution unit 电源转换配电器
power switching over panel 电源转换屏

power swivel 动力水龙头;旋转喷水钻头
power system 动力系统;电力系统;电力网
power system operation 动力系统运行
power system parallel operation 电力系统并联运行
power system separation 电力系统解列
power system stabilizer 电力系统稳定器
power tail stock 动力顶尖座
power takeoff 动力切断;动力输出装置;动力输出轴;动力输出;动力分出;功率输出;分出功率
power takeoff baler mechanism 动力输出驱动的压捆机构
power takeoff bearing support 动力输出轴支持轴承
power takeoff bushing 脱力衬套
power takeoff clutch 功率输出轴离合器;动力输出轴传动离合器
power takeoff clutch lever 动力输出轴离合器操纵杆
power takeoff control lever 动力输出轴操纵杆
power takeoff coupling 动力输出轴联轴节
power takeoff cover 动力输出轴罩
power takeoff device 分动箱
power takeoff drive 动力输出轴驱动
power takeoff drive gear 动力输出传动齿轮
power takeoff driven finger-wheel rake 动力输出轴传动的指轮式搂草机
power takeoff driven hammer mill 动力输出轴传动锤式粉碎机
power takeoff driven hay loader 动力输出轴驱动的装干草机
power takeoff-driven pump 动力输出驱动泵
power takeoff driven tipping device 动力输出轴驱动的拖车倾翻装置
power takeoff driven trailer sprayer 动力输出轴驱动拖车式喷雾机
power takeoff drive shaft 动力输出传动轴
power takeoff dynamometer 动力输出轴测功仪
power takeoff gear 去力机齿轮
power takeoff lever 动力输出轴接合杆
power takeoff lever bearing 动力输出轴轴承
power takeoff mechanism 动力输出机构
power takeoff operating lever 动力输出轴操纵杆
power takeoff power 动力输出轴功率
power takeoff protective device 动力输出轴防护装置
power takeoff pulley 动力输出轴皮带轮
power takeoff shaft 动力输出轴
power takeoff shaft bearing 动力输出轴万向传动支座
power takeoff shaft center line 动力输出轴轴心线
power takeoff (shaft) guard 动力输出轴防护罩
power takeoff shield 动力输出轴外罩
power takeoff shifter ball spring 动力输出移动球簧
power takeoff shifter shaft 脱力移动轴
power takeoff shift lever 动力输出轴接合杆
power takeoff speed 动力输出轴转数
power tamper 机动夯;动力夯;打夯机
power tamper machine 电动捣固机
power tariff 电费
power termination 吸收头;终端功率负载;功率负荷
power test 能力型测验;动力试验
power test codes 动力试验规程
power tester 功率测试器
power test rig 动力试验装置;动力试验台
power thresher 机动脱粒机;电动脱粒机
power tie tamper 动力枕木夯泊机
power-tight 强力拧紧(螺钉)
power tiller 动力耕作机
power tire pump 动力泵
power to engage in import and export trade 进出口经营权
power tongs 动力大钳;功率钳
power tonnage 马力吨位;功率吨位
power tool 重型机床;动力工具;电动工具
power to sign 签字权
power to tax 征税权;课税权力
power-to-volume ratio 动力容积比;单位工作容积功率
power-to-weight ratio 动力重量比
power trace 功率曲线;功率记录
power traction 机动牵引;动力牵引
power traction tamper 动力牵引式打桩机;动力牵

引式打夯机
power train 动力列车；动力传动系；动力传动轮系；动力齿轮系；传动系；动力系
power transfer 功率传输；动力分配装置
power-transfer relay 故障继电器；电源切换继电器；电力传输继电器
power transfer theorem 功率转移定理
power transformation 变电
power transformer 变压器；功率变换器；动力变压器；电源变压器；电力变压器
power transformer gas protection 主变压器气体保护
power transformer mast 架空变压器塔
power transformer zero phase sequence impedance protection 主变压器零序阻抗保护
power transistor 功率晶体管；动力功率管
power transistor switching unit 功率晶体管开关单元
power transmission 送电；输电；动力输送；动力传递；电能输送；电力输送；电力传动
power transmission chain 动力链
power transmission device 传动装置；传动设备；传动机构
power transmission fluid 动力传递液
power transmission line 主输电线；输电线路；输电线；电力传输线
power transmission loss 输电损失；输电损耗
power transmission network 输电网
power transmission screw 动力传动螺杆
power transmission sequence 送电线路；传送线
power transmission shafr 动力传动轴
power transmission system 输电系统；电力输送系统；电力传动系统
power transmission tower 输电塔桅；输电塔架；动力输电铁塔
power traveling wave tube 功率行波管
power traverse 动力回旋
power triangle 功率三角形
power trip lever 自动起落器离合操纵杆
power trowel 机械铲；电铲；机动镘刀；机动抹子；电动镘刀
power-trowelled finish 机动镘刀抹光
power truck 动力转向架；动力搬运车
power tube 功率管；功率电子管；动力管道
power tubing 动力管道
power tunnel 压力隧洞；压力隧道；水电站引水隧道；发电隧洞
power tunnel intake 发电隧洞进水口
power turbine 动力涡轮机；涡轮发电机
power turbine room 涡轮发电机舱
power turn （履带拖拉机的）伺服转向
power turn-on 上电
power-type venting system 动力式通风系统
power unit 执行部件；功率单位；动力组；动力装置；动力头；动力设备；动力机组；动力单元；动力厂；电源部件；发电机组；电源设备
power-up 升高
power user 特权用户【计】
power user end 受电端
power utility 发电站
power utilization index 功率利用指数
power valve 增力阀
power variable 权力变数
power variation rate 功率变化率
power vehicle 动力车
power ventilation 机械通风；动力通风
power vessel 机动船
power vibrator 电动激振器；电动振动器
power voltage input 电源电压输入
power voltage line 电源电压输送线
power voltage output 电源电压输出
power wagon （牵引试验用的）测功车；动力车
power waste 功率损耗；动力消耗
power-wasting 耗能的
power water 压力水；做功的水；动力水；发电用水
power-weight ratio 功率重量比；每单位重量的功率；单位重量功率
power wheel 动力轮
power wheelbarrow sprayer 手推车式机动喷雾机
power wheel drive 动力驱动
power winch 卷扬机；机动绞车
power winding 功率绕组
power wood bits 木工扁钻头
power wrench 机械扳手；动力扳手；电动扳手；电动板钳
power yield 功率产额
Poxboard 帕克斯隔热板
Pox felt 帕克斯隔热(毡)板
Poyang-Dagu interglacial epoch 鄱阳—大姑间冰期【地】
Poyang glacial epoch 鄱阳冰期【地】
Poyang glaciation 翻阳冰期【地】
poyntel 对角斜铺；菱形小砖铺面；菱形铺砌
pozzolan(a) 火山灰
pozzolan(a) admixture 石灰灰掺和料
pozzolan(a) aggregate 火山灰质集料；火山灰质骨料
pozzolan(a) cement 火山灰(质)水泥；硅酸盐水泥与火山灰水泥的混合物
pozzolan(a) cement mortar 火山灰水泥砂浆
pozzolan(a) concrete 火山灰混凝土
pozzolan(a) material 火山灰质材料
pozzolan(a) reaction 火山反应
pozzolanic 火山灰的
pozzolanic action （水泥的）凝硬作用；火山灰(质)作用
pozzolanic activity 火山灰活性
pozzolanic addition 火山灰质混合(材)料
pozzolanic admixture 火山灰质掺和料；火山灰掺和料
pozzolanic aggregate 火山质集料；火山质骨料
pozzolanic blast-furnace cement 火山灰矿渣水泥
pozzolanic cement 火山灰(质)水泥
pozzolanic-cementation 灰结作用
pozzolanic-cementation reaction 灰结反应
pozzolanic concrete 白榴火山灰混凝土
pozzolanic effect 火山灰效应
pozzolanicity 火山灰活性
pozzolanic lime 火山灰(质)石灰
pozzolanic material 火山灰质材料
pozzolanic mortar 火山灰砂浆
pozzolanic Portland cement 火山灰硅酸盐水泥；火山灰波特兰水泥
pozzolanic reaction 火山灰反应；水泥凝硬作用
pozzolanic silicate cement 火山灰硅酸盐水泥
pozzolan-lime cement 火山灰石灰水泥
pozzuolana 白榴火山灰
pozz(u)olanic 凝硬性的；火山灰质的
pozz(u)olanic cement 白榴火山灰质水泥
pozz(u)olanic concrete 白榴火山灰混凝土
PPG-Pennvernon process 无槽引上法
p-phenylazoaniline 对苯偶氮基苯胺
p-phenylene diamine 对苯二胺
p-phthalic acid wastewater 对苯二甲酸废水
practicability 可行性；实用性
practicable 实际的
practical analysis 实用分析
practical anomaly 实异常
practical application 实际应用
practical astronomy 实用天文学
practical athodyd 实用冲压式空气喷气发动机
practical capacity 实际交通量；实际(生产)能力；实际(交通)容量
practical capacity point 现实的生产限度
practical cascade 实用级联
practical cataloguing 实用编目
practical ceiling 实用升限
practical ceiling height 实用上升限度
practical chemistry 实用化学
practical circle-setting table 实用度盘表
practical classification 实用分类
practical collocation 实用列类法
practical column temperature 实用柱温
practical computability 实用计算性
practical computer 实用计算机
practical concentration of measurement 实测浓度
practical consideration 实际问题；实际情况
practical convergence 实际会聚
practical decarburized depth 脱碳层厚度
practical degree of saturation 实用饱和度
practical drying process 实际干燥过程
practical duty 实际能率
practical econometric models 实用经济计量模型
practical efficiency 实际效率
practical elelctric(al) unit 实用电单位
practical envelope demodulator 实际包络线解调器
practical equivalent circuit 实用等效电路
practical error 实际误差
practical experience 实践经验；实际经验
practical factors 实际能力因素
practical fault diagonostic circuit 实用故障诊断电路
practical flow rate 实际排量
practical formula 实用公式
practical formulation 实用配方；生产配方
practical geology 应用地质学
practical grading 实用级配
practical hydraulics 实用水力学
practical hydrocarbon productivity curve 实用烃产率曲线
practicality 实用性
practicality of information 信息的实用性
practical knowledge 实用知识；实际知识
practical level 实际水平
practical life 实用寿命；实际寿命；实际使用寿命；实际使用年限
practical limit 实际极限
practical load 实用荷载
practically minded 有实际经验的
practical material figure 实际材料图
practical material figure of mise-a-la-masse 充电法实际材料图
practical measured profile 实测剖面图
practical operator 有实际经验者
practical optical system 实际的光学系统
practical optics 实用光学
practical plant capacity 工厂实际生产能力
practical porosity 有效孔隙度；实际孔隙率
practical pressure limitations 实际压力限度
practical problem 实际问题
practical production capability 实际生产能力
practical quarry-stone yield 实际成荒率
practical question 实际问题
practical ramjet 实用冲压式空气喷气发动机
practical residue limit 实际残留极限；残留允许量
practical result 实践结果；实效
practical room 实习教室
practical room block 实习楼
practical room building 实习楼
practical salinity 实用盐度
practical salinity scale 实用盐标
practical salinity unit 实用盐度计
practical seismology 实用地震学
practical shot 试验爆破
practical significance 现实意义
practical situation 实际情况
practical specific capacity 实用电容率
practical spreading rate 实际涂布率
practical standard 技术定额；实用标准
practical standard cost 实际的标准成本；实际标准成本
practical stratigraphy 实用地层学
practical superelevation 实用超高度
practical survey 实测
practical system 实用单位制
practical system of units 实用单位制
practical threshold 实际阈
practical throughput capacity 实际通过能力
practical tips 实用操作小品
practical tonnage capacity 实际通过能力
practical traffic capacity 实用通行能力
practical training 实地训练
practical unit 实用单位
practical usage of inclusion study 包裹体研究实际应用
practical value 实际价值
practical viscosity 实际黏度
practical viscosity measurement 实际黏度测定
practice 开业；实行；实习
practice area buoy 演习区浮筒
practice barre 练功扶手
practice basic 基本操作
practice ground 实验用地
practice hall 训练馆
practice of architecture 建筑业务
practice of frost protection 防霜措施
practice recovery 实际回收率
practices prevailing 习惯做法
practice weld 试验焊缝
practice workshop 实习车间
practicing engineer 开业工程师

practicing system engineer 实践的系统工程师
practise economy 实行节约
practise usury 放高利贷
practising certificate 执业证书
practitioner 专业人员;执业者
praeocinctio (古罗马具有上下级排座间的)步道
praetersonics 高超声波学;特超声学
praetorium (古罗马省级官员的)官方住宅
Prager model 帕拉格模型
Pragian 帕拉格阶
pragiogranite group 斜长花岗岩类
praglit 插字
pragma 编译指示
pragmatic economic criterion 实用经济准则
pragmaticism 实用主义
pragmatic solution 实用解法
pragment 插话
Prague red 铁红;(主要含红铁氧的色剂)布拉格红
praguite 莫来石
prain 白色软黏土
prairie 高草原;大草原;草原
prairie breaker 草原犁
prairie-buster plow 草原开沟犁
prairie combine 高草原联合收获机
prairie community 高禾草原群落
prairie dropseed 草原鼠尼草
prairie ecosystem forecasting 草原生态预测
prairie environment 草原环境
prairie fire 草原火灾
prairie-forest soil 湿草原土
prairie house 草原式住宅
prairie land 高草地
prairie landscape 草原景观
prairie soil 普列利群落土壤;湿草原土;草原土(壤)
prairie-steppe brown soil 棕色湿草原
Prairie style 大草原风格;草原式风格
prairie vegetation 普列利群落植被
Prall mill 冲击式粉碎机;冲击磨
pram 婴儿车;童车;平底船;手推车;(装混凝土用的)独轮车;方驳
Pramaxwell 波拉麦克斯韦
pram bow 平头前倾船首
pram dinghy 平头方尼圆底小艇
pram storage room 童车储[贮]藏室
Prandtl bearing capacity theory 普郎特尔承载力理论
Prandtl boundary layer theory 普郎特尔边界层理论
Prandtl number 普朗特尔(准)数
Prandtl plastic equilibrium theory 普朗特尔塑性平衡理论
Prandtl's theory of mixing length 普朗特尔混合长理论
prang (13~18世纪泰国建筑中的)神殿
prank 点缀
prase 绿石英;葱绿玉髓;葱绿色
praseodymium chloride 氯化镨
praseodymium doped fibre amplifier 掺镨光纤放大器
praseodymium doped yttrium alumin(i)um tetraborate single crystal 掺镨四硼酸铝钇单晶体
praseodymium hydride 氢化镨
praseodymium ores 镨矿
praseodymium sulfate 硫酸镨
praseodymium sulfide 硫化镨
praseo salt 绿色盐
prasinous 草绿色
prata 草甸植被
pratique 检疫证(书);无疫通行证
pratique boat 检疫船
Pratt-Hayford theory of isostasy 普拉特·海福特均衡理论
Pratt's hypothesis 普拉特假说
Pratt's mechanism 普拉特机制
Pratt theory of isostasy 普拉特均衡理论
Pratt truss 普拉特(式)桁架
Prausnitz-Kustner anti-body 反应素
prawn and fish farming 海产养殖场
prawn boat 虾艇
prawning ship 捕对虾船队
prayer book 小砂砖(擦甲板转角用)
prayer niche 祈祷用壁龛;圣龛
prayer rug 祈祷地毯(穆斯林教徒祈祷用)

prayer-tower 回教寺院的尖塔
praying chamber 祈祷室
praziquantel 吡喹酮
preabsorption 预吸收
preaccelerated polyester 预促进聚酯树脂
preacceleration 预加速
preaccentuation 预加重
preaccentuator 预增强器
preacceptance 事先承兑
preacher 砖石工砌墙用U形木块
pre-acidification alkalinity 酸以前的碱性
preaction sprinkler system 自动喷水消防系统;预作用喷水灭火系统
preaction system 火警预报的机械装置系统;预作用灭火系统
preaction valve 预作用阀
preadditive category 预加性范畴
preaddressed mode 预定址方式
preadduct 预加成聚合物
pre-adjusted 预先调整;预调
preadjustment 预先调整;预调节
preadjustment trial balance 调整前试算表
preadmission 预进(气);提前进气
preadult 成年期前的
pre-aerated burner 机械预混燃烧器
pre-aerated combustion 完全预混(式)燃烧;无焰燃烧
preaeration 预曝气
preaeration of sewage 污水预曝气
preaeration tank 预曝气箱;预曝气池
preaerator 预曝气箱;预曝气池
preag(e)ing 人工老化
preagitation 预搅动
pre-air filter 前置过滤器
pre-alarm state 预报警状态
prealignment 预校正;预先对准
prealloy 预制合金
prealloyed ingot 预制合金锭料
prealpine facies 阿尔卑斯前缘相
preamble 前言;导言
preamble bit 附加信息位
preamble clause 船舶概况条款
preamplification 前置放大作用;前置放大
preamplifier 前置放大器;前期放大器
preamplifier disable 前置放大器阻塞
preamplifier input 前置放大器输入
preamplifier stage 前置放大级
preanalysis 预先分析;预分析;事前分析
preanchored 预先锚碇的
preand post-dredging 浚前浚后
pre and post-processing 前处理与后处理
preannealing 预先退火
pre-Appalachia ocean 前阿巴拉契亚海洋
preapplied force 预加力
preapplied water nodulizing system 预加水成球系统
pre-approved subcontractor 事先批准的分包人
prearc curve 预燃曲线
pre-arch method 预拱法
prearc period 预燃时间
prearrange 预先安排计划;预先安排
prearranged aircraft maneuver 预先约定的机动动作
prearranged altitude 预定飞行高度
prearranged assignment 预分配
prearranged heading 预定航向
prearranged maneuver 预先约定的机动动作
pre arranged planning 预案
prearranged position 预定船位
prearranged signals 预定信号
preassemble 预装配
preassembled 预装的;预先装配的
preassembled bridge 预装配桥
preassembled lock 预组装锁;预装式止动器
preassembled member 预装构件
preassembled section 预装构件
preassembled system 预装配装置
preassemble time 预汇编时间
preassembling 预装配;预组装;预先安装
preassembly 预配;预装(配);预先组装
preassembly in factory 厂内预装
preassembly method 预制装配法;预制法;预先配安装法

preassembly selection 装配零件预选分组
preassembly shop 预组装车间
preassign 预先指定
preassigned multiple access 预分配多址
preassigned multiple access satellite system 预分多址卫星系统
preassigned pattern 预定模式
preassignment 预指定;预赋值
preassignment system 预分配系统
preauction trading 拍卖前交易
preaudit 事前审计
preaugering 预钻
preauthorization 批准前
preaxial 轴前的
prebai-u rainfall 梅雨前兆雨
prebake 预烘干;预焙
prebaked-anode type cell 预焙阳极电解槽
prebaked carbon slab 预焙炭板
prebaked cell for alumin(i)um-reduction 预焙阳极铝电解槽
prebaking 预先烘干
prebaratic chart 天气预报图;地面天气形势预报图
prebatched 预先配料的;预先分批配料的
prebatching bin 预称量斗仓;预称量料斗
prebedding 先铺法
prebend(ing) 预弯;预先弯曲
prebending girder 预弯梁
prebending girder bridge 预弯梁桥
prebeveling 预开坡口
prebias 预偏置
prebid conference 投标预备会议;标前会(议)
pre-bid meeting 投标预备会;招标前会议;标前会(议)
prebin 磨头仓
prebleaching 预漂
preblend 预先混合;预拌混合料;预拌;预掺混料
preblend dome 圆屋顶预均化堆场
preblending 预混合;预均化;预掺混料
preblending action 预捏合作用;预拌作用;预搅拌作用;预混合作用;预掺和作用
preblending pile 预均化料堆
prebodying 预executes炼
preboiler 预热锅炉
preboreal epoch 前北方期
prebored hole 预钻孔
prebored pile driving method 预钻孔注桩法
preboring 预钻钉眼;预钻;初步探钻;钻导孔;钻孔;打导洞
preboring for drainage 钻孔引水
preboring for nails 预钻钉眼
preboring of spike holes 木枕预钻孔
preboring of wooden tie 木枕预钻孔
prebrachium 四叠体上臂
prebreakdown 预击穿
prebreakdown characteristic 预击穿特性
prebreakdown state 预击穿态
prebreaker 预碎机;预粉碎的破碎机
prebreak-in 磨合前
prebriefed point 预先航路点
prebrightening 预增白
prebuckling 预弯曲
prebuilt 预建的;预制的
prebuilt in anchor and nut 预埋螺母和螺栓
prebuilt-in parts 预埋件(管子或电线等)
prebuilt in U ring 预埋U形环
prebuncher 预聚束器
prebunch(ing) 预聚束
preburn curve 预燃曲线
preburning 预烧法;预烧;预燃
preburn(ing) time 预燃时间
preburnish 磨合前
preburnish check 磨合前制动检查试验
precalcination 预煅烧
precalcinator 分解炉
precalcinator with fluidised bed 流化床预分解炉
precalcine 预烧
precalcined 初步煅烧的
precalciner kiln system 预分解窑系统
precalciner tower 预热器塔架
precalcining technology 预分解技术
precalcining zone 初步煅烧区
precalculated 预先计算好的
precalculated position 预定位置
precalculation 概算

Pre-Caledonian ocean 前加里东海洋【地】
Pre-Caledonian period 前加里东期【地】
precalibration 测前检定
precamber 预拱度
Pre-Cambrian 前寒武系的;先寒武【地】
Pre-Cambrian basement 前寒武纪基底【地】
Pre-Cambrian era 前寒武纪【地】
Pre-Cambrian geochronologic(al) scale 前寒武纪地质年代表【地】
Pre-Cambrian period 前寒武纪【地】
Pre-Cambrian series 前寒武统【地】
Pre-Cambrian shield 前寒武纪地盾【地】
Pre-Cambrian system 前寒武系【地】
pre-carbonation (treatment) 预先碳化处理
precarburization 预先碳化
pre-carcinogen 前致癌物
pre-Carpathian depression 喀尔巴阡山前凹陷
precarry 预进位
precast aerated concrete 预制加气混凝土
precast air-raid sheltering bunker 预制防空庇护掩体
precast anchor 锚固铁件;预制锚固铁件
precast apartment tower 预制塔式公寓
precast arch bridge 预浇拱桥
precast architectural concrete member 预浇混凝土浇注构件
precast articles 预制制品
precast artistic terrazzo flooring 预制美术磨石地面
precast assembled bridge 预制装配桥
precast balancing cantilever method 平衡悬拼法
precast bathroom and toilet 预制卫生间
precast beam 预制梁
precast beam and hollow-tile floor 小梁空心砖楼盖;预制小梁空心砖楼盖
precast beams moved rollers 滚筒托板法
precast beam unit 预制梁构件
precast block 预制块;预制砌块
precast block and lintels 预制块及过梁构造
precast blockwork 预制砌块
precast bomb shelter 预制的防空掩体
precast-boxed cofferdam 套箱围堰
precast bridge 预装配桥;预制构件桥;装配式桥(梁)
precast cable tunnel 预浇电缆坑道
precast cement hollow block 预制水泥空心砌块
precast cenamiste concrete 预制陶粒混凝土
precast cladding panel 预制面板
precast cladding slab 预制面板
precast column unit 预制柱构件
precast component 预制构件
precast concrete 混凝土预制件
precast concrete anchor 预制混凝土锚碇
precast concrete armo(u)r unit 预制混凝土护面块体
precast concrete beam 预制混凝土梁
precast concrete block 预制混凝土砌块;预制混凝土块(体);预制混凝土铺块
precast concrete block flue 预制混凝土块烟道
precast concrete blockwork 预制混凝土砌体
precast concrete caisson 预制混凝土沉箱
precast concrete ceiling 预制混凝土顶板
precast concrete chimney 预制混凝土烟囱
precast concrete cill 预制混凝土窗台
precast concrete cladding 预制混凝土外挂板;预制混凝土饰面
precast concrete component 预制水泥构件
precast concrete compound unit 预制混凝土组合单元
precast concrete connection details 预制混凝土节点详图
precast concrete cover 预制混凝土盖板
precast concrete crib cofferdam 预制混凝土木笼围堰
precast concrete culvert 预浇混凝土涵洞
precast concrete cupola 预拌混凝土穹顶
precast concrete curb 预制混凝土路缘石
precast concrete curtain 预制混凝土围护板
precast concrete curtain wall 预制混凝土护墙
precast concrete deck unit 预制混凝土面板构件
precast concrete dome 预制混凝土穹隆
precast concrete edging 预制混凝土镶边
precast concrete element 混凝土预制构件
precast concrete end-block 预制混凝土端块
precast concrete exposed aggregate slab 混凝土预浇设备
precast concrete facade 预制水刷石混凝土板
precast concrete face member 预浇混凝土贴面构件
precast concrete facing 预制混凝土饰面
precast concrete (facing) form 预制混凝土模板
precast concrete facing panel 预制混凝土面板
precast concrete factory 预制混凝土工厂
precast concrete fence 装配式混凝土围墙
precast concrete filler slab 预制混凝土立面填充板
precast concrete flag 预制混凝土人行道石板
precast concrete flight 预浇混凝土梯级
precast concrete flight of stair(case) 预制混凝土梯段
precast concrete floor 预制混凝土楼板;预制混凝土楼盖
precast concrete floor unit 预制混凝土桥面板;预浇混凝土楼面板
precast concrete flue 预制混凝土烟道;预制混凝土管道
precast concrete frame 预浇混凝土框架
precast concrete framework 预浇混凝土框架(工作)
precast concrete gable beam 预制混凝土山墙梁
precast concrete gallery 预制混凝土廊道
precast concrete girder 预制混凝土梁
precast concrete goods 预制混凝土制品
precast concrete grandstand 预制混凝土看台
precast concrete home 预制混凝土住宅
precast concrete house 预制混凝土构件房屋;装配式混凝土房屋;预制混凝土房屋
precast concrete kerb 预制混凝土路缘石;预制混凝土路边石
precast concrete lighting column 预浇混凝土照明杆;预制混凝土电灯杆
precast concrete lining 预制混凝土衬砌
precast concrete lintel 预制混凝土过梁
precast concrete manhole 预制混凝土检查井
precast concrete manhole shaft 预制混凝土检查井井筒
precast concrete manufacturing 混凝土预浇工场
precast concrete member 预制混凝土构件
precast concrete part 预制混凝土构件
precast concrete pavement 预制混凝土铺面;预制混凝土路面
precast concrete paving block 预制混凝土铺块
precast concrete penstock 预制混凝土压力水管;预制混凝土进水管
precast concrete perimeter frame 预制混凝土外围框架
precast concrete pilaster tile 预制混凝土墙垛块材
precast concrete pile 预制混凝土桩
precast concrete piles wharf with reinforced concrete whole sole deck 整片式码头
precast concrete plank 预制混凝土板
precast concrete plant 预制混凝土车间;预制混凝土厂
precast concrete purlin(e) 预制混凝土檩条
precast concrete raking strut 预制混凝土斜撑
precast concrete roof truss 预制混凝土屋架
precast concrete segment 预制混凝土弧形块;预制混凝土管片
precast concrete septic tank 预制混凝土化粪池
precast concrete shaft ring 预制混凝土井圈
precast concrete sheet pile 预制混凝土板桩
precast concrete sill 预制混凝土窗台
precast concrete silo 预制混凝土筒仓
precast concrete skeleton 预制混凝土骨架
precast concrete slab 预制混凝土板
precast concrete slab pavement 预制混凝土板铺面
precast concrete stave 预制混凝土桶板;预制混凝土环状板
precast concrete string 预制混凝土楼梯斜梁
precast concrete T Pees 预制混凝土三面棱柱
precast concrete tube segment 预浇混凝土巷道筒壁弧形砌块
precast concrete umbrella 预制混凝土隔水墙
precast concrete unit 预制混凝土单元;预制混凝土构件
precast concrete vault 预制拱
precast concrete wall panel 预制混凝土墙板
precast concrete wall structure 预浇混凝土墙板结构
precast concrete wall unit 预制混凝土墙板单元
precast concrete weight carrying skeleton 预制混凝土承重骨架
precast concrete work 预浇混凝土工作
precast concrete worker 预制混凝土工
precast construction 装配式结构;预制构造;预制装配式建筑(物);预制结构;预制构造物;预制工程;预制法施工;预填骨料结构
precast construction system 预制结构体系
precast curb 预制侧石;预制(路)缘石
precast dwellin tower 预制塔式住宅
precasted concrete 预制混凝土
precast element 预制构件
precast expanded concrete building component 预制泡沫混凝土建筑构件
precast exposed aggregate panel 预浇外露骨料板
precast factory 预制厂(混凝土构件)
precast floor segment 预制楼板构件
precast floor slab 预制楼板
precast floor unit 预制楼板构件
precast foam concrete 预制泡沫混凝土
precast-forging 液态锻造;铸后锻造;半液态模锻
precast frame 预制框架
precast framework 预制构架
precast gable 预制山墙
precast gas concrete building unit 预制泡沫混凝土建筑构件;预制加气混凝土建筑构件
precast gypsum product 预制石膏制品
precast haydite concrete 预制陶粒混凝土
precast highrise block of flats 预制高层公寓建筑
precast high(rise) floor 预制高层建筑楼板
precast hole 预铸孔
precast hollow concrete block 预制空心混凝土块;预制混凝土空心砌块
precast hollow concrete plank 预制空心楼板
precast hollow floor unit 预制空心楼板
precast immersed tunnel section 预制沉放管段
precast industry 预制制造业;预浇工业
precasting 预制;预铸;预制的
precasting block 预制块
precasting insulating pie 预制装配式保温管道
precasting of concrete units 混凝土构件的预浇
precasting slab 预制板
precasting table 预制台
precasting unit 混凝土预制件
precast joint 预制接缝
precast joint filler 预制填缝(材)料;预塑嵌缝条;预塑嵌缝板
precast jointing plate 预制填缝板
precast jointing slab 预制填缝板
precast kerb 预浇路缘石
precast light concrete slab 预制轻混凝土板
precast lightweight concrete unit 预制轻混凝土构件
precast lining 拼装式衬砌;预制衬砌
precast lintel 预制过梁
precast marble tile 预制大理石块
precast member 预制构件
precast nailable concrete plank 预制可受钉混凝土厚板
precast panel 预制墙板
precast panel construction 预制板材结构
precast panel house 预制板式房屋
precast panel shipbuilding 预制建造船法
precast parachute 预制伞形结构
precast permanent light(weight) concrete formwork 预制永久性的轻质混凝土模板
precast pile 预制桩
precast prestressed 预制预应力的
precast prestressed beam 预制预应力梁
precast prestressed concrete 预制预应力混凝土
precast prestressed concrete box girder bridge 预制预应力混凝土箱梁桥
precast prestressed concrete manufacturing yard 预浇预应力钢筋混凝土工场
precast prestressed concrete member 预浇预应力混凝土构件
precast prestressed concrete T-beam bridge 预制预应力混凝土 T 形梁桥
precast prestressed concrete unit 预制预应力混凝土构件
precast prestressed concrete voided slab bridge 预制预应力混凝土空心板桥
precast prestressed light(weight) aggregate concrete 预制预应力轻质骨料混凝土

precast product 预制品;预制件
precast pumice concrete 预制浮石混凝土
precast reinforced concrete 预制钢筋混凝土
precast reinforced concrete bridge 装配式钢筋混凝土桥梁
precast reinforced concrete building 装配式钢筋混凝土房屋
precast reinforced concrete column 预制钢筋混凝土柱
precast reinforced concrete construction 预制钢筋混凝土结构
precast reinforced concrete frame 预制钢筋混凝土框架
precast reinforced concrete frame and slab platform type wharf 装配式框架梁板码头
precast reinforced concrete framed support 钢筋混凝土支架
precast reinforced concrete garage 预浇钢筋混凝土汽车库;预制钢筋混凝土车棚
precast reinforced concrete house 装配式钢筋混凝土房屋
precast reinforced concrete lintel 预浇钢筋混凝土门窗过梁;预浇钢筋混凝土门楣
precast reinforced concrete manhole 预浇钢筋混凝土探查孔;预浇钢筋混凝土探井
precast reinforced concrete member 预制钢筋混凝土房屋构件;钢筋混凝土预制构件
precast reinforced concrete panel 预制钢筋混凝土板材
precast reinforced concrete pile 预制钢筋混凝土柱
precast reinforced concrete pipe 预制钢筋混凝土管
precast reinforced concrete retaining unit 预制钢筋混凝土挡块
precast reinforced concrete rib 预制钢筋混凝土肋
precast reinforced concrete sheet pile 预制钢筋混凝土板桩
precast reinforced concrete slab bridge 预制钢筋混凝土板桥
precast reinforced concrete structure 装配式钢筋混凝土结构;预制钢筋混凝土结构
precast reinforced gypsum slab 预制加筋石膏板
precast residence tower 预制高层公寓;预制混凝土高层住宅;预制塔式住宅
precast ribbed roof slab 大型屋面板
precast ribbed slab 预制肋形板
precast roof slab 预制屋顶板
precast roof unit 预制屋顶构件
precast section 预制构件
precast sectional pile 预制拼装柱
precast segment 预制段;预制沟管段
precast-segmental 预制弓形块;装配式预制的
precast segmental bridge 分段预制拼装桥(梁)
precast-segmental sewer 预制沟管段;预制装配式沟管;预制下水管;预制卵形管;预制污水管渠;预制沟渠管节
precast slab 预制板;预制混凝土铺板
precast slab floor 预制板楼盖
precast slab track 预制板式轨道
precast-slab-type construction 预制板式结构
precast stair(way) 预制楼梯
precast stone 人造石材;再造石
precast structural concrete member 预浇混凝土结构件
precast structural member 预制构件
precast subaqueous tunnel 预制管水底隧道
precast system 预制体系
precast tall flats 预制高层公寓
precast terrazzo 预制水磨石墙板;预制水磨石
precast terrazzo flooring 预制水磨石地面
precast terrazzo skirting 预制水磨石踢脚
precast terrazzo slab 预制水磨石板
precast terrazzo tile 预制水磨石板
precast terrazzo tiling 预制水磨石铺贴
precast tetrapod block 预制四角方块
precast tetrapod hollow block 预制四角空心方块
precast trough shaped floor unit 预制槽形楼板
precast tunnel tube 预浇隧道衬管(节)
precast unit 预制构件
precast vermiculite concrete 预制蛭石混凝土
precast wall panel 预制墙板
precast wall panel structure 预制墙板结构
precast wall structure 预制墙板结构
precast wall unit 预制墙构件

precast wood concrete block 预制木纤维混凝土砌块
precast wood concrete tile 预制木混凝土块
precatalyzed polyester 预催化聚酯树脂
precaution 预防;防护措施
precaution against cracking 预防开裂;预防裂纹
precaution against fire 小心防火;防火措施
precautionary 预防的
precautionary area 谨慎航行区
precautionary demand for money 货币的预防性需求
precautionary landing pattern 预防性着陆航线
precautionary measures 预防措施;预防性措施
precautionary motive 谨慎动机
precautionary pattern 预防性着陆航线
precautionary range 预先警告范围
precautionary underpinning 预防性托换
precaution device 预警器
precautioning measure 预防措施
precautioning measurement 预防措施
precautions against earthquakes 防震措施
precautions before welding 焊接前的预防措施
precautions during building operations 建筑施工的预防措施
precede 优先;领先
precedence 优先权;优先次序
precedence diagram 前导网络图;顺序图
precedence function 优先函数
precedence in running 优先运行
precedence language 优先语言
precedence matrices 上位矩阵
precedence matrix 优先矩阵
precedence method 优先法
precedence network 先趋网络图;作业次序网络图;前导网络
precedence network diagram 单代号网络图
precedence relation 优先关系
precedence relationship 先后关系
precedence table 先后顺序表
precedence technique 优先技术
precedence technology 优先技术
precedent 先例的
precedential 先例的
preceding cutter 超前切刀
preceding dividend 前期股利
preceding limb 前导边缘
preceding line 前测边;导线前测边
preceding seismic activity 以往地震活动性
preceding settlement 前期决算;前期结算
preceding stage 前级
preceding train 先行列车
preceding value 先前值
preceding vehicle 先行车
preceding wave 先头波;先导波
precell 置置炉
precelly 普里希利页岩(产于南威尔士)
precementation borehole 预灌浆钻孔
precentagewise 按百分比
precentral Asia Mongolia ocean 前中亚蒙古海洋
precept 技术方案
preceramics 预制陶瓷
precesional motion 进动
precession 旋进;进动
precessional constant 岁差常数
precessional motion 岁差运动
precessional period 岁差周期
precession and nutation correction 岁差与章动改正量
precession angle 进动角
precession axis 进动轴
precession camera 旋进照相机
precession circle 岁差圈
precession coefficient 岁差系数
precession cone 进动锥;岁差锥
precession gyro 进动陀螺仪
precession in declination 赤纬岁差
precession in longitude 黄经岁差
precession in right ascension 赤经岁差
precession of equinoxes 分点岁差;岁差
precession of perihelion 近日点进动
precession of the earth 地球进动
precession photograph 旋进照相图
precession photograph method 旋进照相法
prechamber 预燃室

prechamber diesel engine 预燃式柴油机
prechannel flow 入槽前径流;波面径流
precharge 预先充电;预加压;充电
precharge gate 预充电栅
precharge pulse 预充电脉冲
precharge time 预充电时间
precheck 预先校验;预先检验;预检验
precheck for drilling 钻前验收
prechlorinating 预氯化消毒;预氯化处理
prechlorination 预氯化;预加氯(化)
prechlorination tank 预氯化池
pre-Christian 公元前的
pre-Christian building style 公元前的建筑风格
pre-Christian cross 基督教以前的十字架
pre-Christian era 公元前
pricicheck 指针式精密校表仪
precinct 所属区;(教堂的)境域;管区
precinctio (古罗马具有上下级排座间的)步道
precious alloy 精密合金
precious animal 珍贵动物
precious deposit 宝藏
precious goods 贵重货(物)
precious metal 贵重金属;贵金属
precious metal chemical complex 贵金属化合物
precious metal commodity 贵金属矿产
precious metal ion 贵金属离子
precious metal powder 贵金属粉
precious metal recovery 贵金属回收
precious metal refiner 贵金属精炼厂
precious opal 贵蛋白石
precious stone 宝石
precious stone deposit 宝石矿床
precious tree 珍贵物种;珍贵树种
precious wood 贵重木材
precipice 绝壁;峭壁;陡崖;陡壁
precipitability 临界沉淀点;可沉淀性;沉淀性;沉淀度
precipitable 可沉析;可沉淀的
precipitable water 可降水分;可降水(量)
precipitable water vapo(u)r 可降水分
precipitant 沉淀剂
precipitate 沉淀物;沉淀凝结;沉淀
precipitate composition 沉淀物组分
precipitate composition concentration 沉淀物组分浓度
precipitated 沉淀的
precipitated amorphous silica 沉淀无定形二氧化硅
precipitated barite 沉淀硫酸钡
precipitated barium carbonate 沉淀碳酸钡
precipitated barium sulfate 沉淀硫酸钡
precipitated barium sulphate 沉淀硫酸钡
precipitated basic dye blue 沉淀碱性染料蓝
precipitated basic dye violet 沉淀碱性染料紫
precipitated calcium carbonate 轻质碳酸钙;沉淀碳酸钙
precipitated calcium superphosphate 沉淀过磷酸钙
precipitated catalyst 沉淀催化剂
precipitated chalk 沉淀碳酸钙
precipitated copper 沉淀铜
precipitated deposit 沉积物矿床
precipitated di-calcium phosphate 沉淀磷酸二钙
precipitated dryer 沉淀干料;沉淀(法)催干剂
precipitated dust 捕集粉尘
precipitated light calcium carbonate 沉淀轻质碳酸钙
precipitated magnesium carbonate 沉淀碳酸镁
precipitated phase 析出相;沉淀相
precipitated phosphate 沉淀磷肥
precipitated pigment 沉淀的颜料
precipitated sedimentary rock 沉积水成岩;沉淀水成岩
precipitated silica 沉淀二氧化硅
precipitated solid 沉淀固体
precipitated sulfur 沉淀硫黄
precipitated water 大气降雨
precipitated white lead 沉淀的铅白
precipitate metal hydroxide 沉淀金属氢氧化物
precipitate water 大气降水
precipitating 导致沉淀的
precipitating action 沉淀(作用)
precipitating agent 沉淀剂
precipitating agitator 沉淀搅拌机
precipitating bath 沉淀浴
precipitating cause of ore-forming elements 成

矿元素沉淀原因
precipitating cloud 降水云
precipitating factor 诱发因素
precipitating solid 沉淀固体
precipitating tank 沉淀罐;沉淀槽
precipitation 雨量沉淀;降雨;降水量;降水;脱溶;沉落;沉积;沉淀反应;沉淀(作用);沉底
precipitation absorption 沉淀物吸收
precipitation acidity 降水酸度
precipitation amount 降水量
precipitation analysis 降水分析(法);沉淀分析;沉淀分离法;沉淀滴定法
precipitation and runoff 降水和溢出;降水和径流
precipitation area 降水区
precipitation at a time 次降水量
precipitation barrier 沉降障;沉淀障
precipitation basin 沉降盆;沉降槽
precipitation bed 沉淀层
precipitation by electrolysis 电解沉淀
precipitation capacity 沉淀量
precipitation centrifuge 沉淀式离心机
precipitation characteristics 降水特征
precipitation chemistry 降水化学
precipitation chromatography 沉淀色谱法
precipitation climatology 降水气候学
precipitation clutter 雨雪杂波
precipitation clutter suppression 雨滴杂波抑制
precipitation coefficient 沉淀系数
precipitation collector 降水收集器
precipitation cone 置换圆锥;沉淀圆锥
precipitation curve 降水曲线;沉淀曲线;降水量曲线
precipitation data 降水数据;沉淀数据
precipitation deficiency 降水量差数;同期降水量差数
precipitation depth 沉积厚度;沉淀厚度;降水深度
precipitation discharge 沉淀放电
precipitation distribution 降水分布
precipitation distribution chart 降水分布图
precipitation duration 降水持续时间
precipitation during calculating period 计算时段的降水量
precipitation effectiveness index 降水有效指数;桑斯威特降水效率指数;沉淀效率指数
precipitation efficiency 降水效率
precipitation efficiency ratio 降水效率比
precipitation enhancement 增加降水;降水增加;人工降雨;人工促进降水
precipitation enhancement project 增加降雨量计划
precipitation equipment 沉淀分离装置
precipitation evapo(u)ration 沉淀蒸发
precipitation-evapo(u)ration index[P/E index] 降水蒸发指数
precipitation-evapo(u)ration ratio[P/E ratio] 降水蒸发比
precipitation event 降水场合
precipitation excess 剩余降水;超渗降水量
precipitation exchange resin 沉淀交换树脂
precipitation facies 沉积相
precipitation factor 沉淀因数
precipitation fallout 降水沉降物
precipitation field 降水场
precipitation form 沉淀形式
precipitation formula 雨量公式
precipitation fractionation 沉淀分级
precipitation from homogeneous solution 均匀沉淀;均相沉淀
precipitation ga(u)ge 雨量筒;雨量计;雨量筒;降水量计
precipitation-ga(u)ging station 雨水量计量站;量雨筒站
precipitation grading 沉析级配
precipitation hardening 弥散硬化;沉积硬化;沉淀硬化;析出硬化;脱溶硬化
precipitation-hardening stainless steel 沉淀硬化(型)不锈钢;弥散硬化不锈钢
precipitation hardening steel 沉淀硬化钢
precipitation hardening type magnet 沉淀硬化型磁铁
precipitation heat 沉淀热
precipitation heat treatment 沉淀硬化处理
precipitation index 降水指数
precipitation indicator 沉淀指示剂
precipitation in early days 前期降水(量)
precipitation in rainy season 雨季降水量

precipitation in ten-day periods 旬降水量
precipitation intensity 降水强度
precipitation-intensity-area curve 降水强度—面积关系曲线
precipitation interception 降水截流(量)
precipitation inversion 降水逆增
precipitation isotopic composition 降水同位素成分
precipitation map 降水量图
precipitation mass curve 雨水量累积曲线;雨量累积过程线;雨量累计曲线;雨量累积曲线;降水量累积曲线;降水量累积过程线;降水量曲线
precipitation mechanism 降水机制
precipitation method 滴定法;沉积法;沉淀法
precipitation moisture 凝结水分
precipitation naphtha 沉淀石脑油
precipitation noise 雨雪噪声
precipitation normal 降水量正常值
precipitation number 沉淀值
precipitation observation 降水观测
precipitation of calcium carbonate 碳酸钙沉淀
precipitation of hydroxide 氢氧化物沉淀
precipitation of salts 盐类沉淀
precipitation of settlement 沉降量推算
precipitation of sewage 污水沉淀
precipitation of sewage sludge 沉淀污泥的污水处理法
precipitation oscillation 降雨量变化;降水量波动;降水量摆动
precipitation particle 析出粒子
precipitation pattern 降水形式
precipitation peptization 沉淀胶溶
precipitation per day 每日降雨量;每日降水量
precipitation pollution 雨水污染;降水污染;大气降水污染
precipitation polymerization 沉淀聚合
precipitation prediction 降水预报
precipitation producing process 造雨过程;降水过程;沉积过程;产雨过程;沉淀过程;沉淀法
precipitation rate 降雨率;降水速率;降水率
precipitation rate parameter 收尘效率参数
precipitation reaction 沉淀反应
precipitation reagent 沉淀剂
precipitation recharge 降水补给
precipitation record 降雨记录;雨量记录
precipitation regime 降水特征;降水情势;降水情况;季节降水特征
precipitation rose 降水频率图
precipitation runoff 降雨径流
precipitation runoff relation(ship) 降水径流关系
precipitation sampler 降水取样器
precipitation scavenging 降水净化
precipitation separation 沉淀分离
precipitation softening 沉淀软化
precipitation statics 雨滴静电干扰;静电干扰;降水静电
precipitation station 雨量站;降水观测站
precipitation stimulation 人工影响降水
precipitation strength 析出强化;沉淀强度
precipitation surface 沉降表面
precipitation tank 沉淀器;沉淀箱;承淀槽;沉淀池
precipitation technique 沉淀技术
precipitation threshold 临界沉淀点;沉淀性;沉淀度
precipitation titration 沉淀滴定
precipitation trend 降水趋势
precipitation value 沉淀值
precipitation variation 降雨量变化
precipitation velocity constant 沉淀速度常数
precipitation water 凝结水
precipitation water quality 降水水质
precipitation with zinc dust 锌粉沉降
precipitative filter 沉凝式过滤器
precipitator 除尘器;集尘器;收尘器;沉降器;沉淀器;沉淀剂
precipitator gas velocity 气流速度
precipited dust 沉降粉尘
precipitin 沉淀素
precipitinogen 沉淀原
precipitin reaction 沉淀素反应
precipitin test 沉淀素试验
precipitometer 沉淀计
precipitous 悬崖狂涛;险峻的
precipitous cliff 陡峭悬崖
precipitous sea 巨浪;怒涛;道格拉斯八级浪;八级风浪

precipitum 沉淀细菌
precipitus 降水性云
precipmatrix 淀杂基
precise 精确(的);精密(的)
precise adjustment 精确的调整
precise alignment 精密准直;精密定线
precise angle indicator 精密角度指示器
precise bar mill with rapid roll change 带快速辊装置的精密棒材轧机
precise benchmark 精密水准点
precise calculation 精确计算
precise casting 精密铸造
precise code 精码
precise coordinates 精密坐标
precise data 精确数据
precise differential leveled height 精密微差水准测量高程
precise distance measurement 精密测距
precise distillation 精密蒸馏
precise drawing 精确图纸
precise engineering control 精密工程控制
precise engineering survey 精密工程测量
precise ephemeris 精确星历表;精密星历
precise examination 精密探查
precise information 精确资料
precise installation point 精密大地点
precise instrument 精密仪器
precise integrated navigation system 精密组合导航系统
precise interrupt 精确中断
precise interruption 确切中断;确定中断
precise level 精密水准仪
precise level(l)ing 精密水准测量
precise level(l)ing rod 精密水准尺
precise level(l)ing staff 精密水准尺
precise mass determination 精密质量测定
precise measurement 精确量度;精确尺寸;精密测量
precise measurement of strata movement 地层移动精密测量
precise measuring instrument 精密量具
precise metal alloy 精密合金
precise method of measuring discharge 测流精密法
precise ocean survey 精密海洋测量
precise orientation 精密定向
precise orientation sensor 精确定向传感器
precise output 精确产量
precise plumbing 精密垂准
precise positioning service 精密定位业务
precise positioning system 精密定位系统
precise prospecting 精查
precise radar navigation system 精确近程导航系统
precise radar significant location 精密雷达有效定位
precise ranging 精密测距
precise setting 精确配置;精确定位;精确安装
precise sounding 精密测深
precise staff 精密标尺
precise staking lines 精密定线
precise standard capacitors 精密标准电容器
precise stereoplotter 精密立体测图仪
precise stereoscope 精密立体镜
precise theodolite 精密经纬仪
precise tilting level 精密微倾水准仪
precise time interval 精密时间间隔
precise timing mechanism 精密计时装置
precise transit 精密经纬仪
precise traverse 精密导线
precise traversing 精密导线测量
precise triangulation 精密三角测量;精度三角测量
precision 准确度;精确性;精确(度);精密(度);精度
precision adjustment 精密度调整;精密调整
precision aerial camera 精密航空测量照相机
precision aid 精密测试仪
precision alignment gyrocompass 精密校准陀螺罗经
precision altimeter 精密测高仪
precision analog(ue) computing equipment 精密模拟计算设备
precision analysis 精度分析
precision and advanced 高精尖
precision and exactness 精确度
precision annealing 精细退火

precision apparatus 精密仪器;精密仪表
precision approach 精密进近
precision approach category I lighting system 精密进近 I 类灯光系统
precision approach radar 精密进场雷达
precision approach runway 精密进近跑道
precision assembly 精密装配;精密装备
precision attachment 精确结合
precision attained 达到的精度
precision attenuator 精密衰减器
precision attribute 精度属性
precision automated tracking system 精密自动跟踪系统
precision backlash measurement 侧隙精密测量
precision balance 精密天平
precision-balanced hybrid circuit 精密平衡混合电路
precision bench lathe 精密台式车床
precision block 精密块规
precision board cutting saw 精密裁板机
precision bolt 精密螺栓
precision cartographic camera 精密制图照相机
precision casting 精密铸造;精密铸件;精密浇注
precision ceramics 精密陶瓷
precision chain 精密链
precision chronometer 记录式精密计时器
precision conductive plastic potentiometer 精密导电塑料电位计
precision contact 精密触点
precision control 精密控制
precision coordinate 精确坐标
precision coordinatograph 精密坐标仪
precision cutting tool 精密刀具
precision depth graphic(al) recorder 精确图示测深仪
precision depth recorder 精确深度测定记录器;精密回声测深仪
precision determination method 精密度确定法
precision dial 精密度盘
precision diamond-wheeled surface grin-ding machine 精密金刚石砂轮平面磨床
precision differential manometer 精密差分压力计
precision distance measurement 精密距离测量
precision drafting table 精密绘图机
precision drawing instrument set 成套精密绘图仪器
precision drilling machine 精密钻床
precision echo sounder 精密回声测深仪
precision encoding 精确编码
precision encoding and pattern recognition 精确编码和模式识别
precision engineered polymer 精细工程聚合物
precision engineering 精密工程
precision engineering survey 精密工程测量
precision equipment 精密设备
precision estimation 精度估计
precision fast cutting nozzle 精密快速割嘴
precision feeding 精确喂料
precision fiducial 精密测点
precision film cutter 精密薄膜切割机
precision filter 精密过滤器
precision fine pitch machine 精密小径节插齿机
precision finishing 精密加工
precision flatnosed pliers 精密平口钳
precision focusing device 精密聚焦器件;精密调焦器件
precision forging 精密锻造
precision forging machine 精密锻压机;精密锻造机;精密锻机
precision form 精密成型
precision fractional distillation 精密分馏
precision frequency meter 精密频率计
precision gas cutting machine 精密切割机
precision gear 精密齿轮
precision geared head lathe 精密普通车床
precision graphic(al) recorder 精密图像记录仪;精密图示记录器
precision grid 精密坐标格网
precision grinder 精密磨床
precision grinding 精密研磨
precision high speed lathe 精密高速车床
precision high speed optic(al) lathe 精密高速光学车床
precision high speed production machine 精密高速专用机床
precision high speed tool maker lathe 精密高速度工具车床
precision horizontal surface grinding machine 精密平面磨床
precision hydrostatic(al) level 精密流体静力水准仪
precision index plate 精密分度盘
precision indicator of meridian 子午线精密指示器
precision in expression 表达式的精度
precision infrared tracking system 精密红外线跟踪系统
precision instrument 精密仪器;精密仪表;精密工具
precision instrument grease 精密仪表脂
precision instrument manufacture 精密仪器制造
precision instrument room 精密仪器室
precision interactive operation 精密迭代操作
precision-investment casting 熔模精密铸造
precision kilowatt-hour meter 精密电度表
precision laser automatic tracking system 精密激光自动跟踪系统
precision laser tracker 精密激光跟踪器
precision laser welding machine 精密激光焊接机
precision lathe 精密车床
precision lead screw 精密丝杠
precision level 精密水准仪
precision level(l)er 精密矫直机
precision level(l)ing 精密水准测量;精密校平
precision lidar 精密激光雷达
precision limiter 精确限幅器
precision linear reduction pantograph 精密线性缩小仪
precision lowering mechanism 微速下降机构
precision-machined 精加工的
precision-machined plate 精加工板
precision-machined surface 精加工平面
precision machinery 精密机械
precision machinery and instruments 精密机器仪器设备制造
precision machine tool 精密机床
precision machining 精密机械加工
precision magnifier 精密放大镜
precision mapping camera 精密测图摄应机
precision measure(ment) 精确度量;精密量测;精密测量
precision measurement of length 精密长度测量
precision measuring instrument 精密量测仪器;精密测量仪器
precision measuring machine 精密量测机器
precision measuring microscope 精密测量显微镜
precision measuring stage 精密测量台
precision mechanism 精密机构
precision metal film resistor 精密金属膜电阻器
precision meter 精密仪表
precision micrometer eyepiece 精密测微目镜
precision mill 定径轧机
precision milling machine 精密铣床
precision monocomparator 精密单象坐标仪
precision narrow-beam sounder 精密狭窄测深仪
precision night viewer 精密夜间观察镜
precision nut tapper 精密自动螺母机
precision of analysis 分析(结果)准确性
precision of format items 格式项精度
precision of horizontal positioning 水平停准度【机】
precision of instrument 仪表精密度
precision of laser ranging 激光测距精度
precision of measurement 量测精度;测量精(确)度
precision of network 控制网精度
precision of ranging for laser range finder 激光测距机测距精度
precision of real arithmetic(al) constants 实算术型常数的精度
precision of reference 引用的精确性
precision of research 研究精确度
precision of root 根的精度
precision of rule 精度规则
precision of sample mean 样本均值的精确度
precision of sample results 样本结果的精确度
precision of screen separation 筛分的准确性
precision of separation 筛分的精细度;选分率;选矿精度;分馏精度;离析精度
precision of the readings 读数精度
precision of vertical positioning 垂直停准确度【机】
precision optic(al) instrument 精密光学仪器
precision optic(al) system 精密光学系统
precision optics 精密光学装置;精密光学部件
precision pantograph 精密缩放仪
precision parameter 精度参数
precision photo-electric(al) potentiometer 精密光电电位计
precision photometer detector 精密光度计检测器
precision planimeter 精密积分仪
precision plotter 精密测图仪
precision positioning 精密定位
precision potentiometer 精密电位差计;精密电位计
precision prescribed 要求精(确)度;精度要求
precision presetting machine 精密预调机床
precision-processed 精加工的
precision processing 精处理
precision processing image 精加工图像
precision profiling 精密地貌测量
precision radar 精密雷达
precision ratio 查准率
precision reaming 精密铰孔;精铰
precision refractometer 精密折射计
precision requirement 精度要求
precision research audiometer 精密研究用测听计
precision resistance 精密电阻
precision resistor 精密电阻器
precision rolling 精轧
precision rotary microtome 精密手摇切片机
precision roving 精密无捻粗纱
precision rule 精度法则
precision saw 精密锯条
precision sawing machine 精密锯床
precision scale 精密线纹尺
precision scanning 精密扫描;精测扫描
precision seamless steel tubes 精密无缝钢管
precision self-locking drill chuck 精密自紧钻夹头
precision settling 精密安置
precision settling camera 精确安置摄影机
precision-sleeve splicer 精密套筒接头器
precision small coaxal microwave connector 精密微波同轴小接头
precision sound level meter 精密声级计
precision spectroscopy 精确光谱光度法
precision spectrograph 精密摄谱仪
precision speed reduction device 精密减速器
precision spherometer 精密球径仪
precision spin 精确螺旋
precision square 检验角尺
precision staff 精密水准(标)尺
precision standard 精密标准;精度标准
precision stereoplotter 精密立体测图仪
precision subgrader 路基精密整平机
precision sweep 精密扫描
precision tape 精密带尺
precision template 精密样板
precision test 精密试验;精度检验
precision testing 精度检查
precision testing audiometer 精密研究用测听计
precision theodolite 精密经纬仪
precision timing 精确时标
precision timing system 精确时间系统
precision tools 精密仪器
precision torsion balance 精密扭力天平
precision traverse 精密导线
precision traverse setting device 精密定程装置
precision traversing 精密导线测量
precision tube 精密钢管
precision-type excavator 精密型挖掘机
precision-type pressure regulator 精密型压力调节器
precision wattmeter 精密瓦特计
precision waveform 正确波形;精确波形
precision wave meter 精密波长计
precision weighing equipment 高精度测重仪
precision weighing gear 精密测量(齿轮)传动装置
precision weights 精细砝码;精密砝码
precision welding 精密焊接
precision welding machine 精密焊机
precision wire-wound resistor 精密线绕电阻器
precision work 精密加工;精加工;精工
pre claim cost 净赔款成本
precleaner 空气初级滤清器;预清机;粗选机;粗选器
precleaner guard 粗选机的防护罩;预清机的防护罩

precleaning 预先清理;预清洗;预清洁
preclima 预顶极
preclosing 成交前;结账前
preclosing trial balance 结账前试算表
precludable 能预防的
preclusive 排除的
preclusive buying 排除性采购
precoagulation 预凝结
pre-coagulation-ultrafiltration 预混凝超滤
precoat 预浇;预敷;预涂布;预涂(层);面料层;上底;打底子;底漆
precoated 预浇的;预涂的
precoated aggregate 预涂骨料;预涂集料
precoated base 预涂基层
precoated chip(ping)s 预涂石屑;预涂(层)沥青碎石
precoated filter 预膜过滤器;预涂(层)过滤器;预涂助滤剂的过滤机;覆盖过滤器
precoated grit 预涂砂粒
precoated metal plate 预涂金属板
precoated plate 预感光板
precoated sand 复模砂
precoated sheet 预制涂膜片;预涂层
precoated steel plate 预涂钢板
precoat filtration 预涂层过滤
precoating 镀锡;预涂(层);(油漆的)上底;预浇面层;熔模涂料;底漆
precoating agent 预混剂;预涂剂
precoating composition 预涂混合剂
precoating compound 预涂材料
precoating layer 预涂层
precoating metal 预涂金属
precoating primer 预涂底漆
precognition 预先审查
precollagen 原成胶
precollected revenue 先前吸取的收入
precollection 预收集
pre-collector 初级除尘器;第一级集尘器
precolo(u)red compound 预着色混合料
precolo(u)red resin 预着色树脂
precolo(u)ring 预着色
pre-Columbian architecture 哥伦比亚前期建筑
pre-column 前置柱;预柱;柱前
precombustion 预燃(烧)
precombustion chamber 预燃室
precombustion chamber igniter 预燃室点火装置
precombustion chamber type diesel engine 预燃室(式)柴油机
precombustion diesel 预燃室式柴油机
precombustion engine 预燃室式柴油机
precombustion reaction 预燃烧反应
precomminution 预粉碎
precommissioning 预调试
precompaction 预压(实);初步压坯
precompact set 预紧集
precompensation 预先补偿;预补偿
precompiler 预编译程序
precompiler program(me) 预编译程序
precompile time 预编译时间
precompose 预先构成
precompounding 预混
precompound stock 预混料
precompress 预加压力
precompressed 预压的
precompressed by desiccation 干燥预压
precompressed clay 预压黏土
precompressed load 预加荷载
precompressed tensile zone 预压受拉区
precompressed wood 预压木材
precompressed zone 预压区
precompression 预压(缩)
precompression chamber 预压室
precompression wave 预压波
precompressor 预压器;填装器
precomputation 预先计算;导航预行计算
precomputed 预先算出的
precomputed altitude 预算高度
precomputed altitude curve 预先计算的高度曲线
precomputed course 预计航线角
precomputed curve 预先计算曲线
precomputed height 预算高度
preconceive 预料
pre-concentration 预浓缩
preconcentration 预选;预先富集;预选;预富集

preconcentrator 预浓缩器
preconcert 预先安排
precondensate 预缩合物
precondensation 预冷凝作用;预缩合(作用);预凝结
precondenser 预凝器;预冷凝器
precondition 预处理;先决条件;预先的条件
preconditioner 预调节器
preconditioning 预老化;预调节;预处理
preconditioning circuit 预调电路
precondition of production rule 产生式规则的先决条件
precondition-prerequisite 必要前提
preconduction current 预传导电流
preconnected line 预先与水泵连接的水带
preconnect handline 预先连接的小口径水带
preconsideration 预先考虑
preconsolidate 预先固结
preconsolidated load 预压固结荷载
preconsolidated pressure 预压固结压力
preconsolidated soil 先期压密土;先期固结土;预压实土;前期固结土
preconsolidation 先期固结;预压实;预压固结;前期固结
preconsolidation load 预压固结荷载
preconsolidation pressure 先期固结压力;前期固结压力
pre-consolidation settlement 先行下沉
preconsolidation soil 前期固结土
preconstruction 施工前
preconstruction estimate 施工前估算;施工前预算;施工前的估算
preconstruction fill 预压填土
preconstruction stage 施工准备阶段;施工前期(阶段);施工前阶段
preconstruction thawing 施工前解冻(法)
precontact equipment check 设备连接前检查
precontamination 初期污染
precontinent 大陆前缘;大陆架
precontract negotiation 签约前谈判
precontract preparation 签订合同前的准备
precontrol 预先控制
precool 预冷
precooled aggregate 预冷骨料
precooler 预冷却器;预冷器;前置冷却器
precooling 预冷却(法);预冷
precooling apparatus 预冷装置
precooling coil 预冷却盘管
precooling concrete material 预冷却混凝土材料
precooling grate 预冷篦板
precooling of perishable freight 易冻货物预冷
precooling of perishable goods 易冻货物预冷
precooling room 预冷间
precooling room concrete material 预冷却混凝土材料
precooling zone 预冷却带
precooooling of refrigerated wagon body 冷藏车车体预冷
precoordinate indexing system 先组式标引法
precoordination 先组式
precoring 预钻孔
precorrection 预校正;预先校正
Precote 普里库特(一种冷铺的黑色碎石机黑色石屑)
precray 前贝壳灰岩的
precredit 信贷期以前的;信贷期以前
precreping 预压花
precrete 压力泵送混凝土
precritical 亚临界;临界(点)前的
precrushing 预破碎
precrushing and grinding system 预破碎粉磨系统
precrushing chamber 预破碎室
precrystalline deformation 晶前变形
precure 预熟化;预硫化;预固化
precured period 养护前期;预养护期
precure finish 预定形整理
precuring 静停(蒸汽养护);预养护;早期养护
precuring conveyer 预硬传送带
precursor 先质;先驱物;先驱;预兆;预报器;母体;前兆;前体;前身;前驱物;前驱波
precursor activity 前兆活动
precursor compound 初始化合物
precursor conversion 母体转化法
precursor effect 前兆效应
precursor fault 前兆断层;初始断层

precursor fiber 母体纤维
precursor gas of ozone 臭氧的气体前体
precursor model 前兆模式
precursor phenomenon 前兆现象
precursor pollutant 前体污染物
precursor polymer 母体聚合物
precursor pulse 先行脉冲;前兆脉冲
precursory phenomenon 前兆现象
precursory velocity of slide 滑坡前兆位移速度
precurved rail 预弯钢轨
precurving rail 预弯钢轨
precut 预切割;预开木材
precut fuze 预先装定的引信
precut lumber 预开木材;预开木材;预开木材料
precut preforming 预制
precut timber 预开木材
pre-cutting method 预拱法
precycle signal 定周期(交通)信号
predate bill of lading 提前日期提单
predation 捕食
predator 捕食者
predatory dumping 恶意倾销
predatory exploitation 掠夺性开发
predatory price cutting 竞争性削价;不顾血本地削价
predazzite 水滑结晶灰岩
predecessor 前任;前人;前辈
predecessor block 先行块
predecessor matrix 先行矩阵
predecomposition 预分解
predecorated gypsum board 预装修石膏板;预装饰石膏板
predecorated wallboard 预装饰墙板
predecorated wallboard panel 预装饰镶面墙板
predecrement 算前减量
predefecation 预澄清
predefine 预定义
predefined colo(u)r representation 预定义的颜色表示法
predefined polyline representation 预定义的析线表示法
predefined procedure 预定好的顺序;预定的过程
predefined process 预定处理(过程)
predefined process symbol 预定的处理符号
predefined text representation 预定义的文本表示法
predeflection 偏转之前
predeformation 预变形
predegassing 预先除气
predelay-line re-insertion of subcarrier 延时线前的副载波再注入
predelivery inspection 交付前检查
predelivery room 待产室
predella 坛前踏步;祭坛油画;祭坛台座;祭坛座上油画;祭坛台阶
predella panel 祭坛基座上的油画板
pre-denitrification pool 前置反硝化池
pre-denitrification process 预脱氮法;预反硝化法
predeposition 预淀积
predepreciation profit 事前折旧利润
predesign 概略设计;设计前的;初步设计
predesigned 预设计的
predesign services 设计前服务
predesign survey 设计前期勘测
predetection 检波前的
predetection bandwidth 预检波带宽;检波前带宽
predetection combining 检波前合并
predetection integration 检波前积分
predetection optic(al) filter 检波前滤光片
predetector 预检波器
predeterminated pressure 预定压力
predetermine 预定
predetermined angle 预定角度
predetermined camber 预先起的拱度;预定起拱;预定拱度;预先起拱
predetermined characteristic 预定性能
predetermined concentration 预定浓度
predetermined cost 预计成本;预定成本
predetermined cost accounting 预定成本计算
predetermined cost system 预计成本制
predetermined counter 预置计数器;预调计数器
predetermined course 预定航线
predetermined cycle 预定周期
predetermined dimension 预定尺寸

predetermined dimensional standards 预定尺寸标准
predetermined factory overhead 预估制造费用
predetermined figure 预定数字
predetermined financial statement 预计财务报告;预计财务报表
predetermined formula 预定公式
predetermined job-cost system 预计分批成本制度
predetermined level 预置电平
predetermined maintenance project 预定维修计划
predetermined motion time 预定动作时间
predetermined operations standard 预定经营标准
predetermined orientation 预先定向
predetermined overhead rate 预定间接费用分配率
predetermined period 预定周期
predetermined policy 预定政策
predetermined scale 成比例尺
predetermined scaler 被控脉冲计数器
predetermined setting data 安置数据
predetermined time 预定时间
predetermined time standard 预定时间标准
predetermined time standard method 预定时间标准法
predetermined time study 预定时间研究
predetermined time work 预计计时工作
predetermined value 预先确定的值;预定值
predetermined variable 先定变量;预定变量
predetermined variate 前定变量
predeterminism 先决论
predevelopment landholder 开发前土地所有者
predevelopment site 待开发场地
prediagenesis 前成岩作用
predicable 可预测的;可断定
predicate 判定;断定
predication variable 预估变量;预报变量
predicative decision model 论断性决策模型
predicative decision theory 论断性决策理论
predictability 预报准确度;预报准确率;预报准确度;可预测性;可推算性
predictability of a variable 变量可预测性
predictable 可推算的
predictable interrupt 可预测中断
predictable pattern 可预测模式
predictably computable function 预测可计算性函数
predictand 预测值;预报值;预报量
predicted and corrected altitude 修正的提前高度
predicted area of danger 预测危险区
predicted attenuation value 预期衰减量
predicted average 预期平均值
predicted barrage 移动拦阻射击
predicted chart 预报(天气)图
predicted conclusion 预测成果
predicted condition 预见条件
predicted coordinate 提前坐标
predicted cost 预测成本
predicted cost goods manufactured 制成品预计成本
predicted data 预计数据
predicted dead time 提前量测定时间
predicted depth 预测深度
predicted district 预测地区
predicted drift and leeway 估算风流压差
predicted drift angle 估算漂流角
predicted error 预计误差
predicted exploitation resources 预计开采资源
predicted finishing date 预计完工日期
predicted freight traffic volume 预测货运量
predicted ground speed 预估地面速度
predicted load 预期荷载
predicted mean 预期平均值
predicted mean value 预期值
predicted mean vote 热感觉平均标度预测值
predicted method 预测法
predicted mineral commodity 预测矿种
predicted non-point source risk 预测非点污染源风险
predicted percentage dissatistifed 预测不满意百分比
predicted performance 预计性能;预测运转成果
predicted point 提前点
predicted-point homing 前置法自导引
predicted position 预测点
predicted position device 位置预测装置
predicted propagation correction 传播改正量预测

predicted prospective area 预测远景区
predicted-pulse-shape network 预测脉冲形状网络
predicted range 预测范围
predicted settlement 预计沉降量
predicted starting and finishing times 预测开始和完成时间
predicted strata performance 预测地层状态
predicted tidal flow 推算潮流
predicted tide 预报潮
predicted tide curve 推算的潮汐曲线
predicted time 提前时间
predicted value 预期值;预计值;预测值;预报值
predicted volume 预测体积
predicted volume of traffic 预计交通量
predicted-wave signal(l)ing 预测波通信
predicted wind angle 提前风向角
predicting 预测
predicting apparatus 预测器
predicting cement strength 预测水泥强度
predicting condition 预测条件
predicting date 预测日期
predicting dead time 提前量测定准备
predicting interval 预测间隔
predicting machine 预测机
predicting organization 预测单位
predicting pollutant loading 预测污染物负荷
predicting scale 预测比例尺
predicting settlement of structure 结构物沉降量预估
predicting the effects of risk 预报风险的影响
predicting water yield of mine 矿坑预测涌水量
prediction 预示;预料;预见;预测;预报;前置量;推算
prediction aim 预测目的
prediction algorithm 预测算法
prediction angle 总修正角
prediction coefficient 预测系数;预报系数
prediction column of drillhole 钻孔预想柱状图
prediction computer 总修正角计算机
prediction cost 预测成本
prediction data 提前诸元
prediction equation 推算公式
prediction error 预计误差;预测误差
prediction error deconvolution 预测误差反褶积
prediction error filtering 预测误差滤波
prediction filter 预测过滤器
prediction formula 推算公式
prediction function 预报函数
prediction interval 预测数的变化范围;预报区间;预报时段
prediction method 预报法;预测的方法
prediction model 预测模式;预报模型;预报模式;推算模型
prediction of back-silting quantity 回淤量预测
prediction of calamitous type of earthquake 震害类型预测
prediction of cement strength 水泥强度预测
prediction of coalfield 煤田预测
prediction of depth of focus 震源深度预测
prediction of drawdown value 预测降深值
prediction of earthquake magnitude 地震强度预测
prediction of earthquake triggering area 发震地区预测
prediction of earthquake triggering belt 发震地带预测
prediction of flash-over 袭燃前预测
prediction of immersion 浸没预测
prediction of intensity of seismic calamity 震害强度预测
prediction of magnitude of settlement 沉降量预估
prediction of mean anomalies 平均异常推估
prediction of mineral deposite 矿产预测
prediction of mineral resources 资源预测
prediction of mining quantity 预测开采量
prediction of nuisance 公害预测
prediction of reservoir characteristic parameter 预测储层特性参数
prediction of runoff 径流预测;径流推算
prediction of seismic activity 地震活动趋势预测
prediction of seismic calamity 震害预测
prediction of seismic frequency 地震频度预测
prediction of settlement 预计沉降;沉降预计;沉降预测;沉降预报
prediction of solar activity 太阳活动预报
prediction of solute distribution 溶质分布预测

prediction of tide 潮汐推算
prediction of time series 时间序列预测
prediction of various weather constituents 气象要素预报
prediction of water bursting possibility 突水可能性预测
prediction of water pollution 水污染预测
prediction of water yield of mine 矿井涌水量预测
prediction operator 预测算子
prediction period 提前时间
prediction principle 预测原理
prediction probability 推算概率
prediction problem 预测问题
prediction region 预测区
prediction reliability 预测可靠性
prediction scale 速度预测尺
prediction signal(l)ing system 预告信号装置
prediction-subtraction coding 预测减法编码
prediction theory 预测理论;预报理论
prediction value of point anomaly 点异常推值
predictive amount of resources 预测资源量
predictive analysis 预期分析;预测分析法
predictive cell 预测单元
predictive coding 预先编码;预测编码
predictive control 预测控制
predictive crash sensor 预测碰撞探测器
predictive deconvolution 预测反褶积
predictive distance 预测距离
predictive equation 预断公式
predictive filtering 预测滤波
predictive index 预测指数
predictive map 预地图
predictive mean vote 预测平均热感觉
predictive method by model 模型预测法
predictive mode 预示方式
predictive model(ling) 预测模型
predictive quantizing system 预测量化系统
predictive recognizer 预示识别算法
predictive reserves 预测储量
predictive risk assessment 预测性风险评价
predictive value 预告值;前兆价值
predictivity test 预测性试验
predict length 预测长度
predictor 预示算子;预测指标;预测因子;预测器;预报装置;预报员;预报因子;预测算子;推算者
predictor circuit 预测电路
predictor control 前置控制;提前量的调节
predictor-corrector 预测校正器;预报校正器
predictor-corrector method 预示校正法;预测校正法;预报校正法
predictor-corrector procedure 预测校正法;预报校正法
predictor equation 预测方程
predictor formula 预估公式;预测公式
predictor result 预报结果
predictor servomechanism 提前随动系统
predictor-type controller 预测式控制系统
predict strata 预测的层位
predict structure unit 预测的构造单元
predifferentiation stage 分化前期
prediffused technique 预扩散工艺
prediffusion 预扩散
predigest 预先消化
predigestion 预消化
predigestion of data 数据预先加工;数据预处理加工;数据的预先加工
prediluted 预稀释的
prediluted oil 预先稀释的润滑油
prediluvian 洪积前的
predip 预浸渍;预浸
pre-disaster activity 灾前活动
pre-disaster planning 防灾规划
pre-disaster planning and preparedness 灾前规划和准备工作
predischarge 预先卸载;预放电
predisperser 预色散器
predispersing 预分散
predispersing grating 预色散光栅
predispersing monochromator 预色散单色仪
predispersor 前置分光器
predispose to damage 易受损害的
predisposing 素因性的
predisposing cause 诱因
predisposition 诱因;素因

predissociation 预离解;预分离
predissolve 预溶解
pre-distillation 初步蒸馏
predistorter 前置补偿器
predistorting network 预矫正网络
predistortion 预失真;预矫正;预畸变
predistortion filter 预失真滤波器
predistribution 预先分配
predominance 优势
predominance area diagram 优势区图
predominance of horizontal movement 水平运动主导性
predominance of one colo(u)r 彩色失真
predominant 占优势;突出的;超优势木
predominant bank 主导河岸
predominant cargo flow 主要货流
predominant constituent 主要成分
predominant current 盛行流;盛行海流
predominant formation 最常钻到的岩层;最常钻到的地层;主要岩层;主要地层
predominant frequency 卓越频率;主频率
predominant goods 大宗货
predominant hue 主色相
predominantly residential area 住宅为主的地区
predominant mean height 超优势木平均高
predominant mode 卓越振型
predominant period 卓越周期;主要周期
predominant species 特优势种;超优势种
predominant traffic 主要运量
predominant tree 超优势树
predominant type 主要类型
predominant wave 盛行浪;盛行波(浪);常涌常风
predominant wind 主要风向;主导风向;盛行风
predominant wood 超优势木
predominate 支配
predominating factor 主要因素
predomination 支配
predose 预剂量
predraft 预拉伸
predrainage 预排水法;提早排水
predraining 预排水
predraining ground 降水地层
predraining method 预先抽水法
predraw 预吸取;预提取;预拉
predrawing 预拉伸;预并条
predrawing process 预拉伸过程
predredging survey 浚前测量
predrier 预烘机;预干室
predrilled 预钻孔的
predrilled hole 预钻孔
predrilling 预钻;超前钻进
predrive 预驱动;前级激励
predrive circuit 前置驱动电路
predryer 预烘干机;预干室
predrying 预先干燥;预烘;预干(燥)
predrying chamber 预烘干室
predrying compartment 预烘干仓
predrying period 预干期
predrying treatment 预干燥处理
predrying unit 预烘燥装置
predual 预对偶
predunk tank 净化脱脂槽
pre-earthquake 前震
pre-earthquake deformation 震前变形
pre-earthquake dilatancy 震前扩容
pre-earthquake distortion 震前畸变
pre-earthquake measure 预防地震措施
pre-earthquake movement 震前活动
pre-earthquake slip 震前滑动
Preece test 镀锌层厚度和均匀度测定试验;普里斯钢丝镀锌层的硫酸铜浸蚀试验
pre-echo 前回声;预试反射波
preedit 预先编辑;预编辑
preedit check program(me) 预编辑检查程序
preedited interpretive system 预编解释系统
preediting 事先编辑
preedition 预先编辑
preedit program(me) 预编辑程序
preejection 弹射前准备
preelaboration 预加工
preelect 预选
preelectrolysis 预电解
pre-embedded 预埋的
pre-embedded foot bolt 预埋地脚螺栓

pre-embedded iron member 预埋铁件
pre-embedded parts 预埋件(管子或电线等)
pre-embedded pieces 预埋件(管子或电线等)
pre-embedded steel plate 预埋钢板
preemergence application 出土前施用
preemergency governor 备急调速器
pre-emergency planning 事故前预案
preemergent herbicides 在萌发前施用的除草剂
preemphasis 预加重
preemphasis circuit 预校正电路;预加重电路
preemphasis filter 预加重滤波器
preemphasis network 频响预校网络;预加重网络
preemphasized signal 预加重信号
preemployment activity 聘前作业
pre-employment examination 雇佣前的体格检查
pre-employment health examination 就业前健康检查
pre-employment physical examination 就业前体会
preempt 优先购买
preemption 先买权;优先权;抢先;收买权
preemption capability 占先能力
preemption service 预占业务
preemptive 先买的;抢先的
preemptive algorithm 抢先算法
preemptive priority 抢先优先权
preemptive right 先买权;优先认购权
preemptive scheduling 抢先调度
preemptive scheduling strategy 抢先调度策略
preemulsion 预制乳状液
preen 修剪;整理;修饰;贴脸;镶边
preengagement condition 啮合前状态
preengineered 使用预制构件建造的;预先筹划的;预设计的;用预制构件建造
preengineered belt conveyer 装配式带式运输机
preengineered conveyer 预制标准化运输机
pre-engineered suppression system 预制标准间的灭火系统
preengineering cost 钻前工程费
preengineering technology 初级工程技术
preenrichment 预富集
preentrained jump 预掺气水跃
preentry 预先报关
preentry losses 工程施工前可减免的开支
preenvelop 预包络
pree-photographer 摄影记者
pre-equalizer 前置均衡器
pre-equilibration column 前置平衡柱
pre-equilibrium composition 预平衡成分
pre-erected grade line 预竖立坡度线
pre-erected grade wire 预竖立坡度控制金属丝
preeruption 萌出前期
preestablish 预先设立
preestablished limit 预定的极限
preestablished magnitude of earthquake 预定震级
preestimate 预测
preesure difference 气压差
preetching 预腐蚀
preeutectic 先共晶的
preevacuate 预排气
preevacuated chamber 预抽真空室
preevacuating 前级真空的
preevaluation test 预备性鉴定试验;初步评价试验
pre-evaporation 初步蒸发
pre-evaporator 初步蒸发器;初步蒸馏器
preevent retrofit method 震前改建措施
preevent storage 触发前储存量
preexamine 预先检查
preexcavation 预开挖
pre-excentric pier 预偏心桥墩
preexcitation wave 预激波
preexhaust 预先排气
preexistent pore 原有孔隙
preexisting crack 原有裂缝
preexisting fabric 先存组构
preexisting fault 先在断层;原有断层
preexisting fractural zone 原有破裂带
preexisting fracture 原有裂缝
preexisting micro-feature 原显微形态
preexisting mountain belts 先在山带
preexisting rock 母岩
preexisting valley 前存河谷
preexpanded 预发泡的
preexpander 预扩展器
pre-expansion saturation 膨胀前的饱和

pre-expansion track 膨胀前的径迹
preexport financing 出口前资金融通
preexposure 预曝光
preexposure setting 预曝光调节
preextracted wood 预浸木材
preextraction 预提取
prefab 活动房屋;预制房屋;预装配式房屋;预制装配式房屋;装配式房屋
prefab clean room 装配式洁净间
prefab-form(work) 预制模板
prefab monolithic structure 预制装配式整体结构;预制装配整体结构
prefabricate 预制;装配预制的分段船体;工厂预制的;分段预制
prefabricated 预制装配的;预制的
prefabricated area wall 预制装配式采光井墙;预制装配采光井墙
prefabricated asphaltic bitumen surfacing 预制沥青铺面;预制沥青路面
prefabricated asphaltic blanket 预制沥青铺盖
prefabricated asphalt plank 预制沥青板
prefabricated asphalt sheet(ing) 预制沥青片
prefabricated assembly 预制组装件
prefabricated bathroom unit 预制(装配式)浴室单元
prefabricated bituminized hessian surfacing 预制沥青粗麻布(铺)面
prefabricated bituminous surfacing 预制沥青混凝土板面层;预制沥青混凝土板路面
prefabricated block 预制装配式建筑
prefabricated building 预制装配式建筑(物);预制安装建筑;装配式结构;装配式建筑;装配式房屋
prefabricated building assembly 预制建筑组件
prefabricated building sheet 预制建筑薄板
prefabricated ceramic tiling 预制陶瓷贴面
prefabricated circuit 印刷电路
prefabricated coil 预制线圈
prefabricated component 预制构件
prefabricated-component factory 预制构件厂
prefabricated compound unit 预制组合构件
prefabricated concrete 预制混凝土
prefabricated concrete block structure 块体工程
prefabricated concrete building 预制装配式混凝土房屋
prefabricated concrete building member 预制装配式混凝土建筑构件
prefabricated concrete cladding panel 预制装配式混凝土外围护墙板
prefabricated concrete column 预制混凝土柱
prefabricated concrete gable 预制混凝土山墙
prefabricated concrete garden building unit 预制装配式混凝土园艺建筑单元
prefabricated concrete green block 新预制的混凝土块
prefabricated concrete iplaster tile 预制混凝土墙垛块材
prefabricated concrete panel fence 预制装配式混凝土板材围墙
prefabricated concrete perimeter frame 预制混凝土周边框
prefabricated concrete pile 预制混凝土桩
prefabricated concrete purlin(e) 预制混凝土檩条
prefabricated concrete sheet pile 预制混凝土板桩
prefabricated concrete slab 预制混凝土铺板
prefabricated concrete string 预制混凝土梯斜梁
prefabricated concrete threshold 预制混凝土门槛
prefabricated concrete wall 预制混凝土墙
prefabricated concrete walling 预制混凝土墙砌体
prefabricated conduit system 预制管道系统
prefabricated construction 预制装配(式)结构;预制结构;预制构件建筑;装配式施工
prefabricated construction method 预制装配施工法
prefabricated drain 预制排水管
prefabricated dwelling house 预制装配式住宅
prefabricated element 预制单元;预制构件
prefabricated expanded concrete member 预制泡沫混凝土构件
prefabricated factory 预制厂
prefabricated feed system 预制投加料系统
prefabricated floor 预制楼板
prefabricated floor member 预制楼板构件
prefabricated flue 预制装配烟道;预制烟道

prefabricated form system 预制模板系统
prefabricated form(work) 预制模板
prefabricated frame 预制框架
prefabricated girder 预制梁
prefabricated gypsum product 预制石膏制品
prefabricated heat insulation 预制绝热材料
prefabricated hotel 装配式旅馆
prefabricated house 活动房屋;预装配式房屋;预制(装配式)房屋;装配式房屋
prefabricated house unit 预制装配式住宅构件
prefabricated joinery 定制木作
prefabricated joint filler 预制填缝条
prefabricated lattice panel 预制花格板
prefabricated lining 拼装式衬砌;预制面层;装配式衬砌
prefabricated masonry panel 预制圬工墙板;预制砌体墙板
prefabricated masonry slab 预制圬工墙板
prefabricated member 预制构件
prefabricated-monolithic shipbuilding 预装配整体造船法
prefabricated office 装配式办公楼
prefabricated or preformed parts 预制及预成型件
prefabricated panel 预制墙板;预制板材
prefabricated panel construction 预制板材施工(法)
prefabricated parallel wire strand 预制平行钢丝束
prefabricated part 预制件;预制(安装)构件;工厂预制部件
prefabricated partition wall(ing) 预制隔墙
prefabricated pavilion 装配式亭子
prefabricated pile 预制桩
prefabricated pipe 预制水管
prefabricated pipe conduit system 预制管道系统
prefabricated pipe system 预制管道系统
prefabricated plant 预制厂
prefabricated platform 预制平台;预制安装平台
prefabricated prestressed brick lintel 装配式预应力砖过梁
prefabricated prestressed compound unit 装配式预应力组合构件
prefabricated pumice concrete 预制浮石混凝土
prefabricated reinforced concrete building 装配式钢筋混凝土房屋
prefabricated reinforced concrete pile 制钢筋混凝土桩
prefabricated reinforced concrete wall 预制配筋混凝土墙
prefabricated reinforced pile 预制配筋桩
prefabricated reinforced wall 预制配筋墙
prefabricated reinforcement 预制钢筋骨架
prefabricated rib vault 预制肋形拱顶
prefabricated roofing 预制屋面
prefabricated shuttering unit 预制模板
prefabricated stack 预制烟囱
prefabricated structural element 预制结构构件
prefabricated structure 预装配式结构;预制装配(式)结构
prefabricated subaqueous pile 预制管水底隧道
prefabricated subassembly 预制部件
prefabricated subqaueous tunnel 预制水底隧道
prefabricated surfacing 预制饰面
prefabricated terrazzo 预制水磨石
prefabricated tie 预制拉杆;预制墙箍;(在砌墙内的)预制拉结网片;预理墙系筋
prefabricated tile partition(wall) 预制板材隔墙
prefabricated tiling 预制板
prefabricated tiling cubicle for clothes 预制装配式板材衣橱
prefabricated timber building 装配式木屋
prefabricated timber house 装配式木建筑
prefabricated timber section 预制木制件;预制木零件
prefabricated truss 装配式屋架
prefabricated unit 预制构件;预制安装组合单元
prefabricated vessel 分段建造船
prefabricated wall 装配式墙;预制装配式墙
prefabricated walling 预制砌体
prefabricated wall panel 预制墙板
prefabricated welding mo(u)ld 预制焊模
prefabricated window-wall unit 预制窗间墙板
prefabricated wood-I-joist 预制木材工字桁条
prefabricated wood truss 预制木桁架
prefabricate house 活动房屋

prefabricate treatment plant 预制处理厂
prefabrication 预制装配;预制分段船体的装配;预制(安装);预加工;工厂预制
prefabrication engineering 装配式建筑
prefabrication primer 预涂底漆
prefabrication shed 预制配间
prefabrication technique 预制工艺
prefabricator 预制配件者
prefabticated member 预制件
prefab unit 预制构件
prefading 预调节
prefeasibility exploration 预可行性勘探
prefeasibility study 可行性初步研究;预可行性研究报告;前期可行性研究;初步可行性研究
prefeasibility study of port project 港口工程预可行性研究
prefect dielectric 纯介质
prefectural edifice 县署大厦
prefectural nature conservation area 地区自然环境保护区
prefectural tax policy 地县税收政策
prefecture 地区;辖区;小组办公楼;官邸;专区
prefecture boundary 地区界(线)
preferability condition 更可取的条件
preference 优先选择;优先权;优先(发送);优待
preference as to assets 财产优先还权
preference changes 选好度变化
preference clause 特惠条款
preference duty 特惠关税
preference for probability 概率优先
preference pattern 选择模型
preference premium 特惠酬金
preference ranking 选择顺序
preference relation 选择关系
preference road 优先通行路
preference stock 优先股
preference temperature 适宜温度
preferential absorption 选择(性)吸收;优先吸收
preferential adsorption 选择性吸附
preferential attack 择优侵蚀
preferential clause 特惠条款;优惠条款
preferential combustion 优先燃烧
preferential conformation 优势构象
preferential crystallization 选择性结晶;优先结晶
preferential duties 优惠关税;特惠关税
preferential duty 特惠税;关税优惠
preferential elimination 选择排除
preferential entitlement 享受优惠的资格
preferential etching 择优蚀刻;优先浸蚀
preferential exchange rate 特惠汇率
preferential export financing 出口优惠融资
preferential flo(a)tation 优先浮选
preferential interest 优惠存款
preferential interest rate 优惠利率
preferential loan 优惠贷款
preferential measure 优惠办法
preferential offset 优先(信号)时差
preferential orientation 最佳取向
preferential oxidation 择优氧化
preferential oxidation of hydrogen 氢的优先氧化
preferential pairing 偏向配对
preferential path 最可能渗径;高渗透通道
preferential payment in bankruptcy 破产优先偿付
preferential policy 优惠政策
preferential precipitation 优先沉淀
preferential price 优惠价(格)
preferential rate 优惠运价;优惠汇率
preferential rate of exchange 优惠汇率
preferential recombination 优先复合;择优复合
preferential relation 择优关系
preferential rights 优先权;优惠权
preferential rights of coastal states 沿海国的优先权利
preferential road 优先通行路
preferential runway 特优跑道(机场);优先使用的跑道
preferential segregation 偏向分离
preferential selectivity 有择溶解度
preferential shop 优惠机构
preferential slip 倾向滑移
preferential solubility 选择性溶解;有择溶解度
preferential solution 优先溶解
preferential species 适宜种
preferential street 优先通行道

preferential system 特惠制(度)
preferential tariff 优惠税率;优惠关税;特惠关税
preferential tariff cut 优惠减税额;特惠减税
preferential tariff system 关税优惠制;特惠关税制
preferential tax of environmental protection 环境保护税收优惠
preferential tax policy 税收优惠政策
preferential temperature 适宜温度
preferential tendering 优惠投标法
preferential trade 特惠贸易
preferential trade agreement 特惠贸易协定
preferential trade arrangement 特惠贸易协定
preferential treatment 优先处理;优惠待遇;优待;特惠待遇
preferential trip 先行脱扣器
preferential wetting 选择性溶湿润;优先润湿
preferred alternative index 优先替换使用的合作系数
preferred angle (楼梯的)可取角度;较优坡度
preferred axis of magnetization 易磁化轴
preferred category 前置类目
preferred circuit 标准电路
preferred coordinates 特定座标(系)
preferred diametral pitch 优选径节
preferred dimension 选用尺寸
preferred direction 优先定向;易磁化方向
preferred direction of magnetization 易磁化方向
preferred fabric 优选组构
preferred flo(a)tation 优先浮选
preferred frequencies for acoustic(al) measurement 声学测量的最佳频率
preferred maritime lien 海事赔偿优先留置权
preferred noise criterion curve 适宜的噪声标准曲线;事先选定的噪声标准曲线
preferred number 从优数;优先数;优先系数;优先数字;优先数系
preferred operating rate 优先开工率
preferred option 首选方案
preferred order 选用顺序
preferred orientation 优选方位;优先定位;择优定向;定向方位
preferred orientation of each generation of lineation 各世代线理优选方位
preferred orientations diagram 优选方位图;定向方位图
preferred parts list 选用零件表
preferred plan 最佳规划;最佳方案;可取方案;较佳方案
preferred power supply 优先电源
preferred reinforcement 选用钢筋
preferred scheme 选用的类表;优先方案;倾向性方案
preferred section 推荐截面
preferred sequence 选用顺序
preferred site 选用位置
preferred size 推荐尺寸;选用尺寸
preferred stock 优先股
preferred thickness 推荐厚度
preferred to be selected 优先选用的
preferred value 选用值;优选值;优先值;优先数字
preferred view 理想投影
prefetch 预取
prefetched operand 预取数
prefetch microinstruction 预取微指令
prefigure 预算
prefill 预装填
prefilled 加胶木屑板
prefilled pattern wheel 预装提花轮
prefilling 预装填
prefilling press 成型预压机
prefill surge valve 满油补充阀;充液补偿阀
prefill valve 满油阀;充液阀
prefilmed alloy 预生氧化膜的合金
prefilming 预膜
prefilming agent 预膜剂
prefilter 粗滤器;预滤池;前置过滤器;预(过)滤器;前置滤波器
prefiltered water 预滤过性水
prefiltering 前置滤波;预(先)过滤;预滤(波)
prefiltration 预过滤;粗滤
prefinish 预装修;预做饰面;预加工
prefinished 预做饰面的;预整修的
prefinished colo(u)r coating 预涂色彩层
prefinished door 预制门

prefinished metal plate 预涂金属板
prefinished particle board 预饰刨花板;预涂刨花板
prefinished plywood 预饰胶合板
prefinished wallboard panel 预涂刨花板;预装饰镶面墙板
prefinished wall panel 预涂护墙板;预装饰镶面墙板
prefinishing 预做饰面的;预装修;预加工
prefinishing gypsum plasterboard 预饰面纸面石膏板
prefinishing wall board 预饰面墙板
prefire 预先点火
prefire information 灾前资料
prefire planning 灭火作战计划
prefire situation inspection 灭火前环境状况的调查
prefiring 预热;预烧;提前点火;点火前的
prefiring check 点火前检验
prefiring cycle 发射前准备周期
prefix area 前置区
prefix code 前置代码
prefixed point 前定点
prefixing 加标头
prefix method synchronization 词头法同步
prefix multiplier 乘数引子;词头乘数
prefix operator 前置运算符
prefix register 前置寄存器
prefix value 标头值
preflame 预燃
preflame oxidation 预燃氧化
preflame reaction 预燃反应
preflash 预闪蒸
preflashing 预闪
pre-flash-over period 轰燃前阶段
pre-flash-over temperature 轰燃前温度
pre-flash-over temperature estimation 轰燃前温度估算
preflex beam 预弯梁
preflex beam bridge 预弯梁桥;顶弯梁桥
preflex girder 预弯大梁
preflex girder bridge 预弯梁桥
preflex(ion) 预弯;预加弯力
prefloat 预浮选
prefloat circuit 预浮流程
prefloat treatment 预浮处理
prefloc 初布絮凝体;预絮凝粒
preflood decrease of storage 汛前预泄;汛前水库放水
preflooding 预液泛
preflush 预冲洗;前置液
preflush flow counter 预先冲洗流通式计数管
preflush fluid 预冲洗液;冲洗液
prefluxed polymer 预熔聚合物
prefluxing 预染熔剂;涂助焊剂
preflysch silica muddy association 前复理石硅泥质组合
prefoaming material 预发泡剂;增孔剂
prefoaming method 预制泡沫法
prefocus 预聚焦;初聚焦
prefocus cap 聚光灯头;定焦灯头
prefocused beam 预聚焦束
prefocusing 预聚焦
prefocusing lens 预聚焦镜头
prefocus lamp 预聚焦灯;定焦灯
prefocus lamp base 预调焦灯座
prefocus lens 预聚焦透镜;前置聚焦透镜
prefogging 预曝光
preforce 预先加力
preforging 预锻
preform 初步加工;预成型;压片;盘料;塑坯预型;初加工的成品
preformat 预先格式
preformation 预先形成;预成型
preformationist 预成说者
preformation theory 先成(学)说
preformatted record 预定格式记录
preform die mo(u)lding 预成型对模模塑法
preformed adduct 预制加合物
preformed architectural compression seal 预制建筑压缩密封料
preformed architectural strip seal 预制建筑密封条
preformed armo(u)r rod 预绞式护线条
preformed asphalt joint filler 预塑嵌缝板;预制沥青嵌缝条
preformed building 预制建筑物
preformed camber 预留上拱度

preformed casse 浊雾
preformed cavity 预留孔
preformed ceramic core 预制陶瓷芯
preformed compact 预压坯
preformed culvert 预制管涵
preformed die 预制冲模;预制钢型;预制模具
preformed duct insulation 成型的管道保温绝缘层
preformed flashing piece 成型泛水片
preformed floc 成型絮凝体;预絮凝体
preformed foam 预制泡沫;预发泡沫
preformed gasket 预制密封垫;预制密封圈
preformed groove 预留槽
preformed gum 燃用树脂
preformed hole 预留孔
preformed insulation 预成型绝缘材料;预成型绝缘层
preformed joint 预制缝;预塑缝
preformed joint filler 预制填缝板;预制填缝(材)料;预塑(式)嵌缝板;预塑嵌缝条
preformed joint ring 预制管子连接环
preformed joint seal 预制接缝填料
preformed joint sealant 预制街坊密封料;预制嵌缝料
preformed lay wire 预搓长股的钢索
preformed liner 预制衬里
preformed mastic 预成型的黏结水泥
preformed matched die 预成型模塑法
preformed pile 预制桩
preformed pipe insulation section for heat protection 预制保温用加热管段
preformed polymer 预聚物;预聚(合)物
preformed precipitate 预先形成的沉淀物;预沉淀
preformed rope 预先变形钢丝绳;预成型绳
preformed sealant 预制密封料;预制密封胶
preformed sealing material 定型密封材料
preformed sealing tape 定型密封带
preformed section 预制(管)段;预制型材
preformed shape 预成型
preformed shaped 预成型制品
preformed sintered liner 预制烧结衬里
preformed stemming 预制成型炮泥
preformed tape 预制密封胶带
preformed Tripoli-powder section 预制硅藻土管段(保温用)
preformed unit 预制件
preformed winding 预制品绕组;成型绕组
preformed wire rope 预塑钢丝绳;预搓丝股的钢索;不松散钢丝绳
preform electrode 预成型电极
preformer 预压机;预成型机;预变形器
preforming 预成型模塑法;制坯工序;预成型
preforming die 预成型模头
preforming machine 预成型机
preforming press 预压机;制锭机;成型预压机
preforming screen 半成品检验分类
preforming technique 预成型工艺
preform machine 预成型机
preform mo(u)lding 塑坯模制法
preformulation 预制剂
prefractionation 初步分离
prefractionator 预分馏塔;初步分馏塔
prefractured rock 先在破裂岩
preframe 预装配
preframed 预拼装的;预装配的
preframed panel 预装配板
prefreezing 预冻结
prefreshet decrease of storage 汛前预泄
prefreshet period 汛前期
prefrontal fog 锋前雾
prefrontal shower 锋前阵雨
prefrothing 预发泡
pre-functional building 前功能主义(者设计的)建筑
prefused eutectic 预熔共晶
prefusion 预熔融
pregasified fuel oil fired kiln 烧气化液体燃料的窑
pregassing 预放气
pregate amplifier 带前置门的放大器
pregel 预凝胶(体);多余胶膜
pregelatinization 预胶凝
pregeologic 地质史前的
pregeosynclinal stage 前地槽阶段
pregeosynclinal structural tectonic layer 前地槽构造层
preglacial 冰期前的

preglacial deposit 前冰期沉积物;冰堰沉积(物);冰外缘沉积(物);冰前沉积(物)
preglacial lake 冰前湖
pregnane 孕甾烷
pregnant antimony solution 母液
pregnant liquor 母液
pregnant solution 含贵重矿物溶液;富液,富集液
pregnant year 丰产年
pre-Gothic 前哥特式的
programmed temperature chromatography 程序升温色谱法
pregranular 颗粒期前的
pregreasing pump 预润滑泵
pregrinder 预磨机
pregrinding 预磨;粗磨
pregrinding chamber 预粉磨室;粗粉磨仓
pregrinding mill 粗磨机
pregrinding paste 预磨削用研磨膏
pregrounding 粗压碎
pregroup 准群;前群
pregrouting 预注浆;预灌浆
pregrouting from earth's surface 地表预注浆
pregrouting from face 工作面预注浆
pre-grouting with micropipe 小导管预注浆
pregummed paper 预涂胶纸
pregwood 浸胶木材;浸胶压缩木材
preharden 预坚膜
prehardened mo(u)ld steel 预淬火模具钢
prehardening 初凝;预硬化
prehardening period 预硬化期
prehardening reaction 预先硬化反应
preharmonic 准调和的
pre-harvest desiccant 收获前干燥剂
preheat and hot spot system 预热与局部加热系统
preheat burner 预热燃烧器
preheat circuit 预热电路
preheat coil 预热线圈;预热盘管
preheat current 预热电流
preheated 预热的
preheated aggregate 预热骨料
preheated air 预热空气
preheated electric(al) arc welding 预热电弧焊接
preheated film 预热薄膜
preheated forehearth 保温前炉
preheater 预热器
preheater adjusting valve 预热器调节阀
preheater bypass system 预热器旁路系统
preheater cyclone 预热器旋风筒
preheater kiln 预热器窑;带预热的窑
preheater section 预热(阶)段
preheater shaft 立筒预热器
preheater stage 预热器级
preheater string 预热器系列
preheater tower 预热器塔架
preheater with precalciner 带分解炉的预热器
preheat fan 预热风扇
preheat flame 预热火焰
preheat fluorescent lamp 预热荧光灯;预热荧光灯
preheat hot-cathode lamp 预热的热阴极灯
preheat(ing) 预(加)热
preheating aggregate 预热骨料
preheating bin 预热料仓;预热料斗
preheating chamber 预热室
preheating current 预热电流
preheating deposit 预热沉积
preheating equipment 预热设备
preheating evaporator 预热蒸发器
preheating flame 预热焰
preheating furnace 预热炉
preheating gate 预热孔
preheating liquid metals 预热液态金属
preheating method 预热法
preheating of fuel 燃料的预热
preheating of oil 油的预热
preheating period 预热时间
preheating pipe 预热管
preheating pot 预热锅
preheating process 预热法
preheating range 预热区
preheating region 预热区
preheating section 预热区;预热段
preheating section of kiln 窑的预热段
preheating temperature 预热温度
preheating time 预热时间

preheating treatment 预热处理
preheating tubes 预热管
preheating tunnel 预热隧道
preheating unit 预热装置
preheating zone 预热区;预热带;初步供暖区
preheat margin 预热裕度
preheat oxygen 预热氧
preheat requirement 预热的要求
preheat roll 预热辊
preheat shroud 预温罩
preheat starter 预热起动器
preheat starting 预热起动
preheat train in distillation unit 蒸馏装置中的预热系统
preheat treatment machining 预热加工
pre-Hellenic 古希腊前建筑
pre-Hellenic architecture 古希腊前建筑
prehensile 缠绕性
preheterodyne by-passing technique 预外差旁通技术
prehistoric age 史前时代
prehistoric earthquake 史前地震
prehistoric naturalized plant 史前归化植物
prehistoric pottery 史前期土器
prehistory 史前学;史前史
prehnite 葡萄石
prehnite-pumpellyite facies 葡萄石—绿纤石相【地】
prehnitic acid 连苯四酸
prehnitite 葡萄石岩
prehomogenizing 预均化
prehomogenizing efficiency 预均化效率
prehomogenous technique 预均化技术
prehung door 预装门
prehydrated bentonite 预水合膨润土
prehydration 预先水化;预先泡水;预水化;预水合
pre-hydrogen technique 前置氢技术
prehydrolysis 预水解;预加水分解
preiection 凸起
preignition 预燃;早燃;早;提前点火
preignition chamber 预点火室
preignition period 预燃期
pre-ignition phase 提前点火阶段
preimbrian 前雨海纪【地】
preimport financing 进口前资金融通
preimpounding seismicity 蓄水前地震活动性
preimpoundment 先期蓄水
preimpregnated 预浸渍的;预先浸渍的
preimpregnated carbon fibre web method 碳纤维网预浸渍法
preimpregnated coil 预浸渍线圈
preimpregnated process 预浸渍工艺;预浸工艺
preimpregnated reinforced plastic 预浸增强塑料
preimpregnated varnish 预浸渍漆
preimpregnating insulation 预浸渍绝缘
preimpregnation 预浸渍;预浸
prein 压平布面方法
pre-incident plan 事前计划
preincubation 预保温
pre-industrial background concentration 先进工业本底污染浓度
preindustrial city 前工业城市
preindustrial country 发展中国家
preindustrial prototype 试制品
preinitiation 提前起爆
pre-injection 预注浆;预先灌浆法
pre-injection clearing hole 预注浆扫孔
pre-injection design 预注浆设计
pre-injection technique for waterproofing 预注浆防水设计
preinjector 预注入器
preinlet valve 预启阀
preinput editing 输入前编辑
preinspection check 预检
preinstall 预设
pre-installation 预安装
preinstallation test 装配前试验;安装前试验
preintegral 准整数
preinvasive 蔓延前的
pre-investment activity 投资前活动
pre-investment assistance 投资前援助
pre-investment fund 投资前基金
pre-investment phase 投资前时期(阶段)
pre-investment program(me) 投资前方案

pre-investment project 投资前项目
pre-investment study 投资前(的)研究
preiodic type 周期型
preionization 先电离
preirradiation 预照射
preirradiation treatment 辐照前处理
pre-islamic architecture 前伊斯兰建筑
prejudge 预先判断;预断
prejudication 预先判断
prejudice 偏见;损害
preknock 预爆震
preknotting 预除节
prelaying 预砌
preleach 预浸出
prelevelled 预先整平的
preliberation 棚户区
prelife operation 正式使用前试用操作
prelime 预加灰
prelimer 石灰混合器
preliminary stratigraphy 初级地层学
preliminaries 准备工作;初步行动
preliminarily crushed 初步破碎的
preliminarily under control 初步得到控制
preliminarty calibration curve 初始校正曲线
preliminary acceptance 初步验收
preliminary acceptance certificate 初步验收证(明)书
preliminary acceptance trial 预备性交接试验
preliminary action 初步行动
preliminary adjustment 预调;初步调整
preliminary aeration of wastewater 污水预曝气
preliminary ag(e)ing 预先老化
preliminary agreement 初步协议;初步协定
preliminary air classification 初步风力预筛;初步风力分选
preliminary alignment 初步对准
preliminary amplifier 前置放大器
preliminary analysis 初步分析
preliminary anger press 螺旋预榨机
preliminary announcement 预告
preliminary appraisal 初步评估
preliminary approval 初步审核;初步批准
preliminary audit 初步审计
preliminary audit survey 初步审计鉴定;初步审核
preliminary balance 初步平衡
preliminary balance computation 初步平衡计算
preliminary balance sheet 初编资产负债表
preliminary beam test 梁的初步检验;梁的初步试验
preliminary blasting 初次爆破
preliminary borehole log 原始钻孔记录表
preliminary bracing 初期支护
preliminary breaker 预先压碎机;粗碎;初碎机
preliminary breaking 初碎
preliminary budget 试编预算;初步预算
preliminary budget estimate 初步概算
preliminary calculation 初算;初步计算
preliminary calibration 初步校准
preliminary cataloging 预编目
preliminary centering 初步对中
preliminary charges 筹备费
preliminary checkout 预备测试
preliminary clarification 初步澄清
preliminary clarification tank 初步分选的容器;初步分离的容器
preliminary clarifier 初步澄清池
preliminary classification 预先分级
preliminary cleaning 预净化;初步清洗;初步净化
preliminary combustion 预燃
preliminary command 预令
preliminary compaction 预压;预捣固;预压实;预压紧
preliminary compilation 蓝图拼贴
preliminary compression curve 初始压缩曲线
preliminary computation 概算;设计概算;初步计算
preliminary concentration 预选
preliminary concept(ion) 初步设想
preliminary conference 预备会议
preliminary consolidation 初期凝结作用;预固结;初期固结
preliminary consultation 事先磋商
preliminary contact 预动触点
preliminary contractural agreement 契约性的初步协定
preliminary cook 预煮

preliminary coordinates 概略坐标
preliminary crusher 一次破碎机;粗压碎机;初轧(碎)机;初碎机
preliminary crushing 粗压碎;初轧;初碎
preliminary-crushing department 一次破碎工段
preliminary cube test 初步立方体强度试验
preliminary data 原始数据;预备数据;初始数据
preliminary data report 原始数据报表;准备数据报告
preliminary definition 预先定义
preliminary demolition 预先爆破
preliminary design 准备设计;预先设计;原始设计;预备设计;打样设计;初步设计
preliminary designation 初步命名
preliminary design drawing 初步设计图纸;初步设计图
preliminary design features 初步设计要点
preliminary design of a capital construction project 基本建设初步设计
preliminary design parameter 初步设计参数
preliminary design phase 初期设计阶段;初步设计阶段
preliminary design report 初步设计报告
preliminary design review 初步设计审查
preliminary design stage 初步设计阶段;初级设计阶段
preliminary determination 初步测定
preliminary determination epicentre 初步确定的震中
preliminary dimension 预定尺寸;初步尺寸
preliminary dip 预浸渍;预浸
preliminary direction(al) hole 初级定向孔【岩】
preliminary discussion 初步讨论
preliminary disposal technique 初步处置技术;初步处理技术
preliminary draft 预牵伸
preliminary drawing 草图;初设图纸;初步设计图
preliminary drying 预烘;预干燥
preliminary drying chamber 预烘干室
preliminary dust collector 初级集尘器
preliminary edition 初版
preliminary effect 预效应
preliminary element 初始要素
preliminary elevation 概略高程
preliminary elongation 预拉伸
preliminary engineering inspection 初步工程检验;初步工程检验
preliminary entry 进口预先申报;初步分录
preliminary environment(al) impact assessment 环境影响初评(价)
preliminary estimate 估算;设计概算;设计概算;初步估算;初步估算;初步估计
preliminary evaluation 预先估计;初评;初步评估;初步鉴评试验
preliminary evaluation of technical economics 初步技术经济评价
preliminary examination 预试验;粗探伤;初步检查;初步调查
preliminary examination of bids 标书预审
preliminary expenditures 开办费支出;初期支出;筹备费
preliminary expenses 开办费;初期支出;筹备费
preliminary exploratory report 初勘报告
preliminary feasibility 预可行性研究
preliminary filter 预滤池;燃料粗滤器;初滤器;初滤池
preliminary finish 预整理
preliminary fire refining 预先火法精炼;初步火法精炼
preliminary fractionation 预先部位分离作用
preliminary fractionator 预分馏塔
preliminary fuel filter 柴油初滤器
preliminary fueling 预先油量
Preliminary General Catalogue 博斯总星表初编
preliminary geotechnical investigation 初步岩土工程勘探;初步岩土工程勘察
preliminary glow 初始辉光
preliminary grinding 预先磨矿;预粗磨;初磨
preliminary grinding chamber 粗磨室;粗磨仓
preliminary grinding compartment 粗磨室;粗磨仓
preliminary groundwork 创建工作
preliminary hazard analysis 初步事故分析
preliminary heating 预热
preliminary heating zone 预热区;预热层;初步供

暖区
preliminary hole 粗钻孔;初试钻孔
preliminary hydro-extraction 初步脱水
preliminary hydrogeologic(al) investigation 水文地质初步勘察
preliminary hydrologic work 水文学准备工作;初步水文工作者
preliminary impregnation 预浸
preliminary information 事前资料
preliminary infusion 预先注水
preliminary inspection 初验
preliminary interpretation 初步解释
preliminary investigation 初步勘察;初期勘测;初勘;初查;初步研究;初步调查;初步查勘
preliminary investigation and study 初步调研
preliminary investigation stage 初步调查阶段
preliminary irradiation 初步辐照
preliminary latitude 概略纬度
preliminary layout 初步方案;草图
preliminary list of items 暂定项目表
preliminary loan 初期贷款(建设项目的)
preliminary location 初步定线
preliminary locking 初步锁闭
preliminary longitude 概略经度
preliminary map 临时版地图;初版地图
preliminary market testing 初步市场试验
preliminary matter 正文前的图文
preliminary measure 初步措施
preliminary mentor 预先蒙导
preliminary mill 粗磨
preliminary mixing 预先混合
preliminary mixing chamber 初混室
preliminary note 首项附注
preliminary notice 预告
preliminary notice to mariners 航海预告
preliminary obsrvations relating structure and functions 结构与动能关系的初步观察
preliminary operation 装备工作;试行运转;现场施工准备;预备操作;准备工序;试验;试运行;初步工序
preliminary orbit 最初轨道;初轨;初步轨道
preliminary order 预令
preliminary orientation 初步定向
preliminary pedological work 土壤学的准备工作;初步的土壤学工作
preliminary phase 前期;初期;初步阶段
preliminary pile 试探(性)桩;预置桩
preliminary pile assembly 实验性反应堆
preliminary plan(ning) 初步规划;初步计划;初步方案
preliminary plot 地段草图;初步地段图;临时版地图
preliminary pollutant limit value 初始污染极值
preliminary position 概略位置
preliminary precipitation 预沉淀
preliminary pressure 初压
preliminary program(me) 预备程序
preliminary project 初步设想;初步设计;初步计划方案;初步计划
preliminary project planning 工程初步规划
preliminary project programming 工程初步规划
preliminary prospecting 踏勘;初步普查;初步勘探;初步勘察
preliminary prospecting report 初查报告
preliminary pulse 初期脉冲
preliminary purification 初步提纯;初步净化
preliminary ray trace 初始光线追迹
preliminary reading 初步读数;预先读出;初读数
preliminary reconnaissance 初步踏勘
preliminary refining 初精炼
preliminary rejection 预选尾矿
preliminary report 预备报告;初步报告
preliminary review 预备调查
preliminary rinsing screen 预冲洗筛分机;预冲洗筛子
preliminary robbing 初步回采
preliminary roller 前辊
preliminary route determination 初步定线
preliminary safety analysis report 初步安全分析报告
preliminary sample 初步样本
preliminary scheme 初步计划;初步规划;初步方案
preliminary screeding 初步刮平
preliminary screening 预筛分;初步筛选
preliminary sedimentation 预沉(作用);沉淀降(作用);初次沉积作用;初次沉积
preliminary sedimentation tank 初次沉淀池;预沉降槽;预沉池;初沉池;初步沉淀池
preliminary series 临时版地图
preliminary settling tank 初级沉淀池
preliminary shaft 锁口
preliminary shaping 预成型
preliminary shearing 初剪
preliminary sheet 试印样
preliminary shock 首震;初期微震
preliminary site selection 位置预选
preliminary sizing 粗筛选
preliminary sketch 初步设计;初步草图;草图;初步设想轮廓图
preliminary software 初级软件
preliminary specification 简要说明书;初步说明书;暂定指标
preliminary stage 原始阶段;初始阶段;初期
preliminary-stage duration 预点火阶段持续时间
preliminary start 预先起动
preliminary steeping 预浸
preliminary step 预备步骤
preliminary study 初步研究;初步审查;初步分析;初步调查;初报
preliminary support 隧道初期支护
preliminary support parameter 初期支护参数
preliminary survey 预测;初始测量;初测;初步勘测;初步调查;草测
preliminary survey stage 草测阶段
preliminary symptom 地震前兆
preliminary technical development plan 初步技术发展计划
preliminary tender enquiry 投标的初步查询
preliminary tension 初张力
preliminary term insurance 初期保险
preliminary test estimator 初步检验估计量
preliminary test(ing) 初步试验;预试验;初试;初级试验
preliminary test of significance 初步显著性检验
preliminary time standards 初步时间标准
preliminary time study 初步时间研究
preliminary transit line 初测导线
preliminary traverse 草测导线
preliminary treatment 预处理;简易处理;前期处理;初步处理
preliminary tremor 初始微震;初期微震
preliminary trial 初步试车
preliminary trial period 预试期
preliminary triangle 概算三角形
preliminary tube mill 粗磨管磨机
preliminary vacuum 初期减压
preliminary value 初值
preliminary value sorting device 初值分选设备
preliminary warning notice 缺陷通告
preliminary washing drum 预洗矿筒
preliminary wave 初至波
preliminary winding 预备卷绕
preliminary work for construction 实际建筑施工的准备工作;施工准备
preliminary working 初加工
preliminary working plan report 施业案初步报告
preliminary works 预备工程;准备工作;准备工程;前期工程;初步工作
preliminator 球磨机(水泥)
preliming 预加灰
preliming tank 预灰槽
preliquefier 初步液化器
preload 预加荷载;预填入;预装入;预加料;预载;预压;预试载;预加压;预负荷;初载
preload compaction 预压加固
preload concrete 预应力混凝土
preload concrete slab 预应力混凝土板
preload consolidation 预压加固
preload consolidation apparatus 预压固结两用仪
preloaded bearing 预紧轴承;装配过紧的轴承
preloaded ion exchange packed bed adsorption reactor 预载离子交换填料床吸附反应器
preloaded rubber bushing 预紧橡胶衬套
preloaded soil 超固结土
preloaded spring 预载弹簧
preloader 预紧器
preloading 预压(法);预加载;预加荷载;预加负荷;预先负载;预压法;预先加料;预加负荷
preloading by material 堆载预压

preloading centering 拱架预压
preloading compaction 预载压实
preloading consolidation 预压固结
preloading fill 前期加载填土
preloading method 预压法
preloading washer 预载垫圈
preload machine 预载机器
preload stress 预应力
preload system 预压法
preload tank 预应力混凝土水池;预应力混凝土柜;预压水箱
prelocalization 前定位
prelocalizing subcategory 预局部化子范畴
prelodgement system 先行申报制度
prelogging 伐前准备伐
prolonged aeration process 延时曝气法
prolonged agitation 长期搅动
prolonged corrosion test 长期腐蚀试验
prolonged erosion test 长期冲刷试验
prolonged exposure 长时间曝光
prolonged heating 长期加热
prolonged station 长期观测站
prolonged storage 长时间存放
prelubricated ball 预润滑球
prelubricated bearing 预加润滑剂的轴承
prelubricated sealed ball bearing 预润滑封闭式球轴承
prelubricated sealed bearing 预润滑密封轴承
prelubrication 预先润滑
prelude 序言;过程标题
premade 预制的
premagnetization 预先磁化;预磁化
premagnetization type current transformer 预磁化式变流器
premagnification 预放大(率)
premain sequence 主序前
premarked point 预设标志点
premarking 预设标志
premastery 预先掌握
premastication 预先捏和作用
premature amniotic rupture 沥浆产
premature beat 过早收缩
premature blast 早爆;过早爆炸
premature block 岩芯过早卡塞
premature bolting 早期抽苔
premature burning 炮眼内炸药过早起爆
premature combustion 早燃
premature contact of incline plane 早期斜面接触
premature contraction 早期收缩
premature cure 早期固化
premature damage 过早损坏
premature delivery 提前交货
premature destroy 早期破坏
premature drying-out 过早干燥
premature explosion 早爆;过早起爆;过早爆炸
premature failure 早期破损;早期破坏;过早失效;过早破坏
premature firing 过早起爆
premature freezing 过早凝固
premature gear failure 齿轮过早损坏
premature gelation 先期胶凝;早期胶凝
premature hardening 早期硬化;过早硬化(混凝土)
premature harvesting 提前收割
premature ignition 先期点火
premature loss 过早缺失
prematurely opened bid 早开的投标
premature polymerization 早期聚合
premature release 预先释放
premature setting 过早凝结;急凝(混凝土的)
premature sharpening 磨得过早
premature sintering 早期烧结
premature stiffening 过早硬化(混凝土);过早凝结;假凝(现象);(混凝土的)急凝
premature systole 过早收缩
premature vulcanization 早期硫化
premature wear 过早磨耗
prematurity payment 预付款;未到期付款;提前付款;到期前付款
prematurity repayment 到期前偿付
premeditation 预先计划
premelt 预熔化
premelter bushing 预熔漏板
premelting 预熔
premelting transition 预熔融转变

premetallised dye 预金属络合的染料
premetamorphic emplacement 变质前侵位
premetro 预筑的地下铁道;半地铁
Premier mill 普雷迈磨
preminator 粗磨机
premineral fault 成矿前的断层
premises 房屋连地基;房屋及其附属建筑物;房屋基地;房产
premises wiring system 房屋布线系统;房屋布线系统
premium 奖励工资;奖金;权利金;贴水;升水;超票面价格;附加费;保险金;保险费
premium and bond issuance expenses 摊提贴水
premium association 保险费联盟
premium bond 超过票面值的证券
premium bonus system 职工奖金制度
premium casting 优质铸件
premium cement 优质水泥;高级水泥
premium due 到期保险费
premium engine oil 高级机油
premium fuel 优质燃料;高值燃料
premium funds 保费准备金
premium grade 高级;优等;优等品;一级
premium-grade material 高级材料
premium in arrears 拖欠的保险费
premium insulation 优质绝缘;高级绝缘
premium insurance 保费保险
premium list 保险率表
premium loan 纳费贷款;保费借款
premium motor fuel 高级车用汽油
premium motor oil 优质发动机油;高级车用汽油;高级车用机油
premium mud 优质泥浆
premium note 纳费通知;保险费付款通知单
premium on capital stock 超面值缴入资本
premium on insurance 保险费
premium product 高级产品
premium rate 保险公司标准费率;保险费率
premium receipt 保险费收据
premium reserves 保险费准备金;保费准备金
premium returns 额外收益;额外利润
premium statement 保险费账单;保费账单
premium surface dressing system 高级表面处理系统
premium system (隧道坑道工程中的)超额奖励工资制度;奖励工资制;奖金制度;分红制
premium tariff 保险费率表
premium tax 保险费税
premium theory of taxation 课税的保险理论
premium to be arranged 保险费另议
premium uses of gas 优先用气
premix 预拌;预先混合;预混合(料);预混
premix burner 预燃器;预混燃烧器;预混合烧嘴;层流燃烧器
premix compound 预混料
premixed 预先拌和的
premixed aggregate 预拌集料;预拌骨料
premixed air and gas burner 预混气燃烧器
premixed bituminous surface 预拌沥青混合料路面;预拌沥青面层;预拌沥青路面
premixed burner 预混燃烧器;预混(空气)烧器
premixed carpet 地毯料层;预混合的铺层
premixed chippings 预拌石屑
premixed colo(u)red exterior plsater 预拌着色的外墙灰浆
premixed colo(u)red rendering 预拌着色的外墙抹灰
premixed colo(u)red stucco 预拌着色的拉毛粉饰
premixed combustion 预混燃烧
premixed concrete 预拌混凝土
premixed elements 预混素材
premixed facility 预拌混凝土摊铺机
premixed flame 预混焰;预混火焰
premixed foam solution 预混泡沫溶液
premixed gas 预混合气
premixed gypsum-lime plaster 预拌石膏石灰灰浆
premixed gypsum plaster 预拌石膏灰浆
premixed gypsum stuff 预拌石膏材料
premixed light(weight) gypsum plaster 预拌轻质石膏粉饰
premixed macadam 预拌碎石;碎石路(面)
premixed material 预拌材料
premixed mixture 预混合料
premixed plaster 预拌灰浆;预拌灰浆

premixed solution 预混溶液
premixed stuff 预拌材料
premixed surfacing 预拌路面材料
premixed truck 预拌混凝土运送车;预拌混凝土拌运车
premixed water-bound macadam 湿拌碎石
premixed wet paste 预拌湿的浆状物
premixer 预拌搅拌器;预拌搅拌机;预先混合器;预混室;预混合器
premix gas burner 预混气体燃烧器
premixing 预先搅拌;预混合;预拌
premixing cavity 预燃室;预混合室
premixing chamber 预混合室
premixing equipment 拌浆机
premixing method 预混合法;预拌法
premixing plant 预拌设备;预拌工厂;预拌车间
premix injector 预先混合喷射器
premix material 预混料
premix mo(u)lding 预混模制;预混料模塑;预混合模塑成型
premix mo(u)lding compound 预混合模塑料
premix-plunger machine 预混柱塞式注模机
premixture 预混合物
premix type burner 预混合型燃烧器
premix type of treatment 预混式表面处治
premode 前方式
premodern 现代化之前
pre-modern architecture 近代建筑
premodification 预先修正;预先修改;预先变质处理
premodulation 预调制
premoistening 预湿(混凝土)
premolar teeth 槽牙
premonition 预兆;前兆
premonitoring phenomenon 前兆现象
premonitoring slip 前兆滑动
premonitoring symptom 预兆
premonitory 先兆
premonitory symptoms 地震前兆
premonochromator 前置单色器
premo(u)lded 预制的;预先制制的;预塑的
premo(u)lded asphalt joint filler 预制沥青嵌缝条
premo(u)lded asphalt panel 预制沥青板
premo(u)lded asphalt plank 预塑地沥青板;预塑(地)沥青板
premo(u)lded asphalt sealing strip 预塑式沥青嵌缝板;预塑嵌缝式沥青嵌缝条;预塑嵌缝沥青嵌缝板;预塑沥青嵌缝板
premo(u)lded bitumen strip 预制沥青条
premo(u)lded bituminous joint filler 预制沥青嵌缝条
premo(u)lded bituminous panel 预制沥青板
premo(u)lded bituminous strip 预制沥青条
premo(u)lded expansion joint filler 预制伸缩缝填料
premo(u)lded filler 预模填缝条
premo(u)lded filler strip 预制塞缝条;预制嵌缝条
premo(u)lded gasket joint 预模制套口接头
premo(u)lded joint 预塑缝
premo(u)lded joint filler 预制填缝(材)料;预塑嵌缝条;预塑嵌缝板
premo(u)lded pile 预制桩;预制混凝土桩;预试桩;模制桩
premo(u)lding 预先铸模;预模塑
premultiplication 左乘
prenectarian 前酒海纪【地】
prennical root 多年生根
prenormalization 预规格化
prenormalize 预规格化
prentice 学徒工
Prentice's rule 勃伦蒂斯定则
preohmic 预欧姆
preohmic contact 预欧姆接触
preohmic etch 预欧姆腐蚀
preoiler 预先加油器
preoiling 预先润滑
preopening expenses 开办准备费
preoperation(al) test 运行前试验;交付使用前试验;试运行;试车;空转试验
preoperative control 预定位控制
preoptive control 预选操纵
preordain 预先规定
preorder 预次序
preorder for binary tree 二叉树的前根次序
Pre-Ordovician 奥陶纪前;前奥陶纪

preoreforming fracture 成矿前断裂
preoreforming structure 成矿前构造
pre-ore halo 矿化前晕
preorientation degree 预取向度
preorogenic 造山前的
preorogenic coalification 造山前期煤化作用
preorogenic phase 造山前期
preoscillation current 振荡前电流;起振前电流
preoval 粗轧椭圆孔型
pre-oxidation 预氧化
pre-ozonization 预加臭氧;预臭氧化
prepack(age) 预(先)包装;预组装的;预先装填
prepackaged concrete 预填骨料混凝土;预压骨料混凝土
prepackaged mix(ture) 预填混合料
prepacked aggregate 预埋集料;预填集料;预填骨料;预埋骨料(后用水泥浆灌注成混凝土)
prepacked aggregate concrete 压力灌浆混凝土;预填混凝土;预填骨料混凝土;预填骨料灌浆混凝土
prepacked bearing 预加润滑剂的轴承
prepacked clay brick 预压实黏土砖;预填黏土砖
prepacked coarse aggregate 预捣实粗集料;预填粗集料;预填粗骨料
prepacked column 预填充柱;预充填柱
prepacked concrete 预制混凝土;预填料混凝土;预填集料混凝土;预填骨料混凝土
prepacked gravel sleeve 预填砾石过筛筒
prepacked liner 填砾石衬筒
prepacked method 预填集料压力灌浆法;预填骨料灌浆法
prepacked support 预填充载体
prepacked with grease 预装填润滑脂
prepacking 预装填;预先包装
prepacking treatment 包装前处理
prepack method 预装法
prepadding 预浸轧
prepaging 预约式页面调度
prepaid 预付的;付讫的
prepaid amount 预付金额
prepaid annuity 预付年金
prepaid expenses 预付费用;垫支资金
prepaid freight 预付运费
prepaid insurance 预付保险费
prepaid interest 预付利息
prepaid items 预付项目
prepaid items of expenses 费用预付项
prepaid materials 预付费用
prepaid operating capital 预付营业资金
prepaid rent 预付租金
prepaid reply 预付回报费
prepaint(ing) 预涂(漆);预描绘;预喷漆;预先油漆
prepainting work 涂漆漆;预漆工作
prepakt concrete 预压骨料混凝土;预填骨料混凝土;压浆混凝土
Prepakt cone 普雷派克锥(测定泥浆稠度)
prepakt method 预填骨料法
prepalletized cargo 预先集装的托盘化货物
preparation 预备;准备;制剂;制备品;制备
preparation and utilization of coal slurry 煤水浆的制备及应用
preparation bay 准备区
preparation by magnetic separation 磁选
preparation by screening 筛选
preparation concentrate 精矿
preparation cost 准备工作费;准备费用
preparation equipment 选矿设备
preparation expenses 筹备费
preparation formulation 剂型
preparation for sea 开航准备工作;出海准备
preparation for starting 起动准备
preparation geology work of mine design 矿山设计准备地质工作计划任务书
preparation in quantity 成批制备
preparation method 制备方法
preparation notice 准备装货通知
preparation of a distribution plan 分配计划的编制
preparation of an instalment distribution plan 分期付款计划的编制
preparation of base metal for welding 焊接基本金属的处理
preparation of batch 配料的制备;配合料制备
preparation of bids 投标准备
preparation of cash distribution plan 现金分配

计划的编制
preparation of ceramic body 陶瓷坯料的制备
preparation of coal samples 煤样设备
preparation of construction 施工准备
preparation of construction plan 施工组织设计
preparation of contract documents 合同文件的准备
preparation of core samples 岩芯样(品)的制备
preparation of documents 编制文件
preparation of drawing 绘制原图
preparation of feed materials 进料准备
preparation of forecast 做预报；编制预报方案
preparation of foundation 基础准备
preparation of geodetic parameters 大地诸元准备
preparation of grouped raw coal 分组入选
preparation of land 整地；平整土地；整理土地；土地平整
preparation of normative documents 标准文件的制订
preparation of plates 板材的制备
preparation of plates for welding 焊接板材的准备
preparation of programs 程序设计
preparation of project 项目准备
preparation of pure water 纯水制备
preparation of reinforcement 钢筋加工
preparation of sample 样品制备
preparation of sections 薄片的制备
preparation of sections for study 研究薄片的制备
preparation of site 现场清理；清理场地
preparation of sized raw coal 分级入选
preparation of soil 整地
preparation of soil sample 土样制备
preparation of source material 编图资料准备
preparation of specimen 制造样品；试样制备；标本制备法
preparation of statement 报表编制
preparation of the route 准备进路
preparation of trench 槽段准备
preparation operation 准备作业
preparation period 准备期
preparation room 预备室
preparation routine 准备程序
preparations 筹备工作
preparations for construction 施工准备
preparation surface for painting (油漆的)底层表面处理
preparation symbol 准备符号
preparative 预备号
preparative chromatography 制备色谱(法)
preparative column 制备柱
preparative column electrophoresis 制备柱电泳
preparative gas chromatography 制备级气相色谱法
preparative layer 制层
preparative layer chromatography 制层色谱法
preparative liquid chromatograph 制备液相色谱仪
preparative paper chromatography 制备级纸色谱法
preparative partition chromatography 预备分配色谱法；制备分配色谱法
preparative photosynthesis 制备性光化合成
preparative radiation chemistry 制备辐射化学
preparative scale chromatography 制备型色谱法
preparative scale plate number 制备级塔板数
preparative scale sample 制备级试样
preparative thin-layer chromatography 制备(级)薄层色谱法
preparative treatment 预加工
preparative ultracentrifuge 制备超速离心机
preparative zone electrophoresis 制备区带电泳法
preparative zone electrophoretic separation 制层电泳分离法
preparator 选矿机；介体
preparatory 准备上需要的；准备的
preparatory command 预令
preparatory cost 预备工作成本
preparatory cutting 预加工；预备伐
preparatory degassing 初步除气
preparatory facility 施工设施
preparatory function 准备机能
preparatory gas freeing 初步除气
preparatory key system 预开钥匙系统
preparatory measure 初步措施
preparatory nursery 预备苗床

preparatory operation 准备作业
preparatory period 预备期
preparatory plan 施工准备计划
preparatory plan of construction 施工准备计划；施工准备
preparatory roasting 预焙烧
preparatory shock 预震
preparatory signal 预备信号
preparatory stage 预备伐；准备阶段
preparatory time 准备时间
preparatory treatment 预处理；预备处理；预加工
preparatory workings 准备巷道
preparatory works 准备工作；施工准备工作；施工设施；筹备工作
preparatum milking 产生挤乳
prepare 准备；制备；编制
prepare aggregate 配拌集料；配拌骨料
prepare budget 编制预算；编预算
prepare channel sensitivity of oriented sampling 定向取样灵敏度
prepared atmosphere 制备气氛
prepared cavity 备填洞
prepared chalk 研细白垩
prepared double-line 预留第二线
prepared edge 坡口加工面
prepared foundation 填筑好的基础
prepared glue 配制胶
prepared gravel 准备好的砾石；筛选砾石；备制的砾石
prepared gypsum plaster 精调的石膏粉刷(灰浆)
prepared hole 预留孔
prepared linseed oil 精制亚麻油
prepared lithopone-zinc white paint 锌钡白—锌白调合漆
prepared mo(u)lding sand 制备好的型砂
preparedness 准备好的状态
preparedness meter 灭火计算尺
preparedness planning 准备工作的规划
prepared paint 调合漆；调和漆；调和涂料；调好的漆
prepared plaster 准备好的灰浆
prepared powder 熟粉
prepared reserves 准备储备量；准备储量；采准矿量；采准储量
prepared roof covering 预制屋面覆盖层
prepared roofing 预制屋面料；预制屋面；卷材屋面
prepared roofing manufacture 预制屋顶的制造
prepared roofing shingle 预制屋顶木瓦
prepared sheet roofing(paper) 预制纸板屋面材料
prepared subgrade 填筑好的路基
prepared tar 精制(煤)焦油
prepare for drilling 钻前准备；钻井准备
preparer 调制机
prepare statement 编制报表
prepare the seedbed 播前整地
preparing cash budget 编制现金预算
preparing salt 锡酸钠
preparing seedbed 整理苗床
preparing shop 准备车间
preparing specification 制订规范
prepasted wallpaper 带胶壁板
prepatent 潜伏期的
pre-patinate 预生铜锈；预生铜绿
pre-patinated clay roof(ing) tile 预涂铜绿的黏土屋面瓦
pre-patinated concrete roof(ing) tile 预涂铜绿的混凝土屋面瓦
pre-patinating agent 预涂铜绿剂
pre-patinating paint 预涂铜绿的油漆
prepattern register 预图案寄存器
prepattern template matching 预图案模板匹配
prepause 预计歇
prepay 预先付款
prepayment 预付项目；预付(款)；提前付款；提前偿付
prepayment clause 提前还款条款
prepayment gas meter 预付式煤气表
prepayment guarantee 预付款保函
prepayment issue 预先付款的物资分发
prepayment meter 预付款仪表；预付式电度表；预付费用表计；投币式煤气表
prepayment money 预付金
prepayment of freight 运费预付
prepayment penalty 提前还款罚金

prepayment privilege 提前偿付特权；预付优惠
prepayment tariff 预付款收费制
prepayment valve 投币阀；自动阀
prepay-set 投币式公用电话机
prepay telephone 投币式公用电话
prepedestal level 预消隐脉冲电平
prepeening 预冷作硬化
prephenic acid 预苯酸
prephoto-phase 先光期
pre-piling 试桩；预桩
prepiped 预装好管道的
preplaced-aggregate concrete 预铺骨料灌浆混凝土；预填骨料混凝土；灌浆混凝土；预置骨料混凝土
preplaced brazing 预加钎料钎焊
preplaced ground control 预先施测的地面控制
preplanned allocation 预先分配
preplanned mission 预定任务
preplanned mission request 预定任务申请
preplanned request 预定的申请
preplanned search 预先计划搜索
preplan(ning) 预先计划；初步计划
preplanting soil fumigant 播前土壤熏蒸剂
preplant treatment 播前施用；播前处理；移栽前处理
preplasma 超前等离子体
preplastication 预塑炼
preplasticizer 预塑化装置
preplasticizing 预塑化
preplate 预镀
preplatform stage 前地台阶段
preplodder 预压条机
preplumbed system 预制导波系统
preply 预制层
prepn 制剂
prepolarization 预偏振
prepolarized material 预极化材料
prepolarizing field 预极化场
prepolish(ing) 预抛光
prepollution 初期污染
prepolycondensate 预缩聚物
prepolymer 预聚物；预聚(合)物
prepolymer gel 预聚物凝胶
prepolymerization 预聚合
prepolymer mo(u)lding 预聚物模塑
prepolyurethan(e) 聚氨酯预聚物
preponderance 超重
preponderant ages 主要年龄
preponderating anode 主优阳极
preposition 预先进入阵地；预先放好；前置；放在前面
prepositioned 预装的
prepositioned stock 预先放置的物资
prepost-tensioned concrete 先后张拉混凝土(预应力的)
prepost-tensioned construction 先后张拉结构(预应力的)
prepost tensioning 先后张拉结合法；先张后张法；半后张法(预应力的)
prepotential 前电位；前电势
preppy 预科生(大学的)
preprecipitation 预沉淀
prepreg 半固化片；聚酯胶片；预浸渍制品；预浸料坯；预浸料；预浸处理；预浸材料
prepreg machine 预浸胶机
prepreg mat 预浸渍毡
prepreg method 预浸渍法
prepreg mo(u)lding 预浸料成型
prepreg tape 预浸渍带
prepreparation 预先处理
prepressing 预压
prepressing-die-float 预压浮沉模
prepressing-die-float control 预压浮模控制杆
prepressing method 预压加固法
prepressing method of stack 堆载预压法
prepressing process of drawdown 降水预压法
prepressing process of jet-drawdown 射流降水预压法
prepressing process of sand filled drainage well 砂井排水预压法
prepressing process vacuum drawdown 真空降水预压法
prepressure compacting method 加压法
prepressurization 预加压
prepressurized water 预增压水
preprimed 预底涂
preprint 打样

preprinted symbol 预印符号
preprinting 预印图案
preprocess(ing) 预加工;预处理;预先加工
preprocessing aid 预操作助剂
preprocessing of plate-fin elements 板翅元件的预处理
preprocessing program(me) 预处理程序
preprocessing subprogram(me) 预处理子程序
preprocessing unit 预处理机
preprocessor 先行处理机;预加工程序;预处理器;预处理程序;前处理程序
preprocessor procedure 预处理程序过程
preprocessor variable 预处理程序变量
preproduction 预制造;试制;试验性生产;试采
preproduction capital expenditures 生产前基建费用
preproduction cost 生产前成本
preproduction engine 试生产发动机
preproduction machine 小批生产机械;试生产机械
preproduction model 样机;试制样品;试制模型
preproduction stage 试采阶段
preproduction test 正式投入运行前的试验;投料前生产试验;试制前试验
pre-production trial 试生产
preproduction type test 生产前试验
preproduction vehicle 预生产车
preproduction version 预生产型
preprofessional grade 就业前的(技术)等级
preprofiling 初成型
preprogram(me) 预编程序
preprogrammed autopilot 预定程序自动驾驶仪
preprogrammed chain 预定程序电路
preprogrammed control system 预定程序控制系统
preprogramming 预先编程序
preproof 预打样
pre-proposal conference 提交建议书预备会
prep stall 准备牛栏
prep stand 准备台
prepulse 前置脉冲
prepulsing 发送超前脉冲
prepulverize 预粉磨
prepump 前级泵
prepunch 预先穿孔
prepunched card 预穿孔卡片
prepunched hole 预穿孔
prepunched reader card 穿孔卡片
prepunching 预穿孔
prepurchase conseling 购买前咨询
prepurging 洗炉;清炉
prepurification 预纯化
preputial 包皮的
prepyramidal 锥体前的
prequalification 资格预审;资格审查
prequalification application 资格预审申报表;资格预审申请
prequalification documents 资格预审文件
prequalification of(prospective) bidders 投保人资格预审;投标者资格预审;信誉承包商合格预选(指工程招标)
prequalified firm 资格预审合格商号;资格预审合格公司
prequalify 预先具有资格
prequenching 预淬火
preradical 预根式
prereacted glass batch 预先烧结的玻璃配合料
prereacted magnetiste-chrome brick 预反应镁铬砖
prereacted raw batch 预反应料
prereacting 预反应
prereaction 预反应
preread disturb pulse 读前打扰脉冲
prerealcomplete uniformity 预实完全一致性
prerecord 预记录
prerecorded signal 预记信号
prereduced burden material 预还原炉料
prereducing 预还原
prereduction 预先还原
prereduction method 预还原法
prefining 预先精炼;预精炼;初步提纯
pre-regeneration cleaning 前更新伐
preregister operation 预记录操作;预信号操纵
preregular linear operator 预正则线性算子
preregulator 预调节器;前置调节器
prerelease 过早揭纸;过早剥离;提前排气
preremained hole 预留孔

prerequisite 先决条件;必要条件
prerift arc 前裂谷期岛弧
prerift arch 裂谷前隆起
prerinse 预清洗
preripening 预熟成
preroast 预先焙烧;预焙烧
preroast furnace 预焙烧炉
preroasting 初步焙烧
prerogative 特权
preroll 预滚动
preroll sequencer 预转定序器
preroll shift 预转移动
pre-Romanesque sculpture 早期罗马风格的雕刻
pre-Romanesque style 早期罗马风格
prerotation 预转;预旋
prerotation vane 预旋叶片
prerotator 预旋器
prerun 预试
prerupture 破坏前的
prerupture flow 破坏前流动
prerupture response 破坏前响应
presag conveyer 预垂弯传送机
presale 预售
presaturation 预饱和
presaturation column 预饱和柱
presaturator 预饱和器
presbyterium (教堂的)圣坛旁的空间;(社寺建筑的)中央大殿;(教堂的)内堂;十字架坛
presbytery (教堂的)圣坛旁的空间;(教堂的)司祭席
prescaler 预定标器
prescaler counter 预标定计数器
prescaling 预定比例因子
prescheduling 预先安排计划
preschool 幼儿园
preschool children's playground 学龄前儿童游戏场地
prescore pile 沉管浇注桩
prescouring 预洗炼
prescouring agent 预精炼剂
prescreener 预筛分器;预筛分装置;预筛分机;空气粗滤器
prescreening 预筛分的;预先筛分
prescribe 限定;指令;规定
prescribed burning 规定用火;规定烧除
prescribed dynamic(al) loading 非随机动荷载
prescribed examination 法定检验
prescribed flow 预定流量
prescribed form 规定格式
prescribed laboratory test 指定实验室试验
prescribed limit 已知范围;规定限度;规定极限;规定范围;给定极限
prescribed load 规定载重
prescribed minimum 下限
prescribed mix 规定混合料
prescribed number of passengers 旅客定额
prescribed orbit satellite 静止卫星
prescribed penalty 法定刑罚
prescribed rights 规定的权
prescribed time 法定期内
prescribed time-limit 规定的期限
prescribed water head 规定水头
prescribed water rights 规定水权
prescribed yield 规定收获量
prescribing trajectory 给定弹道
prescription 获取权利;质量要求;规定;起诉期限;诉讼时效
prescription of rights 权利的消灭时效
prescription water rights 法定用水权;法定水权
prescriptive code 指令性规范
prescriptive easement 因长期使用而获得的通行权
prescriptive nature 指令性
prescriptive period 时效期内
prescriptive rights 依时效而取得的权利
prescriptive rules 法定规则
prescrubber 预洗涤装置;预涤气装置
Presdwood 普列斯德层压式板材;层压板材
presedimentary ages 沉积前年龄
presedimentation 预先沉淀;沉沉降(作用);预积;预沉淀
presedimentation tank 预沉池;预沉淀池
preseed 预制晶种
presegmentation 预先分节
preseismic 震前的
preseismic activity 震前活动性

preseismic deformation 震前形变;震前变形
preseismic displacement 震前位移
preseismic slip 震前滑动
preselect 预选
preselect controls 预定调节参数
preselected depth 预定深度
preselected masses 预选质量数
preselected site 预定取样点
preselected stop 预选光栏
preselected temperature control 预选温度控制
preselected value 预选值
preselection 预选性;预选择;预选(送);预先选择
preselection of concrete 混凝土预选
preselection of shutter speed 快门速度预选
preselection stage 预选级
preselection switch 预选开关
preselective 预选的
preselective control 预选操纵
preselective gearbox 预选换挡变速器;预选变速器
preselector 前置选择器;预选装置;预选器;高频预选滤波器
preselector control 预选控制;预定位控制
preselector gearbox 预选式齿轮变速器;预选式变速箱
preselector iris system 预调彩虹光栏系统
preselector mechanism 预定位控制机构
preselector of information 信息预选器
preselector valve 预选阀
presence 现实感;存在度;存在
presence bit 内存联系指示位;存在位
presence chamber 接见厅;接见室
presence detection 存在检测
presence detector 存在型检测器
presence equalizer 单频音调补偿器
presence of animal life in a forest 森林动物的存在
presence room 接见室
presensitized plate 预制感光板
presensitized resist 预敏化抗蚀剂
presensitizing 预敏化
present 现存的;目前的;出席;呈现
present advanced stage 目前先进水平
present angular coordinate 现时角坐标
present angular height 现时高低角
present a ship's manifest 提供装货详单
presentation 显像;显示;影像;增送;交单;图像;提交;呈现;产式;表象
presentational fashion of results 成果提交方式
presentation and analysis of field data 野外数据的描述和分析
presentation class 表示分类
presentation drawing 草图;示意图
presentation format 表示格式
presentation function 表示功能
presentation layer 表示层
presentation medium 表示媒体
presentation of budget 预算编制方式
presentation of information 信息显示;数据输出
presentation of number 数的显示
presentation of reliability data 可靠性数据表示法
presentation protocol data unit 表示协议数据单元
presentation quality 外观质量
presentation services 表示服务
presentation services command processor 表示服务命令处理程序
presentation surface 表示表面
presentation switch 图像开关
presentation time 显现时间;显斑时间
present azimuth 现时方位
present building code 现行建筑规范
present capital value 现时资本价值;资本现值
present contract 本合同
present coordinate 现在坐标
present cost 现在成本
present-day 现代的
present-day fuel 现用燃料
present-day hydrocarbon pollution 现时烃污染
present-day mill 现代化轧机
present-day pollution of bottom sediments 现时底泥污染
present-day pollution of water 现时水污染
present-day regime of natural water 现时天然水状况
present-day state of water bodies 现时水体状况
present-day state of waters 现时水体状况

present discounted value 现在贴现值
present discount value 贴现现值
present distance 现时距离
presented for payment 凭单付款
present elevation 现在高低角；现实仰角
present employment 现职
present environment 现时环境
presenter 压结前烧结
present experience 现有经验
present generation 现代
present geothermal gradient 今地温梯度
present ground range 现在水平距离
present ground stress field 现今地应力场
present gum in gasoline 汽油中的显胶
present heading 现在航向
present ice 现代冰川
present incidence 目前税项与价格之比率
present income 现在收入
presenting bank 提示银行
presenting date of reserves 储量提交日期
present instruction 当前指令
present in the upper 15 cm of the soil 出现在土壤上层15cm内
present land-use 土地利用现状
present meteorological condition 现在气象条件
present net worth 纯现值
present pattern method 现状法（交通预测）；增长系数法
present performance rating 现时性能评级
present period 现代；本期
present policy 现行政策
present position 现在位置
present-position coordinate 现在坐标
present price 现行价格；现价；时价
present range 现在距离；瞬间距离
present salt precipitation 现代成盐作用
present serviceability 现有服务能力；现有路况服务能力
present serviceability index 现有功能指标；现时耐用性指数
present serviceability rating 现有功能评定；现时耐用性分类
present shagreen surface 有糙面
present situation 现状
present situation map of station 站现状图
present slant range 现在斜距离
present tectonic stress field 现今构造应力场
present tectonic system 现今构造体系
present the conclusion and recommendations 提出结论和建议
present value 现值；当前值
present value analysis 现值分析
present value approach 现值分析法
present value depreciation 现值折旧
present value index 现值指数
present value method 现值法
present value of a deferred annuity 延期年金的现值
present value of an annuity 年金的现值
present value of an annuity due 到期年金的现值
present value of annuity 年金现值
present value of benefit 效益现值
present value of cost 成本现值
present value of income from investment 投资收入现值
present value of income investment 投资收入的现值
present value of investment 投资现值
present value of one dollar per period 每期一元现值
present value of one Yuan 一元现值
present value of ordinary annuity 普通年金现值
present value of the additional investment 追加投资的现值
present value rate 现值率
present value standard 现值标准
present water demand 目前需水量
present weather 现在天气
present worth 现值
present worth analysis 现值分析
present worth comparison method 现值比较法
present worth factor 现值因子
present worth method 现值法
present worth of an annuity 年金现值

present worth of an annuity factor 年金现值系数
present worth of income 收入现值
present worth of one factor 一元现值系数
present worth of salvage 残值折现值
present worth of total cost 总成本现值法
present yield 现在蓄积
pre-separation 预离析；预分离
preservation 储藏；防腐作用；防腐；保存；保持；保藏
preservation coat 保护涂层；防腐涂层
preservation in core box 在岩芯盒中保存
preservationist 自然资源保护者
preservation life 油封期限；封存期
preservation of ancient capitals 古都保护
preservation of buildings 房屋维护；建筑物维护
preservation of cultural relics 文物保护
preservation of ecology 生态的保持
preservation of environment 环境保护
preservation of fertility 保持土壤肥力
preservation of mark 测量标志保护
preservation of monuments 纪念碑的保护
preservation of physical evidence 保存物证
preservation of samples 试样保存
preservation of sludge 污泥保存
preservation of soil fertility 土壤肥力保持
preservation of soil moisture 保墒
preservation of structures 结构的维护
preservation of the natural environment 保护自然环境
preservation of timber 木材防腐
preservation of urban historic relics and sites 城市文物古迹保护
preservation of water samples 水样保存
preservation of wooden tie 木枕防腐
preservation plastic film 保护用塑料薄膜
preservation plastic sheeting 保护用塑料薄板
preservative 防腐的；预防的；防腐剂；保护膜；保护料
preservative against blue stain 防止蓝色斑点（木材的一种疵病）
preservative agent 保存剂；防腐剂；保护剂
preservative chemicals 防护化学品
preservative combination 防腐合剂
preservative engine oil 防护机器油
preservative fluid 防腐液；保藏液
preservative for structures and buildings 防护建筑和结构
preservative liquid 防腐液
preservative liquor 防腐液
preservative lubricating oil 防护润滑油
preservative oil 油类防腐剂；防护油
preservative paint 防腐油漆；防腐涂料
preservative plywood 防腐胶合板
preservative process 防腐工序；防腐法
preservative reflex 保存反射
preservative salt 防腐盐
preservative substance 防腐料；防腐物质；防腐剂
preservative treated timber 防腐处理木材
preservative treated wood 防腐处理木材
preservative treatment 防腐处理
preservative treatment of timber 木材防腐处理
preservative treatment technology 防腐技术
preservatize 加防腐剂
preservatory 储藏所；保藏所
preserve 猎物养殖区；禁猎地；保留；保存区
preserve amplitude deconvolution 保持振幅反褶积
preserve area 禁猎区
preserve bunker 备用燃料舱
preserved context index system 保留上下文索引系统
preserved food 加工的储[贮]藏食物
preserved hole 预留孔
preserved latex 储[贮]备胶乳；储备胶乳
preserved live load increasing factor 预留活载发展系数
preserved plywood 防腐胶合板
preserved skin 备皮
preserved timber 防腐木材
preserved tree 保护树
preserve forest 保护森林
preserve moisture and fertility 保水保肥
preserver 防腐物；保存者；安全装置；保护人
preserver solution 防腐液；保护液体
preserve time 保存时间
preservice 预处理
preserving action against blue stain 抗蓝斑的保护作用（木材的一种疵病）
preserving of water sample 水样的保存
preserving oil 防腐油
preserving timber 木材防腐；防腐木材；防腐材
preserving value for resources 资源的保存价值
preset 初凝；预置；预调；初调；安装程序；预先调整
preset adjustment 预调整；预调谐
preset air meter 预定气压充气装置
preset angle 预装角；预置角
preset aperture ring 光圈调节环
preset apparatus 预调设备
preset attenuator 预调衰减器
preset automatic equalizer 预置式自动均衡器
preset bearing 预定方位
preset capacitor 微调电容器
preset code 预置码
preset control 预调控制钮
preset controller 预置控制器
preset count 预调计数；预定计数
preset course 预定航向；预定航线
preset decimal counter 预置计数器；前置装置计数器
preset device 预定装置
preset diaphragm 预置光阑
preset digit layout 预先给定的数位配置
preset distance 预定距离
preset enlargement 预定放大
preset fader 预调灯光衰减器
preset flight path 预定航迹
preset flow limit valve 前置流量限制阀
preset frequency 预调频率
preset function 预置功能
preset guidance 预置制导
preset instruction 程序预置指令；程序机构信号
preset iris 预置型光圈
preset jammer 预调干扰机
preset knob 预调旋钮
preset lens 预调物镜；预调透镜
preset level 预置级；预定能级
preset limit 预定极限
preset measuring instrument 预调式仪器
preset mechanism 预调机构；程序机构
preset mode 预置方式
preset oscillator 预调振荡器
preset output 预定产量
preset parameter 预置参数；预定参数；固定参数
preset period 预凝期；静停期（蒸汽养护的）；预养护期
preset position 预定位置；预调位置
preset potentiometer 预调电位计
preset pressure 预调定压力
preset prism 预调棱镜
preset program(me) 程序调整；预定计划；预定程序表；预定程序
preset regulation 预选装置调节
preset scale 成比例尺
preset screwdown 程序调整的压下装置
preset sequence 预定程序；给定程序
preset shrinkage 预定形收缩
preset signal 预调信号
preset stop 预调光阑
preset system 半自动化控制系统
presettable counter 可预置计数器
preset temperature 预定温度
presetter 预校正器；机外对刀装置
preset time 预定时间；预调时间；规定时间；给定时间；提前时间
preset timer 预置定时器
preset timing bracket 预调定时器托架
presetting 预凝；预调节；预测；初凝；预先调整
presetting apparatus 预整定装置；程序控制装置
presetting bit 预置位
presetting circuit 预调谐电路；预调电路
presetting machine 预调机床
presetting mechanism 预调器
presetting of water temperature 预调水温
presetting period 预养期；预凝时间（混凝土）；预凝（结）时期；静停期；凝前期；初凝时期；初凝期
presetting system 预选制
presettling 预澄清；预沉淀
presettling chamber 预沉井
presettling tank 预沉池；预沉淀池
preset tool 预调刀具
preset torque-wrench 预定扭矩扳手

preset tuning 预调谐
preset type switcher 预置型开关
preset underpinning 预试打支撑桩
preset value 预置值
preset velocity 预定速度
preset yield load 预调初始荷载
preshaping 预先形成；预先成型
preshaving 剃前
preshaving cutter 剃前刀具(齿轮)
preshaving hob 剃前滚刀
preshaving pinion cutter 剃前插齿刀
presheaf 预层
presheaf of regular functions 正则函数的预层
preshear hole 预裂炮孔
preshear(ing) 预剪
preshearing blasting 预裂爆破法；预裂爆破
preshimmed sealant 预衬垫密封膏
preshimmed tape 垫补条(装配玻璃复合体)；预衬垫(密封)带
preshipment financing 装运前资金通融
preshipment inspection 装船前检查；初验
preshipment quality terms 装运前品质条件
preshock noise test 震前音响试验
preshoot 前置尖头信号
preshrinkage expansion 收缩前膨胀
preshrinking 预收缩
pre-shrink mixing 预收缩搅拌
preshrunk 缩拌；预收缩
preshrunk concrete 预收缩混凝土
preshrunk textiles 防缩整理产品
preside 主持；负责
president 总裁
presidential palace 总统府
President press 总统型压砖机(商品名)
president's palace 总统宫
presidio 城堡；炮台；前沿要塞；前线堡垒
presignal 预信号
pre-signal delay 报警延迟
pre-Sinian basement 前震旦纪基底【地】
pre-Sinian period 前震旦纪【地】
pre-Sinian system 前震旦系【地】
presintered bar 预烧结棒
presintered compact 预烧坯
pre-sinter(ing) 预烧结；初步烧结
presizing 预涂胶水；预先筛分；预上胶
presizing screen 预筛筛分机
Presky-Martens flash-point 潘马氏闪火点
pre-slag 前期渣
preslaker 预消化器
presling 预扎吊索
pre-slit method 预开切沟槽法
preslotting 预开槽
presmelting 预熔炼
presoaked 预先浸湿的
presoaked aggregate 预浸集料；预浸骨料
presoak(ing) 预浸；预浸洗涤
presoaking of light(weight) aggregate 轻质集料的预浸；轻质骨料的预浸
presoil 预染污
presomet 普里素密特(一种涂刷铸铁的沥青油漆)
pre-sort(ing) 预分选；预分(类)
presotim 普里素提恩(一种木材装饰防腐剂)
presowing irrigation 播前灌水；播前灌溉
presowing treatment 播前整地；播前处理
presowing water 播前灌溉
prespark curve 预燃曲线
presparking 预放电
prespark period 预燃时间
prespheroidizing 预球化
presphygmic 脉波前的
pre-spill planning 溢漏应急计划
presplit aberration 分裂前异常
presplit cut 预裂掏槽
presplitting 预裂法；预裂
presplitting blasting 预裂爆破法；预裂爆破
presplitting method 预裂法
presplitting technique 预裂技术；预裂爆破法
prespondylolisthesis 初期脊柱前移
prespray 提前喷药
prespring 预弯
prespringing 预弯(焊件)；预弯
press 印刷(厂)；印刷；压力机；压机；床床；通讯社；叠压机；冲压；冲床
pressability 压榨性能；可压制性

press adjustment 上版
press-and-blow machine 压吹机
press-and-blow process 压吹法(玻璃)
press arrangement 压机构造
press atomization 压力喷雾
press automation 压力机自动装置
press-baler 压捆机
press bar 压力棒
press bed 压机座；水压机下横梁
press bending 压弯
pressboard 压纸板；压榨纸板；压制板；绝缘用合成纤维板
pressboard padding 纸板垫
press bolster 压模承台(压床的)；冲床台面
press bond 压力接合；压端连接
press box 记者席
press brake 弯边机；压弯机；折板机；弯板机
press brake dies 压弯机模
press building 出版大楼；印刷大楼
press button 按钮
press-button box 按钮匣；按钮箱
press-button control 按钮控制；按钮操纵
press-button controlled mixer 按钮控制搅拌机
press-button controlled mixing 按钮控制搅拌
press-button ignition 按钮引燃；按钮点火
press-button key 非锁定按钮
press button sensing system 按钮感应系统
press-button starter 按钮启动器
press-button station 按钮试验台；按钮站
press-button switch 按钮式开关
press-button system 按钮操纵系统
press-button warfare 按钮战争
press cake 压滤饼；压滤(泥)饼；滤饼压
press capacity 印刷能力；冲压能力；冲压机吨位
press casting 压力铸造
press channel 压捆室
press coating 压榨施胶
press cold welding 冷压焊；加压冷焊
press compacting 压力机压制
press conference 记者招待会
press contact 压力接点
press controller 按钮控制器
press cooling time 压机冷却时间；压机闭合时间
press crown 压力机上横梁
press cupboard 双层碗橱
press cure 压力硫化
press cylinder 压榨筒
press dewaxing 压滤脱蜡
press dewaxing process 压榨脱蜡过程
press die casting 压力铸造
press distillate 压滤馏分
press down 压下
press down system 压下系统
press drill 压力钻
press drip 压力机油滴
press drying 压滤干化
pressductor 湿纸幅张力测定仪
press dumper 压滤机卸料装置
pressed 压制的
pressed active carbon 成型活性炭
pressed bar 压制棒坯
pressed-base seal 加压基座密封；冲压平底封接
pressed blocks lining 压制成的刹车块
pressed body 压型体
pressed brick 压制砖
pressed carbon cartridge 成型炭滤芯
pressed cement tile roofing 厚石棉水泥瓦屋面
pressed charge 压制燃料
pressed clay brick 压制黏土砖
pressed clay roof(ing) tile 压制黏土屋面瓦
pressed clay-tile 压制黏土瓦
pressed coal 构造挤压煤
pressed colo(u)r coated steel sheet 彩色压型钢板
pressed compact 压坯；压块
pressed concrete 压制混凝土管；压制混凝土
pressed container glass ware 压制的玻璃容器
pressed-core cable 压心电缆；压力电缆
pressed crystal ware 压制的晶серые玻璃器皿
pressed density 压制密度
pressed disc method 压片法
pressed-disc technique 压片技术
pressed distillate 冷榨去蜡油；石蜡馏分；压滤去蜡油
pressed edge 受压边缘；最大土压力边(基础倾覆时)

pressed explosive 压制炸药
pressed felt 压制毡
pressed felt manufacture 压制毡的制造
pressed-fiber board 壁板；压制纤维板
pressed film 压模
pressed finish 压光；压光面饰
pressed fuel 压制燃料
pressed girder 压制梁；模压梁
pressed glass 压制玻璃
pressed glass base 压制玻璃基底
pressed glassware 压制的玻璃制品
pressed green compact 压坯；生坯
pressed halide disk technique method 卤化物压片法
pressed hollow glassware 压制的玻璃容器
pressed laminated wood 压实的材料
pressed lens 压制透镜
pressed liquor 压榨液
pressed loading 加压装载
pressed machine-brick 机压砖；机制砖
pressed magnet 压制磁铁
pressed material 压制薄片木；压制材料
pressed metal 压制金属
pressed metal door 模压金属门
pressed metal node 压制金属接头
pressed metal section 压制金属型材
pressed mica 压制云母
pressed-on 套压的
pressed-on iron tyre 压合铁轮箍
pressed part 冲压制品
pressed pellet 压饼
pressed permalloy powder 压制坡莫合金粉
pressed pile 压入桩
pressed pipe 压型管
pressed plutonium oxide 压制过的氧化钚
pressed polish 压力磨光；压力擦亮
pressed powder process 压粉过程
pressed product 压型制品
pressed profile 压成的轮廓；压制外形；模压轮廓
pressed raw clay brick 压成的黏土砖坯；压制粗黏土砖
pressed recess 压槽
pressed residue 压缩残渣
pressed sampler 压入式取样器；压入式取土器
pressed scrap 压缩废钢
pressed section 压制件；压制截面
pressed shackle 压制钩环
pressed shape 压成的形状；压坯
pressed sheet 压制的板材
pressed sheet metal 型钢板；压制金属薄板
pressed sheet steel 压制钢板
pressed shield 压制防护板
pressed steel 压制型钢；压制钢
pressed steel bogie 型钢转向架
pressed steel boot 模压钢开沟器
pressed steel building component 压制钢结构建筑构件
pressed steel door frame 钢板压制的门框
pressed steel duct 钢板压制的风道
pressed steel eaves gutter 钢板压制的檐沟
pressed steel eaves trough 钢板压制的檐槽；钢板压制的檐沟
pressed steel flange(d) shear plate 钢板压制的法兰剪切板；钢板压制的突缘剪切板
pressed steel hook chain 模压钢钩头链
pressed steel liner plate 挤压式钢支撑板
pressed steel lintel 钢板压制的过梁
pressed steel manhole cover 钢板压制的检查孔盖；钢板压制的人孔盖
pressed steel manway cover 钢板压制的检查孔盖
pressed steel panel 钢板压成的镶板
pressed steel partition(wall) 钢板压成的分隔(墙)
pressed steel pipe 钢板压制的管道
pressed steel plate 压制钢板；压延钢板
pressed steel rainwater pipe 钢板压制的雨水管
pressed steel section 钢板压制的部件
pressed steel sectional tank 压制型钢水箱；压制型钢水槽
pressed steel shape 压制型钢
pressed steel stair(case) 钢板压制的楼梯
pressed steel trim 钢板压制的线脚；钢板压制的镶板
pressed steel unit 钢板压制的构件
pressed steel valley gutter 钢板压制的雨水沟；钢

板压制的斜沟;钢板压制的天沟
pressed straw 压缩稻草
pressed tile 压制的屋面瓦;压制板
pressed tumbler with enamelled decorations 彩花机压杯
pressed type bond 压入式轨隙连接器;压入式轨缝连接器
pressed ware 压制的制品;压制器皿
pressed wood 压制木材
pressed work(ing) 压力加工
pressed zircon block 压制锆(英石)砖
pressel 悬挂式电铃按钮
pressel switch 悬吊开关;悬垂式按钮站
presse-pate 湿抄机
presser 压制工;压榨器;压榨机;加压器;承压滚筒
presser bar lifter 压紧杆扳手
presser bar spring bracket 压杆导架
presser bit 压片
presser blade 压板片
presser cam 压片凸轮
presser device 压板装置
press erection 压力机安装
presser foot 压脚
presser foot lifter 压脚提升器
presser lifting amount 压脚提升高度
presser motion 压板运动
presser-plate hub 压盘毂
presser slide 压板式滑片
presser wheel 压片轮
Press-Ewing seismograph 普雷斯—尤因地震仪
Press-Ewing seismometer 普雷斯—尤因地震计
Press-Ewing vertical seismometer 普雷斯—尤因垂直向地震计
press fabrics 压榨辊的合成网套
press felt 压毡
press figure 印刷数字
press filter 压滤器;压滤机
press filtration 压滤作用;压滤
press finger 压指
press finish(ing) 推压精加工;压光
press fit 压入配合;压配合;压力装配
press fit case 压入配合表壳
press fit package 压嵌组件
press fit test 压入配合试验
press fit type 压入型
press-fit wearing bushings 压配耐磨导套
press for fitting watch glass 装表玻璃压力器
press forging 压力锻造;压锻
press forming 压制成型;模压成型
press for mo(u)ld extrusion 铸型落砂冲锤机
press for payment of debt 催债
press for removing tyres 去箍压力机
press for repayment of a loan 要账
press for the repayment of debts 逼债
press gallery 记者席
press gilder 烫金机
press head 压板
press heater 热压硫化锅;热压锅
press house 压榨间
press-in connector 压入榫;压入接头(木材联结器)
press-in dowel 压入榫
pressing 冲压;压制;压榨;压型毛坯;压型;压成声盘;加压;迫切的;施压;按压法
pressing aquifer 承压含水层
pressing back 模印
pressing bend method 压力弯曲方法
pressing blank 压制雏形
pressing board 压书板;挡纸板
pressing condition 压制条件
pressing crack 压制开裂
pressing creep 抗压蠕变
pressing debts 即还债务
pressing die 压模;压接模
pressing factor 压缩系数
pressing force 压紧力
pressing forceps 挤压钳
pressing in 压入;压进
pressing in and pulling out of bearing 轴承的压入和取出
pressing in method 压入法
pressing lap 冲压折叠
pressing lens blank 压制透镜毛坯
pressing lubricant 压制润滑剂
pressing machine 压力机
pressing machinery 压制机械
pressing mo(u)ld(ing) 压模
pressing of glass 玻璃压制
pressing of powder 粉末压制
pressing on of wheels 轮对压装
pressing on the tyre 轮箍压装
pressing paper 粗面滤纸
pressing piece 模压制品
pressing position 压制位置
pressing powder 压制用粉末
pressing pressure 模压压力
pressing printer for gilding 镀金印刷机
pressing process 压制过程
pressing process of heat exchange plate or fin 换热板片或翅片的冲压
pressing product 冲压制品
pressing ring 压制的模环;压圈
pressing section 压紧部分
pressing shaft 针床水平运动轴
pressing sintering 加压烧结
pressing-sintering-melting machine 压制烧结熔化机
pressing-sintering-melting unit 压制烧结熔炼设备
pressing speed 压制速度
pressing strength 压制强度
pressing time 加压时间
press-in pile driver 压桩机
pressiometer 横压仪;旁压仪
pression 压力
pressional member 受压构件
press juice 压出液
press-key control 键控
Pressley tester 卜氏纤维强力检验仪
press machine 压床
pressman 模压工;冲压工人
press mandrel 压入心轴;压进心轴
press manifold 液压油路阀箱
press marked pottery 印纹陶
press mark(ing) 压痕
press misalignment 压力机调整不良
press mo(u)lding 模压成型
press mo(u)lding machine 压模机;压型机
press mo(u)lding machine with lift-off 顶升卸模的压模机
press mo(u)ld method 压铸法
press mo(u)ld ware 模压制品
press mounting of car wheel 车轮压装
press mud 压滤泥浆
press number 印刷数字
press-off bolt 压开螺钉;压开螺栓;顶开螺钉
press-off cam 脱圈三角
press-off detector 脱套自停装置
press office 报社
press of fluidity 液性压力
press-off position 脱套位置
pressofiner 螺旋压榨机
press of weather 风暴天气
press oil 陶瓷滑模油
pressolution 压溶
pressometer 压力测量器
press-on cap 压上的盖
pressor 加压剂;加压的
pressor areas 加压区
pressoreceptor 压力感受器
pressoreceptor reflex 压力感受器反射
pressor effect 加压效应
pressor reflex 加压反射
pressosensitivity 压力感受性
pressostat 抵押继电器;压制控制器;稳压器;恒压器
pressotransducer 压力换能器
press over system 溢流法(鞣液)
press packed bales 紧压包装
presspahn 木浆压制板;纸板;压板
presspahn sleeve 纸板套管
press paper 薄纸板
press part 压榨部
press pile driving method 压桩法
press plate 压板
press-plate machine 压榨机
press plunger 压力机柱塞
press polish 压力抛光;轧光;高度光泽;加压抛光
press pouring furnace 气压浇注炉
press-powder 压粉
press process 压制工艺
press proof 开印样;机样;清样
press pull 样图;开印样
press pump 压榨泵;压力泵;压泵
press quenching 模压淬火
press ram 压力机压头;压力机滑块
press ratio 压榨比
press residue 压榨渣
press-resin 压缩树脂
press ring 压力环
press roll 压滚;压辊;轧辊;压榨辊
press roller 镇压器
press roll lever 压辊杠杆
press roll weight 压辊砝码
pressroom 冲压车间;印刷间;印刷机房;印刷车间
pressrun 耐印力
press seal 压封;冲压密封圈
press seat 记者席
press section 压榨部
press shell 压榨辊外壳
press shrinking 压力紧箍
press slide 压力机滑块
press split blasting 挤压微差爆破
press switch assembly 压力开关组件
press table 压架;压台
press-talk switch 按键通话方式
press-talk system 按讲制;按键通话方式
press temperature 压制温度
press tempering 加压回火;模压回火
press tenter 压烫拉幅定形机
press-through die 深拉模
press time 压制时间
press tonnage 压力机吨位
press tool 冲压工具
press-to-talk switch 按讲开关
press-twist cap 热塑螺纹盖
press-type resistance welder 预压式接触焊机
pressurant 受压物体
pressure 压强
pressure above bubble point 高于饱和点压力
pressure above the atmosphere 正压(力)
pressure absorption tower 压力吸收塔
pressure accumulator 蓄压器;压缩空气箱;稳压装置
pressure acidizing 压力酸化;压力酸处理
pressure action 压力作用
pressure-actuated 压力传动的
pressure-actuated alarm switch 压力驱动的报警开关
pressure-actuated ga(u)ge 液压水位计
pressure-actuated valve 压力作动阀
pressure adjusting blade 压力调整叶片
pressure-adjusting screw 压力调节螺丝;压力调整螺钉
pressure-adjusting spring 压力调整弹簧
pressure adjustment 压力调整
pressure adjustment relief valve 调压溢流阀
pressure aerator 加压曝气器
pressure aging 加压老化;加压陈化
pressure-air drilling 压缩空气钻进
pressure-air monte-jus 压缩空气升液器
pressure airship 软式飞艇
pressure alarm 压力报警器
pressure-alarm indicator 压力报警器
pressure alarm system 压力警报系统
pressure all around 各向压力;多向压力
pressure altimeter 气压测高表
pressure altitude 压力高度;气压高度
pressure-altitude indicator 气压高度表
pressure-altitude sensor 气压高度传感器
pressure-altitude variation 气压高度变差
pressure amplification 压力放大
pressure amplification factor 压力放大系数
pressure amplifier 增压器
pressure amplitude 压力幅度
pressure and suction effect 压吸效应
pressure and suction hose 耐压吸引胶管
pressure-and-vacuum breather valve 压力真空呼吸阀
pressure-and-vacuum release valve 压力和真空释放阀
pressure and velocity stages 压力速度联合分级法
pressure anemometer 压力式风速计;风速计
pressure angle 压力角
pressure anomaly 压力异常;气压异常

pressure apparatus 压力装置
pressure application 压力法
pressure applicator 压力浇铸机
pressure applied at the surface 井口压力；地面压强；地面压力
pressure approach onto wire 压力上网
pressure aqueduct 压力输水道
pressure aquifer 承压含水层
pressure arch 压力拱
pressure area 压力区
pressure artificial breathing 加压呼吸
pressure at down stream of choke flow line 油嘴下流压力
pressure atomization 压力喷雾；加压雾化；加压喷雾
pressure atomized fog jet 压力雾化喷嘴
pressure atomizer 压送式喷雾器；压力弥雾机；喷燃器
pressure atomizing burner 压力喷雾燃烧室；喷燃器
pressure atomizing lance 压力雾化喷枪
pressure atomizing oil burner 压力雾化油烧嘴
pressure at rest 静压力；静力合
pressure at right angles 垂直压力；正交压力
pressure attachment 加压装置附件
pressure at the well bore 孔底压力；井底压力
pressure axis 压力轴
pressure back 压平板；压板
pressure bag 压力袋
pressure bag compacting 气袋压实
pressure bag mo(u)lding 压力袋模塑成型；压力袋成型
pressure balance 压力平衡
pressure balance adapter 压力平衡装置
pressure balance method 压力提拉法
pressure balance workover rig 不压井修井机
pressure balancing 压力均衡法
pressure balancing device 压力平衡装置
pressure band level 频带压强级
pressure bar 压梁；压杆；夹紧棒
pressure bar gap 压尺间距
pressure bar lead 压尺垂距
pressure-base factor 基压系数
pressure-bearing surface 承压面
pressure bell 压力钟
pressure below atmosphere 负压
pressure bias load 压力偏载荷载
pressure bleeder 压力式排气器
pressure block 压下垫块；承压块；承受地块
pressure blower 压力鼓风机；压力吹气机；压力吹风器；高压鼓风机；空气压缩机
pressure boiler 压力沸水器；压力锅炉；液体高热器
pressure bomb 火花塞试验器
pressure bonding 压力黏结；加压焊(接)
pressure boost 压力放大；增压
pressure booster 增压器
pressure bottle 高压气瓶；耐压瓶
pressure bottoms 蒸馏釜残渣
pressure boundary 压力边界
pressure bowl 压力滚筒；承压滚筒
pressure box 压力箱；压力水箱
pressure box of the classifier 分级器的压箱
pressure-break 压裂差
pressure breakdown 压力下降
pressure breathing oxygen regulator 增压供氧调节器
pressure breathing regulator 加压呼吸调节器
pressure-breathing system 增压式呼吸设备
pressure breccia 构造角砾岩
pressure broadening 压致增宽；压力作用下加宽；压力致宽；压力变宽
pressure build-up 压力升高；压力增大；压力增长；压力上升
pressure build-up curve 压力增长曲线
pressure build-up technique 升压技术
pressure build-up test 升压试验
pressure build-up vapo(u)rized 汽化升压
pressure build-up vapo(u)rized valve 汽化升压阀
pressure built-up test 恢复压力测试
pressure bulb 压力泡；球形土压力分布曲线；等压泡
pressure bulkhead 耐压舱壁；气密隔板
pressure bump 岩爆；受压突出；冲击地压
pressure burner 压力燃烧器；压力喷油燃烧器
pressure burst 岩石破裂
pressure button cover 按钮罩
pressure button spring 压力按钮弹簧

pressure by alongside wharf (due to wind, wave) 挤靠力【船】
pressure cabin 增压舱；加压舱；气密座舱
pressure cabin examination 气压舱检查法
pressure cable 加压电缆；充气维护电缆；充气电缆
pressure calibration 压力校准；声压校准
pressure calibration curve of inclusion 包裹体压力校正曲线
pressure capacity 加压能力；承压能力
pressure capsule 功率计；压力计；压力盒；测压仪；测压力计
pressure capsule system 压力盒系统；压力感传器系统
pressure car 气罐车
pressure carburet(t)or 喷射化油器
pressure casing 耐压外壳
pressure-cast 加压铸成的
pressure casting 压铸件；压铸法；压铸；压力铸造；压力注浆；压力浇铸
pressure-casting die steel 压铸模具合金钢
pressure casting machine 压力铸造器
pressure casting process 压铸法
pressure cell 压力传感器；压应力计；压力盒
pressure cell installation 压力盒装置
pressure cell test 压力盒试验
pressure cement grouting 压力灌浇水泥浆
pressure cementing 压力灌浆；加压注水泥
pressure center 压强中心；压力中心
pressure chamber 压力室；压力腔；压力舱；高压舱；耐压舱
pressure chamber test 压力洞室试验；水压法洞室试验；沉箱试验
pressure change 压力变化
pressure change chart 气压变化图
pressure change of hydrothermal system 热液系统压力变化
pressure change rate 压力变化率
pressure characteristic 压力特性曲线
pressure charge 压缩化学试剂
pressure charge compressor 增压压气机
pressure-charged 增压式
pressure-charged engine 增压式发动机
pressure charging 增压
pressure charging/scavenging system 增压扫气系统
pressure charging system 增压系统
pressure chart 压力图；气压图；等压线圈
pressure check 压制裂纹；压力检验；气压系统检查
pressure chlorinator 压力加氯机
pressure circuit 压力回路；正压回路
pressure circulation 压力循环
pressure circulation lubricating system 压力循环润滑系统
pressure circulation system 压力循环系统
pressure clack 压力瓣
pressure cleat 压力裂隙；压力节理
pressure cloth 压平布
pressure coat 压力沉附物
pressure coefficient 压力系数
pressure coefficient of mine 矿山压力系数
pressure coefficient of viscosity 黏度的压力系数
pressure coil 电压线圈
pressure colling system 加压冷却系统
pressure combustion 正压燃烧；加压燃烧
pressure combustion boiler 增压式锅炉
pressure combustion motor fuel 压燃式发动机燃料
pressure-compensated control 压力补偿控制
pressure-compensated flowmeter 压力补偿式流量计
pressure-compensated pump 压力补偿式泵
pressure-compensated valve 压力补偿阀
pressure-compensating control 压力补偿控制
pressure compensating system 压力补偿系统
pressure compensation 压力补偿
pressure-compensation bottle 压力补偿瓶
pressure-compensation control 压力补偿调节
pressure compensation device 压力补偿装置
pressure compensator 压力补偿器
pressure component 压力分量；分压(力)
pressure-compounded impulse turbine 复式压力级冲动式涡轮机
pressure-compounded turbine 高压复式涡轮机；高压复式透平；高压复式汽轮机；多级式汽轮机；复式压力涡轮机
pressure conduit 压力水道；压力管渠；压力管(道)
pressure cone 压力锥印(坯块缺陷)
pressure connection 压力管接头；加压连接
pressure connector 压力连接器；拉线夹；加压接头
pressure contact 压力接点；压力触点；力触点
pressure contact board 压力接触板
pressure contact method 压力接触法
pressure contact welding 压力接触焊
pressure container 压力容器；耐压容器
pressure-containing member 承压部件
pressure-containing parts 受压部件
pressure containment 压力安全壳
pressure contour 等压线；气压等高线；等压高度线
pressure control 压力控制；压力调节
pressure control circuit 压力控制回路
pressure control device 压力控制装置
pressure control equipment 压力控制设备
pressure controlled downhole test valve 环空压力控制井底测试阀
pressure-controlled plant 压力控制式装置
pressure controlled valve 压力控制阀；压力调节阀
pressure-controlled weld 压力控制焊接
pressure controller 压力控制器；压力调节器；电压调整器
pressure-control loop 压力控制回路
pressure-control relay 压力控制继电器
pressure-control servovalve 压力控制伺服阀
pressure control valve 压力调节阀
pressure converter 压力变换器
pressure conveyer 压紧式输送器
pressure cooker 压力式蒸压器；压力锅；高压锅；高压釜
pressure cooling 压流冷却；加压冷却
pressure core 人口稠密的中心
pressure core barrel 恒压取芯器；保压取心筒
pressure corer 压力式钻芯取样机；压力式钻芯采样机
pressure correction 压力校正
pressure correction factor 压力校正因子
pressure corrosion 压力腐蚀
pressure crack 压裂
pressure creosoted 加压浸油的；防腐剂加压渗透
pressure creosoted timber 加压油浸木材；加压浸油木材；加压防腐木材
pressure creosoting 压力油浸；防腐剂加压渗渗；压注油浸防腐法；杂酚油加压油浸
pressure culvert 压力式涵洞；压力涵洞
pressure curing 加压养护
pressure curtain coating 压力幕涂法
pressure curve 压力曲线
pressure cushion 压力枕
pressure cycle 压力周期
pressure cycle water 压力循环水
pressure cylinder 压力筒；压力缸
pressure damping 压力消减
pressured coring equipment 压力取芯工具
pressured cowling 加压罩
pressure deaeration 压力除气
pressure decay 压力下降
pressure decline 压力衰减；压力递减
pressure decrease 压力降低；压力降落
pressure decreasing coefficient 压力降系数
pressure-defined chamber 压力调制室
pressure deflection 压力偏转
pressure deflection test 压力挠曲试验
pressure dehydrator 加压脱水机
pressure delay 压力滞后
pressure deluge valve 加压式雨淋阀
pressure density 压块密度
pressure dependence 压力依存性
pressure-dependent 随压力变化的
pressure-dependent control 压力控制
pressure depletion 压力递减
pressure depth 压强深度
pressure detector 压力探测器；压力检测器；压力检波器；压电式检波器；水下地震检波器；水听器
pressure determination 压力测定
pressure developed by concrete on formwork 模板压力
pressure developed by concrete on framework 新浇混凝土对模板的压力
pressure device 增压装置；加压装置

pressured hydrocyclone separator 压力式水力旋流分离器
pressured hydrolysis 加压水解
pressure diagram 水压图;压力图
pressure dialysis 压力渗析法
pressure diazo 压重氮;加压重氮法
pressure die 压紧模
pressure-die-casting 压力模铸;压铸件
pressure-die-casting machine 压铸机
pressure-die-casting pirn 压铸金属纡管
pressure difference 压强差;压力差;压差;压差
pressure difference bed-load sampler 压差式底沙采样器
pressure difference bed material sampler 压差式底河床材料采样器
pressure difference hydrophone 压差水听器
pressure difference indicating controller 压力差指示控制器
pressure difference of production 生产压差
pressure difference sensor 压差传感器
pressure difference sub-area 压差分区
pressure difference switch 压差开关
pressure difference transmitter 压差变送器
pressure difference type bed-load sampler 压差式底沙采样器
pressure difference under production 生产压差
pressure difference valve 压差调节阀
pressure differential 压力差;压差
pressure differential hydrophone 压差水听器
pressure differential level controller 压差料位控制装置
pressure differential meter 差压式流量计;压差式流量计
pressure differential navigation 压差领航
pressure differential range 压差范围
pressure differential transmitter 压差变送器
pressure differential warning valve 前后轮制动压差警报阀
pressure disc 压力垫圈;止推垫圈
pressure discharge 压力释放
pressure discontinuity 压力突跃
pressure dispersion 压力分散
pressure-displacement curve 压力—位移曲线;压力—变形曲线
pressure-displacement transducer 压力位移变换器
pressure dissipation 压力消散
pressure dissolved air flo(a)tation 加压溶气气浮法
pressure distillate 压力蒸馏物;裂化溜出油;压滤馏分;加压馏出物
pressure distillation 加压蒸馏
pressure distributing valve 压力分配阀
pressure distribution 压力分布
pressure distribution calculation 压力分布计算
pressure distribution chart 压力分布图
pressure distribution measurement 压力分布测定
pressure distribution of foundation 基础压力分布
pressure distribution on pavement 路面压力分布
pressure distribution pattern 压力分布图
pressure distribution ring 压力分布环
pressure distributor 压力洒布机;压力喷洒机;压力分配器;压力分布器;沥青撒布机;沥青喷洒机;沥青喷布机
pressure distributor truck 沥青压力喷车
pressure disturbance 压力故障;压力干扰;压力扰动
pressure diversion system 压力引水系统
pressure divider 分压器
pressure dividing 压力分配
pressure dividing valve 压力分配阀
pressure dome 胀裂丘;压力穹隆
pressure dosing 压力投配
pressure drag 压制阻力;压力阻力
pressure drag coefficient 压差阻力系数
pressure drainage 压力排水
pressure drainage system 压力排水系统
pressure drilling 压力钻进;加压钻进
pressure driven instability 压力驱动不稳定性
pressure driven membrane separation technology 压力驱动膜分离技术
pressure drop 压力损失;压力下降;压力落差;压力降;压降值;压降;压差
pressure drop across the core 沿岩芯方向压力降
pressure drop coefficient 压损系数;压力损失系数;压降系数

pressure drop effect 压降效应
pressure drop in collar 钻铤内压降
pressure drop increase 压降增加
pressure drop in drill pipe 钻杆内压降
pressure drop manometer 压降压力计
pressure drop method 压降法
pressure drop on bit 钻头压降
pressure drop oscillation 压降振荡
pressured sequencing batch reactor activated sludge process 加压序批间歇式反应活性污泥法
pressure duct 高压管路
pressure due to breaking wave 破波压力
pressure due to wind 风压力
pressured water fire extinguisher 高压水灭火器
pressured-water nuclear reactor 压水核反应堆
pressured-water reactor 压水反应堆
pressure dwelling brake 保压制动
pressure dy(e)ing machine 高压染色机
pressure dye test 压染试验;压力着色试验;加压吸附染料试验
pressure eddy 压力旋涡
pressure effect 压力作用;压力效应
pressure efficiency 压力效率
pressure element 压力元件;测压元件
pressure ellipse 椭圆形压力圈
pressure-energized seal 受压自紧密封件
pressure energy 压能
pressure energy of bottom and edge water 底水和边水的压能
pressure engraving machine 轧钢芯机
pressure-enthalpy chart 压焓图
pressure-enthalpy diagram 压力—焓图
pressure environment 压力环境
pressure equalization 压力平衡;压力均匀化
pressure equalized fuel element 压力平衡的燃料元件
pressure equalizer 压力补偿器;均压器;均压管
pressure equalizer for twin tyres 双轮胎胎压平衡器
pressure equalizing device 均压装置
pressure equalizing layer 松解层;等压层
pressure equalizing passage 平衡压力的通道
pressure equalizing passageway 压力平衡管道;均压管道
pressure equalizing pipe 压力平衡管
pressure equalizing tank 均压箱
pressure equalizing tube 压力平衡管;均压器;均压管
pressure equation 压力方程
pressure equilibrium 压力平衡
pressure equilibrium constant 压力平衡常数
pressure error 压力误差
pressure evaluation 压力评定
pressure evaporator 压力蒸发器
pressure excess 压力增量
pressure excursion 压力偏离额定值
pressure exerted by massed 惯性压力
pressure exerted by the stowing material 充填材料产生的压力;堆积材料产生的压力
pressure expanded joint 压力膨胀接头
pressure expansion curve 压力膨胀曲线
pressure exploration 压力分布测定
pressure face 压力面;承压面
pressure face machine 压切掘进机
pressure factor 压力系数
pressure fall 压力降
pressure fall center 降压中心
pressure fan 压力通风机;压力风扇;压风机
pressure fault 逆断层
pressure fed carburetor 压力给油汽化器
pressure fed lubrication 压流润滑法
pressure feed 压力喂料;压力送料;压力进给;加压装料
pressure-feed air bearing 静压空气轴承
pressure feedback control 压力回授控制
pressure-feed container 增压桶
pressure feeder 压力喂料机;加压冒口
pressure-feed filler 压力加油器
pressure-feed gasoline 压力输送汽油
pressure-feed lubrication 压力润滑;压力送料润滑;压力给油润滑
pressure-feed oiling 压力加油法
pressure-feed system 压力进给系统
pressure-feed tank 压力燃料箱

pressure-feed type spray gun 压料式喷枪
pressure field 压力场
pressure field structure 压力分布
pressure filling 压力加油
pressure film instrument 压力薄膜仪
pressure filter 压力滤罐;压滤机;压力式过滤器;压力滤油;压力滤器;压力滤池;压力过滤机
pressure-filter-bulb method 球压法
pressure filter tank 压力过滤缸
pressure-filter-tube method 管压滤法
pressure filter with cycloid filter leaves 圆形滤叶加压叶滤机
pressure filtration 压滤;压力过滤;加压过滤
pressure filtration process 压力过滤法
pressure finger 压指
pressure firing device 压发装置
pressure fissure 压缩裂缝
pressure fixation 压力固定
pressure fixing 压力固定;加压定影
pressure flange 压力凸缘
pressure flap 压力瓣;均压拍击气门;均压拍击气门
pressure flask 耐压烧瓶
pressure float 压力浮子
pressure flo(a)tation 压力浮选(法)
pressure flow 有压流;压力流
pressure flowmeter 压流计
pressure flow pipeline 压力管线
pressure flow system 压力输送系统
pressure fluctuation 压力起伏;压力脉动;压力波动
pressure fluid 压力油;工作液;受压液体;受压流体;承压液
pressure flush 压力冲洗
pressure flush valve 压力冲洗阀
pressure fluviograph 压力水位计
pressure foot 压力脚
pressure force 压力
pressure force on submerged surface 潜体表面的压力
pressure for dredge pump 泥泵压力表
pressure forging 热模压
pressure former 压力成型装置
pressure forming 压力成形
pressure-free 无压的
pressure front 激震前沿
pressure fuel feed 压力加油
pressure fueling system 加压供燃料系统
pressure fuel system 压力输送燃料系统
pressure function 压力函数
pressure fusing 压力溶化;加压定影
pressure fuze 气压式引信
pressure gain 压力放大
pressure gallery 压力坑道
pressure-gas burner 正压煤气燃烧器
pressure gasification 压力气化
pressure gasifier 压力气化器
pressure gasoline 热裂化汽油
pressure gas plant 稳压煤气工厂
pressure gas tank 压缩气体罐
pressure gas vessel 压力气体瓶
pressure gas welding 加压气焊
pressure gate 缝隙内浇口
pressure ga(u)ge 压力校准装置;压力限位装置;压强器;压强计;压力仪;压力器;压力计;压力表;增压表;气压计;膛压计;水压验潮仪;测压仪;测压计
pressure ga(u)ge connection 压力表接口
pressure ga(u)ge damper 压力表缓冲器
pressure ga(u)ge dash unit 压力表板总成
pressure ga(u)ge snubber 压力表减振器
pressure ga(u)ge switch 压力表开关
pressure ga(u)ge tester 压力表检验器
pressure ga(u)ge thermometer 示压温度计
pressure ga(u)ge with electric(al) contact point 电接点压力表
pressure ga(u)ge with iron case 铁壳压力表
pressure ga(u)ging 压力测定
pressure gear 增压齿轮(油泵的);加压齿轮
pressure gear pump 压力齿轮泵
pressure geophone 压力式地音计
pressure-glued 压力上胶
pressure governing 压力调节
pressure governor 压力控制器;压力调节器
pressure grade line 自由水面线;压头线;压缩头线
pressure gradient 气压坡度;水头梯度;压力梯度;

压力坡降;压力陡度;压力下降;气压梯度;气压记录器;水头损失
pressure gradient analogy 压力梯度相似
pressure gradient correction factor 压力梯度校正因子
pressure gradient effect 压力梯度效应
pressure gradient force 气压梯度力
pressure gradient in tubing 井筒内压力梯度
pressure gradient microphone 压差传声器
pressure gradient transducer 压差换能器
pressure-graph 气压记录器;压力自记仪
pressure grease 压力润滑脂
pressure grease gun 黄油枪
pressure grip throttle 压力夹钳风阀;压力夹钳风门
pressure group 压力团体;施压集团
pressure-grouted 有压注浆;压力灌浆的
pressure grouted aggregate concrete 压力灌浆混凝土
pressure grout(ing) 压力注浆;压力灌浆(法);压浆
pressure grouting machine 压力灌浆机
pressure grouting pan 压力灌浆筒
pressure grouting pipe 压力灌浆管
pressure grouting system 加压灌浆法
pressure grouting tube 压力灌浆管
pressure grout penetration 压力灌浆深度
pressure gun 挤压枪;压力枪;压力注射枪;灌缝喷枪;黄油枪;润滑油枪;堵缝喷枪
pressure-gun fitting 黄油枪油嘴
pressure-gun grease 压力枪(注入的)润滑脂;黄油枪用润滑脂
pressure-gun type asphalt 压力枪喷射型沥青
pressure-gun type cement 压力枪喷射型水泥浆
pressure-gun type cork 压力枪喷射型软木塞
pressure-gun type exterior plaster(ing) 压力枪喷洒型墙外粉刷
pressure-gun type mastic 压力喷射型玛琋脂
pressure-gun type putty 压力喷射腻子;压力喷射油灰
pressure-gun type vermiculite 压力喷洒蛭石(轻骨料)
pressure handhole 受压手孔
pressure hardness 压印硬度
pressure harness 组合吊综装置
pressure head 扬程(泵的);压位差;压头;压力头;压力水头;压差;水位差;承压水位
pressure header 压力总管
pressure headphone 压强耳机
pressure head switch 压力保护开关;触点式压力计
pressure head tank 压力水箱
pressure height 气压高度
pressure-height corrector 航高校正数测定器
pressure-height curve 气压—高度曲线
pressure-helmet 压力盔形图
pressure history 压力变化曲线;压力变动情况
pressure hoisting gear 耐压升降装置
pressure holder 高压储气罐
pressure holding circuit 压力保持回路
pressure hole 测压孔
pressure hose 压力软管;压力胶管;高压软管;耐压软管
pressure house 压力套筒
pressure hull 压力舰体;内壳;耐压壳(体)
pressure-hull structure 耐压艇体结构
pressure hydration 加压水化
pressure hydrophone 压强水听器;声压水听器
pressure ice 压力冰;挤压冰;起伏冰
pressure ice area 冰块拥塞区
pressure ice foot 沿岸冰麓;挤压冰壁
pressure imposing element 加压元件
pressure impregnation 压力注入;加压浸渍(木材防腐);加压浸透
pressure impregnation carbonation 加压浸渍碳化
pressure in 压合
pressure in bubbles 气泡压力
pressure inclination ga(u)ge 倾斜压力计
pressure increasing valve 增压阀
pressure increment 压力增量
pressure increment ratio 压力增量比
pressure indicating controller 压力指示控制器
pressure indicator 压力表;压力指示器
pressure indicator and recorder 压力指示记录器
pressure indicator light 压力指示灯
pressure-induced band 压力感生带

pressure-induced spectrum 压力感生光谱
pressure in excess of atmospheric pressure 超过大气压的压力
pressure-injected footing 压力喷注桩基法;压力喷射造桩基法
pressure injection 压力注入;压力注法;压力注;压力喷射;压力灌注
pressure inlet 增压管;接管嘴
pressure in main 干管压力
pressure in money market 银根紧
pressure-in pile 压入桩
pressure in pores 孔隙压力
pressure instrumentation 测压仪表
pressure insulation 压力绝缘
pressure intensifier 增压器;增压剂
pressure intensity 压强;压力强度
pressure interfere 压力干扰
pressure in the tank 油罐内部压力
pressure in vacuum system 真空系统压强
pressure in vacuum tank 真空室压强
pressure investigation 压力分布研究
pressure ionization 压致电离
pressure iron bar 顶铁
pressure jet 压力射流
pressure jet apparatus 压力喷嘴
pressure jet atomizer 压力喷射雾化器
pressure jet burner 压力喷嘴燃烧器;压力喷油燃烧器
pressure jet nozzle 压力式射流喷嘴
pressure jet(oil) burner 压力射油燃烧器
pressure jig 高压卷染机
pressure joint 压力节理(岩石的)
pressure jump 压力跃变;压力突增;压力突变;压力剧变
pressure jump line 气压涌升线
pressure knotter 压力除节机
pressure law 压力分布规律
pressure leaching 加压浸出
pressure leaching train 高压浸出系统
pressure leaf filter 叶片式压滤机;加压叶片过滤机;加压叶滤器
pressure lenses 压力透镜体
pressureless compacting 无压成形
pressureless sinter(ing) 常压烧结;无压烧结
pressureless steam 无压(力)蒸汽
pressureless type evapo(u)rative cooling 低压汽化冷却
pressure level 声压级
pressure level-off 压力稳定
pressure lever 压杆
pressure lifting 压力升(高)
pressure limit 压力极限;压力范围
pressure limitation 压力限度
pressure limitation valve 限压阀
pressure limiting device 压力安全装置;限压装置
pressure limiting valve 限压阀
pressure line 压送管路;压力线;压力管线;压力管路;啮合线(齿轮传动);耐压管路
pressure line arch 压力光曲拱;压力线拱
pressure line method 压力管道法
pressure line of position 压力位置线
pressure liner 高压缸套(水泵)
pressure line vault 压力线拱顶
pressure link 加压导杆
pressure liquid 液压液体
pressure liquid tank 液压液体容器
pressure-loaded prefilling tank 充液罐
pressure load(ing) 压力负荷;压力荷载;抗压应力
pressure lock 压力栓;空气阀;气压阀
pressure-locked grating 压力锁结格栅;压接栅板
pressure log 压力测井;水压计程仪
pressure loss 压力损失;压力耗;压降
pressure loss coefficient 压力损失系数;压力损耗系数
pressure-loss conversion chart 压力损失换算表
pressure losses through surface equipment 地面压降
pressure loss factor 压力损失系数;压力损耗系数
pressure loss in bit 钻头泵压损失
pressure loss in drilling rod 钻杆泵压损失
pressure loss in flowing 沿程压力损失
pressure loss of heat-transporting fluid 热流体压损
pressure-loss operated deluge valve 压力损失动

作的雨淋阀
pressure-low warning light 压力过低警告灯
pressure-lubricated 压力润滑的
pressure lubricating 压力润滑
pressure lubricating system 强制循环润滑系统
pressure lubrication 压力润滑(作用);压力润滑法;加压润滑
pressure lubrication system 压力润滑系统
pressure lubricator 压力润滑器
pressure main(pipe) 压力主管;压力总管;压力干管
pressure maintaining 保压
pressure maintaining pump 供压泵;压力供给泵
pressure maintaining valve 压力控制顺序阀;稳压阀
pressure maintenance 压力维持;压力保持
pressure maintenance method 压力保持方法(油层)
pressure maintenance process 油层压力保持法
pressure manhole 受压人孔
pressure manifold 液压油路阀箱;集流管;汇管
pressure manometer 压力计;差动(式)压力计
pressure mark(ing) 压痕
pressure matchplate process 石膏型加压整铸铝模板法
pressure measurement 压力测量;压力测定
pressure measurement pump 压力测量泵
pressure-measuring controller 测压控制器
pressure-measuring hole 测压孔
pressure-measuring instrument 压力测量仪器
pressure-measuring microphone 声压测量传声器
pressure-measuring set 压强测量仪
pressure-measuring system 压力测量系统
pressure mechanism 机械压力装置;水压引信
pressure medium 液压介质;压力介质
pressure membrane apparatus 负压测定仪(用于测定孔隙水压力);压力膜析装置
pressure membrane method 压力膜法
pressure metamorphism 压力变质(作用)
pressure meter 横压仪;旁压仪;压力计;压力表;气压计;测压计
pressuremeter limit pressure 旁压仪极限压力;旁压极限压力
pressuremeter method 侧压仪法
pressuremeter modulus 旁压仪模量;旁压模量
pressuremeter test 横压仪试验;旁压试验
pressuremeter type 旁压仪类型
pressure meter type of wave recorder 压力计式波高计;水压式波高仪;水压式波高计
pressure method for air content 含气量压力测定法
pressure microphone 压强式传声器;压力传声器;声压式传声器
pressure mixing 揉捏混合(作用)
pressure modulated radiometer 压力调制辐射仪
pressure modulation valve 调压阀;压力调整阀
pressure modulus 压力模数
pressure-momentum curve 压力动量曲线
pressure monitor 压力监视器;压力监测器
pressure mortar 压力喷浆
pressure mottling 斑迹
pressure mottling resistance 耐压力斑点性
pressure mo(u)lding 压铸模铸造;加压模塑;加压成形
pressure mounting 压力支撑
pressure multiplication factor 压力倍数
pressure multiplier 压力倍增器;增压器
pressure naphtha 裂化石脑油;热裂化粗汽油
pressure nozzle 压力喷嘴
pressure of air 空气压力
pressure of application 压力法;压力施加法
pressure of crystallization 结晶压力
pressure of distribution 分布压力
pressure of explosion 爆炸压力
pressure of foundation 基础压力
pressure of ground 山体压力
pressure of ice 冰压力
pressure of impounded water (筑坝的)静水压力
pressure of inspiration 吸气压(力)
pressure of laser light 激光光压
pressure of light 光压
pressure of liquid 液体压强
pressure of moving water 流水压力
pressure of overlying strata 上覆地层压力
pressure of press 压机压力

pressure of rolling 轧制压力
pressure of saturation 饱和压力
pressure of sound wave 声波压强
pressure of strata 岩层压力;地层压力
pressure of surge chamber 空气包压力
pressure of surrounding rock 围岩压力
pressure of top roll 上辊压力
pressure of wind 风压力;风压
pressure oil 压力油
pressure oil engine 增压重油机;增压柴油机
pressure oiling 压力加油
pressure oil jack 液压千斤顶;压力油千斤顶
pressure oil lift 油压起重机
pressure oil line 压力油管线;压力油管道
pressure oil pipe 压力油管
pressure oil pipeline 压力油管线
pressure oil pump 液压(油)泵;压油泵;增压油泵
pressure oil ram 液压(油)缸
pressure oil tank 压力油箱;压力油柜;高压油箱
pressure on capacity 扩大生产能力的迫切需要
pressure on forms 模壳压力
pressure on formwork 模板工程压力
pressure on foundation 基础承压;基础承受的压力;地基土压力
pressure on foundation soil 基土压力;基土承受的压力
pressure on ground 地基压力
pressure on imposts 拱脚压力;拱基压力
pressure on shuttering 模壳受到的压力
pressure on the bit (钻进时的)钻头压力;钻头负荷
pressure on the bosh 炉腹压力
pressure on the guideway 导轨压力
pressure on the subgrade 路基面压力
pressure on transport 交通运输紧张
pressure-operated 气动的;压力操纵的
pressure-operated device 压力操作装置
pressure-operated microphone 压强传声器
pressure operated siren 气动笛
pressure-operated switch 压力作用开关;压力操纵开关
pressure-operated switch water flow alarm 压力动作开关水流报警器
pressure-operated switch water flow initiating device 压力动作开关水流启动装置
pressure-operated valve 压力操纵阀
pressure operation 增压运行
pressure oscillation 压力振荡;压力变动
pressure oscillation damper 压力脉冲阻尼器
pressure override 压力超增
pressure oxidation 加压氧化
pressure packer 膨胀性封隔器
pressure packing 加压填充法
pressure pad 压料垫;压力垫
pressure-pad-force 压边力
pressure-pad lift-out 压力垫顶出
pressure pair 压接对偶
pressure pan 高压釜
pressure parting 压裂
pressure pattern 压力模式
pressure pattern flying 测高术航行
pressure pattern metamorphism 压力变质(作用)
pressure pattern theory 声压图像说
pressure peak 压力峰值
pressure penetration 压注;压力贯入;压力穿透
pressure penetration test 压力贯入试验
pressure percussion 压力冲击
pressure period 加压阶段;受压期间
pressure per stone (每颗金刚石所受的)钻压
pressure per unit area 压强;单位面积压力
pressure per unit of area 单位压力
pressure pick-up 压力传感器
pressure pile 压力灌注桩;压力灌浆管桩;承压桩
pressure pillow 压力垫枕
pressure pin 气垫上压力顶杆
pressure pipe 压力小管;压力管;增压管
pressure pipeline 压力管渠;压力管道;压力管线;加压管路;耐压管线
pressure piping(-line) 压缩空气管道;压送管路;压力管线;压力管道(路);增压管路;耐压管线
pressure piston 施压活塞;压力活塞
pressure plate 压力试验机座板;压力板;压板
pressure-plate anemometer 压板风速表;风板压速表

pressure plate roller 压板滚柱
pressure plate stud 压板销
pressure-plotting 压力分布图绘制
pressure-plotting model 压力分布模型
pressure plunger 加压柱塞
pressure pneumatic conveyer 压力气流式输送器
pressure point 压力点;压点
pressure polishing 压力抛光
pressure polymer 加压聚合物
pressure polymerization pot 加压聚合器
pressure pond 压力水池
pressure pool 压力水池
pressure port 压气入口;压力孔;取压孔
pressure pot 压力桶;压力罐
pressure potash process 加压钾碱法
pressure potential 压力势
pressure pouring 压力铸锭;压力浇注;气压浇注
pressure pouring furnace 气压浇注炉
pressure preservative treatment 加压防腐处理(木材);加压(防腐)处理
pressure preserved 压力防腐木料;加压防腐的
pressure probe 压力探头
pressure process 加压法;加压处理法
pressure process equipment 加压防腐处理设备
pressure process of timber preservation 压力木材防腐法;压力防腐木材保存法
pressure profile 压力分布;多压廓线
pressure programming 压力程序设计
pressure-proof 耐压的
pressure-proof housing 耐压外壳
pressure-proof layer 耐压层
pressure-proof pipe 耐压管
pressure-proof protecting housing 耐压保护壳
pressure-proof tank 耐压油箱
pressure-proof television camera 耐压电视摄影机
pressure propagation 压力传播
pressure proportioner 压力式比例混合器
pressure proportioning tank 压力比例混合器
pressure protection valve 压力保护阀
pressure proximity fuze 气压式近炸引信
pressure-pull firing device 压发拉发混合式发火装置
pressure pulsation 压力脉动;压力波动
pressure pulsation damper 压力脉冲阻尼器
pressure pulse 压力脉冲
pressure pulse wave 压波;压力脉动波
pressure pump 压气泵;压力(水)泵;增压泵;加压泵
pressure pump delivery pipe 压力泵输送管
pressure pumped down 压力降(因孔内油水而造成的)
pressure pump suction pipe 压力泵吸油管
pressure quench(ing) 加压淬火
pressure radius 压力半径
pressure ram 挤压杆
pressure range 加压范围;压力级制;压力范围
pressure rating 承压率(管子可承受的稳定水压力);压力率;压力定额;压力等级
pressure ratio 压力比;增压比
pressure ratio control 压(力)比控制
pressure ratio controller 压力比控制器
pressure ratio control system 压比控制系统
pressure ratio detector 压力比检测器
pressure ratio in cylinder 汽缸压力比
pressure reading 压力读数;气压计读数
pressure receiver 压力接收器
pressure recorder 液体压力记录器;压力自记计;压力记录仪;压力记录仪;记压器
pressure recorder for blower 鼓风机压力记录器
pressure recording controller 压力记录控制器
pressure-recording system 压力记录系统
pressure recovery 压力恢复
pressure-recovery characteristic 压力恢复特性
pressure-recovery factor 压力恢复系数
pressure redistribution 压力再分布
pressure reducer 减压器
pressure reducer valve 减压阀
pressure reducing chamber 减压室
pressure reducing fill 减压棱体
pressure-reducing regulator 减压调节器
pressure-reducing valve 安全阀;减压阀
pressure reduction 压降;减压;气压归正;气压订正
pressure reflection coefficient 声反射系数
pressure reflection ratio 声反射率
pressure reforming 加压重整

pressure refueling 压力加油
pressure register 压力自记仪;记压器
pressure regulating box 调压箱
pressure regulating device 调压装置;调压机构
pressure regulating lock-in and lock-out submersible 调压进出式潜水器
pressure regulating roller level(l)er 调压式多辊矫直机
pressure regulating thumb screw 压脚调节翼形螺钉
pressure regulating valve 压力调节阀;调压阀
pressure regulation 压力调节;调压
pressure regulator 压力调整器;压力调节器;减压安全阀;调压器
pressure regulator valve 压力调节阀
pressure relay 压力继电器
pressure release 断气
pressure-release action fuze 松发引信
pressure-release assembly 释压管
pressure-release coupling 释压接口
pressure-release surface 释压面
pressure-release zone 免压带
pressure relief 卸压;泄压
pressure-relief and sustaining valve 减压和持续压力阀
pressure-relief cone 压力降低漏斗;锥形压力降低区(地下水)
pressure-relief damper 减压调节风门;泄压风门;降压风挡;减压阻尼器
pressure-relief device 降压装置;泄压装置;减压装置;压力安全装置
pressure-relief governor 超压泄放调压器;超压泄放调节器
pressure-relief hatch 减压通风口
pressure-relief opening 泄压口
pressure-relief pipe 消压管
pressure-relief provision 泄压装置
pressure-relief tank 减压舱
pressure-relief test 释压试验
pressure-relief valve 降压阀;减压安全阀;释压阀;安全阀;卸压活门;卸闭阀;卸压安全阀;泄压阀;压力安全阀;减压阀;安全减压阀
pressure-relief valve ball 限压阀钢球
pressure-relief valve plunger 减压阀柱塞
pressure-relief valve unit 减压阀装置
pressure-relief vent 减压孔;安全气道;卸压孔
pressure-relief well 减压井
pressure relieve 去荷
pressure-relieving device 压力泄放装置
pressure-relieving joint 压力分散层;减压缝
pressure remanent magnetization 压剩余磁化;压加剩余磁化强度
pressure rendering 压力炼油
pressure repeater 压力中继器
pressure replacement 压力交替(作用)
pressure reservoir 蓄势器;压力水箱;压力容器;压力风缸;储压器
pressure resistance 压力阻力;加压阻力;耐压性
pressure-resistant 承受住压力
pressure resonance 压强共振
pressure response 声压响应;声压灵敏度
pressure-response data 压力响应数据
pressure responsive device 压敏装置
pressure restoration 压力恢复
pressure restoration method 压力恢复方法
pressure-retaining member 压力保持部件
pressure retaining parts 保压部件
pressure retaining valve 保压阀
pressure return water 压力回水
pressure ribbon microphone 声压带式传声器
pressure ridge 压脊;胀裂脊;挤压脊;熔岩脊;冰脊;冰丘脉;冰脊
pressure rigid airship 压力刚性飞艇
pressure ring 压缩环;压力环;压紧环;压环;耐压环
pressure rise 压力增长;压力增加;压力升(高);气压上升
pressure rise center 升压中心
pressure riser 压力帽口
pressure rising curve 升压曲线
pressure rod 加压杆
pressure roll(er) 压力辊;压紧辊;压辊;卷压辊;开坯轧辊
pressure roller cam 压纸滚轮凸轮
pressure rubber pipe 耐压橡皮管

pressure rubber piping 耐压橡皮管
pressure rubber tubing 耐压橡皮管
pressure safeguard 压力防护装置
pressure safety switch 液压系统保险开关;压力保险开关
pressure safety valve 压力安全阀
pressure sand filter 压力砂滤池;加压式砂滤池;加压砂滤池;加压滤池;压力砂滤器
pressure sand filtration method 压力砂滤法
pressure-scanned laser 压力扫描激光器
pressure schedule 压力图表;压力范围
pressure scheduler 压力调整器
pressure schistosity 压力片理
pressure screen 压网
pressure screw 压下螺丝;压力螺钉
pressure screw check nut 压力螺钉防松螺母
pressure screw extractor 螺旋压榨机
pressure screw for tank block 池壁压力螺钉;池壁顶丝
pressure seal 压力密封;压力封口;加压密封
pressure seal bonnet 压力密封阀盖
pressure seal cap 压力密封帽
pressure seal cover 压力封盖
pressure-sealed car 压力密封车;气密车辆
pressure-sensing device 压力传感器;压敏元件;压力敏感元件;灵敏压力计
pressure-sensing line 感压管
pressure-sensing resistor 压敏电阻
pressure-sensing tide ga(u)ge 压敏潮位计
pressure-sensitive 压敏材料;压敏的;压敏
pressure-sensitive adhesion 压敏黏合
pressure-sensitive adhesive 压敏(性)黏合剂;压敏黏结剂;压敏胶黏剂;压合胶黏剂
pressure-sensitive adhesive lettering 压敏注记
pressure-sensitive adhesive tape 压敏性黏结带;压敏胶带;轻质黏着带
pressure sensitive-cell 压力传感盒
pressure-sensitive cement 压敏黏合剂
pressure-sensitive device 压力敏感元件
pressure-sensitive diaphragm 压力感压传感膜片
pressure-sensitive direction probe 压敏式方向测针;压敏式测向管;压敏式侧针;压敏流向测定管
pressure sensitive dye 压敏染料
pressure-sensitive effect 压力电敏效应;加压制度;压敏效应
pressure-sensitive element 压敏元件
pressure-sensitive film 压敏薄膜
pressure-sensitive geophone 压敏检波器
pressure-sensitive hydrophone 压敏水听器
pressure-sensitive keyboard 压敏键盘
pressure-sensitive pad 压力敏感块;压力传感块
pressure-sensitive paper 压敏纸
pressure-sensitive pendant type level ga(u)ge 垂球压敏式液位计
pressure-sensitive relay 压力继电器
pressure-sensitive sketch pad 压敏绘图垫板
pressure-sensitive switch 压力感应开关
Pressure Sensitive Tape Council of American 美国压敏胶带委员会
pressure-sensitive tape test 压敏胶带试验
pressure-sensitive transparency lettering 压敏透明注记
pressure-sensitive valve 压力敏感阀
pressure-sensitive vehicle detector 压力感知检数器
pressure sensitivity 声压灵敏度;压力灵敏度
pressure-sensitivity characteristic 压力敏感性能
pressure-sensitivity to temperature 对温度的压力灵敏度
pressure sensor 压力传感器
pressure sequence(-controlled) valve 压力顺序控制阀;定压阀
pressure-set ink 受压快干油墨
pressure setting 压力装置;压力设定;压力调整;压力调节
pressure-setting plug 压力调定螺塞
pressure-setting spring 压力可调节弹簧
pressure shadow 压力影
pressure shaft 压力(竖)井;压杆轴
pressure shell 压力壳层;受压壳层
pressure shift 压致移动
pressure shock 压力波
pressure shock wave 压力冲击波
pressure shower screen 压力水幕

pressure side 压力面;加压侧;受压面;承压侧
pressure side of the blade 叶片工作面;叶片承压面
pressure side proportioner 压力式比例混合器
pressure silicification method 压力硅化加固法
pressure silicified process 压力硅化法
pressure sink 压力降;压力差
pressure sinkage 压力下陷
pressure sintering 加压烧结;热压法
pressure skin 承压蒙皮
pressure sleeve 压力套筒;高压缸套(水泵)
pressure slip casting 压力泥浆浇注
pressure slope 压力斜面
pressure slot 压力缝
pressure sludge pipe 压力输泥管
pressure sluice 压力水闸;压力闸门
pressure snubber 压力缓冲器;压力波阻尼器
pressure solution 压溶;压溶作用
pressure-solution cleavage 压溶劈理
pressure sounder 水压探测器;压力(测)深针;压力测深仪
pressure span 压力区域;压力范围
pressure spectrum level 压谱级
pressure spike 压力尖峰
pressure spot 压力点
pressure spray 压力喷洗;压力喷雾;压力喷水;压力喷洒
pressure sprayer 压力喷洒器
pressure spray tank 压力喷雾箱
pressure spray tanker 沥青喷洒(罐)车
pressure spray-washing method 加压喷射洗涤法
pressure spring 压缩弹簧;压力弹簧;自流泉
pressure spring cap 压力弹簧帽
pressure spring insulating washer 压力弹簧隔热垫圈
pressure spring seat 压力簧座
pressure spring thermometer 弹簧管式压力温度计;弹簧管压力式温度计
pressure stability 压力稳定性;压力稳定
pressure stabilization 压力稳定
pressure-stabilized 压力稳定的
pressure stage 压力(阶)段;压力级
pressure-stage impulse turbine 压力级冲动式汽轮机
pressure starter 压力启动器
pressure stat 自动调压器;恒压器
pressure steel pipe 压力钢管
pressure stem 挤压杆
pressure stencil 压力模版
pressure sterilization 压力灭菌
pressure sterilizer 热压杀菌器
pressure still 裂化炉;加压蒸馏器
pressure-still distillate 热裂化馏出物
pressure-still tar 裂化炉焦油
pressure strainer 压力筛;压力滤网
pressure strength 压强;压力强度
pressure stress 压胁强
pressure strip 压紧板
pressure stroke 压缩行程;压力冲程
pressure suit 增压服
pressure supply 压缩空气供应;压缩空气供给;压力供应
pressure suppression 弛压系统
pressure suppression system 压力抑制系统
pressure suppression type containment 压力抑制型安全壳
pressure surface 压力面;推进面
pressure surface map 承压面图;压力面图
pressure surge 压力冲击波;压力波动
pressure surge limit 压力波动极限
pressure surge tank 压力缓冲器
pressure survey 压力检漏(测量);压力测量;通风压力测量
pressure sustaining 压力持续
pressure-sustaining valve 恒压阀;定压阀
pressure-swing absorber 变压吸附塔
pressure-swing adsorption 变压吸附
pressure-swing adsorption system 压力转换吸附法
pressure-swing cycle 压力转换循环
pressure switch 按钮开关;压力开关;压力继电器
pressure system 压力系统;压力系数;气压系统
pressure system of ventilation 压入式通风系统
pressure tank 压力油箱;压力容器;压力箱;压力(式)水箱;压力柜;压力罐;压力槽;水槽;高压

箱;高压水箱;耐压水舱;耐压容器;传压罐;承压储油啼
pressure tank container 压力箱容器
pressure-tank conveyer 压力箱式输送机
pressure tank filter 压力箱式过滤机;加压箱式过滤机
pressure tank lorry 高压水箱汽车
pressure tank water closet 压力水箱冲洗大便器
pressure tank water supplying 压力水箱供水方式
pressure tap 压力计接口;压力表接口;承压分接管;测压孔;测压接嘴
pressure tapping 压力表接头
pressure tap(ping) hole 压力表接口;取压分接管;测压孔
pressure tar 压裂焦油(沥青);裂化焦油
pressure technique 高压技术
pressure temperature 加压温度
pressure-temperature coefficient 压力温度系数
pressure-temperature compensated valve 压力温度补偿式阀
pressure-temperature rating 额定热力参数
pressure-temperature region 压力温度范围
pressure tendency 气压趋势;气压倾向
pressure tendency chart 气压变化图
pressure terminal 无焊接端头
pressure test fittings 压力试验接头
pressure test ga(u)ge 压力试验计
pressure test(ing) 强度试验;压力试验;加压试验;密封试验;试压
pressure testing device 压力试验设备
pressure testing instrument 压力试验设备
pressure testing machine 压力试验机
pressure testing oscillograph 测压示波器
pressure testing room for car doors 车门压力试验室
pressure-testing unit 压力试验装置
pressure test of distribution system 配水系统压力试验
pressure test of field permeability 现场压水试验
pressure texture 压缩结构;压碎结构;压力结构
pressure thermit(e) welding 加压热剂焊;压力熔焊;加压铸焊;加压铝热剂焊接
pressure thermometer 压力温度仪表;压力温度计
pressure thief 孔底取样器
pressure thrown by holder 储气罐压力
pressure thrust 压力推力
pressure tide ga(u)ge 压力潮位计;压力潮位表
pressure-tight 密闭的;压力密封的;耐压密封的;气密;受压不漏气的
pressure-tight box 气密盒
pressuretightness 加压密封性;气密性
pressure-tight test 耐压试验
pressure-tight weld 密封焊道;气密焊
pressure-time diagram 压力时间图;压力时间曲线图
pressure-time history 压力时间关系曲线
pressure-time indicator diagram 压力时间示功图
pressure-time method 压力时间(流流)法;吉普逊法;水锤测流法;压力时间法
pressure topography 高度形式;等压面形势
pressure transducer 压力换能器;压力传感器;压力变换器
pressure transfer 压力转移;压力传递
pressure transformation 声压转换
pressure transformer 电压互感器
pressure transmission 压力传播
pressure transmission medium 压力传递介质;传压介质
pressure transmitter 压力传送器;压力传感器;压力变送器
pressure trapping 人工增压
pressure-travel curve 压程曲线
pressure traverse 压力剖面;压力横向分布
pressure-treated wood 经加压处理的木材;加压处理木材
pressure treater 加压处理设备
pressure treating 加压处理
pressure treating method 加压防腐处理法
pressure treatment 压力蒸炼;压力处理(法)
pressure trip 压力断开装置
pressuretrol 自动调压器
pressure trough 气压槽
pressure tube 承压管;压力管
pressure tube anemometer 压管式风速仪;压管

式风速计;压管式风速表;压管风速计;压力管风速表
pressure tube gland 测压表管压盖
pressure tube reactor 压力管反应器;压力管式反应堆
pressure tubing 压力输送管道;压力管
pressure tunnel 有压隧洞;有压隧道;压力隧洞;压力隧道;压力水道
pressure turbine 压力叶轮机;压力涡轮机;高压涡轮机;反击式涡轮机;反击式透平;反击式水轮机;反击式汽轮机
pressure type 压力式布滤集尘器;压入式;挤压型
pressure-type apparatus 压力式装药器
pressure-type backwater system 压力式流量计
pressure-type baghouse 压力袋集尘器;正压袋集尘器
pressure-type carburetor 压力型化油器
pressure-type cloth filter dust collector 高压储气罐
pressure-type core barrel 压力式取样筒
pressure-type electrostatic(al) accelerator 压力式静电加速器
pressure-type evapo(u)rative cooling 高压汽化冷剂
pressure-type filter 压力砂滤池
pressure-type flowmeter 压力式流量计
pressure-type glass baghouse 压力玻璃纤维袋集尘器;正压玻璃纤维袋集尘器
pressure type head box 压力流浆箱
pressure-type holder 压力式支持器
pressure-type hydrant 压力式消火栓;压力式消防栓
pressure-type mercury-filled thermometer 压力式水银温度计
pressure-type retarder 压力式缓行器
pressure-type return system 压力式流量计
pressure-type sand filter 压力砂滤池
pressure-type spirit master 加压型原版
pressure-type storage cylinder 高压筒;压力式储存筒
pressure-type system 气压式系统
pressure-type thermometer 压力式温度计
pressure-type tide ga(u)ge 压力式验潮仪
pressure-type type of recording ga(u)ge 自记压力表
pressure-type valve lifter 压力式起阀器
pressure-type water heater 压力式水加热器
pressure-type waterproof lubricating device 压力式防水注油器
pressure-type wave ga(u)ge 压力型测波仪
pressure-type wave meter 压力式波高计;压力式波高计
pressure unit 压强单位;压力装置;压力单位;增压装置;膜片型模;气压单位
pressure unloading 压力降低
pressure upkeep 保持压力
pressure vacuometer 压力真空表
pressure vacuum ga(u)ge 真空压力计
pressure vacuum relief valve 真空安全阀
pressure vacuum vent valve 真空呼吸阀
pressure value 压力值
pressure value of well cleaning liquid 洗井液压力值
pressure valve 压力阀;高压阀(门)
pressure valve spring 限压阀弹簧
pressure vapo(u)rizer lamp 汽灯
pressure vapo(u)r lamp 煤油蒸汽灯
pressure variation 压力变化
pressure variograph 气压变量计
pressure variometer 气压变量表
pressure velocity compounded turbine 复式压力速度级涡轮机;复式压力速度级透平;复式压力速度级汽轮机;压力速度级复合式汽轮机
pressure vent 表面裂纹;表面裂缝
pressure ventilation 压力通风
pressure ventilation pipe 压风管
pressure ventilator 压风机
pressure vent valve 压力放空阀;通风阀
pressure vessel 压力容器;高压容器;耐压容器;受压容器
pressure vessel code 压力容器规范
pressure vessel construction code 压力容器结构规范
pressure vessel conveying system 压力仓式输送机系统

pressure vessel conveyer 压力仓式输送机
pressure vessel head 压力壳顶盖
pressure vessel manufacturer 压力容器制造厂
pressure vessel reactor 压力容器式(反应)堆
pressure vessels accessibility 压力容器可达性
pressure vessels exempted from inspection 免检压力容器
pressure vessels for chemical industry 化工压力容器
pressure vessels for the chemical industry 化工容器
pressure vessel tower reactor 压力容器塔式反应器
pressure viscosimeter 压力黏度计
pressure viscosity 压力黏度;受压黏度
pressure viscosity characteristics 压力黏度特性
pressure viscosity exponent 压力黏度指数
pressure-viscosity ratio 压力—黏度比
pressure-void ratio 压力—孔隙比
pressure-void ratio curve 压缩曲线;压力—孔隙比(关系)曲线
pressure-void ratio curve for sand 砂土压力与孔隙比曲线
pressure-void ratio diagram 压力—孔隙比图(表)
pressure voltage protection 高压防护
pressure-volume chart 压—容图
pressure-volume diagram 压力—体积曲线;压力容积图
pressure-volume relation 压力—体积关系
pressure-volume relationship 压力—容积关系
pressure-volume-temperature-relation 压力—体积—温度关系(曲线);压力—容积—温度关系(曲线)
pressure-volume-temperature superposition 压力—体积—温度叠加
pressure ware 耐压器皿
pressure warning unit 压力报警装置
pressure washer 压力清洗装置
pressure wash(ing) 压力冲洗;压洗
pressure washing operation 压力清洗作业;压力冲洗作业
pressure water 压力水;加压水;承压水
pressure water jet 压力水喷射器
pressure water line 压力水管路
pressure water pipe 压力水管;高压水管
pressure water pipeline 压力水管路
pressure water piping 压力水管
pressure water-spraying system 压力喷水系统
pressure water supply 压力供水
pressure water surface 压力水面
pressure water tube 压力水管
pressure water tubing 压力水管
pressure wave 压缩波;压力波;气压波
pressure wave antinode 压力波腹
pressure wave emission 压力波传播
pressure wavefront 压力波前
pressure wave node 压力波节点
pressure wave propagation 压力波传播
pressure wave reflection 压力波反射
pressure wave supercharge 压力波增压
pressure-wave supercharging 压力波增压
pressure-wave travel time 压力波传播时间
pressure wave valve 压力波动阀
pressure wave velocity 压力波动速度
pressure weather stripping 压力挡风雨条
pressure wedge 油压楔
pressure-welded junction 热压焊结
pressure welder 压焊机
pressure welding 压力焊;压力焊接;压接;压焊;加压焊(接)
pressure well 压注井
pressure-wetting equipment 湿压仪
pressure winding 电压线圈
pressure wind tunnel 高压风洞
pressure-wire 电压线
pressure wire connector 加压电线连接器;加压电线接头
pressure zone 压力区
pressuring medium 加压介质
pressurization 压力输送;加压密封;加压(法);加压状态;耐压;填压;增压(过程);密封
pressurization bag 增压袋
pressurization blower 加压鼓风机
pressurization energy 增压能量

pressurization-gas cascade 压缩气体容器
pressurization gas connection 加压气体接头
pressurization installation 定压装置
pressurization method 定压方式
pressurization of liquid oxygen 液氧的压送
pressurization point 定压点
pressurization pressure 压送压力;增压压力
pressurization process 加压(防腐)处理
pressurization pump 加压泵
pressurization supply fan 增压供气风机
pressurization system 增压系统
pressurization time 增压时间
pressurize 增压;产生压力;保压
pressurized access chamber 高压密封出入舱
pressurized air 压缩空气
pressurized air brakign 压缩空气制动
pressurized aircraft 座舱增压式飞机
pressurized aqueous combustion 压水燃烧法;湿空气氧化(法)
pressurized area 承压面积;增压室
pressurized bay 增压舱
pressurized bio-contact oxidation process 压力式生物接触氧化法
pressurized bladder 加压橡皮囊
pressurized blast furnance 加压鼓风炉
pressurized boiler 正压锅炉
pressurized bottle 汽水瓶
pressurized cabin 增压舱
pressurized canopy 增压座舱盖
pressurized cement lorry 密闭水泥罐卡车;增压水泥罐车
pressurized chamber 加压室
pressurized combustion 微增压燃烧
pressurized connector 充气密封连接器
pressurized consistometer 增压稠度仪
pressurized construction 增压结构
pressurized container 压缩空气箱;增压容器
pressurized container spray 喷雾剂
pressurized crankcase 增压曲轴箱;密封曲轴箱
pressurized curing chamber 增压养护室
pressurized-enclosure motor 增压防爆型电动机
pressurized environment 增压环境
pressurized escape route 正压疏散路线
pressurized exvacuation route 正压疏散路线
pressurized feeding system 压力输送系统
pressurized flammable liquid 加压的易燃液体
pressurized flo(a)tation 加压浮选
pressurized flo(a)tation method 加压浮选法
pressurized flo(a)tation process 加压浮选法
pressurized fluid insulation 加压绝缘液
pressurized fluidized bed combustion 加压硫化燃烧
pressurized food 气溶胶包装食品
pressurized gas 压力氧化煤气;压缩煤气
pressurized gas cylinder 增压气瓶
pressurized gas fire 加压气体火灾
pressurized gas pumping system 压气抽运系统
pressurized heavy and light water reactor 密封式重水和轻水反应堆
pressurized heavy water reactor 加压重水堆
pressurized hopper 正压料斗
pressurized hot water 承压热水
pressurized ion exchange 加压离子交换(法)
pressurized ionization chamber 增压电离室
pressurized motor 充高压气体的密封电动机
pressurized oil 液压油;加压油
pressurized oil braking 油压制动
pressurized oil engine 增压柴油发动机
pressurized oil jack 液压(油)千斤顶
pressurized oil pump 增压油泵
pressurized oil ram 液压缸
pressurized oil system 液压(油)系统;油压系统
pressurized package 气溶胶包装
pressurized refiner 压力磨浆机
pressurized refining 压力磨浆
pressurized reservoir 加压油箱;充压油箱
pressurized sample 压缩试样
pressurized screen 压力筛
pressurized seal 气压密封
pressurized shaft 压气竖井
pressurized slurry 泥水加压盾构
pressurized smoke control system 加压送风烟气控制系统
pressurized spray 喷雾剂

pressurized spreader 压力式液肥洒布机
pressurized stair(case) 增压楼梯间
pressurized staiway 加压楼梯
pressurized stopping 封闭灾区加压隔火墙
pressurized stored water unit 加压的储水容器
pressurized structure 加压式结构;加压结构
pressurized suit 气衣
pressurized system 耐压系统
pressurized system of ventilation 加压换气法
pressurized tunnel 有压隧洞
pressurized type water supply unit 气压给水装置
pressurized water 压力水;加压水
pressurized-water distributor 压水喷水机
pressurized-water fire extinguisher 高压水灭火器
pressurized-water nuclear reactor 压水核反应堆
pressurized-water reactor 压水堆;压力水冷反应堆;加压水反应堆;压水反应堆
pressurizer 压力保持装置;增压装置;加压器;体积补偿器;保持压力装置
pressurizer tank 稳压箱
pressurizing 增压;压力升(高);加压
pressurizing and ventilating-fan system 强制进气排气通风系统
pressurizing and ventilation equipment 增压和通风设备
pressurizing cable 气密电缆
pressurizing duct 增压系统导管
pressurizing non-return valve 增压单向阀门
pressurizing pump 预先加压用油泵
pressurizing recharge 加压回灌
pressurizing stand 增压检验台
pressurizing system 增压系统
pressurizing tank 增压箱
pressurizing unit 增压器
pressurizing valve 压力控制阀
pressurizing vessel 加压容器
pressurizing water test 压水试验
pressurizing water test packer 压水试验封(闭)隔器
pressurizing window 密封窗口;气密口
press vulcanization 加压硫化
press vulcanizer 平压硫化机
pressware 压制玻璃器皿
press water 压力水
press welding 加压焊(接)
press wheel 压土机
presswood 压制板
presswork 印刷作业;压制品;压制成品;压力加工;冲压成品
press-working cost 压力加工费用
press-working method 压力加工法
press-working spring 冲压弹簧
presswre roll 挤压辊
prestack deconvolution 迭前反褶积
prestack dip-filter 叠前倾角滤波
prestack G-log processing 迭前 G-log 处理
prestack migration 前偏移
prestack partial migration 叠前部分偏移
prestage 预级;前置级
prestall detector 失速探测器
prestandard 预标准
prestart 起动前
prestart check 起动前检查
prestart checklist 起动前检查单
prestarting 起动前的
prestarting inspection 起动前检查
prestarting instructions 启动前指南;启动前须知
prestart operation 起动前操作
Prestcore 普里斯特打桩法(一种打桩方法适用于有限净空高度和防震)
prestcore pile 沉管浇注桩(钢筋混凝土管)
presteady state kinetics 稳态前动力学
presteaming 预汽蒸
presteaming chamber 预汽蒸室
presteaming period 静置期;静停期;(混凝土蒸汽养护前的)预养期;预养护期;(混凝土的)蒸汽养护前期
presteaming vessel 预汽蒸室
prestessing under stress control 预应力控制张拉
prestex trap 厕所阻气具;厕所存水弯
prestige building 著名建筑物
prestige lighting 强力照明
prestocking 预储备
prestore 预先存储;预存(储)

prestoring 预先存储;预存(储)
prestowage plan 计划积载图;配载图
prestowage plan for container 集装箱配载图
prestowage planning 配载计划
pre-straightened 预调直的
prestrain(ing) 预应变;预加反向变形;钢化;预隔滤;预钢化处理
prestraining glass 钢化玻璃
prestratigraphy 基础地层学;初级地层学
prestreak 原条
prestreaming period 静停期
prestress 施加预应力;预拉伸;预应力;预加应力于
prestress casting yard 预应力浇注工厂
prestress design 预应力设计
prestress eccentricity 预应力偏心距
prestressed 预拉伸的;预受力的;预应力的
prestressed abutted assemblies 预应力对接拼装
prestressed aerated concrete 预应加气混凝土
prestressed airport concrete pavement 预应力混凝土机场跑道面
prestressed anchor 预应力锚杆;预应力锚
prestressed and automated shotcrete machinery 预应力自动喷浆机
prestressed beam 预应力梁
prestressed block-beam 预应力混凝土分块拼装梁
prestressed bolt 预应力锚杆
prestressed brickwork 预应力砖工
prestressed cable 预应力钢索;预应力钢丝绳
prestressed cable roof 预应力悬索屋盖
prestressed cast-concrete member 预应力浇注混凝土构件
prestressed casting 预应力浇筑(混凝土)件
prestressed centrifugally cast concrete pipe 离心浇制预应力混凝土管(道)
prestressed clay 预应力黏土
prestressed clay beam 预应力黏土梁
prestressed clay factory 预应力黏土工厂
prestressed clay lintel 预应力黏土过梁
prestressed clay plank 预应力黏土厚板
prestressed clay plant 预应力黏土工厂
prestressed clay roof 预应力黏土屋顶
prestressed clay roof slab 预应力黏土屋顶板
prestressed clay wall panel 预应力黏土墙面板
prestressed clay window lintel 预应力黏土窗过梁
prestressed component 预应力构件
prestressed composite tank 预应力组合水池
prestressed concrete 预应力混凝土
prestressed concrete bar 预应力混凝土钢筋;预应力混凝土芯棒
prestressed concrete barrel vault 预应力混凝土筒形拱壳体
prestressed concrete beam 预应力混凝土梁
prestressed concrete beam bridge 预应力混凝土梁(式)桥
prestressed concrete beam bridge by precast balanced cantilever segmental construction 预应力混凝土梁桥用预制分段平衡旋臂施工
prestressed concrete bearer 预应力混凝土托座;预应力混凝土支座
prestressed concrete bearing skeleton 预应力混凝土支承骨架
prestressed concrete binder joist 预应力混凝土梁搁栅
prestressed concrete block beam 预应力混凝土嵌空心砖梁
prestressed concrete bowstring 预应力混凝土铰索;预应力混凝土弓弦
prestressed concrete box beam 预应力混凝土匣形梁;预应力混凝土箱形(截面)梁
prestressed concrete box girder 预应力混凝土箱形大梁
prestressed concrete bracket 预应力混凝土托架;预应力混凝土牛腿
prestressed concrete bridge 预应力混凝土桥
prestressed concrete bridge member 预应力混凝土梁构件
prestressed concrete broad sleeper 预应力混凝宽枕【铁】
prestressed concrete building 预应力混凝土房屋;预应力混凝土建筑
prestressed concrete cable-stayed bridge 预应力混凝土斜拉桥
prestressed concrete caisson 预应力混凝土沉箱
prestressed concrete cassette plate 预应力混凝土箱板

prestressed concrete cast(ing) 预应力混凝土浇注
prestressed concrete ceiling joist 预应力混凝土顶棚搁栅
prestressed concrete center spiral 预应力混凝土中心螺旋筋
prestressed concrete central tube 预应力混凝土芯管
prestressed concrete centrifugally cast pipe 预应力混凝土离心浇注管
prestressed concrete coffer palte 预应力混凝土方格天花板
prestressed concrete column 预应力混凝土柱;预应力管柱
prestressed concrete composite beam 预应力混凝土合成梁;预应力混凝土混成梁
prestressed concrete compound unit 预应力混凝土复合构件
prestressed concrete construction 预应力混凝土建筑;预应力混凝土施工;预应力混凝土构造
prestressed concrete continuous beam 预应力混凝土连续梁
prestressed concrete cross wall 预应力混凝土桥墙;预应力混凝土隔板;预应力混凝土横墙
prestressed concrete cupola 预应力混凝土圆屋顶
prestressed concrete cylinder 预应力混凝土管柱
prestressed concrete cylindrical shell 预应力混凝土圆筒形壳体
prestressed concrete dam 预应力混凝土坝
prestressed concrete dome 预应力混凝土穹隆
prestressed concrete door lintel 预应力混凝土门过梁
prestressed concrete drilled caisson 预应力混凝土管柱;预应力管柱
prestressed concrete element 预应力混凝土构件
prestressed concrete elevated road 预应力混凝土高架道路
prestressed concrete embedded cylinder pipe 预应力混凝土埋藏式筒芯管
prestressed concrete engineering 预应力混凝土工程
prestressed concrete equipment 预应力混凝土设备
prestressed concrete fence 预应力混凝土栅栏
prestressed concrete flexture test beam 预应力混凝土挠曲试验梁
prestressed concrete floor 预应力混凝土梁楼板
prestressed concrete floor beam 预应力混凝土楼板梁;预应力混凝土地板梁
prestressed concrete floor slab unit 预应力混凝土楼面板构件
prestressed concrete folded-plate structure 预应力混凝土折板结构
prestressed concrete for civil engineering structure 预应力混凝土土木工程结构
prestressed concrete foul water pipe 预应力混凝土污水管
prestressed concrete frame 预应力混凝土框架
prestressed concrete girder floor 预应力混凝土带大梁楼板
prestressed concrete header 预应力混凝土横梁
prestressed concrete high-pressure pipe 预应力混凝土高压管
prestressed concrete highway 预应力混凝土公路
prestressed concrete highway bridge 预应力混凝土公路桥梁
prestressed concrete hipped plate roof 预应力混凝土折板屋顶
prestressed concrete hollow beam floor 预应力混凝土空心梁地板;预应力混凝土空心梁楼板
prestressed concrete hollow-core plank floor 预应力混凝土空心毛地板
prestressed concrete hollow-core slab floor 预应力混凝土空心楼板
prestressed concrete horizontal tie back 预应力混凝土水平拉杆
prestressed concrete I-beam 预应力混凝土工字梁
prestressed concrete insert 预应力混凝土插入件
prestressed concrete insertion 预应力混凝土嵌入物;预应力混凝土插入物
prestressed concrete intermediate floor 预应力混凝土中间楼面
prestressed concrete joist 预应力混凝土地板梁;预应力混凝土搁栅

prestressed concrete lattice girder 预应力混凝土花格大梁
prestressed concrete lift-slab construction 预应力混凝土顶升楼板构造
prestressed concrete lighting column 预应力混凝土照明灯柱
prestressed concrete lighting mast 预应力混凝土照明桅杆
prestressed concrete lined cylinder 预应力混凝土衬里钢(筒)芯；预应力混凝土衬里管柱
prestressed concrete lined pile 预应力混凝土衬里桩
prestressed concrete long-span slab 预应力混凝土长跨距板
prestressed concrete non-cylinder pipe 预应力混凝土非钢筒心管
prestressed concrete panel slab 预应力混凝土凸边板
prestressed concrete parachute 预应力混凝土伞形结构
prestressed concrete parallel truss 预应力混凝土平行桁架
prestressed concrete parking deck 预应力混凝土停车楼层；预应力混凝土停车甲板
prestressed concrete parking structure 预应力混凝土多层停车场结构
prestressed concrete pavement 预应力混凝土路面
prestressed concrete pavement for airport 预应力混凝土跑道
prestressed concrete pile 预应力混凝土桩
prestressed concrete pipe 预应力混凝土管
prestressed concrete pipeline 预应力混凝土管线；预应力混凝土管道
prestressed concrete pipe penstock 预应力混凝土给水管龙头；预应力混凝土给水管栓
prestressed concrete piping 预应力混凝土管
prestressed concrete plain web(bed) beam 预应力混凝土平腹梁
prestressed concrete plant 预应力混凝土车间；预应力混凝土工厂
prestressed concrete plate base 预应力混凝土板基
prestressed concrete plate foundation 预应力混凝土平板基础
prestressed concrete platform 预应力混凝土平台
prestressed concrete portal frame 预应力混凝土门式支架；预应力混凝土龙门架
prestressed concrete power pole 预应力混凝土电(力)杆
prestressed concrete pressure pipe 预应力混凝土受压管；预应力混凝土压力管
prestressed concrete pressure vessel 预应力混凝土压力容器
prestressed concrete prismatic shell 预应力混凝土棱柱壳体屋顶
prestressed concrete purlin(e) 预应力混凝土桁条；预应力混凝土檩条
prestressed concrete railroad sleeper 预应力混凝土铁道轨枕；预应力混凝土铁路轨枕
prestressed concrete railway sleeper 预应力混凝土铁道轨枕；预应力混凝土铁路轨枕
prestressed concrete reactor vessel 预应力混凝土反应堆容器
prestressed concrete rectangular grid slab 预应力混凝土矩形网络板
prestressed concrete refuse water pipe 预应力混凝土废水管
prestressed concrete reservoir 预应力混凝土储液器；预应力混凝土蓄水池
prestressed concrete ribbed slab 预应力混凝土有肋楼板；预应力混凝土肋板
prestressed concrete rigid frame 预应力混凝土刚(性)构架
prestressed concrete ring beam 预应力混凝土环梁；预应力混凝土圈梁
prestressed concrete road 预应力混凝土道路
prestressed concrete road bridge 预应力混凝土公路桥
prestressed concrete road pavement 预应力混凝土路面
prestressed concrete roof 预应力混凝土屋面
prestressed concrete roof beam 预应力混凝土屋顶梁
prestressed concrete roof-girder 预应力混凝土屋顶大梁
prestressed concrete roofing slab 预应力混凝土屋顶板
prestressed concrete roof sheathing 预应力混凝土屋顶望板
prestressed concrete runway 预应力混凝土跑道
prestressed concrete sales engineer 预应力混凝土(制件)销售工程师
prestressed concrete section 预应力混凝土构件
prestressed concrete segment 预应力混凝土管件
prestressed concrete segmental bridge 分段式预应力混凝土桥
prestressed concrete series beam 预应力混凝土串联梁
prestressed concrete sewage pipe 预应力混凝土污水管
prestressed concrete sheet pile 预应力混凝土板桩
prestressed concrete sheet pile wall 预应力混凝土板桩墙
prestressed concrete shell 预应力混凝土壳体；预应力混凝土薄壳
prestressed concrete shell roof 预应力混凝土薄壳屋顶；预应力混凝土壳体构造
prestressed concrete silo 预应力混凝土筒仓
prestressed concrete simple supported beam 预应力混凝土简支梁
prestressed concrete skeleton 预应力混凝土构架；预应力混凝土骨架
prestressed concrete slab 预应力混凝土板
prestressed concrete slab-and-beam 预应力混凝土梁板结构
prestressed concrete slab bridge 预应力混凝土板桥
prestressed concrete slab foundation 预应力混凝土平板基础
prestressed concrete slab unit 预应力混凝土板构件
prestressed concrete sleeper 预应力混凝土枕(木)
prestressed concrete solid girder 预应力混凝土实体大梁
prestressed concrete solid web(bed) beam 预应力混凝土实心腹梁
prestressed concrete spun pipe 预应力混凝土旋转浇制管；预应力混凝土离心浇制管
prestressed concrete steel wire strand 预应力混凝土结构用钢绞线
prestressed concrete storage tank 预应力混凝土储[贮]罐；预应力混凝土储[贮]液槽；预应力混凝土储[贮]水箱
prestressed concrete string(er) 预应力混凝土楼梯斜梁；预应力混凝土桁条
prestressed concrete structure 预应力混凝土结构
prestressed concrete structure system 预应力混凝土结构体系
prestressed concrete support 预应力混凝土支架；预应力混凝土支撑；预应力混凝土支座
prestressed concrete supporting skeleton 预应力混凝土骨架
prestressed concrete suspended floor 预应力混凝土架空悬挑面；预应力混凝土架空地板
prestressed concrete suspended roof 预应力混凝土悬挂式屋顶
prestressed concrete system 预应力混凝土系统
prestressed concrete tank 预应力混凝土储[贮]液池；预应力混凝土蓄液池
prestressed concrete T(ee)-beam 预应力混凝土T形梁
prestressed concrete T-frame bridge 预应力混凝土T形刚构桥；预应力钢筋混凝土T形刚构架
prestressed concrete T-frame bridge with suspended beam 预应力混凝土T形刚构加吊梁桥
prestressed concrete through girder 预应力混凝土槽形梁
prestressed concrete tie 预应力混凝土轨枕
prestressed concrete tie-rod 预应力混凝土系杆
prestressed concrete tilted-slab roof 预应力混凝土斜板式屋顶
prestressed concrete tower 预应力混凝土塔
prestressed concrete triangular truss 预应力混凝土三角孔桁架；预应力混凝土三角形构架
prestressed concrete trimmer plank 预应力混凝土装饰板
prestressed concrete truss 预应力混凝土桁架
prestressed concrete truss bridge 预应力混凝土桁架桥
prestressed concrete truss bridge lining segment 预应力混凝土桁架梁
prestressed concrete tube 预应力混凝土管
prestressed concrete tubing 预应力混凝土管
prestressed concrete tunnel vault 预应力混凝土圆筒拱顶
prestressed concrete umbrella 预应力混凝土伞形物
prestressed concrete unit 预应力混凝土构件
prestressed concrete vessel 预应力混凝土容器
prestressed concrete vierendeel 预应力混凝土佛伦第尔大梁；预应力混凝土带上下桁条的梁；预应力混凝土空腹梁
prestressed concrete voided slab bridge 预应力混凝土空心板桥
prestressed concrete waffle slab 预应力混凝土双向格子板；预应力混凝土双向密肋板
prestressed concrete wagon vault 预应力混凝土筒形拱顶
prestressed concrete wall 预应力混凝土墙
prestressed concrete wall panel 预应力混凝土壁板；预应力混凝土镶板；预应力混凝土护墙板
prestressed concrete water tower 预应力混凝土水塔
prestressed concrete window lintel 预应力混凝土窗过梁
prestressed concrete wire 预应力混凝土钢丝；预应力钢丝混凝土
prestressed concrete with fine wires 用细钢丝的预应力混凝土
prestressed concrete with wires 预应力钢丝混凝土
prestressed concrete work 预应力混凝土工作；预应力混凝土工程
prestressed connection 预应力连接
prestressed construction 预应力施工；预应力工程
prestressed continuous beam 预应力连续梁
prestressed die insert 预应力模具镶块
prestressed double tee 预应力双T形构件
prestressed elastic neoprene tube expansion joint 预应力弹性胶管伸缩缝
prestressed-ferro-cement ship 预应力钢丝网水泥船
prestressed ferro-concrete ship 预应力钢丝网混凝土船
prestressed folded plate 预应力折板
prestressed girder 预应力梁
prestressed glass 预应力玻璃
prestressed ground anchor 预应力地锚
prestressed H beam column 预应力H形截面梁柱
prestressed hollow beam 预应力空心梁
prestressed hollow core slab 预应力空心板
prestressed in pairs 成对地施加预应力(的)
prestressed layer 预应力层
prestressed lightweight concrete 预应力轻质混凝土
prestressed lightweight concrete roof slab 预应力轻质混凝土屋面板
prestressed loss 预应力损失
prestressed masonry tank 预应力圬工水罐；预应力圬工水箱；预应力圬工水池子
prestressed medium 预应介质
prestressed member 预应力构件
prestressed membrane 预应力薄膜
prestressed monolithic concrete wall 预应力整体混凝土墙
prestressed mortar 预应力灰浆；预应力砂浆
prestressed normal concrete 预应力普通混凝土
prestressed pile 预应力桩；预应力钢桩
prestressed plank 预应力厚板
prestressed precast beam 预应力预浇梁
prestressed precast concrete construction 预应力预浇混凝土结构；预应力预浇混凝土建筑
prestressed precast concrete member 预应力混凝土预制构件
prestressed precast concrete slab 预应力混凝土预制板
prestressed precast concrete system building 预应力预浇混凝土装配式建筑
prestressed prefabricated concrete floor 预应力预制装配式混凝土地板；预应力预制装配式混凝土楼板
prestressed redistribution 预应力重分布
prestressed reinforced concrete 预应力钢筋混凝土

prestressed reinforced concrete beam 预应力钢筋混凝土梁
prestressed reinforced concrete bridge 预应力钢筋混凝土桥
prestressed reinforced concrete girder 预应力钢筋混凝土大梁
prestressed reinforced concrete pipe 预应力钢筋混凝土管
prestressed reinforced concrete sheet pile 预应力钢筋混凝土板桩
prestressed reinforced concrete sleeper 预应力钢筋混凝土轨枕
prestressed reinforced concrete tie 预应力钢筋混凝土轨枕
prestressed reinforcement 预应力(钢)筋
prestressed road 预应力道路;预应力(混凝土)路
prestressed rock 预应力岩石
prestressed rock anchor 预张紧的岩石锚碇;预应力岩石锚栓
prestressed rock bolt 预应力锚杆
prestressed roof skin 预应力屋面
prestressed shell 预应力薄壳
prestressed stand 预应力机架
prestressed steel 预应力钢筋
prestressed steel bar drawing jack 预应力钢筋张拉千斤顶;张拉预应力钢筋千斤顶
prestressed steel bar tensioning machine 预应力钢筋张拉机
prestressed (steel piston) ring 预应力活塞环
prestressed steel profile 预应力钢筋弧线
prestressed (steel) wire 预应力钢丝
prestressed strand 预应力钢索;预应力钢丝绳
prestressed stranded steel wire 预应力钢绞线
prestressed structure 预应力结构
prestressed tendon 预应力钢丝束;预应力钢筋束
prestressed tieback 预张紧的牵索;预张紧的拉条;预张紧的拉杆
prestressed tie rod 预应力式拉杆
prestressed tower 预应力井架
prestressed unit 预应力构件
prestressed wall 预应力墙
prestressed wire 预应力锚索
prestress fabricator 预应力制作者
prestress forming 预应力成型;预模压加热蠕变成型
prestress-free without bond 无黏结预应力
prestress(ing) 预应力;预拉伸
prestressing anchor 预应力锚杆
prestressing anchorage fixture 预紧的锚固件;预紧的锚固装置
prestressing area 预加应力的面积
prestressing bar 预应力钢筋
prestressing bed 预应力张拉台;预应力台座;预加应力座;预加应力台
prestressing block 预加应力块体
prestressing by electric(al) heating 电热法预加应力
prestressing by heat 加热法预加应力
prestressing by winding 绕线预应力
prestressing cable 预应力钢索;预应力钢丝绳
prestressing camber 预应力反拱
prestressing ceramics 预应力陶瓷
prestressing distance 预加应力的距离
prestressing distance increment 预加应力的距离增量
prestressing duct 预加应力的(通风)管道
prestressing element 预应力构件
prestressing engineering 预加应力工程
prestressing equipment 预应力设备;预应力张拉设备
prestressing factory 预应力工厂
prestressing force 预应力;预加应力
prestressing gang 预应力工作队
prestressing iplaster strip 预应力壁柱板条
prestressing jack 预应力千斤顶;预加应力用千斤顶
prestressing lesene 预加应力壁柱
prestressing line 预应力张拉线
prestressing load 预张紧荷载;预张紧力
prestressing loss 预应力损失
prestressing loss due to creep 徐变引起的预应力损失;蠕变引起的预应力损失;预应力徐变损失
prestressing loss due to friction 摩擦引起的预应力损失;预应力摩擦损失
prestressing loss due to shrinkage 收缩引起的预应力损失;收缩引起的预应力损失;预应力收缩损失

prestressing loss due to slip 滑动引起的预应力损失
prestressing loss due to slip at anchorage 锚固滑动引起的预应力损失;预应力锚固滑动损失
prestressing material 预应力材料
prestressing mechanism 预应力机构
prestressing method 预加应力方法;预加应力法
prestressing moment 预加应力力矩
prestressing mo(u)ld 预加应力模具
prestressing order 预加应力次序
prestressing plate 预应力板
prestressing process 预加应力工艺过程;预加应力工艺步骤
prestressing reinforcement 预应力筋腱;预应力钢筋
prestressing rod 预应力杆
prestressing steel 预应力钢筋;预应力形钢;预变形钢
prestressing steel tendon 预应力钢筋
prestressing strand 预应力钢缆线
prestressing stress 预先拉伸应力
prestressing system 预张系统;预应力系统
prestressing table 预应力张拉台(钢筋)
prestressing technique 预应力工艺;预应力技术
prestressing tendon 预应力钢丝束;预应力筋束;预应力筋腱;预应力钢筋
prestressing tool 预应力工具
prestressing tube 预制管段
prestressing value 预应力值
prestressing wedge 预加应力用楔块
prestressing wire 预应力索线;预应力钢丝
prestressing wires in pairs 成对的预应力钢绞线
prestressing with bond 内传力法预加应力
prestressing without bond 无黏结预应力;外传力法预加应力
prestressing with subsequent bond 复传力法预加应力;后成黏结预加应力
prestressing yard 预加应力场地
prestressing zone 预加应力区
prestress with bond 有握裹力预应力;有黏结预应力
prestress without bond 无握裹力预应力;无黏结预应力
prestretched strand 预拉钢(绞)索
prestretched wire 预拉钢丝
prestretching 先张拉;预先拉伸;预伸;预拉;预加应力
prestretching cable 预伸索缆
prestretching process 预伸过程
prestripped bench 超前剥离平台
prestripper 前剥离器
prestripping 预剥离
prestripping acceleration 预剥离加速
prestrobe delay 预选通脉冲延迟
presubcript 前下标
presume 推断
presumed total loss 推定全损
presumption 假设;假定;或然率;推定;推测
presumption bearing value 假定承载力
presumption of value 价值推定
presumptive address 假定地址;基准地址;基地址【计】;基本地址
presumptive area 预定区域
presumptive bearing pressure 推断承载力
presumptive epidermis 预定表皮
presumptive error 预定误差;假定误差;设想误差
presumptive evidence 推定的证据
presumptive income 推定收入
presumptive instruction 原始指令;假定指令;基本指令
presumptive test(ing) 假定试验;推定试验
presumptive waste discharge 估算排污量
presupercript 前上标
presuperheater 预过热器
presupernova 爆前超新星
presuppression 预抑制
presweeping 工前扫测
preswirl 进气预旋
presynchronization 预同步
presynthesis phase 合成前期
pretan(ning) 预鞣
pretanning agent 预鞣剂
pretax accounting income 税前账面收益
pre-tax allowance 课税前津贴

pre-tax income 税前收入
pretax profits 税前利润
pretectonic crystallization 构造前结晶
pretelescope astronomy 望远镜前天文学
pretence of colo(u)r 保色性
pretender 招标前;标底
pretender estimate 标底
pre-tender meeting 标前会(议)
pretend to be famous brand 假充名牌
pretension 先张(拉);预张力;预拉力;预紧力;预加张力;预加载;预拉伸
pretensioned 先张的
pretensioned bar 预张拉钢筋
pretensioned beam 先张梁
pretensioned bed 先张法(预应力)台座
pretensioned binding 预张绑线(电机)
pretensioned bolt 预张拉锚杆
pretensioned concrete 先张预应力混凝土;先张法混凝土
pretensioned concrete cylinder pipe 预张拉混凝土钢心管
pretensioned concrete pipe 预张法混凝土管道
pretensioned construction 先张拉结构
pretensioned force 预应力
pretensioned member 先张法预应力构件
pretensioned pipe 预张法预应力混凝土管道;先张管道
pretensioned prestress concrete 先张法预应力混凝土
pretensioned prestressed concrete girder 预张拉的预应力混凝土大梁
pretensioned prestress(ing) 先张法预(加)应力
pretensioned prestressing concrete 先张法预应力混凝土
pretensioned reinforcement 预张拉钢筋
pretensioned steel 预张钢筋;预拉钢筋;预张拉钢筋
pretensioned system 先张法
pretensioned tendon 先张钢筋束;先张拉线束;先张拉钢筋束;预应力钢丝索
pretensioned wire 先张拉钢丝;先张钢丝;预张钢丝;预拉钢丝;预张伸
pretensioning 先张法(钢筋);先张技术;预拉;钢筋先张(法)
pretensioning bed 预张台;先张法台座;先张法池座;预应力张拉台
pretensioning bench 预张台;先张法(预应力)台座
pretensioning(method) 先张法
pretensioning prestressed concrete 预张法预应力混凝土
pretensioning prestressed pile 先张法预应力桩
pretensioning process 先张法
pretensioning strand 先张拉的钢绞索
pretensioning system 预应力法先张法
pretensioning type 先张法
preterminal 终端前的
pretersonics 微声电子学
pretest column 预试柱
pretest conditioning 试验前的调整
pretested packing 预制填充物
pretest(ing) 预试;预检;预先试验;预先检验;事先试验
pretesting footing 预试基础
pretest inspection 试验前检查
pretest pile 预试桩
pretest pile underpinning 预试桩托换
pretest production 生产上试用
pretest risk function 验前风险函数
pretest treatment 试验前准备;试验前处理
pretest underpinning 预试打支撑桩;预试托换基础
prethawing 预溶化
prethickening 预浓缩
prethreshold 阈前
prethreshold phenomenon 阈的现象
pretightening 预浓缩
pretightening force 预紧力
pretightening load 预紧力
pretightening unit sealing load 预紧密封比压
pretightening up force 预紧力
Pretiglian cold epoch 前梯格林寒冷期
pretil 土坯女儿墙
pre-tiled partition 先贴好瓷砖的隔墙
pretimed mainline control 定时式主线控制
pretimed metering mode 定时限流法
pretimed signal 定时信号

pretopping 挖前切顶;提前打顶
pretorium 官方住宅(古罗马省级官员的)
pretravel 预行程
pretreat 预处理;预先处理
pretreated ore 预处理过的矿石
pretreated rock 预处理过的岩石
pretreated seed 提前处理种子
pretreated surface water 预处理过的地表水
pretreated water 预处理水;预处理过的水
pretreater 预处理塔;预处理器
pretreating 预加工;初加工
pretreatment 预加工;预处置;预处理;前处理;预先处理
pretreatment agent 预处理剂
pretreatment coating 预处理涂料
pretreatment cost 预处理费用
pretreatment filming agent 预膜剂
pretreatment fluosilicate 预处理氟硅酸盐
pretreatment of hot metal 铁水预处理
pretreatment of samples 试样前处理
pretreatment pontoon 预处理的平底船
pretreatment primer 预处理涂底料;前处理底漆
pretreatment procedure 预处理方法
pretreatment process 预处理工艺
pretreatment river water 预处理过的河水
pretreatment tank 预处理池;预处理槽
pretreatment unit 预处理装置
pretreatment wallpaper 预整理壁纸
pretrigger 预置触发器;预触发器;预触发;前置触发器
pretrigger data 前置触发数据
pretriggering 预先触发
pretriggering pulse 预触发脉冲
pretrigger time selector 预触发时间选择器
pretrimmed wallpaper 预整理壁纸
pre-trip inspection 开航前检查;运输前检查;启程前检查
pretty cold 挺冷
pretty good privacy 加密程序
pretuning 预先调谐
pretuning control 预调控制机构
preupset 预镦粗
pre-Ural ocean 前乌拉尔洋
prevacuated sampling tube 预真空样品瓶
prevail 流行
prevailing condition 主导条件
prevailing current 主导水流;盛行流
prevailing direction of current 主流向
prevailing direction of wind 主导风向;常风向
prevailing price 现行价格;普通价格;通行价格
prevailing rate 现行汇率;通行汇率
prevailing rate of interest 现行利率
prevailing (rock) formation 主要岩层
prevailing stem 支配木
prevailing torque 最常作用的扭矩
prevailing value 一般数值
prevailing visibility 主导能见度;盛行能见度
prevailing wage 现行工资
prevailing wave 盛行波(浪);常浪
prevailing wave direction 常浪向
prevailing weather condition 主要气候条件;经常气候条件
prevailing westerlies 西风带;盛行西风带
prevailing wheel load 经常轮压;经常车轮荷载
prevailing wind 主风;主导风;盛行风;常风
prevailing wind direction 主导风向;盛行风向;常风向
prevailing wind runway 主导风跑道
prevailing wood species 主要材种
prevail price 时价
prevalence-duration-intensity index 优势—持续—强度指数
prevalence rate 流行率
prevalence survey 现况调查;现患调查
prevalent pollutant 一般污染物;盛行性污染物
prevaluation test 预评价试验
prevalue 预置值
prevalution test 模拟试验
prevarication 不相关性
prevent 预防;阻挡;防碍
preventability 可制止性;可预防性
preventable 可预防的
preventable accident 可防止的事故
prevent and control pollution 防治污染

prevent and kill off insect pests 防除害虫
preventation of corrosion by metallic coating 金属敷层防蚀法
preventative maintenance 预防性维护;定期修理
preventative measures of sea water intrusion 防止海水入侵的措施
preventative remedy 预防性修补;预防性补救;预防性医疗
preventative resistance 防护电阻
prevent condensation 防止冷凝
prevent contamination 防止污染
prevent corrosion 防止腐蚀
prevent disease 防病
preventer 阻止物;止动索;警告装置;辅助索;防喷器;防护设备;保险索;安全索
preventer guy 辅助稳索
preventer hawser 解链挂索
preventer lift 辅助千斤索
preventer of double ram type 双柱塞式防喷器;双闸板型防喷器;双闸板式防喷器
prevent erosion and leaching 防止侵蚀和淋溶
preventer pin 止动销;保险销;安全销(钉)
preventer shroud 加强护索
preventer stay 辅助支索
preventer winch 保险索绞车
prevent frostbite 防冻
prevent heatstroke 防暑
preventing and curing disease 防病治病
preventing casehardening 防止外干里湿
preventing groundwater contamination 防止地下水污染
preventing groundwater continuously drawdown 防止地下水位持续下降
preventing land subsidence 防止地面沉降
preventing precipitation 防止沉淀
prevent intensive washing 防止强烈冲刷
prevention 预防;防止;防护
prevention against electromagnetic radiation 电磁辐射防治
prevention against mice 防鼠
prevention against rats 防鼠
prevention against water-logging 防渍
prevention and control of pollution 污染防治
prevention and control of pollution with afforestation 绿化防治污染
prevention and control of pollution with organism 生物防治污染
prevention and control of soil pollution 土壤污染防治
prevention and treatment of damage to water 防治水害
prevention first 预防为主
prevention for falling stone 落石防护设施;防止落石
prevention measure 安全技术
prevention measurement 安全措施
prevention mesh for falling stone 拦石网
prevention of accidents 安全技术;事故预防;防止事故
prevention of alkali 防碱
prevention of burglary 防盗
prevention of corrosion 防腐(蚀)
prevention of cycling 循环防止
prevention of damage 预防破坏;预防损坏
prevention of damage to salty foundation soil 盐渍土地基危害的预防
prevention of decay 防止腐烂;防朽
prevention of deviation drilling 定向钻进;防偏钻进
prevention of disasters 灾害预防
prevention of fire spreading 火灾蔓延预防
prevention of infectious disease 传染病预防
prevention of losses 预防损失
prevention of loss of material 防止材料流失
prevention of marine pollution 防止海洋污染
prevention of misappropriation of funds 防止滥用基金
prevention of overspending of appropriations 防止超支
prevention of repetitive clear of a signal 防止重复
prevention of resonance 防止共振
prevention of risks 风险的预防
prevention of skin formation 防止表皮形成
prevention of slide 滑坡(的)防治
prevention of static electricity 防静电

prevention of toxicants 防毒
prevention of unauthorized access to information 对信息的未授权访问的预防
prevention of unwise or inappropriate expenditures 防止不当支出
prevention of weathering 防止风化的措施
preventive 预防剂;预防法;防止的
preventive belt 预防(林)带
preventive cost 预防费用
preventive cost of environment(al) disruption 环境破坏的预防费用
preventive depot maintenance 预防性后方保养
preventive device 防护设备
preventive expenditure method 防护费用法
preventive inspection 预防性检查
preventive maintenance 预检(修);预防性维修;预防性检修;预防性(技术)保养;预防性安排;预防维修;预防检修;再生修理;再生保养;定期修理
preventive maintenance contract 预防维护合同
preventive maintenance inspection 预防性维修检查;预防性维修检查;预防性技术保养检查
preventive-maintenance overhaul 预防性大修
preventive maintenance program(me) 预防性保养计划
preventive maintenance roster 预修项目表
preventive maintenance schedule 预防性维修计划
preventive maintenance time 预防性维修时间;预防性维护时间;预防性安排时间;预防维护时间
preventive management 事前预防管理
preventive measure 防范措施;预防(性)措施;防护(性)措施
preventive measures against rats 防鼠措施
preventive medicine 预防医学
preventive officer 海岸缉私队
preventive overhaul 预防修理;预防性修理;定期修理
preventive patrolman 护林员
preventive repair(ing) 预防(性)修理
preventive safety 预防性安全措施
preventive safety measures 预防性措施
preventive sanitary supervision 预防性卫生监督
Preventive Service 沿岸海关缉私队
preventive servicing 预防保修
preventive solution 预防溶液
preventive stratum 预防层
preventive treatment 预防法;防治
preventive wood protection 预防性的木材防护
prevent noise 防噪声
prevent of accidents 技术保安措施
preventor 防护器
prevent oxidation 防止氧化
prevent water creep 防止渗水
prevent waterlogging 防涝
prevernal 早春的
prevernal aspect 早春季相
preview 预检;预观;预测
preview control 预看控制
preview gain 预看增益
previewing 预视;预观察
preview matrix 预看混合
preview monitor 预检监视器
preview monitoring 预看监视
preview selector 预看选择器
preview switch 预看开关
preview switching 预检接通
previous balance method 前期余额法
previous calcining 预煅烧
previous church 先期教会;先期教堂
previous cleaning 初洗;初清理
previous consolidation 先期固结;预先固结
previous decade 前十年
previous fracture network 先存断裂网络
previous input state 先前输入态
previous line prediction 扫描前行预测
previously 预先
previously arranged signal 约定信号
previously designated 预先指定的
previously digested sludge 预消化污泥
previously vaporized charge 预先汽化进料
previous mode 前方式
previous output state 先前输出态
previous pass 预轧道次
previous quotation and date 先前报价和日期
previous total 前班累计

prevision for expenses 预提费用
prevue 预告片
prevulcanization 预先硫化
prevulcanize 早期硫化
prevulcanized latex 预硫化胶乳
pre-war housing 战前公寓；战前住房
prewarming 预热
prewarning 预警
prewasher 预洗机
prewash(ing) 预先水洗；预洗；预冲洗
prewashing machine 预洗机
prewash phase 预洗期
prewash tank 预洗槽
prewatering 预先湿润；预先浸湿；提前灌水
prewatering granulating technology 预加水成球技术
prewatering nodulizing technique 预加水成球技术
prewave 前置波
preweld 格构屋面(一种钢制的)；预焊；焊接前
pre-weld heat treatment 烧焊前热处理；焊接前处理
prewelding 预焊
preweld interval 焊前间隙
prewet 预先湿润
prewet screen 预湿筛
prewetting 预先浸湿；预湿；预浸水
prewetting agent 预湿剂
prewhirl 预旋转；预旋
prewhirler 预旋器
prewhirl vane 预旋叶片
prewhitening 预白化
prewhitening filter 预白噪声化滤波器；前置白噪声化滤波器；白化滤波器
prewilted silage 经过调萎的青储[贮]料
prewire 预制线
prewired 预配线的
prewired option 预配选择
prewired program(me) 保留程序
prewood 浸脂木材；浸胶压缩材料；浸胶木材
preworking a block 预办闭塞
pre-work of railway construction 铁路建设前期工作
prey 被捕食者
Prey anomaly 普赖异常
Prezagros basin 扎格罗斯山前盆地
prezenta 黏胶纤维素
prezone 前区
prian 软白黏土
price adaptation 价格适应
price adjustment 价格调整；调整价格；调价
price adjustment clause 价格调整条款
price adjustment formula 调价公式
price adjustment lag 价格调整上的时滞
price adjustment mechanism of financial statement 财务报表物价调整机制
price advancing 价格上涨
price after tax 纳税后价格
price-aggregate 价格累计；价格积数
price agreed upon 协定价格
price agreement 价格协定
price analysis 价格分析
price and income freeze 冻结物价和收入
price and output effect of external diseconomy 外部不经济性的价格和产出效应
price appeal 索取价格
price arbitrage 价格套购
price arbitration 议价；公断价格
price asked offer 开价
price at factory 工厂价
price at the time of order 订货时价格
price bargain 讨价还价
price basis 运费认价标准
price before tax 税前价格
price behavior 价格变化情况
price beneficiation 价格补偿
price bonification 价格补贴
price calculation 价格计算
price card 价标；标价
price cartel 价格协定；价格卡特尔
price catalog(ue) 定价表；价目单；价目表；价格表
price ceiling 限定价格；最高限价；最高价(格)；价格上限
price change 价格变动；调整价格
price change approach 价格变动法

price change clause 价格变动条款
price change ticket 改价单；价格变化标签
price check 价格检查
price comparability 价格可比性
price competition 价格竞争
price concession 价格核准；让价；特许价格
price condition 价格条件
price-consumption curve 价格—消费曲线
price contract 价格契约；价格合同；定价合同
price control 控制物价；控制价格；价格控制；价格监督；价格管制；价格管理；物价控制；物价管制
price control of export commodities 出口商品价格管理
price control of import commodities 进口商品价格管理
price control of material and fuel 材料燃料价格的控制
price converting index 物价折算指数
price coordination method 价格协调法
price current 价目表；定价表；市价表
Price current meter 旋杯式流速仪；旋杯式海流计；普赖斯海流计；普赖斯(旋转式)流速仪
price cutting 削价；减价
price-cutting competition 削价竞争
price cycle 价格周期；价格循环
priced bid 标价的投标
priced bill 标价清单；单价表；工程概算表；标价工程数量清单
priced bill of labo(u)r and material 工料定价单
priced bill of material and labo(u)r 工料定价单
priced bill of quantities 工料定价单；工程量标价清单；工程概算表；投标单价；标有价格工料单；标价工程量清单
priced catalog of labo(u)r and material 工料定价单
priced current 定价表
price-delivered terms 价格交货条件
price description record 价格说明记录；价格描述记录
price determination 决定价格
price difference 价格差异；价差；差价
price difference of material 材料差价
price differentials between commodities 商品差价
price differentials of commodities 商品差价
price differentiation 分化价格
price discrimination 价格歧视；价格差别；区别对待的价格
priced list of labo(u)r and material 工料定价单
priced list of material and labo(u)r 工料定价单
price downslide 价格剧降
price dumping 价格倾销
price duty paid 完税货价
price earning ratio 高变动率；价格—收益比(率)
price elasticity 价格弹性
price elasticity of demand 需求价格灵活性；需求的价格弹性
price elasticity of demand for import 进口需求的价格弹性
price elasticity of import and export 进出口价格弹性
price equilibrium 价格平衡；价格均衡
price equity 价格权益
price escalation 价格升级
price escalation clause 价格自动调整条款；价格逐步上涨的条款
price escalation rate 价格上涨率
price escalator clause 价格自动调整条款；价格伸缩条款；价格调整条款
price evaluation and classification according to quality 按质分等论价
price expectation 预期价格；价格预测
price-fixer 卡特尔定价者
price fixing 价格协商(投标人之间的)；限价；价格垄断；默许订价
price fixing agreement 固定价格协议
price flexibility 价格伸缩性
price floor 最低限价；最低价(格)；价格下限
price fluctuation 价格涨落；价格浮动；价格波动；价格变动；物价波动
price fluctuation reserves 物价变动准备
price for account 代销价格
price for factory 厂盘
price for forward delivery 预付价格
price for tax assessment 完税价格

price for the indicated year 当年价格
price freeze 价格冻结
price gain 价格收益
price gap 价差；差价
price gap between different localities 地区差价
price generally charged 一般加法定律；一般价格
price group 物价组；价格组
price guideline 价格限度
price has skyrocketed 价格猛涨
price has tobogganed 价格突然下降
price hike 价格增加；价格暴涨；价格上涨
price idea 价格意见
price including commission 含佣金价格；包含手续的价格
price including tax 税金在内的价格
price index 价格指数；物价指数
price index number of primary commodities 初级商品价格的指数
price index number of service expenditures 服务性支出价格指数
price index of service items 服务项目价格指数
price indication 价格表示；参考价格
price inelasticity 价格无弹性
price inflation 价格膨胀；物价膨胀
price involving foreign countries 涉外价格
price is on high side 价格偏高
price is rather stiff 价格相当高
priceite 白硼钙石
price leader 先导价格；领头定价者
price leadership 先导价格制；领头定价；价格上的领先地位
price level 价格水准；价格水平；物价水准；物价水平
price level adjusted statement 按价格水平调整的报告
price level shifts 按物价水平调整
price limit 限价
price line 同一定价的一批货物
price lining 底价
price list 价目单；价目表；价格表；报价单
price loco 当地交货价格
price loss 价格损失
price lower 矿产价格偏低
price making according to average cost 平均成本定价
price making according to burden capability 负担能力定价
price making according to marginal cost 边际成本定价
price making according to supply and demand relations 供求关系定价
price margin 价格差额
price mark 价格记号
price mark-down 价格折扣
price markdown percentage 价格降低率
price markup equation 提高标价的方程
price measure 价格尺度
price mechanism 价格结构；价格机能；价格构成
price memory 价格存储器
price movement limit 价格移动限额
price movement trend 价格动向
price not-competitive 无竞争力的价格；非公开招标的价格
price of an hour's leisure 一小时空闲时间的价格
price of award 决标价
price of construction and installation 建筑安装工程造价
price of day 日牌价
price of delivery to destination 目的地交货价(格)
price of equipment should be set at the normal allocation price 按国家调拨价格作价
price of factory 厂盘
price of land 土地价格
price of luggage and parcel traffic 行包运价
price of money 延期日息；利率；贷款利率
price of production 生产价格
price of products 产品价格
price of share 股票价格
price of the selling party 卖方价格
price of transaction 交易价格
price of waiting 等待价格
price on a uniform basis 统一价(格)
price oneself out of the market 漫天要价
price on spot 现货价(格)

price ordinance 价格法令
price out of the market 价高无销路
price-output determination 价格—产量的决定
price-output policy 最宜价格产量政策
price parity of commodities 商品比价
price parity scissors 比价剪刀差
price partiy in the exchange 交换比价
price per Bushel 每普式耳价格
price performance 性能价格比;价格成绩
price per kilowatt-hour 电度单价;电度单价
price policy 价格政策;物价政策
price preference 价格优惠
price proportion 单价比;价格比值
price-push inflation 价格推动型通货膨胀
price quotation 估价单;报价(单);价格行情
price range 价格变动幅度
price rates system 价格比率制
price ratio 比价
price ratios between commodities 商品比价
price reform 价格改革
price relative 价比
price restraint 矿产价格稳定
price revision clause 价格修正条款
price rigidity 价格稳定性;价格固定性
price rigidity in the downward 价格稳定趋势
price ring 价格控制集团
Price rotary current meter 普赖斯旋转式流速仪
Price rotary excavator 普赖斯旋转掘进机
prices and incomes policy 物价与所得政策
prices at a discount 折扣价格
price schedule 价格表
price scissors 价格剪刀差
price-sensitive 价格敏感性
prices fluctuating within a certain range 幅度价格
prices for specialization and cooperation 专业化协作价格
price sheet 报价单
price signal 价格信号
prices inflated 物价上涨
prices in local currencies 以本国货币表示的价格
prices of commodities 物价
prices of miscellaneous goods 小商品价格
price spiral 轮番提价;物价螺旋上升
price spread 价格差距
price spread effect 矿产价格刺激
price stabilization 稳定物价
price-stabilizing 价格竞争
price standard 价格标准
price strategy of mineral commodities 矿产价格对策
price structure 物价结构;价格结构
price subsidy 价格补贴
price support 价格支持;价格补助金
price support program(me) 价格支持方案
price tag 价格标签;标价条
price tendered 投标价格
price terms 价格条款;价格条件
price ticket 标价条
price trend 价格动向
price variance 价格差异
price war 削价竞争
price-wealth pair 价格—资源对
pricing 价格形成;计价;定价;标价
pricing department 定价部门;物价局
pricing entire product package 全部产品综合订价
pricing information 询问定价
pricing of by-product inventory 副产品存货计价
pricing policies of resources 资源的定价政策
pricing policy 物价政策
pricing structure 价格体系
pricing suboptimization 多价算法优化
pricing system 价格体系
pricing the inventory 库存计价
pricing vector 评价矢量
prick 钉伤;刺痕
prick ear 竖耳
pricked rendering 穿孔抹灰
pricker 锥子;火钩;通气针;刺针;刺孔针;触针;冲子
pricker microscope 刺点显微镜
pricker point 刺针尖
pricker staff 探测标杆
pricking 刺孔;试ань;探测;刺伤
pricking device 刺点装置
pricking microscope 刺点显微镜

pricking needle 刺针
pricking needle point 刺针尖
pricking pin 刺针
pricking resistance 抗起毛性
pricking up 抹灰底层;粗涂(抹);括粗层;板条抹灰打底;抹灰打底;灰浆层划痕
pricking-up coat 划痕打底层;粗涂层;靠骨灰【建】;板条抹灰打底层;抹灰打底
prickly sago palm 刺细谷
prick off 划出
prick out 划出;点出
prick photograph 刺点片
prick point 刺点
prick post 桁架中间杆;主次双柱式;框架侧柱;间柱;双柱桁架立杆;穿柱
prick punch 冲孔錾;中心冲头;针孔冲;刺孔器;冲心凿;冲心錾;冲孔器
prick punch mark 定中心点
prick-through type lampholder 插入式灯座
prick transfer 转刺(点)【测】
prick up 打底子
pride of the morning 晨雾
priderite 柱红石
pridinol 二苯哌丙醇
prie-dieu 祷告椅;祷告台
priest's choir 牧师唱诗班
priest's door 教堂中祭出入的门
priest's house 牧师住宅
prifinium bromide 吡芬溴铵
prill 造粒;金属颗粒;富矿颗粒
prilling 造粒工艺;造粒;成球
prilling spry 造粒喷头
prilling tower 造粒塔
prillion 渣锡
prima 初波
prima beneficiary 人寿保险赔偿的第一受益人
primacord 起爆引线;引爆索;导火索;导爆索;传爆索
primacord explosive 引爆炸药
primacy 首位(度)
primadet 起爆体
prima evidence 表面证据
prima facies case of damage 初步推定有损害
prima facies evidence 初步证据
primage 运费回扣;运费补贴;货主送给船长的酬金;水分诱出量;水分带出量;额外酬金
primal 最初的
primal algorithm 原始对偶算法
primal constraint 原有约束
primal dual algorithm 原有对偶算法
primal dual type 主对偶形
primal environment 原始环境
primal feasible condition 主容许条件
prima liability 表面责任
primal objective function 原始目标函数
primal problem 主要问题;主问题
primal simplex method 主单纯形法
primal system 原来系统
primaquine 伯氨喹啉;伯氨喹
primaquine phosphate 伯氨喹啉磷酸盐
primaries 比较原色
primarily crushed 初步破碎的;初始破碎的
primary 基本的;坯料;第一位的;第一阶;第一级;初生;初期的;初等项;初等式;初等量;初等的;初次
primary aberration 第一级像差;初级像差
primary aberration contribution 初级像差贡献量
primary absorption factor 初始吸收因子;初始吸收因数
primary access 直接存取
primary account 初级账户;主要账目
primary accumulation 原始积累
primary accumulator 主累加器
primary accuracy 一等精度
primary acids 伯酸
primary action 直接作用
primary additive colo(u)rs 相加合成基色
primary adsorption 一级吸附
primary aerodynamic characteristic 原始气动力特性
primary afferent fiber 初级传入纤维
primary aftershock 第一期余震
primary agricultural products 初级农产品
primary air 原空气;一次助燃空气;一次空气;一次风;主空气;初级空气

primary air duct 主空气通路
primary air enter 主空气进口
primary(air) fan 一次风机;主送风机
primary air fan coil system 风机盘管加新风系统
primary air inlet 主进气口
primary air pollution 同源空气污染质
primary air regulator 一次调风门
primary air system 新风系统
primary air temperature 一次风温度
primary air velocity 一次风速
primary alarm circuit 主报警电路
primary alarm system 主报警系统
primary alcohol 一级醇;伯醇
primary algebra 准则代数
primary alkyl peroxide 伯烷基过氧化合物
primary allergen 原生变应素
primary allochthonous coal 原始异地生成煤
primary alumin(i)um 原生铝
primary alumin(i)um pig 初次铝块
primary amine 一级胺;伯胺
primary amount of pollutant 初始污染物量
primary amplitude 原始辐角
primary amyl alcohol 伯戊醇
primary annulus 初级齿圈
primary anomaly 原生异常
primary antenna 照明天线
primary application 初始应用
primary application block 初始应用块
primary application program(me) 初级应用程序
primary approximate value 初始近似值
primary approximation 初始近似
primary aquatic plant 原始水生植物
primary arc 主山岛弧;主岛弧【地】
primary area 主区
primary assets 主要资产
primary association 主始群丛
primary austenite 初生奥氏体
primary autopilot 主要自动驾驶仪
primary axis 主轴
primary backing 背衬基材;基底布(簇绒地毯的)
primary back reaction 初级反作用
primary bainite 初生贝氏体
primary ball mill 一次球磨机
primary basalt 原生玄武岩
primary battery 原电池组;一次电池组
primary beam 原始束流;原射线束;有主束流;梁;照明束;初级射束
primary bedroom 主卧室
primary benchmark 一级水准点;基本水准点;基本水准标石;水准基点
primary bending stress 初次弯曲应力;一次弯曲应力
primary benefit 直接受益;一级受益收入
primary bin 原矿仓
primary binder 主黏合剂
primary blasting 岩石移动错位爆破;岩石初爆;原体爆破;一次爆破;初始爆破
primary blasting explosive 一次爆破炸药;基本炸药;主要炸药
primary blast injury 直接的冲击波伤害
primary block 一次群;基群
primary blower 一次风鼓风机
primary blowhole 一次气孔
primary body 主星;主体
primary bond 主价键
primary bow 主虹
primary boycott 初级抵制
primary brake-shoe 自紧制动蹄片
primary branch 第一级支管;主支管
primary branch drain 主分支排水管
primary break 一次断头
primary breaker 一次破碎机;总电路断流器;粗碎机;初轧机;初碎机;初切碎机;初级破碎机
primary breaking 一次爆破;粗碎;初碎
primary brown earth 原生棕壤
primary cable access 主干电缆接入
primary calcium phosphate 一代磷酸钙
primary calibration method 一级校准法
primary calorifier 基本加热器
primary canal-distributor 干渠配水管;干管配水管
primary capacitance 初级线圈电容
primary capacitor 高压电容器
primary carbon 原始有机碳;伯碳
primary carbon atom 伯碳原子

primary carburet(t)or 起动汽化器
primary carrier 原载波
primary cavitation 初ális蚀;初始空化
primary cell 原电池;一次电池;初选浮选机
primary cell supply 一次电池供电
primary cementing 初次注水泥
primary cementite 一次渗碳体;初生渗碳体
primary center 原始中心;基层中心
primary chain 一级类链
primary chamber 主燃烧室
primary characteristic 原有特性
primary characteristic curve of hydraulic coupling 液力耦合器原始联合特性曲线
primary charging 初次充电
primary choke 主节流口
primary circle 主圈;基圆;基本圈
primary circuit 原电路;一次回路;一次电路;初级电路
primary circuit diagram 一次电路原理图;主要接线图
primary circuit routing 主要路由
primary circulating pipe 第一循环管
primary circulating pipe system 第一循环管系
primary circulation 一级环流;一次循环;初级环流;初次循环
primary circulation pipe 一次水循环管路
primary clarification 初级澄清;初次澄清
primary clarifier 初级澄清池;初次澄清池
primary clarifier effluent 初级澄清池出水
primary classification of kerogen 干酪根原始分类
primary classifying screen 粗筛;粗分级筛
primary clay 原生黏土;一次黏土;残余黏土;残留黏土
primary cleaner 一次精选槽;初清室
primary cleaner tailing 一次精选尾矿
primary cleaning 一次精选;主选;初选;初级澄清;初次净化;初步清除;初步净化
primary clearance 主后角;初次间隙
primary clock 主钟;主时钟;母钟
primary cluster 一次簇
primary clustering 首次分簇
primary coagulant 主混凝剂
primary coagulating agent 主混凝剂
primary coast 原生海岸;幼年海岸
primary coat(ing) 沥青透层;结合层;首涂层;打底(指油漆);一次涂层;涂底层;抹灰底层;底层;底釉;底涂;底灰(指油漆);初次镀层
primary coating fiber 一次被覆光纤;一次涂覆光纤
primary coefficient of permeability 主渗透系数
primary cohomology operation 一阶上同调运算
primary coil 原线圈;原边线圈;一次线圈;一次绕组;初级线圈
primary cold trap 主冷阱
primary collecting ducts 初级集合管
primary colloidal material 原生胶体物质
primary colo(u)r 底色;本色;原色;基色;基本色
primary-colo(u)r filter 原色滤色镜
primary colo(u)r signal drive 原色激励方式
primary colo(u)r theory 元色学说
primary colo(u)r unit 基色单元
primary combustion 预燃;初次燃烧;初级燃烧
primary combustion air 初级燃烧空气
primary combustion zone 一次燃烧区;主燃区
primary comminution 初步粉碎
primary commodity 初级货物
primary communication 一次情报
primary community 原生群落
primary compensation 初次补偿
primary compensator 初级补偿器
primary compilation 编绘原图
primary component 主星;主构件
primary component of error 误差的主要成分
primary component space 准素分支空间
primary compression failure 初始压缩破坏;主要受压破坏;受压先破坏;初始压缩破坏;初始受压破坏
primary compression failure of concrete 混凝土的初始压缩破坏
primary compression failure of soil 土的初始压缩破坏
primary concentration 原始浓度;初始浓度
primary condenser 初次冷凝器;一次冷凝器;初级电容器
primary conditioning 初调节

primary console 主控制台
primary consolidation 主固结;初始固结;初期固结;初次固结
primary constant 主要常数;起跑常数;初始常数
primary constituent 主要组分
primary constriction 初级缢痕
primary construction 原始结构;主结构
primary consumer 初级消费者
primary containment 内层安全壳
primary contaminant 原生污染物
primary content 初含量
primary contour 主等高线
primary contrast 原始衬度
primary control 一等控制;主要控制;主干控制;初期控制;初步控制
primary control circuit 主操纵系统
primary control element 灵敏元件;初级检测元件
primary control network 一等控制网
primary control point 一级控制点;一等控制点
primary control program(me) 基本控制程序;主控制程序;初级控制程序
primary control ship 主控船
primary control station 主引导站
primary converter 一段转化炉
primary coolant 一次载热剂;一次冷却剂
primary coolant circuit 一次冷却剂回路
primary coolant inventory 一次冷却剂总量
primary coolant pump 主冷却剂循环泵
primary cooler 一次冷床;初级器;一次冷却器
primary cooling system 初次冷却系统;一次冷却系统
primary cooling water 一次冷却水
primary corona assembly 初次电晕充电装置
primary cortex 初级皮质
primary cosmic radiation 原宇宙辐射
primary cosmic ray 初级宇宙射线
primary cover 初次覆盖
primary coverage 主要内容
primary cover layer 主护面层
primary crack 主要裂缝
primary cracking 初级裂化
primary crankshaft 初级曲轴
primary creep 主蠕变;第一阶段蠕变;初始徐变;初始蠕变;初期蠕变
primary crusher 初轧碎机;一级破碎机;一次破碎机;重型压碎机;粗碎机;初轧机;初碎机;重型碎石机
primary crushing 初步破碎;一次破碎;粗碎;初轧;初碎;初次破碎
primary crushing plant 粗碎设备;粗碎车间
primary crystal 初析晶;初凝晶;初晶;初次结晶(体)
primary crystal field 初晶区
primary crystalline phase 初析晶相
primary crystallization 一次结晶;初结晶;初次结晶
primary cultivation 初耕
primary culture 初级培养物
primary cup 密封皮碗;初级皮碗
primary current 一次电流;主要水流;主流;初级电流
primary current modulation 原电流调制
primary curve 原始曲线
primary cut-out 高压断路器;初级断路器
primary cutting angle 主切削角
primary cycle 主循环
primary cyclone 主气旋
primary damage 财产直接损失
primary dark space 初级暗区
primary data 原始资料;原始数据;主要资料;第一手资料;初始数据;初级资料
primary data matrix 原始数据矩阵
primary data set group 初级数据集组
primary datum 主基(准)面
primary debtor 主要负债人
primary decomposition 原始分解
primary deformation 初次变形
primary degeneration 初级退变
primary demand deposit 初级活期存款
primary demineralization system 一级除盐系统
primary departure runway 主要用于起飞的跑道
primary deposit 现金存款;原生矿床;原生沉积;初始沉积;初级存款
primary design 初步设计
primary detecting element 初级检验元件

primary detector 一次探测器
primary deuteron 初始氘核
primary dewatering 预先脱水;一次脱水
primary digital group 数字基群
primary dimensionality 原始维数
primary dip 原始倾斜
primary discontinuity 原生结构面
primary disintegrator 一级破碎机;一次破碎机
primary dispersion 原生分散;初次分散;主分散
primary dispersion halo 原生分散晕
primary dispersion pattern 原生分散模式
primary displacement 主位移
primary display 一次显示
primary distillation 初级蒸馏
primary distillation of crude oil 原油初馏
primary distillation tower 初馏塔
primary distress alerting network 主遇险报警网络
primary distribution 一次配电
primary distribution main 一次配电干线
primary distribution network 一次配电网
primary distribution of the national income 初次分配
primary distribution road 主要分流路;主要分流道
primary distribution system 一次配电系统
primary distribution trunk line 一次主配电线路
primary distributor road 主要分流路;主要分流道
primary diversion field 主备降(机)场
primary divert field 主备降(机)场
primary division 主要渔区;(比例尺的)主尺
primary documents 原始文件
primary document file 一次文献库
primary dolomite 原生白云岩
primary drilling 一次凿岩
primary drinking water regulation 饮用水基本规程
primary drinking water standards 饮用水基本标准
primary drive 初始传动
primary drying 头冲
primary drying chamber 上干燥室
primary dump system 初级转储系统
primary earnings per share 每股主要盈利额;每股基本收益
primary earthquake 初震;初始地震
primary economic activity 直接人类经济活动
primary effect 一次效果;主要影响
primary effect process 原初作用过程
primary effluent 原生污水;原次出流物;初沉池出水
primary effluent chlorine contact tank 初级污水氯气处理接触槽
primary effluent filtration 原生污水过滤
primary electron 原电子
primary element 原生要素;一次元件;基本元件
primary elevator 第一升运器
primary emergency escape hatch 主要应急离机口
primary emission 原排放空气污染物;一次排放;初级发射;初次排放
primary energy 原始能量;初始能量;初级能源
primary energy consumption 初级能量消耗
primary energy fuel 一次能源燃料
primary energy sources 一次能源
primary environment 原生环境;第一环境
Primary era 古生代
primary evidence 主要证据
primary excavation 原土开挖;一次开挖;一次回采;初挖;初期开挖;初次开挖;原土石层开挖;原始洞穴
primary exception vector 第一异常向量
primary excitation 初级激发
primary explosive 引炸药;起爆炸药
primary explosive ratio 一次爆破单位炸药消耗量
primary exposure 一次接触
primary exposure limiting value 一次接触限值
primary external nodal point 主外结点
primary extinction 初级消光;初次消光
primary fabric 同生组构
primary fabrication 初加工
primary factor of production 初级生产要素
primary failure 自身故障;首次损坏
primary fan 主扇;主风机
primary fault(ing) 原始故障;原断裂
primary fault test 一次电故障试验
primary feed 一次辐射器;主馈电器;主辐射器
primary feedback 主回授
primary feeder lateral 一次进水支管
primary fermentation 一级发酵

primary ferrite 初生铁素体
primary fiber 一次纤维
primary field 一次场
primary filament 一次纤维
primary file 原始文件
primary filling 初装填
primary filter 粗滤池;原色滤色镜;粗过滤;初效过滤器;初滤池;初级滤清器;初级滤池;初级过滤器;初次滤池
primary filter element 基色滤色器
primary filter element filling 初滤件
primary filtration 初滤
primary financing 初期投资;基本投资
primary fission fragment 原始裂片
primary fission product 裂变碎片;初始裂变产物
primary flight line 本航线
primary flight line viewfinder 前方线瞄准器
primary flo(a)tation cell 粗选用浮选机
primary float feed 初次浮选给料
primary flow 原始流;主流
primary flow-and-return 主流与回流;原始流与回流
primary flow-and-return pipe 进出主管道;一次水循环管路
primary flow-and-return piping 一次水循环管路
primary flow pipe 主流管
primary flow structure 原生流动构造
primary focus 主焦点
primary foliation 原生叶理
primary forest 原始森林;原生林
primary form 原形
primary formation 原始结构;原始地层;一级结构;基底岩
primary fossil 原生化石
primary fouling film 原生污着膜
primary fractionator 初馏塔
primary fracture 原生裂隙
primary framing 主框架
primary free face 原生自由面
primary frequency 主频率
primary frequency microseism 主频微震
primary frequency regulation 一次频率调节
primary frequency standard 原始频率标准;一级频率标准;基本频率标准
primary fuel cell 原燃料电池
primary function 原函数;主要机能;基函数;基本功能
primary furcation 一级分叉
primary fuse 一次测熔断器
primary fusinite 原生丝质体
primary gas circulator 初级煤气循环泵
primary gas-oil ratio 原始油气比
primary general membrane stress 一次总体薄膜应力
primary geodetic control network 一等大地控制网
primary geologic(al) map 地质草图
primary geosyncline 原生地槽;原地槽
primary girder 主梁;立梁
primary glider 初级滑翔机
primary gluing 一次胶合
primary gneiss 原生片麻岩
primary gneissic banding 原生片麻状条带
primary gneissosity 原生片麻理
primary gold-bearing sulfide 原生含金硫化物
primary goods 初级货物
primary Gothic 早期哥特式【建】
primary governor control 直接作用调节器;直接调节
primary graphic(al) element 基本图形元素
primary graphite 一次石墨;初生石墨
primary graphitization 一次石墨化
primary graphitizing 初次石墨化
primary gravity network 一等重力网
primary great circle 主大圆;基准大圆;基本大圆图
primary grid voltage 第一栅极电压
primary grinding 初碾;初磨;初级研磨;初次磨矿
primary grinding mill 初粉碎碾磨机
primary grinding stone 粗磨石
primary grinding unit 第一段磨矿设备
primary group 一次群;前古生界;初期组
primary grouping 初级分组
primary group rate 一次群速率
primary gyratory crusher 回转式初碎机;回旋式初碎机

primary half-session 主对话端
primary halo 原生晕
primary hardwoods 原始硬材
primary harmonic 一次谐波
primary harmonic wave 基本谐波
primary haulage 一次运输;回采工作面运输
primary hazard 主要危险
primary headbox 主流浆箱
primary health 初级卫生保健
primary heater 主加热器;第一级加热器
primary heater unit 预热装置
primary heat transport system 主回路系统
primary high explosive 起爆药
primary high polymer 一次性高聚物
primary high-pressure turbine 主高压涡轮机;主高压透平;一级高压涡轮机;一级高压透平
primary highway 主要公路;公路干线;干路
primary highway system 一级公路网;公路干线系统;公路干线网;干道网
primary hue 基色调
primary humic acid 原生腐殖酸
primary hydrocyclone 主选水力旋流器
primary hydrogeologic(al) map 水文地质草图
primary image 原像;主像;初像
primary impedance 原边阻抗;初级线圈阻抗
primary inclusion 原生包裹体
primary income 主要收益
primary index block 初级索引块
primary indexing 一次标引
primary inductance 一次绕组电感
primary industrial sectors 一次产业部门;初级产业部门
primary industry 一级产业;一次产业;第一产业;初级工业
primary inertia force 原惯性力
primary infection 初生侵染;初次感染;初次传染
primary information 原始资料;原始信息;原始数据;一次情报;初次情报
primary ingredient ore 一次配矿
primary(initial) production 初产量
primary inlet port 主进气口
primary input 主输入;初级输入
primary input assignment technique 初始输入赋值法
primary input state 初始输入态
primary input table 初始输入表
primary insect 初期害虫
primary instrument 一次仪表;一次测量仪表;主要仪器;初级仪表
primary interference 核副反应
primary interference reaction 初级干扰反应
primary interfering reaction 核副反应;初级干扰反应
primary interstice 原生间隙;初始间隙
primary investigation 初步调查
primary ion beam 一次离子束
primary ionization 原电离;一次电离;初级电离;初步电离
primary ionizing event 初级电离事例
primary ion pair 初步电离对
primary ion-processes 初级电离过程
primary isotope effect 一级同位素效应
primary iteration 基本迭代;初始迭代
primary jaw crusher 一级颚式破碎机;初碎颚式破碎机;颚式初碎机
primary joint 原生裂隙;原生节理
primary junctional 初级接头褶
primary key 主码;基本键
primary launder 初洗槽
primary layer 初生层
primary lead 原生铅
primary leakage 一次测漏泄
primary leakage reactance 原边漏抗
primary lest of mineral dressing ability 初步可选性试验
primary level index block 初级索引块
primary level line 水准干线
primary level(l)ing 一等水准(测量)
primary level network 一等水准网
primary level of prevention 第一级预防
primary liability 主要债务
primary lifting elevator 挖掘升运器;第一升运器
primary light 基色光
primary lighting 常用照明

primary light source 常用光源;原始光源;原光源;主要光源;主光源;初始光源
primary line 一次线
primary line of sight 主观测线;主视线
primary line switch 第一级寻线机
primary lining 一次衬砌;临时衬砌;底层衬砌;初始衬砌;初衬板
primary link station 主链路站
primary liquid 原液体
primary load 主要荷载;主荷载
primary local membrane stress 一次局部膜应力
primary loess 原生黄土
primary loop 一次活套;一次回路;主回路(件)
primary low 主气旋;主低压
primary luminous standard 发光源标准
primary magma 原始岩浆;原生岩浆
primary magnet 原磁铁
primary magnetic field 一次磁场
primary magnetic standard 主磁点
primary main 一次干线
primary market 主要市场;初级市场;初次市场
primary master clock system 一级母钟系统
primary materials 原材料;主要材料
primary matte 原生冰铜
primary matter 引物
primary maximum 主极大
primary maximum allowable concentration 一次最高容许浓度
primary maximum contaminant level 初始最大污染水平
primary meander 原生曲流
primary means of escape 主要避难通道
primary measuring element 调整器测量机构
primary measuring instrument 一次测量仪表
primary medium 初级广告媒介
primary melting zone 初熔区
primary member 主要构件;主要杆件;主构件
primary memory 主存储器;主储存器
primary memory space 主存空间
primary meridian 起始经线
primary metabolite 初级代谢产物
primary metal 原生金属;初级金属
primary metal industry 基础金属工业
primary metalworking 初次成型加工
primary metamorphosis 初级变态
primary meter 煤气总表;一次测量仪表
primary migration 一次迁移(石油);原始迁移;初次运移
primary mill 一级磨机;粗磨机;初磨机
primary mill in open circuit 开路系统中的粗磨机
primary mineral 原生矿物
primary mineral mater 原生矿物质
primary minerals of rutile 金红石原矿
primary minimum 主极小
primary mining 一次回采
primary mining area 首采区
primary mirror 主镜
primary mirror cell 主镜室
primary mission 主要任务
primary molecule 原始分子
primary moment 主弯矩;主力矩;首力矩
primary morphologic(al) map 地貌地质草图
primary mortgage market 初级抵押市场;初级抵押贷款市场
primary motion 主运动
primary motor 主电动机
primary mullite 一次莫来石
primary multiplex equipment 一次辐射器
primary mycelium 初生菌丝体;初期菌丝体
primary navigation 沿航线导航
primary network 一次电力网
primary nitrocompounds 伯硝基化合物
primary nitroparaffin 伯硝基烷
primary node 初始结点
primary nodule 初级小结
primary no-load tap 一次侧无载抽头
primary nondisjunction 初级不离开
primary nozzle 一次喷嘴
primary observation 一等观测
primary oil 原油
primary opening 原生孔隙;原生孔洞
primary optic(al) axis 主光轴
primary optic(al) vesicle 初级视泡
primary order longitudinal aberration 初级轴向

像差
primary ore 原生矿石;原生矿
primary ore deposit 原生矿藏
primary ore zone 原生矿带
primary organic matter 原始有机质;原生有机质
primary organic matter of lake 湖泊原生有机质
primary orientation 基本定向
primary orogeny 原生造山作用;主造山运动
primary oscillation 主振荡
primary oscillator 原始振荡器;初级振荡器
primary outlet 中心局;初级引出线
primary output 初期出力;主要输出;主要出力;主输出
primary oxide 原氧化物
primary paging device 初级分页装置
primary panel 镶板的直观面
primary paper 原始论文
primary parameter 一次参数
primary particle 原始粒子;一次颗粒;基本颗粒;初始粒子;初生粒子
primary passive potential 初始钝化电位
primary path 主通路
primary path satellite 主通道卫星
primary pattern 照明方向图
primary pearlite 初生珠光体
primary period 主周期;主要时期;主要阶段;初始周期
primary permeability 原生渗透性
primary pest 主要害虫
primary phase 初相;初生相;初晶相
primary photoelectric current 原光电流;一次光电流
primary photoelectric effect 初级光电效应;原光电效应;一次光电效应
primary photometric standard 光度基准
primary pigment 原色素
primary pile 初级堆
primary pipe 初次缩孔
primary piping 一次侧配管;一次侧管路
primary pit-field 初生纹孔场
primary plane 主面
primary planet 主行星
primary planet carrier 一级行星架
primary planet pinion 一级行星齿轮
primary planet ring 一级行星齿轮轮座圈
primary plant (核反应堆的)蒸汽发生装置
primary plant body 初生植物体
primary plasticizer 主增塑剂
primary plate movement 初始板块运动
primary platform 原地台
primary point 一等(测)点
primary pole 主极
primary pollutant 原生污染物;原发性污染物;一级污染物;一次污染物;初始污染物;初级污染物
primary pollutant mass 初始污染物质量
primary pollution 初级污染;原生污染;原发性污染;一次污染;初级污染;初级污染
primary pollution effect 原生污染影响
primary pollution source 主要污染源;原始污染源;原生污染源;原发性污染源;一级污染源;一次污染源
primary population 基本人口
primary pore 原生孔隙
primary porosity 原始孔隙度;原生孔隙度
primary port 主要港口
primary position 初期位置
primary power 可靠电力;主要功率;主要电力;主功率;基本功率
primary power cable 主电力电缆
primary power distribution 一次配电
primary power plant 主动力装置
primary power source 一次电源;主动力源
primary prescriptive 初次指示
primary pressure 原生水压
primary process 初级处理过程(污水);基本过程;初级过程;初步处理
primary producer 初级生产者
primary producer country 初级生产国
primary product 初次产物;原生产物;主要产物;初级产品
primary production 一次产品;初级生产;初次生产
primary production volume 初级生产量
primary productivity 初级生产(能)力
primary project benefit 工程基本收益

primary protection 一次保护装置
primary protection standard 首要保护标准
primary protective barriers 主防护墙
primary publication 原始文献出版物
primary pulp stream 原始矿浆流
primary pump 一次泵;主泵;起动注油泵;起动注水泵;初始引液泵;初始泵
primary pumping station 一泵站
primary purification 一次提纯;初次净化
primary quality casting 一级铸件
primary quantum number 主量子数
primary quenching 初次淬火
primary rachis branch 第一次枝梗
primary radar 一次雷达;雷达导航系统始发射台;初级雷达
primary radial 原辐板
primary radiation 原辐射;主辐射;初级辐射;初辐射
primary radiator 原辐射器;初级辐射体
primary radioisotope 原始放射性同位素;初级放射同位素
primary rainbow 一次虹
primary rate 基群速率;基群
primary rate access 一次速率接入;基群速率接入
primary rated voltage 一次侧额定电压
primary rate interface 主速率接口
primary raw materials 初级原料
primary ray 原射线;初生射线;初级射线
primary reaction 一次反应;主要反应
primary reading 原始读数
primary reception 原始接收
primary record 原始记录
primary recovery 一次回收率;一次采油;最初回收率;首次可采销量;初次开采
primary recovery method 一次开采方法(石油);一次回收方法(石油)
primary recrystallization 一次再结晶;初级再结晶;初次重结晶
primary reducer 一级减压器
primary reduction 初次破碎
primary reference fuel 正标准燃料
primary reflection 原始反射;一次反射
primary reflection light 原始反射光
primary reformer 一段转化炉
primary reformer tubes 一段转化炉管
primary regenerator 主蓄热室
primary regulation 初次调节
primary reject elevator 排矸提升机
primary relay 一次继电器;初级继电器
primary relief 钻尖后角
primary relief system 一次回路泄放系统;一次回路释放系统
primary relief tank 一回路释压凝汽箱
primary remanent magnetization 原生剩余磁化
primary remanent magnetization strength 原生剩余磁化强度
primary remix 初级飞羽
primary repair 一期修补术
primary request 初级请求
primary reserves 一次采油可采储量;首要准备金;原始储[贮]量
primary residential street 居住区主要街道
primary resin 原树脂
primary resistance 一次电路电阻
primary responsibility 主要责任
primary retarder 主缓行器
primary return 主要回程;一次回收
primary return air 一次回风
primary ring 准质环
primary road 干线道路;主要路;主公路;主要道路;干路;干道
primary road network 主要道路网;干道网
primary rock 原生岩
primary rock formation 原岩建造
primary rocks 原生岩类
primary root 初生根
primary route 汽车干道;一级线路;主路由;公路干线;干路
primary row 初始行
primary runway 主跑道
primary safety control 主安全控制;总安全控制
primary safety mechanism 主安全机构
primary safety valve for inner container 一级安全阀
primary saline soil 原生盐土

primary salinization 原生盐渍化
primary salinized soil 原生盐渍化
primary salt 一代盐
primary salt effect 原盐效应
primary sample 原始试样;原矿样;一级样本;一次取样;初级试样;初次试样;初次取样
primary sample feeder 一次试样给料机
primary sample unit 初级抽样单位
primary sampling 一级抽样
primary sampling fraction 一级抽样比
primary sampling unit 一级抽样单位;一次取样单位;一次取样单;一次抽样单位
primary satellite 主用卫星
primary saturation pressure 原始饱和压力
primary scale 主要比例(尺);基本比例(尺);初次氧化皮
primary scale breaker 粗轧破鳞机
primary scattering 初级散射
primary school 小学
primary scour 初始冲刷
primary screen 头道筛;初级筛;初分筛
primary screening plant 初筛机
primary screening test 初步筛选试验
primary seal 主座封;主密封;初级封密
primary search 第一次寻查
primary-secondary alcohol 伯仲醇
primary-secondary clock 子母钟
primary sector 第一产业;采掘业和农业
primary securities 初级证券
primary securities market 初级证券市场
primary sedimentary structure 原始沉积构造
primary sedimentation 初次沉淀
primary sedimentation basin 一次沉淀池
primary sedimentation pond 初次沉淀池
primary sedimentation tank 初次沉淀池;初次沉淀槽;初级沉淀池;初次沉淀池
primary seepage cave 原生渗流洞
primary segment 原体节;初始段
primary segregation 原生偏析
primary seismic wave 地震initial波
primary semifusinite 原生半丝质体
primary sensing device 一次敏感装置
primary separation 一次分离
primary sere 原生演替系列
primary series circuit 初级串联电路
primary service 主要业务
primary service area 主服务区;主播送区;基本服务区
primary set 主台
primary setting 一次沉陷
primary setting tank 初次澄清池
primary settlement 初次沉降
primary settling 初次沉降;初次沉淀
primary settling tank 一次沉降池;一次沉降槽;初级沉淀池;初次沉降槽;初次沉淀池;初步沉淀池
primary sewage sedimentation tank 初级污水沉淀池
primary sewage settling tank 初级污水沉淀池
primary sewage sludge 原下水污泥
primary sewage treatment 一级污水处理;污水一级处理;初级污水处理
primary sewage treatment capacity 初级污水处理能力
primary sewage treatment method 初级污水处理法
primary sewage treatment plant 初级污水处理装置
primary sewage treatment unit 初级污水处理装置
primary sex ratio 原始性比率
primary shaft 原动轴;初动轴
primary shear plane 主剪力面
primary shear wire 主剪力线
primary shield(ing) 一次遮蔽;主屏蔽层;主屏蔽;初级屏蔽
primary shredding 一次破碎
primary side 原始侧;原边;主侧;初级端;初级侧
primary signal 原始信号
primary signal face 主信号
primary simple drawing 初步简图
primary sinusoid 原正弦波;正弦形基波
primary skip zone 起始跳越区
primary slime 原生矿泥
primary sludge 原污泥;原始污泥;初次污泥;初沉污泥;初沉池污泥
primary sludge digestion 初级污泥消化

primary sludge hydrolysis 初级污泥水解
primary smelting 一次冶炼；初步熔炼
primary soil 原生土(壤)
primary soil structure 原生土结构
primary soil type 初级土壤类型
primary sol 初生溶胶
primary solid solution 原生固溶体；原固溶体
primary solution 初始解
primary solvent 一次(稀释)溶剂；主要溶剂；初次(稀释)溶剂
primary sorbite 初生索氏体
primary source 一次光源；一次电源；主电源
primary sources of nutrient 原始营养源
primary space allocation 初始空间分配
primary space target 主要空间目标
primary space vehicle 主航天飞船
primary specific ionization 初生加厚
primary spectroscopic standard 原光谱标准
primary spectrum 一级光谱；第一级光谱
primary speed control 原速调整
primary spool 主滑阀
primary stage 主要阶段；初始阶段；初期
primary stage port 主进气口
primary standard 原始标准；原标准；一级标准；一级标样；主基准；基准；基础标准
primary standard fuel 主要参比燃料；正标准燃料
primary standard material 一级标准物质
primary standard pyrheliometer 基准直接日射表
primary standard sample 一级标样；一级标样
primary standard sea-water 原始标准海水
primary standard sound source 原标准声源
primary standard substance 基准物质；基本标准物(质)
primary star 主星
primary state 始态
primary statement 本原命题
primary state of stress 初始应力状态
primary station 一等(测)站；一等(测)点；主站；主要观测站；主测站；基本站
primary station peg 主测站标桩
primary statistics 原始统计
primary steam 原蒸汽；一次蒸汽
primary steel 通用钢
primary steering instrument 主要驾驶仪表【船】
primary stimuli 原刺激
primary stoping 一次回采
primary storage 主存储器；基本主存储器
primary store 主存储器；主储存器
primary strainer 粗滤网
primary strain stage 初始蠕变阶段
primary stratification 原始层理；原生层理
primary stratigraphic trap 原生地层圈闭
primary strength 初始强度
primary stress 一次应力；主应力；主要应力；初应力；初始应力
primary stress field 初始应力场；环境应力场
primary structural component 主结构件
primary structural elements 初级构造
primary structural map 构造草图
primary structure 基本体系；原生碎屑构造；原生构造；一级结构；一次组织；主要结构；主结构；基本结构；初始构造；初生构造；初级结构
primary structure of igneous rock 岩浆岩原生构造
primary subset 简单子集
primary substation 一次变电站；一次变电所
primary subtractive colo(u)r 相减合成基色；减法混合基色
primary succession 原生演替；正常演替系列
primary sun gear 初级太阳齿轮
primary sun wheel 初级太阳轮；初级太阳齿轮
primary supply point 主要补给点
primary support 初期支护
primary surface 主表面
primary surfacing hopper 屋面铺砂器
primary survey network 初始测量网络
primary survey stage 主测阶段
primary suspension 一次悬挂；第一系悬挂
primary suture 初期缝合
primary switch 一次转换
primary(switching) center 初次转换中心
primary system 初回路；主要系统；主系统
primary table 原始表
primary tabulation 原始表列
primary tangent 主切线；主切曲线

primary tank 主容器
primary tapetal layer 初生绒毡层
primary tar 原沥青；原焦油；原柏油
primary target 主要目的；主要目标
primary target area 主要目标区
primary target line 主要目标线
primary technical maintenance 一级技术保养
primary tectonite 原生构造岩
primary telescope 主望远镜
primary tension failure 初始筋拉伸破坏
primary term 一级主题词
primary terminal 一次电路终端；一次电路端子
primary terminal voltage 初级端电压
primary-tertiary alcohol 伯叔醇
primary test board 主测试台；基本测试台
primary test(ing) 初步试验；预初试验
primary thickening 初生加厚
primary throttle valve 主节流阀；初次节流阀
primary tide 原潮汐
primary tide station 一级验潮站；一级潮位站；一级测潮站；主要潮汐观测站；主要潮位站；主要潮流观测站；主验潮站
primary tillage equipment 初耕机具
primary timber 主要木材
primary time effect 主时间效应；初始固结效应
primary tower 初馏塔；初级塔
primary toxicity 一相毒性
primary toxin 初期毒素
primary train diagram 基本运行图
primary transaction 初级交易
primary transducer 一次换能器；一次传感器；一次变换器
primary transition 初级转变
primary translation 初基平移
primary transmission 一级传动(装置)
primary transmission feeder 一次输电的馈电线
primary transmission line 一次输电线
primary traverse 一等导线；主干导线；主导线【测】；基本导线
primary traverse point 一等导线站；一等导线点
primary traverse station 一等导线站；一等导线点
primary treatment 一级处理；一次处理；初级处理；初步处理
primary treatment of sewage 污水一级处理
primary treatment plant 初级处理厂
primary treatment unit 初级处理装置
primary triangle 一等三角测量三角形
primary triangulation 一级三角测量；一等三角测量
primary triangulation net 一等三角网；一等三角测量网
primary triangulation point 一等三角点；一等三角测量点
primary troostite 初生屈氏体；初次屈氏体
primary truss 主桁架
primary tundra 原生冻原
primary tunnel liner 隧道初试衬砌
primary turbine 第一轮汽轮机；初压汽轮机
primary turns 一次线匝
primary twin 原生双晶
primary twister 初捻机
primary twister for glass fibre 玻璃纤维缠绕机
primary twist(ing) 初捻
primary tympanic cavity 原鼓室
primary type glider 初级滑翔机
primary unit 主单位；基层单位；初级单位
primary unit cell 初生晶胞；初级晶胞
primary uranium ore 原生铀矿
primary user-network interface 一次群用户接口；基群速率用户—网络接口
primary valence 主价(键)【化】
primary valence bond 主原子价键；主价键
primary valve 主阀
primary variable 基本变数
primary vector 第一向量
primary vegetation 原生植被
primary vision area 主视区
primary vitreous 原始玻璃(状)体
primary voltage 原电压；一次电压；初级电压
primary vulcanization 定型硫化；第一次硫化
primary washbox 主选跳汰机；初选跳态机
primary waste heat boiler 第一废热锅炉
primary wastewater treatment 污水一级处理；废水初级处理
primary water 原始水；原生水；一次回水；离子近

层水；近层水；初生水
primary water circulator 初级水循环泵
primary water lifting machinery 初级提水机械
primary wave 原始波；一次电波；地震纵波；初始波；初生波
primary winding 原绕组；原边线圈；一次绕组；初级线圈；初级绕组
primary winding voltage 原绕组电压
primary wire 原电路；初级电线；初级电路
primary wood 初生材
primary workers 一等工人
primary X-ray 一次X射线
primary yarn 初捻纱
primary zone 原生带；一次燃烧室
primate city 初级城市；首位城市
Primatol 阿特拉通
prime 最初的；装信管；质数；首要
prime advance 最大优越性
prime airlift 实际空运飞机数
prime amplifier 主放大器；前置放大器；前级放大器
prime-and-dot notation 点撇符号
prime a pump 泵充水
prime area 初级区域
prime attribute 主属性
prime bank 一级大银行
prime bill 上等汇票
prime business income 主营业务收入
prime cargo 主要货物
prime cartridge 过滤器第一层滤芯；主要卡盘
prime-coat enamel 底涂珐琅
prime coat(ing) 沥青透层；抹灰打底；结合层；首涂层；底涂层；底漆；涂底漆；涂底层；头道涂层；抹灰打底；透层；打底层；头道涂层；第一道抹灰；底沥青；打底
prime coating road tar 底涂柏油(道路)
prime coating with linseed oil 亚麻籽油底涂
prime coat operation 沥青透层施工；第一层涂抹操作
prime commercial paper 一流商业票据；最佳商业票据
prime compatible set 本原相容集合
prime containerizable cargo 最佳集装箱货
prime contract 基本合同；大包合同；原合同；总承包合同；主要合同；直接承包合同
prime contractor 总承包人；总包人；总承人；主要订货商；主要订约人；主要承包者；主要承包商；主承包人；直接承包人
prime contract termination 原合同终止；原合同期满
prime cost 原始成本；原价；最初成本；主要成本；直接费；直接成本(包括材料工时在内)；进货价格；成本
prime cost contract 成本合同
prime cost item 主要成本项目；总成本
prime cost method 主要成本法(指负荷分配的方法)
prime cost sum 营造商支付给分包商的总价；总成本；成本总额
prime couple 质数偶；素数偶
primed 灌水引动水泵；待发的；已上漆；已打内涂层
primed charge 起爆炸药
prime decomposition 素数分解
prime direction 起始方向
primed surface 浇过透层的路面；浇过透层的路面
prime element 质元素
prime energy 原始能
prime engine 主发动机
prime entry 进口预先申报
prime factor 质因数
prime fictitious meridian 虚拟首子午线；虚本初子午线
prime field 质域；素域
prime fish 上等鱼
prime flow table 本原流程表
prime focus 主焦点
prime focus cage 主焦笼
prime focus cell 主焦点光电管
prime frame 主帧
prime-grade oak 优质橡木
prime grid meridian 格网首子午线；网格本初子午线
prime half page 主半页
prime implicant 质蕴涵；素项
prime information 素信息
prime interest rate 优惠利率

prime in the spots 在擦净点上涂底漆
prime investment 优等投资
prime lacquer 打底漆；上底漆
prime lending rate 最优惠短期贷款利率；优惠放款利率
prime material 底漆；透层材料；首涂材料；底层材料；打底材料
prime matrix 素矩阵
prime membrane 首涂层；底涂层
prime meridian 零度经线；起始子午线；本初子午线；本初子午圈线
prime modulus 素模
prime mover 起动机；原动力；原动机；牵引机
prime-mover and semi-trailer 铰接拖挂车
prime-mover cycle 原动机的循环
prime-mover governor 原动机的调速器
prime-mover truck 载重牵引车
prime navaids facility 主要导航设施
prime number 质数；素数【数】
prime number generation 素数生成
prime number theorem 质数定理
prime oblique meridian 本初斜子午线
prime paint 底漆；底层涂料；底层漆
prime painting 上底漆；底漆
prime pair 素数对
prime pigment 遮盖颜料
prime piling box 优质板垛箱
prime plate 一级镀锡薄钢板
prime polynomial 素多项式
prime power 牵引车；原动机；原动力；基本功率
prime professional 与业主订约的业务单位
prime pump 启动泵；起动泵
prime quality 上等货
prime quotient 素商
primer 预注装置；油漆打底层；雷管；浇透层结合料；起动注油器，透油层，始爆器，底涂料；底层涂料；底层漆；打底涂料；初给器；初步；发火药；发火极
prime rate 优惠利率；最优惠利率；最低利率；基础利息率；基本费率；头等利率
prime rate of interest 优惠利率；最优惠利率
primer base 底层涂料；底漆；底漆成分；底漆涂料；打底涂料
primer boiled oil 底漆用熟油
primer cap 雷管帽；起爆雷管
primer cartridge 起爆炸筒；起爆药包
primer charge 起爆药；点火药
primer coated board 底涂板
primer coat(ing) 底涂层；涂底漆；首涂油漆；涂料底层；打底涂层
primer composition 点火剂；起爆剂
primer cup 底火帽
primer-detonator 始爆剂；底火雷管
primer detonator unit 起爆雷管
prime redemption privilege 提前赎回押品的权利
primer-filler 底涂板料
primer fluid 启动液
primer fuel tube 起动燃料管
primer(gilding) brass 雷管黄铜
primer handle 初动手柄
primer house 起爆器材储存室；起爆材料储[贮]存室
prime rib 主肋
prime ring 素环
prime risk 最低风险；基本风险
primer jus 上等脂
primer knife 底漆刮刀
primer knob 起动加油器捏手柄
primer leak 底火漏泄
primerless 无底漆
primer line 起动管路；启动管路
primer lock 起动注油器的保险器
primer membrane 底漆
primer mixed paint 木材底层用调合漆
primer mixture 击发药
primer oil 涂底用熟炼油；桐油底涂料；熟桐油打底漆
primer pump 起动泵；启动泵
primers and fillers 底漆及木材填孔漆
primer-sealer 密封涂料；打底密封涂料；封闭底漆；封闭涂料
primer seat 底火座
primer setback 底火后退
primer strand 元股
primer surfacer 打底面层底漆；头二道混合底漆

底漆二道浆；底层面层两用漆
primer-top coat coating 底面合一涂料
primer treatment 防水层底面处理
primer valve 起动注油阀
primes 一级品
prime sheet 优质薄钢板；优质薄板
prime singular cube 素奇异立方
prime site 贵重地皮
prime spot 素点
prime standby power source 备用电源
prime steam 湿蒸汽
prime stratum 原始地层
prime system 素系
prime target 主要目的；主要目标
prime tenant 最优承租人；主要租户
prime the fuel system 燃油系统充油驱气
prime tide 初潮
prime transverse meridian 横轴首子午线；本初横子午线
primeval forest 原始森林；原生林
primeval isotope 原生同位素
primeval lead 原始铅
primeval vegetation 原始植被
prime variables 带撇号的变量
prime vertical 真卯酉圈
prime vertical circle 卯酉圈；东西圈
prime-white kerosine 上等白色煤油
prime window 主窗
primin 樱草素
priming 首次灌满渠道；浇透层；刷底子油（沥青）；引火药；装填物；装雷管；高潮间隙缩减；雷管接引线；汽水共腾；起爆；偏压点火；涂底漆；刷底漆；底子；底层漆；打底（指底漆）；充水起动；充水，沸溅
priming action 激发作用
priming adapter 起爆接头
priming apparatus 引爆器；加水装置（水泵）；起爆器
priming application 涂底漆；第一遍浇油
priming a pump 泵充水
priming arrangement 起动注水器
priming by vacuum 真空启动
priming can 注油器
priming carburet(t)or 起动汽化器
priming chamber 起动室
priming charge 灌入的水（水泵的）；起爆药包；雷管；起动注水；起动加注燃油
priming circuit 起爆网络；起爆网路
priming coat 底层；底釉；首涂层；底涂层；油漆打底层；头道漆；底漆；底面涂层；底层漆；打底层
priming coating 底涂层
priming cock 起动注油开关
priming colo(u)r 首涂色浆；底涂颜料；底色；底层油漆；初漆色
priming composition 起动加注剂；起爆剂；点火剂；底漆料
priming compound 起爆药包；雷管药包
priming connection 起动注油管
priming control 打注控制；初给控制
priming cup 起动注油旋塞
priming depth 起动水深；起动深度
priming device 注液装置；起动给油装置；起爆装置；涂底漆设备
priming discharge 引火放电
priming dose 初次剂量
priming effect 激发效应
priming emulsion 启动乳化剂；启动乳化液
priming explosive 起爆药
priming fuel 启动燃料；起动用燃油
priming funnel 注油漏斗；加油漏斗
priming handle 起动手柄
priming hole plug 起动喷口螺塞
priming illumination 固定照明
priming impoundment 初次蓄水
priming inductor 启动感应器
priming jet 加注喷嘴
priming level 虹吸启动水位；虹吸水位；起动水位
priming lever 起动杆；初给杆
priming line 灌注管路
priming material 起爆器材；透层材料；打底材料
priming method 起动方法；起爆方法
priming mixture 引火药；点火混合剂
priming model 引动模式
priming mover 原动机
priming nose (溢洪道的)跳水凸缘

priming nozzle 喷布器
priming of battery 蓄电池加注酸液
priming of centrifugal pump 离心泵启动
priming of explosive 炸药起爆
priming of neap 小潮提前
priming of pump 水泵充水；泵起动；泵的起动
priming of the tide 潮汐提前
priming of tide 先潮（日潮先于月潮）；潮时提前
priming oil 透发油；打底用油
priming operation 浇沥青透层；涂底层
priming paint 头道漆；底漆；底层漆
priming pipe 起动注油管
priming plug 注油口塞；加油口旋塞；起动注油塞
priming plug hole 起动喷口螺塞孔
priming point 起爆点；点火点
priming potential 引燃电压；着火电位
priming powder 起爆粉
priming pump 引液泵；灌注泵；灌液泵；起动注油泵；起动注水泵；起动泵
priming reservoir 吸水池；抽水池
priming sensibility 起爆灵敏度
priming siphon 起动虹吸
priming siphon pipe 起动虹吸管
priming siphon valve 起动虹吸阀
priming solution 启动溶液
priming speed 起动速度
priming syphon 起动虹吸
priming syphon pipe 起动虹吸管
priming system 起爆系统
priming the fuel system 燃油系统充油驱气
priming the pump 泵充水；使泵起动
priming the water 泵前注水
priming treatment 透层处理（沥青）
priming tube 导爆管；爆破管；雷管
priming valve 加油阀；起动注水阀；起动阀；初给阀；沸腾安全阀
priming varnish 底清漆
priming wire 撞针
priming with linseed oil 用亚麻籽油启动
primitive 原语；基元；本原
primitive area 原始林区；原始地区
primitive axis 原轴；初始轴
primitive body segment 原体节
primitive campground 原宿营地
primitive cell 初基胞
primitive character 原始性质
primitive circle 原圆
primitive colo(u)r 原色；基本色
primitive component 原始成分
primitive computer 原始计算机
primitive connection matrix 原始联络矩阵
primitive cube 原始立方
primitive curve 原始曲线
primitive data map 实际材料图
primitive D-cube of failure 故障的原始 D 立方
primitive D-cube of fault 故障的原始 D 立方
primitive decision 基本判定
primitive decisive 基本判定
primitive desert landscape 原始荒漠景观
primitive digraph 本原有向图
primitive dislocation 初期脱位
primitive elastic limit 原始弹性极限
primitive element 基本元素；素元；本原元素
primitive environment 原始环境；原生环境
primitive equation 原始方程
primitive equation model 原始方程模式始气方程组；原始方程模式
primitive eyepiece 一般目镜；单透目镜
primitive factor 本原因子
primitive fiber 微纤维
primitive flow table 原始流程表
primitive fold 原褶
primitive forest 原始森林；原生林
primitive form 基本形(式)
primitive function 原函数
primitive glass 原始玻璃
primitive green 原绿色
primitive groove 原沟
primitive knot 原结
primitive land animal 原始陆生动物
primitive lattice 原始晶格；原始格子
primitive map 原始地图
primitive map for compilation 编辑原图
primitive matrix 素阵；素矩阵

primitive nebula 原云
primitive node 原结
primitive ocean 原始海洋
primitive of concept(ual) dependency 概念从属性原语
primitive operation 原始操作
primitive orbit 初轨
primitive organism 原始生物
primitive parallelogram of period 周期的原始平行四边形
primitive period 原始周期;基波周期
primitive period parallelogram 原始周期平行四边形
primitive pit 原窝
primitive polynomial 本原多项式
primitive pottery 原始陶瓷器
primitive projection 原始射影
primitive reaction 基本反应
primitive recursion 原始递归
primitive recursive function 原始递归函数
primitive recursiveness 原始递归性
primitive recursive remainder function 原始递归余数函数
primitive region 原始地区
primitive road 天然土路;原始土路
primitive rock 原岩;原始岩;原生岩(石)
primitive rocks 原始岩类;原生岩类
primitive rock stratum 原岩层
primitive root 原根;本原根
primitive rule 基本规则
primitive soil 未开垦土地;未开垦地;生荒地
primitive solution 原始解
primitive streak 原线;原条
primitive term 原始项
primitive test of container board 箱板性能测定
primitive translation 素平移
primitive variety 原始变种
primitive water 原始水;原生水;初生水
primo 首先
primordial 原生的
primordial argon 原始氩
primordial cosmic(al) abundance 原始宇宙丰度
primordial ecosystem 原始生态系统
primordial helium 原始氦
primordial lead 原生铅
primordial leaf 初叶
primordial plant 原始植物
primordial radioelement 原生放射性元素
primordial shaft 原柱期
primrose 报春花;淡黄色
primrose chrome 淡铬黄
primrose(chrome) yellow 樱草色;樱草黄;柠檬铬黄
primrose green 樱草绿
primula garden 樱草园
primulin 樱草素
primuline yellow 樱草灵黄
primus stove 燃烧汽化油的炉子;手持煤油炉
prince albert fir 冷杉
princes-metal 普林士黄铜
princess 屋桁架小柱;屋面石板瓦
princess mineral 土红
princess pine 加拿大短叶松
princess post 小柱;加劲支柱;桁架两侧小直(腹)杆;辅助柱
princess post truss 小柱屋架
princess seam 高背缝
principal 负责人;主要的;屋架;委托人;本金、利息、税金和保险费
principal absorption wavelength 主吸收波长
principal account 主要账户
principal advantage 主要优点
principal-agent relationship 当事人与代理人关系
principal airway beacon 主航路信标
principal alloying constituent 主要合金成分
principal altar 主神座;主台阶
principal amortization 摊还本金
principal amount 本金
principal analytic set 主解析集
principal and income cash 本金及收益现金
principal and interest 本息;本利
principal angle 主角
principal angle of incidence 主入射角
principal architect 主要建筑师

principal arterial highway 公路主干道
principal assembly 主要总成
principal assistant secretary 主任助理秘书
principal audit 控股公司核数师
principal automorphism 主自同构
principal axes of a body 刚体的主轴
principal axes of strain 主应变轴
principal axis 主轴(线);光轴
principal axis of camera 摄影主光距
principal axis of centroid 形心主轴;质心主轴;矩心主轴
principal axis of compliance 柔性主轴
principal axis of ellipse 椭圆主轴
principal axis of inertia 惯性主轴;惯量主轴
principal axis of quadric 二次曲面的主轴
principal axis of section 截面主轴
principal axis of strain 应变主轴;主应变轴
principal axis of stress 应力主轴;主应力轴
principal axis of stress tensor 应力张量主轴
principal axis of symmetry 主对称轴
principal axis transformation 主轴变换
principal azimuth 主方位角
principal bar 主圆棒
principal base 主要基面
principal beam 主梁
principal bearing 主像限角
principal bearing structure 主承重结构
principal benefits 主要利益
principal block 主体(建筑)区
principal body axes 主体坐标系
principal bond 主键
principal book 主要账簿
principal boring 主孔
principal brace 屋架斜撑;主要拉条;屋架主撑
principal broker 总经纪人
principal building 主体建筑;主要建筑;主楼;主要建筑物
principal building bay 主体建筑开间;主体建筑跨度
principal building vault 主建筑拱顶
principal business 主要经销商
principal cable 主缆索;主电缆
principal canal 主渠;干渠
principal center 主中心
principal centers of curvature 主曲率中心
principal central axis 中心主轴
principal chain 主链
principal character 主要特征;主特征标
principal characteristics of navigation(al) lights 灯标主要特征
principal characteristics of ship 船舶主要要素
principal charge 主要装药
principal choir (唱诗班的)主要席位;唱诗班主要席位
principal church 主中心教堂
principal circle 主圆
principal climatological station 基本气候站
principal cock 主旋塞;主开关
principal coincidence 主叠合素
principal collector 主下水道;主集尘器;主收尘器
principal column 主柱;金柱
principal-column foundation 主柱基础
principal commodity 主要货物;主要货件
principal component 主组分;主分量;主成分
principal component analysis 主组分分析;主元件分析;主要成分分析;主成分分析
principal component image 主组分图像
principal component regression 主成分回归
principal component transformation 主分量变换;主成分变换
principal composition series 主组成列
principal composition series of subgroups 主组成子群列
principal compression stress 主压应力
principal compressive stress 主压应力
principal computer 主计算机
principal conduit 主渠道
principal consolidation settlement 主固结沉降
principal constituent 主要素;主成分
principal construction time 主体(建筑)工程时间
principal content of contract 合同的主要内容
principal contour 主要等高线;主等高线;首曲线
principal contractor 主要承包者;主要承包商;主要承包人
principal contradiction 主要矛盾

principal controller 主控制器
principal controlling dimension 主要控制尺寸
principal control point 主要控制点
principal convergent 主收敛的
principal coordinate plane 主坐标(平)面
principal coordinates 主坐标
principal cornice 屋檐;顶部挑檐;主檐(古典建筑的)
principal coupling 主联结器
principal course 主航向
principal crest 主脊
principal crop 主林木
principal(cross-)section 主截面;主横截面
principal crystallographic axis 主要结晶轴
principal cubic root 主立方根
principal currency 主要货币
principal current 主电流
principal curvature 沿长度方向弯曲;主弯曲;主曲率
principal curvature line 主曲率线
principal curvature of a surface 曲面的主曲率
principal curvature radius 主曲率半径
principal curve 主曲线
principal customer 开证申请人(信用证);开证申请人(信用证)
principal cutting movement 主切削运动
principal data 主要记录;主要资料;主要数据
principal debtor 主债务人
principal defecation 最后澄清
principal deformation 主形变
principal degree 主次数
principal demandable nations 主要需求国
principal derivative 主导数
principal design section 主设计截面;设计控制截面
principal diagonal 主斜杆;主斜撑;主对角线
principal diagonal line of matrix 矩阵的主对角线
principal diagonal polynomial 主对角多项式
principal diagonal rib 主对角(线)加劲肋
principal dielectric(al) constants 主介电常数
principal dimension 主要尺寸;主尺度
principal direction 主方向
principal direction of curvature 主曲率方向
principal direction of distortion 主变形方向
principal displacement 主点位移
principal distance 像主距;主距
principal-distance error 主距误差
principal-distance focusing 主距调焦
principal distance of camera 摄影机主距
principal distance of photo 像片主距
principal distance of projector 投影器主距
principal distortion 主变形
principal distribution function 主分布函数
principal distribution line 主干线;主配电线路
principal distribution panel 主配电盘
principal ditch 干渠
principal dock 主要港池;主要船坞
principal drawing 主要图纸
principal earthquake 主震
principal edition 正本
principal element of oil 石油的主要元素
principal elliptic term 主椭圆项
principal elongation 主伸长
principal engineer 主任工程师
principal entrance 主进口;主入口;主要入口
principal entrance hall 主门厅
principal epipolar ray 主核线
principal equation of gravimetric geodesy 重力大地测量基本方程
principal etched zone 主蚀带
principal exponent 主指数
principal extension ratio 主拉伸比
principal facade 主门面;主立面;主正面
principal facility 主要设备
principal factors of consumer decreasing progressively 消耗递减主要因素
principal factors of consumer increasing progressively 消耗递增主要因素
principal factors of demand contraction 需求减缩主要因素
principal factors of demand increase 需求增长主要因素
principal factors of marketing decreasing 销售减衰主要因素
principal factors of marketing increasing 销售增长主要因素

principal factors of ore reserve deplete 储量耗竭主要因素
principal factors of ore reserve increase 储量增长主要因素
principal factors of price decreasing 价格下降主要因素
principal factors of price increasing 价格增长主要因素
principal factors of productive decreasing 生产衰减主要因素
principal factors of productive increasing 生产增长主要因素
principal factors of slope failure 斜坡失稳主导因素
principal factor solution 初始因子解
principal fault 主断层
principal fault belt 主断层带
principal fault zone 主断层带
principal feature 主要要素
principal felling 主伐
principal fiber 主纤维
principal fiber bundle 主纤维丛
principal floor 主楼层;主楼板;主层楼面
principal focal distance 主焦距
principal focus 主焦点
principal fold 主褶皱
principal force main 主压力管
principal fractional ideal 主分式理想
principal frame 主结构;主框架;主构架
principal front 主锋【气】
principal fuel supply 主要燃料供给
principal fuse 主保险丝
principal ga(u)ge 主量规
principal generating rule 主生成规则
principal genus 主亏格
principal girder 横梁;大梁;主梁
principal grid 主格栅
principal group 主群
principal half-space 主半空间
principal hall 主厅
principal harbo(u)r 主要海港;主要港口
principal horizontal line 主子午线;主水平线
principal horizontal stress 主水平应力
principal hydrolysis 主水解
principal hydrometric station 一等水文站;基本水文站
principal ideal 主理想
principal ideal domain 主理想整环
principal ideal integral domain 主理想整环
principal ideal minor ring 主理想子环
principal ideal ring 主理想环
principal ideal theorem 主理想定理
principal image components 图像主要参量
principal image plane 主像面
principal-in-charge 主要负责人;主管人
principal indicator of maturity 成熟度的主要指标
principal industrial products 主要工业产品
principal ingredient 主要成分
principal in professional practice 专业负责人
principal inspection chamer 主检验井;主检验室
principal installation 主要装置;主要设备
principal invariant of tensor 张量主不变量
principal jet 主射流
principal joint 主节理
principal lateral aberration 主横像差
principal length 主体长度
principal level 主楼层;主平面
principal level(l)ing 主要水准测量
principal line 干线;主要路线;主纵线;主要谱线;主线
principal line scale 主纵线比例尺
principal linkage 主键
principal lithologic unit 主要岩性单位
principal living area 主生活区
principal load 主要荷载;主要负载;主要负荷;主力;主荷载
principal(load) bearing structure 主要的承载结构
principal load-carrying parts 主要承重部件
principal load-carrying structure 主要承重结构
principal loaded structure 主要的受载结构
principal loading case 主要的加载情况
principal lobe 主要裂片
principal longitudinal aberration 主纵像差
principal longitudinal force 主轴向力;主纵向力
principal lunar diurnal component 太阴主要日分潮
principal lunar diurnal constituent 太阴主要日分潮
principal lunar semidiurnal component 太阴主要半日分潮
principal lunar semidiurnal constituent 太阴主要半日分潮
principal machine 主机
principal main 主干管
principal mall 主林阴大道
principal map 基本地图
principal mapping 主映射
principal mark 主标志;主标
principal matrix 主矩阵
principal maximum 主最大(值);主极大
principal mechanical components 主机组
principal member 主要构件;主要杆件
principal merdian 基准线
principal meridian 主子午线;首子午线
principal meridian of gnomonic chart 心射投影基准子午线
principal mineral 主要矿物
principal mines 主要矿山
principal minimum 主极小
principal mining level 主要开采水平
principal mining nations 主要矿产国
principal mining regions 主要矿产地区
principal minor 主子式;主要(行列)式
principal mode 主振型;主型;主模;基谐模式
principal-mode excitation 主模激励
principal mode of vibration 主振动形式
principal moment 主力矩;主惯量
principal moment of inertia 主转动惯量;主惯性矩;惯性主矩
principal motif 主要特色;主题花纹
principal motion 主运动
principal mo(u)lding 基本线条;主要线条;主要线脚
principal network 主网(络)
principal norm 主范数
principal normal 主法线
principal normal direction 主法线方向
principal normal form 主范式
principal normal section 主正截面
principal note 保证期票
principal nozzle 主喷嘴
principal object plane 主物面
principal office 主办事处;主办公室
principal optic(al) axis 主光轴
principal orbit 主轨道
principal order 主序
principal order module 主序模
principal orebody 主要矿体
principal organic hazardous constituent 主要有机危险成分
principal orientation direction 主取向方向
principal orthogonal steel-plate bridge 主正交各向异性钢板桥
principal orthogonal system 主正交系(统)
principal oscillation 主振荡
principal parallel 像片主横线
principal parallel line 主平行线
principal parameter 主参数
principal parameter of generation hydrocarbon model scheme 成烃模式图的主要参数
principal part 主震相
principal particulars 主要项目;主要规范
principal particulars of ship 船舶主要技术参数
principal part of a differential operator 微分算子的主部
principal path 主要路线
principal pathway 主要途径
principal pedestrian zone 主要行人区域
principal period 主周期
principal phase 主相
principal picture components 图像主要分量
principal pipe 总管;主管道
principal place of business 业务总部所在地
principal plan 规划总图
principal plane 主平面
principal plane of bending 主弯曲(平)面;弯曲主平面
principal plane of inertia 主惯性平面
principal plane of strain 主应变面
principal plane of stress 应力主平面;应力的主平面;主应力面
principal plane of symmetry 主对称面
principal plane pattern 主平面方向图
principal planet 主行星
principal point 要点;主点;光心;基点
principal-point error 主点误差
principal-point location 像主点位置
principal point method 主点法
principal-point of assumption 假定主点
principal point of autocollimation 自准直点
principal point of circular curve 圆曲线主点
principal point of curvature 曲线主点
principal point of optimum symmetry 最佳对称主点
principal-point of photograph 像主点
principal-point of photograph on water 像主点落水
principal point radial line plotting 主点辐射线绘图
principal point radial ray 主点辐射线
principal point traverse 主点导线
principal polarizability 主极化率
principal polarization plane 主极化面
principal pollutant 主要污染物
principal polygonometric network 主导线网
principal port 主要港口
principal portal 入口(尤指建筑物的正门);主门
principal portion 主相
principal post 主桩;主柱;门梃;窗梃
principal pressure 主压力
principal problem 主要问题
principal product 主产物
principal productive commerce 主要生产商
principal productive nations 主要生产国
principal professional adviser 首席专业顾问
principal project 主投影
principal projection property 主射影性质
principal purlin(e) 金檩;金桁
principal quantum number 主量子数
principal radius of curvature 主曲率半径
principal radius of inertia 主惯性半径
principal rafter 上弦木;人字木;主椽木;主椽;屋架上弦杆;背顶支撑
principal rafter beam 人字木之间的系梁
principal railroad station 总站【铁】;主干车站【铁】
principal raw material 主要原(材)料
principal ray 主射线;主光线
principal reactance 主电抗
principal reaction 主要反应
principal refraction[refractive] index 主要折光率;主折射率
principal reinforcement 主钢筋
principal reinforcing bar 主钢筋
principal repayment of foreign loan 国外借款本金偿还
principal representation 主表现
principal restaurant 主要饭店;主要餐馆
principal riser 主立管;主提升器
principal risk 主要风险
principal river 主流;干流
principal road 主要路
principal road line 主要道路干线
principal root 主根
principal rotating motion 主旋转运动
principal round bar 主钢筋;主圆棒
principal route 主要路线
principal saddle 主座板
principal scale 主比例尺;基本比例(尺)
principal search direction 基要搜索方向
principal section 主截面;主剖面;主断面
principal sector 主分脉
principal series 主线系;主级数
principal series line 主系线
principal services center 主要服务中心
principal sets of fault 主断裂组
principal sewer 主下水道
principal shareholder 主要股东
principal shear(ing) stress 主切应力;主剪应力
principal shipyard 主要船厂
principal shock 主震;主要震动
principal slope 主坡
principal solar diurnal component 太阳主要日分潮
principal solar diurnal constituent 太阳主要日分潮
principal solar semidiurnal component 太阳主

principal solar semidiurnal constituent 太阳主要半日分潮
principal solar semidiurnal tide 太阳半日主潮
principal solution 主解
principal source of light 主要光源
principal space 主空间
principal spectrum 主光谱
principal square root 主平方根
principal square submatrix 主子方阵
principal stair(case) 主楼梯
principal star 主星形
principal station 总厂
principal steel beam 主钢梁
principal steel girder 主钢梁
principal stimulated seismic activity 主要激发地震活动
principal stor(ey) 主楼层
principal straight motion 主直线运动
principal strain 主应变
principal strain directions 主应变方向
principal strain space 主应变空间
principal stream 干流
principal stress 主应力
principal stress axis 应力主轴
principal stress-carrying part 主应力承重部件;主要受力部件
principal stress circle 主应力圆
principal stress line 主应力线
principal stress plane 主应力面
principal stress ratio 主应力比
principal stress space 主应力空间
principal stress trajectory 主应力络网;主应力迹线;主应力轨迹;主要应力迹线
principal structure 主结构
principal stud 主要自动关门器;主螺栓;主榫头;主立柱
principal stupa(mound) 主神龛塔(印度)
principal subspace 主子空间
principal sulcus 主钩
principal sum 基金;资金;本金
principal supply pipe 主要给水管
principal supply road 主要供应线路
principal supporting structure 主固定架;主下部结构;主支承结构
principal surface of accumulation 主埋面积;主堆积面
principal switch board 电话主交换机;主交换台;主配电盘
principal switchroom 主配电室
principal symmetry plane 主对称面
principal synoptic(al) observation 基本气象观测
principal synoptic(al) station 主要气象站
principal system (结构的)基本体系
principal tangent 主切线;主切曲线
principal tapping 主分接头
principal telescope 主望远镜
principal temple 主要圣殿;主要圣坛;主要庙宇
principal tensile reinforcement 主要受拉的钢筋
principal tensile stress 主张应力;主拉应力
principal tensile stress limit 主拉伸应力的极限
principal tensor 主章动;主张量
principal term 主项
principal terminal 主端子
principal terminal building 主候机室;终点站建筑(铁路);主港口房屋
principal test section 主测试段
principal tetrahedron 主四面体
principal theorem 基本定理
principal tide 主潮
principal tope(mound) 主要圆顶塔(印度)
principal trace 主迹
principal trade partness 主要贸易区
principal traffic roads network 交通干道网
principal traffic route 主要通信路由;主要交通线
principal transept 建筑主要翼部
principal transversal 主横断面
principal traverse 基本导线
principal triangle 主三角形
principal triangulation 一等三角测量
principal triangulation point 一等三角测量点
principal trihedral 主要三面形
principal trunk 主干道
principal type of organism 有机体主要类型

principal underwriter 主要担保人
principal unit assembling 主要总成装配
principal unit stress 单位主应力
principal use 主要用途
principal useful component 主要有用组分
principal useful mineral 主要有用矿物
principal use of mineral commodities 矿产品主要用途
principal use of mineral commodities at percentage 主要用量占百分比
principal use tonnage 主要用量
principal valence 主价【化】
principal valley 主溪谷;主山谷;主斜沟坡谷;主屋谷;主屋面天沟
principal value 主值
principal value of stress tensor 应力张量主值
principal vanishing point 主灭火点;主合点
principal variable 主变数
principal variance 主方差
principal vault 主拱顶
principal vector 主向量
principal velocity 主速度
principal vent 主喷发口
principal vertical circle 子午圈;观测者子午线;天球子午圈
principal vertical(line) 主垂线;像片主总线
principal vibration 主振动
principal view 主视图
principal virgin stress 原始主应力
principal visual ray 主视线
principal voltage 主电压
principal walk 主要走道;主要散步之处
principal water-associated disease 主要水媒疾病
principal water ga(u)ge 基本水尺
principal water supply pipe 主给水管
principal wave 主波;基波
principal wave length 主波长
principal weld 主要焊缝
principal wind direction 主导风向
principal windings 主绕组
principal wire bundle 主线束
principal wire feeder 主电源线
principal workable coal seam 主要可采煤层
principal work item 主体工程项目
principal yield 主产量
principle 原则;原理;法则;本质;被代理人
principle addition canonic(al) expansion theorem 主加法标准展开定理
principle audit 控股公司审计
principle axis 主轴
principle component method of probability matrix 概率矩阵主分量法
principle component method of product matrix 乘积矩阵主分量法
principle connection diagram 原理接线图
principle controller 主控器
principle(cross-)section 主截面
principle drawing 原理图
principle earthquake 主震
principle evolutionary diagram 展开式原理图
principle layout 总布置图
principle of absolute advantage 绝对利益原则
principle of abstraction 抽象原则
principle of accounting 会计原理
principle of action and reaction 作用与反作用相等原理
principle of all inclusiveness 全包原则
principle of anti-fraud 制止背信原则
principle of Archimedes 阿基米德原理
principle of argument 辐角原理
principle of berthing 靠泊原理
principle of causa proxima 近因原则
principle of circulation 环流原理
principle of clarity 明晰清楚原则
principle of classification 分组原则;分类原则
principle of Clausius 克劳修斯原理
principle of comparative advantage 比较利益原则
principle of comparative cost 比较成本原理
principle of complementarity 互补性原理;并协原理
principle of complementary virtual work 余虚功原理
principle of comprehensive utilization of turning harm into good 综合利用、化害为利的原则;化害为利综合利用原则
principle of conservation 守恒原理
principle of conservation of matter 物质守恒定律;物质不灭定律
principle of conservation of momentum 动量守恒原理
principle of consistency 连续一致原则
principle of construction of energy 能量守恒原理;能量守恒定律
principle of continuity 连续原理;连续性原理
principle of continuity of electric(al) current 电流连续性原理
principle of continuity of magnetic flux 磁通连续性原理
principle of correspondence 对应原则
principle of corresponding state 对应状态原理
principle of covariance 协方差原理
principle of cross liability 交叉责任原则
principle of debit and credit 借贷原理
principle of delegation 授权原则
principle of design 设计原则;设计原理
principle of Deville 德维勒原理
principle of double-entry bookkeeping 复式簿记原则
principle of duality 对偶原理;二象性原理
principle of dynamic(al) virtual work 动态虚功原理
principle of dynamics 动力学原理
principle of economic rationality 经济合理原则
principle of effective stress 有效应力原理
principle of electromagnetic inertia 电磁惯性原理
principle of electroneutrality 电中和原理
principle of energy conservation 能量守恒原理;能量守恒定律
principle of environment(al) sciences domination 环境科学主导原则
principle of equal effects 等效原理
principle of equality and mutual benefit 平等互利原则
principle of equality of treatment 平等待遇原则
principle of equal net advantage 纯效益均等原理
principle of equilibrium 平衡原理;借贷平衡原则
principle of equipartition 均分原则
principle of equipollent loads 等效荷载原则;等效荷载原理
principle of equivalence 等效原理;等效性原理;当量原则(路面厚度的)
principle of equivalent loads 等效荷载原理
principle of exception 例外原则
principle of exploration geochemistry 勘查地球化学原理
principle of extreme convergence 极端趋同原理
principle of first-in first-served 先到先做原则
principle of flow 流量原则
principle of free energy minimum 自由能最小原理
principle of full disclosure 完全公开原则
principle of geometric(al) reverse 几何反转原理
principle of good mooring 良好系泊的原则
principle of grouting 灌浆原理
principle of holocoenotic environment 综合环境原理
principle of holography 全息学原理;全息摄影原理
principle of hydraulic similitude 水力学相似原理
principle of independent assortment 独立分配定律
principle of indeterminacy 测不准原理
principle of keeping pollution controlled by whomever making 谁污染谁治理原则
principle of law 法理
principle of least action 最小作用原理;最小作用量原理
principle of least constraint 最小约束原理
principle of least effort 费力最小原则
principle of least squares 最小二乘原理
principle of least work 最小功原理
principle of least work of deformation 最小功变形原理
principle of lever 杠杆原理
principle of limitation on marginal benefit 边际效应有限原则
principle of linear superposition 线性叠加原理
principle of locality 局部性原理
principle of machinery 机械原理
principle of management 管理原则;处理原则

principle of marginal value 边际价值原则
principle of mass conservation 质量守恒原理
principle of massive support 大质量支撑原理
principle of matching 配合对应原则
principle of matching cost with revenues 费用收益配比原则
principle of maximum entropy 最大熵原理
principle of mediocrity 折衷原则
principle of minimal action 最小限度活动的原则
principle of minimal redundancy 最小冗余原则
principle of minimization risks 风险最小化原则
principle of minimum complementary energy 最小余能原理
principle of minimum energy 最小能原理;最小能量原理
principle of minimum potential energy 最小势能原理;最小位能原理
principle of minimum strain-energy 最小应变能原则
principle of minimum voids content 最少孔隙含量原则
principle of moment of momentum 动量矩原理
principle of momentum and energy 动量与能量原理
principle of momentum conservation 动量守恒原理
principle of non-reciprocity 非互惠原则
principle of operation 操作原理;运转原理;工作原理
principle of optic(al) equivalence 光学等效原理
principle of optimality 优化原理;最优性原理;最优化原则;最优化原理;佳性原理
principle of organization 组织原则
principle of orthogonality 直交原理;正交原理
principle of overall planning and rational layout 全面规划、合理布局原则
principle of parallel vertical gaseous flow 气体分流定则
principle of payment for the use of funds 占用资金付费原则
principle of periodic(al) allocation of cost 费用按期分摊原则
principle of permanent residence 永久居所原则
principle of point association 相配取穴
principle of possible displacement 可能位移原理
principle of probability 概率原理
principle of radioactivity 放射性原理
principle of rationalization in economy 经济合理原则
principle of reaction 反作用原理
principle of reasoning 论证原理
principle of reciprocity 互易原理;互惠原则;互反(性)原理
principle of reflection 反映原理;反射原理
principle of regulation 治理原则
principle of relativity 相对性原理
principle of resistance line 阻力线原理
principle of respondent superior 委托人对受托人行为负责的原则
principle of Saint Venant 圣维南原理(即等力载原理)
principle of scientific management 科学管理原则
principle of similitude 相似性原理;相似原理
principle of single liability 单一责任制原则
principle of smaller variance 较小方差原则
principle of solidification 固化原理
principle of specification 规定原理
principle of statics 静力学原理
principle of stationary phase 稳相原理
principle of stationary potential energy 恒定势能原理
principle of stratificaition 分层原理
principle of style 风格原理
principle of substitution 替代原理;代入原理
principle of superimposed stresses 应力叠加原理
principle of superposition 叠加原理;叠覆原理;层序原理
principle of the equipartition of energy 能量均分原理
principle of the maximum 极大原理
principle of the minimum 极小值原理;极小原理
principle of time invariance 时间不变原则
principle of unattainability of absolute zero 绝对零度不能达到原理

principle of undistortion-free model 无扭曲模型法原理
principle of uniform boundness 一致有界原理
principle of utmost good faith 最大诚信原则
principle of valuation 估价原则
principle of virtual complementary work 虚补功原理
principle of virtual displacement 虚位移原理
principle of virtual forces 虚力原理
principle of virtual work 虚功原理
principle of winding 缠绕规律
principle of work 工作原理
principle of work and energy 功能原理
principle of work and power 功能原理
principle section 主截面
principles of drainage 排水原理
principles of electric(al) and electronic engineering 电工原理
principles of electric(al) engineering 电工原理
principles of food composition 食品结构原则
principles of good organization 良好组织原则
principles of hydrogeology 水文地质学基础
principles of instruction 教学原则
principles of integrity 统一原则
principles of line 线条原理
principles of measurement 计量原则
principle strain 主应变
principle stress 主应力
principle structural elements 主要构造要素
print 印痕;样片;晒图;印刷
printability 印刷适性
print a seal 盖章;盖印
printback 缩影版的放大照片
print bar 印刷杆
print bonding 压印黏接
print clearance 芯头间隙
print command 印刷命令;打印指令
print data 打印数据
print density 墨色浓度
print dryer 烘相器;印刷干燥器;晒印干燥器;像片烘干机
print dyeing 印染
printed antenna 印刷天线
printed apparatus 印制设备
printed board 印刷电路板
printed board for spud carriage 定位桩台车印刷电路板
printed book 印刷本
printed cable 印刷电缆
printed card 印刷目录卡
printed card form 印制卡格式
printed card format 印制卡片格式
printed carpet 印花地毯
printed casement 印花窗布
printed character 印刷字母
printed chassis 印刷电路底板
printed circuit 印刷电路
printed circuit assembly 印刷电路组件
printed circuit backplane 印制电路底板
printed circuit board 印刷电路板
printed circuit broad connector 印刷板边部插头
printed circuit cable 印刷导线
printed circuit epoxy-type coating 印刷电路环氧树脂层
printed coil 印制线圈
printed component 印制元件
printed conductor 印制导线;印刷导线
printed contact 印刷接触
printed decoration 印刷装饰
printed desing 印刷设计;印染设计
printed electrical conductor 印刷导线
printed element 印制元件
printed image 行打图像
printed inductance switch 印刷电感开关
printed inductor 印刷电感线圈
printed linoleum 印花油(地)毡;印花漆布
printed mark 印刷标志
printed matrix wiring 印制矩阵布线
printed matter 印刷品
printed motor 印刷(式)电动机
printed original 印刷原稿
printed original map 印刷原图
printed output 印刷品
printed panel 印刷板

printed parts 印刷元件
printed pattern 印刷的风格;印刷的设计;印刷图案
printed picture tile 印花瓦;印花砖
printed plate 印刷板
printed plywood 印花胶合板
printed sheet 印张
printed sheet capacity 印张容量
printed subject indexes 印刷本主题索引
printed substrate 印刷衬底
printed times 印刷次数
printed unit 印刷单位
printed wallpaper 印花墙纸
printed wiring 印刷电路;印刷布线
printed wiring assembly 印刷电路装配
printed wiring substrate 印制导线衬底
printer 印字机;印相机;印刷器;印刷机;打印机
printer control 印字机的控制
printer controller 打印机控制器
printer control unit 光强控制器
printergram 印字电报
printer interface 打字机接口
printer lens 复照镜头
printer page 打印页面
printer perforator 印字凿孔机
printer plotter 打印绘图仪;打印绘图机
printer's bit 耐火隔离片(彩烧装窑用);耐火隔块
printer sharing 共享打印机
printer's ink 印刷油墨;印刷墨
printer's liquor 乙酸亚铁溶液
printer's varnish 印刷漆;调色料;调墨料
printer waste 印工消耗
printer with multicolo(u)r copper plate 彩色铜版印刷机
printery (棉布的)印花厂;印刷所
print film 印片用胶片
print format 打印格式
print gingham 印花格子布
print glazer 图片烘干上光机
print hand 印刷体
print head 印刷头;打印头
printing 印刷;晒印;印花;印刷
printing ability 印刷适性
printing and dyeing 印染
printing and dyeing assistant 印染助剂
printing and dyeing industry 印染工业
printing and dyeing industry wastewater 印染工业废水
printing and dyeing mill 印染厂
printing and dyeing sludge 印染污泥
printing and dyeing wastewater 印染废水
printing apparatus 印制设备;印相设备
printing areas 印刷的图文部分
printing art 印刷术
printing blanket 石印毡
printing block 手工印花木模;凸印板
printing box 晒像柜
printing building 印刷用房屋
printing by stencil 油印
printing calculator 打印计算器
printing capacity 印刷容量
printing carriage 网版行车
printing chronograph 印字记时仪;印字记时器;印字计时仪
printing circuit 印制电路
printing counter 晒印计数器
printing current meter 印刷海流计
printing cylinder 印刷筒
printing density 印相密度;印刷密度
printing diameter 印版直径
printing dimension 图幅大小
printing down 晒印
printing-down machine 连晒机
printing dye 印染染料
printing elements 印版的图文部分
printing equipment 印刷设备
printing filter 复制用滤光镜
printing forme 印板
printing frame 晒像框;晒图架;晒版架
printing grade 像纸反差等级
printing head 打印头
printing house 印刷厂
printing house craftsman 印刷技工
printing industry 印刷工业
printing ink 油墨;印刷油墨;印刷墨;打印墨水

printing input 打印输入
printing keyboard perforator 印字键盘凿孔机
printing keypunch 印刷式键控穿孔机；打印键控穿孔机
printing lamp 晒像灯
printing line 印铁线
printing machine 晒图机；印刷机
printing machine room 印刷机房
printing material 印刷材料
printing mechanism 印刷机器；印刷机构；印刷机械
printing number 印数
printing office 印刷所
printing oil 油墨
printing-out paper 即时显影相片；像纸；印像纸；直晒纸
printing paper 印图纸
printing paste 印浆
printing plant 印刷厂
printing plate 印刷板
printing precision 测点精度
printing press 印刷机
printing process 印刷方法
printing program(me) 打印程序
printing pseudo-colo(u)r composition 假彩色合成印片法
printing punch 印刷穿孔机；打印穿孔机
printing-punching 印字穿孔
printing quota 印数
printing range 打印范围
printing rate recorder 打点记录式校表仪
printing reperforator 印字复凿孔机
printing room 晒图间；印刷间；印刷车间；印花间；晒图室
printing routine 打印程序
printing server 打印服务器
printing shop 印刷所
printing smoothness 印刷平滑度
printing speed 印象速度
printing surface 印刷面
printing table 印染台
printing technology 印刷工艺学
printing telegraph 印字电报
printing troll(e)y 筛网印花框机
printing tube 晒像管
printing unit 印刷机
printing work 打印工作
printing work order 打印工作通知单
print-marking machine 标composition机
print-member 印刷构件
print-meter 调节晒片时间测时计
print motor 微电机
print name 印刷名
print-on-alarm 报警打印
print operator 打印操作员
print order of map 地图印数
printout 打印输出
print-out equipment 印刷输出设备
printout paper 印相纸
printout task 印出任务
print overlay 晒图涂层
print pad 印轧
print plotter 晒图绘图机
print reference 照片索引
print resistance 抗印刷性；耐印痕性
print restore code 恢复打印码
print room 墙面用印刷品装饰的房间
print routine 印刷程序
print run 全印数
printshop 印刷厂
print speed printing speed 印刷速度
print system 打印系统
printthrough 印透
print totals only 制表
print trimmer 切边机
print value 打印值
print washer 洗片罐
printwork 印花厂；印花机
prionotron 调速管
prior accuracy analysis 先验精度分析
prior activity 先行活动
prior analysis 优先分析
prior appropriation doctrine 优先专用权原则；优先用水权原则
prior approval 预先核准；事先批准

prior-art search 事前技术调研
prior building survey 施工前测量
prior claim 优先索赔权
prior consultation 事先磋商
prior distribution 先验分布
prior endorser 前手背书人
prior equity 前期公平
prior estimate 先验估计
prior examination 预先检查
prior glass 早期玻璃
prior information 先验信息；初始信息
prior informed consent 事先知情同意
prior informed consent procedure 事先通知制度
priorite 钇易解石
prioritization of debt 债务的优先考虑
priority 优先权；优先级；优先
priority algorithm 优先算法
priority-arbitration circuit 优先级仲裁线路
priority at traffic signal 指示先行信号
priority built-in function 优先内函数
priority call 优先呼叫
priority check 优先检查
priority chemical 优先化学品；首要化学品
priority clause 优先条款
priority communication multiplexer 优先级通信多路传输器
priority concordance 优先权相关索引
priority concordance index 优先权相关索引
priority considered pollutant 优先控制污染物
priority construction 优先建筑；优先建设；重点建筑；重点工程；近期建筑；首期建筑；首期建设
priority control 优先控制；优先级控制
priority control unit 优先控制部件
priority definition 优先定义
priority designator 优先等级符号
priority development area 优先发展区
priority discipline 优先规定
priority documents 优先权文件
priority ecosystem 优先生态系统；首要生态系统
priority encoder 优先编码器
priority encoding 优先编码
priority error dump 优先错误转储
priority extension override 优先用户插入
priority facility 优先业务；设备优先级
priority flag 优先标记
priority guidelines of industrial policy 倾斜产业政策
priority indicator 优先指示信息；优先指示符；优先权指示符
priority in running 优先运行
priority installation 优先掩护的设施
priority interrupt 优先中断
priority interrupt basis 优先中断制
priority interrupt channel 优先中断通道
priority interrupt control 优先中断控制
priority interrupt controller 优先中断控制器
priority interrupt control module 优先中断控制模件
priority interrupt control unit 优先中断控制器
priority interruption inhibit switch 优先中断禁止开关
priority interrupt scheme 优先中断方案
priority interrupt table 优先中断表
priority in transmission 优先传递
priority job 优先作业
priority lane 优先车道
priority level 优先级；优先等级
priority lien 优先留置权
priority logic 优先逻辑
priority material 优先考虑的材料
priority memory access 优先存取
priority method 优先法
priority mode 优先方式
priority monitoring 优先监测
priority monitoring pollutant 优先监测污染物
priority network 优先网络
priority number 优先级；优先号
priority of a process 进程的优先权
priority of contract documents 合同文件的优先次序
priority of delivery 投送等级
priority offset 优先(信号)时差
priority of investment 投资重点
priority of lien 法定优先的留置权

priority of operator 运算符优先权
priority of partition 分区优先权
priority of starting 优先通行权
priority of subtask 子任务优先级
priority of system task 系统任务优先数
priority of task 任务优先(权)
priority of water use 用水优先权
priority order 优先次序
priority ordered interrupt 优先有序中断
priority parameter 优先参数
priority parking 优先停车
priority permit sign 特许通行标志
priority phase 优先级相位
priority pollutant 优先污染物；最先污染物；重点污染物；首要污染物
priority pollution rights system 优先污染权制度
priority processing 优先处理
priority program(me) flip-flop 优先程序触发器
priority project 重点项目；重点建设项目；重点工程
priority pseudo-variable 优先伪变量
priority queue 优先排队
priority ranking unit 优先分级单位
priority rating 评定的优先次序；优先检定
priority register 优先寄存器
priority requirement objective list 优先需求目标单
priority research areas 重点研究领域
priority ring 优先振铃
priority route 优先进路；优先放行线
priority routine 优先程序
priority rule 优先规则；顺序法则
priority scheduler 优先调度程序
priority scheduling 优先调度
priority scheduling system 优先调度系统
priority selection 优先选择
priority selection interrupt 优先选择中断
priority sequence table 优先顺序表
priority service 优先服务；优先操作
priority signal 优先信号
priority structure 优先结构
priority substances list 首要物质清单
priority suppliers 优先供应单位
priority switch 优先次序开关
priority symbol 优先标识
priority system 优先系统
priority target 优先目标
priority use of water 用水优先权
priority value 优先显示度(交通标志设计质量指标之一)
priority value of task 任务优先数
prior lien 优先留置权
prior loading 先期加荷
prior notification for hazardous waste transport 危险废物运输的事先通告
prior operation 预加工
prior period 前期
prior-period adjustment 前期损益调整；前期调整项目
prior posterior analysis 先验后验分析
prior pressure application 在升压之前
prior probability 先验概率；预先概率
prior processable 可提前处理的
prior processing 初次加工
prior service cost 前期服务成本
prior to planting 播前
prior value 初值
prior weight 先验权
priory (小的)修道院
prisere 正常演替系列
prising 利用杠杆移动的
prism 斜侧面；棱柱(体)；棱镜(体)；棱晶；三棱形
prisma 棱镜；棱晶
prism alignment 棱镜校准；棱镜校正
prism anamorphic attachment 棱镜变形附件
prism and lens interferometer 棱镜透镜干涉仪
prism angle 棱镜角
prism-angle error 棱镜角差误
prism antenna 棱柱形天线
prism array 棱镜排列
prism astrolabe 棱镜等高仪；天体高度测量仪
prismate astrolabe 棱镜
prismatic 棱柱状的；棱柱；棱形；棱镜；棱晶形的；三棱形；等截面的
prismatic action 棱镜作用
prismatical 柱状

prismatical strength 棱柱(体)强度;长直(抗压)强度
prismatic astrolabe 棱镜等高仪
prismatic bar 等截面杆
prismatic barrel vault 棱柱筒穹顶【建】;角柱筒形穹顶
prismatic beam 柱状梁;棱柱状梁;等截面梁
prismatic beam bending test 棱柱形梁弯曲试验
prismatic beam bridge 等截面梁桥
prismatic beam compressive strength 棱柱形梁抗压强度
prismatic beam crushing strength 棱柱形梁的破坏强度
prismatic beam strength 棱柱形梁强度
prismatic bearing structure 棱柱形承载结构
prismatic bed 棱柱状基础;棱柱状路基;棱柱状河床;棱柱状底基;等截面河床
prismatic billet mo(u)lding 棱镜错齿线脚;棱形长方带线脚;棱柱错齿线脚
prismatic binoculars 棱镜双筒望远镜;棱镜式双目望远镜
prismatic binocular telescope 棱镜双筒望远镜
prismatic blade 直叶片
prismatic blastic texture 柱状变晶结构
prismatic bottom level 用棱镜底部的水准仪
prismatic box girder bridge 等截面箱梁桥
prismatic camera 棱镜照相机;棱镜取景式照相机
prismatic channel 棱柱形河槽;棱镜河槽;等截面河槽
prismatic coefficient 圆柱形系数;棱柱系数;棱形系数
prismatic colo(u)rs 光谱七色
prismatic compact 棱柱形坯块
prismatic compass 棱镜罗盘;棱镜罗经;手提罗经
prismatic compensator 棱镜补偿器
prismatic compressive strength 棱柱体抗压强度
prismatic condenser 棱形聚光器
prismatic-cone mixed cut 角柱与锥形混合掏槽
prismatic crystal 斜方晶;棱晶;方晶
prismatic cupola 棱柱形圆屋顶
prismatic curve 棱形曲线
prismatic cut 角柱掏槽
prismatic cutplate 棱柱切片
prismatic decomposition 棱镜分光
prismatic diffuser 棱柱形扩散器
prismatic disc 三棱镜盘
prismatic dislocation 棱位错;棱镜位错
prismatic dispersion system 棱镜色散系统
prismatic dome 棱柱形穹隆
prismatic drum 棱镜鼓
prismatic effect 棱镜效应
prismatic error 楔形误差;棱性差;棱镜误差
prismatic eye 棱镜搜索镜
prismatic eye piece 棱镜目镜
prismatic face 棱镜曲面;棱柱面
prismatic fibrous texture 柱纤状结构
prismatic focusing magnifier 棱镜聚焦放大镜
prismatic folded plate 棱镜折叠板
prismatic folded plate structure 棱柱状折板结构
prismatic foundation pier 棱柱状基础墩
prismatic glass 一面有锯齿形凹凸花纹(折光)的玻璃;棱形花纹玻璃;棱形玻璃窗;棱形压花玻璃;棱面玻璃
prismatic guide 棱柱形导轨;三面导槽
prismatic habit 棱柱习性
prismatic ingot 棱柱形铸锭
prismatic joint(ing) 柱状节理
prismatic layer 棱柱层;角柱层
prismatic lead spar 棱柱形铬铅矿
prismatic lens 棱镜透镜
prismatic level 棱镜水准器
prismatic light 棱镜光
prismatic light guide 棱镜光导
prismatic loop 棱柱环
prismatic member 棱柱形构件;棱形构件;等截面杆;棱柱构件
prismatic pitched-slab structure 琢有斜边板块的结构
prismatic plane 棱柱曲面
prismatic quartz 棱柱状石英
prismatic reflector 棱镜反射镜;棱镜反光镜
prismatic refraction 棱镜折射
prismatic roof 棱柱状屋顶
prismatic roundness 棱圆度

prismatic rustication 棱突粗面石砌体;棱状粗琢;棱柱状粗琢
prismatic sextant 棱镜六分仪
prismatic shell 棱柱状薄壳;棱柱状壳(体);棱柱壳
prismatic shell cupola 棱柱形折板圆屋顶;棱柱形薄壳圆屋顶
prismatic shell segment 棱柱形折板扇形拱;棱柱形薄壳扇形拱
prismatic slab 琢有斜边的板块
prismatic soil structure 棱柱状土壤结构
prismatic space truss 棱柱形空间桁架
prismatic spectrograph 棱镜摄谱仪
prismatic spectrometer 棱镜分光计
prismatic spectroscope 棱镜分光镜
prismatic spectrum 棱镜光谱;棱镜分光谱
prismatic square 直角镜
prismatic stereoscope 棱镜体视镜
prismatic stride level 棱镜跨式水准仪
prismatic structure 棱柱形结构;柱状构造;棱柱状结构
prismatic substructure 棱柱式亚结构
prismatic sulphur 棱形硫
prismatic supporting structure 棱柱形固定结构;棱柱形下部结构;棱柱形支承结构
prismatic surface 棱柱曲面
prismatic tank 棱柱形储罐
prismatic telemeter 棱镜测距仪
prismatic telescope 棱镜望远镜
prismatic texture 柱状结构
prismatic transit instrument 折轴中星仪
prismatic trommel 棱形滚筒筛
prismatite 柱晶石
prismatoid 旁面三角台
prism attachment 前置棱镜
prism axis 棱镜轴
prism bending test 棱柱体弯曲试验
prism block 棱镜体;棱镜胶盘
prism chromatic resolving power 棱镜色分辨率
prism clamp 棱镜夹
prism combination 棱镜组(合)
prism compensator 棱镜补偿器
prism correction 棱镜校正
prism coupler 棱镜耦合器
prism coupler mount 棱镜耦合器装置
prism cruising 棱镜清查
prism demagnification 棱镜缩小率
prism design 棱镜设计
prism diopter 棱镜折光度;棱镜屈光度
prism dispersing system 棱镜色散系统
prism drum 棱镜鼓
prism-erecting system 棱镜转像系统
prism-erecting telescope 棱镜转像望远镜
prism eyecap 陡视棱镜
prism fabrication tolerance 棱镜制造公差
prism face 棱镜面
prism fill for absorbing earth pressure 减压棱体
prism ghost 棱镜幻像
prism glass 棱纹压花玻璃;棱镜玻璃;三棱镜
prism glasses 棱镜望远镜
prism holder 棱镜框;棱镜架
prism imperfection 棱镜缺陷
prism infrared spectrometer 棱镜红外分光光度计
prism instrument 棱镜仪
prism interferometer 棱镜干涉仪
prism inversion system 棱镜正像系统
prism level 棱镜水准器
prism light 道路棱块;路面采光
prism-like medium 似棱镜媒质
prism-like structure 似棱柱状结构
prism magnification 棱镜放大率
prism method 棱柱法;棱镜法
prism mirror 棱镜反光镜
prism mirror system 棱镜反射镜系统
prism monochromator 棱镜单色仪
prism monocular 棱镜单目镜
prism mo(u)ld 棱柱试模
prism offset 棱镜偏置
prism of sand 砂棱(状)堆
prism of second order 第二柱
prismograph 棱镜仪
prismoid 截头棱柱体;棱柱体;棱台;棱晶样的;平截头棱锥体
prismoidal correction 似棱体校正;似棱体改正
prismoidal formula 似棱柱体公式;棱柱体土方公式;棱柱体土方公式;棱台公式;似棱体公式
prismoidal rule 棱柱体法则
prismoid(al) slab roof 棱柱体形折板屋顶
prismoidal specimen 棱柱试件
prismoidal structure 柱状构造
prismoptric 棱镜屈光度
prismosphere 棱球镜
prism photometer 棱镜光度计
prism plate glass 棱镜板玻璃
prism positioning micrometer 棱镜定位测微仪
prism power 棱镜作用
prism predisperser 棱镜前置色散器
prism probe 棱柱型测针
prism reflector 棱镜反射器;棱镜反光镜
prism refraction 棱镜折射
prism refractometer 棱镜折射计
prism rotation 棱镜旋转
prism sextant 棱镜六分仪
prism shape 棱镜形状
prism spectrograph 棱镜摄谱仪;棱镜光谱仪
prism spectrometer 棱镜光谱仪;棱镜分光仪;棱镜分光计
prism spectrophotometer 棱镜分光光度计
prism spectroscope 棱镜分光镜
prism spectrum 棱镜色散光谱
prism square 直角棱镜;直角镜;棱镜直角器
prism stereoscope 棱镜立体镜
prism storage 柱蓄(量);棱柱蓄水体;河槽蓄水量
prism substrate 棱镜衬底
prism surface 棱镜面
prism table 棱镜座
prism telecope 棱镜望远镜
prism test 棱柱体试验
prism test mo(u)ld 棱柱体试模
prism tile 棱面玻璃砖
prism tile glass 棱形玻璃砖
prism transit 折轴经纬仪
prism tunnel diagram 棱镜展开图
prism turntable 棱镜转台;棱镜转盘
prism-type anamorphotic attachment 棱镜变形装置
prism-type beam splitter 棱镜型分束器
prism-type spectrometer 棱镜型分光计
prism wedge 楔形棱镜
prismy 棱柱;棱镜
prison 看守所;监狱
prison ashlar 琢孔石石
prison building 监狱建筑物
prison construction 监狱结构;监狱建筑
prisoner 固定销;锁紧销
prison safety glass 监狱安全玻璃
pristane 姥鲛烷
pristine area 处女地
pristine beauty 原始美
pristine fiber 原始纤维
pristine/n-17 alkane ratio 姥鲛烷/正十七烷
pristine/phytane ratio 姥鲛烷/植烷
pristine strength 原始强度
privacy 隐密;机密;私密性;保密性;保密
privacy device 隐蔽设备;保密设备
privacy equipment 保密设备
privacy fence 民用栅栏
privacy glass 隐密玻璃
privacy issue 内部发行
privacy key 机密键;保密键
privacy landscape screen 活动隔断
privacy lock 机密锁;保密锁
privacy lock procedure 机密锁过程
privacy problem 保密问题
privacy procedure 保密过程
privacy protection 秘密保护;保密保护
privacy system 保密通信制
privacy telephone set 保密电话机
privacy transformation 保密通信制
private 专用的;民营的;民间的;私营的;非公开的
private access 私人通道;私人入口
private account 个人账号;私人账目
private accountant 专任会计人员
private accounting 民间会计
private address 专用通信处;私人通信处
private-aid public housing 公建民助住房
private aid-to-navigation 私营助航设施;私营导航设备(美国)
private airplane hangar 私人飞机库

private alarm system 专用报警系统
private and confidential 机密(的)
private and secure call 保密安全呼叫
private architect 私人开业的建筑师
private area 私人场地;私用面积(室内外为每户保留的面积)
private armed vessel 民用自卫武装船
private audit 民间审计;不公开审计
private automatic branch exchange 专用自动交换分机;专用自动(电话)小交换机
private automatic branch exchange user 程控交换机用户
private automatic exchange 自动小交换机;专用自动交换机;专用自动电话交换总机
private automatic message accounting 用户自动计费
private automatic telephone system 专用自动电话系统
private balcony 私人阳台;(剧院的)私人楼座
private bank 专线触排;私营银行
private bank contact 专线触排接点
private bathroom 私人浴室
private boat 私家商船
private box 专用火警箱
private branch exchange 用户小交换机;专用支线交换机;专用小交换机;专用交换机;专用交换;专用电话交换机;内部分线交换机
private branch exchange switchboard 内部通信小交换机
private branch switchboard 用户交换台;专用交换台
private brands 私人厂牌
private building 私人建筑;私人房屋
private cable 专用电缆
private call library 专用调用程序库
private canal 专用运河
private car 专车;专用车;企业自备车;私人汽车
private cargo 私人货物
private car ownership 私人车主
private car park 私人汽车停车地;私人汽车停车处;私人汽车停车场
private carriage 专用运输工具
private carrier 专用运输工具;私人运输业
private cartography 民办地图制图业
private channel 专用信道;专用航道
private chapel 私人小教堂
private charter party form 非公认租船契约
private circuit 专用线路
private club 私人俱乐部
private code 自编密码;专用码;专用代码;私人密码
private colo(u)r 秘色
private communication channel 专用通信信道
private communication network 专用通讯网
private communication technology protocol 私有通信技术协议
private company 私营公司;私人公司;非公开公司
private conduit-type sewer 私有管道式下水道
private confidential 机密的
private consumption 单户给水量
private contract 私人契约
private corporation 私营公司
private corridor 私人通道
private cost 私人费用
private councillor 私人顾问
private crossing 专用道路与铁路平交道口
private data 专用数据;私人数据;保密数据
private data network 专用数据网
private data set 专用数据组
private debt 私人债务
private development 私人发展;私人开发
private dining room 私人餐厅;私人餐室;小餐厅
private direct investment 私人直接投资
private dock 私营码头
private door 便门
private driveway 专用支路;专业支路
private dwelling 私人住宅
private dwelling house 私人住房
private enterprise 民营企业;民间企业;私营企业;私人企业
private entity 私人机构
privateer 武装民船;私掠船
private exchange 用户交换机;专用交换机
private facility 专用设施
private favorite 个人收藏夹

private file 专用文件
private fire-fighting force 私人灭火力量
private fire hydrant 自备消火栓
private fire protection system 专用消防系统
private fire service system 专用消防系统
private fund 私人资金
private garage 专用车库;私人汽车库
private garden 私家花园;私家园林
private ground 私有土地
private home 个人家庭
private hospital 私人医院
private house 私有房屋;私房
private housing 私人建房屋;私产房;私房
private housing estate 私人房产业
private hydrant 专用消防栓
private identification sign 私有房产标志牌
private incinerator 专用焚化炉;专用焚化炉
private insured mortgage 保过险的私人抵押财产
private interest 私营企业
private investment 私人投资
private land grant 将公有土地授予私人
private law 私法
private ledger 秘密总账;暗账
private letter-box 专用信箱
private library 专用程序库
private line 专用线路;专用线;专用铁路线;专线;民营铁路
private line arrangement 专线连线
private line car 厂商自备车
private line interconnection service 专线互联业务
private line service 专用线业务;专用线服务;专线业务
private line two-way trunk circuit 专线双向中继电路
private loan 私人贷款
private local memory 专用局部存储器
private lot 宅基地
privately aided public housing 公建民助房屋
privately operated automatic alarm system 民用自动火灾报警系统
privately-owned civil aerodrome 私人民航机场
privately-owned container 私有集装箱
privately owned industrial and commercial enterprises 私营工商业
privately-owned line 民营铁路
privately-owned railway 民营铁路
privately-owned ship 私有船泊
privately owned water-supplies 自备水源
private manual branch exchange 专用人工支线交换机
private manual exchange 专用人工交换机
private memory 专用存储器
private monopoly 私人专利
private mortgage insurance 个人抵押保险
private network 专用网(络)
private occupancy 私人占用
private office layout 单间式布局
private open area 专用旷地;私人旷地
private open space 私人游憩用地;私人空地
private-owned corporation 非公众性公司
private owned railway 私有铁路;私营铁路
private ownership 私有制
private ownership of land 土地私有制
private packet network 专用包网
private parking area 私人停车场;私人停车区
private parking garage 专用车库;私人车库
private parking lot 私人停车地区;私人停车地段
private parking place 私人停车地(点)
private path 私人用小路
private person 私人
private phototelegraph station 专用传真台
private port 工业港(口);私有港口;私人商港;厂用码头;厂矿自设港口
private process stack 专用进程栈
private property 私有财产;私产
private property tax 私人财产税
private property week 私有产业周
private provision of water 私人储备的水
private quay 私营码头;厂用码头
private railway 专用铁路线;民营铁路;私有铁路;私营铁路
private real-time relay 保密实时转播
private regulation 民间准则
private relocatable library 专用浮动式存储库

private rented housing 私人出租住宅
private residence 私人住所
private residence elevator 私人住宅电梯
private residence inclined lift 私人住宅斜坡升降机
private residential swimming pool 私宅游泳池;私宅游泳池
private right 专用权
private river 夺流河
private road 私道路;专用道路
private room 个人用房间;私人室;私室
privates 第三导线
private school 私立学校
private seal 私人签署
private section 专用节
private sector 私有部分;私营企业界;私营部门
private semaphore 专用信号量
private sewage disposal 专用污水处理厂;私有污水处理厂
private sewer 地方性污水管;自用污水管;自备污水管;专用下水道;专用污水管;内部污水管
private shares 私人股份
private short-term capital movement 私人短期资本流动
private siding 专用铁路支线;专用线;民营岔线
private signal 轮船公司自订信号
private space 个人占用面积;个人占用空间;私用空间
private stable 专用马桶;小型马桶
private stairway 专用楼梯;独用楼梯间;私用楼梯
private station 专用电台
private storage 专用存储器
private storage garage 私人储藏室
private stream 夺流河
private swimming pool 私人游泳池
private switched telephone network 专用电话交换网
private switching network 专用电话交换网
private system 私有系统
private tap 专用水栓
private telegraph and telephone service 专用电报和电话服务
private telephone 专用电话;专线电话
private telephone line 专用电话线
private telephone network 专用电话网
private telephone set 自用电话机
private telephone system 私人电话系统
private tender 指定投标
private terminal 专用码头;厂用码头
private terms 专用条款;私人条款
private track 民营岔线
private transit 私人交通
private transportation 专用客车运输
private treaty 私人协议;财产出让契约
private trust-fund 私人信托款
private use of water 私用水
private utility 私有公用事业;私家公用事业
private vessel 私营船舶
private view 预展
private village housing 私人乡村屋宇
private volume 专用卷宗;专用卷帘;专用存储媒体
private volumn 专用量
private ward 单间病室
private water 私有水
private water distribution system 私有配水系统
private water supply 私有给水工程
private water supply equipment 专用供水设备
private water supply system 非公用给水系统;专用供水系统
private waterway 非公用水道
private waterworks 专用给水
private well 私用井;自备水井
private wharf 专用码头;私营码头
private wiper 测试电刷
private wire 专用路线;专线;塞套引线;测试线
private wire connection 专用线路连接
private wire network 专用通讯网
private wire service 专用线业务
private-wire teletypewriter system 专线电传系统
privatization 私有化
privatizing ownership of the land 土地私有化
privet 女贞属【植】
privileged access 特许存取
privileged direction 优先方向;优惠方位
privileged expenses 特殊费用

privileged input 特许输入
privileged instruction 特许指令;特权指令
privileged memory operation 特许存储器操作
privileged module 优先模块
privileged norm 特许规范
privileged operation 特许操作
privileged operation exception 特殊操作异常;特权指令异常
privileged price 优待价
privileged processor state word 处理器特权状态字
privileged processor status 处理器特权状态
privileged program(me) state 优先程序状态
privileged ship 直航船(被让路船)
privileged sites 特殊部位
privileged state 特许状态;特惠状态
privileged vessel 权利船;被让路船
privilege of access 存取特许
priving bogie 主动转向架
privity of contract 合约当事人;合同默契
privy 室外厕所
privy pit 私人矿井;私人用灰坑;坑厕
privy vault 茅坑;通气式便桶
prixite 砷铅矿
prize 撬开;撬棍;撬杠;被捕船
prize competition design 有奖设计竞赛
prize for architecture 建筑设计奖
prize law 捕获法
prize master 捕获船押送官
prize ring 拳击场
prize-winning 获奖
proactivator 激活剂前体
pro-active fiscal policy 积极的财政政策
proactive inhibition 经验的阻碍作用
proagglutinoid 强力类凝素
pro and con 赞成和反对
proaulion 教堂门厅;教堂门廊
probabilistic 概率性的;不确定性的
probabilistic algorithm 概率算法
probabilistic analysis 概率分析
probabilistic analysis of risk 风险概率分析
probabilistic approach 或然机遇论方法
probabilistic assessment of groundwater contamination 概率统计法地下水污染评价
probabilistic automatic pattern analyzer 概率自动模式分析器
probabilistic automaton 随机自动机
probabilistic automator 概率自动机
probabilistic base 或然机遇基础
probabilistic budget 依据概率预算
probabilistic budgeting 概率预算
probabilistic demand 或然需要;随机需要
probabilistic design 概率设计
probabilistic design method 概率设计(方)法
probabilistic dominance 概率支配;概率超优
probabilistic equation 概率方程
probabilistic error estimation 概率性误差估计
probabilistic fire model 概率火灾模型
probabilistic grammer 概率文法
probabilistic graph 概率图
probabilistic hydrology 概率水文学
probabilistic indefiniteness 概率不明确性
probabilistic logic 概率逻辑
probabilistic long range planning 可能性的远景规划
probabilistic machine 随机元件计算机
probabilistic method 概率统计(方)法;概率(方)法
probabilistic model 概率模型;随机模型
probabilistic monotonicity 概率的单调性
probabilistic multiroute traffic assignment 多路线概率分配法
probabilistic multistage model 概率多级模式
probabilistic nature 概率性;随机性
probabilistic network evaluation technique 概率网络估算技术
probabilistic optimality method 概率优化(方)法
probabilistic plot 概率图
probabilistic programming 概率规划
probabilistic sampling 概率抽样
probabilistic search 概率搜索法
probabilistic simulation 概率仿真
probabilistic stratigraphy 概率地层学
probabilistic system 概率系统;随机系统
probabilistic theory 概率理论
probabilistic time 概率时间

probabilistic uncertainty 概率(的)不确定性
probabilistic vibration 概率振动
probabilistic water quality control policy 概率水质控制法
probabilities of advantage 可能牟利的机会
probability 概率;可能性;几率;机率;或然率
probability after effect 后效概率
probability amplitude 概率振幅;几率幅
probability analyser 概率分析仪
probability analysis 概率分析;可能性分析
probability analysis compaction 概率分析精简数据法
probability analyser 概率分析仪
probability and statistic(al) analysis program (me) 概率统计分析程序
probability a posteriori 后验概率
probability assessment with simulation 模拟概率评估
probability calculus 概率演算
probability concept 概率概念
probability cumulative curve 概率累积曲线
probability current 概率流量;几率流量
probability current density 几率流密度
probability curve 概率曲线;几率曲线;或然率曲线
probability damage matrix 概率破坏矩阵
probability decision function 概率判决函数
probability density 概率密度;几率密度;机率密度
probability density analyser 概率密度分析器
probability density distribution 频率分布
probability density function 概率密度函数
probability description 概率描述
probability description of random processes 随机过程的概率描述
probability design method 概率设计(方)法
probability deviation 概率偏差;或然误差;或然偏差
probability differential 概率微分
probability distribution 概率分布;几率分布;随机变量分布
probability distribution analyser 概率分布分析器
probability distribution curve 概率分布曲线
probability distribution function 概率分布函数
probability distribution of a continuous variable 连续变量的概率分布
probability distribution of a discrete variable 离散变量的概率分布
probability distribution of random variable 随机变量的概率分布
probability event 概率事件
probability factor 概率因数;几率因数
probability forecast(ing) 概率预报
probability function 概率函数;机率函数
probability generating function 概率母函数
probability graph method 概率图形法
probability graph paper 概率坐标纸
probability independence 概率性独立;随机独立
probability index 概率指数
probability in discrete space 离散空间概率
probability interpretation 概率解释
probability law 概率律;几率律;机率律
probability level 概率水平
probability limit 概率界限;概率极限
probability limits for control chart 控制图的概率极限
probability line 概率线
probability measure 概率测度
probability method 概率统计(方)法;概率法
probability model 概率模型;随机模型
probability nature 概率特性
probability of adapting strategy 可采用策略的概率
probability of a non-zero family error rate 非零族概率误差率
probability of construction 施工可能性;施工概率
probability of damage 破坏概率
probability of energy loss 失电量概率
probability of equipment failure 设备故障概率
probability of error 误差概率
probability of error per digit 每位错误概率
probability of events 机率事件
probability of exceedance 超值概率;超越概率;超过概率
probability of failure 破坏概率;失效概率;失败概率
probability of fatigue failure 疲劳破坏概率
probability of happening 事件概率
probability of large deviation 大偏差概率

probability of malfunction 故障概率;失灵概率
probability of misjudgment 判断错误的概率
probability of missing bit 漏码概率
probability of new marketing 开发新市场概率
probability of new mining 增建矿山概率
probability of new usage develop 新用途开发概率
probability of none earthquake 无震概率
probability of none event 无震概率;无事件概率
probability of non-exceedance 不及概率;不超过概率
probability of nonfailure 无故障概率
probability of obstacle-clearing 越障概率
probability of occurrence 事件概率;出现机率;出现概率;发生概率
probability of occurrence of wave height 波高出现概率
probability of outcome 结果概率
probability of performance failure 性能故障概率
probability of probit 概率单位变换
probability of radar detection 雷达发现概率
probability of recombination 复合概率
probability of repeated trials 重复试验的概率;重复试验的概率
probability of ruin 破产概率
probability of scattering 散射概率
probability of stability 稳态概率;稳定概率
probability of substitutional develop 代用品开发概率
probability of success 成功概率
probability of successful operation 有效运转概率
probability of survival 可靠概率;幸存概率;或然概率;残存概率
probability of type I error 第一类错误的概率
probability of type II error 第二类错误的概率
probability of yielding 屈服概率
probability optimization 概率优化法
probability paper 概率坐标纸;概率纸;概率格纸
probability paper test 概率纸检验
probability parameter 概率参数
probability prediction 概率预测
probability-preserving transformation 同概率转换
probability proportional sampling 概率比例抽样(法)
probability quota sampling technique 概率配额抽样法
probability ratio sequential detector 概率比顺序探测器
probability regression method 概率回归法
probability sample 概率样本;随机样品;随机样本
probability sampling 概率抽样
probability scale 概率比例尺
probability screen 概率筛
probability space 概率空间
probability statistics 概率统计
probability system 概率系统
probability table 概率表
probability theory 概率论;概率定理
probability unit 概率单位
probability unit method 概率单位法
probability unit transformation 概率单位变换
probability value 概值;概率值
probability vector 概率向量
probability weight 概率加权
probable amplitude 概率幅值
probable analysis 概率分析(法)
probable analyzer 概率分析仪
probable approach 概率方法
probable coefficient 概率系数
probable construction cost 可能建筑费用
probable correlation 概率相关
probable current 可能的海流
probable curve 概率曲线
probable density 概率密度
probable design 概率设计
probable deviation 概差;可能偏差;或然偏差
probable discharge 可能流量;可能(发生)流量
probable distribution 概率分布;可能分配
probable distribution analyser 概率分布分析器
probable distribution function 概率分布函数
probable effect limit 概率有效极限
probable equation 概率方程
probable error 概然误差;概念误差;概率误差;概差;或然误差;可几误差;近真误差;几率误差;或是误差;或然误差;大概误差

probable error circle 概率误差圆
probable error of mean 平均值的概差
probable error of standard deviation 标准差的概差;标准变差概差
probable errors based upon different samples 样本性概差
probable estimate 概率估计(值)
probable estimation 概率评定
probable evaluation method 概率估算方法
probable flood 可能洪水;可能发生洪水
probable flow 可能(发生)流量
probable generating function 概率母函数
probable independence 概率独立
probable integral 概率积分
probable life 预计使用期限;预计使用年限;预测使用期限;预测使用年限
probable line 概率线
probable maximum 概极大值
probable maximum flood 概率最大洪水;可能最大洪水
probable maximum loss 可能发生的最大损失额
probable maximum precipitation 可能最大降水(量)
probable maximum storm 可能最大暴雨
probable maximum water level 可能最高水位
probable mineral 可能的矿物
probable minimum 概极小值
probable minimum water level 可能最低水位
probable model 概率模型
probable nature 概率性质
probable ore 推定储存矿量
probable parallelogram 或然误差平行四边形
probable period 或然周期
probable position 最或然船位
probable precipitation 可能降水量
probable range 概率范围
probable reserves 估计储量;概略储量;控制储量;可能储(藏)量
probable seismic design 概率抗震设计
probable sizing 预先筛分
probable stability 可几稳定性;大概稳定性
probable strength 似真强度;大概强度
probable theory 概率论
probable value 概值;概然值;概率值;可能值;可几值
probable velocity 概然速度;概念速度;概率速度;可能速度;近真速度
probang 除鲠器
probate court 遗嘱检验法庭
probate sale 在遗嘱检验法庭监督下出售财产
probation 见习;试用
probation agricultural insecticide 试用农药杀虫剂
probationary period 预备期;试用期
probationary ward 观察病室
probationer 见习生;试用人员;实习生
probation period 见习期;试用期
probation report 检定报告
probe 横销;探子;探头;探试;探视器;探空火箭;探查;探测针;探测剂;探棒;附加器
probe aircraft 试验机(有实验设备的飞机)
probe and drogue 插头与锥管
probe antenna 探针天线
probe body 探测器外壳
probe boring 试钻
probe card 探针板
probe carriage 探针架
probe coil 探针线圈;探头线圈;探测器线圈
probe compartment 探测设备舱
probe conduction 探针电导
probe counter 探头计数器
probe coupling 探针耦合
probe current 探针电流;探测电流
probe diameter 探针直径
probe diffraction 探针衍射;探针绕射
probe drilling 勘探钻井
probe-drogue 探管锥套式
probe-drogue system 探管锥套系统
probe electrode 试探电极
probe-forming system 探针形成系统
probe gas 探测气(体);试验性的气体;示漏气体探针
probe handle 探柄
probe housing 探测器外壳
probe inclinometer 探头测斜仪
probe inlet 试探孔

probe insertion 探头插孔
probe inspection 取样检验
probe lamp 探查灯
probe machine 探针仪
probe material 示踪物质
probe measurement 探针测量
probe method 针入法;探针(测试)法;探头法
probe-microanalyzer 探针显微分析仪
probe microphone 探管传声器
probe particle 试探粒子
probe positioner 探头定位器
prober 探针器
probe radius 探头半径
prober interface 探示器接口
probe rocket 探测火箭
probe rod 探杆
probertite 斜硼钠钙石;基性硼钠钙石
probe sampling 探针取样;探针采样
probe shield 探头屏蔽
probe-shift 探头移位
probe signal 试探信号
probe spacing 探针间距
probe station 探针台
probe-strength 探针触探强度(混凝土);探针触探强度
probe target 试探靶
probe technique 探针技术
probe temperature 仪表指示的温度
probe test 探试器测试
probe tube 取样管
probe type leak detector 探针型检漏器
probe type level indicator 探针式料位指示器
probe-type liquidlevel meter 探针式液位计
probe-type pyrometer 探针式高温计
probe-type vacuum tube voltmeter 探针式电子管电压表
probe unit 检测器;测试装置
probe vehicle 探测火箭
probe window 探头窗口
probing 探针探查;探索;探测;触探
probing distance 探测距离
probing staff 探测标杆
probing step 探测阶段
probing technique 探测技术
probing water and draining water in the mine 矿井探水与放水
probit 概率单位;几率单位
probit analysis 概率单位分析
probit-logarithm drawing 机率对数图纸
probit method 概率单位法;机率单位法
probit unit 机率单位
problem analysis 问题分析
problem analysis by logic(al) approach 逻辑逼近分析法
problem area 研究地区;难题地区
problematic social situation 不能预测的社会情况
problem-board 解题插接板
problem characteristics and fire types 问题特征与火灾类型
problem data 问题数据;题目数据
problem definition 问题说明;问题定义;题目说明;定义问题
problem description 问题说明;题目说明
problem determination 问题决定
problem diagnosis 问题诊断
problem file 题目文件
problem formulation 问题定式化
problem input tape 题目输入带
problem job 疑难工程
problem language 题目语言
problem mode 解题状态;问题方式
problem of accretion 淤积问题
problem of clean bill of lading 清洁提单问题
problem of cost balance 成本平衡问题
problem of deformation of foundation 地基变形问题
problem of elimination 消元问题
problem of fixed detachment 船迹问题
problem of ground subsidence 地面沉降问题
problem of immersion of groundwater 地下水浸没问题
problem of karst 岩溶稳定问题
problem of many bodies 多体问题
problem of model selection 模型选择问题

problem of policy 政治问题
problem of random walk 随机走动问题
problem of seepage and permeation deformation 渗漏与渗透变形问题
problem of stability foundation 地基稳定问题
problem of stability of mountain bodies 山体稳定问题
problem of stability of regional earths crust 区域地壳稳定问题
problem of stability of rock mass 岩体稳定问题
problem of stability of slope 边坡稳定问题
problem of stability of soil mass 土体稳定问题
problem of stability underground excavation 地下洞室稳定问题
problem of three-bodies 三体问题
problem of two bodies 二体问题
problem of two fixed centers 双不动中心问题
problem oriented 面向问题的
problem-oriented language 面向问题的语言
problem-oriented software 面向问题的软件
problem program(me) 解题程序;问题程序
problem program(me) partition 解题程序分区
problem section 难题地段
problems of common concern 共同关心的问题
problem solving 问题求解;问题解决
problem-solving ability 解决问题的能力
problem-solving behavio(u)r 问题解决行为
problem-solving capability 解决问题的能力
problem solving mode 解题法
problem-solving theory 解题理论
problem system 问题系统
problem time 解题时间
problem waste 问题废物
problem with equality constraint 等式约束问题
problem with inequality constraint 不等式约束问题
probolog 电子探伤仪;电测定器
proboscis 吻突
procathedral 临时教堂
proceding crop 轮作中的前作物
procedural 过程的
procedural and exception test 例行和异常检查
procedural attachment 过程的附加段
procedural audit 程序审计
procedural block 过程块;程序块
procedural command language 过程命令语言
procedural control 程序控制
procedural form 程序图
procedural grammar 过程文法
procedural knowledge 过程知识
procedural knowledge representation 过程知识表示法
procedural language 过程语言;面向过程的语言
procedural law 仲裁使用的程序法;诉讼法;程序法
procedural law of environmental 环境程序法
procedural model 过程模型;程序模型
procedural semantic model 过程语义模型
procedural standard 过程标准
procedural test(ing) 过程测试
procedure 作业;规范;手续;程序(框图);方法过程;步骤
procedure activation 过程激励;过程活动
procedure analysis 过程分析;工序分析
procedure analysis chart 程序分析图
procedure as parameter 参数过程
procedure block 过程块;过程分程序
procedure body 过程体
procedure call 过程调用
procedure call information 过程调用信息
procedure call request 过程调用请求
procedure chart 过程;程序图
procedure command 过程命令
procedure component 过程成分
procedure control 工作技术监督;工序技术检查
procedure data area 过程数据区
procedure declaration 过程说明
procedure definition 过程定义
procedure descriptor 过程描述符
procedure design 过程设计
procedure designer 过程设计员
procedure division 过程划分;过程部分
procedure entry 过程入口
procedure entry mask 过程入口屏蔽码
procedure execution 过程执行

procedure exit 过程出口
procedure for arranging a contract 合同洽订程序
procedure for claims 索赔程序
procedure formation 过程生成
procedure for preparation 备料工序
procedure for refund 退费手续
procedure function 过程函数
procedure grammar 过程文法
procedure head 过程首部
procedure heading 过程导引;过程标题
procedure identification 过程标识符
procedure identifier 程序识别器
procedure in bankruptcy 破产程序
procedure incarnation 过程实体
procedure initialization 过程预置
procedure in production 生产过程
procedure input 过程输入
procedure invocation 过程引用
procedure language 程序语言
procedure level 过程级;过程层次
procedure library 过程库;过程程序库
procedure manual 作业程序手册
procedure map 过程图;处理图
procedure member 过程成员
procedure name 过程名字
procedure of analysis 分析程序
procedure of approval 批准程序
procedure of capital construction 铁路基本建设程序
procedure of cargo handling 装卸工序
procedure of extended reach drilling 延伸钻井程序
procedure of geologic(al) surveying 地质调查程序
procedure of mining exploitation 矿山开发程序
procedure of permit system of pollutant discharged 排污许可证实行程序
procedure of repairing 修理程序
procedure of site investigation 勘察工作程序
procedure of stable production 稳产措施
procedure of test 分析手续;分析步骤
procedure of trial production of new products 新产品试制程序
procedure of urban planning 城市规划编制程序
procedure of using land for state construction 国家建设用地程序
procedure of verification 检验程序
procedure-oriented 面向过程
procedure package 过程包;程序包
procedure parameter 过程参数
procedure pointer 过程指针;过程指示字
procedure qualification 工艺评定;程序评定
procedure reference 过程引用
procedure return 过程返回
procedures and arrangements manual 程序与布置手册
procedure schema 过程模式
procedure section 过程节;过程段
procedure segment 过程段
procedures for transfer payments 转移支付办法
procedures for transition 变换过程
procedures manual 过程手册
procedures of new product development 新产品开发程序
procedure specification 过程说明
procedure statement 过程语句
procedure step 过程步
procedure structure 过程结构
procedure subprogram(me) 过程子程序(段);过程辅程序
procedure subroutine 过程(型)子程序
procedure symbol 过程符号
procedure to apply 申请步骤
procedure to follow 遵守规定;遵守程序
procedure-tons by cargo-handling machines 机械操作工序吨
procedure-tons of cargo-handling 装卸工序吨
procedure track 程序航迹
procedure turn 程序转向
procedure value input 过程值输入
procedure work log system 程序工作记录系统
proceed 开始;进行;继续进行;起诉
proceed at restricted speed 限速运行
proceed down the river 下水航驶;沿江下驶

proceeding measurement 顺序测量
proceedings 科研报告集;纪要;会刊;学报;会议论文集;会议录;事项
proceedings from production 生产收益
proceedings of a conferences 会刊
Proceedings of American Society for Testing and Materials 美国材料试验学会会刊
Proceedings of American Society of Civil Engineers 美国土木工程师学会会刊
Proceedings of Institution of Mechanic(al) Engineers 机械工程师学会会报(英国)
Proceedings of National Academy of Sciences of the USA 美国国家科学院院报
Proceedings of Royal Society 英国皇家学会会刊
Proceedings of the Institution of Civil Engineers 土木工程师学会会刊(英国)
Proceedings of the Institution of Electric(al) Engineers 电气工程师学会会刊(英国)
proceed into harbo(u)r 进港
proceed out of harbo(u)r 离港
proceeds 收入
proceeds at the meeting 会议事项
proceed signal 进行信号【铁】
proceeds in cash 货币收入
proceeds of instalment sales 分期付款销售收入
proceeds of loan 贷款所得
proceeds of sale 销售收入;销货收入;实收款项
proceed technical and economic appraisal 进行技术经济论证
proceed-to-select 开始选择
proceed up the channel 沿水道上驶;上水行驶;上水航驶
procellarian 风暴洋纪
procenium 舞台防火幕(框);舞台防火帘
process 过程;工序;进程;加工;手续;处理过程;处理;分支过程;方法
processability 制备性能;可加工性;加工性能;成型性能
processability of product structure 结构工艺性
processable card 可处理卡片
processable scored card 可加工的得分卡片;可处理型记分卡;可处理缺口卡片
processable task graph 可处理任务图
process adaptability 工艺适应性
process air-conditioning 工艺性空气调节;工业用空气调节
process allowance 空裕时间;空裕
process a loan 办理贷款手续
process amplifier 处理放大器
process analysis 过程分析;流程分析;生产过程分析
process annealing 中间退火;工序间退火;低温退火
process art 概念艺术
process attachment table 进程连接表
process automation 过程自动化;工艺过程自动化;工艺程序自动化;工序自动化;加工自动化;生产过程自动化
process average 过程平均
process blister 轧制泡沫表面
process box 工作框
process building 工艺过程厂房
process camera 制版照相机;复写仪;合成照相机;双片照相机
process capability 加工能力
process chain 流程链
process characteristic 过程特性;工艺流程特性
process chart 工作程序图;工艺图;工艺流程图;工艺卡(片);工艺程序图;工艺作业图;工序图;流程图表;加工流程图;施工程序图;生产指示图表;程序框图
process check 过程检查;进程检查
process chemistry 过程化学;工艺化学
process chromatograph 流程色谱仪
process code 过程码
process coding mode 处理编码方式
process-colo(u)r relief presentation 彩色地貌图
process communication 进程通信
process communication system 过程通信系统;进程通信系统
process computer 过程控制计算机;过程计算机;工艺程序计算机;处理计算机;程序计算机
process condition 工艺条件(操作)条件
process configuration 工艺过程排列
process connection 工艺连接件
process console 过程控制台

process construction 过程生成
process context 进程关联
process context stack pointer 进程关联堆栈指示字
process context switching 进程关联转换
process control 过程控制;工艺控制;工艺过程控制;工艺管理;工艺程序控制;工序控制;连续调整;进程控制;加工程序控制;生产过程控制
process control analog(ue) modules 过程控制模拟组件
process control block 过程控制块;过程控制程序
process control center 工艺控制中心
process-control chart 工艺控制图
process control compiler 过程控制编译器;过程控制编译程序
process control computer 过程控制计算机
process control computer system 处理控制计算机系统
process control engineering 程序控制工程
process control equipment 过程控制设备
process control instrument 过程控制仪表;工艺程序控制装置
process control interface 过程控制接口
process control language 过程控制语言
process controller 过程控制器;工艺过程控制装置;工艺操作控制器;程序控制装置
process control loop 过程控制回路;过程控制环
process control micro processor 过程控制微处理机
process control model 过程控制模型
process control of cost 成本过程控制
process-control setting 程序控制调定
process control software 过程控制软件
process control system 过程控制系统;处理控制系统;程序控制系统
process control technique 工艺过程控制技术
process control unit 过程控制单元
process-control viscometer 程序控制黏度计
process cost 工序成本
process cost accounting 分步成本会计
process costing 分步成本计算法;分步成本计算
process cost sheet 分步成本单
process cost system 分步成本计算;分步成本制度
process coupling unit 过程耦合单元
process criterion 加工标准
process curve 过程线;过程曲线;发展曲线
process data 整理资料;过程数据;分理数据
process decision program(me) chart 过程决策程序图
process design 过程设计;工艺(流程)设计;流程设计;生产流程设计;生产流程的设计
process development 工艺过程开发
process diagram 工艺图
process dispatcher 进程调度程序
process distinguish stack 进程区分栈
process distinguish stack number 进程区分栈号
process disturbance 过程扰动
process drawing 工艺图;工艺过程图
processed aggregate 加工集料
processed bentonite 加工膨润土
processed bit 工厂生产的钻头;成批生产的钻头
processed black-and-white negative film 处理黑白负片
processed black-and white positive film 处理黑白正片
processed black-and-white prints 处理黑白相片
processed colo(u)r positive film 处理彩色正片
processed colo(u)r prints 处理彩色相片
processed diamond 经处理的金刚石
processed drilling fluid 经处理的钻探泥浆
processed excrement 处理过的粪便
processed filament yarn 变形长丝
processed film 已冲洗胶片;处理胶片
processed gas 精制过的气体;加工过的气体;脱硫气体
processed glass 加工玻璃;精制玻璃
processed goods 加工品
processed human excreta 处理过的粪便
processed information 修正数据
processed juice 无菌汁
processed material 加工材料;已加工(原)材料
processed negative film 处理彩色负片
processed prints 处理相片
processed quicklime 磨细生石灰
processed rock 加工石料

processed rubber 加工过的橡胶
processed shake 锯成的木瓦；工艺环裂；装饰裂纹；加工裂缝
processed sheet 精整薄板
processed steel products 钢铁制品
processed water 加工过的水
process energy consumption 工艺耗能
process engineer 工艺工程师；程序工程师
process engineering 制造技术；制造工程；加工技术；过程工程；工艺过程(管理)；加工工程；生产过程工程；程序工艺工程；程序工程
process engraving 照相制版
process engraving ink 凸图版腐蚀墨
process entry 过程入口；过程表目；进程表目
process equipment 工艺设备；处理设备；工艺流程；加工设备
process equipment and machine 工艺设备与机器
process equipment design 化工设备设计
process evaluation 过程评价；过程评估；工艺过程评价
process event 过程事件
process exception 过程异常；进程异常
process express 过程表示
process flow 工艺流程；生产流程
process flowchart 进程流图
process flow diagram 工艺流程图
process flow diagram and layout 工艺流程及布置图
process flow sheet 生产流程图
process flowsheet features 工艺流程特点
process fluid 工艺流体
process fluid test 工艺流体试车
process for drying tobacco 烟叶干燥法
process fuel 加工用燃料
process furnace 化工过程加热炉
process gain 过程增益
process gas 工艺气体；工业废气；生产气体；生产废气
process gas scrubber 气体回收洗涤器
process goods locally 就地加工
process header 进程标题
process header slot 进程标题槽
process heat 生产过程用热
process heating 工艺加热；加工加热；程序加热
process heating load 生产工艺热负荷
process hold up time 操作中断时间
process identification 过程辨识；进程标识
process imput table 进程输入表
process-induced defect 处理引起的缺陷
process industry 制造工业
processing 作业；加工；处理加工
processing agent 加工助剂
processing aid 操作助剂；加工助剂
processing alloy 待熔合金
processing amplifier 程序放大器；信号处理放大器
processing and arranging of railway transport statistics 铁路运输统计加工整理
processing and control element 处理控制单元；处理和控制单元
processing array wafer 处理阵列圆片
processing bank 押汇银行
processing block 程序方块图
processing by computer 计算机处理；电算整理
processing capacity 处理能力；加工能力
processing cell 生产操作室
processing chain 加工链
processing check 处理校验
processing condition 处理条件；加工条件
processing control 处理控制
processing cost 加工成本；安装费；装配费；加工费；选矿综合费用
processing criterion 加工标准
processing cycle 加工周期；处理循环
processing data 处理数据
processing deal for export 委托加工外销；委托加工出口
processing device 处理设备
processing effect 显影效应
processing element 处理元件；处理部件
processing elementary memory 处理单元存储器
processing ensemble 处理集合机
processing entry 处理项目
processing equipment 处理设备；工艺设备；工艺流水作业线设备

processing error 车辆误差
processing expenses 加工费用
processing facility 装配设备；加工设备
processing factory 加工厂
processing furnace 加热炉
processing gain 处理增益
processing gas turbine 生产用燃气轮机
processing imported materials 进料加工
processing industry 加工工业
processing industry area of aquatic products 水产加工业区
processing installation 工艺装置
processing instrument 加工设备
processing intent 处理意图；处理目的
processing interrupt 处理中断
processing interrupt status word 处理中断状态字
processing interrupt time sharing 处理中断分时
processing know-how 操作秘密
processing limit 处理限制；处理极限
processing line 生产线
processing list 处理表
processing load 处理量
processing machine 洗片机
processing machinery 工艺设备
processing material for export 加工出口用料
processing materials supplied by clients 来料加工
processing method 处理方法；加工方法；加工法
processing methods of water analytical data 水分析资料整理方法
processing method with coherent/partial coherent light 相干光处理方法
processing method with incoherent light 非相干光处理方法
processing mill 加工作坊
processing mode 处理方式
processing modularity 处理模式化
processing module 处理模块
processing natural gas 加工的天然气
processing of aggregate 骨料制备
processing of auxiliary work 辅助作业过程
processing of coherent noise 不正常噪声处理
processing of crude oil 原油(的)加工
processing of data 资料整理；资料整编；资料处理；数据处理
processing of drinking water 饮用水的处理
processing of hydrologic(al) data 水文资料整理；水文资料整编
processing of passengers 旅客进出口站手续
processing of radioactive data 放射性数据处理
processing of read request 读请求的处理
processing of statistic(al) material 统计资料的加工整理
processing of swimming pool water 游泳池水的处理
processing oil 加工油
processing organic liquid waste 加工有机废液
processing parameter 工艺参数
processing parts 处理部件
processing passengers and baggage 运输旅客和行旅
processing plant 处理设备；处理厂；炼油厂；加工装置；加工处理厂；加工厂；石油加工厂
processing platform 处理平台
processing procedure 加工方法
processing program(me) 处理程序；加工程序
processing program(me) table 处理程序表
processing property 工艺特性
processing quality 加工品质
processing repetitive transactions 重复事项处理
processing result 治理效果
processing safety 操作安全(性)
processing schedule of bars 钢筋加工表
processing scheme 工艺方案；加工流程(图)
processing season 加工季节
processing section 处理段
processing sequence 处理顺序；加工顺序
processing service 处理业务；处理服务
processing set-up 工艺布置
processing speed 处理速率；加工速率
processing support 处理支持；加工支援(程序)
processing system 处理系统
processing tank 显影槽；洗片桶
processing tax 加工税
processing technic 处理技术；加工技术；加工工艺

processing technique 制造工艺；加工技术
processing technologic(al) sampling 加工技术取样
processing technology 加工工艺
processing temperature 处理温度；操作温度；作业温度；加工温度
processing time 处理时间
processing tower 加工塔；处理塔
processing trade 加工贸易
processing train 加工程序；加顺时；加工序列
processing uncoiler 机组开卷机
processing unit 处理装置；处理设备；处理机；处理单元；处理单位；处理部件；处理器；运算器；加工装置；加工设备
processing waste 加工废料；生产废料
processing water 加工水
processing window 工艺窗
processing with customer's material 来料加工
processing with supplied drawings 来图加工
processing with supplied materials 来料加工
processing workshop 加工车间
process input image 过程输入映像
process input-output 过程输入输出；进程输入和输出
process inspection 工序检验
process instrumentation 生产过程用检测仪
process instrumentation drawing 工艺仪表流程图
process interface module 过程接口模块
process interrupt 过程中断；进程中断
procession 旋进
processional moment 转动惯量；惯性矩
processional path 队列行进的道路
processional temple 进行列队仪式的庙宇
process lag 过程延迟
process lapse rate 过程直减率
process layout 分类布置
process lens 复照镜头；分色镜头
process limited 处理时限
process line 工作顺序；工艺管道；生产流水线
process loading criterion 工艺负荷标准
process-loading factor 过程负荷系数
process location plan 过程位置图
process loop test 循环进程测试
process machinery 加工机械
process management 工艺程序管理
process map 进程图；进程代换信息
process mark 加工符号
process material 复照材料
process metallurgy 过程冶金学；生产冶金(学)
process method of cost-finding 分步计算成本法；分步成本计算法
process mix 生产过程组合
process model 过程模型
process monitoring 过程控制；过程监测；工艺流程监控；工艺流程监测；工艺流程调节
process monitoring alarm system 工艺过程监督报警系统
process monitoring supervisory system 工艺过程监视系统
process multiplexing 进程多路转换
process name 进程名
process naming 进程命名
process node 过程节点
process of accumulation overtime 跨时累积过程
process of adaptation 调整过程；适应性的变化过程
process of behavio(u)r change 行为变革的过程
process of cargo-handling 装卸过程
process of chopping 切断法
process of compaction 压实过程
process of concreting 混凝土浇注过程
process of consolidation 凝固过程；固结过程
process of contact reduction 接触还原法
process of cooling 冷却过程
process of coordination 协调过程
process of crustal movement 地壳运动过程
process of curing 硬化过程；烘焙过程；养护过程(混凝土)
process of depolarization 除极过程
process of deposition 沉积过程
process of field shear test 野外剪切试验方法
process of growth 生长过程
process of hardening 硬化过程；淬火过程
process of hydration 水合过程；水化过程
process of import 进口手续；输入手续

process of initial setting 初凝过程(混凝土)
process of iteration 迭代法
process of loading 加料过程;装载过程
process of malleus 锤凸
process of manufacture 制造过程
process of measurement 测量程序
process of nitrification 硝化处理
process of passenger casualty accident indemnity 旅客伤亡事故赔偿程序
process of precast member 预制构件工艺工程
process of precipitation 沉积过程;沉淀过程
process of prestressing 预加应力过程
process of production 生产过程
process of purification 净化过程
process of repolarization 复极过程
process of reproduction 再生产过程
process of rock wastage 风化过程;岩石耗蚀过程
process of salt exclusion 脱盐过程
process of salt return 返盐过程
process of separation 分选程序
process of setting 凝结过程;调整过程;沉降过程;装配过程;安装过程;凝固过程
process of settlement 沉降过程
process of soil formation by phreatic water 潜水成壤作用过程
process of solidification 固结过程
process of stressing 产生应力
process of stretching 拉伸过程
process of successive improvement 逐次改良过程
process of super-critical oxidation 超临界水氧化法
process of surveying data on hole deviation 钻孔偏斜测量资料的整理
process of tectonic movement 构造运动程式
process of tensioning 张拉过程
process of the duplication 复制过程
process of unconventional heat treatment of mo(u)ld 模具非常规热处理工艺
process of wave propagation 波浪传播过程
process of weathering 风化过程
process operation 过程操作
process operator console 处理操作员控制台
process optimization 过程最佳化;过程优化;工艺过程优化
process optimization program(me) 最佳操作程序
processor 旋进磁铁;信息处理机;加工者;加工机械;处理系统;处理器;处理机;处理程序
processor address space 处理机地址空间
processor-bound 处理器限制;处理机限制
processor-camera 照相显影两用机
processor clock 处理机时钟
processor configuration 处理机配置
processor control 处理机控制
processor control cards 处理机控制卡
processor cycle 处理机周期
processor element 处理机部件
processor error interrupt 处理机出错中断
processor evaluation module 处理器评价组件;处理机评价模块
processor flag 处理器标识位
processor identification register 处理标识寄存器
processor input interface 处理机输入接口
processor intercommunication 处理机内部通信
processor interface 处理机接口
processor interface module 处理器接口组件;处理机接口模块
processor interface routine 处理机接口程序
processor interrupt 处理机中断
processor-limited 受处理机限制;处理机限制
processor link 处理机连线
processor management 处理机管理
processor-memory-switch 处理机—存储器开关
processor module 处理器组件;处理机模块
processor multiplexing 处理机多路连接;处理机多路复用
processor multiplexing mechanism 处理机多路转换机构
processor operation 处理机工作;处理机操作
processor register 处理机寄存器
processor register space 处理机寄存器空间
processor scheduling 处理机调度
processor sharing 处理机共享
processor slice 处理机芯片
processor stack pointer 处理机堆栈指示器
processor stall 处理机故障

processor state bit 处理机状态位
processor state register 处理机状态寄存器
processor state word 处理机状态字
processor status long word 处理机状态长字
processor status word 处理器状态字;处理机状态字;程序状态字
processor storage 处理机存储器
processor storage relocation 处理机存储器再定位
processor table 处理机表
processor time 处理机时间
processor transfer time 处理器传送时间
processor verb 处理程序动词
processor word cell 处理机字单元
process output image 过程输出映像
process page 分页处理
process page table 进程页表
process parameter 进程参数
process partition 进程划分
process patent 方法专利
process payment 施工分期付款
process performance guarantees 工艺性能保证
process phase 加工阶段
process picture sheet 过程图表;工序图
process pipe 工艺管道
process pipe line 工艺管线
process piping 加工管线
process planning 过程设计
process plant 制炼厂
process plate 照相版;复制硬板
process printing 多色印刷;彩色套印
process priority 进程优先级
process privilege 进程特权
process production 分步生产
process pump 运行泵;作用泵;中间泵;工艺过程用泵;工艺泵;化工工艺用泵;维持泵
process queue 过程队列;进程队列
process reaction 过程反应
process redundance[redundancy] 加工余量
process regulation 过程调节
process reliability 工艺可靠性
process remote control 远距离过程控制
process requirement 工艺要求
process residue 工艺残渣;加工残渣
process safety limit 工艺安全限度
processs average 生产过程次品率
process schedule 进度时间表
process scheduling 进程调度
process section 进程段
process section table 进程段表
process segment 加工阶段
process selfregulation 过程自调节
process service 工艺管理
process sheet 过程处理单;工艺过程卡(片);程序图表
process shrinkage 收缩率总校正;多次收缩
process-side fouling factor 工艺侧污垢系数
process-side fouling resistance 工艺侧污垢热阻
process simulation 过程模拟;过程仿真
process simulator 过程模拟器;程序模拟器
process space 进程空间
process specification 工艺说明书;工艺规程;工艺标准;加工规格说明书;加工标准;程序说明书
process stability 工艺稳定性
process stage 加工阶段
process standards 工艺标准
process state vector 进程状态向量
process state word 进程状态字
process status 过程状态;进程状态
process steam 过程用汽;供热与供水的蒸汽;工艺用汽;工业生产用汽;生产用汽;生产用汽
process step 工艺步聚;加工阶段
process stipulation 工艺规定
process stream 工业生产液流
process structure 过程结构;进程结构
process study 过程研究
process switching 进程切换;程序控制开关
process symbol 处理符号
process synchronization 进程同步
process tag 进程标志
process tank 作业池;处理罐
process technique 程序加工技术
process technology 过程工艺;加工技术;加工工艺学;生产工艺学
process termination racks 过程终端网络

process time 加工时间
process-time curve 进度时间曲线
process time limit 加工时限
process to process flow control 进程间流程控制
process to process level communication 进程级间通信
process train 工艺程序链;流程链
process transponder 加工转发器
process unit 工艺设备;加工单位;处理单位;生产装置
process utilities 工艺过程公用事业;工艺公用工程
process variable 过程变量;工艺变数;工艺变量;可调变量
process-verbal of awarding 判标纪要
process wastes 工艺废物
process waste(water) treatment plant 生产污水处理装置
process water 工艺用水;生产用水;操作中用水
process water quality monitor 过程水质监测仪
process water reclamation and reuse 工艺用水回收和回用
process wire 中间钢丝
process with electroplated 电镀(铸)法
process with independent increment 具有独立增量的过程
process with non-zero mean 非零平均过程
process yellow 正黄
procetane 柴油添加剂
prochlorite 蠕绿泥石
proclaim one's stand 表态
proclamation 公告;公布
procoelous 前凹的
procoeton 古希腊或罗马有前厅的房间
procollagen 原胶原;溶胶原
Proctor compaction curve 普氏击实曲线;普罗克托击实曲线
Proctor compaction test 轻型击实试验;原位击实试验;普氏击实试验;普罗克托击实试验
Proctor compactive effort 普氏压实效力;普罗克托压实效力
Proctor cylinder 普氏击实筒;普罗克托击实筒
Proctor density test 普氏密实度试验;普罗克托密实度试验
Proctor dynamic(al) test 普氏动力试验;普罗克托动力试验
Proctor hardening test 贯入阻力试验
Proctor hardness 普氏硬度;普罗克托硬度
Proctor hardness test 普氏硬度试验;普罗克托硬度试验
Proctor maximum dry density 普氏最大干密度;普罗克托最大干密度
Proctor method 普氏压实法;普氏方法;普罗克托压实法;普罗克托方法
Proctor needle moisture test 普氏针测含水量试验;普罗克托针测含水量试验
Proctor penetration curve 普氏贯入曲线;普罗克托贯入曲线
Proctor penetration needle 普氏密实度测定针;普氏贯入仪;普罗克托密实度测定针
Proctor penetration resistance 标准贯入阻力;普氏贯入阻力;普罗克托贯入阻力
Proctor penetration test 普氏贯入试验;普罗克托贯入试验
Proctor plasticity needle 普氏塑性指针;普氏贯入锤;普氏塑性测定锤;普氏塑性针;普罗克托塑性针
Proctor resistance needle 普氏贯入针;普罗克托贯入针
Proctor test 普氏试验;普罗克托试验
Proctor-type compaction tester 普氏密实度计
procumbent 平铺的
procurability 可获得性
procurate 检察院
procuration 借贷业务佣金;代理权
procurational work 检察
procuration money 借贷业务佣金
procuratorate 检察院
procuratorate of forest region 林区检察院
procure 预养护;采办
procure agreement 达成协议
procured period 预固化期
procurement 征购;采购
procurement agency 收购机构;采购代办所
procurement authorization 采购授权书

procurement contract 采办合同
procurement cost 采购成本
procurement cycle 采购工作周期
procurement division 采购部门
procurement guidelines 采购指南
procurement lead time 预定提前时间;采办期
procurement manager 工程采购经理
procurement method 采购方法
procurement of capital investment 集资
procurement of materials 购买材料
procurement of sources 货源
procurement planning 采购规划
procurement price 收购价格;收购价
procurement price index 收购价格指数
procurement repair parts list 采购备件清单
procurement regulations 采购规程
procurement request 采购申请
procurement schedule 采购进度计划;采购进程计划;采购程序表
procurement scheme 采购方案
procurement volume 采购量
procuring 过早固化;预固化
procuring cause 招揽途径(指促使房地产成交过程)
procyclidine 丙环定
procymate 丙环氨酯
prod 测试棒
prod cast 椎模
prodeconium bromide 丙癸溴铵
prodelta 前三角洲;三角洲前沉积
prodelta facies 前三角洲相
prodeltaic deposit 前三角洲沉积
prodie against contingencies 以防万一
prodigiosin 灵菌素
prod(magnetizing) method 双磁头通电磁化法(磁粉探伤);双头通电磁化法
prod mark 戳痕
prodomus 门厅;堂前内柱廊
prodophytium 先锋群落
Prodorite 普罗多莱(一种防酸、耐火或沥青敷面水泥)
prodromal stage 前驱期
prod-type pyrometer 锥式高温计
producd fibers 生产纤维
produce 产生
produce animal products useful to man 生产对人类有用的畜产品
produce crude oil 采出的原油
produced dissolved organic carbon 生产的溶解有机碳
produced water-oil ratio 采出水油比
produced water treatment 采出水处理
produce electricity 发电
produce fire by friction 摩擦起火
produce lathes 制造车床
produce petroleum 生产石油
producer 制造者;制气炉;生产者;生产器;发生器;发生炉
producer capital 生产资金;生产资本
producer-city 生产城市
producer control desk 导演控制台
producer cover 发生炉盖
produce results 出成果
producer gas 发生炉气;发生炉煤气
producer-gas brown-coal tar 煤气发生炉褐煤焦油
producer-gas coal tar 发生炉煤焦油;发生炉煤沥青
producer gas lignite tar 煤气发生炉褐煤焦油
producer gas plant 煤气厂
producer gas process 煤气发生过程
producer gas tar 煤气发生炉焦油;发生炉焦油
producer gas tar pitch 煤气发生炉焦油沥青
producer gas tractor 煤气拖拉机
producer goods 产品;生产货物;生产资料;生产物资
producer goods under unified control 统配物资
producer of boundary condition function 边界条件函数发生器
producer of commodity 商品生产者
producer of flow function 流量函数发生器
producer of initial condition function 初始条件函数发生器
producer plant 生产厂;制造厂
producer price 生产者价格;生产价格
producer responsibility 生产者责任
Producers Council 生产者协会

producers' durable goods 耐用生产资料
producer's risk 生产者的风险
producers stock 原料
producer's surplus 生产者剩余
produce warehouse 农产品仓库
produce with low cost 廉价生产
producibility 可制造性;可延长性;可生产性;可伸长
producible 可延长
producing area 开采面积;产区;产地
producing department cost 生产部门成本
producing depth 矿层深度;生产层深度
producing formation 产气层
producing gas-oil ratio 生产油气比
producing gas trap 产气圈闭
producing gas well 沼气生产井;天然气生产井
producing homogeneous magnetic field 产生均匀磁场的马鞍形线圈
producing horizon 生产层;产油层位
producing interval 生产井段
producing method 生产方法
producing oil and gas trap 产油气圈闭
producing oil trap 产油圈闭
producing period 生产周期
producing place 产出部位
producing practice 生产技术;生产实践
producing reserves 动用可开采储量
producing test 试生产
producing well 开采井;生产井;出油井
product 生成物;生产物;生产品;乘积;乘;成果;产物;产品
product accounting 产品会计
product accumulation 乘积存储;产品存储
product accumulator 乘积累加器
product analysis 成品分析;产品分析
product and support requirements request 产品支持条件的请求;产品和支持条件请求
product appraisal certificate 产品鉴定证书
product approval 产品批准
product area 乘积区
product assessment test 产品评定试验
productbility 生产性
product break-even chart 产品盈亏平衡图
product buyback 产品返购
product calculation circuit 乘积计算电路
product catalog(ue) 产品目录
product certification 生产合格证
product charges 产品价格
product code 乘积码
product collector 成品收集器
product comparison 商品比较法
product concentration 产品浓度
product control(ling) 产品控制
product copy 产品样本
product cost 产品成本
product costing 产品成本计算
product cycle 乘积循环
product data 产品数据;产品说明书
product demodulator 乘积解调器
product density method 产生密度法
product departure 产品起输
product description 产品说明;产品阐述
product design 产品设计
product-design development shop 产品试制车间
product design outline 产品设计草图
product detection 乘积检波
product detector 乘积解调器;乘积检波器
product development 产品研制;产品开发;产品发展
product discharge door 成品卸料口
product dislocation 生成位错
product disposal charges 产品垃圾成本费
product distribution 产品分配
product ecobalance 产品生态平衡
product ensemble 乘积概率空间
product event 积事件
product examination 产品检查
product flow 产品流通
product form algorithm 积矩阵算法
product form of specification 标准的产品形式
product form recurrence algorithm 乘积形式递归算法
product from failing 尾矿产品
product function 产品功能
product-fuzzy decision 乘积模糊判决

product gas 煤气化煤气;产品气(体)
product generator 乘积发生器
productibility of soil 土壤生产力
product identification number 产品编号
productimeter 求积仪
product improvement 产品改进
product information 产品资料
product in hand contract 交产品合同
product inhibition 产物抑制
product inspection 成品检验
product instruction 产品说明书
product integrator 乘积积分器
product intermediate storage tank 产品中间储[贮]槽
product in transit 输送中的产品
production 制作;生产;产生式;产量
production acetate 原版胶片
production adjustment 生产调整
production agency 摄制机构
production age of planning 设计生产年限
productional sheet 成果图板
production analyzer 产品缺陷记录仪
production and construction funds 生产建设资金
production and efficiency 生产和效率
production and finance plan 生产财务计划
production and technical guidance 生产技术指导
production and technical indication 生产技术指标;生产和技术指标
production and testing ship 生产试验船
production approval inspection 生产定型检验
production aquifer 生产含水层
production area 摄制项
production automation 生产自动化
production average-weighted pressure method 产量的加权平均压力法
production base 生产基地
production batch 产品批次
production bays 生产区
production bottle neck 生产薄弱环节
production break 生产中断
production breakdown 生产故障
production budget 生产预算
production building 生产(用)建筑
production burst 生成脉冲
production capability 生产能力
production capacity 生产能力;生产量;生产规模;产量
production capacity of coal pits 煤矿矿井生产能力
production capacity of dredge(r) productivity 挖泥船生产能力
production capacity of mine 矿山生产能力
production capacity of mining field 采场生产能力
production capacity of ore block 矿块生产能力
production capital 生产资金;生产资本
production capital quota 生产资金定额
production casing 生产套管
production casing string 生产套管线(石油)
production certificate 生产许可证
production chain 生产线
production charges 生产费(用)
production coal sample number 生产煤样编号
production coefficient 生产率;产率系数
production comes before capital construction 先生产后基建
production concept of orientation 生产概念
production condition 生产条件
production configuration 生产型
production constant 生产常数
production constraint 生产限制因素
production contract 生产合同
production control 生产控制;生产管制;生产管理
production control graph 生产控制图表
production control process 生产控制程序
production cost 制造成本;生产费(用);生产成本
production cost account 生产成本账户
production cost and freight 生产费用和总运输费
production cost budget 生产成本预算
production cost center 生产成本中心
production cost for the period 本期生产成本
production cost ratio 生产成本比率
production cost sheet 生产成本表
production cost variance 生产成本差异
production current 作业流;生产流
production curve 产量曲线;采油曲线

production cycle 生产周期
production cycle of domestic fowl and livestock 畜禽生产周期
production data 产品指标
production date of bit 钻头出厂日期
production-decline curve 生产下降曲线；产量递减曲线
production density 生产密度
production department 生产科；生产单位
production department expenses 生产车间费用；生产部门费用
production derrick 采油井架
production design 生产设计
production despatching 生产调度
production diagram 生产图解
production difficulty 生产故障
production director 制作指导
production dispatching 生产调度
production drawing 制作详图；施工详图
production drilling 开采钻进；生产钻井
production drive 生产竞赛
production ecology 生产生态学
production efficiency 生产效率
production-element allocating market regulation 要素配置市场调节
production elements market 要素市场
production emulation 生产竞赛
production engine 正常系列发动机；成批生产的发动机
production engineer 制造工程师；生产工程师；采油工程师
production engineering 生产工程
production enterprise 生产企业
production environment 生产环境
production equipment 生产设备
production error failure 生产误差故障
production evaluation test 产品鉴定实验
production expenses 生产费（用）
production exploration 生产勘探
production facility 生产设备
production factor 生产要素；采注比
production field 生产现场
production figure 生产数字
production flow 生产流程
production flowchart 生产流程图
production flow diagram 生产流程图
production flowline 生产流水线
production flow(line) analysis 生产流程分析
production flow pattern 生产流程模型
production fluid 可以开采的流体
production forest 经济林；生产性森林；生产林
production formation 生产层
production formula 生产配方
production foundry 成批生产铸造厂
production fracture 产水汽裂隙
production function 生产职能；生产函数
production fund 生产基金
production funds put to a better use 提高生产资金利用效果
production goal 生产目标
production goods 生产资料
production-grade 工业品位
production halts 停产
production hoist 主井提升机
production holdup 生产故障
production hole 生产孔
production index 生产指数；生产指标
production index unit cost 生产指数单位成本
production indicator 产量指标；产量计
production information 生产情报
production information and control system 生产信息控制系统
production information control system 生产信息控制系统
production information processing system 生产信息处理系统
production-injection ratio 采注比
production in series 流水作业的生产；顺序生产
production inspection 生产检验
production intensification 生产集约化
production intensity 生产强度
production intensity of simple well 单井采油强度
production language 产生式语言
production language compiler 产生式语言编译

程序
production level 生产水平
production library 生表程序库
production license 生产许可证
production life 生产设备有效使用寿命；生产设备有效使用寿命
production line 装配线；流水作业线；生产作业线；生产线；生产流水线；出油管；流水作业生产线；流水线
production line evaluation 生产线的评价
production line for dry cell 干电池生产线
production line for manufacturing battery 电池生产线
production line for masaic glass 马赛克玻璃生产线
production line for rice noodle making 米线生产线
production line performance test 生产线性能考核
production line profitability 生产线盈利能力
production line profit and loss statement 生产线损益计算书
production-line technique 生产线工艺
production log 生产日记
production logic tester 生产性逻辑测试机
production lot 生产批量
production machine 专用机械；专用机床
production management 生产管理；生产调度
production management in cutting areas 伐区生产管理
production management of passenger traffic 旅客列车运输生产管理
production management of steel-making 炼钢生产管理
production management system 生产管理制度
production manager 生产经理
production(manufacture) plant 生产工厂
production-manufacturing 生产制造
production-manufacturing engineering 生产制造工程
production mask 生产掩模
production mechanism 生产机制
production metallurgy 生产冶金(学)
production meter 产量计
production method 生产方法
production method of high pure metals 高纯金属制取
production miller 专用铣床
production model 生产型号；生产形式；制造样机；产品模型；生产模型；成批生产模型
production model paver 产品型路面机械
production of concrete 混凝土制品；混凝土生产
production of consumer goods 消费资料的生产
production of digester gas 消化池气(体)生产
production of explosive 炸药生产
production of heat 产热
production of neutron 中子产生
production operational capability 生产操作能力
production order 生产凭单；生产命令；生产定单
production order cost sheet 生产成本单
production packer 产品装填机；采油封隔器
production patent 产品专利
production pattern 开采方式；产量曲线图
production people 生产人员
production performance 生产性能
production performance method 生产动态法
production per hour 每小时产量；每小时产量
production period 生产周期
production period cost 生产期费用
production per man-hour 每人每小时产量；人时产量
production permit 生产许可证
production per point 每个钻头进尺
production per unit area 单位面积产量
production phase 生产阶段
production plan(ning) 生产计划；生产规划
production planning and control 生产计划和控制
production planning control system 生产计划控制系统
production planning problem 生产计划问题
production plan of industry 工业生产计划
production plateau 高产稳产期
production platform 制造岛(指一组独立完整的制造体系)；钻探平台；采油平台
production point of view 生产观点

production poll 生产轮询
production possibility 生产可能性
production preparation fee 生产准备费
production preparation of coal pits 矿井生产准备
production press 生产用压机
production pressure 生产压力
production price 生产价格
production process 生产流程；生产（工艺）过程；生产方法
production process equilibrium 生产过程衡性
production process of coal pits 煤矿矿井生产流程
production profile 采油曲线
production profit sharing 生产利润分配
production program(me) 生产大纲；开采计划；生产说明书；生产计划；生产程序
production programming 生产规划
production prototype 投产原型
production quality control 产品质量控制；产品质量管理
production quality test 生产质量检验；产品质量检验
production quantity 生产量
production quantity guarantee 产量保证
production quarter 生产用房屋；生产区
production quota 生产指标；生产定额
production range 生产井段
production rate 生产率；采油速度
production rate of dredge(r) 挖泥船生产率
production rate of fixed capital 固定资产值率
production rate of geothermal fluid 地热流体的产率
production rate per hour 每小时生产率
production record 开采记录；生产记录；标准产乳记录
production registry 生产性能登记
production registry system 生产性能登记制
production requirement 生产要求
production reserves 生产储备；生产矿量
production reservoir delineation 生产储层边界圈定
production residues 生产残留物；生产废渣；生产残余物
production responsibility system 生产责任制
production rig 采油平台
production routine 生产性程序；生产程序
production rule 形成规则；生成规则；产生式规则
production rule based on implications 基于蕴涵式的产生式规则
production rule for units 对单元的产生式规则
production run 正式运转；全印数；生产运行；生产性运行；生产过程
production run equipment 流水生产线设备；成批生产设备
production run number 生产批号
production sample 生产样品
production-scale 生产规模的
production scale cell 生产用电解槽；大型电解槽
production schedule 开采表；生产进度表
production scheduling 生产进度；生产调度
production sector 生产部门
production sequence 生产程序
production sewage 生产污水
production shaft survey 生产矿井测量
production skills 生产技能
production specifications 生产技术条件
production stages 生产阶段
production standard 生产指标；生产标准；标准时间
production statistics of mineral industry 矿业生产统计
production string 生产套管
production string of casing 生产套管线(石油)
production stripping ratio 生产剥采比
production structure diagram 生产结构图
production study 生产研究
production-supply-marketing relation 产供销关系
production surface 生产面
production system 运行系统；生产系统；产生式系统
production target 生产指标
production technique 生产技术
production technique of high grade zinc oxide 高级氧化锌生产技术
production technology 生产工艺
production technology of high grade zinc oxide 高级氧化锌制取
production temperature 生产温度

production test(ing) 产品试验；产品检验；生产测定；生产试验
production test plan 产品测试计划
production time 工作时间；生产时间
production timer 记件秒表；计产品用计时器
production tool and equipment 生产工具和设备
production track 生产声道
production trait 生产性状
production tree 产生式树；采油树
production-type test 生产定型试验；产品形式试验
production unit 生产单位
production view of ecologic(al) economics 生态经济生产观
production volume 开采量；产量
production waste 生产废水；生产废渣；生产废气；生产中的废料；生产污水；生产垃圾；生产废料
production water supply 生产用水；生产供水
production water supply system 生产用水系统；生产供水系统
production well 生产抽水井；生产矿井；生产(油)井
production well drilling 生产井的钻探
production well for conjunctive use 兼用的生产水井
production well testing 生产井测试
production work 生产工作
productive 生产的；产生性的
productive accumulation 生产性积累
productive and operating cost audit 生产经营成本支出审计
productive and technical preparation 生产技术准备
productive arterial road 生产干线【道】
productive assets 生产性资产
productive branch road 生产支线【道】
productive building 生产性建筑
productive capacity 生产容量；生产能力
productive capacity cost 生产能量成本
productive capacity of well 油井生产能力
productive capital 生产资金；生产资本
productive cold store 生产性冷库
productive construction funds 生产建设投资
productive cooperation 生产合作
productive downtime 生产性停歇
productive downtime of dredge(r) 挖泥船生产性停歇
productive energy 生产能
productive expenditures 生产支出
productive exploration 矿山开发勘探
productive flow analysis 生产流程分析
productive force 生产力
productive forces criterion 生产力标准
productive head 有效压头；发电水头
productive investment 生产投资
productive labor in its general sense 一般意义的生产劳动
productive labo(u)r 生产性劳动
productive labo(u)r hours method 生产人工时数法
productive land 生产用地；地力保持
productive life 生产年限
productive life length 生产年限
productive line 作业线
productive loan 生产放款
productive machine-hour 生产台时
productive maintenance 生产维修；生产保全
productive management 生产管理
productive metamember 产生性元成员
productiveness 生产效能；生产力；产量
productive output 生产量
productive-output method of depreciation 按产量计算折旧法
productive plantation area 生产绿地
productive poisons 生产性毒物
productive poll 生产性查询
productive power 生产能力；生产力
productive practice 生产实践
productive process 生产过程
productive program(me) 生产程序
productive rate 产率
productive rating index 生产指数
productive reservoir 生产性热储
productive routine 生产性程序
productive sand 可采砂层
productive set 积集

productive structure 生产结构
productive task 生产(性)任务
productive technology management 生产技术管理
productive time 作业时间；生产时间
productive value 生产价值
productive wealth 生产财富
productive workings 采区
productive working time 生产性工作时间
productive year 生产年份
productivity 入出库能力【机】；生产率；产率；生产能力；生产量
productivity accounting 生产率会计
productivity coefficient 出铁比
productivity difference 生产率差异
productivity difference 生产率差异
productivity factor 生产率系数；采油指数
productivity index 采油指数
productivity labo(u)r hours method 生产工时法
productivity level 生产力水平
productivity measurement 生产率测定；生产力计量；生产力测定
productivity of added value 增值的生产能力
productivity of capital goods 资本财的生产力
productivity of furnace 炉子生产率
productivity of gas 煤气发生率
productivity of handling machinery 港口装卸机械生产率
productivity of land 土地生产力
productivity of the soil 地力
productivity of transpiration 蒸腾生产率
productivity of workman 劳动者生产率
productivity pattern 产量曲线图
productivity per worker 每个工人的产量；人均劳动生产率
productivity quantitative scale 生产力数量规模
productivity rating index 生产力控制指数
productivity ratio 生产能力比率；生产力比率；产率比
productivity test 产率试验
productivity yield 产量
productized 煤的产品；按产品分类的
product label(l)ing 产品标签
product length operand 积长度操作数
product liability 产品责任问题
product liability prevention 防产品责任问题
product life 产品寿命
product life cycle 产品寿命周期；产品生命循环
product life extension 产品寿命的扩展
product line 产品品种；产品排出管；产品花色品种
product-line profit and loss statement 类别产品损益表
product liquid 液体产率
product literature 产品说明书
product maker 产品制作者；产品制造商
product management standard 产品管理标准
product manufacturer 产品制造者；产品制造厂
product mark 产品标志
product matrix 积矩阵
product measure 乘积测度
product mix 产品结构
product model 乘积模型
product modulator 乘积调制器
product moment 积矩
product-moment correlation 积矩相关
product-moment correlation coefficient 积距相关系数
product number 产品编号
product of abrasive 磨料制品
product of areas 面积乘积
product of centrifuge 离心过滤产物
product of classification 产品分类
product of combustion 燃烧产物
product of concentration and time 浓度时间乘积
product of destruction 破坏的产物；毁坏的产物
product of electrolysis 电解产物
product of handiwork 手工制品
product of inertia 惯性积
product of matrices 矩阵积
product of numbers 数字乘积
product of passenger traffic 旅客运输产品；客运产品
product of railway goods transport 铁路货物运输产品
product oil 成品油

product oil carrier 成品油轮
product outlet 饲料口；成品出口
product overflow 积溢出
product patent 产品专利
product-pay-back scheme 以产品偿还的计划
product percent of pass 产品合格率
product performance evaluation 产品性能评价
product phase 子相
product pipe 产品管
product pipe line 产品管道(石油)
product plus 标准以上的产品
product price 产品价格
product price differentials 产品差价
product property 制品性能；产品性能
product protection 产品保护
product purity 产品纯度
product quality 产品质量
product quality guarantee 产品质量保证
product quality inspection 产品质量检验
product quality liability 产品质量责任制
product quality management 产品质量管理
product quality mark 产品质量标志
product quality standard 产品质量标准
product quality supervision and inspection 产品质量监督检验
product rack 成品台架
product record 生产记录
product register 乘积寄存器
product relay 乘积继电器
product representation 乘积表示
product research committee 产品研究委员会
product rule 积法则；乘法定则
products 制品；产成品
product sampling test 产品抽验
products become hot on the market 热门货(品)
product selector 乘积选择器
product separation 产品的分离
product separator 产品分离器
product seriation 产品系列化
product series 产品系列
product set 乘积集合
product sign 乘号
products index 产品目录
product size 制品尺寸；产品粒度；产品尺寸
products liability insurance 产品赔偿责任险
products name 产物名称
products of inertia 惯量乘积
products of key projects 基础产品
products of photosynthesis 光合作用产物
product sold 销售的产品
product solution 产品溶液
product solution concentration 产品溶液浓度
products outlet 产品出口
product space 乘积空间
product specification 产品说明书；产品规格；产品规范
products sent back for repair 返修品
product standard 产品规格；产品标准；产品质量标准；产品技术标准；产品规格标准
products that have a ready market 适销对路的产品
products that yield a high added value 附加值高
product stored 成品保管；产品存储[贮]；库存货物
product store house 成品库
product string 积串
product structure 产品结构
product styling 产口外形设计
product summary 产品目录；产品概览
product supply task control 产品供应检查
product system 生产系统
product tank 成品油罐
product tanker 成品油轮
product tax 产品税
product term 积项
product test(ing) 产品试验；研究市场对产品的反应
product thickness recovery 回粗；产品厚度回缩
product topology 积拓扑
product type of coal 煤的产品类型；煤产品类型
product-use 制品用途
product validation 对产品质量的评定
product variable 乘积变量
product verification test 产品评价试验；产品检验试验
product warranty 产品保证书

product water 产品水
product withdraw 收回已售产品
product without a material form 非物体形态的产品
product word 乘积字
product yield 成品收率
proection switching duration 保护交换时间
proeutectic 先共晶
pro-eutectic austenite 先共晶奥氏体
pro-eutectic cementite 先共晶渗碳体
pro-eutectoid 先共析体;先共析
proeutectoid cementite 先共析渗碳体
proeutectoid ferrite 先共析铁素体
profane architecture 世俗建筑学;非宗教性建筑
profane Gothic style 非宗教哥特式;世俗哥特式风格
profane monument 非宗教墓穴;非宗教纪念碑
profane structure 非宗教结构
profession 专业;职业;工种;行业
professional 专业的;专门的;职业(的)
professional ability 业务能力
professional accountant 专业会计师;职业会计师
professional adviser 专业顾问
professional agency for building procedure 建筑业务代理处
professional architect 职业建筑师
professional auditor 职业查账员
professional cancer 职业癌
professional competence 专业资格
professional component 专用元件
professional contingent 专业队伍
professional dermatosis 职业性皮肤病
professional diving 职业潜水
Professional Drivers' Council 职业驾驶员协会
professional education 职业教育
professional element 专用元件
professional engineer 职业工程师;有执照的专业工程师;专业工程师;注册工程师
professional ethics 职业道德
professional fee 业务收费;专业费用
professional film 专用胶片
professional fire brigade 专职消防队;职业消防队
professional grade 业务等级;职业等级;专业等级
professional group on automatic control 自动控制专用组件
professional image 职业概念;职业映像
professional indemnity insurance 职业保障保险
professional instrument 工厂制仪器
professional intership 职业实习
professionalization 专业化
professionalization of management 管理专业化
professionalize 职业化
professional judgment 职业性判断
professional knowledge 业务知识;专业知识
professional liability insurance 职业责任保险;专业业务保险
professional licensing 职业执照
professional literature 专业著作;专业文献
professional management 专业管理
professional manager 专业经理人
professional paper 专门论文;专业论文;专题报告
professional personnel 专业人员
professional practice 工程业务;业务实践;专业实践;专业操作;专业业务
professional proficiency 业务能力
professional programmer 专业程序设计员
professional qualification 业务水平;业务能力;专业熟练程度;职称等级;技术水平;从业资格
professional rank and titles 职称
professional real estate executive 职业房地产经理
professional school 专业学校
professional service 专业服务
professional skill 专门技术;专门技能;一技之长
professional society 专业团体
professional specialty 业务专长
professional standard 专业标准
professional team 专业队
professional terminology 专业术语
professional title 职称
professional training 专业培训
professional transport 专业运输
professional work 业务(工作)
profession distribution of disease 疾病职业分布
profession skill 专业技能

proficiency 精通;潜能
proficiency testing 熟练水平测试
proficient 精通的;熟练的
profile 纵剖面;剖面(图);土壤断面;正交剖面;概貌;轮廓;龙门板;廓线;剖面(图);皮数杆【测】;外形;特征测验图;唇面形状【地】;侧影(像);侧形;侧面像;侧断面图;仿形
profile accuracy 齿形精度
profile and cross-section(al) survey for river 河流断面测量
profile and thread rolling die 滚轧型材与螺纹模具
profile angle 齿形角;齿廓角
profile-blasting shop 型钢喷砂车间
profile board 模板;侧板;样板规;剖面样板;剖面模板
profile book 剖面记录本
profile broach 成型拉刀
profile calculator 剖面计算器
profile cavitation 叶型汽蚀;叶片断面汽蚀;断面空蚀
profile chain link 模制链节
profile checking 限界检测
profile computer 剖面计算机
profile contact ratio 齿廓接触系数
profile coordinate 型线坐标
profile copy grinding 靠模磨削;仿形磨削
profile correction 廓型修正;齿廓修正
profile curved grooving cutter 成型曲槽铣刀
profile curve of amplitude ratio 振幅比剖面曲线
profile curve of dip angle 倾角剖面曲线
profile curve of phase difference 相位差剖面曲线
profile curve of relative anomaly 相对异常剖面曲线
profile cutter 成型刀具;成型铣刀
profile cutting 成型切削
profiled alumin(i)um panel 异形铝板
profiled alumin(i)um roof(ing) sheet 异形铝片屋面材料
profile data 剖面数据;断面数据
profile data collecting 断面数据采集
profile date 剖面资料
profiled bar 异形钢材;异形钢棒;异形钢
profiled blade 定形叶(涡轮的)
profiled board 异形木板
profiled bucket 成型铲斗
profiled cast concrete panel 异形浇注混凝土镶板
profiled chisel 外形凿刀;轮廓凿刀
profiled coping 墙帽轮廓
profiled curb 异形路肩石;异形路缘石
profiled curved grooving cutter 成型曲槽铣刀
profiled cutting machine 成型切割机
profiled edge 倒圆
profiled element 异形构件
profile depth 外形深度
profile design 纵断面设计
profile device 仿形装置
profiled face 成型采掘面;成型工作面
profiled grooving cutter 成型槽铣刀
profile diagram 纵断面图;轮廓图
profiled iron 型铁;型钢
profile dispersion 折射率分布色散
profile displacement 纵断面位移
profiled kerb 异形路肩石;异形路缘石
profiled keyway 成型键槽
profiled nosing 突沿的轮廓线;突缘
profiled panel 异形镶板
profiled parts 齿形零件
profiled pipe 异形管
profile draft 侧面图
profile drag 形面阻力;断面阻力;船型阻力
profile dredge(r) 整形挖泥船
profiled roof 齿形屋顶
profiled shaft 靠模轴;仿形轴
profiled sheet iron 成型薄钢板
profiled steel 型钢
profiled(steel) sheet 压型钢板
profiled strip 异型钢带;异形钢带
profiled tube 管剖面图;模压筒管
profiled wire 异形钢丝
profiled wire slotted screen 异形钢丝编成的长孔筛子
profile echograph 纵剖面回声测深仪
profile error 廓型误差;齿形误差
profile error in the tool 刀具廓形误差
profile exec 开工文件

profile exponent 折射率分布指数;分布指数
profile fabricator 轮廓修整工
profile facing 靠模端面车削;仿形端面车削
profile figure of amplitude ratio and phase difference 振幅比相位差剖面图
profile figure of potential induced in transient field method 瞬变场法感应电势剖面图
profile figure of relative imaginary component 相对虚分量剖面图
profile figure of relative magnetic anomaly of horizontal loop method 水平线框法相对磁异常剖面图
profile figure of results of dip-angle method 倾角法测量剖面图
profile figure of results very low frequency method 甚低频法测量剖面图
profile flange 凸缘断面图
profile flow 翼型绕流
profile for interior work 室内修整装饰工作
profile gasket 异形垫圈
profile ga(u)ge 样板;轮廓量规;曲线板
profile glass 浮雕玻璃;异形截面玻璃
profile grade 纵坡(度);纵断面坡度
profile grade of dike crest 坝顶纵坡
profile grade of yard 调车场纵坡
profile graph 体型图
profile gravity survey 剖面重力测量
profile grinder 光学曲线磨床;曲线磨床
profile grinding 成型磨削
profile grinding machine 仿形磨床
profile ground tooth 齿形磨光的齿
profile height 剖面高度
profile improvement 纵断面改善
profile index[prl] 路面平整度指标
profile in elevation 立剖面(图);有高程注记的地形剖面图;纵剖面;注有标高的纵断面图;高程纵断面;平剖图;水平剖面(图);水平断面
profile in plan 平剖面图;平剖图;平剖面
profile inspection 齿形检验
profile interval 剖面间隔;断面间距
profile iron 型铁;型钢;异形钢
profile irregularity 轮廓微观不平度;轮廓不均匀度
profile length 剖面长度
profile length of photo taken in a well 井下摄影剖面长度
profile level(l)ing 纵剖面水准测量
profile lift 翼型升力;叶型升力
profile line 纵断面线
profile line method 断面线法
profile-log of water injection 注水井受水剖面
profile loss 叶型损失
profile maker 外形制作者
profile map 纵断面图
profile measuring aid 断面量测设备
profile measuring plant 断面量测设备
profile meter 轮廓仪;表面光洁度测定仪;表面测量仪
profile microscope 轮廓显微镜
profile milling cutter 成型铣刀
profile mismatch loss 分布失配损耗
profile model 剖面模型;断面模型
profile modelling 靠模;仿形
profile model test 断面模型试验
profile modification 齿形修整;齿形修缘
profile modified gear 变位齿轮
profile modulus 剖面模数
profile of any surface 面的轮廓度
profile of a well 钻井记录;井的轮廓;钻井剖面
profile of canal 渠道纵断面
profile of contour of self-potential method 自然电位等值线平面图
profile of engineering geology 工程地质纵剖面图
profile of equilibrium (河流的)夷平剖面;均衡面;平衡剖面
profile of excavation 隧道轮廓
profile of exploratory 探槽剖面图
profile of fillet weld 填角焊缝轮廓
profile of geochemistry anomaly 地球化学异常剖面
profile of geophysics anomaly 地球物理异常剖面图
profile of groundwater level 地下水位纵剖面
profile of highway 道路横断面(图)
profile of magnetic anomaly 磁异常剖面图

profile of non-equilibrium 非均衡剖面
profile of pipe-line route 管线的立面图
profile of pollution hydrogeology 污染水文地质剖面图
profile of pumping well construction 抽水孔结构剖面图
profile of quaternary system 第四系剖面图
profile of recent valley 现代河谷剖面图
profile of rib 拱肋外形;肋外形
profile of riverbed 河底纵断面;河床纵剖面;河床纵断面
profile of road 道路横截面
profile of sandy loam 沙壤土剖面
profile of sedimentary structure of delta 三角洲沉积削面结构
profile of slope 坡道(纵)剖面(图);斜坡纵剖面(图);斜坡线;边坡纵剖面(图)
profile of soil 土壤剖面
profile of terrace 阶地剖面
profile of tooth 齿形
profile of water head distribution 剖面水头分布图
profile of water table 地下水位纵剖面;地下水位纵断面;水位截面图;地下水位剖面图;地下水位截面图
profile of wave 波剖面
profile on plane of alpha intensity 阿尔法强度平剖图
profile on plane of Eman 埃曼平剖图
profile on plane of gamma intensity 伽马强度平剖图
profile on plane of helium content 氦含量平剖图
profile on plane of polonium content 钋含量平剖图
profile on plane of potassium content 钾含量平剖图
profile on plane of Th/K ratio 钍钾比值平剖图
profile on plane of thorium content 钍含量平剖图
profile on plane of track density 径迹密度平剖图
profile on plane of U/K ratio 铀钾比值平剖图
profile on plane of uranium content 铀含量平剖图
profile on plane of U/Th ratio 铀钍比值平剖图
profile paper 坐标纸;纵断面图纸;方格纸;格子纸;断面图纸
profile parameter 折射率分布参数
profile plan 侧面图
profile plane 侧面
profile plane of projection 侧投影面
profile-plan figure of amplitude ratio and phase difference 振幅比相位差剖面平面图
profile-plan figure of relative imaginary component 相对虚分量剖面平面图
profile-plan figure of results of dip-angle method 倾角法测量剖面平面图
profile-plan figure of results very low frequency method 甚低频法测量剖面平面图
profile plate 靠模样板;仿形样板
profile plotter 断面绘图设备
profile plotting 断面测图
profile plotting attachment 断面绘图设备
profile plotting device 断面绘图设备
profile point 剖面点;断面点
profile position 剖面位置
profile precast concrete panel 成型预浇混凝土镶板;成型预浇混凝土嵌板
profile pressure distribution 翼型压力分布
profile program(me) 剖面图程序
profile projection 侧面投影
profile projection plane 侧立投影面
profile projector 轮廓投影仪;轮廓测定投影器;断面投影器
profiler 靠模铣床;断面仪;型刨
profile recorder 剖面记录器;地表不平度记录仪
profile recording chart 齿形记录卡
profile relieved hob 铲齿滚刀
profile relieved tooth 成型铲齿
profile report 轮廓报告
profile resistance 翼剖面阻力
profile revision 纵断面改进
profile rib 翼肋
profile sampling 剖面采样
profile scanning 断面扫描
profile scheme 纵断面略图
profile set 翼型系列
profile shaft 特型轴

profile shaping machine 刨模机
profile sheet(ing) 纵断面图(纸);异型板;纵断面图幅
profile shifted gear 变位齿轮
profile shooting 剖面爆破法;地震剖面勘探(法);地震测线勘探(法)
profile spacing 断面间距
profile steel 型钢
profile strip 异形钢带;异形带钢
profile surface 成型面
profile survey 纵断面测量;剖面测量
profile surveying 剖面测量法
profile symbol 断面符号
profile tangent 纵向切线;竖曲线切线
profile testing instrument 轮廓仪
profile tolerance 轮廓度
profile tolerance of a surface 面的轮廓度
profile tracer 轮廓仿形;靠模仿形
profile-turning lathe 仿形车床
profile variable 仿形变量
profile view 侧视图
profile wire 异型钢丝;异形钢丝;成型钢丝
profiling 压型;靠模加工;靠模;绘制剖面图(剖面勘探);仿形切削
profiling attachment 靠模附件;仿形附件
profiling bar 靠模杆;仿形杆
profiling calender 胎面机
profiling copy 仿形
profiling cutter 靠模刀具
profiling error 断面测量误差
profiling figure of gradient curve 梯度曲线剖面图
profiling figure of induced polarization combined array 激电联剖剖面图
profiling figure of induced polarization dipole array 激电偶极剖面图
profiling figure of induced polarization gradient array 激电中梯剖面图
profiling figure of potential curve 电位曲线剖面图
profiling figure of self-potential measurement 自然电位测量剖面图
profiling figure of SP gradient measurement 自电梯度测量剖面图
profiling line 剖面线
profiling machine 靠模机床;仿形机床;仿形机
profiling mechanism 仿形机械;仿形机构
profiling method with EM dipole 偶极电磁剖面法
profiling-plan figure of gradient curve 梯度曲线剖面平面图
profiling-plan figure of imaginary component 虚分量剖面平面图
profiling-plan figure of induced polarization combined array 激电联剖剖面平面图
profiling-plan figure of induced polarization dipole array 激电偶极剖面平面图
profiling-plan figure of induced polarization gradient array 激电中梯剖面平面图
profiling-plan figure of potential curve 电位曲线剖面平面图
profiling-plan figure of real component 实分量剖面平面图
profiling-plan figure of second channel response 第二道响应剖面平面图
profiling-plan figure of self-potential measurement 自然电位测量剖面平面图
profiling-plan figure of SP gradient measurement 自电梯度测量剖面平面图
profiling-section figure of induced polarization dipole array 激电偶极拟剖面图
profiling slide 仿形滑板
profiling systems 特征描述制
profiling vibro-tamper 成型振动夯实机
profilograph 轮廓曲线仪;验平仪;纵剖面测绘器;纵剖面测绘器;自记纵剖面测绘器;轮廓测定仪;面形描记器;面部轮廓描记器;断面测绘器;表面光度仪
profilograpy 显微光波干涉仪
profilometer 表面测量仪;表面光度仪;表面光度计;表面平坦度仪;显微光波干涉仪;纵剖面测绘器;纵剖面测绘器;轮廓仪;轮廓测定仪;轮廓测定器;路面验平器;路面平整度测定仪;面形测定器;断面测绘器;粗糙度测定计
profilometry 轮廓测定法
profiloscope 纵剖面测绘仪;拉模孔光洁度光学检查仪;模孔光学检查仪

profit 盈利;利润;收益;得益
profitability 有用性;经济性;有利性;有利程度;盈利能力;获利指数
profitability analysis 盈利分析
profitability guideline 获利能力指标
profitability index 盈利能力指数
profitability investigation 有利性调查;经济性调查
profitability of operating units 经营单位获利能力
profitability of the import and export enterprises 进出口企业的盈利性
profitability ratio 获利能力比率
profitable firm 盈利企业
profitable haul 经济运距
profitable thickness 有用厚度
profit account 利润账户
profit accounting 利润核算
profit accrued from evaluation 估价增益
profit accrued from insurance 保险差益
profit after tax 纳税后利润
profit allowance 利润扣除额
profit analysis 利润分析
profit and loss 损益;收益和亏损;得失
profit and loss account 盈亏计算;损益账户
profit and loss adjustment account 损益调整账户
profit and loss appropriation account 损益分配账户
profit and loss contribution 损益份额
profit and loss for preceding term 前期损益
profit and loss for the previous period 前期损益
profit and loss from warehouses 货栈损益
profit and loss method 损益法
profit and loss of warehouses or godowns 货栈损益
profit and loss on capital and interest 本利的盈亏
profit and loss on realization account 财产变现损益账户;变产损益账户
profit and loss on valuation of real estate 不动产估价损益
profit and loss ratio 损益比率
profit and loss sharing ratio 损益分配率
profit and loss statement 盈亏报表;损益清单;损益表
profit and loss summary account 损益汇总表
profit and loss transaction 损益转账事项
profit and loss transfer 损益转让
profit and tax rates for every hundred Yuan of output value 百元产值的利税率
profit area 盈利区;利润区
profit available for dividend 可作为支付股息的利润;可供分配股利的利润;可分配利益
profit before tax 纳税前利润
profit brought forward 前期结转利润
profit budget 利润预算
profit calculation 利润核算
profit center 利润中心
profit change into taxes system 利改税
profit clearing 利润结算
profit commission 盈利佣金;纯益手续费(保险)
profit commission statement 纯益手续费清单
profit contribution 利润贡献
profit control 利润管理
profit criterion 利润准则
profit deduction and reserving 提留
profit distribution decision 利润分配决策
profit distribution quota 利润分配定额法
profit drawing 利润提取
profit-duty rate of investment 投资利税率
profit earned 已获盈利;已得利润
profiteer 投机商(人);暴发户
profiteering 投机活动;投机倒把;抬高利润
profit estimating 利润预算
profit function 利润函数
profit gap 利润差额
profit improvement plan 利润改善计划
profit index 利润指标
profit-investment criterion 利润投资准则
profit-loss breakeven point chart 利润亏损分界图
profit-making 盈利性的
profit management 利润管理
profit manager 利润经理
profit margin 利润限度;利润率;利润幅度;利润边际;边际利润
profit margin of current capital 流动资金利润率
profit margin of fixed capital 固定资金利润率

profit margin of wage 工资利润率
profit margin per person 人均利润率
profit margin rate 边际利润率
profit maximization 最高利润点；最大限度利润
profit measurement 利润计量
profit model 利润模式
profit norm 利润指标
profit of accrued from insurance 保险增益
profit on disposal of property 财产处理利益
profit on environment 有益于环境
profit on sale of scrap equipment 报废设备出售利益
profit on sals 销售利润
profit on working capital 流动资本利润率
profit or loss frominstallment sales 分期付款销售损益
profit or loss on exchange 外汇买卖损益
profit or loss on installment sales 分期付款销售损益
profit participating preferred stock 参与分红优先股
profit payable audit 应付利润审计
profit perannum 每年利润
profit planning 利润规划；编制利润计划
profit plus taxes ratio of investment 投资利税率
profit-proved 经济的
profit rate 盈利率；利润率
profit rate on funds 资金利润率
profit ratio 盈利比率
profit ratio of investment 投资利润率
profit ratio of net worth 净值利润率
profit ratio of paid-up capital 实收资本利润率
profit ratio of total liabilities and net worth 负债总额与净值的利益率
profit remaining system 利润留成制
profit retention 利润留成
profits discounting method 刚润贴观法
profit-seeking 追逐利润
profit sensitivity ratio 盈利敏感率
profit sharing 利润分享；利润分成；分红
profit-sharing bond 分红债券
profit-sharing security 分红证券
profit-sharing system 分红制
profits in excess of the norm 超目标利润
profits realized 实现利益
profit structure 利润结构；利润构成
profits turned over to the state 利润上缴
profit system 利润制
profit taking 见利抛售
profit target 目标利润
profit tax 利润税；利得税
profit through circulation 流通利润
profit transferred 已转让利润
profit vector 利润向量
profit volume 利润量
profit-volume analysis 利润量分析
profit-volume graph 利润量图表；利量图
profit-volume ratio 利润量比率；利量比率（利润销货额比率）
profondometer 深部异物定位器
profore 矿胎
pro forma 形式上
pro forma bill 估算单；估价发票
pro forma charter party 租约格式
pro forma cost 估计费用
pro forma disbursement 预估使费
pro forma finance 财务预计
pro forma interview sheet 访问表格
pro forma invoice 预计装货清单；暂定的发货单；暂定发票；估价发票；估价单；估算单；形式发票
pro forma statement 形式决算表；定式决算表
pro forma transaction 初步提议；初步交易
pro-form income statement 预算损益表
profound 深刻
profound and lasting 深远的
profound effect 深刻影响
profound fault 深断裂；深断层
profound groundwater 深层地下水
profound hypothermia 深低温
profound intervention 深入干预
profound significance 深远的意义
profundal 湖底的
profundal zone 深水区；深水带；深水层；深海底深底带

profusely enriched 非常浓缩的；极其丰富的
profusely gilt 丰满的金色涂层
progagation of population 人口增殖
progenitor 正本
progeny 次级粒子
proglacial 冰前的
proglacial delta 冰前三角洲
proglacial lake 冰堰湖；冰川前缘湖泊
prognosis 预后；预测；预报
prognosis formula 预测公式
prognostic chart 预报图(表)；预报(天气)图
prognostic contour chart 等压面预报图
prognostic equation 预报方程
prognostic formula 预测公式
prognostic index 预后指数
prognostic map 预图，预地图；预测图
prognostic value 预测值
prognostic wave chart 波浪预报图
progradation 延伸作用；进积作用；进变作用
progradational beach 进积滩
progradation of coast 海岸推进；岸进
progaded delta 进积三角洲；推进三角洲
prograde orbit 顺行轨道
prograding growth sequence 进积型生长层序
prograding pattern 进积型
prograding sequence 进积层序
prograding shoreline 推进滨线
program (施工进度)计划；纲要；进度计划；节目；日程；大纲；程序；方案；时间表
programable film reader 程序胶卷读出器
program-based budgeting 方案基础预算法
programmability 可编程序性
programmable 可编程序
programmable amplifier 可编程序放大器
programmable array 可编程序阵列
programmable attenuator 程控衰减器
programmable calculator 可编程序计算器
programmable clock 可编程序时钟
programmable communication interface 可编程序通信接口
programmable concentrator 可编程序集中器；可编程集线器
programmable controller 可编程序控制器；可编程控制器；编程控制器
programmable counter 可编程序计数器；程控计数器
programmable data generator 程控数据信号发生器
programmable data mover 可编程序数据传送器
programmable data rate 可编程序数据速率
programmable data selector 可编程序数据选择器
programmable desk computer 可编程序公式计算机
programmable digital controller 可编程序数字控制机
programmable divider 可编程序除法器
programmable fire detection zone 可编程序的的火灾探测区
programmable graph generator 可编程序图表生成器
programmable input, output 可编程序输入输出
programmable interface 可编程序接口
programmable interval 程序控制时间间隔
programmable logic 可编程序逻辑
programmable logic(al) array 可编程序逻辑阵列；场致程序逻辑阵列
programmable logic(al) control 可编程序控制
programmable logic(al) controller 可编程序逻辑控制器
programmable logic(al) controller control 可编程序逻辑控制器控制
programmable logic(al) control system 可编程序逻辑控制装置
programmable logic(al) spectrum 可编程序逻辑系列；可编程序逻辑流
programmable logic(al) system 可编程序逻辑系统
programmable logic(al) type 可编程序逻辑类型
programmable mable operational amplifier 可编程序运算放大器
programmable maintenance procedure 可编程序维护过程
programmable operation control 可编程序作业控制器
programmable oscilloscope 程控示波器
programmable peripheral interface 可编程序外围接口
programmable point-of-sale terminal 可编程序销售点终端
programmable power supply 程序控制供电系统；程控电源
programmable pulse generator 程序可控脉冲发生器
programmable read-only memory 可编程序只读存储器；程序可控只读存储器
programmable sequencer 可编程序定序器
programmable system interface 可编程序系统接口
programmable terminal 可编程终端
programmable timer 可编程序计时器；可编程定时器
programmable vertical timing generator 可编程垂直定时发生器
programmatic management of ground water environment 地下水环境规划管理
programmatics 程序设计学；程序设计技术；程序编制学
programme = program
program(me) abort system call 程序中止系统调用
program(me) access code 程序存取码；程序存取代码
program(me) access key 程序访问键；程序存取键
program(me) activation vector 程序启用向量；程序激活向量；程序活动向量
Programme Activity Center for Environmental Education and Training 环境教育和训练方案活动中心
program(me) address 程序地址
program(me) addressable clock 程序可寻址时钟
program(me) address counter 程序地址计数器
program(me) address error 程序地址错误
program(me) address register 程序地址寄存器
program(me) address storage 程序地址存储器
program(me) allocation 程序分配
program(me) alternatives 方案抉择
program(me) analysis 程序分析；方案分析
program(me) analysis and review 规划分析和检查
program(me) analysis method 程序分析方法
program(me) analysis technique 程序分析技术
program(me) analyzer 程序分析器；程序分析程序
program(me) and data management unit 程序及数据管理部件
program(me) and papers 规划与论文
program(me) appropriation 方案拨款
program(me) area 程序区
program(me) area block 程序区块
program(me) assembly 程序组件
program(me) attention key 程序注意键；程序引起注意键
program(me) audit 方案审计
program(me) authority 程序审定；程序权限
program(me) authorization 程序授权；程序审定
program(me) authorized credentials 程序特许凭证；程序审定信任书
program(me) batch process 程序批处理
program(me) behaviou(u)r 程序动作；程序功效
program(me) behaviou(u)r model 程序功效模型
program(me) block 程序块
program(me) block address 程序块地址
program(me) block loading 程序块存入
program(me) board 程序控制台；程序控制盘；程控台
program(me) body 程序体
program(me) budget 项目预算；方案预算
program(me) budgeting 项目预算编制
program(me) bus 程序总线
program(me) capacity 程序控制的容量
program(me) card 程序控制卡；程序卡(片)
program(me) certification 程序检验
program(me) chaining 程序链接
program(me) change command 程序更换命令
program(me) change control system 程序改变控制系统
program(me) channel 控制信息；传输程序通路；程序通道
program(me) chart 进度图；程序图

program(me) chart of subdivisional work 分项工程进度表
program(me) check interrupt 程序校验中断
program(me) check(out) 程序检查;程序校验;程序测试;程序检验
program(me) checkout condition 程序校验条件;程序检查条件
program(me) checkout on-conditions 程序校验条件
program(me) check run 程序校验操作;程序检验操作
program(me) check subsystem 程序校验子系统
program(me) chooser 程序选择器
program(me) circuit 节目电路;程序电路
program(me) clock 程序钟;程序时钟
program(me) clockwork 程序钟表机构
program(me) comment 程序注解
program(me) communication block 程序通信信息块;程序通信块
program(me) communication block mask 程序通信块特征码
program(me) compaction 程序压缩
program(me) comparison 程序比较
program(me) comparison index 程序比较指数
program(me) compatibility 程序兼容(性)
program(me) competition multiprogramming 程序争用的多道程序设计
program(me) compilation 程序编译
program(me) compiler 程序的编译程序;程序编制器;编译程序
program(me) compiling 程序编译
program(me) complexity 程序复杂性
program(me) component 程序成分
program(me) composition 程序设计;程序编制;编制程序
program(me) comprehension 程序理解
program(me) computer 程序计算机
program(me) connectivity 程序内连关系
program(me) control 程序控制
program(me)-control computer 程序控制计算机
program(me) control data 程序控制数据
program(me) control data attribute 程序控制数据属性
program(me)-control device 程序控制装置
program(me) control die casting machine 程序控制压铸机
program(me) control execution 程序控制执行
program(me) control flow 程序控制流
program(me) control hardware 程控硬件
program(me) control instruction 程序控制指令
program(me)-controlled 程序控制的;程序控制的
program(me)-controlled computer 程序控制计算机
program(me)-controlled coordinate boring machine 程序控制坐标镗床
program(me)-controlled copying trimming lathe 程序控制仿型修坯机
program(me)-controlled drilling 程控钻进
program(me)-controlled exchanger 程控电话交换机
program(me) controlled high precision broacher 程控高精度拉床
program(me)-controlled input-output 程序控制输入输出
program(me)-controlled interrupt 程序控制中断
program(me) controlled interruption 程序控制中断
program(me)-controlled lathe 程序控制机床
program(me)-controlled liquidoid extensional furnace 程序控制液相外延炉
program(me)-controlled machine 程序控制机床
program(me) controlled piston ring mo(u)lding machine 程控活塞环造型机
program(me)-controlled sequential computer 程序控制的时序计算机
program(me)-controlled software interrupt 程序控制的软件中断
program(me) controlled telephone 程控电话
program(me) controlled telephone switching 程控电话交换机
program(me) controlled telephone switching system 程控电话交换装置
program(me) controller 程序控制器;程序器
program(me) controlling element 规划控制元件;程序控制器;程序控制部件
program(me) control machine tool 程序控制机床
program(me) control milling machine 程序控制铣床
programme control of technologic(al) process 工艺过程序控制
program(me) control precision hobbing machine 程控精密滚齿机
program(me) control real-time clock 程序控制实时时钟
program(me) control register 程序控制寄存器
program(me) control sheet 程序控制表
program(me) control statement 程序控制语句
program(me) control switch 程序控制开关
program(me) control system 程序控制系统
program(me) control table 程序控制表
program(me)-control test 程序控制试验
program(me) control testing vehicle for preliminary equipment of substation 变电站一次程控测试车
program(me) control testing vehicle for secondary equipment of substation 变电站二次程控测试车
program(me) control transfer 程序控制转移
program(me) control turret lathe 程控转塔车床
program(me)-control unit 程序控制装置
program(me) conversion 程序转换
program(me) copy 程序复制
program(me) correctness 程序正确性;程序的正确性
program(me) cost 方案费用
program(me) counter 指令计数器;程序计数器
program(me) counter address 程序计数器地址
program(me) counter addressing 程序计数器寻址
program(me) counter register 程序计数寄存器
program(me) counter relative 程序计数器相关
program(me) counter stack 程序计数器栈
program(me) counter store 程序计数存储器
program(me) current pulser 程序电流脉冲发生器
programme(d) 可程式化
programmed address modification 程序地址修改;程序地址修改
programmed algorithm 程序算法
program(me) damage assessment and repair 程序故障的固定和恢复
programmed application library 程序应用库
program(me) database information system 程序库信息系统
program(me) data set 程序数据集
programmed automatic circuit tester 程序自动电路测试装置
programmed automatic communication equipment 自动程序控制通信设备;程序控制自动通信设备
programmed automatic press system 程序自动压力系统
programme(d) calculator 卡片分析(计算)机
programmed channel 程序控制通道
programmed check 程序校验;程序检验;编程检验
programmed checking 程序校验;按程序检查
programmed computer 程序计算机;编程计算机
programmed control 程序操纵
programme(d) cost 可控制的成本;计划成本
programmed cryptographic facility 程序加密设备
programmed data processor 程控数据处理机
programmed data transfer 程序数据传送
programmed decision 程序判定
programmed decision-making 规划决策
programme(d) digital automatic control 数字程序自动控制
programmed digital logic 程序数字逻辑
programmed divider 程序除法器
programmed drill 程序控制式的播种机
programmed dump 程序转储;程序控制转储
program(me) debug(ging) 程序调试;程序调整;程序排除错误;程序查错
program(me) debugging tool 程序调试工具
program(me) decision 方案决策
program(me) deck 程序卡片叠
program(me) definition language 程序定义语言
programmed elements 程序单元
program(me) delivery 方案执行
program(me) description 程序说明;程序描述
program(me) descriptor 程序描述符
program(me) design 程序设计
program(me) design data 程序设计数据
program(me) determination 程序终止
program(me) development 程序展开;程序开发;程序发展;程序编制
program(me) development aid 程序扩展辅助设备
program(me) development computer 程序开发计算机
program(me) development software 程序开发软件
program(me) development system 程序开发系统;程序发展系统;程序编制系统
program(me) development time 程序开发时间;程序编制时间;编制程序时间
program(me) device 程序装置
programmed film reader system 程序胶片阅读系统
programmed flow 程序流速
programmed flow chromatography 程序变流色谱法
programmed flow gas chromatography 程序变流色谱法
programmed gain control 增益程序控制;程序增益控制
programmed grammar 程序设计文法;程序化文法
programmed guidance 程序制导
programmed halt 程序停机
programmed heating 程序加热
programmed input-output 程序(控制)输入输出
programmed input-output address 程序输入输出地址
programmed input-output command 程序输入输出命令
programmed input-output microcomputer 程序控制的输入输出微型计算机
programmed input-output operation 程序输入输出操作
programmed instruction 程序指令;程序控制教学;程序教学
programmed integrated control equipment 积分程序控制设备
programmed interrupt request vector 程序中断请求向量;编程中断请求向量
program(me) display 程序显示
program(me) divider 程序除法器
programmed keyboard 程序键盘
programmed leaning 程序控制的学习;计算机辅助教学;程序学习;程序(化)教学
programmed load control 负载程序控制
programmed logic 可编程逻辑;程序控制逻辑;程序逻辑;程序控制的逻辑;编程逻辑
programmed logic array 程序逻辑阵列;程序(控制)逻辑阵列;程控逻辑阵列;编程逻辑阵列
programmed maintenance 预防性维修
programmed maneuver 程序机动
programmed marginal check 程序监督的边缘校验
programme(d) mixture-ratio control 混合比程序控制
program(me) documents 程序文件
program(me) documentation 程序文件(编制);程序文档
programmed operating system 程序操作系统
programmed operator 程序控制操作指令
programmed pressure 程序变压(力)
programmed protection 程序保护
programmed route control 进路程序控制
program(me) drum 程序鼓;程序磁鼓
programmed shutter 程序快门;程序控制快门
programmed solvent 程序溶剂;程序变溶剂
programmed sonar 程序控制声呐
programmed stop 程序停机指令;定点停车;程序停机
programmed switch 程序开关
programmed temperature 程序控制温度;程序升温
programmed temperature gas chromatography 程序升温气相色谱法
programmed temperature works 程序升温机件
programmed time scale 程序时间比例尺
programmed turn 程序转弯
program(me) effectiveness 方案成效
program(me) efficiency 程序效率
program(me) element 程序元素;程序因素;程序单元

program(me) emulator 程序仿真器
program(me) end flag 程序结束标志
program(me) entry 程序进入
program(me) error 程序误差;程序错误
program(me) error control 程序错误控制
program(me) error correction report 程序误差校正报告
program(me) error dump 程序错误转储
program(me) error interrupt 程序错误中断
program(me) evaluation and review 计划评审
program(me) evaluation and review technique 计划评审法;计划协调技术;计划评审技术;计划评审法;计划评估法;计划估评法;统筹方法;统筹法;程序评价和审查技术;程序鉴定技术;程序鉴定和检查技术
program(me) evaluation and review technique-cost 成本计划协调技术
program(me) evaluation and review technique schedule 计划评估情况表
program(me) evaluation procedure 程序鉴定过程
program(me) evaluation process 计划评审过程
program(me) event 程序事件
program(me) event recording 程序事件记录
program(me) evolution and review technique 网络法
program(me) exception 程序异常
program(me) exception code 程序异常代码
program(me) exception interrupt 程序异常中断
program(me) execution 程序执行
program(me) execution control 程序执行控制
program(me) execution control sequence 程序执行控制顺序
program(me) execution monitor 程序执行监督
program(me) execution services 程序执行服务(程序)
program(me) execution speed 程序执行速度
program(me) execution time 程序执行时间
program(me) exit 程序出口
program(me)-exit hub 程序输出插孔
program(me) expression 程序表达式
program(me) extension 程序扩展
program(me) failure alarm 程序故障警报
program(me) fetch 取程序;程序取出
program(me) fetch time 程序取出时间;程序读取时刻
program(me) file 程序文件
program(me) filing 程序文件生成
program(me) fix 固定程序
program(me) flexibility 程序灵活性
program(me) flow 程序流程
program(me) flow analyzer 程序流程分析器
program(me)(flow) chart 程序框图;程序流程图
program(me) flow diagram 程序流程图;程序框图
program(me) for agriculture 农业计划方案
program(me) for approximation of function 函数逼近程序
program(me) for automatic coding technique 自动编码技术程序
program(me) for cross-sectioning 横断面测量程序
program(me) for maintenance 维修程序
program(me) for planning population growth 人口规划
program(me) frame 程序框架
program(me) function key 程序功能键
program(me) generated parameter 程序生成参数
program(me) generation system 程序生成系统
program(me) generator 程序的生成程序;程序编制器;程序编辑程序
program(me) graph 程序图
program(me) halt 程序停止;程序固有停机指令
program(me) header 程序头
program(me) header block 程序头块;程序标题块
program(me) identification 程序标识
program(me) identification code 程序识别码
program(me) identifier 程序标识符
program(me) image 映像程序;程序映像
program(me) in basic-plus 基本加法程序
program(me)-independent modularity 独立程序模型化
program(me) indicator code 程序指令码
program(me) information 程序信息
program(me) information block 程序信息块
program(me) information block extension 程序信息块扩充
program(me) information code 程序信息(代)码
program(me) initialization 程序初始化
program(me) in machine code 机器代码程序
program(me) installation 程序机构
program(me) instruction 程序指令
program(me) interaction 程序相互作用
program(me) interchange 程序互换
program(me) interrupt 程序中断
program(me) interrupt active 程序中断工作区
program(me) interrupt condition 程序中断条件
program(me) interrupt control 程序中断控制
program(me) interrupt exit routine 程序中断出口程序
program(me) interruption control area 程序中断控制区
program(me) interruption element 程序中断元素
program(me) interrupt message 程序中断信息
program(me) interrupt request 程序中断请求
program(me) interrupt signal 程序中断信号
program(me) interrupt signal sequence 程序中断信号序列
program(me) interrupt transfer 程序中断转移
program(me) isolation 程序隔离
program(me) key 程序键
program(me) language 程序语言
program(me) layout 程序框图
program(me) lending 计划借贷
program(me) level 程序级;程序电平
program(me) level change tape 程序级修改带
program(me) level diagram 程序电平图
program(me) library 程序库
program(me) library and directory maintenance 程序库和目录的维护
program(me) library release 程序库发行版次;程序库版次
program(me) limit monitoring 程序界限监督;程序范围监督
program(me) line 广播中继线;程序行
program(me) linkage 程序连接
program(me) linking 程序连接
program(me) list 程序表
program(me) list control block 程序清单控制表;程序表控制块
program(me) listing 程序列表
program(me) listing editor 程序编目编辑
program(me) listing output 程序列表输出
program(me) loader 程序装入装置;程序装入器;程序的装入程序
program(me) load(ing) 程序装入;程序调入
program(me) loading operation 程序装入操作
program(me) loading routine 程序装入的例程;程序装入程序
program(me) loan 计划贷款
program(me) locality 程序局部性
program(me) location counter 程序单元计数器
program(me) location directive 程序单元指令
program(me) location table 程序存储位置表
program(me) lock register 程序锁定寄存器
programme logic 程序逻辑
program(me) logical design 程序逻辑设计
program(me) logic array 程序逻辑阵列
program(me) logic flowchart 程序逻辑流图
program(me) loop 程序周期;程序循环
program(me) maintenance 程序维修;程序维护
program(me) maintenance procedure 程序维护过程
program(me) management 程序管理
program(me) management control table 程序管理控制表
program(me) management system 程序管理系统
program(me) manager 程序主管员;程序管理人员
program(me) manipulation 程序处理
program(me) mask 程序掩码;程序屏蔽
program(me) master-file update 主件程序修改
program(me) master-file updating 程序主文件更新
program(me) mechanism 程序机构
program(me) memory 程序存储器
program(me) memory system 程序存储系统;程序存储方式
program(me) message 程序信息
program(me) meter 响度计
program(me) miss 程序错误
program(me) mode 程序状态;程序方式
program(me) mode field 程序方式字段
program(me) model 程序模型
program(me) modelling 程序模型化
program(me) mode remote control system 程序方式远程控制系统
program(me) modification 程序修改
program(me) module 程序模块
program(me) module dictionary 程序模块字典
program(me) monitor 程序监控器;程序监督程序
program(me) name 程序名字;程序名称
program(me) notation 程序化记数法
program(me) objective 规划目标
program(me) of capital work 基础工程计划
program(me) of investigation 勘探程序
program(me) of loading and unloading operation in section 区间装卸车计划
program(me) of meeting 会议日程安排
programme of N-layered system 多层体系计算程序
program(me) of pollution control 污染控制规划
program(me) of preparatory works 施工准备工作计划
programme of project 工程计划进度
program(me) of prospecting 勘探程序
program(me) of salvage operation 打捞施工计划
program(me) of surveying 勘探程序
program(me) of topographic(al) analysis 地形分析程序
program(me) of variety test 品种试验方案
program(me) of water conservancy of farmland 农田水利规划方案
program(me) of water resources development 水资源开发方案
program(me) of works 工作日程;工作程序;工序;工程计划进度
program(me) on greenhouse gases 温室效应气体计划
program(me) on long-range transport of pollutants over Europe 欧洲污染物长途运输方案
program(me) on ocean dynamics and climate 海洋动力学和气候计划
Program(me) on Ocean Mapping 绘制海洋图计划
program(me) operator 程序化操作员;程序操作员
program(me) operator interface 程序操作员接口
program(me) optimization 程序(最)优化
program(me) order 程序指令
program(me) order address 程序指令地址
program(me) organization 程序组织;程序结构
program(me) origin 程序设计起始地址
program(me) over-lay 程序重叠;程序覆盖
program(me) over-lay capability 程序覆盖能力
program(me) over-lay structure 程序覆盖结构
program(me) package 程序组;程序库;程序包
program(me) packing 程序包
program(me) paging function 程序分页功能
program(me) parameter 程序参数;程序参量
program(me) part 程序部分
program(me) patch 程序修改;程序补片
program(me) patching plug 程序修改插接板
program(me) phase 程序阶段
program(me) portability 程序可移植性
program(me) portable 程序移植
program(me) post-edit 程序后编辑
program(me) preparation 程序准备
program(me) preparation aids 程序准备的工具程序
program(me) preparation facility 程序准备用软件;程序准备设备
program(me) priority 程序优先级
program(me) process(ing) 程序执行过程;程序过程;程序处理
program(me) processor 程序处理机
program(me) product 程序产品
program(me) pulse 程序脉冲
programmer 定值器;程序装置;程序写入器;程序设计员;程序设计师;程序(设计)器;程序机构;程序编制器;编程器
programmer analyst 程序机构分析员;程序设计员;程序设计和系统分析员
programmer and probability analyser 程序设计器与概率分析器
programmer card 程序编制器插件
programmer check 程序员校验;程序员检查

programmer console 程序员工作台
programmer control panel 程序员控制板
programmer-defined macro 程序员定义宏指令
programmer-defined on-conditions 程序员定义的条件
programmer-defined variable 程序员定义的变量
program(me) reader 程序阅读器;程序读出器
program(me) read-in 程序读入
program(me) recording control instrument 程序记录控制仪
program(me) recycling 程序再循环
program(me) reference 程序引用
program(me) reference behaviour 程序引用功效
program(me) reference table 程序参照表;程序参考表
program(me) region 程序区
program(me) region base register 程序区基址寄存器
program(me) region length register 程序区长度寄存器
program(me) register 程序寄存器
program(me) relative paging 程序相对分页
program(me) relocatability 程序再定位性
program(me) relocate 程序再定位;程序浮动
program(me) relocation 程序再定位;程序浮动
program(me) request 程序请求
program(me) request count 程序请求计数
program(me) required credentials 程序使用凭证;程序请求凭证
programmer error correction 程序员改错
program(me) reserves 方案储备金
program(me) restructuring 程序重新构成
program(me) ring 程序环
programmer job 程序员作业
programmer logical unit 程序员逻辑部件
programmer message 程序设计员信息
programmer module 程序编制组件;程序编制微型组件
programmer named condition 程序员命名的条件
program(me) roll-back 程序重算;程序卷回
program(me) route control 程序进路控制
program(me) routine 程序的例行程序;程序的例程
programmer peripheral interface 程序装置机构外围接口
programmer's console diagnostic 程序员控制台诊断
programmer's macroinstruction 程序员宏指令
programmer's support 程序员的支持程序
programmer's template 程序员用模板
programmer's workbench 程序员工作台
programmer system 程序编制器系统
programmer tool 程序员工具
program(me) run 程序运行
program(me) run mode 程序运行方式
program(me) scheduler 程序调度;程序的调度程序
program(me) scheme 程序图式
program(me) segment 程序段
program(me) segmentation 段式程序;程序分段;程序段
program(me) segmenting 程序分段
program(me) segment size 程序段长度
program(me) segment table 程序段表
program(me) selection 程序选择
program(me) selector 程序选择器
program(me) selector switch 程序选择开关
program(me) sensitive error 特定程序错误
program(me) sensitive malfunction 特定程序错误;程序检测误差
program(me) sensor 感应选台装置
program(me) sequence 程序序列
program(me) sequencer 程序序列发生器
program(me) sequencing 程序次序
program(me) server 程序编制服务器
program(me) service 程序服务
program(me) sharing 程序共享
program(me) sheet 程序用纸
program(me) shutter 程序快门
program(me) signal generator 程序信号发生器
program(me) size 程序大小
program(me) skeletal coding 程序纲要
program(me) skip 程序空区;程序空白指令
program(me) space 程序空间
program(me) space counter 程序空间计数器

program(me) specification 程序说明书;程序规格;程序规范说明书;程序规范
program(me) specification block 程序说明块
program(me) stack 程序栈;程序机;程序堆栈
program(me) standard 程序标准
program(me) state 程序状态
program(me) statement 程序语句
program(me) statement analysis 程序语句分析
program(me) status 程序状态
program(me) status double-word 程序状态双字
program(me) status register 程序状态寄存器
program(me) status vector 程序状态向量
program(me) status word 程序状态字
program(me) status word register 程序状态字寄存器
program(me) step 程序步骤;程序步长
program(me) stop 程序停止
program(me) stop instruction 程序停机指令
program(me) storage 程序存储器
program(me) storage unit 程序存储单元
program(me) store 程序存储
program(me) store enable 允许程序存储
program(me) store unit 程序存储器
program(me) structure 程序结构
program(me) structure design 程序结构设计
program(me) structure statement 程序结构语句
program(me) stub 程序余段
program(me) subroutine 程序的子例行程序
program(me) supervision 程序管理
program(me) support cost 方案支持费用
program(me) support representative 程序支持表示
program(me) switch 程序开关
program(me) switching 程序转接;程序转换
program(me) switching interrupt 程序转接中断
program(me) switch list 程序变换表
program(me) synchronization 程序同步
program(me) syntax 程序文法
program(me) synthesis 程序综合
program(me) system 程序系统
program(me) system testing 程序系统测试
program(me) tansformation 程序变换
program(me) tape 程序带
program(me) tape unit 程序带装置
program(me) technic 程序技术;程序方法
program(me) temporary fix 程序暂时固定;程序临时固定;程序的临时修改
program(me) terminal 程序终端
program(me) termination 程序终止;程序的终止
program(me) termination processing 程序最终加工;程序最后加工
program(me) tester 程序试验器
program(me) test(ing) 程序检查;程序检验;程序测试;程序校验
program(me) testing software 程序测试软件
program(me) testing time 程序检验时间;程序检查时间;程序测试时间
program(me) test system 程序检验系统
program(me) test tape 程序检验带
program(me) text 程序原文;程序文档;程序文本
program(me) time 程序时间
program(me) timer 计划调节器;程序计时器;程序定时器
program(me) timing matrix 程序定时矩阵
program(me) to be furnished 提供进度计划
program(me) trace 程序追踪;程序跟踪
program(me) tracking antenna 程序跟踪天线
program(me) transferability 程序可移植性;程序可移植性
program(me) translation 程序翻译
program(me) transmitter 程序发送机
program(me) tree 程序树
program(me) unit 程序装置;程序单元;程序单位
program(me) update tape 程序修改带
program(me) validation 程序正确性检查
program(me) validation services 程序证实服务程序;程序合法性业务
program(me) variable 程序变数;程序变量
program(me) verb 程序动词
program(me) verification 程序验证;程序检验
program(me) welding 程序点焊
programme with progress 工程计划进度图
program(me) word 程序字
programming 用程序规划;规划;程序设计方法

论;程序设计;程序编制;编程序;编制程序
programming aid 程序设计工具;程序设计辅助程序
programming approach 程序设计方法
programming check 程序设计检验
programming control card 程序设计控制卡
programming controller 自动顺序控制器;程序控制器
programming control panel 程序设计控制器;程序设计控制板
programming control unit 程序控制装置
programming convention 程序设计约定
programming cool 程序冷却
programming cost 程序设计代价
programming counter 程序设计计数器
programming device 程序设计装置;程序编制机
programming environment 程序设计条件;程序设计环境
programming error 程序设计错误
programming flexibility 程序设计灵活性
programming flow chart 程序设计流程图;程序设计框图;程序编制流程图;编程序流程图
programming flow diagram 程序设计流程图;编程序流程图
programming in logic 逻辑程序设计;逻辑编程
programming job 程序设计作业
programming language 程序设计语言
programming language grammar 程序设计语言文法
programming linguistics 程序设计语言学
programming machine 编程机
programming manual 程序设计手册
programming method 程序法
programming methodology 程序设计方法学
programming module 程序设计方法编制程序法;程序设计模块;程序模块;编程序模块
programming of construction 施工程序设计
programming panel 程序设计板;程序编制板
programming phase 计划阶段
programming practice 程序设计实践
programming pressure 程序变压(力)
programming primitive 程序设计的基本指令
programming procedure 程序设计过程
programming program(me) 编程序的程序
programming redesign 程序重新设计;程序的再设计
programming request for price quotation 编制市场报告要求程序
programming semantics 程序设计语义学
programming statement 程序设计语句
programming style 程序设计风格
programming system 程序编制装置;程序系统;程序设计系统;编程序系统
programming technique 程序设计技术
programming template 程序设计样板
programming tool 程序设计工具
programming transformation 程序转换
programming transparency 程序透明性;程序设计透明性
programming unit 程序装置
programming utility 程序设计应用
program-output hub 程序输出插孔;程序输出插孔
program-selected terminal 程序选择终端;程序选用终端
program-sensitive error 程序过敏错误
program-sensitive fault 特定程序故障;程序过敏故障
program-sensitive malfunction 程序过敏故障;程序过敏错误
program-storage manager 程序及存储器管理程序
program-suppress hub 程序插孔
program-synthesizing algorithm 程序综合算法
program-to-program(me) communication 程序间通信
progress 进度;创造性
progress and status reporting 进展和现状报告
progress biling 工程进度款账单
progress certificate 进度证书
progress chart 工艺卡(片);工序图;工程进度表;进度图;进度表;时间进度表;施工进度线(图)
progress chart for pipe-laying 铺管进度表
progress chart of cargo work 装卸进度表
progress chart of for concrete work 混凝土工作进度表
progress chart of loading/discharging 装卸进度表

progress chaser 材料检查员(英国);进度检查员
progress clerk 进度记录员
progress control 进度控制;进度管理
progress controlling of construction 工程建设进度控制
progress estimate 进度估计
progressing-cavity pump 腔进式污泥泵;多腔螺旋泵
progressing of excavation 开挖进尺
progression 级数
progression of failure 故障升级
progression saturation method 累进饱和法
progressive 累进利息;累进的;渐进;顺序的
progressive action 连续作用;均匀作用
progressive ag(e)ing 连续升温时效;连续加热时效;分段时效;分级时效
progressive approximation 逐次近似;逐次逼近法
progressive architecture 先进建筑
progressive assembly 流水装配;流水线装配法;传送带式装配
progressive assembly line 流水作业输送线;流水装配线;流水线
progressive austemper 分级等温热处理
progressive austempering 分级等温淬火;奥氏体不完全等温退火
progressive average 移动平均(数);累加平均值;累加平均数;累计平均;累计平均;累积平均(数);滑动平均(数)
progressive average lag method 累进平均滞后法
progressive block method 连续块焊法;分段多层焊
progressive block sequence 分段多层焊工序
progressive bonding 逐渐黏合
progressive brake 渐进式制动器
progressive braking 进行式制动
progressive burning 增推力燃烧;增面燃烧;随堆随烧
progressive burning charge 增面性燃烧火药柱
progressive burning powder 递进燃烧药
progressive cast-in-place method 顺序渐进现浇法
progressive caving 连续崩落
progressive cavity pump 螺杆抽油泵
progressive classified grinding 顺序分级粉磨
progressive clear strip system 连续带状皆伐式渐伐作业法
progressive collapse 渐坍(房屋)
progressive combustion 增面燃烧
progressive consolidation 渐进固结作用
progressive construction method 顺序渐进施工法
progressive contact 渐进接触;顺序接触
progressive contracting 逐渐缩090
progressive contracting of a stream 河流渐缩
progressive conversion model 渐进转化模型
progressive copy turning 顺序仿形车削
progressive crack 渐进裂缝
progressive cutting 累进切削法
progressive cycle 海侵旋回
progressive cylindrical plug ga(u)ge 分极柱形测孔器;分级柱形测孔器
progressive damage 累进破坏
progressive defecation 逐步澄清
progressive deformation 累积变形;递增变形;递进变形
progressive demagnetization 逐步退磁
progressive deposition 沿程淤积
progressive depression 逐渐沉陷
progressive derivative 右导数
progressive deviation 渐进偏差
progressive die 连续模(冲压用);连续冲模;跳步模;顺序冲模
progressive-die press 连续模压力机;连续冲压压力机
progressive diminution 累进递减
progressive dryer 逐步干燥器;分段烘干机
progressive dry kiln 连续干燥窑
progressive erection 逐跨施工(法);逐孔施工(法);顺序拼装(法)
progressive erosion 沿程冲刷
progressive error 行程差;累进误差;积累误差
progressive evolution 前进演化;前进式演化;前进进化
progressive exercise 渐进性运动
progressive failure 逐渐破坏;进展性破坏;渐进破坏;渐进破坏
progressive fault 渐进性断层

progressive flow 行进水流
progressive folding 递增褶皱作用
progressive forming 连续成型;级进成型;步进成型
progressive fracture 扩展裂缝
progressive freezing 逐步冷冻法
progressive ga(u)ge 分级量规
progressive gear 无级变速;顺序变速装置
progressive geosyncline 渐进式地槽
progressive glide 逐渐滑动
progressive gravity wave 前进重力波
progressive hardening 依次淬火;顺序淬火;二次硬化
progressive hose lay 逐步分流式水带铺设
progressive illumination 图像逐点照明
progressive improvement 逐步改善;分期改善
progressive income tax 累进所得税
progressive induction seam welding 连续感应滚焊
progressive inequality 行程差
progressive interlace 逐行交错
progressive inventory method 累计盘存法
progressive kiln 干燥木材的连续(作业)窑;连续(运转)窑;连续式干燥窑;渐进干燥窑
progressive lap winding 右行叠绕组
progressive ledger 累计分类账
progressive loading 递增荷载
progressive loading trial 逐步加载试验
progressive massive fibrosis 进行性大块纤维化
progressive mean 移动平均(数);累加平均值;累加平均(数);滑动平均(数)
progressive metallogenesis 递进成矿说
progressive metamorphism 前进变质作用
progressive method 循序前进法;(线路架设的)前进法
progressive motion 前进运动
progressive movement 推进式运行
progressive narrowing 逐渐缩窄
progressive narrowing of a stream 河流渐窄
progressive narrowing of river 河流逐渐变窄
progressive narrowing of stream 河流逐渐变窄
progressive nature 发展特征
progressiveness of income tax 所得税的累进度
progressive offlap 渐进退覆
progressive operation 连续操作;分级操作
progressive oscillatory 行进振幅波
progressive overflow 依次溢出;顺序溢出
progressive overflow method 顺次溢出法
progressive overlap 渐进超覆;前进超覆
progressive-part method 分解—累进学习法
progressive payment 分阶段付款;按进度(表)付款
progressive performance curve 增速特性曲线
progressive pitting 进展性点蚀
progressive placement in one-direction 单向前进安装
progressive plane strain 渐进面应变
progressive plane wave 前进平面波;平面行波;平面前进波
progressive plastic yield 渐进塑性屈服
progressive press tool 连续冲模
progressive procession 循序渐进;顺序渐进
progressive production work 连续作业
progressive proofs 分色打样
progressive provisioning 累进供应
progressive quenching 分级顺序淬火;分级淬火;分段淬火
progressive ratio 速比
progressive resistance exercise 渐进抗阻运动
progressive rotation 往复旋转
progressive rupture 渐进破坏;渐进断裂
progressive safety clamp 渐进式安全钳
progressive sampling 渐进取样
progressive sand wave 向下游推移的沙波;前进砂波
progressive scaling 逐渐剥落;顺序鳞剥
progressive scan(ning) 顺序扫描;顺次扫描;逐行扫描
progressive schedule 累计表
progressive scour(ing) 沿程冲刷;渐进性冲刷
progressive settlement 逐渐沉降;累积性沉降;逐渐沉降;累进沉降;渐进沉降
progressive shear 渐进剪切;步进剪
progressive sheet 套色打样
progressive shuttering 逐次启闭快门

progressive slide 逐渐滑动;连续滑动;累进性滑动;渐进性滑动;递增滑动
progressive sliding 渐进性滑动
progressive solidification 顺序凝固
progressive sorting 渐进分选
progressive speed trials 逐点加速试航;标准试航
progressive splitting 前进式分岔
progressive stage 进展期;进行期;发展期
progressive stage construction 渐进分散施工法;渐进分期施工法;前进分期施工;分期渐进施工法
progressive strain 渐进应变;递增应变
progressive subsidence 逐渐沉陷;累积沉陷(量)
progressive successive 前进演替
progressive surface wave 前进表面波
progressive system of traffic control 先进交通管理系统;先进的交通管制系统;推进式交通控制信号联动系统
progressive tax(ation) 累进税
progressive tax in excess of specific amounts 超额累进税
progressive tax on excess earning 超额收益累进税
progressive tax rate 累进税率
progressive tax structure 累进税结构
progressive tax system 累进税制
progressive tidal wave 前进潮波
progressive total 累计额
progressive trial 逐步加载试验
progressive type 进行型
progressive type of tidal oscillation 前进式潮汐波动
progressive(type) transmission 级进式传动
progressive upsetting 步进式顶锻
progressive wave 行进波;渐进波;推进波
progressive-wave antenna 行波天线
progressive wave theory 前进波说
progressive wave winding 行波式绕组
progressive welding 前进焊;分段焊接
progressive widening 逐渐放宽;逐渐变宽
progressive widening of a river 河流逐渐变宽
progressive widening of a stream 河流逐渐变宽
progressive winding 右旋绕组;前进绕组
progressivity 进展性
progress map 进度图
progress method 进度拟订
progress of cargo work 装卸进度
progress of condolidation 凝固进程
progress of(construction) work 施工进程
progress of excavation 开挖进度
progress of hardening 硬化进程
progress of loading/discharging 装卸进度
progress of material wear 材料磨损过程
progress of project 工程进度
progress of wave 波的演进;波的行进;波的传播
progress of wear 磨损过程
progress of weld 焊向
progress of works 工作进程;工程进度;工程进程
progress payment 按进度付款;工程进度款;进度(付款);进度付款;交货前分期付款;施工分期付款;分期付款;分阶段付款;分段付款制
progress payment for a project 工程进度款
progress plan 施工进度计划
progress plastic flow 渐进性流塑
progress record 工程进度记录;进度记录
progress report 工作进展报告;工程进度报告;进展报告;进度报告;进度报表;进度记录
progress report for project 工程进度报告
progress reporting system 进度报告制度
progress schedule 施工进度计划;工程进度表;进度时间表;进度计划;进度规划图表;进度工作日程
progress schedule of project 工程计划进度
progress sheet 进度表;进度报表
progress sketch 作业进展略图
progress succession 进度演替
progress taxation 累进税
progress wave 行波;进行波
prohibit 禁止
prohibited area 禁止航行区域;禁区;禁航区;禁航地区
prohibited article 违禁品
prohibited flight area 禁飞行区
prohibited goods 禁运品;违禁品
prohibited import articles 禁止进口货物
prohibited list 拒绝承保险表

prohibited materials for fences 篱笆禁用材料(指脆性的、易燃材料)
prohibited pass area 禁止通行区
prohibited risks 拒绝承保的风险
prohibited road 禁止通行道路
prohibited to traffic 禁止交通;禁止车辆通行
prohibition 禁止;禁令
prohibition of direct transaction 禁止直接交易
prohibition of export 禁止出口
prohibition of import 禁止进口
prohibition tariff 禁止性关税税则
prohibitive duty 高额关税;禁止性关税
prohibitive import duties 禁止性进口税
prohibitive inhibition 超限抑制
prohibitive tariff 禁止性关税
prohibitive tax 禁止性税收
prohibitory and restrictive sign 禁令标志
prohibitory duty 高额关税
prohibitory road marking 禁止路面标记
prohibitory sign 禁令标志;禁止标志
prohibitory turn marking 禁止(左右)转弯标记
prohumic substance 原腐殖质
proidonite 氟硅石
proiferation 层出现象
proiosystole 过早收缩
project 项目;规划;工程;计划项目;计划;突出;设计项目;设计;方案;板影
project accountant 工程会计
project age 项目年限
project agreement 工程协议
project alternative 比较方案;项目方案;项目备选方案
project analysis 项目分析;工程分析
project analysis report 工程分析报告
project application for payment 工程付款申请书
project appraisal 工程项目评价;项目评估;工程评价;工程评估;计划评估;项目审查;计划审查
project appraisal approach 计划评估方法
project approval 项目批准;项目计划核批;项目的审定;项目的批准;项目的核准;计划批准
project architect 项目建筑师
project area 规划区;工程面积
project assessment 工程评定
project assignment 项目转让
project audit of construction enterprise 施工企业工程项目审计
project authority 方案职权
project benefit 工程效益;工程受益
project boundaries 建设项目用地界线;建设项目地界线;项目边界
project brief 项目简介
project budget 建设项目投资预算;项目预算;工程预算
project budget change 工程预算更改通知
project budget change notice 工程预算更改通知
project budgeting 按项目编制预算
project budget requirement 项目预算方案
project capital 项目资本金
project certificate for payment 工程付款证明书
project charts and layouts 项目图与布置图
project chart and layout 工程图及布置图
project circuit 投影电路
project classification 项目分类
project compass 反射罗盘;反射罗经
project completion report 工程竣工报告
project components 项目组成部分
project comprehensive evaluation 项目综合评价
project configuration 工程规模
project construction 工程施工
project construction contract 工程施工承包合同;建设工程施工合同
project construction owning cost 建设工程成本
project construction price 建设工程价格
project contract 工程合同;工程承包合同
project contracting 工程承包
project contractor 工程承包者;工程承包商;工程承包人
project control 项目控制;方案控制
project coordinate system 投影坐标系统
project coordination 项目协调
project corporation responsibility system 项目法人责任制
project cost 项目费用;项目成本;工程总造价;工程造价;工程项目成本;建设项目总费用;施工费用指数;施工费(用);设计总造价;工程费(用)
project cost-benefit assessments 项目利润评价
project crashing 应急计划
project criterion 项目准则
project cycle 项目周期
project data 设计资料;设计数据
project database 项目数据库
project definition 确定任务;确定工程建设任务;编制计划任务书
project definition and secrecy 设计定义和保密
project definition and survey 设计定义和评述
project definition phase 方案论证阶段
project-dependent 随项目不同的
project depth 计划水深;设计水深;设计孔深
project description 工程说明书;工程简介
project design 项目设计;工程设计;方案设计
project designation 工程名称
project design documents 项目设计文件
project designer 项目设计师
project design flood 工程设计洪水;设计洪水
project development cycle 项目发展周期
project diary 项目日志
project director 工程总负责人;工程负责人;计划项目负责人;设计总负责人;项目负责人
project documentation 工程计划文件
project draft environmental impact report 工程环境报告影响报告书草案
project drawing 设计图
project-due date 指定完工日期
project economic appraisal 建设项目经济评价
project economic effect 工程经济效益
project economic evaluation 项目经济估价
projected 计划中;投影的;伸出的
projected angle 投影角
projected area 投影面积;投射面积
projected area mo(u)lding 柱塞注压机
projected area ratio 投射面积比
projected background 放映背景
projected baseline 投影基线
projected beam-type detector 射束型探测器
projected bearing area 轴承投影面积
projected bundle 射影丛
projected capacity 计划能力;设计能力
projected charge 喷射爆破缆
projected charge demonition kit 喷射大爆破缆
projected concrete 喷射混凝土
projected contrast ratio 投影对比度系数
projected-core spark plug 芯部伸出的火花塞
projected cost 预计造价;设计成本;预计成本;预定造价
projected current 引出电流
projected curve 投影曲线
projected cut-off 投射截止点;投影截止点
projected delta 突出三角洲
projected density 投影密度
projected depth 设计水深;设计深度
projected diameter 投影径
projected dimension 设计尺寸
projected dimensions of channel 计划航道尺度
projected display 投影显示(器)
projected dredge-cut 设计挖槽
projected expenditures 预计费用
projected expenditure audit 预提费用审计
projected financial statement 预计财务报告;计财务报表;预测财务报表
projected flood 设计洪水
projected flow 计划流量
projected goal 预定目标
projected image 投影影像;投影图像
projected income statement 预计收益表
projected inner bit (双层岩芯管突出的)内管钻头
projected lantern slide 幻灯片
projected length 投影长度
projected light beam 投影光束
projected line 投影字行
projected linear foam smoke detector 投射光束式烟感火灾探测器
projected map display 投影地图显示器;飞行位置投影
projected map display system 投影地图显示系统
projected mean length 投影平均长度
projected meridian 投影经线
projected micrograph 投影显微照片
projected mine 设计的矿井
projected object 投影目标
projected parallel ridges 突起的平行脊
projected path 预定的轨迹
projected pipe 预置于地面的管子
projected planform 投影平面图
projected point 预计(定)点
projected profit 预期利润
projected propeller area 螺旋桨投影面积
projected railroad 计划铁路线;计划铁道线
projected railway 计划铁路线;计划铁道线
projected route 预定路线;计划铁路线;计划铁道线
projected (sash) window 滑开窗
projected scale 投影比例尺;投影标尺
projected-scale balance 投影标尺天平
projected-scale instrument 投影刻度仪;投影标度尺仪表
projected scenery 幻灯景
projected tolerance 突出公差
projected tolerance zone 突出公差带
projected top hung window 滑轴窗
projected traffic volume 计划交通量;设计交通量
projected transfer 喷射过渡
projected undertaking 计划项目;设计项目
projected-vertical-plane method of orienting 按给定垂直方向钻杆定向法
projected window 挑窗;有旋转窗框的窗;滑轴窗;凸窗
projected workload 预定工作量
project engineer 项目工程师;主管工程师;规划工程师;设计工程师;主任工程师
project engineering 项目设计
project engineer responsibility system 项目设计工程师负责制
project engineer system 项目工程师制度;项目工程师负责制
project environmental impact assessment 工程环境影响评价
project equipment 项目设备
project estimate 工程估算;工程概算
project estimation and budget 工程概预算
project estimator 项目估算师
project evaluation 项目评价;工程评价;工程评定
project evaluation and review technique 项目评审技术;项目评估法;工程评审技术
project execution 项目实施
project feasibility 工程可行性
project finance 项目贷款
project financial appraisal 建设项目财务评价
project financial evaluation 项目财务估价
project financial payments 工程财务支付
project financing 项目资金筹措
project financing evaluation 项目财务评价
project-financing loan 项目贷款
project flood 计划洪水值
project formulation 项目拟订;项目的拟订;制订项目
project for sedimentation purposes 放淤工程
project for upgrading urban and rural power grid 城乡电网改造
project gang 工程小组
project group 项目工作组;项目(研究)小组;工程小组;特别小组;设计组
project hole 勘探点
project identification 确定项目
project identification notes 项目识别附注
projectile 抛射体;投射体;塌缩隔离器;弹丸
projectile body 弹体
projectile energy 入射能(量)
projectile expelling chamber 抛射体发射室
projectile filling 装药
projectile motion 抛物运动;抛体运动
projectile particle 抛射粒子
projectile penetration 发射触探
projectile sampler 抛射式取样机;射弹式取样器;冲击式取样器
project impementation schedule 项目实施表
project implementation 项目执行;项目实施;建设项目实施
project implementation schedule 项目建设进度表
project implementation scheduling 项目建设进度表
project improvement 项目环境改善
project indicator 项目指标
project-induced 工程引起的

projecting 突出(的);投影;伸出
projecting abutment 突出式桥台;突出桥台;伸出式桥墩
projecting angle 投射角
projecting apparatus 幻灯装置;投影仪
projecting bar 启动杆;投影圆棒
projecting beam 悬臂梁;投影光束;投射光束;挑梁
projecting belt course 凸出带层;挑砌腰线
projecting bluff 突出陡岸
projecting breaks 预测破裂
projecting brick 凸出(墙面的)砖;突出墙面砖;外突式砖;突砖
projecting brick course 挑砖层
projecting camera 幻灯;投影机
projecting conduit 露出地面管道;凸埋式管道
projecting cone 射影锥
projecting consumption 预测耗水量
projecting coping 挑出式压顶;悬挑压顶;挑出的压顶
projecting corner 投影角
projecting cover 压顶;突出覆盖层
projecting curvature 投影曲率
projecting cylinder 投影筒;射影柱
projecting diamond 斜井字
projecting diffuser 挑出的扬声器;挑出的扩散器
projecting eaves 突檐;挑檐
projecting end 外突的桥端;突出桥端
projecting figure 投影图形
projecting flange 突出的法兰;突出的翼缘
projecting foot 突出底部;(夯土机的)夯足
projecting groove 凸出沟槽
projecting illuminator 投光灯具
projecting inlet 外突式进水口
projecting instrument 投影器
projecting keel 倒挂龙骨
projecting lamp 投影灯
projecting lantern 映画器;幻灯(机)
projecting ledge rock 突出石梁
projecting leg 挑出的腿
projecting length 凸出长度;突出长度
projecting lens 投影透镜
projecting lever 引长杆
projecting light 投影光
projecting line 投影(射)线;投射线
projecting lug 凸耳
projecting mapping 射影映射
projecting mirror 投射镜
projecting part 突出部分
projecting pier 突堤
projecting pipe 凸埋式管道;安全管
projecting plane 投影面
projecting point 突嘴
projecting pole 凸极
projecting pole rotor 凸极转子
projecting portion 凸出部分
projecting quoins 突隅石;突出隅石
projecting reinforcement 伸出钢筋;设计钢筋量
projecting scaffold 悬伸脚手平台;悬挑式脚手架;挑脚手架;挑出脚手架
projecting scoop 凸出风斗
projecting shaft 伸出轴
projecting sign 挑出的招牌;悬挑标志
projecting slab 悬臂板
projecting spray irrigation 喷灌
projecting table 突边桌
projecting tile of eaves 滴水瓦
projecting top-hung window 上悬窗;突出式上旋窗
projecting tube 凸埋式管道;安全管
projecting unit 投影装置
project in preparation 筹建项目
project in set 成套项目
project insurance 工程保险
project-in vent 内凸通风口
project investment 工程投资
project investment evaluation 项目投资评价
project invitation for bid 工程招标
projection 规划;估算;突出部分;突出部;凸块;凸出(部分);投影图;投影;投射;射
projection amplifier 投影放大器
projection angle 投影角;投射角
projection apparatus 投影仪
projection area 投影面积;投射区
projection balance 投影天平
projection booth 放映室;放映间

projection by suspended plumbing 吊锤投影
projection by suspended weight 吊锤投影
projection calculation 投影计算
projection camera 投影仪;复照仪
projection cathode-ray tube 投射式阴极射线管;投射式显像管
projection center 投影中心
projection chamber 放映室;放映腔;投影火花室
projection change 投影变换
projection communication system 投影通信系统
projection comparator 光学投影比较仪;投影式坐标量测仪;投影比较仪;投影比较器
projection compass 投视式罗经;反映式磁罗经
projection computation 投影计算
projection conduit 露出地面管道;地面管道;出地面管道
projection conversion program(me) 投影换算程序
projection conversion routine 投影换算程序
projection copier 放大印像机
projection copying 设计复制投影
projection course 挑出层
projection cular 投射目镜
projection cupola 放映圆屋顶
projection curvature 投影曲率
projection datum with compensation effect 抵偿高程面
projection detector 投影检验器
projection device 投影装置
projection diameter 凸起直径
projection diapositive 幻灯片
projection direction error 投向误差
projection display 投影显示(器)
projection distance 投影距离;投射距离
projection dome 放映穹隆;突出圆顶
projection drawing 投影图
projection effect 投影效应
projection efficiency 投射效率
projection Ektar lens 艾克塔摄影透镜
projection equation 投影方程(式)
projection equipment 放映设备
projection eyepiece 投影目镜
projection fiber 投射纤维
projection figure 投影图
projection focus depth 投影焦深
projection function 投影函数
projection geometry 射影几何(学)
projection grid 投影格网
projection grinder 光学曲线磨
projection homomorphism 射影同态
projection initiation 计划开始
projection instrument 映像器
projection interferometer 映射干涉仪;投影干涉仪
projection into a street 投射到街道
projection in zone 分带投影
projectionist 投影图绘制者
projection kinescope 投影显像管
projection knot 射影纽结
projection lamp 投射灯
projection lantern 映画器;幻灯
projection lens 投影透镜;放映镜头
projection lens jacket 投影透镜套
projection level of surface of mining area 矿区投影水准面
projection light 投射光;投射灯
projection light source 投影光源
projection line 投影线
projection machine 投影机;放映机
projection manometer 投影压力计
projection map of coal seam and strata 歪斜钻孔投影图
projection matrix 投影矩阵;射影矩阵
projection method 计化划;投射法;投影法
projection microradio graphy 投影显微射线照相术
projection microscope 投影显微镜
projection mode 投影方式
projection nucleus 投影环
projection objective 投影物镜
projection of a masonry wall 圬工墙的突出部分
projection of cornice 檐口突出部;飞檐突出部;飞机分级号码
projection of diamonds 出刃(金刚石)
projection of image 投影
projection of orebody 矿体投影图
projection of pole 极投影

projection of talus 坡脚堆石
projection of the celestial sphere 天球投影图
projection on a plane 平面投影
projection on horizon 地平平面图
projection on meridian 子午线平面图
projection operator 射影算子
projection optic(al) system 投影光学系统
projection optics 投影光学;投射光学
projection optimeter 投影光学比较仪
projection oscillograph 投影示波器;投射示波器
projection panoramic display 投影全景显示器
projection period 评估阶段;(投资的)规划阶段
projection picture size 投影图像尺寸
projection plane 投影面
projection plan position indicator 平面位置投影显示器
projection plaster 喷涂石膏;喷涂灰泥
projection plastering 喷涂抹灰
projection plastwork 喷涂装饰
projection plotter 投影测图仪
projection point 投影点
projection point error 投点误差
projection port 放映孔;放映口
projection position 投影位置
projection print 投影相片;投影晒像
projection printer 照片放大器;放大机
projection printing 光学印刷;投影晒印;投影翻印
projection ratio 凸影比
projection ratio-printer 投影晒印器
projection ratio-printing 投影晒印
projection receiver 投影式电视接收器
projection room 放映室;放映间
projections 预算规划项目
projection sampling 投影取样
projection scale factor 投影比例系数
projection screen 映像屏;投影屏;放映银幕
projection source 投射光源
projection spark chamber 投影火花室
projection stereoplotter 投影立体测图仪
projection suite 放映用房
projection surface 投影面;放映面
projection system 投影系统
projection table 投影表
projection technique 投射技术
projection theorem 投影定理
projection thermography 投影测温术
projection tick 投影网延伸短线
projection transformation 投影转换;投影变换
projection transformation parameter 投影转换参数
projection tube 映射管;投影管;投射管
projection type 投影型;投影式
projection type construction 突出式结构
projection-valued measure 射影值测度
projection-view-display 投影图像显示器
projection viewer 投影图像显示器
projection wall sign 凸出墙面式标示牌
projection weld 凸焊(焊)缝;凸焊
projection welded cap 凸焊帽
projection welded joint 凸焊接头
projection welder 凸焊机
projection welding 凸焊;凸出焊接;多点凸焊
projection weld machine 凸点电阻焊机
projection with heights 标高投影
projection zone 投影度带;投影带
projection zone number 投影带号
project item 工程项目
projective 投影的
projective abundance 投影多度
projective algebraic 射影代数的
projective algebraic curve 射影代数曲线
projective algebraic variety 射影代数簇
projective area 投影面积
projective axioms of order 射影次序公理
projective broken chain 投影断链
projective change 射影变换
projective class group 射影类群
projective collineation 射影直射变换
projective cone 射影锥
projective conformal geometry 射影保形几何
projective connection 射影联络
projective coordinates 射影坐标
projective coordinate system 射影坐标系
projective correspondence 射影对应
projective cover 射影覆盖

projective cover degree 投影盖度
projective curvature tensor 射影曲率张量
projective deformation 射影变形
projective differential geometry 射影微分几何学
projective dimension 射影维数
projective drawing 射影画
projective figure 射影图形
projective frame 射影标架
projective general linear group 一般射影线性群
projective geometry 投影几何(学)
projective geometry code 投影几何码
projective grid 投影格网
projective group 射影群
projective homology group 射影同调群
projective interval 射影区间
projective invariance 射影不变性
projective invariant 射影不变式
projective lemniscate 射影双纽线
projective limit 射影极限
projective limit group 射影极限群
projective limit space 射影极限空间
projective line 射影直线
projective line bundle 射影线丛
projective line element 射影线元素
projectively 投影地
projectively complete 射影完全的
projectively flat 射影平坦的
projectively flat space 射影平坦空间
projectively identical 射影地恒等射影不变性原理
projective measurement 射影度量;射影测度
projective method 投影法;射影法
projective metric space 射影度量空间
projective minimal surface 射影极小曲面
projective module 投影模
projective module A 射影 A 模
projective net 投影网
projective normal 射影法线
projective object 射影对象
projective operator matrix 射影算子矩阵
projective parameter 射影参数
projective pencils of lines 射影线束
projective pencils of planes 射影面束
projective plane 投影面;射影平面
projective point 投影点
projective property 射影性质
projective ranges of points 射影点列
projective rejection 射影干扰
projective relation 射影关系
projective representation 射影表示
projective representative 现场代表
projective resolution 射影分解
projective sampling 投影取样
projective scale 射影尺度
projective scheme 射影标架
projective set 射影集
projective space 射影空间
projective special linear group 特殊射影线性群
projective straight line 射影直线
projective subspace 射影子空间
projective symplectic group 射影辛群
projective system 射影系
projective system of topologic(al) group 拓扑的射影系
projective technique 投射技术
projective tensor product 射影张量积
projective test 投射测验
projective theory of curves 曲线的射影理论
projective topology 射影拓扑
projective total cover degree 投影总盖度
projective transformation 投影变换;射影变换
projective transformation group 射影变换群
projective transformation parameter 投影转换参数
projective variety 射影簇
projectivity 直射
projectized organization 项目化组织
project justification 项目论证;工程论证
project layout 设计方案;工程方案;枢纽布置(图)
project layout chart 方案陈列图
project leader 工程指导者
project lending 计划贷款
project level 方案水准
project life 工程寿命
project line 项目贷款额度
project loan 项目贷款;工程贷款

project management 项目管理;专案管理;工程项目管理;工程管理;建设项目管理;方案管理
project management by objectives 项目目标管理
project management information system 工程项目管理信息系统
project management organization 工程管理机构
project management tool 项目管理工具
project manager 现场经理;工地主任;项目经理;项目管理人;项目负责人;工程主任;工程经理;工程负责人;设计负责人
project manual 设计手册;工程项目文件(不可工程说明书、招标文件、合同条件等);工程手册;设计纪要手册
project mark 项目标记
project master plan 工程总体规划
project matrix organization 项目矩阵组织
project Mercury 水星计划
project method 项目法
project mortgage 建设项目抵押贷款
project name 工程名称
project national economy appraisal 建设项目国民经济评价
project network 工程网络图;计划网络;施工程序网络图
project network analysis 项目网络分析;计划网络分析;设计管网分析
project network technique 项目网络法
project norm 工程定额
project notification and review system 工程项目通知与检查制度
project number 项目号
project office 项目办公室;工程项目办公室
project of high calorie coal gasification 高热量煤气化工程
project of regional geologic(al) survey 区域地质调查工作设计
project of stratigraphic(al) code 地层规范草案
projectometer 投影比长
projector 幻灯机;投影转绘仪;投影仪;投影系统;投影器;投影机;投影灯;探照灯;放映机
projector compass 投视式罗经;反映式磁罗经
projector controller 投影控制器
projector distance 投影距离
projector efficiency 投影机效率
projector equipment 投影设备
projector frame 投影器座架
project order 工程任务书
project organization 设计机构;工程组织机构;设计机构
projector head 射声头;放映机头
projector lamp 探照灯灯泡;聚光灯灯泡;投影机灯泡;放映机灯泡;投射灯
projector lens 投影透镜
projector light 探照灯光
projector measurement 射声器测量
projector mechanism 投影机构
projector position 投影器位置
projector power response 射声器功率响应
projector principal distance 投影器主距
projector scope 投影显示仪
projector screen 投影屏
projector-type filament lamp 集光型白炽灯;投射式白炽灯
projector type unit heater 投射式单位散热器
projectoscope 投射器
project outputs 方案计划产出
project-out vent 外凸通风口
project owner 建设项目业主
project-packaging 全工程统包
project paper 项目报告
project parameter 项目参数
project participants 项目的参与方
project performance audit report 工程性能审查报告;项目执行情况审计报告
project plan 项目计划;自给计划;工程计划
project planner 工程设计人员
project planning 工程设计;工程计划;工程规划;课题计划
project planning and scheduling 工程项目计划与进度安排
project planning chart 项目计划设计图
project planning for machine planting 造林设计;机械造林设计
project position 工程位置

project precipitation 设计降雨量;设计降水量
project preparation facility 项目准备融通资金;项目筹备融通资金
project preparation phase 项目规划编制阶段;项目编制阶段
project preparation section 筹建处
project price settlement audit 工程价款结算的审计
project procedure 项目程序
project production 项目生产
project program(me) 工程计划
project programming 项目计划;制订项目计划;工程计划;工程规划;编制项目方案
project progress 工程进度
project progress chart in physical shape 工程形象进度图
project progress in physical shape 工程形象进度
project promoter 项目发起人;工程主办人
project proposal(report) 项目建议书;立项报告
project purchasing 项目一次采购法;项目采购
project quality 工程质量
project quality audit 工程质量的审计
project quality test 项目质量考核
project rain 设计雨量
project rate of return 项目资金回收率
project rehab 重建项目
project report 工程报告;项目报告
project representative 工程代理人;设计建造师的现场代表;工程代表;项目代表;驻场代表
project requirement 工程要求
project revisit 工程回访
project risk 工程风险
projects approval authority 项目核准权
project schedule 工程计划
project scheduler 项目进度计划师
project scheduling 工程进度计划
projects completed 已完工程
project scope 建设规模
project selection 项目选择
project selection in capital investment 资本投资中项目的选择
projects in capital construction 基本建设项目
project site 工地;工程现场;工程地点;设计现场
projects meeting all quality standards 全优工程
project social assessment 建设项目社会评价
projects of complete plants 成套工程
project specification 工程说明书;工程规格(书);工程规范;设计规格;设计规范
project speed 计划车速
project sponsor 项目发起人;工程主办人
project's social value appraisal 项目社会评价
project staff 项目人员
project stage 设计阶段
project status 工程状态
project storm 设计暴雪(量);设计暴雨(量)
project superintendent 项目监督人
project supervision 工程项目监督;工程监理;项目监理;工程质量管理
project supervision system 工程监理制
project supervisor 项目监督人;现场监工员(美国)
project support equipment 工程辅助设备
project supporting services 项目资助业务;项目支助业务
project support plant 工程辅助设备
project survey 工程测量
project surveying, design contract 建设工程勘察设计合同
project suspension 项目暂停
projects were suspended or deferred 停建缓建项目
project systems engineering 工程系统工程
project team 项目工作组;项目(研究)小组;工程小组;设计组;设计方案小组;方案小组
project team organization 项目小组
project technical specification 工程技术规格书
project technology evaluation 项目技术评价
project time-cost trade-off graph 工程工时成本权衡图
project to be accomplished on schedule 计划内要完成的项目
project training 项目训练(法)
project type assistance 项目型援助
project under construction 在建项目;在建工程;未完工程;施工项目;施工的建设项目
projecture 凸出物;突出物;投影

project variable 工程变量
project volume in physical units 实物工程量
project water-level 设计水位
project works 主要工程
projtectite 热衷庞大计划者
prokaolin 高岭石无定形中间体
pro-knock compound 促爆剂
prolapse 脱垂
prolapse of uterus 阴挺
prolate 扁长的
prolate axis 长轴
prolate cycloid 长辐摆线；长幅旋轮线
prolate ellipsoid 长型椭球体；长椭圆面；长椭球（面）；长球面
prolate ellipsoid of rotation 旋转长椭球
prolate epicycloid 长幅外摆线
prolate hypocycloid 长幅内摆线
prolate sphere 长球体
prolate spheroid 长轴旋转椭圆体；长椭球（面）；长球面；长球
prolate spheroidal coordinates 长球面坐标
prolate spheroidal wave function 长球波函数
prolate tractrix 长曳物线
proliferation 再育；扩散化；快速扩大；层出
proliferation inhibiting factor 增殖抑制因子
proliferation of funds 分散资金
proliferous polymer 增生性聚合物；增生聚合物
prolificacy 丰产
prolification 再育；层出现象
prolific water zone 富水层
prolific well 高产油井；高产井
proline 脯氨酸
prolog(ue) 导引
prologue 序幕；程序块的开始处理
prolong 拉长
prolong a contract 延长合同有效期
prolongation 延期；延长部分；延长
prolongation clause 续阻条款
prolongation of effect 延效
prolongation of latency 潜伏期延长
prolongation of river mouth 河口向海延伸部分
prolongation of scouring duration 延长冲刷历时
prolongation of working day 工作日延长
prolong core barrel 加长岩芯管
prolong core tube 加长岩芯管
prolonged aerator 延时曝气器
prolonged agitation 延时搅拌；持续搅动
prolonged anti-fouling power 长效防污能力
prolonged blast 汽笛长声；长声
prolonged can stability 长期罐藏稳定性
prolonged corrosion test 耐蚀试验；耐蚀试验；长期耐蚀试验
prolonged delay 进一步的误期
prolonged exposure 迁延照射；长期暴晒
prolonged flooding 持久灌水
prolonged frontal rain 持续锋面雨
prolonged heating 长期加热
prolonged inhibition 延长性抑制
prolonged inundation 长期淹没
prolonged junction terminal 伸长式枢纽
prolonged labo(u)r 滞产
prolonged lenticular 延伸双凸透镜
prolonged secondary stage of labo(u)r 二程延长
prolonged shrinkage 长期收缩；长期收缩性
prolonged succession 长期演替
prolonged suspension 持续的暂时停工
prolonging time-distance system 延长时距系统
prolong the period of validity 延长有效期
proluvial fan 洪积扇
proluvial plain 洪积平原
proluvium 洪积物；洪积层
Promal 特殊高强度铸铁
promenad deck 天幕甲板；上层轻甲板
promenade 散步场；运动场；娱乐场；宽廊；散步场所；(防波堤上的)堤顶道路；(防波堤上的)堤顶大路
promenade deck 顶层屋盖板；小艇甲板；外走廊；外廊道；散步甲板
promenade gallery 散步游廊；散步长廊
promenade lane 游览路；游览观赏慢车道
promenade slab roof(ing) 平板铺砌的屋顶散步道
promenade tile 铺面缸砖；高硬度地砖；铺面缸砖
prometacenter 假稳心
prominence 日珥；起伏度

prominence of lateral semicircular canal 外半规管凸
prominent 突起的；凸出的；突出的
prominent fold 主褶皱
prominentia mallearis 锤凸
prominentia spiralis 螺旋凸
prominent natural feature 明置天然地物
prominent peak 主峰
prominent resonance 显著共振；突出共振
prominent surveyed tree 独立树
prominent trees 突出树
promise 承诺
promisee 受约人
promise to pay 承付
promising deposit 有开采价值的矿藏
promising result 满意效果
promisor 立约者；订约人；立约人
promissory note 债务人的期票；期票；本票
promnard 散步器；散步小路
promnesia 记忆错误
promo 宣传品
promontory 高角；岬；海角
promote 促进
promote home products 提倡国货
promoter 助聚剂；助催化剂；加速剂；激发器；启动区；推销商；促进剂；出资人；发起人；发包人
promoter action 促进作用
promoter site 启动子部位
promoter's profit 发起人利润
promoter's share 发起人股
promoter's stock 发起人股
promote sales 推销
promote the sale of products 推销产品
promoting agent 促成物；促长剂
promotion 宣传推销；升级；促进作用
promotional activity 发起活动
promotional allowance 业务推广贴；推广津贴
promotional allowance discount 业务推广奖励折扣；业务推广奖励奖
promotional department store 大众性百货店
promotional literature 宣传推广资料
promotional practice 晋升制度
promotional pricing 推销定价
promotional study 推销研究
promotion and transfer 提升和调动
promotion and transferring of employee 职工升调
promotion by seniority 依据资历提升
promotion expenses 推销费用；推广费用；发起费用
promotion media 推广媒介
promotion mix 推广组合
promotion of binder adhesion 胶合剂中掺加添加剂
promotion of market development 市场发展与培育
promotion of personnel 人员提升
promotion plot trial 跳级小区试验
promotion policy 推销政策
promotion sales 访销
promotion share 发起人股份
promotion workers 推销员
promotor 助催化剂；启动子；促进者
promotor of initial resistance of cement 水泥早强促进剂
promotor's stock 发起人股份
prompt 提示符；随到随发
prompt beam current 瞬时束流
prompt beta 瞬发贝塔
prompt biphasic reaction 迅速二相反应
prompt box 剧院舞台提词人窗口；提示席
prompt cash discount 付现折扣；现金付款的折扣
prompt cash(payment) 当场付款；即时付现；即期付现；立即付现；立即付款
prompt check 即期支票
prompt coincidence curve 瞬间符合曲线
prompt critical 即发临界；瞬发临界
prompt criticality 即发临界；瞬间临界性；瞬发临界性
prompt critical reactor 瞬发临界堆
prompt critical transient 瞬发临界瞬变过程
prompt day 延期交易的交割日；交货日；交割日
prompt decay 瞬发衰变
prompt decay rate 瞬发衰变率
prompt delivery 立即送交；即时交货；即期交货
prompt delivery of goods 早交货
prompt dose 瞬时剂量

prompt drop 速降
prompt drop of reactivity 反应性速降
prompt energy 瞬时能量
prompter's box 剧院舞台提示人窗口；提示席
prompt fission gammas 瞬发伽马
prompt fission product 瞬发裂变产物
prompt flux tilt 瞬发通量倾斜变
prompt gamma 瞬发伽马射线
prompt gamma energy 瞬发伽马能量
prompt gamma radiation 瞬发伽马辐射
prompt gamma ray 瞬发伽马射线
prompt gamma-ray analysis 瞬发伽马射线分析
prompt generation time 瞬发每代时间
prompt goods 现货
prompt harmonic 瞬发谐波
prompt hood 提示罩
prompt in answering payment 付款迅速
prompting 催促
prompting message 催促信息
prompt inversion 快转变
prompt jump 瞬跳变
prompt jump approximation 瞬跳变近似
prompt jump of reactivity 反应性瞬发跳变
prompt-mode 瞬发态
prompt negative temperature coefficient 瞬发负温度系数
prompt neutron 瞬发中子
prompt neutron activation analysis 瞬发中子活化分析
prompt neutron area 瞬发中子面积
prompt neutron burst irradiation 瞬发中子短时间闪光辐照
prompt neutron decay constant 瞬发中子衰变常数
prompt neutron fraction 瞬发中子份额
prompt(neutron) lifetime 瞬发中子寿期
prompt(neutron) multiplication 瞬发倍增
prompt neutron spectrum 瞬发中子能谱
prompt note 期货通知单；交割日期通知单；期货金额及交付日期通知
prompt NOx 瞬发氮氧化物
prompt nuclear radiation 瞬间的核辐射
prompt of fission energy 迅发裂变能
prompt payment 立即付款
prompt period 瞬发周期
prompt period accident 瞬发周期事故
prompt poisoning 瞬发中毒
prompt pulse 瞬发脉冲
prompt radiation 瞬发辐射
prompt reactivity 迅发反应性；瞬发反应性
prompt reactor 瞬发中子堆
prompt runoff 快速径流；瞬发径流
prompt sale 即时出售
prompt sample system 瞬时取样系统
prompt ship 即期船；随到随装船
prompt shipment 立刻装运；即期装运；即期装船；即刻装运
prompt shut-down coefficient 瞬发停堆系数
prompt subsurface runoff 快速地下径流；壤中流；表层流
prompt(-super) critical 瞬时临界的；瞬发临界的
prompt tectonic effect 迅发构造效应
prompt temperature coefficient 瞬发温度系数
prompt tempering 直接回火；快速回火
prompt test 快速试验
promulgate 公布；颁布
promulgated depth 公布的水深
promulgation 颁发
promurit 灭鼠舟
pronaos 前厅(古建筑)
pronate 倾斜的
prone 倾斜的；俯的；伏的
proneness to clogging 容易堵塞(筛子)；容易堵孔(筛子)
prone position 伏卧位
prone position headrest 俯卧位头靠
prone pressure method 俯伏加压法
prone to pitch 趋向于纵向颠簸
prong 凸钩爪；射线；插脚；叉子
prong brake 抓钩制动器；夹子制动器
pronghorn 分叉原木
prong reef 尖头礁
prongs (岩芯采取器的)弹簧
prong-shaped 叉状的
prongs hook 爪钩

prong test 叉齿试验
pronounced node 明显节点
pronounced staining 明显色渍
pronounced valley 深谷
pronouncement 公告
pronouncement of judgment 宣告判决
pronucleus 原始核
Prony brake 波朗尼测功器;摩擦测功器
proof 证据;考验;校样;校模铸件;坚牢性;耐力;合乎标准的;打样
proof bar 校正杆
proof box 保险箱
proof by contradiction 反证法
proof by induction 归纳证明
proof checking 样图审校
proof checking paper 审校纸
proof control 印样控制;验证控制
proof-copy mode 校样方式
proof drawing 校样图
proofed breathable fabric 防水透气织物
proofed cloth 防水胶布;防水布
proof edition 试用版
proofed sleeve 浸胶软管
proofed tape 涂上橡胶的带
proof fabric 胶布(指上胶的布)
proof gallon 标准加仑;标准伽仑
proof gold 纯金;标准金(试金用)
proof ground 试验场
proofing 刮胶;校对;审校;防护
proofing ink 打样油墨
proofing liquid 防水液剂
proofing machine 打样机
proofing material 防护材料
proofing press 打样机
proofing sheeting 标准坯布
proofing unit 校正器;打样机
proof limit 实用弹性极限
proof line 检核基线
proof list(ing) 验证表;验证列表;校对表
proof load 允许荷载;验证荷载;验证荷载;检验荷载;弹性极限荷载;试验负载;试验负荷;标准负荷;保证负载;安全负荷
proof machine 验证机
proof mark 验证记号;校对符号;试验标志
proof of cash 现金核证
proof of cost 成本证明
proof of journal 日记账检查
proof of loss 损失证明
proof of performance 合同与履行的证明
proof of program(me) correctness 程序正确性证明
proof of the contrary 反面证据
proof-plane 验电盘;验电板
proof plate 打样板
proof plotter 校样绘图机
proof plotting 印样绘制
proof pressure 检验压力
proof-proof water glass mortar 耐酸水玻璃砂浆
proofreader 校对员;校对者;校对人;审校员;审校人
proof reader's mark 校对符号
proof read(ing) 校对
proofreading and correcting 校正
proof resilience 总弹(回)性;标准回弹能
proof rod 料钎
proof rolling 检验碾压;复验碾压
proof room 校对室
proof sample 样品;验证样品;验证样本;试样
proof scheme 验证方案
proof sheet 校样
proof silver 纯银;标准银
proof sphere 球形样板
proof-spirit 标准酒精
proof staff 金属直规
proof stick 探针;探测杆;试验棒
proof strength 弹性极限强度;允许强度;弹限强度;保证强度
proof stress 校验应力;实用弹性极限应力;屈服点;许用应力;控制应力;弹性极限应力;试验应力;实用弹限应力;保证应力
proof test 验证性试验;验证试验;验收试验;校验;检验;试用试验;复核试验;安全试验
proof to swelling 抗膨胀化;防止隆起;防止膨胀
proof total 总计验证
proof voltage 耐电压

pro-oxidant 助氧化剂
prooxygenic agent 氧化强化剂
prop 支台;支器;支持物;道具;撑材
propadiene 丙二烯
propaedeutic prospecting stage 初步勘探阶段
propaedeutic stratigraphy 初步地层学
propaedeutic survey stage 初步普查阶段
propafenone 丙胺苯丙酮
propagandist 宣传员
propagated blast 诱导爆破;传播法爆破
propagated blasting 传爆
propagated error 延伸误差;扩散误差;传播误差
propagating burning 传播燃烧
propagating crack 传播裂纹
propagating flame 传播中的火焰
propagating mode 传播模
propagating source 传播震源
propagating velocity 传播速度
propagating wave 传递波;行波;扩散波;推进波;传播波;播波
propagating wavefront 波前
propagation 准横向传播;蔓延;传播;繁殖
propagational distortion 传播畸变
propagational reliability 传播可靠度
propagation anomaly 传播异常
propagation characteristic 传播特性
propagation coefficient 繁殖率;传播系数
propagation condition 传播条件
propagation constant 传播常数;传播常量
propagation constant of a semi-conductor 半导体传播常数
propagation correction 传播改正量
propagation delay 传送延迟;传输延迟时间;传播滞后;传播延时;传播延迟
propagation delay path 传送延时通道
propagation delay time 传输延迟时间;时延;传播延时时间;传播延迟时间
propagation distance 传播距离
propagation distortion 传播失真
propagation effect 传播效应
propagation energy 扩展(裂纹)的扩展能量
propagation error 延încer误差;扩散误差;传播误差
propagation factor 传播因素;传播因数
propagation forecasting 传播预报
propagation function 传播函数
propagation length 传播距离
propagation light 光传播
propagation loss 传播损失;传播损耗
propagation medium 传播介质
propagation mode 传播方式
propagation notice 传播通知
propagation of cracks 裂纹扩展
propagation of detonation 起爆传播
propagation of disturbance 干扰传播;扰动传播
propagation of errors 误差传播
propagation of explosion 传爆
propagation of flame 火焰传播
propagation of flood wave 洪水波传播
propagation of ground wave 地波传播
propagation of heat 热传播
propagation of high frequency radio wave 高频电波传播
propagation of light 光传播
propagation of low-frequency radio wave 低频电波传播
propagation of microwave 微波传播
propagation of sound 声音传播;传声
propagation of surges 涌浪传播;涌波传播
propagation of tea cuttings 茶树插条繁殖
propagation of very high frequency 甚高频电波传播
propagation of very low frequency 甚低频电波传播
propagation of vibration 振动传播
propagation of wave 波浪传播;波的传播
propagation path 传播途径;传播通路;传播路径
propagation power 传播功率
propagation power loss 传播功率损失
propagation process 传播过程
propagation range 传播距离
propagation rate 传播速率
propagation ratio 传播比
propagation reaction 传播反应
propagation rule 繁衍规则

propagation source 传播源
propagation speed 传播速率;波浪传播速度
propagation structure 繁衍结构
propagation time 传输时间;传播时间
propagation time delay 传播时间延迟
propagation transfer 误差传递
propagation vector 传播向量;传播矢量
propagation velocity 传播速度
propagation velocity of flame 火焰传播速度
propagation velocity of flood wave 洪水波传播速度
propagator 传播者;传播函数
propalanine 氨基丁酸
propandioic acid 丙二酸
prop-and-sill 柱梁式棚子;支柱与地梁
propane 丙烷
propane-air mixture 丙烷—空气混合气
propane asphalt 丙烷脱沥青;丙烷地沥青
propane bacteria 产丙烷菌
propane blowlamp 丙烷喷灯
propane bottle 丙烷瓶
propane bubble chamber 丙烷气泡室
propane crop drier 丙烷加热农产品干燥机
propane deasphalted asphalt 丙烷脱沥青的沥青
propane deasphalted oil 丙烷脱沥清油
propane deasphalting 丙烷脱沥青
propane decarbonizing 丙烷脱碳法
propane dewaxing 丙烷脱蜡
propane dewaxing process 丙烷脱蜡过程
propane diacid 丙二酸
propane diamine 丙二酰胺
propane dicarboxylic acid 丙二羧酸
propanedione 丙二酮
propane dryer 丙烷干燥器
propane drying 丙烷干燥
propane evapo(u)rator 丙烷蒸发器
propane fractionation 丙烷分馏;丙烷萃取分离
propane fractionator 丙烷分馏器
propane gas cutting 丙烷切割
propane gas lighted buoy 丙烷(气)灯光浮标;丙烷(气)灯光浮筒;燃丙烷气的灯浮标
propane-heated drier unit 丙烷加热干燥设备
propane-propylene fraction 丙烷丙烯馏分
propane refrigeration unit 丙烷冷冻设备
propane soldering iron 丙烷加热烙铁
propane sultone 丙磺酸内酯
propane tetracarboxylic acid 丙烷四羧酸
propanolon acid 丙醇酮酸
propanone 丙酮
propants 支承剂
propargyl alcohol 炔丙醇
propargyl chloride 氯丙炔
propargylic acid 丙炔酸
propatylnitrate 丙帕硝酯
propcopter 螺旋桨直升机
prop-crib timbering 隔离井框支架
prop density 支柱密度;支撑密度
prop drawer 支柱拉拔机;支柱拉拔器;支撑拉拔机;支承拉拔器
prop-drawing gang 支柱拆除工作队;支撑拆除工作队
prop-drawing shift 拆支柱工作班;支柱拔出移位;拆支撑工作班;支撑拔出移位
propel 推进
propellant 液体发泡剂;火箭燃料;挥发剂;喷气燃料;抛射剂;推进剂;发生剂;发射剂
propellant action 推进作用
propellant-actuated device 推力校正装置;发射剂驱动装置
propellant additive 推进剂添加物
propellant atomization 推进剂雾化
propellant binder 推进剂黏合剂
propellant bottle 火箭燃料筒;燃料筒
propellant burning velocity 推进剂燃烧速度
propellant capacity 燃料箱的容量
propellant chemical 推进剂化学原料
propellant chemistry 推进剂化学
propellant combustion 推进剂燃烧
propellant composition 燃料成分
propellant constituent 推进剂成分
propellant consumption 推进剂消耗量
propellant container 推进剂箱
propellant-cooled 推进剂冷却的
propellant cut-off 停止输送推进剂

propellant cut-off mechanism 推进剂止送装置	诱导屋顶通风装置	propeller-type agitator 螺旋浆式搅拌机;螺旋浆搅拌机
propellant expenditures 推进剂消耗	propeller feathering solenoid 顺桨螺线管	propeller-type current meter 旋桨式海流计;旋桨式流速仪;螺旋桨式流速仪
propellant explosion 推进剂爆炸	propeller feed 螺旋桨进料	
propellant explosive 抛掷炸药	propeller feeder 螺旋桨进料机	propeller-type dredge(r) 旋桨式挖泥船
propellant feed system 燃料输送系统;推进剂供给系统	propeller flowmeter 螺旋桨式水表;螺旋桨水表;螺旋桨式水表	propeller-type fan 螺旋(桨)式风扇;螺旋桨式通风机;螺旋桨式鼓风机;螺旋桨式通风机;螺旋桨式送风机
propellant fire 推进剂点火	propeller frame 螺旋桨框架;推进器框架	
propellant flow 燃烧消耗量	propeller generator 螺旋桨式(风力)发电机	
propellant flow rate 燃料消耗率	propeller governor 螺旋桨调速器	propeller-type flowmeter 螺旋桨式流量计
propellant gasifier 燃料气化器	propeller guard 推进器护栏	propeller-type impeller 旋桨式转轮;螺旋桨(式)叶轮
propellant handling 推进剂的使用保管	propeller horsepower 螺旋桨马力;推进器马力	
propellant injection 喷气燃料注入	propeller hub 螺旋桨毂	propeller-type mixer 螺旋桨式搅拌机;螺旋桨搅拌机;螺旋(桨)式拌和机;旋桨式拌和机
propellant injector 燃料喷嘴;推进剂喷头	propeller jet 螺旋桨式喷气发动机;螺旋桨发动机	
propellant-loading operation 推进剂加注操作	propeller-jet engine 螺桨喷气发动机	propeller-type speed meter 螺旋桨式速度计
propellant mass fraction 推进剂质量比	propeller lathe 螺旋桨车床	propeller-type stirrer 螺旋桨(式)搅拌机
propellant mass ratio 推进剂质量比	propeller load 螺旋桨负荷	propeller-type thruster 螺旋桨式器
propellant module 推进剂舱	propeller load curve 螺旋桨负荷载曲线	propeller-type turbine 螺旋桨式叶轮(机);螺旋式涡轮(机);螺旋式涡轮机;螺旋桨式水轮机;旋桨式水轮机
propellant nature 推进剂性质	propeller lock nut 推进器紧锁螺帽	
propellant orifice 火箭燃料喷嘴;燃料喷嘴	propeller log 转轮式计程仪	
propellant performance 推进剂性能	propeller mechanism 螺桨机构	propeller(-type) water turbine 螺桨式水轮机
propellant residual 推进剂残余物	propeller meter 螺旋桨式水表;推进器式流量计	propeller wake 推进器伴流
propellant servicing area 推进剂加注区	propeller mixer 螺旋桨式混合器;旋桨式搅拌器;螺旋桨式搅拌混合器;推进器搅拌器	propeller well 推进器框穴
propellant-servicing vehicle 燃料加注车		propeller whirling test 螺旋桨旋转试验
propellant shutoff valve 推进剂断路开关	propeller mixing chest 螺旋桨式混合池	propeller windmill 螺桨式风车
propellant silo 推进剂井	propeller mud mixer 叶型泥浆搅拌机	propeller with ring guard 带护环的螺旋桨
propellant spray 燃料喷注	propeller noise 螺旋桨噪声	propellor = propeller
propellant storage area 推进剂储存区	propeller nut 螺旋桨桨母	propelling agent 推进剂
propellant structure 推进剂结构	propeller pitch 螺旋桨螺距;桨距;推进器螺距	propelling booster 助推器
propellant supply 推进剂的输送	propeller pitch control 螺旋桨桨距调节机构	propelling chain 运转链;主动链
propellant terminal 推进剂仓库	propeller port 推进器框穴	propelling effort 推进力
propellant trailer 运送火箭燃料的油罐车	propeller post 螺旋桨柱;推进器柱	propelling energy 驱动能
propellant transfer equipment 推进剂输送设备	propeller power 推进功率	propelling explosive 抛掷炸药
propellant volume 推进剂容积	propeller power coefficient 螺旋桨功率系数	propelling hours 航行时数
propellant waste 推进剂消耗量	propeller protractor 推进式量角器	propelling machine 牵引机
propellant weight fraction 推进剂重量比	propeller pump 螺桨泵;轴流式叶片泵;轴流泵;螺旋式水泵;螺旋桨式水泵;螺旋桨泵;桨叶泵	propelling machinery 推进机械
propeller 螺旋桨;螺桨;搅拌叶(片);推进器;螺旋辊		propelling motor (自行钻机的)移动式电动机;推进发动机
propeller advance 螺旋桨进程	propeller race 推进器尼流	
propeller aerator 旋桨式曝气器;螺旋桨式曝气池	propeller racing 螺旋桨飞车	propelling movement 推送
propeller agitator 螺旋桨式搅拌机;螺旋桨式搅拌器	propeller rake 螺旋桨斜度	propelling nozzle 推力喷管;推进喷管
propeller anemometer 螺旋桨式风速仪;螺旋桨式风速计;螺旋桨式风速表	propeller ramp 螺旋桨式坡道	propelling pencil 活动铅笔
	propeller reinforcing girder 螺旋桨增强梁;螺旋桨加力梁	propelling plant 推进机械
propeller anti-icer 螺旋桨防冰器		propelling power 推进力
propeller aperture 螺旋桨框穴;推进器框穴	propeller release 旋桨式排气装置	propelling rod 推力杆
propeller arc 螺旋桨弧	propeller repair shop 螺旋桨修理车间	propelling screw 螺桨推进器
propeller arch 尼框顶部;船尼拱架	propeller root 螺旋桨根	propelling sheave 往复滑轮;推动滑轮;导引轮
propeller area 螺旋桨总面积	propeller runner 螺旋桨式转子	propelling torque 推进扭矩;推进转矩
propeller area ratio 螺旋桨盘面比	propeller scoop 螺旋桨后进气口	propelling unit 推进装置
propeller axis 螺旋桨轴线	propeller section 螺旋桨剖面	propeiling worm 压榨螺旋
propeller balance 螺旋桨平衡	propeller set 螺旋桨系	propel shaft 传动轴
propeller balancing machine 螺旋桨平衡试验机	propeller shaft 螺旋桨轴;螺桨轴;驱动轴;尼轴推进器轴;尼轴填料涵;尼轴;动力输出万向传动轴;传动轴	propel wheel 推进轮;推动轮
		propene 丙烯
propeller balancing stand 螺旋桨平衡架		propene dicarboxylic acid 丙烯二羧酸
propeller blade 螺旋桨叶片;螺旋桨叶;推进器叶;车叶	propeller shaft bearing 尼轴轴承;推进器轴轴承	propene oxide 氧化丙烯
	propeller shaft bossing 轴包架;螺旋桨轴孔膨出部	propene thiol 丙烯基硫醇
propeller-blade angle 螺桨叶角	propeller shaft bracket 人字架	propenol 烯丙醇
propeller-blade area 螺旋桨叶面积	propeller shaft hole 推进器轴孔	propensity to barter 交易的倾向
propeller blade flange 螺旋桨叶凸缘	propeller shaft lining 尼轴衬套	propensity to consume 消费倾向
propeller bolt 螺旋桨螺栓	propeller shaft nut 尼轴螺帽	propensity-to-consume schedule 消费倾向图表
propeller boom 推进器护杆	propeller shaft pinion flange 传动轴主动齿轮凸缘	propensity to import 进口倾向
propeller boss 螺旋桨毂;桨毂;推进器毂;车叶毂	propeller shaft slip 螺旋桨轴伸缩节	propensity to invest 投资倾向
propeller brake 螺旋桨制动器	propeller shaft slip joint 传动轴滑动接头	propensity to save 储蓄倾向
propeller cap 螺旋桨帽;推进器帽	propeller shaft splined yoke 螺旋桨轴槽轭	propenyl 丙烯基
propeller cavitation 螺旋桨旋转真空;螺旋桨气穴;螺旋桨气蚀	propeller shaft stay 人字架	propenylbenzene 丙烯苯
	propeller shaft strut 人字架	propenyl cyanide 丙烯基腈
propeller coefficient 螺旋桨系数	propeller shaft thrust spring 螺旋桨轴推力弹簧	propenyl ethyl ether 丙烯基乙基醚
propeller cone 螺旋桨帽;推进器帽		propenyl phenol 丙烯基酚
propeller controller 螺桨调节器	propeller shaft tube 传动轴套	proper alignment 正确线向;合理线路设计;同轴度;同心性;适当线向
propeller cuff 螺桨根套	propeller shaft turning wrench 螺旋桨轴扳手	
propeller current 螺旋桨水流;推进器尾流	propeller shaft yoke 传动轴叉	proper application 合理施药
propeller current meter 旋桨式流速计;螺桨式流速仪	propeller sled 螺旋桨式滑橇	proper arrangement 适当的安排
	propeller slip 推进器滑失率;推进器滑距	proper bearing 紧密接触轴承
propeller dead stick 螺旋桨停转	propeller spacing 螺旋桨间距	proper card 主要梳理机
propeller disk area 螺旋桨旋转面积	propeller spinner 螺旋桨毂盖;螺旋桨毂盖;螺桨毂盖	proper channel 正槽
propeller-driven 螺旋桨带动的;螺旋桨带动	propeller starter 螺旋桨起动器	proper channeling 固有沟道效应
propeller-driven plane 螺旋桨飞机	propeller stirrer 螺旋桨式搅拌器	proper circle 真圆;常态圆
propeller drive shaft 螺旋桨传动轴;搅拌器驱动轴	propeller test stand 螺旋桨试验架	proper class 真类
propeller efficiency 螺旋桨效率;螺桨效率;推进器效率	propeller thrust 螺旋桨推(进)力;推进器推力	proper complex orthogonal group 正常复正交群
	propeller tip 螺旋桨叶尖	proper conduction 固有传导
propeller fan 螺旋桨式通风机;旋浆式扇风机;轴流通风机;轴流式(通)风机;螺旋式风扇;螺旋桨式风扇;螺旋桨式风扇	propeller tip speed 螺桨梢速	proper conics 常态二次曲线
	propeller torque 螺旋桨转矩;螺桨扭矩	proper crossover 正向换向
	propeller tube 传动轴轴套	proper deep seeding 深播适宜
propeller-fan induced draught system 螺旋桨通风机强制通风系统;螺旋桨通风机诱导通风系统	propeller turbine 桨式水轮机;轴流定桨式水轮机;螺旋桨式涡轮机;螺旋桨式涡轮	proper distance 真距离
		proper distribution 正常分布
propeller-fan induced powered roof ventilator 螺旋桨通风机强制屋顶通风装置;螺旋桨通风机	propeller twist 螺旋桨式扭曲;麻花状扭曲	proper drag 固有阻力;固有曳力
	propeller-type 螺旋桨式	

proper drag coefficient 固有阻力系数
proper duration 日照长度适宜
proper ellipticity 真椭圆性
proper energy 原能
proper equation 特征方程
proper factor group 真剩余群
proper flag manifold 真旗流形
proper fraction 真分数
proper fuel 合格的燃料
proper function 正常函数;固有功能;特征函数;常义函数;本征函数
propergol 推进剂
proper hardware design 适宜的硬件设计
proper homotopy 真同伦【数】
proper hypersphere 正常超球面
proper implements 适当的工具
proper inflated tire 正常充气轮胎
proper inflation 适度内压
proper integral 正常积分;常义积分
proper isomorphism 正常同构
proper journal 普通日记账
proper length 真长度;静长度
proper Lorentz group 正常洛仑兹群
proper Lorentz transformation 本征洛仑兹变换
proper lubrication 可靠润滑;适当润滑
properly designed 精心设计的
properly deterioration indicator 合理连续作业
properly discontinuous 纯不连续
properly divergent 正常发散的
properly divergent sequence 正常发散序列
properly divergent series 正常发散级数
properly improved strain of tree 合理改良树种
properly n-dimensional 特征 n 维
properly primitive binary quadratic form 真本原二元二次型
properly primitive quadratic form 真本原二次型
proper maneuvering 适当的机动灵活
proper mapping 真映射
proper map-place 相应图面位置
proper mass 原质量;静质量
proper maximum 正常极大
proper meromorphic mapping 真亚纯映射
proper minimum 正常极小
proper mixture ratio 正常混合比;合适的混合比
proper mode of membrane vibration 膜振动本征模式
proper morphism 真射
proper motion 自行(指天体自行);自然运动;固有运动;实际运行
proper motion group 自行群
proper motion in declination 赤纬自行
proper motion in right ascension 赤经自行
proper motion membership 自行成员
proper name 固有名字
proper network 常态网络
proper node 正常结点
proper operating condition 正常操作条件
proper operation 正确操作
proper order 正确次序
proper orthogonal group 正常正交群
proper orthogonal matrix 正常正交阵;正常正交矩阵
properous 富裕的
proper packing 适宜包装
proper path 真路
proper pedigree culture 合理谱系栽培法
proper penetration 合适的切入
proper period 固有周期
proper phase 固有相位;适宜相位
proper phasor 特征相量
proper polynomial 特征多项式
proper prefix 真前缀
proper program(me) 真程序
proper property 适当性能
proper pruning 合理修枝
proper quadric 正常二次曲面;真二次曲面
proper quadric surface 真二次曲面
proper quantity 适量
proper redistribution 合理布局
proper redistribution of industry 工业合理重安排
proper reference frame 固有参考架
proper rotation 真转动;真旋转
proper scale 适合标度;成比例尺
proper seeding 播期适宜

proper selection 合理性选择
proper selection method 合理择伐作业
proper semi-conductor 本征半导体;固有半导体
proper sequence 固有顺序
proper set 正常集
proper set system 固有装置系统
proper shock 本震
proper solution 正常解
proper space 正常空间
proper star 基准星
proper station 原驻地
proper stocking 适度放牧量;适当放养度
proper string 固有字符串
proper symbol 正常符号
proper tail 真尾;固有尾
proper termination 正常终止
properties of matter 物性
proper time 恰当时刻;本征时间
proper tone 常态音
proper tree 正常树;固有树;特有树
property 性质;资产;所有物;财物;财产
property accessory to immovable property 不动产的附属财产
property account 财产账(户)
property accountability 财产责任
property accounting 财产清查;财产会计
property accumulation 财产积累
property accumulation saving contract 财产积累储蓄契约
property and liability insurance 财产和责任保险
property and risk pass on shipment 财产权和风险在货物装时转移
property asset 房地产;不动产
property balance sheet 财产资本负债表;财产平衡表
property-basic form insurance 基本型资产保险
property betterment expenses 房地产改善费用
property bond 财产债券
property boundary 地产界
property boundary survey 地产界测量
property capital 财产资本
property chargeable with duty 必须缴税的财产
property claim 产权要求;财产请求权
property company 房地产公司
property-comprehensive form insurance 综合型资产保险
property conservation 财产保护
property corner 地界标石;地界
property currency 财产通货
property damage 资产损失;财产损失;器材损坏;动迁损失
property damage accident 财产损害事故
property damage insurance 财产损失保险
property damage liability insurance 财产毁损责任保险
property damage only 只负责财产损失
property deed 房契
property detector 属性检测器
property distribution 财产分配
property dividend 财产股利
property exempt form attachment and execution 免于扣押和执行的财产
property improvement loan insurance 修缮房产的贷款保险
property in action 无体物权
property income 财产收益
property information 地产信息;财产信息
property inspection 财产清查
property insurance 资产保险;财产保险;产业保险
property inventory 财产盘存;财产目录
property leasing contract 财产租赁合同
property ledger 财产分类账
property line 地权线;用地线;建筑红线;地盘界线;地界线
property line of planned road 规划道路红线
property line of road 道路红线
property line post 地界标柱;地界标(志)
property line right of way line 用地线
property line survey 建筑物红线测量;建筑红线测量
property line wall 地界(围)墙;界墙
property loss 财产损失
property man 煤矿装备管理员;工具管理员
property management 地产管理;财产管理;物业管理

property management agreement 物业管理协议
property manager 物业管理
property map 地籍图;大比例尺平面图
property mapping 地籍图测图
property of accumulation of harbor 港湾堆积物岩性
property of air-mass 气团性质
property of estimator of a model 模型的估计量性质
property of fluid 冲洗液性能
property of graphs 图件性质
property of initial setting (混凝土的)初凝特性
property of intersecting point in combined profiling method 联剖交点性质
property of material 材料的性质
property of optimal plan 最优方案的性质
property of plane area 几何性质
property of refraction 折射特性
property of section 截面性质;断面性质;断面特性
property of self-confinement 自限性
property of setting 硬化特性;硬化性质
property of wave transmission 波传播特点
property owner 土地所有人;业主;财产所有者
property protection 财产保护
property register 财产登记簿
property rehabilitation standard 产更房地新标准;房地产更新标准
property rental tax 财产出租税
property reserved 累积折旧;财产准备
property residual technique 财产剩余技巧
property rights 产权;财产权
property rights in general 一般产权
property room 舞台储藏室;道具室
property selective detector 特性灵敏检测器
property settlement 财产协议书
property settlement agreement 财产授予协议
property soil 特种土
property sort 特性分类
property-special policy insurance 特种政策型资产保险
property speculator 地产投机商人;地产经纪人
property stone 界碑
property subject to attachment 遭受扣押的财产
property suit 财产诉讼
property surplus 财产盈余
property survey 土地丈量;测绘图;地界;土地勘测;地界测量;地产测量
property tax 地产税;物业税;财产税;房地产税
property tax liens 财产税留置权
property test of lower coalification degree 低煤化程度的煤特性试验
property used in business 营业用财产
property valuation 财产估价
property value 属性值;地价;地产价值;地产价格
property value method 资产价值法
property well 地产上的井;家井
proper use 适度利用
proper-use factor 适用系数
proper value 特征值;本征值
proper vector 特征矢量
proper velocity 本征速度
proper vibration 固有振动
proper wedge graft 合理楔接
proper working load 正常工作荷载
prop-free front 无支柱采掘面;无支柱工作面;无支撑采掘面;无支撑工作面
prophase 前期;分裂前期
prophylactic 预防药;预防剂
prophylactic disinfection 预防性消毒
prophylactic health measure 预防保健措施
prophylactic measure 预防性检修
prophylactico-therapeutic institution 防治机构
prophylactic repair 预防检修;定期检修
prophylactic treatment 预防法
propicillin 苯丙西林
propine 丙炔
propine acid 丙炔酸
propinyl 丙炔基
propiolactone 丙内酯
propiolic acid 丙炔酸
propiolic acid series 丙炔酸系
propiolic alcohol 丙炔醇
propioloyl 丙炔酰
propionaldehyde 丙醛

propionaldoxime 丙醛肟
propionamide 丙酰胺
propionamido- 丙酰胺基
propionate 丙酸酯;丙酸盐
propionic acid 丙酸
propionic aldehyde 丙醛
propionic andydride 丙酸酐
propionic ether 丙酸乙酯
propiono 丙酰基
propionyl chloride 丙酰氯
propionylcholine 丙酰胆碱
propionyl group 丙酰基
propionyl iodide 丙酰碘
propionyloxy 丙酸基
propionyl peroxide 过氯化丙酰
proplasm 选型;首division模型;铸型
proplatinum 镍银铋合金
prop leg 支承式支腿
prop method 窑衬砌筑支撑法
prop of tunnel support 支撑大立柱
propolis 植物树脂;蜂巢蜡胶
propolis oral cavity membrane 蜂胶口腔膜
proponent 提倡者
proporting system 比例混合系统
proportion 相对数;颗粒形状均齐度;调和;比率;比例;按比例
proportionable 可均衡的
proportional 数字比例量;成比例的;比例的
proportional action 比例作用
proportional action coefficient 比例作用系数
proportional action factor 比例作用系数
proportional allocation 按比例分配
proportional allocation of sample size 样本量的比例分配
proportional amplifier 比例放大器
proportional analysis 比例分析
proportional analysis approach 比重分析法
proportional area 成比例区域
proportional band 线性范围;正比区;成比例范围;比例区;比例频带;比例范围(控制计量仪表);比例度;比例带
proportional band potentiometer 比例带电位计
proportional bandwidth filter 比例带宽滤波器
proportional burner 配比燃烧器;比例调节烧嘴
proportional change 按比例变化
proportional charges 比例费用;按比例收费
proportional coefficient 比例系数
proportional compasses 比例两脚规;比例(分)规
proportional component 比例环节
proportional constant 比例常数
proportional control 比例控制;比例调节;比例操纵
proportional control action 比例调节作用
proportional control factor 比例调节系数
proportional controller 比例控制器;比例调节器
proportional controlling means 比例控制法
proportional control valve 比例控制阀
proportional cost 比例费用
proportion(al) counter 正比计数器;比例计数器
proportional counter tube 正比计数管;比例计数管
proportional data 比例(型)数据
proportional deduction 按比例缩减;按比例扣除
proportional depth 充满度
proportional detector 正比探测器;正比检测器
proportional distribution 比例分配;比例分布;按比例分配
proportional divider 比例配水建筑物;比例分配器;比例分流器;比例两脚规;比例分规
proportional division box 比例分水口
proportional division water 比例分水
proportional elastic limit 比例弹性限;比例弹限;弹性比例极限
proportional error 相对比例误差;比例误差
proportional expenses 比例费用
proportional factor 比例因子;比例因数;比例系数
proportional feedback 比例反馈
proportional flow control unit 比例流量调节器
proportional flow filter 比例流量过滤
proportional flow weir 比例流量堰
proportional fluid blending system 比例流体掺和系统
proportional format 比例格式
proportional gain 比例增益
proportional gas meter 比例气体计量器
proportional governor 比例调速器;比例调节器

proportional graduated symbol 等级符号
proportional illumination 匀称照明
proportional integral control 比例积分控制
proportional integral controller 比例积分控制器;比例积分调节器
proportional integral derivative controller 比例积分微商控制器;比例积分微分调节器
proportional integral differential controller 比例积分微分控制器
proportional intensification 比例加强;比例加厚
proportional interpolation 线性内插(法);比例内插
proportional ionization chamber 正比电离室
proportionality 相称;均衡性;比值;比例性
proportionality constant 常系数;比例恒量;比例常数
proportionality factor 比例因子;比例因数;比例系数
proportionality law 比例定律
proportionality limit 配比限度;比例限度;比例极限
proportionality of production process 生产过程比例性
proportionality range 比例范围
proportional law 比例规律
proportional limit 比限;比例限度;比限;比例界限;比例极限
proportional limit intensity 比例极限强度
proportional limit of shear 剪切比例极限
proportional liquid sampler 比例液体采样器
proportional loading 比例加载;比例加荷;比例荷载
proportionally 按比例
proportional map 等积投影地图
proportional mean 比例中项
proportional meter 比例式水表
proportional mixer 比例混合器
proportional navigation course 比例导航
proportional navigation laser seeker 比例导引式激光寻的器
proportional number 比例数
proportional parts 按比例的部分;按比例部分;内插因子;比例部分
proportional personnel allocation method 比例定员法
proportional-plus-derivative action 比例加微商作用
proportional-plus-derivative control 比例加微分控制
proportional plus derivative controller 比例微分控制器;比例加微商控制器
proportional-plus-floating action 比例加无静差作用
proportional-plus-floating control 比例加浮动控制;均衡调节;重定调节
proportional-plus-integral control 比例加积分控制
proportional-plus-integral control action 比例加积分控制作用
proportional point symbol 等级点状符号
proportional position action type servomotor 比例位置式伺服电动机
proportional power unit 比例执行机构
proportional pressure reducing valve 定比减压阀;比例减压阀
proportional pump 定量泵;比例泵
proportional quantity 比例量
proportional range 线性范围;成比例范围;按比例部分
proportional rate of tax 比例税率
proportional reducer 比例减薄液
proportional reduction 按比例的减低
proportional region 正比区;成比例区段
proportional regulator 比例调节器
proportional relation 比例关系
proportional reserve system 比例准备制
proportional reset controller 比例重定调节器
proportional resolver 比例分相器
proportional sample 比例样品
proportional sampler 比例取样器;按比例取样器;按比例采样器
proportional sampling 比例取样;比例抽样(法);比例采样;比例抽样;按比例采样
proportional sampling device 比例取样器;按比例取样器
proportional scale 比例尺
proportional scale coefficient 比标系数
proportional selection 比例选择
proportional share-cropping of net income 净收入按比例分成

proportional spacing 变距;比例协调间隙;比例间隔
proportional spacing machine 比例间隔打字机
proportional spacing mechanism 比例间隔机构
proportional stamp tax 比例印花税
proportional stratified sampling 按比例分层取样
proportional subclass numbers 按比例的子类数目
proportional taxation 比例税制;比例税
proportional tax rate 比例税率
proportional test piece 成比例试件
proportional time-limit distance relay 比例限时远距离继电器
proportional type meter 正比例水表
proportional valve 比例阀
proportional viscous damping 比例黏滞阻尼
proportional volume gradient 比例体积梯度
proportional weight method 比例加权法;按比例加权法
proportional weir 比例(式)堰;比例式量水堰;比例分流堰
proportionate and speedy development 按比例高速发展
proportionate burden-sharing 按比例分担
proportionate measurement 比例量测法
proportionate mortality 比例死亡率
proportionate mortality ratio 比例死亡比
proportionate part 按比例的部分;按比例部分
proportionate sampling 按比例抽样
proportionate sharing of the revenue 比例包干
proportion by addition 合比
proportion by addition and subtraction 合分比
proportion by alternation 更比
proportion by assignment 按经验配比
proportion by inversion 反比
proportion by subtraction 分比
proportion by weight 重量比;重量配合法
proportion calculator 比例计算器
proportioned 均称的;成比例的
proportioned mix 配好混合料
proportion elastic limit 弹性比例极限
proportion electro-magnet 比例电磁铁
proportioner 定量给料斗;配合加料(漏)斗;输送量调节装置;定量器;比例器;比例配料器;比例加料器;比例调节器
proportion in a general average 共同海损分担额
proportioning 配合比选择;设计配合比;配料;配量;成比例;废水比例排放;按比例配合
proportioning and mixing 配比和搅拌
proportioning belt 配料皮带
proportioning belt balance 配料用胶带秤
proportioning belt scale 配料用胶带秤
proportioning bin 配料称量斗;分批称量斗;配料仓
proportioning box 配料称量盒;分批称量盒;配料称量箱;分批称量箱
proportioning burner 比例调节燃烧器
proportioning by absolute volume 按绝对体积配合(法)
proportioning by arbitrary 经验配料法;经验配比法
proportioning by assignment 经验配料(法);经验配合法;经验配比法
proportioning by conveyor belt 输送带配料称量;输送带分批称量
proportioning by grading charts 照级配曲线配料;按级配曲线配料;按级配曲线配合
proportioning by mortar-voids method 灰浆空隙配料法;灰浆空隙配合法;灰浆孔隙配合法
proportioning by trial method 试验配料法;试验配合法;试验法配料;试配法(配料);试配法配合混凝土成分;按试配法配合
proportioning by void cement ratio 用隙灰比法配料(混凝土混合料);隙灰比法配料;隙灰比法配合混凝土成分
proportioning by volume 容积配料;容积配合;体积配合比;体积比配料法;体积比配合法;按体积配料;按体积配合;按体积比例调配;按容积配合
proportioning by water-cement ratio 水灰比配料(法);水灰比配合(混凝土);按水灰比配合;水灰比法配料;水灰比法配合混凝土
proportioning by weight 重量配合(法);重量比配合法;质量配料;按重量配料(法);按重量配合;按重量比例调配
proportioning cart 配料称量车;分批称量车
proportioning concentration 比例混合液
proportioning container 配料称量容器;分批称量

容器
proportioning control system 中间位置控制系统
proportioning conveyor belt scale 配料称量输送带秤;分批称量输送带秤
proportioning cycle 配料称量循环;分批称量循环
proportioning damper 配比档板
proportioning device 计量装置;计量器;计量配料装置;比例装置;比例配料装置;比例配料设备;配料设备
proportioning drum 配料称量鼓筒;分批称量鼓筒
proportioning feeder 计量配料喂料机
proportioning formula 配比公式
proportioning frame 配料称量架;配料称量箱;分批称量架;分批称量箱
proportioning grid 比例格
proportioning hygrometer 配料湿度计
proportioning installation 配料设备;配料装置
proportioning lock 配料阀门;配料闸门
proportioning machine 配比计量机
proportioning meter 配料计;配量计
proportioning mixer 比例混合器
proportioning motor 比例电动机
proportioning of cement concrete 水泥混凝土配合比
proportioning of concrete 混凝土配合(比)
proportioning of mortar 灰浆配合(比)
proportioning of the bridge spans 桥梁分孔
proportioning of the span 跨径划分
proportioning of waste 废料配(合)比
proportioning on a cement-paste basis 灰浆法配料;灰浆法配合混凝土成分
proportioning plant 比量投料器;配料车间;配料装置;配料线;投配器;比例配料装置;比例配料设备
proportioning principle 比例混合原理
proportioning probe 配比探头
proportioning pump 配量泵;配合泵;定量泵;分配泵
proportioning reacto 比例电抗器
proportion(ing) scale 配料称量传动装置;配料秤
proportioning screw 配料称量螺旋输送机
proportioning screw conveyer 配料螺旋输送机;配料绞刀
proportioning screw feeder 配料螺旋喂料机;配料绞刀
proportioning section 配料部分;配料区段
proportioning selector 配料程序分选器
proportioning silo 定量配料仓;配料筒仓
proportioning star 配料星形架
proportioning tower 配料塔
proportioning trough conveyer 配料槽式输送机
proportioning unit 配料称量单元;分批称量机组;分批称量单元;配料称量机组
proportioning valve 配量阀;比例分配阀
proportioning waste 配合废料
proportioning weigh gear 分批称量传动装置
proportioning worm conveyer 配料称量螺杆输送机;配料螺旋喂料机;配料绞刀
proportionment 相称;按比例划分;按比例分配
proportion meter 比例计
proportion model 比例模型
proportion of critical damping 临界阻尼比
proportion of death cause 死因构成比
proportion of equipment in good condition 设备完好率
proportion of gradient 含量比例
proportion of ingredients 成分比例
proportion of mixture 混合比例
proportion of outsize 非标准尺寸的大小
proportion of recoverable block 可采块段比例
proportion of resin present 树脂含量
proportion of size 尺寸比例
proportion of total value methods of weighting 按总值比例加权法
proportion of water supply reaching sewers 排水量与给水量之比
proportion set screw 比例度设定螺丝
proposal 议案;建议;提案;推荐方案
proposal for insurance 投保;保险申请
proposal form 投标表格;标单格式;投保单
proposal item 投标条款;投标项目
proposal of insurance 要保书
proposal request (建筑师、工程师发出的)招标书
proposals for the projects 项目建议书

proposal stage 建议阶段
proposal system 提案制
propose 提出建议
proposed access road 拟建入口道路
proposed alignment 假定路线;拟用路线
proposed building 拟建房屋
proposed common name 建议通用名称
proposed dividend 拟派股利
proposed dock 拟建码头
proposed flowchart 建议采用的流程
proposed grade 建议坡度;拟用坡度;推荐坡度
proposed highway subgrade 拟建公路路基
proposed layout 推荐平面布置方案
proposed life 拟用年限
proposed maximum contaminant level 建议最大污染水平
proposed monitoring item 拟定监测项目
proposed project 建议方案;拟建工程
proposed quay 拟建码头
proposed quay wall 拟建岸壁
proposed regulation 试行条例
proposed scheme 推荐方案
proposed second(ary) track 拟建第二线【铁】
proposed site 推荐地点;待选场址;待选坝址
proposed structure 计划采用的建筑物;计划建造的建筑物;拟用建筑物
proposed task of project 项目建议书
proposed terminal 拟建码头
proposed test method 建议试验方法
proposed traverse net 图上设计的导线网
proposed wharf 拟建码头
proposer 要保人
proposition 论题;建议;命题
propositional algebra 命题代数
propositional calculus 命题演算
propositional constant 命题常数
propositional formula 命题公式
propositional function 命题函数
propositional law 命题定律
propositional logic 命题逻辑
propositional variable 命题变量
proposition letter 命题字母
proposition matrix 命题矩阵
proposition tree 命题树
propotion of gross to net load 毛重与净重之比
propoxide 丙氧化物
propoxylate 丙氧基化物
propoxylation 丙氧基化作用
propped beam (临时的)支顶梁;临时支撑梁;加撑梁
propped cantilever 一端简支的悬臂梁
propped cantilever beam 有支悬臂梁;临时悬臂斜撑;加撑悬臂梁;支撑悬臂梁
propped single-wall 支撑单排板墙
propping 用支柱支撑;纵支梁;横支梁;支柱支撑
propping of floors 楼板支撑
prop post 临时支柱
proprietary 专买的;所有人的;所有人;财产权
proprietary account 业主账户
proprietary alarm system 私有专用报警系统
proprietary appliance 专用设备
proprietary article 专利品
proprietary capital 业主资本
proprietary chemical 专利化学品
proprietary company 独资公司;控股公司;土地兴业公司
proprietary concentrate 商品材料
proprietary construction material 专利建筑材料
proprietary consumables 专有消耗品
proprietary device 专利设备
proprietary door 专利门
proprietary environmental protection technology 有产权的环境保护技术
proprietary equipment 专利设备
proprietary equity 业主权益;业主产权
proprietary formula 专用配方
proprietary formulation 有产权的配方;专利配方
proprietary hospital 私人医院
proprietary information 有产权的情报
proprietary interest 业主权益
proprietary lease 业主租约;集资建房房客契约;产权租赁
proprietary limited 控股有限公司
proprietary material 专买材料;专利品;专利材料

proprietary mix 专卖混合物;专卖混合料;专利混合料
proprietary name 专用商标名;商品名
proprietary plant 专利厂
proprietary product 专利产品
proprietary program(me) 专有程序
proprietary protective signaling system 私有保护信号系统
proprietary rights 所有权(权利);产权
proprietary signaling system 私有信号系统
proprietary software 专利软件
proprietary solvent 专用溶剂
proprietary specification 有专利权的说明书;专利说明书
proprietary system 专用系统
proprietary technique 产权技术
proprietary technology 专有技术;专利技术
proprietor 业主;经营者;所有者;独资业主
proprietor account 业主账户
proprietorship 独资企业;所有权
proprietorship account 业主账户
proprietorship accounting 业主会计
proprietorship of land parcels 小块土地所有制
proprietor's income 业主收入
proprius 背侧固有束
proproduction area 试采区
prop room 道具室
prop root 支柱根
prop-setters 支柱定位器;支柱装定器;支撑定位器;支撑装定器
prop spacing 支柱间距;支撑间距
prop stay 支柱撑杆;支柱;撑杆
prop support 支承式支座
prop test rig 支柱试验装置
propulsion 驱动;气力推动;推进;顶进
propulsion and electric (al) operating system 推进和电气操纵系统
propulsion arm 顶推杆
propulsion assisted 辅助推进
propulsion auxiliary machinery control equipment 推进辅机控制设备
propulsion bay 推进装置舱
propulsion branch 主动段
propulsion capability 推进能力
propulsion characteristic 推进特征
propulsion component 推进分量
propulsion control panel 动力装置控制台
propulsion control unit 牵引控制单元
propulsion current 牵引电流
propulsion data 推进数据
propulsion data display panel 推进动力数据显示板
propulsion drive 行走部分传动装置;推进传动
propulsion efficiency 推进效率
propulsion engine 推进发动机
propulsion force per unit area 单位面积推力
propulsion gear 行走部分传动系
propulsion generator 推进发电机(船用)
propulsion inverter 牵引逆变器
propulsion inverter test equipment 牵引逆变器试验装置
propulsion invertor 牵引逆变器
propulsion local control console 局部推进控制台
propulsion motor 推进电动机
propulsion noise 推进器噪声
propulsion nozzle 推进喷管
propulsion phase 推进阶段
propulsion plant 驱动装置
propulsion pontoon 平底船;机动趸船
propulsion power 驱动功率;推进功率
propulsion reactor 推进用反应堆;推进用堆
propulsion simulator 推进装置模拟器
propulsion subsystem 推进分系统
propulsion system 牵引系统;推进装置;推进系统;动力装置
propulsion-system braking 用推进装置减速
propulsion test facility 推进装置试验设备
propulsion transmission 推进传动
propulsion type 推进动力类型
propulsion unit 驱动装置;推进装置
propulsion wind tunnel 推进式风洞;推进机风洞
propulsive 推进的
propulsive arrangement 推进装置
propulsive coefficient 推进系数;推进力系数
propulsive control 推进操纵

propulsive device 推进装置
propulsive duct 喷气发动机;无压气机式喷气发动机
propulsive efficiency 推进效率
propulsive effort 驱动力
propulsive engine 推进发动机
propulsive force 推力;推进力
propulsive force theory of gas 天然气推动力说
propulsive gas 喷射推进气体
propulsive horsepower 推进马力
propulsive jet 推进射流;冲压式空气喷气发动机
propulsive machinery 推进装置
propulsive output 推进功率
propulsive parameter 推进参数
propulsive performance 推进性能
propulsive unit 推进装置
propulsor 喷气发动机
prop-up and dismountable window 支摘窗
prop wall 密集支柱
propyl 丙基
propyl acetate 乙酸丙酯
propyl-acetic acid 丙基乙酸
propyl acetoacetate 乙酰乙酸丙酯
propylaeum 建筑物前门厅;雅典卫城精致的门道;围墙前的入口;寺庙境域的入口
propyl alcohol 丙醇
propyl amine 丙胺
propyl aminoben-zoate 氨基苯甲酸丙酯
propylate 丙醇盐
propyl benzene 丙基苯;苯基丙烷
propyl benzene phenylpropane 丙苯
propyl benzoate 苯甲酸丙酯
propyl benzyl cellulose 丙基苄基纤维素
propyl benzyl ketone 丙基苄基(甲)酮
propyl bromide 丙基溴
propyl butyrate 丁酸丙酯
propyl cellosolve 乙二醇—丙醚;丙基溶纤剂
propyl cellulose 丙基纤维素
propyl chloride 丙基氯
propyl cyanide 丙基氰
propyl cyclohexane 丙基环己烷
propyldioctylamine 丙基二辛基胺
propyldisulfide thiamine 优硫胺
propylene 丙烯
propylene carbonate 碳酸丙烯酯;碳酸丙烯;碳酸丙二酯;丙烯碳酸脂
propylene diamine 丙二胺
propylene dichloride 氯化丙烯
propylene dimer 二聚丙烯
propylene epoxide saponated sewage 环氧丙烷皂化污水
propylene-glycol 丙二醇
propylene glycol diacetate 丙二醇二乙酸酯
propylene glycol diricinoleate 丙二醇二蓖麻油酸酯
propylene glycol laurate 月桂酸丙二醇酯;十二烷酸丙二醇酯
propylene glycol monoacrylate 丙二醇单丙烯酸酯
propylene glycol monostearate 丙二醇一硬脂酸酯
propylene glycol phenyl ether 苯基丙二醇醚
propylene liguid 液态丙烯
propylene oxide 氧化丙烯;环氧丙烷
propylene oxide adduct 环氧丙烷加成物
propylene oxide glycidyl ether 环氧丙烷—烯丙基缩水甘油醚共聚物
propylene plastics 丙烯塑料
propylene product 丙烯成品
propylene resin 丙烯类树脂
propylene steam 气态丙烯
propylene sulfide 硫化丙烯;硫丁环
propyl ester 丙酯
propyl ether 丙醚;丙基醚
propyl ethylene 丙基乙烯
propylformic acid 丁酸
propyl glycol 丙基乙二醇
propyl halide 丙基卤
propylidyne 次丙基
propyl iodide 丙基碘
propyl iodone 丙碘酮
propyl isocyanate 异氰酸丙酯
propylisome 丙基增效剂
propylite 绿盘岩;绿盘石;青盘岩;变安山岩
propylitization 青盘岩化
propylization 绿盘岩化(作用)
propyl ketone 庚酮
propylmalonic acid 丙基丙二酸

propyl mustard oil 丙基芥子油
propylnitramine 丙基硝胺
propyl nitrolic acid 丙基硝肟酸
propylon 古埃及寺庙进口处及前方(与寺庙不相连)之大门;埃及式门楼;入口;普罗皮伦(聚丙烯纤维)
propylparaben 丙对苯
propyl para-hydroxybenzoate 丙对苯
propyl propionate 丙酸丙酯
propyl ricinoleate 蓖麻油酸丙酯
propyl succinic acid 丙基丁二酸
propyl sulfide 二丙硫醚;二丙硫
propyl thioether 二丙硫醚
propylthiopyrophosphate 丙基硫代焦磷酸酯
propyl thiourea 丙硫脲
propyl-trichlorosilicane 丙基三氯硅
propyl-triethoxysilicane 丙基三乙氧基硅
propyl-triethylsilicane 丙基三乙基硅
propyl-trimethylsilicane 丙基三甲基硅
propyl urethane 丙氨基甲酸乙酯
propyne 丙炔
prop yoke 叉形架;支柱轭架
pro-rata 按比例
pro-rata freight 比例运费;按比例计算运费
pro rata sinking fund 按比例的偿债基金
prorate 摊派;按照比率;按比例分配
prorate average 按比例平均;按比例均摊
prorate cancellation 比例注销
prorate distribution 比例分摊
prorated section single busbar configuration 分段单母线接线
prorate itineraries 按预定的旅程
prorate rebate 按比例回扣
pro ratio 按照比例
pro ratio average 按比例平均
proration 摊派;按比例分配
prororoca 南美亚马逊河口暴涨潮
prosaic industrialist 平凡的工业家
prosapogenins 次皂苷元
proscenium 舞台台唇;舞台前部装置;舞台前部;舞台;台唇;[复]proscenia
proscenium arch 台景的框架;舞台台口;舞台口;台口
proscenium box 剧场前侧包厢;前侧包厢;舞台前侧包厢;舞台包厢;特别包厢
proscenium bridge 舞台口上部灯光廊
proscenium door 舞台口门
proscenium frame stage 箱形舞台
proscenium opening 舞台(台)口;舞台门孔;台口
proscenium stage 镜框式舞台;有台口的舞台
proscenium stage type 台口式剧场布置
proscenium theater 镜框舞台剧院;有台口的剧院
proscenium wall 舞台台口侧墙;舞台侧墙;舞台侧墙
proscription 不予法律保护;剥夺所有权
prosecute 起诉
proseptazine 苄基磺胺
proshenion 古希腊剧场后台房屋
prosign 同一符号
proskenion 后台房屋(古希腊剧场的);舞台台口
prosopite 水铝氟石
prosopoanoschisis 面斜裂
prospect 展望;景色;前景
prospect area 勘探远景区
prospect drilling 普查钻探
prospect evaluation 远景评价
prospect evaluation of regional mineral resource 区域矿产远景评价
prospect hole 探井;试孔;试坑;探孔
prospecting 探查的,勘察的;钻探;勘探;探矿
prospecting adit 勘探平硐
prospecting and exploration of mineral resources 矿产勘查
prospecting apparatus 探测设备
prospecting audio-indicator 探矿用音响指示器
prospecting baseline 勘探基线
prospecting bore bit 勘探钻头;勘探钻;探钻
prospecting borehole 探测井;探测钻孔;勘探钻孔
prospecting by boring 钻孔勘探
prospecting by boring or trial pits 钻孔或试坑勘探
prospecting by trial pit 试坑勘探
prospecting counter 电测计数器
prospecting disk 淘金盘
prospecting drift 探矿巷道

prospecting drill 勘探钻;探钻;试钻
prospecting drill-hole 勘探钻孔
prospecting drilling 试钻;探钻;勘探钻井
prospecting drilling machine 探钻机
prospecting engineering planimetric(al) map 探矿工程平面图
prospecting entry 钻探巷道
prospecting for construct mine 普查最终
prospecting gallery 勘探坑道;眺望长廊;眺望游廊
prospecting guide 勘探标志
prospecting hammer 地质锤;地质铁锤;地质锤
prospect(ing) hole 勘探孔;探孔;普查钻孔
prospecting indications and guides 找矿标志
prospecting instrument 勘探仪器;探测仪(器)
prospecting interval 勘探网度
prospecting license 勘探许可证
prospecting line 勘探线
prospecting line profile map 勘探线剖面图
prospecting line survey 勘探线测量
prospecting map 勘探图
prospecting meter 探矿仪
prospecting method 勘探方法;普查(方)法
prospecting net(work) layout 勘探网测设
prospecting net(work) survey 勘探网测量
prospecting of mineral resources 矿产普查
prospecting pit 勘探试坑
prospecting potential customer 辨认潜在顾客
prospecting project 勘探工程
prospecting report 普查报告
prospecting scheme 勘探计划;探索计划
prospecting seismology 勘探地震学
prospecting shaft 勘探竖井;探井;探察井;试钻井
prospecting stage 勘探阶段
prospecting study 远景研究
prospecting team 勘探队
prospecting trench 探槽
prospecting tunnel 探查坑道
prospecting type of ore deposits 矿床勘探类型
prospecting well 勘探井
prospecting work 勘探工作;探矿工程
prospection 预期
prospective acreage 预期有石油的(地)面积
prospective area of engineering construction 工程建设规划区
prospective area of mineral deposits 成矿远景区
prospective bibliography 预告目录
prospective bidder 有希望得标的投标者;未来投标者
prospective buyer 可能的买主
prospective control 可预见性控制
prospective current 远景电流;预期电流;期望电流
prospective design 勘察设计
prospective earnings 预期收入
prospective glass 小型轻便望远镜
prospective investigation line 远景勘探线
prospective oil 石油储量
prospective ore 可采矿;远景储量;推测矿量
prospective path 预计航迹
prospective peak value 预期峰值
prospective plan 远景规划
prospective port site 可能选用港址
prospective reserves 推断储量;远景储量
prospective resource 远景资源量
prospective significance 预定意义
prospective site 勘探坝址;可能坝址
prospective study 远景调查;前瞻性研究
prospective tender 未来投标者
prospective traffic 远景运量
prospective traffic volume 远景交通量
prospective value 远景价值;预期值
prospective yield 预期收入
prospect-mining hole 勘探开采孔
prospect network 勘探网
prospector 勘探人员;探险者航天器
prospector and/or explorer of reserves 储量提交单位
prospector's pick 地质勘探者用的丁字镐
prospect pit 探坑
prospect pit documentation 探井编录
prospect traffic volume 远景交通量
prospect tunnel 探矿平巷
prospectus 意见书;计划(任务)书;任务书
prosperite 羟神钙锌石
prosperity 景气;繁荣

prosperity in farming 农业兴旺
prosperity peak 繁荣高峰
prosperity phase 繁荣阶段
prosperity through productivity 发展生产力以促进繁荣
prosperous and strong 富强
prospite 氟铝钙石
prostas 门厅;前厅
prostasis 门廊壁柱之间
prosthesis 修复体;修复术;修补物;附加物
prosthetic 修复的;弥补性的
prosthetic ceramics 修复用陶瓷
prosthetic group 修复组;辅基
prosthetic robot 关节式机器人
prosthetics 修复学
prostoon 门廊
prostratigraphy 原始地层学
prostylar 前柱式的;柱廊式的
prostyle 前柱式;列柱式;柱廊式建筑;柱廊式的;柱廊式
prostyle building 柱廊式建筑
prostyle column 前柱廊式建筑物的柱
prostyle temple 前柱廊式庙宇
prostylos 前柱式建筑;多柱建筑
prosultiamine 丙硫硫胺
protable document format[pdf] 可移植文档格式
protable moisture titrator 轻便式湿度滴定器
protable turbidimeter 携带式浊度计
protactinium-231 method 镤231法
protangiospermae 原始被子植物
protated emission charges on air pollutants 按比例摊派的空气污染物排放费
protean stone 石膏制造石;人造石膏石;石膏(质人造)石;石膏人造石
protectant 防护剂;保护的;保护剂
protect bit 保护位
protect coastal zone from irreversible degradation 防止沿海地带不可逆转的恶化
protect coral reefs, mangrove and fishing resource 保护珊瑚礁、红树林和渔业资源
protected 保护性成色剂;保护的
protected apron 防护板;防护墙
protected area 保护区;保护地
protected band 安全带;防护带
protected bear 抛补空头
protected bulkhead 防护舱壁
protected cable 包皮电缆
protected cap 安全帽
protected check 防涂改支票;保护性校核;保护性检查;保护性核查
protected coal seam 被保护煤层
protected code 保护码
protected compartment 消防区图
protected construction 防护结构
protected corner (混凝土路面的)加筋角隅;护隅;护角;有增强钢筋的板角(混凝土路面);受保护的混凝土板角
protected corridor 设防走廊
protected cover 保护罩
protected cut-out 封闭式熔断器
protected data 保护数据
protected data set 被保护数据集;保护数据集
protected device 保护器;防护设备
protected dynamic(al) storage 保护动态存储器
protected electric(al) machine 防护式电机
protected enclosure 防护罩壳
protected escape route 防火疏散路线;安全疏散路线
protected exterior 防火外围;防火外墙
protected field 受保护区域;保护字段;保护域
protected file 被保护文件;保护文件
protected flame lamp 火焰安全灯
protected forest 封禁林
protected formatting 被保护格式
protected free storage 保护自由存储器
protected fuel tank 密闭的汽油桶
protected harbo(u)r 有掩护港口;有屏障海湾;有屏障港湾;保护港
protected key 保护键
protected landscape 受保护的自然景观
protected landscape area 风景保护区
protected level 保护级
protected(level) crossing 有防护道口
protected location 受保护单元;被保护单元;保护单元
protected lowland 护堤内低地;堤内低地
protected machine 受保护机器
protected membrane 屋面保护层
protected metal sheeting 防锈金属薄板
protected mode 保护模式
protected motor 防护型电动机;防护式电动机
protected natural area 自然保护区
protected non-combustible construction 耐火建筑(物)
protected object 保护目标
protected opening 防火出口;耐火通道口;设防洞口;防火洞口
protected ordinary construction 一般耐火建筑
protected paste volume 防冻水泥砂浆体
protected port 保护港
protected queue area 保护排队区;保护队列区
protected reversing thermometer 闭管海水温度表;闭端颠倒温度计
protected roofing 防护屋面材料
protected route 防护进路
protected section 防护区段
protected shaft 安全通路;安全通道
protected shelf 受护陆架
protected shoulder 有保护层的路肩
protected space 受保护空间
protected species 受保护的物种
protected state 保护国
protected storage 受保护存储;被保护存储区;保护存储器
protected storage area 保护存储区
protected strike 防护锁舌
protected subsystem 保护子系统
protected switch 防护式开关;盒装开关
protected thermometer 闭管温度表;闭端颠倒温度计
protected trade 保护贸易
protected tree 保护树
protected type 屏蔽式;保护方式
protected usage mode 保护使用方式
protected wall 防护墙
protected waste pipe 半封闭的污水管;室内排水管;防护污水管
protected wild animals 受保护的野生动物
protected wood frame construction 耐火木框架建筑;防护型木架建筑物
protected zone 防护区;被保护区;保护区
protect forests from overexploitation 防止过度利用森林
protect form light 避光
protect from moisture 不可受潮
protecting 防护
protecting agent 保护剂
protecting aperture 放映口
protecting apron 防护板;防护墙
protecting bag 保护袋
protecting band 防护带;防护带;安全带
protecting belt 防护带
protecting bush 护套
protecting cap 防护帽;保险帽;安全帽
protecting casing 保险罩;保护套管
protecting circuit 保护线路;保护电路
protecting cladding 保护包装
protecting coat 保护涂料;保护涂层
protecting colloid 保护胶体
protecting concrete against corrosion 防护混凝土免被侵蚀
protecting cover 护罩;护盖;防护罩;保护罩
protecting crust 防护壳
protecting cultivated soil 耕作土壤保护
protecting device 防护设备;防险器;保险器;保护装置
protecting embankment 护堤
protecting equipment 保护装置
protecting film 防护膜;保护层
protecting furnace 保护窑
protecting glass 防护玻璃;玻璃保护罩
protecting glasses 防护眼镜
protecting glass plate 防护玻璃片;防护玻璃板
protecting grate 防护格栅
protecting grill 护罩
protecting grounding network 保安接地网
protecting group 保护基
protecting hood 护盖
protecting housing 保护管
protecting jacket 防护套;保护套
protecting layer 保护层
protecting levee 护堤
protecting lid 护盖
protecting mask 防护面具;保护面具
protecting masonry wall 防护圬工墙
protecting means 保护装置
protecting measure 保护手段
protecting net 保护网;安全网
protecting piece 护板
protecting pillar 保安矿柱
protecting pipe 保护管
protecting planting 防护造林
protecting plastic film 防护用塑料膜
protecting primer 保护底漆
protecting rack 保护格栅
protecting railing 保护栏杆;保护扶手
protecting roof 御坍棚;护棚
protecting salt 防护用盐
protecting scaffold(ing) 保护脚手架
protecting screed 保护带
protecting screen 保护屏;防护屏
protecting shield 护罩;防护板
protecting skin 保护层
protecting sleeve for pipe 导管防护套
protecting spectacle 防护眼镜
protecting spray 防护喷射
protecting tube 护管;热电偶保护管;保护套管;护管
protecting wall 护墙板;护墙【港】;挡土墙;挡水墙;防护墙
protecting zone 防护带
protection 安全装置;保护装置;防护;保护(作用)
protection against acid 防酸
protection against acid and fumes 防酸和毒烟
protection against atomic bacteria chemical weapon 对原子化学细菌武器的防护
protection against atomic weapon 防原子武器
protection against avalanches 防雪崩;防崩塌
protection against biological weapon 防细菌武器
protection against cold 防寒
protection against cold and freezing condition 防寒防冻
protection against condensation 防冷凝
protection against corrosion 防腐蚀;防锈保护;防蚀
protection against damage claims 损失赔偿的预防
protection against earthquake 防震
protection against electromagnetic radiation 电磁辐射防护
protection against erosion 冲刷防护;防冲刷
protection against fire 防火
protection against floods 防洪(水)
protection against freezing 防止冻冰
protection against frost 防霜(冻);防冻(措施)
protection against ice run 防凌
protection against insects 防虫
protection against light pollution 光污染防护
protection against mining subsidence 防矿穴沉陷;防坑道下陷
protection against moisture 防潮
protection against noise 防噪声
protection against occupational poisoning 防止职业中毒
protection against oscillation 防振荡;防摆动
protection against over-pressure 超压防护
protection against radiation 辐射保护;放射防护
protection against rain 防雨
protection against sand drifting 防沙
protection against scour(ing) 防擦伤;防冲刷(措施)
protection against silicosis 防止矽尘危害
protection against snow 防雪
protection against sound 防音响
protection against termites 防白蚁
protection against unsymmetric(al) load 不对称负载保护装置
protection against vibration 防振
protection against weather 防风雨措施
protection agent 保护剂;抗氧化剂;防冻剂
protection analysis project 保护分析计划
protection and indemnity 保障与赔偿;保赔协会保险;保赔;保护与赔偿

protection and indemnity association (船舶所有人的)保赔协会
protection and indemnity clause 保赔保险条款
protection and indemnity club (承租人的)保赔协会
protection and indemnity insurance 保障与赔偿保险;保赔保险
protection and indemnity model bill of lading 保赔标准提单格式
protection and indemnity risks 保赔(责任)险
protection and treatment of frozen soil 冻土地基的防治
protection and treatment of swelling foundation soil 膨胀土地基的防治
protection apron 护坦;防护板;保护墙
protection architecture 保护结构
protection area 保护面积
protection area length of catch siding 安全线保护区段长度【铁】
protection bar 护杆;保护杆
protection barrier 保护栏
protection belt of trees 防护林带;防风林带
protection bit 保护位
protection blanket 反压填土;保护层
protection boundary 护林边界
protection building 防护建筑(物)
protection by dikes 用堤坝防护;堤坝防护
protection by oxide films 氧化物膜保护法
protection capability 防护能力
protection casing 技术套管;护壁套管
protection casing depth 技术套管深
protection character 保护符
protection check 保护检查
protection circuit 保护电路
protection circuit for heavy down grade approaching 下坡道防护电路
protection circuit for switch lying in midway of receiving-departure track 到发线出岔电路
protection coat (油漆的)防护层;保护涂剂
protection code 保护码
protection countermeasure 保护对策
protection course 保护层;防护层
protection course of reinforced concrete 钢筋混凝土的保护层
protection covering 护面
protection design 保护设计
protection device 防护装置;保护装置;保护设备
protection device against boom overturn 吊杆防倒装置
protection device for boom overturn 吊杆防倒装置
protection during construction 建筑施工防护
protection duty 保护关税
protection embankment 护堤;防护围墙;防护堤;防洪堤
protection environment 环境保护
protection equipment for transformer 变压器保护装置
protection exception 保护异常;保护事故
protection facility 防护设施
protection failure detector 保护故障探测器
protection feature 保护特性;保护机构
protection fence 护篱;护栏
protection film 防护膜
protection film for heat treatment of metal 金属热处理保护膜
protection forest 禁伐林;防护林;保护林
protection forest fire-proof road 护林防火道路
protection forest for erosion control 防冲林
protection forest for flood hazard 洪害防护林;防洪林
protection forest for landslide prevention 滑坡防护林
protection forest for shifting sand 固沙防护林;流沙防护林
protection forest for water conservation 水源涵养林;水源保护林
protection from frost 防霜冻设施
protection from glare 防眩光设施
protection from radiation 放射性防护
protection from the wind 防风设施(防风林、挡风墙等);防风
protection gap 保护间隙
protection glass 保护等级

protection grade 防护等级;保护等级
protection group 保护组
protection helmet 防护头盔
protection hole 坑道超前钻孔;防护孔
protection in buried zone 埋入区的防蚀
protection installation 防护设施
protectionism 保护(贸易)主义;保护贸易制;保护贸易政策;保护贸易条例;保护关税主义
protectionist 保护贸易主义者
protection key 护符;保护键
protection level 防护水平;保护水平
protection location of segregated ballast tank 专用压载舱保护位置
protection measurement 安全措施
protection measures of gully head 沟头防护工程
protection mechanism 保护机构
protection membrane 保护膜
protection method 保护措施
protection method of soil-hungry 防止土壤瘠薄措施
protection mode 保护方式
protection monitoring area 防护监测区
protection net 保护网
protection network 保护网
protection of bank 护岸
protection of beach dunes 海滩沙丘保护
protection of biologic(al) resources 生物资源保护
protection of biotic resources 生物资源保护
protection of child labo(u)r 童工的保护
protection of closure gap bed 截流龙口护底
protection of coastline 海岸防护
protection of distance 距离防护
protection of drinking water 饮用水保护
protection of ecologic(al) environment 生态环境保护
protection of environment 环境保护
protection of exhaust fan 排气风机的保护
protection of face 护面
protection of frost 防冻
protection of grassland 草原保护
protection of green concrete by curing overlay 用养护罩保护新拌混凝土
protection of groundwater environment 地下水环保
protection of historic(al) relics 文物保护
protection of historic site 保护古迹
protection of home industries 民族工业保护;保护民族工业
protection of inner radiation 内照射防护
protection of islands 岛屿保护
protection of land 土地保护
protection of land resources 土地资源保护
protection of molybdenum electrode 钼电极保护
protection of monitoring points 监测点的保护
protection of monuments 保护纪念碑
protection of nature 自然(界)保护
protection of nursery ground 索饵场保护
protection of occupant 占用人保护
protection of outer radiation 外照射防护
protection of pedestrians 行人(交通)安全措施
protection of power direction 功率方向保护
protection of time 时间防护
protection of trade and commerce 保护贸易
protection of traffic 交通安全措施
protection of tree 树木保护
protection of tunnel 隧道防护
protection of waters 水资源保护
protection painting(work) 涂漆保护
protection plate 防护板
protection policy 保护政策
protection range 保护范围
protection ratio 保护比
protection reactor 保护电抗器
protection requirement 保护要求
protection ring 保护环
protection scheme 保护方案
protection screen 防护屏;保护屏
protection screen deck 防护筛板
protection shield 导坑护板;防护屏;防护板
protection signal 防护信号
protection string 技术套管
protection structure 防护建筑(物)
protection suit 防护服(装)
protection switch 保护开关

protection switching count 保护交换计数;保护倒换计数
protection system 防护系统;保护系统
protection system design 保护系统设计
protection tariff 保护税率
protection tent 防护帐篷
protection test(ing) 保护试验;防护试验
protection unit 保护装置
protection valve 安全阀
protection wagon 防护车
protection wall 挡土墙;防护墙
protection wood pile head 木桩顶的保护
protection works 防护(品)工厂;防护工程;防护设施
protection writ 保护性令状
protection zinc 防护锌板
protection zone 安全区
protection zone within the liquefied gas tank farm 液化气储罐站安全区
protective 防护的;保护的
protective ability 防护能力
protective action 保护作用
protective adaptation 保护性适应;保护适应
protective agent 防护剂;防辐射剂;保护剂
protective anode 保护阳极
protective antigen 保护性抗原
protective apron 防护围裙;铅橡皮围裙
protective arch 保护拱
protective arch rib 拱架支护
protective armo(u)r of breakwater 防波堤护面块体
protective articles 防护用品
protective atmosphere 保护气层;保护气氛
protective atmosphere equipment 保护气氛供应设备
protective bandage 保护绷带
protective bank of earth 防护土堤
protective bar 保护棍
protective barrier 防御屏障;防护栅障;防护栅栏;防护栏(杆);防御阻障物;防御栅栏
protective bedding 防毒被垫
protective belt 护岸林;防护地带;防护带;保护林带
protective bevel 保护性倒边
protective blanket 防护铺盖;防护面层;遮热板;保护层
protective body 保护体
protective boot 防护腿靴
protective bulkhead 防护舱壁
protective cable 保护电缆
protective calcite coating 保护方解石涂层
protective camouflage coating 保护迷彩涂料
protective cap 保护帽;保护盖;安全帽
protective capability 防护能力
protective cap watch 带防护罩的表(测斜仪)
protective case 护壳
protective casing 技术套管;防护罩;防护套;保护外壳;保护套管
protective center 拱架支护
protective choke 防干扰扼流圈;扼流线圈
protective choke coil 保护性扼流圈;保护扼流圈
protective circuit 保护电路
protective(circuit) breaker 保护断路器
protective clothing 防护工作服;防护衣(服);防护服(装);保护服
protective clothing for hazardous chemical operation 危险化学品作业防护服
protective clothing for proximity fire fighting 隔热防护服
protective coat 保护涂层;保护层
protective coat by penetration 浸渍形成的防护面层
protective coating 护膜;防护涂料;防护涂料;防护涂层;保护层;保护性涂料;保护涂层;保护膜
protective coating and lining 保护涂层和衬理
protective coating for the radiation 防辐射线涂料
protective coating material 保护涂层材料
protective coat of concrete 混凝土保护(面)层
protective colloid 保护胶体
protective colo(u)ration 保护色
protective column 保护柱
protective concrete layer 混凝土保护膜;混凝土保护层
protective condenser 保护电容器
protective construction 防护工事

protective course 保护层
protective course of cement mortar 水泥砂浆保护层
protective covenant 保护性契约(中的限制条款);土地保护性契约
protective cover 涂料;护罩;护盖;覆盖(保护)层;防护外壳
protective coverall 防护连衣裤
protective cover(ing) 保护层;防护罩;护面
protective cover of vegetation 植被防护层;防护性植被
protective cover on a coat 涂层外的保护罩层
protective cowl 防护罩
protective cream 防护膏;保护性乳剂
protective current 防护电流
protective customs duty 保护性关税
protective deck 防护甲板
protective design 防护设计
protective device 安全装置;防护设备;保护设备
protective diffusion 保护扩散
protective door 防护门
protective duty 保护关税
protective earth 保护地
protective earthing 保护接地
protective earthing device 保护接地装置
protective earth wall 保护土堤
protective eco system 保护性质生态系统
protective effect 保护效应
protective enclosure 屏蔽室
protective equipment 防护装置;防护设备;保护装置;保护设施;保护设备
protective export duty 保护性出口税
protective facing 护面;防护面层;保护层
protective factor(of an anti-oxidant) 抗氧剂保护系数
protective fence 防护篱
protective fender 防撞装置;防护装置;保护性舷
protective film 钝化膜;防护膜;保护膜
protective filter 防护滤层;反滤层;保护性过滤器
protective finish 防护处理;表面处理;保护镀层;保护性饰面;保护涂层
protective finish mass 面层保护物质
protective finish of metals 金属面漆油漆;金属面涂层
protective fire fighter's clothing 消防员防护服
protective foil 防护金属片;保护膜
protective footwear 防护鞋
protective forest 防护林
protective forest belt 护田林带
protective fungicide 保护性杀菌剂
protective furnace atmosphere 保护锅炉介质气层;炉腔保护空气层
protective gap 保险隙;保护放电器;安全隙
protective garment 防护服(装)
protective gas 保护气体
protective gaseous envelope 气体保护层
protective gear 保护装置
protective glass 防紫外线玻璃;保护镜;安全玻璃
protective glasses 护目镜
protective glass plate 防护玻璃片;防护玻璃板
protective gloves 铅橡皮手套;防护手套
protective goggles 护目镜
protective greening-area 防护绿地
protective groin 护岸丁坝
protective ground(ing) 保护(性)接地;保护地
protective guard 护角;防护屏板
protective harbo(u)r structure 港口防护结构;港口防护建筑物
protective head-cover 防护面罩
protective helmet 安全头盔;防护钢盔
protective hood 防护罩;保护罩
protective hose 防护软管;防护水龙带
protective housing 防护箱;保护壳
protective import duty 保护性进口税
protective inhibition 保护性抑制
protective installation 防护设备
protective insulation 保护绝缘
protective interlock 保险丝联锁装置;保安互锁装置
protective isolation 隔离式防护
protective layer 钢架保护层;护膜;防护层;保护层;膜;保护层
protective layer of fine-grained concrete 细粒混凝土保护层
protective lead-glass 防护铅玻璃

protective legislation 保护立法
protective lighting 安全照明;防扩式照明;防护式照明
protective lining 保护衬里
protective lug 防护手柄
protective maintenance 预防性维修
protective mark 商品包装标志
protective mask 防护面罩
protective mask filter 防护面罩滤器
protective mask for workman 工人保护面罩
protective masking 保护挡板
protective material 防护材料;保护剂
protective measure 防护措施;防腐措施;保全措施;保护措施
protective medium 保护介质
protective membrane 保护膜
protective mimicry 保护性拟态
protective mittens 防护手套
protective mortar 防护砂浆
protective mulch 覆盖保护层
protective multiple earthing 多线保护接地;保护性多点接地
protective netting 防护网;保护网
protective neutralization 保护接零
protective output norm 保护性产量定额
protective ozone shield 臭氧防护屏;保护线臭氧屏蔽
protective package 保护性包装
protective packing 安全包装;保障包装;保护包装
protective paint 保护涂层;保护涂料;防护漆;保护漆
protective paint for radioactive contamination 防放射性污染涂料
protective paper 防护纸
protective pigment 保护颜色;保护颜料
protective pipe 防护管
protective pitching 砌石护面;堆石护坡;堆石护面
protective planting 保护性栽植;保护性栽植
protective plot 保护小区
protective practice 保护性措施
protective price 保护价格
protective provisions 保护条款
protective railing 护栏;保护用栏栅
protective rate 保护率
protective reactance 保护电抗
protective reaction 保护反应
protective reactor 防护扼流圈;保护电抗器
protective redundancy 保护冗余
protective rejector circuit 保护性抑制器电路
protective relay 保护继电器
protective relay device 继电保护装置
protective relay system 保护继电方式
protective relay test set 保护继电器测试仪
protective resistance 保护电阻
protective resistor 保护电阻器
protective revetment 护坡;护岸
protective ring 保护环;护环
protective rock blanket 块石护层
protective rubber gloves 防护橡皮手套
protective sanitary zone 卫生防护区
protective screen 保护屏蔽
protective screening 保护网
protective sealant 保护封面料
protective sealing door 防护密封门
protective sea wall 防护海墙;护岸海堤
protective sheet 防爆板
protective shelf of tree along a road 护路林
protective shield 防护屏(蔽);防护(面)罩;防护板
protective shipping policy 保护海运政策
protective signal(l)ing system 保护信号系统;保安信号系统
protective site clothing 场地保护罩
protective sleeve 吊弦保护套;保护套
protective source housing 辐射源防护罩
protective space 防护区
protective spacing 防护间距
protective spark gap 保护火花隙;安全火花隙
protective spectacles 护目镜;保护眼镜
protective spur 护岸丁坝
protective steam heating 有保护的蒸汽供热
protective stock 保险存货
protective stone wall 保护性石墙
protective storage capacity 防洪库容
protective string 技术套管
protective strip 防护条

protective structure 防护建筑物;防护结构;防护建筑(物)
protective suit 防护服(装)
protective suit cooling system 防护服冷却系统
protective surfacing mass 保护面层物质
protective survey 防护监测;防护测量
protective switch 防护式开关
protective system 保护制度;保护系统;保护贸易制度;保护贸易条例;保护关税制;安全系统
protective system of trade 保护贸易条例
protective tape 保护带
protective tariff 保护性税率;保护性关税;保护税则
protective tariff policy 保护性关税政策
protective threshold 保护始效值
protective time 防护时间
protective tissue 保护组织
protective trade 保护性贸易;保护贸易
protective trade policy 保护贸易政策
protective trade system 保护贸易制度
protective treatment 防护处理
protective trust 保护信托
protective tube 护管;保险管;保护管
protective tunnel 防护性隧道
protective turnout 防护道岔
protective uniform pant 防护制服裤
protective unit 保护装置
protective value 防护效力;保护性能
protective varnish 防护漆;防护清漆
protective varnish for construction 结构防护漆
protective wall 防护墙
protective wire 保护线
protective(working) clothing 防护工作服
protective works 保护性工程
protective zone 防护区;保护区(域);安全区
protect meter 投影比长计
protector 劳教所;防护装置;保险丝;保护装置;保护物;保护设备;保护器;保安器
protector and handrail 挡板与扶手
protector apparatus 保护设备;防护设备;保护装置;防护装置
protector base 保安器座
protector block 保安器组件
protector casing 技术套管
protector feeler 保护装置探头
protector for heading 导坑护板
protector ga(u)ge glass 水位玻璃管保护装置
protector lamp 保险灯
protector plate 护板;防护板
protector rack 保护器架
protector sleeve 保护套
protector string 技术套管
protector tube 保护管
protectory 少年感化院;流浪儿收容所;少年教养院
protectoscope 掩望镜;测望镜
protect-range 草原保护区
protect-range of grassland 草原保护区
protect relay 保护继电器
protect stratum of waterproofing 防水保护层
protect switch 盒装开关
protect the environment against pollution 保护环境
protect the sample distortion and erosion 防止岩样变形和冲蚀
proteetive gap 安全隙
protein adhesive 蛋白质黏合剂;蛋白质胶黏剂
protein-aldehyde resin 蛋白甲醛树脂
protein colloid 蛋白胶体
protein gel 蛋白凝胶
protein glue 蛋白质水胶;蛋白质动物胶
protein paint 蛋白涂料
protein resin 蛋白树脂
protein retarder 蛋白质缓凝胶
proteolite 英云红柱角岩
proterobase 次闪辉绿岩
Proterozoic basement 元古代基底
Proterozoic(era) 原生代;元古代;阿尔冈纪
Proterozoic erathem 元古界
protest 抗议(书);拒付;海难报告书
protest a bill 拒付期票
protest for nonacceptance 拒收承兑证书;拒绝承兑通知书
protest for non-payment 拒付证书;拒付申明书;拒绝支付抗议书;拒绝付款通知书
protesting party 拒付人

protest note 拒付通知书
protest of bill 票据拒付证书
protestor 拒付人
protest signal 抗议信号
protest waived 免除作成拒绝证书;免除作成拒绝书
Protex 普罗特斯(一种防潮乳剂)
protfolio effect 投资搭配效果
prothesis 古希腊教堂圣堂;修复术;圣餐室
prothetely 先成现象
prothetic 修复的
prothiofos 低毒硫磷
prothipendyl 丙硫喷地
prothoate 发呋
prothyride (砖石砌体上突挑的)牛腿;(砖石砌体上突挑的)托座
prothyron 游廊(古希腊建筑门前)
protist 原生生物
protista 原生生物界
protizinic acid 丙替嗪酸
proto-arctic 原北极
proto-Atlantic ocean 原大西洋
protoatmosphere 原始大气
proto-Baroque architecture 原始变态式建筑;原始巴洛克建筑
protobastite 顽火辉石;铁顽火辉石
protobiochemistry 原始生物化学
protobiology 嗜菌体学
protobiont 原生物
protobitumen 原石油质
protocalcite 原方解石
protoccooperation 互生
protochrome 原色素
protoclase 原生解理
protoclasis way 初级碎裂作用方式
protoclastic structure 原生碎屑构造
protoclastic texture 原生碎屑结构
protocol 协议(书);议定书;规约;外交礼节;草约;草案
protocol engineering 协议工程
protocol level 规约层次
protocol of frontier railways between two neighbo(u)ring countries 国境铁路协定
Protocol on Chorofluorocarbons 含氯氟烃议定书
protocol stack 协议栈
proto-cooperation 原始合作
protoderm 原表皮层
Protodiakonov's hardness coefficient 普氏硬度系数
proto-dolomite 原白云岩
protodoloresite 原氧钒石
proto-Doric 原始陶立克柱式;古希腊建筑之形式
proto-Doric column 原始陶立克(柱)型
proto-Doric order 陶立克柱式的原型
Protodyakonov's coefficient of rock strength 普氏岩石坚固性系数
Protodyakonov's classification of rock 岩石坚固性分类
Protodyakonov's number 普氏系数;普罗托吉雅可诺夫数
protoenstatite (人造的)原顽火辉石
prot of discharge 起货港
protofluorine 原氟
protogene gneiss 原生片麻岩
protogenesis 原生
protogenetic 原生的
protogenic rock 原生岩
protogenic yield 给质子溶剂
protogeosyncline 原地槽
protogeosyncline stage 原地槽阶段
protogine 原生岩
protograph 原模图
protohistoric age 原史时代
protohistory 原史学;史前时期
proto-Ionic capital 原始爱奥尼克柱头
protolignin 原木素;天然木素
protolith ages 原岩年龄
protolithionite 黑磷云母
protolog 原记述
protomagma 原始岩浆
protomai capital 半兽饰柱头;四角兽像柱头
protomatrix 原杂基
protome (古典建筑中半凸出的)人兽雕像装饰
protomer 原体
protomere 微胶粒

protometamorphism 初级变质作用
protomontronite 原蒙脱石
protomylonite 原生糜棱岩
proton 质子
proton absorption capacity 质子吸收能力
proton acceptor 质子接受体
proton acceptor solvent 质子受体溶剂;受质子溶剂
proton activation 质子活化(法)
proton activation analysis 质子活化分析
proton activity 质子活度
protonatomic mass 质子质量
proton balance 质子平衡
proton donor solute 给质子溶质
proton donor solvent 质子给体溶剂
proton-electron spectrometer 质子中子分光仪;质子电子谱仪;质子电子分光仪
proton excited X-ray spectrometry 质子激发X射线光谱法
proton free precession magnetometer 质子自由旋进磁力仪
proton hydrate 水合质子
protonic acid 质子性酸
protonic medium 质子性介质
protonic solvent 质子性溶剂
proton induced-X-ray emission fluorescent analysis 质子激发X射线荧光分析
proton magnetic resonance spectroscopy 质子磁共振光谱法
proton magnetometer 质子地磁仪;质子磁强计
proton mass 质子质量
proton microscope 质子显微镜
proton-neutral species 中性质子物种
proton number 质子序数
protonogram 质子衍射图
proton precession magnetometer 质子旋进磁力仪
proton probe microanalyzer 质子探针显微分析
proton pump 质子泵
proton-recoil counter 反冲质子计数器
proton-recoil detector 反冲质子探测器
proton-recoil ionization chamber 反冲质子电离室
proton-recoil method 反冲质子法
proton-recoil telescope 反冲质子望远镜
proton-stability constant 质子稳定常数
proton telescope 质子望远镜
proton theory of acid 酸的质子理论
proton transfer 质子传递
proton uptake capacity 质子吸收能力
protonym 原用名
proto-ocean 原始大洋
proto-Pangaea 始泛大陆【地】
protoparaffin 原石蜡
protopepsia 初步消化
protopetroleum 原油
protophase 前期
protophile 质子亲和物
protophobic 疏质子的
protophyte 原生植物
protoplasm 原生质;首次模型
protoplatform stage 原地台阶段
proto-porcelain 原始瓷器
protoprism 原棱镜
protopyramid 初级棱锥体
proto-Qinling old land 原秦岭古陆
protoquartzite 原石英岩
protore 无开采价值的脉石
proto-Renaissance architecture 原始文艺复兴时期建筑
protosalt 低价金属盐
protospecies 原种
protosun 原太阳
prototpye container 原型集装箱
prototpye data 原型试验数据
prototpye measurement 原型观测;原型测量;原体观测;原体测量
prototpye observation 原型观测
prototroph 原营养型微生物;原养生物;矿质寄生物
prototrophic 矿质寄生的
prototrophic bacteria 原养型细菌;原始营养细菌
prototype 原型样机;原型(的);原体;原生型;样板;足尺模型;主型;模型原模;模型(机);试验模型;设计原型;标准
prototype aircraft 原始飞机
prototype bit 样品钻头
prototype board 样机板;样板机;模型板;模件板;

模板件
prototype bridge 原型桥
prototype bridge model 原桥模型
prototype carbon fund 原型基金
prototype casting 试生产铸件
prototype core 原型堆芯
prototype cost limits 建设项目的最高贷款额;原型成本限额
prototype data 原型资料;原型数据;原体试验数据
prototype debug 样机调试
prototype development 原型机研制
prototype development phase 原型机研制阶段
prototype development system 样机开发系统;模块开发系统
prototype development system module 样机开发系统模块
prototype drawing 原型图
prototype drill 样品钻机
prototype environment(al) buoy 原型环境浮标
prototype equipment 原型设备;定型设备
prototype experiment 原型试验
prototype fast reactor 原型快速反应堆
prototype filter 原型滤波器;模型过滤器
prototype foundation test 原型基础试验
prototype gear 测量齿轮
prototype gradation 天然级配
prototype grating 母光栅
prototype hardware 原型样品
prototype information 原型资料
prototype installation 原型试验装置
prototype instrument 样机
prototype liquid 原型液体
prototype logiic(al) operation 样机逻辑操作
prototype machine 原型电机
prototype macro-model 原始宏观模型
prototype macrostatement 模型宏语句
prototype meter 标准米原尺
prototype micro model 原始微观模型
prototype model 原始模型
prototype network 模型网络
prototype pattern 原型模式
prototype plant 原型装置;样机试验工厂;样机工厂
prototype problem 典型问题
prototype project 样板工程
prototype reactor 原型反应堆;模式反应堆
prototype Reynolds number 真机雷诺数
prototype room 样板间
prototype scour 典型冲刷
prototype software 原型软件
prototype stage 原型阶段
prototype standard 标准原器
prototype structure 原型结构;足尺结构
prototype system 原型系统
prototype test(ing) 原型试验;原体试验;实船试验;抽样试验;样机试验
prototype transport 原型运输机
prototype treatment 原型处理
prototype turbine 原型水轮机
prototype unit 原型单元
prototype variable 原型变量
prototype vehicle 原型车
prototype vehicle for test purposes 试验用车辆样车
prototype verification 原型验证
prototype workpiece 样件
prototypical ligand 原型配位体
prototypic(al) layout 足尺放样
prototyping 原型机制造
prototyping card 样机插件
prototyping card system 样机插件系统
prototyping component 样机部件
prototyping kit 配套样机部件
prototyping operating system 原型操作系统;样机操作系统
prototyping system 样机系统
prototyping testing 样机测试
protour chart 高空预报图
protovermiculite 原蛭石
protoxide 氧化亚物;低(价)氧化物
Protozoa 原生动物
protozoan insecticide 原生动物杀虫剂
protphilic solvent 亲质子溶剂
protract 突出;延长;用比例尺制图;绘平面图;前伸

protracted 拖长的
protracted irradiation 迁延照射
protracted labo(u)r 滞产
protracted stagnation 长期徘徊不前
protracted test 疲劳试验;持续试验
protracted test machine 疲劳试验机
protractile 可外伸的;可伸出
protracting map 原图
protraction 制图;突出;延长;伸长
protraction dose 迁延剂量
protractor 角规;量角器;钳取器;测角器;分度器;分度规;半圆规
protractor head 量角器头
protractor scale 分角器标度;分角器标度
protractor screen 测角投影屏
protractor set square 分度三角板
protractor tool guide 磨刀斜角样板;车刀定角导板
protruded packing 多孔填料;冲压填料
protruding bar 伸出的钢筋;伸出的钢杆;伸出的钢档
protruding point 突嘴
protrusion 装货突出部分(集装箱);固态侵入体;挤进;突起;突出物;凸出;伸出
protrusion allowance 凸出容差
protrusion of surface 表面突起
protrusive 固态侵入的
protrusive balance 前伸平衡
protrusive movement 前伸移动
protrusive occlusion 前伸
protuberance 节疤;突起;凸起形
protuberance type hob 凸角式剃前滚刀
protuberance type shaper cutter 凸角式剃前插齿刀
protyle 不可分原质
proud 凸出部
proud exposure 大量出刃
proudite 硒硫铋铅铜矿
proud seam 合缝凸齿
proustite 硫砷银矿;淡红银矿
proved acreage 证明有石油的地面积
proved area 探明面积
proved carcinogen 确认致癌原;确认致癌物
proved developed reserves 已开发探明储量
proved field 已勘探矿区
proved ore 肯定储量
proved point 标定点
proved recipe 验方
proved reserves 证明有工程价值的储量;已探明藏量;证实储量;肯定储量
proved undeveloped reserves 未开发探明储量
provender mill 饲料加工厂
proven electrostatic moisture meter 静电探测水分仪;静电含水量计量表
proven fill reserves 探明的填料储量
proven life 可靠使用年限
proven reliability 经证实可靠性
proven reserves 高级储量;探明储量
proven software 成熟软件
proven technology 已证明的技术
proven territory 已探明地区
prover 校准仪;校验仪(器);校验仪
prover tank 校准罐
prove the subsidiary ledger 核对明细分类账
provide 规定;配备;提供
provide amenities for the people 方便生活
provided by the owner 物主提供的;业主提供的
provided groove 预留槽
provided hole 预留孔
provided school 地方公立学校
provide financing services 举办融资业务;融资业务
provide in the margin 图廓整饰样图
provident fund 准备基金
provide people with door-to-door service 方便生活
provider 供应者;供应商;提供者
provide shelter from the wind 防风
provide the ticks 绘晕线
provide tilth to a soil 改善土壤耕性
providing a complete set of services 提供成套服务
providing detailed soil map 提供土壤详图
providing fundamental basis for 提供理论基础
providing labor service 提供劳务
providing land planning 提供土地规划
providing soil investigation conditions 提供土壤调查情况
providing tilth to a soil 提供土壤耕性
province (bore) hole 构造钻孔;前探钻孔
province (drill) hole 构造钻孔
province of activities 业务范围;活动范围
provincial administration building 州行政(管理)办公大楼;省行政(管理)办公大楼
provincial alternation 沉积岩交互出现区
provincial boundary 省界
provincial capital 省会
provincial characteristic 区域特征
provincial exhibition 州展览会;省展览会
provincial government 省政府
provincial government building 省政府大楼
provincial highway 区域性公路;省道
provincial highway commission 省公路局
provincial highway system 省公路系统;省道网
provincial hospital 省医院
provincially administered municipality 省辖市
provincial map 省地图
provincial planning 州的规划;地方规划;省的规划
provincial road 省道
provincial standard 地方标准
provincial standardization 地方标准化
provincial trunk highway 省干线公路(省道)
proving 校验
proving circuit 校验线路
proving flowmeter 标定流量计
proving frame 应力(环)架;校正架;试验架;测力环试验架
proving ga(u)ge 校验计
proving ground 考验场;汽车性能检验场;检验场;试验基地;试验场;射击场
proving-ground facility 试验场设施
proving ground test 试验场试验;试验性试验;实车试验场试验
proving hole 探孔
proving instrument 校验仪器
proving lever 校验杆;测力杆
proving oven 发面炉
proving point 标定点
proving press 手动小打样机;打样机
proving program(me) 验证程序
proving range 实验靶场
proving ring 应力环;量力杯;校验环;检验环;试验环;测应(力)环
proving road 试车道
proving stand 试验架
proving tank 校准罐;标定罐
proving test 校核试验;复核试验;验证性试验
proving the subsidiary ledger 明细分类账核对
proving time 验证时间
proving trial 验证性试验;复核试验
provision 对工程提供方向性的估算;准备;规定;供给;条款
provisional 暂时(性)的;暂定;临时(的)
provisional acceptance 临时验收
provisional acceptance certificate 临时验收证书
provisional account 临时账户
provisional agreement 临时协议;临时协定
provisional budget 临时预算
provisional certificate 临时证书
provisional certificate of class 临时船级证书
provisional chart 应急图
provisional classification certificate 临时船级证书
provisional cofferdam 临时围堰
provisional compilation 监到拼贴
provisional contract 临时契约;临时合同;草约
provisional coordinate 假用坐标;临时坐标
provisional coordinate system 假定坐标系(统)
provisional cost 暂定成本
provisional dike 临时围堰
provisional edition 临时版
provisional estimate 估算;概算;暂时估算;暂时估价;临时估计
provisional evaluation 暂时估算;暂时估价;暂时估计
provisional hazard area 临时危险区
provisinal index 预定指数
provisional invoice 临时发票
provisional issue 临时版
provisional item 暂定项目;临时项目
provisionality 临时性
provisional license 临时许可证
provisional light 临时灯标
provisional lining 临时衬砌
provisional list of tenders 临时标单
provisional loan 临时贷款
provisional map 临时版地图
provisional marker 临时标志
provisional measure 临时措施
provisional method 暂定方法
provisional note 临时保险单
provisional payment 临时付款
provisional point 临时点
provisional pratique 临时检疫证书
provisional regulations 暂行条例
provisional regulations for exchange control 外汇管理暂行条例
Provisional Regulations for Reporting Accidents of Environmental Pollution and Damage 报告环境污染与破坏事故的暂行办法
Provisional Regulations on Administration of Energy Saving 节约能源隔离暂行条例
Provisional Regulations on Environmental Administration in Open Areas 对外经济开发区环境管理暂行规定
provisional release 临时许可
provisional report 临时报告
Provisional Rules on Administration of Civil Airports 民用机场管理暂行规定
provisional shiplock 临时船闸
provisional sum 预备费;准备费;暂定金额;临时金额;非计划工程补偿费;不可预见费(用)
provisional survey 初测
provisional treaty 临时条约
provisional value 初值
provisional variation 暂时的变化
provisional weekly tolerance intake 暂行一星期容许摄取量
provisional weight 暂定权;临时荷载;临时负荷
provisional works 临时工程
provision as to payment 付款规定
provision fee for the growth of project costs 工程造价增长预留费
provision for bad debt 坏账准备
provision for contingency 不可预见费
provision for depreciation 预提折旧;折旧准备
provision for direct scribing 直接刻图设备
provision for income tax 备付所得税款
provision for leakage of control 防渗措施
provision for plotting along the edges of adjacent map sheets simultaneously 地图抄接边设备
provision for repairs 修理准备
provision for taxes 备付税款
provision hoy 运粮大驳船
provision ice chamber 冷藏食品室
provisionnement 粮食供应
provision of barrages and locks in a river 河流拦河坝和水闸设备
provision of employment 提供职业
provision of facility 设施的补充;工具的储备
provision of ford 渡口设备
provision of gas 供应煤气
provision of heat 热的补充;热的供给
provision of labo(u)r 提供劳务
provision of parking spaces 可供停车车位数
provision of water 水的补充;水的供给
provision room 食品储[贮]藏室;食品室
provisions already agreed upon 商定的条款
provisions of contract 合同条款
provisions room 粮食储[贮]藏室;食品室
provisions shop 食品商店
provision store 食品库
provision sum 暂定总金额;备用金
proviso 限制性条款;限制条款;但书;附文;附带条款;附带条件
proviso clause 保留条款
provisory 附带条件
provisory clause 附带条款;保留条款
pro-vitrain 结构镜煤;类镜煤
provitamin 维生素原
provitrinite 类镜质组亚组
provitrite 微类镜煤
provocation 诱发
provocative 激发的
provocative dose 激发剂量
provocative reaction 激发反应

provocative test 诱发试验;激发试验
provoke 挑拨
prow 船头;船首水面部分;防卫设施头部
prowersite 橄辉云煌岩
prowl car 警备车
prow slope 船头波
proximal 近端的
proximal cause 近因
proximal end 中枢端;近端
proximal environment 近源环境
proximal face 邻面;近极面
proximal fan deposit 扇近端沉积
proximal phase 近物源相
proximal scanning system 近似扫描系统
proximal sedimentary deposit 近源沉积
proximal sensing 近感
proximal storm deposit 近端风暴沉积
proximal surface 邻面
proximal turbidite 近源浊积岩
proximatal turbidite 近端浊积岩
proximate 紧接的
proximate analysis 显性分析;组分分析;近似分析;实用分析
proximate analysis of coal 煤工业分析
proximate calculation 近似计算
proximate carcinogen 近致癌物
proximate coal analysis 煤的工业分析
proximate component 近似成分
proximate composition 近似组成
proximate grade 次级
proximity 邻近;接近
proximity clothing 隔热服
proximity detector 临近探测器
proximity effect 邻近效应;近场效应
proximity exploder(fuse)近爆信管;近炸雷管;接近引炸雷管
proximity fire fighting 近火灭火战斗
proximity focus 近贴聚焦
proximity fuse 近炸信管
proximity gloves 隔热手套
proximity helmet 隔热头盔
proximity log 邻近侧向测井
proximity log curve 邻近侧向测井曲线
proximity meter 邻近探测仪;接近探测仪
proximity space 邻近空间
proximity suit 隔热服
proximity switch 引发开关;接近开关
proximity theorem 邻近定律
proximity transducer 接近传感器
proximity warning indicator 防撞报警指示器;防撞报警指示器;避撞报警显示器
proximity warning laser radar 防撞报警激光雷达
proximity warning radar 近程警戒雷达
proxy 委托书;代理(人)
proxy attribute 间接属性;代理(人)品质
proxy data 代用资料
proxy fight 委托竞争
proxy measure 间接测定
proxy server 代理服务器
proxy statement 委托书
proxy system 委托书制度
prozone 前区
PR-thermocouple 铂铑热电偶
prudential committee 咨询委员会
prudential policy 稳妥的政策
prudent investment 审慎(的)投资
prudent limit of endurance 航时可靠限度
pruina 粉霜
prune drier 黑李干燥机
prune products 梅制品
pruner 剪枝剪
pruning 修剪;整形修剪;剪裁;天然整枝;删改
pruning hook 剪枝镰
pruning implement 修枝工具;修剪工具
pruning knife 修枝刀
pruning peaches 修剪桃树
pruning range 修剪高度
pruning saw 修剪锯;截锯;修剪锯
pruning scissors 修枝剪
pruning trees 修剪树木
prunning shears 剪枝刀;剪枝剪;修剪剪
prunt decoration 粘花装饰
Prunus persica 桃树
Prussian blue 贡蓝色(普鲁士蓝);青光铁蓝;普鲁士蓝;深蓝色;柏林蓝;亚铁氰化铁
Prussian blue colo(u)r 普鲁士蓝
Prussian blue pigment 普鲁士蓝颜料
Prussian(cap) vault 普鲁士蓝拱顶
prussiate 氰化物
prussiate of potash 亚铁氰化钾
prussic salt 氰铁酸盐
pry 杠杆;杠秤;撬具;撬动;起货钩
pryan 白色细脆黏土;白色软黏土
pry bar 拔钉器;撬杠;撬棍;撬杆
prying force 杠杆力
prying tool 撬具
pry-out 撬出
prytaneion (古希腊城市的)公共礼仪大厦;宾馆;(古希腊的)市政厅
prytaneum (古希腊城镇官员的)接待公民大厅;公共建筑(物)
przhevalskite 铅铀云母
psamathium 海滨砂生群落
psammite 砂质岩;砂屑(砂)岩
psammitic schists 砂屑片岩
psammitic texture 砂状结构;砂屑结构
psammobont 沙栖生物
psammo-littoral zone 砂质海滨地带
psammoma bodies 沙粒小体
psammon 沙栖生物
psammonsere 沙生演替系列
psammophile 喜砂植物;适砂植物
psammophilous plant 好砂植物
psammophyte 沙生植物
psammyte 砂质岩
psamnite-gneiss 砂屑片麻岩
pscudoleucite 假白榴石
psephicity 磨圆度;磨圆的
psephite 砾质岩;碎屑质岩;碎砾岩
psephitic 砾状的
psephitic structure 砾状构造
psephitic texture 砾状结构
Psephyrus gladius 白鲟
psephyte 砾质岩;碎砾岩
P-set of cycle 回路 P 集
pseudisodomum 块石砌墙法
pseudo-aborption 闪突起
pseudo-absolute velocity section 虚绝对速度剖面
pseudo-acceleration 拟加速度
pseudo-acceleration spectrum 拟加速度谱
pseudo-accidental error 伪偶然误差
pseudo-acid 假酸
pseudo-acid form 假酸式
pseudo-acidity 假酸性
pseudo-activation energy 视活化能
pseudo-adiabat 假绝热线
pseudo-adiabatic atmosphere 假绝热大气
pseudo-adiabatic chart 假绝热图
pseudo-adiabatic diagram 假绝热图
pseudo-adiabatic expansion 假绝热膨胀
pseudo-adiabatic process 假绝热过程
pseudo-adsorption 假吸附(作用)
pseudo-agglutination 假凝聚;假凝集
pseudo-albite 中长石
pseudo-allelism 拟等位性
pseudo-alloy 假合金
pseudo-anisotropy 假各向异性
pseudo-anomaly 虚异常
pseudo-anticline 假背斜
pseudo-artesian groundwater 假自流地下水
pseudo-asymmetry 假不对称
pseudo-asynchronous 伪异步
pseudo-autunite 假钙铀云母
pseudo-azimuthal projection 伪方位投影
pseudo-basicity 假碱性;假碱度
pseudo-basilica(church)假的早期基督教教堂建筑
pseudo-bed 假层
pseudo-bedded 假层状的
pseudo-bedding 假层理
pseudo-bercciated texture 假角砾结构
pseudo-binary chalcogenide glass 赝二元硫族化合物玻璃
pseudo-binary diagram 伪二元图
pseudo-boiling 似沸腾
pseudo-bolite 水氯铜铅矿
pseudo-bomb 假火山弹
pseudo-box deck 假箱式桥面
pseudo-breccia 假角砾岩
pseudo-Brewster angle 赝布儒斯特角
pseudo-brookite 假板钛矿;铁板钛矿
pseudo-bulk compressibility 视总体压缩系数
pseudo-cannel coal 假烛煤
pseudo-capsule 假包膜
pseudo-carburizing 假渗碳
pseudo-cast 假管型
pseudo-cationic polymerization 假正离子聚合;假阳离子聚合
pseudo-channel 伪通道
pseudo-chrysolite 莫尔道熔融石
pseudo-circulation 假回流
pseudo-classical 假标准的;假传统的;假经典的
pseudo-classic architecture 仿古(典)建筑
pseudo-classicism 拟古典主义;伪古典主义
pseudo-classic style 仿古式
pseudo-clathrate texture 假网格状结构
pseudo-coarctation 假缩窄
pseudo-code 伪码
pseudo-cohesion 准黏聚力
pseudo-cold front 假锋【气】
pseudo-colloidal 假胶态的
pseudo-colony 假菌落
pseudo-colo(u)r composition 假彩色合成银片组合法
pseudo-colo(u)r composition with dyeing 假彩色合成染印法
pseudo-colo(u)r composition with separate exposing 假彩色合成分层曝光法
pseudo-colo(u)r density encoding through phase modulation 位相调制密度假彩色编码
pseudo-colo(u)r display 伪彩色显示
pseudo-colo(u)r image 假彩色片
pseudo-colo(u)r satellite image graph 假彩色卫星照片图
pseudo-component 假组分
pseudo-compound 假化合物
pseudo-concave 伪凹
pseudo-concretion 假结核
pseudo-conditioning 假条件作用
pseudo-cone 伪晶体
pseudo-conformity 假整合;准整合
pseudo-conglomerate 假砾岩
pseudo-conglomeratic texture 假圆砾结构
pseudo-conic(al) projection 伪圆锥投影
pseudo-consolidated pressure 准固结压力【岩】
pseudo-consolidation pressure 准固结压力
pseudo-continuous zone refining 假连续区熔提纯
pseudo-convex 伪凸
pseudo-convexity 伪凸性
pseudo-corposclerotinite 假团块菌质体
pseudo-cotunnite 钾氯铅矿
pseudo-coupled bending-torsion vibration 伪弯曲—扭转耦联振动
pseudo-critical 假临界的
pseudo-critical constant 假临界常数
pseudo-critical point 假临界点
pseudo-critical pressure 假临界压力
pseudo-critical property 伪临界性质
pseudo-critical temperature 假临界温度
pseudo-crocidolite 虎睛石
pseudo-cross-bedding 假交错层理
pseudo-crossing 假交叉
pseudo-cruciform 假十字纹的;假十字形的
pseudo-crystal 假晶;赝晶(体)
pseudo-crystalline 伪晶的
pseudo-crystalline material 赝晶体物质
pseudo-crystalline state 假结晶态
pseudocursor 伪光标
pseudo-cycle 伪循环
pseudo-cycloid 伪旋轮线
pseudo-cycloidal curve 伪旋轮类曲线
pseudo-cylinder 伪圆柱体
pseudo-cylindric(al) equivalent projection 等面积伪圆柱投影
pseudo-cylindric(al) projection 伪圆柱投影;拟圆柱投影法
pseudo-cylindric(al) arbitrary projection 任意伪圆柱投影
pseudo-cyst 假孢囊;伪胞
pseudo-decimal digit 伪十进制数字
pseudo-deltidium 假柄孔盖
pseudo-demand schedule 拟需求表
pseudo-dielectric 假电介质

pseudo-differential operator 伪微分算子
pseudo-diffusion 伪扩散
pseudo-dipole 赝偶极子
pseudo-dipteral 仿双柱廊式;仿双廊式
pseudo-dipteral building 仿双排柱廊式建筑;仿双廊式建筑
pseudo-dipteral temple 仿双廊式庙宇
pseudo-dipteros 仿双排柱廊式建筑物;(古希腊神庙的)仿双重周柱式
pseudo-directed family of points 伪有向点族
pseudo-directed set 伪有向集
pseudo-distance 伪距离
pseudo-dont 假齿
pseudo-ductility 假延性
pseudo-dynamic(al) method 拟动力法
pseudo-dynamic(al) test 拟动力试验
pseudo-eclogite 假榴辉岩
pseudo-elastic approximation 拟弹性近似
pseudo-ellipse 三心椭圆(仿椭圆)
pseudo-elliptic(al) integral 伪椭圆积分
pseudo-entity 虚拟实体
pseudo-equalizing pulses 伪均衡脉冲
pseudo-equilibrium 假平衡
pseudo-Euclidean metric 伪欧几里得度量
pseudo-Euclidean space 伪欧几里得空间
pseudo-eutectic 伪共晶
pseudo-fibre space 纤维空间
pseudo-field 赝场
pseudo-field-sync pulses 伪场同步脉冲
pseudo-flat 赝平面
pseudo-force 伪力
pseudo-form texture 假像结构
pseudo-four-tensor 伪四维张量
pseudo-free ideal ring 伪自由理想环
pseudo-frequency 伪频率
pseudo-front 假锋【气】;伪锋【气】
pseudo-function 伪函数
pseudo-galena 闪锌矿
pseudo-gamy 假配合
pseudo-geometric net 伪几何网
pseudo-geometric ring 伪几何环
pseudo-gley soil 伪潜育土
pseudo-globular texture 伪球状结构
pseudo-Gothic style 仿哥特式
pseudo-gravity chart 假重力异常图
pseudo-group of transformations 伪变换群
pseudo-guarantee 假担保
pseudo-hardness 假硬度
pseudo-header 半砖;假丁头砌块;假丁砖
pseudo-hexagonal crystal 假六方晶体
pseudo-holographic technique 赝全息术
pseudo-holography 赝全息术
pseudo-homogeneous view 假均相见解
pseudo-host 伪主机
pseudo-hot isostatic pressure 准热等静压
pseudo-hydrologic sequence 伪水文序列;伪水文系列
pseudo-hyperbolic distance 伪双曲距离
pseudo-hyperelliptic(al) integral 伪超椭圆积分
pseudo-hypha 假菌丝
pseudo-impedance section 虚波阻抗剖面
pseudo-instruction 拟指令;伪指令【计】
pseudo-instruction form 伪指令形式
pseudo-inversion matrix 伪逆矩阵
pseudo-isochromatic diagram 伪等色图
pseudo-isochromatic plate 假同色板
pseudo-isochromatic test 伪同色试验
pseudo-isochromat plate 假等色板
pseudo-isochromat plate test 假等色板试验
pseudo-isochron 假等时线
pseudo-isoline map 伪等值线地图
pseudo-isomeric crystal 假等轴晶体
pseudo-job scheduling 伪作业调度
pseudo-karst 假岩溶;假喀斯特
pseudo-lateral gradient log 接地电阻梯度测井
pseudo-lattice 伪格
pseudo-laueite 伪劳埃石
pseudo-length 伪长度
pseudo-lens 幻视透镜
pseudo-leucite microsyenite 伪白榴石微晶正长岩
pseudo-leucite syenite 假白榴石正长岩
pseudo-linear 伪线性
pseudo-line-sync pulses 伪行同步脉冲
pseudo-linkage 拟连锁

pseudo-liquid density 拟液态密度
pseudo-load method 假载法
pseudo-lock 赝锁相
pseudo-logging well technique 拟测井技术
pseudo-loop 假环
pseudo-luxation 假脱位
pseudo-malachite 假孔雀石
pseudo-matrix 假杂基
pseudo-matrix isolation 伪矩阵隔离
pseudo-membrane 假膜
pseudo-membranous conjunctivitis 暴风客热
pseudo-merism 假节构造
pseudo-metamerism 假分节
pseudo-metric 半度量;伪度量
pseudo-metric space 伪度量空间
pseudo-metric uniformity 伪度量一致性
pseudo-mica schist 假云母片岩
pseudo-micrite 假微晶
pseudo-microsparite 伪微亮晶
pseudo-monas 假单胞菌
pseudo-monochromator 赝单色仪
pseudo-monotropy 假单变性
pseudo-moraine 假碛
pseudo-morph 假同晶
pseudo-morphic crystal 假晶
pseudo-morphic image 假像
pseudo-morphism 假像;赝像
pseudo-morphosis 假同品(现象)
pseudo-morphy 假同晶
pseudo-morphy by chemical alteration 化学蚀变假象
pseudo-motion 伪运动
pseudo-mountain 分割高原山
pseudo-mycelium 假菌丝体
pseudo-mycorrhiza 假菌根
pseudo-net graph 伪网图
pseudo-noise 伪噪声;伪随机噪声
pseudo-normal 伪法线
pseudo-object-language 伪对象语言
pseudo-off-line input-output 伪脱机输入输出
pseudo-offline working 伪脱机工作
pseudo-olite texture 假鲕粒结构
pseudo-ooid 假鲕
pseudo-oolitic chert 假鲕状燧石
pseudo-operation 伪运算;伪操作;虚拟操作
pseudo-operation code 伪操作码
pseudo-optic(al) band 赝光波段(毫米波);准光学波段
pseudo-order 伪指令【计】
pseudo-order form 伪指令形式
pseudo-parallel lines 伪平行线
pseudo-parallel plane 伪平行面
pseudo-pearlite 伪珠光体
pseudo-peridium 拟包被
pseudo-periodic(al) arc 伪周期弧
pseudo-periodic(al) function 假周期函数;伪周期函数
pseudo-periodic(al) problem 伪周期问题
pseudo-periodicity 伪周期性
pseudo-peripteral 仿单(柱)廊式
pseudo-peripteral building 仿柱廊式建筑;仿单廊式建筑
pseudo-peripteral temple 仿单廊式庙宇
pseudo-peripteros 仿周柱式(古希腊神庙)
pseudo-phase 伪相
pseudo-phase-conjugate 赝相位共轭
pseudo-phenocryst 假斑晶
pseudo-phite 似蛇纹石
pseudo-phlobaphinite 假鞣质体
pseudo-pin hole 假针孔
pseudo-pisenmatic character 拟态性
pseudo-plane 赝平面
pseudo-plankton 假浮游生物
pseudo-plastic 假塑性体
pseudo-plastic behaviou(u)r 假塑性行为
pseudo-plastic consistency 假塑性稠度
pseudo-plastic deformation 假塑性变形
pseudo-plastic flow 假塑性流动
pseudo-plastic fluid 准塑性流体;假塑液(体);假塑性流体
pseudo-plasticity 假塑性
pseudo-polygonal 伪多边形的
pseudo-porphyritic 假斑状的
pseudo-porphyritic texture 假斑状结构

pseudo-porphyroblastic structure 假斑状变晶构造
pseudo-potential 假势能
pseudo-potential temperature 假位温度
pseudo-preconsolidation pressure 似前(先)期固结压力
pseudo-prepreg material 准预浸料
pseudo-primeval condition 模拟原始条件
pseudo-principal solution 伪主解
pseudo-productivity of gas 视煤气发生率
pseudo-program(me) 伪程序
pseudo-prostyle 半壁柱列柱式;假列柱式;假廊式;仿柱廊式
pseudo-prostyle building 仿柱廊式建筑
pseudo-Raman technique 赝拉曼技术
pseudo-random 伪随机的;虚拟随机;赝随机
pseudo-random binary sequence 伪随机二进制序列
pseudo-random binding signal 拟随机二值信号
pseudo-random code 伪随机码
pseudo-random code stream 伪随机码流
pseudo-random data 伪随机数据
pseudo-random digital signal generator 伪随机数字信号发生器
pseudo-random impulse response technique 伪随机脉冲响应技术
pseudo-random noise 准随机噪声
pseudo-random noise generator 伪随机噪声发生器
pseudo-random noise jammer 伪随机杂波干扰机
pseudo-random noise signal 伪随机噪声信号
pseudo-random number 准随机数;拟随机数目;拟随机数量;伪随机数
pseudo-random number sequence 准随机数序列;拟随机数序列;伪随机数序列
pseudo-random perturbation 伪随机扰动
pseudo-random phase reversal 伪随机反相
pseudo-random process 拟随机过程
pseudo-random pulse coded source 伪随机脉冲编码震源
pseudo-random radar 伪随机调制雷达
pseudo-random sequence 伪随机序列
pseudo-random sequence of numbers 准随机序列;拟随机数序列;伪随机数序列
pseudo-random system 伪随机系统
pseudo-range 伪距离
pseudo-range correction 伪距校正
pseudo-range measurement 伪距测量
pseudo-rate constant 拟速率常数
pseudo-ray tracing 赝光线追迹
pseudo-recurrence 伪递推
pseudo-reduced property 伪折减性质
pseudo-reflection coefficient section 虚反射速度剖面
pseudo-register 伪寄存器
pseudo-regular function 伪正则函数
pseudo-relative velocity section 虚相对速度剖面
pseudo-relative velocity spectrum 拟相对速度谱
pseudo-resolvent 伪预解式
pseudo-resonance 拟共振;伪共振
pseudo-rhyolitic structure 假流纹构造
pseudo-Riemannian bundle 伪黎曼丛
pseudo-Riemannian manifold 伪黎曼流形
pseudo-Riemannian metric 伪黎曼度量
pseudo-Riemannian vector bundle 伪黎曼向量丛
pseudo-ripple mark 假波痕
pseudo-rutole 假金红石
pseudo-salt 假盐
pseudo-scalar field 伪纯量场
pseudo-scalar(quantity) 假标量;伪标量;赝标量;伪纯量
pseudo-science 伪科学
pseudo-sclerote 假菌类体
pseudo-scope 幻视镜;反影镜
pseudo-scopic effect 伪立体效果
pseudo-scopic holographic(al) stereomodel 幻视全息粒体模型
pseudo-scopic image 反视立体像;假立体像;赝像
pseudo-scopic imagery 假立体象
pseudo-scopic stereo 反立体;假立体
pseudo-scopic stereopair 反立体象对
pseudo-scopic view 反立体图
pseudo-scopic viewing 反立体观察
pseudo-scopic vision 反立体视觉
pseudo-scopy 幻视术

pseudo-scroll 假旋涡形装饰
pseudo-scroll sexpartite 假旋涡形六部分(一种建筑形式)
pseudo-scroll vault 假旋涡形拱顶
pseudo-secondary inclusion 假次生包裹体
pseudo-second order kinetic model 准二级动力动力学模型
pseudo-second order reaction kinetic model 准二级反应动力学模型
pseudo-second order reaction kinetics 准一级反应动力学;准二级反应动力学
pseudo-selectivity 准选择性
pseudo-shear 似剪切
pseudo-shock 伪冲激波
pseudo-side 伪边
pseudo-single crystal 伪单晶
pseudo-single eye 伪单眼
pseudo-skew symmetric 伪斜对称的
pseudo-solid body 假固体
pseudo-solution 假溶液;胶体溶液
pseudo-spar 假亮晶
pseudo-spar fabric 假亮晶组构
pseudo-sparite 假亮晶灰岩
pseudo-sphere 假球面
pseudo-spheric(al) helicoid 伪球形螺旋面
pseudo-static approach 准静力法
pseudo-static displacement 拟静力位移;伪静位移
pseudo-static earthquake analysis 拟静力地震分析
pseudo-static failure mechanism 拟静力破坏机制
pseudo-static force 拟静力;伪静力
pseudo-static influence coefficient 拟静力影响系数
pseudo-static influence factor 拟静力影响系数
pseudo-static method 拟静力法
pseudo-static process 准静态过程;准静力过程;拟静态过程;拟静力过程
pseudo-static spontaneous potential 假静自然电位;拟静自然电位
pseudo-static test 拟静力试验;伪静力试验
pseudo-static transmission 拟静力传递
pseudo-stationary 假稳
pseudo-stationary flow 假稳定流;假恒定流;拟稳定流
pseudo-steady hypothesis 假稳态假定
pseudo-steady state 假稳态
pseudo-steady-state pressure distribution 拟稳定状态压力分布
pseudo-steel 假钢;烧结钢
pseudo-stereoscopy 反立体
pseudo-stoichiometry 假化学计算法
pseudo-stratification 假层理
pseudo-structure 伪结构
pseudo-style 假柱状建筑物
pseudo-sulum 假出水孔
pseudo-supernormal conduction 伪超常传导
pseudo-symbol 伪符号
pseudo-symmetry 假对称;赝对称;伪对称
pseudo-syncline 伪向斜
pseudo-tachylite 假玄武玻璃
pseudo-tangent line 伪切线
pseudo-tangent plane 伪切面
pseudo-tensor 赝张量;伪张量
pseudo-tensorial form 伪张量形式
pseudo-termination 假终止
pseudo-ternary 伪三进制
pseudo-ternary signal 伪三进制信号
pseudo-terrace 假阶地
pseudo-tetrad 伪四位二进制
pseudo-thickness 假厚度
pseudo-three-dimensional computation 准三维计算
pseudo-three-dimensional consolidation theory 准三向固结理论
pseudo-three-dimensional solution 准三元解
pseudo-thrill 假震颤
pseudo-titration 假滴定
pseudo-topaz 黄石英;假黄玉
pseudo-tracheae 拟气管
pseudo-tractrix 伪曳物线
pseudo-trajectory 虚假弹道
pseudo-transfer law 伪传递律
pseudo-transposition 伪转置
pseudo-triangle 伪三角形

pseudo-unconformity 假不整合
pseudo-vacuole 假空泡
pseudo-valuation 伪赋值
pseudo-valuation ring 伪赋值环
pseudo-value 伪值
pseudo-valve 假瓣膜
pseudo-variable 准变量;伪变量
pseudo-variable method 伪变量法
pseudo-vector 赝矢量;伪向量;伪矢量;伪变量
pseudo-vector field 伪向量场
pseudo-velocity 拟速度;虚速度;虚拟速度
pseudo-velocity logging section 伪速度测井剖面
pseudo-velocity spectrum 拟速度谱
pseudo-viscosity 假黏性;假黏度;人工黏性;非牛顿黏度
pseudo-viscosity method 人工黏性方法
pseudo-viscous flow 假黏性流(动);拟黏滞流
pseudo-viscous fluid 假黏性流体
pseudo-viscous pressure 人工黏性压力
pseudo-viscous system 假黏性系统
pseudo-vitrinite 假镜质组;假镜煤
pseudo-vitrinoid 假质组
pseudo-vitrinoid group 假质煤组;假镜质组
pseudo-volcano 假火山
pseudo-volute 假涡旋饰
pseudo-wax 假石蜡
pseudo-wet-bulb potential temperature 假湿球位温
pseudo-wet-bulb temperature 假湿球温度;虚拟湿球温度
pseudo-witch 伪箕舌线
pseudo-wollastonite 假硅灰石
psf[pounds per square foot] 每平方英尺磅(数)
psi[pounds per square inch] 每平方英寸磅(数)
Psi function 斐函数
psilomelane 硬锰矿
psilomelane ore 硬锰矿矿石
psilophyte 草原植物
psittacinite 钒铜铅矿
psj[planed and square-jointed] 刨光并平头接合的
psophometer 噪声测量仪;噪声表;噪压计;噪声计;声级计
psophometric chart 焓湿图
psophometric electromotive force 噪声电动势;估量噪声电动势
psophometric voltage 噪声电压计;噪声电压
psophometric weight 噪声评价系数
psychada alternate 生物滤池灰绳
psychiatric bed areas 精神病住院区
psychiatric clinic 精神病诊所;精神病院
psychiatric hospital 精神病院
psychiatric window 精神病房的窗
psycho-acoustic(al) criterion 音质评价(标准)
Psychoda 毛蠓属
psychological test 心理检查室
psychological testing center 心理检查中心
psychological theory of business cycle 经济周期心理学
psychology of art 艺术心理学
psychology of design 设计心理(学)
psychophysical colo(u)r specification 色刺激值
psychotechnology 心理技术学;工程心理学
psychrograph 自记干湿球湿度计;干湿计
psychrolusia 冷水浴
psychrometer 干湿仪;干湿温度计;干湿球温度表;干湿表;湿度计;测湿计
psychrometer constant 湿度计常数
psychrometer effect 干湿效应
psychrometric calculator 湿度计算器
psychrometric chart 空气湿度图;湿度图;湿度计算图
psychrometric difference 干湿温差;干湿球(温度)差;湿度差
psychrometric formula 湿度计算式;湿度公式
psychrometric method 湿度测定法
psychrometric ratio 湿度比
psychrometric regulator 湿度调节器
psychrometric room 人工气候室;人为气候室
psychrometric table 干湿球湿度表;湿度查算表
psychrometry 湿度测定法;测湿度法
psychrophile 喜湿菌;低温微生物
psychrophilic anaerobic wastewater treatment 嗜冷厌氧废水处理
psychrophilic bacteria 嗜冷细菌

psychrophilic microorganism 嗜冷微生物
psychrophilic plant 喜冷植物
psychrophilic sludge digestion 污泥低温消化
psychrophore 冷却导管
psychrophyte 高寒植物
Psycoda alternata 北美毛蠓
PT boat 鱼雷舰
Pteridophyta 蕨类植物
pterion 翼区;翼点
Pterochore 翅分布植物
pteroma 柱墙间距;柱廊空间
pteromata 希腊神殿的边墙
pteron 侧翼附属建筑房屋(古建筑);外露柱廊
pteropod ooze 翼足虫软泥
p-tert-butyltoluene 对叔丁基甲苯
p-tertiary butyl phenol 对一叔丁基酚
p-tertiary octyl phenol 对一叔锌基酚
pterygopolymorphosis 翅多型
pteryla 羽管
pterylosis 羽序
Pt-group element ore 综合铂族矿石
Pt-hydroger electrode 铂氢电极
ptilolite 发光沸石
Ptilota filicina 羽状翼藻
Ptolemaic temple 托勒玫庙宇
ptoleruian 托勒玫纪
p-toluene sulfonamide 对甲苯磺酰胺
p-toluenesulfonamido-formaldehyde resin 对甲苯磺酰胺(缩)甲醛树脂
p-toluene sulphonyl chloride 对苯磺酰氯
ptototype statement 模型语句
P-trap P 形存水弯;单弯头水封
PtSi detector 铂硅探测器
PtSi infrared focal plane array 铂硅红外焦平面阵列
PtSi infrared sensor 铂硅红外传感器
ptygma 肠状褶皱;肠形脉
ptygmatic fold 肠状褶皱
ptygmatic injection 肠状侵入体;肠状贯入体
ptygmatic migmatite 肠状混合岩
ptygmatic structure 肠曲构造
P type highway 客运公路
P-type semi-conductor P 型半导体
ptyxis 个叶卷叠式
Pt-Zn couple 铂锌电偶
puaplane 滑水板
pub 小旅馆;小酒店;客栈
pubertal inflection 成熟转折点
public access site 公共入口地点
public access to land 公众对土地的获取
public accountant 职业会计师
public accumulation 公共积累
public accumulation fund 公积金
public address broadcasting system 有线广播系统
public-address loudspeaker system 广播扩音系统
public address receiver 公共接收机
public address system 广播系统;扩声系统;扩音装置;公共广播系统;扩声装置
public administration 公共行政机构
public adminstrator 公定遗产管理人
public agency 政府机构;官方机构
public airplane hangar 共用飞机棚
public airport 公共机场
public air-raid shelter 公共防空避护所
publican 赋税的征收员
public arcade 公共拱廊步道;廊道
public arch 公用拱门
public area 公用面积;公共用地;公共广场;公共场所
public assembly hall 公共会堂;公共会场;礼堂
public-assisted authority housing 公助住房
public-assisted bath 公共浴室
public-assisted block 公助住宅街坊
public-assisted boarding school 公助寄宿学校
public-assisted bomb shelter 公共防空避护所
public-assisted building 公助建筑
public-assisted collector 公共垃圾收集器
public-assisted conduit-type sewer 公共管路式阴沟
public-assisted convenience 公共厕所
public-assisted corridor 公共通道;公共走廊
public-assisted dwelling construction 公助住宅建设;公助住房建设
public-assisted electric current grid 公助电流网

public-assisted funds 公助基金
public-assisted green area 公共绿地
public-assisted health engineering 公共卫生工程
public-assisted heath authority 公助卫生管理部门
public-assisted library 公共图书馆
public-assisted open area 公共空旷地
public-assisted phone 公用电话
public-assisted sewer 公共阴沟;公共下水道
public-assisted structure 公共结构;公共建筑
public-assisted telephone coin box 公用电话投币箱
public-assisted toilet 公共厕所
public-assisted traffic 公共交通
public-assisted transport 公共运输
public-assisted utility 公用事业
public-assisted water consumption 公助水耗量;公助用水量
public-assisted water supply 公助水库;公助自来水
publication 出版物;发表
publication and distribution 出版和发行
publication date 出版日期
publication date note 出版日期说明
publication note 出版说明
publication original 出版原图;出版原版
publication scale 出版比例尺
public auction 公开拍卖;拍卖
public authority 建筑科;建筑部门
public authority 地方当局
public automobile service 自动租借汽车
public aviation service 公用航空通信
public bath 公共浴室
public bath house 公共浴室
public bath room 公共浴室
public beacon 公海航标
public bench for washing 公共厕所
public berth 公用码头
public bid(ding) 公开招标;公开投标
public bidding corporation 公开招标公司
public bidding system 招标制
public bid opening 公开开标
public body 公共团体
public bond 公债
public broadcasting center 公众广播中心
public broadcasting system 公共广播系统
public building 公共建筑(物);公共建筑;公共房屋
public building groups 公共建筑群
public building planting 公共建筑绿化
public call office 公用电话通话所
public canteen 公共食堂
public car park 公共停车场
public center 公共中心
public cleaning 公共清洁;公共净化
public cleansing 公共清洁
public comfort room 公共厕所;公共盥洗室
public comfort station 公共厕所
public communication service 公用通信业务
public company 公开公司;开放公司
public comsumption 公共消耗量;公共消费
public construction agency 公营建筑公司
public contract 公共合同
public convenience 公共厕所
public corporations 开放公司;公共事业公司
public correspondence 公用通信
public corridor 公共走廊
public cost 公费代价
public crossing 公用道路与铁路平交道口
public damage 公害
public dance hall 公共舞厅
public data 公开数据
public data network 公众数据网(络);公用数据网;公共数据网络;公共数据网
public data transmission service 公用数据传输服务
public debt 公债
public debt transaction 政府债务交易
public decision making 公共事业决策;公共决策
public development 公共工程;公办工程
public disaster 公害
public domain 共有土地;国有土地;公产
public domain land 国有地;公地
public-domain software 公共流通软件
public drinking fountain 公共饮水处
public driveway 公用车道
public easement 公共用地

public economy 财政
public education 公众教育;公共教育
public(electricity) supply 公用电气事业
public employment program(me) 社会就业纲领
public engineering survey 市政工程测量
public enterprise 公营企业;公共企业
public entrance 公共入口
public errand 公差(指办事)
public escape 公共疏散口
public evacuation door 公共疏散门
public exchange 公用交换局;公用交换
public facility 公共建筑;公用装置;公共设施;公共设施;市政设施
public facsimile bureau 公众传真局
public finance 财政
public finance and finance 财政金融
public-fire alarm address system 火警广播系统
public(fire) brigade 公共消防队
public fire education 公众消防教育
public fire protection 公共消防
public fire reporting system 公共火灾报告系统
public fire safety education 公众防火安全教育
public fire supply 公共消防设施
public float 公用码头
public forces 社会劳动力;公共劳动力
public gallery 旁听席
public garage 公共汽车库;公共车库
public garden 公共花园
public garden plot 公共绿地
public goods 无需付偿可享受之事物(社会中的);公共物品;公共财产
public green area 公共绿地
public green area ratio 公共绿地率
public green space 公共绿地
public green space norm 公共绿地指标
public hack 出租汽车
public hall 广分配厅【铁】;广(大)厅【铁】;会堂;大会堂
public hazard incident 公害事件
public hazards 公害
public health 公共卫生(学)
public health acts 公共保健法规;公共卫生法规
public health authority 公共卫生管理部门;公共卫生管理机构
public health bacteriology 公共卫生细菌学
public health bureau 卫生局
public health clinic 公共保健门诊
public health doctor 公共卫生医师
public health engineering 公共卫生工程
public health hazard 公共卫生危害
public health insect 环境卫生害虫
public health laboratory 公共卫生检验室
public health nurse 公共卫生护士
public health pest 环境卫生害虫
public health practice 公共卫生实施
Public Health Service 公共卫生署(美国);公共卫生服务;公共卫生机构;公费医疗
public health standard 公共防护标准
public health worker 公共卫生工作者
public hearing 公共意见听证会;公开听证会
public hearing meeting 民意听取会
public highway 公路
public holiday and vocation 公共假日
public holidays 公共假日
public house 小旅馆;客栈;酒馆;酒店
public housing 国民住宅建设;公共房屋;公房;公产住房;公共住房
public housing acquisition 为公共住房建设置地
public housing agency 公共住房机构
public housing project 公共住房建设项目
public hydrant 公共消火栓
public hygiene 公共卫生
public illumination 公共照明
public improvement 公共设施的改进
public information 公共信息
public information assist team 公共信息援助队
public information officer 公共信息官员
public infrastructure 市政基础设施
public institution 公共机构
public interest 公营企业;公共利益
public international law 国际公法
public investment 政府投资;公共投资
public investment program(me) 政府投资计划
public involvement 公众参与

publicity 公共性
publicity agent 宣传员;广告员
publicity expenses 宣传费
publicize company policy 宣传公司政策
public key infrastructure 公共密钥基础结构
public land 公有土地;公用地;公共用地
public land mobile network 公众陆地移动网
public land ownership 土地公有制
public land survey 公用地测量
public land system 公用土地系统
public latrine 公共厕所
public lavatory 公共盥洗室
public law 公法
public liability 公共责任
public liability insurance 公共责任保险
public liaison office 公众联络处;公众接待处
public library 公共图书馆;公共库
public lighting 公用照明;公共照明;街道照明
public lodging house 简易宿舍
public lounge 公共休息室
publicly aid private housing 自建公助住房
publicly assisted 官方援助的;公共资助的
publicly assisted house-building 公助房屋建设
publicly owned 公用的
publicly posted price 通告价格
public market 公共市场
public medical care 公费医疗制度
public message service 公用信息业务
public money 公款
public monument 公共纪念碑;公共纪念物;公共纪念馆
public museum 公共博物馆
public music channel 公用音响通道
public navigation 民航
public network 公用网(络);公共网络
public notary 公证员;公证人
public notary organization 公证机构
public notice 通告
public nuisance 公害
public nuisance analysis 公害分析
public nuisance disease 公害病
public nuisance event 公害事件
public nuisance law 公害法
public nursery 公共托儿所;托儿所
public offer 公开报价
public offering statement 公开报告
public office 办事处
public officer 公职人员;公务员
public official 公共机构代表;公职人员;政府工作人员
public off-site sanitary sewage disposal 市政污水系统
public open space 公享空地;公共旷地;共享绿地
public operated house 公营住宅
public opinion 舆论;群众意见;民意
public opinion game 民意对策
public opinion poll 民意调查;民意测验
public opinion survey 民意测验
public order 治安;公共秩序
public(owned) rental house 公营租房
public owned treatment works 公共废水处理厂
public ownership 公有制
public ownership of land 土地公有制
public ownership with a popularly-run nature 民营性质的公有制形式
public packet switched network 公共分组交换网
public packet switching network 公共包交换网络
public park 公园
public parking area 公共停车区
public parking decks 公共停车台
public parking garage 公共车库
public parking lot 公共停车场地
public parking place 公共停车地;公共停车处;公共停车场
public participation 公众参与;民众参与;群众参与
public passage(way) 公共通道
public passenger transport 公共旅客运输
public phone 公用电话
public place 公共场所;公共广场
public plumbing fixture 公共卫生设备
public policy 国家政策;公共政策
public pollution 公共污染
public pollution control work 公害防治工作
public pool 公共游泳池

public port 公用港口;公营商港
public port facility 公用港口设施
public power station 公用发电厂
public-private partnership 公私合作
public program(me) library 公用程序库
public property 公共财产;公产
public property land 公有财产土地
public purse 国库
public queue 公用队列;公共队列
public radio communications service 公用无线电通信业务
public railway 国家铁路;公营铁路
public recreation 公共休养;公共娱乐
public recreation facility 公共娱乐设施
public relation 公共关系
public relation and secretary office 公关文秘室
public report 公开报告
public requirements 公众要求
public reserve fund 公积金
public reservoir 公用水库
public restaurant 公共食堂
public rights of way 公众通行权
public road 公路;公用道路
Public Road Administration 公路管理局(美国)
Public Roads Administration Classification of Soil 美国公路局土的分类法
public room 公用舱室
public safety 公共安全
public safety frequency bands 公共安全频带
public safety radio frequency 公共安全无线电频率
public safety radio service 公共安全无线电服务
public sale 公开出售;拍卖
public saler 拍卖商;拍卖人
public sanitary engineering 公共卫生工程
public sea area 公共海域
public seas 公海
public sector 公共部分;公有部分
public sector of the economy 公共经济中的公营成分
public security agency 公安机关
public security bureau of forestry 林业公安局
public security fire brigade 公安消防队
public security office 公安办公室
public security organ 公安机关
public segment 公用段
public service 公用事业;公益服务;公共事业;公共服务;公职
public service announcement 公益广告
public service building 公共建筑;公共服务楼
public service facility 公共服务设施
public service trades 服务性行业
public service vehicle 公共车辆;公共汽车
public sewage disposal 公共厕所
public sewage work 公共污水工程
public sewer 公用下水道;公共下水道;公共污水管;公共污水道;公共排水管道;市政排水管道
public sewerage disposal 公共厕所
public sewerage work 公共污水工程
public shelter 公共隐蔽处
public space 公共用地;公用舱室;公共空间;公共场所
public space light(ing) 公共场所照明
public space waste 公共场所废水
public square 广场;公共活动广场;公共广场
public square planting 广场绿化
public stable 公用马棚
public stair(case) 公用楼梯
public station 公共电话亭
public storage garage 公共存车库
public store 海关仓库
public store and general order clerk 货品验关估值登记员
public street 公路;大路
public supply 公共给水
public supply mains 城市给水管道;城市供应管网;城市配水网;城市供应干线;城市供水管网;城市给水干线;给水干管
public(swimming) pool 公共游泳池
public switched network 公用交换网(络)
public switched telephone network 公众交换电话网;公用交换电话网;公用电话交换网
public system 公共给排水系统
public telegram service 公用电报业务
public telephone 公用电话
public telephone booth 公用电话间

public telephone network 公众电话局;公用电话网
public telephone room 公用电话室
public telephone system 公用电话系统
public television 公用电视
public tender 公开招标;公开投标
public tender bid 公开竞标
public terminal 公用码头
public toilet 公用厕所;公共厕所
public toilet rooms 公共盥洗室
public traffic 公共交通
public traffic system 公共交通系统
public transit 公共运输(工具);公共交通
public transit facility 公共交通设施
public transit planning 公共交通规划
public transit system 公共交通系统
public transport 公共运输;公共交通
public transportation system 公共交通系统
public transportation usage study 公共交通客流调查
public transport company 公共交通公司
public transport information system 公共交通信息系统
public transport junction 公共交通枢纽
public transport line 公共交通线路
public transport line facility 公共交通线路设施
public transport line network distribution 公共交通线路布局
public transport means 公共交通工具
public transport network 公共交通线路网
public transport parking place 公共交通停车场
public transport planning 公共交通规划
public transport priority 公共交通优先
public transport service 公共交通公司
public transport station 公共交通车站
public transport stop 公共交通车站
public transport yard and station arrangement 公共交通场站布局
public treasury 国库
public trustee 公设信托人;公共受托人
public undertaking 公共事业公司
public use 公用事业;公共事业;公用
public-use civil aerodrome 公用民航机场
public use of water 公共用水
public utility 公用设施;公用事业;公共事业;公共设施;公用设备;市政公用设施
public utility and service 公用事业服务机构
public utility building 公用设施用房
public utility commission 公用事业委员会
public utility corporation 公用事业公司
public utility easement 公用设施通过权
public utility facility 公用事业设施
public utility industry 公用工业
public utility project 公用事业项目;公用事业工程
public utility system 公用事业系统
public utility undertaking 公共企业;公用事业设备
public vehicle 公共交通工具
public vehicle traffic 公共车辆交通
public vessel 政府船舶;公务船(舶)
public view (建筑物的)敞开部分;公开展示;公众意见
public volume 共用卷宗;公用卷(宗)
public volume table 公用容量表
public walk 散步场所
public wall 公用隔墙
public warning 公开警告
public washroom 公共盥洗室
public waste 公共场所垃圾
public wastewater treatment facility 公有污水处理厂
public water 公用水
public water area 公共水域
public(water) consumption 公用耗水量;公共用水量
public water main 公用供水干管
public water project 公共给水工程
public waters 公用水域;公海;公共海域
public water supply 公用供水;公共给水;公共供水
public water supply demand 公共供水需求量
public water supply equipment 公共供水设备
public water supply-station 公共给水站
public water supply system 公共给水系统
public water system 公共用水系统;公用自来水系统
public waterway 公用水道
public water works 自来水厂;自来水工程;公共给水工程;公共给水厂

public water works system 自来水工程系统
public way 公用通道;公用道路
public weigher 公证行;公证衡量员
public welfare 公用福利设施;公共福利;社会福利
public welfare establishment 公益事业
public welfare facility 公用福利设施;公共福利设施;生物福利设施
public welfare fund 公益金
public wharf 公用码头
public wire 公用电话线
public works 公用工程;公共设施;公共建筑(物);公共工程;市政工程
Public Works Administration 市政工程管理局;公共事业管理局
public works contractor 公共建筑承包者;公共工程承包者;市政工程承包者;公共建筑承包商;公共工程承包商;市政工程承包商;公共建筑承包人;公共工程承包人;市政工程承包人
public works department 市政工程局
public works duties 民工建勤
public works project 公用建筑设计;市政工程设计
public works survey 市政工程测量
published 公开发表
published date 出版时间
published depth 公布的水深
published symbol 图式符号
published symbolism 图式符号
published tariff 公布的运价表;公布的价目表;公布的关税表;发布的税率
published times 出版次数
published tolerance 名义公差;公称公差
publisher 出版者;出版单位;发行者;发行人
publisher's binding 特装本
publisher's note 出版者说明
publishing house 出版社
puce 紫褐色
pucherite 钒铋矿
puck 弹力盘;矸石墙;石垛
pucked barbed spindle 冲压倒刺纺锭
pucker 皱褶;皱纹;皱起;摺皱缝;摺叠;折叠
puckered slate 皱纹板岩
pucker-free 无皱缩
puckering 起褶;皱纹;小褶皱;焊缝皱皮;起皱
puckers 拉延件侧壁皱纹;拉延壁上皱纹
pucking (煤矿层的)隆起
pucking cutter 铲平工
pudding 艇护舷;捣密;船尾碰垫;船首碰垫(拖船上)
pudding ball 贴沙砾泥球
pudding bin 化石灰池;石灰消和池
pudding fender 护舷靠垫;碰垫;艇护舷;艇(首)碰垫;船尾碰垫
pudding granite 球粒花岗岩
pudding rock 圆砾岩
pudding stone 圆砾岩;蛮石;布丁岩
puddle 直浇口窝;胶土;夯实体;熔池;水坑;捣拌;稠黏土浆
puddle ball 搅炼铁块;搅拌铁球
puddle clay 黏土浆;胶黏土;黏土胶泥
puddle clay core wall 夯土心墙
puddle clay cut-off 夯土截水墙
puddle clay lining 捣密黏土涂层
puddle cofferdam 胶土围堰;黏土围堰;夯土围堰;夯土堆筑围堰;揉泥围堰
puddle cofferdam weir 黏土围堰
puddle core wall 黏土心墙;夯土心墙
puddled 夯实的
puddle dam 胶土墙
puddled backfill 夯实回填(土);捣实回填(土)
puddled ball 搅炼炉熟铁块
puddled bar 熟铁棒材
puddled billet 搅炼炉坯铁
puddled bloom 搅炼炉坯铁
puddled clay 夯实黏土;捣实黏土;胶土
puddled clay core 夯土心墙
puddled concrete 捣实的混凝土
puddled core 胶土心墙;夯实胶土心墙
puddled cut-off 夯土截水墙
puddled dam 夯实堤;胶土坝
puddled earth 夯实土(壤)
puddle dike 胶土心墙堤;有胶土填心的堤坝;黏土堤;黏土坝
puddled iron 搅炼熟铁;锻铁
puddled land 黏闭土

puddled pig iron 搅炼炉用生铁
puddled runway 积水的跑道
puddled soil 黏闭土壤；浆土；夯实土(壤)；捣实土(壤)
puddle(d) steel 熟铁；搅炼钢
puddled wall 夯实胶土墙
puddle furnace 搅炼炉
puddle iron 熟铁；锻铁
puddle mixer 搅泥机；混砂机
puddle quick lime 捣碎的生石灰
puddler 搅炼炉；搅捣机；搅实器；搅实机；捣密机；捣拌泥浆工；捣拌机；表面振捣器
puddle(rolling) mill 熟铁轧机
puddle structure 搅泥结构
puddle trench 夯土沟；水坑沟槽
puddle wall 胶土墙；黏土心墙；夯土墙
puddle weld(ing) 熔焊；堆焊
puddling 搅炼；黏土质矿石洗矿；黏土夯实；捣成泥浆；成浆
puddling basin 搅炼炉膛
puddling bin 化灰池
puddling cinder 搅炼炉渣
puddling furnace 搅炼炉
puddling phenomenon 黏闭现象；黏闭现象
puddling treatment 拌土处理
puddly-plain coast 淤泥质平原海岸
puddly soil 淤泥质土(壤)
pudendal canal 阿尔科克氏管
pudling machine 捣泥浆机
pudlink 脚手架短横木
Pudlo 普得洛(一种粉状防水剂)
pueblo 集体住所；(印第安人的)村庄
Pueblo sandstone 普伟布洛砂岩(一种产于美国科罗拉多州的浅灰色砂岩)
puelche 帕尔希风
Puerto Limon 利蒙港(哥斯达黎加)
Puerto Rico cotton 波多黎各棉
Puerto Rico trench 波多黎各海沟
puff 喷焰；爆音
puff blowing 吹洗；吹去灰屑
puff blowing device 吹气冲尘装置(凿岩机)
puff blowing head 反向风风嘴
puff blowing piston 吹气冲尘活塞(凿岩机)
puff blowing piston rock drill 吹气冲尘活塞式凿岩机
puff cake 粉饼
puff cone 泥火山锥
puff diffusion 烟团扩散
puff duster 手动风箱式喷粉器
puffed bar 气胀棒坯
puffed compact 气胀压坯
puffed grain 膨化谷物
puffed up concrete 膨胀混凝土；加气混凝土
puffer 哄抬价格者；小绞车；小型发动机
puffer fish 河豚
puffer-pipe 中心喷管
puffing 晶胀现象；急喷出；气浪
puffing activity 膨松活性
puffing advertising 口头广告
puffing agent 增稠剂
puffing drying 膨胀干燥
puff of wind 阵风
puff pipe 防虹吸管；膨突管；排气管；反虹吸管
puff port 喷气口
puff sub-model 烟团子模型
puflerite 钙辉沸石
pug 泥料；搅087机；断层泥；拌土机；练泥；捏和
pug(concrete) mixer 强制式混凝土搅拌机
Puget Sound pine 红松；美国松；洋松
pugged clay 窖泥；捏练过的黏土；捣塑黏土
pugging 捣捏黏土；隔音层；楼板隔声材料；黏土捣塑；填塞泥土隔墙；练泥；捏和
pugging mill 捏和碾磨机
pugging mullering 捏练
puglianite 云白辉长岩
pug lining 捣实(黏土)衬垫
pug mill 叶片式搅拌机；叶片式拌和机；练泥机；搅泥机；搅拌捏和机；捏拌机；碾泥机；黏土拌和机；拌土机；拌泥机；捣土机
pugmill-mixed 捏拌机拌和的；捏拌机搅拌的
pug(-mill) mixer 窖泥机；泥土搅拌机；泥土捏和机；捏土机
pugmill mixing 捏拌机拌和；捏拌机搅拌
pugmill mortar mixer 砂浆搅拌机

pug mill type mixer 捏拌机
pugmill type soil stabilization mixer 稳定土搅拌机
pug mixer 叶片式搅拌机；叶片式混料机；叶片式拌和机；捏拌机
pug pile 鸠尾堆；鸠尾桩；燕尾桩；黏土隔水桩
puking 冒顶
pulae duration coder 脉冲宽度编码器
Pulaski shales and sandstones 普拉斯基页岩和砂岩
pulaskite 斑霞正长岩
pulau-pulau 群岛
pulcom 小型晶体管式测微指示表
pulegium 长叶薄荷
pulegium oil 海地油
pulegol 长叶薄荷醇
pulegone 长叶薄荷酮
pulic system engineering 公用事业系统工程
pull 吸引；样张；拉手【建】；牵引；拖曳；拖；出料量；爆破进尺
pull(action) fuze 拉发引信
Pull-and-out block 牵引复滑车
pull-apart 拉裂；扯断
pull-apart basin 拉分盆地
pull-apart connector 拉脱接头
pull-apart downwarped extracontinental basin 拉离型陆外下陷盆地
pull-apart of rail 断轨拉开轨缝
pull-apart structure 拉分构造
pull a print 印图样
pull a proof 印试样
pull a well 从孔内起拔套管柱
pull a well in 拆卸钻塔
pullback 牵引式安全装置
pullback blade 反向铲
pullback cable 复拉钢索；反拉钢丝绳
pullback cam disk 后退凸轮盘
pullback cylinder 回程缸
pullback draw bridge 后拉式开合桥；后曳式仰开桥；拖拉桥；便桥
pullback line 拉返绳索；复拉绳索
pullback mechanism 后撤机构
pullback spring 拉簧
pull bar 拉杆
pull-behind 牵引式
pull-behind cylinder 外置式油缸
pull bituminous concrete and aggregate spreader 牵引式沥青混凝土和骨料摊铺机
pull blade 牵引板
pullboat 平底拖船；(集运木材的)浮式绞木机
pull box 火警报警器；引线箱(消防用)；拉线箱；拉线盒；穿线盒；分线箱
pull-broaching 牵引拉削
pullbutton 拉钮
pull cable (拉铲的)牵引缆；(拉铲的)牵引索
pull casing 拔去套管；起拔套管；提升套管
pull chain 牵引链；拉链
pull-chain operator 拉链开关；拉链操作器
pull-chord 拉线；拉桁弦；拉弦；拉索
pull-chord type switch 拉线开关
pull contact 拉钮开关
pull-cord 牵引绳
pull crack 拉裂裂纹；拉裂；横向不规则裂缝；热裂
pull cut 拉切
pull-cut shaper 拉切式牛头刨
pulldevil 拖钩
pull direction 提拉方向
pulldown 拉下；向下拉；拆除；拆除；油墨(刮涂)膜；拉开距离；轴压力；拉片距；间歇拉片；间隙拉片
pulldown bed 活动床【给】
pulldown claw 输片爪
pulldown crystal 拉晶
pulldown current 反偏电流
pulldown equipment 强制给进机构(转盘钻机)
pulldown handle 下拉把手
pulldown hook 窗钩
pulldown menu 按落式窗口菜单；下拉式菜单
pulldown roller 慢拉辊
pulldown test 冷却试验；降温试验
pulldown time 倒片时间
pulldown utility shelf 下拉式搁板
pull dozer 推土机
pull dynamometer 拉力表
pulled crystal 拉制的单晶；拉晶；抽晶
pulled out of its binding 脱壳

pulled pattern 花纹伸长；图案花纹重复；图案花纹重叠
pulled-type coil 穿入式线圈
pulled wire 横丝错位
puller 拉轮器；拉单晶机；拉出器；牵引器；提引器；拆装器；拔具；拔出器；拔板机
puller adapter 拆卸工具接头
puller bolt 预开螺栓；顶紧螺栓
puller cylinder 张拉液压缸；拉拔缸
puller cylinder range 拉拔缸行程；张拉液压缸行程
puller-deseeder 亚麻提取脱粒机
puller detacher 拆卸器
puller device 拉紧设备；牵拉设备
puller finger 拆卸工具夹头
puller grip 拔拔夹
puller nut 拆卸工具螺帽
puller rod 连杆
puller section 牵引段
puller set 一套拉出器
puller sheave 拔取轮
puller strap 拉紧带
puller unit module 牵引机组合模块
pulley 托辊；滑轮
pulley-and-finer feeder 刮板转轮式给料机
pulley and loose pulley 固定轮和游滑轮；固定轮和动滑轮
pulley-and-scraper feeder 刮板转轮式给料机
pulley assembly 滑轮总成
pulley assembly for tensioning 补偿滑轮装置；补偿滑轮
pulley axle 滑车轴
pulley block 手拉吊挂；滑轮组；滑车(组)；滑轮
pulley block for a gain in force 增速滑车(轮)组；省力滑车(轮)
pulley block luffing gear 起重机变幅滑轮组；(起重机的)升降支臂用滑轮组
pulley boss 滑车轮毂；滑车毂
pulley box 天车轴承外壳
pulley bracket 滑轮托(架)
pulley casing 滑轮箱
pulley casting 滑轮铸件；皮带轮铸坯
pulley center 皮带轮中心距
pulley chain 轮带链系
pulley cheek 滑轮颊板；滑轮侧板
pulley class gate lifting device 滑车类闸门启闭机
pulley cone 锥轮；塔轮
pulley conical disk 三角皮带轮
pulley cover 皮带轮罩
pulley drive 滑轮传动；皮带轮传动；皮带传动
pulley fork 滑轮轭
pulley for main grain auger 谷粒推运螺旋皮带轮
pulley frame 天轮架
pulley gear 滑轮装置；皮带轮传动装置
pulley groove 滑轮槽；皮带轮槽
pulley guard 皮带轮安全装置
pulley guide 滑轮导向器
pulley head 窗头板；滑轮安装端(吊窗)；提拉窗上口槽
pulley holder 滑轮托(架)；皮带轮托架
pulley holding forceps 滑车钳
pulley idler 滚筒托辊；空转轮；惰轮
pulleying 过卷
pulley key 皮带轮键
pulley lathe 皮带轮车床
pulley magnet 磁轮(选矿用)
pulley mortise 滑轮榫槽；滑槽榫接；一侧加长并凿斜的榫眼
pulley motor 电机皮带轮；电动机皮带轮
pulley mo(u)lding machine 皮带轮造型机
pulley nut 皮带轮螺母
pulley ratio 滑轮组倍率
pulley rig 悬吊滑车式支架
pulley room 滑轮间
pulley separator 滑轮式分离器；磁轮(选矿用)
pulley shaft 提升井
pulley sheave 滑轮；滑车滑轮；滑车
pulley shell 皮带轮罩
pulley speed 皮带轮转速
pulley spindle 滑轮轴
pulley stile 滑轮竖框板；滑轮框架；滑车槽；提拉窗滑车槽；吊窗滑车槽
pulley style 滑车槽
pulley support 皮带轮支架
pulley tackle 起重滑车；复滑车；辘轳；滑车组

pulley tap 皮带轮丝锥;皮带轮螺丝攻
pulley tightener 紧带机
pulley(-type) feeder 转筒给料机;滑轮式给料器
pulley type wire-drawing machine 滑轮式拔丝机;滑轮拔钢筋机
pulley wheel 滑轮
pulley with clevis 补偿滑轮
pulley with suspension 悬吊滑轮
pulley yoke 滑轮架;滑车夹板
pull forming 拉挤与摩压联合工艺
pull grader 拖式平地机;拖拉式平地机;拖拉机平地机;牵引式刮土平地机
pull handle 拉手【建】;拖曳柄
pull hardware 拉手;门用五金
pull head 拖管头
pull hook 牵引钩
pull in 引入;吸合
pull-in acquisition 捕捉
pull-in-and-slide window 驾驶舱或车厢的边窗
pull-in bandwidth 捕捉带宽
pull-in box 引入箱
pull-in collet 内拉夹套
pull-in frequency 引入频率
pulling 拉制;拖力;图像伸长
pulling ability 牵引能力
pulling and running the drill pipe 提下钻杆
pulling and turning 扳法
pulling an oar 荡桨
pulling attachment 拔取装置
pulling belt 拔取皮带
pulling boat 划桨船
pulling capacity 拉力;拉拔力;牵引能力
pulling chain 拉链;牵引链
pulling channel 拔取导槽
pulling chuck 拉刀夹头
pulling clrcult 牵引电路
pulling core 抽芯
pulling device 拉紧器;牵引装置
pulling dog 提引器
pulling down 拆毁;推倒;拉缩
pulling down of bank 边坡切削
pulling down work 拆如工作
pulling drum 拔取皮带轮
pulling effect 同步效应
pulling equipment 提升设备(升降钻具)
pulling eye 引线孔;拉孔
pulling force 拉力;切削能力;拔力;牵引力
pulling form 拉模
pulling form mo(u)lding 拉模成型
pulling furnace 拉晶炉
pulling head 拖管头
pulling-in dogs 夹料钳;带头线钳;牵引链钳
pulling-in machine 牵引穿模机;穿模带头机
pulling in parallel operation 牵入并联运行
pulling in step 引入同步;拉入同步;牵入同步
pulling into synchronism 牵入同步
pulling into tune 拉引调谐;强制调谐
pulling jack 拉力千斤顶;手动葫芦
pulling line 拉绳
pulling machine 拉丝机;拔桩机;套管起拔机;拔管机
pulling material 牵引材料
pulling mechanism 拔取机构
pulling method 引上法;提拉法
pulling motor 牵引电机
pulling motor speed control 牵引电机速度控制
pulling off device 拉出装置
pulling-off silk end 拖头丝
pulling of seedlings 采苗
pulling on whites 白像拖延;白像后延带线条;白拖黑尼
pulling operation 拉晶操作
pulling-out force 拉脱力
pulling out of step 牵出同步
pulling out of synchronism 牵出同步;失步
pulling out the hole 提出钻孔
pulling over (漆面的)溶剂抛光
pulling over of nitrate paint 硝基漆漆膜擦平
pulling pillars 回收矿柱法
pulling plug 拉栓;牵栓
pulling point 岩芯提断点;岩芯提断部位(孔内)
pulling power 牵引力;拉力
pulling rate 拉速;拉晶速率
pull in green 过早提钻(为更换钻头)

pulling resistance 桩抗拔力;抗拔力
pulling rope clamp 牵引绳夹紧器;绑绳夹
pulling scraper 刮土拖拉机;拉铲;拖拉刮土机;拖铲
pulling sheeting 拔板桩
pulling speed 牵引速度
pulling technique 拉引技术;拉单晶技术
pulling tension 拉伸应力
pulling test 拉伸试验;拉力试验;张拉试验;拔桩试验;(桩的)拔出试验
pulling test on pile 桩的拉力试验;拔桩试验
pulling tool 拆卸工具;拉伸器
pulling to place 拉曳就位
pulling tops 拖梢护树
pulling type 牵引式;拖挂型;拖挂式
pulling under the brush 抗漆刷拉力
pulling unit 管子起拔机;拔取装置
pulling up 掘路;提起;拔起;表面软化;脱层(旧漆层的软化);咬底(涂料涂刷缺陷);拉毛
pulling-wheel 牵引轮
pulling yoke 提升架
pull-in jack 拉合千斤顶;牵引千斤顶
pull-in line 拖曳线
pull-in pliers 夹料钳
pull-in range 牵引范围;同步范围;捕捉范围;捕捉带
pull-insert 锥体拉出试验插入物
pull-in step 拉入同步
pull instruction 拉指令
pull-in test 整步试验;牵入试验
pull-in time 拉片时间;回场时间;捕捉时间
pull into derrick 拉管子到钻塔中
pull-in torque 牵入转矩;启动转矩
pull-in value 吸入值
pull-in winding 拉入绕组;穿入式绕组;插入式绕组
pull-iron 牵钢
pull-jack 拉合千斤顶
pull length 一次钻进深度;四次提升岩芯长度;四次进尺深度
pull lever 拖杆
pull lift 轻便起吊具;拉链式起重器;轻便提升设备;手绞车;神仙葫芦
pull line 拉绳;牵引索;牵引绳;牵引钢索
pull line method 拉线法
Pullman 普尔门式火车卧车
Pullman balance 普尔门式弹簧
Pullman berth 可折叠卧铺(铁路车厢);船上床铺
pull mandrel rivet 抽心轴铆钉
Pullman train 普尔门式列车
pull-off 脱掉;正定位;曲线定位装置
pull-off cantilever 正定位腕臂
pull-off clamp 双耳线夹
pull-off clamp for strand wire 绞线固定定位线夹
pull-off connection 脱落连接
pull-off lane 停车车道(路上)
pull-off plug 脱落插头
pull-off pole 锚柱;系定柱;系定杆;锚定柱;双撑杆
pull-off spring 放松弹簧
pull-off strength 拉脱强度;扯发强度
pull-off strip 停车带(路上);停车道(路上)
pull off test of adhesion 附着力拉脱试验
pull off the bottom 提离孔底(钻局)
pull of the brake band 拉紧制动带
pull of vacuum 真空吸力
pull-on section 进料区段
pull-on value 吸合值
pull open 拉开
pull out 拉长;开出;起钻;脱出;提钻;使失步;拔出;脱落;退出俯冲
pull-out and reducing stand 拉轧机
pull-out bond test(for reinforcement) (钢筋的)拔出(黏结力)试验;(钢筋的)拔拉握裹试验;拔拉结合力试验;抗拔力试验;拔拉黏结力试验
pull-out capacity 固着力;抗拔力
pull-out distance 挡阻距离
pull-out force 拉拔力;拔力;拔拉力;拔出力
pull-out fuse 插入式熔断器;插入式保险丝
pull-out guard 拉出式保护装置;牵手式安全装置;推出安全装置(冲床用)
pull-out hand brake 手拉式手制动器
pull-out hook 抬圈钩
pull-out lead 牵出线【铁】
pull-out load 拔拉荷载
pull-out map 插图
pull-out oven 抽屉烤箱
pull-out piece spring screw 压簧螺钉

pull-out power 牵出功率;失步功率
pull-out resistance 拉出阻力;固着力;拔拉阻力;拔拉阻力
pull-out resistance of pile 桩的抗拔力;桩的表面阻力
pull-out roll 拉料辊;拉出辊
pull-out shear test 拉锚剪切试验
pull-out specimen 拔出试验用试体(钢筋);抗拔力试验用试件
pull-out spot 路侧停车坪
pull-out strength 拔出强度;抗拔拉强度;拔拉强度
pull-out strength of nails 钉子的拔拉强度
pull-out test 拔出试验;拔桩试验;固着力试验;拉拔试验;拔拉试验;抗拔试验;牵出试验;失步试验;拔拉试验
pull-out test for pile (桩的)拔出试验
pull-out test of bars 钢筋拔出试验
pull out the tubing string 起油管
pull-out time 吸动时间;出场时间
pull-out torque 临界过载力矩;输出转矩;牵出扭矩;失步转矩;拔出转矩
pull-out-type crucible furnace 可拆式坩埚炉
pull out type fracture 剥落破坏
pull-out type test 拔拉结合力试验
pull-out value 固着力值
pull-out vehicle 出场车
pull-over 递回;拨送器
pull-over gear 移送机;拖运机
pull-over mill 递回式轧机;二辊周期式(薄板)轧机
pull-over roll 递回轧辊
pull phenol 提取酚
pull plunger press 压捆机;拖式柱塞压捆机
pullrake 指轮式搂草机;牵引式搂草机
pull rate 制成率;出料率
pull ratio 出料率
pull reel 张力卷取机
pull ring 延伸圈;拉手环
pull-rise curve 拔升曲线
pull rod 拉杆;拉棒;牵引杆
pull-rod adjusting nut 拉杆调整螺母
pull-rod gasket 拉杆衬垫
pull rod nut 拉杆螺母
pull-rod operator 拉杆开关
pull rods carrier (抽屉的)拉杆座架;拉杆式工具;拉杆式搬运器
pull rods connection 拉杆连接
pull rods hanger 吊杆;栏杆悬挂物
pull-rod spacing 拉杆间距
pull roll 引出杆
pull-root 下曳根
pull rope 信号绳;牵引索;牵引绳;牵索
pull-rope winch 牵引绞车;牵引卷扬机;牵引绞车;牵引卷扬机
pull scraper 拉铲;手工刮具
pull shovel 拉铲挖掘机;拉铲;索铲
pull side 拉紧端
pull socket 拉线灯口;抽拉插座;插头
pull-station 拉式火警箱
pull strategy 前拉策略
pull strength 拉拔强度;抗拔强度
pull-stretch fuel pump 拉伸机油泵
pull strings form behind the scenes 暗中操纵
pullstroke trenching machine 反铲挖掘机;反铲挖沟机
pull switch 拉线开关;拉线电门
pull tab or button drown 易开口包装
pull tackle 起重滑车
pull tap 拔出断丝锥用的四爪工具
pull tension ga(u)ge 张力器;张力计;牵引力计
pull tension on casing 提紧套管柱
pull test 抗拔力试验;牵引试验
pull-test of rock bolts 锚杆拉拔试验
pull through machine 穿料装置
pull-through winding 拉入绕组;穿入式绕组
pull-tows 拖驳船队
pull-tows in series 一列式拖带驳船队
pull tube 拉管
pull-type 拖挂型;拖挂式;牵引式
pull type broach 拉刀
pull-type go-devil 拉引式清管器
pull-type pig 拉引式清管器
pull-type recoiler 张力式卷取机
pull-type roller 拖式压路机;拖式碾压机;拖挂式

压路机;拖式路碾
pull-type scraper 牵引式铲运机;拖式铲运机;拉引式清管器
pull-type wire feeder 拉丝式送丝机构
pull-up 拉起;拉高;牵力;咬底(涂料涂刷缺陷);脱层(旧涂层的软化);拉毛;掘路;表层软化;拔起;停车处;路边休息所;路边咖啡馆;休息所;张力
pull-up circuit 工作电路
pull-up cold (补偿器的)冷态拉紧
pull-up curve 拉力曲线
pull-up into a roll 跃升进入横滚
pull-up method 选取法
pull-up position 吸持位置
pull-up resistor 上拉电阻箱;工作电阻;正偏电阻
pull-up riser 穿插式升气管
pull-up test 最低起动试验
pull-up the casing 起拔套管
pull-up time 吸引时间;动作时间
pull-up torque 最小起动扭矩;最低起动转矩
pull-up weeds 拔掉草
pull wedge 调整斜铁;调整楔
pull well 从孔内起拔套管柱
pull well in 拆卸钻塔
pull wheel 天轮
pull wheel process 拉轮法
pull winding 拉挤缠绕工艺
pull wire 牵引索;牵引绳;牵引线;拉线
pull yoke 起管卡;起管夹
pulmonary asbestosis 石棉肺
pulmonary barotraumas 肺气压伤【救】
pulmonary silicosis 矽肺
pulmonary talcosis 滑石肺
pulons 塔门
pul-out piece spring 压簧
pulp 矿泥;矿浆;浆状物;浆化;浆;泥浆;制浆
pulp aeration 矿浆充气
pulp and paper industry 纸浆造纸工业
pulp and paper mill 纸浆造纸厂
pulp and paper mill black liquor 纸浆造纸厂黑液
pulp and paper mill wastewater 造纸工业废水
pulp assay 矿浆分析;纸浆的试验
pulp balance 矿浆天平
pulp bin 木浆储[贮]仓
pulp black liquor 纸浆黑液
pulp bleaching agent 制浆漂白剂
pulpboard 纸浆(纸)板
pulpboard liner 硬纸板衬垫
pulp-body froth flo(a)tation 矿浆体泡沫浮选;内界面泡沫浮选
pulp-body machine 矿浆充气式浮选机
pulp canal drill 根管钻
pulp catcher 受浆器
pulp cement board 纸浆水泥板
pulp chest 纸浆池
pulp clarification 矿浆澄清
pulp cleaner 筛浆
pulp climate 矿浆物理和化学状态;调和矿浆
pulp colo(u)r 水浆颜料;水调色浆
pulp concentration 矿浆浓度
pulp density 纸浆密度;矿浆密度
pulp density recorder 矿浆密度记录器
pulp dewatering 矿浆脱水
pulp digester 蒸煮器
pulp digger 掏料机
pulp dilution 矿浆稀释
pulp divider 矿浆分配器
pulp-dresser 纸浆洗涤机
pulped cloth 纸浆布
pulp elevator 甜菜废丝提升器
pulper 浆粕机;制浆机;搅碎浆机;碎浆机;水力碎浆机
pulp flow graph 矿浆流程图
pulp flowsheet 矿浆流程图
pulp grinder 木浆研磨机
pulphate method 硫酸盐法
pulp hook 制浆材抓钩;木浆材搬钩
pulphurous acid gas 亚硫酸气
pulp industry 纸浆工业
pulpiness 柔软性
pulping 打浆;成浆;制浆
pulping and paper-making effluent 制浆造纸废水
pulping device 浆板撕碎机
pulping engine 浆板撕碎机;碎浆机
pulping grinder 磨浆机

pulping liquor 纸浆液
pulping machine 研碎机;研磨机;磨浆机;捣碎机
pulping process 制浆过程;制浆方法
pulping tank 制浆槽
pulpit 机器操纵台;讲坛;讲经台;船首突出台;布道坛
pulpit baldachin 祭坛华盖
pulpit canopy 讲坛柱顶
pulpit-operated crane 控制室操纵的起重机
pulpit over sarcophagus 石棺上面的祭坛
pulpit stair(case) 讲坛扶梯
pulpitum (教堂的)唱诗班隔屏;教堂石屏栏;演讲讲坛;(罗马剧院里与乐队席相邻的)舞台
pulp kneader 纸浆揉混机;碎浆机
pulp liquor from pulping mill 纸浆厂废水
pulp machine 浆粕抄造机
pulp making intermediate wastewater 制浆中段废水
pulp mill 制浆厂;纸浆厂
pulp mill digester 纸浆蒸煮锅;蒸煮锅
pulp mill effluent 纸浆厂废水
pulp mill pollution 纸浆厂污染
pulp mill screenings 纸浆筛余物
pulp mill waste 纸浆厂废物
pulp mill wastewater 纸浆厂废水
pulp mo(u)lding 纸浆浸渍模塑
pulp of finger 指垫
pulp-preparation plant 砂浆制备车间
pulp press 废粕压榨机
pulp press water 压榨水;甜菜废丝压榨水
pulp process paper 纸浆法纸
pulp pump 纸浆泵;渣泵
pulp saver 纸浆回收机
pulp screen 纸浆筛;浆筛;筛浆机
pulp screening 筛浆
pulp separator 渣分离器
pulp sifting 纸浆筛选
pulp sluice 纸浆堰
pulp slurry 液体浆
pulpstone 磨浆石;磨制矿浆用磨石;磨木石;髓石
pulp strainer 纸浆筛滤器
pulp thckener 纸浆浓缩机
pulp washer 洗浆机;洗浆池
pulp wastewater 纸浆废水
pulpwater 浆水
pulp web 浆幅
pulpwood 造纸木材;制浆(木)材;纸浆原材;木浆材料
pulp wood fork 装卸制纸浆木材的叉车
pulp wood grapple 抓制浆圆木的抓斗
pulp yarn 木浆纱
pulsactor 磁脉冲发生器
pulsafeeder 脉动给料机
pulsar 脉冲星
pulsar receiver 脉冲星接收机
pulsar signal 脉冲星信号
pulsar synchronizer 脉冲星同步器
pulsar time scale 脉冲星时标
pulsar time transfer system 脉冲星时间传递系统
pulsated air conveying 脉动式空气输送
pulsated bed 鼓动床
pulsating 脉动的
pulsating air 脉冲空气
pulsating air classifier 脉动空气分级机
pulsating air intake 脉动进气孔
pulsating air jig 风力脉动跳汰机
pulsating amber light 脉动琥珀色灯
pulsating arc 脉动弧;脉动电弧
pulsating aurora 脉动极光
pulsating auroral radar echoes 脉动极光雷达回波
pulsating auroral zone 脉动极光带
pulsating bubble 脉动气泡
pulsating cavity 脉动空蚀
pulsating combustion 脉动燃烧;脉冲燃烧
pulsating compressive loading 脉动压缩荷载
pulsating correction factor 脉动校正系数
pulsating current 脉冲电流;脉动水流;脉动电流
pulsating current factor 脉动电流因数
pulsating current motor 脉动电流电动机
pulsating cyclic load 脉动循环荷载
pulsating deicer 脉动除冰设备
pulsating direct current 脉动直流
pulsating dye bath 脉动染浴
pulsating effect 脉动效应
pulsating electric(al) source 脉动电源

pulsating electromotive force 脉动电动势
pulsating energy 脉动能源
pulsating equipment 脉动装置
pulsating error 脉动误差
pulsating fatigue strength under bending stress 脉动抗弯应力疲劳强度
pulsating feed 脉动供给;脉动给进
pulsating field 脉动磁场;脉动场
pulsating firing 脉冲烧成
pulsating flow 脉动水流;脉动流
pulsating flow gas turbine 脉动式燃气轮机
pulsating flow measurement 脉动流量测量;脉动流测定
pulsating fluidized bed 脉动流化床
pulsating flux 脉动通量
pulsating force 脉动力
pulsating laser action 脉动激光作用
pulsating light 脉动光
pulsating light source 脉动光源
pulsating load(ing) 脉冲荷载;脉动荷载;脉动负载;脉动负荷
pulsating machine 脉动机
pulsating magnetic field 脉动磁场
pulsating medium 脉冲介质
pulsating motor 往复运动电动机
pulsating oxidative pyrolysis 脉动氧化裂解
pulsating plunger pump 脉动柱塞泵
pulsating power 脉动功率
pulsating power control 脉动功率控制
pulsating press 脉动压机
pulsating pressure 脉动压力;脉动电压;脉冲压力
pulsating pump 吸水泵
pulsating quantity 脉动值;脉动量
pulsating radio source 脉动射电源
pulsating rectified current 整流过的脉动电流
pulsating relay 脉动继电器
pulsating rotating field 脉动旋转磁场
pulsating sampler 脉动采样器;脉动取样机
pulsating scintillation 脉动闪光
pulsating screen 脉动筛;摆动筛;振动筛
pulsating sphere 脉动球体;脉冲范围
pulsating sphere source 脉动球体声源
pulsating spring 间断泉
pulsating star 脉动星;闪光星
pulsating steam pump 蒸汽吸水机
pulsating stress 循环应力;脉动应力;脉冲应力
pulsating surface 脉动表面
pulsating surface heat-flow 脉动式地表热流
pulsating test 脉动试验
pulsating torque 脉动转矩
pulsating traction motor 脉动牵引电机
pulsating variable 脉动变星
pulsating voltage 脉动电压
pulsating water 脉冲水
pulsating wave 脉动波
pulsating weld 脉动接触焊
pulsating welder 脉冲焊机
pulsation 造山脉动期;交流电角频率;间断;脉动
pulsation acceleration 脉动加速
pulsation action 脉动作用
pulsation but in which the essential trend is contracting 收缩为主要趋势的脉动
pulsation chamber 脉动室
pulsation chamber vacuum record curve 脉动室真空记录曲线
pulsation combustion 跳动燃烧
pulsation constant 脉动常数
pulsation current velocity 脉动流速
pulsation cycle 脉动周期;脉动循环
pulsation dampener 脉动消除器
pulsation dampening 脉动衰减器
pulsation damper 脉动阻尼器;脉动节制器;脉冲消除装置
pulsation damping 脉动阻尼
pulsation damping diaphragm 脉动阻尼薄膜
pulsation effect 脉动效应
pulsation equipment 脉动设备;脉动器
pulsation hypothesis 脉动说
pulsation-induced vibration 脉动引起的振动
pulsation instability 脉动不稳定性
pulsation instability strip 脉动不稳定带
pulsation interference 脉动干扰
pulsation loss 脉动损耗;脉动耗损
pulsation mass 脉动质量

pulsation mode 脉动模式
pulsation of current 电流脉动
pulsation period 脉动周期
pulsation phenomenon 脉动现象
pulsation point 脉动点
pulsation rate 脉动(频)率;脉冲率
pulsation ratio 脉动节拍比
pulsations and noise recorders 磁脉动和噪声记录
pulsations continuous 连续型
pulsations irregular 非连续型
pulsation source 脉动源
pulsation spot welding 脉冲点焊
pulsation system 脉动系统
pulsation tachometer 脉冲转速表
pulsation theory 脉动学说;脉动理论
pulsation variable 脉动变星
pulsation wave 脉动波
pulsation welding 脉动焊接;脉冲焊;多脉冲接触焊;多脉冲焊接
pulsation welding timer 脉冲焊接计时器
pulsation wind velocity 脉动风速
pulsative movement 脉动
pulsative oscillation 脉动振荡;脉冲式振荡
pulsative zoning 脉动分带
pulsator 液压拉伸压缩疲劳试验机;蒸汽双缸泵;振动筛;脉动器;脉冲试验机
pulsator air jig 风力脉动跳汰机
pulsator clarifier 脉动澄清池
pulsator classifier 脉动分级器;脉冲分极器
pulsator controller 脉动控制器
pulsator deduster 脉动吸尘器
pulsator dyeing 脉动染色
pulsator dyeing machine 脉动染色机
pulsator jig 脉动跳汰机
pulsator line 脉动器真空管路
pulsator rig 凿岩机
pulsator screen 脉动筛
pulsator vaccum line 脉动器真空管路
pulsatory 脉动的
pulsatory current 脉动电流
pulsatory oscillation 脉动(波)
pulsatory pressure 脉动压力
pulsatory ratio 脉动比率
pulsatron 脉冲加速器;脉冲管
pulscope 脉冲示波器
pulse 脉冲;半周
pulse-absorbed 脉冲吸收的
pulse absorber 脉冲吸收器
pulse absorption 脉冲吸收
pulse accumulator 脉冲累加器;脉冲存储器
pulse acquisition radar coverage 脉冲搜索雷达的探测范围
pulse address multiple access 脉冲寻址多址连接
pulse advancing 脉冲提前
pulse air 脉冲空气
pulse airbore electromagnetic instrument 脉冲电磁仪
pulse airbore electronic magnetic system 脉冲航电仪
pulse air cleaning 脉冲空气清洗
pulse air induction reactor 脉冲负压空气喷射净化器
pulse air vibration 脉冲空气振动
pulse altimeter 脉冲测高计
pulse ambiguity 脉冲模糊性;脉冲二重性
pulse amplification 脉冲放大
pulse amplifier 脉冲放大器
pulse amplitube 脉冲放大管
pulse amplitude 脉冲振幅;脉冲幅度
pulse amplitude analysis 脉冲幅分析;脉冲幅度分析;脉幅变分析
pulse amplitude analyzer 脉冲幅度分析器
pulse amplitude code modulation 脉幅编码调制
pulse amplitude code modulation telemetry 脉冲幅度编码调制遥测技术
pulse amplitude discrimination 脉冲振幅鉴别
pulse amplitude discriminator 脉冲幅度甄别器
pulse amplitude distribution 脉冲幅度分布
pulse amplitude-modulated altimeter 脉冲调幅式高度表
pulse amplitude-modulated carrier 脉冲幅度调制载波
pulse amplitude-modulated sample 脉幅调制样值
pulse amplitude modulation 脉冲调制;脉冲振幅调制;脉冲幅(度)调制
pulse amplitude modulation-frequency modulation 脉幅调制调频
pulse amplitude modulation switching system 脉幅调制交换机
pulse amplitude modulator 脉冲调幅器
pulse amplitude spectrum 脉幅谱;脉冲振幅谱;脉冲幅谱
pulse amplutude-to-time converter 脉冲幅度—时间变换器
pulse analysis 脉冲分析
pulse analyzer 脉冲分析仪
pulse annealing 脉冲退火
pulse antenna 脉冲天线
pulse apex 脉冲顶点
pulse arc starter 脉冲引弧器
pulse arc welding 脉冲电弧焊(接)
pulse arrival 脉冲波至
pulse attenuation 脉冲衰减
pulse attenuator 脉冲衰减器
pulse average time 脉冲平均时间
pulse averaging circuit 脉冲平均电路
pulse back edge 脉冲后沿
pulse band 脉冲频带
pulse bandwidth 脉冲频带宽度;脉冲带宽
pulse bandwidth filtering 脉冲滤波;脉冲带宽滤波
pulse base 脉冲基线;脉冲底
pulse beating 脉冲差拍
pulse biased photicon 脉冲偏压式移像光电摄像管
pulse biasing 脉冲偏置法
pulse bleaching 脉动漂白
pulse blocking 脉动阻塞;脉冲阻塞
pulse blocking counter 脉冲阻隔计算器
pulse broadening 脉冲展宽
pulse broadening circuit 脉冲展宽电路
pulse bucking adder 脉冲补偿加法器
pulse build-up time 脉冲形成时间;脉冲建立时间
pulse burst 脉冲群;脉冲冲击
pulse bus 脉冲总线
pulse cable 脉冲电缆
pulse calorimetry 脉冲量热法
pulse cam 脉冲凸轮
pulse carrier 脉冲载波
pulse chamber 脉冲室
pulse characteristic 脉冲特性
pulse charging precipitator 脉冲荷电除尘器
pulse chase 脉冲追踪
pulse chopper 脉冲斩波器;脉冲断续器
pulse circuit 脉冲电路
pulse circuit theory 脉冲电路理论
pulse clarifier 脉冲澄清池
pulse clipper 脉冲削波器
pulse clipping 脉冲削波
pulse clock 脉率计
pulse clock generator 脉冲时钟发生器
pulse code 脉码;脉冲电码;脉冲代码;脉冲编码
pulse code control system 脉码控制系统
pulse code demodulator 脉(冲)码解调器;脉冲解码器
pulse coded source 脉冲编码震源
pulse code mdoulation exchange 脉码调制交换机
pulse code modulated system 脉码编码调制系统
pulse code modulating system 脉码调制系统
pulse code modulation 脉码调制;脉宽调制;脉冲(代)码调制;脉冲编码调制;脉冲编码调节;编码调制
pulse code modulation cable 脉码调制电缆
pulse code modulation converter 脉码调制变换器
pulse code modulation digital 脉码调制数字
pulse code modulation-frequency modulation 脉码调制—调频
pulse code modulation integrative measuring set 脉冲编码调制综合测量仪
pulse code modulation multiplex equipment 脉码调制多路复用设备;脉宽调制复用设备
pulse code modulation multiplexer 脉码调制多路复用器
pulse code modulation recorder 脉码调制记录器
pulse code modulation simulator 脉码调制模拟器
pulse code modulation telemetry system 脉码调制遥测系统;脉冲编码遥测系统;脉冲编码调制遥测系统
pulse code modulation terminal 脉码调制终端机
pulse code modulation transmission tester 脉冲码调制传输测试仪
pulse code modulator 脉码调制器;脉冲码调制器
pulse coder 脉冲编码器
pulse code receiver 脉码接收器
pulse code reticle 脉冲编码整盘
pulse code signalling system 脉码信号系统
pulse code system 脉冲码系统
pulse code transmission 脉码传输
pulse code tube 脉码管
pulse coding and correlation 脉冲编码和相互关系
pulse coding system 脉冲编码制
pulse coding tube 脉冲编码管
pulse coefficient 脉冲系数
pulse coincidence 脉冲符合
pulse coincidence detector 脉冲重合检测器;脉冲符合计数器
pulse coincidence effect 脉冲重合效应
pulse column 脉动柱;脉冲塔
pulse combustion 脉冲燃烧
pulse communication 脉冲通信
pulse communication system 脉冲通信制
pulse compaction technique 脉冲压缩技术
pulse comparator 脉冲比较器
pulse compression 脉冲压缩
pulse compression filter 脉冲压缩滤波器
pulse compression method 脉冲压缩法
pulse compression radar 脉冲压缩雷达
pulse compression ratio 脉冲压缩比
pulse compressive receiver 脉冲压缩接收机
pulse concentration 脉冲聚焦
pulse condition 脉冲状态
pulse connector 脉冲连接器;脉冲接插件
pulse control 脉冲控制
pulse controlled generator 脉冲控制发生器
pulse controlled tube oscillator 脉冲控制电子管振荡器
pulse controller 脉冲控制器;脉冲控制器
pulse control modulation 脉冲控制调制
pulse control system 脉冲控制系统
pulse converter 脉冲变换器
pulse-converter system of turbocharging 涡轮增压的脉冲变换系统
pulse copper vapour laser 脉冲铜蒸气激光器
pulse corrector 脉冲校正器;脉冲边沿修整器
pulse correlation 脉冲波对比
pulse count 脉冲数
pulse count converter 脉冲计数换算器
pulse count encoder 脉冲计数编码器
pulse counter 脉冲计数器;脉冲分频计数器;脉冲定标器
pulse counter detector 脉冲计数器—检波器;脉冲计数检测器
pulse counter discriminator 脉冲计数式鉴频器
pulse counting 脉冲计数
pulse counting discriminator 脉冲计数鉴别器
pulse counting equipment 脉冲计数装置
pulse counting method 脉冲计数方法
pulse counting module 脉冲计数组件;脉冲计数模件
pulse counting photometer 脉冲计数光度计
pulse counting system 脉冲计数制;脉冲计数系统
pulse count modulation 脉冲计数调制
pulse coupling 脉冲耦合
pulse crest 脉峰;脉冲顶峰
pulse crest factor 脉冲波峰因数
pulse cross-spectrum 脉动交叉光谱
pulse crowding 脉冲拥挤
pulse current 脉冲电流
pulsecutting 脉冲削波
pulse cycle 脉冲周期
pulsed 脉冲激发的
pulsed accelerator 脉冲加速器
pulsed adsorption bed 脉冲吸附床
pulsed adsorption bed process 脉冲吸附床过程;脉冲吸附床法
pulse damper 脉冲阻尼器
pulsed amplitude-modulated carrier 脉冲调幅载波
pulsed anaerobic baffled reactor 脉冲厌氧折流板反应器
pulsed anaerobic filter 脉冲厌氧滤池
pulsed arc 脉冲电弧
pulsed arc-exciter 脉冲激弧器
pulsed arc stabilizer 脉冲稳弧器
pulsed arc starter 脉冲引弧器

pulsed arc-welding power source 脉冲弧焊电源
pulsed arc welding rectifier 脉冲弧焊整流器
pulsed attenuator 脉冲衰减器;脉冲分压器
pulsed automatic gain control 脉冲自动增益控制
pulsed beam 脉冲束
pulsed beam current 脉冲束流
pulsed beam transport system 脉冲束流输送系统
pulsed bed 脉冲床;脉动床
pulsed bed filter 脉动床滤池
pulsed bed ion exchanger 脉动床离子交换器
pulsed bed sorption 脉冲床吸着
pulsed bending magnet 脉冲偏转磁铁
pulsed biased photicon 脉冲偏压式辐帖康
pulsed biasing 脉冲偏压法
pulsed bombardment 脉冲袭击
pulsed carrier-current system 脉冲载波系统
pulsed cavitation 脉动空蚀
pulsed coil 脉冲线圈
pulse(d) column 脉动塔;脉冲塔;脉冲柱
pulsed compression wave 脉冲压缩波
pulsed condition 脉冲状态
pulsed continuous wave 脉冲连续波
pulsed cooling temperature control system 脉冲冷却温度控制系统
pulsed corona discharge 脉冲电晕放电
pulsed counter 脉冲计数器
pulsed cyclotron 脉冲式回旋加速器;脉冲回旋加速器
pulsed deflection 脉冲偏转
pulsed deflector 脉冲偏转板
pulsed dielectric(al) laser 脉冲式电介质激光器
pulsed diffusion 脉冲扩散
pulsed dipole 脉冲偶极
pulsed discharge 脉冲式放电;脉冲放电
pulsed discharge lamp 脉冲式放电灯
pulsed discharge tube 脉冲放电管
pulsed Doppler lider 脉冲多普勒激光雷达
pulsed Doppler radar 脉冲调制多普勒雷达系统
pulsed Doppler system 脉冲式多普勒制
pulsed Doppler trackier 脉冲多普勒跟踪雷达
pulsed Doppler tracking 脉冲多普勒雷达跟踪
pulsed dosage 脉冲剂量
pulsed drive 脉冲驱动
pulsed dust collector 脉冲除尘器
pulsed dye laser densitometry 脉冲染料激光密度测定法
pulsed dye technique 脉冲染料技术
pulse decay 脉冲衰变
pulse decay time 脉冲下降时间;脉冲衰减时间;脉冲后沿持续时间
pulsed echo ultrasonoscope 回波反射式超声诊断仪
pulse decoder 脉冲译码器
pulse deconvolution 脉冲反褶积
pulsed eddy 脉动涡流
pulsed eddy current instrument 脉冲涡流测量仪
pulsed ejector 脉冲喷射器
pulse delay 脉冲延迟
pulse delay fall time 脉冲下降延迟时间
pulse delay line 脉冲延迟线
pulse delay mode 脉冲延迟模
pulse delay network 脉冲延迟网络
pulse delay nomogram 脉冲延迟时间列线图
pulse delay time 脉冲延迟时间
pulse delay unit 脉冲延时器;脉冲延迟器
pulsed electron accelerator 脉冲电子加速器
pulsed electron capture detector 脉冲电子俘获检测器
pulsed electron electrostatic accelerator 脉冲电子静electric加速器
pulsed electron impaction source 脉冲电子碰撞离子源
pulsed electrostatic deflector 脉冲静电偏转器
pulse delta modulator 脉冲增量调制器
pulse demagnetization 脉冲退磁
pulsed emission 脉动放射
pulsed emission current 脉动发射电流
pulse demoder 脉冲译码器;脉冲解码器
pulse demodulation 脉冲解调
pulse demodulator 脉冲解调器
pulse density 脉冲密度
pulse density modulation 脉冲密度调制
pulse density envelope detector 脉冲包络检波器
pulse envelope principle 脉冲包络原理

pulse deopiking 脉冲钝化
pulse detection 脉冲探测;脉冲检波
pulse detector 脉冲探测器;脉冲检测器;脉冲检波器
pulsed expansion joint 脉冲胀接
pulsed extraction 脉冲引出
pulsed extraction column 脉动式萃取塔;脉冲萃取塔
pulsed feed 脉动加料
pulsed feeding 脉动馈给
pulsed field 脉冲场
pulsed field accelerator 脉冲场加速器
pulsed field alternating-gradient accelerator 脉冲场交变陡度加速器
pulsed field constant-gradient accelerator 脉冲场恒陡度加速器
pulsed field constant-gradient synchrotron 脉冲场恒陡度同步加速器
pulsed field gel electrophoresis 脉冲凝胶电泳
pulsed filter 脉冲过滤器;脉冲滤528器
pulsed flash gas discharge lamp 脉冲闪光气体放电灯
pulsed flow hydraulics 交流液压技术
pulsed fluorescence hydrogen sulfide monitor 脉冲荧光法硫化氢监测仪
pulsed fluorescence sulfur dioxide monitor 脉冲荧光法二氧化碳监测仪
pulsed Fourier transform 脉冲傅立叶转换
pulsed frequency modulated carrier 脉冲调频载波
pulsed frequencymodulated signal 脉冲调频信号
pulsed frequent flash lamp 脉冲频闪灯
pulsed fusion system 脉冲聚变系统
pulsed gas analyzer 脉冲气体分析仪
pulsed gas cleaning 脉冲气体清洗
pulsed gasdynamic(al) laser 脉冲气动激光器
pulsed gas fluidization 脉动气体流态化
pulsed gas laser 脉冲气体激光器
pulsed glide path 脉冲滑翔道
pulsed gravity wave 脉冲引力波
pulsed guidance beam 脉冲制导波束
pulsed guidance laser 脉冲制导激光器
pulsed high-voltage discharge 脉冲高压放电
pulsed high-voltage generator 脉冲高压发生器
pulsed hydrogen thyratron 脉冲氢闸流管
pulse dialer 脉冲拨号器
pulse dialing 脉冲式拨号
pulse diathermy apparatus 脉冲透热机
pulse differentiator 脉冲微分器
pulse diffraction 脉冲衍射
pulse digit 脉冲数位
pulse digit spacing 脉冲数字间隔
pulsed individual tone system 脉冲单音制
pulsed inflector 脉冲偏转器
pulsed inflector voltage 脉冲偏转器电压
pulsed infusion 脉动浸渗;断续注入;脉动注水
pulsed infusion blasting 脉动注水爆破
pulsed infusion explosive 脉冲导燃炸药
pulsed infusion shot 高压注水爆破
pulsed infusion shot-firing 脉冲导燃装药放炮;脉动注水爆破
pulsed intensity 脉冲强度
pulsed ion chamber 脉冲电离室
pulsed ion cyclotron resonance spectrometer 脉冲离子回旋振波谱仪
pulsed ionizing beam 脉冲离子束
pulse direct current motor 脉冲式直流电动机
pulse direction finder 脉冲探向器
pulse direction finding 脉冲探向
pulse discharge 脉冲放电
pulse discriminator 脉冲甄别器;脉冲鉴别器
pulse dispersion 脉冲色散
pulse displacement modulation 脉冲位调制
pulse display 脉冲显示
pulse dissipation 脉冲耗散
pulse distance 脉冲间距
pulse distance finder 脉冲测距计
pulse distance system 脉冲测距制
pulse distortion 脉冲失真;脉冲畸变
pulse distribution amplifier 脉冲分配放大器
pulse distributor 脉冲分配器
pulse divider 脉冲分频器;脉冲分配器
pulsed jet cloth filter 脉冲布袋除尘器
pulsed jet intensifier 脉冲射流增强器
pulsed klystorn 脉冲速调管
pulsed lamp 脉冲灯

pulsed landing system 盲目着陆脉冲系统
pulsed laser 脉冲激光器;脉冲激光
pulsed laser action 脉冲激光作用;脉冲激光振荡
pulsed laser beacon 脉冲激光指向标;脉冲激光信标
pulsed laser beam 脉冲激光束
pulsed laser bonding 脉冲激光焊接
pulsed laser hologram 脉冲激光全息照相;脉冲激光全息图
pulsed laser holography 脉冲激光全息术
pulsed laser illuminator 脉冲激光照明器
pulsed laser power 脉冲激光功率
pulsed laser rangefinder 脉冲激光测距仪
pulsed laser ranging equation 激光脉冲测距方程
pulsed laser welder 脉冲激光焊机
pulsed light 光脉冲;脉动光;脉冲光
pulsed-light ceilometer 脉冲光束云高计
pulsed light cloud height indicator 脉冲光束云高计
pulsed light generator 脉动照明光源;脉冲光发生器
pulsed light range finder 脉冲光线测距仪
pulsed magnet 脉冲磁铁
pulsed magnetic extractor 脉冲磁引出装置
pulsed magnetic field apparatus 脉冲磁场装置
pulsed magnetic mirror 脉冲磁镜
pulsed magnetic shutter 脉冲磁快门
pulsed magnetic slit 脉冲磁(场)狭缝
pulsed magnet power 脉冲磁功率;脉冲磁场功率
pulsed magnetron 脉冲磁控管
pulsed matrix isolation 脉冲矩阵隔离
pulsed metal argon-arc-welding 熔化极脉冲氩弧焊
pulsed metal vapo(u)r laser 脉冲金属蒸气激光器
pulsed microwave radar 脉冲微波雷达
pulsed mixer-settler 脉冲混合澄清器
pulsed mode 脉冲工作方式
pulsed modulated carrier 脉冲调制载波
pulsed modulator tube 脉冲调制管
pulsed momentum 脉冲动量
pulsed multi-cavity klystron 脉冲多腔速调管
pulsed neutron 脉冲中子
pulsed neutron activation 脉冲中子活化
pulsed neutron die-away method 脉冲中子消失法
pulsed neutron generator 脉冲中子发生器
pulsed neutron interrogation 脉冲中子探询法
pulsed neutron log 脉冲中子测井
pulsed neutron method 脉冲中子法
pulsedneutron reactor 脉冲中子堆
pulsed neutron source method 脉冲中子源法
pulsed neutron technique 脉冲中子技术
pulsed nuclear magneitc resonance spectrometer 脉冲核磁共振分光仪;脉冲核磁共振分光计
pulsed nuclear reaction 脉冲核反应
pulsed off signal 脉冲断路信号
pulsed off signal generator 脉冲切断式信号发生器;脉冲断路信号发生器
pulsed operation 脉冲状态工作
pulse Doppler 脉冲式多普勒系统
pulse Doppler clutter 脉冲多普勒杂波
pulse Doppler early warning radar 脉冲多普勒预警雷达
pulse Doppler radar 脉冲多普勒雷达
pulse Doppler search and acquisition radar 脉冲多普勒搜索与捕获雷达
pulse Doppler spectrum 脉冲多普勒频谱
pulse Doppler surveillance radar 脉冲多普勒监视雷达
pulse Doppler system 脉冲多普勒系统
pulsed oscillator 脉控振荡器
pulse dot soldering iron 脉冲点钎焊烙铁
pulse doubling 脉冲加倍
pulsed packed column 脉冲填料柱
pulsed packed tower 脉冲填料塔
pulsed perforated-plate tower 脉冲筛板塔
pulsed perturbation 脉冲微扰
pulsed phase-modulated carrier 脉冲调相载波
pulsed photomultiplier 脉冲光电倍增器;脉冲光电倍增管
pulsed plasma 脉冲等离子体
pulsed plasma arc welding 脉冲等离子弧焊
pulsed plate column 脉冲盘柱
pulsed pneumatic dryer 脉冲式气流干燥器
pulsed power technology 脉冲能量技术
pulsed power travelling-wave tube 脉冲功率行波管
pulsed pyrolysis unit 脉冲热解单元

pulsed quenching 脉冲淬火
pulsed radar altimeter 脉冲式雷达测高仪
pulse(d) radar impulse radar 脉冲雷达
pulsed radar set 脉冲雷达设备
pulsed radar system 脉冲雷达系统
pulsed radioactive source 脉冲放射源
pulsed reactor 脉冲反应堆;脉冲堆
pulsed rectifier 脉冲整流器
pulsed reset 脉冲重调;脉冲复位
pulsed resonance spectroscopy 脉冲共振光谱学
pulsed ring 脉冲场环
pulsed ring discharge 脉冲环形放电
pulse-driven loop reactor 脉冲驱动循环反应器
pulse driver 脉冲驱动器
pulse droop 脉顶倾斜;脉冲衰减;脉冲顶部倾斜
pulse dropout 脉冲脱落
pulsed ruby laser 脉冲式红宝石激光器;脉冲红宝石激光器;红宝石脉冲激光器
pulsed ruby optic(al) laser 脉冲红宝石激光器;脉冲红宝石光学激光器
pulsed sandwich hologram 脉冲夹层全息图
pulsed sandwich holography 脉冲分层全息照相术
pulsed sequencing batch reactor process 脉冲序批间歇式反应器
pulsed servo(system) 脉冲随动系统
pulsed short wave 脉短波
pulsed sieve-plate column 脉冲筛板柱;脉冲筛板塔
pulsed sieve-plate extraction column 脉冲筛板萃取柱;脉冲筛板抽提塔
pulsed sound level 脉冲声级
pulsed sound wave 脉冲声波
pulsed spot welding 脉冲点焊
pulsed spray arc 脉冲喷射电弧
pulsed spray arc welding 脉冲射流电弧焊
pulsed spray transfer 脉冲喷射过渡
pulsed strong-current accelerator 脉冲强流加速器
pulsed submillimeter laser interferometer 脉冲亚毫米波激光干涉仪
pulsed superconducting magentic field 脉冲超导磁场
pulsed superconducting magnet 脉动超导磁铁
pulsed synchrotron 脉冲同步加速器
pulsed temperature gradient 脉冲温度陡度
pulsed thyratron 脉冲闸流管
pulsed transfer function 脉冲传输函数;脉冲传递函数
pulsed transformer accelerator 脉冲变压加速器
pulsed triaxial test 脉冲三轴试验
pulsed tungsten argon arc weld(ing) 脉冲钨极氩弧焊;钨极脉冲氩弧焊
pulsed tungsten gas arc welding 脉冲钨极气体保护焊
pulsed tungsten inert gas arc welding 脉冲钨极惰性气体保护电弧焊
pulse duct 脉动式喷气发动机;冲压管
pulse duct engine 脉动式空气喷气发动机
pulsed ultraviolet laser rangefinder 脉冲紫外激光测距仪
pulse duplicator 脉冲复制器
pulse duration 脉冲时间;脉冲宽度;脉冲持续时间
pulse duration coder 脉宽编码器
pulse duration control 脉冲宽度控制
pulse duration counter 脉冲时间计数器
pulse duration discrimination 脉宽鉴别
pulse duration discriminator 脉冲宽度鉴别器
pulse duration modulation 脉宽调制;脉幅调制;脉冲宽度调制;脉冲持续时间调制
pulse duration modulation-frequency modulation 脉冲持续时间调制调频
pulse duration ratio 占空系数;脉宽周期比;脉冲占空系数
pulse duration system 脉冲宽度制
pulse dust collector 脉冲除尘器
pulse duty factor 脉冲占空因子;脉冲占空因数;脉冲占空系数;脉冲占空比
pulsed voltage 脉冲电压
pulsed water jet 脉冲水力喷射
pulsed wave Doppler ultrasonic flow meter 脉冲波多普勒超声流量计
pulsed xenon lamp 脉冲氙灯
pulsed xenon light 脉动氙灯;脉冲氙灯
pulse echo 脉冲回波
pulse echo method 脉冲回波法;脉冲反射法
pulse echo receiver 脉冲回波接收器

pulse echo tester 脉冲回波测试仪
pulse echo ultrasonics 脉冲回波超声学
pulse echo ultrasonogram 脉冲反射超声图
pulse edge 脉冲沿;脉冲边缘;脉冲边沿
pulse edge sense network 脉冲边沿检测网络
pulse electrolysis 脉冲电解
pulse electromagnetic energy apparatus 脉冲电磁能机
pulse electromagnetic force 脉冲电磁力
pulse electromagnetic kinetic energy generator 脉冲电磁动能发电机
pulse eletromagnetic method 脉冲电磁法
pulse emission 脉冲发射
pulse emitter 脉冲发射器
pulse emitter load 脉冲发射器负载
pulse encoder 脉冲编码器
pulse energized electrostatic precipitator 脉冲激能集尘器
pulse energy 脉冲能量
pulse envelope 脉冲波包;脉冲包络
pulse envelope method 脉冲包线法
pulse equalizer 脉冲均衡器
pulse excitation 脉冲激励;脉冲激发
pulse excited antenna 脉冲激励天线
pulse experiment 脉冲实验
pulse extender 脉冲扩展器
pulse extraction column 脉动萃取塔;脉动抽提柱
pulse extra high-tension generator 脉冲超高压发生器;超高压脉冲发生器
pulse factor 脉冲系数
pulse fali delay 脉冲下降延迟
pulse fall time 脉冲下降时间;脉冲衰减时间
pulse fatigue machine 脉冲疲劳试验机
pulse fatigue testing machine 脉动疲劳试验机
pulse fault indicator 脉冲故障指示器
pulse feeder 脉动式给矿机
pulse finding accuracy 脉冲探测精确度
pulse firing 脉冲点火
pulse fission chamber 脉冲裂变箱
pulse flow performance 脉冲流特性
pulse follower 脉冲跟踪器
pulse form 脉冲形式;脉冲形
pulse formation 脉冲形成
pulse former 脉冲形成器
pulse forming amplifier 脉冲形成放大器
pulse forming coil 脉冲形成线圈
pulse forming delay line 脉冲形成延迟线
pulse forming line 脉冲形成线
pulse forming network 脉冲形成网络
pulse forming panel 脉冲形成电路板
pulse-free chromatographic pump 无脉冲色谱泵
pulse-free flow 无脉动流;无脉冲流
pulse-free pump 无脉冲泵
pulse-free signal 无脉冲信号
pulse frequency 脉冲频率
pulse frequency control 脉冲频率控制
pulse frequency discriminator 脉冲鉴频器
pulse frequency divider 脉冲分频器
pulse frequency dividing 脉冲分频
pulse frequency modulating 脉冲频率调制
pulse frequency modulation 脉冲调频
pulse frequency modulation telemetry 脉冲调频遥测术
pulse frequency nuclear magnetic resonance flowmeter 脉冲频率核磁共振流量计
pulse frequency spectrum 脉冲频谱
pulse front edge 脉冲前沿
pulse fuel rod 脉冲燃料棒
pulse gas shutter 脉冲气体断路器
pulse gate 选通脉冲;闸波脉冲门;脉冲门
pulse gated system 脉冲选通系统
pulse generating 脉冲发生
pulse generating means 脉冲振荡装置
pulse generating tube 脉冲形成管
pulse generator 门脉冲发生器;脉冲振荡器;发生器
pulse generator clock 脉冲发生器时标
pulse generator disc 脉冲发生器圆盘
pulse generator system 脉冲发生器系统
pulse glide display 脉冲滑翔道显示图形
pulse group 脉冲群;脉冲组;脉冲群;脉冲列
pulse guidance laser radar 脉冲制导激光雷达
pulse guidance radar 脉冲制导雷达
pulse half wave 脉冲半波

pulse hand magnet 脉冲式手持吸铁器
pulse height 脉冲高度;脉冲幅值;脉冲幅度;冲击高度
pulse height analysis 脉冲高度分析
pulse height analyzer 脉冲高度分析仪;脉冲高度分析器;脉冲幅度分析器;脉冲幅度分析器
pulse height discriminator 脉冲振幅鉴别器;脉冲幅度鉴别器;波高鉴别器;幅度甄别器;幅度鉴别器
pulse height distribution 脉冲振幅分布;脉冲高度分布
pulse height measurement 脉冲高度测定
pulse height resolution 脉冲振幅清晰度;脉冲高度分辨率
pulse height selector 脉幅选择器;脉冲振幅选择器;脉冲高度选择器;脉冲幅度选择器
pulse height spectrum 脉冲高度谱
pulse height-to-time converter 脉冲振幅时间转换器
pulse high frequency apparatus 脉冲高频机
pulse hologram 脉冲全息图
pulse improvement threshold 脉冲信噪比改善阈值
pulse induction 脉冲感应
pulse inhibit 脉冲禁止;脉冲封锁
pulse initiating rotor 脉冲起动转子
pulse injection 脉动注射;脉冲注入
pulse input 脉冲输入
pulse input asynchronous network 脉冲输入异步网络
pulse inserting circuit 脉冲引入电路
pulse integrating detector 脉冲积分检波器
pulse integrating divider 脉冲积分选择器
pulse integration 脉冲积分法
pulse integrator 脉冲积分器
pulse interference 脉冲干扰
pulse interference blanker 脉冲干扰消隐器
pulse interference eliminator 脉冲干扰消除器
pulse interference separator 脉冲干扰分离器
pulse interference separator and blanker 脉冲干扰分离器和消除器
pulse interferometer 脉冲干扰仪
pulse interlacing 脉冲交织;脉冲交错
pulse interleaving 脉冲交织
pulse interrogation 脉冲询问
pulse interval 脉冲周期;脉冲间隔
pulse interval jitter 脉冲间隔跳动
pulse interval modulation 脉冲间隔调制
pulse interval time 脉冲间隔时间
pulse inverter 脉冲倒相器
pulse ionization chamber 脉冲电离箱;脉冲电离室
pulse ionizing device 脉冲电离器
pulse irradiation 脉冲辐照
pulse jet 脉动喷射;脉动式喷气发动机;脉冲式喷气发动机
pulse-jet baghouse 脉冲喷吹袋式除尘器
pulse-jet engine 脉动式空气喷气发动机;脉冲式喷气发动机
pulse jet filter 脉冲喷气式袋集尘器
pulsejet missile 脉动式喷气发动机导弹
pulse jitter 脉冲跳动;脉冲抖动;脉冲颤动
pulse justification 脉冲码速调整
pulse keyer 脉冲键控器
pulse label(l)ing 脉冲标记
pulse label(l)ing technique 脉冲标记技术
pulse laser amplifier 脉冲激光放大器
pulse laser beam 脉冲激光束
pulse laser induced epitaxy 脉冲激光外延
pulse laser system 脉冲激光系统
pulse laser welding 脉冲激光焊接
pulse lead 脉冲提前
pulse leading edge time 脉冲前沿时间
pulse length 脉宽;脉冲宽度;脉冲持续时间;脉冲长度
pulse lengthener 脉冲延长器
pulse length error 脉冲宽度误差
pulse length impulse 脉冲持续时间
pulse length modulation 脉幅调制;脉冲宽度调制;脉冲长度调制;脉冲持续时间调制
pulseless pump 无脉动泵
pulse limiting rate 脉冲限制率;脉冲的极限重复频率
pulse line 脉冲管
pulse link repeater 脉冲线路增音机;脉冲链重发器;脉冲链路转发器;脉冲链路增音机
pulselite 脉光灯

pulse load(ing) 脉冲荷载;脉冲负载
pulse logic system 脉冲逻辑系统
pulse machine 脉冲发生器
pulse mass-spectrometer 脉冲质谱仪
pulse matching 脉冲协调;脉冲均调
pulse matching technique 脉冲匹配法
pulse measuring oscillograph 脉冲测示波器
pulse meter 脉冲计
pulse method 脉冲法
pulse method calibration 脉冲法校准
pulse method of measuring sound velocity 测量声速的脉冲法
pulse/minute/cm 计数率/厘米
pulse mixer 脉冲混频器
pulse mixer-settler 脉动混合沉降器
pulse mixing 脉冲混合
pulse mode 脉冲型;脉冲形式;脉冲方式
pulse mode multiplex(ing) 脉冲式多路传输;脉冲型多路传输
pulse moder 脉冲编码装置
pulse modulated 脉冲调制的
pulse modulated carrier 脉调载波
pulse modulated communications system 脉冲调制通信系统
pulse modulated jamming 脉冲调制干扰;调制脉冲干扰
pulse modulated magnetron 脉冲调制磁控管
pulse modulated oscillator 脉冲调制振荡器
pulse modulated radar 脉冲调制雷达
pulse modulated signal 脉冲调制信号
pulse modulated sonar 脉冲调制声呐
pulse modulated wave 脉冲调制波
pulse modulation 脉冲调制
pulse modulation infrared system 脉冲调制红外系统
pulse modulation method 脉冲调制法
pulse modulation receiver 脉冲调制接收机
pulse modulation system 脉冲调制系统
pulse modulation technique 脉冲调制技术
pulse modulation transmission 脉冲调制传输
pulse modulation transmitter 脉冲调制发射机
pulse modulator 脉冲调制器
pulse modulator radar 脉冲调制器雷达
pulse monitor 脉冲监控器
pulse motor 脉冲电动机
pulse multiplex 脉冲调制多路传输
pulse multiplex coding 脉冲多路编码
pulse multiplex distance-measuring equipment 脉冲多路式测距装置
pulse narrowing 脉冲变窄
pulse narrowing device 脉冲窄化器
pulse narrowing system 脉冲压缩系统
pulse network 脉冲网络
pulse network modulator 脉冲网络调制器
pulse noise 脉冲噪声
pulse noise generator 脉冲噪声发生器
pulse normalization 脉冲整形;脉冲标准化
pulse normalizer unit 脉动规度装置
pulse nulling phenomenon 脉冲消失现象
pulse number 脉动数;脉波数
pulse number modulation 脉冲数调制;脉冲密度调制
pulse of air 大气压脉冲;气压脉冲
pulse off period 脉冲停止周期
pulse of sea-floor spreading 海底扩张的脉动
pulse of sound 声脉冲
pulse of the earth 地球脉动
pulse on 开启
pulse operated chamber 脉冲电离室
pulse operated cyclotron 脉冲运行回旋加速器;脉冲回旋加速器;脉冲操作回旋加速器
pulse operater chamber 脉冲操作电离室
pulse operating time 脉冲运行时间
pulse operation 脉冲运用;脉冲操作
pulse optic(al) radiation 光脉冲辐射
pulse oscillator 脉冲振荡器
pulse oscillator radar 脉冲振荡器雷达
pulse oscilloscope 脉冲示波器
pulse output 脉冲输出
pulse output power 脉冲输出功率
pulse output voltage 脉冲输出电压
pulse overlap 脉冲重叠
pulse packet 脉冲系列;脉冲群;脉冲链;脉冲块
pulse pairing 脉冲成对性

pulse parameter 脉冲参数
pulse passage 脉冲通道
pulse pattern 脉冲图形
pulse pattern generator 脉冲波形发生器;标准脉冲发生器
pulse peak 脉峰;脉冲峰值
pulse peak detector 脉冲峰值检波器;脉冲巅值检波器
pulse peak height 脉冲峰高
pulse peaking 脉冲修尖;脉冲峰化
pulse peak power 脉冲峰值功率;脉冲峰功率
pulse peak width 脉冲峰宽度
pulse period 脉冲周期
pulse per second 秒脉冲
pulse persistence 脉冲持续;脉冲宽度
pulse phase 脉相
pulse phase conveying 脉冲固相输送
pulse phase-locked loop 脉冲锁相环
pulse phase modulation 脉相调制;脉冲相位调制;脉冲调相;脉冲位置调制
pulse phase modulation-amplitude modulation system 脉冲调相—调幅制
pulse phase system 脉相系统
pulse phasing 脉冲相位调整;脉冲定相
pulse pile-up 脉冲聚积;脉冲堆积
pulse polaraulic selector 脉冲极性选择器
pulse polarity 脉冲极性
pulse polarity selector 脉冲极性选择器
pulse polarizer 脉动极化器;脉冲极化器
pulse polarograph 脉冲极谱仪
pulse polarographic determination 脉冲极谱测定(法)
pulse polarography 脉冲极谱法
pulse porch 脉冲边沿
pulse position 脉冲位置
pulse position indicator 脉冲位置指示器
pulse position modulation 脉位调制;脉宽调制;脉冲位置调制
pulse position modulation code 脉位调制码;脉冲位置调制码
pulse position modulation opical tracker 脉位调制光学跟踪仪
pulse position modulation system 脉冲位置调制系统
pulse positon modulator 脉冲位置调制器
pulse power 脉冲功率
pulse power breakdown 脉冲击穿功率;脉冲功击穿
pulse power driver 脉冲功率激励器
pulse powering cycle 脉冲供电周期
pulse power output 脉冲功率输出
pulse power source 脉冲电源
pulse power supply 脉冲功率源
pulse pressure-charging 脉冲增压
pulse process 脉冲过程
pulse profile 脉冲轮廓
pulse propagation 脉冲传播
pulse propagation meter 脉冲传播测定计
pulse propagation time 脉冲传播时间
pulse propagation velocity 脉冲传播速度
pulse pump 脉冲泵
pulse punching 脉冲穿孔
pulse pyrolysis 脉冲高温分解
pulse quadrupling 脉冲成四倍
pulse quenching press 脉冲淬火压床
pulser 脉冲器
pulse radar 脉冲式雷达
pulse radar altimeter 脉冲雷达高度表
pulse radar distance-measuring system 脉冲雷达测距系统
pulse radar transmitter 脉冲雷达发射机
pulse radiation 脉冲辐射
pulse radiation frequency 脉冲射频
pulse radiolysis 脉冲射解作用;脉冲射解
pulse ranging 脉冲测距
pulse ranging navigation 脉冲测距导航
pulse rate divider 脉冲重复频率分频器
pulse rate division 脉冲重复频率分隔
pulse rate factor 脉冲占空比
pulse rate meter 脉冲率表
pulse rate multiplication 脉冲重复频率倍增;脉冲重复倍频
pulse rate multiplier 脉冲速率乘法器
pulse rate of rise 脉冲上升率

pulse rate telemetering 脉率式遥测术
pulse ratio 脉冲比
pulse ratio encoding 脉冲比编码
pulse ratio modulation 脉冲比调制
pulse receiver 脉冲接收器;脉冲接收机
pulse recording 脉冲记录
pulse rectifier 脉冲整流器
pulse recurrence interval 脉冲周期;脉冲重复间隔;脉冲重复
pulse recurrence period 脉冲重复周期
pulse recurrence rate 脉冲重复率
pulse recurrence time 脉冲重复时间
pulse reflection fault locator 脉冲反射式故障定位器
pulse reflection principle 脉冲反射原理
pulse regeneration 脉冲正反馈;脉冲再生;脉冲恢复
pulse regeneration unit 脉冲再生器
pulse regenerative amplifier 脉冲再生式放大器
pulse regenerative oscillator 脉冲再生振荡器
pulse regenerator 脉冲再生器;脉冲再发器
pulse regime 脉冲范围
pulse registering device 脉冲记录线路
pulse rejector 脉冲排除器
pulse relaxation amplifier 脉冲张弛放大器
pulse relay 脉冲继电器;脉冲继电器
pulse remote control system 脉冲遥控系统
pulse repeater 脉冲转发器;脉冲重发器
pulse repeating relay 脉冲重复继电器
pulse repetition frequency 脉冲重复频率
pulse repetition frequency generator 脉冲重复频率发生器
pulse repetition frequency limitation 脉冲重复频率极限
pulse repetition frequency switch 脉冲重复频率开关
pulse repetition interval 脉冲重复间隔
pulse repetition period 脉冲重复周期
pulse repetition rate 脉冲重复频率
pulse repetition-rate modulation 脉冲频率调制
pulse reply 脉冲应答
pulse reproducibility 脉冲重复性
pulse reshaping 脉冲整形
pulse resistance 脉冲电阻
pulse resolution 脉冲分辨力
pulse resonance 脉冲共振
pulse resonance method 脉冲共振法
pulse response 脉冲反应;脉冲响应
pulse response adaptive 脉冲响应自适应
pulse response analysis 脉冲响应分析
pulse response analyzer 脉冲特性分析仪
pulse response curve 脉冲响应特性曲线
pulse response duration 脉冲响应宽度
pulse response time 脉冲响应时间
pulse response width 脉冲响应宽度
pulse ringing 脉冲振铃;脉冲瞬变
pulse ripple 脉冲纹波;脉冲涟波;脉冲波纹;脉冲波动
pulse rise delay 脉冲上升延迟
pulse rise time 脉冲上升时间;脉冲前沿时间
pulse rise time correction 脉冲前沿校正
pulse rising edge 脉冲上升边
pulser rack 脉冲发生器架
pulser timer 标记定时器
pulse sample-and-hold circuit 脉冲取样和保持电路
pulse sampling 脉冲取样
pulse sampling oscilloscope 脉冲取样示波器
pulse sampling technique 脉冲取样技术
pulse scaling circuit 脉冲计数电路
pulse scatterometer 脉冲散射计
pulse sealing 脉冲热合
pulse selector 脉冲选择器
pulse sending 脉冲发送
pulse separation 脉冲分离
pulse separator 脉冲分离器
pulse separator unit 脉冲分离装置
pulse sequence 脉冲序列;脉冲系列
pulse sequence detector 脉冲次序检测器
pulse series 脉冲序列
pulse series generator 脉冲组发生器;脉冲序列发生器;脉冲串发生器
pulse servo 脉冲伺服机构
pulse servo mechanism 有脉动信号的伺服机构
pulse servosystem 脉冲伺服系统
pulse setting controller 脉冲给定调节器

pulse shape 脉冲波形
pulse shape analyzer 脉冲形状分析器
pulse shape discriminator 脉冲形状甄别器
pulse shaper 脉冲整形器;脉冲形成器
pulse shaper-divider 脉冲整形分频器
pulse shaping 脉冲整形;脉冲成型
pulse shaping and re-shaping 脉冲整形和再整形;脉冲的成型与再成型
pulse shaping line 脉冲形成线
pulse shaping stage 脉冲形成级;脉冲成型级
pulse sharpening 脉冲锐化
pulse shifter 脉冲定相装置
pulse shortening 脉冲缩短
pulse shrinkage 脉冲变狭
pulse signal 脉冲信号
pulse signal generator 脉冲信号发生器
pulse signal source 脉冲信号源
pulse sintering 脉冲烧结
pulse site 脉冲产生点
pulse size 脉冲振幅大小
pulse slicer 脉冲限制器;脉冲限幅器
pulse smearing 脉冲拖影
pulse sonar 脉冲声呐
pulse sound 冲击声
pulse source 脉冲源
pulse spacing 脉冲间距;脉冲间隔
pulse spacing code 脉冲间隔编码
pulse spacing coding 脉冲间隔编码
pulse spacing modulation 脉冲间隔调制
pulse spectrograph 脉冲摄谱仪
pulse spectrometry 脉冲能谱学
pulse spectrum 脉冲谱;脉冲频谱
pulse spectrum bandwidth 脉冲频谱带宽
pulse spectrum envelope 脉冲频谱包络
pulse speed 脉冲速度
pulses per second 每秒脉冲数
pulse spike 脉冲尖锋(信号);脉冲波尖
pulse spike amplitude 脉冲尖峰幅度;脉冲巅值振幅
pulse spreading 脉冲展开
pulse spreading specification 脉冲展宽鉴定
pulse spring 脉动簧;脉冲触簧
pulse sprinkle head 脉冲式喷头
pulse stability 脉冲稳定性
pulse stacker 脉冲堆积器
pulse standardization 脉冲标准化
pulse starting stability 脉冲起始稳定性
pulse steepness 脉冲陡度
pulse steering 脉冲指引
pulse stepper 脉冲步进器
pulse storage time 脉冲存储时间
pulse strength 脉冲强度
pulse stress 脉冲应力
pulse stretcher 脉冲展宽器;脉冲伸张器;脉冲伸张管;脉冲扩展器
pulse stretching 脉冲展宽;脉冲拖尼
pulse string 脉冲链
pulse stripping analysis 脉冲解吸分析
pulse strong magnetic assembly 脉冲强磁装置
pulse stuffing 脉冲填充;脉冲塞入
pulse stuffing technique 脉冲塞入技术
pulse subcarrier 脉冲副载波
pulse superposition 脉冲叠加
pulse suppressor 脉冲抑制器
pulse switch 脉冲开关
pulse switching optic(al) circuit 脉冲开关光电路
pulse switch out 脉冲断开
pulse symmetrical phase modulation 脉冲对称相位调制
pulse synchronization 脉冲同步
pulse synchroscope 脉冲同步示波器
pulse synthesis 脉冲合成
pulse synthesizer 脉冲合成器
pulse system 脉系;脉冲系统
pulse system differential 脉冲系统微分
pulse system differential analyzer 脉冲系统微分分析器
pulse system simulation 脉冲系统仿真
pulse system turbo-charging 脉冲式涡轮增压
pulse tachometer 脉冲转速计
pulse tail 脉冲尾部
pulse tail clipping 脉冲后沿斩断
pulse target acquisition radar 脉冲目标捕捉雷达
pulse technique 脉冲技术
pulse telemetering method 脉冲遥测方法

pulse telemetry 脉冲遥测术
pulse temperature adjusting control board 脉冲温度调节控制板
pulse test(ing) 脉冲试验;脉冲测试
pulse threshold energy 脉冲阈能
pulse time 脉冲时间;脉冲间隔时间;脉冲间隔
pulse time code 脉冲时码
pulse time demodulator 脉时解调器;脉冲时间解调器
pulse time distribution 脉冲时间分布
pulse time division multiplex 脉冲时分多路复用
pulse time division system 时间分隔脉冲通信制
pulse time jitter 脉冲宽度波动
pulse time-modulated radiosonde 脉冲时间调制无线电(探空)仪
pulse time modulation 脉时调制;脉冲时间调制
pulse time modulation system 脉时调制系统
pulse time multiplex 脉冲定时多路传输
pulse timer 脉冲计时器;脉冲定时器
pulse time ratio 脉宽周期比;脉冲占空系数;占空系数
pulse timing 脉冲同步;脉冲计时;脉冲定时
pulse timing marker oscillator 脉冲时标发生器;脉冲时标振荡器
pulse to bar ratio 脉冲条信号比;脉冲对条信号比
pulse-tone audiometer 音频脉冲测听计
pulse top 脉冲顶部
pulse top clipper 脉冲顶部限幅器
pulse top compensation 脉冲顶部补偿
pulse top droop 脉冲顶部倾斜
pulse to-pulse correlation 脉冲间相关性;脉冲间的相互关系
pulse tracing 脉冲描记法;脉波描记法
pulse tracking system 脉冲跟踪系统
pulse trailing edge 脉冲下降边;脉冲后沿
pulse trailing-edge time 脉冲后沿时间
pulse train 脉冲序列;脉冲系列;脉冲列;脉冲链;脉冲串;脉冲波列
pulse train analysis 脉冲序列分析
pulse train frequency spectrum 脉冲序列频谱
pulse train generator 脉冲序列发生器;脉冲链发生器;脉冲串发生器
pulse train relay 脉冲继电器
pulse train spectrum 脉冲序列频谱
pulse transducer 脉冲变换器
pulse transformation 脉冲变换
pulse transformer 脉冲变压器;脉冲变量器
pulse transformer box 脉冲变压器室
pulse transit-time 脉冲传播时间
pulse transit-time method 脉冲传播时间法
pulse transmission 脉冲传输
pulse transmission circuit 脉冲传送线路
pulse transmission method 脉冲透射法
pulse transmission mode 脉冲传播模;脉冲传播模式
pulse transmission mode operation 脉冲透射运转
pulse transmission-reflection method 脉冲传输回波法;脉冲传输反射波法
pulse transmission technique 脉冲传输技术
pulse transmission tester 脉冲传播试验仪
pulse transmitter 脉冲发送器;脉冲发送机;脉冲发射机
pulse transponding 脉冲转发
pulse triaxial test 脉冲三轴试验
pulse trigger 脉冲触发器
pulse trigger action 脉冲触发
pulse triggered binary 脉冲触发二进元
pulse triggering 脉冲触发
pulse tube 脉冲管
pulse tube oscillator 脉冲电子管振荡器
pulse tube refrigerator 脉管制冷器
pulse type airborne radar 脉冲式机载雷达
pulse type altimeter 脉冲式测高计
pulse type correlator 脉冲式相关器
pulse type detector 脉冲型探测器
pulse type distance-measuring system 脉冲测距系统
pulse type disturbance 脉冲扰动
pulse type electric(al) clock 脉冲式电钟
pulse type ionization chamber 脉冲型电离箱
pulse type laser diode 脉冲型激光二极管
pulse type magnetohydrodynamic(al) generator 脉动式磁流体发电机
pulse type reactor 脉冲反应堆

pulse type speedometer 脉冲式速度计
pulse type telemeter 脉冲型遥测计;脉冲式遥测计
pulse type telemetering 脉冲型遥测;脉冲式遥测
pulse unit 脉冲部件
pulse vacuum steam disinfector 脉冲真空蒸汽消毒器
pulse vacuum tube voltmeter 脉冲真空管电压表
pulse valley 脉冲谷
pulse value 脉冲值
pulse valve 脉冲管;脉冲阀
pulse velocity 脉动速度
pulse velocity determination 脉冲速度测定
pulse velocity measurement 脉冲速度测定
pulse velocity test 脉冲速度试验;脉冲波速试验;超声波传播速度试验
pulse video thermography 脉冲电视热成像
pulse visibility 脉冲可见度
pulse visibility factor 脉冲可见率
pulse voltage 脉冲电压
pulse voltage divider 脉冲电压分压器
pulse voltage multiplier 脉冲倍压发生器
pulse voltage recorder 脉冲电压记录器
pulse volume 脉量
pulse wave 脉冲波;脉波
pulse wave Doppler apparatus 脉冲式多普勒仪
pulse waveform 脉冲波形
pulse wave of inducing field source in electromagnetic method 电磁法的激励场源脉冲波
pulse wave packet 脉冲波包
pulse welding interrupter 脉冲焊接断续器
pulse welding timer 脉冲焊接计时器
pulse widescope 脉冲宽度显示器
pulse width 脉宽;脉冲时宽;脉冲持续时间
pulse width coding 脉宽编码;脉冲宽度编码
pulse width control 脉宽调整;脉冲宽度控制;脉冲宽度调整
pulse width decoder 脉宽译码器;脉冲宽度译码器
pulse width discriminator 脉宽鉴别器;脉冲宽度鉴别器
pulse width distortion 脉宽失真
pulse width encoder 脉冲宽度编码器
pulse width encoding 脉冲宽度编码
pulse width keyer 脉宽键控器;脉冲宽度键控器
pulse width method 脉冲宽度法
pulse width-modulated analog/digital converter 脉宽调制模—数转换器
pulse width-modulated chopper 脉冲宽度调制的斩波器
pulse width-modulated inverter 脉宽调制变换器
pulse width-modulated power signal 脉冲宽度调制功率信号
pulse width-modulated wave 脉宽调制波
pulse width modulation 脉宽调制;脉幅调制;脉冲宽度调制;脉冲持续时间调制
pulse width modulation amplifier 脉宽调制放大器
pulse width-modulation chopper 脉宽调制斩波器
pulse width modulation-frequency modulation 脉冲持续时间调制调频
pulse width modulation principle 脉冲宽度调制原理
pulse width modulator 脉宽调制器;脉冲宽度调制器
pulse width mulitivibrator 脉冲宽度可控多谐振荡器
pulse width multiplexer 脉冲宽度多路调制器
pulse width multiplier 脉宽倍增器
pulse width multivibrator 脉冲多谐振荡器;脉冲宽度主控多谐振荡器
pulse width phase detector 脉宽相位检波器
pulsewidth recorder 脉宽记录器
pulse width recording 脉宽记录
pulse width selector 脉冲宽度选择器
pulse width shrinkage 脉宽收缩
pulse width signal(ling) 脉宽调制信号
pulse width standardization 脉宽标准化
pulse window 脉冲窗
pulse wise 脉冲法
pulse wound cleaner 脉冲式伤口清洗器
pulsimeter 脉力计;脉力计
pulsing 脉冲调制;脉冲(的)发生;脉冲(的)产生;发生脉冲
pulsing circuit 脉冲电路
pulsing device 脉冲装置
pulsing feed 脉动推进

pulsing form 脉冲形状
pulsing key 脉冲控制键;脉冲键控
pulsing laser 脉冲激光器
pulsing mechanism 脉动机构
pulsing operation 脉冲运行
pulsing rate 脉冲速率
pulsing signal 脉冲信号
pulsing system 脉冲制;脉冲系统
pulsing unit 脉冲设备
pulsing voltmeter 脉冲电压表
pulsion 推出
pulsion phase 脉冲相
pulsion stroke 脉动冲程
pulsion-suction cycle 脉动吸入周期
pulsive noise 脉冲性噪声
pulsometer 脉搏表;蒸汽抽水机;蒸气双缸泵;蒸气抽水机;气压扬水机;气压抽水机
pulsometer dredger 气压抽水挖掘机
pulsometer plant 蒸汽抽水机
pulsometer pump 蒸汽吸水泵;蒸汽双缸泵;蒸气吸水泵;汽压泵;气压抽水泵
pulsto-fatigue machine 脉冲疲劳机
pulszkyite 铜镁矾
pultrudate 挤拉产品
pultruded profile 挤拉型材
pultruder 挤拉机
pultrusion 挤挤成型;挤拉法;挤拉成型
pultrusion-compounding 挤拉复合
pultrusion machinery 挤拉机械
pultrusion product 拉挤产品;挤拉产品
pultrusion roving 挤拉用玻纱无捻粗纱
pulvation distance 抛散距离
pulverability 可粉化性;喷雾性能
pulverator 振动器;冲击器;轧机机;粉碎器
pulverator plough 松土犁
pulverescent 粉末状的
pulverin 草灰
pulverised anthracite 无烟煤粉
pulverised asbestos 石棉粉
pulverised ash cement 粉煤灰水泥
pulverised ash wall panel 粉煤灰墙板
pulverised bulk material 粉末状散装物料
pulverised burner 微粉喷燃器
pulverised coal 粉煤;煤粉;煤末
pulverised coal ash 粉煤灰
pulverised coal blower 煤粉鼓风机
pulverised-coal burner 粉煤燃烧炉
pulverised coal classifier 煤粉分离器
pulverised coal collector 细粉分离器;煤粉收集器
pulverised coal combustion chamber 煤粉燃烧室
pulverised coal feeder 给煤粉机
pulverised coal fired 烧煤粉的;煤粉加热的
pulverised coal fired boiler 煤粉锅炉
pulverised coal firing boiler 煤粉锅炉
pulverised coal firing plant 煤粉燃烧设备
pulverised coal fuel 煤粉燃料
pulverised coal nozzle 煤粉喷嘴;煤粉喷嘴
pulverised coal preparation 煤粉制备
pulverised coal(storage) bin 煤粉仓
pulverised coal-storage system 仓储[贮]式制粉煤系统
pulverised coal system 煤粉系统
pulverised corundum 金刚砂粉;刚玉砂
pulverised fuel 粉煤;粉状燃料;粉状燃料;雾状燃料
pulverised fuel ash 粉煤灰;粉状灰烬;粉状燃料粉;粉状燃料灰
pulverised-fuel-ash aggregate 粉煤灰轻骨料;粉煤灰轻集料
pulverised-fuel-ash cement 粉煤灰水泥
pulverised fuel boiler 煤粉锅炉
pulverised fuel feeder 给粉机
pulverised fuel system 燃料粉化系统
pulverised lime 粉石灰;粉状石灰;石灰粉
pulverised magnesian lime 镁质石灰粉
pulverised mixer 粉碎拌和机
pulverised powder 粉碎的粉末
pulverised sample 粉磨样品
pulverised slag 粒化渣;炉渣粉
pulverised solder 粉状焊锡;粉状焊剂;粉状焊料
pulverised stone 石粉
pulverised sulfur 粉末硫磺
pulverised talc 粉状滑石
pulverised texture 粉末状结构
pulveriser 粉磨机

pulverising mill 粉磨机
pulverite 细粒沉积岩
pulverizable 可粉化;可成粉末的
pulverization 研末(作用);研成细末;金属喷镀;喷雾;雾化(法);粉碎(作用)
pulverization carburet(t)or 喷雾式化油器
pulverization test 粉碎试验
pulverizator 粉碎器
pulverize 研末;磨碎;磨成粉状的
pulverized anthracite 无烟煤粉
pulverized asbestos 无石棉粉
pulverized ash cement 粉煤灰水泥
pulverized ash collector 粉状灰收集器
pulverized ash wall panel 粉煤灰墙板
pulverized bulk material 粉末状散装物料
pulverized burner 微粉喷燃器
pulverized coal 煤末;煤粉;粉煤
pulverized coal ash 粉煤灰
pulverized coal blower 煤粉鼓风机
pulverized-coal burner 粉煤燃烧炉
pulverized coal classifier 煤粉分离器
pulverized coal collector 煤粉收集器;细粉分离器
pulverized coal combustion chamber 煤粉燃烧室
pulverized coal feeder 给煤粉机
pulverized coal fired 煤粉加热的;烧煤粉的
pulverized coal fired boiler 煤粉锅炉
pulverized coal firing boiler 煤粉锅炉
pulverized coal firing plant 煤粉燃烧设备
pulverized coal fuel 煤粉燃料
pulverized coal nozzle 煤粉喷嘴;喷煤嘴
pulverized coal preparation 煤粉制备
pulverized coal(storage) bin 煤粉仓
pulverized coal-storage system 仓储[贮]式制粉煤系统
pulverized coal system 煤粉系统
pulverized coal transport through pipes 粉煤管道输送
pulverized corundum 刚玉砂;金刚砂粉
pulverized fuel 雾状燃料;粉状燃料;粉末燃料;粉状燃料
pulverized fuel ash 粉状燃料灰;粉状燃料粉;粉状灰烬;粉煤灰
pulverized-fuel-ash aggregate 粉煤灰轻集料;粉煤灰轻骨料
pulverized-fuel-ash cement 粉煤灰水泥
pulverized fuel ash-lime block 粉煤灰和石灰制的砌块
pulverized fuel boiler 煤粉锅炉
pulverized fuel feeder 给粉机
pulverized fuel plant 煤的粉磨设备
pulverized fuel system 燃料粉化系统
pulverized lime 石灰粉;粉状石灰;石灰粉
pulverized limestone 粉化的石灰石;磨碎的石灰石
pulverized magnesian lime 镁质石灰粉
pulverized material 粉化的材料;磨碎的材料
pulverized mixer 粉碎拌和机
pulverized powder 粉碎的粉末
pulverized sample 粉磨样品
pulverized slag 炉渣粉;粒化渣
pulverized solder 粉状焊锡;粉状焊剂;粉状焊料
pulverized stone 石粉
pulverized sulfur 粉末硫磺
pulverized talc 粉状滑石
pulverized texture 粉末状结构
pulverizer 碎土器;喷射器;研碎器;碾碎;磨粉机;雾化器;微粉机;粉碎器;粉碎机;粉磨机
pulverizer exhauster 排粉机
pulverizer feeder with magnetic separator 装有磁铁分离器的磨煤给煤机
pulverizer for lubricating spring 簧板喷油器
pulverizer plate 雾化板
pulverizer reducing machine 粉碎机
pulverizing 粉碎(的);粉磨
pulverizing element 研磨机件
pulverizing installation 粉磨设备
pulverizing jet 喷烧器
pulverizing mill 细粉机;微粉机
pulverizing mixer 粉碎拌和机;松土拌和机
pulverizing plant 粉磨设备;粉磨工厂
pulverous 粉状的;粉末状的
pulverulence 粉末状态
pulverulent 自匀货;粉状的
pulverulent material 粉末材料;粉状物料;粉化材料
pulverulent product 粉末制品
pulverulite 细尖塔凝灰岩

pulveryte 细粒沉积岩;细尘岩
pulvimixer 松土搅拌机;松土拌和机;打松拌和机;粉碎拌和机
pulvimix(ing) 经粉碎拌和的混合料;松土拌和
pulvin 柱顶垫石
pulvinante colony 垫形高塔【给】
pulvinarium 古罗马庙宇内举行圣餐的房间
pulvinated 凸弯形的;垫子形的;鼓突的
pulvinated frieze 稍凸的檐壁
pulvinated mo(u)lding 鼓突线脚
pulvination 鼓胀形
pulvinis 毛垫
pulvinus 圆栏杆形饰(爱奥尼亚柱头侧面的);柱头垫石;拱端托块
pumeconcrete 轻质混凝土;浮石混凝土
pumecrete 轻质混凝土;浮石混凝土
pumicate 轻石
pumice 泡沫岩;浮石;浮岩;轻石
pumice agglomerate 浮岩集块岩
pumice aggregate 浮石集料;浮石骨料
pumice aggregate concrete 浮石集料混凝土;浮石骨料混凝土
pumice block 浮石砌块
pumice breaker 浮石破碎机
pumice breccia 浮岩火山角砾岩
pumice breeze block 敷设焦渣混凝土砌块
pumice building block 浮石混凝土砌块
pumice building material 浮石建筑材料
pumice building materials factory 浮石建筑材料厂
pumice cast(ing) 浮石混凝土浇筑;浮石混凝土浇制
pumice chip(ping)s 浮石小块;浮石碎片
pumice concrete 浮石混凝土砌块;轻质混凝土;浮石混凝土
pumice concrete cast(ing) 浇捣浮石混凝土杆件
pumice concrete cavity block 浮石混凝土空心(砌)块
pumice concrete hollow beam 空心浮石混凝土加筋梁
pumice concrete plank 浮石混凝土厚板
pumice concrete roller box 装置活动百叶窗的浮石混凝土匣
pumice concrete solid slab 浮石混凝土实心板
pumice concrete unit 浮石混凝土建筑构件;浮石混凝土建筑单元
pumice constructional material 浮石建筑材料
pumice constructional materials factory 浮石建筑材料厂
pumice deposit 浮岩矿床
pumice dust 浮石粉末
pumice expanded concrete 浮石膨胀混凝土
pumice fall 浮岩雨
pumice gas concrete 浮石气体混凝土
pumice grain 浮石颗粒
pumice gravel 浮砾石
pumice gravel fill(ing) 浮砾石填料
pumice grinding grease 浮石研磨膏
pumice hollow(core) plank 浮石混凝土空心板
pumice industry 浮石工业
pumice meal 浮石粉
pumiceous 泡沫状的
pumiceous concrete 轻集料混凝土;轻骨料混凝土;浮石混凝土
pumiceous fragment 浮岩碎屑
pumiceous structure 泡沫构造;浮岩构造
pumiceous tuff 浮石凝灰岩
pumice polishing grease 浮石抛光膏
pumice roll crusher 浮石滚压破碎机
pumice sand 轻石砂;浮岩砂;浮石砂
pumice-slag brick 浮石渣砖
pumice soap 去油垢皂(含浮石粉);去油垢洗手皂(含熔岩粉)
pumice soil 轻石土
pumice stone 轻石;浮岩;浮石
pumice stone concrete 浮石混凝土
pumice-stone paper 浮石纸
pumice-stone powder 浮石粉
pumice structural material 浮石结构材料
pumice structural materials factory 浮石结构材料厂
pumice tile 浮石(混凝土)平砖
pumice tuff 浮岩凝灰岩
pumicing 用浮石摩擦;浮石粉抛光
pumicite 火山灰;细浮石;火山尘埃;浮岩;浮石粉
pummel 夯;捣实机;球端;碾

pump 泵；唧筒；机泵；抽运
pumpability 可抽运性；输送量；供给量；可泵性；唧量；抽送量；泵送能力；泵唧性
pumpability of concrete 混凝土的易泵性
pumpability test 可抽送性试验；泵抽试验
pumpability time 可泵送时间
pumpable 可泵送的
pumpable concrete 塑性混凝土；泵送混凝土；可泵(送的)混凝土
pumpable condition 泵送状态
pumpable mix 可泵(送的)混合物
pump accessory 泵附件
pump adapter 水泵接合器；泵接合器；泵的接合器
pump adjustment 抽运调节；泵的调节
pumpage 水泵抽水量；抽送能力；抽水能力；抽水量；泵送量；泵的能力；泵的抽运量；泵的抽送率；泵的抽送量
pumpage cost 抽水费用
pumpage of nodal well 节点井抽水量
pumpage of non-nodal well 非节点井抽水量
pumpage rate 扬水率
pump agitator 液压泵搅拌器
pump air chamber 泵的气(压)室
pumparound 泵唧循环
pump arrangement 水泵安置；泵装置
pump ashore unit 排泥上岸装置；吹泥装置；泵送上岸装置
pump assembly 泵机组
pump assistant 司泵工
pump axis 水泵轴；泵轴
pump back 汲送回；泵回；侧线循环回流
pump back test 反馈试验
pump barrel 泵筒；泵体
pump base 泵座
pump basket 水泵吸水管滤网；水泵莲蓬头
pump bay 给水泵区；泵区
pump bearing 泵轴承
pump blade 泵叶片
pump block 泵座
pump bob 泵摆锤
pump body 水泵体；泵体；泵壳
pump bonnet 泵盖
pump-bow compasses 弹簧点圆规
pump bowl 水泵缸(体)；泵缸
pump box 泵房；水泵间
pump bracket 泵架
pump brake 唧筒把手
pump brake windlass 摇臂起锚机；手摇式锚机
pump bucket 活塞泵的活塞；水泵皮碗；泵筒；泵缸
pump bucket dredge(r) 自扬链斗挖泥船
pump building 泵房
pump bushing 泵衬套
pump by heads 间歇性抽吸
pump bypass 泵的旁通路
pump bypassing 回油路泵油
pump cage 泵盖
pump capacity 输出量；抽水量；泵流量；水泵流量；泵抽出水率；抽水机出水率；泵容量；泵的能力
pump capacity head curve 水泵流量扬程曲线；泵的容量压头曲线
pump cartridge 泵支架；泵座
pump case 水泵外壳；泵外壳；泵壳
pump case filter 泵上过滤器
pump casing 泵壳
pump casing cover 泵体盖
pump cavitation 泵气蚀；泵空穴作用
pump cavity 抽运谐振腔
pump centrifugal type 离心泵
pump certification 消防泵检验合格证
pump chamber 水泵空气室；泵内腔室
pump characteristic 水泵特性；泵的特性；泵特性
pump characteristic curve 水泵特性曲线；泵特征曲线；泵特性曲线
pump check ball 泵的闭锁球
pump check spring 泵止回簧
pump check valve 泵止逆阀；泵的止回阀
pump check-valve ball 止回阀珠；止回阀球
pump circuit 泵的电路
pump circulation 强制循环；泵压循环；泵循环；泵送循环
pump column 水泵底座
pump compartment 泵舱
pump concrete 泵浇混凝土；泵送混凝土
pump connecting link 泵连接杆
pump control mechanism 泵调节器
pump control valve 泵控制阀
pump conveyance 泵送
pump cooling 强制冷却；泵冷却
pump cover 泵盖
pumpcrete 泵注混凝土；泵送混凝土；泵浇混凝土；混凝土泵
pumpcrete machine 混凝土泵；泵送混凝土机械；泵浇混凝土机
pump cup(leather) 水泵皮碗；泵轴环；泵皮碗
pump curve 泵特性曲线；水泵特性曲线
pump cylinder 泵筒；泵缸体；泵缸
pump data 泵数据
pump delivery 抽水量；水泵排(出)量；水泵抽水量；泵输送量；泵输出量；泵排(出)量；泵量；泵供水量；泵的排量；水泵扬水量；水泵抽运量
pump delivery head 扬水水头；泵的扬程；泵的输出水位
pump delivery test 泵排量试验
pump diaphragm 泵膜片
pump diffuser 泵扩散管
pump direct driven press 泵直接传动液压机
pump disc 泵盘
pump discharge 水泵流量；抽水量；泵排送量；泵的流量；泵抽送量
pump discharge control valve 泵出口控制阀
pump discharge head curve 水泵流量扬程曲线
pump discharge line 高压管线；高压管道；泵排水管路
pump discharge pressure 泵出口压力
pump displacement 泵排出量；泵排送量；泵排水量；泵排量；泵工作容积；泵的排水量
pumpdown 抽空
pump down cycle 抽气循环
pump-down time 抽成真空的时间；抽空时间；抽真空时间
pump drain 水泵排水沟；抽水沟
pump drainage 机泵排水；水泵排水量；抽排；泵的排水量
pump drain cock 泵放水旋塞
pump drain equipment 排水设备
pump drain line 油泵排油道
pump dredge(r) 吸泥船；吸扬(式)挖泥船；吸扬(式)挖泥机；吸力挖泥机；汲泥机；抽泥机；泵吸泥机；泵式挖泥机；泵式挖泥船
pump drill 泵钻
pump drive 泵驱动；泵(的)传动
pump drive assembly 泵的传动组；泵传动组件；泵传动装置
pump drive gear 泵传动齿轮
pump driven by power 动力驱动泵
pump dry 抽干
pump duty 泵输送量；泵量
pumped 腐心的
pumped circulation 泵循环
pumped concrete 泵浇混凝土；泵送混凝土
pumped-fed power station 抽水蓄能发电站；抽水蓄能发电厂
pumped fluid characteristics 输送液体性质
pumped gas 抽运气体
pumped hydroelectric(al) storage 抽水蓄能水力发电
pumped level 抽运能级
pumped medium 输送介质
pumped-out dock 排空压舱水的浮船坞
pumped pipe 泵管
pumped rectifier 抽气式整流阀
pumped storage 抽水蓄存；抽水蓄能；扬水蓄能；泵蓄水池
pumped-storage aggregate 抽水蓄能机组
pumped-storage development 抽水蓄能开发
pumped-storage groups 抽水蓄能机组
pumped-storage hydro-electric(al) plant 抽蓄式水力发电站；抽蓄式水力发电站设备；抽水蓄能水电站；抽水蓄能水电厂
pumped-storage hydro-electric(al) power station 抽水蓄能式水力发电厂
pumped-storage hydroplant 抽水蓄能电站
pumped-storage hydro station 抽水蓄能发电站；抽水蓄能发电厂
pumped-storage plant 抽水蓄能厂
pumped-storage power generation 抽水蓄能发电
pumped storage power plant 抽水蓄能电站
pumped-storage power station 抽水蓄能发电站；抽水蓄能发电厂；抽水蓄能电站
pumped-storage pump station 抽水蓄能泵站
pumped-storage scheme 抽水蓄能计划
pumped-storage set 抽水蓄能机组
pumped-storage water 抽汲蓄能水库
pumped tube 泵管
pumped vacuum system 抽真空系统
pumped well 抽水井
pumped well drain 水井抽排
pump efficiency 水泵效率；泵效率
pump element 泵芯子
pumpellyite 绿纤石
pumpellyite lawsonite glaucophane schist 绿纤硬柱蓝闪石片岩
pump energy 泵的能量
pump energy density 泵的能密度
pump engagement 泵的衔接
pump engine company 泵车中队
pumper 司泵员；司泵工；抽水机
pumper apparatus 泵消防车
pumper connection 水泵接合器
pumper-decanter 泵送倾注洗涤器
pumper inlet valve 泵车进水阀
pumper ladder truck 泵消防梯车
pumper outlet 消火栓接消防泵的出口
pumper outlet nozzle 消防大接头；消防大接头
pumper pit 试泵箱
pumper test pit 泵车试验槽
pump escape 泵摇梯消防车
pump excitation 泵的激发
pump exhausting 泵抽
pump exhaust inlet 泵排气口
pump-fed 泵输给
pump feed evapo(u)rator 泵供液蒸器
pump feed system 泵给水系统
pump field 泵场
pump filter 泵压过滤器
pump flow 水泵流量；泵的流量
pump-foot 泵底座
pump for gradient elution 梯度洗脱泵
pump frame 泵座
pump frequency 泵的频率；泵激频率；激励频率
pump gear 泵齿轮
pump-generator plant 抽水蓄能电站
pump gland 泵密封圈
pump governor 水泵调节器；泵调整器；泵调节器
pump governor case cover 泵调节器盒盖
pump governor cover 泵调节器盖
pump grease 唧筒润滑脂
pump handle 泵的摇杆；泵的手把
pumphandle footing 泵柄式基础
pump head 水泵扬程；水泵水头；泵水(压)头；泵的压头
pump hold 泵舱
pump hopper (泵送混凝土的)受料斗
pump(horse)power 泵功率；泵送马力；泵马力
pump house 水泵间；水泵房；泵房
pump housing 水泵壳；泵壳
pump impeller 水泵叶轮；泵叶轮；泵轮
pump impeller blade 水泵叶轮片
pumping 道路翻浆；截补；唧泥(刚性路面的喷泥现象)；抽水；抽汲；抽动；泵送；抽运
pumping ability 泵送能力
pumping action 增压泵动作；抽吸作用；抽水作用；抽汲作用；泵吸作用
pumping action of slab 混凝土板的抽吸作用
pumping adjuster 泵杆调整器
pumping admixture 泵送剂
pumping aid 泵送设备；泵送剂
pumping and drainage plan 进排水系统图；泵系和泄水布置图
pumping appliance 泵车；泵唧装置；泵唧设备
pumping aquifer 抽水含水层
pumping area 泵区
pumping ashore 泵送上岸
pumping back method 反泵法
pumping band 泵带
pumping beam 泵的手柄；泵的摇柄
pumping box 泵房
pumping capacity 扬水能力；抽水容量；抽水能力；泵送能力；泵流量；泵唧本领
pumping capacity number 排出流量数
pumping chamber 泵室；压油室；增压室；泵腔
pumping charges 抽水费用

pumping circuit 激励电路；抽运电路
pumping clearance 水泵的活塞余隙；泵活塞余隙
pumping columm 抽气塔
pumping compasses 啷筒式圆规
pumping concrete 泵送混凝土
pumping cone 降落漏斗；抽水漏斗
pumping cone profile 抽水漏斗剖面图
pumping cost 抽水成本
pumping cycle 抽运周期；泵的周期
pumping delivery limit point 泵送极限
pumping depletion 泵的损耗
pumping depression area 抽水降落范围；抽降面积
pumping depression cone 抽水降落锥面；抽降漏斗
pumping depression well 抽水减压井
pumping derrick 采油井架
pumping device 抽水设备
pumping diaphragm 泵膜
pumping down 抽空
pumping draft 泵水量
pumping drainage 抽水排水
pumping duration 抽水时间；抽水历时；泵的持续时间
pumping during construction 施工抽水
pumping effect 泵的效应
pumping efficiency 泵送效率；泵的效率
pumping element 有010水单元
pumping energy 泵的能量
pumping energy storage 泵的能量存储
pumping engine 蒸汽泵；救火车；泵发动机
pumping equipment 泵送设备；排液装置；排水装置；排气设备；抽水设备
pumping equipment depreciation apportion and overhaul charges 排水设备折旧摊销及大修费
pumping flow 泵流量
pumping flow field 抽水流场
pumping fluid 真空泵油
pumping frequency 激励频率；泵频
pumping from well point 井点抽水
pumping gear 泵送传动装置；抽油机减速装置
pumping generating set 抽水储[贮]存发电机组
pumping grinding mill 泵送碾磨机
pumping ground water 抽吸地下水
pumping head 扬水水头；扬程；抽水扬程；抽水头；泵送水头；扬水水头扬程；泵扬头；泵的扬程
pumping head matter 扬水水头物质
pumping head of capillarity 毛细啷送压头
pumping helix 螺旋型泵
pumping hole 抽水孔；泵孔
pumping hole clean 孔眼抽汲干净
pump(ing) house 泵站；泵房
pumping index 溅浆指数；啷泥指数
pumping in-line 水带线泵送
pumping installation 泵送装置；抽油装置；抽水装置；抽水站；抽水机械；水啷装置；啷送装置
pumping intake piping 泵入口管
pumping intensity 抽运强度
pumping-in test 压水试验；泵入试验；注水试验
pumping irrigation 扬水灌溉；机械提水灌溉；提水灌溉；抽水灌溉
pumping jack 泵四脚支架；泵支架；抽油机
pumping joint 汲泥接缝；翻浆冒泥接头
pumping lamp 抽运灯
pumping level 抽水水位；抽水高程；泵吸高程
pumping lift 抽吸提升机；泵的扬程；抽水扬程
pumping light 激光
pumping limit 抽吸极限
pumping limit of mud 泥浆泵送极限
pumping line 泵送管线；泵送管(道)；扬水管道；水泵出水管道
pumping load 泵啷负荷
pumping loop 排炭线圈
pumping loss 泵损失；泵送损耗；泵气损失
pumping machine 泵
pumping machinery 抽运机械；抽水机械；抽水机；泵
pumping main 增压总管；增压管路；泵送干管
pumping mechanism 抽水机理
pumping method 抽水法
pumping mode 泵送方式
pumping of concrete 混凝土泵送
pumping of mortar 泵送砂浆；砂浆的泵送
pumping of pavement 泵送铺(路)面材料；铺(路)面材料的泵送
pumping of sludge 污泥泵送
pumping of the barometer 气压表的振荡
pumping oil 抽油
pumping operation 水泵工况运行
pumping out 排水；抽水；抽空；抽汲；排出
pumping outfit 泵吸成套设备；泵送成套设备；水泵设备
pumping output 抽水量；泵流量
pumping(-out) system 抽水系统；抽空系统；排水系统；扬水系统
pumping out water from a distressed vessel 难船抽水
pumping over 泵送
pumping pit 吸水坑；抽水坑
pumping plan 泵系布置图
pumping plant 泵房；水泵站；抽水站；抽水设备；泵送设备
pumping plant capacity 抽水站抽水能力
pumping plant discharge capacity 泵站排水能力
pumping point 啷动点；抽动点；泵出点
pumping polluted water body 抽出已污染水体
pumping pontoon 泵船
pumping power 泵的功率；抽output功率；抽汲功率；泵动力
pumping power level 抽运功率级；泵的功率级
pumping power requirement 泵送流体所需功率
pumping pressure 泵压
pumping process 抽运过程
pumping pulse 抽运脉冲冲
pumping pulse length 抽运脉冲长度
pumping radiation 抽运辐射
pumping rate 抽运速率；抽气速率；泵送速度；泵送率；泵啷速度
pumping rate of acidizing 注酸泵量
pumping rate of fracturing fluid 压裂液泵排量
pumping relief well 抽水减压井
pumping rig 采油井架
pumping rod 泵杆；扬油杆；泵拉杆
pumping room 泵房；抽水(机)房
pumping saline water and replenish fresh 抽咸换淡
pumping sampler 抽吸采器
pumping schedule 泵送进度表
pumping scheme 泵示意图
pumping set 泵组；抽运机组；抽气泵组
pumping shaft 扬水竖井；排水井；抽水竖井
pumping shed 泵房
pumping signal 激励信号；抽运信号；参数激励频率信号；泵频信号
pumping sleeper 松动枕木
pumping source 抽运源；泵源
pumping speed 抽吸速度；泵速；抽速
pumping station 水泵站；抽水站；泵站；扬水站
pumping station house 泵站房
pumping station of long distance oil transmission pipeline 长距输油管线泵站
pumping suction temperature 泵吸入温度
pumping system 水泵系统；水泵网
pumping temperature 泵送温度
pumping test 扬水试验；抽水试验；泵送试验；注水试验
pumping test from above to below 正向抽水
pumping test from bottom to top 反向抽水试验
pumping testing 扬水试验
pumping test in-situ 原位抽水渗透性试验
pumping test of field permeability 现场渗透抽水试验
pumping threshold 抽运阈值
pumping track 翻浆冒泥线路
pumping trench 抽水坑
pumping trough 抽水槽
pumping tube 泵管
pumping type suspension load sampler 抽吸式悬移质采样器
pumping unit 水泵机组；抽油机；抽水设备；抽水机组；泵设备
pumping-up drainage method 扬升式排水法
pumping-up power plant 扬水式水力(蓄能)发电厂
pumping value 泵送值；泵啷值
pumping wave 泵波
pumping well 抽油井；抽水井；泵抽井
pumping well array 抽水井的布局
pumping windlass 手摇起锚机
pumping works 抽水工程

pump inlet 泵的进油口；泵的进水口；泵进口
pump inlet check valve 泵进口止回阀
pump inlet head 泵的进口水头
pump inlet strainer 泵入口过滤器
pump inlet valve 泵吸入阀
pump in pipeline 管道泵
pump input 水泵输入功率
pump in recirculation 循环泵
pump in series 泵串联；串联泵
pump installation 泵装置；泵送设备；泵的装配；泵的安装；啷送装置
pump insulation 水泵绝缘装置
pump intake 水泵吸水孔
pump intake piping 泵的入口管
pump intensity 泵的强度
pump-in test 压水试验
pump jet type 射流泵
pump kettle 泵吸入管的过滤龙头
pumpkin 连续处加强套管；南瓜
pump lead 泵引水
pump leather 水泵皮碗
pumpless drilling 无泵钻进
pump lever 泵杆；泵操纵杆
pump lift 泵的扬程；水泵扬程
pump lighting 抽运光照明
pump line 抽气管；泵电缆控制线
pump liner 缸体衬套；泵套；泵的衬里
pump lining 泵缸衬里
pump link 泵连接杆
pump liquid end 泵的液体端
pump load 水泵荷载；水泵负荷
pump load-drop cavitation 泵减载气蚀
pump loading 泵的加载
pump low down type 卧式手摇泵
pump lubrication 强制润滑；泵润滑
pump main 泵输出管；泵排出管
pumpman 油船上管理泵的船员；司泵工
pump manifold 水泵歧管；泵管复式接头
pump manufacturer 泵制造厂
pump-mixing system 泵搅拌系统
pump-mix type 泵混合型
pump motor 泵用电动机；泵的电动机
pump nozzle 喷吸泵
pump number 泵号
pump of constant delivery type 定量泵
pump off 泵出
pump oil 泵油
pump oiler with flexible nozzle 带软喷嘴的油泵
pump oil seal 泵的油封
pump operated sprayer 液力喷雾器
pump operating status 泵操作状态指示器
pump oscillator 泵激振荡器
pump out 泵出；汲出
pump-out bean 从岩芯管中压出岩芯用管接头
pump-out(hopper) dredge(r) 排泥船
pump outlet 泵的出油口；泵的出水口；泵出水口
pump outlet valve 泵排放阀；泵放出阀
pump output 抽水量；水泵排(出)量；水泵流量；水泵抽水量；泵送量；泵排量；泵啷本领；泵出量
pump-out reversing valve 泵开式反循环阀
pump out valve 抽空阀
pumpover 泵送
pump overcapacity 泵的超负荷
pump package 泵组
pump packing 水泵活塞杆的密封；泵密封垫
pump performance 水泵性能；泵性能；泵的性能
pump performance characteristic 泵的工作特性
pump performance data 泵的效能数据
pump pinion 泵小齿轮
pump(pipe)line 泵管
pump piping 水泵输送管；泵压管；泵啷管线
pump piston 泵活塞
pump piston clearance 泵活塞余隙
pump piston collar stud 泵活塞环柱螺栓
pump pit 水泵坑；抽水机坑；泵坑
pump plant 水泵房
pump platform 泵(平)台
pump pliers 泵钳
pump plunger 泵的柱塞；泵柱塞
pump plunger shaft 泵柱
pump power 泵耗电力
pump-power chart 泵的功率图
pump power end 泵的动力端
pump-power modulation 泵的功率调制

pump pressure 泵的压力;泵压力
pump pressure ga(u)ge 泵压力计
pump pressure meter 汞压表
pump pressure of acidizing 注酸泵压
pump pressure of cementing unit 水泥车泵压
pump pressure of fracturing 压裂泵压
pump pressure regulator 泵压调节器
pump pressurizing unit 泵压装置
pump primer 起动真空泵;泵起动器
pump priming 自力发展经济的政府政策;泵充水启动;引泵;水泵充水;泵起动注水;泵的起动注水;泵充水
pump program(me) 泵送进度表
pump proportioner 环泵式比例混合器
pump pulley 泵皮带轮;泵轮
pump pulper 泵式打浆机
pump pulse duration 泵的脉冲持续时间
pump push rod 泵推杆
pump ram 泵柱
pump rate distribution 泵的速率分布
pump reciprocating type 往复式水泵
pump reflector 抽运反射器;泵反射器
pump regulator 泵量调节器
pump relief valve 泵安全阀
pump ring 水泵管垫圈
pump rod 水泵轴;水泵杆;捣冒口棒
pump-rod fishing socket 抽油杆打捞器
pump rod joint 抽油杆接头;抽油杆接撼
pump room 水泵间;水泵房;泵舱
pump rotary type 旋转泵
pump rotor 泵的转子;泵的转轮;泵转子;泵工作轮
pump runner 离心轮;泵叶轮;泵的工作轮
pump running dry 泵空运转
pump running free 泵空运转
pump screen 泵滤网;泵过滤器
pump screen retainer 泵滤网护圈
pump screw 泵组
pump screwdriver 伸缩螺丝旋凿
pump seal 水泵的水封
pump set 泵组
pump setting 井泵总扬程;水泵深度;水泵安装
pump setup 泵机组
pumps for special liquids 特种液体泵
pump shaft 泵轴
pump shaft seal 泵轴密封
pump shell 泵壳
pump ship 消防船
pump side 泵侧
pumps in parallel 并联泵;泵并联
pumps in series 串联泵
pump size 泵的规格
pump sleeve 泵套筒
pump slip 泵滑水率
pump slippage 泵滑转
pump snoring box 水泵吸水龙头
pump speed 泵速
pump speed regulator 泵速调节器
pump spindle 泵轴
pump spring 泵弹簧
pump stage 水泵级;泵级
pump station 泵送场;水泵站;抽水站;泵房
pump station design 泵站设计
pump station header piping installation 泵站管路安装
pump-station of recirculation water 循环水泵站
pump status contact 水泵状态触点
pump steam inlet 泵进气口
pump strainer 水泵进水管口滤网;泵滤网;泵滤头;泵口拦污网;泵粗滤器
pump stroke 泵行程;泵冲程
pump stroke counter 泵冲程计数器
pump structure 泵的结构
pump submergence 水泵淹没深度
pump suction 泵吸
pump suction chamber 泵吸收水池;泵吸水池(船坞充水)
pump suction drilling 泵吸反循环钻进
pump suction ga(u)ge 泵的进水表
pump suction head 水泵吸水高度;泵吸收下头;泵吸升高度
pump suction hose 吸水胶管;泵的吸入胶管
pump suction line 泵吸入管道
pump suction pipe 泵吸水管;泵吸入管
pump suction pit 泥浆泵吸池

pump suction supply pressure 泵进水压力
pump suction valve seat 泵吸水阀座
pump suction well 水泵集水井
pump sump 泵池;泵集水坑;水泵集水井;抽水集水坑
pump sun gear 泵太阳齿轮;泵的太阳齿轮
pump supply oil flow 压力油流量;液压油流量
pump surge 泵脉动
pump suspended in well 水井吊泵
pump suspension 泵悬挂装置
pump switch 泵开关
pump tank extinguisher 小型带泵灭火器
pump tank fire extinguisher 小型带泵灭火器
pump test(ing) 水泵试验;泵试验
pump test machine 泵试验器
pump the bilge 抽干舱底水
pump thrust 泵的压头
pump transition 泵的跃迁
pump transition probability 泵的跃迁概率
pump transmission 泵传动
pump transport system 泵送系统
pump-turbine 可逆式水轮机;水泵水轮机
pump type 泵型号;泵分类;泵的类型
pump-type circulation lubrication system 泵送循环润滑系统
pump unit 抽水机成套装置;泵装置;泵机组
pump universal joint drive 泵的万向节驱动(装置)
pump up 唧送
pump valve 泵阀
pump-valve cage 泵阀体;泵阀盖座
pump valve pressing spring 泵阀压紧簧
pump valve seat 泵阀座
pump voltage 泵电压
pump volume 水泵排(出)量;泵量
pump water conductor 水泵排出管
pump water conduit 泵排出管;泵的导水管
pump waveguide 抽运波导
pumpway 水泵间格;排水间
pump well 抽水井;泵井
pump wheel 泵轮;泵工作轮
pump with frequency conversion and speed governor 变频调速水泵
pump with helicoidal piston 螺旋活塞式泵
pump working pressure 工作泵压
pump works 水泵工程;水泵站;泵站
pun 捣
puna 高山荒原;高寒气候;山间高原
punch 阳模冲子;压力机;钻孔;凿孔(器);卡片穿孔;戳子;穿孔;穿孔器;穿孔机;穿卡机;冲子;冲压模;冲头;冲孔;冲杆;打孔机;冲孔工具
punch adapter 凸模接头;冲杆连接器
punch advance 冲杆冲程
punch and anvil 冲眼与铁砧
punch-and-die process 穿孔冲模法
punch and profile shaping machine 刨模机
punch and shear 冲剪两用机;冲孔剪切机
punch, bind machine 冲孔装订两用机
punch block 凸模坯料;穿孔区
punch block mechanism 穿孔机构
punch boring 冲击钻孔
punch bowl 成卷机;山坳
punch brush 穿孔电刷
punch card 穿孔卡;冲孔卡
punch card computer 穿孔卡片计算机
punch card equipment 穿孔卡片
punch card key 穿孔卡索引;穿孔卡标志
punch card machine 纸板打孔器;卡片穿孔机;洞卡;穿孔卡片机;凿孔机
punch-card machinery 卡片穿孔机械
punch card reader 穿孔卡片阅读机
punch card system 穿孔卡片系统
punch-card transcriber 穿孔卡转录器
punch carrier 凸模接头
punch check 穿孔校验
punch column 穿孔栏
punch combination 孔组
punch connection 冲压接合
punch core 孔壁取心;井壁取心
punch corer 冲击式取样管
punch cutting machine 阳模雕刻机
punch die 冲模;凸凹模
punch displacement 凸模位移
punch-dressed masonry 凿面圬工
punched acoustic(al) tile 冲孔天花板

punched-barb spindle 冲齿摘锭
punched bend 冲压弯头;穿孔板
punched card 凿孔卡;穿孔卡(片);已穿孔卡片;打孔卡片
punched-card accounting machines 穿孔卡片会计机械设备
punched-card control 冲孔纸板控制裁
punched-card devices 冲孔卡片法装置
punched-card equipment 凿孔卡设备;打孔机;穿孔卡片设备
punched-card field 穿孔卡片区;穿孔卡片字段
punched-card input device 穿孔卡片输入装置
punched-card jacquard mechanism 冲孔卡片提花机构
punched-card machine 穿孔卡片机;穿孔卡机;穿卡机
punched-card method 卡片穿孔法
punched-card reader 卡片输入机;穿孔卡片阅读器;穿孔卡(片)输入机;穿孔卡片读入机
punched-card recorder 穿孔卡片记录器
punched-card sorter 穿孔卡分类机
punched-card system 穿孔卡片系统;穿孔卡控制系统
punched-card terminal 穿孔卡片终端机
punched carpet 针刺毡;针刺地毯
punched drain pipe head 冲压排水管弯头
punch edge 冲头镶块
punched head 冲孔机头
punched hole 穿孔;冲孔
punched-hole sieve 冲孔筛
punched louver 门的镶板中设置的百叶
punched matrix 冲压字模
punched needle 冲刺钢针
punched(paper) tape 穿孔纸带
punched-plate 眼子板;冲孔板
punched-plate screen 冲孔板筛
punched-plate screen deck 冲孔板筛网
punched screen 冲孔金属板筛分机;冲孔金属板筛子;冲孔钢筛板
punched shear rupture 冲剪破坏
punched square nut 冲压方形螺母
punched steel plate 穿孔钢板
punched steel plate screen 冲孔钢板筛分机;冲孔钢板筛子
punched tape 凿孔纸带;穿孔带;冲孔带
punched tape code 穿孔纸带代码;穿孔纸带代码;穿孔带码
punched tape collator 穿孔纸带校对机
punched tape machine 穿孔纸带机
punched-tape program(me) 穿孔带程序
punched tape programming 穿孔带程序设计
punched tape reader 穿孔纸带输入机;穿孔纸带读入装置
punched-tape storage 穿孔带存储
punched washer 冲制垫圈
punched work 斜纹缀面;琢痕石板;石面刻线槽;宽凿的石料;刻对角线槽石面
puncheon 架柱;替灯;短柱;短支柱;大桶(容量单位);粗刨木料;半圆木料
puncher 凿孔机;泵员;穿孔机操作员;冲压工;冲孔器;冲孔机;冲床;报务员;打孔机;穿孔机
punch error 穿孔错误
punch flange 凸模固定凸缘
punch format 穿孔形式;穿孔格式
punch forming 冲压成型
punch hammer 冲孔锤
punch heel 凸模背靠块
punch-holder 凸模固定板
punch-holder shank 冲头柄
punch inclosure 凸模围栅
punching 漆膜穿孔;冲制;冲压;冲板;凿孔;穿孔;冲孔工具
punching and forming automatic 冲剪成型自动机
punching and shearing machine 冲剪机;联合冲剪机;冲剪两用机
punching area 冲剪面积
punching bear 打孔机;打孔器
punching block 钢模(板)
punching column 冲压机柱
punching device 打孔装置
punching die 冲孔模
punching disc 冲孔提花盘
punching drilling 冲击钻井
punching driven-in-place piles 冲孔灌注桩

punching effect 冲孔效应
punching failure 刺入破坏;冲压破裂面
punching field 穿孔区
punching force 冲剪力
punching head 模压机头;冲头
punching hole diameter 冲孔直径
punching machine 轧切机;冲床;打孔机;穿孔装置;穿孔机;冲压机;冲孔机;冲床机
punching machine auto-feeder 冲床自动送料装置
punching method 冲击钻探法
punching nippers 冲孔钳
punching of screen 筛子打孔
punching oil 模锻机油;冲孔机油
punching over-capacity 穿孔过载容量
punching pin 冲子;冲孔器
punching pliers 冲孔钳
punching position 穿孔位置
punching press 冲床;冲孔压机
punching-process die 冲模
punching process of heat exchange plate or fin 板片或翅片的冲压
punching program(me) 穿孔程序
punching rate 穿孔速率;穿孔率
punching routine 穿孔程序
punchings 冲孔屑
punching shear 冲剪破坏;冲切力;冲压剪切;冲剪力;冲力
punching shear area 冲剪面积
punching shear failure 冲剪破坏
punching(shearing) machine 剪板刀纵放带模剪联合冲剪机
punching shear stability test 冲剪应力稳定性试验
punching shear stress 冲剪应力
punching shear test 冲剪试验
punching speed 穿孔速度
punching station 穿孔站;穿孔台
punching stress 穿孔应力
punching test 冲压试验;冲孔试验;冲剪试验
punching the wad (锻后的)冲连皮
punching tongs 穿孔夹具;冲孔钳
punching tool 冲孔工具
punching track 穿孔导轨
punching type 冲压铅字
punching velocity 穿孔速度
punch knife 穿孔刀
punch land 凸模面刃口宽度
punchless drawing 凸模拉延
punch light 穿孔指示器
punch list 检查项目表;建设缺陷清单;建设工程缺陷清单;穿孔表;承包工程项目清单
punch load 冲剪荷载;冲击荷载
punch locator plate 冲模导向板
punch magnet 穿孔磁铁
punch mark 钢印;冲压记号;原点;起标点;打标记;冲压标志;冲孔标记;冲标记
punch material 冲杆材料
punch mica 冲压云母片
punch nose 冲模头
punch-nose angle 凸模圆角
punch-nose radius 凸模圆角
punch off 凿孔停止;穿孔结束;穿孔偏离
punch on 穿孔开始
punch operator 穿孔员;穿孔机操作员
punch out 穿孔输出;冲掉芯料
punch-out equipment 捅箱机
punch pad 凸模压板
punch path 穿孔通路
punch perforator 三柱凿孔机
punch pin 冲子;尖冲头;冲头
punch pipe 直开浇口管
punch plate 凸模接头;凸模固定板
punch pliers 打眼钻;轧孔钳
punch press 冲床;片材造型机;冲孔压力机;冲裁压力机;冲压压力机
punch press for popsicle 冲压冰棒模
punch pressure 冲杆压力
punch profile radius 凸模圆角半径
punch prop 冲孔机支柱;冲孔机支座
punch radius 冲头半径
punch register 纸带式火警记录器
punch register system 穿孔套合系统
punch retainer 凸模固定板;冲头护圈
punch rivet hole 冲铆钉孔
punch row 穿孔一行;穿孔行

punch shank 模柄
punch shears 冲孔剪割机;冲剪床
punch sign 冲压记号;冲压标志
punch station 穿孔站
punch tape 穿孔带
punch tape code 穿孔纸带代码;穿孔带代码
punch tape controlled programming system 穿孔纸带控制程序系统
punch tape recording 打点记录
punch tape storage 穿孔带存储器
punch test 冲击断裂试验
punch through 穿通(现象)
punch-through breakdown 穿通击穿
punch-through effect 穿通效应;击穿效应
punch-through voltage 穿通电压
punch travel 凸模行程
punch trimming 冲模修边;冲具修整
punch typewriter 打字穿孔机
punch with corrugated nose 波形冲头
punch with fluted nose 波形冲头
punch work 冲孔工具
puncll cam 穿孔凸轮
punctate calcification 点状钙化
punctuality 正点
punctuality of freight train 货物列车正点率
punctuality of passenger train 旅客列车正点率
punctuality ratio of train moving 列车运行正点率
punctuality ratio of train starting 列车出发正点率
punctuated facade 分段立面
punctuation 分段;点标点;标点
punctuation bit 标示长度位;标示位;标长位
punctuation character 标点字符
punctuation marks 标点符号
punctuation symbol 标点符号
punctum 刻点
punctum remotum 远点
puncturatio 穿刺
puncture 击穿;漆膜穿孔;缩凹;刺扎;刺破;刺孔;刺穿;戳穿;刺伤
puncture and irrigation 穿刺冲洗法
puncture and shears 冲孔剪切机
puncture core 通气芯
puncture counter 击穿计数器
punctured 凿孔的;冲孔的
punctured codes 收缩码
punctured element 针孔元件
punctured plane 有孔平面
puncture instrument 穿刺器械
puncture lightning arrester 击穿保险器
puncture needle 穿刺针
puncture-proof tire 自动封漏(口)轮胎;防穿孔轮胎;防刺轮胎;自封轮胎
puncture reaction 穿刺反应
puncture resistance 抗冲穿强度;击穿阻力;冲穿阻力
puncture seal 补洞胶片
puncture strength 击穿强度;冲穿强度
puncture test 击穿试验;耐压试验;破坏性试验;电击穿试验;戳穿试验;冲孔试验
puncture transducer with center channel 中孔型穿刺用探头
puncture tire 漏气轮胎
puncture voltage 击穿电压
puncture wound 刺伤
puncturing 穿孔
puncturing lever 穿刺杆
puncturing machine 冲孔压力机
puncturing membrane catheter 刺胎膜导管
punding up machine 锤打机
pungent odo(u)r 刺激性臭味
puniceous 鲜红的;紫赤色;暗红紫色
punicic acid 石榴酸
punish(ing) stress 疲劳破坏应力;疲劳强度
punitive 惩罚性赔款
punitive clause 罚款条款
punitive coefficient 惩罚因子
punitive damages 惩罚性赔偿费
punitive factor 惩罚因子
punitive provision 罚则;罚款条款
punka 大吊扇
punkah 拉风;顶棚吊扇;布风扇;压入逸出式(通风);大吊扇
punkah louvre 大风扇;布条百叶遮阳;布风扇
punk knot 半朽节

punky 软韧的(木料)
punned concrete 捣实混凝土
punner 夯(具)
punning 夯实;夯捣;打夯
pun silk yarn 绢丝
punt 整修平底船;瓶底;平底船
punt code 瓶底标记
punter 平底船船员
punt ferry 撑篙渡船
punt glass 调焦玻璃
puntied base 研磨的瓶底
puntist 平底船船员
punt pole 篙
punty 瓶底;吹玻璃用的铁棒
puntying 研磨底座
puoztas 无林草地
pup 低功率干扰发射机;插缝砖
pup boat 小型消防艇
pupils dining room 学生食堂
pupils reading room 学生阅览室
pupin coil 加感线圈
pupinization 加感
pup jack 小型塞孔
pup joint 嵌入短管接头
puppet 下水梁;木偶;垫架
puppet head center 随转尼座顶尖(车床的)
puppet head tail stock 随转尼座
puppet play 木偶戏
puppet theatre 木偶剧院;木偶剧场
puppet valve 随转阀
pura 筑成台地的圣坛
puratized agricultural spray 苯汞铵
Purbeck bed 普贝克层【地】
Purbeckian 普贝克阶【地】
Purbeckian marble 普贝克大理石
Purbeckian stage 普贝克层【地】
Purcell method 普塞尔气孔率测定法
purchase 购买(品);绞辘;起重装置;收买;收购;定单;采办;保证金购进
purchase, storage and transportation 收购、仓储及运输
purchase agreement 购买协议;购货协议
purchase allowance 购买折扣;进货折让
purchase an anchor 起锚
purchase and installment saleback 购买后分期收款售出
purchase and leaseback 购买后租出
purchase and sale contract 购销合同
purchase average cost 进货平均价
purchase beyond state quota 超购
purchase block 起重滑车
purchase book 进货簿
purchase budget 进货预算
purchase by grade 分等收购
purchase by order 定购
purchase by state quotas 派购
purchase by tender 通过招标方式购买
purchase cable 滑车索
purchase capital 采购资金
purchase cardinals 收购基数
purchase commitment 约定付款数
purchase confirmation 购货确认书
purchase contract 购货合同;进货合同;订货合同
purchase cost 购货成本;进货成本
purchase design 购买的设计图样
purchased finished materials 采购的成品材料
purchase discount 购货折扣;进货折让
purchase discounts and allowances 进货折让
purchased life annuity 已购人寿保险年金
purchased on a volume basis 按体积采办
purchased parts 购置的零件
purchase estimate 采购估计
purchase expenses 采办费用
purchase fall 辘绳;滑车绳
purchase farm produce and sideline prducts 收购农副产品
purchase from the market 市场采购
purchase goods on a selective basis 择优选购
purchase in advance 预购
purchase investigation 采购调查
purchase invoice 购货发票;进货发票
purchase journal 进货日记账(簿);进货簿
purchase ledger 进货分类账
purchase list 购货单;订货单

purchase-loan storage program(me) 收购借贷储存方案
purchase money 买价;定价金
purchase money bond 购房款抵押债券
purchase money mortgage 置产人以抵押款付给出售人作为部分产价;购货款抵押;押金;购买财产担保;买人财产抵押
purchase money trust deed 购房款信托证券
purchase of a subsidiary 附属公司之买入
purchase of land 土地购置
purchase of merchandise 进货
purchase on installment 分期付款购入
purchase on margin 差额购买
purchase option 进货选择权
purchase order 购货订单;购货单;进货(订)单;定货单;定购单;订货单
purchase order procedure chart 采购定货程序图
purchase order quantity 采购订货量
purchase permit 准购证
purchase power 绞辘功率;滑车组功率
purchase price 购买价格;买价;收购价
purchase quantity 购货量
purchaser 发包人;买主;购买人
purchase rebates and allowances 进货回扣及折让
purchase records 进货记录;采购记录
purchase requisition 请购书;进货申请书;请购单
purchase return and allowance journal 进货退回及折让日记账
purchase return book 进货退回簿
purchase returns and allowances 进货退回及折让
purchases flow chart 购货核算流程图
purchase shears 扩力剪
purchases journal 进货日记账(簿)
purchases ledger 进货总账
purchase tackle 滑轮组;滑车;起重复滑车;提升滑轮;提升滑车
purchase tax 消费品零售税
purchase the land 征购土地
purchase-to-lease ratio 购买与租赁价格比
purchase value 购入价
purchase with cash 现金购买
purchasing 进货;采购
purchasing agency 采购代理
purchasing agent 经纪人;进货纯价账;物资技术供应代理人;采购代理商;采购代理人;采购部门负责人
purchasing analysis 进货分析
purchasing and constructing using capital construction allocated funds 基建拨款构建
purchasing and constructing using capital construction loans 基建借款构建
purchasing and stock budget 进货及存货预算;采购和存储预算
purchasing and store department 供应科
purchasing and supply station 采购供应船
purchasing by invitation to bid 招标采购
purchasing clerk 采购办事员;采购员
purchasing commission 购货佣金;采购代理业务
purchasing cost 进货成本
purchasing delivery receipt 收到材料
purchasing department 采购部门
purchasing department expenses 进货部门费用
purchasing inspection 购入检验;购货检验;采购检验
purchasing management 采购管理
purchasing office 采办处
purchasing order 进货订单
purchasing pattern 采购式样
purchasing power 购买力
purchasing price 买入价(格);买价;收购价格;收购价
purchasing procedure 采购程序
purchasing records 材料采购的核算
purchasing requisition 进货申请书;请购单;采购申请
purchasing right of locomotive and train 机车购置权
purchasing right of materials 物资采购权
purchasing schedule 采购程序表
purchasing work load 采购工作负荷
Purde method 普度法
pure γ radioactive nuclide 纯伽马放射性核素
pure act 纯行动
pure aerodynamic(al) coupling 纯气动连接

pure agglomerated cork 结块的纯软木
pure air 纯空气;纯洁空气
pure alumina 纯氧化铝
pure alumin(i)um 纯铝
pure alumin(i)um gas welding rod 纯铝气焊条
pure alumin(i)um foil 纯铝箔
pure alumin(i)um plate 纯铝板
pure alumin(i)um wire 纯铝线
pure asphalt 纯地沥青
pure bacterial bacteria 纯菌性菌
pure bending 单纯弯曲;纯弯曲
pure bending fatigue testing machine 纯弯曲疲劳试验机
pure benzene 纯苯
pure benzol 纯苯
pure binary 普通二进制;纯二进制
pure binary notation 纯二进制记法;纯二进记法
pure binary number 纯二进制数
pure binary number base 简单二进计数制
pure binary numeration system 纯二进制数制;纯二进制记数制;纯二进记数制
pure birth process 纯增过程;纯生过程
pure bitumen 纯沥青
pure bitumen index 纯沥青指数
pure black stone 纯黑石
pure cable system 纯缆索体系
pure calcium carbonate 纯碳酸钙;方解石
pure carbon 纯碳
pure car carrier 汽车专用运输船
pure carrier network 纯载波网络
pure car-truck carrier 小车—卡车专用运输船
pure cast iron 纯铸铁
pure caustic potash 纯苛性钾
pure caustic soda 纯碱
pure cement clinker 纯水泥熟料
pure cement compound 纯水泥化合物
pure change rate of solute in volume element 体积元内溶质质量的纯变化率
pure chemical corrosion 纯化学腐蚀
pure chemical laser 纯化学激光器
pure chemistry 理论化学;纯化学
pure circulating decimal 纯循环小数
pure citral 纯净柠檬醛
pureclarite 纯微亮煤
pure clay 纯白黏土;纯土;纯黏土;不合格砖泥
pure clay deposit 纯黏土沉积物
pure coal 纯煤
pure coal basis 纯煤基
pure coal substance 煤可燃质
pure code 纯代码;纯编码
pure colo(u)r 纯色
pure colo(u)ration 纯色
pure coma 纯彗形象差
pure commercial network 纯商业网
pure commodity table 纯商品表
pure competition 完全竞争;纯粹竞争
pure compound 纯净化合物;纯化合物
pure compression 纯压缩
pure compression method 纯压缩法
pure conflict game 纯斗争对策;纯冲突对策
pure conic(al) rolling 纯圆锥滚动
pure continuous waves 纯连续波
pure coordinated time 纯粹协协调时
pure copper 红铜;纯铜
pure copper starting sheet 纯铜始极片
pure copper tube 纯铜管
pure correlation 纯相关
pure cost 净成本;纯粹成本
pure cost of circulation 纯流通费用
pure cotton fabric 纯棉织物
pure crop 纯林
pure culture 纯培养(物);纯培养;纯粹培养
pure cuticodurite 纯微角质暗煤
pure death process 纯消过程;纯灭过程
pure delay 纯延迟
pure delay river model 纯延时河流模型
pure deviatoric shear 纯偏剪力
pured germanium crystal spectrometer 高纯锗谱仪
pure diffraction breadth 纯衍射宽度
pure dilatational deformation 纯膨胀变形
pure dilatational strain 纯膨胀应变
pure distance system 纯距离导航制
pure dominance 纯支配;纯优势

pure drilling cost 钻井直接成本
pure economic profit 纯经济利润
pure economic rent 纯粹经济地租
pure economics 纯经济学
pure electrolytic copper 纯电解铜
pure element 纯元素
pure error 纯误差
pure fan system 纯扇形体系
pure flax fabric 纯亚麻布
pure flexture 纯弯曲
pure forest 单纯林;纯林
pure form 纯洁形式
pure functional organization system 纯功能组织系统
purefusite 纯微丝煤
pure gas 纯气体
pure gaseous material 气体纯物质
pure gas turbine cycle 纯燃气轮机循环
pure generator 生产程序的程序;纯生成程序;产生程序的程序
pure geometry 纯粹几何学
pure germanium radiation detector 纯锗辐射探测器
pure gold 纯金;赤金
pure gypsum 纯石膏;生石膏
pure heteroscedasticity 纯异方差
pure hiding strategy 纯隐藏策略;纯陷藏策略
pure holding company 纯持股公司
pure horsepower 纯马力
pure humic acid 纯腐殖酸
pure imaginary 纯虚数
pure inductance 纯电感
pure industrial table 纯产业部门表
pure integer programming 纯整数规划
pure interest 纯利息
pure interest rate 纯利润率
pure-internal pressure test 纯内压试验
pure international trade 纯国际贸易
pure interpreter 纯解释程序
pure iron crucible method 纯铁坩埚法
pure kaolinite flake 纯高岭石薄片
pure kinematics 纯运动学
pure kraft paper 无光牛皮纸
pure labor-cost theory of value 纯劳动成本价值说
pure laminar flow 纯片流;纯层流
pure lead 纯铅
pure lime 纯石灰
pure lime mortar 纯石灰浆
pure lime paste 纯石灰黏土
pure lime putty 纯石灰黏料
pure line 纯系
pure linen grey cloth 纯亚麻坯布
pure line organization system 纯直线组织系统
pure line selection 纯系选择
pure liquid 纯液体
pure literal rule 纯基本式规则
pure load 纯外荷载
pure local time 纯粹地方时(原子时)
pure lump lime 纯石灰块
purely chemical parameter 纯化学参数
purely discontinuous distribution 纯不连续分布
purely functional structure 只供使用而不住人的结构
purely imaginary number 纯虚数
purely mathematical probability 纯数学概率
purely quantitative measure 纯数量计量
purely random error 纯随机误差
purely random model 纯随机模型
purely random process 纯随机过程
purely scattering atmosphere 纯散射大气
purely state-owned enterprises 纯国有企业
purely state-owned stock company 纯国有股份公司
purely viscous flow 纯黏性流动
purely viscous fluid 纯黏性流体
pure made land 素填土
pure magnetic iron 纯磁铁
pure marble 白大理石
pure Markov process 纯马尔可夫过程
pure material 纯材料
pure mathematics 纯数学;纯粹数学
pure mechanism 纯粹机构学
pure metal 纯金属;精金属
pure metallic cathode 纯金属阴极

pure metallic crystal 纯金属结晶
pure metallic hydroxide 纯金属氢氧化物
pure metallic powder 纯金属粉末
pure mild soap 纯肥皂
pure monetary theroy of business cycle 纯粹货币周期论
pure neon discharge 纯氖放电
pureness 纯净;纯度
pure-neutron shielding 纯中子屏蔽
pure nickel 纯镍
pure nickel electrode 纯镍焊条
pure nickel welding electrode 纯镍电焊条
pure nitrogen generator 纯氮发生器
pure no par stock 纯无面值股票
pure normal mode 纯正则振型
pure nugget effect model 纯块金效应模型
pure number 纯数
pure oscillation 正弦震荡;正弦振荡;纯正弦振荡;纯摆动
pure oxygen 净气;纯氧
pure-oxygen aeration 纯氧曝气
pure-oxygen aeration activated sludge process 纯氧曝气活性污泥法
pure-oxygen aeration method 纯氧曝气法
pure-oxygen aeration system 纯氧曝气系统
pure-oxygen aeration tank 纯氧曝气池
pure-oxygen aerobic digestion 纯氧好氧消化
pure oxygen top-blown converter 纯氧顶吹转炉
pure phase 纯相
pure phase filter 纯相位滤波器
pure phenolic baking electro-deposition primer 纯酚醛烘干电泳底漆
pure phenolic boat topping paint 纯酚醛水线船顶漆
pure phenolic resin 纯酚醛树脂
pure pitch 纯音调
pure plane shear wave 纯平面剪切波
pure plastic bending 纯塑性弯曲
pure plastic flow 纯柔性流动
pure plastic state 纯塑性状态
pure Portland cement 纯硅酸盐水泥;纯波特兰水泥;不掺混合料的波特兰水泥
pure precipitate 纯沉淀物
pure precipitated adsorbate 纯沉淀吸附质
pure precipitated solid 纯沉淀固体
pure premium 纯保险费
pure procedure 纯过程
pure product 纯净产品
pure profit 纯利润;净利(润);纯利
pure projective geometry 纯射影几何
pure project structure 纯粹方案结构
pure public goods 纯公共物品
pure pursuit 单纯追踪;单纯追击;纯跟踪方式
pure quantum number 纯量子数
pure quicklime 纯生石灰
purer 浇筑工(人)
pure rain 净雨;纯雨
pure random process 纯随机过程
pure raster 洁净光栅
pure reactance network 纯电抗网络
pure reaction turbine 纯反动式汽轮机
pure recall method 纯粹回想法
pure repeating decimal 纯循环小数
pure research 纯理论研究
pure resistance 纯电阻
pure retail network 纯零售商业网
pure risk 纯危险性;纯风险
pure rolling 纯滚动
pure rolling contact 纯滚动接触;纯粹滚动接触
pure rotational strain 纯旋转应变
pure rotational transition 纯转动跃迁
pure rubber hose 纯橡胶管
pure rubber sheet 纯胶板
pure salvage 纯救助
pure-sample standard method 纯样品标准法
pure sand 纯砂
pure sandstone 纯砂岩
pure scattering 纯散射
pure science 纯科学
pure seawater 纯海水
pure selenium 纯硒
pure selenium ingot 精硒锭
pure selenium powder 精硒粉
pure semiconductor 纯半导体

pure shale shaker 纯振动筛
pure shear 单纯剪力;纯剪;纯剪切;纯剪力
pure shearing strength 纯剪强度
pure shearing stress 纯切应力
pure shear strain 纯剪应变
pure shear test 纯剪试验
pure silica core single-mode fiber 纯石英纤芯单模光纤
pure silica fiber 纯石英纤维
pure silicon 纯硅
pure silicon polycrystal 纯多晶硅
pure silk 纯丝绸
pure silver 纯银
pure sine wave 纯正弦波
pure sinusoidal oscillation 纯正弦振动
pure slip 纯滑(动)
pure sociology 纯理论社会学
pure solid 纯固体
pure soluble blue 纯溶蓝
pure sound 纯声
pure source of monochromatic light 纯单色光源
pure spectrum 纯光谱
pure state 纯态
pure statistical estimation 纯统计估计
pure strain 单纯应变;纯应变;纯系;纯品系
pure strategy 纯策略
pure stress 单向应力;单纯应变;纯应力
pure stretch levelling 全拉伸矫直
pure style 纯正的风格
pure substance 纯物质;纯剂
pure substitution elasticity 纯代替弹性
pure swap transaction 纯换汇交易;纯对冲交易
pure talc 纯滑石
pure temperature radiation 纯温度辐射
pure tension 纯张;纯拉(伸)
pure terephthalic acid 纯对苯二甲酸
pure terephthalic acid wastewater 纯对苯二甲酸废水
pure thallium 纯铊
pure time-delay factor 纯时后因子
pure titanium spectacle frame 纯钛金属眼镜架
pure tone 正弦波音;纯音
pure tone audiometer 纯音听力计
pure tone generator 纯音发生器
pure-tone pulse 单频脉冲
pure tone screen audiometer 纯音筛选测听仪
pure torsion 纯扭转;纯扭力
pure tungsten 纯钨极
pure undamped wave 纯非衰减波;纯等幅度;纯等波幅
pure-vapour counter 纯汽计数器
pure-vapour-filled counter 充纯蒸汽计数管
pure variable-pressure turbine 纯冲动涡轮机;纯冲动透平
pure vessel 专用船
pure vibrational transition 纯振动跃迁
pure viscosity 纯黏性
pure water 纯水;纯净水
pure water attack 纯水腐蚀
pure water cooling unit 纯水冷却器
pure wave 正弦波;纯波
pure wet clay 纯湿黏土
pure white 纯白
pure white stone 纯白石
Purex process 普里克斯法
pure zinc ingot 纯锌锭
purfle 美化;装饰;镶边;边缘饰
purfled 镶边的
purfled work 镶边细工
purfling 边饰;镶边;花边;边缘饰
purga 布冷风;拨格风
purgation 清洗
purgative gas 清洗气
purgative prescription 泻剂
purge 清洗
purgeable organic chloride 可清除有机氯化物
purgeable organic halide 可清除有机卤化物
purge and trap capillary column gas chromatography 驱气俘获毛细管柱气相色谱法
purge and wash system 清除和洗涤系统
purge chamber 清洗室
purge date 消除年月日;清除日期
purge ducting 吹扫风管
purge gas 净化气体;吹扫用的气体;吹扫气体

purge gas recovery 释放气回收
purgemeter 曲线测长仪;转子流量计;转子式测流速仪
purge meter interlock 内侧联锁送气表
purge oil 冲洗油;冲洗用油
purge period 钻孔冲洗周期
purge pipe 排气管
purge process 纯化过程
purger 清洗装置;净化器;清洗器;排除管
purge stream 清除气流
purge tank 清洗室
purge unit 清洗装置
purge valve 清除阀;清洗阀;放泄阀;放空气阀
purging 吹扫;清除;放泄;管道吹扫;吹洗
purging air fan 清扫风机
purging by crystallization 结晶净化
purging cock 放汽旋塞
purging enclosure 清吹封闭空间
purging medium 吹扫介质
purging method 冲洗方法
purging of pipe 管道冲洗
purging pipe 清洗管
purging piping 清洗管
purging process 净化过程
purging system 冲净系统
purging the holder with gas 储气罐气体置换
purging tube 清洗管
purging tubing 清洗管
puridryer 净化干燥器
purification 净化(作用);净化法;清洗;提纯(作用);纯化(作用)
purification agent 油田水净化剂
purification amount 净化量
purification by chromatography 色谱提纯
purification by diffusion 扩散净化
purification by heat-treatment 热处理净化
purification by liquid extraction 萃取提纯
purification can 净化罐
purification capacity 净化能力
purification coefficient 净化系数
purification column 净化蒸馏塔
purification constant 净化常数
purification demineralizer 净化脱矿质器;脱矿质净化器
purification device 净化设备
purification efficiency 净化效率
purification equipment 净水设备;净化装置
purification index 净化指数
purification mechanism 净化机理
purification mechanism of sand filter 矿滤池净化机理;砂滤池净化机理
purification mechanism of stabilization pond 稳定塘净化机理
purification method 净化法
purification ocular estimate method 提纯目估法
purification of automobile exhaust gas 汽车废气净化
purification of drinking water 饮用水净化
purification of kerogen 干酪根提纯
purification of paper-making effluent 造纸污水净化
purification of sewage 污水净化
purification of sewerage 污水净化
purification of soil-plant system 土壤—植物系统的净化
purification of waste water 废水净化
purification plant 净水工厂;净水厂;净化装置;净化场;净化厂
purification plant with trickling filters 有滴滤池的净水场
purification process 净水过程
purification rate 净化速率
purification ratio 净化率;净化比;提纯比
purification residual section 残渣净化工段
purification run 净化操作
purification section 净化工段
purification structure 净化构筑物;净水构筑物
purification system 净化系统
purification tank 净化池;净化槽
purification tank for liquid waste 废水处理池
purification technique 净化技术
purification tower 净化塔;提纯塔
purification treatment 净化处理
purification water 净化水;净水

purification works 净水工程
purified 纯化的
purified air 净化的空气
purified animal charcoal 精制动物炭
purified bit speed 完全净化钻速
purified brine head tank 精制盐水高位槽
purified clay 精陶瓷黏土
purified cotton 纯棉
purified gas 净煤气;净化气(体)
purified hydraulic horsepower 完全净化水马力
purified liquor outlet 清液出口
purified petroleum benzin 纯苯
purified product 精制产品
purified rubber 纯化橡胶
purified salt 精制盐
purified signal 静噪信号;提纯信号
purified solution 净化溶液
purified standard material 标准纯物质
purified stand oil 纯熟油;纯聚合油;纯定油
purified steel 精炼钢
purified talc 净化滑石粉
purified terephthalic acid wastewater 纯对苯二甲酸废水
purified vector 净化向量;净化量
purified water 净水;净化水
purified water gallery 净水集水道
purified water tank 净水箱
purifier 滤清器;净水装置;净化器;精炼器;清洗装置;清洗器;提纯器;纯水器;净化器;净化器
purifier centrifuge 离心式油水精分机
purifier of flue gas 烟气净化设备
purify(ing) 纯化;净化;提纯
purifying agent 净水剂;净化剂;提纯剂;纯化剂
purifying apparatus 净化仪器;提纯仪器
purifying box 净化箱
purifying capacity 净化能力
purifying column 净化塔
purifying device 净水装置;净化装置;净化设备
purifying effect 净化效果
purifying method 净化法;精炼法
purifying of sewage 污水净化
purifying of wastewater 废水净化
purifying pond 净化池;净水池;净化塘
purifying reaction 精炼反应
purifying tank 净水池
purifying unit 净化装置;净水装置
purifying waste 废物净化
purifying with air 用空气净化
Purilon 普里纶医用黏胶短纤维
Purimachos 普里马克斯(一种耐火水泥)
Puriri 普里里灰褐色硬木(即新西兰柚木)
purism 纯洁主义
purist house 纯粹派艺术家的房屋
purity 洁净;品位;纯正;纯净;纯度
purity check(ing) 纯度检验;纯度校核
purity circuit 彩色电路
purity coil 纯度控制线圈;纯度调整线圈
purity control 色纯度调节;纯度控制
purity grade 纯度级别
purity magnet 纯化磁铁;彩色纯度调节磁铁
purity of caustic soda 苛性钠纯度
purity of core 岩芯的纯洁性
purity of electrode 电极纯度
purity of hue 色相纯度
purity of quotient 纯度
purity of sand 泥沙纯度
purity of state 态的纯度
purity requirement 纯度要求
purity style 纯净式样;式样的纯净
purity test 纯度试验
purl 刺绣;潺潺流水声
purl fabrics 双反面针织物
purlieu 近郊;贫民区;森林边缘
purlin 檩条支托楔块;檩条;檩
purline = purlin
purlin(e) anchoring 檩条锚固;檩条锚碇
purlin(e) arrangement 檩条布置
purlin(e) brace 屋顶桁条;檩香条撑;檩条撑;檩撑
purlin(e) cleat 檩条扣件;檩条支托的楔块;支托檩条的楔块;檩垫
purlin(e) connection 檩条联结
purlin(e) hanger 檩托;檩托
purlin(e) head 檩条端部
purlin(e) hinge 檩条铰接

purlin(e) joint 檩条接合
purlin(e) layout 檩条布置
purlin(e) level 檩条标高
purlin(e) load 檩条上荷载
purlin(e) nail 檩条钉
purlin(e) on hypostyle 老檐桁
purlin(e) plate 复斜屋顶转折处的檩条;檩条板;复斜屋顶转折处的檩条;檩条板
purlin(e) post 檩柱;檩条支柱;檩条下柱
purlin(e) projection 檩条挑头;檩条伸出部分
purlin(e) roof 檩条屋架;檩支屋顶;檩支屋面;檩香屋架;硬山搁檩屋顶
purlin(e) roof with inclined struts and wind filling 有斜支撑和填充墙檩支屋顶
purlin(e) roof with king post 单柱式檩支屋顶;单柱檩屋架
purlin(e) roof with king post and slanting studs 有单柱和斜支撑的檩支屋顶
purlin(e) roof with king post and wind filling 有正立柱和填充墙的檩支屋顶
purlin(e) roof with queen post 双柱式檩支屋顶;双柱檩屋架
purlin(e) roof with slope studs 有倾斜支撑的檩支屋顶
purlin(e) roof with struts inclined away from the center 支撑从中心向外倾斜的檩支屋顶
purlin(e) roof with struts inclined towards the center 支撑向中心倾斜的檩支屋顶
purlin(e) roof with three posts 三柱式檩支屋顶
purlin(e) roof with three posts and wind filling 有填充墙的三柱檩支屋顶
purlin(e) roof with two posts and wind filling 有填充墙的两柱檩支屋顶
purlin(e) spacing 檩条间距
purlin(e) with king post 单柱檩屋架
purlin(e) with queen post 双柱檩屋架
purl loop 反面线圈
purlpe convex lens 紫色凸透镜
purolator 滤油器
puron 高纯度铁
purpil capacity 学生容量
purple 紫色;红紫色;红紫
purple black 紫黑色
purple blende 红锑矿
purple boundary 紫色分界线
purple brown 紫棕色;紫褐色
purple canker 斑溃疡
purple copper ore 斑铜矿
purple dye 紫色染料
purple heart 紫红木;紫心木(产于圭亚那)
purple lake 紫色淀
purple light 紫霞;紫光
purple of cassius 桂皮紫;金紫红
purple ore 紫矿石
purple osier 紫柳
purple oxide 紫色氧化物颜料;赤色氧化铁
purple pigment 紫色颜料
purple slate 紫色板岩
purple soil 紫土;紫色土
purple willow 紫柳
purple yulan 紫玉兰
purpling 变紫
purplish 带紫色的
purplish-black 紫黑色;红紫黑
purplish brown 紫褐色
purplish grey 紫灰色
purplish red 紫红色;枣红(色)
purpose 用途;目的
purpose and scope of reconnaissance 踏勘的目的和范围
purpose and scope of the study 研究的目的与范围
purpose-built 专门建造的;特制的
purpose-built box 特制盒子
purpose-built rack 特制架子
purpose-built slope cleansing jet box 特制清坡射流箱;特制的斜坡清洗射流箱
purpose-built vehicle 特殊用途车辆
purpose-built washing equipment 专用清洗设备
purpose-designed 专门设计的
purpose-designed terminal 专门设计的码头
purpose loan 目的贷款;目的贷款
purpose(ly) made 特制(的)
purpose-made(bonding) adhesive 特制的黏合剂

purpose-made bonding agent 特制黏结剂
purpose-made box(roof) gutter 特制的箱形天沟
purpose-made brick 特征形状的砖;特种用途的砖;特制砖;特型砖
purpose-made bucket 特制铲斗;特制斗
purpose-made builders fitting 特制的施工用器具
purpose-made builders hardware 特制的施工用的小五金
purpose-made composition 特制的混合物
purpose-made core extruder 专门岩芯挤出器
purpose-made lift 特制的升降机
purpose-made ornamental block 特制的装饰材料
purpose-made protecting foil 特制的防护薄片
purpose-made reinforcement steel 特制的配筋钢
purpose-made section 特制的型材
purpose-made slope cleansing jet box 特制清坡射流箱;特制的斜坡清洗射流箱
purpose-made structural profile 特制的结构型材
purpose-made wire 特制的金属线
purpose money 定金
purpose of artificial recharge 人工补给目的
purpose of grouting 灌浆目的
purpose of model 模型目的
purpose of operation 经营目的
purpose of reconnaissance survey 踏勘目的
purpose of use 使用目的
purpose of water sealing 止水目的
purposes of 的用途
purposes of development 开发目标
purposes of reservoir 水库用途
purposes stated in detail 详细的目的
purposive behavio(u)r 有目的的行为
purposive sampling 立意抽样法;目的抽样;取样
purpure 紫色
purpureal 红紫色的
purpurin red 蔥紫红素
purpurite 紫磷铁锰矿;磷锰石
purree 印度黄
purse boat 围网渔船;围拉网渔船
purseful 满袋
purse net 围网
purser 业务主任【船】;管事;事务长【船】
purser department 事务部
purser room 事务长室【船】
purser's stocking 卧具等物
purse seine (双船拖曳的)大型拖网;围网
purse seine design 围网设计
purse seiner 围网渔船;大型围网
pursuance method 追赶法
pursuant to 根据
pursue a claim 进行索赔
pursuit display 追踪式显示
pursuit-evasion problem 跟踪逃避问题
pursuring speed to match others 攀比速度
purves lignin 高碘酸盐木素
purveyance 供伙伙食;承办伙食
purveyor 供应者;供应商;伙食供应商
Pusan Port 釜山港(朝鲜)
push action cylinder 推力油缸
push ahead 向前推进
push a kiln 强化窑工作
push and guard bar 推梗
push-and-guide 推入
push and pull 推拉
push and pull bar 推拉把手
push and pull brace 推拉联杆;(桁架的)推拉撑;推拉把手;推拉把撑;推力杆
push and pull handle 推拉把手
push and pull method 顶进法
push and pull pumping power 中央泵送功率;推拉泵送功率
push and pull rod 推拉杆
push and pull switch 推拉开关
push arms 推杆;铲刀主推臂;推臂
push-away buggy 运输小车;带辊架管子
push-back 推回
push-back blank 向上顶出坯料;重压入条料坯料
push-back cam 退回三角
push bar 推门横杆;门推手;推门横条
push bar conveyer 链板运输机;推杆传送器
push-bar drawing frame 推排式并条机
push-barge unit 顶推驳船队
push-bar gill 推针排式针梳机
push bat kiln 推板式窑

push bench 推拔钢管机；推拔床；顶管机
push-bicycle 自行车
push-bike 自行车
push binder 手推割捆机
push block 顶推座（铲运机）；顶推块
push board 推出板
push boat 推轮【船】；顶推船；推船
push bolt 推销锁；推闩
push boring 推式镗孔
push-bottom control 揿钮操纵
push-bottom drilling 按钮操纵钻进
push-bottom drilling machine 按钮操纵钻机
push-bottom-operated alarm 揿钮报警器
push-bottom switch 揿钮开关
push brace 桁架压杆；推撑；桁架压杆；推力系杆；推杆
push bracing 推杆
push broom 推式路帚
push-broom scanning 推扫式扫描
push-broom sensor 推扫式传感器
push bulldozer 手扶拖拉机
push-buttnon-starter 按钮开关起动器
push-button 接通按钮；电钮；按钮（开关）
push-button actuator 按钮开关
push-button alarm 按钮报警器
push-button array 按钮开关阵列
push-button box 按钮开关箱；按钮盒
push-button contact 按钮触点
push-button control 按钮控制；按钮操纵
push-button controlled concrete proportioning and mixing 混凝土混合料自动配料与拌和
push-button controlled mixer 按钮控制搅拌机
push-button controlled mixing 按钮控制搅拌
push-button controlled proportioning 按钮控制配料
push-button control system 按钮控制系统
push-button dialing 按键式拨号；按钮拨号
push-button dialing pad 按钮拨号盘；按钮式拨号板
push-button drill 按钮遥控钻机
push-button flame ignitor 按钮式火焰点火器
push-button ignition 按钮点火
push-button lock 按钮锁
push-button menu 按键菜单
push-buttnon coal mining 遥控采煤法
push-button operation 按钮操作
push-button receiver 按钮接收机
push-button reset 按钮复位
push-button rotary switch 按钮旋转开关
push-button safety 按钮式保险
push-button safety-cap 按钮保险罩
push-button selector 按钮式选择器；按钮波段开关
push-button ship 按钮化船
push-button signal 按钮式信号
push-button starter 按钮启动器
push-button station 按钮台；控钮站
push-button switch 圆头开关；按钮式开关；按钮开关
push-button timer 按钮定时器
push-button tuner 按钮式调谐器；按钮工调谐装置；按键式调谐装置
push-button tuning 按钮式调谐；按钮调谐
push-button valve 按钮开关阀；按钮阀
push-button winding control 按钮操纵半自动绞车控制
push-button yard 机械化调车场；按钮控制式调车场
push cam 顶推凸轮
push car 居间车；运料车（铁路）；间隔车（机车推上轮渡用）；推送连接车；手推车；铁道工程运料车
push cart 推动式试验车；手推车
pushcart for mounting-dismounting of coupling cushion system 钩缓装置拆装小车
push cat 履带拖拉机
push conveyer 推送式输送机；推送式运输机
push-cut 推削
push-cut shaper 推切式牛头刨
push cycle 自行车
push detonator 压发引信
push-down 叠加；后进先出方式；推下
push-down automata 后进先出自动机
push-down automaton 叠加自动机；排队式自动机
push-down list 下推表；后进先出表
push-down machine 电力起爆器
push-down register 后进先出寄存器
push-down stack 下堆栈；后进先出栈

push down storage 下推式存储器；叠式存储器
push-down transformer 后进先出变换机
push-down wheel 退圈轮
push dozer 后部作压实用
push drill 冲击钻；手推钻
pushed 推挤现象的
pushed barge 顶推驳（船）
pushed-bat kiln 推杆窑；推板窑；低推式窑
pushed convey 顶推驳船队
pushed-down stack 后进先出堆栈
pushed grade term of reserves 推荐的储量分级术语
pushed length 推进长度
pushed punt 凸形瓶底；凹瓶底
pushed slab kiln 推板窑
pushed tow 顶推驳船队
pushed-up bottom 凸形瓶底
pusher 推轮【船】；推料机构；推料机；推器；推进机；推钢机；推出器；推车机；顶推器；顶推机；顶推船；顶车机
pusher airplane 推进式飞机
pusher airscrew 推进式螺旋桨
pusher arm 推动杆
pusher autogiro 推进式旋翼机
pusher bar 推杆；推出器
pusher-bar booster 推杆提升机
pusher bar conveyer 推杆运输机
pusher barge 顶推驳（船）
pusher barge navigation 顶推驳船航行
pusher barge pair 推驳搭配
pusher barge system 顶推驳运方式；顶推驳运法
pusher barge tows 顶推驳船队
pusher blade 推刀；推土刀；推进叶片
pusher bow 顶推船首
pusher bracket 推进杆托架
pusher carriage 推料机小车
pusher cleat 止推器
pusher-click banking stop 推爪止挡
pusher-click driver 推爪驱动轮
pusher-click spring 推杆爪簧
pusher collar 移动调节环
pusher dog 拨爪
pusher end 推焦机侧；推机焦侧
pusher engine 推进式发动机
pusher feed 推杆式送料
pusher-feed rock drill 气腿式凿岩机
pusher furnace 推杆式炉
pusher gear 推车装置；推车机
pusher grade 加力牵引坡度；推挽坡度；多机坡度；辅助坡（度）；补机坡度
pusher installation 自动推杆器
pusher jack 推力千斤顶
pusher kiln 推板窑
pusher leg 凿岩机腿架；自动推式气腿；风动钻架
pusher leg drill 气腿式凿岩机
pusher locomotive 推送机车
pusher machine 推焦车
pusher mechanism 推料机构装置
pusher operation 补机作业
pusher propelller 推进式螺旋桨
pusher pump 压气泵
pusher ram 推钢机的推杆
pusher relay 推杆式继电器
pusher screw 推进式螺旋桨
pusher side 机侧；推焦面
pusher tow unit 顶推驳船队
pusher tractor 助推拖拉机；助铲拖拉机；后推机；推式拖拉机
pusher train 顶推船队
pusher tray 推杆式料盘
pusher tray furnace 推盘式炉
pusher tug 推拖两用船；推轮【船】；顶推轮
pusher-tug unit 顶推驳船队
pusher-type 强压式
pusher-type airplane 推进式飞机
pusher-type combine 推式联合收获机
pusher-type fan 强压通风机
pusher-type furnace 推杆式加热炉；推送式炉
pusher-type kiln 推送式隧道窑
pusher-type shaping 推式插削
pusher-type stock guide 推杆式送料导向装置
pusher vehicle 推进式螺旋桨飞行器
push feed 推送送料
push-feed drill 自动推进式钻机

push-feed drilling 自动给进钻进
push-feed separator 推送喂料分选机；推送喂料分离机
pushfiller 回填机；填土机；推土回填机
push fit 推入配合；推合座
push fleet 顶推船队
push frame 顶推架（铲运机）；推耙
pushfulness 冲劲
push grease gun 手压式牛油枪；手推式牛油枪
push gun 推式焊枪
push handle 推手
push hardware 门推手；推门五金
push head 压入端
push hoe 手推中耕器
push-in 按上
push-in fitting 推入式配件
pushing 推挤；变薄挤压
pushing action 推进作用
pushing body 推进体
pushing carriage 牵引小车
pushing-collecting machine 推耙机
push(ing) conveyer 刮板式运送机；推进式运送机；推进式输送机
pushing device 推钢机；送料装置
pushing down the cullet 推下破玻璃
pushing figure 推出系数
pushing force 推力
pushing gradient 推送坡
pushing handle 推手柄
pushing-in wheel 套圈轮
pushing jack 推进千斤顶
pushing landslide 推动式滑坡
pushing machine 推焦车
pushing off the slag 挡渣
pushing off the wagons or cars 冲溜法的调车
pushing of the receiver 受油机接近加油机
pushing open action 撑开作用
pushing pipe 顶管
pushing pit 推进的矿井
pushing process 推挤成型法
pushing ram 推焦杆
pushing rod jack 挺针片
pushing schedule 推焦顺序
pushing section 推送部分【铁】
pushing section of hump 驼峰推送部分
pushing site 顶管现场
pushing speed 推送速度
pushing system 顶推运方式
pushing track 推送线
pushing track of hump 驼峰推送线
pushing-type dung channel cleaner 推式粪沟清理器
pushing-type flat plane 推式平刨
pushing-up board 推毛板
pushing vessel 推轮【船】；顶推船
push in of cutting edge 贯入推进
push-in-type soap dispenser 按入式肥皂配出器
push-jack 推压千斤顶
push joint 挤浆接缝；挤接；挤浆缝（用挤浆法砌砖形成的砖缝）
push-joint end 推接端
push-key 按锹锁；按钮；按键开关
push lever 推动控制杆
push-loading 推式装载
push-loading scraper 推装式铲运机；推式铲运机
push-loading tractor 推装式拖拉机；推式拖拉机
push money 推销鼓励佣金；推销员佣金
push moraine 伸展的冰碛层；推进冰碛
push net 推网
push nipple 柱形连接口；锥形连接口
push-off 推出器；推出；反定位；推换
push-off arm 推料杆
push-off buck rake 带推草板的集草器
push-off cantilever 反定位悬臂
push-off mo(u)lding machine 带顶杆的造型机
push-off pin 托杆；推出杆
push-off stack 推卸式堆垛机
push-on 进栈；推进
push-on filter 附加滤光镜
push-on gland 推入密封套
push-on joint 推入式接头
push-on PVC pipe 推入式PVC管
push-on starter 按钮起动器；按钮启动器
push operation 压栈操作；进栈操作

push-out 推出;顶出
push-out case 外推式表壳
push-out chuck 外推夹套
push-out collet 外推夹套
push-out (collet) chuck 外推弹簧夹头
push-out cylinder (现场从未硬化混凝土中取出的)圆柱试体
push out test 推出试验
push-out windshield 推出式挡风玻璃
pushover 推出器
pushover wipe 推换
push pawl (与棘轮相配合的)推爪
push pedal 推板
push penny (电线导管口上遮挡异物掉入的)圆片
push phone 按钮式电话机
push-piece 按销;推销
push pin 推针
push plate 执手挡板;门推板;推土铲;推门板;推拉板;推板;顶推板(平地机);撒板
push-plate conveyer 高链板输送机
push plunger baler 推式柱塞压捆机
push plunger press 推式柱塞压捆机
push pointer 压尖机
push pointer bench 强迫送料的拉拔机
push pointing 强制压尖
push pole 推杆;撑船篙
push pole pocket 推杆插口
push-pop control 压弹控制
push-pop stack 压弹式堆栈
push-pull 推挽
push-pull amplification 推挽放大
push-pull amplification system 推挽放大方式
push-pull amplifier 推挽放大器;推挽放大器
push-pull arrangement 推拉装置;推挽装置
push-pull audio amplifier 声频推挽放大器
push-pull cascade 推挽级联;推挽级
push-pull centring 推挽定心
push-pull circuit 推挽电路
push-pull computer amplifier 推挽式运算放大器
push-pull connection 推挽连接
push-pull connection method 推挽接法
push-pull control cable 推拉操纵缆
push-pull-critical experiment 推挽式临界实验
push-pull deflection 推挽偏转
push-pull detection 推挽检波
push-pull detector 推挽检波器
push-pull distribution 推挽分布
push-pull drive 推挽激励
push-pull driver 推挽式驱动器;推挽激励器
push-pull effort 推拉力
push-pull electret transducer 推挽驻极体换能器
push-pull equipment 推拉设备;推挽设备
push-pull error 推挽误差
push-pull feed arrangement 推拉式送进装置
push-pull forklift 推挽叉式装卸车
push-pull inverter 推挽变流器
push-pull jack 推拉千斤顶
push-pull jacking rig 推拉式顶进装置
push-pull knob 推拉手把;推拉钮
push-pull microphone 推挽式传声器
push-pull mode 推挽式
push-pull modulation 推挽调制
push-pull multiplier 推挽式倍频器
push-pull neutrodyne method 推挽中和法
push-pull operation 推拉运行
push-pull oscillation 推挽振荡
push-pull oscillator 推挽振荡器
push-pull output amplifier 推挽输出放大器;推挽式输出放大器
push-pull output stage 推挽输出级
push-pull photocell 推挽式光电管
push-pull port tug 推拖两用港作船
push-pull power amplification 推挽功率放大
push-pull process 推拉作用
push-pull propeller stirrer 推挽式螺浆搅拌器
push-pull radio frequency amplifier 推挽射频放大器
push-pull reactor 推挽式堆
push-pull rod 推拉杆
push-pull rod linkage 硬式传动机构
push-pull rule 钢卷尺
push-pull running 可逆运行
push-pull sawtooth wave 推挽锯齿波
push-pull sawtooth wave form 推挽式锯齿波形

push-pull scraper 推挽式铲运机
push-pull screw 推拉螺杆
push-pull shunting 推送调车(法)
push-pull signal 推挽信号
push-pull sound-track 推挽声迹
push-pull stage 推挽级
push-pull support system 液压支柱推拉移动法
push-pull sweep 推挽扫描
push-pull switch 推拉开关
push-pull system 推挽制;推挽系统
push-pull tape 卷尺
push-pull train 推拉式列车
push-pull transformer 推挽式变压器;推挽变压器
push-pull type 推挽式
push-pull wave 疏密波
push-push amplifier 推挽放大器;双推放大器
push-push circuit 双推电路
push-push operation 同相工作
pushrake 推集机
push rod 推杆;推棒;顶杆
push rod actuator 推杆执行机构;推杆传动装置
push rod brake cylinder 推杆闸缸
push rod bushing 推杆衬套
push rod clearance 推杆间隙
push rod cup 推杆头
push-rod deluge valve 推杆型雨淋阀
push rod guide 推杆导杆
push rod holder 推杆托架
push rod holder pin 推杆托销
push rod housing 推杆套筒
push rod pin 推杆销
push rod spring 推杆弹簧
push rod system 推杆操纵系统
push rod tube 推杆套管
push roller 推滚
push rolls 推送辊
push section car 手推平车;手推车
push sets 顶推船队
push socket 推入式插座
push starting 推车起动
push stem power unit 直进式执行部件;直推式执行机构
push straight on 一直向前推动
push tap 手推杠杆式放水龙头
push the heater 强化炉子的操作
push the speed 增大速度
push through 推过去;冲过
push-through deep drawing die 推落式深拉延模
push-through die 下漏式模;压入式模;推落式模
push-through geyser 通过式热水器
push-through loading system 推换装料系统
push-through pinch-trim die 推落式挤切修边模
push-through presentation 满屏显示;全显示
push-through tie 推落式连杆;推送式拉杆;压入式连杆
push-through winding 拉入绕组;插入绕组
push toggle clamp 推式肘板夹紧装置
push to talk 按键通话
push-to-talk button 通话按钮
push-to-type operation 推键式工作;打键型操作;按钮启动打印操作
pushtow boat 推拖两用船
pushtowing 顶推
pushtow navigation 顶推航行
pushtow train 顶推驳船队
pushtow vessel 顶推航行
push-turn key 按下旋钮
push type 推进式
push-type broach 推刀;压刀
push-type centrifuge 推送式离心机;推挤式离心机
push-type conveyer 推式输送机
push-type duster 推车式喷粉器
push-type grader 推式平地机
push-type grease gun 推压式滑脂枪
push-type lubricating fitting 压力油嘴;挤压式润滑器
push-type machine 推式机具
push-type motor grader 推式平地机;推板式平路机
push-type puller 前置式拔销器
push-type scraper 推压式铲运机;推土机
push-type slab kiln 推入式扁平窑
push-type sprayer 手推车式喷雾器
push-type wire feed 推式送丝
push-type wire feeder 推式送丝机构

push-up 上推;上凹底
push-up block 立式推线拉丝卷筒
push-up bottom 凸形瓶底
push-up door 上推的门
push-up engine 推峰机车
push-up leaf 上推页
push-up list 上推表
push-up operation 上推操作
push-up queue 上推队列
push-up speed 推送速度
push-up storage 上推式存储器;上推存储器
push-up store 上推存储
push-up-type soap dispenser 上压式液体肥皂配出器
push water 推动水
push wave 推波
push wedge 调整块;调整楔
push weld(ing) 钳点焊;手(压)点焊;挤焊;手动挤焊
push well 平推管井;推进井
pusle response characteristics 脉冲响应特性
pussolane 白榴火山灰
Pussur 布苏尔酒红色硬木(印度产)
put…into circulation 投放;存放记录;存放;拨动
put about 掉转航向
put across 渡过(河、湾等)
put a fence across the middle of the fields 地中间隔上一条栅栏
put a head on a stem 冲击杆损坏
put a high price on 高度评价
put a machine into service 使用机器
put and call dealer 期货抛出和买进自营商
put and call option 卖出买入选择权;期货抛出和买进选择权
put a signal at stop 关闭信号
put back 回原地;退回;送回
put back to port 返回港口
put down 削减;笔录
put down a borehole 向下钻进
put down expenditures 削减支出
put down in colo(u)r 色环检查
put down the price exworks 压低出厂价格
puteal 井栏(古罗马)
put elevator on links 挂接引器
puteus (古罗马沟渠的)洞口或进人井
put forth 离港;伸出
put forward a preliminary plan for test 提出初步试验方案
put forward for discussion 提出讨论
put goods on sale with the prices clearly marked 明码售货
puthole 墙上脚手架(横木)孔
put in 加入;放入
put in a skylight 开天窗
put in a tender for 投标承包
put in circuit 接入电路
put in force 施行
put in repair 修理
put in series 串联接入
put in stalk 接钻杆
put into commission 开始投产;投产
put into effect 实施
put into gear 挂档;啮合
put into motion 开动
put into operation 开通;开始投产;开工;交付使用;投入运转;投入运行;投入生产;投产
put into overhaul 交付检修
put into practice 实行
put into practice freezing method 实行冷冻法
put into production 开始投产;投入生产
put into service 投入营运;交付使用;投入运行;投产;使运行
put into stalk 挂档
putizze 硫化氢气孔
putlock 脚手架跳板横木;脚手架小横木;排木
putlog 脚手架小横杆;脚手架跳板横木;横格材;排木
putlog hole 脚手架孔;架眼;脚手眼
putlog holes of wall 墙上脚手架(横木)孔
putlog scaffold 单排脚手架
put-off clamp for twin catenary wire 双承力索双耳线夹
put-on 安装
put on a demonstration of aero-spraying 示范表演空中喷药

put on production 投产
put on pump 井口安装泵
put on sale 发售
put on stamp duty 贴印花税
put on stream 投入生产
put on the market 卖出
put on trial 试验
put on weight 体重增加
put option 卖期货选择权;卖出选择权;卖出期权
putoranite 波硫铁铜矿
put out 熄灭;驶出
put out of circuit 切断电路
put out to contract 给人承包
put out to lease 出租
put out to pasture 放牧
put over 延期
put part on product 配置于产品上
put-phase 相位不重合
putrefaction 腐烂作用;腐化作用;腐败作用;腐败
putrefaction of paints 涂料的腐败
putrefactive 腐烂的
putrefactive bacteria 腐败菌
putrefactive earth bacteria 腐败性土壤菌
putrefactive fermentation 腐败发酵
putrefactive odo(u)r 腐臭味;腐败臭气
putrefactive organism 腐败性有机物
putrefactive process 腐败过程
putrefactive substance 腐败性物质
putrefy 腐化;腐败
putrefying bacteria 腐败菌
putrefying blebs 腐败气泡
putrescence 腐化作用;腐败作用
putrescibility 可腐败性;腐熟度;腐败稳定性
putrescible 易腐烂的;可腐败的
putrescible goods 易腐烂商品;易腐烂货物
putrescible matter 易腐烂物质
putrescible waste 可腐烂废物;腐烂废物;腐烂的废物
putrescine 腐胺
putrid 腐烂的;腐败性的;腐败的;腐臭的
putrid fermentation 腐烂发酵
putrid fetid 恶臭的
putrid food 腐败食物
putridity 腐烂
putrid mud 腐殖泥;腐泥
putrid odo(u)r 恶臭味
putrid weather 讨厌的天气
putromaine 腐败毒
put statement 放置语句
put the ring on her 量规检查钻头直径
put the rudder to port/starboard 向左/右舵
put the sludge on before cultivate 耕地前施上淤泥
put through by line 管线输送
put-through time 中转时间
puttied split 用灰泥塞裂隙
puttier 油灰工
putting 刮油灰
putting back into operation 推迟运行
putting-down machine 澄清机;沉降机;沉积机
putting-down method 沉降方法;钻井方法;钻孔方法;扑灭方法
putting-down of borehole 加深钻孔
putting green 小高尔夫球场
putting into effect the policy of trade protection 实现保护贸易政策
putting into operation 投入运转;投入运行;实施;实行;启动;开动
putting into port on account of bad weather 入港避风
putting in working order 投入正常工作状态
putting leaf 贴金属箔;贴金
putting on the stocks 上墩
putting out 挤出
putto 裸儿雕饰;爱神裸体雕像饰;裸体小儿雕像饰;裸体神童雕像饰;[复]putti
put together 装配;拼
put to sea 出海
put to test 做试验;经受试验;试验
putty 油漆腻子;油灰;用油灰镶嵌;氧化锡抛光粉;腻子;泥子;水泥封涂
putty adhesive force 油灰附着力
putty and plaster 装饰石膏
putty chap cracking test 油灰龟裂试验

putty chaser 油灰磨;油灰刮刀
putty-chaser-type mixer 研磨混合机
putty coat 石灰膏罩面层;抹灰压光面;最后抹面
putty consistency apparatus 腻子稠度计
putty filling 填腻子
putty finishability 油灰操作性
putty for glazing 玻璃腻子
putty for laminated glass 叠层玻璃的油灰
putty for steel(plate) windows 钢窗腻子;钢窗油灰
putty glazing 括油灰安装玻璃;用油灰安装玻璃;油灰装玻璃;油灰镶玻璃法
putty gun 红丹油泥喷枪
puttying 抹刮油灰;嵌油灰
putty in plastering 抹灰嵌缝
putty joint 油灰缝
putty knife 油灰(刮)刀;刮腻子刀;开刀
puttyless 无油灰的
puttyless alumin(i)um glazing bar 无油灰装玻璃用的铝条
puttyless glazing 无腻子玻璃;无油灰镶玻璃法;无灰装玻璃
puttyless glazing bar 不用油灰的玻璃条
puttyless glazing roof 无油灰的玻璃屋顶
puttyless steel bar 无油灰装玻璃用钢条
putty-like property 灰泥状性质
putty method 刮腻子方法;打腻子方法
putty oil 油灰油
putty pointing 油灰勾缝
putty powder 擦光粉(氧化锡);油灰粉;氧化锡粉;氧化锡;去污粉
putty pump 油灰唧筒
putty removal 油灰清除
putty retouch 找补腻子
putty sand 石灰质土
putty seal(ing) 油灰嵌缝;油灰密封
putty soil 油灰土
putty up 刮底漆腻子
put under separate control 分口管理
put up 建造;挂起;搭架
put up a tent 支帐篷
put up for auction 付之拍卖
putward passage 出口航道
Putzmeister pump 双缸液压混凝土泵
puy 死火山锥;布义型火山
puzzle 难题
puzzle lock 转子组合锁
puzzolana 白榴火山灰
puzzolana cement 火山灰水泥
puzzolanic capability 火山灰性能
puzzolano 白榴火山灰
PVA emulsion 聚乙烯醇乳液
PVC conduit 聚氯乙烯管
PVC duct 聚氯乙烯线槽
PVC gravity sewer pie PVC 管重力污水管
PVC pipe 聚氯乙烯管;PVC管
PVC pipe joint restraint PVC 管制约管接头
PVC plastic pipe 聚氯乙烯塑料暗管
PVC resin 聚氯乙烯树脂
P-veatchite 副水硼锶石
PVF 聚氟乙烯
P wave 地震纵波;P 波
p-wave velocity 纵波波速
P-wire 测试线
p-xylene resin 对二甲苯树脂
pycastyle 列柱式柱廊
pycnocline 密跃层;密度跃层
pycnometer 液体比重计;密度瓶;比重瓶;比重计;比重管;比重表(指仪器仪表)
pycnometer method (测定液体和骨料比重的)比重瓶法;比重瓶测定比重法
pycnometer specific gravity factor 比重瓶法测定的比重因数
pycnometry 测比重术
pycnonuclear reaction 超密态核反应
pycnosis 固缩
pycnostyle 列柱廊(柱边间距为1.5倍柱径的柱廊);一个半柱径式
pycnoxylic wood 密木材
pygidium 尾板
pygmy current meter 微型流速仪;微型电流表
pygmy lamp 小型灯
pygmy meter 微型流速仪
pyimithate 嗒啶磷

pyinkado 派恩卡多硬木(一种产于印度的红褐色至纯红色的硬木)
pyinma 派恩玛硬木(一种产于印度的淡红至红褐色的硬木)
pyknometer 密度瓶
pyknosis 致密化
pyknostylos 密柱式(古希腊、古罗马神庙的,柱间为1.5倍柱径的形式)
pyll 涨潮时形成的小(海)湾
pyller 塔门;标塔
pyllonite 千糜岩
pylon 桥台;柱台;高压电缆塔;机场标塔;外支架;铁塔;梯形门框;塔状物;塔柱;塔形门;塔架;电缆铁塔;电缆塔;标塔;埃及式塔门
pylon antenna 圆筒隙缝天线;铁塔(式)天线
pylon bearing 墩支座
pylon bent 塔架
pylon bridge tower 桥塔
pylon head 桥塔顶端;支架顶部
pylon shape 支架形状
pylon tower 耐张力铁塔;桥塔
pylumin process 铝合金涂漆前铬酸盐浸渍处理;铬酸盐浸渍处理
pyod 热(电)偶
Pyongyang marine trough 平壤海槽
pyophyte 耐火植物
pyorrhea alveolaris 齿槽脓溢
pyrabrom 吡拉布隆
pyragyrite 深红银矿
pyralarm 烟火报警器
pyralin 珀拉玲
pyralmandite 镁铁榴石
pyralspite 铝榴石
pyramal 吡拉明
pyramid 锥形;棱锥体;金字塔;经营或连续投机;角锥体
pyramidal 锥状;金字塔形的;金字塔形的
pyramidal antenna 棱锥形天线;角锥天线
pyramidal cleavage 锥状解理
pyramidal crown of stone slabs 石板的棱锥顶
pyramid(al) cut 角锥形掏眼;角锥形掏槽
pyramidal error 尖塔差
pyramidal error of prism 棱镜尖塔差
pyramidal face 锥面
pyramidal folded plate 棱锥形折板
pyramidal folded structure system 棱锥形折结构系统
pyramidal form 棱锥形式
pyramidal formula 棱锥体土方公式
pyramidal hipped roof 金字塔形四坡屋顶
pyramidal horn 棱锥形号筒
pyramidal iceberg 锥状冰山;峰形冰山
pyramidal light 屋顶灯光照明;棱锥天窗
pyramidal lobe 锥体叶
pyramidal roof 棱锥屋顶;棱锥形屋顶
pyramidal sedimentation tank 倒锥形沉淀池
pyramidal slewing crane 锥形架旋臂起重机
pyramidal space truss 棱锥形空间桁架
pyramidal structure 棱锥结构
pyramidal surface 棱锥面
pyramidal system 锥体系;正方晶系;四方晶系
pyramidal tomb 金字塔;棱锥形坟墓
pyramidal toothed crushing roll 带角锥形齿的辊式破碎机
pyramidal volcano 锥状火山
pyramidal wave 锥形波;三角浪;三角波;大三角浪
pyramid building 金字塔建筑
pyramid carry 锥形进位
pyramid column 角锥柱
pyramid complex 金字塔建筑群
pyramid construction 金字塔式建筑
pyramid control 金字塔式控制
pyramid crane 角锥架起重机
pyramid cut 锥体切割
pyramid dam 金字塔坝
pyramid feet roller 有棱锥形(羊)足的压路碾;棱锥形(羊)足的压路机
pyramid formation 锥体形成
pyramid glass 棱锥形玻璃(一种装饰玻璃)
pyramid hardness 角锥硬度(金刚石)
pyramidic space frame 四角锥空间构架
pyramiding 金字塔式交易法;金字塔式加价;积累利润的使用
pyramidion 角锥体;小金字塔;小角锥体;小方尖塔

pyramid matrix 锥形矩阵
pyramid method 角锥形法;角锥法
pyramid of biomass 生物量金字塔
pyramid of credit 信贷的层层加大
pyramid of numbers 数量锥体;数量金字塔
pyramid of polluted population 受污染人口金字塔
pyramid of productivity 生产力金字塔
pyramid of the first order 初级棱锥体
Pyramid of the Moon 月亮金字塔(位于墨西哥特奥蒂瓦坎)
Pyramid of the Sun 太阳金字塔(位于墨西哥特奥蒂瓦坎)
pyramidon (高尖碑顶端的)尖锥头
pyramidonosis 氨基比林中毒
pyramidon test 氨基比林试验
pyramid ratio 塔式比率;金字塔形比率(进行财务比率分析的一种方法)
pyramid roller 宝塔式传墨辊
pyramid roof (亭,台,阁,榭的)多边形屋顶;棱锥屋顶;攒尖顶;四角攒尖顶;方攒尖顶
pyramid selling 金字塔式推销
pyramid-set bit 角锥钻头
pyramid-shaped dune 金字塔形沙丘
pyramid-shaped roof 角锥形屋顶
pyramid slewing crane 棱锥架旋臂起重机;角锥架旋臂起重机
pyramid structure 金字塔形结构(即纵向结构)
pyramid temperature tester 测温角锥
pyramid tower 方锥形塔
pyramine 嘧胺
pyramis roof 攒尖顶【建】
pyrane 吡喃
pyranium salt 吡喃盐
pyranoid form 吡喃型
pyranoid ring 吡喃环
pyranometer 总日射表;日(辐)射强度计;测辐射计;辐射强度计
pyranon 电介浸渍器
pyranose 吡喃糖
pyranthrone 阴丹士林金黄;皮蒽酮
pyranyl 吡喃基
pyranylation 吡喃基化
pyrargyrite 浓红银矿
pyrasteel 铬镍耐蚀耐热钢
Pyrator 煤粉烘干粉磨机(商品名)
pyrauxite 叶蜡石
pyrazinamide 对二氮苯酰胺;吡嗪酰胺
pyrazine carboxamide 对二氮苯酰胺;吡嗪羧酰胺
pyrazine carboxylamide 吡嗪酰胺
pyrazines 吡嗪
pyrazinoic acid amide 对二氮苯酰胺;吡嗪酰胺
pyrazole 吡唑
pyrazole blue 吡唑蓝
pyrazoles 吡唑类
pyrazolidone 吡唑烷酮
pyrazolidyl 吡唑烷基
pyrazoline 吡唑啉
pyrazolon 吡唑啉酮
pyrazolone 吡唑啉酮
pyrazolone dye 吡唑啉酮染料
pyrazolone red 吡唑啉酮红
pyrazolone spectrophotometry 吡唑啉酮分光光度法
pyrazolone yellow 吡唑啉酮黄
pyrazolyl 吡唑基
pyrazophos 定菌磷
pyrazothion 彼硫磷
pyrazoxon 彼氧磷;吡唑磷
pyre 大堆薪材;柴堆;火葬柴堆
Pyrelide 彼拉里德(一种防火门或百叶窗)
pyrene (金属制成的)厨房用具
Pyrenean orogeny 比利牛斯造山运动;比利牛斯运动
Pyrenees Mountains 比利牛斯山脉
Pyrene process 派来恩方法(一种制造泡沫砂浆的方法)
pyretic sulfur 硫铁矿硫
pyretogen 致热物
pyretogenesis 热发生;热产生
pyretogenetic 热发生的;热产生的
pyretogenic 致热的;热源的
pyretogenous 致热的
pyretolysis 热消退
Pyrex 派勒克斯玻璃(商标名称);耐热玻璃;硼硅酸玻璃

Pyrex bulb 硬质玻璃瓶;耐热玻璃瓶
Pyrex cooling jacket 硼硅玻璃冷却套;派勒克斯玻璃冷却套
Pyrex glass 高硬玻璃;高耐火玻璃;高硅玻璃;派勒克斯耐热玻璃
Pyrex unit 耐热玻璃设备
pyrgeometer 地面辐射强度计;地面大气辐射强度计;大气辐射强度计
pyrheliograph 直接日照强度计;日光辐射强度仪
pyrheliometer 直接日照强度表;直接日射强度计;直接日射强度表;日照仪;日温计;日射强度计;日光辐射强度计;太阳热量计
pyrheliometry 太阳热量测量学
pyrhibole granular poprhyrite 辉石安山玢岩
pyric 与燃烧有关的;由燃烧引起的
pyric climax 火烧顶级
pyric factor 火灾因素
pyridina poisoning 吡啶中毒
pyridine 吡啶
pyridine acid 吡啶酸
pyridine-barbituric acid colo(u)rimetry 吡啶—巴比妥酸比色法
pyridine bases 吡啶碱类
pyridine carboxylic acid 吡啶羧酸
pyridine chloride 氯化吡啶
pyridine dye 氮苯染料;吡啶染料
pyridine-insulable matter 吡啶不溶解物质
pyridine monocarboxylic acid 吡啶一羧酸
pyridine monosulfonic acid 吡啶一磺酸
pyridine-N-oxide 吡啶氧化物
pyridine poisoning 吡啶中毒
pyridine tricarboxylic acid 吡啶三甲酸
pyridinium salt 吡啶盐
pyridino 吡啶并
pyridium 苯基偶氮二氨基吡啶
pyridol 吡啶酚
pyridone 吡啶酮
pyridopyridine 吡啶并吡啶
pyridostigmin(e) 吡啶斯的明
pyridoxal 吡哆醛
pyridoxal thiosemicarbazide fluorophotometry 吡哆醛氨基硫脲荧光光度法
pyridoxamine 吡哆胺
pyridoxic acid 吡哆酸
pyridoxine 吡多辛
pyridyl 吡啶基
pyridylaldehyde 吡啶甲醛
pyriform 梨形线脚;梨状屋顶;梨形
pyriform aperture 梨状孔
pyriform profile 拱梨形轮廓
pyrilamine 吡拉明
pyrimidine 间二氮杂苯
pyrinuron 灭鼠优
pyrisuccideanol maleate 马来酸
pyrite 愚人金;硫铁矿;黄铁矿
pyrite cement 黄铁矿胶结物
pyrite cinder 硫铁矿渣
pyrite concretion 黄铁矿结核
pyrite copper ore 黄铜矿
pyrite deposit in carbonate rock and sandstones hale 碳酸盐岩及砂页中硫铁矿床
pyrite deposit in volcanic series 火山岩系中硫铁矿床
pyrite dust 黄铁矿粉
pyrite marcaite 二硫化铁
pyrite ore 黄铁矿石
pyrite roasting 黄铁矿焙烧
pyrite roastor 黄铁矿焙烧炉
pyrite rock 黄铁矿岩
pyrites 硫化铁矿类;硫化矿类
pyrite slag 硫铁矿渣
pyrite smelting 黄铁矿冶炼
pyrites plow 黄铁矿刮板
pyrithiamine 吡啶硫胺
pyrithyldione 吡乙二酮
pyritic action 自热脱硫作用
pyritic copper deposit 黄铁矿型铜矿床
pyritic oolite 黄铁矿鲕状岩
pyritic rock 黄铁矿石
pyritic smelting 黄铁矿冶炼;自热熔炼
pyritic sulfur 黄铁矿硫
pyritic-type-polymetallic deposit 黄铁矿型多金属矿床

pyritinol 吡硫醇
pyritization 黄铁矿化
pyritobitumen 火成沥青
pyritohedron 五角十二面体
pyroabietic acid 焦松香酸
pyro acid 焦酸
pyroantimonic acid 焦锑酸
pyroarsenic acid 焦砷酸
pyroarsenous acid 焦亚砷酸
pyroaurite 菱水碳铁镁石;鳞镁铁矿
pyrobelonite 钒锰铅矿
pyrobiolite 焦性生物岩
pyrobitumen 焦性沥青;焦沥青;火成沥青
pyrobituminous shale 焦性沥青页岩
Pyroc 派洛克(一种轻质砂浆,可用喷枪喷涂)
pyrocarbon 高温炭;高温石墨
pyrocatechol 邻苯二酚
pyrocatechol violet 邻苯二酚紫
pyrocellulose 高氮硝化纤维素;焦纤维素
pyroceram 高温瓷;耐高温陶瓷;耐高温玻璃
pyroceramic 玻璃结晶材料
pyrochemical processing 高温化学处理
pyrochemistry 高温化学
pyrochlore content 烧绿石含量
pyrochlore ore 黄绿石矿石
pyrochlorite 焦绿石;烧绿石
pyrochlor structure 烧绿石结构
pyrochlor type structure 烧绿石型结构
pyrochore-bearing carbonatite deposit 烧绿石碳酸岩矿床
pyroclast 火山碎屑;火成碎屑
pyroclastic 火成碎屑的;火成的
pyroclastic cone 集块火山锥;火山碎屑锥
pyroclastic eruption 火山碎屑喷发
pyroclastic fall phase 降落火山碎屑相
pyroclastic flow 火碎流;火成碎屑流
pyroclastic flow phase 火山碎屑相
pyroclastic flow plateau 火碎流台地
pyroclastic fragmental facies 火成碎屑岩相
pyroclastic ground surge 火成碎屑岩涌
pyroclastic lava 火山碎屑熔岩
pyroclastic lava clan 火山碎屑熔岩类
pyroclastic lava texture 碎屑熔岩结构
pyroclastic material 火成碎屑物
pyroclastic rock 火山岩;火山碎屑岩;火成碎屑岩
pyroclastics 火成碎屑沉积
pyroclastic sedimentary rock 火山碎屑沉积岩
pyroclastic texture 火山碎屑结构
pyroclean 热浸脱脂;热浸净化
pyrocoal coating process 高温溶胶镀膜法
Pyrococcus furiosus 火球菌
pyrocondensation 热缩(作用)
pyroconductivity 高温导电性;热电导性;热电导
pyrocone 塞格锥;测温锥
Pyro-Control 煤粉计量系统(德国洪堡公司产品名)
pyrocxene monzogneiss 辉石二长片麻岩
pyrodynamics 爆发力学;爆发动力学
pyroelectric 热电的
pyroelectric(al) bolometer 热释测辐射热器
pyroelectric(al) ceramics 热释电陶瓷
pyroelectric(al) charging 热电充电
pyroelectric(al) constant 热电常数
pyroelectric(al) copying process 热释复制法
pyroelectric(al) crystal 热释晶体;热释电晶体;热电晶体
pyroelectric(al) detector 热释探测器
pyroelectric(al) device 热释装置
pyroelectric(al) display 热释显示
pyroelectric(al) effect 热释效应;热释电效应;热电效应
pyroelectric(al) energy meter 热释能量计
pyroelectric(al) imaging system 热电成像系统
pyroelectric(al) material 热释电材料;热电材料
pyroelectric(al) optic(al) sensor 热释电式光传感器
pyroelectric(al) optic(al) transducer 热释电式光传感器
pyroelectric(al) production 热电联合生产
pyroelectric(al) radiation detector 热释辐射探测器
pyroelectric(al) relay 热电继电器
pyroelectric(al) sensor 热释传感器
pyroelectric(al) signal 热释信号
pyroelectric(al) sputtered thin film detector 溅

射热电薄膜探测器
pyroelectric(al) technology 热释技术
pyroelectric(al) temperature sensor 热释电式温度传感器
pyroelectric(al) temperature transducer 热释电式温度传感器
pyroelectric(al) thermal detector 热释热探测器
pyroelectric(al) vidicon 热释视像管;热电式光导摄像管
pyroelectricity 热释电学;热释电性;热释电;热电学;热电性;热电现象;热电物质;热电
pyroelectrics 热释电体;热电体
pyroferrite 热电铁氧(化)体
Pyrofill gypsum concrete 派洛石膏混凝土
pyrofoam 泡沫焦性石墨
pyrofusinite 火焚丝质体
pyrofusites 微火焚丝煤
pyrogallic acid 焦棓酸
pyrogallol 连苯三酚;顺苯三酚
pyrogen 致热原;致热物;热原(质)
pyrogenesis 火成作用
pyrogenetic 火成的;热发生的
pyrogenetic decomposition 热解作用
pyrogenetic deposit 火成矿床
pyrogenetic mineral 火成矿物
pyrogenetic rock 火成岩
pyrogenetic stage 发热期
pyrogen-free 无热原的
pyrogenic 致热的;焦化;火成的
pyrogenic action 致热作用
pyrogenic attack 火法处理
pyrogenic decomposition 高温分解
pyrogenic deposit 火成矿床
pyrogenic effect 发热效应
pyrogenic process 火法
pyrogenic reaction 高温反应;焦化反应;生热反应
pyrogenic rock 火成岩
pyrogenic silica 热解(法)二氧化硅
pyrogenous 干馏的;火成的
pyrogenous asphalt 火成沥青
pyrogenous bitumen 火成沥青
pyrogenous wax 火成蜡
pyrogensis 火成作用
pyrogen test 热原检查法
pyrogeometer 地面辐射计;大气辐射表
pyrogram 裂解色谱图;热解图;热解色谱图
pyrograph 裂解色谱;烙画;热谱;热解色谱;烫画
pyrographalloy 热解石墨合金
pyrographic analysis 热解色层分析;热解分析
pyrographic method 热解色层法
pyrographite 高温石墨;高温石墨;热解定向石墨
pyrography 裂解色谱法;烙画法;热法;热解色谱法;热解色层
pyrogravure 烙画法;烫画
pyrohroite 羟锰矿
pyrohydrolysis 高温水解;热水解
Pyro-Jet-burner 高速燃烧器
pyrolic alloy 镍铬合金
pyroligneous 焦木的;干燥木材而得的;干馏木材而得的
pyroligneous acid 焦木酸
pyroligneous alcohol 焦木酒精
pyrolin plastics 硝化纤维素塑料
pyrolite 地幔岩
pyrolith 火成岩
pyrolithic acid 三聚氰酸
pyrology 高温热学;热工学
pyrolsis product 热解产物
pyrolusite 软铁矿;软锰矿
pyrolusite ore 软锰矿矿石
pyrolysate 热分解物
pyrolysis 高温分解;加热分解;热解作用;热解离;热分解作用
pyrolysis analysis 热解分析
pyrolysis apparatus 热解器
pyrolysis chamber 热解室
pyrolysis-chromatogram gas logging 热解色谱气测井
pyrolysis coils 热解旋管
pyrolysis curve 热解曲线
pyrolysis diagram 热解图
pyrolysis gas 裂解气
pyrolysis gas chromatograph 裂解气相色谱
pyrolysis gas chromatography 裂解气相色谱法;热解气相色谱(法)
pyrolysis-gas chromatography combination 热解气相色谱联机
pyrolysis gas oil 裂解柴油
pyrolysis gasoline 裂解汽油;热解汽油
pyrolysis genetic hydrocarbon of kerogen 干酪根中热解生成烃
pyrolysis in tubular furnace 管式炉裂解
pyrolysis liquids 热解液体产物
pyrolysis-mass spectrometry 热解质谱测量
pyrolysis naphtha 高温裂解石脑油;热解汽油馏分
pyrolysis of coal 煤热解
pyrolysis of refuse 垃圾热解;垃圾的热解
pyrolysis of wastes 废物热解
pyrolysis oven 热解炉
pyrolysis rate 热解速度
pyrolysis reactor 热解反应器
pyrolysis spectrum 热解光谱
pyrolysis temperature 热解温度
pyrolysis time 热解时间
pyrolysis tube furnace 裂解管式炉
pyrolysis unit 热解组件;热解单元
pyrolysis zone 热分解区
pyrolytic black 热解石墨
pyrolytic boron nitride 热解氮化硼
pyrolytic carbon 焦化石墨;热解石墨
pyrolytic carbon-coated particle 热解碳涂敷颗粒
pyrolytic carbon film resistor 热解碳膜电阻器
pyrolytic chromatography 热解色层分析;热解色谱法;裂解色谱法
pyrolytic coating 高温喷涂;热沉积涂层
pyrolytic coating technique 高温沉积技术
pyrolytic conversion 高温转化
pyrolytic conversion of hydrocarbons 碳氢化合物的高温转化
pyrolytic copolymer 热聚物
pyrolytic cracking 热裂(作用);热裂解
pyrolytic damage 高温分解损伤;热损害
pyrolytic decomposition 热解;热分解
pyrolytic deposition process 热解沉积法
pyrolytic elimination 热解消除
pyrolytic gas chromatography 裂解气相色谱分析;热解气相色谱(法)
pyrolytic graphite 热解分解石墨;热解石墨
pyrolytic infiltration 热解渗滤
pyrolytic plating 高温喷涂;热沉积涂层
pyrolytic polymer 热解聚合物
pyrolytic process 热解过程;热分解法
pyrolytic product 热解产物
pyrolytic reaction 热解反应
pyrolytic silicon carbide 热解碳化硅
pyrolytic spectrum 热解图;热解光谱
pyrolytic stability 耐高温分解性
pyrolytic technique 热解技术
pyrolyzate 干馏物;热解产物
pyrolyzate spectrum 热解光谱
pyrolyze 热解
pyrolyzed-polymer 热解聚合物
pyrolyzed substance 热解物
pyrolyzer 裂解炉;热解器
pyrolyzing apparatus 热解器
pyrolyzing furnace 热解炉
pyromagma 浅源岩浆
pyromagnetic 热磁
pyromagnetic detection 热磁探测
pyromagnetic detector 热磁探测器
pyromagnetic effect 热磁效应
pyromagnetic generator 热磁发电机
pyromagnetic substance 热磁物质
pyromagnetism 高温磁学;热致内磁性;热致磁性;热磁性
pyromaniac 纵火狂;放火癖
pyromellitic acid 均苯四酸;苯均四酸
pyromellitic anhydride 均苯四酸酐
pyromellitic dianhydride 均苯四酸二酐;苯均四酸二酐
pyromellitic diimide 均苯四酸二酰亚胺
pyromellitic ester 均苯四酸酯
pyromeride 球泡霏细岩
pyrometallurgical method 高温冶金法;火法冶金
pyrometallurgical plant 高温冶金设备
pyrometallurgical process 高温冶金过程
pyrometallurgical refining 火法精炼
pyrometallurgy 高温冶炼;火法冶金学;火法冶金;热冶金学
pyrometamorphism 高温变质(作用);高温变相;高热变质(作用);热力变质(作用);热变相
pyrometasomatic 热液交代的
pyrometasomatic deposit 热液交代矿床
pyrometasomatism 高温交代变质作用;高热交代;热液交代变质
pyrometer 高温仪表;高温温度计;高温计;高温表
pyrometer block 热电偶砖
pyrometer cone equivalent 三角锥等值
pyrometer couple 高温热电偶;高温计热电偶;计热电偶;热电偶
pyrometer fire-end 高温计热端
pyrometer probe 高温计探头
pyrometer-protecting tube 高温计保护管
pyrometer recorder 高温计记录器
pyrometer sighting tube 高温计窥视管
pyrometer tube 高温计管;高温计保护管
pyrometric 高温测量的;测温的
pyrometrical photography 高温摄影学
pyrometric cone 西格示温锥;西格测温锥;高温三角锥;示温熔锥;三角锥;测温锥;测高温熔锥
pyrometric cone equivalent 耐火度测温锥当量值;耐火度;熔锥当量;熔锥比值;示温熔锥当量;示温熔锥比值;三角锥等值;测温锥耐火度
pyrometric cone equivalent value 热锥比值
pyrometric cone evaluation 示温熔锥估价
pyrometric cone number 测温锥号码
pyrometric ga(u)ge 高温规
pyrometric indicator 高温测定指示器
pyrometric probe 高温探针
pyrometric reference cone 标准测温锥
pyrometric scale 高温计刻度;高温计表;高温表
pyrometric Seger cone equivalent 西格三角锥耐火板(高温测量用)
pyrometric telescope 高温计望远镜
pyrometry 高温学;高温测量学;高温测量法;高温测定学;高温测定法;温度测定学;测高温学;测高温法
pyromike 微型高温计
pyromorphism 热变法
pyromorphite 火成晶石;磷氯铅矿;磷绿矿石
pyromotor 热动机
pyromucic acid 糠酸;焦黏液酸
pyromucyl 焦黏酰
pyromucyl acid 焦黏酸
pyromucylanhydrin 焦黏酐
pyron 拍隆
pyronaphtha 热解粗汽油
pyrone carboxylic acid 吡喃酮羧酸
pyrones 吡喃酮
Pyron iron powder 派朗海绵铁粉
pyro-oxidation-reduction 高温氧化还原
pyrope 镁铝榴石;镁硅铝榴石
pyrope kimberlite 美铝榴石金伯利岩
pyrophane 火蛋白石
pyrophanite 红钛锰矿
pyrophobia 火焰恐怖
pyrophore 引火物;撞燃的引火物
pyrophoric alloy 引火合金;米氏合金;发火合金;发光合金
pyrophoric iron 引火铁
pyrophoricity 自燃性
pyrophoricity accident 燃烧事故
pyrophoric material 自燃物
pyrophoric metal 引火合金;自燃金属
pyrophoric powder 引火粉
pyrophoric propellant 自燃推进剂
pyrophorus 引火物;自燃物
pyrophosphate 焦磷酸盐
pyrophosphate exchange reaction 焦磷酸交换反应
pyrophosphate method of electroplating 焦磷酸盐电镀法
pyrophosphodiamic acid 二氨基焦磷酸
pyrophosphoric acid 焦磷酸
pyrophosphorolysis 焦磷酸解作用
pyrophosphorous acid 焦亚磷酸
pyrophosphoryl 焦磷酰
pyrophyllite 叶蜡石
pyrophyllite asbestos 叶蜡石石棉
pyrophyllite ceramics 叶蜡石陶瓷
pyrophyllite clay 叶蜡石黏土
pyrophyllite deposit 叶蜡石矿床

pyrophyllite fire brick 叶蜡石耐火砖
pyrophyllite-kaolinite ore 叶蜡石高岭石岩矿石
pyrophyllite sleeve 叶蜡石套
pyrophyllitite 叶蜡石岩
pyrophyte 防火植物
pyropissite 蜡煤
pyropissite brown coal 蜡质褐煤
pyroplastic deformation 高温塑性变形;热塑性变形
pyroplasticity 高温塑性;热塑性塑性;热塑性
pyroplastic state 高温塑性状态
pyropolymer 焦化聚合物
pyro-probe-ribbon 热解取样带
pyroprocess 高温过程
pyroprocessing 高温冶金处理回收;高温冶金加工;高温冶金处理;高温加工;高温处理;热工
pyroprocessing plant 高温处理设备;热工设备
pyropruf 具有较强耐火性能的轻质灰浆
pyroracemamide 丙酮酰胺
pyroracemic aldehyde 丙酮醛
Pyrorapid 超短窑(德国洪堡公司产品名)
pyroscan 一种红外线探测器;红外(线)探测器
pyroscheerite 超碳地蜡
pyroschist 沥青质页岩;沥青页岩
pyroschist bitumen shale 焦性沥青页岩
pyroscope 测温锥;高温仪;高温计;测温熔锥;测热辐射器;辐射式高温计;辐射高温计
pyrosemifusinite 火焚半丝质体
pyroshale 可燃性有母页岩;可燃页岩;焦页岩
pyrosmalite 热臭石
pyrosol 高温溶液;高温溶片;熔溶胶
pyrosphere 岩浆圈;火界
pyrostat 高温恒温器;恒温槽;高温控制器;高温调节器;高温传感器;高温保持器
Pyrostep 篦冷机
pyrostibnite 红锑矿
pyrostipnite 火红银矿
pyrosulfate 焦硫酸盐
pyrosulfite 焦亚硫酸盐
pyrosulfuric acid 焦硫酸
pyrosulfuryl chloride 胶硫酰氯
pyrotechnic 烟火的
pyrotechnic blow-off charge 烟火喷射装药
pyrotechnic code 烟火信号码
pyrotechnic communication 烟火信号通信
pyrotechnic composition 烟火剂
pyrotechnic compound 烟火剂
pyrotechnic delay train 烟火信号延时线
pyrotechnic dispenser 引爆式投放器
pyrotechnic effect material 烟火效果物质
pyrotechnic gas generator 烟火气体发生器
pyrotechnic generator 火焰发生器

pyrotechnic igniter 烟火发火剂;烟火点火器
pyrotechnic inflator 烟火式充气器
pyrotechnic installation 烟火信号装置
pyrotechnic light 烟火信号
pyrotechnic locker 烟火器材储藏室
pyrotechnic pistol 信号枪
pyrotechnic reaction 烟灰反应
pyrotechnics 烟火制造术;烟火使用技术
pyrotechnic shock 烟火冲击
pyrotechnic signal 烟火信号
pyrotechnic-type igniter 烟火点火器
pyrotechnite 无水芒硝
pyrotector 烟火探测器
Pyrotenax (一种高韧性、不燃、耐高温的)矿物绝缘低压电缆
pyrotic 腐蚀的
pyrotitration 热滴定(法)
pyrotitration analysis 热滴定分析
pyrotomaleanic acid 黑酸腐殖酸
Pyrotop 分解炉顶部混合室
Pyrotrol 回转窑燃烧控制系统
pyrotron 高温器;磁镜热核装置
pyrovalerone 吡咯戊酮
pyrovanadic acid 焦钒酸
pyrovanadium 焦钒
pyroxehe group 辉石组
pyroxene 辉石
pyroxene alvikite 辉石方解石碳酸岩
pyroxene amphibole peridotite 辉石角闪石橄榄岩
pyroxene andesite 辉石安山岩;辉安山岩
pyroxene anorthosite 辉石斜长岩
pyroxene ariegite 辉石尖榴辉岩
pyroxene beforsite 辉石镁云碳酸岩
pyroxene dacite 辉英安岩
pyroxene diorite 辉石闪长岩
pyroxene glass 辉石玻璃
pyroxene granulite 辉石麻粒岩
pyroxene hornblende plagioclase gneiss 辉石角闪斜长片麻岩
pyroxene hornblendite 辉石角闪石岩
pyroxene-hornfels facies 辉石角岩相;辉角页岩相
pyroxene olivine kersantite 辉石橄榄石云斜煌岩
pyroxene olivine minette 辉石橄榄云蝗岩
pyroxene peridotite 辉石橄榄岩
pyroxene plagioclase gneiss 辉石斜长片麻岩
pyroxene rauhaugite 辉石白云(石)碳酸岩
pyroxene sovite 辉石黑云碳酸岩
pyroxene spinifex 辉石鬣刺岩
pyroxenite 辉岩;辉石岩
pyroxenite group 辉石岩类
pyroxenoids 似辉石

pyroxenolite 辉石岩;辉岩
pyroxferroite 三斜铁辉石
pyroxmangite 锰三斜辉石;三斜锰辉石
pyroxylin(e) 可溶硝棉;焦木素;漆用硝基纤维素;低氮硝化纤维;低氮硝化棉;火棉;硝化棉;绵火药
pyroxylin(e) cement 火棉胶
pyroxylin(e) lacquer 焦木素漆;硝基漆;硝化棉涂料;火棉漆
pyroxylin(e) plastics 胶棉塑料;硝化纤维素塑料
pyroxylin(e) sheeting (窗户上的)挡阳光的板
pyroxylin(e) stopper 火棉制动器;硝棉制动器
pyrozolin 常咯林
pyrradio 高温射电
pyrrhite 烧绿石;烧绿岩
pyrrholite 钙块云母
pyrrhotine 磁黄铁矿
pyrrhotine ore 磁黄铁矿矿石
pyrrhotite 磁黄铁矿
pyrrocoline 中氮茚系(颜料)
pyrrole 吡咯
pyrrole ring 吡咯环
pyrrolidine 四氢化吡咯;吡咯烷
pyrrolidine alkaloids 吡咯烷生物碱类
pyrrolidine carboxylic acid 吡咯烷羧酸
pyrrolidinetrione 吡咯烷三酮
pyrrolidone 吡咯烷酮
pyrrolidyl 吡咯烷基
pyrrolo-indole 吡咯并吲哚
pyrrolylcarbonyl 吡咯基甲酰
pyrromonazole 吡唑
pyrrone 吡咙
pyrrones 吡酮类
pyrroporphyrin 焦卟啉
pyrryl 吡咯基
pyruma 一种耐火黏土水泥
pyruvate 丙酮酸盐
pyruvic acid 丙酮酸
pyruvonitrile 丙酮腈
pyruvoyl 丙酮酰
pyrvinium chloride 吡维氯铵
pyrvinium embonate 吡维铵恩波酸盐
pysuma 耐火黏土水泥
Pythagoras's theorem 勾股定理
Pythagorean number 毕达哥拉期数
Pythagorean proposition 勾股定理
Pythagorean right-angle inversor 毕达哥拉斯直角控制器
Pythagorean test 毕达哥拉斯验算
Pythagorean theorem 毕达哥拉斯定理;勾股定理
pyx 硬币样品箱;小盒子;小保险箱
P. Z. method P. Z. 法【道】

Q

Q address 传送数据地址;源地址
Qahar tectonic segment 察哈尔构造段【地】
qala'a 筑于山上的阿拉伯要塞或据点
Q alloy 镍铬合金
qanat (伊朗的)灌溉暗渠
qasr 阿拉伯宫殿;阿拉伯公馆;阿拉伯城堡
qaudricycle 脚踏四轮车
Q-band Q 波段
Q block 特殊质量砌块
Q department 军需部
Q disk 横盘
Q-factor 品质因素;品质因数
q-feel 动压荷载感觉器
Q-flag Q 旗
Q-floor 格形桥面板;格形楼板
Q-floor unit 多孔钢楼板单元
Q-grade gasoline 正规汽油
qibla 朝向(穆斯林礼拜的方向);伊斯兰教中要求祈祷壁龛朝向麦加的方位
Q-incoming neutron flux 进入地球表面的中子通量
qingbai glaze moon palace pillow 青白釉广寒宫枕(瓷器名)
qingbai glaze with brown mottles 青白釉褐斑
qingbai mottled porcelain 青白花瓷器
qingheiite 青河石
Qing stone 青石
Q-inverse filtering Q 值反演滤波
qitianlingite 骑田岭石
Q-joint 横节理;Q 节理
Q-loaded 加载品质因数
Q-meter 品质因数计;优值计
Q-mode cluster analysis Q 型聚类分析
Q-mode factor analysis Q 型因子分析
Q-pattern analysis Q 式分析
q-pot 动压荷载感觉器
Q-scale Q-表
Q-switch 光量开关;Q 开关
Q-tempering 淬火回火
Q-test (土的)快剪试验
quack effect 反向回声效应
quad 样方;空铅;嵌块;四重;四芯线组;四芯导线;四合院;四边缆;四倍重的;四分仪
quad alloy 四元合金
quad (cable) 四心(扭绞)电缆
quad clock driver 四路钟驱动器
quad crown 全开图纸
quadded cable 四芯电缆;四线电缆
quadded flip-flop 四线触发器
quadded logic 四线逻辑
quadergy 无功能量
quad flat package 四列扁平封装
quad in-line 四行排齐封装
quad multiplexer 四路调制器
quad pair cable 扭绞八芯电缆;四线八芯电缆
quad pairing 四线对绞
quad pairing cable 四线对绞电缆
quadplex type cable 四芯型电缆
quadra (图置像的)方框;平台基座;踢脚板;勒脚
quadrajet 四喉管式(化油器)
quadramaran (浮式钻台的)四体浮座
quadrangle 四角形;四边形
quadrangle-centered photograph 图幅中心相片
quadrangle frame 四边形机座
quadrangle-houses 四合院
quadrangle map 梯形图幅;标准地形图幅
quadrangle name 图名;图幅名称
quadrangle report 图历簿
quadrangle roof 四坡屋顶;四边形屋顶;四边形层顶;方形屋顶
quadrangle sheet 正方角图幅;标准地形图幅
quadrangle truss 四角形桁架
quadrangular 四棱柱
quadrangular jar 方壶
quadrangular lobule 方形小叶
quadrangular method 方格法
quadrangular number 四角数
quadrangular prism 四方柱;四方鼓轮
quadrangular pyramid 四角锥

quadrangular roof 四边形层顶
quadrangular set of six points 四点型六点
quadrangular truss 四角形桁架
quadrangular tumbler 四方鼓杯
quadrant 象限(仪);象限四分仪;象角;直角换向器;国际亨利;无线电导航信号区;四分之一圆;四分圆线脚;四分体;扇形体;扇形舵柄;扇形齿轮;扇形板;四分仪
quadrant aeration unit 扇形分区充气装置
quadrant aerial 正方形天线
quadrant(al) angle 象限角
quadrantal arc 象限弧
quadrantal bearing 方位角;象限方位(角);罗盘方位
quadrantal component (误差的)平方分量
quadrantal component of error 象限误差成分
quadrantal correction 象限修正
quadrantal davit 弧齿式吊艇柱;俯仰式吊艇柱
quadrantal deviation 象限自差;象限偏差;象差
quadrantal deviation corrector 象限偏差校正器
quadrantal diagram 象限图
quadrantal distribution 象限分布
quadrantal error 象限差
quadrantal error corrector 象限误差修正器
quadrantal heading 象限方向
quadrantal point 中间基本方位;象限点
quadrantal sphere 软铁球;象限球
quadrantal spherical triangle 直边三角形;象限球面三角形
quadrant angle 射角;象限角
quadrant angle of fall 落地俯角
quadrant antenna 正方形天线
quadrant aquifer 象限含水层
quadrant arm 象限仪水准器臂
quadrant balance 扇形天平
quadrant blending method 四分搅拌法
quadrant blending system 四分搅拌系统;扇形分区搅拌系统
quadrant blocks 扇形座
quadrant bracket 扇形托架
quadrant coil 象限线圈
quadrant compass 象限罗盘仪
quadrant dependence 象限相关
quadrant divider 象限分规;象限分线规
quadrant doffing motion 扇形落纱装置
quadrant drive 扇形齿轮传动
quadrant electrometer 象限静电计
quadrant elevation 仰角;水平射角;射角
quadrant elevation coarse synchrodata 同步传送的概略射角
quadrant elevation fine synchrodata 同步传送的精确射角
quadrant error 象限误差
quadrant formula 平方根分式
quadrant gear 扇形齿轮
quadrant homogenizing 扇形均化
quadrant intercardinal rolling error 象限横摇误差
quadrant iron 方钢
quadrant latch 犁子弧形板
quadrant method 象限法;四分之一矢高法
quadrant mo(u)ld 四开木压条
quadrant mount 象限仪座
quadrant multiplier 象限乘法器
quadrant of truncated cone 桥台锥坡;四分之一截头锥体
quadrant photomultiplier 象限光电倍增器
quadrant plate 挂轮板
quadrant reflector 四分之一圆形反射器
quadrant scale 扇形天平
quadrant sight 象限瞄准具
quadrant spheric(al) triangle 球面直边三角形;球面象限三角形
quadrant stay 扇形空窗撑杆;弧形撑杆;窗扇支撑
quadrant steering gear 舵扇式转动装置
quadrant step 扇形踏步
quadrant style Shore durometer 扇形肖氏硬度计
quadrant system of aeration 扇形分区充气系统
quadrant tap bolt 齿节弧形板

quadrant test 象限检核
quadrant tiller 舵扇;舵柄弧
quadrant tooth 扇形轮齿
quadrant type 象限类型
quadrant-type shiploader 扇形装船机
quadranture 象限值
quadrant warhead 扇形弹头
quadraplex cable 四芯电缆
quadrat 样方;嵌块;铅块
quadrat chart 样方图
quadrate 正方形;四等分;成正方形;方块物;方材
quadrate algebra 方代数
quadrate bone 方骨
quadrate ligament 方形韧带
quadrate lobe 方叶
quadrate part 方部
quadrate pillar tungsten carbide bit 方柱状硬合金钻头
quadrat-free number 无平方因子数
quadratic 正方形;二次方程式;二次的
quadratically integrable function 平方可积函数
quadratic approximation 二次近似;二次逼近
quadratic assignment problem 二次分配问题
quadratic average 二次平均(值)
quadratic code 二次码
quadratic component 矩形成分;平方项;二次方项;二次方分量
quadratic cone 二次锥面
quadratic congruence 二次线汇
quadratic constraint 二次约束
quadratic convergence 二次收敛
quadratic curve 二次曲线
quadratic cylinder 二次柱面
quadratic dependence 二次关系(式)
quadratic diagram 二次图
quadratic differential equation 二次微分方程
quadratic differential form 二次微分形式
quadratic differential method 二次微商法
quadratic discriminant 二次方程判别式
quadratic dual problem 二次对偶问题
quadratic effect 二次效应
quadratic electro-optic(al) effect 平方光电效应;平方电光效应
quadratic equation 二次方程(式)
quadratic error 均方根误差
quadratic Euclidean domain 二次欧几里得整环
quadratic filter theory 二次滤光理论
quadratic fit 二次拟合
quadratic forecast model 二次预测模型
quadratic form 方形;二次型
quadratic frequency magnitude relation 频率震级二次关系式
quadratic function 二次函数
quadratic functional 二次泛函
quadratic hill-climbing 二次登山法【数】
quadratic homogeneous polynomial 二次齐次多项式
quadratic index profile 二次平方折射率分布
quadratic inequality 二次不等式
quadratic interpolation 二次内插法;二次插值(法)
quadratic law effect 平方律效应
quadratic loss function 二次损失函数
quadratic magnitude frequency law 震级频率的平方定律
quadratic mean 均方值;二次平均(值);二次方均值
quadratic mean deviation 中误差;均方差;平均误差;二次平均偏差;二次平均离差;标准误差
quadratic mean error 中误差;均方误差;二次平均误差
quadratic method 平方区法
quadratic minimization problem 二次极小化问题
quadratic non-linearity 二次非线性
quadratic non-residue 二次非剩余
quadratic performance index 二次性能指标;二次型性能指标
quadratic phase mismatch 二次相位失配

quadratic planning 平方规划
quadratic polynomial 二次多项式
quadratic potentiometer 平方电位计
quadratic pressure drop 压力平方差
quadratic prism 方棱柱
quadratic probing 二次探测
quadratic program(me) 二次计划
quadratic programming 平方规划；二次计划；二次规划(法)
quadratic programming problem 二次规划问题
quadratic pulse function 二次脉冲函数
quadratic reciprocity law 二次倒数律
quadratic residue 二次剩余
quadratic resistance 二次阻力
quadratic response 二次响应
quadratic root formula 二次根公式
quadratic root transformation 平方根变换
quadratics 二次式；二次方程式论
quadratic search 二次检查
quadratic selection 二次选择方(式)
quadratic slack variable 二次松弛变量
quadratic smoothing with five point 五点二次曲线平滑
quadratic smoothing with nine point 九点二次曲线平滑
quadratic smoothing with seven point 七点二次曲线平滑
quadratic spline 二次条样；二次仿样
quadratic Stark effect 二次斯塔克效应
quadratic sum 平方和
quadratic surface 二次曲面
quadratic system 正方晶系
quadratic term 平方项；二次项
quadratic termination 二次限定
quadratic termination property 二效终结性质
quadratic transformation 二次变换
quadratic trend model 二次趋势模型
quadratic utility function 二次效用函数
quadratic variation 二次变分
quadratic variational problem 二次变分问题
quadratic Zeeman effect 二次塞曼效应
quadrat list 样方表
quadrat method 样方法
quadratrix 割圆曲线
quadrat sampling 方样
quadrat sampling method 样方法
quadratura 透视画
quadrature 转像差；求面积；求几何面积法；求积分；平方面积；方照
quadrature-adjust 正交调节
quadrature amplitude modulation 正交调幅
quadrature axis 象限轴；正交轴(线)；交轴
quadrature axis reactance 正交轴电抗
quadrature-axis synchronous internal voltage 横轴同步电动势
quadrature axis synchronous reactance 横轴同步阻抗
quadrature balanced modulation 正交平衡调制
quadrature booster 正交增压器
quadrature brushes 正交电刷
quadrature carrier 正交载波
quadrature chrominance signal 正交彩色信号
quadrature circuit 交轴电路；交磁电路
quadrature coherent demodulator 正交相干解调器
quadrature component 转像差成分；正交分量；正交分力；正交部分；横轴分量；无功分量；电抗性电流；电抗分量
quadrature current 电抗性电流
quadrature demodulation 正交解调
quadrature demodulator 直角相位调制；正交解调器
quadrature detector 积分检波器
quadrature differential phase shift keying 四相差分相移键控
quadrature displacement 正交位移
quadrature distortion 正交失真
quadrature error 正交误差
quadrature factor 无功因数
quadrature field 正交场
quadrature filtering 正交滤波
quadrature formula 求积公式
quadrature formula with central differences 中心差分求积公式
quadrature frequency hopping 正交跳频

quadrature impedance 交轴阻抗
quadrature instrument for IP 积分式激发电位仪
quadrature kilovoltampere 无功千伏安
quadrature-lagging 滞后九十度
quadrature-modulated carrier 正交调制载波
quadrature modulated chrominance signal 正交调制彩色信号
quadrature modulation 直角相位调制；正交调制
quadrature modulator 正交调制器
quadrature motion 正交运动
quadrature network 正交网络；正交电路
quadrature of a circle 求圆的面积
quadrature of a conic 求二次曲线的面积
quadrature of the circle 圆求方问题；方圆法
quadrature oscillator 正交振荡器
quadrature phase 正交相位
quadrature phase component 正交相位分量
quadrature phase shift key 正交相移键控
quadrature phase shift keying 四相相移键控
quadrature phase splitting circuit 正交相位分离电路
quadrature phase subcarrier signal 正交相位副载波信号
quadrature phasing 正交定相
quadrature potentiometer 正交电位计
quadrature-sampled 正交取样
quadrature sensitivity 正交灵敏度
quadrature spectrum 转像谱；正交谱
quadrature subroutine 平方子程序
quadrature subtransient reactance 正交初始瞬态电抗
quadrature theory 求积理论
quadrature tide 弦月潮水；弦潮；方照潮
quadrature tube 直角管；电抗管
quadrature voltage 正交电压
quadrature winding 正绕组；交磁绕组；横向绕组
quadravalence 四阶
quadrel 四方形瓦；方瓷砖；方砖；方瓦；方块石
quadrennial cargo gear survey 起货设备每四年的检验
quadriad 包括四个一组
quadribasic acid 四元酸；四价酸
quadric 二次曲面的；二次的
quadri-carriageway road 四幅(式)路
quadric cone 二次锥体
quadric crank mechanism 四联杆机构；摆杆曲柄连杆机构
quadric curve 二次曲线
quadric cylinder 二次柱面
quadric dilution ratio 二次贫化率
quadric discriminator 二次判别器
quadricentennial 四百周年；四百年的
quadric form 二次型；二次形式
quadric-hinged pipe culvert 四铰管涵
quadric mean deviation 均方偏差
quadric of deformation 形变二次式
quadric of revolution 回转二次曲面
quadricorrelator 自动调节相位线路
quadrics 二次型
quadric stress 曲面应力
quadric surface 二次曲面；二次面
quadric transformation 二次变换
quadricycle 特省油汽车；四轮车
quadridentate 四配位体
quadridentate chelate 四配位体螯合物
quadrifid 分成四部分的
quadrifid petal 四裂花瓣
quadrifilar torsion suspension 四边形扭转悬挂
quadrifoil 四叶
quadrifoliate 有四叶的；四叶的
quadrifores 折门(古建筑)
quadrifrontal 有四个正面的
quadriga 四马战车雕饰；四马拖车雕饰；[复]quadrigae
quadrigeminal body 四叠体
quadrigeminal plate 四叠板
quadrigeminus 四叠体的
quadrilateral 四边的；四角形；四级侧生的；四方面
quadrilateral characteristics 四边形特性
quadrilateral element 四边形单元体；四边单元体
quadrilateral element characteristic 四边形单元体特性
quadrilateral form 四线性形式
quadrilateral jib 蝴蝶三角帆

quadrilateral peristyle 四边形列柱(廊)式
quadrilateral planning machine 四面刨床
quadrilateral prism 四边形棱柱
quadrilateral pyramid 四棱锥
quadrilateral tension crack stake 四边形裂隙桩；【测】
quadrilateral torsion suspension 四边形扭转悬挂
quadrillage (地图上的)经纬线
quadrille 方眼的
quadrille paper 方格纸
quadrille pattern 方格图样
quadrillion 一千万亿(10^{15})
quad ring 方形密封环
quadrinomial 四项式；四项的
quadrinomial distribution 四项分布
quadripartite 四分的；四部分组成的；四部分组合体；四深裂的；四分构造体系
quadripartite cross-rib(bed) vault 四根交叉肋拱顶
quadripartite rib vault 四肋拱顶
quadripartite shutter 分成四部分的百叶窗
quadripartite tracery 四分花窗格
quadripartite vault 四区穹窿；四分穹顶
quadripartition 分成四部分
quadriphase system 四相制
quadriplanar plane coordinates 四点面坐标
quadriplane 四翼飞机
quadriply belt 四层带
quadripod 四角锥体(防波堤用)；平底四角锥体；四足鼎；四脚架
quadripolar 四端
quadripole 四端网格；双偶极
quadripole attenuation factor 四端网络衰减因数；衰减因数
quadripole of near source 近场源四极
quadripolymer 四元聚合物
quadriporticus 方庭；古罗马建筑中环形柱廊且近似方形的中梃；方厅
quadripuntal 四孔的；穿四孔
quadrisect 分为相等的四部分
quadrisection 四分切
quadrivalve 四折门扇之一
quadrivial 四条道路交会于一点的；四路交叉的
quadro 方形住宅区
quadrode 四级管
quadrominium 四户自有公寓
quadrosilan 二苯硅烷
quadroxide 四氧化物
quadruped 有四足的
quadrupedal extensor reflex 布雷恩氏反射
quadruped mast 四脚桅
quadruplane 四翼机；四翼飞机；反平行四边形
quadruple 四重；四伏；四元组；四路；四工；四倍频；四倍(的)；翻两翻
quadruple address 四地址
quadruple block 四滑轮组
quadruple board platform 二层台(四根钻杆组成立根高度)
quadruple bolt cutter 四轴螺钉机
quadruple boom 四台凿岩机的伸梁(钻车上)
quadruple burner 四头灯；四连闪光
quadruple chain drive 四排链条传动
quadruple charge 四柱装药
quadruple cloth 四层织物
quadruple compound turbine 四缸复式涡轮机
quadruple curve 四相曲线
quadruple degeneracy 四重简并
quadruple distributor 四路分配器
quadruple diversity operation 四重分集运用
quadruple-diversity system 四路分集传输系统
quadruple effect 四效
quadruple effect evapo(u)rator 四效蒸发器
quadruple error detection 四重误差检测
quadruple expansion engine 四胀式蒸汽机
quadruple flashing light 四连闪光
quadruple-flow turbine 四排汽口涡轮机；四排汽口汽轮机
quadruple form 四元组形式
quadruple-frequency system 四频制
quadruple ion 四重离子
quadruple ladder truck 四节云梯车
quadruple lattice 四重格
quadruple line 四线铁路
quadrupleness 四重性
quadruple notation 四元组表示

quadruple operator 四元组算符
quadruple orthogonal 四重正交
quadruple pass 四程
quadruple peak 四重峰
quadruple point 四重点;四相点
quadrupler 四频器;四倍压器;四倍(频)器;四倍乘数;乘四装置
quadruple railway 四线铁路
quadruple recorder 四用自记计
quadruple rig 四台凿岩机钻车;四机台车
quadruple riveted joint 四行铆钉接合
quadruple riveting 四行铆接
quadruple root 四重根
quadruple screw 四叶螺旋桨;四线螺旋;四纹螺旋;四头螺旋
quadruple screw motor ship 四螺旋桨船
quadruple screw steamer 四螺旋桨船
quadruple serial air survey camera 四镜头连续航测摄影机
quadruple serial photogrammetric camera 四重联拍测绘照相机
quadruple star 四合星
quadruple system 四路传制;四路传输系统
quadruplet 四联体;四件一套的
quadruple table 四项表
quadruple thread 四线螺纹
quadruplet lens 四透镜物镜
quadruple track 四线铁路
quadruple type 四元组类型
quadruple unit attachment 四元附件
quadruple vector product 四重向量积;四重矢量积
quadruple well completion 四井采掘;四井开采
quadruple-working switches 四动道岔
quadruplex 四重(线号);四路多工系统;四工;四倍的
quadruplex circuit 四工电路
quadruplex house 四联式住宅
quadruplex rake classifier 四耙分级机
quadruplex receptacle outlet 四孔插座出线口;四孔插座出口线
quadruplex system 四路多工制;四工电路系统
quadruplicate 一式四份;四次方
quadruplicated plot 重复四次小区
quadruplication 增到四倍;乘以四;反复四次
quadrupling 四倍
quadruply 四倍地
quadrupod 四角架
quadrupolarity 四极性
quadrupole 四极子
quadrupole adsorption line 四极吸收谱线
quadrupole aperture 四极透镜孔径
quadrupole broadening 四极展宽
quadrupole coincidence set 四重符合记录设备
quadrupole component 四极分量
quadrupole coupling 四极耦合
quadrupole coupling constant 四极偶合常数
quadrupole-double peak 四极双峰
quadrupole doublet 四极磁透镜对
quadrupole-doublet lens 四极双合透镜
quadrupole effect 四极效应
quadrupole electromagnet 四极磁铁
quadrupole focusing lens 四极调焦透镜
quadrupole force 四极力
quadrupole lens 四极透镜
quadrupole lenses pair 四极透镜对
quadrupole-lens focusing system 四极透镜聚焦系统
quadrupole-lens triplet 四极三合透镜
quadrupole magnet 四极磁铁
quadrupole mass analyser 四极质量分析器
quadrupole mass filter 四极滤质器
quadrupole mass spectrometer 四极质谱仪;四极质谱计
quadrupole mass spectrometry 四极矩质谱法
quadrupole matching 四极匹配
quadrupole residual gas analyzer 四极剩余气体分析器
quadrupole RF field 四极射频场
quadrupole source 四极源
quadrupole spectrometer 四极谱仪
quadrupole spectrum 四极能谱
quadrupole split 四极分裂
quadrupole strength 四极子强度
quadrupole system 四极系统

quadrupole transition 四极跃迁
quadrupole-triplet lens 四极三透镜组
quadrupole vibration 四极振荡
quadruquaternion 四极四元数
Quad-track 联结式履带拖拉机
quad word 四倍长字
quag 沼泽地
quagginess 泥塘性;沼地性
quaggy 沮洳的;泥泞的;泥沼的
quaggy timber 环裂木材;心裂木材
quagmire 沼泽地;泥炭沼泽;软泥地;颤沼
quagmiry 泥泞的
quaint brick 异形砖
quake center 震中
quake ooze 泥沼
quake-prone 易于发生地震的
quake-proof 抗震;耐地震的
quake-proof structure 抗震结构;耐震结构
quaker drab 单调的;淡褐色的
quake resistant structure 抗震结构
quake-resistant wall 抗震墙
quake-sensitivity 地震感觉
quake sheet 地震岩席
quaking aspen 颤杨
quaking bog 震沼;跳动沼;颤沼;浮酸沼
quaking concrete 软混凝土;塑性混凝土;大塌落度混凝土
quaking concrete column 塑性混凝土柱
quaking mat 浮生植物层
quale 可感受的特性
qualification 限制条件;限定;资格;质量评定;执照;工程试验鉴定;技术指标;技能;合格性;条件;熟练程度;赋予资格;附加保留条款;附带申明
qualification affirmative 资格证件;资格确认
qualification approval test 技术合格考试
qualification certificate 资格证(明)书;合格证(书)
qualification documents 资格(预审)文件
qualification joint 合格接缝
qualification of procedure and welder 程序及焊工的评定
qualification of welding procedure 焊接工艺评定
qualification (on test) 合格
qualifications of the guarantors 保人地位
qualification statement 资格审查文件
qualification stock 资格股(就任重要职务时的必要股份)
qualification test 资格考核;质量鉴定试验;鉴定试验;检定试验;合格(性)试验;合格检验;合格检查;评定试验
qualification test procedure 质量鉴定试验程序
qualification test specification 质量鉴定试验规范
qualification to the form 格式说明
qualification welder 合格焊工
qualification welding operator 合格焊工
qualified 具有资格的;有资格的;合乎要求的;合格的
qualified acceptance 修改性承兑;限制承兑;条件承兑(票据);附条件认支汇票;附条件的认支汇票;附条件的签署;附条件承兑;附带条件承兑汇票;不单纯承兑
qualified acceptance as to time 修改到期日承兑
qualified acceptance of a bill of exchange 附条件承兑汇票
qualified agency 有资格的机构
qualified apparatus 合格设备
qualified approval 有条件的批准
qualified audit certificate 限定查账证明书
qualified bill of lading 加批注的提单
qualified buyer 有条件的买主;合格买主
qualified captain 合法有些的物权
qualified certificate 资格证书;附条件证明书
qualified coating 合格涂料
qualified component 合格元件
qualified condition 合格条件
qualified contributions 附条件的交款
qualified diver 合格潜水员
qualified driver 合格驾驶员
qualified endorsement 限制性背书;限定背书;限制资格的背书;附条件的背书
qualified engineer 合格轮机员
qualified fee estate 有限制条件的房地产继承权
qualified institution 主管机关;主管机构;主管部门
qualified majority 有条件的多数;特定多数
qualified manufacturer 合格制造者;合格制造

商;合格制造厂家
qualified name 限定名;受限名;当限名称
qualified notary 合格律师;合格公证人
qualified officer 合格驾驶员(指船)
qualified operator 合格操作工
qualified opinion report 限定意见报告书
qualified paragraph name 限定段名
qualified parts list 零件一览表;零件目录(表);合格商品目录;合格零件一览表;合格产品一览表
qualified person 合格人员
qualified personnel 专门人才
qualified pilot 合格领航员
qualified product 合格品
qualified products list 经检定产品清单;合格商品目录;合格零件一览表;合格产品一览表;商品目录
qualified project 合格的建设项目
qualified property 限定产权财产
qualified rate 合格率
qualified raw sheet 合格原版
qualified report 附条件报告
qualified size 合格尺寸
qualified steel 合格钢材
qualified stock option 附条件认股权;附条件的股份认购权
qualified ultimatum 附条件的最后通牒
qualified worker 合格工人
qualifier 限定符号
qualify 证明合格;渗淡
qualify for 考核
qualifying expenditures 符合规定的支出
qualifying reserves 限定准备金
qualifying reserve account 评价账户
qualimeter 硬度测量仪(X射线);X光硬度计
qualimetry 质量计量学
qualitative 定性的
qualitative analysis 定性分析
qualitative analysis for two times 两次定性分析
qualitative assay 定性分析
qualitative assessment 质量评价;质量评定
qualitative attribute 质量属性;品质属性
qualitative attribution 质的属性
qualitative background method 质底法
qualitative carbon steel 定性碳钢
qualitative change 性质变化;质变
qualitative character 质量性状
qualitative characteristic 定性特性;品质特性
qualitative choice 定性抉择
qualitative classification 质分类;品质分类
qualitative collection 定性采集
qualitative composition of samples 试样的定性组成
qualitative control 品质管理
qualitative correlation 质相关
qualitative credit control 信用质量控制
qualitative credit restriction 品质信用限制
qualitative data 质量数据;定性资料;定性数据
qualitative demography 质量人口学
qualitative dependent variable 质量反应变量
qualitative determination 定性测定
qualitative economic forecast 定性经济预测
qualitative elementary analysis 元素定性分析
qualitative evaluation standard 定性评价标准
qualitative examination 定性测定;定性检验;定性检查;定性分析
qualitative examination of ash 灰分定性测定
qualitative experiment 定性实验
qualitative factor 质的因素;定质量因素
qualitative facts 定性论据
qualitative filter paper 定性滤纸
qualitative forecast(ing) 质量预报;定性预报
qualitative geothermometer 定性地温计;定性地热温标
qualitative governing 变质调节
qualitative grouping 品质分组
qualitative identification 定性鉴定
qualitative index 质量指标
qualitative indicator 质量指标
qualitative indices of locomotive utilization 机车运用质量指标
qualitative information 定性信息
qualitative interpretation 定性判读;定性解释
qualitative investigation 质量集中分析;定性研究;定性分析
qualitative layer interpretation 定性地层解释

qualitative limitation 质的限制
qualitatively identical labor 同质的劳动
qualitative materiel development objective 物资质量发展目标
qualitative method 定性(方)法
qualitative observation 质的观察;定性观察;定性观测
qualitative parameter 定性参数
qualitative perception 质量感
qualitative photogeology 定性摄影地质学
qualitative prediction 定性预测
qualitative prescription of 质的规定性
qualitative property 品质属性
qualitative reaction 定性反应
qualitative regulation 定性调节
qualitative reliability consumption data 品质可靠性消耗数据
qualitative representation 质量表示法
qualitative response 质反应
qualitative sampler 定性取样器
qualitative sampling equipment 定性取样器
qualitative similarity 定性相似
qualitative spectroanalysis 光谱定性分析
qualitative study 定性分析
qualitative temperature 定性温度
qualitative term 定性术语
qualitative test 定性试验;定性测定
qualitative test of lead monoxide 一氧化二铅定性试验
qualitative variable 定性变量
qualitative variables model 品质变量模型
qualities of the female parent 显出母本的品质
quality 质量;合格;品质;品位
quality acceptance standard of maritime engineering 水运工程质量验收标准
quality agreement 质量契约;质量合同;质量协议
quality amplifier 高品质放大器
quality analysis 质量分析
quality and measurement office 质量计量室
quality and reliability 质量和可靠性
quality and reliability objective 质量和可靠性目标
quality appraisal 质量鉴定
quality arbitration 质量检定
quality as per buyer's sample 质量以买方样品为准;凭买方样品质量交货;凭买方样品交货;品质按买方样品
quality as per seller's sample 质量以卖方样品为准;凭卖方样品交货
quality as seller's sample 凭卖方样品
quality assurance 质量担保;质量保证;品质保证
quality assurance acceptance standard 质量保证验收标准
quality assurance (control) system 质量保证体系
quality assurance data system 质量保证数据系统
quality assurance directive 质量保证规定
quality assurance for environment(al) monitoring 环境监测质量保护
quality assurance for monitoring 监测质量保证
quality assurance inspection 质量保证检查
quality assurance laboratory 质量保证实验室
quality assurance operating procedure 质量保证控制程序;质量保证操作程序
quality assurance personnel 质量保证验收人员;保证产品质量的人员
quality assurance plan 质量保证计划
quality assurance provision 质量保证条例
quality assurance system 质量保险系统
quality assurance technical publications 质量保证技术出版物
quality assurance test 质量保证试验
quality audit 质量审查;质量检查;质量审核
quality auditing system 质量审查制度
quality audit observation 质量审查观察
quality auditor 质量审核员
quality base method 质底法
quality bonus 质量奖;品质奖金
quality booster 质量改善剂
quality car 质量优良的小客车
quality card 质量卡片
quality center 质量中心
quality certificate 质量证(明)书;质理证明书;品质证(明)书
quality certificate issued by manufacturers 制造商出具的质量证明

quality certification 质量鉴定
quality certification system 质量保证制度
quality change 质量变化
quality character 质量性状
quality characteristic 质量特征;质量特性;品质属性
quality characteristic and parameter 质量特征和参数
quality check 质量监视
quality checker 质量检验器
quality check(ing) 质量检验;质量检查;质量评价
quality claim 品质索赔
quality class 质量级;质级
quality classification of rock mass 岩体的质量分级
quality clauses 质量规定;品质条款
quality coefficient 精度系数;强度系数
quality concrete 优质混凝土;高质量混凝土;高级混凝土;合格混凝土
quality concrete production 优质混凝土生产
quality consciousness 质量意识
quality control 质量控制;质量检查;质量管理;品质管理;商品检验;产品质量控制;产品质量检查
quality control activity 质量控制活动
quality control analysis 质量控制分析
quality control and basis of acceptance 质控制和验收依据
quality control (assurance) manual 质量保证手册
quality control by attributes 按固定指标检验质量
quality control center 质量控制中心
quality-control chart 质量控制图(表);质量评估图
quality control check(ing) 质量控制检查;质量监控;质量管理检验;质量管理检查
quality control circle 质量管理小组;品质管理圈
quality control circle activity 质量管理小组活动
quality control committee 质量控制委员会
quality control cycle 品质管理过程
quality control data 质量控制资料;质量控制数据
quality control division 质量检查科
quality control engineer 质量控制工程师;质量管理工程师
quality control engineering 质量管理工程
quality control expert 质量控制专家
quality control form 质量控制表格
quality controlled concrete 质量控制的混凝土
quality controlled investing 质量控制投资
quality control level 质量控制等级
quality control manual 质量控制手册
quality control of coal 煤质管理
quality control officer 质量控制人员
quality control of food 食品质量管理
quality control of life 使用期限质量控制
quality control of oil products 石油产品质量管理
quality control of railway transportation statistical data 铁路运输统计数据质量控制
quality control organization 质量控制机构
quality control package 质量控制装置;品质控制装置
quality control personnel 质量控制人员
quality control pick-up 质量控制传感器
quality control plan(ning) 质量保证计划;质量控制计划
quality control reliability 质量控制可靠性
quality control representative 质量控制样品
quality control selector 质量控制取样器
quality control specification 质量控制标准
quality control staff 质量控制员
quality control standard 质量控制标准;质量检验标准;质量管理标准
quality control strip 质量控制条
quality control surveillance 质量检验
quality control system 质量控制系统;质量控制体系;质量检查系统;质量管理制度;质量管理系统;质量保证计划
quality control table 质量检验表
quality control technical report 质量控制技术报告
quality cost 质量费用;质量成本
quality cost control 质量控制成本
quality cost ratio 质量成本比率
quality-crafted 优质手工制品
quality criterion 质量判据;质量标准
quality criterion for water 水质评定标准
quality curve method 质量平均线法
quality data system 质量资料系统
quality description 质量说明(书)
quality design 优质设计

quality detection 质量鉴定
quality deterioration limit 品质降低极限
quality determination 质量鉴定;质量检验;质量测定
quality diagnosis 质量诊断
quality discrepancy record 质量不符规定记录
quality drinking water resource 优质饮用水源
quality engineering 质量管理;质量工程(学)
quality entity 质量实质
quality equation 品质方程
quality evaluation 质量评价;质量评定
quality evaluation of groundwater 地下水水质评价
quality examination 质量检验
quality examination program(me) 质量检查计划
quality expert 质量管理专家
quality factor 线质系数;质量因素;质量因数;质量系数;品质因子;品质因素;品质因数
quality factor ratio 质量因素比
quality factor value 质量系数值
quality feedback 质量反馈
quality first 质量第一
quality forging 优质锻件
quality franchise 品质免赔限度
quality function 质量函数
quality goals 质量目标
quality goods 优质产品;正品
quality governor 变质调节器
quality grade 质量品级;质量等级
quality grade description 质量品级说明(书)
quality grade of concrete 混凝土的质量品级
quality guarantee 质量担保函;品质保证
quality guideline 质量指南
quality improvement 质量改进
quality index for inspector 检验质量指数
quality index (number) 质量指数;形式指数;质量指标;品质指数
quality index of a channel 信道质量指数
quality index of rock mass 岩体质量指标
quality information feedback 消费者对产品质量的反应
quality inspection 质量检验;质量检查;品质检验
quality inspection certificate 质量检查证明书
quality inspection criterion 质量检查标准
quality inspection of finished products 成品质量检查
quality inspection phase 质量检验阶段
quality inspector 质量检验员
quality latitude 质量幅度;品质机动幅度;品质差幅
quality leadership 品质领先
quality level 质量水平
quality level certificate 质量水平证明书
quality level certification agreement 质量水平鉴定合同
quality loop 质量环
quality loss 质量损失
quality lubricant 优质润滑剂
quality management 质量管理
quality management department 质量管理部门
quality management of groundwater environment 地下水环境质量管理
quality management plan 质量管理计划
quality management policies 质量管理政策
quality management program(me) 质量管理方针
quality management specialist 质量管理专家
quality manger 质量经理
quality manual 质量手册
quality map 质量分析图
quality map of urban environment 城市环境质量图
quality mark 质量标志;质量标号;品质标志
quality market 重质量的市场
quality material 质量好的材料
quality measure 质量度量
quality measurement system 质量测定系统
quality merchandise 高级商业;高级商品
quality meter 质量因数计;品质因数计
quality method 高质法
quality method of life 生活质量法
quality mind 质量意识;品质意识
quality monitor console 质量监视控制台;质量监控台
quality monitoring 质量控制;质量监控;质量监督
quality neutralization 质量中和;品位中和
quality number 品质号数;特性指数

quality objective 质量目标
quality ocntrolling of construction 工程建设质量控制
quality of air environment 空气环境质量
quality of bag filter cloth 袋滤器袋布质量
quality of balance 平衡质量;平衡度
quality of beam 束流品质
quality of boring 钻探质量
quality of boring coal seam 煤层质量
quality of bottom 海底性质
quality of coal sample 煤样质量
quality of colo(u)r 颜色品质
quality of combined sewage 合流污水质量;合流污水性质
quality of commutation 整流质量
quality of comprehensive boring 综合质量
quality of concrete 混凝土质量
quality of construction 工程质量
quality of construction works 建筑工程质量
quality of credit 信贷质量
quality of design 设计质量;设计品质
quality of dredge-cut 挖槽质量
quality of effluent 出水水质
quality of estate 房地产的特质;不动产占有方式
quality of finish 抛光质量
quality of fit 配合等级
quality of flow 流量等级(交通工程);车辆流量等级
quality-of-flow index 交通品质指数
quality of fuel 燃料的质量
quality of gasoline 汽油的质量
quality of geologic(al) environment 地质环境质量
quality of geologic(al) map 地质图性质
quality of goods transport 货物运输质量
quality of grit 沉沙质量
quality of groundwater 地下水水质
quality of heat supply 供热质量
quality of information transfer 信息传输性质
quality of initial set(ting) 初凝质量
quality of irrigation water 灌溉水质
quality of labor 劳动质量
quality of life 社区生活质量;生活质量;生活水平
quality of lighting 照明质量
quality of line drafting 线划质量
quality of logging 测井质量
quality of lot 批(量)质量
quality of materials 材料质量
quality of mineral 矿产质量
quality of ore 矿石质量
quality of passenger transport 旅客运输质量
quality of rainfall 雨水性质
quality of reception 接收质量
quality of recharging water 回灌水质
quality of reclamation 吹填质量
quality of regulation 调节质量
quality of reproduction 重现质量;保真度
quality of river water 河水质量
quality of sample analysis 样品分析质量
quality of screen cloth 筛分机织物质量;筛分机织物品质
quality of service 服务质量
quality of set(ting) 凝结质量
quality of sewage 排水水质
quality of snow 雪质
quality of soil 土质
quality of sounds 音质
quality of steam 蒸汽质量;蒸汽的干燥度
quality of textile products 纺织品质量
quality of the natural water 天然水质(量)
quality of the work 工作质量
quality of tobacco 烟叶质量
quality of tolerance 公差等级
quality of track 线路质量
quality of urban life 城市生活质量
quality of vibration 振动品质
quality of water 水质
quality of water sealing 止水质量
quality of water supply 供水质量;水供应质量
quality of works 工作质量;工程质量
quality parameters 质量参数
quality part 高级部件合格品;合格品
quality plan 质量计划
quality policy 质量方针
quality pre-controlling of works 工程质量预控
quality price differential 质量价差;品质差价

quality process 质量工序
quality producer 优质生产厂
quality product 优质产品
quality production 优质生产
quality program(me) 质量保证计划
quality range 质量范围
quality rank of drilling engineering 钻探工程质量
quality rating 质量分等
quality rating system 质量评价系统
quality reduction 质量下降
quality reference index 质量参考指数
quality regulations 质量控制条例
quality-related cost 质量成本
quality reliability assurance 质量可靠性保证
quality requirement 质量要求;质量规格;规格
quality research 质量研究
quality restriction 品质管制
quality retention 质量保持率
quality retention rating 质量稳定性评价标准
quality review 质量审查
quality sample 品质样品
quality seed 优质种子;优良种子
quality selection 按质选择
quality series 品质数列
quality shipped 装船品质
quality specification 质量规格(说明书);质量说明书;质量技术条件;质量技术说明;质量规范;质量标准;技术规范;技术安全规程
quality standard 质量说明书;质量规格;质量标准;技术规范;品质标准
quality standard for drinking water 饮用水水质标准
quality standard for water 水质标准
quality standard of agricultural water 农业用水水质标准
quality standard of boiler water 锅炉用水水质标准
quality standard of cooling water 冷却用水水质标准
quality standard of drinking water 生活饮用水水质标准
quality standard of effluent 出水水质标准
quality standard of mineral springs 矿泉水水质标准
quality standard of mineral spring water for drinking 饮料矿泉水标准
quality standard of mineral spring water for industry 工业矿泉水标准
quality standard of mineral spring water for medical treatment 医疗矿泉水标准
quality standard of rich water 肥水水质标准
quality standard of sewage irrigation 污水灌溉水质标准
quality standard of soil environment 土壤环境质量标准
quality standard of water for building industry 建筑业用水水质标准
quality standard of water for cotton industry 棉纺业用水水质标准
quality standard shake 有木裂纹的板的质量标准
quality steel 优质钢
quality supervision 质量监督
quality supervision for approval 评价型质量监督
quality supervisor 质量管理
quality surveillance 质量监督
quality survey 品质调查
quality symbol 品质标记
quality system 质量系统;质量体系
quality tester 质量试验器
quality test(ing) 质量试验;质量检验;质量检查
quality test of concrete 混凝土质量检验
quality test of water 水质试验
quality test schedule 质量试验规范
quality to be considered being about equal to the sample 交货品质与样品大体相符
quality tolerance 品质公差
quality tolerance of boiler water 锅炉水质允许限度
quality transplants 品质移植
quality treated wood 高级防腐木材
quality trends 质量趋势
quality value 品质因素;品质因数
quality variation 质量变化
quality variation from sample 品质与样品不符
quality verification test 质量验证试验

quality warranty clause 品质担保条款
quant 撑篙
quanta 坎儿井
quantameter 光量子能量测定器
quantasome 量子转化体
quanative attribute 数量特征;数量符号
quantic 齐次多项式;代数形式
quantifiable data 可量化数据
quantifiable function 可量化函数
quantification 量化;确定数量;定量评价;定量化
quantification of uncertainty 不确定的量化
quantification symbol 量化符号
quantified economic data 经济数量资料
quantified statement 量化命题
quantified system analysis 量化系统分析;定量系统分析
quantifier 计量器;计量计;配量计;配量斗
quantify 量化
quantify activity 量化活度
quantifying drainage 量化污染物排放
quantifying pollutant discharge 量化污染物排放
quantifying predictive uncertainty 量化预测不确定性
quantify pollution level 量化污染水平
quantile 分位数(的一位);分位点;分数位
quantile fractile 分位点
Quantimet 定量电视显微镜
quanting pole 撑篙
quantiser 量化器
quantitaes 数量表示
quantitative 定量的
quantitative absorption 定量吸收
quantitative adsorption 定量吸附
quantitative analysis 量的分析;数量分析;定量分析
quantitative analysis method 定量分析(方)法
quantitative analysis of agricultural chemicals 农用化学物定量分析;农药定量分析
quantitative analysis of CO gas 一氧化碳定量分析
quantitative analysis of hydrogen sulphide 硫化氢的定量分析
quantitative analysis of suspension 悬浮物质定量分析
quantitative analyzer of residual chlorine 残氯定量分析仪
quantitative approach 数量方法
quantitative assessment 数量查定;定量评价
quantitative attribute 数量属性
quantitative automatic balance 定量自动秤
quantitative autoradiography 定量自体放射照相术
quantitative change 量变;数量变化
quantitative character 数量性状;数量特征
quantitative characteristic 定量特性
quantitative check 定量检查
quantitative chemical analysis 定量化学分析
quantitative classification 定量分组;定量分类
quantitative classification system 定量分类系统
quantitative collection 定量收集;定量采集
quantitative composition of samples 试样的定量组成
quantitative correlation 定量相互关系
quantitative criterion 数量标准;定量准则;定量判据;定量标准
quantitative data 数据;定量资料;定量数据
quantitative decision analysis 定量决策分析
quantitative delivery pump 定量输送泵
quantitative determination 定量测定
quantitative diagnosis 计量诊断
quantitative dilution 定量稀释剂
quantitative distribution 定量分布
quantitative economic forecast 定量经济预测
quantitative effect 定量效应
quantitative electron-probe micro-analysis 电子探针定量微区分析
quantitative elemental analysis 定量元素分析
quantitative elementary analysis 元素定量分析
quantitative estimation 定量估算
quantitative estimation of plankton 定量浮游生物估算;浮游生物定量估测
quantitative evaluation 定量评价;定量计算
quantitative evaluation standard 定量评价标准
quantitative examination 定量检验;定量检查;定量测定
quantitative experiment 定量实验
quantitative expression 数量表达式

quantitative filter paper 定量滤纸
quantitative forecast(ing) 定量预报；数值预报；数量预报
quantitative gel diffusion tests 定量凝胶扩散试验
quantitative geothermometer 定量地热温标
quantitative governing 变量调节
quantitative governor 变量调节器
quantitative housing shortage 住宅数量短缺
quantitative hydroecology 定量水生生态学
quantitative hydrology 定量水文学
quantitative image analyzer 定量图像分析仪
quantitative index number 数量指数；数量指标
quantitative indices of locomotive utilization 机车运用数量指标
quantitative indices of passenger transportation statistics 旅客运输统计数量指标
quantitative indices of wagon utilization 货车运用数量指标
quantitative information 定量资料；定量信息
quantitative interpretation 定量判读；定量解译；定量解释
quantitative investigation 定量研究
quantitative layer interpretation 定量地层解释
quantitative limitation 量的限制
quantitative limits 数量界限
quantitative lithofacies analysis 岩相定量分析
quantitatively definite social need 特定数量的社会需要
quantitative maturity 数量成熟期
quantitative measurement 定量测量
quantitative measuring 定量测定
quantitative metallogeny 定量矿床学
quantitative metallography 定量金相学
quantitative method 数量方法；定量(方)法
quantitative methods of ecological control 定量生态控制法
quantitative microscope 定量显微镜
quantitative mineralogic(al) analysis 矿物定量分析
quantitative model 定量模型
quantitative modeling 定量模拟
quantitative monitoring 定量监控；定量监测
quantitative morphology 定量形态学
quantitative objective 定量目标
quantitative observation 计量观测；定量观察；定量观测
quantitative paleoecology 定量古生态学
quantitative paper 定量滤纸
quantitative parameter 定量参数
quantitative plankton sampler 定量浮游生物采样器；浮游生物定量采样器
quantitative polarizing microscope 定量偏光显微镜
quantitative precipitation test 定量沉淀试验
quantitative prediction 定量预测；定量预报
quantitative prediction of minerogenetic prospective region 成矿远景区定量预报
quantitative prediction of ore-content of geologic(al) body 地质体含矿性定量预测
quantitative pyramid 数量金字塔
quantitative-qualitative flow graph 数量质量流程图
quantitative rating 数量等级
quantitative reaction 定量反应
quantitative regulation 定量调节
quantitative regulation of imports 进口定额
quantitative relation 定量关系
quantitative relationship 数量关系；定量关系
quantitative representation 数量表示法
quantitative response 量反应
quantitative restriction 数量管制
quantitative result 定量结果
quantitative rheology 定量流变学
quantitative sampling equipment 定量取样仪
quantitative school 计量学派
quantitative sedimentology 定量沉积学
quantitative seismicity map 定量地震活动分布图
quantitative shock 定量振动
quantitative similarity 定量相似
quantitative spectral classification 定量光谱分类
quantitative spectrochemical analysis 光谱化学定量分析
quantitative standard 数量标准
quantitative statistics 量的统计

quantitative stratigraphy 定量地层学
quantitative structure activity relationship 定量结构活性关系；定量构效关系法
quantitative structure-biodegradability relationship 定量结构与生物降解性关系
quantitative structure-property relationship 定量结构与性质相关关系
quantitative study 定量研究；定量分析
quantitative technique 定量技术
quantitative television microscope 定量电视显微镜
quantitative tester 定量测试器
quantitative test(ing) 定量试验；定量测定
quantitative theory 定量理论
quantitative three-dimensional information 三维定量信息
quantitative tracer 定量示踪器
quantitative tracing 定量示踪(法)
quantitative value 定量数值
quantitative variable 量变项；数量变量；定量变量
quantitative variation 数量变化
quantitative variation index of orebody 矿体变化性定量指标
quantitative X-ray diffraction 定量X射线衍射
quantities bill 工程量表
quantities of bars 钢筋数量表
quantities of earthwork 土方工程量
quantities of high quality water 优质水量
quantities of labo(u)r and material 工料数
quantities of major works 主要工程数量表
quantities of material and labo(u)r 工料数
quantities placed 灌筑的数量；放置的数量
quantities uplifted 数量增加
quantity 数量；定量
quantity added to 加入量
quantity allocation 数量分摊；数量分配
quantity allotted 分配数量
quantity and productivity of organic matter 有机质的量和产率
quantity at one's option 数量有某人决定
quantity change 数量变化；数量增减
quantity claim 数量索赔
quantity combusted 燃烧量
quantity consumed 消耗数量
quantity contract 数量合同
quantity control 数量控制；数量调节
quantity controller 数量控制器
quantity control of pollutant 污染物量控制
quantity delivered 交付数量
quantity demanded 需求量
quantity deposit 沉积物量
quantity diagram 积星图；积量图；土积图；土方图
quantity difference 量差
quantity discharged 排出量
quantity discount 大宗交易的折扣；数量折扣
quantity discount agreement 大宗交易折扣合同
quantity distance tables 炸药存储[贮]量安全距离表；炸药储存安全距离表
quantity distribution 数量分布
quantity equation of exchange 费歇尔氏交换方程式
quantity estimate sheet 工作量估算表(建筑成本)；工程量估计表
quantity for reference 参考工程量
quantity for spray 喷涂量
quantity forwarded 发送数量
quantity galvanometer 测量电荷量的电流计
quantity governor 变量调节器
quantity index number 数量指标
quantity in the exponent 指数函数
quantity list 工程量表
quantity mark 数量标志
quantity measure 实物量度
quantity meter 总流量表
quantity method 变量法
quantity not sufficient 量不足；数量不足
quantity of allocation trains 配车数【铁】
quantity of artificial exported water 人工引出水量
quantity of artificial inputted water 人工引入水量
quantity of back-pumping 回扬量；补给工程总投资
quantity of borrow 取土量
quantity of cargo 船货量
quantity of charge 负载量
quantity of clean-up 清舱量
quantity of current 流量
quantity of deposit 沉积量

quantity of discharge 排泄量；污染物排放量
quantity of dissolution 溶蚀量
quantity of dust 尘量；粉尘量
quantity of earthwork 土石方数量
quantity of electricity 电量
quantity of employment 就业人数；就业量
quantity of excavation 开挖量；挖方量
quantity of excess sludge 剩余污泥量
quantity of explosive charge 炸药量
quantity of filler present 填充剂配合量
quantity of fixed assets 固定资产数量
quantity of floating ice 冰量
quantity of flowing water at well mouth 孔口自流水量
quantity of flow of non-point source pollutant 非点源污染流量
quantity of flushing wastewater 冲洗废水量
quantity of goods available for sale 社会商品可供量
quantity of goods produced 实物量
quantity of grit 杂粒量；沉沙数量
quantity of heat 热量
quantity of heat dissipation in workshop 车间散热量
quantity of heat given up 输出热量
quantity of heat supplied 输入热量
quantity of illumination 曝光量
quantity of incoming detritus from mountain torrent 山洪来石量
quantity of information 信息量；资料量
quantity of information method 信息量法
quantity of leaky recharge 越流补给量
quantity of light 光量
quantity of liquid phase 液相量
quantity of magnetism 磁量
quantity of mass 物质的量
quantity of materials 材料数量
quantity of mineral resources 资源量
quantity of motion 动量
quantity of order 订货量
quantity of overtopping 漫顶(流)量
quantity of oxygen deficit 缺氧量
quantity of oxygen lack 缺氧量
quantity of percolation 渗漏量；渗流量；深层渗入量；地面渗入量
quantity of pollutant discharged 污染物排放量
quantity of pollutant flow 污染物流量
quantity of pollutant reduced 污染物削减量
quantity of pollution 污染量
quantity of precipitation infiltration 降水入渗量
quantity of production 生产量
quantity of pumping water 抽水数量
quantity of radiant energy 辐射能量
quantity of radiation 辐射量
quantity of rainfall 降雨量
quantity of reflux 回流量
quantity of refuse 垃圾量
quantity of reinforced concrete members 钢筋混凝土构件数量表
quantity of relative subsidence 相对下沉量
quantity of residual sludge 剩余污泥量
quantity of river leakage 河水漏失量
quantity of rock powder 岩粉用量
quantity of screenings 筛渣数量
quantity of section steel 型钢数量表
quantity of sediment 泥沙量；沉积量
quantity of seepage 渗透量；渗流量
quantity of selection 选择量
quantity of sewage effluent 污水处理后出水量
quantity of shaped steel 型钢数量表
quantity of shipment 装运的数量
quantity of slag 渣量
quantity of snow 雪的蕴藏量
quantity of spring overflow 泉的溢出量
quantity of stabilizer 稳定剂加入数量
quantity of state 状态量
quantity of suspended solid 悬浮固体量
quantity of tracer injection 注水量；示踪剂注入量
quantity of tracer used 示踪剂用量
quantity of using trains 运用车数
quantity of waste flow 污水流量；废水流量
quantity of wastes 废料量
quantity of wastewater 废水量
quantity of wastewater disposal 污水处理量

quantity of wastewater effluent 污水处理后出水量
quantity of wastewater treatment 污水处理量
quantity of water 水量
quantity of water bursting 突水量
quantity of water bursting in the early days 初期突水量
quantity of water pouring 灌水量
quantity of water recharge 回灌量
quantity of work 施工量;作业量;工作量;劳动量;工程量
quantity of wound used 铂用量
quantity on hand 现存数量
quantity originated 发送数量
quantity overrun 超出估计的量;工程量差额
quantity overrun/underrun 估计工程量与实际竣工量之差
quantity per capita 人均量
quantity performed 完成数量
quantity pricing 工程量估价
quantity production 大量生产
quantity rate 高货量频率
quantity recording 数量记录
quantity reproduction 大量复制
quantity required 需要数量
quantity sensitivity 电量灵敏度;冲击灵敏度
quantity sheet 工作量表;工程数量表;工程量清单;工程量表;土方量;土方表
quantity sludge 污泥量;淤泥量
quantity sold 销售量
quantity standard used in cutting 切削用量标准
quantity survey(ing) 土方量估计;数量估算;数量估计;数量概算;空地或空房调查表;估料;工程用料与设备清单;工程量估算;工程量估计;工程量评估;数量调查
quantity survey method 数量概算法(用于房地产评估)
quantity surveyor 概算员;预算师;估算师;估算师;估计员;工程量统计员;工程量估算师;工程量测算师;测试师
quantity take off 工程量计算
quantity to be conveyed 运输量
quantity underrun 低于估计的量;工程量差额
quantity unit 数量单位
quantity value 量值
quantity variance 数理差异
quantity weighed out for testing 称取的试样量
quantivalence 原子价;化合价
quantivalency = quantivalence
quantization 量化;取离散值;数字转换;变量分区法
quantization accuracy range 分层精确度范围
quantization distortion 量化噪声;量化误差;定量失真
quantization error 量化误差
quantization law 量化律
quantization level 量化电平
quantization noise 量子(化)噪声;分层噪声
quantization of amplitude 振幅量化;脉冲调制
quantization step 分层隔距
quantize 量化;取离散值
quantized collector 分层信号集电极
quantized computer 数字转换型计算机
quantized frequency modulation 量化调频
quantized interaction 分层互作用
quantized point data 数字化点数据
quantized pulse modulation 分层脉冲调制
quantized signal 量化信号
quantized spin wave 磁振子
quantized symbol 数字符号
quantized visual signal 量化图像信号
quantizer 量化器;脉冲调制器;数字化器;读数器;分层器
quantizer input quantization increment 量化器输入端量化增量
quantizing distortion 量化失真
quantizing distortion power 量化失真功率
quantizing encoder 数字转换器
quantizing frequency modulation 量化调频
quantizing interval 量化间隔
quantizing noise 量化噪声
quantizing structure 量化结构;分层结构
quantometer 光量计;剂量计;冲击电流计;测电量器;光谱分析仪
quantorecorder 光量计
quantum 量子;时间片;定额

quantum biology 量子生物学
quantum biophysics 量子生物物理学
quantum chemistry 量子化学
quantum clock 量子钟;时阴计
quantum condition 量化条件
quantum constant 量子常数
quantum contract 数量合同
quantum detector 量子探测器
quantum efficiency 量子效率
quantum electrodynamics 量子电动力学
quantum electronics 量子电子学
quantum energy 光子能量;量子辐射能量
quantum field theory 量子场论
quantum index of imports 进口物量指数
quantum libet 随意量
quantum mechanical state 量子态
quantum mechanics 量子力学
quantum mechanics of wave field 波场量子力学
quantum nature of light 光的量子性质
quantum number 量子数
quantum of goods sending-out 货物交出量
quantum of rainfall 总降雨量;降雨总量
quantum of world trade 世界贸易总量
quantum-optic(al) generator 光量子振荡器
quantum physics 量子物理学
quantum placet 随意量
quantum state 量子状态;能态
quantum statistics 量子统计学
quantum step 量化阶梯
quantum theory 量子论
quantum theory of heat capacity 热容量的量子理论
quantum theory of radiation 辐射的量子理论
quantum theory of spectra 光谱的量子理论
quantum theory of valence 价的量子理论
quantum value 量子值
quantum well material 量子阱材料
quantum yield 量子产额
quaquaversal 穹状的;穹形的
quaquaversal dip 穹倾斜
quaquaversal fold 穹褶皱
quaquaversal structure 穹状构造;穹形构造
quar 坚实砂岩
quarantine 交通检疫;检疫申报
quarantine act 检疫法
quarantine anchorage 检疫锚地
quarantine authority 检疫部门
quarantine boat 检疫船
quarantine buoy 检疫锚地浮标;检疫浮筒;检疫浮标
quarantine certificate 船舶检疫证书
quarantine disease 检疫疾病
quarantine dues 检疫费
quarantine duration 检疫停船期间
quarantine fee 检疫费
quarantine flag 检疫旗
quarantine harbo(u)r 检疫港
quarantine inspection 检疫
quarantine law 检疫法
quarantine mark 检疫标
quarantine measures 检疫措施
quarantine message 检疫通信
quarantine object 检疫对象
quarantine of animals and plants 动植物检疫
quarantine office 检疫所
quarantine officer 交通检疫员
quarantine period 检疫期间
quarantine port 检疫港
quarantine range 检疫范围
quarantine regulations 检疫规则;检疫规定
quarantine service 检疫站;检疫机关;检疫隔离工作
quarantine ship 检疫船
quarantine signal 检疫信号
quarantine station 交通检疫站;检疫站;卫生防疫站
quarantine treatment 检疫处理
quarantine unit 不可分的数据块
quarantine unit identifier 不可分数据块识别器
quarantine vessel 检疫船
quarantine weeds 检疫性杂草
quarantine wharf 检疫码头
quarded cable 扭绞四芯电缆
quardrel 四方形瓦
quadruple block 四饼滑车
Quarella sandstone 夸雷拉砂岩
quark 夸克(理论上一种比原子更小的基本粒子)

quarl block 喷嘴砖;喷火口砖
quarl(e) 异形耐火砖;大块耐火砖
quarrel 小板方形玻璃板;菱板;四边形建筑构件;方形瓦
quarrel pane 菱形玻璃片;方形玻璃片
quarrel panel 方形玻璃片;菱形玻璃片
quarried block 粗石砌块;粗面方块
quarrier 采石工(人)
quarry 露天采石场;矿场;厚地砖;石矿;石碴场;采石区;采石场
quarry bar 露天矿管柱;机钻架
quarry barge 采石驳船
quarry bed 天然石层;天然层;石矿层
quarry bench 台背
quarry blast-hole rig 钻眼设备(采石场);炮孔钻机(采石场)
quarry blast(ing) 露天爆破;采石(场)爆破;采石爆炸
quarry body 采石场谷地;采石场凹地
quarry bucket 采石场铲斗
quarry car 采石场斗车;采石场用车;采石场斗车
quarry chips 碎石片
quarry cord 矿场起爆点火线;采石点火索;采石场用点火线
quarry cut 边裂
quarry damp 原石水分
quarry drainage 采石场排水
quarry dressed 修琢方石
quarry drill 采石场钻机;采石钻(孔)机;采石穿孔机
quarry drilling 采石场钻进
quarry drill machine 矿场钻孔机
quarry dumper 采石用自动倾卸汽车
quarry dust 细石屑;石屑;石粉;采石场粉尘
quarry dynamites 采石场用硝甘炸药
quarry engineering 采石工程
quarry equipment 采石设备;采石场设备
quarry face 粗面(指石材天然面);粗琢石面;采石开凿面;采石工作面;采石场采掘面
quarry-faced 粗面的;毛面(的);表面未经琢磨的(建筑石块)
quarry-faced stone 粗琢石;毛面石;粗面石;凿石面;石材天然面;原开石(面)
quarry fines 细石屑;采石场石屑
quarry floor 料场底;采石场工作面;采石场底盘
quarry for animal shelters 动物笼舍用的地砖
quarry glass 小块方形玻璃;菱形玻璃(片);方形玻璃
quarry gravel 粗砾石
quarry hammer 采石锤
quarrying 开采;石料开采;采石(工程);采掘;拔蚀作用
quarrying action of the waves 冲击波的采石作用
quarrying by hand 人工开采
quarrying by the use of explosives 爆破采石
quarrying machine 采石机(械)
quarrying method 采石方法
quarrying operation 采石工作
quarrying rock breakwater 块石防堤堤;石砌防波堤;石块防波堤
quarry light 菱形玻璃窗;菱格玻璃窗
quarry lorry 采石场载重卡车;采石场载重汽车;矿石运输卡车
quarryman 采石工(人)
quarryman's pick 石工镐
quarry master 大型履带凿岩机;露天矿钻机;露天打眼设备
quarry master piston drill 往复式凿岩机
quarry material 采石材料;采石场石料
quarry method 采矿方法;采石方法
quarry-mine shovel 采矿铲
quarry operation 矿山开采;采矿作业
quarry pane 菱形玻璃片;菱形玻璃块
quarry pavement 片石路面;粗石铺砌;粗石路面
quarry pick 石工镐
quarry-pitched 粗琢的;粗凿的;粗凿
quarry-pitched stone 原开石;粗琢石
quarry plant 采石设备;采矿设备
quarry powder 采石炸药
quarry refuse 石屑;采石场弃石;采石场废石;采石场废料;矿场弃石
quarry rock 毛石(料);粗石;采石场岩石;采石毛料
quarry rubbish 弃料;尾料;石屑;石碴
quarry run 毛石踏步;(料坑的)杂乱石料;采石场

毛料;毛石料;混合石料;石场未分选料;粗毛石
quarry-run core 混合石料堆心;混合石料坝心
quarry-run rock 毛石;粗石;混合石料
quarry-run rockfill 乱石堆筑;毛石填筑
quarry-run sand 未筛选砂料;天然砂料
quarry-run stone 采石场石料;采石场原石;毛石料;未筛(分)石料
quarry sampling 采场采样
quarry sap 建筑石料的天然含水量;天然石含水量;石坑岩层含水
quarry shovel 采矿型挖掘机;采石场挖掘机;采场挖土机
quarry site 采石山场;采石场地
quarry spall 采石场碎石;采石场碎片
quarry spoil 采石场弃方
quarry stone 毛石;荒料;石块;粗石
quarry-stone bond 毛石砌筑;毛石工程;毛石砌合;粗石砌体
quarry stone masonry 粗石圬工;粗石砌体
quarry-stone yield 成荒率
quarry tile 粘土地板方砖;缸砖;无釉地砖;大铺地砖;铺地砖
quarry tile base 方砖基座
quarry tile floor 缸砖楼地面;缸砖地面
quarry tile flooring 缸砖楼地面
quarry tile room 方砖房间
quarry tile skirting 缸砖踢脚
quarry transportation 露天矿运输
quarry tripod 采石场上三脚架
quarry truck 采石场载重汽车;采石场卡车
quarry waste 采石场碎块;石屑;采石场废块;采石场尾料;采石场弃石;采石场废石;采石场废料;废石
quarry water 矿泉(水);石窝水;石坑水
quarry work 采石场
quarry yard 采石场
quart 夸(脱)(容量单位,1夸脱=1/4加仑)
quartation 析银法;四分取样法
quarter 住所;中部舷侧;罗盘基点;夸特;刻钟;季度;每季;尾舷;四分之一;(军分;船舶后部;船尾部
quarterage 季度津贴;四分之一每季付款
quarterage day 季度结账日;每季付款日
quarter bat 四分砖;四分之一块砖;二五砖
quarter beam 四开木梁;四开材梁
quarter belt 直角联动皮带
quarter bend 四分之一弯头;直角弯头;直角弯管;矩管
quarter bending 直角弯头
quarter bill 岗位表;部署表
quarter bitt 船尾系缆桩
quarter block 转弯滑轮;转角滑轮
quarter boat 住宿船;舰尾小艇;船尾救生艇
quarter bollard 船尾系缆桩
quarter bond 一顺一丁砌(砖)法
quarter-bound 布脊纸面装订
quarter breast (line) 后横缆
quarter brick 四分之一砖
quarter-chord point 四分之一翼弦点
quarter circle 四分之一圆;四分圆
quarter closer 二寸头;四开填塞砖;四分砖;四分之一砖;小磙(四分之一砖)
quarter closure 四分之一砖;二寸头
quarter cone filling 桥台填堆体;桥台四分之一锥体填坡;岸墩筑锥体;岸墩堆体;岸墩四分之一锥体填坡
quarter-cut 圆木四开锯法;径面刨切的;径锯;四开木材
quarter davit 尾舷吊艇柱
quarter day 季度的第一天;季度结账日
quarter deck 后甲板;艉甲板;尾甲板;船尾甲板;后部甲板
quarter decker 后甲板船
quarter deck vessel 船尾甲板船
quarter district 居住区
quarter-diurnal constituent 四分之一日分潮
quarter-diurnal tide 四分之一日间潮;四分之一日潮
quarter drain 区间排水沟;区间沟
quartered 径锯;纵隔墙;方隔墙;四开的;四等分的
quartered log 四分之一圆木;四开圆木
quartered lumber 四开木材
quartered partition 木筋隔墙;立楞隔墙;方木隔墙;方隔墙
quartered point loading 四分之一跨度荷载
quartered timber 径向锯材;四开锯木
quartered veneer 四开单板;四开薄木板

quarter fairlead 船尾导缆器
quarter fast 后倒缆;尾缆
quarterfoil 四瓣花饰;四叶饰
quarterfoil crossing 四叶式交叉
quarterfold 四叶饰
quarter gallery 船尾跳台
quarter girth 四分周干围
quarter-girth measurement 圆木折合立方数量
quarter-girth rule 圆木折合立方法则;圆木围长法
quarter-girth tape 四分周卷尺
quarter grain 四开木材纹
quarter-hard annealing 低硬度退火
quarter-hard temper sheet 软回火的薄钢板
quarter hatch 后甲板舱口
quarter height-to-span ratio [1/4 height-to-span ratio] 四分之一高跨比
quarter hollow 小凹圆饰;四分之一凹圆线脚
quarter hoop 木桶箍
quarter hour 一刻钟
quarter-inch map 四分之一英寸地图
quartering 锥堆四分法;立筋;尾舷方向;四分(取样)法;四等分(取样)法;成直角(的)
quartering attachment 车削曲柄轴装置
quartering hammer 破碎锤
quartering machine 曲柄轴钻孔机
quartering process 四分法工艺过程
quartering system 一刻钟制度
quartering way 四分法案件
quartering wind 尾舷风;船尾风
quarter iron 桁腰铁箍
quarter joint 四分之一接头
quarter ladder 船尾楼梯
quarter landing (楼梯的)直角转弯平台;直角转弯楼梯过渡平台
quarter light 边窗;(车的)侧窗
quarter line 后倒缆;船尾缆
quarter-line offset 四分之一行频置
quarterly 季度
quarterly dividend 季度股利;每季付一次的股息
quarterly edition 季度版
quarterly estimate 每季估计
quarterly interest 季度息
quarterly output 每季产量
quarterly payment 按季付款
quarterly premium 每季保险费
quarterly progress report 季度进展报告
quarterly report 季节报告;季度报告;季报
quarterly settlement days 分季清算日
quarterman 造船厂工段长
quarter master 舵手;舵工
quarter master depot 军需补给仓库
quartermaster truck 军需车
quarter-matching 四分之一波长匹配
quarternary 四元(的)
quarternary alloy 四元合金
quarternary eutectic alloy 四元易熔合金
quarternary form 四元型;四元形式;四元齐式
quarternary phase shift keying 四相相移键控
quarternary quantic 四次式
quarternary steel 四元素钢
quarternary structure 四级结构
quarternary system 四合组
quarter newelled stair (case) 直角转弯楼梯
quarter octagonal 缺角四边形的
quarter octagon drill rods 带圆角方断面钻探用钢材
quarter octagon steel 四角方钢
quarter of a mast 桅的鼓出部分
quarter pace (直角转弯的)梯台;直角转弯的楼梯平台;直角转弯梯台;曲尺楼梯平台
quarter pace landing 曲尺楼梯平台
quarter pace stair 直角转弯梯台;直角楼梯台
quarter panel 四角板
quarter partition 立楞隔断;立楞隔板;方木融断;木筋隔墙
quarter peg 四分之一桩(放在道路横断面的四分之一处)
quarter phase 两相(的);双相位;两相的
quarter-phase heating 两相加热;二相加热
quarter phaseshift-keying modulator 二相移相键控调制器
quarter-phase system 两相制
quarter pillar 侧支柱;边支柱
quarter pipe 管子直角弯头;船尾导缆孔

quarter pitch 高跨一与四比;四分之一高跨比(斜坡)
quarter plate 11cm×8cm相片
quarter point 四分之一间距点;四分点
quarter point loading 四分之一跨度(点)荷载;四分点点荷载
quarter point moment 四分之一跨度点处弯矩
quarter-quarter section 土地测量地块
quarter rack 报刻齿板
quarter rail 后甲板栏杆
quarter ramp 船尾引桥;船尾引道
quarter rampway 尾角跳板
quarter red partition 用正方形制的分隔墙
quarter rip saw 手锯
quarter rope 后倒缆
quarter round 扇形线脚;象鼻圆饰;四开木压条;四分之一圆缘饰;四分圆线脚
quarter-round mo(u)lding 四分之一圆线脚;四分圆线脚
quarter sampling method 方形取样法
quarter saw 纵向锯成四块;四开(指矛圆木)
quarter-sawed 四开锯木;径切的;四开的;四分之一径向锯
quarter-sawed clear oak 四开无节疤橡木材;四开精选橡木材
quarter-sawed grain 四开径切木料
quarter-sawed lumber 径锯板
quarter-sawed select oak 四开精选橡木材
quarter sawed (timber) 四开圆木;四开木材
quarter sawing 四开锯木;径向锯木
quarter-sawn 径锯面;四分之一径向锯木;四开的
quarter-sawn conversion (of log) 原木径截;锯制直纹木材
quarter-sawn grain 木材径面纹理;(木材的)直行纹理;径面纹理
quarter-sawn timber 直锯四开材;径向锯材
quarter screw 调节螺钉;四等分螺钉
quarter section 象限线;四开截面;四开断面;四分之一图幅;四分之一平方英里;四分之一波长线段;半方英里(等于160英亩)
quarters for the single 单身宿舍
quarter size 四分之一缩尺
quarter small column 四分之一小柱
quarter snail 报刻涡形轮
quarter space 梯台(直角转弯处);楼梯转变处之梯台
quarter-space landing (楼梯的)直角转弯平台;曲尺楼梯平台
quarter-space stair (case) 90°转角式楼梯
quarter spring 后倒缆;船尾倒缆
quarter-square multiplier 四分之一平方乘法器
quarter staff 铁头木棍
quarter-stake 四分之一桩
quarter stanchion 侧支柱;边支柱
quarter stuff 四分之一英寸厚木板;四分之一寸厚木板;二分板
quarter surface 径切面
quarter tie 四分枕木
quarter timber 四开圆材
quarter turn 直角转弯;直角回转
quarter-turn belt 直角回转带;直角挂轮皮带
quarter-turn drive 直角回转传动;皮带直角转动传动;十字轴传动;半交叉传动
quarter-turn stair (case) with winders 带扇步的直角式楼梯
quarter-turn stair (case) 曲尺楼梯;转弯楼梯;直角转弯楼梯
quarter-turn with winders 带扇步的直角转弯
quarter-twist belt 直角回转带
quarter watch 四班轮值
quarter wave 尾侧浪;四分之一波长
quarter-wave antenna 四分之一波长天线
quarter-wave attenuator 四分之一波长衰减器
quarter-wave criterion 四分之一波长判断标准
quarter-wave film 四分之一波长膜
quarter-wave filter 四分之一波长滤波器
quarter-wave layer 四分之一波长层
quarter wavelength 四分之一波长
quarter wavelength plate 四分之一波片
quarter-wave limit 四分之一波限
quarter-wave line 四分之一波长线;四分之一波线
quarter-wave plate 四分之一波晶片;四分之一波片
quarter-wave plate compensator 四分之一波长补偿器

quarter-wave radiator 四分之一波长辐射器
quarter-wave resonance 四分之一波长谐振
quarter-wave stack 四分之一波长膜系
quarter-wave transformer 四分之一波长变换器
quarter-wave voltage 四分之一波电压
quarter width 四分之一码宽;夸特宽;跨特宽
quarter wind 尾舷风;船尾风
quartet(te) 四重线;四重组;四等体;四个一组;四个一套
quartferrous rock 石英质岩
quartic 四次;双二次
quartic curve 四次曲线
quartic equation 四次方程(式)
quartic surd 四次不尽根
quartic surface 四次曲面
quartiere 放射形装饰
quartile 四分值;四分位数;四分点
quartile coefficient 四分位系数
quartile deviation 四分位偏差;四分位差
quartile diameter 四分直径
quartile dispersion 四分位散度
quartile division 四分位分法
quartile measurement 四分位量度
quarto paper 四开纸
quarts rod strain seismometer 石英杆应变地震仪
quartz 水晶;石英(质)
quartz accelerometer 石英加速度计
quartz aggregate 石英骨料
quartz alarm clock 石英闹钟
quartz-albite phyllite 石英钠长千枚岩
quartz-albitite 石英钠长岩
quartz ampoule 石英管
quartz analog(ue) clock 模拟石英钟
quartz analog(ue) watch 模拟石英表
quartz andesite 英安岩;石英安山岩
quartz-anorthosite 石英斜长岩
quartz apparatus 石英仪器
quartz arenite 石英砂屑岩
quartzarenite 正石英岩;石英砂屑岩
quartz-banakite 石英粗面粒玄岩
quartz-banded ore 石英条带矿
quartz bar 石英条
quartz bar extensometer 石英棒伸缩仪
quartz basalt 石英玄武岩
quartz beaker 石英烧杯
quartz-bearing diorite 石英闪长岩
quartz block 石英砖
quartz boat 石英舟
quartz-bostonite 英淡歪细晶岩
quartz bottle 石英玻璃瓶
quartz brick 石英砖
quartz bulb 石英灯泡
quartz burner 石英燃烧器;石英(喷)灯
quartz-bytownite 石英倍长岩
quartz capillary 石英毛细管
quartz capsule 石英管
quartz cat's-eye 石英猫眼石
quartz cement 石英水泥;石英胶接剂
quartz chamotte 硅质黏土熟料
quartz chip(ping)s 石英碎屑
quartz-chlorite phyllite 石英绿泥千枚岩
quartz chlorite schist 石英绿泥片石岩
quartz chronometer 石英钟
quartz claim 脉矿采矿用地
quartz clock 石英钟
quartz clock tester 石英钟检验仪
quartz cloth 石英玻璃布;玻璃布
quartz compound 石英化合漆
quartz condensing-lens 石英聚光镜
quartz conglomerate 石英砾石
quartz container 石英外壳;石英容器
quartz container tube 石英储存管
quartz control 石英变换控制
quartz controlled oscillator 石英控制振荡器
quartz controlled transmittter 石英控制发射机
quartz core 石英管芯(熔模中)
quartz coupler 石英耦合器
quartz crucible 石英坩埚
quartz crystal 水晶;石英晶体
quartz-crystal clock 石英晶体时钟
quartz-crystal control 石英晶体控制
quartz crystal controlled blocking oscillator 石英晶体控制间歇振荡器
quartz-crystal controlled FM receiver 石英晶体控制调频接收机
quartz-crystal controlled receiver 石英晶体控制接收机
quartz crystal cutter 石英晶体切割机
quartz crystal cutting 水晶切型
quartz-crystal electric(al) clock 石英电钟
quartz crystal electronic watch 石英晶体电子表
quartz crystal filter 石英晶体滤波器
quartz crystal for artifacts 工艺水晶
quartz crystal frequency oscillator 石英晶体频率振荡器
quartz crystallization 石英晶体;石英结晶(作用)
quartz crystal monitor 石英晶体监视器
quartz crystal monochromator 石英晶体单色仪
quartz-crystal movement 石英表机心
quartz crystal oscillator 石英晶体振荡器
quartz crystal oscillator side 石英晶体振荡器装配面
quartz-crystal pressure transducer 石英晶体压力变换器
quartz-crystal resonator 石英晶体谐振子
quartz-crystal stabilizer 石英晶体稳频器
quartz-crystal subaqueous microphone 石英晶体水下传声器
quartz crystal thermometer 石英温度计
quartz crystal thin film monitor 石英晶体薄膜监控器
quartz-crystal unit 晶体振子;石英晶体单元
quartz-crystal vibrator 石英振子
quartz-crystal watch 石英电子手表
quartz crystal wave filter 石英晶体滤波器
quartz cutting 晶体切割;石英切割
quartz day-date watch 石英双历表
quartz delay line 石英延迟线
quartz delay-line analyzer 石英延迟线分析器
quartz delay-line memory 石英玻璃延迟线存储器
quartz diabase 石英辉绿岩
quartz digital car clock 石英岩汽车钟
quartz diorite 石英闪长岩
quartz-diorite intrusion 石英闪长岩侵入
quartz diorite porphyrite 石英闪长纷岩
quartz disk 石英盘
quartz dolerite 石英粒玄岩
quartz double monochromator 石英双单色仪
quartz-dust phagocytosis 石英粉尘吞噬作用
quartz electronic modules 石英电子组件
quartz electronic watch 石英电子表
quartz element 石英元件
quartz enriched porcelain 高石英瓷
quartz feldspar mica porphyry 石英长石云母斑岩
quartz feldspar porphyry 石英长石斑岩
quartz-feldspathic gneiss 长英片麻岩
quartzfels 石英岩
quartz felsite 石英霏细岩
quartz fiber 石英纤维;石英丝
quartz fibre array 石英纤维阵列
quartz-fibre balance 石英纤维秤;石英丝弹簧秤
quartz-fibre dose meter 石英丝剂量计
quartz-fibre dosimeter 石英丝剂量计
quartz-fibre electrometer 石英丝静电计
quartz-fibre electroscope 石英丝验电器
quartz fibre ga(u)ge 石英丝黏滞规
quartz-fibre manometer 石英丝压力计
quartz-fibre microbalance 石英丝微量天平
quartz fibre pendulums 石英丝摆
quartz fibre reinforced phenolics 石英纤维增强酚醛树脂
quartz fibre suspension 石英纤维悬浮液
quartz field flattener 石英视场致平器
quartz film 石英膜
quartz filter 石英滤波器
quartz filter sand 石英过滤砂
quartz firebrick 石英耐火砖
quartz flash tube 石英闪光管
quartz flour 石英粉
quartz-fluorite achromat 石英萤石透镜
quartz-fluorite lens 石英萤石透镜
quartz-fluorite ore 石英—萤石矿石
quartz flux 石英熔剂
quartz foam glass 石英泡沫玻璃
quartz for artifacts 工艺石英
quartz force transducer 石英测力感传器
quartz frame 石英架
quartz-free diorite 无石英的闪绿岩
quartz-free porphyry 无石英质斑岩
quartz frequency 石英频率
quartz frequency stabilizer 石英稳频器
quartz gabbro 石英辉长岩
quartz glass 石英玻璃
quartz glass brick 石英玻璃砖
quartz glass fibre 石英玻璃纤维
quartz glass lamp 石英玻璃灯
quartz glass pot 石英玻璃坩埚
quartz glass spring 石英玻璃弹簧
quartz-glass thermometer 石英玻璃温度计
quartz glass ware 石英玻璃器皿
quartz grain 石英砂;石英颗粒
quartz gravel 石英砾(石)
quartz graywacke 石英杂砂岩
quartz halogen lamp 石英卤灯;石英卤气灯
quartz-halogen-tungsten lamp 石英卤素钨灯
quartz heater 石英加热器
quartz holder 石英支架
quartz holder for boat 石英舟架
quartz hook 石英钩
quartz horizontal magnetometer 石英水平磁强计
quartz hornblende schist 石英角闪片岩
quartziferous 石英质的
quartziferous rocks 石英质岩
quartzification 石英化
quartzific sandstone 泡砂岩;泡砂石
quartz index 石英指数
quartz indicator 石英指示器;晶体指示器
quartzine 正玉髓
quartz insulator 石英绝缘子
quartz inversion 石英转变
quartz-iodine lamp 石英碘灯
quartz iodine tungsten lamp 石英碘钨灯
quartzite 石英岩
quartzite ballast 石英岩道砟
quartzite brick 硅砖;石英岩砖
quartzite conglomerate 石英岩砾岩
quartzite deposit 石英岩矿床
quartzite fireproofing material 硅耐火材料
quartzite for chemical firtilizer 化肥用石英岩
quartzite for flux 溶剂用石英岩
quartzite for glass 玻璃用石英岩
quartzite gneiss 石英片麻岩
quartzite jade 石英岩玉石
quartzite material 石英岩材料
quartzite oscillator 石英振荡器
quartzite sandstone 石英质砂岩;石英岩状砂岩
quartzite-schist 耐火石
quartzite slab 石英岩薄板
quartzite slate 石英岩状板岩
quartzite tile 石英面砖;石英岩花砖
quartzitic intrusion 石英侵入体
quartzitic sandstone 正石英砂岩;石英质砂岩;石英岩状砂岩;白泡石
quartzitic soil 石英质土
quartzitic stone 白泡石
quartz keratophyre 石英角斑岩;石英角斑玢岩
quartz keratophyre glass 石英角斑岩质玻璃
quartz knife edge 石英刃口
quartz ladder 石英梯
quartz lamp 水晶灯;石英灯
quartz latite 石英安粗岩
quartz lattice 流纹安粗岩
quartz lens 石英透镜
quartz-lens method 石英透镜方法
quartz magnetometer 石英磁力仪
quartz marble 石英大理岩
quartz master oscillator 晶体主振荡器;石英主控振荡器
quartz membrane ga(u)ge 石英膜真空计
quartz mengwacke 石英蒙瓦克岩
quartz mercury arc lamp 石英水银弧光灯;石英银电弧灯
quartz mercury lamp 石英水银灯;石英汞灯
quartz mercury vapo(u)r lamp 石英水银气灯
quartz-mica gneiss 石云片麻岩
quartz microbiorite 石英微晶闪长岩
quartz microsyenite 石英微晶正长岩
quartz mill 石英磨机
quartz mine 石英脉矿
quartz mining 石英金矿脉
quartz moisture 石英砂水分

quartz monzonite 石英二长岩
quartz-monzonite-aplite 石英二长细晶岩
quartz monzonite porphyry 石英二长斑岩
quartzoid bulb detector 玻璃球火灾探测器
quartzoid bulb pendant type sprinkler 石英球悬吊型喷淋头
quartzoid sprinkler system 玻璃球喷水灭火系统
quartzolite glass 透紫外线玻璃
quartz olivine dolerite 石英橄榄粒玄岩
quartz optic(al) flat 石英光学平面
quartz optic(al) phonon maser 石英光(学)声子激射器
quartz oscillator 石英振子;石英晶体振荡器
quartzose 含石英的;石英质的
quartzose conglomerate 石英质砾岩
quartzose limestone 石英质灰岩
quartzose phyllite 石英千枚岩
quartzose sandstone 石英质砂岩;石英砂岩
quartzose shale 石英页岩
quartzose subgraywacke 原石英岩;石英次杂砂岩
quartzous 含石英的
quartz paint 石英纹理漆
quartz pantellerite 石英碱流岩
quartz pebbles 石英小卵石
quartz pendulum 石英摆
quartz pendulum tiltmeter 石英摆倾斜仪
quartz-piezoelectric(al) transducer 石英压电变送器
quartz pilot 石英控制
quartz pit sand 石英质砂坑
quartz plate 水晶片;石英片;石英板
quartz plate holder 晶体片支架
quartz-polaroid monochromator 石英偏振单色仪
quartz porphyrite 石英玢岩
quartz porphyry 水晶斑岩;石英斑岩
quartz powder 石英粉
quartz powder insulation 石英粉绝缘
quartz pressure sensor 石英压力传感器
quartz pressure transducer 石英压力传感器
quartz-printer 石英频率检查仪
quartz prism 石英棱镜
quartz prism filter 石英棱镜滤波器
quartz prism instrument 石英棱镜仪
quartz-propyry 石英青盘岩
quartz protecting tube 石英保护管
quartz pushrod 石英棒
quartz reaction chamber 石英反应室
quartz reef 石英脉;石英矿脉
quartz regrowth cement 石英再生生长胶结物
quartz resonator 石英振子;石英谐振器
quartz retaining spring 石英夹持簧
quartz-rich domain 富石英域
quartz rock 石英岩
quartz rock crystal 石英石晶体
quartz rod 石英杆;石英棒
quartz-rod illumination system 石英棒照明系统
quartz rod technique 石英棒检术
quartz sample 石英进样管
quartz sand 白硅砂;石英砂
quartz sand deposit 石英砂矿床
quartz sandstone 石英质砂岩
quartz sandstone deposit 石英砂岩矿床
quartz sandstone formation 石英砂岩建造
quartz-schist 石英片岩
quartz sediment 石英质泥沙;石英质沉积物
quartz-sericite phyllite 石英绢云母千枚岩
quartz slate 石英板(岩);水晶片
quartz sleeve 石英套筒
quartz spectrograph 石英摄谱仪
quartz spectro-oltaphmeter 石英分光测光仪
quartz spectrophotometer 石英分光光度计
quartz spectroscope 石英分光仪
quartz spiral balance 石英螺旋线秤;石英螺旋线天平
quartz spring 石英弹簧
quartz spring balance 石英弹簧秤
quartz spring granimeter 石英弹簧式重力仪
quartz stabilization 石英稳定性
quartz stabilizer 石英稳定器
quartz-steel resonator 石英钢片谐振器
quartz still 石英蒸馏器
quartz strain ga(u)ge 石英应变仪
quartz stringer 石英细脉
quartz substrate 石英基片;石英衬底

quartz syenite 石英正长岩
quartz tester 石英手表测试仪
quartz testing instrument 石英试验仪表
quartz thermometer 石英温度计
quartz tholeiite 石英拉斑玄武岩
quartz tile 石英面砖
quartz-timer 晶体定时器;数字式石英校表仪;石英表校仪
quartz topaz 仿造黄晶;黄晶
quartz trachyte 流纹岩;石英粗面岩
quartz trachyte porphyry 石英粗面斑岩
quartz transducer 石英换能器;石英感传器
quartz transmitter 石英发送机
quartz tube 石英管
quartz-tube infrared heater 石英管红外线加热器
quartz-tube radiator 石英管辐射器
quartz tube reactor 石英管反应器
quartz tube test for sulfur 石英管定硫试验
quartz-tube thermometer 石英(管)温度计
quartz ultraviolet laser 石英紫外激光器
quartz ultraviolet light 石英紫外光
quartz ultraviolet light weatherometer 石英紫外光人工老化试验机
quartz vacuum microbalance 石英真空微量天平
quartz vein 石英脉
quartzwacke 石英瓦克岩
quartz ware 石英制品
quartz watch 石英表
quartz watch tester 石英表校仪
quartz watch with liquid crystal display 液晶显示石英手表
quartz wavemeter 石英波长计
quartz wedge 水晶劈;石英楔(子);石英劈;石英光楔
quartz wedge analyser 石英楔片检偏器
quartz wedge compensator 石英光楔补偿器
quartz Wollaston prism 渥拉斯顿石英棱镜
quartz wool 石英棉;石英毛
quartz wool sheet 石英纤维板
quartzy 石英质的
quartzy sandstone 石英质砂石;石英砂岩
quartz yarn 石英纱
Quarzal 夸尔扎耳铝基轴承合金
quash 压制
quash an indictment 撤消起诉
quashy 软湿似沼泽的
quashy ground 软湿地
quasi-absolute method 准绝对方法
quasi-adiabatic 准绝热
quasi-adiabatic convection 准绝热对流
quasi-affine algebraic variety 拟仿射代数簇
quasi-algebra 拟代数
quasi-algebraically closed field 拟代数闭域
quasi-algebroidal 拟代数体的
quasi-analogue model 准模拟模型
quasi-analogue system 准模拟系统
quasi-analytical method 拟分析方法
quasi-analytic function 拟解析函数
quasi-analyticity 拟解析性
quasi-anisotropic layered medium 准各向异性分层媒质
quasi-aperiodic(al) interference 准非周期干扰
quasi-arc welding 潜弧自动焊
quasi-aromatic compound 似芳族化合物
quasi-associated reserve signal(l)ing link 准直联备用信号线路
quasi-asymptote 拟渐近线
quasi-asymptote method 拟渐近线法
quasi-asymptotical stability 拟渐近稳定
quasi-attenuance 似衰减光度
quasi-autocorrelation 拟自相关
quasi-axial electrode configuration 准轴电极结构
quasi-axiomatizable 拟可公理化的
quasi-Besse merizing 锭模吹氧铸造法
quasi-biennial oscillation 准两年振荡;赤道地区平流层风的两年振荡
quasi-bistable circuit 准双稳态电路
quasi-bound state 准结合态
quasi-capital goods 准资本货物
quasi-center of action 准活动中心
quasi-character 拟特征标
quasi-characteristic function 拟特征函数
quasi-chemical approximation 准化学近似
quasi-chemical equilibrium 准化学平衡

quasi-chemical equilibrium of defect 缺陷的准化学平衡
quasi-chemical method 准化学方法;拟化学方法;似化学方法
quasi-chemical solution model 准化学溶液模型
quasi-circular 准圆形
quasi-circular orbit 准圆形轨道
quasi-classical 准经典的
quasi-cleavage crack 准解理断裂
quasi-cleavage fracture 准解理断口
quasi-climax 拟顶极植物群落
quasi-coherent beam 准相干束
quasi-coherent sheaves 拟凝聚层
quasi-coincidence 拟重合
quasi-compact cluster 拟紧聚集
quasi-compactness 拟紧性
quasi-compact set 拟紧集
quasi-compact space 拟紧空间
quasicompensation 半补偿
quasi-complement 拟补
quasi-complemented lattice 拟补格
quasi-complete locally convex space 拟完备局部凸空间
quasi-complete space 拟完备空间
quasi-complexity 拟复性
quasi-complex lattice 拟复格
quasi-complex manifold 拟复流形
quasi-component 拟分量
quasi-component space 拟分支空间
quasi-concave function 拟凹函数
quasi-concave programming 拟凹规划
quasi-concavity 拟凹性
quasi-conductor 准导体;半导体
quasi-conformality 拟保形
quasi-conformal mapping 拟保角映射
quasi-conjugation 似共轭效应
quasi-conjunctive equality matrix 拟合取等值母式
quasi-continental crust 准陆壳
quasi-continous action of a threelevel controller 三级控制器的准连续动作
quasi-continuity 拟连续性
quasi-continuity principle 拟连续性原理
quasi-continuous 准连续的
quasi-continuous function 拟连续函数
quasi-continuous functional 拟连续泛函
quasi-continuous measure space 拟连续测度空间
quasi-continuous process 拟连续过程
quasi-continuous variation 似连续变化
quasi-continuum of level 准连续能级
quasi-contract 准契约;准合同
quasi-contraction semi-group 拟收缩半群
quasi-convergence 拟收敛
quasi-convergent 拟收敛的
quasi-convex function 拟凸函数
quasi-convex functional 拟凸泛函
quasi-convexity 拟凸性
quasi-convex sequence 拟凸序列
quasi-coordinates 准坐标
quasi-critical damping 准临界阻尼
quasi-crosslink 准交联
quasi-crystal 准晶体;次晶
quasi-crystalline lattice 准晶格
quasi-crystalline structure 似晶结构
quasi-crystal model 准晶模型
quasi-cyclic(al) code 准循环码
quasi-cyclic(al) group 拟循环群
quasi-cylinder 拟圆柱
quasi-cylindric(al) coordinates 准柱面坐标
quasi-cylindric(al) documents 准圆柱形文件
quasi-cylindric(al) orthogonal coordinates 准柱面形正交坐标
quasi-decimal 拟小数
quasi-definite 拟定义
quasi-degenerated mode 准简并模式
quasi-derelict 准弃船
quasi-deterministic model 准确定性模型
quasi-diagonal 拟对角的
quasi-diagonal matrix 准对角矩阵;拟对角(线)矩阵
quasi-dielectric(al) 准电介质
quasi-differential 拟微分
quasi-digit 准数位
quasi-direct sum 拟直和
quasi-discontinuity 准不连续体
quasi-discrete spectrum 拟离散谱

quasi-disjunction 拟析取式
quasi-disjunctive equality matrix 拟析取等值母式
quasi-distance 拟距离
quasi-distortion 准畸变
quasi-diurnal variation 准周日变化
quasi-divisor 拟因子
quasi-dominant diagonal 拟强对角线
quasi-dual space 拟对偶空间
quasi-dynamic(al) height 似动力高
quasi-dynamic(al) model 拟动态模型
quasi-elastic 准弹性的
quasi-elastic dipole 准弹性偶极子
quasi-elastic force 准弹性力
quasi-elastic law 准弹性定律
quasi-elastic light scattering 准弹性光散射
quasi-elastic oscillator 准弹性振子
quasi-elastic rock 准弹性岩石
quasi-elastic scattering 准弹性散射；似弹性散射
quasi-elastic vibration 准弹性振动
quasi-elastoplastic flow 准弹塑性流(动)
quasi-electric(al) field 准电场
quasi-electroneutrality 准电中性
quasi-elliptic 准椭圆形的
quasi-elliptic function 拟椭圆函数
quasi-elliptic geometry 拟椭圆几何学
quasi-elliptic space 拟椭圆空间
quasi-energy gap 准能隙
quasi-equality 拟等值
quasi-equality function 拟等值函数
quasi-equilibrium 准平衡；准均衡
quasi-equilibrium state 准平衡(状)态；准平衡过程
quasi-equilibrium theory 准平衡理论
quasi-equilibrium theory of mass spectrum 质谱的似平衡原理
quasi-equivalence classes 拟等价类
quasi-equivalent unitary representation 拟等价酉表示
quasi-ergodic 准遍历性的
quasi-ergodic hypothesis 准各态经历假说；准遍历假设；拟遍历性假设
quasi-ergodic system 拟各态历经系统
quasi-eruptive prominence 准爆发日珥
quasi-eutectic 伪共晶
quasi-eutectoid 伪共析
quasi-expectation 拟期望值
quasi-experimental design 拟实验设计
quasi-exponential attenuation 准指数衰减
quasi-extremal compactum 准极值紧致
quasi-factor 拟因子
quasi-factorical design 拟析因设计
quasi-Fermi levels 准费密能级
quasi-fibrous 准纤维的
quasi-field 拟域
quasi-filtered 拟滤子化的
quasi-filtered category 拟滤子化范畴
quasi-fission 准裂变；拟裂变
quasi-flexural fold 准弯曲褶皱
quasi-flow 曲线流(动)；准流动；半流动
quasi-free ion 半自由离子
quasi-full 准完全
quasi-full employment 接近成分就业
quasi-function 拟函数
quasi-fundamental mode 准基波型
quasi-gelifusinite 次胶丝质类
quasi-gelifusinite-collinite 次胶丝质无结构体
quasi-gelifusinite-posttelinite 次胶丝质次结构体；次胶丝质似无结构体
quasi-gelifusinite-telinite 次胶丝质结构体
quasi-geodal height 似大地水准面高程
quasi-geodesic 拟测地线
quasi-geoid 准大地水平面；似大地水准面
quasi-geostrophic approximation 准地转近似
quasi-geostrophic solitary wave 准地转孤立波
quasi-goods 准货物
quasi-gravity 准重力；人造重力
quasi-group 拟群
quasi-harmonic oscilation 准谐波振荡；准简谐振荡
quasi-harmonic system 准谐波系统；准简谐系统；拟简谐系统
quasi-hexagon 似六角形；似六边形
quasi-high-speed train 准高速列车
quasi-historical 带有历史性质的
quasi-holographic 准全息的
quasi-homogeneous 准齐次；准均匀的

quasi-homogeneous material 准均匀材料
quasi-homogeneous mixture 准均匀混合物
quasi-homogeneous radiation 准均匀辐射
quasi-homogeneous reactor 准均匀反应堆
quasi-homogeneous vibration 准均匀振动
quasi-homogeneous waves 准均匀波
quasi-horizontal motion 准水平运动
quasi-hydrodynamic(al) lubrication 似流体动力学的润滑；半流体润滑
quasi-hydrostatic(al) analysis 准静力分析
quasi-hydrostatic(al) approximations 准静水压力近似；准静力近似
quasi-hydrostatic(al) assumption 准静力近似
quasi-hyperbolic orbit 准双曲轨道
quasi-hyperbolic reflector 准双曲形反射器
quasi-idempotency 拟幂等
quasi-idempotent matrices 拟幂等阵
quasi-identities 拟恒等式
quasi-indecomposable 拟不可分解的
quasi-industrial zone 准工业区
quasi-instantaneous 准同时的
quasi-instantaneous change 准瞬时变化
quasi-instruction 拟指令
quasi-instruction form 准指令形式；拟指令形式
quasi-insulator 准绝缘体
quasi-integral 拟积分的
quasi-integral lattice-ordered semigroup 拟整格序半群
quasi-interior point 拟内点；拓扑单位
quasi-interpenetrating polymer network 似互穿聚合物网络
quasi-invariant measure 拟不变测度
quasi-invariant measure space 拟不变测度空间
quasi-inverse element 拟逆元
quasi-inverse matrix 拟逆阵
quasi-invertible element 拟可逆元
quasi-investment banker 半投资银行家
quasi-isloate idea 准细目概念
quasi-isloates 准细目
quasi-isobaric process 准等压过程
quasi-isostatic displacement 准均衡位移
quasi-isothermal layer 准等热层
quasi-isotropic 准各向同性的
quasi-isotropic laminate 准各向同性层合板
quasi-isotropic material 类无向性材料
quasi-isotropy 准无向性；类无向性
quasi-Lagrangean minimization problem 拟拉格朗日极小化问题
quasi-Latin square 准拉丁方
quasi-lattice 准点阵；似晶格
quasi-lattice model 准晶格模型
quasi-lattice theory 准晶格理论
quasi-left-continuity 拟左连续性
quasi-length 拟长度
quasi-linear 准线性的；拟线性的
quasi-linear amplifier 准线性放大器
quasi-linear differential equation 拟线性微分方程
quasi-linear elasticity 准线性弹
quasi-linear elliptic equation 拟线性椭圆型方程
quasi-linear equation 拟线性方程
quasi-linear feedback control 反馈控制系统
quasi-linear feedback control system 准线性反馈控制系统
quasi-linear functional 拟线性泛函
quasi-linear hyperbolic equation 拟线性双曲型方程
quasi-linear hyperbolic systems 拟线性双曲型组；拟线性双曲型方程组
quasi-linearity 拟线性
quasi-linearization 似线性化；准线性化；拟线性化
quasi-linearization method 准线性化方法；拟线性化法
quasi-linear model 准线性模型
quasi-linear operator 拟线性算子
quasi-linear partial differential equation 拟线性偏微分方程
quasi-linear planning 拟线性化规划
quasi-linear programming 拟线性化规划
quasi-linear programming algorithm 准线性程序算法
quasi-linear system 准线性系统；拟线性系统
quasi-linear system of equations 拟线性方程组
quasi-linear theory 准线性理论
quasi-liquid 似液体；半液体

quasi-local operator 拟局部算子
quasi-local ring 拟局部环
quasi-longitudinal propagation 准纵向传播
quasi-longitudinal wave 准纵波
quasi-Markov chain 拟马尔可夫链
quasi-maximum likelihood estimation 拟极大似然估计
quasi-maximum value 准最大值
quasi-membrane method 薄膜比拟法
quasi-metric 拟度量的
quasi-minimal set 拟极小集
quasi-moment 拟矩
quasi-monochromatic beam 准单色束
quasi-monochromatic light 准单色光
quasi-monochromatic pulse 准单频脉冲
quasi-monochromatic radiation 准单色辐射
quasi-monostable circuit 准单稳态电路
quasi-monotone 拟单调的
quasi-monotone function 拟单调函数
quasi-monotonic 拟单调的
quasi-Monte-Carlo method 拟蒙特·卡罗法
quasi-multiple 拟倍数
quasi-multiplication 拟乘法
quasi-negative operator 拟负算子
quasi-neutral cluster 准中性凝聚
quasi-neutrality 准中性
quasi-Newtonian flow 准牛顿型流动
quasi-Newton's method 拟牛顿法
quasi-nilgroup 拟零群
quasi-nilpotent element 拟幂零元
quasi-nilpotent operator 拟幂零算子
quasi-nilpotent ring 拟幂零环
quasi-nil ring 拟诣零环
quasi-nonlinear characteristic 准非线性特征；准非线性特性
quasi-norm 拟范数
quasi-normal family 拟正规族
quasi-normal flow 准正常(水)流
quasi-normality 拟正规性
quasi-normal operator 拟正规算子
quasi-normed linear space 拟赋范线性空间
quasi-nuclear mapping 拟核映射
quasi-observable 虚拟观测量
quasi-observation 拟观测值
quasi-official 半官方的
quasi-off-line-digitizing 半脱机数字化
quasi-open mapping 拟开映射
quasi-open transformation 拟开变换
quasi-operator 拟算子
quasi-optic(al) 准光学的；似光学性
quasi-optic(al) feed system 准光学馈源系统
quasi-optic(al) frequencies 准光频
quasi-optic(al) technique 准光学技术
quasi-optic(al) wave 准光波；类光波
quasi-optimal 近似最佳；拟最优的
quasi-optimal solution 准优解
quasi-ordering 拟序
quasi-orthogonal 准正交集；准正交的
quasi-orthogonal polynomial 拟正交多项式
quasi-orthographic(al) projection form 准正射投影形式
quasi-orthotropic 准正交各向异性的
quasi-orthotropic plate method 比拟正交各向异性板法
quasi-orthotropic slab method 准正交各向异性板法
quasi-overconsolidation[QOC]准超固结
quasi-parabolic reflector 准抛物形反射器
quasi-paracomplex 拟仿复的
quasi-parahermitian 拟仿埃尔米特的
quasi-parallel processing 拟并行处理
quasi-particle 准粒子
quasi-particle model 似粒子模型
quasi-partner 准合伙人
quasi-peak 准峰值
quasi-peak detector 准峰值检测器
quasi-peak meter 准峰值测量仪
quasi-peak output 准峰值输出
quasi-peak-to-peak amplitude 准峰至峰幅度
quasi-peak-to-peak voltage 准峰至峰电压
quasi-peak type voltmeter 准峰型伏特计
quasi-peak value 准峰值；类峰值
quasi-perfect code 准完备码
quasi-perfect mapping 拟完全映射

quasi-period 准周期
quasi-periodic(al) 准周期的
quasi-periodic(al) function 拟周期函数
quasi-periodic(al) motion 拟周期运动
quasi-periodic(al) orbit 拟周期轨道
quasi-periodic(al) oscillation 拟周期振荡；拟周期波动
quasi-periodic(al) solution 拟周期解
quasi-periodic(al) variation 拟周期变化；准周期性变化
quasi-periodic(al) vibration 准周期振动
quasi-periodicity 准周期性；拟周期
quasi-permanent deformation 似永久变形
quasi-permanent load 准永久荷载
quasi-permanent low 准常定低压
quasi-permanent value of action 作用准永久值
quasi-phase space 准相位空间
quasi-plane 拟平面
quasi-plane curve 拟平面曲线
quasi-plane wave 准平面波
quasi-plastic flow 准柔性流动；半柔性流动
quasi-plasticity 似塑性；似塑度
quasi-point radio source 类点射电源
quasi-polar sun-synchronous orbit 准极地太阳同步轨道
quasi-polynomial 拟多项式
quasi-potential 准势
quasi-potential equation 准势方程
quasi-potential mapping 拟势映射
quasi-potential operator 拟位势算子
quasi-power series 拟幂级数
quasi-preconsolidation pressure（土的）准先期固结压力；准前期固结压力
quasi-primitive 拟本原的
quasi-private revenue 半私有经济收入
quasi-probability 拟概率
quasi-probable error 拟或然误差；似概差
quasi-project 准方案
quasi-projective 拟射影的
quasi-projective algebraic variety 拟射影代数簇
quasi-projective scheme 拟射影概型
quasi-projective variety 拟射影簇
quasi-proprietary 准所有的
quasi-proprietary right 半所有权
quasi-propulsive coefficient 准推进系数
quasi-public 私营公用事业的
quasi-public goods 准公共物品
quasi-quaternary 伪四元
quasi-quaternionic 拟四元数的
quasi-radical ring 拟根环
quasi-random 拟随机的
quasi-random access 准随机存取
quasi-random continuous wave 准随机连续波
quasi-random number 准随机数；拟随机数目；拟随机数(量)；伪随机数
quasi-random points 拟随机点
quasi-random sampling 拟随机抽样
quasi-random sequence of numbers 准随机序列；拟随机数序列；伪随机数序列
quasi-ratio scale 拟比率尺度
quasi-real simulation 半实时仿真
quasi-real time 准实时
quasi-reconsolidation pressure 似前(先)期固结压力
quasi-rectilinear stress circle 拟直线应力图
quasi-rectilinear stress diagram 似直线应力图
quasi-reentrant 准可重入；拟重入人
quasi-reflectoin 准反射
quasi-regular 拟正规的
quasi-regular element 拟正则元素
quasi-regular function 准正则函数
quasi-regularity 拟正则性
quasi-regular left ideal 拟正则左理想
quasi-regular point 拟正则点
quasi-regular sequence 拟正则序列
quasi-rent 准租金；类似地租
quasi-reorganization（指对公司的）准改组
quasi-residual 拟剩余的
quasi-resonance 准共振
quasi-ring 拟环
quasi-rotation 拟旋转
quasi-safe area 准安全区
quasi-sample 准样本；拟样本
quasi-sample size 拟样本容量

quasi-satellite city 准卫星城
quasi-saturated soil 准饱和土
quasi-seasonal fluctuation 拟季节波动
quasi-secretive illuminator 半隐蔽照明器
quasi-self-conjugate extension 拟自共轭扩张
quasi-semigelifusinite 次半胶质类
quasi-semigelifusinite-collinite 次半胶丝质无结构体
quasi-semigelifusinite-posttelinite 次半胶丝质次结构体
quasi-semigelifusinite-precollinite 次半胶丝质似无结构体
quasi-semigelifusinite-telinite 次半胶丝质结构体
quasi-semi-local ring 拟半局部环
quasi-sidereal time 准定量时
quasi-simple 拟单的
quasi-simple ring 拟单环
quasi-simple wave 拟简波
quasi-sine-wave 准正弦波
quasi-single sideband transmission system 准单边带发送系统
quasi-singular vertex 拟奇异顶点
quasi-sinusoid 准正弦曲线
quasi-slab 比拟板
quasi-slab method 拟板法；似板法；比拟板法
quasi-soft 亚软质
quasi-solid 准固体的
quasi-spin group 拟自旋群
quasi-spiral 拟螺线
quasi-spiral orbit 准螺旋形轨道
quasi-split 拟分裂
quasi-split algebraic group 拟分裂代数群
quasi-stability 准稳定性；似稳态
quasi-stable 似稳定的
quasi-stable adjustment 拟稳平差
quasi-stable distribution 拟稳定分布
quasi-stable distribution function 拟稳定分布函数
quasi-stable distribution law 拟稳定分布律
quasi-stable element 准稳元素
quasi-stable flow 准稳定流
quasi-stable island 准稳岛
quasi-stable isotope 准稳定同位素
quasi-stable law 拟稳定分布律
quasi-stable state 准稳态；准定态；拟稳态
quasi-standardized distribution 拟标准化分布
quasi-static 准静态的；准静力的；似稳态的
quasi-static approximation 准静态近似
quasi-static calculation 准静态计算
quasi-static change 似静态变化
quasi-static component of strain 应变准静态分量
quasi-static cone penetration 准静力触探贯入度
quasi-static deformation 准静态形变
quasi-static earthquake mechanics 准稳态地震力学；准静态地震力学
quasi-static electric(al) field 准静态电场
quasi-static lateral force 准静横向力
quasi-static load 准静载；准静荷载
quasi-static model 准静力模型
quasi-static power coefficient 准静态功率系数
quasi-static problem 拟静态问题
quasi-static process 可逆过程；准静态过程；准静力过程；拟静力过程
quasi-static solution 准静态解
quasi-static sounding 准静态探测
quasi-static strength test 准静态强度试验
quasi-static stress field 准静态应力场
quasi-static test 准静态试验；准静力试验
quasi-stationary 准稳态的；似稳态的；似稳定的
quasi-stationary current 准稳定流；似稳电流
quasi-stationary (energy) level 准稳能级
quasi-stationary field equations 准稳态场方程
quasi-stationary flow 准稳流；准稳定流；似稳的流动准稳流
quasi-stationary front 准静止锋
quasi-stationary level 准稳级
quasi-stationary motion 准稳定运动
quasi-stationary oscillation 准稳振荡
quasi-stationary phenomenon 准平稳现象
quasi-stationary process 拟平稳过程；似稳过程
quasi-stationary processing 准稳定过程
quasi-stationary random process 准稳随机过程
quasi-stationary sound pressure 准稳态声压
quasi-stationary spectrum 准稳定谱
quasi-stationary state 准稳定状态；拟稳状态；似稳态
quasi-steady 准稳定的
quasi-steady flow 似稳定流；准稳流；准定常流
quasi-steady flow analysis 准稳态流量分析
quasi-steady impulsive noise 准稳态脉冲噪声
quasi-steady method 准稳定法
quasi-steady operation 准稳态工作
quasi-steady process 准稳定过程
quasi-steady state 准定常态；似稳状态
quasi-steady state solution 似稳定状态解
quasi-stochastic 拟随机的
quasi-sufficiency 拟充分性
quasi-sun-synchronous orbit 太阳同步准极轨道
quasi-superheavy element 准超重元素
quasi-symmetric(al) pulse 准对称脉冲
quasi-synchronization 准同步
quasi-synchronous 准同步
quasi-systematic error 准系统误差
quasi-tensor 准张量；拟张量
quasi-thixotropy 准触变性
quasi-three dimensinal solution 准三元解
quasi-three dimensional computation 准三维计算；拟三维计算
quasi-toroidal coordinates 准圆环坐标
quasi-traditional 半传统的
quasi-translation 拟平移
quasi-transverse 准横向传播
quasi-transverse wave 准横波
quasi-two-dimensional flow 准二元流
quasi-unemployment 准失业
quasi-uniform 拟一致
quasi-uniform convergence 拟均匀收敛
quasi-uniform distribution 拟一致分布；拟均匀分布
quasi-uniform flow 准均匀流量；准均匀流
quasi-uniformity 拟一致性
quasi-uniformly continuous 拟一致连续的
quasi-uniform space 拟一致空间
quasi-uniform time 准均匀时
quasi-unitary 拟酉的
quasi-unmixed local ring 拟纯粹局部环
quasi-variable 准变数；拟变量
quasi-viscous 半黏性的
quasi-viscous creep 准黏滞蠕变；准黏性蠕变
quasi-viscous effect 黏性效果
quasi-viscous flow 半黏性流；准黏滞性流变；准黏滞流(动)；准黏性流；半黏滞性流(动)；半黏度流(动)
quasi-viscous fluid 准黏滞流体；准黏性流体；半黏滞性流体；半黏滞度流体
quasrupole multiplet 四极多重线
quassation 压碎
quaterdenary 十四进制的
quater dividing method 四分取样法
quaterfoil 四叶式
quaterfoil crossing 四叶式道路交叉
quaterisation 季铵化反应
quaternaries 季铵盐类
quaternarization 季铵化作用
quaternarized polysulfone 季铵化聚砜
quaternarized polysulfone membrane 季铵化聚砜膜
quaternarized polysulfone reverse osmosis membrane 季铵化聚砜反渗透膜
quaternarized polysulfone ultrafiltration membrane 季铵化聚砜超滤膜
quaternary 四部组成；四变数
quaternary adduct 四元加成物
quaternary alloy 四元素合金
quaternary alloy steel 四元素合金钢
Quaternary alluvium 第四纪冲积物【地】；第四纪冲积层【地】
quaternary amines 季铵类
quaternary ammonium base 季铵碱
quaternary ammonium compound 季铵化合物
quaternary ammonium salt 季铵盐
quaternary ammonium salt catinic surfactant 季铵盐阳离子型表面活性剂
quaternary ammonium salt cationic polyelectrolyte 季铵阳离子型聚电解质
quaternary ammonium surfactant 季铵表面活性剂
quaternary ammonium type resin 季铵型树脂
quaternary carbon atom 李碳原子
Quaternary climate 第四纪气候
quaternary code 四进制码

quaternary compound 四元化合物
Quaternary deposit 第四纪沉积(物)
quaternary diagram 四相图
Quaternary eustatic movement 第四纪海面升降运动
quaternary eutectic alloy 四元共晶合金；铋锡铅镉易熔四元合金
quaternary facies analysis 第四系相分析法
Quaternary fault 第四纪断层
quaternary geochronologic(al) scale 第四纪地质年代表
quaternary geologic(al) profile 第四纪地质剖面图
Quaternary geology 第四纪地质学
Quaternary geology map 第四纪地质图
Quaternary geology mapping 第四纪地质测绘
quaternary geomorpho-geologic(al) map 第四纪地貌地质图
Quaternary glacial stage 第四纪冰期
Quaternary glaciation 第四纪冰川(作用)
quaternary halide 卤化季盐
Quaternary ice age 第四纪冰期
quaternary ionisation 四级电离
quaternary link 四件运动链
quaternary liquid system 四元液体系统
quaternary mix 四元配料
quaternary mixture 四元混合物
quaternary monocrystalline diffractometer 四圆单晶衍射仪
quaternary monocrystalline diffractometer method 四圆单晶衍射仪法
quaternary multiplication 四进制乘法
quaternary number 四进制数
quaternary number system 四进制数系
quaternary parasite 四重寄生物
Quaternary period 第四纪
quaternary phase equilibria 四相平衡
quaternary salt 季盐
quaternary sampling 四分采样法
Quaternary sediment 第四纪沉积(物)
quaternary sedimentary type 第四系沉积类型
Quaternary sedimentation 第四纪沉积作用
quaternary signal 四元通信
quaternary steel 四元合金钢
Quaternary system 第四系【地】
quaternary system 四元制；四元(体)系；四进制；四个一组系统
quaternary system phase diagram 四元相图
quaternary thickness analysis 第四系厚度分析法
quaternary treatment 四级处理
quaternate 四个一组的
quaternion 四元数
quaternion algebra 四元数代数
quaternion crane 港口门式起重机
quaternion elliptic space 四元数椭圆空间
quaternion field 四元数域；四元数体
quaternion function 四元数函数
quaternion Grassmann manifold 四元数格拉斯曼流形
quaternion group 四元群
quaternion hyperbolic space 四元数双曲空间
quaternionic 四元
quaternionic projective space 四元射影空间
quaternion numbers 四元数
quaternion ring 四元数环
quaternions 四元法
quaternion vector bundle 四元数向量丛
quaternion wall 堤岸墙
quaternization 碱化反应
quaterphenyl 四联苯
quater-square multiplier 开四次方运算器
quatrefoil 四叶饰；四叶式；四叶形；瓣形花饰
quatrefoil crossing 四叶式道路交叉
quatrefoil oculus window 四瓣眼形窗
quatrefoil tracery 四叶式花窗棂
quatter's rights 土地所有权
Quattrocento (15世纪的)文艺复兴初期；文艺复兴时期
Quattrocento architecture 文艺复兴建筑
quaver 震颤
quay 码头(驳岸)；平行式码头；贴岸码头；顺岸码头；岸壁码头
quayage 码头总长；码头用地；码头税；码头使用税；码头使用费；码头面积；码头空货位；码头费；码头长度；泊位费

quay and pier construction 码头结构
quay appliance 码头设施；码头设备
quay apron 码头凸堤；码头前沿(地带)
quay apron space 码头前沿区
quay area pavement 码头路面
quay berth 码头(停)泊位；顺岸(码头)泊位
quay breast wall 堤岸前沿岸墙
quay breastwork 码头前沿岸墙；堤岸码头前沿岸墙；顺岸码头不紧靠水池
quay conduit 码头管沟
quay crane 港岸起重机；码头桥式起重机；码头前沿起重机
quay edge 码头边缘
quay extension 码头扩建
quay floor 码头地面；码头面板；码头地板
quay for goods-carrying barges 载货驳船的停泊处；载货驳船的停泊码头；载货驳货码头
quay for propeller thrust trial 试车码头
quay furniture 码头用设备
quay-gallery 码头廊道
quay level 码头水位；堤岸水位；停泊处水位；码头面高程
quay line 码头铁路(线)；码头前沿线；码头边线
quayman 码头带缆工人；码头搬运工
quay of sloping blockwork 斜砌块式方块码头
quay of steel sheet piling 钢板桩码头
quay on piles 桩基码头；桩基承台码头
quay operation 码头业务
quay pattern 码头型式
quay pier 凸堤码头；防波堤；码头墩柱；突堤(式)码头；单边码头；岸壁式突码头
quay pontoon 码头驳船
quay rate 码头提货费率(不经仓库的卸货)
quay run rockfill 块石堆筑
quay sensor 岸壁传感器
quay shed 码头前方货栈；码头前方仓库；码头货棚；码头仓库；前方仓库；前方仓库
quayside 码头区；码头边
quayside building 码头旁建筑(物)；码头建筑(物)
quayside container 岸边集装箱
quayside container crane 码头区集装箱起重机；岸边集装箱起重机
quayside crane 码头沿岸起重机；码头前沿起重机；岸边起重机；岸壁起重机
quayside portal crane 岸壁门吊
quayside railway 码头前沿铁路
quayside railway track 码头铁路线
quayside shed and yard 前方库场
quayside track 码头前沿铁路
quayside travel(l)ing crane 码头前沿移动式起重机
quay siding 码头线
quay space 码头地域；码头前沿地面；码头附属区
quay staircase 码头阶梯
quay surface 堤岸面；码头面；码头地面；(码头的)前沿；码头面积
quay surface level 码头面高程
quay system 码头系统
quay-to-quay 码头到码头
quay-to-quay traffic 码头到码头运输
quay-to-quay transportation 码头之间的运输
quay-to-ship stevedoring 码头到船舶的装卸搬运
quay track 码头前沿轨道
quay trestle 码头栈桥
quay wall 堤岸码头墙(沿岸边线挡土墙)；直立式码头；码头驳岸；码头墙；码头岸墙；岸壁；岸壁型码头
quay wall foundation 岸壁基础
quay wall of piles 桩式岸壁
quay wall of precast reinforced concrete units 预制钢筋混凝土块砌筑的岸壁；预制钢筋混凝土块砌筑的岸墙
quay wall of set blocks 方砌块砌成的岸壁；方砌块砌成的岸墙；方块码头；方块岸墙
quay wall of sliced blockwork 斜砌式方块码头
quay wall of sloping blockwork 斜砌式方块岸壁
quay wall of steel sheet pile 钢板桩岸壁
quay wall of steel sheet piling 钢板桩岸壁；钢板桩岸墙；钢板桩码头岸墙
quay wall on a pile foundation 桩基岸壁
quay wall on arches 拱顶岸墙；拱座码头；拱座岸墙
quay wall on caissons 沉箱码头
quay wall on piles 桩承式岸壁；桩基承台岸壁；桩

承台(式)岸壁
quay wall on pile type 桩承式岸壁
quay wall pf sliced blockwork 削波砌块的岸壁；削波砌块的岸墙
quay wall structure 岸壁结构；岸墙结构
quay wall with stone facing 石块面层的岸壁；石块面层的岸墙；镶石护面码头
quay wall wit platform on piles 桩承台(式)岸壁
quay with battered face 内切倾岸壁；岸壁内倾的码头
quay with half tide landing 半潮登岸码头
quay with sloping face 内切倾岸壁；岸壁内倾的码头
quay with vertical face 直立岸壁
queasy stock market 不稳定股票市场
Quebec Port 魁北克港(加拿大)
Quebec Spruce 魁北克云杉
quebrachitol 白雀木醇
quebracho 白雀玫瑰红硬木；竖木；破斧木；铁木；极坚硬的木材
quebracho extract 白雀树皮萃取物
quebrachomine 白雀碱
quebrada 高低不平的地形；(美国西部的)峡谷
Queen Anna's summer house 安娜王后夏宫(建在布拉格的观景楼)
Queen Anne arch 安娜女王拱
Queen Anne style 英国住宅式；英国18世纪安娜女王时代的建筑家具式样
queen bolt 双螺杆；盔形螺栓；双柱桁架螺栓；桁架竖杆；双螺栓杆
queen closer 小接砖；纵半砖【建】；小半砖接头；长向半砖
queen closure 半头砖；纵剖砖【建】
queen duck 遥控船舰射击靶
queen girder 双柱梁
queen mast 双柱桅
queen metal 锡锑铅焊料
queen post 瓜柱【建】；桁架双柱；双柱桅；双柱式；双柱架；双柱桁架立杆；双柱屋架
queen-post and wind filling 双柱架与填充墙
queen-post beam 双柱上撑式梁
queen-post girder 双柱托梁；双柱上撑式大梁
queen post of truss 桁架副柱
queen-post purlin(e) 双柱式檩条
queen-post roof 双柱(式)屋顶；双柱桁架屋顶
queen-post roof truss 双柱式屋顶桁架
queen post timber truss bridge 双柱木桁架桥
queen-post truss 双竖杆桁架；双柱上撑式桁架；双柱托架
queen-post trussed beam bridge 双柱上撑式梁桥
queen-post truss roof 双柱上撑式桁架屋顶；双柱上撑式屋架
queen post wooden 双柱木桁架
queen rod 吊杆柱；双柱桁架竖杆
queen rule 排队规则
queens 大石板
Queen's chamber 王后卧室
queen-size 大号
Queensland walnut 昆士兰黑胡桃壳色硬木(产于澳大利亚)
Queen's megaron 王后寝宫
queen's metal 锡基合金
Queen's suite 王后套房
queen's ware 精华陶器；奶油色陶器；乳酪色陶瓷器皿；上乳白色釉的英国瓷器
queen theory 排队理论
queen truss 双柱式桁架；双柱桁架
queen trussed beam 双柱上撑式梁；双柱桁架撑梁
queen-trussed beam 双柱桁架加强梁；双柱桁构加强梁
queit aircraft 低噪声飞机
queitite 硫硅锌铅石
quence check word 顺序检验字
quench 熄灭；灭弧；平息；骤冷；淬火
quench ag(e)ing 冷淬时效；马氏体时效处理
quench alloy steel 淬硬合金钢
quench annealing 急冷退火
quenchant 淬火介质
quench area 挤气面积；挤流面积
quench bath 淬火液
quench bend test 淬火弯曲试验
quench blanking 猝灭
quench blast furnace slag 淬火高炉渣
quench carbon steel 淬硬碳钢

quench circuit 熄灭电路
quench cooler 急冷器
quench crack(ing) 淬裂
quenched and tempered 调质过的
quenched and tempered high tensile strength plate 调质高强度钢板
quenched and tempered steel 调质钢
quenched bit 淬火钻头
quenched blast furnace slag 水淬炉渣
quenched cullet 淬火法形成的碎玻璃；水淬(法)碎玻璃
quenched in water 水中淬火的
quenched martensite 淬火马氏体
quenched region 猝灭区
quenched sorbite 淬火索氏体
quenched spark converter 猝熄火花变频器
quenched spark gap 猝熄火花隙
quenched spark gap converter 猝熄火花隙变频器
quenched steel 淬硬钢
quenched structure 淬硬组织
quenched troosite 淬火屈氏体
quenched zone 淬硬区
quencher 熄灭器；骤冷器；急冷器；灭弧器；淬火器(具)；淬火冷却器；淬火工；猝熄物；猝灭器；猝灭剂
quench fan 急冷风机；淬火风机
quench frequency 歇振频率
quench gas 骤冷气；急冷气
quench grate 急冷炉篦
quench-hardened case 淬硬层
quench hardening 急冷硬化；淬火硬化；淬硬
quench head 淬火喷头；淬火机机头
quench heating and cooling 急冷急热
quench hot 高温淬火
quench immersion thermometry 淬火油浸测量法
quench index 淬火指数
quenching 因骤冷引起的漆膜失光或起雾；急冷；激冷；淬硬；猝熄；猝灭；骤冷；淬火
quenching of a flama 熄火
quenching ag(e)ing 淬火时效
quenching agent 猝熄剂；冷却剂；猝灭剂
quenching and age-hardening 完全时效硬化
quenching and tempering 调质处理；淬火及回火
quenching annealing 淬火退火
quenching apparatus 淬火器
quenching arrangement 淬火装置
quenching bath 急冷槽；淬火池；淬火槽；淬化浴
quenching boiler 裂解气急急锅炉
quenching centre 猝灭中心
quenching circuit 猝熄电路
quenching clinker cooler 熟料骤冷冷却机
quenching collision 猝灭碰撞
quenching compound 冷却剂；骤冷剂；冷却化合物；淬火剂
quenching concentration agent 猝熄浓剂
quenching condensation 骤冷凝
quenching constant 淬火常数
quenching correction 猝灭校正
quenching crack 骤冷裂纹；淬冷裂纹；淬致裂纹；淬火裂纹
quenching crack susceptibility 淬裂敏感性
quenching cross-section 猝灭截面
quenching cup 淬火盂
quenching defect 淬致缺陷；淬火缺陷
quenching degree 急冷度
quenching delay 淬火延迟
quenching diagram 淬火变化图
quenching diameter 熄火直径
quenching distance 淬熄距离；猝熄距离
quenching distortion 淬火变形
quenching effect 骤冷效应；淬火作用；淬火效应；猝灭效应
quenching fissure 淬火裂纹
quenching-frequency oscillation 猝熄振荡
quenching-frequency oscillator 猝熄振荡器
quenching from forging heat 锻件余热淬火
quenching furnace 淬火炉
quenching gas 猝灭气体
quenching grate 骤冷炉床；淬冷篦子
quenching hardness 淬火硬度
quenching head 喷水头
quenching heat-exchanger 急冷热交换器
quenching intensity 淬火烈度
quenching-in water 水中淬火
quenching machine 淬火机(床)

quenching mechanism 猝灭机理
quenching medium 骤冷剂；淬火媒介；淬火介质；淬火剂
quenching method 淬火法
quenching moment 灭弧时刻；猝熄时刻
quenching nozzle 喷射冷却器
quenching of arc 熄弧
quenching of clinker 熟料急冷；熟料骤冷
quenching of explosive condition 爆炸状态的骤熄
quenching of frit 熔块淬火
quenching oil 急冷油；淬火油
quenching oil bath 淬油槽
quenching-oil-immersion thermometry 淬火油浸测量法
quenching oscillator 猝熄振荡器
quenching particle 猝灭粒子
quenching pot 灭弧箱；灭弧室开关；猝熄室
quenching press 成型淬火机
quenching process 骤冷法
quenching rate 淬冷率
quenching steel 淬火钢
quenching strain 淬火应变；淬火变形
quenching strength 淬火强度
quenching stress 骤冷应力；淬火应力
quenching system 冷却通风系统；通风系统
quenching tank 骤冷槽；淬火槽
quenching temperature 淬火温度
quenching-tempering 淬火回火
quenching test 淬火试验
quenching time 灭弧时间
quenching tower 骤冷塔；冷却气体塔；冷却塔
quenching transformer 灭弧变压器
quenching trough 急冷槽；淬火槽
quenching unit 灭火器；淬火装置；猝熄设备；猝灭装置
quenching water 冷却水；淬火水
quenching water bath 淬火水槽
quench instrumentation panel 淬火控制台
quench in water 水淬
quench-modulated radiofrequency 歇振调制射频
quench number 淬火指数
quench oil 高闪点精炼油
quenchometer 冷却速度试验器
quench oscillator 猝灭振荡器
quench photocurrent 淬火光电流
quench pulse 消隐脉冲
quench pump 骤冷泵；急冷泵
quench ring 淬冷环
quench steel 淬钢
quench stock 骤冷原料
quench-tank extrusion 淬火槽挤压
quench tempering 等温回火
quench-temper process 淬火回火过程
quench time 间歇时间；灭弧时间
quench tower 急冷塔；骤冷塔
quenselite 基性锰铅矿；羟锰铅矿
quenstedtite 紫铁矾
quercetin 栎精
quercetinic acid 栎精
quercitron 栎皮粉
quercitron lake 栎皮粉黄色淀
Quercus 栎属橡木
Quercus robur 英国栎；欧洲栎
queried access 查询访问
quern 手推磨；小手磨；手动碾磨机
quern-stone 磨石
Quer-wave 奎尔波
query 询问；查询
query by example 举例询问
query enhancement 请求扩充
query evaluation algorithm 询问判定算法
query facility 查询工具
query language 询问语言；查询语言
query-random blockwork 不规则阻挡物【港】；乱抛块体工程【港】；抛筑块体工程【港】
query station 查询台
query strategy 查询策略
quest for profit 追逐利润
questionable 有争议的
questionable figures 虚假数字
question answering system 问题回答系统
question blank 查询表；访查表格
question failure 提问失败
question format 提问格式

question formulation 提问方式
question input phase 提问输入阶段
question length 问题应答长度
question logic table 提问逻辑表
questionnaire 意见调查表；征询意见表；征信书；问卷法；调查表；查询表
questionnaire data 调查表资料
questionnaire mail 邮件调查
questionnaire method 质凝法；调查表法
questionnaire post card 调查用明信片
question of common interest 共同关心的问题
question of substance 实质性问题
question profile 提问式
question signal 询问语言
questions of classification 分类问题
question tag 附加疑问
question term 提问术语
question term table 提问术语表
question validity table 提问有效性表
question vector 提问向量
quetsch 轧浆辊
quetta bond 砖石和混凝土组合砌体；配竖筋砖砌体；砖石加筋砂浆平胎砌合；砖石和钢筋混凝土组合砌体(在垂直墙中的)；奎达砌式(巴基斯坦的一个城市建筑形式)
quetzalcoatlite 羟碲铜锌石
queue 队列
queue 1 第一类队列
queue 2 第二类队列
queue analysis 队列分析
queue anticline 背斜辫
queue arrangement 停车站台设备
queue back chain 队列返回链
queue barrier 队列用栅栏
queue control 排队控制；队列控制
queue control block 排队控制块；队列控制块；队列控制分程序
queue control block extension 队列控制块扩充
queued access 排队存取
queued access method 排队存取法
queue data set 队列数据组；队列数据集
queued entry 队列项目
queue descriptor 队列说明符
queue detector 排队检测器
queued for connection 等待连接排队；等待连接
queued for log-on 等待注册；等待登入排队
queue discipline 队列规定
queue distance 排队距离
queue(d) request 排队请求
queue-driven task 队列驱动任务
queued sequential access 排队顺序存取
queued sequential access method 排队顺序存取法
queue element 队列元(素)
queue element control block 队列元素控制块；队列元控制块
queue file 排队文件
queue header 队列首标
queueing time 等待时间
queue length 排队长度；队列长度
queue linkage entry 队列链接入口
queue link word 队列链接字
queue management 队列管理
queue name 队列名
queue overflow 排队溢出；队列溢出
queue priority 队列优先级
queue request 队列请求
queue search 队列搜索；队列检索
queue shelter (排列的)候车棚；排队候车棚
queue size 列的长短；队列大小
queue study 队列研究
queue system with loss 损失排队系统
queue table 队列表
queue traffic 排队通信量
queuing 排队
queuing analysis 排队分析
queuing area 排队区
queuing buffer 排队缓冲器
queuing delay 排队延时
queuing discipline 排列规则；排队原则；排队规则
queuing model 排队模型；排队(理)论
queuing network 排队网络
queuing principle 排队原则
queuing process 排队过程
queuing register 队列寄存器

queuing routine 排队程序
queuing rule 排队规则
queuing system 排队系统
queuing system with waiting 等待排队系统
queuing theory 排队(理)论
queuing theory and waiting time problems 排队论和等待时间课题
quiangulator 等高线仪
quicentenary 第五百周年(的)
quick access data 快速存取数据
quick access memory 快速存取存储器
quick access report 立即可取报告
quick access storage 快速存取存储器
quick acting 快动的;速动(的);快作用的
quick acting accelerator 快速促进剂
quick-acting automatic switch 高速自动开关;高速自动断路器
quick acting binder 快干黏结剂
quick acting brake 快动闸
quick-acting brake valve 快速制动阀
quick-acting braking 快速制动
quick acting charging 快速充电
quick-acting chuck 瞬时夹紧卡盘
quick acting clamp 快动夹具
quick-acting clamp 快速夹具;快速夹钳;快作用夹具;快作用夹钳
quick-acting coupling 快作用联轴器;快速联轴器
quick-acting equipment 快速行动装备
quick-acting explosive 快作用炸药
quick-acting folding shutter door 快速折叠百叶门
quick-acting fuse 速效保险丝
quick acting mechanical briquetting press 快动机械压坯机
quick acting recorder 快速记录器
quick acting regulator 速动调节器
quick-acting relay 速动继电器;快继电器
quick-acting reverse-current circuit-breaker 速动逆电流断路器
quick acting scuttle 快速关闭式舱口盖
quick acting shutoff 快速关闭
quick-acting spring switch 速动弹簧开关
quick-acting starter 速动起动器
quick-acting switching 速动合闸
quick acting switching-off 快速切断
quick-acting switching-over 速动切接
quick-acting switch machine 快速转辙器;快速转辙机
quick-acting voltage regulator 速动电压调整器
quick action 快速作用;速动
quick-action ball valve 快动作球阀
quick action clamp 快作用夹器;快作用夹紧装置;速动夹具
quick action cylinder cap 快动汽缸帽
quick-action distributor valve 快速分配阀
quick action fuse 速动保险丝
quick action relay 速度继电器
quick action slide tool 快动滑动刀架;快动刀架
quick action switch 速动开关
quick-action triple valve 快速三通阀
quick-action valve 快动阀;速动阀
quick-action valve spring 速动阀簧
quick-action water heater 快速热水器
quick active vise 快动虎钳
quick actuating closure 快开(封头)盖
quick additive 快凝剂
quick adjusting 快速调整的;快速调整;迅速调整
quick-adjusting micrometer calliper 速调测微仪
quick-adjusting nut 自动销紧弹簧螺母
quick ag(e)ing 快速老化;快速陈腐
quick air drying paint 常温快干漆
quick-and-direct prediction 快速并直接预测
quick-and-dirty evaluation 近似估计
quick anneal oven 快速退火炉
quick ash 烟道尘
quick asphaltic emulsion 速凝乳化沥青
quick-assembly method 快速组装法
quick assets 流动资产;可兑现的资产;能迅速变现的资产;往来资产;速动资产
quick attachable coupling 快速联轴节;快速联轴器
quick balance 杆秤
quick bead 齐平圆线脚
quick bending test 高变形速度抗弯试验;快速弯曲试验;快速抗弯试验
quick biofilter 快速生物过滤器

quick bioreactor 快速生物反应器
quick bitumen emulsion 速凝乳化沥青
quick-blow 急速冲击
quick blow drilling 高频冲击钻进
quick brake application 快制动;快上闸
quick break 快速断路;速断;迅速熔断
quick break cut-out 快速熔丝保险器;快速断路器
quick breaker 高速断路器
quick-break feeder fuse 速断馈电保险丝
quick-break fuse 速断熔断器;速断保险丝
quick-breaking 快速断裂
quick-breaking emulsion 快裂乳液(英国);快解乳浊液;速裂乳胶体
quick-breaking foams 速断泡沫
quick-break knifeswitch 速断闸刀开关
quick-break switch 速断开关;高速断路器;急断开关;速断开关;速动开关
quick burning 速燃
quick-burning composition 速燃成分
quick burning fuse 速燃引信;速燃导火线;速燃导火索
quick burst pressure 快速崩裂压力
quick burst strength 快速崩裂强度
quick burst test 快速崩裂试验
quick capital 流动资金;流动资本
quick cell 快速单元;快速保留单元
quick cell facility 快速单元设施;快速保留设备
quick cement 快凝水泥;速凝水泥
quick census 快速普查
quick change 速换;可快速调换
quick change adapter 快速调换器
quick change attachment 快速更换附加装置;快速变换装置
quick change chuck 速换夹头
quick change drilling chuck in process 不停车快换钻攻夹具
quick-change engine unit 速换发动机部件
quick-change gear 快速变换装置;速度箱;走刀箱
quick-change gearbox 快速变换齿轮箱;速变齿轮箱
quick change gearing 快速变换装置
quick change gear lathe 快换轮车床
quick change pin 快换销轴;快换销钉;快换销子;速换销
quick-change room 剧院后台快速换装室
quick change share 快换式犁铧
quick change spray boom (喷雾机的)快换式喷杆
quick change tool holder 快换刀架
quick change tool rest 快换刀架
quick changing 快速更换的;快速变换的;快换的;速变
quick charge 快速充电
quick charger 快速充电机
quick chilling 快速冷却的;急速冷却的
quick circulating percentage 速动比率
quick clay 过敏(性)黏土;高灵敏度黏土;流(性)黏土;流动性黏土;超灵敏黏土;不稳黏土
quick closedown 快速停闭;快速停送;快速关闭
quick-closing door 速闭门
quick-closing emergency valve 应急速闭阀
quick-closing gate 快速关闭闸门
quick-closing gear 急闭装置;舱壁门控制装置
quick-closing lock 快速关闭闸门;快速关闭装置
quick-closing stop valve 速闭截止阀
quick-closing valve 快速关闭阀;急闭阀;速闭阀
quick composting 快速堆肥
quick condition 流动状态;(土的)悬浮状态;流沙现象;流沙条件;快速条件;浮动状态
quick condition boil 地基管涌现象
quick condition failure 快速条件破坏
quick connect 快速接头
quick connect coupling 快式接口
quick connection 快速接头
quick connector 快速联接器
quick consolidated compression test 快速固结压缩试验
quick consolidated shear test 快固结剪力试验
quick consolidated test 快速固结试验;快固结试验
quick cooking 快速蒸煮
quick cooling 快速冷却
quick cooling zone 快速冷却带
quick coupler 快速松脱离合器;快速松脱挂钩;快接头
quick coupler pipe 快速联接管
quick coupling 快速联接器;快速管接头;速接联轴节

quick cure 快速硫化法
quick curing 快凝的(液体沥青)
quick-curing bitumen 速凝沥青
quick curve 急转弯
quick cutting 高速切削
quick cutting steel 高速钢
quick dependable communication 快速可靠通信
quick despatch 快装快卸;快递件
quick-detachable 易拆卸的;可迅速拆卸
quick-detachable rim 速卸式轮辋
quick-detachable semitrailer 速脱钩式半拖车
quick-detechable share 快换式犁铧
quick die change system 快速换模方式
quick-disconnecting arrangement 速闻接合
quick-disconnecting coupling 速闻接合;快解接头
quick-disconnecting fitting 快卸接头
quick disconnecting nozzle 快速分离喷嘴
quick dispatch 快装快卸;快运;快速装卸;快递件
quick dispatch freight 快运货物;快货运
quick dissolving 速溶的
quick dive 速潜
quick-diving tank 速潜水舱
quick draw press 快速拉伸压力机
quick drop catch 快速下降档
quick drop valve 速降阀
quick dry facing wash 快干涂料
quick drying 快干性的
quick drying adhesive 快干胶黏剂
quick drying black lacquer 快燥黑凡立水
quick drying boiled oil 快干熟油
quick drying enamel 瓷釉;磁釉
quick drying liquid product 快干液剂
quick drying paint 快干油漆;快干涂料;快干漆
quick drying resin 快干树脂
quick drying varnish 快干清漆;快干凡立水
quick effect 反向回声现象;反射回声效应
quick effective 速效的
quicken 锐折曲线
quickening 加速
quickening admixture 促凝外加剂
quickening agent 促凝外加剂
quickening liquid 催镀液
quick erection method 快速架设法;快速安装法
quick estimate 快速估算法
quickest route 最快路线
quickest route computer 最短路线计算机
quick exhaust valve 快速排气阀
quick feed 快速进给
quick fence 树篱
quick fire 快烧
quick fire camera 快拍照相机
quick firing 快速煅烧
quickfit joint 快速装配接合
quick fixing 快速安装
quick flashing light 快速闪光;快闪光;急闪光;闪光灯
quick flashing light with blinks 联断急闪光;长间歇急闪光
quick flow 快速径流;瞬发径流
quick freeze 快冻;速冻
quick-freezen fresh fruit 速冻水果
quick freezer 速冻器;急冻冷库;速冻冷库
quick freezing 迅速冻结;快速冻结;急冻;速冻
quick-freezing plant 速冻装置
quick-freezing room 速冻间
quick-frozen solution 速冻溶液
quick fuze 瞬发引信
quick gearbox 速度变速箱
quick goods 家畜
quick gouge 小圆凿
quick-gowing tree 快速生长树木
quick gravel 流沙(也称流砂)
quick gripping 快速卡头
quick grip take in 快速压紧装置
quick ground 流砂土;流沙土;流沙地层
quick-growing plantation 速生人工林
quick-growing species of trees 速生树种
quick hardening 快硬;快速硬化;快速硬固;速硬
quick-hardening cement 快硬水泥;早强水泥
quick-hardening gypsum plaster 速凝石膏粉刷
quick-hardening lime 快硬石灰
quick hardening of concrete (混凝土)快硬剂
quick hedge 插树篱;树篱;活篱

quick hitch arm 快速联接拉杆
quickie strike 非法罢工
quick index(ing) 快速分度
quicking 快镀
quicking bog 活动沼泽地
quick initialize 立即预置
quick-jellying property 速凝性
quick knock down 快速拆卸
quick level(l)ing head 快速调平装置;快速调平头;快速调平水准仪;快速调平球座;快速安平装置;速调水平装置
quick liability 速动负债
quick lift attachment 快速提升装置
quick-lift cam 急升凸轮;速升凸轮
quick light 快闪光
quicklime 氧化钙;生石灰;轻烧氧化镁
quicklime bunker 生石灰仓库
quicklime grease 钙基润滑脂
quicklime manufacturing plant 生石灰厂
quicklime mill 生石灰研磨机;生石灰磨
quicklime pile 生石灰柱;生石灰桩;石灰桩;石灰桩
quick line 加压水带线
quick liquid limit test 快速液限试验
quick load change capability 快速变负荷能力
quick loading 快速卷卷;快速加载
quick loading system 快速加载系统
quick loan 快速收回的贷款
quick-lock bayonet type head 错齿式机头
quick-look method 快速解析法
quick-look recording 速视记录
quick-look system 速视系统
quick lowering mechanism 快速下降机构
quick lunch counter 快餐部
quick lunch stand 快餐食堂
quickly drying colo(u)r 立干漆;快干油漆
quickly drying paint 立干漆
quickly frozen goods 速冻货物
quickly heating and cooling test 急热急冷试验
quickly taking cement 快凝水泥;早凝水泥
quick make 快速闭合;瞬时开关
quick-make switch 快速闭合开关;速通开关
quick malleable iron 快速可锻铁
quick marking method 快速划线法
quick match 速燃引信条;速燃导火线;速燃导火索
quick measurement 快速量测
quick money 速动资金
quick-motion apparatus 降速拍摄装置
quick-motion movement 粗装定机构
quick-motion screw 速动螺旋
quick-mounting 快装
quick movement 快速运动
quickness 火药燃烧率
quickness mineral prospecting 加速矿产普查
quick open flow characteristic 快开流量特性
quick-opening device 速开装置
quick-opening escape hatch 速开逃脱舱口
quick-opening gate valve 速启闸阀;速启闸门
quick-opening level 速启杠杆
quick opening manhole 快开人孔
quick opening rising stem gate valve 快开杆式闸阀
quick opening sprinkler system device 快开喷水灭火系统装置
quick-opening valve 快开阀;速开阀;速开阀
quick operating valve 快动阀;快动闭阀
quick operation on station 车站快速作业
quick order 快速订单
quick payoff 快速盈利
quick pickup relay 快吸继电器
quick-pitch thread 粗扣螺纹
quick planting works 速成绿化
quick priming siphon spillway 快速起动的虹吸式溢洪道
quick printer 快速打印机
quick profit 暴利
quick ratio 速动比率
quick reaction capability 快速反应能力;敏感度
quick reaction system 快速反应系统
quick recovery fall 快收辘轳
quick reduction series 快速压缩孔型系统
quick release 快速断路;速释
quick release catch 快速释放动片
quick-release clamp 快松夹器;快松夹紧装置
quick-release cock 快启龙头

quick release coupling 快速偶合;快释偶合;快卸接头;快脱联轴节;快速松脱离合器
quick release device 速开装置
quick release fitting 速断装置
quick-release gear 瞬时分离机构
quick-release grip 快速式夹具;快卸夹具;速释卡栓
quick-release handle 速开手柄
quick-release joint 速卸接头
quick release latch 快速释放销
quick release mechanism 快卸机构
quick-release mooring hook 快速解缆钩
quick-release parachute 速开式降落伞
quick-release pin 速开锁销
quick-release pitman 快速松放连杆
quick-release stopper 易开塞子
quick-release valve 快泄阀;快速泄气阀;速动阀
quick release valve spring seat 快泄阀簧座
quick releasing hook 快速脱缆沟
quick-releasing relay 速释继电器
quick repair 快修
quick repair washer 快速修理垫圈;迅速修垫圈
quick replaceable 快速变换的;快速更换的
quick report of goods transport accident 货运事故速报
quick response 快速响应;快速反应
quick-response control system 高灵敏度控制系统
quick-response excitation 快速励磁
quick response voltage control 快速电压控制
quick result 速效
quick return 快速回行;快速返回;急回装置;急回运动;急回机构
quick-return belt 急回皮带
quick-return flow 快速回归水流;快速回归径流
quick-return mechanism 急回机构
quick-return motion 急回运动
quick-return stroke 急回冲程
quick revolution engine 快转发动机
quick runner 高速叶轮
quick sand 流沙(也称流砂);流土;流沙现象;漂沙;浮沙
quicksand area 流沙区
quick saturation converter 速饱和变压器
quick seal coupling 速封耦合;速密耦合
quickserved noodle 方便面
quick service 快修
quick set 插条;速凝
quickset hedge 插树篱(笆)
quickset level 速调水准仪
quickset reamer 快调铰刀
quick-setting additive 快凝剂;快凝添加剂;快凝掺和剂;速凝添加剂;速凝剂
quick-setting admix(ture) 速凝掺和料;速凝剂
quick-setting agent 快凝剂;速凝剂
quick-setting grout 速凝灌浆
quick-setting mansic 速凝胶泥
quick-setting material 速凝料
quick-setting mixture 速凝混合料
quick-setting particles 速沉质点
quick settlement assembly 快速稳定装置
quick settling 快速沉淀的
quick-settting and rapid hardening fluoaluminate cement 快凝快硬氟铝酸盐水泥
quick-setttlng and rapid hardening Portland cement 快凝快硬硅酸盐水泥
quick-settting asphalt 快凝沥青
quick-setttting binder 快凝胶接剂;快凝胶结料
quick-setting cement 快凝水泥;快凝角化材料;快干水泥;速凝水泥
quick-setttting concrete 快凝混凝土;速凝混凝土
quick-setttting emulsion 快裂乳液(美国)
quick-setttting glass 快凝玻璃;短性玻璃
quick-setttting grouting 及时注浆
quick-settting ink 快干油墨;快凝油墨
quick-setttting instrument 快速整定仪表;快速安放仪器;快速安放仪表
quick-setttting level 快调水准仪;速调水准仪;速调水平仪
quick-setttting mortar 快凝砂浆;快凝灰浆;速凝灰浆
quick-setttting Portland cement 快凝硅酸盐水泥;快凝波特兰水泥
quick-setttting quick hardening cement 快凝快硬水泥
quick shear (土的)快剪
quick shear strength 快剪强度

quick shear test 快速压缩试验;(土的)快剪(试验)
quick shift hoist bucket 快速提升机;快速移动提升机
quick silver 水银
quick silver lamp 水银灯
quick slaking lime 快消石灰;快熟石灰;快化石灰
quick soft clay 高灵敏度软黏土
quick soil 土流;浮土
quick soil classification 土的简易分类法
quick solder 易熔焊料
quick solution 快速解
quick sort 快速分类
quick-speed 高速
quick-speed steel 高速钢
quick spindle adjustment (钻床的)快速主轴调整
quick spiral 螺旋角大的螺线;陡峭的螺旋;快速盘旋
quick start 快速启动
quick start cathode 速动阴极
quick start fluorescent lamp 快速启动日光灯
quick start lamp 快速起动灯;快速启动灯;瞬时起动灯
quick steaming 快速蒸发
quick steaming unit 快速蒸发装置
quickstick 快粘
quick stick test 快粘试验
quickstone 流积岩
quick stoppage 骤止
quickstryp 快速除漆剂(一种清除油漆剂)
quick survey 快速测量
quick sweep 小半径的环形工程;有急弯的木工制品;建筑圆形作业;小半径作业;急转
quick switch 快速开关
quick switching 快速转换
quick system 快速系统
quick-take 抢购;快速购置建设用地
quick taking cement 快凝水泥
quick taper 快速拔梢
quick test 快速不排水剪切试验;快速压缩试验;快速试验;快速检验;速测法
quick test for soil 迅速测定土壤法
quick testing method of strength 快速强度试验法
quick test of soil 土壤快速试验
quick tests for soils 土壤速测法;土壤快速检验法
quick text mode 快文本方式
quick throw 急掷
quick transport 快速运输
quick transport lever 快速卷片拨杆
quick travel 快速移动
quick triaxial shear test 三轴快剪试验
quick triaxial test 三轴快剪试验
quick triple valve 快动三通阀
quick tunnel(l)ing 快速掘进
quick turnaround system 立即折返制【铁】
quick turn bottom 快翻式犁体
quick union 直速连接;快速接头
quick valve 速闭阀;快闭阀
quick volatile organic chemicals load equalizer 快速挥发性有机化学物负荷缓冲器
quick walk 疾行
quick warmup (发动机的)快速加温
quick water 急流;激流;水银液;水流湍急处
quick-water in water channel 水道中的湍急水流
quick-wear parts 易碎件;易损零件
quid 扩孔器
quiesce-at-end-of-chain indicator 链端静态指示器;串指示器末端为静态
quiesce communication 静态通信
quiesce-completed indicator 静止完成指示器
quiescence condition 静止状态
quiescence settling 静止沉淀
quiescence stage 静止期
quiescence tank 静止池
quiescence water 静止水
quiescency 静止状态
quiescent angle 寂静角
quiescent antenna 沉静天线
quiescent arc 静弧
quiescent area 静止区
quiescent carrier 抑制载波
quiescent-carrier modulation 抑制载波调制;静态载波调制
quiescent-carrier transmission 静态载波传输
quiescent chamber 静止室;无旋涡室

quiescent condition 静止状态
quiescent current 静态电流
quiescent dissipation 静态功耗
quiescent fluid 静止流体
quiescent fluidized bed 平稳流化床
quiescent load 本底荷载;恒载;固定荷载;静应力;静荷载;静负载;底负载;底负荷
quiescent luminosity 宁静态光度
quiescently loaded 静载的
quiescent operating point 静态工作点
quiescent operation 静态运用;静态运行;静态工作;稳态运转;稳态工作
quiescent period 静止期(间)
quiescent point 静态工作点
quiescent prominence 宁静日珥
quiescent push-pull 静态推挽
quiescent push-pull circuit 低板流推挽电路
quiescent sedimentation 静置
quiescent settling 静止沉降;静止沉淀;静土沉降
quiescent spectrum 宁静光谱
quiescent stage 休眠期;静止期
quiescent state 静止状态;静态
quiescent tank 静水沉淀池
quiescent time 停止时间
quiescent value 开路值;静态值
quiescent water 静水
quiescent waters 静水水域;静水区域
quiescent work 静态工作
quiesce protocol 静态协议;静态规程
quiescing 禁止操作
quiet air 无风
quiet alloy 低碳合金;稳定的合金
quiet at approach of earthquake 震前的平静
quiet automatic volume control 延迟式自动增益控制
quiet battery 无噪声电池
quiet circuit 无噪声电路
quiet component 宁静分量
quiet component of radio radiation 无线电辐射宁静部分
quiet condition three 三级静止状态
quiet corona 宁静日冕
quiet day 平静日
quiet-day variation 宁静日变化
quiet engine 低噪声发动机,低噪声发动机
quieter 消音器;内燃机的消音装置
quieter material 消音材料
quieter motor 无噪声马达
quiet fast boat 无噪声快艇
quiet-form aurora 宁静极光
quiet gear 无声换档;无声齿轮
quiet homogeneous arc 平稳均匀弧
quiet hour 静态时间
quieting 静噪;静态特性
quieting curve 无声运转曲线
quieting ramp 消音锥度;消声锥面
quiet internal gear pump 低噪声内啮合齿轮泵
quietive 镇静剂
quiet lines 无噪声线路
quiet location 少地震活动地区
quiet mode 静止模式
quiet motor 无噪声电动机;低噪声电动机;低噪声电动机
quietness 无噪声
quiet period 平静期
quiet piling rig 低噪打桩机
quiet pouring 平静浇铸
quiet power supply 无噪声电源
quiet radio sun 射电宁静太阳
quiet reach 静水河段;静水段(河流);河流静水段
quiet reach of river 稳定河段
quiet-riding 安静行驶
quiet rivet 临时定位铆钉
quiet room 安静的房间
quiet run 均匀运转;平稳运转;平稳操作;无声运转
quiet running 无噪声运转
quiet shade 素净色
quiet shifting 无声变速
quiet solar radio radiation 宁静太阳射电
quiet speed 无声换挡
quiet steel 软钢;全镇静钢;全脱氧钢
quiet sun 宁静太阳
quiet sun condition 宁静太阳条件
quiet sun noise 宁静太阳噪声

quiet take-off and landing 无声起飞与着陆
quiet thermal emission 宁静热发射
quiet third 无声(的第)三档
quiet-time magnetosphere 宁静磁层
quiet-time solar wind 宁静太阳风
quiet title 判决产权归属
quiet tuning 无噪调谐
quietude 平静
quietus 债务的清偿
quiet volcano 静火山
quiet water 静水
quiet water region 静水区
quiet zone 安静地区
quiet zone of oil 静止油区
quill 羽毛状花纹;空心轴弹性传动;空心轴传动;活动套筒;纬管;导火线;衬套
quill bearing 滚针轴承;(钻机立轴的)导管轴承
quill bit 有纵槽的长钻头
quill drive 空心轴弹性传动;空心套管传动;套管传动
quill embroidery 毛饰刺绣
quiller 卷纬机
quill feather 大羽毛
quill gear 套管齿轮;背齿轮
quilling 卷纬;网眼纱褶裥边饰
quill master 套式校对规
quill roller 滚针
quill shaft 传动轴内的传动杆;中空轴;套筒轴;套管轴
quill spindle 套筒主轴
quill-type drilling unit 可移动轴组合钻削头
quilot fiber 马尼拉麻
quilt 缝合
quilted eelgrass 大叶藻填褥
quilted fabric 踏花被;被褥织物
quilted figure 丘陵木纹;板面中的褶皱纹;褶裥花纹(木板的);泡状花纹;被絮纹
quilted insulation blanket 保温被
quilted plastic panel 塑料芯板
quilting 船首防冰垫
quilting rivet 紧缝铆钉;定位铆钉
quilting weld 钉焊
quilt insulation 纸面绝缘材料;被状绝缘材料
quilt radiator 被褥形采暖器;被形采暖器
Quimby pump 双螺杆泵
quinacridone pigment 喹吖啶酮系颜料
quinacridone red 喹吖啶酮红
quinacridone violet 喹吖啶酮紫
quinalbarbitone 司可巴比妥
quinaldine red 喹哪啶红;甲基氮萘红
quinalizarin 醌茜素
quinary 五进制
quinary alloy 五元合金
quinary code 五进码
quinary digit 五进制数
quinary notation 五进制记数法
quinary steel 四元合金钢
quinary system 五元系
quinazoline 喹唑啉;间二氮杂萘
quince 温桲树
quincentenary 五百年
quincuncial 五点形的
quincuncial aestivation 双盖覆瓦状
quincuncial dredging method 梅花式挖泥法
quincuncial piles 梅花状交错桩;梅花桩
quincunx 梅花形;梅花点式
quincunx system of planting 梅花形栽植制度
quincy cutter 对称弯犁刀
Quincy granite 昆西花岗岩(产于美国)
quindenary 十五进制的
quinhydrone 醌氢醌;对苯醌合对苯二酚
quinhydrone electrode 氢醌电极
quinidine 康奎宁
quinine profile 醌图谱
quinisatin 奎靛红
quinite 对环己二醇
quinitol 对环己二醇
quinizarin 醌茜
quinochrome 醌色素
quinocyanine 喹啉蓝
quinol 醌醇
quinoline 喹啉
quinoline blue 喹啉蓝;花青染料;氮萘蓝
quinoline colo(u)ring matter 喹啉染料
quinoline dye 喹啉染料;氮萘染料

quinoline ethiodide 碘化 N-乙基喹啉
quinoline yellow 酸性喹啉黄
quinoline yellow spirit-soluble 醇溶喹啉黄
quinone 醌;苯醌
quinone pigment 醌系颜料
quinonimine dye 醌亚胺染料
quinophthalone 喹啉酮
quinquagenary 五十周年纪念
quinquangular 五角形的;五边形的
quinquefoil 梅花饰;五叶饰
quinquennial 持续五年的
quinquepartite 有五部分;由五部分组成的;分为五部分
quinquevalent 五价的
quint 五件一套
quintal 公担(等于一百千克);百千克;百公升
quintant 五分仪
quintessence 精华;浓萃
quintet(te) 五重奏;五件一套
quintic 五次的
quintic curve 五次曲线
quintic electrode longitudinal axis sounding 五极纵轴测深
quintic equation 五次方程
quintic splines 五次样条函数
quintic surd 五次不尽根
quintile 五分之一对座
quintozene 五氯硝基苯
quint-point arch 五点拱
quintuple 以五乘之;五倍(量);成五倍
quintuple form 五元组形式
quintuple harmonics 五次谐波
quintuple point 五相点
quintupler 五倍器;乘五装置
quintuple riveting 五行铆接;五行铆钉
quintuple space 五维空间
quintuple Turing computable function 五元组图灵可计算函数
quintuple Turing machine 五元组图灵机
quintuplicate 一式五份;五倍的
quintuplication 乘以五
quintupling 成五倍
quire 歌唱队席位;教堂中唱诗班的席位;一刀
quire aisle 唱诗班走廊;教堂侧厢走廊
quire arcade 唱诗班拱廊;教堂拱廊
quire arch 教堂拱圈
quire architecture 唱诗班席建筑;教堂建筑艺术
quire bay 唱诗班部位
quire chapel 唱诗班小教堂
quire gallery 唱诗班楼座
quire loft 唱诗班阁楼;教堂楼座
quire screen 唱诗班屏风;围隔唱诗班的栏杆
quire stall 唱诗班席位
quirk 斜角镶条;狭凹槽;水道;深槽
quirk bead 圆形木条板;深槽凹圆线脚;刻槽串珠线脚;角上刻槽圆线脚;槽隔小圆;凹槽圆线脚
quirk board 刻槽线脚板
quirk cutter (开挖圆形地沟的)专用工具
quirked bead 半圆形凸缘
quirk float 深槽镘刀
quirk-mitered corner 斜接海棠角【建】
quirk mo(u)lding 深凹圆线脚;方突角线脚;小槽线脚;鸟啄形线脚
quirk router 槽刨;线脚槽刨;狭槽刨
quirt 槽;企口
quisqueite 钒镍沥青矿;硫沥青(即高硫钒沥青)
quit 解数
quit and deposed rate of fixed assets 固定资产退废率
quitch grass 茅草;野草
quit claim 放弃要求;放弃权利
quit claim deed 放弃索偿契约;放弃权利的契约;放弃合法权利转让契据
quite bad rock mass 较坏的岩体
quite circular in outline 外形十分圆
quite cold 挺冷
quite entire 完全全缘的
quite good rock mass 较好的岩体
quite tensile testing machine 静拉力试验机
quite useful 很有用的
quit office 退职
quit radiator 棉被形采暖器
quit rent 偿还地租
quit score 结清账目

quit score with 和……结清账目
quittance 免除债务证书
quitting-time 下班时间
quit work 停止工作;停工
quivalance value 等效值
quiver 震颤;箭筒
Qumarleb tectonic knot 曲麻莱构造结
Q-unloaded 无载品质因数
quoin 角石;墙角石;屋角隅石;屋角构件;桶堆楔块;堆装楔块;房屋端部墙角;版楔
quoin bar 闸门墩铸钢垫座
quoin bead 隅限隔条;隅角隔条
quoin bonder 屋角顶砖
quoin bonding 墙角砌合;屋角石块墙砌合;尼角石块墙砌合;隅石块砌合;墙角砌法
quoin bondstone 屋角束石;屋角顶石
quoin brick 屋角砖;隅砖;墙角砖
quoin end 闸扇枢轴端;门扇枢轴端
quoin header 屋角桩;屋角丁头砖;突角露头石
quoining 装屋角石;突角构件;外角构件
quoining up jambs 门窗洞边砌石
quoin key 版楔键
quoin plate 闸门墩铸钢垫座
quoin post 屋隅角柱;房外角柱;突角支柱;屋外角柱;闸隅柱;闸门轴柱;隅柱;门轴柱(边梃);外角柱;枢轴柱
quoin reaction casting 闸门墩铸钢垫座
quoin stone 楔形石;隅石;角石;墙角石;屋角石
quoin veneer split tile 隅石表层面石;楔形表层面砖
quoit 抓圈;金属圈;铁圈;绳圈
quoit terminal 卷尼状扶手末端
quonset hut 半圆形活动房屋;半圆柱体临时房屋;半圆瓦楞铁临时房屋
quorum 法定人数
quorum sensing 定感感应
quota 限额;指标;控制额;配额;摊额;定量;定额;份额;分配额
quota administered 配额管理
quota allocation 配额分记;定额分配
quota allowed to somebody 配给的份额
quota and nonquota immigrants 限额与不限额移民
quota clause 销售配额条款
quota control of energy consumption 能源消耗定额管理
quota current asset days outstanding 定额流动资产周转天数
quota current funds checked days outstanding 定额流动资金考核周转天数
quota design 限额设计

quota for repair labor 修理劳动量定额
quota foundation 定额分配依据
quota game 限额对策
quota management 定额管理
quota management system 定额管理制度
quota management with contract for different part of job 定额管理
quota od budgetary estimate 概算定额
quota of consumption 消耗定额
quota of expenditures 消耗标准
quota of materials in building and installation 建筑安装材料消耗定额
quota of sales contribution margin 业务贡献率责任额
quota periods 配额期限
quota policy 限额政策
quota profit 定额利润
quota quantity 配额数量
quota restriction 比额限制
quota rights 配额权利
quota sample 限额样本;按比例抽样样品
quota sampling 配额抽样;定额抽样
quota scheme 比额计划
quota system 限额制;限额进出口制;配额制;定额分配制
quota system for distributing materials 限额发放物资制度
quota ticket 定额票
quotation 估价单;行情;报价;填空铅
quotation for building a workshop 建造一所车间的估价单
quotation for foreign currency 外汇牌价
quotation mark 引用符号;引号
quotation of prices 报价单
quotation on foreign market 国外市场报价
quotation (sheet) 报价单
quotations list 行情表
quota wage system 定额工资制
quote 要价;开盘;开列;开价;开估价单;报价
quoted government subsidies receivable 应收政府定额补贴
quoted market price 开列的市价
quoted price 牌价;报价
quoted radius of big aperture well 大井引用半径
quoted standard 引用标准
quoted string 引证串;引用串
quoted width 引用宽度
quote favorable terms 报优惠价
quote image 引证映象

quote symbol 引证符号
quote-worthy 有引证价值的
quotient 系数;商数;份额;比值
quotient address operand 商地址操作数
quotient algebra 商代数
quotient bundle 商丛
quotient category 商范畴
quotient chain complex 商链复形
quotient counter 商计数器
quotient extension 商扩张
quotient field 商域
quotient function 商函数
quotient graph 商图
quotient group 商群
quotient lattice 商格
quotient law 商定律
quotient length operand 商长度操作数
quotient Lie algebra 商李代数
quotient Lie group 商李群
quotient linear space 商线性空间
quotient manifold 商流形
quotient mapping 商映射
quotient measure 商测度
quotient meter 流比计
quotient module 商模
quotient-multiplier register 乘数商数寄存器
quotient object 商对象
quotient of amplitudes 振幅比值
quotient of difference 增量比
quotient of increase 增长系数
quotient of real numbers 实数商
quotient proximity 商邻近
quotient register 商数寄存器
quotient relay 商数继电器;比例继电器
quotient representation 商表示
quotient ring 商环
quotient rule 商法则
quotient semigroup 商半群
quotient set 商集(合)
quotient sheaf 商层
quotient space 商空间
quotient string 商串
quotient symbol 商符号
quotient system 商系;份额制度
quotient variety 商簇
quotiety 系数
quoting 引用
Q-value 核反应能量值;质量系数值;品质因素;品质因数;Q值
Q wave Q波

R

R-1 zone 独门独户住宅区
R-2 zone 二～四单元住宅区
R-3 zone 公寓楼住宅区
rab 拌砂浆棒
rabal 探空气球测风
rabat 拉巴特磨料
rabatment 沿迹线旋转而展出的面
rabbet 嵌接；榫头；插孔；槽舌接合；槽口；裁口；半槽企口；凹凸缝；凹部
rabbet bead 槽口凹角中的压条；转角圆线；槽口珠缘
rabbet depth 嵌接深度
rabbeted butt joint 企口对接；槽舌对接
rabbeted corner pile 企口角桩；带槽角桩
rabbeted door 槽口接合门
rabbeted doorjamb 裁口门框；有槽的门樘；截口门框边樘
rabbeted frame 裁口门框
rabbeted jamb 企口窗框；嵌接窗框
rabbeted lock 裁口锁；凹槽锁；门边梃锁；接缝锁；槽口门锁
rabbeted siding 榫接外墙板；披叠墙板；互搭墙板；铲口接合墙板
rabbeted stop 门窗档架脚；门窗止条
rabbeting bit 凹槽铣床
rabbeting machine 刨榫槽机；制凹凸榫接机；开槽机；榫槽刨机；车边机
rabbet joint 铲口缝；嵌接；槽舌接合；半舌接合；企口接合；裁口接头；半槽接合；企口接头
rabbet ledge 榫接横档
rabbet line 嵌接线
rabbet plane 企口刨；开槽刨；凸边刨；槽刨；裁口刨；边刨；半槽边刨；木工板刨
rabbet size 裁口尺寸；嵌玻璃的槽口尺寸
rabbit 气动速送器；单级再循环
rabbit ear faucet 兔耳形水龙头；兔耳式龙头
rabbit's ear 夹丝上的气泡
rabbittite 针钙镁铀矿；水碳钙镁铀矿；水菱镁钙石
rabbit warren 拥挤住宅；养兔场；过度拥挤的共同住宅
rabble 搅拌棍；搅炼棒；搅拌耙；机械搅拌器；长柄耙
rabble arm 耙柄
rabble furnace 搅拌炉；机械多层焙烧炉
rabbler 刮刀；铲子；刮器；搅拌器；搅拌刀；加煤工
rabbling 搅炼；搅拌
rabbling hoe 搅拌铲；搅拌扒
rabbling iron 搅拌棒
rabbling mechanism 搅拌机构
rabbling roaster 搅拌焙烧炉
rabdionite 铜铝锰土
rabicide 四氯苯酞
rabies 恐水病
Rabitz ceiling plaster 拉比兹式顶棚抹灰
Rabitz construction (钢丝网建筑的)拉比兹构造
Rabitz finish 拉比兹式最后加工；拉比兹灰浆饰面
Rabitz type 拉比兹式
Rabitz type board 拉比兹式平板
Rabitz type box 拉比兹式箱匣
Rabitz type casing 拉比兹式抹灰管道外套
Rabitz type ceiling paster 拉比兹式顶蓬抹灰
Rabitz type cornice 拉比兹式挑檐
Rabitz type gypsum 拉比兹式石膏
Rabitz type lining 拉比兹式衬砌；拉比兹式衬里
Rabitz type reed lath(ing) 拉比兹式芦苇板条
Rabitz type steel-wire plaster fabric 拉比兹式钢丝灰泥纤维
Rabitz type vault 拉比兹式圆顶
Rabitz type work 拉比兹式操作
Rabitz vault 拉比兹式拱顶
Rabitz wall 拉比兹墙
Racah's coefficient 拉卡系数
Racah's notation 拉卡符号
Racah's symbol 拉卡符号
Racal expert system 雷开尔专家系统
race 轴承环；种族；海峡水道；全速运转；品种；碳酸钙小结核；梭道；水力机械的水道；石灰石屑；潮路；座圈
race an engine 使发动机空转(确定发动机的工作规范)

race condition 竞争状态；竞态条件；紊乱情况
racecourse 跑马场；赛跑道；赛马场
racecourse bend 导向弯管
racecourse grand stand 跑马场大看台
racecourse stand 跑马场看台
race grinder 轴承滚道磨床
raceline (水车的)水渠；导水路
race lip 座圈缘
racemation 消旋作用
racemic mixture 消旋混合物；外消旋混合物
racemic solid solution 外消旋固体溶液
racemization 消旋作用
racemose 蔓状的
race of screw 尾流；推进器流
race pulverizer 中速球磨机
race rotation 逸转；空转
race seal 座圈密封装置
race track 跑道形电磁分离器；跑道形放电管；水道；电缆管
racetrack microtron 跑道型微波加速器
racetrack orbit 跑道型轨道
racetrack stability 跑道稳定性
racetrack synchrotron 跑道型同步加速器
racetrack torus 跑道型加速器的环形室
racetrack type winder 轨道式绳绕机
raceway 座圈；导水道；线槽；输水管(道)；电缆管(道)
raceway floor 线槽楼面
raceway for wiring 电槽路；电线管路
raceway grinder 轴承滚道磨床
raceway grinding machine 轴承滚道磨床
raceway track 轴承座圈滚道
raceway wiring 槽板布线
racewinite 杂变柱岩
rach car 装ово双层汽车的货车
rachero 牧场主；牧场工人
rachion 湖岸线
racing 急转
racing body 流线型车体；流线型车身；赛车车身
racing car 赛快车
racing-car engine 赛车发动机
racing-car tire 赛车轮胎
racing course 赛车路线
racing driver 赛车运动员
racing engine 赛车发动机
racing fuel 赛车燃料
racing of engine 发动机空转
racing of the screw 打空车
racing skiff 双人双桨赛艇
racing test 空转试验；空转调试
racing-type foam-filled rubber fuel cell 赛车型泡沫橡皮燃油盒
racing yacht 赛艇
rack 小集材道；支架盘；挂物架；挂具；挡泥物栅；结绳架；交叉扎架；架(子)；滑轨；强摇；台架；堆架；粗帘格；齿棒；齿条
rack and gear 齿条齿轮
rack and gear drive 齿条齿轮传动
rack and gear jack 棘轮千斤顶；棘轮起重器；齿轮(式)千斤顶；齿条齿轮起重器
rack and gear swing mechanism 齿条齿轮回转机构
rack-and-lever jack 棘轮千斤顶；棘轮起重器
rack and panel connector 抽屉机架接插件；转接插头座；矩形插头座
rack and panel construction 架式结构
rack and pinion 齿条齿轮(传动)；齿条和小齿轮；齿轮齿条副
rack-and-pinion drive 齿条小齿轮传动；齿条齿轮传动
rack and pinion elevator 齿条齿轮电梯
rack and pinion gate lifting device 齿条齿轮式闸门启闭机；齿条齿轮式启门机
rack and pinion gearing 齿条齿轮传动装置
rack and pinion hoist 齿条齿轮吊轮；齿条齿轮吊车
rack and pinion jack 齿条齿轮千斤顶；齿条齿轮起重器；齿杆式千斤顶
rack and pinion loom 齿条齿轮式织带机

rack-and-pinion luffing 齿条变幅
rack and pinion meter 齿条齿轮传动仪
rack and pinion railway 齿轨和齿轨铁路
rack-and-pinion steering-gear 齿条齿轮式转向器
rack and pinion transfer 齿条齿轮传送
rack and snail 齿条和蜗形轮
rack-and-tank processing 挂架浴槽显影
rack a tackle 绞住绞绳
rack back 用齿条退回；齿条齿轮退离装置
rack bar 绞索棒；齿条；齿杆
rack bar operating machine 齿条式启闭机
rack bar operating machinery 齿条式启闭机
rack (bar) screen 格筛；格栅
rack bar sluice valve 齿条齿轮装置的阀
rack block 多饼滑车
rack building hoist 齿条提升机；齿条升降机；齿杆升降机
rack car 框架平台车；架式干燥车
rack carriage 齿条导轨
rack chamber 格栅室
rack circle 扇形齿轮；齿圈；齿弧
rack cleaning machine 拦污栅清理机
rack clock 齿轨钟
rack clogging 拦污栅阻塞
rack compasses 齿轮圆规；齿轮罗盘
rack container 框架货柜
rack course 斜砌砖层
rack cutter 齿条铣刀；齿条刀(具)
rack cutting 排刀切削；齿条切削
rack cutting machine 齿条切削机
rack division 齿轨段
rack driven planer 齿条式龙门刨床
rack driving 齿条传动
rack driving planer 齿条传动刨床；齿条传动车床
rack drum 牵引卷筒
rack dryer 干燥架
rack dryer car 台架式烘干小车
rack earth 机架接地
racked 搭临时木架的；临时架的
racked drill pipe 装在支架上的钻管
racked joint 留齿缝；深凹接缝
racked stitch 波纹组织
racked timbering 加撑支架；木支撑架；加撑木架；斜木撑；斜角支撑；临时木架
racked wale 波纹纵行
racket 非法买卖
racket court 网球场
rack feed gear 齿条进给装置
rack for curing concrete (混凝土块的)养护格栅
rack for curing concrete block 混凝土块的养护架；混凝土块的养护格栅
rack for rods 钻杆架
rack gear(ing) 齿条齿轮传动装置；齿条传动
rack guide 齿条导轨
rack hoist 卷扬机；牵引绞车
rack hook 齿条钩
rack housing 齿条罩
racking 斜茬；阶梯形砌墙；(墙的)阶梯形砌合；阶梯形边接；退步茬；推压载荷；推压动作；船体横断面结构变形；齿条传动；阶梯形砌接
racking back 预留阶梯形齿接；预留阶梯形接头；留齿缝；墙壁接齿缝；斜茬；留齿缝待接；砌墙时齿缝待接；退步茬；阶梯形砌接
racking bond 齿咬接；咬齿砌合；斜茬；接茬
racking capacity of derrick 钻塔储(贮)放钻杆容量
racking course 斜砌砖层；找平垫层；铺砌层
racking for milk churns and buckets 奶桶吊挂架
racking load 振动荷载；挤压荷载；船体横歪引起的负载；变形荷载
racking machine 分装机
racking moment 推压弯曲
racking of drill pipe 摆放钻杆(钻塔内)
racking of rubber 橡胶整理
racking out 勾缝准备；准备勾缝
racking platform (钻塔的)工作台
rackings 格栅渣
racking seizing 交叉绑扎
racking stopper 绞辘制动索

racking strength 剥离强度
racking stress 挤压应力;船体横断面结构扭变应力
racking turns 交叉扎绳的八字形绕数
racking wire 系紧钢丝(模板)
rack insulator 架式绝缘子
rack intake 河流入口
rack jack 齿杆千斤顶;齿条千斤顶
rack jobber 超级市场批发商
rack level(l)ing staff 支架式水准尺;带支架水准尺
rack lever 齿条操纵杆
rack lever escapement 齿条杠杆擒纵机构
rack lift 爬梯式自动起落机构;齿条式自动起落机构
rack limiter 齿条限位器;齿条限制器
rack line 牵引索;牵引绳
rack loader and unloader (混凝土块的)齿条装垛卸载器
rack locomotive 齿轨机车
rack loss 装笼损失;耙平损失
rack marks 折痕;链痕
rack mechanism 齿条机构
rack miller 齿条铣床
rack milling cutter 齿条铣刀
rack milling machine 齿条铣床
rack mixer 耙式混合器
rack-mounted 安装在架上的;安装在机架上的
rack-mounted power unit 电源架
rack mount type 机架安装型
rack off 佣金
rack of fusion 焊接不良
rack of track circuit 机架(轨道电路用)
rack-operated jack 齿条小齿轮起重器
rack panel 机架仪表屏
rack pinion 齿条小齿轮;齿轮齿条
rack-pinion jack 齿条齿轮千斤顶;齿条齿轮起重器
rack pipe 摆管架
rack post 支杆
rack press 齿轮齿条传动压力机
rack presser 木棍压榨器
rack pusher for transfer car 齿条传送顶车机
rack rail 齿轨
rack rail locomotive 齿轮机车
rack railroad 齿条铁路;齿轮铁路;齿轨铁路;齿轨铁道
rack rake 栅耙;格栅耙;拦污栅耙
rack rent (大于年产值的三分之二的)租金;苛酷租金;高额租金;苛刻的租金
rack resistance 装笼阻力;耙平阻力
rack road 齿轨铁道
rack rod 绞索棒
rack saw 粗齿锯;阔齿锯;宽齿锯;齿条锯
rackscope 支架式示波器
rack seat 拦污栅座
rack section 机架部分
rack setting ga(u)ge 齿条安装定位机
rack shaft 齿条轴
rack shaper 梳齿机;齿条刀插齿机
rack shaping machine 梳齿机;齿条插齿机
rack shaving 齿条刀剃齿
rack shaving tool 齿条形剃齿刀
rack sleeve 齿条套筒
rack soot blower 齿轨式吹灰器
rack stacker 退火炉送料装置;退火窑送料装置
rack stake 栅栏杆;栏栅杆
rack supported S/R machine 货架支承型有轨堆垛起重机
rack switch 齿轮转辙机;齿轨转辙器
rack system 箱架堆装方式;托架系统
rack tail 抬闸止杆
rack tooth 直边齿
rack track effect 轨道效应
rack-type car 棚加车
rack type cutter 梳齿刀;齿条铣刀
rack-type differential 齿条(式)差动装置
rack-type-drag 耙式路刮
rack-type drawbench 齿条式拉拔机
rack-type gear cutter 齿条式齿轮刀具
rack-type gear shaving cutter 齿条式剃齿刀
rack-type jack 齿条式起重器
rack-type pusher 齿条式推钢机
rack-type shaving 齿条式剃刀剃齿法
rack-type wagon 栅栏式拖车;高栏板拖车
rack warehouse 以台架存放物品的仓库
rack well 栅格人孔;栅格竖井
rack wheel (电路的)开关设备转轮;(电路的)转换转轮;齿条传动齿轮;齿轮
rack winch 牵引卷扬机
rack work 齿条(加工);调位装置;齿条机构
racon 雷康;雷达信标;雷达控制
rad 拉德(辐射吸收剂量单位)
radameter 预防碰撞的雷达仪
radan 雷达导航
rad and dab 篱笆抹泥墙
radappertization 辐射消毒
radar 雷达
radar absorption characteristic 吸收电磁波的特性
radar advisory 雷达报告
radar aerial 雷达天线
radar aerosurveying 雷达航空测量
radar aids 雷达助航(系统)
radar aids to navigation 雷达助航(系统)
radar altimeter 雷达高度计;雷达高度表;雷达测高仪
radar-altimeter elevation 雷达测高仪
radar altimeter profile 雷达测高仪剖面图
radar altimeter surveying 雷达高程测量
radar altimetry 雷达测高法
radar altimetry area 雷达测高区
radar altitude 雷达高度
radar and television aid to navigation 港口短程助航仪器;雷达和电视导航设备
radar and television aid to navigation system 雷达及电视导航系统
radar and television navigation 雷达电视导航
radar antenna 雷达天线
radar approach control 雷达临场指挥;雷达进场控制
radar astronomy 雷达天文学
radar automated data transmission assembly 雷达自动数据传输装置
radar azimuth resolution 雷达方位分辨率;雷达方位分辨力
radar band 雷达波段
radar barrier ship 雷达哨舰
radar beacon 雷康;雷达指向标;雷达信标;雷达发射装置;雷达导航信号反射台;雷达导航标
radar beam 雷达射束;雷达波束
radar beam marking 雷达射束图;雷达辐射图
radar beam pattern 雷达波束图
radar bearing 雷达方位
radar board 雷达测绘板
radar Braun tube 雷达显示管;雷达布朗管
radar bright display 雷达明亮显示器
radar buoy 雷达浮标
radar cell 雷达波包
radar chain 雷达链;雷达防线
radar chart 雷达图表;雷达海图;雷达导航图
radar chart projector 雷达海图投影器
radar climatology 雷达气候学
radar clutter 雷达杂乱回波;地物回波
radar collision 雷达防撞装置
radar collision warning device 雷达碰撞警报装置
radar complex 全套雷达设备
radar confusion reflector 扰乱反射体
radar console 雷达台座
radar conspicuous object 显著雷达目标;雷达波反射力强的物标
radar constant 雷达常数
radar contact 雷达捕捉
radar control 雷达控制
radar controller 雷达控制器;雷达操纵员
radar control plate 雷达控制板
radar converter 雷达信息变换器
radar countermeasures 雷达干扰措施
radar coverage 雷达探测区;雷达覆盖区
radar coverage-diagram 雷达威力图
radar coverage indicator 雷达有效区域显示器
radar cross-section 雷达截面;雷达目标有效截面;雷达横截面;目标等效反射面;回波面积
radar data computer 雷达数据计算机
radar data processing 雷达数据处理
radar data processing system 雷达数据处理系统
radar detection geo-radar exploration 雷达探测
radar detector 雷达探测器;雷达侦察器
radar diagram method 雷达图法
radar-directed 雷达操纵的
radar direction-finder 雷达测向仪
radar display 雷达荧光屏显示器;雷达显示(器)
radar display console 雷达显示器支座
radar display system 雷达显示系统
radar display tube 雷达显示管
radar display unit 雷达显示器
radar distance warning system 雷达距离警报系统
radar distribution switchboard 雷达配电盘
radar dome 雷达罩;雷达天线整流罩
radar duplexer 雷达转接头
radar early-warning 雷达预警
radar early-warning equipment 雷达预警设备
radar echo 雷达回波
radar environment 有空中交通管制雷达的环境
radar equation 雷达方程
radar-equipped inertial navigation system 装有雷达的惯性导航系统;雷达惯性导航系统
radar fence 雷达网
radar fire 雷达火力
radar fix 雷达船位
radar fixing 雷达定位
radar fixing accuracy 雷达定位精度
radar for collision avoidance 避撞雷达
radar for ground traffic 地面交通管制雷达
radar frequency band 雷达频带;雷达波段
radar geophysical survey 雷达物探法
radargrammetry 雷达图像测量(术);雷达摄影学
radar guidance 雷达制导;雷达导航
radar helm 装有雷达定向仪的头盔
radar-homer 雷达自动瞄准头
radar homing 雷达自动寻的
radar hood 雷达荧光屏遮草
radar horizon 雷达作用距离;雷达地平(线)
radar illumination strip 雷达照射条带
radar illumination swath 雷达照射条带
radar imagery 雷达图像
radar imaging sensor 雷达成像传感器
radar indicated face 雷达荧光屏;雷达显示表面
radar indicator 雷达指示器;雷达显示器
radar information 雷达信息
radar information network 雷达情报网
radar infrastructure 永久性雷达基地
radar inspection 雷达检测
radar installations 雷达装置
radar interface equipment 雷达接口设备
radar interference 雷达干扰
radar interferometer 雷达干涉仪
radar interswitch 雷达互换装置
radar island 海洋雷达岛;海洋雷达站;海上雷达站
radar jammer 雷达干扰台
radar jamming 雷达干扰
radar jamming device 雷达干扰装置
radarkymography 雷达计波摄影
radar layover 雷达图像摺叠
radar line of position 雷达位置线
radar link 雷达中继站
radar lunar landing device 月球雷达着陆装置
radar man 雷达员;雷达操纵员
radar map 雷达图;雷达地形显示图;雷达地图
radar mapping satellite 雷达测图卫星
radar marker 主动雷达航标;雷达指向标;雷达航标;雷达导航信号反射台;雷达航标信标
radar marker beacon 标志雷达信标
radar mark 雷达信标
radar mast 雷达天线杆
radar measurement 雷达量测;雷达测量
radar measurement accuracy 雷达量测精度;雷达测量精度
radar meteorologic(al) observation 雷达气象观测
radar meteorology 雷达气象学
radar meter 雷达测速计;雷达表
radar microwave link 雷达微波中继装置
radar mile 雷达英里
radar monitoring 雷达监控
radar mosaic 雷达图像镶嵌图
radar nautical mile 雷达海里
radar navigation 雷达导航
radar navigation aid 雷达导航设备
radar navigation system 雷达导航系统
radar navigator 雷达导航仪;雷达导航系统
radar netting 雷达联络
radar netting station 雷达联络中心站
radar observation 雷达观测
radar observation of precipitation 雷达测雨
radar observation system 雷达监测系统
radar operator 雷达手
radar outpost 雷达站

radar patrol submarine 雷达哨潜艇
radar pencil beam 雷达锐方向性射束
radar performance 雷达性能
radar performance figure 雷达性能指标;雷达效率
radar photogrammetry 雷达摄影测量
radar photograph 雷达像片;雷达传真
radar photography 雷达摄影学;雷达摄影术;雷达摄影
radar picket escort ship 雷达哨护卫舰
radar picket ship 雷达哨舰;雷达警戒舰
radar picket submarine 雷达哨潜艇;雷达警戒潜艇
radar platform 雷达站;雷达海上平台
radar plot 雷达目标显示图;雷达测绘板
radar plotter 雷达作图器;雷达描绘器;雷达测图仪
radar plotting 雷达标绘
radar plotting sheet 雷达绘迹图
radar point 雷达点
radar position line 雷达位置线
radar precipitation echoes 雷达降水回波
radar prediction format 雷达预测形式
radar prediction type 雷达预测方式
radar presentation 雷达显示(器)
radar prism 雷达棱镜
radar profile 雷达剖面图;雷达断面图
radar-proof 防雷达的
radar pulse 雷达脉冲
radar rainfall equation 雷达雨量方程
radar range 由雷达测出的物标距离;雷达作用距离;雷达作用范围
radar range-finder 雷达测距仪;雷达测距器
radar range indicating system 雷达距离指示系统
radar range marker 雷达距标;雷达测距基线
radar range resolution 雷达距离分辨率;雷达距离分辨力
radar range scale 雷达距离标尺
radar range unit 雷达测距装置
radar ranging 雷达测距
radar ranging set 雷达测距仪;雷达测高仪
radar receiver 雷达接收机
radar receivertransmitter 雷达收发两用机
radar reconnaissance 雷达勘测
radar recording camera 雷达记录照相机
radar reflected buoy 雷达反射浮标
radar reflecting buoy 雷达反射浮标
radar-reflecting target 装有雷达反射器的目标
radar reflection 雷达反射
radar reflection data 雷达反射信息
radar reflection plotter 雷达反射作图器
radar reflectivity 雷达反射率
radar reflectivity plate 雷达反射板
radar reflector 雷达反射器
radar reflector buoy 雷达反射浮标
radar reflector rocket 雷达反射火箭(救生艇筏用)
radar reflector tape 雷达反射带(救生艇筏用)
radar refraction 雷达折射
radar relay 雷达中继站
radar relay(ing) 雷达中继发送
radar repeater 雷达转发器;雷达复示器
radar repetition interval 雷达间隔周期
radar representation 雷达显示地形
radar resolution 雷达分辨率;雷达分辨力
radar responder 雷达应答器
radar responder beacon 雷达应答标;雷达响应信标
radar responding beacon 雷达应答标
radar responser 雷达应答器接收机
radar return 雷达回波;雷达反射
radar return analysis 雷达回波分析
radar rise gradient 雷达作用范围
radar room 雷达室
radar safety beacon 雷达安全航标
radar safety beacon system 雷达安全信标系统
radar same frequency interference 雷达同步干扰
radarsat 雷达卫星
radar satellite 雷达卫星
radar satellite oceanograph 雷达卫星海洋学
radar saturation and overload 雷达饱和与过载
radar scanner 雷达扫描器
radar scan(ning) 雷达扫描
radar scanning antenna 雷达扫描天线
radar scanning technic 雷达扫描技术
radar scatterometry 雷达散射测量技术
radarscope 雷达相片;雷达显示器;雷达显示屏;雷达示波器

radarscope overlay 雷达显示器透明图
radarscope photograph 雷达照片
radar(scope) speedometer 雷达测速器
radar screen 雷达荧光屏
radar screen picture 雷达屏幕图像
radar search unit 搜索雷达
radar sea state analyzer[analyser] 雷达海况分析器
radar section 雷达剖面图;雷达断面图
radar selector switch 雷达选择开关
radar sensor 雷达传感器
radar set 雷达装置;雷达设备
radar shade 雷达静影
radar shadow 雷达阴影;雷达盲区
radar sidelap 雷达旁向重叠
radar signal detection 雷达信号检测
radar signal film 雷达信号胶片
radar signal processing 雷达信号处理
radar signal spectrograph 雷达信号摄谱仪
radar silence 雷达静止时间
radar simulation plate 雷达模拟板
radar simulator 雷达模拟器
radar sonde 雷达测风仪
radar sonobuoy 雷达声呐浮标
radar sounder 雷达探测器
radar sounding 雷达探空;雷达探测
radar speed estimator 雷达测速器
radar speed meter 雷达测速仪;雷达测速计;雷达测速器
radar speedometer 雷达测速器
radar speed survey area 雷达测速区
radar spinner 雷达旋转天线装置
radar station 雷达站;雷达点
radar station ship 雷达舰;防空雷达哨舰
radar storm detection 雷达风暴探测
radar stormdetection equation 雷达风暴探测方程
radar strip 雷达摄影条带
radar sub-modulator 雷达预调制器
radar surveillance station 雷达监视站
radar surveying 雷达测量
radar system parameter 雷达系统参数
radar target 雷达目标;雷达物标
radar target discrimination 雷达上物标鉴别能力
radar telescope 雷达望远镜
radar television 雷达电视
radar theodolite 雷达经纬仪
radar time delay 雷达时延
radar tower 雷达天线塔
radar tracking 雷达跟踪
radar tracking information 雷达跟踪情报
radar tracking station 跟踪雷达站
radar track speed 雷达跟踪速率
radar traffic control 雷达空中交通管理
radar trainer plate 雷达训练板
radar transceiver 雷达收发器;雷达收发机
radar transmission 雷达数据传输
radar transmitted pulse 雷达发射的脉冲
radar transmitter 雷达发射机
radar transponder 雷达应答器;雷达信标
radar transponder beacon 雷达反射信号台
radar triangulation 雷达三角测量
radar trilateration 雷达三边测量
radar tube 雷达显示管;定位管
radar tuning indicator 雷达调谐指示器
radar unit 雷达装置
radar vector 雷达向量
radar video face 雷达显示表面;雷达图像表面
radar video indicator 雷达图像指示器
radar video signal 雷达视频信号
radar view 雷达视野;雷达视界;雷达对景图
radar volume 雷达区域
radar watch 雷达监视
radar wave packet 雷达波包
radar weather observation 雷达天气观测;雷达气象观测
radar wind 雷达测风
radar window 雷达窗
radar wind sounding 雷达测风仪
radar wind system 雷达测风装置;雷达测风系统
radation 灰度
Radburn 人车分流设计;人车分离设计(首先在美国拉德伯恩使用)
Radburn method 人车分隔设计法
Radburn plan 拉德柏恩规划【道】
Radburn system 拉德柏恩方式设计法

radcrete 喷射(处理)混凝土;辐射处理混凝土
raddle 圆木;交织;树干
raddle pick-up 链板式捡拾器
raddle rakes 输送带式逐稿器
raddle straw rack 输送带式逐稿器
Radechon 雷德康management
rade conveyer 耙式输送机
radgas 放射性气体
radiabiliy 透辐射能力
radiac 放射性检测
radiac instrument 辐射仪
radiac instrumentation 放射性探测仪器
radiacmeter 核辐射测定器;辐射仪
radiac set 辐射仪
radiactive phosphorus 放射性磷
radiactivity meter 放射性强度计
radiacwash 放射性去污液
radial 星形;径向的;辐射状的;半径的
radial acceleration 径向加速度
radial activity inspection 放射性穿透检测
radial adjustment 径向调节
radial admission 径向供给
radial aeroengine 星形航空发动机
radial air-cooled engine 星形空冷式发动机
radial amplitude 径向振幅
radial and checker-board street system 方格放射混合式道路网
radial and ring road network 放射加环形线路网
radial and ring road system 环形放射道路系统
radial angular contact ball bearing 径向角面接触滚珠轴承
radial anti-friction bearing 径向抗摩轴承
radial arch 拱形弧
radial arch design 弧形拱架设计
radial arch roof (由中柱旋转成的)穹屋顶;辐射拱形屋顶
radial arm 旋臂;径向臂;臂形拐肘
radial armature 凸极电枢
radial arm bearing 旋臂支承;横力臂支承
radial arm boring machine 旋臂钻床
radial arm drill 旋臂钻床;摇臂钻床
radial arm routing 旋臂式定线机
radial-arm saw 多功能圆锯;摆锯;转向锯;旋臂锯
radial arrangement 间隔排列
radial artery 辐射(式)干线
radial assumption 假定辐射中心
radial astigmatism 径向像散
radial average frequency spectrum 径向平均频谱
radial average logarithmic energy spectrum curve 径向平均对数能量谱曲线
radial axial flow turbine 轴向轴流式水轮机;轴向轴流式气轮机;混流式涡轮机;混流式透平;混流式水轮机;混流式汽轮机;辐向轴流式水轮机
radial axle 转向轴;径向轴
radial axle bogie 径向转向架
radial backlash 径向侧隙
radial bar 圆规尺;旋转棒;径向钢筋;辐射形钢筋
radial basin 扇状流域;辐形流域;辐射流域
radial beam 径向梁(指弧形闸门)
radial beam position 束流径向位置
radial-beam tube 径向聚束管
radial bearing 向心轴承;径向轴承
radial bending moment 径向弯矩;径向弯曲弯矩
radial bending stress 径向弯矩应力
radial blade 径向叶片
radial blade blower 径向叶片吹风机
radial bladed impeller 径向式叶轮
radial blanket 径向再生区
radial blastic texture 放射状变晶结构
radial bonding header 径向砌合顶砖
radial borehole 辐射状钻孔
radial box 辐射箱(切割异形砖用)
radial bracing 径向支撑
radial brick 楔形砖;弧形砖;扇形砖;大小头砖;辐向砖;辐射型砖;辐射形砖
radial brick chimney 扇形砖烟囱
radial brick industrial chimney 扇形砖工业烟囱
radial brush 径向电刷
radial buffer 球形缓冲器
radial bundle 辐射维管束
radial burner 辐射灯
radial cable crane 辐射式缆索起重机
radial cam 径向凸轮
radial canal 辐水管;辐管;放射沟

radial capacity 辐射能力
radial center 径向中心
radial chaser 径向螺纹梳刀
radial check gate 弧形节制闸门;弧形节制阀
radial chimney brick 扇形烟囱砖
radial chromatic displacement 径向色位移
radial chromatic error 径向色误差
radial chromatography 径向色谱法;径向层析
radial city 辐射形城市
radial clearance 径向净空;径向间隙;环向间隙
radial clutch 径向离合器
radial coefficient 辐射系数
radial collector well 辐射状集水井
radial commutator 径向排列换向器;辐射式整流子
radial component 径向分量;径向部分
radial component of load 荷载的径向分力
radial composite error 径向综合误差;双啮误差
radial compression 径向压缩
radial compression stress 径向压应力
radial compressor 离心式压缩机
radial cone 辐射锥(体)
radial consolidation 径向固结
radial contraction 径向收缩
radial control of beam 束流径向控制
radial control point 辐射控制点
radial convergence 径向聚焦;径向会聚
radial convergence magnet 径向聚焦磁铁;径向会聚磁铁
radial conveyer 径向输送机
radial coordinate 径向坐标
radial crack 径向裂缝;径向破裂;辐射形裂缝;辐射式裂缝;放射状裂缝;放射(形)裂隙;放射(形)裂缝
radial cracking 径向开裂
radial crane 辐射式起重机
radial cross-vein 横射脉
radial crushing strength 径向压溃强度;径向破碎强度;径向抗压强度;中心破碎强度
radial cut 径向切口;径向开锯;径切面;原木四开据法;径向锯木
radial cut ring 径向切口环
radial cutter 侧面铣刀
radial cylinder 星形汽缸
radial davit 旋转式吊艇柱;转动式吊艇柱;转动式吊艇杆
radial deflecting electrode 径向偏转电极
radial deflection 径向偏转;径向挠度
radial deflection terminal 中心电极;径偏转电极
radial deflection type 径向偏转式
radial defocusing 径向偏焦
radial deformation 径向变形
radial derivative 径向导数
radial development 径向展开(法)
radial deviation 径向偏差
radial diffuser 径向扩压器;径向扩散器;径向扩散廊道(船闸泄水);径向导叶
radial diffusion 径向扩散
radial dike 放射状岩墙;辐射岩墙
radial dilation 径向膨胀;径向扩容
radial dilution effect 径向稀释效应
radial direction 径向;辐射方向
radial dispersion 径向散布
radial displacement 径向位移
radial displacement of the joints 连接点的径向位移
radial distance 辐射距离;径向距离
radial distortion 径向畸变
radial distortion of lens 透镜径向畸变
radial distribution 径向分布
radial distribution analysis 径向分布分析
radial distribution feeder 径向配电馈路
radial distribution function 径向分布函数
radial distribution system 径向配电制
radial drainage 辐射形排水(系统);径向泄流;径向排液;径向排水;辐射(状)水系;辐射状排水
radial drainage pattern 辐射状水系型;辐射型水系【地】;辐射水系型
radial drainage system 辐射形水系排水系统;辐射式排水系统
radial-draw deformation 径向拉伸变形
radial drawing 径向拉伸
radial drill 摇臂钻(机)
radial drill hole 辐射状钻孔
radial drilling 径向钻进
radial drilling machine 旋臂钻床;摇臂钻机;摇臂钻床;扇形孔凿岩机
radial drilling pattern 扇形布孔;辐射状布孔;辐射状布井
radial ducts 径向通风道
radial-elastic modulus 径向弹性模量
radial electric(al) oil pump 径向式电动油泵
radial electrostatic field 径向静电场
radial energy distribution 径向能量分布
radial engine 星形发动机;径向配置活塞发动机;径向排列发动机
radial engineering brick 工程上用扇形砖
radial error 径向误差
radial error of satellite orbit 卫星轨道径向误差
radial extension vibration mode 径向伸缩振动模式
radial facing 径向进给车端面;车端面
radial fan 离心式通风机;离心扇;径流式风机
radial fault 辐射(状)断层;放射(状)断层;放射形断层
radial feed screw 径向进给丝杠
radial feed servo 径向馈送伺服
radial-fibrous fabric 放射纤维的组构
radial field 径向场
radial field cable 分析屏蔽电缆
radial filament 径向暗条;辐射细丝
radial filter 辐射形滤电
radial fin filter 辐射翅形滤器
radial fission 辐射分裂
radial fissure 径向裂缝;径向沟
radial flank 径向齿面
radial flat jack 径向液压枕;径向钢枕;径向扁千斤顶
radial flat jack test 径向液压枕法试验;径向扁千斤顶试验
radial flow 径向流(动);径流;辐(状)流;辐向(射)流;辐射流
radial-flow back pressure turbine 辐射式背压汽轮机
radial-flow basin 辐射式水池;辐流式水池
radial-flow blower 径向式鼓风机;径向鼓风机
radial-flow classifier 径向式分级机
radial-flow compressor 离心式空(气)压(缩)机;离心式压缩机;径流式压气机
radial-flow consolidation coefficient 径向流水固结系数
radial-flow double-rotation turbine 辐流式正反双转子汽轮机
radial-flow fan 径流式风扇;轴流式扇风机
radial-flow gas turbine 辐流式燃汽轮机
radial-flow impulse turbine 辐流背压式汽轮机
radial-flow noncondensing turbine 辐流背压式汽轮机
radial-flow pump 径流泵;辐流泵
radial-flow reaction turbine 辐流反冲透平;辐流反冲涡轮机
radial-flow reactor 径向反应器
radial-flow sedimentation tank 辐流式沉淀池
radial-flow settling basin 辐射式沉淀池;辐射流沉淀池
radial-flow settling tank 辐流式沉淀池
radial-flow steam turbine 辐流式汽轮机
radial-flow system 辐射流系统
radial-flow tank 径流式水池;辐流式水池;辐射流水池
radial-flow tray 辐流盘
radial-flow turbine 径流涡轮;径流式汽轮机;射流涡轮机;辐向轴流式涡轮机;辐流(式)涡轮机;辐流式水轮机;辐流透平
radial-flow turbocharger 径流式涡轮增压器
radial flux 辐射流(量)
radial flux density 辐射流密度
radial force 径向力;辐射力
radial forging machine 径向锻机
radial forging process 径向锻造法
radial format 辐射形格式;扇形格式
radial Fourier analysis 径向傅立叶分析
radial fracture 放射状断裂;辐射断裂
radial function 径向函数
radial gantry crane 辐射式门式起重机
radial gap 径向缝隙
radial gap of double cone ring 双锥环的最大半径间隙
radial gate 扇形闸门;弧形闸门
radial gear 偏心传动装置
radial girder 径向大梁
radial gland 径向汽封
radial glass 散热玻璃;热辐射玻璃
radial grating 径向光栅;径向光栅;扇形排架;辐向光栅;辐射状格栅
radial grinding 径向磨削
radial grooved filter plate 辐射凹纹滤板
radial guide apparatus 径向式导水机构
radial hard brick 扇形烧透砖;辐射形硬砖
radial header 扇形顶筒
radial height of blade 辐向叶片长度
radial highway 辐射式公路;公路放射线
radial hobbing 径向滚削
radial impeller 径向叶轮
radial incision 放射状切口
radial inflow 向心式;径向流入
radial inflow compressor 向心式压气机
radial inflow turbine 径向内流式水轮机
radial inflow wheel 径向内流式水轮机
radial inlet 径向进口
radial inlet impeller 径向进口式泵;径向入口叶轮
radial interferogram 径向干涉图
radial intersection method 辐射交会法
radial inversion 径向反演
radial inward 向心式的
radial-inward admission 向心进汽
radial inward flow 径向内流;内向径流;辐射内向流
radial inward-flow gas turbine 向心式燃气轮机
radial-inward flow turbine 径向内流式水轮机;向心式涡轮机
radial-inward turbine 向心式涡轮机
radialization 辐射
radial jack test 径向千斤顶试验
radial joint 径向连接;径向接缝;径向缝
radial journal bearing 支持轴承
radial-knife cutterhead 盘刀式切屑器
radial ladle 回转吊包(配合转台)
radial layout of streets 辐射式街道布置;放射形街道布置;放射街道网
radial lead 径向引线
radial line 径(向)线;径向沟;辐射形线路;辐射线;放射线【铁】
radial line control 辐射三角测量
radial line intersection 辐射线交会
radial line load 径向线荷载
radial line method 辐射线法
radial line of position 辐射位置线
radial line plot 辐射线图;辐射三角变形锁
radial line plotter 辐射线绘仪;辐射线平面绘图仪
radial line plotting 辐射线转绘成图
radial line strip 辐射三角带
radial line triangulation 辐射三角测量
radial link 放射键丝
radial link head 放射键丝头
radial loader 弧线装船机
radial load(ing) 径向荷载
radial location 径向定位
radial lock gate 弧形闸门
radial locomotive 转向机车【铁】
radial loop 反箕
radial loop cable 径向(循)环电缆
radially cut grating 扇状格栅
radially increasing pitch 径向递增螺距
radially non-uniform inflow 径向非均匀流
radially split casing pump 径向剖分壳式泵;径向分段壳式泵
radially stratified fiber 径向分层光纤
radially symmetric(al) 径向对称的
radially unbounded 径向无界;放射无界
radially unbounded function 径向无界函数;放射无界函数
radially varying inflow 径向非均匀流
radial magnetic anomaly 放射状磁异常
radial method 光线法;辐射法
radial mode 辐射状水系模式
radial momentum 径向动量
radial motion 径向运动
radial motor 星形马达;星形发动机
radial movement 径向运动;径向移动;放射式运动
radial multiplepiston type of pump 径向多活塞式泵
radial navigation 辐射信号导航
radial nodal plane 径向节面
radial nozzle 弧形喷嘴
radial oscillation 径向振荡
radial outrigger 径向悬臂梁;辐射式外伸支腿;辐

射式外伸支架;辐射式外伸叉架
radial outward admission 离心进气
radial outward flow 径向外流;径向水量;径向流量;辐射外(向)流
radial outward flow turbine 离心式涡轮机;离心式水轮机;径向外流式水轮机
radial outwash flow 径向外流;径向水量;径向流量
radial oxygen loss 径向耗氧量
radial packing 轴封;径向密封
radial panel 辐射板
radial parallel search 平行线族搜索
radial particle velocity 径向质点速度
radial passage 径向通道;径向风道
radial pattern 辐射型;放射型
radial penetration 沿径向穿云
radial percolation test 径向渗透试验;辐向渗透试验
radial permeability coefficient 径向渗透系数
radial photometer 径向光度计
radial piston motor 径向柱塞马达
radial piston pump 径向柱塞泵;径向活塞泵
radial piston pump with exterior admission 外配流径向柱塞泵
radial piston pump with interior admission 内配流径向柱塞泵
radial pitch variation 沿半径变化的螺距
radial plan 辐射式布置;放射形布局
radial planimeter 径向求积仪
radial planimetric(al) plotter 辐射线平面绘图仪
radial play 径向游隙;径向间隙
radial plot 辐射三角测量
radial plotter 辐射线平面绘图仪
radial plunger oil pump 旋转柱塞油泵;旋转往复油泵;径向柱塞油泵
radial plunger pump 径向柱塞泵
radial plunger spinning pump 旋转往复油泵
radial ply construction 子午合股结构
radial pole piece 径向极靴;径向磁极块
radial polynomial 径向多项式
radial pore multiple 径向复管孔
radial porous wood 辐射孔状木材
radial positioning grid 辐射线格网
radial power 辐射能力
radial pressure 径向压力;辐射压力
radial pressure stress 径向压应力
radial principal point 辐射中心
radial principal stress 径向主应力
radial probability density 径向几率密度
radial probability function 径向几率函数
radial profile 辐射状剖面
radial projection 轴测透视(图);放射(型)投影
radial pull 径向拉力
radial pulling scraper 径向拉铲
radial pulsation 径向脉动
radial pump 径向泵
radial rake 径向前角;刀面径向角
radial rate 径向速率
radial ray 径向射线;辐射线
radial reaction turbine 径向反击式水轮
radial refraction 径向折射
radial resolving power 径向分辨率
radial reticle 辐射式调制盘
radial retrieval 半径检索
radial rib 径向加强肋;径向肋
radial-rib cupola 径向肋圆(屋)顶
radial rift 放射状裂隙;放射状裂缝;放射状断裂
radial rigidity 径向刚度
radial road 径向放射式道路;辐射式公路;辐射式道路;辐射路;放射形道路;放射线【铁】;放射(式道)路
radial road (line) 半径线
radial road system 放射式道路系统
radial roller bearing 向心滚子轴承
radial roof fan 旋叶式屋顶风扇;径向吊扇
radial root mean square error 径向均方差
radial rotor machine 凸极电机
radial runout 径向跳动
radial saw 悬臂(圆)锯;转向锯;不定向圆锯;摆锯
radial sawing 径向锯(木)法
radial scan 径向扫描
radial scatter 径向散射器
radial scattering 径向散布
radial scheme 辐射式系统布置(道路管路线路等的);辐射式布置(道路,结构等)
radial scraper arms 径向刮泥机臂

radial screen 径向筛分机;径向遮光板
radial screen tank 辐射流格网池
radial seal(ing) 轴封;径向密封
radial search 半径检索
radial section 径向切片;径(向)切面
radial section of coal 煤的辐射剖面
radial sector 径分级
radial sedimentation tank 辐射形沉淀槽
radial selector 全方位选择器;无线电定向标选择器
radial self seal 径向自紧密封
radial seriation 径向排列
radial sett paving 扇形石块铺砌;扇形石块路面
radial sewer network 径向污水管网
radial shake 径向开裂;径裂;辐(射)裂;辐式裂缝
radial shaped city 放射形城市
radial shear 径向剪切;径向剪力
radial shearing interferometer 径向切变干涉仪
radial-shear interferometer 径向剪切干涉仪
radial shear zone 径向剪切带
radial shear zone of city 城市径向剪切带
radial ship-loader 弧线装船机
radial shrinkage 径向收缩;径向干缩量(木材)
radial slot 沿径槽;径向切口;径向缝
radial small stone sett paving 小方石辐射形路面;小方石辐射形地面;小方石辐射形铺面
radial solid block 扇形实心砌块
radial solid brick 辐射形实心砖;扇形实心砖
radial spark drilling 辐射(电)火花钻进
radial speed 径向速率;径向速度
radial speed field 径向速度场
radial spiral 辐射螺旋线
radial spreader (conveyor) 旋转布料胶带运输机
radial spreading effect 放射状传播效应
radial square 辐向角尺
radial stacker 环形堆垛机;辐射形堆料机;辐射式堆料机
radial stacker-conveyer 径向堆料输送机
radial steps 辐射梯阶;扇形踏步
radial strain 径向应变;径向变形
radial strand 辐射束
radial stress 径向应力;辐射应力
radial stress in flange 法兰中的径向应力
radial stretcher 扇形顺砖
radial striae 放射状线纹
radial structure 辐射状构造
radial strut 径向支柱
radial strut in tunnel support 扇形支撑
radial style 放射式
radial support 径向支撑
radial surface shingle 扇形面板;径向面板
radial swing 径向摆动
radial-symmetric(al) distortion 辐射对称畸变
radial symmetry 径向对称;辐射对称
radial system 辐射式排水系统;排水系统布置-放射式系统;放射供电系统
radial system of sewerage 辐射式下水道系统
radial system of street layout 辐射式街道布置
radial system of streets 放射式街道网
radial temperature 辐射温度
radial template 方向模片
radial tension 径向张力
radial test 径向取样试验;射线检验
radial texture 放射状影纹
radial thrust 径向压力;径向推力
radial-thrust ball bearing 向心推力球轴承
radial-thrust bearing 径向止推轴承;向心推力滚动轴承;径向推力轴承
radial-thrust force 径向抗力
radial tile line 辐射(式)缸瓦管线
radial-time-base display 径向时基显示器
radial tire 子午线轮胎
radial tool 圆角切刀
radial tool holder 径向刀杆
radial top-slicing 下向分层崩落采矿法
radial transfer 径向传送;内外传送
radial transfer-type mo(u)lding press 径向转移型模压机
radial transformation 辐射变换
radial transporter 径向运送机;径向输送机
radial travel(l)ing cableway 辐射移动式架空索道;径向移动式缆道;辐射移动式缆索起重机
radial triangulation 辐射三角测量
radial triangulation net 辐射三角网
radial triangulator 辐射三角仪

radial truck 向心卡车
radial truck sliding seat 单轴转向架滑座
radial turbine 辐射式涡轮机;辐流式水轮机
radial turbulent jet 径向紊动射流
radial two-dimensional model 径向二维模型
radial-type fill valve 环转式装料活门
radial-type mottling 放射性斑纹
radial-type of cable stayed bridge 辐射形斜拉桥
radial type oil seal 径向式油封
radial type road network 放射型线网
radial type transmission line 辐式传输线
radial universal copying milling machine 仿形摇臂万能铣床
radial universal milling machine 摇臂万能铣床
radial valve gear 辐射式阀动装置
radial vane 径向叶片;辐射式叶片
radial vane meter 径向叶轮表
radial vane pump 径向叶轮泵
radial varying pitch 径向变螺距
radial vascular 辐射维管束
radial vascular bundle 辐射维管束
radial vector 径向向量;径向矢量
radial velocity 径向速度;径向流速;视向速度
radial velocity component 径向分速度
radial velocity curve 视向速度曲线
radial velocity field 径向速度场
radial velocity meter 视向速度仪
radial velocity scanner 光电视向速度仪;视向速度扫描仪
radial velocity spectrometer 视向速度光谱仪
radial ventilation 辐向通风;辐式通风
radial vibration 径向振动
radial wall 径向壁
radial wave resonator 径向波共振腔
radial weir 辐流堰
radial well 辐射井;放射井
radial well-burnt brick 扇形烧透砖
radial whirl 径向涡流;辐向旋涡
radial winding 辐(射)式绕组
radial wiring 径向拉线
radian 弳(弧度法的角度单位,约等于57°17′44.8″);弧度;光度
radiance 面辐射强度;辐射率;辐亮度
radiance conservation principle 辐射率守恒原理
radiance contour map 辐射外形图
radiance contrast 辐射度对比;辐射度差
radiance factor 辐射亮度因数
radiance function 辐射度函数
radiance intensity resolution 辐射强度分辨率
radiance relationship 辐射率关系
radiance theorem 辐射率定理
radiancy 辐射率;辐射度
radian frequency 角频率
radian frequency domain 角频率域
radian frequency of vibration 振动弧度频率
radian length 弧度波长
radian measure 弧度法
radiant 热源;辐射的;放射能
radiant absorptance 辐射吸收系数
radiant absorption 辐射吸收
radiant arc furnace 辐射电弧炉
radiant artery 辐射干线
radiant band 辐射频带
radiant boiler 辐射热锅炉
radiant burner 辐射(式)燃烧器
radiant ceiling 辐射采暖顶棚
radiant ceiling heating 辐射顶棚采暖
radiant chapel 扇形平面礼拜堂
radiant city 辐射形城市
radiant colo(u)r 辐射颜色
radiant convector gas fire 辐射对流采暖炉
radiant cooling 辐射致冷;辐射冷却
radiant cooling method 辐射冷却法
radiant correction 辐射量校正
radiant cup burner 杯状辐射燃烧器
radiant densifier 辐射式浓缩剂
radiant density 辐射能密度;辐射点密度;辐密度
radiant drying 辐射法干燥
radiant efficiency 辐射效率
radiant electric(al) heater 辐射电热器
radiant electric(al) heating 辐射电供热
radiant element 辐射元件
radiant emissivity 辐射通量密度
radiant emittance 照射通量密度;辐射率;辐射度;

发能能流密度;发射率
radiant energy 辐射能量;辐射能
radiant energy density 辐射能量密度
radiant-energy thermometer 辐射高温仪表;辐射高温计
radiant energy type analyzer 辐射能式分析仪
radiant excitation 辐射激发
radiant exitance 辐射发散度;辐射度;辐射出射度
radiant exposure 曝辐量;辐照量;辐射曝露量
radiant floor 辐射热楼面;辐射采暖地面
radiant flow 扇形流
radiant flux 辐(射)通量;辐射流量
radiant gas burner 辐射式气体喷灯
radiant gas fire 辐射式采暖炉
radiant glass 散热玻璃;热辐射玻璃
radiant heat 辐射热
radiant heat boiler 辐射热锅炉
radiant heat density 辐射热强度
radiant heat dryer 辐射热干燥器
radiant heat drying 辐射热烘干;辐射热干燥
radiant-heated 辐射加热的
radiant heat energy source 辐射热能源
radiant heater 红外线辐射器;辐射式加热炉;辐射式供暖器;辐射加热器;辐射发热管
radiant heat exchange 辐射热交换
radiant heat flux 辐射热流量
radiant heating 热辐射烘干;辐射供暖;辐射采暖
radiant heating ceiling 辐射采暖顶棚
radiant heating elements 辐射加热元件
radiant heating floor 辐射采暖地面
radiant heating installation 热辐射装置
radiant heating of street layout 辐射式供暖系统
radiant heating panel 热辐射板;辐射供暖板
radiant heating stoving 红外辐射热烘干
radiant heating system 辐射式供热系统;辐射加热系统;辐射供暖系统
radiant heat lamp 辐射热灯
radiant heat reflector 辐射热反射器
radiant heat transfer coefficient 辐射传热系数
radiant heat transmission 辐射传热
radiant heat zone 辐射热区
radiant illumination 辐射照度
radiant intensity 辐射强度
radiant interchange 辐射交换
radiant loss 辐射热损失
radiant marmth 辐射热
radiant matter 辐射物(质)
radiant mode 辐射模式
radiant oven 辐射烘干炉
radiant panel 红外线辐射器;辐射采暖炉;辐射板
radiant panel burner 辐射板式燃烧器
radiant panel heater 辐射板供暖器
radiant panel heating 辐射板供暖(法)
radiant panel test 辐射板试验
radiant power 辐射通量;辐射功率
radiant-power density 辐射功率密度
radiant power detector 辐射功率探测器
radiant quantity 辐射量
radiant ray 辐射线
radiant reflectance 辐射反射率
radiant resistance 辐射电阻
radiant resistance furnace 辐射电阻炉
radiant sensor 辐射感应器;辐射传感器
radiant source 辐射源
radiant source distribution 辐射源能量分布
radiant supercharger 辐射式增压器
radiant superheater 辐射式过热器
radiant supply of electricity 辐射式供电
radiant surface 辐射面
radiant surface absorptivity 辐射表面吸收率;发光表面吸收能力;发光表面吸收率
radiant temperature 辐射温度
radiant track 辐射轨道
radiant transmittance 辐射透射率
radiant tube 辐射管
radiant tube annealing 辐射加热退火;辐射管加热式退火
radiant tube-fired cover-type furnace 辐射管加热的罩式退火炉
radiant tube furnace 辐射管加热炉;辐射管式炉;装有辐射管的炉;辐射管炉膛
radiant tube heated furnace 辐射管加热管式炉
radiant tube heating 辐射管加热
radiant tube recuperator 辐射管换热器

radiant tubular heater 辐射型管式炉
radiant-type boiler 辐射式锅炉
radiant-type furnace 辐射型管式炉
radiant-type pipe still 辐射型管式炉
radiant umbel 辐射伞形花序
radiant wall burner 侧壁燃烧器
radiant wall heating 墙面辐射供暖
radiant wall type heater 辐射墙式炉
radiate 放射
radiate crown 辐射冠
radiated 放射状
radiated acoustic(al) field 辐射声场
radiated characteristic 辐射特性
radiated electric(al) field 辐射电场
radiated electro-magnetic field 辐射电磁场
radiated element 发射元件
radiated energy 辐射能
radiated flange 散热凸缘;散热片
radiated heat 辐射热
radiated interference 辐射干扰
radiated light 辐射光
radiated part 辐状部
radiated power 辐射功率
radiated power output 辐射功率输出
radiated pyrite 白铁矿
radiated rib 散热肋片
radiated structure 放射构造
radiated television 电视辐射发送
radiated wave 辐射波
radiated zeolite 束沸石
radiate fibers 辐状纤维
radiate green space 放射状绿地
radiate layer 放射状层
radiate shutter 散热器风门片
radiate vein 辐射脉
radiating 散热(的);辐射(状)的
radiating antenna 辐射天线
radiating area 散热面积
radiating block 扇形块
radiating body 辐射体
radiating bonder 扇形顶砖
radiating brick 楔形砖;辐射型砖;辐射形砖
radiating bridge 扇形桥
radiating burner 辐射烧嘴;辐射燃烧器
radiating cable 辐射电缆
radiating canal rays 辐射极隧射线
radiating canals 辐射水沟
radiating capacity 辐射能力
radiating chapels 辐射状小教堂
radiating church 辐射形小教堂
radiating circuit 辐射电路
radiating collar 冷却套管
radiating configuration 辐射形
radiating crack 放射(状)裂隙;放射裂缝
radiating curtain 辐射帘
radiating element 辐射单元
radiating engineering brick 扇形工程用砖
radiating fault 放射状断层
radiating fiber 放射纤维
radiating fin 散热片
radiating flange 散热片;散热凸缘
radiating format 辐射形格式
radiating gas 辐射气体
radiating glacier 放射状冰川
radiating guide 辐射波导;波导天线
radiating header 扇形露头砖
radiating heat 辐射热
radiating joint 径向缝
radiating layer 辐射层
radiating material 放射性材料
radiating medium 放射性介质
radiating mirror 辐射反射镜
radiating panel coil 散热盘管
radiating pattern 辐射状
radiating pipe 散热管
radiating power 辐射能力;辐射本领
radiating principal 辐射主体
radiating process 散热过程
radiating rib 散热(筋)片
radiating ridge 辐射纹
radiating scattering 辐射散射
radiating solid block 扇形实心块
radiating stretcher 扇形顺砖
radiating structure 放射状组织;放射状构造

radiating surface (蒸汽采暖散热器的)散热面(积);辐射面
radiating tracery 辐射形花格窗
radiating track 辐射状轨道
radiating valve 散热阀
radiating vein 辐射状(岩)脉;放射状矿脉
radiating well-burnt brick 扇形烧透砖
radiation 照射(作用);辐射;放射
radiation ablation 辐射消融;辐射烧蚀
radiation absorbed dose 辐射吸收量
radiation absorber 放射性物质吸收剂
radiation-absorbing glass 防辐射玻璃
radiation absorption 辐射吸收
radiation absorption analysis 辐射吸收分析
radiation absorption analyzer 辐射吸收分析仪
radiation absorption factor 辐射吸收因数
radiation absorptivity 辐射吸收率
radiation accident 辐射事故
radiation-activated sensor 辐射活化传感器
radiation alarm 辐射警报信号;放射性警报
radiation alarm network 辐射警报网
radiational tide 辐射潮
radiation amount 辐射量
radiation and radapertization and radurization 辐射消毒、杀菌及防腐法
radiation angle 辐射角
radiation annealing 辐射退火
radiation area 辐射区
radiation auto-oxidation 辐射自氧化
radiation background 辐射背景
radiation balance 辐射天平;辐射平衡;辐射差额
radiation balance meter 辐射平衡表
radiation band 辐射带
radiation barrier 防辐射屏;辐射屏蔽
radiation barrier system 辐射隔离系统
radiation beam 辐射束
radiation belt 辐射带
radiation biochemistry 辐射生物化学;放射生物化学
radiation biology 辐射生物学
radiation biophysics 辐射生物物理学
radiation boiler 辐射型锅炉
radiation burden 辐射负荷
radiation burn 辐射烧伤;放射烧伤
radiation burner 辐射烧嘴
radiation calorimeter 辐射热量计
radiation carcinogenesis 放射致癌作用
radiation cascade 辐射串级
radiation catalysis 辐射催化
radiation cataract 辐射缓冲
radiation characteristic 辐射特性
radiation chart 辐射图
radiation-chemical chromatography 放射化学色谱法
radiation chemical engineering 辐射化学工程;辐射化工
radiation-chemical reduction 辐射化学还原
radiation chemistry 辐射化学;放射化学
radiation chimera 辐射嵌合体
radiation cleavage 辐射裂解
radiation climate 辐射气候
radiation coefficient 辐射系数;放射系数
radiation concentration guide 辐射浓度指导限值
radiation condition 辐射条件
radiation conductivity 辐射传导
radiation cone 辐射锥(体)
radiation constant 辐射常数;放射常数
radiation control 辐射管理
radiation-convection temperature 辐射对流温度
radiation cooled 辐射致冷
radiation-cooled structure 辐射冷却结构
radiation-cooled tube 冷法管;气冷管;辐射冷却管
radiation cooling 辐射致冷;辐射冷却
radiation core 散热器型心
radiation correction 冷却改正
radiation counter 辐射计数器
radiation counter tube 计数管;辐射(线)计数管
radiation coupling 辐射耦合
radiation crosslinking 辐射交联法
radiation cross-section 辐射截面
radiation curable paint 辐射固化涂料
radiation curable polymer 可辐射固化聚合物
radiation-curable resin 可辐射固化树脂
radiation cure 辐射熟化

radiation curing 辐射线固化
radiation curve 辐射曲线
radiation damage 辐照损伤;辐射性损伤;辐射线损失;辐射损伤;辐射伤害;放射性损伤;放射线损害
radiation damage inhibitor 防放射线剂
radiation-damage reaction 辐射损伤反应
radiation-damage susceptibility 辐射损伤敏感性
radiation-damage threshold 辐射损伤阈
radiation damping 辐射阻尼
radiation damping coefficient 辐射阻尼系数
radiation damping force 辐射阻尼力
radiation danger zone 辐射危险区
radiation data 辐射数据
radiation death 辐射死亡
radiation decay 辐射衰变
radiation decomposition 辐射分解
radiation decontamination 辐射去污
radiation defect 辐射缺损;放射性缺损
radiation degradation 辐射降解
radiation degradation of polymer 聚合物辐射降解
radiation densitometer 放射性密度计
radiation density 辐射密度;放射(性)密度
radiation density constant 辐射密度常数
radiation density matrix 辐射密度矩阵
radiation dermatitis 放射性皮炎
radiation detecting element 辐射高温计元件
radiation detecting instrument 辐射探测器
radiation detection 辐射探测
radiation detection instrument 辐射探测仪器
radiation detector 粒子探测器;辐照探测器;辐射探测器;辐射测定器
radiation device 放射仪
radiation diagram 辐射图
radiation disease 放射性疾病;放射病
radiation dominated gas 辐射占优气体
radiation dosage 辐射剂量
radiation dosage indicator 辐射剂量指示器
radiation dose 辐射量
radiation dose distribution 辐射剂量分布
radiation dose measurement 辐射剂量测量
radiation dose meter 辐射剂量计
radiation dose rate 辐射剂量率
radiation dose rate meter 放射性剂量仪
radiation dose sensor 射线剂量传感器
radiation dose standard 辐射剂量标准
radiation dose transducer 射线剂量传感器
radiation dosimeter 辐射剂量计;辐射计量仪
radiation dosimetry 辐射剂量学;辐射剂量测定法
radiation dryer 辐射干燥器
radiation drying 辐射干燥
radiation effect 照射效应;辐射效应;放射效应
radiation efficiency 辐射效率;放射效率
radiation electrochemistry 辐射电化学
radiation energetic parallax 辐射能量视差
radiation energy 辐射能
radiation energy density 辐射能密度
radiation energy thermometer 辐射能温度计
radiation environment 辐射环境
radiation equilibrium 辐射平衡;放射平衡
radiation-equivalent manikin calibration 生物标定辐射当量
radiation era 辐射时代
radiation error 辐射误差
radiation estimation 辐射估定
radiation estimator 辐射剂量计
radiation exchange 辐射交换
radiation excitation 辐射激发
radiation exposure 射线照射;辐射接触;辐射监测;辐射暴露
radiation factor 辐射系数
radiation fatigue 辐射疲劳
radiation field 照射野;辐射场
radiation field statistics 辐射场统计
radiation filter 辐射滤波器
radiation fin 散热片
radiation fin filter 辐射翅形滤器
radiation fire detector 辐射火感器;辐射式火警探测器
radiation flux 辐射流量;辐射通量
radiation flux density 辐射通量密度;辐射流密度
radiation fluxmeter 辐(射)通量计
radiation fog 辐射雾;放射雾
radiation for disinfection 放射性消毒
radiation-free glass 无放射性玻璃

radiation frequency 辐射频率
radiation frost 辐射霜冻
radiation function 辐射函数
radiation furnace 辐射(电)炉
radiation gain 辐射增益
radiation ga(u)ge 辐射仪
radiation-generated current 辐射产生电流
radiation gradient 辐射梯度
radiation grafting 辐射接枝
radiation hardening 辐射硬化
radiation hazard 辐射危害;放射(性)危害
radiation heat 辐射热
radiation heating 辐照加热;辐射增温;辐射加热;辐射采暖
radiation heating-surface 辐射受热面
radiation heat loss 辐射热损失
radiation heat transfer 辐射热传递;辐射传热
radiation heat transfer coefficient 辐射换热系数;辐射传热系数
radiation high-pressure area 辐射高压区
radiation hygiene 辐射卫生学;辐射卫生
radiation illness 放射性疾病
radiation image pickup tube 辐射图像摄像管
radiation impedance 辐射阻抗
radiation indicator 辐射指示器
radiation-induced 照射诱导
radiation-induced copolymerization 辐射引发共聚
radiation-induced crosslinking 辐射诱导交联
radiation-induced decomposition 照射分解
radiation-induced grafting 辐射诱导接枝
radiation-induced graft polymer 辐射诱导接枝聚合物
radiation-induced ionic polymerization 辐射诱导离子聚合
radiation-induced pollutant decomposition 辐射诱导污染物分解
radiation-induced polymerization 辐射诱导聚合
radiation-initiated 照射引发
radiation-initiated polymerization 辐射引发聚合
radiation injury 辐射(性)损伤;辐射(性)伤害;放射(性)损伤;放射(性)杀伤
radiation intensity 辐射通量密度;辐射强度;放射强度
radiation inversion 辐射逆温
radiation ionization 辐射电离
radiation ion polymerization 辐射离子聚合(作用)
radiation isnult 放射损伤
radiation law 辐射定律
radiation leaktightness 防止辐射能漏出设备
radiation length 辐射距离;辐射长度
radiationless decay process 无辐射衰变过程
radiationless generation 无辐射产生
radiationless process 无辐射过程
radiationless resonance 无辐射共振
radiationless transition 无辐射跃迁
radiation level 辐射水平
radiation-level indicator 辐射强度指示器
radiation lifetime 辐射寿命
radiation load 放射荷载
radiation lobe 波瓣
radiation logging 辐射测井
radiation longitudinal cable arrangement of cable-stayed bridge 斜拉桥辐射形纵向布置
radiation loss 辐射损失;辐射损耗
radiation material contamination 放射性物质污染
radiation matrix 辐射矩阵
radiation measurement 辐射(性)测量
radiation measurement equipment 辐射能测量设备;辐射剂量测量设备
radiation measuring assembly 辐射测量装置
radiation measuring instrument 辐射测量仪
radiation mechanism 辐射机制
radiation medicine 放射医学
radiation melt 辐射融化
radiation meter 伦琴计;射线计;辐射计;辐射测量装置
radiation meter spectral analyzer 辐射计频谱分析仪
radiation method 辐射法;放射法;放射测量法(平板测量用的)
radiation mode 辐射模式
radiation moisture ga(u)ge 辐射线湿度计
radiation monitor 辐射监测器;放射线检验器;放射监测器;放射探测器
radiation monitoring 辐射检查;辐射监控测定法;辐射监测
radiation-monitoring film 辐射监察片
radiation-monitoring instrument 辐射监察仪器
radiation monitoring system 辐射监测系统
radiation night 强辐射夜
radiation noise 电磁噪声;辐射噪声
radiation of 106Rh 铑106辐射
radiation of 130Ru 钌130辐射
radiation of 131I 碘131辐射
radiation of 137Cs 铯137辐射
radiation of 140Ba 钡140辐射
radiation of 140La 镧140辐射
radiation of 95Nb 铌95辐射
radiation of 95Zr 锆95辐射
radiation of energy 能量辐射
radiation of heat 热辐射;热的辐射
radiation of light 光辐射
radiation of lightening discharge 雷电辐射
radiation oven 辐射炉
radiation pasteurization 辐射消毒
radiation pathology 放射病理学
radiation pattern 辐射型;辐射图;辐射方向图;不同方向的相对辐射能量
radiation pattern plotter 辐射图绘图器;辐射式绘图器
radiation peak 辐射峰值
radiation physics 射线物理学;辐射物理学
radiation plasma turbulence 辐射等离子体湍动
radiation point 辐射点
radiation poisoning 辐射中毒
radiation polymerization 辐射聚合(作用)
radiation power of sound source 声源辐射功率
radiation precaution sign 辐射警告标志
radiation preservation 辐射保藏
radiation pressure 辐射压力
radiation pressure balance 辐射压力天平
radiation probe 辐射探测器
radiation process 辐射过程
radiation processing 辐射加工
radiation-producing 辐射的
radiation proof 防辐射的
radiation-proof container 耐辐射容器
radiation proof design 防辐射设计
radiation-proof glass fiber 防辐照玻璃纤维
radiation proofing test 耐光性试验
radiation property 辐射性质;辐射特性
radiation protection 辐射防护;放射防护;防辐射
radiation protection criterion 辐射防护基标;辐防护标准
radiation protection device 防辐射装置
radiation protection equipment 放射防护设备
radiation protection guide 辐射防护指南
radiation protection material 防辐射材料
radiation protection standard 辐射防护标准
radiation protection work clothes 放射性防护工作服
radiation-protective glass 防辐射玻璃
radiation protective paint 防辐射涂料
radiation pyrometer 辐射高温仪表;辐射高温计
radiation pyrometer temperature measurement 辐射高温计温度测量系统
radiation quality 辐射质量
radiation range 辐射范围
radiation recuperator 辐射(式)换热器
radiation reflection 辐射反射
radiation refrigerator 辐射制冷器
radiation relaxation time 辐射弛豫时间
radiation resistance 抗放射性;辐射阻力;辐射电阻
radiation resistance furnace 辐射电阻炉
radiation resistance gear oil 耐辐射齿轮油
radiation resistant 防辐射的;抗辐射性;抗放射的;抗辐射的
radiation resistant ceramic 耐辐射陶瓷
radiation-resistant electric(al) glass fiber 耐辐射电工玻璃纤维
radiation-resistant fiber 耐辐射玻璃纤维
radiation-resistant glass 耐辐射玻璃
radiation-resistant glass wool 耐辐射玻璃棉
radiation resisting paint 防辐线涂料
radiation resolution 辐射分辨率
radiation-retarding door 防射线门;防辐射门;辐射阻滞门;保温门

radiation-retarding frame 防射线门框;辐射阻滞门框;包铅皮樘子
radiation rib 散热筋片
radiation risk 辐射危险
radiation safety 辐射安全
radiation safety control 辐射安全控制
radiation sand shoal deposit 辐射沙洲沉积
radiation scattering 辐射散射
radiation scattering analyzer 辐射散射分析仪
radiation screen 辐射屏;辐射屏;防辐射屏
radiation section 辐射段
radiation self-decomposition 放射性自分解
radiation selling 放射式销售法
radiation sensitive center 辐射敏感中心
radiation sensor 辐射感应器;辐射传感器;射线传感器
radiation setting coating 射线固化涂料
radiation shelter 放射性物体掩蔽室
radiation shield block 防辐射单元;防辐射装置;防辐射砌块
radiation shield design 防辐射设计
radiation shield door 防辐射门
radiation shield(ing) 辐射屏蔽;辐射防护;防放射掩护;防辐射屏(蔽);辐射防护屏;辐射防护罩
radiation-shielding concrete (医院放射科墙壁使用的)防射线混凝土;防辐射混凝土;屏蔽混凝土
radiation-shielding design 辐射屏蔽设计
radiation-shielding designer 辐射屏蔽设计者
radiation-shielding door 辐射防护门;防辐射门;辐射屏蔽门
radiation-shielding expert 防辐射专家
radiation-shielding glass 辐射屏蔽玻璃;防辐射玻璃
radiation-shielding material 辐射屏蔽材料
radiation-shielding window 防射线窗
radiation shield wall 防辐射墙
radiation sickness 辐射病;放射性疾病;放射病
radiation skin 辐射包壳
radiation snowmelt 辐射融雪
radiation somatic effect 辐射躯体效应
radiation sound 辐射声
radiation source 照射光源;辐射源;放射源
radiation source in environment 环境中的放射源
radiation spectrometer 辐射谱仪
radiation spectrum 辐射(光)谱
radiation stabile 耐辐照度
radiation stabilizer 抗辐射稳定剂
radiation stable 对辐射作用稳定的
radiation standard 辐射标准
radiation station 辐射观测站
radiation sterilization 辐射消毒;辐射灭菌(法)
radiation sterilization of water 水的辐射消毒
radiation strength 放射性强度
radiation stress 辐射应力
radiation superheater 辐射过热器
radiation supervisor 剂量员
radiation surface 辐射面(积);辐射表面
radiation survey(ing) 辐射测量
radiation survey meter 辐射检测计;辐射测量仪;辐射测量计
radiation symmetry 辐射对称
radiation syndrome 放射综合征
radiation synthesis 辐射合成
radiation system 放射式传送制
radiation temperature 辐射温度
radiation test 辐射试验
radiation therapy department 防射线治疗部
radiation thermal exchange 辐射热交换
radiation thermocouple 辐射温差电偶;辐射热电偶
radiation thermometer 辐射温度计
radiation thermometry 辐射测温法
radiation thickness ga(u)ge 放射性元素测厚仪
radiation tolerance 容许照射剂量
radiation transducers 射线传感器
radiation transfer 辐射传递;放射传递
radiation transfer index 辐射传递率
radiation transfer rate 辐射转移率
radiation transformer 辐射变压器
radiation transmission 无线电发送
radiation transmitting glass 透辐射的玻璃
radiation transport and storage 辐射储运
radiation trap 辐射迷宫;辐射阱
radiation trapping 辐射俘获
radiation-trapping process 辐射俘获过程

radiation treatment 照射处理;辐射处理
radiation-triggered 辐射起动的
radiation type density meter 辐射式密度计
radiation type electric(al) supply 辐射供电
radiation type level detector 辐射式液位检测器
radiation-type pyrometer 辐射式高温计
radiation-type timbering 辐射式支撑
radiation unit 辐射单位;辐射长度
radiation vacuum ga(u)ge 放射真空计
radiation value 辐射值
radiation value of air 214Bi 大气铋214辐射值
radiation value of geologic(al) body 地质体辐射值
radiation value of radioactive body 放射性矿体辐射值
radiation value of radon daughter 氢气体辐射值
radiation vulcanization 辐射硫化
radiation wall 防辐射墙
radiation warning assembly 辐射报警装置
radiation warning symbol 放射性警告标志
radiation waste 辐射废物
radiation wasteheat boiler 辐射式废热锅炉
radiation weapon 辐射武器
radiation width 辐射宽度
radiation window 辐射窗
radiation work 辐射工作
radiation zone 夫琅和费区
radiative 辐射的
radiative braking 辐射制动
radiative broading 辐射加宽
radiative capture 辐射俘获
radiative collision 辐射碰撞
radiative cooling bolometer 辐射致冷测辐射热计
radiative correction 辐射(场)改正
radiative data code 辐射数据编码
radiative detonation model 辐射引爆模型
radiative diffusivity 辐射扩散率
radiative dissociation 辐射离解
radiative effect 辐射效应
radiative efficiency 辐射效率
radiative envelope 辐射壳
radiative gear 放射齿轮
radiative heat transfer 辐射热传动
radiatively active gas 影响辐射效应气体
radiatively important gas 对辐射产生重要效应的气体
radiative mode 辐射模式
radiative properties of gas 气体的辐射特性
radiative recombination 辐射复合
radiative recombination coefficient 辐射复合系数
radiative recombination lifetime 辐射复合寿命
radiative relaxation 辐射弛豫
radiative temperature 辐射温度
radiative temperature gradient 辐射温度梯度
radiative transfer 辐射转移
radiative transition 辐射跃迁
radiative transition probability 辐射跃迁概率
radiator 铸铁散热器;取暖器;散热器;辐射子;辐射体;辐射器
radiator air baffle 散热(器)阻流板
radiator air deflector 散热器挡风板
radiator air passage 散热器空气通道
radiator and core testing stand 散热器及芯子试验台
radiator antifreeze solution 散热器防冻剂;散热器防冻溶液
radiator apron 散热器保温罩
radiator baffle 散热器阻板;散热器挡板
radiator bar 散热器支杆
radiator base 散热器座
radiator bearer 散热器支架
radiator blanket 散热器毡帘(防冻帘)
radiator bleeder tap 暖气片的排出孔
radiator blind 散热器百叶窗
radiator blind plank 散热器百叶窗板条
radiator block 散热器组
radiator bonnet 散(热器)罩
radiator bottom header 散热器下水箱
radiator bottom plate 散势器底板
radiator bottom tank 散热器下集水槽
radiator box 暖气罩;散热(器)罩
radiator brace 散热器支架角撑
radiator brace rod 散热器拉杆
radiator bracket 散热器托架

radiator brush guard 散热器护栅
radiator bumper 散热器弹性支架
radiator bumper rod 散热器防冲杆
radiator cap 水箱盖;散热器加水盖
radiator cap tester 散热器水箱盖测试器
radiator-cap thermometer 散热器盖温度计
radiator case 散热器箱
radiator case frame 散热器箱形框
radiator center 散热器芯子
radiator chaplet 螺旋心撑;盘香式芯撑
radiator clamping strip 散热器夹条
radiator cleaner 冷却器洗涤剂;散热器洗涤剂;散热器清洁器
radiator cleaning compound 散热器清洗剂
radiator clip 散热器软管夹箍
radiator closing nut 散热器螺母
radiator coat 散热器上涂层
radiator coil 散热装置蛇管;散热器蛇形管
radiator connection 散热器连接管
radiator control valve 散热器上控制阀
radiator coolant heater 散热器冷却水加热器
radiator cooling fin 散热器冷却片
radiator cooling tube 散热器冷却管
radiator core 散热器芯(子)
radiator core adapter 散热器心子接头
radiator core cleaner gun 散热器芯部清洗枪
radiator core fin 散热器芯片
radiator core guard 散热器芯护罩
radiator core section 散热器芯子单元
radiator core support 散热器芯子支架
radiator cosy 散热器保温罩
radiator cover 散热器罩;散热器保温罩
radiator cowling 散热器罩壳
radiator cross member 散热器横条
radiator damper 散热器百叶窗
radiator drain 散热器放水口
radiator drainage valve 散热器放水开关
radiator drain cock 散热器放水旋塞
radiator drain outlet elbow 散热器放水弯头
radiator drain plug cover 散热器放水塞盖
radiator draw-off (cock) 散热器放水塞
radiator draw-off plug 散热器放水塞
radiator duct 散势器导管;散热器导管
radiator element 散热器单片;散热器元件
radiator enclosure 散热器罩
radiator false front 散热器前护栅
radiator fan 散热(器)风扇
radiator-fan belt 散热扇带
radiator fan shroud 散热器风扇罩
radiator fender 散热器护网
radiator filler 散热器加水口
radiator (filler) cap 散热器盖;散热器加水盖
radiator filler cap gasket 散热器加水口密封垫
radiator filler gun 散热器注液枪
radiator filler with cap 散热器加水口
radiator fin 散热器散热片;散热(器)片
radiator flap 鱼鳞板;散热器风门片;散热风门
radiator flap handle 散热器百叶窗操纵手柄
radiator flush 散热器洗涤剂;散热器清洗剂
radiator frame 散热器框架
radiator gasket 散热器密封垫
radiator grate 炉片格栅;散热器护栅
radiator grid 散热器(格)栅
radiator grill(e) 炉片格栅;散热器(前)护栅;散热器百叶窗
radiator grill molding 散热器护栅嵌条
radiator guard 暖气片的防护罩;散热罩;散热器罩;散热器护罩;散热器护网
radiator header 散热器上水箱
radiator heating 散热器供暖
radiator heating system 散热器采暖系统
radiator hood ledge 散热器罩支架
radiator hose 散热器软管;散热器皮管
radiator hose connection 散热器软管接头
radiator inlet and outlet hose 散热器进出水软管
radiator inlet fitting 散热器进水管配件
radiator inlet hose clamp 散热器进水软管夹
radiator inlet strainer 散热器进水滤清器
radiator leak stop 散热器防漏剂
radiator liquid 散热器用液
radiator locking cap 散热器加水盖
radiator louver 散热器放气孔
radiator louver antisqueak 散热器百叶窗减声片
radiator louver frame 散热器百叶窗框架

radiator louver shaft 散热器百叶窗轴
radiator lower tank 散热器下水箱
radiator mounted fan 散热器所安装的风扇
radiator mounting 散热器固定架
radiator mounting pad 散热器座垫
radiator mounting spring 散热器座弹簧
radiator muff 散热器保温罩
radiator niche 放暖气片的壁龛
radiator of sound 声辐射器
radiator ornament 散热器饰件
radiator outlet fitting 散热器出水管配件
radiator outlet hose 散热器出水软管
radiator outlet hose clamp 散热器出水软管夹
radiator outlet pipe 散热器出水管
radiator overflow tank 散热器溢流箱
radiator overflow tube 散热器溢流管
radiator pad 散热器衬垫
radiator paint 散热器油漆;散热器用漆
radiator panel 散热器组
radiator priming paint 散热器底漆
radiator protection 散热器防护装置;防辐射保护
radiator relief valve 散热器安全阀
radiator repair cement 散热器堵漏油灰
radiator return fitting 暖气片的回水装置
radiator rust preventive 散热器防锈液
radiator rust resister 散热器防锈液
radiator screen 散热器保护栅;散热器保温帘
radiator screw-cap 散热器加水盖
radiator section 散热片;暖气片
radiator shell 散热器(外)壳
radiator shell apron 散热器罩的挡板
radiator shell lacing 散热器壳衬带
radiator shell molding 散热器罩嵌条
radiator shell support 散热器壳支架
radiator shield 散热器保温罩
radiator shroud 散热器护罩
radiator shutter 散热器(百叶)气窗;散热器百叶窗
radiator shutter control 散热器百叶窗操纵
radiator sreeen 散热器护栅
radiator stand 散热器(支)架
radiator stay rod bracket 散热器撑杆固定架
radiator stay rod pin 散热器撑杆固定销
radiator stoneguard 散热器前护栅
radiator strainer 散热器粗滤器
radiator strut 散热器支柱
radiator support 散热器支架
radiator support bracket 散热器支架;散热器托架
radiator support cushion 散热器支架垫
radiator surge tank 散热器平衡水箱
radiator tank 散热器水箱
radiator temperature drop 散热器温度降落
radiator thermometer 散热器温度计
radiator thermostat 冷却系恒温器
radiator tie rod 散热器拉杆
radiator top frame 散热器上支架
radiator top tank 散热器上水箱
radiator trap 散热器疏水阀
radiator trunnion 散热器耳轴
radiator tube 散热器管
radiator-type cooling unit 散热式冷却装置
radiator-type transformer 散热器型变压器;散热器式变压器
radiator upper tank 散热器上水箱
radiator valve 散热器(调节)阀
radiator vent line 散热器通气管路
radiator vent pipe 散热器通风透气管;散热器通风管
radiator water-level 散热器水位
radiatus 辐辏状云
radiaxial turbine 混流式涡轮机;混流式透平;混流式水轮机;混流式汽轮机;辐向轴流式涡轮机
radical 根数;根式;根本的
radical axis 根轴;幂线
radical beat eccentricity 径向摆动量
radical center 等幂心
radical change 根本变化;彻底改变
radical circle 根圆
radical copolymerization 游离基共聚合
radical cross linking 自由基交联
radical diffusion model 射ραδ扩散模型
radical drainage 放射状水系
radical drainage pattern 放射状水系形式
Radical Eclecticism 激进折衷主义
radical equation 无理方程
radical expression 根式

radical logging 放射状测井
radical of a Jordan algebra 约当代数的根
radical of a vector lattice 向量格的根
radical outward flow 外向径向流
radical play eccentricity 径向跳动量
radical polymerization 游离基聚合;自由基聚合
radical principle 基本原理
radical ray 放射线
radical reduction 大减价
radical scavenger 游离基清除剂
radical sign 根号
radical survey 放射状测量
radical telomerization 游离基型调聚反应
radical texture 放射轴结构
radical transfer 基团转移
radicand 被开方数
radication 开方
radices 根值数
radician 远程预警线操作员
radicidation 辐射杀菌
radicle 根端部
radico functional name 根基官能名称
radicula 根端部
radii lenti 晶状体辐射线
radii loss 半径损耗
radiiophotoluminescent dose glass 辐射光致法光剂量玻璃
radii vectors 矢径
radio 无线电;射电;放射性的
radio absorption measurement 无线电吸收测量
radio acoustic(al) method 无线电声学法;电声法
radio acoustic(al) position finding 无线电声响定位;无线电声定位;声呐定位;电声测距(法)
radio acoustic(al) range finding equipment 无线电声波测距仪
radio acoustic(al) (sound) ranging 电声测距(法);无线电回声测距法;无线电声(响)测距
radio acoustics 无线电声学
radioactinium 放射性锕
radioactivation 辐射活化;辐射激活;放射性活化
radioactivation analysis 放射(性)活化分析;放射化分析
radioactive 放射性的
radioactive aerosol 放射性气溶胶
radioactive aerosol monitor 放射性悬浮微粒监测仪
radioactive age determination 放射性(碳)年代测定法;放射性测定年代
radioactive air monitor 放射性空气监测仪
radioactive air pollution 放射性空气污染
radioactive air sampler 放射性空气取样器
radioactive analysis survey meter model 放射分析测量仪型号
radioactive analyzer 放射性分析器
radioactive apparatus 放射性仪器
radioactive ash 放射性尘埃
radioactive assay 放射性验定
radioactive atom 放射性原子;放射同位素原子
radioactive background 放射性本底
radioactive beam monitor 放射线监测仪
radioactive breakdown 放射性损伤
radioactive brick 放射性砖
radioactive calcium 放射性钙
radioactive carbon 放射性碳
radioactive carbon dating 放射性碳测定年龄
radioactive cement 放射性水泥(测定水泥在管外上升高度用)
radioactive cemetery 倾倒放射性废物场所
radioactive chain 放射性系
radioactive change 放射性变化
radioactive cleaning wastewater 清除放射性废水
radioactive cleaning wastewater treatment 放射性洗涤废水处理
radioactive clock 放射性时计
radioactive cloud 放射云;放射性烟云
radioactive cobalt 放射性钴
radioactive colloid 放射性胶体
radioactive compound 放射性化合物
radioactive concentration 放射性浓度
radioactive concentration indicator 放射性浓度计
radioactive conductance 辐射电导
radioactive constant 放射性常数
radioactive contaminant 放射性污染物
radioactive contamination 放射性沾污;放射性沾染;放射性污染
radioactive contamination monitoring 放射性污染监测
radioactive contamination of soils 土壤的放射性污染
radioactive content 放射性含量
radioactive counting 放射性计数
radioactive damage 放射性物质引起的损害;放射性损伤;放射性损害
radioactive dating 放射性(碳)年代测定法;放射性测定年代
radioactive daughter 放射性子体
radioactive daughter isotope 放射性子同位素
radioactive debris 放射性碎屑;放射性碎片
radioactive decay 放射性衰变
radioactive decay chain 放射性衰变链
radioactive decay constant 放射性衰变常数
radioactive decay law 放射性衰变律
radioactive decay phenomenon 放射衰变现象
radioactive decay product 放射性衰变产物
radioactive decay rate 放射性衰变率
radioactive decay rule 放射衰变规律
radioactive decay scheme 放射性衰变图式
radioactive decay series 放射性系;放射性衰变系
radioactive decontamination 消除放射性沾染;辐射去污;放射性污染清除;放射性去污;放射性净化
radioactive decontamination agent 放射性去污剂
radioactive decontamination of soils 土壤的放射性净化
radioactive densiometer 放射性同位素密度计
radioactive density meter 放射性密度计
radioactive deposit 放射性淀积物;放射性沉降(物);放射性沉积(物)
radioactive detector 放射性探测器;放射性检测器
radioactive determination 放射性测定
radioactive discharge 放射性排弃物
radioactive disintegration 放射性蜕变;放射性衰变;放射性分解
radioactive displacement law 放射性位移定律
radioactive distribution 放射性分布图谱
radioactive dust 放射性微尘;放射性尘埃
radioactive effect 放射性效应
radioactive effluent 放射性排出物;放射性流出物;放射性废水
radioactive effluent disposal 放射性流出物处置
radioactive electron capture detector 放射性电子俘获检测器
radioactive element 放射性元素
radioactive element analysis 放射性元素分析
radioactive element decay 放射元素衰变
radioactive emanation 放射性射气
radioactive energy 辐射能;放射能
radioactive equilibrium 辐射平衡;放射性平衡
radioactive fallout 沉降物;放射状沉降;放射性下落灰;放射性微粒回降;放射性散落物;放射性落下灰;放射性落尘;放射性回降物;放射性沉降灰;放射性沉降(物);放射性降落物
radioactive filling 放射性填充
radioactive fission product 放射性裂变产物
radioactive flow analyzer 放射性流分析仪
radioactive flowmeter 放射性同位素流量计
radioactive gas 放射性气体
radioactive gas decay tank 放射性气体衰变箱
radioactive gas meter 放射性气体仪
radioactive goods 放射性物品
radioactive half-life 放射性半衰期;半衰期
radioactive hazard 放射性公害
radioactive heat 放射性脱变热
radioactive heating surface 放射性辐射加热面
radioactive heat production 放射性热生产
radioactive heat source 放射性蜕变热源
radioactive-hydrogeologic(al) survey 放射性水文地质调查
radioactive impurity 反射杂质
radioactive index 放射性指数
radioactive indicating organism 放射性指示种
radioactive indicator 放射性指示剂;放射性示踪剂
radioactive indicator method 放射性指示剂法
radioactive ink 放射性油墨
radioactive instrumentation 放射性监测仪表
radioactive intensity of tracer 示踪剂放射性强度
radioactive ionization ga(u)ge 放射性电离计
radioactive isotope 放射性同位素
radioactive isotope dilution method 放射性同位

素稀释法(测流方法之一)
radioactive isotope radio-graphy 放射性同位素射线照相检验术
radioactive isotope survey 放射性同位素监测仪
radioactive isotope tracer 放射性同位素示踪物;放射性同位素示踪剂
radioactive leak 放射性泄露;放射性漏泄
radioactive level 放射性水平
radioactive level ga(u)ge 放射性液位计;放射性液面计
radioactive level indicator 放射性同位素料位计
radioactive level-meter 放射性同位素料位计
radioactive lighting 荧光照明
radioactive liquid waste 放射性液体废弃物;放射性废液
radioactive logger 放射性测井仪
radioactive logging 放射性测井
radioactive logging equipment 放射性测井设备
radioactive logging method 放射性测井法
radioactive luminescent paint 放射性发光涂漆
radioactive luminous paint 放射性夜光漆
radioactive material 放射性物质;放射性材料
radioactive material contamination 放射性物质污染
radioactive metal 放射性金属
radioactive metal commodities 放射性金属矿产
radioactive method 放射性方法
radioactive micrometer 放射性测微计
radioactive mineral 放射性矿物
radioactive mixed waste 放射性混合废液
radioactive nature 放射特性
radioactive nitrogen 射氮
radioactive nuclear species 放射性核类
radioactive nucleus 放射性核
radioactive nuclide 放射性核素
radioactive-ore detector 放射性矿物探测器
radioactive origin 放射成因
radioactive paint 放射性涂料;放射性发光颜料
radioactive parent 放射性母体
radioactive parent isotope 放射性母同位素
radioactive particle 放射性微粒
radioactive particle protection 放射性微粒保护
radioactive pebble 放射性卵石
radioactive phosphorus 标志磷
radioactive photography 放射性照相法
radioactive pigment 放射性(发光)颜料
radioactive piston ring 放射性活塞环
radioactive poisoning 放射性中毒
radioactive pollutant 放射性污染物
radioactive pollution 放射性(的)污染;反射沾染
radioactive pollution control 放射性污染控制
radioactive pollution of water body 水体放射性污染
radioactive pollution sources 放射性污染源
radioactive pottery 放射性陶器
radioactive precipitation analysis 放射性沉淀分析法
radioactive probe 放射性探测器
radioactive process 辐射过程
radioactive processing 放射性(物质)处理
radioactive prospecting 放射性找矿;放射性勘探;放射能测井
radioactive purity 放射纯度
radioactive radiation 放射性照射;放射性辐射
radioactive rain 放射性雨
radioactive rare metal 放射性稀有金属
radioactive ray 放射线
radioactive release 放射性释放;放射性释出
radioactive resistance 辐射阻器
radioactive salt 放射性盐
radioactive sample 放射性样品
radioactive sample analyzer[analyser] 放射性样品分析仪
radioactive sand 放射性砂
radioactive secular equilibrium 放射性长期平衡
radioactive sediment tracer test 泥沙放射性示踪试验
radioactive selfabsorption 放射性自吸收
radioactive sensing device 放射性传感器
radioactive sensor 放射性传感器
radioactive series 放射系列
radioactive settlement 放射状沉降
radioactive single radial diffusion 放射单幅射状扩散

radioactive snow ga(u)ge 放射性量雪计
radioactive solid wastes 放射性固体废物
radioactive solid waste treatment 放射性固体废物处理
radioactive solution ga(u)ging 放射性溶液测流速法;放射性溶液测流;放射性溶液测量
radioactive source 放射源
radioactive source of water pollution 水污染放射性源
radioactive source shielding 屏蔽辐射源
radioactive sources of water pollution 水放射性污染源
radioactive source term 放射性源顶
radioactive spring 放射性泉
radioactive standard 放射性标准
radioactive standard source 放射性标准源
radioactive substance 放射性物质
radioactive surface contamination 放射性表面污染
radioactive test 放射性试验
radio-active testing 同位素试验
radioactive thickness ga(u)ge 放射性同位素厚度仪;放射性厚度计;放射性测厚计
radioactive trace 放射性示踪原子
radioactive trace method 放射示踪法
radioactive tracer 放射性示踪器;放射性示踪物;放射性示踪剂;示踪原子
radioactive tracer element 放射性示踪元素
radioactive tracer log 放射性示踪剂测井
radioactive tracer method 放射性示踪法
radioactive tracer technique 放射性示踪技术
radioactive tracing 放射性示踪法
radioactive transfer 辐射转移;辐射传输;放射性迁移
radioactive transfer code 辐射传输规律
radioactive transformation 放射性变化
radioactive transient equilibrium 放射性暂时平衡
radioactive tritium source 放射性氚源
radioactive variation 放射性变化
radioactive waste 放射性废(弃)物;放射性废物玻璃固化;放射性废料
radioactive waste concentration 放射性废物浓度
radioactive waste conditioning 放射性废物调理
radioactive waste cooling tank 放射性废物冷却槽
radioactive waste disposal 放射性废物处置
radioactive waste disposal method 放射性废物处置方法
radioactive waste facility 放射性废物设施
radioactive waste final disposal 放射性废物最终处置
radioactive waste liquid 放射性废液
radioactive waste management 放射性废物管理
radioactive waste processing and disposal 放射性废物处理和处置
radioactive waste release 放射性废物释放
radioactive waste repository 放射性废物库
radioactive waste storage 放射性废物储[贮]存
radioactive waste treatment 放射性废物处理
radioactive waste treatment facility 放射性废物处理装置
radioactive waste water 放射性废水
radioactive wastewater treatment 放射性废水处理
radioactive water 放射性水
radioactive wave 放射波
radioactive wear measuring 放射性同位素法测量磨损
radioactivity 辐射性;放射学;放射性渠道;放射性强度;放射性(活度);放射现象
radioactivity absorber 放射线物质吸收器;放射性物质吸收剂
radioactivity analysis 活化分析
radioactivity anomaly 放射性异常
radioactivity build-up 放射性积累
radioactivity concentration 放射性浓度
radioactivity concentration guide 放射性浓度指导限值
radioactivity decay 放射性衰变
radioactivity density 放射性比度
radioactivity detection 放射性探测
radioactivity determination 放射性测定;放射法测定
radioactivity index 放射性指数
radioactivity in the ocean 大洋放射性
radioactivity level 放射性程度
radioactivity limits of release 放射性排放标准

radioactivity log(ging) 放射性测井
radioactivity logging (method) 放射(性)测井法
radioactivity meter 放射性测量计
radioactivity monitor 放射性监测器
radioactivity of minerals 矿物放射性
radioactivity of precipitation 降落物放射性
radioactivity of water 水的放射性
radioactivity property of rock and soil 岩土的放射性
radioactivity prospecting 放射性勘探
radioactivity protection 放射性防护;放射性保护
radioactivity ratio 放射性比(值)
radioactivity release 放射性产物排放
radioactivity seeker 放射性探测器
radioactivity shielding 放射性屏蔽
radioactivity spectrum analyzer 放射性能谱分析仪
radioactivity strength 放射性强度
radioactivity survey 放射性勘探
radioactivity testing 放射性测试
radioactivity (well) logging 放射性测井
radio adaptation 无线电操作
radio aerial 无线电天线
radioaerosol 放射性气溶胶
radio aids 无线电航标;无线电助航设施;无线电导航设备
radio aids to navigation 无线电导航设备;无线电助航设备
radio aid to air navigation 无线电空中导航
radio aiming device 无线电瞄准装置
radio alarm and control system 无线电报警和控制系统
radio alarm signal 无线电求救信号;无线电警报信号
radioallergo sorbent test 放射变应原吸附试验
radio altimeter 无线电测高仪;无线电测高计
radio altimeter antenna 无线电高度计天线
radio altimeter indicator 无线电高度计指示器
radio altimetry 无线电测高法
radio altitude 雷达高度;无线电高度
radio altitude indicator 无线电高度指示器
radio amplifier 无线电放大器
radioanalysis 放射性分析
radioanalytic(al) chemistry 放分析化学
radioanalytic(al) technique 放射性分析技术
radio and chart room 无线电海图室
radio and electric(al) pliers 电信钳
radio and panel section 无线电与布板通信分排
radio and telecommunication 电台和电信通讯
radio and telephone control 无线电及电话管制
radio and television aid to navigation 无线电与电视导航工具
radio and television engineering 无线电与电视工程
radio and wire integration 无线与有线综合系统设施
radio angle 无线电方位角
radio antenna 无线电天线
radio antenna installation 无线通信天线安装
radio antenna truck 无线电天线车
radio apparatus 无线电设备;无线电机
radio appliance factory 无线电器材厂
radio approach aids 无线电进场导航设备
radio-arm box 电阻电桥箱
radioassay 放射性鉴定;放射性检验;放射分析;放射(性)测量
radioassay detector 放射鉴定检测器
radio astrometry 射电天体测量学
radio astronomer 射电天文学家
radio astronomical observatory 射电天文台
radio astronomical station 射电天文站
radio astronomy 射电天文(学)
radio astronomy measuring methods 射电天文学测量法
radio astronomy observation 射电天文观察
radio astronomy satellite 射电天文卫星
radio astronomy service 射电天文业务
radio astronomy station 射电天文台
radio atmometer 辐射蒸发计
radio atmosphere 无线电大气
radio attenuation 射电衰减
radio attenuation measurement 无线电衰减测量
radio aurora 模拟无线电极光;无线电极光
radio autocontrol 无线电自动控制;无线电自动调整
radio autogram 无线电传真;自动射线照相;放射

自显影图
radioautograph 自体放射照片;射线显迹图;放射自显影(照片);放射显迹图
radioautography 自体放射照相术;放射自显影法
radioaxial mosaic 放射轴嵌晶
radiobarite 放射性重晶石
radio base station 无线基台
radio bay 无线电设备舱
radio B battery 无线电用乙电池组
radio beacon 无线电指向标;无线电信标;无线电警标;无线电航标;电指向标
radio beacon buoy 无线电导航信号浮标;无线电导航测标
radio beacon course 无线电信标航向
radio beacon diagram 无线电指向标分区图
radio beacon facility 无线电信标设备
radio beacon identification 无线电信标识别
radiobeacon in general 一般无线电指向标
radio beacon marker 无线电指点信标台
radio beacon monitor station 无线电指向标监视站;无线电信标监视站
radio beacon network 无线电信标网
radio beacon range 无线电指向标射程
radio beacon receiver 无线电信标接收机
radio beacon set 无线电信标机
radio beacon station 无线电指向标站;无线电导航信号台
radio beacon system 无线电标系信统
radio beam deviation indicator 无线电波束偏离指示器
radio beam guidance 无线电波束制导
radio beam transmitting 定向无线电发送
radio bearing 无线电方位;无线电定向
radio bearing chart 无线电方位图
radio bearing conversion table 无线电方位换算表
radio bearing deviation 无线电方位偏差
radio bearing installation 无线电定向仪
radiobiochemistry 放射生物化学
radiobiologic(al) action 放射生物作用
radiobiologic(al) dosimetry 放射生物剂量学
radiobiologic(al) effect 辐射生物效应;放射生物效应
radiobiologic(al) laboratory 放射生物实验室
radiobiologic(al) protection 放射线防护
radiobiology 辐射生物学;放射生物学;放射生态学
radiobiology effect 放射生物学效应
radio blackout 无线电中断
radio blackout period 无线电中断时间
radio blind landing equipment 无线电盲目着陆设备
radio bomb fuze 无线电引信
radio bridge 射电桥
radio brightness 射电亮度
radio brightness temperature 射电亮温度
radio broadcasting 无线电广播;无线电播发
radio broadcasting channel 无线电广播通道
radio broadcasting equipment 无线电广播设备
radio broadcasting station 广播电台;无线电广播站
radio buoy 无线电浮标
radio cable 高频电缆
radiocalcium 放射性钙
radio call letters 无线电呼号
radio call sign 无线电呼号
radio call signal 无线电呼号
radio camera 无线电摄像机
radio canal 无线电波道
radio capsule 微型电子内诊器
radio car 警务车;无线电通信车;无线电汽车
radiocarbon 炭同位素;放射性碳
radiocarbon age 放射性碳龄
radiocarbon chronology 放射性碳年代学;放射性碳测定年代法
radiocarbon dating 碳同位素年龄测定;放射性碳年代测定法;放射性碳测定年龄
radiocarbon method 放射性碳方法
radiocarcinogenesis 辐射致癌;放射致癌作用
radio cast 无线电广播
radio celestial navigation system 无线电天文导航系统;射电天文导航系统
radio center 无线电中心
radio central office 无线电中央台;无线电总局
radio ceramics 高频陶瓷;无线电陶瓷
radiocesium 放射性色
radio channel 无线电信道;无线电波道
radio channel jammer 无线电通信干扰台

radio chart 无线电信标导航图;无线电定向图
radio chassis 无线电仪器底板
radio check 无线电通信检查
radiochemical analysis 放射化学分析
radiochemical analyzer 放射化学分析仪
radiochemical behavio(u)r 放射化学行为
radiochemical contamination 放射化学污染
radiochemical engineering 放射化学工程
radiochemical laboratory 放射化学实验室
radiochemical method 放射(性)化学法
radiochemical neutron activation analysis 放化中子活化分析
radiochemical polarography 放射化学极谱法
radiochemical processing plant 放射化学处理工厂
radiochemical pure 放射化学纯
radiochemical purity 放射化学纯度
radiochemicals 放射化学试剂
radiochemical separation 放射化学分离
radiochemical synthesis 放射化学合成
radiochemical test 放射化学检定
radiochemical yield 放射化学产额
radiochemistry 放射(性)化学
radiochlorine 放射性氯
radiochroism 放射吸收性
radiochromatic dose glass 变色剂量玻璃
radiochromatogram 放射色谱;辐射色谱图
radiochromatograph 放射色谱仪
radiochromatographic separation 辐射色谱分离法
radiochromatographic technique 放射性色谱技术
radio chromatography 放射色谱法
radiochromatography 辐射色谱法;放射层析法
radiochrometer 射线硬度测定计
radiochronology 放射测年学
radio circuit 高频电路;无线电通路;无线电电路
radio circuit discipline 无线电通信纪律
radio climatology 无线电气候学
radiocobalt 放射性钴
radio code 无线电通信电码
radio code aptitude area 无线电通信简语适用范围
radio coil 无线电线圈
radiocolloid 放射性胶体;放射性胶质
radio command 无线电指令
radio command channel 无线电指挥波道
radio command control system 无线电指挥控制系统
radio command guidance 无线电指令制导
radio command guidance system 无线电指令制导系统
radio command indicator 无线电指令指示器
radio command link 无线电指令链路
radio command system 无线电指令系统
radio command unit 无线电指令控制装置
radio common channel 共用无线电信道
radio communication 无线电通信;无线传输
radio communication beyond the horizon 视距外无线电通信
radio communication diagram 无线电通信线路
radio communication engineering 无线电通信工程
radio communication equipment 无线电通信设备
radio communication frequency 无线电通信频率
radio communication guard 无线电警戒
radio communication line of centimetric wave 微波无线电通信线路
radio communication net 无线电通信网
radio communication range 无线电通信距离
radio communication set 无线电通信设备
radio communication statistics 无线电通信统计
radio communication system 无线通信系统
radio communication transmitter 无线电通信发射机
radio compartment 无线电舱
radio compass 自动测向器;无线电罗盘;无线电罗
radio compass bearing indicator 无线电罗盘方位指示器
radio compass calibration data 无线电罗盘校准数据
radio compass error 无线电罗盘误差
radio compass fix 无线电罗盘定位
radio compass indicator 无线电罗盘指示器
radio compass receiver 无线电罗盘接收机
radio compass sensor 无线电罗盘传感器
radio compass station 无线电定向台;无线电定向站
radio-competitive assay 放射竞争分析

radio connection 无线电连接
radio contact 无线电联络
radiocontamination 放射性污染
radio continuum 射电连续辐射
radio contour 射电轮廓图
radio control 无线电控制
radio control and metering service 无线电遥控遥测站
radio control box 无线电操纵台
radio control bulldozer 无线电操纵推土机
radio controlled 无线电控制的
radio controlled aerial-navigation 无线电控制空中航行区
radio controlled aerial target 无线电控制的空中目标
radio controlled buoy 无线电操纵的浮标
radio controlled center 无线电控制中心
radio controlled locomotive 无线电控制机车
radio controlled pump station 无线电操纵泵站
radio controlled radar 无线电控制雷达
radio controlled station 无线电控制站
radio controlled switch engine 无线电控制的调车机车
radio controlled target boat 无线电遥控靶艇
radio controlled traffic 无线电控制的交通
radio controlled train operation 无线电控制的列车运行
radio control operation 无线电控制运行
radio control panel in driver's desk 驾驶室无线电控制屏
radio control receiver 无线电对讲接收机
radio control relay 无线电控制继电器
radio control system 无线电监控系统
radio copying telegraph 无线电传真电报
radio corona 射电晕
radio countermeasures 无线电干扰
radiocounting 放射性计数
radio course 无线电指向标波束
radio coverage 无线电设备作用范围
radiocrystallography 射线晶体学;辐射晶体学
radiocurable binder 可辐射固化基料
radio current meter 无线电海流计
radio data 无线电数据
radio data link 无线电数据传输线路
radio data link set 无线电数据通信设备
radio data set 无线电数据设备
radiode 镭植入管
radiodermatitis 放射性皮炎
radio detecting and ranging 无线电探测
radio detection 无线电探测;无线电警戒
radio detection and location 无线电探测与定位
radio detection and ranging 无线电定向与测距;无线电定位
radio detector 无线电探测器
radio detector equipment 雷达设备;无线电探测设备
radio determination 无线电判位;无线电测定
radio determination satellite service 卫星无线电测量业务
radio determination station 无线电定位电台
radio detonator 无线电信管
radiodiagnosis 放射性诊断;反射诊断
radio diameter 射电直径
radio diffusion 无线电散射
radio diffusion van 广播车
radio digital system 无线电数字系统
radio digital terminal 无线电数字终端
radio directing and ranging 雷达
radio directional bearing 无线电定位
radio direction(al) finder 无线电测向仪;无线电定向仪;无线电测向台;无线电测向器;无线电向计
radio direction finder error 无线电测向仪误差
radio direction finder method 无线电测向法;无线电定向法
radio direction finder station 无线电测向台
radio direction finding 无线电定向;无线电测向
radio direction finding chart 无线电定向图
radio direction finding control station 无线电测向控制台
radio direction finding data base 无线电测向数据库
radio direction finding network 无线电测向网
radio direction finding receiver 无线电测向定向

仪;无线电测向接收机
radio direction finding station 无线电定向台;无线电测向站;无线电导航站;无线电导航台;无线电测向台
radio direction finding system 无线电导航系统;无线电测向系统
radio directive device 无线电定向设备
radio director indicator 无线电指引仪
radio dish 射电抛物面天线
radio dispatching system for train 无线电调度防灾系统【铁】
radio distance finder 无线电测距器
radio distance -finding set 无线电测距仪
radio distance-finding station 无线电测距站
radio distance-measuring 无线电测距
radio distance-measuring set 无线电测距仪
radio distress frequency 无线电求救信号频率
radio distress signal 无线电求救信号
radio distress signal watcher 监听无线电求救信号
radio disturbance 无线电干扰
radio division 无线电通信处
radio Doppler 多普勒无线电技术
radio Doppler effect 射电多普勒效应
radio duct 无线电波导
radio echo 雷达反射;无线电回波
radio echo detector 无线电回波探测器
radio echo method 无线电回波法
radio echo observation 无线电回波探测(法)
radio echo sounding 无线电回波探测(法)
radioeclipse 射电食
radioecological concentration 放射生态学浓集
radioecology 辐射生态学;放射生态学
radioelectrochemical analysis 放射电化学分析
radioelectrochemistry 放射电化学
radioelectrophoresis 放射电泳
radio electrophoretic method 放射电泳法
radioelement 放射(性)元素
radio emergency installation 应急无线电装置
radio emission 无线电辐射;无线电发射;射电辐射
radio energy 射电能量
radio engineering 无线电工程
radioenvironmental chemistry 放射环境化学
radio equipment 无线电装备;无线电设备
radio equipment bay 无线电设备舱
radio equipment factory 无线电器材厂
radio equipped inertial navigation system 无线电惯性导航系统
radio-equipped inertial navigation system 装有无线电的惯性导航系统
radio escutcheon panel 无线电仪表板
radio evapo(u)rimeter 辐射蒸发计
radioexamination 伦琴射线透视法;射线检验(法)
radio exchange 无线电交换台
radio facility 无线电设备
radio facility chart 无线电导航图
radio facsimile 无线电传真
radio facsimile equipment 无线电传真设备
radio facsimile system 无线电传真系统
radio factory 无线电厂
radio fade-out 电波消失;无线电波消失
radio fall-out 放射尘埃
radio fan marked beacon 无线电扇面标识浮标
radio fan marker beacon 扇形无线电指向标;扇形指点标
radioferrikinetics 放射性铁动态
radio field intensity 高频场强度;无线电场强
radio field-to-noise ratio 射电量噪声比
radio firing device 无线电发火装置
radio fix 无线电定位;无线电台定位;无线电船位
radio fixing aids 无线电定位设备
radiofluorescence 射电荧光;辐射荧光;放射荧光
radio flux 射电辐射流量
radio flying 无线电导航飞行
radio fog signal 无线电雾号
radio forecast 无线电波传播情况预报
radio-frequency 射频;无线电频率;高周波;射电频率
radio-frequency accelerating potential 射频加速电压
radio-frequency accelerating structure 射频加速结构
radio-frequency acceleration 射频加速
radio-frequency accelerator 射频加速器
radio-frequency adapter 高频适配器;射频转接器;射频适配器

radio-frequency alternator 高频同步发电机;射频交流发电机;射频发电机
radio-frequency ammeter 射频电流计
radio-frequency amplification 射频放大
radio-frequency amplifier 射频放大器
radio-frequency amplifier gain 射频放大器增益
radio frequency analyzer 无线电频率分析仪;射频分析器
radio-frequency arcing 射频击穿
radio-frequency attenuation 射频衰减
radio-frequency bandwidth 射频带宽度
radio-frequency beacon 射频信标机
radio-frequency beam 无线电射束
radio-frequency bridge 高频电桥
radio-frequency buncher 射频聚束器
radio-frequency bunching 射频聚束
radio-frequency cable 射频电缆
radio-frequency capture 射频俘获
radio-frequency carrier 射频载波
radio-frequency carrier shift 射频载波漂移
radio-frequency carrying surface 射频电流传导面
radio-frequency cavity 射频空腔;射频共振腔
radio-frequency channel 无线电频道;射频通道;射频波道;射频波导
radio-frequency channel synchronization 射频信道同步
radio-frequency characteristic measuring set 射频特性测试仪
radio-frequency chart 射频图
radio-frequency check receiver 射频监测接收机
radio-frequency choke 高频阻流圈;射频扼流圈
radio-frequency coil 高频线圈
radio-frequency cold tests 射频冷却试验
radio-frequency component 高频成分;射频分量;射频成分
radio-frequency confinement 射频抑止;射频限制
radio-frequency control system 射频控制系统
radio-frequency converter 射频变换器;射频变换器
radio-frequency coupling loop 射频耦合线圈
radio-frequency coupling ring 射频耦合环
radio-frequency coupling technique 射频耦合技术
radio-frequency cure 射频固化
radio-frequency curing 射频固化
radio-frequency current 射频电流
radio-frequency current transformer 射频变流器
radio-frequency deflector 射频偏转器
radio-frequency deflector system 射频偏转器系统
radio-frequency detector 射频检测器
radio-frequency dielectric(al) heating 射频介电加热
radio-frequency discharge 射频放电
radio-frequency discharge detector 射频放电检测器
radio-frequency distribution system 射频分配系统
radio-frequency drive 射频激发
radio-frequency driver 射频激发器
radio-frequency dropout compensator 射频信号失落补偿器
radio-frequency dropout killer 射频信号失落消除器
radio-frequency drying 高频干燥;射频烘燥;射频干燥
radio-frequency electrode 射频电极
radio-frequency electromagnetic field 射频电磁场
radio-frequency electromagnetic radiation 射频电磁辐射
radio-frequency emission 无线电发射;射频发射
radio-frequency energy 射频能量
radio-frequency engineering 频射无线电工程
radio-frequency equipment 射频设备
radio-frequency exciter 射频振荡激发器
radio-frequency fault detection 射频故障探测
radio-frequency field 高频场;射频场
radio-frequency field intensity 射频场强度
radio-frequency filter 射频滤波器
radio-frequency gain control 射频增益控制
radio-frequency gap 射频加速隙;射频放电器
radio-frequency generator 射频信号发生器;射频发生器
radio-frequency harmonic 射频谐波
radio-frequency harmonic number 射频谐波序数
radio-frequency hazard 射频危害
radio-frequency head 高频端;高频头;射频头;射频端
radio-frequency heating 高频加热;射频加热;射

钉枪
radio-frequency holography 射频全息照相术;射频全息术
radio-frequency homing guidance 射频自导引
radio-frequency impulse 射频脉冲
radio-frequency induction brazing 射频焊接
radio-frequency induction coil 高频感应线圈
radio-frequency induction furnace 射频感应炉
radio-frequency induction heater 射频感应加热器
radio-frequency induction heating 射频感应加热
radio-frequency influence 射频感应
radio-frequency input signals 射频输入信号
radio-frequency interference 无线电频率干扰;射频干扰
radio-frequency interference suppression equipment 射频干扰抑制装置
radio-frequency interference test 射频干扰测试
radio-frequency intermodulation distortion 射频互调失真
radio-frequency knockout 射频击出
radio-frequency knockout probe 射频击出探针
radio-frequency knockout resonance 射频击出共振
radio-frequency knockout technique 射频击出技术
radio-frequency lamp 射频灯
radio-frequency line 射频电缆;射频传输线
radio-frequency linear accelerator 射频直线加速器
radio-frequency liner 射频共振腔
radio frequency location 无线电频率定位(法)
radio-frequency loss 射频损失
radio-frequency loss-angle technique 射频损耗角技术
radio-frequency magnetic field 射频磁场
radio-frequency maser 射频微波激射器;射频脉塞
radio-frequency mass spectrometer 射频质谱仪;射频质谱计
radio-frequency microphone 射频传声器
radio-frequency microstructure 射频微观结构
radio-frequency mine detector 射频探雷器
radio-frequency modulation 射频调制
radio-frequency modulator 射频调制器
radio-frequency monitor 射频监控器
radio-frequency mute 射频杂波消除
radio-frequency noise 射频噪声
radio-frequency operation 射频运用
radio-frequency oscillation 射频振荡
radio-frequency oscillator 射频振荡器
radio-frequency oscillator high-voltage supply 射频振荡器型高压电源
radio-frequency output probe 射频输出探头
radio-frequency passband 射频通带
radio frequency pattern 交变脉冲图形
radio-frequency performance 射频性能
radio-frequency period 射频周期
radio-frequency permeameter 射频磁导计
radio-frequency pick-up 射频传感器
radio-frequency plumbing 射频波导(管)
radio-frequency polarography 射频极谱(法)
radio-frequency potential 射频电位
radio-frequency power 射频功率
radio-frequency power amplifier 射频功率放大器
radio-frequency power dissipation 射频功率耗散
radio-frequency power loss 射频功率损失
radio-frequency power supply 射频高压电源;射频(辐射)电源
radio-frequency preheating 射频预热
radio-frequency preionization 射频预电离
radio-frequency preselector 射频预选器
radio-frequency probe 射频探针;射频探测器
radio-frequency protection ratio 射频保护率
radio-frequency pulse 射频脉冲
radio-frequency pulse generator 射频脉冲发生器
radio-frequency pulse shape 射频脉冲波形
radio-frequency radiation 射频辐射
radio-frequency radiation power 射频辐射功率
radio-frequency radiation recorder set 射频辐射记录器
radio-frequency range 射频波段
radio-frequency reactor 高频电抗器
radio-frequency reading 射频读出
radio-frequency receiver 射频接收器;射频接收机
radio-frequency receiving tube 射频接收管
radio-frequency region 射频范围
radio-frequency repeating system 射频接转制

radio-frequency resistance 高频电阻;射频电阻
radio-frequency resonance transformer 射频共振变压器
radio-frequency resonator 射频共振腔
radio frequency resource 无线电频率资源
radio-frequency sealing 射频密封;射频封接
radio-frequency sealing technique 射频封接技术
radio-frequency seasoning 射频干燥
radio-frequency selectivity 射频选择性
radio-frequency sensitivity 射频灵敏度
radio-frequency sensor 射频传感器
radio-frequency separator 射频分离器
radio-frequency shielding fence 射频屏蔽装置
radio-frequency shift 射频漂移
radio-frequency sideband equalizer 射频边带均衡器
radio-frequency signal 射频信号
radio-frequency signal generator 射频(信号)发生器
radio-frequency signal source 射频信号源
radio-frequency spark 射频火花
radio-frequency spark discharge source 射频火花放电源
radio-frequency spark ion source 射频火花离子源
radio-frequency spark tester 高频火花检漏仪
radio-frequency spectrometer 高频谱计;射频分光计
radio-frequency spectroscopy 射频(波)谱学
radio-frequency spectrum 无线电频谱;射频谱
radio-frequency spectrum analyzer 射频谱分析仪
radio-frequency sputtering 射频溅射
radio-frequency stability 射频稳定度
radio-frequency stability region 射频稳定区
radio-frequency stacking 射频聚积
radio-frequency stage 射频级
radio-frequency standard signal generator 射频标准信号发生器
radio-frequency steady state discharge 射频稳态放电
radio-frequency structure 射频结构
radio-frequency suppressor 高频抑制器;射频抑制器
radio-frequency switch 射频开关
radio-frequency switching pulse inversion 射频开关脉冲倒置
radio-frequency switching relay 射频转换继电器
radio-frequency system 射频系统
radio-frequency tank 射频室
radio-frequency technique 射频技术
radio-frequency thermonuclear 射频热核装置
radio-frequency torch 射频(火)炬
radio-frequency transformer 射频变压器;射频变量器
radio-frequency transition 射频相变
radio-frequency transmission line 射频传输线
radio-frequency transmitter 射频放大器
radio-frequency tuner 射频调谐器
radio-frequency tuning 射频调谐
radio-frequency unit 射频部件
radio-frequency voltage 射频电压
radio-frequency voltmeter 射频伏特计
radio-frequency wanted-to-interfering signal ratio 射频信扰比
radio-frequency wave 射频波
radio-frequency waveguide 射频波导(管)
radio-frequency welding 高频焊接;射频熔接
radio gain 无线电增益
radiogallium 放射性镓
radio gas chromatography 放射气体色谱法
radiogen 放射物质
radiogenic 放射所致的;放射成因的;放射产生的
radiogenic age determination 放射性测定年代
radiogenic argon 放射成因氩
radiogenic calcium 放射成因钙
radiogenic dating 放射性测定年代
radiogenic heat source system 放射性蜕变热源系统
radiogenic helium 放射成因氦
radiogenic isotope 放射性生成同位素;放射成因同位素
radiogenic lead 放射成因铅
radiogenic strontium 放射成因锶
radiogeochemistry 放射生物地球化学
radio geodesy 无线电大地测量学;无线电测量

radio geology 无线电测地学;放射(性)地质学
radio-geophysical chemistry 放射地球物理化学
radiogold 放射性金
radio goniograph 无线电定向计
radio goniometer 无线电方位测定器;无线电定向器;无线电测向仪;无线电测向器;无线电测向计;无线电测距仪;无线电测角器
radio goniometer control 测角器旋钮
radio goniometer error 无线电测向仪误差
radio goniometry 无线电测向术
radio gonio station 无线电方向探测站
radiogram 无线电报;射线照相;放射照片;X射线照片
radiograph 伦琴射线照相;射线照片;射线照相;射频照相;放射照片;X射线照片;X光照片
radiographer 放射照相技术员
radiograph factor 射线照相系数
radiographic(al) 放射线照相的
radiographic(al) analysis 放射线照相分析
radiographic(al) apparatus 放射线照相设备
radiographic(al) contrast 射线底片对比度
radiographic(al) effect 放射线照相效应
radiographic(al) emulsion 射线照相乳胶
radiographic(al) equivalence factor 射线照相等效因子
radiographic(al) examination 射线照相检验
radiographic(al) exmination 射线照相
radiographic(al) film 射线(照相)胶片;射线底片
radiographic(al) inspection 透视;辐射照相检查;放射性照相检验;放射性探伤;放射线探伤;射线照相探伤法;射线探伤;射线检验;射线故障检验(法)
radiographic(al) lens 射线透镜
radiographic(al) paper 射线照相纸
radiographic(al) plate 射线照相底版
radiographic(al) search of weld 焊缝照相检验
radiographic(al) sensitivity 射线照相灵敏度
radiographic(al) set-up 射线照相装置
radiographic(al) source 射线源
radiographic(al) stereometry 立体X线片测定法;射线立体测量学
radiographic(al) stereoscopy 射线体视术
radiographic(al) technique X射线照相技术
radiographic(al) test 放射线照相试验;X射线检验
radiographic(al) testing 射频照相检查;射线照相试验
radiographic(al) testing equipment 放射线检查器
radiographic(al) testing machine 放射线探伤机
radiographic(al) thickness ga(u)ge 放射线摄影测厚仪
radiographic(al) thickness ga(u)ging 放射线照相测厚法
radiographing requirements 射线照相要求
radiography 射线照相术;放射照相(法);射线摄影;辐射照相术;放射性探伤;X光照相术
radiography department 射线照相部门
radiography examination 射线照相探伤检查
radiography with molybdenum target tube 钼靶X线摄影
radiogravimetry 放射性重量分析法
radio ground-wave propagation 无线电地波传播
radio guard 无线电警戒
radio guidance 无线电导航
radio-haloes 放射晕
radio-haloes of combination gas field 油气田放射晕
radio handset 无线电手机
radio hazard 射线伤害危险;射线危害;射线危害
radio heater 高频加热器;射频加热器
radioheating 射钉枪;射频加热
radio heat meter 辐射量热计
radio heliograph 射电日象仪
radiohm 雷电欧
radio hm alloy 铁铬铝电阻合金
radio hold 无线电舱
radio hole 射电洞
radio holography 射电全息照相术
radio homing 无线电归航
radio homing aid 无线电归航辅助设备
radio homing beacon 归航信标;无线电归航信标
radio horizon 无线电地平(线);电波水平线
radio horizon distance 无线电地平线距离
radiohydrochemical index 放射水文化学指数

radio hydrometry 放射性水文测量
radiohygience 放射卫生(学);辐射卫生学
radioimmunoassay 放射免疫分析;放射免疫测定
radioimmunochemistry 放射免疫化学
radioimmunoelectrophoresis 放射免疫电泳
radioimmunology 放射免疫学
radioimmunosorbent test 放射免疫吸附试验
radioimmunossay 免疫放射测定
radioimpurity 放射性杂质
radio index 无线电指数;射电指数
radioindicator 放射性指示剂
radio-induced 放射所致的
radio inertial guidance 无线电惯性制导
radio inertial guidance system 无线电惯性制导系统
radio inertial monitoring equipment 无线电惯性监控设备
radio influence 高频感应;无线电干扰
radio-influence field 高频干扰场
radio input 无线电信号输入
radio installation 无线电设备
radio instrument 无线电测量仪器
radio intercept control set 无线电截听控制设备
radio intercepting post 无线电截听站
radio interception or intelligence analysis post 无线电截听或情报分析站
radio intercept station 无线电截获站
radio intercept unit 无线电截听单位
radio interference 无线电干扰;射电干扰
radio interference eliminator 无线电干扰消除器
radio interference field 无线电干扰场
radio interference field intensity meter 无线电干扰波场强仪
radio interference filter 无线电干扰滤波器
radio interference generator set 无线电干扰信号发生器
radio interference level 无线电干扰水平;无线电干扰电平
radio interference locator 无线电干扰探测器
radio interference measuring instrument 无线电干扰测量仪表
radio interference measuring set 无线电干扰测量仪
radio interference suppression 无线电干扰的排除
radio interference suppressor 无线电干扰抑制器
radio interference test 无线电干扰试验
radio interference voltage 无线电干扰电压
radio interferometer 无线电干涉仪;射(电)干涉仪;射电干扰仪
radio interferometry 无线电干涉测量;射电干涉测量(法)
radioiodine 碘碘;放射性碘
radioiodine scintigraphy 放射性碘闪烁照相法
radio ionospheric propagation 无线电电离层传播
radioiron kinetics 放射性铁动态
radioisomerization 放射异构现象
radio-isophote 射电等强线;射电等辐透
radioisotope 放射性同位素
radioisotope analysis 放射性同位素分析
radioisotope assay 放射性同位素检验
radioisotope battery 原子电池;核电池
radioisotope deep therapy equipment 放射性同位素深部治疗机
radioisotope detector 放射性同位素检测器
radioisotope examination 放射性同位素检查
radioisotope flowmeter 放射性同位素流量计
radioisotope fluoroanalyzer 放射性同位素荧光分析仪
radioisotope ga(u)ge 放射性同位素测量计
radioisotope generator 放射性同位素发生器
radioisotope heater unit 放射性同位素加热装置
radioisotope heat source 放射性同位素热源
radioisotope level meter 放射性同位素料位计
radioisotope log 放射性同位素测井
radioisotope log curve 放射性同位素测井曲线
radioisotope method 放射性同位素测定法
radioisotope monitor 放射性同位素监测仪
radioisotope pollution 放射性同位素污染
radioisotope release 放射性同位素的排放
radioisotope resistant paint 防放射性同位素漆
radioisotope scanner 同位素扫描仪;放射性同位素扫描仪
radioisotope scan(ning) 放射性同位素扫描
radioisotope sediment concentration meter 放

射性同位素含沙量测定
radioisotope sediment probe 放射性同位素测沙仪
radioisotope smoke alarm 放射性同位素烟雾报警器
radioisotope snow ga(u)ge 放射性同位素测雪仪;放射性同位素测雪计
radioisotope thermoelectric(al) generator 放射性同位素温差电堆
radioisotope tracer 放射性同位素示踪物;放射性同位素示踪剂
radioisotope transmission ga(u)ge (测厚度、密度、液面等的)放射性同位素透过测量计;按放射性同位素透射测量计
radioisotope X-ray fluorescence analyzer 放射性同位素X射线荧光分析仪
radioisotopic generator 核电池
radioisotopic purity 放射性同位素纯度
radio jammer 无线电干扰发生器
radio jamming 无线电干扰
radio jeep 有无线电台的吉普车
radio jet 射电喷流
radio knife 射电刀
radio-label(l)ed 放射性同位素示踪的;放射性同位素标记的
radio-label(l)ed compound 放射性标记化合物
radio-label(l)ed counting rate isotope 放射性同位素的示踪计算率
radio-label(l)ed molecule 放射性标记分子
Radiolaria 放射虫目;放射虫类
radiolarian chert 放射虫燧石
radiolarian earth 放射虫土
radiolarian ooze 深海淤泥;放射虫软泥
radiolarian rock 放射虫岩
radiolarite 硬化放射虫软土;放射器;放射虫岩;放射虫壳化石
radio launch control system 无线电发射控制系统
radiole 辐棘
radiolead 放射性铅
radio lens 射电透镜
radiolesion 放射性损害
radioligand 放射性配体
radio light vessel 无线电浮标;无线电灯船;浮动无线电信标船
radio line 无线电播送;射电谱线
radio line of position 无线电位置线;无线电定位线
radio line-of-sight 无线电瞄准线
radio linked seismometer network 无线电地震台网
radio linked tide ga(u)ge 无线电传感潮位仪
radio link interferometer 无线电中继干涉仪;无线电连接干涉仪
radio link jammer 无线电通信干扰台
radio link network 无线电通信网
radio link protection 无线电保护装置
radio listening silence 无线电静默
radiolite 钠沸石
radiolite compass 放射性(涂料)罗盘
radiolite survey instrument 放射性测量仪
radiolitic texture 放射煳状结构
radio local loop 无线电本地环路
radio location 无线电定位
radio locational astrometry 射电定位天文学
radio locational astronomy 无线电定位天文学
radio location chart 无线电定位图
radio location equipment 无线电定位设备
radio location fixing 无线电定位
radio location land station 无线电定位地面电台
radio location mobile station 无线电测位行动电台
radio location station 无线电定位站
radio locator 无线电定位器;雷达无线电定位器
radio log 无线电通信记录本;报房日志
radiological agent 放射剂
radiological chemistry 放射化学
radiological clinic 放射科
radiological contamination 放射性污染
radiological contamination decontaminating agent 放射性沾污消除剂;放射性污染消除剂
radiological defense 辐射防御
radiological department 放射科
radiological dose 放射剂量
radiological examination 放射学检查
radiological exclusion area 放射性禁区
radiological hazard 放射性危害
radiological health 放射卫生

radiological program(me) 放射学规划
radiological safety 放射安全
radiological safety control 放射性安全管理
radiological safety protection 辐射安全防护;放射安全防护
radiological training manual 放射学训练指南
radiological work 放射工作
radiological worker 放射工作人员
radiology 应用辐射学;射电学;放射学;放射科
radio long call system 无线电长途电话系统
radiolucency 射线透过性
radio lucent 透射的伦琴射线
radioluminescence 射线致发光;辐射致发光;辐射光;放射(线)发光
radioluminescent watch 夜光表
radio luminosity 射电光学;射电光度
radio luminous material 放射发光材料
radio luminous paint 辐射发光涂料
radio luminous pigment 辐射发光颜料
radiolysis 射解;辐照分解;射解;辐解作用
radiolysis of alkanes 石蜡烃的射解作用
radiolysis of water 水的射解
radiolytic decomposition 辐射分解
radiolytic degradation 辐解降解
radiolytic oxidation 辐射氧化
radio magnetic indicator 无线电磁航向指示器
radio marker 无线电信标;无线电标志
radio marker beacon 无线电(导航)信号发射台;无线电标志信标;标志无线电信标
radio marker station 无线电指点标
radio mast 无线电天线杆;无线电天线桅
radio mast antenna 无线电柱形天线
radiomateriology 材料辐射探伤
radiomaximograph 大气天电干扰场强仪
radio measuring equipment 无线电测量器材
radio message transmission 无线电报传输
radio metal 无线电高导磁性合金;射电金属
radio metal locator 无线电金属探测器
radiometallography 金属射照相术;射线金相学;射电金相学;放射金相学
radiometallurgy 辐射冶金学
radio meteorogram 无线电气象图解
radio meteorograph 无线电气象仪;无线电气象计;无线电高空仪
radio meteorograph observation 无线电气象观测
radio meteorological service 无线电气象报告业务
radio meteorology 无线电气象学
radiometeorrograph 无线电气象计
radiometer 射线探测仪;辐射仪;射线计;放射量测定器;比例规
radiometer calibration 辐射计校准
radiometer channel 辐射计通道
radiometer frequency response 辐射计频率响应
radiometer ga(u)ge 辐射式真空计;辐射测压计
radiometer model 辐射仪型号
radiometer sensitivity 辐射计灵敏度
radiometer telescope 辐射计式望远镜
radio method 无线电勘探法
radiometric 辐射度的;放射性测量的
radiometric age 放射性年龄;放射性年代;放射测年代
radiometric age dating 地质年代的放射性测定
radiometric analysis 射量分析法;辐射度分析;放射分析法
radiometric conversion chart 辐射换算图
radiometric correction 辐射(度)校正
radiometric data 辐射测量数据
radiometric dating 放射性碳年代测定法;放射性测定年代
radiometric detector 辐射度探测器
radiometric distortion 辐射失真;辐射畸变
radiometric documentary contour map 辐射编录等值图
radiometric error 辐射误差
radiometric force 辐射力
radiometric magnitude 辐射星等
radiometric manometer 辐射测量压力计
radiometric mapper 热辐射成像仪;辐射度测绘仪
radiometric measurement 辐射测量
radiometric photometry 辐射测量光度学
radiometric polarography 放射极谱法
radiometric property 辐射测量性
radiometric prospecting 放射性勘探
radiometric reflectance 辐射度反射比

radiometric resolution 辐射分辨率
radiometric sampling volume 辐射取样体积
radiometric sextant 无线电六分仪;射电六分仪
radiometric survey 放射性测量
radiometric survey for petroleum 放射性方法
radiometric technique 射量分析技术;放射分析技术
radiometric temperature 辐射温度
radiometric titration 放射性滴定法;放射滴定
radiometric unit 辐射度单位
radiometry 辐射度学;辐射度测量;辐射测量术;辐射测量法;辐射测定法;放射性测量(学);放射量(测定)学;放射分析法
radiomicrobiological assay 放射微生物分析
radiomicrometer 微辐射计;辐射微热计;放射显微镜;显微辐射计;辐射微热量计
radio microphono 无线电传声器
radio microware link 无线电微波线路
radio microwave search set 无线电微波搜索仪
radiomimesis 拟辐射
radiomimetic activity 拟辐射活性
radiomimetic agent 拟辐射剂
radiomimetic chemical 拟辐射化合物
radiomimetic effect 拟辐射效应;模拟放射效应
radiomimetic substance 拟辐射物质
radio mobile telephone 无线电移动电话
radio molecular effect 辐射分子效应
radio monitoring 无线电监控
radio monitoring equipment 无线电监听设备
radio monitoring set 无线电监控器
radio monitoring system 无线电监控系统
radio multisource 多重射电源
radiomutation 辐射致变变;放射突变
radion 射粒;放射微粒;放射粒子
radionautography 自动射线照相术
radio navigation 无线电导航
radio navigation aid instrument 无线电导航仪
radio navigational aids 无线电导航设备
radio navigational device 无线电导航装置
radio navigational lattice 无线电导航图网
radio navigational warning 无线电航行警告;无线电航海警告
radio navigation buoy 无线电导航测标
radio navigation chart 无线电导航图
radio navigation facility 无线电助航设施
radio navigation guidance 无线电导航式制导
radio navigation land 无线电导航岸台
radio navigation land station 无线电导航陆地电台
radio navigation mobile station 移动无线电导航台;无线电导航移动电台
radio navigation receiver 无线电导航系统接收机
radio navigation satellite 无线电导航卫星
radio navigation set 无线电导航仪
radio navigation silence 无线电导航静默
radio navigation system 无线电导航系统
radionecrosis 放射性坏死
radio net 无线电网
radio network 无线电通信网
radionics 射电电子学;电子学
radionitrogen 放射性氮
radio noise 射频噪声;射电噪声
radio noise burst 无线电辐射激扰
radio-noise field strength 射频噪声场强度
radio noise figure 无线电噪声指数
radio noise flux 射电噪声通量
radio noise from the sun 太阳射电噪声干扰
radio noise map 无线电噪声图
radio noise meter 无线电噪声计
radio noise source 无线电干扰源
radio non-thermal brightness temperature 非热射电亮温度
radio note 通知性电传打字电报
radionuclide 放射性核素;放射性核毒;反射原子核
radionuclide battery 放射性核电池
radionuclide concentration 放射性核素浓度
radionuclide contamination 放射性核素污染
radionuclide generator 同位素发生器;放射性核素发生器
radionuclide kinetics 放射性核素动力学
radionuclide migration 放射性核素迁移
radionuclide toxicity classification 放射性核素毒性分组
radio observation 无线电观测
radio observatory 射电天文台
radio office log book 电台日志

radio-opacity 辐射不透明度
radio optics 辐射光学
radio oscillator 无线电振荡器
radio out procedure 无线电联络中断时的处置办法
radio output light 无线电发射机工作信号灯
radio overlay network 无线重叠网
radio-oxidation wastewater treatment 辐射氧化废水处理
radiopacity 射线不透性
radio pager station 无线寻呼台
radio paging 无线电寻呼
radio paging service 无线寻呼电话
radio paging station 无线寻呼台
radio paging system 无线寻呼系统；无线电联络指挥系统；无线电呼叫系统
radiopaque 辐射透不过的；辐射不透明的；不透X线（的）；不透放射线的；不透X线的
radioparency 射线可透性
radio parent 透射线
radiopasteurization 射线和加热混合杀菌法；辐射杀菌作用；辐射灭菌（法）
radio path 无线电波路径
radio phare 无线电指示台；无线电航标
radiopharmaceutical 放射性化学药品
radio phase 无线电定位仪
radio phone 无线电话
radio phone auto-alarm 无线电话自动报警器
radio phone transmitter 无线电话发射机
radiophosphorus 放射性磷
radio photogram 无线电传真照片
radio photo(graph) 无线电传真
radio photography 无线电摄影术
radiophotoluminescence 辐射光致发光；放射光致发光（现象）
radiophotoluminescent dosimeter 放射光致发光剂量计
radio photon 射电光子
radiophotoscanning 放射照相扫描术
radiophotostimulation 辐射光致激发；辐射光刺激
radio phototelegraph 无线电传真电报
radio phototelegraphic(al) receiver 无线电传真接收机
radio photo-transmission 无线电传真
radiophyllite 球硅钙石；水硅灰石
radio picture 无线电传真照片
radio picture signal 射频图像信号
radio pilot 无线电领航；无线电测风气球
radio pilot balloon 无线电控制气球；无线电导航气球
radio plage 射电谱斑
radio plant 无线电装置
radio platoon 无线电排
radio pocket 无线电静区
radio polarimeter 射电偏振计
radio polarimetry 射电偏振测量
radiopolarography 放射极谱法
radiopolymerization 辐射聚合（作用）
radio position finder 无线电定位仪；无线电测仪
radio position finding 无线电测位
radio position finding land station 无线电测位陆地电台
radio position finding station 无线电测位电台
radio position fixing instrument 无线电定位仪
radio position fixing (method) 无线电定位法
radio positioning land station 无线电定位陆地电台
radio positioning method 无线电定位法
radio positioning mobile station 无线电移动式定位电台；无线电定位移动电台
radio position line 无线电船位线
radio position plotter 无线电定位量测仪
radio post 无线电台
radiopotassium 放射性钾
radio power 无线电传送功率
radio print 缩放晒印
radio prism 无线电棱镜
radio procedure 无线电通信工作程序
radio program(me) 广播节目
radio propagation 无线电波传播
radio propagation forecast 无线电波传播预报
radio propagation prediction 无线电波传播预测
radio prospecting 射线探矿
radio prospection 无线电勘探
radioprotectant 辐射防护剂；放射防护剂
radioprotection 放射防护

radio protection channel 无线电保护通路
radioprotective 放射防护的
radio-protective substance 辐射防护物质
radioprotector 辐射防护装置；辐射防护剂；放射防护剂
radio proximity fuse 无线电近发引信
radio pulse 射频脉冲；电磁脉冲
radio pure 放射纯
radiopurity 放射性纯度；放射化学纯度；放射纯
radio quarantine reports from ship at sea 海上船舶无线电检疫报告
radio quiescence 无线电静默
radio radar absorber 无线电雷达电磁波吸收剂
radio radiation 射电辐射
radio range 无线电导航仪；无线电测得的距离；射电轨；等信号区无线电信标
radio range beacon 无线电导向标；无线电导航信标
radio range beam 无线电测距波束
radio range filter 无线电信标用滤波器
radio range finding 无线电测距
radio range fix 无线电波导航定位
radio range leg 无线电测距射束
radio range network 无线电信标网
radio range orientation 无线电法导航定向；无线电导航定向法
radio range receiver 无线电测距接收机
radio range station 无线电导航台
radio range transmitter 无线电测距发射机
radio ranging measurement 无线电测距
radio ranging system 无线电导航系统
radio ray path 无线电波传播路径
radioreaction 辐射反应；放射反应
radio re-broadcast system 无线电转播系统
radio receiver 无线电收信机；无线电接收机
radio receiver parts 接收机零件
radio receiver set 无线电收发机
radio receiver-transmitter 无线电收发机
radio receiving room 无线电接收室
radio receiving set 无线电接收装置
radio receiving station 无线电接收站
radio reception 无线电接收
radio reception and transmitting installation 无线电接收发送装置
radio reception condition 无线电接收情形
radioreceptor 辐射感受器；放射感受器
radioreceptor assay 放射性受体分析
radio recognition 无线电识别
radio recognition and identification 无线电识别
radio recognition equipment 无线电识别设备
radio recombination line 射电复合线
radio reconnaissance monitoring system 无线电侦察监听系统
radio refraction correction 大气折射改正；大气折射订正
radio regulation 无线电(通信)规则
radio relay 无线电中继
radio relay line 无线电中继线路
radio relay repeater set 无线电中继增音器
radio relay satellite 无线电中继卫星
radio relay station 无线电中继站
radio relay terminal section 无线电中继终端分排
radio-release analysis 放射—释放分析
radio-release determination 放射性释放测定（法）
radio remote-recording system 无线电遥测系统
radio remote-sensing fluviograph 无线电遥测水位计
radio remote-sensing system 无线电遥测系统
radio reporting river ga(u)ge 无线电发报水文站；无线电报汛站
radio repository for wastes 放射性废料储[贮]藏器
radio research ship 无线电通信试验船
radio-resistance 抗辐射(性)；耐辐射性；辐射抗性
radio-resistant coating 耐辐射涂料
radio-resistant material 耐辐射材料
radio-resistant plant 耐辐射植物
radioresonance method 射频共振法
radio robot 遥控机器人
radio robot level-meter 无线电自动遥控水位计
radio robot ombrometer 无线电自动遥控雨量计
radiorocket electrophoresis 放射火箭电泳
radio room 无线电室；报房
radio safety signal 无线电安全信号
radioscan(ning) 放射性同位素扫描

radio scattering 无线电波散射；射电散射
radio school 广播学校
radioscintigraphy 放射性闪烁摄影法
radio scintillation 射电闪烁
radioscope 剂量测定用验电器；γ射线透镜；放射镜
radioscopy X射线透视检查法；射线检验法；γ射线检查法；放射检查
radio search 无线电搜索
radio security 无线电保密
radiosensitivity 辐射敏感性；辐射灵敏度；放射敏感性
radiosensitization 辐射敏化
radio sensitizer 辐射敏化剂
radio sensor 无线电接收机；无线电传感器
radio service 无线电业务
radio service code 无线电公电密语
radio set 接收机；无线电收发信机；无线电收发机；无线电设备；无线电接收机
radio setting apparatus 无线电测向仪
radio-shielded 防高频感应的
radio shielding 无线电屏蔽；射频屏蔽；高频感应屏蔽
radio ship 无线电通信舰；无线电通信船
radio sight 无线电瞄准器
radio signal 无线电信号
radio signal carrier service area 无线载波服务区
radio signal interceptor 无线电信号截听器
radio signal reporting code 无线电信号报告码
radio silence 无线电静默
radiosilicon 放射性硅
radio skip zone 无线电盲区
radio sky temperature 天空射电温度
radio sondage 无线电探空
radio sondage technique 无线电探测术
radiosonde 无线电探空仪；无线电高空测候器；γ线电探测器；无线电测器
radio sonde and radar wind sounding 无线电探空仪和雷达测风（联合装置）
radio sonde and radio wind station 无线电探测风站
radio sonde balloon 无线电探空气球；探空仪气球
radio sonde commutator 无线电探空仪转换开关
radio sonde data 无线电探测资料
radio sonde flight equipment 无线电控空仪信号发送装置
radio sonde ground equipment 无线电探空仪信号地面接收装置
radio sonde observation 无线电探空观测；探空仪观测
radio sonde receptor 无线电探空接收机
radio sonde recorder 无线电探空记录器
radio sonde-rocketsonde set 无线电火箭探空仪
radio sonde station 无线电探空站
radio sonde wind sounding 无线电探空仪测风
radio sonic buoy 无线电声纳浮标
radio sonic ranging 无线电声呐测距
radio sonic ranging system 无线电声呐测距系统
radio sonobuoy 无线电声呐浮标；声呐浮标；防潜浮标
radio sonobuoy and hydrophone 无线电声呐浮标与水下听音器
radiosorption luminescence 辐射吸收发光
radio sounding 无线电探空
radio sound ranging 无线电回声测距法；电声测距（法）
radio source 射电源
radio source count 射电源计数
radio spark suppressor 无线电火花制止器
radio speaker 无线电扩音器
radio spectral index 射电谱指数
radio spectral line 射电谱线
radio spectrograph 射电谱仪
radio spectrography 无线电频谱学
radio spectroheliogram 射电太阳单色图
radio spectroheliograph 射电太阳单色仪
radio spectroscopy 射频频谱学；辐射光谱学
radio spectrum 无线电频谱；射频频谱；射电谱
radio spectrum allocation 无线电频谱分配
radiostat 中放晶体滤波式超外差接收机
radio station 广播电台；无线电台
radio station call sign 无线电台呼号
radio station for long distance 远程无线电台
radio station interference 无线电台干扰
radiostereoassay 放射立体化学分析法

radio stereophony 无线电立体声
radiosterilization 辐射消毒(作用);辐射杀菌;放射性消毒
radio storm 无线电扰动;射电暴
radiostrontium 放射性锶
radio subsolar temperature 日下点射电温度
radio sun 射电太阳
radio suppression 无线电干扰
radio suppressor 无线电抑制器;无线电阻尼器
radio surveying 无线电勘测
radiosusceptibility 辐射敏感性
radio switching centre 无线电切换中心
radiosynthesis 放射合成
radio system 无线电系统
radio system for depot 基地无线电系统
Radio Technical Commission for Maritime Services 海事无线电技术委员会
radio technical means of communication 无线电技术通信工具
radio technics 无线电技术
radio technology 无线电技术;无线电工艺
radio telecontrol 无线电遥控
radio telegraph 无线电报
radio telegraph alarm signal 无线电报报警信号
radio telegraph auto-alarm 无线电报自动报警器
radio telegraph communication 无线电报通信
radio telegraphic(al) call 无线电报呼叫
radio telegraphic(al) transmission 无线电报发送
radio telegraphic installation 无线电报设备
radio telegraph operation room 无线电报室
radio telegraph procedure 无线电报通信程序
radio telegraph receiver 收报机
radio telegraph ship station 无线电报船舶电台
radio telegraph transmitter 无线电发射机;无线电报发射机
radio telegraphy 无线电报学;无线电报术
radio telemetering 无线电遥测
radio telemetering buoy 无线电遥测浮标
radio telemetering equipment 无线电遥测设备
radio telemetering station 无线电遥测站
radio telemeter receiver 无线电遥测接收机
radio telemeter transmitter 无线电遥测发射机
radio telemetry 遥测技术;无线电遥测学
radio telemetry and remote control 无线电遥测与遥控
radio telemetry network 无线遥测台网
radiotelemtry 无线电遥测法
radio telephone 无线电话
radio telephone alarm signal 无线电话报警信号
radio telephone auto-alarm 无线电话自动报警器
radio telephone call 无线电话呼叫
radio telephone communication 无线电话通信
radio telephone distress procedure 遇险无线电话操作程序;遇难无线电话操作程序
radio telephone signal procedure 无线电话信号程序
radio telephone system in station and yard 站场无线电话系统
radio telephone transmitter 无线电话发射机
radio telephony 无线电话术;无线电电标
radio telephony network 无线电话网
radio teleprinter 无线电传真打印机
radio telerecording seismograph 无线遥测地震仪
radio telescope 射电望远镜
radio teletype writer 无线电电传打字机
radiotelevisor 电视接收机
radiotellurium 放射性碲
radio temperature 无线电辐射温度
radio terminal 无线电终端设备
radio terminal assembly 无线电终端组件
radio terminal box 无线电接线盒
radio terminal equipment 无线电终端设备
radio terminal set 无线电终端设备
radio tester 无线电检验器
radio test set 无线电测试设备
radio theodolite 无线电经纬仪
radiother 放射(性)指示剂
radiothermics 射频加热(技)术
radio-thermoluminescence 射线热发光;辐射热致发光
radio thermoluminescent dose glass 热致发光剂量玻璃
radio thin layer chromatography 放射性薄层色谱法

radio-thorium 放射性钍
radio tick 无线电时间信号
radio tide meter 无线电验潮仪
radio time code 无线电时间码
radio time signal 无线电时间信号;无线电时号
radiotine 星蛇纹石
radio tolerance 辐射容量;耐辐射性
radio tower 无线电(铁)塔;无线电天线塔;无线电天线杆;无线电发射塔;天线塔
radiotoxicity 辐射毒性;放射毒性;放射毒理学
radiotoxin 放射性毒素
radiotracer 放射(性)示踪物;放射指示物;放射指示剂;放射性示踪剂
radiotracer compound 放射性示踪化合物
radiotracer element 放射性示踪元素
radiotracer leak detector 放射针示踪检漏仪
radiotracer technique 放射性示踪技术
radio tracking 无线电追踪;无线电跟踪
radio tracking station 无线电跟踪站
radio tracking technique 无线电追踪技术
radio transceiver 无线电收发信机;无线电收发报机
radio transformer 射频变压器
radio transit 无线电发射
radio transmission 无线电发射;无线电传送;无线电波传播
radio transmission center 无线电发射中心
radio transmission formula 无线电波传播公式
radio transmission frequency measuring system 无线电发射频率测量系统
radio transmitter 无线电发送器;无线电发射机;无线电传送机
radio transmitter and receiver 无线电收发报机
radio transmitter equipment 无线电发射机
radio transmitting center 无线电发射中心
radio transmitting room 无线电发射室
radio transmitting set 无线电发射机
radio transmitting station 广播电台;无线电(发射)台
radio transparency 射线穿透能力
radio transparent 透放射线
radiotreatment 放射处理
radio triangulation 无线电大地测量
radiotropism 向辐射性
radio true bearing 无线电真方位
radio trunk call system 无线电长途电话系统
radio trunk telephone system 无线电长途电话系统
radio tube 无线电真空管;电子管;电磁阀
radio two way communication 无线电双向通信
radio uranium 放射性铀
radio urgency signal 无线电紧急信号
radio valve 整流器真空管;无线电真空管
radiovision 电视
radiovisor 光电监视器;光电继电器装置
radio voice transmission 无线电话报传输
radio warning 无线电警报;无线电报警
radiowaste 放射性废物
radio watch 值班
radio watcher 值班报务员
radio wave 无线电波;辐射波
radio wave absorbing material 无线电波吸收材料
radio wave absorption 无线电波吸收
radio wave channel 无线电波道
radio wave frequency 射频
radio waveguide 无线电波导
radio wave interference method 无线电波干涉法
radio wave maser 射频激射器
radio wave method 无线电波法
radio wave path 无线电波路径
radio wave penetration method 无线电波透视法;阴影法
radio wave propagation 无线电波传播;射频波传播
radio wave propagation prediction 无线电波传播预测
radio wave radiation 无线电波辐射
radio wave reception 射频波接收
radio wave reflection method 无线电波反射法
radio wave sounding 无线电波探测
radio wave spectrum 无线电频谱
radio wave transmission 无线电波传播
radio weather message 无线电气象通信
radio-well logging 放射性测井
radio wind 无线电测风(仪);测风雷达
radio wind observation 无线电测风风观测
radio window 射电窗口

radio wind sounding 无线电高空测候仪
radio wind station 无线电测风站
radio wire 绞合天线
radio wire integration 无线电有线电综合通信
radio wire integration station 无线电有线电综合通信台
radist 空中目标速度测量装置
radium age 镭龄
radium-bearing 含镭
radium bomb 镭炮
radium cache 藏镭器
radium carbonate 碳酸镭
radium cell 镭管
radium chloride 氯化镭
radium contamination 镭污染
radium content determined by cycle method 循环法测镭含量
radium content determined by Ra (B+C) method 镭(B+C)测镭含量
radium content determined by vacuum method 真空法测镭含量
radium dating 镭年代测定法
radium electroscope 镭检电器
radium emanation 镭射气
radium equivalence 镭当量
radium ionium method 镭镤法
radium irradiation equipment 镭辐射设备
radiumize 受镭作用
radium needle 镭针
radium new thorium water 镭—新钍水
radium off-set of joint 接头错位半径
radium plaque 镭板
radium plaque adaptometer 镭板适应计
radium poisoning 镭中毒
radium pollution 镭污染
radium thermal water 镭热水
radium tube 镭管
radium-type vacuum ga(u)ge 放射真空计
radium vermilion 镭朱红
radium water 镭水
radius 径向射线;半径(范围)
radius acceleration 半径加速度
radius-and-safe-load indicator 吊杆仰角与安全荷载指示器
radius angle 圆心角
radius at bend 曲率半径;曲流率半径;弯曲半径;弯道半径
radius attachment 圆角磨削装置
radius bar 极臂;曲拐臂;半径杆
radius bar brace 径向杆拉条
radius beam pitman 摇摆梁枢轴;辐梁枢轴
radius block 半径杆块
radius brick 楔形砖;径形砖;扇形砖;大小头砖;辐射形砖;半圆砖
radius bucket 圆底铲斗
radius-changing mechanism 变辐机构
radius-changing speed 变辐速度
radius compensation 径差补偿;半径补偿
radius correction 半径校正;半径改正
radius curvature in a normal section 法截线曲率半径
radius cutter 半径铣刀
radius die 弧形成型模具
radius dimensioning 半径尺寸标注
radiused punch nose 凸模端部圆角
radius former 圆角样板;圆弧样板
radius gate 弧形闸门;扇开闸门
radius ga(u)ge 圆角规;曲度规;半径样板;半径规
radius grinding attachment 半径磨削附件
radius indicator 曲率指示器;曲率表
radius influence of well 井的影响半径
radiusing 舍去零数化成整数;弄圆;使成圆形;使成圆角
radiusing machine 去毛刺机;齿轮倒角机
radius kerb mo(u)ld 转弯处缘石的模板
radius link 摇杆;半径杆
radius-luminosity relation 半径光度关系
radius-mass relation 质径关系
radius mill cutter 半径铣刀
radius mounting 半径装配法
radius of a circle 圆周半径
radius of action 有效距离;作用半径;工作半径;活动半径

radius of addendum 齿顶圆半径
radius of arched observation line 弧形观测线半径
radius of bending 转弯半径
radius of central line 中心线半径
radius of clean-up 工作半径;(挖土机的)切削半径;挖掘半径
radius of clean-up at floor 清底半径
radius of connecting curve 连接曲线半径
radius of convergence 收敛半径
radius of corner 圆角半径;转角半径
radius of crank 曲柄半径
radius of curvature 圆尖半径;曲率半径;弯曲半径
radius of curvature for turn-off (飞机场出租汽车车道出口处的)曲率半径
radius of curvature in meridian 子午圆曲率半径
radius of curvature in prime vertical circle 卯酉圈曲率半径
radius of curvature method 曲率半径法
radius of curvature of tendon 钢筋束的曲率半径
radius of curve 曲线半径;弯道半径
radius of curve of turnout 道岔导曲线半径
radius of cut 挖掘半径
radius of damage 毁伤半径
radius of deformation 变形半径
radius of destruction 破坏半径
radius of drainage 油井的汲油半径;油井的排油半径
radius of elbow 弯管半径
radius of excavation 挖掘半径
radius of extent 扩散半径;分布半径
radius of extrados 拱外弧半径;外拱圈半径
radius of flange 弯边半径
radius of flange mouth arc 法兰口圆弧半径
radius of flute 沟底半径
radius of flute of tooth 齿底半径
radius of generating circle 生波圆的半径
radius of graph 图的半径
radius of gyration 转动半径;惯性半径;回转半径;环动半径
radius of gyration of area 断面二次半径
radius of gyration of ship 船舶回转半径
radius of horizontal curve 平曲线半径
radius of inertia 惯性半径;惯量半径
radius of influence 诱导半径;影响半径
radius of influence circle 影响圆半径;影响圈半径
radius of influence of well 水井影响半径
radius of interpolation circle 内插圆半径
radius of intrados 拱内(弧)半径;拱腹半径;内拱圈半径
radius of lead curve 导曲线半径
radius of lens 晶状体半径
radius of maximum wind 最大风速半径
radius of meridional section 子午面曲率半径
radius of molecular action 分子作用半径
radius of operation 作用半径
radius of particle 粉粒半径
radius of perceptibility 震感半径;可感半径
radius of plane curve 平曲线半径
radius of principal curvature 主曲率半径
radius of protection 保护半径
radius of railway curve 铁路曲线半径
radius of relative stiffness 相对劲度半径;相对刚性半径;相对刚度半径
radius of rolling circle 滚圆半径
radius of root 根部半径
radius of rotation 旋转半径
radius of rounded angle 圆角半径
radius of rounded corner 圆角半径
radius of rupture 毁伤半径;破坏半径;断裂半径
radius of safety 安全半径
radius of service 服务半径
radius of soffit 拱内弧半径;拱腹半径
radius of sphere 球半径
radius of sphere of equal volume 等体积球的半径
radius of sphericity 球体半径
radius of spin 螺旋半径
radius of steaming range 续航距离
radius of stereoscopic vision 体视半径
radius of stiffness 刚度半径
radius of swing 回船半径;掉头半径
radius of switch point 尖轨半径
radius of terrain correction 地改半径
radius of the core 铁芯半径;磁芯半径;岩芯半径;芯子半径

radius of torsion 扭转半径;挠率半径
radius of turn 旋转半径;转弯半径
radius of turning circle 旋回圆半径;回转圈半径
radius of vector 向量径
radius of vertical curve 竖曲线半径
radius of visibility 能见度半径;视界半径;视地平距离
radius parameter 半径参数
radius planer 曲面刨床
radius ratio 半径比(率)
radius ratio effect 半径比效应
radius rod 大半径臂杆;旋转半径;半径杆
radius rod bracket 半径杆托
radius rod drive 推杆驱动
radius rod segment 圆角切刀
radius rod shim 半径杆垫片
radius safety 安全半径
radius segment 圆角切刀
radius segment tool 圆角切刀
radius served 服务半径
radius shoe 旋转棒的刮板
radius template 半径样板
radius test (油毛毡的)弯折试验
radius-thickness ratio 径厚比;半径与厚度比
radius tip 球面电极头
radius tool 圆角切刀
radius truing device 半径修整装置
radius vector 向(量)径;位置向量;矢径;动径;辐向矢径;辐
radius vector method 矢径法
radivection 辐射对流加热
radix 根值;词根
radix complement 基数补数;基数补码;底数补数;补码
radix conversion 基数转换
radix converter 基数变换器
radix digit 基数数字
radix exchange sort 基数交换分类法;基数调换分类法
Radix Inulae Racemosae 藏木香
radix marker 基数标记符
radix-minus-one complement 基数减1补码;反码【计】
radix nasi 鼻根
radix notation 根值记数法;基数记数法;基数表示法
radix point 小数点;基数小数点
radix pointer 基数指示字
radix scale 基数记数法;基数计数制
radix sorting 基数分类(法)
radix sorting algorithm 基数排序算法
radix system 基数制
radix transformation 基数转换
radix transformation method 基数变换法
radix two computer 多位二进制计算机;二进制计算机
radix two counter 多位二进制计数器
Radke-Prausnitz equation 拉德克─普劳斯尼茨方程
radlux 辐射勒克司
radome 雷达罩;微波天线屏蔽器;天线罩(雷达)
radome of sandwich type 多层天线罩
radon 镭射气
radon chloride 氯化淀粉
radon consistence 氡浓度
radon emanation 氡析出
radon fluoride 氟化氡
radon gas 氡气
radon measurement method 氡气测量法
radon method 氡测法
radon-radium water 氡镭水
radon routine measurement 常规氡测量
radon thermal water 氡热水
radon water 氡水
radphot 辐射透(照度单位)
Rad-stockian (stage) 拉德斯托克阶【地】
radula 齿舌
rad unit 拉德单位
radux 远距离双曲线低频导航系统;计数制基数
radwaste 放射性废物
radwaste final disposal 放射性废物最终处置
radwaste management 放射性废物管理
radwood 辐射处理木料
Raeberry castle group 拉柏里堡岩群
rafaelite 钒地沥青;阿根廷地沥青
raff 大量;粗矿石
raffinal 高纯度铝

raffinate 提余液;萃取残液;抽提液;残油;残液
raffinate layer 残油层
raffinate splitter column 残液分离塔
raffinator 单动盘磨机
raffle 三角顶帆;杂物;船上绳索杂物;抽彩售货
raff wheel 转轮提斗机
rafraichometer 体温表
raft 木筏
raft apron 筏道末端护坦
raft body 救生筏体;平行舯体
raft bridge 浮桥;木筏(浮)桥;筏桥
raft canal 放筏运河
raft chute 过筏斜槽;木材流放槽;浮运水槽;筏道
raft construction 筏基结构
raft contact pressure 承台接触压力
rafted cover 筏状冰
rafted ice 多层冰;叠冰;重叠冰;筏冰;阀状多层冰
rafted pack ice 重叠流冰群
rafter 木筏运工;椽子;椽(条)
rafter-and-strut framed dam 木框架式坝;椽撑框架式坝
rafter bearer 承椽板
rafter bearer notch 承椽槽口
rafter cleat 椽子防滑木
rafter connection 椽子连接;椽接
rafter cross-section 椽子截面
rafter dam 木板坝;人字坝
rafter end 椽子头
rafter fill(ing) 梁与椽子间填充物;椽间空隙填塞
rafter harbo(u)r 木筏港
rafter-head entasis treatment 椽头卷杀
rafter interval 椽子间距
rafter joist plate 支承梁板;支承椽子板
rafter nail 椽木钉
rafter plate 檐檩(支承椽下端的檩条);椽板
rafter roof 人字木屋顶(椽子承重的);人字椽屋顶;椽子屋顶;椽木屋顶
rafter set 屋架;人字(形)木屋架
rafter slope 椽子斜度
rafter spacing 椽子间距
rafter span 椽子跨度
raftersupporting beam 承椽枋
raftersupporting purlin(e) 支承椽子的檩条
rafter system 椽子系统(布置)
rafter table 椽子下料表;椽座
rafter tail 椽尾
rafter timber 椽木
rafter timbering 拱式支撑;屋椽式支护法
rafter with ornamental end 有端部装饰的椽子
raft foundation 片筏基础;满堂基础;浮筏基础;筏形基础;筏(式)基础;筏基
rafting 木筏流放;木材流放;成筏过程;浮运木材;筏流;伐运;冰块重叠;冰块叠挤
rafting canal 放筏运河;木材运河;木材流放渠;放排河;筏木排运河
rafting channel 浮运水槽;筏流水路
rafting ice 浮筏冰
rafting operation 筏运作业;木材输送工作
rafting reservoir 筏运储[贮]水池;筏运水库
raft lake 木材拥塞积水湖
raft lock 过筏闸;木材流放闸
raft log 木筏
raftman 木排工(人);木筏工人;木材流放工(人);筏运工;[复]raftmen
raft mat foundation 筏形基础
raft mounted aerator 筏式曝气器
raft of foundation 片筏基础
raft of land vegetation 沿岸植物漂流块;浮生植物层(湖沼的)
raft of pontoons 舟桥筏
raft of vegetation (河、湖上的)沿浮生植物层
raft pass 木材流放渠;筏道
raft passway 过筏道
raft path 木材流放渠;筏道
raft port 木材装船港
raft-slab floor 拼板地面
raft sluice 过筏道;木材流放渠;筏道
raft stiffening 筏基加强(法)
raft tectonics 板块构造
raft test 放排试验
raft tow 木排
raft way 筏道
raft-wood 木排流送材
rag 刻纹;毛口;毛刺;坚硬石灰岩;硬质岩石;抹布

破布;条石;石板瓦;擦布;钢板等切断处不光滑整齐的边
rag and bone man 废品收购者
rag and stick work 抹纤维性灰泥工作
rag armo(u)ring 条石护面
rag boiler 破布蒸煮器
rag bolt 大头螺栓;地脚螺栓;棘螺栓
rag breaker 破布拌浆机
rag content paper 含破布纸
rag cutter 破布切断机
rag cutter with disc knives 盘刀式切布机
rag cuttings 碎布
rag devil 碎布开松机;扯碎机
rag dike 条石堤
rag doll 湿布卷条
rag-doll tester 布卷发芽器
rag dust 破布尘屑
rag duster 破布除尘机
rageas 拉格特风
rag engine 碎布打浆机
rag fair 估衣市场
rag felt 棉毡胎油毡;沥青油毡;粗制油毡;粗制毡(建筑用)
rag felt floor cover(ing) 粗毡地毯
rag felt with lead foil 有铅箔油毡
rag fiber 破布纤维
rag-frame 洗矿台
ragged 外形参差不齐的;粗糙(的)
ragged canvas 破烂的帆布
ragged ceiling 不定云幕高度
ragged coast 凹凸不平的海岸
ragged cut 高低不平的切割
ragged edge 曲折边沿;粗糙的边缘;不平边缘
ragged hole 孔壁不平的钻孔;井壁不平的钻井
ragged margin 留边不齐的
ragged roll 刻痕轧辊
ragging 跳汰机床层;压花;击碎;滚花
ragging frame 倾斜洗矿台
ragging mark 刻纹
ragging off 撮落松煤
raggle 墙上槽口;披水槽(砌);石水槽;承水槽
raggle block 盖板;盖块(防漏用);披水槽砌块;防漏盖块
rag grinder 碎呢机
rag iron spring 压板弹簧
rag knife 切布刀
raglan 泛水石槽;开槽;凿槽;墙上凹槽;挡水板槽;平顶格栅
raglanite 奥霞正长岩
raglet 墙上披水槽;墙上槽口;墙上槽口;披水槽
raglin 楼板格栅;披水槽;凿槽;开槽;泛水石槽
rag line 大绳索
ragman 埋管涂油工人
rag mix 碎布胶料
rag nail 棘螺栓;棘钉;折钉
rag paper 优质纸
rag paving 石板铺面
rag pitching 条石护面
rag pulp 破布浆
rag pulping waste 破布纸浆废水
rag-rolled finish 乱滚花抹面;碎布滚花涂装;滚涂花饰面;滚划抹面
rag rope waste 麻绳厂废水
rag rubble 硬质岩毛石工程;粗型块石;粗型石
rag rubble wall 粗面毛石墙
rag rubble walling 砌毛乱石墙;片石墙
rag rug 碎呢地毯;织物条毛毯
rags calender 碎布胶料压延机
rag shaker 碎呢除尘机
rag slitter 破布条切割机
rag sorting roll 破布拣选辊
rag sorting room 选布间
rag stone 硬质岩灰质岩;硬岩;硬灰石;铺路石板;硬质岩石;石板瓦
ragstone concrete 片石混凝土
ragstone facing 硬质粗石贴面
rag thrasher 破布乱散机;碎布敲打机
ragtop 折叠式车顶
raguinite 硫铁铊矿
Ragusa 拉古萨沥青
ragut 拉格特风
rag washer 半料洗涤机
rag wheel 磨(光)轮;抛光轮;布抛光轮;布轮

rag willow 破布敲打机;破布除尘器
rag work 砌扁石工;片石砌筑;片石墙;小块石砌体;毛石工程;毛石贴面;砌石板(圬工);石板砌法
rahmen 环形天线;框架结构
rahtite 铜闪绿矿
raider company 劫夺性公司
raider firm 劫夺性厂商
raid patrol 袭击性巡逻
raid period 袭击时间
raid report 袭击报告
raid the market 扰乱市场;冲击市场
raid the sinking fund 挪用偿债基金
raies ultimes 元素光谱特征线条
rail 钢轨;路轨;栏杆横档;抹头;冒头;横木;铁轨;围栏;栏杆
rail abrasion measuring instrument 钢轨磨耗测量器
rail advice 到达通知单
railage 铁路运输;铁路运费;铁道运费
rail anchor 轨卡;钢轨锁碇器;钢轨防爬器;防爬器
rail anchor device 钢轨锁碇装置;钢轨锚定装置;钢轨防爬装置
rail and air 铁路及空运
rail and ocean 铁路运输及海运
rail and road traffic 铁路公路联合运输
rail and structural steel mill 钢轨及钢梁轧机;轨梁轧机
rail and truck 铁路及卡车运输
rail and water 铁路运输及水运
rail and water terminal 铁路水路联运码头;水陆联运站
rail and water traffic 铁路水道联合运输
rail approach 进站引线【铁】
rail arcn 轨拱;钢轨拱形支架
rail attachment 铁轨附件
rail bar 轨条
rail base 轨底(座);铁路轨座
rail batter 钢轨打塌
rail bead 凸面线脚
rail beam 轨束梁;钢轨梁
rail beam and section mill 轨梁轧机
rail bearer 轨托纵梁;轨道梁;托轨梁
rail bearing 轴枕
rail bed 钢轨底座
rail belt conveyer 轨道皮带运输机
rail bender 弯轨器;弯轨机
rail bending machine 弯轨机
railboat 石英舟
rail bogie 有轨小车;轨道式运料台车;轨行式运料矿车
rail bolt 扶手螺栓
rail bolt hole 钢轨螺栓孔
rail bond 轨端电气连接;轨道夹紧器;钢轨接续线;钢轨导(电)接线;导轨夹紧器
rail bond grinder 钢轨接头研磨机
rail bonding drill 钢轨钻孔机
rail bond resistance tester 钢轨导接线电阻测试器
rail borer 轨条钻
rail bottom 轨底
rail bound manganese steel frog 钢轨组合锰钢叉心辙叉
rail bound vehicle 轨道车
rail brace 轨撑
rail brake 轨制动装置;轨道制动(器)
rail brake equipment 车辆减速设备【铁】
rail-brake system 轨道制动系统
rail brand 钢轨印记;钢轨标志
rail breaker 钢轨落锤实验机
rail bucket ladder excavator 有轨链斗挖掘机
rail buckle 轨道鼓曲;钢轨挠曲
rail buckling 钢轨屈曲
rail buggy 有轨小车;轨行式运料台车;轨行式运料矿车
railbus 轨道汽车
rail butt 钢轨(的)切头
rail cambering machine 钢轨弯曲机
rail cant 轨底坡
rail canting 钢轨内倾
rail cap 钢轨顶梁
rail capacity 线路通过能力【铁】;铁路通过能力
rail car 有轨(自动)车;轨道车;铁路轻油车;铁路车皮;铁路车厢;热力动车;动车;单节动力有轨车
railcar dumper 翻车机

rail carriage 轨道手推
rail carrier 钢轨搬运机;铁路公司;铁路承运人
rail carrying capacity 铁路输送能力
rail-carrying wagon 运轨车
railcar set 动车组
railcar sway 车辆摆动【铁】
railcar tank 铁路散装水泥罐车
railcar tanker 铁路罐车
rail-cement lining 铁轨混凝土衬砌
rail center 轨道交会点;轨道交叉中心;钢轨中心
rail chair 轨座;铁路轨座
rail clamp 轨头座栓;轨夹;夹轨器
rail clamping device 轨头座栓;轨夹;夹轨器
rail classification 钢轨分类
rail claw hammer 道岔羊角锤
rail clearance point 铁路卸车点
rail clips 轨卡;轨夹;钢轨钳;扣轨弹条;夹轨器;单趾弹簧
rail concrete mixer 轨行式混凝土搅拌车
rail conditioning shop 钢轨整修工厂
rail conditioning unit 钢轨校直精整机组
rail connecting lines 铁路联络线
rail contact 轨条接触点
rail container crane 轨道式集装箱起重机;集装箱轨道起重机
rail conveyer 轨式输送机
rail cooler 钢轨冷却器
rail corrosion 钢轨锈蚀
rail corrugation 钢轨波纹磨耗;钢轨波磨
rail coupling 轨联接
rail crack 钢轨裂缝
rail crane 钢轨起重机;轨道起重机;轨道吊车
rail creep 钢轨爬行
rail creep indication posts 钢轨位移观测桩
rail creep(ing) 轨道蠕动;钢轨爬行
rail cropping method 钢轨截断方法
rail crossing 铁路(平交)道口;轨道交叉
rail cross-sectional area 钢轨断面积
rail curve 铁路曲线
rail cutting machine 锯轨机;切轨机
rail defect detector 轨道探伤器;轨道探伤车;钢轨探伤器
rail defect detector car 钢轨探伤车
rail defects and failure 钢轨伤损
rail deflection 钢轨挠曲;钢轨挠度
rail delivery 铁路运送;铁路发送
rail destressing 钢轨应力放款
rail detecting car 钢轨探伤车
rail detection 钢轨探伤
rail detection by portable ultrasonic equipment 移动式钢轨探伤设备
rail detector 钢轨探伤仪
rail dolly 单轨小车
rail drill 钢轨钻孔机
rail drilling machine 轨道钻孔机;钢轨(端)钻孔机
railed floor 铺钢轨的底板
railed side 栏杆式侧板;横条栏板
rail effective mass 钢轨有效质量
rail end 轨端;铁轨端部
rail-end batter 钢轨接头马鞍形磨损
rail-end chamfering 轨端削角
rail-end chipping 轨端断裂
rail-end fitting 横杆末端配件
rail-end flow 轨端流坍
rail-end-hardening 轨端淬火
rail-end-hardening machine 轨端淬火机
rail-ending machine 轨端淬火机
rail-end milling machine 钢轨端面铣床
rail-end rebuilding 轨端焊补
rail ends unevenness in line or surface 错牙接头
rail-end welding 轨端补焊
rail ex 快运
rail examination 钢轨检验
rail expansion adjuster 钢轨伸缩调节器
rail expansion device 钢轨伸缩调节器
rail expansion joint 钢轨伸缩接缝
rail expansion rollers 钢轨滚筒
rail failure 钢轨伤损;钢轨损伤
rail failure rate 钢轨损伤率
rail fastener 钢轨紧固件;钢轨扣件;扣件
rail fastening 装钢轨配件;钢条扣件;轨道固定;钢轨扣件;钢轨固紧件;扣件【铁】
rail-fastening coach screw 钢轨扣件的方头螺钉;钢轨扣件的方头螺栓

rail fastening down 钢轨锁定
rail fastening system 钢轨扣件系统
rail fatigue 轨轨疲劳；钢轨疲劳
rail fence 围栏；栏杆栅栏；栅栏；篱笆
rail-fence jammer 连续波干扰器
rail ferry 火车渡轮；火车渡口
rail field weld 钢轨工地焊接
rail finishing mill 轨梁精轧机；钢轨轧机；轨梁轧机精轧机座
rail fissure 钢轨裂纹
rail fitting 钢轨配件
rail flaw 钢轨缺陷；钢轨裂缝；钢轨核伤
rail flaw detecting car 钢轨探伤车
rail flaw detection vehicle 钢轨探伤车
rail flaw detector 钢轨探伤仪
rail foot 轨底(座)；铁路轨座
rail foot surface 钢轨底面
rail foot width 钢轨底宽
rail for fixing elevator buckets 提升机斗的固定钢轨
rail fork 轨道分岔
rail-form cement concrete paver 轨模式水泥混凝土摊铺机
rail form concrete paver 轨模式混凝土摊铺机
rail fracture 钢轨断裂
rail freight 铁路运输
rail gap 轨隙，轨缝
rail gap adjuster 轨缝调整器
rail gap adjusting 轨缝调整
rail ga(u)ge 铁道轨距；道尺
rail ga(u)ge template 轨距规
rail girder 扣轨梁
rail grabbing crane 有轨抓斗起重机
rail grinding 磨轨；钢轨研磨；钢轨打磨
rail grinding coach 磨轨车
rail grinding machine 钢轨打磨机；磨轨机
rail grinding train 轨道研磨车；钢轨打磨列车
rail grinding wagon 钢轨打磨车；磨轨车
rail grip 钢轨夹牢；夹轨
rail grizzly 轨条筛
rail guard 挡条；排障器；排障架
rail guide 轨道导向器；钢轨罐道
rail-guided 安装在轨道上的
rail-guided machine 轨模式摊铺机
rail-guided transport system 轨道运输系统
rail guide type rock drill 导轨式凿岩机
rail haulage 有轨运输(方式)；轨道运输；轨道式
rail haulage system 铁道运输系统
railhead 垂直刀架(刨床)；轨顶；轨头；铁路线终点；铁路转运点；铁路尽头站；钢轨端部；钢轨头；临时终点站；尽头式车站
railhead calipers 测轨头卡钳
railhead elevation 轨顶标高
railhead level 轨顶标高
railhead lubricator 轨头涂油机
railhead reprofiling 轨头整形
railhead reserves 铁路终点储备的物品
rail heater 钢轨加热器
rail height 钢轨高度
rail-highway crossing 铁路公路平交道口
rail impact 纵梁碰撞
rail impedance 钢轨阻抗
rail inclination 轨底坡
railing 硬木镶轨条；栅形干扰；栅栏干扰；铁路装运；扶梯栏杆；栏杆；围栏
railing barrel 栏杆管子
railing fitting 栏栅装修
railing for timber bridge deck 木桥面栏杆
railing height 栏杆高度
railing ladder 舷梯
railing material 栏杆(建筑)材料
railing of stainless steel tubes 不锈钢管栏杆
railing operating expenses 铁路运营费用
railing operating fixed assets 铁路运营固定资产
railing operating fixed capital 铁路运营固定资金
railing panel 栏板
railing post 栏杆立柱
railing removable 活动栏杆
railing stair(case) 有栏杆扶手的楼梯
railing system 栏杆系统；扶手系统
railing track 轨道【铁】
rail-inspection vehicle 轨道检测车
rail insulation 钢轨绝缘
rail insulation fault 钢轨绝缘不良

rail insulation wire 钢轨绝缘线
rail internal stress 钢轨内应力
rail iron 铁路钢
rail jack 起道机
rail joint accessories 接头联结零件【铁】
rail joint and point grinding machine 轨道接头及道岔打磨机
rail joint bar 接连钢轨用鱼尾钣
rail joint base plate 轨节垫板
rail joint bond 轨缝连接器
rail joint clearance ga(u)ge 轨隙规；轨缝尺
rail joint expander 轨缝调整器
rail joint fastenings 钢轨接头配件；接头联结零件
rail joint filling compound 钢轨接头填缝剂；轨缝填料
rail joint gap 轨缝
rail joint gap of structure 构造轨缝
rail joint grinder 钢轨接头研磨机
rail joint heater 钢轨接头加热器
rail joint impedance 轨道接头阻抗
rail joint(ing) 接头【铁】；钢轨接头
rail joint welding 钢轨接头焊接
rail key 钢轨固定楔；轨条固定楔
rail ladder 扶手梯
rail layer 铺轨机
rail-laying car 铺轨车
rail-laying crane 铺轨吊车；紧螺帽机
rail-laying equipment 铺轨机械
rail-laying gang 铺轨队
rail leakage resistance 钢轨泄露电阻
rail length 钢轨长度
railless earthmoving 无轨土方(作业)；无轨土方运输
railless line 公路；无轨道路
railless tram 无轨电车
railless (trolley) car 无轨电车
rail level 轨顶高程；铁轨面；轨面；轨顶标高；测轨水准仪
rail level(l)er 平轨机；钢轨整平机；钢轨平路机
rail life 钢轨寿命
rail lift 升轨
rail lifter 起轨器；抬轨器；升轨器；吊轨机
rail link 铁路联络线；铁路连接线
rail lip 钢轨肥边
rail loading plant 装运钢轨设备
rail log 船尾舷侧拖曳式计程仪
rail lowering motion 钢领板下降装置
rail lubrication 钢轨润滑
rail lubricator 钢轨涂油器；曲线涂油器
raill-water terminal 水—铁联运码头；水陆枢纽
railman 码头工人；铁路职工
rail mileage 铁路(英里)里程
rail mileage open to traffic 铁路通车里程
rail mill 轧轨机；轧钢轨设备
rail-motor 铁路公路联运的
rail motor car 有轨机动车；轨道车；铁路机动车；铁道机动车
rail-mounted 安装在轨道上的；轨道式
rail-mounted belt conveyer 轨道式带式输送机
rail-mounted bucket ladder excavator 轨道式链斗挖掘机
rail-mounted bucket wheel excavator 轨道式斗轮挖掘机
rail-mounted continuous excavator 轨道式连续挖掘机
rail-mounted crane 装在轨道上的起重机；轨行(式)起重机；轨道起重机；铁路起重机；铁道起重机
rail-mounted crane with fixed jib 固定吊臂的轨道起重机
rail-mounted diesel-drives grabbing crane 柴油机驱动的轨道式抓斗起重机
rail-mounted electric(al) grabbing crane 电驱动的轨道式抓斗起重机
rail-mounted equipment 有轨设备
rail-mounted excavator 轨道式挖掘机；轨行式挖土机；轨行式挖掘机；轨行式挖沟机
rail-mounted gantry crane 轨行式门式起重机；轨道龙门吊
rail-mounted hoisting gear 轨行式提升绞车；轨道提升机；轨道式起重设备；轨道式起重机
rail-mounted mobile crushing plant 有轨移动式破碎机组
rail-mounted overhead crane 轨行高架起重机
rail-mounted power shovel 轨道式单斗动力铲；

轨道式单斗挖掘机
rail-mounted slewing crane 轨道式悬臂起重机；轨道式回转起重机
rail-mounted stacker 轨道式堆垛机
rail-mounted steam grabbing crane 轨道式蒸汽抓斗起重机
rail-mounted steam shovel 轨道式蒸汽动力铲；轨道式蒸汽单斗挖掘机
rail-mounted tipping concrete skip 轨道式混凝土翻斗车
rail-mounted tip wagon 轨道式倾斜货车
rail-mounted traffic 轨道式车辆；轨道式交通
rail-mounted transtainer 轨行式门式集装箱装卸机
rail-mounted trenching machine 轨道式挖沟机
rail-mounted vehicle 轨道式车辆
rail-mounted wagon drill 轨道式钻车
rail neutral temperature 中和轨温
rail of self-hardening steel 自淬硬钢钢轨
rail of square steel 方钢轨
rail oiling 钢轨涂油
railophone 火车电话机
rail overlapping device 钢轨伸缩调节器
rail overturning 钢轨倾翻
rail pad 轨下胶垫；轨下垫板
rail pass 钢轨孔型
rail pile 钢轨肚柱
rail pincers 轨道钳
rail pit 铁路运输露天矿
rail plane 轨刨；刨轨机
rail planing machine 刨轨机
rail plate 主轴箱滑动导轨；钢轨垫板；滑动导轨
rail platform car 铁路平(板货)车
rail port 火车与货船联运的港口
rail post 扶手；栏杆；栏杆(立)柱
rail potential 钢轨电位
rail potential limiting device 轨电位限制设备
rail press 压轨机
rail pressure 轨压；钢轨压力
rail profile ga(u)ge 钢轨断面磨耗测量器
rail profile measuring coach 钢轨磨耗检查车
rail property 钢轨性能
rail puller 曳轨器；轨缝调整器；钢轨拉伸器
rail punch 钢轨冲压机
rail rapid transit 快速铁路
rail rate 铁路费率；铁道运费
rail reconditioning 钢轨整修
rail relaying machine 更换钢轨机
rail removing device 搬移钢轨机
rail renewal 钢轨更新
rail renewal machine 更换钢轨机
rail replacement 钢轨更换
rail replacer 换轨器
rail rest 轨座；钢轨支架
rail-retarder 车辆轨条缓行器
rail return 轨道电流回路
rail return current 钢轨回流
railroad 火车轨道；铁路；铁道(路线)
railroad accident 铁路事故
railroad accouting 铁路会计核算
railroad administration 铁路管理局
railroad advance warning sign 铁路道口预警标志
railroad advice 铁路货物通知书
railroad apartment 列车式单元房；车厢式住宅单元
railroad architecture 铁道建筑学
railroad atlas 铁路地图集
railroad ballast 铁路道砟
railroad bed 铁路路基
railroad bill of lading 铁路货运提单
railroad bond 铁路债券
railroad boom crane 铁路吊臂起重机；机车起重机
railroad boxcar 铁路运输货车
railroad branch 铁道支线
railroad bridge 铁路桥(梁)；铁道桥
railroad building 铁路房屋；铁路用房；铁路建造
railroad car 有轨电车；火车车厢；铁路车辆；铁道车辆
railroad car axle 火车车轴
railroad car tank 铁路水泥罐车
railroad chronometer 铁路计时器
railroad clearance 铁路限界；铁路净空；铁道界限
railroad connection 铁路到港口的终点；铁路连接
railroad construction 铁路建筑；铁路建造
railroad container 铁路散装水泥车
railroad crossbuck sign 铁路道口警告标志
railroad crossing 道岔锁闭器；平交道口【铁】；铁路平

交道(口);铁路交叉口;铁路交叉点;铁路道口
railroad crossing angle 铁路道口交叉角
railroad crossing gate 平交道口栅门;铁路道口交叉栅门
railroad crossing lock 道岔锁闭器
railroad crossing sign 铁路(道口)交叉标志;铁路道口标志
railroad crossing signal 平交道口信号;铁路道口交叉信号
railroad curve 铁路弯道;铁路曲线
railroad curve survey 铁路曲线测量
railroad cuts 铁路路堑
railroad cutting 铁路挖方
railroad data system 铁路数据系统
railroad depot 铁路机务段;铁路车站
railroad embankment 铁路路基;铁路路堤
railroad employee 铁路雇员
railroad engineering 铁路工程学
railroader 铁路职工
railroad excavation 铁路开挖
railroad export bill of lading 铁路外贸提单
railroad ferry 火车轮渡;火车渡轮;火车渡口
railroad ferry terminal 火车轮渡码头
railroad flat 车厢式住宅单元
railroad freight car 铁路货运车
railroad freight charges 铁路货运费
railroad freight rates 铁路运费率
railroad freight yard 铁路货运场
railroad ga(u)ge 轨距;铁路轨距;铁道轨距
railroad grade crossing 道路与铁路的平交叉道(口);平交道口【铁】;铁路与轨道平交道口;铁路平面交叉;铁路道口;道口【铁】
railroad greening 铁路绿化
railroad hauling car 轨道牵引车
railroad hopper car 铁路运输活底车
railroad infrastructure 铁路设施
railroading 铁路业务;铁路建设
railroad interline traffic 铁路联运
railroad intermodal yard 铁路换装场
railroad intersection 铁路公路交叉
railroad jack 有轨起重器;移动式起重器;铁路举重器
railroad journal box 铁路轴箱
railroad journal-box grease 货车轴用润滑脂
railroad level-crossing 铁路平交道口
railroad line 铁路线
railroad lines opened to traffic 铁路营业里程
railroad location survey 铁路定线测量
railroad locomotive engineer 机车机械师
railroad logging 铁路运材;铁道运材
railroad map 铁路路线图
railroad motor car 铁道机动车
railroad network 铁路网
railroad noise 铁路(交通)噪声
railroad opening 铁路道口
railroad operation 铁路运营
railroad overcrossing 跨线桥;铁路天桥
railroad pen 双(曲)线笔
railroad platform 铁路站台
railroad point 铁路岔尖;铁路辙尖;铁道辙尖
railroad pool 铁路联营
railroad powder 铁路炸药
railroad profile 钢轨断面
railroad rail 铁路轨道;钢轨;铁轨
railroad ramp 铁路斜坡道
railroad rate schedule 铁路运费表
railroad repair shop 铁路修理车间;铁路车辆检修车间
railroad right-of-way 铁路筑路权;铁路用地
railroad roller 钢轨滚机
railroad rolling sequence 钢轨轧制系统
railroad saw 切轨锯;切钢锯
railroad sawing machine 钢轨锯切机
railroad section 钢轨断面;线路区间
railroad service 铁路公路联运业务
railroad shoe 轨道座
railroad shovel 轨道挖掘机
railroad showshed 铁路防雪棚
railroad siding 铁路盒线;铁路调车线;铁路岔线;铁路岔线;铁路侧线;轨道支线
railroad sleeper 铁路枕木;铁路轨枕
railroad slewer 移轨器
railroad span 轨距
railroad spike 铁路道钉
railroad spur 铁路支线;铁路调车线;铁路岔道

railroad station 火车站
railroad survey(ing) 铁路勘测;铁路测量
railroad switch 铁路道岔;道岔【铁】
railroad system 铁路系统
railroad terminal 铁路公路联合枢纽;铁路终点(站);铁路枢纽(站);枢纽站
railroad test track 试验滑轨
railroad thermit 钢轨铝热剂
railroad through transport 铁路联运
railroad tie 铁路枕木;铁路轨枕
railroad tire 火车轮箍
railroad track 轨道【铁】;铁路轨道
railroad track scale 轨道衡
railroad tracks of coating textiles 涂层织物的轨状压花
railroad track-work 铺轨工作
railroad train 铁路列车;火车
railroad transportation cost 铁路运输成本
railroad transportation map 铁路运输图
railroad transport(ion) 铁道运输
railroad tunnel 铁路隧道
railroad tunnel kiln 窑车式隧道窑
railroad turnout 铁路道岔
railroad underbridge 跨铁路桥;铁路在下交叉;铁路跨线桥
railroad underconstruction 建筑中铁路
railroad warning sign 铁路警告标志;铁道警告标志
railroad wear tolerance 钢轨磨耗余量
railroad wharf 铁路码头
railroad wheel lathe 车轮车床
railroad yard 铁路调车场
rail rolling mill 钢轨轧机
rail rope 承载索(架空索道的)
rail running surface 钢轨运行表面
rail rupture 钢轨断裂
rails 铁路股票
rails and accessories 钢轨与配件轨料
rail saw 钢轨锯;切轨锯
rail saw frame 钢轨锯架
rail sawing machine 锯轨机
rail scrap 废钢轨
rail screen 栏杆围幕
rail seat 承轨槽
rail section 钢轨截面
rail section drawing instrument 钢轨断面测绘仪
rails for transfer platform 转向平台轨道
rail shear 钢轨推凸机
rail shift 铺轨班
rail shifting machine 移道机
rail-ship interface 车船交接;车船换装
rail shipment 钢轨发送;钢轨装运
rail shovel 有轨挖土机
rail shunter 扳道工;转辙器;转器
rail siding 铁路支线;铁路侧线
rail skip 有轨箕斗
rail slewer 移轨器
rail-slitting mill 钢轨切分轧机;钢轧切分轧机
rail socket 栏杆插座
rail specification 钢轨规格
rail spike 道钉
rail splice 连接板【铁】;鱼尾板;钢轨鱼尾板
rail spur 铁路支线;铁路专用线;短支线
rail square 准轨尺;钢轨角尺
rail stanchion 栏杆(支)柱
rail steel 轨钢;轨道钢;钢轨
rail steel reinforcement 轨用钢钢筋;钢轨钢钢筋;T形钢筋;轨形钢筋
rails tie plate 垫板
rail straightener 直轨器;直轨机;钢轨矫直机
rail straightening machine 直轨机;轨道矫直机;钢轨校直机;钢轨矫直机
rail-straightening tool 直轨器
rail straighter 钢轨校直器
rail strength 钢轨强度
rail stress 钢轨应力
rail stretcher 轨缝调整器
rail string 长轨节
rail support 轨座
rail support at discrete points 点支承钢轨
rail surface 轨面
rail tank car 铁路(油)槽车
rail tanker 火车油槽车;铁路(油)槽车
rail temperature 钢轨温度
rail tensor 钢轨拉伸器

rail terminal 铁路终点(站)
rail test 钢轨试验
rail tester 钢轨检验器
rail testing wagon 路轨检查车
rail thermit 钢轨焊接用铝热剂
rail thermometer 轨温测量表;钢轨温度计;路轨测温计
rail thread 钢轨踏面
rail tie bar 钢轨系杆
rail tie plate 轨道垫钣
rail tilting 钢轨倾翻
rail to earth resistance 钢轨过渡电阻
rail tongs 轨夹;钢轨(夹)钳;夹轨钳
rail top 钢轨顶部;轨顶
rail-to-rail 铁路间中转运输
rail-to-water 铁路转水路运输
rail track 铁路轨道;铁路线路
rail-track product 钢轨制品
rail traction 铁路牵引
rail traffic 铁路运输
railtrain dispatch system 铁路列车调度系统
rail transit 有轨运输;有轨交通;轨道交通;铁路直达运输
rail transit corridor 轨道交通走廊
rail transit network planning 轨道交通线网规划
rail transit passenger turnover 轨道交通乘客周转量
rail transit system 轨道交通系统
rail transit train 直达列车
rail transport(ion) 铁道运输
rail transposing 左右钢轨互换
rail travel 有轨运输
rail trimming machine 钢轨研磨平整机
rail trolley 轨道自动车
rail truck 铁路车辆
rail turner 翻轨机
rail type drawing frame 平台式并条机
rail-type hot bed 轨式温床
rail tyre 钢丝轮胎
rail unloading point 铁路闸车点
rail upright 钢轨支柱;栏杆立柱
rail vibration 钢轨振动
rail voltage 干线电压
rail-water facility 铁路水路中转设备
rail-water rate 铁路水路联运运价
rail-water terminal 铁路水路联运码头
rail-water through goods transport plan 铁水货物联运计划;水陆联运计划
railway = railroad
railway accident 铁路事故
railway accommodation 铁路客货运容量
railway accounting 铁路会计核算;铁路会计
railway accounting pattern 铁路会计核算方式
railway accounting statement 铁路会计报表
railway accounting system 铁路会计制度
railway actual capability 铁路实际运输能力
railway aerosurveying 铁路航(空)测(量);铁道航空测量
railway and highway combined bridge 铁路公路两用桥
railway annual plan 铁路年度计划
railway annual report audit 铁路年度报告审计
railway assets operating 铁路资产经营
railway atlas 铁路地图集
railway auditing 铁路审计
railway authority 铁路管理机构
railway axis 铁路轴线;铁路中线
railway axle box 铁路轴箱
railway basic construction 铁路基本建设
railway basic construction management 铁路基本建设管理
railway basic construction management agency 铁路基本建设管理机构
railway basic construction management content 铁路基本建设管理内容
railway basic construction management system 铁路基本建设管理制度
railway bed 铁路路基
railway bed settling 路基下沉【铁】
railway bed transect 路基横断面【铁】
railway berth 通火车的码头泊位
railway bill of lading 铁路提货单
railway bogie 轨道转向架
railway bond 铁路债券

railway boundary 铁路界标;铁路边界
railway box wagon 铁路箱车;铁路棚车
railway brake 铁路用制动器
railway brake hose 机车制动胶管
railway branch administration 铁路分局
railway bridge 铁路桥(梁)
railway bridge load 铁路桥梁荷载
railway bureau paid cash in transit 路局回缴途中款
railway bureau transport profit 路局运输利润
railway bureau turnover 铁路局资金周转额
railway capacity 铁路能力
railway capital structure 铁路资本结构
railway car 铁路车辆
railway car accessories plant 车辆附件厂【铁】
railway car parts plant 车辆附件厂【铁】
railway carriage 铁路客车;铁道车
railway carrying capacity 铁路通过能力
railway cars plant 车辆制造厂
railway catheter 槽式导管
railway center 铁路枢纽;铁路中心
railway centerline 铁路中心线
railway chair 铁路辙枕
railway charges 铁路费用
railway charter 铁路铺设权
railway classification 铁路等级
railway classification yard 铁路编组场
railway clearance 铁路限界;铁路净空;铁道界限
railway coachbuilding 铁路车辆制造
railway colony 铁道聚居地(靠港口)
railway communication 铁路通信
railway commutation 铁路通讯;铁路通勤
railway company 铁路公司
railway connection 铁路联络线
railway consignment note 铁路托运单
railway construction 铁路建筑;铁路建造;铁路建设;铁道建筑;铁道工程
railway construction battalion 铁路修建营
railway construction clearance 建筑限界
railway construction enterprise 铁路施工企业
railway construction enterprise audit 铁路施工单位审计
railway construction enterprise management 铁路施工企业管理
railway construction enterprise quality 铁路施工企业素质
railway construction enterprise standard 铁路施工企业标准
railway construction funds payable 应缴铁路建设基金
railway construction ga(u)ge 建筑限界
railway construction organization design 铁路施工组织设计
railway construction project 铁路建设项目
railway construction supervision 铁路建设监理
railway container terminal 铁路集装箱站
railway-controlled 铁路管辖的
railway convertor 铁路用变流机
railway corporation 铁路公司
railway crane 轨道式起重机;铁路起重机
railway cross-cut saw 横割轨锯
railway crossing 平交道口【铁】;铁路交叉口;铁路交叉点;铁道交叉
railway crossing bar 铁路公路交叉口拦路杆
railway culvert 铁路涵洞
railway current 铁路电流
railway current assets management 铁路流动资产管理
railway curve 铁路曲线(板);铁道曲线(板);铁道弯尺
railway curve rule 铁路曲线尺
railway cuts 铁路路堑
railway cutting crest 铁路挖方顶部
railway cutting on sloping ground 坡地上铁路挖方
railway data-processing 铁路运输数据处理
railway data-processing center 铁路运输数据处理中心
railway deck 铁路桥面
railway depot 火车站
railway design criterion 铁路设计准则;铁路设计标准
railway despatch concentration 铁路调度集中
railway development planning 铁路发展规划
railway diagram 铁路图

railway directory 铁路指南
railway distribution system 铁路配电系统
railway district 铁路区段
railway diversified economy 铁路多元经营
railway division 铁路(区)段
railway division communication system 铁路区段通信系统
railway dry dock 铁道干坞;修船台;铁路干船坞;铁道干船坞;升船滑道;船排;船台滑道
railway duty 铁路税
railway economic accounting 铁路经济核算
railway economic benefit audit 铁路经济效益审计
railway economic efficiency audit method 铁路经济效益审计方法
railway economic efficiency audit procedure 铁路经济效益审计程序
railway economics 铁道经济
railway electrification 铁路电气化;铁道电气化
railway embankment 铁路路基;铁道路堤
railway end-loading ramp 铁路端式装车斜坡台
railway engineering 铁路工程(学);铁道工程
railway engineering survey 铁路工程测量
railway engineering surveying 铁路工程测量学
railway engineer troops 铁道工兵部队
railway equipment 铁路设备;铁道设备
railwayese 铁路用语
railway estate 铁路地产
railway export bill of lading 铁道出口提单
railway express and fast freight 特快服务(运)
railway facility 铁路装备;铁路(装车)设施
railway fare 铁道运费
railway fast parcel transport agency 铁路快运包裹代理人
railway feeder 铁路支线
railway ferry 铁路轮渡;火车轮渡;火车渡轮;火车渡口
railway ferry route 铁路轮渡线路
railway ferry terminal 火车轮渡码头;铁路轮渡码头
railway ferry vessel 列车渡轮
railway fill 铁路填土;铁路路堤
railway filling point 铁路装车点
railway finance 铁路财务
railway finance and accounting information library 铁道财务会计信息库
railway finance and accounting management information system 铁路财务会计管理信息系统
railway financial analysis 铁路财务分析
railway financial budget 铁路财务预算
railway financial control 铁路财务控制
railway financial decision 铁路财务决策
railway financial management system 铁路财务管理体制
railway fixed assets management 铁路固定资产管理
railway for mixed passenger and freight traffic 客货运混合线路
railway freight 铁路运费;铁道运费
railway freight car oil 铁路用粗润滑剂
railway freight price 铁路货物运价
railway freight transport statistics 铁路货物运输统计
railway freight transport tariff system 铁路货物运价体系
railway ga(u)ge 铁路轨距;铁道轨距
railway general accounting and management information system 全路通用会计核算及管理信息系统
railway general investigation 铁路普查
railway goods transport 铁路货物运输
railway grade crossing 铁路道口;铁路与轨道平交道口;铁路公路水平交叉
railway grand division 铁路总段管理局
railway grease 铁路用润滑脂;齿轮油
railway "green passage" 铁路绿色通道
railway guide 铁路旅行指南
railway haulage capacity 铁路牵引力
railway heavy haul traffic 铁路重载运输
railway high speed traffic 铁路高速运输
railway hub 铁路枢纽
railway in a cut 铁路路堑;路堑上的铁路
railway income capital 铁路收入资金
railway income management 铁路损益管理
railway-industrial joint marshalling station 路厂联合编组站

railway industrial statistics 铁路行业统计
railway information project 铁路信息化工程
railway infrastructural construction 铁路基本建设
railway infrastructure 铁路设施
railway infrastructure capacity 铁路设施能力
railway injury 火车损伤
railway in operation 运营铁路
railway in service 运营铁路
railway inspection car 检测车;铁路验道车
railway inspection trolley 检道车;铁道巡查车
railway institute 铁道学院
railway integration mechanism 铁路联锁设备
railway intelligent transportation system 铁路智能运输系统
railway interest 铁路业务;铁路部门
railway intermodal yard 铁路换装站场
railway internal labo(u)r market 铁路内部劳动力市场
railway investment system reform 铁路投资体制改革
railway jib crane 铁路悬臂起重机
railway journal (box) grease 铁路货车轴承润滑脂
railway junction 铁路联轨站;铁路交叉点
railway junction terminal 铁路枢纽
railway keep-price transport 铁路保价运输
railway labo(u)r organization 铁路劳动组织
railway labo(u)r quota 铁路劳动定额
railway lamp fitting 铁路照明器附件
railway land 铁路用地
railway launching platform 铁轨式发射台
railway level-crossing 铁路平交道(口)
railway lighting equipment 铁路照明设备
railway line 线路【铁】;铁路线
railway line for freight traffic 货运专线
railway line for passenger traffic 客运专线【铁】
railway load 铁路用电负荷
railway loading 火车装车处
railway loading facility 铁路装车设施
railway loading point 铁路装车点
railway locating engineer 铁路选线工程师
railway location 选线【铁】;铁路选线;铁路定线
railway logging 铁道运材
railway long plan 铁路长期计划
railway long-term program(me) 铁路长期计划
railway luggage and parcel transport statistics 铁路行包运输统计
railway luggage and traffic contract 铁路行包运输合同
railway mail service 铁路邮政
railway main line length 铁路正线长度
railway maintenance car 铁路养护车
railway maintenance gang 铁路道班;铁路养护队
railway man 铁路职工
railway management information system 铁路运营管理信息系统
railway map 铁路路线图
railway material circulation enterprise 铁路物资流通企业
railway material consumption 铁路物资消耗量
railway material contracting supply management 铁路物资合同制供应管理
railway material incomes 铁路物资收入量
railway material stocks 铁路物资库存量
railway mid-term program(me) 铁路中期计划
railway-mining joint marshalling station 路矿联合编组站
railway motive power 铁路牵引动力
railway motor 铁路用电动机
railway motor car 轨道式自动车;内燃机车
railway museum 铁路博物馆
railway nationalization 铁路国有化
railway network 铁路(路线)网
railway network capacity 铁路网通过能力
railway network density 铁路(路)网密度
railway network marshalling station 路网性编组站【铁】
railway network planning 铁路(路)网规划
railway noise 铁路噪声
railway non-operating state-owned capital 铁路非经营性国有资产
railway of national network 路网铁路;国家路网铁路
railway on trestle 栈桥高架线
railway operating income 铁路运营进款

railway operating kilometrage 铁路营业里程
railway operating length of railway 铁路运营里程
railway operating management theory 铁路运营管理理论
railway operating plan 铁路作业计划;铁路运营计划
railway operating revenue 铁路运营收入
railway operating state-owned capital 铁路经营性国有资产
railway operation 铁路运营
railway overbridge 铁路跨线桥
railway overcrossing 铁路天桥
railway-owned 铁路所属的;铁路所属的
railway parcel transportation statistics 铁路行包运输统计
railway passenger 旅客
railway passenger company 铁路客运公司
railway passenger contract 铁路客运合同
railway passenger fare 铁路旅客运价
railway passenger fare kilometreage district 旅客票价里程区段
railway passenger tariff kilometrage table 铁路客运价里程表
railway passenger traffic 铁路旅客运输
railway passenger traffic control 铁路客运调度工作
railway passenger traffic record 客运记录
railway passenger traffic statistics 铁路旅客运输统计
railway passenger traffic tariff 铁路旅客运输价格
railway passenger transportation plan 铁路旅客运输计划
railway passenger transport tariff 铁路旅客运输价格
railway pen 双曲线笔
railway performance 铁路绩效
railway permitted velocity 线路允许速度
railway pile driver 有轨打桩机
railway plan(ning) 铁路计划
railway planting 铁路绿化
railway platform 月台
railway point 铁路辙尖
railway point of frogs 铁路道叉转辙器
railway police 铁路警察
railway post appraisal 铁路岗位测评
railway post office 铁路邮局
railway potential capability 铁路潜在运输能力
railway power supply network 铁路供电网
railway precincts 铁路管界
railway premises 铁路房产
railway private 铁路民营化;铁路公司化
railway project investment 铁路建设项目总投资
railway project quality check 铁路工程质量检查
railway property 铁路资产;铁路财产
railway protection forest 铁路防护林
railway rail 铁轨
railway rate 铁路运价
railway rate of exchange 铁路兑换价率
railway rate structure 铁路运价结构
railway reconnaissance 铁路勘测
railway refrigerated wagon 铁路冷藏车
railway renewing, transforming project 铁路更新改造项目
railway reorganization 铁路重组
railway repair shop 铁路修配厂;铁路修理车间;铁路车辆检修车间
railway right-of-way 铁路用地
railway rolling stock 铁路车辆
railway roof 月台棚;站台棚
railway roundhouse 铁路保养场
railway route kilometrage 铁路线路公里数;铁路线路公里里程
railway route map 铁路线路图
railway sack 粗麻袋
railway sennit 双千线编索
railway service 铁路业务;铁路服务
railway service car 路用车【铁】
railway service train 路用列车
railway service transport kilometers 路用车辆公里【铁】
railway shop 铁路保养场
railway showshed 铁路防雪棚
railway shunting yard 调车场
railway siding 专用铁路线;铁路专用线;铁路支

线;铁路盒线;铁路侧线;铁道支线
railway signal(ling) 铁路信号
railway signals equipment 铁路信号设备
railway sleeper 铁路枕木;铁路轨枕
railway slip 轨道滑船台;船排滑道
railway span 轨距
railway special line 铁路专用线
railway spike 铁路道钉
railway spur 铁路岔道
railway's right of lien 铁路留置权
railway staff service pass 铁路乘车证
railway station 火车站;铁路车站
railway station communication system 铁路站内通信系统
railway station hall 铁路车站候车大厅
railway statistic(al) analysis 铁路统计分析
railway statistic(al) consultation 铁路统计咨询
railway statistic(al) educational training 铁路统计教育培训
railway statistic(al) factory 铁路统计工厂
railway statistic(al) management information system 铁路统计信息管理系统
railway statistic(al) scientific research 铁路统计科学研究
railway statistic(al) survey 铁路统计调查
railway statistics 铁路统计
railway statistics arrange 铁路统计整理
railway statistics supervision 铁路统计监察
railway structure 铁路建筑物
railway subgrade 铁路路基
railway substation 铁路变电所
railway surroundings 铁路周围环境
railway survey and design 铁路勘测设计
railway survey(ing) 铁路勘测;铁路测量
railway survey in selected items from totality at random 铁路抽样调查
railway system 铁路系统;铁路网;铁道系统
railway system control 铁路体系控制
railway tank car 铁路槽车
railway tariff 铁路运价(表);铁路运费率
railway technical regulation 铁路技术管理规程
railway technics 铁路技术
railway telegraph ordinary 铁路普通电报
railway telephone 铁路电话
railway terminal 铁路终点(站);铁路枢纽(站)
railway terminal with one station 一站枢纽
railway territory 铁路地产;铁路地区;铁路土地
railway test vehicle 铁路试验车
railway track 铁路线路;铁路轨道
railway track in use 通车里程
railway track line length 铁路延展长度
railway-track scale 铁路过车磅;铁路过车秤;铁路轨道地磅;铁路车辆秤
railway traction power 铁路牵引动力
railway traffic enterprise responsibiltiy 铁路运输企业责任
railway traffic revenue 铁路运营收入;铁路运输收入
railway traffic turnover 铁路运输周转量
railway's trail-car 铁路拖车
railway train 火车
railway train ferry 铁路轮渡
railway train load norm 列车质量标准
railway transit 铁路用纬仪
railway transport 铁路运输
railway transport action receipts check 铁路运输收入稽查
railway transportation 铁路运输
railway transportation cost 铁路运输成本
railway transportation cost analysis 铁路运输成本分析
railway transportation cost and expense calculation 铁路运输成本费用计划
railway transportation cost and expense control 铁路运输成本费用控制
railway transportation cost and expense decision 铁路运输成本费用决策
railway transportation cost and expense forecast 铁路运输成本费用预测
railway transportation cost calculating 铁路运输成本计算方法
railway transportation cost calculation 铁路运输成本计算
railway transportation cost planning 铁路运输成本计划

railway transportation map 铁路运输图
railway transportation noise 铁道交通噪声
railway transportation regulation 铁路运输规程
railway transportation revenue and expense audit 铁路运输收入与支出审计
railway transportation revenue management 铁路运输进款管理
railway transportation statistics information network 铁路运输统计信息网络
railway transportation volume forecasting 铁路运输量预测
railway transport company 铁路运输公司
railway transport density 铁路运输密度
railway transport enterprise reform 铁路运输企业改革
railway transport enterprise responsibility 铁路运输企业责任
railway transport equipment statistics 铁路运输设备统计
railway transport expense entry 铁路运输支出科目
railway transport labo(u)r productivity 铁路运输劳动生产率
railway transport management system reform 铁路运输管理体制改革
railway transport marketing 铁路运输营销
railway transport of passengers 铁路客运
railway transport operation income management 铁路运输营业收入管理
railway transport rate 铁路运输价格
railway transport rate index 铁路运输价格指数
railway transport revenue 铁路运输收入(进款)
railway transport revenue account 铁路运输进款科目
railway transport revenue accounting 铁路运输进款会计核算;铁路运输进款会计
railway transport revenue financial statement 铁路运输进款会计报表
railway transport revenue impropriate days calculation sheet 运输进款占用天数计算表
railway transport revenue receipts 铁路运输进款凭证
railway transport statistics 铁路运输统计
railway transport statistics theories and methods 铁路运输统计理论与方法
railway transport timber 铁路运输木材
railway truck 铁路车皮;车皮
railway tunnel 铁路隧道
railway turbine 机车燃气轮机
railway turntable 转车盘
railway tyre 火车轮箍
railway underbridge 铁路在下交叉;铁路下穿桥;铁路跨线桥
railway under construction 在建铁路;建设中铁路
railway utilities 铁路公用事业
railway valuation 铁路估定价值;铁路估(定)值
railway van 车皮
railway vehicle 铁路车辆
railway wagon 铁路货车;车皮
railway wagon detergent 火车车皮洗涤剂
railway warning sign 铁路警告标志
railway wear tolerance 钢轨磨耗余量
railway wharf 铁路码头
railway wheel and type mill 车轮和轮箍轧机
railway wheel lathe 轮箍车床
railway wheel mill 火车车轮轧机;车轮轧机
railway wheel tyre 车轮轮箍
railway work 铁路工程
railway working capital 铁路营业资金
railway work shop 铁路工厂
railway yard 编车场【铁】;铁路站场;铁路调车场
railway yard and hub designing standards 铁路车站及枢纽设计规范
railway yard designing 铁路车站设计
railway yearly plan 铁路年度计划
rail wear 钢轨磨耗
rail wear tolerance 轨道磨耗容限
rail web 轨腹(板);轨腰
rail-web bond 轨腹接续线
rail weighbridge 轨道衡
rail weight 钢轨重量
rail welding 钢轨焊接;钢轨焊缝
rail welding machine 钢轨焊接机;焊轨机
rail welding plant 焊轨厂
rail welding train 焊轨列车

rail welding trimming machine 轨道焊接修整机
rail weld seam shearing machine 钢轨推凸机
rail wheel 铁路车轮;轨轮
rail-wheel contact stress 轮轨接触应力
rail winch 起轨(绞)车
rail wing 辙叉翼轨
rail with blind ends 两端无螺栓孔的钢轨
rail yard 铁路站场
rain and/or freshwater damage 雨淋和/或淡水货损
rain and snow ga(u)ge 雨雪测量器;降水量计
rain arc-over test 湿飞弧试验
rain area 雨水面积;雨区
rain attenuation 雨致衰减
rain awning 舱口雨篷;舱口天幕
rain backscatter 雨滴后向散射
rain band 雨带
rain barrel effect 过补偿线路上的响应
rain barrier 雨水屏障
rain basin 雨水坑
rain-bearing cloud 致雨云;造雨云
rain-bearing system 降水天气系统
rain belt 雨区;雨带
rainbow 虹;彩色带信号;彩虹
rainbow arch 虹拱;同心圆拱
rainbow dressing 挂满旗【船】
rainbow effect 虹彩效应
rainbow film 彩虹膜
rainbow generator 彩条信号发生器
rainbow granite 虹状花岗岩
rainbow holography 彩虹全息术
rainbow light 彩虹灯具
rainbow on the flush pit 沉淀池中彩色油花
rainbow pattern 彩色带条信号图
rainbow roof 虹形或鲸背形坡屋顶;拱形坡顶;弓形屋顶;凸弧形坡屋顶
rainbow scattering 霓虹散射
rainbow sprinkler 彩虹喷水器
rainburst 降雨阵发
rain cap 雨帽;雨罩;雨盖
rain capacity 降雨量
rain cape 雨披
rain catch 承雨(量)器
rain catchment 降雨集水
rain catchment basin 降雨集水盆地
rain chamber 喷水吸尘室;喷水水塔
rain channel 雨水流痕(岩石表面);雨水沟;雨水道;水蚀沟
rain chart 雨量图;降雨曲线;降水分布图
rain cistern 雨水储[贮]池
rain-climatic zone 雨区
rain cloth 雨布;舱口雨篷;舱口天幕;防雨布
rain cloud 雨云
rain clutter 雨滴杂乱回波
rain clutter suppression 雨雪干扰抑制
raincoat 雨衣
rain collector 集雨器
rain conductor 雨水管;水落管
rain cone 金属伞
rain crust 雨斑壳
rain curtain 雨幕
rain damage 淡水濡损
rain day 雨日;降雨日
rain depth-duration curve 雨量—历时关系曲线
rain detection radar 测雨雷达
rain diagram 雨量图
rain discharge 排雨量
rain down-pipe 落水管
rain drainage inlet 雨水排入口
raindrop 雨点;雨滴
raindrop echo intensity 雨滴反射信号强度
raindrop erosion 雨水冲刷;雨点冲刷;雨滴侵蚀
raindrop fall velocity 雨滴下落速度
raindrop figure 斑点与条带相间的木纹;雨滴状花纹;雨滴图案;雨点图案
raindrop glaze 雨点釉
raindrop impact 雨痕;雨滴冲击
raindrop impression 雨点印痕;雨痕;雨点痕迹
raindrop imprint 雨痕;雨滴痕
raindrop interference 雨滴干扰
raindroplet collector 雨滴谱仪
raindrop pattern 雨点图案
raindrop size distribution 雨滴谱
raindrop spectrograph 雨滴谱仪

raindrop spectrometer 雨滴谱仪
raindrop splash on soil surface 降雨土表溅水
raindrop test 雨淋试验
rain duration 降雨历时
rain echo 雨状回波
rain-echo intensity 雨滴反射信号强度
rained nucleus 雨滴核
rain ending 降雨终止
rainer 雨枪;人工降雨装置;人工降雨器;喷灌装置
rain erosion 雨水浸蚀;雨蚀(作用)
rain-erosion damage 雨蚀损伤
rain exclusion of walling 墙体防雨
rain factor 雨量因素;降雨系数
rainfall 雨量;降雨(量);降水(量)
rainfall amount 降雨量;降雨量
rainfall anomaly 降水偏距
rainfall area 降雨面积;雨区;降雨地区
rainfall at point 点雨量
rainfall catchment 降雨集水
rainfall characteristics 降雨特征(值)
rainfall chart 雨量图
rainfall collection 汇集雨水
rainfall curve 雨量曲线;降雨曲线
rainfall data 雨量资料;降雨资料
rainfall day 雨天
rainfall density 雨量强度;降雨密度
rainfall depth 降雨量;降雨深度
rainfall-depth contour 降雨等深线;等雨深线
rainfall depth duration curve 雨量深度历时关系曲线
rainfall desert 降雨沙漠
rainfall distribution 雨量分布;降雨分布;降水分布
rainfall distribution chart 降雨分布图
rainfall distribution coefficient 降雨分布系数
rainfall distribution density 降雨分布密度
rainfall distribution function 降雨分布函数
rainfall dump box machine 百叶窗式翻斗壳型机
rainfall duration 降雨持续时间
rainfall-duration curve 雨量—历时关系曲线
rainfall effectiveness 雨量效率;有效雨量;降雨效率
rainfall energy-intensity factor 降雨能量强度系数
rainfall erosion 雨失;降雨侵蚀(作用);降雨冲蚀
rainfall erosion index 降雨侵蚀指数
rainfall erosivity 降雨侵蚀能力
rainfall erosivity factor 降雨冲刷能力系数;降雨冲蚀指数
rainfall excess 有效雨量;降雨量过剩
rainfall expectancy 降雨概率
rainfall factor 降雨因素;降雨因数
rainfall-frequency 降雨频率
rainfall-frequency analysis 雨量频率分析
rainfall-frequency atlas 降雨频率图集
rainfall-frequency diagram 降雨频率图
rainfall ga(u)ge 雨量计
rainfall grandient 降雨量梯度
rainfall gutter 檐沟
rainfall hood 雨罩
rainfall hours 降雨时数
rainfall index 雨量指数;降雨指数
rainfall indicator 降雨指示器
rainfall infiltration 雨水渗入;降雨渗透(量);降雨渗入过程;降雨入渗
rainfall infiltration coefficient 降水渗入系数
rainfall infiltration coefficient analogy method 降水渗入系数比拟法
rainfall influence 降雨影响
rainfall intensity 雨量强度;降雨强度;降水强度;暴雨强度
rainfall-intensity curve 降雨强度曲线
rainfall-intensity-duration curve 降雨强度—历时曲线;降雨强度—持续时间曲线
rainfall-intensity frequency 降雨强度频率
rainfall-intensity occurrence 降雨强度重现期
rainfall-intensity recurrence interval 降雨强度重现期
rainfall interception 降雨截溜(量)
rainfall inversion 雨量逆增
rainfall leader 水落管
rainfall map 等—降雨量曲线图;雨量图
rainfall mass curve 雨量累计曲线;雨量累积曲线
rainfall observation 降水量观测
rainfall occurrence interval 降雨重现时段
rainfall on area 面雨量
rainfall pattern 雨型

rainfall penetration 雨水渗透;降雨渗入过程
rainfall penetration depth 降雨渗入深度
rainfall percentage 降雨百分率
rainfall period 降雨周期
rainfall pipe 雨水管
rainfall point of concentration 集中降雨点
rainfall pollution 大气降水污染
rainfall precipitation 雨量;降雨
rainfall probability 降雨几率;下雨可能性;降雨概率
rainfall province 雨量区;降雨区域
rainfall purification theory 雨洗理论
rainfall rate 降雨强度;降雨率
rainfall reappearing period 重现期(指降雨)
rainfall recharge 大气降水补给
rainfall recorder 雨量记录器;自记雨量计
rainfall record(ing) 雨量记录;降雨记录
rainfall recording instrument 雨量记录仪
rainfall recording station 雨量站
rainfall recurrent interval 雨量重现期
rainfall regime 雨情;雨量型;(一个地区一年中的)雨量分布情况;降雨型;降雨方式
rainfall runoff 地面径流;暴雨径流;雨水径流;降雨径流
rainfall-runoff correlation 降雨径流相关关系
rainfall-runoff method 雨量径流法;暴雨径流法
rainfall-runoff model 降雨径流模型
rainfall-runoff relation(ship) 雨量径流关系;降雨径流关系(曲线)
rainfall self-recorder 自记雨量计
rainfall simulation 降雨人工模拟
rainfall simulator 人工降雨装置;降雨模拟装置;降雨模拟器
rainfall standard intensity-duration curves 降雨标准强度历时曲线
rainfall station 雨量站
rainfall storm 暴雨
rainfall subsequent to spraying 喷洒后降雨
rainfall volume 降雨(总)量
rainfall water 雨水
rainfall water head 水落头
rainfall watering 雨水的
rainfall year 降雨量水文年
rain fastness 耐雨性;耐阴性
rain-fed agriculture 靠雨水灌溉的农业
rain-fed crop 旱作物
rain-fed farming 靠雨水灌溉的农业
rain-fed river 雨水源河;雨水补给河
rain-fed stream 雨水补给河(流);雨水源河
rain field 雨区;雨量场
rain fog attenuation 雨雾衰减
rain forest 雨林;热带雨林
rain forest climate 雨林气候;常湿气候
rain-free days 无雨天数
rain-free period 无雨期
rain frequency 降雨频率
rain from warm clouds 暖云雨
rain front 雨锋
rain furrow 雨沟
rain ga(u)ge 雨量器;雨量计;测雨计;雨量表(测量用)
rain ga(u)ge bucket 雨量筒;雨量桶;量雨筒;量雨桶;储雨筒;盛雨筒;盛雨斗
rain ga(u)ge data 雨量计资料
rain ga(u)ge glass 雨量杯
rain ga(u)ge network 雨量站网;雨量测量网
rain ga(u)ge shield 雨量器(防)护罩;雨量计风挡
rain ga(u)ge station 雨量站
rain ga(u)ging 雨量测量;雨量计;雨量测量
rain ga(u)ging station 雨量计量站
rainger 喷灌机
rain glass 气压表;雨量计;量雨杯;晴雨计;晴雨表
raingraph 雨量图
rain graph (indicator) 列车运行图
rain-green forest 雨绿林
rain gun 雨枪;人工降雨器;远射程喷灌器
rain gun irrigation 雨枪灌溉
rain gush 骤雨;暴雨
rain gutter 雨水(排水)沟;檐沟;天沟
rainhat 雨帽
rain height 雨深
rainhill 雨山红砂石(产于英国兰开夏)
rain hood 雨套
rain ice 雨凇(下降时呈液态,着地后就冰冻);雨冰;冻雨

rain impact erosion 雨水冲刷；雨水冲蚀；雨水冲击侵蚀
rain impermeability 雨水不渗透性
raininess 雨强；雨量强度
raining 人工降雨机；淋降现象
raining intensity 降雨强度
raining solid reactor 淋粒反应器
raining test 喷淋试验
rain intensity 降雨强度
rain intensity curve 雨量曲线
rain intensity duration curve 降雨强度历时曲线
rain intensity ga(u)ge 雨量强度计
rain leader 雨水管；水落管
rain leader downspout 雨水立管
rainless 无雨的
rainless desert 无雨沙漠
rainless period 旱季；无雨期
rainless region 无雨区
rain-like condensation 雨滴状凝结
rainmaker 制雨器；人工造雨器；人工降雨装置；人工降雨设备；人工降雨工作人员；喷灌设备
rain-making 人工降雨；人工造雨
rain making machine 人工降雨装置
rain map 雨量图
rain-on-snow flood 雨融雪洪
rain-out 雨除；云中的尘埃被形成的水滴清除；冲洗
rain outlet 雨水出口；雨水泄水管
rain-out washout 雨水冲洗
rain-patch 雨季
rain patches 小片雨区；小块雨区
rain penetration 雨水浸透；雨水渗透
rain penetration depth 雨水入渗深度；降雨渗入深度
rain pillar 雨柱
rain pipe 水落管
rain pitting 雨坑
rain pool 雨水塘；雨水坑
rain precipitation 降水；降雨(量)
rain print 雨痕；雨点痕迹
rain producing 造雨；产雨
rain producing cloud 致雨云；造雨云；产雨云
rainproof 雨具；防雨用具；雨衣；防雨的；不透雨的
rainproof hood 防雨罩；防雨外壳；防雨盖；防雨套
rainproof masonry wall 防雨圬工墙
rainproof material 防雨材料
rain recorder 自记雨量计
rain repellent 排雨剂；防雨的
rain-repellent polymer coatings 透明防雨聚合物涂层
rain-repelling 防雨水的
rain resistance 抗雨力
rain return 雨反射
rain rill 雨水沟；雨冲沟
rain ripper 雨水沟沟槽
rain rot 雨斑病
rain sampler 雨水取样器；降雨收集器
rain scavenging of semi-volatile organics 半挥发有机物雨水清除
rain screen 防雨屏
rain screen cladding 防雨遮挡板
rain sculpture 雨水冲蚀
rain sculpturing 雨水蚀痕
rain seepage 雨水渗透
rain shade 遮雨罩
rain shadow 雨影；雨痕；无雨干旱带；未着雨地带
rain shed 雨棚
rain shell 高炮降雨弹
rain-shield insulator 防雨隔电子
rain shower 阵雨
rain shower glaze 雨淋墙釉
rain simulator 人工降雨装置
rain soaked runway 雨水浸湿的跑道
rain spell 雨期；阴雨期；连绵雨期；多雨期
rain splash 雨水飞溅
rain splash erosion 雨水溅击侵蚀；雨滴侵蚀
rain spot 雨斑
rain spotting 雨水沾污的
rain spout 水落管；排水口
rain squall 雨飑；狂风暴雨；暴风雨；暴风骤雨；雨前暴风
rain stage 成雨阶段
rain stimulation by artificial means 人工致雨；人工影响降水；人工催雨；人工催化降雨
rainstorm 雨爆；暴(风)雨

rainstorm centre 暴(风)雨中心
rainstorm frequency 暴雨频率
rainstorm intensity 暴雨强度
rainstorm magnification 暴雨放大
rainstorm pattern 暴雨类型
rainstorm runoff 暴雨径流
rainstorm transposition 风暴移置；暴雨移置
rainstorm type mud-stone flow 暴雨型泥石流
rain strainer 雨水斗
rain switch 篷顶操纵开关
rain test 淋雨试验；降雨试验；人工淋雨试验
rain-tight 防雨性；防雨的；不漏雨
raintight material 不漏油材料
rain tire 雨季用轮胎
rain tract 降雨区
rain trap 集雨器
rain tree 雨树；雨林
rain utilization system 雨水利用系统
rain visor 遮雨板
rain wash 雨水洗刷；雨水洗涤；雨水冲洗；雨水冲刷(物)；雨水冲蚀；流积土；冲刷流失；雨致蠕动；雨运物；雨水的冲刷
rain wash-out 降雨冲刷
rainwater 雨水；软水
rainwater acidity 雨水酸度
rainwater articles 雨水排泄构件
rainwater basin 雨水池
rainwater bend 雨水弯管
rainwater channel 雨水明沟
rainwater cistern 雨水箱；雨水槽；雨水蓄池；储水池
rainwater drain 雨水沟
rainwater drainage 雨水排水
rainwater goods 雨水管件
rainwater gutter 雨水沟
rainwater gutter board 天沟板
rainwater gutter heating 加热天沟
rainwater gutter hook 天沟钩钉
rainwater gutter sheet 天沟薄片
rainwater head 水落头；雨水斗；水落斗
rainwater hopper 水落斗；雨水斗
rainwater junction 雨水管分叉
rainwater junction pipe 雨水管分叉
rainwater lath 顺水条
rainwater leader 雨水管；水落管
rainwater offset 雨水弯管
rainwater outlet 雨水出(水)口
rainwater pipe 落水管；雨水管；雨水
rainwater pipe hanger 雨水管挂钩
rainwater removal 雨水排除
rainwater retention basin 滞洪区；蓄雨水池
rainwater runoff 雨水径流
rainwater settling tank 雨水沉淀池
rainwater sewer 雨水下水道；排雨水管
rainwater shoe 落水斜口
rainwater stain 雨水污迹
rainwater storage 雨水量
rainwater storage tank 储[贮]雨水槽；储[贮]雨水箱
rainwater system 雨水排出系统
rainwater tank 雨水箱
rainwater treatment 雨水处理
rain wear 防雨布；雨衣
rain-weather 雨天
rainwet 雨淋污损铺面材料；雨湿木料
rain with snow 雨雪；雨夹雪
rain work 雨中作业；雨中工作；雨天作业
rainworm 蚯蚓
rainy 阴雨的；多雨的
rainy climate 多雨气候
rainy day 下雨天
rain year 降雨水文年
rainy period 雨季；多雨期；丰水期
rainy season 雨季；黄梅季
rainy season construction 雨季施工
rainy spell 下雨时间
rainy tropics 多雨热带
rainy weather 多雨天气；多雨气候；雨天天气；雨天(气候)；雨季
rainy year 多雨年(份)；丰水年
raio brightness distribution 射电亮度分布
rairoad engineering 铁路工程
raisable jib 变幅吊臂；可举升的吊臂
raise 扬起；天井；上山

raise a boom 升起吊杆
raise adequate finance 筹足资金
raise advance 上山掘进
raise a light 灯光进入视界(驶到一个灯塔能见范围)
raise a loan 举债；借债
raise anchor 起锚
raise animal 畜牧
raise a price 加价；抬价
raise a purchase 安装绞辘
raise block 解除封锁
raise-bore advance 天井钻孔进尺
raise-bore hole 天井钻孔
raise borer 天井钻机
raise borer machine 天井钻机
raise boring 天井钻进；提升钻进；(自下而上的)上向凿井；反井钻进
raise boring machine 天井钻机
raise by filling 灌浆提高水位
raise cam 起针三角
raise capital by floating shares 招股
raise chute 天井溜口
raise climber 爬行式升降台；爬罐；天井爬罐
raise crops 栽培作物
raised and fielded panel 突面镶板；中间鼓起的镶板
raised and sunken plating 内外叠板法
raised apex lands 转子角顶凸岸
raised approach 斜坡道
raised arch 突起拱
raised bands 书脊棱带
raised bank 自然堤
raised bars 突起的齿条(防车辆进入路面)
raised beach 海滩高地；海滨台地；上升(海)滩
raised beach plain 上升滨岸平原
raised beach platform 上升滨岸平原
raised bench 上升阶地
raised block 加层房屋
raised bog 高位沼泽；高地沼泽
raised cable 架空缆索
raised-cable technique 挠曲钢筋束工艺
raised check 涂改支票；提高金客支票
raised cheese head screw 突起的圆头螺钉
raised chequer pattern 凸方格图案
raised choir 唱诗班高台席位
raised chord 折线下弦
raised-chord truss 起拱桁架
raised coast(al) barrier 兴建的沿岸挡水堤
raised control tower 交通指挥塔
raised copper 铜箔
raised cosine wave 余弦平方波
raised cottage 高架村舍小屋；架空住房
raised countersunk head screw 突起的埋头螺钉
raised crossing 高架道口；高起交叉道；高起道口
raised crossover rib 十字凸花纹
raised curb 高起路缘；突起路缘
raised curve 外缘加高的拐弯；超高曲线
raised deck 升高甲板
raised diamond checker plate 斜方格钢板
raised effect 凸纹效应
raised fabric 起绒织物
raised face 高起表面；突起面；凸起表面；凸面
raised face flange 突面法兰；凸面法兰
raised face tool 凸面刀具
raised face welding neck flange 凸面对焊法烂
raised fiber 凸起的(木)纤维
raised field 隆起的部分；隆起的场地；台田
raised flange 突面法兰
raised floor(ing) 双层地板；高架地板；楼面板；架空地板；活(动)地板；上升面；上升架
raised flooring system 活动地板系统；高架地板系统；架空活动地板；活动地板体系
raised flower bed 高设花台
raised forecastle 升高首(楼)甲板
raised foredeck 前半升高甲板
raised foredeck ship 升高首甲板船
raised girt(h) 抬高的圈梁；角柱两侧略高的横档；活动地板边撑
raised grain 隆起的木纹；木纹突起；木纹隆起；起棱纹理；突(起)纹理；波浪纹(理)；凹凸纹理
raised growth 大量生长
raised head 凸头
raised head bolt 凸头螺栓
raised head rivet 凸头铆钉
raised head screw 凸头螺钉

raised hotbed 高设温床
raised left side 反面拉绒
raised-letter printing 热式复制术
raised line 凸条
raised line design 弦纹
raised moss 高温泥炭沼泽
raised mo(u)lding 隆起式线脚；隆起式线条；突线条
raise domestic animals 畜养
raised panel 尖顶隆起镶板；中间较厚的镶板；鼓面的镶板；鼓面镶板；突面镶板；凸嵌板
raised part 增高部分
raised pattern 浮雕花纹
raised piece 承梁短板；浮雕制品
raised pile 回升桩
raised platform 高站台；升高平台
raised power takeoff 高置式动力输出轴
raised pressure 扬压力
raised printing 压凸印刷
raised quarter-deck 升高后甲板；后升高甲板
raised quarter-deck ship 升高尾甲板船
raised quarter-deck vessel 后半升高甲板船
raised quarter vessel 后升高甲板船
raised rear fender 翘尾护板
raised reef 上升礁
raised reef coast 抬升礁海岸
raise drilling 天井钻进；提升钻进；(自下向上的)上向凿井；反井钻进
raise driving 天井掘进；提升掘进；反井掘进
raised separator 高架式分隔带；突起式分隔带
raised shaft 天井
raised skylight 隆起天窗
raised sludge 上升污泥
raised stomata 拱起气孔
raised strake 外列板；外包板列板
raised strands 架空缆；架空的钢绞线；架空的钢绞线；钢丝外露；夹丝外露
raised stripe 形成凸凹的条纹
raised style enamel 凹凸珐琅
raised surface structure 起绒表面结构
raised table 高出周围的水平顶面
raised tank top 升高二层底铺板；从中线向船侧倾斜的油舱
raised traffic control tower 交通指挥塔
raised umbrageous bog 凸起高位沼泽
raised wage 提高(的)工资
raised walkway 高置行道
raised water level 壅高水位
raised welding method 上升焊接法
raised wheel seat axle 无内凸肩车轴(火车)
raised work 浮雕细工
raise face 天井工作面
raise fascine road 柴排路堤
raise funds 集资；筹款
raise gold content 提高含金量
raise interlock relay 上升联锁继电器
raise lift 天井掘进工作台；上向凿井提升机；吊罐
raise livestock 畜牧
raise loan 贷款
raise mining 上向掘进
raise money 筹款
raise opening 天井井口；上山开石
raise price 提价；提高物价
raise productivity 提高生产率
raiser 经纬线浮点；提升器；坝内竖井
raiser block 垫块
raise scaffold 升降式脚手架
raise set 天井支架
raise steam 蒸汽升压
raise stope 逆倾斜工作面
raise tax 征税
raise the current price 提高时价；抬高时价
raise the land 陆地进入视界
raise the multiple cropping index 提高复种指数
raise the output of products 增加产量
raise the per unit yield 提高单位面积产量
raise the price 提价
raise the value of money 提高货币价值
raise the wind 起风浪
raise to a power 自乘
raise to third power 立方
raise wages 提升工资
raise water table 地下水位上升
raise work 竖井反掘工作；沉排工(作)

raising 向上开挖施工法；起伏成型；上挖
raising against the hair 反向起绒
raising-and-depression 举折【建】
raising and lowering of the bit 钻头上下晃动
raising a sunk vessel by inserting floats 船内充塞浮具打捞法
raising a wreck 打捞沉船；沉船起捞
raising block 垫块
raising body 挖掘犁体
raising by inserting floats 充塞浮具打捞法
raising capabilities for multi-production 提高多种生产的能力
raising cord 拉力用绳索
raising crank 提升曲柄；提升机构曲柄
raising cribber 放样间划线笔刀
raising directly 直接打捞【救】
raising force 提升力；升举力
raising fork 举杆叉
raising form 顶升模板
raising fund 筹资
raising gear 起落机构(齿轮)；提升机构
raising gig 拉绒机；起绒机
raising hammer 圆头大木槌；大木锤
raising in section 分段打捞
raising in steps 逐步打捞
raising iron 刮缝凿；刮缝刀
raising knife 刮缝凿；刮缝刀；括缝刀；划线笔刀；放样间划线笔刀
raising labour productivity 提高劳动生产率
raising legs 提升架
raising lever 提升杆
raising machinery 起落机构；提升机械
raising mechanism 提升机构
raising method 竖井反掘法
raising middle flask 箱圈；中(砂)箱
raising of an area 地面隆起
raising of dam 坝顶加高
raising of dust 扬尘
raising of fill 路堤填高；路堤加高
raising of founds 筹集资金
raising of ground water 潜水位上升；地下水升高
raising of groundwater table 地下水位上升表
raising of indices 升标
raising of pool 抬高水库水位
raising of river bottom 河底淤高
raising of sags 使凹处隆起(筑路)
raising of sewage 下水道抬高
raising of steam 蒸汽产生
raising of strands 弯起缆索
raising of sunken concrete slabs 已沉陷的混凝土板抬高
raising of the stream floor 河床抬高
raising of threshold 阈域上升
raising of track 起道
raising-of-truss method 举架【建】
raising of water 水面提高；水位提高；水的提升
raising of water level 增高水位；提高水位；抬高水位；水位壅高
raising on new ballast 新渣起道
raising operation 沉船起浮
raising operator 上升算符
raising piece 柱顶垫木；墙顶垫木
raising plate 垫板；椽端板；墙顶承梁板；檐檩；高架托板；联系基础构件的横梁；顶升板；顶梁板
raising platform 提升式平台；升降台；下降(式)平台
raising pusher leg (凿岩机的)推压腿架；举升腿架
raising rate 上升速率
raising rate of water-level 水位上升速率
raising rod 提升拉杆
raising scaffold 升降式脚手架
raising scriber 刮缝凿；刮缝刀；括缝刀；划线笔刀
raising shaft 提升井
raising ship 打捞船
raising sling 提升钢丝绳
raising speed 升速；上升速度
raising spring 升弓弹簧
raising temperature crystallization 升温结晶法
raising the level of impoundage 提高蓄水位
raising the nitrate level 提高硝酸盐水平
raising the purlin(e) 举析；举架
raising the wing 增大机翼安装角
raising velocity 上升速度
raising water storage 提高蓄水量
raising wire 抬起杆

raising with beams supported by pair of barges 大梁对驳打捞
raising worm 提升机构蜗杆
raite 水硅钠锰石
rajite 亚碲铜矿
rake 搂草机；磨削伤；倾斜度；倾入；前倾面；不规则褐铁矿脉
rake aft 向后倾斜
rake angle 洗流角；(钻头的)割刀角；翘角；前角；刀面角
rake attachment (for graders) 松土耙；松土装置
rake blade 刮板
rake board 挡风板；(山墙的)封檐板；倾斜板
rake bond 曲折砌合；斜纹接合；对角接合
rake-bowl classifier 浮槽耙式分级机
rake classifier 刮板式分选机；耙式分级器；刮板式选分机
rake classifier for dewatering 脱水用的耙式分级器；脱水用的耙式分级机
rake conveyer 耙式输送机；链板运输机；刮板运输机
rake cooling bed 斜坡式冷床
rake cornice 斜檐；斜挑檐
rake cylinder 搂集滚筒
rake damper 倾斜闸板
raked blade 后倾桨叶
raked floor 地面升起；坡形地板
rake dimension 斜跑尺寸；梯段标杆
raked joint 捋缝；刮缝；清缝；剔缝
raked mast 倾斜桅
raked mo(u)lding 山墙装饰线；斜面线条；山墙线脚；倾斜的线脚；斜线脚
rake dozer 刮板推土机；除根器；齿板推土机
raked stem 斜船首
raked stern 斜船尾
rake dune 齿形沙丘
rake face 倾斜面
rake-form reef 耙状似
rake gantry 清污耙起重台架
rake gate 倾斜闸门
rake grab 扒集式抓斗
rake hoist platform 清污耙起重平台
rake in enormous profits from 攫取高额利润
rake machine 耙式洗毛机
rake mixer 耙式混合器
rake mo(u)lding 斜面线条；斜面线脚；山墙装饰线条
rake of bow 斜柱船首
rake-off 耙出；佣金；回扣
rake of the stem 船首柱倾斜
rake of tubes 管斜度
rake-out (of joint) 清理；刮缝；勾出；清理灰缝
rake probe 梳状探针；梳状测针
raker 斜支撑木；斜撑；搂草机；火耙；耙子；耙路机
rake radius 斜耙首柱底部曲率半径
rake ratio 后倾率；倾斜比
raker-braced wall 设斜撑的墙
raker dolphin 斜撑簇船墩；斜撑系船柱；高架系船柱
raker dump car 斜撑自卸车
rake reception 瑞克接地；分离多径接收
raker ga(u)ge 刨齿规
raker pile 斜桩
raker set tooth 交错偏挠锯齿
raker stirrer 刮板式搅拌器；搅拌耙
raker stripper 搂耙清理器
rake scraper 梳形耙斗
rake teeth 耙形牙；耙齿；间齿
rake tool 组合刀具
rake tooth 耙齿
rake-type cooling bank 倾斜式冷床
rake-type cooling bed 倾斜式冷床
rake-type drag 耙式路刮；耙式刮路机
rake-type duplex wet magnetic cobber 双耙湿式粗选磁选机
rake-up ballast remover 蟹耙式清岩机
rake vein 竖脉
raking 耙(除)；侧面进针
raking anchor pile 斜向锚桩
raking aperture block 斜孔预制块
raking arch 高低脚拱；跛拱
raking back 预留踏步接头；阶梯形砌块；墙壁接齿缝
raking baluster 斜面栏杆
raking balustrade 斜面栏杆
raking beam 斜梁
raking bond 斜(纹)接合；人字砌合；对角砌合

raking capping 斜面压顶(石)
raking coping 斜(面)压顶;倾斜的墙帽
raking cornice 倾斜的檐板;斜边装饰;跳檐;斜挑檐
raking course 斜砖砌层;斜砖砌合
raking cutting 斜切
raking element 斜撑
raking equipment 清耙机
raking flashing (piece)斜挡水板;斜(向)泛水;斜遮雨板
raking frame 斜拉框架
raking machine 搂草机
raking mast 倾斜桅
raking mechanism 机械耙
raking mo(u)lding 斜面线条;斜面线脚
raking off the slag 撇渣
raking out 勾出
raking out joint 挦出缝;勾突缝
raking out of joint 清理灰缝;剔清接缝;清缝
raking pile 斜桩;叉桩
raking-piles anchorage 锚碇叉桩
raking prop 斜撑;斜桩;斜支柱
raking reel 搂集轮;搂集滚筒
raking riser 倾斜台阶;楼梯踏步斜竖板;斜踢面;踢面
rakings 耙除物
rakings and screenings 耙出物和筛余物
raking shore 斜支柱;临时顶撑;斜支撑木;斜撑(木)
raking shoring 斜支撑法
raking shoring of building pit 基坑的(临时)斜支撑
raking stem 斜船首
raking stretcher bond 斜纹顺砖砌法;斜条砌合;斜错缝满条砌法
raking strut 斜支撑
raking support 斜支承
raking surface 倾斜面
raking timber 斜撑木
raking trestle 人字(形)排架
raking-up effect 集段效应
rakish 后倾的(指桅、烟囱等)
Rako 未上釉的白瓦载体
Raku ware 厚壁茶具
Raleigh fractionation 瑞利分馏
ralstonite 氟钠镁铝石
ram 撞头;撞角;推钢机;挑杆;水下冰山;捣固(修船斜道);锤(体);串杆货叉
ram accumulator 柱塞式蓄力器
ramada 敞开的亭子;盖凉棚构架;凉棚(构架)
ram-air 冲压空气
ram-air condenser 冲压空气冷凝器
ram-air cooling 冲压空气冷却
ram-air pipe 全压管;冲压空气管
ram-air turbine 冲压空气涡轮
ram-air turbopump 冲压式空气涡轮泵
Ramalina denticulata 细齿树花
Ramalina pacifica 太平洋树花
Raman activity 拉曼活性
Raman band 拉曼谱带
ram anchor 油缸连结销
Raman dephasing processes 拉曼相移过程
Raman diffusion 拉曼散射
ram-and-inner frame machine 锤头内架式高速锤
Raman displacement 拉曼位移
ram and piston pump 活塞泵
Raman echo 拉曼回波
Raman effect 拉曼效应
Raman intensity 拉曼强度
Raman laser 拉曼激光器
Raman laser action 拉曼激光作用
Raman laser material 拉曼激光材料
Raman laser source 拉曼激光源
Raman line 拉曼谱线
Raman mixing 拉曼混频
Raman non-activity 拉曼非活性
Raman oscillator 拉曼振荡器
Raman probe analysis 拉曼探针分析
Raman rotational spectrum 拉曼转动光谱
Raman scattered light 拉曼散射光
Raman scattering 拉曼散射
Raman scattering coefficient 拉曼散射系数
Raman scattering laser radar 拉曼散射激光雷达
Raman-shifted light 拉曼位移光
Raman shifting 拉曼频移
Raman spectrogram 拉曼光谱图
Raman spectrograph 拉曼摄谱仪

Raman spectrometer 拉曼光谱计
Raman spectroscopy 拉曼光谱学;拉曼光谱分析;拉曼光谱法
Raman spectrum 拉曼(光)谱
Raman transition 拉曼跃迁
Raman-type scattering 拉曼式散射
ram area 柱塞面积
ramark 主动雷达航标;雷达航标;雷达信标
ram arm 液压油缸杠杆
ram attachment 夯锤附属装置
ram attack 撞击式攻击
ramaway 落砂;模型捣动
ram-baler 柱塞式压捆机
Ramberg-Osgood force-deformation relation 伦贝格-奥斯古特力与变形关系
Ramberg-Osgood skeleton curve 伦贝格-奥斯古特滞变曲线
ramble 煤层顶板页岩
rambler 农场平房;牧场平房;单层住房
rambler rose 蔓性种蔷薇;攀缘蔷薇
ramble strip 搓板带
ram block 锤砧;冲头垫块
ram bolster 下托板;活动平台
ram bow 撞角型船首
Ramb's noise silencer 长波式静噪器
ram chisel 冲钻
ram clamping handle 锤头夹持手柄
ram compression 冲压;速压;速度头压缩
ram compressor 冲压缩机;冲压式压气机
ram controls 油缸操纵杆
ram cooling 增压风冷
ram cylinder 压头油缸;柱塞油缸;水压机汽缸
ramdohrite 辉锑银铅矿
ramdom shingles 宽度不等的木板
ram down 夯实
ram drag 冲压阻力
ram drive 柱塞传动装置
ram drive cylinder 柱塞传动油缸
ram drive release 柱塞传动松脱器
ram drive release mechanism 柱塞传动装置分离机构
ram driving mechanism 冲压驱动机构
rameauite 黄钾钙铀矿
ramee 苎麻(纤维)
ramee retting wastewater 苎麻纤维浸解废水
ramee wastewater 苎麻废水
ram effect 冲压效应
ram efficiency 冲压效率
ram engine 打桩机
Rames(s)eum 拉美西姆(古埃及国王拉美西斯二世陵庙)
ramet 同系株;碳化钽
Ramet alloy 碳化钽硬质合金
ram extruder 柱塞式压出机
Ramey type curve matching 瑞米图版法
ram filter 压滤器;冲压过滤器
ram framework 打桩架
ram guide 桩锤导柱;桩锤导架;桩锤导杆;滑枕导轨
ram guiding 桩锤导架
ram head 机头;侧刀架
ram-headed sphinx 公羊头狮身像
ram home 夯实
ramie 苎麻(纤维)
ramie cotton fabric 棉麻织物
ramie fabric 纯麻布
ramie hemp 苎麻纤维
ramie stripes 麻条
ramification 分枝(式);分支机构
ramification exponent 分歧指数
ramification field 分歧域
ramification group 分歧群
ramification index 分歧指数
ramification migmatite 树枝状混合岩
ramification number 分歧数
ramification order 分歧阶
ramification point 分歧点
ramification valve 分歧值
ramified pipe system 树枝状管网
ramified system 树枝式(水管)网
ramified theory of type 分歧类型论
ramiform 分枝纹孔
ramiform pit 分枝壁孔
ramiform pitting 分枝壁孔式

ramify 成网眼
ramifying 分支的过程
ram impact machine 打夯机;锤击机;捣锤冲击机;冲击(式)打夯机
ramin 棱柱木(东南亚)
ram injection machine 柱塞式注压机
ram intake 迎风进气口;全压受感器;全压接收管;冲压(式)进气口
ram intake manifold 冲击式进气歧管
rami parietales 壁支
rami perforantes 穿支
ramirite 铜钒铅锌矿
rami spinales 脊支
rami ventrales 腹侧支
ramjet 冲压(式)喷气发动机
ramjet cycle 冲压发动机循环
ramjet duct 冲压发电机
ramjet engine 冲压式喷射发动机;冲压式喷气发动机
ramjet exhaust nozzle 冲压式喷气发动机排气喷管
ramjet helicopter 冲压喷气式直升飞机
ramjet performance 冲压喷气发动机性能
ramjet propulsion 冲压式喷气发动机;冲压式发动机;冲压喷气推进
ramjet speed 冲压发动机速率
ramjet stage 冲压发动机级
ramjet sustainer 冲压式主发动机
ramjet-turbojet combination 冲压涡轮喷气发动机组合
ramjet-turbojet power plant 冲压式涡轮喷气动力装置
ram jolt 震实
ram knife 柱塞剪切刀
ramleh 海岸风成岩
ramless extrusion 静水压挤压
ram lift 柱塞升程;冲压升力
rammability 春实性能
ram machine 打桩机;锤击机
rammed 夯实的;冲压成的
rammed 3∶7 lime earth 三七灰土夯实
rammed bottom 捣筑炉底;捣固电解槽底;捣固槽底;冲压成的炉底
rammed bottom lining 捣筑的衬底
rammed brick 捣打砖
rammed clay 捣实黏土;夯实黏土
rammed concrete 夯筑混凝土;夯实混凝土;捣实混凝土
rammed dolomite 捣实的白云石
rammed earth 夯实土(壤);素土夯实;捣实土(壤)
rammed earth construction 干打垒;夯土结构;夯土建筑;夯土构造
rammed earth foundation 夯土基础
rammed earth house 干打垒房屋
rammed earth subgrade 夯实土基
rammed earth wall 夯土墙
rammed energy 夯击能
rammed foundation 夯实地基
rammed ground 夯实地基
rammed in layers 分层夯实
rammed ladle 捣制钢包;打结盛钢桶
rammed lime 夯实的石灰
rammed lime earth 灰土夯实;夯实灰土
rammed lining 捣筑的炉衬;捣打炉衬
rammed loam and lime 捣实的黏土与石灰
rammed loam construction 壤土的夯实施工;垆姆的夯实施工;壤土的夯实工程;垆姆的夯实工程
rammed refractory 捣实的耐火材料
rammed soil 夯实土(壤)
rammed soil foundation 夯土基础
rammed soil pile 夯实土桩
rammed soil type of foundation 夯土基础;夯实土基(类型)
rammed three-to-seven lime earth [3∶7 lime earth rammed] 三七灰土夯实
rammed wall(ing) 夯筑(墙)
rammed width 夯击宽度
rammel 松石
rammell 页岩层
rammelsbergite 斜方砷镍矿
rammer 撞锤;火力夯;夯实器;夯(具);夯击机;退弹器;捣锤;春砂器
rammer board 压榨板
rammer compacter 夯土机;夯实机

rammer man 夯土工人
rammer method 撞捶法
rammer number 夯击数
rammer process 捣打成型法
rammer the sand 舂砂
ramming 压实;夯筑;夯实;夯打;抛砂;捣打成型法;捣打;打夯;舂砂
ramming arm 抛砂头横臂
ramming attack 撞击式攻击
ramming bar 捣固杆;夯实杆;夯棍;夯杆;捣实杆
ramming board 压榨机;压榨板
ramming clay 捣实黏土
ramming compactor 夯实机
ramming compound 捣打混合物
ramming concrete 捣打混凝土
ramming depth 打(夯)入深度
ramming device 捣实装置
ramming dolomite 捣实白云石;捣固白云石
ramming engine 人力打夯机
ramming hammer 夯锤
ramming head 抛头;抛砂器
ramming in layers 分层夯实
ramming iron 铁夯
ramming machine 捣固机;打夯机;蛤蟆夯;锤击机
ramming mass 捣打料
ramming material 冲压物料;捣打料
ramming mix(ture) 捣固材料;密封用料;打结用料;捣打混合物;搪炉料
ramming motion 冲压运动
ramming mo(u)lding machine 夯实造型机
ramming of the sand 捣实型砂
ramming piston 栓塞
ramming process 捣打成型法
ramming refractory 捣打成型耐火材料
ramming row 排夯
ramming space 夯距
ramming speed 抛砂速度;捣实速度;打夯速度
ramming up 楔起
ramming weight 夯锤;桩锤
ram mounted extrusion 活塞挤出
ram mounted extrusion system 带压盘挤压机
ramnant 零头
ramoff 落砂
ram of shaper 牛头刨的刨头
ramollescence 软化作用
Ramond's point 拉蒙氏点
Ramon flocculation test 拉蒙絮状试验
ramose 分支的
ramp 雪坡;斜坡道;装卸货跳板;匝道;坑线;接线夹;接线端钮;滑行台;倾斜端头;坡路;坡道;断坡;等变率
rampactor 跳击夯
ramp and twist 面的倾斜与扭曲;面的提升与扭曲
ramp angle 上坡角
rampant 具有一个比一个高的拱座的
rampant arch 跷拱;高低脚拱;跛拱
rampant arch bridge 跛拱桥
rampant barrel vault 高低座的筒拱顶
rampant center 高低脚拱的弧心;层层高的中心
ramp anticline 冲断层背斜
rampant inflation 恶性通货膨胀
rampant mo(u)ld concrete 连续模捣混凝土
rampant vault 跛穹窿;穹肩高低的穹顶;斜拱顶;跛拱顶
ramp approach 引道坡
rampart 筑垒;防御物;环壁;防御墙;波蚀残丘;壁垒;堡垒
rampart walk 城墙通道;城墙上人行道
ramp beam 斜梁
ramp between floor 楼面间的斜坡道
ramp bow 有斜坡出入道的船首;登陆舰船首
ramp brake 峰加速坡缓行器
ramp breakover angle 纵向通过角
ramp bridge 引桥;匝道桥;坡道桥
ramp capacity 匝道通过能力;立交入口通行能力
ramp carriage 斜梁
ramp control 匝道控制
ramp crossing 有接坡的交叉口
ramp design speed 匝道设计车速
ramp door 登陆舌门
ramped 倾斜的
ramped approach 斜坡引道;斜坡进口道
ramped cargo berth 活动板泊位
ramped dump barge 开底垃圾驳

ramped floor 倾斜地板;坡形地板
ramped pile (集料的)斜坡料堆
ramped steps 阶形坡道;斜坡踏步板
ram penetrometer 锤式贯入硬度计
ramp entrance 斜坡进口;匝道进口
ramp equipment 停机坪设备
ramper 捣棒;夯具
ramp error voltage 斜坡误差电压
ramp fault 对冲断层
ramp for climbing 上滑坡道
ramp former 斜坡形成器
ramp function 斜坡函数
ramp function generator 斜坡函数发生器
ramp function response 斜坡函数响应
ramp generator 斜坡发生器
ramp grade 斜坡坡度
ramp incline 坡道的斜面
ramping 修整斜坡
ramp input 斜坡输入;扫掠输入
ramp input signal 斜坡输入信号;瞬变输入信号
ram pipe 增压管;冲压空气管
ram piston 压力机活塞
ramp platen 水压机压板
ramp loader 箕斗提升机
ramp merge 匝道口车辆汇流;立交口车辆汇流
ramp metering 匝道限流;匝道交通调节;匝道车流调节
ramp metering system 匝道车流控制设施
ramp method 斜坡法
ramp mining 斜坡道开采;缓坡开采法
ramp of freeway 高速公路匝道
ramp of rerailing 复轨斜坡台
ram pot 液压挺液压筒
ramp rail 坡接轨
ramp rate 缓变率
ramp region 冲断层陡倾区
ram press 冲压机
ram pressing 塑压法;湿压法
ram pressure 全压力;速压头;冲压
ram pressure confinement 冲压约束
ram pressure recovery 冲压压力恢复
ram preventer 闸板式防喷器
ramp road 斜坡道
ramp sight 斜板式瞄准具
ramp signal 斜坡信号
ramp slope 坡道坡度
ramp step 倾斜阶跃波
ramp step wave 斜阶跃波
ramp surface 坡道面
ramp terminal 匝道枢纽;匝道终点;接坡终点
ramp thrusts 对冲式冲断层
ramp tower 螺旋形坡道
ramp traffic volume 匝道交通量
ramp turnout 匝道让车道;坡道道岔
ramp type (高速干道的)坡道形式
ramp-type garage 坡道式车库
ramp-type sight 斜板式瞄准具
ram pump 柱塞泵;水锤泵;冲击泵
ramp valley 对冲断层谷
ramp voltage 斜坡电压;斜坡电压
ramp wall 翼墙
ramp waveform 斜坡波形
ramp way 跳板;斜道;引道;倾斜通道
ramp weignt 停机重量
ramp with a slight pitch 小坡度坡道
ramp with non-slip surface 防滑坡道
ramp with serrated surface 礓磜坡道
ramp with slip-resistant surface 防滑坡道
ram ratio 压缩比(率);冲压比
ram recovery 冲压恢复
ram recovery pressure ratio 冲压恢复压力比
ram-retraction piston 锤头返回活塞
ram rocket 冲压式火箭发动机;复合式冲压火箭发动机
ramrod 推杆;洗杆;通条;直挺;捣棒
ram rubber 防喷器的活动胶皮衬
ramsaite 褐硅钠钛矿
Ramsay problem 拉姆齐问题
Ramsay Sound Series 拉姆齐松德统【地】
Ramsbottom safety valve 拉姆斯博顿安全阀
ram schooner 无顶帆三桅纵帆船
ramsdellite 斜方锰矿;拉锰矿
Ramsden's circle 冉斯登圈;冉斯登环
Ramsden's disc 冉斯登圈

Ramsden's eyepieces 冉斯登目镜
Ramsden's ocular 冉斯登目镜
ram seal 金属陶瓷封接;陶瓷金属撞入封接
Ramsey fringes 冉赛条纹
ramshackle 似塌建筑;要倒塌状建筑
ram shaft 动力油缸活塞杆
ram's horn figure 羊角图案
ramshorn hook 山字钩
Ramshorn test 拉姆逊冲孔法可锻性试验
ram side 机侧
ramson 熊葱(一种植物)
ramsonde 冰雪硬度器
ram speed 冲击速度
ram steam pile driver 汽锤打桩机
ram steering gear 撞杆式舵机
ram stem 撞角型船首
ram tensioned highline system 柱塞式高架(海上补给用)
ram test 撞击试验
ram tester 撞击试验仪;撞击试验机;冲击试验机
ram travel 柱塞行程
ram truck 挑杆式装卸机
ram turbine 冲压式空气涡轮
ram turret lathe 滑板式转塔六角车床
ram-type bending machine 压头式弯机
ram-type blow-out preventer 液压缸式防喷器
ram-type continuous extruder 活塞式连续挤出机
ram-type cylinder 柱塞式动力油缸
ram-type pump 柱塞泵
ram-type pusher 杆式推钢机
ramular 小枝的
ramulus 小枝;副枝
ramulus et uncus uncariae 勾藤
ramup 预填
ramup core 舂入芯
ramup wing core 舂入侧芯
ram weight 桩锤重量
ram wing surface effect ship 冲翼艇
ram-wing vehicle 升降翼汽车
ram with ram type 双重活塞型
Ramzin boiler 拉姆辛(直流)锅炉
ran 绳段
rance 支柱;窄矿柱;比利时大理石
ranch 大农场;专业农场;果园;大牧场
rancher 牧场主;牧场工人
Rancheria 大牧场工人住屋
ranchero 小农场主
ranchette 小牧场
ranch house 平房建筑;农场住房;牧场住房;农舍;农场房屋;牧场房屋
ranchman 牧场主;牧场工人
rancho 大牧场工人住屋
ranch shape 矩形坡式(屋顶);农场式;大牧场式
ranch style house 农场式住宅
ranch style window 牧场住宅型窗
ranch waggon 旅行汽车;客货两用车
rancid flavor 酸败气味
rancidity 酸败
rancieite 钙硬锰石;钙锰石
rand 地边;卡圈;田埂
randan(n)ite 硅藻石
randing 浅井勘探
Rando-bonder 兰多黏合机
random acceleration 随机加速度
random access 记忆装置;随机存取
random access addressing 随机存取寻址
random access auxiliary storage 随机存取辅助存储器
random access beam deflection system 随机存取电子束偏转装置
random access beam positioning 随机存取电子束定位
random access beam positioning system 随机存取电子束定位系统
random access card equipment 随机存取卡片装置
random access compater equipment 随机存取计算机设备
random access computer 随机存取计算机
random access controller 随机存取控制器
random access control unit 随机存取控制器
random access delta modulation 任意选址增量调制
random access device 随机存取装置;随机存取器
random access discrete address 任意选址;无规

存取分立地址;离散用户的随机使用
random access file 随机访问文件;随机存取文件
random access input-output 随机存取输入输出
random access I/O routine 随机存取输入输出程序
random access machine 随机存取机
random access memory 随机存取存储[贮]器
random access memory computer 随机存取存储计算机
random access memory digital/analog(ue) converter 随机存取存储器数模转换器
random access message storage 随机存取消息存储器
random access method of accounting and control 计算和控制随机存储法
random access module 随机存取功能块
random access optic(al) memory 随机存取光存储器
random-access programming 随机存取程序设计
random access programming and checkout equipment 随机存取程序设计与检查设备
random access software 随机存取软件
random access sort 随机访问分类;随机存取分类(程序)
random access stationary satellite 任意选址同步卫星
random access storage 随机存取存储[贮]器;随机存取存储器
random access telemetry 随机存取遥测
random access time 随机存取时间
random access X-Y positioning unit 随机存取X-Y 定位装置
random accumulation 随机累积
random acess sorter 随机存取分类程序
random allocation 随机分配
random amplitude 随机振幅
random amplitude distribution 随机振幅分布
random analysis 随机分析
random analysis of variance 随机方差分析
random angle domain 任意角域
random antenna 代用天线
random arrangement 概率配置法;无规则排列;随机排列;不规则排列
random array 无规排列;随机阵;不规则排列
random ashlar 乱砌料石;乱砌方石;不规则方石;乱砌琢石;层砌粗琢石毛石
random ashlar facing 乱砌琢石(面);乱砌琢石墙面
random ashlar work 乱砌方石工程
random assembly 随机系集
random assignment 随机分配
random assignment multiple access 随机分配多址
random assortment 随机分酸;随机分配
random background object 无规分布背景天体
random bar 游荡性洲滩;游荡性沙洲
random base motion 随机基底运动
random "beat" characteristic 随机偏置"拍"特性
random-bias hologram 随机偏置全息图
random block 乱抛块体;抛筑块体;抛筑的块体
random bond 不规则砌合;乱砌体;乱砌合;乱砌法
random bonding 乱砌
random borehole 无规则钻孔
random chain 无规链
random chain model 随机链模型
random change 随机变化
random check 任意抽查;随机校核;随机抽查;抽样
random claim amount 意外的赔偿数额
random clumped 随机成群的
random code 随机码
random-coding bound 随机编码限
random coil 乱绕线圈;无规绕(制)线圈;无规卷曲
random coil conformation 无规卷曲构象
random coiling polymer 无规卷曲聚合物
random coil model 无规线团模型
random coincidence 随机重合;随机符合
random collision 无规碰撞
random colo(u)r 杂乱色;任意色
random communication satellite 随机通信卫星
random communication satellite system 随机通信卫星系统
random component 随机分量;随机成分;随机部分
random connection 任意连接;不规则连接
random contraction operator 随机收缩算子
random convergence 随机收敛
random-coordinate method 随机配位法
random copolymer 任意异分子聚合物;无共聚体

random copolymerization 无规共聚(作用)
random coupling 随机耦合
random course 乱砌层
random coursed rubble 不规则成层毛石;乱砌成层毛石
random coursed wall 毛石砌墙
random coursed work 乱砌层
random crack(ing) 杂乱裂纹;无规则裂纹;不规则裂纹;不规则裂缝;随机开裂
random criterion 随机判定准则
random crosslinking 无规交联
random cross winder 不规则交叉卷绕络筒机
random current 随动流动
random data 随机数据
random data processing 随机数据处理
random defect 杂乱缺陷;不规则缺陷
random degradation 无规则衰变;无规则降解
random dense arrangement 无规则紧密排列
random dependence 随机相依
random depletion 偶然损耗
random deviate 随机偏差
random deviation 随机偏差;随机离差;不规则井斜
random device 随机策略
random difference 随机差(数)
random difference equation 随机差分方程
random diffusion chamber 混响室
random digit 随机数字;随机数位
random digitizer 任意数字转化器
random direction 随机方向
random displacement 随机位移
random disposition 任意排列
random distribution 任意分布;无序分布;分布;随机分配;随机分布;无规则分布
random disturbance 偶然干扰;随机扰动;随机干扰
random diversification strategies 随机分散策略
random dot grain screen 任意大小网点目板
random dot stereogram 随机点体视图
random draft 任意牵伸;不规则牵伸
random drift 随机漂变
random drift rate 随机漂移率
random drill hole 无规则钻孔
random drilling 无规则推进
random dumping 倾倒;乱石抛填;抛堆土石料
random dynamic(al) loading 随机动荷载
random early bite 偶然早期放牧
random earthquake motion 随机地震运动
random eccentricity 随机偏心
random effect 随机效应;随机效果
random ejection 随机出料
random electron 无规电子
random electron current 随机电子流
random element 随机元素;随机成分
random endomorphism 随机自同态
random entity 随机量
random entry 随机项
random entry method 随机输入法
random environment 随机环境
random ergodic theorem 随机遍历定理
random error 偶然误差;无规误差;随机误差;随机错误;随机差错
random-error-correcting convolutional code 随机误差校正卷码
random error protection 随机误差防护
random error term 随机误差项
random event 偶然事件;随机事件
random excitation 随机激振
random experiment 随机试验;随机实验
random fabric 随机组构
random failure 偶然失效;偶然故障;偶发故障;随机失效;随机故障
random failure period 偶然故障期;随机故障间隔期
random fashion 随机形式;随机方式
random fatigue 随机疲劳
random fault 随机性故障
random fiber array 纤维随机排列
random fiber prepreg sheet 纤维随机分布的预浸板材
random field 随机场
random file 随机文件
random fill 任意填筑
random flashes 杂乱闪光
random flow 不规则流动
random fluctuating data 杂乱起伏数据
random fluctuating system 随机起伏系统

random fluctuation 偶然升降;偶然波动;偶然变动;随机涨落;随机起伏;随机波动;随机变动;不规则起伏
random fluctuation system 随机起伏系统
random forcing function 随机扰动函数
random forecast 随机预报
random fragmentation model 无规碎裂模型
random fringe 不规则条纹
random function 随机函数
random gate signal 随机选通信号
random generation 随机生成
random geochemical model 随机性地球化学模型
random geologic(al) geophysical model 随机性地质地球物理模型
random geologic(al) model 随机性地质模型
random geometry 任意几何状态;任意几何形状;随机几何学;不规则几何形状
random geometry technique 杂乱形状技术;不规则形状技术
random grab 随机取样
random grain 混乱纹理;不规则纹理
random ground point 任意地面点
random harmonic 无规谐波
random incidence 随机入射声系数;无规入射
random-incidence response 无规入射响应
random-incidence sensitivity 无规入射灵敏度;散乱入射灵敏度
random-incidence sound absorption coefficient 无规入射吸声系数
random-incidence sound field 无规入射声场
random-incidence transmission loss 无规入射隔声量
random increment 随机增量
random indication 不规则显示
randoming scheme 随机化方案
random input 散乱输入
random inspection 随机抽样检验;随机抽样检查;抽检;抽查
random integer 随机整数
random integral equation 随机积分方程
random interference 随机干涉效应
random interference pattern 无规则干涉图样
random interlaced scanning 无规隔行扫描
random interrogation 无规则询问
random interval 随机区间
random inventory 随机性存储
random irregularity 随机不匀率
random isolation method 随机分离法
randomization 无规则分布;随机化;不规则化;不规则分布的形成
randomization method 随机化方法
randomization of energy 能量无规则化
randomization test 随机性试验;随机性检验
randomize 随机选择
randomized block 随机化区组;随机分组法
randomized block design 随机化区组设计
randomized block experiment 配伍组试验
randomized complete-block design 随机化完全区组设计
randomized decision function 随机化判决函数
randomized estimator 随机化估计值
randomized jitter 随机(图像)跳动
randomized orthographic(al) coordinate 随机正射坐标
randomized policy 随机策略
randomized process 随机法
randomized program(me) test 随机程序试验
randomized replication 随机重复
randomized sampling 随机化抽样
randomized table 随机化表
randomized test 随机检验;随机化试验;随机化检定
randomizer 随机函数发生器
randomizing 使不规则化
randomizing scheme 随机图表;随机式计划;随机方案
randomizing technique 随机技术
random jamming 随机干扰
random jump 随机跳动;无规则跳动
random Kriging 随机克立格法
random labelling 随机标记
random lack-of-fit structure 随即不贴合性结构;随机不贴合性结构
random-laid web 无定向纤维网
random lamination 非同向纹理层压

random lateral force 随机横向力
random length 任意长度(材);随机长度;长度不齐;不规则长度;不定尺;不定长
random line 试测线;辅助测线
random linear copolymer 无规线型共聚物
random load 随机荷载
random loading 无规负载;随机荷载
random logic 随机逻辑
random logic design 随机逻辑设计
random logic device 随机逻辑设备;随机逻辑部件
random logic testing 随机逻辑测试
random-loop transferring 不规则移圈
random loose arrangement 无规则松散排列
randomly coiled chain 无规旋卷链
randomly coiled macromolecule 无规旋卷大分子
randomly fluctuating data 随机起伏数据
randomly inhomogeneous medium 无规不均匀媒质
randomly modulated carrier 随机调制载波
randomly oriented fiber 无规定向纤维
randomly packed column 无规填料塔
randomly phased eccentricity 随机相位偏心
randomly time-varying linear system 随机时变线性系统
randomly traceable graph 随意可溯图
random manual mode 任意手工方法
random masonry 乱石圬工
random masonry bond 圬工随意砌法
random masonry work 圬工随意砌筑
random matching 随机拼合
random material 任意料;未筛选(的)材料;不规则料
random matrix 随机矩阵
random meander 游荡性曲流;游荡性河弯
random measure 随机测度
random measurement error 随机测量误差
random medium 随机介质;散乱介质
random mismatch 随机失配;随机失配
random model 随机(性)模型
random mosaic 不规则的马赛克;随机型马赛克
random motion 无规则运动;无规划运动;随机运动;不规则运动
random multi-level process 随机多值过程;随机多级过程
random multiple access 随机多路存取
random mutation 随机突变
randomness 无序性;随机性;不规则性
randomness component 随机性成分
randomness test 随机性试验;随机性检验
random network 无规则结构网
random network hypothesis 无序网络结构假说
random network simulation technique 随机网络模拟技术
random network theory 无规则网络学说;不规则网络学说
random noise 杂乱噪声;杂乱噪声;无规则噪声;随机噪声;随机噪声;不规则噪声
random noise correction 杂乱噪声消除
random noise generator 随机噪声发生器
random noise interference 随机噪声干扰
random noise level 随机噪声级
random noise power 随机噪声功率
random noise voltage 随机噪声电压
random-noise-weighting network 随机噪声加权网络
random number 随机数
random number generation 随机数发生
random number generator 随机数(生成)程序;随机数发生器
random number program(me) 随机数(生成)程序
random number sequence 随机数序列
random number series 随机数序列
random observation 随机观察(在工作抽样中所作的随机观察)
random occurrence 偶然事件;随机事件
random operator 随机算子
random optimalizer 随机最佳器
random orbit 随机轨道
random-orbit satellite system 随机轨道卫星系统
random order 任意顺序;随机位;随机顺序
random ordered sample 随机有序样本
random organization 随机编制
random orientation 无规取向;随机定向
random origin interval 随机原点间隔

random-packed column 任意堆积填料塔
random packet length 随机包长度
random packing 无规则填充;随机填料;散堆填料;堆堆填料
random pairs method 随机成对法
random pairs technique 随机成对技术
random paneling 不规则镶板
random parameter 随机参数
random path 随机路线
random pattern 乱铺纹样;任意排列;任意布置;随机模式;不规则形式;不规则镶嵌;不规则图案
random paving 乱石铺砌;乱石路面;乱推路面;简易路面;不整齐小方石铺砌路面;不规则砌铺路面;不规则毛石铺面;不规则毛石路面
random perforated 任意穿孔的
random perforated cellulose fiber tile 无规穿孔纤维板
random permutation 随机排列
random-perturbation optimization 随机扰动最优化
random phase 无规相位;随机相位
random phase angle 随机相角
random phase approximation 随机相位近似
random phase distribution 周相的无规分布
random phase error 随机相位误差
random phase factor 不定位相因子
random phase shift 随机相移;随机相变动
random phenomenon 随机现象
random-placed 乱抛的;抛筑的
random-placed concrete blocks 方块抛填
random-placed riprap 乱石工程;抛石工程
random placement 随意布局
random placing 抛筑
random plotting error 随机绘图误差
random point process 随机点过程
random polarization 随机偏振
random pollution 随机性污染
random polymer 无规聚合物
random population 随机种群;随机群体;随机母体
random pore model 无规孔隙模型
random-position scattering 无规散射
random posterior probability 随机后验概率
random posting 随意过账
random prediction 随机预测
random problem 随机问题
random process 无规过程;随过程;随机程序
random processing 随机处理
random pulse 随机脉冲
random pulse generator 随机脉冲发生器
random pulse jamming 杂乱脉冲干扰
random pulse train 随机脉冲串
random pulse train process 随机脉冲过程
random quadrature formula 随机求积公式
random quantity 随机(自变)量
random range ashlar 不规则粗料石工程;不成层方整石砌体;乱方石工程
random range masonry 随机砌圬工;乱石圬工
random range work 乱砌方石工程
random reactivity function 随机反应性函数
random reading error 随机读数误差
random reflectance of vitrinite 镜质体随机反射率
random reflectivity 随机反射率
random rehash method 随机再散列法
random rehash technique 随机再散列技术
random response 随机响应;随机反应
random retransmission interval 随机重发间隔
random riprap 乱石护坡;乱石堆;抛石工程;乱石工程
random rockfill 乱石堆填;任意石料填筑;任意石料抛筑;任意石料堆筑
random routing 随机路径选择
random rubber fill 乱石堆筑
random rubble 虎纹毛石墙;虎纹荒石墙;虎皮墙;乱石;乱砌毛石;粗石砌石;不规则毛石
random rubble facing 乱毛石饰面
random rubble fill 乱石堆筑(体);抛填乱石;房屋渣堆填;房碴堆填;乱石堆填
random rubble finish 乱毛石饰面
random rubble masonry (work) 乱石砌体;乱砌毛石圬工;不规则毛石砌体;乱石圬工;乱石墙
random rubble paving 不规则毛石铺面;不规则毛石路面
random rubble walling in course 分层砌筑乱石墙
random sample 随机试件;任意试件;随意抽样;随机样品;随机样本;随机试样;散装试样

random sampled parts 任意抽取件
random sampling 任意取样;随机取样;随机抽样;随机采样;抽样;抽查取样
random sampling by classification 分类随机抽样
random sampling distribution 随机抽样分布
random sampling error 随机抽样误差
random sampling method 随机原样法
random sampling oscilloscope 无规取样示波器;随机取样示波器
random sampling performance test 生产性能随机测定
random sampling voltmeter 随机取样伏特计
random sampling without replacement 不重复随机抽样;不放回随机抽样
random satellite system 随机卫星系统
random scan(ning) 散乱扫描;随机扫描
random scatter(ing) 无规散射;乱散射;漫射;随机散射;不规则的反射
random scission 无规断链
random sea 随机波;不规则波
random search(ing) 随机探测;随机搜索;随机检索;随机调查;随机探寻
random search method 随机搜索法
random seeding 随机引晶;多晶籽晶引单晶技术
random select 随机选择
random selection 随机选择
random sensitivity 无规灵敏度
random sequence 随机序列;随机顺序;随机时序
random sequential access 随机顺序存储
random sequential memory 随机时序存储器;随机按序存储器
random series 随机序列;随机级数
random service 随机服务
random service system 随机服务系统
random service system theory 随机服务系统理论
random set bit 不规则镶嵌的钻头(金刚石)
random setting 无定向摆放
random-shaped stone 形状不规则金刚石
random-sheared carpet 不规则剪线地毯
random-sheared rug 不规则剪线地毯
random shingle 不定型屋顶板;不定型层顶板;不规则屋面板(等长不等宽);不规则木瓦
random shock 随机震动;随机冲量;随机冲击
random signal 杂色信号;偶然信号;随机信号;不规则信号
random simulation method 随机模拟法
random sine wave 随机正弦波
random skip position 随机跳位
random slate 粗石饰面;不规则石板;不定型石板
random soil sample 散装土柱
random solid solution 无序固体体;无规固溶体
random sound 无规声
random source 无规源;随机源
random spacing 不规则排列;不规则间距
random specimen 随意试样
random splicing 无序联接
random Stark effect 无规斯塔克效应
random start 随机起动
random stone work 不规则毛石砌体;乱砌石工
random strain 随机应变
random strategy 随机策略
random stress 随机应力
random stripe 不规则条纹
random subsample 随机次级样品
random sum peak 随机叠加峰
random superposition 随机叠加
random support acceleration 随机支承加速度
random surface profile 随机表面形状
random switching function 随机开关函数
random synchronizing 自同期;不规则同步
random table 乱数表
random target course 任意的目标航线
random tensor field 随机张量场
random test 任意试验;随机试验;随机抽样检验;随机抽样检查;随机测试
random time 任意时间
random time distribution 杂乱时间分布
random time spacing 随机时间间隔
random time-varying channel 随机时变信道;随机时变通道
random tooled ashlar 乱琢方石;不规则修琢的粗料石;乱凿纹方石
random tooled finish 乱凿纹面
random traverse 任意联测导线;辅助测线

random trial-and-error test 随机边试边改试验
random trial method 随机试验法
random turbulence 不规则湍动;不规则紊流
random type 随机型
random uncertainty 随机不确定度
random value variance 随机值变化
random variable 随机应变;随机变数;随机变量
random variable of continuous type 连续型随机变量
random variation 偶然性变化;随机变异;随机变化
random variation parameter system 随机变参数系统
random vector 随机向量
random-vector generator 随机向量生成程序
random velocity 无规速度;随机速度
random vibration 随机振动;随动振动;不规则振动;不规则振荡
random vibration environment 随机振动环境
random walk 乱行;随机走向;随机走动;随机游动
random walk diffusion 无规则游动扩散;无规则移动扩散
random walk method 随机游动法;随机移动法
random walk problem 随机走动问题
random walk process 无规行走过程
random walk simulation 随机游走扩散模拟
random walk theory 随机走动理论
random water-killing 偶然冻死
random wave 随机波;不规则波;无规波
random waveform 任意波形
random wave shape 随机波形
random web 无定向纤维网
random web-laying equipment 气流成网机;无定向成网机
random white noise 无规白噪声
random width 任意宽度
random winding 无规则线圈;散下式绕组;不规则线圈
random work 不成层石砌体;随意划分;随意排列;断层石砌体;乱石墙;杂纹方石工程;乱砌工事;乱砌(方)石工程;不规则石工
random wound coil 散绕线圈
Randupson process 伦道普森水泥砂造型法
Raney nickel 镍催化剂
Raney nickel film 热催化活性镍合金片
Raney's alloy 拉尼镍铝合金
Raney's nickel 拉尼镍
Raney's nickel wastewater 拉尼镍废水
range 续航力;航程;行程;域;营林区;值域;涨落差;量程;粒子射程;方程;距离;火格子;全距;山区绵亘;支盘标度;单位草场;测程;幅射;幅度;分类;变动范围;变程;靶场
rangeability 可调节范围;可调范围;幅度变化范围;变化幅度;适应范围;运行距离;量程范围
range accelerometer 二重积分加速度表;二重积分加速表
range accuracy 距离精度;测距准确度;测距精度
range adjustment 量程调整;距离修正;距离校正;距离调整
range allotment 放牧面积
range ambiguity 轮廓模糊;距离模糊
range amplifier 测距放大器
range analysis 值域分析;量程分析;极差分析
range and bearing marker 距离和方位标志
range aperture 沿距离射角;距离孔径
range appraisal 草场评定
range area 断面面积;放牧区
range arm 距离尺
range at cruising speed 巡航速度航程
range at future position 提前位置距离
range at maximum speed 最大速度航程
range at present position 现在位置距离
range attenuation 距离衰减;途程衰减
range-azimuth 距离一方位
range-azimuth display 距离一方位显示器
range-azimuth indicator 距离一方位指示器
range-azimuth presentation 距离一方位显示
range-azimuth tube 距离一方位管
range beacon 航线导航无线电信标;导向标;叠标(导航)
range bearing 叠标方位
range-bearing display 距离一方位显示器
range blanking 距离照明
range blind hole 测距盲区
range boiler 灶旁热水锅炉;炉灶热水器

range calibration 固定距标距离校准;量程校准;距离校准;范围校准
range-calibrator target 距离校正靶;校距标
range capability 可达距离;活动半径;测距能力
range card 方位物略图
range carrier 星形齿轮架
range-changing method 距离变更法
range channel 测距信道
range-channel selector 测距通道选择钮
range chart 极差控制图
range check 量程检查;区域校核;区域检查
range checking 区域检查
range circuit 测距电路
range cleat 系索耳;大羊角
range closet 多蹲位厕所
range clutch 极限行星离合器
range coal 块煤
range coding 距离信号编码
range coefficient 测程系数
range compensator 距离补偿器
range conservation 牧场保持
range constraint 限程约束;区域约束
range control 距离控制;航程控制
range control chart 航程检查图
range control switch 测量范围转换开关
range conversion 量程变换
range converter 量程变换器;距离变换器
range correction 距离校正
range corrector 距离修正器
range counter 距离计数器;距离测定器
range coverage 可达区域
range cruise 最大航程巡航
range cubes 草原用块状饲料
range-cutoff function 有效距离截止
range-data telemetry 射程遥测术
range deflection fan 扇形分划板
range delay 距离延迟
range determination 距离测定
range difference 亮度差
range difference measurement of satellite fixing 卫星距离差定位
range digital indicator 距离数字显示
range discrimination 距离分辨(力)
range display 距离显示器
range-distance indicator 距离高度指示器
range district 放牧区;保护区
range drive 联合驱动;多动力驱动
ranged rubble 毛石工程
range drum 距离分划筒
range ecology 草原生态学
range-elevation display 距离一仰角显示器
range end 书架右侧板
range equation 距离方程
range error 量程误差;距离误差;测量误差
range estimation 距离判定;距离估测
range expander 量程扩展器
range expansion 分布区扩大
range finder 行程定位器;测远仪;测远计;测距仪;测距器;测距机
range finder camera 测距照相机
range finder filter 测距仪滤光镜;距离滤波器
range finder interval 测距仪间隔
range finder prism 测距棱镜
range finder scope 自动测距镜筒;测距瞄准器
range finder window 测距器孔;测距窗
range finding 距离测定;测距
range finding test 预备实验;预试验;预备测试
range fire 草场火灾
range flag 靶场警戒旗
range-follow-up unit 距离跟踪装置
range formula 距离公式
range for pulsating compressive stresses 脉冲压缩应力范围
range for pulsating tensile stresses 脉冲拉伸应力范围
range front 书架左侧板
range gate 距门;距离选通脉冲;距离开关;测距选通门
range gated capture 假目标俘获
range-gated detection 距离选通探测
range-gated imaging technique 距离选通成像技术
range-gated laser 距离选通激光器
range gating 距离选通
range guide 书架导标

range gun 射流枪
range heads 起锚机(系)柱
range-height converter 距离高度变换器
range-height display 距离高度显示器
range hole 测距盲区
range hood 炉用排气罩;炉灶罩;厨房拔气罩
range-in 试测和误差修正定线法
range index 距离指数;距离指标
range indicating system 距离指示系统
range indicator 距离指示器
range interval 间距
range into line 排列成行
range-ionization telescope 射程电离望远镜
rangekeeper 射程计算仪
range ladder 梯级试射
rangeland 牧场;牧地;草场;放牧地
range laser radar 激光测距雷达
range lavatory basin 成列盥洗盆
range light 后桅灯;航迹灯;叠路灯(之一);导向灯标;导(航)灯
range light column 叠灯柱
range light installation 导航(灯)设施;导航设备
range light post 叠灯柱
range light tower 叠灯塔
range line 镇区;延线;边界;国境线;限距;市镇范围界;射程线;定位线;叠标线
range line profile 断面面积(水库某一断面线上的);断面轮廓线
range management 牧地管理;草原管理
range mark 距离标记;方位标志;方位标(记);距离标志;叠标(导航);导标
range marker 距离指点标;距离刻度指示器;距离刻度校准器;距离(刻度)标识器;距离标志
range marker circuit 距离定标电路
range marking 距离校准标记
range mark offset 距离标志偏移
range masonry (work) 成层砌体;成层圬工;层砌琢石;毛石整砌层;垒石工程;成层石工;每皮等厚的圬工
range maximum 最大续航力;最大射程
range measurement 距离测定
range measurement of satellite navigation system 卫星测距导航系统
range measurement system 测距系统
range-measuring 测距的
range meter 测距仪
range method 断面法
range-mode selector 距离方向选择器
range multiplier 量程扩大器;扩程器;倍率器
range multivibrator 距离扫描多谐振荡器;按距扫描多谐振动
range network 测距网
range noise 距离噪声
range normalization 距离校正;距离归一化
range notch 距离选择器标尺
range nozzle 多斗旋转喷嘴
range number 断面号数
range observation 距离观测
range of accommodation 调节幅度
range of action 有效范围;作用距离;作用范围
range of active seress zone 营力活动带范围
range of a default specification 缺省说明范围
range of adjustment 调节幅度;校正范围;调整范围;调节阀
range of administered price 行政定价范围
range of allowable error 容许误差范围
range of altitude 高度范围
range of a mapping 映射的值域
range of (an) instrument 仪表量程
range of applicability 适用范围
range of application 应用范围
range of a relation 关系的值域
range of a signal 显示范围
range of atomization 雾化区
range of a transformation 变换的量程
range of attenuation 衰减范围
range of audibility 可听度范围;可闻距离;可听范围;声频范围
range of balance error 对称误差范围
range of boiling 沸腾范围
range of brightness 亮度范围
range of building 一排房子;房屋的排列
range of cable 一排锚链;排链
range of capacity 容量范围

range of colo(u)rs 颜色的范围;颜色的分布
range of columns 柱列
range of concentration 浓度范围
range of conditional simulation 条件模拟变程
range of conics 二次曲线列
range of contrast 反差范围
range of decrepitation temperature 爆裂温度范围
range of definition 定义域;定义范围
range of deflection 偏转范围
range of density 密度范围
range of deposition and degradation 冲淤幅度
range of detection 探测距离;搜索距离
range of detector 探测距离
range of distribution 分配区域;分配范围;分布域;分布范围
range of disturbance 干扰范围;扰动范围
range of effectiveness 有效范围
range of equipment 设备系列;成套设备;成套机具
range of erosion and siltation 冲淤幅度
range of error 误差范围
range of execution 执行范围
range of exposures 曝光宽容度;照射范围
range of feeds 进给量范围
range of fire 射程
range of flood and ebb 潮幅;潮差
range of flow 流量范围
range of fluctuation 涨落范围;起伏范围;波动范围;变动范围
range of focus setting 调焦幅度
range of furrow width 沟宽范围
range of ground swing 地面反射误差范围
range of groundwater withdrawal 地下水抽取范围
range of hardness 硬度范围
range of head 落差范围;压头范围;水头(变化)范围
range of headlamp 前灯照射距离
range of hearing 听觉范围
range of hills 多丘陵地区;丘陵(山脉)
range of homogenization temperature 均一温度范围
range of indication 指示范围
range of infinitely variable speeds 无级变速范围
range of influence 影响区域;影响范围
range of instability 不稳定区
range of integration 积分限
range of inundation 淹没范围
range of irradiation 照射范围
range of jet 射流射程;射流流程
range of lay 捻距范围
range of light 光程
range of light-variation 光变幅
range of linearity 线性区域
range of linearity control 线性控制范围
range of load combination 荷载组合范围
range of load fluctuation 荷载变化范围
range of longitudinal feeds 纵向进刀量范围
range of luminous reflectance 发光折射范围
range of Mach's numbers 马赫数范围
range of magnification 放大限度
range of maturity 成熟期的范围
range of maximum speed 最大速度航程
range of maximum tension area 最大拉应力区范围
range of measurement (仪器的)量距;量程
range of metric thread cut 公制螺纹螺距范围
range of models 模型范围;型号系列
range of module thread cut 模数螺纹螺距范围
range of movement 移动范围;活动范围;动程
range of noises 噪声幅度
range of non-conditional simulation 非条件模拟变程
range of normal audibility 正常听觉范围
range of normal hearing 正常听觉范围
range of normal value 正常值范围
range of nuclear force 核力范围
range of ocular accomodation 目镜调节范围
range of operation 运行范围;动作范围
range of oscillation 振幅;振动范围;摆动范围
range of particle 粒子行程
range of points 点数
range of possibility 可能性范围
range of power levels 功率级范围
range of price 价格幅度;价格范围
range of price changes 调价浮动
range of products 产品种类

range of quadrant 象限尺;表尺
range of quadrics 二次曲面列
range of regulation 调节范围
range of relaxed zone 松弛带范围
range of response 视场
range of responsibility 责任范围
range of revolution 转数范围
range of sample 样本范围
range of scatter 扩散范围;散乱范围;散布区;分散范围
range of screening 颗粒大小范围
range of setting 安置幅度
range of set value 给定值范围;设定值范围
range of sight 视线范围;视界
range of silence 哑点宽度
range of sizes 尺寸范围
range of society-wide wage increases 全社会工资增长的幅度
range of speed control 转速控制范围;速度上下范围
range of speeds 转速范围
range of speeds of rotation 转速范围
range of spring 弹簧变形限度
range of spring tides 子午潮幅度;大潮潮差
range of stability 稳定阶段;稳性范围;稳定范围
range of stage 水位差;水位变幅
range of standard grading for aggregate 集料标准级配范围
range of strength 强度变化范围
range of stress 应力限程;应力幅度;应力(变化)范围;应力变幅
range of study 研究范围
range of super-humid zone 过湿带范围
range of survey 测图范围
range of survey area 测绘区范围
range of temperature 温度较差;温度范围;温度变化幅度;温度变化范围
range of thermal neutrality 温度适中区
range of thermometer scale 温度计读数范围
range of the sample 样本的范围
range of throughput 输送范围
range of tide 潮汐幅度;潮位变幅;潮幅;潮差
range of time-lag settings 时延调整范围
range of trajectory 抛射距离
range of transgression 海侵范围
range of unclear force 核力力程
range of urinal stalls 小便斗分隔间的排列
range of use 应用范围;使用范围
range of validity 有效范围
range of values 值的范围;评价范围;数值范围
range of variables 变数范围;变量间隔;变量范围;变量的范围
range of variation 变异范围;变化范围
range of viscosity 黏度的变化范围
range of visibility 可见距离,能见(度)距离;能见度;视野;视距
range of vision 视野;视力范围;视界;视程
range of vitrification 玻化范围
range of wave height 波高范围
range of wave length 波长范围
range of well distribution 水井分布范围
range of Whitworth's thread cut 惠氏螺纹螺距范围
range of wind tide 风潮差
range of work 加工范围
range of working temperature 作业温度范围
range-only radar 测距雷达
range operator 距离测定员;距离操作手
range out 定位
range out line 定线
range photomultiplier 距离光电倍增管
range pile 导桩
range plant cover 牧区植被;草场植物覆盖
range pole 测杆;花杆【测】;视距尺;标杆
range positioning system 圆—圆定位系统;距离定位系统;测距定位系统
range potentiometer 量程电位器;距离分压器
range power supply 测距电源
range prediction 作用距离预报
range preparation ground 射击预习场
range presentation 距离显示
range pulse 距离脉冲
range quadrant 象限仪;表尺
ranger 营林员;模板横撑;测力计;测距仪;板桩横挡

range radar 测距雷达
range rake T 形距离尺
range-range navigation 圆—圆航法
range rate 临近速度;接近速度
range-rate correction 距离扫描校正
range-rated appliance 热负荷可调节的燃具
range-rate data 距离变化率数据
range rate error 距离变化率误差
range-rate observation 距离变化率观测
range readiness 草场条件具备
range reconnaissance 草场踏查
range recorder 距离记录器
range region 放牧区
range re-seeding 草场补播
range resolution 空间分辨率;距离分辨率;距离分辨(力)
range resolution rate 距离分辨率
range ring 距离圈;距离刻度圈;距离刻度环;距离比例圈;距离比例尺
range rod 花杆【测】;视距尺;测距尺;测杆;标杆;标尺
ranger program(me) 测距者计划
range safety 安全范围
range safety system 靶场安全系统
range scale 高低分划;量程指示;量程刻度;距离刻度(盘);距离标度
range scale and range ring interval indicator 量程及距圈间隔指示器
range scale selector 量程指示转换开关;距离标度选择开关;距离比例尺转换器
range scanning 按距扫描
range scanning rate 按距扫描速度
range scattering 纵向散布
range science 草原学
range scope 距离指示器
range selector 量程转换开关;量程选择器;量程选择钮;距离转换开关
range selector lever 速比范围选择杆;变速杆;高低开关杆
range selector valve 行程控制阀;行程操纵阀
range selsyn 距离自动同步机
range sensor 距离传感器
range set 区域集
range setter 瞄准装定器
range setting 装定距离
range shelter 草地畜棚;草场畜棚
range signal 距离标志
range site 草场地生境
range socket 电灶插座
range sowing 草场补播
range space 列空间
range spacer 横杆;支撑;横撑;隔片;隔块;隔撑;测距仪
range spotting 炸点距离偏差观测
range spread 射程散布
range station 无线电指向标;导航台
range step 距离阶梯
range straggling 射程距离
range strobe 距离范围选通脉冲
range surveillance 距离监视
range survey 牧场调查;断面测量;草场调查
range swath 航带;地带
range sweep 距离扫描
range-sweep gate generator 距离扫描选通脉冲发生器
range switch 量程开关;距离转换开关;波段(选择)开关
range switching 量程转换;波段变换
range system 测距系统;草原利用制度;方位线系统
range table 射程表
rangetaker 测距员
range the cable 盘或排列锚链
range tie 方向系线
range-time display 距离时间显示器
range-to-go 剩余航程;到目标的距离
range-to-go display 剩余距离显示器
range tracker 距离跟踪仪
range-tracking 距离跟踪
range transformation 极差变换
range transmission 传动系统
range transmission hydraulic control valve 行星传动装置液压控制阀
range transmitter 测距发射机
range-transmitting potentiometer 距离数据输出

分压器
range-transmitting system 距离数据传输系统
range tube 距离显示管;测距管
range unit 测距装置
range utilization 草场利用
range velocity component 纵向速度分量
range vertical 距离垂线
range-viewfinder 测距探视仪;检影器;测距取景仪
range wall(ing) 分区墙
range wind 纵风
range wind error 纵风测定误差
range with maximum tankage 最大载油量航程
range work 方石砌体;垒石工程;整层砌筑;成层琢石;成层圬工;成层石工;层砌工程
range zero 距离零位
range-zero calibration 零距校准
range-zero control 距离零位调节
range zone 分布区带;延限带
range zone of organism 生物地带
ranging 距离调整;测距;测程
ranging accuracy 测距精度
ranging a curve (路线测量的)曲线放线;(路线测量的)曲线放样
ranging amplifier 测距放大器
ranging bond 排列式砌合;石缝握裹木条;排列式砌筑
ranging bridge 排列式砌筑
ranging computer 距离测定仪;测距计算机
ranging data 测距数据
ranging gating 测距选通
ranging into line 瞄准;将……定位;给……定线;照准
ranging laser 测距激光器
ranging line 测距线;轮廓引线;基准线;定位线;测程线;标距线
ranging method by radio and sound wave 无线电电声测距法
ranging method of sound wave 声波测距法
ranging of curve 曲线测设
ranging of line 定线
ranging oscillator 测距振荡器
ranging pole 视距尺;标杆
ranging pole support 标杆支架
ranging pole tripod 标杆三脚架
ranging prism 测距棱镜
ranging radar 测距雷达
ranging radio positioning 无线电测距定位
ranging rod 花杆;视距尺;测距尺;测杆;标杆
ranging set 声呐站测距分机
ranging sonobuoy 测距声呐浮标
ranging theodolite 测距经纬仪
ranging transit 测距经纬仪
ranging unit 测距计
rang land conservation 牧场土地保护
rang management 牧场管理;草场管理
ranite 水霞石
ranitidine 甲胺呋硫
rank 职位;级别;横排;行列;位次;等级;次序;层号
rankachite 莱卡石
rankamaite 羟碱铌钽矿;钽钠石
rank and file 一般职工;普通成员
rank correlation 秩相关;等级相关
rank correlation coefficient 秩相关系数
rank defect 秩亏
rank difference 等级差
ranked air 污浊空气
ranked area of mice sheet 云母片型号面积
ranked data 等级资料;分级资料;分级数据;分等级的数据
ranked data model 分级数据模型
ranked payoff 序列支付
rank factorization 因子分解;因式分解;秩分解
rank force element 秩力元素
rank fusinite 煤化丝质体
rank growth 杂草丛生
rank indigo 腐致靛蓝
Rankine active failure zone 主动朗肯破坏区
Rankine active state equilibrium 朗肯主动土压力平衡状态
Rankine-Hugoniot equation 朗肯-胡贡纽方程
Rankine's body 朗肯体
Rankine's cycle 朗肯循环
Rankine's cycle engine 朗肯循环发动机
Rankine's design method 朗肯设计方法

Rankine's earth pressure theory 朗肯压力理论
Rankine's efficiency 朗肯效率
Rankine's formula 朗肯设计立柱和挡土墙公式;朗肯公式
Rankine's scale 朗肯温度标;朗肯刻度
Rankine's solution 朗肯解
Rankine's states of stresses 朗肯应力状态
Rankine's state 朗肯状态
Rankine's temperature 朗肯温度
Rankine's temperature scale 绝对华氏温标;朗肯温标
Rankine's theory 朗肯土(压力)理论
Rankine's vortex 朗肯涡流
Rankine's zone 朗肯区
ranking 秩评定;评序;分类
ranking alternatives 备选方案排序
ranking bar 手推车
ranking criterion 排序判据
ranking form 投资方案排队格式
ranking function 评级函数
ranking method 评级方法;排序方法;分等级方法
ranking mo(u)lding 斜檐脚
ranking of alternatives 方案排队
ranking of grade 品位分级
ranking of waterway 航道定级
ranking or rating of landfill site 垃圾处理地点顺序或等级
ranking problem of capital investment 投资方案排队问题
ranking test 评价试验
ranking work joint 斜茬施工缝
rankinite 硅钙石
rank macrinite 煤化粗粒体
rank of a bilinear form 双线性形式的秩
rank of a congruence 线汇的秩
rank of a group 群的秩
rank of a weakly stationary process 弱平稳过程的秩
rank of commensurability 通约秩
rank of determinant 行列式的秩
rank of geothermal hazard area 热害区级别
rank of matrix 矩阵秩;矩阵的秩
rank of selectors 选择器列
rank of tensor 张量的阶
rank order 排位次序;等级序列;等级位次;等级次序
rank order statistics 秩次统计量
rank program(me) 评序程序
rank score 评级
rank semifusinite 煤化半丝质体
rank set 粗装
rank sheer 船首突然转向
rank-size rule of cities 城市顺位-规模法则
ranks of observation 观测值的秩
rank soil 十分肥沃的土地;肥沃的土壤
rank space 秩空间
rank statistics 秩统计量
rank-sum test 秩和检验
rank sum test method 秩和检验法
rank system 序级制
rank technique 秩法
rank test 秩检验;秩和检验;等级测试
rank weeds 丛生的杂草
Ranney collector 辐射管井;兰尼集水管
Ranney oil-mining system 兰尼石油采矿系统
Ranney well 兰尼式井
Ranney well collector 兰尼井内集水管
ransack 搜查
ransatite 杂镁榴石
Ransburg electro-air gun 兰斯堡电离空气喷枪
Ransburg electrostatic blade coater 兰斯堡叶片式静电涂漆机
Ransome iron 螺纹钢筋
Ransome mixer 拉斯姆搅拌机
Ransome's process 拉斯姆工艺
ransomite 铜铁矾
Ransom type dam 兰松式坝(1908年左右推荐的平板式坝)
rantidine hydrochloride 甲胺呋硫
rant loans in accordance with applicant's assets-liabilities ratios 资产负债比例管理
ranunculite 纤磷铝铀矿
Ranvier's gold chloride staining method 郎飞氏氯化金染色法
Ranvier's membrane 郎飞基础膜

raob 无线电探空观测;探空(仪)观测
Rao formula 拉奥公式
Raoult's law 拉乌尔定律
Raoult's method 拉乌尔法
rap (模型的)松动
rapakivi 环斑状;奥环状花岗岩
rapakivi aplite 环斑细晶岩
rapakivi granite 环斑花岗岩
rapakivi syenite 环斑正长岩
rapakivite granite 更长环斑花岗岩
rapakivi texture 环斑结构;奥环状花岗岩结构
Rapaloid 拉帕劳(一种纤维素塑料)
rapatelle 马毛筛绢
rapcon 雷达临场指挥
rape of forest 违犯森林规则
rape oil 菜籽油;菜油
rape seed oil 菜籽油
rape weed 油菜子
raphide 针状晶束
raphidine 针状纤维
raphilite 透闪石
raphioid fiber 针形纤维
Raphiolepis indica 石斑木
raphisiderite 针赤铁矿
Raphoneis amphiceros 双角缝舟藻
Raphoneis elliptica 椭圆缝舟藻
rapic acid 菜籽酸
rapid access 高速存取;快速取数
rapid access loop 高速存取环;快速循环取数区;快速访问循环;快速访问环;快速访问道;快速存取环
rapid acting coupling 快速接头
rapid acting folding shutter door 速动折叠百叶门
rapid acting fuse 速断保险丝
rapid action brake servo 快速制动伺服机构
rapid action hammer 速动锤
rapid action mixer 高速强制搅拌机
rapid action valve 快动阀;速动阀
rapid activated carbon filtration 快速活性炭过滤
rapid adjustment 快速调整
rapid ag(e)ing 快速老化
rapid ager 蒸发机
rapid agitating 快速搅拌
rapid agricultural growth 农业迅速增长
rapid air change 快速换气
rapid amortization 快速摊销;快速摊提
rapid analysis 快速分析
rapid analysis machine 快速分析仪
rapid analysis method 迅速分析法;快速分析法
rapid analysis of fresh concrete 新拌混凝土快速分析
rapid and sensitive feed 快速灵敏进给
rapid annealing 快速退火
rapid answer 迅速应答
rapid-application position 快速制动位
rapid approach 快速进给
rapid approach pump 快动泵
rapid ascent 陡坡
rapid assembly method 快速装配法;快速拼接法
rapid assessment 快速现场评定
rapid assessment contaminant transport model 快速评估污染物输移模型
rapid automatic checkout equipment 快速自动检查仪;快速自动检查设备
rapid automatic press 高速自动冲床
rapid ball grinder 快速球磨机
rapid blow drilling 快速冲击钻进
rapid brake application 应用快速制动
rapid break-in 快速磨合
rapid breaking emulsified asphalt 快裂乳化(地)沥青
rapid building hoist 快速施工升降机;快速施工提升机
rapid burning clinker 急烧熟料
rapid burning process 快速烧成法
rapid calculation 快速计算
rapid cement 快硬水泥;快速水泥
rapid ceramic firing 陶瓷快速烧成
rapid change 急剧变化
rapid charge rate 快速充电率
rapid charging 快速充电法
rapid chopper tester 高速断路器试验装置
rapid clamping fixture 快速夹紧夹具
rapid cleaning agent 快速清洗剂

rapid clinker cooler 水泥熟料快速冷却器
rapid clinker cooling 熟料快速冷却
rapid closing 迅速闭合
rapid closing stop valve 快速截止阀
rapid closing valve 速闭阀
rapid clutch 快动离合器
rapid coagulation 快速混凝
rapid combustion 迅速燃烧
rapid combustion method 快速燃烧法
rapid combustion period 迅速燃烧期
rapid compression ignition of fuel 燃料快速压燃
rapid condensation 快速凝结；快速冷凝
rapid construction 快速施工
rapid contactor 快速接触装置
rapid contractor 快速萃取装置
rapid convergence 快速收敛；快速聚焦
rapid cooling 快速冷却
rapid cooling of clinker 熟料的快速冷却
rapid copy 快速复制
rapid core drill anchor 快速岩芯钻固定器
rapid coupler 快速连接装置；快速接头；快速联轴器
rapid coupling pipe 快速连接管
rapid crack propagation 裂缝迅速扩展
rapid cure adhesive 速定胶粘剂
rapid-curing 冷底子油；快速养护；快速湿治；快凝的（液体沥青）
rapid-curing asphalt 快固沥青；快干沥青；速凝沥青；快凝沥青
rapid-curing cutback 快干；快凝；速凝
rapid-curing cutback asphalt 快凝轻制（地）沥青；速凝沥青
rapid-curing cutback bitumen 快凝轻制（地）沥青
rapid-curing liquid asphalt 快凝液体（地）沥青
rapid-curing material 快凝材料；速凝材料
rapid-curing mortar 速凝灰浆
rapid cut-off valve 快速切断阀
rapid cycle telemetry 脉动循环遥测
rapid damage assessment 损坏情况快速判定
rapid data base management system 快速数据库管理系统
rapid data processing 快速数据处理
rapid decay 快速衰减
rapid dechlorination 快速脱氯
rapid decompression 迅速失压；急剧减压；突然失压
rapid deployment 迅速展开
rapid depressurization 迅速降压
rapid design method 快速确定尺寸方法；快速设计（方）法
rapid determination 快速确定；快速测定
rapid digital automatic computing 快速数字自动计算
rapid-direct-reading spectrographic determination 快速直读光谱测定法
rapid downward flow sand filter 下流式快滤砂池；快速下流砂滤池
rapid drawdown 水位骤降
rapid-drying 快干的
rapid-drying material 快干材料
rapid dy(e)ing 快速染色
rapid earth flow 快速泥石流
rapid erection 快速安装
rapid erection crane 快速架设的起重机；快速竖立起重机
rapid erosion 快速侵蚀
rapid escape valve 快速排气阀
rapid examination 快速检验
rapid excavation 快速开挖；快速掘进
rapid exit taxiway 高速脱离滑行道
rapid exposure photography 快速曝光照相
rapid fading 急剧衰落
rapid fast colo(u)r 快速坚牢颜料
rapid fatigue test 快速疲劳试验
rapid feed 快速进给
rapid field assessment 快速现场检定
rapid-filing device 快速锉磨器
rapid filter 高速过滤器；快（速）滤池；快速过滤器
rapid filtering funnel 快速过滤漏斗
rapid filtration 快速过滤；快滤
rapid firing 快烧；快速烧成
rapid firing kiln 快速烧成窑
rapid firing stain 高温快烧颜料
rapid fixing 快速装配
rapid flood passage 泄洪（道）
rapid flow 急流；激流；湍流；超临界流

rapid flow kinetics 速流动力学
rapid flow technique 速流技术
rapid fluctuation 急剧波动（水面）
rapid fluctuation of water-level 水面急剧涨落
rapid fluorometric method 快速荧光测定法
rapid formation 迅速形成
rapid fuse 快速熔断丝
rapid gas water heater 煤气快速热水器
rapid geodetic surveying system 快速测地系统
rapid gravel filter 快速砾石滤池
rapid gravity filter 快速重力滤池
rapid gravity sand filter 重力式快砂滤池；快速重力砂滤池
rapid hardener 速凝剂；快速硬化剂
rapid hardening 快硬；快速硬化；快速硬固；速凝
rapid hardening cement 早强水泥；快硬水泥；快凝水泥；特快硬水泥
rapid hardening concrete 快硬混凝土
rapid hardening lime 快硬石灰
rapid hardening Portland cement 快硬硅酸盐水泥；快硬波特兰水泥
rapid heater 快速加热器
rapid heating 快速加热
rapid index(ing) 快速分度
rapid infiltration 渗滤；快速渗滤
rapid infiltration basin 快速渗滤池
rapid infiltration land treatment system 快速渗滤土地处理系统
rapid infiltration system 快速渗滤系统
rapid information 有效情报
rapid information technique for evaluation 评核加速资料（技）术
rapid inversion method 快速转换法；快速反转法
rapid inversion system 快速反转系统
rapidity of changeover 变换快速性
rapidity of convergence 收敛速度
rapidity of fading 衰落速度
rapidity of ignition 发火快速性
Rapidity separator 拉比迪蒂磁选机
rapid lens 快速透镜；大相对孔径物镜
rapid lining support 快速衬砌
rapid loading 快速加载
rapid look technique 快速检视
rapid loss of load 快速卸载；快速失重
rapidly available phosphorus 速效磷
rapidly developing disasters 急剧恶化的灾害
rapidly drying paint 快干油漆
rapidly rotating shift system 快速轮班制
rapidly soluble material 易溶物质
rapidly strengthen 急骤增强
rapidly switched tube 快速开关管
rapidly varied flow 急变流
rapidly varied unsteady flow 急变不稳定流
rapid measurement 快速测量
rapid method 速测法
rapid mix and flocculation basin 快速混合絮凝池
rapid mixer 快速混合器
rapid mixing 快速搅拌；快速混合
rapid mix unit 快速混合设备
rapid monitoring 快速测定
rapid movement 快速运动
rapidograph 快速绘图头
rapidoprint 双液冲洗
rapid orientation 快速定向
rapid original survey 快速初测
rapid percussive drilling 快速冲击钻孔
rapid permeability 迅速渗透；快速渗滤；速渗（透）
rapid picture manipulation 快速图像处理；快速图像变换
rapid polymerization 快速聚合
rapid pressure fluctuation 压力突然波动
rapid printer 快速打印机
rapid processor unit 快速信息处理机
rapid quenching 快速淬火
rapid quick coupling 速接联轴器
rapid radioactive contamination indicator 放射性污染快速指示器
rapid rail transit 快速轨道交通
rapid-reader sextant 鼓轮六分仪
rapid reconnaissance 遥感；快速勘测
rapid record oscillograph 快录示波器
rapid record oscilloscope 快速记录示波器
rapid rectilinear lens 快速直线透镜；快速消畸透镜；快速消畸变镜头；快速无线性畸变镜头

rapid release 快速缓解
rapid repair method 快速修理法
rapid repair sidings 旁线快速修理
rapid retrieval 快速检索
rapid return 快速回程；快速返回
rapid return motion 快速返回运动
rapid return stroke 快速返回行程
rapid revision 快速修测
rapid rise in prices 价格飞涨
rapid rise test 快速升压试验
rapid river 湍急河流；陡峻河流
rapid road 高速公路
rapid road towing 路上高度拖行
rapid rotary braiding machine 高速编带机
rapid rotary dryer 快速转筒烘干机
rapids 险滩；急湍；急滩；湍滩
rapid sand filter 快速砂滤池；快（砂）滤池
rapid sand filter bed 快速滤清器
rapid sand filter rating 快速砂滤池定额；快滤池设计滤速；快速池定额
rapid sand filter strainer 快速砂滤池滤器；快滤砂滤器；快滤池滤器
rapid sand filtration 快速砂（过）滤
rapids and shoals 滩险
rapids-ascending ability of ship 船舶过滩能力
rapid scan monochromator 快速扫描单色仪
rapid scanning 快速扫描
rapid scanning interferometer 快速扫描干涉仪
rapid scanning spectrometer 快速扫描分光（光度）计
rapid scanning spectrophotometry 快速扫描分光光度法
rapid scan spectrometer 快速扫描分光计
rapid screening machine 快速筛分机
rapid screening toxicity 快速筛分毒性
rapid screw 大节距螺钉
rapid seating 座面快速磨合
rapid selector 快速选择器
rapid separator 感应盘式磁选机
rapid sequence camera 快速序列照相机
rapid setting 快裂；快凝的；快结
rapid setting admixture 快凝剂
rapid setting agent for cement 水泥速凝剂
rapid setting cement 早强水泥；快凝水泥；速凝水泥
rapid setting concrete 快速凝固混凝土
rapid setting emulsified asphalt 快凝乳化（地）沥青；快裂乳化（地）沥青
rapid setting emulsion 快凝乳化液
rapid setting mortar 快凝凝固砂浆；快凝凝固灰浆
rapid setting mo(u)lding plaster 快凝装饰线条石膏
rapids-forming period 急滩成滩期
rapids/hazard-abating stage 急、险滩消滩水位
rapids/hazard-forming stage 急、险滩成滩水位
rapids/hazard-regulation stage 急、险滩整治水位
rapids head（急滩的）滩头
rapids heaving 绞滩
rapids heaving barge 绞滩驳船
rapids heaving station 绞滩站
rapids heaving winch 绞滩机
rapid-shifting bridle arrangement 快换拉紧装置
rapid-shifting drag scraper machine 快变换拖式铲运机
rapid-shiter 快速挂钩脱钩的拖式铲运机；快速开关
rapid sintering 快速烧结
rapid slaking lime 快化石灰；快消石灰
rapid sludge return mechanism 快速污泥回流机理
rapids of face-to-face pattern 对口型急滩
rapids of narrow channel pattern 窄槽型急滩
rapids of narrow channel type 窄槽型急滩
rapids of opposite protruding points type 对口型急滩
rapids of protruding point pattern 突嘴型急滩
rapids of protruding point type 突嘴型急滩
rapids of staggered protruding point pattern 错口型急滩
rapids of staggered protruding point type 错口型急滩
rapids of submerged ridge pattern 潜埂型急滩
rapid solidification 快速固化
rapid speech transmission index 快速语言传输指数
rapid speed drilling technology 快速钻井技术
rapid splitting 急速分岔

rapid spoil removal 快速除渣法
rapids regulation 治滩;理滩;急滩整治
rapid start fluorescent lamp 快速启动日光灯;快速启动荧光灯
rapid start lamp 快速启动灯
rapid state 急态
rapid steam ager 熟化器
rapid steamer 蒸发机
rapid steel 高速切削钢;高速钢;风钢
rapid steel tool 高速钢工具
rapids throat 滩口
rapid stirring 快速搅拌
rapid stock-removal 快速切削
rapids tongue 滩舌
rapid stream 湍急河流
rapid street railway 快速城市铁路
rapid stripping concrete 快速脱模混凝土
rapid survey 快速测量
rapids warping 绞滩
rapids warping barge 绞滩驳船
rapids warping by machine 机械绞滩
rapids warping by manpower 人力绞滩
rapids warping by water power 水力绞滩
rapids warping station 绞滩站
rapids warping winch 绞滩机
rapid swing hammer 快摆桩锤
rapid switching movement 快速开关运动
rapid symmetrical lens 快速对称透镜
rapids zone 急流区
rapid test 迅速测定;快速试验;速测
rapid tester 快速测定器
rapid test for soil 土壤速测法;迅速测定土壤法
rapid test method 快速测定法
rapid the pattern 起模
rapid tool 高速工具
rapid tool steel 高速工具钢
rapid traffic 快速交通
rapid trailing 高速拖行(道路上)
rapid transit 高速通行;高速交通;快速运输;快速移动;快速通行;快速交通;捷运系统
rapid transit bridge 轻轨交通桥
rapid transit car 高速通行车
rapid transit fall 快递吊货索
rapid transit line 高速交通线;快速交通线路
rapid transit network 高速通行交通网;快速交通网
rapid transit railway 高速铁路;快速铁路;城市高速铁路
rapid transit self-propelled car 高速通行自行式车辆
rapid transit subway 高速通行地下铁道
rapid transit system 高速交通系统;快速(交通)运输系统;快速交通系统;捷运系统
rapid transit tunnel 高速运输隧道;快速通行隧道
rapid transmission data system 高速传输数据系统
rapid traverse 快移;快速横动
rapid traverse drive 快速横动
rapid traverse gears (横刀架和工作台的)快速行程齿轮传动
rapid traverse piston 快速行程活塞
rapid trough 急流槽
rapid tunnel driving 隧道快速掘进
rapid tunnel(l)ing 快速挖掘隧洞;隧道快速施工;隧洞快速开挖
rapid vehicle lane 高速车道;快车道
rapid water quality monitor 水质快速检测仪
rapid wear 易损
rapid wearing 迅速磨损
Rapier 拉皮尔(一种混凝土搅拌机)
rapier clamp machine 夹紧器
rapier silage rake 杆条式青储[贮]料集运器
raplot 等点绘图法
rappage 起模胀砂
rapper 门环;敲顶锤;松模工具
rapping 起横棒;松模;松动模型
rapping allowance 铸型尺寸增量;起模松动量
rapping bar 敲模棒;起模棒;拍杆;松模棒
rapping device 振动装置
rapping equipment 振打装置(清灰)
rapping gear 振动装置
rapping hole 敲模孔
rapping intensity 振打强度
rapping iron 敲模棒;起模棒;拍杆;松模杆
rapping pin 松模棒
rapping plate 敲模垫板;起模板

rapping puff 振打冒灰
rapping system 振打系统
rapport 协调关系;相关
Rap-rig 利普里格脚手架(商品名);速立脚手架(商品名)
rapture 断裂
rare 稀有的;稀疏的;稀少的;名贵的
rare air 稀薄空气
rare alkaline earth metal 稀有碱土金属
rare alkaline metals 稀碱金属元素
rare animal 稀有动物
rare base 稀有碱基
rare book library 稀有图书馆;珍本图书馆;善本图书馆
rare book storeroom 珍本书库
rare commodity worth boarding to corner the market 奇货可居
rare concrete mixture 贫混凝土混合料
rare degree figure 稀有性程度图
rare-dependent 与速率有关的
rare earth 稀土
rare earth addition 稀土添加剂
rare earth alloy 稀土合金
rare earth betafite 稀土贝塔石
rare earth catalysis-H_2O_2 oxidation process 稀土催化双氧水氧化工艺
rare earth chelate 螯合稀土金属
rare earth chloride 氯化稀土
rare earth cobalt magnet 稀土钴磁钢
rare earth commodities 稀有金属矿产;稀土金属矿产
rare earth complex 稀土络合物
rare earth compound 稀土金属化合物
rare earth compounds 稀土化合物类
rare earth concentrate decomposition wastewater 稀土精矿分解废水
rare earth dopant 稀土掺杂剂
rare earth doped calcium fluoride 掺稀土氟化钙
rare earth doped crystal 掺稀土晶体
rare earth doped fiber 掺稀土光纤
rare earth doped optic(al) fiber 稀土掺杂光纤
rare earth drier 稀土(金属)催干剂
rare earth element 稀土元素
rare earth element analysis 稀土元素分析
rare earth fluoride 氟化稀土
rare earth garnet 稀土石榴石
rare earth glass 稀土玻璃
rare earth ion 稀土离子
rare earth ion doped nonlinear crystal 掺稀土离子非线性晶体
rare earth iron garnet 稀土铁石榴子石
rare earth luminescent material 稀土发光材料
rare earth magnet 稀土磁体
rare earth metal 稀土元素;稀土金属
rare earth mineral 稀土(金属)矿物
rare earth miserite 稀土钠硬硅钙石
rare earth optic(al) glass 稀土光学玻璃
rare earth ore 稀土矿(石)
rare earth orthoferrite 稀土正铁氧体
rare earth oxidation 稀土金属氧化物
rare earth oxide 稀土氧化物
rare earth oxide-coated cathode 稀土金属氧化物阴极
rare earth oxide phosphor 稀土氧化物荧光粉
rare earth phosphor 稀土类荧光粉
rare earth removal system 稀土除去系统
rare earths 稀族
rare earth salts 稀土盐
rare element 稀有元素
rare element analysis 稀有元素分析
rare element chemistry 稀有元素化学
rarefaction 稀疏作用;稀疏区;稀疏化;稀薄;疏松;纯净
rarefaction quadrant 稀疏像限
rarefaction state 稀疏状态
rarefaction wave 稀疏波;稀薄波;膨胀波
rarefied air 稀薄空气
rarefied atmosphere 稀薄大气
rarefied gas 稀疏气体;稀薄气体
rarefied gas dynamics 稀薄气体动力学
rarefied mixture 稀疏混合物
rarefied mudflow 稀化泥
rarefied zone of wave 波疏区
rare flood 稀遇洪水

rare gas 稀有气体;稀薄气体
rare gas clathrate 稀有气体笼形包合物
rare gas clathrate compound 稀有气体笼形物
rare gas method 稀有气体法
rare gas plasma 稀薄气体等离子体
rare gas polymerization method 稀有气体聚合法
rare gas removal pilot plant 分离稀有气体中间工厂
rare issue 稀有本
rare light metal 稀有轻金属
rarely occurred earthquake 罕遇地震
rare materials 稀有文献资料
rare metal 稀有金属;稀土金属
rare metal couple 稀有金属温差电偶
rare metal ore 稀有金属矿石
rare metals industry 稀有金属工业
rare-metal thermocouple 稀有金属温差电偶
rare mineral 稀有矿物
rare minute echoes 稀疏微波型
rare mixture 稀有混合物;贫混凝土拌和物;贫混合料
rare nobble metal 稀有贵金属
rare oxide 稀有氧化物
rare refractory metal 难熔稀有金属
rare reinforcement concrete 少筋混凝土
rare species 稀有(物)种;稀见品种;珍稀物种
rare stable isotope 稀有稳定同位素
rare storm 罕有暴雨
rare taxa 稀有分类群
rare variety 稀有品种
rare veneer 稀有花饰贴面
rarity 稀有性;稀薄
Raschig ring 拉歇环;填充圈
Raschig-ring cooler 拉歇环冷却器
Raschig tubes 填充圈
rascle 灰岩参差蚀面[地]
rasher 薄片
rashings 煤层底板的软碳质页岩;薄软炭质页岩
rashly launching new projects 盲目上项目
rashly putting up establishments 铺摊子
rasidual pearlite 残余珠光体
rasorite 斜方硼砂;斜方锰矿;四水硼砂
rasp 木锉(刀);粗木锉;粗锉(刀);粗处理
rasp bar drum 纹杆式脱粒滚筒
raspador 剥麻机
rasp away 锉掉
rasp bar 纹杆
rasp bar cylinder 纹杆式滚筒
rasp cut 粗锉纹
rasp-cut file 粗锉;木锉(刀)
rasper 锉刀;锉床;粗锉
rasp file 粗锉
rasping 切边
rasping file 木锉(刀)
rasping machine 磨光机
raspings 锉屑
rasping sucking mouthparts 锉吸口器
raspite 斜钨铅矿
rasp mill 湿磨
rastellus 小耙;耙齿
raster 窄平行光栅;光栅;屏面
raster algorithm 光栅算法
raster aperture 光栅孔径
raster area 光栅区域;光栅面积
raster burn 光栅烧伤;光栅破坏
raster count 光栅数(目)
raster data 栅格数据;光栅数据
raster density 光栅密度
raster dial 光栅度盘
raster digitizer 光栅扫描数字化器
raster display 光栅显示(器)
raster display device 光栅显示设备
raster element 光栅元素
raster equation 光栅方程
raster excitation 图案激励
raster flyback time 扫描逆程时间
raster font 光栅字型
raster form 光栅形式
raster formatted picture 光栅扫描图
raster generator 光栅发生器
raster glass measuring device 光栅玻璃测量装置
raster graphic 光栅图形
raster graphic extension 光栅图形扩充
raster graphics 光栅图形学
raster grid 光栅网络;光栅格

raster image 光栅图像
rasterization 光栅化
rasterization scheme 光栅化图案
raster laser plotter 光栅激光绘图机
raster line 光栅线
raster matrix 光栅矩阵
raster microscope 光栅显微镜
raster mode plot 光栅方式绘图
raster operation method 光栅操作法
raster pattern 光栅图样;光栅图
raster plotted cartographic(al) data 光栅扫描绘图数据
raster plotter 光栅绘图仪;光栅图形显示器
raster plotting 栅格绘图
raster point 光栅点
raster principle 光栅原理
raster scan computer graphics 光栅扫描计算机图形
raster scan display 光栅扫描显示
raster scan display system 光栅扫描显示系统
raster scanned image 光栅扫描图像
raster scanner 光栅扫描机
raster scan(ning) 光栅扫描
raster scanning model 光栅扫描方式
raster scanning plotter 光栅扫描绘图机
raster scanning program(me) 光栅扫描程序
raster scanning recorder 光栅扫描记录
raster-scanning system 光栅扫描系统
raster-scan plotter 光栅扫描绘图仪
raster-scan technique 光栅扫描技术
raster segment generator 光栅段发生器
raster shape 光栅形状;光栅类型
raster timing 光栅定时
raster to vector 光栅到矢量
raster unit 栅距;光栅单元
Rastrites 耙笔石属
Rast's method 拉斯特法
rasura 碎片,锉屑
rasvumite 硫铁钾矿
Rata 拉塔暗红色装饰硬木(产于新西兰)
ratable 应征税的
ratable charge method 比例计费法
ratable distribution 比例分配
ratable estate 应纳税的房地产
ratable property 应纳税财产
ratable value 课税价值
ratal 纳税额
ratan 雷达和电视导航设备
ratative diffuser 旋转扩散体;旋转扩散器
rat barrier 防鼠栅栏
rat cellar 地下室饮食店
ratch 制轮;齿杆
ratchel 大石块;毛石;砾石
ratcher brace 扳钻
ratchet 棘齿;单向齿轮
ratchet adjustment clutch 棘轮调整式离合器
ratchet-and-pawl 棘轮和掣爪;棘轮机构
ratchet-and-pawl drive 棘轮推爪传动装置
ratchet arrangement 棘轮装置
ratchet bar 棘齿条
ratchet bit 棘轮摇钻
ratchet bit brace 手摇钻机;弓摇钻;棘轮摇钻;扳钻
ratchet blocking system 棘轮闭锁装置
ratchet box 棘轮油箱(凿岩机)
ratchet brace 棘轮(手)摇钻;手摇旋钻机;手摇旋转机;板钻
ratchet brace bit 棘轮摇钻钻头
ratchet casing 棘轮罩
ratchet casing bushing 棘轮罩套
ratchet chain 棘轮链条
ratchet clutch 棘爪离合器
ratchet coupling 爪形联轴器;爪形联轴节;爪形接器;爪形离合器;爪式联轴器;爪式联轴节;爪式联接器;爪式离合器;闸轮联轴节;棘轮联轴节;逆制套
ratchet cover plate 棘轮盘
ratchet cyclometer 带有棘轮的转数表
ratchet cylinder 棘轮滚筒
ratchet dial feed 棘轮式转盘送料
ratchet diestock 快割咬口钳
ratchet drill 棘轮凿岩机;棘轮摇钻;棘轮手钻;棘轮式手摇钻;手摇钻机;手摇旋转机;手扳钻;扳钻
ratchet drive 棘轮驱动;棘轮传动

ratchet driver 棘轮传动装置
ratchet feed 棘轮给进;棘轮进给
ratchet feeder 棘轮给料机
ratchet feed mechanism 棘轮进给机构
ratchet gear(ing) 棘轮(传动)装置,棘轮机构,棘轮齿轮
ratchet gear wheel 棘齿轮
ratchet handle 棘轮手柄
ratchet handle spanner 棘轮单头扳手
ratchet head 棘轮头(风钻)
ratchet hob 棘轮滚刀
ratchet hoist 棘轮起重葫芦;环链手拉葫芦;环链手扳葫芦;扳链手扳葫芦
ratcheting device 棘轮装置
ratcheting effect 棘轮效应
ratchet jack 棘轮千斤顶;棘轮起重器
ratchet lever 棘轮柄
ratchet lever jack 轮杠千斤顶;棘轮杠杆千斤顶;蜗轮蜗杆千斤顶;涡轮涡杆千斤顶
ratchet lift 棘轮式起落机构
ratchet magnetic pulse oscillator 磁性脉冲振荡器
ratchet nail 倒刺钉
ratchet nut 棘轮螺母
ratchet pawl 止回棘爪;棘爪;棘轮挚子;棘轮卡子;棘轮掣子
ratchet pin 棘轮销
ratchet punch 棘轮冲头
ratchet purchase 棘齿起重设备
ratchet rate 锯齿价格
ratchet relay 棘轮式继电器
ratchet reversing lever 棘轮回动装置
ratchet screwdriver 棘轮螺丝起子;棘轮公锥;棘轮改锥;棘轮旋凿
ratchet screwdriver with wood(en) handle 木柄棘轮式螺丝起子
ratchet screw jack 棘轮螺旋千斤顶;棘轮螺旋起重器
ratchet screwstock 快割螺丝板
ratchet sleeve 棘轮套
ratchet spanner 棘轮扳手
ratchet spring 棘轮弹簧
ratchet stock 棘轮架
ratchet stop 棘轮式挡料装置
ratchet threader 棘轮螺纹铣床;棘轮螺纹磨床
ratchet time-base 滞后时基;棘轮时基
ratchetting 松脱振动
ratchet to eliminate running back 防止倒转的棘轮
ratchet tool 棘轮工具
ratchet tooth 棘轮齿
ratchet-tooth lever escapement 尖齿式擒纵机构
ratchet type grease cup 棘轮式注油杯
ratchet type jack 棘轮式千斤顶;伞齿轮式起重器
ratchet type packer 棘轮封隔器
ratchet type roll feed 棘轮式滚轴送料
ratchet washer 棘齿垫圈
ratchet wheel 爪轮;闸轮;棘轮;擎逆轮
ratchet wheel pawl bracket 棘轮爪托
ratchet wheel pawl pin 棘轮爪销
ratchet wheel pawl weight 棘轮掣子锤
ratchet winch 棘轮绞车;棘轮传动绞车
ratchet winding wheel 上条大钢轮
ratchet winding wheel spring 上条大钢轮簧
ratchet winding wheel swing lever 上条大钢轮摆杆
ratchet winding wheel swing lever ring 上条大钢轮摆杆衬圈
ratchet winding wheel swing lever spring 上条大钢轮摆杆簧
ratchet wrench 棘轮钳;棘爪扳手;棘轮扳手;套筒棘轮扳手
ratchet wrench extension 手摇螺丝扳手接长节
ratching 作棘齿
rat destruction 灭鼠
rate 价格;速率;等级;费(率);比率
rateability 纳税义务
rateable 应征税的;可估价的
rat(e)able property 抽税产业
rateable value 征税价值;纳税的价值;财产应课税价值
rate accelerating material 增速剂;速率加速剂
rate accuracy 速率准确度;测速准确度;测速精度
rate act 比率作用
rate action 速率作用
rate action control 优先率调整;微分控制
rate action controller 速率动作控制器

rate actually paid 实际支付率
rate adjustment 时差调整
rate agreement 运费协定
rate analysis 价格分析;价格分解;比率分析
rate animal 珍稀动物
rate arrangement 费率协定
Rateau formula 拉多公式
Rateau turbine 压力外级冲动式汽轮机;拉托涡轮机
rate base 费用基数;价格基数;费率基础
ratebuster 抢定额者
rate by volume 容积比(率)
rate card 价目牌
rate center 测量中心
rate change 走时变化
rate checking 走时检验
rate-command system 速率指令系统
rate constant 速率常数;速度常数;反应速度常数
rate construction unit 费率结构单位
rate control 速率控制
rate controller 流率控制器;速率控制器
rate controller action 速率控制器作用
rate-controlling 速率控制;速度控制
rate-controlling step 速率限制阶段
rate control pattern 速率控制型
rate control to filter 滤池滤速控制
rate conversion 兑换率
rate correction 钟速校正
rate counter 计数率计数管
rate curve 速率曲线;速度曲线
rate cutting 削减工资;降低费率
rated 公称;(额定的);标称的
rated accuracy 额定准确度;额定精(确)度
rated age 标准年龄(保险)
rated air pressure 额定风压
rate damping 速率阻尼
rated apparent power 额定视在功率
rated boost 额定升压
rated boost pressure 额定吸入压力
rated (bore)hole diameter 设计钻孔直径
rated breaking capacity 额定遮断功率
rated breaking current 额定馈式电流;额定开断电流;额定断路电流
rated burden 额定装载量;额定负载
rated bursting pressure 额定破碎压力
rated busbar current 额定母线电流
rated capability 额定出力
rated capacity 设计能力;额定生产率;额定(容)量;额定起重量;额定(排放)能力;额定功率;额定电容
rated capacity of boiler 锅炉额定蒸发量;锅炉额定容量
rated carrying current 额定荷载电流
rated cell (or room) capacity 额定房间居住人数
rated chamber condition 燃烧室计算条件
rated chamber pressure 燃烧室额定压力
rated compression deflection 设计压缩变形
rated compression ratio 额定压缩比
rated condition 规定条件;计算条件;额定状态;额定条件;额定工况
rated consumed power 额定需要功率
rated consumption 定额消耗;额定消耗(量)
rated continuous running time 额定连续运转时间
rated continuous working voltage 额定连续工作电压
rated current 额定电流
rated current of generator 发电机额定电流
rated demand 额定需量
rated differential head 额定压差
rated discharge 额定流量
rated discharge pressure 额定排气压力
rated dissipation 额定耗散
rated duration 额定状态工作时间
rated duty 额定装载量;额定工作方式;额定负载
rated effective head 额定有效水头
rated efficiency 额定效率
rated energy 额定能量
rated engine speed 发动机额定转速
rate-dependent character of corrosion 腐蚀速度特性
rate-determining step 速率控制步骤;律速阶段;控速阶段;速率决定阶段;定速步骤
rate-determining step of reaction 反应速率决定步骤
rated evapo(u)ration capacity 额定蒸发量

rated excitation 额定励磁
rated fatigue limit 额定疲劳极限
rated filling weight 容许充装量
rated flow 设计流量;额定流量
rated frequency 输出频率;额定频率
rated frequency range 额定频率范围
rated head 额定扬程;额定水头
rated heat input 额定热负荷;额定热输入;额定热荷载
rated horse-power 额定马力
rate diffusion 扩散速率
rated impedance 额定阻抗
rated incinerator capacity 焚化炉额定容量
rated input 额定输入量;额定输入
rated input power 额定输入功率
rated input voltage 额定输入电压
rated insulation voltage 额定绝缘电压
rate distortion 速率失真
rate-distortion coding 速率失真编码
rated kiln output 窑的额定产量
rated kilovolt-ampere 额定千伏安
rated lamp life 额定的灯泡寿命
rated life of lamp 灯泡额定寿命
rated life time 额定寿命
rated lightning impulse withstand voltage of line to earth 额定雷电式脉冲反压一线对地
rated load-bearing coefficient for bridge 桥梁检定承载系数
rated load current 额定负荷电流
rated load(ing) 额定荷载;额定负荷;额定装载量;额定荷载;标准荷载
rated loading capacity 额定装载能力;额定载重量
rated load(ing) operation 额定负荷运作;额定负荷运行
rated load marking 额定荷载标志
rated load test 标准荷载试验
rated measure 额定尺度
rated nozzle characteristic 额定喷管特性
rated octane number 实测辛烷值
rated operating specification 额定工作规程
rated operational voltage 额定工作电压
rated output 额定输出(功率);额定生产率;额定出力;额定产量;标准产量
rated output power 额定输出功率
rated partition 耐火等级隔墙
rated passenger capacity 额定载客量
rated payload 额定有效荷载
rated people on the train 列车定员
rated performance 额定性能
rated pipeline capacity 管线额定输送能力
rated point 额定点
rated power 计算功率;设计功率;额定功率;额定动力;不失真功率
rated power frequency withstand 额定工频耐受电压
rated pressure 名义吨位;额定压力
rated primary current 一次测额定电流
rated primary voltage 一次测额定电压
rated productivity 额定生产率
rated quantity 额定量
rated resolving power 额定分辨能力
rated revolution 额定转速;额定转数
rated rimpull 额定轮牵引力
rated room capacity 额定房间居住人数
rated secondary voltage 二次侧额定电压
rated short circuit breaking current 额定短路开断电流
rated size 额定尺寸
rated slip 额定转差率;额定滑失;额定滑差
rated specific speed 额定比速
rated speed 额定转速;额定速率;额定速度
rated speed of pump 泵的额定转速
rated speed of rotation 额定转速
rated spring capacity 弹簧额定负载量
rated steam 额定耗汽量
rated suction pressure 额定吸入压力
rated tax 定率税;额定税率
rated temperature-rise 容许温升
rated temperature-rise current 容许温升电流
rated thrust 标准推力
rated time 额定时间
rated tonnage 额定吨位;额定吨数
rated torque 额定转矩;额定扭矩
rated tractive effort 测定牵引力

rated tractive force 计算牵引力;额定牵引力
rated turbine flow 额定透平流量;额定涡轮机流量
rated value 额定(价)值
rated velocity 额定速度
rated voltage 额定电压;标称电压;设计电压
rated wall 耐火等级墙
rated weight 额定重(量)
rated wind pressure 计算风压
rated working condition 额定工作条件
rated working pressure 额定工作压力
rate effect 速率效应;变速效应
rate elongation 延长率
rate equation 速率公式;速率方程
rate error 速率误差
rate established 所规定的比价
rate feedback 速率反馈;速度反馈
rate feedback system 速度反馈系统
rate fixer 速度调节器;定额员
rate fixing 规定定额
rate follow-up unit 速率随动部件
rate generator 速率发生器;比率发电机
rate governor 流量调节器;调速器
rate growing method 变速生长法
rate-grown transistor 变速生长晶体管
rate growth junction 变速生长法
rate gyro(scope) 速度陀螺仪;速率陀螺(仪);阻尼陀螺仪;微分陀螺仪;二自由度陀螺(仪)
rate if melting recovery 冶炼回收率
rate impact study 价格影响研究
rate increase 速率增长
rate index 利率表;频率指标
rate index of intermediate input in railway transport 铁路运输中间的投入价格指数
rate indicator 速度计;速率指示器;速度指示器
rate injury accident of hitting from flying stone 飞石打伤
rate instrumentation 速率测量装备
rate integrating gyro(scope) 速率积分陀螺
rate law method 时率法
rate limitation 速率限制;速度限制
rate-limiting factor 限速因子;限速因素
rate-limiting process 速率限制过程
rate-limiting reaction 限速反应
rate-limiting step 限制步骤;限速步(骤)
rate-limiting substance 限制速率基质
rate making 制定运价;制定费率;确定保险费率
rate marking 制定价格
Ratematic 自动校正快慢仪
rate matrix 比率矩阵
rate measuring instrument 走时测定器
ratemeter 计数率计;计数测量计;强度测量仪;速率计;测速计
rate multiplier 比率乘数;比例乘数;比例乘法器
rate of ablation 消融率
rate of abrasion 磨蚀率
rate of absorption 吸附速率;吸收速率
rate of accumulation 蓄积率;增长率(冰川);累积率;积累率;积聚率;堆积率
rate of activation 激活速度
rate of actual use of available equipment 现有设备实际使用率
rate of adsorption 吸收率;吸附速率
rate of adsorption grout 吸浆率
rate of advance 钻进速度;掘进速度(隧道);前进速度
rate of advancement 前移速度
rate of aeration 充气程度
rate of after culture 补种量
rate of air circulation 换气次数
rate of allowable aquifer dewatering 含水层年允许疏干率
rate of angular motion 转动速率
rate of annual depletion 年耗水率;年耗亏减率;年耗尽率
rate of apparent polar-wandering 视极移速率
rate of approach 逼近速度
rate of arrival 到达(车)率;到车率
rate of aspiration 提吸速率
rate of assembly 装配速率
rate of assessment 分摊比率
rate of astronomic(al) clock 天文钟误差率
rate of attendance 出勤率
rate of attendance in loading/discharging 装卸出勤率

rate of back silting 回淤率
rate of bed load discharge 推移质输沙率
rate of bed load transport 底质输送搬运率;底砂搬运率;底沙输移率
rate of bed load transportation 河床质运送速率
rate of best quality engineering 优质工程率
rate of bicycle riding 骑车率
rate of bio-degradation 生物降解(速)率;生物降解速度
rate of bituminous application 洒油率
rate of body weight gain 增重率
rate of both profits and taxes on entire funds 全部资金利税率
rate of brain retention 人才存留率
rate of braking 制动强烈程度
rate of break 离析速度
rate of breakage 破碎率
rate of broken stowage space 亏舱系数
rate of burning 燃烧速率
rate-of-burning test 燃烧速度试验
rate of call 通知利率;通知存款利率
rate of camber change 前轮外倾角变化率
rate of capacity (设备的)容量利用率
rate of capacity utilization 开工率
rate of capillary rise 毛细管上升(速)率;毛管水上升率
rate of capital accumulation 资本积累率
rate of capital net profit 资金利润率
rate of capital return 资本利润率
rate of capital turnover 资本周转率;投资周转率
rate of change 变化(速)率;变动率
rate of change circuit 一次导数调节电路
rate of change curve chart 变(化)率曲线图
rate of change in azimuth 方位变化率
rate of change indicator 速率变化指示器;变率指示器
rate of change map 变率图
rate of change of acceleration 加速度的时间变率
rate of change of capacitance 电容量变化率
rate of change of phase 相位变化率
rate of change relay 变化率继电器;微分继电器;变率继电器
rate of change sensor 瞬态变量感传器
rate of charge 充电时间;充电率
rate of charging 装料速率
rate of chronometer 日差率;表速
rate of circulating flow 循环流速
rate of circulation 循环速度
rate of climate change 气候变化速度;气候变化率
rate of climb 爬高速度;上升速度
rate-of-climb control 上升率控制
rate-of-climb indicator 爬升速率指示器
rate-of-climb meter 爬升速度表
rate-of-climb sensor 上升率传感器
rate of clock 钟速
rate of closure 逼近速度
rate of coagulation 凝固速度
rate of coal consumption 煤耗率
rate of coal contaning waste rock 煤炭含矸石率
rate of coal liquefaction 煤液化速率
rate of coast(al) erosion 海岸侵蚀速率
rate of collapse 倒塌率
rate of colo(u)r removal 色度去除率
rate of combined package 拼件率
rate of combustion 燃烧速度;燃烧(速)率
rate of completion of on-line construction projects 建设项目投产率
rate of comprehensive utilization of equipment 设备综合利用率
rate of compression 压缩率
rate of computation 计算速率
rate of concrete placement 混凝土浇筑速度;混凝土浇筑速率
rate of conductivity for sheet 板体电导比
rate of consolidation 固结率
rate of construction 施工速度
rate of consumption 消耗率;耗水率;耗电率
rate of contraction 收缩率
rate of convergence 收敛速率;收敛速度
rate of conversion 换算率;转化率
rate of conveyance loss 输水损失率
rate of cooling 冷却速率
rate of correction 校正率
rate of corrosion 腐蚀速率;腐蚀速度;腐蚀率

rate of cost net profit 成本利润率
rate of cost reduction 成本降低率
rate of cost reduction to total assets 固定资产总折价率
rate of crack growth 裂纹增长率
rate of cracking 裂化速率
rate of crack propagation 裂隙扩展速度;裂缝扩展速度
rate of creep 蠕变速度;徐变速度;徐变(速)率;蠕动速率;蠕变(速)率;爬行速度
rate of cross linking 交联速率
rate of crystallization 结晶速率
rate of cure 养护速率;凝固速率;硬化速率
rate of curing 养护速率
rate of current 流速
rate of curves 曲线斜率
rate of curving 弯曲率
rate of customs duty 海关税率
rate of cutting 切削速度
rate of cutting and sizing 切裁率
rate of damaged cargo 货损率
rate of damping 阻尼率
rate of data signal(l)ing 数据信号传输速率;数据传送率
rate of death 死亡(速)率
rate of debarkation 卸载速度
rate of decay 破碎程度(岩石);损坏速度;衰减速率;衰变(速)率;腐烂(速)率;风化率;分解率
rate of decay attenuation 衰减率
rate of deceleration 减速率
rate of decline 下降率
rate of decomposition 分解速率
rate of decosition 熔敷速度
rate of decrease 递降率
rate of defect 缺陷率
rate of deflection 挠曲速率
rate of deforestation 森林砍伐率
rate of deformation 形变(速)率;变形速率
rate of degradation 降解速率;河底刷深速率
rate of delayed report 迟报率
rate of delay shift 前移延迟系数
rate of delivery 交付率(货物);输送率
rate of departure 离开速度
rate of depletion allowance 资源耗竭补偿税率
rate of deposit 沉积率
rate of deposition 焊着率;沉积速率;沉积率;沉淀率
rate of depreciation 折旧率;贬值率
rate of depreciation charges 折旧费率
rate of descend 下降率
rate of descent 下降(速)率(飞机);垂直下降速度
rate-of-descent control 下降率控制
rate-of-descent hold 下降率保持
rate-of-descent indicator 下降率指示器;爬升率指示器
rate-of-descent sensor 下降率传感器
rate of despatch money 速遣费率
rate of detonation 爆炸速率;爆轰速率
rate of development 显影速度;开发速度;发展速度;发育速度
rate of development of strength 强度增长率
rate of deviation change 偏差率
rate of diffusion 扩散速率
rate of digestion 消化率
rate of digging for exploiting and lengthening tunnel and shaft 可掘进率
rate of dilution 淡化率;稀释率
rate of dirty removal dy washing 冲洗去污率
rate of discharge 驶出率;流量率;倾卸速度;排出(速)率;出料速率;出口流量;放电率
rate of discharge of ground water 地下水流率;地下水流出量
rate of discharging 卸货效率;卸货速率;卸货速度
rate of discount 折扣率;贴现(利)率
rate of displacement 位移速率
rate of dissipation 消失率
rate of dissociation 分解率
rate of distillation 蒸馏速率
rate of dividend 股利率;分红率
rate of doing work 作功率
rate of draft 吸出率;引出率;开采率(地下水);取样率;取水量;吃水率(船舶)
rate of drainage 排水(速)率
rate of drift method 漂移率测量法
rate of drilling 钻进速度

rate of driving 冶炼强度;打桩速率
rate of drop 下落速度
rate of drying 干燥速率
rate of drying curve 干燥速率曲线
rate of duty 税率
rate of earnings on total capital employed 总资本利益率;使用总资本收益率
rate of earthquake damage 震害率
rate of earthquake occurrence 地震发生率
rate of easement curvature 缓和曲线变更率
rate of economizing gasoline 节油率
rate of economizing petrol 节油率
rate of elastic wave propagation 弹性波传播速度
rate of electricity loss from transmission line 线路损失率
rate of elongation 延伸率;伸长率
rate of energy consumption 能量消耗率
rate of energy dissipation 消能率;耗损率
rate of energy expenditures 能量消耗率
rate of energy flow 能流速度;能流率
rate of energy loss 能量损失率;能量损耗率
rate of energy transformation 能量转换率
rate of enrichment 选矿系数
rate of erosion 侵蚀速率;冲刷速率
rate of evapo(u)ration 蒸发(速)率;蒸发速度;汽化速率
rate of exceed the flight elevation 超高率
rate of excess air 过剩空气率
rate of exchange 汇率;汇兑牌价;汇兑率;汇兑价;汇兑换算表;兑换率
rate of exhaust 抽气率
rate of expansion 扩展率;膨胀率
rate of expenditures 消耗定额
rate of explosion 压力升高比
rate of extension 拉伸速率
rate of external work 外功率
rate of extinction 绝灭率
rate of extraction 提取率
rate of extrusion 挤出速率
rate of face advance 工作面开采速率;工作面推进速率
rate-of-failure curve 破坏率曲线
rate of failures 失效率
rate of fall 降落率;沉降速率;沉淀速度
rate of feathering 出羽速度
rate of feed 供料速度;进料(速)率;喂料速度;送丝速度;送料速率;上水率
rate of feeding 投料速率
rate of fiber flying 飞丝率
rate of filtration 过滤速率;滤速
rate of finished products 成品率
rate of firing travel 火行速率
rate of flame propagation 火焰传播速度
rate of flattening of flood wave 洪水波坦化率;洪水波平化率;洪水波(的)平复率
rate of floor space completed 房屋建筑面积竣工率
rate of flow 移动率;给水定额流量;流速;流率;流量(率);流动速率;交通流率;水流速率;车流率
rate-of-flow controller 流速控制装置;流率控制器;流量控制器
rate-of-flow control valve 流速控制阀;流量控制阀
rate-of-flow indicator 流量表;流速指示器;流量指示速率
rate-of-flow meter 流量计;流量表
rate-of-flow of fluid 流体流速
rate-of-flow of sewage 污水流速
rate-of-flow recorder 流量自记仪;流率记录器;流量记录仪;流量记录器
rate-of-flow-stress curve 流速应力曲线
rate of following 跟踪速度
rate of foreign exchange 外汇兑换率
rate of formwork turnover 模板周转速度;模板周转率
rate of forward shift 前移速度
rate of foundation settlement 基础沉陷速率;基础沉降速率
rate of freight 运费率
rate of freight (passenger) traffic by water 水运比重
rate of frequency drift 频率漂移速率
rate of frost heave 冻胀速率
rate of fuel change 燃料变化率
rate of fuel consumption 燃料消耗率
rate of fuel flow indicator 燃料耗量表

rate of full package 满筒率
rate of funds used by project under construction 未完工程资金占用率
rate of gasification 气化效率
rate of gas transfer 气体传输(速)率
rate of good-roads 好路率
rate of grade 斜率;坡度变化率
rate of gross investment 总投资率
rate of gross profit 毛利率
rate of ground settlement 地面沉陷速率
rate of groundwater discharge 地下水流量率
rate of groundwater flow 地下水流量率;地下水流量
rate of grout-acceptance 水泥吸收速度;吃浆率;吃浆量
rate of growth 增长率;生长(速)率;生长速度
rate of growth of benefits 年收益增长率
rate of hardening 硬化速率;硬化速度;固化率
rate of harmonic 谐波含有率
rate of head loss 水头损失速率
rate of head movement 压头移动速率;施加外力速度
rate of heat addition 加热速率
rate of heat development 热增长速度
rate of heat dispation 散热速率
rate of heat exchange 热交换率
rate of heat flow 热流率
rate of heat generation 热量生成速率
rate of heating 加热速度
rate of heat liberation 放热速率
rate of heat loss 热损失率;热损耗率;热耗
rate of heat production 热量生成速率
rate of heat release 热释放率;散热速率;放热速率
rate of heat transfer 热传递速率;传热速率
rate of heave 隆起速率
rate of hindered settling 受阻沉降速率
rate of horizontal shear 水平剪切速率
rate of husked rise 出糖率
rate of hydration 水化(作用)速率
rate of incidence 入射率
rate of income from investment 投资收益率
rate of income from the funds 资金收入率
rate of increase 增长率
rate of increase of angle 角度增长率(定向钻进时)
rate of increasing reserves accomplishment 增力储量完成率
rate of increasing returns 报酬递增率
rate of increment 增长率
rate of infiltration 入渗(速)率;入流率
rate of inflation 通货膨胀年率;通货膨胀率
rate of inflow 流入速度
rate of information throughput 信息周转率;信息吞吐率;信息吞吐量;信息传送速率;信息传送速度
rate of information transmission 信息传送率;信息传输速率
rate of inherent regulation 自调节速度
rate of injection 注入速度;注入率
rate of ink penetration 油墨渗透率
rate of interest 利率
rate of interest per annual 年利率
rate of internal energy combustion dissipation 内能消耗率
rate of internal return 内部回收率
rate of inventory turnover 库存周转率;存货周转率
rate of inventory usage 库存耗用率
rate of investment 简单收益率;投资率
rate of irrigation 灌水率;灌溉用水定额
rate of joint fill 接头填充率
rate-of-keying error 键错率
rate of land reduction 用地递减率
rate of laying 铺管速度
rate of leakage 漏水率
rate of level rise 水位上涨率
rate of liquid aspiration 吸液速度
rate of littoral transport 沿岸漂沙率;沿岸泥沙输移率;漂沙率
rate of live weight growth 增重率
rate of load growth 荷载增长率
rate of loading 装载速度;装货速度;荷载速率;加载速率;加荷速率;载重率
rate of loading and discharge 装卸效率
rate of loading and discharging 装卸(速)率
rate of locomotive under repair 机车检修率

rate of locomotive under shed repair 断修率
rate of loss 损耗率
rate-of-loss of coolant 失水速率
rate of lost circulation 漏失速度
rate of lubricant flow 润滑剂流速
rate of lubrication oil consumption 润滑油消耗率
rate of mail transfer 信汇汇率
rate of making mud 造浆率
rate of margin 毛利率
rate of marginal profit 边际利润率
rate of mass transfer 传质速率
rate of melting 熔融速率;熔化(速)率
rate of metal consumption 金属消耗率
rate of metal removal 切屑率;出屑率
rate of migration 移动速率
rate of migration flow 人口迁移流动率
rate of mineral dissolution 无机物溶解速率
rate of mineral precipitation 无机物沉积速率
rate of missed diagnosis 漏诊率
rate of missing report 漏报率
rate of modulus variation 径流模比系数
rate of momentum transport 动量传递率;冲量传递率
rate of mortality of aquatic life 水生生物死亡率
rate of movement 移动速率
rate of mutation 突变率;变化率
rate of national savings 国民储蓄率
rate of national taxes 国家税率
rate of natural increase 自然增长率
rate of natural increase of population 人口自然增长率
rate of naturalized plants 归化植物率
rate of natural loss of commodity 商品自然损耗率
rate of natural population growth 人口自然增长率
rate of net drilling 纯钻进速度
rate of net output 净产出率
rate of net present value 净现值率
rate of nitrification 硝化速率
rate of nominal wages 名义工资率
rate of non-commodity expenditures 非商品支出比例
rate of nucleation 成核速率
rate of occupancy of available beds 可用床位占用率
rate of occurrence of errors 误差出现率
rate of oil production 采油速度
rate of opening 离开速度
rate of operating rig 钻机利用率
rate of operation 开工率
rate of ore dilution 矿石贫化率
rate of oscillation 振动速率;摆动率
rate of outflow 流出速度;流出率
rate of overtopping 漫顶率
rate of oxidation 氧化速率
rate of oxygenation 充氧速率
rate of ozone layer depletion 臭氧层的消耗速率
rate of passage 运费率(船舶)
rate of passenger fare 旅客票价;票价率
rate of passenger satisfaction 旅客满意率
rate of pay 报酬率
rate of penetration (桩的)打入速度;钻进速度;机械钻速;(泥浆的)渗漏速度;穿透速度
rate-of-penetration log 钻速记录
rate of percolation 渗吸速度;渗透速度;渗滤(速)率;滴漏滤速
rate of perforation 穿孔速率
rate of performance 施工速度
rate of phosphorus release 释磷速率
rate of photosynthesis 光合强度
rate of pickup 悬浮率
rate of placement 浇注速率(混凝土)
rate of placing (建筑混凝土的)升高速率;(混凝土每小时的)浇注高度;铺管速度
rate of plant coverage 植物覆盖率
rate of polymerization 聚合速率
rate of population 总体率
rate of population growth 人口增长率
rate of population growth form migration 人口机械增长率
rate of port dues 港口税率
rate of pour 浇注速度
rate of precession of equinox 岁差变化率
rate of precipitation 降雨强度
rate of precipitation losses 降雨损失率;降水损失率

rate of premium 升水率;保险费率
rate of pressure rise 压力升高(速)率
rate of pressurization 增压速率
rate of product 生产率
rate of production 生产率;采出速度
rate of production test 变产量测试
rate of profit 利润率
rate of progress 进展速度;进度;施工进度;发展速度
rate of progressively decrease of spring discharge 泉水流量的递减速度
rate of propagation 扩展率;开发速度
rate of pulse repetition 脉冲重复率
rate of pumping 泵送速度
rate of purity 纯度
rate of radiographic(al) examination 射线探伤率
rate of rainfall 降雨率
rate of rainfall ga(u)ge 雨量强度计
rate of rainfall losses 降雨损失率
rate of rate 加速率;速率导数
rate of reactions 反应率
rate of reading 读数速度
rate of reaeration 复氧速率
rate of real loading capacity 实载率
rate of rebound 回弹率
rate of recapture 重捕率
rate of recovery 回收率
rate of recovery of reagent 还原剂收回率
rate of redemption 偿还率
rate of redox 氧化还原率
rate of reduction 破碎率;缩微率;锻造比
rate of reinforcement 含钢率
rate of reinjection-extraction 回灌量—开采量之比
rate of release 释放速率
rate of replenishment 补给速度
rate of reproduction 再生产率
rate of residual volumetric(al) deformation 容积残余变形率
rate of retardation 减速率
rate of retention 滞洪率;拦洪率;持水率;分洪率;保水率
rate of return 利润率;回归率;投资盈余率;投资效益率;投资收入率;收益率;投资返回率
rate of return function 返回函数率
rate of return method 资金回收率评估法;回收利用率法
rate of return on common stock equity 普通股收益报酬率
rate of return on incremental investment 附加投资收益率
rate of return on invested capital 投入资本所得率
rate of return on invested capital method 投资收益率法
rate of return on investment 投资收益率;投资(收入)报酬率;投资利得率;投资回收率
rate of return on new investment 新资本所得率;新投资所得率;新投资收益报酬率
rate of return on sales 销货收益率
rate of return on the use of foreign capital 使用外资收益率
rate of return on total assets 总资产报酬率;资产总额利润报酬率
rate of returns of operating assets 经营性资产收益率
rate of returns on invested capital 投入资本利润率
rate of returns pricing 投资报酬率订价法
rate of revolution 旋转速率
rate of rework losses 返工损失率
rate of rise 增长速率;曲线上升斜率;升压速度;上涨率;上升速率
rate of rise of water 水位上涨速率
rate of rise restriking voltage 再击穿电压上升率
rate of river bank recession 河岸崩退率
rate of river bank retrogression 河岸崩退率
rate of road in green area 道路绿化率;道路绿地率
rate of roasting 焙烧速度
rate of rock absorption 岩石吸水率
rate of rock decay 岩石风化程度
rate of roll indicator 倾斜角指示器
rate of rotation 转率
rate of royalty 使用费率
rate of ruling grade 限坡坡率
rate of runoff 径流流量;径流率;径流模数
rate of safety margin 安全界限率

rate of sample 样本率
rate of sand motion 输沙率
rate of sand movement 输沙率
rate of saturated of rock test 岩石饱水率试验
rate of saturation 饱水率
rate of scale formation 生垢率
rate of scan(ning) 扫描频率
rate of secondary consolidation 次固结速率
rate of sedimentation 沉降(速)率;沉淀速率;沉积(速)率
rate of sediment discharge 输沙率
rate of sediment production 产沙率
rate of sediment transport 泥沙输移率;底质推移率
rate of sediment yield 产沙率
rate of self-absorption 自吸值
rate of self-purification 自净(速)率
rate of self-regulation 自调节速度
rate of set 凝结速度
rate of setting 凝结速率;水泥凝固速度
rate of settlement 沉降(速)率;沉淀速率
rate of settling 沉淀速率;沉降速率
rate of severe injury 负伤严重率
rate of shear 剪切速率;剪切率;切变率
rate of shore(line) erosion 海岸侵蚀速率
rate of shorts 缺货率
rate of shrinkage 收缩速度;收缩率
rate of siltation 回淤率
rate of silting 淤填率;淤积率
rate of sinking 下沉率;钻进速度;沉陷速度;沉降速率
rate of sintering 烧结速率
rate of slaking 消解速率
rate of slide 滑移速率
rate of slip 滑动速率
rate of slope 斜率;陡度;边坡系数
rate of snowfall 降雪强度;降雪率
rate of social increase of population 人口机械增长率
rate of soil preconsolidation 土的前期固结比
rate of sonic wave 声波速率
rate of specific gravity 比重值
rate of specific surface area 比表面积率
rate of speed 速率;速度变化率
rate of speed change 变嫌率
rate of spoilage 腐败度
rate of spoiled 废品率
rate of spoiled products 废品率
rate of spread 扩展率;蔓延速率
rate of spreading 扩散速率
rate-of-spread meter 蔓延速率计算器
rate of standard coal consumption 标准煤耗率
rate of stick-slip 黏滑速率
rate of stocking 载畜率;载畜量
rate of stock turnover 存货周转率
rate of storage change 蓄水量变率
rate of strain hardening 应变硬化速率
rate of strain(ing) 应变率;应变速率
rate of strength development 强度增长率
rate of strength gain 强度增加速率
rate of stress application 应力施加速率
rate of stretch 伸展速度
rate of subsidence 沉降速率;沉陷(速)率;消落率(洪水)
rate of substitution 替换率
rate of sulphur expulsion 脱硫速率
rate of supercharging 增压比
rate of super-elevation 超高率
rate of surfaced roads 路面铺装率
rate of surface runoff 地面径流率;地表径流率
rate of survival 成活率
rate of table traverse 工作台横动率
rate of taxation 税率
rate of taxes 税率
rate of temperature change 温度变化速率
rate of temperature rise 温升速率
rate of tension 张紧度
rate of throughput 流过率;物料通过速率
rate of tidal stream 潮流速率
rate of time preference 时间优惠率
rate of titration 滴定速度
rate of traffic flow 行车密度;交通强度
rate of transfer of suspended particles 悬浮质移动率
rate of transmission 传导率;透光率

rate of transpiration 蒸腾率;散发率
rate of transport 输送率
rate of transportation of bed load 推移质输沙率;底质推移率
rate of transportation of suspended load 悬移质输沙率
rate of transportion of suspended load 悬移质输送率
rate of travel 行速;移动速率
rate of travel of flood wave 洪水波移动速率;洪水波行进速率;洪水波传播速率;洪峰移动速度
rate of tread wear 轮胎磨耗率
rate of turbidity removal 浊度去除速率
rate of turn 转弯速率;转动速度;匝数比
rate-of-turn gyroscope 角速度陀螺仪
rate-of-turn indicator 转弯速度指示器
rate-of-turn on investment 投资回收率
rate of turnover 周转速度;周转率
rate-of-turn record 转速记录
rate-of-turn recorder 转速记录器
rate of twist 来复线缠度
rate of unemployment 失业率
rate of unloading 减负荷速度
rate of unserviceable locomotive 机车不良率
rate of upward gradient 上升梯度变化率
rate of use of invested funds 投资资金交付使用率
rate of use of maintenance equipment 维修设备使用率
rate of using passenger seat 客座占用率
rate of utilization 利用率;使用率
rate of utilization of equipment's capacity 设备能力利用率
rate of utilization of working time of equipment 设备时间利用率
rate of valve closure 阀闭合速率;阀截流速率
rate of vapo(u)r content 含汽率
rate of vegetation 植物覆盖率
rate of ventilation 换气次数
rate of vertical displacement 垂直位移速率
rate of vitrification 玻璃化速率;玻化速率
rate of voltage rise 电压增长率;电压上升速度
rate of volume flow 体积流速;体积流动速率
rate of wage 工资率
rate of warming 加热速率
rate of washing 洗涤速率
rate of water absorption 水分吸收率
rate of water circulation and water utilization 水的循环利用率
rate of water demand 用水量标准
rate of water discharge 流量
rate of water infiltration 水分渗透率
rate of water injection 水注入速度
rate of water loss 失水率
rate of water supply 供水率;给水率;给水定额
rate of water temperature progressively decreasing 水温递减速度
rate of water temperature progressively increasing 水温递增速度
rate of wave decay 波浪衰减速率
rate of wear(ing) 磨损率;磨耗率;磨耗程度
rate of weathering 风化速率
rate of withdraw 引出速率;退出速度
rate of withdrawal 回收率;回采率;出水量;采出速度
rate of work 功率;劳动速率
rate of working 运营比率;工作效率
rate payer 纳税人;纳地方税人(英国)
rate per annum 年率
rate per working hour 计时工资;工作小时工资
rate plate 铭牌
rate-plus-displacement control 速率
rate process 累进法;速率过程
rate process curve 速率过程曲线
rate process theory 速率过程理论;速度过程理论
rater 估价人;收费器
rate receiver 速率信号接收器;速率接收机
rate-recognition circuit 扫描频率测定电路
rate regulator 速率调节
rate response 微商响应;速率响应;导数响应;按速度作用
rate risk 利率风险
rates and taxes 地方税和国家税
rate scale 费率表
rate-sensitive 对速度灵敏的

rate-sensitivity 速度灵敏度
rate sensor signal 速率传感器信号
rate servo 速度跟踪系统
rate servosystem 速度伺服系统;按速控制的伺服系统
rate-setter 定额员
rate setting 规定定额
rate sheet 收费表
rate signal 速度信号;比率信号
rates of capacity utilization 制造业开工率
rates of exchange 兑换率;汇率
rates of executing work 工程单位造价
rates of gas transfer 气体传输速率
rates of wages 工资标准
rate stabilization 速率稳定
rate station 主控台
rate system 收费制
rate tariff 运费价目表
rate test 速率测试
rate theory 速率学说
rate the process 工艺过程的评定;评定流程
rate time 微分时间;比率时间
rate to be arranged 费率待定
rate transmitter 速率发射机
rate transparency 速率的透明性
rate tree 珍贵树种
rate type detector 比例式检测器
rate up 征收比规定费率高的保险费
rate variable 比例变动
rate variance 工资率差异;费率差异;比率的差异
rate-versus-concentration information 速率浓度数据
rate war 价格竞争
rate worker 计件工(人)
rat guard 伞形防鼠板;挡鼠板;防鼠隔板
rather cold 挺冷
rather level areas 较平坦地
rathite 双砷硫铅矿
rat-holding factor 料仓管状卸料系数
rat hole 放备用钻杆的浅孔;料仓内物料形成的管状漏斗;大钻压导孔;储钻杆的浅孔
rat hole discharge 管状卸料
rat hole drilling 细孔钻
rat holing 钻斜孔
ratholite 针钠钙石
rathskelter 地下餐厅;地下啤酒店
ratification 追认的代理;核准;认可;批准
ratification of agent's contract 追认代理人所订合同;批准代理人签订的合同
ratified reserves 已审批储量
ratifier of available reserves 保有储量审批单位
ratifier of report 报告审批部门
ratifying data of reserves 储量审批日期
ratifying date 报告审批日期
ratifying date available reserves 保有储量审批日期
rating 率定;校准;检定;级别;确定税率;评级;评定等级;定额;等级;测定;分等级;额定性能;额定能力
rating a ga(u)ge station 建立测站水位-流量关系;测站率定
rating canal 率定池;率定槽
rating car 率定车;校准车
rating chart 定额表;流量表(格)
rating current 额定电流
rating curve 关系曲线;率定曲线;流量曲线;水位-流量(关系)曲线;定级曲线;标定曲线
rating curve of spillway 溢流堰流量与堰顶水深关系曲线;溢洪堰关系曲线
rating data 额定数据;标定数据
rating diagram 检定图
rating factor 额定值系数;额定因数
rating filtration 过滤速率
rating flume 率定(水)槽;率定池
rating form 定额表
rating formula 功率计算公式
rating life 额定寿命
rating-limiting factor 限制因素
rating-limiting step 限制步骤
rating load(ing) 额定负荷;额定荷载
rating loop 额定绳套;环形率定曲线;水位-流量关系环线
rating matrix 分等矩阵
rating method 率定法;评价法
rating number 评级;分等级
rating nut 钟摆调节螺母;调节螺母

rating of aeration 曝气速率
rating of backwash(ing) 反洗速度;反冲速度;反冲速率
rating of blowdown 排污率
rating of bridge 桥梁鉴定;桥梁检定
rating of chronometer 日差率测定
rating of consumption 耗量
rating of covering of tunnel 隧道覆盖率
rating of current meter 流速仪定;流速仪检定
rating of deposits 储量评价;储量评估
rating of engine 发动机额定功率
rating of evolution 转速
rating of exchange 系数
rating of feeding (water) 上水率
rating of filtration 过滤速度
rating of freight container 集装箱额定质量
rating of fuel 燃料的评价
rating of gear 传动比;齿轮比
rating of ground water 地下水流量
rating of hindered settling 后凝速度
rating of machine 机械规格;机械定额;机器定额;电机规格
rating of merit 性能评价
rating of output 额定产量
rating of oxygen consumption 耗氧速率
rating of reaction 反应速率
rating of return 回收期限
rating of sound level 声级评定
rating of water consumption 用水率
rating of water demand 用水量标准
rating of water supply 供水率
rating of well 油井产油率;钻井估计产量
rating parameter curve 额定参量曲线
rating plate 铭牌;定额牌;标牌
rating power 额定功率
ratings 评定等级
rating schedule 检定程序表
rating speed 额定速度
rating system 收费制;定额制度
rating table 率定表;比率表
rating tank 率定水槽;检定槽
rating test 评定试验;检定试验;评级试验
rating tolerance 额定快慢公差
rating ton 检定吨
rating tonnage 率定吨位;检定吨位;额定吨位;额定吨数
rating under working condition 作业时的定额
rating unit 评定部分
rating value 额定值;标准值
rating voltage 额定电压
rating weight 额定重(量)
ratio 比值;比率
ratio adjuster 匝比调节器;比率调整器
ratio amplifier 比率放大器
ratio analysis 相对含量分析;比率分析
ratio analysis of net profits to net sales 净利与净销售额比率分析
ratio analysis of net sales to net working capital 净销售额与流动资本净额比率分析
ratio analyzer 相对含量分析器
ratio and difference estimation sampling 比率差异推算抽样法
ratio and phase meter 振幅比相位差计
ratio arm 比率杆;比率臂;比例臂
ratio arm box 电桥用电阻箱;比例臂(电阻)箱
ratio arm circuit 比例臂电路
ratio average method 比率平均法
ratio-balance relay 差动继电器;比值平衡继电器;比率平衡继电器;比例平衡继电器;百分率差动继电器
ratio-balance type 比率平衡式
ratio between assets and liabilities 资产负债率
ratio between investments 投资比例
ratio between liabilities and assets 资产负债率
ratio between water and soil 水土比例
ratio by public traffic 公交出行比例
ratio by weight 重量比
ratio calculator 比例计算器;比例计算机
ratio change gears 滚切交换挂轮架
ratiocination 推理
ratio code 恒比代码
ratio-computation sheet 比例计算模片
ratio control 关系调节;比值控制;比率控制;比例调制;比例调配;比例调节

ratio controller 关系调整器；比值控制器；比值调节器；比例控制器
ratio control system 比值调节系统；比率控制系统；比例控制系统
ratio correction factor 比例校正系数
ratio counter 比率计数器；比值计数器
ratio coverage 速比范围
ratio data 比例数据
ratio delay study 工作的抽样检验
ratio detector 比值检波器；比率鉴频器；比例检波器
ratio deviation 调制指数
ratio-differential relay 比率差动继电器；百分率差动继电器
ratio differential relaying system 比率差动继电系统
ratio divider 比例分配器
ratio electromagnet 比例电磁铁
ratio enhancement 比值增强
ratio error 变比误差；比值误差；比误差；比(率)误差；比例尺误差
ratio estimate 比率推算(法)；比率估计；比例估计
ratio estimates in stratified random sampling 分层随机抽样的比率估计值
ratio estimation 比率估计
ratio estimation sampling 比率推算抽样法
ratio estimator 比率估计量；比例估值
ratio factor 比率因数
ratio feeder 配量给料器
ratio flow control 比率流量控制
ratio for weight per axle 轴重比率
ratio-frequency welding 高频焊
ratio gear 变速轮
ratio governing 质量调节；改变成分的调节
ratiograph 比例图解图
ratio hopper loader 比例料斗加料器
ratioing 比值法
ratioing image 比值影像；比值图像
ratio kill 射杀比例
ratio law 比率定律
ratioless 无比率
ratio meter 流比计；比值计；比率表；比率计
ratio method 比值法；比例法
ratio method of balancing 平衡比值法；比率平衡法
ratio method of Fourier spectrometry 傅立叶光谱测量比例法
ratiometric conversion 比例换算
ratio microzonation method 比值小区划方法
ration 限额分配；日定量；配额；定量
rational 合理的
rational action 有理作用
rational activity coefficient 有理活度系数
rational algebraic form 有理代数式
rational algebraic fraction 有理代数分数
rational algebraic variety 有理代数簇
rational analysis 有理分析；理论分析；合理分析；推理分析；示性分析；示构分析
rational approximation 有理近似；有理逼近
rational behavio(u)r 合理性能
rational canonical form 有理标准型
rational canonical matrix 有理标准矩阵
rational canonical reduction 有理曲型简化
rational character 有理特征标
rational construction 合理施工
rational correspondence 有理对应
rational curve 有理曲线
rational cycle 有理循环
rational depth of well 水井合理深度
rational design 合理设计
rational differential equation 有理微分方程
rational differential system 有理微分组
rational division algebra 有理可除代数
rationale 原理；理论(基础)
rational element 有理元素
rational entire function 有理整函数
rational equation 有理方程式
rational equivalence class group 有理等价类群
rational equivalent weight 合理当量
rational estimates of runoff from rainfall 降雨记录推理估计
rational exploitation 合理开发
rational expression 有理式
rational filter length 滤水管合理长度
rational fishing 合理捕鱼
rational form 有理式

rational formula 有理化式；推理公式；示性式；示构式
rational formula of rainfall 降雨推理公式
rational fraction 有理分数；有理分式
rational fractional function 有理分数函数
rational fraction interpolation 有理分式插值
rational frequency curve 理论频率曲线
rational function 有理函数
rational function approximation 有理函数近似；有理函数逼近
rational function field 有理函数域
rational grazing 限量放牧
rational homomorphism 有理同态
rational horizon 真水平线；真地平(圈)；理论地平；天球地平(圈)；地心地平(面)
rational index 有理指数
rational injectivity 有理内射性
rational integer 有理整数
rational integral expression 有理积分表示式
rational integral function 有理积分函数
rational integral solution 有理整数解
rational interpolating function 有理插值函数
rational interval 有理区间
rational interval function 有理区间函数
rational invariant 有理不变式
rational involution 有理对合
rational involutorial algebra 有理对合性代数
rationalism 理想主义
rationalist period 理性主义时代；理性主义建筑
rationality 有理性；合理性
rationality analysis 合理性分析
rationality assessment of water consumption in industry 工业用水合理性评价
rationalizable integral 可有理化积分
rationalization 合理化
rationalization method 合理化方法
rationalization of goods transport 货物运输合理化
rationalization of insulation 绝缘配合
rationalization proposal 合理化建议
rationalize 有理化；理化
rationalized making of bituminous paving 流水作业铺沥青路面；合理化铺沥青路面
rationalized system 有理单位系统；合理单位系统
rationalized system of units 有理化单位系统
rationalized unit 合理单位
rationalizing denominator 有理化分母
rationalizing factor 有理化因子
rational knowledge 理性认识
rational layout of station 车站合理布置
rational length of water filtering sections 滤水管合理长度
rational magnetic surface 有理磁面
rational mapping 有理映射
rational material flow chart 货物合理流向图
rational mechanics 理性力学
rational mechanization 合理机械化
rational method 有利方法；合理方法
rational method of runoff calculation 径流计算推理法
rational method of runoff estimates 径流估算推理方法
rational mining of groundwater 地下水合理开采
rational mix design 合理配合比
rational mix proportioning 合理配合比
rational number 有理数
rational number field 有理数域
rational operation 有理(化)运算
rational point 有理点
rational polynomial 有理多项式
rational prime number 有理素数
rational proper fraction 有理真分式
rational rank 有理秩
rational readjustment 合理调整
rational real number 有理实数
rational representation 有理化表示法；有理表示
rational rolling schedule 合理的轧制制度
rational root 有理根
rational runoff formula 雨量理论公式；有理化径流公式；径流推算理论公式；推理径流公式
rational sampling number 合理采样数
rational series 有理级数
rational skinning surface 有理蒙皮曲面
rational space 有理空间
rational spacing between wells 合理井距

rational structure of industrial sectors 工业部门结构合理化
rational subgroup 有理子群
rational surface 有理曲面
rational swinging surface 有理回转型曲面
rational synthesis 示构合成
rational system of units 有理单位制
rational tillage 合理耕作
rational traffic 合理运输
rational traffic lines 合理运输线路
rational traffic routes 合理运输线路
rational transformation 有理变换
rational transformation group 有理变换群
rational transport 合理运输
rational transport flow diagram for major products in the country 全国主要物资合理运输
rational unit 合理单位
rational untilization of warehouse 仓库合理使用
rational utilization of groundwater 地下水合理利用
rational-valued nests 有理值巢
rational vane hydraulic motor 定量叶片式液压马达
rational variety 有理簇
rational well depth 合理井深
rational work 合理功
rational zero points 有理零点
ration book 配额供应本
ration cycle 日给养
ration depot 粮秣库
rationed credit 限额信贷
rationed exchange 限额交易
rationed export 限额输出
rationed grazing 限量放牧
ration export 出口限额
rationing 强制控制调节；定量给给；限额配给；顾客配给；定量供应
rationing equipment 定量配料设备
rationing exchange 配额外汇
rationing of exchange 外汇配给
rationing price 限额价格
rationing procedures 配给程序
ration of floating particles in coarse aggregate 粗集料中浮粒率；粗骨料中浮粒率
ration of goods 货物配给
ration of mud/sand 泥沙比值
ration price 配给价格
ration storage 定额储[贮]量
ration system 配给制
ratio of absorption 吸收率
ratio of activity 比活度
ratio of a parallel projection 平行射影的比例
ratio of area 面积比(率)
ratio of asset value to sales 资产价值对销货额比率
ratio of attenuation 衰减率；衰减比
ratio of average gross profit 平均毛利率
ratio of brazing area 钎着率
ratio of building volume to lot 建筑容积率
ratio of burnt area 烧损率(建筑物烧损面积与总面积之比)
ratio of capital to current liabilities 资本对流动负债比率
ratio of capital to fixed liabilities 资本对固定负债比率
ratio of capital to liabilities 资本负债比率
ratio of carbon 定碳率
ratio of carrier-to-interfering signal 载频信号对干扰信号比
ratio of cars under repair 车辆检修率
ratio of cash to current liabilities 现金对流动负债比率
ratio of cash to income 现金对收入的比率
ratio of cations 阳离子比率
ratio of cement 混凝土比
ratio of cement of aggregate 水泥骨料比率
ratio of clearance volume 余隙容积比
ratio of climbing 爬升比
ratio of closing error 闭合比
ratio of closure 相对闭合差
ratio of closure of traverse 导线闭合比
ratio of coefficient of variance against precision demanded 标志变化系数和精度要求比
ratio of collapsed building 房屋倒塌率
ratio of combustion 燃烧比
ratio of commodities' supply and demand 商品供求比率

ratio of completion under productive capacity 生产能力建成率
ratio of components 组成比;配料比
ratio of compression 压缩比(率)
ratio of compressive reinforcement 抗压钢筋比率
ratio of concentration 选矿比
ratio of constituents 成分比
ratio of consumer decreasing progressively 消耗递减率
ratio of consumer increasing progressively 消耗递增率
ratio of crushing 破碎比
ratio of current assets to fixed assets 流动资产与固定资产比率
ratio of current assets to total assets 流动资产与总资产比率
ratio of curvature 曲率比
ratio of cut to fill 挖填(方)比
ratio of damping 阻尼率;阻尼比;减幅比
ratio of day and night 昼夜比
ratio of daytime to nighttime population 昼夜人口率
ratio of debt to net worth 负债率
ratio of demand contraction 需求减缩率
ratio of demand full 需求满足率
ratio of demand increase 需求增长率
ratio of dependence on foreign trade 外贸依存度
ratio of depot repairing 检修率
ratio of development meters to ten thousands tons of mined ore 万吨采掘比
ratio of diameters of flange 法兰径比
ratio of diameter to high 径高比
ratio of discounted net benefit to cost 净收益贴现值对成本比率
ratio of division 分割比
ratio of drawing 牵伸比
ratio of dynamic(al) to plastic viscosity 动塑比
ratio of elongation 延伸率;延伸比;伸长率;伸长比(值)
ratio of enlargement 放大率;放大比
ratio of enrichment 富集比
ratio of equality 等比
ratio of exaggeration 夸大率
ratio of exchange 交易比例
ratio of expansion 膨胀率;膨胀比(率)
ratio of expense to sales 费用对销售额的比率
ratio of exploitation and reserves 采储比
ratio of external standard 外标道比
ratio of extraction 萃取率
ratio of fire danger 火灾危险率
ratio of first-grade products 一级品率
ratio of fixed assets to fixed liability 固定资产对资产总额比率
ratio of flattening 压扁率
ratio of floating particles in coarse aggregate 浮粒率
ratio of floor 地基系数
ratio of flow to mean flow 径流模数
ratio of fluid pressure to overburden pressure 流体压力/积土压力
ratio of forging reduction 锻压比;锻比
ratio of freight traffic 运输系数
ratio of full consumption 消耗满足率
ratio of gain 增益比
ratio of gate-to-pin 门针比;门引线比
ratio of gear 齿轮速比;齿轮传动比;变速比
ratio of geometric(al) anisotropy 几何异向比
ratio of gold and silver 金银比价
ratio of greater inequality 优比
ratio of green space 绿地率
ratio of grinding media to material 球料比
ratio of gross profit to net sales 毛利对销货净额比率
ratio of gross to tare weight 总重皮重比
ratio of gun currents 束电流比
ratio of heat capacities 热容量比
ratio of heat transfer to skin friction 热传导与表面摩擦比
ratio of height to thickness 高厚比
ratio of hole size enlargement 井径扩大率
ratio of inertia 惯性比
ratio of installed equipment available 现有设备安装率
ratio of interest expenses to sales 利用费用与销货比率
ratio of internal interval 内间隙比;内间距比
ratio of internal reserves to net profits 内部准备对净利润比率
ratio of investment to output 投资产出率
ratio of labo(u)r to cost 劳动成本比率
ratio of lake surface to drainage 湖泊率;湖泊度
ratio of land for public utilization 公共用地率
ratio of land use 大地利用率
ratio of land utilization 土地利用率
ratio of lay 捻比系数(钢丝绳)
ratio of length to diameter 长径比
ratio of lens aperture 透镜孔径比
ratio of less inequality 劣比
ratio of linear shrinkage 线收缩率
ratio of living area 居住面积比
ratio of major and minor axis of ellipse 椭圆长短轴比
ratio of margin on sales to net sales 销货毛利与销货比率
ratio of marketing decreasing 销售减衰率
ratio of marketing divided by output 销采比
ratio of marketing increasing 销售增长率
ratio of mechanized handling 机械化比重
ratio of melting 冶炼比
ratio of mixture 混合物配比;混合(料配)比;配料比
ratio of modal split 交通方式分配率
ratio of modulus elasticity 弹性模量比
ratio of molecular parameter 分子参数的比值
ratio of moment of inertia 惯性矩比
ratio of mountain runoff to tidal volume 山潮比
ratio of net earnings to dividend 净收益对股息比率
ratio of net income to net sales 净收益对销售比率;净收入与净销货额比率;纯益与销货总额比率
ratio of net income to net worth 净收益对资本净值比率
ratio of net income to outstanding stock 净收益对现发股本比率;净收入对在外股票比率
ratio of net income to total assets 净收益对资本总额比率
ratio of net profit 净利比率
ratio of net profit to capital 净利润对资本比率
ratio of net profit to net worth 净利对净值的比率
ratio of net sales to net worth 净销货额对资本值比率
ratio of net sales to receivables 净销售额对应收款项比率
ratio of net sales to total assets 销货净额对资产总额的比率
ratio of net worth to total capital 净值对资产总额比率
ratio of net worth to total debt 净值对负债总额的比率
ratio of non-volatile carbon to total organic carbon 定碳比
ratio of number of particles 粒子数比
ratio of opening to total surface 筛孔与矿粒总表面比
ratio of operation(al) tons to physical tons 操作系数(指装卸)
ratio of operational tons to ports throughput 装卸工作系数
ratio of ore 矿比
ratio of ore reserve depleting 储量增长率;储量耗渴率
ratio of ore reserve divided by annual output 储采比
ratio of organ to body weight 藏器系数
ratio of outside interval 外间距比
ratio of parking for peak phase 高峰停放比率
ratio of population in crease 人口增长率
ratio of precision 精度比
ratio of price decreasing 价格下降率
ratio of price increasing 价格增长率
ratio of principal stresses 主应力比
ratio of productive increasing 生产增长率
ratio of profit payments and tax turnover 投资利税率
ratio of putting coke into furnace 入炉焦比
ratio of radiation 比辐射率
ratio of radioactive content 辐射量比值
ratio of range against precision demanded 标志极差和精度要求比
ratio of ranges 潮幅比;潮差比;变幅比
ratio of reduction 破碎比;缩小率
ratio of reed divided by supply 供需比
ratio of reed divided by supply in country 矿产自给率
ratio of refraction 折射比
ratio of refractory to crude steel 耐钢比
ratio of reinforcement 钢筋比(率);配筋率
ratio of relative stiffness 相对刚性比;相对刚度比
ratio of relative width of cleavage domain 劈理域相对宽度比
ratio of rentable area 出租面积比
ratio of repaired wagon 车辆检修率
ratio of repair to fixed assets 维修费对固定资产比率
ratio of reserves 储量比例
ratio of return on resources employed 使用资源收益率
ratio of return sludge 回流污泥比
ratio of revolutions 转数比
ratio of rise (两地的)潮升比
ratio of rise to span 高跨比;矢跨比
ratio of roll gears 滚比挂轮
ratio of runoff 径流率
ratio of settlement 沉陷率
ratio of silica to alumin(i)um 硅铝比率
ratio of similitude 相似比
ratio of size reduction 破碎比
ratio of slenderness 细长比(例);柔性系数;长细比
ratio of slip 滑动比
ratio of slope 坡率;坡度比;边坡斜度;边坡系数;边坡比率
ratio of soil pressure to settlement 基础系数
ratio of solvent 溶剂比
ratio of space allocated to buildings 建筑楼比
ratio of specific heat 绝热指数;热容比;比热比
ratio of standing wave (吸声系数用的)驻波比
ratio of stream discharge obtainable 河道可得流量比
ratio of strength to weight 强重比;强度—重量比;强度与重量比率;比强(度)
ratio of stroke to diameter 冲程与直径比;冲程缸径比
ratio of tare weight to gross 皮重与总重比
ratio of the circumference of a circle to its diameter 圆周率
ratio of the investment contributions 出资数额比例
ratio of the three-colo(u)r primary signals 三基色信号比
ratio of Th/K 钍钾比值
ratio of tidal ranges 潮差比
ratio of tide (两地的)潮升比
ratio of total liability to net worth 负债总额对净值的比率
ratio of total nitrogen to total phosphorus 总氮与总磷比
ratio of transformation 变压器变压系数;变压比
ratio of transformer capacity to load 变电容载比
ratio of transmission 转速比;传动比
ratio of turned over income from investment 投资收益上交率
ratio of two ratio estimators 两个比率估计量的比率
ratio of U/K 铀钾比值
ratio of useful compression 有效压缩比
ratio of U/Th 铀—钍比值
ratio of value added to sales 附加价值对销售额百分比
ratio of vanadium to nickel 钒镍比
ratio of varying capacitance 电容变化比
ratio of viscosities 黏度比
ratio of void 孔隙比
ratio of volume 体积比
ratio of wastewater discharge capacity to runoff 污水排放量与径流量比;污径比
ratio of weight 重量比
ratio of wet and dry strength 干湿强度比(混凝土)
ratio of width 宽度比
ratio of winding 匝数比;绕组比;绕圈比
ratio of yield point to tensile strength 屈服点与推力强度之比
ratio oxygen cutting machine 比例氧切割机

ratio plug valve 比率调节阀
ratio print 缩放相片；投影晒印
ratio printer 比例缩放仪；比率晒印器
ratio productive decreasing 生产衰减率
ratio recorder 比记录器
ratio regulation 比例调节
ratio reinforcement 含筋率
ratio resistor 比率电阻箱
ratio scale 比率量表；比率尺度；比例尺（度）；比例标尺
ratio separator 离心比分离机；比率分离机
ratio set 比率装置
ratio spacer 隔离液配比
ratio spectrum-derivative spectrophotometry 吸光度比值—导数分光光度法
ratio squelch 比率啸声抑制电路
ratio switch 比例换转开关
ratio table 比率表
ratio table for scale focusing 焦距表
ratio test 比率检验法
ratio testing 比率测试
ratio to material used 材料利用率
ratio-to-moving-average method 比率移动平均法；移动平均比率法
ratio to the primary unit 对立单位之比
ratio tracing receiver 比例跟踪接收机
ratio transformation 比值变换
ratio transformer 比率变压器
ratio turbidimeter 比率浊度计
ratio-turn 匝数比
ratio use of resources 资源合理利用
ratio value 比移值
ratio value of stable isotope 稳定同位素比值
ratio variance against precision demanded 标志方差和精度要求比
ratio-voltage 电压比；比例电压
ratio-weight blender 比例重量掺合机
ratio wheel 减速轮
Ratisbo Cathedral 雷根斯堡大教堂
rat joint runner 石棉绳（用于嵌缝）
ratkiller 灭鼠优
ratlin(e) 桅梯横绳；梯索；绳索梯
ratlin(e) down 扎桅梯绳
ratlin(e) stuff 桅梯绳
ratofkite 萤石
ratproof 挡鼠板；防鼠的；防鼠隔板；防鼠措施
ratproof construction 防鼠构造
rat-proofing 防鼠
rat race 环形波导
rat-race mixer 环形波导混频器
ratran 三雷达台接收系统
rat repellent 鼠忌避剂；避鼠剂
rat's nest type bolometer 鼠巢式测辐射热计
rat stop 防鼠挡板
rat stoppage 拒鼠设施
rat tail 天线升高部分和引下线的连接线束；铸铁鼠尾
rat-tail burner 针状火焰燃烧器；鼠尾喷灯；单射流喷燃器；单喷焊炬
rat-tail file 细圆锉；圆锉；鼠尾锉
rat-tail fringe 边饰小穗
rat-tail splice 尖尾插接
rattan 藤条；藤料
rattan articles 藤器
rattan bar 藤条
rattan chair 藤椅
rattan fender 藤条碰垫；藤料防冲器
rattan net bridge 藤网桥
rattan product 藤制品
rattan rope 藤索
rattan work 藤制品
ratteening 毛绒成珠工艺
rattening machine 毛绒搓磨成珠机
ratter 往复式振动筛；大孔选矿筛；动筛
rattle 急响器；拍击；脆响
rattle barrel 滚筒式磨机
rattle free 无吵闹声的
rattlehead 吸水管
rattle jack 炭质页岩
rattler 滚筒；滚筒机；磨损试验机；倾盆大雨
rattler loss 磨损；磨耗率
rattler test 洛杉矶试验；磨耗（机）试验；滚磨试验；磨砖试验；磨损试验
rattle stone concretion 同心层结核
rattle trap 旧车辆

rattling 滚筒清理法；脆梯横绳；拍击
rattling machine 抗磨试验机
rat trap 捕鼠器；捕鼠式
rat-trap bond 竖砖空斗墙；空斗墙砌合；空斗间砌合；花墙砌法；咬合砌砖
rat-trap pedal 空心踏板；锯齿踏板
rat-trap wall 空斗墙
rauchwacke 糙面白云岩
raucous breathing 息粗
rauhaugite 白云石碳酸岩
rauhaugite group 白云石碳酸岩类
raummeter 堆积立方米
raunchy 不够标准的
rauvite 红钒钙铀矿；水钙钒铀矿
Rauvolfia sumatrana 苏门答腊萝芙木
rave （运货车四周的）栏板
rave hook 刮缝钩
rave iron 刮缝凿；刮缝刀
ravelin 半月形城堡
raveling cord 分离横列线
raveling course 分离横列
ravelling (of pavement) 路面松散；松散路面
ravelly ground 碎屑岩质；松散岩层
Raven 雷纹石膏墙板
raven 深墨色
raven black 墨黑色
raven gray 墨灰色
Ravenite 拉文奈（一种覆盖屋面用的压缩油毛毡）
ravine 细谷；深谷；雏谷
ravine afforestation 沟壑造林
ravine erosion 峡谷侵蚀；冲沟浸蚀
ravine head protection 沟头防护
ravinement 沟蚀
ravine stream 溪流；山溪；山涧
ravine wind 峡谷风
ravison oil 劣质菜子油
raw 未加工的；湿寒
raw acid 原酸
raw acid settling tank 原酸澄清槽
raw aggregate 原状集料；原状骨料；生集料；生骨料
raw alluvium 粗冲积物
raw analysis 未灼烧分析
raw anchor 粗锚
raw and processed materials 母料
raw and semi-finished materials industry 原材料工业
raw asbestos 纯石棉；生石棉
raw asphalt 原柏油
raw auto exhaust 未净化的汽车排气
raw bank 天然河岸；天然岸
raw-bank system 木桩护岸法
raw bar horsepower 拖钩牵引马力
raw basis 未灼烧基
raw batch 粉料配合料；配合料；不含碎玻璃的配合料
raw bauxite 高铝生料
raw blending fluid 粗混合油
raw blending silo 生料搅拌库
raw bolt 粗制螺栓；圬工锚栓
raw brick 砖坯；生砖
raw casting 未经清理的铸件；粗铸件
raw catalyst 新催化剂
raw clay 纯黏土；未加工(的)黏土
raw cloth 未处理的织物
raw coal 原煤
raw coal amounts to 原煤合计
raw coal desulfurization 原煤脱硫
raw coal inlet 原煤进口
raw coal moisture 原煤水分
raw coal silo 原煤仓
raw condition 原状；未加工状态；粗糙状态
raw constituents 原料组分
raw copal 天然柯巴脂
raw copper 泡铜；粗铜
raw cord 粗糙绳索
raw cork 未加工的软木
raw cotton 原棉
raw cullet 全熟料配合料；碎玻璃配合料
raw data 原始资料；原始数据；素材；待处理的数据
raw data form 原始数据格式
raw-data map 原始数据图
raw department 原料备料车间
raw diatomite 生硅藻土
raw digital data 原始数据
raw dirt trail 荒野小路

raw drinking water 原饮用水
raw edge 裂边；毛边
raw effluent 原污水；原废水；未经处理的废水
raw emery paper 粗砂纸
raw energy consumption 毛电能消耗量
raw exhaust 未净化废气；未处理废气
raw fall head 原降落水头
raw feed 喂入的生料
raw felt unroller 胎éhub开卷机
raw file 原始材料
raw film 原胶片
raw fireclay 火泥生料
raw-food material 食品原料
raw foundation 粗筑路基
raw-fuel feeder 原煤给煤机
raw gas 原料气；未净化气体；粗煤气
raw gas dust content 含尘气体粉尘浓度
raw gasoline 不纯汽油
raw glass 毛坯玻璃
raw glaze 窑土；生釉；生铅釉；生料釉
raw glycerine 未加工的甘油；粗甘油
raw goods 生货
raw gravel 粗砾石；粗糙的砾石
raw grinding 初磨；粗磨
raw grinding dust 粗磨粉尘
raw gypsum 生石膏；冰石膏
raw hand 新手
raw head 原水头
raw hide 原料皮；生皮料
rawhide gear 硬皮革齿轮
rawhide hammer 生皮手锤
rawhide wheel 生皮轮
raw humus 粗腐殖质
raw hydrocarbons 原状碳氢化合物
rawin 雷达气球；雷达测风；无线电测风（仪）
rawin balloon 无线电测风气球
raw information 原始资料；原始信息；原始数据
raw ingot 粗锭
rawinsonde 无线电探空测风仪
rawinsonde observation 无线电探空测风观测
rawin target 雷达气球目标
raw jute 生黄麻
raw lacquer 生漆
raw limestone 石灰石
raw linseed oil 毛亚麻油；生亚麻籽油；生漆油
raw liquor 原液
raw log 原始日志；原始测井曲线；粗原木
Rawlplug 罗乐塞；罗尔纤维管胀塞
raw manuscript data 原始数据
raw marble 未加工的大理石
raw material 原料；材料；母料
raw material bin 原料仓
raw material commodities for cast stone 铸石原料矿产
raw material commodities for chemical firtilizer 化肥原料矿产
raw material commodities for chemical industry 化工原料矿产
raw material consumption 原料消耗
raw material deposit 原料储存（量）
raw material drying 原料烘干
raw material drying and grinding mill 烘干原料磨
raw material feed pipe 给料管
raw material for building and construction 建筑材料矿产
raw material for glass industry 玻璃原料
raw material for optic(al) use 光学原料
raw material grinding 原材料研磨；原料粉磨
raw material grinding control 原料粉磨控制
raw material inlet 原料入口
raw material of wood-charcoal 炭材
raw material preparation 原料准备；原料制备
raw material property 原料的性质
raw materials for industry 工业原材料
raw material shop 原料车间
raw materials inventory 库存原料
raw material sizing 原料粒度分级
raw materials storage 原材料储[贮]藏
raw materials taked from the soil and air 在土壤和空气中得到的原料
raw materials to produce food and fiber 生产食品和纤维的原料
raw material storage 原料库；原料储仓
raw material storage building 原料库房

raw material storage shed 原料棚
raw material store 原材料仓库
raw material storing facility 原料储[贮]藏设备
raw material to clinker ratio 生料/熟料比
raw matte 粗冰铜
raw meal 生料(粉)
raw meal and anthracite nodule 黑料球;含无烟煤的生料球
raw meal and coke nodule 含焦碳的生料球
raw meal blending 生料搅拌;生料调配
raw meal blending silo 生料搅拌库
raw meal composition control 生料成分控制
raw meal granule 生料球
raw meal grinding aid 生料助磨剂
raw meal grinding plant 生料粉磨设备;生料粉磨车间
raw meal homogeneity 生料均匀性
raw meal nodule 生料球
raw meal nodulizing 生料成球
raw meal overflow 生料溢流
raw meal pellet 生料球
raw meal preheater 生料预热器
raw meal prepared from lime 石灰配料
raw meal proportioning 生料配料
raw meal sampling 生料取样
raw meal silo 生粉筒仓;生料库
raw meal with blast furnace slag 矿渣配料
raw metal 金属原料;粗金属
raw mica 生料云母
raw mill 原生料磨;生料磨;粗磨
raw mineral materials 矿物原料
raw mix 生料拌和;生混合料;未加工的混合料
raw mix bin 生料库
raw mix composition control 配料成分控制
raw mix preheater 生料预热器
raw mix preparation 生料制备
raw mix proportioning 生料配料
raw mix slurry 料浆
raw mixture 原料混合物;配合料;粉磨前混合料
raw natural gas 未精制的天然气;未加工的天然气
raw natural gasoline 未加工的天然汽油
raw oil 原油;原料油;未精制油料;粗制油
raw ore 原矿(石)
raw organic soil 粗有机质土壤
raw pelletizing 原料成球
raw peppermint oil 薄荷原油
raw pig iron 铸造用生铁
raw pine oil 松脂油
raw pitch 原生沥青
rawplug 罗尔塞;罗尔纤维管胀垫
rawplug anchor 粗锚塞
raw polymer 原始聚合物;原料高聚物
raw pot glass 光学玻璃坯块
raw primary sludge 原初级污泥
raw producer gas 未净化的发生炉煤气;发生炉煤气
raw product 原产地;粗制品;粗制产物
raw pull 拖钩牵引力
raw quality control system 生料质量控制系统
raw radiance 原始辐射率
raw raise 无支架天井
raw refuse 原始垃圾;未经任何处理的垃圾
raw rubber 生橡胶
raw rubber block 生胶块
raw rubber emulsion 生橡胶乳化液
raw rubber emulsion paint 生橡胶乳化涂料
raw rubber powder 生橡胶粉
raw sand 粗糙砂;原砂
raw sanitary sewage 原生活污水
raw sanitary waste 原生活废水
raw sanitary wastewater 原生活污水
raw sanitary water 未经处理的生活废水
raw score 原记分
raw seam 毛缝;粗缝
raw sewage 原污水;未净化(的)污水;未经任何处理的污水;未经净化污水
raw sewage analysis 原污水分析
raw sewage concentration 原污水浓度
raw sewage discharge 原污水排放
raw sewage sludge 原污水污泥;未经处理的污水污泥
raw shale 生页岩
raw sheet 原板;粗加工薄钢板
raw sheet horse 原板木架
raw ship 新服役船

raw sienna 富铁黄土
raw silk production process 制丝工程
raw silk rings 直丝
raw sludge 原(生)污泥;未净化污泥;未经处理的污泥;未处理的淤渣;生污泥
raw slurry 生料浆;粗制泥浆
raw slurry mill 湿法生料磨;生料浆磨
raw slurry mixer 纯泥浆搅拌机
raw soil 原土;原始土;未处理土;生土
raw spirit 无水酒精;原酒精
raw state 原状态;原始状态
raw steel 原钢;粗钢
raw stock 原材料;毛坯
raw stone 未加工过石料;原石
raw stone store 原石库
raw storage 原始数据存储器
raw storage pile 原料堆(场)
raw stuff 原料
raw subgrade 粗筑路基;原土路基;原土地基;天然路基
raw sugar 原糖;粗制糖;粗糖
raw tall oil 粗妥尔油
raw tape 净带
raw timber 未加工的木料
raw tin 原料锡
raw traffic capacity 原始交通量;毛交通量
raw tung oil 生桐油
raw umber 棕土棕;哈巴粉;富锰棕土;富锰赭土
raw value 原始数值
raw video picture 未经处理的图像
raw waste load 原废负荷
raw wastewater 原污水;原废水;未处理的废水;生废水
raw wastewater irrigation 原废水灌溉
raw water 未经处理的水;原水;未净化的水;未经净化水;生水
raw water chlorination 原水氯化;原水氯消毒
raw water influent 原水来水管
raw water inlet 原水进水管
raw water network 生水水网;未净化水水网;原水水网
raw water pump 原水泵;水泵
raw water pumping station 原水泵站
raw water quality 原水水质
raw water sample 原水样本
raw water source 原水水源
raw water storage basin 原水储[贮]存池;未净化的蓄水池;未经净化储[贮]水池;生水储[贮]水池
raw water supply 未净化水供应;原水供应;原水供水
raw water tank 未净化水箱;原水箱
raw water tunnel 未净化水的隧道;引原水的隧道
raw water turbidity 原水浊度
raw wool 原棉
raw yarn 未加工纤维
ray 线;射线;辐射纹;辐肋
ray aberration 光行差;光线像差
ray acoustics 几何声学;射线声学
rayage 射线用量规定
ray approximation 射线近似法
ray approximation method 射线近似法
ray axis 光轴
ray bundle 射线束
ray center[centre] 射线中心
ray check 射线开裂
ray cone 光锥(体)
ray coordinates 射线坐标
ray crossing 射线交叉场
ray curvature 射线曲率
ray deflection 射线偏移
ray deflection type storage tube 射线偏转存储管
ray depth migration 射线深度偏移
ray detection 射线探伤
ray diagram 射线图
ray displacement 光线位移;射束位移
Raydist 雷迪斯特双曲线定位系统
Raydist series 雷迪斯特系统
ray divergence 射线发散
ray effect 射线效应
ray ellipsoid 光线椭球;菲涅尔椭球
ray envelop 光束包络;射线包迹
ray equation 射线方程
ray fan 光线扇面

ray filter 滤光器;滤光镜;射线过滤器
ray filter holder 滤光片框
ray fleck 射线径纹
ray floret 边花
ray flower 边花
ray focusing 射线聚焦
ray gun 射线枪
Rayhead 瑞海德煤气红外线加热器
ray index 光束指数
ray intercept curve 光线交点轨迹
ray intersection 光线交会;射线交点
ray intersection point 射线交点
rayite 银板硫锑铅矿
Raykin (fender) buffer 瑞金(式)消能垫;瑞金缓冲层
rayl 瑞(耳)(声阻抗率单位)
Rayleigh-Jeans formula 瑞利-杰恩斯公式
Rayleigh-Jeans law 瑞利-杰恩斯定律
Rayleigh-Lamb frequency equation 瑞利-兰姆频方程
Rayleigh-Lamb frequency spectrum 瑞利-兰姆频谱
Rayleigh number 1 第一瑞利数
Rayleigh number 2 第二瑞利数
Rayleigh number 3 第三瑞利数
Rayleigh-Ritz method 瑞利-里茨法
Rayleigh's approximation 瑞利近似
Rayleigh's atmosphere 瑞利大气
Rayleigh's balance 瑞利天平
Rayleigh's condition 瑞利条件
Rayleigh's constant 瑞利常数
Rayleigh's criterion 瑞利准则;瑞利判据
Rayleigh's criterion for resolution 瑞利分辨率判据
Rayleigh's cross-section 瑞利(散射)截面
Rayleigh's current balance 瑞利电流天平(安培秤)
Rayleigh's cycle 瑞利周期;瑞利循环
Rayleigh's disc method 瑞利声盘法
Rayleigh's dish 瑞利声盘(测声强用)
Rayleigh's disk 瑞利盘;瑞利斑
Rayleigh's dissipation function 瑞利散逸函数;瑞利耗散函数
Rayleigh's distillation 瑞利蒸馏
Rayleigh's distribution 瑞利分布
Rayleigh's distribution of peaks 瑞利峰值分布
Rayleigh's effect 瑞利效应
Rayleigh's elastic quality factor 瑞利波弹性品质因子
Rayleigh's energy method 瑞利能量方法
Rayleigh's equation 瑞利公式;瑞利方程
Rayleigh's fading 瑞利衰落
Rayleigh's flow 瑞利流(动)
Rayleigh's function 瑞利函数
Rayleigh's group velocity 瑞利波群速度
Rayleigh's interferometer 瑞利干涉计
Rayleigh's law 瑞利定律
Rayleigh's law of scattering 瑞利散射定律
Rayleigh's limit 瑞利限度;瑞利极限
Rayleigh's line 瑞利线
Rayleigh's loop 瑞利回线
Rayleigh's method 瑞利法
Rayleigh's number 瑞利数
Rayleigh's phase velocity 瑞利波相速度
Rayleigh's principle 瑞利原理
Rayleigh's prism 瑞利棱镜
Rayleigh's process 瑞利过程
Rayleigh's quotient 瑞利商数
Rayleigh's ratio 瑞利比值;瑞利比(率)
Rayleigh's reciprocity theorem 瑞利倒易定理
Rayleigh's refractometer 瑞利折射计
Rayleigh's resolution 瑞利分辨率
Rayleigh's resolution limit 瑞利分辨极限
Rayleigh's scatterer 瑞利散射器
Rayleigh's scatter(ing) 瑞利散射
Rayleigh's scattering function 瑞利散射函数
Rayleigh's scattering method 瑞利散射法(X射线分析)
Rayleigh's scattering spectrum 瑞利散射光谱
Rayleigh's spectrometer 瑞利分光计
Rayleigh's speed 瑞利速度
Rayleigh's wave 瑞利波
Rayleigh's wave singlet transfer function 瑞利波单一传递函数
Rayleigh-Willis relation 瑞利威利斯关系
Rayleigh-wing scattering 瑞利翼状散射

rayless 无光线的
ray-locking device 射线锁定装置
ray-matrix method 光线矩阵法
ray method 射线(方)法
ray modulation 射线调制
Raymond cast-in-place concrete pile 雷蒙德现场灌注混凝土桩
Raymond concrete pile 雷蒙德式桩;雷蒙德混凝土桩;雷蒙德带套管钻孔灌注桩
Raymond granite 雷蒙德石(产于美国)
Raymond impact mill 雷蒙德型锤磨机;雷蒙德型冲击式磨机
Raymond mill 雷蒙德研磨机;雷蒙德磨;雷蒙德粉碎机
Raymond pile 雷蒙德桩
Raymond ring roller mill 环辊式磨粉机;摆式磨粉机
Raymond standard concrete pile 雷蒙德标准混凝土桩
Raymond system 雷蒙德污泥干化系统
Raymond vertical mill 雷蒙德竖式粉碎机
Raynaud's disease 雷诺病
ray normal 射线法线
ray of light 光射线
ray of refraction 折射线
ray of wood (木材的)射髓
rayon 粘胶(纤维);人造纤维;人造丝
rayon-based carbon fiber 黏胶纤维基碳纤维
rayon belt 人造丝芯运输带
rayon coning oil 人造丝络筒油
rayon cord 人造丝绳
rayon fabric 人造丝织物
rayon factory 人造纤维厂
rayon fiber 人造丝纤维
rayon filament 人造丝
rayon manufacturer 人造丝制造厂
rayon map 丝绸地图
rayon mill 人造丝厂
rayon mill wastes 人造丝厂废水
rayon monofil 人造纤维单丝
rayonnant 辐射状条的;辐射式装饰法
Rayonnant period 辐射式时期
Rayonnant style 辐射式(风格)(指十四世纪法国哥特式以辐射式窗为特色的建筑)
rayon palace checked 彩格纺
rayon pulp 人造丝浆
rayon reject 人造丝浆筛渣
rayon silk works 人造丝厂
rayon staple 黏胶短纤维
rayon tow 黏胶丝束
rayon tyre yard 人造丝轮胎线
rayon wastewater 人造纤维废水
rayon works 人造丝厂
ray optics 射线光学
ray oscilloscope 光线示波器
rayotube 光电高温计
rayotube pyrometer 全辐射高温计
ray paper 废纸
ray parameter 射线参数
ray-parameter method 射线参数方法
ray path 光路;光程;射线途径;射线路径
ray picture 声线图
ray plotter 光线绘迹器
ray projector 光线足迹器
rayproof 辐射保护
ray proofing 防辐射的
ray protecting glass 防射线玻璃
ray pulse 射线脉冲
Rayrad 拉拉德散热器(一种隐蔽式散热器)
ray radiation 光辐射
ray ratio 射线比
ray representation 射线表示法
ray sieve tube 射线筛管
Raysistor 光控变阻器;全波整流管
ray spectrum 射线谱
ray stopper 射线防护塞
ray structure 射线结构
ray surface 射线曲面;射线面
ray system 辐射纹系统
Raytheon tube 全波整流管
ray theoretic(al) method 射线理论方法
ray theoretic(al) seismogram 射线理论地震图
ray theory 射线理论
ray trace 光线追迹;射线径迹

ray trace technique 射线追踪技术
ray tracing 光线跟踪;射线示踪;射线径迹;射线轨迹
ray tracing apparatus 光线跟踪仪
ray tracing calculation 光路追迹计算
ray tracing devic 光线跟踪装置
ray tracing diagram 射线轨迹图
ray tracing formulas 光线迹迹公式
ray tracing migration 射线追踪偏移
ray tracing model of zero offset VSP 零偏移距VSP 射线追踪模型
ray trajectory 光线径迹;射线轨迹
ray transformer 射线变压器
ray tube 射线管
ray type 射线类别
Rayvecto 拉维克托(一种采暖板)
ray vector 光矢(量)
ray velocity 光速;射线速度
ray velocity surface 光线速度面
ray-vessel 射线导管的
ray-vessel pitting 射线导管间纹孔式
ray wheel 射线轮
raze 拆毁;消除;刮去
raze a vessel 取消船级
raze to the ground 夷平(房屋);推平
Razin effect 拉金效应
Razin-Tsytovitch effect 拉金效应
razor 剃刀
razor back 刃脊
razor blade 刀片
razor blade cutter head 刀片切割头
razor-blade share 薄片梯形犁铧
razor blade slitter 刀片式薄膜分切机
razor-edge liner 楔形垫板;楔形衬垫
razor-edge stone 均密砂岩
razor saw 剃锯
razor steel 剃刀钢;刀片钢
razor stone 细磨石;均密石英岩
razor strop 磨刀皮带
Rb-Sr dating method 铷—锶测年法
R chart R 控制图
RC oscillator 电阻电容振荡器
R curve R 型曲线
R day 预计供应日
rderal strategy 杂草策略
reabsorbed cost 再吸收成本
reabsorption 再吸收;重吸收(作用)
reabsorption rate 再吸收率
reaccelerate 再加速
reacceleration 再加速
reaccess 再存取;重存取
reacclimatization 再驯化
reaccommodation 再调节
reach 直线流域;直水道;横风行驶;湾段;射程;地段;到达;达到;范围
reachability 可达(到)性;能达到性
reachability graph 可达图
reachability map 可达到映射
reachability matrix 可达性矩阵
reachability set 可达到集
reachable accessible 可达到的
reachable point 可达点
reachable set 可达集
reachable state 可达状态
reach and burden 容积与载重量
reach an identity of views 取得一致看法
reach forklift crane 伸缩叉式起重机
reach forklift truck 伸缩叉式装卸车
reach full growth 达到完全长成了的
reach-in freezer 冻结柜
reaching-in 穿经
reach-in refrigerator 大型冷库;柜台冷柜;大型冷柜
reach me 请与我联系
reach of channel 河段;渠段;槽段
reach of crane 起重机臂伸出长度;起重机臂工作半径
reach of river 河区;河流流程;河流的上下游;河段;河的上下游;河道
reach of stream 河区;河流流程;河段;河的上下游
reach post 下端桩
reach range 所及范围;达到的宽广度;作用半径
reach rod 联结杆
reach stacker 正面吊运机
reach targeted levels 达到指标水平
reachthrough 透过

reach travel curve 河段汇流曲线;河段回流曲线
reach truck 前伸式叉车
reach with divided flow 汊流河段
reacquired bond 取回债券;重得债券;重购进债券
reacquired stock 入库股票;取回股份;重购进股票
react 起反作用;反应
react acid 呈酸性反应
reactance 阻抗;电抗;反应性
reactance amplifier 电抗耦合放大器
reactance arm 电抗臂
reactance attenuator 电抗衰减器
reactance blade 反作用叶片
reactance bond 接合扼流圈;电抗耦合
reactance bridge method 电抗桥法
reactance capacity 无效功率;无功功率
reactance characteristic 电抗特性
reactance chart 电抗图
reactance coil 电抗线圈;扼流(线)圈
reactance component 电抗元件;电抗分量
reactance-controlled circuit 电抗控制电路
reactance coupling 电抗耦合
reactance drop 电抗电压(下)降
reactance element 电抗元件
reactance equalizer 电抗均衡器
reactance factor 电抗因数
reactance filter 电抗滤波器
reactance frequency multiplier 电抗式倍频器;电抗频率倍增器
reactance function 电抗函数
reactance grounded 电抗接地
reactance leakage 电抗漏泄
reactance load 电抗荷载
reactance meter 电抗测定计
reactance modulation 电抗调制
reactance modulation system 直接调制方式
reactance modulator 电抗调制器;变电抗调制器
reactance network 电抗网络
reactance of armature reaction 电枢反应电抗
reactance of opposite phase sequence 反相序电抗
reactance protection 电抗保护装置
reactance relay 电抗继电器
reactance resistance ratio 电抗电阻比
reactance resonance indicator 电抗谐振指示器
reactance stabilized oscillator 电抗稳定振荡器
reactance starter 电抗起动器
reactance triangle 电抗三角形
reactance tube 电抗管
reactance-tube frequency modulation 电抗管调频
reactance-tube modulator 电抗管调制器
reactance valve 电抗管
reactance variation method 变抗法
reactance voltage 电抗电压
reactant 作用物;成分;分应物;反应物;反应体;反应剂
reactant concentration 反应剂浓度
reactant cross-linker 反应性交联剂
reactant fixing dyes 反应性固色染料
reactant gas 反应气体
reactant mixture 反应混合物
reactant ratio 推进剂组分比;反应物比例
reactant resin 活性树脂
reactatron 低噪声微波放大器
react basic 呈碱性反应
reacting 起反应;反应(的)
reacting dose 反应量
reacting field 反作用场;反应场
reacting fluid 液体燃料的组成部分;反应流体
reacting force 反作用力;反推力;反力
reacting furnace 反应炉
reacting furnance 反应堆燃炉
reacting gas 反应气体
reacting interphase 作用相界
reacting phase 作用相;反应相
reacting species 反应物种
reacting steel 再结晶钢
reacting substance 起反应物;反应物质
reacting time 吸动时间
reacting wall 支顶墙;后座墙
reaction 反应;反(作用)力;反动力
reaction acceleration 反应加速
reaction accelerator 反应加速器;反应促进剂
reactional mineral 反应矿物
reaction alternator 反应交流发电机
reaction arrester 反应阻止剂

reactionary 反应(的)
reactionary torque 反作用力矩
reaction bath monitor 反应槽监测器
reaction beam 反力梁
reaction bed 反应床;反应池
reaction blade 反作用叶片
reaction blading 反击式叶片配置
reaction bonded silicon nitride 反应结合氮化硅
reaction bonding 反应结合
reaction bulb 反应瓶
reaction buttress 反力扶壁
reaction capacity 反应能力
reaction casting 铸钢枕座
reaction cavity 反馈谐振腔
reaction cement 活性粘结剂;反应粘结剂
reaction center[centre] 反应中心
reaction chamber 反应箱;反应室
reaction channel 反应道
reaction characterize equilibrium 反应表征平衡
reaction coefficient 反应系数;反馈系数
reaction coefficient of rock 岩石抗力系数
reaction coil 回授线圈;反馈线圈
reaction column 反应柱;反应塔
reaction component 无功分量;无功部分;电抗部分
reaction constant 反应常数
reaction control 回授控制;反作用控制;反应控制;反馈控制;反馈调整
reaction control agent 反应调节剂
reaction control aircraft 喷气操纵的飞机
reaction control and stabilizaton system 反应控制与稳定系统
reaction control correction 反应控制修正
reaction control data 反应控制数据
reaction control system 反作用控制系统;反应控制系统
reaction coordinate 反应坐标
reaction coupled 电抗耦合的
reaction coupling 回授耦合;电抗耦合;反馈耦合
reaction cross-section 反应截面
reaction current 反作用流
reaction curve 反作用曲线
reaction cylinder 反作用缸
reaction debris 反应屑
reaction differentiation 反应分异作用
reaction diffusion equation 反应扩散方程
reaction distance 反应距离
reaction distillation 反应蒸馏
reaction dynamometer 反作用测力计
reaction earth pressure 被动土压力
reaction effect 反应作用
reaction engine 喷气发动机;反作用式发动机
reaction enthalpy 反应焓
reaction enthalpy number 反应焓数
reaction equation 反应方程(式)
reaction equilibrium 反力平衡;反应平衡
reaction equilibrium constant 反应平衡常数
reaction factor 反应因数
reaction feedback of feel 反馈系数
reaction field 反应野
reaction flask 反应烧瓶
reaction floor 反力床
reaction flux 反应焊药
reaction force 反作用力
reaction formation 反应形成
reaction frame 反力框架;反力架
reaction frame for loading 加载反力架
reaction function 反应函数
reaction furnace 反应炉
reaction gas 喷射推进气体;工艺废气
reaction gas chromatography 反应气相色谱法
reaction gear 反转齿轮;反动齿轮
reaction generator 反馈发电机
reaction getter 反应型吸气器;反应收气剂
reaction heat 反应热
reaction hot-pressing 反应热压
reaction hydraulic turbine 反击式涡轮机;反击式透平;反击式水轮机;反击式汽轮机;反动水轮机
reaction hysteresis synchronous motor 反应式磁滞同步电动机
reaction indicator 反应指示剂
reaction-inhibiting 反应抑制
reaction initiator 引发剂
reaction injection mo(u)lding 反应注塑;反应注射模塑成型;喷射成型

reaction injection mo(u)lding machinery 反应注模机
reaction installation 反应装置
reaction intensity 反力强度
reaction interface 反应界面
reaction intermediate 反应中间体
reaction isochore 反应等容线
reaction isotherm 反应等温线
reaction jet 喷射流
reaction kettle 反应釜
reaction key 反应键
reaction kinetics 反应动力学
reaction-kinetic spectrometer 动力反应谱仪
reactionless 无反应的
reactionlessness 反应惰性
reactionless servo-mechanism 无反作用的伺服机构
reaction limiting current 反应极限电流
reaction liquid 反应液体
reaction locus 反应轨迹
reaction machine 反作用电机
reaction mass 反应物料
reaction mechanism 反应历程;反应机制;反应机理;反应规范
reaction medium 反应介质
reaction mineral 反应矿物
reaction motor 喷气发动机;反作用式发动机;反作用(式)电动机;反应式发动机;反应(式)电动机
reaction network 反应网络
reaction nozzle 反作用喷油嘴
reaction of bearing 轴承反作用(力);轴承反力;支承反力
reaction of clinker formation 熟料形成反应
reaction of combustion 燃烧反应
reaction of degeneration 变性反应
reaction of disappearance 消失反应
reaction of earthquake 地震反应
reaction of formation 生成反应
reaction of hardening 硬化反应
reaction of imposts 拱墩反力
reaction of non-identity 非一致性反应
reaction of nozzle 喷嘴反作用
reaction of propagation 增长反应
reaction of set(ting) 凝结反应
reaction of supports 支座反力;支点反力
reaction of the second order 双分子反应
reactionogen 反应原
reaction oil 反应油
reaction order 反应级(数)
reaction pad 反力衬垫
reaction pair 反应对;反力对
reaction pan 反应锅
reaction particle 反应粒子
reaction path 反应途径
reaction pattern 反应模式
reaction period 反应(周)期
reaction point 反力点
reaction polarization 反应极化
reaction pot 反应锅
reaction potential 反应电位
reaction power 反应能力
reaction-powered helicopter 喷气式直升机
reaction power plant 喷气式动力装置
reaction pressure 反压力;反应压力;反压
reaction primer 反应型底漆
reaction principle 反作用原理;反应原理;反动原理
reaction probability 反应概率
reaction process 反应过程
reaction product 反应生成物;反应产物
reaction profile 反应剖面图
reaction promoter 反应促进剂
reaction propelled 反作用力推进的
reaction propeller 反作用式螺旋桨
reaction propulsion 反作用力推进;反冲推进
reaction quality 反应质量
reaction quotient 反应商
reaction raft foundation 反梁式浮筏基础
reaction rail (高架高速公路的)反应护栏
reaction range 反应范围
reaction rate 反应(速)率
reaction rate analyser 反应率分析器
reaction rate (analysis) method 反应速率分析法
reaction rate constant 反应速率常数;表面反应速率常数

reaction rate equation 反应速率方程
reaction rate laws 反应速率公式
reaction rate theory 反应速率理论
reaction regeneration 正反馈
reaction relationship 反应关系
reaction resin 反应树脂
reaction rim 反应池;反应边
reaction ring 内凸轮环;导向环;反作用环
reaction scheme 反应图解
reaction sedimentation basin 反应沉淀池
reaction sedimentation tank 反应沉淀池
reaction selectivity 反应选择性
reaction sensitive adhesive 反应灵敏性粘结剂;反应灵敏性粘合剂
reaction sensitivity 反应灵敏度
reaction series 反应系列
reaction servo-mechanism 有负荷感觉的伺服机构;反作用伺服机构
reaction sintering 反应烧结
reaction soldering 反应钎焊;反应焊接
reaction speed 反应速度
reaction spring 反作用弹簧;反力弹簧
reaction stage 反应期;反应阶段
reaction stand 反力墙
reaction steam turbine 反作用汽轮机;反击式蒸汽轮机
reaction step 反应阶段
reaction step(ping) motor 反应式步进电机
reaction still 反应锅
reaction stress 约束应力;反作用力;反应应力
reaction structure 反应结构
reaction substance 反应质
reaction surface 反应面
reaction system 阻力装置(桩试验);反应系统;反应体系
reaction tank 反应池
reaction temperature 反应温度
reaction test 反应试验
reaction test for grease 润滑脂反应试验
reaction testing plates 反应测试板
reaction texture 反应边结构
reaction thrust 反推力
reaction time 反应周期;反应时间;反馈时间
reaction time method 反应时法
reaction timer 反应速度测量器
reaction time tester 反应时间测试仪
reaction torque 反抗扭矩;反作用扭矩;反作用力矩;反转力矩;反扭矩;反抗转矩
reaction torque arm 反扭矩杆
reaction tower 反应塔
reaction trap 止回阀;防逆瓣;止回活门
reaction turbine 反作用(式)涡轮(机);反力(式)涡轮(机);反动式透平;反动式汽轮机
reaction type 反馈式;反作用式
reaction-type brake tester 反力式制动试验台
reaction-type brush holder 反应式电刷架
reaction-type cavity wavemeter 反馈式空腔波长计
reaction-type cooler 涡轮式冷却器
reaction-type engine 反作用式发动机
reaction-type frequency meter 反馈式频率计
reaction-type stage 反动级
reaction-type wheel 反击式叶轮;反击式涡轮机;反击式水轮(机);反击式汽轮机
reaction vane 反应式叶片
reaction velocity 反应速度
reaction velocity constant 反应速度常数
reaction vessel 树脂反应锅;反应锅
reaction wall 反力墙
reaction water pump 反作用式水泵;反动式水泵;反冲式水泵
reaction water turbine 反击式水轮机;反动水涡轮
reaction water wheel 反击式水轮
reaction wave 反应曲线;反射波
reaction wheel 燃气涡轮转子;反作用轮;反力轮;反动轮;反冲式叶轮
reaction winding 反作用绕组;反作用绕法
reaction wood 应力木;反应木材
reaction yield 反应产率
reaction zone 反应区;反应带
reactivate 再活化;再生;活性恢复;重激活
reactivated mountains 复活山地
reactivated old slide 重新活动的老滑坡
reactivating temperature 复活温度
reactivation 再生(过程);再激活;再活化(作用);

活性化;复原作用
reactivation cycle 复活周期
reactivation line 再活化管路
reactivation of partition 分区再启动
reactivation of sludge 活性污泥再生
reactivation surface 复活面
reactivation surface structure 再作用面层理构造
reactivator 再生器
reactive 活性的;电抗性的;反应性的;反应
reactive acrylate adhesive 反应性丙烯酸盐粘合剂
reactive adhesive 反应性黏合剂
reactive aggregate 与水泥起反应的骨料;活性集料;活性骨料
reactive aggregate concrete 活性集料混凝土;活性骨料混凝土
reactive aluminium in the Vermont soil test 测定佛蒙特土壤中的活动性铝
reactive atmospheric(al) processing 反应气氛法
reactive azo dye 活性偶氮染料
reactive black 活性黑
reactive blue 活性蓝
reactive bond 活性键
reactive brilliant red 活性艳红
reactive brilliant red dye wastewater 活性艳红染料废水
reactive brilliant yellow 活性嫩黄
reactive characteristic 反应特性
reactive circuit 电抗电路;反馈电路
reactive coil 电抗线圈
reactive component 无功分量;无功部分;电抗分量;电抗部分
reactive concrete aggregate 活性混凝土集料;活性混凝土骨料
reactive curing agent 反应性固化剂
reactive current 无功电流;电抗性电流
reactive dark green 活性墨绿
reactive disperse dye 活性分散染料
reactive drawing 反力拉伸;反拉力拉丝
reactive drilling 反冲钻进
reactive drop 无功电压降
reactive due hydrolysate 活性染料水解产物
reactive dye 活性染料
reactive effect 电抗效应
reactive equalizer 电抗均衡器
reactive etching 反应侵蚀
reactive evapo(u)ration 反应蒸发
reactive factor meter 电抗系数测量计
reactive fiber 活性纤维
reactive fixing agent 反应性固色剂
reactive flame retardant 反应性阻燃剂;反应性阻燃剂
reactive flavine 活性嫩黄
reactive force 反作用力;反应力;反冲力
reactive golden yellow 活性金黄
reactive green 活性绿
reactive group 反应基团
reactive haulage system 反力牵引系统
reactive high molecular weight phosphorus 化学高分子量磷
reactive hot isostatic pressing 反应热等静压
reactive hot pressing 反应热成型;反应热压
reactive hydrocarbon 活性烃
reactive hydrogen 活性氢;活泼氢
reactive intermediate 活性中间体;活性中间产物;活泼中间体
reactive ion beam etching 反应离子束刻蚀
reactive ion etching 反应离子刻蚀
reactive ion exchange 活性离子交换
reactive iron 电抗铁(心)
reactive light yellow 活性嫩黄
reactive line 无功线路
reactive liquid polymer 反应性液态聚合物
reactive liquid sintering 反应液相烧结
reactive-load 无功负载;无功负荷;无功性负载
reactive-load compensation 无功补偿
reactive-load compensation equipment 无功补偿设备
reactive material 反应物质
reactive metal 活性金属
reactive mode 反应式(工作制);反应方式
reactive monomer 活性单体;反应性单体
reactive muffler 抗性消声器;反作用消声器
reactive navy blue 活性藏青
reactiveness 反应性
reactive orange 活性橙

reactive orientation 反应式行动倾向
reactive particle 反冲粒子
reactive-peak limiter 无功峰值限制器
reactive phosphorus 反应磷
reactive pigment 活性颜料
reactive plasticizer 活性增塑剂
reactive polymer 活性聚合物;反应性聚合物
reactive powder 活性粉末;活性粉末
reactive powder concrete 活性粉末混凝土
reactive power 无效功率;无功功率
reactive power compensation equipment 无功功率补偿装置
reactive power control 无功功率控制
reactive red 活性红
reactive red violet 活性红紫
reactive resin 活性树脂;反应型树脂
reactive resistance 电抗
reactive scarlet 活性大红
reactive scattering 反应散射
reactive silica material 活性硅(石)材料
reactive silicate 活性硅酸盐
reactive sintering 活性烧结;反应烧结
reactive site 活化点
reactive solid waste 反应性固体废物
reactive solute 活性溶质;反应溶质
reactive solute concentration 活性溶质浓度
reactive solute transport 活性溶质运移
reactive source 电抗源
reactive species 活性种;活性反应组分
reactive sputtering 反应溅射
reactive state 反应状态;反应状况
reactive step(ping) motor 反应式步进电机
reactive still 反应釜
reactive suspended carrier 反应悬浮载体
reactive tank 反应槽
reactive texture of aggregate 集料活性质地
reactive thrust 反冲力
reactive turquoise blue 活性翠蓝
reactive-type flame retardant 反应型阻燃剂
reactive vapour deposition 反应气相沉积
reactive voltage 无功电压
reactive voltage-ampere 无功伏安
reactive voltage component 无功电压分量
reactive volt-ampere hour 乏时
reactive wiredrawing 反拉力拉丝
reactive yellow 活性黄
reactivity 再生性;活性;反作用性;反应作用;反应性;反应度
reactivity control system 反应性控制系统
reactivity disturbance 反应性扰动;反应性偏差
reactivity excursion 反应性急剧上升
reactivity feedback 反应性反馈
reactivity gain 反应性增益
reactivity index 反应活性指数
reactivity insertion rate 反应性插入率
reactivity lifetime 反应堆连续运行时间
reactivity meter 反应性测量计
reactivity monitor 反应性监测仪
reactivity noise 反应性噪声
reactivity of clinker 熟料活性
reactivity oscillator 反应性振荡器
reactivity power coefficient 反应性功率系数
reactivity pressure coefficient 反应性压力系数
reactivity rate 反应性变化率
reactivity ratio 反应率;反应竞形率
reactivity spectral density 反应性谱密度
reactivity-to-power transfer function 反应性功率传递函数
reactivity transient 反应性瞬变
reactivity worths 反应性价值
reactology 反应学
reactor 稳定器;电抗器;反应器;反应剂;反应锅;反应釜;反应堆
reactor arrangement 反应堆外形轮廓
reactor auxiliary systems 反应堆辅助系统
reactor behavio(u)r 反应堆性能
reactor biodegradation system 反应器生物降解系统
reactor biological tunnel 反应堆生物实验孔道
reactor blanket 反应堆再生区
reactor block 反应堆体;反应堆室;反应堆区;反应堆舱
reactor breakdown 反应堆损坏
reactor by-product 反应堆副产品

reactor calculation 反应堆计算
reactor capability 反应堆功率
reactor cavity 反应堆腔
reactor chemistry 反应堆化学
reactor circulation 反应器循环
reactor code 反应堆计算程序
reactor-colo(u)red polymer 反应器着色聚合物
reactor commissioning 反应堆试运行
reactor compartment 反应堆室;反应堆舱
reactor composition 反应堆成分
reactor containment 反应堆安全壳;安全壳
reactor containment structure 反应堆(密封)外壳结构
reactor control 反应器控制;反应堆控制
reactor control board 反应堆控制台
reactor control materials 反应堆控制材料
reactor control system 反应堆控制系统
reactor-converter 反应堆转换器
reactor cooling 反应器冷却
reactor cooling water 反应器冷却水
reactor core 堆芯
reactor core radiation detector 反应堆堆芯辐射检测仪
reactor cross section 反应堆截面
reactor cubicle 反应堆室;反应堆舱
reactor debris 反应堆裂变物
reactor design 反应器设计
reactor-down 反应堆功率下降
reactor dynamics 反应堆动力学
reactor effluent 反应器流出物
reactor engineering 反应器工程学
reactor experiment 实验反应堆
reactor feed 反应器进料
reactor fission 反应堆裂变;反应堆分裂
reactor fouling 反应器积垢
reactor fuel 核燃料;反应堆燃料
reactor fuel cycle 反应堆燃料循环
reactor fuel failure monitor 反应堆燃料事故监测仪
reactor fuel meltdown 反应堆燃料熔化
reactor fuel pellet 核燃料芯块
reactor-grade graphite 适合反应堆用的石墨
reactor grade metal 反应堆级金属
reactor grounded neutral system 中心点电抗接地系统;电抗线圈中点接地系统;电抗接地中性系统
reactor group 电抗器组
reactor hazard 反应堆灾害
reactor heating 反应堆放热
reactor height 反应堆高度
reactor housing 反应堆外壳建筑;反应堆外壳结构
reactor hydrodynamics 反应器水动力学
reactor installation 反应堆装置
reactor-irradiator 反应堆辐照器
reactor kinetics 反应堆动力学
reactor kinetics equation 反应堆动力学方程
reactor make-up water 反应堆补给水
reactor materials 反应堆材料
reactor metallurgy 反应堆材料冶金学
reactor network 反应器网络
reactor neutron dosimeter 反应堆中子剂量计
reactor noise 反应器噪声;反应堆噪声
reactor off-gas system 反应堆除气系统
reactor optimization 反应堆最佳化
reactor oscillator 反应堆振荡器
reactor period 反应堆周期
reactor physicist 反应堆物理学家
reactor physics 反应堆物理学
reactor poison skirt 反应堆毒物围筒
reactor polison removal 反应堆毒物排除
reactor power level 反应堆功值
reactor pressure vessel 反应堆压力壳
reactor reliability 反应堆可靠性
reactor resistance 电抗器电阻
reactor restart 反应堆再起动
reactor riser 提升管反应器
reactor room 原子锅炉舱
reactor safety 反应堆安全
reactor shell 反应器筒体
reactor shield 反应堆屏蔽层
reactor shimming 反应堆补偿
reactor shut-off system 停堆系统
reactor simulator 反应堆模拟装置;反应堆仿真器
reactor source 反应堆启动源
reactor stability 反应堆稳定性
reactor starter 电抗起动器

reactor starting 电抗器起动
reactor start motor 电抗起动电机;电抗线圈起动式电动机
reactor start-up 反应器启动;反应堆启动
reactor start-up accident 反应堆启动事故
reactor start-up source 反应堆启动源
reactor statics 反应堆静力学
reactor structure 反应堆结构
reactor technology 反应堆工艺学
reactor transfer function 反应堆传递函数
reactor trip 事故保护停堆;反应堆事故停堆
reactor-up 反应堆功率上升
reactor variables 反应堆变量
reactor vessel 反应罐;反应堆压力外壳;反应堆容器;反应壶
reactor voltage 电抗器电压
reactor waste 反应器废料;反应堆废物;反应堆废料
reactor water cleanup system 反应堆水净化系统
read 读写与计算;判读;读出
readability 易读性;可读性;清晰度;便读性
readability of map 地图易读性
readable 清楚的;易读的
readable line 可读行
read access 读访问
read access time 读取数时间
read access type 读访问类型
read address counter 读地址计数器;读出地址计数器
read-after-write check 写后读检查
read-ahead queue 读前队列
read all data 读出全数据
read analog input 读模拟量输入
read analog(ue) input 模拟量输入
read and write 直读式记录
read-around number 整个阅读数字;连续读出比(率);环读次数;读出次数
read-around ratio 连续读出比(率);连读次数;读数比;读出比
read-back 回读;读回
read-back check 回读校验;回送检测;读回检查;读后校验;读出回校
read-back signal 读回信号
read backward 反向读出;反读
read bit 读位
read brush 读出刷
read bus 读出总线
read circuit 读数线路
read clock 显示时钟
read command 读命令
read compare check 读出比较校验
read current 读出电流
read cursor 读出光标
read cycle 读出周期
read data latch signal 读出数据自锁信号
read data strobe 读数据选通
read detector 读出检测器
read digital input 数字量读入;读数字输入
readdressing 再寻址
readdressing routine 变地址程序
read driver 读数驱动器;读驱动器
reader 阅读机;读数器;读出器;读程序;分类读器
reader check 读出校验
reader for microfilm 缩微胶片阅读器
reader interpreter 读入解释程序;读入翻译程序
reader-interpreter routine 读入解释程序
reader light 读出器指示灯
reader-out unit 读出装置
reader printer 读数打印机
reader-punch 读数穿孔装置
read error 读错误;读出错
reader's comment form 读者注解形式
reader's interested 图书发行对象
reader-sorter 读出分类器
reader's set 内校样
reader stop 读出器停止
reader typer 读数打印机;读数打字机;读数打印装置
reader unit 阅读器
read exit option 读出出口选择
read forward 正向读出
read frequency 读出频率
read frequency input 读频率输入
read gate 读选通
read half-pulse 半选脉冲;半读脉冲
read head 读数头;读出(磁)头

readily accepted 便于接受的
readily accessible 易通过的
readily available fertilizer 速效肥料
readily available water 有效水分;易效水;最大有效含水量;速效水
readily biodegradable organics 易生物降解物质
readily carbonizable substance 易炭化物
readily cash 现金
readily dissociate 易溶解
readily removable pack 易卸包
read-in 读入
read-in readout 读入与读出
read-in data 读入数据;写入数据
readiness area 报警信号台;战备区
readiness for operation 运行状态
readiness for sea period 准备出航时期(船大修后)
readiness monitoring operation 准备情况检查操作
readiness posture 战备状态
readiness reserves 应急补给品与装备
readiness review 就绪复审;启用前检验;备用状态观察;安装后检验
readiness time 预备时间;准备时间
reading 量测读数;示度;读数;传感读出;判读;读出
reading access time 读出时间
reading accuracy 观读精度;读数精度
reading and writing amplifier circuit 读写放大电路
reading apparatus 读数装置
reading beam 显示电子束
Reading beds 雷丁层【地】
reading by estimation 估读
reading by reflection 反射式读数
reading consistency 读数符合度
reading desk 斜面书桌;读书桌
reading device 阅读器;刻度盘;示读装置;度盘;读数装置
reading dial 读数盘
reading drawings 理解和解释图纸;读图(纸)
reading drum (自记仪的)读数卷纸筒
reading duration 读数时间;读出时间
reading electrode 读数电极
reading error 记录误差;判读误差;读数误差;读出误差
reading error of parallax 读数视差
reading for press 校对清样
reading forward 正向阅读;正读
reading function button 读入按钮
reading glass 读数(放大)镜;放大镜
reading gun 读出电子枪
reading hall 阅览厅
reading head 读头;读数头;读出头
reading-illuminated lamp 读数照明灯
reading index 读数指示
reading instrument 指示仪器;指示仪表
reading in thousandths of an inch 读数0.001英寸的
reading jump 读数突变
reading lamp 台灯
reading lens 读数镜
reading light 读数照明灯
reading list 读数表
reading machine 阅读器;阅读机;读出装置
reading magnifier 读数(放大)镜
reading method 读出法
reading micrometer 读数测微器;读数测微计
reading microscope 读图显微镜;读数显微镜;读数测微镜
reading mirror 读数镜
reading number 读出数字
reading of a pointer 指针读数
reading of density log 密度测井读数
reading off device 示值读数装置;阅览设备
reading of master ground-wave and slave sky-wave match 主台地波重复副台天波的改正量
reading of piezometer 测压计读数;测压管读数
reading of pointer 指针读数器
reading of record(ing) instrument 记录仪表读数
reading of tracking error 跟踪误差读数
reading on rod 标尺读数
reading optic(al) beam 读出光束
reading plotter 记录转换系统记录装置;读数装置
reading point 读数点
reading position 读数位置
reading precision 读数精度

reading prism 读数棱镜
reading process 读数过程
reading rate 读出速度
reading report 书面报告
reading room 阅览室
reading room library 阅览室藏书
reading scale 读数比例尺
reading scan 显示扫描;读出扫描
reading spectacles 读数镜
reading speed 读数速度;读出速度
readings per second 每秒读数
readings pointer 读数指针
reading station 阅读台;读数装置;读数台
reading system 校对系统;读数系统
reading table 读数台;读效桌;读书桌
reading telescope 读数望远镜
reading tolerance 读数公差
reading track 读出道;读出导向装置;读出磁道
reading type of instrument 读数方式
reading unit 读数装置
reading value of inflection point 拐点读值
reading value per division of vernier 游标每格读数值
reading velocity 读速度;读出速度
reading-writing amplifier 读写放大器
read initial program(me) 读出初始程序
read-in process 读入过程
read-in program(me) 读入程序
read in unit 读入部件
read-jump protection 读出转移保护
readjust 重调整;重安装
readjust current 重调电流
readjusting entry 重调整分录
readjusting gear 重调整装置
readjusting industrial structure 调整产业结构
readjusting industry and commerce 调整工商业
readjusting operation 整理作业
readjustment 再调整;重新平差;重新调整;重校准;重调
readjustment of farm boundaries 补整田地
readjustment of field plot 地块整理
readjustment of foreign exchange rate 调整外汇汇率
readjustment of interest rate 调整利率
readjustment of investment structure 投资结构调整
readjustment of zero 重新调零
readjustment plan 修正方案;调整方案
readjust price 调整物价;调整价格
readjust supply and demand 调整供求关系
read-lock 封锁读
read-made 制成品
readmit 重新接纳
read-modified operation 读改操作
read modify 读出修正
read number 读出数
read off 读取;读出
read only file 只读文件
read only memory 只读存储器
read only procedure 只读过程
read only storage 只读存储器
read operation 读出操作
read operator 读出操作员
read or write speed 读写速度
readout 数字显示装置;直读式的;结果输出值;结果传达;示值读数;读出
readout amplifier 读出放大器
readout apparatus 读出装置
readout circuit 读出电路
readout command 显示指令
readout connector 测读连接器
readout counter 读出计数器
readout coupling 测读连接头
readout cryotron 读出冷子管
readout desk 读出台
readout device 显示装置;读出装置;测读装置
readout display 读数显示器;读出显示器
readout error 读出误差;读出错误
readout gate 读出门
readout head 读出头
readout information 读出信息
readout instrument 直读式仪表;读数仪表
readout memory 永久存储器;固定存储器
readout meter 直读式仪表;直读式频率计

readout mode 读出方式
readout noise level 读出噪声电平
readout plotter 读数装置
readout print 读出打印机
readoutput signal 读出信号
readout signal 读出信号
readout station 读出装置
readout system 读出系统
readout-tube 读出管
readout unit 读数装置;读出装置
read path 读通路
read preamplifier 读出前置放大器
read proof 初校样
read-protected 读保护的
read protection 读保护
read pulse 读数脉冲;读出脉冲
read pulse switch 读脉冲开关
read-pulse synchronization logic 读脉冲同步逻辑
read-punch 读孔穿孔机
read punch unit 卡片输入穿孔机;读卡穿孔机;读取穿孔机
read rate 读出速度
read release 读机构断开
read request 读请求
read request cycle 读请求周期
read response 读出响应
read-response cycle 读响应周期
read reverse 倒转读出
read routine 读数程序
read saturation 读数饱和度
read signal 读数信号
readsorption 再吸附
read specially paper tape 读取特殊纸带
read statement 读语句
read station 读出台
read status microinstruction 读状态微指令
read stop 读出停止
read store 读出存储
read strobe 读出选通
read through model 读模块
read time 读出时间;存取时间
read-to-read crosstalk 读间串扰
read track 读出磁道
read vector instruction 读向量指令
read-while-write 边写边读
read-while-writing 写时读
read winding 读绕组;读出绕组
read-write access mode 读写访问方式;读写存取方式
read-write amplifier 读写放大器;读出记录放大器
read-write channel 读写通道
read-write check 读写校验;读写检验;读写检查
read-write check indicator 读写校验指示器;读写检验位
read-write command 读写指令
read-write compatible head 读写兼用头
read-write control 读写控制
read-write counter 读写计数器
read-write cycle 读写周期
read-write deck 读写卡片组
read-write drive 读写驱动器
read-write head 读写磁头;读写头
read-write memory 读写存储器
read-write memory module 读写存储器插板
read-write operation 读写操作
read-write process 读写处理
read-write protection 读写保护
read-write speed 读写速度
read-write station 读写站
read-write storage tape 读写存储带
read-write subpool 读写子池
read-write waiting time 读写等待时间
ready 准备就绪的;读就绪
ready berth clause 准备停泊条款
ready cash 现金
ready chain 就绪链
ready coating 预制涂料;快速罩面;简易浇面
ready condition 准备状态;可算条件;就绪(条件)
ready cut house 预制木结构房屋;装配式木屋
ready delivery 即期交货
ready disposition 准备部署
ready flow 流动性
ready foods 方便食品;现场食品
ready for data 就绪接收数据

ready for delivery 待发运
ready for installation 准备好安装
ready for mounting 准备好安装
ready for occupation 做好居住准备
ready for operation 运行准备就绪;准备运行;待运转
ready for sail 准备航行
ready for sea 准备出海
ready for sending 准备发送;发送就绪
ready for service 准备投入业务
ready for towing 备拖
ready-for-use 备用(的)
ready indicator 准备就绪指示灯;就绪指示符;就绪显示器
ready-issue room 日用品供应舱
ready lifeboat 值勤救生艇
ready light 就绪指示灯;操作准备指示灯
ready line 准备线;就绪引线
ready list 就绪表
ready list as a queue 就绪表队列
ready-made 预先准备的;现成的;预制的;制成的
ready-made articles 成品
ready-made clothes production line 成衣生产线
ready-made goods 现成制品
ready-made hinge 普通铰链
ready-made mixture 现成配料;备好配料
ready-made mud 预制泥浆
ready-made partition 预制隔墙
ready-made tiled partition 现场贴面砖隔墙
ready-made unit 预制部件;预制构件;工厂预制部件
ready message 准备信息
ready-mix 厂拌的
ready-mix cement truck 预拌水泥输送车
ready-mix concrete 预拌混凝土;已搅拌混凝土
ready-mix distribution facility 商品混凝土摊铺机
ready-mixed 混合好的;搅拌好的;预拌的
ready-mixed alumin(i)um paint 铝粉调合漆
ready-mixed concrete 商品混凝土;预拌混凝土;厂拌混凝土
ready-mixed concrete lorry 预拌混凝土输送车
ready-mixed concrete operator 预拌混凝土操作者;预拌混凝土生产者
ready-mixed concrete plant 预拌混凝土站;预拌混凝土(工)厂
ready-mixed concrete spreader 预拌混凝土摊铺机
ready-mixed concrete tower 预拌混凝土站;预拌混凝土楼
ready-mixed distribution facility 预拌混凝土摊铺机
ready-mixed glue 商品混合胶;混成胶
ready-mixed gypsum-lime plaster 预拌石膏石灰灰浆
ready-mixed joint compound 预拌嵌缝混合料
ready-mixed lean concrete 预拌弱混凝土
ready-mixed lime-sand mortar 预拌好的石灰砂浆
ready-mixed mortar 预拌砂浆
ready-mixed oil-base(d) paint 预拌油基涂料
ready-mixed paint 调合漆;调和漆;调和油漆
ready mixed paint of synthetic(al) resin 合成树脂调合漆
ready-mixed plaster 预拌粉刷;预拌石膏灰
ready-mixed stuff 预拌填料;预拌涂料;预拌材料
ready-mixed truck 成品混凝土运送车;预拌混凝土运送车;预拌混凝土拌运车
ready-mixed wet paste 预拌湿浆糊
ready mixed zinc white paint 锌白调合漆
ready-mix paint 预调合油漆
ready-mix plant 预拌工厂;预拌车间;混凝土(工)厂
ready-mix truck 商品混凝土拌运车;成品混凝土运送车
ready mode 预备式
ready money 现款;现金
ready money business 现款交易
ready-package 快装式
ready payment 现款支付;付现
ready program(me) 就绪态程序
ready queue 就绪队列
ready-read card 可读卡片;备读卡片
ready reckoner 简便计算表;计算便览
ready reference 现成参考资料
ready reserves 现成备用容量;容量(能立即投入的)一级备用容量
ready roofing 预制屋面料;商品屋面料;组合屋面料
ready roofing manufacture 生产预制屋面料
ready roofing shingle 沥青屋面板

ready sheet roofing paper 预制屋面油毛毡
ready signal 准备信号
ready state of a process 进程的就绪状态
ready status word 准备状态字
ready task 就绪任务
ready task queue 就绪任务排队
ready-to-assemble 准备装配
ready to be picked up 准备装运
ready-to-erect 准备好架设
ready-to-mix 准备好拌和
ready-to-mix alumin(i)um paint 双包装铝粉漆
ready to mount 随时可以安装
ready to operate 随时可使用的;时刻准备好的
ready-to-paint 准备上漆
ready-to-place 准备安放;准备安装
ready to put-off 准备出海
ready-to-receive signal 准备接收信号
ready to sail 准备出海
ready-to-servo requirement 备用设备
ready-to-spray 准备喷射
ready to start 准备起动
ready-to-use 随时可用;可供使用的
ready-to-use material 随时可用的材料
ready-to-use paste 随时可用的胶
ready-to-wear 现成的
re-aerate 复氧;再曝气
reaeration 再充气;重新充气;复氧;再曝气
reaeration by atmosphere 大气复氧
reaeration coefficient 复氧系数
reaeration constant 再曝气常数;复氧常数
reaeration in river 河流复氧作用
reaeration of aeration tank 曝气池再曝气
reaeration rate 再曝气率
reaeration sludge 复氧污泥
reaeration tank 再曝气池;再生池
reaeration test 再曝气试验
reafforestation 再造林;重新造林
reag(e)ing 再老化;重试线;反复时效
reagent 反应物;试药;试剂;反应力
reagent blank 试剂空白
reagent bottle 试剂瓶
reagent chemicals 试剂类化学药品
reagent concentration 还原剂浓度
reagent distributor 试剂分配器
reagent feeder 给药机;试剂加入器
reagent-feeder battery 给药机组
reagent for chromatography 色层试剂
reagent for nickel 试镍剂
reagent grade 试剂级别;试剂等级
reagent identification 试剂鉴定
reagent method of water softening 化学试剂软化法
reagent quantity 还原剂用量
reagent resistance 抗试剂性
reagents of high purity 高纯试剂
reagent solution 试剂溶液
reagent's purity 试剂纯
reagent type 试剂种类
reagglomeration 再结块;再附聚
reaggregation 重聚合
reaggregation agent 再聚集剂
reagin 反应素
reaging 反复老化
real 真实的;实数【数】
real absorption 有效吸收(量)
real account 实物账户;实账户(即资产账户负债账户及资本账户);实账;实体账户
real accumulator 实数累加器
real action 产权诉讼
real add 实数加
real address 实(际)地址
real address area 实址区;实(际)地址区
real air temperature 实际气温
real allowance 实际容许量
real amplitude preservation stack 真振幅保持叠加
real analysis 实分析
real aperture 有效孔径;真实孔径;实际孔径
real aperture antenna 真实孔径天线
real aperture radar 真实孔径雷达;实孔径雷达
real area 实际面积
real argument 实变元
real arithmetic 实(数)运算
real arithmetic(al) constant 实运算常数
real arithmetic(al) data 实际计算数据

real assets 不动产;实物资产;实体资产
real assets depreciation 固定资产折旧
real axis 实轴
real balance 实际平衡
real-balance-effect theory 实际余额效应论
real bill 真实票据;实质票据
real bitter spring 硫酸镁泉
real brightness 真亮度
real capital 实物资本;实际资本
real capital flow 实际资本流动
real capital maintenance 保持实际资本
real capital stock 全部实际资本
real cash balance 实际现金余额
real channel 真实信道
real colo(u)r 天然色
real component 有功分量;实分量;时分量
real component of secondary field 二次场实分量
real constant 有效常数;实常数
real construction time 实际建造时间
real correlation 真相关
real cost 真实成本;实际成本
real cost of production 实际生产成本
real cost terms of trade 贸易实际成本条件;实际成本贸易条件
real credit 实物信用;实际信用
real cross-section 实际截面
real crystal 全晶玻璃;实际晶体;不完美晶体
real curves 实际曲线
real data 实型数据
real data item 实数据项
real debt 实际债务
real definition 真实定义
real demi-gross weight 实际半毛重
real denotation 实(数)标志
real density 真密度;干у重
real dimension 实际尺寸
real displacement 实际位移
real disposable per capita income 人均实际可支配收入
real-dollar value 真实美元价值
real domain 开域;实域
real domestic wastewaters post-biotreatment 真实生活污水后生物处理
real drive 实驱动
real driver 实驱动器
real earning 实质收益
real earth 实地球
real earthquake 真实地震
real effective exchange rate 实际有效汇率
real effective exchange rate index 实质有效汇率指数
real electrolyte 真正电解质
real error 真误差
real estate 物业不动产;地产;房地契;房地产;不动产
real estate acquisition 不动产购置
real estate agency 房地产公司
real estate agent 房地产经纪人;地产商
real estate analyst 房地产分析员
real estate and property 不动产与其他财产
real estate appraising 房地产估价
Real Estate Board 房地产商会;房地产董事会
real estate broker 房地产经纪人
real estate bureau 房地产管理局
real estate cadaster 房地产地籍
Real Estate Commission 房地产委员会
real estate contract 不动产合同;不动产契约;房地产合同;房地产契约
real estate counseling 房地产交易顾问
real estate credits 不动产信贷
real estate debt 不动产负债
real estate development 房地产开发;物业开发
real estate equity trust 房地产信托公司
real estate finance 不动产财务
real estate financing 购置不动产借款
real estate investment trust 房地产投资信托
real estate license 房地产经营执照
real estate lien 不动产留置权
Real Estate Limited Partnership 房地产投资合作有限公司
real estate loans 不动产放款
real estate market 房地产市场
real estate matter 房地产问题
real estate mortgage investment conduit 房地产抵押投资渠道

real estate mortgage trust 房地产抵押信托
real estate office 房地产事务所
real estate owned 持有的房地产
real estate pool 不动产投资共同基金
real estate professional 不动产专业人员;房地产专业人员
real estate property 土地资产
real estate research 房地产调查;房地产研究
real estate salesman 房地产经销商
real estate salesperson 房地产经销商
Real Estate Settlement Procedures Act 房地产处置程序法
real estate tax 房地税;不动产税
real estate trust 不动产托拉斯;房地产托拉斯
real evapotranspiration 实际蒸腾
real evapo(u)ration 实际蒸发(作用)
real expansion 实膨胀
real experience 实际工作经验
real exponent 实指数
real expression 实型表达式
real extra 上等产品
real fault displacement 实际断层移距
real field 实域;实场
real field of view 实视场
real figure 实际数字
real file 实文件
real flattening 真扁率
real flow 实物流动
real fluid 真实流体;实际液体;实际流体
real fluid flow 真实流体流
real flying height 真航高
real focal plane 实焦面
real focus 实焦点
real forest 现实林
real format item 实格式项
real frequency axis 实频率轴
real function 实函数
realgar 雄黄;鸡冠石;二硫化二砷
real gas 真实气体;实在气体;实际气体
real geometry 实素几何学
real gradient 实际坡度
real growth rate 实际增长率
real guarantee 实物担保
real hardening curve 实硬化曲线
real head 实际水头
real hinge 实铰
real hologram spread function 实像全息图的扩散函数
real horizon 真地平(圈)
real horsepower 指示马力;实际马力
realign(ing) 改线;重新定线;重新校准;拨道
realignment 改线;调线;重新组合;重新排列;重新定线;重定(路)线;公路中心线调直
realignment escalation 重组升级
realignment of curve 曲线调整
realignment of exchange rates 重整汇率
real image 实像
real image holographic(al) stereomodel 实像全息立体模型
real image plane 实像平面
real income 房地产所得;实际收益;实际收入
real income index 实际收入指数
real income per capita 人均实际收入
real indicator 实际示意图
realinement 改线;重新定线
realining 改线
real installation capacity 实装机容量
real interest rate 真实利率;实质利率;实际利率
real interval 实区间
real investment 实际投资
realism 现实主义
realist 房地产经纪人
realistic accuracy 真实精度
realistic colo(u)r 逼真彩色
realistic design 切合实际的设计
realistic display 真实显示
realistic object 实物
realistic rendering 逼真渲染
realistic scale 实用水平
realistic simulation 实际模拟
realistic testing specimen 仿真试件;仿真样品
realistic two-phase soil 天然三相土
reality condition 现实性条件
reality criterion 逼真度准则

reality of roots 根的实性
reality principle 现实原则
realizable assets 可变现资产
realizable function 可实现函数
realizable limit 鉴定极限;确认极限
realizable utilization of heat energy stage by stage 逐级合理利用热能
realizable value 可实现价值;可变现价值
realizable value of accumulated net cash flow 累计净现金流量现值
realizable value of net cash flow 净现金流量现值
realization 变现
realization and liquidation 变产清算
realization and liquidation profit and loss 变产清算损益
realization convention 变现常规
realization of appropriation 占有的实现
realization of loss 亏损实现
realization of profit 利润实现
realization of property 变卖财产
realization response 实现反应
realization value 实际变现值;变现价值
realize a profit 赚得利润
realized appreciation 实际增值;已实现增殖;实得增值
realized capital 变现资本
realized depreciation 实际折旧;已实现折旧;变卖折旧
realized gains 已实现收入
realized gains and losses 实际损益
realized gains or losses 已实现损益
realized holding gain 已实现资产持有利得
realized investment 已实现投资
realized natality 实际出生率
realized niche 实际生态位
realized price 实现价格;实售价格;实际价格
realized profit 已实现利润;实现的利润;实际利润;变卖利益
realized property 实际财产
realized revenue 已实现收入;变现收入
realized surplus 已实现盈余;实现盈余;实际盈余
realized value 市场价值;实现价值
realize property 变产
realize returns on an investment 发挥投资效果
real labo(u)r cost 实际劳动成本
real lead (打桩时的)实际导程
real liability 实际负债
real-life 有效寿命
real line 实线
real liquor 实际液体
real load 实际荷载;有效荷载;有效负载;有效负荷
real loading capacity 实际载重量
reallocate sequentially stored table 对顺序存储表进行再分配
real location 实位置;重新分配
reallocation of land 土地规划
reallocation of partition 分区重新分配
real loss 实际损失
reall-valued function 实值函数
realm 领域;区域
real machine 实机
real magnetism 实磁性
real material figure in induced polarization method 激电法实际材料图
real matrix 实矩阵
realm currency indicator 域当前指示符;领域现行指示器
realm data item 领域数据项
realm description entry 域描述项;区域描述体
real memory 实(在)存储器
realm identifier 域标识符
real mining yield 实际开采量
real mixture standard cycle 实际混合气标准循环
realm name 域名;区名
real mode 实(址)方式;实模式
realm of art 艺术领域
realm of freedom 自由王国
realm of literature and art 文艺领域
realm of nature 自然界
realm of necessity 必然王国
realm of superstructure 上层建筑领域
real money 现金;实际货币
real money balance 实际现金余额;实际货币余额
real mould 实模型

real national income per capita 人均实际国民收入
real net profit 实际净利润
real number 实数【数】
real number axis 实数轴
real number field 实数域
real number of semi-cushioned seats on passenger train 硬座实际定员
real object 实物
real observation condition 实际观测条件
real observation time 实际观测时间
real orthogonal group 实正交群
real output 实际输出功率;实际产量
real oxygen requirement 实际需氧量
real parameter 实参数
real part 实数部分;实部
real-part condition 实部条件
real partition 实分区
real part of symbol 实部符号
real part operator 实部算子
real pattern 实模式
real personnel 实际人员数量
real point 实点
real position 实际位置
real potential 真实电位
real power 有效功率;有功功率;实际功率
real power analysis 有效功率分析
real precession 真实进动
real pressure 实际压力
real price 实价;实际价格
real process 实际过程
real product 实际产值
real productive activity 现实的产生活动
real property 遗赠财产;实际产业;不动产
real property loan law 不动产贷款法规
real property sales contract 不动产出售合同
real pseudo-variable 实伪变量
real purchasing power 实际购买力
real quadratic form 实二次型
real quantity 实数【数】;实量
real quantity of water consumption 实际用水量
real quantity of water supply 实际供水量
real radius 实际半径
real rate of exchange 实际汇率
real resources 实际资源
real restriction 实域局限
real result 实际结果
real right 物权
real river 真实河流
real root 实根
real saving 实际储蓄
real scale 真比例尺;实际比例尺
real security 不动产抵押保证;实物担保
real signal 实信号
real singularity 实奇点
real situation 真相
real size 实际规模;实际尺寸
real slip 实滑距
real solubility 真溶(解)度
real solution 实际溶液
real source 实声源
real space 实物空间;实空间
real specific gravity 真比重
real spread function 实扩散函数
real stack height 实际烟囱高度
real staff 实际工作人员
real stake number of receiver point 检波点实际桩号
real standard of living 实际生活水平
real state 实际状态;实际状况
real stock 实股
real storage 实在存储器;实存
real stream 真实河流
real structure 实际结构
real stuff 真材实料;上等货
real substract 实数减
real sun 真太阳
real surplus 实际剩余
real symbol 实符号
real system 真系统
real tare 实际皮重;包皮净重
real tax 实物税
real thickness 实际厚度
real thickness of aquifuge 隔水层实际厚度
real thing 上等产品;真材实料

real-time 实时(的);实际时间
real-time acquisition and processing system 实时采集处理系统
real-time address 实时地址
real-time addressing 直接选址;高速寻址;零级定址;实时定址
real-time algorithm 实时算法
real-time algorithmic tester 实时算法试验器
real-time all-digital spectrum analyzer 实时全数字频谱分析仪
real-time ampacity 实时安培容量
real-time-analog-digital computation 实时模拟数字计算
real-time analogue computer 实时模拟计算机
real-time analysis 快速分析;实时分析
real-time analysis system 实时分析系统
real-time analyzer 实时分析仪;实时分析器
real-time and in-situ observation 原位实时观察法
real-time and service software routine 实时服务程序
real-time application 实时应用
real-time assignment 实时指定
real-time automatic digital optic(al)-tracker 实时自动数字光学跟踪器
real-time batch processing 实时成批处理
real-time buffer 实时缓冲器
real-time central processing 实时中央处理
real-time central processor 实时中央处理机
real-time channel 实时通道
real-time clock 实时装置;实时(时)钟;实时计时器
real-time clock count word 实时时钟计数字
real-time clock diagnostics 实时时钟诊断程序
real-time clock interrupt 实时时钟中断
real-time clock log 实时时钟记录
real-time clock module 实时时钟模件
real-time clock time-sharing 实时时钟分时
real-time coding 实时编码
real-time command 实时指令
real-time communication 实时通信
real-time communication processing 实时通信处理
real-time communication system 实时通信系统
real-time computation 实时计算
real-time computer 实时计算机
real-time computer center 实时计算机中心
real-time computer complex 实时计算综合装置;实时复合计算机
real-time computer program(me) 实时计算机程序
real-time computer system 实时计算机系统
real-time computing centre 实时计算中心
real-time computing complex 实时计算复合体
real-time computing control 实时计算控制
real-time computing technique 实时计算方法
real-time concurrence 实时并行性
real-time concurrency operation 实时并行操作
real-time control 实时控制
real-time control application 实时控制应用
real-time control input/output 实时控制输入输出
real-time controller 实时控制器
real-time control of combined sewer system 混合下水道系统实时控制
real-time control of urban drainage system 城市排水系统实时控制
real-time control routine 实时控制(例行)程序
real-time control system 实时控制系统
real-time coordination 快速配位;实时配位
real-time correction 实时校正
real-time correlation 实时相关
real-time data 实时数据
real-time data collection 实时资料收集
real-time data control system 实时数据控制系统
real-time data processing 实时数据处理
real-time data reduction 实时数据处理
real-time data transmission 实时资料传输;实时数据传输
real-time debug program(me) 实时调试程序
real-time debug routine 实时调试例行程序
real-time decoder 实时译码器
real-time decompression 实时减压法
real-time detection 实时探测
real-time digital governor 实时数字控制器
real-time digital spectrum analyzer 实时数字频谱分析器
real-time display 实时显示

real-time display device 实时显示器
real-time display program(me) 实时显示程序
real-time Doppler imaging system 实时多普勒成像系统
real-time dynamic(al) projection display 实时动态放映显示器
real-time electronic surveillance 实时电子监视
real-time executive 实时管理
real-time executive routine 实时执行程序
real-time executive system 实时执行系统
real-time extrapolation 实时外推法
real-time fail 实时故障
real-time filter 实时滤波器
real-time flow routing 实时流量演算
real-time forecast(ing) 实时预测;实时预报
real-time foreground task 实时前台任务
real-time frequency analysis 实时频率分析
real-time frequency analyzer 实时频率分析仪
real-time fringe 实时条纹
real-time grey level display 实时灰阶显示
real-time guard 实时保护
real-time guard mode 实时保护方式
real-time hologram 实时全息图
real-time holographic interferometry 实时全息干涉测量(术)
real-time holographic reconstruction 实时全息照相再现
real-time holography 实时全息术
real-time identification 实时辨识
real-time information 实时信息
real-time information-processing system 即时信息处理系统
real-time information system 意见交换制度;实时信息系统;实时情报系统
real-time information theory 实时信息理论
real-time input 实时输入
real-time input-output 实时输入输出
real-time input-output control 实时输入输出控制
real-time input-output translator 实时输入输出译码器
real-time interactive reference retrieval system 实时相作用情报检索系统
real-time interface 实时接口
real-time interferometer 实时干涉仪
real-time interferometry 实时干涉量度学;实时干涉测量法
real-time job 实时作业
real-time language 实时语言
real-time laser link 实时激光线路
real-time linear detection system 实时线性探测系统
real-time link 实时传输线
real-time management information system 即时管理情报系统
real-time measurement 实时测量
real-time microcomputer 实时微计算机
real-time microcomputer system 实时微计算机系统
real-time mode 实时方式
real-time monitor 实时监控器;实时监控程序;实时监督程序;实时监测仪
real-time multicomputing 实时多重计算
real-time multispectral viewing 实时多光谱观察
real-time multitape Turing machine 实时多带图灵机
real-time on-line operation 实时联机操作
real-time on-line system 实时联机系统
real-time operating system 实时操作系统
real-time operation 快速操作;实时运算;实时工作;实时操作
real-time operational mode 实时操作方式
real-time operational system 实时操作系统
real-time operational on-line 实时操作联机
real-time optic(al) correlation 实时光学相关
real-time optic(al) matched filter 实时光学补偿滤光器
real-time optic(al) signal processor 实时光学信号处理器
real-time option board 实时选择板;实时控制板
real-time oscilloscope 实时示波器
real-time output 实时输出(信号)
real-time pattern recognition 实时模式识别
real-time photogrammetry 实时摄影测量
real-time photography 实时摄影术

real-time point of view 实时观点
real-time position 实时位置
real-time precision level measurement 实时精确声级测量
real-time prediction 实时预测;实时预报
real-time process 实时进程
real-time process control system 实时过程控制系统
real-time processing 实时处理
real-time processing communication 实时处理通信
real-time processing control system 实时处理控制系统
real-time processing system 实时处理系统
real-time process language 实时过程语言
real-time processor 实时处理机
real-time processor design 实时处理机设计
real-time production monitoring system 实时生产监视系统
real-time program(me) 实时程序
real-time program(me) development system 实时程序开发系统;实时程序编制系统
real-time program(me) simulation 实时程序仿真
real-time programming 实时程序设计
real-time protocol 实时协议
real-time prototype analyzer 实时样机分析器
real-time quality control 实时质量控制
real-time queue 实时排队
real-time range 实时范围
real-time range processing 实时距离处理
real-time readout 实时读出
real-time recording 实时录音;实时记录
real-time relative addressing 实时相对寻址
real-time remote inquiry 联机询问(站);实时远程询问(站);实时查询
real-time response 实时响应
real-time revision and correction system 实时更新系统
real-time satellite computer 实时卫星计算机
real-time scale 实时度标
real-time signal analysis 实时信号分析
real-time signal processing 实时信号处理
real-time simulation 实时模拟;实时仿真
real-time simulator 实时模拟装置;实时模拟器;实时仿真器
real-time software 实时软件
real-time spectral analysis 实时谱分析
real-time spectrum analyzer[analyser] 实时频谱分析器
real-time survey 实时测量
real-time system 实时系统
real-time system characteristic 实时系统特性
real-time system microprocessor 实时系统微处理机
real-time system software 实时系统软件
real-time task 实时任务
real-time telemeter transmitter 实时遥测发射机
real-time telemetry 实时遥测
real-time telemetry link 实时遥测线
real-time telemetry processing system 实时遥测处理系统
real-time television 实时电视
real-time three-dimensional display 实时三维显示
real-time time-sharing 实时分时
real-time traffic 实时通信
real-time traffic responsive control system 实时交通感应控制系统
real-time transmission 实时传输
real-time transmission and non-real-time transmission 实时传输和非实时传输
real-time transmission system 实时传送系统;实时传输系统;实时传递系统
real-time visual information 可见实时信息
real-time waveform 实时波形
real-time working 实时工作
real-time working ratio 实时工作比
realtist 房地产经纪人
realtor 房地产经纪人
real-to-random ratio 真实计数对偶然计数比
realtor-associate 房地产助理经纪人
Realtors Land Institute 房地产经纪人协会
real traffic capacity 实际通行能力
real transaction 实际交易
real transition 实跃迁

real triaxial process 实际荷重法
realty 房地产
realty business 房地产业
real type 实型
real type specification 实型说明
real unimodular group 实单模群
real unit weight process 实容重法
real value 实值;实际值;实际价值
real-value assets 实值资产
real valued intensity 实数值场强度
real-valued stochastic process 实值随机过程
real-value item 实值项目
real value of drawdown 实际水位降深值
real value of safe water pressure 实际静水压力值
real variable 自由变数;自由变量;实变数;实变量
real variable data 实变量数据
real variable function 实变(数)函数;实变量函数
real vector space 实向量空间
real vegetation 实际植被
real velocity 实际流速
real velocity of pore water 地下水实际流速
real velocity of seepage 渗流实际流速
real wage 实得工资;实质工资;实际工资
real wage income 实际工资收入
real wage index 实际工资指数
real wage rate 实际工资率
real waste activated sludge 实废活性污泥
real water-level decrease 实际水位下降值
real well 实井
real work 实际工作
real work time 实际工作时间
real-world coordinate 实物坐标
real-world operation 实际运转
real-world system 现实界系统
ream 扩缝;扩大铆钉孔;绞孔(用绞刀);波筋
ream a rivet hole 铰锥铆钉孔
ream back 从孔底向孔口扩孔(在坑道中钻仰孔时)
ream cutter 闸刀式切纸机
reamed bolt 精制螺栓;铰制螺栓;密配合螺栓;穿钉
reamed extrusion ingot 中空挤压锭
reamed hole 扩大的孔眼;铰制孔;铰孔
reamer 整孔钻;根管扩大针;扩眼器;扩孔钻;扩孔锥;扩孔工人;铰刀;铰床
reamer and tap fluting cutter 铰刀及丝锥槽铣刀
reamer bar 绞刀杆
reamer bit 扩张钻(头);扩孔钻;铰刀
reamer blade 扩孔器切削刃
reamer body 扩孔器体
reamer bolt 密配合螺栓;镶嵌螺栓;铰(刀)螺栓
reamer bushing 铰孔导向套;铰刀导向套
reamer chuck 铰刀夹套
reamer cutter 铰刀槽铣刀;铰刀(槽铣刀);铰铣刀
reamer drill 铰孔前钻头
reamer fluting milling cutter 铰刀开槽铣刀
reamer for camshaft aligning 凸轮轴孔校正铰刀
reamer holder 铰刀刀杆
reamer set 成套铰刀
reamer shell 扩孔筒;扩孔器
reamers in tandem 分级扩孔器
reamer stone (钻头外表面上的)扩孔金刚石
reamer tap 铰孔攻丝复合刀具
reamer tap with handle 带柄铰刀丝锥
reamer with pilot and guide 带前后导向部的铰刀
reamer with staggered straight flutes 不等分齿直槽铰刀
reamer with straight flutes 直沟铰刀
ream grab 抓土机;抓泥机
ream hole 扩孔炮眼
ream indexer 闸刀切纸机
reaming 扩孔(钻进);铰孔
reaming-after-boring method 先钻后扩法
reaming angle 展开角;扩孔角;刷大角
reaming barrel 管状扩孔钻具;管状扩孔器;扩孔器稳定管;扩孔器导管
reaming bench 铰孔机
reaming bit 扩孔钻头;铰孔锥;铰孔钻;铰孔锤
reaming bottom drill 矿底钻机
reaming diamond (钻头外表面上的)扩孔金刚石
reaming drill press 铰孔钻床
reaming edge 扩孔刃
reaming edge taper 锥形扩孔器
reaming enlarge hole 扩眼
reaming face 扩孔刃面

reaming holes for screw stays 为固定螺钉用的铰孔
reaming iron 扩孔锥;扩孔凿;铆钉孔扩孔钻
reaming jig 铰孔夹具
reaming lathe 铰孔车床
reaming machine 钻探机;扩孔机;铰孔机;铰床
reaming of a borehole 钻孔扩大
reaming of hole 扩孔
reaming pilot 扩孔钻头导杆;扩孔器导向部分
reaming pilot adapter (套管与扩孔器之间的)扩孔导向钻头
reaming pilot assembly 扩孔导向器
reaming pilot horn 扩孔导向器连接管
reaming ring 扩孔环
reaming shell 扩孔筒;扩孔器
reaming shell height 扩孔器高度
reaming shell outside diameter 扩孔器外径
reaming stand 铰孔架
reaming stone (钻头外表面上的)扩孔金刚石
reaming surface 扩孔件表面(钻头端部)
reaming teeth (铰刀的)扩孔齿
reaming time 划眼时间
reaming wheel 扩孔牙轮
ream out 扩孔
re-amplification 再放大
re-amplify(ing) 再放大;重复放大
ream weight (木、砖的)垛重
ream weight-ring 令重
reamy glass 表面不平整的玻璃
reamy yarn 单双花线
reanneal 再退火
reannealing 重退火
reaped ripeness 可采成熟度
reaper 收割机
reaper-and-binder 割捆机
reaper-binder 割捆机
reaper for hay 干草收割机
reaping attachment for motormower 动力割草机的谷物收割附加装置
reaping hook 镰刀
reappear 再现
reappearance 重现
reappoint 重新指定;重新委任
reappointment 重新委任
reapportion 再分配;重新分配
reapportionment of cost 费用再分配
reapportionment of service charges 服务费用的再分摊
reappraisal 再鉴定;再估价;重新评价;重新估价;重估
reappraisal lease 重新估价租约
reappraisal of fixed assets 固定资产再估价
reappraisal surplus 重估价盈余
reappraise the stocks and assets of enterprises 清产核资
reappropriation 再次拨款
rea-projection display device 背投式显示器
rear 后(端);背部(的)
rear abutment pressure 后墩座的压力
rear access 后面入口
rear airport 后方机场
rear and front axle stand 前后车轴架
rear anomaly 尾部异常
rear aperture 后孔
rear apron 后挡板;防冲扩底;闸后护坦;坝后防冲护底
rear arch 后拱;背拱
rear arrangement 背列
rear axle 后轴;后桥
rear axle bearing lubricant nipple 后轴轴承加油嘴
rear axle bearing oil seal 后轴轴承油封
rear axle bevel pinion flange 后桥主动伞齿轮法兰
rear axle bogie 后桥转向架
rear axle bumper 后轴挡
rear axle casing 后轴箱
rear axle casing cover 后轴箱盖
rear axle cone bearing 后轴锥轴承
rear axle differential 后桥差速器
rear axle differential cross 后桥差速器十字头
rear axle differential side gear 后桥差速器半轴齿轮
rear axle differential spider pinion 后桥差速器行星齿轮
rear axle disconnecting 后桥脱开;脱开后桥;后轴

脱开
rear axle drive 后桥驱动;后轴驱动;后轮传动
rear axle drive gear 后轴变速齿轮
rear axle dual drive 后桥双重传动
rear axle filler plug 后轴壳注油孔塞
rear axle gear 后轴齿轮
rear axle gear drive 后轴齿轮传动
rear axle housing 后轴套;后轴壳
rear axle housing cover 后轴壳盖
rear axle housing filter plug 后桥壳注油孔塞
rear axle hub key 后轴毂键
rear axle load 后轴荷载;后桥荷载
rear axle nut 后轴螺母
rear axle nut washer 后轴螺母垫圈
rear axle overhang of a towing vehiche 牵引车后悬
rear axle propeller shaft 后桥传动轴
rear axle protection 后轴护板
rear axle radius rod 后轴半径杆
rear axle ratio 后轴比
rear axle reductor 后桥减速器
rear axle shaft 后轮轴
rear axle shaft key 后轴轴键
rear axle spring seat 后轴弹簧座
rear axle stabilizer 后轴稳定器
rear axle stand 后桥架
rear axle steering 后轴转向
rear axle tie-rod 后轴系杆
rear axle tube 后桥半轴套管
rear axle universal joint 后轴万向节
rear axle universal joint jaw 后轴万向节爪
rear balcony 剧场后边楼厅;后阳台
rear base 后方基地
rear beam 后部梁
rear bearing 后轴承
rear bearing bushing 后轴承衬套
rear bed (bunk) 后挂车底架
rear bellcrank steering 后桥双臂曲柄操纵方向;后桥双臂曲柄转向
rear blade (plate) 机尾护板;后护板
rear board 后拦板
rear body offset 车厢偏距
rear bogie 尾部转向架
rear bottom rail 下端梁(集装箱)
rear brake 后轮制动器
rear brake backing disc 后闸底板
rear brake bracket 后闸托架
rear brake extension lever 后闸伸杆
rear brake lever 后闸杆
rear brake operating shaft 后闸操作轴
rear bridle chains 后部链条悬挂装置
rear bucket 后置挖掘机;后铲斗
rear building 后侧建筑物
rear bumper 后挡
rear bumper guard 后保险杆
rear car 两轮拖车;两轮挂车;手推两轮货车
rear carrier 后行星齿轮架;后输送带
rear center 后顶尖
rear chainman 后点测工
rear clearance 尾端离地净空(汽车车身)
rear coil spring 后身绕圈弹簧
rear connection 盘后接线;背面接线
rear connection diagram 盘后接线图;背面接线图
rear control 尾舵
rear control surface 尾部操纵面
rear conversion equipment item 后可拆换设备品种
rear conveyer 尾端运输机
rear cord 内部塞绳
rear court 后院
rear cover 后盖
rear crane rail beam 起重机后轨道梁
rear cross beam 后横梁
rear cross conveyer 后部横向输送器
rear cross-feed lever 后横向进给手柄
rear cross member 后横梁
rear cut-off distance 后截距
rear deck drain 车身后部排泄孔
rear deflector door 后挡板
rear-delivery apron 后送式输送器
rear-delivery chute 后送式卸料槽
rear digging attachment 后挖掘装置
rear door 后门
rear dorse 外墙无突出烟囱的壁炉

rear-draft fulcrum leverage (轮式铲运机的)后部牵引一转动杠杆机构
rear drive 后驱动
rear drive chain 后传送链
rear-drive grader 后轴驱动平地机
rear-drive machine 后轴驱动机械
rear drive shaft 后驱动轴
rear-drive vehicle 后轴驱动车辆
rear-drive wheel 后驱动轮
rear driving gear 后驱动装置
rear-dump body 后翻倾车厢
rear dumping 向后倾卸;后翻
rear-dump lorry 后倾卸自卸车;后斗自卸车;后卸车
rear-dump off-road hauler 后翻斗越野承运者;后翻斗越野运输机
rear-dump rocker 装运石块的后翻斗卡车
rear-dump semi-trailer 后翻斗半挂车
rear-dump tipper 后卸式车
rear-dump trailer 后卸式拖车
rear-dump truck 后翻斗自卸卡车;后卸式卡车;后卸式货车;后翻斗式自卸汽车;尾卸式自卸(汽)车;翻斗式自卸汽车
rear-dump wagon 后卸(式)卡车;后卸(式)货车;尾卸式自卸(汽)车;翻斗式自卸汽车;后卸(式)车;倾卸货车
rear eccentric 回程偏心轮
rear edge 后刃
rear edge contact 后端接点
rear element 后组;尾部元素
rear elevation 后立面;背视图;背面立视图;后立(面)图
rear encroachment 后轮内偏
rear end 后部;后端
rear-end collision 尾端冲撞
rear-end delivery spreader 后送式撒布机
rear-end of spindle 主轴后端
rear-end plate 后盖板;后端盖板
rear-end radius 尾部回转半径
rear-end stop lamp 后端停车灯
rear-end transmission gear 后端传动装置
rear-end winch 尾部绞车;尾部卷扬机;后端卷扬机
rear engine 带动水泵的发动机;后置(式)发动机
rear engine bus 后置发动机公共汽车
rear engine car 后置发动机汽车
rear-engined 后部装发动机的;发动机置于尾部的
rear engine support 后置发动机支架
rear entrance 后门;背入口
rear-entry access road 后门对外交通路
rearer 中耕机
rear excavator bucket 后部(装的)挖掘机
rear exit 后安全门;后太平门;后出口
rear face 背面
rear face armo(u)r 后坡护面
rear face fail 背面失效
rear-facing seat 向后座
rear farm wastewater 饲养场废水
rear-feed horn 螺形喇叭
rear fender 后翼子板;后挡板
rear finder 指示器
rear fitting radius 后回转半径
rear fitting radius of semi-trailer 半拖车后回转半径
rear fitting radius of semi-trailer tractor 半牵引车后回转半径
rear flank 牙轮的背面
rear focus 像方焦点;后焦点
rear frame 后框梁
rear frame member 后横梁
rear furrow wheel 尾轮
rear furrow wheel adjustment screw 尾轮调节螺钉
rear furrow wheel scraper 尾轮刮土板
rear fuselage fairing 机尾整流罩
rear grab 装于后部的抓斗
rear guard 后轮挡泥板
rear halo 尾晕
rear head 后端罩
rear header 后上端梁(集装箱);后管箱
rear hinge pin 后铰链销
rear-hitch tool 牵引式农机具
rear hodpital 后方医院
rear house 后房
rearhouse of silkworms 养蚕室
rear idler eccentric wheel 后空转偏心轮

rear idler hub 后空转轮壳
rear idler wheel 后空转轮
rearing 翘头(车辆);竖码入窑;竖码(盘类制品竖着装窑);侧装(入钵)
rearing house 蚕室
rearing pond 后池;饲养塘
rearing stage 培育期
rear interchangeable equipment item 装于后部可拆换装置
rear lamp 尾灯;后灯
rear lamp bracket 后灯托架
rear land 背面建筑基础;不临街的建设基地
rear leading mark 后导标
rear leaf-spring 后身钢板弹簧
rear lens 后透镜
rear lifting 后方起吊
rear lift linkage 尾轮起落机构
rear light 后灯(标)
rear linkage 后悬挂装置
rear lip tile 后唇砖
rearloader 后装载机;后悬挂式装载机
rear-loading facility 尾部装货设备
rear-loading freighter 尾部上货式货机
rear-loading vehicle 后装载车
rear lorry 后开式车
rear main bearing 后部主轴承
rear main bearing oil seal 主轴承后油衬
rear mill table 后工作辊道
rearming 重新装定;重新装备
rear mirror extension arm 望后镜伸长臂
rear-most 最后面的
rear-most axle 后轴
rearmost end 最后端
rearmost position 最后位置
rear-mounted 后置(式)的;装于后部的;后悬挂式的
rear-mounted buckrake 后悬挂集草器
rear mounted engine 后置发动机
rear-mounted finger-wheel rake 后悬挂指轮式搂草机
rear-mounted flywheel 后置飞轮
rear-mounted grader 后悬挂式平地机
rear-mounted grader attachment for tractors 拖拉机后部装的平地装置
rear-mounted hydraulic digger attachment 装于后部的液压挖掘装置
rear-mounted installation 尾部装置
rear mounted loader 后缀装土机
rear-mounted mower 后悬挂式割草机
rear-mounted three-point linkage 后三点悬挂装置
rear-mounted toolbar 后悬挂通用机架
rear navigation lamp 尾部航行灯
rear nodal point 后节点
rear node 后方节点
rear of a house 房屋背部
rear of concave 凹板后段
rear of panel mounting 在面板后面的装配
rear of queue 队列尾;队列的尾部
rear-of-shed 码头仓库后面的
rear operating aperture 后工作孔径
rear outreach 后伸距
rear overhang 后悬长度
rear panel 后面板
rear part 后部
rear part actual length of turnout 道岔后部实际长度
rear part theoretical length of turnout 道岔后部理论长度
rear piled platform 后方承台
rear pilot 后导杆
rear-pivoting boom 后枢接式转臂
rear plot line 后地界线
rear porch 后门廊
rear portico 后柱廊
rear position 后方位置
rear position light 尾部航行灯
rear poultry 畜牧
rear power take-off shaft 后动力输出轴
rear power unit 后动力装置
rear program (me) picture recording 背景节目录像
rear projection 背面投影;背景放映
rear-projection reader 背观投影阅读器
rear-projection readout 后投影显示器
rear projection screen 背景屏幕

rear propeller shaft 后螺旋桨轴
rear pump 输出泵
rear pump driven gear 后泵主动齿轮
rear pump driving gear 后泵从动齿轮
rear push block 后推座
rear rack 后升运器
rear radius rod 后半径杆
rear rail 后导轨
rear raking pile 后斜桩
rear raking pile driving 仰打(桩)
rear raking pile force 后斜桩力
rearrange 整理;重新排序;重新布置;重排
rearrangeable switch 可重新排序的交换设备
rearranged sterane content 重排甾烷含量
rearranged steranen 重排甾烷
rearranged twills 重排斜纹
rear range light 后导(向)灯
rearrangement 调整;调配;重新整理;重新配置;重新排列;重新编队;重新安排;重排(列)
rearrangement catalyst 重排催化剂
rearrangement cost 重新安置成本
rearrangement ion 重排离子
rearrangement of crossings 交叉口重新布置
rearrangement of interior partitioning 内部分隔重新布置
rearrangement reaction 重排反应
rearranging 重新规划;重新安排
rear rapid traverse lever 后快速横向移动手柄
rear receiver 尾部接收装置
rear red reflex reflector 车后红色反光器
rear rig 装于后部的装置
rear roll 后滚轮
rear roll center (汽车的)后转动中心
rear row velocity stage 复速级的后列
rear scarifier attachment 装在后部的松土装置
rear screed 后夯刮板
rear-screen projection 后屏幕投影;背景放映
rear screw shaft 后螺旋桨
rear screw shaft bushing 后螺轴衬套
rear service 后勤
rear service road 屋后辅助道路
rear-set type speed variator 后置式变速器
rear shaft 后轴
rear shaft-extension type tubular turbine 后轴伸贯流式水轮机
rear shaft seal 后轴封
rear sheet-piling platform 后板桩式高桩码头
rear-shooting camera 向后拍摄航摄仪
rear-side elevation 后视图
rear-signal system 尾部灯光信号系统
rear-silvered mirror 背面镀银(反射)镜
rear slagging 炉后出渣
rear slagging spout 炉后出渣槽
rear slide rail 后座滑轨
rear smoke box 后烟道室
rear span 外形后宽
rear spar 后翼梁;后梁;后部梁
rear split ring 后隔环
rear spring 后弹簧
rear spring clamp 后弹簧夹
rear spring clip 后弹簧夹
rear spring clip bar 后簧夹杆
rear spring pivot seat 后弹簧支枢座
rear spring seat 后弹簧座
rear spring shackle 后弹簧钩环
rear staffman 后司尺员
rear stage area 后台区
rear steer(ing) 后轮转向
rear stop 后挡
rear street 后街
rear strut 后支柱
rear support 后支撑
rear support bracket 后支架
rear surface 后表面
rear suspension shock absorber 后悬挂减震器
rear suspension spring 后悬架弹簧
rear suspension wrench 后悬挂螺母扳手
rear tape man 后尺手【测】;后司尺员
rear tipper 后倾自卸车
rear tool rest 后刀架
rear torus 涡轮
rear track 后轮距;后方铁路线(码头)
rear-traction drive 后行走轮驱动
rear transmission gear box 后变速箱

rear turbine bearing 透平后轴承
rear turret 尾部炮塔
rear udder 后翻
rear unit assembly 后部件总成
rear-unloading vehicle 后卸汽车
rear up 竖立;举起;抬起
rear vault 后穹顶;深拱;背穹隆;背拱顶
rear vibrator 装在后部的振动器
rear view 后视(图);后观;后视图;背面立视图
rearview mirror 倒车镜;后视镜;后方反射镜;反视镜
rearview mirror arm 望后镜臂
rearview mirror bracket 望后镜伸长臂
rear vision mirror 后视镜
rear wall 驾驶室与客室隔墙;后墙
rear wall arrangement 驾驶室与客室隔墙布置
rearward 向后;在后背
rearward conveyer 尾部运输机
rearward extend 向后延伸的
rearward face 背弧面;凹面
rearward perception 司机后方视野
rearward shift 后移
rear warp-round booster 后置环列助推器
rear wheel 后轮
rear-wheel ballast 后轮配重
rear-wheel bearing nut wrench 后轮轴承螺母扳手
rear-wheel brake 后轮闸
rear-wheel cylinder 后轮制动缸
rear-wheel depth adjustment 尾轮深浅调节器
rear-wheel drive 后轮驱动;后轮传动
rear-wheel hub cap 后轮毂帽
rear-wheel lifting link 尾轮起落拉杆
rear-wheel spindle 后轮轴
rear wheel steering 后轮转向
rear-wheel suspension 后轮悬挂(系统)
rear-wheel torque 后轮转矩
rear window 后(车)窗
rear window wiper 后(车)窗雨刷;后(车)窗刮水器
rear windscreen 后风挡
rear wing 后保护板
rear yard 后院
reason 理性
reasonability 合理程度
reasonable 合理的
reasonable arch axis 合理拱轴线
reasonable assumption 合理假设
reasonable beam 适中束
reasonable care and skill 应具备的能力和技艺;合格质量验收
reasonable comparability 合理的可比性
reasonable condition 合理条件
reasonable cutting slope 合理削坡
reasonable departure time range of passenger trains 旅客列车合理开车范围
reasonable deviation 合理绕航
reasonable dispatch 合理调遣
reasonable distribution of burden 合理负担
reasonable evaluation process 合理评估法
reasonable image 合理图像
reasonable minimum volume of sample 合理的最小试样体积
reasonable mixing ratio 合理配合比
reasonableness 合理性
reasonableness analysis 合理分析
reasonableness check 合理性检测
reasonableness check table 合理性检验表
reasonableness test 合理性检测
reasonable notice of abandonment 适当的委付通知
reasonable price 公平价格;价格公道;合理价格;售价公道
reasonable profit 适当利润
reasonable quality 合理质量
reasonable reward 合理报酬
reasonable size 适当尺寸
reasonable starting date 合理开工日期
reasonable time path tree 合理时间路径树
reasonable use of water 合理用水
reasonable use rule 合理利用准则
reasonable use theory 合理利用理论
reasonable value 合理(价)值
reasonable wear and tear 合理磨损
reasonably assured resources 确实可靠资源
reasonably foreseen 合理预见
reason for outage 线路中断原因

reasoning 论证;推理
reasoning backward 倒推论
reason leading unreasonable exploratory degree 勘探程度不合理的原因
reason of fire 火灾原因
reasonpiece 顶梁板
reasons of impunity 免责理由
reassemble 重装;重新装配;重编
reassembling 重新组合
reassembling mechanism 重新组合机构
reassembly 再装配;重(新)组合;重(新)装配
reassembly block 重编块
reassembly deadlock 重装封锁
reassembly lock-up 再汇编封锁;重装封锁
reassert 再主张
reassessment 重新估价;重新评定;重估
reassign 再分派;重新指定;重赋值
reassignment 再赋值;再分配;重赋值
reassociation 重新组合
re-assort 再分类
reassurance 再保
reasting island 靠船基;承冲护船桩;岸壁系缆桩
reather valve 弹子阀
reattachment 再附着;重附着
reattachment flow 再附着流
Reaumur alloy 列奥米尔锑铁合金
Reaumur scale 列氏温标
Reaumur temperature scale 列氏温度表
Reaumur thermometer 列氏温度计;列奥米尔氏温度计
reaustenitizing 重新奥氏体化
reawaking of a volcano 火山的复活
rebabbit 重铸巴氏合金;重浇巴氏合金
rebabbiting 重新浇注巴氏合金
rebabbiting of bearing 重新浇注轴承
rebabbiting of lead bronze 重新浇注铅铜合金
re-ballasting machine 道砟整补机
re-bar 螺纹钢筋;配筋
re-bar bundle 钢筋束
re-bar mat 钢筋筏;钢筋网
re-bar spacing 钢筋间距
re-bar tie encased in concrete 混凝土内的钢筋拉杆
re-bar work area 钢筋加工场
rebasing 基托垫底;垫低术
rebat 里巴风
rebatching 重卷绕;重复添加;重复配料
rebate 减少;门窗侧壁槽;回扣;切口;槽口;裁口;半槽(企)口;凹凸榫
rebate bead 转角圆线;槽舌圆线;槽口圆线
rebated acceptance 扣息贴现承兑;提前付款
rebated and filleted joint 槽舌圆贴角接缝;槽舌贴角接缝
rebated boarding 企口板条
rebated cement slab 槽口水泥板;半槽口水泥板;凹凸榫(接)水泥板
rebated concrete slab 凹凸榫接混凝土板
rebated door 有槽门框
rebated floor 企口地板
rebated grooved and tongued joint(ing) 企口接合
rebated joint 凹凸榫搭接;槽舌榫接合;裁口接头;半槽接合;凹凸榫接合;企口接合
rebated jointing 企口接头
rebated lock 槽口门锁;裁口锁
rebated miter joint 凹凸榫斜接
rebated pipe 有凹凸榫的管子;凸凹套接管
rebated plane 企口刨;凹凸榫刨
rebated sheet metal sheath 有凹凸榫的金属板
rebated siding 凹凸榫盖板
rebated weather-boarding 裁口鱼鳞板;企口风雨板;槽舌风雨板
rebated wood(en) floor(ing) 木企口楼面;木企口楼板
rebate for putty 油灰槽口
rebate joint 搭接拼接
rebate ledge 企口挡条
rebate of glazing 镶嵌玻璃的槽口;镶玻璃裁口;镶板玻璃槽口
rebate plane 企口刨;窄槽木刨;裁口刨;凹凸榫刨
rebate siding 凹凸榫墙板
rebate size 凹槽尺寸;裁口尺寸
rebate system 回扣制度;回扣方式
rebate tariff 折扣计费率
rebating 开裁口
rebating cutter 切削刀具

rebatron 高能电子聚束加速器
Rebecca 雷别卡导航系统
Rebecca-Eureka system 雷别卡—尤利卡系统
rebed 浇注轴承(修理时)
rebend 倒车转弯角;折弯;重新弯曲
rebending 重加弯曲;反复弯曲
rebending test 反复弯曲试验
rebend test 倒车转弯考核;倒车转弯试验
rebevelling 重开坡口
rebid 再次投标;再招标;重新招标
rebirth 重生;复兴式
reblade 重复整型
reblending 再拌和;重复拌和;重混合;再(次)混合;重新搅拌
reblowing 后吹
reblunge 重混合
reboil 再煮;再沸;重新煮沸;重热
reboil bubble 再生气泡;二次气泡
reboiled absorber 吸收精馏塔
reboil effect 再生气泡效应
reboiler 再蒸馏锅;再沸器;加热再生器;重沸(腾)器
reboiler circulation pump 再沸器循环泵
reboiler furnace 再沸炉;重沸炉
reboiler section 再煮旋管
reboil heat 重煮热
reboiling 再生气泡
reboiling temperature of glass marble 玻璃球二次起泡温度
reboil ratio 再沸比
rebonded electrically fused magnetic-chrome brick 电熔再结合镁铬砖
rebonded fused-grain refractory 熔粒再结合耐火材料
rebonded sand 加黏土回用砂
rebonding clay 复用黏土
rebonding of mo(u)lding sand 型砂翻新
rebore 再镗孔;重镗孔
reborer 重镗孔钻
reboring 再镗;重钻;重镗孔
reboring of worn cylinder 磨损汽缸重镗
Reboul effect 雷步耳效应
rebound 回跳;回缩;回弹(值);弹回;反跳
rebound action 弹回
rebound apparatus 回弹仪
rebound blow 回跳冲击;回弹冲击
rebound check 弹簧减振器
rebound clearance 回跳间隙
rebound clip 钢板弹簧骑马螺栓;回跳夹;回弹夹箍
rebound coefficient of elasticity 弹性回弹系数
rebound curve 回跳曲线;反弹曲线
rebound deflection 弹性挠度;回弹弯沉
rebound degree 回弹率
rebound effect 反跳效应
rebound elasticity 回弹性;弹回性
rebound ga(u)ge 回弹计
rebound hammer 混凝土回弹仪;回弹仪;回弹锤;回弹测强仪;混凝土试验锤
rebound hardness 回跳硬度;回弹硬度;反跳硬度;反弹硬度;肖氏硬度
rebound hardness test 回弹硬度试验
rebound hardness tester 回弹硬度试验仪
rebound hardometer 回跳硬度计
rebound height 回弹高度
rebound index 回弹指数
rebounding lock 击锤的反回锁定
rebound leaf 回跳弹簧片
rebound leaf of the spring 辅助弹簧片
rebound method 回弹法
rebound modulus 回弹模量
rebound number 回弹数
rebound of pile 桩的回跳;桩的回跃;桩的回弹
rebound of shotcrete 喷射混凝土回弹
rebound phenomenon 回缩现象;反跳现象
rebound procedure deflection 回弹过程挠曲;回弹法弯沉
rebound resiliometer 回弹性测定计
rebound resistance 伸张行程阻尼力
rebound resistance of shock absorber 减震器回跳阻力
rebound Shore hardness test 肖氏弹跳硬度试验
rebound stop 悬架回弹限位器
rebound strain 回弹应变
rebound stress 回弹应力
rebound test 肖氏硬度试验;回跳试验;回缩试验;回弹试验
rebound tester 混凝土回弹仪;回弹仪;测强仪
rebound valve 回跳阀;伸张阀
reboyo 里也约风
rebreaker 再割机
rebreaker rolls 第二次破碎用辊式破碎机
rebreaking zone 再割区
rebreather 再生式氧气系统;换气器
rebreather-type mask 再生式面罩
rebrick 重砌;(内衬的)改砌
rebricking 换容衬;换衬里;重新砌衬
rebricking a kiln 重砌窑砖
rebroadcast 转播;重播
rebroadcasting station 转播台
rebroadcasting transmitter 转播发射机
rebroadcasting receiver 转播接收机
rebroadcasting station 转播台
rebroadcast transmitter 转播发射机
rebrushing 再次挑顶卧底
rebudgeting 重审预算;重编预算
rebuild 再建;改造;重新组装;重新装配;重建
rebuildability 修复性
rebuild a tyre 修复轮胎
rebuilding 再造作用;改建;翻修;重建
rebuilding instruction 修复说明
rebuilding work 翻修工程
rebuild kit 修复机器用的成套零部件
rebuilt truck 整修货车
reburn 再烧
reburner 转化炉
reburning 再烧(过程);重烧
reburning chamber 再燃烧室
reburnishing 再磨合
reburring 再刻石
rebush 换衬
rebuttal evidence 反驳证据
rebutting evidence 反驳证据
rebval 门窗洞口;门窗帮
recalcification 再钙化
recalcification time test 再钙化时间试验
recalcitrant compound 难降解化合物
recalcitrant substance 抗生物降解物质;难生物降解物质
recalculate 重新计算;重算
recalculated A. P. W. curve 再计算视极移曲线
recalculation 再计算;再核算
recalescence 再辉;金属骤然放热;复辉
recalescence curve 复辉曲线
recalescence point 再炽热点;再辉点;突放热量临界温度;复辉点
recalescent point 实放热量临界温度
recalibrate 再校准;重新刻度
recalibration 再校准;微校;重新校准;重新率定;重校(准);重率定;重刻度;重核准;重标定
recalibration screw 重新校准螺钉
recalibrator 重新校准器
recall 取回;重复呼叫;撤换;二次呼叫
recall factor 再现率
recall-fallout plot 查全率—错检率曲线
recall field 阻尼场
recall image 清除图像
recall level average 平均查全率
recall loan 收回贷款
recall ratio 检索率;查全率
recall switch 信号复原开关
recall thrust 反冲力
recamber 重新翘起
recanalisation 管道再造
recap 胎面翻新的轮胎
recapitalization 资本重定;调整资本;重整
recapitalize 调整资本结构
recapitulation 公式汇编;汇总表;重演
recapitulation statement 摘要表
recap job 翻新(轮胎)
recapped tire 胎面翻新的轮胎
recapper 轮胎翻新器
recapping 重修面层
recapture 再俘获;取回;收复;重俘获;重捕
recapture of chorofluorocarbons 含氯氟烃的回收
recapture of depreciation 折旧回收
recapture of overall foreign loss 全部国外损失冲减
recapture rate 资金回收率
recapture tax 超定额税收
recarbonation 再碳酸化
recarbonation feed 再碳酸化料液
recarbonation tower 再碳酸化塔
recarbonization 再碳化
recarbonize 再碳化
recarburation 增碳作用;再碳化;再渗碳;二次增碳作用
recarburization 再碳化;再渗碳
recarburize 增碳
recarburizer 增碳剂;再增碳剂;再碳化剂
recarburizing 增碳作用
recarburizing agent 增碳剂
recarding 重复梳理
rec-assay method 重组缺陷型测定法
recast 另算;重作;重铸;重新计算;重算
reca(u)lking 重凿缝;重(新)嵌缝
recce pack 侦察设备舱
recce pod 侦察设备吊舱
recce radar 侦察雷达
receded disk impeller 离心式叶轮
recedence time 退缩时间
receding branch 退缩分枝
receding chaser collapsing tap 可调径丝锥
receding colo(u)r 远感色;后移色;后退色;褪色;似远色
receding cornice 后缩檐口;退缩檐口
receding dental 后斜齿板
receding difference 后向差分;后退差分;退差(值)
receding glacier 后退冰川
receding hemicycle 冷却半周期阶段(冰川)
receding line (建筑物的)退缩线
receding log 倾斜航行
receding metal 缩金属
receding prices 逐渐下跌的物价;价格跌落
receding shock wave 后退激波;后退冲击波
receding side 退出侧
receding side of belt 皮带退出侧;皮带松边
receding water table 退落潜水面;退落地下水位
receding wave 后退波
receibable on demand 催收账款
receipt 接收;签收;凭证;收条;收到;发票
receipt account 应收账款
receipt and disbursement statement 收支明细账
receipt and expenditures on capital account 资本收支账
receipt communication method 回执通信方式
receipt documents 收入凭单
receipted account 应收账目;已收账目
receipt for a loan 借据
receipt in full 全数收清(条据)
receipt in part 部分收清
receipt invoice 收讫发票
receipt ledger 收入分类账
receipt note 收据
receipt of goods 货已收到
receipt of license clause 收到许可证条款
receipt of payment 收款
receipt of transport income 运输收入票据
receipt on accounts 入账
receipt period 进货周期
receipt rate 进货率
receipts 进款;入款
receipts and disbursements breakeven point 收支平衡点
receipts and expenditures out of budget 预算外收入与支出
receipts and payment documents 收支凭证
receipts basis 实收基础
receipts from enterprise 企业收入
receipts from retirement of fixed assets 固定资产清理收入
receipts from services 服务收入
receipt side 收方
receipts of general accounts 一般账户收入
receipt stamp 收讫章
receipt voucher 收条;收入传票;收款传票;收货凭证
receivable account 应收账款
receivable and advanced payment audit 应收及预付款项的审计
receivable from subordinary unit 应收下级款
receivable testimony 可接受的证据
receivable turn over 应收款周转率
receive, send keyboard set 收发键盘装置

receive and depart 接发车
receive clock 接收时钟
received 收讫;收到
received and forwarded 收到和发出
receive data buffer 接收数据缓冲器
receive data register 接收数据寄存器
received cash book 现金收入簿
received current 输入电流
received data 接收数据
received energy 接收能量
received for shipment 收货представ;收货备运
received for shipment bill of lading 收讫待运提单;备运提(货)单
received in full 全数收清;收清;收讫
receive dividend 分红
received payment 款已收到
received power 接收功率
received pulse 接收脉冲
received report 收货报告单
received shipment 备运
received signal level 接收信号电平
received stamp 收讫章
received version 标准译本
received view 公认的观点
receive effluent 受纳污水
receive in advance 预支;预收
receive interruption 接收中断
receive leg 接收引线
receive mode 接收方式
receive not ready 不能接收
receive only 只接收;只收
receive-only device 只收器
receive-only earth station 只收地面站
receive-only equipment 只收设备
receive only page printer 页式打印机;纸页式印字机;纸页式打字机
receive-only terminal 只收终端
receive-only typing reprerforator 接收用打字穿孔机
receive pacing 接收定步;接收调步
receive payment on the basis of their shares of the land 土地入股分红
receiver 验收人;转化炉;储[贮]液罐;储[贮]槽;接受器;接收器;接收机;接卡器;接管人;清算人;委托管理人;受话器;收集器;电话听筒;存储器;串行数据发送机接收机;船舶收发报话机;储液器
receiver, transmitter communication controller 接收发送通信控制器
receiver alignment 接收机调整
receiver amplifier 接收机放大器
receiver and transmitter 送受话器;送收话器
receiver antenna 接收机天线
receiver bandwidth 接收机带宽
receiver bin 受料仓
receiver cabin 接收机室;无线电接收机壳
receiver cabinet 无线电接收机盒
receiver case 接收机箱;受话器盒;电话室
receiver circuit 接收机电路
receiver clause 抵押产业管理人条款
receiver compass 分罗经
receiver cone 装料漏斗
receiver cord 受话器软线
receiver cupola 带前炉冲天炉
receiver current sensitivity 自由场电流灵敏度
receiver diaphragm 受话器膜片
receiver drum 集汽包;集气包
receive ready 接收准备
receiver earpiece 受话器耳承
receiver element 容器
receiver field of view selector 接收机视场选择器
receiver gain control 接收机增益控制
receiver gating 接收机选通
receiver ga(u)ge 验收规;综合量规;外形检验样板
receiver hopper 受料仓
receiver in bankruptcy 破产财产管理人
receiver incremental tuning 接收机增量调谐
receiver inlet 接收机输入端
receiver input 接收机输入
receiver inspection 验收检查
receiver interference 接收机干扰
receiver isolation 接收机隔离
receiver ladle 移动式前炉
receiver language 转写语言
receiver limiter 接收机限幅器

receiver line 深度记录线(海洋)
receiver machine 收信机
receiver matrix 接收机矩阵变换电路;接收机矩阵
receiver mechanism 接收机制
receiver noise 接收机噪声
receiver noise power level 接收机噪声功率电平
receiver noise threshold 接收机噪声阈
receiver of main engine revolution 主机转数表
receiver of remote control sytem 遥控系统
receiver of the refrigerator 冰箱储[贮]液器
receiver of ultrasonics 超声波受波器
receiver optic(al) system 接收光学系统
receiver oscillator 接收机振荡器
receiver pass-band 接收机通带
receiver pipe 油罐;接受管
receiver pipe connection 接受管接头;储柜管接头
receiver point elevation 检波点高程
receiver point sit 检波器位置
receiver pressure 受器压力
receiver primary 显像三基色;接收机原色
receiver protective device[reprod] 接收机保护装置
receiver recorder 二级记录器
receiver response 受话器响应
receiver retrace 接收机回扫线
receiver's cash account 收款人现金账
receiver's certificate 借款证
receiver sensitivity 接收器灵敏度;接收机灵敏度
receiver shell 受话器外壳
receivership 破产产业的接管;财务清算;财产接管人身份
receiver's office 收款处
receiver synchro 自同步接收机
receiver system 客式
receiver tag 接收机标记
receiver tank 蓄汽桶;储[贮]藏柜;收汽箱
receiver telescope 接收望远镜
receiver test 接收机检测
receiver-transmitter 收发报机;接收发射机;送受话器;送话器;收发两用机
receiver-transmitter switch 收发转换开关
receiver tube 接收管
receiver unit 受话装置
receiver vessel 储油罐;储气罐
receive status 接受状态
receive tax 收税
receive time-out 接收超时
receive trains in humping direction 顺向接车
receive window 接收窗口
receiving 验收;接收
receiving account 收入账户
receiving aerial 接收天线
receiving amplifier 接收放大器;收信放大管
receiving and delivery 收货与交货
receiving and delivery at junction station cars loadings 分界站交接重车数
receiving and dispatching room 收发室
receiving antenna 接收天线
receiving aperture 接收孔径
receiving apparatus 接收设备;接收机
receiving apron 验收单
receiving area 卸货台;接收区(域);接收面积;承受面积;安置区
receiving array 接收阵(列)
receiving attenuation 净衰减;接收衰减
receiving auger 接受螺旋
receiving barrel 接受阱
receiving basin 储[贮]水池;进水池;受水池
receiving beam 接收光束
receiving belt conveyer 受料胶带输送机
receiving bin 受料仓
receiving body of water 承泄水体;受纳水体
receiving box 接受筒
receiving bucket 受料罐;承雨筒
receiving bunker 受料仓;受矿(煤)仓
receiving center 接收中心
receiving chamber 接待室;进水室;接收室
receiving channel 接收频道
receiving chute 装料溜槽
receiving clerk 受货员;受料员;收货员
receiving coil 接收(用)线圈
receiving compartment 接受阱
receiving compensation piston 补偿反应活塞
receiving condenser 接收电容器
receiving cone 受料漏斗

receiving container 受料容器
receiving conversion 接收变换
receiving conveyer 受料输送机
receiving crystal 接收晶体
receiving cup 传力杆帽套
receiving department 验收部门
receiving-departure track 到发线【铁】
receiving-departure yard 到发场【铁】;到达线
receiving device 接收装置
receiving diode 接收二极管
receiving direction 接收方向
receiving distributor 接收分配器
receiving electrode 积尘极;沉积电极;沉淀极
receiving element 接收单元
receiving elevator 接受升运器
receiving encoding 接收端编码
receiving end 接收端;受电端;到货终点
receiving-end crossfire 接收端串扰;接收端串报
receiving-end impedance 接收端阻抗;受电端阻抗
receiving equipment 接收设备
receiving error 接收机误差
receiving expenses 收货费(用)
receiving facility 接收设施
receiving filter 接收滤波器
receiving flask 接受瓶;长颈玻璃容器;测量杯
receiving funnel 承雨漏斗;承雨(量)器
receiving ga(u)ge 轮廓量规
receiving hood 接收式排风罩
receiving hopper 受料(漏)斗;承接斗
receiving hopper of bucket elevator 斗式提升机喂料机
receiving house 收油车间
receiving individual project 单项工程验收
receiving inductance 接收电感
receiving inspection 验收检查;进料检验;接受检查
receiving inspection and maintenance 验收与维护
receiving installation 接收设施;接收设备
receiving instrument 受信机
receiving item 接受项
receiving jigger 接收用高频振荡变压器
receiving ladle 储[贮]铁包
receiving line 来液管线;深度记录线(海洋);接收线路
receiving line to depot 入段线【铁】
receiving loop 接收线圈
receiving loop loss 接收环路损耗
receiving machine 接收机
receiving magazine 接收箱
receiving manifold 来液汇管;采油管汇
receiving margin 接收裕度;接收余量
receiving mechanism 接收机制
receiving modem 接收调制解调器
receiving mouth 承接口;受料口
receiving note 收货通知;收货单
receiving nozzle 承接管嘴
receiving objective 接收物镜
receiving office 接收站;收件局;收货处;收报处
receiving only 只接收
receiving opening 受料孔;受料口;装料口;装料孔
receiving optics 接收光学装置;接收光学系统
receiving order 接管令;法院发出的收款指令
receiving oscillator 接收振荡器
receiving parabola 抛物面接收天线
receiving paraboloid 抛物面接收天线
receiving perforator 接收凿孔机;接收穿孔机;穿孔机
receiving pipe 承接管
receiving piston 接受活塞
receiving pit 接受箱;接受坑;受料浅井
receiving plane 接收面
receiving plant 接收设施
receiving pocket 受料仓
receiving point 受料地点;受料场所;受振点
receiving power 输入功率
receiving punch 输入穿孔机
receiving quotation 应收汇价;收货估价单
receiving range 接收范围
receiving register 接收寄存器
receiving relay 接收机继电器
receiving reperforating apparatus 收报复凿机
receiving report 验收报告(单)
receiving river 纳污河流;调入河流
receiving room 接声室;收发室
receiving route 接车进路

receiving scale 接收秤
receiving sector 接受部门
receiving selsyn 自动同步接收机；同步系统接收装置
receiving sequence 接受序列；接收序列
receiving set 接收机
receiving side 接收端
receiving singal 接车信号
receiving skid 受料台架；受料格栅
receiving slip 收入传票
receiving slit 接收槽
receiving spectrum 接收光谱
receiving station 接受点；接收站；接收台；接料站；受料站；收信台
receiving station terminal 接收站码头
receiving storage delivery 卸货费用及栈租
receiving stream 纳污河流
receiving stylus （仪器的）记录笔
receiving substation 降压变压器；降压变电所
receiving surface 接收表面
receiving synchro 同步接收机
receiving system 接收系统；接收装置；接收制式
receiving tank 储[贮]水槽；储[贮]槽；中间罐；接受罐；接收振荡回路；倾入槽；收集槽
receiving telescope 接收望远镜
receiving terminal 接受终端；接收站；接受端；收货码头；到货终点
receiving terminal station 终端接收站
receiving track 接车线【铁】；进车轨道【铁】；到达站；到达线
receiving transducer 接收探头；接收换能器
receiving transducer of ultrasonics 超声波受波器
receiving-transmitting branch equalizer 收发信支路均衡器
receiving-transmitting level difference 收发信防卫度
receiving tray 接收支架
receiving tube 接受管；接收管；收信（放大）管
receiving unit 接收装置
receiving valve 接收阀
receiving voltage sensitivity 自由场电压灵敏度
receiving water body 接受水体；纳污水域；纳污水体
receiving watercourse 受纳水道
receiving water ecosystem 纳污水体生态系统
receiving water impact 纳污水体影响
receiving water pollutant concentration 纳污水体污染物浓度
receiving water quality 纳污水体水质
receiving water quality model 纳污水体水质模型
receiving waters 纳污水域；纳污水体；受（纳）水体；承受水体
receiving water segment 扇形纳污水域
receiving water standard 受纳水体标准
receiving wave range 接收波段
receiving weigher 储[贮]料称量装置
receiving well 进水井
receiving wire 接收天线
receiving yard 接收场；到达场【铁】
recemented glacier 再生冰川；重新固结冰川；复活冰川
recementing 补注水泥
recemose branching 总状分枝式
recent 近期的
recent activity of fault 断层现代活动性
recent advance 新发展
recent alluvium 近代冲积层
recent bottom sediments 现代底部沉积（物）
recent channel 现代河床
recent cost 最近成本
recent crust movement 近代地壳运动
recent deposit 近代沉积（物）
recent deposited cohesive soil 新近沉积黏性土
recent deposited loess 新近堆积的黄土
recent deposits of alluvium 新冲积物
recent development 新发展；最新进展
recent drainage map 现代水系图
recent drizzle 观测时有细雨
recent epoch 现代
recentering 重新置中
recentering calibration system 再定中心标定系统
recentering mechanism 再定中心机构
recent estimates indicate 最近的估计指出
recent fill 新填土
recent fluvial sediments 现代河流泥沙；现代河流沉积物；现代冲积物

recent fracturing pattern in the crust 地壳现代破裂网络
recent geosyncline 近代地槽
recent glaciation 现代冰川作用
recent information 最新资料
recent isoanabase map of earth crust 现代地壳等升线图
recent landform 现代地貌
recently deposited soil 新近堆积土
recent model 现代形式
recent period 现世纪
recent rain 观察时有雨；观测时有雨
recentralizing 恢复到中心位置
recentre 重新置中
recentrifuge 再次离心
recent sediment 新沉积物；现代沉积物；近代沉积（物）
recent snow 观测时有雪
recent snow shower 观测时有阵雪
recent soil 新生土
recent studies 近代研究
recent studies show that 近代研究指出
recent tectonic movement 现代构造运动；近代构造运动
recent tectonics 近代构造
recent thunderstorm 观测有雷暴
recent time 现时
recent trend 新动向
recent vegetation 现代植被
recent volcanic activity 近期火山活动
recent volcanism 近代火山活动
recent volcano 近期火山
recent years 近年
receptacle 插孔板；接线盒；仓库；储藏所；接受器；容器；电源插座；电器插座；插座
receptacle box 同步传动装置的受信仪箱
receptacle contact 插座接点
receptacle outlet 输出插孔；插座输出
receptacle plug 插头；电插头
receptacle trough 受料槽
receptance 响应；敏感性；接受率
recept(ion) 接收；接受
reception and departure yard 到发场
reception area 接收区（域）；受水区；安置区
reception basin 储[贮]水池；进水池；集水区；受水盆地；储水池
reception basin of river 河流补给区
reception chamber 会客室；接待室
reception congestion 接收拥塞；接收挤塞
reception desk 接待处；总服务台
reception distributor 收报分配器
reception file read-out 接收文件读出
reception good 接收良好
reception hall 宾馆接待厅；接待厅；门厅
reception lounge 会客休息厅
reception of double polarization 双偏振接收
reception of echoes 接收回声
reception office 接待办公室；传达室
reception of impulse 脉冲接收
reception of maps 图幅验收
reception points on the earth 地球上的接收站
reception poor 接收不良
reception room 接待室；会客室；会客室；候诊室
reception saloon 接待厅
reception sampling plan 验收取样方案；验收抽样计划
reception set of siding 到达线群
reception sidings 到达线；接车叉线
reception stall 接待柜台
reception tank 储[贮]水池
reception terminal 收货码头；到货终点
reception test 验收试验
reception time 验收时间
reception wall 防潮层的保护墙
reception yard 到达场【铁】
receptive phase 可接受状态
receptivity 吸收性能；接收能力；可接收度；接收性能
receptor 感受器；淋浴槽；接受器；接收器；受体
receptor area 接受面积；受体面积
receptorium （与古罗马方形会堂相连的）接待室
receptor limestone 输入灰岩
receptor model 受体模型
recess 建筑后退；海岸凹进处；切槽；壁龛；凹座；凹穴；凹口；凹坑；凹进处；凹进（部分）；凹挡；凹处；凹部；凹壁
recess-action gear 啮出齿轮；全齿根加高齿轮
recess angle 渐远角
recess basin 三面嵌入式浴缸
recess bed （折入凹墙或壁柜内的）折叠床；壁床
recess contact 渐离接触；啮出接触
recess continuous row fluorescent fixture 镶入式荧光灯带
recess drawing 预留孔洞图
recessed abrasive wheel 槽式砂轮
recessed arch 层叠拱；套碹；双碹；叠内拱；凹进叠层拱；凹叠拱
recessed bath tub 嵌入式浴缸
recessed bead 凹入的圆条线脚
recessed bollard 闸墙凹槽内的系船柱
recessed bolt 凹面栓
recessed bulkhead 阶形舱壁；槽形舱壁；凹形舱壁
recessed cathode 隐藏式阴极；退藏式阴极
recessed ceiling 凹顶棚
recessed drinking fountain 嵌入式饮水器；砌入式饮水器
recessed dry joint 凹槽干缝
recessed element 凹入构件
recessed entrance 凹形入口
recessed fitter press plate 安框滤板
recessed fitting 下水管道承插式内平弯头；下水管道承插式内平弯路
recessed fixture 埋安装；凹入式顶灯；隐藏式照明装置；吸顶灯（具）
recessed flanged joint 埋头法兰连接
recessed fluorescent fixture 镶入式荧光灯
recessed flush cistern 嵌入式的冲水水箱；凹进的冲水水箱
recessed heading tool 空心头镦凸模
recessed head screw 凹头螺钉；槽头螺丝；埋头螺钉；十字头螺钉
recessed-head screw-driver 槽头螺丝刀
recessed heater 隐藏式暖器；暗装取暖器
recessed jewel 带油槽直孔钻
recessed jig 挖入式钻模
recessed joint 方槽（灰）缝；凹（灰）缝；凹槽灰缝
recessed joint pointing 勾凹缝
recessed key 平头埋入键
recessed ledge 凹进墙架；凹进壁架
recessed light fixture 凹入式顶灯
recessed lighting fitting 凹槽照明装置；埋入式照明器具
recessed lightings 嵌入式照明设施；嵌入式照明设备
recessed locker 嵌墙存衣柜
recessed loop aerial 隐藏式环形天线
recessed luminaire 嵌装式照明设备
recessed margin 凹陷边缘
recessed mounting 凹入设置
recessed order of arch 壁渐收进式的拱
recessed paper towel dispenser 砌入式纸巾配出器
recessed plate filter 凹版式压滤机
recessed plate filter press 凹板式压滤机
recessed plate press 凹板式压滤机
recessed pointing 凹灰缝；凹槽嵌填；凹槽勾缝；勾凹缝
recessed portal 凹形入口；凹形门；缩后的入口；门洞内的门
recessed powdered soap dispenser 砌入式肥皂粉配出器
recessed radiator 嵌墙散热器
recessed room 凹室
recessed soap dish 嵌入式肥皂盆
recessed soap dispenser 砌入式肥皂配出器
recessed soap holder 凹进的肥皂盒
recessed surge tank 埋藏式调压井
recessed thread 加深螺纹；深扣螺纹
recessed toilet paper holder 嵌入式手纸架
recessed tub 嵌入式浴盆；壁龛式浴盆
recessed tubewall 缩入管式炉墙
recessed vanity shelve 凹进式梳妆架；凹入式梳妆架
recessed vee-joint 尖槽（灰）缝
recessed veranda 凹阳台
recessed V-joint 尖槽缝
recessed wall heater 轻便取暖设备；墙内暖气设备
recessed wall urn ash tray 墙上凹进式烟火盒
recessed waste receptacle 凹进式废物箱

recessed wheel 带中心凹槽的砂轮；槽式轮
recessed wide-ledge (bath) tub 嵌入式宽边浴盆
recess engraving 凹版雕刻
recess for concealed lighting 暗灯凹槽
recess for housing end frames 端架护罩的凹座
recess for main spring 主簧凹缺
recess for stop log 叠梁槽
recess ga(u)ge 槽规
recess grinding 凹槽磨削
recess hole 凹槽孔
recess incandescent lamp 镶入式白炽灯
recessing 开凹槽
recessing bar 内槽镗杆
recessing bit 切口刀；切槽车刀
recessing machine 切槽机；刨槽机
recessing tool 圆根切刀；切口刀；凹槽车刀
recess in rotor 转子工作面凹坑
recession 减弱处；海退；退行；退水；退离；衰退；凹处
recessional 后退的
recessional moraine 后退(冰)碛
recession analysis for snowmelt 融雪后退分析
recession constant 后退常数
recession curve 减退曲线；退水曲线；亏水曲线
recession discharge 消退排放量
recession equation 亏水方程；退水方程
recession flow 后退水流
recession gas generator 置换系统气体发生器
recession hydrograph 后退水文曲线；退水曲线；退水过程线
recession in demand 需求下降
recession limb 过程线下降段；退水曲线
recession line 后退线
recession of beach 海滩后退
recession of coast 岸线后退
recession of flood 洪水消落；洪水退落
recession of level 水位下降
recession of sea level 海面下降
recession of shoreline 岸线后退
recession of the groundwater level 地下水位降落
recession of valley sides 谷岸后退
recession of water-level 水位退落
recession of water stage 水位退落
recession period 退水期
recession phase of flood 洪水位后退
recession segment 后退部分
recession time 退水时间；地面水消退时间
recession velocity 退水速度；退水流速
recessive 劣势；倒退的
recess of press ram 压力机滑块上的模柄孔
recess path 啮出轨迹
recess phase 渐离相位；啮出相位
recess tub 凹入墙内的浴盆
recessus sphenoeth-moidalis 蝶筛隐窝
rechaimed water 再生水
rechain 重新丈量(指用测链重新丈量)
rechange 应急吊杆索具
recharge 再(次)装料；再充电；灌注(量)；二次充电；补给(量)
rechargeable 可再充电的
rechargeable battery 蓄电池组；可(再)充电电池
rechargeable battery cell 可充电蓄电池组
rechargeable power drill 可充电电钻
rechargeable sealed lead-acid battery 封闭式可充电铅酸电池
recharge aquifer for spring 泉的补给含水层
recharge area 地下水补给区；补给区；补给面积
recharge area of aquifer 含水层补给区
recharge basin 补给池；回灌池；补给流域
recharge by rainfall penetration 雨水入渗补给
recharge by seepage fo stream 河水渗漏补给
recharge canal 补给渠
recharge capacity 补给量
recharge coefficient 补给系数
recharge condition of groundwater 地下水补给条件
recharge cone 回灌锥
recharge current 再充电电流
recharge during ground water extraction 开采补给量
recharged water level 回灌水位
recharge form 补给方式
recharge in summer and utilizing in winter 夏灌冬用
recharge intensity 补给强度

recharge in winter and utilizing in summer 冬灌夏用
recharge line 补给(曲)线
recharge line of wells 井的补给曲率
recharge modulus 补给模数
recharge of an aquifer 含水层回灌
recharge of aquifer 含水(地)层回灌；含水层补给
recharge of condensation water 凝结水补给
recharge of groundwater 地下水回灌(量)；地下水补给(量)
recharge of melted snow 融雪水补给
recharge per single well 单井回灌量
recharge per unit pressure head 单位回灌量
recharge pit 补给竖井；补给孔
recharge quantity of phreatic water replenished by surface water 地表水对潜水的补给量
recharge quantity of water table condensation 潜水凝结补给量
recharger 再装填器
recharge rate 回灌率；补给率
recharge sources 补给来源
recharge sources of river 河水补给来源
recharge through skylight 天窗补给
recharge water 回充水；回灌水；补给水
recharge well 回灌井；受补井；复修井；补给井
recharge with irrigation water 灌溉回渗法
recharge with method of decreasing temperature and increasing pressure 降温加压回灌
recharge zone 回灌区
recharging 再装料；再充电；回灌；二次充电
recharging a storage reservoir 回灌蓄水库
recharging basin 灌水池；补给流域；再充灌水池
recharging phase 回灌期
recharging time 再充电时间
recharging water quantity after back-pumping 回扬后的回灌量
recharging water quantity before back-pumping 回扬前的回灌量
recharging well 回灌井
recharter 重发执照
recheck 再检验；再检查；再核对；再调查；重新检查；复验；复检；复核；复查
re-check observation 测回【测】
recheck oil level 再检查油位
rechipper 精加工机；木片再碎机；复切机
rechlorination 再氯化作用
rechuck 再夹紧；倒杆【岩】；重新卡紧；重新夹紧
recidivity 再发性
recipe-type specification 配方式规范
recipher 译成密码
recipience[recipiency] 容纳
recipient 信息接收器；接受者；接受方；技术引进方；收受人；收货人
recipient country 接受国；受援国
recipient of goods 受货人
recipient of the technology transferred 接受技术转让方
recipient processor 接收处理机
recipient river 排污河
reciprocable motor 往复运动电动机
reciprocal 相互的；相反的；互易；互逆；互换的；倒易的；倒数
reciprocal account 来往账户；相互账户；相对往来账户
reciprocal action 交互作用
reciprocal affiliations 相互投资联营公司
reciprocal agreement 互惠协定
reciprocal algebra 反代数
reciprocal apportionment 相互分摊额
reciprocal arrangements 相互安排
reciprocal assimilation 相互同化
reciprocal automorphism 反自同构
reciprocal axis 倒易轴
reciprocal bearing 对向方位角；倒方位；反方位；反方位(角)
reciprocal bond 相互约束
reciprocal buying 相互购买
reciprocal cancellation 相互注销
reciprocal capacitance 反向电容
reciprocal centimetre 厘米倒数
reciprocal circuit 可逆电路
reciprocal coefficient 反系数
reciprocal colo(u)r temperature 倒色温
reciprocal combustion 往复燃烧

reciprocal communication 相互间通信
reciprocal compressor 活塞式压缩机；往复式压气机
reciprocal cone 配极锥面
reciprocal consignment 相互委托销售
reciprocal contract 互惠合同
reciprocal correlated colo(u)r temperature 倒相关色温
reciprocal correspondence 反对应
reciprocal counter 倒计数器
reciprocal credit agreement 互惠信用协议
reciprocal cross-section 截面值倒数；倒截面
reciprocal currency agreement 互惠货币协议；通货互换协定
reciprocal curve 倒数曲线；倒角曲线
reciprocal curve method 倒数曲线法
reciprocal deflection 互等变位
reciprocal demand 相互需求
reciprocal depth 对应水深
reciprocal diagram 受力图；作用力示意图；力线图
reciprocal dielectric(al) constant 倒介电常数
reciprocal difference 倒(数)差分；反商差
reciprocal dispersion 相互分散；色散率倒数；反分散
reciprocal dispersive power 色散系数；色散率倒数
reciprocal distribution 对应分布
reciprocal duties 互惠关税
reciprocal effect 交互影响
reciprocal eigenvalue 逆本征值
reciprocal energy theorem 能量互易定理
reciprocal engine 往复(式)发动机
reciprocal expression 相反表达式；倒数式
reciprocal feed pump 往复式给水泵
reciprocal ferrite switch 互易性铁氧体开关
reciprocal figure 可易图形
reciprocal force polygon 力多边形
reciprocal friction test 磨损试验
reciprocal function 倒数函数
reciprocal holdings 相互控股
reciprocal identity 互等恒等式
reciprocal impedance 互易阻抗
reciprocal inductance 反向电感
reciprocal induction 相互诱导
reciprocal inhibition 相互抑制；交互抑制
reciprocal inhibition method 交互抑制法
reciprocal innervation 交互支配
reciprocal interchange 相互交换
reciprocal interlock 可逆联锁；互相联锁
reciprocal inverse matrix 倒逆矩阵
reciprocal junction 互易性接头
reciprocal kernel 逆核；互易核
reciprocal knitting 往复编织
reciprocal lattice 逆点阵；互易点阵晶格；互逆点阵；倒易格子；倒易点阵晶格；倒易点阵
reciprocal lattice coordinates 倒易点阵坐标
reciprocal lattice point 倒易阵点
reciprocal letter of credit 对开信用证
reciprocal level(l)ing 往复水准测量；双程水准测量；对向水准测量
reciprocal linear dispersion 倒线色散率
reciprocal linear representation 反线性表示
reciprocal load 往复荷载
reciprocal logarithmic curve 反商对数曲线
reciprocally proportional 成反比(例)的
reciprocal mass 质量倒数
reciprocal matrix 逆矩阵
reciprocal mixing 相互混频
reciprocal motion 往复运动
reciprocal multiplication 增殖率倒数；倒数相乘；反乘(法)
reciprocal natural selection 相互自然选择
reciprocal network 可逆网络
reciprocal normal sections 相对正截面
reciprocal observation 对向观测
reciprocal observation angle 对向观测角
reciprocal observation vertical angle 对向观测垂直角
reciprocal of amplification factor 放大系数倒数
reciprocal of base-altitude ratio 基线航高比倒数
reciprocal of base-height ratio 基线航高比倒数
reciprocal of oil mobility 油类流动性的倒数
reciprocal of the base height ratio 基线航高比倒数
reciprocal one-to-one correspondence 一一反对应
reciprocal orientation 相对方位
reciprocal parasite 相互寄生物
reciprocal permutation 反排列

reciprocal piezoelectric(al) effect 反电压效应
reciprocal points 互易点
reciprocal polar 反极线
reciprocal pole 相反的极;反极
reciprocal polynomial 反多项式
reciprocal precipitation 相互沉淀(作用)
reciprocal projection 反方向投影
reciprocal proportion 反比例
reciprocal proportion law 倒数比定律
reciprocal purchase 相互采购
reciprocal quantity 倒数量
reciprocal quotient 反商(系数)
reciprocal radius 倒半径;反向量径;半径倒数
reciprocal rate 相关汇率;反数汇率
reciprocal ratio 反比(例)
reciprocal reaction 往复反应
reciprocal recombinant 相互重组
reciprocal recurrent selection 交互反复选择
reciprocal reference 相关参照
reciprocal relation 互易关系;互反关系;倒易关系;倒数关系
reciprocal relative dispersion 相对色散的倒数
reciprocal representation 逆表示
reciprocal restraint 互反制约;互反约束
reciprocal sale 相互销售
reciprocal salt pair 平衡盐对
reciprocal sec 秒倒数
reciprocal sight 对向照准
reciprocal sight line 双向视线;双向测线;对向视线
reciprocal space 倒易空间;倒晶格空间;波矢空间
reciprocal space method 倒易空间法
reciprocal spacing 倒易间距
reciprocal specific turbidity 倒数比混浊度
reciprocal spectrum 逆谱
reciprocal spiral 双曲螺线;倒螺线
reciprocal square distance method 距离平方倒数法
reciprocal square root 反平方根
reciprocal strain ellipse 逆应变椭圆
reciprocal stress 往复应力
reciprocal substitution 倒数代换
reciprocal symbiosis 互惠共生
reciprocal system 相反系统
reciprocal tariff 互惠税则
reciprocal theorem 逆定理;互易定理;倒易定理
reciprocal theory 互易理论;互换理论
reciprocal time 往复次数
reciprocal trade 经惠贸易
reciprocal trade agreement 互惠贸易协定
Reciprocal Trade Agreement Act 相互贸易协定法
reciprocal transducer 互易换能器;倒易换能器
reciprocal transfer of property 财产互换
reciprocal transformation 相互转化;反向变换
reciprocal transformation model 反向变换模型
reciprocal translation 倒易平移
reciprocal translocation 相互易位
reciprocal treatment 互惠待遇;平等待遇
reciprocal treaty 相互条约;互惠条约
reciprocal value 互易值;倒数(值)
reciprocal variational problem 反变分问题
reciprocal vector 互易矢量;倒易向量;互易矢量
reciprocal vectors 倒易矢量组
reciprocal velocity 倒易速度
reciprocal velocity region 速度倒数区
reciprocal virtual work theorem 虚功互等定理
reciprocal viscosity 黏度倒数;倒易黏度
reciprocal zenith distance 对向天顶距
reciprocant 微分不等式;微分不变式
reciprocate 交替;互换;前后转动;往复移动;上下移动
reciprocate the casing 上下滑动套管
reciprocating 前后转动
reciprocating action 往复作用;往复动作
reciprocating air compressor 活塞式空气压缩机
reciprocating air engine 直动风马达
reciprocating air motor 空气马达;往复式气动发动机
reciprocating air pump 往复式空气压缩机
reciprocating and low pressure turbine 往复和低压涡轮机;往复和低压透平;往复和低压汽轮机
reciprocating and rotating rollers 往复回转罗拉
reciprocating apparatus 往复(运动)装置
reciprocating-arm distributor 往复式布水器;往复臂式配水器
reciprocating-arm grit washer 往复臂式洗砂器

reciprocating bar 连杆
reciprocating beam conveyer 往复式刮板运输机
reciprocating bearing 往复运动零件用的轴承
reciprocating blade spreader 往复式叶桨铺摊机;往复式叶桨分布机;倒刮式(混凝土)摊铺机
reciprocating block slider crank mechanism 往复滑块曲柄机构
reciprocating blower 往复式压气机;往复式鼓风机
reciprocating cacuum pump 往复式真空泵
reciprocating calibration 互易校准
reciprocating cam carriage 往复三角座滑架
reciprocating charger 往复式加料机
reciprocating chute 往复摆动式溜槽
reciprocating cold oil pump 往复式冷油泵
reciprocating column 往复塔
reciprocating compressed-air hammer 往复式压缩气锤
reciprocating compressor 往复式压缩机;活塞式压缩机;往复式压气机;往复式空气压缩机
reciprocating compressor with radial cylinders 星形往复式压缩机
reciprocating condensing unit 往复式制冷机;往复式冷凝装置;往复式冷凝器
reciprocating conveyer 往复式运输机;往复式输送机;往复摆式输送机
reciprocating-conveyor continuous centrifuge 往复式连续推料离心机
reciprocating crusher 往复式破碎机
reciprocating cutter 往复式破碎机
reciprocating diaphragm pump 往复式隔膜泵
reciprocating discharge grate 往复式卸料箅子
reciprocating displacement pump 往复式排代泵
reciprocating distributor 往复式撒布机
reciprocating drill 往复式凿岩机;往复式手钻;往复式风钻
reciprocating duplex pump 往复式双缸泵
reciprocating engine 活塞式发动机;往复式引擎;往复式发动机
reciprocating expansion-engine 活塞式膨胀机
reciprocating feeder 往复式加料器;往复式加料机;往复式供料器;往复式供料机;往复式给料器;往复(摆)式给料机
reciprocating feeder pump 往复式加料泵
reciprocating feed pump 往复式进料泵
reciprocating filing machine 往复式锉床
reciprocating flight conveyer 往复刮板输送机
reciprocating force 往复力
reciprocating furnace process 移动炉法
reciprocating gas compressor 往复活塞式气体压缩机
reciprocating grate 往复式炉排;往复式炉箅;摆动式炉箅子
reciprocating grate cooler 推动箅式冷却器;摆动箅式冷却器
reciprocating grate stoker 往复炉箅式加煤机
reciprocating grate stoker boiler 往复推动炉箅炉
reciprocating grid 摇筛
reciprocating grinder 往复式磨床
reciprocating heater 往返互换加热器
reciprocating hedge cutter 往复式树篱修剪机
reciprocating impedance 倒易阻比
reciprocating inertia generator 往复惯性发生器
reciprocating intensifier pump 往复式增压泵
reciprocating internal combustion engine 往复式内燃机
reciprocating knife 往复式割刀
reciprocating lever 往复杆
reciprocating liquid chiller 往复式液体冷却器
reciprocating machine 往复式机器
reciprocating main motion 往复主运动
reciprocating mass 往复质量
reciprocating mechanism 曲拐机构;往复运动机构
reciprocating motion 往复运动;往返运动
reciprocating-motion vertical conveyer 往复式垂直输送机
reciprocating number of mud pump 泥浆泵往复次数
reciprocating piezoelectric(al) effect 往复压电效应
reciprocating piston engine 活塞式发动机;往复活塞式发动机
reciprocating piston internal combustion engine 往复活塞式内燃机
reciprocating piston pump 往复活塞泵

reciprocating piston tool 往复作用活塞式工具
reciprocating plate and worm type distributor 摆动板和螺旋式排肥器
reciprocating plate extractor 往复式孔板萃取器
reciprocating plate feeder 往复式平板给料机;往复片式给料机;抽板喂料机;往复板式给料机
reciprocating power-driven tine harrow 动力驱动摆动式钉齿耙
reciprocating pressure-type crusher 往复压挤式破碎机
reciprocating pump 活塞(式)泵;往复(式水)泵
reciprocating pump compressor 往复压缩泵
reciprocating rake 往复式路耙
reciprocating ram pump 往复式水锤泵
reciprocating refrigerating machine 往复式制冷机
reciprocating rock drill 往复式凿岩机
reciprocating roller press 往复式滚压机;往复式辊压机
reciprocating rolling 往复滚动;往复碾压
reciprocating rolling mill 往复式轧机
reciprocating rolling process 往复滚压法
reciprocating rotary piston engine 往复式转子发动机
reciprocating saw 往复(式)电锯;往复锯
reciprocating scraper 往复式刮板输送器
reciprocating screed (整压混凝土的)往复式整平板
reciprocating screen 往复式振动筛;往复(式)筛;摆动筛;震动筛
reciprocating-screw extrusion machine 往复式螺杆挤出机
reciprocating shaft 往复轴
reciprocating share 摆动犁铧
reciprocating sieve 震动筛;往复(式)筛
reciprocating single-screw compounder 往复式单螺杆配料机
reciprocating spading machine 往复式铲掘机
reciprocating spout 摆动撒布管
reciprocating spraying machine 往复式喷涂机
reciprocating steam engine 往复式蒸汽机
reciprocating steam pump 往复式蒸汽泵
reciprocating swashplate type engine 往复斜盘式发动机
reciprocating table 往复式工作台
reciprocating theorem 互等定理
reciprocating through conveyer 往复式运输机;簸动送料机;簸动运输机
reciprocating time 纯钻进时间(冲击钻)
reciprocating type 往复式
reciprocating type air compressor 往复式空气压缩机
reciprocating type compressed-air hammer 往复式压缩气锤
reciprocating type compressor 往复式压缩机
reciprocating-type milling fixture 往复式铣削夹具
reciprocating-type supercharger 往复式增压器
reciprocating unbalance 往复运动不平衡(切割器等)
reciprocating vertical conveyer 往复式提升机
reciprocating water chiller 活塞式冷水机组
reciprocating water wheel distributor 往复式水轮配水器;往复式水轮配水池
reciprocating wedge wire screen 往复式振动筛(筛面用截面金属丝编成)
reciprocating welding electrode 往复焊接电极
reciprocation chiller 活塞式冷水机
reciprocation screen 往复筛
reciprocator 抖料机;往复(运动)机件;往复式引擎;往复式发动机;往复喷涂机;倒数器
reciprocity 相互状态;可逆性;交互作用;互易(性)互利;倒易(性);反比
reciprocity calibration 互易校准;互易定标;互易标准
reciprocity clause 互惠条款
reciprocity coefficient 互易系数;互异常数
reciprocity constant 互异常数
reciprocity curve 倒易曲线
reciprocity effect 倒易效应
reciprocity failure 倒数率失效
reciprocity in trade 互惠贸易
reciprocity law 互易律;互易定律;倒易律;倒易定律;反比定律
reciprocity parameter 互易参数
reciprocity principle 可逆性原理;互易原理
reciprocity property 互易特性

reciprocity relation 可逆关系;互易关系;互反关系
reciprocity rule 倒易定律
reciprocity theorem 倒易理论
reciprocity transaction 互惠交易
recirculate 再循环;逆环流
recirculated air 循环空气;循环风
recirculated air by-pass 循环空气旁管
recirculated air intake 循环空气进口
recirculated air operation 空气回流运行
recirculated gas 回流气体
recirculated iron 回炉铁
recirculated seal 循环式密封
recirculated ultrafiltration membrane bioreactor 循环超滤膜生物反应器
recirculated water 循环水
recirculating 再循环
recirculating air 再循环空气
recirculating air separator 循环选粉机
recirculating aquaculture system 循环水养殖系统
recirculating ball guide 循环滚珠导管
recirculating ball type steering gear 循环球形转向器
recirculating ball type steering wheel 循环滚珠式转向盘
recirculating cooling 循环冷却
recirculating cooling system 循环冷却系统
recirculating cooling water 循环冷却水;再循环冷却水
recirculating cooling water system 循环冷却水系统
recirculating dip tank 回流施釉槽
recirculating dissolver 环流式溶解器
recirculating dry kiln 多次循环干燥窑
recirculating fan 再循环风机
recirculating flume 循环水槽
recirculating heater 环流加热器
recirculating heating system 循环热水系统;循环供暖系统
recirculating high pass velocity preheater 高速再循环预热器
recirculating linear accelerator 回注式直线加速器
recirculating load 循环负荷
recirculating loop 再循环程序区;回路
recirculating loop frequency 闭合回路频率
recirculating loop memory 循环存储器
recirculating mass 再循环流量
recirculating oil 再循环(润滑)油;再循流(润滑)油
recirculating pipe 再循环管
recirculating (pipe) line 再循环管线
recirculating pool 循环水池
recirculating pump 循环泵;再循环泵
recirculating ratio 循环系数;循环比
recirculating sand filter 循环砂滤池
recirculating scrap 回炉废钢;返回废钢
recirculating skimmer 重复循环撇沫器
recirculating store 重复循环存储器
recirculating system 再循环系统
recirculating tube 再循环管
recirculating water 循环水;再循环水
recirculating water cooling system 循环冷却水系统
recirculating water portion 再循环水分配
recirculation 信息重复循环;再循环;重复循环;二次循环
recirculation air 再循环空气
recirculation cavity 空气动力阴影区
recirculation current 循环水流
recirculation drying stove 热风循环烘干炉
recirculation duct 回流管道
recirculation flow 循环水流
recirculation flow in biologic(al) filter 生物滤池回流量
recirculation flume 循环水槽
recirculation furnace 热风循环炉
recirculation gel trap 循环胶体捕集
recirculation heater 循环加热器
recirculation heating system 循环供暖系统
recirculation lands 回流台肩
recirculation of air 通风再循环
recirculation of gas 气体的回流
recirculation of over-size particles 粗粉循环
recirculation of sewage 污水循环
recirculation of sewage in high rate (biological) filters 高负荷生物滤池污水循环

recirculation of sludge 污泥循环
recirculation of wet air 湿空气回流
recirculation operation 循环操作
recirculation pipe 循环管
recirculation piping 循环管
recirculation piping line 循环管线
recirculation process 循环法
recirculation pump 循环泵;重复循环泵
recirculation rate in biologic(al) filter 生物滤池回流比
recirculation ratio 循环系数;循环比;回流比
recirculation reactor 循环反应器
recirculation system 循环(水)系统
recirculation water 再循环水;循环水
recirculation water flow 循环水流
recirculation water portion 再循环水分配
recirculator 再循环器
recirculatory pipe (水、汽、油的)回流管
recision 交易取消
recision diffusion 废弃扩散
recital 详细说明
recitation room 课室;教室
reckless bonus paying 滥发奖金
reckless conduct 轻率的行为
reckless investment 冒险投资
reckon 估计
reckon account 结账
reckoned from 计算起自
reckoner 计算员;计算手册
reckoning 结算;计算;推算船位;推算
reckoning book 账簿;出纳簿
reckoning by piece 计件
reckoning fix position 船位推算(位置)
reckoning in every item of income or expenditures 满打满算
reckoning of useful life 估定使用年限
reckon up the bill 合计账单
reclad 在砖、石上再做上一层贴面
reclaim 再生胶;土地的开垦;补釉;回收
reclaimable 可回收的
reclaimable land 可利用土地;可垦地
reclaimable virgin soil 生荒地
reclaimable waste 可再利用废物;可回收废物
reclaimable wasteland 宜耕荒地
reclaim area 开垦区;开垦区
reclaimation 回收利用
reclaimed alkali wastes 碱回收
reclaimed asphalt mixture 再生沥青混合料;复拌沥青混合料
reclaimed bitumen pavement 再生沥青铺面;再生沥青路面
reclaimed bituminous pavement 再生路面
reclaimed concrete 再生水泥混凝土
reclaimed copper 再生铜
reclaimed dam 填筑坝
reclaimed earth 吹填土
reclaimed effluent 再生污水
reclaimed fiber 回收纤维
reclaimed filler 沥青填料
reclaimed ground 围垦的土地;填筑地
reclaimed gypsum 再生石膏
reclaimed land 土地复垦;新垦地;垦拓地;开垦地;围垦地;填筑地;吹填地
reclaimed leather 再生革
reclaimed lubricating oil 再生润滑油
reclaimed marsh 垦殖沼地
reclaimed material 修旧利废材料;再生材料;回收材料
reclaimed municipal wastewater 回收城市污水
reclaimed oil 再生油
reclaimed quarry 可再使用碎石;可回收碎石
reclaimed rosin 再生松香
reclaimed rubber 再生(橡)胶
reclaimed rubber dust 再生橡胶粉
reclaimed rubber modified asphalt 再生胶改性沥青
reclaimed rubber solution 再生的橡胶溶液
reclaimed rubber tyre 再生橡胶轮胎
reclaimed sand 再生砂
reclaimed sewage 再生污水;回污水
reclaimed soil 已改良的土壤
reclaimed waste glass 复用废玻璃
reclaimed wastewater 再生污水;回收废水
reclaimed water 再生水;回收水

reclaimed water reuse 再生水回用
reclaimed water system 循环水系统
reclaimer 旧料复拌机;储[贮]存场装载输送机;再生装置;(旧料复拌机;回收设备;回收器;取料机;取矿机;脱硫剂;填筑机;输送机;出场给料漏斗
reclaim extrusion system 再生挤出系统
reclaim floodland and plant it to crops 河滩造田
reclaim hopper (向地下输送机给料的)出场给料斗;出场给料漏斗
reclaiming 开垦荒地;翻造;回收
reclaiming action 再生作用
reclaiming aid 再生助剂
reclaiming barrow 填筑耙
reclaiming belt conveyer 取料胶带输送机
reclaiming bucket wheel 斗轮取料机
reclaiming by centrifuge 离心法回收
reclaiming by filtration 过滤法回收
reclaiming by plastification 再塑造
reclaiming conveyer 取料输送机;转载运输机;出场输送机
reclaiming facility 回采设备;回收设备
reclaiming hopper 出场给料漏斗
reclaiming installation 回采装置;回收装置
reclaiming line 废板修整线
reclaiming of gullies 开垦沟地
reclaiming of lead 废铅重炼
reclaiming paddle 取料叶桨;出料叶桨
reclaiming plant 回采设备;回收设备
reclaiming rate 取料量
reclaiming salt area 盐渍地改良
reclaiming saw cylinder 回收锯齿滚筒
reclaiming scraper 回收刮刀;回收铲运机
reclaiming system 回收系统
reclaiming tunnel (散料堆场的)出料坑道;取料坑道
reclaiming tunnel conveyer (散料堆场的)出料坑道输送机
reclaiming unit 取料机械设备;出料机械设备
reclaiming waste 回收废料
reclaim land from the lake 围湖造田
reclaim land from the sea 围海造田
reclaim mix 再生胶拌料
reclaim of land 土地复垦
reclaimor 回收器
reclaim rinse 漂洗回收
reclaim tunnel 出料坑道
reclaim lake bottom land and plant in to crops 围湖造田
reclamation 资源化;再次分选;改良;利用废料;垦石;开拓(采矿);纠正银行清算错误;回收;荒地垦拓;围垦;围海造陆;填筑;收回的请求;吹填;复原工程
reclamation activities 修复措施
reclamation and utilization of sewage 污水回收利用
reclamation area 垦区;填筑区;吹填区
reclamation by enclosure 围垦
reclamation by filling 填筑
reclamation dam 垦堤;垦拓围堤;填筑坝
reclamation dike 垦拓滩地用围堤;吹填围埝
reclamation district 围垦区;填筑区;吹填区
reclamation dredge(r) 吸扬(式)挖泥船;吹泥船;开垦挖泥机
reclamation earth 吹填土
reclamation engineering 填筑工程
reclamation expenses 垦殖费用
reclamation ground 开垦地
reclamation innovation 回收革新
reclamation land 围垦地;填筑地;吹填地
reclamation law 回收律
reclamation of eroded area 侵蚀地改良
reclamation of geotechnique 土质改良学
reclamation of land 土地填筑;土地垦拓;土地开垦
reclamation of land by enclosure 围垦造田;围垦造地
reclamation of land by filling 填土造地
reclamation of material 废料利用
reclamation of paper 纸回收
reclamation of saline-alkali soils 盐碱土改良
reclamation of saltern 盐碱土改良
reclamation of sewage 污水回收
reclamation of soils 粪水回收
reclamation of solid resources 固体资源回收
reclamation of solid wastes 固体废物资源化
reclamation of waste water 废水利用;废水回收

（利用）
reclamation project 垦拓计划；围垦工程；填筑工程；吹填工程
reclamation sand 再生砂
reclamation scheme 围垦工程
reclamation service 垦拓事务；收回使用
reclamation site 围垦区；填筑区；吹填区
reclamation soil layer 填土层
reclamation unit 回收装置
reclamation volume 吹填方
reclamation wall 垦区围墙
reclamation works 垦殖工程；开垦工程；围垦工程；填筑工程；填土工程；吹填工程；修复工作
reclamation works presently underway 现在进行中的吹填工程
reclamation zone 吹填区
reclamp 再夹(住)
reclassification 再入级；再分类；再(次)分级；重新分类
reclassification entry 分录的重新分类
reclassification of post 重新定职定级
reclassification track 改编线
reclassifier line 重选作业线
reclassify 二级分组；二级分类
recleaner 再选机；二次洗涤机
recleaner cell 精选槽
recleaner flo(a)tation 再精选
recleaner screen 精选筛
recleaning 再精选
recleaning screen 精选筛
reclined Budda 卧佛
reclined fold 斜卧褶皱；平卧褶皱；伏褶曲
reclined woman pillow 卧女枕
reclining chair 可调靠背椅；活动靠背椅
reclining posture 半卧式
reclining radio antenna 折倒天线
reclocking 重新计时；重复计时
reclose 再次接通；重新接通
recloser 反发光电设备；自动重接器；自动开关(装置)
reclosing 再投入；重合(闸)
reclosing device 重合装置
reclosing fuse 重合熔断器
reclosing relay 自复继电器；重合闸继电器；重闭继电器
reclosing time 再闭合时间；重合时间
reclosing type circuit breaker 复合式断路器；复合断路器
reclosure 自动接入；再次接入
reclothed stone 包石
recoal 再供给煤
recoat 再涂；重涂
recoatability 可再涂性；再涂性；再涂抗力
recoat adhesion 再涂附着力
recoating 重新涂(膜)
recoating test 罩漆试验；再涂试验
recoat time 再涂时间
recockling cylinder 麦郎选除筒
recodification 重编码
recogizer 测定器
recognition algorithm 识别算法
recognition and approval arrangement 承认和批准协议
recognition and control processor 识别和控制处理机
recognition arrangement 承认和批准协议
recognition capability 识别能力
recognition code 识别密码
recognition differential 认可差别；识别差；分辨差
recognition dispute 认可的争执
recognition function 识别函数；识别功能
recognition gate 识别门
recognition hazard 识别危险
recognition hazardous diamond 识别危险菱形图
recognition image 识别图像
recognition intercept 识别性截击
recognition lag 决策时间滞差；认识时间差滞
recognition light 识别信号灯；识别灯标
recognition mark 识别标志
recognition matrix 识别矩阵
recognition method 识别方法
recognition network 识别网络
recognition of end-point 终点辨认
recognition of expenses 费用(的)确定
recognition procedure 识别程序

recognition process 识别过程
recognition program(me) 识别程序
recognition rate 识别速率
recognition rule 识别规则
recognition sequence 识别顺序
recognition service 认可服务处
recognition signal 识别信号
recognition site 识别部位
recognition system 认识系统；识别系统
recognizable kerogen classification at present in China 我国目前采用的干酪根分类
recognizable pattern 可识别模式
recognizable set 可识别范围
recognization part 识别部位
recognization phase 识别期
recognization site 识别部位
recognization system 识别系统
recognized 公认的
recognized bank 认可银行
recognized carcinogen 确认致癌原；确认致癌物
recognized component index 认可元件索引
recognized custom of trade 公认航运惯例
recognized dealer 认可的交易商
recognized sum 识别总数
recognized target 识别目标
recognize economic law 遵重经济规律
recognizer 识别算法；识别器
recognizes mycolic acid 确认霉菌酸
recognizing ability 认识功能
recognizing machine 识别机
recoil 再绕；后座力；跳回；反绕；反撞；反跳；反冲
recoil absorber 制退机
recoil adapter 后坐力接合器
recoil-and-compression stress-wave 回弹压缩应力波
recoil angle 反冲角
recoil apparatus 制退机
recoil atom 反冲原子
recoil booster 后座增压器
recoil brak 驻退机
recoil broadening 反冲加宽
recoil check spring 复进弹簧
recoil check spring bolt 后座弹簧栓
recoil chemistry 反冲化学
recoil counter 反冲粒子计数器
recoil cylinder 驻退筒；制退筒
recoil cylinder closing plug 驻退筒密封塞
recoil cylinder replenisher 驻退筒调节器
recoil depletion 反冲消耗
recoil detector 反冲探测器；反冲核探测器
recoil distribution 反冲分布
recoil effect 反冲效应
recoil energy 反冲能(量)
recoiler 卷取机；重绕机
recoil escapement 锚式擒纵机构；反冲擒纵机构
recoil experiment 反冲实验
recoil factor 反冲因数
recoil fault 反冲断层
recoil fluid 驻退液
recoil fragment 反冲碎片
recoil impulse 反冲脉冲
recoiling 反冲；再绕
recoiling click 带退程的掣子
recoiling machine 重卷机；卷取机
recoiling nucleon 反冲核子
recoiling ratchet going barrel 条盒反冲棘爪
recoil ion 反冲离子
recoil ionization chamber 反冲粒子电离室
recoil label(l)ing 反冲标记
recoil test 反跳试验
recoilless 无后坐力的
recoilless emission 无反冲发射
recoilless resonance absorption 无反冲共振吸收
recoil lug 后座突耳
recoil mechanism 驻退机
recoil movement 后座运动
recoil nucleus 反冲原子；反冲核
recoil oil 反冲油
recoil-operated 受反冲力而动作的
recoil pad 反冲垫；后座(缓冲)垫
recoil particle 反冲粒子
recoil particle counter 反冲粒子计数器
recoil piston 驻退活塞
recoil piston rod 驻退(活塞)杆

recoil piston rod lug 驻退杆固定突座
recoil pit 后退槽
recoil-proton counter 反冲质子计数器
recoil radiation 反冲辐射
recoil reaction 回击
recoil separation 反冲分离
recoil shift 反冲位移；反冲漂移
recoil spectrometer 反冲粒子分光仪
recoil spring 驻退簧；回位弹簧；复进簧
recoil spring pilot 反冲弹簧导向架
recoil spring stop 反冲弹簧停止块
recoil starter 反冲启动器
recoil streamer 反冲电子流
recoil strength 反冲力
recoil system 驻退系统
recoil tagging 反冲标记
recoil test 反跳试验
recoil throttling rod 驻退机调节杆
recoil valve 驻退机阀；反冲阀
recoil velocity 后坐速度
recoil wave 反冲波
recollapse 再坍缩
recollimate 重新照准
recolo(u)r 重新着色
recombinant 重组体
recombinated luminescence bacteria 重组发光菌
recombination 再组合；再化合；重组作用；重组(合)；重新组合；重新结合；复合
recombination action 复合作用
recombinational mapping 重组作图法
recombination center 复合中心
recombination coefficient 复合系数
recombination continuum 复合连续区
recombination cross section 复合截面
recombination current 复合电流
recombination deficient mutant 重组缺陷型
recombination electroluminescence 注入式电致发光
recombination energy 复合能
recombination epoch 复合期
recombination frequency 重组频率
recombination H-line 复合氢线
recombination laser 复合激光器
recombination lifetime 复合寿命
recombination mechanism 复合机理
recombination mosaic 重组型嵌合体
recombination nodule 重组节
recombination noise 复合噪声
recombination of gaseous ions 气体离子的复合
recombination probability 复合概率
recombination process 重组过程
recombination radiation 复合辐射
recombination ramjet 复合式冲压空气喷气发动机
recombination ramjet engine 复合冲压式发动机
recombination rate 复合率
recombination-rate surface 表面复合速率
recombination spectrum 复合谱
recombination time 复合时间
recombination value 重组值
recombination velocity 复合速度
recombination with double excitation 双激发复合
recombinative repair 重组修复
recombine 重新组合；重新结合
recombined 调配
recombined imagery 合成图像
recombined materials 重新组合的材料
recombiner 复合器
recombing 复精梳
recommendal figure of plan 计划建议数字
recommendation 建议；介绍；推举；建筑及维护规则；推荐
recommendation route 推荐航线；推荐航路
recommendatory annotation 推荐性简介
recommended allowable settlement 建议容许沉降量
recommended alternative 推荐项目；推荐工程；推荐方案
recommended course indicator 最佳航向指示器；最佳航向指示
recommended daylight factor 推荐天然采光系数；标准天然采光系数
recommended exposure limit 建议的暴露限值
recommended flowsheet 推荐流程
recommended illumination 推荐照度

recommended limit 建议极限;推荐标准
recommended maximum concentration 建议最大浓度
recommended maximum contaminant level 建议最大污染水平
recommended maximum impurity content 建议最大污染物含量
recommended parameter 推荐参数
recommended plan 推荐方案
recommended practice 推荐方法
recommended project 推荐项目;推荐工程;推荐方案
recommended retail price 推荐零售价格
recommended route 推荐航线;推荐航路
recommended spare parts list 推荐的备件单
recommended standard 推荐性标准;推荐标准
recommended temperature ranges for heat treatment 推荐的热处理温度范围
recommended tender 被推荐的投标书
recommended tolerance 推荐规范
recommended track 推荐航线;推荐航路
recommended track for deep draught vessels 深水船推荐航道
recommended value 推荐值
recommended variety 推荐品种
recommend special tools 推荐专用工具
recommend substitute 推荐代用器
recompact 重复压实;再压实
recompaction 再压制;再压实;重复压实
recompense 赔偿(金);报酬
recompilation 重新编译
recompile 重新编译
recompility 重新编译性
recomplementation 再求补;再补法
recompletion 重新完井
recompose 重新组合
recomposed granite 再组合花岗岩
recomposed rock 再胶结岩石
recomposition 再组成
recompounding 重配料
recompression 再压(缩);重复压缩
recompression chamber 加压舱;重新加压舱
recompression curve 再压缩曲线
recompression index 再压缩指数
recompression modulus 再压缩模量;回弹模量
recompression pressure 再压力
recompression therapeutic scheme 加压治疗方案
recompression treatment 加压治疗
recomputation 重新计算;复核计算
recompute 重算
recomputed point of turn 重新算定的转向点
recon 交换子;重组子
reconcentration 再浓缩;再精选
reconcile 使一致
reconciliation account 调节账户
reconciliation agreement 和解协议;调解书
reconciliation item 调整项目
reconciliation of bank accounts 银行账户核对
reconciliation of depreciation 折旧核对
reconciliation of inventory 物料平衡;调整物料
reconciliation of reciprocal accounts 往来账户的调节;相互账户的协调
reconciliation procedure 对账程序
reconciliation statement 调节表
reconciling bank account 调节银行存款账户
reconciling inventory 调节存货
recondite map 难读地图
recondition 修理;重车;检修;恢复;还原;重磨轧辊;重激活;重调节
re-conditional 翻新的
reconditioned (moulding) sand 再生砂
reconditioned mud 再生泥浆
reconditioned vehicle 修复车辆
reconditioned wood 改善的木材;二次处理木材;矫形木材
reconditioner 油净化装置;再生装置;净化装置;调整机;重调湿室
reconditioner-carrier reception 调整载波接收
reconditioning 蒸汽处理(用于硬木);(木材的)调直处理;重车磨;整修;整理改装;再(生)处理;模孔修磨;重调湿;暴露剂量
reconditioning charges 整理费
reconditioning dimensions 重调尺寸
reconditioning equipment 修配的设备

reconditioning fee 整理费
reconditioning of collapsed timber 干缩变形木材(的蒸汽)处理
reconditioning of rail joint 钢轨接头整修
reconditioning of road 道路修复
reconditioning of stone 再刻石
reconditioning of timber 蒸汽处理木材;木材的蒸汽处理
reconditioning plant 泥浆处理设备
reconditioning press 复榨机
reconditioning system 回收系统
reconditioning work 修理作业
reconfiguration 再组合;改变外形;结构变换;重新组合;重(新)配置;重构
reconfiguration program (me) 重新组合程序;重构程序
reconfiguration system 再组合系统;重构系统
reconfigure itself 自我重组
reconfirm 重新证实;重新确认
reconfirmation 再确认
reconnaissance 选线;选点;勘察;勘查;勘测(选点);踏勘(调查);查勘;采测
reconnaissance and underwater demolition group 侦察与水下爆破队
reconnaissance by direct observation 直接观测侦察
reconnaissance camera 勘测摄影机
reconnaissance diagram 选点略图;勘测草图
reconnaissance drill 轻型普查钻机(钻15～20m深)
reconnaissance/establishment of interpretation 踏勘解译标志建立
reconnaissance examination 踏勘检查
reconnaissance flight 勘测飞行
reconnaissance flying boat 侦察飞船
reconnaissance for control point selection 控制网选点
reconnaissance geologic(al) report 普查地质报告
reconnaissance in field 地面踏勘
reconnaissance instrument 普查用仪器
reconnaissance ladder 选点梯
reconnaissance level 勘测阶段
reconnaissance map 勘测(草)图;踏勘(地)图;草(测)图
reconnaissance method 路线踏勘法
reconnaissance module 侦察仪器舱
reconnaissance objective 侦察目标
reconnaissance of mineralization point 矿点踏勘
reconnaissance of route 路线测绘
reconnaissance operations center [centre] 侦察活动中心
reconnaissance orbiting stabilized platform 侦察用轨道式稳定平台
reconnaissance pallet 侦察仪器箱
reconnaissance party 踏勘队
reconnaissance phase 踏勘阶段
reconnaissance photography 勘测摄影
reconnaissance pod 侦察设备吊舱
reconnaissance point 选点
reconnaissance radar 侦察雷达
reconnaissance recorder 侦察记录装置
reconnaissance reference point 侦察基准点
reconnaissance report 勘测报告;踏勘报告
reconnaissance satellite 侦察卫星;勘测卫星
reconnaissance ship 侦察船
reconnaissance sketch 勘测草图
reconnaissance soil map 土壤概图;土壤调查图
reconnaissance soil survey 土壤约测;土壤概测
reconnaissance sonar 侦察声呐
reconnaissance stage exploration 踏勘阶段的勘察
reconnaissance strip 条幅式侦察照片
reconnaissance survey 选线;线路普查;概测;概查;路线测绘;勘测;普查;踏勘(测量);查勘测量;草测
reconnaissance survey in field 野外踏勘
reconnaissance team 勘测队
reconnaissance trip 现场踏勘
reconnaissance well 普查井
reconnaissance work 踏勘作业
reconnect 重接
reconnection 再连接;恢复(原先)接线;重新接入;重连
reconnoiter [reconnoitre] 踏勘
reconnoitring boat 侦察艇

reconsequent drainage 再顺水系
reconsider 重新考虑;复议
reconsideration 复议
reconsideration of the case 复议案件
reconsolidate 再巩固;再加强;重新巩固
reconsolidation 再固结
reconstitute 重建;重新制定;重新构成
reconstituted emulsion 再生乳状液
reconstituted feed 计算入料;重组给料
reconstituted marble tile 复制的大理石面砖
reconstituted mica 复制云母
reconstituted oil 翻造油;再造油
reconstituted specimen 再制试样;再制试件
reconstituted stone 再造石;再加工宝石;铸石;人造石
reconstituted stone shop 复制石工场
reconstituted stone work 复制石作业
reconstituted wood 再造木材
reconstitution 重组成;重新组成;重新制定;重构;补充头寸
reconstruct 翻修;重建;改建;按原样修复;再建
reconstructed coal 加工煤
reconstructed cost 重计成本
reconstructed granite 再生花岗岩
reconstructed image 重现像;重建像;重构象
reconstructed item 改建项目
reconstructed line 改建线路
reconstructed model 复制模型
reconstructed project 改建工程
reconstructed railway 改建铁路
reconstructed sample 重构样
reconstructed specimen 再制试样;再制试件
reconstructed stone 再造石;人造石
reconstructed wave front 再现波前
reconstructed wood panel 再造木板
reconstructing beam 再现光束
reconstructing wave 重建波
reconstruction 再造;再现;再建;改建;重新构建;重显;重塑;重建;重构
reconstructional measures 改建措施
reconstruction beam 再现光束
reconstruction beam angle 重建光束角
reconstruction brightness 再现亮度
reconstruction cost 改建费用;重建费
reconstruction effect 重建效应
reconstruction engineering 改建工程
reconstruction item 改建项目
reconstruction of acoustic(al) hologram 声全息图的再现
reconstruction of bridge 桥梁改建
reconstruction of existing line 旧线改建;既有线改建
reconstruction of face contour 复原面容
reconstruction of image 图像重显;图像再现
reconstruction of old port 老港改建
reconstruction of pavement 路面改建;路面翻修;翻修路面
reconstruction of ship-yard 船厂改建
reconstruction of tunnel 隧道改建
reconstruction park 大修场
reconstruction phase 重建期
reconstruction plan(ning) 改建规划;改进规划;重建规划;重建计划
reconstruction sample 重建样值
reconstruction system 重建系统
reconstruction wave 重现波
reconstruction work 改建工程;重建工程
reconstructive inversion 重建转变;重建性转换
reconstructive policy 改建政策
reconstructive surgery 重建外科
reconstructive transformation 重建型转变
reconstructor 再现器
Reconstuction Finance Corporation 财政复兴公司(美国)
re-contamination 再污染
recontrol 再检校;复查
reconvergent node 再收敛节点
reconverging fan-out path 再收敛扇出通路
reconversion 再转换;恢复原状
reconversion maneuver (垂直起落飞机的)逆过渡机动
reconvert 再转变
reconverted analog picture 再恢复模拟图像
reconveyance deed 产权归还证书;取回文据;归

还文据
recooking 重蒸煮
recooler 重冷却器;二次冷却器
recooling 再冷却;二次冷却
recooling tower 二次冷却塔
record 记事;地层资料
record-based system 以记录为基础的系统
record block 实际记录
record-breaking 破纪录的
record button 录像按钮
record-card number of geologic(al) observation point 地质点记录卡片号
record cast 记存模型
record changer 自动换盘放送器
record chart 自记图;自动记录仪表用的记录带;自动记录带;记录线路图
record copy 存档底图
record drawings 竣工图;修改记录图;施工记录图
recorded cost 账面成本;记录成本
recorded delivery 记录的交割;保价邮件
recorded depth 记录深度
recorded fringe 记录条纹
recorded hologram 记录全息图
recorded hydrograph 实测过程线
recorded key 记录键
recorded plat 注册地界图
recorded position 记录位置
recorded room 记录室
recorded text 记录正文
recorded valuation 账面价值
recorded value 观测值;记录值;实测资料;实测值
recorder 记录者;记录员;记录仪(表);记簿员;记簿员;数据记录者
recorder adjustment 自动记录仪校准
recorder apparatus 记录室
recorder carrier 压力计托筒
recorder chart 自动记录器(记录纸);记录线路图
recorder control console 记录器操纵台
recorder-controller 记录控制器
recorder counter 记录计数管
recorder driver amplifier 记录装置放大器
recorder drum 记录(辊)筒
recorder file 记录器文件
recorder finder 记录器式寻线机
recorder float 自记浮子
recorder motor 记录器电动机
recorder of volume 容积描记器
recorder paper 记录(仪)纸
recorder pen 自动记录器笔;记录笔
recorder reception 记录接收
recorder shelter 自记仪仪器箱;自记仪仪器房
recorder strip 记录仪卷带;记录带
recorder switch 记录器转换
recorder tape 记录器纸带
recorder unit 记录器;记装置;记录装置
recorder well 自记仪器测井;观测井
recorder with paper disc 纸盘记录器
record flood 特大洪水;记录洪水;历史记载特大洪水;实测洪水;创记录的洪水
record for surveying in parts 碎部测量记录
record head 检频头
record holdings 文献馆藏
record hole 构造孔
recording 记录(式);登记
recording accelerometer 自记加速表;自动加速表;可记录加速仪
Recording Acts 登记法(契据)
recording address file 记录地址文件
recording aggregate 记录聚集
recording airspeed indicator 自记风速仪;自记风速表;自记风力计;自记风力表
recording altimeter 记录式高度计;自动记录高程计;自动记录测高计
recording ammeter 记录式电流表;自记电流表
recording amplifier 记录放大器
recording analysis method 记录式分析法
recording anemometer 自记风速仪;自记风速表;自记风力计;自记风力表
recording aneroid barometer 自记无液气压计;自记空盒气压计
recording apparatus 记录设备;记录装置;记录室;记录器;自记仪器
recording appraisal 记录鉴定
recording area 记录区(域)

recording arm 记录杆;记录臂
recording attribute 记录属性
recording avapo(u)rimeter 自记蒸发器;自记蒸发计
recording balance 记录天平;自动记录秤
recording bar 记录杆
recording barometer 自记气压计;自记气压表
recording beam 记录电子束
recording block 记录块
recording blocking 记录块化
recording board 记录台
recording book 记录簿
recording bridge 记录电桥
recording bus 自动记录仪表车
recording button 记录按钮
recording cabinet 记录箱;记录柜
recording calorimeter 记录式量热器;自动记录热量计
recording camera 记录照相机;记录摄影机;光转筒记录器;电视胶片录像机
recording camera device 记录照相装置
recording car 自动记录仪表车
recording card 记录卡(片);自动记录卡片;图表记录
recording casing head pressure 记录套管压力
recording cell 记录隔室;记录单元
recording chain 记录链
recording characteristic 记录特性
recording chart 记录图表;(自动记录仪记录带上的)记录曲线表
recording check time 记录检查时间
recording circuit 记录电路
recording clockwork 记录用的钟表机构;记录仪表的计时机构
recording code 记录码
recording colo(u)rimeter 记录式色度计;自记比色计
recording communication 记录通信
recording compensator 记录均衡器;记录补偿器
recording-completing trunk 记录通话长途线
recording component 记录成分
recording condition 记录条件
recording contains clause 记录包含子句
recording controller 记录式调节器;记录控制;自动记录调节器;数据控制器
recording control schedule 记录控制调度
recording count 记录数目;记录计数
recording current 记录电流
recording current meter 自记流速仪;自记流速表;自记海流计;直读式海流计
recording current optimizer 记录电流优化器
recording curve 记录曲线
recording cylinder 记录鼓;记录(辊)筒
recording declaration 记录说明
recording definition field 记录定义段
recording deflection 记录器偏转度
recording demand meter 记录式需量计
recording densitometer 记录式密度计;记录式显像密度计
recording density 记录密度
recording description 记录说明;记录描述
recording description entry 记录描述项
recording description entry format 记录描述项格式
recording description word 记录说明字
recording descriptor 记录说明符;记录描述符
recording descriptor word 记录描述字
recording device 记录转换系统装置;记录器;自动装置;自记装置
recording differential manometer 记录差动压力计
recording disc (测斜仪的)记录纸盒
recording Doppler comparator 记录型多普勒比长仪
recording drawing 记录图纸
recording drive 记录驱动
recording driver amplifier 记录装置放大器
recording drum 记录(转)鼓;记录滚筒;记录(辊)筒
recording duration 记录(持续)时间
recording dynamometer 自记式拉力表
recording echo sounder 记录式回声测深仪;自记回声测深仪
recording electrode 记录电极;引导电极
recording electrometer 记录式静电计
recording element 记录元件

recording end flag 记录结束标志
recording entry 记录项目;记录描述项
recording equipment 记录转换系统记录装置;记录设备;自记设备
recording extensometer 自动记录伸长试验仪
recording fee 契据登记费
recording file 记录文件
recording film 记录胶片
recording film plane 记录膜面
recording firedamp indicating detector 甲烷自动记录探测计;沼气自动探测计
recording float ga(u)ge 记录式浮子水位计;记录式浮标水尺;自记式浮子水位计;浮子自记水位仪
recording flood 记录洪水
recording flow controller 自记式流量控制器
recording flow-meter 流量记录仪;记录式流量计;自记流量计
recording fluxmeter 自动记录磁通计
recording form 记录形式
recording format 记录格式
recording format descriptor 记录格式描述文件
recording format message 记录格式消息
recording format option 记录格式选择
recording form descriptor 记录格式说明
recording frequency response 录放幅频响应
recording galvanometer 记录式检流计
recording gap 记录间隙;记录间隔;记录间断
recording gas analysis 记录气体分析器
recording gas analyzer 自记气体分析仪;自记气体分析器;记录式气体分析器
recording ga(u)ge 记录(式)计;液面记录器;自记水位计;自记水位计;自记水尺;自计器
recording gear 记录装置
recording generator 记录生成程序
recording geometry 记录图形
recording gravimeter 比重记录仪
recording group 记录组
recording head 记录(磁)头;刻纹头
recording header 记录目标;记录标头
recording heading 记录标题
recording holder 记录保持者
recording house 记录室
recording hydrometer 自记(液体)比重计;自记流速计
recording hygrometer 记录式温度计;记录式湿度计;自记湿度计;自记湿度表;湿度记录器
recording identification code 记录识别码
recording identification indicator 记录标识
recording identifier 记录标识
recording image 录像
recording impulse 记录脉冲
recording indicator 记录式指示器
recording infrared spectrometer 自记红外分光(度)计
recording infrared spectrophotometer 记录式红外分光光度计
recording infrared tracking instrument 记录式红外跟踪仪;红外(线)跟踪记录仪器
recording ink 记录墨水
recording insert 记录插入
recording in situ 井场记录
recording instrument 记录仪(表);记录(式)仪器;自记装置
recording interval 记录范围
recording key 记录键;记录关键码
recording key clause 记录键子句
recording key position 记录关键码位置
recording kilo watthour meter 记录电度表
recording lamp 记录灯
recording layout 记录结构;记录格式;记录布局;记录安排
recording length 记录长度
recording length indicator 记录长度指示符
recording lens 记录图像
recording level 记录水平;记录电平
recording level access 记录级存取
recording level control 记录级控制(器)
recording level indicator 记录级指示器
recording lever 记录杆;记录臂
recording link 记录杆
recording liquid level ga(u)ge 液面记录仪
recording location 记录位置
recording location counter 记录位置计数器
recording lysimeter 自记液度计;自记渗漏仪;自

记溶度计;自记测渗仪;自记测渗计
recording management 记录管理程序;记录管理
recording management service 记录管理服务
recording manometer 记录压力表;自记压力计;压力自记器
recording mark 记录标记
recording material 记录材料
recording material shrink 记录材料的体缩
recording mechanism 记录机械装置;记录机构
recording medium 记录转换系统记录装置;记录媒体;记录介质
recording memory 记录存储器
recording message 记录信息
recording meter 记录仪(表);自记计数器
recording method 记录方法
recording micrometer 记录测微器;自记测微计
recording microphotometer 记录测微光度计;自记测微光度计
recording microtitrater 记录式微量滴定器
recording mode 记录方式
recording modification 记录修改
recording name 记录名
recording needle 记录针
recording network 自记站网
recording number 记录数;记录(编)号
recording observation 记录观测
recording occurrence 记录具体值
recording of digital result 数字结果记录
recording of sound distribution 声分布记录
recording of subdivision load lines 分舱载重线记载
recording of the noise level 噪声电平记录
recording operation 记录操作
recording optic(al) spectrum analyzer 记录光谱分析器;记录式光学跟踪仪
recording optic(al) tracking instrument 光学跟踪记录仪
recording option 记录选择;记录任选
recording oscillograph 记录示波器;自记示波仪;自记示波器
recording oscillometer 记录式示波器
recording paper 记录纸
recording paper of sounder 测深仪记录纸
recording paper roller 记录纸卷筒;记录纸滚筒
recording peak flood 记录最大洪峰
recording pen 记录笔;自记笔
recording pen linkage 记录笔尖的传动装置
recording period 记录周期
recording phase hologram 记录相位全息图
recording photometer 自记光度计
recording placement strategy 记录布局方式
recording-playback robot 记录—再现机器人
recording pluviometer 记录雨量计
recording point 记录点
recording pointer 记录指针;记录笔
recording position 记录台
recording potentiometer 自记(式)电位计
recording precipitation station 雨量记录站;自记雨量站
recording precision 记录精度
recording pressure ga(u)ge 记录式压力表;压力记录仪;自记压力计;自记压力计
recording principle 记账原则
recording profile 记录剖面
recording program(me) 记录程序
recording psychrometer 记录式干湿球湿度计;自记湿度计;自记湿度计
recording pyrometer 自记高温计
recording radiation level 记录辐射水平
recording rainfall 记录特大暴雨
recording rainfall ga(u)ge 雨水记录仪;自记雨量器;自记(录)雨量计
recording range 记录范围
recording receiver 记录接收器
recording reel 记录卷筒
recording reference 记录引用;记录参考表
recording release 记录释放
recording replacement 记录替换
recording retention schedule 记录留存调度
recording room 记录室;录音室
recording rotameter 记录式转子流量计
recording scale 自记(录)式称重计;自记(录)式秤
recording section 记录剖面
recording segment 记录段

recording selection expression 记录选择表达式
recording selector 记录选择器
recording sensing machine 记录读出机
recording sensitivity 记录灵敏度
recording separator 记录分隔符
recording separator pressure 记录分离器压力
recording sequence own code 记录顺序扩充工作码
recording serial number 记录顺序号
recording set 记录集
recording sheet 记录纸;记录单;记录(图)表
recording ship 考察船;探测船
recording site 记录场地
recording size 记录大小
recording sounder 自记测深仪
recording spectrometer 记录式分光计
recording spectrophotometer 记录式分光光度计
recording spectrum analyzer 光谱分析记录器
recording speed 记录速度
recording speed indicator 自记速度表;速度记录器;自记速率表
recording speed(o)meter 自记速率计;可记录速度计
recording spot 记录光点
recording stage 记录阶段
recording station 记录台站
recording statoscope 自记高差仪
recording storage area 记录存储区
recording storage mark 记录存储标记
recording storage tube 记录存储管
recording strain ga(u)ge 记录式应变仪;自记应变仪
recording stream ga(u)ge 自记流量仪
recording studio 录音室
recording stylus 记录(触)针;描记笔
recording surface 记录面
recording system 记录系统
recording tach(e)ometer 记录式速度计
recording tachograph 自记速度计图;自记速度计;自动(记录)转速计
recording tachometer 自动记录转速计;自动式转速计;自动记录流速计
recording tambour 描记气鼓
recording test method 记录试验法
recording theodolite 自记经纬仪
recording thermometer 记录(式)温度计;自记温度表;自记式温度计;记录温度计;温度记录器;温度记录计
recording tide ga(u)ge 自记验潮仪;自记潮位计;自动潮位仪;自动潮位计
recording time 记录时间
recording traffic 记录通信量
recording transmission 记录传输
recording transmission measuring set 自记传输测试器
recording truck 自记测试车
recording trunk 记录中继线;记录信息通路
recording type 记录类型
recording typewriter 记录打字机
recording unit 记录装置;记录器
recording updating 记录修改;记录更新
recording van 录像车
recording velocity 记录速度
recording vessel 考察船;探测船
recording visco(si)meter 自记黏度计;记录黏度计;记录式黏度计
recording voltimeter 自记电压表
recording volumeter 记录式体积计;记录式容积计
recording water ga(u)ge 自记水位计
recording watt and varmeter 自动记录式瓦特乏尔计
recording wattmeter 记录瓦特计;自记电力表
recording wavelength 记录波长
recording weighing machine 自记(录)式称重计;自记(录)式秤
recording well head pressure 记录井口压力
recording Wheatstone bridge 记录式惠斯通电桥
recording wheel 记录轮
recording wind vane 自记风向计;自记风向标
recordist 录像员
recordkeeping 记录保持;记录保存
record keeping system 档案系统
record key 录像按键
record least depth 历史最小水深

record magnetism flaw detector 录磁探伤仪
record of acceptance 验收记录
record of acceptance of concealed works 隐蔽工程验收记录
record of analog(ue) transform 模拟变换式记录
record of concealed works 隐蔽工程记录
record of decision 决策报告
record of differential spectrum 微分谱记录
record of discharge 排放记录
record of earnings 收入记录
record of geologic(al) observation point 地质点记录
record of goods sold 销货记录
record of inspection 检查记录
record of integral spectrum 积分谱记录
record of measurement 测量记录
record of non-destructive detection 无损探伤记录
record of number of passengers on train by districts 列车旅客密度表
record of pen 笔式记录
record of penalty charges 罚款记录
record of performance 性能记载;生产性能测定
record of production 生产记录
record of product test 产品检验记录
record of radiographic(al) examination 射线照相检验记录
record of route survey 路线记录
record of test 试验记录
record of work 工时记录
record on spot 现场记录
record-oriented data 面向记录的数据
record-oriented data transmission 记录数据传输
record-oriented device 面向记录的设备
record-oriented input, output 记录式输入输出
record-oriented transmission 记录式传输;面向记录的传输
record owner 登记注册的业主
record peak flood 历史最大洪峰
record player 声盘放送器
record preparation 录像准备
record receiver 自记接受器
records 档案;案卷
records blocking 记录封锁
records deblocking 记录封锁启封
records holding area 无用档案保管所
records made by radio altimeter and statoscope 无线电测高仪和高差仪记录
records management 档案管理
records of actual floods 实际洪水记录;实际洪峰记录
records of cost 费用记录
records of performance 经营成果报表
records review 记录检查
records room 病历室
record storage 数据存储(器)
record storage system 数据存储系统
record table 重型粗砂摇床
record table of drilling hole 钻孔编录表
record table of exploring 坑探编录表
record table of exposure 露头记录卡片
record table of geophysics 地球物理记录表
record table of hydrogeologic(al) observation point 水文观测点记录表
record table of morphologic(al) observation point 地貌观测点记录表
record table of photograph 照相记录表
record table of profile 剖面记录表
record table of samples 标本编录表
record table of structural observation point 构造观测点记录表
record time 创记录时间
record title 注册产权
recount 重新计算
recoup air 三次风;补偿风
recoup duct 三次风管
recouping sales income 取回销售收入
recouping the capital outlay 收回投资
recoupling 机械闭合与开断
recoupment period 回收期
recoupment period of equipment investment 设备投资回收期
recoup oneself 收回费用
recourse action 追索诉讼
recourse against the responsible third party 向

责任第三方追偿
recourse agreement 追索协议;追索权协定
recourse arrangement 追偿条款;追偿合约;追索合约
recourse back 申请偿还
recourse basis 追索权基础;保留追索权基础
recourse claim 追索请求
recourse for non-acceptance 拒绝承兑的追索权
recourse for non-payment 对拒绝付款的追索权
recourse function 偿付函数
recourse loan 有追索权的贷款
recourse note 追索通知;索债文件
recourse repudiation 拒绝偿还(债务);否定追索权
recover 收回;回浆;复原
recoverability 复原性;修复性;可修复性;可恢复性;可复原性;恢复性
recoverable 可恢复的
recoverable abnormal end 可恢复的异常结束;可恢复的非正常结束
recoverable catalog 可恢复目录
recoverable coal reserves 煤可采储量
recoverable control point 可恢复控制点
recoverable cost 可收回成本
recoverable deformation 可回复的变形;可恢复(的)变形;弹性变形
recoverable deposits 保有储量
recoverable error 可恢复的误差;可恢复错误
recoverable expenditures 可回收支出;可回收费用
recoverable form of energy 可恢复能量形式
recoverable heat 可回收的热量
recoverable level(l)ing point 可恢复水准点
recoverable loss 可回收损失;可回收损失;可补偿损失
recoverable mean grade 可回采平均品位
recoverable mine age 可供开采年限
recoverable natural resources 可恢复自然资源
recoverable oil reserves 可(开)采石油储量
recoverable point 可恢复点
recoverable quantity of metal 可回采金属量
recoverable reference point 可恢复控制点
recoverable reserves 可(开)采储量;可储[贮]量
recoverable resources 可回收的资源;可恢复资源;可复用资源
recoverable shear 可回复剪切;可回变剪切
recoverable station 可恢复测站
recoverable strain work 可恢复变形功;弹性变形功
recoverable synchronization 可恢复同步
recoverable transaction 可恢复事务处理;可恢复的事项处理
recoverable value 可回收价值;可回收(的)价值
recover damage 取得赔偿;赔偿损毁
recover damaged drawing 修复受损的图形
recover debt 追偿债务;追债
recovered acid 再生酸;回收酸
recovered brine 回收盐水
recovered carbon 回收活性炭
recovered casing 回收的套管(钻井)
recovered circulating funds 回收流动资金
recovered civil works capital rest value 回收土建资产余值
recovered control 恢复的控制点
recovered dense medium 再生重介质
recovered dust 回收粉尘
recovered enamel glaze 回收釉
recovered materials 回收的材料;回收物料
recovered oil 再生(润滑)油;回收油
recovered pressure 恢复压力
recovered resources 回收的资源
recovered rolling stock residual value 回收机车车辆残余值
recovered solvent 回收(的)溶剂
recovered steam 回收蒸汽
recovered temperature 留б温度;回收温度
recovered treated water 回收改良水
recovered water quality 回收水水质
recovered water quantity 回收水量
recovered water table 恢复水位
recoverer 回收器
recoverer of waste heat 废热回收器
recover funds in full 追回全部款项
recoveries 追回款
recovering 重新覆盖
recovering expansion energy 膨胀能量回收
recovering of vapo(u)rized hydrocarbons 轻质油回收
recovering tap 打捞公锥
recover loan 收回贷款
recover loss 弥补损失
recover losses 追偿损失
recover value 收回(价)值
recovery 追回;追偿;改出;弥补;回收;回采(率);回浆;恢复;合金过渡系数;清除应力退火;收获率;采收率;财产收回;复原;复苏
recovery action 追索诉讼
recovery airfield 收容机场
recovery apparatus 复原装置
recovery area 回收区
recovery capability 恢复能力
recovery capacity 恢复能力
recovery capacity on gas heater 气体加热回收量
recovery capsule 回收舱
recovery carbon 回收碳
recovery circle 回升周期(指地下水位)
recovery circuit 恢复电路
recovery coefficient 回收系数;恢复系数;采收率;复采系数
recovery coefficient of investment 投资回收系数
recovery collecting point 损坏车辆收集站
recovery column 回收塔
recovery constant 恢复常数
recovery control unit 恢复控制部件
recovery creep 回返徐变;回复蠕变
recovery current 恢复电流
recovery curve 暂态上升曲线;过渡过程曲线;回复曲线;恢复(过程)曲线;采油曲线
recovery curve of water table 水位恢复曲线
recovery cycle 补偿周期(地下水);回升周期
recovery debts 追回欠债
recovery device 回收装置
recovery efficiency 回收效率;回收率;实收率;采收率
recovery factor 回收系数;恢复系数;采收率
recovery field 着陆回收场
recovery file 恢复文件
recovery force 恢复力
recovery from a disturbance 退出扰动区
recovery from creasing 折皱回复性
recovery from drought stress 受旱后的复原
recovery from submergence 受浸后的复原
recovery from system failure 系统故障排除
recovery from the blade stall 消除叶片气流分离
recovery from transplanting 返青
recovery fund 补偿基金
recovery furnace 回收炉
recovery guideline 收回资金的时间标准
recovery heat 恢复热
recovery hook 打捞钩
recovery information set 恢复信息组
recovery in kind 实物补偿
recovery interrupt 可恢复中断;恢复中断
recovery length 取样长度
recovery locator beacon 失事救援方位信标
recovery management 恢复管理
recovery method 回收法
recovery nursing unit 恢复室
recovery of bundle of rays 光束复原
recovery of casing 回收套管
recovery of chemicals 化学药品回收
recovery of colo(u)r 色泽复原
recovery of core 岩芯采收率;取岩芯
recovery of elasticity 弹性恢复
recovery of energy 能量恢复
recovery of form 形状的恢复
recovery of grease 油脂回收
recovery of heat 热的回收
recovery of information 信息重获
recovery of instruments 仪器回收
recovery of interfered recorder paper 记录纸带干扰恢复
recovery of interfered record(ing) tape 记录带干扰恢复
recovery of late payment 要求支付欠款
recovery of loss 补偿损失
recovery of nature 自然复原
recovery of oil 油的回收
recovery of platinum-rhodium 铂铑合金回收
recovery of potable water 饮用水的回收
recovery of price 价格回升
recovery of renovated water 回收再生水
recovery of sand 采砂
recovery of service station vapo(u)r 加油站蒸汽的回收
recovery of shape 形状恢复
recovery of soil-equilibrated potassium 恢复土壤全部的平衡
recovery of station 测量标志恢复
recovery of structure form 恢复构造形态
recovery of the film 胶片回收
recovery of the ozone hole 臭氧层空洞的复原
recovery of the performance 恢复执行合同
recovery of thermal equilibrium 热平衡恢复
recovery of waste heat 废热回收
recovery of water 水的回收
recovery oxygen system 氧气再生系统
recovery package 回收装置;回收舱;回收标识器箱
recovery peg 复原测标;重复用测标;参考桩【测】;参考标桩
recovery percent 开采程度;回收率;回收百分数
recovery percent of reserves 采出程度
recovery period 复原期间
recovery period of added investment 附加投资回收年限
recovery period of investment 投资回收期
recovery phase 恢复期;复原期间;复原阶段
recovery plant 再生工厂;回收设备;回收车间
recovery point 恢复点
recovery potential 回收潜力
recovery procedure 恢复过程
recovery processing 故障处理;回收处理;重新处理;回收加工
recovery program(me) 恢复程序
recovery quotient 复原商
recovery rate 恢复速度;恢复率;复得率
recovery rating 额定回收率
recovery ratio 岩芯获得率;岩芯回采率;回采率;回收率;恢复率;提取率;采样比;采收率;采取率
recovery record 恢复记录
recovery room (手术后的)康复室;休养室;恢复室;复原室
recovery scrubber 回收洗涤器
recovery separator 回收分离器
recovery service 抢修
recovery ship 宇航器回收舰
recovery slip 修复标签
recovery standard 回收标准
recovery strategy 重生策略
recovery sump pit 集液(器)回收井;集水回收井;回灌补给坑
recovery system 回收系统
recovery tap 打捞公锥
recovery temperature 恢复温度
recovery termination manager 复原终止管理程序
recovery test 性能恢复试验;更新试验;回收(率)试验;复原试验
recovery time 还原时间;再现时间;再生时间;恢复时间;恢复期
recovery timer 恢复计时器
recovery tool 打捞工具
recovery tower 回收塔
recovery tunnel 出料隧道(输送机);出料坑道(输送机)
recovery unit 回收装置
recovery utility 恢复实用程序
recovery vacuum pump 回收真空泵
recovery value 更新价值;回收价值;残值;复原价值
recovery value of water-level 水位恢复值
recovery vehicle 救险车;紧急修理车;回收飞行器
recovery voltage 恢复电压;还原电压
recovery volume 开采量;复原卷
recovery waste heat 废热回收;余热回收;回收余热
recovery water 回收水
recovery weight 回收重量
recovery well 回采井;回收井
recovery work 恢复工作;破坏车辆之抢修工作
recovery zone of river 河流复氧区
recracking 再裂化
recreatable data base 可重建数据库
recreational activity 消遣活动
recreational area 文娱场地;休养区;休养娱乐区;娱乐区;休养娱乐用地;游览区
recreational boat 游艇
recreational boat engine 游艇发动机

recreational boating 驾驶游艇
recreational center 娱乐中心;游憩中心
recreational district 娱乐区
recreational environment 休养环境;游览环境
recreational facility 娱乐设施;娱乐设备;游憩设施;再生设施;再生设备
recreational (focal) area 休养区
recreational forest 游息林
recreational ground 娱乐场
recreational impoundment 娱乐蓄水
recreational park 游息公园
recreational reuse 娱乐性再用
recreational room 娱乐室
recreational traffic 旅游交通
recreational use of water 休养用水;消遣用水
recreational use 供游览休养用
recreational vehicle 游览车;游乐性车辆;周末旅游车
recreational water 娱乐用水;再生加工用水;文娱场所用水
recreational waters 游乐水域
recreation area 休养娱乐用地
recreation beach 娱乐海滩
recreation building 文娱建筑
recreation center 旅游中心;游憩中心
recreation complex 综合旅游区
recreation development 专设户外娱乐区
recreation engineering 造园工程
recreation enterprise 旅游业
recreation facility loans 游憩设施的贷款
recreation ground 娱乐园地(公共户外娱乐绿地);娱乐场;游乐场
recreation park 游息公园
recreation priority area 康乐优先区
recreation resources 旅游资源;天然山水旅游区
recreation room 休息室;文娱室
recrement 余渣;回吸液;废物;废料
recriminate 反控
recroom 休息室;娱乐室
recrudescence 再燃
recruit 招募
recruit and use work force 用工
recruitment 增添量;增加量;进界木;补充鱼群;补充木
recruitment period 进级期
recruitment test 重振试验
recruit technical forces 补充技术力量
recruit training center 新兵训练中心
recruit worker 招工
recrusher 次(破)碎机;二次破碎机
recrushing 二次破碎;再破碎;重轧碎;重扎碎;二次粉碎
recrystal grain 再结晶晶粒
recrystalline dolomite 重结晶白云岩
recrystalline limestone 重结晶灰岩
recrystallization 重结晶(作用);再结晶(作用)
recrystallization annealing 再结晶退火
recrystallization behavio(u)r 再结晶性质
recrystallization boundary 再结晶边界
recrystallization breccia 假角砾岩
recrystallization replacement way 重结晶替代方式
recrystallization stage 重结晶阶段
recrystallization temperature 析晶温度;重结晶温度
recrystallization texture 再结晶组织;再结晶织构;再结晶结构
recrystallization twin 再结晶孪晶
recrystallization way 重结晶作用方式
recrystallization welding 再结晶焊接
recrystallization zone 再结晶区
recrystallize 再结晶
recrystallized 再结晶的
recrystallized graphite 再结晶石墨;重结晶石墨
recrystallized layer 再结晶层
recrystallized limestone 纯结晶大理石;再结晶石灰岩
recrystallized quartzite 再结晶石英岩
recrystallized snow 再结晶雪
recrystallized structure 再结晶构造
recrystallized threshold 再结晶阈值
recrystallized tungsten wire 再结晶钨丝
recrystallized zone 再结晶区
recrystallizing heat treatment 再结晶热处理
recta 上裳【建】
rectangle 矩形;长方形
rectangle boudin 矩形石吞肠

rectangle conduit 矩形通道
rectangle connector for industry 工业矩形插头座
rectangle grid 矩形网格
rectangle of dispersion 全数必中界
rectangle ruler 直角尺
rectangle scriber 矩阵符号刻图仪
rectangle section steel 矩形截面钢板
rectangular 矩形的;长方形的
rectangular acceleration pulse 矩形加速度脉冲
rectangular air distributor 矩形散流器
rectangular air supply opening 矩形送风口
rectangular alignment 直角校正
rectangular aperture 矩形孔径
rectangular approach 直角航线进场着陆;矩形航线进场
rectangular apse (教堂端部的)长方形凸室
rectangular aquifer 矩形含水层
rectangular area 长方形面积;矩形面积
rectangular area grating 长方形栅板
rectangular array 巨阵列;矩阵列;矩形数组;长方阵列
rectangular association scheme 矩阵相连法
rectangular axes 直角坐标轴线
rectangular axis 直角坐标系轴;直角轴;直交轴
rectangular baler 长方形压捆机
rectangular bar 矩形棒;矩形条;长方形杆
rectangular bar blade 矩形条叶片
rectangular bar of steel 长方形钢
rectangular basin 矩形盆(地);矩形池
rectangular bay 长方形开间
rectangular bay window 长方形凸窗
rectangular beam 矩形梁
rectangular bent 矩形排架
rectangular block 矩形砌块;矩形方块;平板砖
rectangular bloom 矩形坯
rectangular board 矩形板
rectangular brick 长方形砖;矩形砖
rectangular bunker 矩形料仓
rectangular caisson 矩形沉箱
rectangular canal 矩形河槽
rectangular cantilever(ed) step 挑出的长方形踏步
rectangular Cartesian coordinates 直角笛卡儿坐标;笛卡儿直角坐标
rectangular Cartesian coordinate system 笛卡儿(直角)坐标系
rectangular cavity 矩形共振空腔
rectangular cell 直角网络;长方形电池;长方形单元
rectangular chamber 长方形房间
rectangular channel 矩形(截面)水道;矩形(截面)水渠;矩形河槽;矩形(断面)水槽
rectangular chapel 长方形小教堂
rectangular chart 正交投影图;矩形波图板
rectangular choir 长方形唱诗班席位
rectangular clarifier 矩形澄清池
rectangular clipping window 矩形裁剪窗口
rectangular collar 矩形套筒
rectangular column 矩形柱
rectangular condensate tank 矩形冷凝水箱
rectangular conductor in semi-closed slot 半闭口槽内的矩形导体
rectangular conduit 矩形输水道
rectangular connector 矩形接插件;矩形插头座
rectangular contracted weir 矩形缩口堰
rectangular controller 长方形控制器
rectangular coordinate axis 直角坐标轴
rectangular coordinate display 有直角坐标的显示
rectangular coordinate grid 直角坐标网
rectangular coordinate method 直角坐标法
rectangular coordinate plotter 直角坐标展点仪
rectangular coordinates 直角坐标
rectangular coordinate system 直角坐标系
rectangular coordinate type potentiometer 直角坐标式电位计
rectangular coordinatograph 直角坐标展点仪
rectangular coordinatometer (直角的)坐标尺
rectangular course 矩形起落航线
rectangular cover 矩形罩盖;矩形罩壳
rectangular crib groin 矩形(断面)木笼式丁坝
rectangular cross flow 垂直交叉流
rectangular cross ripple mark 直角交错波痕
rectangular cross-section 矩形断面
rectangular culvert 矩形涵洞
rectangular curvilinear coordinate 直角曲线坐标
rectangular cylindric(al) projection 方块投影;方格圆柱投影

rectangular design 长方形设计;矩形设计
rectangular distance 矩形距离
rectangular distribution 均匀分布;矩形分布
rectangular distributor 矩形布水池;矩形配水器;矩形布水器
rectangular dome 矩形穹顶
rectangular dowel 钢筋混凝土的长方形连接筋
rectangular downpipe 矩形水落管;矩形落水管
rectangular drainage 矩形排水系统;长方式水系;方格式水系
rectangular drainage pattern 方格式排水;直角式排水;矩形排水;长方水系型;长方式排水
rectangular drainage system 矩形排水系统;方格式排水系统
rectangular drawing 矩形图纸;矩形拉延
rectangular drop shaft 矩形开口沉井
rectangular duct 矩形管道;矩形风道
rectangular element 长方形构件;矩形单元
rectangular engine house 矩形机车库
rectangular equation 直角坐标方程
rectangular expansion tank 矩形膨胀水箱
rectangular exploration grid 矩形勘探网
rectangular face hammer 长方平锤
rectangular fetch 矩形风区
rectangular figure 矩形
rectangular flag 长方形旗
rectangular flask 矩形砂箱
rectangular flat-plate 矩形板;矩形平板
rectangular flue 矩形烟道
rectangular flume 矩形(引)水槽;矩形渡槽
rectangular footing 矩形基础
rectangular forecourt 长方形前院
rectangular foundation 矩形基础
rectangular foundation pier 矩形基础墩
rectangular frame 长方形框架
rectangular frequency diagram 矩形频率图;频率直方图
rectangular frequency distribution 矩形频率分布
rectangular full-width weir 矩形全宽堰;矩形满堰宽堰
rectangular function 直角函数
rectangular game 矩阵对策;矩形对策
rectangular ga(u)ging weir 矩形量水堰
rectangular grating 矩形光栅
rectangular grid 长方坐标格;长方形格子;长方形交叉梁;直角坐标网;矩形格网;井字格法;方格网
rectangular grid ceiling 长方形交叉梁式顶棚
rectangular grid city 方格状城市
rectangular grid floor 长方形交叉梁式楼板
rectangular grid pattern 长方形交叉梁式模式
rectangular grid soffit 长方形交叉梁式挑檐底面
rectangular groyne 矩形(断面)木笼式丁坝
rectangular guide bar 矩形导杆
rectangular guideway 矩形导轨
rectangular gutter 箱式排水沟;箱式檐沟;箱式天沟
rectangular head bolt 方头螺栓
rectangular-headed loophole 长方形顶盖的观察孔
rectangular histogram 矩形直方图
rectangular hollow section 矩形空心断面(构件);矩形空腹型钢
rectangular hooping 矩形箍
rectangular horn 矩形喇叭
rectangular hyperbola 等轴双曲线
rectangular hyperbolic coordinates 直角双曲线坐标
rectangular hysteresis loop 矩形滞后回线
rectangular hysteresis material 矩形磁滞材料
rectangular image 矩形图像
rectangular ingot 扁钢锭
rectangular integration 直交积分
rectangularity 矩形性
rectangular kiln 矩形窑;长方窑;方窑
rectangular lag structure 矩形滞后结构
rectangular lattice 矩形点阵;长方点阵
rectangular lattice design 矩形格子设计
rectangular layout of streets 方格式街道网;棋盘式街道(网)
rectangular lining 矩形衬砌
rectangular loaded area 矩形荷载面积
rectangular loop 矩形回线;矩形箍
rectangular loop hysteresis 矩形磁滞曲线
rectangular magnetic chuck 矩形磁性吸盘
rectangular manhole 长方形检查井

rectangular manway 长方形孔口;长方形人孔
rectangular map format 矩形图廓
rectangular map projection 正交地图投影
rectangular map-subdivision 矩形分幅
rectangular masonry wall crossing 圬工墙直角交叉
rectangular matrix 矩阵;矩形矩阵;长方形矩阵
rectangular measuring weir 矩形水堰;矩形测量堰
rectangular membrane 矩形膜
rectangular mesh 矩形筛眼;矩形筛孔;矩形网格;长方形网格
rectangular mesh grating 炉筛条;矩形孔眼格栅
rectangular mesh grid 长方形筛格
rectangular mesh screen 长方孔筛
rectangular mirror 矩形反射镜;长方镜
rectangular mosaic 锦砖镶嵌细工;长方形马赛克;矩形马赛克
rectangular multiple compartment bin 矩形多格斗仓;矩形多分隔斗仓;矩形多仓室谷仓
rectangular net(work) 矩形网(络);矩形格网
rectangular notch 矩形凹槽;矩形缺口;矩形切口;矩形槽口
rectangular notch weir 矩形缺口堰;矩形切口堰;矩形槽口堰
rectangular open caisson 矩形开口沉箱
rectangular opening 矩形孔(口)
rectangular orifice 矩形小孔;矩形孔(口)
rectangular or square section solid strip and D-hollow fender 矩形或方形断面实心条带和 D 型空心护舷
rectangular otter board 矩形网板
rectangular parallelepiped 直角平行六面体;矩体;长方体
rectangular paralleling 矩体
rectangular patch budding 矩形芽接
rectangular pattern 正常航线;矩形型
rectangular picture tube 矩形屏显像管
rectangular pier 矩形桥墩
rectangular pierhead 矩形墩头;矩形堤头;矩形堤块
rectangular pile 矩形桩
rectangular pipe 矩形管
rectangular plane coordinates 平面直角坐标
rectangular planting 方形穴播
rectangular polyconic(al) projection 正交多圆锥投影
rectangular port 矩形孔(口)
rectangular prism 直角棱镜;矩形棱柱(体);矩形棱镜
rectangular profile 长方形轮廓;矩形剖面(图)
rectangular projection 正交投影;矩形投影;等渐长率圆柱投影;方格投影
rectangular projection point 正交投影点
rectangular property of saddle-point 鞍点的矩形特性
rectangular pulse 矩形脉冲
rectangular pulse stimulator 方波刺激器
rectangular pyramid 四棱锥
rectangular rainwater gutters 矩形雨水沟
rectangular raster 矩形光栅;长方形光栅
rectangular ratio 矩形比
rectangular refractory brick 长方形耐火砖
rectangular rod 直角扁形杆件
rectangular sandstone 砂条石
rectangular scanning 矩形扫描
rectangular search 矩形搜索
rectangular section 矩形截面;长方形剖面
rectangular section beam 矩形截面梁
rectangular section nozzle 矩形截面喷管
rectangular section ring 方断面密封圈
rectangular section volute spring 方体锥形螺簧
rectangular sedimentation tank 矩形沉淀池
rectangular set 矩形集
rectangular settling tank 矩形沉淀池
rectangular setup 直角坐标系
rectangular shaft 矩形断面轴
rectangular shaft set 矩形井筒支护
rectangular shaped tunnel 箱形隧道
rectangular sheet pile 矩形板桩
rectangular shell 矩形壳体
rectangular shield 矩形盾构
rectangular shock pulse 矩形冲击脉冲
rectangular single base 矩形独立基础
rectangular single wall caisson 矩形单墙沉井

rectangular skip 长方形箕斗
rectangular skylight 矩形天窗
rectangular slot 矩形槽;长方形槽
rectangular solid 长方体
rectangular source 矩形源
rectangular space coordinates 空间直角坐标
rectangular spheric(al) coordinates 球面直角坐标
rectangular spirals 回纹
rectangular steel shaft set 矩形井筒钢支护
rectangular step 长方形踏步
rectangular stone slab 压阑石;阶条石
rectangular street 风格式街道;棋盘式街道
rectangular street system 棋盘式街道系统;棋盘式街道体系;方格形道路系统;棋盘式街道网;方格式道路系统
rectangular survey 矩形分区测量;矩形调查法
rectangular surveying 经纬测量
rectangular system 直角系;长方系
rectangular system of city planning 棋盘式城市规划体系
rectangular system of street layout 棋盘式街道网;棋盘式街道布置(网)
rectangular table 矩形工作台
rectangular tank 矩形罐;长方形储槽
rectangular thread 矩形螺纹
rectangular three-dimensional coordinate system 三维直角坐标系
rectangular tie 矩形墙箍;矩形拉结筋;矩形系铁;矩形拉杆;矩形轨枕
rectangular tile 矩形瓦;长方形面砖
rectangular timber 矩形锯材;方木材;矩木材;长方形木材;长方矩材
rectangular tolerance zone 矩形公差带
rectangular triangle 直角三角形
rectangular tube 矩形断面管;矩形(沉)管
rectangular twin-track underground tunnel with supporting web 矩形中间支柱双向地下隧道
rectangular underground fire hydrant 矩形地下消防栓
rectangular vault bay 长方形拱顶间
rectangular vessel 矩形容器
rectangular wave 矩形波
rectangular wave generator 矩形波发生器
rectangular weir 矩形堰;矩形量水堰;矩形(断面)堰
rectangular weir meter 矩形堰量计
rectangular weir with double side contraction 两侧收缩矩形堰
rectangular window 长方形窗;矩形窗口;特殊滤波器;矩门窗口
rectangular wire 直角扁形金属线
rectascension 赤经
rectiblock 整流片
rectifed current 已整电流
rectifiability 可矫正性
rectifiable 可整流;可矫正的
rectifiable curve 可求长曲线
rectifiable work piece 可修件
rectification 整流;整顿;再生作用;精馏;校正;矫频;检波;订正
rectification base 纠正底图
rectification by facets 分块纠正
rectification characteristic 整流特性
rectification circuit 整流电路
rectification coefficient 整流系数
rectification efficiency 整流效率
rectification error 整流误差
rectification factor 整流因数
rectification mechanism 整流机构
rectification method 整流法
rectification of a conic 锥线求长
rectification of aerial photograph 像片改正;航摄像片纠正
rectification of alignment 定线调整
rectification of bankline 修整岸线
rectification of boundaries 地界修正
rectification of channel 河道取直;河槽整治
rectification of distortion 变形纠正
rectification of harbo(u)r approach 港湾进口水道改直
rectification of photograph 像片纠正
rectification of projection 投影纠正
rectification of river 河道整治;河道改直
rectification packed column 填充精馏塔
rectification picture 纠正像片

rectification plane 纠正仪承影面
rectification point 纠正点
rectification ratio 整流比
rectification section 整流工段
rectification switchgear 整流变压器
rectification tower 精馏塔
rectification under vacuum 真空精馏;减压精馏
rectification with control points 用控制点纠正
rectification with predetermined settings 用安置值纠正
rectified 改正的
rectified action 整流作用
rectified air speed 仪表修整速度(飞机)
rectified altitude 视高度;视高程
rectified channel 整治了的水道;经过整治的水道;经过整治的河槽;经过整治的航道;河槽整治
rectified current 整流电
rectified feedback 整流反馈
rectified index number 调整的指数
rectified oil of vitriol 精馏硫酸
rectified output 整流输出
rectified photograph 纠正像片
rectified photographic(al) mosaic 纠正像片镶嵌图
rectified power 整流功率
rectified print 纠正晒印像片
rectified recording 整流信号接收
rectified scan 校正扫描
rectified signal 整流信号
rectified spirit 精馏酒精
rectified value 整流值
rectified voltage 整流电压
rectified waveform 整流波形
rectifier 整流装置;整流器;整流管;纠正仪;精馏器;矫正器;检波器;变流器
rectifier anode 整流管阳极
rectifier assembly 整流器装配
rectifier block 整流器关闭
rectifier cascade 整流极
rectifier cell 整流元件;整流片
rectifier characteristic 整流器特性
rectifier charging 整流器充电
rectifier circuit 整流电路
rectifier detector 整流检波器
rectifier device 整流装置
rectifier disk 整流器片
rectifier doubler 整流倍压器;倍压整流器;倍压整流
rectifier-driven motor 整流器馈电的电动机
rectifier efficiency 整流器效率
rectifier element 整流器元件
rectifier equation 整流方程
rectifier equipment 整流设备;整流器设备
rectifier filter 整流器滤波器
rectifier for charging battery 充电整流器
rectifier for contact protection 消除火花用整流器
rectifier for daylight lamp 日光灯整流器
rectifier for mountainous terrain 山地纠正仪
rectifier heater 整流器热丝
rectifier instrument 整流式仪表
rectifier inverter 整流换流器
rectifier leakage current 整流器漏电流
rectifier load 整流器负载
rectifier locomotive 整流电力机车
rectifier nets 伸长网络
rectifier output voltage regulator 整流器输出电压调整器
rectifier phase comparator 整流型相位比较器
rectifier phenomenon 整流现象
rectifier photocell 整流光电管;障层光电管
rectifier plate 整流器片
rectifier power supply 整流器供电
rectifier protection 整流器保护
rectifier rating 整流器容量
rectifier ratio 整流比
rectifier relay 整流器继电器
rectifier ripple factor 整流脉动系数
rectifier roll 整流辊
rectifier set 整流器组
rectifier stack 整流器组;整流片组;整流堆
rectifier stage 整流极
rectifier storage characteristic 整流器存储特性
rectifier substation 整流变电所
rectifier system 整流方式
rectifier tank 整流槽
rectifier transformer 整流器变压器

rectifier transformer feeder cubicle 整流变压器馈电柜
rectifier tube 整流管
rectifier type 整流式
rectifier type arc welder 整流式直流弧焊机
rectifier type differential relay 整流式差动继电器
rectifier type distance relay 整流式距离继电器
rectifier type echo suppressor 整流器式回波抑制器
rectifier type instrument 整流式仪表
rectifier-type meter 整流式仪表
rectifier type relay 整流继电器
rectifier unit 整流装置;整流机组
rectifier voltmeter 整流电压表
rectifier wavemeter 整流式波长计
rectifier welding machine 整流焊机
rectifier welding set 整流焊机
rectiflow system 直流法(勘探)
rectiformer 整流变压器
rectify 纠正;精馏
rectify a deviation 纠偏
rectifying 整流;精馏过程;检波;求曲线长
rectifying action 整流作用;精馏作用
rectifying arc welding machine 整流弧焊机
rectifying characteristic function 整流特性函数
rectifying coefficient of ore grade 矿石品位校正系数
rectifying column 精馏塔
rectifying contact 整流接触
rectifying detector 整流检波器
rectifying device 整流装置
rectifying efficiency 整流效率
rectifying hydraulic press with single pole 单柱校正液压压力机
rectifying installation 整流设备
rectifying instrument 纠正仪
rectifying plane 从切平面
rectifying printer 纠正仪
rectifying probe 整流探针
rectifying screw 调整螺旋
rectifying section 精馏段;提纯段
rectifying still 精馏釜
rectifying surface 伸长曲面;伸长表面
rectifying tank 精馏柜;沉淀柜
rectifying tower 精馏塔
rectifying tube 整流管
rectifying valve 整流阀
rectigradation 直线渐变
rectilineal asymptote 渐近直线
rectilineal flow 直线流动
rectilineal hoop 环箍筋
rectilinear 直线(性)的
rectilinear accelerator 直线加速器
rectilinear air classifier 直线型空气分级机
rectilinear angle 直线角
rectilinear building (以垂直线条为特征的)哥特式建筑
rectilinear cascade 直线叶栅
rectilinear chart 方格纸;方格图
rectilinear composition 直线组成
rectilinear congruence 直线线汇
rectilinear coordinates 直线坐标;直角坐标
rectilinear correlation 直线相关
rectilinear creeping 直线爬行
rectilinear cross-stratification structure 直线交错层理构造
rectilinear current 线性电流;往复流
rectilinear direction 直线方向
rectilinear distributor 直线布水器
rectilinear drainage 直线形水系;直线形排水系统
rectilinear evolution 直线演化
rectilinear face tooth 直面轮齿
rectilinear figure 直线图(形)
rectilinear flight 直线飞行
rectilinear flow 直线水流
rectilinear generator 直线生成元;直纹母线
rectilinearity 直线性
rectilinearity of particle motion in three-dimension 三维质点运动的直线度
rectilinear jet 直线喷流
rectilinear lens 直线性透镜;无线性畸变镜头
rectilinearly polarized light 直线偏振光
rectilinear manipulator 直线式控制器;直线式机械手
rectilinear motion 直线运动

rectilinear movement 直线运动
rectilinear oscillation 线性振荡;线性摆动;周期振荡
rectilinear path 直线行程
rectilinear period 直线式时代
rectilinear polygon 直线多边形
rectilinear potentiometer 线性变化电位器
rectilinear prestressing 直线预应力法
rectilinear propagation 直线传播;直接传播
rectilinear propagation of light 光的直线传播
rectilinear relation 直线关系
rectilinear sampling 长方格采样
rectilinear scale 直尺
rectilinear scanner 线性扫描器
rectilinear scan(ning) 直线扫描;线性扫描
rectilinear screw-thread reluctance motor 直线螺纹磁阻电动机
rectilinear shoreline 直岸线
rectilinear solution 直线解
rectilinear stream 往复流;往复潮流
rectilinear stress diagram 直线应力图
rectilinear style 直线条式建筑;直线式建筑风格
rectilinear system 直线系;无畸变系统
rectilinear system of coordinates 直线坐标系统
rectilinear tidal current 直线往复潮汐流;直线潮流
rectilinear tracery 直线式窗花格
rectilinear transducer 线位移传感器;直线运动传感器;平移传感器
rectilinear translation 直线运动
rectilinear transmissivity 直线透射比
rectilinear transmittance 直线透射比
rectilinear trend 直线趋势
rectilinear triangle 直线三角形
rectilinear vibration 直线振动
rectilinear vortex 直线涡流;直线涡
rectilinear wire 直线导线
rectimarginate 直缘型
rectipetaly 直向性
Rectisol process 低温甲醇法
rectisorption 整流吸收
rectistack 整流堆
rectitude 正直
recto 奇数页
rectoblique plotter 改斜绘图仪;变角器(倾斜面上测得的角度变成平面上相应角度的仪器)
rectograver 半自动刻图仪
rectometer 精馏计
rectorite 累托石;钠板石
rectorite clay 累托石黏土
rectorite deposit 累托石矿床
rectory 教区长住宅
rectron 电子管整流器
recued fuel oil 重油
re-cultivation 再植(被);恢复植被;恢复生态
recumbent anticline 伏卧背背斜(层)【地】
recumbent effigy 横卧雕像
recumbent fold 平卧褶皱;伏卧褶;伏卧褶皱【地】
recumbent position 斜卧位
recumbent statue 卧像
recumbent-upright test 卧倒直立试验
recuperability 可回收性;回收的可能性;恢复能力
recuperate 余热利用
recuperate and multiply 休养生息
recuperated 回收的
recuperated dust silo 回收窑灰库;回收粉尘库
recuperated rubber 再生胶
recuperation 休养;余热利用;间壁;换热;复原
recuperation lock 回流船闸
recuperation of deformation 变形恢复
recuperation of heat 同流换热法
recuperative 反馈的
recuperative airheater 间壁式空气预热器
recuperative air preheater 蓄热式空气预热器;同流换热式空气预热器
recuperative burner 同流换热炉
recuperative cooler 换热式冷却器;换热式冷却机
recuperative furnace 回热式电炉;换热炉;热交换窑
recuperative gas turbine 间壁回热式燃气轮机;回热式燃气轮机;同流换热式燃气轮机
recuperative heater 再生式热风机;同流换热器
recuperative heat exchange 同流换热
recuperative heat exchanger 同流换热器;再生式热交换器;间壁式换热器;回流式热交换器
recuperative heat exchange system 同流换热系统

recuperative heating 再生式加热
recuperative incineration 复热焚烧
recuperative oven 换热炉;同流换热炉
recuperative pot furnace 换热式坩埚窑;换热玻璃熔炉
recuperative power 恢复能力
recuperative regenerator 间壁式回热炉
recuperative soaking pit 换热式均热炉
recuperative system 同流换热系统;同流换法
recuperative tank furnace 换热式池窑
recuperator 间壁(式)换热器;隔道换热器;回流换热室;能量回收器;换热器;同流换热器
recuperator cylinder 复进调节筒
recuperator tube 热交换管
recure 再次硫化
recuring 再固化
recurrence 再现;再发生;再发;递推;递归;重现;复现
recurrence code 递归码
recurrence curve 重现曲线
recurrence equation 递推方程
recurrence factor of public transport line 公共交通线路重复系数
recurrence formula 循环公式;递演公式;递推公式;递归公式
recurrence frequency 脉冲重复频率;重复频率;重发频率;重现频率
recurrence horizon 重复地层【地】
recurrence interval 重现期;重现间隔;重复间隔
recurrence interval of designed highest stage 设计最高水位重现期
recurrence interval of designed maximum discharge 设计最大流量重现期
recurrence interval of designed water-level 设计水位重现期
recurrence interval of design wave 设计波浪重现期
recurrence method 递归法
recurrence of tectonic movement 构造运动新生性
recurrence period 重现周期;重复周期
recurrence rate 再发率;重现率;重复率
recurrence relation 递演关系;递推关系;递归关系;重现关系
recurrence risk 再显危险率
recurrence risk rate 再发危险率
recurrence theorem 循环定理;回归定理
recurrence time 重现时间
recurrence time of earthquake 地震重复时间
recurrence type 递归类型
recurrent 复现的;反回的
recurrent algorithm 递演算法;递推算法;递归算法
recurrent appropriation 周期性拨款;经常性拨款
recurrent backcross 相互回交
recurrent code 循环码;连环码;卷积码
recurrent community 重现群落
recurrent congestion 周期性拥挤;常态拥挤
recurrent cost 经常性开支;经常性费用;经常性成本
recurrent deformation 重复变形
recurrent deposition 迭次沉积;重复沉积
recurrent determinant 循环行列式;循环小数
recurrent disturbance 复现扰动
recurrent event 循环事件
recurrent expenditures 经常性支出;经常性费用
recurrent failures 周期性发生的故障
recurrent fever 回归热
recurrent flare 再现耀斑
recurrent fluctuations 重复波动;反复波动
recurrent folding 重复褶皱作用
recurrent formula 循环公式;递归公式
recurrent frequency 重发频率
recurrent interval 重复间隔
recurrent lap 反复折叠;波状重皮;重复折迭
recurrent network 链形网络;重复网络
recurrent number 循环数
recurrent orogenesis 重复造山作用
recurrent profit and loss 经常性损益;重复性损益
recurrent pulse 重复脉冲;周期(性)脉冲
recurrent reaction 再发反应
recurrent reciprocal selection 相互反复选择
recurrent requency 脉冲重复频率
recurrent selection 轮回选择;反复选择
recurrent shoaling 重复变浅
recurrent signal 重发信号
recurrent state 循环状态;递归状态

recurrent surge oscillograph 重复脉冲示波器
recurrent time 重现时间
recurring audit 经常性审计
recurring continued fraction 循环连分数
recurring cost 续生成本
recurring decimal 循环小数
recurring frequency 再生频率
recurring group 重复基团
recurring issue 再补给
recurring load 循环荷载;反复荷载
recurring maintenance 计划维修
recurring period 循环小数循环节;循环节;小数的循环节
recurring polynya 反复出现的冰期
recurring series 循环级数
recurring structural unit 聚合物重复结构单元
recurring unit 聚合物重复链节;重复单位
recursion 递推;递归式
recursion algorithm 循环算法
recursion and iteration 递演与迭代
recursion formula 递盘公式;递推公式;递归公式
recursion function 递归函数
recursion instruction 循环指令
recursion length 递归长度
recursion relation 循环关系
recursion theorem 递归定理
recursion theory 递归理论;递归理论
recursive 递推的;递归的
recursive analysis 递归分析
recursive arithmetic 递归算术
recursive attribute 递归属性
recursive call 循环呼叫;循环调用;递归调用;递归调入
recursive contour coding 周线递归编码
recursive control algorithm 递归控制算法
recursive data flow 递归数据流
recursive deconvolution 递归反褶积
recursive definition 递归定义
recursive digital filter 循环数字滤波器
recursive equation 递归(型)方程
recursive estimation 递演估计;递推估计;递归估计
recursive estimator 递归估算子
recursive facility 递归能力
recursive fashion 递归方式
recursive filter 递归滤波器
recursive filtering 递归滤波
recursive function 递归函数
recursive function theory 递归函数论
recursive game 递归对策
recursive grammar 递归文法
recursive in 递归于
recursive instruction 递归指令
recursive language 递归语言
recursive least square 递归最小二乘方
recursive limit 递归极限
recursive linear regression 递演线性滤波;递推线性滤波;递归线性滤波
recursive list 递归表
recursive list processing 递归表处理
recursively countable set 递归可数集合
recursively defined sequence 递归定义的序列
recursively defined variable 递归定义变量
recursively enumerable 递归可枚举
recursively enumerable language 递归列举语言
recursively enumerable set 递归可列举集
recursive macro call 递归宏调用
recursive method 循环消去回代法;递归法
recursive nature 递归性
recursiveness 递归性
recursiveness problem 递归性问题
recursive predicate 递归谓语
recursive procedure 递归过程
recursive process 递演过程;递推过程;递归过程;递归处理
recursive program(me) 递归程序
recursive program(me) scheme 递归程序模式
recursive programming 递归程序设计
recursive relation 递演关系;递推关系;递归关系
recursive routine 递归例程
recursive rule 递归规则
recursive sequence 递归序列;递归顺序
recursive set 递归集(合)
recursive solution 递归解法
recursive structure 递归结构

recursive subroutine 递归子程序
recursive unsolvability 递归不可解性
recurvate 反弯曲(的)
recurvation 反向弯曲;反弯;反屈
recurvature 转向;反向弯曲;反曲
recurvature of storm 风暴移动轨迹的转向点
recurvature of wind 风的转向
recurvature point 转向点
recurve 向后弯
recurved occlusion 后曲锢囚【气】
recurved parapet 挡水墙;防浪(胸)墙
recurved spit 钩状岬;钩形沙嘴;钩形沙咀;弯曲沙嘴;沙钩
recurving typhoon 转向台风
re-cut 再切削
recut block 复切石块
recut stone 复切石块
recutter 第二次切碎装置
rec-vehicle 周末旅游车
recyclable coagulant 可再循环凝结剂
recycle 再循环;再回收;重复循环;重复旋回;重复利用;复循环;反复循环
recycle back 反向循环
recycle brine 循环海水
recycle chromatography 循环色谱法
recycle circuit 循环线路
recycle compressor 循环压缩机
recycle cooler 再循环冷却器
recycle cracking operation 循环裂化操作
recycled-aggregate concrete 再生集料混凝土;再生骨料混凝土
recycled backwash water 循环(的)回洗水
recycled coagulant 循环凝结剂
recycled concrete 再生混凝土;重复利用混凝土
recycled concrete aggregate 重复利用混凝土集料;重复利用混凝土骨料
recycled fabric 再生织物
recycled fludlized bed 循环流化床
recycled fuel 再循环燃料
recycled-gas mixing method 再循环气体混合法
recycled material 循环物质;循环使用的材料;回收的材料
recycled mixture 再生混料
recycled nutrient 再循环养分
recycled off-gas 循环尾气
recycled organic matter 再循环的有机质
recycled sludge 酸渣回收;污泥回收
recycled waste-process 废水回流过程
recycled wastewater 重复利用废水
recycled wastewater-process 废水回流过程
recycled water 再循环水
recycled water portion 再循环水分配
recycled wood 再生木材
recycle fraction 循环馏分
recycle gas 循环气
recycle gas blower 循环气(气)鼓风机
recycle gas furnace 循环气体炉
recycle gas oil 循环瓦斯油
recycle gasoline 循环汽油
recycle gel permeation chromatography 循环凝胶渗透色谱
recycle leaching 循环浸出
recycle liquid pump 循环液泵
recycle mixing 循环混合
recycle monomer 循环单体
recycle of chorofluorocarbons 含氯氟烃的重复使用
recycle of effluent 出水回流
recycle oil 循环油
recycle pump 循环泵
recycler 反复循环器
recycle ratio 循环系数;循环比
recycle ratio of gaseous constituents 气态组分的循环比
recycle ratio of sludge 污泥回流率
recycle slurry 循环悬浮体;循环浆
recycle stock 循环油;再循环物料
recycle stream 环流气体
recycle system 循环系;再循环系统
recycle unit 循环设备
recycle valve 循环阀;再循环阀
recycle water 循环水;回用水;回收水
recycling 回收;回灌(地下水);再循环;再(生)利用;重复循环;循环利用

recycling agent 回收剂
recycling binder 复拌结合料
recycling chromatography 循环色谱法
recycling continuous 循环连续
recycling course 复拌层
recycling elution 再循环洗脱
recycling glass 回用玻璃
recycling loading test 重复荷载试验
recycling of end gas 废气循环
recycling of pavement 路面复拌
recycling of petrodollars 石油美元再循环
recycling of waste gas 废气循环
recycling of wastes 废物循环使用
recycling of water 水的循环
recycling of water or sewage 水域污水循环
recycling process 复拌法
recycling program(me) 再循环程序
recycling pump 回流泵
recycling relay 重合式继电器
recycling stock 再周转原料;再周转备料
recycling surface 复拌面层;复拌道路
recycling surface treatment 复拌式表面处治
recycling system 再循环系统
recycling trays 循环塔盘
recycling tub 再混合池
recycling wastewater after treatment 处理后的废水重复利用
red 指北;红的
red acetate 媒染剂红液
redact 编校;编辑
redaction 校订;编校
redactor 审订人
red alder 红桤木;赤杨
red alert 紧急警报
red algae 红藻
red alizarin(e) lake 红茜素湖
Redalon 利达隆缓凝剂
redan 突角堡;凸角堡
red-and-black chequered buoy 红黑方格浮标
red-and-black horizontal stripes buoy 红黑横纹浮标
red-and-black vertical stripes buoy 红黑直纹浮标
red-and brown compilation 红版编绘原图
red-and-white beacon 红白立标
red-and-white chequered buoy 红白方格浮标
red-and yellow basket 小物品打捞篮(钻孔内);小物件打捞篮(钻孔内)
red-and yellow podzolic soil 红黄色灰化土
red antimony 红锑矿
red antimony ore 红锑矿石
redar 红外(线)光测距仪;红外(线)测距仪
red arc 红弧
red azimuth table 太阳方位角表
red balance 赤字(结余)
red ball 特快列车;快运货车
red balmoral 陆佑红花岗石
red band 转速禁区
red bauxite 铁铝土;铁矾土
red beam 红色光束;红光束
red bed 红砂岩层;红色岩层;红层
red bed deposit 红层矿床
red bed facies 红层相
red beds copper deposit 红层铜矿床
red beech 红假山毛榉
red bill of lading 红色提单
red birch 红桦;赤杨
red board 停车牌
red body 红坯
red bole 土状赤铁矿
red book 红皮书
red branch 红支
red brass 锡锌合金;红铜;红色黄铜(含铜大于80%的);红黄铜
red brass alloy 红铜合金;红色黄铜合金
red brick 红砖
red bridge washer valve 红色桥形气门嘴
red brittleness 高温脆性;热脆性
red bronze 红铜;红黄铜
red-brown 棕红色;红褐色
red-brown alkyd transparent paint 红棕醇酸透明漆
red-brown soil 红棕土(壤)
red cabbage 红(球)甘蓝
red calcareous soil 红色石灰性土壤;石灰性红土

red carpet 铺红地毯的;红地毯
red cedar 北美圆柏;大侧柏;铅笔柏;美国南方红桧;红杉木;红松
red cedar pigment 铬红(颜料)
red cedar shingle 红松屋面板
red cedar wood oil 红雪松木油
red chalk 红铅粉;红垩;红色铁矿
red charcoal 红木炭
red check 红液渗透探伤法
red clay 红土;红黏土;红色黏土
red clay square floor tile 红泥方阶砖
red cobalt 钴华
red colo(u)r 红彩
red colo(u)ration 红色
red colo(u)r difference modulator 红色差调制器
red colo(u)r difference signal axis 红色差信号轴
red colo(u)r glass 红玻璃
red colo(u)r weakness 红色弱
red compressed asbestos fibre jointing in sheet 红色石棉橡胶板
red compressed asbestos sheet 红石棉纸板;石棉红纸板
red concrete 红色混凝土
red convex lens 红色凸透镜
red copper 赤铜;紫铜
red copper ore 赤铜矿
red copper oxide 氧化亚铜;红色氧化铜
red coral 红珊瑚
red coronal line 日冕红线
red cross health unit 红十字卫生站
red cross hospital 红十字医院
Red Cross Society 红十字会
Red Cross Society of China 中国红十字会
red crowberry 岩高兰
red-crowned crane 丹顶鹤
red crystalline glaze 红结晶釉
red current 赤潮
red cut-off 红色截止
red cypress 红丝柏;红柏木
red deal 红松材;红木;洋杉
red decometer 红台卡计
red deep sea earth 深海红土
red degeneration 红色变性
reddendum (授权人、出租人的)权益保留条款
reddening 红化
reddening line 红化曲线
reddening ratio 红化比
red desert soil 红色沙漠土;红漠土
red dextrin 红色糊精
reddingite 磷锰矿;水磷锰石
reddish 微红的
reddish black 红黑色;红黑;带红相黑色
reddish blue 红蓝色;品蓝(色);带红相蓝色
reddish brown 红棕(色);带红相棕色;红褐色
reddish-brown chestnut soil 淡红褐色栗钙土
reddish-brown lateritic soil 红棕色红土;淡红褐色土壤
reddish-brown soil 淡红褐色土壤
reddish colo(u)r 微红色
reddish dark brown 暗红棕色
reddish orange 带红相橙色
reddish prairie soil 淡红色草原土
reddish purple 紫红色
reddish violet 红紫(色);带红相紫色
reddish yellow 红黄(色);带红相黄色
reddle 代赭石;红土;红丹粉
red dog 灰面;粗面粉
red drive control 红色激励控制
red dysentery 赤痢
red earth 红土;红壤;红染
redeck 翻修路面;重修平屋顶
redecking 重修屋顶;翻修路面;修复路面;翻修面层
redecorate 重新油漆;重新装饰
redecoration 重新装饰
redecoration work 重新装饰作业
red edge 红边
redeem 弥补;缓冲;赎回;偿还;变卖证券为现款
redeemable bond 可赎回债券;可通知偿还的债券;可提前偿还的债券
redeemable debenture 可提前偿还公司债;可赎回公司债;可赎公司债券
redeemable paper money 可兑换纸币
redeemable rent 可冲销租金

redeemable share 可赎回股份;可偿还份额
redeemable stock 可赎回股票;可换现股票
redeemed 债券收回溢价
redeem mortgage 赎回抵押品
redefine 重定义;重新定义
redefining 重新定义
redefinition 再定义
redelivery 再装船
redelivery area 租还地
redelivery clause 还船条款
redelivery date 租环期
redelivery notice 还船通知
redelivery of chartered vessel 还船
redelivery survey 退租检验
red elm 红榆
red emitting phosphor 红色磷光体
redemption 押品的赎回;缓冲;赎回;被补偿
redemption annuities 补偿年金
redemption by drawing 抽签偿还
redemption by year installment 分年偿还
redemption date 偿还贷款日期
redemption fund 折旧费;折旧基金;赎债基金;偿债基金
redemption loan 偿还贷款
redemption of mortgage 赎回典契
redemption of stock by distribution of net profit 分配净利润赎回股票
redemption penalty 补偿赎金
redemption period 赎回期;偿还期
redemption premium 赎回溢价
redemption price 清偿价格
redemption rate 兑换率
redemption value 赎回价值
red end 红端(磁铁指北极);磁针指北端
red engobe 红色化妆土
redented 齿状的
redeploy 调遣
redeployment 调整措施
redeployment airfield 转场备用机场;战备机场
redeployment of personnel 人员重新配置
redeposit 再沉积
redeposited clay 次生黏土
redeposited loess 再积黄土;次生黄土
redeposition 再沉积作用;再沉淀(作用)
rederal plant 杂草植物
redesign 再设计;重新设计;重算
redesigned 重新设计的
redesigned berth 重新设计的码头
redesigned device 改进型装置
redesigned model 改进模型
redesigned quay 重新设计的码头
redesigned strategy 改进型策略
redesigned terminal 重新设计的码头
redesigned wharf 重新设计的码头
redesigning 重新设计
redetermination 新定义;重新确定;重新测定
redetermine 重新定出
redeveloped area 重新开发区
redeveloper's statement of public disclosure 城市改建地区开发商的公开计划报告
redevelopment 改造;照相加厚;再显影;再开发;再发展;二次显影
redevelopment company 土地开发公司
redevelopment district 再开发区
redevelopment of a well 井的修复
redevelopment of landfill site 垃圾填埋地的重新整修
redevelopment plan 重建计划;复兴计划;复兴规划
redevelopment scheme 重建计划
redevidees silk 复摇丝
red eye (平板玻璃边缘的)橙色斑点;红色信号灯
red fiber sheet 红色硬化底板
Redfield efflux viscometer 雷德菲尔德流出型黏度计
red-figure style 朱红图像式样
red-figure vase 朱红图像瓷瓶
red filter 红色滤光镜;红色过滤片
red fir 红松【植】;花旗松;红杉
red fir strip floor(ing) 红冷杉木条地板
red flare signal 红色火焰信号
red flea 红虫
red flooring tile 红地砖
Redford's alloy 红福特合金
red for lake 色淀性红

Red Fox 雷德福克斯镍铬耐热耐蚀钢
red fox alloy 一类不锈钢
red gain 红色信号增益
red glass 玉红玻璃
red glassine paper 红色半透明玻璃纸
red globe lamp 红光球灯
red globular shape 红色球型
red gold 纯金
red gold alloy 红金合金
red green-blue colo(u)r decoder 红绿蓝三色解码器
red green-blue system 红绿蓝色系
red green colo(u)r charts 红绿色测试环
red green crossover 红色交叉
red grey 红灰色
red ground tile 红地砖
red gum 糖胶树;美国枫香;赤桉;桉树胶
Red Halloysite 赤石脂
Red Hand 红手(各种屋面毡的注册商标)
red hardness 红硬性;红热硬度;热硬性;热硬度
red-hard steel 红硬钢;高速工具钢;热硬钢
Redhead ga(u)ge 雷德黑德真空计
red head series 红崖统【地】
red heart 腐朽心材
red heat 灼热;炽热;赤热
red-heat temperature 赤热温度
red-heat test 赤热试验
red helium-neon light 红氦氖灯
red hematite 红赤铁矿;赤铁矿
red herring issue 非正式上市的股票
red herring prospectus 非正式招股章程
Redhill beds 红丘层
red hot 炽热的;赤热的
red hot component 红热零件
red hot dust particles 炽热窑灰粒子;炽热粉尘粒子
rediactive change 放射变化
re-dialing function 重新拨号功能
rediation-noise-limited system 辐射噪声限制系统
rediaul modulation 残留调制
redifferentiation 再分化
rediffusion on wire 有线广播
rediffusion station 有线广播站;转播站;广播站;广播台
redilution 再稀释
red incandescent electric(al) lamp 红色炽光灯
red indigo 红靛
red induration 红色硬结
redingtonite 铁明矾;水铬镁矾
red ink 亏本
red ink paste used for seals 印泥
re-dipping unit 二次镀锡机组
redir 雨后储水区
redirected call 改发呼叫
redirecting the use of investment 调整投资使用方向
redirection 改址;改寄新址;后视方向;重新定向
redirection of call 调用的重新编址
red iron ochre 赤铁赭石
red iron ore 红铁矿;赤铁矿
red iron oxide 氧化铁红;红土子;红(色)氧化铁
red iron oxide in lump 红铁氧块
red iron oxide in powder 红铁氧粉
red iron oxide pigment 氧化铁红颜料
red ironwood 红铁木
rediscount 再折扣;再贴现;重新贴现
rediscounted bill 再贴现票据
rediscount rate 再贴现率;重贴现利率
rediscover 再发现
redislocation 再脱位
redispersibility 再分散性
redispersion 再弥散;再分散法;再分散(作用);重分散
redissolution 再溶解
redissolve 再溶解
redistill 再蒸馏
redistillation 再蒸馏;重蒸馏
redistilled water 再蒸馏水
redistilled zinc 精制锌
redistribute 再分配;再分布
redistribution 再分配;再分布(作用);重(新)分配;重新分布
redistribution at surface 表面再分布
redistribution baffle 再分布挡板
redistribution by fiscal process 财政再分配

redistribution current 重新分配电流
redistribution effect 再分布效应
redistribution of creep 蠕变再分布;蠕变再分配
redistribution of head 水头再分配
redistribution of load 荷载重分布
redistribution of moments 弯矩重分配;弯矩再分布;弯矩再分配
redistribution of momentum 动量再分配
redistribution of shear 剪力再分布
redistribution of stress 应力重分布
redistribution of stress under foundation 基底应力重分布
redistribution of water head 水头再分配
redistribution reading 重新分配读数法;重分配读数
redistribution writing 再分配记录;重新分布记录
redistributor 再分布器
redistrict 把……重新划区
redivide 再分配
redix 环氧类树脂
red japan lacquer 朱漆;日本红漆
red knot 红节疤;活节
red label 红标签
red lacquer 朱漆
red lake 红色淀
red lake C pigment 色淀 C 颜料红
red lake pigment 红色淀颜料;深红色颜料
red lamp 红灯;危险信号
red lantern 红灯
red lauan 菲律宾柳桉木;南太平洋红木;红柳安
red lead 红铅粉;铅丹;四氧化三铅
red lead anti-corrosive paint 红丹防锈漆;铅丹防锈漆
red lead-based building mastic 铅丹底房屋胶粘料
red lead-based oxide 铅丹底氧化物
red lead-based paint 铅丹底涂料;铅丹油漆
red lead-based spar 铅丹打底的橡子
red lead cement 红铅油灰;红丹油灰
red lead-coated pigment 包核红丹颜料
red leaded zinc 铅丹铬酸锌
red leaded zinc chromate anti-corrosive paint 铅丹铬酸锌涂料;铅丹铬酸锌防锈漆
red lead epoxy coat 红丹环氧涂层
red lead gun 红丹油泥喷枪
red lead injector 红丹油泥喷枪
red lead oil 红丹清油
red lead oxide 红铅粉
red lead paint 红丹涂料;红丹防锈漆;铅丹(油)漆
red lead powder 红铅粉;红丹粉
red lead primer 红丹底漆
red lead putty 红铅油灰;红丹油灰;红油灰;铅油
red lead putty gun 红丹油泥喷枪
red lead ready-mixed paint 红丹调合漆
red lead rust-proof paint 红丹防锈漆
red lead varnish 红丹漆
redledgeite 硅镁铬钛矿
red lens 红透镜
Redler conveyer 雷德勒输送机(一种运送粉料的槽式链刮板输送机)
Redler en masse conveyer 雷德勒埋刮链输送机
red light 红灯(光);危险信号
red light district 红灯区
red light relay 红灯继电器
red lime mud 红灰基泥浆
red limit 红光极限
red line 红线
redlining 安全极限;红线注销
red liquor 媒染剂汇液;红碱液
red litharge 红相黄丹
red litmus paper 红石蕊纸
red loam 红壤
redlocatable program(me) 可再排序
red locust 红刺槐木
red machine oil 红机油
red magnetism 红(色)磁性;地磁南极性
red manganese 红锰矿
red marl 红泥灰岩
red media filter 红色填料过滤器
red mercanti 热containerfd硬料
red mercuric oxide 红色氧化汞
red mercury sulfide 硫化汞
red mercury sulphide 硫化汞红
red metal 红铜;(含铜大于80%的)红色黄铜
red mineral colo(u)r powder 红色土粉
red minium 红丹;红铅(丹)

red monoclinic crystal 红色单斜晶体
red mud 红土泥浆;红泥(浆);赤泥
red-mud sulphated cement 赤泥硫酸盐水泥
redness 红色
redness of the skin or complexion 赤色
red nitric acid 红色硝酸
Redo 雷度(一种乙烯树脂涂胶织物)
redo 再循环
red oak 红橡木;红栎木
red ocher 红赭石颜料;红赭石;赤铁矿
red oil 油酸;工业级油酸;红油;不纯的油酸
red ooze 红软泥
red organic pigment 红有机颜料
red orpiment 雄黄
redoubling 再折叠
redoubt 棱堡;方形堡
redox 氧化还原(作用);还原氧化(作用)
redox-activated emulsion polymerization 氧化还原活性乳化聚合
redox-active couple 氧化还原活性电对
redox-active media 氧化还原活性介质
redox-active species 氧化还原活性物种
redox-active system 氧化还原活性系统
redox analyser 氧化还原分析器
redox analysis 氧化还原分析
redox buffering 氧化还原缓冲
redox catalyst 氧化还原催化剂
redox-catalyst system 氧化还原催化剂体系
redox cell 氧化还原电池
redox chemistry 氧化还原(过程)化学
redox condition 氧化还原条件
redox control (in bleaching) 氧化还原势控制
redox couple 氧化还原偶;氧化还原对
redox equilibrium 氧化还原平衡
redox equilibrium constant 氧化还原常数
redox half-cell reaction 氧化还原半电池反应
redox-hypothesis 氧化还原说
red oxide 氧化红物;红色氧化物;铁丹
red oxide hold paint 紫红货舱调和漆
red oxide of iron pigment 氧化铁红颜料
red oxide of zinc 红锌矿
red oxide paint 紫红漆
redox indicator 氧化还原指示剂
redox initiate polymerization 氧化还原引发聚合
redox initiation 氧化还原引发(作用)
redox initiator 氧化还原引发剂
redox ion exchange 氧化还原离子交换
redox ion exchanger 氧化还原离子交换树脂
redoxite 氧化还原树脂
redox-mediated decolo(u)rization 氧化还原间接脱色
redox mediator 氧化还原介体
redox method 氧化还原(滴定)法
redox of soil 土壤氧化还原(作用)
redoxomorphic stage 氧化还原阶段;氧化还原成岩变化阶段
redoxostat 氧还电位稳定器
redox pairs 氧化还原对
redox polymer 氧化还原聚合物
redox polymerization 氧化还原聚合
redox potential 氧化还原电位;氧化还原电势
redox potential detector 氧化还原电位检测器
redox potential discontinuity 氧化还原电势突变
redox potentiometry 氧化还原电位法
redox precipitation 氧化还原沉淀
redox process 氧化还原过程;氧化还原法
redox pump 氧化还原泵
redox rate constant 氧化还原速率常数
redox reaction 氧化还原反应
redox resin 氧化还原(型)树脂
redox series 氧化还原系列
redox state 氧化还原态
redox system 氧化还原系统;氧化还原体系;氧化还原(引发)系统
redox thermochemistry 氧化还原热化学
redox titrant 氧化还原滴定剂
redox titration 氧化还原滴定(法)
redox zone 氧化还原带
red paint 朱漆;红丹(又名红铅)
red painted steel pipe 红油涂面钢管
red pairing tile 红地砖
red pecan 红山核桃木
red-pencil 改正
red phoenix in morning sun 丹凤朝阳

red phosphorus 红磷;赤磷
red pigment 红色颜料
red pine 红松【植】;红粉;赤松
red podzolic laterite 灰化砖红壤
red podzolic soil 灰化红土;灰化红壤;红色灰化土
red polzolic soil 红土化
red porcelain 紫砂
red pottery 红陶
red pottery bowl with textile impression 布纹红陶碗(陶器名)
red precipitate 氧化汞;红色沉淀物
red printing 套红
red prussiate 赤血盐
red putty cement 红丹油灰
red quartz 红水晶
redraft 逆转票据
red ramie 红麻
red raspberry 覆盆子
redraw 重拉拔
redrawing 重画;再拉拔;重拉伸
redrawing die 续次拉延模;重复拉模
redrawing test 再拉延试验
redraw rammed piles 拔桩
redraw rod 再拉条
red reflecting dichroic 红色反射镜
red resin 红树脂
red response 红色信号响应
redress 重车;再平衡;纠正;重新修整;重磨轧辊;补救
redress damage 赔偿损失
redress disequilibrium 纠正不平衡
redresser 高频检波器
redressing coefficient of rolls 轧辊重车系数
redressing of setts 重新修整小方石铺砌
redressing shop 修锉间
redress life 重修寿命
redrier 二次干燥机
re-drill 再钻
re-drill bit 扩眼钻头;扩孔钻头
re-drilling 重新钻孔
red ripe 红热的
Red River 红河
redriving (桩工的)复打
redriving test (桩的)重钻试验
red rock 红层;全晶质霏细斑岩
red rock formation 红色岩建造
red roof tile 红瓦;素烧瓦
red rot 红朽斑(木材缺陷);腐木变红;腐朽变红
red roundel 红色表示镜
red route 危险的道路
red rubber acetylene hose 红色乙炔胶管
red rubber oxygen hose 红色氧气胶管
red rubber sheet 红色橡胶板
red rust 红锈
redruthite 辉铜矿
redry 再烘干
redrying 二次干燥
red sand 红砂
redsandal 紫檀香木
red sandal wood 紫檀
red sanders 紫檀
red sandstone 红砂岩
red saturation scale 红色饱和度
Red Sea 红海
Red Sea Bay 红海湾
Red Sea convergence zone 红海辐合区
Red Sea graben 红海地堑
red seal 红色填缝灰泥
red sea lined snapper 线纹笛鲷
Red Sea rift seismotectonic zone 红海地堑地震构造带
Red Sea rift system 红海裂谷带
red sector 红区;红光扇形区
red seraya 马来红柳桉;红杉木
red shade 带红光的
red shift 红移;红色光栅偏移
red shift effect 红移效应
red-short 热脆的
red-short iron 热脆铁
red shortness 红脆性;热脆性
red short steel 热脆钢
red-side light 左舷灯;红色舷灯
red signal light 红灯(光)
red signal repeater 红灯信号复示器

red signal repeater relay 红灯信号复示继电器
red silver fir 红银杉
red silver ore 红银矿
red slag 烧红的铁渣
red slate 红色板岩
red snow 红雪
red softening 红色软化
red soil 红土；红壤
red soil area 红壤区
red spruce 红(果)云杉【植】
red squill 红海葱
red stain 红色着色剂；红色彩料
red staining 铜红扩散着色
red star 红星
red stern light 红尾灯；(船)
red stone 氧化铁；红石；红色岩
red stoneware 红炻器
red stringybark 大钩刺桉(产于澳大利亚)
red stuff 红料
red-tape operation 整理操作；程序加工操作；辅助操作
red tide 红潮；赤潮
red tide alga 赤潮藻
red tide control 赤潮控制
red tide microalga 赤潮微藻
red tide plankton 赤潮生物
red-time delay 灯照时间
red top 路基标桩；牧草；小糠草
red tulip oak 银达理木；郁金香栎木
red turbid urine 赤油
redtwig dragon spruce 异鳞云杉
reduce 折算；折合；紧缩；降低；简化；减缩；减少；还原
reduceability 再现性
reduce a fraction to its lowest terms 约分
reduce air pollution 减少空气污染
reduce budget expenditures 紧缩开支
reduce by half 减少一半
reduce casing 缩小套管直径
reduce cost 降低成本
reduced 归约
reduced Abelian group 约化阿贝耳群
reduced abutment 简式支座；简式墩座
reduced acceleration spectrum 折减加速度谱
reduced admittance 正规化导纳
reduced air 再循环风
reduced algebraic scheme 约化代数概型
reduced amount of cost of comparable products 可比产品成本降低额
reduced area 改算面积
reduced astronomical latitude 归化海面的天文纬度
reduced astronomic latitude 天文归化纬度
reduced automaton 归约自动机
reduced bit load drilling 减压钻进
reduced bore 缩小孔径
reduced calcine 还原焙砂
reduced capacity 减低(输出)容量
reduced capacity tap 减载抽头
reduced channel 简约信道
reduced character 约化特征标
reduced charge 减量装药；减费
reduced colo(u)r 还原色(料)
reduced comfort boundary 不快感界限
reduced complex 约化复形
reduced compliance 折合柔量
reduced Compton wavelength 折合康普顿波长
reduced condition equation 约化条件方程
reduced cone 约化锥
reduced coordinate directions 简化坐标方向
reduced correlation matrix 约相关矩阵；约化相关矩阵
reduced cost coefficients 缩减成本系数
reduced cost matrix 减值费用矩阵
reduced cost of comparable products 可比产品成本降低额
reduced crude 常压渣油
reduced current 折算电流
reduced cycle matrix 缩减圈矩阵
reduced de Broglie wavelength 约化德布罗意波长
reduced degree 简化次数；递减次数
reduced density 折算密度；折合密度；对比密度
reduced density effect 约化密度效应
reduced dependence 简化的依赖性
reduced depth 折算深度
reduced dimensionality system 低维系统

reduced discharge 折算流量；折减流量
reduced distance 改化距离
reduced distribution function 约化分布函数
reduced doping concentration 折合掺杂浓度
reduced effective mass 折合有效质量
reduced elastic modulus 折减弹性模量
reduced enamel epithelium 缩余釉上皮
reduced energy 折合能量
reduced equation 约简方程；约化方程；简化式；简化方程；对比方程
reduced extremal distance 约化极值距离
reduced eye 简约眼；简化眼；模型眼
reduced eye lens 简化目镜
reduced factor 折合因子；折合系数
reduced-fare ticket 减价票
reduced file size 缩减的文件长度
reduced filling period 减慢充盈期；缓慢充盈期
reduced flange 异径法兰
reduced flow 简化流动；收敛(水)流
reduced flush device 节水冲洗装置
reduced form 简约型
reduced form of zero-sum n-person game 简约型 n 人零和对策
reduced frequency 折合频率；换算频率
reduced fuel oil 重质重油
reduced gain 抑低增益
reduced grade 换算坡道
reduced gradient method 简化梯度法
reduced graph 简化图；约化图
reduced gravity field 弱重力场
reduced gravity value 改化重力值
reduced ground motion 折减地面运动
reduced growth in trade volume 贸易量增长率减低
reduced hardness 折合硬度
reduced head 折算水头；换算水头
reduced heat flow 简化热流
reduced heat input 折算热负荷
reduced height 折算高度；折合高度；换算高度
reduced height of soil 土的骨架净高
reduced Hessian matrix 简化海森矩阵
reduced homology group 约化同调群
reduced image 缩小像
reduced impedance 折合阻抗；归一化阻抗
reduced incidence matrix 约化关联矩阵
reduced influence diagram 简化影响线图
reduced instruction set computer 精简指令系统计算机
reduced interest cost 减低利息成本
reduced isothermal 对比等温线
reduced join 约化联接
reduced Jordan algebra 约化当代数
reduced kilovolt-ampere tap 低负荷抽头
reduced latitude 规化纬度；归心纬度；归化纬度
reduced lattice 化约格
reduced law 递减律
reduced laying length 折算铺设长度
reduced length 约简长度；折算长度；折合长度；简约长度；计算尺度
reduced length of the geodetic line 大地线归化长度；大地线改化长度
reduced level 降落水准；折算高度；折算高程；折算标高；折合水准；归化高度；简化水准
reduced level of water 降落水位
reduced liability tax 赋税义务减轻
reduced lift of sleeve 套筒减速升程
reduced lighting 折照灯光；掩饰照明；简化照明
reduced linear differential equation 约简线性微分方程
reduced line of loading 缩减的受荷载线
reduced loading line 缩减的受荷载线
reduced longitude 改化经度
reduced mapping cone 约化映射锥
reduced mapping cyclinder 缩减映射圆柱
reduced mass 约化质量；折算质量；折合质量
reduced mass effect 约化质量效应
reduced mass-flow 折合质量流量
reduced matrix 约简矩阵
reduced matrix element 约化矩阵元
reduced model 缩小模型；简化模型；缩微模型
reduced modulus 换算模数；化成的模量
reduced modulus of elasticity 弹性折合模量；折算弹性模量
reduced moment 缩减的力矩

reduced moment of inertia 折算惯性距
reduced mordant dye 还原煤染料
reduced natural frequency 折算固有频率；折减固有频率
reduced nipple 异径螺纹接套
reduced nitrogen species 被还原氮化物
reduced normal equation 约化正则方程；约化正规方程；约化法方程
reduced oil 还原油；拔顶(原)油
reduced optic(al) length 折合光程
reduced ordered pair group 约化有序对群
reduced-order model 降阶模型
reduced osmotic pressure 比浓渗透压
reduced output 降额输出；降低出力
reduced parameter 折算参数；折合参数；简约参数；简化参数；换算参数
reduced partial width 约化分宽度
reduced pendulum length 折合摆长；改化摆长
reduced photograph 缩微(像)片
reduced pigment 冲淡颜料
reduced plan 缩小平面
reduced Planck constant 约化普朗克常数
reduced point 缩减点
reduced point of satellite 卫星归算点
reduced potential energy 约化位能
reduced powder 还原粉
reduced power 折算功率；降低功率
reduced premium 降低保险费
reduced pressure 折合压力；减压；换算压力；对比压强；对比压力
reduced pressure backflow preventor 减压回流防止器
reduced pressure coefficient 换算压力系数
reduced pressure distillate 减压馏分
reduced pressure distillation 减压蒸馏；真空蒸馏
reduced pressure drilling 降压钻进
reduced price 已降低的价格；降价
reduced principal stress 缩减的主应力
reduced print 缩小影印图；缩微像片
reduced product 归纳积
reduced product space 约化积空间
reduced proper motion 简化自行
reduced property 折合性质；对比值
reduced pumping 减量泵送
reduced quadratic form 约化二次形式
reduced quadratic function 简化二次函数
reduced quantity 约化量；折合量；减缩量
reduced radiation 减弱辐射
reduced rank tensor 约化张量
reduced ratio 折合比
reduced representation 约化表示
reduced residual equation 约化残余方程；约化残差方程
reduced resistance 折算电阻
reduced root 退化根
reduced sampling inspection 缩减抽样检查；分层抽样检查
reduced scale 缩小规模；缩小的比例(尺)；缩小例尺；缩尺
reduced scale prototype model test 缩尺真型试验
reduced scheme 约化格式
reduced section 折算断面
reduced set 缩减集
reduced source 简约信源
reduced space 约化空间
reduced specific viscosity 比浓黏度
reduced specimen 缩尺试片
reduced speed sign 减速地点标
reduced speed signal 减速信号
reduced state 约化态
reduced state classification 简化状态类别
reduced state diagram 简化状态图
reduced steam 减压水蒸汽
reduced stone 碎石
reduced-strength design (路基、地基霜冻期的)减低强度设计
reduced stress 约化应力；折算应力；简化应力
reduced subgrade strength 减低地基强度；减低路基强度
reduced sulphur species 被还原硫化物
reduced suspension 约化双角锥
reduced system of equations 归化方程组
reduced T 异径三通
reduced tee 异径三通(管)；缩径三通(管)

reduced tee pipe coupling 异径T形管接管
reduced telemetry 处理遥测
reduced temperature 折算温度;降低温度;对比温度;订正温度
reduced the size 缩小体积
reduced to vacuum 折合到真空
reduced trace 缩减迹
reduced transition probability 约化跃迁几率
reduced transparency 对比透明度;比透明度
reduced travel time 折合走时
reduced tree height 简化树结构
reduced turbidity 比浓浊度
reduced unit 折合单位;换算单位
reduced unitized table 约简表
reduced value 折算值;折价;对比值
reduced variable 约减变量;简化变量
reduced variate 转化变量
reduced velocity 折合速度;换算速度
reduced viscosity 折合黏度;对比黏度;比浓黏度
reduced visibility 减弱能见度
reduced voltage 折算电压
reduced voltage method 降压法(检查集装箱气密性)
reduced voltage starter 降压起动器
reduced volume 缩小体积;对比体积
reduced wave vector 简约波矢
reduced white 加填料铅白
reduced width 约化宽度;折合宽度
reduced winding diagram 缩小绕组图
reduced zone scheme 还原区域图
reduce expenditures 节流
reduce expenses 压缩开支
reduce fractions to a common 通分
reduce in scale 缩小比例
reduce investment outlay 节省投资
reduce price 减低价格
reduce production 紧缩生产
reducer 练条机;节流器;简约器;简化器;减压器;减速机;减薄液;还原者;还原剂;切碎器;缩节管;分解者;变径段
reducer adapter 减压活门联接器
reducer and increaser 大小头
reducer angle 直角异径弯管
reducer casing 减速箱
reducer coupling 缩径接管;缩径(管)接头;缩颈接头;缩接
reducer for conveyer 输送机用减速机
reducer for doffer 滚筒减速机
reducer for finned drum feeder 翅片转鼓给料机用减速机
reducer gas brown coal tar 煤气发生器用褐煤的焦油
reducer jointing 缩接
reduce roll machine 锻轧机
reducer output shaft 减速器输出轴
reducer (pipe) 变径管
reducer ratio printer 比例缩小仪
reducer section 渐缩管截面
reducer spacer 减速箱调整垫
reducer thinner 冲淡剂
reducer union 过渡接头;变径接头
reduce soil acidity and improve soil texture 减轻土壤酸性和改良土壤结构
reduce speed sign 慢行标志
reduce staff 紧缩编制
reduce stocks 压缩库存
reduce the effect of acid in a soil 减弱土壤酸性
reduce the menace of ice run 防凌
reduce the number of rejects and seconds 降低废品
reduce the number of vehicle 减少机动车辆
reduce the price 降低价格
reduce the risks of loss 分散发生损失的危险
reduce to scrap 降为废品;敲成碎片
reduce waste 减少浪费
reducibility 可缩减性;可约性;还原性
reducibility axiom 可约性公理
reducibility of a transformation 变换的可约性
reducibility of problem 可约性问题
reducibility test 稀释性能试验
reducible 可缩小的;可简化的;可减少的;可还原的;可复位的
reducible correspondence 可约对应
reducible curve 可约曲线

reducible fraction 可约分数
reducible graph 可约图形
reducible linear representation 可约的线性表示
reducible matrix 可约矩阵;可简化的矩阵
reducibleness 可折合性;可约性;可简化性;可减少性;可还原性;可粉碎性
reducible operator 可约算子
reducible polynomial 可约多项式
reducible procedure 可约过程
reducible representation 可约表示
reducine 还原碱
reducing 拉细;减低;还原作用;还原的
reducing admixture 促凝型减水剂
reducing agent 减薄剂;还原剂
reducing and enlarging machine 缩放机
reducing and sustaining valve 减压和持续阀
reducing apparatus 减压装置;缩影仪
reducing atmosphere 还原空气
reducing balance depreciation method 余额递减折旧法
reducing-balance form 余额递减式
reducing-balance method 余额递减法
reducing barrier 还原障
reducing bath 还原浴
reducing bend 异径弯头;缩径弯头;变径弯头
reducing bend pipe 变径弯管
reducing bend piping 变径弯管
reducing bend tube 变径弯管
reducing bend tubing 变径弯管
reducing box 制砖框架(专门制壁龛需要的砖)
reducing branch 异径支管
reducing bushing 缩孔垫圈;缩径轴套;过渡套管;过渡衬套;变径衬套
reducing calculus 归算
reducing capacity 还原能力
reducing catalyst reactor 还原催化反应器
reducing condition 还原条件
reducing conduit 变径管
reducing cost balance method 降低成本余额法
reducing coupling 缩径管接头;异径管节;缩小接管;缩口接头;缩径联轴节;变径管节
reducing cross 异径四通;异径十字(形接)头;变径四通
reducing die 拉线模板;拉模;缩径模
reducing elbow 异径弯头;异径弯(接头);渐缩弯管;大小头弯管;变径弯头
reducing electrode 还原电极
reducing end group 还原性(末)端基
reducing environment 还原环境
reducing extension piece 异径内外螺纹管接头
reducing factor 折合因数;简化因数;实积形数
reducing finder adapter lens 简化深测器接合器透镜
reducing firing 还原烤色;还原燃烧
reducing fittings 缩减装置;异径管件
reducing flame 还原性火焰;还原(火)焰
reducing flange 异径法兰;渐缩法兰;渐缩突缘;缩口法兰;变径法兰
reducing force 还原力
reducing furnace 还原炉
reducing gear 减速齿轮
reducing gear box 变速箱
reducing gear indicator 减速齿轮示功器
reducing glass 缩小透镜
reducing grade 折减坡度;减少坡度
reducing hose coupler 减径软管管套
reducing in a single pass 开路粉碎;单程粉碎
reducing instalment method 分期递减(摊销)法;分期递减偿付法
reducing joint 变径接头;缩径接头;缩管接头;变径接合
reducing jointing 缩口接头
reducing lamp holder 灯头缩节
reducing lateral 缩径支管;缩接(分)支管
reducing lens 简化透镜;缩短管筒透镜
reducing machine 磨碎机
reducing mechanism 减速机构
reducing medium 还原介质
reducing metal 还原金属
reducing method 缩分方法
reducing mill 粉碎磨(机);减径机
reducing nipple 异径(螺纹)管接头;渐缩联支管;缩接连接管;变内径内螺纹接头
reducing nozzle 异径喷嘴;渐缩喷嘴

reducing number 缩分次数
reducing of crude oil 石油直馏
reducing of flood 减少行洪量;减洪;洪水消落;洪水位下降;洪水退落
reducing of sample 样品缩分
reducing pass 开坯孔型;减径孔型
reducing piece 异径接头;异径接管;渐缩套管;渐缩管箍;缩小接管;缩小管接头
reducing pipe 渐缩管;变径管
reducing pipe elbow 异径弯头
reducing pipe fitting 异径管配件;渐缩管配件
reducing pipe joint 异径管节;大小头(承插管接头);变径管节
reducing piping 渐缩管
reducing plane table 简化平板仪
reducing potential 还原电位;还原电势
reducing power 消色力;减色能力;还原能力;还原本领
reducing press 缩口用压力机
reducing pretreatment 预还原处理
reducing printer 缩印仪;缩像仪
reducing process 还原过程
reducing quantity 减量化
reducing resister 减流电阻器
reducing roast 还原焙烧
reducing roll 开坯轧辊
reducing screen 减光屏
reducing sizing mill 减径—定径机
reducing sleeve 转接套;缩径衬套;变径套
reducing socket 缩径套节;异径套节;异径管接头;异径承插管;缩径承窝;大小头承插管接头;变径管接头
reducing solution 还原性溶液
reducing solvent 稀释溶剂
reducing space 还原空间
reducing still 轻质油蒸馏锅
reducing substance 还原性物质
reducing surface 简约表面
reducing tacheometer 简化测速仪
reducing taper elbow 渐缩斜弯头
reducing tee[T] 缩径丁字管节;异径三通(接头);渐缩三通管;缩径三通(管);大小头三通;变径三通;异径T形接头
reducing test 缩口试验
reducing theodolite 简化经纬仪
reducing the row-spacing 缩短行距
reducing to scale 按比例尺归算
reducing transformer 降压变压器
reducing transitive grammar 还原传递文法
reducing tube 渐缩管;变径管
reducing tubing 渐缩管
reducing twin elbow 异径三通双弯接头
reducing value 还原值
reducing value of water-level 水位削减值
reducing valve 减压阀
reducing valve strain 减压活门滤器
reducing valve strainer 减压阀滤器
reducing waste 还原废料
reducing zone 还原带;还原层
reduct 小块料;边脚料
reductant 还原剂
reductant character 冗余字符
reductant constraint 冗余约束
reducted form of general game 简约型一般对策
reductible Markov chain 可减缩的马尔可夫链
reduction 修正;整理编辑;折扣;归正;归约;降低;简约;还原(作用);缩小;缩减;复位术
reduction activation 还原活化(作用)
reduction activator 还原活化剂
reduction algorithm 归约算法
reduction and exemption 减免
reduction and exemption of tax 税收减免
reduction ascending 向上折算
reduction at source 在源头减少污染
reduction-azo colo(u)rimetry 还原偶氮比色法
reduction-azo spectrophotometry 还原偶氮分光光度法
reduction baffle 还原隔板
reduction box 减速箱
reduction by carbon 用碳还原
reduction by continuous traction 持续牵引复位
reduction by impact 冲击破碎
reduction by lithium aluminium hydride 氢化铝锂还原

reduction by-product 还原副产品
reduction by pulling and kneading 拔伸捏正
reduction camera 缩小摄像机
reduction carrier 还原载体
reduction cascading 收缩级联
reduction case 减速器箱体
reduction cell 还原槽;电解还原槽
reduction cell sidewall 电解槽侧壁
reduction clear 稀释用印浆
reduction coefficient 转换系数;折算系数;折减系数;折合系数;折合率;归算系数;减少系数;换算系数;缩分系数
reduction coefficient of allowable stress 许用应力折减系数
reduction colo(u)r test 还原色试验
reduction compasses 比例规;缩比两脚规
reduction cone 减速锥;铁炮式粗砂机
reduction cracking 还原开裂
reduction crusher 碎减式破碎机;次轧碎石机;次轧破碎机;次碎机;碎石机
reduction curve 缩减曲线;衰减曲线
reduction descending 向下折算
reduction device 减速装置
reduction division 减数分裂
reduction efficiency 粉碎效率
reduction environment 还原环境
reduction environment of hydrogen sulfide containing 含硫化氢的还原环境
reduction environment of non-hydrogen sulfide 无硫化氢的还原环境
reduction factor 折减因素;折减因数;折减系数;降低因数;减摩因数;缩小系数;缩小复制图;缩减系数
reduction factor of dredge(r) productivity 挖泥船生产率折减系数
reduction factor of fissure water 裂隙水折减系数
reduction factor of pile group 群桩折减系数
reduction factor of rock modulus 岩石模量折减系数
reduction flame 还原焰
reduction for distance 距离缩小
reduction formula 折算公式;归约公式;换算公式;缩分公式
reduction for polar motion 极移减少
reduction for principal-distance 主距改正
reduction for the height of sighting point 照准点高程校正
reduction furnace 还原炉
reduction gear 减速装置;减速齿轮;齿轮减速机
reduction gear box 减速箱
reduction gear ratio 减速比;传动比
reduction gear train 减速齿轮系
reduction gear unit 齿轮减速设备
reduction Gothic (style) 缩减哥特式
reduction gyratory breaker 旋回破碎机
reduction gyratory crusher 碎减式旋回破碎机
reduction hub 齿轮减速的轮毂
reduction-in-alkality test 碱性下降试验(混凝土)
reduction in area (拉力试件的)截面缩小;断面减缩率;断面减少
reduction in a single pass 开路粉碎;单程粉碎
reduction in bulk 容续减小;体积缩减;体积减少
reduction-in-expansion test 胀缩试验
reduction in force 裁减人员
reduction in height 高度压下量
reduction in load intensity 荷载的强度组合
reduction in pass 道次压下量
reduction in rates 费率降低
reduction in rigidity 刚性减小
reduction-in-risk approach 减少危害法
reduction in stiffness 劲性减小
reduction in subgrade strength design 地基减低强度设计;路基减低强度设计
reduction in tariff 削减关税
reduction in thickness 厚度压下量
reduction in tooth thickness 齿厚减薄量
reduction in visibility 能见度减低
reductionism 简化论;简化法
reductionizer 气雾器;粉碎机
reduction jaw crusher 次碎颚式破碎机
reduction kinetics 还原动力学
reduction law 约化律
reduction lens 缩影镜头
reduction mechanism 还原机理

reduction method 氧化还原滴定法;还原法
reduction mill 冷轧板材轧机
reduction motor 单独驱动马达
reduction moving-bed reactor 动态床还原反应器
reduction nucleus
reduction of adhesion at small radius curve 小半径曲线黏降【铁】
reduction of adhesion on minimal radius curve 小半径曲线黏降【铁】
reduction of a fraction 约分
reduction of altitude to another place of observation 位置线移线订正
reduction of area 断面缩减率;截面缩小;截面收缩;面缩;面积缩小
reduction of area at fracture 断裂收缩;断裂颈缩;断口收缩率
reduction of articulation 清晰度降低系数
reduction of a transformation 变换的简约化
reduction of baseline 基线缩小
reduction of basin size 流域大小化算
reduction of bleeding 减少泌水(现象)
reduction of bulk 体积减缩
reduction of construction time 缩短工期
reduction of cross-section(al) area 断面收缩(率);断面减少量;截面(面积)缩小
reduction of data 信息简缩变换;信息简化;资料整理;资料折算;资料归算;资料归约;数据简化;数据换算;数据还原;数据归纳;数据处理
reduction of detection efficiency 检效率降低
reduction of dissolved oxygen 溶解氧的还原
reduction of drill hole 钻孔缩径
reduction of drying shrinkage 干缩折减
reduction of efficiency 效率减少
reduction of erection time 缩短工期
reduction of expenditures 节减经费
reduction of fiber bundle 纤维丛的约化
reduction of fractions to a common denominator 通分(母)
reduction of fused salts 熔盐电解还原
reduction of gradient 坡度折减
reduction of gravity value 重力值归算
reduction of heat 减热
reduction of image 像片减薄
reduction of iron ores 铁矿石还原
reduction of latitude 纬度归算;纬度订正
reduction of legal capital 法定资本的减少
reduction of levels 水准折算;高程折算;高程归算
reduction of moments 力矩减小
reduction of mortgage certificate 欠款余额证书
reduction of nitrogen in wastewater 废水中氮的减少
reduction of observation 观测订正
reduction of ore 矿石还原(法)
reduction of porosity 孔隙度降低
reduction of pressure head 压头减小
reduction of rates 降低税率
reduction of rhythmic(al) signal 科学时号归算
reduction of schedules 压缩分类表
reduction of service life 缩短使用寿命
reduction of singularity 奇点还原
reduction of soundings 水深值的改正;水深归算;测深归算;测深订正
reduction of spatial distances 球面距离改化
reduction of speed 降速
reduction of temperature to mean sea level 平均海平面温度订正
reduction of the microscope 显微镜分辨率
reduction of the normal equations 法方程组的解
reduction of tidal current 潮流资料整理
reduction of tide 潮汐(的)归算
reduction of time signals 时号归算
reduction of wages 克扣工资
reduction of wastewater flow 废水流量减少
reduction of wave 波浪减容
reduction operation 归约操作
reduction or cancellation of debts 减轻债务负担
reduction oxidant 还原氧化剂
reduction oxidation 还原氧化(作用);氧化还原
reduction-oxidation polymerization 氧化还原聚合
reduction-oxidation potential 氧化还原电位
reduction parameter 归算参数
reduction paste 稀释浆
reduction per area 每道次压下量
reduction period 还原期

reduction per pass 每道次压下量
reduction planning of pollutant at river outfall 河流排污口污染物削减量规划
reduction plant 还原车间
reduction potential 还原电势
reduction print 缩小复制;缩微像片
reduction printer 像片缩小仪;缩印机;缩小仪;缩微仪;缩图仪;缩片机
reduction printer lens 缩小仪镜头
reduction printing 缩印
reduction procedure 简约过程;简化方案
reduction process 还原过程
reduction program(me) 压缩程序
reduction range 压缩范围;破碎度;缩微幅度;缩微范围
reduction rate 转速比;减速比;还原率
reduction ratio 折缩比;磨碎程度;还原比;破碎比;缩小比(例);缩微率;缩减比(例);冲淡比;粉碎度;粉碎(比)率
reduction reaction 还原反应
reduction reactor 还原反应器
reduction relay valve 减压中继阀
reduction roaster 还原焙烧炉
reduction roasting 还原焙烧
reduction roll 压延轧辊;轧缩辊;弯形辊
reduction roller mill 辊轧机
reduction rule 简约规则
reduction run 还原操作
reduction sailing 减速航行
reduction scale 缩微率;缩尺
reduction sintering 还原烧结
reduction sleeve 异径连接套
reduction stage 磨碎段
reduction stand 压轧机架
reduction state 还原态
reduction step 破碎步骤;粉碎步骤
reduction stone crusher 石料破碎机
reduction subsystem 数据压缩子系统
reduction table 归算表;高度方位表
reduction tacheometer 归算视距仪;自动归约测量仪
reduction technique 简化技术
reduction test 还原试验
reduction theory 约化理论
reduction thickening 稀释糊
reduction to average sea level 平均海平面订正
reduction to center 归心(计算)
reduction to centre correction 归心校正;归心改正
reduction to chord length 弦长归算
reduction to common denominator 通分
reduction to ellipsoid 椭球面归算
reduction to equator 赤道归算
reduction to flux 直蒸馏到沥青
reduction to group mean 组内改正
reduction to half-frame area 缩摄一半
reduction to mean sea level 平均海平面订正
reduction to sea level 折合为海平面值;海平面归算;海平面订正
reduction to standard condition 归算至标准条件
reduction to station center 测站归心
reduction to target centre 照准点归心
reduction to the centre of signal 照准点归心校正
reduction to the centre of station 测站归心校正
reduction to the equator 赤道订正
reduction to the horizon 倾斜改正
reduction to the meridian 子午订正;中天高度修正
reduction to the pole 化到地磁极
reduction to the sun 太阳订正
reduction transformer 降压变压器
reduction value 还原值
reduction valve 减压阀;减速阀
reduction voltage 降压
reduction welding 还原焊接
reduction with carbon 用碳还原
reduction with equal marginal cost 等边际费用削减
reduction with equal proportion 等比例削减
reduction work 粉磨功
reduction zone 还原作用带
reductive 还原的
reductive agent 还原剂
reductive algebra 约化代数
reductive ammonolysis 还原性氨分解
reductive dechlorination 还原脱氯

reductive dehalogenation 还原性脱卤(作用);还原脱卤化
reductive desulfuration 还原性脱硫(作用)
reductive firing 还原烧成
reductive homogeneous space 约化齐性空间
reductive operator 约化算子
reductive ozonolysis 臭氧物还原分解
reductive pair 约化偶
reductive rate of injuries and deaths 伤亡事故降低率
reductive smog 还原性烟雾
reductive sulfonation 还原性硫化(作用)
reductometric titration 还原滴定法
reductometry 还原滴定法
reductometry solution 参比溶液
reductor 变径管;缩位仪;减压器;减速器;归正仪;减速机;还原器;电压表附加电阻;复位器
red ultramarine 带红深蓝色;红群青
redundance 冗余(度);冗余性;剩余度;多余(信息);多余度;超静定(性)
redundance bit 冗余位
redundance check 冗余检验
redundance code 冗余码
redundance encoding 冗余编码
redundance selection 余度选定
redundance unit 冗余设备
redundancy = redundance
redundancy automaton 冗余自动机
redundancy bit 冗余位
redundancy character 冗余字符
redundancy check 可靠性余度校验;冗余校验;冗余检验;剩余校验
redundancy design 冗余设计
redundancy determination 多余测定
redundancy digit 冗余位
redundancy of attribute 多余属性
redundancy of information 信息剩余度
redundancy payment 冗余人员解雇费;编余费
redundancy rate 冗余率
redundancy reduction 冗余度压缩
redundant 超静定杆件;多余的;超静定的
redundant analysis 冗余分析
redundant autopilot 余度自动驾驶仪
redundant bar 多余杆件
redundant bit 冗余位
redundant capacity 过剩生产能力
redundant channel 冗余通道
redundant check bit 冗余检查位
redundant check(ing) 冗余校验;冗余检验;多余位数检验
redundant circuit 备用电路
redundant code 冗余码;多余电码
redundant coding 冗余编码
redundant constraint 赘余约束;多余约束;多余的约束条件
redundant data 冗余数据;多余数据
redundant digit 冗余位
redundant displacement 多余位移
redundant equation 冗余方程;多余方程
redundant equipment 多余设备
redundant fault 冗余故障
redundant force 多余(未知)力;赘余力
redundant frame 赘余构架;超稳定框架;超定框架;超静定构架
redundant in determination 多余观测
redundant indexing 冗余标引
redundant information 冗余信息
redundant Maitra cascade 多余梅特拉串级连接
redundant member 冗余杆;多余支撑杆;超静定杆件;赘余杆件;赘杆;多余杆件
redundant mesh 附加回路
redundant moment 多余力矩;冗余力矩
redundant node 冗余节点
redundant number 冗余数
redundant observation 多余观测
redundant operation elimination 冗余运算消除
redundant phase record 冗余相位记录
redundant plant 多余设备
redundant quantity 超静定量
redundant reaction 赘余反力;超静定反力
redundant rod 多余杆件
redundant state 冗余状态
redundant structure 静不定结构;赘余结构;超稳定结构;超静定结构

redundant support 超静定支承;赘余支承;多余支座
redundant symbol 冗余符号
redundant tie 冗余系杆
redundant unknown 多余未知数
reduplicate 再复制;重复
reduplication 重复
reduster 再除尘器
Redux binder 雷杜克斯粘合料;雷杜克斯粘合剂
Redux bonding 雷杜克斯胶接法
reduzate 还原沉积物
red violet monoclinic crystal 红紫色单斜晶体
red vitriol 赤矾
red water 赤潮;红水;红潮;铁锈水
red water bloom 红水花;赤潮
red water fever 特克萨斯热
red water trouble 铁锈水污浊;红水问题
red wax mesh for retention 固位用红蜡网
red white compound jewel 红白人造宝石
red willow 红柳(木);平滑柳
redwood 欧洲赤松;红杉[植];红木
redwood bark 红木树皮(绝缘层材料)
redwood carving 红木刻
redwood fiber 红木纤维
redwood furniture 红木家具
redwood grade 红木等级;红杉等级
Redwood orifice viscosimeter 雷德伍德锐孔黏度计(又称雷氏锐孔黏度计)
redwood plywood 红杉胶合板
redwood plywood standard 红杉胶合板标准;红木胶合板标准
Redwood second 雷德伍德黏度计流出秒数
Redwood visco(si)meter 雷德伍德黏度计
Redwood viscosity 雷德伍德黏度
redye 复染
red yellow podzolic soil 灰化红黄壤;红黄色灰化土
red yellow soil 红黄色的层化土壤
red zinc chromate primer 红色锌铬打底漆
reecho 回声反射;再回音;再反射;回声的回响
reechoed wave 重回声波
reed 钢筘;钢箍;芦苇;箝;簧片;舌簧
reed and bead 芦珠饰【建】
reed and harness polishing machine 总筘抛光机
reedback 凸嵌线脚背(椅子)
reed basket 小物品打捞篮(钻孔内);小物件打捞篮(钻孔内)
reed baulk 扎筘杆
reed beach 芦苇滩
reed bed 苇地
reed bed dewatering and treatment system 针床脱水处理系统
reed bed treatment system 针床处理系统
reed blade 舌簧叶片
reed board 芦苇板
reed bog 芦苇沼泽;芦苇酸沼
reed brass 簧片黄铜
reed clip 簧片式夹紧装置
reed comparator 振簧式比较仪
reed contact 簧片触点
reed control system 衔铁控制装置
reed cutter 割苇机
reed dent 筘齿
reeded 芦苇盖的;平行的半圆波纹装饰
reeded glass 有肋片和凸纹的毛玻璃;小凸嵌线玻璃;瓦楞玻璃;槽纹玻璃;芦苇形玻璃
reeded pattern 横纹图案
reeded roof 芦苇屋顶
reeded tile 防滑踏步砖
reed field 芦苇塘
reed fog horn 舌簧雾笛
reed fog signal 低音雾号
reed frequency 簧片振动频率
reed frequency meter 振动簧片式频率计;簧片式频率计
reed ga(u)ge 筘齿隔距;舌簧频率计
reed glass 芦苇形玻璃
reed horn 簧喇叭
reed indicator 簧振指示器;簧片指示器
reeding 芦苇束状线脚;小凸缘线脚;褶皱;凸缘;防滑条
reeding machine 穿簧机
reeding tool 刮平刀
Reed inkometer 里德油墨计
re-editing 再编辑
reed lath(ing) 芦苇板条

reed lathing plaster 芦苇板条灰泥
reed loudspeaker 磁簧式扬声器
reed mace 宽叶香蒲;宽芽香蒲
reed-making machine 扎筘机
reed marsh 芦苇草沼;芦苇荡
reed mat 芦席;苇席
reed mat factory 芦苇垫层厂
reed mat pattern 芦席纹图形
reedmergnerite 硅硼钠石;硼钠长石
reed method 簧片振动法
reed mo(u)ld 小凸嵌线饰;小凸嵌线脚
reed nail 筘钉
reed of buzzer 蜂鸣器簧片
reed panel 芦苇板
reed peat 芦苇泥炭
reed pipe 簧管
reed planting 芦苇种植
reed pond 苇塘
reed relay 簧式继电器;簧片继电器;舌簧(式)继电器;笛簧继电器
Reed roller bit 里德牙轮钻头
reed roof 茅草屋顶;芦苇屋顶
reeds 小凸嵌线饰;钢坯梳状裂纹;梳状裂纹
reed sand-break 苇巴子沙障
reed screen with flower 草花苇帘
reed sheet 芦苇薄板
reed signal 簧片音响雾号
reed slab wall 芦苇加筋土墙
reed spike 宽叶香蒲
reed spring 笛片式簧
reed swamp 芦苇沼泽(地);芦苇渍水沼泽;芦苇林沼
reed switch 簧片开关;舌簧接点元件;笛簧接点开关
reed tachometer 簧片式转速表
reed thatch 芦苇屋面
reed-type comparator 振簧比较仪;扭簧比较仪
reed-type glass 槽形毛玻璃
reed-type suspension valve 片簧式悬置阀
reed valve 针阀;簧片阀
reed vegetation 芦苇植被
reed wax 针茅蜡
reed width 筘幅
reedy 芦苇丛生的
reedy marshes 芦苇丛生的沼泽地
reef 礁;环礁;砂矿底岩
reef a sail 缩帆
reef atoll 礁脊
reef band 礁带
reef bank 浅滩珊瑚礁
reef barrier 堡礁(指与海岸平行的珊瑚礁)
reef bin 矿石仓
reef blasting 炸礁
reef blasting ship 炸礁船
reef blasting under ice cover 冰下炸礁
reef blasting underwater 水下炸礁
reef blasting with enclosure 围堵炸礁
reef breccia 礁角砾岩
reef-building coral 造礁珊瑚
reef canal 礁沟
reef cap 礁盖
reef cape 礁帽
reef cave 礁洞
reef cluster 礁丛
reef complex 杂礁灰岩;礁组合
reef conglomerate 礁砾岩
reef core 礁核
reef core deposit 礁心沉积
reef crest 礁脊
reef-crest deposit 礁脊沉积
reef debris 礁岩屑
reef drift 脉内平巷
reef drive 砂矿底岩中开凿的巷道
reef edge 礁缘
reefer 冷藏车;冷藏员;冷藏室;冷藏(货)船
reefer cargo 冷藏货
reefer cargo carrier 冷藏船
reefer chamber 冷藏柜
reefer connection 冷气管线接头
reefer container 冷藏集装箱;冷藏货柜(集装箱)
reefer container management system 冷藏集装箱管理系统
reefer ship 冷藏船
reefer space 冷藏货舱位
reef explosion 炸礁

reef facies 礁相
reef farthest from a bank 自岸伸出最远的礁
reef flank 礁翼
reef flat 礁坪;浅石滩
reef flat deposit 礁坪沉积
reef-forming algae 造礁藻类
reef frame 礁架
reef front 礁前
reef front deposit 礁前沉积
reef-front slope deposit 礁前斜坡沉积
reef-front talus deposit 礁前塌新沉积
reef front terrace 礁前阶地
reef gold 石英脉金
reef hook 刮缝钩
reef-induced traffic-hazard 礁石险滩
reefing 缩帆
reefing becket 帆眼绳
reefing float 栓定浮筏
reefing iron 刮缝凿;刮缝刀
reef island 珊瑚岛
reef island karst 礁岛岩溶
reef island power supply equipment 岛礁供电设备
reef knoll 圆礁丘
reef knot 方结;平结;缩帆结
reef lake 礁湖
reef lake deposit 礁湖沉积
reef level karst 礁坪岩溶
reef limestone 礁灰岩
reef limestone reservoir 礁灰岩储集层
reef line 缩帆索
reefoid rock 类礁岩石
reef patch 补丁礁
reef picking 手拣金砾
reef pinnacle 尖礁
reef pool 礁塘
reef proper 礁本部
reefs 暗礁群
reef segment 礁块
reef slope 礁坡
reef stone 礁石
reef tackle 缩帆绞辘
reef talus 礁砾岩
reef type oil-gas field 礁型油气田
reef type pool 礁型油藏
reefy ground 多暗礁的海底
reefy outer edge bar 礁外缘坝
reek 蒸气;冒烟;薄雾
reel 线轴;均整;卷轴;卷(线)车;卷筒;绞车;绕线架;绕线管;盘架;丝框;绳车;纱框;拨禾轮
reelability 可绕性
reelable 可绕的;可卷的
reel alternation 带盘交换
reel and bead 串珠卷饰线脚;珠饰
reel and bead enrichment 卷带串珠装饰
reel-and-cable locomotive 电缆式电机车
reel-and-cable trolley locomotive 电缆架线两用电机车
reel block 卷取机卷筒
reel brake 盘盘制动器
reel carriage 电缆盘拖车;凸圆线脚;电缆车
reel cart 绞车
reel case 卷带盒
reel chain 长链条
reel control mechanism 带盘控制机构
reel core oven 立式回转烘芯炉
reel crab 卷扬机;卷筒绞车;绞车
reel cutter 转轴切砖器;转筒切坯机;钢丝切泥坯器
reel drier 卷筒热风烘煤机
reel drive 绕线盘传动装置
re-electrolysis 再电解;二次电解
reeled off 退绕
reeled riveting 之字铆接;交错铆接;曲折形铆钉
reeler 均整机(管材);卷取机;卷盘机;矫正机;拆卷机
reel fed automatic sack applicator 袋卷自动插袋器
reel gripper jaws 卷取机卷筒夹紧爪
reel hoist 绞轮提升机
reel horn 雾笛
reeling 矫直
reeling condition 卷绕状态
reeling furnace 卷取炉
reeling hook 捻缝钩
reeling machine 整径机(管材);整经机;均整机
(管材);卷取机;绕线机;绕丝机
reeling mill 制丝厂;整经机;均整机(管材)
reeling silk 摇绞
reel label 带卷标签
reel lift 拨禾轮升降机构
reel lifting tackle 升举卷带的滑车
reel locomotive 带钢绳卷盘电机车
reel machine 卷纸机
reel motor 卷带电动机;带盘电动机
reel number (钢丝绳、电线等的)卷号;盘号;带号
reel off 绕出;抽出
reel-off gear 松卷装置
reel-off stand 退卷机
reel of overlapping sacks 叠接纸袋卷盘
reel of sacks 纸袋卷盘
reel oven 转炉
reel pipe 卷取导管;盘条卷取导管
reel purchase 手摇起重绞车
reel slitter 卷筒分切机
reel spindle 供带轴
reel stand 纱框架
reel stop 抗剥离件
reel stopper 带卷盘制动器
reel support 拨禾轮支架
reel suspension 卷索(悬挂法)
reel swap 带盘交换
reel tape 卷尺;纺车式卷尺
reel tension regulator 卷取机张力调节器
reel-to-reel 卷盘到卷盘
reel-twist crimping 扭转加捻卷曲
reel-type cutting unit 滚筒式切碎装置
reel-type feed mechanism 卷筒式进料机构
reel-type reader 带盘型输入机
reel-type relay 旋转式割草机
reel-type sack magazine 卷盘型集袋器
reel up 绕成;绕起
reel welded tube 卷焊管
reel winder 绞轮提升机
Reema 瑞玛建筑体系
reeming beetle 木板捻缝锤
re-emission 二次辐射;二次发射;再发射
re-emitted radiation 重发辐射
re-employment service center 再就业服务中心
reemulsification 再乳化(作用)
reemulsified 再乳化的
reenabiling 重新开放;重新赋能
reenergize 重新激励;重供能
reenforce 重新实施
re-enforcing development effort 强化发展计划(工作);大力发展
re-engage 重新啮合;重新接入
re-engagement 再联锁;再次起动;再次接入
reengined 换过发动机
re-engineering 再设计;重建
reenrichment 再浓缩;更浓缩
reenterabilty 可重入性
reenterable 可重新进入的
reenterable attribute 可重入属性
reenterable load module 可重入装配模块
reenterable module 可重入模块
reenterable program(me) 可重用程序;可重入程序
reenterable routine 可重入(例行)程序
reenterable subroutine 可重入子程序
re-entering 再返;重新输入
reentering abutment 凹进桥台
reentering angle 入射角;凹入角;重入角
re-entering the original data 重新输入原始数据
reentrainment 二次扬尘;再夹带;粉尘二次飞扬
reentrance 重入
reentrant 内曲(的);凹入的;再进入的;可再入;凹腔;凹港
reentrant acousto-optic light modulator 凹腔声-光学光调制器
reentrant angle 缺陷角;重入角;凹入角;凹角
reentrant beam 重入注
reentrant beam crossedfield amplifier 重入型正交场放大管
reentrant bushing 缩短引入线;缩短套管
reentrant cathode 凹形阴极
reentrant cavity 再入式谐振腔;重入腔;凹状空腔
reentrant cavity oscillator 凹状振荡器
reentrant cavity resonator 凹型空腔谐振器
reentrant circuit 重入式电路
reentrant code 重入(代)码
reentrant corner 内角;内隅角;凹角;阴角
reentrant deflector 凹腔偏转器
reentrant gas cooling 再入式气体冷却
reentrant horn 折叠式喇叭
reentrant inlet 凹入(式)进水口
reentrant intake 凹入式进水口
reentrant loop 再入式回路
reentrant mold 侧凹模
reentrant molding 凹穴造型
reentrant orifice 内狭泄水孔;内缩排水孔;凹入式泄水孔;凹入孔口
reentrant part 内角;凹角
reentrant pipe 凹入管
reentrant polygon 凹多边形
reentrant program(me) 可重入程序;重入程序
reentrant programming 重入程序设计
reentrant resonator 凹腔谐振器
reentrant routine 可重入(例行)程序;重入程序
reentrant shift register 重入式移位寄存器
reentrant slow-wave structure 重入式慢波结构
reentrant subroutine 重入子例;重入子程序
reentrant syncline 伸入向斜
reentrant tube 凹入式接管;凹入管
reentrant-type frequency meter 凹腔式频率计;半同轴频率计
reentrant winding 闭路绕组;闭合绕组
re-entry 收回权;再记入;收回租借;重入;重返(大气层)
re-entry ablation 再入烧蚀
re-entry aid 再入辅助设备
re-entry angle 再入角
re-entry arc length fuze 再入弧长引信
re-entry body 再入体
re-entry capsule 回收密封暗盒
re-entry communication 重返大气通信
re-entry condition 再入条件
re-entry control 再入控制
re-entry decoy 再入假目标
re-entry drilling 再入钻探
re-entry effect 折回效应
re-entry environment 再入环境
re-entry environment and system technology 再入环境与系统工艺
re-entry module 回返舱
re-entry nose cone 再入头锥
re-entry observable prediction and experiments 再入观察预报与试验
re-entry on leased land 回收已批租土地
re-entry phenomenon 重返现象
re-entry point 再入点;重入点
re-entry positioning 再入定位
re-entry recovery system 再入回
re-entry system 重入系统;重入导流;重返系统
re-entry system environmental protection 再入系统环境防护
re-entry test facility 再入大气层试验设备
re-entry trajectory 再入轨道
re-entry turbine 回流式汽轮机;回流式涡轮机
re-entry velocity 再入速度
re-entry window 再入通过区
re-equip 重新装备
re-equipment 改装;重新装备;再新设备
re-equipped ship 改装船
re-equipping a house 增设住宅辅助设施
re-erect 再立;重建
reerection 重新组装
re-establish communication 重建通信
re-establishment 重建
re-establishment of livelihood 重建家园
re-establishment of natural flood 天然径流还原;天然洪水还原
re-establishment of station 重新设站
reesterification 再酯化(作用)
re-evacuate 再抽空
re-evacuation 再汲出;再抽空
reevaluate the remaining assets 作价剩余资产
reevaluation 重估计
reevaluation of capital 资金重估
re-evapo(u)ration 再蒸发;再汽化
reeve 用绳穿过;缚紧;穿索;绳穿入孔
reeved fall (复滑轮中的)缆绳
reeverse wound geophone 反绕地震检波器
reevesite 陨菱铁镍矿;水碳铁镍矿
reeving 绕线;(滑轮中)穿绳;支索;绳索绕缠法

reeving beetle 填絮锤
reeving line bend 互穿两个半结
reeving of pulley block 滑车组倍率;滑车(轮)组倍率
reeving of ropes 钢丝绳缠绕
reeving system 绳索绕缠系统;穿索方式
reeving thimble 穿绳套环
reexamination 复试;补充试验
reexamination of anomalies 异常检查
reexamination of feedback devices 反馈方法的重验
reexamine 再检验;再检查
reexchange 要求退汇的金额;再交换;退汇(要求)
reexchange bill 退汇汇票
reexecution 重新执行
re-expand 再膨胀
re-expansion 再膨胀;二次膨胀(使用时膨胀);重复膨胀
re-expedition 再装运
re-experiment 再试验
reexport 再输出;复出口
re-exportation 再出口;复出口
reexports of imported goods 进口商品的再出口
re-expose 二次曝光
reextraction 再抽提;反萃取
reextractor 反萃取器
refable 圣坛后高架
refabricate 再制(备)
refabrication 再制造;再制(备)
refabrication plant 再加工装置
reface 整修表面;再车削
reface clutch disc 更换离合器片
refacer 表面修整器;整修工(具);光面器
refacing 修整表面;整修门面;重磨;表面重磨;整修墙面
refacing a valve 研磨阀面
refdanskite 水硅镍矿
refectorium (教堂,神学院的)食堂
refectory 食堂(又称职工饭堂);(寺院神学院中的)餐厅;(指寺院或学院中者的)饭厅
refectory for lay brethren 杂役僧侣食堂(寺院中);平信徒食堂(寺院中)
refeed 重新喂料
refeed line 返料管
referable 可涉及的;可交付的
referance leak 标准漏孔
referee (受法庭委托的)鉴定员;公断人;裁判员
referee in bankruptcy 破产公断人
referee in case of need 需要时进行公断者
referee method 仲裁法
referee method test 仲裁法试验
referee test 仲裁试验
referee whistle 裁判哨子
reference 引示波;证明书;基准;起始位置;参照;参考资料;参考物;参考文献;参考书目(录);参考电源;参比;查询人;保证人
reference acoustic(al) pressure 基准声压
reference address 转换地址;基本地址;参考地址
reference address zero point 基地址【计】
reference adjustment 基准调整;参考值调整;参考调整
reference amplifier 参考放大器;标准放大器
reference anchor 基准锚头
reference angle 基准角;基准方位角;入射角;参考角
reference arc voltage 基准弧压
reference area 起始面;参考面积;参考面;比较面积
reference arm 基准臂;参考臂
reference atlas 参考地图集
reference audio level 声频基准磁平
reference axis 相关轴;坐标轴;组构轴;计算轴;基准轴(线);基准线;参照轴;参考轴(线);参考线
reference azimuth 原始方位;基准方位角;基准方位;参考方位(角)
reference bank 参考银行
reference base period 标准基期
reference basin 参证流域
reference beam 基准梁;备用梁;基准光束;参考光束
reference beam angle 参考光束角
reference beam method 参考光束法
reference beam polarization 参考光束极振
reference bearing 基准方位;参考轴承;参考方位角
reference bias 基准偏磁
reference bit 参考位;访问位
reference block 标定参考块;基准块;对比试块;参

考试块;参考程序块;标准块;比较试块
reference body 基准物件
reference book 工具书;参考用工具书;参考书
reference burst 基准副载波群;色同步信号
reference bushing 参照漏板;标准漏板
reference capacitor 标准电容器
reference capillary 参比毛细管
reference carrier 基准载波;参考载波
reference carrier burst 基准载波脉冲群
reference cavity 基准空腔谐振器;标准谐振腔;标准空腔谐振器
reference cell 参考试池;参比电池;参比池
reference cell cavity 参比池腔
reference center distance 标准中心距
reference central meridian 基准中央子午线
reference chart folios 参考海图册
reference check button 基准检查按钮
reference chromatogram 参比色谱(图)
reference circle 参照圆;参考圆
reference circuit 基准电路;参考圆
reference class 参考类
reference clause 仲裁条款
reference climatological station 基准气候站
reference clock 基准时钟;参考时钟
reference coefficient 参照系数
reference colo(u)r 参考色;标准色
reference column 定位柱;参考柱;参比柱;参比电柱;比较柱;备注栏;备考栏
reference compound 参比物
reference computer 基准计算机
reference condition 参考条件;参比条件
reference cone 参考圆锥;分锥
reference cone angle 分锥角
reference configuration 参考配置
reference control amplifier 基准控制放大器
reference coordinate 基准坐标系;基准坐标
reference coordinate axis 参考坐标轴
reference coordinates 参考坐标
reference coordinate system 参照坐标系;参考坐标系
reference corner 参考角
reference count 引用计数;检验读数;基准计数;参考计数;参考读数
reference counter 引用计数器
reference coupler 基准耦合腔
reference craft 基准飞机
reference culture 标准菌株
reference currencies 参考货币
reference currency 参考通货
reference current meter 校验流速仪;基准流速仪;标准流速仪
reference current metre 标准流速仪
reference cycle 参照周期;参考周期
reference cylinder 参考圆柱
reference damping 参考阻尼
reference data 基准数据;参考数据;参考数据;参比数据
reference data book 参考资料手册
reference data for anchoring 锚地参考资料
reference date 标准
reference datum 基准零点;参考基准(面);参考基(准)面
reference debugging aids 调试辅助程序
reference defect 参考缺陷
reference delay frame 基准延时帧
reference design 参考设计
reference dew point 参考露点
reference diagram 参考图
reference dimension 基准尺寸;参考尺寸
reference diode 参考二极管
reference dipole 基准偶极子
reference direction 基准方向;参照方向;参考方向
reference disc 校对盘;标准圆盘量规
reference distance 参考距离
reference dose 参考剂量
reference drawing 参考图
reference drawing number 参考图号
reference edge 基准缘;基准边;参考边缘;参考边
reference electrode 参考电极;参比电极
reference electrode filling solution 参比电极充填液
reference element 基准元件;基准要素;基准成分;参考元素;参比元素
reference ellipsoid 参照椭球(体);参考椭球(体)
reference equivalent 基准等效值;基准当量;参考

等效值;参考当量
reference error ratio 参考差错率
reference external symbol 外部参考符号
reference eyepiece 参考目镜
reference feedback 起始回授;起始反馈
reference filter 标准滤光片
reference flowchart 参考流程
reference form 参考式
reference format 基准格式;标准格式
reference frame 空间坐标;计算系统;基准标架;参照系;参照构架;参考坐标系;参考系(统);参考架;参考构架;参考轴网;参考标架
reference frame center 框架中心
reference frequency 基准频率;参考频率
reference frequency multiplier 基准倍频器;基频倍加器
reference fuel 参考燃料;标准燃料
reference gas 基准煤气;基准(燃)气
reference ga(u)ge 考证规;校检计;检验计;基准水位尺;基准标尺(校正用);参考水尺;参考量规;标准水尺;标准量规
reference generator 基准信号发生器;参考信号发生器
reference glass 参比玻璃
reference grid 坐标网格;基准栅极;参考坐标网;参考栅板;参考网板;标格
reference group 参考群体
reference half-cell 基准半电池
reference harbo(u)r 参考港
reference height 参考标高
reference helix 分度圆螺旋线
reference holder 参比器
reference horizon 对比层;标准层
reference illumination 标准照度
reference image height 参考帧高
reference image plane 基准像面
reference index 参考指标
reference input 基准输入;参考输入;额定输入;标准输入
reference input variable 基准输入变量;参考输入变量
reference instruction 基准指令;参考指令
reference instrument 参考仪器;参比仪表;标准仪器;标准仪表
reference ion 比较离子
reference issue 特辑
reference junction 支撑焊接处;基准结;参考结
reference junction compensation 冷端补偿
reference keyway 参考键槽
reference kilowatt hourmeter 标准电度表
reference lamina 标志层(位)
reference language 参考语言
reference layer 参考层
reference length 基准长度;参考长度
reference letter 参考字母
reference level 引用层;假定水位;基准水准面;基准水平面;基准级;基准电平;海图基准(面);参照级;参考(水准)面;参考水位;参考级;参考电平;参考层;参考标高;参考水平;标准电平
reference level species 参比水平物种
reference library 参考图书馆
reference light 参考光
reference light-water reactor 标准轻水反应堆
reference line 零位线;基线;起读线;参证线;参照线;参考线
reference list 引用表
reference listing 对照表;参考列表;备用号码表
reference load calibration machine 基准测力机;标准测力机
reference lobe 基准瓣
reference location 基准位置
reference magnet 参照磁体
reference manual 参考手册
reference map 参考图
reference mark 假定水准点;控制点;基准标志;基准标记;基准(标)点;基线标志;起点读数;参证点;参照符号;参考刻度;参考标志;参考标记;参考点;标准点;标记刻度;标高刻度
reference mark on observation hole 观测孔基准点
reference mark on observation well 观测井基准点
reference master ga(u)ge 校对规
reference material 参考资料;参考材料;参比物质;标准物质;参考物质
reference material temperature 参比物温度

reference measurement 基准测量
reference member 参考部分;参考部件
reference men and women 参考方位
reference meridian 基准子午线
reference method 参考方法
reference method of analysis 参照分析方法
reference mirror 后视镜;参考反射镜
reference model 基准模型;参考模型
reference modulation 基准调制
reference monitor 标准监测器;标准监测机
reference monument 参考标石
reference net 基准网;参考网
reference node 参考结点
reference noise 基准噪声;参考噪声
reference number 引得号数;基准数;文号;参考（号）数
reference object 基准目标
reference object ratio 参物比
reference offset 基准偏置
reference of reference 基准的基准
reference operator 标准算符
reference orbit 参考轨道
reference order 转接指令;参考指令
reference orifice 标准孔口
reference oscillator 基准振荡器;参考振荡器
reference oscillator muting 基准振荡器抑制
reference oscillator phase 振荡器基准相位
reference page 参考页
reference pair method 双点法
reference pairs 参考像对
reference parallel 基准纬度
reference path 参考光路
reference peak location 参比峰位;参比单位
reference peg 参考标桩;基准桩;参照桩【测】;参考桩【测】
reference performance 参考性能
reference period 对照期;参考期
reference person 参考方位
reference phase 基准相;参考相位
reference-phase information 基相信息
reference piece 参考试件
reference plane 基准平面;基准面;参证面;参考平面;参考面
reference plane of rectification 纠正自由度
reference plate 基准板
reference point 控制点;基准点;起算点;水准点;测量参考点;参证点;参照点;参考点;标准点
reference point method 共同点法
reference point of picture 相片参考点
reference point source 参考点光源
reference position 原始位置;基准位置;起始位置;参考位置
reference potential 基准电势;参考电位;标准电位
reference power 基准功率;参考功率;标准功率
reference power meter 标准功率计
reference power supply 基准动力源;基准电源;参考电源
reference preparation 参考制品
reference pressure 基准压力;参考压强
reference price 参考价格
reference price mechanism 参考价格制度
reference price system 参考价格制度
reference prism 参考棱镜
reference profile 基准齿廓;参考剖面
reference program(me) 引用程序
reference program(me) table 引用程序表
reference projection 参照延伸
reference pulse 基准脉冲;参考脉冲;标志脉冲
reference pyrometer 基准高温计
reference quantity 基准量;参考量;标准量
reference radiation 基准辐射;参考辐射
reference radiation level 参考辐射水平
reference radiographs 参考底片
reference range 参考距离
reference range marker 参考距离标记器
reference rate 参考汇率
reference rated output 标准额定产量
reference rate zone 参考汇率区域
reference ray 参考光(线)
reference reading 参考读数
reference reagent 参考试剂
reference receiver 基准受话器;基准接收机;标准接收机
reference record 引用记录;参数记录;参考记录;编辑记录
reference recording 监视录声;基准记录;备查录音
reference reflector of Michelson 迈克尔逊基准反射器
reference relation 参照关系
reference rod 标杆;参考标杆;测杆
reference room 资料室;参考书室
reference rounds 标准弹
reference sample 对比样品;参考样品;参考样本;参考试样;参比样品
reference scale 基准标度;参照标尺;参考标尺;参考比例
reference sealant 基准密封膏
reference section 计算断面;基准断面;参考剖面;参考断面(图)
reference seismometer 参考检波器
reference sensitivity 基准灵敏度;参考灵敏度
reference share 标准犁铧
reference ship 基准船;设计船舶;参考船
reference signal 基准信号;参考信号
reference signal generator 标准信号发生器
reference signal level 基准级
reference sketch 参考草图
reference skid-resistant material 参考抗滑材料;标准抗滑材料
reference soil system 参考土体系
reference solution 参考溶液;参比溶液
reference sound 基准声
reference sound absorbing power 基准吸声力
reference sound intensity 基准声强;参考声强
reference sound level 基准声级
reference sound power 基准声功率;参考声功率
reference sound pressure 基准声压;参考声压
reference sound source 标准噪声(声)源
reference source 基准源;定标源;参考源
reference space grid 基准网格(空间)
reference species 参考物种
reference speed 基准速度;参考速度
reference sphere 基准球;参考球
reference spheroid 参考椭球(体)
reference stake 基准桩;基准柱;护桩【道】;参考桩【测】
reference standard 基准;参考标准;参比标准;标准规;比照标准
reference standard fuel 准燃料
reference star 参考星
reference state 参考态;参比态
reference state composition 参考态成分
reference state concentration 参考态浓度
reference state condition 参考态条件
reference state environment 参考态环境
reference state environmental condition 参考态环境条件
reference statement 参考语句
reference station 主潮港;基准台;基准检潮站;基准潮位站;参证站;参照站【测】;参考港
reference stiffness 基准劲度
reference strain 参考应变
reference stream 引用流
reference stress 基准应力;标准应力
reference substance test 参考物实验
reference supply 基准电源;参考电源
reference surface 基准面;参考面
reference surface center 基准面中心
reference surface deviation 基准面偏差
reference surface diameter 基准面直径
reference surface induction 基准磁平
reference surface induction for the sound track 声迹基准磁平
reference symbol 参考符号
reference system 基准制;参照系统;参考系;参考系(统)
reference system convertor 参照系转换器
reference table 换算表;参引表;参考表;备查表
reference table for thermocouples 热电偶参考表
reference tape 基准带;参考带;标准带
reference temperature 基准温度;对比温度;参照温度;参考温度;参比温度;标准温度
reference termination 参考终端负载;标准终端负载
reference test bar 测试杆;测试棒
reference test block 标准试块
reference test piece 参考试块;对比试块
reference test rod 测试杆;测试棒
reference tide station 基准潮位站
reference time 基准时间;参考时间;参考期
reference time scale 时间参考尺度;参考时标
reference to an element of structure 对结构元素的引用
reference to a structure or an array element 对结构元素或数组元素的引用
reference tone 基准音调
reference-to-object energy ratio 参考对物光能量比
reference toroid 分度超环面
reference to standard 引用标准;参考标准
reference to standard in regulation 在条例中引用标准
reference to storage 访问存储器
reference to symbol 引用符号
reference to variable 引用变量
reference track 参考道
reference transit station 中途测站;经纬仪参考测站
reference trigger pulse 基准触发脉冲
reference turning dates 参考性的转折日期
reference unit 参考装置
reference value 参考(价)值;基准值;参照值
reference value convertor 参考价值转换
reference variable 基准变量;基准变量;参考变数;参考变量
reference vector 参考向量
reference vehicle 基准飞行器
reference velocity 参考速度
reference video 基准视频
reference voltage 参考电压;标准电压
reference volume 基准声量;参考(音)量;参考卷
reference water 参考水
reference water ga(u)ge 辅助水尺
reference water level 基准水位
reference water table 参考水位
reference wave 基准波;参考波
reference wavefront 参考波前
reference wavelength 参考波长
reference wave polarization 参考偏振性
reference weight 参考重量
reference well 参数井
reference white 参考白(色);标准白色
reference white level 白色信号基准电平
reference works 参考著作
reference world market price 世界市场参考价格
reference zone 参证区域;参照区
referencing 核对位置;定位;参照
referent 涉及的对象
referential 参考用的
referential experience 可借鉴的经验
referential standard of water quality 参考水质标准
referential transparency 对透明性有关的
refer expression 引用表达式
referral information 供查询的信息
referral information system 信息查询系统
referral leads 推销线索
referred inertia 折算惯量
referred value 折算值
referring factor 折合系数
referring impedance 折算阻抗
referring object 参照符号;基准标点
referring of rotor quantity 转子量的折算
referring point 参考点;参照点
refer to 参看
refer to acceptor 请询问承兑人
refer to arbitration 提交仲裁
refer to drawer 请与出票人洽询;查问出票人;洽询出票人;查问开票人
refer to endorser 请询背书人
refer to reference 引用
refery test 仲裁试验
refetch logic 重取逻辑
refetch state 重取状态
refigure 重新塑造;重新描绘
refikite 海松酸石
refile 重锉;接力传送
refill 还土;再装填;再加料;再充填;再充水;再充满;回填
refill check ball 进口止回球
refiller 注水器;注入装置
refilling 再填(充);再充填;再充水;再充满;回填(的);还土;重新蓄水
refilling crankcase 向曲轴箱加满油
refilling of oil tank 油桶再装油
refilling of the brake 制动机再充气

refilling of trench 沟槽还土;回填(壕)沟;堑壕回填
refilling process 再充水过程;回填过程
refilling zone 回填区
refill of fuel 燃油再加注
refill opening 加油口
refilter 再过滤
refiltered oil 再生油;再度过滤的油
refiltration 再过滤
refinance 重筹资金;重新筹集资金
refinancing 发行新债券;再筹资金;重新筹集资金
refinancing mortgage 重筹资金抵押
refinancing of debt service payments 为偿付债务开支再筹资;为偿付债务筹集资金
refinancing of outstanding borrowings 为未偿还借款重新筹集的资金
refine 精炼;清理;细化
refined 精炼的
refined analysis 严密(的)分析;精细分析
refined antimony 精炼的锑
refined asphalt 细沥青;精制石油沥青;精制(地)沥青
refined bitumen 精制(地)沥青
refined bleached lac 脱蜡漂白紫胶
refined bolt 精制螺栓
refined Bouguer gravity value 精化布格重力值
refined calculation 精确计算
refined compost 精炼肥料
refined copper 精炼铜
refined design 优化设计
refined grade solvent naphtha 精制级溶剂石脑油
refined iron 精炼铁
refined kerosine 脱臭煤油
refined lake asphalt 精制湖沥青
refined lead 精炼铅;软铅
refined linseed oil 精制亚麻油
refined magnesium 精炼镁
refined manganese dioxide 活化锰
refined metal 精炼金属
refined molybdenum trioxide 精制三氧化钼
refined net 细网格;加细格网;加密网络
refined nutritious substances 精微
refined oil 精制油;精炼油
refined oil product 成品油
refined paraffin wax 精制石蜡
refined petroleum 精制石油
refined pig iron 再制生铁;精炼生铁
refined porcelain 细瓷
refined processing 精处理
refined salt 精制盐
refined salt (iodined) 加碘精制盐
refined sand 精砂
refined-smelting ladle furnace 钢包精炼炉
refined steel 精制钢;精炼钢
refined stock 已磨浆料
refined sugar 精制糖
refined tail oil 精炼的妥尔油
refined tar 精制焦油;精制柏油
refined white paraffin wax 精白蜡
refined zone 再结晶区
refinement 精炼;加细;提炼;提纯;细化
refinement of tariff classifications 税则的精细分类
refiner 均料机;精制机;精研机;精选机;精碎机;精磨机;精炼机;澄清带
refiner roll 精磨辊;精炼辊
refinery 炼油厂;炼制厂;精制厂;精炼厂;提炼厂
refinery and petrochemical plant 石油精炼和石油化工厂
refinery (asphaltic) bitumen 炼油厂沥青
refinery (asphaltic) bitumen of penetration-grade 膏体级精制石油沥青
refinery building program(me) 炼油厂建设计划
refinery catchment lake 炼油厂集污池
refinery cinder 精炼炉渣
refinery coke 石油焦
refinery connection 精炼厂接丝
refinery construction 炼油厂建筑;炼油厂结构
refinery effluent 炼油厂废水;精炼厂废水
refinery furnace 精炼炉
refinery gas 炼厂气;精炼气;精炼厂气体
refinery hydrocarbons 炼油烃类
refinery loading rack 炼油厂起重机
refinery output 提炼产品产量
refinery petroleum producls 石油加工产品
refinery pit 炼制槽;精制槽

refinery plant 炼制装置
refinery process unit 炼油厂工艺设备
refinery products 石油加工产品
refinery sludge 炼油厂污渣;炼厂酸渣
refinery tank 炼油厂油罐
refinery waste 炼油厂下脚;炼油厂废弃物;炼油厂废料;精炼厂废水;精炼厂废料
refinery waste sulfuric acid 炼制厂废硫酸
refinery wastewater 炼油废水
refinery wastewater reuse 炼油厂废水回用
refining 精炼;细化;匀料;匀浆;冶炼;精制;清除伐;提纯;使成熟
refining agent 纯化剂
refining by liquation 熔析精炼
refining cell 电解精炼槽
refining earth 精炼用白土
refining equipment 炼油设备;精炼设备
refining equipments in sets 炼油成套设备
refining furnace 精炼炉
refining in the liquid phase 液相精炼
refining mill 精研机;磨浆机
refining of gasoline 汽油的精制
refining of lead bullion 粗铅精炼
refining of metals 金属精炼
refining of petroleum product 石油产品的精制
refining of the twice metal 二次金属精炼
refining period 精炼期
refining plant 精炼车间
refining plant for solvents 溶剂的精制设备
refining process 炼制过程;精制加工;精制过程;提炼过程
refining slag 精炼渣
refining steel 精炼钢
refining tank 澄清池
refining techniques 精炼技术
refining temperature 细化温度;精炼温度;晶粒细化温度;调质温度
refining with adsorbents 吸附精制
refinish 再修整;重新磨光;返工修光
refinish enamel 重涂用瓷漆
refinishing 在修饰;重新装修;重涂;整修表面;再修饰
refinishing a flat section 整修平截面
refinishing paint 修补漆;维修漆
refire capability 再起动能力;再发射能力
refiring 重烧;复烧
re-first in first out 重新进行先进先出
refit 整修;改装;重(新)装配
refitment of a ship 船舶改装
refitment of a vessel 船舶改装
refit requirements 改装要求
refitted ship 改装船
refitting quay 整修码头
refixation 再定位
refixation of station 重新设站;测站重建
reflect 弹回;折转;反光;反射;反映;折回
reflectafloat glass 反射涂膜玻璃
Reflectal 反射铝合金
reflectal 锻造铝合金
reflectalyte 反光玻璃(一种灯具玻璃)
reflectance 反射能力;反射比;反光度
reflectance anomaly 反射异常
reflectance attachment 反射装置;反射附加器
reflectance curve 反射率曲线
reflectance dispersion curve 反射率色散曲线
reflectance factor 反射因素;反射率因数
reflectance ink 反射油墨;反光墨水
reflectance measurement 反射系数测量;反射能力测定;反射测定(法)
reflectance microdensitometer 反射微密度测定仪
reflectance of vitrinite 镜质体反射率
reflectance photometer 反射式光度计;反射度计
reflectance radiometer-photometer 反射率测定用辐射—光度计
reflectance ratio 反射率;反射系数比
reflectance signature 反射特征
reflectance spectrometer 反射分光计
reflectance spectrometry 反射光谱学
reflectance spectrophotometer 反射光谱光度计
reflectance spectrophotometry 反射(比)分光光度学
reflectance spectroscopy 反射光谱学;反射光谱法
reflectance spectrum 反射(光)谱
reflectance target 反射目标;反射靶

reflectance value 反射(率)值
reflect compass 反射罗盘;反射罗经
reflected 反射的
reflected amplitude 反射振幅;反射信号振幅
reflected beam 反射束
reflected beam kinescope 反射式显像管
reflected binary 反射二进制码
reflected binary code 反射二进制码
reflected binary number system 反射二进(位)制
reflected body wave 反射体波
reflected code 循环码;反射码;反射二进制码
reflected colo(u)r 荧光;反射色
reflected component 反射的部分
reflected echo pattern 反射波型
reflected electron detector 反射电子探测器
reflected electron image 反射电子图像
reflected energy 反射能(量);反射波能量
reflected flux 反射通量
reflected galvanometer 反射镜式电流计
reflected glare 反射强光;反射眩光
reflected global solar radiation 地球反射太阳辐射
reflected grating 反射光栅
reflected halo 反射晕圈
reflected harmonics 反射谐波
reflected head 反折头
reflected high energy electron diffraction 反射式高能电子衍射
reflected high-pressure mercurylamp 反射型高压汞灯
reflected image 反映像;反射(映)像;反射图像
reflected impedance 反射阻抗
reflected infrared 反射红外(线)的
reflected infrared ray 反射红外线
reflected intensity 反射强度
reflected light 反射光;反光
reflected light beam 反射光束
reflected-light halo 反射光晕
reflected-light luminaire 反射光照明器
reflected light meter 反射光式曝光表;反射光测光表
reflected light microscope 反(射)光显微镜
reflected light microscopy 反射光显微学
reflected light reading 反射光测定读数
reflected light scanning 反射光扫描
reflected load 反映负载
reflected near-end crosstalk 反射近端串音
reflected number 反射码
reflected object wave 反射物波
reflected plan 仰视图;反射平面图;反镜平面;投影图
reflected plane wave 反射平面波
reflected power 反射功率
reflected pressure 反射压力
reflected pulse 反射脉冲
reflected P-wave 反射纵波
reflected radiation 反射辐射
reflected ray 电离层反射波;反射(射)线
reflected reactance 反射电抗
reflected refraction 反射折射
reflected-refraction wave 反射折射波
reflected resistance 反映电阻;反射电阻
reflected secondary wave 次生反射波
reflected seismic information 反射地震资料
reflected shock front 冲击波(的)反射波面
reflected shock wave 反射震波;反射冲击波
reflected signal 反射信号
reflected signal indicator 反射信号指示器
reflected signal of echo sounder 回声测深仪回波信号
reflected sky wave 反射天波
reflected solar radiation 太阳反射辐射
reflected sound 反射声
reflected surface 反光表面
reflected tension wave 反射张力波
reflected traffic 反射话务
reflected ultraviolet method 反射紫外线摄影法
reflected value 反射值
reflected wave 电离层反射波;反射波
reflected wave coupler 反射波耦合器
reflected wave field 反射波场
reflectible 可反射
reflecting ability 反射能力
reflecting absorption grating 反射吸收光栅
reflecting acoustic(al) pressure 反射声压
reflecting anamorphotic optic(al) system 反射

变形光学系统
reflecting angle 反射角
reflecting antenna 无源天线;反射天线
reflecting background 反射背景
reflecting barrier 反射障;反射屏蔽;反射壁
reflecting beacon 反射标
reflecting bead 反光串珠
reflecting beam 反射光束
reflecting beam waveguide 反射束波导
reflecting-block folding 反射块状褶皱
reflecting body 反射体
reflecting boundary 反射边界
reflecting bundle 反射束
reflecting button 反光路钮
reflecting camera 反光式照相机
reflecting chopper 反射截光器
reflecting coating 反光涂面;反光涂层
reflecting collimator 反射式准直仪
reflecting compass 反射罗盘;反射罗经
reflecting component 反射分量
reflecting concrete 反光混凝土
reflecting condenser 反射聚光镜
reflecting curb 反光缘石;反光路缘(石);反光侧石
reflecting curtain 反射幕;反射帘
reflecting density 反射密度
reflecting dichroic mirror 反射分色镜
reflecting dielectric(al) grating 反射电介质光栅
reflecting diffraction grating 反射衍射光栅
reflecting disc 反射盘
reflecting electrode 反射电极
reflecting element 反射元件
reflecting elements in series 串联反射元件
reflecting film 反射膜
reflecting filter 反射滤光器
reflecting finder 反射取景器
reflecting force 反射力
reflecting galvanometer 镜式电流计;反射(式)检流计;反射镜式电流计;反射镜面电流计;反射镜电流;反射电流计
reflecting glass type 反射式玻璃液面表
reflecting glazing 反射窗玻璃
reflecting goniometer 反射测向计;反射测角仪
reflecting grating 反射栅
reflecting hemisphere type pyrometer 反射半球式高温计
reflecting hologram 反射全息图
reflecting holographic(al) optic(al) element 反射式全息光学元件
reflecting image 反射像
reflecting instrument 镜面仪表
reflecting insulating cover 反射绝热盖
reflecting interference microscope 反射光干涉显微镜
reflecting kerb 反光路缘(石);反光侧石
reflecting lamp for photography 反射型照像灯
reflecting layer 反射层
reflecting level 反照镜水准仪;反射水准仪;反光(式)水准仪
reflecting lobes 反射波瓣
reflecting loss 反射损耗
reflecting mark(er) 反射标记
reflecting material 反射材料;反光材料
reflecting medium 反射媒质;反射介质
reflecting membrane reflectance coating 反射膜
reflecting metallic film 反射金属膜
reflecting microscope 反射(式)显微镜;反光显微镜
reflecting microscope identification 反光显微镜鉴定
reflecting microscopy 反射显微镜检术
reflecting mirror 反射镜;反光镜
reflecting nephoscope 测云镜;反射测云器
reflecting object 反射物体
reflecting objective 反射物镜
reflecting optics 反射光学
reflecting paint 反光涂料
reflecting pavement marker 反光路面标志
reflecting point 反应点
reflecting pool 镜池;倒影池
reflecting power 反射功率;反射能力;反射比;反射本领;反光能力;发射能力
reflecting prism 反射棱镜
reflecting projector 反射投影器
reflecting prospecting 反射法勘探
reflecting pyrometer 反射(式)高温计

reflecting reduction 反光缩小
reflecting region 反射区
reflecting (road) stud 反光路钮;反光的指路标;反射路钮;反射路标
reflecting satellite 反光卫星
reflecting satellite communication antenna 卫星通信(用)发射天线
reflecting scan 反射扫描
reflecting screen 反射屏
reflecting segment 反射段
reflecting seismograph 反射地震仪
reflecting seismology 反射地震学
reflecting shade 反光罩
reflecting sign 反射标志;反光标志
reflecting signal 反射信号
reflecting sign board 反光标志板
reflecting spectrograph 反射式光谱仪
reflecting sphere 反射球(面)
reflecting square 反光直角器
reflecting stereoscope 反射立体镜;反光立体镜
reflecting stratum 反射层
reflecting surface 反射面;反光路面;反光表面
reflecting sweep 反射扫描
reflecting system 反射系统
reflecting telescope 反射(型)望远镜;反射式望远镜
reflecting-type microscope objective 反射型显微物镜
reflecting-type water level indicator 反射式水位指示器
reflecting viewfinder 反射式取景器
reflecting wave 反射波
reflecting zenith tube 反射天顶筒
reflection 折转;倒影;反映物;反作用;反射热;反射;反映
reflectional coefficient 反射系数
reflection altimeter 无线电测高计;反射测高计
reflectional well 反映井
reflection amplifier 反射放大器
reflection amplitude 反射振幅
reflection and refraction loss 反射与折射损失
reflection angle 反射角
reflection apparent rotation angle 反射视旋转角
reflection arrival 反射波至
reflection at critical 临界反射
reflection at critical angle 临界反射角
reflection at plane mirror 平面反射
reflection axis 反射轴
reflection behavio(u)r of surfaces 表面反射特性
reflection case 反射情况
reflection character analysis 反射波特征分析
reflection characteristic 反射特性
reflection coating 反射涂层
reflection coefficient 振幅反射率;反射系数;反射率
reflection coefficient circle 反射系数圆
reflection coefficient measuring bridge 反射系数测量电桥
reflection coefficient meter 反射系数计;反射系数测试仪
reflection colo(u)r 反射色
reflection colo(u)r determination method 反射色测定法
reflection colo(u)r luminance 反射色亮度
reflection colo(u)r saturability 反射色饱和度
reflection colo(u)r tone 反射色色调
reflection colo(u)r tube 反射式彩色显像管
reflection condition 反射条件
reflection configuration 反射结构
reflection constant 反射常数
reflection copy 反射原稿;反光晒印
reflection crack 对应裂缝【道】;反射裂缝
reflection cracking 反射裂缝;对应裂缝【道】
reflection crosstalk 反射串话
reflection curve 反射曲线
reflection cylinder 反射圆柱面
reflection densitometer 反射显像密度计;反射(光)密度计
reflection density 反射(光)密度
reflection diffraction 反射衍射
reflection digital densitometer 反射式数显密度计
reflection echelon grating 反射阶梯光栅
reflection effect 反射效应
reflection efficiency 反射效率
reflection-electroabsorption modulator 反射电吸收调制器

reflection error 反射误差
reflection external form 反射外形
reflection factor 反射因素;反射因数;反射系数;反射率
reflection factor meter 反射系数测试计
reflection filament 反射暗条
reflection finder 反射取景器
reflection flaw detector 反射式探伤器
reflection flaw method 反射式探伤法
reflection-free areas 无反射区
reflection free wall 无反射墙壁
reflection frquency influence-immune 不受射频干扰的
reflection gain 反射增益
reflection ga(u)ge 反向散射测量计;反射测量计
reflection goniometer 反射测角仪;反射天线方向调整器;反射测角器
reflection halation 反射光晕
reflection hologram 反射全息图
reflection horizon 地震反射地层;反射层
reflection image 反射像
reflection index 反射率
reflection interaction 反射光相互作用;反射波相互作用
reflection interference filter 反射干涉滤光片
reflection interval 反射时间间隔
reflection invariance 反射不变性
reflection invariant 反射不变量
reflection kernel 反射中心
reflection law 反射定律
reflection lens 反射透镜;反射镜头
reflection logging 反测井
reflection loss 反射损失;反射损耗
reflection loss measuring instrument 反射损耗测量仪
reflection marker 反射标准层
reflection matrix 反射矩阵
reflection measuring set 反射测量仪
reflection method 映射法;反法
reflection model 反射方式
reflection modulated optic(al) fiber sensor 反射体调制光纤传感器
reflection modulation 反射系数调制
reflection objective 反射物镜
reflection of electromagnetic wave 电磁波反射
reflection of light 光的反射;反光
reflection of ocean waves 海浪反射
reflection of P-waves 初波反射
reflection of radiation 射线反射;辐射反射
reflection of sea-surface 海面反射
reflection of shallow earthquakes 浅震反射
reflection of SH-wave 水平横波反射
reflection of sound 回声;声(音)反射
reflection of wave 波的反射
reflection operator 反射算符
reflection oscillator 反射(速调管)振荡器
reflection painting 反射涂料
reflection path 反射路径
reflection peak 反射峰
reflection phase transformation 反射相变
reflection photometer 反射式光度计
reflection plane 反射平面;反射面
reflection plane model 反射平面模型
reflection plane of symmetry 反射对称面
reflection pleochroism 反射多色性
reflection plotter 反射作图器;反射描绘器;反射绘图器(雷达附件);反射标绘器
reflection point 反射点
reflection-point trajectory 反射点轨迹
reflection polarization character 反射偏光性
reflection pool 倒影池
reflection power 反射功率
reflection principle 反射原理
reflection prism 棱镜反射镜;棱镜反光镜;反射棱镜
reflection probability 反射概率
reflection profile 反射剖面
reflection prospecting 反射法勘探
reflection rainbow 反射虹
reflection ray 反射光线
reflection reaction 反射反应
reflection reducing coating 增透膜;减反射膜;防反射涂层
reflection rotation 反射旋转
reflection rotation angle 反射旋转角

reflection rotation angle dispersion 反射旋转角色散
reflection rotation dispersion 反射旋转色散
reflection rule 反射规则
reflection-seismics 反射地震测量学
reflection seismograph 反射式地震仪
reflection seismology 反射地震学;反射波勘探法
reflections from phosphor particles 荧光粉粒的反射
reflection shield 反射屏
reflection shooting 反射波勘探法
reflection sounding 回声探测;回波探测;反射探测;反射(声)探测
reflection space transit angle 反射空间渡越角
reflection spectrum 反射光谱
reflection sphere 反射球(面)
reflection stereoscope 反光立体镜
reflection survey 反射勘探法;反射法勘探
reflection symmetry 反射对称
reflection target 反射靶
reflection termination 反射终端
reflection thickness ga(u)ge 反向(散射)厚度计
reflection time of penstock 压力水管中压力波往复时间;压力水管中压力波反射时间
reflection twin 反射双晶
reflection type 反射式
reflection-type cavity 反射式共振腔
reflection-type colo(u)r tube 反射式彩色显像管
reflection type electron microscope 反射式电子显微镜
reflection-type infrared polarizer 反射式红外偏振镜;反射型红外偏光镜
reflection-type kinescope 反射式显像管
reflection-type laser protective spectacles 反射式激光防护眼镜
reflection-type overhead projector 反射教学投影仪
reflection-type teaching projector 反射式教学投影仪
reflection unit 反射装置
reflection wave method 反射波法
reflection X-ray microscopy 反射(式)X 射线显微术
reflective 反射的
reflective array 反射阵
reflective attenuation 反射衰减
reflective blind 反射帘幕
reflective bulk hologram 反射体积全息图
reflective button 反射路钮;反光路钮
reflective coat(ing) 反射涂膜;反射涂料;反射膜;反射镜涂膜;反射镜涂层;反射层
reflective coefficient 反射系数
reflective cracking 反映裂缝
reflective crystal coating 晶体反射膜
reflective fabric 反光织物
reflective film 反射薄膜
reflective glass 反射玻璃
reflective high-energy electron diffraction 反射高能电子衍射
reflective index 反射率
reflective index of liquid 液体反射指数
reflective infrared collimator 反射式红外平行光管
reflective insulant 反射绝缘材料
reflective insulation 反射绝热材料;反射(型)绝热;反射绝缘材料;反射绝热体;反射隔热
reflective insulation material 反射保温材料;反射绝缘材料
reflective liquid level ga(u)ge 反射式液面计
reflective mark(er) 反射标记;反光标志;反光标记
reflective material 反光材料
reflective network 反射网络
reflective optic(al) system 光反射器;反射光学系统
reflective paint 发光漆;反射涂料
reflective pavement marker 反光路标志
reflective photoelectric(al) switch 反射式光电开关
reflective pigment 反光颜料
reflective placard 反光性标志牌
reflective power 反射(能)力
reflective radiation from earth's surface 地面反射辐射
reflective scan 反射扫描
reflective sensor 反射传感器
reflective sheeting 反光挡板

reflective sight 反射瞄准器
reflective sign 反光标志
reflective spectrophotometer 反射分光光度计
reflective spot 反射点
reflective starcoupler 反射星形耦合器
reflective surface 反射面
reflective surface area 反射面积
reflective tail indication plate 车尾反光灯
reflective tracking 反射式跟踪
reflective traffic paint 交通标志反光漆
reflective-type display 反射型显示
reflective-type variable lens 反射式变形镜头
reflective viewing screen 反射式银幕;反射式视屏;反射式屏幕
reflective vinyl film 反射乙烯膜
reflective wall 反射墙
reflectivity 反射性;反射率;反射力;反射比;反差系数
reflectivity curve 反射曲线
reflectivity determination method 反射率测定法
reflectivity dispersion 反射率色散
reflectivity factor sequences 反射系数序列
reflectivity model 反射系数模型
reflectivity modulation 反射率调制
reflectivity of volume scattering 体散反射率
reflectivity test 反射性试验;反射比试验
reflecto-ellipsometer 反射式椭圆率测定仪
reflectoga(u)ge (金属片的)厚度测定器;厚度测量器
reflectogram 探伤图;反射图
reflectogram of vitrinite 镜质体反射率分布图
reflectometer 反射仪;反射系数计;反射计;反光白度计
reflectometry 反射法;反射测量术
reflecto-polarimeter 反射偏振计
reflector 折射器;反射物;反射体;反射镜;反射极;反射层;反射标;反射板;反光罩
reflector alignment 反射器准直
reflector antenna 无源天线;反射器天线
reflector backup structure 反射器架结构
reflector board 反光板;反射板
reflector buoy 反射浮标
reflector button 反光路钮(交通标线用)
reflector carrier 反光镜座
reflector characteristic 反射器特性曲线
reflector collector 反射集热器
reflector compass 反射罗盘;反射罗经
reflector constant 反射器常数
reflector control 反射控制;反射层控制
reflector curvature 反射界面曲率
reflector dipole 反射偶极子
reflector economy 反射层节省
reflector element 反射器元件;反射单元
reflector flood 反射镜泛光灯
reflector frequency cutoff 反射器频率截止
reflectorised paint 反光涂料;反光漆;反光涂料
reflectorization 反射作用;反光处理;反光(作用)
reflectorized lamp 装有反射器的灯
reflectorized marker 发射器标志
reflectorized paint 反光涂料;反光漆;反光涂料
reflectorized route marker 反光路标;反光路标
reflectorized sign 反射标志;反光标志
reflectorizing concrete 反光混凝土
reflectorizing curbstone 能反光的路缘石
reflectorizing foil 反光(金属)薄片;反光箔
reflectorizing glass bead 反光玻璃珠
reflectorizing of paint 涂料的反光
reflectorizing road stud 反光的指路标
reflector lamp 聚光灯;泛光灯;反射灯;反光灯
reflector lamp bulb 反光灯泡
reflector-layer lamp 反射层灯
reflector lens 反射器透镜;反射镜镜头
reflector light fitting 反光灯装置
reflector marker 反光路标;反光路标
reflector material 反光材料
reflector microphone 反射器传声器
reflector mount 反射镜架
reflector oven 反射式加热炉
reflector panel 反射器板
reflector path 镜路
reflector plate 反射片;反射板
reflector potential 反射极电位
reflector prism 反射棱镜
reflector road button 反光路钮
reflector satellite 无源转发卫星;反射卫星

reflector saving 反射层节省
reflector sight 反射式瞄准具;反射瞄准镜
reflector sign 反射标志;反光标志
reflector spectral frequency slope 反射镜光频斜率;反射镜光频特性斜率
reflector stud 反光钉
reflector telescope 反射式望远镜
reflector tracker 反射跟踪装置
reflector type 反射器式
reflector-type antenna 反射器式天线
reflector volt 反射器电压
reflector voltage 反射极电压
reflectoscope 投影灯;声探伤仪;超声波探伤仪;反射仪;反射系数测试仪;反射探伤仪;反射镜;反射检测仪;反射仪;反射测试仪;反光镜
reflect sun rays 反射太阳光线
reflecxion hologram 反射全息图
reflet 陶瓷器表面的光泽
reflex 映像;折回;来复;回复;倒影
reflex action 反射作用;反射动作
reflex amplification 来复(式)放大;回复放大
reflex amplifier 来复式放大器
reflex angle 优角;优弧角
reflex arc 反射弧
reflex attachment 反射装置
reflex baffle 反音障
reflex boresight 反光小孔瞄准
reflex bunching 反射聚束
reflex camera 镜面摄影机;反射(式)照相机;反射摄影机;反射式照相机
reflex center 反射中枢
reflex chain 反射链
reflex circuit 回复电路;反射(回)路
reflex coefficient cross section map of magnetotelluric method 大地电磁法反射系数断面图
reflex condenser 回流冷凝器;回流冷凝管
reflex conduction 反射传导
reflex contraction 反射性收缩
reflex copy 反光晒印
reflex copying 反射复制;反射法复印;反光晒像法
reflex current 反流
reflex detector 来复式检波器;反射(式)检波器
reflexed 反折
reflex electro-optic display device 反射电光学显示装置
reflex enclosure 反音箱;反音匣
reflex exposure 反射曝光
reflex exposure photoreceptor 反射曝光感光体
reflex feed system 反射馈源系统
reflex finder 反射瞄准器;反光取景器
reflex finder system 反射取景系统
reflex galvanometer 反射式电流计
reflex glass 液面显示玻璃
reflex-hammer 反射锤
reflex horn 反射喇叭
reflex horn speaker 反射式喇叭扬声器
reflex housing 反射罩
reflexible 可折转的;可反射
reflex inhibition 反射性抑制
reflexion 折转;反射
reflexion electron microscope 反射电子显微镜
reflexion of wave 波浪反射
reflexion twin 反射孪晶
reflexive contraction 反射性收缩
reflexive locally convex space 自反局部凸空间
reflexive module 自反模
reflexive motion 反射性运动
reflexive normed space 自反赋范空间
reflexive order 转移指令
reflexive regulation 反射性调节
reflexive relation 自反关系
reflexive rotary dryer 回流式转筒烘干机
reflexive space 自反空间
reflexivity of a fuzzy relation 模糊关系的自反性
reflexivity of strategic equivalence 策略等价的自反性
reflex klystron 反射速调管;反射式速调管
reflex-klystron amplifier 反射速调管放大器
reflex lamp trough 反射灯槽
reflexless terminal 匹配终端;无反射终端
reflex level ga(u)ge 反射式液面计
reflex ligament 反转韧带
reflex-like reaction 反射样反应
reflex mirror locking lever 反光镜锁紧扳手

reflex movement 反射(性)运动
reflexograph 反射描记器
reflexology 反射学
reflex oscillator 反射速调管
reflex paper 反光印像纸
reflex pathway 反射径路
reflex pickup 反光拾音器
reflex printer 反射复制器
reflex print(ing) 反光晒印;贴印法
reflex prism 反射(三)棱镜
reflex process 反射光复印法
reflex reception 来复式接收法
reflex regulation 反射性调节
reflex reinforcement 反射性增强
reflex sight 反射式瞄准器;反射镜式瞄准镜
reflex strength 反射强度
reflex sweep high speed camera 反射扫描高速照相机
reflex time 反射时间
reflex-type two-dimensional steady water quality model of river 边界反射二维河流水质模型
reflex viewer 反射式视片器
reflex viewfinder 反射型录像器;反射型取景器;反射式指示器;反射式寻像器;反射式取景器
reflex water ga(u)ge 反光水位表
reflight 再次试飞
refloat a stranded vessel with drag 拖绞出搁浅船
reflo(a)tation 再浮选
refloated concentrate 再浮选精矿
refloat(ing) 脱浅;浮起
refloating operation 打捞作业;打捞工程
refloating with slipway process 滑墩脱浅法
reflocculation 再絮凝
refloor 重铺地板;重新铺面
reflooring 重铺地板
reflow 逆流;回流;再济平;反流
reflow coating 再流平涂料
reflowing 软熔发亮处理
reflow soldering 回流焊接
reflow unit 软熔机组
refluence 退潮;回流
refluent tide 退潮
reflux 回流(的);反流
reflux accumulator 回流液收集器;回流液储器;回流储液器
reflux coil 回流盘管
reflux column 回流(蒸馏)柱;回流(蒸馏)塔
reflux condenser 回流冷凝器;回流冷凝管
reflux condition 回流条件
reflux conduit 回水渠;回流管
reflux distillation 回流蒸馏
reflux drum 回流液收集器;回流液储器
reflux duty 回流热消耗
reflux exchanger 回流冷凝器;回流冷凝管;回流换热器
refluxing 回流
reflux(ing) coil 回流蛇管
reflux line 回流管道(从冷凝器通到塔中)
reflux pipe 回流管
reflux process 回流过程
reflux pump 回流泵
reflux rate 回流速率
reflux ratio 回流系数;回流比
reflux splitter 回流分配器
reflux tower 回流塔
reflux tube 回流管
reflux tubing 回流管
reflux unit 回流装置
reflux valve 单向阀;止回阀;逆止阀;回流阀
refocused revolving beam 再聚焦旋转光束;二次聚焦旋转光束
refolded fold 重(复)褶皱
refolder 重折叠器
refolding 重褶皱作用
refoliation 再生页理;重叶理
reforest 再植林木
reforestation 再造林;更新造林;迹地更新;人工造林(法);重新造林
reforestation with exotic speices 引种造林
reforested area statistics 重新造林面积统计
reform 改造;改革;重新组成
reform, opening to the outside world and enlivening 改革、开放、搞活
reformat 重定格式

reformate 重整油
reformation 修正;改造作用;改革;重新组成
reformation of contract 重订合同
reformatory 教养院;教养所
reformatting 重定格式
reformatting transformation 格式变换
reform building 改建
reformed bar 再生鱼尼板
reformed gas 改制气;转化气;重整气
reformed gasoline 重整汽油
reformer 转化器;转化炉;改制炉;改革者;重整装置;重整器;重整炉
reformer desuperheater 转化炉预热器
reformer effluent gases 改制气
reformer pipe 重整炉管
reformer pretreating 重整预处理
reformer tube 转化炉管
reforming 重整(作用)
reforming catalyst 转化催化剂;重整催化剂
reforming economic relations 调整经济关系
reforming furnace 转化炉
reforming gas 转化气
reforming gas boiler 转化气废热锅炉
reforming plant 转化装置
reforming process 重整过程
reforming the methods of operation 经营机制改革
reforming tube 重整炉管
reforming unit 转化装置
reforming with recycle 循环重整
reform lantern 改制灯(一种顶棚灯)
reform of farmtools 农具改革
reform of financial structure 财务体制改革
reform of insurance business 保险业的改革
reform of school administrative system 学校管理体制改革
reform of the financial system 财政体制改革
reform of the organizational structure 机构改革
reform of wage system 工资改革
reform school 少年教养院
reformulate 再简述;重新制定
reformulation 料方配重
reform weave 变换组织
reforwarding 续运
reforwarding of luggage 转运
reforwarding station 重新发送站
refrachor 等折比容
refract 折射
refractability 可折射性
refractable 可折射的
refractary 休复的
refracted angle 折射角
refracted axis 折射轴
refracted drawing 折射图
refracted flow 折向流;偏转流
refracted image 折射成像
refracted law 折射定律
refracted layer 折射层
refracted light 折射光
refracted plane 折射面
refracted prospecting 折射勘探法
refracted pump 折射脉冲
refracted ray 折射线
refracted seismology 折射地震学
refracted wave 折射波
refracted wave correlation 折射波对比
refracting ability 折射能力
refracting angle 折射(棱)角;棱角
refracting glass 折射玻璃;折光玻璃
refracting layer 折射层
refracting medium 折射介质
refracting meter 折射计
refracting power 折射率;折光力;屈折力
refracting prisms 折射棱镜
refracting sphere 折射球体
refracting surface 折射面
refracting telescope 折射(型)望远镜;折光望远镜
refracting wedge 折射楔
refraction 折射(度测定);折光(差)
refraction action 折射作用
refraction(al) coefficient 折射系数;折光系数
refraction analysis 折射计法
refraction angle 折射角
refraction anisotropy 折射各向异性
refraction at plane boundary face 边界面折射

refraction at spheric(al) surface 球面折射
refraction cleavage 折射劈理
refraction constant 折射常数
refraction correction 折射改正;折光差改正;蒙气差改正
refraction diagram 折射图
refraction diagram of wave 波(浪)折射图
refraction effect 折射作用
refraction error 折射误差
refraction factor 折光系数
refraction filter 折射滤光片
refraction index 折光指数
refraction in level(l)ing 水准测量折光差
refraction law of seepage 渗流折射定律
refraction light 折射光
refraction line 折射光线
refraction loss 折射损失;折光损耗
refraction method 折射(波)法;折光率法
refraction objective 折射物镜
refraction of convection 对流折射
refraction of light 光(的)折射
refraction of ocean wave 海浪折射
refraction of sound 声折射;声音折射
refraction of wave 波浪折射
refraction parameter 折射参数
refraction prism 折射棱镜
refraction process 折射法
refraction property 折射特性
refraction prospection 折射勘探法
refraction ray 折射光线
refraction-reflected wave 折射反射波
refraction seismic prospecting 折射地震勘探
refraction seismograph 折射地震仪
refraction seismology 折射地震学
refraction shooting 折射爆破
refraction surface 折射面
refraction survey 折射勘探法
refraction table 折光表
refraction term 折光差改正项
refraction test 折射试验
refraction velocity 折射波速
refraction wave 折射波
refractive 折射的
refractive exponent 折光指数
refractive index 折射率;折光指数;折光率
refractive index correction 折射率校正
refractive index detector 折光率检验器;折光检测器
refractive index determination 折射率测定
refractive index dispersion 折射率色散
refractive index ellipsoid 折射率椭球
refractive index for petroleum 石油的折光指数
refractive index increment 折光指数增值
refractive index interface 折射率界面
refractive index of liquid 液体的折光率
refractive index profile coefficient 折射率分布系数
refractive index table 折光指数表;折光率表
refractive law 折射定律
refractive medium 折射介质
refractive modulus 修正折射指数;折射模数
refractive optics 折射光学
refractive power 折射能力;折射率;折射本领
refractive projector 折射式投影机
refractive property 折射性能
refractive quality 折射功效
refractive technique 折射技术
refractive wave 折射波
refractivity 折射率;折射性;折射系数;折射能力;折射本领;折光性
Refractoloy alloy 赖弗拉托洛伊合金
refractometer 折射指数;折射仪;折射率仪;折射计;折光仪
refractometric analysis 折射分析
refractometric titration 折射(率)滴定
refractometry 折射率测量;折射计法;折射(分析)法;折光法;量折射术
refractor 折射器;折射镜;等析比容
refractoriness 耐熔性;耐熔度;耐热度;耐火性;耐火度;不应态
refractoriness pins 耐火烧针
refractoriness under load 荷重软化温度
refractory 难除物;耐熔物质;耐熔(的);耐火的;耐火材料
refractory aggregate 耐高温集料;耐高温骨料;耐热集料;耐热骨料;耐火集料;耐火骨料

refractory alloy 高温合金;难熔合金;耐火合金
refractory alumina cement 耐火铝水泥;耐火高铝水泥
refractory anchor 耐火材料锚固件
refractory and insulating fire brick 耐火隔热砖
refractory behavio(u)r 难熔性
refractory biochemical degradation organic wastewater 难生化降解有机废水
refractory biodegradation organic pollutant 难生物降解有机污染物
refractory block 耐火砌块;耐火垫
refractory body 耐火物
refractory bottom ring 耐火底环
refractory brick 耐火砖
refractory "bubble" 耐火空心砖
refractory bubble product 耐火空心球制品
refractory building 耐火建筑(物)
refractory building material 耐火建筑材料
refractory cable 耐火型电缆
refractory carbide 难熔金属碳化物
refractory castable 耐火浇注料;耐火浇灌料
refractory casting 耐火浇注料;耐火浇灌料
refractory casting tube 耐热铸管
refractory cathode 耐热阴极
refractory cement 铝酸钙水泥;耐火水泥;耐火胶结材料
refractory ceramics 高温陶瓷;耐火陶瓷
refractory chrome ore 耐火铬矿石
refractory clay 耐火(黏)土;耐火泥
refractory clay ore 耐火黏土矿石
refractory coat 耐火面层
refractory coating 耐火涂料;耐热(保护)层;耐火涂层;防火隔热涂料
refractory cobalt chromium carbon alloy 耐熔钴铬碳合金
refractory composition 耐火制品
refractory compound 难熔化合物;耐火化合物;耐处理化合物
refractory concrete 耐热混凝土;耐火混凝土;耐高温混凝土
refractory concrete aggregate 耐火混凝土骨料;耐火混凝土集料
refractory concrete block 耐火混凝土砌块
refractory construction material 耐火建筑材料
refractory corrosion 耐火材料侵蚀
refractory cover 耐火盖板
refractory crucible 耐火材料坩埚
refractory damage 耐损害
refractory dam ring 耐火砖挡圈
refractory degradation 难降解
refractory dress 耐火涂料
refractory dressing 耐火涂料
refractory elements 难熔元素
refractory enamel 耐火漆
refractory erosion 炉衬熔蚀
refractory expansion joint 耐火材料伸缩缝
refractory faced 涂复耐火材料的
refractory fiber 耐火纤维
refractory fiber product 耐火纤维制品
refractory fibre reinforced plastic 耐火纤维增强塑料
refractory fireclay block 耐火黏土砌块;黏土耐火砖
refractory furnace 耐火炉
refractory fused cement 耐火水泥熔融水泥料
refractory gas oil 耐热瓦斯油
refractory glass 耐火玻璃
refractory glass fiber 耐火玻璃纤维
refractory gold 难选金
refractory gun 喷浆枪;喷镀枪
refractory gunite 喷射耐火混凝土
refractory gun mix 喷浆
refractory gunning equipment 喷浆装置
refractory gunning material 耐火喷涂材料
refractory hard metal 高熔点硬质合金;耐火硬金属
refractory hollow brick 耐火空心砖
refractory hot-gas grate 耐火热风炉栅
refractory industry 耐火工业材料
refractory insulating brick 隔热耐火砖;绝热耐火砖
refractory insulating concrete 耐火绝热混凝土;耐火隔热混凝土
refractory insulation 耐火材料绝缘
refractory insulator 耐火绝缘体
refractory ionic compound 难降解离子化合物
refractory iron ore 耐火的铁矿石

refractory lifter 凸棱窑衬
refractory lime 耐火石灰
refractory lined 耐火材料衬里的
refractorylined chamber 衬有耐火材料的加热室
refractory-lined copper bowl 耐火材料衬里的铜炉体
refractory-lined oven 耐火炉
refractory liner 耐火材料衬里
refractory lining 耐火炉衬;耐火衬套;耐火衬砌
refractory lining mixture 耐火炉衬混合料
refractory lining of kiln 耐火窑衬
refractory magnesite ore 耐火菱镁矿矿石
refractory masonry 耐火圬工
refractory material 耐熔材料;耐火材料
refractory matter 耐火料
refractory metal 高熔点金属;难熔金属;耐熔金属;耐火金属;耐高温金属
refractory metal alloy 难熔金属合金
refractory metal base composite material 难熔金属复合材料
refractory metal carbide 难熔金属碳化物
refractory metal-ceramics laminate 高熔点金属陶瓷层压材料
refractory metal contacts 难熔金属电接触器材
refractory metal fibre reinforced ceramics 高熔点金属纤维增强陶瓷
refractory mineral 难熔矿石
refractory mineral wool 耐高温棉
refractory mix(ture) 耐火泥
refractory mortar 耐火灰浆;耐火砂浆;耐火泥(浆);耐火胶泥
refractory nozzle 耐火材料喷管
refractory ore 难选矿石;难熔矿石
refractory organic matter 难降解有机物
refractory organic micro-pollutant 难降解有机微污染物
refractory organics 难降解有机物;难分解有机物
refractory organic silt 难降解有机污泥
refractory organic wastewater 难降解有机废水
refractory organic wastewater treatment 难降解有机废水处理
refractory oxide 耐火氧化物
refractory oxide crucible 耐火氧化物坩埚
refractory oxide fiber[fibre] 耐高温氧化物纤维
refractory paint 耐火涂料
refractory patching cement 填补用耐火水泥
refractory patching mixture 耐火的补炉料
refractory patch material 补炉材料
refractory pebbles 耐火石
refractory peridotite 耐火用橄榄岩
refractory period 不应期
refractory plaster facing 防火隔热涂料
refractory plastics 耐火塑料
refractory pollutant 难降解污染物
refractory porcelain 耐热瓷器;耐火瓷
refractory powder 耐火粉剂
refractory product 耐火制品;耐火材料制件
refractory proof 耐火
refractory protection 耐热保护层;耐火保护层;防火装置
refractory quality 难熔性
refractory quotient 耐火率
refractory ramming material 捣打料
refractory raw material 耐火原料
refractory raw material commodities 耐火材料矿产
refractory repairing mass 耐火补炉料
refractory rock 耐火岩
refractory rocks 耐火岩类
refractory sand 耐火砂
refractory seal 耐高温密封
refractory shell mo(u)ld 耐火材料壳型
refractory shop 耐火材料车间
refractory sill 玻璃液流料槽
refractory slab 耐火板
refractory slag 难熔矿渣
refractory solid 耐火固体
refractory stage 不应期
refractory state 不应状态
refractory steel 耐热钢;热强钢
refractory stone 耐火材料结石
refractory stopper 耐火塞
refractory structural material 耐火结构材料
refractory substance 耐处理物质
refractory sulfide 难选硫化物

refractory support 耐火垫板
refractory surface 耐火面;钝热表面
refractory technology 耐火材料工艺
refractory temperature 耐热温度
refractory thermal insulator 耐火绝缘体
refractory tube 耐火管
refractory value 耐火率
refractory volatile organic matter 难降解挥发性有机物
refractory volatile solid 难降解挥发性固体
refractory wad 耐火泥条
refractory wall 绝热墙;隔热墙;耐火墙
refractory ware 耐火器材
refractory wash 耐火涂抹(材料)
refractory wastewater 难降解废水
refractoscope 折射测定仪
refrain from 避免
reframe 安装新框架;再组织;再构造
refrangibility 折射能力;折射度;可折射性;屈折度
refrangible 可折射的
Refrasil 石英玻璃纤维材料
refrax 金刚砂砖;碳化硅耐火材料
refreeze 再冻;再结冰
refreezing 再冷冻;再冻结
refresh 刷新
refreshable 可刷新的
refreshable load module 可刷新装入模块
refreshable program(me) 可刷新程序;可复制程序
refresh address 更新地址
refresh amplifier 更新放大器
refresh controller 刷新控制器
refresh counter 更新计数器
refresh cycle 刷新周期
refresher display 重复显示器
refresher driving 刷新驱动
refresh grant 重新响应
refresh information 更新信息
refreshing irrigation 降温灌溉
refreshing the water periodically 定期换水
refresh memory 显示重复存储器;更新存储器;刷新存储器
refreshment kiosk 小食亭
refreshment room (车站或餐车中的)餐室;小吃部;茶点室
refreshment saloon 餐厅;饮食店
refreshment stand 点心摊;小吃摊
refresh mode 更新态
refresh overhead time 更新整理操作时间
refresh page 更新页
refresh rate 更新速度;刷新速率
refresh request 更新请求
refresh routine 更新程序
refresh scheduler 更新调度程序
refresh system 更新系统
refresh test 更新测试
refresh time 更新时间
refex 碳化硅耐火材料
refrigerant 致冷剂;制冷剂;冷却物;冷冻剂
refrigerant bypass 冷媒旁路
refrigerant capacity 制冷能力;制冷量
refrigerant charge 系统内制冷剂量;制冷剂容量
refrigerant coil 制冷剂盘管
refrigerant compressor 制冷压缩机;冷冻剂压缩机
refrigerant compressor unit 制冷压缩机组;制冷压缩装置
refrigerant condenser 制冷冷凝器
refrigerant controller 制冷剂控制器
refrigerant filter 制冷剂过滤器
refrigerant fluid 致冷液;制冷液
refrigerant gas 气态制冷剂;制冷气体
refrigerant gas for cooling motor 冷却电动机用冷煤气
refrigerant heater 制冷剂预热器
refrigerant latitudes 寒带地
refrigerant line 冷冻线;制冷剂管路
refrigerant medium 制冷介质
refrigerant metering device 制冷剂计量装置
refrigerant method 冷冻法
refrigerant oil 冷冻机油
refrigerant pressure vessel 制冷剂压力容器
refrigerant pump 冷媒泵
refrigerant reducing orifice 冷媒节流孔板
refrigerant return pipe 冷媒回流管

refrigerant system 冷冻系统
refrigerant temperature 制冷(剂)温度
refrigerant vapo(u)r 制冷剂蒸气
refrigerant velocity 制冷剂流速
refrigerant vessel 制冷剂容槽;制冷剂导管
refrigerate 制冷
refrigerated 冷藏的
refrigerated and heated container 冷藏和加热集装箱
refrigerated baffle 冷却挡板
refrigerated body 冷藏车身
refrigerated brine 冻结盐水
refrigerated cargo 冷冻货(物)(冷藏船装运华氏30度以下货物);冷冻船货;冷藏货
refrigerated cargo hold 冷藏货舱
refrigerated cargo ship 冷藏运输船
refrigerated carrier 冷货运输船;冷藏船
refrigerated centrifugation 冷冻离心法
refrigerated centrifuge 冷冻离心机
refrigerated centrifuge separator 冷冻离心分离机
refrigerated compartment 冷藏舱
refrigerated container 冷冻集装箱;冷藏集装箱;冷藏货柜(集装箱)
refrigerated cycle 循环制冷
refrigerated enclosed container 冷冻货柜
refrigerated fish ship 冷藏鱼船
refrigerated food storage 冷藏食品库
refrigerated fruit ship 冷藏水果船
refrigerated goods 冷藏货
refrigerated hold 冷藏舱
refrigerated lorry 冷藏汽车
refrigerated plant 冷藏厂
refrigerated plate freezing method 平板式冻结法
refrigerated space 冷藏舱位
refrigerated storage 冷冻储[贮]存;冷藏
refrigerated storage room 冷藏间
refrigerated storage tank 低温储[贮]罐
refrigerated stowage 冷藏库
refrigerated traffic 冷藏运输
refrigerated train 冷藏列车
refrigerated transport 冷藏运输
refrigerated truck 无机械致冷的冷藏车皮
refrigerated van 冷藏车
refrigerated warehouse 冷库
refrigerated window 冷藏橱窗
refrigerating 冷藏的
refrigerating accumulator 冷冻压缩机
Refrigerating and Air-conditioning Engineers 美国供暖、制冷及空气调节工程师学会
refrigerating and cannery ship 水产冷藏制罐船
refrigerating cabinet 冰柜
refrigerating capacity 制冷能力;制冷量;冷却能力;冷冻能力;产冷量
refrigerating cargo 冷藏货(物)
refrigerating carrier 冷藏船
refrigerating chamber 冷库;冷冻室;冷藏室
refrigerating circuit 制冷电路;制冷循环;制冷回路
refrigerating circulating pump 制冷装置泵
refrigerating coefficient of performance 制冷性能系数
refrigerating coil 冷冻盘管
refrigerating compressor 制冷压缩机
refrigerating condenser pump 制冷装置冷凝器泵
refrigerating cycle 制冷循环;冷冻循环
refrigerating effect 制冷效应;制冷效果;制冷能力;制冷量;冷冻作用;产冷量
refrigerating effect per brake horsepower 单位轴马力制冷量
refrigerating effect per unit of swept volume 单位容积产冷量
refrigerating engine 制冷机
refrigerating engineer 冷藏员
refrigerating engineering 制冷工程
refrigerating equipment 冷冻设备
refrigerating factory ship 冷藏加工船
refrigerating fluid 制冷液;冷却液;冷冻液
refrigerating industry 制冷工业
refrigerating installation 制冷装置;制冷设备;冷冻装置;冷冻设备;冷藏设备
refrigerating machine 冷气机;制冷机;冷冻机;冷藏设备;单筒型溴化锂吸收式制冷机
refrigerating machine certificate 冷冻机证书
refrigerating machinery 冷藏机械
refrigerating meat ship 冷藏肉船

refrigerating medium 制冷介质;制冷剂;载冷剂
refrigerating output 制冷(产)量
refrigerating output under air conditioning condition 空调工况制冷量
refrigerating plant 冷气机房;冷藏库;制冷装置;制冷设备;制冷厂;冷冻装置;冷冻设备;冷冻厂;冷藏设备
refrigerating plant room 制冷机房
refrigerating plant wagon 冷冻机械车
refrigerating room 冷藏间
refrigerating ship 冷藏船
refrigerating station 制冷机房
refrigerating system 制冷系统;致冷装置;冷却系统;冷冻系统
refrigerating temperature 制冷温度;冻结温度
refrigerating treatment 冷处理
refrigerating unit 制冷装置;制冷设备;制冷机组
refrigerating work 制冷工程
refrigeration 致冷;制冷(学);冷冻(作用);冷藏(法)
refrigeration apparatus 制冷装置;制冷设备;制冷机
refrigeration capacity 冷冻能力;制冷能力
refrigeration clause 冷藏条款
refrigeration coil 制冷盘管
refrigeration component 制冷构件;制冷部件
refrigeration compressor 制冷压缩机;制冷压气机;冷冻机
refrigeration compressor unit 制冷压缩机组
refrigeration condenser 制冷凝汽器;制冷冷凝器
refrigeration cooler 制冷冷却器
refrigeration cycle 致冷循环;制冷循环
refrigeration dehumidifying 冷冻除湿
refrigeration duty 制冷负荷;冷负荷;热负荷
refrigeration engineering 制冷工程
refrigeration equipment 致冷装置
refrigeration installation 致冷装置
refrigeration load 制冷负荷;冷负荷
refrigeration machine 冷冻机;制冷机
refrigeration machinery building 制冷机房
refrigeration method 冷冻法;冰冻法
refrigeration oil 冷冻油
refrigeration output 制冷能力
refrigeration pipe 冷却管(道)
refrigeration piping 制冷工管道
refrigeration plant 制冷装置;制冷设备;冷却装置;冷冻装置
refrigeration plant room 冷冻机间
refrigeration room 冷冻间
refrigeration ship 冷藏船
refrigeration station 冷冻站
refrigeration surface 冷却面
refrigeration system 制冷系统;冷却系统;冷冻系统
refrigeration ton 冷吨;冷冻吨
refrigeration truck 冷藏车
refrigeration tube 冷却管(道)
refrigeration unit 制冷单元;冷藏装置
refrigerator 致冷器;制冷器;制冷机;制冰机;冷柜;冷冻器;冷冻机;冷藏箱;冷藏器;冷藏库;家用冰箱;电冰箱;冰箱
refrigerator car 冷藏车;保温车
refrigerator cleaner 冰箱清洗剂
refrigerator compressor 电冰箱压缩机
refrigerator compressor motor 冰箱压缩机电机
refrigerator express car 冷藏特快车
refrigerator lorry 冷藏车
refrigerator oil 冷冻机油
refrigerator paper 冰箱用纸
refrigerator room 冷冻机房
refrigerator system rating 冷冻机参数;冷冻系统额定功率
refrigerator thermometer 冰箱温度计
refrigerator truck 冷藏车
refrigerator unit 致冷装置
refrigerator van 冷藏车
refrigerator wagon 冷藏车
refringence 折射(系数)
refringence index of coal 煤的折射率
refringency 折射系数
refuel 增添燃料;加燃料
refueled aircraft 受油机
refueling 加油;装燃料;中途加油;续加燃料;换油
refueling aerodrome 加油机场
refueling airport 加油机场
refueling apron 加油坪(航空港)

refueling boat 燃料加注艇
refueling hydrant 加油龙头;加油栓
refueling machine 装燃料机
refueling machine of nuclear power plant 核电厂装卸料机
refueling nozzle 加注嘴
refueling satellite station 人造卫星加注燃料站
refueling ship 加油船
refueling station 加油站
refueling submarine 输油潜水艇
refueling tank 给油箱
refueling unit 加油设备
refueller 燃料器;加油器;汽油加油车
refuel stop 添加燃料停车
refuel system 加燃油系统
refuge 禁猎区;避难所;避车台;安全地带
refuge aerodrome 停避机场
refuge airport 停避机场
refuge area 避难区(域)
refuge capital 逃避资金
refuge center 避难中心;避难所
refugee and exvacuee 难民和灾民
refugee camp 难民营
refugee capital 逃避资本
refugee shelter 难民收容所
refuge for fish 鱼类安全地带
refuge harbo(u)r 避难港;避风港
refuge hole 避车洞
refuge hut 避难小屋
refuge island 安全岛;避车岛
refuge manhole 行人避车处;让车道;道路避车处
refuge passage(way) 避难通道
refuge platform 避车台;安全台(分洪区)
refuge recess 避车洞
refuge shelter 避难所
refuge siding 避难线【铁】
refuge siting 越行线
refuge track 待避线;避车线
refugium 避难所
refund 退款;退回款;退还
refundable 可偿还的
refundable deposit 可退还的定金
refundable taxes 可退税款
refund and rebate 回扣
refund bond 偿付保函
refund by carrying a loss back to a prior year 亏损转回退税
refund check 退款票据
refunding 以新债券取代旧债券;偿还
refunding bonds 偿债证券;替换债券;调换债务
refundment 资金偿还;付清
refundment bond 偿还担保书
refund of a part of the income taxes already paid 退还缴纳的部分所得税
refund of duty 退税
refund offset 退税补偿
refund of overpayment of tax 退还溢缴税款
refund of tax 退税
refund retroactive to the past taxable years 退还以前年度税款
refunds and rebates 退款和回扣
refund the price 照原价偿还
refurbish 再刷新;重新擦亮
refurbishment 翻修;整修;改建
refurbishment of existing building 旧房整修
refurnish 再供给
refurnishment 再生修理;再生保养
refusal (钻探中的)未穿透层;(桩的)抗沉(点);(桩的)止点;桩阻力
refusal increasing factor 阻力增长系数(打桩)
refusal of building license 建造许可证拒发
refusal of consent 拒绝同意
refusal of payment 拒绝付款;拒付
refusal of pile 桩的止点(桩的最后打入深度);桩的抗沉
refusal of registration 拒绝登记
refusal point 桩的止点;拒受点
refusal pressure 极限压力;回抗压力
refusal to accept 拒绝承兑
refusal to deliver 拒绝发运
refusal to sell 拒绝出售
refusal to start 不能起动
refusal to treatment 拒受处理
refuse 垃圾;拒绝;尾矿;重新熔化;废品;废料

refuse and night soil composting 垃圾粪便堆肥化
refuse and sludge composting 垃圾污泥堆肥化
refuse an order 拒绝订货
refuse baling plant 垃圾打包厂
refuse barge 垃圾输送船
refuse bin 矸石仓;废渣仓;废石仓
refuse boiler 垃圾燃烧锅炉
refuse box 废物箱;垃圾箱
refuse bulk density 垃圾体积密度
refuse burner 垃圾燃烧炉;垃圾焚化炉
refuse burning 垃圾燃烧;垃圾焚烧
refuse burning equipment 废物燃烧设备;垃圾焚毁设备
refuse burning generation 焚烧垃圾发电
refuse-burning plant 燃烧垃圾发电厂;燃烧废料发电厂
refuse cart 手推垃圾车
refuse cartage 废物运费;废物运输
refuse channel 垃圾管道
refuse chemical and distributing center 垃圾站
refuse chemical composition 垃圾的化学成分
refuse chute 垃圾滑道;垃圾管道;垃圾坑;垃圾井筒;垃圾(井)道;垃圾槽
refuse classification 垃圾分类
refuse cleaning 垃圾清除
refuse clinker 废炉渣
refuse clinker block 废炉渣块
refuse collecting chamber 垃圾堆放室
refuse collecting station 垃圾收集点
refuse collection 垃圾收集
refuse collection lorry 垃圾运输(汽)车;垃圾收集运送车
refuse collection point 垃圾收集站
refuse collection truck 垃圾收集(运送汽)车;垃圾集运汽车
refuse collection vehicle 垃圾运输(汽)车;垃圾收集运送车;垃圾(集运)车
refuse collector 垃圾运输(汽)车;垃圾收集器
refuse combustion gas 垃圾燃烧气
refuse compacting 垃圾压缩
refuse compaction ratio 垃圾压缩比
refuse compartment 废料间
refuse composition 垃圾组成;垃圾成分
refuse composting 垃圾堆肥
refuse compressibility 垃圾压缩比
refuse conduit 垃圾(输送)管道
refuse constituent 垃圾组成
refuse container 垃圾袋;垃圾桶;垃圾箱;垃圾容器
refuse container chamber 垃圾间
refuse conveyer 废石运输机
refuse conveying 垃圾运送
refuse crane 垃圾吊车
refuse crusher 垃圾轧碎机;垃圾研碎机;垃圾破碎机;垃圾磨碎机
refuse density 垃圾密度
refuse-derived 由垃圾产生的
refuse derived fuels 从废料中提取的燃料
refuse derive energy 垃圾能
refuse destructor 垃圾焚化炉
refuse destructor furnace 废渣焚烧炉
refuse destructor plant 垃圾焚烧厂;垃圾焚化场;垃圾焚化厂
refuse difficult to burn 难燃垃圾;不易燃垃圾
refuse disposal 废物处理;垃圾处理
refuse disposal method 垃圾处理法;废物处置法
refuse disposal plant 垃圾处理厂
refuse disposal site 垃圾处理场
refused landing 中断着陆
refused on delivery 货到拒收
refused payment 拒付
refuse dump 垃圾倾弃地;废料堆;垃圾堆
refuse dumping 垃圾翻倒
refuse dumping at sea 垃圾倾入大海
refuse dump seepage 垃圾堆渗出液
refuse ejector 排渣器;废渣排出器
refuse extraction chamber 排渣室
refuse feeder 垃圾给料机
refuse feed mechanism 垃圾送料装置
refuse fermentation treatment 垃圾发酵处理
refuse filter press 尾矿压滤机
refuse for combustion 燃烧处置垃圾;燃气处置垃圾
refuse fuel 垃圾燃料
refuse furnace 垃圾焚化炉
refuse grinder 垃圾研磨机;垃圾破碎机;垃圾磨碎机

refuse handling 垃圾处理
refuse heap 废物堆;垃圾堆
refuse hopper 垃圾漏斗;垃圾斗
refuse incineration 垃圾焚化
refuse incineration plant 垃圾焚烧厂
refuse incinerator 垃圾焚化装置;垃圾焚化炉;焚秽炉
refuse incinerator plant 废物燃烧厂
refuse island 安全岛
refuse landfill 垃圾填埋场
refuse moisture 垃圾水分
refuse moisture content 垃圾含气率
refuse odo(u)r 垃圾臭气
refuse oil 废油;不合标准的润滑油
refuse ore grade 尾矿品位
refuse particle density 垃圾颗粒密度
refuse physical composition 垃圾的物理组成
refuse pit 垃圾坑
refuse pollution 垃圾污染
refuse porosity 垃圾空隙度
refuse porosity ratio 垃圾空隙比
refuse pre-treatment 垃圾预处理
refuse processing 垃圾处理
refuse processing plant 废物处理厂;垃圾处理厂
refuse production peak 产生垃圾高潮季节
refuse property 垃圾特性
refuse pulping unit 垃圾研磨装置
refuse pyrolysis plant 垃圾热解装置
refuse quantity 垃圾量
refuse receptacle 垃圾箱;垃圾桶;垃圾容器
refuse reclamation 垃圾利用;垃圾回收;废物再利用
refuse regain 垃圾回收
refuse removal 垃圾清除;清除垃圾
refuses baler 垃圾打捆机
refuse shredding 垃圾撕碎;撕碎垃圾
refuse shredding compactor 垃圾撕碎压实机
refuse shredding compressor 垃圾撕碎压缩机
refuse siding 尽头线【铁】
refuse storage chamber 垃圾储藏室
refuse storage container 垃圾储[贮]存容器
refuse storage container chamber 垃圾箱;垃圾储[贮]存桶
refuse temperature 垃圾场温度
refuse tip 垃圾(堆放)场
refuse transfer station 垃圾转运站
refuse transport channel 垃圾输送管道
refuse transport system 垃圾输送系统
refuse treatment 垃圾处理
refuse truck 垃圾(汽)车
refuse turning 垃圾翻倒
refuse utilization 废物利用
refuse volume weight 垃圾容重
refuse wagon 垃圾运输(汽)车;垃圾车
refuse water 污水;废水
refuse water disposal facility 污水处理设施
refuse water gallery 地下污水道
refuse water pipe 污水管
refuse water pipe trench 污水管沟
refuse water purification 污水净化
refuse water tunnel 地下污水道
refuse wood 废木料;废材
refuse yard 矸石场;尾矿场
refuse yarn 废纱
refusing conduit 垃圾管道
refusion 再熔(化);重熔
refutation 反演
refutation process 反演过程
refutation tree 反演树
refute the testimony 反驳证据
reg 砾质沙漠
regain 再增长;收回;标准回潮率
regaining contact with the well 重返井口
regalite (无缝的)菱苦土地面
regarded alkali soil 复原盐土
regarding 整修(圬工表面);重整坡度
regardless of the level of the productive forces 离开生产力发展要求
regards 致意
regas 再加油
regasification 再蒸发;再气化
regasify 再气化
regatta 水上节日赛艇运动;赛船会
regauged mortar 反复打出的灰浆
regelation 再凝;再冻;重新凝结;重凝作用;复冰(现象)

Regel metal 瑞格尔合金
Regency ornament 摄政时期风格装饰
Regency style 摄政时期风格;摄政王式建筑
regenerant 再生物;再生剂;回收物
regenerant consumption 再生剂耗量
regenerant level 再生剂量
regenerant measurement 再生剂计量
regenerant metering 再生剂计量
regenerate 再生的
regenerate cycle 再生周期
regenerated acid storage tank 再生酸储[贮]罐
regenerated activated carbon 活性炭再生(作用)
regenerated activated charcoal 活性炭再生(作用)
regenerated anomaly 再生异常
regenerated capacity 再生容量
regenerated carbon 再生炭
regenerated cell 再生电池
regenerated cellulose 再生纤维素
regenerated cellulose fiber 再生纤维素纤维
regenerated clay 再生陶土
regenerated crustal 再生晶体
regenerated dense medium 再生重介质
regenerated deposit 再生矿床
regenerated drainage 再生水系
regenerated energy 再生能量
regenerated fiber 再生纤维
regenerated flow 回归水流
regenerated fuel 再生燃料
regenerated glacier 再生冰川
regenerated material 再生物质
regenerated nodule 再生结节
regenerated noise 再生噪声
regenerated polymer 再生聚合物
regenerated protein fiber 再生蛋白质纤维
regenerated river 更生河
regenerated rock 再生岩
regenerated rubber 再生橡胶;再生胶
regenerated signal 再生信号
regenerated stream 更生河
regenerated water 再生水;更生水;回水
regenerate in situs 原地再生
regenerating activated carbon 再生活性炭
regenerating circuit 再生电路
regenerating column 再生塔
regenerating forest land 更新林地
regenerating furnace 交流换热炉;再生炉
regenerating medium 再生介质
regenerating oxygen apparatus 再生式氧气设备
regenerating section 再生部分
regenerating type 反馈型
regenerating used foundry sand 旧砂再生
regenerating valve 再生阀
regeneration 正反馈放大;再生(作用);再生过程;更新幼林;交流换热;复兴
regeneration area 更新面积
regeneration block 更新分区
regeneration brake 再生制动器
regeneration capacity 再生能力
regeneration circuit 再生电路
regeneration column 再生塔
regeneration connection 再生连接
regeneration counter 再生计数器
regeneration crucible furnace 蓄热坩埚炉
regeneration curve 水位回升曲线
regeneration-cutting 主伐
regeneration cycle 再生循环
regeneration efficiency 再生效率
regeneration facility 再生设施;再生设备
regeneration factor 再生因子
regeneration felling 更新采伐
regeneration field 再生区
regeneration furnace 再生炉
regeneration gas 再生气体
regeneration interval 更新间隔期
regeneration law 再生律
regeneration level 再生水平
regeneration link 再生链路
regeneration memory 再生存储器
regeneration of doctor solution 铅酸钠溶液的再生
regeneration of heat 交流换热法;热量回收
regeneration of ion(ic) exchange softener 离子交换软化剂再生

regeneration of vegetation 植被再生
regeneration oscillator 再生振荡器
regeneration period 再生周期；更新期
regeneration phenomenon 再生现象
regeneration point 再生点
regeneration pulse 再生脉冲
regeneration rate 再生速率
regeneration reuse 再生回用
regeneration signal 再生信号
regeneration sludge 再生污泥
regeneration system 再生系统
regeneration technology 再生工艺
regeneration temperature 再生温度
regeneration test 再生试验
regeneration through one's own efforts 自力更生
regeneration time 再生时间；更新期
regeneration tower 再生塔
regeneration treatment 再生处理
regeneration under cleat cutting 皆伐更新
regeneration under selection system 择伐更新
regeneration under shelterwood system 伞伐更新
regeneration unit 再生装置
regeneration valve 快速动作阀
regeneration wastewater 再生污水
regeneration water 回渗水；回ushed水
regeneration zone 再生区
regenerative 再生的
regenerative accumulator 回热储存器；反馈累加器
regenerative action 再生作用
regenerative actuator 快速回程缸
regenerative adiabatic operation 蓄热式绝热操作
regenerative afterburner 再生式后燃器
regenerative agent 再生剂
regenerative agriculture 再生农业
regenerative air heater 再生式热风机
regenerative air preheater 蓄热式空气预热器；再生式空气预热器
regenerative amplifier 再生式放大器
regenerative apparatus 再生装置
regenerative bed demineralizer 再生床净化器
regenerative brake 再生制动器
regenerative braking 再生制动(作用)；能量回收式制动；电力再生制动；反馈制动
regenerative capacity 再生能力
regenerative cell 再生电池
regenerative chamber 蓄热室；回热室
regenerative circuit 正反馈电路；再生电路
regenerative clipper 再生限幅器
regenerative coil 再生线圈
regenerative colo(u)r killer 再生消色器
regenerative condenser 回热式冷凝器
regenerative converter 再生转换堆
regenerative conveyer 反馈式输送机
regenerative cooling 再生冷却；回热冷却
regenerative coupling 再生耦合
regenerative cycle 回热循环
regenerative cycle gas turbine engine 再生回热式燃气轮机
regenerative cylinder 再生式汽缸；快速回程缸
regenerative deflection 再生式偏转
regenerative deflector 再生偏转器
regenerative detection 再生检波
regenerative detector 再生式检波器
regenerative device 再生装置
regenerative efficiency 再生效率
regenerative energy 可再生能源
regenerative engine 再生式发动机
regenerative extraction 再生引出
regenerative feedback 正反馈；再生反馈
regenerative feedback amplifier 正反馈放大器
regenerative feedback loop 正反馈电路
regenerative feed system 回热给水系统
regenerative feed-water heater 回热给水加热器
regenerative firing 蓄热式加热法
regenerative flue 蓄热烟道；回热烟道
regenerative frequency divider 再生式分频器
regenerative fuel 再生燃料
regenerative fuel cell 再生燃料电池
regenerative fuel cell system 再生燃料电池系统
regenerative furnace 蓄热(式窑)炉；再生炉；交流换热炉；交换热量炉；回热炉
regenerative gas turbine 再生式燃气轮机；回热式燃气轮机
regenerative grid detection circuit 再生栅极检波电路
regenerative heat exchange 蓄热式换热
regenerative heat exchanger 蓄热式换热器；再生(式)热交换器；再生式换热器；交流(热)交换器
regenerative heating 交流换热法采暖；再生加热；回热加热
regenerative hot blast stove 蓄热式热风炉
regenerative image intensifier 再生式像增强器
regenerative impediment 再生干扰
regenerative integrator 正反馈积分器
regenerative interlock valve 再生联锁阀
regenerative isotopes 再生同位素
regenerative laser amplifier 再生激光放大器
regenerative life support system 再生式生命保障系统
regenerative link 再生(式)链路
regenerative liquid-fuel rocket 再生式液体燃料火箭
regenerative loop 正反馈环路；再生环路；再生电路
regenerative method 复演法
regenerative modulator 再生式调制器
regenerative motor 再生发动机
regenerative noise 再生噪声
regenerative operational amplifier 再生运算放大器
regenerative oven 蓄热式炉
regenerative power energy 再生电能
regenerative preheating system 回热式预热系统
regenerative process 再生式过程
regenerative process of fume gas desulfur 回收法烟气脱硫
regenerative pump 再生泵；涡流泵
regenerative quenching 再生淬火；二次淬火
regenerative reactor 再生(反应)堆
regenerative receiver 再生(式)接收机
regenerative reception 再生式接收法
regenerative reflector 再生反射器
regenerative regenerator 再生式交流换热器
regenerative reheat cycle 回热再热循环
regenerative repeater 再生转发器
regenerative repeater for intermediate station 中间站再生中继机
regenerative repeater for terminal station 终端站再生中继机
regenerative section 再生中继段
regenerative skid steering 再生式滑动转向
regenerative soaking pit 蓄热式均热炉
regenerative solution 再生溶液
regenerative squaring amplifier 再生式方波放大器
regenerative steam turbine 回热式(蒸)汽轮机
regenerative storage 再生存储器
regenerative store 再生存储器
regenerative switch 再生式开关
regenerative system 再生系统；交流回热法；交流换热法；回热制
regenerative system of matter 物质再生系统
regenerative tank furnace 蓄热式池窑
regenerative technology 再生技术
regenerative track 快速访问道
regenerative tracking 再生跟踪
regenerative trigger circuit 再生触发电路
regenerative turbine 回热式涡轮机；回热式透平
regenerative-turbine pump 再生式涡轮泵
regenerative turboprop engine 再生式涡轮螺桨发动机
regenerative voltage 再生电压
regenerative water 再生水
regenerative zone 再生区
regenerator 蓄热器；再生装置；再生器；再生炉；再生剂；再生换热器；再生电路；交流换热器；回热器
regenerator effectiveness 回热度
regenerator flue 蓄热室烟道
regenerator kiln 蓄热式窑；再生炉
regenerator packing 蓄热室格子砖堆砌
regenerator reflux drum 再生塔回流槽
regenerator reflux pump 再生塔回流泵
regenerator section 预热部分；再生段
regenerator section overhead 再生段开销
regenerator section termination 再生段终端
regenerator setting 蓄热室布置
regenerator timing generator 再生器定时发生器
regenerator top space 蓄热室顶部空间
regent specification 试剂规格
regeration efficiency of ion(ic) exchange resin 离子交换树脂再生效率
regeration of ion(ic) exchange agent 离子交换剂再生
regeration of ion(ic) exchange resin 离子交换树脂再生
Regge pole 雷吉极点
Regge trajectory 雷吉轨迹
regia 古罗马剧院舞台中门
regime 正常状况；体制；体系
regime analysis 情况分析
regime and balance of groundwater 地下水动态与均衡
regime approach 冲淤平衡法
regime behavior 河槽演变特性
regime channel 缓变平衡河槽；稳定河道；稳定河槽；冲淤平衡的稳定河槽；不冲不淤(的)渠道
regime channel dimension 稳定航道尺度
regime concept 准衡概念
regime curve of groundwater development 地下水开采动态曲线
regime curve of groundwater level 地下水水位动态曲线
regime curve of groundwater quality 地下水水质动态曲线
regime curve of groundwater temperature 地下水水温动态曲线
regime curve of mine drainage 矿井排水量动态曲线
regime curve of single peak 单峰动态曲线
regime depth 稳定水深；冲淤平衡水深
regime discharge 河道缓变平衡流量；河道造床流量；平衡流量；不冲不淤流量
regime element change profile 动态要素变化剖面
regime equation 准衡方程
regime flow 河道缓变平衡水流；平衡水流；平衡流量；稳定流量；冲淤平衡流量
regime gradient 正常比降；冲淤平衡比降
regime method 冲淤平衡(水)法
regime(n) 格律；季节变化特征；河道缓变平衡；作息制度；生活制度；变律
regime(n) channel 缓变平衡河槽
regime(n) data 河性数据
regime(n) discharge 河道缓变平衡水流；河道缓变平衡流(量)；河道缓变平衡量
regime(n) equation 河相关系式
regirne(n) flow 河道缓变平衡水流；河道缓变平衡流(量)
regime(n) formula 河相关系式
regime(n) of river 河性；河流状况；河流情态；河流情势；河流情况
regime(n) of runoff 径流状况
regime(n) of slope 河道坡降状况
regime(n) of slope equation 河道坡降状况方程
regime(n) of stream 河性；河流状况；河流情态；河流情势；河流情况
regime(n) relation formula 河相关系式
regime(n) relationship 河相关系
regime(n) relationship formula 河相关系式
regiment 系统化；编制
regimentation 统一管理；编制
regime(n) theory 河性理论
regime(n) value 河性数值
regime observation point number 动态观测点数目
regime of confined water 承压水动态
regime of elastic deformation 弹性变形范围
regime of flow 流态；水流状态；水流状况；水流情势；水流情况
regime of groundwater development 地下水开采动态
regime of groundwater-level 水位动态
regime of high seas 公海制度
regime of international inflation 通货膨胀的国际结构
regime of phreatic water 潜水动态
regime of river 河流变迁情况
regime of runoff 径流状况
regime of subterranean river 地下暗河动态
regime of underground water 地下水动态
regime of water quality 水质动态
regime of water quantity 水量动态
regime river 平衡河流；稳定河流；冲淤平衡河流
regime stream 稳定河流；冲淤平衡河流
regime theory 准衡理论
regime value 均衡值；冲淤平衡数值；正常数值
regime width 均衡宽度；稳定宽度；冲淤平衡宽度
regina purple 苯胺紫色

reginic anti-body 反应素
reginned cotton 再轧棉
region 领域;区(域);地域;地区;地带
regional 区域(性)的;区的;地区(性)的;部位的
regional account 地区账户
regional address 区地址;同区相邻地址
regional addressing 分区访问
regional adjustment 地区审查
regional aeronautic(al) chart 区域性航图
regional air chart 区域航空图
regional air navigation 区域领航
regional alluvial coast 区域冲积海岸
regional analysis 区域分析
regional anomaly 区域(性)异常
regional anomaly of geochemistry 区域地球化学异常
regional anticline 区域背斜
regional aquifuge 区域性隔水层
regional architecture 地区建筑
regional arrangements 区域性安排
regional array 区域台阵
regional assessment 区域评价
regional assignment 划定栽培区;区域划定
regional atlas 区域地图集
regional automatic circuit exchange 地区自动线路交换机
regional background 区域背景
regional basic synoptic network 区域基本天气观测站网
regional benchmark 地下水准(基)点
regional biota 区域性生物群
regional block 区域阻滞
regional boiler plant 区域锅炉房
regional broadcast 区域性广播
regional centralized heat supply 分区集中供热
regional center 分区中心;区域中心;区域电话中心局;地区中心;地方门
regional centre of seismology for South America 南美地震学区域中心
regional change 区域性变化
regional channel 区域(性)波道;区域通道
regional chorographic(al) map 地区一览图
regional chorography 地区编制
regional clearing 冰穴;冰湖个
regional cleavage 区域性劈理
regional climate 区域(性)气候
regional climatology 区域气候学
regional code 区域编码
regional coding 区域编码
regional community 区域群落
regional compensation 区域补偿
regional computer network 区间性网络
regional constitution 区域结构
regional convention 区域性公约
regional cooperation 区域性合作
Regional Cooperation Organization for Development 区域合作开发组织
regional corner 区域边界角
regional correction 区域校正
regional correlation 区域对比
regional cycle 区域周期
regional deep tectonic background 区域深部构造背景
regional demographic characteristic 区域性人口特征
regional depot 地区储配站
regional development 区域开发;区域发展;地区(性)开发
regional development bank 区域开发银行
regional development plan 区域开发计划
regional development strategy 区域性开发战略;区域性发展战略
regional differentiation 局部分化
regional dip 区域倾斜
regional disparities 地区差别
regional dispatching 地区间调度
regional distribution 区域(性)分布
regional distribution center 区域性储存中心
regional distribution network 区域配电网
regional distribution of elements 元素区域分布
regional distributor 区域性输送线;区域性分流道路;区域干路
regional disturbance 区域扰动
regional dynamometamorphism 区域动力变质作用

regional dynamo-thermal metamorphism 区域动热变质作用
regional earthquake 区域地震
regional economic cooperation 区域(性的)经济合作
regional economic impact 区域经济影响
regional economic integration 区域经济一体化
regional economics 区域经济学;地区间经济学
regional economic structure 区域经济结构
regional economy 地区经济
regional engineering geologic(al) map 区域工程地质图
regional engineering geology 区域工程地质(学)
regional environment 区域环境
regional environmental anomaly 区域性环境异常
regional environment(al) assessment 区域环境评价
regional environmental monitoring 区域环境监测
regional environmental noise 区域环境噪声
regional environmental planning 区域环境规划
regional environment(al) water quality standards 地区环境质量标准
regional epidemicological analysis 地区流行病学分析
regional evapo(u)ration 区域蒸发
regional exhibition 地方性展览
regional express railway 地区快运铁路
regional facility 区域设施
regional factor 地区(性)系数;区域性因素
regional flight 地区航线
regional flood 区域性洪水;地区性洪水;地方性洪水
regional flood forecast 区域洪水预估;区域洪水预测;区域洪水预报
regional forecast 区域预报;区域报告
regional gains and losses 地区性的获得和损失
regional gains of nitrogen 地区性氮获得
regional gap 地区差价
regional generalization 区域综合;区域简化;区域概化;地区综合;地区简化;地区概化
regional geochemical anomaly 区域性地球化学异常
regional geochemical process 区域地球化学作用
regional geochemical survey 区域地球化学调查;区域地球化学测量
regional geochemistry 区域地球化学
regional geography 区域地理(学);地质学
regional geologic(al) map 区域地质图
regional geologic(al) map based on remote sensing 遥感区域地质图
regional geologic(al) mapping 区域地质填图
regional geologic(al) survey(ing) 区域地质调查
regional geology 区域地质(学)
regional geology-hydrogeologic(al) survey 区域性地质—水文地质调查
regional geomorphology 区域地貌(学)
regional geophysical anomaly 区域地球物理异常
regional geophysical prospecting 区域地球物理勘探
regional geophysical survey 区域地球物理调查
regional geotechnique 区域岩土力学;区域土质学
regional gradient 区域梯度
regional gravity 区域重力
regional gravity anomaly 区域重力异常
regional gravity anomaly map 区域重力异常图
regional gravity survey 区域重力调查
regional gross survey stage 区域概查阶段
regional ground water system 区域地下水流系统
regional harbo(u)r 区域性港口;地区性港口
regional heat flow 区域热流量
regional heating 区域(集中)供热
regional heating plant 区域供热锅炉房
regional heat source 区域热源
regional highway 区域性公路
regional hub city 区域中心城市;地方核心城市
regional hydrochemical abnormality 区域水化学异常
regional hydrogeologic(al) investigation 区域水文地质调查
regional hydrogeologic(al) map 区域水文地质图
regional hydrogeologic(al) profile 区域水文地质剖面图
regional hydrogeologic(al) survey 区域性水文地质调查

regional hydrogeology 区域水文地质学
regional hydrogeology history method 区域水文地质历史法
regional hydrologic(al) cycle 区域水文循环
regional hydrology 区域水文学
regional hypothesis 区域假定
regional identify code 区域标识码
regional impact analysis 区域影响分析
regional influence 地区影响
regional information 区域性信息
regional infrastructure 区域基础设施;地区基础设施
regionalism 区域划分;地区主义;地区特性
regional isostasy 区域均衡
regional isostatic anomaly 区域均衡异常
regional isotope equilibrium 区域同位素平衡
regionality 地区性
regionalization 区域划分;区域化;区划;地区化
regionalization map 区划地图
regionalization of water quality parameters 水质参数区别
regionalized variable 区域化变量
regional key 地区判读样片
regional level 区域水平
regional level(l)ing survey 区域性水准测量
regionally harmonized standard 区域协调标准
regional magnetic anomaly 区域磁异常
regional magnetic anomaly chart 区域磁异常图
regional map 区域地质(图);区域地图
regional map of assemblage of the heavy minerals in the superficial sediments 表层沉积物中分区重矿物组合图
regional maps 区域性图件
regional market 区域性市场
regional marshalling station 区域性编组站【铁】
regional metamorphic rock 区域变质岩
regional metamorphic rocks 区域变质岩类
regional metamorphism 局部变质(作用);区域混合岩作用;区域变质(作用)
regional metasomatism 区域交代作用
regional meteorologic(al) center 区域气象中心
regional meteorologic(al) office 区域气象中心
regional metro 地区性地下铁道
regional migmatization 区域性混合岩化
regional migration 区域性迁移
regional movement 区域性运动
regional multiplier 区域乘数
regional nations 地区国家
regional natural park 区域自然公园
regional network 区内电力网
regional nitrogen budgets 地区性氮的预算
regional observation net 地区性观测网
regional oceanography 区域海洋学
regional office 地区办事处
Regional Organization for the Protection of the Marine Environment 海上环境保护区域组织
Regional Organization of Protective Marine Environment 保护海洋环境区域组织
regional overstep 区域超覆
regional paracycle 区域准周期
regional park 区公园
regional petrology 区域岩石学
regional picture 区域像片
regional plan 区域(平面)图;区域规划;区划图;地区计划
regional planning 区域规划;地区规划
regional planning commission 区域规划委员会
regional planning in metropolitan areas 都市区域规划
regional planning program(me) 区域规划方案
regional planning stage 区域规划阶段
regional pollution 区域(性)污染;地区(性)污染
regional pollution model 区域污染模式
regional pollution resources 地区性污染源
regional pollution sources 区域污染源
regional port 区域性港口;地区性港口
regional port authority 地区性港务局
regional power station 区域性发电厂
regional precipitation 区域雨量;区域降水(量)
regional prediction 区域预测
regional preliminary trials 初级区域试验
regional price differences 地区差价
regional price differential 地区差价
regional products 区域性产品
regional profile 区域性剖面;地区性剖面

regional pumping station 区域泵站
regional quota 区域配额制;按国家分配的输入配额
regional rail network 地区铁路网
regional railway 地方铁路
regional rate 地区级
regional reconnaissance 区域勘探;区域勘察;区域勘测
regional reflex 区域反射;分节反射
regional regularity of groundwater 地下水的区域性规律
regional representative station 区域代表站
regional resettlement strategy 库区移民安置方针;地区移民安置方针
regional-residual correction value 区域剩余校正值
regional resource inventory 区域资源调查
regional resource investigation 区域资源调查
regional road 区域性道路
regional road network 区域道路网
regional runoff coefficient 区域径流系数
regional scale 区域规模
regional science 区域科学
regional sea 区域海洋
regional search 区域探查
regional securites exchanges 区域性证券交易所
regional seismic coefficient 区域地震系数
regional-seismic geology 区域地震地质
regional seismicity 地区基本烈度;区域烈度
regional seismic network 区域地震台网
regional seismic risk 区域地震危险性
regional seismology 区域地震学
regional setting 区域环境
regional settlement 区域性沉降
regional sewage disposal 区域污水处置;区域污水处理
regional sewage system 区域污水系统
regional sewage treatment 区域污水处理
regional sewage treatment facility 区域污水处理设施
regional sewerage system 区域排水系统
regional shopping center 区域性购物中心;区域性商业中心
regional sketch 地区性略图
regional slope 区域坡度
regional slope deposit 区域坡面沉积
regional smoothing of isohyets 等雨量区域修匀
regional soil 地区性土(壤);地带性土(壤)
Regional Soil Resources Development Projects for Africa 非洲区域性土壤资源开发计划
regional specialization 区域性专业
regional specialty meteorological center 区域专业气象中心
regional stability of mine 矿山区域稳定性
regional standard 区域标准;地区标准
regional standardization 地区标准化
regional standardization organization 地区标准化组织
regional standard of water quality 地方性水质标准
regional standards organization 地区标准组织
regional station 区域站;地方无线电台
regional stock exchange 地区证券交易所
regional stratigraphy 区域地层学
regional stress 区域应力
regional strike 区域走向
regional structural map in time domain 时域区域构造图
regional structural position of the ore deposit 矿区尽域构造位置
regional structure map 区域构造图
regional subsurface heat flow 区域地下热流
regional supercycle 区域超周期
regional survey 区域勘探;区域勘察;区域勘测;区域调查;区域测量
regional survey stage 区域普查阶段
regional syncline 地槽;大地向斜
regional-synthesis 区域综合;地区综合;分区合成
regional system 局部的系统;城区系统
regional teaching materials 乡土教材
regional tectonic movement 区域构造运动
regional tectonics 区域构造地质学
regional tectonic stress field 区域构造应力场
regional telecommunication hub 区域电信枢纽
regional test 区域(性)试验;区域(性)检验
regional testing program(me) 区域试验方案
regional threshold 区域异常下限

regional town planning 区域城市规划
regional transportation planning 区域运输计划;区域运输规划
regional transport building 外部运输建设
regional trend 区域性走向
regional trial 区域试验
regional type distribution 区域形分布
regional unconformity 区域不整合
regional unit 区域单位
regional utility 区域公用事业
regional variation 区域性变化
regional verification office 区域检定所
regional volcanic complex 区域火山组合
regional volcanic geology 区域火山地质
regional volume table 区域材积表
regional water 区域水
regional water balance 区域水量平衡
regional water intercepting 全区性截流
regional water pollution in the Gulf 区域性海湾水污染
regional water supply 区域供水;区域给水
regional water supply and waste project 区域给水和废水项目
regional water supply planning 地区给水规划
regional water supply system 区域给水系统;区域供水系统
regional water system 区域水系;地水水系
regional weather office 区域气象站
regional wire 区域线
regional wireless beacon 区域无线电信标
region class 区域级;区分类
region code 区域码
region control task 区域控制任务
region-dependent segmentation 区域分割
region elimination 区域消去
region enclosed by a curve 曲线所围包的区域
region estimation 区域估计
region extraction 区域抽取
regionite 大都市区居民
region job pack area 区域操作装配区;分区作业组合区域
region list 区域表
region of acceptance 接受域;接受区(域)
region of alimentation 补给区
region of binary glass formation 二元玻璃形成区
region of connective stability 联结稳定域
region of construction 建筑区;建设区
region of convergence 收敛区(域)
region of defocusing 散焦区
region of disease 发病区
region of disorder 无序区;不规则区
region of dispersed water 水分散区
region of dissipation 散热区
region of disturbance 干扰区(域)
region of elongation 延长区
region of fracture 断裂区域
region of heat accumulation 聚热区
region of heating 加热区
region of high activity 强放射性区
region of high barometric(al) pressure 高气压区
region of high-pressure 高压区
region of high stress 高应力区
region of humid tropics 湿润热带地区
region of hydrothermal alteration 水热蚀变区
region of indeterminacy 不肯定区域
region of initiation 原始区;创始区
region of intake 补给区
region of integration 积水区
region of interest 研究区;感兴趣区;评价区
region of limited proportionality 有限正比区(域);比例有限区
region of little relief 小地形区域
region of low-pressure 低压区
region of maximum shear 最大剪力区
region of melting 熔融区
region of non-operation 不动作范围
region of non-rejection 非拒绝区域
region of no pressure 真空区
region of no relief 平原区;平坦地区
region of operation 运行区
region of outflow 排泄区
region of perpetual snow 永久积雪区
region of power supply 供电范围
region of proportionality 正比区

region of rationality 有理数域
region of rejection 拒绝(区)域;否定(区)域
region of search 搜索区(域);查索区(域)
region of separation 分离区
region of stable values 稳定值区
region of streaking 拖尾区域
region of target 目标区
region of ternary glass formation 三元玻璃形成区
region of thunderstorm activity 雷暴活动区
region of transitional stress 转移应力地区
region of ultimate boundedness 毕竟有界区域
region of vorticity initiation 旋涡起始区
region of water pollution 水污染区
region-oriented 面向地区需要的
region parameter 区域参数
region sheltered form winds 避风处
regions inhabited by ethnic groups 民族地区
regions of flow 流域
region static 区域静校正
region strain 区域应变
region structure of population 人口地区构成
region studied 研究地区
region with abundant light supply 光照充足的地区
regioselective reaction 区域选择性反应
regioselectivity 区域选择性
register 自动记录器;注册员;累加器;节气阀;寄入器;寄存器;记忆装置;计数器;记录员;记录器;分发器;名册;调温装置;调风器;套准;示度;登记员;登记表
register bed 记录牌;记录板
register block 寄存器组
register book 船舶登记簿
register box (管子通入室内的)接管箱
register button 记录按钮;计数按钮
register calipers 指示卡钳
register capacity 寄存器容度
register chooser stage 记选级
register circuit 记忆电路
register constant 记录器常数;计量器常数
register contention 抢占寄存器
register control 定位控制器
register controlled system 记录器控制系统
register corner 规矩线
register designator 寄存器指示符
register device 套准装置
register difference 套合差
registered 已注册的;登记的
registered architect 注册建筑师
registered assets 登记资本
registered bonds 记名债券
registered book 登记证书
registered brand 注册商标
registered breadth 登记宽度
registered capital 注册资本;已注册资本;注册股本
registered certificate of shares 注册股票
registered certification 记录在案的证明;记录在案的鉴定
registered certification trade mark scheme 注册商标示意图
registered company 法人公司
registered corporation 已注册公司
registered depth 登记深度
registered design 已注册设计;注册图案;设计商标
registered dimension 登记尺寸
registered engineer 注册工程师
registered firm 记名公司
registered gross tonnage 登记总吨位
registered horsepower 注册马力;登记马力
registered information provider 注册信息提供当局
registered land 在册地址
registered length 寄存器可容符号数;登记长度
registered letter 挂号信
registered luggage 托运的行李
registered mail 挂号信
registered net tonnage 注册净吨位;登记净吨位
registered number 登记号(码);注册号码;注册编号
registered office 注册地址
registered paper 记名证券
registered permanent residence 户口
registered plot 注册地块;登记的地块
registered population 注册人口;定居人口;常住人口
registered resident 户口

registered seed 原种
registered shares 记名股票
registered ship 注册船
registered stock 记名股票
registered ton 注册吨
registered tonnage 注册吨位;登记吨位【船】
registered trademark 注册商标
registered trader 注册交易人
registered unemployment rate in cities and towns 城镇登记失业率
registered user 注册用户
registered warrant 已登记支付书
registered water stage 记载水位;记录水位;登记水位
register error 套合误差
register face 计数器面板
register finder 记发器选择器
register galley 定位活版盘
register gate 调节闸门
register glass 定位测微镜;成像平面玻璃片
register-glass reseau 压平玻璃检查格网
register group 寄存器组
register guide 定位指示器;导向记录器
register hole 记录孔;套印孔
register hole punch 套合定位孔
register image 寄存器映像;重合影像
register indexed address 寄存器索引地址
registering 配准;记录;套版
registering apparatus 自记器;记录仪器;记录设备;记录器
registering arrangement 记录设备
registering balloon 探测气球
registering calipers 指示卡规
registering chronograph 自记计时器
registering equipment 记录设备
registering instrument 自记器
registering manometer 自记录压力计
registering micrometer 记录测微器
registering pen 记录笔
registering pluviometer 自记雨量计
registering rain ga(u)ge 雨量器;雨量计
registering ribbon 记录纸带
registering statoscope 自记高差仪
registering surface tolerance 定位面公差
registering thermometer 自记温度计;自记温度表
registering tide gauge 自动记录潮位计
registering voltmeter 记录伏特计
registering weather vane 风向计;风向标
register instrument 记录仪(表)
registerion certificate 注册执照;注册证(书)
register key 记数器按钮;记数键;记发器按钮;计数器按钮
register lamp 记录器指示灯
register length 寄存器长度
register lock-up 记录定位装置
register mark 规矩线;十字规矩线;登记标志
register observation device 定位检视器
register of collections 捐税收入簿
register of expenses 经费收支登记簿;费用(支出)登记簿
Register of International Conventions and Protocols 国际公约和议定书登记簿
register of luggage and parcel accident 行包事故立案
register of ownership of state-owned assets 企业国有企业资产产权登记
register of real estates 不动产登记;房地产登记
register of shipping 船舶检验局
Register of shipping of the People's Republic of China 中华人民共和国船舶检验局
register of toxic chemicals 有毒化学品登记
register pin 套合栓;定位针;定位销
register plate 承片玻璃板
register pointer 寄存器指示器
register proof 套合样图
register punch 套合定位孔
register rack 记发器架
register ratio 机械速度比;机械传动比;机械齿轮比;齿轮比
register reading 寄存器读数;计数器读数
register relay 记录继电器
register renaming 寄存器重命名
register rollers 套准调节辊
register rotation 寄存器循环移位

register saving operation 寄存器存储操作
register-sender 记录发送机
register set 寄存器组;对版装置
register sheet 对版台纸
register signal 记录信号
register stack 寄存器栈
register sticks 十字规矩线
register stove 居室火炉
register stud 打孔定位销栓
register system 寄存系统;记录系统;记发制
register table 拼版台
register tax 登记税
register the changes 变更登记
register trail 套合试验
register-translator 记发器—转发器
register tube 定位测微镜
register valve 调节阀
register work 套印
registrant 登记员
registrar 挂号员;记录员;登记员
registrar general shipping 船舶登记局
registrar of deeds 契据登记员
registrar's office 注册处
registrating frame 承片框
registration 注册;记录;配准度;对准;对齐;登记;重合;重叠
registration accuracy 记录精度;配准精度;套合精度;重合精度
registration arm 定位管;定位杆
registration arrangement 配准系统;重合装置
registration authorities 登记部门
registration bar 套合定位杆
registration certificate of import 进口登记证明书
registration chart 光栅配准测视图;配准测试卡;重合测试卡
registration clamp 定位线夹
registration control 配准控制;重合控制
registration device 定位装置
registration drift 配准漂移
registration error 配准误差;套合误差;重合误差
registration fault 配准缺陷;重合缺陷
registration fee 注册费
registration files 注记文件
registration law 注册法
registration mark 注册号;规矩线;记录标记
registration mark light 牌照灯
registration mast 定位柱
registration number 登记码;登记号(码)
registration office 注册处;挂号处
registration of juristic persons 法人登记
registration of patent 专利权登记
registration paper 记录纸
registration problem 注册问题;登记问题
registration sticks 十字规矩线
registration unit 记录装置
registration wire 定位索【铁】
registrator 记录装置
Registro Italiano Navale 意大利船级社
Registro Italiano 意大利船级社
registry 注册处;挂号处;记录处;配准;登记处
registry agency 船舶登记局
registry of shipping 船级社
reglaze 重配玻璃
reglazing 重新装配玻璃
regle 滑槽
reglet (混凝土中的)狭长槽;墙上槽口;嵌料缝;平嵌(饰)线【建】
reglette 读数尺;(基线尺两端的)端尺;精密短尺;精密端尺;基线尺端点分划尺
reglowing 再辉;再炽热
regmagenesis 破裂作用;大断裂作用
regnant fashion 流行样式
Regnault hygrometer 雷诺德湿度计
regnum vegetabile 植物界
regolith 表土;土被;表岩屑;基岩上松软层;浮土;风化层
Regon exhaust gas treatment process 雷特法
regosity 皱褶状态
regosol 岩成土;岩层土;粗骨土;非固结屑土
regosolic soil 粗骨土
regradation 再均夷(作用)
regradation of soil 土壤复原
regrade 再分级
regraded alkali soil 再生碱土;复原碱土

regraded saline soil 复原盐土
regrading 重叠坡度;二次筛分;二次分级
regrading of line 线路坡度整修;线路改坡
regrading skin 修整表面
regrading stream 均夷流
regrate 修琢;石面琢新;重轧
regrating 用斧修整石料表面
regrating skin 石面琢新
regressed luminaire 缩入式照明设备;退缩照明装置(装在顶棚线之上的照明装置);泛光照明
regression 逆回演替;回归(作用);退行;退化作用;退步;衰退;倒退;返回
regression analysis 回归分析
regression analysis forecasting 回归分析预测法
regression analysis model 回归分析模型
regression and correlation analysis 回归和相关分析
regression coefficient 回归系数
regression conglomerate 海退砾岩
regression curve 回归曲线;回归截取曲线
regression equation 回归方程(式)
regression equation model 回归方程模型
regression equation of deverisity 变化度的回归方程
regression estimate 回归估算;回归估计
regression estimate value 回归估值
regression estimation of event probability 事件概率回归估计
regression estimator 回归估计量
regression factor 回归因子
regression function 回归函数
regression line 脊线;回归(直)线
regression matrix 回归矩阵
regression model 回归模型
regression of glacial tongue 冰川舌退缩
regression of sea 海退
regression of the nodes 交点西退;交点退行
regression of the orbit 轨道退行
regression parameter 回归参数
regression phase 海退相
regression point 回归点
regression program(me) 回归程序
regression relationship 回归关系式
regression species 退化种
regression speed 海退速率
regression surface 回归(曲)面
regression technique 回归技术
regression test 退化试验
regression theory 回归理论
regression-type statistic(al) analysis 回归型统计分析
regression variable 回归变量
regressive 海退的;退化的
regressive adaptability 退化适应
regressive burning 减推力燃烧;减面燃烧
regressive burning charge 减面性燃烧火药柱
regressive coast 后退海岸
regressive combustion 减面燃烧
regressive cycle 海退旋回
regressive depreciation 递减折旧
regressive derivative 右导数
regressive differentiation 逆行分化
regressive erosion 向源冲刷;逆向侵蚀;海退侵蚀
regressive evolution 逆行演化
regressive facies 海退相
regressive interpolation 回归插值法
regressive metamorphosis 退化变态
regressive overlap 海退退覆;海退超覆;海滩退覆;退覆
regressive performance curve 减推力特性曲线
regressive reef 海退礁
regressive rhythm 水退韵律
regressive ripple 逆行流痕;退流波痕
regressive ripple mark 逆行波痕
regressive sand ripple 逆行沙痕
regressive sand wave 逆行沙波
regressive sediment 海退冲积(物);海退沉积物
regressive sequence 海退层序
regressive series 海退岩系;海滩岩系
regressive shear thickening 递降剪切稠化
regressive shear thinning 递降剪切稀化
regressive succession 海退序列
regressive supply 逆进供应
regressive tax 累退税;递减税

regressive tax rate 递减税率
regressive type 消退型
regressivity 燃烧面递减性
regret criterion 塞韦奇原则
regret critical value 遗憾临界值
regrind 重新研磨;重磨的;二次研磨;再磨;再次研磨
regrindability 磨锐性
regrind bit 重磨钻头
regrinding 再次粉磨;再磨;再次研磨;重磨削;复磨;二次粉碎物料
regrinding circuit 再磨流程;再磨回路
regrinding degree 再粉碎程度
regrinding mill 细磨机;第二级磨机
regrinding of die 模具的再研磨
regrinding section 再磨部分
regrinding tool 重磨削工具
regrinding unit 再磨机
regrind mill 细粉磨;再磨机
regroover 再次刻纹机;胎面开槽机
regrooving 再次刻纹
regrooving of worn tyre 胎面花纹翻新
reground 再次粉磨的;重新研磨的;重新粉磨的;再磨的
reground floor 重铺地板
reground material 再磨物料
regroup 重新组合
regrouping data item 数据项再分组
regrouping switch 重新组合开关
regrown material 再生材料
regrown region 再生区
regrowth 再生长
regula(e) 扁饰带;三角板下短条线脚;方嵌条
Regulaitons of the People's Republic of China on Administration for Environmental Protection of Offshore Oil Exploration and Exploitation 中华人民共和国海洋石油勘探开发环境保护管理条例
regular 正规的;正常的;整齐的;规则状;规则的;单层的
regular affine transformation 正则仿射变换
regular aid 定期补助
regular airport 正规航空站
regular air service 定期班机
regular algebraic equation 正则代数方程
regular alumina 重结晶氧化铝
regular analysis 普通分析
regular analytical function 正则分析函数
regular analytic curve 正则解析曲线
regular angle-type step 普通角型梯级
regular arc 正则弧;正弧
regular arm 箱拱;令拱
regular army 常备军
regular asteroid 正(则)星形线
regular astigmatism 正则像散性;规则散光
regular band 有序谱带
regular bank account 定期银行存款账户
regular barrel 正规桶;标准桶
regular bed 规则层;稳定层
regular bedding 普通层理;整合层理;规则层理
regular bedset sequence 规律层组式层序
regular binary 正规二进制;普通二进制
regular bond 正规砌合;普通砌合
regular bonus 经常性奖金
regular boundary point 正则边界点
regular branch office service 经常性分行服务
regular budget 经常预算
regular burden(ing) 正常料;正常负荷
regular bus 班车
regular calling 正常停靠
regular calomel cell 正常甘汞电池
regular calomel electrode 正常甘汞电极
regular cardinal number 正则基数
regular Cartan space 正则嘉当空间
regular cast method 普通浇注法
regular cement 普通水泥;正常水泥
regular center 普通顶尖
regular chain 正则链
regular change 正常变化
regular channels 正常渠道
regular check 定时检查
regular check and repair of locomotive 机车定期检修
regular checking 合格检验;合格检查;定期检查
regular classroom 正规的教室

regular client 常年委托人
regular closed set 正则闭集
regular closed subset 正则闭子集
regular coast 规则海岸;平直海墙;平直岸;顺直(海)岸
regular coastline 顺直海岸线;顺直岸线
regular college course 本科
regular column 正则列
regular command 规则命令
regular commutator 正则换位子
regular complement matrix 正则补阵
regular complex function 正则复函数
regular concrete sand 正规混凝土用砂
regular connector 通用连接器
regular content 正则容度
regular continued fraction 正则连分数
regular contour 规则轮廓
regular convergence 正则收敛
regular convex solids 正凸多面体
regular cost computation 定期成本计算
regular coupling 普通联轴器;常见联轴器
regular course 整层砌的
regular coursed ashlar masonry 普通成层琢石砌体;普通成层琢石圬工;整层砌筑琢石圬工
regular coursed ashlar stone work 整齐方石砌筑
regular coursed rubble 成层乱砌毛石;成层琢石;成层砌筑毛石;正规砌筑毛石;整(齐)层砌的毛石;毛石整层砌
regular covering space 正则覆盖空间
regular covering surface 正则覆盖面
regular cracking 规则裂缝
regular crack 规则裂缝
regular crystal 正方晶
regular cultivation 常规耕作
regular curve 正弧;正则曲线
regular-cut share 正常切幅犁铧;通用犁铧
regular cutting 预定的采伐
regular cyclic(al) group 正规循环群
regular decagon 正十边形
regular demand 固定需求
regular diamond 规则金刚石
regular directed-graph 正则有向图
regular discharge 正常放电
regular distortion 正规失真;规则畸变
regular distribution 正规分布;定长分布
regular diurnal tidal current 全日潮流
regular dividend 固定股息;经常性股息
regular diving dress 常规潜水服
regular divisor 正则分母
regular dodecahedron 正十二面体
regular dome 正规球形穹顶
regular dot pattern 规则点图样
regular doublet 正常双线
regular down time 定期停歇时间
regular drawing rights 一般提款权
regular element (数学中的)正规元素;正常的因素
regular embedding 正则嵌入
regular employee 正式职员
regular energy consumption 经常能耗
regular equipment 正规设备;标准设备
regular error 系统误差;规律误差
regular estimate 正规估计
regular expenses 经常费(用)
regular expression 正则表达式;正规表达式
regular extension 正则扩张
regular extinction 规则消光
regular eye needle 穿线孔缝合针
regular false interaction 试位迭代法
regular false method 试位法
regular fee 正常费用
regular feed 正常钻进;正常进给;正常给进
regular five-hedron 正五面体
regular flight 正常航标
regular flow 正常流
regular flower 整齐花
regular forecast 定期预报
regular frustum 正平截头体
regular fuel super car 普通燃油超级车
regular function 正规函数;正则函数;正常函数
regular function algebra 正则函数代数
regular function of complex variable 复变量正则函数
regular gas 普通汽油
regular grade 正规坡度;正规品级;正常坡度;普通

品类
regular graph 正则图
regular growing 正常生长
regular growth 正常生长
regular growth patterns 规则的生长方式
regular guard 标准型切割器护刃器
regular gypsum wallboard 普通石膏墙板
regular hand feed 正常手动进给
regular hexagon 正六边形
regular hexahedron 立方体;正六面体
regular icosohedron 正二十面体
regular inclusion 规则包体
regular income 经常性收入
regular inspect 定期检验
regular inspection 定期检查;常规检查
regular integral 正则积分
regular integral element 正则积分元
regular integral manifold 正则积分流形
regular interrupt 正规中断
regular intersection 一般交叉口;普通交叉口
regular interval 正则区间;等值等距(离)
regular inverse matrix 正则逆阵
regular investigation 常规调查
regularity 正则性;规则性;规律性
regularity assumption 正则假设
regularity attenuation 正规衰减
regularity condition 正则性条件;正则条件
regularity meter 均匀度试验器
regularity of aggradation and degradation 冲淤规律
regularity of distribution 分布规律
regularity of fluctuation in stage 水位涨落规律
regularity of oscillations 振荡规律性
regularity of outbreak 发生规律
regularity of scour and silting 冲淤规律
regularity of shoal process 浅滩演变规律
regularity property 正则性质;正规性质
regularity rule 规范性法规
regularization 正则化;正规化;重整化
regularization of land tenure 调整土地使用权
regularization of the earth 地球质量调整
regularized-photon propagator 规则化光子传播函数
regularizing 木板刨平;木板规整化
regularizing transposition 正则化转置;正规化转置
regularizing variable 正则化变数
regularizing variogram 正则化变差函数
regular jaw 固定钳口
regular lay 正规捻;交叉捻(绕);普通捻;标准捻绕
regular lay left lay 钢丝绳左向交叉捻绕
regular lay right lay 钢丝绳右向交叉捻绕
regular lay rope 正搓绳;交叉捻钢丝绳;普通捻钢丝绳
regular lightship 正规灯船
regular line 正常航线;定期航线
regular linear mapping 正则线性映射
regular linear service 定期航运业务
regular liner 定期班轮;班轮
regular line speed 正常生产线速度
regular link 正常线路
regular load 正则荷载
regular local equation 正则局部方程
regular local ring 正则局部环
regular longwall along strike 沿走向规则长壁法
regular lot 经常批量
regularly hyperbolic operator 正则双曲算子
regularly perforated 有规则穿孔的
regularly pulsating laser 规则脉动激光器
regularly spaced signal 纯周期性信号
regular maintenance 例行检修;定期检修
regular maintenance of buildings and structures 房建维修
regular manifold 正则流形
regular map 正规图幅
regular mapping 正则映射
regular matrix 正则矩阵;正规矩阵;规则矩阵
regular maximal ideal 正规极大理想
regular measure 正则测度
regular meeting 例会
regular mesh point 正常网点
regular mirror 规则镜
regular motor oil 正规的车用机油
regular net 规则网格;规则格网
regular net region 规则网格区域

regular network 规则网
regular node 规则节点
regular noise 规则干扰
regular normed space 正则赋范空间
regular observation 定时观测；常规观测
regular octahedron 正八面体
regular of space distribution 空间分布规律性
regular order 正规定单
regular ordinal 正则序数
regular ordinal number 正则序数
regular outer measure 正则外测度
regular overhaul(ing) 定期检修
regular packing 规则填料
regular parameter 正则参数
regular parcel post 一般包裹
regular pattern 正常航线；规律
regular pattern composite 规则型
regular pentagon 正五边形
regular permutation group 正则置换群
regular perturbation 正则摄动
regular perturbation system 正则摄动系统
regular pillar file 普通柱锉
regular pitch chain 标准节距链
regular pitch roller chain 标准节距套筒滚子链
regular point 正则点
regular point of spectrum 谱的正则点
regular polygon 正多角形；正多边形【数】
regular polyhedral angle 正多面角
regular polyhedral group 正多面体群
regular polyhedron 正多面体；规则多面体
regular polymer 规整聚合物
regular Portland cement 普通波特兰水泥
regular position 正则位置
regular practice 习惯做法
regular price 正常价格
regular prism 正棱柱；规则棱镜
regular procedure 正规手续
regular procession 有规则进程
regular production 正常生产
regular program(me) 正常程序
regular progression 有规则进程
regular projection 正则射影
regular projective transformation 正则射影变换
regular pulse excitation-long term prediction 规则码激励长期预测
regular pyramid 正棱锥(体)
regular ramification 正则分歧
regular rational mapping 正则有理映射
regular reflectance 正常反射率；单向反射率
regular reflection 正反射；规则反射；单向反射；单相反射
regular reflection factor 平面反射因数
regular regeneration 正常再生
regular reinforcement 普通钢筋
regular representation 正则表示
regular right ideal 正则右理想
regular ring 正则环
regular runway 常用跑道
regular sampling 按常规取样
regular sanitary supervision 经常性卫生监督
regular satellite 规则卫星
regular screw thread 基本螺纹；普通螺纹
regular sea 规则波
regular selector 普通选择器
regular sequence 正则序列
regular service 正常运行；正常使用；定期快速交通；定期航线
regular service condition 正常(的)使用条件
regular set 正则集
regular shackle 正装卸扣
regular shackle block 带卸扣滑车；带斜扣滑车
regular shape 规则现状
regular ship 定期船
regular shore 顺直(海)岸
regular shoreline 顺直(海)岸线
regular simplex 正则单形
regular single-mo(u)ldboard plow 普通单铧犁
regular singularity 正则奇点
regular singular point 正则奇点
regular size 正规尺寸；标准尺寸
regular sleeper 普通轨枕
regular slope 均匀坡
regular smooth bottom steel bench plane 普通平底钢刨

regular soft-center steel 正心软心钢
regular solution 正规溶液；规则溶液
regular solution model 正规溶液模型
regular solution theory 正规溶液理论
regular space 正则空间
regular spacing 整齐间距
regular spectral sequence 正则谱序列
regular speed 正常速度
regular spike 规则尖峰
regular stone 方正石；整齐石(料)；条石料
regular streaks 规则条纹
regular stream pattern 正常水系；正常河型
regular strength 普通强度
regular structure 正规结构
regular subdivision 正则剖分
regular submanifold 正则子流形
regular substitution group 正则代换群
regular supply 正常供应
regular surface element 正则曲面元素
regular surface ga(u)ge 普通平面规
regular survey 正规测量
regular survey station 固定测站
regular survey station point 固定测站点
regular system 正则系统；等轴晶系
regular system of parameters 正则参数系
regular tenacity 普通强度
regular tetrahedral orientation 正四面体取向
regular tetrahedron 正四面体
regular thread 正规扣
regular thumb screw 对称翼形螺钉
regular tidal-cyclic sequence 规律间隙式潮后周期层序
regular tie 普通木枕；普通轨枕
regular time scanning 定时扫描
regular tone variation 色调变化规则
regular tool-grinding machine 标准工具磨床
regular tool joint 正规接头
regular topographic(al) map 标准地形图
regular transition probabilities 正则转移概率
regular transmission 正透射；正常传输；规则透射
regular triangle 正三角形
regular tube 正则管
regular turn 正常次序；按序靠泊
regular turning tool 普通车刀
regular twill 正则斜纹
regular twist 反手捻
regularty of epidemic 流行规律
regular-type biparting gate 普通对开闸门
regular-type gear lubricant 标准齿轮润滑油
regular unbia(s)sed critical regions 正则无偏临界域
regular unit matrix 正则单位矩阵
regular variable 规则变星
regular variation 规则变化
regular variety 正则簇
regular ventilating circuit 正常的通风线路
regular wave 规则波
regular wave train 规则波列
regular wholesaler 正规批发商
regular work 正规工作；正常工作；日常工作
regular worker 正式工人；固定工人
regular work hours 正规工作时间
regulated amplifier 控制放大器
regulated attenuator 调节衰减器
regulated channel 整治了的水道；经过整治的水道；经过整治的河槽；经过整治的航道
regulated company 控制投资公司
regulated cost 节制性成本
regulated crossing 管制交叉口
regulated discharge 经过调节的流量；调节流量
regulated discharge during dry season 枯水期调节流量
regulated fee 正规费用率
regulated flow 经过调节的流量；调节水流；调节流量；调节径流
regulated geoid 调整大地水准面
regulated investment company credit 投资公司税收抵免
regulated level 经过调节的水位；调节水位
regulated polymer 有规聚合物
regulated power supply 稳压电源；稳定电源
regulated quantity 调节量；被调量
regulated receiver 稳定接收机
regulated rectifier 稳压整流器

regulated river 已整治河流；经过整治的河流
regulated-set cement 控凝水泥；调凝水泥
regulated-set cement and jet cement 调凝水泥和喷射水泥
regulated-set concrete 调凝凝固混凝土
regulated split 风流调节分路
regulated start 稳定起动
regulated state of track circuit 轨道电路调整状态
regulated stream 已整治河流；整治河流；经过整治的河流；稳定河流；调节河流
regulated stream flow 调整流量；调节水流；调节河流流量
regulated value 调节值
regulated variable 受控变量；被调节量
regulated voltage 稳压；稳定电压
regulated water stage 整治水位；调节水位
regulated width 整治宽度
regulate expenditures 控制用费
regulate in a particular manner 一定形式的调节作用
regulate price 调整物价
regulate supplies 调剂余缺
regulate supply and demand 调整供求关系；调节余缺；调剂余缺
regulate the exchange of commodities 调节商品交换
regulate the green house temperature 调节温室温度
regulate the market 管理市场；控制市面
regulate the speed 调整速度
regulate the velocity 调整速度
regulate watercourses 治水
regulating 控制；校正
regulating action 调节作用
regulating additive 调节剂
regulating agent 调整剂；调节剂
regulating amplifier 稳定放大器；调节放大器
regulating apparatus 调整装置；调节装置；调节设备；调节器
regulating barrage 调节坝
regulating box 调节箱
regulating brake 调节制动器
regulating capacity 调节容量
regulating carpet of sand 调整砂垫层
regulating cell 调节电池
regulating characteristic 调节特性
regulating chest 调节箱
regulating circuit 调节电路
regulating cock 调节栓；调整旋塞；调节旋塞
regulating coil 调节线圈
regulating command 调节主令；调节指令；调节命令
regulating constant 调节常数
regulating contact 调整接点；调节触点
regulating control rod 调节棒
regulating course 整平层；调节层
regulating crank 拐臂；调整杆；导叶臂；操纵杆
regulating current 调节电流
regulating dam 调节坝；分水坝；分流坝
regulating damper 调节风门；调节挡板
regulating device 调整装置；调节装置；调节设备
regulating diagram 调节图；调度曲线
regulating economic activities 调节经济活动
regulating economy 调节经济
regulating effect 节制作用；调节作用
regulating effect of lake 湖泊调节作用
regulating element 调节元件
regulating equipment 调节设备
regulating error 调节误差
regulating fitting 调节配件
regulating flap valve 调节瓣阀
regulating frequency 调节频率
regulating gate 控制闸(门)；节制闸门；调节闸门
regulating gear 调节机构；调节齿轮
regulating globe valve 调节球阀
regulating gradient 调节梯度
regulating handle 调整手柄；调节手柄
regulating heat extraction 散热调整
regulating hole for discharging interval 落料间隙调节孔
regulating impulse 调节脉冲
regulating instrument 调整仪器
regulating knob 调整捏手；调整按钮
regulating lever 拐臂；导叶臂；操纵杆
regulating line 整治线

regulating link 调整联杆
regulating linkage 调节联动装置
regulating load 调节负载
regulating lock 节制闸;节制阀;节流闸
regulating lock lights 节制闸灯
regulating loop 调节回路
regulating magnet 调节(电)磁铁
regulating measures 整治措施
regulating mechanism 调节机制;调节机构
regulating motor 调节电动机
regulating multi-zone production rate 分层配产
regulating network 调节网络;调节管网
regulating nut 调整螺母
regulating organ 调整机构
regulating pilot 调节用的导频
regulating piston 调节活塞
regulating plate 调节板
regulating plunger 调节柱塞
regulating pole 调节极
regulating pond 调节水库;调节池
regulating pondage 调节(蓄)水量;调节库容;调节池
regulating port 调节孔
regulating precipitation records 调节降水记录
regulating printed circuit board 调整印刷电路板
regulating processing 调节过程
regulating property 调节特性
regulating range 调节范围
regulating reflection 可调反射
regulating relay 调节继电器
regulating reserves 调节储量
regulating reservoir 调节水池;调节(蓄水)池;调节水库
regulating resistor 调节电阻器
regulating rheostat 调节变阻器
regulating ring 调整环;调节环;调节圈;调速环
regulating rod 微调杆;口调节棒
regulating screw 调整螺旋;调整螺丝;调整螺钉
regulating separate stratum oil production rate 分层配产
regulating set 调节机组
regulating shaft 调节轴
regulating sign 指导标志
regulating sleeve 调节套(筒)
regulating slide valve 调节滑阀
regulating sluice 节制闸;节制阀
regulating speed motor 调速电动机
regulating spindle 调节轴
regulating spring 调节弹簧
regulating stage 调节级
regulating starting rheostat 起动调节变阻器
regulating station 调节发电厂;备用电厂;运输调度站
regulating stem 调节杆
regulating stem set screw 调节杆止动螺钉
regulating stock 调节库存
regulating storage 调节储存;调节库[贮]仓;调节水池;调节库容;调节库存量
regulating structure 整治建筑物;调治构筑物;调治建筑物
regulating subunit 调节亚单位
regulating switch 可调变阻器;调节开关
regulating system 自动控制系统;调整系统;调节系统
regulating tank 调节水箱;调节池;调节槽
regulating temperature 调节温度
regulating transformer 调压变压器;调节变压器
regulating underlay 调节垫层
regulating unit 调节装置
regulating valve 调节阀;分配阀
regulating valve bonnet 调节阀帽
regulating valve cap nut 调节阀螺母
regulating valve gland 调节阀压盖
regulating valve seat 调节阀座
regulating valve seat bush 调节阀座衬套
regulating valve stem 调节阀杆
regulating water source 调节型水源地
regulating weir 调节堰
regulating wheel 调整轮;调节轮
regulating winding 调整绕组;调节线圈;调节绕组
regulating works 调节工程;整治建筑物;整治工程
regulation 整治;章程;规章;规则;规范;规程;调整(率);条例
regulation alternative 整治方案

regulation and control 调控
regulation by market 市场调节
regulation by the market mechanism 市场机制调节作用
regulation chamber 调节室
regulation characteristic 调节特性
regulation chip 调整片
regulation cock 调节栓
regulation constant 调节常数
regulation curve 调节曲线;负载特性曲线
regulation damper 调节挡板
regulation design 整治设计
regulation diagram 调节图;调度曲线
regulation discharge 整治流量
regulation drop-out 失调负荷
regulation facility 整治设施
regulation factor 调整因素;调整系数;调整率
regulation for safety 安全条例
regulation for technical operation 技术操作规程
regulation hole 调节孔
regulation in rotor circuit 转子电路调节
regulation in steps 分级调节
regulation light 规定号灯
regulation limit 受规章限制
regulation line 整治线;河流治导线
regulation loss 调节损失
regulation means 调整方法
regulation meter 稳定度测量器;调节(式)仪表
regulation nylon 正规耐纶
regulation of capital 节制资本
regulation of cascade reservoirs 梯级水库调节
regulation of channel 河道整治;河槽整治;航道整治;渠道整治
regulation of circulation 循环调节
regulation of classification 分等规定
regulation of derricks 改变吊杆位置
regulation of estuary 河口整治
regulation of exhaust gas 排气管理规则
regulation of fellings 采伐调节
regulation of flow 水流整治
regulation of forest 森林调整
regulation of fork river section 分叉河段整治
regulation of heavily sediment-laden river 多沙河流整治
regulation of heavily sediment-laden stream 多沙河流整治
regulation of inspection 验收规章
regulation of inspection and repair 检修规则
regulation of level 水准仪调整
regulation of load 荷载规定
regulation of local river section 局部河段整治
regulation of low water flows 枯水流量调节;低水位流量调节
regulation of luminous intensity 光强调节
regulation of meandering reach 蜿蜒性河段整治
regulation of meandering river section 蜿蜒性河段整治
regulation of motor 电动机调整
regulation of navigation channel 航道整治
regulation of pressure 压力调节
regulation of protection 防护规定
regulation of reservoir cascade 梯级水库调节
regulation of river 河流整治;河道治理;河道整治
regulation of river confluence section 河渠交汇段整治
regulation of river mouth 河口整治
regulation of sand bar 沙洲整治
regulation of shoal 浅滩治理;浅滩整治
regulation of step reservoirs 梯级水库调节
regulation of stream 河川整治;河流整治
regulation of streamflow by forest 森林治水;森林调节径流
regulation of temperature 温度调节
regulation of the contact wire tension 接触线拉力调整
regulation of the number of locomotive in services 运用机车调整
regulation of tidal reaches 潮汐河段治理;潮汐河段整治
regulation of tidal river 潮汐河流整治
regulation of tidal water 潮汐控制
regulation of torrent 山溪整治
regulation of total emission 总排放量管理规则
regulation of water and soil conservation work 水土保持工作条例

regulation of waterway 航道整治
regulation on adminsitration for environmental protection of construction projects 建设项目环境保护管理办法
regulation on engineering design for environmental protection of construction projects 建设项目环境保护设计规定
regulation on foreign investment 外来投资条例
regulation on forest fire-fighting 森林防火条例
regulation pullout 失调负荷
regulation range 调节范围
regulation resistance 调节电阻
regulation rod 调整杆
regulations 守则
regulation scheme 整治方案
regulations for diving 潜水条例
regulations for preventing collisions 避碰规则
Regulations for Preventing Collisions at Sea 海上避碰规章;海上避碰规则
regulations for radiation protection 辐射防护规定
regulations for reconstruction 改建规程
regulations for safe use of pesticide 农药安全使用规定
regulations governing foreign vessels 外轮管理规则
regulation sign 管制标志;道路标志
regulations in force 现行规章;现行规则
regulations of blast 爆破规程
regulations of breeding and conservation of aquatic resources 水产资源繁殖保护条例
regulations of company 公司章程
regulations of inspection 验收规则
regulations of level 水位调节
regulations of output 输出调整
regulations of railway technical operation 铁路技术管理规程
Regulations of the People's Republic of China on Administration of Wastes Dumping to Ocean 中华人民共和国海洋倾废管理条例
Regulations of the People's Republic of China on Administration Penalty against Public Security 中华人民共和国治安管理处罚条例
Regulations of the People's Republic of China on Adminsitration for Prevention of Pollution in Sea Areas by Vessels 中华人民共和国防止船舶污染海域管理条例
Regulations of the People's Republic of China on Prevention and Control of Environmental Noise Pollution 中华人民共和国环境噪声污染防治条例
Regulations of the People's Republic of China on Qurantine of Animals and Plants' Import and Export 中华人民共和国进出口动植物检疫条例
regulations on administration for environmental protection of capital construction of nuclear power plant 核电站基本建设环境保护管理法规
Regulations on Administration for Preventing Environmental Pollution from Dismantling Ships 防止拆船污染环境管理条例
Regulations on Administration of Environmental Protection Standards of the People's Republic of China 中华人民共和国环境保护标准管理条例
Regulations on Administration of National Environmental Monitoring 全国环境监测条例
Regulations on Administration of Pollution Prevention and Control in Drinking Water Source Area 饮用水水源区污染防治管理条例
regulations on adminsitration for radioactive wastes of cities 城市放射性腐物管理办法
regulations on defence against radioactive emissions 放射防护规定
Regulations on Protection of Upper Reaches of Water Head 上游水源保护条例
regulations on reclamation of land 土地复垦规定
regulation speed 限制车速
regulations regarding loads 加载规程;加荷规程
regulations relating to carriage of dangerous goods 危险货物运输规则
regulation stage 整治水位
regulation standard 整治标准
regulation storage 调节容量;调节库容

regulation structure 节制建筑物;调节建筑物;整治建筑物
regulations waste 调节损失
regulations tank 调蓄池;储[贮]水池;调节池
regulation tax of state enterprise 国营企业调节税
regulation thermostat 调节恒温器
regulation through the market 市场调节
regulation voltage 调节电压
regulation whistle 规定汽笛
regulation width 整治线宽度
regulation work of estuary 河口整治工程
regulation works 整治建筑物;整治工程;治河工程
regulative cleavage 调整裂
regulator 节制阀;稳压器;控制器;校准器;调节闸;调节器;调节结构;调节剂
regulator base 调节器座
regulator by ferroresonance 磁铁谐振式调压器
regulator bypass switch 调节器旁通开关
regulator canal 调节渠
regulator check nut 调节器防松螺母
regulator clock 标准钟
regulator coil 调节器线圈
regulator cover 调节器盖
regulator cover gasket 调节器盖衬
regulator crank 调节器曲柄
regulator cubicle 调速器控制盘;调节器(操作)柜;调节柜
regulator drive 调速传动
regulator for gas concrete 加气混凝土调节剂
regulator gate 调节闸门
regulator generator 调节发电机;电机调整器
regulator handle 调节器手柄;调节杆
regulator in steps 分级调压器
regulator limiter 调节器限制单元
regulator link 调节器连接杆
regulator of fan 电扇调速器
regulator of industrial waste handling practices 工业废液处理法的管理单位
regulator of level 水平调节器
regulator of mitogenic signals 分裂信号调节者
regulator of the market 市场调节器
regulator opening 调节孔
regulator pendulum 调整摆
regulator pin 调速器销
regulator potentiometer 调节电位器
regulator pressure 调节器压力
regulator quadrant 调节器扇形齿轮
regulator rod 调节棒
regulator shaft 调节器轴
regulator shaft jaw 调节轴爪
regulator site 调节部位
regulator solution 调节液
regulator spindle 调速器轴
regulator spring screw 调节簧螺钉
regulator station 调压站
regulator storage 日调节蓄水量;日调节库容;调节容量
regulator subunit 调节亚单位
regulator system 调节系统
regulator tap 控制旋塞
regulator tester 调节器试验装置
regulator tube 稳压管;稳流管
regulatory action 管理的行为
regulatory agency 管理机关
regulatory authority 制定规章的机构;法规机构
regulatory body 管理单位;管理机构
regulatory business tax 调节税
regulatory control 法律限制;法规防治
regulatory factor 调节因子
regulatory fishery bodies 渔业管理机构
regulatory framework 规章总纲
regulatory income tax 收入调节税
regulatory inhibition 调节性抑制
regulatory mechanism 调节机制
regulatory metabolite 调节代谢物
regulatory method of analysis 常规分析法
regulator yoke 调节器轭
regulatory operation 按章操作
regulatory products 管制物资
regulatory regime(n) 管理制度
regulatory regime(n) of river 河道管理制度
regulatory sign 管制标志;禁令标志
regulatory signal 调节信号
regulatory site 调节部位

regulatory system 调节系统
regulatory tax 调节税
regulatory work 调节功;整治设施;整治工程
regulex 电子放大器;电机调节器;磁饱和放大器
reguline metal 块状金属
reguline of lead 铅熔块
regulon 调节子
regulus 金属渣;熔块;[复]reguli 或 reguluses
regulus falsi 虚位法
regulus antimony 金属锑;锑渣
regulus lead 含锑铅;铅块;铅锑合金
regulus metal 锑铅合金
regulus metal alloy 抗酸金属合金
regulus of antimony 锑熔块
regulus of Venus 耐腐蚀合金
regurgitate 反喷
regurgitation 反逆;反流
regur soil 黑棉土
rehabilitating of tank 油罐的修理
rehabilitation 修建;修复;休养生息;恢复(作用);地区修复;重建;复原
rehabilitation area 休整地区
rehabilitation center 康复中心;恢复中心
rehabilitation effectiveness testing 修复有效度测试
rehabilitation expenditures 重建开支;重建费用;复兴开支;复兴费用
rehabilitation expenses 修复费用;复兴费
rehabilitation loan 复兴贷款
rehabilitation of bridge 桥梁修复
rehabilitation of dollar 美元复兴
rehabilitation of housing 住宅的翻新
rehabilitation of the environment 恢复环境
rehabilitation of victims of a disaster 灾害受害者的复原
rehabilitation of water environment 水环境修复
rehabilitation program(me) 休整计划
rehabilitation shop 修理间;修理厂
rehabilitation work 复兴工作;改建工程;重建工程;修复工作
rehabilitative measures 善后措施
rehalogenation 再卤化
rehandle 再处理;重(新)处理
rehandle facility 转载设备
rehandling 再处理;改铸;捣载作业;捣垛;重新处理;重复劳动;翻舱;予整顿
rehandling device 转载装置
rehandling facility 转载装置
rehandling of container 集装箱捣箱;集装箱倒箱
rehandling plant 转运站;转运设备
rehandling scraper 再处理铲运机;回填土铲运机
rehandling unit 再处理机械;再处理设备
reharden 再硬化;重淬火
rehardened streak 再度硬化条纹
rehardening 重新硬化;再硬化
rehashing 重散列;重散列
rehashing procedure 再散列过程
Rehbock dental 雷伯克齿形挡板
Rehbock dentated apron 雷伯克齿槛式护坦
Rehbock dentated sill 雷伯克齿形消力槛
Rehbock measuring weir 雷伯克量水堰
Rehbock weir flow formula 雷伯克堰流公式
Rehbock weir formula 雷伯克水堰公式
reheader 二次成型凸缘件镦锻机
rehearing 重新审理
rehearsal hall 排演厅
rehearsal room 排演场;排练室
rehearsal studio 排演厅
reheat 再热;加力燃烧;级间加热;再加热;二次加热;重加热;重热;重烧
reheat air conditioning system 再热式空气调节系统
reheat boiler 再热锅炉
reheat coil 二次加热盘管;再热盘管
reheat condensing turbine 再热凝汽式汽轮机
reheat control valve 中压调节阀;再热调节阀
reheat cycle 再热循环
reheat cycle gas turbine 再热循环燃气轮机
reheat emergency valve 中压汽门;再热汽门
reheat engine 再热式发动机;补燃式发动机
reheater 中间加热器;中间过热器;再热器;再热炉;灯丝加热器;重热器
reheater engine 中间再热式发动机
reheater-regenerative plant 再热回热式装置
reheat factor 再热系数;热量回收系数

reheat gas turbine 再热式燃气轮机
reheating 再加热;二次加热;重加热;重烧
reheating and regenerative cycle 再热再生循环
reheating bath 再热槽
reheating crack(ing) 再热裂纹
reheating cycle 再热循环;重热循环
reheating expansion 再加热膨胀
reheating furnace 二次加热炉;加热炉;再热(锅)炉;重热炉
reheating furnace with movable hearth 移底式加热炉
reheating hotwell 回热式热井
reheating of flue gas 烟道气再热
reheating of joint after welding 焊接接头再热
reheating oven 再热炉
reheating regenerative cycle 再热回热循环
reheating regenerative vapo(u)r cycle 重热回热汽循环
reheating regulator 再加热调节器
reheating shrinkage 再加热收缩
reheating stage 再热级
reheating test 二次加热试验;再热试验
reheating turbine 中间再热式涡轮机;重热式汽轮机
reheating vapo(u)r cycle 重热汽循环
reheat interceptor valve 中压调节阀
reheat machine 再热装置
reheator 再热器;重热器
reheat plant 蒸汽再热装置;再热装置
reheat pressure 再热汽压
reheat regenerative steam cycle 再热回热蒸汽循环
reheat-return point 再热蒸汽引入点
reheat shrinkage 再热收缩
reheat staged expansion 中间再热多级膨胀
reheat steam 再热蒸汽
reheat stop interceptor valve 中压联合汽门
reheat stop valve 回热停止阀
reheat stretch-type stretch-blow mo(u)lding machine 再加热拉伸型拉吹模塑机
reheat system 再加热系统
reheat test 热稳定性试验;反复加热试验
reheat testing furnace 重烧试验炉
reheat test of fireclay brick 火砖重热试验
reheat thrust 加力燃烧室发出的推力
reheat turbine 中间再热式涡轮机;再热式汽轮机
rehouse 重新分配住房
rehousing 再建房屋;旧房翻新
re-hump operation 重复驼峰作业【铁】
rehydration 再水合;再水化
re-hypothecate 再抵押
Reichdrill 瑞奇型旋转钻机
Reichert-Meissl value 赖克特—迈斯耳值
Reichert number 赖克特数
Reichert value 赖克特值
reichite 方解石
reidentification 重新识别;重新鉴定
Reid equation 瑞德方程式
Reid vapo(u)r pressure 瑞德蒸汽压力
reignier work 镶嵌细木工
reigning winds direction 盛行风向;当地主要风向
reignite 再次点火
re-igniter chamber 二次起动室
reignition 再次离子化;重复放电;二次起动;二次电离;二次点燃
reignition capability 二次点火能力
reignition of arc 再起弧
reignition voltage 再引弧电压;再起弧电压
reignition voltage of arc 重燃电压
reimaging 重新成像
reimbursable expenditures 可收回的支出
reimbursable expenses 可偿还的支付;工程补偿费;赔偿费用;补偿费
reimbursable financing 可偿还的资助
reimbursable service 可收回费用的事务;可收回的事务费
reimburse bank 清算银行
reimbursed leave 补假;补偿假期
reimbursed time 租用时间
reimbursement 赔偿;偿还;补回;报销
reimbursement credit 偿还信贷
reimbursement draft 求偿汇票
reimbursement expenses 补偿费用
reimbursement for expenses 补偿费用
reimbursement of customs duties 补偿关税
reimbursement period of loan 贷款偿还期

reimbursible expenditures 可偿还的支出
reimbursing agent 补偿银行
reimbursing bank 补偿银行
reimpaction 重复压实
re-import 再进口(货物)
reimportation 再进口(货物)
re-imposition 重新拼版
re-impregnating 再渗透
rein 拱底石;手柄;黏因(黏滞度单位)
rein chord bridge 单弦式桥梁
rein chord system 单弦体系
reindex 重订指标
reindexing 改变符号
reinduration 再硬结;再硬化
reinerite 砷锌矿
reinfection of newer portions 新生部分的侵染
reinfestation 再次蔓延
reinfm type 补强型式
reinforce 增强(物);补强
reinforce concrete dock 钢筋混凝土船坞
reinforce concrete vessel 水泥船
reinforced 增强的;加强的;加筋的;加固的
reinforced aerated concrete 充气的钢筋混凝土;加气钢筋混凝土
reinforced aerated concrete lintel 钢筋加气混凝土过梁
reinforced apron 加筋的护墙
reinforced arch 钢筋混凝土拱
reinforced arched bridge 钢筋混凝土拱桥
reinforced arched girder 钢筋混凝土曲形大梁
reinforced asbestos-cement middle corrugated tile 钢丝网石棉水泥中波瓦
reinforced asbestos rubber sheet 增强石棉橡胶板
reinforced asphalt 加筋沥青
reinforced asphalt mattress 加筋沥青排
reinforced bar 钢筋;螺纹钢筋;加筋杆
reinforced barrel vault 钢筋混凝土筒形圆顶室
reinforced barrel-vault shell 钢筋混凝土筒形圆顶壳体
reinforced beam 增力梁架;钢筋混凝土梁
reinforced beam floor 钢筋混凝土梁楼面
reinforced bearer 钢筋混凝土联结格栅
reinforced belt 高强度输送带
reinforced bitumen felt 增强沥青油毡;加强沥青油毡
reinforced bituminous coat 增强沥青涂层
reinforced bituminous coating 加强沥青涂层;加筋的沥青涂层
reinforced block 钢筋混凝土房屋
reinforced block beam 混凝土及钢材混合梁
reinforced block floor 混凝土及钢梁混合楼面
reinforced block partition (wall)钢筋混凝土隔墙
reinforced block roof floor 钢筋混凝土屋顶楼面
reinforced block work 加筋砌块工程
reinforced board 增强板
reinforced box caisson 钢筋混凝土箱形沉箱
reinforced box girder 钢筋混凝土箱形大梁
reinforced bracing 加固支撑
reinforced bracket 钢筋混凝土托梁
reinforced breast 钢筋混凝土拱肩墙
reinforced brick 钢筋砖
reinforced brick beam 加筋砖砌过梁
reinforced brick construction 砖配筋结构;钢筋砖结构;配筋砖结构
reinforced brick lintel 钢筋过梁
reinforced brick masonry 加筋砖圬工;钢筋砖圬工;钢筋砖砌体;加筋砖砌(体);配筋砖砌体
reinforced brick structure 钢筋砖结构
reinforced brickwork 配筋砌体;增强砌体;钢筋砖砌(体);加筋砌砌(体);加筋砖砌工
reinforced brick(work) lintel 加钢筋的砖过梁
reinforced brickwork pier 加钢筋的砖墩
reinforced brickwork wall 配筋砖墙
reinforced bridge 钢筋混凝土桥
reinforced bridge construction 钢筋混凝土桥梁建筑
reinforced building 钢筋混凝土房屋
reinforced bulkhead (沉箱或沉井的)钢筋混凝土隔墙
reinforced bunker 钢筋混凝土筒仓
reinforced butt weld 补强式对接焊缝;补强的对接焊缝
reinforced caisson 钢筋混凝土沉箱
reinforced cames 钢芯增强铅条;钢芯有槽铅条;加强铅棂
reinforced cast floor 预制的钢筋混凝土楼面
reinforced cast(ing) 钢筋混凝土浇制;加筋铸件;预制的钢筋混凝土单元
reinforced cast in-situ rib 现浇钢筋混凝土肋
reinforced cellular concrete 泡沫加强混凝土;泡沫加筋混凝土
reinforced cement mortar 加筋水泥砂浆
reinforced cement mortar board 水泥砂浆板;砂浆板
reinforced ceramic coating 加固陶瓷层
reinforced coat 铠装层;加强面层
reinforced coating 钢丝网抹灰(面)
reinforced column 组合柱
reinforced composite 增强复合材料
reinforced compression ring 加劲受压环
reinforced concrete 钢筋水泥;钢筋混凝土;加筋混凝土
reinforced concrete aerated with foam 泡沫加气钢筋混凝土
reinforced concrete arch 钢筋混凝土拱
reinforced concrete arch bridge by segmental construction without scaffolding 钢筋混凝土拱桥用预制构件无支架分段施工
reinforced concrete beam 钢筋混凝土梁
reinforced concrete beam and slab 钢筋混凝土梁板
reinforced concrete beam floor 钢筋混凝土带梁楼板
reinforced concrete bearer 钢筋混凝土联结格栅
reinforced concrete binder 钢筋混凝土联结格栅
reinforced concrete block 钢筋混凝土房屋;钢筋混凝土块
reinforced concrete bolt 钢筋砂浆锚杆
reinforced concrete bored cast-in-place pile 钢筋混凝土钻孔灌注桩
reinforced concrete box arch bridge 钢筋混凝土箱形拱桥
reinforced concrete box caisson 钢筋混凝土箱式沉箱
reinforced concrete box girder bridge 钢筋混凝土箱形梁桥
reinforced concrete bridge 钢筋混凝土桥
reinforced concrete buttressed dam 钢筋混凝土支墩坝
reinforced concrete buttressed retaining wall 钢筋混凝土前扶垛式挡土墙
reinforced concrete caisson 钢筋混凝土沉箱
reinforced concrete cantilever abutment 钢筋混凝土悬臂式桥台
reinforced concrete cantilever bridge 钢筋混凝土悬臂桥
reinforced concrete cantilever retaining wall 钢筋混凝土悬臂式挡土墙
reinforced concrete cap 钢筋混凝土帽梁
reinforced concrete cast floor 钢筋混凝土浇制楼板
reinforced concrete cast on site 现场浇制钢筋混凝土
reinforced concrete chimney 钢筋混凝土烟囱
reinforced concrete circular water storage tank 钢筋混凝土圆形蓄水池
reinforced concrete clump 钢筋混凝土重块
reinforced concrete collar 钢筋混凝土套环
reinforced concrete column 钢筋混凝土柱
reinforced concrete compound unit 钢筋混凝土建筑部件
reinforced concrete construction 钢筋混凝土结构;钢筋混凝土建筑;钢筋混凝土构造
reinforced concrete construction type 钢筋混凝土类(建筑物);钢筋混凝土构造类型
reinforced concrete coping course of bridge pier 桥墩钢筋混凝土顶板
reinforced concrete core 钢筋混凝土机械井体;钢筋混凝土芯体
reinforced concrete counterforted retaining wall 钢筋混凝土后扶垛式挡土墙
reinforced concrete cover and frame of manholes 检查井钢筋混凝土盖座
reinforced concrete culvert 钢筋混凝土涵洞
reinforced concrete culvert pipe 钢筋混凝土涵管
reinforced concrete cupola 钢筋混凝土圆屋顶
reinforced concrete dam 钢筋混凝土坝
reinforced concrete deck 钢筋混凝土面板
reinforced concrete dock wall 钢筋混凝土坞墙
reinforced concrete double-shaft bridge pier body 钢筋混凝土双程式桥墩身
reinforced concrete draught tube 钢筋混凝土尾水管
reinforced concrete drill 钢筋混凝土钻
reinforced concrete dry-dock 钢筋混凝土干船坞
reinforced concrete element 钢筋混凝土构件
reinforced concrete engineering 钢筋混凝土工程(学)
reinforced concrete-faced rockfill dam 钢筋混凝土面板堆石坝
reinforced concrete facing 钢筋混凝土护面层
reinforced concrete facing dam 钢筋混凝土面板堆石坝
reinforced concrete fence 钢筋混凝土护栏
reinforced concrete flat slab floor 钢筋混凝土无梁楼盖
reinforced concrete flexible foundation beam 钢筋混凝土柔性地基梁
reinforced concrete floating dock 钢筋混凝土浮船坞
reinforced concrete floating tank type graving dock 钢筋混凝土浮箱式干船坞
reinforced concrete floor 钢筋混凝土楼盖;钢筋混凝土楼板;钢筋混凝土桥面
reinforced concrete flume 钢筋混凝土渡槽
reinforced concrete footing course of bridge pier 桥墩钢筋混凝土底板
reinforced concrete foundation 钢筋混凝土基础
reinforced concrete frame 钢筋混凝土框架;钢筋混凝土构架
reinforced concrete frame structure 钢筋混凝土框架结构
reinforced concrete framing 钢筋混凝土构架
reinforced concrete gate 钢筋混凝土闸门
reinforced concrete girder 钢筋混凝土大梁
reinforced concrete graving-dock from floating tank 钢筋混凝土浮箱式干船坞
reinforced concrete grill 钢筋混凝土格子;钢筋混凝土格栅
reinforced concrete hollow block floor 钢筋混凝土空心砖地板
reinforced concrete hollow pile 钢筋混凝土空心桩
reinforced concrete hollow tile floor 钢筋混凝土空心砖地板
reinforced concrete lift-slab construction 钢筋混凝土升板施工
reinforced concrete lined tunnel 钢筋混凝土衬砌隧洞;钢筋混凝土衬砌隧道
reinforced concrete lintel 钢筋混凝土过梁;混凝土过梁
reinforced concrete masonry 钢筋混凝土砌体;钢筋混凝土砌体
reinforced concrete masonry construction 配筋混凝土砌块结构
reinforced concrete mattress 钢筋混凝土垫层
reinforced concrete member 钢筋混凝土构件
reinforced concrete mine support 钢筋混凝土矿井巷道支架
reinforced concrete moment resisting frame 钢筋混凝土抗弯框架
reinforced concrete parachute 钢筋混凝土伞形结构
reinforced concrete pavement 钢筋混凝土路面;钢筋混凝土铺面
reinforced concrete penstock 钢筋混凝土压力水管
reinforced concrete pier 钢筋混凝土闸墩;钢筋混凝土桥墩
reinforced concrete pile 钢筋混凝土桩
reinforced concrete piling 钢筋混凝土板桩
reinforced concrete pipe 钢筋水泥管;钢筋混凝土管
reinforced concrete pipe column 钢筋混凝土管柱
reinforced concrete piping 钢筋混凝土管
reinforced concrete plate 钢筋混凝土板
reinforced concrete pole 钢筋混凝土电杆
reinforced concrete pontoon 钢筋混凝土浮船;钢筋混凝土趸船
reinforced concrete precast frame 钢筋混凝土预制构架
reinforced concrete pressure pipe 钢筋混凝土压力(水)管
reinforced concrete problem 钢筋混凝土负荷
reinforced concrete radial gate 钢筋混凝土弧形闸门
reinforced concrete rail 钢筋混凝土轨道;钢筋混凝

凝土栏杆;钢筋混凝土横档
reinforced concrete railroad tie 钢筋混凝土轨枕
reinforced concrete relieving platform 钢筋混凝土卸荷平台
reinforced concrete reservoir 钢筋混凝土水库
reinforced concrete retaining wall 钢筋混凝土挡土墙
reinforced concrete rib 钢筋混凝土肋(板)
reinforced concrete rib arch bridge 钢筋混凝土肋拱桥
reinforced concrete rib(bed) floor 钢筋混凝土带肋楼板
reinforced concrete rigid frame 钢筋混凝土刚架
reinforced concrete road 钢筋混凝土路
reinforced concrete road slab 钢筋混凝土路面板
reinforced concrete roof 钢筋混凝土屋顶
reinforced concrete sandwich slab 钢筋混凝土夹层板
reinforced concrete screed 钢筋混凝土整平板;钢筋混凝土括平板
reinforced concrete sector gate 钢筋混凝土扇形闸门
reinforced concrete segment 钢筋混凝土管片;钢筋混凝土砌块
reinforced concrete service boat 钢筋混凝土工程船
reinforced concrete sewer pipe 钢筋混凝土排水管
reinforced concrete shear wall 钢筋混凝土剪力墙
reinforced concrete sheet pile 钢筋混凝土板桩
reinforced concrete sheet pile wall 钢筋混凝土板桩墙
reinforced concrete sheet piling cofferdam 钢筋混凝土板桩围堰
reinforced concrete shell 钢筋混凝土薄壳
reinforced concrete skeleton construction 钢筋混凝土骨架结构
reinforced concrete skeleton frame 钢筋混凝土骨架
reinforced concrete slab 钢筋混凝土(平)板
reinforced concrete slab bridge 钢筋混凝土板桥
reinforced concrete slab mattress 钢筋混凝土板沉排
reinforced concrete sleeper 钢筋混凝土轨枕
reinforced concrete space structure 钢筋混凝土空间结构
reinforced concrete spiral casing 钢筋混凝土蜗壳
reinforced concrete stair(case) 钢筋混凝土楼梯
reinforced concrete storage 钢筋混凝土油罐
reinforced concrete structural topping 钢筋混凝土结构覆盖层
reinforced concrete structure 钢筋混凝土结构
reinforced concrete structure regulation 钢筋混凝土结构规范
reinforced concrete sun breaker 钢筋混凝土遮阳板
reinforced concrete (sur)facing 钢筋混凝土面层
reinforced concrete surge tank 钢筋混凝土调压塔
reinforced concrete system 钢筋混凝土体系
reinforced concrete tension member 钢筋混凝土受拉件
reinforced concrete tie 钢筋混凝土轨枕
reinforced concrete tie rod 钢筋混凝土拉杆
reinforced concrete truss 钢筋混凝土桁架
reinforced concrete truss bridge 钢筋混凝土桁架桥
reinforced concrete trussed arch bridge 钢筋混凝土桁架拱桥
reinforced concrete tube 钢筋混凝土管
reinforced concrete tubing 钢筋混凝土管
reinforced concrete vault 钢筋混凝土地下室;钢筋混凝土拱顶
reinforced concrete vessel 钢筋(混凝土)水泥船
reinforced concrete wall 钢筋混凝土墙
reinforced concrete wall panel 钢筋混凝土墙板
reinforced concrete with glass tiled fillers 有玻璃砖填充的钢筋混凝土
reinforced concrete works 钢筋混凝土工程
reinforced construction 加固结构
reinforced continuous frame 钢筋混凝土连接框架
reinforced continuous pavement slab 连续配件路面板;连续配筋混凝土(路面)板
reinforced core 加强线芯;强化线芯
reinforced cored block floor 钢筋混凝土空心砖楼板
reinforced core wall 钢筋混凝土心墙
reinforced cross member 加固的横向构件

reinforced cross wall 钢筋混凝土剪力墙;加强的横向(剪力)墙
reinforced culvert pipe 钢筋混凝土涵管
reinforced cushion course 加筋垫层
reinforced cutting curb 钢筋混凝土沉箱刃脚
reinforced dead lock 加强的单栓锁
reinforced deck 钢筋混凝土楼板;钢筋混凝土地板;钢筋混凝土桥面
reinforced dense concrete 密实钢筋混凝土
reinforced design 钢筋混凝土设计
reinforced dikes and dams 加固堤坝
reinforced dorman lock 加强的暗榫锁
reinforced earth 加劲土;加筋土;加固土;加固地层
reinforced earth dam 加筋土坝
reinforced earth retaining wall 加筋(混凝土)挡土墙
reinforced effect of birefringent coatings 双折射贴片的加强效应
reinforced element 钢筋混凝土构件
reinforced epoxy resin gas pressure fitting 增强环氧树脂气体压力管件
reinforced epoxy resin gas pressure pipe 增强环氧树脂气体压力管
reinforced expanded concrete 加气钢筋混凝土;膨胀钢筋混凝土
reinforced exposed concrete 不加饰面的钢筋混凝土
reinforced facing 加固的饰面;钢筋混凝土护面层
reinforced factory 钢筋混凝土(构件)工厂
reinforced fair-faced concrete 光面钢筋混凝土
reinforced fiber and polymer 增强纤维与聚合物
reinforced fiber composite glass 玻璃纤维增强复合材料
reinforced filled masonry cell construction 加筋砌体网格结构
reinforced flange fitting 加强的法兰(管)接件
reinforced foam concrete 加气钢筋混凝土
reinforced foam concrete panel 钢筋泡沫混凝土镶板
reinforced for compression 抗压加筋
reinforced for handling 吊装加劲
reinforced foundation 钢筋混凝土基础(加强的基础)
reinforced foundation pier 钢筋混凝土基础墩
reinforced frame 加强的框架;增强架;加固架
reinforced framed construction 钢筋混凝土框架建筑
reinforced framing 钢筋混凝土构架
reinforced gasholder 钢筋混凝土煤气储罐
reinforced girder 增强梁;加强的大梁
reinforced glass 增强玻璃;加筋玻璃;钢化玻璃
reinforced glass concrete 玻璃纤维增强混凝土
reinforced glass fiber mo(u)ld 加筋玻璃纤维模型;加筋玻璃纤维模板
reinforced glazing bar 镶玻璃的钢芯铅条
reinforced graphite sheet 增强石墨板材
reinforced grillage 钢筋网(格);网状钢筋
reinforced grinding wheel 加筋砂轮;补强砂轮
reinforced grouted brick masonry 加筋灌浆砌体;配筋(灌浆)砖砌体
reinforced grouted masonry 配筋灌浆砖石砌体;钢筋砖圬工
reinforced gunite layer 加筋喷(浆)层
reinforced hatchway side coaming 加强舱口侧围板
reinforced hinge 强固销;强固铰
reinforced hollow brick masonry 加筋空心砖石砌体
reinforced hollow unit masonry 配件空心砌块砌体;配筋灌浆砌体
reinforced horizontal member 加固横杆
reinforced hose 强化皮管
reinforced hot mix asphalt 加筋热拌地沥青(路面)
reinforced in-situ concrete 现浇钢筋混凝土
reinforced in-situ screw pile 现浇钢筋混凝土螺旋桩
reinforced jetty 钢筋混凝土突堤;钢筋混凝土码头
reinforced joint 增强接缝;加筋接合;加筋(接)缝;加固接头;补强接合
reinforced joist 加强的格栅
reinforced lattice girder 钢筋混凝土格构大梁
reinforced layer 加固层
reinforced lighting (隧道的)加强照明

reinforced lighting column 钢筋混凝土照明灯杆;钢筋混凝土照明灯柱
reinforced line 加强线
reinforced liner 加固支架
reinforced lining 加固的衬筑;加筋的衬砌
reinforced lintel 钢筋混凝土横楣;钢筋混凝土过梁;加筋砖横楣;加筋砖过梁
reinforced liquid coating encapsulation product 增强液体涂料包封产品
reinforced masonry 配筋砌体;加筋圬工;加筋砌体;配筋圬工
reinforced masonry construction 加筋砌体结构
reinforced masonry lintel 加筋砖过梁
reinforced masonry structure 加筋砌体结构
reinforced masonry wall 加筋砖石墙
reinforced mast 加强桅杆
reinforced mat 钢筋混凝土垫层;加筋毡
reinforced material 增强材料
reinforced mat foundation supported on piles 支承在桩上的钢筋混凝土垫层基础
reinforced membrane 钢筋混凝土隔板;钢筋混凝土护面层
reinforced molding compound 加固的造型混合物
reinforced mortar 加筋的砂浆
reinforced mo(u)lding compound 增强模塑料
reinforced non-vulcanized polymeric sheet 增强型非硫化聚合物片材
reinforced open caisson 钢筋混凝土开口沉箱
reinforced opening 加筋洞口;补强的开孔
reinforced panel 加固(墙)板
reinforced paper tape 增强纸带
reinforced partition 增强的隔板
reinforced partition wall 增强的隔墙
reinforced pavement 钢筋混凝土路面
reinforced perforated block floor 加筋(混凝土)多孔砖楼板
reinforced pier 群柱;集墩
reinforced piled pier 钢筋混凝土桩墩
reinforced pipe penstock 钢筋混凝土管闸门
reinforced plaster 加筋墁灰
reinforced plastic boat 增强塑料艇
reinforced plastic case 加固塑料壳体
reinforced plastic corrosion-resistance equipment 耐蚀增强塑料设备
reinforced plastic drain sheet 增强塑料排水板
reinforced plastic mortar 加筋塑料灰浆;强化塑性砂浆
reinforced plastics 增强塑料;加强塑料;加筋塑料
reinforced plywood 强化胶合板
reinforced pole 加强杆;钢筋混凝土电杆;加固杆
reinforced-polyester laminate 增强聚酯层压材料
reinforced polypropylene 增强聚丙烯塑料
reinforced polystyrene 增强聚苯乙烯塑料
reinforced polytetrafluoro ethylene 增强聚四氟乙烯
reinforced precast concrete beam 钢筋混凝土预制梁
reinforced precast element 预制钢筋混凝土构件
reinforced precast floor 预制钢筋混凝土楼板
reinforced precast frame 预制钢筋混凝土构架
reinforced precast wall 预制钢筋混凝土墙
reinforced prism 钢筋混凝土棱柱体
reinforced purlin(e) 加强的桁条
reinforced pylon 钢筋混凝土塔(架)
reinforced raft 钢筋混凝土筏形基础
reinforced rail 补强轨道
reinforced reactor shield 钢筋混凝土反应堆屏蔽外壳
reinforced refractory brick 配筋耐火砖
reinforced resin 增强树脂
reinforced retaining wall 钢筋混凝土挡土墙
reinforced rib 加强肋
reinforced rigid frame 钢筋混凝土刚架
reinforced road pavement 钢筋混凝土路面
reinforced rockfill 加筋堆石体
reinforced roller bearing 钢筋混凝土滚柱支承
reinforced rubber 补强橡胶
reinforced rubber belt 加强橡胶带
reinforced rubber suction hose 吸管橡胶套
reinforced screw pile 钢筋混凝土螺旋桩
reinforced seal 骨架密封
reinforced segment 钢筋混凝土(矿)井壁
reinforced shaft 钢筋混凝土矿井;钢筋混凝土竖井
reinforced shear wall 加固剪力墙;钢筋混凝土剪

力墙
reinforced sheet pile 钢筋混凝土板桩
reinforced sheet pile wall 钢筋混凝土板桩墙
reinforced shell 钢筋混凝土薄壳
reinforced shotcrete 加强喷射混凝土
reinforced single base 钢筋混凝土单独基础
reinforced skeleton construction 钢筋混凝土骨架构造
reinforced skewed rigid frame bridge 钢筋混凝土斜交单跨刚构桥
reinforced soil 加劲土;加筋土
reinforced solid pile 钢筋混凝土实体桩
reinforced span 钢筋混凝土柱的跨度;钢筋混凝土桥的跨度
reinforced spiral flume 钢筋混凝土螺旋形渡槽
reinforced spun pipe 钢筋混凝土离心旋制管
reinforced square set 加固方框
reinforced steel 钢骨架
reinforced steel bar round 混凝土圆钢筋;圆钢筋
reinforced stock 加固橡胶
reinforced stress 增强应力
reinforced string 加固的楼梯梁
reinforced structure 钢筋混凝土结构;加固结构
reinforced stull 斜撑加强支架;加强的斜撑支架;加强的横撑支架
reinforced suspension preheater 强化悬浮预热器
reinforced T-beam 钢筋混凝土 T 形梁;加强 T 形梁
reinforced thermoplastic resin 热塑性增强树脂
reinforced thermoplastics 增强热塑性塑料;加强热塑材
reinforced thermoplastic sheet 增强热塑性片材
reinforced thermosetting plastics 增强热固性塑料
reinforced thermoset(ting) resin 增强热固性树脂
reinforced tile 加固瓦
reinforced tile lintel 加筋(空心)砖过梁
reinforced tire 加强外胎
reinforced topping 钢筋混凝土面层;加固的面层
reinforced truss bridge 钢筋混凝土桁架桥
reinforced type 加强型
reinforced umbrella 钢筋混凝土蘑菇顶;钢筋混凝土伞形顶
reinforced unit 钢筋混凝土构件
reinforced vault 钢筋混凝土拱顶;加固的拱顶
reinforced vessel 钢筋混凝土船
reinforced waling 钢筋混凝土横撑
reinforced wall 钢筋混凝土墙;配筋墙;加固墙
reinforced water tank 钢筋混凝土水箱
reinforced weld 加强焊缝;加固焊接
reinforced wheel 加强砂轮;加固砂轮
reinforced wood wool 加强木丝板;增强刨花板
reinforced wood wool slab 加强木丝板
reinforced works 钢筋混凝土工厂;钢筋混凝土工程
reinforcement 余高;增强(物);增力;拉筋;加筋物;加强筋;加强件;加强法;加强(材);加固(作用);强化;配筋;补强
reinforcement against tensile splitting 抗拉裂加固;抗拉裂钢筋
reinforcement area 钢筋截面积
reinforcement arrangement 钢筋排列;钢筋布置
reinforcement assembly shop 钢筋装配工场;钢筋装配车间
reinforcement bar 钢筋(条)
reinforcement bar-bender 弯筋机;钢筋弯曲机
reinforcement bar-bending machine 弯筋机;钢筋弯曲机
reinforcement bar-cutting machine 钢筋剪切机
reinforcement bar shear cutter 断筋机
reinforcement bar shears 钢筋剪刀
reinforcement bar spacing 钢筋间距
reinforcement bending 弯筋;钢筋弯折
reinforcement bending yard 弯钢筋工场
reinforcement binding 钢筋绑扎
reinforcement blockwork 配筋砌块工程
reinforcement bumper 强化保险
reinforcement butt welding 钢筋对焊
reinforcement butt weld 补强对接焊;对焊的补强
reinforcement by flange 凸缘补强
reinforcement by integrated forging piece 整体锻件补强
reinforcement by outside panel 外侧补强圈补强
reinforcement by thickened nozzle 接管加厚补强
reinforcement cage 钢筋笼;钢筋骨架
reinforcement coating 钢筋涂层
reinforcement cohesion 加固内聚力

reinforcement concrete 钢筋混凝土
reinforcement concrete of support legs 护基混凝土
reinforcement concrete oval arch culvert 钢筋混凝土卵形拱涵
reinforcement concrete pipe with bell and spigot 承插式钢筋混凝土管
reinforcement conditioning 钢筋调整反射
reinforcement corrosion 钢筋腐蚀
reinforcement cover 钢筋保护层
reinforcement cross-section 钢筋截面
reinforcement cutter shear 钢筋切割机
reinforcement cutting and bending 断弯钢筋
reinforcement cutting and bending plant 钢筋剪切弯曲设备
reinforcement cutting shear 钢筋切割机
reinforcement cylinder 钢筋筒箍
reinforcement density 钢筋密度;配筋密度
reinforcement detail 钢筋细部;配筋大样
reinforcement detail drawing 配筋细部图
reinforcement detector 钢筋检测仪
reinforcement displacement 钢筋移动;钢筋变位;钢筋移置;钢筋位移
reinforcement distance piece 钢筋定距块
reinforcement drawing 配筋图
reinforcement effect 补强效果
reinforcement engineering 加固工程
reinforcement erection 钢筋架设
reinforcement fabric 钢筋网
reinforcement filler 增强填料
reinforcement for adjacent openings 邻近开孔的补强
reinforcement for compression 抗压加固;抗压钢筋
reinforcement for handling 搬运钢筋
reinforcement for shearing 抗剪加固
reinforcement for stresses in erection 抗施工应力加固;抗施工应力钢筋
reinforcement for tension 抗拉加固;抗拉钢筋
reinforcement geometry 增强材料几何形状
reinforcement in compression 压力区钢筋
reinforcement index 配筋指数
reinforcement in ribs 肋内钢筋
reinforcement layer 钢筋加固层
reinforcement layout 钢筋布置
reinforcement limitation 加筋率上下限;配筋限度
reinforcement loop 箍筋;钢筋环;加强环
reinforcement machine 钢筋加工机械
reinforcement mat 钢筋网
reinforcement material 增强材料;加强材料;加固材料
reinforcement member 加固构件
reinforcement metal 钢筋;加强金属
reinforcement meter 钢筋测力仪;钢筋测力计
reinforcement method 加筋法;增量法【计】
reinforcement notes 配筋说明
reinforcement of additional elements 另附加元件的补强
reinforcement of concrete 混凝土加筋
reinforcement of content 含(钢)筋率
reinforcement of large opening 大孔径补强
reinforcement of multi-openings 并联孔开孔补强
reinforcement of ramp 道路接坡加固
reinforcement of reflex 反射增进
reinforcement of weld 加强焊缝;焊接补强;焊缝加厚;补强金属;焊接加强
reinforcement on the reverse side 反面补强
reinforcement pad 补强圈
reinforcement percentage 配筋率
reinforcement placement 钢筋配置;配筋
reinforcement plate 加筋板;增强板;钢筋锚碇垫板;加劲板;加固板;补强板
reinforcement prefabrication 钢筋预加工
reinforcement ratio 钢筋比(率);含(钢)筋率;配筋率
reinforcement rib 加劲肋(条)
reinforcement ring 箍筋;钢筋环;加强环
reinforcement rod 钢筋
reinforcement sheet 加强薄板
reinforcement shop 钢筋加工场;钢筋加工工厂
reinforcement spacer 钢筋定位器;钢筋定位件
reinforcement steel 加劲钢板
reinforcement steel bar 钢筋
reinforcement steel rod 钢筋
reinforcement-steel shear machine 钢筋剪切机

reinforcement stirrup 钢箍
reinforcement storage yard 钢筋堆场
reinforcement store 钢筋仓库
reinforcement strength 补强强度
reinforcement stress 钢筋应力
reinforcement stress detector 钢筋应力检测器;钢筋应力测试仪
reinforcement summary 钢筋明细表
reinforcement system 加固系统;钢筋分类
reinforcement theory 强化理论
reinforcement tying 钢筋绑扎;扎铁
reinforcement type 补强形式
reinforcement wire 预应力钢丝
reinforcement work 钢筋工作
reinforcement worker 钢筋工
reinforcement yard 钢筋加工场地
reinforce provision 增加供应
reinforcer 加强件;增强填料;增强剂;增强材料;加固物;激励物;强化剂
reinforce set 加强支撑
reinforce the conservation of water and soil 加强水土保持
reinforce the hole wall 加固孔壁;保护孔壁
reinforcing 增强方法;补强
reinforcing agent 增强剂;补强剂
reinforcing angle 加劲角铁;加劲角钢
reinforcing band 增强带;加强带;补强带
reinforcing bar 增强筋;钢筋;加强用圆钢;混凝土内的配筋;配筋
reinforcing bar mill 周期断面型钢轧机;钢筋轧机
reinforcing bar of switch rail 尖轨补强板
reinforcing bar spacer 钢筋隔块
reinforcing bar spacing 钢筋间距
reinforcing beam 加固用梁
reinforcing cage 钢筋骨架;钢筋笼
reinforcing cage welding machine 钢筋骨构架焊接机
reinforcing coat 硬化层;增强层
reinforcing concrete bar 混凝土配筋;钢筋
reinforcing course 加固层
reinforcing dam 护堤
reinforcing diaphragm 增强膜
reinforcing efficiency of asbestos 石棉增强效率
reinforcing element 加强构件
reinforcing fabric 钢丝网;钢筋网;增强用织物
reinforcing fiber 强化纤维
reinforcing filler 增强填(充)料;活性填充剂;补强(填充)剂
reinforcing fillet 凸焊填角
reinforcing for compression 抗压加强
reinforcing foundation 加固地基
reinforcing girder 加劲大梁;加强梁;加力梁
reinforcing glass fiber 增强用玻璃纤维
reinforcing gussets 增强结点板
reinforcing iron worker 钢筋工人
reinforcing layer 加固层
reinforcing legs 加固钻塔支腿
reinforcing limit 加劲限度
reinforcing mat 增强毡;加劲排;钢筋网
reinforcing material 增强材料;加固(材)料
reinforcing member 加强杆件;加固件
reinforcing mesh 钢筋网
reinforcing mirror 增强(发射)镜
reinforcing pad 加固托座
reinforcing panel 加强板;加劲板
reinforcing pigment 增强性颜料;补强性颜料
reinforcing plate 加筋板;加强板;加劲板
reinforcing post 加固支柱;辅助支柱;加强支柱
reinforcing pull rod 钢拉杆
reinforcing rail 加强钢轨
reinforcing relay 加强继电器
reinforcing rib 加强肋;加劲筋;加劲肋(条)
reinforcing ring 加劲环;补强环
reinforcing rod 钢条;钢筋条;钢筋(棍)
reinforcing rod bender 弯筋机;钢筋弯曲机
reinforcing rod-cropping machine 钢筋剪切机
reinforcing rod distance piece 钢筋分距块;钢筋定距块
reinforcing rod separator 钢筋分距块;钢筋定距块
reinforcing rod spacer 钢筋分距块;钢筋定距块
reinforcing round steel 圆钢筋
reinforcing screen 钢筋网;增强用网布
reinforcing shear 钢筋切断机;钢筋切割机
reinforcing sheet 增强薄板

reinforcing spiral 增强弹簧;螺旋形钢筋
reinforcing stay 加固支撑;加固拉条
reinforcing steel 钢筋
reinforcing steel adjusting cutter 钢筋调直切断机
reinforcing steel area 钢筋截面积
reinforcing steel bar 钢筋
reinforcing steel bender 钢筋弯曲机
reinforcing steel contractor 钢筋承包商
reinforcing steel-crooking machine 钢筋弯曲机
reinforcing steel cutter 钢筋切断机
reinforcing steel hook 钢筋钩
reinforcing steel mesh 钢筋网
reinforcing steel plate 加劲钢板
reinforcing steel rod 钢筋条
reinforcing steel rust 钢筋锈蚀
reinforcing system 钢筋系
reinforcing top soil 地面补加
reinforcing unit 加固件
reinforcing wall 起加劲作用的墙;加固墙
reinforcing wire 加劲钢弦;加劲铁丝;钢筋(钢)丝;加劲钢丝
reinforcing works 钢筋工程
reinfusion 再输注
reinhardbraunsite 莱拉硅钙石
reinitialize 重新预置
reinject 回灌
reinjection 回注;再喷入
reinjection control 回灌控制
reinjection fluid 回灌流体
reinjection line of gas 天然气回注管线
reinjection program(me) 回灌程序
reinjection test 回灌试验
reinjection well 回灌井
Reinschia-alginite 伦奇藻类体
reinsert 重新引入;重新插入;重拉复入
reinserted subcarrier 还原副载波;重置副载波
reinsertion of carrier 恢复载波
re-insert transmitter 再插发射机
reins of vault 支承拱顶的墙或边
reinspection 再检验;再检查;再次检查
reinspector 复验工
re-inspiration 再吸气
reinstall 装回原位;顶替
reinstatement 恢复原状
reinstatement cost 恢复价值
reinstatement of structure 结构的恢复;结构的修复
reinstatement value 重建价值
reinstatement work 修复工作;恢复工作;修整工作
reinsurance 分保;再保险
reinsurance agreement 再保险协定
reinsurance arrangement 分保安排
reinsurance certificate 分保单
reinsurance clause 分保条款;再保险条款
reinsurance commission 再保险佣金;分保手续费
reinsurance covering excess losses 超额赔款再保险
reinsurance offer 分保建议
reinsurance open cover 预定保险
reinsurance policy 再保险单;分保单
reinsurance reserves 分保责任准备金
reinsurance slip 分保条(单)
reinsurance treaty 分保合同
reinsurer 分保接受人
reintegration 再整合
re-interlocking 再联锁
reinterpret 重新解释
reintubation 重插管法
reinundation 再淹没
reinversion 重新求逆;重反演
reinvest 再投资;重新投资
reinvested earning 再投资的收益
reinvested profit 再投资利润
reinvestigation 重新调查
reinvestment of earnings 收益再投资
reinvestment rate 再投资率;重新投资率
reinvest profit 利润再投资;净利润再投资
reionize 再电离
reiphenylmethane 三苯甲烷
reisner work 彩色镶嵌木工(17世纪德国)
reissacherite 锰土
reissite 柱沸石
Reiss microphone 赖斯传声器
reissue rate 续保收费率
reiteration 重申;反复迭代(法);复测(法);正倒镜复测
reiteration control 复测控制装置
reiterative method 反复逼近法
Reith's alloy 锡锑铅青铜;莱斯铅青铜
reja (西班牙教堂中保护墓地的)钢栏杆
reject 筛余粗料;废料;下脚料;拒绝;粗粉;不合格品;报废;选矿废渣;筛余粗料
rejectable quality level 不合格质量水平;可拒绝的质量水准;拒收品质水准
rejectamenta 漂浮物;排泄物
rejectance number 拒绝数
reject an offer 不接受报价
reject bid 拒绝投标
reject bin 废品箱;废料仓
reject chute 节余粗料槽;筛余粗料槽;废料槽
reject code 作废码
reject control 选片控制器
rejected casting 不合格铸件
rejected concrete 废弃的混凝土;混凝土下脚料
rejected footage 报废进尺
rejected heat 排出的热
rejected material 节余粗料;筛余粗料;废弃材料;废品;不合格材料
rejected parts 拒收的部件或零件;不合格零件;不合格部件
rejected works 拒收工程
reject failure rate 抑制故障率
reject heat 排出的热
reject indicator 废品指示器
rejection 拒收;拒绝采用;截留率;退回;剔除;废弃
rejection amplifier 带除放大器
rejection and waste 返工和浪费
rejection band 抑制频带
rejection chute 废品排除斜道
rejection clause 退货特约条款
rejection efficiency 粗选出效率;粗粉率
rejection factor 抑制因数;排除因数
rejection filter 拒波滤波器
rejection frequency 抑制频率
rejection image 抑制图像
rejection iris 抑制窗孔
rejection level 排斥水平
rejection mask 金属膜掩蔽
rejection method 拒绝法
rejection number 拒绝数;不合格判定数;不合格个数
rejection of accompanying sound 伴音抑制
rejection of all bids 拒绝全部投标
rejection of full load 甩满负荷
rejection of goods 退货
rejection of heat 散热
rejection of load 弃荷;甩负荷
rejection of order with explanation 拒绝订单并加以说明
rejection of plans 否定设计图;否定计划
rejection of tender 投标书的拒绝;拒收投标书;拒绝投标(书)
rejection of work 工程的拒收
rejection or condemnation clause 废弃或没收条款
rejection point 抑制点
rejection quality level 不合格质量水平
rejection rate 舍弃率;废品率;不合格率;报废率
rejection ratio 抑制比;衰减比;舍弃率;废品率;减弱系数
rejection region 拒绝域;否定区域
rejection risks 拒收风险
rejection technique 舍选技术
rejection trap 拒波器;带阻陷波器
rejection zone 截止带
rejective amplifier 带除放大器
rejective sampling 拒绝抽样
reject light 废品指示灯
rejector 抑制器;拒波器;带除滤波器;除波器
rejector circuit 抑制器电路
rejector resonance 衰减器谐振
rejector unit 拒波部件
reject overtime 拒绝加班
reject piler 废品堆积机;废板堆垛机
reject pocket 废弃卡片袋
reject rate 拒识率;拒绝率;废品率
reject rate of raw sheet 原板废品率
reject refiner 筛渣磨浆机
reject region 抑制区;拒绝区
rejects 选矿废渣;筛余粗料
reject sheet conveyer 废薄板运输机
reject signals at nearby frequencies 抑制邻近频率信号
rejects outlet 粗粉出口
reject stacker 废奔收卡箱;废板堆垛机
reject tender 拒绝投标
reject tension reel 废品卷取机
reject test 拒绝检验
reject the null hypothesis 拒绝虚无假设
rejoin 再接合
rejoining 重勾(灰)缝
rejoint 再填缝;重接
rejointing 填缝;灌浆;勾缝;重新勾缝;重接(合)
rejointing of sett paving 块料铺砌的再填缝
rejuvenate 恢复;更新;再生;回春;复初
rejuvenated 更生的
rejuvenated fault 再生型断层
rejuvenated fault scarp 再生断层崖;更生断层崖;回春断层崖;复活断层崖
rejuvenated mountain 回春性山地
rejuvenated mountain land 再生型山区
rejuvenated ore deposit 再生矿床
rejuvenated river 回春河
rejuvenated rubber 再生胶
rejuvenated stream 更生河;回春河
rejuvenated water 再生水
rejuvenate geosyncline 新生地槽
rejuvenating pruning 更新修剪
rejuvenating stage 回春期
rejuvenation 回春(作用);恢复过程
rejuvenation stage of geosyncline 再生地槽阶段
rejuvenator 再生器;破壳机
rekindle 重新燃起
rel 勒耳(磁阻单位)
rel[unit of reluctance] 利尔(磁阻单位)
relaid of tracks 重新铺轨
relaining key 固定键
relamping 更换损坏的灯具
relapsable 可复发的
relapse 再发
relapsing fever 回归热
relaskop 测高仪
relatching 再接合
related activity 相关活度
related angle 相关角
related channel program(me) 相关通道程序
related coefficient 相关系数
related cost 相关成本;有关成本
related curve of balance element-influence factor 均衡要素—影响因素关系曲线
related curve of regime element-influence factor 动态要素—影响因素关系曲线
related data 相关资料;相关数据
related department 有关部门
related differential equation 相关微分方程
related expression 相关式
related feature 相关形体
related function 相关函数
related interpolator 相关内插器
related parties 有关单位
related (perceived) color 相关知觉色;相关感色
related product 相关产品;有关产品;关联产品;相关商品
related rate 相关变率
related ratio 相关比(率)
related relation 相关关系
related shades 相似色调的色泽
related subject 相关主题
related term 相关项
related terminal 相关线端
related terminals of a transformer 变压器的相关线端
related topics 有关主题
related trades 相关工艺;有关行业;有关工种
relational algebra 关系代数
relational approach 关系方法
relational automaton 关系自动机
relational calculus 关系运算;关系演算
relational coiling 相关螺旋
relational complete language 关系完备语言
relational completeness 关系的完备性
relational composite 关系的合成
relational database 相关数据库;关系型数据库;关系(式)数据库

relational data-base management system 关系型数据库管理系统
relational data-base schema 关系数据库模式
relational data file 有关数据文件
relational direct access method 关系直接存取法
relational expression 相关式;关系(表达)式;比例式
relational file 关系文件
relation algebra 关系代数
relational indexing 相关标引
relational matrix 关系矩阵;联系矩阵
relational model 关系模型
relational model of database 关系式数据库模型
relational model of data for large shared data 大型共享数据的关系模型
relational operator 关系运算子;关系(运)算符;关系操作符
relational programming 关系程序设计
relational query 关系查询
relational query language 关系查询语言
relational record 关系记录
relational spiral 相关螺旋
relational storage 相关存储器
relational symbol 关系符号
relational system 关系系统
relational T-algebra 关系 T 代数
relational theory 关系理论
relational tree 关系树
relation among well alignment and topography tectonic line 井排与地形构造线关系
relation analysis 关系分析
relation back 追溯效力
relation between bank and enterprise 银企关系
relation between ground river and surface water 暗河与地表水联系
relation between seismic grade and seismic degree 地震烈度和地震震级关系
relation between storage water and reservoir earthquake 水库地震与蓄水关系
relation between tectonic layers 构造层层间关系
relation between well alignment and groundwater flow 井排与地下水流向关系
relation calculus 关系演算
relation character 关系符(号)
relation condition 关系状态;关系条件
relation curve 相关曲线;关系曲线
relation curve between mud sandstone and interval velocity 泥沙岩与层速度关系曲线
relation curve of water-level 水位相关曲线
relation graph 关系图;相关图形
relation model 关系模型
relation of distribution 分配关系
relation of diversity 相异关系
relation of equivalent 等价关系
relation of homotopy 同伦关系
relation of identity 恒等性关系
relation of inclusion 包含关系
relation of location to drainage 水系对选线的关系
relation of non-linear stress-strain 非线性应力—应变关系
relation of plane 面的关系
relation of results to the originals 将结果与原来对比
relation on attributes 属性关系
relation operator 关系运算符
relation phase 相关相
relation property 相关特性
relation record 关系记录
relation schema 关系图;关系模式
relation-sensitive adhesive 反应性胶粘
relation(ship) 关系;媒质
relationship between foliation and main structure 面理与主构造关系
relationship between lineation and fold axis 线理与褶皱轴关系
relationship between stratigraphic(al) attitude and engineering work 地层产状与工程关系
relationship expression 关系式
relationship of arguments and parameters 变元和参数的关系
relationship of production 生产关系
relationship price index 比价指数
relations of dependence 依附关系
relation stage 相关水位

relative absorbance 相对吸收物
relative absorption 相对吸收
relative-absorption term 相对吸收项
relative abundance 相对丰度;相对多度
relative abundance of counterclockwise foram and clockwise foram 左旋有孔虫与右旋有孔虫相对丰度
relative abundance of foram in warm water and foram in cola water 冷水有孔虫与暖水有孔虫相对丰度
relative acceleration 相对加速度
relative accuracy 相对准确度;相对精度
relative acid content 相对硫酸含量
relative acidity 相对酸度
relative activity 相对活性;相对活度
relative activity determination 相对活度测量(放射性);放射性活度相对测量
relative activity level 相对作业水准
relative activity of soil water 土壤中水相对流动性;土壤中水相对活动性
relative adaptability 相对适应性
relative address 相对地址
relative addressing 相对认址【机】
relative addressing time-sharing 相对编址分时
relative address label 相对地址标号
relative adsorbed amount 相对吸附量
relative adsorption coefficient 相对吸附系数
relative advantage 相对利益
relative age 相对时代;相对年龄;相对年代
relative age of landform 地貌相对年龄
relative age of soil 土壤的相对年龄
relative air density 相对空气密度
relative air speed 相对气流速度
relative algebraic number field 相对代数数域
relative altitude 相对高度
relative altitude position 相对高度位置
relative amounts of reactants 相对反应物量
relative amplitude 相对幅值
relative amplitude conserved processing 保持相对振幅处理
relative-amplitude processing 相对振幅处理
relative amplitude section 相对振幅剖面
relative and absolute returns 相对收益和绝对收益
relative anomaly 相对异常
relative anomaly curve of imaginary component 虚分量相对异常曲线
relative anomaly curve of real component 实分量相对异常曲线;实分量相对异常曲线
relative antiskid characteristics 相对抗滑性
relative aperture 相对孔径
relative aquifuge 相对隔水层
relative area of peak 峰的相对面积
relative area response 相对面积响应
relative articulation 相对清晰度
relative asymmetry 相对不对称(性)
relative attenuation 相对衰减
relative average difference of second order 相对平均二级差
relative azimuth 相对方位角;相对方位
relative balance 相对平衡
relative balance between acid and alkali 酸碱相对平衡
relative band 相对波段
relative bandwidth 相对带宽
relative bearing 相关方位;相对方位(角);舷角
relative bearing indicator 相对方位角指示器
relative bearing of the current 水流舷角
relative bell 继电器铃
relative bending 相对弯曲
relative bending radius 相对弯曲半径
relative biologic(al) effect 相对的生物效应
relative biologic(al) effectiveness 相对生物有效性;相对生物学效能;相对生物效应
relative biologic(al) efficiency 相对生物效率
relative bireflectivity 相对双反射率
relative blackness 相对黑度;比较黑度
relative blowing rate 相对鼓风强度
relative boundary 相对边界
relative brightness 相对亮度
relative burnup 相对燃耗
relative capacity 相对容量;相对能力
relative capacity to pay 相对支付能力
relative catalytic activity 相对催化活性
relative chain complex 相对链复形

relative change 相对变化
relative change rate of precipitation 降水相对变率
relative character 相对性状
relative charge 相对充气量
relative chroma level 相对色度电平
relative chroma time 相对色度时间
relative chronology 相对年代学
relative circle 相对圆
relative classification 相关分类法
relative closed set 相对闭集
relative closing error 相对闭合差
relative coarseness 相对粗度
relative cochain complex 相对上链复形
relative code 相对代码;相对程序
relative coding 相对编码
relative coefficient 相对系数
relative coefficient of migration 迁移相对系数
relative cohomology 相对上同调
relative cohomology group 相对上同调群
relative cohomology theory 相对上同调论
relative colo(u)r 关联色
relative colo(u)r strength 相对着色力
relative command 相对命令
relative compaction 相对压实作用;相对密实度;压实系数
relative compactness 相对压实度;相对密实度
relative comparison 相对比较
relative comparison figures 比较相对数
relative complement of a set 集合的相对补
relative component 相对分量
relative compressibility 相对沉降量;相对压缩性;相对压缩量;相对压缩度;单位压缩量
relative concentration 相对浓度
relative concentration of surface area 相对表面积浓度
relative confidence interval 相对置信区间
relative configuration 相对构型
relative consistency 稠度指数;相对相容性;相对稠度;稠度指标
relative consistency curve 相对稠度曲线
relative consumption 相对消耗量
relative contamination level 相对污染水平
relative content 相对含量
relative content analyzer 相对含量分析器
relative content of ice test 冻土相对含冰量试验
relative contour 相对等高线;厚度线
relative contrast 相对对比率;相对对比度;百分对比度
relative control 相对控制
relative conversion ratio 相对转换比
relative coordinates 相对坐标
relative coordinate system 相对坐标系
relative correction factor 相对校正因子
relative cost 相对费用;有关费用;比较成本
relative cost coefficient 相对成本系数
relative cost control 相对成本控制
relative cost vector 相对成本向量
relative course 相对航向
relative covariance 相对协方差
relative cover 相对覆盖度
relative coverage ratio 相对覆盖率
relative cumulative frequency curve 相对累积频率曲线
relative current 相对流
relative current level 相对电流电平
relative current speed 相对流速
relative curvature 相对曲率;相对曲度(弯道曲度与间宽之比)
relative curve 相关曲线
relative curve between value aromatic versus saturated hydrocarbons fraction 芳烃—饱和烃馏分的值关系曲线
relative curve of total solids with depth 总矿化度—深度关系曲线
relative cutting speed 相对切削速度
relative cycle 相对循环
relative damage factor 相对损伤因数
relative damping factor 相对阻尼系数
relative damping ratio 阻尼比
relative data 相对数据
relative dating 相对年代测定
relative deficit 相对亏损
relative deflection 相对垂线偏差;天文大地偏差;相对形变

relative deformation 相对变形
relative deformation gradient 相对形变梯度
relative degree 相对次数
relative degree of humidity 相对湿度
relative delay 相对延迟
relative delay in transit 传输相对延迟
relative density 相对密(实)度;比重
relative density of cohesiveless soil 无黏性土相对密度
relative density of sand 砂土相对密度;砂的相对密度
relative density of sands and gravel 砂砾的相对密度
relative density test 相对密度试验
relative density tests 相对密实度试验
relative departure 相对偏离
relative depth 相对水深;相对深度;相对厚度
relative derived functor 相对导函子
relative determination 相对测定
relative detuning 相对失调
relative deviation 相对偏差
relative deviation of vertical 相对竖直线偏差;相对垂线偏差
relative dielectric(al) constant 相对介电常数
relative difference limen 相对差阈
relative differential geometry 相对微分几何学
relative diffusion 相对扩散
relative dimension 相对尺寸
relative direction 相对方向
relative discharge 相对排污量;相对排放量;相对流量
relative discrete space 相对离散空间
relative discriminant 相对判别式
relative dispersion 相对色散
relative displacement 相对位移;相对变位
relative displacement of fault walls 断盘相对位移
relative distance 相对距离
relative distribution 相对分布
relative distribution function 相对分布函数
relative distribution of income 相对收入分配
relative divergence 发展指数
relative diversity 相对多样性
relative dormancy 相对休眠
relative dormant period 相对休眠期
relative dry hiding power 相对干遮盖力
relative dullness 比较浊音
relative eccentricity 相对偏心率
relative eddy 相对旋涡;相对涡流
relative effect ratio 相对效应比
relative efficiency 相对效能;相对效率
relative efficiency inspection 相对效率检验
relative elastic recovery 相对弹性回弹;相对弹性恢复
relative electric(al) susceptibility 相对电极化率
relative electron diffraction 相应电子衍射
relative elevation 相对高程;相对标高
relative elongation 相对延伸率;相对伸长
relative embedment 相对埋深
relative emission index 相对排放量指数
relative enhancement coefficient 相对浓缩系数
relative enrichment factor 相对富集系数
relative entropy 相对熵;相对平均信息量
relative environment index 相对环境指数
relative environment value 相对环境价值
relative equilibrium 相对平衡;相对均衡
relative equivalent 相对当量
relative error 相对误差;比例差率;比较误差
relative error ellipse 相对误差椭圆
relative error of aero-survey 航测相对误差
relative error of the frequency 相对频率误差
relative error of the period 相对周期误差
relative evapo(u)ration 相对蒸发
relative exit angle 相对流出角
relative expansion 相对膨胀
relative extremal 相对极线
relative extremal curve 相对极值曲线
relative extreme (value) 相对极值
relative factor cost 相对要素成本
relative factor price 相对要素价格;要素比价
relative fall 压头相对落差
relative field drop 电场下降率
relative firmness 相对紧度
relative fitting error of water table 水位相对拟合误差

relative fixed cost 相对固定费用
relative fixed expenses 相对固定费用
relative flammability index 相对性燃烧指标
relative flight height 相对航高
relative flow 相对流动
relative fluctuation 相对涨落;相对波动
relative flying altitude 相对航高
relative focal length 相对焦距
relative force 相对推力
relative forest land 相对林地
relative form of value 相对价值形式
relative fractionation coefficient 相对分馏系数
relative frequency 相对频数;相对频率
relative frequency deviation 相对频率偏差
relative frequency diagram 相对频率图
relative frequency distribution 相对频率分布
relative frequency draft 相对频率变差
relative frequency histogram 相对频率直方图
relative frequency variation rate 相对频率变差率
relative friction coefficient 相对摩擦系数
relative gain 相对增益
relative gamma field map 相对伽马场图
relative geochronometry 地貌相对年龄研究方法
relative geologic time 相对地质时期
relative glare 相对眩光
relative glossmeter 相对光泽计
relative gradient 相对梯度
relative gravity 相对重力
relative gravity data 相对重力数据
relative gravity determination 相对重力测定
relative gravity measurement 相对重力测量
relative ground stress measurement 相对地应力测量
relative growth 相对生长
relative growth rate 相对生长率
relative hardness 相对硬度
relative harmonic content 相对谐波含量
relative harmonic measure 相对调和测度
relative hazard 相对危害性
relative hazard factor 相对危害因数
relative heading indicator 相对航向指示器
relative height 相对高度
relative height difference 相对高差
relative height of hydraulic jump 水跃的相对高度
relative homogeneity 相对均匀性
relative homological algebra 相对同调代数
relative homology group 相对同调群
relative homology theory 相对同调理论
relative homotopy group 相对同伦群
relative hoop force 相对环向(压)力
relative hue 相对色彩
relative humidity 相对湿度
relative humidity of the air 空气相对湿度
relative hydrophobicity 相对疏水性
relative ice content 相对含冰量
relative impact energy 相对冲击能量
relative importance 相对重要性
relative inclination 相对倾角
relative inclinometer 相对倾斜仪;相对倾斜计
relative income 相对所得;相对收入
relative income difference 相对收入差别
relative income hypothesis 相对收入假设
relative increase of speed 转速相对升高;速率上升
relative increment 相对增量
relative indefatigability 相对不易疲劳性
relative independence 相对独立
relative independentability 相对独立性
relative index 相对指数
relative index of refraction 相对折射率
relative index weight 相对标志权重
relative indicators 相对指标
relative indicatrix 相对指标
relative indices 相对指标
relative induction 相对感应
relative inductivity 相对介电常数
relative inertness 相对惯性
relative insertion loss 相对插入损耗
relative instantaneous center 相对瞬心
relative integral invariant 相对积分不变式
relative intensity 相对强度;相对烈度
relative intensity I/I0 value 相对强度输入/输出值
relative intensity of pressure 相对压强
relative interference effect 相对干涉作用
relative invariant 相对不变式

relative invariant measure 相对不变测度
relative ionospheric opacity meter 电离层相对不透明度计
relative isohypse 厚度线
relative isotope abundance determination 相对同位素丰度测量
relative length 相对长度
relative length closing error of traverse 导线相对闭合差
relative level 相对电平;相对标高
relative level(l)ing 相对流平性
relative level point 相对电平点
relative light efficiency 相对发光效率
relative limit 相对极限
relative limit of error 误差相对极限
relative line of flow 相对流线
relative load capacity 相对负荷容量
relative loader 相对装入程序
relative location 相对排架法
relative location system 相对排架法
relative log exposure 相对对数曝光时间
relative log H(E) 相对曝光量对数
relative longitudinal slope 相对纵坡
relative luminance curve 相对亮度曲线
relative luminosity 相对亮度;相对发光度
relative luminosity factor 相对亮度系数;光度函数
relative luminous efficiency 相对发光效率;明视觉光谱光视效率
relatively closed set 相对闭集
relatively comfortable life 小康
relatively comfortable standard of living 小康
relatively compact 相对紧致
relatively compact domain 相对紧域
relatively compact semi-group 相对紧半群
relatively compact set 条件紧集
relatively complemented 互补
relatively complemented lattice 相对补格
relatively erosive intensity index of watershed 流域相对侵蚀强度指标
relatively hard wood 比较硬的树木
relatively insoluble solid 相对不可溶性固体
relatively invariant functionals 相对不变泛函
relatively invariant measure 相对不变测度
relatively minimal model 相对极小模型
relatively noncompact subregion 相对非紧子区域
relatively open set 相对开集
relatively rotating elements 相对旋转元件
relatively short sailing distance 较短航距
relatively steady period 相对稳定期
relatively steady range 相对稳定地带
relative machinability 相对切削性
relative magnet 比较磁铁
relative magnetic attractability 相对磁力吸引性
relative magnetic susceptibility 相对磁化率;比磁化率
relative magnetometer 相对磁强计
relative magnitude 相对值;相对等级;相对大小;相对程度
relative manifold 相对流形
relative marginal value 相对边际价值
relative market values method 市价比较法
relative mass defect 相对质量亏损
relative mass density of rock 块石相对重度
relative mass density of rock armo(u)r 护面块石相对重度
relative mass excess 相对质量过剩
relative maximum 相对极大值
relative maximum or minimum 相对极大或极小
relative mean deviation 相对平均离差;相对均差
relative mean error 相对中误差
relative mean square error of adjustment points 相邻点间相对中误差
relative mean square error of baseline 基线测量相对中误差
relative mean square error of extended side 扩大边相对中误差
relative mean square error of side length 边长(测定)相对中误差
relative mean square error of starting side 起始边相对中误差
relative mean square error of the weakest side 最弱边相对中误差
relative measurement 相对测量
relative merit 相对优点;优缺点

relative metabolic rate 相对代谢率;能量代谢率
relative method 比较法
relative minimum 相对极小值
relative minimum of area 相对最小面积
relative minimum point 相对极小点
relative mobility of element 元素相对活动性
relative moisture 相对湿度
relative moisture content of soil 土的相对含量
relative moisture of soil 土壤相对湿度
relative molar heat content 相对摩尔热含
relative molar response 相对摩尔响应
relative molecular mass 相对分子质量
relative molecular mass distribution 相对分子质量分布
relative molecular weight 相对分子质量
relative momentum 相对动量
relative mortality index 相对死亡率指数
relative motion 相对运动;视运动;表观运动
relative motion display 相对运动显示
relative movement 相对运动
relative movement line 相对运动线
relative name in a link definition 连元定义的相对名字
relative neighbo(u)rhood 相对邻域
relative net precision 相对净精确度
relative noiseness 相对噪度
relative noise power 相对噪声功率
relative noise temperature 相对噪声温度
relative norm 相对范数
relative number 相对数;相对编号
relative opening 相对孔径
relative operating characteristic 相对运转特性
relative orbit 相对轨道
relative orientation 相对方位;相对定向
relative orientation of conjunction of successive photographs 连续像对相对定向
relative orientation of independent photo pair 单独像对相对定向
relative oriented photo 相对定向照片
relative oriented picture 相对定向照片
relative osculating circle 相对密切圆
relative outflow of gas 瓦斯相对涌出量
relative outlet velocity 相对排气速度
relative output value 相对输出量
relative overpopulation 相对过剩人口
relative overshoot 相对过调
relative parallax 相对视差
relative parameter sensitivity function 相对参数灵敏度函数
relative partial free energy 相对微分自由能
relative pauperization 相对贫困化
relative peak area 相对峰面积
relative percentage 相对百分数
relative percentage error in magnetic survey 磁测百分相对误差
relative performance 相对性能
relative periodic(al) movement 相对周期运动
relative permeability 相对透气率;相对渗透性;相对渗透率;相对磁导率
relative permeability ratio 相对渗透率之比;相对渗透率与饱和率的关系(曲线)
relative permeability to oil 原油的相对渗透率
relative permeability to water 水的相对渗透率
relative permittivity 相对电容率
relative persistence 相对持久性
relative perviousness 相对渗透性;相对可透性
relative phase 相对相(位)
relative photographic(al) gap 相对漏洞
relative photometer 比较光度计
relative photometry 相对光度测量;相对测光
relative pitch 相对螺距;相对跨距
relative pitch-shortening value 相对短距系数
relative plasticity index 稠度指数;相对塑性指数
relative plateau slope 相对坪斜率
relative plate motion 相对板块运动
relative plot 相对运动作图;相对标绘
relative poisoning 相对中毒
relative polarity 相对极性
relative pole 相对极
relative pollutant index 相对污染指数
relative pollutant loading 相对污染负荷
relative pollution index 相对污染指数
relative pore space 相对孔隙
relative porosity 相对孔隙率;相对孔隙度

relative position 相对位置
relative positioning 相对定位
relative position movement 相对位移量
relative potency 相对功效
relative power gain 相对功率增益
relative precision 相对精确性
relative presentation 相对表示法
relative pressure 相对压力
relative price 相对价格;比价
relative price difference 相对价格差别
relative price effect 相对价格效应
relative price of commodities 商品比价
relative product 相关乘积
relative production density 相对生产密度
relative productivity 相对生产率;相对生产量
relative program(me) 相关程序;相对程序
relative programming 相对程序设计
relative proper motion 相对自行【天】
relative property 相对性
relative proximity 相对邻近
relative pulse height 相对脉冲高度
relative purchasing power 相对购买力
relative quantity 相对值;相对量
relative quickness 相对灵敏度
relative quiet 相对平静
relative radial velocity 相对视向速度;相对径向速度
relative radial wear 相对径向磨损
relative radiance strength signature 相对辐射强度特征
relative radius (拉伸时的)相对转角半径
relative radius of curvature 相对曲率半径
relative ramification index 相对分歧指数
relative range 相对全距
relative range of control 相对控制范围
relative rate 相对率
relative rates of change in concentration 浓度变化相对速度
relative ratio 相对比
relative ratio method 相对比值法
relative ratio with fixed base 定基比
relative reaction rate 相对反应速率
relative recall 相对查全率
relative receiving response 相对接收响应
relative reduction 相对压下量;相对缩短
relative redundancy 相对多余度
relative refraction index 相对折射率
relative refractory period 相对不应期
relative regulation 相对调节;调整率
relative reliability 相对可靠性
relative relief 相对起伏;相对地势
relative relief map 相对地形图;分层设色图
relative relief method 分层设色法
relative remission 相对缓解
relative resident time 相对停留时间
relative resistance 相对阻力;相对电阻
relative resistance value 相对抗滑值
relative resonance integral 相对共振积分
relative response 相对响应(值);相对灵敏度
relative response factor 相对响应因子
relative response value 相对响应值
relative restraint 相对制约;相对约束
relative result 相对结果
relative retardation 相对阻滞
relative retention 相对保留;相对保持
relative retention ratio 相对保持率
relative retention time 相对停留时间;相对保留时间
relative retention value 相对保留值
relative retention volume 相对保留体积
relative reverberation level 相对混响级;混响级
relative rigidity 相对刚性;相对刚度
relative rise of sea level 海面相对上升
relative risk 相对危险(性);相对危险(度);相对风险
relative risk factor 相对风险系数
relative rolling 相对横摇
relative rotative efficiency 螺旋桨效率比
relative rotor displacement 转子相对移动
relative roughness 相对粗糙(度);相对糙率;相对糙度
relative roughness factor 相对粗糙度系数
relative sales value method 相对销售价值法;相对销价法;比较售价法
relative saturation 相对饱和
relative saturation deficit 相对饱和差
relative saturation degree 相对饱和度

relative saving loss 相对补救损失
relative scalar 相对尺度
relative scale 相对比例(尺)
relative scatter 相对离差
relative scattering power 相对散射力
relative scatter intensity 相对散射强度
relative search 相对搜索
relative sector search 相对扇形搜索
relative sediment charge 泥沙相对含量
relative sediment size 泥沙相对粒径
relative seismic velocity 相对振速
relative selectivity 相对选择性
relative-sensitive 相对灵敏
relative sensitivity 相对敏感性
relative sensitivity coefficient 相对灵敏度系数
relative sensitivity factor 相对灵敏度系数
relative setting angle 相对安置角
relative settlement 相对沉陷;相对沉降(量);不均匀沉降
relative share 相对份额
relative shear 相对剪切
relative shift 相对移动
relative size 相对值;相对大小;相对尺寸
relative skewness 相对偏度
relative sliding movement 相对滑移运动;相对滑动
relative slip 相对滑动
relative slip theory 相对滑移理论
relative solubilizing power 相对加溶能力
relative solvation 相对溶剂化作用
relative spacing 相对间距
relative specific ionization 相对电离比度;相对比电离
relative specificity 相对专一性;相对特异性
relative spectral directional reflectance 相对光谱定向反射比
relative spectral energy 相对光谱能量
relative spectral energy distribution 相对(光)谱能量分布
relative spectral power distribution 相对光谱功率分布
relative spectral sensitivity 相对光谱灵敏度
relative spectrum curve 相对频谱曲线
relative specular glossiness 相对镜面光泽度
relative speed 相对速度
relative speed approach 相对接近速度
relative speed drop 转速降落率
relative speed rise 转速升高度
relative speed variation 转速变化率
relative square search 相对正方形搜索
relative stability 相对稳定性;相对稳定度
relative stability of earth crust 地壳相对稳定
relative stability stage 相对稳定阶段
relative stable area 相对稳定地块
relative standard 相对标准
relative standard deviation 相对标准误差;相对标准偏差
relative standard deviation ellipse 相对标准偏差椭圆
relative standard error 相对标准误
relative steam entrance velocity 蒸汽相对入口速度
relative steam exit velocity 蒸汽相对出口速度
relative step 相对梯级
relative stiffness 相对劲度;相对刚性;相对刚度
relative stiffness of slab 水泥混凝土板相对转动刚度;水泥混凝土板相对转动劲度
relative stiffness ratio 相对劲度比;相对刚性比;相对刚度比
relative stopping power 相对阻止能力
relative strength 相对强度
relative strength index 强弱指标
relative strong adsorption 相对强吸附
relative subsidence 相对下沉量
relative sunshine 相对日照
relative sunshine duration 日照百分率
relative sunspot number 相对日斑数;太阳黑子相对数
relative surface activity 相对表面活度
relative surface area 相对表面积
relative surface fitting 相对曲面拟合
relative surface wear 相对表面磨损
relative surplus 相对剩余
relative surplus-population 相对过剩人口
relative swing 相对旋角
relative tack 相对黏性

relative temperature 相对温度
relative tensor 相对张量
relative term 相对项(目)
relative thermal stress modulus 相对热应力模数
relative thickness 相对厚度
relative thickness of orebody 矿体相对厚度
relative thinning intensity 相对疏伐强度
relative tilt 相对倾斜
relative tilt angle 相对倾角
relative time delay 相对延时
relative time difference 相对时差
relative time scale 相对地质年代表
relative tinting strength 相对着色力
relative to geologic(al) design work 与地质有关的设计工作
relative topography 相关地形;相对地形
relative total loss 相对全损
relative toxicity 相对毒性;相对毒度
relative toxic ratio 相对毒力比
relative track 相对磁道
relative translation 相对平移
relative transmission level 相对传送电平;相对传输电平
relative transmittance 相对透射率
relative transmitting response 相对传输响应
relative transpiration 相对蒸腾
relative travel time 相对走时
relative triple precision 相对三倍精度
relative twist 相对捻度
relative uniform converge 相对一致收敛
relative uniform convergence 相对均匀收敛性
relative uniformity 相对一致性
relative uniform star convergence 相对一致星形收敛【数】
relative value 相对值
relative value approximation 相对近似值
relative value form 相对价值形式
relative value of consequence 结果的相对值
relative value of outcome 结果的相对值;产量的相对值
relative value of strategies 策略的相对值
relative vapo(u)r density 相对蒸气密度
relative vapo(u)r pressure 相对蒸气压
relative variability 相对变动
relative variance 相对方差
relative-variance for stratified sampling 分层抽样的相对方差
relative variation 相对变化
relative variogram 相对变差函数
relative vector 相对矢量
relative velocity 相对速度;相对流速
relative visco(si)meter 对比黏度计
relative viscosity 相对黏(滞)度
relative viscosity temperature number 相对黏度温度数(值)
relative visibility 相对能见度
relative visibility curve 相对可见度曲线
relative visibility factor 相对可见度因数
relative volatility 相对挥发度;相对发挥度
relative voltage drop 相对电压降
relative voltage level 相对电压电平
relative voltage response 相对电压反应
relative volume 相对体积
relative volume method 相对体积法
relative volumetric deformation 相对体(积)变形
relative vorticity 相对涡度
relative wage 相对工资
relative water content 相对含水量;液性指数
relative water depth 相对水深
relative water vessel speed 船对水的相对速度
relative wear of the tool flank 后刀面相对磨损率
relative wear rate 相对磨损率
relative wear resistance 相对耐磨性
relative weight 相对重量;相对权重
relative wettability 相对湿润性
relative wind 相对气流;相对风;视风
relative wind velocity 相对风速
relative worth 相对价值
relative yield 相对产额
relativism 相对论
relativistic astrophysics 相对天体物理学
relativistic case 相对论性情况
relativistic change in mass 质量的相对论性变化
relativistic correction 相对论性修正;相对论性改正

relativistic covariant 相对论性协变量
relativistic deflection 相对论偏折
relativistic deflection of light 光线相对论偏折
relativistic degeneracy 相对论性简并
relativistic dilation of time-scale 时标的相对论性扩展
relativistic effect 相对论性效应
relativistic electrodynamics 相对论电动力学
relativistic electronic beam accelerator 相对电子束加速器
relativistic energy 相对论性能量
relativistic energy range 相对论性能量范围
relativistic factor 相对论性因子
relativistic field 相对论性场
relativistic increase of mass 质量的相对论性增加
relativistic kinematics 相对论运动学
relativistic linac 相对论性直线加速器
relativistic mass 相对论性质量
relativistic mass equation 相对论性质量方程
relativistic mechanics 相对论力学
relativistic particle 相对论性粒子
relativistic plasma 相对论等离子体
relativistic rate of energy loss 相对论性能量损失率
relativistic thermodynamics 相对论性热力学
relativistic velocity 相对论(性)速度
relativistic wave equation 相对论性波动方程
relativity 相对性
relativity correction 相对论修正
relativity principle 相对性原理
relativity shift 相对论位移
relativity theory 相对论
relativization 相对化
relativization of a topology 拓扑的相对化
relativization principle 相对性原则
Relatyrex 回转轮带滑移监测系统
relax 松弛;弛豫;放宽
relaxant 弛缓剂;弛缓的
relaxation 应力释放;弛张;减轻;松弛(作用);弛张;弛豫;弛缓;放松
relaxation allowance 休息减计时间
relaxation balance 松弛平衡
relaxation broadening 弛豫加宽
relaxation circuit 张弛电路;弛张电路
relaxation constant 弛张常数
relaxation curve 松弛曲线
relaxation dispersion 弛豫色散
relaxation effect 松弛效应
relaxation factor 松弛因子
relaxation frequency 张弛频率
relaxation function 松弛函数
relaxation generator 张弛振荡器;弛张振荡器
relaxation in concrete 混凝土的松弛
relaxation index 松弛指数
relaxation integral law 松弛积分定律
relaxation inverter 弛张换流器
relaxation length 张弛长度;松弛长度;衰减长度;弛豫长度
relaxation mechanism 松弛机理
relaxation method 逐次近似法;张弛法;渐近法;松弛法;迭豫法;弛豫法
relaxation modulus 松弛模数;松弛模量
relaxation of power series 幂级数的松弛
relaxation of prestress tendons 松弛损失(预应力筋)
relaxation of residual stress 残余应力消除
relaxation of rock 岩石松弛
relaxation of selection 选择放松
relaxation of steel 钢筋的松弛
relaxation of stress 消除应力;应力松弛;应力缓冲;应力弛豫
relaxation oscillation 张弛振荡;弛豫振荡
relaxation oscillator 张弛振荡器;张弛电路;弛豫振荡器
relaxation-oscillator alarm 张弛振荡器报警器
relaxation parameter 松弛参数;弛豫参数
relaxation pattern 松弛格式
relaxation period 张弛周期
relaxation phase 松弛相
relaxation phenomenon 松弛现象
relaxation polarization 松弛极化;张弛极化
relaxation pressure of surroundings 松动围岩压力
relaxation procedure 松弛方案
relaxation process 逐次近似过程;弛豫过程
relaxation pulse 弛豫脉冲

relaxation rate 弛豫率
relaxation reagent 松弛试剂
relaxation shrinkage 松弛回缩
relaxation shrinkage on wetting 湿后松弛回缩
relaxation spectrum 松弛时谱;弛豫光谱
relaxation strain 松弛应变
relaxation strength 弛豫强度
relaxation suture 减张缝合法
relaxation table 松弛表
relaxation technique 松弛法;迭豫技术
relaxation term 弛豫项
relaxation test 松弛试验;弛豫试验
relaxation tester 松弛试验机
relaxation testing machine 松弛试验机
relaxation time 阻尼时间;张弛时间;缓和时间;松弛时间;弛豫时间
relaxation time measurement 弛豫时间测量
relaxation transition 弛豫跃迁
relaxation type absorption 松弛型吸收
relaxation type dispersion 松弛型色散
relaxation variable 松弛变量
relaxation velocity 松弛速度
relaxation viscometer 松弛型粘度计
Relax cylinder drier 赖拉克斯辊筒烘燥机
relaxed and loose structure 舒适型结构
relaxed fracture 卸荷裂隙
relaxed market 宽松的市场
relaxed modulus of elasticity 松弛弹性模量
relaxed peak process 拟裂变
relaxed replication control 松弛复制调控
relaxed rock 卸载岩体;消除应力岩石;弛豫岩体
relaxed stress 松弛应力
relaxed synthesis 松弛合成
relaxing bulking treatment 膨松处理
relaxing exercise 放松运动
relaxing factor 松弛因素
relaxing medium 松弛介质
relaxing nonlinearity 弛豫非线性
relaxing suture of spring 弹簧的松弛
relaxometer 应力松弛仪;应力松弛计;张弛测量器;松弛测量器
relaxor 张弛振荡器;弛豫振荡器
relaxor ferroelectrics 弛豫性铁电体
relay 重铺;转播;中继;继电器;替续器;伺服机;分程传递
relayable rails 重新铺设用钢轨
relay act trip 继电器操作跳闸
relay actuator 继电器起动装置;继电器操作机构
relay amplifier relay panel 继电器放大器
relay armature hesitation 衔铁瞬时延迟
relay armature travel 衔铁行程
relay assembly test room experiment 接力装配试验
relay base 转播站
relay bay 继电器台;继电器架
relay board 继电器盘
relay box 中继变速器;继电器箱
relay broadcasting 转播
relay broadcast station 广播转播电台
relay bulk plant 中继油库
relay bulk station 中继油库
relay bungalow 继电器室
relay cabinet 继电器箱
relay case 继电器箱
relay center 中继中心;交换中心;转接中心;中转中心
relay chamber method 中继间法
relay channel 转播信号通道
relay characteristic 继电特性
relay check-in check-out register 继电检入检出寄存器
relay chest 继电器箱
relay clutch 继动离合器
relay communication 中继通信
relay comparator 继电器比较仪
relay computer 继电器计算机
relay contact 继电器触点
relay-contactor 继电器接触器
relay-contactor control 继电器接触器控制
relay contact welder 继电器接点焊接机
relay controller 继电控制器
relay control system 继电器控制系统
relay core 继电器铁芯
relay cropping 轮种法

relay cubicle 继电器柜
relay cycle timer 周期定时继电器
relay cylinder 继动油缸;伺服机缸;传递油缸
relay decoder 继动器译码器
relay dep[英]中继油库
relay depot 中继油库
relay drive 继电器驱动
relay driver 继电器驱动器
relay duty cycle 继电器工作周期
relay earthquake 递发地震
relay earth station 中继地面站
relayed call 转接呼叫
relayed capacity 继电保护容许功率
relayed rail 再铺钢轨
relayed surveillance network 远程(遥信)网络
relayed surveillance subsection 远程遥信分区
relay electromagnet 继电器电磁铁
relay end voltage 继电器端电压
relay energized 继电器吸起
relay equipment 继电设备
relay fan 接力风机
relay for industry control 工业控制继电器
relay for position indication 指示位置的继电器
relay gate 选通继电器;继电器闸门
relay gear 继动装置
relay governing 继动调节
relay governor 继动调节器
relay governor gear 继动调速器
relaying 重铺块料;继电保护;重新铺砌
relaying action 中继作用
relaying platform 中继平台
relaying rail 再用轨;更换钢轨;重铺轨
relaying station 批发站;转输站
relay interlocking 继电联锁(装置);继电联动装置;继电集中联锁;继电联锁器
relay interlocking for large station 大站电气集中联锁
relay interrupter 继电器式断续器
relay just-release value 继电器始可释放值
relay key 继电器式电键
relay lens 旋转透镜;中继透镜
relay-lens system 透镜中继系统
relay lever bushing 继电器杆衬套
relay magnet 继电器磁铁
relay matrix 继电器矩阵;插栓矩阵
relay mechanics 继电器力学
relay mechanism 继电器机构
relay must-operate value 继电器保证起动值
relay network system 继电器网络系统;继电器电路方式
relay non-operate value 继电器未动作值
relay-operated accumulator 继电(器)累加器;继电器控制的累加器
relay-operated controller 间接作用调节器;继电器操作的控制器
relay-operated type 继电器控制形式;继电控制型
relay-operating time 继电器吸合时间;继电器动作时间
relay operation 继电器动作
relay overrun 继电器超限运行
relay pad 继电器垫圈
relay picked up 吸起的继电器
relay piston 自动转换活塞;继动活塞;调节器一次滑阀;从动活塞
relay piston valve 继动活塞阀
relay plugboard 继电器插座板
relay point 中继站;转播站
relay protection 继电保护
relay pump 中转泵;接力泵
relay pumping 中继泵转输;接力泵送;分程泵送
relay pump station 中继泵站
relay pump station of heating network 热网中间泵站
relay rack 继电器架
relay racks bay 继电器架
relay receiver 转播用接收机;接力接收机
relay regulator 继电调节系统;断续式调节器
relay released 继电器释放
relay reservoir 中间储[贮]罐
relay retractile spring 继电器回缩弹簧;继电器复位弹簧
relay ring 继电环
relay room 继电器室
relay route storage 继电(器)存储器

relay satellite 中继卫星
relay seating 继电器衔铁入位
relay seating time 继电器入位时间
relay selector 继电器(式)选择器
relay selsyn 中断自动同步机
relay semi-automatic block system 继电半自动闭塞系统
relay sensitivity 继电器灵敏度
relay sequence 继电器序列
relay-set 继电器装置;继电保护屏
relay slug 继电器延时套管
relay spring 继电器弹簧
relay stagger time 继电器参差时间
relay station 转运站;转播站;转播台;转报台;中转站;中继站
relay storage 中继储存
relay switch 延时开关;继电器开关
relay switchboard 继电器盘
relay switching circuit 继电器开关电路
relay tank 中间油箱
relay tank farm 中继油库
relay tank truck 油槽可替换运油车
relay test set 继电器测试设备
relay timing 继电器时延;继电器定时
relay tip 继电器接点
relay track 中转线
relay transmitter 中继发射机;接力发射机
relay tube 替续管
relay-type recorder 继电式自动仪表;继电器式记录器
relay-type recording instrument 继电式自记仪表
relay types 继电器类型
relay underreaching protection 继电器失灵保护装置
relay valve 继动阀
relay valve body 继动阀体
relay valve diaphragm spring 继动阀隔膜簧
relay valve supply valve 继支阀供给阀
relay valve supply valve spring 继动供给阀簧
relay valve top cover 继动阀顶盖
releasability letter 解密代码字母
releasable casing spear 可以松开的打捞工具(钻探用);可以松开的套管夹(钻探用)
releasable connection 可拆连接
releasable energy 可释放能量
releasable shank 带安全器的铲柄
releasable standard 带松脱装置的铲柄
release 解锁【铁】;解除;缓解;让渡;操纵装置;排放;脱聚;投放市场;提单限额超装许可证;释放;放松;放出
release action 释放作用
release agent 防粘剂;隔离剂;脱模剂;分离剂
release agent for fair-face concrete 光面混凝土脱模剂
release air pressure 释放风压
release alarm 释放报警
release altitude 投放高度
release analysis 优选分析
release and fixation of potassium 钾的释放与固定
release arm 松脱臂
release bar 释放闩
release bearing 分离器分离轴承
release bearing of clutch 离合器分离轴承
release (bore)hole 泄水钻孔
release brake position 制动器松开位置
release buoy 失事浮标
release busy 释放占用
release button 缓解按钮;放松钮
release by 投放
release by two sections 两点检查
release cam 释放凸轮
release catch 摘钩;摘纵器;脱钩
release circuit 投放电路;释放电路
release clause 让与条款;解除条款
release clutch 松脱安全离合器
release coating 脱模涂料
release cock 放气旋塞;放气活栓;放气活塞;放气活门(门)
release command 释放指令;释放命令;复原指令
release condition 释放状态
release connection 断开连接
release control retainer 缓解控制保持阀
release course 豁免条款
release current 释放电流

release curve 泄放曲线;松放曲线
release curve of strain 应变释放曲线
release cut-off valve 缓解截断阀
release cycle 缓解循环
released bill of lading 放行提单
released by three sections 三点检查【铁】
released energy 释出能量
release device 快门装置;脱钩装置
released flare 放空火炬
released fracture 解压破裂;解压裂隙
released gate 泄水闸门
release diagram 泄放曲线
released ion 获释离子
released joint 解压节理
released liquor 排出液
released mineral 释出的矿物;解放矿物;释放矿物
released mortgage 解除抵押
released neutron 释出中子
release dredge(r)s from a project to assist in another project 从某工程抽出挖泥船参与另一工程
released thinning 透光伐
releasee 让受人;财产受让人
release electromagnetic valve 缓解电磁阀
release emulsion 脱模乳剂(混凝土)
release factor 释放因子;返还系数
release film 分离膜;剥离膜
release force 脱模力
release fracture 减压破裂;释压裂隙
release-free 自由释放
release-free relay 自由释放继电器
release-free test 自由释放试验
release from debt 免除债务
release from liability 解除债务
release from performance 解除履约
release fund for 抽出资金
release fuze 松发引信
release guard 放松警戒;放松监护
release guard signal 释放保护信号
release handle 释放柄
release histories of contaminants 污染源排放史
release hook 释放钩
release hook bracket 释放钩架
release if in order after examination 验后放行
release insuring valve 缓解保证阀
release joint 释重节理
release key 释放键
release lanyard 脱钩拉索
release layer 脱模层
release lever 离合器压盘分离杆;快门杆;脱解子;缓解手柄;脱扣杆;脱钩杆;释放杆;断开杆;放松杆
release lever yoke 脱扣杆叉
release limiting valve 缓解限制阀
release line 松解缆
release link 释放杆;复位杆
release load 脱开力;断开力
release lube 脱模润滑剂
release magnet 释放电磁铁
release mechanism 脱钩机构;松开机构;释放机构;分离机构;安全机构
release mesh 合格粒度
release moment 释放力矩
release mooring hook 脱缆钩
release note 版本注释
release of bank account 解冻银行存款
release of basic data 释放基本数据
release of brakes 制动缓解;松闸
release of contract 合同解除
release of debt 免偿债务
release of distress 解除扣押
release of forms 脱模
release of guarantee 解除担保
release of heat 热量释放;释放热(量)
release of humidity 去湿
release of industrial contaminants 工业污染物排放
release of industrial pollutants 工业污染物排放
release of jaws 夹钳松开(制动器的)
release of liability 债务解除;责任解除
release of lien 解除财产留置权
release of mortgage 解除抵押
release of pollutant 排出污染物;污染物(的)释放
release of pressure 压力释放;压力减弱
release of record 销账(解除抵押品的留置权)
release of stored water 蓄水泄放;蓄水开放;蓄水

放出
release of stress 应力释放
release of suspended solid 悬移质泄放
release of tension 张力释放
release of work hardening 加固硬化消除
release oil 脱模(润滑)油
release order 释放指令
release paper 松脱纸;防护薄膜;黏性处理纸;剥离纸
release paste 脱模(润滑)膏
release pattern 释放规律
release pay 遣散费
release pedal 松放踏板;分离踏板
release pheromone 释放信息素
release piece 释放爪
release pipe 缓解管;放泄管
release point 排泄点;排污点;释放点;出逸点
release position 关闭位置;离开位置;缓解位(置);松开位置;放松位置
release power 释放力
release pressure 释放压力
release price 赎回扣押权的费用
release property 脱模性质
release-push 释放开关;释放按钮
release-quiesce indicator 释放停止指示符;释放禁止操作指示器
releaser 释放装置;释放器;释放剂
releaser rate 松弛速率;释放速率
release rate of chlorofluorocarbons 放出氯氟烃的速率
release ratio 释放比
release relay 释放继电器;复归继电器
release requisition 解锁条件;释放条件
release restraining 解除约束
releaser stimulus 排放刺激物
release sensitivity 释放灵敏度
release sequence 投放顺序
release shackle 释放挂钩
release shaft 分离轴
release signal 释放信号
release sleeve 分离套筒
release sluice 泄水闸
release speed 缓解速度
release speed at a retarder 减速器出口速度
release spring 缓解弹簧;松开弹簧;松放弹簧;复原弹簧
release spring bracket 缓解弹簧座
release spring of clutch 离合器分离弹簧
release stability 释放稳定性
release strength 脱模强度(混凝土)
release structure 泄水构筑物
release the brake 松开制动踏板
release the drive 切断传动
release the pedal 释放踏板
release thermal energy 释放热能
release the shutter 开动快门
release time 缓解时间;释放时间
release timer 释放定时器
release to birth ratio 释放比
release trunk 释放中继线
release-type launcher 投出式发射装置
release type 应变释放方式
release unit 释放装置
release value 释放值
release valve 降压阀;泄水阀(门);泄放阀;缓解阀;排除阀;排出阀;放油阀;放气阀(门);泄气阀
release valve ball (滤清器的)旁通球阀
release valve gasket 放油阀垫密片
release valve spring 放泄阀弹簧
release valve stud 放松阀柱螺栓
release voltage 释放电压
release wave 泄水波;放水波
release wave below dam 大坝泄水坡
release wave below-power station 电站泄水坡
release wave below ship lock 船闸泄水波
release wave below spillway 大坝泄水坡
release wax 脱模蜡
release wire 释放线
release without payment of duty 免税放行
release work 泄水建筑(物);泄水构筑物
release yoke 释放轭;分离叉
releasing 解脱;拉扣
releasing agent 脱模剂;分型剂;防粘剂
releasing and circulation overshot 能松开和冲洗液循环的打捞筒

releasing area 排气管截面积
releasing arrangement 脱扣装置;释放装置
releasing coil 释放线圈
releasing contact 释放触点
releasing coupling 释放联轴节
releasing current 释放电流
releasing device 释放装置;断电装置;分离装置;分离杠杆
releasing factor 释放因子
releasing force 解锁力;释放力
releasing gear 解脱装置;解脱机构;脱离齿轮;脱钩装置;释放装置;释放机构
releasing gear of hooks 艇钩联动脱钩装置
releasing handle 释放柄
releasing hook 松绳钩;释缆钩;脱扣钩
releasing hormone 释放因子
releasing key 脱膜楔;脱模楔;松模楔
releasing lag 释放延迟
releasing lever 分离杠杆;放松杆
releasing line 脱钩索
releasing liquid 释放剂
releasing magnet 释放电磁铁
releasing mechanism 释放装置
releasing moment 放松力矩
releasing of a route 解锁进路;释放进路
releasing of brake 松开制动器
releasing oil 松弛模楔
releasing overshot 松开的打捞筒
releasing state of retarder 缓行器缓解状态
releasing temperature (自动消防器的)引喷温度;引放温度;释放温度
releasing the packer 介封打隔器
releasing the pull 切断拉线;放松拉力
releasing track 线路开通
releasing value 释放值
releasor 让渡人(权利或财产)
releg a derrick 加固钻塔大腿
relegs 加固金属构件(塔腿)
re-lending 再贷款
relevance 关联性
relevance analysis 相关分析
relevance factor 适用度
relevance feedback 关联性反馈
relevance grade 适用度
relevance ratio 相关比(率);检出率
relevance tree 相关树;树状分析图
relevance weights 适用度
relevant action 相关动作
relevant benefits and cost 相关受益与成本
relevant character 并联字符
relevant clauses 有关条文
relevant control 相关控制
relevant cost 相关成本;有关成本
relevant costing 相关成本计算
relevant data 有关数据
relevant documents 并联文件
relevant equilibrium 相应平衡
relevant failure 关联失效
relevant identifier 相关标识符
relevant information 适用资料
relevant item 相关项目
relevant nodal displacement 相应节点位移
relevant parameter 关联参数
relevant parties 有关方(面)
relevant point 相应点
relevant portion 相关部分
relevant position 相应位置
relevant pressure 相应压力
relevant range 相关范围
relevant reaction 相应反应
relevant risk 相关风险
relevant segment 相关线段
relevant stability constant 相应稳定性常数
relevant value 相应值
relevant vocoder 相关声码器
relevant volume range 相关产量范围
relevant wavelength 有关波长
releve 原建筑图
relevel(l)ing 再平整;重新调平【测】;重复水准测量(法);复测水准;重新整化;再平层
reliability 性能最佳性;信赖度;可靠性;可靠度;确实性;安全性
reliability, availability, serviceability 可靠性、可用性与可维修性

reliability activity coordination 可靠性业务协调
reliability administration 可靠性管理
reliability allocation 可靠性分配
reliability analysis 可靠性分析
reliability apportionment 可靠性分配
reliability assessment 可靠性估计
reliability assignment 可靠性分配
reliability assurance 可靠性保证
reliability assurance program(me) 可靠性保证规划
reliability-based criterion 根据可靠性准则
reliability block diagram 可靠性框图
reliability certification 可靠性认证
reliability coefficient 可靠性系数
reliability compliance test 可靠性验证实验;可靠性鉴定试验
reliability consideration 可靠性研究;可靠性考虑
reliability constraint 可靠性约束
reliability control 可靠性控制
reliability control algorithm 可靠性控制算法
reliability critical problem 可靠性临界问题
reliability datum period 可度性基准期
reliability design 可靠性设计;可靠性理论
reliability design analysis 可靠性设计分析
reliability design analysis report 可靠性设计分析报告
reliability determination test 可靠性测定试验
reliability engineering 信赖度工程;可靠性技术;可靠性工程
reliability evaluation 可靠性评审
reliability evaluation test 可靠性鉴定试验
reliability factor 可靠系数
reliability failure analysis 可靠性破坏分析
reliability figures of merit 可靠性灵敏值
reliability function 可靠性函数
reliability index 可靠性指标;可靠系数;可靠度指数
reliability index of spring 弹簧的可靠指标
reliability information coordination 可靠性信息协调
reliability level 可靠性水平;可靠度等级;可靠程度
reliability logic 可靠性逻辑
reliability management 可靠性管理
reliability mathematics 可靠性数学
reliability measurement 可靠性测量
reliability mechanism 可靠性机理
reliability method 可靠性方法
reliability of aids 航标可靠性
reliability of contact 触点(的)可靠性
reliability of correlation coefficient 相关系数的可靠性
reliability of design 设计可靠性;设计可度性
reliability of financial records 财务记录可靠性
reliability of forecast 预报可靠性
reliability of material 材料的可靠性
reliability of operation 运转安全度;运行(的)可靠性
reliability of safety 安全可靠性;安全可靠度
reliability of service 服务的可靠性
reliability of source of supply 供应来源的可靠性
reliability of standard deviation 标准差的可靠性
reliability of statistics 统计可靠性
reliability of structure 结构可靠度
reliability of the instrument 仪器的可靠性
reliability of transmission 传输可靠性
reliability optimization 可靠性最优化
reliability performance 可靠性能
reliability performance measure 可靠性性能测定技术
reliability physics 可靠性物理
reliability policy 可靠性方针
reliability principle 可靠性原理
reliability program(me) 可靠性规划;可靠性计划
reliability proof cycle 可靠性验证周期
reliability reassignment 可靠性再分配
reliability report 可靠性报告
reliability screening 可靠性筛选
reliability service 可靠运行
reliability standard 可靠性程度;可靠性标准
reliability stress analysis 可靠性应力分析
reliability test 可靠性试验;可度性试验
reliability test assembly 可靠性测试装置
reliability testing 可靠性测试
reliability test road 可靠性试验路

reliability theory 可靠性理论
reliability trade-off 可靠性综合标准
reliability trial 长距离耐久试验(汽车等);可靠性试验;可靠度试验;强度试验
reliable 可靠的,稳妥的
reliable account 可靠账户
reliable analysis 可靠性分析
reliable assurance 可靠性保证
reliable coefficient 可靠性系数
reliable compiled map 校订版地图
reliable contour 精确等高线
reliable demonstration 可靠性验证
reliable design 可靠性设计
reliable documents 可靠证件
reliable estimate 可靠估计
reliable estimation 可靠性估计
reliable evidence 可靠证据
reliable fit 可靠拟合
reliable index 可靠(性)指标;可靠性指数;可靠度指数
reliable information 确实可靠的情报资料
reliable interval 可靠区间
reliable marker 指示层;可靠标准层
reliable observation 可靠观测
reliable operating system 可靠的操作系统
reliable operating value 可靠工作值
reliable pick-up value 可靠吸取值
reliable position keeping 可靠定位
reliable precision 可靠精确度
reliable probability 可靠性概率
reliable result 可靠成果
reliable seal 密封可靠
reliable service 可靠服务
reliable software 高可靠性软件;可靠软件
reliable source 可靠来源
reliable standard 可靠性标准
reliable system 可靠系统
reliable test 可靠性试验;可靠度试验
reliable theory 可靠性理论
reliable weight of sample 样品可靠重量
reliable witness 可靠证人
reliable yield 可靠产水量
reliablistic design methodology 可靠性设计方法论
relic 遗迹;残余物;残遗体;残遗地貌;残片;残留物;残块
relic bar 残留沙洲;残留沙坝
relic bioclastic texture 残余生物碎屑结构
relic coil 残留螺旋
relic fauna 残遗动物区系
relic mineral 残余的矿物;残留的矿物
relic mountain 蚀余山
relics 文物;残迹
relic scar 残留痕
relic sea 残海
Relics of Yuan Ming Yuan 圆明园遗址
relic species 残遗种
relict 残余种;残余物;残余(的);残遗群落;残留(生物)
relict age 残余年龄
relict allochem texture 残余异化颗粒结构
relict alteration 残余蚀变
relict bedding 残留层理
relict bittern 残留卤水
relict dyke 残余岩脉
relict dune 残遗沙丘
relic texture of carbonate rocks 碳酸盐岩残余结构
relict gypseous solonchak 残余石膏盐土
relict intraclastic texture 残余内碎屑结构
reliction 退水土地的所有权;海退;出水土地
reliction sediment 残留沉积(物)
relict island 残岛
relict lake 残余湖;残湖
relict lump texture 残余团块结构
relict micritic texture 残余微晶结构
relict mineral 残余矿物
relict mountain 残山
relict oolitic texture 残余鲕粒结构
relict pellet texture 残余球粒结构
relict permafrost 残遗永久冻土
relict porphyitic texture 残余斑状结构
relict sediment 残余沉积(物);残遗沉积物;残留堆积物;残留沉积(物)
relict soil 残余土
relict structure 残余构造;残存构造

relict texture 残余结构;残存结构
relict texture dolomite 残余结构白云岩
relict tuffaceous texture 残余凝灰结构
relict water 残余水
relief 卸荷;离隙;救助;减荷;免责;缓减;后角;退刀槽;突起;凸纹;顶尖缺口;地势起伏;地形;浮花雕饰;浮雕
relief against forfeiture 免除罚金
relief angle 后让角;起伏角
relief annealing 消除残余应力退火;去应力退火
relief arch 卸荷拱;反砌
relief arrangement 卸荷装置;减荷措施
relief ball 球阀
relief blasting 卸载爆破(法);剥离爆破(法)
relief blasting method 辅助炮眼
relief block 凸版
relief board 地貌版
relief bridge 泄洪桥;溢流桥;排涝桥;辅助桥
relief bridge openings 辅助桥孔
relief bronze door 青铜雕饰门
relief bypass 降压支管;减压旁通管
relief cable 应急电缆;救援电缆
relief cam 减压凸轮;退动凸轮
relief car 救援车
relief carving 浮花雕饰
relief carving knife 浮雕刀
relief center 救援中心
relief chamber 减压室
relief channel 辅道(指航道);分洪渠(道);分洪道;放水渠(道)
relief circuit 卸荷回路
relief coating 凸纹涂料;浮雕涂料
relief cock 卸荷旋塞;泄压旋塞;溢流开关;降压管门;减压龙头;减压开关;安全(放泄)旋塞
relief cone 卸料锥(体)
relief cover 保险盖
relief crank 臂形拐肘
relief culvert 辅助涵洞;边沟泄水涵洞
relief cut 防咬锯的先行锯
relief cylinder 保险汽缸;安全汽缸
relief damper 泄压风门;减压风门;自动气阀;压力调整器
relief data 地貌资料;地貌数据
relief decoration 浮雕装饰
relief device 安全装置;泄放装置;减压装置;保险装置
relief difference 起伏差
relief displacement 投影差;地形起伏位移
relief distortion 投影差
relief ditch 泄水沟;放水沟
relief diver 救助潜水员
relief drain 降压排水;降低地下水位的排水
relief drawing 地形原图;地貌描绘
relief duct 安全通气管;安全通风管
relief effect 立体作用;立体效应;浮雕效应
relief element 地貌因素;地貌要素;地貌成分
relief enamel 堆釉
relief engraving 凸纹雕刻
relief engraving machine 浮凸雕刻机
relief expenses 救济费
relief feature 地形特征;地势特征;地貌因素;地貌要素;地貌特征;地形要素
relief fitting 溢流塞;减压装置
relief flank 铲背
relief for pioneer industry 先行性行业减税
relief frame 减压架;辅助肋骨
relief frieze 浮雕檐壁
relief from obligation 免除义务
relief fund 救济金
relief fund reserves 救济基金准备
relief gas 排出气;废气;吹气
relief gate 放水闸(门)
relief globe 立体地球仪
relief grating 浮雕光栅
relief grinder 铲齿机
relief grinding 铲磨
relief grinding machine 铲齿机
relief groove 压力平衡槽
relief ground 铲磨的
relief height 地面高度
relief holder 缓冲储气罐;平衡气柜
relief hole 辅助炮眼;减载炮眼;衬托孔洞;放水孔
relief image 立体像;浮雕像
relief incision 减张切开

relief intensity 地形强度
relief irons 解脱式提引环;解脱式提引钩
relief joint 伸缩缝
relief lever 卸荷手柄;释放操作手柄
relief lightship 替用灯船
relief lines 切片刀痕
relief load 卸载;甩负荷
relief locomotive 救援机车
relief manager 救灾工作负责人
relief map 模塑地图;立体地图;立体地形图;模型地图;地形图;地势图;浮雕地形图
relief map of excavation levels 开挖层立体图
relief mechanism 减荷装置;释放装置;保险装置;安全装置
relief method 卸载爆破(法);剥离爆破(法)
relief milling 铲背铣削
relief-milling device 石膏地形模型塑造装置
relief model 立体模型;地形模型
relief modelling 浮雕塑像
relief model panel 地形模型板
relief mo(u)lding 浮雕成形
relief needle nozzle 保险针形管嘴;安全针形喷嘴
relief of customs 关税豁免
relief of pressure 释压
relief of stress 应力释放
relief of traffic 减轻交通负荷;疏散交通
relief opening 减压阀门;溢流口;排气口;减压口;安全排放口
relief ornament 浮雕装饰
relief outlet 放泄口;放水口
relief overlay 等高线透明片
relief painting 浮雕画
relief passage 卸载通道
relief passageway 安全通路;安全通道
relief pattern 阳纹样本;阳纹花样;阳纹图案(立体图案)
relief pattern tile 凸纹花样瓦
relief payment 救济金
relief perception 立体感
relief phase hologram 浮雕型相位全息图
relief pipe 排气管;溢流排水管;减压管;放泄管;放水管;安全排出管
relief (pipe) line 泄压管线
relief piping 溢流排水管
relief piston 阻气活塞;平衡活塞;调压活塞;辅助活塞
relief plan 地形图
relief platform 卸荷(平)台
relief plug 放气塞
relief polishing 显微切片浮雕抛光;凸磨光
relief port 放气口
relief pressure 表压力;超过大气的压力;减压
relief pressure control valve 释放压力控制阀
relief pressure valve 安全阀;减压阀
relief pressure valve cartridge 减压阀芯
relief printing 凸版印刷
relief printing machine 凸版印刷机
relief printing plate 凸印版;凸版
relief pump 辅助泵
relief ratio 流域高宽比
relief relay 交替继电器;辅助继电器
relief representation 地形表示(法);地貌表示法
relief road 间道;辅助道路;辅道(指道路);疏散车辆道路
relief roadway 卸载巷道
relief service car 救援车
relief sewer 泄流排水管;溢流排污管;溢流污水管;救济污渠;减流污水道;污水泄放管
relief shade 地貌晕渲
relief siphon 放水虹吸管;安全虹吸管
relief spring 保险弹簧
relief sprue 减压直浇口;冒渣口;除渣减压冒口;补助浇口
relief stretching 立体夸大
relief survey 地形观察
relief tax 宽减税项
relief telescope 体视望远镜
relief television 立体电视
relief track 临时替用线;避让线;避车线;避车道
relief train 救援列车
relief treatment of stress 应力解除处理
relief tube 溢流排水管;排放管
relief tubing 溢流排水管
relief tunnel 放水隧洞

relief valve 卸压阀;泄压阀;溢流阀;自动气阀;开放阀;保险阀;安全活门;安全阀
relief valve ball 保险阀球;安全阀球
relief valve connection 安全阀接口
relief valve connector 安全阀接口
relief valve spring 安全阀簧
relief valve unit 减压阀装置;保险活门装置
relief valve with dashpot plunger 带缓冲柱塞的安全阀
relief variation 地貌变化
relief vent 辅助排气孔;辅助排气管;辅助排气口;辅助透气管;安全通气管;安全通风管
relief welding 凸焊
relief well 减压井;排水井
relief worker 救灾工作人员
relief works 失业救济工程
relieve 凸起;卸载;降压;减压;解开;减轻;脱开
relieve dark appearance 改为淡色
relieved compression ring 卸荷气环
relieved cutter 铲齿铣刀
relieve distortion 残余变形挠曲
relieved land 后刀面棱宽
relieved shank 缩小的螺栓杆身
relieved steel plate 花纹钢板
relieved work 浮雕饰品;浮雕装饰
relieved zone 释压带
relieve from 使解除
relieve from obligations 解除义务
reliever 解脱器;辅助炮眼
reliever relay 接替者
relieving 分离;使凸起;显露;解脱;减轻;衬托;铲齿
relieving amount 让刀量
relieving anode 释荷阳极
relieving arch 卸载拱;卸力拱;载重拱;减压拱;辅助拱;反碹;暗拱
relieving attachment 铲齿装置;拆卸装置
relieving block 减载块
relieving board 舱底货上面的垫板
relieving capacity of safety valve 安全阀的泄放容量
relieving cutter 后角铣刀;铲齿铣刀
relieving device 减载装置;铲齿装置
relieving effect of overhanging element 悬出的构件卸荷作用
relieving floor 卸荷板
relieving gear 解脱机构;解扣装置
relieving hyperacidity 制酸
relieving lathe 铲齿车床;铲背车床
relieving layer 减压层
relieving machine 铲齿车床
relieving of internal stress 消除内应力
relieving pipe 减压管
relieving piping 减压管
relieving platform 卸载式平台;卸荷板;卸荷(平)台;减载(平)台
relieving-platform quay 减荷台式码头
relieving platform type of quaywall 桩承台(式)岸壁
relieving platform type wharf supported on bearing piles 后板桩式高桩码头
relieving-platform wall 减压平台式墙
relieving-platform wharf 卸荷台式码头;减荷台式码头
relieving prism 松弛棱镜
relieving quay 卸载式码头
relieving ring 保险环
relieving rubble backfill 减压棱体
relieving rudder 应急舵
relieving set 衬托装置
relieving sewer 溢流排污管
relieving shot 卸载爆破;卸荷爆破;应力解除爆破;清除拒爆的炮眼
relieving slab 卸荷板
relieving stress 消除应力
relieving summer-heat 解暑
relieving tackle 缓冲滑车组
relieving tiller 应急舵柄
relieving timber 辅助支撑;辅助支柱
relieving tool 铲齿刀(具)
relieving tool holder 铲齿刀座
relieving tube 减压管
relieving tubing 减压管
relieving-type platform supported on piles 桩承台
relieving valve 卸荷阀;减压阀;溢流阀
relieving variable 辅助变数
relieving vault 减载拱顶
relieving wall 辅助墙
relieving well 减压排水井
relieving wharf 卸载式码头
relievo 浮花雕饰;浮雕
relift station 污水提升站
relight button 重复起动按钮
relighter flame safety lamp 自点火焰安全灯;装有再点火器的火焰安全灯
relight(ing)再次起动;重新点火;重复点火
religion 宗教信仰
religion institute 宗教机构
religious architecture 宗教建筑
religious building 宗教建筑
religious customs 宗教习惯
religious ecology 宗教生态学
religious Gothic (style) 宗教哥特式
religious holidays 宗教节日
religious house 寺院;修道院
religious monument 宗教纪念建筑
religious painting 宗教绘画
religious symbolism 宗教象征手法
reline 换衬里;重新划线;重砌内衬
reliner 换衬器
relining 换衬;更换衬套;重衬;重挂白合金(轴承)
relining a kiln 更换窑衬
relining of bearing 轴承重挂白合金
relining tool 摩擦衬片装合工具;换衬工具
relinking 再连接
relinquish 撤回
relinquishing of business 停办企业
relinquishment clause 让出条款
reliquary 遗物箱;古迹;圣物匣
reliquefaction 再液化
relish 榫头加腋宽度;凸肩;吞肩;榫肩
Relizian (stage) 雷利兹阶(北美)
reload 重加荷载;再装载;再装;再放电;重装入;重新装载;重新加载;半装入
reload cycle 再装药周期;再加载循环
reloader 转载机;装卸机;换装机;重新装载机;复载机
reloading 倒装;重复加载;再放电
reloading and rearrangement of goods 货物换装整理
reloading curve 再加荷曲线;重新加载曲线
reloading machine 装卸机;换装机
reloading modulus 再加荷模量
reloading point 转运点
reload model 再加荷模型
reload modulus 再加荷模量
relocatability 再定位;可定位性;可浮动性
relocatability attribute 浮动属性
relocatable 可重定位的;可定位的;浮动形式;浮动的;浮点形式
relocatable address 可重定位地址;可浮动地址;可再定位地址;可浮动地址;浮动地址
relocatable area 可再定位区;可浮动区
relocatable assembler 浮动汇编程序
relocatable binary 可再定位二进制程序;浮动二进制
relocatable block of instruction 可再定位指令块
relocatable building 可搬迁的建筑
relocatable code 相对程序;可再定位代码;浮动码
relocatable coding 可重新配置编码
relocatable emulator 浮动仿真程序
relocatable entry address 可浮动入口地址
relocatable expression 可重定位表达式;可浮动表达式;浮动表达式
relocatable file 浮动文件
relocatable form 可浮动形式
relocatable library 可定位库;可浮动库;浮动(程序)库
relocatable library module 浮动库模块
relocatable linking loader 浮动连接装配程序
relocatable loader 浮动装配程序
relocatable load module 浮动装入模块
relocatable mode 浮动方式
relocatable module 可再定位模块;可浮动模块;浮动(程序)模块
relocatable name 可浮动名;浮动名称
relocatable object program(me) 浮动目标程序
relocatable partition 可拆装的隔断;装卸式隔断;活动隔墙;活动隔断
relocatable phase 可再定位阶段
relocatable program(me)可再定位程序;可重定位程序;可浮动程序;浮动(输入)程序
relocatable program(me) library 可再定位程序库;浮动程序库
relocatable program(me) loader 浮动输入程序
relocatable routine 浮动程序
relocatable subroutine 可浮动子程序;浮动子程序
relocatable symbol 可浮动符号;浮动符号
relocatable symbolic address 浮动符号地址
relocatable task set 可再定位任务组;浮动任务集
relocatable term 可重定位项;可再定位项;可浮动项;浮动项
relocate 改线;重新选址;重新定线
relocated address 置换地址
relocated channel 旧航道
relocate hardware 浮动硬件
relocating object loader 浮动目标装配程序
relocation 再分配;再定位;改线;重新定线;重新安置;重定位置;搬迁
relocation binary program(me) 再定位二进制程序
relocation bit 重新(分)配位
relocation criterion 迁建准则;迁建标准
relocation dictionary 重新配位表;浮动地址表
relocation grant payment 搬迁补助付款
relocation household 拆迁户
relocation housing 周转房
relocation mechanism 浮动机构
relocation of building 建筑物易位
relocation of control point 控制点重新定位
relocation of control station 控制点重新定位
relocation of road 改线;路线重定;道路改线;重定路线
relocation of storage pile 转垛
relocation of track 改移轨道
relocation payment 搬迁费
relocation plan(ning)迁建规划
relocation register 浮动寄存器
relocation settlement cost 搬迁安置费(用)
relocation workload 搬迁任务;搬迁对象登记
relogging 回头采
re-lubrication 再润滑;重新润滑
reluctance 磁阻
reluctance coefficient 磁阻系数
reluctance force 磁阻力
reluctance generator 磁阻发电机
reluctance machine 磁阻电机;反应式同步电机
reluctance microphone 电磁传声器
reluctance motor 磁阻电动机;反应式同步电动机
reluctance pick-up 磁阻拾音器
reluctance pressure transducer 磁阻压力传感器
reluctance stepper 磁阻步进电动机;反应式步进电动机
reluctance synchronous machine 反应式同步电机;磁阻同步电机
reluctance torque 磁阻转矩;反应转矩
reluctance-type pressure ga(u)ge 磁阻压力表
reluctancy 磁阻系数
reluctivity 磁阻系数;磁阻率
reluctometer 磁阻鉴质仪
relume 重新点燃
reluster 无光泽
reluxation 再脱位
rem 雷姆(人体伦琴当量)
remachine 再加工;重新加工
remachining 再加工
remagnetization 再磁化
remagnetize 再磁化;重新磁化
remain 余下;剩余;剩下;残留
remain arc 残留(岛)弧
remainder 剩余产权;余数;零位;净余;继承权;剩余数;剩余;残余;残料
remainder depreciation 筛余折旧
remainder error 剩余误差
remainder estate 地产未指定继承权
remainder formula 余项公式
remainder function 余项函数
remainder index 温度余值指数
remainderman 有房产未来指定权的人
remainder mountain 环蚀山;残山
remainder of a series 级数的余部
remainder of beam 剩余束流
remainder of distribution 分布的剩余
remainder of exhaust gas 剩余废气
remainder of exhaust gases 排气余气

remainder radio deviation 剩余无线电自差
remainder term 余项;剩余项
remainder theorem 余数定理
remained oil 残余油
remain function of absorption 保持吸收作用
remain inclination 剩余倾角
remaining 残留的;残存
remaining arch 残余拱
remaining arrears 拖欠
remaining at post without wage 停薪留职
remaining-benefit method 剩余效益法;剩余收益法
remaining deviation 剩余自差
remaining durable years (建筑物的)剩余使用年限
remaining error 剩余误差
remaining force 剩余力
remaining liquid of fertilizer 化肥残液
remaining liquid of pesticide 农药残液
remaining load 遗留负载
remaining oil saturation 保持油饱和
remaining on board 未卸;船上存有
remaining operational life 剩余使用寿命
remaining period 残存期间
remaining pore 残留气孔
remaining runway 跑道剩余段;跑道剩余长度
remaining service life (建筑物的)剩余使用年限
remaining star 残留星
remaining strategy 剩余策略
remaining time 剩余时间
remaining trace 残迹
remaining unbalanced friction factor 剩余未平衡摩擦系数
remaining value 剩余价值
remaining value of house 房屋现价
remaining variable 剩余变量
remaining variation 残余波动
remaining velocity 剩余速度
remaining video information 剩余图像信息
remain in operation 继续运转
remains 剩余物;残余;残骸
remains of animal and plant 动植物残骸
remains of wreck 沉船残骸
remain stationary 保持固定
remain unfulfilled 仍未完成
remain unperformed 仍未执行
remain unshipped 仍未装船
remake 重做;重制
remake a plan 修订计划
remaking of nature 改造自然
remaking surface 地面改造
remalloy 勒马罗伊铁钴钼合金;磁性合金
remalloy permanent magnet alloy 勒马罗伊永磁合金
remand home 青少年拘留所
remanence 剩余磁化强度;剩余磁感应;剩磁感应
remanence effect 剩磁效应
remanence field profile on plane 剩磁平剖图
remanence magnetization 剩余磁化
remanence relay 剩磁继电器
remanence type relay 剩磁感应式继电器
remanent 残余
remanent core 带剩磁铁芯
remanent elongation 残余伸长
remanent field 剩余(磁)场
remanent flux 剩余磁通
remanent flux density 剩余磁密
remanent induction 剩余(磁)感应
remanent induction coefficient 残余感应系数
remanent magnetic moment 剩余磁矩
remanent magnetism 剩余磁性;剩磁;剩余磁性
remanent magnetization 剩余磁化(强度);残余磁化(强度)
remanent magnetometer 剩余磁力仪;剩磁仪
remanent median surface 剩磁场中平面
remanent ocean basin 残余洋盆
remanent permeability 残余磁导率
remanent point 残留点
remanent polarization 剩余极化强度
remanent stage 残留阶段
remanent strain 残余应变;残余胁变
remanent stress 残余应力
remanent value of equipment 设备残值
remanent volume 残留容积
remanent work 修补工作;修补工程
remanufactured lumber 再加工木材

remapping 地图重测;重测图
remark 注视;雷达指向;雷达光标;附记;摘要栏;备注
remarkable local irregularity 明显的局部不匀度
remark(ing) 改造
remark on a contract 合同说明
remarks board 备注栏
re-marshalling of restoring the trains 列车改编
rematch 再匹配
rematching relay 连续匹配继电器
rembrandt bonds 瑞士债券
re-measure 重新量度
re-measuring 再测量
remedial acid job 补救性酸化作业
remedial action 补救行动;补救措施
remedial action plan 补救活动计划
remedial cementing 补注水泥
remedial dredging 补救疏浚
remedial dredging technique 补救性挖泥技术
remedial facility 补救设施
remedial frames 纠正图像
remedial inspection 补救检验
remedial maintenance 出错维修;补救维修
remedial management 事后对策管理;补救管理
remedial measure 矫正措施;补救措施;补救办法
remedial operation 修理作业;补救操作
remedial pathway 补救途径
remedial solution 补救方法
remedial time 补救时间
remedial treatment 补强处理
remedial underpinning 补救托换
remedial works 补强工事;修理工作;整改;补救(性)工程
remediation of aquatic ecology 水生生态修复
remediation of polluted underground water 污染地下水修复
remediation of water body 水体修复
remediation of water pollution 水污染修复
remediation technology 修复技术
remedied treatment 补救处理
remedies precede rights 救济先于权利
remedy 弥补;补救方(法)
remedy a leak 修补漏缝
remedy allowance 公差
remedying defect 修补缺陷
remedy measure 补救措施
remedy the trouble 排除故障
remelt crucible 再熔坩埚
remelted alloy 再熔合金;重熔合金
remelted iron 再熔铁
remelter 再熔器;再熔炉
remelt(ing) 再熔(化);重熔(化);重新熔化;复熔
remelting alloy 再熔合金
remelting furnace 再熔炉;重熔炉
remelting hardness 熔融硬度
remelting machine 火抛光机
remelting technology 再熔工艺
remelt iron 再熔铁
remelt junction 再熔结
remelt process 再熔法
remelt technique 再熔技术
remelt zinc 再熔锌
remembering 存储
rememberment 土地登记
remembrance 纪念物
remesh 重啮合
remetal 更换金属挡板
remethylation 再次甲基化
remaining property liquidation and distribution 资产清理分配
remigration 再运移;再迁移
reminder 催付通知书
reminder letter 催函
remineralization 再矿化作用;补充矿质
remingtonite 土红钴矿
remint 重铸
remise 立契出让
remission 减免;缓和;罚金等
remission form 缓解型
remission of debts 免除债务
remission of penalty 免除罚款
remission of tax 免税
remission photometer 分光光度镜
remission rate 缓解率
remission stage 缓解期

remit 减轻;减免;免于处罚
remit account 划账
remit a debt 免债;免除债务
remit money 划拨款项
remit rent 免租
remittance 汇款;汇兑
remittance abroad 往国外汇款
remittance advice 汇款通知单
remittance and draft outstanding 未提取的汇出汇款
remittance bill 汇(款)条(单)
remittance by banker's demand draft 票汇
remittance by migrants 侨汇
remittance charges 汇费
remittance check 汇款支票
remittance fee 汇费
remittance order 汇款单
remittance permit 汇款核准书
remittance rate 汇费率
remittance slip 汇款通知单
remittee 领取汇款人;取款人;收款人
remittent 弛张的
remittent fever 弛张热
remittent temperature 弛张温度
remitter 免除者;汇款人;出票人
remitting bank 汇款银行;托收银行
remitting funds 恢复费用
remix 二次搅拌;再混合
remixed concrete 再拌混凝土;复拌混凝土;二次搅拌混凝土
remixer 再拌机;重复搅拌机;复拌机
remixing 再拌和;再混合;(混凝土的)重拌和;复拌;二次搅拌
rem-jet 炭黑防光晕层
remnant 遗留物;侵蚀残余;残余的;残迹;残存
remnant arc 残余岛弧;残留(岛)弧
remnant island 残岛
remnant magnetism 残磁(性)
remnant ocean 残留大洋
remnant ocean basin 残留洋盆
remnant pillar 残留矿柱
remnant ridge 残留脊
remnant sea 残余海
remnants of glacial abrasion 冰溜通痕
remnants of glaciations 冰川遗迹
remnant strain release 剩余应变释放
remnant value 残值
remobilization 再活化(作用)
remobilization of a chemical 化学品的再流动
remobilized platform type formation 地台活化型沉积建造
remobilized stage of platform 地台活化阶段
remodel 重制;重新塑造
remodel(l)ing 改型;改造;重塑;城市或街坊重建
remodel(l)ing a house 住宅翻建
remodel(l)ing of city centre 市中心区重新规划
remodel(l)ing of nature 改造自然
remodulation 再调制;重复调制;二次调制
remodulator 再调制器;二次调制器
remoistening adhesive 再湿性黏合剂;再湿性胶黏剂
remold 重铸
remolded soil 扰动土(壤)
remolten 再度熔化
remote-access and control 远程存取和控制
remote-access computing system 远程访问计算系统
remote-access data processing network 远程存取数据处理网络
remote-access service 远程访问服务
remote-access terminal 远程存取终端;遥控存取终端
remote action 远程作用
remote-action system 远程作用系统
remote adjustment 遥调
remote aerial reconnaissance 航空遥感
remote aftereffect 远程后效
remote aggregation contact 远凝聚接触
remote aiming 远距离瞄准
remote airport 远方机场
remote alarm 远程报警;遥警
remote anchorage 远隔锚地
remote area 偏僻地区;边远地区
remote area channel 遥远地区信道

remote automatic calibration system 远距离自动校准系统
remote automatic telemetry equipment 远距离自动遥测设备;自动遥测设备;自动遥测仪
remote azimuth selsyn 遥控方位角自动同步机
remote back-up 远距离后备保护(装置)
remote backup system 远程备份系统
remote balance 远距操作天平
remote batch 远程成批处理
remote batch access 远程信息存取;按键式信息存取
remote batch computing 远程成批计算
remote batch entry 远程成批输入
remote batch operation 远程成批操作
remote batch process 遥控批量处理
remote batch processing 远程成批处理
remote batch station 远程成批站
remote batch terminal 远程成批处理终端
remote batch terminal system 远程成批处理终端系统
remote bulb thermostat 遥控恒温器
remote buoy 遥控浮标
remote calculator 远程计算器
remote calibration 远距离校准
remote casting 遥控浇注
remote cause 远因;诱因
remote center compliance 间接中心顺从性
remote communication 远距离通信
remote communications complex 远距离通信全套设备
remote communications console 远程通信控制台
remote compass 遥控罗盘
remote computer 远程计算机
remote computer networking 远程计算机连网
remote computing monitor system 远程计算监督系统
remote computing system 远程计算系统
remote computing system consistency error 远程计算系统的一致性误差
remote computing system exchange 远程计算系统交换器
remote computing system exchange device 远程计算系统交换机
remote computing system language 远程计算机系统语言
remote computing system log 远程计算系统记录表
remote computing system monitor 远程计算系统监督程序
remote concentrator 远方集中器;远程集中器;远程集线器
remote connector 远操连接器
remote console 远距离控制台;远程终端;远程控制台
remote control 远距离控制;远程控制;远程操纵;遥远控制;遥控
remote control apparatus 远距离控制设备;遥控设备;遥控操作装置
remote control assembly 遥感装置
remote control automation board 遥控自动化仪表板
remote control balance 遥控天平
remote control board 遥控盘
remote control box 遥控盒
remote control bulldozer 遥控推土机
remote control cable 遥控电缆
remote control center 遥控中心
remote control channel 遥控通道
remote control circuit breaker 遥控断路器
remote control coal mining 遥控采煤法
remote control coder 遥控编码器
remote control coding 遥控编码
remote control command 遥测指令
remote control connector 遥控连接器
remote control cylinder 分置式控制油缸
remote control device 遥控设备
remote control dial unit 遥控拨号装置
remote control distributor 遥控分配器
remote control door 遥控风门
remote control electronic switching system 遥控电子交换系统
remote control engine 遥控发动机
remote control equipment 遥控(监视)设备
remote control excavator 遥控挖掘机

remote control facility 遥控设施
remote control focusing 遥控调焦
remote control foot pedal 遥控脚踏板
remote control for automatic transmission 自动变速器遥控装置
remote control gate 遥控闸板
remote control gear 远程操纵机构;遥控(操作)机构
remote control handle 遥控手柄
remote control high tension switch 遥控高压开关
remote control ignition 遥控点火
remote control installation 遥控装备
remote control-instruction 遥控指令
remote control iris 遥控可变光阑
remote controlled 遥控的
remote controlled anchor windlass 遥控起锚机
remote controlled boiler 遥控锅炉
remote controlled demolition carrier 遥控爆破车
remote controlled device 遥控装置
remote controlled directional well drilling 遥控定向钻进
remote controlled drier 遥控式干燥机
remote controlled injector 遥控喷射器
remote controlled instrument 遥控记录仪表
remote controlled interlocking 遥控联锁
remote controlled motorized submersible 遥控动力潜水器
remote controlled mount 遥控装置
remote controlled operation 远程控制操作;遥控运行;遥控操作;远距离操作
remote controlled piping stations 远距离控制的泵站
remote controlled point 遥控道岔【铁】
remote controlled priming 遥控起爆
remote controlled pump 遥控泵
remote controlled pump station 遥控泵站
remote controlled rail retarder 遥控轨道缓行器
remote controlled receiver 遥控收信机
remote controlled robot 遥控自动装置
remote controlled solenoid 遥控螺线管
remote controlled station 远程泵站;远程控制站;遥远受控站
remote controlled switch 遥控道岔【铁】
remote control(led) system 遥控系统
remote controlled underwater manipulator 遥控水下机械手
remote controlled underwater robot 遥控水下机器人
remote controlled unit 遥控装置
remote controlled valve 遥控控制阀
remote controlled X-ray unit 遥控 X 线诊断机
remote controller 远距离控制器;遥控器
remote controller with infrared ray 红外(线)遥控器
remote control lever 遥控杆
remote controlling telephone exchange of central traffic control 调度集中遥控总机
remote control manipulator 远距离操纵器;遥控操作机
remote control microscope 遥控显微镜
remote control modulation 遥控调制
remote control monitoring system 遥控监视系统
remote control of an area 区域遥控
remote control office 遥控室
remote control of hump engines 驼峰机车遥控
remote control of transmission 遥控传动装置
remote control operation 遥控操作
remote control panel 远程遥控制台;远距离操纵盘
remote control pipet(te) 远控吸移管
remote control pipetting device 远控吸移装置
remote control rack 远距离控制起重机
remote control radio system 遥控无线电系统
remote control reading 遥测读数
remote control receiver 遥控接收机
remote control receiving unit 遥控接收装置
remote control releaser 遥控释放机构
remote control rover 遥控自行装置
remote controls 远程操纵机构
remote control sampler 遥控取样器
remote control scriber 遥控刻图器
remote control signal 遥控信号;远程控制信号
remote control signals selection 遥控信号选择
remote control starter 遥控起动器
remote control station 遥控站

remote control steering gear 遥控转向机构
remote control substation 遥控变电所
remote control switch 遥控开关
remote control switching system 遥控开关系统
remote control system for distribution 分散目标控制系统
remote control system in station 车站遥控系统;车站遥控
remote control system testing 遥控系统测试
remote control target-boat 遥控靶艇
remote control thermostat 遥控恒温器
remote control valve 遥控阀
remote control vehicle 遥控船
remote coupling constant 远程耦合常数
remote cut-off 遥控开关;遥截止
remote cut-off characteristics 遥截止特性
remote cut-off tube 遥截管
remote cycle change 遥控周期变更装置
remote damage 间接损害
remote data 远距离数据
remote data base 远程数据库
remote data box 远距离数据传递部件
remote data capture 远程数据收集
remote data collection 远程数据收集;遥控数据收集
remote data concentration 远程数据集中
remote data concentrator 远程数据集中器
remote data entry 远程数据输入
remote data indicator 数据遥示器;远距离数据指示器
remote data input 遥测数据输入
remote data processing 远距离数据处理;远程指示系统;远程数据处理
remote-data station 远端数据站
remote data telemetry 数据遥测(术)
remote data terminal 远距数据终端
remote data transmission 远距离数据传输
remote data transmitter 遥控数据发射机
remote debugging 远程调试;远程测试
remote-detector 遥测器
remote device 远程设备
remote diagnosis 远程诊断
remote diagnosis system 远程诊断系统
remote dial flowmeter 远程数字流量计;远距离表盘流量计
remote disconnect 远距离操作活接头
remote display system 远程显示系统
remote display unit 远距离显示器
remote district 偏僻地区
remote drive 远距离驱动
remote effect 远期影响;远期效应;远隔作用;间接影响
remote electro-optic(al) sensor 光电遥感仪
remote elevation selsyn 遥控仰角自动同步机
remote end 远端
remote end feeding circuit 送受分开电路
remote entry network 远程输入网络
remote entry unit 远程输入装置
remote equipment 远距离设备;遥控装置
remote error sensing 远距离误差传感;远程误差读出
remote exchange concentrator 远地交换局集中器
remote failure indication 远端失效指示
remote-fed 远程馈电
remote feed control 远距离进给控制
remote filling point 偏僻地区灌瓶站
remote firing panel 遥控发射指挥台
remote float 遥控浮子
remote float system 浮子遥控系统
remote focus control 遥控调焦
remote format item 间接格式项
remote format specification 间接格式指明部分
remote frame 远程帧
remote gain amplifier 遥控增益放大器
remote gain control 远距离增益控制
remote gain control dial 远距离增益控制度盘
remote gauging of tanks 油罐远距离计量
remote gearbox control 远程操纵变速器机构
remote gear control 遥控变速;变速器遥控
remote handling 远距离操纵;遥控
remote handling device 远距离操作装置
remote handling equipment 远距离操作设备
remote-handling gear 远距离操作装置
remote handling system 远距离操作系统

remote handling tongs 遥控操作钳
remote handling tool 远距离操纵设备
remote head 遥控泵头
remote head pump 远距给料泵
remote hydraulic cylinder 分置式液压油缸
remote identifier 远程识别符
remote index value adjustment 远距离指标值调整
remote-indicating compass 罗盘遥示器;远读罗盘
remote-indicating device 遥示装置
remote-indicating instrument 遥控记录仪表;遥测仪器
remote-indicating pressure ga(u)ges 遥控指示压力表
remote-indicating rotameter 遥示旋转流量计
remote-indicating thermometer 遥控温度计
remote indication 遥测;遥示;远距(离)指示
remote indication instrument 遥测仪表
remote indication manometer 远距离指示压力计;远距离压力计
remote indication system 远程液面指示器
remote indication telemeter 远距离指示遥测计
remote indicator 遥控(指示)器
remote-indicator servomechanism 遥示器伺服机构
remote information 远程信息
remote information processing system 遥控图像处理系统
remote injector 遥控喷射器
remote input 遥控输入
remote input-output station 远程输入输出站
remote input units 遥控输入装置
remote inquiry 远程询问
remote inquiry function 远程询问功能
remote inspection 远距监视
remote intelligent terminal 远距离智能终端
remote interactive computing 远程交互计算
remote interlock connector 遥控联锁连接器
remote job entry 远程作业输入
remote job entry network 远程作业输入网络
remote job entry system 远程作业输入系统
remote job entry terminal system 远程作业输入终端系统
remote job output 远程作业输出
remote job receiving and dissemination 远程作业的收发
remote jump 远驱水跃
remote keying 遥控按键
remote keypad 遥控键盘
remote launch 遥控发射
remote launch control 远距离发射控制
remote level control 远距离水平控制
remote level indicating system 远距离液位指示系统
remote level indicator 远距液面指示器
remote line 远距离传输线;遥控线路
remote line concentrator 远程线路集中器
remote loading 远距离装料;远程装入
remote location 远距离位置;边区
remote login 远程登录
remote loss 间接损失
remotely adjustable collimator 远距离可调准直器
remotely computer-controlled tetherless submersible 计算机遥控无缆式潜水器
remotely controlled camera 遥控式摄像机
remotely controlled float 遥控浮标
remotely controlled level crossing 遥控道口
remotely controlled midwater trawl 遥控中层拖网
remotely controlled monitor 遥控的监控装置
remotely controlled object 遥控对象
remotely controlled operation 远距离控制操作
remotely controlled railway trains 遥控铁路火车
remotely controlled repeater station 遥控中继站
remotely controlled section 遥控区段
remotely controlled vehicle 远距控制运载工具
remotely maintained plant 遥控维修装置
remotely monitored sensor 遥控传感器
remotely operated camera 遥控式摄像机
remotely operated polarograph 遥控极谱仪
remotely operated relief valve 遥控安全阀
remotely operated telescope 遥控望远镜
remotely operated work arm 遥控作业工作臂
remotely sensed drainage pattern 遥感水系类型
remotely sensed geomorphologic(al) types 遥感地貌类型

remotely sensed image processing 遥感图像处理
remotely sensed imagery 遥感成像
remotely sensed lithologic(al) kinds 遥感岩石类型
remotely sensed scanner 遥感扫描器
remotely sensed stratigraphic(al) units 遥感地层单位
remotely sensed structural elements 遥感构造要素
remotely supervised hydroelectric(al) station 遥控水电站
remotely surveillance object 遥信对象
remotely surveillance section 遥信区段
remotely surveillance subsection 遥信分区
remote Mach number 远方马赫数
remote maintenance 遥控维修
remote manipulating equipment 远距离操作设备
remote manipulation 远距离操作;遥控
remote manipulator 遥控机械手;遥控操作器
remote manual control 手动遥控
remote mass-balance weight 远置重量平衡配重
remote measurement 远距离测量;遥测
remote measurement by carrier system 载波制遥测
remote measurement device 遥测装置
remote measurement of discharge 排污遥测
remote measurement seismograph 遥测地震仪
remote measuring 远距离测量;遥测
remote measuring device for cargo tank 货油舱位遥测装置
remote measuring element 远距离测量元件
remote measuring equipment 遥测设备
remote measuring system 远距离测量系统;遥测系统
remote-measuring unit 遥测装置
remote message input, output 远距离信息输入输出
remote message processing 远程信息处理
remote metering 远距离测量;遥测
remote metering system 遥测系统
remote meter reader 远程水表读数计
remote mode 远程传送方式
remote modem 远程调制解调器
remote monitor 远距离监视器;遥控监视仪;遥测装置
remote monitoring 远距离监测;遥控监测
remote monitoring equipment 远距离监视设备
remote monitoring station 遥控监测站
remote monitoring system 远距离监控系统
remote mounting capability 遥置能力
remote multiplexer 远程多路转换器
remote multiplexing system 遥测多路传输系统
remote navigation 远距离导航
remoteness of community 居民点的偏僻性
remote network 远程网络
remote networking system 远程网络服务系统
remote node 远端节点
remote observation 远距离观测
remote observation system automation 自动化遥测系统
remote observatory 遥测台
remote operated balance 遥控操作天平
remote-operated controller 远距遥控制器
remote-operated crane 遥控挖掘机
remote-operated valve 遥控阀
remote-operated vehicle 遥控潜水器
remote operating and signal(l)ing device 遥控遥信装置
remote operating equipment 远距离操作设备;遥控设备
remote operating system 远程操作系统
remote operation 远距离操纵;远程操作;远程操纵;遥远操作;遥控操作
remote operation cylinder lever 分置式液压油缸操纵杆;外置式液压油缸操纵杆
remote operation service element 远端操作服务单元
remote operation vehicle 遥控船
remote operator panel 远程操作面板
remote oxygen plasma generator 远程氧等离子体反应器
remote packet concentrator 远程包集中器
remote patch system 遥控插入码系统;遥控补码系统
remote pickup 远距离拾波

remote pickup equipment 远距中继设备
remote pickup unit 遥控摄像设备
remote pilot control valve 远程先导控制阀
remote pipe connector 遥控接管器
remote pipet 遥控吸移管
remote pipetter 遥控移液器
remote pipetting 远程移液
remote plan (position) indicator 平面位置遥示器;外置平面位置显示器
remote platform 遥控平台;遥感平台
remote plug 远距离转换开关
remote point 偏僻地点
remote polling 远程探询
remote-position control 位置遥控
remote position indicator 远距离位置指示器;远程位置指示器
remote possibility 极小的可能性
remote power control system 远距动力控制系统
remote power cylinder 分置式动力油缸
remote power-off 遥控断电
remote pressure control circuit 压力遥控系统
remote printer 远程打印机
remote probability 低或然率
remote probe 远程探测
remote probing techniques 远距离探测技术
remote processing 远程(信息)处理;遥感处理
remote processing computer 遥控处理计算机
remote program(me) entry 远程程序输入
remote program(me) translation 远程程序翻译
remote programming control 远距离程序控制
remote pumping unit 远距离控制的泵装置
remote radio beacon 远程无线电信标
remote Raman spectroscopy 远距喇曼光谱法
remote reader 远程读出器
remote reading 遥示;远距读数;遥测读数
remote-reading compass 罗盘遥示器
remote-reading current meter 遥测海洋流速仪;遥测海流计
remote-reading ga(u)ge 遥测仪
remote-reading indicator 读数遥示器;远距指示器
remote-reading instrument 远距离读数仪器
remote-reading tach(e)ometer 远程读数转数计
remote-reading tank ga(u)ge 远距离液位计;远程读数液位计;遥测储罐液位计
remote-reading thermometer 遥控温度计
remote-reading water level indicator 远距离显示水位表
remote readout 远距离显示;远程示值读数;远距读出;远程读数
remote readout system 遥读系统
remote-read register 遥读记录器
remote-ready strainometer 遥测式应变计
remote real-time branch controller 远程实时分支控制器
remote real time terminal 远程实时终端
remote recorder 远读记录器;远距计数器;遥控记录仪
remote recording 远距离遥测记录;遥控记录
remote recording apparatus 远程记录仪
remote recording seismograph 遥测记录地震仪
remote region 偏僻地区
remote regulating 远距离调节;遥调
remote regulation 远程调整;遥调
remote reporting 遥测报告
remote reservoir 偏僻地区水库
remote restart 远程再启动
remote restart process 远程再启动过程
remote sampler 遥控取样器;遥感采样器
remote sampling 远距离取样;遥控取样;遥感采样
remote sampling probe 远程采样探头
remote satellite 远距卫星
remote satellite computer 远程卫星机
remote selector 远程选择器
remote self-sealing coupling 遥控自封接头
remote sense 遥感
remote sensing 遥感技术;远距离显示;远距离读出;遥感(学)
remote-sensing aircraft 遥感飞机
remote-sensing analysis of oil pollution 石油污染的遥感分析
remote-sensing and measuring analysis 遥感遥测分析法
remote-sensing applications for coal field pros-

pecting 遥感煤田勘查应用
remote-sensing applications for comprehensive environmental surveying 遥感环境综合调查应用
remote-sensing applications for energy exploration 遥感能源勘查应用
remote-sensing applications for engineering geologic(al) survey 遥感工程地质调查应用
remote-sensing applications for environmental pollution monitoring 遥感环境污染监测应用
remote-sensing applications for geothermal field prospecting 遥感地热勘查应用
remote-sensing applications for hydrogeologic (al) survey 遥感水文地质调查应用
remote-sensing applications for mineral exploration 遥感矿产勘查应用
remote-sensing applications for oil-gas field prospecting 遥感油气勘查应用
remote-sensing applications for seismogeologic(al) survey 遥感地震地质调查应用
remote-sensing applications for structural geology and geologic(al) mapping 遥感构造地质及地质制图应用
remote-sensing applications in geology 遥感地质应用
remote-sensing balloons 遥感气球
remote-sensing cartography 遥感制图
remote-sensing center 遥感中心
remote-sensing data 遥感数据
remote-sensing data processing 遥控数据处理
remote-sensing data processing centre 遥控数据处理中心
remote-sensing device 遥感装置
remote-sensing equipment 遥感设备
remote-sensing experiment 遥感试验
remote-sensing for agriculture 农业遥感
remote-sensing for atmospheric pollution 大气污染遥感
remote-sensing for environment(al) engineering 环境工程遥感
remote-sensing for forest environment 森林环境遥感
remote-sensing for water pollution 水污染遥感
remote-sensing geometry 遥感几何学
remote-sensing image 遥感图像
remote-sensing image interpretative results 遥感图像解译成果
remote-sensing imagery 遥感成像
remote-sensing indication 遥感标志
remote-sensing information 遥感信息
remote-sensing information acquisition 遥感信息获取
remote-sensing information processing 遥感信息处理
remote-sensing information transmission 遥感信息传输
remote-sensing in geology 遥感地质
remote-sensing in geomorphology 地貌学的遥感技术
remote-sensing in hydrology 水文遥感
remote-sensing instrument 遥感器;遥感(测量)仪
remote-sensing instrumentation for oceanography 海洋学遥感仪器
remote-sensing instruments and equipments 遥感仪器设备
remote-sensing line 遥感线
remote-sensing mapping 遥感制图
remote-sensing material 遥感资料
remote-sensing method 遥感方法
remote-sensing of atmosphere 大气遥感
remote-sensing of microclimatic stress 小气候异常的遥感
remote-sensing of water resources 水资源遥感
remote-sensing platform 遥感平台
remote-sensing prospecting 遥感找矿
remote-sensing rocket 遥感火箭
remote-sensing survey 遥感测量
remote sensing system 遥感系统
remote-sensing technique 遥感技术
remote-sensing with audio frequency pulse 音频脉冲遥感
remote-sensing with radio-frequency pulse 射频脉冲遥感
remote sensor 远距离传感器;遥感(传感)器;遥感设备;遥测(传感)器

remote sensor package 遥感器组件
remote servicing 远距离操作
remote set 副台
remote set point adjustment 远距离设定值调整;远距离给定值调整
remote set point unit 远距离设定点调整器
remote setting 遥调
remote setup 远程设置
remote shift linkage 远距离换挡杆系
remote signal 遥测信号
remote signal(l)ing 远距离发信号;遥信;远距信号装置
remote signalling plant 远距离信号装置
remote slave display 远程从属显示器
remote software 远程软件
remote sounding 遥测探空
remote source 远距离电源
remote speed adjustment 远距转速调节
remote-spotting tube 遥示管
remote start(ing) 遥控起动;遥感发动;遥控启动
remote station 远距离端机装置;遥控台;对方站
remote station alarm 遥控台报警器
remote steering gear 遥控操舵装置
remote subscriber 远端模块
remote subscriber 远距离用户;对方用户
remote subscriber concentrator 远端用户集中器
remote supervision plant 遥感监测设备
remote supervision system 遥测系统
remote supervisory and control(ling) equipment 远程监控设备
remote supervisory and control system 远程监视与控制设备
remote supervisory control 远距离监视控制;遥控监测
remote supervisory equipment 遥控监视设备
remote surveillance 远距离监视;遥控监视
remote surveillance and telemetering for highway crossing 道口通信遥测设备
remote surveillance for section 区段遥信
remote surveying of liquid level 遥测液位
remote switch 遥控开关
remote switching control 远距离开关控制
remote switch unit 远端交换模块
remote-synchronizing system 遥控同步系统
remote system 分置式系统;远程系统;遥控系统
remote target 远端目标
remote terminal 远距(离)终端;远端
remote terminal console 远程终端控制台
remote terminal display 远程终端显示器
remote terminal plotter 远程终端绘图仪
remote terminal processing 远程终端处理
remote terminal support 远程终端辅助设备
remote terminal type 远程终端类型
remote terminal unit 远程终端装置;远程终端设备
remote test equipment 远程试验设备
remote testing 远程测试
remote thermal map 遥感热异常图
remote throttle control 远距离节流控制
remote transducer 远距离传感器
remote transmission 远程输送;远程传送;远程传输
remote transmitter 远距离发射机
remote transmitter-receiver 遥控收发报机
remote transmitting ga(u)ge 远传送测量仪
remote triggering 遥控触发
remote trip control 远动跳闸控制
remote tripping device 远方脱扣装置;远方跳闸装置
remote trip protection 远方跳闸保护(装置)
remote tuning 遥控调谐
remote-type coupling 遥控接头
remote underwater detection device 水下遥测装置
remote underwater manipulation 水下遥控操纵
remote underwater manipulator 遥测水下操纵器
remote unit 遥控装置
remote unit interface 远程单元接口
remote unit monitor 遥控装置控制器
remote unmanned work system 遥控无人作业系统;遥控无人管理的工作系统
remote vacuum-pump loader 遥控真空泵加料机
remote valve 远距离操纵阀;遥控阀
remote valve control 远程阀门控制;遥控阀控制
remote vehicle 遥控飞行器
remote vertical gyroscope 远距垂直陀螺仪
remote video display unit 遥控视频显示单元

remote video source 遥控图像信号源
remote-viewing equipment 远距离观察设备
remote-viewing system 远距窥视系统
remote visual display 远距离可见显示
remote visual display unit 远程可见显示器
remote volumetric titrimetry 遥控容量滴定
remote water-level controller 水位遥控器
remote water-level indicator 远距水面指示器;遥示水尺;遥读水位表;遥测水位计;遥感水位表;远距水位指示计
remote wave and tide meter systems 波浪潮位遥测装置
remote window control equipment 遥控开窗设备
remote work station 远程工作端
remotion 切削
remotored 电动机重新投入
remo(u)ld 改铸;改塑;重铸
remo(u)ldability 重塑性(能)
remo(u)lded clay 扰动黏土;重塑黏土
remo(u)lded curve 重塑曲线
remo(u)lded disturbed soil sample 重塑扰动土样
remo(u)lded fat clay 重塑肥黏土
remo(u)lded sample 扰动试件;重塑试件;重塑(土)样
remo(u)lded soil 改塑土;重塑土(壤)
remo(u)lded state 重塑状态
remo(u)lded strength 扰动试件强度;重塑(试件)强度
remo(u)lded undrained shear strength 重塑不排水抗剪强度
remo(u)lded undrained shear test 重塑不排水抗剪试验
remo(u)lding 重塑;重新造型
remo(u)lding apparatus 重塑仪
remo(u)lding degree 重塑度
remo(u)lding effect 重塑效应
remo(u)lding effort 和易性;易浇注性;易灌注性;重塑数
remo(u)lding enhancement 重塑增强
remo(u)lding gain 重塑增益;重塑增强
remo(u)lding index 扰动指数;重塑(指)数
remo(u)lding loss 扰动减弱;重塑损失;重塑减弱
remo(u)lding sample 重塑试样;扰动试样
remo(u)lding sensitivity 重塑灵敏度
remo(u)lding soil 重塑土
remo(u)lding test 重塑试验
remo(u)ld piece 重塑试件
remo(u)ld sample 重塑试样
remo(u)ld tire 翻新轮胎
removability 可除去性;可拆性
removability of slag 脱渣性
removable 可拆的;活动的;轻便的
removable agitator 移动式搅拌机;移动式拌和机;可拆装式搅拌机
removable axle 可拆卸轴;活动轴
removable bit 活动钎子;活动钻头;可拆卸钻头
removable bridge 活动桥;跳板
removable casement 可装卸窗扇
removable ceiling 易拆卸顶棚
removable cementer 可取出的水泥塞
removable contaminant 可去除的污染物;可除去的污染物
removable controls 可卸控制机构
removable core bit 活动空心钻头;活动岩芯钻头
removable cross section 分出截面
removable cutter 可卸式钻头;可卸式切削具;可卸式钎头
removable discontinuity 可去不连续点
removable draw bar 可拆式联结装置
removable drill bit 可卸式钻头;可卸式钎头;移动式钻头
removable elements 可更换零部件;活动零件
removable end 可拆卸端部
removable equipment 可卸设备
removable flask mo(u)lding 脱箱造型
removable glass roof 滑动玻璃顶棚
removable graphite blocks 可移动石墨块
removable grate 可拆式格栅
removable grating 可卸格栅;可卸炉条
removable handle 可拆手柄
removable isolated singularity 可去孤立奇点
removable jaw (钻杆夹持器的)可卸式卡瓦
removable jumper 可移跳线
removable lid 活动盖子

removable lifting handle 活络箱把手
removable liner 活动衬套
removable louvers 活动百叶窗
removable magnetic concentrator 活动磁性选矿机
removable monorail beam 可移动单轨吊车梁
removable mullion 活动中梃
removable muntin 可拆卸窗梃;可拆卸窗棂
removable packer 可拆卸式封隔墩;可卸式堵塞器
removable paint 可清除的涂料
removable parts 更换部分;可换零件;可更换零部件;活动零件
removable plugboard 可卸插线板
removable pollutant 可去除的污染物;可除污染物
removable rail 活动栏杆
removable random access 可拆装的随机存取;可拆卸的随机访问
removable retainer 活动保持器
removable rim 可拆轮辋
removable rock bit 可卸式钻头;可卸式钎头
removable sampling tube 可拆卸式取样管
removable shoe 活动闸瓦
removable singularity 可去奇异点
removable sluice pillar 活动水闸墩(柱)
removable snow fence 移动式防雪栏
removable stop 可卸止动器;活动挡条
removable stopping 移动式挡墙
removable strap cam 变曲线凸轮板
removable subdie 移动式小冲模
removable substructure 可拆的下部结构
removable support 可移动支架;可收回支架;活动支架;移动式支架;可拆卸的支架;玻璃压条
removable tackle 移动滑车
removable traction device 可拆卸牵引装置
removable type die set guide pin 可拆卸式模架导柱
removable-type whipstock 活楔子;拆卸式造斜楔
removable varnish kettle 可移动式清漆热炼锅
removable viewfinder 可卸取景器
removable volume 可拆装的卷宗
removable wind crank 卷片扳手
removable window 支摘窗
removal 去除;清除;除去;拆迁;拆除;搬移;搬迁
removal age 采伐龄
removal bond 搬移保证书
removal by filter 滤除
removal by ligature 扎除
removal by suction 抽除
removal capacity 排除容量
removal circuit 接插电路
removal class 砍去类别;删除种类
removal coefficient 去除系数
removal coil type alternating current welder 可动线圈交流焊机
removal constant 移出常数
removal contractors 搬运公司
removal cost 拆迁成本;拆除费(用)
removal cross-section 移出截面
removal cutting 后伐;受光伐
removal device 清除装置
removal diffusion method 移出扩散法
removal efficiency 去除效率
removal expenses 搬运费(用)
removal factor 转移系数
removal felling 后伐
removal flux 移出通量
removal method 去除方法
removal of adsorbed impurities 清除被吸附的杂草
removal of ash 除灰
removal of atmospheric pollutants 大气污染物质去除
removal of bridge 拆桥
removal of cake 下筒
removal of chromium 除铬
removal of coat 除漆;面层剥离
removal of contamination 清除杂质
removal of contractor's equipment 承包者设备的撤离;承包商设备的撤离;承包人设备的撤离
removal of core 取出岩芯
removal of cuttings 排除岩粉(从孔内)
removal of defects 缺陷的清除;清除缺陷
removal of demolition waste 清理废墟
removal of dissolved inorganic substance 溶解无机物的去除
removal of dissolved salt 溶解盐的去除

removal of dominance 优势转移
removal of engine 拆除发动机
removal of export ban 解除出口禁令
removal of faults 排除故障;故障的排除
removal of float stone 清除浮石
removal of fluorides 除氟
removal of form (work) 拆膜;脱模(混凝土)
removal of fractured root by flap operation 翻瓣取根法
removal of grease 脱脂
removal of gypsum from soil samples 除去土样中石膏
removal of heat 排热
removal of iodide 除碘
removal of iron 除铁
removal of iron from water 水质除铁
removal of load 卸载;去除荷载
removal of locomotive rods 机车连杆拆卸
removal of material 材料移动
removal of mercury 除汞
removal of mo(u)ld 拆模(板)
removal of nitrate 除硝酸盐
removal of nitrogen 除氮
removal of nitrogen in wastewater 废水中氮的去除
removal of non-carbonate hardness by soda-ash 苏打灰除非碳酸盐硬度法
removal of odor from water 除水臭
removal of odo(u)r from water 水质除臭
removal of oil and grease 除掉油脂
removal of overburden 覆岩剥离
removal of peat 移走泥炭
removal of pipe 提升钻杆;提升管子
removal of pollutants 消除污染物;去除污染物
removal of putty 去除油灰
removal of rainwater 雨水排除
removal of rock cuttings 清除钻孔石尘;清除钻孔石屑
removal of ruins 清除废墟(瓦砾)
removal of salt by chemicals 化学除盐
removal of salt by heating 热除盐法
removal of scaffold(ing) 拆除脚手架
removal of scale 锈鳞铲除;除垢
removal of sediment 清淤
removal of shuttering 拆模(板)
removal of silica 除硅
removal of silt 淤泥免除
removal of stress 除掉应力
removal of support 拆除支撑
removal of surplus soils 残土处理;余土处理
removal of taste and odo(u)r 除臭味
removal of timbering 拆除(木)支撑
removal of top soil 清除表土
removal of track 线路拆移;线路拆除
removal of tree stumps 清除树根
removal of water in gas 天然气脱水
removal of weeds 除草
removal organism 除生物
removal permit for duty-paid goods 已上税物品搬运证
removal plant 清除植物
removal process 消除过程;消除步骤
removal process of atmospheric pollutants 大气污染物质去除过程
removal rate 去除率
removal ratio (沉砂池的)除砂比
removal sampling 去除取样法
removal value of central value 中心值移值
removal work 拆除工程
remove 移置;移动;去除;取去;除去
remove and replace 拆卸与置换
remove-and-replace method 换土法
remove barriers and blockades 消除壁垒和堵塞
remove burrs 清理毛刺
remove coating 撕膜
remove control panel 移动式控制台
removed 拆除的
remove debris 除去杂物
remove debris from water 清除水中残留物
removed heat 散热
removed mean value 移动平均值
removed rail 移轨;拆卸钢轨
removed section 移出断面;移出剖面;移出剖视
remove fair and refit 取下修复再装上

remove feeder 去除冒口;除冒口
remove flash 去除披缝;清除披缝;除披缝;除毛刺
remove flaw 清除缺陷
remove forms 拆除模板
remove foul air 排出污染的空气
remove from an account 销账
remove from structure 从结构中撤销
remove from the line 关闭
remove pollutant 除污染物
remover 消除剂;洗净剂;去除器;清除装置;清器;排除装置;脱涂膜剂;拆卸工具;搬运工
remove redundant operation 消除多余运算
remove riser 去除冒口;除冒口
remove scaffold(ing) 拆除脚手架
remove scale 除水垢
remove the bark in strips 条状剥皮
remove the burr 整平;去除凸纹
remove the defects of distribution 消除分配的缺点
remove the moisture for easy storage 去掉水分以便于储藏
remove the wheels from an axle 从轴上拆下车轮
removing 移动
removing burrs 清除毛刺
removing carbon 消除积炭
removing device 拆卸装置
removing engineering 拆迁工程
removing gear 滑动齿轮;推移齿轮
removing indemnity 迁移赔偿
removing of cake 滤饼脱落
removing of mud around a wreck 沉船周围除泥
removing of mud inside a wreck 沉船内除泥
removing of piston ring 拆卸活塞环
removing of radiological agents 消除放射性质点
removing of supports 拆除支架
removing old paint 铲去旧漆
removing organism contaminant 去除有机污染物
removing paraffin with xylol 用二甲苯脱蜡
removing phenol in wastewater 废水脱酚
removing plow 转动犁
removing restraint 清除限制因素
removing shaft 风井
removing surface 表面清理
removing tank gas by air 用空气驱除油罐气
removing trace organism 去除痕量有机物
removing trace organism contaminant 去除痕量有机污染物
removing track from service 封锁轨道
remreed relay 剩簧继电器
Remscope 长余辉同步示波器
remuneration 待遇(指物质报酬);报酬
remuneration and bonus 报酬和红利
remuneration for work 劳动报酬
remunerative transfer of new technology 新技术有偿转让
Remy floor 雷密楼板(一种轻质混凝土密肋楼板)
renail 再钉
Renaissance architecture 文艺复兴时期建筑
Renaissance church 文艺复兴式教堂
Renaissance dome 文艺复兴式穹隆
Renaissance palace 文艺复兴式宫殿
Renaissance style 文艺复兴式
Renaissance style villa 文艺复兴庄园
rename 更名;改名;重命名
renardite 黄磷铅铀矿
Renard number 工业标准数
Renard series 雷纳数列
renascent 草本多年生植物
renascent herb 多年生草木
renaturation 复性
renature 恢复原状
rend 割裂
render 三道抹灰;三层抹灰;放松穿眼绳;反映
render, float and set 打底及中层和罩面抹灰
render a bill 开账单
render account 开账目;开列账单;报送账单
render and set 两道抹灰;两层抹灰;打底和结硬;二层抹灰
render a tackle 绕起绞辘绳索
render bill 开列账单
render coat 防水砂浆层;抹灰底层
rendered 粉刷;打底
rendered brickwork 已抹底灰的砖砌体
rendered building 抹灰面房屋
rendered facade 抹灰立面

rendered finish 涂抹底层灰
rendered lath 抹过灰的板条
render-float-and-set 三道抹灰；三层涂抹
rendering 复制图；透视图；建筑示意图；炼油；抹灰（打底）；砂浆粉面；打底（指油漆）；初涂
rendering aggregate 粉刷集料；粉刷骨料
rendering base 粉刷底涂
rendering coat 抹灰底（涂）层；头道灰；底涂层；底灰（指油漆）；打灰；打底灰泥层
rendering industry 炼油工业
rendering lath 抹灰板条
rendering mix(ture) 抹灰拌料
rendering mortar 抹灰砂浆
rendering of account 编送账单
rendering of detail 细部再现
rendering of services 提供劳务
rendering plant 炼油厂；炼动物油油厂；熬油厂
rendering sand 抹灰（用）砂
rendering scheme 粉刷方案
rendering stuff 粉刷材料
rendering technique 渲染技术；铺砂浆技术；抹灰技术
rendering waste(water) 油脂废水
rendering with water and ink 水墨渲染
rendering work 抹灰工程
render services 提供服务
renderset 二道抹灰；二层抹灰
rendezvous 公共集会场所；会合点
rendezvous method 对接法
rendezvous sensor laser 会合传感激光器
rendezvous value 集合值
rending action 爆破作用
rending effect 破裂效应
rending explosive 高猛度炸药
rendition 再现；重显
rend of main survey line 主测线走向
rendoll 黑色石灰软土
rendu 已渲染设计；(渲染了的)建筑学设计图；建筑渲染
Rendulic's plot 兰杜勒克图
Rendulic's surface 兰杜勒克面
rendu price 到达目的地价格
rendzina 黑色石灰土
rendzina soil 黑色灰土
renegotiation 重新议价；重新协商；重新谈判
renegotiation reserves 再议价准备；重议价准备
renew 续保；更新；重新开始
renew a bill 将期票延期
renewable 可复制的；可再用的；可以更换的；可更新的
renewable brake shoe 可换闸瓦
renewable bushing 可换导套
renewable contract 可继续的契约
renewable elements 更新部件
renewable energy certificate 可再生能源核实证
renewable energy resource 再生能源；可再生能源
renewable energy (source) 可再生能源
renewable filter media 再生式滤材
renewable fuse 再用熔断丝；可再用的熔断丝
renewable letter of credit 可继续的信用证
renewable natural resources 可再生的自然资源；可恢复的自然资源；可更新资源
renewable obligation certificate 可再生义务证书
renewable of oil 换油
renewable parts 更新部件；可更新部件
renewable point 可换尖
renewable resource 再生资源；可再生资源；可恢复资源；可恢复的资源；可更新资源
renewable sill 可更新的门槛
renewable skid 可换滑板
renewable sources of energy 可再生能源；可更新能源
renewable steel 磨坏的钢板
renewable wearing bushing 可换耐磨导套
renew a contract 合同展期；重订合同
renewal 续保；更换；换新；期限延长
renewal a contract 展期合同
renewal and reconstruction fund 更新改造基金
renewal and reform of fixed asset 固定资产的更新改造
renewal capital of fixed assets 固定资产更新资金
renewal charges 更新费(用)；更换费
renewal corrective maintenance 更新校正维修
renewal cost 更换费；更新成本；换新成本；更新费

(用)
renewal equation 更新方程
renewal expenses 更新费(用)
renewal fees 更新费(用)
renewal function 再生函数
renewal fund 更新基金
renewal notice 到期通知书
renewal of air 换气
renewal of authorization to use 继续委任使用
renewal of contract 续订合同
renewal of equipment 设备更新
renewal of facing 换衬片
renewal of furnace bottom 更换炉底
renewal of geologic(al) information 地质资料更新
renewal of geologic(al) theoretics 地质科学理论的更新
renewal of insurance 续保
renewal of licensed patents 专利权的维持
renewal of parts 零件修复；零件更新；旧件修复
renewal of pavement 路面翻新；路面大修
renewal of registration 续展注册
renewal of rolling stock 机车车辆的更新
renewal of track 线路更新
renewal of vault 窑拱修复
renewal option 续订权
renewal parts 替换零件；更换零件
renewal premium 续保(险)费；展延保险费
renewal process 更新过程
renewal pruning 更新修剪
renewal rate of mine field 矿产地突破率
renewal receipt 续保收据
renewal repair 换件修理
renewal theory 更新(理)论
renewal toxicity test 换液式毒性实验
renewal value 更新价值
renewal works 改建工作；更新工作；翻修工程
renew bid 重新递盘
renew contract 重订合同
renewed fault 复活断层
renewed offer 重新发盘
renewing 整新
renewing leaking joint 更新漏水接头
renewing rail 抽换钢轨
renewing the oil 换油
renewing tie 抽换枕木
renew lease 重订租约
renew offer 更新发盘；恢复发盘
renew order 续订货
renew premium 续保费
renew the water in the tank 把水箱再灌满水
renforce 船帆布
Rengas 婆罗洲玫瑰木(产于东南亚一带)
renierite 硫锗铁铜矿
reniform ore 肾状矿石
reniform slate 豆形石板；肾形石板
reniform structure 肾状构造
Re-Ni wastewater 铼镍废水
rennet 皱层；粗制凝乳酶
renormalizability 可重整性
renormalization 再归一化；重正火；重正化；重整化
renormalization group methods 重整群方法
renormalization technique 重正化技术；重整化技术
renounce claim to an inheritance 放弃遗产立继承权
renounce right 放弃权利
renovated pasture 更新牧地
renovation 修复；整修；更换；改进；翻新
renovation and reformation project 更新改造项目
renovation cost 修缮经费
renovation fund of fixed assets 固定资产更新改造资金
renovation of old buildings 旧房整修
renovation of wastewater 废水再生
renovation process 再生方法
renovation project 改造项目
renovation work 修缮工作；翻修工作
renovator 更新器
rent 租赁；裂口；断口；出租；房租
rentable 可(出)租的
rentable area 出租面积
rentable room 出租房间
rentable space 可租面积；出租土地
rentable space ratio 可出租面积系数；出租面积比

rentage 租金
rental 租金；租费；可租的
rental agent 租赁代理人
rental agreement 租赁协议
rental allowance 房租税收优惠；房租津贴
rental apartment 出租公寓
rental area 阻力面积
rental basis 出租方式
rental charges 租费；租金
rental concession 阻力特许权
rental expense 阻力费用
rental fee 租赁费
rental housing 租用住房
rental housing to cooperative housing 多户出租住房改造为合作住房
rental income 租金收入
rentalism 租赁制度
rental library 租书处
rental living unit 出租居住单元
rental market value 租金市价；租金和维修费
rental office building 出租办公楼
rental payment 租费
rental period 租赁期
rental pool 集体出租
rental purchase 分期租金购买
rental rate 租额
rental receipts 租金收入
rental requirement 阻力必要条件；租费需要
rental room 出租房间
rental space 租用面积；出租面积
rental system 租赁制
rent-car 租赁汽车
rent charges 租费；地租税；租金负担
rent control 房租控制
rent deposit 押租
rented area 租用面积
rented space 租用地方；租用场所；租用面积；租用场地
renter 出租人；租赁人
renter's insurance 租户保险
rentes 统一公债
rent escalation 租金调整
rent expenses 租赁费；租金费用
rent-free 免租金的；不收租金
rent-free period 免租期
rent house management 经营出租房屋业务
renting 租金
renting of construction equipment 租赁建筑设备
renting with furniture 带家具出租；全套出租(住宅和家具设备)
rent in kind 实物地租
rent insurance 出租保险；房租险
rent lath 劈开的灰板条
rentline of displacement 位移裂线
rent multiplier 租金乘数
rent of displacement 移位裂缝；断层线
Rentokil 雷托杀(一种杀虫剂)
rent on movable estate 动产租用费
rent on real estate 不动产租金
rent pale 劈开的窄木条
rent prepayment 预付租金
rent rebate 租金退还
rent roll 租金账单
rentschlerization 紫外线灭菌作用
rent service 以工代税
rent-sharing system 分成制
rent subsidies 租金补助；房租补助
rent-up 房产的出租率
rent-up period 起租期间
renunciation 放弃权利声明书
renunciation of succession 放弃继承权
re-occupation 重新定居
reoccupied 重新占据
re-occurrence period 重现期
reoil 再加油
reoiling 重泼油
reometer 流变仪；流速计；电流计
re-open 重开；再开始
reopener clause 重新谈判条款
reopening clause 重开谈判条款
reopening function 重开门
reoperate 重新运转；重新运行
re-operation 重新操作
re-operational 翻新的；再使用的；重新运转的

reorder 再订货;重新安排;按序排列
reordering point 再订购点
reorder point 订货点
reorganization 整理(活动);整顿;再编制;改组;结构改造;重新组织
reorganization bond 重整债券
reorganization of assets 资产重组
reorganize 改编;重新组织
reorganized data 重组数据
reorientation 再定位;重新确定方向;重(新)取向;重新定向;重定方位
reorientation energy 重取向能量
reoriented insulation 保温带
reorienting production 调整产品方向
reoxidation 再次氧化
reoxygenation 再氧化作用;再充氧;重新充氧作用;复氧
reoxygenation by discharging 泄水复氧
reoxygenation coefficient 再曝气系数
reoxygenation in rivers 河流的复氧
reoxygenation of river 河流的再充氧作用
reoxygenation of stream 河流的再充氧作用
repack 再装配;换填料;重新包装;拆修
repackage 再装配;改装
repacking 改组;改装
repack with grease 填充润滑脂
repaintable 可重新涂漆的
repainted period 重涂周期
repaint(ing) 重新涂漆;重施涂料;重漆;再油漆;重新油漆
repair 修理(工作);修复(作用);修补(术);返修
repairability 可修(理)性
repairable 可修理的;可修复的
repairable damage 可修复的破坏
repairable data base 可修复数据库
repairable defective products 可修复的废品
repairable items 可修复部件
repairable loss 可补偿损失
repairable material 可修材料
repairable parts 可修复部件
repair according to status 状态修
repair and dismantling yard 修船拆船厂
repair and maintenance 修理和维护
repair and maintenance workshop 维修车间
repair and service trades 修理服务行业
repair and servicing time 修理和技术保养时间;维修及服务时间
repair and spare parts workshop 修配车间
repair and utilize old or discarded things 修旧利废
repair as required 按需要修理
repair assignment 修理任务
repair barge 修理驳
repair base 修理基地
repair bay 修理间;维修间
repair bench 修理台
repair berth 修船泊位
repair beyond the scope of repairing course 超范围修理
repair by renewal of parts 互换修【铁】
repair by welding 焊接修补
repair capacity 检修能力
repair car 流动修理车
repair charges 修理费(用)
repair complex coefficient 修理复杂系数
repair concrete 修补用混凝土
repair cost 修缮成本;修理价格;修理费(用)
repair cycle 修理周期;检修周期
repair cycle structure 修理周期结构
repair defects in welded vessel 修理焊接容器中的缺陷
repair depot 修理基地;修理厂
repair drawing 修理图
repair drill hole assembly 修井设备
repaired biscuit 修坯
repaired mouth 补口
repair emulsion 修理用乳剂;修理用乳胶
repair equipment 修理设备
repairer 修理工人
repair expectancy 预计修理量
repair expenses 修理费(用)
repair facility 修理装置;修理设施;修理设备
repair factory 修理厂
repair for 1st time 一次返修

repair forecast 修理预测;修理计划
repair for secondary time 二次返修
repair funds 修理基金
repair garage 汽车修理(车)间
repair gate 检修闸门
repair grafting 修补嫁接
repair group 修配工组
repair guide 修理守则
repair in depot 断修【铁】
repairing 整修
repairing account 修理账目
repairing and keeping 修理和保养
repairing and maintaining work 修理工程
repairing base 修船基地
repairing basin 修船坞池
repairing berth 修船位
repairing budget 修理预算
repairing cost 修理成本;修缮经费
repairing dock 修船(船)坞;修船厂
repairing drydock 修船(干)船坞
repairing equipment 修理设备
repairing expenses 修缮经费
repairing expenses of house 房屋维修费
repairing glazier 修配玻璃工人
repairing item 修理项目
repairing quay 修船码头
repairing room 修理间
repairing technology 修理工艺
repairing welding 补焊;修补焊;焊修
repairing work 修理工作
repairing yard 修船厂
repair in-situ 现场修理
repair installation 修理设备
repair interval 修理间隔期
repair jetty 修理码头
repair kit 修理用整套器具;修理工具包
repair label 修理标签
repair link 备修链节
repair list 修理单
repair load 检修荷载
repair machinery 修理用机床
repairman 安装工;装配工;修配工;修理人员;修理工(人);维修工
repair manual 修理手册
repair manual for alumin(i)um freight container 铝质集装箱修理手册
repair manual steel freight container 钢制集装箱修理手册
repair mortar 修补用砂浆;修补用灰泥;填补砂浆
repair of blemishes 瑕疵修补
repair of defects 缺陷修理
repair of information 信息的恢复
repair of lining 衬砌修理
repair on schedule 定期修
repair order 修理通知单;修理任务单
repair outfit 修理用具;修理工具;修补工具
repair parts 修理(用)零件;修理(用)备件;修理部分;修理部件;配件;备用零件;备件
repair parts line item 备件名称
repair parts stock 备品仓库
repair period 检修期
repair piece 备份;备件;备用零件;备品;配件
repair pit 检修坑(汽车)
repair plant 修配厂
repair plant for handling machineries of port 港区机械修理厂
repair planting 补植
repair plaster 修补用粉刷
repair platform 修理台
repair port 修理港口;修船港
repair position 检修台位
repair production order 修理(工作)通知单
repair quay 修理码头;修船码头
repair rate 返修率
repair risk insurance 修缮期间保险;修理保险
repair rod 修理拉杆
repairs and betterment 修理和改良
repairs and replacement 修理和更换
repair schedule 修理计划;修程;检修计划
repairs done while you wait 即刻修妥
repair section 修理工段
repairs equalization reserves 修理费均衡准备
repair services 修理工作;修配服务;修理业
repair sheets 修补胶

repair ship 修理工作船;修理船
repair shipyard 修船厂
repair shipyard shop 修配船坞厂
repair shop 修配车间;修理厂;修理所;修理工场;修理店;修理车间;修理厂;维修车间
repair siding 修车线【铁】;检修线
repair space vehicle 修理用航天器
repair standard value 修理标准值
repair synthesis 修复合成
repair tag 修理单
repair the facility 修复水毁设施
repair the injury 赔偿损害
repair the old farm tools 废农具修理
repair the ozone hole 修复臭氧层空洞
repair time 修理时间;修复时间;检修时间
repair tool 修理工具
repair to original shape 修复原状
repair track 修车线【铁】;检修线
repair track at station 站修线
repair valve 检修阀(门)
repair ventilation 检修通风
repair wagon 修理车辆
repair welding 焊修;焊补
repair work 修理间;修理工程;定期小修
repair work norm 修理工作量定额
repair workshop 修理车间
repair yard 修车场
repaper 用纸复贴;用纸复铺;用纸重新包装
reparation 修复;赔偿
reparation agreement 赔偿协议
reparation duty 赔偿率
reparation payments 赔款支付
reparations 赔款
reparation's agreement 赔偿协定
reparations in cash 现金赔款;现金赔偿
reparations in kind 实物赔偿
reparative 修复的
reparative surgery 成型外科
repartition bar 分配钢筋
repaste 再涂
repatching 修补
repatriate 遣送回国;遣返
repatriation 遣返
repatriation of labour 劳务人员的遣返
repave 重铺路面
repaver 重铺路面机
repaving 路面重铺;重新铺面;重铺路面
repay 偿还;付还
repayable loan 可收回的贷款
repayable on demand 立即付款
repayable to either 可偿付任何一方;偿还任何权人
repay capital with interest 还本付息
repaying capability 偿还能力
repayment 还款;贷款偿还;偿还
repayment ability 清偿能力
repayment ability analysis 清偿能力分析
repayment contract 偿还合同
repayment of advance 预付款偿还
repayment of bank loan 归还贷款
repayment of loan 债务支出;清还借款
repayment of principal 还本
repayment of shares 股款付还
repayment period 偿还期;偿还年限
repayment with penalty 附带罚金的提前还款
repay principal and interest within the current year 当年还本付息
repeal 作废;撤销;废止
repeal by implication 相互抵消;默认作废
repeat 重复;重发;反复
repeatability 再现性;可重复性;重现性;重合性;重复性
repeatability error 重复性误差
repeatability measure 可重复率度量;重合性度量
repeatability of results 结构的重复性
repeatability of the instrument 仪器的重复性
repeatability precision 重复精度
repeatability range 重复性的幅度;重复性的限度
repeatable accuracy 重复准确度
repeated action 重复作用
repeatable robot 重复型机器人
repeat action key 重复作用键
repeat-back 指令应答装置;指令应答发射机
repeat base unit 重复基本单元

repeat change device 重复变换装置
repeat character 重复字符;反复字符
repeat circuit 重复电路
repeat count 重复计数
repeat counter 重复(次数)计数器
repeat coverage 重复覆盖
repeat customer 重购顾主
repeat cycle 重复周期
repeat demand 重复需求
repeat design 重复设计;重复图案;循环图案
repeat dialing 重复拨号
repeated accuracy 重复精度
repeated addition 叠加;重复相加;重复加法
repeated admission 重复进气
repeated application 重复操作
repeated averaging 累次平均;反复平均
repeated bending 重复弯曲;反复弯曲
repeated bending fatigue 重复弯曲疲劳;反复弯曲疲劳
repeated bending fatigue test 重复弯曲疲劳试验;反复弯曲疲劳试验
repeated bending stress test 弯曲应力疲劳试验;重复弯曲应力试验;反复弯曲应力试验
repeated bending test 反复弯曲试验
repeated bend test 弯曲疲劳试验
repeated blading 重复整型;重复整形;反复整形
repeated block input-output 重复分程序输入输出
repeated breaches 多次违约
repeated call attempt 重复试呼
repeated check 重复检查
repeated combination 重复组合
repeated compression fatigue test 多次压缩疲劳试验
repeated compression test 压缩疲劳试验
repeated crystallization 重复结晶(作用);反复结晶(作用)
repeated cycle 重复循环
repeated cycles of freezing and thawing 重复的冻融循环
repeated deformation 重复变形
repeated difference 累差分
repeated direct stress test 重复拉伸应力试验;反复拉伸应力试验
repeated dosage 重复添加;重复配料
repeated dose 重复剂量
repeated dose toxicity test 多次性毒性实验
repeated dragging 重复拖刮;反复拖刮
repeated dynamic(al) stress test 重复动应力试验;反复动应力试验
repeated emergence 重复出现
repeated examinations 反复检验
repeated fault 重叠断层
repeated flexural strength 受弯耐力
repeated flexural stress 重复挠曲应力
repeated flow turbine 回流式汽轮机
repeated fold(ing) 重复褶皱
repeated geodetic surveying 重复大地测量
repeated hardening 多次淬火
repeated impact 反复冲击
repeated impact bending strength 重复冲击弯曲强度;反复冲击弯曲强度;反复冲击抗弯强度
repeated impact tension test 重复冲击拉伸试验;冲击拉伸疲劳试验;冲击拉伸试验;冲击拉伸疲劳试验
repeated impact test 重复冲击试验;冲击疲劳试验;反复冲击试验
repeated impact tester 重复冲击试验机;冲击疲劳试验机
repeated index 重复指标
repeated integral 累积分;重积分;叠积分
repeated layer soil 复层土
repeated level(l)ing (survey) 重复水准测量(法)
repeated load(ing) 重复荷载;多种交变荷载;循环负载;反复荷载
repeated load(ing) test 重复荷载试验;重复加载试验
repeated measurement 重复测量;重测;反复测量
repeated measurement design 重复观测设计
repeated measurements model 重复测量模型
repeated melting 重复熔化
repeated midpoint formula 合成的中矩形公式
repeated motion 反复放像
repeated observation 重复观测(值)

repeated operation 重复操作
repeated optic(al) link 光中继
repeated permutation 重复排列
repeated practice 重复实践;反复实践
repeated precipitation 多次沉淀法
repeated puncture 反复针刺
repeated quenching hardening test 重复淬火硬化试验
repeated rate 重复率
repeated reflection 重复反射
repeated robot 重型工序机器人
repeated roots 重根;复根
repeated sample 反复的抽样;反复抽得的样本
repeated sampling 重复取样
repeated seam 重复缝
repeated selection 多次选择
repeated selection sort 重复选择分类
repeated shock 反复冲击
repeated simultaneous force 重复联合作用力
repeated solution 重复求解
repeated strain 疲劳应变
repeated stress 交变应力;重复应力;反复应力
repeated stress cycle 应力反复周期;应力重复循环
repeated stress failure (交变应力引起的)疲劳损坏;(交变应力引起的)疲劳断裂
repeated stress test 疲劳试验;重复应力试验;反复应力试验;反复胁强试验
repeated stress testing machine 反复应力试验机
repeated stress variation 重复应力变换;重复应力变化
repeated study 重复研究;反复研究
repeated summation 累次和;多级求和;叠次求和
repeated surfacing 重复加铺面层
repeated survey 反复进行的调查
repeated tempering 多次回火
repeated tensile strength 反复抗张强度
repeated tensile stress test 脉动荷载下拉伸试验;反复拉伸应力试验
repeated tensile test 拉伸疲劳试验
repeated tension and compression test 拉压疲劳试验
repeated tension test 重复拉伸试验;反复拉伸试验
repeated test 重复试验
repeated theodolite survey 经纬度重复测定法
repeated tillage 多次耕耘
repeated torsion 重复扭转;反复扭转
repeated torsion(al) test 扭曲疲劳试验;反复扭力试验
repeated traffic 重复运输
repeated transformation 重复变换
repeated transmission 再传
repeated transverse impact test 反复弯曲冲击试验
repeated transverse stress strength 重复弯曲下应力强度;反复弯曲下应力强度
repeated trauma 重复损伤
repeated trials 重复试验;反复试验
repeated twining 重复双晶;反复双晶
repeated use 重复使用
repeated use of form 模板周转
repeated utilization 重复利用
repeated Y types intersection 麦穗状交叉
repeated zeroes of a characteristic function 特征函数的重零点
repeater 转发器;中继器;增音器;增音机;单线增音器;重发器;复示器;帮电机
repeater circuit 转接电路
repeater compass 罗盘转发器;罗盘复示器;复示罗盘;分罗经
repeater-gain measurement 增音机增益测量
repeater gyrocompass 陀螺罗经复示器
repeater-indicator 转发指示器
repeater lamp 指令应答灯
repeater lamp switch 分罗经照明灯开关
repeater optic(al) transit 复测光学经纬仪
repeater panel 指令显示屏
repeater pennant 代字旗
repeater scope 加接显示器
repeater section 中继段
repeater station 转发站;中插站;中继站;中继台;增音站
repeater system 增音系统
repeater theodolite 复测经纬仪
repeater-transmitter 转发机
repeater valve 增音管

repeater without drop 直通中继站
repeat feed 重复进给;分级进给
repeat formation tester 重复式地层测试器
repeat frequency 重复频率
repeat glass 复制镜头
repeating 转发;转播;中继;重复
repeating amplifier 增音放大器
repeating audit 经常审计
repeating center 转播中心
repeating circle 复测经纬仪度盘
repeating coil 转续线圈
repeating-coil bridge cord 转电线圈桥接软线
repeating coil rack 转电线圈架
repeating cone 复测轴
repeating current meter 多层联测流速仪;多层联测海流计;串联流速仪;串联海流计;复测流速仪
repeating data 重复数据
repeating decimal 循环小数
repeating display terminal 复示终端
repeating flash tube 反复闪光管
repeating group 重复组
repeating indication system 复示式显示制度
repeating instrument 中继设备;复测仪
repeating key 重复键
repeating load(ing) 重复荷载;反复荷载
repeating mechanism 转播装置;连打机构;打簧机构
repeating motion 重复运动
repeating motor 步进电机
repeating optic(al) theodolite 复测光学经纬仪
repeating optic(al) transit 复测光学经纬仪
repeating part 重复部分
repeating pattern 周期性形式;重复形式
repeating point 重复调谐点
repeating selector 复述选择器
repeating ship 信号中继船
repeating signal 重复信号(器);复示信号
repeating slide 连打滑片
repeating station 增音站;广播转播电台
repeating structural unit 重复结构单元
repeating theodolite 复测(式光学)经纬仪
repeating tie 循环穿吊法
repeating timer 反复计时器
repeating turning point 重复调谐点
repeating unit 重复链段;重复单位
repeating watch 打簧表
repeating work 打簧机件;打簧机构;重复机构
repeat integral 反复积分
repeat key 重复键
repeat key stroke 重复键行程;重复击键
repeat knob 重复按钮
repeat number 重复数
repeat offer 再发盘;再报价;重复开价;重复发盘;重复发价
repeat operator 重复操作符
repeat order 添定;补定购
repeat phasing 重复相位
repeat point 双点调谐
repeat point number of equivalent point system 等效点系重复点数
repeat position number of equivalent place group 等效位置组重复位置数
repeat print 重复印刷
repeat-rolling 反复轧制
repeat specification 重复说明
repeat statement 重复语句
repeat station 重复观测点
repeat station position 重复观测点位置
repeat test 重复试验
repeat to address 重复地址;重复编址
repeat transmission 重复传输
repeat transmission system 重复传输方式
repellant 排斥力
repelled plate 斥板
repellence 相斥性;抵抗性
repellency = repellence
repellent 相斥的;忌避剂;驱除的;防水布
repellent building paper 油毛毡
repellent capacity 防水能力;排斥能力
repellent cement 防水水泥
repellent coat 避潮层;防水层
repellent concrete 防水混凝土
repellent emulsion 防水乳化剂;防水浆
repellent escapement 反叉式擒纵机构

repellent finish 防水表面处理
repellent grouting compound 防水灌浆材料
repellent liquid 防水浆
repellent material 防水材料
repellent membrane 防水膜
repellent plaster 防水抹灰
repellent property 防水性能;排斥性能
repellent solution 防水剂
repeller 排斥极;弹回装置
repeller bias 反射极偏压
repeller oscillation mode 反射极振荡模
repeller plate 反射屏极
repeller-type oscillator 反射型振荡器
repeller-type tube 反射式电子管
repeller voltage 反射极电压
repelling agent 防水剂
repelling board 挡板;防水板
repelling force 排斥力;推斥力
repelling groin 挡水丁坝
repelling groyne 挡水防波堤;挡水丁坝
repelling lug 推销铁
repelling odour 除臭
repeptization 再胶溶
repercolation 再渗入(地下水);再渗透;再渗滤(作用)
repercussion 消退法;弹回;反撞;反响;反冲效应
repercussion effect 反冲效果;波及效果
repercussion study 相互作用研究;反应研究
repercussion type 反击式
Repere matic 自动校正偏执仪
reperforator 接收穿孔机;收报凿孔机;收报穿孔机;带式自动纸带穿孔机;复凿(孔)机;复孔机
reperforator switching center 自动纸带穿孔机转换中心
reperforator/transmitter 复式穿孔发送机
repertoire of computer 计算机指令系统
repertory 储藏所;仓库;栈
repetition 再现;重现;重复;副本;反复
repetitional loading 反复荷载
repetition clamp screw 复测制动螺旋
repetition count 重复计数
repetition cycle 重复周期
repetition cycle of satellite 卫星重复周期
repetition drive screw 复测微动螺旋
repetition equivalent 再现当量;重复当量
repetition factor 迭代因子;重复因子
repetition frequency 重复(频)率
repetition impulse 重复脉冲
repetition instruction 重复指令
repetition interval 重复周期;重复间距
repetition lathe 仿形车床
repetition line flying 重复线飞行
repetition measurement 重复测量;复测角法;复测(法)
repetition method 重复法;重测法;反复法
repetition method of observation 复测法
repetition number 复测次数
repetition of angles 复测法测角
repetition of beds 地层重复
repetition of symbol 符号重复
repetition optic(al) theodolite 复测光学经纬仪
repetition period 重复周期
repetition pulse 重复脉冲
repetition rate 重现率;重复(速)率
repetition rate divider 重复频率分频器
repetition rate generator 重复频率发生器
repetition survey of existing 线路复测
repetition test 反复试验
repetition theodolite 复测经纬仪
repetition transit 复测经纬仪
repetition work 复制工作;按形仿制;成批生产;重复工作
repetitious stimulation 反复刺激
repetitive 重复单元
repetitive accuracy 重复准确度
repetitive addition 重复(相)加
repetitive addressing 重复寻址
repetitive air dive 反复空气潜水
repetitive analog computer 重复模拟计算机
repetitive check 反复核对
repetitive computer 重复运算计算机;重复计算器
repetitive construct 重复构造
repetitive cycle 重复循环
repetitive decision process 重复决策过程;反复决策过程

repetitive direct memory access 重复直接存储器存取
repetitive dive time table 反复潜水时间
repetitive element 正常的因素
repetitive error 重复误差;重复错误
repetitive fault 重复断层
repetitive fixed-level diving 反复固定深度潜水
repetitive forms 反复使用的模板;重复使用模板
repetitive frequency 重复频率
repetitive frequency laser 重复频率激光器
repetitive group designation table 反复潜水分组表
repetitive group sampling plan 重复群抽样方案
repetitive heating 重复加热
repetitive housing 重复性住房
repetitive irradiation 重复辐照
repetitive load(ing) 重复荷载;反复加载;反复荷载
repetitive loading test 重复荷载试验
repetitively pulsed laser 重复脉冲激光器
repetitively pulsed tunable 重复脉冲可调谐的
repetitive manufacturing process 重复的制造程序
repetitive member 反复使用构件;重复构件
repetitiveness 重复性
repetitive operation 反复操作;反复运转;重复动作;重复操作
repetitive operations in building 施工中重复操作;施工中有节拍的操作
repetitive peak reverse voltage 反向重复峰值电压
repetitive process 迭代法;重复过程;重复程序;反复过程
repetitive projector 重复式投影仪;重复放映机
repetitive pulse 重复脉冲
repetitive routine 重复(例行)程序
repetitive scrubbing 反复擦洗
repetitive separation 重复分离
repetitive sequence 重复顺序
repetitive shear test 单点剪切试验;重复剪切试验
repetitive short pulses 重复短脉冲
repetitive specification 重复说明(部分);重复描述
repetitive statement 重复语句
repetitive stress 重复应力;反复应力
repetitive structure 重复结构
repetitive time estimates 重复性作业时间估计
repetitive time method 反复计时法
repetitive timing 反复测时法
repetitive training 重复训练
repetitive type analog(ue) computer 周期运算式模拟计算机
repetitive unit 重复单元
repetitive(-use) form(work) 重复使用的模板
repetitive work 重复工作
repetive addressing 重复定址
rephase 重相位
repiercing mill 扩径机
repiling 再打桩
repipe 更换管子
repiping 更换管子
repivoting 轴尖修理
replace 置换;接替;替换;代替;顶替;代换;重置;撤换;放回原处
replaceability 置换能力;更换性;可代换性;替换性
replaceable 可置换的;可取代的;可互换的;可代换
replaceable base 置换性盐基
replaceable bit 可替换钻头;可替换刀具;可替换刀架
replaceable blade bit 可置换切削翼片钻头
replaceable blade knife 多用刀
replaceable cartridge 可置换芯体
replaceable (central) pilot (细粒金刚石钻头的)可替换的中心凸块
replaceable chuck bushings 可替换式卡盘轴套
replaceable cutter 可换刀具
replaceable cutter tooth 可更换刀齿
replaceable ferrule 可替换稳定圈
replaceable hydrogen 可置换氢
replaceable insert 可替换镶嵌块
replaceable rock bit 可替换式牙轮钻头
replaceable shin plate 可换式犁胫板;可换犁胫刀板
replaceable sleeve 可换套筒
replaceable stem seal 可更换阀
replaceable teeth-cutter 可换齿铰刀
replace borrowed money 归还贷款
replace clause 重置条款
replace concretion 置换结核
replaced mineral 交代矿物

replaced position line 移线位置线;转移船位线
replaced row 被替换行
replace in the supports 纵转望远镜
replacement 置换;更换;交替;交代(作用);换新;换位;取代;替换(件);替代;代替;代换;重置;复置;放回
replacement accounting 重置会计
replacement agent 赎换代理
replacement air 新风
replacement algorithm 置换算法
replacement along cleavage 沿劈理交代作用【地】
replacement along contacts 沿接触交代作用
replacement along crystallographic(al) planes 沿晶面交代作用
replacement along grain boundaries 粒周交代作用
replacement analysis 置换分析;更新分析;替换分析
replacement and maintenance 更新与维修
replacement array 置换阵列;替换数组;替代阵列
replacement asset 重置资产
replacement bearing 替换轴承;备用轴承
replacement bit 替换钻头;重嵌钻头
replacement bogie used during repairs 修理期间用的替换转向架
replacement by purchase 购入重置
replacement capital 重置资产;重置资本
replacement changes way 置换变化方式
replacement chart 替换图;继任图
replacement chemical 代用化学品;代替化学品
replacement circuit 替代电路
replacement clause 修复条款;重置条款;保险的重置条款
replacement coefficient 替换系数
replacement coefficient of groundwater 地下水交替系数
replacement control 置换控制
replacement corner fitting 更换角件
replacement corner post 更换角柱
replacement cost 更换成本;设备更新费(用);重置成本;重建费;更新费(用)
replacement cost accounting 重置成本会计
replacement cost approach 重置成本法
replacement cost basis 重置成本制
replacement cost method 恢复费用法;重置成本法
replacement cushion 换土垫层
replacement cushion method 换土垫层法
replacement cycle 更新周期
replacement demand 替换需求
replacement deposit 交代矿床
replacement depreciation 重置折旧;按重置成本折旧
replacement dike 交代脉
replacement equipment 更换设备
replacement expenses 重置费用
replacement factor 更换因数
replacement film 备用胶卷(片)
replacement fork pocket 更换叉槽(集装箱等)
replacement formula 替换公式
replacement fund 更新基金;重置基金
replacement housing 拆迁户住房
replacement insurance 重置或修复保险
replacement intensity groundwater 地下水交替强度
replacement investment 更新投资;更换投资
replacement item 调换零件;调换项目;备件;更换零件目录
replacement length 置换长度
replacement material 置换材料;替换材料;代用材料
replacement mechanism 置换机理
replacement method 置换施工法;换土(施工)法
replacement method of depreciation 重置成本折旧法
replacement model 置换模型
replacement of capital 资本抵补
replacement of foundation 基础替换
replacement of oil 换油
replacement of parts 替换零件;替代零件
replacement of property 财产的重置
replacement of rail 更换钢轨
replacement of retired and scrapped assets 报废资产的重置
replacement of service laterals 更换用户接管
replacement of the old structure with the new 体制转换

replacement outfit 更换设备
replacement part of a reduction rule 归约规则的代元部分
replacement parts 配件;调换零件;替换零件;代用零件;备用零件;备件;代元部分
replacement pavement 重铺路面
replacement period 更新期限;可用年限;替换期
replacement pile 复打桩
replacement pipe 换管
replacement policy 淘汰策略;代换策略
replacement pressure 排驱压力
replacement price method 重置价格法
replacement problem 置换问题;更新问题;替换问题
replacement property 重置财产
replacement quantity 置换量
replacement rate 替换率
replacement rate of 210Po 钋置换率
replacement ratio 置换率;置换比;补偿比率
replacement reaction 置换反应;复分解反应
replacement resources 代用资源
replacement roof bow 更换顶梁(集装箱等)
replacement root 替代根
replacement rule 置换规则
replacement section 调换段
replacement selection technique 替换选择技术
replacement series 置换(次)序;代替序列
replacement set 更新装设
replacement shift impulse method 交替位移脉冲法
replacement sorting technique 置换分类技术
replacement steel door assembly 钢门更换
replacement substoichiometry 置换不足当量法
replacement tanning agent 取代鞣剂
replacement technique 取代技术
replacement temperature 置换温度
replacement theorem 替换定理
replacement theory 置换理论;更新理论;替换论
replacement time 置换时间
replacement titration 置换滴定
replacement tool 装配工具;安装工具
replacement traffic 顶运输
replacement trap 交代圈闭
replacement tree 补植树
replacement tube 换管
replacement unit 重置单位
replacement value 更新价值;再建价;重置价值
replacement value accounting 替代价值会计
replacement value insurance 更新价值保险
replacement value of fixed assets 固定资产重置价值
replacement velocity of groundwater 地下水交替速度
replacement water 换水
replacement wheel 备用轮
replacement with sand 砂代
replacer 置换器;换装器;拆装器
replace stress 置换应力
replace taxes by charges 以费挤税
replace tube 换管
replace worn part 更换磨损零件
replacing 复位
replacing acid 取代酸
replacing bar 拆卸工具的杆
replacing member 替换杆件
replacing of support 扩换支撑
replacing of timbering 扩换支撑
replacing of timber support 更换木支撑
replacing panel assembly 整顶板更换;整壁板更换(集装箱)
replacing power 置换(能)力;替打动力
replacing top soil 替换表(层)土
replanning 重新计划;重新规划
replanting 再种;改种;重栽;补植
replanting soil fumigant 移栽前土壤熏蒸剂
replastering 重新抹灰
replat 谷肩;山肩
replating 金属堆焊
replay 重放;重播
replay head 重放磁头
re-pledge 转抵押
replenish 再装满;再加料;再补给
replenisher 感应起电机;反后座装置调节器;补充液;补充物;补充器;补充剂

replenisher tank 补充浴液槽
replenishing 再充电
replenishing basin 补给池
replenishing period 灌注期;满水时期;补给期
replenishing soimoisture reserves 补给土壤储水量
replenishing tank 补给油柜
replenishing valve 充液阀;补给阀
replenishing water 补给水
replenish manpower 补充人力
replenishment 补给;补充(法)
replenishment-based ordering system 以补充点为基准的订货方式
replenishment cost 补充费用
replenishment developer 补充显影剂
replenishment intensity of precipitation during peak season 峰期的降雨补给强度
replenishment of groundwater 地下室的补给
replenishment of inventory 补充存货
replenishment order 补充订货
replenishment point 补充点(即最大库存量)
replenishment pump 补给泵
replenishment spare parts 补充备件
replenishment supply 补充
replenishment water 补充水;补给水
replenish stock 进货
replenish the thoroughfare and conception vessels 补益冲任
replete state of soil 土壤充水状态;土圈充水状态
replevin 扣押财产物发还;收回非法扣留动产
replevy 取回被扣押物
replica 拷贝;模仿复制品;复制试样;复制品;复制片;仿制品;仿造品
replica beam 复制梁
replicability 复制性
replica girder 复制大梁
replica grating 重摹(衍射)光栅;复制光栅
replica impedance 重复阻抗
replica mirror 复制反射镜
replica of grating 复制光栅
replica plant 中间工厂;复制厂
replica planting 反复平面培养法
replica plating 复制平板法
replica technique 复型技术
replicated data 复制资料;复制数据
replicate determination 平行测定
replicated field test 重复现场实验;重复田间试验
replicated literal 重复文字
replicated plot 重复小区
replicated tree-plot experiment 重复单株小区试验
replicate septum 反迭隔膜
replica test 重复性试验
replicate sub-sample 复分样品
replicate/transfer switch 复制转移开关
replicatile 重折的
replicating fork 复制叉
replication 制模;重作;重现;重复;复制品;仿作;反响
replication count 重复计数
replication cycle 复制循环
replication factor 复复因子
replication mistake 复制错误
replication pattern 复制型
replication process 复型作用;复型过程
replication unit 复制品
replicative function 重叠函数
replicator 重复符(号);复制员
replicon 复制子
replicon hypothesis 复制子假说
replotting 重划用地;重复描绘
replotting plan 换地规划
replum 门芯板
reply as soon as possible 尽快答复
reply code 应答密码
reply delay 延迟应答
reply efficiency 应答效率
reply-frequency coding 应答频率编码
reply paid 回电费付讫
reply pulse 应答脉冲;回答脉冲
reply pulse spacing 应答脉冲间隔
reply queue element 应答队列元素
reply request 应答请求
reply signal 应答信号
repointed masonry 重嵌坛工;重勾(灰)缝坛工

repoint(ing) 锻伸铧尖;重嵌灰缝;重新勾缝;重勾(灰)缝
repointing masonry 重嵌坛工;勾缝砌体;重勾(灰)缝坛工
repolarization 重极化;复极化
repolish 再抛光;再磨光
repolisher 复碾机
repolishing 重新抛光
re-pollution 再污染;二次污染
repopulation 再引种;再放养;粒子数再增
report 通报;表报;报告(书);报表书
reportable quantity 需报量
report calculating reserves 储量计算报告
report call 通报呼叫
report center 通信中心站
report clause 报告子句;报表子句
Report Customizer Program(me) Product 报告定制人程序产品
report definition mode 报表定义方式
report delay 报告延迟
report description 报表描述
report description entry 报表描述项
report development 拟定报告
reported 据报
reported date 报告日期
reported taxable income 课税所得申报额
report entry 报表款
reporter 报告人
reporter general 总报告人
report file 报告文件;报表文件
report footing 报告总计;报表尾部
report form of balance sheet 报告式资产负债表
report form profit and loss statement 报告式损益表
report forms 报表
report for variety test 品种试验报告
report generating 报告生成
report generation 报告形成;报告生成;报告产生(法);报表形成;报表生成;报表产生
report generation parameter 报告生成参数;报表生成参数
report generator 报告文件处理机;报告生成程序;报告产生程序;报表生成程序;报表生产程序
report generator language 报告生成语言
report group 报告群;报表组;报表栏
report group description entry 报表组描述项;报表栏描述款
report head(ing) 报表头;报表提要;报表表头
reporting chain 通报系列
reporting contract 申报式保险契约
reporting date of report 报告提交日期
reporting day 通知日;报告日
reporting form 报告格式;报告形式;报告格式
reporting format 报告格式
reporting framework 报告格式
reporting lag 报告时滞差
reporting network 水情站网;报汛站网
reporting of accidents 事故报告
reporting of an accident 事故报告
reporting of data 资料的报告
reporting of secretariat 报道秘书处的情况
reporting pay 签到工资
reporting period 报告周期;报表周期
reporting point 报告点
reporting policy 按保险对象价值变动调整的保单
reporting room 报到室;申报办公室
reporting standard 审计报告标准
reporting station 水情站;报汛站
reporting system 通信系统;报告制度;报告系统
reporting unit 报告单位
reporting year 报告年度
report interval 报告时间间隔;报表时间间隔
report item 报表项目
report line 报告行;报表行
report name 报告名;报表名
report number 报告编号
report of analysis chemistry 化学分析报告
report of analysis spectrum 光谱分析报告
report of cargo damage survey 货物损害鉴定书
report of clearance 出口报告书
report of deviation 绕航报告
report of disclaimer of opinion 放弃表达意见报告
report of entry for foreign-going ship 国际航行船舶进口报告书

report of geologic(al) surveying 地质调查报告
report of investigation 研究报告;调查报告
report of mine environment quality 矿山环境质量报告书
report of oil analysis 油样化验报告
report of passenger number on train by accident 旅客区段密度报告
report of passenger number on train by junction station 分界站报告
report of reconnaissance 勘察报告
report of regional geologic(al) survey 区域地质调查报告
report of scrap materials 废料报告
report of shipment 装运报告
report of undelivered container 集装箱催提单
report of voting 表决报告
report of wreck surveying 沉船测量报告
report on a special topic 专题报告
report on exception to cargo discharged 卸货事故报告书
report on feasibility 可行性研究报告
report on fossil identification 化石鉴定报告
report on identification heavy mineral sand 重砂鉴定报告
report on inspection of port state control 港口国管理检验报告
report on mineral identification 矿物鉴定报告
report on ore identification 矿石鉴定报告
report on rock identification 岩石鉴定报告
report on sea casualty 海事报告
report on tenders 投标审查报告
reportor 报表生成程序
report package 一套财务报表
report page format 报告页面格式
report period 报告期
report program(me) 报告程序;报表程序
report program(me) generator 报告程序的生成程序;报告程序编制器;报表程序发生器;报表的生成程序;报表产生器;报表编程语言
report ratifier 报告审批单位
report section 报告部分;报表节
reports-status 现状报表
report table of viscosity result 黏度结果报表
report tape 报告带
report testing separability of ore 矿石可选性试验报告
report title 报告名称
report type 报告种类
report write control system 报表书写控制系统
report write logical record 报表书写逻辑记录
report writer 报告编写人员;报表书写程序;报表生成;报表程序的生成程序
report writer control system 报表编制控制系统
report writer feature 报表编辑功能
report writer logical record 报表编制逻辑记录
report writing 报告编写
report-writing date 报告编写日期
report-writing unit 报告编写单位
repose 静态
repose angle 休止角;安息角
repose angle of particle 粒子安息角
repose period 休眠期
repose soil 静止土
reposition 改变地位;重置;储[贮]藏;复原位;复位术
repositor 复位器
repository 陈列馆(美术品);仓库;储[贮]藏室;处置库;储存所;博物馆
repository for radioactive wastes 放射性废水储[贮]藏器
repository of radioactive solid wastes 放射性固体废物库
repository system 处置库系统
repossessed goods 重新占有的货物;重新拥有的货物;重新收回的物资
repossessed property 收回产业;重置财产
repossession 收回分期付款货物
reposting 重新支撑;替换(模板)支撑
repour 重浇;重新浇筑
repoussage 锤版
repousse 金属浮雕花饰;敲花细工;凸纹制作;凸纹锤制;冲压花纹
repousse lace 凸纹花边
repousse work 敲花细工
repower 改建动力装置

rep(p) 棱纹平布;有纹理的丝或毛织品
Reppe process 雷珀(合成)法
reppling instability 皱纹不稳定性
reprecipitation 再沉降;再度沉淀;再沉淀(作用)
representatin of a group 群的表示法
representation 再现;投保声言;说明;申述;重现;重显;表示(法);表达(法)
representation allowance 开会津贴
representational symbol 象形符号
representation error of gravity-anomaly 重力异常代表误差
representation formula 表示公式
representation fraction 数字比例尺
representation language 表示语言
representation method 表示法
representation of a game 对策的表示
representation of a group 群表示
representation of algebraic(al) formulas 代数公式的表示
representation of a matrix 矩阵的表示法
representation of arrays 数组的表示
representation of a table 表格表示法
representation of binary tree 二叉树的表示
representation of contour lines 等高线表示法
representation of forests 森林的表示
representation of ground 地貌表示法
representation of hill features 山形表示法;山丘特征图
representation of list 列表的表示
representation of polynomial 多项式的表示
representation of queues 队列的表示法
representation of stack 堆栈的表示法
representation of surface 曲面表示法
representation of the shape of ground 地貌表示法
representation point 相点
representation position 代表部位
representation specification 表示法指明
representation system 表示系统
representation theorem 表示定理
representation theory 表象理论
representative 有代表性的;典型的;代理人;代表物,代表(的);表示的
representative area 示范区;典型地区
representative basin 典型流域;代表性流域;代表流域
representative calculating operation 典型计算工作
representative calculating time 典型工作周期
representative coefficient 表象系数
representative computation time 典型计算时间
representative computing time 典型计算时间
representative condition 特征条件;典型条件
representative core 典型岩芯;代表性岩芯
representative cost 代表成本
representative data 代表资料
representative director 代表董事
representative element 代表元(素);代表性元素;表现元素
representative engine 典型发动机
representative error 代表性误差
representative example 典型实例
representative firm 代表厂商
representative formula 典型配方
representative fraction 自然比例尺;缩尺;分数比例尺;比例分数
representative goods 财富代表物
representative method of sampling 典型抽样法;代表性抽样法;代表性采样法;抽样表示法
representative model 表现模
representative module 表示模
representativeness 典型性
representativeness of core 岩芯的代表性
representativeness of sample 样本代表性
representative observation 典型观测;代表观测
representative of contractor 承包者代表;承包商代表;承包人代表
representative of sampling 代表性抽样法
representative pattern 典型样图
representative point 代表性点
representative property 典型性质;代表性特性
representative region 典型区域
representative ring 表现环
representative river 代表性河流
representative river section 典型河段

representative sample 土样;典型样品;典型样本;典型土样;典型试样;典型试件;代表性样品;代表性样本;代表性试样;标准样(本);试样
representative sampling 典型抽样;代表性取样
representative scale 惯用比例尺;典型比例尺;代表性比例尺
representative section 代表性剖面;典型截面;等效截面;代表性截面
representative short-term period 短系列代表性
representative simulation 代表模拟
representative soil sample 代表性土样
representative specimen 典型试件
representative station 代表性测站
representative stream 代表性河流
representative temperature 特性温度;代表性温度
representative test 代表性试验
representative valley 代表流域
representative value 典型值;代表值
representative work 代表作
represent dynamically 动力模拟
represent graphically 以图表说明
represent wave parameter 波浪特征值
repressed brick 加压砖;压制的砖
repressed lug brick 加压凸边砖
repressed plain tile 压制的平瓦
repressed tile 压制的瓦
repressible system 阻遏系统
repress(ing) 再压;再加压;补加压力;重压;补充加压
repressing line of gas 天然气回注管线
repression 压抑;阻遏
repress oil 陶瓷脱模油
repressor 阻遏物;阻断剂
repress pile 加压桩
repressuring 加压合成处理;再加压;天然气回注
repressuring coefficient 压力恢复系数
repressuring gasoline 加压汽油
repressuring medium 驱油试剂
repressurize 再增压
reprint 翻印;翻版
reprint copy 翻印本
reprint edition 再版
reprise 租金;年金;墙上线脚转延侧面;门窗框缘转角线条
repro 清样
reprocessed 后处理过的
reprocessed material 再生材料
reprocessed plastics 再生塑料
reprocessed products 返修品
reprocessed waste 后处理废物
reprocessing 再加工;再处理;后处理;重制
reprocessing cost 再加工成本
reprocessing cycle 再生循环;再加工循环
reprocessing plant 再加工厂
reproduce 再制(备);复制;翻印
reproduce by pantograph 缩放仪复制
reproduced image 再现图像;收像;重显图像
reproduced image contrast 显像对比度;收像对比度
reproduced image fineness 图像重显精细度;收像细节
reproduced image resolution 显像清晰度;显像分辨力;收像清晰度;收像分辨力
reproduced model mo(u)ld(ing) 实物造型;实物模铸型
reproduced paper-making wastewater 再生造纸废水
reproduced pulse 输出端脉冲
reproduce head 重放磁头
reproducer 扬声器;再现设备;再生器;再生程序;重现装置;复制机;复制程序;复印机;复穿孔机
reproduce the population 人口再生产
reproduce to scale 按比例复制
reproducibility 增殖率;再现性;再生性;再生产能力;可重现性;可再现性;可复现性;重现性;重复(再现)性;重复能力;复验性;复演性
reproducibility of tests 试验结果再现性;平行试验;试验结果的复验性;试验复现性
reproducibility range 还原范围;重复范围;再现范围;再生范围
reproducible 可重复的;可再生的;可复制
reproducible assets 可再生产的资产
reproducible copy 可复制副本
reproducible data 能再现数据

reproducible fixed assets 可再生产的固定资产
reproducible result 重复结果
reproducible tangible assets 可再生产的有形资产
reproducible tracing 能再现的示踪
reproducible value 复制价值;再生产价值
reproducing 卡片复穿
reproducing amplifier 再生放大器;重现放大器
reproducing area 造林地
reproducing brush 复制电刷
reproducing camera 复制照相机;复制仪;复照机
reproducing channel 重放通道
reproducing characteristic 再现特性;重现特性
reproducing head 重放头
reproducing image 复制像
reproducing lathe 仿形车床
reproducing output 重现输出
reproducing picture 复制像
reproducing punch 复制穿孔机
reproducing seismic signal 回放地震信号
reproducing stylus 放音计
reproducing system 翻印法;翻译法;重放系统;复制法;放声系统
reproducing unit 复制装置;复制穿孔机;复穿孔装置
reproduction 增殖;再现;再生产;重现;重显;重放;复制(品);繁殖;翻版
reproduction camera 制版照相机;复照仪
reproduction chain 重放通路
reproduction coefficient 再生产系数
reproduction constant 再生常数
reproduction cost 资产再生成本;再生产费用;再生产成本;复制成本
reproduction curve 繁殖曲线
reproduction cutting 更新伐
reproduction equipment 复照设备
reproduction factor 转换因子;增殖因素;重现因素;重现因数
reproduction felling 更新伐
reproduction fidelity 重现保真度;重放保真度
reproduction fund 再生产基金
reproduction gamma 还原伽马
reproduction lens 复照镜头
reproduction method 更新作业法;复制法
reproduction method of deprecation 重置法折旧
reproduction negative 复照用底片
reproduction of ancient bronze chariot and horse 仿古铜车马
reproduction of antique pottery ware 仿古陶瓦器
reproduction of capital 资本再生产
reproduction of drawing 图纸复制
reproduction of image 图像重显
reproduction of labo(u)r power 劳动力再生产
reproduction of sound 声的重ží
reproduction on an enlarged scale 扩大再生产
reproduction on a progressively increasing scale 扩大再生产
reproduction positive mo(u)ld 阳模
reproduction potentiality 增殖潜力
reproduction process 再生产过程;复制过程;复制法
reproduction proof 清样
reproduction quality 重现性
reproduction rate 增殖率;再生产率
reproduction ratio 重现比;复制比
reproduction replica 复制副本
reproduction room 复印室;晒图室
reproduction scale 复制定标;复制比例尺;翻印比例
reproduction speed 复制速度
reproduction test 繁殖试验
reproduction value 再生产价值
reproductive 再生(产)的;复现的
reproductive investment of capital 再生产投资
reproductive isolation 生殖隔离
reproductive law 再现定律
reproductive stenohaline 生殖狭盐性
reproductive stenothemy 生殖狭温性
reproductive system 繁重方式
reproductivity 复现性
re-profiling of rail 钢轨翻新
reprogram(me) 程序重调
reprogrammable 可改程序的
reprogramming 改变方案;改变程序;改编程序;重新规划;重新程序设计;重编程序;程序再设计
reprographic light source 复照光源
reprographic original 复照原图

reprography 复制术
reprojection 二次投影
reproportion 再定比率;改变比例;比率改变;再定比例
re-proving 再标定
reptation 表层塌滑
reptile 爬行动物
Repttian (stage) 雷佩蒂阶(北美)【地】
repudiate 拒付债务
repudiate a contract 否定合同;否认合同有效
repudiate a debt 赖债;拒付欠款
repudiate obligation 拒绝履行义务
repudiation 拒绝;拒付债务
repudiation of claims 拒绝赔付
repugnant substance 恶臭物质
repulpator 再浆机
repulper 再浆机槽
repulp filter feed tank 再压滤器的进料罐
repulse excitation 碰撞激励;推斥激励;冲击激励
repulsion 排斥力;斥力
repulsion and induction type motor 推斥感应式电动机
repulsion and induction type single phase motor 推斥感应型单相电动机
repulsion coil 上托线圈
repulsion force 推斥力;反斥力
repulsion-induction motor 推斥感应型电动机;推斥感应电动机
repulsion instrument 推斥型仪表
repulsion motor 推斥(式)电动机
repulsion motor with damper winding 有阻尼绕组的推斥式电动机
repulsion phase 相斥相
repulsion-start induction motor 推斥起动感应电动机
repulsion type meter 推斥型仪表
repulsion type motor 推斥型电动机
repulsive effect 排斥效应
repulsive energy 排斥能量
repulsive force 排斥力;推斥力;斥力
repulsive interaction 互相推斥作用;互斥力;排斥作用
repulsive potential 排斥势
repulsive state 推斥态
repulverize 再(次)粉碎;重新粉碎
repumping 重新泵回;再抽汲;重抽汲;再汲取;重汲取
repumping house 二级泵房
repumping installation 泵回设施
repumping plant 泵回设施
repurchase 买回;重购买
repurchased stock 重购入股份;重购买股票
repurchase price 重购买价格
repurification 再净化;再纯化
repurified 重提纯
repurifier 再纯器
repurify 再纯化
reputable bank 信誉良好的银行
reputation 信誉;名誉
reputation of firm 厂商信誉
re-putty 找补腻子
requalification 再审定(资格);重审定(资格)
request allocate overhead 请求分配开销
request a radar vector 请求雷达导航
request block 请求块;请求分程序
request call 请求呼叫
request control block 请求控制块
request data transfer 请求传输数据
request enable 请求赋能
request entry key 请求输入键
request for accident indemnity 赔偿要求书
request for an offer 要求报价
request for bid 要求承包;邀请投标(书)
request for bid documents 要求承包文件
request for comments 请求评议
request for estimation 成本估计申请书
request for price quotation 请求价格引证
request for proposal 征询方案;请提建议
request for quotation 报价申请
request for references 征询要求
request for tenders 组织投标
request for transfer 请求调职;申请调离
request header 请求报头
request level 请求级

request monitor 请求监控器
request note 先卸批准单;申请单
request of utilization 使用申请
request on-line 请求接通线路
request packet 请求包
request parameter list 请求参数表;申请参数目录
request parameter list exit routine 请求参数表出口例行程序
request parameter list string 请求参数表串;申请参数目录串
request pending light 请求待处理灯
request queue 请求排队;请求队列
request reexamination 申请复审
request repeat system by interference detection 干扰检测请求重发方式
requests by contractor 承包者的要求;承包商的要求;承包人的要求
request-send circuit 请求发送线路
requests for alternations 改装要求
request signal 询问信号;请求信号
request stacking 请求加工;请求堆积;请求处理
request stop 请求停机;随意停机
request stop of traffic 公共交通的招呼站
request system 咨询系统
request to extend time-limit of claim 要求延长索赔期
request to receive 要求接收
request to send 请求发送
request unit 请求单元
request unit chain 请求单元链
request unit portion 请求单元部分
required accuracy 要求精(确)度
required air amount 空气需要量
required air volume 需要风量
required amount of water 需要水量
required area 需要面积;所需面积
required chemical characteristics of drinking water 饮用水所需化学性质
required date 需用日期
required delivery date 要求交货日期
required depth 要求深度
required design load 必要的设计荷载
required excavation 必要开挖
required field intensity 所需场强
required fire flow 消防需水流量
required fund 所需经费
required gear 需要挡位
required (horse) power 需用功率;需用马力
required hyphen character 必要连字符
required labor 所需劳动力
required length of life 耐用年限
required live load 要求的活荷载;规定的活荷载
required net positive suction head 必要净吸入扬程
required power 需要功率
required precision 精度要求
required pressure 要求的压力
required rate of return 应得收益率;要求报酬率
required reservoir capacity 所需库容
required scale 规定比例尺
required space character 所需空格符
required strength 要求强度
required take-off distance 起飞需用跑道长度
required take-off run 起飞需用滑跑距离
required test time 试验需用时间
required thickness 要求厚度
required value 预期值;目标值;求定值;期待值;待定值
required window 需用窗户面积
require equable moderate temperature 要求均恒而适中的温度
requirement 需要量;需要订立协议;保证金比率
requirement contract 附有限制条款的合同
requirement description language 要求描述语言
requirement factor 需要因素;需要因数
requirement for anti-corrosion 防腐要求
requirement for map introduction 图幅说明书编制要求
requirement for raw materials 对原料的需求
requirement for tax exemption 免税规定
requirement of air 需气量
requirement of backwash water 反冲水需要量
requirement of building code 建筑规范要求
requirement of drain 排水沟要求

requirement of gas 需气量
requirement of motion of motor vehicle 汽车行驶的必要条件
requirement of tolerable settlement 容许沉降量要求
requirements 调整需要量
requirements clearance symbol 需求批准符号
requirements definition 要求定义
requirements engineering 要求定义技术
requirements of artificial illumination 人工照明条件
requirements of assembling 装配要求
requirements of process 流程需要
requirements vector 要求向量；要求矢量
requirement tracer 要求跟踪程序
require performance of an obligation 要求履行义务
requiring additional reinforcement 需要附加补强
requisite depth 必要水深
requisition 要求；征用(土地)；领取申请书；领料单；请求书；请料单；申请书；申请(单)
requisition by purchase 征购
requisition channel 申请系统
requisition cycle 申请周期
requisitioned figures of plan 计划原提数字
requisitioned land for project 工程征地
requisition for materials 请料单
requisition for payment 申请付款书；付款通知书；付款通知单；付款申请(书)
requisitioning of goods 征用财产
requisition in kind 实物征用
requisition of land 征用土地
requisitions on title 业权调查书
requisition unfilled 未发材料的领料单
reradiate 再辐射
reradiating surface 再辐射面
reradiation 反向辐射；再辐射
reradiation error 再辐射误差
reradiation factor 再辐射系数；重辐射系数
reradiation pattern 二次辐射图样
reradiation power 再辐射功率
rerail 重铺轨
rerailer 复轨器
rerailing 铺复轨
rerailing crane 多轨起重机；复轨起重机
rerailing device 复轨器
rerailing equipment 复轨设备
rerailing frog 复轨辙叉
rerailing ramp 复轨斜坡台
Rerardo 雷塔都(一种水泥缓凝剂)
rerating contract 重新估价合同
reread(ing) 重读
rere arch 背拱；扁腹拱；内拱；扇面背拱
rerecording 重新登记；重录
rerecording device 转录装置
rerecording system 转录系统
reredorter 寺庙后的厕所
reredos 祭坛背壁(上的屏风)；祭坛背后的饰物；壁炉背面
rereeled silk 扬返丝；复摇丝
re-reeler 重卷机
rereeling machine 扬返机；复摇机
rerefined oil 再生润滑油
rerefining 再精制
reregister 再配准；再对准；再定位
reregistered emitter 再对位射极
reregistration 再定位；重复对准
reregulating reservoir 反(复)调节水库；平衡水库；反(复)调节库容
reregulating storage 反调节水库；反调节库容
reregulation 反调节
rere vault 后背拱顶
rering locked in 重发振铃信号镇定
rering signal 再呼叫信号；重发振铃信号
reroll 再轧；二次滚扎
rerollable 可重新再轧；重新再轧的
rerolled bar 重轧钢筋；重扎钢筋
rerolled rail 可重新再扎的钢轨；重轧钢轨
rerolled steel 半成品钢
rerolling 二次混料
rerolling feed 再轧坯
rerolling mill 再轧机
rerolling quality blooms 优质方坯
rerolling rail 再轧钢轨

re-roofing 重铺屋面
rerouting 绕行；重新定线；改道；重选路由；改线
rerouting for slow vehicles 加设慢车专用路线
rerouting of channel 河槽改道
rerouting of lines 改变管路布置；重新铺设管线；重新布线
rerouting of river 河流改道
rerouting of road 道路重新定线；道路改线；重定路线
rerun 再运行；再启动；再开动；再度蒸馏；再处理；重新执行；重新运行
rerun a bit 使用修磨过的钻头
rerun bottom 再蒸馏后的残油
rerun check point 再运行检查点
rerun mode 再运行方式
rerunning 再蒸馏；再度运行
rerunning control survey 控制测量复测；重复精密控制测量法
rerunning level(l)ing 重复水准测量(法)
rerunning o f level(l)ing 水准测量复测
rerunning plant 再蒸馏设备
rerunning still 再蒸馏锅
rerunning tower 再蒸馏塔
rerunning unit 再蒸馏装置
rerun oil 再处理油；再蒸馏油
rerun point 再运行点；重新运行点；重新起动点；重算点
rerun roll-back 重算
rerun routine 再生程序；恢复程序；重新执行程序；重新运行(例行)程序；重算程序；重复运行程序；重复程序
rerun routine restart routine 重算例行程序
rerun still 再蒸馏锅；再蒸馏釜
rerun time 重算时间；重复运行时间
rerun tower bottoms 再蒸馏塔接底产物
rerun yield 再蒸馏产率
Resac 雷萨克深褐色硬木(产于马来西亚)
resaca 暴风浪
resail 归航；驶回
resale 转售；转卖；再出售(进口商无力付款提货，只好在进口地点减价出售)
resale contract 转卖合同
resale price 转售价格；再售价格
resale price maintenance 维持转卖价格
resale proceeds 转售收入
resample 重复取样；重采样
resample method 再取样法
resampling 再取样；重新取样；重取样；重采样
resampling interval 重采样间隔
resanding 再铺砂
resaturated 重新浸透的；再使饱和的
resaturating 重新浸透；重使饱和
resaturation 再饱和(作用)
resaturator 再饱和器
resaw 顺纹锯；纵锯；带锯；再锯；解锯
resawed 再锯；粗锯
resawed lumber 顺纹锯材
resawed shale 锯板头
resawed tapered shake 斜锯板
resaw lumber 改锯材
resaw mill 再锯车间
resawn board 顺纹锯的木板
resawn lumber 再锯材；顺纹锯材
resawn tapered shake 斜锯板
resazurin-rennet test 刃天青凝乳试验
resbenzophenone 苯酰间苯二酚
Resbon block 不透性石墨块
Resbon tube 不透性石墨管
rescaling 改变尺度；改变比例；重标度；尺度改变
rescan 重复扫描
rescanning 重扫描；二次扫描
rescap 封装阻容
rescatter 再扩散；重新扩散
rescattering 重散射；二次散射
rescattering light 重散射光
reschedule 修订计划；重(新)调度
reschedule the debt 推迟偿债期
rescheduling 重安排
rescheduling of debt 债务重议
rescind 废除；撤销
rescind a contract 取消合同
rescind contract 废除合同；取消合同
rescission 解约
rescission of contract 取消合同

rescission of dividends 退回股利；取消股利
rescission of original plan 撤销原计划
rescissory action 废约行为
rescrape 刮研
re-screener 二次筛分机；再筛
re-screen(ing) 再筛分；重新筛分；二次筛分
re-screening plant 二次筛分设备
re-screening unit 二次筛分装置
re-screening provision 重新筛分装置；二次筛分装置；二次筛分设备
rescue 营救；救助；救援；救生；抢救；挽救
rescue apparatus 救护用具
rescue authority 营救当局
rescue beacon 呼救信号器
rescue bell 救生钟
rescue boat 救助艇；救生船；救护船；海上救助船
rescue borehole 救援钻孔
rescue buoy 救生浮标
rescue by stages 分级救治
rescue capsule 救生舱
rescue car 救生车
rescue chamber 救生钟；救生舱
rescue clause 救助条款
rescue co-ordination center 救助协调中心；营救协调中心
rescue craft 救助艇
rescue crew 打捞队
rescue cruiser 巡逻救助艇
rescue diver 援救潜水员
rescue dump 重入点信息转储
rescue equipment 救援设备
rescue equipment locker 救助设备柜
rescue group 救援组
rescue hawser 救助拖缆
rescue helicopter 救援直升机
rescue message 遇险呼救信号；遇难呼救信号
rescue mission 营救任务
rescue motor launch 救助机动艇
rescue of survivals 救生
rescue operation 营救作业；救生作业
rescue orange 营救橙；救生橙颜料
rescue outfit 抢救装备
rescue party 抢救队；抢救班；抢救组
rescue path 人员紧急撤离飞机的路线
rescue personnel 抢救人员
rescue point 重新运行点；重入点
rescue post 救护站
rescuer 救助人(员)；抢救人员
rescue radio beacon 无线电救援信标
rescue rope 救援索
rescue seaplane 救助水上飞机
rescue service 海上救助业务
rescue ship 营救船；救助船；救援船；救难船；海难救助船
rescue signal light 救生火号
rescue station 抢救站
rescue sub-center 救助分中心
rescue submersible 救生潜水器
rescue system 救生系统
rescue train 救援列车
rescue truck 紧急救助用救援车
rescue tug 救助拖轮
rescue unit 救助设备；救助单位
rescue vessel 救助船；打捞船
rescue work 救援工作；抢救工作
reseal(ing) 重新填缝；再浇封层；再封(死)；再密封
resealing of joint 接缝重封；重新填缝
resealing pressure 再密封压力
resealing trap 再密封存水弯
reseal voltage 灭弧电压
research 研究；考查；调查研究；探索
research achievement 研究成果
research activity 研究活动；科学研究工作
research agency 研究机构
research and development personnel 研究和发展人才
research and development expenses 研究发展费用
research and development 研究与开发；研究和开发；研究和发展；开发研制
research and development center 研究与发展中心
research and development cost 研究与开发费用；研究和开发成本；研究和发展成本；科研费用
research and development efforts 研究试制工作

research and development equipment 研究和发展设备
research and development expenditures 研究和开发支出
research and development function 研究和发展机能
research and development plan 研究和开发计划
research and development program(me) 研究与发展计划;研究试制方案
research and development project 研究与发展项目;研究与发展计划;研究和开发项目
research-and-development reactor 研究和发展用堆
research and development silo 研究与发展用发射井
research and development investment 研究和开发投资
research and evaluation of pool 油藏研究和评价
research and evaluation of stratum 地层研究和评价
research and evaluation of structure 构造研究和评价
research and exploration vessel 勘探船
research and extension 研究与推广;研究和推广
research and produce 研制
research and technique intension 研究和技术密集;研究和技术集约
research and technology 研究与工艺
research budget 研究支出预算
research bureau 调查局
research center[centre] 研究中心
research class 最优级(材料);重要样品;重要试件;头等样品;头等试件
research conclusion 研究结论;调查结论
research contract 科学研究合同
research data 研究资料
research department 研究室;研究部门
research design 研究设计
research device coupler 研究设备耦合
research diver 科研潜水员
research effort 研究力量;研究工作
research engineer 研究工程师
research environment(al) satellite 环境科研卫星
research equipment 研究设备
researcher 研究(人)员;研究工作者
researches 研究工作
research expenditures 研究费支出;研究费用
research facility 研究设备
research fellow 研究员
research findings 探测研究;研究成果
research foundation 研究基金
research funds 研究费;研究基金
research grade water 研究级水
research grant 研究补助金;科研补助金
research idea 研究思考
research information service 科研情报服务
researching test 研制试验
research in man's development in an all-round way 人的全面发展研究
research institute 研究院;研究所
Research Institute of Chemical Machinery 化工机械研究院
research institute of nutrition 营养卫生研究所
research institution 研究机构;研究单位
research instrument 研究设备
researchist 研究(人)员;研究工作者
research laboratory 研究试验室;研究工作实验室;科研实验室;科学实验室
research laboratory waste 研究实验室废物
research man 研究人员
research master blueprint 研究总蓝图
research memorandum 研究备忘录
research method 研究方法
research method of neotectonic movement 新构造研究法
research method of sediment sequence 沉积物层序研究方法
research microscope 研究用显微镜;研究显微镜
research model 研究模型
research monitoring 研究性监测
research observation 调查观测
research of transport capacity 运输能力的调查
research on climate 气候研究
research on man's all-round way development 人的全面发展研究
research-on-research 关于研制过程的研究
research-on-research corporation 研究的研究公司
research opportunity 研究课题
research paper 学术论文;研究论文
research party 勘探队
research personnel 研究人员
research plan 研究计划
research potential 研究潜能
research problem 研究课题
research production 研究成果
research program(me) 研究计划;研究规划
research project 研究项目;研究计划
research report 研究报告
research results of ground spectrum 地物波谱研究成果
research results of image processing 遥感图像处理研究成果
research ship 研究船;科学调查船;考察船;勘查船;海洋研究船;调研船;调查船
research staff 研究人员
research stage 研究阶段
research subject 研究课题
research target and test 研究目标与试验
research task 研究任务
research team 研究(小)组
research technique 研究方法
research topic 研究题目
research trial 研究试验
research unit expenditures allowance 科研单位经费
research vessel 科学调查船;考察船;调研船;调查船
research vessel unit 考察船队
research work 研究工作
research worker 研究人员;研究工作者
reseat 研磨;更换底座
reseater 阀座;修整器
reseating 座的修整
reseau (彩色照相用的)滤屏;网格;测候网
reseau camera 网目摄影机
reseau cross 格网十字
reseau grid 格网
reseau photograph 格网相片
reseau register plate 格网承片板
reseau stereoscopy 格网立体观测
resected air station 后方交会空中摄影站
resected duct 切口式导管
resected point 后方交会点
resected station 后方交会点
resection 截点法;后方交会;切除(术);反切法
resection by inverted triangle of error 示误三角形后方交会法
resection by plane-table 平板仪后方交会
resection by tracing paper 透明纸后方交会法
resection in space 空间后方交会
resection method 后方交会法
resection point 后方交会点
resection process 后方交会法
resection station 后方交会点
resedimentation 再沉降;再沉淀;再沉积(作用);回淤
resedimented rock 浊流沉积岩;再沉积岩
reseed 补种
reseeder 补种机
reselection 再精选;再分离
resell 转售
resemblance 相似性;类似
resene 氧化树脂;碱不溶树脂;含氧树脂
resequent drainage 再顺水系
resequent fault line scarp 再顺断线崖
resequent river 再顺(向)河;复活顺向河
resequent stream 再顺(向)河;复活顺向河
resequent-subsequent river 再顺后成河
reservation 禁猎区;权益保留;储备;附加保留条款;备用;保留意见;保留权益;保留(地)
reservation area 预留地;保留地
reservation clause 保留条款
reservation demand 保留需求
reservation land 备用地
reservation of landscape 风景保护
reservation of opinion 保留意见
reservation of right 权利保留;保留权利
reservation price 保留价格
reservation water 储存水
reserve 埋藏量;蕴藏量;储藏量;后备(力量);储量;储备物;储备量;储备(力);变换极性;备用的;保留木;保留林
reserve account 公积金账户
reserve accumulator 保留累加器
reserve accuracy 储量精度
reserve aerodrome 备用机场
reserve against decline in prices 价格下降准备
reserve air 储气;补呼气
reserve aircraft 备用飞机
reserve an entry 冲账
reserve area of wild medical herbs resources 野生药材资源保护区
reserve as natural farmland 保留为自然耕地
reserve assets 备用资产
reserve awaiting ratification 已审核储量
reserve bank 准备银行
Reserve Bank of Malavi 马拉维储备银行
reserve base 辅助基地;备用基地
reserve batch hopper 备用配合料料斗
reserve battery 备用电池组
reserve bunker 备用燃料舱
reserve buoyancy 后备浮力;储备浮力;保留浮力
reserve buoyancy for operating depth 工作深度储备浮力
reserve buoyancy tank 浮力调整舱室
reserve bus-bar 备用母线;备用汇流线
reserve-calculating plan on mining level 中段储量计算平面图
reserve-calculating section on exploratory line 勘探线储量计算剖面图
reserve capacity 潜力;储备能力;储备功率;备用容量;备用能力;备用能量;保存量
reserve circuit 备用电路
reserve clause 保留子句;保留条款
reserve coach 备用客车
reserve coal bunker 备用煤舱
reserve currency 准备货币
reserve cutting 保留作业法
reserved 后备的;保留的
reserved addressing mode fault 保留寻址方式故障
reserved area 保留区
reserved as spare 备用
reserved authority 保留权限
reserved bandwidth 保留带宽
reserved boiler 备用锅炉
reserved buoyancy 预备浮力
reserved cable 备用电缆
reserved capital 保留资金
reserved car 备用(货)车
reserved cars parking track 备用车停留线
reserved copy 正本
reserved deformation 预留变形量
reserved energy 备用能
reserve depot 备用仓库
reserved file type 保留文件类型
reserved foreign currencies 备用外汇资金
reserved forest 保留林
reserved groove 预留槽
reserved instruction 保留指令
reserved land 预留土地;保留地
reserved lanes 保留车道
reserved locomotive 备用机车【铁】
reserved lounge 贵宾休息室
reserved market 保留市场
reserved mining borehole 预留开采孔
reserved observation borehole 预留观测孔
reserved operand 保留操作数
reserved operand abort 保留操作数失败
reserved operand exception 保留操作数异常
reserved operand fault 保留操作数故障
reserved page frame performance option 保留页坐标性能选择;保留页框性能可选项
reserved page option 保留页选择;保留页面可选项
reserved pattern operator 保留模式操作符
reserved procurement 保留采购
reserved property of formation 岩层储集参数
reserved pump 备用泵
reserved quota 保留份额
reserve drive 备用传动装置
reserved road 保留道路;专用道路;备用泵
reserved route 专用路线
reserved seaplane area 水上飞机专用水域
reserved second line 预留第二线

reserved space 保留空间
reserved stock 保留股票
reserved strength 预留强度;保留强度
reserved surplus 保留盈余
reserved turning lane 转弯专用车道
reserved valve 备用阀
reserved vehicle 备用车
reserved volume 预约卷宗;保留卷
reserved water 备用水
reserved water pipe 备用水管
reserved water piping 备用水管
reserved water pump 备用水泵
reserved water tube 备用水管
reserved water tubing 备用水管
reserved word 保留字
reserve electric(al) source automatic connection device 备用电源自动投入装置
reserve energy of capacity 能量储备能力
reserve energy reduction factor 能量储备降低系数
reserve equipment 备用设备
reserve error 储量误差
reserve estimate 储量估计
reserve estimation 储量估计
reserve estimation package 储量计算程序包
reserve factor 储备因数;储备系数;备用系数
reserve feed tank 备用给水箱
reserve feed water 储备供水
reserve feed water tank 备用给水柜
reserve for amortization 摊销准备;摊提准备
reserve for appreciation 涨价准备
reserve for bonded debts 债券债务准备
reserve for business expansion 扩大经营公积金;扩充营业准备
reserve for construction 基建准备
reserve for depletion 折耗准备;耗减准备
reserve for depreciation 折旧准备;贬值准备金;备抵折旧
reserve for encumbrances 保留数准备
reserve forest 禁伐区
reserve for expansion 业务扩充准备;扩充准备;发展公积金
reserve for extensions 业务扩充准备;扩展公积;扩充准备
reserve for improvement on property 财产改良准备
reserve for insurance 保险准备
reserve for interdepartmental encumbrance 内部保留数准备
reserve for losses 亏损补偿准备
reserve for market fluctuation 市场价格变动准备金
reserve for outstanding losses 未付清赔款准备金;未决赔款准备金
reserve for overhead 间接费用准备
reserve for pensions 恤金准备
reserve for realized profits and losses 实现损益准备
reserve for redemption of preferred stock 收回优先股票准备
reserve for repairs 修缮准备;修理准备
reserve for replacements 重置准备
reserve for retirement allowance 退休金准备
reserve for unrealized increment in assets 未实现资产增价准备
reserve funds 公积金;预备费;准备金;储备基金;备用(基)金;备用款项
reserve fund for major repair 大修准备金
reserve gate 备用闸门
reserve generator 应急发电机
reserve in ore 矿石储量
reserve instruments 备用仪器
reserve lamp 备用灯
reserve level 准备金水平;备用量
reserve liabilities 准备性负债;负债准备
reserve lifting buoyancy 储备浮力
reserve light 备用灯
reserve line 备线路
reserve loop 备用回路
reserve lubrication oil tank 滑油储存箱
reserve machine 备用机(器)
reserve maintenance period 保养间隔期
reserve management 储量管理
reserve motion 逆动
reserve of construction material 建材储备
reserve of materials 库存材料

reserve of strength 力量保存
reserve of taxable capacity 可课税潜力
reserve oil bunker 备用燃油舱
reserve oil tank 储油罐;辅助油箱;备用油箱
reserve parachute 备份伞
reserve parts 备品;备件
reserve pit 备用泥浆坑
reserve plant 备用装置
reserve policy 公积金政策
reserve pool 蓄水池
reserve power 储备功率;备用功率;保留权利
reserve power circuit 备用电力电路
reserve power source 备用能源;备用电源
reserve-power source installing system 备用电源设备系统
reserve power station 应变发电站;辅助发电站;备用发电站;备用发电厂
reserve power supply 备用电源
reserve preferences 反优惠问题
reserve price 最低价(格);卖主的最低售价;底价
reserve protection 后备保护装置
reserve pump 备用(水)泵
reserve quantity 储备量;备用量
reserve ratifier 储量审批单位
reserve-ratifying file 储量审批文据
reserve ratio 法定准备金比率
reserve receiver 备用收信机
reserve requirement 准备金要求;法定(存款)准备金
reserve requirements on deposit 法定存款准备金
reserve return circuit 备用回路
reserve rimpull 储备轮缘牵引力;备用轮缘牵引力
reserves according to classifications 分级储量
reserve salt 防染盐
reserve sample 保留试样
reserves control 矿量管理
reserve set 备用机组
reserve settlement 预留沉落量;预留沉降量
reserve ship 后备船
reserve shutdown 备用停机
reserve side 背面
reserve siding 备用(侧)线
reserve signaling link 备用信号链路
reserves increment 矿量增加
reserves in metal 金属储量
reserves in mineral 矿物储量
reserves in useful component 有用组分储量
reserve sleeper 备用轨枕
reserves lifting promotion of reserves 储量升级
reserves of associated element 伴生元素储量
reserves of natural gas 天然气储量
reserves of ore block 块段储量
reserves of ore deposit 矿床储量
reserves of paragenic ore 共生矿产储量
reserve source of cash 现金的储备来源
reserve space 保留舱空
reserve speed (应急的)备用速度;保留航速
reserves period of production 保有矿量
reserve-sprout forest 中林
reserve-sprout system 中林作业
reserve statement 保留指令
reserve-statistic(al) date 储量统计日期
reserve stock 储备库存
reserve stockpile 备用料堆;备用储存物
reserve storage 备用仓库;保留存储器
reserve strength 强度储备;储备强度
reserve substance 储藏物质
reserve tank 储槽;储油罐;备用槽
reserve the right 保留权利
reserve the right to claim 保留索赔权
reserve the right to recourse 保留追索权
reserve to reduce inventory value to market 盘存估价准备
reserve tractive ability 储备牵引能力
reserve tractive force 储备牵引力
reserve transmitter 备用发信机
reserve trunk 备用中继线
reserve without ratification 未审批储量
reserving agent 防染剂
reserving triangle 折返三角线
reservoir 蓄水池;水库;钢瓶;容器;吸收库;储油器;储气器;储存库;储槽;储油池;储液器;储液槽;储器;储油窟;储存器;层眼;风缸
reservoir accretion survey 水库淤积测量

reservoir action 调节作用(水库)
reservoir and roof rock 储集层和盖层
reservoir area 库区
reservoir area-capacity curve 水库面积-库容(关系)曲线
reservoir area survey 库区调查
reservoir backwater 水库回水
reservoir bag 储存袋
reservoir bank 库岸
reservoir bank caving 水库坍岸
reservoir bank change 库岸演变
reservoir bank erosion 库岸侵蚀;库岸冲刷
reservoir bank failure 水库坍岸
reservoir bank protection forest 水库岸防护林
reservoir bank variation 库岸演变
reservoir basin 库区;水库区
reservoir bed 储油层;储水层;储气层
reservoir bitumen 储层沥青
reservoir bottom 油箱底
reservoir boundary 储层边界
reservoir by-wash channel 水库减淤积渠
reservoir cap 储蓄器盖
reservoir capacitor 储存电容器
reservoir capacity 储罐容量;水库容量;水库库容
reservoir capacity curve 水库容量曲线;水库库容曲线
reservoir capacity profile 水库容量纵剖面
reservoir characteristics 库区特征;水库特性
reservoir cistern 储水池
reservoir clearance 清库(水库蓄水前)
reservoir clearing 库区清理;水库清基
reservoir climate 库区气候
reservoir compartment 储油室
reservoir comprehensive utilization 水库综合利用
reservoir configuration 热储形状
reservoir-cooling 水库冷却
reservoir country rock 热储围岩
reservoir cupola 带前炉冲天炉
reservoir cycling 油藏循环法
reservoir dam system 水库坝系统
reservoir delta 库首三角洲;水库三角洲
reservoir depleting curve 水库放水曲线
reservoir deposit 水库淤积物;水库淤沙;水库沉积(物)
reservoir deposition 水库淤填;水库淤积
reservoir deposition observation 水库淤积观测
reservoir deposition survey 水库淤积测量
reservoir design capacity 水库设计容量
reservoir design level 水库设计水位
reservoir desilting 水库清淤
reservoir detention 水库拦蓄
reservoir draft 水库提取水量;水库供水量
reservoir drain plug 储筒放泄塞
reservoir drawdown 库水位下降;库水位下降;库水位降落;水库消落;水库工作深度
reservoir drive mechanism 油藏驱动机制
reservoir dynamics 油藏动力学
reservoir earthquake 水库地震
reservoir ecology 水库生态学
reservoir element 油池式折皱滤清元件
reservoir embankment 水库堤
reservoir empty 空库
reservoir-empty condition 库空情况
reservoir emptying 水库放空
reservoir engineer 油层工程师
reservoir engineering 油层工程;水库工程;采油工程
reservoir environmental quality assessment 水度环境质量评价
reservoir evapo(u)ration 库区蒸发(量);水库蒸发
reservoir experiment station 水库实验(水文)站
reservoir filling 水库蓄水;水库充水
reservoir filling-triggering earthquake 水库诱发地震
reservoir filter 油箱用滤油器
reservoir fishery 水库渔业
reservoir flooded line survey 水库淹没线测量
reservoir flood routing 水库调洪演算
reservoir flowage 水库淹没
reservoir flowage area 水库淹没区
reservoir fluid 热储流体;储层流体
reservoir fluid recharge 热储流体补给
reservoir for condensation 凝结液储槽
reservoir for irrigation 灌溉蓄水池;灌溉(用)水

库;灌溉蓄水库
reservoir for low-flow augmentation 调节水库
reservoir formation characteristic parameters analysis 储集层特征参数分析
reservoir formation temperature 热储建造温度
reservoir for mud 泥浆池
reservoir for overyear storage 多年调节水库
reservoir for power generation 发电水库
reservoir foundation 水库基础
reservoir fracturing 油层水力压裂
reservoir gas-oil ratio 油层油气比
reservoir habitat 水库生境
reservoir head 库尾;水库末端
reservoir holdout 水库拦蓄量
reservoir host 储[贮]存宿主;存储宿主
reservoir impounding induced seismicity 水库诱发地震
reservoir-induced earthquake 水库地震
reservoir-induced quake 水库诱发地震
reservoir inflow 入库水流;水库入流
reservoir inflow flood 入库洪水
reservoir inflow hydrograph 水库入流过程线;入库流量过程线;水库进水过程线
reservoir inflow rate 入库流量
reservoir inflow sediment 入库泥沙
reservoir in series 串联水库
reservoir integrated regulation 水库综合调度
reservoir interface 热储界面
reservoir inundation 水库淹没
reservoir inundation area 水库淹没区
reservoir inundation damage 水库淹没损失
reservoir inundation loss 库区淹没损失;水库淹没损失
reservoir investigation 水库调查
reservoir ladle 铁水混合桶;铁水混合包
reservoir leakage 水库渗漏;水库渗漏(量)
reservoir level 库水位;水库水位
reservoir life 水库寿命;水库使用年限
reservoir limit test 油层边界测试
reservoir lining 水库衬里
reservoir live storage 水库有效容量;水库有效库容;有效库容
reservoir load 水库荷载
reservoir loading 水库负载
reservoir mechanics 热储机制
reservoir observation 水库观测
reservoir of chlorine 氯吸收库
reservoir of clean water 清水池
reservoir of natural gas 天然气储备
reservoir operating policy 水库经营政策
reservoir operation 水库运作;水库运用;水库管理;水库调度;水库操作
reservoir operation chart 水库运作表;水库运用(图)表;水库调度表
reservoir operation guide curve 水库运作指导曲线;水库运用指导曲线;水库调度指导曲线
reservoir operation procedure 水库运作表;水库运用表;水库运行表
reservoir outflow 水库流出(水)
reservoir outlet works 水库排出口工程
reservoir overdraft 热储开发过度
reservoir parking spaces 暂时性停车处
reservoir pen 管形储水绘图笔
reservoir performance 热储工况
reservoir perimeter 水库周边
reservoir permeability 热储渗透性
reservoir plug 储筒塞
reservoir pollution 水库污染
reservoir pollution control 水库水质控制
reservoir pore space 储集层孔隙空间
reservoir port 油箱口;水库港
reservoir power plant 蓄水式水电站;库式水电站
reservoir pressure 水库压力;油层压力
reservoir pressure gradient 水库压力梯度
reservoir priming 增压加油
reservoir protection forest 水库防护林
reservoir raise 库水位抬高;水库水位抬高
reservoir range 水库断面
reservoir recovery 油层采收率
reservoir recreation 水库旅游业
reservoir region 库区
reservoir region afforestation 库区绿化
reservoir regulation 水库调蓄;水库调动;水库调度
reservoir release rate 水库泄水速率;水库放水量

reservoir rest level 热储静水位
reservoir rock 多孔存储[贮]岩;储集岩;热储岩石;储油层岩;储水岩层
reservoir roof 蓄水池顶盖
reservoir routing 水库洪水演算;水库调洪演算
reservoir safety provision act 水库安全规定法
reservoir sand 储油砂层
reservoir sandstone 储油砂岩
reservoir section 水库断面
reservoir sediment 水库淤积物;水库泥沙;水库沉积(物)
reservoir sediment accumulation rate 水库淤积速度;水库泥沙淤积速率
reservoir sedimentation 水库淤填;水库淤积;水库沉积(作用)
reservoir sedimentation observation 水库淤积观测
reservoir sedimentation survey 水库淤积测量
reservoir sediment flushing 水库冲沙
reservoir sediment washout 水库冲沙
reservoir seepage 水库渗漏(量)
reservoir seismic effect 水库地震效应
reservoir shore 库岸
reservoir shoreline 水库岸线
reservoir silt 水库泥沙
reservoir silting 水库淤积;水库沉泥
reservoirs in parallel 并联水库
reservoir site 库址;水库库址;水库位置
reservoir site selection 库址选择
reservoir species 吸收库中的各种物质
reservoir spillway 水库溢洪道
reservoir stage 库水位;水库水位
reservoir static pressure 地层静压力
reservoir station 水库(水文)站
reservoir statistics 水库统计学
reservoir steam pipe 储汽气管道
reservoir storage 库容;水库储存;水库蓄水量;水库储水(量)
reservoir storage capacity 水库蓄水量
reservoir storage curve 库容曲线
reservoir storage quality 水库蓄水水质
reservoir storage survey 库容测量
reservoir stripping 水库清基;水库表土清除;水库表土剥离
reservoir surface 库水面
reservoir survey 水库测量
reservoir sweep 排油
reservoir temperature 油层温度;热储温度
reservoir trap 储油圈闭
reservoir trap efficiency 水库拦沙效率;水库截沙效率;水库截留(泥沙)效率
reservoir-type 热储类型
reservoir-type manometer 带容器式压力计
reservoir type power plant 蓄水式水电厂;蓄水池式水力发电站;水库式发电厂
reservoir utilization 水库利用
reservoir volume 水库(总)容积;总库容;储[贮]气容积;库容;水库总容积;水库容量
reservoir wall 水库墙
reservoir water 水库贮水
reservoir water balance 水库水量平衡
reservoir water quality 水库水质
reservoir water quality condition 水库水质条件
reservoir water quality model 水库水质模型
reservoir water shaft 水库水井
reservoir water tower 水库水塔
reservoir with dam 有坝式
reservoir with damless 无坝式
reservoir yield 水库供水量
reset 转接;置零;计数器复零;回零;回到原状;回到零位;清除;微调;重置;重新设置;重镶;重放;重调整;重调(节);重安装;重定
reset action 重新调整动作;重调动作
reset adjustment 复位调整
reset and add 清加
reset and subtract 清减
reset a theodolite 重新安置经纬仪
reset a transit 重新安置经纬仪
reset attachment 再调附件
reset bias circuit 复位偏置电路
reset bit 回收钻头;重嵌钻头
reset button 回零按钮;清除按钮;重镶球齿;重复起动按钮;复位按钮
reset chuck 倒立轴
reset circuit 复位电路;复接电路

reset code 复位码
reset condition 原始状态;复位条件
reset contactor 复原接触器
reset control 预置控制
reset controller 复位控制器
reset counter 零位计数器;复位计数器
reset cycle 复位周期
reset driver for external synchronization 外同步复位驱动器
reset error 复原误差;复位误差;复位错误
reset flip-flop 复位触发器
reset gate 复位门
reset handle 重调把手
reset information data set 重置信息数据集;复位信息数据集
reset inhibit 禁复位
reset-input 复位输入
reset instruction 复位指令
reset key 回原键;恢复键;清除键;复位键;复归键
reset knob 可调按钮;重复起动按钮
reset line 复位线
reset magnet 复位磁铁
reset mechanism 重复定位机构
reset mode 回零方式;恢复方式;清除方式;重置方式;复位方式
reset operation 积分作用;重调工作
reset pressure 原始压力
reset procedure 清除过程
reset pulse 清除脉冲;复位脉冲
reset pulse generator 复原脉冲发生器
reset rate 置零速度;恢复速度;复位频率
reset reference bit 清除参考位
reset relay 跳返继电器;复位继电器
reset response 无静差作用
reset sequential machine 复位时序机
reset-set flip-flop 置位复位触发器
reset signal 清除信号;复位信号
reset spool valve 复位柱塞阀
reset state 复位状态
reset switch 转换开关;清零开关;重复启动开关
resettability 可重调性;可复位性;重调性
resettable 可重调的
resettable clock module 可清除时钟模块
resettable data base 可重置数据库;可恢复数据库;可复位数据库
resetter 恢复元件;复归机构
reset terminal 零输入端;恢复端
resetter-out 再平衡
reset time 复归时间
resetting 重新设置;重新铺砌;重新接入;重新给定;重复定位;复位
resetting age 再生年龄
resetting cam 回动凸轮
resetting device 复归机构
resetting function 复位功能
resetting magnet 复归磁铁
resetting of zero 零复位;回到零位
resetting period 凝前期
resetting price 重新定价
resetting shaft 回授轴;反馈轴
resetting spring 复归弹簧
resetting switch 复位开关
resetting time 复位时间
resetting value 复归值
resettlement 移民安居;重新安装;安置
resettlement area 移民安置区
resettlement budget 移民投资预算;移民安置费预算
resettlement bureau 移民局
resettlement compensation cost 移民安置费补偿
resettlement cost estimate 移民安置费估算
resettlement effort 移民计划;移民工作
resettlement expenditures 移民安置经费
resettlement feasibility 移民安置可行性
resettlement level 移民水位
resettlement of the residents 迁移居民
resettlement planning 移民安置规划
resettlement policy 移民安置政策
resettlement regulation 移民法;移民安置条例
resettlement scenario 移民安置实施方案
resettlement strategy 移民安置策略
resettler 移民;移居者
resettling 二次沉降;二次沉淀
reset to n 复位到 n

reset to zero 零复位
reset-type 嵌入式
reset unit 复位单元
reset valve 重调阀
reshabar 黑风(土耳其克尔迪斯坦一带的干冷东北风)
reshape 再成形
reshaper 整形器
reshaping 整形;重新修整;城市或街坊重建
reshaping of trees 树木修整
reshaping signal 整形信号
resharpen 再磨锐;再磨快;重新磨快;重新变锋利
resharpened 重磨锐的
resharpening 重磨锐
reshear 重剪机;精剪机
re-shearing 再剪切
reshears 钢板重剪机
reship 再次装船
reshipment 进口物再出口;转载;再装船;重新装运
reshore 重新加的支撑
re-shoring 临时支撑;重新支撑
re-shrinkage 再收缩
reshuffle 改组;重配置;重安排
resid 残油
reside 居留;居住
residence 住所;住处;滞留;居住;居民;常驻
residence area 住宅区
residence building 居住房屋;住宅;住房
residence casement 住宅窗扉
residence certificate 户口簿
residence district 居住区;住宅区
residence half-time 停留半衰期
residence hall 住房厅堂;住宅厅堂
residence housing 住房
residence lot frontage 住宅土地临街界线
residence of the president 总统府
residence quarter 住宅区;居住区
residence range 住宅区
residence region 住宅区
residence section 住宅区
residence street 居住区街道;住宅区街道
residence telephone 住宅电话
residence time 阻滞时间;滞留时间;停留时间;逗留时间;残留时间;闭模时间
residence time distribution 停留时间分布
residence time indicator 停留时间显示器
residence tower 塔式住宅
residence zone 住宅区
residency (英国住印度各邦的)官员的官邸;住处
resident 驻留的;住户;常驻的
resident architect 驻场建筑师;工地建筑师
resident area 常驻区
resident assembler 常驻汇编程序
resident bird 留鸟
resident buyer 常住采购人员
resident community 居住社区
resident compiler 常驻编译程序
resident control executive area 常驻控制执行程序区
resident control program(me) 驻留控制程序;常驻控制程序
resident density 居住密度
resident development system program(me) 常驻编制系统程序
resident district 居住区(域);住宅区
resident element 驻留单元
resident engineer 驻工段工程师;驻工地工程师;驻地(盘)工程师;驻场工程师;(代表业主的)工地工程师;常驻工程师代表
resident executive program(me) 常驻执行程序;常驻管理程序
resident fish 常栖鱼
resident flora 聚居植物群
resident resource 驻留资源
residential 住宅的;居住的
residential accommodation 住宅设施
residential air-conditioning 柱塞空气调节
residential allotment 住宅分配;住宅建设拨地
residential-apartment pool 公寓游泳池
residential appliance type chimney 适用于住宅炉灶的烟囱
residential architecture 居住建筑
residential area 居住区;居民区;居民点
residential awning 住房遮篷

residential belt 居住区
residential block 居住街坊;居住房屋
residential block green belt 街坊绿地
residential building 居住建筑(物);居住房屋;居民住宅
residential building permits 居住用建筑许可
residential building rate 住房定额;住房利率;住房等级;住房价格
residential camp 居住营地
residential care 家庭式疗养所;家庭康复所
residential care institution 家庭式疗养院
residential club 居民俱乐部
residential college 寄宿学校
residential community 居住社区;居住小区
residential construction 住宅建设;住宅建筑;住宅建造;民用建筑
residential construction industry 民用建筑业
residential custodial care facility 供膳宿的居民护理所;住房管理试验设施
residential density 居住密度
residential developer 住房建设者;住宅开发商
resident(ial) district 居住区(域);生活区;住宅区
residential district planning 居住区规划
residential drainage piping 住宅排水管
residential dwelling unit 住宅单元
residential electrical wiring 住宅布线
residential entrance 住宅入口
residential floor 居住层
residential floor area 居住建筑面积
residential frontage 住宅前;居住区街道
residential function 居住功能
residential garage 住宅车房
residential heating 住宅采暖;住宅区供热
residential heating system 住宅采暖系统
residential hinge 住宅用铰链
residential holdings 房产
residential home 住家式旅店;住宅
residential hotel 住宅旅馆
residential hot-water demand 住宅热水需要量
residential houses 民用住宅
residential hygiene 住宅卫生
residential illumination 住宅照明
residential incinerator 住宅用焚化炉
residential kindergarden 寄宿幼儿园
residential land 居住用地
residential lighting 住宅采光
residential local street 住宅区街道;居住区街道
residential location 住宅位置
residential maintenance equipment 住宅维护设备
residential mortgage 住宅抵押贷款
residential occupancy 住房占用率;住宅占用率;居住用途
residential park 居住区公园
residential permit 居住证
residential planned community 有规划的住宅地段建设
residential planning hygiene 住宅设计卫生
residential plot 住宅地块;住宅用地
residential population 居住人口
residential portion 住宅用地;土地的居住用部分;住宅部分
residential property 居住的房地产
residential quadrangle 四合院(住宅)
residential quarter 住宅区;居住(小)区
residential range 居住区
residential rental 房租
residential rental property 住房租赁房地产
residential sanitation 住宅卫生
residential school 寄宿学校
residential service 家庭供水;家庭供气;家庭供电
residential sewage discharge 住宅污水流量
residential sewage flows 生活污水流量
residential sewage pipe 住宅排水管
residential sewage treatment 生活污水处理
residential sewage treatment plant 生活污水处理厂
residential standard 住宅标准;居住标准
residential stor(e)y 居住层
residential street 居住区街道;住宅区街道;居住区道路
residential structure 住宅建筑;住宅用建筑物
residential suburb 城郊住宅区;近郊居住区;居住城郊;郊外住宅区
residential swimming pool 家庭游泳池

residential tower 塔式住宅
residential town 卧镇
residential-type equipment 住宅用设备
residential-type hotel 公寓式旅馆
residential unit 居住单元
residential usable floor area 住宅使用面积
residential use 供居住用;家庭用水;家庭用气;家庭用电
residential waste 家庭废水;住宅废物
residential water consumption 生活用水量
residential water pipe size 住宅水管尺寸
residential window 住宅型窗
residential zone 居住区;住宅区;居住带
resident inspector 驻场监察员;驻地监工员
resident load module 常驻装入模块
resident macroassembler 常驻宏汇编程序
resident mission 驻地代表团
resident module 常驻模块
resident monitor 常驻监督程序;常驻管理程序
resident monitoring system 常驻管理系统
resident nucleus program(me) 常驻核心程序
resident office 常驻办事机构
resident on-line hardware 常驻联机硬件
resident population 常住人口
resident program(me) 常驻程序
resident program(me) select list 常驻程序选择表
resident program(me) storage 常驻程序存储器
resident representative 常驻代表
resident road 居住区道路
resident routine 常驻例行程序
resident's committee 居民委员会
resident segment 常驻程序段
resident software 常驻软件
resident supervisor 常驻管理程序
resident system 常驻系统
resident text editor 常驻文本编辑程序
resident time 停留时间
residing time in furnace 炉能停留时间
residium 残渣;残余物(石油蒸馏)
residua 剩余物
residual 留余;剩余的;残余的;残数;残留;残积层;残差
residual aberration 剩余像差;残余像差
residual absorption 剩余吸收
residual achromatic aberration 剩余色差
residual acid 残留酸
residual acidity 剩余酸度;残余性酸度
residual action 后效;残效;残留作用
residual activity 剩余活度;剩余放射性;残余活性;残效;残留活性
residual activity level 剩余放射性强度
residual affinity 残余亲和力
residual air 余气;剩余空气
residual air volume 余气量
residual alkali content 残余碱含量
residual altitude 剩余高度
residual alum 残留明矾
residual-ammonia wastewater 残氨废水
residual amplitude 剩余振幅
residual amplitude modulation 残余调幅
residual analysis 残留(量)分析;残差分析
residual angle of internal friction 残余内摩擦角
residual anomaly 剩余异常
residual anomaly method 剩余异常法
residual anticline 残余背斜【地】
residual arc 残留电弧
residual arkose 残余长石砂岩
residual asphalt 残余(地)沥青;残留(地)沥青
residual assets 剩余资产
residual astigmatism 剩余像散
residual atmospheric drag 剩余大气阻力
residual attenuation 剩余衰减
residual attenuation distortion 剩余衰减畸变
residual austenite 残余奥氏体
residual austenite content 残余奥氏体含量
residual background field 残余背景电场
residual basin 残留盆地
residual beach cusp 残余滩尖嘴
residual betatron amplitude 终端自由振荡幅度
residual bioclast dolomite 残余生物屑白云岩
residual bioherm dolomite 残余生物礁白云岩
residual biostromic dolomite 残余生物层白云岩
residual bitumen 残留沥青
residual block 残余岩块;残余陆块

residual body 残质体;残余小体;残存小体
residual bond 残留键;残键
residual boulder 残余漂砾
residual breccia 残余角砾岩
residual bubble 剩余气泡
residual buoyancy 剩余浮力
residual burning equipment 残余燃烧设备
residual cake valve 排渣阀;排油阀
residual calculation 残差计算
residual capacitance 剩余电容;残余电容
residual capacity 剩余容量
residual carbon 残余碳
residual cavity 残腔
residual cementite 残余渗碳体
residual charge 剩余电荷;残余电荷;残留电荷
residual check 剩余校验
residual chlorine 余氯;剩余氯
residual chlorine analyzer 残余氯分析器;残氯分析仪
residual chlorine comparator 余氯比色器;残余氯比较器
residual chlorine removal 除残氯
residual circuit 剩余电路
residual claim 剩余财产索偿权
residual class ring 剩余类环【计】
residual clay 残余黏土;残积黏土
residual coating 残余覆盖层;残留膜层
residual cohesion (strength) 残余黏聚力;残余凝聚力
residual coldwork 残余冷作
residual collapse settlement 剩余湿陷量
residual communality 剩余公因子方差
residual compaction 残余压密
residual compression 剩余压力
residual compressive stress 残余压应力
residual conduction 剩余电导
residual conductivity 残余导电性
residual contact poison 残效接触毒
residual contaminant 残余污染物
residual contamination 残余污染
residual contribution 残余贡献
residual core 残留岩芯
residual core length 残留岩芯长度
residual correlation 剩余相关;残余关联
residual cost 残余成本
residual count 剩余计数
residual coupling 剩余耦合;残余耦合
residual crown 残冠
residual current 余流;零序电流;剩余电流;残余电流
residual current law 残余电流定律
residual current relay 零序电流继电器;剩余电流继电器
residual current state 初速电流状态
residual curve of energy percentage for low-frequency bandpass 低通能量百分比剩余曲线
residual curve of interval velocity 层速度剩余曲线
residual curve of middle frequency 中心频率剩余曲线
residual curve of peak frequency 峰值频率剩余曲线
residual cylinder stock 残渣汽缸油料
residual damage coefficient 残余破坏系数
residual damage mechanism 残余破坏机理
residual defect 残余缺陷
residual deflection 残余挠度;残余弯沉;残留挠度
residual deformation 剩余变形;残余变形;残余形变;残留变形
residual deposit 残余矿床;残余沉积(物);残剩矿床;残留矿床;残留沉积(物);残积物
residual depth 残余深度
residual deviation 剩余自差;剩余偏差;残差
residual diameter 残留直径
residual discharge 剩余流量;剩余放电;残余流量
residual discrepancy 剩余偏差
residual dislocation loop 残余位错环
residual dispersion 剩余色散
residual displacement 剩余位移
residual dissolved oxygen concentration 残溶解氧浓度
residual distortion 剩余失真;剩余畸变;残余畸变
residual disturbance 余扰;剩余扰动;剩余干扰;剩余磁扰
residual dolomicrite 残余微晶白云岩
residual dome 残余背斜【地】

residual drawdown 剩余降深;残余降深
residual dumidity 剩余湿度
residual dust 残留粉量
residual earth 沉积土壤;残余泥土;剩余耕地
residual effect 滞效;剩余效应;剩余效果;残余效率;残效
residual effect of pesticide 农药残效
residual effect spray 滞效喷洒
residual elasticity 弹性后效;剩余弹性;残余磁性
residual electric(al) charge 剩余电荷
residual electric(al) power 过剩电力
residual electricity 剩电
residual elements 残余元素;残积元素
residual elongation 剩余伸长
residual emission 剩余发射
residual energy 残余能量
residual enthalpy 残余焓
residual entropy 残余熵
residual equation 剩余方程;残差方程
residual equity 剩余资本主权
residual error 漏检故障;剩余误差;残余误差;残留误差;残差
residual error rate 漏检故障率;差错漏检率;残留错误率;残差率
residual error ratio 漏检故障率;漏检错误比例;剩余误差比;残留错误率
residual excitation 剩余激发;剩磁激励
residual expansion 残余膨胀
residual expansion of tank volume 罐体容积残余变形值
residual expansion rate of tank volume 罐体容积残余变形率
residual extension 剩余伸长
residual external angle 余外角
residual fabric 残余组构
residual factor 剩余因子;残余因素;残余系数
residual failure envelope 残余破坏包络线
residual field 剩余(磁)场;残场
residual field curvature 剩余场曲
residual field method 剩磁法;残余场法
residual flame 余焰
residual flicker 余闪;剩余闪烁
residual flow 余流;剩余流量;残流
residual flux 剩余通
residual flux density 剩余磁通(量)密度;残余磁通(量)密度
residual force 残余力
residual form 残渣态
residual fraction 尾馏分;残余馏分;残余分数;残余部分
residual free gas 剩余自由气
residual frequency modulation 寄生频率调制;残余调频
residual friction 残余摩擦力
residual frictional angle 残余摩擦角
residual fuel 剩余燃料;残渣燃料油
residual fuel oil 残渣燃料油;残余燃料油
residual fuel oil contamination 残余燃料油污染
residual function 微平功能;补差功能
residual gap 剩余气隙
residual gas 剩余气(体);剩余废气;残余气(体);残气
residual gas analyzer 剩余气分析器;残余气体分析仪
residual gas atom 剩余气体原子
residual gas composition 剩余气体成分
residual gas pressure 残余气压
residual gas rate 残气率
residual gas saturation 残余(天然)气饱和度;残余气饱和率
residual gas scattering 剩余气体散射
residual generation 残留物产生
residual geosyncline 残余地槽
residual glass phase 残余玻璃相
residual gravitational field 剩余引力场
residual gravity 剩余重力值
residual gravity anomaly 剩余重力异常;残余重力异常
residual gravity anomaly map 剩余重力异常图
residual halo 残积晕
residual hardness 剩余硬度;残余硬度;残留硬度
residual head 剩余水头
residual heat 余热;残余热
residual heat of workshop 车间余热量

residual heat removal system 余热排除系统
residual hill 残丘
residual homology 残余同源性
residual hum 残留哼声
residual hydrocarbon saturation 残余油气饱和度
residual hydrogen 残余氢
residual image 余像;残留影像;残留图像
residual image charge 残留图像电荷
residual impact 残存影响
residual impairment 残余损伤;残留损伤
residual impulse 剩余脉冲;残留脉冲
residual impurity 剩余杂质;残留杂质
residual income 残余收益
residual indentation 残余凹陷
residual induction 剩余电感;剩余磁感应;剩磁电感;残余感应
residual infiltration 残余下渗
residual insecticide 滞留杀虫剂
residual intelligibility 剩余可理解度
residual interaction 剩余相互作用;残余互作用
residual interionic force of attraction 残余离子间引力
residual internal stress corrosion 残余内应力腐蚀
residual intraclast dolomite 残余内碎屑白云岩
residual ion 剩余离子
residual ionization 剩余电离
residual land 剩余土地
residual landform 残余地形
residual landscape 剩余景观;残余景观
residual lattice absorption 残余点阵吸收
residual level 残毒含量
residual life 残效期;残留期
residual light 余光
residual lime 剩余石灰
residual linear change 残余线性变化
residual liquid 岩浆残余挥发分;残余液体;残余岩浆;残液
residual liquor 岩溶残液;残余岩浆;残浆;残留药液
residual load 后期荷载
residual loss 剩余损耗;残余损失
residual lump dolomite 残余团块白云岩
residual magma 残余岩浆
residual magnesium content 残余镁量
residual magnetic anomaly 剩余磁异常
residual magnetic field 残余磁场
residual magnetic flux 剩余磁通
residual magnetic flux density 剩余磁感应强度
residual magnetic induction 剩磁通量密度
residual magnetism 余磁;剩磁;残留磁性;残磁(性)
residual magnetism effect 剩磁效应
residual magnetism excitation 剩磁励磁
residual magnetism measurement 剩磁测定
residual magnetization 余磁化强度;剩余磁化强度;剩磁化;残余磁化
residual map 剩余图
residual mass curve 距平积曲线;剩余累积曲线;差积曲线;残余曲线
residual mass diagram 剩余累积(曲线)图;差积曲线
residual material 残积土
residual matrix 剩余矩阵
residual mean 残余平均
residual mean square 剩余均方;残差均方
residual melt 残余熔体
residual metal elements 残余金属元素
residual method 剩磁法
residual migration 剩余偏移
residual mine age 尚可服务年限
residual mineral 残余矿物;残留矿物
residual-mobility period 余动期
residual mode 剩余模
residual modulation 载波噪声;寄生调制;残余调制
residual moisture 剩余水分;残余水分;残留水分
residual moisture content 残余含水量
residual molten process 残余熔融作用
residual moment capacity 残余受弯能力
residual monomer 残余单体;残留单体
residual mountains 残余山地
residual net income 剩余纯收入;残余纯收入
residual nitrogen time 剩余氮时间
residual noise spectrum 剩余噪声谱
residual normal movement 残余正常运动

residual normal moveout 残余正常时差
residual nuclear radiation 剩余核辐射
residual nucleus 剩余核
residual nutrient load 残营养物负荷
residual ochre 残遗赭石
residual odo(u)r 残余恶臭
residual offset 剩余偏移
residual oil 渣油;残余油;油;废料渣油
residual oil binder 渣油黏结剂
residual oil blob 残余油滴
residual oil pavement 渣油路面
residual oil road 渣油路
residual oil saturation 裂余油饱和度;残余油饱和度
residual oil stabilized sand pavement 渣油稳定砂路
residual oolitic dolomite 残余鲕粒白云岩
residual organic compound 残有机化合物
residual organic load 残有机负荷
residualorganic matter 残有机物
residual organic matter in source rock 残留母岩中有机质
residual organic phosphorus 残有机磷
residual origin 残积成因
residual outcrop 残余露头
residual over-burden 残余覆盖层;残积覆盖层
residual over-burden anomaly 残积物异常
residual oxidant 残氧化剂
residual oxygen 残余溶解氧
residual parallax 剩余视差;残余视差;残存视差
residual parent material 残积母质
residual part 残余部分
residual payment 剩余付款
residual peak 剩余最大量
residual pellet dolomite 残余球粒白云岩
residual percentage 剩余百分率
residual period 残效期
residual permeability 剩余磁导率
residual pesticide 残留(性)农药;残留杀虫剂
residual phase error 剩余相位误差
residual phenomenon 残留现象
residual phosphate 残余磷酸盐
residual pin 防磁钉
residual placer 残积砂矿
residual plate 防粘片
residual plot 残差图
residual point-group 剩余点群
residual polarization 剩余极化
residual pollutant 残余污染物
residual pollution 残余污染
residual pollution product 残余污染产物
residual population 剩余种群
residual pore 残余孔
residual pore pressure 剩余孔隙压力;残余孔隙压力
residual pore water pressure 剩余孔隙水压力;残余孔隙水压力
residual porosity 残余孔隙率;残余孔隙度
residual potential 剩余位;残余电位
residual powder 残余粉末
residual pressure 剩余压力;残余压力;残压
residual pressure drop 剩余压力降低
residual product 副产品;剩余产品;残余(产)物;残余产品;残留产物
residual-prone agricultural chemicals 残留性农用化学物
residual property 残余性质;残留性能
residual quantity 残留量
residual radial velocity 剩余视向速度;剩余径向速度
residual radiation 剩余辐射;残余辐射;残留辐射
residual radioactive contamination 剩余放射性污染
residual radioactive dust 残余放射尘;残余放射性尘埃
residual rail stress 钢轨残余应力
residual rain(fall) 残余降雨;残余雨量
residual random variable 剩余随机变量;残余随机变量
residual range 剩余射程
residual ratio 残值比率
residual ray 剩余射线;剩余光线;残余光线
residual recoverable reserves 剩余可采储量
residual recovered drawdown 剩余恢复降深
residual red clay 残余红黏土

residual reflectance 剩余反射率
residual reflection 剩余反射;残余反射
residual refraction correction 剩余折射改正
residual regain 剩余回潮率
residual regression method 剩余回归法
residual resistance 余阻力;剩余阻力;剩余电阻;残余电阻
residual resistance factor 残余阻力系数
residual resistance ratio 剩余电阻比
residual response 残留响应区;剩余响应
residual rock 残余岩石
residual root 残根
residual rust 锈屑;未除尽锈斑
residual rust coating 带锈涂料
residuals 残余
residual salt content 残余含盐量
residual sand 残留砂
residual saturation 剩余饱和度;残余饱和度
residual sea water pump 排污泵【船】
residual secondary emission 剩余二次发射
residual sector 剩余地区
residual sediment 抗蚀残积;残余沉积(物);残积物
residual settlement 残余沉降;残留沉降
residual share 残余份额
residual shear resistance 剩余剪抗
residual shear strength 剩余抗剪强度;残余抗剪强度;残余剪切强度
residual shear stress criterion 残余剪切应力准则
residual shock response spectrum 剩余冲击响应谱
residual shock spectrum 剩余冲击谱
residual shrinkage 剩余缩率;残余收缩;残余收缩
residual side 残余边带
residual sideband transmission system 残余边带传输制
residual signal 剩余信号
residual silica constant 余硅常量
residual silica index method 余硅指数法
residual single-sideband filter 残余单边带滤波器
residual skeletal texture 残余骸晶结构
residual slag valve 排油阀
residual slat feature 残余盐丘构造
residual slip 残留泥浆
residual sludge 残渣
residual sludge concentration 剩余污泥浓度
residual sodium carbonate 残余(性)碳酸钠
residual soil 残余污物;原积土;残余土;残积土
residual soil anomaly 残余土异常
residual solution 残余溶液
residual solvent 残留溶剂
residual spectrum 剩余谱;残谱;残留谱
residual spheric(al) aberration 残余球差
residual spray 滞留性喷洒;滞留喷雾
residual spray mixture 残留液量
residual stability 残余稳定性
residual-stability period 余定期
residual stage 残留阶段
residual stand 主林分;伐后残余林木
residual standard error 剩余标准差
residual state of stresses 应力剩余状态
residual static correction 剩余静校正
residual strain 残余变形;剩余应变;残(余)应变
residual strain by welding 焊接残余应变
residual strain energy 残余应变能
residuals stray light 剩余杂散光
residual strength 剩余强度;残余强度
residual stress 剩余应力;残(余)应力;剩余应力
residual stress by welding 焊接残余应力
residual stress distribution 残余应力分布
residual stress due to welding 焊接残余应力
residual stress field 环境应力场;残余应力场
residual stressing force 残余内张力
residual stress in rocks 岩石残余应力
residual stress measurement 残余应力测量
residual stretch 剩余拉伸
residual stretching force 残余伸张力
residual structure 残余构造;残留体构造
residual structure stress 残余构造应力
residual stud 防粘螺栓
residual subluxation 残留半脱位
residual sulfur 残硫量
residual sulfur content 残硫量
residual sulphate 残余硫酸盐
residual sum of squares 剩余平方和;残差平方和
residual swelling 残余膨胀

residual tack 残余黏性;固化不良;发黏(漆膜);油漆过黏
residual tackiness 残余黏性
residual tar 残余焦油
residual technique 余值法(用于房地产价值评估)
residual tectonic stress 剩余构造应力
residual temperature 残余温度
residual tensioning force 剩余拉力
residual thaw layer 残融化层
residual thaw zone 残融层
residual theory 残值理论
residual thermal radiation 残余热辐射
residual thickness 剩余厚度
residual thickness of source rock 残余母岩厚度
residual thiosulphate content 剩余硫代硫酸盐含量
residual thread 丝扣余扣
residual time 残余时间;残留时间
residual time distribution 残留时间分布
residual tolerance 残留容量;残留耐量
residual toner 残余色调剂
residual toxic dose 残毒量
residual toxic effect 残留毒效应
residual toxicity 农药残毒;后期毒性;残余毒性;残留毒性;残毒
residual transport capacity 富余运输能力
residual unbalance 剩余失衡
residual valence 残余化合价
residual valence force 剩余价力
residual valley 残谷
residual value 余值;剩余(价)值;残值
residual value in capital investment 投资残值
residual value of conditional simulation 条件模拟剩余值
residual value of lime 石灰的残效
residual value of non-conditional simulation 非条件模拟剩余值
residual variance 剩余方差
residual variation 剩余变化
residual vector 剩余向量
residual velocity 剩余速度
residual-velocity differential 余速差
residual velocity head 剩余流速水头
residual velocity loss 余速(水头)损失
residual vertical parallax 残余上下视差【测】
residual vibration 零点振动;剩余振动;残余振动
residual viscosity 剩余黏度;残余黏度
residual volatile matter 残余挥发物
residual voltage 零序电压;剩余电压;残余电压;残留电压
residual voltage at switching impulse 操作波残压
residual voltage of lighting arrester 避雷器残压
residual volume 余气量;剩余容积;残余体积;残气量;残留容积;残留量
residual waste 家庭废水
residual water 残余水;残留水
residual water head 剩余水头
residual water height 剩余水位
residual water-level 剩余水位
residual water mass 剩余水体
residual water pressure 残余水压力;残余水压
residual water saturation 裂余水饱和度;残余水饱和度
residual water stage 剩余水位
residual water use 家庭用水
residual water vapo(u)r 剩余水蒸气
residual wave 残波
residual wavelet cutback 剩余子波消除
residual-wax soil 残积蜡土
residual weathering crust 残积型风化壳
residual wedge 剩余楔
residuary 留数
residuary earth 残留泥土
residuary estate 剩余资产
residuary product 残留产品
residuary resistance 剩余阻力;残余阻力
residuary soil 残留泥土
residue 渣滓;留数;扣除捐税后的剩余财产;剩余(物);残渣;残余物;残数;残留物;残基;不溶残渣
residue accumulation 残毒积累
residue analysis 残留物分析;残留量分析
residue black body radiation 残余黑体辐射
residue by distillation 蒸馏残渣
residue by evapo(u)ration 蒸发残渣
residue by weight 按重量计的残余物;残渣重量;

residue cargo 剩余货物
residue character 剩余特征
residue check 余量检查;剩余校验;剩余检查;残数校验
residue class 剩余类
residue class algebra 剩余类代数
residue class field 剩余类域
residue class group 剩余类群
residue class ring 剩余类环【计】
residue class space 剩余类空间
residue code 剩余(代)码
residue coke 残留焦炭
residue condition 留数条件
residue decomposition 残渣分解作用
residue disposal system 残渣处理系统;残余处理系统
residue distribution of particle size 残留粒子分布
residue extraction mechanism 排渣设备
residue field 剩余域
residue gas 残留的天然气(石油中);残余(煤)气;残留气
residue half life 残留半衰期
residue heat dryer 余热干燥器
residue heat drying 余热干燥
residue length 剩余长度
residue number system 剩余数系
residue of pesticide 农药残余;残留农药
residue of sieve 筛余;筛剩残渣
residue oil 残渣油
residue on evapo(u)ration 蒸发残渣;蒸发残余;蒸发残留(物)
residue on-evapo(u)ration method 蒸发余渣法;蒸发残留物法
residue on ignition 灼烧残渣
residue on sieve 筛余物;筛剩残渣
residue-on-sieving test 筛余试验
residue on the sieve 筛剩余渣;残渣物
residue output 剩余输出
residue pool 残渣水池
residue problem 残留问题
residue prone 残留性农药
residue-prone agricultural chemical in soil 土壤中残留性农药
residue quench system 残渣淬火系统
residue sliding 残余滑动
residues of previous fertilizer application 残积肥
residue stress 余应力
residue system 剩余系;剩余数系
residue term 残项
residue test for liquefied petroleum gas 液化气蒸发残余试验
residue theorem 留数定理;剩余定理;残数定理
residue valence 余价
residue weight 残基量
residuum 重残油;渣油;残渣;残余物;残油;残坡积物;残留物;残积层;风化壳;不透明微粒体
residuum coking 渣油焦化
residuum conversion process 渣油转化过程
residuum cracking process 渣油裂化过程
residuum hydroconversion 渣油加氢转化
residuum hydrocracking 渣油加氢转化
residuum hydrodesulfurization 渣油加氢脱硫
residuum lodge 残留物沉淀
residuum of carbonate type 碳酸盐型风化壳
residuum of kaolin type 高岭土型风化壳
residuum of laterite type 砖红土型风化壳
residuum tar 渣油焦油;残渣焦油(沥青);残留焦油
resign 辞职
resilicification 复硅作用
resilience 回弹性;回弹(能);恢复力;弹力;弹回;冲击韧性;斥力
resilience connector 弹性接头
resilience design procedure 回弹设计方法
resilience energy 回弹能
resilience factor 回弹系数
resilience force 回弹力
resilience meter 回弹计;弹性计;弹性测量仪
resilience of beam 梁的回能
resilience of roadbed 道床弹性
resilience of track 线路弹性
resilience test 回弹试验;弹性试验
resilience testing machine 夏氏冲击试验机
resilience under stress 应力之下的回弹力

resilience work of deformation 弹性变形功
resiliency 回弹力
resilient 有回弹力的;有弹性的;能恢复原状
resilient anchor 弹簧防爬器
resilient anti-creeper 弹簧防爬器
resilient beacon 活节式灯桩
resilient bearing 反弹轴承
resilient board 弹性板
resilient bollard 弹性系缆桩
resilient bushing 弹性衬套
resilient chaff 弹性干扰物
resilient chair 弹簧钢椅;弹性垫座
resilient channel 弹性槽形夹
resilient clip 弹性扣板;弹性卡子;有弹性的卡箍;弹性夹
resilient coating 弹性涂层;弹性覆层
resilient connector 挠曲性联轴节;弹性接头
resilient coupling 弹性耦连;弹性联轴器;弹性连接
resilient cushion member 弹性减振垫
resilient damping board 弹性减振板
resilient deformation 回弹变形;弹性变形
resilient drive 弹性传动
resilient-elasticity recovery 弹性变形恢复
resilient energy 回弹能
resilient escapement 弹性摆轮
resilient felt 弹性毡
resilient fender 弹性碰垫;弹性防撞装置
resilient fiberboard 弹性纤维板
resilient fixing 弹性连接
resilient floating floor 弹性地板
resilient floor 弹性地板
resilient floor covering 弹性地面铺料;弹性地板面
resilient flooring 弹性地面材料;弹性地板(材料)
resilient furring channel 弹性槽形龙骨
resilient gear 弹性齿轮
resilient gear ring 弹性齿圈
resilient gear wheel 弹性齿轮
resilient hanger 弹性悬挂装置;弹性吊挂
resilient hanger clip 弹性吊卡
resilient isolator 防振材料
resilient joint 弹性接缝
resilient joint material 弹性接缝料
resilient mastic 弹性玛琋脂
resilient material 弹性材料
resilient modulus 弹性模数;弹性模量
resilient modulus of base material 基层材料回弹模量
resilient mooring 弹性锚链系泊
resilient mount 隔振基座;隔振底座
resilient mounting 减振器;弹性支座;柔性支撑
resilient packing 弹性填料
resilient pad 缓冲垫
resilient quilt 减振垫;弹性毡
resilient rail fastenings 弹性扣件
resilient resin 弹性树脂
resilient response 回弹反应;弹性反应
resilient ring 弹性环
resilient seal 弹性密封;弹性封口
resilient sealant 弹性密封料
resilient seal valve 带弹性密封的阀
resilient seated gate valve 弹簧座闸阀
resilient sheet 弹性薄板;弹性模片
resilient shock absorbing suspension 弹性减振悬挂装置
resilient sleeper bearing 弹性垫板
resilient sleeve 弹性套筒
resilient spacer 弹性隔片
resilient stirrup 弹性套环扣件
resilient strength 弹性强度;复原强度
resilient strip 减振条
resilient support 弹性支承
resilient-supported track 弹性支承轨道
resilient-supporting member 弹性支座
resilient supporting unit 弹性支承装置
resilient support plate 弹性支承板;防振支座
resilient tie pad 弹性垫板
resilient tile 弹性面砖
resilient transmission 弹性传动
resilient trip 弹性片
resilient wheel 弹性车轮
resiliometer 回弹仪;弹性计
resillage 网状裂纹
resilver 重新挂银

resin 精制松香;树脂状沉淀物;树脂(松香);树胶松香
resin acceptor 偶联剂
resinaceous 树脂的
resin acid 树脂酸
resin acid lake 树脂酸色淀
resin/additive blender 树脂/添加剂掺和机
resin adhesive 树脂型黏合剂;树脂胶黏剂
resin adsorbent 树脂吸附剂
resin adsorbent-Fenton reagent oxidation process 树脂吸附剂—芬顿试剂氧化法
resin adsorption process 树脂吸附工艺
resin alcohol 树脂醇
resin alcohols 树脂醇类
resin alloy 树脂合金
resin-anchored bolt 树脂固定销;树脂锚固锚杆法
resin anchored bolting method 树脂锚固锚杆法
resin anchor 树脂锚固栓;树脂锚固杆
resin anion 阴离子树脂
resin applicator roller 树脂涂覆辊
resinat 聚胺甲烯树脂
resinate 树脂酸盐
resinate lustre 树脂酸盐光泽彩
resinate of cobalt 钴的树脂酸盐
resinate of lead 铅的树脂酸盐
resinate of manganese 锰的树脂酸盐
resinates 树脂酸盐类
resin base 树脂基
resin-base ca(u)lking compound 树脂基嵌缝料
resin-base curing compound 树脂基养护剂
resin-based 树脂基的
resin-based adhesive 树脂基黏合剂;树脂胶黏剂
resin-based binder 树脂基黏合剂
resin-based cement 树脂胶黏剂
resin-based concrete 树脂基混凝土
resin-based concrete curing agent 树脂基混凝土养护剂
resin-based concrete curing compound 树脂基混凝土养护剂
resin-based liquid membrane 树脂基涂膜
resin-based membrane curing 树脂基薄膜养护
resin-based mortar 树脂基灰浆
resin-based putty 树脂基灰泥;树脂基腻子
resin bed 树脂层
resin belt 树脂结合剂砂带
resin binder 树脂黏合剂
resin blaze 采脂裂痕
resin blushing 树脂致白
resin bolt 树脂锚杆
resin bond 树脂黏结
resin-bonded 树脂胶合的
resin-bonded board 树脂棉板
resin-bonded chipboard 树脂胶木屑板;树脂胶结碎木板
resin-bonded coil 树脂黏合线圈
resin-bonded felt 树脂棉毡
resin-bonded magnet 树脂黏结的磁铁
resin-bonded pigment 彩色树脂型颜料
resin-bonded pigment printing 涂料印花
resin-bonded plywood 合成树脂黏结层压板;合成树脂胶合板;树脂胶合板
resin-bonded refractory 树脂结合耐火材料
resin-bonded sand 树脂砂
resin-bonded slab 树脂胶合板
resin-bonded wheel 树脂结合砂轮
resin bonding 树脂胶结;树脂结合的;树脂胶合
resin bonding bolt 树脂黏合螺栓
resin bond wheel 树脂结合剂砂轮
resin-bound 树脂黏合的
resin-bound cork filler 树脂胶结软木填缝材料
resin canal 树脂道
resin capacity 树脂容量
resin cast component 树脂密封元件
resin cation 阳离子树脂
resin cavity 树脂腔
resin cement 树脂水泥;树脂腻子;树脂胶合剂
resin channel 树脂沟道
resin chip board 树脂碎木板;树脂木屑板;树脂胶结碎木板
resin clad transformer 环氧树脂变压器
resin clear varnish 树脂凡立水;树脂清漆
resin coat 树脂涂层
resin-coated paper 涂塑相纸
resin coated sand 树枝覆膜砂

resin coating 树脂覆膜
resin coating of vinyl chloride 氯乙烯树脂涂料
resin colo(u)r chip 颜料树脂片；彩色树脂片
resin-column 树脂交换柱
resin compound 树脂混合料；树脂化合物
resin compounding 树脂混配
resin concentrate 树脂提浓物；树脂母料
resin concrete 树脂混凝土；树脂改性混凝土
resin constituent 树脂成分
resin content 含脂率；树脂含量
resin core binder 树脂型芯黏结剂
resin-cored solder 松香芯焊锡条；松脂芯软钎料
resin-cored solder wire 松脂芯焊条
resin cure 树脂硫化
resin curing machine 树脂烘焙机
resin degradation 树脂降解
resin depletion 树脂变质
resinder 砂带磨光机
resin dispersion 树脂分散体
resin dropping point 树脂滴点
resin duck 树脂道
resin duct 树脂囊
resin emulsion 树脂乳(状)液
resin emulsion cement 乳化树脂黏结料；乳化树脂粘接料
resin-emulsion paint 树脂乳液涂料；树脂乳胶漆；树脂涂料；树脂漆
resin-encapsulate 树脂包封
resin encased pigment particle 树脂包裹颜料颗粒
resinene 中性树脂
resineon 树脂油
resin ester 酸化树脂；树脂酯
resin exchange capacity 树脂交换容量
resin extender 树脂增量剂
resin-faced building paper 树脂表面墙纸
resin filling 树脂填充
resin film 树脂薄膜
resin finish(ing) 树脂整理
resin finishing agent with super-low formaldehyde 超低甲醛树脂整理剂
resin-fixed pigment printing 涂料印花
resin flake 树脂薄片
resin-forming alcohol 树脂表面形成的乙醇
resin fouling 树脂污染
resin-free 不含树脂质
resin gall 树脂瘤
resin glass 塑料玻璃
resin-glass interface 树脂—玻璃界面
resin-glass lifeboat 树脂玻璃救生船
resin glue 树脂(黏)胶；树脂黏结剂；树脂黏接剂
resin grout 树脂浆液；树脂灌浆
resin grouted bolt 灌树脂锚杆
resin grouting 树脂灌浆
resin gun 树脂喷枪
resin-gypsum plaster 树脂配合石膏灰浆
resinification 含有树脂的；树脂化(作用)
resinification concrete 树脂化混凝土
resinifying agent 树脂化剂
resinifying solution for root canal 根管塑化液
resinifying therapy of root canal 根管塑化术
resin-impregnated 树脂浸渍的
resin-impregnated ceramics 树脂浸制陶瓷
resin-impregnated coil 树脂浸渍线圈
resin-impregnated fabric 浸树脂织物
resin-impregnated graphine cloth 浸树脂石墨布
resin-impregnated paper-faced plywood 树脂浸透纸面胶合板
resin-impregnated porous ceramics 树脂浸制多孔陶瓷
resin-impregnated wood 树脂浸渍木材；树脂浸注木材
resin impregnation 树脂填充
resin impregnation bath 树脂浸渍洗槽
resin-in-column 树脂柱式吸附法
resin injection machine 树脂压力注射成型机
resin injection mo(u)lding 树脂注射模塑
resin injection process 树脂压力注射成型
resin ink 树脂油墨
resin-in-pulp ion(ic) exchange 矿浆树脂离子交换(法)
resin-in-pulp process 树脂矿浆吸附法；树脂分批入浆法
resin-insulated winding 树脂绝缘绕组
resinite 树脂体；树脂煤素质

resin kettle 树脂锅；树脂反应釜
resin lacquer 树脂漆
resin latex 树脂乳液
resin layer 树脂层
resin leading 树脂浸析
resin life 树脂寿命
resin-ligand 树脂配合体
resin-lining 树脂衬里
resin-lint 树脂屑
resin liptobiolith 树脂残植煤
resin-loaded paper 饱和树脂纸
resin manufacturing wastewater 树脂生产废水
resin-membrane electrode 树脂膜电极
resin-modified cement 树脂改性水泥；改性树脂水泥
resin-modified cement concrete 树脂改性的水泥混凝土
resin molded transformer 树脂浇注型变压器
resin mortar 树脂砂浆
resin mortar core 树脂胶泥芯
resinoclarite 微树脂亮煤
resinoclarodurite 微树脂亮暗煤
resinocyst 树脂囊
resinodurite 微树脂暗煤
resinoduroclarite 微树脂暗亮煤
resin of asphalt 沥青胶质
resin of sweetgum 枫香脂
resinogen 树脂原
resinography 显微树脂学
resinoic acid 树脂型酸
resinoid 热固性黏合剂；树脂型物；树脂类；熟树脂
resinoid and vitrified bond 树脂陶瓷结合剂
resinoid bond 树脂黏结剂；树脂胶合；树胶黏结剂；树胶结合剂
resinoid group 树脂组
resinoid wheel 树脂黏结砂轮；树脂结合砂轮；树脂胶结砂轮
resin oil 松香油；树脂油
resinol 树脂酚；树脂醇
resinolic acid resin 树脂酸树脂
resinophore 易树脂化
resinophore groups 成树脂基
resino-sclerotinite 树脂—菌类体
resinosis 泌脂；树脂渗溢
resinotannol 树脂丹宁醇
resinous 树脂的
resinous acid 树脂酸
resinous bond 树脂结合剂；树脂胶合
resinous cement 树脂结合剂
resinous coal 树脂煤
resinous composition 树脂成分
resinous electricity 树脂电
resinous emulsion 树脂乳液
resinous exchanger 树脂交换剂
resinous fracture 树脂状断口
resinous impurities 树脂状杂质
resinous lignite 树脂褐煤
resinous luster 松脂光泽；树脂(状)光泽
resinous material 树脂状物质
resinous matter 树脂状物质
resinousness 树脂性
resinous odo(u)r 树脂味
resinous oil 树脂质油
resinous plasticizer 树脂状增塑剂；树脂(型)增塑剂
resinous polymer 树脂状聚合物
resinous polyol 树脂状多元醇
resinous putty 树脂腻子；树脂油灰
resinous shale 油页岩
resinous stabilizer 树脂类稳定剂
resinous substance 树脂状物质
resinous surface treating method 树脂表面处治法
resinous surface treatment 树脂表面处治；树脂表面处理
resinous tar 树脂柏油(沥青)
resinous timber 含树脂木材；充树脂木材
resinous trap 树脂阱
resinous trapper 树脂捕捉器
resinous varnish 树脂系涂料；树脂清漆
resinous waterproof adhesive 树脂防水胶黏剂
resinous wood 含脂材；多脂材；充脂材；明子
resinox 酚—甲醛塑料(酚—甲醛树脂)
resin paint 树脂漆
resin panel 树脂板
resin particle 颗粒树脂；树脂颗粒

resin passage 树脂道
resin paste 树脂糊
resin permalloy 树脂坡莫合金
resin-pigment ratio 树脂颜料比
resin plaster 树脂膏
resin plasticizer 树脂增塑剂
resin plate 树脂板；树胶板
resin plug 树脂胶合塞
resin pocket 油眼；树脂淤积；树脂囊
resin poison 树脂中毒
resin pot life 树脂适应期
resin powder 树脂粉末；粉状树脂；干托粉
resin precoating 树脂覆膜
resin putty 树脂灰泥；树脂腻子
resin reaction vessel 树脂反应器
resin regeneration 树脂再生
resin-rich area 富树脂区
resin-rich layer 富树脂层
resin-rich spot 树脂集聚
resin-rich surface 富树脂表面
resin-rock bolt 树脂型锚杆
resin roof bolting 灌注树脂固定锚杆法
resin saturated 树脂饱和的；饱和树脂
resin seam 树脂缝
resin sheet 树脂薄膜
resin sheet overlay 树脂薄膜罩面层
resin shelf life 树脂储[贮]存期
resin shell mold casting 树脂壳模铸造
resin-soaked wood 充脂木材
resin soap 松香皂；树脂皂
resin solution 树脂溶液
resin solvent 树脂溶剂
resin spirit 树脂精
resin spot test 树脂点滴试验
resin spot test method 树脂点滴试验法
resin-starved resin 贫树脂区
resin storage life 树脂储[贮]存期
resin streak 木材渗出的树脂；树脂条纹；树脂渗出条痕
resin sulfur 树脂硫
resin support 树脂支承
resin surge pot 树脂搅动罐
resin suspension 树脂悬浮液；树脂悬融
resin swelling life 树脂溶胀
resin system 树脂系统
resin tapper 树脂采集器
resin tapping 树脂采收；采脂
resin technology 树脂工艺学
resinter 酯化树脂
resin transfer mo(u)lding 树脂传递模塑成型
resin trap 树脂收集器
resin trapper 树脂捕捉器
resin-treated wood 树脂浸渍木；经树脂处理的木材；树脂处理木材；树脂表面木材
resin treatment 树脂加工
resin tube 树脂管
resin type curing agent 树脂固化剂
resin varnish 树脂清漆
resin viscosity 树脂黏度
resin wax 树脂蜡
resin wearing 树脂磨损
resin working life 树脂适应期
resist 抵抗；防蚀剂
resist agent 防燃剂
Resistal 耐蚀硅砖
resistance 阻抗；抗力；电阻；抵抗力
resistance against damage 抗损强度；抗损力
resistance against pelting rain 抗暴雨
resistance against rust 抗锈蚀
resistance alloy 电阻合金
resistance alloy-powder 电阻合金粉
resistance amplification 电阻放大
resistance amplifier 电阻放大器
resistance angle 抗滑角
resistance angle of shearing 剪阻角
resistance arrester 电阻避雷器
resistance attenuator 电阻衰减器
resistance at ventilation pipe joint 风筒接头阻力
resistance at ventilation pipe turning 风筒转弯阻力
resistance bank 抗蚀河岸
resistance box 电阻箱
resistance braking 电阻制动
resistance brazing 接触钎焊；电阻(加热)钎焊

resistance bridge 电阻(电)桥
resistance bulb 测温电阻器；变阻泡
resistance butt joint 接触焊对接接头；电阻对接接头
resistance butt-seam welding 电阻对缝焊接
resistance butt-welding 对接接焊；电阻对焊；接触对焊
resistance capacitance 电阻电容
resistance-capacitance circuit 阻容电路
resistance-capacitance constant 电阻容常数
resistance-capacitance coupled amplifier 电阻电容耦合放大器；阻容耦合放大器
resistance-capacitance coupling 阻容耦合；电阻耦合
resistance-capacitance divider 阻容分压器
resistance-capacitance network 电阻电容网络；阻容网络
resistance-capacitance network analog(ue) machine 电阻电容网络模拟机
resistance-capacitance oscillator 阻容振荡器
resistance-capacitance time constant 阻容时间常数
resistance card 电阻卡；电阻表
resistance coefficient 阻力系数；电阻系数
resistance coefficient 1 第一阻力系数；第一达西数
resistance coefficient of partial penetrating well 非完整井阻力系数
resistance coefficient of pipe 管道阻力系数
resistance coil 线绕电阻；电阻线圈
resistance commutation 电阻换向
resistance comparator 电阻比较器
resistance control 电阻控制
resistance controller 电阻控制器
resistance converter 电阻变换器
resistance cord 电阻线
resistance-coupled amplifier 电阻耦合放大器
resistance coupling 电阻耦合
resistance coupling amplifier 电阻耦合放大器
resistance curve 阻力曲线
resistance cutting 电阻加热切割
resistance deflecting force 防斜力
resistance deflector 防斜器
resistance degradation by sunlight 抗阳光降解性；抗曝晒降解性；耐阳光降解性
resistance degree 阻抗程度
resistance deterioration on weathering 耐室外曝晒老化性
resistance diagram 阻力图
resistance divider 电阻分压器
resistance drag 阻力
resistance drop 电阻(电)压降
resistance due to running 运行阻力
resistance dynamometer 测阻力仪
resistance electrode 电阻焊条
resistance electroslag welding 接触电渣焊
resistance element 电阻元件
resistance energy head 阻力能高
resistance environment 耐环境条件性
resistance factor 阻力因数；阻力系数；抵抗因子
resistance feedback 电阻反馈
resistance flash-butt welding 闪光对焊头；闪光对焊
resistance-flash welding 电阻闪光焊接；电阻弧花压焊接
resistance flowmeter 电阻式流量计
resistance for car rolling 车辆溜放阻力
resistance force 阻力；抗力
resistance frame 电阻框
resistance function 阻力函数
resistance furnace 电阻(电)炉
resistance ga(u)ge 电阻计
resistance gradient 电阻梯度
resistance grid 电阻栅
resistance grounded 经电阻接地
resistance grounded neutral system 中心点电阻接地方式
resistance grounded system 电阻接地系统
resistance grounding 电阻接地
resistance hard soldering 电阻加热硬钎焊
resistance head 阻力水头；摩阻水头
resistance-heated evapo(u)ration coating 电阻加热蒸镀
resistance-heated furnace 电阻加热炉
resistance heater 加热电阻丝；电阻加热器
resistance-heating 电阻加热

resistance-heating cutting 电阻加热切割
resistance-heating element 电阻发热元件
resistance-heating furnace 电阻加热电炉
resistance hybrid 电阻桥接岔路
resistance hygrometer 电阻湿度计
resistance indicator 阻力指示器
resistance industrial fume 抗工业烟雾性
resistance industrial solvent 耐工业溶剂性
resistance in series 串联电阻
resistance in the dark 暗阻
resistance lamp 电阻灯
resistance law 阻力法则
resistance lead 电阻竖片
resistance load 电阻负荷
resistance loss 水流水头损失；阻力损失；电阻损失；电阻损耗
resistance magnetometer 电阻式磁强计
resistance manometer 热线真空计；电阻压力计；电阻式真空计
resistance mass 抗力体
resistance-matching transformer 电阻匹配变换器
resistance material 电阻材料
resistance measurement 电阻测量；电阻测定
resistance meter 电阻计
resistance methanometer 电阻式甲烷指示器
resistance method of temperature determination 电阻测温法
resistance modulus 抗力模量
resistance moment 抗外力内弯曲；抗滑力矩；抵抗力矩；抗外力弯矩
resistance moment diagram 抵抗弯矩图
resistance movement 抵抗力矩
resistance network 电阻网络
resistance network analog(ue) 电阻网络模拟
resistance noise 热噪声
resistance of bolt 锚杆抗力
resistance of conductor 导体电阻
resistance of conduit 管道阻力
resistance of deformation 变形抗力
resistance of ducting 管道阻力；管道压力损失
resistance of filtration 过滤阻力
resistance of flow 流动阻力
resistance of heat-transfer 传热阻
resistance of masonry 砌体强度
resistance of materials 材料力学；材料抗力
resistance of motion 运动阻力
resistance of overall heat transmission 总传热阻
resistance of parallel 并联电阻
resistance of pile brooming during driving 桩顶抗裂能力
resistance of piles 桩群抗力
resistance of river channel 河槽阻力
resistance of rivet 铆钉阻力
resistance of shock absorber 减震器抗振强度
resistance of soil 土壤强度
resistance of suction 吸入阻力
resistance of tearing 撕裂强度
resistance of trailing dredging 耙吸(式)挖泥阻力
resistance of ventilation 通风阻力
resistance on oxidation at high temperature 抗高温氧化性
resistance pad 电阻衰减器
resistance paint 电阻漆
resistance percussion welding 接触冲击焊
resistance percussive-welding 电阻冲击焊；电阻锻接
resistance plug 附电阻的插塞
resistance potentiometer 电阻电位计
resistance power 抗断裂强度
resistance pressure 抗压性
resistance probe 电阻温度计探头
resistance projection welding 电阻凸焊
resistance property 牢固特性；阻抗性能
resistance pyrometer 电阻温度计；电阻温度表；电阻高温计；电测温度计
resistance ratio 电阻比
resistance reinforcement 阻力钢筋
resistance-resistance network analog(ue) machine 电阻电阻网络模拟机
resistance roller-spot-welding 电阻滚动点焊
resistance seam welding 接触缝焊；电阻滚焊；电阻缝焊
resistance shunt method 分流电阻法
resistance silencer 阻性消声器；电阻式消声器

resistance soldering 电阻钎焊
resistance spiral 螺旋形电阻丝
resistance spot weld(ing) 接触点焊；电阻点焊
resistance-stabilized oscillator 电阻稳定振荡器
resistance starter 电阻起动器
resistance starting 变阻器起动
resistance-start motor 电阻起动电(动)机
resistance stitch welding 电阻点缝焊接
resistance stopping 阻力停车
resistance strain ga(u)ge 电阻应变计；电阻应变仪
resistance stress 抵抗应力
resistance switch 带电阻的两级开关
resistance switchgroup 切断电阻组合开关
resistance temperature coefficient 电阻温度系数
resistance temperature detector 热电阻温度计；电阻式温度检测器
resistance test 电阻试验；阻力试验
resistance tester 电阻测试器
resistance thermometer 铂丝温度计；电阻温度器；电阻温度计；电阻温度表；电阻式温度计
resistance tide ga(u)ge 电阻式验潮仪
resistance time 滞留时间；停留时间
resistance to abrasion 抗磨性；抗磨(力)；耐磨强度；磨蚀阻力；磨耗阻力
resistance-to-abrasion test 耐磨性试验
resistance to acid (attack) 耐酸(性)；耐酸碱性
resistance to ag(e)ing 抗老化性(能)；抗老化；耐老化(性)
resistance to agglomeration 抗团聚力
resistance to aggressive agencies 对腐蚀介质的阻抗；对腐蚀剂的阻抗
resistance to aggressive influences 抵抗侵蚀性能
resistance to alcohol 抗醇性
resistance to alkali 耐碱性
resistance to alternate freezing and thawing 抗冻融交替
resistance to alternating mechanical stress 耐交变力学应力性
resistance to alttrition 耐磨性
resistance to atmosphere 抗大气性
resistance to atmospheric corrosion 耐空气腐蚀性
resistance to axial compression 抗轴向压缩力
resistance-to-bake test 耐烘烤性试验
resistance to bascule action 平钩阻力
resistance to bend(ing) 耐弯曲性；抗弯强度；坎弯性；抗弯能力
resistance to bending strain 抗弯曲应变力；抗弯(曲)性
resistance to bleeding 抗变色性；抗泌水性；抗色料扩散性
resistance to blush 抗发白性
resistance to boiling water 耐沸水性
resistance to bond 抗黏着；抗黏性；抗黏结；黏着阻力
resistance to breakage 抗碎强度
resistance to breakage graduation 抗碎强度分级
resistance to breaking up 抗破裂能力；抗崩裂能力
resistance to bronzing 抗青铜斑性
resistance to broom 桩顶抗裂能力；桩顶开花抗力
resistance to case 对机壳电阻
resistance to cavitation 抗气蚀性能；抗气蚀能力
resistance to chemical attack 抗化学侵蚀性；抗化学腐蚀能力；耐化学腐蚀
resistance to chemical deterioration 抗化学变质性
resistance to chemical reagents test 抗化学剂试验
resistance to chipping 抗碎落性；抗石击性
resistance to cleaning 抗去污剂性
resistance to cold 耐冷性(能)；耐寒性；耐寒强度；低温不脆化性
resistance to colo(u)r change by ultraviolet 抗紫外线变色性；耐紫外线变色性
resistance to compression 抗压强度；抗压性；压缩变形阻力；耐压强度
resistance to compressive strain 抗压缩应变力；抗压缩性
resistance to control 难控(制)性
resistance to corona 耐电晕性
resistance to corrosion 抗腐蚀能力；耐腐蚀性；耐蚀(性)
resistance to corrosion from seawater 抗海水腐蚀性；耐海水浸蚀性
resistance to cracking 抗裂强度；抗裂纹性；抗龟裂性
resistance to crack propagation 抗裂缝扩展性

resistance to crazing 耐裂纹性
resistance to creep 耐蠕变
resistance to crippling 压屈阻力
resistance to crocking 抗脱色性;抗染污性
resistance to crushing 抗压碎性;抗碾compactness;耐压碎性
resistance to cutting 抗切削性;切削抗力
resistance to cutting damage 抗切削破坏能力
resistance to cycle movement 耐周期位移性
resistance to darkening 抗变黑性;抗变暗性
resistance to decomposition 分解阻力
resistance to deformation 对变形的阻力;抵抗变形;变形阻力;变形抗力;抗变形性
resistance to deterioration on weathering 耐气候老化性
resistance to determination on ag(e)ing 抗老化变质性
resistance to dirt 抗脏污(水)性
resistance to displacement 位移阻力
resistance to distortion 抵抗变形
resistance to driving 抗锤入力;击入阻力;打桩阻力
resistance to dynamic(al) indentation 耐动态压痕性
resistance to effect of heat 耐热性能
resistance to efflorescence 耐风化性
resistance to elements 耐候化性能
resistance to elevated temperatures 耐高温性能
resistance to emulsion 抗乳化强度
resistance to emulsion number 抗乳化值
resistance to environment 耐周围条件性
resistance to environment(al) cracking 耐环境(条件)龟裂性
resistance to environment(al) stress cracking 抗环境条件应力龟裂性
resistance to erosion 耐侵蚀性
resistance to etching solution 抗腐蚀液性
resistance to expansion and contraction 体积恒定性
resistance to exposure 抗照射;抗曝光
resistance to extinction 对消退的抗性
resistance to failure 抗破坏强度
resistance to fire 耐火性
resistance to flame erosion 耐火焰侵蚀性
resistance to flexing 耐反复挠曲性
resistance to flow 流动阻力
resistance to flow output 出口阻力
resistance to flow through valve 阀流阻力
resistance to foaming 抗起泡沫能力
resistance to former oils 耐变压器油性
resistance to fouling 抗侵蚀性;防污着性
resistance to freezing 抗冻能力;抗冻性
resistance to freezing and thawing 抗冻融性(能)
resistance to friction 摩擦阻力
resistance to frost 抗冻性;抗霜(冻)性;防结霜性
resistance to frost action 耐霜冻
resistance to frost attack 防冻害性能
resistance to ground 对地电阻
resistance to head 水头阻力
resistance to heat 抗热性;抗热度;耐热性;热电阻
resistance to heat and humidity 耐湿热性
resistance to heat distortion 耐热变形性
resistance to heat emission 耐热辐射性
resistance to heat shocks 抗热冲击(性);耐热振性
resistance to heat transmission 阻热传动性能
resistance to high-temperature 耐高温性
resistance to-high-temperature test 耐热性能试验
resistance to humid ag(e)ing exposure 抗吸潮老化性;耐湿老化性
resistance to hydraulic shock 抗液冲性;耐液压冲击性
resistance to impact 抗冲击(性);抗冲击强度;耐碰击性;耐冲击(性);碰撞阻力;冲击阻力;抗冲击性能
resistance to impact fracture 抗冲击破坏强度;耐冲击破坏性
resistance to indendation 抗压入强度
resistance to indentation 压痕阻力
resistance to industrial corrosion 耐工业腐蚀性
resistance to insecticide 对杀虫剂的抗药性
resistance to intergranular corrosion 抗晶间腐蚀性
resistance to lateral bend(ing) 抗侧弯强度;抗侧弯能力;侧弯阻力;抗侧弯
resistance to light and weather test 抗光及抗风

化试验
resistance to lime 抗石灰性能
resistance to longitudinal displacement 纵向位移阻力
resistance to lubricating oil 耐润滑油性
resistance to moisture 防潮性;耐潮
resistance to moisture test 抗湿度试验
resistance to nitrocellulose lacquer 耐硝基漆性
resistance to oil 抗油性
resistance to organic solvent 耐有机溶剂性
resistance to oscillation 抗振荡阻力
resistance to outflow 流出阻力
resistance to overbake 抗过烘烤性
resistance to overhead flooding injury 耐涝性
resistance to overturning 整体稳定性;抗倾覆(性)
resistance to ozone 耐臭氧性
resistance to ozone cracking 抗臭氧裂纹性;耐臭氧龟裂性
resistance-to-penetration test 抗穿透性试验
resistance to pesticide 耐农药性;对农业的抵抗性
resistance to pit corrosion 抗坑状腐蚀能力;抗点蚀性
resistance to plant roots 耐植物生根性
resistance to poison 耐毒性
resistance to polishing 抗耐磨光能力;抗磨光性
resistance to pressure 耐压性;耐压力
resistance to propulsion 推进阻力
resistance to pull-off 抗拔性;耐拔性
resistance to punching shear 抗冲剪能力
resistance to radiative contaminants 耐辐射污染性
resistance to radioactive contaminants 耐放射性污染物
resistance to rolling 滚动阻力
resistance to rupture 抗破裂性;抗破坏性;抗裂强度;抗裂能力;破裂阻力
resistance to salt 耐盐性
resistance to salt spray fog 抗盐雾性
resistance to scrubbing 耐擦性
resistance to scuffing 耐擦伤性
resistance to seawater 耐海水的
resistance to separation 阻裂;抗分裂
resistance to settlement 沉陷阻力
resistance to severe heat 耐高温性
resistance to shear 抗剪性能;抗拉强度;抗剪强度;抗剪力
resistance to shock 抗振动;抗冲击(性);耐振性;耐振荡;耐冲击性
resistance to simulated weathering test 模拟耐候试验
resistance to skid(ding) 抗滑溜能力;抗滑溜性
resistance to slag erosion 抗渣性
resistance to sliding 抗滑稳定性;抗滑(动)能力;耐滑移性;滑移阻力;滑动阻力
resistance to softening at high temperature 耐高温软化性
resistance to soiling 耐污性
resistance to solvent 耐溶剂性
resistance to spalling 抗破裂性
resistance to sparking 击穿电阻;耐击穿性
resistance to spirit 耐酒精性能
resistance to spittle 耐唾沫性
resistance to splash water 抗溅水性
resistance to staining 抗污染性
resistance to static indentation 耐静态压痕性
resistance to stone-chipping 抗石击性
resistance to stream flow 耐河川径流能力
resistance to stress crazing 耐应力碎裂性
resistance to sulfate attack 抗硫酸盐侵蚀性
resistance to sulfide tarnishing 抗硫蚀
resistance to sweat 耐汗性
resistance to swelling 耐(溶)胀性
resistance to swelling by solvent 耐溶剂溶胀性
resistance to swelling by water 耐水泡胀性
resistance to swing action 立转阻力
resistance to tack development 抗显黏性
resistance to tearing 耐撕裂性
resistance to temperature change 耐温度变化能力
resistance to temperature change test 耐温度变化性试验
resistance to temperature test 抗温度试验
resistance to temperature variation 耐温差性
resistance to tempering 回火抗力
resistance to tensile stress 抗拉强度;抗拉力

resistance to tension 抗拉(伸)能力;抗拉力
resistance to termite 抗白蚁性
resistance to termite attack 防止白蚁侵蚀
resistance to thermal cycling 热循环阻力
resistance to thermal decomposition 耐热裂解性;耐热分解性
resistance to thermal shock 耐热振性;耐热冲击性
resistance to torsion 抗扭转强度;抗扭性
resistance to tracking 爬电电阻
resistance to traction 牵引阻力
resistance to traffic abrasion 车辆磨耗阻力
resistance to turnover 抗倾覆
resistance to ultraviolet 耐紫外线性
resistance to ultraviolet discoloration 耐紫外线变色性
resistance to uplift 上托力的抗力
resistance to vibration 抗振性
resistance to washing 耐冲刷能力;耐洗(涤)性
resistance to water 抗水阻力;抗水性;耐水性
resistance to water-flow 水流阻力
resistance to water pressure 耐水压性
resistance to water vapo(u)r permeability 蒸汽渗透阻力
resistance to water vapo(u)r transmission 蒸汽气密性
resistance to wear 耐磨(性);抗磨损;耐磨性;耐磨损性;磨损度;磨耗阻力
resistance to wearing 耐磨损性
resistance to weather 耐受天气自然作用的能力;耐气候性;耐候性;气候稳定性;全天候的;抗风化(能力);抗风化性
resistance to weathering 抗风化作用;抗风化(能力);抗候化性;耐候化性能;耐风蚀性(能);耐风蚀强度;抗风化性能;抗风化强度
resistance to wind flap 耐风抽性
resistance to yellowing 抗泛黄性;抗发黄性
resistance to yield 抗流塑能力;抗屈服能力;沉陷阻力
resistance transducer 电阻传感器
resistance transformer 电阻变换器
resistance tube 管状电阻
resistance-tuned oscillator 电阻调谐振荡器
resistance type flowmeter 电阻式流量计
resistance type furnace 电阻式熔炉
resistance type high voltage meter 电阻式高压表
resistance type liquidometer 电阻式液位计
resistance-type temperature detector 电阻式温度检测器
resistance unit 阻力单位;电阻件
resistance value 阻抗值;热阻值;土基或路面材料抗力(值)
resistance valve 阻流阀;电阻阀
resistance variation method 变阻法
resistance voltage divider 电阻分压器
resistance water head 阻力水头
resistance welder 电阻焊接机;电阻焊机
resistance weld(ing) 接触焊;电阻焊(接)
resistance welding die 电阻焊模
resistance welding electrode 电阻焊电极
resistance welding machine 电阻(焊)接机
resistance welding time 电阻焊通电时间;接触焊通电时间
resistance weld mill 电阻焊管机
resistance weld pipe 电阻焊管
resistance winding 线绕电阻;欧姆线圈
resistance wire 电阻丝
resistance wire extensometer 电阻丝引伸仪
resistance-wire strain ga(u)ge 电阻丝应变仪
resistance-wire wave ga(u)ge 电阻丝测波仪
resistance wire wave probe 电阻丝式浪高仪
resistance-yield curve 荷载屈服曲线
resistance zone 抗拔区
resistant 抗滑力
resistant asbestos packing sheet 石棉耐油橡胶板
resistant bank 抗冲(刷)河岸
resistant block 阻力地块
resistant component 耐蚀成分
resistant crystalline region 稳定的结晶区
resistant earth pressure 土抗力
resistant form 抗污染型
resistant insect 抗性害虫
resistant lining 防腐衬里
resistant material 耐磨蚀材料;耐腐蚀材料
resistant metal 耐蚀金属

resistant mineral 耐风蚀矿物;稳定矿物
resistant plant 抗性植物
resistant proof 抗性试验
resistant rock 难钻岩石;耐蚀岩石
resistant spring 耐久弹簧;长寿命弹簧
resistant structure 稳定结构
resistant to alkali 抗碱的
resistant to bending 抗弯曲的;抗弯刚度的
resistant to chemicals 耐化学制品性;耐化学药品性
resistant to corrosion 不锈的;耐腐蚀的
resistant to elevated temperatures 耐高温的
resistant to hydrostatic pressure 抗静水压力的
resistant to oxidation 抗氧化的
resistant to pollution 抗污染的
resistant to rust 抗锈(蚀的);耐蚀的
resistant to saponification 耐皂化作用的
resistant to sunlight 耐日晒的
resistant to tarnishing 抗锈(蚀)的
resistant to wheel-chairs 防轮椅冲撞的;耐轮椅磨损的
resistate 抗蚀残积
resist compression 阻力
resist-dye 套染
resist-dyeing 防染色工艺
resisted rolling 几横摇
resist-etchant 保护膜抗蚀剂
resist film 抗蚀膜
resist heat 耐热
resistibility 抵抗力
resisting 耐久的;有抗力的;抵抗的
resisting force 阻力;抗力
resisting function 阻力函数
resisting medium 阻尼媒质;阻尼介质;黏性介质
resisting moment 阻(抗)力矩;抗倾覆性;抗扭(力)矩;抗(力)矩;抵抗矩
resisting pollution by green plants 绿化抗污
resisting power 阻力;抵抗力
resisting pulleys type cold-drawing machine 阻力轮式冷拉机
resisting shear 抗剪切;抗剪力
resisting steel 耐热钢
resisting strength 抗强;抵抗力
resisting torque 抗力矩;抗转矩;抗扭转;抗扭力;抗力力矩
resisting wall system 反力墙系统
resisting weathering 抗风化(能力)
resisting wind force 抗风力;抗风化(能力)
resistive attenuator 电阻衰减器
resistive bed 高电阻层
resistive component 电阻分量
resistive conductor 高阻导线
resistive coupling 电阻耦合
resistive cut-off frequency 负阻截止频率
resistive exercise 抗阻运动
resistive flowmeter 电阻流量计
resistive formation 高(电)阻层
resistive hearth furnace 炉床电阻炉
resistive heater 加热电阻丝;电阻加热器
resistive load 有功负载;阻性负载;电阻性负载
resistive muffler 阻性消声器
resistive network modeling 电阻网络模型
resistive rippling instability 电阻性皱纹不稳定性
resistive side 阻力边
resistive unbalance 电阻性不平衡
resistive-wall amplifier tube 吸收壁放大管
resistivity 电阻系数;抵抗性力;抵抗能力;比阻;电阻率
resistivity against water 耐水性
resistivity computerized tomography 电阻率层析成像
resistivity contour 等电阻线
resistivity curve 阻力曲线
resistivity device 电阻测井装置
resistivity dipmeter 电阻式测斜仪
resistivity exploration 电阻勘探法;电阻查勘法
resistivity for second layer 第二层电阻率
resistivity index 电阻率指数;电阻率
resistivity log(ging) 电阻率测井;电阻测探;电(阻)法测井
resistivity log plot 电阻率测井图
resistivity measurement 电阻率测量
resistivity method 电阻(率)法;比电阻法
resistivity of adjacent formation 围岩电阻率
resistivity of different materials 不同土层的电阻率

resistivity of flushed zone 冲洗带电阻率
resistivity of liquid in the flushed zone 冲洗带液体的电阻率
resistivity of mud 泥浆电阻率
resistivity of mud cake 泥饼电阻率
resistivity of mud filtrate 泥浆滤液电阻率
resistivity of petroleum 石油的电阻率
resistivity profile 电阻率剖面;电阻测量剖面
resistivity prospecting 电阻率法勘探
resistivity reconnaissance 电阻勘测
resistivity survey(ing) 电阻勘探(法);电阻率测量;电阻(法)测量
resistivity survey method 电阻勘探法
resistivity test device 电阻率测量装置
resistivity well log(ging) 电阻率测井
resist lustre 露花光泽
resistor 电阻器
resistor assembly tester 电阻检验器
resistor bulb 电阻测温包
resistor-capacitor oscillator 阻容振荡器;电阻电容振荡器
resistor-capacitor unit 阻容元件
resistor coating 电阻器涂料
resistor colo(u)r code 电阻器色标
resistor core 电阻芯子
resistor coupled transistor logic 电阻耦合晶体管逻辑电路
resistordivider 电阻分压器
resistor dryer 电阻干燥器
resistor furnace 电炉
resistor house 舱面电机操纵室
resistor microelement 微型电阻元件
resistor network 电阻网络
resistor oven 电阻烘箱
resistor starting 变阻器起动
resistor termination 电阻端接法
resistor transistor logic 电阻晶体管逻辑
resist pattern 抗蚀图形
resist permalloy 高电阻坡莫合金;强磁性铁镍合金
resist printing 防染印花
resistron 光阻摄像管
resist sagging 防流挂
resist technique 掩蔽涂装法
resist valve arrester 阀电阻避雷器
resist ware 露花器皿
resite 微树脂煤【地】;不熔酚醛树脂;丙阶(酚醛)树脂
resite layer 微树脂煤分层【地】
resite stage B B 阶酚醛树脂
resitol 乙阶(酚醛)树脂;半溶酚醛树脂
resiweld 环氧树脂类黏结剂;环氧树脂类黏合剂
resize 改变尺寸;重定尺寸
resizing 重定尺寸
resleeve 换汽缸套;换衬套
resmooth 重新弄光
resmoothing 重新弄光;再镀;复镀
resmousness 树脂度
resocollinite 树脂镜质体
resoiling 再污染;还土(工程);(河渠库底植草防冲的)重铺表土
resojet 脉动式喷气发动机
resol 甲阶酚醛树脂;可溶酚醛树脂
resold grain 返销粮
resolidification 再凝结
resolite 乙阶酚醛树脂;半溶酚醛树脂
resolving power test target 分辨率测试图
resoluble method 再溶解接着法
resolute 分矢量
resolution 消解;离析;解决;鉴别率;清晰度;重新溶解;拆分;分辨作用;分辨力;分解度;分辨率;分辨度;分辨本领;分辨(力)
resolution ability 分辨能力
resolution angle 鉴别角;分辨角
resolution bars 清晰度测试条
resolution block 分辨力楔;清晰度楔
resolution capability 鉴别力;分辨能力;分辨本领
resolution capacity 分辨能力
resolution cell 分辨单元
resolution characteristic 分辨特性
resolution chart 鉴别率板;测试图(像);分辨率图;分辨率测试卡;分辨力图表;分辨测试图
resolution detail 细节清晰度;细节分辨力
resolution diagram of multipopulation 多重总体分解图
resolution distance 解像距离
resolution element 分离元件;分辨(单)元
resolution ellipse 分辨率椭圆
resolution error 解算误差;分辨误差
resolution estimation 分辨估计
resolution factor 分辨系数;分辨因子
resolution filtering 分辨率滤波
resolution function 分辨函数
resolution graph 分解图
resolution in azimuth 方位角分辨能力;方位角分辨率
resolution in bearing 方向分辨率;方位分辨率;方位分辨力
resolution in line direction 水平方向清晰度
resolution in range 距离分辨率;距离分辨(力)
resolution into components 分解成分量
resolution into factors 因数分解
resolution into factors of the resources situation 资源形势因素分析
resolution limit 分辨限度;分辨极限
resolution limiting 分辨率限度
resolution line 解像线;析像线
resolution loss 分辨损失
resolution meter for photosensitive materials 光敏材料分辨仪
resolution noise 分圈噪声
resolution of a vector 向量的分解;矢量解算
resolution of board 董事会决议
resolution of camera 摄影机分辨率
resolution of composite crystal 组合晶体分辨率
resolution of displacement 位移分解
resolution of down measured crystal 下测晶体分辨率
resolution of emulsion 乳剂解像力
resolution of eye 人眼分辨能力
resolution of forces 力的分解;求分力
resolution of laser ranging 激光测距分辨率
resolution of lens 镜头解像力
resolution of operator 运算子的分解
resolution of photographic(al) material 感光材料分辨力
resolution of photography 摄影分辨率
resolution of polar to Cartesian 极坐标—直角坐标转换;极坐标—直角坐标换算;极坐标—直角坐标变换
resolution of single crystal 单晶体分辨率
resolution of strain 应变分解;应变的分解
resolution of stress 应力分解
resolution of tensor 张量分解
resolution of the identity 单位分解
resolution of unity 单元分析;单元分解
resolution of up measured crystal 上测晶体分辨率
resolution of vectors 矢量分析
resolution of velocity 速度分解
resolution on a design 一个设计决定;决定设计
resolution passed 通过的议案
resolution pattern 清晰度测试图;分解力测试图
resolution performance 分解性能;分辨性能
resolution power 析像能力;分解力;分辨(能)力;分辨本领
resolution principle 分解原理;分解法则
resolution proof graph 分解证明图
resolution proof tree 分解证明树
resolution raster pattern 分辨率测试图
resolution ratio 析像系数;分解系数;分辨率;分辨系数;分辨率
resolution reading 分辨力读数
resolution refutation 归结反驳
resolution-refutation graph 分解反驳图
resolution requirement in the primary image 三基色析像能力
resolution response 析像系数;清晰度响应;分辨力响应;分辨力响应
resolution-retrieving 分辨率恢复
resolution sensitivity 分辨灵敏度
resolution square 分辨方块
resolution stability testing method 分辨率稳定度测试法
resolution stage 溶解期
resolution standard 清晰度标准
resolution target evaluation 分辨率测试板评定
resolution test 分辨率测试;分辨检验
resolution test bar 分辨力测试条

resolution test chart 分解力测试卡
resolution testing method 分辨率测试法
resolution test pattern 鉴别率板
resolution theorem 分解定理
resolution theorem-proving 解析定理证明
resolution theory 分解理论
resolution threshold 分解力阈;分辨阈
resolution time 消散时间
resolution-time correction 分辨时间校正
resolution to borrow 借款决定
resolution to ground 对地分辨率
resolution unit 分辨单位
resolution view 分辨图
resolution wedge 清晰度楔;分解力楔形图;分辨力楔形图
resolution width 分辨宽度
resolvable balanced incomplete block design 可解析平衡不完全区组设计
resolvable number 可分解单元数
resolvable picture element 可析像素;可辨像素
resolvable spots 可分辨点(数)
resolvant equation 预解方程
resolved by negotiation 协商解决
resolved echo 清晰回声
resolved motion rate control 解析运动速率控制
resolved shear stress 分解切应力;分解剪应力
resolved-time migration 分解时间偏移法
resolvent 消散剂;消散的;预解式;分物
resolvent kernel 预解核
resolvent operator 预解算子
resolvent set 预解集
resolver 快存区;解析器;解算装置;解算器;溶剂;求解仪;分析器;分解区;分解器
resolver feedback system 分解反馈系统
resolver potentiometer 分解器电位计
resolver rotor 解算仪转子;求解仪转子
resolving 解析过程
resolving ability 解算能力;鉴别能力;分解力;分辨;辨别能力
resolving agent 拆解试剂
resolving block 分解力楔
resolving capacity 分辨
resolving cell 分解单元;分辨波胞
resolving chromatic power grating 色分辨率光栅
resolving coefficient 分解系数
resolving device 分解设备
resolving function 分辨函数
resolving index 清晰度指数;分解力指数;分辨指数
resolving limit 分解力极限;分辨限度;分辨极限
resolving mechanism 解算机构
resolving potentiometer 解算电位计
resolving potention meter 分解电位计
resolving power 析像能力;解像能力;分辨;鉴别本领;分离本领;分解力;分辨率;分辨本领
resolving power chart 解像能力测视图;分辨率测试卡
resolving-power curve 分辨率曲线
resolving power limit 分辨能力极限
resolving power multiplier 分辨力倍增器
resolving power of emulsion 乳剂分辨率
resolving power of image 影像分辨率
resolving power of lens 物镜分解力
resolving power of objective lens 物镜分辨率
resolving power response 分辨力响应
resolving power spectrogram 分辨率谱图
resolving power target 分辨率检验图
resolving power test 清晰度测试;分辨力测试
resolving power test target 分辨率板
resolving shear stress 分切应力
resolving stage 好转期
resolving system 分解系
resolving time 解算时间;分解时间;分辨时间
resonance 谐振;共振(吸收);叩响
resonance absorber 共振吸收器
resonance absorber effect 共振吸收体效应
resonance absorption 谐振吸收;共振吸收
resonance absorption peak 共振吸收线峰
resonance action 共振作用;谐振作用;共鸣作用
resonance amplifier 共振放大器;调节放大器
resonance amplitude 共振振幅
resonance and ultrasonic tests 共振和超声试验
resonance angular frequency 谐振角频率
resonance-antiresonance method 谐振反谐振法
resonance area 共振区

resonance arm 谐振臂
resonance bandwidth 共振带宽
resonance bond 共振键
resonance box 共振箱;共鸣箱
resonance braking 变阻器制动
resonance brazing 接触黄铜钎焊
resonance bridge 谐振电桥
resonance broadening 共振展宽;共振加宽
resonance butt weld 谐振式对接焊
resonance capture 共振俘获
resonance chamber 谐振空腔;共振室
resonance characteristic 共振特性(曲线);谐振特征
resonance circuit 谐振电路
resonance coil 谐振线圈;共振线圈
resonance coil antenna 谐振圈天线
resonance condition 谐振状态;共振状态;共振条件
resonance constant 谐振电路常数;振荡回路常数
resonance conveyer 共振输送机
resonance cross-section 谐振截面;共振截面
resonance current 谐振电流
resonance curve 谐振曲线;共振曲线
resonance curve area method 谐振曲线面积法
resonance damper 动力吸振器
resonance detector 共振检测器
resonance device 谐振装置
resonance dispersion 共振色散
resonance domain 共振域
resonance effect 谐振效应;共振作用;共振效应
resonance efficiency 谐振效率
resonance efficiency meter 谐振频率计
resonance elastometer 共振弹性计
resonance energy 共振能
resonance energy transfer 共振能量传递;能量共振转移
resonance enhancement 共振增强
resonance escape factor 逃脱共振因子
resonance escape probability 逃脱共振几率
resonance excitation of betatron oscillations 自由振荡共振激发
resonance fatigue characteristic 共振疲劳特性
resonance field 谐振场
resonance fluorescence 共振荧光
resonance frequency 谐振频率;谐振模式;共振频率;共鸣频率
resonance frequency indicator 振动式频率计
resonance frequency meter 谐振式频率计;共振频率仪
resonance hummer 谐振蜂鸣器
resonance hump 共振峰
resonance impedance 谐振阻抗
resonance indicator 谐振指示器
resonance inductor 谐振感应器
resonance inspection of or pile 共振法测桩
resonance instrument 谐振式仪表
resonance integral 共振积分
resonance intensifier 谐振增强器
resonance interference 谐振干扰
resonance lamp 共振灯
resonance level 共振(能)级
resonance-level spacing 谐振能级间距
resonance line 谐振(传输)线;共振线
resonance line satellite 共振伴线
resonance matching 谐振匹配
resonance measurement 谐振测量法
resonance method 谐振法;共振法;(测混凝土强度的)共振测定法
resonance model 共振模式
resonance modulus 共振模量;动力模量
resonance neutron activation 共振中子活化
resonance neutron detector 共振中子探测器
resonance of spectrum 共振光谱
resonance orbit 共振轨道
resonance oscillation 谐振荡;共振荡;共振摆动
resonance oscillatory circuit 谐振振荡电路
resonance overlap 共振叠加
resonance peak 谐振曲线峰值;谐振峰(值);共振峰
resonance pendulum 共振摆
resonance phenomenon 谐振现象;共振现象
resonance plastic 共振塑性
resonance point 谐振点
resonance potential 谐振电压;谐振电位;共振电位;共振电势
resonance probe 谐振探测;共振探针

resonance problem 共振问题
resonance pulsator 共振脉动器
resonance quality 共振性
resonance radiation 共振辐射
resonance radiometer 谐振辐射计
resonance Raman spectrometer 共振拉曼光谱仪
resonance Raman spectroscopy 共振拉曼光谱学
resonance range 共振范围;谐振区;谐振范围
resonance ratio 谐振放大系数;谐振放大率;谐振(比)率
resonance reaction 共振反应
resonance region 谐振区;共振区
resonance resistance 谐振电阻
resonance response 谐振响应
resonance response principle 共振反应原理
resonances 共振态
resonance satellite 共振卫星
resonance scattering 共振散射
resonance screen 共振筛
resonance screening 共振筛分
resonance self-focusing 共振自聚焦
resonance sharpness 谐振锐度;共振锐度
resonance sieve 共振筛
resonance spectrometer 共振谱仪;共振分光计
resonance spectrum 谐振频谱;共振谱
resonance speed 共振转速
resonance state 共振状态
resonance step-up 谐振升压
resonance table vibrator 共振振动台
resonance technique 共振测试技术
resonance test 共振试验
resonance test method 谐振测试法;(弹性模量的)共振试验法
resonance theory 共振学说;共振理论;共鸣学说
resonance transfer 共振转移
resonance transformer 谐振变压器;共振变压器
resonance trough 共振槽
resonance tube 共鸣管
resonance type capacitor resistance oscillator 谐振式阻容振荡器
resonance type voltage regulator 谐振式调压器
resonance variation method 谐振变化法
resonance vibrating screen 共振筛分机;共振筛
resonance vibration 谐振振动;共振振动
resonance voltage 谐振电压
resonance wave 谐振波(浪);共振波
resonance wave coil 谐振波线圈
resonance wave length 谐波长;共振波长
resonance wavemeter 谐振式波长计
resonance width 共振宽度
resonance wood 共振材
resonant 共振的;共鸣的;叩响的;反响的
resonant absorbent material 共振吸收材料;共振吸音材料;共振吸声材料
resonant absorber 共振吸收器;谐振吸减器
resonant aerial 谐振天线
resonant air space 共振腔
resonant amplifier 谐振放大器
resonant amplitude 共振振幅
resonant antenna 谐振天线;共振天线
resonant-bar transducer 棒状共振换能器
resonant capacitor 谐振电容器
resonant cavity 谐振(空)腔;共振腔;空腔谐振器;空腔共振器
resonant chamber 谐振腔;空腔共振器
resonant circuit 共振电路
resonant circuit impedance 谐振电路阻抗
resonant circuit type frequency indicator 谐振式频率指示器
resonant coefficient 共振系数
resonant column 共振柱
resonant column method 共振柱法
resonant column test 共振柱试验
resonant column triaxial test apparatus 共振柱三轴(试验)仪
resonant combustion 脉冲燃烧
resonant coupling 谐振耦合
resonant diaphragm 共振膜片
resonant discharger 谐振放电器
resonant earthed system 谐振接地系统
resonant electric(al) spark 共鸣火花
resonant element 空腔共振器
resonant excitation 谐振激励;共振激发
resonant factor 共振系数

resonant feeder 谐振馈线
resonant flip-flop 谐振触发器
resonant four-terminal network 谐振四端网络
resonant frequency 谐振频率;共振频率
resonant frequency meter 共振频率仪
resonant frequency modulation 谐振频率调制
resonant frequency response 谐振频率响应
resonant gap 谐振隙;谐振空隙
resonant-gate transistor 谐振控制式晶体管
resonant grounded system 谐振接地系统
resonant insulating apparatus 谐振式绝缘装置
resonant interference 共振干扰
resonant internal reflection 共振内反射
resonant iris 谐振膜片
resonant-iris switch 谐振膜转换开关
resonant jet 脉动式(空气)喷气发动机
resonant line 谐振线;共振线
resonant line oscillator 谐振线振荡器
resonant-line tuner 谐振线调谐器
resonant load 谐振荷载
resonant mode 谐振模;共振模态;共振模;共振方式;共振波模
resonant-mode analysis 共振模分析
resonant model 共振模式
resonant-mode number 谐振模数
resonant mode structure 共振模结构
resonant modulus 共振模量
resonant optic(al) Faraday's rotator 法拉第光谐振旋转器
resonant optic(al) schlieren system 共振光学纹影仪系统
resonant orbit 共振轨道
resonant oscillation 共振振荡
resonant oscillation in harbo(u)r 港内谐振
resonant particle 共振粒子
resonant pile driver 共振打桩机
resonant pile driving method 共振打桩法
resonant point 谐振点;共振点
resonant radiation sound 共振辐射声
resonant Raman scattering 共振拉曼散射
resonant-reed relay 谐振簧片继电器
resonant-reed tachometer 谐振舌片式测速发电机;谐振簧片式转速计;谐振簧片式转速测量计
resonant resistance 谐振电阻
resonant ring 谐振环
resonant satellite orbit 调谐卫星轨道
resonant shunt 谐振分路
resonant slit 谐振缝
resonant sound absorber 谐振吸声体;共振吸声器;共鸣消声器
resonant spectrum 共振光谱
resonant speed 共振速度
resonant structure 共振结构
resonant test 共振试验
resonant tester 谐振试验机
resonant transfer 共振转移
resonant transformation 谐振变换
resonant transformer 谐振变压器
resonant transition 共振跃迁
resonant type instrument 谐振式仪表
resonant type muffler 共振型消声器
resonant type vibrating pile driver 共振式沉桩机
resonant vibration frequency 谐振频率
resonant vibrometer 共振示波仪
resonant voltage step-up 谐振式的电压升高
resonant whip 共振抖动;共振拍打
resonant window 谐振窗;共振窗口
resonated impedance bond 谐振式阻抗连接变压器
resonating cavity 共振空腔
resonating circuit 谐振电路
resonating group 共振群
resonating piezoid 谐振用石英晶体片;谐振压电石英片
resonation 共鸣作用
resonator 谐振腔;共振腔;谐振器;共振器;共鸣器
resonator banking stop 谐振器止块
resonator damper 谐振阻尼器
resonator entropy 共振器熵
resonator fixing clamp 谐振器固定夹
resonator gap 谐振器间隙
resonator grid 谐振器栅(极)
resonator guard 谐振器护罩
resonator insulator 谐振器绝缘板
resonator magnet 谐振器磁铁

resonator mirror 共振腔反射镜
resonator mode 共振器模
resonator mode of oscillator 振荡器的谐振模
resonator-type silencer 谐振式消声器
resonator wall 共振墙
resonator wavemeter 共振器波长计
resonatron 谐振管
resonoscope 谐振示波器;共振示波器
resorb 再吸收;重吸收
resorcin 雷琐辛
resorcin blue 间苯二酚蓝
resorcin brown 间苯二酚棕
resorcin dark brown 间苯二酚深棕
resorcine-azobenzene sulfonic acid 间二酚偶氮苯磺酸
resorcine yellow 间苯二黄
resorcinism 间苯二酚中毒
resorcinol 间苯二酚
resorcinol acetate 单乙酸间苯二酚酯
resorcinol adhesive 间苯二酚黏合剂;间苯二酚树脂胶;间苯二酚胶黏剂
resorcinol diglycidyl ether 间苯二酚二缩水甘油醚
resorcinol formaldehyde 间苯二酚甲醛
resorcinol formaldehyde adhesive 间苯二酚甲醛胶
resorcinol formaldehyde latex 间苯二酚甲醛配合胶乳
resorcinol formaldehyde resin 间苯二酚甲醛树脂
resorcinol formaldehyde resin adhesive 间苯二酚树脂胶
resorcinol glue 间苯二酚胶
resorcinol monoacetate 单乙酸间苯二酚酯
resorcinol monobenzoate 间苯二酚单苯甲酸酯
resorcinol resin 雷锁辛树脂;间苯二酚树脂
resorcinol resin adhesive 间苯二酚树脂胶黏剂
resorcinol resin glue 间苯二酚树脂胶
resorcinol test 间苯二酚试验
resorcinol-type synthetic adhesive 间苯二酚类合成黏合剂
resorption 再吸作用;再吸收;回吸;熔蚀
resorption border 熔蚀边;熔蚀边
resorption refrigerating system 再吸收式制冷系统
resorption rim 熔蚀池
resorption schlieren 熔丝异离体
resorptive effect 吸回效应
resort city 游览城市;观光城市
resort condominium 名胜地区的共有公寓房屋
resort hotel 游览旅馆
resorting capacity 改编能力(调车场)
resorting capacity of lead track 牵出线改编能力
resorting yard 改编车场
resort place 休养胜地
resort town 休养城镇;休养城市;疗养城镇
resort waste 旅游地废物
resound(ing) 反响;回声
resource 资源
resource accounting 资源使用费用
resource allocation 资源分配;资源分布
resource allocation microprocessor 资源分配微处理机
resource analysis 资源分析
resource application controller 资源应用控制程序
resource appraisal 资源评价;资源估价
resource area of land 土地资源区域
resource base 资源基础
resource-based planning 资源规划
resource centralized management 资源集中管理
resource conservation 资源保护
Resource Conservation and Recovery Act 资源保护与恢复法案
resource contention 资源冲突
resource cost 资源成本
resource deallocation 资源再分配
resource depletion allowance 资源耗竭补偿税
resource depletion by mining 资源开发正常枯竭
resource development 资源开发
resource economics 资源经济学
resource efficiency 资源效率
resource enhancement 提高资源质量
resource environmental hydrogeology 资源环境水文地质学
resource equalization tax 资源平衡税
resource exploitation 资源开采
resource exploration 资源勘探
resource flow 资金流量

resource function 资源函数
resource information 资源信息
resource-intensive industry 资源密集型工业
resource inventory 资源清单
resource investigation 资源调查
resourcelization of wastewater 污水资源化
resource management 资源管理
resource management information system 资源管理情报系统
resource management program(me) 资源管理程序
resource manager 资源管理程序
resource map 资源(分布)图
resource of control region 控制区资源量
resource of dust 尘埃来源
resource of information 情报资源
resource-oriented industries 面向原料产业
resource out of place 不适当的资源
resource partitioning 资源分配
resource pool 资源库
resource protection and recovery act 资源保护和回收法案
resource quality enhancement 资源质量提高
resource queue 资源排队
resource reclamation 资源回收利用
resource reconnaissance 资源勘探
resource recovery 资源回收(利用)
resource recovery of sludge 污泥资源回收
resource recovery plant 资源回收厂;回收车间
resource recovery system 资源回收系统
resource recovery system of solid waste 固体废物资源化系统
resource recovery technique of solid waste 固体废物资源化技术
resource recovery utilization 资源回收利用
resource repudiation 拒绝偿还债务
resource reserves 资源保护区
resources allocation 资源配置
resources allocation and multiple project scheduling 资源分配及多计划调度
resource scheduling method 限制资源
resource science 资源科学
resources control block 资源控制块
resources crisis 资源危机
resources environment(al) planning 环境资源规划
resources environmental policy 资源—环境政策
resources exploitation planning 资源开发规划
resources for production 生产资源
resource sharing 资源共享
resource sharing executive 资源共享执行程序
resource sharing network 资源共享网络
resources marine investigation 海洋资源调查
resources of manpower 人力资源
resources of solid wastes 固体废物资源化
resources of the continent 大陆架资源
resources-providing country 提供资源的国家
resources reserved to be used 储备待用
resources reserve gap 有缺口
resource status modification 资源状况修改
resources tax 资源税
resources trade 资源贸易
resource survey 资源调查
resource survey satellite 资源考察卫星;资源调查卫星
resources war 资源战
resource utilization 资源利用
resource waste 资源浪费
resource-wasting 浪费资源
resow 补种
resowgrass 补播草
respace 重新隔开
respacing of cross ties 调整轨枕间距
respective error 人差【测】
respect the reputation 讲究信誉
respirable aerosol mass monitor 呼吸式粉尘监测仪
respirable dust 可吸入性粉尘;可吸气的尘埃
respirable dust content 可呼吸的粉尘含量
respirable suspended particulates 可因呼吸进入人体的悬浮微粒
respiration action 呼吸作用
respiration anaerobic 厌氧呼吸
respiration calorimeter 呼吸测热计

respiration heat 呼吸热
respiration intensity 呼吸强度
respirator 口罩;呼吸(保护)器;防毒面具;防尘面罩
respiratory acidosis 呼吸性酸中毒
respiratory alkalosis 呼吸性碱中毒
respiratory device 吸氧装置;吸空气装置;防毒设备
respiratory exercise 呼吸操练
respiratory protective device 呼吸防护器
respiratory quotient 呼吸比
respiratory ratio 呼吸比
respiratory tube 呼吸管
respiratory wave 呼吸波
respirometer 呼吸仪;透气性测定仪
respite 停付
respond 应答
respond beacon 应答信标
respondentia 冒险借款;货物抵押借款
respondentia bond 货物抵押借款契约
respondent superior 主要被告
responder 应答器;应答机(器);回答器
responder action 应动作
responder assisted surveilance radar 应答式监视雷达
responder beacon 应答式无线电指向标;应答器信标;应答(式无线电)信标;应答(器)信标
responder coding 应答器编码
responder link 应答信道
responder loss 应答器损耗
respond in damages 赔偿损失
responding curve 响应曲线
responding factor 响应因子
responding range 响应范围
responding surface 响应表面
respond opportunity 可回答状态
respond time 响应时间
response 响应;应答;反应
response adequacy 反应适当性
response alternate program(me) communication block 应答交替程序通信块
response amplitude 反应振幅;反应幅度
response analysis program(me) 应答分析程序
response analyzer 反应分析器
response budgeting 反应预算法
response characteristic 响应特性;灵敏度特性曲线
response circuit 反应回路
response coefficient 响应系数;反应系数
response colo(u)r relation 色感度
response curve 响应曲线;应答曲线;灵敏度(特性)曲线;反应曲线
response curve of controlled plant 调节对象响应系数
response data 响应数据
response detector 响应探测器
response diagram 响应图;反应图;博德图
response envelope spectrum 反应包线谱
response equation 反应方程(式)
response excursion 扰动振幅
response factor 响应因子;响应因数
response field 响应场
response frequency 响应频率
response frequency characteristic 响应频率特性
response function 响应函数;反应函数
response header 应答头;应答首部;应答标题
response instruction 响应指令
response key 反应键
response lag 响应滞后
response latency 反应潜伏期
response limit 反应极限;灵敏限
response magnitude 反应量
response measurement 反应测量
response message 响应信息;应答信息;应答信号;应答消息
response method 响应法
response mode 响应模式;应答方式
response modification factor 反应修正系数
response of harmonic excitation 简谐激振反应
response orientation 响应调整能力
response parameter 反应参数
response period 反应期
response point 反应点
response pulse 回答脉冲
response quantity 反应量
responser 询问机应答器;响应器;应答机(器)
response range 响应范围

response rate 响应速率;应答率;回答率
response ratio 响应比;反应系数;反应比
response set 反应定势
response shape 响应形状
response signal 响应信号
response space 反应空间
response spectrum 响应谱;反响谱
response spectrum envelope 反应谱包线
response spectrum intensity 反应谱烈度
response spectrum method 反应谱法
response spectrum with double peaks 双峰反应谱
response spectrum with multiple peaks 多峰反应谱
response spectrum with single peak 单峰反应谱
response speed 响应速度
response strategy 回应战略
response strength 反应力
response surface 响应(曲)面;感应面;反应面
response surface analysis 反应面分析
response surface experiment 反应面实验
response surface method 响应面法
response surface methodology 反应曲面分类研究法
response time 响应时间;吸合时间;应答时间;反应时间;暴雨集流时间
response time distribution 反应时间分布
response time history 反应时程
response time of a measuring instrument 测量仪器的响应时间
response to inhibitors 对抑制剂的感应性
response to initial excitation 初始激振反应
response to irrigation 灌溉效应;灌溉反应
response to non-periodic(al) excitation 非周期激振反应
response to open flame exposure 明火试验反应
response to periodic(al) excitation 周期激振反应
response to random excitation 随机激振反应
response transform 响应变换;反应变换
response type 反应型
response unit 应答单元;应答部件
response value 响应值
response variable 响应变数;反应变量
response variance 响应方差
responsibilities and interests of an enterprise 权责利
responsibility 职责;责任能力;可信赖性
responsibility accountant 责任会计
responsibility accounting 责任会计
responsibility accounting system 责任会计制度
responsibility area 责任者项
responsibility audit 责任审计
responsibility budget 责任预算
responsibility cost 责任成本
responsibility crew system of passenger train 旅客列车包车制
responsibility for a crime 刑事责任
responsibility for crossing accidents 道口事故责任
responsibility of manufacturer 制造者职责
responsibility range 职责范围
responsibility system 责任制(度);岗位责任制
responsibility system for operation 经营责任制
responsibility system for production 生产责任制
responsibility system in economy management 经济管理的责任制
responsibility system of assets management of railway bureau 铁路局资产经营责任制
responsibility system of cost and fee management 成本费用管理责任制
responsibility system of environmental protection 环境保护责任制度
responsibility system of hierarchical object cost management 分级目标成本管理责任制
responsibility system with the aim of balancing depleting and growing forest resources 森林资源消长目标责任制
responsibility zone 责任区
responsible 负责的
responsible accident 责任事故
responsible bidder 驰誉投标商;可靠投标者;可靠投标商;可靠投标人
responsible institution 主管部门
responsible operator position 责任话务员座席
responsive 有反应的
responsive algorithm 应答算法

responsive bandwidth 响应带宽
responsive bid 响应性投标
responsive design 响应设计
responsive device 响应器
responsiveness 响应性;易起反应
responsive planning 响应设计;反应计划法
responsive plans for unexpectedly oil spilling 溢油应急计划
responsive tender 应答标
responsive time constant 响应时间常数
responsivity 响应性;响应率;反应性
responsor 应答机(器)
respool 重绕
respooler 复绕机
re-spooling machine 再绕机;复绕机
resprayed body 重复喷漆过的车身
resquaring 剪切成方形
resquaring shears 精确剪板机
ressaut 墙面突出部分;墙面凸出部分;圆棒形线脚;双弯凹凸弧形线脚
rest 盈余;架;剩余(部分)
restabilization 再稳定(作用);重新稳定(作用)
restacking 捣垛
restaff 重新配备人员
rest and build up strength 休养生息
rest and recuperation 休整
rest area 休息场所
restart 重新启动;重新开始;再(次)启动
restartable 可重新起动的
restart address 再启动地址
restart and recovery technique 重新启动和恢复技术
restart call 重新启动调用
restart capability 再启动能力;重新启动能力
restart checkpoint 再启动检查点
restart condition 再启动条件
restart control 再启动控制
restart data set 重新启动数据集
restart facility 重新启动功能
restarting 再引弧;重新启动;再(次)启动
restarting ability 重新启动能力
restarting a weld 再引弧
restarting capability 再次启动能力
restarting injector 再生喷射器
restarting procedure 再启动程序
restarting routine 再启动程序
restart instruction 再启动指令;再启动命令
restart key 重新启动键
restart point 再启动点;重新启动点;重入点;程序再启动点
restart procedure 再启动过程;重新启动过程
restart routine 重新启动程序;重算程序;重启动例行程序
restart sorting 重新排序
restart sorting point 再排序点
rest assured 深信不疑
restatement 重编报表
restaurant 餐馆;饭馆;饭店
restaurant car 餐车
restauranteria 自助餐馆
restaurant floor drive system 餐厅楼板驱动系统
restaurant level 餐厅层
rest bar 支承梁
rest base 架底;刀架基面
rest bend 支架弯头;端弯头;直角弯头;直角弯管(接头);托座弯头
rest button 支承销;搁钉
rest center 疗养中心
rest condition 原始条件;静止条件
rest-day 休息日
rest density 静止密度
rest diamond bit 回收金刚石钻头
rest duration 静止持续时间
rested reduction gear 套接式减速齿轮
rest energy 静能
rest exponent integer 余留指整数
rest frame 静止坐标系;静止框架
rest frequency 静止频率
rest home 休养院;疗养院
rest house 小客栈;旅途旅馆;休养所;招待所;途中旅馆
restiform body 绳状体
rest image 静止图像
resting area 休息区

resting barrel 固定滚筒
resting basin 静止池
resting contact 静止接点;静止触点
resting form 休眠型;潜伏型
resting frequency 载频
resting guide 固定导板
resting mark 停息痕
resting membrane potential 膜静电位
resting nucleus 静止核
resting order 取决于市价订单
resting oxygen uptake 安静氧吸取量
resting period 静止期;间歇周期;存活时间;放置期
resting phase 休止期;休静
resting place 休息处
resting plate 支承板
resting point 支持点
resting pool 休息池
resting position 静止位置;静止状态
resting potential 静态电位
resting rung 爬梯的休息梯级
resting stage 静止期
resting state 休眠状态;静止状态;安静状态
resting time 静止时间;静止期
resting trace 停息迹
resting value 安静值
restitute 退赔
restitution 要求恢复原状的诉讼;归还;解调;等时调制;重原
restitution base 测图基线
restitution coefficient 回弹系数;弹力恢复系数
restitution coordinatograph 立体坐标测仪
restitution nucleus 重建核
restitution of advance payment 退回预付款
restitution of model 模型重建
restitution of property 物件归还;财物归还
restitution of skid resistance 抗滑能力的恢复
restitutor 恢复器
rest magma 残余岩浆
rest mass 静(止)质量
rest metabolism 安静代谢
restock 重新进货
restocking 再引种;再放养;立木重建
restocking forest land 重建林地
restorability 可恢复性;恢复性能
restorable system 可恢复系数
restoration 修复;再造;建筑复原;恢复;复原
restoration after failure 故障复原
restoration characteristic 恢复力特性
restoration drawing 修复图
restoration due to a failure 故障复原
restoration map of pal(a)eodrainage 古水系复原图
restoration map of pal(a)eomorphology 古地貌复原图
restoration of aquatic ecology 水生生态修复
restoration of a weld 焊缝修补
restoration of a weld by reheating 焊缝再加热修复
restoration of balance 平衡恢复
restoration of hydrologic(al) data 水文资料修复;水文资料还原;水文资料修复
restoration of site 恢复现场
restoration of soil 土壤的恢复
restoration of water 水的复原
restoration of water quality 水质修复
restoration of waters 水体复原
restoration premium 补额保险费
restoration priority 恢复线路的优先等级
restoration ratio 恢复率
restoration stage 恢复期
restoration to the original condition 恢复原状
restoration work 复原工作;修复工作;修缮工程
restore 修复;恢复
restore circuit 恢复电路
restore circulation 恢复循环
restore control 复原控制
restored acid 再生酸;回收的酸
restored building 重新修复建筑物;重新翻修建筑
restore deacidize 还原
restored energy 回收能量;复原能量
restored rubber 再生胶
restored taxa 复原分类群
restore-pulse generator 时钟脉冲发生器
restorer 迁原设备;回复者;修补物;修建者;恢复设备;恢复器;复位器
restorer line 恢复系
restore screen 恢复屏幕
restore to traffic 恢复行车
restore voltage 重建电压
restore wetland 修复湿地
restoring clogged filter 恢复阻塞滤池
restoring component 复原构件
restoring couple 回复力偶
restoring current 恢复电流
restoring force 回复力;恢复力;稳定力;复原力
restoring force behavio(u)r 恢复力特性
restoring function 恢复力函数
restoring lever 复位杠杆
restoring lever length 复位杠杆长度
restoring link 复位连杆
restoring lost circulation 恢复冲洗液循环
restoring method 恢复法;复原法
restoring moment 回复力矩;恢复力矩;稳定力矩;复原力矩
restoring of topsoil 表土复原
restoring pipeline capacity 恢复管道输送能力
restoring relay 回复继电器
restoring signal 恢复信号
restoring spring 复位弹簧
restoring to normal position 恢复定位;返回定位
restoring torque 回原位力矩
restoring traffic 恢复通车
restoring voltage 恢复电压
restoring water 回收水
restor sleep during winter 冬季休息或睡眠
restowage 捣载作业;捣载【港】;翻舱
restowing 翻舱
rest pad 支承块
rest parameter 剩余参数
rest period 休息时间
rest pier 支墩;支架墩
rest pin 支承销;支承销
rest plate 支承板
rest-pocket 袖珍
rest point 静止点
rest position 息止位;不工作位置
rest position of plummet 垂球静止位置
rest potential 静止电位;残余电位
rest precipitation tank 绝对静止沉淀池(废水处理)
rest quarter 生活间;生活区
restrain 约束
restrainable key 色带控制键
restrain block 防振挡块
restrained beam 约束梁;固端梁;控制梁
restrained cooling shrinkage 限制冷缩
restrained end 约束端;固定端
restrained expansion 有限扩展;固端板;约束膨胀
restrained factor 约束条件
restrained from rotating 不能转动的
restrained gyroscope 框架陀螺仪
restrained joint 固定接头
restrained joint and prestressed non-cylinder pipe 固定接头与预应力非圆筒管
restrained maximization(minimization) problem 约束最大(最小)化问题
restrained motion 受限运动
restrained pile 约束桩;钳制桩
restrained post 约束柱;传力柱
restrained shrinkage 约束收缩
restrained slab 端固板;约束板
restrained structure 约束结构
restrained support 约束支承
restrainer 限制器;酸洗缓蚀剂;阻尼器
restraining anchor cable 约束锚索
restraining coil 阻尼线圈;制动线圈
restraining dolphin 限制墩
restraining effect 约束效应;牵制效应;牵引效应
restraining force 约束力
restraining moment 固端力矩;约束弯矩;约束力矩
restraining order 禁令
restraining order to control pollution within prescribed time 责令限期治理
restraining pressure 约束(压)力
restraining spring 限制弹簧;保险弹簧
restraining structure 约束结构
restraining system for parallel suspension cable 平行吊索的加劲系统

restrain of displacement 位移约束
restraint capability 栓固能力
restraint coefficient 约束系数
restraint column 约束柱;传力柱
restraint condition 固定条件;约束条件
restraint crack 约束裂缝
restraint deformation 限制变形;约束变形
restraint device 制约装置
restraint fixing 约束固定装置;非承重固定件
restraint forces 约束力
restraint joint 约束节点
restraint layer 约束层
restraint modulus 约束模量
restraint moment 约束力矩;固(定)端力矩
restraint of alienation 转让限制
restraint of concrete 混凝土的约束
restraint of loads 负载的约束
restraint of princes 入港的禁止;出港的禁止(海上保险)
restraint of subgrade 地基约束;路基约束;地基限制;路基限制
restraint of trade 贸易约束;贸易限制
restrain trade 限制贸易
restraint relay 制约式继电器;牵制式继电器
restraints on financial budgets 财务预算约束
restraint stiffness 约束劲度
restraint system 约束系统;监督系统
restraint system on deck 甲板集装箱绑扎系统
restraint test 固定试验;栓固试验
restraint to slab warping 板翘曲约束
restraint welding 约束(下)焊接;抑制焊接
restress corrosion 再应力下锈蚀
restressing 副应力;次应力
restretch 重新拉直;重铺
restrict 限制;限定;约束
restricted adhesive 限制使用的胶黏剂;限用胶黏剂
restricted air-bled 限制空气喷口
restricted air bled carburetor 受制渗气式汽化器
restricted anchorage 特许锚地
restricted and unrestricted random walk 限制与非限制随机走动
restricted area 限航区;禁地;受限制区域;缩径孔数
restricted articles 限制物品
restricted assets 限制用途的资产;受限制的资产
restricted basin 隔离盆地;受限盆地
restricted bit 小水眼钻头
restricted cash 限用途的现金;限定用途的现金;受限制的现金
restricted channel 限制性航道;限制航道;狭窄河槽;狭浅河槽;窄浅航道;束狭河道
restricted clearance 限制净空;限制净距
restricted credit 限制议付信用证
restricted current funds 限定流动基金
restricted data 内部资料
restricted depth water area 有限水深水域
restricted diffusion 限制扩散
restricted diffusion chromatography 限制扩散层析法
restricted dimensional analysis 狭义量纲分析
restricted direct product group 限制直积群
restricted discharge 节制排料
restricted distribution 限制性分配
restricted-draft ship 限制吃水船
restricted earth fault protection 限制接地故障保护
restricted earth protection 限制接地保护
restricted endorsement 限定背书
restricted entry type of surge tank 束口式调压塔
restricted extension line 受限分机线
restricted flow 限止流
restricted format 限制格式
restricted freight 限运货物
restricted fund 限定用途基金
restricted fuzzy automaton 受限模糊自动机
restricted height 限制高度
restricted holonomy group 限制完整群
restricted homogeneous holonomy group 限制齐次完整群
restricted hour maximum demand indicator 限时式最高需量指示器
restricted industrial district 工业专用(地)区
restricted internal rotation 被阻内旋转
restricted isomorphism 有限类同象;不完全类质同象
restricted jet 受限射流

restricted list of bidders 投标者的限价单；拟订投标人名单；推选标商名单；拟订投标人清单
restricted market 限制市场
restricted mode 限速模式
restricted navigation zone 禁航区
restricted negotiable letter of credit 限制议付信用证
restricted negotiation 限制议付
restricted occupancy 限制占有
restricted operation 限制性运转
restricted orifice surge tank 束口式调压塔；阻力孔口式调压塔
restricted passage 限制通过
restricted passing sight distance 限制停车视距；制约超车视距
restricted platform deposit 局限台地沉积
restricted platform facies 局限台地相
restricted proper motion 限制性自行
restricted quarter 限制空间
restricted random sampling 限制性随机抽样；限制的随机抽样法；限定随机抽样法
restricted reach 狭窄河段；窄浅河段
restricted residential district 居住专用地区；住宅专用区；居住限用地区
restricted retained earnings (or surplus) 受限制的留存收益
restricted root 限制根
restricted sequential procedure 限制序贯程序
restricted service 有限航区
restricted share 附限股制
restricted space 有限空间；局限空间
restricted speed 限制速度；限速
restricted speed signal 速度限制信号
restricted speed way 限制行车速度的道路
restricted speed zone 限速区
restricted step method 约束步长法
restricted stock 受限制的股票
restricted stopping sight distance 制约停车视距
restricted surplus 限定用途的公积金
restricted tender 受限制的投标；限制性投标；限制投标；特约比价
restricted traffic 受管制的车流
restricted Turing machine 受限图灵机
restricted turns 受制转弯
restricted use 限制使用
restricted use cable 专用电缆
restricted variable length code 限制可变长代码
restricted vibration 限制振动；有限振动
restricted view 受限制的视野
restricted visibility 能见度受限
restricted voyage 限制航行
restricted water depth 有限水深
restricted water resistance 限制航道阻力
restricted waters 限制水域；狭窄水域；狭浅水域；窄浅水域；受限制水域
restricted waterway 束狭水道；限制水道；狭窄水道；窄浅水道
restricted zone 禁区
restricter 限制产量者
restricting 限燃层；表面覆层
restricting catch 限定捕获量
restricting graphitization 限制石墨化
restricting signal 限制信号；速度限制信号
restricting speed indication sign 限速指示器
restriction 限制；限度；节流(口)
restriction amplifier 节流放大器
restriction arc 限制弧
restriction area 节流面积
restriction area of industry 工业专用地区
restriction board 限速牌；限速标
restriction characteristic 节流特性
restriction chock 节流器
restriction control 节流控制
restriction controller 限制控制器
restriction crack 阻碍型裂缝；拉裂
restriction efficiency 限定效率
restriction factor 限制因子
restriction in wagon loading 限制装车
restrictionism 限制主义
restriction jet 缩孔喷嘴
restriction loss 节流损失
restriction lost circulation 恢复循环
restriction modification system 限制修正系统
restriction of credit 信贷管制

restriction of draught 吃水限度
restriction of expenditures 限制费用
restriction of flow 液流限制
restriction of luxury imports 限制奢侈品进口
restriction of output 产量限制
restriction of trade 贸易限制
restriction of visibility 能见度限制
restriction orifice 限制孔口；节流孔板
restriction ratio 限制比
restriction screw 节流螺旋杆；节流尖头螺钉；带螺纹针阀
restriction section 阻力区
restriction section of carrying capacity 通过能力限制区间
restriction sign 限制标志
restriction signal 减速信号
restriction site 限制性内切位点；限制部位
restriction sleeve 节流短管
restrictions on authority 授权的限制
Restrictions on Hazardous Substances 欧盟关于危险物质限定的指令
restrictions on length 长度限制
restrictions on shipping space clause 限制配船部位的条款
restrictions on sulfur dioxide emission 二氧化硫排放限制
restrictions on sulphur dioxide emission 二氧化硫排放限制
restrictions on the parameters 参数约束
restriction versus opposition to restriction 限制和反限制
restrictive business practices 限制性商业惯例；限制性经营办法
restrictive business tariff 限制营业税率
restrictive clause 限制(性)条款
restrictive coating 铠装；涂料；护套层
restrictive commercial practice 限制性商业措施
restrictive condition 限制条件
restrictive covenant 限制性契约；限制性条款；土地保护性契约；限制使用财产协定
restrictive credit policy 限制性信贷政策
restrictive effect 限制性效应
restrictive endorsement 限制性背书
restrictive flow-regulator 节流式流量调节阀
restrictive host 限制性寄生
restrictive labo(u)r practice 限制性劳工措施
restrictive length 节流长度
restrictive license 限制性许可证
restrictive normality assumption 限制正态性假定
restrictive policy 限制性政策
restrictive policy of a discriminative nature 歧视性的限制政策
restrictive position 限制位置
restrictive practice 限制性措施
Restrictive Practices Court 限制贸易实施法庭
restrictive ring seal 阻流环密封
restrictive sight distance 限视距；约束视距
restrictive signal 限速信号
restrictive speed 限制速度
restrictive temperature 限制性温度
restrictive trade 限制贸易
restrictive trade practices 限制性交易措施
restrictive zone 限速区
restrictor 限制器；限流器；闸板；节汽门；节气门；节气阀；节流器；节流阀
restrictor check-valve 止回节流阀
restrictor ring 限流环
restrictor screw 节制孔螺钉
restrictor size 节流面积
restrict power 约束力
restrict sale 限制销售
restrike 再触发；再点火；重燃
restrike of arc 再点火；电弧再触发；重燃电弧
restriking 整形锻压；再闪击；矫正锻压
restriking die 校正模；修整模
restriking voltage 再闪击电压；再起弧电压
restringing 更换导线
rest room 休养所；休息室；洗手间；值班休息室
rest-rotation grazing 休闲转牧；休闲轮牧
restructing algorithm 重构算法
re-structuralization of trade 贸易结构的调整
restructured debt 调整后债务
restructuring 城市或街坊重建；重组
restructuring graph 重构图

restructuring of enterprises 企业改制
restructuring program(me) 重构程序
restructuring technique 再构成技术
restructuring the international economic framework 重建国际经济结构
rest shoe 爪块
rest state 安静状态
reststrahlen band 剩余射线(谱)带
reststrahlen filter 剩余射线滤光器
reststrahlen plate 剩余射线滤光板
reststrahlen ray 剩余射线
reststrahlen region 剩余射线区
reststrahlen spectrum 余辉带光谱
reststrahlung plate 剩余辐射滤光板
rest switch 静止开关
rest time control 停机时间控制
restuff 重新填塞
rest value 安静值
rest water level 静水平面
resubdivide 重新分区
resubdivision 再分区
resubgrade 重筑路基
resublimation 再升华
resublimed iodine 再升华碘
resue method of mining 选别采矿法
resue stoping 选别回采
resuing 剥离围岩回采法
resuing development 削壁采准
resuing method 选别采矿法
resulfurize 再用硫处理；再硫化
resulphurize 再硫化
resulphurized steel 加硫钢
result 效果；结果；成效；成果；产物
result address 结果地址
resultant 消元式；终结式；结式；结果的；合量；合成物；生成物
resultant acceleration 合加速度
resultant accuracy 最后精度；总准确度；总精度
resultant action 总作用
resultant action potential 综合动作电位
resultant admittance 总导纳
resultant amplitude 合成振幅
resultant cohesion 合成的黏结力
resultant colo(u)r 总的色位移
resultant compressive stress 合成压应力
resultant couple 总力偶；总力矩；合(成)力偶
resultant current 合成流
resultant curve 综合曲线；合力曲线
resultant cutting movement 合成切削运动
resultant cutting pressure 合成切削压力
resultant cutting speed 合成切削速度
resultant displacement 总位移；总变位；合成位移
resultant draft 实际牵伸
resultant drag 总阻力
resultant error 总误差；综合误差；真误差；结果误差；合成误差
resultant fault 综合障碍
resultant field 结果域
resultant force 合力
resultant force coefficient 合力系数
resultant frequency characteristic 总频率特性
resultant gear ratio 总传动比
resultant gradient 综合坡度；合成坡度
resultant ground displacement 合成地面位移
resultant identifier 结果标识符
resultant impedance 合成阻抗
resultant intensity 总强度
resultant law 结合分布律
resultant line capacity 线路最终通过能力
resultant load 总荷重；总荷载；合成荷载；合成负荷
resultant magnification 总放大率
resultant maps of geophysical exploration 物探成果图
resultant matrix 结式矩阵
resultant metal 产品金属
resultant moment 合力矩
resultant motion 合成运动
resultant movement 合成运动
resultant of an algebraic(al) equation 代数方程的结式
resultant of forces 合力；力的合成
resultant of reaction 反应生成物；反应产物
resultant of velocities 速度的合成
resultant pitch 总节距；合成节距

resultant plot of metallo-metric survey 金属量测量结果图
resultant pressure 总压力;合压力;合成压力
resultant prism 综合棱镜
resultant quantity 合成量
resultant radial force 辐向合力
resultant rake 合成前角
resultant shock wave 合成冲击波
resultant signal 总信号
resultant strain 合成应变
resultant stress 合成应力;合应力
resultant temperature 合成温度
resultant tool force 切削合力
resultant tool force on the tool flank 后刀面上的合力
resultant unbalance force 合成不平衡力
resultant unbalance moment 合成不平衡力矩
resultant vector 综合向量;合矢量;合成向量;合矢量
resultant velocity 合(成)速度
resultant wind 合成风
resultant wind force 合成风力
resultant wind force coefficient 合成风力系数
resultant wind velocity 合成风速
result buffer 结果缓存
result data item 结果数据项
result element 结果元素
result figure of electric(al) method 电法勘探成果图件
result function 结果函数;目标函数
resulting 结果的
resulting beat 组合音拍
resulting curve 综合曲线;合力曲线
resulting dimension 最终尺寸
resulting display 最终显示
resulting force 合力
resulting from lack of tolerance 因缺乏耐性所致
resulting indicator 运算结果指示符;结果指示符
resulting instantaneous rotational vector 有效瞬时旋转矢量
resulting mixture 合成混合物
resulting moment 合力矩
resulting record of reserve ratification 储量审批结果
resulting stress 合应力;合成应力
resulting tape 结果带
resulting two-dimensional stress state 合成二维应力状态
result map 成果图
result maps of geochemical prospecting 化探成果图
result maps of geophysical prospecting 物探成果图
result matrix of mapping 成果矩阵
result of decision 决策产物
result of detection 探伤结果
result of experiment 实验结果
result of hydraulic test 水压试验结果
result of hydrogeologic(al) investigation 水文地质调查成果
result of interpretation of ground EM method 地面电磁法解释结果
result of investigation 勘察工作成果
result of operation concept 经营成果概念
result of reserve calculations 储量计算结果
result of scientific research 科研成果
result of test 检验结果
result phase 结果状态
result reduction 结果处理;测量结果归化
result register 结果寄存器
results and material of handing 上交成果资料
results management 业绩管理法
results of a poll 投票结果
results of inbred variety cross 顶交结果
results of rating 考核成绩
results of regional geologic(al) survey 区域地质调查成果
results of remote sensing applications for comprehensive environmental investigation 遥感环境综合调查应用成果
results of remote sensing applications for energy exploration 遥感水文地质调查应用成果
results of remote sensing applications for engineering geologic(al) survey 遥感地震地质调查应用成果

results of remote sensing applications for exogenic geologic(al) hazard survey 遥感外生地质灾害调查应用成果
results of remote sensing applications for hydrogeologic(al) survey 遥感工程地质调查应用成果
results of remote sensing applications for mineral exploration 遥感矿产勘查应用成果
results of remote sensing applications for structural geology and geologic(al) mapping 遥感构造地质及地质制图应用成果
results of surveying and mapping 测绘成果
results of tests 试验结果
results of water analysis 水分析成果
result success 成绩
result value 实验值
result verification 实验证明
resume 履历(表);重新开始
resume load 重新输入
resuming about hours 大约继续进行小时
resumption 再取回
resumption of navigation after river thawing 开江
resumption of payment of suspended funds 停付资金的恢复支付
resumption of work 复工
resuperheat 再过热
resuperheater 再(过)热器
resuperheating 重加热;再加热
resupply 再补给
resupply voucher 再补给单
resurface 修整工具;铺新路面;检修炉衬;重做面层;重新车削
resurfacer 表面修整器
resurfacing 修复路面;路面重铺;面层修补;重做面层;重修面层;重修面;重铺路面;表面修整
resurfacing agent 再敷面剂
resurfacing by addition 加料翻修
resurfacing by replacement 补料翻修
resurfacing by welding 再次堆焊
resurfacing material 表面修整材料;路面翻新材料
resurfacing of rail 钢轨焊补;焊修钢轨
resurfacing of tyres 轮胎重旋
resurgence 复流;再现;风暴潮余振
resurgence of pest 虫害回升
resurgent caldera 复活破火口
resurgent water 再现水;再生水
resurge phase of water hammer 水锤的再现期
resurrected peneplain shoreline 复露准平原岸线
resurrection glass fibre 高硅氧玻璃纤维
resurvey 再勘查;再勘测;再测量;重测
resurvey of line 线路复测
resuscitator 复苏器
resuspending 再悬浮
resuspension 再悬移(作用);再悬浮(作用);再悬
resweld 树脂黏结的层压板
reswitching 改换投资
resynchronization 再同步;重新同步;二次同步
resynthesis 再合成
retable 祭坛后部的高屏;祭坛后的壁板
retackiness 回黏性
retail banking market 小额银行业务市场
retail business 零售业务
retail business district street 零售商业街(道)
retail center 零售商业中心
retail department 小卖部;门市部
retail deposits 小额零星存款
retail depot 换瓶点;零售供应站
retail district 零售商业区
retailing business district 零售商业区
retailing sales area 零售场地
retailing sphere 零售区
retail market 零售市场
retail network 零售网
retail nursery 零售苗圃
retail price 销售价格;零售价格
retail rate of exchange 小额交易汇价
retail sale 零售;门市
retail sales area 零售商品场地
retail sales department 门市部
retail sales office 分销处
retail salesroom 零售处
retail service 提供生活服务的小型设施
retail shipment 零星运输
retail shop 零售商店;分销店

retail shopping district 零售区
retail store 零售(商)店
retail trade 零售商业
retain 夹持;保留
retainable drill pipe 可保留的钻管
retainage 滞留量;保留金(合约条款);进程支付的扣留总额
retain a portion of foreign exchange 留成外汇
retain a portion of profit to be placed at the disposal of the enterprise 利润留成
retained amount 筛余量;保留金
retained austenite 残余奥氏体;残留奥氏体
retained correspondence 保留信件
retained data 保留数据
retained dose 持留剂量;保留剂量
retained earning 留存盈余;留存收益;保留收益
retained earnings statement 营业盈余表
retained fund 自留资金
retained gasket system 滞留衬垫法
retained heat 保存的热量
retained height of structure 结构物的挡高度
retained import 保留输入
retained income 留存收入;保留收益
retained material 筛余粗料
retained material on the sieve 筛上剩余物料
retained neutral polarized relay with heavy-duty contacts 有极加强继电器
retained ownership 保留所有权
retained percentage 保留金百分比;留筛百分率;筛余率;筛留百分数;筛留百分率;残留百分率;保留百分率
retained percent revenue on over-revenue 超收加成收入
retained porosity 残留孔隙度
retained profit 留存利润;留成利润;待分配利润
retained profit per capita 人均留利
retained saving 保留储蓄
retained strength 焦砂强度;残留强度
retained thing 留置物
retained water 存留水;吸着水;阻滞水;滞留水;持面水;持留水
retained water level 存留水位;集水水位;蓄水水位
retained water pressure 存留压力;集水压力;蓄水压力
retained weight 筛余重;保留重
retainer 限位器;轴承保持架;止动装置;止动器;固位体;集料(又称骨料);拦阻物;护圈(拦阻物);定位器;抵住物;不透水岩;保持器;保持架
retainer belt 夹送皮带;保持带
retainer bolt 止动螺钉
retainer clip (打结器的)绳夹的夹头
retainer collar 轴环;钎子挡环
retainer cup 套杯
retainer disc 压绳盘
retainer flange 固定凸缘;挡住凸缘
retainer gland 止动密封套
retainer neutral polarized combination relay 组合保持继电器
retainer nut 防松螺母
retainer oil seal groove 挡油环油槽
retainer pliers 挡圈钳
retainer production packer 悬挂采油封隔器
retainer ring 固定环;夹持环;护圈
retainer screw 护圈螺钉
retainer seal 护圈密封
retainer spring 压紧弹簧
retain funds for depreciation 留成折旧基金
retaining arm 制动臂
retaining backwall 子墙;堆墙
retaining bar 止动杆;(转动的)插入件
retaining basin 蓄水池;拦洪池
retaining block 支块;挡土坎
retaining board 衬板;(冷压机的)垫板;临时挡板
retaining bolt 固定螺栓;留挂螺栓;扣紧螺栓
retaining box 挡水闸
retaining cable 定位缆索
retaining cap 抵盖
retaining capacity of reservoir 热储存能力
retaining catch 闭锁犁子
retaining circuit 保持电路
retaining click 止爪
retaining clip 固定夹
retaining coil 吸动线圈;保持线圈
retaining collar 保持环

retaining contact 保持接点
retaining crib wall 木垛填石挡土墙
retaining current 吸持电流
retaining dam 蓄水坝;拦水坝;挡水坝
retaining device 止动装置;止动器;锁定装置;挡住装置
retaining device of ship chamber 承船厢栓锁装置
retaining differential mechanism 制动差动机构
retaining dike 围堤;挡水堤
retaining grillage 挡土格床
retaining ledge 止动台肩
retaining lever catch 把手挡
retaining lock 蓄水闸;挡水闸
retaining log 圆挡木
retaining mechanism 制动机构;锁紧机构
retaining member 锁紧件
retaining nut 制动螺母;制动螺帽;固定螺母;锁紧螺母;锁紧螺帽
retaining part of the extra profit 超额利润分成
retaining pawl 制动爪;止爪;止动爪
retaining pay 保留工资
retaining piece 制动片
retaining pin 固定销;插入销;制动销(钉);制动箱;制动栓;止动销
retaining plate 定位板;制板;支承板;固定板
retaining possession 保留占用权
retaining ring 护圈;定位环;止推环;支撑环;扣金;扣环;抗磨环;卡环;护环;锁紧环;挡圈;车轮扣环;绑环
retaining ring bar 扣环钢条
retaining ring bolt 扣环螺钉
retaining ring for bearing 轴承挡圈
retaining ring for bore 钻孔扣环
retaining ring for shaft 柱身扣环
retaining ring for shaft shoulder 轴身挡圈
retaining ring for tyre 轮箍扣环
retaining ring groove (轮箍的)卡环(沟)槽;扣环槽
retaining ring lip 扣环唇
retaining rope 系留缆
retaining screen 阻滞筛;留粒筛
retaining screw 固定螺栓;固定螺钉
retaining shield 锁紧式防尘罩
retaining shoulders 护肋
retaining sieve 留粒筛
retaining size 筛余粒径
retaining snap ring 制动弹簧垫圈;固定开口环
retaining spring 止动弹簧
retaining structure 挡土结构;支挡结构
retaining structure with anchor 锚碇式挡土结构
retaining structure with bracing 支撑挡土结构
retaining suture fixator 留置缝线固定夹
retaining time 保留时间
retaining valve 止逆阀;止回阀;单向阀
retaining variety 保留品种
retaining wage 保留工资
retaining wall 挡土墙;支承墙;拥壁;拦水坝;挡土墙
retaining wall design 挡土墙设计
retaining wall drain 挡土墙排水
retaining wall drain pipe 挡土墙排水管
retaining wall for falling stone 拦石墙
retaining wall for rock-fall 栏石墙
retaining wall shear failure 挡土墙剪切破坏
retaining wall transition 挡土墙过渡段
retaining wall with anchored bulkhead 锚碇板式挡土墙
retaining wall with anchored tie-rod 锚杆式挡土墙
retaining wall with stepped back 踏步式挡土墙;步式挡土墙;背面台阶式挡土墙
retaining wall with surcharge 上有土坡的挡土墙;超载挡土墙
retaining wing 挡土翼墙
retaining works 蓄水工程;支挡;拦水工程;挡土构筑物;挡土工程;挡水工程
retaining zone 同步区域;同步区
retain varying portions of profit 超额利润分成
retally 重理货
retally charges 重新理货费;翻仓费
retally note 重理单
retamp 再夯实
retangular scale 直尺
retan(ning) 复鞣
retanning agent 复鞣剂
retapering 修钎
retapping (桩工的)复打

retardance 延迟性
retardance coefficient 阻滞系数;滞流系数
retardant 阻滞剂;迟凝剂
retardant combustion treatment 迟延燃烧处理
retardant die-away curve 缓凝剂消失曲线
retardant pattern simulation model 按地面图阻滞的模拟方法
retardation 延迟量;阻滞;滞后;制动;减速时期;减速度;缓凝作用;缓聚作用;推迟量;迟滞差;迟延;迟差
retardation angle 滞后角
retardation area 减速面积
retardation axis 延迟轴
retardation basin 滞水池;滞洪水库
retardation coefficient 滞留系数;扼流系数
retardation coil 迟滞线圈;扼流圈
retardation controller 减速控制器
retardation device 减速装置
retardation efficiency 制动效率;减速效率
retardation factor 延迟因数;阻滞因数;制动系数;减速因数;迟后因子
retardation field 减速场
retardation formula 制动公式
retardation function 推迟作用
retardation method 阻尼法
retardation-modulated light wave 延迟调制光波
retardation motion 阻止运动
retardation network 推迟网络
retardation of discharge 阻滞系数;排水阻滞
retardation of mean lunar time 平太阴时迟滞差
retardation of mean solar time 平太阳时迟滞差
retardation of phase 相位滞后
retardation of set(ting) 凝结迟延
retardation of solar on sidereal time 恒星时换算平时的减量
retardation of standing 立迟
retardation of tide 潮汐迟滞;潮时滞后;潮时后延
retardation period 减速期
retardation phase 增殖减退期;缓长期
retardation plasticized plaster 缓塑化石膏
retardation plate 延迟板
retardation sheet 波片
retardation spectrum 推迟谱
retardation stage 减速期
retardation test 制动试验;减速试验
retardation theory 延迟理论
retardation time 制动时间;减速时间;延时;延迟时间;滞后时间;慢化时间;推迟时间
retardation to target speed 制动到规定的速度
retardation wedge 减速光楔;减速光劈
retard combustion 延迟燃烧
retarded acid 阻滞酸液;钝化酸
retarded action 推迟作用
retarded action fuse 延发电雷管;延发导火线
retarded admission 延迟进气
retarded baseline 推迟基线
retarded caving 滞后崩落
retarded cement 缓凝水泥
retarded closing 延迟闭合
retarded coagulation 延迟凝固
retarded combustion 延缓燃烧
retarded concrete 缓凝混凝土
retarded control 延迟控制;延迟调节;推迟控制
retarded creep 延迟蠕动
retarded distance 推迟距离
retarded effect 推迟效应
retarded elastic deformation 推迟弹性形变
retarded elasticity 延迟弹性;阻滞弹性;推迟弹性
retarded elasticity function 推迟弹性函数
retarded electrode 抑制电极;阻滞电极
retarded field 推迟场
retarded field and retarded potential 迟滞场与迟滞势
retarded flow 减速流(动);缓慢流动;推迟流
retarded force 制动力;减力
retarded hemihydrate gypsum 缓凝半水合石膏
retarded hemihydrate gypsum plaster 缓凝半水合石膏灰泥
retarded hemihydrate plaster 缓凝半水化熟石膏
retarded intraparticle diffusion 迟滞颗粒内扩散
retarded layer 滞止层
retarded motion 减速运动
retarded offset 推迟偏置
retarded oil-well cement 缓凝油井水泥

retarded oxidation 缓慢氧化
retarded potential 推迟势;推迟电位
retarded reaction 阻滞反应
retarded scalar potential 推迟标量位
retarded set(ting) 缓凝
retarded spark timing 延迟点火正时
retarded spontaneous recovery 推迟自发回复
retarded type equations 推迟方程
retarded velocity 减速度
retarder 延时器;迟延线圈;延迟器;阻滞剂;阻聚剂;减速(制动)器;减速顶;缓行器;缓凝剂;推迟线圈;迟凝剂;车辆减速器
retarder booster 减速辅助器
retarder capacity 缓行器制动能力
retarder classification yard 有缓行器的编组场【铁】
retarder controller 缓行器控制机
retarder control machine 缓行器控制机
retarder-equipped 装有减速器的
retarder-equipped hump 有缓行器装备的驼峰
retarder-equipped yard 有缓行器的编组场【铁】
retarder exit speed 车辆从缓行器出口速度
retarder for alkali-aggregate reaction 碱—集料反应抑制剂;碱—骨料反应抑制剂
retarder for cement 水泥缓凝剂
retarder-impregnated paper 缓凝剂浸渍纸
retarder in closed state 减速器制动状态
retarder in open state 缓行器在缓解状态
retarder in release position 缓行器在缓解位置
retarder in release state 缓行器在缓解状态
retarder in retarding position 缓行器在制动位置
retarder in working condition 缓行器在工作状态
retarder in working state 减速器工作状态;缓行器在工作状态
retarder jaw 缓行器钳;减速器夹钳
retarder location 制动位;缓行器位置
retarder motor 减速马达
retarder operation 缓行器操作
retarder plate 延迟板
retarder pressure 缓行器压力
retarder released 减速器缓解状态
retarder rotor 减速器转子
retarder solvent 阻滞溶剂
retarder spring 阻滞弹簧
retarder stage of retardation 缓行器制动级
retarder thinner 稀释溶剂;阻滞稀释剂
retarder under pressure 缓行器制动状态
retarder valve 减速器阀
retardin 阻滞素
retarding action 阻滞作用;缓凝作用;迟延动作
retarding admixture 缓慢掺和剂;缓凝外加剂;缓凝剂;延缓剂
retarding agent 延缓剂;阻滞剂;缓凝剂
retarding and water reducing admixture 延缓减水剂;缓凝型减水剂
retarding and water reducing agent 延缓减水剂;缓凝型减水剂
retarding basin 滞洪水库;滞洪区;滞洪池;拦洪池;分洪区
retarding braking 减速制动
retarding chain 减速锚链
retarding chain conveyer 阻滞式链板运输机
retarding characteristic 滞后特性
retarding coefficient 阻滞系数
retarding conveyer 限速输送机
retarding corrosion 抑制腐蚀
retarding device 延时装置
retarding disc 制动板
retarding disk conveyer 阻滞式圆盘运输机
retarding effect 抑制作用;阻滞效应;滞洪作用;滞洪效应;减速作用;减速效应;推迟效应
retarding effort 制动力
retarding elasticity 滞弹性
retarding electrode 抑制电极;抑止电极;减速电极
retarding factor 延滞因素
retarding field 迟滞电场
retarding field generator 减速场振荡器
retarding-field oscillator 正栅振荡器
retarding field tube 减速场管
retarding flow 减速流(动)
retarding force 延迟力;制动力;减速力
retarding ignition 延迟点火
retarding ion mass spectrometer 迟滞离子质谱仪
retarding line 延迟线

retarding loss 制动损耗
retarding magnet 制动磁铁
retarding mechanism 减速机构
retarding moment 减速力矩
retarding of growing 延缓生长;迟缓生长
retarding period 减速时期
retarding phase 滞后相位;推迟相位
retarding pool 滞洪水库
retarding potential 抑止电极;减速势;减速电势
retarding potential method 滞后电位法
retarding reservoir 滞洪水库;拦洪水库
retarding setting 缓凝
retarding spark 迟火花
retarding stage 减速级
retarding stage of drying 降速干燥阶段
retarding system 延滞系统;阻尼系统;减速系统
retarding thermal decomposition 延迟热分解
retarding time 延滞时间
retarding torque 制动转矩;制动力矩
retard of tide 迟潮现象;迟潮时间;潮迟
retard position 延迟点火装置
retard transmitter 迟延发送机
retarring 重浇煤沥青;重浇煤沥青
retemper 改变稠度;加水重拌(砂浆等);二次回火
retempered concrete 重拌混凝土;重塑混凝土
retempering 再拌和(砂浆或混凝土);加水重塑;重新回火;(混凝土的)重拌和;二次搅拌;再次回火
retempering of concrete 混凝土加水改变稠度
retempering of mortar 砂浆重新拌和
re-tender 再招标;重新招标
re-tendering for conveyance of goods unloaded 重复运输
retene 惹烯
retening temperature 保留温度
retention 滞留降水;固位法;留存;保留值;保留金;押金
retention ability 滞洪能力;拦洪能力;持水力;分洪能力
retention ages 保存年龄
retention aid 助留剂
retention analysis 持着分析法;残留(量)分析;保留分析(法)
retention barrier 拦油栅
retention basin 拦蓄区;蓄洪水库;储水池;贮留池;调节池;调节槽
retention capacity 滞水容量;保洪能力;拦蓄能力;持水能力;分洪能力
retention capacity of soil 土壤保水能力;土的保水能力
retention characteristic 保留特性
retention class 留置类别;保存类别
retention coefficient 阻滞系数;保持系数
retention colo(u)r 保色性
retention curve 滞留曲线
retention dam 挡水坝
retention dike 壅水丁坝
retention effect 滞留作用;滞洪作用;调节作用;持水作用
retention effect of lake 滞洪作用;湖泊滞洪作用;湖泊调节作用
retention efficiency 去除效率
retention elevation 壅水高程
retention factor 滞留系数;持水系数;保持系数
retention flange 固定法兰(盘)
retention form 固位形
retention fraction 滞留份额
retention fund 保留金;留存金;保留费;留成资金
retentiongram 保留分析图
retention groin 壅水丁坝
retention groyne 壅水丁坝
retention index 滞留指数;持着指数;保留指数
retention lagoon 储[贮]留池
retention level 壅水(水)位
retention mechanism 保留机制
retention mesh 固位网
retention modulus 保留模型
retention money 留存款项;保证金;保留款项;保留金;保留费;押金
retention money bond 保留金保证
retention of color 着色稳定性
retention of lubricity 润滑性能的保持
retention of memory 记忆保持
retention of oil on board 船上保留油量额
retention of ownership 所有权保留

retention of phosphates 磷酸盐固定作用
retention of records recorder 记录的保管
retention of sodium and water 钠水潴留
retention of title clause 物权保留条款
retention of viscosity 黏度的保持
retention parameter 保留参数
retention period 阻滞期;停留周期;停留时期;保留;保存(周)期
retention period check 保存期检查
retention period of grit chamber 沉沙池停留时间
retention phase 阻滞相位
retention pin 止动销
retention pool 拦蓄池;溢液池;澄清池
retention pound 溢液池
retention range 陷落范围;同步区;保持同步范围
retention rate 滞留率;滞洪率;拦洪率;持水率;分洪率;保水率;保留率
retention ratio 留存比率;截留率
retention reservoir 洪水调节池
retention right 扣留权
retention ring 挡料圈
retention screw 止动螺钉;定位螺钉
retention stage 阻滞阶段
retention storage 滞水库
retention strategy 保留策略
retention temperature 保留温度
retention time 阻滞时间;持水时间;保留时间;保持时间
retention time in kiln 窑内停留时间
retention toxicosis 蓄积性中毒
retention valve 止回阀
retention volume 滞留容积;滞留量;保留体积
retention wage 保留工资
retention wall 挡水墙;防水隔墙;外墙防水隔墙
retention water level 壅水(水)位
retentive activity 保留活性
retentive alloy 硬磁性合金
retentive magnetism 剩磁
retentive material 硬磁性材料
retentiveness 剩磁;持水性
retentive power 持水能力
retentive soil 能保持水分的土壤;保水土壤;持水土壤;保持水分的土壤
retentivity 顽磁性;剩磁;保持性;保持力
retest 再试验;重新试验;复试
retesting 重复检验
retesting period 检定周期
retest of materials 材料复验
retest of vessel test plate 容器试板的复试
retest reliability 再测信度
retextured sediment 变形结构沉积物
retgersite 镍矾
rethickening 再次增稠
rethread 修理螺纹;再过丝;重新喂料;重新穿进
rethreading 修理螺纹
rethreading die 螺纹修整板牙
rethresher 二次脱粒装置
rethresher concave 复脱器凹板
reticle 调制盘;丝网;十字线;分度线;标线
reticle adjusting ring 瞄准具调整环;十字丝校正环
reticle alignment 十字线对准
reticle axis 调制盘轴
reticle camera 制版照相机;制板摄影机;网线照相机
reticle chop 调制盘调制
reticle configuration 调制盘结构
reticle illumination 十字丝照明
reticle image 标线影像
reticle infrared system 光栅红外系统
reticle-mosaic system 光栅镶嵌系统
reticle-on-target 十字分划对准目标
reticle pattern 分划板图形
reticle plate 分划板
reticular 网状的
reticular activating system 网状激活系统
reticular density 网(眼)密度;标线密度
reticular drainage 网状排水系
reticular lamina 网状板
reticular layer 网状层
reticular magma 网状填充组织
reticular membrane 网状膜
reticular plate 网状板
reticular polymer 网状聚合体
reticular structure 网状结构;网形结构
reticular substance 网状物质

reticular system 网状系统
reticular zone 网状带
reticulate 似网状的
reticulated 网状的
reticulated ashlar 网状纹琢石;网状纹石面
reticulated bar (与海岸线成交角的)斜沙洲;网状沙洲
reticulated bond 网状砌合
reticulated conveyer 网状输送器
reticulated cracks 网状裂隙
reticulated dressing 网状纹琢面;网状纹修琢(石面)
reticulated duct 网纹管
reticulated element 网纹分子
reticulated glass 网玻璃
reticulated masonry 网状纹砖石工;网状圬工
reticulated mo(u)lding 网状线脚;有网状饰的线脚
reticulated network 环状管网
reticulated pattern 网状图案
reticulate drainage 网状水系
reticulated root piles 网状树根桩
reticulated shell 网壳;网状薄壳
reticulated structure 网(脉)状构造
reticulated texture 网状组织;网状结构
reticulated tracery 网状窗花格
reticulate duct 网纹导管
reticulated vault 网格状拱顶;网状拱
reticulated vein 网状脉
reticulated vitreous carbon 网状玻璃态碳
reticulated work 网状铺砌;斜方砌法;嵌砖混凝土墙;网状物
reticulate evolution 网状进化
reticulate pattern 网纹(状)
reticulate perforation 网状穿孔
reticulate thickening 网状加厚
reticulate vessel 网纹导管
reticulation 小皱纹;网状形成;网状物;方格网
reticulatum opus 嵌方石块的混凝土墙
retic(u)le 焦点板;交叉线;中间掩膜;十字丝;分划板;分度线;标线(片);标度线
retic(u)le adjusting ring 十字丝校正环
retic(u)le adjustment screw 十字丝校正螺旋
retic(u)le alignment 标度线对准
retic(u)le holder 标度线保持器
retic(u)le ring 十字丝环
retic(u)le screen 十字丝网目板
reticuline bar 支承杆间的弯曲连杆;波状弯曲钢筋
Reticulitermes arenincola 沙地散白蚁
Reticulitermes virginicus 南方散白蚁
retiform 网状的
retighten(ing) (螺)重新拉紧;重新固定
retimber 修理木支架;重新支撑
retimbering 重新支撑
retime 重新定时
retime ignition 重调点火定时
retimer 重新定时器
retiming 再定时
retina [复]视网膜;retinas 或 retinae
retina character reader 网膜字符阅读器
retinacs insert 石棉塑料镶块
retinal cone 视锥
retinal illumination 网膜照度
retinal rod 视杆
retinite 树脂石
retinol 松香油
retinyl 视黄基
retip a bit 修整钻头
retipping 修磨
retire 退役;退休;付清本息
retire bonds 收回债券
retired bill 已付讫票据;收回的票据
retired bond 已赎回债券;已收回债券;已回收债券
retired employee 退休雇员
retired line of position 后移位置线
retired pay 退休金
retired stock 已收回注销的股票;已收回股份
retired worker 退休工人
retiree 歇业者
retirement 指令引退;赎回
retirement accounting 退废会计
retirement age 退休年龄
retirement allowance 养老金;退休金;退积金
retirement allowance reserves 退休金准备
retirement and liquidation 报废清理

retirement annuity 退休金
retirement by instalment 分期拨付汇票
retirement communities 退休者居住区
retirement curve 报废曲线
retirement fund 资产置换基金；退休金基金
retirement home 退休住处；退休家庭
retirement hotel 退役旅社；退役旅馆
retirement income 退休收入
retirement income insurance 退休收入保险
retirement life 报废寿命
retirement method 废弃法
retirement of a partner 退伙
retirement of fixed assets 固定资产报废
retirement of outdated processes, techonologies and equipment 淘汰落后工艺、技术和设备
retirement of property 财产退废
retirement pay 退休金
retirement pension 退休金
retirement pension plan 退休金计划
retirement period 收回期
retirement policy 退休方针；退废政策(指折旧)
retirement price 报废价格
retirement rate 赎票率
retirement room 厕所；休息室
retirement system 退休制度
retirement table 退休表
retirement through accident 事故报废
retire shipping documents 赎单
retiring a bill 赎票
retiring partner 退伙人
retiring room 休息室
retool 改进工具；重装；重新装备
retooling 再加工；重组
retort 蒸馏瓶；蒸馏罐；曲径瓶；曲颈蒸馏器；曲颈瓶；炭化炉
retort basket 杀菌篮
retort bench 煤气厂
retort brown coal tar 干馏的褐煤焦油
retort carbon 蒸馏碳；蒸馏罐碳精
retort clamps 曲颈夹
retort coke 蒸馏焦
retort for processing of oil shale 页岩干馏炉
retort furnace 干馏炉；反射炉
retort gas 炭化炉煤气；蒸馏气(体)；干馏煤气
retort glaze 干馏罐内表面的釉层
retort grate 下饲式炉排
retort holder 曲颈甑架
retort house 碳化炉室
retorting 蒸馏法
retorting of shales 页岩干馏
retort lignite 干馏褐煤
retort pine tar 干馏松焦油
retort press 蒸馏罐压制机
retort process 甑馏法；干馏过程
retort resistance 抗蒸馏性
retort sponge 蒸馏罐海绵金
retort tar 干馏的焦油
retort-type incinerator 多室串烧焚化炉；甑式焚化炉
retort-type stoker 下饲式炉排
retort with ground-in glass stopper 带磨砂玻璃塞的曲颈甑
retouch 润色
retouch colo(u)rs 修版颜料
retouched photograph 修饰照相
retouching 修饰；修版；手修；补釉；润色
retouching machine 修版机
retouching operation 修改操作
retouching stand 修饰台
retrace 逆程；回扫描；回扫
retraceable 可缩回的
retraceable belt conveyer 可缩传送带
retrace blanking 逆程消隐；回扫消隐
retrace characteristic 逆程特性；回扫特性
retrace course 沿原路返回的航线
retrace interval 回程间隔
retrace line 回扫线
retracement line 后视线
retrace period 回描周期
retrace ratio 逆程率；回描率
retrace scanning 回描
retrace time 回描时间
retracker 复轨器
retract 拉回；回缩；退刀；缩回；收缩核
retractable 可收缩的；可伸缩的

retractable air brake 收放式减速板
retractable antenna 收缩式天线
retractable awning 可折回遮篷
retractable bollard 收缩式系缆柱
retractable bridge 可缩桥
retractable cable reel 可缩的缆索卷筒；收放式电缆卷筒
retractable chute 伸缩式卸槽；伸缩式倾斜槽；伸缩式溜槽
retractable clothes line 可回缩式衣绳
retractable core barrel 可收回岩芯管；活动岩芯管
retractable core bit 可收回钻头；可换式岩芯钻头
retractable direction finder loop 收缩式测向器环形天线
retractable drawbar 伸缩式联结装置
retractable erase head 可动抹磁头
retractable fender system 伸缩式防撞装置
retractable fiber 收缩性纤维
retractable finger pick-up 伸缩指式拾捡器
retractable floor (轮式铲运机的)可缩斗底
retractable floor jack (轮式铲运机的)可缩斗底液压缸
retractable floor sequence valve (轮式铲运机的)可缩斗底顺序阀
retractable fork 伸缩式货叉
retractable guide 可撤导板
retractable jib slewing crane 伸缩臂式旋转起重机
retractable landing gear 可伸缩起落架；收缩起落架
retractable landing lamp 收放式着陆灯
retractable landing light 收放式着陆灯
retractable launcher 收放式发射装置
retractable launching box 收放式发射器
retractable launching device 收缩式发射装置
retractable mast 伸缩式桅杆
retractable periscope 伸缩式潜望镜
retractable pilot-house 伸缩式驾驶台
retractable plug sampler (带可提出式的矛式钻头的)振击式取土器；可缩式活塞取土器
retractable pusher-leg 伸缩式气腿
retractable radiator 可伸缩散热器；收放式散热器
retractable ramp 收放式跳板
retractable rock bit 可换式牙轮钻头；活动钎头
retractable roof 可伸缩屋顶
retractable scaffold bracket 伸缩式吊盘卡
retractable shuttle 伸缩货叉
retractable shuttle belt conveyer 可缩梭动式输送机；可缩的梭动式输送带
retractable soot blower 伸缩式吹灰器
retractable spear 伸缩杆
retractable stairway 收放式扶梯
retractable steel joist 可缩钢梁；伸缩式钢梁
retractable steel pocket tape 伸缩式袖珍钢卷尺
retractable steel rule
retractable tail wheel 收放式尾轮
retractable telescopic chute 可回缩脊椎节式溜斗(装船用)
retractable time 收放支腿时间
retractable under carriage 可伸缩起落架；伸缩起落架
retractable wedge 可回收的偏斜楔
retractable wheel 伸缩轮
retractable wrapper rolls 伸缩式助卷机辊
retract a confession 翻供
retract bolt 伸缩螺栓
retract clutch 回缩离合器(挖掘机斗杆)
retracted film 回缩膜
retracted position 内缩位置
retracted yarn 收缩纱
retracter 移镗器；拉钩；牵开器；收缩器
retractible 可缩回的
retractile 可回缩的；能缩回的；伸缩自由的
retractile (draw-)bridge 可缩进的开启桥；可缩吊桥；拖曳桥
retractile force 收缩力
retractility 可缩进性；退缩性；伸缩性
retracting 收起
retracting cylinder 伸缩筒
retracting cylinder stroke 移动油缸冲程
retracting finger auger 带伸缩指的运禾螺旋
retracting inflatable seal 可伸缩充气密封
retracting motion 回缩运动
retracting spring 复位弹簧；回程弹簧；释放弹簧
retracting spring-tooth pick-up cylinder 伸缩弹指式捡拾滚筒

retracting stroke 倒转
retracting transformation 收缩变换
retraction 回缩；缩进；缩回
retraction bulb 回缩球
retraction device 钻具自动提升装置(缩孔时)；退缩装置
retraction jack 收放机构
retraction lock 防收锁
retraction rate 回缩率
retraction speed 内缩速度
retraction stress 回缩应力；收缩应力
retraction stroke 内缩行程
retractometer 收缩仪
retractor 拉钩；收缩装置；收缩器
retractor device 伸缩装置
retractor launcher 收缩式发射器
retractor support 牵开器支撑器
retract rearward 向后收起
retract stroke 内缩行程
retransformation 逆变换
retransliteration 文字再转换
retransmission 转发；中继
retransmission counter 重复传输计数器
retransmission interval 重发间隔
retransmission receiver 转播接收机
retransmission unit 转播装置
retransmit 中继站发送；重(复)传输
retransmitter 转发发射机；转播发射机；中继发射机
retransmitting station 转播电台
retransmsstion 重复传输
retransposing 再转置；重易位；重交叉
retread 修补轮胎；重装新轮胎；复拌处治层；路拌处治层；重复处治层；路面复拌层
retread binder 复拌结合料
retread course 复拌层
retreading 路面修补；路面翻新
retreading machine 翻新机
retreading road 道路翻新
retread mixer 处治层路拌机；处治层复拌机；路面复拌层材料搅拌机
retread process 复拌法
retread road 复拌道路
retread surface 复拌面层
retread surface treatment 复拌式表面处治
retreat 后退式回采；重新处理；再精制；再加工；再处理
retreated tire 翻新轮胎
retreater 退降温度计；翻新器
retreat front 后退线
retreating 再处理；再精制；再加工
retreating angle 后倾角
retreating bank 后退河岸
retreating bankline 后退岸线
retreating coast 后退海岸
retreating glacier 后退冰川
retreating sea 海退
retreating system 再提炼系统；再加工系统；再处理系统；后退式开采法
retreatment 再精制；再处理；再加工；重(新)处理；重复处治
retreatment cell 再选浮选槽；浮选精选机
retreatment mill 再磨机
retreatment of tailings 尾矿再处理
retreat mining 后退式开采(法)
retreat of beach 海滩后退
retreat of monsoon 季风后退
retreat of river bank 河岸后退
retreat velocity 退水速度
retree 次等纸
retrench 压缩开支；紧缩
retrenching 缩减
retrenchment 节约；紧缩；节省
retrenchment measures 紧缩措施
retrenchment of expenditures 紧缩开支
retrenchment policy 紧缩政策
retrial 再试验
retriangulation 重新布设三角网
retrievability 可回收性
retrievability of solid wastes 固体废物的可回取性
retrievable 可恢复的
retrievable bit drilling without drill string lifting 不提钻可换钻头钻进
retrievable bridge plug 可回收式桥塞
retrievable choke 可更换阻流嘴；可拆管堵

retrievable deformation 可恢复的变形
retrievable drill bit 可换式钻头
retrievable form of energy 可恢复能量形式
retrievable information 可检索的信息
retrievable inner barrel 可回收的内筒;可换式内管
retrievable natural resources 可恢复自然资源
retrievable packer 可拆改的包装机;可卸式封隔器;可卸式堵塞器;可回收式封隔器
retrievable resources 可恢复资源
retrievable storage 可收回储[贮]存;可回收储[贮]存
retrievable tool 可卸式工具
retrievable valve 可卸阀
retrievable wedge 可提式偏斜楔;活动楔子
retrieval 检索
retrieval and dissemination system 信息管理检索和传播系统
retrieval by header 标题检索
retrieval by window 开窗检索;定位检索
retrieval facility 分类数据的检索设施
retrieval language 检索语言
retrieval oil 回收油
retrieval performance 检索性能
retrieval subsystem 检索子系统
retrieval system 检索系统
retrieve 检索;收回;出库(取货)
retriever 抢险车
retriever conveyer 取料运输机
retriever end 补正器端
retriever's hoist 抢救车起重机
retrieve vessel 回收打捞船
retrieving a bridge plug 起桥塞
retrieving anchor 起锚
retrieving disturbance sample 取样扰动样
retrieving line 回收缆索;取回绳
retrieving ring 提升环
retrieving rod 捞杆
retrieving sub 可取出接头
retrieving system 回收系统
retriggable 可预触发的
retrigger 再触发器
retriggerable 可再触发的
retrim 再平衡;再调整;再调平;重新调整
retrimming 再配平;再调平
retroact 倒行
retroaction sequence 制动顺序
retroactive adjustment 追溯调整
retroactive admission 追计加入
retroactive amplification 反馈放大
retroactive bank 可追溯的资金供应
retroactive contributions 追溯交款
retroactive effect 有追溯效力;追溯效力
retroactive financing 追补资金
retroactive method 追溯法
retroactive notation 追溯标记
retroactive pay 补发的增加工资
retroactive receiver 再生接收机
retroactive reception 再生式接收法
retroactivity of payments 款项的追领;补付款项的追领
retroarc basin 弧后盆地
retroazimuthal projection 反方位投影
retrocede 转分保
retroceding company 转分保分出公司
retrocession 再分保;退滩
retrocessionaire 转分保接受人
retrocession treaty 转分保合同
retrocession welding method 后退焊接法
retro-choir 教堂唱诗班席位后方;高祭坛后的礼拜堂;教堂主祭坛后面的空间
retroclusion 逆压法
retrodiffusion 反扩散作用
retrodirected ray 反向光线
retrodirective array 反向天线阵
retrodirective component 反向元件
retrodirective mirror 反向镜
retrodirective prism 反向棱镜
retrodirective reflection 逆向反射
retrodirective reflector 后向反射器
retrodirector 反向导向器
retroengine 制动发动机
retrofire 制动发动机点火
retrofire maneuver 制动发动机起动时的机动
retrofire time 制动点火时间

retrofit 更新改进;改进;改装;改型
retrofit device 翻新改进后设备;改进型装置
retrofit of a ship 船舶改装
retrofit of a vessel 船舶改装
retrofit strategy 改进型策略
retrofit technology 改装技术
retrofitted low-pollution vehicle 经改装的低污染车辆
retrofitting 翻新(改进);改装;改型
retrofitting existing service 在现有用户上补装
retroflection 折回;反曲
retroflexion 折回;回射
retrofocus 焦点后移
retrofocus lens 后焦点镜头;反远距透镜
retrogradation 逆行;后退;海蚀后退(作用);海蚀变狭作用;海岸线后退;海减(作用);返原
retrogradation condensate pool 逆凝析油池
retrograde 逆行的;后退的
retrograde condensate pool 反凝析气藏
retrograde condensation 逆冷凝;反转凝析;反缩合;反凝析
retrograde conduction 逆行性传导
retrograde conduction block 逆向传导阻滞
retrograde degeneration 逆行性变性
retrograde dissociation 退减离解
retrograde embolism 逆行栓塞
retrograde embolus 逆行栓子
retrograde evapo(u)ration 逆蒸发
retrograde evolution 逆行演化
retrograde fall out 反凝析
retrograde gas condensate reservoir 反凝析气藏
retrograde maneuver 制动机动
retrograde materiel action 物资倒流
retrograde metamorphism 逆变质作用
retrograde metastasis 逆行转移
retrograde motion 逆行;后退运动;反向运动
retrograde orbit 逆行轨道
retrograde precession 反向旋进
retrograde ray 逆行射线
retrograde reentry 逆向再入
retrograde reservoir 逆向储层
retrograde return 逆行返回
retrograde rocket 减速火箭
retrograde rotation 逆向旋转
retrograde solid solubility 退缩固溶度
retrograde solubility 退缩性溶解度;倒溶解度
retrograde stationary 逆留
retrograde trajectory 逆行轨道
retrograde vernier 逆游标;反读游标
retrograde-vision Daubresse prism 逆视四面体棱镜;逆视杜布雷斯棱镜
retrograde-vision tetrahedral prism 逆视四面棱镜
retrograde wave 退行波
retrograding beach 后退海滩
retrograding coast 后退河岸;退缩海岸
retrograding growth sequence 退积型生长层序
retrograding pattern 退积型
retrograding sequence 退积层序
retrograding shoreline 冲蚀岸线;后退岸线
retrogression 逆行;逆流运动;后退;退步;倒退作用;反向运动
retrogression of beach 海滩后退
retrogression of coast 海岸后退
retrogression of flame front 前部火焰的消退
retrogression of level 水位消退;水位消涨
retrogression of riverbed 河床后退
retrogression of strength 强度退化
retrogression of succession 逆行演替
retrogression of water stage 水位后退
retrogressive 后退的;退化的;退化变质作用
retrogressive accumulation aggradation 向源堆积
retrogressive deposition 溯源淤积
retrogressive erosion 向源侵蚀;溯源侵蚀;溯源冲刷
retrogressive evolution 后退演化
retrogressive flowslide 逆行流滑流
retrogressive landslide 牵引式滑坡
retrogressive metamorphism 退向变质;退化变质(作用)
retrogressive metamorphosis 逆行变态
retrogressive movement 逆动
retrogressive mutation 退化突变
retrogressive phenomenon 退化现象
retrogressive slide 向源滑坡;逆滑动
retrogressive slope 倒坡;反坡

retrogressive succession 逆行演替;倒显演替;倒退演替
retrogressive wave 逆行波;退行波;退缩波;反波
retrogressive wave motion 后退波动
retrogressive winding 倒退绕组
retroignition 制动发动机点火
retroillumination 后部反光照相法
retro maneuver 制动机动
retrometamorphism 退向变质作用
retro module 制动发动机轮
retromotor 制动发动机
retropack 制动发动机
retro pack jettison 投弃减速发动机
retroposition 后移
retro-ray reflector 回光反射器
retro-reflecting electron microscope 背向反射电子显微镜
retro-reflecting material 定向反光材料
retro-reflecting sheet 定向反光片
retro-reflecting sign 定向反光标志
retroreflection 回射;指向性反射;逆向反射;回复反射;后向反射
retroreflective finishing 反光整理
retroreflective placard 后向反光标志牌
retroreflector 后向反射器;定向反光路钮;折回反射镜;反应镜
retro-rocket 制动发动机;减速火箭
retrorse leaf 倒向叶
retrorsine 惹卓碱
retrosection 纽形剖线
retrosequence 制动顺序
retrospect and prospect 回顾和展望
retrospective analysis study 回顾性分析研究
retrospective application 追溯效力
retrospective assessment 回顾性评价
retrospective cohort study 回顾性群组调研;回顾性队列研究
retrospective information 追溯检索情报资料
retrospective paired observation 回顾性配对调查
retrospective-prospective study 回顾前瞻性调研
retrospective rating 追溯费率
retrospective retrieval method 逆检索法
retrospective risk assessment 回顾性风险分析
retrospective search 追溯检索(法);逆检索(法)
retrospective search method 逆检索法
retrospective search system 逆检系统
retrospective study 回顾性研究;回顾性调查
retrosynthesis 逆合成
retrosynthetic analysis 逆合成分析
retrosystem 制动系统
retrothrust 制动推力
retroversioflexion 后倾后屈
retroversion 后倾;倒退
retroverted 后倾的
retrusion 后移
retry 重试;复执;复算
retry register 重算寄存器
retting 浸解
retting clam 沤麻地
retting liquid 浸解液体
retting liquor 浸解液
retting poud 纤维浸解坑
retube 更换管子
retubing 更换管子;更换管件
retuning 重调谐;重调
retun trace 回扫描
return 折回;回复;回程;回车;申报表;重返;返回;反程
returnability 可收回性
returnable 可回收使用的
returnable bottle 周转使用的瓶子;可回收瓶;回收瓶
returnable container 可回收集装箱;多次使用容器
returnable containers 可退容器
returnable glass container 可以退回的玻璃容器
returnable pallet 回收式货板
return-acting pump 双动式(水)泵
return action 回复作用
return activated region 回流区
return activated sludge pumping 泵抽回流活性污泥
return activated zone 回流区
return address 返退地址;返回地址
return address instruction 返回地址指令
return address link 返回地址连接

return a deposit 退押
return after all taxes 纳税后收益
return air 循环空气;循环风;循环充气;回风;回气
return-air condenser 回气冷凝器
return air course 回风道;回风巷道
return air duct 回风管道;回气通道;回气管道
return air duct system 回风管道系统
return air fan 送气(回)空气的风机;回风机
return air grill(e) 回风算子;回气(管)花格栅;转角风格栅;回气格栅
return air inlet 回风口
return air intake 回气入口;回风口
return air register 回风节气门
return air system 回风系统
return airway 回流风道;回风巷道
return anchor wall 转角锚墙
return and allowance 退货及折让
return and allowance book 退货及折让簿
return a sum 归还款项
return a ticket 退票
return attribute 返回属性
return back 回流管
return bar 离合杆
return bead 串珠饰;转角凸圆线;转角处的联珠线脚;联珠角线脚;曲角线;转延侧面的压条
return bead and rebate joint 联珠线脚(木框转角凹凸榫上);木框转角凹凸榫上联珠线脚
return beam 返回射束;反射束
return beam vidicon 反束视像管;反束光导管摄像管
return beam vidicon camera 反束光导管摄像机
return belt 回程皮带
return bend 180 度弯头;回转弯头;回管;返向弯管;反向弯头;半环形套;U 形弯头;U 形管
return bill 退汇汇票
return blank 空白所得税申报表
return blanking die 向上出料落料模
return-blank type blanking die 带顶件机构的落料模
return block 开口滑车;导回滑轮
return branch of radiator 散热器回水支管
return branch to radiator 散热器供水支管
return brine 从冻结钻孔中返回的盐水
return brush 回流电刷
return buffer 返回缓冲器
return cable 回流电缆;返回电缆
return call 回答符号
return cam 回动凸轮;复归凸轮
return cargo 回运货(物);回航货(物);回程货(物)
return cargo freight 退货运费
return-carrying coefficient of passenger flow 客流回运系数
return catch 返回挡
return chamber 回流室
return channel 返回通道
return check valve 回油单向阀
return chute 回油槽;回料斜槽;回料溜子
return circuit 回流管路;回路;返回电路;反向回路
return-circuit rig 反向导流器
return circulation 冲洗液循环
return clause 退回保险费条款
return-cocked 突圆(转)角的
return code 返回(代)码
return code register 返回(代)码寄存器
return coefficient 复原系数
return commission 回扣;退还佣金
return condenser 回流冷凝器;回流冷凝管
return conduit 回流管(渠)
return connecting rod 回行连杆;回头连杆
return convection 回返对流
return conveyer 返料皮带
return conveyor for caulphater 型板回送机
return corner block 转角砌块
return crank 回行曲柄;偏心曲拐
return cullet 回炉碎玻璃
return current 逆流;回流;返回电流;反向流
return current rail 回流轨
return current zone 回流区
return curve 回复曲线
return cylinder 回程缸
return deposits to tenants in the land reform 退押
return-drain pump 疏水泵

return duct 回风管道
returned acid 回流酸
returned activated sludge 回流活性污泥
returned cargo 退回货物;返运货物
returned check 退回支票
returned corner (墙的)端头墩
returned end 复原端头
returned flood of five hundred years 五百年一遇洪水
returned flood of one hundred years 百年一遇洪水
return edge 回路边
returned goods 退回货物;退赔品
returned goods allowances 备抵退货
returned material 回料;退回材料
returned mo(u)lding 转弯的线脚;迂回线脚;转延装饰嵌线;回转线脚;回旋延伸式线脚
returned purchase 进货退回
returned purchase account 进货退回账户
returned purchase invoice 进货退回凭单;退货单据
returned quantity of broken glass 碎玻璃返回量
returned sales account 销货退回账户
returned sales report 销货退回报告单
returned sand 回用砂
returned shipment rate 回程费率
returned sludge 回流污泥
returned sludge ratio 污泥回流比
returned value specification 返回值说明
return elbow 回弯头
return ell 线脚端部;回弯头
return/exhaust fan 回排风机
return fan 回风机
return farmland to forests 退耕还林
return feed 回行进给
return feeder 回路馈线;回流馈路
return filler 返回填充料
return filter 回流过滤器
return flame boiler 回焰式(火管)锅炉;回焰锅炉
return flange 回油板
return flanging die 折叠式凸缘成型模;鹅颈模
return flow 回水;逆流;回流;回归水(流);反向流
return flow atomizer 回流雾化器
return flow burner 回油喷嘴
return flow combustion chamber 回流式燃烧室
return flow compressor 回流式压缩机
return flow line 回流通道
return flow range 回流区
return flow scavenging 回流换气法
return flow system 回流系统
return flow (wind) tunnel 回流式风洞
return flow zone 回流区
return flue boiler 回焰式(火管)锅炉;回焰锅炉
return fluid 返回的液体
return flume 回水槽
return flushing 反循环冲洗
return for no claim 无赔款退货
return freight 回程运费
return from final clearing 主伐收益
return from interrupt 中断返回
return from seal 从密封返回
return from subroutine 从子程序返回
return function 返回函数
return gear 回行装置
return gear box 回行齿轮箱
return ghost 逆程重影
return grill(e) 回气管格栅;回风格子窗;回转格栅;回风算子
return guide vane 反导叶
return head 线脚转弯处;线脚端头;屋角石镶边
return hose 回水软管;回油软管;回流软管
return idle 运输机回程皮带惰轮
return idler 回行惰轮;从动滚轮;输送机回程皮带惰轮
returning activated sludge 回流活性污泥
returning conduit 回流管渠
returning curve 回复曲线
returning handle 回动把手
returning iron 回炉铁
returning land for farming to forestry 退耕还林
returning main 回流总管
returning security money 退押
returning spring 复位弹簧
returning system 回水系统
returning to service 重配种

returning tube 回水管
returning water conduit 回收管渠
returning wave 回波
return in invested capital 投入资本收益率
return instruction 返回指令
return interval 回扫时间;回扫间隙;回程时间
return inward 销货退回
return item 退回的拒付票据
return journey 回程;返航
return jump 返回跳转
return key 回车键;返回键
return laser beam 激光反射光束;激光反回光束
return launcher 返回流槽
return light 应答灯光信号;回光
return line 回汽管;回油管路;回水(总)管;回水管线;回描线;回流管;返回线路
return line filter 回油管滤油器;回油管过滤器
return link 返回连接
return liquor 回流液
return load 回载
return loan 退回贷款
return loss 逆程损耗;回程损耗;回波损耗;反射波损耗
return loss measuring set 失配损耗测量器;反射波损耗测量器
return louvers 回流放气孔
return macro 返回宏指令
return main 溢流总管;回水总管;回水干管;回流总管;回流干管
return material 回磨;返料;材料退回
return materials journal 退料账
return mechanism 回行机构;返回机构
return motion 回行运动;回复运动;反转运动
return motion linkage 摇摆拐臂
return movement 回动
return mud 返回泥浆
return nosings 踏步转角延伸的小突沿
return of control 控制返回
return of drilling mud 泥浆返回
return offset 迂回管;连续迂回管;回头乙字管;归偏置管;绕行管
return of goods 退货
return of investment 投资回报
return of material 退料
return of movement 回复运动;运动的返回
return of net worth 净值的盈利
return of points 辙叉返回原位
return of premium 退回保险费
return of slurry 泥渣回流
return of stroke 行程换向
return oil 回油
return oil pipe 回油管
return oil system 废油的回流系统
return on assets 资产收益率;资产利润率
return on assets managed 可控资产投资报酬率
return on capital 资本收益率
return one has unlawfully taken 退赔
return on equity 资产净值的盈利;产权收益率
return on investment 投资收入率;投资收回期;资利润;投资利得率;投资收回(率)
return on investment ratio 投资收入比率
return on net worth 净值收益(率)
return on post 请即答复
return on sales 销货收益率
return opening 回风洞
return orifice 回流孔
return orifice check valve 单向节流阀
return outward 进货退回
return outwards invoice 返运货物清单
return pass 返回道次
return passage 回路
return period 逆程期;回复期;重现期(间);重复周期
return period of design wave 设计波浪重现期
return period of earthquake 地震重复周期
return period of investment 投资回收期
return phase 重现相位
return pipe 回油管;回水管;回气管;再循环管;回行管道;回流管;回浆管;返水管
return pipeline 回水管线;回水管路
return pipe of re-usage 重复利用回水管
return pipe of re-utilization 重复利用回水管
return piping 回水管线
return plate 返料叶片
return platt 回油板

return point 山形穿综;返回点
return point in calling procedure 调用过程的返回点
return port 回流孔
return premium 退回保险费;退还保险费
return pressure 回压;回流压力
return pressure null shift 回油压力零漂
return pressure of safety valve 安全阀回座压力
return pressure wave 反射压力波
return propagation factor 反传因子
return pulley 反向滑轮
return pump 回流泵;抽空泵;回压油泵
return pumping 回流泵送
return purchase 进货退回
return push-button 返回按钮
return rate on capital employed 资本利润率
return recording 数字间有间隔的记录
return refrigerant velocity 制冷吸入速度
return riser 再循环上升管;回水立管;回流竖管
return roller 支承轮;托带轮
return rope 尾绳
return route concealment 隐蔽返航路线
return run 逆行段;回程折流段;无载分支
returns 回用料;回炉物;回炉料;返回料;报告
return sales 销货退回
return sales memo 销货退回通知单
return sample 回样
return sand 回用砂
returns book inward 销货退回账
return scanning beam 回扫电子束
return scrap 回炉废钢;返回废钢
return seepage 回归渗流;渗透回水
return selling 返销
return-service rate 复配率
return sheave 尾绳滑轮;导向滑轮
return shipment 回程装运;返程装运
return shipping order 退货发运单
return shock 反冲
return side 回水部分;回气部分
return signal 返回信号;反馈信号
return signal strength 返回信号强度
returns in capital construction investment 基建投资收益
returns in real term 实际收益
return siphon 回水虹吸管
return siphon 倒虹吸
return slag 回炉渣
return sludge 回流污泥(污水处理曝气池用)
return sludge concentration 回流污泥浓度
return sludge flow 回流污泥流量
return sludge requirement 回流污泥的要求
returns of capital investment 投资报酬率
returns of contractor's equipment 承包人配备报告
returns of investment 投资利润
returns of labo(u)r 有关劳务的统计;劳务报告
returns of trade 贸易报告
return solution 重新解算
returns on investment 效果
return speed 回程速度
return spring 回位弹簧;回动弹簧;回动发条;复位弹簧;复进簧
return spring casing 回动簧箱
return spring guide 回动弹簧导承
return spring pressure 复位弹簧压力
return spring stop 回动簧止杆
return state 返回状态
return streamer 回返闪流
return stroke 逆程;回行冲程;回返闪流;回程;反回冲程;反冲程
return-stroke time 返回行程时间
return subroutine 返回子程序
return swing arm drip 摆动溢流管
return system 回复系统;回风系统
return tabular steam boiler 回焰蒸汽锅炉
return tank 回水水箱
return temperature limiter 回水温度限制器
return temperature of water 回水温度
return ticket 来回票
return time 回描时间;倒程时间
return time of vision 眩光恢复时间
return-to-bias 归偏制;归零(制)
return to bias recording 归偏制记录
return top of slurry 水泥返浆
return-to-reference recording 归基准记录;归参考记录;返回基准记录法
return to scale 与生产规模成比例的收益
return-to-zero 归零(制)
return-to-zero code 归零(代)码
return-to-zero coding 归零编码
return-to-zero logic 归零制逻辑
return-to-zero method 归零制;归零法
return-to-zero mode 归零方式
return-to-zero position 归零位
return-to-zero record(ing) 归零(制)记录;复零记录
return-to-zero representation 归零表示
return-to-zero system 归零(制)系统
return trace 反行程;逆程;回描
return-trace blanking 回扫消隐;回描熄灭;回程消隐
return track 反信风
return transfer function 返回传送操作
return transport 回运
return trap 回水弯(管);回水隔气弯管;回水隔气具
return travel 反向运动;反向行程
return tray 回油盘
return trip 回程;返航
return trip C/P 往返航次船租合同;往返航程船租合同
return tube 再循环管;回油管;回浆管;回管
return tube boiler 回焰锅炉;回水管式锅炉
return tubular boiler 回焰式(火管)锅炉;回流锅炉
return type blanking die 带顶件机构的落料模
return value 回收价值;返回值
return valve 止回阀;回油阀;回水阀;回路阀;回流阀
return valve plug 回流阀塞
return valve seat 回流阀座
return valve spring 回流阀簧;回动阀弹簧
return velocity 返回速度
return voyage 回程;返航
return voyage charter 来回程租船契约
return wall 翼墙;扶壁;墙垛;尽端墙墩;迁回墙;转角延伸侧墙;转角墙;连续迁回墙
return water 回注污水;回水;回流钻液;回收水;回归水;返回的水
return water main 回水干管
return water pipe 回水管
return water piping 回水管
return water tank 回水箱;回水池
return water temperature 回水温度
return water tube 回水管
return water tubing 回水管
return water waves 回收水波
return wave 回波;反波
return what one has unlawfully taken or pay compensation for it 经济退赔
return wing (建筑物的)转角侧房;(建筑物的)拐角侧房
return wire 回线;回流线【电】
return wire leading roll 转向导网辊
retwister 复捻机
retyred wheel 再组合车轮
retzbanyite 铜辉铅铋矿
retzian 羟砷钇锰矿
retzianite 砷钙钇锰矿
retzian-(La) 羟砷镧锰石
retzian-(Nd) 羟砷钕锰石
reunion 再合并;再结合;重叠式皮带运输机
Reunion normal polarity subchron 留尼汪正向极性亚时
Reunion normal polarity subchronzone 留尼汪正向极性亚时间带
Reunion normal polarity subzone 留尼汪正向极性亚带
reuptake 再摄取
reuring bay 避险处
reusability 重复使用(可能)性;复用性;再用性
reusable 可再用的
reusable attribute 可再用属性
reusable container 可再用容器;可反复试验的集装箱
reusable file 可重复使用文件;可复用文件
reusable module 可再用模块
reusable program(me) 可重复使用程序;可再用程序
reusable resource 可重用资源
reusable routine 可重用例行程序;可重复使用(例行)程序;可再用例程
reuse 再使用;重新使用;再用;重复使用;重复使用
reused numbers 启用旧号
reuse of forms 模板重复使用
reuse of materials 材料再利用
reuse of old and discarded things 废旧物资回收利用
reuse of plastic waste 废塑料利用
reuse of shuttering 模板重复使用
reuse of wastewater 废水重复利用
reuse rate 再用率
reuse system 复用水系统
reuse water 再用水;再生水;回用水
reusing of industrial effluent 工业废水再使用
reusing sample 复用试样
reutilization 再用;重新利用;重复利用;二次利用
revale 现场雕凿的石材台脚
revalidation sticker 重新确认签条
revalorization 币值重估
revaluate 重新评估;再估价
revaluation 再估价;升值;重新评价;重新估价;重新定值;重估
revaluation account 重估价账户
revaluation clause (租约的)重新估价条款
revaluation excess 重估价溢额
revaluation lease 定期重新估价租约
revaluation of assets 资产重估价
revaluation of data 数据换算
revaluation process 重估计价程序;重估价程序
revaluation profit 重估价利益
revaluation reserves 重估价准备金
revaluation surplus 重计价盈余;重估价盈余
revaluation surplus reserves 重估价盈余准备
revalue 升值
revalving 更换电子管
revamped mill 改装过的轧机;改造过的轧机
revamping 重修;翻新;部分改进
revamping of equipment 设备改造
revamping of exiting enterprises 改组现有企业
re-vapo(u)rization 再气化;二次蒸发
re-vapo(u)rize 再蒸发
re-vapo(u)rizer 二次蒸发器
rev counter 转数计
revdite 雷水硅钠石
reveal 门窗侧墙;门窗侧边;门窗侧壁
reveal lining 门窗口(侧面)筒子板;筒子板
revealment 显露
reveal pin 墙洞上的肘钉;窗口脚手销钉
reveal tie 窗口脚手拉杆
revegetation 植被重建;植被再造;植被恢复;再(生)植被;重新绿化
revegetation of land 土地再植被
reveiving of train 接车【铁】
revelation 展现
reventing 复管排气(室内排水系统)
revent pipe 重装新通气管
revenue 钱粮;岁收;岁入;税收;收益;收入
revenue accounting 收入核算
revenue act 岁入法案
revenue and expenditures 收支;收入和支出
revenue assets 收益资产
revenue balance sheet 年度资产负债表
revenue boat 巡逻船
revenue bond 岁入债券;税收合同
revenue budget 收入预算
revenue charges 业务支出
revenue cruiser 海关巡逻船
revenue cutter 海关巡逻艇
revenue depreciation method 营业收入折旧法
revenue duty 财政关税
revenue-earning 有利可图
revenue effort 取得收入的工作
revenue elasticity 岁入弹性
revenue expenditures 营业支出;收益支出
revenue financing 收入财政
revenue form enterprise and undertaking 企业和事业收入
revenue from financial subsidy 财政补助收入
revenue from goods sold 销售收入
revenue from senior administrative agency subsidy 上级补助收入
revenue map 税务地图
revenue office 税务所;税收机构
revenue officer 税务官员
revenue of per ton-kilometre 吨/千米收入
revenue of railway state-owned assets 铁路国有资产收益

revenue paid by subordinary unit 附属单位上缴收入
revenue passenger-kilometer 人·千米收入
revenue position 收入额
revenue-producing activities 盈利机构
revenue-producing water 年总产水量
revenuer 缉私员；税务员
revenuer cutter 缉私艇
revenue realization 收益之实现；实现收入
revenue requirement 收入需要
revenue reserves 保留盈余；保留收入
revenues and expenditures are in balance 收支相抵
revenues and expenditures balanced 平衡收支
revenues are over expenditures 收大于支
revenue sharing 收益分享
revenue stamp 印花税票；税收印花
revenue stream 收入流程
revenue tariff 岁入关税（为财政收入而征收的关税）；财政性关税税率
revenue tax 岁入税；财政税收；财政税；财政收入税
revenue ton 载货吨；计费吨
revenue vector 收入向量
revenue vessel 海关巡逻艇
reverb 余响
reverberant camera 反射摄影机
reverberant noise level 混响声级
reverberant room 混响室
reverberant sound 回声；混响声
reverberant sound absorption coefficient 混响声系数
reverberant sound field 混响声场
reverberant steady-state sound 混响声
reverberant test room 混响试验室
reverberant unit 混响器
reverberate 反冲
reverberating 反焰
reverberating burner 返焰炉
reverberating echo 混响回波
reverberating effect 混响效应
reverberating furnace 倒焰炉；返焰炉；反射炉
reverberation 交混回响；混响；回响；反响；反射物；反射热
reverberation absorption coefficient 混响吸收系数
reverberation bridge 测定混响电桥
reverberation chacteristic 混响特性
reverberation chamber 混响室
reverberation circuit 混响回路；反响回路
reverberation control 混响控制
reverberation damping 混响减幅；混响阻尼
reverberation degree 混响度
reverberation field 混响声场
reverberation level 混响级
reverberation meter 混响仪；混响测定器
reverberation method 混响法
reverberation of sound 声音的混响
reverberation period 交混回响时间；混响期
reverberation radius 扩散地距离；混响半径
reverberation response 混响响应
reverberation response curve 混响响应曲线
reverberation room 混响室
reverberation room method 混响室法
reverberation room technique 反射室法
reverberation-suppression filter 混响抑制滤波器
reverberation time 余响时间；混响时间
reverberation-time meter 混响时间（测量）计
reverberator 反射器；反射灯；反射镜；反射灯
reverberatory 反射的；交混回响的；返焰的
reverberatory burner 返焰炉
reverberatory burning 返焰灼烧
reverberatory calcination 返焰煅烧；反焰煅烧
reverberatory calciner 返焰煅烧炉；反焰煅烧炉
reverberatory forehearth 反射前炉
reverberatory frit kiln 反射式熔块炉
reverberatory furnace 反焰炉；反射炉
reverberatory furnace smelting 反射炉熔炼
reverberatory furnace refining 反射炉精炼
reverberatory roaster 返焰炉；反射焙烧炉
reverberatory roasting 返焰焙烧
reverberatory slag 反射炉炉渣
reverberometer 混响时间测量计；混响仪；混响计
reversa 下裂
reversal 逆转；换向；颠倒；反向（型）；反变；变号
reversal colo(u)r film 反射彩色胶片

reversal condition 可逆条件；倒易条件
reversal cost 反向成本
reversal cost method 逆向成本法；反向成本法
reversal design 可逆设计
reversal-developed image 反转显影图像
reversal developer 反转显影剂
reversal development 逆转显影；反转显影
reversal device 反向装置
reversal dip（地层的）逆倾斜
reversal effect 反转效应
reversal emulsion 反转型乳胶；反转乳剂
reversal exposure 反转曝光
reversal film 反转片
reversal flow 反向流
reversal flue 换向烟道
reversal gear（wheel）回动齿轮
reversal inhibition 回返抑制
reversal in loading 反复加载
reversal inter locking relay 反向联锁继电器
reversal interval 换向间隔；换火周期
reversal knob 换向钮
reversal load 变向荷载
reversal loading 交变荷载；交变负载；反复荷载
reversal material 反转（感光）材料
reversal nozzle 反向推力喷管
reversal of a machine 机械的反向
reversal of a spectral line 光谱线的自蚀
reversal of control 操纵反效
reversal of current 反流；电流反向；逆电流
reversal of curvature 反曲率
reversal of dip 倾斜逆转
reversal of flame direction 火焰换向；换火
reversal of flow 倒流
reversal of geomagnetic field 地磁场反转；地磁场倒转
reversal of judgment 取消判决
reversal of load 反向荷载；变号荷载；荷载反向；荷载变号；变换荷载
reversal of magnetic field 磁场反向
reversal of magnetism 磁性反转；磁极变换；反（复）磁化
reversal of phase 相反转；倒相
reversal of photography 反转摄影
reversal of polarity 极性反转
reversal of polarization 反极化
reversal of poles 极性颠倒；极性变换
reversal of season 季节转换
reversal of spectral line 光谱线的自蚀
reversal of stress 应力逆转；应力交变；应力反向；反向应力
reversal of stroke 反冲程
reversal of the head 压头反向
reversal of the pressure 压力逆变
reversal of vibration 振动反向；反向振动
reversal ogee mo(u)lding 反 S 形线脚；反双曲线线脚
reversal paper 反转相片
reversal per minute（柱塞的）每分钟往复次数
reversal phase coding 反相编码
reversal phenomenon 反转现象
reversal photographic(al) processing 反转摄影处理
reversal point 峰点；反向点
reversal point method 逆转点法
reversal position 反转位置
reversal process(ing) 反转显影法；反相处理；反转（成像）法
reversal river 反向河
reversal rotation 反向旋转
reversal run 返测【测】
reversal scattering 反转散射
reversal side 背面；反面
reversal spectrum 自蚀光谱
reversal stream 反向河
reversal stress 重复应力；反向应力；反复应力
reversal tainter gate 反向弧形闸门
reversal tainter valve 反向弧形阀门
reversal temperature 反转温度
reversal test 倒转检验
reversal tidal current 反向潮流
reversal time 换向时间
reversal-type emulsion 反转型乳剂
reversal valve 可逆阀；换向阀
reversal vibration 反复振动

reversal zone 反转区
reverse 相反的；颠倒的；倒转；倒体；反向的；反模；回动
reverse a-c resistance 反向交流电阻
reverse acting 回动传动
reverse acting control system 反作用控制系统
reverse acting control valve 反作用控制阀
reverse acting solenoid valve 反向动作螺管阀
reverse acting spiral 反向螺旋
reverse action 倒动作
reverse address resolution protocol 逆地址解析协议
reverse after moving back out of the station 站前折返
reverse after moving forward out of the station 站后折返
reverse air baghouse 逆气流反吹袋式除尘器
reverse air blast 反向鼓风
reverse air cleaned unit 反吹风清扫装置
reverse air cleaning 反吹风清扫
reverse air flow cleaning 反向吹风清扫
reverse anchor 反向锁闭
reverse anchor jam nut 反向锁闭机构紧固螺帽
reverse an entry 冲销
reverse angle-shot 侧角反摄镜头
reverse-annuity mortgage 反向年金抵押贷款
reverse a procedure 颠倒程序
reverse arching 倒拱作用
reverse arm 回动杆
reverse armature 反极性磁铁
reverse a telescope 纵转望远镜
reverse azimuth 反方位角
reverse azimuth sweep 反向方位扫掠；反方位扫描
reverse band 反向带
reverse bar 反肋骨
reverse beam vidicon camera 反束光导管摄像机
reverse bearing 后象限角；反象限角；反方向角；反方位（角）
reverse behind station according to the proceeding direction 站后折返
reverse belt 反向带
reverse bend 反向弯头；反向弯曲；S 形弯曲
reverse bend test 弯曲疲劳试验；反向弯曲试验
reverse bevel 反向弹簧锁闩；反向斜面
reverse bevel door lock 反向门锁
reverse bias 逆向偏压；反向偏置；反偏压
reverse-biased diode 反（向）偏压二极管
reverse biased junction 反向偏压结；反向偏压结；反偏结
reverse biased junction source 反向偏压结辐射源
reverse bias voltage 反向偏压
reverse blocking 反向阻断
reverse blocking current 反向阻断电流
reverse blocking diode thyristor 反向阻断二极晶闸管
reverse blocking interval 反向阻断间隔；反向关断期间
reverse blocking state 反向阻断状态；反向闭锁状态
reverse blocking tetrode thyristor 可控硅开关
reverse blocking triode thyristor 可控硅整流器
reverse board and batten 反钉宽板与板条
reverse bonded-phase chromatography 反键合相色谱法
reverse book fashion 从左往右摆放（岩芯）
reverse box 换向箱
reverse brake shoe 后向制动蹄
reverse Brayton cycle 逆布雷顿循环
reverse break 反向中断；反向断开
reverse breakdown 反向击穿
reverse burning 反向燃烧
reverse capacity 折返能力
reverse capstan 倒转主动轮；反行主动轮
reverse card 反向（罗盘）刻度盘
reverse Carnot cycle 逆卡诺循环
reverse-cascaded flow control 倒联级式流量控制
reverse casehardening 逆表层硬化
reverse caster 主销后倾；（地层的）逆倾斜；负倾斜
reverse cell 逆环流圈
reverse channel 反向信道；反向通道
reverse characteristic 反向特性
reverse charge call 对方付款呼叫
reverse charging 反向充电
reverse chill 反白口
reverse circuit 反作用电路

reverse circulating and killing the well 反循环压井
reverse circulating sub 反循环短节
reverse circulation 逆(转)循环;反循环
reverse circulation boring machine 反循环钻孔机
reverse circulation boring method 反循环钻孔法
reverse circulation continuous core drilling 反循环连续取芯钻进
reverse circulation core barrel 反循环钻进岩芯管;逆循环钻进岩心管
reverse circulation core barrel without pump 无泵反循环取芯钻具
reverse circulation drilling 钻进反循环;反循环(洗孔)钻进;反循环(洗井)钻进
reverse circulation drilling without pump 无泵反循环钻进
reverse circulation gravel pack 反循环砾石充填
reverse circulation gravel pack technique 反循环砾石充填技术
reverse circulation junk basket 反循环打捞篮
reverse circulation method 逆循环法;反向循环法
reverse circulation portadrill 反循环轻便钻机
reverse circulation rotary drilling 反循环旋转钻进
reverse circulation valve 反循环阀
reverse circulation washing 反循环冲洗
reverse classification 反分级
reverse clipping 反向削波;反向限制
reverse clutch 换向离合器;反转离合器
reverse code 逆向码
reverse cold bend test 反复冷弯试验
reverse collection 反托收
reverse combustion 反向燃烧
reverse compatibility 逆兼容性
reverse compression 反向压力
reverse cone grate 倒锥形格条
reverse connection 反接
reverse contact 反转接点;反向接点;反位接点
reverse contactor 换向接触器
reverse control 换向控制
reverse control handle 反向控制手柄
reverse control hodograph 相遇时曲线系统
reverse convex program(me) 反向凸规划
reverse conveyer pawl 输送器回动爪
reverse copy 反像晒印
reverse copying 反像直接晒印
reverse counter gear 回动对齿轮
reverse counting 反向计数
reverse-coupling direction(al) coupler 反向激励定向耦合器;反相激励定向耦合器
reverse cover 回动盖
reverse creep 反向蠕变
reverse current 逆流;逆电流;倒流;反转电流;反向电流;反流
reverse-current braking 逆流制动;反向电流制动
reverse-current breaker 逆流断路器
reverse-current circuit breaker 逆电流断路器;反向电流断路器
reverse-current cut-out 反向电流自动开关
reverse-current dryer 回流式干燥器
reverse-current filter 双流滤池;逆流过滤器
reverse-current metering 反向电流测量;反流计量;逆流测量
reverse-current protecting equipment 逆电流保护设备
reverse-current protection 逆(电)流保护;反向电流保护
reverse-current relay 逆流继电器;逆电流继电器
reverse-current release 逆电流释放器
reverse-current switch 逆电流开关
reverse-current time-lag relay 逆流延时继电器
reverse-current trip 逆(电)流自动切断
reverse-current working 反向电流工作法
reverse curvature 反曲率
reverse curve 反向曲线;反转曲线;反莫深曲线
reverse curve sign 之字形曲线标志
reverse cut-off current 反向截止电流
reverse cutting needle 外弯割口缝合针;反三角缝合针
reverse cycle 反转循环
reverse cycle heating 逆循环供热
reverse cylinder 回动汽缸
reverse cylinder valve arm 回动汽缸阀臂
reverse cylinder valve body cap 回动汽缸阀体盖
reverse cylinder valve stem 回动汽缸阀杆
reversed 被撤销的;颠倒的;逆转的;反向的;倒转的

reversed access 反向存取
reversed airblast process 反向鼓风制气法
reversed air classifier 闭式空气分离器
reversed arc 反向岛弧
reversed arch 倒发券;倒拱;反拱
reversed arrangement 反向排列
reversed barb 逆刺;反向刺
reversed beam 反弯梁;反梁
reversed bending 反向弯曲
reversed bending strength 交变弯曲强度
reversed bend test 冷弯试验;弯曲试验
reversed-bias diode 反向偏低二极管
reversed blanking 反向落料
reversed brass 锌基压铸合金;高锌黄铜
reversed budding 逆芽接
reversed charges 对方付款的电话簿;逆充电
reversed compression 反向压力
reversed cone nozzle 倒锥形喷嘴
reversed control 反向控制
reversed current air classifier 逆流分级机;逆流空气分离器
reversed current dryer 逆流干燥器;回流干燥机;对流式干燥机
reversed current relay 逆流继电器;反流继电器
reversed curve 反向曲线;反曲线
reversed cyclic action 反复作用
reversed damping 逆阻尼
reversed de Laval nozzle 拉瓦尔型倒喷嘴
reversed dip 逆倾斜(地层的)
reversed direct stress 反复直接应力;拉压交替压力
reversed discharge 反转出料(混凝土搅拌机)
reversed dished head 凹形封头
reversed door 反向(开启)门;反门;外开门
reversed draft 山形穿法
reversed drag 逆牵引
reversed drag anticline 逆牵引背斜
reversed drag structure 逆牵引构造
reversed drainage 反向水系
reversed eddy 反向涡流
reversed electric(al) field 反向电场
reversed electrostatic stencil 反转静电模版
reversed emulsion drilling fluid 反相乳化钻井液
reversed epoch 逆转期;反相期
reversed erosion fault scarp 逆向侵蚀断层崖
reverse detergent 逆性洗涤剂
reversed fault 逆(掩)断层;上冲断层
reversed field 倒性场
reversed-field configuration 反场位形
reversed filter 反滤层
reversed filtration layer 反滤层
reversed fishtail bit 反鱼尾钻头
reversed flame kiln 倒焰窑
reversed flow 反向流;变向流动
reversed-flow condenser 逆流式凝气器
reversed-flow type viscometer 逆流式黏度计
reversed flush boring 反循环冲洗钻进
reversed flush drilling 反循环冲洗钻进
reversed fold 逆褶皱
reversed fold-fault 逆褶断层
reversed francis turbine 轴向辐流式水泵—水轮机
reversed frontage lot 街角地块
reversed grading 反向分品法
reversed head 反向水头
reversed head method 开挖掉头(开挖)法
reverse diffusion 逆行扩散
reversed digit sorting method 逆向数位排序法;反向数字排序法
reverse dilution method 逆稀释法
reversed image 转像;倒像;颠倒相影像;倒立像;负像;反像
reversed inclined impervious bottom bed 逆坡倾斜隔水底板
reverse direction 反(视)方向
reverse direction(al) angle 反(视)方向角
reverse direction(al) element 反向元件;倒相单向元件
reverse direction(al) flow 反向流程图;反相流;反流向;反方向流
reversed jeanette 细斜纹布
reversed king-post beam 倒单柱梁
reversed king-post girder 反向单柱上撑式大梁;倒单柱大梁
reversed king post truss 倒单柱桁架

reversed left to right 左右翻转的
reversed limb 倒转翼
reversed line 自蚀(光谱)线;反转线
reversed loader 反转载机;转载机
reversed load(ing) 反向荷载;反复荷载
reversed logarithmic potentiometer 反转对数式电位器
reversed loop winding 逆行叠绕组;反环绕组
reversed L-shaped retaining wall 反L形挡土墙
reversed magnetic field 反向磁场
reversed magnetization 逆向磁化;逆向磁化;反磁化
reversed matching 对纹拼装;对纹拼板;反向拼装
reversed migration 逆偏移
reversed nozzle 反喷嘴
reverse dog 反位锁簧
reverse domain 逆磁畴
reverse door 反向门
reversed order 逆序;倒转层序;反顺序
reversed pendulum 倒垂线
reversed phase 反相
reversed phase chromatography 反相色谱法
reversed phase coil 反相线圈
reversed phase paper chromatography 反相纸色谱法
reversed phase (partition) chromatography 反相分配色谱法
reversed phase sequence 逆转相序
reversed piezoelcetric effect 负压电效应
reversed polarity 异极性(晶体);反转(极)性
reversed polarity direct current 反接直流
reversed position of telescope 倒镜
reversed price scissors 反剪刀差
reversed profile 反向剖面
reversed queen-post beam 倒双柱梁
reversed queen-post girder 反向双柱(上撑式大)梁;倒双柱大梁
reversed queen-post truss 倒双柱桁架
reverse drag 反牵引
reversed drawing 逆向拉制;反压延;反拉伸
reversed relay 逆电流继电器
reversed return scheme 同程式布置
reversed return system 反向回水系统;可逆回水系统;同程式系统
reversed rhythm 反节律
reverse drive 逆行程;回程;换向传动;反向驱动
reversed river 反河;反电河
reverse drum 反向鼓轮
reversed run 返工
reversed shear 反向剪力
reversed shooting 反向爆炸
reversed sign 相反符号
reversed silhouette 逆剪影;反射的轮廓
reversed single post beam 反向单柱梁
reversed spin 反向自旋
reversed spiral test 逆螺线操纵试验
reversed S-shaped structure 反S状构造
reversed stratigraphic(al) sequence 倒转层序
reversed stream 逆流河;反河;反电河
reversed stress 反向应力;交变应力;重复应力;反复应力
reversed system 反向体系
reversed tainer gate 倒弧形闸门
reversed tainter valve 反向弧形闸门
reversed telephoto objective 反向望远物镜
reversed tension 反向拉力
reversed thrust 逆掩断层;反向推力
reversed tidal current 反向潮流
reversed tide 逆潮(流)
reversed time 反向时间
reverse dual currency bond 可调换二元通货债券
reverse dumping 逆(向)倾销
reversed-current air classifier 粗粉分离器
reversed value 反向值
reversed vault 倒拱顶
reversed zigzag mo(u)lding 反向锯齿线脚;反锯齿形线脚
reverse eddy 反向涡流
reverse efficiency 逆效率
reverse electrodialysis 反电渗析
reverse electromotive force regeneration 反电动势再生
reverse electro-pneumatic valve 回动电气阀;回动电控阀
reverse elevator pawl 升降机回动爪

reverse emission 反向放射；反向发射
reverse end for end 掉头
reverse engine 反转机
reverse epicyclic gear train 倒车行星齿轮系
reverse error 回复误差
reverse eye hook 转眼钩
reverse face 倒镜位置【测】
reverse fault 逆断层；冲断层
reverse fault trap 逆断层圈闭
reverse feed 反向给进
reverse feed gear 反转给进齿轮
reverse feed lever 反向进给手柄
reverse field process 反转电场法
reverse filter 反滤层；倒滤层
reverse fishtail bit 反向鱼尾钻头
reverse flank 迂回侧翼
reverse-flat baghouse 回转反吹扁袋除尘器
reverse-flighted screw 双向螺杆
reverse-flight screw 反螺纹杆
reverse flow 逆流；回流；反向水流；变向交通；变向车流
reverse flow baffle 回流挡板
reverse flow check valve 止回阀
reverse flow cooling system 逆流冷却（水）系统
reverse flow cyclone 逆流旋流器；回流式旋风除尘器
reverse flow gas process 逆流发生气体过程
reverse flow glass capillary viscometer 回流玻璃毛细管黏度计
reverse flow lane 反向交通车道；变向车流车道
reverse flow muffler 反流式消声器
reverse flow nozzle 制动喷嘴；回流喷嘴；反向水流喷嘴；反向流动型喷嘴
reverse flow operation 变向车流运行；变向车流交通
reverse flow roadway 变向车流道路
reverse flow system 回流系统
reverse flow technique 逆流技术
reverse flow theorem 反流定理
reverse flow tray 回流型塔板
reverse flow type combustor 回流（式）燃烧室
reverse flow valve 逆流阀；回流阀
reverse flushing tool 散热器洗涤装置
reverse fold 倒转褶皱
reverse fold in rock beds 岩层反向褶皱
reverse frame 副肋骨；反肋骨
reverse-frequency 相反频率
reverse funds 预备金
reverse gas flow cleaning 反吹清灰
reverse gate 反向门
reverse gear 逆转装置；回动装置；回动电控阀；倒挡齿轮；反向齿轮；回行机构；倒车轮；换向齿轮
reverse gear assembly 倒车齿轮总成
reverse gear block 倒车齿轮组
reverse gearbox 可逆变速箱；换向齿轮箱；双向变速箱
reverse gear clutch 回动齿轮离合器；反转齿轮离合器
reverse gear lever 换向齿轮操纵杆；倒挡操纵杆
reverse gear mechanism 倒车机构
reverse gear operating lever 换向齿轮操纵杆
reverse gear ratio 回动齿轮速比
reverse gear shift 倒车（调）挡
reverse gear shifter shaft 反转齿轮变速轴
reverse gear shifting fork 倒车换挡叉
reverse gear train 换向传动机构；换向齿轮系；回动电控阀
reverse grade 反向坡度
reverse graded bedding structure 反粒序层理构造
reverse graded structure 反粒序构造
reverse gradient 反坡；逆坡；反向梯度；反向坡度；反比降
reverse grading 反向递变
reverse graft 倒接
reverse gun 反向行走；倒车开行
reverse head 回动头
reverse head gasket 回动头衬
reverse hodograph 相遇时距曲线
reverse horn gate 倒牛角浇口
reverse idler 倒车用空套齿轮；反转空转齿轮；反向惰轮
reverse idler gear 换向空转（惰齿）轮；回动空转齿轮；倒挡中间齿轮；反向空转齿轮
reverse idler gear shaft 回动空转齿轮轴

reverse idler gear shaft bearing 回动空转齿轮轴轴承
reverse idler shaft 回动空转轮轴
reverse idler synchromesh 回动空转轮同步配合
reverse impact test 反面冲击试验
reverse impeller 反向叶轮
reverse inching turning 反转点动
reverse income tax 反所得税
reverse indexing 反向转位；反向索引
reverse in front of the station according to the proceeding direction 站前折返
reverse interlocker 回动联锁器
reverse interrupt 可逆中断；反向中断符；反向中断
reverse intersecting 逆向交叉【铁】
reverse interval 折返间隔时间
reverse isotope dilution analysis 逆同位素稀释分析
reverse isotopic dilution 逆同位素稀释
reverse isotopic dilution method 逆同位素稀释法
reverse isotopic exchange method 逆同位素交换法
reverse jato 制动用喷气发动机；喷气刹车；反向助飞器
reverse jet 除灰吹管
reverse jet bag filter 反吹风袋集尘器
reverse jet type duster 反喷式振动落料器
reverse J-distribution 反J形分布
reverse key 换向电键；反向开关
reverse key head bolt 回动键头螺钉
reverse key shackle 扭向卸扣
reverse laid rope 交叉捻钢丝绳；逆捻钢丝绳
reverse lane 变向车道
reverse latch 倒车保险器
reverse lay 反向捻；反捻
reverse leach process 反浸法
reverse leading 反向引导
reverse leakage 逆向杠杆作用
reverse leakage current 反向漏泄电流
reverse lever 回动杆；换向杠杆；反转杠杆
reverse lever catch 回动杆挡
reverse lever catch cover 回动杆挡盖
reverse lever foot rest 回动杆踏板
reverse lever fulcrum pin 回动杆支销
reverse lever handle 回动杆柄
reverse lever latch link pin 回动杆弹键连杆销
reverse lever spring 回动杆簧
reverse limiting valve 回动限制阀
reverse link 回动杆
reverse load 反向负载
reverse locking 反位锁闭
reverse lock magnet 反位锁钩电磁铁
reverse loop 折返环线；反面线圈
reverse loop curve 回头弯道；回头曲线；回头环线；发针形曲线
reversely curved cable net 反曲缆网
reverse mainspring 反转发条
reverse manner 逆序
reverse matching 对纹拼板
reverse matrix 逆矩阵
reverse measurement 返测【测】
reverse mechanism 换向机构
reverse merge 反向归并
reverse micelle extraction 反胶团萃取
reverse motion 逆转
reverse mo(u)ld 倒型；倒模
reverse movement 逆动；反回运动
reverse mutation 回复突变
reverse of direction 换向
reverse of motor 电动机反转
reverse ogee mo(u)lding 反枭混线【建】
reverse one-way street 可变单向交通街道
reverse operating contact 反向动作接点
reverse osmosis 污水的反渗透净化；逆（向）渗透；反向渗透；反渗析；反渗（透）
reverse osmosis and ultrafiltration 反渗透和超过滤法
reverse osmosis apparatus 反渗透设备
reverse osmosis desalination 反渗透（法）脱盐
reverse osmosis device 反渗透渗漏装置
reverse osmosis filter 反渗透过滤器
reverse osmosis machine 反渗透机
reverse osmosis membrane 反渗透膜
reverse osmosis membrane technique 反渗透膜技术

reverse osmosis module 反渗透组件
reverse osmosis permeator 逆渗透渗漏计
reverse osmosis plant 反渗透净化水装置
reverse osmosis pressure 反渗透压力
reverse osmosis process 反渗透过程；反渗透法
reverse osmosis process system 反渗透工艺系统
reverse osmosis purification 逆渗透净化
reverse osmosis separation 逆渗透分离
reverse osmosis system 反渗透系统
reverse osmosis system design 反渗透系统设计
reverse osmosis system optimization 反渗透系统优化
reverse osmosis technology 反渗透技术
reverse osmosis test unit 反渗透测试仪
reverse osmosis treatment 反渗透处理
reverse osmosis ultrafiltration 反渗透超滤
reverse osmosis ultrafiltration membrane 反渗透超滤膜
reverse osmosis unit 逆渗透渗漏装置；反渗透装置；反渗透器
reverse osmosis water 反渗透脱水
reverse osmosis water purification unit 反渗透净化水装置
reverse osmotic treatment for sewage 污水反渗透处理法
reverse over-current relay 逆过载继电器；反向过载继电器
reverse particle tracking 反向颗粒跟踪
reverse peak voltage 反向峰（值电）压
reverse pedal 回动踏板
reverse penetration method 反贯入法
reverse-phase chromatography 反向色谱法
reverse phase fractionation method 反相分馏法
reverse phase high-performance liquid chromatography 反相高效液相色谱法
reverse phase layer chromatography 反相层析
reverse phase liquid chromatography 反相液相色谱法
reverse phase liquid-liquid chromatography 反相液液色谱法；反相液液层析
reverse phase protection 反相保护
reverse phase relay 反相继电器
reverse phase thin layer chromatography 反相薄层色谱法
reverse photogrammetry 逆反摄影测量
reverse photoproduction 逆光性
reverse pickup current 反极性（下的）启动电流
reverse pitch 反螺距
reverse pitch airscrew 反距螺旋桨
reverse plunger spring 倒挡定位柱销弹簧
reverse plunjet 回流式气动量塞
reverse point 反位道岔
reverse polarity 反接；反极性（接法）；反极性
reverse polarity charging 反极性充电
reverse polarity section 反极性剖面
reverse polarity smear 反极性曳尾
reverse Pollish notation 反向波兰表示法
reverse polymerization 反聚合；反向聚合
reverse position 反转位置；反位；反常位置
reverse potential 反向电势
reverse-power circuit breaker 逆功率断路器
reverse-power protection 逆功率保护
reverse-power protector 逆功率保护装置
reverse-power relay 反向功率继电器；逆功率继电器
reverse-power tripping 逆功率动作
reverse-power tripping device 逆功率脱扣装置
reverse preference 逆优惠；反向优惠
reverse pressure 反压力
reverse price-fixing 反向定价
reverse printing 背面印刷
reverse procedure 相反步骤
reverse process 反转过程
reverse projection control 倒影製
reverse pulse cleaning 反向脉冲清袋
reverse-pumped Raman laser 反馈泵拉曼激光器
reverser 自动反转轮；换向器；换向开关；反演机构；反向器；反向机构
reverse rate constant 逆反应速率常数
reverse reaction 逆反应
reverse read 反读
reverse reaming 反向扩孔
reverse recovery current 反向恢复电流
reverse recovery time 反向恢复时间
reverse redrawing 反向再拉延

reverse regulation 反调节
reverse remittance 逆汇
reverse repeater 反围盘
reverse repurchase agreement 逆重购协议
reverse repurchases 反转再购回
reverse resistance 反(向)电阻
reverse rig 反向回转钻机
reverse ripper 后挂松土耙
reverser lock relay 换向开关锁定继电器;反向器锁定继电器
reverser mounted 自动导轮
reverse rod 回动杆
reverse roll 逆向辊
reverse roll coater 逆向辊涂机
reverse-roll coating 逆辊涂布
reverse roller coating 逆向辊涂
reverse rotary drilling 反向旋转式(岩芯)钻进;反循环回转钻进;反向旋转钻进(法)
reverse rotary rig 反循环钻机
reverse rotation 反向旋转;反向回转
reverse rotation drive solenoid 反转驱动电磁线圈
reverse rotation solenoid 反转电磁线圈
reverse running 逆向运行;回程;倒转;回动
reverse sampler 颠倒式采样器
reverse saturation analysis 逆饱和分析
reverse saturation current 反向饱和电流
reverse scan 撤销扫描;反向扫描
reverse scissors gap 反剪刀差
reverse screw 回动丝杠
reverse selection 反向选择
reverse selector 回动选择器
reverse-separation fault 反向错位断层
reverse servo 回动伺服机构;反向伺服机构
reverse shaft 回动轴
reverse shift fork 回动挡叉
reverse shooting 逆向爆炸
reverse side 反面
reverse side of tooth 轮齿背侧
reverse side welding electrode 根部施焊用焊条
reverse sight 反视
reverse sign 反号;变向减号指令;变号指令
reverse signal 反转信号;反向信号
reverse similar fold 反相似褶皱
reverse sliding gear 倒车滑动齿轮
reverse slip fault 逆(滑)断层;冲断层
reverse slope 脊坡;反向斜坡;反比降
reverse slow motion 慢倒动作
reverse speed 倒挡速度
reverse speed operation 逆转运行;反向运行
reverse split 聚集股份;反向析股
reverse spring 还原弹簧
reverse stereo 反立体
reverse stock split-up 反向析股
reverse stop device 逆止器
reverse strain 反向应变;反向形变
reverse stress 反向应力;逆应力
reverse stroke 反向行程;反回行程
reverse superelevation 反超高
reverse surface 反面
reverse-swing door 反向(摇)门
reverse switching loss 反向开关损耗
reverse switch position 道岔反位
reverse symmetry 反向对称
reverse takeover 反向接管企业;反向兼并
reverse tape feed 反向送带
reverse taper ream 倒锥形钻孔;倒锥形铰孔;倒锤形钻孔
reverse telescope 倒镜
reverse temperature 反转温度
reverse tension 反向张力
reverse thrust 逆推力;反推力
reverse thrust device 推力反向器
reverse thrust nozzle 反向推力喷管
reverse thrust unit 反向推力装置
reverse tiller 反装柄舵
reverse time 折返时间
reverse-time migration 逆时间偏移
reverse timing 提前成圈(俗称前吃)
reverse toe 反缝头
reverse tool thrust 反向刀具推力
reverse torque arm 反力矩杆
reverse torsion 反复扭转
reverse torsion(al) machine 扭转疲劳试验机
reverse torsion fatigue testing machine 反向扭转疲劳试验机
reverse torsion ring 反扭曲环
reverse torsion test 反复扭转试验
reverse track 折返线
reverse transfer impedance 反向转移阻抗
reverse transformation 逆相变;逆变换
reverse transformation of martensite 马氏体可逆转变
reverse transmitter 反向发射器
reverse trap 水厕
reverse trap water closet (带存水弯的)水厕
reverse trend adjustment 反向趋势调整
reverse turning bed 翻板
reverse turn sign 短反向急弯标志;之字形急弯标志
reverse twill 山形斜纹
reverse valve 可逆阀;回动阀;反向阀
reverse veneer 反面胶合板
reverse video 可逆视频终端;反向显示;反相显示
reverse wave 逆行波;反向波
reverse-wetting additive 改变湿润添加剂
reverse wheel 换向轮
reverse work(ing) 反向工作法
reverse working value 反极性工作值
reverse-wrench fault 逆平移断层
reverse yield gap 逆收益差额;反收益率差距;反(向)收益差距
reverse yielding 反向屈服
reverse zoning 逆向分带
reversibility 两面可用性;可塑性;可逆性;反转性
reversibility condition 可逆性条件
reversibility of adsorption 吸附的可逆性
reversibility of states 物态逆变
reversibility principle 可逆性原理;微观可逆性
reversible 可逆(性)的;可反转的;回行的;双向的
reversible absorption current 反向吸收电流
reversible addition reaction 可加成反应
reversible adiabatic curve 可逆绝热曲线
reversible airscrew 可逆式螺旋桨
reversible amplifier 可逆放大器
reversible axial flow fan 可逆轴流式风扇
reversible ballasting system 可逆压载系统
reversible belt 可逆转皮带运输机
reversible belt conveyer 可逆运转的带式输送机
reversible billet mill 可逆式钢坯轧机
reversible binary counter 可逆二进位制计数器
reversible blade 翻换式锄铲
reversible bleachable dye solution 可逆漂白染料液
reversible booster 可逆增压机;可逆升压器
reversible bucket 正铲反铲通用铲斗
reversible cell 可逆电池
reversible center path 可变向的中央车道
reversible chain block 倒链滑车
reversible chain tong 双面链钳
reversible change 可塑变化;可逆变化
reversible chemical reaction 可逆的化学反应
reversible circulation valve 双(向)循环阀
reversible close-quarter pneumatic drill 狭小地方用的风钻;双向风钻
reversible cloth 双面织物
reversible coagulation 可逆凝聚
reversible colloid 可逆胶体
reversible compression 可逆压缩
reversible control 反向操纵
reversible controller 双向控制器
reversible conversion 可逆变换
reversible copying unit 可逆转的印刷装置
reversible counter 可逆计数器;双向计数器
reversible creep 可逆徐变
reversible cutter 反面用切刀;可逆切刀
reversible cutter bit 可转刀片
reversible cutting insert 可转动刀片;可换刀片衬垫
reversible cycle 可逆循环
reversible decade counter 可逆十进位计数器
reversible deflection 可恢复性弯沉
reversible deflector 换向挡板;反折板
reversible deformation 可逆变形;反变形;变形恢复
reversible diaphragm piston type pneumatic spring 双向膜片活塞式空气弹簧
reversible diffusion 可逆扩散
reversible disk plow 圆盘翻转双向犁
reversible drive 可逆传动
reversible electric(al) servomotor 正反转伺服电动机
reversible electrode 可逆电极;逆向电极
reversible electrolysis 可逆电解
reversible emulsion 可逆乳剂
reversible encoding 可逆编码
reversible endless-rope haulage 往复式无极绳运输
reversible endless-rope system 往复式无极绳运输系统
reversible engine 可逆(式)发动机;可逆机;可反转发动机
reversible expansion 抗逆性膨胀;可逆膨胀
reversible fan 可反转的鼓风机;可逆转风扇;可反转的扇风机;换向风扇
reversible film 两面可用软片;可逆膜片;反转片;反转薄膜
reversible flat band conveyer 可逆带式输送机
reversible flow 可逆流
reversible flow sign 改变车道通行方向标志
reversible flue 可换式烟道;双向车道
reversible fore-carriage plow 前导轮架式双向犁
reversible function 可逆功能
reversible gas turbine 可逆转式燃气轮机
reversible gate 双向闸机
reversible gel 可逆性凝胶
reversible generator motor 可逆式发电电动机
reversible grating 可翻转的格栅
reversible grease 可润滑脂
reversible hammer breaker 可逆锤式破碎机
reversible hammer crusher 可逆锤式破碎机
reversible hammer mill 可逆锤式破碎机
reversible handle 可把扶手;双向把手
reversible heat engine 可逆热机
reversible high-voltage direct current system 双向输电高压直流系统
reversible hinge 双面可折铰链;可反合页
reversible hydraulic machine 可逆式水力机械
reversible hydrolysis 可逆水解
reversible impactor 可逆型反击式破碎机
reversible impact type hammer crusher 可逆击式锤碎机
reversible insert 可换镶块
reversible jaw 可逆爪
reversible jib 可反向吊机臂;能反向吊机臂
reversible key 预热;两向键
reversible knife 两刃刀
reversible lane 变向车道;可反向车道;可逆行车道
reversible lay days 可互补碇泊期限;可调剂用的装卸日;可调剂试验的装卸日数;并用装卸期
reversible level 可逆电平;可倒水准器;活整水准仪;回转式水准仪;回转式水平仪;翻转式水准仪
reversible level tube 可倒水准管
reversible lifeboat 可翻转救生筏
reversible line 双向传输线
reversible link 双向传输线路
reversible lock 双向锁
reversible magnetic ater effect 可逆磁性后效
reversible magnetic process 可逆磁化过程
reversible magnetization 可逆性磁化
reversible Markov chain 可逆马尔可夫链
reversible medium 可逆媒质
reversible mill 可逆式轧机
reversible mirror 反像平面镜
reversible mortise lock 可反向插锁;反向插锁
reversible motor 可逆电动机;可反转电动机;可反向电动机
reversible motorcoach train 双向运行动车列车
reversible one-way street 可变单向交通街道;往复变向的单向交通街道
reversible operation 可逆运转
reversible optic(al) damage 可逆光损伤
reversible optic(al) recording medium 可逆光学记录媒质
reversible pallet 可反转用的平板架;可反转用的抹子
reversible pan conveyer 可逆裙板输送机;可逆盘式输送机
reversible path 可逆路程
reversible pattern plate 可逆模板;双面型板
reversible pawl 可逆爪
reversible pendulum 可逆摆(锤);可倒摆
reversible permeability 可逆性磁导率;可逆透磁率
reversible pitch propeller 反距螺旋桨
reversible plow 翻转犁
reversible precipitation 可逆沉淀(作用)
reversible pressure curve 可逆压力曲线
reversible process 可逆过程;可逆程序

reversible propeller 活叶螺旋桨
reversible pump 可逆泵；双向旋转泵
reversible pump-turbine 可逆式（水）泵—水轮机
reversible ratchet 可逆转的棘轮；可逆棘轮机构
reversible reaction 可逆反应
reversible recording 可逆记录
reversible ring counter 可逆环形计数器
reversible ring spanner chrome plated 镀铬棘轮梅花扳手
reversible rod 双面标尺
reversible rotation 可逆转动
reversible saturation-adiabatic process 可逆饱和绝热过程
reversible saw 两刃锯；双刃锯
reversible seat 翻椅
reversible series starting rheostat 可逆串联起动变阻器
reversible share 转壁式双向犁桦
reversible shovel 双尖翻换松土铲；翻换松土铲
reversible shrinkage 可逆的收缩
reversible side plough 双向可逆侧犁
reversible siding 可翻转护墙板；可翻转用的壁板
reversible single revolution clutch 单向反转离合器
reversible sol 可逆性溶胶
reversible sprocket 可逆链轮
reversible staff 双面标尺
reversible steel 双尖松土锄铲
reversible steering 可逆转向
reversible steering gear 可逆转向装置；可逆（式）转向器；可逆换向装置
reversible swelling 可逆溶胀
reversible switch 可逆转的开关
reversible television channel 双向发射电视信道
reversible temperature-indicating coating 可逆性示温涂料
reversible temperature-indicating pigment 可逆性示温颜料
reversible thermopaint 可逆性示温涂料
reversible toxic effect 可逆性毒作用
reversible track 可逆向运行线路
reversible tractor 双向拖拉机
reversible traffic lane 双向车行道
reversible tramway 往复（式架空）索道
reversible transducer 可逆换能器；双向变换器
reversible transformation 可逆转变；可逆变换；双向转变
reversible transit circle 可反转式子午仪
reversible turbine 可逆涡轮机；水泵水轮机；可逆式水轮机
reversible TV channel 双向电视通道
reversible unit 周转总成
reversible valve 可逆阀
reversible variation 可逆变化
reversible warp backed weave 双面经二重织物
reversible wave 可逆波
reversible winch 可逆绞车
reversible window 可反转的窗；立转窗
reversing 回动；换向
reversing air cylinder 回动风缸
reversing airway 反风道
reversing amplifier 倒相放大器
reversing arm 回动臂
reversing arrangement 换向装置
reversing bar 换向杆
reversing blade 转向导片
reversing bloomer 可逆式初轧机
reversing blooming mill 可逆式初轧机
reversing carburetor 倒置气化器
reversing chain tong 双面链钳
reversing chamber 回火室
reversing chassis 回转底盘
reversing clutch 逆转离合器；倒出离合器；正倒车离合器；可逆离合器；换向离合器；反转离合器；双向离合器
reversing cock 回动旋塞
reversing cogging mill 可逆式初轧机
reversing cold mill 可逆式冷轧机
reversing colo(u)r sequence 变换彩色相序
reversing commutator 电流方向转换器；反转换开关
reversing contact 可逆触点
reversing contactor 转换接触器；换向接触器
reversing controller 可逆控制器；可逆调节器；反向控制器

reversing correspondence 反序对应
reversing current 可逆水流；逆流；往复（潮）流
reversing damper 换向闸板
reversing device 反向装置；反风设备
reversing dog 反向触件
reversing drill 反转钻孔机
reversing drive 反向传动
reversing drum 回动鼓轮
reversing drum discharge 搅拌筒反转出料
reversing drum mixer 转向滚筒搅拌机
reversing drum switch 倒顺鼓形开关
reversing engine 可逆发动机；倒转发动机
reversing entry 转回分录；反向分录
reversing exchanger 可逆交换器
reversing eyepiece 反像目镜
reversing facility 反风设备
reversing feed 反向进给
reversing field 反向场
reversing flow 可逆水流
reversing flue 回转火道
reversing four way valve 四通回动阀
reversing frame for truss frames 构架翻转机
reversing gear 逆转装置；逆转机构；回动装置；回动机构；回动齿轮；换向机构；换向齿轮；反转装置；反向齿轮
reversing gearbox 逆转机构；换向齿轮箱
reversing gear catch 倒车挡
reversing gear control bracket 换向控制架
reversing gear mechanism 换向齿轮机构
reversing gear spindle 倒向齿轮轴
reversing gravity die 双面金属型
reversing handle 反转手柄
reversing index 反转标牌
reversing interlocker 换向互锁器
reversing key 换向键
reversing layer 反变层
reversing lever 换向手柄；反转杠杆；反向拉杆
reversing light 倒车灯
reversing line 掉头线
reversing link 换向月牙板；换向联杆
reversing loop 迂回环线
reversing machine 换向机
reversing mechanism 回动机构；换向机构；反向机构
reversing mill 可逆式轧机
reversing mirror 反像镜
reversing motion 反向运动
reversing motor 双向旋转马达；双向（旋转）电动机
reversing movement 反向运动
reversing nozzle 二次供给喷嘴
reversing piston 换向活塞
reversing plate 回动板
reversing plate mill 可逆式钢板轧机
reversing plug 反向开关
reversing preference 可逆偏好
reversing prism 转像棱镜；倒像棱镜；反像棱镜
reversing pump 逆动泵；换向泵；反向泵
reversing rod 换向杆
reversing rolling stand 可逆式轧机座
reversing rotation 反向转动
reversing rotation of motors 改变电动机的旋转方向
reversing rougher 可逆式粗轧机座
reversing roughing stand 可逆式粗轧机座
reversing rudder 倒车舵
reversing screw 回动螺旋
reversing screw bolt 回动螺栓
reversing screw bottom guide 回动螺杆下导板
reversing screw gudgeon 回动螺旋
reversing screw top guide 回动螺旋上导板
reversing shaft 倒挡轴；反转轴
reversing shape mill 可逆式型钢轧机
reversing shear box test 往复直剪试验
reversing shifting fork 倒挡变速叉
reversing slide valve 换向滑阀
reversing solenoid valve 换向电磁阀
reversing starter 可逆起动器；双向起动器
reversing station 终点站；折返站
reversing stroke 返回行程
reversing stud 换像组
reversing switch 换向开关；换接开关；倒向开关；反向开关
reversing switch contact 换接开关触点
reversing switch lever 回动开关柄

reversing test 反转试验
reversing thermometer 颠倒温度计；颠倒温度表；倒转（自记）温度计
reversing thermometer method 颠倒温度计法
reversing tidal current 往复潮流
reversing tide 往复潮
reversing time 掉头时间；反转时间
reversing tool 反扣打捞工具
reversing track 折返轨
reversing transmission 反向传动
reversing triangle 转向三角线
reversing turbine 可逆转式涡轮机；倒车透平；倒车汽轮机
reversing two-high mill 可逆式二辊轧机
reversing valve 可逆阀；逆流阀；回动阀；换向阀；反向阀
reversing valve buffer gasket 换向阀缓冲垫
reversing vibrating circular pipeline 振动环形管道可逆输送（系统）
reversing water bottle 南森瓶；水样颠倒采样瓶；颠倒采水器
reversing wheel 回动轮；反转手轮
reversing winch 提升绞车
reversion 逆转；回复变异；回复；复归；返租；返原【给】；返流；反演
reversionary bonus 复归的红利
reversionary interest 可归回收益
reversionary lease 未来租契
reversion chamber 转流室
reversion factor 资金返还系数
reversion of kerosene 煤油的储存变质
reversion of series 级数反演
reversion of test 检查性试验
reversion reaction 逆变作用
reversion type mixer 反转出料搅拌机
reversion value 资金返还价值
revert 下脚料；返料
revertant 逆转株
reverted gear train 回归轮系
reverted image 镜像
reverted train 双向传动轮系
reverted true bearing 逆真方位
revertex 蒸浓胶乳
revertible flue 可换向烟道；双向烟道
revertive control 反控制
revertive control of impulsing 脉冲反控制
revertive control system 反控制系统
revertive impulse 回送脉冲；反脉冲
revertive impulse control 反脉冲控制
revertive pulses 回送脉冲
revert metallics 金属物返料
revert scrap 回炉废钢；回收废物；返回废钢
revesal 反行程
revesing permanent mould 双面金属型
revest 重新投资
revesting 物归原主
revestry 祭具室；祭服室
revetment 护堤壁；护岸；砌石面；土堡
revetment dike 护堤
revetment of slope 护坡
revetment wall 护墙【港】；护坡墙；护岸墙
revetment works 护岸工程
revetted shelter trench 掩蔽沟
revetting 护岸设施；护岸（工程）
Revex 直齿锥齿轮粗拉法
revibration 再振捣；再振捣；重复振动；重复振捣
review 考查；检查；评估；审查；重复观察；复验；复审
review appraiser 评估复审员
reviewing ground 阅兵场
reviewing stand 观礼台；检阅台
reviewing the performances of all workers 人事考核
review meeting of standard 标准审查会
review of extension research 推广研究评论
review of laboratory tests 试验室试验述评
review of tenders 投标审查
review on special information 专业文献综述
review panel 审查小组
review paper 综合介绍
review research 评论研究
revise 修订；更正；改样；改编；校阅
revise a contract 修改合同
revise a schedule 调整进度表
revised 改进的

revised appropriation 修订后的拨款
revised budget 修正(后的)预算
revised cash balance equation 修正的现金余额方程式
revised contract 修正合同
revised death-rate 修正死亡率
revised design 修正的设计;修改(了的)设计;修改了的设计;已修改的设计;改进设计
revised documents 修正文件
revised drawing 修正图
revised edition 修订版
revised estimate 修整预算;订正概算
revised index 修正指数
revised map 修测图
revised minimum standard model 修订最低限度标准模型
revised optimum allocation 修订的最优分配
revised parsing algorithm 修正分析算法
revised price 修正后价格
revised proof 审校参考图
revised rates 修正率
revised recognizer 修正识别算法
revised simplex method 修正的单纯方法;修改单纯形法;修订单体法
revised standard version 修订标准版
revised statutes 修改后的法规(美国)
revised treaty 修正条约
revise in field 实地检测
revise on ground 实地检测
revise production plans 调整生产计划
revision 修改;修订本;修订版;改版;校订;订正
revision cycle 更新周期
revision directive 修正指令
revision in layouts 修正布置
revision note 修订说明
revision notice 更正通知
revision of contract 修改合同
revision of line maps 线划地图修测
revision of map 图纸校正
revision of plan 修改计划
revision of test 校核试验;复核试验
revision of topographic(al) map 地形图修订
revision of topographic(al) map maintenance of topographic(al) map 地形图修测
revision on drawing 修改图纸
revision point list 修正的三角形成果表;订正的三角点成果表
revisions in layouts 修改布置
revision survey 修正测量;修测;再次校测;再次检测
revision test 校核试验;重复试验
revisit interval 重复观测周期
revitalize 使新生;恢复活力;休养生息
revival 复苏
revival architecture 复旧建筑;复古建筑
revivalism 复古主义
Revival of Learning 文艺复兴
revival of policies 保险单的再生效
revival of trade 贸易复苏
revival sunspot group 重现黑子群
revive 复苏
revived fault 复活断层
revived fault scarp 更生断层崖;回春断层崖;复活断层崖
revived fold(ing) 复活褶皱
revived Gothic style 哥特复兴式
revived range 更新范围
revived river 更生河;复苏河;复活河
revived stream 更生河;回春河;复苏河;复活河
revived zone 更新地带
revive geosyncline 再生地槽
reviver 再生铅
revivification 再生(作用);复苏
revivification of solution 溶液的再生
revivifier 叶片式松砂机;再生器;交流换热器;复活剂
revivify 再活化;活性恢复
reviving action 返青作用
rev limit 转速极限
revocable 可废除的;非固定的
revocable credit 可取消的信用证;可取消的信贷;可撤销的信用证
revocable letter of credit 可取消的信用证;可撤销的信用证
revocable offer 可撤销的报价
revocable transfer 可撤销财产转移

revocable trust 可撤销信托
revocable unconfirmed banker's credit 可撤销无保兑银行信用证;可撤销的银行未保兑信用证
revocation of contract 解除合同
revocation of licence 撤销执照
revoke 收回;撤回
revoke an offer 收回报价
revoliving arm 回转臂
revoluting gate 转动式闸门
revoluting loan 循环贷款;自动展期贷款
revoluting speed 转速
revolution 旋转;周转;公转;回转;绕转;变革
revolution alarm 转速警报器
revolutional shell 旋转薄壳
revolutional slip 转差率
revolutionary geosyncline 激进式地槽
revolutionary shell 旋转壳
revolution axis 旋转轴;转轴;回转轴
revolution body 回转体
revolution coefficient 旋转系数
revolution core type relay 旋转铁芯式继电器
revolution counter 计转数表;转速计;转速表;转数器;转数计;转数表
revolution counter drive quill 转数表传动轴
revolution detector 转速检测器
revolution door 旋转门;转门
revolution drop 转速下降;转数降低
revolution frequency 绕转频率
revolution indicator 转速指示器;转速计;转速表;转数指示器;转数计;转数表
revolution mark 切削痕迹
revolution meter 旋转计数器;转速计;转数表
revolution of plane 平面旋转
revolution of the earth 地球公转
revolution period 公转周期;绕转周期
revolution per minute 每分钟转数
revolution per second 每秒钟转数
revolution per unit of time 单位时间转数
revolution ratio 回转比
revolution recorder 转数记录器
revolutions of twisting 扭转圈数
revolution speed of screw 螺旋转速
revolution-speed table 转数航速表
revolution speed transducer 转速传感器
revolutions per hour 转/小时
revolutions per minute 转数/分;转/分;每分钟转动次数
revolutions per second 转数/秒;转/秒;每秒转数;每秒旋转次数
revolution stabilizer 转数稳定器
revolution stop 转速限制器
revolution surface 回转面
revolution value 周值
revolution window 转窗
revolve around an axis 绕轴旋转
revolved section 旋转剖视;旋转断面
revolver 旋转器;不定位置圆
revolver camera 转筒式摄影机
revolver dryer 转筒式烘干机;转筒式干燥机;转筒式干燥机
revolver gantry crane 旋转移动门式起重机
revolver press 旋转压砖机;回转式压砖机
revolver track 快速循环取数区
revolver type multiple punch 复式冲模;多凸模冲模;多滑块压力机;手枪式多穿孔
revolving 旋转;绕转
revolving air plane hangar 旋转式飞机库
revolving antenna 旋转天线
revolving-armature alternator 旋转电枢交流发电机
revolving-armature type 旋转电枢区
revolving-armature type alternator 旋转电枢式发电机
revolving-armature type machine 旋转电枢式电机
revolving arm mixer 叶片式混砂机
revolving ash pan 旋转灰盘
revolving ash table 旋转灰板
revolving barrel 转筒
revolving batch charger 旋转式加供料器;旋转式加给料器
revolving beacon light 旋转信标灯
revolving beam station 旋转波束无线电信标台
revolving bed 旋转床
revolving bedplate 转盘;转台
revolving bell mouth 旋转式喇叭口

revolving black ash furnace 旋转黑灰炉
revolving blade mixer 旋转桨叶搅拌机;转叶式搅拌机;转叶片搅拌机
revolving-block engine 旋转缸体式发动机
revolving blow-out preventer 旋转式防喷器
revolving body 回转体
revolving boiler 旋转锅炉
revolving bookstand 旋转书架
revolving boom crane 旋转伸臂起重机
revolving brush 旋转刷;旋转路刷;旋转路刷
revolving bucket 转斗;回转斗
revolving bundle holder 盘条打捆机
revolving burner 旋转炉
revolving card file 旋转式卡片档
revolving centre 活顶尖
revolving chain locker 旋转式锚链舱
revolving chair 转椅
revolving chalkboard 翻转式黑板
revolving charge account 限额赊销账户
revolving coil 旋转线圈
revolving conveyer boom 旋转输送机臂
revolving-core type 旋转铁芯型
revolving counter 转数计数器
revolving crane 旋转(式)起重机;旋转桁车;立柱式旋臂起重机;回转起重机
revolving credit 循环贷款;限额循环周转信贷;周转信用卡
revolving credit account 限额循环周转信贷账户
revolving credit agreement 限额循环周转信贷合同
revolving credit plan 限额循环周转信贷计划;周转信贷计划
revolving credit technique 转期信贷技术
revolving cumulative credit 可累积定期循环信用证
revolving cupola 冲天转炉
revolving cutter 转刀;回转刀片
revolving cutter head 旋转铣刀头
revolving cylinder 旋转汽缸
revolving cylinder engine 转缸式发动机
revolving damper 转动式风门
revolving derrick 旋转式起重机
revolving diaphragm 回转光圈
revolving die 转模
revolving die-hammer 旋转打号槌
revolving directional radio beacon 旋转定向无线电导向台
revolving disc 旋转圆盘(刀)
revolving disc crane 旋盘式起重机
revolving disc grizzly 圆盘(滚轴)筛
revolving disc shutter 旋转叶片式快门
revolving distributor 旋转布水器;旋转喷水器;旋围喷水器
revolving dome 旋转穹窿;旋转圆屋顶
revolving door 铰接门;旋门;转门;回门;暗室防光转门
revolving door casing 转门门套
revolving door enclosure 转门门套
revolving door wing 转门门扇
revolving draw bridge 旋转吊桥
revolving dryer 卧式烘砂滚筒;转筒烘干机;转筒干燥器
revolving drum 转筒;转鼓;滚筒;回转筒
revolving drum concrete mixer 转式混凝土搅拌机;转筒式混凝土拌和机
revolving drum drier[dryer] 转筒式烘干机;转筒式干燥机;转筒式干燥机
revolving-drum hopper 转鼓式肥料斗
revolving drum mixer 转筒式搅拌机;转筒式拌和机;转鼓式拌和机
revolving drum press 转鼓式压机
revolving drum screen 转筒筛(算)
revolving drum stroboscope 旋转轮闪频仪
revolving drum truck 转筒式搅拌车;转筒式输送车
revolving drum type mixer 转筒式搅拌机
revolving drum washer 旋转鼓式洗涤器;转筒式洗涤器
revolving dump car 旋转卸料车;翻转卸载式矿车
revolving dumper 旋转式翻车机
revolving dynamo 旋转发动机
revolving electro-motive force 旋转电动势
revolving escapement 旋转擒纵机构
revolving excavating processing 转罐挖土法
revolving excavator 旋转挖土机;旋转式挖掘机
revolving expander 旋转式扩幅辊
revolving expeller 旋转发射器

revolving feeder 回转喂料器;回转喂料机
revolving feed table 旋转供料台;转盘加料器
revolving field 旋转场
revolving-field alternator 旋转磁场交流发电机
revolving field theory 旋转场理论
revolving field type machine 转场式电机
revolving field type motor 转场式电动机
revolving field type synchronous alternator 旋转磁场式同步发电机
revolving filter 回转过滤器
revolving flat card 圆转盖板梳棉板
revolving floor restaurant 旋转餐厅
revolving flowmeter 涡轮流量计
revolving force 旋转力
revolving frame 转架;回转框架
revolving fund 循环基金;周转(资)金;周转基金
revolving furnace 旋转炉
revolving gantry crane 旋转式移动门式起重机
revolving gate 旋转式闸门;转门
revolving gear 旋转机构
revolving grate 旋转炉箅
revolving grate producer 旋转炉箅发生炉
revolving grid 旋转筛
revolving grizzlies feeder 转筛式给料机
revolving grizzly 滚轴筛
revolving hammer 回动锤
revolving hatch 旋转舱口
revolving head 六角刀架;回转磁头
revolving head punch 有六角刀架的冲床
revolving hexagonal drum test 六角形滚筒回转试验
revolving hollow spindle 旋转空心轴
revolving hopper 旋转式漏斗
revolving horizontal drum 旋转卧式鼓轮
revolving joint 旋转接头
revolving knife 旋转切刀
revolving ladder 旋梯
revolving leaf 转叶;转门门扇
revolving-leaf type filter 转盘过滤机
revolving light 旋转灯(光);转动灯标
revolving light screen 旋转式遮光罩
revolving light shade 旋转式遮光罩
revolving line of credit 循环信贷限额;限额循环周转信贷额度;周转信贷额度
revolving loading and unloading crane 旋转式装卸起重机
revolving loading floor restaurant 旋转楼板餐厅
revolving loan 自动延期贷款;循环贷款法;周转次款
revolving machine 旋转机
revolving magnetic field 旋转磁场
revolving mast-type jib crane 定臂转柱起重机
revolving member 转动件;回转构件
revolving metal table 旋转圆盘;研磨台
revolving microprobe stage 旋转式显微探针台架
revolving mirror 旋转镜
revolving mirror stroboscope 旋转镜闪频仪
revolving mixer 旋转式混合机
revolving nosepiece 显微镜换镜旋座
revolving nosepiece of microscope 显微镜的换镜旋座
revolving nozzle 旋转喷头
revolving packerhead 旋转打包机
revolving paddle finisher 旋桨修整机
revolving paddle mixer 叶片式混砂机
revolving panel 旋转板
revolving part 旋转件
revolving pillar jib crane 立柱式旋臂起重机;回旋立柱臂杆起重机
revolving planisphere 旋转星座图
revolving plate 回转板
revolving plate feeder 转盘给料机;盘式加料机;盘式给料器;盘式给料
revolving platform 回转台;转盘;旋转台
revolving plow reclaimer 旋转式叶轮给料机;叶轮给料机
revolving position 旋转位置
revolving pot 回转均料盘
revolving press 旋转式压制机
revolving pulp distributor 旋转式矿浆分配器
revolving punch plier 六角冲孔钳
revolving radio beacon 旋转无线电指向标;旋转(无线电)航标
revolving restaurant 旋转餐厅
revolving rheostat 旋桨变阻器

revolving ring 油环;转环
revolving ripper 旋转式松土器;回转式松土机
revolving roll 旋转式辊筒
revolving screen 旋转式栅;旋转(式)筛;旋转格网;旋筒筛;转筛;回转筛;平面旋回筛;多边筛
revolving screen washer 转筒筛式洗矿机
revolving scrubber 旋转式洗涤器;旋转式擦洗机
revolving section view 旋转剖视图
revolving shaft 旋转轴;回转轴
revolving shelf 旋转架;旋转搁板
revolving shovel 回转式挖掘机;旋转挖土机;旋转式挖掘机;旋转式铲;旋转铲(土机)
revolving shutter 旋转遮光器;卷筒百叶窗;卷帘窗
revolving sickle bar 旋转式切割器
revolving sieve 回转筛
revolving sleeve 匀料筒
revolving sludge scraper 旋转清淤机;旋转式污泥清除机
revolving snow plow 掘雪犁
revolving solid 旋转体;转成体
revolving sorting table 转动拣选台
revolving spade 旋转锄
revolving spindle 旋转轴
revolving spiral classifier 转动螺旋分级机
revolving sprinkler 旋转式洒水器;旋转式洒水器;旋转式喷水器;旋转式喷灌机
revolving sprinkler head 回转式喷头
revolving stacker 旋转式堆料机
revolving stage 旋转(式)舞台
revolving stage with disc and outer ring 双圈转台
revolving steam shovel 旋转式蒸汽挖土机;回转式蒸汽挖土机;回转式汽铲
revolving storm 旋转风暴;热带风暴
revolving superstructure (起重机的)上部转台
revolving swap 循环交换
revolving table 旋转工作台
revolving table concrete block press 转台式混凝土砌块压制机
revolving table feeder 圆盘给料机;转盘加料器
revolving table press 转台压制机;转盘式压(团)机
revolving tackboard 旋转式布告板
revolving times of circulating capital 流动资金周转次数
revolving tool box saddle 转塔刀架
revolving tool holder 回转刀架
revolving top 旋转布料器
revolving track crane 回转式轨道起重机
revolving tube carrier 旋转套管架
revolving tubular kiln 回转窑
revolving turret 转盘式换筒装置
revolving type of cutter head 旋转式铣头;旋转式切削头;旋转式刀头
revolving underwriting facility 循环式包销协议;循环承销融资;周转性包销便利
revolving valve 旋转滑阀;回转阀
revolving valve mechanism 旋转阀分配机构
revolving vane 旋转叶片
revolving vase 套环瓶(转心瓶)
revolving vibrating screen 振动式转筒筛分机
revolving washing screen 洗矿转筒筛
revolving wheel 旋转砂轮
revolving window 旋转窗;转窗
revolving wing 转门门扇
revolving/wing crane 摇臂起锚机
revulcanization 再次硫化
revulcanize 再硫化
revulsion 诱导法
revulsion of capital 资本抽回
revulsiva 诱导剂
revultex 浓缩硫化乳胶
rev up 增大转速
rev up the engine 起动发动机
revved up 提高转数
reward 奖金;酬金;报答;报酬
reward and bonus 奖金
reward and penalty system of energy consumption 能源消耗奖惩制
reward and punishment 奖惩
reward and punishment by economic means 经济奖惩制
reward distribution 奖金分配
reward for go-between 好处费
rewarding by merit 按劳付酬
reward system on time-rate basis 计时奖励工资

reward-to-variability ratio 收益对风险比率
rewarming 复温
re-wash box 再洗跳汰机
re-wash jig 再选跳汰机
re-wash launder 再洗槽
re-wash overflow 再洗溢流
re-wash recirculation 再洗循环
re-weigh 再行衡量;再行称量;重新衡量;重新称量
re-weighing 复称
reweld(ing) 返修焊;重焊;重新焊接;缺陷焊补;重复焊接
re-wetable 能回潮的;可再润湿的
rewetting 重新浸湿;再润湿
rewetting ability 再润湿能力
rewetting agent 再润湿剂
rewind 再绕;倒片;倒带装置;重绕装置;重绕;反绕
rewind bench 倒台
rewind button 倒带按钮
rewind crank 倒片手柄
rewind drive 倒带驱动
rewinder 再卷装置;倒筒机;倒片装置;倒片器;倒带装置;倒带机;重绕机;复卷机;反轴机
rewind film processor 胶片冲洗机
rewinding 卷带;反绕;二次卷绕
rewinding coil 重绕线圈
rewinding control 倒片犁
rewinding frame 倒筒机
rewinding motor 倒带电机;重绕电动机
rewinding roller 重卷辊
rewinding room 卷片室
rewind key 反绕键
rewind reel 重卷机
rewind solenoid 倒带螺线管
rewind switch 反绕开关
rewind tank 双轴胶片显影槽
rewind time 倒带时间;反绕时间
rewiring 重新接线;重新布线【电】
rework 修改;再加工;再制(备);返修;返工;二次加工
reworkable waste 再用回花
rework cost 加工综合费用;返工费用;返工成本
reworked deposit 再造沉积物;二次沉积物
reworked tuff 再沉积凝灰岩
reworked units 返工单位数;翻工单位数
rework goods 加工品
reworking 再造(作用);再(次)加工;重新开采;二次开采
reworking of spent catalyst 废催化剂再生
rework plastics 再加工塑料
rework solution 再生溶液;返回溶液
rewrite operation 重写操作
rewrite statement 重写语句
rewriting rule 改写规则;重写规则
rewriting statement 重写语句
Rex 雷克斯(一种油漆洗涤剂)
rexangle 平行四边形屋面材料
Rexilite 雷克西莱特(一种屋面油毛毡)
rexine 强力镀层墙布
Rexoid 雷克西奥德(一种屋面油毛毡)
rextite 平行四边形屋面材料
reyerite 水硅钙钾石
Reyleigh's damping matrix 瑞利阻尼矩阵
reyn 雷恩(动力黏度单位)
Reynaud's phenomenon of occupational origin 职业性雷诺现象
Reynier's isolator 雷尼耳隔离箱
Reynolds 雷诺数
Reynold's alloy 雷诺合金(一种压铸铝合金)
Reynold's analogy 雷诺相似(性);雷诺模拟;雷诺比拟
Reynold's analogy parameter 雷诺数
Reynold's criterion 雷诺准则;雷诺准数
Reynold's critical velocity 雷诺临界速度;雷诺临界流速
Reynold's effect 雷诺效应
Reynold's equation 雷诺方程
Reynold's law of similarity 雷诺相似(定)律
Reynold's model 雷诺模型
Reynold's model law 雷诺模型律
Reynold's number 雷诺数
Reynold's number expression 雷诺数表达式
Reynold's number of fluid 流体雷诺德数
Reynold's number of particles 颗粒雷诺数
Reynold's plastic 雷诺塑性

Reynold's reducing power 用雷诺数表示的消色力
Reynold's similarity law 雷诺相似(定)律
Reynold's stress 雷诺应力
Reynold's stress tensor 雷诺应力张量
Reynold's transport theorem 雷诺运输定理；雷诺推移定理
rezbanyite 块辉铋铅矿
rez-de-chaussee 房子的第一层；建筑物底层
rezeroing 重新调零
rezone 重新分区
rezoning 重新区划
rezoning plan 重新区划图
Réaumur thermometer 列氏温度计
RF-pulsed beam 射频脉冲束
RF-squid magnetometer 射频超导磁力仪
rhabdionite 铜钴锰土
rhabdite 陨磷铁矿
rhabdolith 棒状晶体
rhabdophane 磷稀土矿；水磷铈矿
rhabdosome 笔石群体
Rhacomitrium aquaticum 水砂藓
rhaegmageny 扭裂运动
rhagite 砷酸铋矿
rhagon 复沟型
rhammite 伪杆状体
rhe 流值(动力流度单位)
rhegma 裂损
rhegmagenesis 断裂运动
rheid 黏流体；固流体
rheid fold 流变褶曲
rheid folding 固流褶皱
rheidity 固流限；流变度性
Rheims Cathedral Church 理姆斯主教堂
Rhenish architecture 莱茵式建筑
Rhenish-Bohemian brachiopod region 莱茵—波希米亚腕足动物地理大区
Rhenish brick 莱茵式砖；轻质硅酸盐砖
Rhenish marine trough 莱茵海槽
Rhenish style 莱茵式建筑
Rhenish trass 莱茵凝灰岩
rhenium 自然铼
rhenium black 铼黑
rhenium deposit 铼矿床
rhenium dioxide 二氧化铼
rhenium halide 卤化铼
rhenium ores 铼矿
rhenium osmium age method 铼—锇年代测定法
rhenium osmium dating method 铼—锇测年法
rhenium osmium method 铼—锇法
rhenium oxide 氧化铼
rhenium sesquioxide 三氧化二铼
rhenium tetrachloride 四氯化铼
rhenium trioxide 三氧化铼
rheobase 基强度；基本电流强度
rheobasis 基强度
rheocasting 流变铸造
rheochord 滑线变阻器
rheodestruction 流变破坏
rheodichroism 流变二色性
rheodynamic(al) lubrication 流变动压润滑
rheoelastic extrusion 流变弹性挤压成型
rheogometer 流变性测定仪
rheogoniometer 流变性测定仪
rheogoni(o)metry 流变测角法；流变性测定法
rheogram 阻抗图；流变图；流变曲线
rheograph 流变记录器
rheologic(al) 流变的
rheologic(al) agent 流变剂
rheologic(al) analysis 流变分析
rheologic(al) behavio(u)r 流变特性；流变性(能)；流变行为；流变学性能
rheologic(al) body 流变体
rheologic(al) change 流变变化
rheologic(al) coefficient 流变系数
rheologic(al) constant 流变常数
rheologic(al) control agent 流变调节剂
rheologic(al) dynamics 流变动力学
rheologic(al) equation of state 流变状态方程
rheologic(al) equilibrium 流变平衡
rheologic(al) ga(u)ge 流变仪
rheologic(al) gram 流变图
rheologic(al) hysteresis 流变滞后
rheologic(al) kinematics 流变运动学；流变动力学
rheologic(al) model 流体模型；流变(学)模型

rheologic(al) modifier 流变改进剂
rheologic(al) parameter 流变参数
rheologic(al) phenomenon 流变现象
rheologic(al) property 流变性(质)；流变特性
rheologic(al) property of rock and soil 岩土的流变性质
rheologic(al) reactor model 反应堆流变模型
rheologic(al) settlement 河流沉降
rheologic(al) settling 河流沉降
rheologic(al) stratification 软流分层
rheologic(al) structure 流变结构
rheologic(al) teat of rock 岩石流变试验
rheologic(al) test curve 流变试验曲线
rheologic(al) yield condition 流变屈服条件
rheology 流变学；河流学；塑流学
rheology of suspension 悬浮物流变
rheology of suspensions 悬浮体流变学
rheology of wood 木材流变学
rheometer 流变仪；流变计；电流计
rheometry 流动度测定；流变性测量法；流变性测定；流变测定法；测流学
rheomorphic fold 流变褶皱；深流褶皱
rheomorphic intrusion 流变侵入体
rheomorphism 软流变质(作用)；柔流变质；深流作用【地】
rheonome 电流强度变换器
rheonomic constraint 不稳定约束
rheo-optics 流变光学
rheopectic flow 震凝流动
rheopectic fluid 震凝流体；抗流变流体；触稠流体
rheopecticity 抗流变性；震凝性；震凝度
rheopectic material 流变材料
rheopexy 触变性；抗流变性；震凝性；震凝(现象)；流凝性；抗流变体；减流性
rheophankton 流水浮游生物
rheophilous bog 流水源沼泽；流水植物沼泽
rheophyte 流水植物；河生植物
rheoplankton 流水植物酸沼
rheoreceptor 趋流感受器
rheoscope 电流检验器；电流检验器；验电器
rheospectrometer 流谱计
rheosphere 软流圈
rheostan 高电阻铜合金；变阻合金
rheostat 可变电阻；滑线电阻器；变阻器
rheostatic alloy 变阻器合金
rheostatic arm 变阻器滑块
rheostatic brake 变阻器制动器
rheostatic braking 再生电阻制动；电阻制动；变阻器制动
rheostatic braking controller 变阻制动器
rheostatic brush 变阻器电刷
rheostatic control 变阻调速；电阻器控制；变阻(器)控制
rheostatic controller 变阻器控制器；变阻控制器
rheostatic control rod 变阻器控制杆
rheostatic heating apparatus 变阻式加热器
rheostatic knob 变阻旋钮
rheostatic slider 变阻器滑动触头
rheostatic starter 起动用变阻器；变阻起动器
rheostatic type automatic power factor regulator 变阻型自动功率因数调整器
rheostatic type regulator 变阻器式调节器
rheostatic type voltage regulator 变阻式调压器
rheostatic voltage regulator 变阻型电压调整器
rheostriction 流变压缩；夹紧效应；捏缩效应
rheotan 雷奥坦合金
rheotaxial 液相外延
rheotaxial growth 液延生长；液相外延生长
rheotaxis 趋流性
rheotaxy 液相外延
rheotome 周期断流器；断流器
rheotron 电子回旋加速器；电子感应加速器
rheotrope 电流转换开关
rheotropic brittleness 流动脆性；低温脆性
rheoviscometer 流变黏度计；流变黏弹计
rhine 引水渠；水道；河道
Rhine graben seismotectonic zone 莱茵地堑地震构造带
Rhine graben system 莱因地堑系
Rhinemetal 铜锡合金
rhino 渡口浮桥
rhinoceros horn 犀角
Rhinoceros unicornis 印度犀
rhino ferry 机动方驳

rhizic water 土壤水
rhizobenthos 海底生根植物
rhizocaul 根茎
rhizocretion concretion 根状核
rhizofiltration 根枝过滤法
rhizoma arisaematis 天南星
rhizoma calami 菖蒲
rhizoma curcumae longae 姜黄
rhizome 根状茎；根茎
rhizomorph 根状菌索
Rhizophoraceae 红树科
rhizophore 根托
rhizosphere 根围；根际；根层区
rhizosphere effect 根际效应
rhizospheric microflera of soil 土壤根际微生物群
rhochrematics 货物流通学
rhodamine 碱性蕊香红；玫瑰红；若丹明
rhodamine B 若丹明 B
rhodamine lake 若丹明色淀
rhodamine pigment 若丹明颜料
rhodaminered 若丹明红
Rhodanian orogeny 路丹尼造山运动
rhodanic acid 硫氰酸
rhodanide 硫氰化物
Rhodesian mahogany 罗得西亚桃花心木
Rhodesian old land 罗得西亚古陆
Rhodesian teak 罗得西亚柚木(产于北非)
rhodesite 纤硅碱钙石
Rhodian Law 古罗得海商法
rhodinic acid 香茅酸
rhodium 自然铑
rhodium electroplating 电镀铑
rhodium hydroxide 氢氧化铑
rhodium monoxide 一氧化铑
rhodium oil 旋花油
rhodium ores 铑矿
rhodium oxide 氧化铑
rhodium plating 镀铑
rhodium-rhodium alloy bushing 铂铑合金坩埚
rhodizite 硼铍铝铯石
rhodoarsenian 砷钙锰矿
rhodochrome 铬绿泥石
rhodochrosite 菱锰矿；锰晶石
rhodochrosite ore 菱锰矿矿石
rhododendron 杜鹃花
rhododendron oil 杜鹃油
rhodo-etio-porphyrin 玫红初卟啉
rhodol 对甲氨基酚
rhodolite 镁铁榴石；红榴石
Rhodomonas lacustris 湖沼红胞藻
rhodonite 蔷薇辉石
rhodonite deposit 蔷薇辉石矿床
rhodonite rock 蔷薇辉石岩
rhodophllite 铬绿泥石
rhodophyta 红藻植物
rhodopin 玫红品
rhodo-porphyrin 玫红卟啉
rhodopsin 光紫质
rhodopurpurin 玫红紫素
rhodostannite 蔷薇黄锡矿
rhodplumsite 硫铅铑矿
Rhoduline blue 若杜林蓝
Rhoduline heliotrope 若杜林紫
rhodusite 镁钠闪石
rhogosol 粗骨土
Rhoiptelea chiliantha 马尼树
rholnite 钙铁非石
rhomb 斜方形；斜方六面体；菱面体
rhomb brick 菱形砖；平行四边形砖
rhomb(en)-porphyry 菱形斑岩
rhombic(al) 斜方形的；斜方晶的；菱形的
rhombic(al) antenna 菱形天线
rhombic(al) antenna of feedback type 回授式菱形天线
rhombic(al) base 斜方底面
rhombic(al) bipyramid 斜方双锥
rhombic(al) bit 菱形钻头
rhombic(al) brachy-prism 斜方短柱
rhombic(al) brachy-pyramid 斜方短锥
rhombic(al) disphenoid 斜方四面体；斜方双楔
rhombic(al) dodecahedron 菱形十二面体
rhombic(al) exploration grid 菱形勘探网
rhombic(al) flange unit 菱形法兰式轴承箱组
rhombic(al) holohedral crystal system 斜方全

面体晶系
rhombic(al) knurling 菱纹滚花
rhombic(al) lattice 正交晶格
rhombic(al) lead chromate 正交晶型铬酸铅
rhombic(al) mica 琥珀云母
rhombic(al) net structure 斜方网状构造
rhombic(al) packing 菱形堆砌
rhombic(al) prism 菱形棱镜;斜方柱
rhombic(al) pyramid 斜方(单)锥
rhombic(al) quartz 长石
rhombic(al) rift-sag basin 菱形裂陷盆地
rhombic(al) spar 斜方白云石
rhombic(al) sphenoid 斜方楔
rhombic(al) sphenoidal class 斜方楔体类
rhombic(al) sulfur 斜方硫;正交硫;正交晶硫
rhombic(al) sulphur 斜方硫
rhombic(al) symmetry 正交对称
rhombic(al) symmetry site 菱形对称格位
rhombic(al) system 斜方晶系;正交晶系
rhombic(al) truss 米字形桁架
rhombic base 斜方底面
rhombicumboctahedron 菱形八面体
rhombiod 长斜方形
rhombo-boudin 菱形石吞肠
rhomboclase 板铁矾;板铁钒
rhombododecahedron 十二菱面体;斜方十二面体
rhombogen 菱形体
rhombohedra 菱面体
rhombohedral 菱形形;菱面体的
rhombohedral class 菱面体晶类
rhombohedral close packing 菱面体的密集装填
rhombohedral crystal 菱晶
rhombohedral (crystal) system 菱形晶系
rhombohedral graphite 菱面体石墨
rhombohedral iron ore 赤铁矿
rhombohedral lattice 菱形晶格;菱面体格子
rhombohedral pore 菱面体孔
rhombohedral prism 菱面体棱镜
rhombohedral system 菱形晶系
rhombohedron 菱形(六面)体;菱面体
rhomboid 斜矩形;菱形体;菱形区;平行四边形;长菱形
rhomboidal 斜方晶体;菱形块状;长斜方形的
rhomboidal lead spar 长斜方晶体白铅
rhomboidal prism 菱形棱镜
rhomboidan 菱形晶体
rhomboid chain 菱形锁;菱形链
rhomboid joint 菱形接合;菱形接头
rhomboid lanceolate 长菱形叶(披针叶形)
rhomboid leaf 菱形叶
rhomboid network 菱形网
rhomboid ripple mark 菱形波痕
rhomboid tongue 菱形舌
rhombomagnojacobsite 斜方镁黑镁铁锰矿
rhomb-porphyry 菱长菱岩
rhom brick 斜方砖;菱形砖
rhombus 等边平行四边形;长斜方形;斜方形;扁菱形
rhombus algorithm 菱形算法
rhombus even-speed tube 菱形均速管
Rhometal 镍铁磁合金;镍铬硅铁磁合金
rhometer 电阻率计
rhone (苏格兰式的)檐槽;排水槽;雨水槽
rhonite 钛硅镁钙石
rhooki 成熟紫胶
rho-rho system navigation 圆—圆航法
rho-theta 距离角度导航
rho-theta computer 航线计算机
rho-theta navigation 极坐标导航;全向导航系统
rhumb 罗盘卡;罗盘方位;罗盘点单位;斜航
rhumbatron 空腔共振器;环状共振器
rhumb bearing 墨卡托方位(角);恒向方位角
rhumb card 罗盘刻度盘;罗经刻度盘
rhumb course 墨卡托航向;恒向线航向
rhumb direction 墨卡托方向;墨卡托方位;恒向线方向
rhumb distance 恒向线航程
rhumb line 罗盘方位线;恒向线;无变形线;等角(航)线;等方位线;航程线
rhumb-line bearing 恒向线方位
rhumb-line correction 等角航线改正
rhumb-line distance 恒向线距离
rhumb-line error 恒向线误差
rhumb-line sailing 墨卡托航迹计算法;恒向线航行法

rhumbow line 再生绳
Rhus aromatica 加拿大漆树
Rhus chinensis 盐肤木
Rhus coriaria 西西里漆树
Rhus delavayi 山漆树
Rhus javanica 盐肤木
Rhus lacquer 天然漆;漆树漆
Rhus metopium 西印度毒漆树
Rhus orientalis Schneid 气根毒藤
Rhus shinensis poisoning 盐肤木中毒
Rhus succedanea 野漆树
Rhus toxicodendron 野葛
Rhus verniciflua 漆树
Rhus verniciflua stokes 日本漆树;漆树
rhyacium 急流群落
Rhyacodrilus sinicus 中华河蚓
rhyacolite 透长石
rhyncholite 有齿石
Rhynie Chert 莱尼埃燧石层
rhynohophylla 勾藤
rhyobasalt 流纹玄武岩
rhyocrystal 流纹斑晶
rhyodacite 流纹英安岩
rhyodacite glass 流纹英安质玻璃
rhyodacite porphyry 流纹英安斑岩
rhyodiabasic texture 流纹辉绿结构
rhyolite 流纹岩
rhyolite-dacite-rhyolite group 流纹岩—英安岩—流纹岩类
rhyolite porphyry 流纹斑岩
rhyolith 流纹岩
rhyolitic agglomerate 流纹质集块岩
rhyolitic flood eruption 流纹岩质泛滥喷发
rhyolitic glass 流纹岩玻璃
rhyolitic lava 流纹岩熔岩
rhyolitic magma 流纹岩岩浆
rhyolitic structure 流纹构造
rhyolitic tuff 流纹(质)凝灰岩
rhyolitic tuff lava 流纹质凝灰熔岩
rhyolitic volcanic breccia 流纹质火山角砾岩
rhyolitic welded agglomerate 流纹质熔结集块岩
rhyolitic welded breccia 流纹质熔结角砾岩
rhyolitic welded tuff 流纹质熔结凝灰岩
rhyotaxitic 流纹状的
rhyotaxitic structure 流纹状构造
rhysimeter 流体流速测定计
rhythm 韵律;周期性变动
rhythm due to climate change 气候变化韵律
rhythm due to seasonal change 季节变化韵律
rhythmer section 节奏发生器
rhythmic(al) bedding 韵律层理
rhythmic(al) change 有规律变化
rhythmic(al) corrugation 规律性搓板;周期性搓板
rhythmic(al) crystallization 韵律结晶
rhythmic(al) deposition 间歇沉积(作用)
rhythmic(al) disturbance 韵律混乱
rhythmic(al) drying 间隙干燥
rhythmic(al) graded bedding structure 韵律粒级层理构造
rhythmic(al) image 韵律感
rhythmic(al) layering 韵律成层
rhythmic(al) light 周期性灯光;节奏断续灯光
rhythmic(al) map 韵律图
rhythmic(al) movement 有规律移动
rhythmic(al) production 节奏生产
rhythmic(al) reaction 间歇反应
rhythmic(al) sedimentation 韵律沉积(作用)
rhythmic(al) succession 韵律层序;周期性重复
rhythmic(al) surge 有节奏涌浪;有节奏涌波
rhythmic(al) time signal 科学时号
rhythmic(al) unit 韵律层
rhythmic(al) variation 韵律性变化;有规律变化
rhythmic(al) volcanic mud ball structure 韵律型火山泥球构造
rhythmicity 韵律性
rhythmics 韵律学
rhythmite 韵律层;带状纹泥层
rhythm of engine 发动机振动频率
rhythm of light 灯光节奏
rhythm selection buttons 节奏选择钮
rhythm start or stop 节奏开关
rhythm synchro 节奏同步
rhythm variation 节奏变化
ria 狭长的海湾;溺湾;溺河

ria bay 里亚式港湾
ria coast 溺河海岸;沉降海岸
rial bond grinder 钢轨导接线研磨机
Rialto 威尼斯岛区;大理石廊桥(威尼斯桥);市场
Rias coast 里亚式海岸
ria shoreline 里亚式(海)岸线;沉降海岸线
ria type coast 里亚式海岸
rib 修边刀;整体矿柱;肋条;肋骨;加强部;凸条;补强材;横槽凸缘
rib across the head 顶部横肋
riband 缎带(装饰用);围篱顶端木条;梁的底模条板;装饰木条
rib and filler block floor 空心砖多肋楼板
rib-and-furrow 脊沟相间
riband jasper 带状碧石
rib-and-lagging 拱肋与背板支护
rib and lagging method 肋条横板法
rib and panel 肋拱支承的
rib and panel arch 肋拱
rib and panel vault 肋及拱板组成的穹顶;肋拱支承穹顶;肋拱穹顶;肋板支承的穹顶
rib arch 拱肋;肋拱
ribband 支材;木桁;饰带
ribband carvel planking 外板平铺内加横镶条舢板船壳
ribband jasper 条纹碧玉
ribband stone 条纹砂岩
ribbed 密肋;带肋的
ribbed arch 弯曲构件;扇形肋拱;有肋拱;菱形拱;扇形拱;肋拱
ribbed arch aqueduct 肋拱渡槽
ribbed arch bridge 肋拱桥
ribbed arched girder 肋拱料
ribbed back 加肋座
ribbed bar 竹节钢筋;带筋钢条
ribbed base plate 肋形垫板
ribbed beam 带肋梁
ribbed beam floor 密肋楼板;肋梁楼板
ribbed brake drum 加肋闸轮
ribbed clamp coupling 纵向有肋扭合联轴器
ribbed (clay) brick floor 填砖密肋楼板
ribbed collars 肋骨钻链
ribbed concrete floor 肋形钢筋混凝土楼盖
ribbed construction 有肋结构;横纹图案
ribbed continuous beam 肋形连续梁
ribbed conveyer belt 有肋的输送机胶带
ribbed cotton elastic braids 螺纹松紧带
ribbed countersunk head bolt 带榫沉头螺栓
ribbed cross-section 带肋剖面;带肋断面
ribbed cupola 带肋圆顶
ribbed cylinder 有肋汽缸
ribbed disc wheel 加肋盘轮
ribbed dome 带肋穹隆
ribbed dowel 带肋销钉
ribbed duplex tube 带肋双管
ribbed expanded metal 带肋钢板网;带肋网纹钢板
ribbed extrusion 肋形挤压
ribbed flat 带肋扁钢
ribbed floor 肋形楼面;肋构楼面;肋构楼板;密肋楼面;密肋楼盖;密肋楼板
ribbed floor slab 带肋楼板;肋形楼板
ribbed floor with slabs 加面板的密肋楼板
ribbed flue 加肋烟道;加肋火管
ribbed fluting 凹弧与平肋相间并用的装饰;肋形柱槽
ribbed form(work) 带肋模板
ribbed frame 有肋框架;带肋框架
ribbed framing 肋形外壳;有肋框架
ribbed funnel 线沟漏斗;浅沟漏斗
ribbed glass 槽纹玻璃;柳条玻璃;肋(形)玻璃;木条固定的玻璃;起肋玻璃
ribbed heater 带肋采暖器;带肋加热器
ribbed heating pipe 肋形暖气管
ribbed heating tube 肋形暖气管
ribbed heating unit 肋式供暖器
ribbed indented bar 竹节钢筋
ribbed lath 带肋网眼钢皮
ribbed liner 瓦棱衬板;凸棱式衬板
ribbed material 肋形材料;防滑材
ribbed metal wall 带肋金属板墙
ribbed motor 散热片型电动机
ribbed neck bolt 棱颈螺栓
ribbed panel 肋条加强板;肋板;带肋板

ribbed pipe 肋管;内螺纹管
ribbed pipe heater 肋管采暖器;肋管加热器
ribbed piping 内螺纹管
ribbed plate 多肋板;带肋板;用肋材加强的钢板;花钢板;网纹板;防滑(钢)板
ribbed plate arch bridge 肋板式拱桥
ribbed plate fastening 肋形垫板扣件
ribbed profile 带肋型材;带肋剖面;带肋断面
ribbed radiator 肋片散热器;翅片散热器
ribbed rag felt 条纹粗制毡
ribbed rebar 轧纹钢筋;轧纹钢条;竹节钢筋
ribbed reinforcing bar 竹节钢筋
ribbed rod 竹节钢筋
ribbed roll 带棱的压延辊筒
ribbed roller 肋形碾
ribbed roof 肋形屋顶;加肋穹顶
ribbed roof cladding 带肋屋面板
ribbed roof deck 加肋屋面板;带肋屋面板
ribbed rope finish 麻绳处理的混凝土表面印花
ribbed rubber flooring 条纹橡胶地面
ribbed rubber matting 带楞橡胶垫
ribbed seam 起肋镶轧缝
ribbed seam roof(ing) 肋形缝屋面板
ribbed section 带肋型材
ribbed sheet metal 肋形铁皮
ribbed shell 带肋(薄)壳;有肋(薄)壳
ribbed shuttering 加肋模板
ribbed slab 有肋楼板;肋形板;肋构楼板;肋板;加肋楼板;密肋楼面;密肋楼盖;密肋楼板;肋板楼板
ribbed-slab filler 肋板嵌块
ribbed-slab floor 肋板楼面
ribbed steel 条纹钢板;起肋钢板;横肋形(螺旋)钻杆钢
ribbed steel pantile 带肋钢槽瓦
ribbed stiffener 加劲肋(条)
ribbed stove 棱纹火炉
ribbed surface 带肋(表)面
ribbed surface machine 散热筋型电机;翅面电机
ribbed tile 肋形瓦
ribbed tray 肋形灶面板
ribbed tube 肋筒;肋管;加肋管;内螺纹管
ribbed tube economizer 肋管节能器
ribbed tube heater 肋管采暖器;肋管加热器
ribbed tube radiator 肋管散热器
ribbed tubing 内螺纹管
ribbed-typed reinforcement concrete segment 肋形钢筋混凝土管片
ribbed vault 扇形拱顶;扇形穹顶;肋骨拱;有肋拱顶;带肋穹(顶);带肋拱顶
ribbed wool felt 起楞羊毛毡;条纹羊毛毡
ribber 罗纹机
ribbet 槽口;企口缝
ribbing 压筋;肋形排列;肋材装配;肋材(构架);加肋;瓦楞绉缩;条纹;横槽凸缘
ribbing plow 双壁犁
ribbing up 肋材拼镶;将薄板胶合成圆拱形;层压圆细木工制品;肋状排列
ribbon 紧带;金属带;墨带;飘带;条带状;条板;饰带;带子;带状物;带状;带形;带;船舷包带
ribbon amphogneiss 条带状混合片麻岩
ribbon antenna 带形天线
ribbon band 饰带
ribbon banding 带状层理
ribbon beam 带状注;带状光束
ribbon bed 条式花坛;带状花土云
ribbon blender 螺条混合器;螺条混合机;螺条掺和机;螺条拌和器;螺带式掺混机;带式搅和器
ribbon board 饰物板;嵌入式撑杆;木桁条;横撑;条板
ribbon bobbin 色带盘
ribbon bond 带式连接器
ribbon brake 带状闸
ribbon building 连片式房屋;带形建筑群;建筑带;干路排屋;条形建筑;带形布置房屋;带形建筑;带式房屋
ribbon burner 带状燃烧器
ribbon cable 带状电缆;带形电缆;扁平线
ribbon cable connector 带状电缆连接器;扁形光缆连接器
ribbon cage 浪形保持架
ribbon carriage 色带盒
ribbon cartridge 色带卷
ribbon checks irrigation 带状畦田灌
ribbon clay 条带状黏土;带状黏土

ribbon clip 带夹
ribbon coil 带形线圈
ribbon conductor 带状导线;带形导体;扁状导线
ribbon contact 带状接触;带状触点
ribbon conveyer 皮带输送机;皮带输送带;带式螺旋输送机;螺条输送机;带式运送机;带式运输机;带式运输机;胶带输送机;带状螺旋叶片输送机
ribbon copper 铜带
ribbon core 带形铁芯
ribbon course 带形瓦层;带层(屋面)
ribbon crusher 带条扎碎机
ribbon crystal 带状晶体
ribbon cut 撕成长带;带状切割
ribbon cutting 剪彩
ribbon development 带形商业区;带状发展;带形发展;带式布置;城市带形发展;带状扩展
ribbon diagram 连续剖面图;带状栅栏图
ribbon drier[dryer] 带条干燥机
ribbon edge 板边
ribboner 碎茎打麻机;打麻机
ribbon feeder 平行传输线;带状馈线;带馈线
ribbon feeding 皮带送料
ribbon feed mechanism 色带输送机械装置;色带馈送机构
ribbon feed ratchet 色带输送棘轮
ribbon filament 带状灯丝
ribbon-flame burner 带状火焰喷燃器
ribbon flight 带状螺旋叶片;带式刮板
ribbon flight spiral conveyer 螺旋带式输送器
ribbon flower bed 带状花坛
ribbon foundation 条形基础
ribbon fuse 带状熔线
ribbon ga(u)ge 花带状应变片
ribbon gneiss 条纹片麻岩
ribbon grain 带状纹理;带状条纹
ribbon granulator 带条切片机
ribbon growth 带状生长
ribbon guide 色带导向
ribbon heater 电热带
ribbon injection 舌状侵入体
ribbon iron 窄带钢;条钢;带钢;扁铁;扁钢
ribbonite 绞成束状的软铅;塞管接缝铅条
ribbonization 放料
ribbon lap machine 并卷机
ribbon lapper 并卷机
ribbon lead 带状引出线
ribbon left guide 色带左导向
ribbon left mechanism 色带左向机械装置
ribbon lenticular 条式双凸透镜
ribbon lift guide 色楞提升导向器
ribbon lift mechanism 色带提升机构
ribbon lightning 带状闪电
ribbon-like fiber 带状纤维
ribbon-like filament 带形丝
ribbon-like lead 带状引出线
ribbon line 带状线
ribbon loading 带状下料法;胶带装料法;同时投料法(混凝土搅拌时)
ribbon loss 掉炉
ribbon loudspeaker 带式扬声器
ribbon machine 带式玻璃成型机
ribbon-mender 修边工
ribbon mending 燎裂子
ribbon microphone 动导体传声器;带式扩音器
ribbon mixer 螺旋叶片式搅拌机;螺旋混合器;螺旋(带式)混合机;螺条混合器;螺条混合机
ribbon optic(al) fiber cable 带状光缆;扁形光缆
ribbon parachute 带式快速降落伞
ribbon polymer 条带聚合体
ribbon position indicator 色带位置指示器
ribbon powder 带状火药
ribbon process 玻璃带成形法
ribbon rail 焊接长钢轨;条形横挡;带状横杆
ribbon reef 线状礁;带状礁
ribbon reel 带材卷取机
ribbon resistance 带状电阻
ribbon resistor 带式电阻器
ribbon retainer 浪形保持架
ribbon reverse control 色带可逆控制;色带反绕控制器
ribbon rock 条纹岩;带状岩(层)
ribbon saw 带锯(机)
ribbon screw 带状叶片螺旋输送机
ribbon sendust head 带式铝硅铁粉磁头

ribbon skylight 采光带
ribbon sowing 带状条播
ribbon spool 色带卷
ribbon steel 带钢;窄带钢;打包带钢
ribbon stirrer 螺条搅拌器;带式螺旋搅拌器
ribbon stock 带状(石)板材
ribbon stone 条纹砂岩
ribbon strip 带状条板;条形板;木桁条;肋梁
ribbon stripe 条状纹理;带式(条)纹
ribbon structure 条纹(状)构造;带状构造;带状构造
ribbon swing 玻璃带跑偏
ribbon tape 窄板材;卷尺;窄带材;皮带尺;皮尺
ribbon test 搓条试验
ribbon texture 条纹结构【地】;条带结构
ribbon threader 穿带针
ribbon turbine 螺旋带涡轮搅拌器
ribbon type mixer 螺旋式搅拌机
ribbon type radiator 带式散热器
ribbon type screw conveyer 带状螺旋运料机;带状螺旋输送机
ribbon type skylight 采光带
ribbon type temperature sensor 螺条温度传感器
ribbon vein 带状矿脉
ribbon width 引上玻璃原板宽度
ribbon winding 带绕组
ribbon window (大厦接连成带状的)连窗;统长窗;带形窗;窗带
ribbon-wound core 钢带铁芯;带绕铁芯
ribbon-wound pole 扁绕磁极
ribbon zone selector 色带区选择器
ribby 皱纹的
rib center 拱肋弧的圆心
rib concrete 带肋混凝土
rib-cooled motor 散热肋冷却型电动机
rib-cutter 断肋器
rib depth 肋高
rib design 肋拱设计
riberbed developement 河床形成(过程)
riberirite 稀土锆石
rib fabric 罗纹针织物
rib face 挡边导引面
rib flange 肋突缘;带肋法兰盘
rib floor 密肋楼面;密肋楼盖;密肋楼板
rib floor slab 肋构楼板
rib footing 拱肋基脚;拱肋基础
rib for base 底座肋板
rib forms 带肋模板
rib heating pipe 带肋暖气管;肋管
rib hole 修边炮眼;整边炮眼;边孔(隧洞的、巷道的);边炮眼
rib intersection 肋的相交点;肋相交
rib interval 翼间距;肋间距
rib lath 有V形肋的钢丝网;肋形钢丝网;肋条钢丝网;加劲钢丝网;带肋网眼钢皮
ribless 无翼翅的;无肋(骨)的
ribless corduroy 灯芯布
rib line 煤壁线
rib loading 肋受荷载
rib machine 螺纹机
rib mark 肋状纹理
rib member 肋杆
rib mesh 钢肋条网;带肋钢板网;带肋金属网格;肋形钢丝网;肋筋钢丝网;金属网格;加劲钢丝网
Ribmet 利伯梅网眼钢筋;膨体混凝土;带筋钢
rib of column 柱肋;柱肋;柱体条棱;柱身楼纹
rib of dome 圆穹顶肋
rib of floor 楼板肋
rib of piston 活塞肋
rib of slab 楼板肋;板肋
rib of valve 阀肋
rib oval wire 带棱椭圆截面金属丝
rib pipe 带肋管
rib-ply 带肋层压板
rib-profile 肋的断面
rib ratio 支撑比例
rib reinforcement T形梁的主筋;肋内受力筋;加强肋
rib rifling 等齐膛线
rib roof 带肋屋顶
rib roof slab 带肋楼板
ribs and wool lagging 肋和木横板
rib seat 凸肩支承座
rib section 带肋型材;带肋断面;肋断面
rib shortening 拱肋缩短;拱肋压缩

rib slope 肋的坡度
rib snubber 外围掏槽眼(爆破)
rib space 肋间隔
rib spacing 肋(的)间距;肋间隔
rib-span forms 带肋模板
rib-span formwork 肋板模板;肋板模壳
rib steel 肋钢
rib stiffener 加劲肋
rib-strengthened 用肋加强的;用肋加劲的
rib strip 肋条
rib system 波形系统;肋系统
rib-thread tire 肋纹轮胎
rib tile 带凸牙的瓦
rib tread 肋形花纹
rib tube 带肋管
rib type 肋拱形式
rib-type bailer 肋骨式提砂筒
rib-type expanded metal 肋型金属网;肋型金属钢板
ribut 里巴特
rib valley 肋间凹槽
rib vault 带肋拱顶;扇形肋穹顶
rib vaulting 有肋拱顶;扇形肋穹顶;肋形拱顶
rib weave 亩组织;螺纹;棱纹;重平组织
ribwort 长叶车前
Riccati equation 黎卡提方程
Ricci equation 里西方程
Ricci identity 里西方程
Riccitensor 缩并的曲率张量
Ricci theorem 里西定理
rice and cereals 粮谷
rice bay 灌水稻田
rice bin 米箱
rice bran oil 糠油;米糠油
rice bran wax 米糠蜡
rice character design 米字纹
rice coal 米粒煤;米级无烟煤
rice combine 水稻联合收割机
rice dust 米尘
rice field 稻田
rice field fish culture 稻田养鱼
rice germ oil 米胚芽油
rice glue 米胶
rice grain decoration 米粒装饰
rice grain pattern 米通(花样)
rice grains 米粒
rice grain tile 玲珑饰瓦;稻粒瓦;颗粒瓦
rice grain valve 超小型电子管
rice grass 落芒草类;稻草
rice hull 稻壳;大米壳
rice hull ash 谷糠灰
rice huller 龙壳机
rice huller screen 米筛
rice husk 稻糠(作轻质填料用);稻壳
rice husk ash 谷壳灰;稻壳灰
rice husk ash cement 稻壳灰水泥
rice husks 谷壳
rice husks aggregate 稻壳集料
rice in sack 袋装米
rice made products 米制品
rice market 米市
rice mill 碾米机;碾米厂
Rice neutralization 赖斯栅极中和
rice noodle 米线
rice nursery 秧田
rice oil 米糠油
rice paddy 水稻田
rice paper 宣纸;米纸;通草纸(米纸)
rice pattern fracture 米粒状碎裂
rice polishings 糠
rice scourer 碾米机
rice seed bed 水稻秧田
rice sheller 碾米机
rice shoots 秧田;禾秧;稻秧
rice sieve 米筛
rice sieve design 米筛纹
rice sieve with sieve mesh 冲孔米筛网
rice starch 米淀粉
rice starch paste 米浆
rice stick 米线
rice stone 斑状岩
rice store 粮舱
rice straw 稻草
rice-straw ash 稻草灰

rice-straw ash glaze 稻草灰釉
rice stub 稻桩
rich acacia 相思树
rich alloy 中间合金;富合金
rich and mighty 富强
Richards column 理查德型干扰沉降分级室
Richards deep-pocket hydraulic classifier 理查德型深槽水力分级机
Richards-Janney classifier 理查德—杰尼型分级机
Richards jig 理查德型脉动跳汰机
Richards launder-type vortex classifier 理查德型槽式涡流分级机
Richardson's alloy 理查逊锡砷银锌合金
Richardson's automatic scale 理查逊自动秤
Richardson's effect 理查逊效应
Richardson's law 理查逊定律
Richardson's number 理查逊数
Richardson's plot 理查逊标绘图
rich-ash coal 富灰煤
rich binding 精装
rich bottom layer 薄浆底层
rich-bound 装帧精美的
rich chromium ball-bearing steel wire 高铬轴承钢丝
rich clay 重黏土;高塑性黏土;沃土;富黏土;肥黏土
rich coal 肥煤;沥青煤
rich coast 增砂岸
rich colo(u)r 浓艳色彩;浓色
rich concrete 高水泥含量混凝土;水泥含量高的混凝土;多水泥混凝土;多灰混凝土;富混凝土
richellite 土氟磷铁矿
richelsderfite 砷锑钙铜石
richetite 黑铅铀矿;水板铅铀矿
rich flame 浓焰
rich fusodurite 微富丝质暗煤
rich gas 富煤气
rich-glittering 金碧辉煌的
rich guy 富翁
rich hand-picked ore 手拣富矿
rich humic acids coal 富腐殖酸煤
rich ice content frozen soil 富冰冻土
rich in alumina 高铝的
rich in contrast 反差层次丰富
rich in harmonics 多谐波的
rich in mortar 多砂浆
rich in natural resources 丰富的资源
rich lake 富湖
rich lime 富石灰;肥石灰
rich lime mortar 稀石灰灰浆;白石灰砂浆;浓石灰砂浆;多石灰砂浆
rich low brass 高铜低锌黄铜;红铜;装饰用高锌黄铜
rich lump lime 白钙块石灰
Richman's Club 富人俱乐部
rich metering 浓混合气调节
rich mica-arenite 富含云母砂屑岩
rich mica-arkose 富含云母长石砂岩
rich mica-graywacke 富含云母杂砂岩
rich mica-quartz sandstone 富含云母石英砂岩
rich mica-wacke 富含云母瓦克岩
rich mineral 富矿
rich mixed concrete 富灰混凝土;浓配合混凝土
rich mix(ture) 多灰混合物;多油混合物;多水混合料;混凝土拌和物;富混合物;富混合料;富混凝土;多灰混合(料);富灰混合(料);浓配合;浓(厚)拌和;浓混合气;稠浆拌和物;稠浆拌和;富配合;富拌和;胶凝材料多的混合料
Richmondian 里查满阶
rich mortar 富黏砂浆;浓砂浆;浓灰浆;多砂浆;富砂浆;富灰浆
rich naphtha 饱和石脑油
richness class 富度级
richness index 富度指数;丰度指数
richness of details 细部详尽
richness of mix 拌和浓度
richness of oil source 油源丰度
rich oil 富油;饱和吸收油
rich ore 富矿(石)
rich phase 富相
rich quicklime 纯钙生石灰
rich shade 浓色相;强色
rich soil 沃土;肥沃土
rich solution 浓溶液
rich-sulfur coal 富硫煤
rich-tar coal 富油煤

richterite 碱镁闪石;钠透闪石
Richter magnitude 里氏震级
Richter (magnitude) scale 里氏震级表
Richter motor 里希特电动机
Richter scale 里克特震级;里氏地震强度分级
rich water 肥水
riciaism 蓖麻子中毒
ricin 蓖麻(毒蛋白)
ricinate 蓖麻油酸盐
ricinic acid 蓖麻酸
ricin oil 蓖麻油
ricinoloil 蓖麻油
ricinoleate acid 蓖麻酸
ricinoleic acid 蓖麻油酸;蓖麻醇酸
ricinolein 蓖麻油脂
ricinoleindin 甘油三蓖麻油酸酯
ricinoleyl alcohol 蓖麻油醇
ricinoleyl glycerine 甘油—蓖麻油酸酯
ricinolic acid 蓖麻油酸
ricinstearolic acid 蓖麻硬脂炔酸
Ricinus 蓖麻
ricinus oil 蓖麻油
ricinus poisoning 蓖麻子中毒
ricinus silk 蓖麻蚕丝
rick 堆积成垛;干草垛;草垛
rickardite 碲铜矿
ricker 圆木料;堆垛工;脚手架立柱;堆垛机
Ricker wavelet 雷克子波
rick lifter 堆垛机;草垛提升机
rick-mow yard 堆草场
rickrack 波状花边
rickshaw 手推双轮混凝土车;人力车
rickstand 堆料架
ricochet 回弹;回跳;跳飞;反跳
ridable tide level 乘潮水位
ridax photo printing paper 大光晒纸
riddings 碎屑;筛余粗料
riddle 清选;盘筛;手动筛;粗眼筛;粗筛;粗矿筛;粗孔筛
riddle drum 圆筒筛;转筒筛;筛筒
riddled sand 筛选砂
riddle map 暗射(地)图
riddler 振动筛;震筛
riddle sifter 粗筛(网)
riddle type digger 格筛式挖掘机
riddlings 筛上物;过筛;筛屑;粗筛余料;粗筛筛上产品
ride 碇泊
rideability 行驶性能
rideability of pavement 路面(的)行驶质量
ride a long peak 远锚泊
ride and handling loop 汽车驾驶试验环路
ride a short peak 近锚泊
rideau 小土堆;土丘
ride characteristic 车辆行驶特性
ride clearance 动力挠度
ride comfort 旅客乘坐舒适性
ride-control 车辆减振器
ride-control shock absorber 可调减振器
ride down 压进;捶进
ride dynamics 汽车驾驶动力学
ride easy 纵摇轻微;锚泊时锚链受力不大
ride-frequency 舒适性频率
ride hard 纵摇强烈;锚泊时锚链受力很大
ride hawse full 锚泊中波浪打进锚链孔
ride height 装载高度
ride index test 运行平稳性试验
ride meter 测路面仪(量测路面行驶质量用);平整度(测定)仪;测震仪
ride of vehicle 汽车的行驶平顺性
rideograph 测震仪;测振仪;平整度(测定)仪(路面)
ride performance 行驶平顺性
ride quality 行驶质量;乘坐质量
rider 勾墨辊;游码;追加条款;重辊;制导雷达;可采煤层之上的薄煤层;木船舱盖板内侧肋骨支架;批改条款;导向套;附加条款
rider arch 炉条硷
rider bar 游码标尺;骑码标尺
rider brick 托砖;撑砖
rider cap 桩承台;桩帽;斜撑帽
rider carrier 游码钩
rider frame 补强肋骨;补加肋骨
rider hook 游码钩

rider keel 副龙骨
rider-of-insurance 保险追加条款
rider ring 导向环
rider roll 接触胶皮辊
rider roller 匀墨辊;浮动辊
ridership 客运率;全体乘客
rider shore 斜撑柱;斜撑木;架空斜撑;支墙斜撑
riders' vertigo 乘车性眩晕
rider to charter party 租约附则
ride sharing 合乘车
ride stabilizer 行驶稳定器;侧向稳定器
RI detector 折光检测器
ride the blocks 滑车升降
ride the brush 用漆刷窄边涂漆(俚语)
ride the clutch 脱挡滑行
ride the tide 乘潮
ride the wind 顶风停泊
ride time 乘行时间
ride transfer characteristic 行驶振动传递特性
ridge 形成脊状;高压脊;屋脊;田垄;田埂;山梁;山脊;粉梗;分水岭山脊
ridge aloft 高空脊
ridge and furrow aeration 垄沟式曝气
ridge and furrow aeration tank 垄沟式曝气池
ridge and furrow air diffusion 垄沟式空气扩散(法)
ridge and furrow air diffusion of aeration tank 曝气池谷沟式鼓风曝气
ridge and furrow diffusion 脊和河槽空气扩散
ridge and furrow irrigation 沟灌;垄沟灌溉
ridge and furrow roof 重脊屋顶
ridge and furrow system 垄沟式灌溉系统
ridge and furrow tank 垄沟式池;畦状曝气池
ridge and furrow type of drainage 排水垄沟形式
ridge-and-ravine beach 脊槽海滩
ridge-and-ruined sand beach 脊槽型海滩
ridge and spur sampling 沿山脊采样
ridge and valley roof M 形屋顶;重脊屋顶
ridge and valley under-drain 脊谷式地下排水管
ridge basalt 海岭玄武岩
ridge batten 屋脊梁上加钉的木条
ridge beach 脊状海滩
ridge beam 脊梁;栋梁;脊檩
ridge binder 屋脊系杆
ridge board 脊板;脊木;脊檩;脊梁
ridge buckles 带材的皱纹
ridgecap 屋脊盖;脊帽
ridge capping 屋脊盖瓦;脊压顶片;脊盖;脊帽
ridge capping piece 屋脊盖
ridge capping tile 脊瓦
ridge channel 脊槽
ridge configuration 纹脉结构
ridge corner tile 脊角瓦
ridge course 屋脊瓦层
ridge covering 脊帽;脊盖
ridge cover tile 脊瓦
ridge crest(ing) 屋脊饰;脊饰;脊顶
ridge crossing line 越岭线;脊岭交叉线
ridge cut 脊梁上部切口;屋脊梁的垂直锯口
ridge dammer 作畦埂器
ridged ashlar 尖凿琢面方石
ridged-back file 条形板锉;平三角锉
ridged brickwork 脊形砖砌体
ridged conveyer 锯齿式输送机
ridge design 凹凸花纹
ridged field 垄田;畦田
ridged ice 多脊冰
ridge digger 垄用挖掘机
ridge distortion 脊形歪痕
ridged melon vase 瓜棱瓶
ridged roof 人字屋顶;有脊屋顶
ridged surface 肋形面;搓板面;搓板状面
ridged surface finish 搓板状饰面;搓板状面
ridged sweep 培土铲;培垄铲
ridged wall 人字屋顶
ridge-end ornament (dragon head) (古代建筑屋脊上的)大吻
ridge extractor 屋脊抽气机
ridge extract ventilation 屋脊抽气通风
ridge fault 脊状断层
ridge fillet 柱槽(间)脊;两凹槽间的凸棱;脊槽
ridge flashing sheet 屋脊泛水片
ridge folding (金属板屋面的)屋脊折叠(缝)
ridge form 屋脊形式

ridge-former 筑埂器;起垄器
ridge forming 起垄
ridge framing 屋架
ridge furrow irrigation 垄沟灌溉;畦沟灌溉
ridge girder 屋脊大梁
ridge groundwater 脊形地下水面
ridge gusset plate 脊连接板
ridge height 屋脊高度;山脊高度
ridge horn 脊形喇叭
ridge irrigation 垅灌
ridge jump 海岭跳动
ridge junction tile 屋脊联结瓦
ridge level(l)er 平垄器;平垄机
ridge line 甲板中线;脊线;屋脊线;山脊线;分水线;分水岭
ridge mountains 有脊丛山
ridge-mounted beast ornament 走兽
ridge-mounting timber 扶脊木
ridge of gutter 天沟分水线
ridge of high-pressure 高压脊
ridge of pile 料堆脊峰
ridge of shoal 滩脊
ridge of wave 波峰
ridge ornament 脊饰
ridge ornamentation of celestial-being and beasts 仙人走兽【建】
ridge piece 背垫板;脊檩;脊梁;脊板;栋木
ridge planting 垅植;山岭造林
ridge plate 屋脊板;脊板
ridge plow 起垄犁;培土犁
ridge point 脊点
ridge pole 脊檩;脊梁;脊板;屋脊梁;天幕纵梁;帐篷顶横杆;栋梁
ridge purlin(e) 屋脊檩条;正梁;脊檩
ridge purlin(e) brace 脊檩撑
ridger 筑埂机;起垄机;培土器;松土器
ridger body 作垄犁体
ridge regression 岭回归
ridger girder 屋桁
ridge rib 脊肋
ridge-ridge transform fault 脊—脊式转换断层
ridge rim 凸起轮辋
ridge road 山脊路
ridge rod 圆脊檩
ridge roll 筒形戗脊构件;屋脊铅皮卷筒;栋木;脊瓦;圆脊檩;金属脊帽;屋脊卷形装饰;半圆脊帽
ridge roller 压垄辊
ridge roof 人字屋顶;有脊屋顶
ridge roof type 山形屋顶
ridge rope 小艇舷侧栏杆顶围绳
ridge route 山脊线;山脊路;分水岭线
ridge-runner 线路巡查员;屋脊水沟
ridge segment 海岭块段;海脊块段
ridge slate 盖脊石板
ridge spar 天幕纵梁
ridge spike 脊尖饰
ridge starting tile 脊端瓦
ridge stone 顶端石;脊石
ridge stop 屋脊披水铅皮;脊泛水;屋脊防水铅皮;屋脊止水铅皮
ridge support 天幕纵梁支柱
ridge terrace 梯式沟渠;梯田土埂;起脊台地;屋面截水条;坡面截水堆;等高线梯田台地;等高线阶梯式沟渠;脊状地埂;起脊台地
ridge tiebeam 脊枋
ridge tile 脊瓦;屋脊瓦;瓦脊筒
ridge tile for saw tooth roof 锯齿形屋顶脊瓦
ridge tillage 垄作
ridge-to-ridge folding 屋脊至屋脊间折叠(缝)
ridge-to-valley folding 屋脊至屋面沟折叠(缝)
ridge tree 栋木;脊木
ridge-trench transform fault 脊—沟式转换断层
ridge turret 屋脊小塔
ridge-type rooflight 屋脊式天窗
ridge-type skylight 屋脊式天窗
ridge-up bed 畦床
ridge ventilator 屋脊通风机
ridge versus valley location 山脊线与河谷线对比
ridge waveguide 脊形波导管
ridge way 山脊路;山顶线;山脊线
ridge welding 筋状点焊
ridging 做垄筑埂;局部隆起;筑垄;筑埂;隆起;脊;围埂;成垄;封垄
ridging bottom 起垄犁体

ridging ground 丘陵
ridging plough 培土犁
riding 马道;锚泊中的;波束运动;按曲线运动
riding academy 马术学校;骑术学校
riding anchor 力锚;受力锚
riding athwart wind 横风锚泊
riding athwart wind and tide 横风流锚泊
riding at single anchor 单锚泊
riding at two anchors 双锚泊
riding between wind and tide 横风锚泊
riding bitt 锚链柱;系缆柱
riding boom 舷侧系艇杆;系艇杆;吊艇杆
riding boots 马靴
riding buckler 锚链孔盖
riding cable 力链
riding chain 系链;力链
riding characteristic 行驶质量(街道路面的车辆)
riding check 随车观察
riding chock 擎链器
riding coefficient 乘坐舒适系数
riding comfort 行驶平顺性;行车舒适性;乘行舒适性
riding confortableness 乘坐舒适度
riding cultivator 乘式中耕机
riding cut-off valve 骑式停气阀
riding grade 路面行车质量等级;行车等级
riding hawse fallen 在颠簸中的锚泊
riding hawse full 在颠簸中的锚泊
riding high tide level 乘潮水位
riding lamps 系留灯
riding lee tide 顶风流锚泊
riding light 锚灯;停泊灯
riding lister 乘座式开沟犁
riding moor 后退抛双锚
riding mower 乘式割草机
riding out a gale 顶住风暴
riding plow 乘式犁
riding position 安装位置;安放位置
riding quality 行车平稳质量;路面行车质量;乘坐平稳质量;车辆行驶性能
riding rate 乘车率
riding roller 匀墨辊
riding saddle 马鞍
riding school 骑术学校
riding seizing 绕扎
riding shore 斜撑
riding slip 钩形掣链器
riding stability 运行稳定性
riding stable 马厩
riding stopper 锚链掣
riding surface 车行道路面
riding time 乘车时间
riding to two anchors 双锚泊
riding trail 骑马的小径;跑马道
riding weather tide 横流锚泊
Ridley-Scholes process 里德利—斯科尔斯的砂浮选法
riebeckite 钠闪石
riebeckite aplite 钠闪细晶岩
riebeckite felsite 钠闪石霏细岩
riebeckite granite-aplite 钠闪花岗细晶岩
riebeckite aplite granite 钠闪细花岗岩
riebekite aplite granite porphyry 钠闪细花岗斑岩
riebeckite granophyre 钠闪花岗岩
riebungsbreccia 褶挤角砾岩
Rieche's law of crystallization 李凯结晶定律【地】
Riecke's principle (重结晶作用的)李凯原理
riedel 河间地
Riegler's test 黎格勒试验
Rieke diagram 瑞基图(即振动器的功率和频率同负载关系曲线图)
Riemann curvature 黎曼曲率
Riemann curvature tensor 黎曼曲率张量
Riemann function 黎曼函数
Riemann hypothesis 黎曼假设
Riemannian geometry 椭圆几何
Riemann integral 黎曼积分
Riemann manifold 黎曼流形
Riemann mapping theorem 黎曼映射定理
Riemann method 黎曼方法
Riemann space 黎曼空间
Riemann space curvature 黎曼空间曲率
Riemann sphere 黎曼球面
Riemann surface 黎曼(曲)面
Riemann tensors 黎曼张量

Riemann zeta function 黎曼ζ函数
riffle 凿沟;(板带材侧缘的)皱纹;砂槽;浅滩;微波
riffle and pool 滩槽
riffle-and-pool sequence 滩槽序列
riffle box 分样器
riffled iron 窄带钢;网纹钢
riffled plate 花纹钢板
riffled sheet 花纹钢;网纹钢板
riffle machine 电光机
riffle nut 螺帽
riffle plate 花纹板
riffle-pool difference 滩槽高差
riffle-pool spacing 滩槽间距
riffler 来福锉;溜砂槽;试样缩分器;除砂器;除砂盘;沉砂槽;分样器;凹锉
riffle range 射击场
riffler file 修模锉
riffle sampler 分格(式)取样器;分样器
riffle steel plate 花纹钢板
riffle-to-riffle spacing 滩与滩的间距
riffling 铺格条;除砂;轧辊刻纹;凿沟
rifle 来福枪;制拉来福线;膛线
rifle bar 带旋槽的柱体
rifle barrel groove cutting broach 炮筒来复线槽拉刀
rifle brush 螺旋形钢丝刷
rifled barrel 膛线炮管
rifled bore 膛线炮管
rifled cannon 膛线炮
rifled gun 膛线炮
rifled pipe 来复线管;内螺纹管;输送黏油的内螺纹管(约有3米长的螺纹);带肋管
rifled sheet 皱纹钢板
rifled sheet iron 皱纹铁(板)
rifled shell 槽钢壳
rifled tube 来复线管;来福线管;内螺纹管
rifle ground 步抢射击场
rifle kiln 蛇窑
rifle nut 来复螺母
rifle-pit 散兵壕
rifler 窄刨;波纹锉
rifle range 步抢射击场
rifle scope 来福探测仪
rifle tie 旋条带
rifling 制膛线;来复线;岩芯上螺纹槽;膛线
rifling bar 切螺旋槽刀杆
rifling head 来复线拉刀;膛线拉刀
rifling machine 拉制来复线机
rift 裂隙;裂口;裂痕;裂谷;断裂【地】;断陷谷;长峡谷
rift and sag mode 裂谷和沉陷模式
rift basin 裂谷盆地
rift-block mountain 断块山;裂谷块山
rift-block valley 裂谷断块谷;断裂地堑谷
rift bulge 裂谷隆起
rift crack 木材干裂
rift cushion 裂谷垫层
rift cut 破开;四开的
rift cutting 斜径侧切
rift dome 裂谷穹隆;裂谷隆起
rifted arch basin 拱顶断陷盆地
rifted continental margin 扩张大陆边缘;断裂大陆边缘
rifted cratonic margin 裂谷化克拉通边缘
rifted margin prism 裂谷边缘沉积柱
rifted-sliced 径向刨切
rift geosyncline 裂谷地槽
rift grain 顺纹
rift-grained 原木四开锯法
rift lake 裂谷湖
rift saw 镶片锯切;四开锯木;径切法;镶片圆锯;裂木锯;剖料锯;板条锯
rift-sawed 径向锯切;四开的
rift-sawed board 长纹板
rift sawn 四开锯木的;四开的;径切面
rift-sawn plank 径向锯的木板
rift-sawn timber 四开木材
rift sea 裂谷海
rift-sliced 径向刨切
rift structure 断裂结构
rift system 裂陷系;裂谷系;断裂系
rift trough 正断层谷;断陷谷;裂谷;裂谷
rift trough fault trough 地堑
rift valley 裂谷;断陷谷;裂谷;裂缝谷;地堑;地谷
rift valley lake 裂谷湖;断陷湖;地堑湖;断谷湖

rift zone 断裂(地)带;裂谷带;破裂带
rig 钻井装置;钻井架;钻架;钻机;沙丘;打井机;成套器械
rig accident 钻井装置事故
rigaree 刻棱花
rig a ship 装备船舶
rig base 钻塔基础;钻架支撑座;钻机
rig boom 钻机臂杆;钻车动臂
rig builder 钻探设备安装师
rig builder outfit 钻机安装工的工具
rig column 钻机柱架;钻机托臂
rig cost 钻机操作费
rig day rate 钻探设备每日成本
rig developer 钻探设备设计者
Rigdon's apparatus 雷顿仪
rig down 拆卸钻探设备
rigesity 糙度
rig fixer 钻探设备安装技师
rig floor 机台板
rig for deep submergence 准备深潜
rig for model test 模型试验台;模型试验塔
rig for test 试验架
rig for testing 试验装备
rig front 钻塔正面
rigged ashlar 粗琢石块
rigged market 被操纵的市场
rigged oar 船桨
rigger 特种刷;理纹刷;长鬃刷;装配工;转盘钻司钻机具装配工;索具装配工;索具工人;束带滑车
rigger's horn 索具工油盒
rigger's scaffold(ing) 升降机脚手架;装配工脚手架
rigger's screw 张缆器
rigging 组装模板;装配;(铸工的)造型工具;索具;帆缆索具
rigging abeam 正正横方向
rigging adjoint functor 右伴随函子
rigging adjoint linear mapping 右伴随线性映射
rigging aft 艉后方
rigging ahead 正前方
rigging-and-left coupling 左右偶联管
rigging-and-left-hand-screw clip 左右向螺杆夹
rigging-and-left nut 左右纹螺母
rigging-angle bend 直角弯管
rigging-angle clamp 直角夹
rigging-angle cone 直角锥
rigging-angle connector 直角连接器
rigging-angle crossing 直角交叉
rigging-angle drive 直角传动
rigging-angle elbow pipe 直角弯管
rigging-angle fire 曲射
rigging-angle friction wheel 直角摩擦轮
rigging-angle gear 正交轴齿轮
rigging angle of incidence 迎角
rigging arrangement 缆具装置
rigging band 装吊带
rigging bar 固定凿岩机支座
rigging bell crank 控制直角杠杆
rigging chain 铐子链
rigging datum line 装配根据线
rigging diagram 水平测量图
rigging end fitting 活钩;索端固定装置
rigging error 装配误差
rigging line 索具绳索;吊伞索
rigging loft 缆具车间;索具传动装置
rigging machine 对折机
rigging pendant 短索
rigging pin 装配销
rigging plan 缆具装置图
rigging position 装配位置
rigging rod 控制杆
rigging screw 装配螺钉;螺旋扣;紧索套;夹索螺旋夹具;花篮螺钉;松紧螺扣;伸缩螺钉
rigging shackle 索具卡环
rigging shop 缆具车间;索具工场;索具车间
rigging up 钻探设备安装
rigging up inspecting rule 安装验收制
rigging-up of rotary tools 安装转盘钻机设备
rigging-up of standard tools 安装钢绳冲击钻进设备
Rigg motor 里格径向柱塞液压电动机
riggot 地面排水沟;雨水槽;地面水沟
right 右方;直角的;正确的;权利;法权
right abeam 右正横(方向)
right-about 反对方向
right-about-face 向后转

right adjoint 右伴随
right ahead 正前方
right and bent 直角弯头
right-and-left-hand chart 左右手程序图
right-and-left nut 连接螺母;牵紧螺母;左右纹螺母
right-and-left screw 左右旋螺栓
right-and-left threaded 左右螺纹的;正反扣的
right and obligation of the borrower 借款人的权利义务
right-and tee 直角三通
right angle 直角
right-angle-angle domain 直角角域
right-angle bend 直角弯管;90°弯头
right-angle bending 直角弯曲
right-angle board 直角板
right-angle branch 直角支管
right-angle connector 直角连接器;直角接头
right-angle crank 直角拐肘
right-angle crossing 直角交会;直角交叉
right-angle curette 直角刮匙
right-angled 成直角的;直角的
right-angled bend 直角转弯
right-angled corner 直角转角
right-angled crossing 直角交叉
right-angled elbow 直角形肘管
right-angled intersection 直角交叉;十字交叉;正交叉
right-angled junction 直角交叉;直角连接
right-angled nomogram 直角诺模图;直角列线图
right-angled tee 直角三通
right-angled triangle 直角三角形
right-angled turn 直角转弯
right-angled valve 直角弯头阀门
right-angle escapement 侧角式擒纵机构
right-angle examining telescope 直角检查窥镜
right-angle facing lathe 直角端面车床
right-angle finder 直角检景器
right-angle friction wheel 直角摩擦轮
right-angle gear 正交轴齿轮
right-angle impulse 直角脉冲
right-angle intersection 正交(叉);直角交会;直角交叉
right-angle inversor 直角控制器
right-angle joint 直角联结
right-angle lead 直角引线;直角引出线
right-angle mirror 直角仪;直角镜
right-angle network 直角网络
right-angle parking 直角停车;横列式停车
right-angle pier 突堤码头
right-angle prism 直角棱镜
right-angle retractor 直角牵开器
right-angle seizing (绳结的)十字合扎
right-angle shaft drive 传动轴与主轴成直角传动
right-angle side 直角边
right-angle speed reducer 正角式减速器
right-angle spheric(al) triangle 球面直角三角形
right-angle tee 直角T形接头;直角三通
right-angle three way valve 直角三通阀
right-angle traverse shaving 切线方向进刀剃齿法
right-angle type 直角形
right-angle viewer 直角指示器
right anterior oblique position 右前斜位
right arch 正拱
right ascension 赤经
right ascension circle 赤经圈
right ascension of a celestial body 天体赤经
right ascension of mean sun 平太阳赤经
right ascension of meridian 子午线赤经
right ascension of satellite ascending node 卫星升交点赤经
right astern 正后方
right athwart 正横方向;正横
right avertence 右偏
right balanced functor 右平衡函子
right bank 右岸;河右岸;右堤
right border 右缘
right branch 右支
right bridge 正桥;正交桥
right bridge deck(ing) 直桥桥面板
right bundle branch 右束支
right bundle branch block 右束支传导阻滞
right circular cone 直立圆锥;正圆锥体
right circular cylinder 直圆柱体;直立圆柱
right circular cylinder coordinate 竖圆柱坐标系
right circularly polarized light 右旋圆偏振光

right component 右分量
right cone 直立圆锥;直锥
right congruence relations 右同余关系
right conoid 正劈锥曲面
right container 密封货柜
right continuous function 右连续函数
right culvert 正交涵洞
right-cut tool 右切(车)刀
right cylinder 正圆柱体
right cylindric(al) 正圆柱形的
right cylindric(al) reactor 正圆柱形堆
right decomposition 右分解
right derivative 右微商
right derived functor 右导出函子
right differential coefficient 右微商
right divisor 右因子
right divisor of zero 右零因子
right drainage 排水权
right driving 靠右(侧)行驶;靠右(侧)行车
right elevation 右视图
right elevator 右升降舵
right end clearance 右余隙
right end marker 右端标记
right endpoint 右端点
right engine 右侧发动机
right ening couple 正位力偶
right entry 右端进入
right exact functor 右正合函子
right exposure 适当曝光
right face of the round 炮眼组的正确布置
right full rudder 右满舵
right gable tile 直角山墙顶盖瓦
right global dimension 右整体维数
right half rudder 右半舵
right-hand 右手(的);右面的;右方;右侧;正搓
right-hand adder 低位加器;右移加法器;右侧数加法器
right-hand auger 右向螺纹螺旋
right-hand axle stop 前桥右支挡
right-hand beam 右旋光束
right-hand bend tool 右削弯头刀
right-hand border 右图廓
right-hand buoy 右行标;右边浮标
right-hand channel 右手信道
right-hand chassis beam 底盘右梁
right-hand circularly polarized wave 右旋圆偏振波
right-hand circular polarization 右旋圆偏振
right-hand column 右边列
right-hand complete residue system 右方完全剩余系
right-hand component 右旋分量;右手坐标分量;右侧数
right-hand continuity 右方连续性
right-hand control car 右座驾驶车辆
right-hand crusher 右式破碎机
right-hand curve 右转弯(道);右旋曲线;右向曲线
right-hand cut 右旋切削
right-hand cutter 正手刀
right-hand-cutting tool 右削车刀
right-hand cylinder bar 脱粒滚筒右向槽纹的纹杆
right-hand derivative 右微商
right-hand designation 右向标志
right-hand door 右旋门;右(手)开门
right-hand drive 右侧驾驶
right-hand drive car 右座驾驶车辆
right-handed 右行的;右手的;右搓;右旋的
right-handed coordinate system 右旋坐标系;右手坐标系
right-handed crystal 右晶
right-handed door 右开门
right-handed elliptical polarization 右旋椭圆偏振
right-handed engine 右转发动机
right-handed form 右形
right-handed helix 右手螺旋
right-handed lang's lay 股丝同向右搓(钢丝绳)
right-handed machine 右转电机;正装
right-handed moment 右转力矩
right-handed multiplication 右乘法
right-handed nut 右旋螺母
right-handed person 惯用右手的人
right-handed polarized light 右(旋)偏振光
right-handed polarized wave 右旋极化波
right-handed propeller 右螺旋桨;顺时针转螺旋桨
right-handed quartz 右旋石英
right-handed rope 右旋缆索
right-handed rotation 右旋(光);右向旋转
right-handed scale 右手坐标
right-handed screw 右旋螺旋桨;右旋螺旋杆;右旋螺钉
right-handed side outlet elbow 右手侧流肘管
right-handed spiral 右旋游丝
right-handed system 右旋坐标系;右旋系统
right-handed thread 右旋螺纹;正扣螺纹
right-handed thread screw 右螺纹螺杆
right-handed trihedral 右手三面形
right-handed window 右开窗
right-hand eigenvector 右(手)特征向量
right-hand elevation 右视图
right-hand end mill 右旋端铣刀
right-hand end side outlet tee 右手侧流三通
right-hand engine 右旋轴发动机
right-hand fusee 右圆锥滑轮
right-hand helical polarization 右旋螺旋偏振
right-hand helicity 右旋性
right-hand helix 右旋螺(旋)线
right-hand hinge 右向铰链
right-hand hob 右旋滚刀
right-hand joint 右螺纹接头;正扣钻杆
right-hand justify 向右对齐
right-hand lang-lay 右向顺捻;同向右捻
right-hand lay 右捻;右搓;右向捻(绳子);右转扭绞
right-hand lay wire rope 右捻钢丝绳
right-hand limit 右极限
right-hand lock 右旋门锁;右手门锁
right-hand loose joint hinge 右边活节铰链
right-hand magneto 右转磁电机;右旋磁电机
right-hand man 得力助手
right-hand mark 右侧标
right-hand member 右端构件
right-hand mill 右旋铣刀
right-hand milling cutter 右旋铣刀
right-hand moment 顺时针方向力矩
right-hand multipliers 右方乘子
right-hand offset 右偏置
right-hand open-end auger 右旋输送螺旋
right-hand opening 右手开关
right-hand opening gate valve 右转起动闸阀
right-hand plough 石翻犁
right-hand polarization 右旋偏振
right-hand polarized electromagnetic wave 右旋偏振电磁波
right-hand polarized receiver 右旋极化波接收机
right-hand polarized wave 右旋偏振波
right-hand propeller 右螺旋桨
right-hand propeller right turn 右旋
right-hand reverse door 右旋外开门
right-hand revolution breaker 右转断电器
right-hand rope 右搓绳;正搓绳
right-hand rotating fan 右转通风机
right-hand rotation 顺时针方向转动
right-hand rotation reamer 右旋铰刀
right-hand rule 右手定则
right-hand running 右侧行车
right-hand screw 右螺旋;右旋螺钉
right-hand screw thread 右(旋)螺纹
right-hand set of axes 右手坐标轴系
right-hand shaving tool 右旋剃齿刀
right-hand side 右边
right-hand side tool 右削偏刀
right-hand signal 右信道信号
right-hand stair(case) 右扶手楼梯
right-hand stairway 右扶手楼梯
right-hand switch 右侧道岔
right-hand thread 右转螺纹【机】;正扣
right-hand threaded drill rod 右螺纹钻杆
right-hand thread tap 右螺纹丝锥
right-hand tool 右削车刀;正手刀
right-hand track chain 右履带
right-hand traffic 右行交通;靠右(侧)行驶的交通
right-hand turn 右转弯;右向旋转
right-hand turning share 右翻犁铧
right-hand turning tool 右削车刀
right-hand turnout 右开道岔【铁】;右侧道岔
right-hand twine 开式锭带缠纱
right-hand unit 右侧装置
right-hand upper lower derivate 右上下导数
right-hand vector system 右手向量系
right-hand winding 右旋绕组
right-hand-wise 顺时针方向
right-hand worm 右旋蜗杆
right heir 合法继承人;法定继承人
right helicoid 右螺旋面;正螺旋面
right home 打到底(打桩)
right hyperbola 正双曲线
right identity element 右幺元
right in civil affairs 民事权利
righting 扶正
righting a capsized vessel 扶正沉船
righting arm 回复(稳定)力臂;稳性力臂;复原力臂
righting arm of stability 复原力臂
righting arms and moments 稳性力臂及力矩
righting couple 正位力偶;回复力偶;稳性力偶;稳定力偶;复原力偶
righting lever 正位力臂;回复力臂;稳性力臂;复原力臂
righting moment 改正力矩;正位力矩;回复力矩;恢复力矩;稳定力矩;复原力矩;扶正力矩
righting reflex 正位反射;翻正反射
right injective resolution 右内射分解
right interpolation 右内插值
right invariant 右不变
right invariant tensor field 右不变张量场
right inverse 右逆元
right inverse element 右逆元
right inverse operator 右逆算子
right justify 右对齐;向右对齐;右整列;右侧调整
right-laid 右捻
right lane 右侧车道;右车道
right lang lay 右同向捻;单向右捻钻探钢丝绳
right-lateral deformation 右旋形变;右旋变形
right-lateral displacement 右旋位移
right-lateral fault 右旋断层;右侧断层
right-lateral motion 向右旋运动
right-lateral position 右侧位
right-lateral slip 右旋错动
right-lateral slip fault 右侧断层
right-lateral strike-slip fault 右行走向滑动断层
right-lateral transcurrent fault 右旋横推断层
right lay 右搓绳;正绳索
right lay rope 右捻钢丝绳
right leaflet 右瓣叶
right-left bearing indicator 左右方位指示器
right-left needle 左右摆动式指针
right-left signal 左右方向指导信号
right-left traverse of the arm (机器人的)臂部左右移动
right limit 右极限
right line 右线;直线
right lobe 右叶
right longitude 右经度
right lower derivative 右下导数
right lower quadrant 右下象限
right majorizing function 右强函数
right margin 右端界限
right marginal bank 右边岸
right marginal branch 右缘支
right moment 稳性力矩
rightmost 最右边的
right multiplication 右乘(法)
right-normal slip 右旋正滑
right-normal-slip fault 右行正向滑动断层
right of abandonment 委付权
right of access 近岸权;接近权
right of anticipation 提前还款权
right of asylum 庇护权
right of authorship 著作权
right of cancellation 解约权
right of claim 求偿权;索赔权
right of coercion 强制权
right of convoy 护航权
right of creditor 债权
right of eminent domain 征用权;土地征用权
right of first refusal 首买权
right of fishery 捕捞权
right of indemnity 要求赔偿权
right of ingress 通行权
right of innocent passage 无害通过权(海洋法)
right of inspection 检查权
right of invention 发明权
right of land usage 土地使用权
right of management 经营权
right of navigation 航行权
right of occupancy 居住权

right of patent 专利权;特许权
right of priority 优先权
right of recourse 追索权;求偿权
right of redemption 赎买权
right of rescission 解约权
right of search 检查权
right of special permission 特许权
right of taxation 征税权
right of trade mark 商标权
right of usage 使用权
right of visit(ation) 临检权
right of water 用水权
right of way 通过权;权界;路线征购地界;地界;优先行驶权;用地;公路用地;路权;通航权;通行权;铁路用地;道路通行权;筑路用地;道路用地
right-of-way appraisal 道路用地评价
right-of-way boundary 铁路用地界;用地界【道】
right-of-way fence 用地栅栏;路界栅栏;道路用地栅栏
right-of-way gang 管带开拓班
right-of-way line 道路用地界;道路红线
right-of-way map 路界图;地界图;道路用地图
right-of-way monument 地界石;道路用地界石;道路用地标
right-of-way operation 管带开拓
right-of-way party 管带开拓班
right-of-way rule 道路用地规划
right-of-way screw 管带开拓班
right of way signal 通行信号;通告信号
right-of-way survey 用地测量【道】;道路用地测量
right-of-way term 管带开拓班
right-of-way vessel 直航船(被让路船);权利船;被让路船
right-of-way width 道路用地宽度
right one-dimensional linear subspaces 右一维线性子空间
right operation 右运算
right opposite 正相反;正对面
right package 合适包装
right parallelepiped 正平行六面体
right parenthesis 右括号
right part 右部
right part of a rule 规则右部
right plummet 正锤线
right price 适当价格
right prime formula 右素逻辑式
right prism 直立棱柱
right projective space 右射影空间
right pyramid 直立棱锥;正棱锥(体)
right quantity 适当数量
right quasi-regular element 右拟正则元
right quotient space 右商空间
right rail 右股钢轨
right reading 正像;正向阅读;正读
right-reading negative 正阴片
right-reading positive 正阳片
right recursive 右递归
right recursive definition 右递归定义
right recursive form 右递归形式
right regular-lay cable 右捻钢丝绳
right relating to civil law 民事权利
right residue class 右剩余类
right residue system 右剩余系
right resolution 右分解
right-reverse-slip fault 右行逆向滑动断层
right ring spanner 右封口扳手
right ring wrench 右封口扳钳
right rudder 右舵
rights 权责利
rights and obligation 权利与义务
right section 右部;右段;正截面;右截面;正截影;剖面
right semicircle 右半圈
right semilunar valve 右半月瓣
rights for forest 林权
right shift 右移
right-shift register 右移位寄存器
right-shift times of multiplier 乘法器右移次数
right side 右侧
right side bank 右岸
right-side engine 右侧发动机
right sideline of dredge-cut 挖槽右边线
right side of the figure 图的右侧
right side up with care 小心正面朝上
right side view 右视图
right sighting system 右瞄准系统

right skewness 右偏斜度
right slab bridge 直板桥
right slip 右滑
right-slip fault 右侧断层
right slope angle 右坡角
rights of beneficiaries 受益者权利
rights of mooring 停泊权
rights of water 水利权
right sphere 正(交)天球;垂直球;垂球
right spheric(al) triangle 直角球面三角形;球面直角三角形
right spin 右螺旋
right-spiral screw 正螺纹
rights reserved 保留权利
right stair(case) 右向楼梯
rights to pollute 污染权
right strophoid 正环索线
right subgraph 右子图
right substitutability 右置换
right subtree in a binary tree 二叉树中的右子树
right superior function 右上函数
right swing line 右向绕绳索
right tapered plate 右转弯楔形衬砌环
right tensor 右张量
right the helm 正舵
right to access 出让权
right to attend shareholder annual meetings 出席股东年会权
right to claim damages 损害索赔权
right to claim for damages 要求赔偿损失的权利
right to demand compensation for losses 要求赔偿损失的权利
right to derive benefit from land 土地收益权
right to dispose land 土地处理权
right to dispose of land 土地处分权
right to exclude all others 排除他人权利
right to grass 杂草采取权
right to handle imports and exports 经营进出口业务权
right to hold land 土地占有权
right to interpret 解释权
right-to-left interchangeability 左右方零件互换性
right-to-left parser 右到左分析算法
right-to-left reading 反写
right to make decision 决策权
right to manage land 土地经营权
right to own and use grassland 草原所有权和使用权
right to ownership of land 土地所有权
right to property owned by the whole people 全民产权
right to use in forestry 森林使用权
right to use land 土地使用权
right to use site 土地使用权
right to watercourses 水流权;水道通行权
right to work 工作权利法
right translation 右平移
right triangle 直角三角形
right truncation 右截断
right-turn ban 禁止右转弯
right-turn(ing) 右转弯
right turning single screw 右旋单螺旋桨
right turning tapering plate 右转弯楔形衬砌环
right-turn lane 右转(专用)车道;右转(弯)车道
right-turn ramp 直接右转匝道
right turn switch 右开道岔【铁】
right upper derivative 右上导数
right upper quadrant 右上象限
right vector space 右向量空间
right verge tile 直角檐瓦
right view 右视图
rightward 向右
rightward rotation 右转
rightward skew slab 右向偏斜板
rightward welding 右焊法
right way round 正像
right wheel base 固定车轴距
right-wing-down attitude 右坡度姿态
right wing tank 右翼油箱
right-wound spiral 右旋
right zero 右零
rigid 刚硬的;刚性的;坚硬
rigid adherence 牢固黏附
rigid adhesive 刚性黏结剂
rigid airlift mud pipe 刚性气升吸泥管

rigid airship 硬式飞艇
Rigidal Mansard 瓦楞复合铝板
rigid angle 刚性角
rigid arch 刚性拱;刚拱;整体拱
rigid arch dam 刚性拱坝
rigid armo(u)ring 劲性钢筋;型钢钢筋;刚性钢筋;刚性布筋
rigid asbestos curtain 刚性石棉幕
rigid attachment 刚性连接;刚性固定
rigid axle 刚性轴
rigid-axle type suspension 刚性轴式悬挂
rigid bar 固定顶梁
rigid base 刚性底座;刚性基层
rigid base bar 杆状基线尺
rigid bearing 固定轴承;刚性支承
rigid bed 硬质河床;固定河床;定床
rigid bed model 定床模型
rigid bed river engineering model 定床河工模型
rigid bed river model 定床河道模型
rigid-bed sedimentation model 定床淤积模型
rigid bed-sweeping 硬式扫泥【疏】
rigid bidirectional repeater 双向刚性增音机
rigid block 刚性块
rigid board 硬质板;刚性物
rigid board of glass wool 玻璃棉硬质板
rigid boat 硬式浮舟
rigid body 刚性体;刚体
rigid body kinematics 刚体运动学
rigid body kinetics 刚体动力学
rigid body mode 刚体振型
rigid body motion 刚体运动
rigid body rotation 刚体转动
rigid bond 刚性连接
rigid bottom hole assembly 刚性钻铤组合
rigid boundary 刚性边界;刚体边界
rigid boundary channel 坚硬岩石河道;坚硬岩石河床
rigid braced derrick crane 斜撑式桅杆起重机
rigid bracing 刚性支撑;刚性风撑
rigid brick 刚性砖
rigid busbar 硬母线;刚性母线
rigid buttress dam 刚性支墩坝;刚性连结支墩坝
rigid catenary 刚性接触网
rigid cellular polystyrene 硬质泡沫聚苯乙烯
rigid cellular polystyrene thermal insulation 硬质泡沫聚苯乙烯绝热
rigid chain polymer 刚性链聚合物
rigid charging 固定装料机
rigid classification 机密
rigid clutch 刚性离合器
rigid coaxial cable 刚性同轴电缆
rigid-column-flexible-beam model 刚柱柔性梁模型
rigid composite material 硬质复合材料
rigid concrete frame 混凝土刚架
rigid conductor rail joint 刚性导轨接头
rigid conduit 刚性涵管;刚性管渠;刚性管道;刚性导管;硬管;刚性套管
rigid connected beam method 刚接梁法
rigid connection 刚性连接;刚接
rigid construction(al) system 刚性构造体系
rigid container 固定式集装箱(非折叠式)
rigid control 硬式操纵
rigid control linkage 硬式传动机构
rigid copper bus 硬铜母线
rigid copper coaxial line 刚性铜同轴线
rigid core 刚性内芯
rigid core wall 刚性心墙
rigid coupling 刚性联轴节;刚性联结;刚性连接;刚接
rigid cover plate 刚性盖板
rigid cradle 刚性支船架
rigid crust 刚性地壳
rigid culvert 刚性涵洞
rigid damp course 刚性防潮层
rigid derricking mechanism 刚性吊臂起落机构;刚性变幅机构
rigid design 刚性设计
rigid diaphragm 刚性隔板
rigid dirigible 硬式飞艇
rigid distortion 刚性扭曲
rigid dolphin 刚性(靠船)墩
rigid double tube 双动双层岩芯管;双动双层取芯器
rigid drawbar 刚性牵引装置
rigid drive 刚性传动
rigid driven gear 刚性从动齿轮
rigid drive shaft 固定主动轴

rigid dynamics 刚体力学
rigid earth 刚性地球
rigid economy 严格节约
rigid elastic condition 刚弹性状态
rigid elastic scheme 刚弹性方案
rigid element 刚性构件
rigid end restraint 刚性端部约束
rigid expanded plastics 硬泡沫塑料
rigid extensible bar 伸缩式刚性梁
rigid extensible prop 刚性可伸长支柱
rigid facing 刚性护面
rigid factor 刚性系数
rigid fastening 刚性联结；刚性连接；刚性扣件；刚性固定
rigid feedback 硬反馈
rigid feedback controller 硬反馈控制器
rigid fibre rod 刚性纤维棒
rigidfix 桥墩边角护板
rigid fixing 刚性固定
rigid flask 整体砂箱
rigid flax 条纹亚麻
rigid foam 硬质泡沫
rigid foam adhesive 硬泡沫塑料黏结剂
rigid foam board 硬泡沫塑料板
rigid foam bonding adhesive 硬泡沫速率胶合剂
rigid foam core(d) wall pane 硬泡沫塑性芯墙板
rigid foam fill(ing) 硬泡沫塑料填料
rigid foam insulant 硬泡沫隔声板；硬泡沫隔热板
rigid foam insulation furring 硬泡沫隔热衬料
rigid foam laminate(d) board 硬泡沫塑料夹层板
rigid foam plastic board 硬泡沫塑料板
rigid foam plastics 硬(质)泡沫塑料
rigid foam urethane 硬泡沫氨基甲酸乙酯泡沫
rigid footing 刚性基底
rigid forming tube 硬质拉丝筒
rigid foundation 刚性基础
rigid foundation structure 刚性基础结构
rigid frame 刚架；刚(性)构架；刚性框架；闭合刚构
rigid frame beam 刚架梁
rigid frame bent 刚性排架；刚性构架；刚构排架
rigid frame bridge 刚架(式)桥；刚架桥；单跨钢构桥
rigid frame bridge with inclined legs 斜腿刚架桥
rigid frame construction 刚架结构；刚架建筑；刚架构造
rigid framedarch bridge 刚架拱桥
rigid framed pier 刚构桥墩
rigid framed structure 刚架结构
rigid frame formula 刚架公式
rigid frame foundation 刚构基础
rigid-frame loader 整体底架装载机
rigid frame of foundation 刚架基础
rigid frame pier 刚架墩
rigid frame slab 刚架上薄板
rigid frame type flexible pier 刚架式柔性墩
rigid framework 刚性构架；刚架；刚构(架)；坚强控制网
rigid frog 固定(式)辙叉
rigid full track 刚性全履带车
rigid game 严格博弈
rigid gel 刚性凝胶；无膨胀性凝胶；非弹性凝胶
rigid granule 硬颗粒
rigid graver 固定刻图仪
rigid grid 刚性网格；刚性格网
rigid-hammer crusher 固定锤碎机
rigid hanger 刚性吊杆
rigid highway pavement 刚性公路路面
rigid highway surfacing 刚性公路路面
rigid hitch 刚性连接装置
rigid hook 无弹性针钩
rigidified rubber 硬化橡胶
rigid image guide 刚性传像束
rigidimeter 刚度计
rigid index 刚性指数
rigid insert 刚性插片
rigid insulating foam 硬质隔声泡沫(材料)；硬质隔热泡沫(材料)
rigid insulation 硬质隔声；硬质隔热；硬质绝缘；刚性绝缘
rigid insulation board 硬质绝缘板；刚性绝热板
rigid insulation (grade) material 硬质隔声材料；硬质隔热材料
rigidity 硬度；刚性；刚度；僵硬；僵化
rigidity anchor 刚性锚固
rigidity bar 刚性杆

rigidity coefficient 刚性系数
rigidity computation 刚度计算
rigidity condition 刚度条件
rigidity connected 刚性连接的
rigidity development 刚度改进
rigidity element 刚性构件
rigidity factor 整步功率；刚度因子；刚度因数
rigidity index 刚性指数；刚度指数
rigidity in torsion 抗扭刚度
rigidity jointing 刚性接头
rigidity-longitudinal test 纵向刚性试验
rigidity loss 刚度降低
rigidity matrix 刚度矩阵
rigidity member 刚构件
rigidity modulus 刚性模数；刚度模量
rigidity number 刚度数
rigidity of concrete pipe 混凝土管刚度
rigidity of flange 法兰刚度
rigidity of hairspring 游丝刚度
rigidity of modulus 模壳刚度；形体的刚度
rigidity of mud 泥浆结构黏度
rigidity of ocular wall 球壁硬度
rigidity of rock 岩石刚度
rigidity of structure 结构劲度；结构刚度
rigidity of the bolted flange system 法兰系统刚度
rigidity of the system of bolted flange connection 法兰系统刚度
rigidity pavement 刚性路面
rigidity ratio 刚度比(例)
rigidity reinforcement 刚性钢筋
rigidity slab 刚性板
rigidity supported system 刚性支承系统
rigidity-transverse test 横向刚性试验
rigidization 强化
rigidized 刚性的；加固的
rigid joint 刚性结合；刚性结点；刚性接合；刚结点；刚节点；刚(接)节点
rigid joint between elements 管段刚性接头
rigid-jointed 刚性连接的；刚接的
rigid-jointed frame 刚构；刚架；刚性接合架
rigid-jointed framework 刚性接合构架
rigid-jointed plane framework 刚接平面骨架；刚接平面框架
rigid-jointed prismatic structure 刚接棱柱结构
rigid-joint frame 刚构
rigid-joint(ing) 刚性接头
rigid lattice 刚性晶格
rigid leg 刚性支腿
rigid lifecraft 刚体救生筏
rigid line 刚性管线
rigid link 刚性链杆
rigid load 刚性荷载
rigid lock 预装锁
rigidly fixed power plant 固定动力装置
rigidly jointed cupola 刚性连接的圆顶
rigidly jointed portal 刚性连接门(形)架；刚构门(形)架
rigidly material 刚性材料
rigid magnet 硬磁铁
rigid material 刚性材料；硬路面
rigid mechanical conveyer 刚性机械输送机
rigid member 刚性构件
rigid membrance 刚性膜片
rigid metal conduit 硬金属管；刚性金属管道
rigid metal girder 刚度大的金属梁；刚性金属大梁
rigid mill 强力铣床
rigid mineral wool 硬质矿棉
rigid model 刚性模型；定床模型
rigid modulus 刚性模量
rigid motion 刚体运动
rigid mount 刚性支座
rigid-mounted axle 刚性安装的轴
rigid mounting 与机架刚性连接；刚性机架
rigid network 刚性控制网
rigid nodal point 刚性节点
rigid non-metal(lic) conduit (不能套螺纹的)非金属硬管
rigid of track panel 轨道框架刚度
rigidometer 刚度测定仪
rigid optic(al) system 刚性光学系统
rigid overlay 刚性盖层；刚性面层
rigid patrol airship 硬式巡逻飞艇
rigid pavement 硬路面；刚性铺高；刚性路面
rigid pavement design 刚性路面设计

rigid pavement subgrade 刚性路面地基
rigid penstock 刚性压力水管
rigid-perfectly plastic body 理想刚塑性体；完全刚塑性
rigid phenolic slab 硬质酚醛塑料板
rigid piece 刚性垫条
rigid pier 刚性墩
rigid pipe 刚性管
rigid pipe culvert 刚性管涵；刚性管式涵洞
rigid pipeline 刚性管线；刚性管道
rigid pipe section 硬质管壳
rigid piston 刚性活塞
rigid-plastic 刚塑性的
rigid-plastic body 刚塑性体
rigid-plastic finite element method 刚塑性有限元法
rigid plastic foam 硬(质)泡沫塑料
rigid plasticity 刚塑性
rigid-plastic material 刚塑性材料
rigid-plastic model 刚塑性模型
rigid-plastic resistance 刚塑性阻力
rigid plastics 硬(质)塑料；硬化塑料
rigid plastic sandwich panel 硬塑料夹心板
rigid plastic sheets 硬塑料板材
rigid plastic structure 刚塑性结构
rigid-plastic system 刚塑性系统
rigid plastic tube 硬塑料管；刚性塑料管
rigid plate 刚性板块；刚性板；熔合板
rigid plate of loading method 刚性板荷载法
rigid platform 刚性承台；刚性平台
rigid platform dolphin 刚性台面靠船墩；刚性台面靠船台
rigid plummet 对中杆
rigid point 高精度控制点
rigid polyester foam 硬泡沫聚酯
rigid polyethylene 硬质聚乙烯
rigid polyisocyanurate foam 硬质聚氰酸酯泡沫塑料
rigid polymer 硬质聚合物；刚性聚合物
rigid polythene 硬质聚乙烯
rigid polyurethane foam 硬质聚氨酯泡沫
rigid polyvinyl 硬质聚乙烯化合物的
rigid polyvinyl chloride 硬质聚氯乙烯
rigid portal 刚性龙门架；刚性构架
rigid portal frame 刚性门式框架
rigid pressure machine 刚性压力机
rigid prestressed concrete frame 预应力钢筋混凝土刚架
rigid price 僵硬价格
rigid PVC 硬质聚氯乙烯
rigid PVC conduit 硬聚氯乙烯管
rigid PVC interior-profile extrusion 硬质聚氯乙烯内异形挤出型材
rigid PVC pipe 硬质聚氯乙烯管
rigid PVC sheet 硬质聚氯乙烯板材
rigid PVC siding 硬质聚氯乙烯护墙板
rigid PVC soffit 硬质聚氯乙烯底面
rigid rail brace 固定式轨撑
rigid reinforcement 刚性钢筋；劲性钢筋
rigid resin 硬树脂；刚性树脂
rigid restrained 刚性约束的
rigid restraint 刚性约束
rigid restraint cracking test 刚性拘束裂纹试验
rigid retaining wall 刚性挡土墙
rigid rib 刚性筋
rigid riverbed 硬质河床
rigid road 刚性路(面)
rigid roadbase 刚性路基
rigid road pavement 刚性路面
rigid rolls 刚性压辊
rigid roofing 刚性屋面
rigid rotation 硬性转动
rigid rotator 刚性转子
rigid rotor 整体转筒
rigid runway design 刚性道面设计
rigid runway pavement 刚性道面
rigid scheme 刚性方案
rigid scaffolding 刚性脚手架
rigid secret 机密
rigid section 硬断面
rigid shaft 刚性轴
rigid-shank applicator 刚性施肥开沟器
rigid sheet 刚性平板
rigid shell 刚性壳体
rigid six-wheeler 刚
rigid slab 刚性板

rigid sleeper clips 刚性扣件
rigid sol 刚性溶胶
rigid solid 刚体
rigid solution 严格的解答(法)
rigid steel 硬钢
rigid steel anchor 刚性钢锚固件
rigid steel conduit 硬钢管
rigid stratum 刚性地层
rigid structural system 刚性结构系统
rigid structure 刚性结构;刚架结构;刚构
rigid subgrade 刚性路基
rigid submarine repeater 固定式海缆增音机
rigid support 刚性支座;刚性支护;刚性支点;刚性支持;刚性支承
rigid surface 刚性路面;刚性面层
rigid surface planning 硬直面滑水现象
rigid suspension 刚性悬架
rigid suspension device 刚性悬挂装置
rigid system 刚性系统
rigid temperature control 严格的温度控制
rigid test 严格试验;刚性试验
rigid theory 严格的理论
rigid tile 刚性砖
rigid tow bar 刚性牵引杆
rigid tower 刚性塔
rigid training airship 硬式训练飞艇
rigid truck 整体式车架载重汽车
rigid truss 刚(性)桁架
rigid tube 刚性管
rigid tubing 刚性管
rigid tunnel lining 刚性隧道支护
rigid two-way repeater 双向刚性增音机
rigid type 美式管钳
rigid type base 刚性基层
rigid type bearing 刚性支承点
rigid-type castor 支架滚子
rigid type clutch plate 硬式离合器板
rigid-type construction 刚性结构
rigid-type core barrel 双动双层岩芯管;双动双层取芯器
rigid type design 刚性设计
rigid-type double tube core barrel 双动双管取芯钻具;双动双层岩芯管;双动双层取芯器
rigid type safety fence 刚性护栏
rigid-type structure 刚性结构
rigid-type vibrator 刚式振动器
rigid universal joint 刚性万向节
rigid urethane 刚性聚氨酯板
rigid urethane foam 刚性聚氨酯泡沫材料;硬质聚氨酯泡沫
rigid vehicle 牵引杆式挂车;全挂车
rigid vinyl 硬质乙烯树脂
rigid vinyl component 刚性乙烯基组件;刚性乙烯基合成物
rigid vinyl film dosimeter 硬乙烯膜剂量计
rigid walker 固定式步行扶架
rigid wall 刚性墙
rigid waterproof(ing) 刚性防水
rigid wedge 刚性楔块
rigid wheel base 固定轴距;固定轮距
rigid wing rail 固定(式)翼轨
rigid zone 寒潮
riginal error 固有误差
rig in bow planes 收进舰首水平舵
rig irons 钻机的铁附件;铁件;钻探工具
rigisol 硬质塑料溶胶
riglet 平条;平嵌(饰)线【建】;小方线条
rig line 钻机钢丝绳
rig-month 钻机台月;台月【岩】
rig-mounted drill 凿岩台车
rigol 遮水楣
rigorous adjustment 严密平差;严格平差
rigorous analysis 精密分析
rigorous error limits 严格误差界限
rigorously enforce rules and regulations 严格规章制度
rigorous method 严格(方)法
rigorous selection 严格选择
rigorous solution 精确解;严格解法;精确分析
rigorous survey 精密测量
rig out bow planes 推出舰首水平舵
rig runner 司钻
rig selection 钻机选配
rig-shift 钻机台班

rig ship for dive 准备下潜
rig test(ing) 台架试验;装配试验;试验台上试验
rig the derrick 在井架上安装设备
rig timber 台板
rig time 钻探时间
rig type 钻机型号
rig up 临时安装;装置好
rig-up man 挂索工
rig-up time 安装时间(钻探设备)
rig weight 坠砣
rijkeboerite 钡细晶石
rilabdite 水硅铬石
rilievo 凸起;浮雕
rill 小溪;小河沟;小沟;细流;细沟;折岸浪回流时冲成的沟;沟纹
rill crater 沟纹环形山
rill cutting 倾斜分层回采
rill drainage 细流排水;毛沟排水
rill erosion 细沟状冲刷;细沟侵蚀;溪流侵蚀;条痕状冲刷
rillet 小河沟;小溪
rill mark 细流痕;流痕
rill mo(u)ld 细流模
rill of loessland 黄土区细沟
rill stage 细沟阶段
rillstone 风棱石
rill wash 小河中冲蚀;细沟冲刷;细沟冲蚀
rill washing 细沟冲刷
Rilsan coating 尼龙-11 粉末涂料
rim 缘;支圈;轮圈;火口沿;凸缘;胎环;齿环;边缘;边沿;边框
rim angle 边界角
rim bearing 周缘支承;轮缘座;环梁支承;环承
rim bearing swing bridge 环承式平旋桥
rim bolt 轮缘螺栓
rim brake 轮缘制动器;轮圈刹车
rim cement 边缘胶结物
rim cement texture 环边缘胶结物结构
rim charge 边料
rim clamp 轮环夹
rim clearance 轮缘间隙
rim clutch (coupling) 带式离合器
rim condition 边缘条件
rim cooling 轮缘冷却
rim degeneration 边缘变性
rim deposit 边沿矿床;边沿沉积
rim diameter 轮辋直径
rim discharge mill 周边卸料碾磨机;盘式碾磨机
rim drive 轮缘传动;边沿传动
rime 雾凇;树挂;树挂
rime fog 冰雾
rime ice 霜形冰;不透明冰;冰凇
rimer 绞刀;整孔钻;扩孔钻;扩孔工人;铰刀;铰床
rimestone 边石
rim etching 边蚀刻
rim exhaust 槽边排风罩
rim expander 卸轮辋器
rim-feed sedimentation basin 周边进水沉淀
rimfire 边缘发火弹
rim fit rib 轮圈安装线
rim gear wheel 齿环齿轮
rim generator turbine 全贯流式水轮机
rim holes 周边炮眼;周边孔
rim horsepower 轮周功率
rim inner flange 轮缘内压圈;轮缘内边
rim inside thickness 轮辋内侧厚度
rim joint 轮缘接合
rim joint bolt 轮缘接合螺栓
rim lamination 磁轭叠片
rim latch 弹簧锁;执手插销;明插销
rimless 没有轮缘的;无底缘的
rim lock 门边锁;明门锁;弹簧锁
rim magnet 沿边磁铁
Rimman's green test 林曼绿试法
rimmed 带底缘的
rimmed ingot 沸腾钢锭
rimmed shelf 镶边陆架
rimmed shovel 带卷边的铲
rimmed solution pool 有缘溶潭
rimmed steel 净面钢;沸腾钢;不脱氧钢
rimmed steel sheet 沸腾钢板
rimmed texture 镶边结构
rimming 套上轮缘
rimming ingot 沸腾钢锭

rimming light 轮廓光
rimming steel 沸腾钢
rim moment 轮缘转矩
rim night latch 弹簧锁;弹弓锁;弹子门锁
rim of a cup 杯口
rim of gear 齿轮缘;轮缘
rim of gear wheel 齿轮轮缘
rim of lake 湖缘;湖边
rim of optic(al) cup 杯缘
rim of pulley 滑车边;皮带轮缘
rim of the flywheel 飞轮缘
rim of the guide blading 导向叶片环
rim of tube 镜筒垫圈
rimose 龟裂的;裂缝的
rimous 龟裂的
rim outer flange 轮缘外压圈;轮缘外边
rim outside thickness 轮辋外侧厚度
rim plating apparatus 边缘电镀器
rim potential function 边缘势函
rim pressure 边缘压力
rim pull 轮缘拉力;轮缘牵引力
rim pulley 绳子盘
rim ray 沿边光线;边缘光线
rim requirements 边缘需要量(运输问题)
rimrock 边岩;高原边沿岩石;砂矿边沿;顶岩
rim saw 圆锯
rim section 轮圈轧材
rim shaft 主轴
rims inverse voltage rating 反向电压均方根额定值
rim spacer 双轮胎可拆式轮辋间隔圈
rim speed 轮缘速度
rim stacking 轮辋堆叠
rimstone 石灰沉积物
rimstone pool 喀斯特洞底池
rim stress 轮缘应力
rim surface zona 表层区
rim syncline 周缘向斜;周围向斜
rim tubeless valve hole 无内胎轮胎轮辋气门嘴孔
rimu 红松(产于新西兰)
rim velocity 轮周速度;轮缘速度
rim ventilation 槽边排风
rim vulcanizer 轮圈硫化器
rim wave 边缘波
rim width 轮辋宽度;轮辋宽度
rim with removable flange 可卸凸缘轮辋
rim work 边部加工;库缘凹填方
rim zone 边缘区;边缘带
rinceau 叶旋涡饰;树枝状装饰;叶饰
rincon 小暗谷;河流弯曲区
rind 树皮
rind gall 木疵;隐伤
ring 匝钵圈(无底匣钵);圆环;摇车钟;箍环;拱圈;镜圈;秘密拍卖集团;马戏场;环状;环(绕);插塞中环
ring action 环作用
ring air foil 环形机翼
ring-and-ball apparatus 环球软化点测定器;环球法测定器;环球仪
ring-and-ball method 环球试验法(用于测定沥青软化点)
ring-and-ball point 沥青软化点
ring-and-ball softening point 环球(法)软化点
ring-and-ball softening point method 环球法软化点试验法
ring-and-ball test 环球试验;环球软化点试验;环球法(软化点)试验(沥青)
ring-and-brush 同心环和十字刷
ring and disc for filtrator 过滤仪环和盘
ring and radical road system 环形辐射式道路系统
ring arch 环拱
ring armature 环形电枢
ring around 回转振铃;环呼
ring auger 活把麻花钻
ring auger bit 活把手麻花钻
ring-back apparatus 振铃装置
ring-back tone 回铃音;接通回音
ring baffle 环形挡板
ring-ball mill 环球磨;环球式磨机
ring band (活塞的)环箍区
ring bar 环形杆件;钢领板;环形母线;环形汇流条
ring-barking 环状剥皮
ring-bar slow wave line 环杆型慢波线
ring beam 环梁;弧形轨道梁;圈梁;通过梁
ring belt 活塞环带(活塞上活塞环分布带)

ring-belt temperature 活塞环带温度
ring bending machine 弯环机
ring bit 活把麻花钻；环状钻头；环形钻头
ring blasting 环形炮孔破碎；扇形炮孔爆炸；扇形炮孔爆破
ring-blocking test 活塞环卡住试验
ring bolt 环(首)螺栓；环端螺栓；环形螺栓(干船坞设备)；带环螺栓
ring breaker 环式破碎机
ring bulkhead 环形隔壁；环形舱壁
ring burner 环焰灯；环形燃烧器
ring bus 环形总线；环形汇流条；环形母线
ring cable 环状缆绳
ring canal 环管
ring cathode 环状阴极
ring cell system 圈梁荷构造体系
ring-centered cage 套圈定心保持架
ring chain tautomerism 环链互变异构现象
ring chamber kiln 轮窑
ring change 换环
ring channel 圆形槽；环形槽
ring circuit 环状电路；环形电路；环路【道】
ring clamp 环状夹头
ring closure 闭环(作用)
ring-closure reaction 闭环反应
ring clothing 针布环套
ring collar 瓶颈环
ring collector 整流子；集电环；环形集电器
ring colo(u)rimetry 环圈比色法
ring complex 环状杂岩
ring compression 环压力
ring compression theory 受压环理论
ring compressor 活塞环压紧器(将活塞装入汽缸时用)
ring configuration 环形结构
ring connection 环结；环状接法；环形接法；环螺件；环接(合)
ring connector 环结件；(木结构中的)连接环
ring control network 环形控制网
ring conveyer 环式运输链
ring counter 环形计数器
ring counting 环状计数
ring course 拱背层；拱圈(层)
ring cowl 整流环
ring crack 轮裂
ring crusher 环形碎石机；环式压碎辊；环式碎石机；环式破碎机
ring crush value 环碎值
ring-cupped oak 青冈(木)
ring current 环形电流；环流
ring cut 扇形孔掏槽
ring cut method 环开挖法
ring dam 环闸；环形闸；环形坝
ring demodulator 环形反调制器；环式解调器
ring design 环形设计
ring die 环形冲模；顶管模
ring dike[dyke] 环形岩墙；环状岩脉；环形岩墙；围堤
ring discharge ion laser 环状放电离子激光器
ring dislocation 位错圈
ring distribution system 环状配水系统；环形配水管网；环形配电网
ring ditch 环形沟
ring door knocker 门环
ring dowel 环形暗销
ringdown signal(l)ing 振铃信号；低频监察信号
ring drawbar 牵引环
ring drier 环形干燥器；环式干燥机
ring drilling 环形钻眼；环向钻孔法；平面放射形钻孔钻进
ringed column 叠合柱；箍柱；分段装饰(凸出)柱；箍形柱
ringed-line barrel sampler 束节式取土器
ringed out bit 形成环状沟槽的钻头(由于金刚石损坏在端部)
ring efflux flotator 环射式浮选机
ring electromagnet 环形电磁铁
Ringelmann chart 林格曼图表(表)(测定烟囱排烟量用)；林格曼烟罕鉴别图；烟色比较图；标准烟色图
Ringelmann concentration 林格曼浓度
Ringelmann concentration table 林格曼浓度表
Ringelmann number 林格曼数
Ringelmann smoke chart 林格曼烟黑图；林格曼烟尘浓度图；林格曼烟气浓度图
ring embankment 环形堤；月牙堤；月堤

ring emitter 环形反射极
ring engraver 刻图环
ringer 圆环形匣钵；振铃信号器；无底匣钵；钵型
ring erector 环装配工；环安装器(活塞)
ringer oscillator 振铃信号振荡器；铃流发生器
Ringer's solution 林格溶液
ring expander 环扩张器(活塞)
ring-expansion polymerization 开环聚合
ring-fanned holes 扇形炮孔组
ring fast 系缆环
ring fault 环状断层
ring feeder 环式馈路
ring fence 围墙；围栏
ring filament 环状灯丝；环形灯丝
ring fire 圆火花；整圈打火
ring fissure 环状裂隙；环状裂缝
ring flange 环形凸缘；法兰盘
ring flat-plate 环板
ring-follower gate 附环滑动闸门；附环(活动高压)闸门
ring foot 圈足
ring footing 环形底脚；环形基础
ring for Anamorphic lens 艾奈莫尔弗透镜圈
ring forging 圆筒件锻造
ring form 环形
ring formation 环纹形成；结圈；成环(作用)
ring formation in mill 磨内结圈
ring forming machine 制环机
ring for safety net 系网环
ring for the sealing liquid water 封水环
ring-forward signal 前向振铃信号
ring foundation 环形基础
ring fracture 环状断层；环形断裂
ring free 无横条痕
ring furnace 环形炉
ring gall 木材环裂；环裂
ring gap 活塞环切口间隙
ring gap in bore 气缸内活塞环切口间隙
ring gasket 圆箍衬垫；环状垫圈；环形垫片；垫板；衬圈
ring gas laser 环形气体激光器
ring gate 环形浇口；圆筒阀；环形阀(门)
ring ga(u)ge 环规
ring gauging device 套圈选配装置
ring gear 回转大齿轮；环形齿圈；环形齿轮；齿圈；齿环
ring gear broach 齿圈拉刀
ring girder 圈梁；С梁；环形主梁
ring glass 分度玻璃盘
ring goniometer 环形测向器
ring green 城市周边绿地带；环形绿地；环城绿带
ring grip 环形夹
ring grizzly 环形格筛；环式格筛
ring groove 涨圈槽；环形槽；环槽；牵索环
ring groove bearing 环槽支承
ring groove bottom 环槽底
ring groove carbon remover 环槽积炭清除器
ring groove hyperboloidal lining 环沟双曲面衬板
ring groove insert 涨圈槽嵌圈
ring groove nail 环槽钉
ring groove pad 环槽垫
ring groove side 环槽侧
ring groove side wear 环槽侧磨损
ring ground 环形接地装置
ring guard 防护环式安全装置
ring guide drawing machine 环形导轨并条机
ring handle needle 环柄针
ring head 环形磁头；整流子
ring header 集电环；环状集流器；整流子
ring heater 套圈加热器
ring heating system 环状热网；环形供暖系统；环式供暖系统
ring heliotrope 环回照器
ring highway 环形公路
ring holder 钢领座
ring hole 挑料口；扇形孔；放射环形钻孔
ring induction furnace 环式感应炉
ring infiltrometer 环式测渗仪；同心环测渗仪；套环式测渗仪
ringing 振鸣；振铃(信号)；加标环；鸣震；瞬时振动
ringing cavity 低阻尼空腔谐振器
ringing circuit 冲击激励电路
ringing code 呼叫信号电码
ringing current 铃流
ringing effect 激振效应

ringing engine 人力打桩机；小型打桩机；人工打桩机
ringing fail alarm 振铃信号故障报警
ringing generator 铃流振荡器；铃流发生器；铃流发电机；电动磁石发电机
ringing interrupter 铃流断续器
ringing junction working 直接通报法
ringing-loft 悬钟山墙
ringing noise 自鸣噪声
ringing of pulse 脉冲振铃
ringing pile engine 人工打桩机
ringing pilot lamp 振铃指示灯；呼唤指示灯
ringing pulse 瞬变脉冲
ringing relay 呼叫继电器
ringing set 振铃装置；铃流机组
ringing shipway 人工浮船道
ringing signal 振铃信号
ringing signal oscillator 振铃信号振荡器
ringing test 环形试验
ringing tester 振铃信号测试器
ringing time 瞬时扰动时间
ringing wire 呼叫线；振铃线；拉铃线
ring inserting tester 环插入测试仪
ring installation 安装活塞环
ring intensity 雷达距离圈亮度
ring interface 环形接口
ring-jet unit 环形喷射器
ring jewel 环宝石
ring joint 环状接头；环状接合；环形接头；环结件；环接
ring joint flange 梯形槽式密封面法兰
ring joint gasket 接合垫圈
ring joint welding neck flange 环形式密封面对焊法兰；梯形槽式密封面对焊法兰
ring kiln 转窑；轮窑；环形窑
ring knife 环形刀
ring knocker 门环
ring laser 环状激光器
ring laser gyro(scope) 环形激光陀螺仪；环形激光陀螺仪
ring laser pick-off 环状激光传感器
ring laser rotation sensor 旋转传感器
ring latch (环形执手启闭的)门闩；(环形执手启闭的)插销；挂锁；环锁
ring layout 环形设计
ring lead 塞环引出线
ringlet 小环
ring levee 月堤；环形堤；围堤；套堤
ring lifebuoy 救生圈
ring lifter 岩芯提取器；开口环提芯器；提断环
ring lifter case 提断环外壳(岩芯)
ring lighting 照明环；照明灯
ring-like ore body 环状矿体
ring line 环形管道
ring liner 环形衬板
ring lock 环锁；暗码锁
ringlock nail 环槽钉
ring louver 环形百叶式灯具；环形灯片(照明用)
ring lubrication 油环润滑(法)；钢领润滑
ring lubrication bearing 油环润滑轴承
ring lubricator 油环润滑器
ring magnet 环形磁铁
ring main (电源插座的)环形布线法；环状干管；环形干线；环形干管；环船干线
ring main layout 环形干管布置
ring main system 环状干线系统
ring main unit 环形供电组
ring manometer 环形压力计；环秤压力计
ring marker 刺点标器
ring matching device 套圈选配装置
ring member 环形组件
ring method 拉环法；环绕法
ring micrometer 环状微器；环状测计器；环形测微计
ring mill 环形材轧机
ring motif (pattern) 环状基型
ring motor 环形电动机
ring mo(u)ld 瓶颈模
ring mulching 环状盖草
ring nebula 环状星云
ring network 环形网络
ring nozzle 圆形喷嘴；环状管嘴；环形喷嘴
ring nut 环形螺母；圆顶螺帽；环形螺帽
ring of admission ports 进气口圈
ring of cast-iron manhole 铸铁井圈
ring of coating 结圈
ring of column 环列群柱

ring of compressed cardboard	纸板环
ring of convergent power series	收敛幂级数环
ring of differential polynomials	微分多项式环
ring of endomorphisms	自同态环
ring of exhaust port	排气口圈
ring-off	拧开（钻头或钻杆）
ring-off engine	主机定速
ring of fibre optics	光导纤维环
ring-off indicator	空线指示器
ring-off lamp	拆线指示灯
ring of flywheel	飞轮轮圈
ring of formal power series	形式幂级数环
ring of holomorphic functions	全纯函数环
ring of integers	整数环
ring of light	光环
ring of numbers	数环
ring of operators	算子环
ring of packing	填料环
ring of plug	塞环
ring of quaternions	四元数环
ring of rational integers	有理整数环
ring of right multiplications	右乘变换环
ring of sand	砂环
ring of sandbags	砂袋圈围
ring of statues	环列雕像
ring-of-ten circuit	十进制环形电路
ringoid	广环
ring-oil	油环润滑
ring-oiled	油环润滑的
ring-oiled bearing	轴承护油圈；油环（润滑）轴承
ring-oiled sleeve bearing	油环润滑式滑动轴承
ring oiler	油环注油器
ring-oiling journal box	油环润滑轴颈箱
ring-opening hydrolysis	开环水解
ring opening polyaddition	开环加成聚合
ring oven	环形炉
ring oven analyzer	环炉分析仪
ring oven method	环炉法
ring oven technique	环炉技术
ring-oven test	环形炉试验
ring packed tower	环状填料塔
ring-pattern	环状排孔炮眼
ring pipe	环形管
ring pipeline	环形管网；环形管线
ring piston	筒形活塞
ring piston meter	环形活塞流量计
ring piston servomotor	环形活塞继动器
ring pit	环形泄水沟
ring plate	垫模板；环形板；带环（眼）板
ring plate lifting pillar	钢领板升降杆
ring plugging	装上活塞环
ring-porous wood	环孔（木）材
ring precipitation tank	环状沉淀池
ring precipitation test	环状沉淀试验
ring protection	环保护
ring-radial city	环状放射状城市
ring rail	钢领板
ring reaction	环状反应
ring reef	环礁
ring reinforcement	环状（布）筋
ring replacement	换环
ring resilience	圆环回跳现象
ring riser	调整垫片
ring road	环形道路；环行（公）路；环线；环路【道】；环城（公）路；环状道路
ring road system	环路系统
ring roll crusher	环辊破碎机
ring roller	带胶圈的辊子
ring-roll(er) mill	辊式磨机；环辊研磨机；环辊磨；滚环细磨机
ring-roller mill flash calciner	环辊磨快速煅烧器
ring-rolling mill	环形材轧机
ring-roll press	环辊压力机
ring roll pulverizer	辊式粉磨机；环辊粉磨机
ring rolls	碾环机
ring rope	锚环系索；收锚索
ring rot	环状腐朽
ring route	环状道路；环形路线；环路【道】
ring sampler	环形取土器；环刀取样器
ring saw	圆锯
ring scaler	环形脉冲计数器
ring scission	环裂开；环的破裂
ring scission polymerization	裂环聚合（作用）
ring scotoma	环形暗点
ring scratch awl	环形刻痕锥；环形画针
ring screen	环的隔板
ring screw	环头螺钉
ring screw ga(u)ge	螺纹环规
ring seal	密封环；环状密封；环形止水；环的封闭
ring seal gate	环形止水阀门；环封闸门；附环闸门；附环直滚闸门
ring seat	活塞环支座面
ring section	环状切片；环形截面
ring section examination	切环检验
ring settling tank	环状沉淀池
ring settling test	环状沉淀试验
ring set type	环状表镶式
ring shake	年轮裂；轮裂；环裂；顺年轮干裂；木心裂
ring-shank nail	羊眼钉；圆环体钉
ring-shaped	环状的；环形的
ring-shaped basin	环形池；圆形水池；环形（水）池
ring-shaped beam	圈梁；环形梁
ring-shape collecting flue	环形集合烟道
ring-shaped gear pump for lubricant	环形齿轮润滑泵
ring-shaped hydrocarbon	环烃
ring-shape main layout	环形干管布置
ring-shaped mo(u)ld	环形铸模
ring-shaped nucleus	环状核
ring-shaped sprinkler	环形喷灌机
ring-shaped structure	环状构造；莲花状构造
ring-shaped tank	环形池
ring-shape twin	环状双晶
ring-shaped valve seats	环形阀座
ring shaped water supply pipe net(work)	环式给水管网；扣环紧压接头（石棉水泥管用）
ring shear	环剪
ring shear apparatus	剪力环；环状剪切仪；环形剪力仪；环剪仪
ring shear machine	环形剪力机
ring shear test	环剪试验
ring shelf kiln	蒸笼窑
ring shift	循环移位
ring shooting	射圈法
ring silicate	环状硅酸岩；环硅酸盐
ring silo	环形分室的多室料仓
ringsings	冲洗池
rings intensity	距离圈亮度
ring slab	环形板
ring slot	涨圈槽
rings of die-molded packing	模压填料环
ring solenoid	环式螺线管
ring spanner	封口扳手；梅花双头扳手；环形扳手
ring species	环种
ring spot	斑纹病
ring spread	环形排列
ring spring	止动环；环形弹簧；弹簧环
ring spring buffer	环簧缓冲器
ring squeezer	压环机
ring stand	圆环架
ring standard ga(u)ge	标准环规
ring standby	主机准备【船】
ring sticking	活塞环黏住
ring sticking test	活塞环卡住试验
ring-stiffened	周边加强的
ring stiffener	环形支撑肋；环形支杆
ring stiffening	周边加强
ringstone	拱（石）楔块；（楔形的）拱面石；砌拱用的楔形砖；砌拱用的楔形块材
ring strain ga(u)ge	环式应变计
ring street	环状街道
ring stress	圆周应力；拱圈应力
ring structure	环形构造
ring sum	环和
ring support	环形支座（钢管）；环形支撑；圈座
ring surface	环面
ring switch	环形开关
ring syncline	环状向斜
ring system	环状系统；环形网；环形管网
ring tank	环形水池
ring tension	周边张力；环（形）张力
ring test	圆环拉强试验（水泥及混凝土）；圆环（试验）法；环形实验；成环试验
ring test for shrinkage cracking	收缩开裂钢环不试验法
ring thermostat tube	恒温盘管
ring thread ga(u)ge	螺纹环规
ring tie	环形地脚螺栓
ring-tile coupling	扣环紧压接头（石棉水泥管用）
ring tone	呼叫声
ring topology	环形拓扑
ring to safety ring	系网环
ring transformation	环变换
ring transformer	环形变压器
ring translator	环形变换器
ring traveller	钢丝圈
ring trouble	结圈故障
ring tube cutter	环形截管器
ring twisting	环锭加捻
ring type	环形
ring-type adder	环式加法器
ring-type azimuth mirror	环形方位镜
ring-type cold-strip mill	环式带材冷轧机
ring-type fan	环形风扇
ring-type filling	填充环
ring-type joint face	环槽式密封面
ring-type joint flange	圆环圈式法兰
ring-type reaming shell	环状金刚石扩孔器
ring-type scraper	环形刮泥器
ring-type scriber	刻图环
ring-type seals	圆环式密封装置
ring-type wedge	上部带安装环的偏斜楔
ring valve	圆环阀；环状阀；环形阀（门）
ring vent	环形通气管
ring vessel	环纹导管；环管
ring village	环形村
ring viscometer	环式黏度计
ring voltage	三角连接法线电压
ring wale	环形横梁（支撑用）；环形围撑；环围木；环撑
ring wall	环墙；围墙
ringwall foundation structure	环形基础结构
ring washer	环形垫圈
ring water main	环状自来水干管；环状给水干管
ringway	环形火车路；环形电车路；环形公路；环形道路
ring wear	活塞环磨损
ring-weeding	环状除草
ring welding	滚焊
ring wheel chuck	三爪卡盘
ring whirl	环形旋涡
ring winder	环形缠绕机
ring wire	塞环线
ring with coil spring expander	螺旋弹簧胀圈活塞环
ring with equal pressure	等压环
ringwoodite	林伍德石；尖晶橄榄石
ring-wrench	环形扳手；环形钳
ring zone	结圈带
rink	滑冰场
rinkite	褐硅铈矿；褐硅铈矿；层硅铈钛矿
rinkolite	褐硅铈矿；褐硅铈矿
Rinmann's green	林曼绿
rinneite	钾铁盐
rinse	清洗；漂洗；漂清
rinse and brush cartridge	可冲洗滤芯
rinse bath	洗涤槽
rinse cycle	漂清周期
rinse device	漂洗设备；冲洗装置
rinse displacement	再生液置换
rinse fill	冲填土
rinser	清洗装置；冲洗器；玻瓶冲洗机
rinse stage	清洗阶段
rinse tank	清洗槽；漂洗池；涮洗池；冲洗池；冲洗槽
rinse time	清洗时间
rinse truck	漂清车
rinse tub	冲洗槽
rinse type air humidifier	淋洒式空气增湿器
rinse water	清洗水；漂洗水；涮洗水
rinsing	漂洗；漂清；冲洗（法）
rinsing bar	冲水柜台
rinsing chute	冲洗溜槽
rinsing cycle	清洗周期；漂清周期
rinsing device	漂洗设备；清洗装置
rinsing equipment	漂洗装置；冲洗装置；清洗装置
rinsing flood time	涨水期间
rinsing line	冲洗管
rinsing machine	冲洗机
rinsing main	清洗干管；冲洗干管
rinsing of surface	表面重洗
rinsing period	清洗周期
rinsing pipe	清洗管；冲洗管
rinsing process	清洗过程
rinsing rate	清洗速率；冲洗速率
rinsing ratio	清洗倍率
rinsings	冲洗水；残渣

rinsing screen 冲洗筛
rinsing sludge 冲洗污泥
rinsing speed 冲洗速度
rinsing-spraying screen 喷水清洗筛
rinsing table 清洗台
rinsing tank 冲洗池;清洗池
rinsing trough 冲洗槽
rinsing tub 漂洗盆;淋洗盆
rinsing tube 清洗管;冲洗管
rinsing unit 清洗装置
rinsing velocity 清洗速度;冲洗速度
rinsing water 冲洗(用)水;清洗水
rinsing water norm 清洗用定额;清洗水定额
Rio grand rise 里奥格兰德海丘
riometer 相对电离层吸收仪;噪声探测仪;电离层相对不透明度计
riopan 水化铝酸镁
Riopone 氢氧化铝
riot gun 短筒防暴枪
riotous welter of ornament 装饰堆砌
riots 暴动、内乱和罢工
riots, civil commotion and strikes 暴动、内乱和罢工险
riots and civil commotions 暴动与内乱
rip 直锯;裂流;裂浪;裂口;裂痕;清洗器;挖掉木甲板缝的旧捻絮
ripa 河岸
riparian 湖滨的;河岸的;水滨生活的;水边的;岸地所有者
riparian cities and towns 沿河城镇
riparian country 沿岸国家;滨河国家
riparian doctrine 河岸学说
riparian environment 沿岸环境;河边环境
riparian floodplain 岸边漫滩
riparian freshwater 岸边淡水
riparian land 沿河(土地);沿岸(土)地;河边地
riparian law 河岸法
riparian legislation 沿岸法规;岸边法
riparian owner 河岸所有者
riparian plant 岸栖植物
riparian proprietor 河岸(土地)所有者
riparian provinces 沿河省份
riparian rights 河流使用权;沿岸使用权;河岸所有权;河岸使用权;岸线权;河(边)水权
riparian state 沿岸国(家)
riparian system 河岸系统
riparian use 河岸利用
riparian vegetation 沿岸植被;岸栖植物;岸边植被
riparian water loss 河水蒸发损失;河岸水量损失
riparian water rights 沿岸用水权;岸边水权
riparian water use 沿岸用水
riparian werland 岸边湿地
riparian work 河湖岸滨工程;治水工程;河岸工程
rip bit 可卸(式)钻头;可卸(式)钎头
rip blade 刮刀
rip blade saw 粗齿锯;裂刃锯
rip bulldozer 松裂推土铲
rip channel 裂流水道;急流航道
ripcord 双股绝缘电线;放气幅拉绳
rip current 裂流;离岸(急)流;激流;岸边回流
rip current channel 离岸流水道
ripe bed 熟床【给】
ripe cotton 成熟棉
ripe experience 成熟经验
ripen 挖除石块
ripened drum 熟成鼓
ripened filter 熟化过滤器
ripener 催熟剂
ripeness for cutting 采伐成熟期
ripening 熟化
ripening accelerator 成熟促凝剂
ripening index 成熟(度)指数
ripening period 成熟期
ripening phase 成熟期
ripening stage 成熟时期
ripening time 成熟时间
ripening zone 成熟区
ripe sludge 消耗污泥;熟污渣;熟污泥
ripe snow 软雪
ripe soil 熟土
ripe wood 熟材;成熟木材
rip feeder current 补流
rip head 离岸流头
ripidolite 蠕绿泥石;铁绿泥石

rip-in 板边撕裂
rippability (岩石的)可凿性
rip panel 裂幅;放气裂幅
ripped lumber 纵剖锯材
ripped material 挖松的材料
ripper 撬棍;平巷掘进机;松土机;粗齿锯
ripperability 可凿性;可破碎性;可劈性
ripper attachment 松土装置
ripper beam 裂土器横梁
ripper bracket 挡土墙托架;松土器撑架
ripper carriage bar 松土器支撑杆
ripper die 修边冲模
ripper kick-out 松土器拔出
ripper line 松土路线
ripper linkage 裂土器架
ripper rooter 犁土机
ripper-scarifier 松土翻土机
ripper step bit 阶梯式钻头;梯形钻头
ripper tooth 松土齿;松土器耙齿
ripping 纵向锯木;纵向锯切;纵解;外围切土;顺纹锯木;顺纹锯开
ripping angle 松土角
ripping bar 夹具;撬棍;撬杠
ripping blasting 挑顶爆破
ripping chisel 细长凿(刀);榫眼去屑凿
ripping circular saw 圆形粗齿锯
ripping cord 开伞索
ripping dirt 挑顶岩石
ripping edge 挑顶边界
ripping edger 裁边锯
ripping face support 挑顶工作面支护
ripping iron 刮缝凿;刮缝刀
ripping lip 松土部位;挑顶面工作端
ripping of soft rock 软石翻松
ripping path 翻松路线;翻松路径
ripping punch 横切冲头
ripping saw 粗木锯;解木锯;粗齿锯
ripping size 木材锯后尺寸
ripping stone 挑顶岩石
ripping tooth 松土齿
ripping-up 剖开(断面);锯开(木材);翻挖(路面)
ripple 鱼鳞纹;皱波;鳞纹;脚手架;毛细流;焊缝波纹;焊波;粗钢梳;波纹
ripple adder 逐位进位加法器
ripple amplifier 脉动放大器;波纹放大器
ripple amplitude 波纹振幅
ripple bed 波纹基床;波纹河床;波动床
ripple bedding 波纹层理;波痕层理
ripple carry 行波进位
ripple carry adder 行波进位加法器;脉动进位加法器
ripple carry binary counter 脉动进位二进制计数器
ripple carry system 行波进位方式
ripple coefficient 波纹系数
ripple column 纹波塔
ripple component 脉动成分
ripple contain factor 涟波因数;脉动系数;脉动因数
ripple control 脉动控制
ripple counter 波纹计数器
ripple-covered bed 有沙纹河床
ripple crest 砂纹峰
ripple cross-bedding structure 波痕交错层理构造
ripple cross-lamellar structure 同相纱纹交错纹理构造
ripple cross-lamination 波形交错纹理
ripple current 脉动电流;弱脉动(电)流;波纹电流
ripple current motor 脉动(电流)电动机
rippled 波形的,波纹的
ripple damper 脉动阻尼器
rippled bed 波纹基床;波纹河床
rippled deck transmission 平滑承座传动
rippled edge 波浪边
rippled glass 波纹玻璃
Ripple diagram 里普图;供水水源净出水量曲线图
ripple disable 无波动
ripple drift 活动波痕
ripple drift bedding 波痕迁移层理
rippled sea 微波
rippled wall 波纹状模壁
ripple electromotive force 轻微脉动电动势
ripple eliminator 灭波器;脉动波阻尼器
ripple factor 脉动系数;纹波系数;脉动因数;纹波系数
ripple figure 卷曲状;皱纹状
ripple filter 脉动滤波器;平滑滤波器;波纹滤波器
ripple finish 皱纹(罩面)漆;皱纹面饰;波纹面饰;

波纹面饰
ripple finish tile 波纹瓦;皱纹瓦
ripple formation 波纹形成
ripple form-set 波痕形态组
ripple frequency 脉动频率;纹波频率;波纹频率
ripple glass 波纹玻璃
ripple hammer 摆动锤
ripple index 波痕指数
ripple in input 输入脉动
ripple in output 输出脉动
ripple lamina 波形纹层;波纹叶理
ripple laminae in-drift 迁移波痕纹层
ripple laminae inphase 同相波痕纹层
ripple lamination 砂纹层理;波状纹理
ripple load cast 波痕荷载模
ripple mark 肋状纹理;砂纹(混凝土跑浆造成的纹痕);波状纹;波纹痕迹;波迹;波痕
ripple mark index 波痕指数
ripple marking 波纹
ripple metal sheet 波形金属板
ripple method 涟波法
ripple noise 波纹电压噪声
ripple pattern 波纹图样
ripple percentage 脉动百分数
ripple projection welding 波纹状凸焊
ripple puckers 波形小皱纹
ripple quantity 脉动值;波纹量
ripple ratio 脉动系数;波纹系数;波纹率
ripple rejection 脉动抑制
ripple-rudder ship 三舵船
ripples 纹波
ripple-shedding loom 波形开口织机
ripples on river bottom 河底波纹
ripple spacing 砂纹波长
ripple steepness 砂纹陡度
ripple surface 波纹面
ripple tank 波纹状储水池;波动箱
ripple tank method 水波箱法
ripple through 行波传送进位
ripple through carry 行波传送进位
ripple through carry unit 行波传送进位装置
ripple till 波状冰碛丘
ripple torque coefficient 转矩波动系数
ripple train 砂纹序列
ripple tray 波楞川穿流塔板;波纹塔盘
ripple varnish 波纹清漆
ripple voltage 脉动电压;波纹电压
ripple weld 鳞状焊缝
rippling 亚麻除籽叶工艺;丝纹;波痕面系列
rippling device 梳刷装置;梳麻装置
rippling instability 皱纹不稳定性
rippling sea 细浪;海面细浪
rippling trough 行波传送
riprap 乱石(堆底的护岸);石堆堤;防冲抛石;抛石
riprap apron 抛石护坦;抛石海漫
riprap bank protection 抛石护岸
riprap breakwater 抛石防坡堤;抛石防波堤
riprap cofferdam 抛石围堰
riprap dam 土石坝
riprap dike 抛石坝
riprap dumping 乱石抛填
riprap foundation 乱石基础;抛石基础;堆石基础
riprap pavement 乱石铺砌;乱石护面
ripraped embankment 乱石路基;抛石路基;乱石堤;抛石路堤
riprapping 抛筑;抛石
riprap prism 抛石棱体
riprap protection 乱石保护(边坡);抛石防护
riprap protection of slope 乱石护坡;抛石护坡
riprap revetment 抛石护岸
riprap sealing compound 抛石止水材料
riprap slope protection 抛石护坡
riprap stone 乱石;抛石
riprap stone rubble 抛填乱石
riprap works 抛石工程
rip rooter 松土筑路机
rip saw 纵剖锯;纵切锯;纵割锯;解木锯;粗齿锯;粗木锯
rip saw blade 纵切锯条
rip sawing 纵锯;顺纹锯解;顺纹锯解
ripsnorter 电动工具;移动式锯
rip surf 裂流;离岸(急)流
rip tide 裂流;浪潮;离岸(急)流
rip tooter 犁土机

rise 终升;增长;隆起;矢高;拱度;海隆;溶岩泉;梯段高度;梯段尺寸;踏步高度;上升
rise amount 潮升量
rise and down of pipe 管子的升降
rise-and-fall 涨落;涨潮和落潮;起落;起伏;升降(高差)
rise-and-fall clause 价格涨落条款;价格变动条款
rise-and-falling saw 升降锯机
rise-and-fall luminaire 升降式吊灯
rise-and-fall milling machine 升降铣床
rise-and-fall pendant 升降吊灯
rise-and-fall-saw 升降锯
rise-and-fall system 高低归算系统
rise-and-fall table 升降工作台
rise and run 高宽比(楼梯踏步)
rise and run ratio 楼梯踏步高宽比;踏步级高与踏步宽之比;踏步高宽比
rise angle 水平舵上浮转角
rise by big margin 大幅度增长
rise crest 隆起脊
rised cosine 上升余弦
rise face 升高(工作)面
rise-fall delay 上升下降延迟
rise flank 隆起翼
rise-free design 无位移花纹
rise in price 涨价
rise in steam temperature 蒸汽温度升高
rise in temperature 温度上升;升温
rise in the net output value 纯产值的增长
rise in wage 工资增长
risen mo(u)lding 凸形线脚;隆起的线脚;隆起(装)饰
risen pile 回升桩
rise of arch 拱的矢高;矢高;矗高;拱矢;拱高
rise of a truss 桁架的矢高
rise of barrel vault 筒拱矢高
rise of bottom 船底斜度
rise of carriageway 车行道的路拱高度
rise of columns 柱升起
rise of concrete strength 混凝土强度提高
rise of flight 梯段高度;梯段尺寸
rise of flood 洪水上涨
rise of floor 船底斜度
rise of groundwater level 地下水位上升高度
rise of high tide 高水位高程
rise of level 水位上涨
rise of mean water-level 水位上升平均高度
rise of pile power 反应堆功率上升
rise of river 水位上涨
rise of sea level 海平面升高;海平面上升
rise of sill 闸槛斜度
rise of span 桥跨结构拱度;起拱
rise of stage 水位上涨
rise of the river 水位上涨高度
rise of tide 高水位高程(提高水位与基准面之差);涨潮;潮升
rise of vault 穹顶矢高
rise of water 水位升高
rise of water-level 水位上涨
rise of water-table 潜水位上升;地下水位上涨
rise of wave centre line 波浪中心高度
rise of wave center line
rise of yield point 屈服点升高
rise per tooth 齿升量
rise pipe 竖管
rise project 洋隆调查计划
riser 舷侧纵板;注浆冒口;隔水导管;楼梯踢板;立管;升点管;甲板下支板;集电器接线叉;冒口;气口;提升器;踢面;梯级竖板;竖片;竖板;升气管;升降器;上升道;(坝内的)垂直井;帮板
riser-and-stack 上下水道立管
rise rate of clarifier 澄清池上升流速
rise rating 上升率
riser base 冒口窝
riser board 踏步竖板;(梯级的)竖板
riser bob 补缩包
riser brick 浇口砖
riser bus 立柱母线
riser bush 冒口圈
riser cable 直立电缆;吊索
riser chain 竖链;浮标牵索
riser circuit 上升回路
riser clamp 立管管夹(子)
riser compound 冒口防缩剂;冒口发热剂
riser conduit 上升管

riser connection 整流片的竖端连接
riser-connector 隔水导管连接器
riser distance 冒口颈长度
riser efficiency 冒口效率
riser gating 冒口浇注
riser-head 冒口
riser heat pipe 供热竖管
riser height 冒口高度;竖板高度;踏步(竖板)高度
riser joint 隔水管
riser knock-off ram 打冒口锤
riser leg 上升管支管
riser less 无楼梯踏板的
riser mandrel 隔水导管紧轴
riser modulus 冒口模数
riser neck 冒口颈
riser outlet 上升管出口
riser pad 易割冒口圈;缩颈泥芯
riser padding 冒口贴边
riser pallet 挡板
riser pin 冒口模棒
riser pipe 室内立管;竖管;上升管;竖管;立管
riser piping 上升管
riser plate 座板
riser pressure 冒口压力
riser runner 出气帽口
riser shaft 立井;提升井;升井
riser sleeve 冒口圈
risers on outfall 出口管上的扩散竖管
riser staircase signal 上升阶梯信号
riser support 竖管支架
riser system 冒口系统
riser tensioner 隔水(导)管张紧器
riser tube 提升管;上升管
riser tubing 上升管
riser turnout 上升岔道
rise/run ratio 梯级高进比
riser valve 竖管阀门
riser vent 透气孔;出气帽口
riser water pipe 竖水管
riser with flo(a)tation modules 有浮力装置隔水管
riser with slip joint 伸缩式隔水管
rise-span ratio 高跨比;矢跨比;桥梁高跨比
rise steam 蒸汽升压
rise table 踏步高计算表(梯级)
rise tholeiite 洋隆拉斑玄武岩
rise time 兴起时间;增长时间;起动时间;生成时间;升压时间;升起时间;上沿时间;上升时间;发泡时间
rise-time correction 上升时间校正
rise-time jitter 上升时间跳动
rise-time response 上升时间响应
rise-time switching 上升时间开关
rise to run ration 级高与级宽比(楼梯)
rise to span ratio 矗跨比;高跨比
rise working 上升开采
rising 涨起;溶岩泉;升高;船侧纵板
rising air current 上向风流
rising amplitude 升出幅角
rising and falling isothermal annealing 起伏等温退火
rising and falling saw 活轴圆锯;升降轴锯
rising and setting 出没
rising and sinking 升沉【船】
rising anvil 活动台砧
rising arch 跋拱;跛拱
rising borehole 上向钻孔
rising bottom 上升海底
rising butt 斜升高门铰;斜升门铰链
rising butt hinge 斜升门铰链;升降门铰链;升降合页
rising characteristic 增长特性;升起特性;上升特性
rising coast 上升岸
rising connector 上升管连接件;上升管连接器;竖管连接器
rising current 增大冲洗液流
rising-current separator 上升水流分离器
rising duct 伸出的竖直管道
rising edge 前缘;前沿;上升沿;上升边
rising end 仰起端
rising-falling tone 升降调
rising feeds 千斤顶
rising field 增强场
rising film condensation 升膜浓缩
rising fire main 直立的消防主管
rising flood period 涨水期
rising flood phase 涨水期
rising flood stage 涨水期;涨水阶段

rising floor 升高肋板
rising flow 上升流动
rising-flow expansion neutrailizing filter with changeable rate 变速生流式膨胀中和滤池
rising force 浮力
rising gradient 上坡道
rising grate 上升箅板
rising ground 高地
rising heating system 上升式采暖系统
rising height of smoke plume 烟羽上升高度;烟缕上升高度
rising hinge 斜升合页;升降合页;升高门铰;斜升门铰链
rising hopper feed 升斗式排肥器
rising level 上涨水位
rising limb 涨水段;涨洪段;过程线上升段;涨水线段(曲线的上升段);水位过程线上升段
rising limb of hydrograph 过程线上升段;水文过程线的涨水段
rising limit 出限
rising line 肋板边线
rising main 室内总管;总立管;直上总管;给水竖管;提升干管;主供水立管
rising masonry 沉箱基础上的圩工
rising method 上浮法
rising mine 上浮水雷
rising motion 上升运动
rising of arch 起拱
rising of a vault 拱顶矢高
rising of single ion content year after year 单项离子含量逐年上升
rising out of synchronism 超出同步
rising part 升起部分;凸起部分
rising period 增长周期
rising pin 活动销
rising pipe 压力管;直竖管;给水竖管;出水(竖)管
rising pipeline 上升管线
rising piping 直竖管
rising point 上升点
rising portion 上涌段(波浪的)
rising pouring 底注
rising prices 涨价
rising prominence 上升日珥
rising rate 上升速率
rising run 楼梯踢板
rising salt brine cone 上升咸水锥体
rising sample rate 升采样率
rising sample rate processing 升采样率处理
rising scaffolding bridge 临时鹰架桥;便桥
rising sea bottom 上升海底
rising section 上升段
rising segment (曲线的)上升段
rising service pipe 给气上竖管;给水上竖管;给水上升管;给水上升管
rising shaft 竖井;上行竖井;反井
rising slope 上升斜率
rising sludge 漂浮污泥;污泥上浮;污泥上翻
rising soil 翻浆冒泥
rising-speed characteristic 速度上升特性;升速特性
rising stage 上涨水位;上升水位
rising steel 沸腾钢
rising stem 明杆
rising stem double-disk parallel slide valve 明杆平行式双闸滑阀
rising stem gate valve 明杆式闸阀
rising stem valve 逆止阀
rising stone drift 上升脉外平巷;上升岩巷
rising strength 强度上升
rising-sun magnetron 旭日型磁控管
rising-sun resonator system 旭日谐振系统
rising surface curve 壅水曲线;水面上升曲线;上升水面曲线
rising tide 涨潮;高潮(位);上涨潮
rising transient 上升瞬态
rising tube 直竖管
rising tubing 直竖管
rising valve wheel 阀门的手轮
rising vault 跛拱顶;斜拱顶
rising volt-ampere characteristic 上升伏安特性
rising warded mechanism 向上挡阻的机械装置
rising water 涨水;上升泉
rising water pipe 上水管
rising wave 上升波
rising whirl 上升旋涡

rising window 倾斜的窗
rising wood 钝材;呆木
risk 危险性;危险率;风险(率);分散风险
risk-adjusted discount rate 调整风险后贴现率;风险调整折现率;按风险额调整的折现率
risk agreement 风险协议
risk allowance 安全津贴
risk altitude 危险高度
risk analysis 危险性分析;风险分析(系统工程)
risk analysis unit 风险分析器
risk arbitrage 风险套汇
risk arbitrager 投机证券商
risk area 危险区(域)
risk assessment 危险性评价;危险性评定;危险率估计;风险评价;风险估计
risk assessment categories 风险评价等级
risk assets 风险性资产
risk assets ratio 风险资产比率
risk average 风险平均值
risk averse 风险反感
risk aversion 厌恶风险;风险预防;风险规避;风险反感;防止发生风险;避开风险
risk aversion function 厌恶风险函数
risk-based decision-making 基于风险做出的决策
risk-based design 以险情为基础设计
risk bearer 风险承担者
risk-bearing 容忍风险;风险负担
risk-benefit analysis 风险效益分析
risk capital 承担风险的投资;冒险资本;投机资本;风险资本
risk characteristics 风险特性
risk characterization 危险性特征评定;风险表征
risk characterization for organism level effect 生物体效应风险特征
risk circle 危险圆
risk comparisons 风险比较
risk congruence 风险一致性
risk-consistent spectrum 与危险性一致的反应谱
risk contract 风险合同
risk cost 风险成本
risk covered 承保范围
risk decision 风险型决策
risk decision-making 风险决策
risk discount factor 风险折扣因素;风险贴现因素
risk dispersion index method 风险分散指标法
risk distance 危险距离
risk economist 风险经济师
risk effect 风险效应
risk estimation 风险评估
risk evaluation 危害度评价;风险评价
risk factor 风险因数;风险系数;危险因素;危险度;风险因素
risk for insurance 保险的风险
risk free environment 风险自由环境
risk free rate 无风险投资利率
risk function 危险性函数;风险函数
risk hedging 风险承担
risk indemnification 风险补偿
risk indicator 风险指标
risk in investment 投资风险
risk insurer 承担风险人
risk investment 风险投资
risk label 危险标记
risk level 危险程度;风险水平;风险度
risk loan 风险贷款
risk management 危险性处理;风险管理
risk management system 事故管理系统
risk manager 危险经理;风险经理
risk marking 危险标记
risk matrix criterion 风险矩阵准则
risk-maturity relationship 风险与到期日关系
risk mechanism 风险机制
risk minimization 风险减至最小限度
risk model 风险性模型
risk neutral 对风险持中性态度;风险中性
risk note 暂保单;保险凭条;保险通知;承保证明
risk of breakage 破损风险;破碎风险;断裂风险
risk of buckling 失稳风险;屈曲危险
risk of capture clause 捕获风险条款
risk of clashing 碰损风险
risk of clashing and breakage 碰损破碎风险
risk of collision 碰撞危险
risk of contamination 污染风险
risk of contamination with other cargo 染污风险

risk of contingent import duty 进关税风险
risk of cracking 裂开危险
risk of damage 损坏危险
risk of disturbance 危险故障
risk of error occurring 发生错误的风险
risk of failure 事故危险;破坏危险
risk of frozen products 冷藏风险
risk of glare 炫目危险;眩光危险
risk of heat 发热风险
risk of hook hole 钩损风险
risk of igniting explosive 引爆危险
risk of leakage 漏损风险;渗漏风险
risk of loss 损失风险
risk of misrecognition 误识别风险
risk of non-delivery 提货不着风险;遗失风险
risk of non-observed malfunctioning 非观测出错的风险
risk of non-payment 拒付风险
risk of oil 油渍风险
risk of problem 风险问题
risk of rust 锈损风险
risk of rusting 锈蚀危险
risk of scour(ing) 侵蚀危险
risk of segregation 分离的危险
risk of shortage 货差风险;短重风险
risk of spontaneous combustion 自燃风险
risk of sweating and-or heating 受潮受热风险
risk of total condemnation 完全无法使用的危险
risk of warehouse to warehouse 仓库至仓库风险
risk-oriented audit 风险导向审计
risk prediction 风险预测
risk premium 重视风险;风险生水;风险补贴;风险保险金;风险差额
risk premium rate 风险报酬率
risk prevention 风险防范
risk prevention mechanism 风险防范机制
risk probability analysis 风险概率分析
risk prone 甘冒风险
risk-prone area 风险大的地区
risk proneness 冒风险;倾向风险
risk prophecy 风险估计
risk rate 投资风险率
risk rating 风险率;风险估计
risk ratio 率比;危险比
risk repugnance 风险反感
risk-return analysis 风险收益分析;风险报酬分析
risk-return trade off function 风险与报酬比益函数
risk securities 风险证券
risk-seeking 风险追求
risk set 风险集
risk sharing 风险分担;风险承担
risks of leakage and breakage 漏损及破损风险
risks of regulatory changes 法规变动风险
risks of rejection or condemnation 废弃及没收风险
risks of sea 海上风险
risk standard 危险标准
risk suspended 未肯定的危险
risk theory 风险论
risk transfer 危险性转移;风险转让
risk transfer date 风险转移日
risk value 危险值
risk-value index 风险价值指数
Risley prism 里斯莱棱镜
Risley prism system 里斯莱棱镜系统
Riss 里斯期【地】
Riss glacial epoch 里斯冰期
Riss glacial stage 里斯冰期
Riss glaciation 里斯冰期
Riss stage 里斯阶【地】
Riss-Yumu interglacial epoch 里斯—玉木间冰期
ritalin 苯哌啶醋酸甲酯
Ritchie prism 里奇棱镜
Ritchie wedge photometer 里奇光楔光度计
Ritecure 雷特养护剂(商品名)
rithron zone 高地河段
Ritolastic 雷托防水剂(一种液体沥青防水剂)
Ritter equation of moments 里特力矩方程
Ritter method 里特法(桁架杆件内力的截面分析法)
Ritter method of dissection 里特截面切割法
Ritter pressure cell 压应力计;里特压力盒
rittingerite 砷硒银矿
Rittinger's law 里丁格定律
ritualization 仪式化

Ritz-Galerkin method 里兹·加勒钦法
Ritzide 里兹兹德塑料(一种无缝铺地板塑料,含沥青和皮革浆状物)
Ritz-plazzo 低价无缝铺地板材料(含有木屑和沥青)
Ritz's combination principle 里兹组合原则;里兹并合定则
Ritz's formula 里兹公式
Ritz's method 里兹(解)法(用最小能量计算结构的方法)
rivadavite 水硼钠镁石
rivage 滨
rival 竞争者;对手
rival business firms 竞争商行
rival commodities 竞争商品
rival demand 竞争需求
rival firms 竞争公司
rive 裂缝
rive-borne material 河流夹砂
rivefront 河边
riveling 皱纹饰面;起皱;条纹
rivel varnish 皱纹清漆
riven lath 劈开的板条
riven slate 劈开的板
riven wood siding shingle 手劈木墙板片
river 河;川
Rivera and Tamayo fracture zone project 里塔计划
river abrasion 河流磨蚀
river abstraction 河流袭夺;抽取河水
river action 河流作用
river activity 河流船舶动态
river adjustment 河道调整
river aeration 河水复氧
river aeration technology 河水复氧技术
river aggradation 河流淤积
riverain 近河区;河川的;河边居民
riverain intake 岸边进水口
riverain survey 河网测量
river alluvium 河流冲积物
river and harbo(u)r aid to navigation system 江河及港口导航系统
river-and-harbo(u)r engineering 河海工程
river and harbo(u)r navigation system 河港导航系统
river and lake fishery 内河渔业
Rivera plate 里维拉板块
river arm 河流分支;河汊
river arm of bascule bridge 仰开桥翼
river association 河流协会
river at grade 均衡河流
river authority 河流管理局;河道局
river avulsion 河流冲裂
river axis 河流轴线;河流中心线
river bank 江岸;河岸
river bank cutting 河岸开挖
river-bank dike 河岸堤
river bank ecology 河岸生态
river bank engineering 河岸工程
river bank erosion 河岸冲刷;河岸侵蚀;河岸冲蚀
river bank erosion control 河岸冲防
river bank intake structure 河岸式取水构筑物
river bank protection 河流护岸
river bank protection forest 河岸防护林
river bank recession 河岸后退
river bank scouring 河岸冲刷
river bank slide 河岸滑坡
river bank spillway 河岸式溢洪道
river bank storage 河岸储水
river bar 河沙洲;拦沙沙;河中潜洲
river barge 河边驳
river-bar placer 阶地砂矿
river barrage 河流坝;拦河堰;拦河建筑物
river basin 流域;江河流域;河域;河流盆地;河流流域;河谷盆地
river basin administration 流域管理局
river basin area 流域面积
river basin balance 流域水量平衡
river basin commission 流域委员会
river basin contracts and treaties 流域合同条约
river basin development 流域开发;河川流域综合开发
river basin development plann(ing) 流域开发规划;流域发展规划
river basin investigation 河流流域勘察
river basin management 流域管理
river basin management of water environment 流域

水环境管理
river basin master plan 流域总体规划
river basin model 流域模型;河床模型
river basin plan(ning) 流域规划
river basin project 流域开发项目;流域开发工程
river beach deposit 河口边滩沉积
riverbed 河底;河床
riverbed accretion 河床淤高
riverbed aggradation 河床抬高
riverbed and meander 河床及曲流
riverbed armoring 河道粗化;河床粗化
riverbed change 河床演变
riverbed deepening 河床加深
riverbed deformation 河床变形
riverbed degradation 河槽冲刷
riverbed degradation at downstream of dam 坝下河床下切
riverbed degradation below dam 坝下河底刷深
riverbed deposit 河床沉积
riverbed development 河床演变
riverbed downcutting below dam 坝下河床下切
riverbed elevation 河底高程
riverbed evolution 河床演变
riverbed experimental station 河床实验站
riverbed facies 河床相
riverbed landform 河床地貌
riverbed level 河床高程
riverbed load 河床负荷
riverbed material 河床(底)质
riverbed morphology 河床形态学
riverbed morphometry 河床形态测量
riverbed pier 河墩;河床桥墩;河底桥墩
riverbed placer 河床砂矿
riverbed process 河床演变(过程);河床形成过程
riverbed process observation 河床演变观测
riverbed profile 河底纵断面;河床纵断面;河床剖面
riverbed protection 河床保护
riverbed roughness 河床粗糙度;河床糙率
riverbed roughness form 河床糙率型
riverbed scour 河床冲刷
riverbed sediment 河床沉积
riverbed sweeping 扫床(指探测河床有无障碍物)
riverbed types 河床类型
riverbed variation 河床变化
river behavior 河性;河流特性
river bend 河弯;河曲;河流弯曲;河流弯道
river bend model test 弯道模型试验
river berth 河中泊位
river betrunking 河流断尾
river bill of lading 内河提单
river biological self-purification 河流生物自净化作用
river biota 河流生物群;河流生物区
river bird 河鸟
river blockage due to ice-cover 封江
river board 河流局;河道局;河边
river boat 江船;内河船(舶);河船
river boil 泡水(指水文现象)
river-borne 河水随带的;河流(带来)的
river-borne material 河流挟沙(也称河流挟砂);河流挟带物;河流搬运物质
river-borne substance 河源物质
river bottom 河底
river bottom control 河底控制
river bottom profile 河底纵断面
river bottom protection 护底(河流);河底保护
river bottom roller 河底回溜
river bottom sediment 河底沉积;河道底泥
river-bound 内河的
river branch 支流;河流支流;河流岔道;河道叉道;河汊;汊河
river branch development 支流开发
river branching 河道分叉
river breathing 河水涨落
river bridge 河流上桥梁;跨河桥
river buoy 内河浮标
river cable 过河电缆;河底电缆
river caisson 河流沉箱
river canalization 河流渠化;河道渠化
river capacity 河流能力
river capture 河水夺流;河流袭夺;河流截夺;河流夺流
river cascade project 河道梯级工程
River Catchment Board 河流流域管理处;河流流域水利局

river center 河心
river centre line 河流中心线
river channel 过河航道;河流航道;河底隧道;河道;河槽
river channel aggradation 河道淤积;河床淤积
river channel change 河道演变;河槽变化
river channel cross-section 河槽横断面
river channel degradation 河道刷深;河道冲刷
river channel form 河槽形状
river channel habitat 河道生境
river channelization 河道渠化
river channel pattern 河流航道模式;河槽类型
river channel resistance 河床阻力;河槽阻力
river channel sedimentation 河道淤积
river channel shape 河槽形状
river channel shifting 河槽迁移
river channel silting 河道淤积
river channel slope 河道比降
river channel straightening 河道裁直;河道裁弯取直
river characteristic 导治结构
river chart 江河图
river clarifying basin 河水澄清池
river classification 河流分类;河道分类
river cliff 沿河陡崖;矶头;河流陡岸;山矶
river closure 河流堵塞;河道截流
river closure gap 截流龙口
river closure sequence 河道截流程序
river confluence 汇流点(河流);河水回流;河流交汇点;河道交叉点;河道汇合点
river-connected lake 连河湖
river conservancy engineer 河流保护工程师;河流管理工程师
river conservation plan 河流保护规划
river construction 治河工程
river construction engineering 河道建设工程
river construction job 河道建设工程
river construction project 河道建设工程
river construction work 河道建设工程
river contamination 河流污染
river continuum concept 河流连续统概念
river control 治河;河流治理;河流治导;河流控制;河道治理;河道整治;河道控制
river control practice 河道管理制度
river control works 河道治理工程;河道控制设施
river cooling 河流冷却
river corner 河角;河拐角
river correction 河道治理;河道整治
river course 河道
River Course Administration Regulations of the People's Republic of China 中华人民共和国河道管理条例
river course work 河道工程
river craft 内河船(舶);河船
river crossing 过河管;过河河段;跨河测量;横穿河流;河流桥渡;河流过渡段;河道过渡段;跨河的
river-crossing facility 过河设施;跨河设施
river-crossing level(l)ing 过江水准测量;过江水准测量;跨河水准测量
river-crossing mark 过河标
river crossing-over 过渡河段
river crossing pipe 过河管(道)
river crossing regulation 河渠交汇段整治
river-crossing structure 过河建筑物
river cross-over 河流过渡段;河道过渡段
river cross-section 河流(横)断面
river current 河道水流;河川水流
river cut-off 河道(裁弯)取直
river-cut plain 河流侧移平原
river cut terrace 河切阶地
river dam 拦河坝;河堤
river dam and sluice 拦河闸坝
river data 河道资料
river deery 河麂
river deflection 河流弯曲;河流偏转
river degradation 河道刷深
river delta 河流三角洲
river density 河流密度;河道密度
river deposit 河道淤积;河道泥沙;河道沉积
river deposition 河流沉积作用;河川沉积
river deposition coast 河流沉积海岸;三角洲海岸
river depot 水上储木场
river depression 河流沉陷
river development 河流演化;河流开发;河流发育;河道发育

river-die away 河流衰减
river dike 河流堤防
river dimension 河流尺度
river direction 河流流向
river discharge 河水流量;河流流量;河道流量
river-discharge record 河道流量记录
river disposal 河流处置
river dissolved load 河流溶解质
river diversion 河流改道;河流分流;河道分流;导流
river diversion arrangement 导流布置
river diversion phase 河流导流阶段;导流阶段
river diversion sequence 河流导流程序
river diversion tunnel 导流隧洞;导流隧道;分水隧洞;分水隧道
river dock 内河船坞
river-dominated delta 河控三角洲
river drainage 河流排泄
river drainage system 河道排水系统
river dredge(r) 内河挖泥船
river dredging 河流疏浚;挖河
river drift 河沙流;河面漂浮物;河流砂;河漂沙;河流漂积物;河流堆积物;河流沉积;河道浅滩;河道漂积物;河流吹积物
river driving 放筏排;运运;木材流运;木材流送;河流漂积物
river drop 河流落差
river dues 内河税
river dynamics 河流动力学
river ecology 河流生态学;河口生态学
river ecosystem 河流生态系统
river ecosystem service 河流生态系统机能
river eel 河鳗
river embankment 河堤;沿江大堤;沿江大堤;岸堤
river emission rights 河流排污权
river ending 河流消失
river energy 河流能量
river engineer 治河工程师
river engineering 治河工程;河流工程;河工(学);河道工程(学);河川工程学;河岸工程
river engineering model 河工模型
river engineering model design 河工模型设计
river engineering model in movable bed 动床河工模型
river engineering model with bed load 推移质河工模型
river engineering test 河工模型试验
river environment 河流环境
river environmental element 河流环境要素
river environmental quality 河流环境质量
river environmental quality assessment 河流环境质量评价
river environment capacity 河流环境容量
river erosion 江河侵蚀;河流侵蚀;河流冲刷(作用);河流冲蚀;河道侵蚀
river erosion basis 河流侵蚀基准面
river estuary 感潮河口;河口湾
river etching 河流侵蚀
river experimental station 河流实验站
river facies 河相;河流相;河积相
river facies relation 河相关系;河槽形态
river fairway 河流航道
river fall 河流(水位)落差;跌水
river fan 河流冲积扇
river feeder system 内河疏运系统;内河集散系统
river ferry 过河轮渡;过河渡船
river flat 河滩;冲积平原;岸滩
river fleet 内河船队;内河船(舶)
river float 河流漂浮物
river flooding 河水泛滥
river flood plain 洪水漫淹平原;洪水漫淹滩地;河滩沼泽地(平原);河滩滩;河滩地
river flood-plain swamp 河流漫滩沼泽
river floor 河床;河底
river flow 河道水量;河流流量;河流流水;河道水流;河流流量;河川径流
riverflow depletion 河水亏耗
river flow ga(u)ging 河流流量测量
river flow regulation 河川径流调节
river-flow sampling 河水采样
river forecast(ing) 河水预报;河情预报;河道预报;水情预报;河流预测
river fork 河流分叉(口);河道岔道;河流叉道;河道分叉口;河汊;汊河
river form 河道形态

riverfrac treatment 清水压裂
river freeze-up 河流封冻;河道封冻
river freight 内河运费
riverfront 沿河地带;河边的
river ga(u)ge 河流水(深标)尺;水(位)尺;(河、溪流的)水标(尺)
river ga(u)ging 河流计量;河流测验;河道流量测量;河道测量
river ga(u)ging station 河道水文测量站;河流水文测量站;河流测站
river geomorphic analysis 河流地貌分析
river geomorphology 河流地貌(学)
river gradient 河面坡度;河流比降;河道坡降
river gradient ratio 河流比降比(率)
river gravel 河卵石;河砾(石)
river habitat 河流生境
river habitat classification 河流生境分类
river harbo(u)r 内河港口;河港
river harbo(u)r dock 河港码头
river head 河源;河流源头;河流发源地;河道源头
riverhead sandy loam 河源亚砂土
riverhood 河流状态
river hopper dredge(r) 内河装舱挖泥船
river hull insurance 内河船舶保险
river hydraulics 河流水力学;河道水力学;河川水力学
river hydrology 河流水文学;河川水文学
river hydrometry 河流水文测验
river ice 河冰
river ice-drifting 河流流冰
river icing condition 河流冰情
river improvement 河道整治;治河;河流治理;河道改善;河道整治
river improvement works 治河工程
riverine 河川的;河边的
riverine factories 沿江河建厂
riverine forest 河滩林
riverine input of pollutants 沿河排入的污染物
riverine lake 河成湖
riverine lock-and-dam installation 河流闸坝设施
riverine traffic 河上交通
riverine warfare craft 内河军用艇
river inflow 河水流入量;河道来水量
river in regime 冲淤平衡河流
river intake 江河取水口;河中取水口;河水取水口;河上取水口;河泥进口;岸边取水口
river inversion 河水转向;河水倒流;河流转向;河流逆向倒流
river-irrigated land 渠浇地
river island 河心岛;江间岛
river island head 江心洲头
river island tail 江心洲尾
river islet 河中小岛;江间小岛
river jam 河道壅塞
river junction 江河交叉点;(河流)汇流点;河流汇合点;河道交叉点;河道汇合点
river lake 河床湖
river-lake facies channel 河湖两相航道
river landform 河流地貌
river landscape 河流景观
river launch 内河小艇
river law 河道管理法
river leakage 河水渗漏
river length 河流长度;河长
river levee 防汛堤;沿河大堤;河流堤防;河堤;防汛墙
river level 河水位;河流水位
river level fluctuation 河流水位变化
river load 河流挟沙(也称河流挟砂);河流挟带物含量;河流输沙量;河流泥沙量;河流荷重;河流负荷;河流含沙量
river lock 河道水闸;内河船闸;河中船闸
river loop 河环
river luff 矶头
river-made lake 河成湖
river maintenance 河道维护
river management 河道管理
river management system 河道管理制度
river marking 河状条纹
river meander 河弯
river meandering 河道的蜿曲
river measurement 河道测量
river mechanics 河流(动)力学
river metamorphosis 河型变化;河流变态
river mine 江河水雷
river mining 河底采矿;河床开采;河床采矿;河床采掘

river model 河流模型;河工模型;河道模型
river model mo(u)lded in concrete 混凝土浇筑的河流模型
river model practice 河工模型试验
river model ratio 河道模型比(尺)
river model scale 河道模型比例;河道模型比(尺)
river model study 河工模型试验
river model test 河工模型试验
river model with all sediment 全沙模型
river model with suspended load 悬移质河工模型
river monitoring 河流监测
river morphology 河态学;河貌学;河流形态(学);河川地形学
river mouth 河口
river mouth and delta survey 河口与三角洲调查
river mouth bar 河口沙洲;河口(沙)坝;河口三角洲;河口潜洲;河口拦门沙;拦门沙
river mouth bar deposit 河口沙坝沉积
river mouth branch 河口汊道
river mouth improvement 河口改善
river mouth jetty 河口导流堤
river mouth pollution 河口污染
river mouth shoal 河口浅滩
river mouth water quality 河口水质
river mouth water quality model 河口水质模型
river movement 河流动态
river mud 河泥
river navigation 江河航行;内河航运;内河航行
River Navigation Administration Regulations of the People's Republic of China 中华人民共和国航道管理条例
river navigation system 内河导航系统
river net(work) 河网;河道网;河流网
river network water quality model 河网水质模型
river node 河流节点
river offing 出海河口;河流出海口
river of flat lands 平原河流
river of movable bed 动床河流
river oil spill simulation model 河流溢油模拟模型
river operation condition 河流营运条件
river order 河流分级;河流等级
river organism 河流生物
river outfall 汇流处;河口;河口出口
river outflow 河水泛滥;河流泛滥;河流(出)流量
river outlet 河出海口;汇流处;河出口;河口;河道出口
river outwash 河流冲积(层)
river overcrossing structure 跨河建筑物
river overflow 河水泛滥
river overflow (spill)way section 河流溢洪道截面
river parting 河道分叉
river patrol boat 江河巡逻艇
river patrol craft 内河巡逻艇
river pattern 河型;河流类型;河流花样;河谷型(式);河流形态;河道类型
river persistence 河流持续性
river pilot 领江
river pipe 过河管(道)
river piping 过河管(道)
river piracy 河流袭夺;河流截夺;河流夺流;河道夺流;河道夺流
river plain 河网平原;河成平原;冲积平原
river plan 河流平面图;河道平面图
river plankton 河水浮游生物
river planning 流域规划
River Plate bag 粗黄麻袋
river platform 河流平面形态
river police 内河治安警察
river police station 内河治安警察分局
river pollutant transport 河流污染物输移
river pollution 河流污染;河道污染;河川污染
river pollution control 河流污染控制
river pollution index 河流污染指数
river pool 河流深潭
river port 内河港口;内河港(埠);河港
river power 河流能力
river power plant 河流水力发电设备;河流水力发电厂
river process 河道演变过程
river profile 河流纵剖面;河道纵剖面;河道纵断面(图)
river project 治理规划;治理工程
river purification 河川净化
river purification board 河流净化作用
river purification capacity 河流净化能力

river pusher 内河推轮
river quality 河流质量
river quay 沿河码头;内河码头
river radar 内河(领航)雷达;内河(导航)雷达
river raft 河流木筏;河流木排
river ranges 河道固定断面线
river reach 河区;河段;上下游河区
river reach of bridge crossing 桥渡河段
river realignment 河流改道
river realignment works 河道改道工程
river reconnaissance 河道查勘
river record 河情记录;水情记录
river recovery 河流复原
river rectification 河道改直
river regime(n) 河流动态;河性;河势;河流状况;河流情态;河流情势;河流情况;河况;河道情况
river regulating reservoir 河流调节水库
river regulating structure 河道整治建筑物;河道整治构筑物;河道整治
river regulation 治河;河流调节;河道治理;河道整治;河道控制
river regulation measure 治河措施
river reoxygenation 河流复氧
river reoxygenation constant 河流复氧常数
river resources 河道资源
river rectification 河道整治
river retting 流水沤麻
river rise 河水上涨;河流涨水
river robber 夺流河
river robbery 河流袭夺;河流截夺
river routing 河道流量演算
river-run gravel 河床砾石
river runoff 河川径流
river runoff regulation 河川径流调节
river-run plant 径流式(发)电站;河床式发电站;引水式水力发电厂;贯流式水力发电厂
river-run power plant 河流水力发电站
rivers 江河
river sand 河砂;河沙;河谷砂
Riversand Harbo(u)r Act 河砂河港法
rivers board 河上交通
riverscape 河流景观;河景
river science 河流学
river scriber 河流刻图仪
river-sea cargo vessel 河海型货轮
river section 河流截面;河流横剖面;河流横断面;河流过水断面;河段;河槽面
river section below dam 坝下河段
river sediment 河流泥沙;河流沉积物
river sedimentation 河流沉积作用
river sedimentation process 河流淤积过程
river segment 河流分支;河段
river self-purification 河流自净(作用)
river self-purification mechanism 河流自净机理
river service launch 内河服务艇
river shield 盾构(江河下水隧道开挖)
river shifting 河道变迁
river ship 河船
river shipping 内河航运
river shoal 河流沙洲;河流浅滩;河道浅滩
riverside 河流临水面;河岸;江边;河边
riverside blanket of dike 河堤侧护坡
riverside bluff 河边陡岸
riverside embankment 河堤;河岸堤
riverside face 临水面;临河面
riverside grape 河岸葡萄
riversideite 纤硅钙石;单硅钙石
riverside jetty 内河码头
riverside land 河边地
riverside marsh 河边沼泽
riverside of tunnel 隧道靠河侧
riverside park 河滨公园;滨江公园
riverside pavilion 水榭
riverside pump station 河边式泵站
riverside quay 河港码头
river side road 沿溪路;沿河路;滨河路
riverside slope 迎水坡;临水坡(度);河岸坡;外坡
riverside soil 河岸土壤
riverside well 河边井
riverside wharf 沿河码头;河港码头
river silt 河泥;河流泥沙
river silting 河流淤沙
river simulator for amphibian test 水陆两用车辆江河模拟试验装置

river size 河流尺度;河道大小
river sleeve 河底(管路)套管
river slope 河面坡度;河流坡降;河流坡度;河流陡降;河道坡降
river source 河源;河流源头
river span 河跨(度);河道宽度
rivers prevention of pollution act 河流防污染法
river stage 江(河)水位;河水位;河流水位
river stage fluctuation 河水涨落
river stage ga(u)ging 河道水位测量
river stage hydrograph 河流水位过程线
river station 江河水文站;河流水文站;河流测站;河道水文站
river steamer 江轮
river steps 梯级河道
river stone 河石
river straightening 河流整直;河道(截弯)取直;河道改直
river stream 河流
river stretch 河流伸长;河段
river structure 河流构筑物;河工建筑物
river study 河道研究
river succession 河流序列
river suction dredge(r) 内河吸泥船
river superstructure 河道上部结构
river surface profile 河流水面线
river surface temperature 河面水温
river surveillance 河流监测
river survey 河流调查;河流测量;河道测量
river surveying 河道测量;河流调查
river survey(ing) map 河道勘测图;河道测量图
river survey(ing) vessel 河道调查船
river suspended drift 江中悬浮物质;河中悬浮物;河流悬浮物质;河流漂浮物质
river system 河系;河网;河流系统;河道网;水系
river system pattern 河系形式
river system survey 河系调查
river terminal 内河石油库;河港码头
river terrace 沿河阶地;河流阶地;河谷阶地;河川台地;河成阶地;河岸台地;河岸阶地
river terrace sand 河坡沙
river thaw 河流解冻
river tide 河流潮汐;河口(段)潮汐;河段潮汐
river tortuosity 河流的弯曲率
river traffic 内河航运;内河运输;内河交通;内河货运量
river training 河道整治;治河;河流治导;河流导治;河道导治
river training structure 河道整治建筑物;河道整治构筑物;河道整治
river training works 河导流工程;治河工程;河道整治工程;河道整治工程
river transition section 河流过渡段
river transport(ation) 河道运输;河流运输;河运;河流搬运(作用)
river transportation insurance 内河运输保险
river transverse gradient 河流横向坡度
river tributary area 河道流域
river tug 内河拖轮
river tunnel 河底隧道
river turbulence 河水紊动
river type pumping plant 河岸式抽水站
river type 河流类型
river types and drainage 河流类型及水系
river utility craft 内河工作船
river valley 河谷
river-valley basin 河谷盆地
river valley deposit 河谷沉积物
river valley development 流域开发
river valley line 溪线
river valley planning 流域规划
river valley project 河谷工程
river valley reclamation project 流域垦殖规划
river valley type 河谷类型
river vertical 河道垂线
river wall 岸壁;河流护堤;河流堤防;河堤;河道堤防;河坝
riverward 朝河方向
riverward slope 向河坡
river warm water dish habitat 河流温水鱼类生境
river wash 河流荒地;河流冲积物;河流冲积层;河床冲积物;河流沉积物;河道冲积物
river water 河水(流)
river water alimentation 河水补给;河流水源;河流补给
river water environment capacity 河流水环境容量
river water environment system 河流水环境系统
river water inflow 河水入流
river water intake 河水进口
river water irrigation 河水灌溉
river water isotopic composition 河水同位素成分
river water level 河流水位
river water plume 河水污染烟羽
river water pollution tracing 河水污染跟踪
river water purification 河水净化
river water quality 河流水质
river water quality assessment 河流水质评价
river water quality control 河流水质控制
river water quality criterion 河流水质基准
river water quality model 河流水质(评价)模型
river water quality modeling 河流水质模拟
river water quality monitor 河水水质监测仪
river water quality objective 河流水质目标
river water quality predication 河流水质预测;河流水质预报
river water quality specification 河流水质规范
river water sampling 河水采样
river water scheme 河水取水计划;河水取水规划;河水取水工程;河水取水方案;河道取水计划;河道取水规划;河道取水工程;河道取水方案
river water sources 河流水源
river water table 河水位
river-water temperature 河水温度
river water treatment 河水处理
river weir 拦河堰;拦河坝
river wharf 内河码头
river width 河宽;河道宽度;河床水面宽度
river width constriction 河宽束狭(段);河宽收缩(段)
river winding 河道的曲折
river with shifting bed 有活动河床的河流
river work 河流工程;河工;河道设施
river-worn 河流冲蚀的
river yard 内河造船厂
river yield 河流产生量
river zone 江河生物净化区;河流生物净化区
rivet 铆钉连接;铆接;铆钉;骑缝钉
rivet allowance 铆钉孔裕量;铆孔留量(计算铆接用)
rivet-back plate 铆钉垫圈
rivet bar 铆钉坯;铆钉钢;铆钉杆
rivet bucker 碾压机
rivet buster 铆钉机;铆钉切断机;铆钉截断器;铆钉铲
rivet catcher 铆钉工人;热铆钉接送工;投铆钉手
rivet ca(u)lking tool 铆钉铆合工具
rivet centers 铆钉中(心)距
rivet clasp 铆固
rivet clipper 铆钉钳
rivet closing tool 铆钉封合工具
rivet cold 冷铆
rivet collar 铆钉套环
rivet cutter 截铆钉器
rivet cutting 铆钉切割
rivet cutting blowpipe 铆钉割炬
rivet cutting nozzle 割铆钉喷嘴
rivet diameter 铆钉直径
rivet driver 铆工;铆钉机
riveted 铆接的
riveted allowable 铆钉孔扣除
riveted beam 铆合梁
riveted bolt 铆螺栓
riveted bond 铆接;铆钉接合
riveted bridge 铆接桥
riveted buster 铆钉机
riveted butt joint 铆钉对接;对接(头)铆接
riveted casing 铆接外罩;铆接(的)套管
riveted casing tube 铆接套管
riveted chimney 铆接烟囱
riveted conduit 铆接槽
riveted connection 铆钉连接;铆接;铆接(接)头
riveted connectional 铆接的
riveted connection sheet pile 铆接板桩
riveted construction 铆接结构;铆接钢结构;铆合结构
riveted dog 铆钉锁簧
riveted double butt strap joint 铆钉双复板对接接头
riveted element 铆接构件
riveted flange 铆合凸缘
riveted frame 铆接框架;铆接架
riveted girder 铆接大梁;铆合大梁
riveted grating 铆接格栅
riveted head 铆钉头
riveted hole 铆钉孔
riveted hole deduction 铆钉孔扣除
riveted hull 铆合车盘
riveted iron tyre 铆接铁轮箍
riveted joint 铆接(缝);铆钉连接;铆接(接)头;铆钉接合
riveted lap joint 铆钉搭接;搭接铆接;铆合互搭接头
riveted member 铆接杆件
riveted mild steel pipe 铆接低碳钢管
riveted pipe 铆接管;铆合管
riveted piping 铆接管
riveted plate-and-angle plate 角钢与钢板铆接构件
riveted plate beam 铆合板梁
riveted plate girder 铆接板梁;铆合梁;铆合板大梁
riveted seam 铆缝;铆接(纵)缝;铆合纵缝
riveted sheet steel 铆接钢板
riveted staybolt 铆钉撑杆
riveted steel 铆结钢
riveted steel conduit 铆接钢管
riveted steel girder or truss 铆接钢梁
riveted steel pipe 铆接钢管;铆合钢管
riveted steel piping 铆接钢管
riveted steel truss bridge 铆接钢桁架桥
riveted steel tube 铆接钢管
riveted steel tubing 铆接钢管
riveted structure 铆接结构
riveted tank 铆接的箱;铆接的槽;铆接的罐;铆合箱;铆合储罐
riveted truss 铆接桁架;铆合桁架
riveted truss bridge 铆接桁架桥
riveted tube 铆接管
riveted tubing 铆接管
riveter 铆钉枪;铆钉接合器;铆枪;铆接机;铆机;铆工;铆钉机
rivet factory 铆钉厂
rivet forge 铆钉炉
rivet frame 铆合架
rivet furnace rivet fire 铆钉炉
rivet girder 铆接桁架;铆钉大梁
rivet grip 铆钉夹
rivet gun 自动铆(钉)枪
rivet head 铆钉头
rivet head carpet 铆固簇绒地毯
rivet header 铆钉镦锻机
rivet hearth 铆钉炉
rivet heater 铆钉加热器;铆钉加热炉
rivet heating 铆钉加热
rivet heating forge 铆钉加热炉
rivet holder 铆顶;铆钉托;铆钉夹具;铆钉挡工;抵座
rivet holder up 铆轧机;铆钻;铆钉梭
rivet hole 铆钉孔
rivet hole deduction 铆钉孔扣除
rivet hopper feeder 铆钉漏斗送料机构
rivet hot 热铆
rivet in double shear 双(面)剪(力)铆钉
riveting 铆接法;铆接;铆合
riveting beam 铆合梁
riveting burr 铆钉毛口
riveting by hand 手工铆钉
riveting by machine 机铆
riveting clamps 铆钉夹具;铆钉钳;铆(钉)夹钳
riveting die 铆接模
riveting dolly 铆顶;铆钉托;顶把;抵座
riveting double shear 双剪(切)铆钉
riveting forge furnace 铆钉锻炉
riveting gang 铆工组;铆工班
riveting gun 铆钉枪;铆枪;风动铆钉枪
riveting hammer 铆钉锤;铆钉枪;铆钉撑锤;铆锤
riveting handle 铆钉夹钳;铆叉
riveting head 铆钉头
riveting hearth 铆钉炉
riveting hole 铆钉孔
riveting horn 铆钉垫铁
riveting in rows 成排铆钉
riveting jack 铆接托架
riveting joint 铆钉搭接
riveting knob 铆钉顶;铆工模
riveting machine 铆钉机;铆接机;铆机
riveting machine hammer 铆钉锤;气动铆钉锤
riveting nut 铆钉头
riveting party 铆工组

riveting pin 杆铆钉
riveting plate beam 铆合板梁
riveting plate girder 铆合板大梁
riveting press 压铆机;压力铆接机;铆钉机
riveting pressure 铆接压力
riveting punch 铆接冲孔机;铆冲器
riveting ram 水压铆钉
riveting set 铆钉用具;铆接工具;铆接用具
riveting stake 钢砧;铆砧;铆接砧
riveting team 铆工组;铆工班
riveting tongs 铆钉(夹)钳
riveting tool 铆接工具
riveting worker 铆工
rivet insert 铆接嵌衬
rivet in shear 受剪铆钉
rivet in single shear 单剪(切)铆钉;单剪(力)铆钉
rivet in tension 拉力铆钉;受拉铆钉
rivet interval 铆钉间距
rivet iron 铆钉铁
rivet joint 铆钉接合;铆接(头)
rivet knob 铆工模
rivet lap joint 铆钉搭接
rivet length 铆钉杆长度;铆钉长度
rivet line 铆接线;铆钉(准)线
rivet list 铆钉表
rivet machine 制铆钉机
rivet making machine 制铆钉机
rivet nail 紫铜钉
rivet nut 铆钉帽
rivet on 铆上;加上铆钉
rivet over 铆上
rivet pattern 铆钉排列形式
rivet pin 铆上销子
rivet pitch 铆距;铆钉行距;铆钉心距;铆钉距;铆钉间距
rivet pitching tongs 铆钉定位钳
rivet plug 膨胀铆钉
rivet point 铆钉尖;铆钉端部
rivet punch 铆钉冲头
rivet removal tool 铆钉拆除工具
rivet rigidity 铆接刚度
rivet rod 铆钉杆
rivet row 铆钉行(数)
rivet seam 铆接缝;铆缝
rivet section 铆钉截面
rivet set 铆钉用具;铆钉冲头;铆钉模
rivet shank 铆钉体;铆钉杆
rivet single 单行铆钉
rivet snap 铆头模;铆钉枪头;铆钉窝子;铆钉(冲)模
rivet snap hammer 铆钉罩锤
rivet space 铆钉间隔
rivet spacing 铆接间距;铆钉(间)距;铆钉间隔
rivet spinner 铆钉旋压机
rivet squad 铆工组
rivet stamp 铆钉模
rivet steel 铆钉钢
rivet steeple head 锥头铆钉
rivet stem 铆钉体;铆钉杆
rivet symbol 铆钉符号
rivet tail 铆钉尖;铆钉徽头;铆钉端部
rivetter 铆钉枪;铆工
rivet test 铆钉试验
rivet tester 铆钉试验工
rivet tight 铆紧
rivetting 铆接(法)
rivet value 铆钉强度
rivet washer 铆钉垫圈
rivet weld 铆焊;电铆焊缝;电铆
rivet welding 塞焊;电铆焊
rivet wire 铆钉钢丝
rivet work 铆工工作
riviera 沿海游憩胜地;滨海疗养区;海滨旅游胜地
riving 岩石破裂
riving hammer 劈锤
riving knife 劈板斧;劈木瓦刀;劈刀;缝楔;分板机
RIV measurement 无线电干扰电压测量
rivulet 小溪;小河;溪(流)
r-meter 伦琴计
R-mode cluster analysis R 型聚类分析
R-mode factor analysis R 型因子分析
rnathyite 钾铀铀云母
roach 岩石;波兰德石灰岩上部优质层;船尾滚浪
roach back 弓背
Roach bed 罗奇层【地】

road 公路;矿山平巷;马路;市区道路;道路
roadability 操作灵便性;行车舒适性;行车稳定性和适应性
road access 进港道路;(进港的)通路
road adherence 路面与轮胎间的黏着力
road adhesion 路面与轮胎间的黏力
road administration 道路管理
road aesthetics 道路美化(学);道路美学
road alignment 道路线形
road and bridge 公路和桥梁
road and railroad crossing 公路铁路(平)交叉(道口)
road appearance 路容
road appreciation 行车升值
road approach 桥头(引)路;引桥路;引坡;桥梁引路;桥梁引道;道路引道
road area index 道路面积指数
road area per capita 人均道路面积;居住道路占有率;居民道路占有率
road area per citizen 人均道路面积(城市)
road area ratio 道路面积率;道路面积密度
road asphalt 路用沥青;道路沥青;筑路(石油)沥青
road asphalt emulsion 筑路沥青乳胶
road auger 手摇钻
road authority 道路工程管理局
road axis 道路轴线;道路中心线
road banks 道路边坡
road base 路面底层;路面承重层;道路基层;路面主层;基层
road bay 路面(板)分块;路面(板)分节;路面(板)分色
road beacon (筑路地点的)警告标灯
road bead 道路用微珠
road bed 路基(表)面;路基;路床;路槽面;路槽底;道床
roadbed body 路基本体
roadbed boiling 翻浆
roadbed cavern 路基洞穴
roadbed defect 路床缺点;路基缺陷
roadbed depth 道床厚度
roadbed excavation 路基开挖
roadbed grouting 路基灌浆
roadbed in especial condition 特殊条件下的路基
roadbed material 路基材料
roadbed modulus 道床系数
roadbed on soft soils 软土路基
roadbed on steep slope 陡坡路堤
roadbed pressure 路基面压力
roadbed profile 道床断面
roadbed resistance 道床阻力
roadbed section 路床断面;道床断面
roadbed settling 路基下沉
roadbed shoulder 路肩
roadbed side-slope 路基边坡
roadbed stability 路基稳定性
roadbed stabilization 路基稳定技术;路基稳定措施;路基稳定(法)
roadbed stiffness 道床刚度
roadbed stress 道床应力
roadbed surface width 路基面宽度
roadbed width 道床宽度
road binder 筑路黏合料;筑路结合料;路面黏结料
road bitumen 路用沥青
road blading 用平地机整平(道)路;路面刮土;道路整型
road block 筑路块料;路栏;路障
road board 道路局;公路局
road book 路程指南
road border 路边;路肩
road borrow pit 路旁借土坑
road boundary 道路界限
road breaker 路面破碎机;路面破坏机;掘路机
road bridge 公路桥(梁)
road broom 路刷;扫路机;扫路车
road builder 筑路机;筑路工作者
road building 道路建筑(物);道路工程;道路施工;筑路
road building bitumen 铺路沥青
road building contract section 筑路合同章节
road building equipment 筑路机械;筑路设备
road building firm 筑路公司
road building machine 筑路机械
road building machinery 筑路机械
road building material 筑路材料

road building site 筑路现场
road building technique 筑路技术
road built on seawall 海塘路
road burner 路面热熨器
road camber 路拱
road capacity 道路通车量;线路通过能力;道路容量
road carpet 路面表层;道路毯层;道路磨耗层;道路面层
road carriageway 行车路面
road carrying capacity 道路运载能力
road casing line 道路边线
road chip(ping)s 筑路石屑
road circuit 道路环线
road classification 公路分类;道路分类;道路分级
road clearance 车身离地距;路面与车身之间的空隙;路面与车身之间的净空;道路净空
road closed sign 禁止通行标志
road closure 道路阻塞
road compaction machine 道路压实机械
road concrete 道路混凝土
road concrete sand 道路混凝土用砂
road condition 路况;道路行车条件
road condition survey 路况调查
road congestion 道路交通拥挤;道路拥挤(度)
road construction 路面结构;道路工程;筑路;道路施工;道路结构;道路建筑;道路建设
road construction budget 道路施工预算
road construction financing 道路建筑资金供给
road construction machine 筑路机械
road construction project 动力建筑设计
road construction site 动力施工现场
road construction soil 筑路土壤
road construction technique 动力施工技术
road construction tester 动力施工检验设备;动力施工检验装置
road construction work 动力施工工程;动力建筑工程;筑路工程
road contact of tire 轮胎与路面的接触面
road contractor 筑路承包者;筑路承包人;筑路承包商
road cooker 筑路加热炉
road core 路面钻样
road crane 道路上运行的起重机
road crossing 交叉口;道路交叉(点)
road crossing design 交叉口设计
road crossing signal 道路交叉口信号
road crust 磨耗层;道路面层
road-cum-rail bridge 公路铁路两用桥
road curb 公路边(缘)石
road curve 道路弯道;道路曲线
road cut 路堑;路基切面
road data 道路资料
road data bank 路况资料库
road deck 路面板
road decking 覆盖路面
road decking beam 路面梁
road decking panel 路面覆盖
road defect 道路缺点
road delineator 指路牌;道路标牌
road delivery 载货汽车运输;道路输送;公路货运
road density 道路密度
road depot 道路施工供应站
road depreciation 行车贬值
road design 道路设计
road-design engineer 道路设计工程师
road ditch 道路边沟
road diversion 迂回(道)路;绕行道路;曲折(道)路;道路改线;分行车路
road divider 公路分界石
roaddle 代赭石
road drag 压路机;刮路器;路刮;刮路机
road drain 路边排水(沟)
road drainage 地面排水;路边排水(沟);道路排水
road drainage pipe 道路排水管
road dust 路尘;道路尘埃
roaded catchment (人工的)道路型集水区
road embankment 路堤
road emulsion 铺路乳化沥青
road engineer 道路工程师
road engineering 道路工程
road engineering progress 道路工程进展
road engineering survey 道路工程测量
road equipment 道路设备
roader 受风流漂动的船

road excavator 筑路用挖掘机；路用挖土机
road exhibition 道路展览(会)
road expenditures 道路费用
road fabric 道路用织料
road fare 道路运费
road feasibility study 道路工程可行性研究
road fill 道路路堤
road filling 路堤
road finisher 路面整修机
road-finishing machine 路面整修机；路面修整机
roadfone 道路电话网
road fork 道路分叉
road form 筑路模板；路边石模板；路面模板
road formation 道路构造
road formwork 筑路模板
road for tank traffic 装甲车交通道路
road friction 道路摩擦
road frost boil 道路翻浆
road furniture 道路(附属)设施；道路(附属)设备
road furrow 车辙
road glare 道路眩光
road grade 道路坡度
road grader 压路机；筑路(平地)机；平路机
road grading 道路整型；道路土方整平
road gravel 筑路砾石
road green area 道路绿域；道路绿地
road greening mileage 公路绿化里程
road grinder 道路路面磨平机
road gritting machine 石屑撒布机
road grooving machine 道路路面刻槽机
road guard 道路护栏；道路防护物
road gull(e)y 道路排水沟；路沟
road harrow 路耙；耙路
road haul 行驶里程
road haulage 道路运输；公路货运；轮胎式运输
road haulage industry 汽车货运业
road haul truck 载货汽车
road haunch 支承路面层梁腋
roadhead 筑路工作面；巷道工作面；平巷的工作面
road heater 路面加热器；路面加热机
road heating 道路路面加热
road heating installation 道路路面加热装置
road-holding 道路行车适应性；道路行车稳定性
road hone 土路整平器；土路平整器
road house 路边(饮食)店；路边旅馆；客栈
road hump 路面隆起
road ice-control 道路冰冻防护措施；防止路面结冰措施
road improvement 道路改良；改善工程【道】；道路改善
road industry 筑路业
road inequality 道路的不平整性
road infill 道路套设
road information transmitted aurally system 道路信息传送预感系统
roading 拖运；道路规划
road inlet 道路进水口；道路入口
road intersection 交叉口；道路交叉(口)
road inventory 路况表；道路清单；道路目录；道路登记表
road joint 道路路面接缝
road joint cutter 切缝(缝)机
road joint machine 路面接缝安设机
road junction 道路枢纽；道路交叉(点)；道口；路口
road kerb 侧石；路缘石
road kettle 道路修筑加热炉
road lamp 路灯
road land 道路用地
road landscape 道路景观；道路环境美化
road laying 铺路；筑路；道路设置；道路铺设
road layout 道路规划；道路放样；道路定线；道路定位线；道路布置
road lengthsman 路权决定人；道路长度测量员
roadless 无路可通的
roadless transport 越野运输
road leveller 道路整平机
road level(l)ing machine 压路机；道路整平机
roadlice 微型汽车
road life 道路使用寿命；道路使用年限
road lighting 公路照明；道路照明
road lighting fixture 道路照明设备
road lighting mast 道路灯杆
road line 路面划线；马路划线涂料
road line coatings 马路划线涂料

road line machine 路面划线机
road line marker 道路划线机
road line paint 道路划线漆；路标油漆
road lining 道路衬垫；道路衬砌
road lining paper 道路防水衬纸
road load 道路荷载
road location 道路放样；道路定线
road locomotive 列车机车
road logger 道路记录器
road lorry 罐车
road louse 微型汽车
road louse 小汽车
road luminaire 道路照明光源
road luminance 路面亮度
road macadam 筑路碎石(料)
road machine 筑路机
road machinery 筑路机械；道路机械
road machining 用机械筑路
road magnet 磁力清路机
road maintainer 养路机；道路养护工
road maintenance 道路养护
road maintenance cost 养路费
road maintenance crew 养路道班人员
road maintenance equipment 养路设备
road making 筑路；道路建设；道路建设
road-making mix(ture) 筑路混合料
road-making plant 成套筑路机；平巷掘进机械
road-man 修路工人；筑路工人；路工
road map 公路(交通)图；道路图；路线图；街道地图
road margin 路段；道路界限
road marker 公路等级编号；路面划线机
road marking 路面划线；路面标线；路面标记；道路标示
road marking composition 路面标线混合料
road marking compound 路面标线混合料
road marking machine 道路划线机
road marking material 路面标线材料
road marking paint 路标(油)漆；路线漆；马路标线漆

roadmaster 路段养路工人；养路局长；公路管理局长；养路领工员；养路工长；养路分段长；工务段线路主任
Roadmasters, Maintenance-of-way Association of America 美国铁路养路学会；美国铁道养路学会
road mat 路面面层
road material 筑路材料；筑路器材
road materials survey 筑路器材测量；筑路材料测量
road mender 养路工人；道路修理工
road mending 修路；道路修补
road mesh 道路钢筋网
road metal 硬质路面；道砟；筑路碎石(料)；公路碎石
road mileage 道路里程
road miles 线路总延长
road milling and grooving machine 路面刻槽机
road-mix bituminous mixture 路拌沥青混合料
road-mix bituminous surface 路拌沥青路面
road-mix construction 路拌法施工
road-mix course 路拌层
road-mixed pavement 路拌路面
road mixer 筑路搅拌机；筑路拌料机；筑路拌和机；路拌拌和机；路拌机
road mix(ing) 路拌；路拌法建筑的路面；路拌混合料
road-mix(ing) method 路拌法
road-mix(ing) plant 路拌厂；筑路混合料拌和设备
road-mix machine 路拌机；筑路拌料机
road-mix material 路拌材料
road-mix pavement 路拌法建筑的路面；路拌材料路面
road-mix soil stabilizer 路拌土壤稳定机
road-mix surface 路拌路面
road-mix surfacing 路拌面层；路拌法铺路(面)
road mixture 筑路混合料
road-mix work 路拌工作
road mobility 行驶灵活性
road narrows sign 路幅缩狭标志
road needs 道路修理需要
road net density of public transportation 公共交通线路网密度
road net flow volume figure 路网流量图
road net node 路网节点
road net(work) 公路网；道路网
road network planning 道路网规划
road noise 道路噪声
road occupancy 道路占用
road octane 道路辛烷值

road octane-value 行车辛烷值
road of bridge 桥面(路)
road of clay-stabilized sand 砂掺黏土稳定的道路
road officials 道路管理员；道公务员
road oil 铺路沥青；液化沥青；筑路用油；筑路柏油；路油；铺路油
road-oil distributor tanker 道路沥青喷洒机；道路沥青喷洒车
road oiler 路油洒布机；沥青洒布机；铺路洒油机
road oiling machine 路面洒布机；沥青洒布机；浇沥青(路)机；洒铺沥青机
road-oil mat top 浇油面层
road-oil surface 浇油面层
road on stilts 高架道路
road operation 道路运营；道路营运
road overseer 筑路监工
roadpacker 夯路机；刮土机；刮路机；路用铲路机；平路机；道路夯击机
road paint 道路划线漆；路标油漆；道路划线涂料
road panel 道路面板
road patch 道路修补；道路补坑
road patrol 道路巡逻；道路巡视
road pattern 路型
road pavement 路面；道路铺面
road pavement bed 路基
road pavement construction 路面施工
road pavement construction machinery 路面机械
road pavement coring machine 路面施工钻取土样机械
road pavement foundation 路面基层；路面基础
road pavement tester 路面检测设备
road pavement work 路面工作
road paver 筑路工；砌路机
road paving 道路铺设
road performance 道路性能
road performance test 行车试验
road petroleum asphalt 道路石油沥青
road plan 道路平面图；道路计划
road planer 平路(面)机；铲路机；刨路机
road planning 道路规划
road planning machine 平路机
road planning section 道路设计断面(图)
road planting 道路植树；道路绿化
road plough 路犁
road pocket 路窝；路床
road portability 道路运移能力(道路容许工程机械靠自身动力行驶转场的能力)
road products 道路工程(混凝土)制品
road project 道路工程(方案图)
road property 道路资产
road pug travel-mix plant 路拌机；移动的路拌设备
road pump 可在道路上运移的泵
road-rail 路边轨模；公路铁路两用的
road-rail bridge 公路铁路两用桥
road/-rail crossing 公路铁路(平)交叉(道口)
road railer 公路铁路两用车
road railway 索道
road-railway tunnel 公路铁路两用隧道
road rater 道路使用性能测定机
road ratio 道路率
road reconstruction 道路重建；道路改建
road reconstruction works 道路改建工程
road reflecting mirror 道路反射镜；道路反光镜
road relocation 道路改线
road repair 修路；道路修补；途中修理
road rerouting 道路改线
road research 道路研究
road research laboratory 道路研究实验室
road resistance 道路(行车)阻力
road revenue 道路进款；道路财务收入
road ripper 耙路机；松土机【道】
road roller 压路碾；压路机；光面压路机；路碾；道路碾压机
road roller wheel 压路机碾轮
road rooter 路面除根孔；除根机；犁路机
road safety 道路(交通)安全；公路(交通)安全
road safety research 道路安全研究
road salt 道路冰冻防护添加剂；道路防冻盐；道路用盐
road scraper 铲运机；筑路铲运机；刮路机；平路机
road scum 路面浮渣
road sealing 道路填缝
road section 路段；道路分段
road section design by same standard 同标准的设计路段

road sense 行车判断能力
road service 道路营运;道路服务
road service level 道路服务水平
road service test 道路使用试验
road shaper 路面整修机;路面整形机
road shelf 路棚
road shocks 行车震颤;道路行车震颤
road shoulder 路肩
road shovel 道路挖掘机
road shovel bucket 道路挖掘机铲斗
road shovel loader 道路挖掘装载机
road show 道路展览(会)
road shuttering 筑路模板
roadside 路肩;路边;路侧地带;路旁(地)带
roadside advertisement 路旁广告
roadside assembly sign 路侧组合标志
roadside channel 路边排水渠
roadside condition 沿途条件
roadside delineator 路边(夜间)反光标志
roadside development 沿线开发;路旁建设;路旁开发;路边建设;路旁设施;路旁发展
roadside ditch 公路边沟;路边水沟;道路边沟
roadside embellishment 路旁美化;路旁绿化
roadside environment 路边环境
roadside facility 路旁设施
roadside features 路旁设施
roadside flower bed 路边花坛
roadside garden 路边公园;街头绿地
roadside gully 路旁进水口;路边小沟
roadside hedge 路边灌木篱;路边栅栏
roadside improvement 路旁改建;路旁改良
roadside installation 路边设施
roadside interference 路边干扰
roadside interview 路边询问
roadside interview method 路旁交通询问法
roadside interview origin-destination survey 路访起讫调查
roadside landscaping 沿路景观;路旁景观;路旁风景
roadside obstruction 路旁障碍物
roadside park 街道停车;街道花园
roadside parking 路旁停车;路边停车
roadside parking apron 路侧停车坪
roadside parking area 路边停车场
road side parking lamp 路旁停车信号灯
roadside planting 路旁植树;路旁绿化;路边种植
roadside platform 路旁车站
roadside protection 路旁防护;路旁保护
roadside radio 路边无线电
roadside reflector sign 路边反光标志
roadside repair 路旁修理
roadside rest (长途车的)路旁休息停车场;(长途车的)路旁休息场所
roadside sign 路侧标志
roadside station 路旁车站
roadside stone packing 路旁石料堆
roadside storage 路旁存放(筑路材料)
roadside survey 路边调查
roadside (tree) planting 行道树;路旁植树
roadside turnout 路边让车道
roadside vegetation 路旁植被
roadside verge trimming 路边修整
road sign 路标;道路标志;路牌
road sign adhesive 道路标志胶
road signal 路标
road sign and signal 路标和信号
road sketch 道路略图
road sketching 道路草测
road-skid property 道路防滑性质
road slab 路面板
road slice 道路片段
road slide 路滑
road slip 道路滑移
road slope 道路坡度
roadsman 筑路工人;路工
road spall 道路碎石(屑)片;道路剥落;道路裂开
road speed 道路车速
road spreader 筑路撒料机;粒料撒布机(路用);道路撒料机
road spring 车用弹簧
road sprinkler 道路洒水机;道路洒水车
road stability 道路稳定性
road standards 道路标准
roadstead 停泊所;港外锚地;开敞锚(泊)地;开敞或半开敞锚地;近岸锚地;(海上敞开的)锚地;锚泊区(域);无遮蔽港口(不能挡风浪);停泊地;泊地;半开敞锚地;碇泊处
roadstead buoy (海上敞开的)锚地浮标
road steel fabric 道路钢筋网
roadster 小型货车;跑车;双座敞篷轿车;双门篷车;双门敞篷轿车;受气流漂动的船
roadster at mooring 在锚地锚泊的船
road stone 筑路石屑;筑路石(料);路用石料;铺路石料
road straight 道路直线段
road stratum 道路地层
road stream 道路车流(量)
road stress 道路应力
road structure 道路路面结构;道路结构物;道路建筑物
road studs 路钉;反光路钉
road supporting work 道路支承工作
road surface 道路表层;修筑路面;路面;道路面层
road surface course 路面面层
road surface drainage 路面排水
road surface dressing 道路表面处治
road surfaceman 修路工
road surface mortality curve 路面损坏率曲线
road surfacer 铺路机;路面整修机
road surface stripping 路面剥离
road surface treatment 路面处治;路面处理;道路表面处治
road surfacing 铺路面
road surfacing construction 路面建筑;路面施工
road surfacing foundation 路面基层;路面基础
road surfacing tester 路面检验器;路面检查设备
road survey 路线测量;道路测量
road surveying 道路测量
road survey system 道路量测系统;道路测量体系
road suspension bolt 道路悬吊螺栓
road sweeper 扫路机;道路清扫车
road sweeper-collector 扫路集尘机;扫路集尘机
road sweeping and clearing work administrating system 道路清扫作业管理系统
road sweeping and clearing work system 道路清扫作业系统
road sweeping magnet 磁力清路机
road sweepning machine 扫路机
road switching locomotive 列车调车机车
road symbol 道路符号
road system 道路系统;道路系数;道路网
road tamping machine 道路夯实机
road tank car 油罐车;油槽车
road tank(er) 公路汽车槽车;油槽汽车;罐车;公路运油车
road tar 筑路焦油(沥青);筑路柏油;道路沥青
road tar-asphalt mix(ture) 路用混合沥青;路用煤沥青和石油混合料
road tar emulsion 煤沥青乳液;路用乳化沥青
road tar type penetration macadam 路用煤沥青贯入式碎石路面
road tar visco(si)meter 路用煤沥青黏度计
road tax 路税
road technical standard 道路技术标准
road temperature 道路温度
road temple 路旁庙宇
road test 试车;行车试验;道路试验
road-test car 道路试验车
road test simulation 模拟道路试验
road to be blasted 待爆破的道路
road to be shot down 待炸毁的道路
road toll 养路费;过路费;通行过路费
road towing 道路上拖曳;道路上牵引
road towing gear 道路上拖曳装置;道路上牵引装置
road traction engine 曳引机;路用牵引机
road traffic 公路运输;公路交通;道路交通
road traffic accident 道路交通事故
road traffic control 公路交通控制;公路交通管制;公路交通管理
road traffic engineering 道路交通工程
road traffic fuel 公路交通用燃料
road traffic marking 道路交通标线
road traffic sign 道路交通标志;道路交通信号
road traffic survey 公路交通观测
road traffic volume 道路交通量
road trailer 道路挂车;道路拖车
road train 道路汽车列车;大型运牛车;大型载重汽车;一支车队;汽车列车
road transit 陆上运输
road transition curve 道路缓和曲线
road transport 汽车运输;道路运输
road transportation 道路运输;公路运输
road transportation commitment 汽车运输业务
road transportation means 道路运输工具
road transportation unit 道路运输设备
road transport network 道路运输网
road transport tanker 油槽车
Road Transport Union 国际陆运联盟
road travel(l)ing gear 道路上移式装置
road tread 车行道
road trough 路槽
road tube 公路隧洞
road tunnel 公路隧洞;公路隧道;道路隧道
road turn 道路转折点;道路转向
road underconstruction 建筑中道路
road upkeep 道路维护
road usable all year round 常年通行公路
road usable only during dry periods 旱季通行道路
road usage 道路运营;道路使用
road user 道路使用者
road user cost 道路使用费
road use toll 公路交通费
road van 沿途零担车
road vehicle 道路运输工具;道路车辆
road vehicle engine 道路车辆发动机
road vehicle scale 道路车辆秤
road vehicle unit 路行车单位;道路交通量计算单位
road verge 路旁草坪
road vibrating and finished machine 筑路振动磨面机
road vibrating machine 筑路振实机
road vibration machine 筑路振实机;道路测振机
road watering 浇洒道路(用水)
road wave 道路搓板(现象);道路波浪(现象)
roadway 铁路路线;桥面;车行道;路幅;平巷;道路;车路;车道
roadway above 桥梁上承行车道;上承(式)行车路
roadway arch 车行道弯拱
roadway below 下承行车路;下承行车道;桥梁下承行车道
roadway bucket 公路边排水沟清理铲斗
roadway capacity 道路通行能力
roadway casting 预制车行道铺面板(桥梁)
roadway condition 道路条件
roadway construction equipment 筑路设备
roadway crossing 平巷交叉道口
roadway deck 车行道板
roadway delineation 路面反光标记显示;车行道反光标记显示
roadway delineation treatment 道路轮廓线处理
roadway design 车道设计
roadway drainage 路面排水系统
roadway excavation 路基开挖;平巷掘进
roadway grouping 道路分类
roadway junction 平巷交叉道口
roadway lantern 车行道信号灯
roadway light 路灯
roadway lighting 平巷照明
roadway load 车道荷载
roadway luminance 路面亮度
roadway maintenance 车行道养护;平巷维护
roadway of dam 坝上道路
roadway on bottom boom 上承桥面
roadway platform 路上站台
roadway pressure arch 平巷压力拱
roadway section 道路断面;车行道断面
roadway side 车行道接口;平巷边帮
roadway slab 路面板
roadway stringer 车道纵梁
roadway structure 路基路面结构;道路结构
roadway support 车行道部分支护;车行道部分改建;平巷支护
roadway system 车行道网;车行道系统
roadway width 车(行)道宽度
road wear 道路磨耗
road wedge 掘路楔块
road weigh bridge 公路地秤;地中衡
road wheel 行走轮;车轮
road wheel contact 轮胎与路面的接触面
road wheel drive 地轮驱动
road wheel torque meter 汽车车轮转矩计;动轮转矩计
road white washing 道路刷白
road widener 扩路机

road widening 道路拓宽;道路加宽;道路展宽
road width 路面宽度;路宽;道路宽度
road with engineering brick paving 高强抗蚀砖铺面道路
road with paving of natural setts 天然石块铺砌的路
road with rising gradient 上升坡度的道路
road works 筑路工作;道路工程
road work ahead sign 前面道路施工标志
road worthiness 车辆行驶优良;车辆行驶卓越
roak 表面缺陷
roaldite 罗氮铁矿
roalkylsiloxane polymer 氟代烷基硅氧烷聚合
roamer 透明量图器
roan 斑色;杂色的
roar 噪声;轰鸣
roaring business 门庭若市
roaring flame 烈焰
roaring forties 中纬度风暴带
roast 焙烧生成物
roast-duck restaurant 烤鸭店
roasted molybdenum oxide 焙烧氧化钼
roasted product 焙烧产品
roasted pyrite 焙烧(的)黄铁矿
roasted work 刻石工作
roasted zinc 焙烧过的锌
roaster 炉算;炉桥;烘烤器;烘烤机;电烤箱;焙烧炉;焙炉;焙茶机
roaster ash 焙烧苏打灰
roaster hearth 熔烧炉膛
roast gas 焙烧发生气
roasting 氧化焙烧;焙烧
roasting apparatus 焙烧设备
roasting bed 焙烧床层
roasting chamber 焙烧室
roasting furnace 焙烧炉
roasting heat 焙烧热
roasting in air 氧化焙烧
roasting in kilns 窑中焙烧
roasting in stalls 涂泥焙烧
roasting kiln 烘烧窑;煅烧窑;焙烧窑
roasting method 焙烧法
roasting oven 烘炉;焙炼炉
roasting regeneration 焙烧再生
roasting space 焙烧空间
roasting test 焙烧试验
roast sand mo(u)ld 干砂型
roast sintering 鼓风焙烧法
roatary design 环形设计
rob 拆除坑道支撑;拆卸移作他用;河水夺流;拆迁;拆除支柱
Roband hitch 鲁班结;天幕结
robber 限流阴极
robber absorber 橡胶减振器
robbery accident 被盗事故
robbery protection alarm 防盗报警器
robbing pillars 矿柱回收;矿柱回采;回收矿柱法
robe hook 挂衣钩
Roberoid roofing 橡胶屋面(一种屋面材料)
Roberoid roofing felt 橡胶屋面油毡
Roberoid slab 橡胶石板瓦
Roberl buoy 罗伯特浮筒
Robert converter 罗伯特转炉
Robert gradient 罗伯特梯度
Robert radio current meter 罗伯特无线电测流仪;罗伯特无线电流计
Roberts evapo(u)rator 短管立式蒸发器
Roberts grinder 罗伯特磨木机
robertsite 水磷钙锰矿
Roberts' linkage 罗伯特连杆机构
Robertson 屋面用瓦楞钢皮
Robertson test 罗伯逊脆断试验
Robert's shotting machine 罗伯特喷粉机
Robert's smoke chart 罗伯特烟色图
Roberval's balance 架盘天平
robing room 化妆室
robin hood 蓝灰色砂石(产于英国约克郡)
Robin Hood's wind 罗宾汉风
robinia 洋槐
robin's egg blue 灰绿蓝色
Robins escapement 罗宾斯擒纵机构
Robin's law 罗宾定律
Robin's method 罗宾法
Robinson cement 鲁宾逊水泥
Robinson direction-finding system 鲁宾逊测向系统

robinsonite 纤硫锑铅矿
Robinson's anemometer 鲁宾逊风速表
roble 加州白栎
robot 自动机;机械手;机器人;通用机械手
robot aircraft 无人驾驶飞机
robot brain 自动计算装置;自动计算机
robot buoy 无人浮标
robot calculator 自动计算器
robot-controlled construction 用机器人施工;自动化施工
robot control system 自动控制系统
robot engineering 机器人工程学
robot equipment 自动设备;无人驾驶的设备
Robot fireman 自动У排;罗伯特加煤机
robot for poor environment 用于恶劣环境的机器人
robot guiding system 自动驾驶系统
robotics 自动控制学;模拟机器人学
roboting machine 机械手
robotization 机器人化
robotized plant 自动化工厂
robot mechanism 遥控机械装置;机器人装置
robot navigator 自动领航仪
robot painting 机器人涂装
robot pilot 自动驾驶仪
robot radar 自动雷达
robot scaler 自动计算装置
robot scaling equipment 定标装置
robot spray 遥控混凝土喷射机
robot station 自动输送站
robot storekeeper 仓库管理机器人
robot submarine listening post 自控水下监听站
robot systern 自动控制系统
robot tractor 遥控拖拉机
robot weather station 自动天文站;自动(化)气象站
robot welding system 自动焊接系统
robust decentralized control 坚强分散控制
robust estimation 稳健估计
robust joint 坚固接头
Roc 罗克(电导率单位)
Rocan copper 含砷铜板
Rocard scatter 罗卡散射
Roche Abley 有乳油色斑点的石灰石(产于英国约克郡)
Rochelle salt 四水合酒石酸钾钠
roche moutonnee 羊状岩;羊背石【地】;卷毛岩
roche moutonnee rock 石英电气石岩
roche separator 湿式带型磁选机
Rochester shale 罗彻斯特页岩
Rochon polarizing prism 罗森偏振棱镜
Rochon prism 罗森棱镜
rock 摇摆;岩石;礁;石头
rock abrasion platform 浪蚀岩石阶地
rock abrasiveness 岩石研磨性
rock abrasivity 岩石研磨性
rock abutment 岩拱脚;石桥台;石桥墩
rockair 机载高空探测火箭
rockallite 钠辉细岗岩
rock alteration simulating 蚀变交代实验
rock alum 石明矾
rock analysis 岩石分析
rock anchor 岩石地锚;岩层锚杆;用岩石锚碇;岩石锚固
rock anchorage 岩石加固;锚杆支护
rock anchor hole 岩石中锚碇孔
rock and mineral analysis 岩矿分析
rock and sand fill 填石和填砂
rock and soil foundation 岩土地基
rock and soil ground 岩土地基
rock and soil improvement 岩土加固
rock and soil mass (geotechnical mass) 岩土体
rock and soil property relating water 岩土的水理性质
rock and soil reinforcement 岩土加固
rock and soil symbols 岩石和土壤符号
rock and water miniature garden 水石盆景
rock anomaly 岩石异常
rock arch 岩拱;石拱
rock arch bridge 石拱桥
rock arch door 石拱门
rock arching 石拱圬工
rock architecture 石窟建筑;岩洞建筑
rock arm 摇臂
rock arm plunger 摇臂柱塞
rock asphalt 岩石(地)沥青;岩沥青;沥青岩;天然地沥青

rock asphalt pavement 岩石(地)沥青路面;岩沥青路面
rock asphalt powder 岩地沥青粉
rock association 岩石组合
rock auger 岩石麻花钻;凿岩螺旋钻
rock avalanche 岩崩
rock awash 洗岩;适淹礁
rock ballast 石渣;石砟;道砟
rock bank 石堤;石岸
rock bar 岩坝;冰狀;石梁
rock base 岩石基础;岩石基底;基岩
rock basin 岩盆;岩盘盆地
rock beach 岩滩
rock beam 岩梁
rock-bearing plant 岩生植物
rock bed 岩床;基岩
rock bedding 岩石层理;岩层
rock behavio(u)r 岩石特性
rock bench 岩台;构造阶地;阶状地
rock bin 废石仓
rock biologic(al) filter 碎石生物滤池
rock bit 岩芯钻头;钻岩钻头;凿岩钻头;凿岩钎头
rock bit cone bearing 牙轮钻头锥形轴承
rock bit cone fishing socket 牙轮钻头打捞套筒
rock bit drilling 牙轮钻头钻进
rock bit insert 牙轮钻头镶嵌块;凿岩钻头用硬质合金片
rock bit structure 牙轮钻头结构
rock bit testing rig 牙轮钻头试验钻机
rock bit type hole opener 牙轮式扩孔器
rock blade 推土机
rock blast 凿岩爆破
rock blasting 岩石爆破
rock block 岩块;大块石
rock block agglomerate 岩块集块岩
rock block field 石海
rock block volcanic breccia 岩块火山角砾岩
rock body 采石场运石槽车
rock bolt 锚杆;岩栓;岩石锚杆;岩层锚栓;炸药爆锚固锚杆法;杆栓
rock-bolt diameter 锚杆直径
rock-bolt dynamometer 岩石锚杆测力计;锚杆测力计
rock bolter 锚杆冲凿机;锚杆冲击机
rock-bolt extensimeter[extensometer] 岩石锚杆伸长仪
rock bolting 岩石锚杆;锚杆支护;岩石锚栓支护;岩石锚栓加固
rock bolting and shotcrete 岩石锚杆结合喷混凝土支护
rock bolting jumbo 锚杆机
rock bolting support 锚杆支护
rock-bolt length 锚杆长度
rock-bolt load carrying capacity 岩石锚杆支承能力
rock boltometer 岩石锚杆检测仪
rockbolts and shotcrete supporting method 喷锚支护法
rockbolts and shotcrete supporting of rock mass 岩体喷锚支护
rock-bolt setter 锚杆安装机
rock-bolt shotcrete 锚杆喷射混凝土
rock-bolt spacing 锚杆间距
rock-bolts shore 岩石锚栓支撑
rock-bolt support 锚杆支护
rock-bolt test 岩石锚杆测试
rock-bolt type 锚杆类型
rock bolt with timber 带垫木的岩栓
rock bore 石质隧洞;岩石孔;钻岩机
rock borer 钻石机;岩芯机;岩石钻;蛀石(海)虫;钻岩机
rock boring 岩石钻探;岩石钻孔
rock boring device 岩石钻孔设备
rock boring machine 钻岩机;凿岩;岩芯机
rock bottom 岩层底部;岩底
rock bottom price 最低价(格)
rock bound 岩石围绕的;被岩石包围的;多岩石
rockbound coast 多岩(石)海岸(无锚地)
rock breakdown 岩石的破碎(在钻头作用下)
rock breaker 岩石破碎器;岩石破碎机;凿岩机;碎石机
rock breaker vessel 石料破碎船
rock breaking 岩体破裂;岩石破碎
rock breaking by water jet 水射流破岩
rock breaking jumbo 凿岩台车

rock breaking parameter 岩石破碎参数
rock breaking plant 石料破碎设备;石料破碎厂
rock breaking vessel 凿岩船;碎石船
rock breaking zone 岩石破碎带
rock breakwater 块石防波堤;堆石防波堤
rockbridageite 基磷重铁矿
rockbridgeite 绿铁矿;铁锰绿铁矿
rock bucket 岩石铲斗;挖石铲斗
rock bulk analysis 岩石全分析
rock-bulk compressibility 岩石容积压缩系数
rock bump 岩石突出
rock burst 岩石破裂;岩石爆破;岩石突裂;岩崩;岩爆;冲击地压
rock burst shock 岩石振动;岩爆振动
rock butter 铁明矾
rock buttress 挡石墙
rock candy 冰糖
rock candy structure 岩石状断口;脆性断口
rock cap 覆盖岩层
rock category 岩石分类
rock causeway 石料填筑的人行道;石料填筑的长堤
rock cave 石窟
rock cavity 岩洞
rock cement 岩石水泥
rock chamber 石室
rock channel 岩石河槽;基岩河道
rock channel(l)er 截岩机
rock character 岩性;岩石特性
rock characteristic 岩性
rock chart 岩石地形图
rock chemistry method 岩石化学法
rock chestnut oak 岩栎
rock chisel 石凿
rock chunk 石块;石砟
rock classification 岩石类别;岩石分类;围岩分类(法)
rock classification on permeability 岩石的渗透性分类
rock cleavage 岩石劈理
rock cliff 悬岩;悬崖;岩崖
rock cofferdam 石围堰
rock cohesion 岩石凝聚强度;岩石凝聚力
rock column of water prevention 防水岩柱
rock compartment 岩石提升格
rock compression 岩石压缩
rock constituent 岩石成分;岩石组分
rock construction types and its mutation 岩石建筑种及亚种
rock contraction 岩石收缩
rock control 岩石控制
rock core 岩芯
rock core bit 岩芯钻头;取(岩)芯钻头
rock core drill 岩芯钻
rock cork 烧结软木(板);淡石棉
rock cover 岩石覆盖(层)
rock creep 岩石蠕动;岩石潜动
rock crevice 岩石裂隙
rock crib 石笼;石筐
rock-crib dam 木笼填石坝
rock crosscut 石门
rock crusher 岩石压碎机;轧石机;碎石机
rock crusher lubricating grease 碎石机的润滑油
rock crushing 压碎岩石
rock crushing machine 岩石破碎器;岩石破碎机;碎石机
rock-crushing plant 碎石厂
rock-crushing vessel 碎石船
rock crystal 石英晶质玻璃;水晶;大水晶;天然的透明石英
rock cullet 炉底大块碎玻璃
rock cut 石方开挖;岩石路堑;岩石掘进
rock-cut building 窟屋;石窟
rock-cut ditch 石渠;石槽
rock-cut job 采石工程;凿石工作;开石工作;采石工作
rock-cut statue 石雕像
rock-cut temple 石窟庙宇;石窟寺
rock cutter 凿石机;割岩机;割石机;碎石机;截石机
rock cutter dredge(r) 凿石船
rock cutter vessel 凿石船
rock cutting 岩体切割;岩石开凿;凿岩;岩石质路堑
rock cutting in formation 凿岩岩石以建造地基;凿挖岩石以建造路基
rock cuttings 岩屑;岩粉
rock-cutting tooth 岩石挖凿齿
rock-cut tomb 石窟(陵)墓

rock-cut work 凿石工程
rock cycle 岩石循环;岩石回旋
rock dam toe 块石坝址
rock dash 细石抹灰饰面;干粘石;抹灰撒粘石子、卵石或壳墙面
rock debris 岩屑;岩石碎屑;碎岩;碎石(堆)
rock debris dam 石砟坝
rock decay 岩石腐化;岩石风化
rock-defended terrace 岩石防蚀阶地
rock deformation 岩石形变;岩石变形
rock deformation and stability measurement 岩石变形及稳定性测量
rock density 岩石密度
rock deposit 岩石堆积;岩漠沉积
rock depression 岩层下沉
rock desert 石质沙漠;石质荒漠;石漠
rock dike 抛石坝;堆石堤
rock dike cofferdam 石块围堰
rock dilatancy 岩石扩容性
rock dipper 石料挖斗
rock discontinuity 岩体构造断裂
rock disintegration 岩石剥落;岩石剥蚀;岩石破碎;岩石崩碎
rock dislocation 错落(岩石)
rock disturbance 岩石扰动
rock ditching 挖岩石管沟
rock drain 填石盲沟;排水暗沟;填石排水沟
rock drawing 石山表示法
rock dredge(r) 凿岩船
rock dredging 岩石疏浚
rock drift 岩石平巷;水平石坑道
rock drill 钻岩设备;钻岩机;岩石钻眼机;凿岩(机);开石钻;石头钻机;冲击钻机
rock drillability 岩石的可钻性
rock drillability index 岩石可钻性指数;岩石可钻性系数
rock drill barge 岩石钻探船;钻岩船;钻船;水上钻机
rock drill bit 岩石钻头;牙轮钻头;凿岩机钻头
rock drill column 凿岩机气腿
rock drill design 凿岩机结构
rock drill drifter 凿岩机
rock drill dust exhauster 凿岩机干式捕尘器
rockdriller 钻岩机
rock drill hammer 手持凿岩机
rock drilling 凿石(工作)
rock drilling and blasting 凿岩爆破
rock drilling crawler 履带式凿岩车
rock drilling jumbo 凿岩台车
rock drilling machine 凿芯机;钻岩机
rock drill insert 钻岩机合金衬片;凿岩机硬质合金衬片
rock drill jumper 凿岩钎子
rock drill mounting 钻架
rock drill on airfeed leg 气腿式凿岩机
rock drill shop 凿岩机修配(车)间;凿岩机修理(车)间
rock drill steel 凿岩钎钢;钎头;钎杆;钎子钢
rock drill test 凿岩试验
rock drill vessel 岩石钻探船;钻岩船;钻船;水上钻机
rock drivage 岩巷掘进
rock drumlin 石鼓丘
rock-dumping barge 抛石驳
rock dumping yard 矸石山
rock dust 岩尘;石屑;石粉
rock-dust distributor 岩粉撒布器
rockdusting 撒岩粉
rock-dust nuisance 岩尘污害
rock-dust testing kit 岩粉测验仪
rock dwelling 岩屋
rocked-bed 有岩石河床
rocked pipe 皮尔格式冷轧管
rock ejector 石块消除器
rock electricity 岩石电性
rock element 岩石单元体
rock elm 木栓榆;石榆木
rock embankment 石堤;堆石堤;堆石坝
rock emulsifying explosive 岩石乳化炸药
rock environment 岩石环境
rocker 摇轴;摇椅;摇床;龙骨弯曲的船;淘砂盘;掏金摇动槽;翻斗车;摆架
rocker-and-masonry-plate bearing 摇座座板固定支座
rocker arm 摇臂;摇杆;弹动杆
rocker-arm drive 摇臂传动
rocker-arm link 摇臂连杆

rocker-arm linkage 摇臂杠杆系统
rocker arm mechanism 摇臂机构
rocker-arm oil trough 摇臂油槽
rocker-arm plunger spring 摇臂柱塞簧
rocker-arm push-rod 摇臂推杆
rocker-arm ratio 摇臂传动比
rocker-arm resistance welding machine 摇臂(式)接触焊机
rocker-arm return spring 摇臂复位弹簧
rocker-arm roller 摇臂滚轮
rocker-arm roller pin 摇臂销
rocker-arm shaft 摇臂轴
rocker-arm shaper 摇臂牛头刨
rocker-arm shovel 摇臂式装载机;旋转式挖掘机
rocker-arm spring 摇臂簧
rocker-arm support 摇臂座
rocker-arm way 摇臂导轨
rocker-bar 一端铰接杆;天平杆
rocker-bar bearing 摆杆支座;摇杆支承;铰链支座
rocker-bar furnace 摇杆推钢炉
rocker-bar heating furnace 摇杆式加热炉
rocker beam 平衡梁
rocker bearing 桁架伸缩支座;桥梁支座;摆式支座;摇座;摇轴支座;摇动支座;摇摆式活动支座;桁架摇动支座;摆(动)支座
rocker bearing for bridge 桥梁弧形支座
rocker bent 摇座排架;摆式排架
rocker bottom 瓶底凸出
rocker box 摇杆箱
rocker cam 摆动凸轮
rocker car 侧翻料车;侧翻矿车;翻转料车
rocker column 一端铰接支柱
rocker compensating gear 间隙补偿齿轮
rocker conveyer[conveyor] 悬链式输送机;悬臂式输送机;摆动式输送机;摇动式传送机
rocker cover 摇杆罩盖;摇杆盖
rocker-dump car 摇动卸料车;翻斗矿车
rocker-dump hand-cart 双轮手推翻斗车;带有翻斗的(混凝土)手推车
rockered keel 弧形龙骨;线形龙骨
rocker gear 摇杆传动装置
rocker gear mechanism 摇杆齿轮机构
rocker grate 摆辊炉箅;摆动箅条筛
rocker guillotine type shears 杠杆摆动式铡刀剪
rocker hardness tester 摆杆硬度计
rocker hinge 摇摆铰链
rocker joint 摆柱接头
rocker joint chain 摇轴链
rocker keeper 摆卡
rocker lever 平衡杆
rocker mechanism 摆杆机构
rocker member 一端铰接杆件
rocker of injector-pump 油泵喷油器摇杆
rocker panel 摇板;嵌板
rocker piece 浮动块
rocker shaft 摇杆轴;摇臂轴
rocker shaft circlip 摇臂轴止动环
rocker shaft diameter 摇臂轴直径
rocker shaft oil pipe 摇臂轴油管
rocker shaft spacer spring 摇臂轴推力弹簧
rocker shoe 滑动柱脚;摆动支座
rocker shovel 翻斗装载机;翻铲装载机;斗式装载机;铲装机;反向铲;翻斗铲
rocker side dump(ing) car 侧倾式翻斗车;翻斗矿车
rocker support 摇杆支承铰;摇座;摇轴支座
rocker swaging machine 摆式轧机
rocker switch 摇臂开关
rocker tool post 摇杆刀架
rocker trunnion block 摇轴座
rocker-type cooling bed 摆动齿条式冷床
rocker-type die 摇杆式模具
rocker unloader 摇式卸车机
rocker valve 摆动阀
rocker vat 摇槽
rocker wagon 摆式货车
rocker wall (门架系统的)摇臂墙
rockery 粗面石工;假山(园林)
rocket 火箭
rocket antenna 火箭天线
rocket apparatus 火箭式投射装置;火箭抛绳器
rocket assist 火箭助推
rocket assisted take off gear 火箭助推起飞装置
rocket astronomy 火箭天文学
rocket-borne infrared mapper 火箭运载红外成像

仪;火箭携带的红外绘图仪
rocket-borne ultraviolet spectrograph 火箭携带的紫外摄谱仪
rocket centring 摇架
rocket electrophoresis 火箭电泳
rocket engine 火箭发动机
rocket field 火箭发射场
rocket-flash triangulation 火箭闪光三角测量
rocket fuel 火箭燃料
rocketing prices 物价飞涨
rocket launcher 火箭筒发射器;火箭发射装置
rocket launching 发射火箭
rocket lightning 火箭状闪电
rocket line 火箭抛绳;火箭救生绳;撒缆
rocket line whip line 抛绳
rocket meteorograph 火箭气象计
rocket motor 火箭发动机
rocket motor assembly 火箭发动机组
rocket-motor case 火箭发电机箱
rocket noise 火箭噪声
rocket nose section 火箭头部
rocket nozzle 火箭喷嘴
rocket parachute flare 火箭降落伞信号
rocket plane 喷气式飞机
rocket plane noise 喷气式飞机噪声
rocket-powered 火箭推动的
rocket projectile 火箭
rocket propellant 火箭推进剂;火箭燃料
rocket-propelled 火箭推动的
rocket propulsion 火箭推进;喷气推进
rocket ramjet 火箭冲压喷气发动机
rocket rope 火箭抛绳;火箭救生绳
rocketry 火箭技术
rocket ship 装备火箭的舰船
rocket signal 火箭信号
rocket sled 火箭滑车
rocket-sled testing 火箭滑车试验
rocket socket 火箭筒
rocket sonde 火箭测候仪;箭探空仪;气象探测火箭
rocket sounder 探空火箭
rocket sounding 火箭探测
rocket station 海事救援站;备有火箭抛绳器的救生站
rocket support landing craft 支援火箭登陆艇
rocket tester 烟火检验器;喷烟测漏器
rocket tube 火箭筒
rocket turbine 火箭用透平
rocket X-ray astronomy 火箭X射线天文学
rock-eval pyrolysis analysis 生油岩分析仪热解分析
rock excavation 岩石挖掘(石方工程);岩石开挖;石方工程;采石工程
rock excavation work 石料采掘工作
rock excavator 岩石挖掘机;挖岩机;挖石机
rock explosive 岩石炸药
rock exposure 岩石露头;裸露岩石
rock extensimeter [extensometer] 岩石应变计;岩石形变仪
rock fabric 岩石地层;岩质构造;岩石组构;岩石构造;组构
rock face 凿石斜面纹理;琢石面;粗石面
rock-faced 琢石面的;粗石面的
rock-faced brick 岩面砖;琢石砖;石面砖
rock-faced dam 堆石护面坝
rock-faced dressing 粗石面修琢;粗石面修整;粗琢石(工);粗琢石面
rock-faced masonry work 粗琢面石工
rock-faced stone 粗琢石;粗面石;琢面砖;天然面石料
rock facies 岩相
rock facies analysis 岩相分析
rock facing 堆石面层;堆石护坡
rock factor 岩石阻力系数
rock failure 岩轴冒落;岩石破碎;岩石破裂;岩石破坏
rock fall 岩崩;岩石崩塌;坠落;落石;崩塌堆积物;岩石塌落
rock-fall alarm device 落石警报装置
rock-fall earthquake 岩崩地震
rock-fall furrow 岩崩沟槽
rock-fall prediction 落石预兆
rock-fall prevention net 拦石网
rock-fall prevention wall 拦石墙
rock-fall rapids 崩岩急滩
rock family 岩石族
rock fan 岩石扇状地;岩质扇形地;石扇;扇形基岩
rock feeder 给石机

rock fence detector 落石围墙检测器
rock fiber 岩棉;岩石纤维
rock-fill 堆石填方;堆石(体);岩石充填;岩石填料;抛石填充;填石
rock-fill and mud displacement by explosion 爆炸抛石挤淤法
rock-fill bamboo basket cofferdam 填石竹笼围堰
rock-fill breakwater 堆石防波堤
rock-fill cofferdam 堆石围堰;填石围堰
rock-fill crib 填石木笼;填石木桑
rock-fill dam 抛石坝;填石坝;堆石堤;堆石坝
rock-fill dam with asphaltic concrete core wall 沥青混凝土芯墙堆石坝
rock-fill dam with concrete facing 混凝土斜墙堆石坝;混凝土面板堆石坝
rock-fill dam with mortar 浆砌堆石坝
rock-fill dam with vertical clay core 垂直黏土芯墙堆石坝
rock-fill diversion weir 堆石引水堰;堆石导流堰
rock-fill drain 堆石排水体;堆石盲沟;填石盲沟;填石排水沟
rock-filled breakwater 填石防波堤
rock-filled crib 填石木笼
rock-filled cribbing 填石桑
rock-filled crib dam 填石木笼坝
rock-filled crib type 格笼填石式;叠木填石式
rock-filled crib weir 木笼(式)填石坝
rock-filled dike 排桩填石堤岸;排桩填石堤
rock-filled double row sheet-pile breakwater 双排板桩桑堤
rock-filled jetty 双排板桩填充突堤;堆石突堤;填石突堤
rock-filled pile dike 排桩填石堤岸;排桩填石堤
rock-filled revetment 堆石护岸
rock-filled sheet pile cell 填石板桩格体;填石板桩格笼
rock-filled slope protection 堆石护坡
rock-filled timber crib 填石木笼
rock-filled timber crib dam 填石木笼填石坝;填石木垛坝
rock-filled trickling filter 碎石充填滴滤池
rock-filled (wire) gabion 石笼;填石铁丝笼
rock-fill embankment 堆石路基;堆石堤(岸)
rock-fill face of dam 坝的堆石面
rock-fill foundation 堆石基础
rock-fill groin 堆石防波堤;堆石丁坝;填石丁坝
rock-fill groyne 堆石丁坝
rock-filling 堆石;抛石体;碎石充填;填石
rock-filling dike 岩石路堤
rock-filling facing 堆石护面;堆石护坡
rock-fill in layers 分层填石;分层抛石
rock-fill material 填石材料;堆石材料
rock-fill platform 堆石平台
rock-fill revetment 堆石护坡;堆石护岸
rock-fill riprap 堆石护坡
rock-fill sheet-pile cell 填石板桩格笼
rock-fill spilling dam 堆石溢流坝;堆石滚水坝
rock-fill spur dike 堆石丁坝
rock-fill timber crib 堆石木笼;木笼填石;叠木石笼
rock-fill toe bund 抛石棱体
rock-fill toe mound 抛石棱体
rock-fill toe prevent scour 堆石以防冲刷
rock-fill toe wall 堆石脚墙;堆石坝趾墙
rock-fill to prevent buckling 堆石以防翘曲
rock filter for algae removal 碎石除藻池
rock fines 细石料
rock fissure 岩石裂隙;岩石裂缝;岩石缝隙;岩面裂缝
rock fissurity 岩石裂隙性
rock flexural strength 岩石弯曲强度
rock floor 岩床
rock flour 粉砂;岩粉;石粉
rock flow 泥石流;岩石流动;岩流;流石;石流
rock flowage 石块流动;岩流
rock flowage and fracture zone 岩流岩裂带;裂流带
rock flowage zone 岩流带
rock fold 岩石褶皱;岩层褶皱
rock force-displacement performance 岩石强制位移特性
rock foreshore 岩石滩
rock fork 夹石耙叉
rock formation 岩系;岩石形成;岩层
rock formation grade in engineering geology map 区域工程地质图上岩层分级
rock formation sketch 岩层素描图
rock formation test 岩体试验

rock-forming 造岩的
rock-forming element 造岩元素
rock-forming mineral 造岩矿物
rock-forming simulating 成岩模拟实验
rock foundation 岩基;岩石地基
rock foundation grouting 岩基灌浆
rock foundation test 岩石基础试验
rock fracture 岩石破裂;岩石裂隙
rock fracture by water-jet impact 水射流破碎岩石
rock fracture zone 岩石破裂带;岩石裂隙带;岩裂带;岩石破碎带
rock fracturing technique 岩石碎裂技术;岩石破碎方法
rock fragment 岩块;岩石碎屑;岩石碎块;碎块岩石
rock fragmentation 岩石破碎
rock fragmentation by electrohydraulic effect 水电效应破岩
rock fragment flow 碎石流
rock fragments 岩屑
rock freezing 岩石冰冻法
rock gallery 岩石坑道
rock gang 落石工班
rock gangway 岩巷
rock garden 岩石园;岩石为主的公园;岩石庭园;岩石花园;岩生植物园;假山庭园;假山(公)园
rock gardening 岩石园艺
rock gas 天然(煤)气
rock generator 岩样微剩磁测量仪;岩样剩磁测量;感应式旋转磁力仪
rock getter 采石工人
rock glacier 岩石流;岩石冰川;岩流;石流;冰川石流;石冰川
rock glass-ceramics 岩石微晶玻璃
rock gorge 峡谷
rock grab 岩石开掘机;块石抓斗
rock gradient temperature 地温梯度
rock grapple 石料抓斗;岩石挖掘机;抓石钳;抓石机;块石抓斗
rock grinder 振动研磨机
rock grinding 岩石碾碎;岩石磨碎
rock grout(ing) 岩石灌浆;岩浆
rock groyne 填石丁坝
rock guard (装石槽车的)挡石板;防石护刃器
rock gun 喷石枪
rock gypsum 天然石膏;石质石膏;石膏块
rock hammer drill 凿岩机
rock hand 采石工人
rock-hard beam 超硬织轴
rock hardening 岩石灌浆
rock hardness 岩石硬度
rock hauling semi-trailer 运石半挂车
rock head 硬岩层;岩石顶层;岩基;岩层顶
rock header 露天石;石工;石匠
rock heading 岩石平巷
rock-hewn 造岩而成的
rock-hewn monastery 石窟寺院
rock-hewn seated statue 石雕坐像
rock-hewn temple 石窟寺
rock hoisting 岩石提升
rock hole 岩层钻孔;脉外溜井
rock hound 喜采集奇石的人;业余矿物学家;石油勘探地质学家
rock house 岩洞;矿石破碎筛分车间
rock identification 岩石鉴定
rock improvement 岩层加固
rock inclusion in coal series 煤系中的岩石包裹体
rock industry 石制品工业;采石工业
rock-infested reach 多岩石河段
rocking 摇动;摇摆;填石块;淘砂盘洗选;摆动
rocking angle 摇摆角度
rocking apparatus 摆动设备
rocking arc furnace 摇摆式电弧炉
rock(ing) arm 摇臂;平衡杆
rocking arm pump 杆式泵
rocking ball bearing 摇臂滚珠轴承
rocking bar 摇杆
rocking bar spring 振荡杆簧
rocking bar stud 振荡杆桩
rocking bar wheel 振荡杆轮
rocking beam 摇梁;振荡梁
rocking bracket 摆动托架
rocking buggy 锭座回转式送锭车
rocking cell 摇摆式电解槽
rocking chair 安乐椅;摇椅

rocking chair induction furnace 可倾式感应炉
rocking-contact speed regulator 振动接触式速度调节器
rocking-contact voltage regulator 摇动接点式电压调整器
rocking couple 摆动力偶
rocking curve 摇摆曲线
rocking die method 摆动锻造法
rocking equipment 振荡装置
rocking fly shears 摆式飞剪
rocking foundation 摇摆基础
rocking frame 振动器
rocking furnace 摇动式炉;摇摆(式)电弧炉;可倾炉;回转炉
rocking grate 摇动炉栅;摇动炉排;摇动炉箅
rocking grating 摆动光栅
rocking horse antenna 摆动天线
rocking iron 摇架铁件;支架铁件;吊架铁件
rocking jaw 摆动爪
rocking lath 石膏纸板抹灰底
rocking lever 摇杆
rocking-lever for ink duct roller 墨斗辊摆动杆
rocking link 转动环节
rocking microtome 摇动切片机
rocking mirror 摆动反光镜
rocking motion 摇摆运动
rocking of foundation 基础摇摆
rocking of vehicle 车辆摇摆
rocking oscillation 摇摆振动;摇摆振荡
rocking pier 摇座;摇式桥墩;摆座;摇座桥墩
rocking pin 摆动销
rocking portal 摆动式桥门
rocking-post type 摇柱式
rocking roll(er) 摆动式辊
rocking runner 移动式铁水沟
rocking-sector regulator 扇形摆动式电压调整器
rocking shaft 摇轴;摇臂轴
rocking shears 摆式剪
rocking sieve 摇动筛;摆动筛
rocking slot 环形槽
rocking stiffness 摇摆刚度
rocking table 镀银摆台
rocking the sextant 左右摆动六分仪;六分仪弧尺的转动
rocking the well 废水填井
rocking trenching 挖岩石管沟
rocking trough 摇槽;振动槽
rocking-type hot saw 摆式热锯
rocking vibration 摇摆振动;平行摇摆振动
rocking whole-stream cutter 摇槽式全流取样机
rocking whole-stream sampler 摇槽式全流取样机
rock in place 岩基;岩盘;本生岩;原岩;原生岩石;原地岩石
rock interstice 岩石裂缝
rock island 离堆山
rock isotope composition 岩石同位素成分
rock jaw crusher 颚式碎石机
rock job 须爆破取材的工程
rock joint 岩石节理
rock ladder 卸石梯;斗式石料输送机
rock land 石质地
rock landslide 岩石滑坡
rock lath(ing) 石膏抹灰板条;抹灰用硬板条;粉刷用石膏硬底;石膏纸板抹灰底
rock layer 岩层
rock leather 石棉;一种石棉;坡缕石
rock ledge 岩礁;岩架;石礁
rock lifting 凿岩
rock-like mass 岩状物质
rock-like material 类岩石材料
rock-like site 类岩石场地
rock-like substance 岩状物质
rock line 岩面线
Rocklin granite 罗克林岩石(产于美国加利福尼亚州的一种浅色花岗岩)
Rocklinizing 罗克林碳化钨表面硬化法
rock load 岩石荷载
rock loader on rail 轨行装岩机
rock loading 围岩压力
rock loading conveyer 装岩运输机
rock lump 岩片
rock machining 岩石钻凿
rock magma 岩浆
rock magnetic measurement 岩石磁测

rock magnetism 岩石磁性
rock mantle 表皮岩;岩石风化层;岩石地幔;岩壳;岩基
rock maple 糖槭
rock mark 岩石标石
rock mass 岩体;岩块
rock mass basics quality 岩体基本质量
rock mass classification 岩石分级
rock mass completeness 岩体完整度
rock mass control 岩体控制
rock massif 岩石山丘;整体岩石
rock mass mechanics 岩体力学
rock mass pressure 岩体压力
rock mass quality designation 岩体质量指标
rock mass quality index 岩体质量指标
rock mass resistibility 岩体抗力
rock mass sliding 岩体滑动
rock mass stability 岩体的稳定性
rock mass strength 岩体强度
rock mass stress redistribution 岩体应力重分布
rock-matching epoxy grout 拼石环氧灌浆
rock material 岩石材料;石料
rock matrix 杂石
rock matrix compressibility 岩石基质压缩系数
rock mattress 块石垫层
rock meal 岩粉
rock mechanics 岩石力学;岩体力学
rock mechanics laboratory studies 岩石室内试验研究
rock-media system 碎石填料系统
rock medium 石质介质
rock-melting penetrator 岩石溶化穿透器
rock metal 白色碳酸钙沉淀物(水中)
rock metallometric survey 岩石金属量测量
rock micro hardness 岩石显微硬度
rock milk 松软方解石;山乳;岩乳
rock mill 石材磨床
rock mincer 振动切碎器
rock mineral 造岩矿物
rock mole 堆石突堤;堆石防波堤
rock motion 基岩运动
rock mound 抛石棱体
rock-mound breakwater 抛石防波堤;堆石防波堤
rock-mound of run-of-quarry 不分级堆石防波堤
rock mound prism 抛石棱体
rock movement 岩石移动;岩层移动
rock mover 集石机;除石机
rock noise 岩石噪声
rock of artificial recharge layer 人工补给层岩石
rock of engine 发动机工作时的摆动
rock of reservoir system 热储系统岩石
rock of water-storing bed 储水层岩石
rockogenin 岩配基
rock oil 石油
rockoon (气球带到高空发射的)高空探测火箭
rockoon sounding 气球火箭探测
rock opening 岩洞
rock ordinary benchmark 岩层普通水准标石
rock outburst 岩爆
rock outcrop(ping) 岩石露头
rock outcrops 露头岩石
rockover 翻台
rockover core making machine 翻转式型心机
rockover draw 翻台起模
rockover draw mo(u)lding machine 翻台起模式造型机
rockover mechanism 翻台机构
rockover mo(u)lding machine 翻台式制模机;翻台造型机
rockover rake 抓斗爪
rockover remover 除石器
rockover table 变向台
rock particles 岩石颗粒
rock patch 石滩
rock pavement vegetation 石戈壁植被
rock pedestal 细颈柱
rock pediment 岩石麓原;山麓基岩侵蚀面
rock penetration performance 凿岩效率
rock permeability 岩石渗透性
rock perviousness 岩石透水性
rock phases 岩相
rock phosphate 磷酸盐岩;磷酸岩;磷钙土
rock phosphate in powder 磷矿石
rock picker (破碎岩石用的)风镐
rock pile 摩天楼(俚语);石柱;岩堆;岩桩
rock pillar 礁柱;岩石柱;岩礁柱

rock pin 岩石(锚固)插筋
rock pinnacle 岩石尖礁;岩石尖峰;岩石尖顶;礁岩尖顶
rock pinning 岩石插筋(锚固)
rock pipe 岩筒
rock plane 山麓基岩侵蚀面
rock plant 采石工场;岩石植物;岩生植物;采石设备
rock plateau 石坪
rock plug 岩栓;岩塞
rock plug blasting 岩塞爆破
rock pocket 矸石仓;岩穴;岩石窝孔;骨料架空(混凝土);蜂窝
rock point 穿岩桩尖
rock point tip 凿岩机齿冠尖头
rock popping 岩石爆裂
rock pore 岩石孔隙
rock porosity 岩石孔隙率;岩石孔隙度;岩块孔隙度
rock post shotcrete 岩柱支撑
rock powder 岩粉
rock pressure 岩压;岩石压力;岩层压力;围岩压力;山岩压力;地压;地面压强;地面压力
rock pressure burst 冲击地压;岩爆
rock pressure indications 岩石压力迹象
rock pressure measurement 岩压测定
rock prestressing 岩石预加应力
rock primary benchmark 岩层基本水准标石
rock product 岩石产品;石制品;岩石制品
rock projecting over water 水凸
rock propelling 爆破岩石抛出
rock property 岩石性质;岩石特性
rock pulverizer 岩石粉碎机;岩石粉磨机
rock quality coefficient 岩石质量系数
rock quality designation 岩性指标;岩石质量指标
rock quality index 岩石质量指标
rock quarry 岩石开采;采石场
rockquarry run stone 毛石料
rock quartzite 石英岩
rock raise 脉外天井
rock rake 抓石耙;耙石爪;(推土机的)重型带齿推板;抓石机;耙岩机
rock ranger 移动式破碎筛分装置
rock rash 用异形碎石、卵石、晶石等镶嵌的图案
rock rating 岩石分类;岩石分级
rock ratio map 岩比图
rock reinforcement 岩石加固
rock remanent magnetization 岩石剩余磁化
rock remanent magnetization stability 岩石剩余磁化稳定性
rock-removal project 岩石开采工程;岩石开采方案
rock resistance 岩石抗力
rock response 岩石特性(曲线)
rock retaining wall 假山石挡土墙
rock revetment 块石护岸;抛石护坡;抛石护岸
rock rheological property 岩石流变性
rock riffle sluice 铺石格条流洗槽
rockright mill 摇摆式冷轧管机
rock ripper 松石器
rock riprap 乱石;抛石;填石
rock riverbed 岩石河床
rock roller bit 岩石齿轮钻头;牙轮钻头
rock rot 岩石风化
rock salt 岩盐;石盐
rock-salt lattice 岩盐晶格
rock-salt prism 岩盐棱镜
rock-salt structure 岩盐型结构
rock sample 岩样;岩石样品;岩石试样;岩石标本
rock sample analysis 岩石试样分析
rock sampler 采石取样机
rock sampling 岩石采样
rock sanctuary 石窟圣堂
rock sand 岩砂
rock saw 切石锯
rock schistosity 岩石片理
rock screen 碎岩筛分机;碎石筛分
rock screening machine 碎石筛分机
rock screenings 石屑
rock seal 岩层封闭
rock sealing 岩层灌浆堵塞
rock sepulchre 石窟墓
rock series 岩系
rock series alkalinity ratio of Read 莱德岩系碱度率
rock series index of Rittmann 里特曼岩系指数
rock shaft 下放矸石井筒;摇臂轴;杠杆轴
rock shake 摇动

rock shear test 岩体剪断试验
rock sheet 岩床
rock shelter 岩洞
rockshield 防石护板
rock shoal 石质浅滩
rock shoe 穿岩桩靴
rock shovel 铲斗式清岩机
rock silk 石棉
rock sill 岩石潜坝;抛石潜坝
rock slice 岩石碎片;岩石薄片;石片
rock slide 岩坍;岩石坍方;岩石滑坡;岩石滑落;岩滑;岩崩;坍方;岩石坍方;(地层结构上的)塌方
rock-slide dam 岩崩坝
rock-slide detector fence 落石坍方检测栅
rock-slide in the catchment area 岩滑
rock-slide protection 落石坍方防护法
rock sliding type of dam foundation 坝基岩体滑动形式
rock slinger 拣去大块石的工人
rock slip 岩石坍方;岩石滑移;岩石滑坡;岩滑
rock slope 岩石边坡
rock slotter 岩石掏槽机
rock slump shear test (岩石的)崩落剪切试验
rock smasher 碎石机
rock socket 岩石承插口(桩底)
rock soil mechanics 岩土力学
rock soil nature heat releasing light method 岩壤天然热释光法
rock specimen 岩石标本
rock sphere 岩石圈
rock splitter 劈石器
rock splitting 劈裂岩石
rock sprinkling 岩石飞散
rock stability 岩石稳定性
rock stack 岩柱
rock stack-formed cave 岩堆洞
rock stairway 岩梯;假山石楼梯
rock statue 石像
rock-steady structure 稳定结构
rock storing of oil 岩石中石油的储藏
rock storm 石流
rock stratification 岩石层理;岩层
rock-stratigraphic(al) unit 岩石地层单位
rock stratigraphy 岩石地层学
rock stratum 岩层
rock stratum movement 岩层移动
rock stratum pollution 岩石地层污染;岩层污染
rock stream 岩石流;岩流;石流;石河
rock strength 岩石强度
rock strengthening 岩石加固
rock strength index 岩石强度指数
rock strength loss 岩石强度损失
rock stress 岩石应力
rock stress measurement 岩石应力量测;岩石应力测量
rock structure 岩石(的)构造
rock structure parameter 岩石结构参数
rock study 岩石试验
rock subgrade 石质路基
rock subsidence 岩石塌陷
rock subsoil 岩石底土
rock substance 岩石物质
rock-support 岩石支架;岩石支护
rock-supported 岩石支承的
rock surface depression 岩表坑凹
rocksy 地质学家
rock system 岩石系统
rock taking-up 卧底
rock tallow 伟晶蜡石
rock tar 原焦油;石焦油
rock target 岩基觇标
rock temperature gradient 岩温率
rock temperature map 岩石温度图
rock temple 石窟
rock tend 岩层倾向
rock terrace 岩石阶地
rock test 岩石试验;岩石检测
rock test rig 岩石试验机
rock texture 岩体结构;岩(的)结构
rock throw 岩石抛射距离;(岩石的)飞散距离
rock thrust 岩石压力
rock toe 堆石坝脚;块石坝趾;熔岩趾;石坝堆趾
rock triaxial apparatus 岩石三轴压力机
rock tunnel 岩石隧洞;岩石隧道;脉外巷道

rock tunnel(ling) machine 隧道掘进机;岩石隧道掘进联合机
rock type 岩石类型
rock type of bank slope 岸坡岩性
rock type of leakage portion 渗漏段岩性
rock type of relative aquifuge 相对隔水层岩性
rock uncovered 明礁
rock unit 岩石地层单位;岩石单元
rock vegetation 岩石植物群落
Rockville 罗克维尔石(产于美国明尼苏达州的一种红灰色粗粒花岗岩)
rock volume 岩石体积
rock wall 崖壁;石垛带
rock wall failure 石壁倒塌;石壁破坏
rock washing machine 吸石机
rock waste 岩渣;废石;岩屑
rock water gel explosive 岩石水胶炸药
rock weathering 岩石风化(作用)
rock weathering index (岩石的)风化系数
rock weir 抛石堰;抛石坝
Rockwell ABC scale 洛氏硬度 ABC 级硬度表
Rockwell apparatus 洛氏硬度计;洛氏硬度仪
Rockwell hardness 洛克威尔硬度;洛氏硬度
Rockwell hardness number 洛氏硬度值
Rockwell hardness scale 洛氏硬度数;洛克威尔硬度数
Rockwell hardness test 洛氏硬度试验
Rockwell hardness tester 洛氏硬度计;洛氏硬度试验仪;洛氏硬度试验器;洛氏硬度试验机
Rockwell machine 洛氏硬度机
Rockwell tester 洛氏硬度机
rock winding 岩石提升
rock windrow 抛石堆;石料堆;石堆
rockwood 褐色紧密石棉
Rockwood oolitic limestone 洛克伍德鲕状石灰石(产于美国拉巴州)
rock wool 岩石棉;岩棉;矿石棉;矿毛绝缘纤维;矿毛绝缘材料;石棉;玻璃纤维
rockwool acoustic(al) ceiling 岩棉吸声吊顶
rockwool asphalt board 沥青岩棉板
rockwool board 岩棉板
rockwool building 矿棉建筑板;(高密度的)矿棉毡;岩棉建筑板;(高密度的)岩棉毡;(低密度的)岩棉毡
rockwool building board 矿棉建筑板材;岩棉建筑板材
rockwool cord covering 矿棉绳罩;矿棉绳套;岩棉绳罩;岩棉绳套
rockwool felt 矿棉毡;岩棉毡
rockwool fiber 矿棉纤维;岩棉纤维
rockwool insulating product 岩棉保温制品
rockwool insulation (grade) material 矿棉隔声材料;矿棉隔热材料;岩棉隔声材料;岩棉隔热材料
rockwool insulation-grade slab 矿棉隔声板;矿棉隔热板;岩棉隔声板;岩棉隔热板
rockwool insulator 矿棉隔热料;岩棉隔热料
rockwool lagging section 矿棉保温套管;岩棉保温套管
rockwool lamella 岩棉带
rockwool mat 矿棉垫;岩棉垫;岩棉卷毡
rockwool mat with glass cloth on one side 岩棉玻璃布缝毡
rockwool mat with wire net on one side 岩棉钢丝网缝毡
rockwool pipe casing 岩棉管壳
rockwool pipe section 岩棉管套
rockwool quilt 矿棉保温层;岩棉保温层
rockwool sheet 矿棉薄板;岩棉薄板
rockwool slab 矿棉板;岩棉板
rock work 假山;采石作业;凿石工作;石工;石方(工程);粗面石工;采石工作;凿石工程
rock work dressing 石工修整;粗石工饰面;粗琢石面
rocky 多岩石的
rocky area 岩石(海底)区
rocky bay 岩海湾;山岩湾;岩湾
rocky beach 岩滩
rocky bed 岩石河床
rocky bottom 岩石海底
rocky coast 岩石海岸;岩岸;基岩海岸;石质(海)岸
rocky core 岩核
rocky desert 岩漠
rocky feef 礁脉
Rocky geosyncline 落基山地槽
rocky ground 岩基;岩石(地面);多暗礁的海底
rocky harbo(u)r 岩滨港

rocky island 礁石岛;基岩岛
rocky ledge 岩脉;岩架;暗礁
Rocky marine trough 落基海槽
rocky mineral 脉石矿物
Rocky Mountain spotted fever 黑色热
rocky outer edge bar 基岩外缘坝
rocky point 石嘴
rocky reef 岩礁
Rocky regression 落基山海退
rocky riverbed 基岩河床
rocky shoal 石质浅滩
rocky shore 岩岸;石质海岸
rocky slope 岩质边坡;岩石边坡
rocky soil 坚石类土;岩性土;岩石类土(壤);石质土
rocky subsoil 石质底土;岩质底土
Rocla process 悬模法
Rococo 洛可可式【建】
Rococo architecture 洛可可式建筑
Rococo style garden 洛可可式园林
rod 杆;钎杆;条;长砾;棒
rodability 可插捣性
rod adapter 钻杆接头;接杆器
rodalquilarite 碲铁石
rod and drop-pull system 钻杆冲击钻进装置法
rod and pole drilling 弹弓钻杆钻进
rod antenna 鞭状天线;棒状天线;棒形天线
rod area 杆面积
rod assembly 杆组合件
rod ball mill 杆球磨机
rod basket 罐笼;吊篮
rod bearing 杆轴承
rod beater 杆式逐搞轮
rod-beating 棒击法
rod bench 锯木台
rod bender 钢筋弯曲机;弯钢筋工
rod bending machine 弯棒机
rod bit (不带岩芯管的)钻杆钻头
rod board 吊篮
rod bolt 闩钢
rod bore 钻杆钻
rod borer 杆式钻;杆钻
rod boring 钻杆钻进;杆钻;铁杆钻进
rod boring machine 杆式镗床
rod breaker 辊式破碎机
rod breaker bottom 杆条式破土犁体
rod bundle 圆钢捆;钢筋盘条;盘条
rod buster 钢筋设置工;钢筋切割机;钢筋(安装)工
rod calibration 标尺校准
rod carrier 杆导架
rod-chain conveyer 链杆式输送器
rod-chain elevator 杆链式升运器;链杆式升运器
rod charger 装棒机
rod check-valve 钻杆逆止阀
rod chuck wrench 钻杆卡盘扳手
rod clamp 钻杆夹头;钻杆夹持器;钻杆夹板
rod clearance 钻杆间隙
rod clevis 杆叉;连接叉
rod coil 盘条;线材卷
rod coil compressor 盘条紧捆机
rod comparison 标尺比长
rod connection 杆件连接
rod-control soft light 杆控聚光灯
rod correction 标尺校正
rod coupler 棒状耦合器
rod coupling 钻杆接头;钻杆接箍;杆连接器;活塞杆连接器
rod crack 纵裂纹;纵裂缝
rod-curtain precipitator 静电沉淀器;条屏沉降器;棒屏沉淀器
rod cutter 钢筋切断机;杆状刀具
rod deck 棒条筛面
rod-deck screen 格栅筛
rodded 捣实的
rodded bulk density 捣实密度;捣实容重
rodded concrete 捣实混凝土
rodded end 钢筋加固端
rodded joint 捣实接头;捣实的接缝
rodded point 标定点【测】
rodded volume 捣实体积
rodding 刮杆刮平;管道通条;机械清扫管子;铁杆清扫;用棒通沟;用棒捣实;捣实;捣棒插捣;棒捣
rodding chamber 喷洗室(车辆)
rodding concrete cubes 插捣混凝土立方试件
rodding dolly 孔道探测杆

rodding eye 清扫口;通渠孔;通管孔盖;通管孔
rodding fitting 夯捣装置;通洗装置
rodding opening 通洗口
rodding pipe 夯捣管;通洗管
rodding pit 夯捣坑
rodding structure 杆状构造
rod divider 杆式分禾器
rod divining 探条
rod dope 钻杆涂料;钻杆润滑脂
rod dowel 接头钢筋;暗销杆
rod drag 钻杆摩擦(力)
rod drawing 棒材拉拔
rod drive 连杆传动
rod-drop method 落棒法
rode 小艇系锚绳
rod electrodrill 有杆电钻
rod electrodrilling 有杆电钻钻进
rod element 杆形构件;棒形元件
rod elevator 杆条式升运器;提篮
rod elevator link 杆条式升运器链节
rod end bearing 杆端轴承
rod end chamber 活塞杆腔
rod end coupling 活塞杆端连接件
rod end pin 杆端销
rod end pressure 活塞杆端压力
rod end yoke 杆端连接叉
rodent and vermin-proofing 防鼠防虫的
rodent barrier 防鼠设施
rodent control 防鼠
rodenticide 灭鼠剂;杀鼠剂
rodent poison bait 灭鼠药
rodent proof 防侵蚀的;防鼠咬的;防蚀的
rodent repellent 鼠忌避剂
rodent resistance 耐鼠咬性
rodent-resistant cable 防鼠电缆
rodeo 圈地
rod error 标尺误差
rod extensometer 杆式应变计
rod extractor 钻杆提引器
rod extrusion 棒料挤压
rod fastening rotor 拉杆转子
rod feed 棒料送进器
rod feeding 捣冒口;棒注法;棒料送进
rod fender 木条碰垫
rod field 杆件加工场地
rod fixer 扎钢筋工
rod float 杆式浮标;测流杆;流速测杆;浮标测流杆;浮杆
rod for anchor 下锚埋入杆
rod friction 钻杆柱摩擦
rod gap 同轴电极间火花隙
rod gap arrester 棒状放电器
rod ga(u)ge 塞规;测杆尺;标准棒
rod gear 杠杆传动装置
Rodgers' anchor 冰锚
rod gland 杆密封盖;活塞杆密封
rod gland sleeve 导套
rod grain 杆状药柱
rod grease 钻杆润滑剂;棒状敷脂
rod grinder 棒式磨碎机;棒式碾磨机;棒磨机
rod guidance 杆导
rod guide 钻杆定心器;杆导(承)
rod gun 起管器
rod handling equipment 棒材搬运设备
rod hanger 钻杆挂钩
rod holder 焊把
rod image 标尺影像
rod in compression 受压杆;受压钢筋
roding hole 磨棒装入口
rodingite 异剥钙榴岩
rod injection 钻杆注浆
rod insulator 棒形绝缘子;棒式绝缘子
rod in tension 受拉杆
rod intercept 标尺截距
rod interval 标尺(分划)间隔
rod-in-tube method 棒管法
rod-in-tube technique 管棒法
rod iron 圆铁;元铁;钢棒;盘条
rodite 奥长石铜无球粒陨石
rodius-to-eccentricity ratio 转子半径偏心距比
rod jaw 叉杆
rod jaw pin 叉杆销
rod joint 杆连接;拉杆连接
rod journal 连杆轴颈

rod lashing system 杆式拉紧系统
rod length error 标尺长度误差
rodless angle(moulding)press 无杆角式压塑机
rodless electrodrill 无杆电钻
rodless reverse circulation rig 无钻杆反循环钻机
rod level 杆尺水准仪;标尺水准器;水准尺上的水准器;杆式水准器
rod level(l)ing 杆尺水准测量
rod lifter 杆尺夹持器
rod-like crystal 棒状晶体;圆形晶体;棒型晶体
rod-like molecule 棒状分子
rod-line 抽油杆
rod-line connecting links 抽油杆吊环
rodliner pump 杆式泵
rod-link chain 杆式链
rod-link conveyer[conveyor]杆链式输送器
rod-link digger 杆链升运式挖掘机
rod-link elevator 杆链式升运器
rod list 钢筋明细表
rod-loaded waveguide 棒荷波导;棒负载波导
rod loading 杆件承受的荷载
rod locking mechanism 挑杆锁紧机构
rod machine 制条机
rod magazine 棒料送进器
rodman 钢筋工;竖尺员;照尺员;司尺员;标尺员;立尺员;标尺员
rod-melting process 棒法拉丝工艺
rod memory 杆式存储器;磁棒存储器;棒式存储器
rod meter 米长尺
rod mill 管筒轧机;棒棍轧制机;盘条轧机;滚轴式破碎机;线材(滚)轧机;滚轴式碎石机;杆式研磨机;盘条轧机;棒磨机
rod-mill feed 棒磨机给料
rod milling 棒磨
rod mill liner 棒磨机内衬
rod moment 杆件弯矩
rod-motion indicator 棒移动指示器
rod multiplexer filter 棒(状)复用器滤波器
rod nodule 条状结核
rod note 水准尺读数
rod number 钢筋标号
rod oil-pump 杆式油泵
rodometer 压电轴向计
rod oscillation mechanism 挑杆摆动机构
rod packing 拉杆填料;推杆密封环
rod plug 钻杆提引塞
rod puller 钻杆提取器;拉管器(坑道用);起管器
rod pulls 钻杆立根数
rod pump 杆式泵;拉杆泵
rod rack 钻杆架
rod radiator 介质棒辐射器;棒状辐射器
rod reading 标尺读数
rod reaming shell 扩孔器
rod reducing bushing 钻杆异径接头(两头一公一母)
rod reducing coupling 钻杆异径接头(两头都是公扣)
rod reinforcement 钢筋
rod repeater mill 活套式线材轧机
rod rigging 铁牵条
Rodriguez fracture zone 罗得里格斯破裂带
rod rolling 线材轧制
rod-rolling mill 线材轧机
rod rotor 抽油杆旋转器
rods 棒条体
rod sag 钻杆柱的弯曲
rods and bars 杆和棒;棒材
rods and bars of alumin(i)um alloy 铝合金棒材
rod schedule 钢筋明细表
rod seal 活塞杆密封
rod-setter 钢筋工人
rod shape code 杆的形状标准
rod-shaped 杆状的
rod-shaped flash lamp 棒状闪光灯
rod-shaped micelle 棒状胶束
rod shearing machine 下料机;棒料剪切机
rod shears 钢筋切切机;棒材剪断机
rod shell 扩孔器
rod shoulder 钎肩
rods in axial vibration 杆的轴向振动
rods intermeshed in squares 方格网钢筋;钢筋方格网
rod size 钢筋尺寸
rod slack 减少钻杆对孔底压力
rod slap 钻杆碰撞

rod slide 钻杆导向滑槽;钻杆导槽
rod slope 钢筋弯斜度
rodsman 标尺员
rod snap 钻杆折断
rod sounding 土壤探测;杆(式)探测;杆测深;(水深探测的)杆测法;钎探;测杆测(水)深
rod spacer 钢筋定位器;钢筋间隔物;钢筋间垫块;钢筋定位件
rod spear 钻杆捞矛
rod spherule 杆小球
rod spraying 金属丝喷涂
rod stabilizer 钻杆稳定器
rod stand 钻杆立根
rod steel 圆钢筋;钢筋
rod stock 圆钢条;线材坯
rod storage 钢棒存放架
rod straightener 棒材矫直机
rod stress 杆中应力
rod stretch 抽油杆柱的拉长
rod string 钻杆串;抽油杆柱
rod string stabilizer 钻杆柱稳定器
rod string vibration 钻杆柱振动
rod stuffing box 钻杆填料函;拉杆填料盒
rod support 杆托;水准尺(尺)垫;尺座;尺垫【测】;标尺尺座
rod-suspended current meter 悬杆式测流仪;悬杆流速仪
rod-suspended hydrometric(al) cableway 悬杆式水文缆道
rod system 杆件系统
rod system boring 钻杆钻进
rod tamping 夯棍;捣棒
rod tap 钻杆螺纹攻
rod target 水准尺上的觇板
rod-tension test 圆棒张力试验
rod test 土探测
rod thermistor 棒状热敏电阻器
rod thief 棒形采样器
rod threaded at both ends 螺柱(两端都有螺纹的螺杆);两端有螺纹的螺栓
rod-tiller 钻杆转动手把
rod tipping mechanism 挑杆回转机构
rod tong 管钳
rod-tool rig 钻杆冲击钻机
rod-toothed crusher 齿辊破碎机
rod trail 连杆拖曳
rod transposition 杆件移动
rod type 杆式的
rod-type borehole extensometer 杆式钻孔应变仪
rod-type elevating conveyer 杆条式升运器
rod-type form tie 固定模版用拉杆
rod uncoupling 卸钻杆
rod vibration 钻杆振动
rod vibrator 杆式振捣器;棒式振捣器
rod walk 移动钻杆
rod weeder 杆式中耕除草机
rod wiper 活塞杆刮垢器
rod withdrawal 提棒
roe 桃花心木中短的断断续续的条纹
Roebling 防火地板和隔墙等的构造
roeblingite 硫硅钙铅矿
roedderite 罗镁大隅石
roe figure 鱼卵状木纹
roemer 大酒杯
roemerite 粒铁矾
Roentgen 伦琴
Roentgen apparatus 伦琴射线装置
Roentgen cinemato-graphy 伦琴射线摄影术
Roentgen detectoscopy 伦琴射线探伤法
Roentgen diagnosis X光诊断
Roentgen equivalent 伦琴当量
Roentgen equivalent physical 物理伦琴当量
roentgenite 伦琴石
roentgenization 伦琴射线照射
Roentgen kymography 伦琴射线记录法
Roentgen machine 伦琴射线机
Roentgen meter 伦琴仪;伦琴计
Roentgen microspectrography 伦琴射线显微分光术
roentgenogram 伦琴射线照相;伦琴射线照片;X光片
roentgenographic emulsion 伦琴射线照相胶
roentgenography 伦琴射线照相术;伦琴射线摄影术

roentgenology 伦琴射线学
Roentgenoluminescence 伦琴射线发光
Roentgenometer 伦琴计
roentgen-opaque 不透 X 线的
roentgenoscope 伦琴射线透视机
roentgenoscopy 伦琴射线透视法
roentgen per hour at one meter 一米处每小时伦琴
Roentgen photogrammetry 伦琴射线照相测量术
Roentgen rate meter 伦琴计
Roentgen ray 伦琴(射)线;X 光线
Roentgen ray photograph 伦琴射线照片
roentgen rays X 光
Roentgen spectrum 伦琴射线谱
Roentgen tube 伦琴管
Roentgen unit 伦琴单位
roepperite 锌锰橄榄石
roesslerite 水重砷镁石;砷氢镁石
roe stone 细石;角卵石;鱼卵石
roestone 鱼卵石;缩状岩;鲕状岩;鲕石
Rog 罗格油漆(用于新的水泥面)
Roga index 罗加指数
Roga index graduation 罗加指数分级
rogascope 测金属线裂纹或直径的仪器
roggianite 水硅铝钙石
rogitine 吩妥胺
rognon 圆形冰原石山
rogueing 拔株去劣
rogue peak 畸峰
rogue product 劣品
Rogue's yarn 识别绳条
roguing 淘汰
roguing the weak 去劣去杂
rohaite 硫锑铜铊矿
roheisenzunder process 生铁喷雾粒化法
Rohn mill 罗恩式多辊轧机
roily oil 乳化油
Rok 罗克毡(屋面防水材料)
Rokalba 罗尔卡尔巴石棉毡
rokuhnite 罗水氯铁石
rokui 橹柾
Rolcam 辊压机辊子表面磨损测试仪
Rolcox 辊压机控制系统
role in cycle of matter 物质循环作用
role indicator 作用指示器;作用指示符;功能指示符;任务显示器
role of city 城市性质
role of profit margins 边际利润的作用
roles in groups 团体中的角色
roling circle 展成圆
roll 轧制;轧机;轧辊;滚翻;缕条;卷动;碾平;名册;马脊岭;横滚;辊子;滚轧;滚动
rollability 滚动度;可轧制性(金属轧制变形的能力)
rollability measurement 测阻
roll-a-car hydraulic jack 液压落地千斤顶
roll acceleration 侧倾加速度;翻滚加速度
roll-adjusting gears 轧辊调整机构
roll-along shooting 逐点爆炸
roll-and-fillet mo(u)lding 卷筒形线脚;嵌边卷筒形饰边;凸方线脚(中世纪建筑)
roll and pitch indicator 摇摆仪
roll and weathered 墙头压顶(哥特式卷筒形装饰的)
roll angle 滚动(倾斜)角;旁向倾角
roll angle test 滚动角试验
roll articulation 滚动铰接
rollaway 活动折叠床
rollaway bed 可收拢的床
rollaway glaze 滚釉
roll-away nest 滚道;斜面鸡笼
roll axis 轧辊轴线;翻滚轴
roll axis reference 滚动轴基准
roll back 重新滚压;重新滚转;重算;小轮车架;滚回;重新执行;反绕;压低到以前水平;重新运行;重绕
roll-back point 重新运行点;重算点;反转点
roll-back policy 压价政策
roll-back price 压价
roll-back routine 重新(进行)程序;重新执行程序;重新运行(例行)程序;重算例行程序
roll-back snapshot system 重算抽点打印系统
roll-back system 重算方式;重算系统
roll balance gear 轧辊平衡装置
roll balance rod 长型换辊套筒

roll balancing counter weight 轧辊平衡锤
roll banding 黏辊
roll bar 碾杆
roll barrel 轧辊筒;辊身
roll bearing 轧辊轴承;滚柱轴承
roll bender 轧辊弯曲压力机
roll bending 轧辊挠度;辊弯
roll billet mo(u)lding 滚筒错齿线脚;短圆棍错齿线脚
roll bite 轧入;轧辊咬入轧件;辊缝
roll body length 辊身长度
roll boiling 煮呢
roll bonding 辊压接合
roll book 考勤登记本
roll-booster 回转加速器;回旋加速器
roll-box 轧辊箱;卷捆室
roll breakage 断辊
roll breaker 辊式破碎机;辊碎机
roll cage 滚柱罩
roll calibrating 轧辊型缝校准
roll calibre 轧辊型缝
roll camber 辊身轮廓
roll campaign 轧辊寿命
roll-capped 旋涡形盖顶瓦;漩涡饰盖顶
roll-capped ridge tile 辊帽脊瓦;漩涡饰盖顶脊瓦
roll-capped system 卷楞瓦屋面做法
roll capping 旋涡饰盖顶
roll-cap system 锌板平屋顶系统(表面无钎焊和钉头)
roll centre 辊心
roll change 换辊
roll change buggy 换辊小车
roll change mechanism 换辊机构
roll change pit 换辊用坑
roll changer 换辊装置
roll changing 换辊
roll changing carriage 换辊车
roll changing gear 换辊装置
roll changing rig 换辊装置
roll changing sledge 换辊滑车
roll changing sleeve 换辊套筒
roll check 对滚检验
roll chock 轧辊轴承座
roll cladding 轧制包覆
roll clamp 回转式货夹
roll clamp truck 夹钳式装再机
roll coater 辊涂机;浸胶辊;挤胶滚筒;涂胶辊
roll coating 滚涂;黏辊
roll cogging 在初轧机上轧制
roll collar 辊环
roll compacting 轧压;轧辊压制
roll control 侧滚控制
roll coolant 轧辊冷却乳液
roll coolant solution 轧辊工作表面冷却液
roll coolant system 轧辊冷却系统
roll cooling device 轧辊冷却装置
rollcore 滚筒芯
roll couple 倾翻力偶
roll crossing 辊筒交叉;辊叉叉
roll crown 中凸起轧辊;轧辊凸表
roll crusher 辊压机;辊式碎石机;滚筒式碎石机;滚筒式碎石机;滚筒破碎机;滚筒式碎石机;辊式破碎机;滚碎机;辊式破碎机
roll crushing 辊式破碎
roll crushing machine 辊筒式压碎机
roll crushing mill 对辊破碎机
roll cumulus 卷轴积云
roll curb 滚式缘石;进门侧石;进口缘石;圆角路缘
roll curving 滚轧曲面
roll cutter 辊式切纸机
roll damper 滚动阻尼器
roll damping 滚动阻尼
roll damping tank 减横摇水舱;横摇稳定舱
roll datum 倾侧角读数起点
roll die 辊锻卸模;搓丝模
roll displacement 滚动角位移;倾侧
roll down 轧下
roll down furnace 斜底式炉
roll down hem 朝下卷边
roll-driving shaft 辊轮驱动轴
Rolle's theorem 第一中值定理
roll eccentricity correction 轧辊偏心的修正
rolled 滚轧的;滚压的
rolled alumin(i)um alloy shape 铝合金轧制型材

rolled angle 轧制角钢;角钢
rolled angle bar 滚轧角钢
rolled angle (section) 滚轧角钢
rolled article 轧制成品
rolled asphalt 滚制沥青;滚压沥青;碾压式(地)沥青;碾压沥青
rolled asphalt pavement 碾压式沥青路
rolled backfill 压实回填
rolled bar 条钢;轧制棒材
rolled bar iron 轧制形钢;轧制条钢
rolled bascule bridge 滚轴开启吊桥
rolled base 压实基层
rolled beam 轧制(钢)梁;滚轧梁;辊轧梁;辊压梁
rolled-beam bridge 型钢梁桥
rolled bearing 滚轴承
rolledl bridge 滚轴座桥
rolled broken stone 碾压碎石
rolled cap roofing 卷棱瓦屋面做法
rolled cast steel 滚轧的铸钢件;滚轧铸钢
rolled cement concrete 碾压(式)水泥混凝土
rolled channel (section) 滚轧槽钢;轧制的 U 形钢;轧制的钢槽
rolled compact 轧压坯块
rolled concrete 碾实光面混凝土路面;滚压混凝土;碾压混凝土;碾实混凝土
rolled concrete dam 卷楞瓦屋面做法
rolled concrete pipe 滚制混凝土管
rolled copper strip 滚轧铜条
rolled core 压轧芯材;碾压填土墙
rolled cross girder 轧制钢横梁
rolled curb 滚式缘石;进门侧石;波形(路)缘石
rolled dam 碾压式土坝;碾压坝
rolled dry 干压;干碾
rolled earth dam 碾压(式)土坝
rolled earthen core 碾压填土墙心
rolled earth fill 碾压填土;压实填土;碾压土堤;碾实填土
rolled earthfill dam 碾压(填)土坝
rolled edge 轧制边;卷边
rolled-embankment method (路堤的)碾压填筑法
rolled figured 辊轧花纹的
rolled fill 碾压填土
rolled fill dam 碾压土坝;碾压坝
rolled-fill embankment 碾压筑堤
rolled film 轧膜
rolled flange 轧制法兰
rolled flaw 轧制发纹
rolled foundation subsoil 碾压基础地基
rolled girder 滚轧钢梁;滚轧钢梁;轧制(钢)梁;滚轧大梁;辊轧梁
rolled glass 辊轧玻璃;滚花玻璃;压延玻璃;轧制玻璃;滚轧玻璃;滚压玻璃
rolled glass canopy 滚压玻璃雨棚
rolled gold 轧制金箔;金箔;包金
rolled gold watch 包金表
rolled groove method 滚槽法
rolled hardening 压延淬火;轧制硬化;轧制淬火;辊压淬火
rolled hollow section 轧制空心钢材
rolled inlay 坯体表面滚压入彩色黏土
rolled-in scale 轧入氧化皮
rolled-in (stone) chip(ping)s 压入石屑
rolled-in treatment 压入式骨料处治法
rolled iron 钢材;轧钢铁;压延铁;轧制钢;滚轧铁
rolledl-section joist 滚轧工字形小钢梁;辊轧工字形小钢梁
rolled joint 滚轧接合
rolled joist 轧制钢龙骨;滚轧小梁
rolled laminated tube 滚压层压管
rolled latex 橡皮芯线
rolled lean concrete 碾压贫混凝土;碾压贫混凝土
rolled L-section 滚轧不等边角钢
rolled metal 轧制的金属;轧材
rolled mild steel section 轧制低碳型钢
rolled nonferrous metal 轧制的有色金属
rolled-on 碾压的;轧制的
rolled on flanges 轧制法兰
rolled-on method 碾压法
rolled-on thread 轧制螺纹
rolled paper 卷像卷纸;卷筒纸
rolled parts 轧制件;轧制部件
rolled pavement 碾压的铺面;碾压的路面
rolled pebbles 河床卵石;河床漂砾

rolled piece 轧件
rolled pipe 轧制管
rolled plate 轧制钢板;轧制板材;辊轧钢板
rolled plate glass 压延平板玻璃
rolled plate high pressure cylinder 卷板式单层高压筒
rolled product 轧制的产品;轧制材;钢材
rolled profile 轧制型材;型钢断面
rolled rockfill 碾压(式)堆石;碾压的堆石
rolled rockfill dam 碾压堆石坝
rolled rod 轧制棒材
rolled roofing 屋面卷材
rolled roofing material 屋面卷材
rolled screen 转筒筛;滚筒筛;回转筛
rolled section 异型钢(材);轧制型钢;碾压断面;滚压截面
rolled section brass 异型黄铜条
rolled section frame 型钢框架
rolled shape 型钢;轧制型钢;轧制型材;滚轧成形
rolled sheet 压延板材
rolled sheet glass 压延玻璃板
rolled sheet iron 轧制钢板;轧制铁皮;辊压钢板;碾压铁板
rolled sheet lead 碾压铅板
rolled sheet material 轧制钢板
rolled sheet metal 轧制金属板;轧制薄板;轧制板钢
rolled sheet metal door 轧制金属门
rolled sheet steel 轧制钢板;轧制钢板;轧压钢板
rolled special shape 轧制特种型材
rolled steel 压延钢材;轧制钢;轧制的钢材;滚轧钢;辊轧钢
rolled steel bar 轧制钢条;轧制钢筋
rolled steel beam 压延钢梁;轧制钢梁;轧轧钢梁;辊压钢梁
rolled steel beam arch 轧压钢梁拱
rolled steel beam bridge 型钢梁桥
rolled steel bearing plate 轧制支承钢板
rolled steel channel 热轧槽钢;轧制(的)槽钢
rolled steel for general structure 一般结构用轧制钢材
rolled steel for welded structure 焊接结构用轧制钢材
rolled steel joist(beam) 型钢梁;轧制钢(托)梁;型钢格栅;轧制工字梁;轧制钢龙骨;辊轧小钢梁;轧制钢格栅
rolled steel member 型钢
rolled steel plate 轧轧钢板
rolled steel product 钢材
rolled steel section 轧制型钢
rolled steel wire 盘条
rolled stock 轧制材
rolled stock grab 卷状物件夹具
rolled stock shape consistency 轧制件形状均匀性
rolled stone 轧压磨石;碾压碎石
rolled-stone base pavement 碾压碎石基层路面
rolled strip roofing 屋面卷材
rolled structural shape 轧制结构型材
rolled structural steel 轧制结构钢
rolled structure 轧结构
rolled surfacing 碾压路面
rolled tap 滚压丝锥
rolled thread 轧制螺纹;滚丝
rolled-thread screw 滚丝螺钉
rolled tin 锌板;锌皮
rolled translucent glass 滚压玻璃
rolled tube 轧制管;滚制管;压压管;卷铁皮管
rolled-up stock 压制件
rolled wire 轧制钢丝;轧制的丝材;辊轧钢丝;辊压钢丝
roll eights 八张薄板叠轧
roll end 轧辊端头
roller 压路机;滚子;滚柱;滚珠;滚轴;轧轧机;滚浪;轧刀;滚碎机;辊磨;辊(子);路碾;卷浪;碾子;碾压机;碾;托辊;长滚;长砾;轧辊
roller actuator 转子开关;滚子开关
roller and outer ring assembly 无内圈滚子轴承
roller application 轧轧机应用;滚镀(法);滚制(法);辊涂施工(法);辊涂
roller applicator 辊式浸润器
roller arm 翻窗撑挡;滑轮杆
roller axle 滚子辊轴;辊皮
roller bandage 绷带卷
roller bar grate 辊棒饱筛
roller bar grizzly 辊轴筛;辊动(格)筛

roller bascule bridge 滚动式竖旋桥;滚转跳开吊桥;滚轴吊桥
roller-bat machine 滚压泥饼机
roller beam 滚动梁
roller bearing 活动铰支座;滚柱轴承;滚子轴承;滚轴支架;滚动支承;滚轴支座;辊轴承
roller bearing axle 滚柱轴承轴
roller-bearing axle-box 滚动轴承轴箱
roller bearing center 滚柱轴承中心
roller bearing cone 滚珠轴承内锥体
roller bearing crankshaft 滚柱轴承曲轴
roller bearing cup 滚柱轴承座;滚珠轴承外圈
roller bearing gate 滚轴式闸门;滚轮闸门;滚动闸门
roller bearing lock ring 滚柱轴承锁环
roller bearing lubrication 滚柱轴承润滑
roller bearing needle 滚针轴承滚针
roller bearing pillow block 滚柱轴承轴台
roller bearing sleeve 滚柱轴承套
roller bearing steel 滚柱轴承钢
roller bearing tester 滚动轴承检测仪
roller-bearing wheel 滚柱轴承轮
roller bearing with wound (spiral) rollers 带螺旋滚柱的轴承
roller beater 打浆棍
roller bed 辊道
roller bend test 滚压弯曲试验;辊承弯曲试验
roller bin gate 滚筒式料仓闭锁器
roller bit 滚轮式钻头;滚子旋锥;齿轮钻头
roller bit drilling 牙轮钻井
roller blind 卷帘;卷窗帘;转轴式窗遮帘
roller blind frame 卷帘式百叶窗框
roller blind of steel 钢制滑轮百叶窗
roller blister 滚压泡
roller block 滚动支座底板;滑块
roller boring 牙轮钻头钻进
roller-bound macadam 碾压碎石(路)
roller-bound surface 碾压碎石路面
roller bowl mill 辊式碗磨
roller box 辊转箱;轮箱;轮架
roller breaker 滚轴式碎石机;滚轴式破碎机
roller breaking machine 辊式碎茎机
roller bridge 滚轴式活动桥;滚轮式活动桥
roller bridge balance 滚动式仰开桥;滚动式竖旋桥;滚动式开合桥
roller brush 滚动电刷;滚筒漆刷
roller-bucket type energy dissipater 滚斗式消能器
roller buffing machine 擦皮辊机
roller bump 辊压痕;辊花
roller bumper 滚轴防冲设备
roller burnish 轧辊磨光
roller burnishing 滚珠抛光
roller bush 滚轮衬套
roller cage 轴承滚子夹圈;轧轧机座;滚子保持架
roller calender 辊式压延机
roller cam 滚子凸轮
roller cam steering 滚柱凸轮转向机构
roller carbon black 滚筒炭黑
roller carriage 滚子支座;滚轴架;托滚架
roller carried plunger 滚道柱塞
roller carrier 托轮
roller casting apparatus 墨辊浇铸机
roller catch 碰珠
roller chain 滚子链(条);滚柱链;滚轴链;滚轮链;滚链
roller chain assembly jig 辊子链装配夹具
roller chain coupling 滚子链联轴器
roller chain dive 滚子链条驱动
roller chain idler 链传动惰轮
roller chain sprocket fraise 滚子链轮铣刀
roller chain twist guide 带滚子链的翻钢导板
roller check 滚碾裂纹;碾压裂纹
roller chock 滚柱导缆钳;滚柱导缆器;滑轮导缆器
roller cleaner 辊子洗涤剂
roller cleaning 滚筒清理
roller cleaning device 辊子清洗机
roller clutch 滚柱离合器
roller coaster (公园中游乐用的)滑行铁道
roller coater 辊式涂镀设备
roller coating 辊涂;滚涂;辊筒涂色法
roller-coating enamel 滚涂搪瓷;辊涂瓷漆;滚涂釉
roller-coating machine 辊涂机
roller coating printing ink 滚涂印刷油墨
roller cold header 滚柱冷镦机
roller compacted concrete 碾压混凝土;滚压混凝

土;碾压混凝土
roller-compacted concrete cofferdam 碾压混凝土围堰
roller-compacted concrete dam 碾压混凝土坝
roller-compacted concrete pavement 碾压混凝土路
roller-compacted concrete quality control system 碾压混凝土质量控制系统;碾压混凝土质量控制体系
roller compaction 碾压法;碾压
roller compaction riveting press 滚辗式压铆机
roller compaction test 碾压试验
roller cone 钻头牙轮
roller-cone core bit 取芯牙轮钻头
roller connecting rod 滚子轴承连杆
roller consolidating 辊压密实
roller consolidation 辊压密实
roller contact timer 滚柱接触式定时器
roller control system 辊压机控制系统
roller conveyer 滚轴运输机;辊子输送机;滚轴(式)输送机;滚道;滚筒传送机;滚筒传送带;辊式运输机;辊式输送机;辊式输送机;移动滚柱式传送机;辊道输送机
roller cooling bed 辊式冷床
roller core 滚筒堰
roller cover 辊套
roller crescent tool 圆盘月形孔抛光工具
roller crusher 滚碎机;滚轴式碎石机;滚轴式破碎机;滚筒式碎石机;滚磨机;辊式破碎机;辊式轧碎机;筒式碎石机
roller crusher conditioner 轧式式干草压扁机
roller crushing machine 罗拉碎茎机
roller-crushing mill 对辊磨
roller cup 滚针轴承衬套
roller current collector 滚柱接电器
roller curtain type filter 滚帘式空气过滤器
roller cutter 转轮割管器;转轮割刀
roller cutter bit 牙轮钻头
roller cutter for pipe 滑轮切割器;滑轮割管器
roller dam 圆辊坝;碾压坝
roller delivery motion 罗拉输出机构
roller die push bench 辊式模顶管机
roller dies 滚轮拉丝模;辊轮拉丝模
roller digital indicator 滚动数字显示器
roller distributor 辊式分配器
roller door 旋转门
roller door stop 滚珠门碰头
roller dryer 辊筒式烘干机;辊(筒)式干燥机;滚筒(式)干燥机
roller drill 带镇轮的播种机
roller drive 辊道传动
roller drive chain 滚柱传动链
roller-driven centrifugal mo(u)lding machine 托轮式离心成形机
roller drum 圆辊;压路机滚轮;滚轮
roller drum gate 圆筒闸门;圆辊闸门;滚筒堰
roller drum gate weir 圆辊闸门堰
roller dryer 辊筒式烘干机;辊筒式干燥机
roller drying 辊筒式干燥
roller electrode 滚子电极;滚轮电极;滚动电极
roller end 辊端
roller expander 滚子扩张器
roller extensometer 滚轮式伸长计
roller fading 摆动衰减
roller fairleader 滚柱导缆钳;滚柱导缆器
roller feed 滚珠进给;滚轮给进
roller feeder 滚筒加料器;辊式喂粮机;辊式送料装置;辊式给料机
roller feed mechanism 辊式排肥装置
roller fender 滚式护舷;滚筒碰垫;滚动碰垫
roller film machine 轧膜机
roller finder 仿形滚轮
roller-fitted spiral gravity shoot 装有滚柱的螺旋重力溜槽
roller flag 卷浪危险旗号
roller flange 辊轴凸缘
roller flattening 摇光;钢板压平;平皱
roller follower 滚子从动件;滚轮从动杆;随动滚柱;随动滚筒
roller forceps 转轴镊;车轮式沙眼镊
roller forming 滚压成形
roller forming machine 滚压成形机
roller for self-opening thread rolling head 滚压式板牙头用圆滚子

roller free wheel 滚子单向轮
roller gate 滚轴门;定轮闸门;圆轮闸门;圆辊闸门;滚筒门;辊式闸门
roller gate shield 圆辊闸门挡板
roller ga(u)ge 辊子卡规;辊子隔距
roller gear 滚子齿齿轮;滚柱齿轮
roller gin 辊式轧花机
roller grader 辊式分级机
roller grate 辊式卸料篦子
roller grille 卷动格栅
roller grinding mill 辊式碾碎机
roller grizzly 滚动筛
roller guide 辊式导板;滚柱导向器;滚柱导轨;滚针导轨
roller guide apron 导辊装置
roller head 滚压头
roller head machine 滚压成形机
roller hearth 辊式炉底;辊道炉床
roller hearth continuous rotaflame furnace 转焰炉
roller hearth conveyer 辊道炉膛运输带
roller hearth furnace 辊底式炉;辊道窑
roller hearth kiln 滚底式隧道窑;辊道窑
roller-hearth type furnace 辊底式炉
roller holder 滚柱轴承保持架
roller hydraulic jack 滚轮式液压千斤顶
roller impeller pump 滚子叶轮泵
roller impeller rotary pump 滚子叶轮泵
roller inner race assembly 无外座圈滚柱轴承
roller in sections 分组镇压器
roller iron 滚轮熨斗
roller jack 滚动千斤顶
roller jalousie 卷帘百页
roller jalousie gate 卷帘百叶门
roller jalousie housing 卷帘百叶箱
roller jaw 滚压颚板
roller jewel ga(u)ge 圆盘规
roller joint 球节(点)
roller Kelly bushing 辊式夹具
roller kiln 辊道窑
roller latch 滚柱碰闩;滚压闩
roller leaf actuator 转子刀形开关
roller leveling 辊子矫平
roller leveling machine 辊式矫直机
roller leveller 钢板压平机
roller levelling 钢板压平
roller lever actuator 转子式杠杆开关
roller lift bascule bridge 滚动式仰开桥;滚动式竖旋桥;滚动式开合桥
roller lifter 滚子式气门挺杆
roller lock 圆环锁
roller lube nozzle 滚柱润滑喷嘴
roller lubrication 滚柱加油;滚柱润滑法
roller machine 滚压机
roller man 压路机驾驶员
roller mark 辊印;辊花;碾压痕迹;滚痕;墨辊杠子
roller measuring ga(u)ge 测皮辊机
roller mechanism 双圆盘机构
roller mill 碾压机;辊压机;滚压机;辊式滚压机;辊式粉磨机;碾压机;辊式研磨机;辊式磨
roller milling 轧制
roller mill with corrugated rollers 带槽辊的对辊磨
roller mill with spring-loaded rollers 磨辊用弹簧加压的辊磨
roller-mounted leaf gate 安装在辊轴上的百叶门
roller mulcher tiller 松土覆盖镇压机
roller nest 滚柱窝
roller of breaking waves 破浪带
roller operated lubricating apparatus 滚柱润滑装置
roller operated valve 滚柱操纵的阀
roller operator 压路机驾驶员
roller outer race assembly 无内座圈滚柱轴承
roller packer 滚子式镇压器
roller painting 滚涂法;辊涂
roller pallet 滚筒墁灰板
roller pan mixer 转盘式搅拌机
Roller particle-size analyzer 罗勒粒径分析仪
roller partition 卷升隔墙
roller path 辊道;辊式输送器;辊式输送机;辊道;定radiated道
roller pattern 压路机模型;滚柱模型
roller pattern paint 辊涂花纹漆
roller pawl 滚筒掣子
roller pendulum discharge device 摆辊式卸料装置

roller pin 圆盘钉;滚针;滚柱轴;滚轮销
roller plate 滚筒板
roller-plus-belt applicator 辊带式浸润器
roller press 轧压机;滚压机;滚动机;辊压机
roller press cake 滚压料饼
roller press-down apparatus 轧辊压下装置
roller pressure compensation 轧压力补偿
roller-printed wallpaper 滚筒印花墙纸
roller printing 辊筒印花
roller pump 滚子泵;滚柱泵
roller pusher 辊式推料装置
roller race 滚珠轴承座圈;滚柱(轴承)座圈;滚柱轴承环
roller rail 滚筒轨
roller ram 滚动导轨柱塞
roller-ratchet socket 绳帽
roller reamer 圆柱牙轮扩孔器
roller reel 辊式卷取机
roller remover 圆盘除脱器
roller rest 靠辊
roller retainer 滚柱保持器
roller rim 辊轴凸缘
roller rock arm 滚轮摇臂
roller rocker 滚轮摇杆
roller rule 滚动平行尺
rollers 长涌群
rollers and cage 滚柱和隔圈
roller sander 砂纸辊打磨机
rollers and races contact 滚柱与滚道间的接触点
roller scratch 辊子引起的擦伤
roller screed 滚筒式整平板
roller screen 滚筛;滚动筛;丝网辊
roller seal 辊封;滚筒封口
roller seat 辊座
roller section straightening machine 辊式型材(矫直)机
roller segment 镇压器环
roller separator 滚柱隔离环
roller setting 墨辊调节
roller shackle 有滑轮的卸扣
roller shades 百叶窗
roller shaft 滚轮轴;托轮轴
roller shaft seat 滚轴轴座
roller shape straightening machine 辊式型材矫直机
roller shelf 滑动书架
roller shell 辊筒(防护)罩
roller shield 圆筒屏蔽;圆辊闸门挡板
roller shoe 滚轴支座;滚轴支承
roller shutter 卷升百叶窗;卷门;卷帘百叶
roller shutter casing 卷帘盒
roller shutter door 卷帘门
roller shutter type door 卷升百叶门
roller shutter window 卷帘窗
roller side bearing 滚子式旁承;辊式侧支座
roller side guard 辊式侧导板
roller sizer 辊式分级机
roller skate rink 旱冰场
roller skating rink 旱冰场;室内溜冰场;溜冰场
roller slewing ring 滚子夹套式回转支承环
roller-smoothing 平皱
roller snap thread ga(u)ge 滚柱螺纹卡规
roller sorter 辊式分级机
roller spacing 辊距;辊缝
roller spindle 滚筒轴
roller spiral conveyer 螺旋滚式输送机;螺旋滚式输送道
roller spot-and-seam welding machine 辊式点缝焊机
roller squeegee 橡胶滚筒
rollers set 滚筒组
roller stamping die 辊子冲压模
roller stand 滚柱架
roller standard 滚柱架
roller stay 滚子跟刀架
roller steel 轧制钢
roller steel section 型钢
roller step bearing 滚柱止推轴承
roller stopper 滚柱停止器
roller straightening (钢管型钢的)辊式矫直
roll(er) straightening machine 辊式矫直机
roller stretcher 辊式拉伸矫直机
roller-stretcher machine 辊式板材拉伸(矫直)机
roller strike 滚柱锁舌

roller stripper 辊式摘穗器
roller subsoil 碾压底土
roller support 滚轴支座;滚轴支承
roller supporting mechanism 托滚支承装置
roller surface wearing tester 辊压机辊子表面磨损测试仪
roller suspension pipemaking machine 滚筒悬置式制管机
roller suspension process 悬辊法
roller suspension system 磨辊悬置系统
roller swage 套管修正器;套管修整器
roller sweling 墨辊膨胀
roller switch 滚轮开关
roller table 大圆盘;辊式摇床;辊式磨的磨盘
roller-table gear 辊道
roller tapper 滚柱挺杆
roller test plant 车辆滚动试验装置
roller thrust bearing 滚柱止推轴承;滚柱推力轴承
roller tilter 辊式翻钢机
roller tool 滚轮工具
roller tooth 滚削齿
roller-top card 罗拉盖板梳理机
roller-towel 油卷
roller track 滚轨
roller track roller way 辊道
roller tractor 拖拉机式压路机;轮式拖拉机
roller train (在闸门两侧的)滚轮链串;滚轴运输机;滚轴输送机;滚轮组
roller transmission chain 滚子链条
roller tray 移动式压延台;滚压台;滚动台
roller trough 墨辊洗涤槽
roller twist guide 辊式扭转导板
roller-type annealing lehr 辊式退火窑
roller-type breaker 辊式破碎机
roller-type bucket loader 轮胎斗式装载机
roller-type conveyer [conveyor] 滚轴式输送器
roller-type corrugator 辊轴式瓦楞轧机;辊轴式波纹轧机
roller type discharge grate 辊式卸料篦子
roller type furnace 辊道窑
roller-type grizzly 滚轴筛
roller-type guide 辊式导板
roller-type ingot buggy 辊底式送锭车
roller-type mill 辊式磨碎机
roller-type moving ramp 滚柱式电动斜道
roller-type moving walk 滚柱式电动走道
roller-type one way clutch 滚柱式单向离合器
roller-type overrunning clutch 滚柱式超越离合器
roller-type pump 滚柱式泵
roller-type ratchet 滚柱式棘轮机构
roller-type rice mill 辊式碾米机
roller type rotary pump 滚轴式回转泵
roller type rubber fender 转动型橡胶护舷
roller-type seam welding machine 辊式缝焊机
roller-type sinterer 辊道烧结机
roller-type slab magazine 辊式板坯装料台
roller-type thickness ga(u)ge 辊式测厚规
roller-type thread 滚柱式螺纹
roller-type venier callipers 滚轮式游标卡尺
roller up in a cluster 集束缠绕
roller vane pump 滚子叶轮泵;滚筒叶轮泵
roller vibrator 滚轴式振动器;辊式振动器
roller washing 墨辊冲洗设备
roller wave 滚压波纹;辊压痕
roller way control case 辊道控制箱
roller-way face 溢流坝面;滚水坝面
roller weir 滚筒堰;圆辊堰;圆辊活动坝;辊堰
roller welding 轧焊
roller wheel 压路机滚轮;滚轮
roller with internal vibrator 带内振动器碾压机
roller with supporting ring 带托环炉辊
roller with trunnions 带颈辊
roller working radius 旋轮圆角半径
roll extrusion 滚压
rolley way 运输道
roll feed 辊式进给装置
roll feeder 旋转式加料机;旋转式供料器;旋转式给料器;转筒给料机;辊式给料机
roll fender 圆辊防撞装置;碰垫球
roll film 照相胶卷;胶卷;软片
roll-film camera 卷片式照相机
roll film dryer 胶卷干燥器
roll finish 滚光
roll flanging 凸缘滚形

roll flattening 轧辊压下装置;轧辊弹性压扁;辊轧校平
roll fluting machine 滚筒切割机;辊开槽机
roll for cold rolling 冷轧轧辊
roll-forged straight shank drill 轧制直柄钻头
roll forging 滚锻;滚镀;辊锻
roll forging machine 轧辊锻压机;辊锻机
roll forging under increasing compression 增压下辊锻
roll for hot-rolling 热轧轧辊
roll form body maker 罐身成圆机
roll formed case 滚压成形壳体
roll formed section 辊轧型材
roll formed shape 辊轧型材;冷弯型钢
roll formed unit 辊轧型材
roll former 辊轧成型机;自动卷条装置
roll forming 轧制成型;轧膜成;轧轧成型;辊形;成形轧制
roll forming die 滚压成型模具;辊锻模
roll forming machine 滚形机;成圆机
roll gallows 前导轮架式单铧犁
rollgang (横行的)输送辊道
rollgang elevator 滚道升器
roll gap 轧辊开口度;辊隙;辊间距离;辊缝
roll gap indicator 辊缝指示器
roll gap set 调整压下装置
roll gap setter 辊缝调整器
roll ga(u)ge 轧辊型缝
roll grinder 轧辊磨床;卷带式抛光机;卷带式磨光机
roll grinding machine 滚筒研磨机;辊子磨床;辊式碾碎机
roll grinding mill 辊磨机;轮碾机
roll grip pipe wrench 链条管(子)扳手
roll gyroscope 倾斜陀螺仪
roll hammer crusher 辊锤破碎机
roll hot stamping machine 热滚压印机
roll housing 轧机机架
roll-in 转入;滚入;换入
roll-in and roll-out 转入转出
roll-in freezer 主冷库;存冷冷库
rolling 压延;轧制;轧钢;滚釉;滚压;滚动的;辊轧;轮碾;溜放【铁】;碾轧;碾压;碾光;碾的;横摇;缩釉;翻滚;滚轧;滚动
rolling acceleration 倾侧角加速度
rolling a joint 碾压接缝
rolling and mixing action 揉拌作用
rolling and peening 滚轧与锤击(齿面塑变形式)
rolling and pitching motion 横摇和纵遥
rolling angle 横摇角
rolling avalanche 滚动雪崩;滚动崩落;卷雪崩
rolling average 移动平均数
rolling axis 滚动轴线
rolling ball planimeter 转球求积仪;滚球式求积仪
rolling ball visco(si)meter 滚球(式)黏度计
rolling barrel 滚光筒
rolling barrier 滑轮栅门;滑动栏木
rolling bascule bridge 滚轮式开合桥;滚动式仰桥;滚动式竖旋桥;滚动式开合桥
rolling beach 宽平海滩;宽缓海滩
rolling beam apparatus 旋转秤仪
rolling bearing 滚动轴承;滚动支座;滚动支承
rolling bearing inspecting and mounting table 滚动轴承检查装配工作台
rolling bearing oil seal 滚动轴承油封
rolling bed loading 推移质
rolling belt type surface treatment machine 滚带式表面处理机
rolling billet 轧制用钢坯
rolling blind 卷帘
rolling block 下转后退闭锁机
rolling bridge 滚轮式活动桥;轻型机械化桥
rolling caisson 滚轮沉箱式闸门;滚动闸门;滚动沉箱式闸门;推拉式坞门
rolling cam 滚子凸轮
rolling capital 经营资金;累积资本;经营成本
rolling car favourable condition 溜车有利条件
rolling car resistance due to wind effects 货车溜放风阻力
rolling car unfavourable condition 溜车不利条件【铁】
rolling chair 轮椅
rolling charateristics 滚动性能
rolling chock 减摇龙骨;防御动垫木
rolling chopper 重型碎土镇压器;重型碎石镇压器

rolling circle 转盘;转车台;滚圆;基圆;齿轮基圆
rolling circle aeration 转盘曝气
rolling circle model 滚环模型
rolling circumference 滚动圆周
rolling cleat 滚轮导缆耳
rolling colter 圆犁刀
rolling comb type seedling puller 滚梳式水稻拔秧机
rolling cone 滚锥
rolling contact 滚动接触
rolling contact bearing 滚动接触轴承
rolling contact gears 滚动接触齿轮系
rolling contact joint 滚动接触缝
rolling contact-type equalizer bar 滚接式平衡横梁;滚接式平衡杆
rolling core veneer 卷心式胶合板
rolling coulter 圆盘犁刀
rolling counters forward 计数器计数
rolling counterweight (桥梁的)仰扒滚轴平衡重
rolling country 准平原;丘陵区;丘陵地(区);波形地
rolling couple 倾动力矩;滚动力矩
rolling crack 轧制裂纹;轧制裂缝
rolling cradle 辊子吊架;(斜孔钻进支撑钻杆的)带辊滑台
rolling curtain 卷帘幕;舞台卷幕
rolling curve 滚线
rolling cutter 牙轮切削具
rolling cutter (rock) bit 牙轮钻头
rolling cycle 轧制周期
rolling cyclorama 卷帘天幕;(能横向卷动的)舞台天幕
rolling cylinder gate 滚动式圆筒闸门
rolling dam 滚筒式闸堰;滚筒式溢流坝;滚筒式活动坝;滚动式活动坝
rolling defect 轧制缺陷
rolling depression 滚压
rolling dermatome 滚轴式切皮刀
rolling diaphragm 滚动隔膜
rolling diaphragm actuator 滚动薄膜式液压缸
rolling direction 轧制方向;溜车方向
rolling direction of hump 驼峰溜车方向
rolling disc planimeter 转盘式求积仪
rolling dog 滚动挚子
rolling door 滚动门;卷升百叶门;卷帘门;推拉门;卷升门;卷门;滑动门
rolling down section 溜放部分
rolling drawbridge 滚轮式吊桥
rolling drum 滚筒
rolling edge 飞翅
rolling edger 轧边机
rolling effluent 压延排水;轧钢排出液
rolling element 滚动体
rolling experiment 横摇试验
rolling face 轧辊工作面
rolling factor 碾压因数
rolling fixture 对滚装置
rolling friction 轧制摩擦;滚动摩擦(力)
rolling friction resistance 滚动摩擦阻力
rolling from the molten condition 无锭轧制
rolling fulcrum 轴动支点
rolling gasket 轧辊垫圈;辊轴垫圈
rolling gate 圆辊闸门;滚筒闸门;滚动式闸门;滑轮栅门
rolling gate of sluice chamber 泄水闸室滚动闸门;泄水闸室滚筒闸门
rolling gear 盘车装置
rolling grade 起伏坡度;包线设计的纵坡度
rolling grade design 包线设计
rolling gradient 最大纵坡;控制坡度;(铁路、道路的)限制坡度
rolling greenhouse 移动式暖房
rolling grille 卷格子窗;格栅卷帘;卷帘格栅
rolling grille door 卷帘格栅门
rolling groove 压槽
rolling ground 丘陵地(带);起伏地面
rolling guide 滚动导轨
rolling hatch beam 滚动舱口梁
rolling head 滚轮冲头(滚光孔用);滚压头
rolling hills 起伏丘陵;波状丘陵地
rolling hitch 轮结
rolling hospital 随军医院
rolling impact 滚动振荡
rolling indicator 横摇指示器
rolling ingot 轧制用钢锭

rolling in pack form 叠板轧制
rolling in rail 钢轨内移
rolling instability 横向不稳定性;横倾不稳定;对纵向轴线的不稳定性
rolling jig 碾压模
rolling keel 减摇龙骨
rolling key clutch 旋转键离合器
rolling land 丘陵地(形);坡地
rolling landside 滚动犁侧板
rolling lattice gate 卷动格栅门
rolling lift bridge 滚动升降桥;滚升式开合桥;滚动式仰开桥;滚动式竖旋桥;滚动(式)开合桥;轮开桥
rolling line 轧制线
rolling liquid sealing apparatus 滚动式液膜密封装置
rolling load 车辆载重;滚动载重;车辆荷载;行动载重;轧制负荷;滚动轮压;滚动荷载
rolling-load fatigue testing machine 滚动承载疲劳试验机
rolling loading 滚动荷载;轧制负荷
rolling lock gate 滚动式闸门;滚动式(船闸)闸门
rolling loss 滚动损失
rolling lubricant 轧制工艺润滑剂
rolling lubrication 压延润滑
rolling machine 轧钢机;滚压机;滚皮机;辊式压板机;卷板机;搓圆机;成圆机;滚轧机
rolling machinery 碾压机械
rolling margin 轧制裕量
rolling material 轧制材料
rolling metal 被轧制的金属
rolling method 转桶法(加气混凝土配料法的一种)
rolling mill 轧钢机;轧钢厂;滚轧机;碾压机;合金片碾压机;辊轧机;轧机
rolling mill assembly 轧机装置
rolling mill control 轧机控制
rolling mill drive apparatus 轧机的传动装置
rolling mill equipment 轧制设备
rolling mill for circular shapes 轧圆型材的轧钢机
rolling mill ga(u)ge 滚辊厚薄计;钢板厚度计
rolling mill guide 轧机导板
rolling mill machine 滚轧机
rolling mill machinery 轧制设备
rolling mill motor 轧钢电动机;滚轧电动机
rolling mill product 辊轧制品
rolling mill scale 轧制鳞皮
rolling mill screw down 轧机的压下装置
rolling moment 横摇力矩
rolling-moment inertia 滚转惯性矩
rolling motion 滚动;横摇;倾侧运动
rolling needle 滚针
rolling noise 滚动噪声
rolling off 摘挂
rolling of sectional iron 型钢轧制
rolling of sections 型钢轧制
rolling of sediment particle 沉积颗粒的滚动
rolling of shield 盾构旋转
rolling of ship 船舶横摇
rolling of soil 土基压实
rolling of vessel 船舶横摇
rolling oil system 轧制润滑系统
rolling on 穿轧
rolling on edge 轧边;立轧
rolling O-ring joint 轧辊垫圈接头
rolling-out collapsed string 矫正压扁的钻杆柱
rolling-out process 碾出法
rolling-out test 滚压试验(测可塑性)
rolling over 翻转
rolling pair 滚动副
rolling partition (wall) 移动(式)隔墙;滚动式隔断;卷帘隔墙;卷帘隔断
rolling pass 轧制道次
rolling path of grinding bodies 磨球滚落轨迹
rolling period 横摇周期
rolling period of buoy 浮标摇摆周期
rolling plan 逐年延展计划;滚越;滚动计划
rolling plane 轧制平面;横倾平面
rolling platform 轧钢机台
rolling pontoon 滚动(式)浮筒
rolling press 滚筒印刷机;滚动印刷机;矫平压力机
rolling pressure 辊压;轧制压力
rolling process 滚压工艺;压延法;轧制过程;轧压成型法;滚压法;滚筒式制法
rolling pulley 滚动滑轮

rolling quenching 旋转淬火
rolling rack 滚动支架;滑动支架
rolling radius 轧机半径;滚动半径
rolling rate 轧机生产率
rolling readjustment 交替调整
rolling recorder 横摇记录器
rolling region 起伏地区
rolling resistance 车辆溜放阻力;滚动阻力
rolling resistance factor 滚动阻力系数
rolling resistance measurement 滚动阻力测定
rolling resistance test 滚动阻力试验
rolling ring 滚轧环;滚圈
rolling ring type crusher 圆环式破碎机
rolling roof 移动式屋顶
rolling rotor compressor 滚动转子式压缩机
rolling rubber 橡胶滚轴
rolling rubber O-ring joint 滚轴橡胶垫圈接头
rolling scaffold(ing) 滚轴(移动)脚手架;滚动脚手架;带滚轴的脚手架
rolling schedule 轧制制度
rolling screen 卷帘;回转筛
rolling sea 滚浪;卷浪;长浪
rolling seam 滚轧裂纹(由于轧制引起金属表面的细小裂纹)
rolling section 轧制型钢
rolling section of hump 驼峰溜放部分
rolling segment 转动承座;转动承梁;转动承架
rolling shaft 滚轴
rolling shapes 钢材品种
rolling shaping 滚压成形
rolling shear 滚剪机;滚动剪切
rolling shutter 卷帘百叶帘;卷升百叶窗;卷闸门;卷帘门;卷帘窗
rolling-shutter door 卷闸门;卷升百叶门
rolling-shutter type door 卷升(式)百叶门
rolling skin 波形屋面板
rolling slab 扁钢坯
rolling slat 卷板条
rolling slate 卷帘板
rolling slurry process 滚浆法
rolling smoke curtain 移动烟幕
rolling solution 轧制工艺润滑冷却液
rolling speed 轧制速度;溜入速度;溜放速度;碾压速度
rolling sphere instrument 转动落球黏度计;落球式仪器;滚球式仪器
rolling sphere planimeter 转球求积仪
rolling sphere visco(si)meter 滚球(式)黏度计
rolling stay 防摇牵条或支索
rolling steel door 滚动式(防火)钢门;钢制转门
rolling steel mill wastewater 轧钢厂废水
rolling stock 车皮(俗称);运输车辆;轧制材料;机车车辆;全部车辆;铁路机车及车辆;铁道车辆
rolling stock amount of loading 装车数
rolling stock amount of using 使用车数
rolling stock and rolling stock accessories 车辆及配件
rolling stock clearance 机车车辆限界
rolling stock damage 机车车辆破损
rolling stock department 车辆厂
rolling stock depot 车辆段
rolling stock facility 车辆设备
rolling stock ga(u)ge 机车车辆限界
rolling stock ga(u)ge frame 机车车辆限界架
rolling stock ga(u)ge line 机车车辆限界线
rolling stock inspection 车辆检测
rolling stock kilometer 车辆千米
rolling stock limit frame 机车车辆限界
rolling stock maintenance shed 车辆检修车间
rolling stock manufacturing plant 机车车辆制造厂
rolling stock operation(al) maintenance 车辆运用维修
rolling stock plant 机车车辆厂
rolling stock repair plant 机车车辆修理厂
rolling stock spare parts plant 机车车辆配件厂
rolling stock strength 车辆强度
rolling stock utilization index 机车车辆运用指标
rolling stock utilization plan 机车车辆运用计划
rolling stock workshop 车辆厂
rolling strata 波状层理;波痕交错纹理
rolling support 滚动支座;滚动支承
rolling surcharge 滚压超载
rolling surface 滚动面;起伏地面

rolling swell 长涌
rolling-table saw 滚动台锯
rolling tackle 防摇绞链
rolling tail gate 路面整平器后部滚筒;滚筒式下闸门;滚筒式尾闸;船闸下游滚筒闸门
rolling take-off 边部(平衡)拉引
rolling temperature 碾压温度;轧制温度
rolling temperature of bituminous mixture 沥青混合料的滚压温度
rolling terrain 丘陵区;丘陵地带;微丘区
rolling test 滚动试验;碾压试验
rolling texture 轧制织构
rolling tolerance 轧制公差
rolling topography 丘陵地形;起伏地形
rolling track of hump 驼峰溜放线
rolling trailer loading 滚装挂车荷载
rolling transport 滚移;滚动运输;滚动输送
rolling transposition 转动换位
rolling tread circle 滚动圆
rolling trough 翻滚水流
rolling truck 碾压卡车;(平板闸门的)定轮小车;辊轴吊车;转轴吊车
rolling-tub screen deck 转桶式筛面
rolling-up curtain weir 卷帘式框架堰;帘(式)堰
rolling-up device of roller shutter 百叶窗的卷动装置
rolling vane motor 旋转式叶片电动机
rolling wastewater 轧钢废水
rolling wave 卷波
rolling weight (打夯机上的)滑锤
rolling-wheel taker 滚轮取料机
rolling width 碾压宽度
rolling with (stock) tension 带张力轧制
rolling Wollaston wire 沃拉斯顿碾压导线
rolling work 碾压工作;辊轧工作
roll in one heat 一火轧成
roll-in refrigerator 主冷库;转存冷库
roll insulation 绝缘毡;保温毡
roll integrator 倾斜角积分器
roll-jaw crusher 滚轴颚式破碎机;转动颚式破碎机;滚爪式碎石机;滚动颚式碎石机;单辊颚式破碎机
roll joint 孔型锁口;卷边式接缝
roll kiss coater 双辊涂漆机
roll kneader 辊式捏炼机
roll laminating 辊压层(压)
roll leveller 辊式矫直机
roll life 轧辊寿命
roll line 轧制线
roll machine 轧轧机
roll magazines 滚动料斗
roll magnetic anisotropy 压延磁各向异性
roll mandrel 组合式轧辊的辊轴
roll mark 滚印;轧痕;滚动痕;碾痕;流动痕
roll microfilm 卷轴式缩微胶卷;缩微胶片卷
roll mill 滚磨机;辊式破碎机;辊碾制粉机;筒式碎石机;辊式研磨机;辊式磨;轧轧机
roll mill granulator 转筒式成球机;辊式成球机
roll mill pulverizer 辊式粉磨机;煤粉压榨机
roll mixer 辊式混合机
roll mixing 辊压捏合
roll motor 辊道电动机
roll mo(u)lding 圆形线脚;旋涡形线脚;漩涡形线脚;卷线脚;卷饰
roll move sprinkler 滚移式喷灌机
roll neck 辊颈
roll neck bearing 轧辊轴承
roll nozzle 滚动喷嘴
roll number 纸带卷号
roll-off 碾轧
roll-off characteristic 频响跌落特性
roll-off frequency 向上转移频率
roll-off screen printing machine 滚动丝网印花机
roll of wall covering 饰墙布用滚筒
roll of wire 盘条;钢筋盘条;铁丝盘
roll oil 轧用油
roll on, roll off facility 水平装卸设施
roll on berth 滚装码头
roll-on decorating machine 辊条装饰机
roll on-off adjustable platform 滚装可调平台
roll on return pass 返回轧制
roll on/roll off 滚动装卸;开上开下
roll on roll off berth 滚装码头
roll on/roll off facility 滚装设施

roll on/roll off method 滚装法
roll on/roll off ramp 滚装引桥;滚装坡道;开上开下坡道
roll on/roll off service 滚装运输业
roll on/roll off ship 开上开下船;机动车辆专运船;滚上滚下船
roll on/roll off system 滚装法;滚装卸方式;水平装卸法
roll on/roll off terminal 滚装码头;滚上滚下码头
roll on/roll off traffic 水平装卸法;滚上滚下运输;开上开下运输
roll on/roll off traffic by trailer and ferry 拖车及渡船驶进驶出运输
roll on/roll off transport 滚装运输
roll on/roll off vessel 滚装船
roll on/roll off yard 滚装堆场
roll on service 滚装运输业
roll on ship 滚装船
roll-on traffic 滚装运输
roll on vessel 滚装船
roll opening 轧辊开度
roll opening indicator 辊缝指示器
roll ore body 卷卷矿体
roll out 转出;车辆出厂;转出主存储器;主存储器转出;辊平
roll-out control 着陆滑跑操纵
roll-out guidance 着陆滑跑操纵
roll-over 倾翻;以新债券取代旧债券;滚改器;倾向倒转;图像跳动;翻层现象
roll-over action 倾翻作用
roll-over anticlinal trap 滚卷背斜圈闭
roll-over anticline 滚卷背斜
roll-over board 垫板
roll-over bucket 倾翻式铲斗
roll-over budget 结转预算
roll-over credit 浮动利率信贷;展期贷款
roll-over draw machine 翻转式起模机
roll-over dump 翻斗撒砂
roll-over form 翻转模
roll-over machine 翻型机
roll-over mortgage 滚动利率抵押款
roll-over mo(u)lding machine 翻台式造型机
roll-over paper 短期债券
roll-over plate 翻箱板
roll-over plow 翻转犁
roll-over protection system 驾驶员舱顶保护罩;翻滚保护设施
roll-over protective structure 滚翻防护装置
roll-over scraper 翻转斗式铲运机
roll-over stand 翻转架
roll-over table 翻转台
roll-over test 翻转试验
roll-over type furnace 斜底加热炉
roll packer 滚动器
roll paper 卷轴式纸;卷轴式记录纸;卷纸
roll parting 辊缝锁口
roll parting line 轧辊孔型中线
roll pass 轧辊型缝
roll path 轧辊型缝
roll peripheral speed 辊轴圆周速度
roll piercing mill 穿轧机
roll piercing process 穿轧法
roll plain 波状平原
roll-polishing tool 滚抛磨具
roll polling 轮询
roll-position control 坡度控制
roll preheating device 轧辊预热器
roll press 辊式压制机
roll presses for paper industry 造纸用压光机
roll process 辊轧法
roll pulverizer 辊式粉磨机
roll recovery 中止滚转
roll release 脱耙
roll release lever 辊子释放杆
roll rim 卷拉边;卷边
roll ring mill 辊环式磨机
roll rockfill 碾压的堆石
roll roofing 卷铺屋面材;卷材屋面
roll roofing felt 卷筒油毛毡
roll roofing winding mandrel 布蓬;油毛毡卷轴
roll scale 氧化皮;轧制铁鳞;(压轧的)鳞皮
roll screen 滚动屏幕
roll sea 卷板

roll seal 滚子接封;辊子闸门(密封缝隙用)
roll seam 轧辊间隙
roll seed separator 辊式种子清选机
roll sensing accelerometer 滚动运动加速表
roll sensor 滚动传感器
roll-separating mechanism 轧辊调整装置;轧辊调整机构;辊缝调整机构
roll servo 滚动伺服机构
roll setting 安置轧辊
roll setting angle 转动安装角
roll setup 调整轧辊位置;安置轧辊
roll shaft 辊轴
roll shape 辊型
roll shears 辊式剪断机
roll shell 滚轮缘
roll sleeve 轧辊辊套
roll slitter 辊式钢板切断机
roll speed 轧辊圆周速度
roll spin 滚旋
roll spinning 滚旋
roll-splitter guillotine 辊切机
roll spot welding 滚点焊
roll spring 轧辊弹起度
roll square 卷材方(屋面卷材面积单位,1卷材方=108平方英尺)
roll squeezer 辊压机
roll stability 转动稳定性
roll stabilized 滚动稳定的
roll stand 轧钢机架
roll steering of rear axle 后轴倾斜转向
roll stiffness 轧辊刚性
roll stiffness of suspension 悬挂抗侧倾刚度
roll stop 钢板压痕
roll straightener 辊式矫直机
roll straightening 轧辊矫直法
roll strength 轧辊强度
roll stripper 辊式摘棉铃机
roll stroke 轧辊升程
roll subsidence mode 衰减横摇运动
roll surface 辊面
roll table annealing lehr 辊道式退火窑
roll template 滚压加工样板;卷板
roll testing 对滚检验法
roll thickness ga(u)ge 滚辗厚度计
roll tile 意大利筒瓦
roll-top desk 带折叠盖子的写字台;折叠式写字台;卷折叠式写字台
roll torque 轧制力矩
roll trailer 滚轴挂车
roll turning lathe 轧辊车床
roll turning templet 车削轧辊用的样板
roll type air filter 自动卷绕式空气过滤器
roll type briquetting machine 对辊制团机
roll-type briquetting press 对辊式压制机
roll-type coiler 辊式卷取机
roll-type crusher 辊式破碎机
roll-type decorticator 辊式脱壳机
roll-type down coiler 辊式地下卷板机
roll-type electro-static separator 滚筒型静电分选机
roll-type feed mechanism 滚柱式进给机构
roll-type grader 辊式分级机
roll-type magnetic separator 辊式磁选机;辊式磁力分离器
roll-type rotary sprayer pump 滚柱式旋转喷雾泵
roll-type stripping unit 辊式摘棉铃装置
roll-type upcoiler 辊式地上卷取机
roll up 卷起;卷叠
roll-up door 卷升门;卷帘门
roll-up fire curtain 防火卷帘
roll-up hatch cover 卷叠式舱口盖
roll-up huge fund 积累大量资金
roll-up mechanism 卷布机构
roll-up screen 卷式银幕
roll-up structure 包卷结构;包卷构造
roll-up type solar array 卷式方阵
roll-up window 卷帘窗
roll wave 滚浪;滚进波;滚动波
rollway 滚木坡;溢流口;滚物道;滚水面;滚道;供料台
rollway bed 活动折叠床
roll-welded (monolayered) cylinder 卷焊式单层圆筒
roll welding 热辊压焊接;滚压焊(接);滚焊
roll wobbler 轧辊梅花头

roll wrapper 卷筒纸包装机
rolock 立砖;丁砌的砖;丁头砌面;竖砌砖
rolock arch 立砖发圈;竖砌砖拱
rolock-back 背面竖砌砖墙
rolock paving 竖砖铺砌
rolok 竖砌砖
Rolyat system 罗尔亚特系统(一种罐热水系统)
Roman acanthus leaf 罗马叶饰
Roman amphitheatre 古罗马角斗场
Roman arch 罗马式券;罗马式拱;半圆拱
Roman architecture 罗马建筑;罗马(式)建筑
Roman balance 提秤
Roman barrel vault 罗马筒形穹顶
Roman basilica 罗马式巴西卡利
Roman bath 罗马公共浴场
Roman beam 提秤
Roman brass 罗马黄铜
Roman brick 罗马(式)砖
Roman bronze 改良型锡青铜
Roman cement 罗马水泥;天然水泥
Romanche fracture zone 罗曼希破裂带
Romanche trench 罗曼希海沟
Roman column 罗马柱式;罗马式柱
Roman concrete 火山灰水泥混凝土;罗马混凝土
Roman Corinthian capital 罗马考林辛柱头
Roman Corinthian order 罗马考林辛柱型
Roman cross 罗马十字架
Roman Doric capital 罗马陶立克式柱头
Roman Doric order 罗马陶立克式柱式
Roman Doric temple 罗马陶立克神殿
Romanesque 罗马式
Romanesque architecture 罗马风格建筑;仿罗马建筑
Romanesque basilica of St. George on the Hradcany Hill at Prague 在布拉格赫拉德察尼山上的圣乔治罗马风格巴西利卡
Romanesque church 罗马风格教堂
Romanesque column base 罗马风格柱础
Romanesque leaves 罗马风格叶饰
Romanesque revival 罗马建筑形式的复兴
Romanesque school 罗马风格建筑学派
Romanesque sculpture 罗马风格雕饰
Romanesque style 仿罗马式;罗马式建筑
Romanesque transept 罗马风格侧厅
Romanesque vault(ing) system 罗马风格拱顶体系
Roman extension character set 带罗马字母的字符组
Roman figure 罗马数字
Roman foot 古罗马尺(相当于296毫米)
Roman fort 罗马要塞
Roman High Renaissance 罗马鼎盛期文艺复兴
Romanian architecture 罗马尼亚建筑
Romanian Register 罗马尼亚船级社
Romanian Register of Shipping 罗马尼亚船舶登记局
Roman imperial architecture 罗马帝国建筑
Roman Ionic capital 罗马爱奥尼亚式柱头
Roman Ionic order 罗马爱奥尼亚式柱式
Romanism 罗马式(建筑)
romanite 杂色琥珀
Romanium 罗曼铝合金;钨铜镍铝合金
Roman Law 古罗马法
Roman letter 罗马字体;罗马字母
Roman mosaic 镶嵌玻璃;大理石拼花地面;镶石铺面;罗马式镶嵌细工;罗马式马赛克;嵌花铺面
Roman mo(u)lding 罗马式线脚
Roman notation 罗马记数法
Roman numerals 罗马数字
Roman ogee bit 回旋双弯曲线铣刀;罗马S形铣头
Roman order 混成柱型;罗马柱型;罗马柱式
Roman order of architecture 罗马建筑柱式
Roman ornament 罗马浮雕装饰
Roman ovolo 罗马镘形饰
Roman ovolo mo(u)lding 罗马式凸圆线脚
Romanowsky's stain 罗曼诺夫斯基染剂
Roman palazzo style 罗马宫廷式
Roman pitch 罗马式屋面坡度(25°)
Roman Renaissance 罗马文艺复兴时期艺术
Roman Revival 罗马复兴式
Roman school 罗马学派
Roman shades 罗马遮光帘
Roman steelyard 提秤
Roman structure 罗马式结构

Roman style 罗马式
Roman theatre 罗马剧院
Roman thermae 罗马公共浴场
Romantic classicism 浪漫古典主义
romanticism architecture 浪漫主义建筑
Roman tile 单搭盖式瓦;罗马(式)瓦(仰瓦为槽形,覆瓦为筒形)
Roman triumphal arch 罗马凯旋门
Roman type 罗马字体
Roman wagon vault 罗马筒形穹顶
romarchite 黑锡矿
romeite 锑钙石
romer 坐标格网尺
romissory note 期票
Ronchigram 朗奇图
Ronchi grating 朗奇光栅
Ronchi ruling 朗奇刻线法
Ronchi test 朗奇测试法
rondada 车辙风
rondavel 圆形茅屋
rondel 圆片台玻璃
rondelle 铅镶窗的玻璃圆片
rondure 圆形;圆形物;优美的弧度
rone 苏格兰式屋顶天沟;檐沟雨水槽
rongalite 雕白块
rontgen equivalent physical 物理当量伦琴
rontgen-hour metre 伦琴小时米
rood 路得(英国面积单位,1路得=1/4英亩);十字架
rood altar (教堂靠在十字架隔屏上的)祭坛
rood arch 十字架隔屏中央的拱
rood beam 支承十字架梁;十字梁;十字架支承梁
rood loft 教堂中十字架坛;教堂阁楼
rood screen 教堂十字架围屏;十字梁
rood spire 教堂十字形平面顶上的屋尖
rood stair(case) 上十字架隔屏高台的阶梯
rood tower (教堂平面十字架交叉处的)塔楼
roof 箱顶(集装箱);炉顶;煤层顶板;屋顶;顶壁
roof accessories 屋面附属配件
roof aerial 屋顶天线
roof aerodrome 屋顶机场
roofage 盖屋顶材料;屋面材料
roof aggregate 屋面层集料;屋顶铺面集料
roof airdrome 屋顶机场
roof alumin(i)um 屋面铝材
roof anchorage 屋顶锚固
roof and wall structure 顶壁构造
roof-angle 脊角
roof-angle prism 脊角棱镜
roof antenna 屋顶天线
roof arch 拱形屋顶;屋顶拱
roof area 屋面面积
roof area cable duct 车顶电缆槽
roof area cable duct installation 车顶电缆槽安装
roof baffle 反射炉障板
roof bar 顶杆
roof bar stay 顶杆撑
roof base sheeting 屋顶基层片材
roof batten 挂瓦条;屋面木条;屋面挂瓦条
roof beam 桁条;屋面梁;屋顶梁
roof beam grillage 屋顶纵梁系
roof bearer 垫块;屋顶托架
roof bed 顶板层
roof bent 屋架
roof bit 顶板钻头
roof block 炉顶砌块;焦炉顶砌块
roof board(ing) 顶蓬;屋面板;屋顶板;屋架板;望板
roof boiler room 屋面锅炉房
roof bolt 顶栓;杆柱;隧道洞顶锚栓;洞顶锚栓;顶板栓杆;护顶锚杆
roof-bolt drill 顶板锚杆钻机
roof-bolter 杆柱机;锚杆机
roof-bolt head 洞顶锚栓头;顶板锚头
roof bolting 隧道顶钎;顶板锚杆支护;隧洞顶栓;隧洞顶钎
roof-bolting jumbo 顶板栓锚(眼)用钻车;杆柱孔钻车;锚杆孔钻车
roof-bolting machine 顶板锚栓安装机;杆柱安装机
roof-bolting method 洞顶锚杆法;岿顶栓法
roof-bolting stopper 杆柱钻孔机;顶板锚杆钻机
roof bolt supporting 顶板锚杆支撑
roof boss 屋顶凸饰
roof bow 顶梁;车篷弓
roof bracing 屋盖支撑;屋面支撑

roof bracing system 屋面支撑系统
roof break 顶板裂隙;顶板破裂;顶板断裂;顶板开裂(隧道)
roof brick for electric(al) arc furnace 电弧炉炉顶砖
roof bushing 屋顶线套管
roof cantilever 屋顶悬挑
roof car 屋顶小车
roof carbon 炉顶石墨
roof car park 屋顶停车场
roof caving 顶板陷落
roof chimney 屋顶烟囱
roof chip(ping)s 绿豆砂;铺屋面石屑
roof cladding 屋面覆盖层;屋面材料
roof cladding element 屋面覆盖层
roof cladding metal sheet 铺屋面金属板
roof cladding with roll roofing 用屋面卷材覆盖的屋顶
roof clay 隔水黏土;不透水黏土
roof cleaning 清理顶板
roof clippy 坡屋顶脚手托架
roof coal 留顶板煤
roof coating 屋面层;屋面涂层;屋面涂料;屋面漆
roof collapse 屋顶塌落
roof column 屋顶支柱
roof comb 条脊(沿屋脊砌墙);屋脊饰
roof component 屋面构件
roof concreting 混凝土浇筑屋顶
roof condenser 顶式冷凝器
roof conductor 屋面避雷装置
roof construction 屋盖构造;屋顶构造
roof continuous over three spans 连续三跨度屋顶
roof contour 屋顶轮廓
roof control 顶板控制;顶板管理;地压控制
roof control by wall-shape pillar 煤柱支撑顶板法
roof control drill 顶板锚栓孔钻机
roof cork 屋顶软木
roof cornice 屋面檐板
roof covering 瓦面;屋面覆盖层;屋面;屋架覆盖;屋顶防水层
roof covering sheet-metal 屋顶的金属覆盖层
roof covering with asphalt ready roofing 屋面盖以预制沥青材料
roof covering with roll roofing 用屋面卷材覆盖的屋顶
roof crack 顶板裂缝
roof cradle 屋板托架
roof crane 屋面吊;屋顶起重机
roof crest(ing) 脊饰;屋脊装饰;屋脊饰;条脊(沿屋脊砌墙)
roof cross-section 屋顶剖面
roof curb 屋面边缘
roof curb tile 屋面边饰瓦;屋顶边缘瓦
roof curtain 屋顶隔火屏
roof deck(ing) 屋顶平台;屋顶层面;屋面板;预制板;预制屋顶板;屋面(盖)板;上人平屋顶(指屋顶的一部分)
roof decking panel 屋顶(平台)板
roof decking sheeting 屋顶平台薄板
roof decoration 屋顶装饰
roof decorative feature 屋顶装饰特点;屋顶装饰部件
roof decorative finish 屋顶装饰加工
roof design 屋顶设计
roof displacement 屋顶位移
roof dome 屋顶穹窿
roof dormer 老虎窗;屋顶窗
roof dormer covering 屋顶天窗盖板
roof drain 雨水斗;屋顶排水口;屋顶排水沟
roof drainage 雨水斗;屋面排水;屋顶排水
roof drain strainer 雨水斗罩;雨水斗筛网
roofed 有屋顶的
roofed berth 顶棚式船台
roof edge 屋顶边缘
roofed ingle 壁炉墙角
roofed in wood 木制屋盖
roofed spectator's stand 有屋顶的观众席
roofed walk 带顶的人行道
roof enrichment 屋顶富丽装饰
roofer 屋面工;望板
roofer felt (铺屋面的)油毛毡;屋面油毡
roofer nail 屋顶工用的钉子
roofer pitch 铺屋顶用硬沥青
roofer's mortar 屋顶用砂浆

roofer's scaffold(ing) 屋顶脚手架
roofer work 屋顶工工作
roof exhaust unit 屋顶排气装置
roof expansion joint 屋顶伸缩缝
roof expansion joint cover 屋顶伸缩缝盖板
roof extension 伸顶通气管
roof extract unit 屋顶排尘器
roof extract ventilator 屋顶抽风机
roof failure 冒顶
roof fall 顶板崩落;屋顶塌落;冒口;冒顶;坍顶;顶板陷落;(隧洞等的)掉顶
roof-fall accident 帽顶事故
roof fan 屋顶(排)风扇
roof felt manufacture 屋顶油毡工厂;屋顶油毡制造
roof fill insulation 屋顶绝热填料
roof filter 修平滤波器
roof fissure 顶板裂缝
roof flange 屋面管套
roof flashing 屋面防雨板;屋面披水板;屋顶泛水
roof flashing block 屋面防雨装置;屋面披水装置
roof flashing method 屋面防雨方法;屋面披水方法
roof flashing strip 屋面防雨片;屋面披水片
roof floor 屋顶面
roof foil 屋面(金属)箔层
roof foundering 岩浆坍顶作用
roof frame 屋架
roof frame strap 屋架的金属环箍
roof framing 屋架;屋盖构造;屋顶框架
roof framing system 屋盖构造体系
roof gallery 屋顶走廊
roof garden 屋顶(小)花园
roof garden bar 屋顶花园酒吧
roof gardening 屋顶花园
roof garden planting 屋顶花园绿化;屋顶花样培植
roof garden restaurant 屋顶花园餐厅
roof gate 人字闸门;屋顶式闸门
roof girder 屋顶大梁;屋面梁;屋顶梁
roof glazing 屋顶玻璃窗;天窗
roof glazing bar 天窗(窗扇)芯子;天窗拉杆
roof gravel 屋顶卵石
roof greening 屋顶绿化
roof guard 屋面防护栏;屋顶防护板;屋顶挡雪板
roof gutter 屋檐水槽;屋顶排水沟;屋顶落水沟;天沟
roof gutter heating 屋顶檐沟采暖
roof hatch 屋顶有盖出口;屋顶出入孔
roof heat 屋顶热
roof-heating 顶部加热
roof heat insulation 屋顶隔热
roof helicopter airport 屋顶直升机停机场
roof heliport 屋顶直升飞机停机场
roof hoist 屋顶起重机
roof hole 顶板炮孔;顶部钻孔
roof hook 屋顶钩状物;屋顶吊钩
roof-hung ductwork 悬挂于屋顶上的管道作业
roof-hung load 屋顶悬吊的荷载
roof hydrant 屋顶消防栓
roof inclination 屋顶斜度
roofing 盖顶;屋面(材料);屋脊型;屋面
roofing accessories 屋面附件
roofing alumin(i)um 铺屋顶铝材
roofing application 屋面铺数;屋面施工;屋面装修
roofing asphalt 屋面柏油;铺屋沥青;屋面沥青
roofing asphaltic bitumen 屋面柏油沥青
roofing batten 屋顶木条
roofing beam 挂瓦条;屋架拉梁
roofing bitumen 屋面沥青
roofing board 望板
roofing bond 屋面维修保证书
roofing bracket 屋面托座;屋面施工用支架
roofing cement 屋面黏结剂;屋面水泥
roofing chip(ping)s 屋面小碎石
roofing contractor 屋面承包商
roofing copper 屋面铜片;复面铜板
roofing cost 屋面工程造价
roofing design 屋顶设计;屋面设计
roofing fabric 屋顶织布;油毡基布;屋面纤维质材料
roofing fastener 屋面固定构件
roofing fastening 屋面固定
roofing felt 油毡;沥青油毡;屋面(油)毡;屋面(油)毛毡;屋面(油)毡卷材;屋面毡;屋面(油)毛毡
roofing felt adhesive 屋面油毡胶黏剂
roofing felt layer 屋面油毡铺设工
roofing fixing accessories 屋面固定附件
roofing flux 屋顶柏油

roofing foil 屋面金属箔;屋面卷材
roofing glass 园艺用玻璃;玻璃瓦
roofing granules 屋面砂砾;屋面粒渣;屋面颗粒材料
roofing hammer 独角锤
roofing hammer with tubular handle 管式柄独角锤
roofing insulation 屋面绝缘;屋面绝热
roofing insulation panel 屋面隔热板
roofing iron 屋面(薄)铁皮;屋顶铁皮;覆盖铁
roofing jib 截顶槽截盘
roofing lead sheet 屋面铅皮
roofing malthoid 屋面油毡
roofing mat 屋面毡
roofing material 屋面材料
roofing membrane 屋面油毡防水层;屋面片材;屋面薄膜
roofing metal 屋面金属材料
roofing millboard 屋面麻丝板
roofing nail 屋面(用)螺钉;油毡钉
roofing paint 屋顶漆
roofing paper 纸底油毛毡;屋面防潮纸;屋面油纸;屋顶油纸
roofing paste 屋面浆料
roofing pitch 屋面坡度;屋面斜坡
roofing plume 屋脊型烟羽
roofing preservative 屋面防腐剂
roofing preserver 屋面保护物
roofing product 屋面材料
roofing putty 屋面油灰
roofing rags 屋面石板瓦
roofing sand 屋面铺砂
roofing saturant 制屋面卷材的沥青
roofing shaft 屋面通风井
roofing sheet 顶板;屋面板材;屋顶薄钢板
roofing sheet-metal 屋面金属薄板
roofing shingle 屋顶木瓦板;屋顶纸板
roofing skin 屋面表层
roofing slab 屋面板
roofing slag 屋面炉渣
roofing slate 铺屋面石板;屋顶板;瓦(用)板岩;石板瓦;屋面石板瓦
roofing slate deposit 瓦板岩矿床
roofing square 木工用钢曲尺;屋顶用直角尺;屋面方块钢板;平方数(指屋面材料)
roofing substructure sheeting 屋面基础片材
roofing system 屋面体系
roofing terracotta 屋面琉璃瓦
roofing tile 屋顶瓦;屋面瓦
roofing tile factory 屋面瓦工厂
roofing tile tester 屋面瓦检验机
roofing trade 屋面行业
roofing washer 瓦形垫圈
roofing waterproofing joint 屋面防水接缝
roofing waterproofing system 屋面防水系统
roofing waterstop 屋面挡水条
roofing work 屋面工程
roof in hollow tile 空心砖屋顶
roof in Stahlton 预应力黏土屋面
roof insulating screed 屋面隔热找平层
roof insulating strip 屋面隔热条
roof insulation 屋面隔热材料;屋面隔热层;屋面保温层;屋面绝缘层;屋面隔热;屋面绝热
roof insulation board 屋面绝缘板;屋顶隔热板
roof intersection 屋顶的交接
roof jack 液压顶柱;临时钢支柱;千斤顶支柱;顶部支柱;顶板螺旋千斤顶
roof jib 屋顶吊杆
roof joint 顶板连接;屋面接缝
roof joist 屋顶格栅;屋顶托梁;屋顶龙骨
roof joists laid to falls 屋顶龙骨找坡
roof lacqer 屋面天然漆
roof lacquer 屋面腊克
roof ladder 人字梯;屋顶爬梯
roof ladder hook 屋顶梯吊钩
roof lagging 顶板背板
roof lath 挂瓦条;屋面木条
roof lathing 屋顶金属网;屋顶板条
roofless 没有屋顶的;无顶的
rooflet 小屋顶
roof level 屋顶高度
roof lift 炉顶升降机
roof light 舱顶灯;顶篷灯;车顶灯;屋顶窗;采光天窗;采光屋顶;天窗
roof-light fittings 屋顶天窗配件

roof-light flashing 天窗泛水
roof lighting 屋顶采光
roof lighting sheet 采光屋面板
roof-light opening 屋顶天窗开口
rooflight sheet 采光屋面构件；采光屋面片材；屋顶透明板；采光屋面板；平天窗采光片；侧天窗石棉水泥薄板
roof limb 顶翼
roof line 屋顶线；屋顶轮廓线；车顶线
roof lining 屋顶内衬
roof live load 屋顶活荷载；屋面活荷载
roof load 试验荷载；弹性极限荷载；屋面荷载；屋顶荷载
roof loading 屋顶荷载
Roofloy 耐蚀铅合金
roof mast 屋顶桅杆
roofmaster 金属顶梁
roof material 屋顶材料
roof member 屋顶构件
roof membrane 屋面薄膜
roof membrane system 屋面薄膜防水系统
roof monitor 屋顶监测器
roof-mounted flagpole 安装在屋顶上的旗杆
roof-mounted pantograph 车顶集电弓
roof mounted window washing system 屋顶擦窗系统
roof nail 屋面板用钉
roof of caving 冒口
roof of fire-bolt 火箱顶板
roof of gallery 矿坑顶板
roof of prefabricated elements 预制构件屋顶
roof of tunnel 隧道顶板
roof of vehicle 车顶
roof openings 油罐顶上的开口
roof ornament 屋顶装饰
roof ornamental finish 屋顶装饰处理
roof ornamental feature 屋顶装饰特征
roof outlet 有隔栅的水落管（平屋顶）；屋面排水口
roof overhang 屋顶悬挑；挑檐
roof overhung door 挑檐
roof paint 屋顶漆
roof pane 屋顶网格
roof panel 屋面板；车顶天面
roof paper 油毡
roof parapet slab 屋顶女儿墙板
roof parapet wall 屋面女儿墙
roof parking 屋顶停车（场）
roof parking area 屋顶停车场
roof parking deck 屋顶停车坪
roof paste 屋顶涂料
roof paulin 篷布
roof pendant 顶棚下垂物；顶棚下垂体；顶垂
roof pergola 屋面藤架；屋面凉亭；屋顶花架
roof periodic(al) caving 顶板周期垮落
roof-pinning jumbo 顶板锚栓钻车；杆柱孔钻车
roof pitch 铺屋顶硬沥青；屋顶坡度；屋顶高跨比
roof plan 屋面平面图；屋顶平面
roof plank 屋顶木板
roof plate 檐檩；檐板；承椽板；屋面盖板；承梁板
roof playground 屋面游戏场
roof plumbing 屋顶管道（系统）
roof plummet 屋顶铅垂器
roof-point 顶板点
roof pole 屋顶天线杆
roof ponding 屋面积水
roof pour(ing) 屋面浇筑
roof-powered platform 屋顶小车驱动的工作平台
roof preservation coat 屋面保护层
roof preserver 屋面保护物；屋面保护装置
roof pressure 屋顶压力；顶板压力
roof pressure arch 冒落拱；顶板压力拱
roof principal 屋架
roof prism 屋脊（形）棱镜
roof projection 摇檐
roof promenade 屋面阳台；屋面散步廊
roof protrusion 屋面突出物
roof purlin(e) 屋顶檩条；屋面桁条
roof rafter 屋面椽条
roof rail 顶部桁材；车顶纵梁
roof railing 屋面栏杆
roof repair 屋顶修理
roof rib 屋顶肋条
roof ridge 屋脊
roof rock 顶盖岩；顶板岩石

roof rock seal 盖层封闭
roof rolls 屋面卷材
roof roofer 屋顶铺盖工
roof saddle 鞍形屋脊
roof sampling 屋面采样
roofscape 屋顶景观
roofscaping 屋顶庭园
roof screed material 屋面找平材料
roof screed topping 屋顶找平面层
roof screen 屋顶隔火山墙
roof scrubber 仓顶除尘器
roof scuttle 屋顶天窗；屋顶出入孔
roof seal(ing) 屋顶封顶
roof sealing strip 屋顶密封条
roof set flagpole 竖立在屋顶上的旗杆
roof shape 屋顶形状
roof-shaped crown 屋顶式路拱
roof sheathing 屋面板；屋顶面；屋顶盖板；望板；承梁板
roof sheathing lumber 屋顶望板木材
roof sheathing metal sheet 屋面金属片材
roof sheathing paper 屋顶油毛毡；屋面柏油纸
roof sheathing plywood 胶合板屋顶望板
roof sheet （集装箱的）顶板；屋面板材
roof shell 屋顶壳体
roof shield 半盾构
roof shingle 屋面瓦
roof shop 屋顶商店
roof side sill 车顶侧梁
roof side wall 顶部池墙
roof sign 屋顶标志；屋顶标牌
roof slab 屋面板；屋架板；屋瓦
roof slab in Stablton 预应力黏土屋顶板
roof slate 屋面石板瓦
roof slater 屋面工
roof slating 铺石板瓦；屋顶铺石板瓦
roof slope 屋顶坡度
roof soffit 屋顶底面；屋顶拱腹
roof sounding 问顶
roof space 屋顶空间
roof spalling 顶板剥落
roof span 屋架间距；屋顶跨度
roof spire 屋顶尖塔
roof staging 屋顶脚手架
roof stanchion 屋顶支柱
roof step 顶板阶梯
roof stilted upon suspension cable 支承在悬索上的屋面
roof stone 屋顶板岩石
roof strainer 雨水斗罩；纵护顶背板
roof strain indicator 顶板应变仪（测矿井坑道用）；顶板应变指示器
roof straw 屋顶稻草；屋顶茅草
roof strength test 箱顶强度试验（集装箱）
roof structure 屋顶结构
roof structures of timber 各式木屋顶结构
roof structure to falls 屋顶结构层坡
roof substrate sheeting 屋顶垫片
roof substructure sheeting 屋面下部结构片材
roof support 顶板支架；屋面支承
roof-supported wire rope conveyor 悬吊在顶板上的钢绳支架输送机
roof surround 屋顶外围层
roof suspension bolt 屋顶悬挂螺栓
roof system 屋顶系统；屋面系统；屋架系统
roof tank 屋顶水箱；屋顶水池
roof tapping 问顶；敲顶
roof terminal 屋面盖板
roof terminal line 屋顶通风口
roof terrace 屋顶阳台；屋顶平台
roof thrust 顶板逆冲断层；顶板冲断层
roof tie 屋架拉杆
roof tile 屋顶瓦, 瓦
roof tiler 屋面瓦工；铺瓦工
roof tiling 屋顶铺瓦
rooftilling work 屋面铺瓦工程
roof timber 屋顶（用）木材
roof-to-floor convergence 顶底板会合
roof-top 屋脊的；屋脊状的
roof-top air conditioner 屋顶（式）空调器
roof-top air conditioning unit 屋顶式空调器
roof-top car park 屋顶停车场
roof-top equipment 屋顶设备
roof-top heliport 屋顶直升机降落场

roof-top promenade 屋顶散步地坪
roof-top swimming pool 屋顶游泳池
roof-top tennis court 屋顶网球场
roof-top terrace garden 屋顶花园平台
roof trackway 屋顶跑道
roof tree 脊杆；脊梁；屋宇梁；栋梁
roof trellis 屋顶花架
roof truss 屋顶桁架；人字桁架；屋架
roof type 屋顶类型
roof valley 屋顶排水沟；天沟
roof valley clay tile 屋谷黏土瓦；斜沟陶土瓦
roof valley post 屋谷柱
roof valley tile 屋谷瓦；屋面天沟瓦
roof vault 拱顶屋顶
roof vent 屋顶排气口；屋顶通气装置
vent vent block 屋顶通风砌块
roof ventilater 温室的天窗
roof ventilating tile 屋顶通风砖
roof ventilation 屋顶通风
roof ventilator 屋顶风机；屋顶通风器；屋顶通风机；屋顶风帽；车顶通风器
roof ventilator block 屋顶通风器砌块
roof void 屋顶孔隙
roof walk 屋顶便道
roof walkway 屋顶走道
roof water 屋面水
roof weir 熊阱堰；屋顶式堰；屋顶式坝
roof weir gate 屋顶式（堰）闸门；进式熊阱堰闸门
roof wetting party 屋顶上梁庆宴
roof wire guard 顶板防护网
roof with valley 带天沟屋顶
roof woodwork 木屋架；屋顶木工活
roof work 屋顶工程；屋顶工作
roof worker 建筑屋顶工人
roof work safety lock 车顶安全作业锁
rookery 破旧的住房；贫民窟
rookery clearance 贫民窟的清理
rook heat of mine 矿井岩温
room 余地；煤房
room absorption 房间吸声（量）
room acoustic(al) assessment 室内音质评价
room acoustic(al) characteristic 房间声学特性
room acoustics 室内声学；房间音质
room air conditioner 室内空气调节器；室内空调器；房间空（气）调（节）器
room air conditioning system 室内空调系统
room air cooler 室内空气冷却器
room air heating 室内采暖炉火
room air heating burning appliance 室内空气加热燃烧器
room air heating furnace 室内采暖炉
room air temperature 房间空气温度
room and pillar caving 房柱式采煤法；柱式崩落开采法
room and pillar mining 房柱开采
room and pillar system 房柱式采煤法
room and space 肋距
room antenna 室内天线
room area 室内面积
room background 室内本底
room blower 室内吹风机
room brightness 室内亮度
room brilliancy 室内亮度
room-by-room 每个房间
room capacity 房间容积
room circuit 室内电路
room climate 室内气候
room constant 室内常数；房间常数
room containing the source （用以判断声级D的）声源室
room conveyer 矿房运输机；煤房运输机
room cooler 室内冷却器；房间冷却器
room cooling 房间制冷
room-cooling unit 房间制冷单元
room count （住宅的）房间数
room crosscut 通风联络小巷
room curing 室温固化
room dew point 室内露点
room divide 房间分隔
room-dividing 房间分隔
room door 房门
room door lock 房门锁
room-dry 室内风干的
room dryer 室式干燥器；室式干燥机

room enclosing area 室内净面积
roomer 家庭房间承租者
roomette 卧铺车厢小包房;小房(间)
room finish 房间装修
room finish schedule 室内做法表;房间做法表
room for branches to building 房屋电力引入线室
room for drinks 酒吧;饮酒间
room for rent 出租房间
roomful 一屋子的人;满房间
room-habitable 适宜居住的房间
room heat 室内热量
room heater 室内暖气;室内采暖器;房间取暖器;小型采暖炉
room heating 房间采暖
room heating appliance 房间采暖装置;房间采暖设备
room heating strip 房间采暖片
room-high 房间等高墙板;房间高度
room-high window 房间等高窗
room humidity 室内湿度
room humidity sensor 房间湿度探测器
room illumination 室内照明
room index 室形指数
rooming house 出租住处;公寓;寄宿房屋;出租宿舍
rooming occupancy 住房;卧室
room insolation 居室日照
room lamp 室内灯
room-layout 房间布置;室内平面布置
room light 室内灯
room lighting 房间照明
room-light loading 室光装入
room model 室内模型
room moisture 室内湿度
room neck 洞室口;巷道;进入硐室的巷道;煤房颈
room noise 室噪声;室温噪声;室内噪声;室内噪声
room noise criterion curve 室内噪声标准曲线
room noise spectrum 室内噪声谱
room occupant 房间住户
room of maneuver 回旋余地
room parameter 室内参数
room response 室内频率响应
room reverberation 室内混响
room-scattered 从房间墙壁散射的
room-scattered neutron 室内散射中子
room scattering 室内散射
room-sealed appliance 封闭式燃具;封闭式取暖装置;室内封闭式燃具
room-sealed boiler 封闭式锅炉
room-sealed space heater 封闭式采暖炉
room separation wall 房间隔墙
room service 客房服务(部);房内服务
room-size 房间大小
room sized unit 房间尺寸单元
room sound attenuation 室内声响衰减
room sound field 室内声场
room sound insulation 室内声响绝缘
room spray-type humidifier 室内喷湿式增湿器
room steam-curing 室内蒸汽养护
room temperature 室温;室内温度;常温;房间温度
room temperature ag(e)ing 常温时效;室温时效
room temperature comfort curve 室温舒适曲线
room temperature conductivity 室温电导率
room temperature control 室温控制;房间温度控制
room temperature control system 房间温度控制系统
room temperature cure 室温固化;常温固化
room temperature gluing 室温黏结;常温黏结
room temperature noise 室温噪声
room temperature property 室温性能
room temperature sensor 室内温度探测器
room temperature setting 室温硬化
room temperature setting adhesive 凝固胶;室温硫化硅橡胶;室温固化黏合剂;室温固化胶黏剂
room temperature setting phenolic resin adhesive 常温固化型酚醛树脂黏合剂
room temperature stability 室温稳定性
room temperature storage 室温储藏
room temperature strength 室温强度
room temperature superconductor 室温超导体
room temperature vulcanising 室温硫化
room temperature vulcanization 室温硫化
room thermostat 室内恒温计;室用恒温器;室内恒温器
room-to-let 出租房间

room unit 房间单元
room unit air conditioner 房间单元空调器
room utilization factor 房间利用系数
room ventilation 室内通风
room wall 室内墙
room warming 房间采暖
roomy 有宽敞空间的;宽敞的
roomy body 大容量车身
roomy middle 中躯宽深
rooseveltite 砷铋石
roo-shor 罗索柱(一种可调的支柱)
roost 喧嚣潮汐急流;鸡舍;避风停泊处
rooster-tail 船尾急流
root 根值;根数;根部;根;实码头接岸端;方根;坝根;开平方
root amplifier 开方放大器
root and rock rake 树根和石块清除器;树根和块石抓机
root angle 伞齿轮底角;齿根锥角
root apiex 根尖
root ball 根团
root band 根带
root bend specimen 根部受拉的弯曲试样
root bend test 弯折试验;反向弯曲试验;(焊缝的)根弯试验;根部弯曲试验;根部受拉的弯曲试验;(焊缝的)背弯试验
root bent test 反面弯曲试验
root canal 根管
root canal cleaner 根管清扫剂
root canal depth ga(u)ge 根管长度测量器
root canal dryer 根管干燥器
root canal file 根管锉
root canal file with long handle 长柄根管锉
root canal filler 根管充填器
root canal filling 根管充填
root canal irrigation 根管冲洗
root canal overfilling 根管超填术
root canal plugger 根管充填器
root canal ramification 根管分支
root canal reamer 根管扩孔钻
root canal resinifying agent 根管塑化剂
root canal sealant 根管封闭剂
root cap 根冠
root carving 根雕
root cause 根本原因
root cellar 储藏根块植物的窖
root circle (齿轮的)齿根圆
root clay 根土岩;底黏土
root climber 根攀植物
root collar 根颈
root concentration 根的集中度
root cone (指伞齿轮的)齿根锥
root constant 根常数
root-cooled blade 根部冷却叶片
root crack 根部裂纹;焊根裂纹;焊缝根部裂纹
root cut-and-cleaner 块根切洗机
root cutter 切根机;锄根机
root cutter with cleaner 带清理器的块根切碎机
root cutter with cleaning cage 带清理笼的块根切碎机;带清理滚笼的切根机
root cutting 根插(条)
root cylinder 齿根圆柱面
root defect 根部缺陷
root deposit 脉石矿床
root diameter 根(部)直径;内螺纹外径;齿根(圆)直径
root digger 块根挖掘机
root directory 根目录
root distance 根距
root distribution 根系分布
rootdozer 除根机;掘土机
root edge 接边焊缝;焊缝根部边缘;底缘
rooted graph 有根图
rooted tree 根树
root elevator 根挺
root end of groyne 丁坝靠岸段
rooter 挖土机;路犁;除根犁;除根机;翻土机;拔根器;拔根机
rooter attachment 翻土机附件;除根机附件
rooter bracket 翻土机架;除根机悬挂装置
rooter carriage bar 翻根机支座
rooter kick-out 除根机出料;翻除根机出料
rooter line 除根机开行路线
rooter linkage 翻土除根机联动装置

rooter plough 掘根犁
rooter plow 掘根犁;挖根犁;除根犁
rooter shank 除根机锚杆;翻土机锚杆
rooter tooth 除根机轮齿;翻土机轮齿
Roote's blower 罗兹式鼓风机
Roote's energy absorbing fender system 罗兹吸能护舷系统
root expansion 根系扩张
root extractor 挖根机;拔树根器
root exudate 根渗出物
root face 焊缝坡口钝边;坡口钝边;齿根面
root face height 钝边高度
root face of weld 焊缝根部钝边
root field 根域;伽罗瓦域
root fillet 齿根圆角
root fixing 叶片连接
root-forming substance 成根物
root furcation 根分叉部
root gap 根部间隙
root gap of weld 焊缝根部间隙
root grafting by approach 根靠接
root groove 焊根坡口;焊根坡口
root-hair zone 根主区
root hole 根孔;树根孔洞
root hook 超挖拔根钩
rooting 加固;求根;使固定;开平方
rooting-in of blades 叶片安装
rooting machine 除根机
rooting medium 扎根介质
rooting mixture 补强混合物
rooting zone 生根带;根际
root key 开方计算键
root laboratory 根实验室
root length 根长
rootless intrafolial folds 层内无根褶皱
rootlet bed 根系层;煤层底黏土
rootlets 支根
root lifter 块根挖掘机
root line 齿根线
root loader 装块根机
root locus 根轨迹
root locus diagram 根轨迹图
root locus method 根轨迹法
root locus plot 根轨迹图示
root mass 根群
root mat 根簇
root mean error 均方误差
root mean intensity 均方烈度
root mean square 均方根
root mean square abnormal 均方根异常
root mean square amplitude 均方根振幅值
root mean square criterion 均方根(误差)准则;均方根判据;均方根判别准则
root mean square current 有效电流
root mean square deviation 均方根偏差;均方差;平均平方数离差;标准差
root mean square displacement 均方根位移
root mean square error 均方根(误)差;均方差;标准差
root mean square height 均方根高度
root mean square horsepower 有效马力;有效功率
root mean square horsepower method 均方根马力法
root mean square intensity 均方根烈度
root mean square load 均方根负载
root mean square meter 均方根测量仪
root mean square noise 有效值噪声
root mean square noise voltage 噪声的有效电压
root mean square percent error 均方根百分误差
root mean square power 方均根功率
root mean square prediction error 均方根异常
root mean square pulse duration 均方根脉冲持续时间
root mean square quantity 有效值
root mean square question 均方根提问
root mean square rectifier 均方根检波器
root mean square residual 均方差
root mean square roughness 均方根粗糙度
root mean square simulation error 均方根模拟误差
root mean square sound pressure 有效值声压
root mean square value 均方(根)值;方均根值
root mean square velocity 均方根速度;方均根速度
root mean square velocity gradient 均方根流速

梯度
root mean square wave height 均方根波高
root menu 主菜单;根菜单
root mesh mark 齿轮齿根啮合印痕
root neck 根颈
root node 根节点
root nodule 根瘤
root of a central tube 中心管的根部
root of blade 叶片根部
root of contract 合同根基
root of equation 方程的根
root of flame 火根
root of groyne 丁坝根部
root of joint 接头根部
root of meck 瓶颈根部
root of mountain 山根
root of needle 针根
root of nose 鼻根
root of thread 螺纹牙底
root of title 业权基础
root of tooth 齿根
root of weld(ing) 焊根;焊接(部);焊缝根部
root opening 根部间隙
root opening of weld 焊缝根部间隙
root out 铲除
root parasite 根寄生物
root pass 根焊;根部焊道
root pass welder 根部焊接机
root pass welding 根部焊接
root path 根路径
root penetration 根部焊透层;根部熔深
root phase 基本阶段
root pile 基桩;树根桩
root plow 挖根机
root plugger 根管充填器
root pressure 根压
root-proof 防树根破坏
root pruning 断根
root radius 缺口半径;坡口半径;齿根半径
root radius of weld 焊缝坡口根部半径
root rake 树根耙;搂根耙;除根齿耙
root reinforcement 根部加固(焊接)
root relief 齿根修正
root resistance 耐生根性
root-resisting 根部防护的(焊接)
root room 根系分布区
root rot 根腐
root run 根部焊缝;根部焊道;焊根焊道
root run welding 根部焊接
Roots blower 罗茨鼓风机;双转子鼓风机
Roots blower pump 罗茨增压泵
root sealer bead 根部焊道
root seat 根土岩
root section 螺纹根部剖面
root segment 主段;根基段;根段;根部;基本段;常驻段
root segment addressable area 根段可寻址区
root shavel 挖根铲
Roots high vacuum pump 罗茨高真空泵
Roots meter 转子式煤气表;罗茨式煤气表
roots of large plant 高大植物的根系
root solver 求根解算器
root spread 根系分布区
Roots pump 罗茨泵
root square deviation 均方根差
root-squaring method 根方法;根的平方法;平方根法
root stock 根状茎
rootstock 砧木;根桩;根砧木
Roots type flowmeter 罗茨流量计
Roots type gas exhauster 罗茨排气机
Roots type super-charger 罗茨增压器
root sucker 根出条
root sum square 和的平方根
root-sum-square of three-dimensional method 三维平方和开方方法
root-sum-square value 方和根值
root support 支柱根
root swelling 根部突起;根部膨大
root system 根系
root system branches 根系分支
root system of plant 植物根系
root test 根值判别法;根检验法
root timber 原木;根材

root-time law 时间平方根定律
root tip 根尖
root topper 树根挖掘器
root top ratio 地上地下比率
root toroid 齿根超环面
root tubercle 根瘤
root van 整体式箱形货车
root vegetation 根菜类蔬菜
rootwedging 根劈作用
root weeds 根生杂草
root window 根窗口
root zone 根系层;根带;山根带
roove clinker nail 紫铜钉
ropac 侧立冰
ropak 竖立冰
rope 钢丝绳;干扰雷达用的金属带;干扰雷达用的长反射器;缆;绳索
rope action 缆绳作用
rope and belt conveyer 绳带式运输机
rope-and-button conveyer 绳和小盘输送机
rope and drop-pull method 钢丝绳冲击钻进法
rope and pulley type 绳索滑轮式
rope angle ga(u)ge 绳索倾斜计
rope-arranging device 排绳器
rope attachment 绳头冲持装置
rope bag paper 麻浆纸袋纸
rope-band screen 索带筛
rope belt 排绳传送带;绳带
rope binding hook 束绳钩
rope block 滑车;滑轮
rope boring 钢绳钻进;钢丝绳冲击钻进
rope braiding machine 编绳机
rope brake 绳索制动器;绳测功器
rope breaking strength test 钢丝绳断裂强度试验
rope breaking strength tester 钢丝绳断裂强度试验计
rope bridge 吊桥;缆索桥;索桥
rope bucket elevator 钢绳斗式提升机
rope cager 绳索装罐机
rope cantilevering roof 用缆绳悬吊的屋顶
rope capacity 钢绳长度;绳索牵引能力;可缠绕的绳索长度
rope cappel 绳卡
rope capping 钢丝绳和提升容器连接装置
rope capping metal 钢丝绳封头合金;缆索封头(用)合金
rope carrier 支索器
rope catching device 绳索夹具
rope caulk 绳状嵌缝条
rope changer 换绳装置
rope chemicking machine 绳状轧漂机
rope chopper 钢丝绳切断器
rope circuit 循环索
rope clamp 钢索夹(子);钢丝绳夹卡;钢绳卡
rope clip 钢索夹(子);钢丝绳夹卡;索夹;绳夹;抱索器
rope clutch 接绳器
rope coil 钢丝绳捆
rope control 绳索操纵
rope-controlled 绳索控制的
rope-controlled mechanical lift 绳索控制的机械式起落机构
rope control unit 索控式装置
rope conveyer[conveyor]索绳输送机;绳索运输机
rope core 钢丝绳绳芯
rope coupling 绳索连接;绳接头
rope crab 钢丝绳起重绞车
rope cracker (冲击钻杆与钢绳之间的)短节棕绳
rope cramp 钢丝绳夹
rope creep compensator 松绳补偿器
rope crossing 溜索
rope crowd(ing) 绳索推进(装置)
rope curve 绳索垂度
rope cutter 斜井上下把钩;钢(丝)绳割刀
rope dead end 绳索(终端)固定
roped hydraulic driving machine 缆式液压传动机械
roped hydraulic elevator 缆式液压升降机
rope drift 绳索运输道
rope drill(er) 钢绳冲击(式)钻机;索钻机;索钻器
rope drilling 缆索打井法;缆式掘进机;缆式打井法;索钻(法)
rope drilling bit 缆索式钻头

rope drilling method 钢绳冲击钻进(方)法
rope drive 钢索传动;钢丝绳传动;缆索传动;绳索传动
rope-driven belt conveyer 钢绳牵引带式输送机
rope driven travel(l)ing crane 钢丝绳移动起重机
rope driving 绳索传动
rope drum 绕绳筒;钢绳卷筒
rope dynamometer 绳索测力计
rope end 索端
rope extension 绳索接长
rope factor 提升钢丝绳系数
rope family 缆索系列;缆索类
rope fender 绳碰垫
rope ferry 缆索浮桥;索拉渡船;绳渡
rope fiber 缆索纤维
rope figure 绳纹
rope flat roof 缆索(吊)平顶
rope fork drive 缆索叉型驱动(装置)
rope frame (belt) conveyer 钢绳吊挂胶带输送机;缆架式皮带运输机
rope furnishing tool 帆缆作业用具
rope ga(u)ge 绳索卡尺
rope gearing 钢绳传动装置;绳传动装置
rope grab 缆索夹具;缆索抓固器;捞绳抓钩
rope grader 绳索式分级机
rope grease 钢绳润滑油;绳油
rope-grip 索夹
rope grommet 索耳
rope groove 缆索闸门导向槽;绳槽
rope guard 缆索防护装置;压绳器;钢(丝)绳护板
rope guide 绳导器;导绳器;钢绳导向器
rope guider 排绳器;导绳器
rope guiding resistance 导绳阻力
rope guy 支索;拉索
rope haulage 缆索运输;绳索运输
rope haulage hoist 钢绳运输绞车
rope haulage machine 缆索运输机
rope-hauled bucket 索式运土机铲斗
rope-hauled scraper blade 绳索牵引式刮泥刀
rope heart 绳芯
rope hoist 钢绳绞车
rope installation 钢绳安装
rope insulation 绝热绳
rope knife 钢绳割刀;切钢丝绳刀
rope knife swivel 钢绳割刀加重钻杆
rope ladder 软梯;绳梯
rope-lay conductor 绳绞导线
rope layer 搓绳机
rope laying machine 搓绳机
rope level 缆索高度;钢绳标准
rope lever block 钢丝绳手扳葫芦
rope life 钢绳使用寿命
rope lubricant 钢丝绳润滑剂;绳索润滑剂
rope luffing 钢丝绳变幅
rope lug 绳耳
rope machine 制绳机
rope Manila sand paper 马尼拉麻浆砂纸
rope mo(u)lding 卷缆花饰;绳索饰嵌条;模压绳索;卷缆线脚;绳索纹饰线脚
rope of parallel wire 平行捻钢丝绳
rope of sand 靠不住的结合
ropeology 钢丝绳数据
rope-operated 绳索操纵的
rope-operated lift 绳索式起落机构
rope-operated scraper 缆索式铲运机
rope packing 麻编填料;垫料绳
rope parting machine 钢丝绳切割器
rope pattern 绳纹
rope polygon 玫瑰线脚;索多边形
rope-powered 绳索拖动的
rope pressure on lagging 衬垫绳压
rope propulsion system 绳索推进系统【铁】
rope pull 绳索拉力
rope pulley 绳索轮;绳索滑轮;绳索滑车
rope-pulley flywheel 绳索飞轮;钢绳滑轮
rope pusher 链索式推车机
rope race 钢道
rope railway 索道
rope reel 绳绞盘;缠绳;线卷
rope reeving 穿绳
rope rider 列车跟车工
rope rig 钢绳钻进钻机
rope roll 钢绳卷筒;绳筒
rope roller 导缆滚子

rope roof 绳索吊的屋顶
rope roof bolt 钢丝绳顶板锚杆
rope runner 绕钢索滚筒；电缆敷设工
rope safety factor 钢绳安全系数
rope sag 缆绳垂度
rope saver 钢丝绳终端保险装置
rope scourer 绳索洗涤机
rope-scouring machine 绳状洗涤机
rope sealing 密封绳；填缝绳
rope shackle 缆绳卸扣；缆绳卡环
rope sheave 索轮；绳索轮；绳索滑车；绳轮
rope sling （起重机用的）网兜；吊链；吊索；绳吊索；吊货索环
rope slip 绳索脱；绳索滑落
rope slipping 绳索打滑
rope socket 绳插座；钢绳连接眼环；钢绳连接套筒；钢索连接眼环；钢索端眼具；钢绳打捞筒；绳头扣紧座；绳套；绳缆端头
rope socket catch 钢丝绳端（钮）扣；钢丝绳端紧扣
rope spear 捞绳钩；钢绳打捞钩；钢绳手；（钢绳冲击钻用）捞绳矛；绳索插脚
rope-spear wadder （连接在绳帽上带捞针的）加重异径钻杆
rope speed 提升钢丝绳速度；绳速；卷绕速度
rope spinning 搓绳
rope splice 钢绳环；绳绞接
rope splicing tools 接绳工具
rope squeezer 绳状轧水机
rope stopper 制动索
rope storage 磁芯线存储器
rope storing drum 储绳卷筒
rope strand 钢绳的股
rope strap 绳索环圈
rope stretch 缆绳拉紧
rope strop 滑车环索；滑车带
rope support 绳架
rope suspended sheave 托索轮
rope suspended cantilever(ed) roof 悬索悬臂屋顶
rope suspended elliptic(al) roof 缆索悬吊椭圆屋顶
rope suspension bridge 悬索桥；缆式悬桥；绳索桥；吊桥
rope suspension equalizer 悬索平衡器
rope system 缆绳体系；钢丝绳系统
rope system drill 钢绳冲击（式）钻机
rope tackle block 绳滑车
rope tensile force 缆索拉力
rope tensile force ga(u)ge 缆索拉力仪
rope testing machine 缆绳检验机；绳索强度试验机
rope thimble 钢索套环；绳索套环
rope thread 圆螺纹
rope to drum ratio 卷筒直径与钢绳直径之比；钢丝绳和卷筒直径比
rope tool drilling equipment 钢缆冲击式钻孔装置；钢缆冲击式钻孔设备
rope tool well 钢缆冲击钻井
rope towing traction 缆索牵引力；钢缆牵引航行
rope transfer 钢丝绳移送机
rope transmission 钢绳传动；绳索传动
rope tread pressure 钢绳压力
rope trolley 索挖式起重小车；索控式起重小车；绳索小车
rope turn 绳索捻距
rope-type head 绳式喷头
rope unhitching device 分绳装置
rope walk 制绳狭长走道
rope wander 缆索漂移；缆索漂动
rope washer 绳式洗涤机
rope washing machine 洗绳机
ropeway 索道；钢绳绳道；架空索道；钢道；缆道
ropeway cable 索道钢索
ropeway car 架空索道矿斗；索道小车；索道吊式
ropeway conveyer 索道式输送机
ropeway measurement 缆道测量
rope wear 绳索磨损；钢绳磨损
rope whipping line 缆扎绳
rope winch 钢丝绞车；钢丝绞盘；缆绳绞盘
rope winch attachment 钢丝绞车附件
rope winder 卷绳器
rope wire 钢绳用钢丝；钢绳钢丝；绞合线；合股线
rope working 绳索传动
rope works 制绳厂
rope worm 断绳打捞矛
ropey 涂料轻度痕迹

rope yard 制绳场地
rope yarn 钢索股线
rope yarn knot 绳条结
rope yarn packing 绳条填料
ropiness 胶黏；黏刷性；绳纹；成丝性；刷痕
roping 浆痕；绳索；帆缘索；帆边绳
roping needle 缝帆粗眼针
roping pin 绕绳索柱
ropy 成丝的；黏稠的
ropy finish 有刷痕涂层
ropy lava 绳状熔岩
ropy lava flow 绳状熔岩流
ropy peat-bog 黏稠泥炭沼泽
roquesite 硫铟铜矿
ro/ro access equipment 滚装通道设备
ro/ro barge carrier 滚浮船
ro/ro equipment 滚装设备
ro/ro ramp 滚装坡道
ro/ro rampway 滚装跳板
ro/ro ship 滚装船；滚装进出船；机动车辆专运船
Ro/Ro system 轮渡式船舶；水平装卸法；滚装法
ro/ro terminal 滚上滚下码头
Rorschach test 墨迹测验
rosace 圆浮雕；车轮窗；圆花窗；蔷薇花饰
rosaceous 蔷薇花形
rosalin 结晶玫瑰
rosaniline 品红（碱）
rosarium 玫瑰园
rosary 玫瑰花坛；玫瑰花圃；玫瑰园；串珠
rosasite 锌孔雀石；纤维绿铜锌矿
roscherite 钙磷铁锰矿；水磷铍锰石
roscherite triclinic 三斜水磷铍锰石
roscoelite 钒云母
Roscoe model 罗斯科模型
Roscoe surface 罗斯科面
rose 莲蓬嘴喷涂；莲蓬式喷嘴；玫瑰饰；玫瑰花型；玫瑰花图；蔷薇；喷雾器喷嘴（莲蓬式）
rose A 玫瑰红刚玉
Rose alloy 罗斯易熔合金
rose-arbour 玫瑰棚架
rose arena 粉红麻石
rose aurora 黄玉石
rosebay 石南；夹竹桃；杜鹃花；柳兰
rose Bengal 孟加拉红；玫瑰红
rose beryl 玫瑰绿宝石
rose bit 菊花钻；扩孔锥；星形钻（头）；梅花钻；玫瑰状钻头；玫瑰形钻头
rose bit pilot 玫瑰状铣头导向器；星形钻头导向器
rose box 舱底水滤盒；蜂巢箱【给】
Rosebrae bed 罗斯布拉层
rose chart 玫瑰图
rose chucking reamer 玫瑰形机用绞刀
rose colour 玫瑰红；蔷薇彩
rose coloured 玫瑰红色的
rose comb 玫瑰冠
rose concrete 玫瑰凝结物
rose copper 红铜
rose countersink 菊花钻
rose curve 玫瑰线
rose cutter 菊花形铣刀；玫瑰形铣刀；球头铣刀；板牙状切削器
rose diagram 玫瑰图
rose diagram of joints 节理玫瑰图
rose diagram of wave frequency 波频玫瑰图
rose diagram of wind direction 风玫瑰图
rose drill 圆花钻；梅花钻；球头钻
rose-eye 玫瑰眼
rose flowered 蔷薇花型的
rose garden 玫瑰园；蔷薇园
rose grain 玫瑰色磨料；玫瑰色颗粒
rosehead 莲蓬头；滤水斗；玫瑰头钉；（吸水管末端的）滤水头
rose hip 刺玫果
Rosein 罗新镍铝合金
roseine 品红
roseite 蛭石
rose lake 茜素玫瑰红（色淀）
rose lashing 交花绑
Roseler reinforced block floor 罗斯勒钢筋混凝土楼板
roselite 玫瑰砷钙石；砷钴钙石；三斜砷钴钙石
rose madder 茜素玫瑰红
rosemary 迷迭香【植】
rose medallion 园花饰瓷器

rose metal 铋锡铅合金
rose moulding 玫瑰线脚饰；玫瑰花线脚
rose nail 圆花钉
Rosenbach formula 罗森巴赫公式
Rosenberg generator 罗森堡发电机
rosenbuschite 钴针钠钙石
Rosendale cement 罗森达水泥
rosenhnite 罗水硅钙石
rose of joints 节理玫瑰图
rose of wave frequency 波向频率图；波频玫瑰图
rose oil 蔷薇油
roseolic acid 玫瑰酸
rose olive 玫瑰榄球
rose opal 玫瑰蛋白石
rose ornament 玫瑰饰
rose pale 淡玫瑰色
rose pattern 玫瑰形
rose-patterned glass 玫瑰图案压花玻璃
rose pink 玫瑰（红）色；淡玫瑰红；淡粉红色
rose pipe 滤吸管
rose porcelain 粉彩瓷
rose purple 玫瑰紫；蔷薇紫
rose quartz 蔷薇水晶；蔷薇石英
rose reamer 星形铰刀；玫瑰形铣刀；倒角刀
rose red 玫瑰红
rose ring 玫瑰状环
rosery 玫瑰园
Rose's alloy 罗斯易熔合金
rose seizing 交花绑
rose shell reamer 星形套筒铰刀
Rose's metal 罗斯易熔合金
rose supreme 玫瑰
rose topaz 玫红黄玉
rose trim 圆花装饰
rosette 玫瑰花形；圆花饰；圆花窗；菊花状共晶组织；玫瑰花饰；玫瑰花结；蔷薇状共晶组织；天花板电线盒；三向应变针
rosette bit 菊花状钻头；梅花钻头
rosette copper 红铜；盘铜
rosette ester 玫瑰状松香甘油酯
rosette ester varnish 玫瑰状松香甘油酯清漆
rosette formation test 玫瑰花环形成试验
rosette fracture 星形断口；菊花状断口；玫瑰状断口；花状断口；蔷薇状断口
rosette ga(u)ge 应变片丛；玫瑰瓣状应变片丛
rosette graphite 玫瑰状石墨；蔷薇状石墨
rosette method 系列法
rosette of needles 针状体束
rosette plate 插座式极板
rosette tier 莲座层
rose type head 蜂窝式喷头
rose vitriol 硫酸钴
rose water 玫瑰水
rose white 白玫瑰
rose window 圆花窗；玫瑰形窗；车轮窗
rose wood 玫瑰木；黄花梨木；花梨木；黄檀木
rosewood cabinet with dragon design 雕龙檀木柜
rose work 滚花
rosickyite 斜自然硫
rosin 松脂；松香
rosin acid derivative 松脂酸衍生物
rosin adduct 松脂加成物
rosinate 松脂酸盐
rosinated pigment 松脂酸酯处理的颜料
rosinate soap 松香皂
rosinate soap foamer 松香皂泡沫剂
rosinate varnish 松脂酸酯制清漆
rosin connection 虚焊接线
rosin content 松香含量
rosin-core solder 松脂芯焊料
rosined joints 涂树脂的钻杆接头
rosiness 玫瑰红
rosin essence 松香精
rosin ester 酯胶；松香酯
rosin ethylene glycol ester 松香乙二醇酯
rosin flux 焊剂
rosin glycerin ester 甘油松香（酯）
rosin grease 松香油膏；树脂润滑膏
rosin jack 闪锌矿
rosin joint 虚焊接头；虚焊
rosin-maleic adduct 顺丁烯二酸酐松香加成物
rosin-maleic anhydride condensation compound 松香—顺丁烯二酸酐缩合物
rosin milk 松香乳液

rosin modified alkyd resin 松香改性醇酸树脂
rosin modified glyptal resin 松香改性醇酸树脂
rosin modified maleic resin 松香改性马来树脂;松香改性顺丁烯二酸树脂;松香改性失水苹果酸酸树脂
rosin modified phenolic resin 松香改性酚醛树脂
rosin oil 松香油;松脂油
Rosinos distribution 罗辛分布
rosin pitch 松香沥青
rosin processing wastewater 松脂价格废水
rosin qualitative test 松香定性试验
rosin resin 松香(基)树脂
rosin size 松香胶
rosin sized sheathing 树脂胶夹衬板;涂以松香胶的木纤维片
rosin smoke 松香烟
rosin soap 松脂皂
rosin spirit 松脂醇;松香精
rosin standard 松香标样
rosin type sample 松香型试样
rosin varnish 松香清漆
rosin wash 松脂合剂
rosita 檍叶
Roslin sandstone 罗斯林砂岩
rosoeresite 磷铝铅铜矿
rosolic acid 蔷薇色酸
Ross Barrier 罗斯冰障
Rossby diagram 罗斯比图
Rossby number 罗斯比数
Rossby parameter 罗斯比参数
Rossby regime 罗斯比形式
Rossby term 罗斯比参数
Rossby wave 罗斯比波
Ross chain feeder 罗斯链式给料机
Ross corrector 罗斯校正器
Rosseland mean absorption coefficient 罗斯兰平均吸收系数
Rosseland mean opacity 罗斯兰平均不透明度
Rosseland theorem 罗斯兰定理
Rossel current 罗塞尔海流
Rossellian (stage) 罗赛超阶
rosser 防台(风)窗;剥皮工
Ross feeder 罗斯给料器
Rossi counter 罗西计数管
Rossi-Forel scale 罗西一福利地震烈度级
Rossi-Foruir intensity scale 罗西一佛瑞尔烈度表
Rossi process 罗西连续铸钢法
rossite 水钒钙石
Rossiter effect 罗西托效应
Ross lens system 罗斯透镜系统
Rosslyn metal 不锈铜一铜一不锈铜复合板;包铜薄钢板;罗斯林包铜薄钢板
Rosslyn number 罗斯林数
Rossman drive 罗斯曼驱动
Ross objective 罗斯物镜;罗斯镜头
ross of mull 红色花岗岩(产于英国)
ross quartz 芙蓉石
Ross roll grizzl(e)y 双轴辊轴筛
Ross Sea 罗斯海
Ross's test 厚膜试验
Ross water quality index 罗斯水质指数
Ross wide angle lens 罗斯宽角透镜
roster 逐项登记表;值勤名单;值班登记表;花名册;人名单
roster scan 值班搜索
roster system 定期轮灌制
rosthornite 棕色树脂
rostral column 船头饰纪念柱【建】;(有船头装饰的)海战纪念柱;船头形饰纪念柱【建】
rostrum 城门楼;活动门窗;主席台;讲台;[复]rostra
rosula 莲座簇
rosulated leaf 莲座状叶
rosy 淡红色的;玫瑰色的
rosy quartz 蔷薇石英
rot 糟朽;枯病;腐朽
rota 值班人员表;花名册
rotable loop 旋转环形天线
rot accident 变质事故
rotacore 软岩层用钻头(直径76mm)
rota-evapo(u)ration 旋转蒸发
rotafoil parachute 旋转伞
rotahoist 运转卷扬机
rotalactor 转盘挤奶台
rotamerism 几何异构现象

rotameter 转子式流速仪;旋转式流量计;转子流速计;转子流量计;曲线测长计;曲率测量器;浮子(式)流量计
rotamower 旋转式割草机
Ro-Tap 罗太普筛分机
rotap 机动筛;振筛机;转击式筛析试验机
rotaplane 旋翼机
rotap mechanical shaker 转动轻敲式机械摇动器
rotap shaker 转动锤击式振筛器
rotary 旋转钻井机;旋转性的;旋转机械;旋转的;旋切的;转动的;交通盘;回转的
rotary abrasive tester 旋转式磨料磨损试验机
rotary abutment motor 旋转隔板电动机
rotary acceleration 旋转加速度
rotary accelerator 回旋加速器
rotary actuator 旋转执行器;旋转式激励器;转动装置;摆动(液压)马达;摆动液压缸
rotary actuator with two pistons 双柱塞摆动缸
rotary agitation 旋转搅拌
rotary agitator 旋转式搅拌机;回转式搅拌机
rotary air compressor 旋转式空气压缩机;旋转式空压机;回转式空气压缩机
rotary air drill 回转式风钻
rotary air engine 旋转式热空气发动机
rotary air lock 回转式气塞
rotary air lock feeder 旋转锁风喂料机;格式喂料机
rotary airplane hangar 旋转式飞机库
rotary air pump 转动(式)气泵;回转式空气泵
rotary air sluice 回转式锁风闸门
rotary air tool 旋转气动工具
rotary air vibrator 旋转式压缩空气振动器
rotary amplifier 旋转式放大器;旋转式磁放大器;回旋放大器
rotary angle 旋转角
rotary annular extractor 旋转式环形萃取器
rotary antenna system 旋转天线系统
rotary apartment tower 旋转式公寓塔楼
rotary atomizer 旋转雾化器;旋转式雾化喷头;旋转式喷雾器;回旋式雾化器;回旋式喷雾器
rotary attenuator 旋转衰减器
rotary awner 旋转式除芒机
rotary axis 旋转轴
rotary bagger 旋转式装袋器
rotary balance 转动平衡
rotary balanced jack 转盘平衡摇臂
rotary balance log 转轮式计量仪
rotary ball mill 旋转式球磨机
rotary balloon 圆球形气球
rotary barrel throttle 旋转式节流滑阀
rotary base 转盘基础;回转底盘
rotary batch charge 滚筒式投料机
rotary batch feeder 滚筒式投料机
rotary batch mixer 滚筒式分批拌和机
Ro-tary beacon 旋转航标;旋转光标
rotary beam 旋转波束
rotary beam antenna 射束旋转天线
rotary beater 旋转式切茎器
rotary belt cleaner 旋转式传送带清洁器
rotary bending fatigue test 旋转弯曲疲劳试验
rotary bending tester 旋转弯曲试验机
rotary billet cooler 旋转坯料冷却器
rotary bit 旋转钻(头);旋转刀片;回转钻进钻头
rotary blade 旋转叶刀;旋转叶片;旋叶铲
rotary blade stabilizer 旋转叶片型稳定器
rotary blasthole drilling 旋转式炮眼钻进
rotary block 游动滑车(组)
rotary block radial pump 旋转缸体径向泵;旋转缸体式径向柱塞泵
rotary blower 旋转(式)通风机;旋转(式)鼓风机;回转(式)鼓风机
rotary body producer 转井式煤气发生炉
rotary boiler 旋转式锅炉
rotary bolt 旋转螺栓
rotary boom crane 旋转吊杆起重机
rotary boring 旋转钻(探);旋转式钻探;旋钻;回转钻进
rotary bottle filler 回转式灌瓶机
rotary branch switch 旋转式分路开关
rotary breaker 旋转压碎机;旋转式破碎机;滚筒破碎机;回转锥形破碎机;回转式碎石机;回转式破碎机
rotary broom 转动路耙;旋转式扫路机;旋转路耙;旋转路扫;滚路刷
rotary Brownian motion 转动布朗运动
rotary brush 旋转式路刷;旋转刷;旋转电刷

rotary brush aerator 转刷曝气装置
rotary brush attachment 旋转式路刷附件
rotary brush cutter 旋转式灌木切除机
rotary brush cutter for walking tractor 手扶拖拉机用旋转式灌木切除机
rotary brush drill 旋转式刷子播种机
rotary bucket 回转取样筒
rotary bucket drill 旋转式勺斗钻机;转盘式钻机(用于钻软岩层或土层)
rotary bucket elevator 转斗式提升机
rotary bucket excavator 旋转斗式挖掘机;转斗电铲
rotary bucket-type flowmeter 旋转叶片式流量计
rotary bucket wheel reclaimer 旋转斗轮式回收设备
rotary burner 旋转式烧嘴;转子式燃烧器;转子式喷燃器
rotary bushing 转盘补芯
rotary bushing drag 转盘补芯阻力
rotary button 旋转按钮
rotary calciner 旋转窑;转筒窑;转筒煅烧炉
rotary calculator 旋转计算器
rotary camera 旋转式摄影机
rotary cam type automatic voltage regulator 旋转凸轮式自动稳压器;旋转凸轮式自动电压调整器
rotary car dumper 转子式翻斗车;旋转式翻车机;转子式翻车机;转笼式翻车机
rotary car tipper 旋转式翻车机
rotary casing scraper 套管刮刀
rotary cement kiln 旋转式水泥窑;水泥回转窑
rotary centrer 旋转中心;转动中心
rotary chain 旋转链;套筒滚子链
rotary chain draw works 旋转链式提升机;旋转链式卷扬机
rotary chain tightener 旋转链式张紧装置
rotary chopper 旋转式切碎机
rotary chopper wheel 旋转式短切机轮
rotary circulating pump 旋转式循环水泵
rotary classifier 旋转式分粒器
rotary cleaner 旋转式清理机
rotary clinker 单筒冷却机
rotary clinker cooler 回转式熟料冷却器
rotary clippers 旋转剪板机;滚切钳
rotary clothes dryer 旋转式干衣机
rotary coil 转动线圈
rotary column 旋转塔;转柱
rotary combustion engine 双缸转子内燃机
rotary combustion engine of trochoidal form 余摆线型旋转式内燃机
rotary compression mo(u)lding press 旋转式压缩成形机
rotary compressor 旋转压缩器;旋转压气机;旋转式(空气)压缩机;回转式压缩机;滑片式空压机
rotary concentrator 旋转式选矿机
rotary condenser 旋转式冷凝器;调相器;调相机
rotary cone crusher 圆锥破碎机
rotary cone penetrometer 旋式锥形贯入计
rotary connection 旋转接触装置
rotary construction crane 旋转式施工起重机
rotary contactor 旋转接触器
rotary continuous filter 旋转连续过滤器;旋转连续过滤机
rotary continuous sterilizer 回转式连续杀菌机
rotary converter 旋转变流器;旋转变流机;旋转变换器;同步换流机;电动机发电机组
rotary converter locomotive 机车回转变频器
rotary cooler 转筒式冷却器;回转冷却器
rotary cooling drum 转筒式冷却机
rotary core 旋转铁芯
rotary core bit 旋转式取芯钻头;回旋式取芯钻头
rotary core boring 旋转式岩芯钻探
rotary core machine 旋转式挤芯棒机
rotary couple 旋转力偶
rotary coupler 旋转车钩
rotary coupon test 旋转挂片试验
rotary crane 旋转(式)起重机;立柱式旋臂起重机;回转起重机;回转吊车
rotary crop shears 滚筒式切头剪
rotary crossed axis shaving 旋转交轴剃齿法
rotary crusher 旋转式破碎机;旋转式碎石机;圆锥碎石机;旋转锥形破碎机;回转式碎石机;回转(式)破碎机
rotary cultivator 旋耕机
rotary cultivator attachment 旋耕装置
rotary cultivator with land wheel 带地轮的旋耕机
rotary cultivator with longitudinal shafts 纵轴旋

耕机
rotary cup burner 转杯喷嘴
rotary-cup oil burner 旋转喷注燃油器
rotary current 旋转流(量);旋转潮流;回转流;多相电流
rotary current vibrator 多相电流振动器
rotary cut 旋刨木片;旋削
rotary cut-off 轮转裁切机
rotary cut-off knife 旋转切断刀
rotary cutter 旋转式掘土器;旋转式绞刀;旋转式割草器;旋转切割机;旋转切断机;旋转切刀;旋转剪切机;转割机
rotary cutting 旋转式切削
rotary cutting and bending machine 对开滚线机
rotary cutting assembly 回转式切割器
rotary cutting shield 旋转式切削盾构
rotary cut veneer 旋切层板;旋切单板
rotary cylinder-block 旋转缸体
rotary cylinder thickening filter belt press 转筒浓缩带式压榨过滤机
rotary cylinder viscometry 转筒浓度测定法
rotary cylindrical kiln 旋转圆筒窑
rotary dairy 圆形自转挤奶台
rotary damping 转动阻尼
rotary davit 旋转式吊艇柱
rotary deflection (风的)旋偏向
rotary dehumidifier 转轮除湿机
rotary design 旋转设计
rotary design speed 环形设计速率;环道设计车速
rotary dewaterer 旋转脱水机
rotary diamond drill 旋转式金刚石钻机
rotary diaphragm 回转光圈
rotary diesel engine 转子柴油机
rotary diffusion 转动扩散
rotary digester 旋转浸煮器
rotary digger 旋转式挖掘机
rotary digging method 旋转钻式挖掘法
rotary digitizer 滚筒式数字化器
rotary discharge grate 回转下料箅
rotary discharger 旋转放电器
rotary disc type gritter 圆盘式铺砂机
rotary disk 旋转盘;回转盘
rotary disk bit 盘式钻头;圆盘形回转钻;圆盘形扁钻;圆盘形扁锥
rotary disk column 转盘塔
rotary-disk contactor 转盘接触器;旋圆盘混合器
rotary disk ditch cleaner 转盘式清沟机
rotary disk extractor 转盘萃取器
rotary disk feeder 转盘送料装置;转盘给料器;转盘给料机;转盘给矿机;回转盘送料装置;盘式喂料器;盘式喂料机;喂料盘
rotary disk filter 转盘(式)过滤器;转盘(式)过滤机
rotary disc grizzly 转盘铁格筛
rotary disk grizzly feeder 转盘式筛式给矿机
rotary disk loader 转盘式装车机
rotary disk meter 转盘式流量计
rotary disk press 转盘式压砖机
rotary disk pulsed extractor 转盘脉冲抽提器
rotary disk shutter 旋转叶片式快门
rotary disk trpe gritter 圆盘式铺砂机
rotary dispensing air lock 回转式布料气封
rotary dispersion 旋光性色散
rotary displacement compressor 旋转型容积式压气机
rotary displacement meter 罗茨式煤气表;转子式煤气表
rotary displacement pump 回转式排量泵;转泵
rotary distributing valve 旋转配流阀
rotary distribution 旋转配水;旋转喷水;轮流配水
rotary distributor 旋转喷水器;旋转配水器;旋转配水池;旋分配器;转动配水器
rotary dividing shears 滚筒式切分剪
rotary doffer 旋转式脱棉器
rotary door 旋转门;旋动门
rotary drain cutter 旋转式排水沟挖掘机
rotary draw bender 旋转式拉弯机
rotary drawing frame 梳棍式并条机
rotary draw works 滚筒式提升机;滚筒式卷扬机;转盘钻机绞车
rotary draw works drum 旋转式提升机滚筒;式卷扬机滚筒
rotary drier 回转窑;旋转干燥炉;回转式烘干机;回转干燥机
rotary drifter 旋转架式钻机

rotary drill 旋转式钻孔机;旋转(式)钻孔机;转盘式钻机;回转式钻机
rotary drill cuttings 回转钻进岩粉
rotary driller 转盘钻司钻
rotary drill for shaft sinking 凿竖井用
rotary drill hammer 旋转式钻锤
rotary drill hole 旋转式钻孔
rotary drilling 旋转式岩芯钻探;旋转钻井;旋转钻(孔);旋转钻进;旋转(式)钻探;转盘钻探;磨盘钻探;回转钻进;回转式凿岩;回转钻进
rotary drilling bit 旋转式钻头
rotary drilling by pumping of mud 正循环回转钻孔
rotary drilling by suction of mud 反循环回转钻孔
rotary drilling line 旋转式钻探绳索
rotary drilling machine 转盘钻机;回转(式)钻机;牙轮钻机
rotary drilling outfit 旋转式钻探装置
rotary drilling rig 旋转式钻(孔)机
rotary drilling system 旋转式钻机系统;旋转式钻井法
rotary drilling unit 旋转式钻探装置
rotary drill jumbo 旋转式钻车;回转式钻机的钻车
rotary drill line 游动钢绳
rotary drill pipe 旋盘钻进用钻杆;旋转钻进钻管
rotary drill rig 转盘钻进设备
rotary drills with hydraulic 油压转盘钻机
rotary drive bushing 转盘衬套
rotary drum 转筒;滚筒;转鼓
rotary drum concrete mixer 转筒式混凝土搅拌机;转筒式混凝土拌合机
rotary drum (cone-element) 锥形鼓
rotary drum discharge unit 转筒式卸料装置
rotary drum dryer 转筒(式)烘干机;转筒式干燥机;转筒干燥器;转鼓(式)干燥器;旋转圆筒干燥器;旋转鼓式干燥器;干燥滚筒;滚筒式干燥器
rotary drum feeder 转筒式喂料机;旋转加料器
rotary drum filter 圆筒式(真空)过滤器;转鼓式(过)滤器;转鼓(式)过滤机;转鼓过滤机;转筒式滤器
rotary drum grader 滚筒分级机
rotary drum mixer 转鼓式拌和机;转鼓搅拌机;转筒(式)搅拌机;转筒式拌和机;回转式搅拌和机
rotary drum pulverizer 转筒粉磨机
rotary drum screen 转鼓式筛网;转筒筛;转筒式滤网
rotary drum type continuous pressure filter 转鼓式连续加压过滤器;转筒式连续加压过滤机
rotary drum unloader 滚筒式卸载机
rotary drum vacuum filter 滚筒式真空吸滤机
rotary drum washer 滚筒式清洗机
rotary dryer 旋转式干燥炉;旋转(式)干燥器;旋转式干燥机;转动干燥器;回旋式干燥器
rotary drying 转筒式干燥(法)
rotary drying and heating process 转筒式干燥与加热方法;转筒式干燥与加热过程
rotary dry vacuum pump 旋转真空干燥泵;旋转式干式真空泵
rotary dump car 旋转式翻卸矿车;翻转式卸矿车
rotary dumper 旋转式翻斗车;旋转式翻矿车;旋转式翻车机;转子翻车机;回转式翻车机
rotary dust collector 旋转式吸尘器
rotary duster 旋转式喷粉器
rotary dwelling tower 旋转式居住塔楼
rotary dy(e)ing machine 滚筒式染色机
rotary eccentric piston pump 偏心转子泵
rotary electric(al) drill 回转电钻
rotary electric(al) servo 旋转电伺服机构
rotary electrode 旋转电极
rotary elevator 转轮提升机
rotary elliptical piston pump 旋转式椭圆活塞泵
rotary encoder 旋转编码器
rotary end-sealing machine 回转式封底盖机
rotary end tip 旋转翻笼;旋转倾卸
rotary engine 转子发动机;旋转(式)发动机;转缸式发动机
rotary engine bearing 转子发动机主轴轴承
rotary engine combustion cycle 转子发动机工作循环
rotary engined car 转子发动机小客车
rotary equipment 回转钻进设备
rotary evapo(u)ration 旋转蒸发
rotary evapo(u)rator 旋转(式)蒸发器;旋转(式)气化器

rotary excavator 盾构;圆形隧道掘进机;旋转式挖掘机;滚切式挖掘机
rotary exhaust machine 旋转排气台;旋转抽气机
rotary exhaust valve 回转式排气阀
rotary expander 滚压扩管机
rotary expansion engine 转子发动机
rotary extrusion 旋压
rotary fan 旋转式风扇;扇风机
rotary fan conveyer [conveyor] 旋转式运送机
rotary fault 旋转故障;旋转断层
rotary feed 圆周进给
rotary feed control 回转给进控制
rotary feeder 旋转式加料器;旋转式给料机;旋转式给料器;星形进料器;转动叶轮定量给料机;旋转喂料器;旋转喂料机;回转式给料器;回转式给料机;回转进料器;回转加料器;筒式给煤机
rotary feed reservoir 旋转给料斗
rotary feed table 转盘给料器;转盘给矿器
rotary felt layer 旋转铺毡机
rotary fettling table 转盘精选机;带平台清砂机
rotary field 旋转磁场
rotary field motor 旋转磁场电动机
rotary file assortment set 成套圆头尖头旋转锉
rotary film contactor 转动薄膜式萃取器
rotary filter 旋转(式)过滤器;旋转滤机;转子式滤清器
rotary fine screen 旋转式细筛
rotary finisher 旋转式修整机
rotary finishing 旋转抛光
rotary fixed displacement motor 旋转式固定排量发动机
rotary flange 移动法兰
rotary flanging machine 回转式翻边机
rotary float 电动抹刀;旋转整平机;旋转磨面机
rotary floor 旋转地板
rotary floor restaurant 旋转地板的餐馆;旋转地板的酒店;旋转地板的饭店
rotary flow divider 旋转分流器
rotary flowmeter 转子式流量计;转子流量计
rotary fluid machine 旋转液压机械
rotary fluid power device 旋转液压传动装置
rotary fluid pressure machine 旋转液压机械
rotary flying crop shears 滚筒式飞剪
rotary flying hot saw 回转式热飞锯
rotary flying shears 滚筒式飞剪
rotary force 旋转力;转动力
rotary forced-air cooler 鼓风环式冷却器
rotary forging machine 回转式锻造机;回旋式锻造机;摆旋碾压机
rotary forging mill 周期式轧管机
rotary forging press 摆碾机
rotary frit furnace 回转熔块炉
rotary frit kiln 回转熔块炉
rotary fuel-feed pump 转子式输油泵
rotary fuel oil pump 旋转燃油泵
rotary fulling mill 滚筒式缩呢机;滚筒式漂洗机
rotary fund 周转资金;循环基金
rotary furnace 旋转加热炉;转炉;回转炉
rotary gang slitter 带材多刀圆盘纵切剪
rotary gap 旋转火花隙;旋转放电器
rotary gate 旋转式闸门
rotary gate feeder 旋转门式送料机
rotary ga(u)ge 旋转液面计
rotary gear pump 旋转齿轮泵
rotary gear shaving cutter 盘形剃齿刀
rotary gill box 旋转针梳延展机
rotary grab 旋转式抓斗
rotary grab crane 旋转式抓斗起重机
rotary grader 旋转式分级机
rotary granulator 旋转式制粒机
rotary grass cutter 旋转式割草机
rotary grate 回转式炉箅;旋转炉箅;回转炉排
rotary grate shaft kiln 回转算式立窑
rotary grinder 圆台平面磨床
rotary grinding machine 转台磨床
rotary grizzly 旋转格筛;辊轴筛
rotary guide 旋转导柱
rotary hammer 旋转式锤;电锤
rotary hammer breaker 回转锤式破碎机
rotary hammer mill 旋转锤磨机
rotary hammer type breaker 旋转锤式破碎机
rotary hand drill 手摇钻;手摇钻
rotary hand pump 手摇回转泵
rotary harrow 旋转耙;旋转靶

rotary head 旋转喷洒头;旋转刀盘;回转钻具(钻井筒用)
rotary header 旋转式掘进机
rotary hearth 回转炉
rotary hearth continuous furnace 转底式连续(加热)炉
rotary hearth furnace 转底炉
rotary hearth incinerator 旋转床式焚化炉;回转床式焚烧炉
rotary hearth kiln 转底(环状隧道)窑;转盘式隧道窑
rotary heat exchanger 转轮式换热器;旋转式热交换器
rotary helical screw compressor 回转式螺杆压缩机
rotary helper 司钻助手;副司钻
rotary hoe 旋转锄;旋锄机;松土搅拌机
rotary honing machine 旋转式研磨机
rotary hook 旋梭缝;旋转钩;大钩
rotary hopper 旋转式漏斗
rotary hopper dewaterer 旋转漏斗脱水机
rotary hose 水龙带;回转高压胶管
rotary hose coupling 旋转式软管接头
rotary house 旋转房屋
rotary hunting connector 寻线机
rotary hydraulic expansion wall scraper 回转液压扩张式井壁刮刀
rotary hydraulic jack 旋转液压千斤顶
rotary hydraulic motor 旋转式液动机;旋转式液压马达
rotary hysteresis 回转磁滞
rotary impact drilling 旋转式冲击钻探;旋转式冲击钻孔
rotary impact drilling machine 旋转式冲击钻机
rotary impeller 转叶轮
rotary impeller exhauster 叶轮式排气机
rotary incinerator 旋转式焚化炉
rotary indexing fixture 旋转式分度夹具
rotary indexing machine 多面回转工作台式组合机床
rotary induction system 旋转输送系统
rotary inductosyn 回转式感应同步器
rotary inertia 转动惯量
rotary injection pump 旋转式喷射泵
rotary interchange 环形立体交叉;环形立交;环形互通式立交
rotary interlocked switch 旋转联锁开关
rotary interlocking block 旋转式联锁闭塞机
rotary internal combustion engine 旋转式内燃机;旋转活塞发动机
rotary interrupter 旋转式断续器
rotary interrupter relay 旋转断续继电器
rotary intersection 旋式交叉;转盘式立体交叉;转盘式交叉(口);环形交叉(道)口;道路环形交叉
rotary inverter 旋转逆变器;旋转反向变流机
rotary ironing calendar 旋转式烫轧机
rotary irradiation apparatus 旋转辐射仪
rotary island 交通环绕岛;环形交通岛;环形交叉中心;环岛;交通转盘
rotary jack hammer 旋转式凿岩机
rotary jamming 旋转干扰
rotary jars 回转振击器
rotary jet cooler 旋转喷口冷却器
rotary jig 旋转式钻模;旋转式固定架
rotary joint 旋转联结;旋转连接;转动接头
rotary kiln 旋转窑;旋窑;转筒窑;回窑;回转炉
rotary kiln brick 回转窑用耐火砖
rotary kiln cement 转窑水泥
rotary kiln chains 回转窑挂链
rotary kiln clinkering zone 回转窑烧成带
rotary kiln cooler 回转窑冷却器
rotary kiln dryer 转窑干燥器;回转窑式干燥机
rotary kiln expanded clay 回转窑膨胀陶粒;回转窑膨胀黏土
rotary kiln flue dust 回转窑窑灰
rotary kiln lining 回转窑窑里;回转窑窑衬
rotary kiln method 转窑法
rotary kiln operating stability 回转窑操作稳定性
rotary kiln plant 回转窑设备
rotary kiln shell 回转窑外壳
rotary kiln stone 回转窑石料
rotary kiln tube 回转窑筒;回转窑管
rotary kiln wedge 回转窑楔形砖
rotary kiln with calcinator 料浆蒸发机窑;带分解器的回转窑
rotary kiln with enlarged calcining and burning zone 分解带和烧成带扩大的回转窑
rotary kiln with enlarged calcining zone 分解带扩大的回转窑
rotary kiln with enlarged drying, calcining and burning zone 干燥带、分解带和烧成带扩大的回转窑
rotary kiln with enlarged drying or preheating zone 干燥带或预热带扩大的回转窑
rotary kiln with waste heat boiler 带余热锅炉的回转窑
rotary knife 圆盘刀
rotary knife cutter 旋转切割机
rotary knife feeder 旋转式刮刀加料器
rotary knob 旋钮
rotary lamp 旋式灯
rotary latch 旋转锁闩
rotary launcher 旋转发射装置
rotary launder 转动流槽
rotary letterpress 凸版轮转印刷机
rotary lifter 旋转挖掘器
rotary lime kiln 旋转石灰窑;石灰回转炉
rotary line 游动钢绳
rotary line switch 旋转式预选器
rotary load 旋转负荷
rotary loader 旋转式装载机
rotary locking spring 旋转锁簧
rotary loop aerial 旋转环状天线
rotary loop antenna 旋转环状天线
rotary louvre drier 旋转百叶式干燥机
rotary machine 回转(式)机械;轮转机;旋转式钻机;旋转式凿岩机;转盘钻机
rotary machinery 旋转式机械
rotary machine tool 回转式机床
rotary magnet 旋转(电)磁铁
rotary magnetic amplifier 旋转磁放大器
rotary magnetic chuck 回转磁性吸盘
rotary magnetic table 回转磁吸盘
rotary magneto 旋转式磁电机
rotary manometer 回转式压力计
rotary mass 转动质量
rotary mechanical grate 回转炉算
rotary mechanical output 旋转机械输出
rotary mechanical pump 回转机械真空泵
rotary mechanism 旋转机构
rotary mercury pump 旋转汞泵
rotary meter 旋转式流速仪;转子流速计
rotary microtome 轮转切片机;手摇切片机
rotary midwifery phantom 旋转式接生模型
rotary mill 滚动球磨机
rotary milling machine 滚筒式缩绒机
rotary mirror 旋转镜
rotary mirror scan system 旋转镜扫描系统
rotary mixer 旋转式搅拌机;旋转式混炼机;旋转混合机;转动搅拌机;滚筒搅拌机;式搅拌机;转筒式拌和机;转鼓式拌和机
rotary mixing 转动搅拌;滚筒式搅拌
rotary moment 旋转力矩;旋转力矩
rotary moment of inertia 转动惯性力矩
rotary motion 旋转运动;转动运动;回转运动
rotary motor pump 回转式马达泵
rotary movement 转动
rotary-movement relay 旋转继电器
rotary moving blade vacuum pump 旋转刮板真空泵
rotary mower 旋转式割草机
rotary mud 回转泥浆;回转钻进用泥浆
rotary mud flush boring 旋转泥浆冲洗钻孔;回转泥浆冲洗钻探
rotary mud flush drilling 旋转式泥浆冲洗钻进
rotary mud screen 转筒泥浆筛
rotary muller 滚筒式(连续)混砂机
rotary multiple-spray nozzle coating head 转动式多喷嘴涂装装置
rotary multivane compressor 多叶回转式压缩机
rotary off-normal contact 旋转离位接触点
rotary off-normal spring 旋转离位弹簧
rotary oil burner 燃油炉;旋转式油烧嘴;旋转石油喷灯
rotary oiler 旋转加油器
rotary oil pump 旋转式油泵
rotary oil seal 旋转式油封
rotary oil sealed mechanical pump 旋转式油封机械泵
rotary on-demand cutter 旋转式切割器;旋转切割机;旋转切刀
rotary oscillating pump 回旋振荡泵
rotary outfit 转盘钻进设备
rotary output 旋转输出
rotary oven 旋转炉
rotary packer 旋转压紧器;旋转式包装机;回转式包装机
rotary-paddle feeder 旋转式加供料器;旋转式加给料器;旋浆式加料器
rotary paint film ga(u)ge 转轮式漆膜测厚仪
rotary pan mixer 转锅式拌和机
rotary parachute 旋转伞
rotary pawl 旋转爪
rotary pelleting machine 旋转压片机
rotary percussion boring rig 旋转冲击钻机
rotary-percussion drill 旋转冲击式钻机
rotary-percussion drilling 冲旋钻井;回转冲击钻进;回转冲击凿岩
rotary-percussion drill jumbo 旋转冲击式钻车;回转冲击式钻车;冲击式回转钻车
rotary percussive tool 旋转冲击工具
rotary phase advancer 旋转相位超前补偿器
rotary phase changer 旋转式变相器
rotary phase converter 旋转式相位变换器;旋转式变相机
rotary phase modifier 旋转式调相机
rotary-pick breaker 旋转针齿破碎机
rotary pickup 旋转式捡拾器
rotary piercing 旋转穿孔;回转冲孔
rotary piercing mill 芯棒旋转穿孔机;回转式强力轧孔机
rotary pilot valve 回转式控制阀
rotary pipe bending machine 旋转式弯管机
rotary piston 旋转活塞;回转式活塞
rotary piston blower 旋转活塞鼓风机
rotary piston compressor 旋转活塞压缩机
rotary piston diesel engine 旋转活塞柴油机
rotary piston drive 旋转活塞传动装置
rotary piston engine 旋转活塞发动机
rotary piston engine of epitrochoid design 外余摆线型转子发动机
rotary piston engine of trochoidal type 余摆线型转子发动机
rotary piston hydraulic motor 旋转柱塞液压马达
rotary piston internal combustion engine 旋转活塞内燃机
rotary piston machine 转子发动机
rotary piston motor 旋转活塞马达;转子发动机
rotary piston pump 旋转(式)活塞泵;回转活塞泵
rotary piston sealing ring 转子密封环
rotary piston steam engine 旋转活塞蒸汽机
rotary piston-type-positive-displacement flowmeter 旋转活塞式正压排代流量计
rotary planning 旋转刨
rotary plate 转阀芯
rotary plate feeder 转盘式给料器
rotary plate filler 旋转板式给料器
rotary plate motion 板块旋转运动
rotary platform 旋转平台
rotary plating barrel 旋转电镀桶
rotary plough 旋转犁
rotary plough feeder 旋转犁式给料机;圆盘给料机
rotary plow 旋转式除雪犁
rotary plow feeder 叶式给料器;叶轮给料器;转动犁式给矿机
rotary plug valve 转阀
rotary-plunger pump 旋转(式)柱塞泵
rotary pneumatic actuator 旋转式气动马达
rotary pneumatic tool 旋转气动工具
rotary polarization 旋转极化;旋光性
rotary polygonal piston engine 三角转子发动机
rotary power broom 旋转式机动路刷
rotary power tool 旋转电动工具
rotary press 滚压机;旋转式压路机;工作(台)回转式压力机;轮转印刷机
rotary prime mover 旋转式原动机
rotary printing machine 滚筒印花机
rotary print washer 旋转洗片机
rotary prism 旋转棱镜;对棱棱镜
rotary prism camera 回转棱镜式照相机
rotary process 旋转法;离心法
rotary-proof press 轮印机
rotary pulse generator 循环脉冲发生器
rotary pump 回转(水)泵;旋转式泵;转子泵;转轮泵

rotary pump with liquid piston 液环泵
rotary rainer 旋转式喷灌器
rotary rain gun 旋转式远射程喷灌装置;喷雨枪
rotary rake 旋转式搂草机
rotary-rake thickener 转耙式浓密机
rotary rectifier 旋转整流机
rotary reflector 旋转角反射器
rotary regenerative air heater 旋转再生空气加热器
rotary regenerative heater 旋转式再生加热器
rotary regenerator 旋转式再生器
rotary relay 旋转式继电器
rotary release 旋转释放
rotary releasing spear 螺旋松紧式捞管器
rotary residence tower 旋转式公寓塔楼
rotary resistance 转动阻力
rotary retort 旋转式干馏炉
rotary retort type electric(al) resistance furnace 滚筒式电阻炉
rotary reverse switch 旋转反向开关
rotary rheometer 转动流变仪
rotary rig 旋转式钻探机;旋转钻井机;旋转式钻孔机;旋转(式)钻机
rotary rigder 旋转式培土机
rotary ring crusher 旋转式碎石机;旋转环破碎机
rotary ringing generator 旋转铃流机
rotary road brush 转轮式扫路机
rotary roadway 环形车道
rotary rock drill 旋转(式)钻岩机;旋转(式)钻石机
rotary rock drilling 旋转凿岩;旋转钻岩
rotary roller pump 回转滚筒泵
rotary rolling 滚压扩管
rotary rolling mill 蘑菇形轧辊的穿孔机
rotary roughening 旋转式粗糙化
rotary runner 转盘钻司钻
rotary sagger machine 匣钵回转成型机
rotary sampler 回转式取样器;回转取土器
rotary sampling switch 旋转转换开关
rotary sand 烘砂机
rotary sand drier 卧式烘砂滚筒
rotary sander 砂轮机
rotary sand screen 旋转式筛砂器
rotary saw 旋转锯;圆盘锯;转锯;回转式锯
rotary scan mirror 旋转扫描镜
rotary scanner 旋转扫描器
rotary scanning spectroscope 机械扫描分光镜
rotary scoop trencher 旋转式铲斗挖沟机;轮式挖沟机
rotary scraper 旋转(式)刮土机
rotary screen 圆筒筛;旋转筛(选机);圆网;转(筒)筛;转动筛;多边筛
rotary screen breaker 带旋转筛的破碎机
rotary screen digger 旋转筛式挖掘机
rotary screen dryer 圆网烘燥机
rotary screen magnetic head 旋转筛磁头
rotary screen printing 圆网印刷法
rotary screen roller printing machine 圆网辊筒印花机
rotary screen type precleaner 旋转圆筒筛式粗选机
rotary screw compressor 旋转螺旋压缩机
rotary screwing chuck 旋转螺纹钢板盘
rotary scrubber 旋转式洗涤机;离心洗涤器
rotary scrub slasher 旋转式灌木铲除机
rotary scythe 旋转式割草机
rotary scythe grass cutter 旋转镰刀割草机
rotary seal 旋转密封;转动密封
rotary search 旋转搜索
rotary seat 旋转座
rotary selector 旋转选择器
rotary servomotor 旋转伺服马达
rotary servo valve 旋转伺服阀
rotary sewing machine 旋转梭子式缝纫机
rotary shaft set bolt 转轴固定螺钉
rotary share 旋转犁掘铲
rotary shear 滚轮剪切机;旋转剪床
rotary shearing motion 旋转剪切运动
rotary shear line 回转式剪切作业线
rotary shears 滚轮剪机;回转式剪切机
rotary shell 旋转壳
rotary shield 旋转式盾构
rotary shifts 倒班制
rotary shoe 铣鞋
rotary shot drill 旋转式炮眼钻;回转钢砂钻具
rotary shunt-type flowmeter 旋转分路式流量计
rotary shutter 旋转遮光器;旋转式遮光器;旋转快门

rotary side trimming shears 圆盘式切边剪
rotary sieve 回转筛;旋转滤网;转筒筛
rotary signal 旋转信标
rotary single-vane compressor 单叶回转压缩机
rotary sintering kiln 旋转烧结窑
rotary skip 旋转式翻斗车
rotary slag-resistance test method 旋转抗渣试验法
rotary slaker 旋转式消石灰器
rotary sleeve inductor type magneto 转套感应式磁电机
rotary sleeve valve 转筒阀
rotary slide tongs 旋转式滑钳
rotary slide valve 回转滑阀
rotary slide valve engine 旋转滑阀式发动机
rotary-sliding-vane flowmeter 旋转滑动叶片式流量计
rotary slip 旋转钻管(卡瓦);卡瓦
rotary snow plough 旋转式雪犁;旋转式扫雪机;旋转式犁雪机
rotary snow plow 旋转式除雪机;转子式除雪型
rotary soil mixer 旋转式土壤拌和机
rotary solenoid 旋转螺线管
rotary solenoid relay 旋转螺管式继电器
rotary spading machine 旋转铲
rotary spark gap 旋转火花隙;旋转火花放电器
rotary speed 旋转速率;转速;回转速度
rotary speed control 转速控制
rotary speed controller 转速控制器
rotary speed exponent 转速指数
rotary speed mixer 转筒式快速拌和机;转筒式拌和机;转鼓式拌和机
rotary spheric(al) digester 蒸球;球形蒸煮器
rotary spraying machine 旋转式喷涂机
rotary sprinkler 旋转式喷淋机;旋转式喷洒机;旋转(式)喷洒器;旋转式喷头;旋转式喷灌器;旋转式喷灌机
rotary sprinkler system 转式浇灌
rotary sprinkling method 旋喷法
rotary spud 旋转式定位桩【疏】
rotary squeegee washer 旋转式加压水洗机
rotary squeeze 回转锻造
rotary squeezer 旋转压榨机;回转压榨机
rotary stacker 旋转式堆积机;旋转式叠料器
rotary stage 旋转阶段;旋转平台
rotary stall 转动式挤奶厅
rotary standpipe 立管
rotary starting air distributor 回转式起动空气分配阀
rotary steam-heated film drier 内部蒸汽加热式转动薄膜干燥机
rotary steam joint 旋转式汽管接头
rotary stepped machine 旋转式步进电机
rotary stepping motor 旋转式步进电动机
rotary stepping relay 旋转步进继电器;步进继电器
rotary stepping switch 步进继电器
rotary sterilizer 旋转杀菌器
rotary stop sign 转动式停车标志
rotary storm 旋转风暴
rotary strainer 旋转式滤尘器;圆筛
rotary stream 旋转潮流
rotary stress 旋转应力
rotary strip casting process 转动铸条法
rotary subsoiler 旋转式深耕铲
rotary supply air outlet 旋转式送风口
rotary support frame 转盘座架
rotary surface washer 旋转表面冲刷器
rotary swager 旋转锤锻机
rotary swaging forging 旋转锻造
rotary swaging machine 旋转锻打机;旋锻机;转动型锻机
rotary sweeper 旋转扫路机
rotary switch 旋转开关;旋转机键
rotary switch board 旋转式交换机
rotary swivel 转头;转环;水龙头【岩】
rotary symmetric(al) diaphragm 旋转式对称膜片
rotary synchronizer 旋转式同步器;指针式同步器
rotary synchroscope 旋转式同步示波器
rotary system 环形道口;环路系统
rotary system of drilling 旋转钻进法
rotary table 旋转台;转台;转盘;转动台;分度台
rotary table bushing 转盘衬套
rotary table charging car 转盘式装煤车
rotary table drill 转盘钻机
rotary table feeder 旋转盘式加料机;旋转进料器;

转盘喂料机;转盘给料机;喂料盘
rotary table power 转盘功率
rotary table press 转盘式压(力)机;转盘式压(团)机
rotary table reverse circulation drill 转盘反循环钻机
rotary table rig 转盘式钻机
rotary table rotations per minute 转盘转数
rotary table sand blasting machine 转台喷砂机
rotary table stabiliser 转台稳定器
rotary tablet machine 旋转(式)压片机
rotary tablet presser 旋转压片机
rotary table transmission counter-shaft 转盘式变传动副轴
rotary table type surface treatment machine 回转盘式表面处理机
rotary target 旋转靶
rotary teeth 旋转齿;回转齿
rotary test rig 旋转试验机
rotary texture 旋转结构
rotary thermal rectifying column 旋转式精馏柱
rotary throttle 回转节流阀;转子式节气门
rotary throttle valve 转动节流阀
rotary thrust 旋转推力
rotary tidal current 旋转潮流;回转潮流
rotary tidal waves system 旋转潮波系统
rotary tide 旋转潮
rotary tile press 回转压瓦机;回转式压砖机
rotary tillage 旋耕
rotary tiller 旋转式松土机;旋转式翻土机
rotary timing valve 回转调时阀
rotary tipper 转子式翻车机
rotary tippler 旋转式翻斗车
rotary tongs 大管子钳;钻杆钳;大钳
rotary tool 转盘钻具
rotary torque 回转力矩
rotary-tower crane 塔式转动起重机;旋转式塔吊;回转塔式起重机;塔式旋动起重机
rotary traffic 环行交通(交叉口);环绕交通
rotary trammel fluidized dryer 转筒式流化干燥器
rotary transducer 转速传感器
rotary transfer machine 回转式输送机
rotary transformer 旋转变压器;旋转变量器;变流机
rotary-tray table 转盘式淘汰盘
rotary trencher 旋转式挖沟机
rotary trimming shears 圆盘式切边剪
rotary triple prism 旋转式三棱镜
rotary trochoidal piston engine 余摆线转子发动机
rotary trochoidal-type engine 旋转式余摆线发动机
rotary truck 转子发动机货车
rotary tube expanding machine 斜轧式扩管机
rotary tube furnace 旋转管式炉
rotary tube straightener 旋转式钢管矫直机
rotary tube type magnetic roasting furnace 旋转筒式磁化焙烧炉
rotary tubular calciner 回转烘干机煅烧机
rotary tubular conveyer 转筒式运输机
rotary tumbler 转筒
rotary turret lathe head 转塔车床六角头
rotary type 旋转式;回转式
rotary type agitator 回转搅拌器;回转搅拌机
rotary type air separator 旋转式选粉机
rotary type cam-driven tableting press 旋转式凸轮驱动压片机
rotary-type cooler 单管冷却机;环式冷却机
rotary type cutter 盘形剃齿刀
rotary type flash trimmer 滚筒式焊缝清理机
rotary type fuel feed pump 旋转式供油泵
rotary type heating furnace 旋转式加热炉
rotary-type mixer 转轴式搅拌机;转轴式拌和机;旋转式拌和机
rotary type pile driving plant 旋转式打桩设备
rotary-type pneumatic tool 回转式风动工具
rotary-type pump 旋转泵
rotary type sampling tube 旋转式采样器
rotary type screen 旋转型格网
rotary type shutter 旋转式快门
rotary type soil stabilizer 转子式稳定土搅拌机
rotary type thermoforming machine 旋转型热成型机
rotary type twin sheet thermoforming machine 旋转型双板热成型机
rotary type vibrator 旋转式振动机;旋转式振动器
rotary underreamer 旋转扩孔器;转盘钻进扩孔钻头

rotary vacuum drum filter 转筒式真空过滤器;转筒式真空过滤机
rotary vacuum dryer 旋转式真空干燥器
rotary vacuum evapo(u)rator 旋转真空蒸发器
rotary vacuum filter 旋转(式)真空过滤器
rotary vacuum filtration 旋转式真空过滤
rotary vacuum pump 旋转(式)真空泵
rotary valve 旋转阀(门);旋动阀;回流阀;球形阀(门)
rotary valve key 回转阀键
rotary valve seat 回转阀座
rotary valve spring 回转阀簧
rotary valve train 回转阀系
rotary-vane and directly-connected vacuum pump 旋片式直联真空泵
rotary-vane arrangement 旋转叶片装置
rotary-vane attenuator 旋片型衰减器
rotary-vane batcher 旋转叶片式配料器;旋转叶片式配料机
rotary-vane compressor 叶轮式压缩机
rotary-vane feeder 旋转叶片给料器;叶片加料器;旋叶给料器;旋叶送料器;分格轮(式)喂料器
rotary-vane meter 旋转翼流量计;叶片式水表
rotary-vane mixing pump 回旋叶片混合泵
rotary-vane motor 转翼式液压马达
rotary-vane pump 旋转轮叶式(水)泵;转动叶片泵
rotary-vane steering engine 回转叶片式舵机
rotary-vane steering gear 转翼式操舵装置
rotary-vane type discharger 分格轮(式)卸料器
rotary-vane-type-positive-displacement flowmeter 旋转叶轮式正压排代流量计
rotary-vane type pump 旋转式叶片泵
rotary-vane vacuum pump 回转叶片式真空泵
rotary vector 旋转向量
rotary velocity 旋转速度;转速
rotary veneer 旋制薄板;旋制层板;旋切单板
rotary veneer cut 单板旋切;单板旋切
rotary vibrating riveter 旋转振动铆钉机
rotary vibrating tippler 振动翻车机
rotary vibratory fine screening 旋转式振动细筛
rotary vise 旋转虎钳
rotary visco(si)meter 旋转(式)黏度计
rotary voltmeter 旋转(式)伏特计;高压静电伏特计
rotary wall furnace 转壁炉
rotary warded mechanism 旋转监视器
rotary wash boring equipment 旋转式水冲钻探设备
rotary washer 旋转式清洗机;旋转式表面冲洗器
rotary washing nozzle 旋转洗涤喷嘴
rotary washing pan 转动式淘洗盘
rotary washing screen 旋转式冲洗机;洗矿滚筒筛
rotary washover shoe 套筒冲洗铣鞋
rotary waste heat dryer 回转式余热烘干机
rotary water meter 旋转式水表
rotary water-spreader 旋转式布水器
rotary water-valve 旋转水门
rotary wave 旋转波
rotary waveguide variable attenuator 旋转波导可变衰减器
rotary weeder 旋转式除草机
rotary welding set 旋转式焊机
rotary welding transformer 旋转式焊接变压器;可旋转焊接变压器
rotary wheel-type rake 指轮式搂草机
rotary winding machine 回转式绕线机
rotary wire brush 电动旋转钢丝刷
rotary working 回转加工
rotascope 转动机观察仪
rotas system 翻转胎架造船法;翻转轮胎造船法
rotatable cabin 旋转舱
rotatable design 旋转设计
rotatable feeding shoe 转动给料式刮板
rotatable immersion device 可旋转的浸渍装置
rotatable parts 转动部件
rotatable phase adjusting transformer 旋转相位调整变压器
rotatable phase-adjusting transformer 调相器
rotatable shield 转台式屏蔽
rotatable spray head 可旋转喷雾头
rotatable transformer 旋转变压器
rotate 转动;轮作;回转;辐状的
rotate days off 轮休
rotated bar air sampler 旋转棒空气采样器
rotated disk electrode 转盘电极
rotated dropping mercury electrode 旋转滴汞电极
rotated plate 旋转板块
rotated platinum microelectrode 旋转微型铂电极
rotated rectangular array 旋转矩形阵列
rotated sampling 循环取样法
rotated vector field 旋转向量场
rotate with a wrench 手搬大钳回转
rotating 旋转(性)的
rotating active diffuser 旋转活动式扩散器
rotating aerial 回旋天线
rotating airplane hangar 可旋转的飞机库
rotating air separator 离心式选粉机
rotating amplifier 旋转式放大机;旋转式磁放大器;旋转放大器;电机放大器
rotating amplifier excitation system 旋转放大器式励磁系统
rotating and circulating 回转并循环冲洗
rotating and sliding system 翻转胎架造船法;翻转轮胎造船法
rotating angle 转角
rotating angle of mandrel 芯模转角
rotating annular cooler 环式冷却机
rotating anode 旋转阳极
rotating-anode tube 旋转阳极管
rotating anode X-ray tube 旋转阳极X射线管
rotating apartment tower 旋转式公寓塔楼
rotating arc 旋转电弧
rotating arc pipe welding machine 旋转电弧焊管机
rotating arc welder 旋转式弧焊机
rotating arc welding 旋转电弧焊
rotating arc welding machine 旋转式弧焊机
rotating arc welding set 旋转式弧焊机
rotating arm 旋转臂;回转臂
rotating armature 旋转电枢
rotating-armature generator 转枢式发电机
rotating-armature machine 旋转电枢式同步电动机;转枢式电机
rotating-armature motor 转枢式电动机
rotating-armature relay 转动衔铁继电器
rotating-armature-type exciter 转枢式励磁机
rotating arm winding machine 立式缠绕机
rotating auger 麻花钻;螺旋钻
rotating-auger test 螺旋钻试验
rotating-auger tester 螺旋钻检验器
rotating axis 自转轴
rotating ballhead 旋转球头
rotating ball viscometer 转求法黏度计
rotating band 转动带;驱动带
rotating bar impact fatigue test 转杆式冲击疲劳试验
rotating barrel 旋转枪管
rotating barrel shutter 转筒快门
rotating barrier 旋转隔板
rotating base 旋转底座
rotating beacon 旋转指向标;旋转信标;旋转式灯塔;旋转灯标;旋转航标
rotating beacon light 旋转式防航信号灯
rotating beacon station 旋转信标台
rotating beam 旋转射束;回旋射束
rotating beam ceilometer 旋转射束云高计
rotating beam chopper 旋转式断束器
rotating beam transmitter 旋转波束发射机
rotating beam type machine 转杆式疲劳试验机
rotating bearing seal 旋转封接;转动轴承密封
rotating bed contactor 旋转床接触器
rotating biologic(al) contactor 旋转式生物接触器
rotating biologic(al) contact reactor 旋转式生物接触反应器
rotating biologic(al) disc process 生物转盘法
rotating biologic(al) disk 生物转盘
rotating biological extended contactor 扩展式旋转生物接触器
rotating biologic(al) filter 旋转式生物滤器
rotating biologic(al) media 旋转式生物培养基
rotating biologic(al) surface process 旋转式生物表面处理
rotating blade 转动叶片;转动刮板
rotating blade column for liquid-solid extraction 转盘式液固萃取塔
rotating blade drive 转盘驱动装置
rotating blade groove 动片叶根槽
rotating-blade propeller 直翼推进器;平旋推进器
rotating blade-type disintegrator 旋转叶片式粉碎机
rotating block linkage 回转滑块联动机构
rotating blowpipe 旋转吹管
rotating body 旋转物
rotating bookcase 旋转书架
rotating boom 旋转式横臂;旋转式悬臂
rotating boring 旋转式钻探
rotating bowl 转筒
rotating-bridge scraper 转动桥式刮泥器
rotating broom 旋转式路刷;旋转式扫路机;旋转式路帚
rotating brush 旋转路刷;旋转电刷;回转刷子;旋转毛刷
rotating brush aeration 转刷曝气
rotating brush attachment 旋转路刷附件
rotating brush separator 转刷式分离器
rotating bucket 旋转式铲斗;转斗
rotating building crane 旋转式建筑施工起重机
rotating cabin 旋转舱
rotating caisson 旋转沉箱
rotating cantilever beam type machine 转梁式疲劳试验机
rotating capacitor 旋转电容器
rotating cardioid pattern 旋转的心脏形方向性图
rotating cascade 旋转叶栅
rotating cathode 旋转阴极
rotating cell 转动隔仓
rotating center 旋转中心;回转顶尖
rotating chaff distributor 转动式颖糠匀布器
rotating choke 旋转扼流圈
rotating choke-coil 旋转扼流圈
rotating chopper 旋转调制盘
rotating clamp 回转压板
rotating classifier 旋转式粗粉分离器
rotating coaxial-cylinder viscometry 旋转同轴圆筒测黏法
rotating coil 转动线圈
rotating-coil fluxmeter 旋转线圈式磁通计
rotating-coil indicator 旋转线圈指示器
rotating-coil method 旋转线圈法
rotating coil plan position indicator 动圈式平面位置指示器
rotating combustion engine 转子发动机
rotating compressor 旋转式压缩机
rotating concentric cylinder extractor 旋转中心柱萃取器
rotating concentric tube distilling column 旋转同心管分馏柱
rotating concrete mixer 自落式混凝土搅拌机
rotating condensation 旋转式絮凝(作用);旋转式冷凝(作用)
rotating conical feeder 旋转圆锥给料机
rotating contactor 转动接触器;生物转盘
rotating control assembly 旋转控制装置
rotating converter 旋转变流器
rotating coordinate system 转动坐标系
rotating core type relay 旋转铁芯式继电器
rotating couple 旋转力偶
rotating crane 旋转式起重机;旋转吊车
rotating crank 转动曲柄
rotating crank gear 转柄装置
rotating crossfeed unit 转动横进刀装置
rotating crystal method 旋转晶体法;转晶法;周转晶体法
rotating crystal pattern 旋晶衍射图;旋转晶体衍射图
rotating crystal X-ray photograph 旋转晶体X射线照相
rotating cup type current meter 旋杯式流速仪;旋杯式海流计
rotating current meter 旋转式流速仪
rotating cutter 旋转切割器;旋转切割机;旋转切刀
rotating cutter head 旋转式刀盘
rotating cutting mechanism 旋转式切割器
rotating-cut veneer 旋转切出层板
rotating cylinder 旋转柱体;旋转筒;旋转缸
rotating cylinder method 转筒法
rotating cylinder vacuum ga(u)ge 转筒真空计
rotating cylinder visco(si)meter 旋转(圆筒)式黏度计;转筒黏度计
rotating cylindric(al) drum 圆柱形滚筒
rotating cylindric(al) kiln 回转圆筒窑
rotating davit 旋转式吊艇柱
rotating derrick 悬臂起重机;旋转式起重机

rotating detector 旋转式传感器
rotating device 旋转装置;回转机构
rotating dial knob 转动刻度盘的手柄
rotating diamond bit 旋转式金刚石钻头
rotating diffuser 旋转扩散体;旋转扩散器
rotating dipper unit 旋转铲斗装置
rotating direction finder 旋转(式)测向仪
rotating disc 转盘
rotating disk analyzer 转盘式扫描器
rotating disk atomizer 转盘雾化器
rotating disk biofilm reactor 转盘式生物膜反应器
rotating disk column 转盘塔
rotating disk contactor 转盘接触器
rotating disk device 转盘装置
rotating disk electrode 旋转圆盘电极
rotating disk extraction tower 转盘萃取塔
rotating disk extractor 转盘萃取器
rotating disk field store 转盘式场存储器
rotating disc filter 转盘式过滤器
rotating disk friction generator 转盘摩擦起电机
rotating disk memory 转盘式存储器
rotating disk mill 转盘研磨机
rotating disk scanner 转盘式扫描器
rotating disc shutter 转盘式快门
rotating disc unit 转盘单元
rotating disk vacuum ga(u)ge 转盘真空计
rotating disk visco(si)meter 旋转圆盘式黏度计
rotating distributor 旋转喷水器;回旋配水器
rotating door 旋转门
rotating double crank 转动双曲柄
rotating drilling 旋转式钻孔;旋转式钻探;回转钻进
rotating drive 旋转传动装置;旋转驱动器
rotating drum 旋转鼓轮;转鼓
rotating drum camera 转鼓式照相
rotating drum furnace 转筒窑
rotating drum heat transfer 转鼓传热
rotating drum image scanning digitizer 鼓形图像扫描数字化器
rotating drum incinerator 转鼓式焚化炉
rotating drum method 转筒法
rotating drum mixer 转筒式搅拌机;转筒混合机
rotating drum multistage crystallizer 转筒多级结晶器
rotating drum paver mixer 转筒式铺路拌和机
rotating drum relay 转鼓式继电器
rotating drum shears 滚筒式剪切机;滚筒式飞剪
rotating drum type variable message sign 转筒式可变指示标志
rotating dumbbell pseudo-gravity station 旋转式哑铃形人造重力航天站
rotating dumper 转子式翻车机
rotating duplex compressor 双联式旋转压气机
rotating dwelling tower 旋转式居住塔楼
rotating eccentric weight exciter 偏心起振机
rotating electric(al) field 旋转电场
rotating electric(al) machine 旋转电机
rotating electrode 转动电极
rotating electron 旋转电子
rotating element 旋转元件;转动元件
rotating environment 旋转环境
rotating equipment 旋转设备;转动设备
rotating equipment data 转动设备数据
rotating evapo(u)rator 旋转式蒸发器
rotating excavating bucket 转动挖斗
rotating exciter 旋转励磁机
rotating extractor 旋转萃取器
rotating face 旋转面
rotating fairing 旋转导流罩
rotating falling slurry photocatalytic reactor 旋转向下流动悬浮液光催化反应器
rotating fender 旋转式护舷
rotating field 旋转场
rotating-field antenna 旋转场天线
rotating-field electromagnetic pump 旋转场电磁泵
rotating-field exciter 转场式励磁机
rotating-field generator 转场式发电机
rotating-field instrument 旋转磁场式仪表
rotating-field machine 转场式电机
rotating-field magnet 旋转场磁极
rotating-field type 旋转场型
rotating-field type alternator 旋转场型交流发电机
rotating-field type engine 旋转场型交流发动机
rotating filter 旋转滤色盘;滤色盘;回转式过滤器
rotating filter drum 旋转滤鼓;旋转过滤鼓筒

rotating filtering drum 转筒式过滤机;转筒式过滤器
rotating flail spreader 旋转甩击式撒布机
rotating flange 转动凸缘
rotating fluidized bed combustion 旋转式流化床燃烧
rotating frame 旋转框
rotating frame antenna 转动框形天线
rotating furnace 旋转炉;旋转焙烧炉
rotating furnace drum 旋转式炉鼓
rotating generating unit rotor 旋转发电装置转子
rotating gland 旋转式填料盖
rotating go-devil 旋转清管器
rotating grab scoop 旋转铲斗
rotating grading screed 旋转式路基整平板
rotating grate shaft kiln 旋转炉篦式立窑
rotating gravity concentrator 旋转式重力浓缩器;重力转动浓缩器
rotating grizzly 滚轴筛;辊轴筛
rotating guide 旋转导板
rotating guide vane 旋转导叶
rotating half-bridge scraper 旋转半桥式刮泥器
rotating head 旋转机头
rotating-head sprinkler 旋转喷头式喷洒器
rotating hexagonal prism 旋转六角棱镜
rotating hours 机械钻进时间;纯钻进时间
rotating house 旋转房屋
rotating housing 旋转式护罩
rotating housing tower 旋转式住宅塔楼
rotating hyperboloid 旋转双曲面
rotating inductor-type magneto 旋转感应式磁电机
rotating iron 旋转衔铁
rotating-iron type over-current relay 旋转衔铁式过流继电器
rotating-iron type over-voltage relay 旋转衔铁式过压继电器
rotating-iron type under-voltage relay 旋转衔铁式欠压继电器
rotating jib 悬臂吊机
rotating jig for welding 旋转式电焊台
rotating joint 旋转连接器;联轴器
rotating joint with multiple flow paths 多通回转接头
rotating journal bearing 转动轴颈轴承
rotating key clutch 转键离合器
rotating kiln 回转窑
rotating launcher 旋转式发射装置
rotating lens beacon 旋转航标灯;旋转导航标
rotating lens drum 旋转透镜筒
rotating level(l)er 旋转式整平器;旋转式平地机
rotating lifting spinner 旋转式挖掘抛掷机
rotating light 旋转(式)灯标;旋转灯(光)
rotating lighthouse system 旋转灯塔仪
rotating link 旋转联杆
rotating liquid 旋转液体
rotating load 转动负荷
rotating loader 转臂式装载机
rotating loop 环形旋转天线;旋转环状天线
rotating loop antenna 旋转环形天线
rotating loop beacon 旋转环状(天线)指向标
rotating loop radio beacon 旋转式无线电导标;旋转环状无线电指向标;旋转环形无线电信标
rotating loop station 旋转环形天线信标台
rotating loop transmitter 旋转环形天线发射机
rotating machine 旋转机;旋转电机
rotating machinery 旋转式机器
rotating magnet 转动式磁铁
rotating magnetic amplifier 旋转式磁放大器
rotating magnetic field 旋转磁场
rotating magnet magneto 转极式电机;回磁铁磁电机
rotating magnetometer 回旋式磁力仪
rotating-magnet separator 旋转磁铁式磁选机
rotating manipulation 摇法
rotating mass 旋转质量
rotating mass shaker 旋转质量起振器;偏心起振机
rotating mast crane 旋转桅杆式起重机
rotating-mat V-type cell 旋转垫衬 V 形压气浮选槽;带旋转多孔面的 V 形槽气压浮选槽
rotating mechanism 旋转机构
rotating member 旋转部件
rotating mercury tachometer 旋转式水银转速计;旋转式水银转数计
rotating meter 速度式流量计
rotating mirror 旋转镜;旋镜

rotating mirror analyzer 旋转镜扫描器
rotating mirrors method 傅科法
rotating mirror system 转镜系统
rotating mixer 转鼓式拌和机;转筒式拌和机;旋转搅拌机;转鼓式搅拌机;转动搅拌机
rotating model 旋转模型
rotating moment 旋转力矩
rotating monochromator 旋转式单色器
rotating mooring boom 旋转式系船臂
rotating mud scraper 旋转式刮泥机
rotating nozzle 旋转式喷嘴;旋转式喷丝嘴;转动喷嘴
rotating off-axis blade 偏轴旋转叶片
rotating off-axis reticle 轴外旋转调制盘
rotating on-axis reticle 轴上旋转调制盘
rotating optical-bar panoramic camera 回转光轴全景照相机
rotating packing machine 回转式包装机
rotating paddle (供料机的)旋转刮板;旋转叶轮;旋转叶片
rotating paddle type level indicator 回转桨叶式料面高度指示器
rotating-paddle type material-level indicator 旋桨叶式料位指示器
rotating pan mixer 转盘混合器;转锅式拌和机
rotating parachute 旋转伞
rotating part 旋转部分;回转部分;转动部件
rotating particle 旋转粒子
rotating pattern radio beacon 旋转型无线电指向标;旋转型无线电信标
rotating phase converter 旋转变相机
rotating phasor 旋转相量
rotating photograph 旋转图
rotating photograph indexing 旋转图指标化
rotating pig 旋转清管器
rotating pilot valve 旋转滑阀
rotating piston 旋转活塞
rotating piston pump 旋转式活塞泵
rotating plane-vector 平面旋转向量
rotating plasma device 转动等离子体装置
rotating plate 旋转台
rotating platform 转台
rotating platinum electrode 旋转铂电极;转动铂电极
rotating plug 旋转屏蔽塞
rotating polarizing filter 旋转式偏振滤光片
rotating pole type magneto 转极式电机;转场式磁电机
rotating potentiometer 旋转电位计
rotating priority scheme 轮流优先权方案
rotating prism 旋转棱镜;转像棱镜
rotating prism panoramic camera 旋转棱镜全景照相机
rotating pulley 转动皮带轮
rotating pump 旋转式水泵;旋转泵
rotating radio beacon 旋转无线电指向标;旋转(无线电)航标;旋转式无线电导标;无线电旋转指向标
rotating radio beacon station 旋转无线电指向标站;旋转无线电指向标台;旋转无线电信标台
rotating rake mechanism 旋转式耙集机构
rotating ram 旋转式撞槌;旋转式夯具
rotating rectifier 旋转整流器
rotating rectifier alternator 旋转整流器发电机
rotating rectifier exciter 旋转整流器励磁机
rotating restaurant 旋转餐厅
rotating reticle 旋转形调制盘
rotating rheostat box 旋转式可变电阻箱
rotating ring disk electrode 旋转环盘电极
rotating rock drilling 旋转式岩石钻孔
rotating rod viscometer 转杆式黏度计
rotating room 旋转室
rotating sample magnetometer 转动样品磁强计
rotating sampler 旋转取样器;回转取土器
rotating sand washer 旋转洗砂机
rotating scanner 旋转扫描器
rotating scanner method 旋转扫描法
rotating scoop 旋转铲斗
rotating scraper 旋转清管器
rotating scrubber 旋转式洗涤器
rotating seal 旋转密封
rotating seat 旋转座;动环座
rotating sector 旋转扇形齿轮;旋转窗
rotating sector method 旋转光闸法

rotating sector wheel 扇形旋转轮
rotating segment feeder 旋转星形给料器
rotating shaft 转轴
rotating shield 旋转屏蔽体
rotating shield plug 旋转屏蔽塞
rotating shift 轮流换班;轮班
rotating shift system 轮班制度
rotating shovel 转式挖土机;转式挖土铲;转式铲土机
rotating shutter 旋转快门;转动光闸
rotating shuttle 旋梭
rotating side peak 旋转边峰
rotating sieve 回旋筛
rotating silicon rectifier 转动式硅整流器
rotating skip 旋转式翻斗
rotating slaker 旋转式消石灰器
rotating sleeve type magneto 旋转套筒式磁电机;转套式磁电机
rotating slider crank mechanism 回转滑块曲柄机构
rotating slide valve 回转滑阀
rotating sludge rake 旋转式泥铲
rotating sludge ranger 旋转式泥耙
rotating sludge scraper 旋转式刮泥机
rotating soil loosener 旋转式松土机
rotating soldering machine 旋转式焊(接)机
rotating space station 旋转航天站
rotating speed 转速;回转速度
rotating sphere 旋转球
rotating spindle visco(si)meter 转轴式黏度计
rotating spray 旋转式喷动雾
rotating sprayer 旋转式喷雾器
rotating spray transfer 旋转喷射过渡
rotating spreader 旋转分流梭
rotating sprinkler 旋转式喷灌器;旋转式喷灌机;旋转洒水器;旋转喷洒器
rotating stage 旋转台
rotating stainless steel cage 回转不锈钢笼(浸出用)
rotating stall 旋转失速
rotating stand 转架
rotating standard 电度表转数标准
rotating stirrer 旋转搅拌器
rotating stop 转动光阑
rotating stop sign 转动式停车标志
rotating storage container 转动式储藏容器
rotating storage rack 旋转储藏架
rotating suction pipe dredge(r)旋转吸管式挖泥船
rotating surface 旋转面
rotating surface type air speed indicator 转面式风速计
rotating surveillance radar 旋转天线监视雷达
rotating switch 旋转开关;转动开关
rotating synchro 旋转同步机
rotating synchronous exciter 旋转同步励磁机
rotating table 旋转台;转台;转盘;磨盘
rotating table press 转盘式压(力)机;转动台式压砖机
rotating talking beacon 旋转对话式信标
rotating tape guide 导带轮
rotating target 旋转靶
rotating temple 旋转模板
rotating tensioning device 转动式张力装置
rotating the needle 摇针;针
rotating thermometer 旋转温度计
rotating thin layer chromatograph 旋转薄层色谱仪
rotating thin layer chromatography 旋转薄层色谱法
rotating tipper 旋转式翻斗车
rotating tool 回转刀具
rotating torque 旋转力矩
rotating trace 旋转扫描线
rotating transformer 旋转变压器
rotating trolley 旋转式起重小车
rotating trowel 旋转抹子
rotating tumbling-barrel-type mixer 旋转圆筒混合品
rotating turbine aerator 旋转式水轮曝气器
rotating turret 旋转炮塔
rotating unbalanced mass 旋转不平衡质量
rotating vacuum drier 旋转真空干燥器
rotating valve 旋转阀(门);转动阀;回转阀
rotating valve gear 回转阀齿轮
rotating vane 转叶

rotating vane pump 旋转叶片泵
rotating variable 自转变星
rotating vector 旋转向量;旋转矢量
rotating velocity 转速
rotating vessel 旋转式搅拌机
rotating vibration test 旋转振动试验
rotating view 旋转视图
rotating visco(si)meter 旋转粘度计
rotating visco(si)meter vacuum ga(u)ge 旋转粘度计式真空计
rotating water distributor 旋转布水器
rotating water wheel distributor 旋转式水轮布水器
rotating wave 旋转波
rotating wedge 旋转光楔;旋光光楔
rotating wedge range-finder 转楔测距仪
rotating welding machine 旋转式焊机
rotating wheel 回转轮
rotating wire brushes 回旋钢丝刷
rotating wire rope 旋转钢丝绳
rotating worm plow 旋转式螺旋犁
rotation 旋量;旋光;旋度;轮流
rotation age 轮伐龄
rotational activator of cement 旋转式水泥活化器
rotational analysis 旋光分析
rotational and translator device 转动平移装置
rotational angular velocity 自转角速率
rotational axis 转动轴
rotational axis of the ellipsoid 椭球旋转轴
rotational axis of the sphere 球体旋转轴
rotational axis precession 自转轴进动
rotational balance 转动平衡
rotational band 转动谱带
rotational broadening 自转致宽
rotational casting 旋转铸塑;离心铸造
rotational center 转动中心
rotational characteristic 转动特性
rotational coaxial-cylinder viscometry 转动同轴圆筒测黏法
rotational component 旋转分量;转动分量
rotational constant 转动常数
rotational contour 转致轮廓
rotational control 旁向倾斜控制
rotational coordinates 旋转坐标;转动坐标
rotational current tray 旋转回流式塔盘
rotational deformation 旋转变形;转动变形
rotational delay 旋转减速
rotational diffusion coefficient 旋转扩散系数
rotational direction 旋转方向
rotational disorder 转动无序
rotational displacement 转动角位移
rotational distortion 旋转畸变
rotational disturbance 旋转扰动
rotational ductility 转动延性
rotational earth slump 圆弧形土层滑移;圆弧形土层滑塌
rotational effect 转动效应
rotational efficiency 转动效率
rotational energy 转动能;转动能(量)
rotational energy barrier 转动能障
rotational energy level 转动能级
rotational equilibrium 转动平衡
rotational error 旋转误差
rotational excitation 转动激发
rotational exposing image 旋转曝光图像
rotational exposing method 旋转曝光法
rotational fault 旋转故障;旋转断层
rotational field 旋转场;涡旋场
rotational fine structure 旋转微细结构
rotational flow 旋流;有旋流;涡流
rotational flow grit chamber 旋流沉砂池
rotational flow of mix 混合料的涡流
rotational forming 旋转成形
rotational frequency 转数;旋转频率;转动频率
rotational grain 旋转晶粒
rotational gravitational instability 转动重力不稳定性
rotational gravity force couple 旋转重力力偶
rotational grazing 轮牧;分区轮牧法
rotational grazing cycle 轮牧周期
rotational guide vane 旋转导向叶片
rotational hardening 旋转硬化
rotational impedance 转动阻抗
rotational inertia 转动惯量;惯性矩
rotational inertia coefficient 转动惯量系数

rotational instability 旋转不稳定性;自转不稳定度;自转不稳定性;转动不稳定性
rotational irrigation 轮灌;轮流灌溉
rotational isomerism 旋光异构
rotational kinetic energy 转动动能
rotational laminar displacement 旋转层状位移
rotational landslide 旋转滑坡;圆弧形滑坡;转动式滑坡
rotational lathe head 回转式车床头
rotational line 旋转线;转动线
rotational loss 旋转损耗;转动损失
rotationally symmetrical 轴对称的;旋转对称的
rotationally symmetrical flow 对称涡流
rotational model 旋转模型
rotation(al) motion 旋转运动;转动(运动)
rotational mo(u)lded 旋转铸塑
rotational mo(u)lding 旋转模塑;离心成型(法)
rotational mo(u)lding machine 旋转成型机
rotational movement 转动
rotational noise 旋转噪声
rotational nucleus 转动核
rotational of beam 射线转动
rotational oscillation 转振
rotational paraboloid 旋转抛物面
rotational position sensing 旋转定位读出
rotational profile 转致轮廓
rotational pseudo gravity 模拟旋转重力
rotational quantum number 转动量子数;振动量子数
rotational relaxation 转动张弛
rotational resistance 转动阻力
rotational restraint 旋转约束
rotational rigidity 旋转刚度
rotational scheduling 旋转结构
rotational shear 旋转剪切
rotational shear slide 旋转剪切滑动
rotational shear structural system 旋扭构造体系
rotational shear tectonic system 旋扭构造体系
rotational shell 旋转(薄)壳
rotational slide 旋转滑坡;旋转滑动;旋滑;圆弧滑动;转动滑座
rotational slip 旋转滑坡;转动滑移;滚动滑坡
rotational spectroscopy 转动光谱学
rotational spectrum 转动能谱
rotational speed 转速;转动速度;角速度;回转速度
rotational speed governor 转速控制器
rotational speed of collet 机头转速
rotational speed of drilling machine 钻机转速
rotational speed of spindle 立轴转速
rotational stability 转动稳定性
rotational state 转动态
rotational stiffness 旋转刚度;转动刚度
rotational stiffness factor 转动刚度系数
rotational stiffness ratio 转动劲度比
rotational strain 转动应变;转动应变
rotational strain ellipsoid 旋转应变椭球体
rotational stress 旋转应力
rotational structure 转动结构
rotational sub-level 旋转亚能级
rotational substitution 旋转代换
rotational sum rule 转动求和定则
rotational swirling flow 旋转流
rotational symetry 轴对称
rotational symmetry 旋转对称(性);对称旋转
rotational torque meter 力矩计;转矩表
rotational transform 转动变换
rotational transition 旋转跃迁;转动跃迁
rotational type of motion 旋转型运动;旋转式运动
rotational variable 自转变星
rotational velocity 旋转速度;转速;转动速度;角速度
rotational velocity component 转动分速度
rotational velocity-spectral type relation 自转速度光谱型关系
rotational vibration spectrum 转振光谱
rotational visco(si)meter 旋转黏度计;转动黏度计;库埃特黏度计
rotational visco(si)meter recorder 转动黏度计记录器
rotational voltage 旋转电压;旋转电势
rotational vortex 旋涡
rotational wave 旋转波;有旋波;切变波
rotation anemometer 旋转风速表
rotation angle 旋转角
rotation angle sensor 旋转角传感器

rotation axiom of perspective 透视旋转定律
rotation axis 旋转轴；转轴
rotation beacon 旋流信标
rotation camera 旋转照相机
rotation card file 旋转式卡片档
rotation center 旋转中心
rotation chuck 转动轴箱；卡盘
rotation coefficient 转动系数
rotation contour 转致轮廓
rotation cosine 旋转余弦
rotation couple 旋转力偶
rotation crop 轮作作物
rotation cropping 轮作栽培
rotation curve 自转曲线
rotation cycle 轮作周期
rotation davit 旋转式吊艇柱
rotation-deferred grazing 轮流延迟放牧
rotation deformity 旋转畸形
rotation detection mechanism 旋转检出机构
rotation detector 旋转检测器
rotation diad 二重转动轴
rotation diagram 旋转图；旋转图解
rotation diffusion constant 转动扩散常数
rotation direction 旋转方向；转动方向
rotation direction(al) of anisotropic mineral 非均质矿物的旋向
rotation direction sensor 转向传感器
rotation-disc unit 转盘装置
rotation displacement 旋转移位
rotation drill 旋转钻机
rotation-drum sign 转筒式标志
rotation dynamometer 回转式测力计
rotation effect 旋转效应；自转效应；涡流效应
rotation ellipsoid 旋转椭球面
rotation facing coupling end 从联轴节端看转向
rotation factor 旋转系数
rotation firing 多排孔间隔爆破；轮流引爆
rotation flow method 轮灌方式；轮灌法
rotation frequency 转动频率
rotation gate 旋转式阀门
rotation gear 转动机构
rotation glass 旋光玻璃
rotation grazing on an area basis 分区轮收；分区轮牧
rotation group 旋转群
rotation indicator 旋转指示器
rotation-inversion axis 旋转倒反轴
rotation jig for welding 旋转式电焊台
rotation matrix 旋转矩阵
rotation mechanical impedance 转动机械阻抗
rotation method 旋转法；转动法
rotation mirror 旋转镜
rotation mixer 行星摆转混砂机；行星摆轮混砂机
rotation moment 转矩
rotation momentum 自转角动量
rotation net 旋转网
rotation number 循环号码
rotation of axes 坐标轴旋转；轴旋转
rotation of beam 射线转动；波束旋转
rotation of binaries 双星绕转
rotation of bloom 轮流开花
rotation of circular polarized waves 圆极化波的旋转
rotation of coordinate axis 坐标轴的转动
rotation of crops 轮作
rotation of data 数据旋转
rotation of ecliptic 黄道转动
rotation of flange 法兰旋转
rotation of fodder crops and agricultural crops 草田轮作
rotation of foundation 基础旋转
rotation of maximum volume production 材积最多轮伐期
rotation of molecule 分子转动
rotation of plate 照准部旋转
rotation of the earth 地球自转
rotation of the highest income 收入最高轮伐期
rotation of the line of apsides 拱线转动
rotation of the plane of polarization 偏振面的旋光
rotation operator 转动算符
rotation parameter 旋转参数
rotation pasture 轮作牧场；轮换牧地；农牧轮作
rotation pattern 转动图样
rotation period 自转周期；轮伐期；绕转周期

rotation plan 轮作计划
rotation platform with ladder 带扶梯回转平台
rotation plough land 轮作田
rotation polarization 旋转偏光
rotation prevention joint 防转接头
rotation priority 循环优先
rotation profile 转致轮廓
rotation property 旋光性
rotation radius 回转半径
rotation rate 转动率
rotation recorder 旋转记录仪
rotation-reflection axis 旋转反映轴
rotation regulation 轮换调节
rotation sampling 轮换抽样
rotation seismograph 旋转地震仪；转动地震仪
rotation sense 旋转方向
rotation sensor 旋转传感器；转动传感器
rotation set screw 转动限位螺钉
rotation sideband testing method 旋转边带测试法
rotation sleeve bushing 旋转衬套
rotation slip 旋转滑动
rotation spectrum 转动谱；转动光谱
rotation speed 旋转速率；抬前轮速度
rotations per minute 转/分；每分钟转数
rotation spindle method 旋转针法
rotation stabilizer 防摇动装置
rotation strength of grinding wheel 砂轮回转强度
rotation surface 旋转面；旋回面
rotation symmetry 旋转对称
rotation synchronization 旋转同步
rotation synthesis bore 自转综合孔径
rotation system of grass and crop 草田轮作制
rotation technique 旋转技术
rotation temperature 自转温度
rotation tensor 转动张量；转动张量
rotation test 旋转试验
rotation torque 旋转（机的）转矩
rotation to the right 向右转
rotation transducer 旋转变换器
rotation transform angle 旋转变换角
rotation transformation 旋转变换
rotation twin 旋转双晶；旋转孪晶
rotation valve 旋转阀（门）
rotation-vibration band 转振谱带；转动振动谱带
rotation-vibration spectrum 转动振动谱
rotation-vibration state 转动振动态
rotation volume of freight transport 货运周转量
rotation wave 旋度波
rotative action 旋转作用
rotative component 转动分量
rotative flow 旋转水流
rotative mechanism for arm 导向翼板翻转机构
rotative moment 转动力矩
rotative power 转动功率
rotative speed 旋转速率；转速；转数
rotative surface 回转面
rotative velocity 旋转速度；转速
rotative velocity transducer 转速传感器
rotator 旋转器；旋转机；旋转反射炉；自转天体；转子；转轮；回转装置（盾构内）
rotator crusher 转子式碎石机
rotator noise 转子噪声
rotatory 循环的；转动的
rotatory axis 旋光轴
rotatory Brownian motion 转动布朗运动
rotatory converter 旋转变流机
rotatory current 旋转电流；多相电流
rotatory diffusion 转动扩散
rotatory dispersion 旋光(性)色散
rotatory dryer 旋转干燥器
rotatory evapo(u)rator 旋转蒸发器
rotatory fault 旋转断层
rotatory finisher 旋转平整机
rotatory force 转动力
rotatory irradiation 旋转辐照
rotatory polarization 旋偏振光；圆偏光
rotatory polarization angle 旋偏振光角
rotatory polarized light 转动偏振光
rotatory power 旋光本领
rotatory pump 转子泵
rotatory rock drill 旋转式钻岩机；旋转式钻石机
rotatory viscometer 回旋黏度计
rotatrol 旋转控制机
rotatruder 旋转输送机

rotavator 旋耕机
rotaversion 反顺转变
Rotdoom 罗特顿（一种抗木材虫害的防蚀剂）
rotenturm wind 红塔风
rotexion 转屈
rotex-screen 转动筛分机；滚球筛
Rothemuhle regenerative air preheater 风罩再生式空气预热器
rotine data flow 常规数据流
rotinoff pile （有混凝土浇注的）管桩
rotisserie 烤肉店
Rotliegende 赤底统
rotobaler 卷捆式捡拾压捆机
rotobaler for hay 干草卷捆式捡拾压捆机
rotobeater 滚刀式切碎机
rotobelt filter 滚筒胶带式过滤机
rotoblast 转筒喷砂；喷丸
rotoblast barrel 喷丸清理滚筒
rotoblasting 喷丸除鳞
rotoblower 离心调速器
rotocap 旋转管帽；旋转盖；气阀旋转机构
Rotocel extractor 洛特赛萃取机
rotocleaner 旋转式清粮筒；滚筒式清选器；滚筒清洗矿石机
roto-clone 旋涡收尘器
rotoclone collector 旋风吸尘器；旋风收尘器；动力除尘器
rotoclone dry-type dust collector 干式旋风收尘器
roto-clone separator 旋涡分离器
rotoconditioner 旋转式给湿机
rotocultivator 旋耕机
rotocycle 旋转切碎机
roto-dipping 旋转式浸渍涂漆
roto-dip process 旋转式浸渍涂漆法
rotodome 旋转天线罩
rotodome antenna 旋转罩型天线
rotodome radar antenna 旋转罩型雷达天线
roto-dynamic(al) pump 转子动力泵
roto-dynamic(al) water meter 转子式水表
rotofeed 旋转式推进装置；（钻机的）自动推进装置
rotoflector 旋转反射器
rotoformer 真空圆网抄纸机
rotoforming 旋转成形法
rot of wood 木材糟朽；木材腐朽
rotograph 旋印照片
rotogravure press 轮转照相凹版印刷机；凹印轮转机
roto-hammer 旋转锤
rotoil sand mixer 立式叶片混砂机
rotoinversion axis 旋转反伸轴
roto-lock closure system 旋锁封盖系统
roto-louvre dryer 旋转百叶式干燥机
rotometer 旋转（式）流量计
rotomotor 转子发动机
roton 旋转能量子
rotonda 圆形大厅
roto operator 回转操作器
rotopark car parking system 循环吊运式车库
rotophore 旋光中心
rotopiler 旋转式堆垛机
rotopiston pump 旋转（式）活塞泵；转子活塞泵
rotoplug pump
rotoplug sludge concentrator 旋塞污泥浓缩器
rotoplug type 转塞式
rotor 旋转圆筒；转子；转轴；工作轮；螺桨效应；动片组
rotor aeration 转子曝气
rotor aeration ditch 转子曝气渠
rotor aircraft 旋翼机
rotor ampere turns 转子安匝
rotor angle 转子角
rotor angle detector 转子角度检测仪
rotor angle indicator 转子角位指示器
rotor angular momentum 转子角动量
rotor arm 转动臂
rotor assembling 转子装配
rotor assembly 转子组合体；转子装配
rotor assembly space 转子安装场地
rotor axis 转子轴
rotor balancing 转子平衡
rotor bandage 转子绑线
rotor banding 转子护环装配
rotor bar 转子线棒；转子铜条
rotor bearing 转子轴承
rotor bearing cock 马达支承板
rotor bearing reservoir 转子轴承油槽

rotor blade 转子叶片;转轮叶片;转动叶片;动叶片
rotor block 转子体
rotor-blocked test 转子止转试验
rotor-body slot 转子本体槽
rotor bow 转子弯曲;转子挠度
rotor brake 转子制动器
rotor bridge support 转子夹板座
rotor-brush aerator 转刷曝气机
rotor bushing 套装转子
rotor cage 鼠笼
rotor cap 转子罩
rotor case 转子箱;转子壳;陀螺房
rotor casing 转子箱
rotor center 转子中心
rotor center part 转子中心体
rotor check washer 转子止退垫圈
rotor circuit 转子电路
rotor coil 转子线圈
rotor compressor 转子压缩机
rotor conductor 转子导体
rotor contraction 转子收缩
rotor cooling 转子冷却
rotor copper 转子铜线
rotor copper loss 转子铜耗
rotor core 转子芯
rotor-core assembly 转子铁芯装配
rotor cover 转鼓外壳
rotorcraft 旋翼机;直升(飞)机
rotor current 转子电流
rotorcycle 轻型单座旋翼机
rotor depression (转子发动机的)转子工作面凹坑
rotor diameter 转子外径
rotor digger 旋转式挖沟机
rotor disc 旋翼叶盘
rotor displacement 转子偏移
rotor displacement angle 转子位移角
rotor drive 滚筒传动装置
rotor driven screen 转子传动筛
rotor drum 转鼓
rotor dynamics 转子动力学
rotor earth fault protection 转子故障保护
rotoreflection axis 旋转反映轴
rotor end-cap 转子端帽;转子端箍
rotor end ring 转子端环
rotor excavator 滚筒式挖掘机
rotor expansion 转子膨胀
rotor exterior ring 转子外环
rotor face cavity 转子端面空腔
rotor face recess 转子工作面凹坑
rotor face tub 转子工作面浴盆状凹坑
rotor feed type motor 转子馈电式电动机
rotor feed type polyphase shunt motor 转子馈电式多相并激电动机
rotor field 转子磁场
rotor field spider 转子支架
rotor fine granulator 转子式细粒造粒机
rotor flank 转子周面
rotor flank contour 转子周边曲线
rotor flowmeter 转子流速仪
rotor forming 离心造型
rotor frequency 转子频率
rotor gear 转子齿轮
rotor gearing ratio 转子发动机的速比
rotor groove 转
rotor ground 转子接地
rotor guidance 转子导动装置
rotor guide cone 转子导流锥体
rotor head 转子头
rotor heating 转子发热
rotor hinge 旋翼铰链
rotor housing 转子发动机缸体
rotor hub 转子中心体;转子轮毂
rotor impeller 转动叶轮;转子叶轮组合件
rotor inertia cooling 转子惯性冷却法
rotor inlet gas angle 转子进气角
rotorjet 喷气式旋翼
rotor journal 偏心轴颈
rotor leakage field 转子漏磁场
rotor length 转子长度
rotorless curemeter 无转子硫化仪
rotor loading 转子负载
rotor loss 转子损耗
rotor magnet 转子磁铁
rotormeter 转子流量计

rotor motor 转子式发动机
rotor nog 转子凸棱
rotor noise 转子噪声
rotorod air sampler 旋转棒空气取样器
rotor of aeration 曝气转子
rotor of angle iron 角铁转子
rotor of condenser 转片
rotor orbit 转子运动轨道
rotor oscillation 转子振动
rotor outer diameter 转子外径
rotor overload protection 转子过载保护
rotor peripheral face 转子工作面
rotor peripheral speed 转子圆周速度
rotor phase 转子相位
rotor pinion 转子齿轴
rotor pitch 转子节距
rotor plane 旋翼机
rotor plate 转动板;动片
rotor plate assembly 转子叠片组
rotor-plug sludge concentrator 转塞污泥脱水机
rotor pocket 转子工作面凹坑
rotor pocket recess 转子周期凹坑
rotor-position sensor 转子位置传感器
rotor pump 旋转泵;转子泵
rotor punching 转子冲片
rotor radius 转子半径
rotor reaction 转子反应
rotor recess 转子工作面凹坑
rotor resistance 转子电阻
rotor-resistance starter 转子电阻起动器
rotor-resistance starting 转子串电阻起动
rotor revolution per minute 搅拌器每分转数
rotor rheostat 转子变阻器
rotor rim 转子磁轭
rotor's ring 转子环;动环
rotor sagging 转子挠度
rotor scale 回转计量秤
rotor seal 转子密封
rotor sealing system 转子密封系统
rotor segment 旋转刮板
rotor set 转子组
rotor-set-hammer 可脱开的旋转锤
rotor shaft 转子轴;转轴
rotor sheet 转子冲片
rotor ship 旋筒式风力推进船
rotor short-circuit 转子短路
rotor side 旋转活塞侧边
rotor side oil seals 转子端面油封装置
rotor's leading apex 转子导前角顶
rotor slip 转子转差率
rotor slot 转子线槽
rotor slot armer 转子线圈主绝缘
rotor slot-pitch 转子齿距
rotor slot wedge 转子槽楔
rotor speed 转子转速
rotor spider 转子辐臂
rotor spindle 转子轴
rotor spoke 转子轮辐
rotor spool 转子线圈
rotor-spray washer 转动喷水式洗涤器
rotor spreader 甩链式撒肥机
rotor sprinkler 圆筒喷灌器
rotor squish action 转子挤压作用
rotor starter 转子起动器
rotor stop lever 转子止动杆
rotor's trailing apex 转子尾后角顶
rotor stray field 转子杂散磁场
rotor stress 转子应力
rotor stress program(me) 转子应力程序
rotor temperature gradient 转子温度梯度
rotor terminal box 转子出线盒
rotor thrust 旋翼拉力
rotor thrust balance piston 转子推力平衡活塞
rotor timing gears 转子相位齿轮机构
rotor-tip lights 旋翼桨尖灯
rotor-to-crankshaft speed ratio 转子对偏心轴的转速比
rotor tooth 转子齿
rotor-torque method 转矩法
rotor turn 转子线匝
rotor turning gear 盘车装置
rotor type 转子类型
rotor-type landing 转子制动着陆
rotor-type magnetic separator 滚筒式磁选机

rotor-type operator 摇柄操作器;旋转型操作器
rotor vane 转子叶片
rotor vent 转子风孔
rotor-vibrated screen 转子振动筛
rotor-vibrating screen 旋转振动筛
rotor voltage 转子电压
rotor volume 转子体积
rotorwash 旋翼洗流
rotor water box 转子水箱
rotor wave resistance 转子波阻
rotor wedge 转子槽楔
rotor weight 转子重量
rotor wheel 工作轮
rotor width 旋转活塞宽度
rotor width-engine generating radius ratio 转子宽度对创成半径的比值
rotor windage loss 转子风阻损耗
rotor winding 转子线圈
rotor wire 转子导线
rotor with non-salient poles 隐极转子
rotor with salient poles 凸极转子
rotoscoop 旋转屏斗式脱水机
roto-sifter 回转筛
roto-sort 旋转式分级
rotospeed 旋转切碎机
rotosyn 旋转同步装置
rototherm 转子温度计(自动发出温度临界点信号的仪器)
rototiller 旋耕机;转轴式松土机;转盘式松土机
rototrol 旋转式自动调节器
rototrowel 转子式镘浆机
rotovator 松土机;转轴式松土拌和机;旋转式松土机
rotovert process 高速旋转的氧气顶吹转炉
rot-proof 防腐(朽)的
rot-proof finish 防腐处理
rot-proof material 不易腐烂的材料
rot-proofness 耐腐性;防腐性
rot-repellent aerosol coating 木材防腐气溶胶涂料
rotted earth 朽壤
rotted knot 腐木节;(木材的)腐节
rotted leaf 腐叶
rotted manure 腐熟厩肥
rotted plant material 腐烂植物
rotted rock 岩屑;风化岩石
rotted wood 糟朽木材
rotten 朽坏的;腐朽的;腐败;风化的
rotten bottom 腐朽基质
rotten clause 解除保险条款
rotten ice 蜂窝冰;蜂巢冰
rotten knot 腐朽节;(木材的)腐节;朽木节
rotten lumber 腐木
rottenness 腐烂
rotten river bottom 腐殖质河底
rotten shale section 风化页岩地段
rotten spot 沼穴
rotten stone 磨石;抛光石;含硅软土;擦亮石;蚀余硅质岩;擦光石
rotten stone powder 磨石粉
rotten wood 朽木;腐朽木(材);腐木
Rotterdam school 鹿特丹学派(亦称风格派)
rotter table 辊子台
rot test 腐朽试验
rotting process 风化过程
rottisite 镍叶绿泥石
rottor 定向干扰自动发射机
rotund 华丽的;圆形的
rotunda 圆形建筑(物);圆形大厅;圆厅;圆顶建筑(物);中央大厅
rotunda arch 圆顶建筑拱
rotundity 圆形物
roubaultite 铜铀矿
r(o)uble 卢布
rouching voltage 接触电压
roude 橹柄
rouge 过氧化铁粉;红铁粉;红粉;铁丹
rouge flame 胭脂红;火焰红
rouge glaze vase with concave lines 凹道亮光胭脂瓶
rouge red 胭脂红
rouge red glaze 胭脂红釉
rough 拉毛(墙面);粗略的;粗糙的;粗糙;不平滑的;不平的;不光滑的;凹凸不平的
rough adjustment 粗调整
roughage 粗草料;粗糙食物;粗材料

roughage cutter 粗饲料切碎机
roughage mill 粗饲料粉碎机
rough air 扰动气流;紊气
rough air filter 空气粗滤器
rough air penetration 通过大气扰流区
rough alignment 粗略找正
rough analysis 实用分析;粗略分析
rough and fine tap changer 疏密(抽头)变换器
rough-and-ready 粗糙但还可以用的;草率;粗糙的
rough-and-tumble 无秩序的
rough approximation 概数;粗略近似
rough arch 毛石拱;粗拱
rough area 切割地区;起伏地区;崎岖山区;崎岖地区
rough arrangement 布置略图
rough ashlar 粗方石块;毛石(块);毛方石;毛长石;粗(琢)方石
rough atmosphere 扰动大气
rough average 粗略平均(值)
rough axed 粗斩的
rough axed brick 斧斩砖砌的拱
rough balance 初步平衡
rough balance sheet 概算借贷对照表
rough bark 粗放剥皮
rough bed 粗糙床面
rough blank 粗坯;粗毛坯
rough block 毛石块;粗糙砌块;粗面方块;粗糙块体
rough board 毛板;粗木板
rough bodywork 车体初步加工
rough bolt 粗制螺栓
rough bore 粗镗孔
rough borehole 不规则井眼
rough boundary 粗糙边界
rough brackets 楼梯踏步木块
rough break 粗糙断口
rough breaking 粗碎
rough brick 粗面砖
rough brick arch 粗砖拱
rough brick work 粗砖工
rough broken land 陡坡地
rough builder hardware 粗建筑五金
rough burning 振荡燃烧;不稳定燃烧;不完全燃烧
rough calculation 粗算;概算
rough cargo 长大货件;包装粗糙货物
rough carpenters' work 粗木作(业)
rough carpentry 粗木作(业)
rough carriage 楼梯支承斜梁;梯基加固材
roughcast 干粘石;粗制;粗涂;粗糙粉刷;拟定初步方案;制毛坯;拉毛面;拉毛(面)【建】;打底粗灰泥;粗灰泥
rough cast concrete 粗面混凝土
rough cast glass 毛玻璃
rough casting 毛坯铸件
rough cast plaster 粗面粉饰
rough cast plastering 粗粉刷面;毛粉刷
rough cast plate 浇注原板玻璃;未加工的浇注玻璃板
rough cast plate glass 建筑平板玻璃
rough channel flow 粗糙明渠流
rough check 近似验算,粗算校核
rough classification 粗分级
rough cleaner 粗除尘器
rough cleaning 粗净化;初洗
rough cleavage 粗劈理
rough coal 原煤
rough coat 划毛底灰;粗糙底涂层;油漆底层;抹灰底层;粗抹灰层;粗灰泥
rough coated 粗糙外层的
rough coating 打底子;毛坯;粗灰泥(表面);毛粉刷
rough coat sealer 粗涂密封材料
rough cobbing 拣选大矿石;粗碎脉石中的矿石
rough combustion 不稳定燃烧
rough concentrate 粗精矿
rough concentration 粗选
rough concrete 毛面混凝土;未修整混凝土
rough cost 概算成本;大概费用;粗略成本;成本概算
rough cost book 成本概算表
rough country 崎岖地区;崎岖地带
rough cross-country track 不平整的越野道路
rough crushing 粗碎
rough cut 粗割
rough-cut arch 粗加工石拱
rough-cut file 粗齿锉;大荒锉
rough-cut joint 平灰缝;粗切开接
rough-cut stone chisel 粗石凿

rough cutter 毛刀
rough cutting 粗加工;粗砍砖
rough dentation 犬牙交错
rough deposit 粗沉积物
rough design 初步设计
rough determination 近似测定;初步测定
rough diamond 粗金刚石
rough discontinuity 粗糙的结构面
rough ditch 未修整边沟;粗挖边沟
rough door frame 毛门框
rough draft 略图;示意图;草图
rough draw 粗绘
rough drawing 草图
rough dressing 粗饰;粗琢(石);粗选
rough drill 初钻
rough drilling 硬岩钻进
roughed aggregate 砾石
roughed bloom 粗轧坯
rough edge 毛边
roughed-out 粗制的;粗加工
roughen 琢毛;变粗糙;凿毛;使粗糙
roughened sett road surface 琢毛(块)石路面
roughened surface 粗糙(表)面;凿毛的表面
roughened surface glass fiber 麻面玻璃纤维;粗糙面玻璃纤维;糙面玻璃纤维
rough engine 运转不稳的发动机
roughening (by picking) 拉毛【建】;凿毛;琢毛;打毛(混凝土);粗加工的
roughening by picking chipping 琢毛
roughening concrete 粗糙混凝土
roughening concrete surface 混凝土面打毛;混凝土毛面;粗糙混凝土(表)面
roughening course 粗加工层;凿毛层
roughening machine 磨毛机;凿毛机
roughening of riverbed 河床粗化
roughening tool 修整锤;凿石锤;鳞齿锤;磨锉工具
roughening transition 粗糙化转变
rougher 粗轧机座;粗选机;粗选池;粗选槽;初选机
rougher cell 粗选机;粗浮选槽
rougher-cleaner flowsheet 粗选—精选流程图
rougher floatation cell 粗选槽
rougher mill 粗轧机
rougher-scavenger flowsheet 粗选—扫选流程图
rougher tailings 粗选尾矿
rougher tap 粗切丝锥
roughess length 动力糙度
rough estimate 概算;毛估;粗算;粗略估算;粗略估计;粗估
rough estimate norm 概算定额
rough estimate of unit project 单位工程概算书
rough estimation 估算价(格)
rough estimation yield 粗估产量
rough etch 粗蚀
rough export ware 粗面出口器皿
rough eye 毛糙眼
rough-faced dressing 粗琢石面
rough-faced form board 粗面模板
rough-faced stone 粗面石
rough face of concrete 混凝土毛面
rough-field take-off 土跑道(上)起飞
rough figure 大约数字;粗略数值
rough file 粗齿锉;粗纹锉
rough filter 粗滤器
rough filtering 粗滤
rough filtration 粗滤
rough finish 粗修整;粗面;粗加工;低光洁度;粗饰;初级修整
rough finished 粗面的
rough-finished stone 粗修石;粗琢石
rough finishing 低级光洁度;粗饰;粗加工
rough finish stone 粗琢石
rough fish 杂鱼
rough fit 粗(精度)配合
rough fitting 粗装配
rough float 粗镘
rough floor 地板的底板;毛地板;毛地面;粗(糙)地面
rough flooring 粗地面;毛地板(材)料;铺粗地面
rough focusing 粗调焦
rough forging 粗锻
rough formula 粗略公式
rough form(work) 粗制制工作;粗模板;粗糙型
rough gas 原煤气
rough gilt edges 毛金边

rough glass 切裁后的压延玻璃;未磨光的原板玻璃
rough grade 四级精度;粗精度
rough grading 初步整型(路基);初步整平(土方);初步土方修整;初步土方平整
rough grain 粗糙纹理
rough grind(ing) 粗磨
rough grinding machine 粗磨床
rough grinding stone 粗磨石
rough groove 圆槽
rough ground 木龙骨;起伏地面;衬背龙骨;不平整地面;不平地(面);(钉墙面板的)木砖;粗地面;不平地面
rough ground edge 粗磨边
rough ground for nailing skirting 踢脚板木龙骨
rough guess 粗略估算;粗略估计;粗估
rough hand 钻工
rough handling 粗(略)处理
rough hardware 隐蔽的配件;大五金;粗五金
roughhew 粗切;粗制(木材);粗雕;造材
rough hewing 粗略削平;粗加工
rough-hewn 毛坯的;粗凿的
rough-hewn marble 粗糙的大理石
rough hewn stone 粗砍石;毛石;粗制石块
rough hewn timber 粗制木材;粗砍木材
rough-hewn wood 粗凿的木材
rough ice 压力冰;起伏冰;糙面冰
roughing 毛刻;粗加工;粗轧;粗选;初步加工
roughing belt 粗刻磨带
roughing bevel gear cutter 粗切刀盘
roughing broach 粗切拉刀
roughing cut 粗切削;粗加工
roughing-cut arch 粗加工石拱
roughing cutter 粗加工用刀具
roughing down 粗轧
roughing end mill with waveform teeth 粗加工波形刃立铣刀
roughing feed 大进给;大进刀
roughing file 粗磨锉
roughing filter 高负荷滴滤池;预滤池;粗滤器;粗滤池
roughing floatation 粗浮选
roughing gear shaper cutter 粗切齿轮插齿刀
roughing hammer 粗加工锤
roughing hand taper reamer 手动粗切锥铰刀
roughing heads 粗磨头
roughing hob 粗切滚刀
roughing-in 初步商谈;辅助性工作;准备工作;室内管道架设;室内卫生设备安装;敷设;粗抹灰;抹灰底层
roughing lathe 粗加工车床
roughing line 粗抽真空线
roughing loam 粗黏土砂
roughing machine 粗磨机;粗化机
roughing mill 粗轧机;开坯机
roughing mill group 粗轧机组
roughing milling cutter 粗齿铣刀
roughing operation 粗略处理
roughing out 粗加工
roughing-out pass 粗轧孔型
roughing-out sieve 粗杂质排出筛
roughing pass 粗轧孔型
roughing planing tool 粗刨刀
roughing plant 粗选车间
roughing pump 前置泵;低(压)真空泵;初级泵
roughing reamer 粗铰刀
roughing roll 粗轧
roughing roller 粗面压除机;糙面滚筒
roughing sand 原砂
roughing scale breaker 粗轧破鳞机
roughing shaper cutter 粗插齿刀
roughing slab 粗面装饰板材;粗面板
roughing slotting tool 粗插刀
roughing special slotting tool 特种粗插刀
roughing stand 粗轧机座
roughing table 粗选摇床
roughing tank 粗滤池;粗沉池
roughing tap 粗丝锥
roughing the angles 磨光棱角
roughing tool 粗磨工具;粗车刀
roughing tooth 粗切齿
roughing turning tool 粗车刀
roughing valve 初级阀
rough interface 粗糙界面
roughish 微糙的

rough landing 硬着陆
rough lapping 粗研磨
rough leaf 糙叶
rough log 未加工(的)圆木
rough log book 航海日志草本
rough lumber 毛料;毛材;未加工木杆;粗原木;粗木料;粗木材;粗锯木板;粗锯材
rough lumber floor 毛板地板
roughly 大体上
roughly dressed 毛面处理
roughly dressed ashlar 毛琢石;粗琢石
roughly processed product 粗加工制成品
roughly set block 粗砌块体;抛筑块体;抛弃块体
roughly squared stone 粗方石
rough machined 粗加工的
rough-machining 粗加工
rough-making 粗加工
rough mast 未加工木桅
rough material 原料;原材料
rough measurement 粗测
roughmeter 粗糙度测定仪;平整度(测定)仪;粗糙度仪;粗糙度计
rough mill 粗铣
rough mosaic 未经修正的航测拼图;粗制马赛克;粗制锦砖;粗糙镶嵌
rough neck 钻机工人
roughness 加糙;粗糙率;粗糙(度);糙率;糙度;不平(整)度;凹凸层;凹凸不平状
roughness across the lay 横向粗糙度
roughness axis 糙度轴
roughness block 粗糙块体
roughness category 糙率分级
roughness cavitation 粗糙性空蚀
roughness coefficient 粗糙系数;糙率系数;糙度系数
roughness coefficient of pipe 管道粗糙系数
roughness coefficient of slope 斜坡粗糙度系数
roughness concentration 粗糙度
roughness condition of ground 地面起伏状况
roughness curve 糙度曲线;粗糙度曲线
roughness diameter 粗糙粒径
roughness elements 增加粗(糙)度的构件
roughness factor 相对粗糙度;粗糙因素;粗糙系数;糙率因数;糙率系数
roughness Froude number 糙率弗老德数
roughness height 粗糙(度)高度
roughness index 粗糙(度)指数;不平整度指数
roughness index of mud-stone flow 泥石流粗糙系数
roughness indicator 粗糙度指示器
roughness integrator 轮廓测定器;路面粗糙度测定器
roughness length 粗糙长度
roughness measurement 粗糙度测量
roughness meter 路面粗糙度测定仪;糙度计
roughness number 粗糙数
roughness of channel 河槽糙率
roughness of discontinuity 结构面粗糙程度
roughness of engine 发动机运转不稳
roughness of polished surface 磨光表面的粗糙度
roughness of soil surface 土面粗糙度
roughness of surface 表面粗糙度
roughness parameter 起伏参数;粗糙度参数
roughness ratio 粗糙度比;糙率比(率)
roughness Reynolds number 粗糙雷诺数;糙率雷诺数
roughness size 糙率尺度
roughness spectrum 粗糙度谱
roughness tester 粗糙度测试仪;粗糙度测定器
roughness value 粗糙度值;粗糙度系数
roughness width 粗糙度宽度
roughness-width cutoff 最大粗糙宽度
rough nut 粗制螺母
rough(o)meter 轮廓仪;路面不平度测定仪;平整度(测定)仪;粗糙度(测定)仪;不平整度(测定)仪;不平度仪;表面粗糙度仪
rough(o)meter of road surface 路面平整度仪
rough opening 未竣工的孔洞;粗开孔;粗加工的开口;门窗框塞口
rough ore 原矿
rough orientation 粗略定向
rough out 打草图;粗制
rough pad 草稿簿
rough paper 糙面纸

rough particle 粗表面颗粒
rough piece 粗器
rough plan 草图;平面示意图;平面草图;设计草图;初步计划;初步规划
rough-plane 粗刨;粗加工平面
rough plank 毛板
rough plate 厚平板玻璃;厚钢板
rough plate blank 平板玻璃半成品;平板玻璃毛坯
rough plate glass 压花玻璃;皱纹玻璃;毛玻璃板;未加工的浇注玻璃板
rough pointed finish 粗凿面修整
rough pointed stone 粗琢石
rough polishing 粗抛光
rough porcelain 糙瓷
rough pump 初步抽气机;辅助泵
rough purification 粗净化;初步净化
rough puttying 刮底油灰;刮底腻子
rough rafter 暗椽
rough reading 近似读数;粗略读数;粗读(数)
rough rendering 粗粉刷;粗涂灰泥;粗抹
rough rice 粗谷
rough riding surface 粗糙的路面
rough road 简易公路;不平整道路
rough road tester 毛路试验台
rough rock 粗石
rough roll 粗面轧辊
rough rolled glass 粗压延玻璃;粗滚压玻璃
rough roller mill 粗辊轧机
rough rolling 粗轧
rough rolling mill 初轧厂
rough rubble 乱毛石;粗毛石
rough rubble facing 乱毛石饰面
rough rubble masonry 粗(毛)石圬工
rough rule 近似法则;粗略近似法
rough running 不平稳运转
rough sawed 粗锯;雨锯(板材表面用锯划线,产生装饰效果)
rough-sawn 未刨的木材;粗锯的
rough-sawn timber 粗锯材
rough screw 毛螺栓;粗制螺栓
rough sea 强浪(海浪五级);五级浪;道格拉斯四级浪;大浪;风涛海面;恶浪
rough sea anchor 镇浪海锚
rough seas 波浪汹涌的海面
rough section 毛断面
rough section size 毛断面尺寸
rough selvedge 毛边
rough separation 粗选
rough service lamp 耐振灯泡;耐振电灯;耐用灯
rough shanner 粗大螺栓扳手
rough-shuttered 粗模板制的
rough shuttering 粗模板
rough side 舱壁有防挠材的一面
rough sill 毛下槛;未加工的窗台板;未加工的门槛;未加工的窗槛
rough single cur file 粗齿单纹锉
rough sizing 粗筛(选)
rough sketch 示意图;草图
rough slope 未修整的斜坡
rough spar 未加工木杆
rough stock 原材;粗锯材;原材料;粗加工材料
rough stone 粗加工石;毛石;粗石
rough stone arch bridge 粗石拱桥
rough stone masonry wall 毛石砌墙
rough stone masonry work 粗石圬工工程
rough stone pitching 毛石铺底;粗石铺底;粗石护坡;毛石铺砌(层);粗石铺砌(层);粗石砌墙
rough stoneware 粗炻器
rough storage 毛坯储存
rough string 楼梯踏步梁;毛楼梯梁;木楼梯中间斜梁;内楼梯基
rough-strip takeoff 土跑道上起飞
rough stucco 粗拉毛粉刷
rough subgrade 未完成的地基;未完成的路基
rough sunk face 粗凹面
rough surface 铸疵;毛面;粗糙的表皮;粗糙表面;糙面
rough-surfaced 表面粗糙的
rough-surfaced blade 表面粗糙的叶片
rough-surfaced endoplasmic reticulum 糙面内质网
rough surfaced pavement 麻面(路面)
rough surface finish 粗糙表面成品
rough surface flow 粗糙面水流

rough surface grinding machine 表面粗磨机
rough surface runway 粗糙道面的跑道
rough surface scale 粗糙表面氧化皮
rough survey 初步勘测;初步调查;草测
rough swell 强涌
rough T and G boarding 粗糙凹凸板
rough teeth taper shank vertical milling cutter 粗齿锥柄立铣刀
rough terrain 崎岖地形;不平整地形;起伏地区;崎岖地区;崎岖地带;不平整地段
rough terrain crane 越野起重机
rough terrain mobile crane 越野移动式起重机
rough terrain truck 越野载货汽车
rough terrain wheeled crane 越野轮式起重机
rough texture 粗糙组织
rough tile 粗面条纹砖
rough timber 未加工原木材;粗锯材
rough timber boarding 粗木板安装
rough tool 平缝刀
rough tooth 粗制齿
rough toothed cutter 粗齿切削刀
rough treatment 草率处理;粗处理
rough tuning 粗调谐
rough tunnel 无衬砌隧洞;无衬砌隧道;无衬砌隧道
rough turbulence 不规则湍流;不规则紊流
rough turn condition 粗加工状态
rough turned 粗车削的
rough turned piece 粗削工件;粗车工件
rough turned rounds 粗加工圆坯
rough turn(ing) 粗车(削)
rough turning lathe 粗切车床
rough turning tool 粗切车刀
rough type 粗糙型
rough vacuum 低真空(度);低度真空
rough wall 乱砖墙;乱石墙;毛石墙;粗砌墙;粗糙墙面
rough wall backing 细料表面层
rough wall boundary layer 粗壁边界层
rough walling 粗砌墙工;毛石砌墙;毛石圬工;毛石砌筑;毛石圬墙
rough washer 粗制垫圈
rough-water behavior 水上飞机的水上性能
rough water effect 波浪效应
rough waters 波涛汹涌的水域;波浪汹涌的海面
rough wave 大浪
rough wave effect 波浪效应
rough wearing fiber 粗糙纤维
rough weather 狂风暴雨天气;坏天气;天气不正常;风暴天气;恶劣天气
rough-weather works 恶劣气候下作业
rough weighing 粗称
rough weight 毛概重;粗估量;粗秤重;毛重
rough wire drawing machine 粗拉拉丝机
rough wood 带皮原木
rough wooded country 丘陵森林地带
rough woodwork 粗木作(业)
rough work 混水墙;粗砌砖墙;粗木活;粗活
rough workpiece surface 工件毛面
rough zone 粗切区;粗糙带;不良运行区
rouging 除劣
roulette 旋轮线;一般旋轮线;压花刀具;转迹线;滚花轮;滚花刀
roulette pattern 螺线图形
round 圆形(装饰);圆线脚;一圈;一轮;周围;掘进长度;环绕;绕过;圈状的;炮眼组;围绕;倒角
roundabout 环形交叉;环道;转盘式交叉(口);环形航道;环行道
roundabout chair 角椅
roundabout crossing 环形交叉(道)口;环形交叉;转盘式立体交叉;路心圆盘式道路交叉;大转盘(道)
roundabout intersection 环行高架桥交叉
roundabout island 转盘式中心岛【道】;中心环岛;环形交通岛;环形交叉中央岛
roundabout line 迂回线
round about line of hump 驼峰迂回线
roundabout or loop heading 迂回导坑
roundabout system 回转式流水作业法
roundabout track 迂回线
roundabout trade 间接贸易
roundabout traffic 环行交通;迂回运输
roundabout trajectory 绕圈轨道
roundabout transport 迂回运输
round-about way 大转盘;迂回道

round a corner 转弯;拐角;拐弯
round air diffuser 圆形雾化器;圆形扩散器
round and hollow plane 圆空刨
round angle 圆周角;圆角;周角【数】
round angled square screwed type classifying lining 圆角方形螺旋分级衬板
round apse 圆形半圆顶室
round arch 圆拱;半圆拱
round-arched 圆拱形的
round-arched barrel vault 圆拱形拱顶
round-arched corbel-table (由一片牛腿支承的)挑出面层;圆拱挑檐
round-arched merlon 圆拱形城堞
round-arched opening 圆拱形开口
round-arched wagon vault 圆拱形拱顶
round-arched window 圆形拱窗
round arc point mesh gear 点啮合圆弧齿轮
round arris edge 圆棱边
round backed angle 圆脊角钢
round bar 圆铁;圆条;圆筛条;圆钢;圆棒
round bar crack test 圆棒抗裂试验
round bar davit 旋转式吊艇柱
round bar handle 圆杆门拉手
round bar steel 圆钢条
round bar web 圆钢箍筋
round bastion 圆形堡垒;圆形棱堡
round beam 圆形梁
round bearing 转动件轴承
round beater 圆形排锤
round bed 圆(形)花坛
round bend 圆形弯管机
round bend screw hook 弯钩螺钉;螺丝钩
round bilge 圆舭
round bilge hull 圆舭型船体
round bilge ship 圆舭船
round billet 小圆材
round billet mo(u)lding 圆辊错齿线脚
round blade screwdriver 圆刀螺丝刀
round block 圆形砌块
round bloom 大圆形钢坯
round bodied needle 圆体针
round-body motor 圆机壳电动机(无突出轴承)
round-body packing ring 圆型轴封
round bottom centrifuge tube 圆底离心管
round-bottomed flask with long neck 长颈圆底烧瓶
round bottomed punch 球底冲头
round bottom flask 圆底烧瓶
round bottom hull 圆底船体
round bottom slot 圆底槽
round bottom tank 椭圆形底储[贮]槽
round brackets 圆括弧
round brick window 圆形砖窗
round broach 圆孔拉刀
round brush 圆电刷
round building 圆形建筑(物)
round bur 圆钻;圆牙钻;球钻
round cable 圆形电缆
round campanile 圆形钟楼
round capital 圆形柱头
round casing coupling thread tensile strength 圆扣套管接箍丝扣抗拉强度
round casing thread tensile strength 圆扣管丝扣抗拉强度
round cella 圆形小房;圆形内殿;圆形板窝
round chapel 圆形小教堂
round chart instrument 圆盘形记录仪表
round chart recorder 圆形图记录器
round chip box 圆木匣
round church 圆形教堂
round clamp 圆压板
round collar pin shackle 眼环销螺钉卸扣
round column 圆柱
round column steel form 圆形柱钢模
round concave mo(u)lding 凹圆线脚
round concrete bar 混凝土用圆钢
round concrete flue block 圆形混凝土烟道砌块
round conductor 圆形导线
round copper bar 圆紫铜棒
round core 圆形核心
round core wire 圆形芯线
round corner 圆角
round cornering machine 切圆角机
round corner square 圆角方料

round corner square billet 圆角方坯
round counter 剩弹计数器
round crested weir 圆顶堰
round cross section gasket 圆截面垫片
round cut 圆切割
round cutter tool holder 圆刀刀架
round deformed bar 异形圆钢筋
round die 圆丝板;圆板牙;辊轮(滚制螺纹用)
round diffuser 圆形喷雾器;圆形扩散器
round diminutive tower 圆形小塔楼
round dome 球状圆顶
round donjon 城堡圆形主塔
round dowel 圆形销钉;圆形木榫
round dowel pin 圆形定位销;圆形暗销
round down-draft kiln 倒焰圆窑
round downpipe 圆水落管
round drill rod 圆形钻杆
round dungeon 圆形地牢
round eaves-tile ornament 圆瓦当
rounded 圆拱的;圆状的;滚圆的;磨圆的;成圆的
rounded aggregate 圆形集料;圆形骨料;圆角骨料;卵石集料;卵石骨料;无棱角集料;无棱角骨料;无棱角砾石
rounded algal limestone 含圆藻灰岩
rounded analysis 全面(的)分析
rounded angle 圆角
rounded bottom 圆形底
rounded bow 肥型船首
rounded concrete aggregate 圆形混凝土集料;圆形混凝土骨料
rounded corner 圆抹角
rounded crest 圆形牙顶
rounded crest measuring weir 圆顶量水堰
rounded cross-section 圆拱式横断面(路面)
rounded cutting edge radius 刃口圆弧半径;切削刃钝圆半径
rounded edge 圆缘;圆形边;倒圆的棱边
rounded edge tile 圆边瓷砖;圆边瓦
rounded end 铰接端(梁的);铰接;铰接端(梁的);回转端;球面底
rounded face 磨圆面
rounded figure 取整数;舍入的整数
rounded flatter 圆头平面锤
rounded fold 圆滑褶皱
rounded forend (双向门的)圆形门梃
rounded front 圆头;圆面;(双向门的)圆形门梃
round edge 圆形边缘;圆角刃口
round edge-arrissing 倒圆
round edge arrissing residue 倒圆残留
round-edged 圆边的
round-edged (steel) flat 圆边扁钢
round-edged tire 圆边轮圈
round edge external corner 二面圆边外角
round edge external reveal 外圆边侧砖
round edge joint file 圆边铰链锉
round edge opposite 对圆边方砖;二面圆方砖
round edge reveal 角侧方砖(釉面砖)
round edge tile 圆边砖;一面圆边(釉面)砖
round edge tile fitting 圆边瓦装修
rounded grain 圆粒砂;圆颗粒;磨圆颗粒;浑圆砂粒
rounded gunwale 圆形舷缘
rounded handrail 圆形扶手
rounded internal angle 阴圆角
rounded interval arithmetic 舍入区间运算
rounded leading edge 圆形导边
rounded-lip curb 圆角缘石;圆角侧石
rounded material 圆形材料
rounded mo(u)lding 圆形线脚
rounded nose 圆形头;回转端
rounded number 约束整数
rounded-off 修圆的;取整的
rounded-off value 凑整值
rounded orifice 圆缘孔;圆形孔口
rounded particle 圆形颗粒
rounded pebble 磨圆卵石;卵石
rounded property corner-yards 转angled庭院
rounded roof 圆(形)屋顶
rounded root contour 圆形齿根轮廓
rounded slope 周边圆边坡
rounded step 圆端踏步;圆楼梯踏步;圆角踏步;踏步板的圆形凸缘
rounded support 铰接支座
rounded system 完整的体系
rounded tail 圆尾

rounded value 凑整值;凑整数
roundel 小圆窗;串珠饰;圆形物;圆玻璃;圆堡
round elbow joint 圆弯头
round element 圆形构件
round end clamp 圆头压板
round-ended cylinder 圆底圆筒
round-ended pier 圆端形桥墩
round-ended shape 圆端形
round-ended step 圆端踏步
round end key 圆头键
round end of a choir 教堂唱诗班的圆端
round-end step 端部为半圆形的踏步
rounder 成圆机
rounder willow 圆耳柳
round exedra 半圆形前廊;半圆形室
round extension rod 圆形接杆
round facade 圆形立面
round-face bit 圆端面钻头;圆底唇钻头
round faced roughing tool 圆面粗车刀;端面粗车刀
round face hammer 圆头锤
round feeler ga(u)ge 带测圆部分的塞尺
round fence picket 圆木柱栅栏
round figure 整位数;整数【数】
round file 圆锉
round flange 圆法兰
round flat-headed nail 圆平头钉
round floor 圆形楼厅
round folded plate roof 圆形折板屋面
round footing 圆形底脚;圆形基础
round foundation 圆形基础
round frame motor 圆机壳电动机(无突出轴承)
round front (双向门的)圆形门梃
round girder 圆形大梁
round glass tower 圆形玻璃塔楼
round grain 圆颗粒
round gravel 圆砾(石)
round guard rail profile 圆形围栏外形
round guard section 圆围栏断面
round guide 圆导轨
round gutter cutter 圆形剡刀
round hanger 圆悬杆;圆挂钩
round-haul netter 围拉网渔船
round head 圆头;球头
round-head bench mark 球头(式)水准点
round-head bolt 圆头螺钉;圆头螺栓
round-head buttress 圆头支墩;圆头扶垛;大头肋墩坝;大头坝
round-head buttress dam 大头坝;圆头支墩坝
round-head construction 圆形堤坝
round-head counter sunk rivet 埋头圆顶铆钉
round-head(ed) corbel-table (由一片牛腿支承的)圆头形挑出的面层
round-head(ed) dormer window 圆顶天窗
round-head(ed) loophole 圆头漏洞;圆头窥孔
round-head(ed) rivet 圆头铆钉
round-head(ed) screw 圆头螺钉;圆头(螺)钉
round-head(ed) window 圆顶窗;圆头(形)窗
round-head(ed) wood screw 圆头木螺丝;圆头木螺钉
round-head nail 圆头钉
round-head piston 圆头活塞
round head square neck bolt 圆头方颈螺栓
round-head stake 圆头桩
round-head type buttressed dam 圆头式支墩坝
round-hipped-plate roof 圆斜脊瓦屋面
round hole 圆孔
round-hole card 圆孔卡片
round-hole diameter 圆孔直径
round hole framework 圆孔骨架
round-hole mesh 圆孔筛
round-hole plate 圆孔排种盘
round-hole plate plate 圆孔板
round-hole punched plate 圆孔筛板
round-hole screen 圆孔筛(板)
round-hole sieve 圆孔筛
round-hole test sieve 圆孔试验筛
round horizon 完整地平线
round horseshoe arch 圆马蹄形拱
round house 甲板室(船);上甲板船首附近厕所;扇形机车库;船尾小室;圆弧机车库;圆形机车库;拘留所
round indication 圆形显示
rounding 旋圆;圆角;磨圆(作用);绕行;使成圆形;舍去;倒圆

rounding angle 倒圆角
rounding bending machine 扒圆机
rounding bit 舍入位;倒圆成型车刀
rounding dial 整数盘
rounding error 修整误差;取整误差;舍入误差;舍零误差
rounding figure 舍入的整数
rounding machine 车圆机
rounding number 舍入的整数
rounding off 化成整数;使成圆形;圆滑过渡;舍入
rounding off dollars 纳税时四舍五入计去的金额
rounding of figures 略去尾数
rounding-off method 整数法;数的舍入法;舍入法
rounding-off milling cutter 倒圆角铣刀
rounding off numbers 舍入数
rounding off procedures 舍入方式
rounding of grains 砂粒滚圆度
rounding over cutter 四分之一圆铣刀
rounding plane 圆刨(床)
rounding procedure 舍入过程;舍入方式
rounding switch 舍入开关【计】
rounding tool 修圆角工具
rounding-up tool 齿轮校正器
round inspection 循环检验
round interlocking pipe 连锁圆管
round iron 圆铁;圆条
round iron anchorage 圆铁锚具
round iron bar 圆铁辊
roundish 浑圆形的
round keep 圆形盒底
round key 圆柄钥匙;圆形键
round kiln 圆窑
round kiosk 圆亭
round knob 圆形门捏手
round knob door furniture 圆形门捏手装置
round knot 圆节;圆树节
round ladder 盘梯;螺旋梯
roundlet 小圆(形物)
round line 三股右旋绳;大环索
round links 圆环链
round locomotive house 圆形机车库【铁】
round-log 圆木
round-log construction 圆木构筑成墙;圆木构造
round-looking scan 环式扫描;四周搜索
round-looking sonar 环视声呐
round lost head nail 曲头钉;角钉;土钉;圆形无头钉
round manhole 圆形检查孔;圆形人孔
round manway 圆形检查孔;圆形人孔
round mast 圆桅杆
round mat 草圆垫
round mesh 圆形细孔(筛网)
round-meshed 带有圆孔的
round-meshed screen 圆孔筛
round mesh screen 圆网孔筛
round mesh sieve 圆网孔筛
round mesh test sieve 圆孔试验网筛
round mirror with handle and stand 带架圆镜
round mo(u)lding 圆线脚
round mouth tongs 圆口钳
round mushroom 蘑菇
round neck nut 圆颈螺母
round needle 圆针
roundness 圆柱度;圆度;磨圆度;球度
roundness ga(u)ge 不圆柱性检验规
roundness index 圆度指数
roundness measuring equipment 圆度仪
roundness measuring instrument 圆形测量仪表
roundness ratio 圆度比
roundness tolerance 圆度公差;不圆度公差
round nose 圆端;圆头
round-nose bit 圆唇钻头
round-nose chipping tool 圆头凿
round-nose chisel 圆孔凿;弧口凿;圆鼻凿
round-nose cutting tool 圆头切刀
round-nosed blade 圆头叶片
round-nosed brick 圆头砖
round-nosed plicers 鱼口钳;圆头钳
round-nose pliers 圆头钳
round-nose spinning tool 圆头旋压工具
round-nose tool 圆鼻车刀;圆头车刀
round-nose turning tool 圆头车刀
round number 整数【数】;取整数
round nut 圆螺母

round nut with grooves 带槽圆螺母
round nut with lock channel 带锁紧槽圆螺母
round of beam 梁拱度
round of bearings 绕行方位群
round off 四舍五入;舍去零数;舍入
round-off accumulating 累ληθ舍入误差;尾数四舍五入的累加数;舍入误差累加
round-off accumulator 舍入误差累加器
round-off constant 舍入常数
round-off corner 圆抹角;磨圆边角;圆角
round-off curve 竖曲线
round-off error 约整误差;取整误差;舍入误差;凑整误差
round-off file 弧形锉
round office block 圆形办公室
round-off number 取整数
round-off order 修整指令;舍入指令
round-off sights 连续测天
round-off symbol 舍入符号
round-off wing 圆角机翼
round of holes 一组炮眼;钻孔组;炮眼组
round of sights 天体观测群
round one turn 绕一圈;绕一道
round-opening openwork screen 圆光罩
round orifice 圆孔
round-over bit 四分之一圆铣刀
round packing machine 回转式包装机
round paint brush 圆漆刷
round particle 圆形微粒
round pass 圆形孔型
round pavilion roof 圆攒尖
round pediment (圆形窗头的)挑檐形线脚;圆山头饰;半圆三角楣饰
round peripteral temple 圆形围柱式庙宇
round piece 圆形工件
round pier 圆柱桥墩;圆柱码头
round pillar 圆支墩
round pin 圆销
round pin shackle 带销子卸扣
round pipe 圆管
round plain head nail 圆的平头钉
round plan 圆形平面
round plane 圆刨
round plane mirror 圆形平面反光镜
round plate 圆形板
round plate feeder 圆盘喂料机
round pliers 圆钳;圆嘴钳;圆钳子;夹圆钳
round plug ga(u)ge 圆柱塞规
round-pointed 圆头的
round-pointed shovel 圆筒铲;圆头锹;圆头铲
round-pointed spade 圆筒铲
round pole 圆形磁极
round punch 圆形凸模
round-punched sheet 圆孔筛板
round punt bevel 贝壳状斜边
round radius 圆角半径
round rasp 圆木锉
round reamer 圆头扩孔器
round rear pin type die set 后导柱式圆形模架
round re-bar (钢筋混凝土的)圆钢筋
round reinforcement bar (钢筋混凝土的)圆钢筋
round ridge 圆屋脊
round ridge roof 卷棚
round rivet head 圆铆钉头
round-robin 时间片轮转
round-robin algorithm 循环算法
round-robin scheduling 循环式安排;循环调度;轮式调度
round-robin test 环形试验
round rod 圆条;圆钢;圆杆
round roof 圆屋顶
round roof-light 圆屋顶天窗
round rotor 整圆转子
round rubber stripe 圆橡皮条
round-rung ladder 圆棒索梯
rounds 圆面包;木梯上踏脚横档
round sand 圆粒沙子
round scraper 圆刮刀
round screen 圆网屏;圆筛
round screw 圆(头)螺钉
round screw pin shackle 环眼销螺钉卸扣
round sculpture 圆雕
round seam 卷缝
round seaming 卷缝法

round search 圆周搜索
round section 绕行地段
round section rubber packing ring 圆截面橡胶垫圈
round section spiral casing 圆形蜗壳;圆断面蜗壳
round seizing 绕扎
round sennit 圆编
round set-hammer 圆边击平锤;圆锤
round shaft 圆形井筒
round shaft chisel 圆形井筒凿
round shake 环裂
round-shaped slot 圆形槽
round shave 圆弧刨
round sheath 圆鞘管
round shed 圆形棚
round shovel 圆铲
round silo 圆形筒仓
round slab 圆形板
round sleek 圆角光子
round sleeker 修圆镘刀
round slotted head bolt 圆槽头螺栓
round small tower 小圆塔座
roundsman 推销员;值班人;采购员
round socket 圆形插座
rounds of ladder 梯子横档;梯级;梯子横木
round sole plane 圆底刨刀
round spade 圆头铲;圆铲
round specimen 圆试样
rounds per shift 每班炮眼组数;每班掘进循环数;每班操作循环
round spike 圆道钉
round spirit level 圆水准仪
round splice 顺纹插接
round spline kelly 圆槽方钻杆
round split die 圆形拼合板牙
round spot 圆点
round spud 圆形打捞母锥(钢绳冲击钻的一种)
round stair(case) 圆形楼梯
round stair(case) well 圆形楼梯井
round steel 圆钢
round steel anchor 圆钢锚
round steel bar 圆钢(筋)
round steel column 圆钢柱
round steel fork 圆钢叉子
round steel wire 圆钢丝
round step 圆楼梯踏步;圆踏步
round stern 圆船尾
round stirrup 圆箍筋
round stock 圆钢材
round-stock anchor 圆杆锚锚固件
round stockpile 环形料堆
round stone 圆石;大卵石
round straightener 圆钢矫直机
round strand 圆股钢丝
round strand cable 圆股钢丝绳
round strand hoisting rope 圆股提升钢丝绳
round strand rope 圆股绳;圆股钢丝绳
round strand wire rope 圆股钢丝绳
round strip lease contract 往返租用契约
round stror(e)y 圆楼层
rounds-type trussed girder 圆形桁架大梁
round sum 整数【数】
round table 协商会议;圆桌;圆台
round tank 圆筒柜;圆罐
round temple 圆形庙宇
round-the-clock 昼夜不停的
round-the-clock job 昼夜施工;日夜施工
round-the-clock process 连续过程
round-the-clock work 昼夜服务;昼夜工(作);昼夜施工
round-the-corner 转角门
round-the-corner auger 转角输送螺旋
round-the-corner door 转角门
round-the-corner screwdriver 转角螺丝刀
round the world 环球
round-the-world echo 环球回波
round thimble 圆形索环眼
round thread 圆形螺纹;圆螺纹
round thread casing 圆扣套管
round thread coupling 圆扣接箍
round-threaded screw 圆螺纹螺钉
round throw 炮孔组爆破
round tile 筒瓦
round tilted-slab roof 圆形斜板屋顶

round timber 原木;圆木(材);待采脂林
round timber bulkhead 圆木隔墙
round timber construction pole 圆木建筑柱
round timber falsework 圆木脚手架
round timber pole 圆木桩
round timber prop 圆木撑柱
round timber support 圆木支柱;圆木支承;圆木支撑
round-toothed plate 圆齿状板
round top 下桅盘
round-top mountain 圆顶山
round-topped 圆顶漩涡状装饰
round-topped roll (金属屋面的)圆顶包卷接头
round tower 圆形塔;圆塔
round traceried window 圆形花格窗
round tree 待采脂树
round trip 来回程;一个回次;交路;回次;环游;环行;全程;往返行程;往返旅行;往返旅程;往返航次;往返航程
round-trip charter party 往返航次船租合同;往返航程船租合同
round-trip cycle 往返行程时间;工作循环时间;往返周期
round-trip delay 确认的延迟;往返延迟
round-trip echoes 往复回波
round-trip fare 来回票价
round-trip flight 往返飞行
round-tripping 先贷后存的利差(英国)
round-trip ticket 往返票
round-trip time 来回一次时间;回次时间;往返时间;电梯升降时间
round tube 圆管
round turn 一个回合
round turn and half hitch 旋圆加半径
round turret 圆形塔楼;圆形炮塔
round type reamer 圆形扩孔器
round underground fire hydrant 圆形地下消防栓
round up 进位舍入;收紧辘绳;上舍入;赶拢;四舍五入
round up of beam 梁拱
round valve 圆阀(门)
round voyage 全航程;往返航次;往返航程
round waveguide 圆形波导管
round wedge 圆楔块
round well 圆井
round wind 跟日风;半圆形风
round window 小圆窗;圆窗
round wing tip 圆形翼梢
round wire 圆电线;圆导线
round wire carpet 钢丝绒地毯;圈绒地毯
round wire rod 圆形线材;圆形钢条
round wire screen cloth 圆钢丝网筛布
round wood 圆木;圆材
round wood screw 圆头木螺钉
round work 圆形工件
roundy 成圆形
routine of work 工作制度;工作规则
Rousay flags 鲁赛板层
rouse ultrasonic viscometer 振子式超声波黏度计
rouse up 用力拉
Rousseau diagram 鲁梭图(解)(求光通量的图解法);卢梭图
Roussin's salt 鲁森盐
roust 喧嚣潮汐急流
roustabout 码头工人;徒工;条工;散工(矿山等)
route 线路;旅路;路由;路线;路径;路程;进路【铁】;通路
route alignment 路线走向
route alternative 道路比较线;方案比选
route and track chart 航路图
route books mailing list 定簿籍邮寄目录单
route card 加工卡片
route change of railway 铁路改线
route chart 路线图;流水线工艺卡片;航线图;道路图
route column-cross profile survey 线路纵横断面测量
route component 航线分速
route confirmation sign 路线名牌
route-control digits 路由控制数字
route crossing 路线交叉
route determination 线路决定
route development 展线;路线延展
route direction 路线方向

route diversion 变径;变更径路;变更经路
route entry 进入进路
route extension 路由扩展件;路径扩充
route forecast 航线预测;航线预报;航路预报
route for locomotive entering and leaving the depot 机车出入库进路
route for out-ga(u)ge freight train 超限货物列车进路
route geometry 线路几何学;线路几何尺寸
route gravity survey 路线重力测量
route growth factor 路线增长系数
route guidance 路线制导
route guidance system 路线导向
route header 路径标头
route identification sign 路名牌
route-indicating bottle 漂流瓶
route-indicating signal 指路信号
route indicator 路由指示器;进路表示器
route indicator circuit 进路表示器电路
route inspection 线路检查;路由检查
route interlocking equipment 进路联锁设备
route intersection type 线路交叉形式
route kilometrage 线路里程
route length 铁道线路长度
route level(l)ing 线路水准测量;线路水准测量
route location 定线
route location on paper 纸上定线【测】
route location survey 定线测量
route locking 路由闭塞;进路锁闭
route locking indication 进路锁闭表示
route maintenance 路径维护
route management process 路径管理进程
route map 线路地图;路线图
route map of soil 土壤路线图
route mapping 路线填图;路线测图
route marker 路线标
route marking 路标;设置路标
route matrix 路由矩阵
route mileage 线路里程
route miles 线路总延长
route non-linear coefficient 线路非直线系数
route of circulation 室内交通线
route of entry 进入途径;侵入途径
route of escape 疏散路线;安全疏散路线
route of infection 传染途径
route of journey 游览路线
route of metastasis 转移途径
route of pipe line 管路;管线路
route of road 道路路线
route of transmission 传播途径
route of water transfer 引水线路;调水线路
route parallel weaving 线路平行交织
route plan 线路平面;路由计划
route plan and profile 线路平纵面
route plane control survey 线路平面控制测量;线路高程控制测量
route planning 线路选定;路线计划;航行计划;航线设计;航线计划;路由计划
route planning computer 航线设计计算机
route presetting 预排进路
route profile 管子纵剖面图
router 路由(选择)器;孔眼钻边冀;开槽刨;缩放刻膜机;刨圆削片联合机;钻头排屑槽;拷刨;开槽机;槽刨
router bit 钻头;垂直切刀刃;角刀刃
route reconnaissance 线路踏看;线路踏勘;路线勘测;路线草测
route relay 线路中继站;管路中继站
route release 进路解锁
route release at once 进路一次解锁
route report 路线图;路线说明
route restriction 路由限制
router ga(u)ge 拷刨器
route right-angle intersection 线路垂直交叉
route round the world 环球航行
router patch 拷刨补片
router plane 沟刨;槽刨;开槽刨
route sampling 路线抽样
route selecting circuit 选路电路
route selection 选线;选路;线路选线;路由选择;路线选择
route selection system 路由选择制
route setting 排列进路
route sheet 路线图;路线卡;路线表;流程工艺卡

(片);进路表
route sign 线路标志;路线标志;路标
route signal 进路信号(机)
route signal(l)er for departure 发车进路信号机
route signal(l)er for receiving-departure 接发车进路信号机
route signal(l)ing 选路信号
routes index 航线索引表
route singal(l)er for receiving 接车进路信号机
route skew intersection 线路斜交
routes of invasion 侵入途径
route standby 后备线路方式
route storage 进路存储(器)
route surveillance radar 航线监视雷达
route survey for power transmission line 输电线路测量
route survey(ing) 线路测量;路线测量
routes with overlapped section in the same direction 顺向重叠进路;顺向重叠线路
route table 进路表
route terminal 路径终端
route terminal sink 路径终端宿
route terminal source 路径终端源
route traverse 线路导线【测】;导线【测】
route type all-relay interlocking 进路继电式电气集中联锁
route type all-relay interlocking for a hump yard 进路式驼峰电气集中
route weather forecast 航线气象预报
routhierite 硫砷汞铊矿
Routh's approximant 劳思近似值法
Routh's array 劳思阵列
Routh's array table 劳思数列表
Routh's rule 卢思定律;劳思法则
Routh's rule of inertia 劳思惯性法则
Routh's table 劳思表
Routh test 劳思法则
routine 例行程序;例程;日常工作惯例;常规;分程序
routine accounting function 日常会计职能
routine adjustment 例行调整;经常性调整;日常调整;定期调整
routine analysis 例行分析;日常分析;常规分析
routine analysis method 常规分析法
routine analyzer 例行程序分析程序;分析程序
routine attention 日常维护
routine basin 程序库
routine calculation 普通计算
routine call 程序调用
routine care 例行管理;例行养护【道】;常规维护
routine check 例行查核;日常检验;日常检查;程序校验;常规检查
routine class sample 常规级样品;例常试样;日常样品
routine compiler 程序编制器
routine conductive maglev train 常导磁悬浮列车
routine control 常规控制
routine culture 常规培养
routine data 常规数据
routine data processing 常规资料整编;常规数据处理
routine denotation 例行程序标志;程序标志
routine determination 常规测定
routine diver 常规潜水员
routine dredging 常规挖泥作业;经常性挖泥
routineer 定期测试装置;定期测试器
routine experiment 例行实验
routine exploration 常规勘察
routine factory test 出厂试验
routine initializer 程序的预置程序;程序的初始程序
routine inspection 例行检查;例行检查;例外检查;定期检查;常规检查
routine interface 程序接口
routine investigation 常规勘察
routine laboratory test 例行实验室试验;常规实验室试验
routine library 程序库
routine list 程序单
routine logical instruction 程序逻辑指令
routine maintenance 小修补;小修保养;例行维护(保养);例行维修;例行保养;经常维修;日常维修;日常检修;定期维修;常规维修;定期维护;常规保养;经常(性)养护;常规养护【道】
routine maintenance time 例行维修时间
routine market 路标

routine measurement 例行测量;定期测量;定期测定;常规测量;常规测定
routine monitoring 常规监测
routine name 例行程序名
routine observation 定期观察;定期观测;常规观测
routine observation thermometer 常规观测温度计
routine operating inspection 日常检查
routine operation 例行作业
routine order 保养规程
routine outage 计划停运
routine over-haul 常规维修
routine package 例行程序包;程序包
routine-plan 例行程序图案
routine practice 操作规程
routine priority 普通优先度
routine procedure 规定程序
routine quantitative determination 常规定量测定
routiner 定期测试装置
routine reconnaissance 常规勘察
routine request 日常要求
routine safety 日常安全
routine sampling 定期采样;常规取样;常规采样
routine schedule 例行测验表
routine select 例程选择
routine shear test 常规单剪试验
routine soil test 常规土工试验
routine statement 程序语句
routine supporting services 日常支助性服务
routine test 例行试验;例行测试;日常试验;定期试验;常规试验;程序检查
routine test and dispatch 例行试验和发货
routine test apparatus 例行测试器
routine test equipment 例行测试器
routine tester 携带式测试器
routine testing 程序检查
routine test schedule 定期试验表
routine test set 例行测试器
routine-text 例行程序正文
routine vane test 常规十字板试验
routine water main cleaning application 给水干管例行清管
routine weighing 定期称重;常规称重
routine work 日常工作;常规作业;常规工作
routine zone sampling apparatus 常规分层取样器
routing 路由选择;路径选择;镂铣;航线设计;航行计划;通路指示;水文演算;程序安排
routing affinity 路径选择亲和力;路径选择亲和性
routing algorithm 路径算法
routing and traffic separation 航路规划与分道通航
routing bit 刳钻(木工用)
routing by key 按键选择路径;按关键字选择路由
routing chart 航路图
routing code 路由选择码;路选代码;路线指示码;路径选择码
routing control 路由选择控制;路选控制;路径选择控制
routing criterion 路线准则
routing curve 演算曲线
routing equation 演算方程
routing expression 路径表达式
routing flood 洪水演算
routing function 演算函数
routing in controlled reservoir 控制水库中的洪水演进推算
routing indicator 路线指示器;路线标记;传送标志符;传递标志符;发信路由指示
routing information 路由选择信息
routing information protocol 路由选择信息协议
routing is investigation 经常调查
routing key 路选键
routing key table 路选键表
routing machine 钻版机;洪水演算机;手动靠模铣床;手动仿形铣床
routing message 例行消息
routing method 定线方法
routing method of determination 常规试验方法
routing of aircrafts 飞机确定航线
routing of route 线路路径由
routing of stream flow 河川水流演进
routing parameter 演算参数
routing path 路由选择通路;路选通路
routing problem 路径问题
routing process (of flood) 洪水演进法;洪水追踪法
routing qualifier 路选限定器

routing select 路由选择
routing table 路由选择表;路选表;路径表
routing though channel 河槽洪水演算;河槽洪水演进
routing tool 靠模铣刀
routinization 程序化
routinize 常规化
rovalising 金属磷酸膜被覆法
rove 金属凹面垫片(用于抓紧铆钉);绳环
rove clinch nail 弯头舷板钉
rove or rove 敲平铆钉尼部的金属板
rover 罗弗利行车线路优先法程序;海上掠夺;采取拱形建筑形式;曲线物件;曲线线脚;三道粗纱机
Rovibond tintometer 罗维邦德色调计
roving 纤维束粗纱;无捻粗纱;粗纱
roving cutter 丝束切断机
roving for chopping 短切用无捻粗纱
roving for filament winding 缠绕用无捻粗纱
roving for pultrusion 拉挤用无捻粗纱
roving frame 三道粗纺机
roving let-off equipment 导纱装置
roving maintenance man 巡回检修工
roving vehicle 探险车
row 划船;排;荡桨;草条
row and column numbering 行列编号法
row binary 行二进制
row-block 竖砖
rowboat 划桨船;划子;划艇;舢板
row bow 领桨
row-by-row reading (卡片的)横读
row cleaner 清垄器
row coefficient 行系数
row-coordinate 行坐标
row crop 中耕作物;条播作物
row-crop equipment 中间作物机具
row cropping 条播
row dislocation 行列位错
row display 行显示
row ditch 垄头沟;条头沟
row dwelling 联排式住宅
roweite 硼锰锌石
roweitwe 硼钙石
row engine 直列式发动机;排式发动机
rower 桨手;划船人;行标
row home 联立式住宅
row house 成排房屋中的一幢;连栋住宅;联排式住宅;联立房屋;三幢以上并联住宅;行列式房屋;排房;成排房屋
row-house development 联立式住宅发展
row-house-type dwelling unit 联立式住宅单元
rowing boat 划子;划艇
rowing box 滚桶
rowing equipment 划船设备
rowing ferry 渡河小船
rowing test 划船试验
row installation 成排设备安装
row iteration 行迭代
row knocker 平垄器
rowl 轻便卸货吊车
Rowland arrangement reflection grating 罗兰排列反射光栅
Rowland circle 罗兰圆
Rowland-circle concave grating 罗兰凹面光栅
Rowland effect 罗兰效应
Rowland ghost 罗兰鬼线
Rowland grating 罗兰光栅
rowlandite 氟硅钇石
Rowland mounting 罗兰(发光光谱仪)装置
Rowland ring 罗兰环
Rowland wavelength system 罗兰波长系
rowler 滑车
rowlock 桨门;桨叉;桨架;侧砌顶砖层;侧砌砖;竖砌砖
rowlock arch 竖砖券;竖砌(砖)拱;侧砌砖拱;砖砌圆拱;墙背面竖砌砖的墙
rowlock-back 墙脊竖砌砖墙
rowlock-back wall 背面竖砌砖墙
rowlock bolster 舢板舷侧桨门板
rowlock bond of arch 竖砌砖拱砌合
rowlock cavity wall 空斗墙;竖砌(砖)空心墙
rowlock cheek 桨门加强板;桨门承板
rowlock course 炉床边缘;竖砌砖层
rowlock cover 舢板舷侧桨门板
rowlock paving 顺砌的砖地坪

rowlock plate 桨叉插板
rowlock wall 空斗墙
row matrix 行矩阵
row middle 行间
row number 列号;集装箱船的行号
row of buildings 一排建筑物
row of chairs 一排座椅
row of contacts 触排
row of holes 炮孔排
row of houses 街区
row of piles 桩板;一排桩;排桩
row-of-piles breakwater 密集桩排防波堤;密集排桩防波堤
row of pump 水泵系列
row of racks 机架排
row of rivets 铆钉行列;铆钉排列;铆行(数)
row of seats 一排支座;一排座椅
row of setts 成排小方石
row of soil lying between two furrows 两沟之间隆起的一条土带
row of stores 一排库房;一排商店
row of tubes 管排
row of wells 井群;井列
row order 行阶
row pitch 孔距;行距
row planter 行播机
row plot 行播小区
row rank 行秩
row scanning 行扫描
row-seeder 行播器
row shedding flowmeter 涡流分离流量计
rowship 划桨船
row shooting 多排孔间隔爆破;排孔爆破
rows of area element mesh 面元网的行数
rows of roof tiles 瓦垄
row spacing 排间距;垄宽;垄距;行距;排间距离;排距
row spacing of bore hole 孔间距
row span 行宽
rows per inch 每英寸行数
row-to-row distance 排间距离
row vector 行向量;行矢量;单行矩阵
row wet 水花四溅地划桨
row-width tests 行距试验
roxenol 氯二甲苯酚
roxite 罗赛特橡木塑料
royal 大木瓦板;西方红杉木瓦
Royal Architectural Institute of Canada 加拿大皇家建筑师学会
Royal Australian Institute of Architecture 澳大利亚皇家建筑师学会
royal blue 晶蓝;贡蓝;红光蓝;品蓝(色)
royal blue pigment 深蓝色颜料;晶蓝颜料;红光蓝颜料;贡蓝颜料
royal castle 皇宫
royal cell 王台
royal chapel 皇宫教堂
royal chuck 钢球式消隙夹头
royal court 宫廷
royal door 高大门
royal garden 皇家花园;皇家园林
royal gardener 皇宫园丁
Royal Institute of British Architects 英国皇家建筑师学会
Royal Institute of Naval Architects 英国造船工程师学会
Royal Institution of Chartered Surveyors 英国皇家特许测量员学会
royal mast 顶上桅
royal Mastabah tomb 马斯塔巴皇陵
royal pair 虫偶
royal palace 皇宫
royal pink 深红色
royal pole 顶上桅
royal purple 皇家金字塔;蓝紫色;皇紫;深紫红色
Royal Square 王家广场(伊朗伊斯法罕)
royal stay 桅最上方的前支索
royal studding-sail 顶翼帆
royal suite 皇家套房
royalties-loaded 负担专利税者
royal tomb 皇陵
royal tree 国有林木;保留林木
royalty 专利权税;矿山使用费;矿区使用费;权利

金;提成支付;提成费;使用税(版权专利权等的使用税);森林使用费;产权使用费;版税
royalty expenses 特权费
royalty interest 矿区特许使用权益;产区使用权益
royalty payment 权利金;特许使用金
royalty petroleum 石油产地使用费
royalty rate 提成率
royal yard 桅最上方帆桁;顶桅桁
royal yellow 雌黄
royer 带式松砂机
Royer sand mixer (and aerator) 带式松砂机
Roylar 聚氨酯弹性塑料
rozenite 四水白铁矾
R-pattern analysis R 式分析
RR layup 聚酯树脂增强层压材料
RRR triple junction RRR 型三向联结构造
R-tectonite R(岩组)构造岩
R-test 固结快剪试验;固结不排水剪切
RTF triple junction RTF 三向联结构造
R-theta navigation 极坐标导航;全向导航系统
Ruabon brick 鲁班砖(类似红玻璃砖)
Ruabon kiln 多室煤气窑(英国)
ruanyl sulfate 硫酸双氧钕
ruarsite 硫砷钌矿
ruaskovite 水磷钡铁矿
rub 磨;刷状痕;搓;擦迹
rub-bar cylinder 搓擦杆式脱粒滚筒
rubbed brick 磨砖;软滑砖;磨面砖
rubbed bricks with pointed joints 磨砖勾缝
rubbed bricks with tight joints 磨砖对缝
rubbed brickwork 规准砖砌体;磨砖对缝(砌筑)
rubbed concrete 磨面混凝土;磨光混凝土;水磨石
rubbed down material 擦亮材料
rubbed effect 磨退效果
rubbed finish 磨退面;磨平修整;磨出面;无光磨面层;磨砂面层;磨轨饰面;磨光(漆)面
rubbed joint 胶结缝;磨牲对缝;木板摩擦挤胶拼缝
rubbed pitched kraft paper 柏油纸
rubbed slab 磨光板
rubbed surface 摩擦面;磨光面
rubbed work 磨面工作;磨面砌体
rubber 橡皮(擦);橡胶(三角皮带);舷围;胶皮;磨具;软锉(指可切削及雕刻的软砖);水平止动装置;粗纹锉
rubber acceleratant 橡胶促进剂
rubber acceleratant wastewater 橡胶促进剂废水
rubber accelerator 橡胶制止器;橡胶促进剂;硫化促进剂
rubber adhesive 橡胶黏合剂;橡胶型胶黏剂
rubber adhesive plaster 橡皮膏
rubber aggregate 橡胶集料;橡胶骨料
rubber alloy 合胶
rubber and canvas article 橡胶帆布制品
rubber antioxidant 橡胶防老剂
rubber apron 橡胶防护板;橡胶防护圈;橡胶防护层;橡胶围裙
rubber asbestos plate 橡胶石棉板
rubber asphalt 掺橡胶沥青;橡胶(地)沥青
rubber asphaltic bitumen emulsion 橡胶沥青乳胶
rubber asphaltic bitumen sealing compound 橡胶沥青密封剂
rubber bag 橡胶包套;橡胶袋
rubber ball 橡胶球
rubber-ball deck 橡胶球台面
rubber balloon 橡胶气囊
rubber balloon method 橡胶气球法
rubber band 橡胶圈
rubber bandage 防潮胶带
rubber-banding 橡胶式聚束
rubber-based adhesive 橡胶质黏合剂;橡胶胶黏剂
rubber-based coating 橡胶底层;橡胶敷面;橡胶漆;塑料敷面
rubber based plate vibrator 橡胶座平板振动器
rubber-based sealant 橡胶基密封剂
rubber-base paint 橡胶基涂料
rubber bearing 橡胶垫块;橡胶承座;橡胶轴承;橡胶支承;橡胶支座(桥梁)
rubber bearing pad 橡胶承垫块
rubber bearing plant 橡胶植物
rubber bellows 橡胶皱纹管;橡胶伸缩管
rubber belt conveyer 橡胶皮带输送机
rubber belt(ing) 橡胶带;传动皮带;胶(布)带
rubber belt track 橡胶履带
rubber-bitumen composition 橡胶沥青混合物;橡胶沥青合成料
rubber-bitumen emulsion 橡胶沥青乳胶
rubber-bitumen joint 橡胶沥青填缝
rubber-bitumen sealing compound 橡胶沥青填缝料
rubber bladder 橡胶胀形模
rubber blade 橡胶叶片
rubber blanket 橡胶垫
rubber-blanket cylinder 橡胶布辊筒
rubber block 橡胶块
rubber-block pavement 橡胶块(铺砌)路面
rubber-block shock absorber 橡胶块减振器
rubber-block support 橡胶减振支座
rubber-block track 橡胶块履带
rubber boat 橡胶艇
rubber bond 橡胶黏结料;橡胶结合剂;橡胶胶合
rubber-bonded-to-metal pad 橡胶金属叠合垫片;橡胶金属叠合垫块
rubber boot 橡胶靴套;防尘罩
rubber-boot deicer 橡胶除冰带
rubber boots 橡胶长靴
rubber-bound cork filler 橡胶结软木嵌缝料
rubber bowl 橡胶碗
rubber bridge bearing 肋玻璃(桥梁);桥梁橡胶支座
rubber buffer 橡皮缓冲器;橡皮缓冲垫;橡胶缓冲装置;橡胶缓冲垫
rubber building mastic 橡胶建筑胶黏水泥;橡胶建筑胶黏剂
rubber bulb 橡胶袋
rubber bumper (码头的)橡胶防撞块;橡胶缓冲装置;橡胶缓冲器;(码头的)橡胶顶撞块
rubber bush 橡胶套管
rubber cable 橡胶(绝缘)电缆;橡胶缆
rubber canvas hose 橡胶管;橡胶帆布软管
rubber cap 橡胶帽
rubber cargo hose 运输石油产品的橡胶管
rubber carpet 橡胶铺垫
rubber case 橡胶套
rubber catheter 橡胶导管
rubber cement 橡胶黏结剂;橡胶(水)泥;橡胶胶合剂;胶浆
rubber check 空头支票
rubber-clad transducer cable 换能器胶皮电缆
rubber clay 橡胶用陶土;橡胶工业用黏土;砂质黏土
rubber cloth 橡胶布
rubber -coated 上橡胶涂层的
rubber-coated fabric 橡胶涂层织物
rubber-coated glass fabric 被覆橡胶的玻璃织物
rubber-coated paper 涂胶纸
rubber-coated pressure bowl 橡胶辊筒
rubber coating 橡胶涂层;涂胶
rubber coating asbestos cloth 胶胶石棉布
rubber collar drill steel 橡胶扁肩钎
rubber-compatible 橡胶相容性
rubber composition 橡胶组分;橡胶混合物
rubber compound 橡胶混合物
rubber compounding 橡胶配方
rubber compounding ingredient 橡胶助剂
rubber compression joint 橡胶压接
rubber compression type seal 橡胶压缩式止水条
rubber concave bar 凹板橡胶脱粒杆
rubber conductor 橡胶绝缘导线
rubber conduit 橡胶(导)管
rubber cone 橡胶接管
rubber cone belt 三角橡胶带
rubber constituent 橡胶成分
rubber container 橡胶容器;橡胶集装袋
rubber conveyer 胶带输送机
rubber conveyor belt 橡胶传送带;皮带运输机;皮带运输带
rubber cord 橡胶索;橡胶绳
rubber-cord start 橡筋弹射起飞
rubber core 胶胶芯
rubber cork 橡胶塞
rubber coupling 橡胶连接器;橡胶联结器;橡胶联轴节
rubber cover 涂胶
rubber-covered 橡胶绝缘的;包橡胶的
rubber-covered braided 橡胶绝缘编包风雨线
rubber-covered braided weather proof wire 橡胶绝缘的编包风雨线
rubber-covered cable 包橡胶电缆
rubber-covered double braided 橡胶绝缘双层编包线
rubber-covered outer casing 橡胶绝缘外罩;橡胶绝缘外壳
rubber-covered roll 橡胶套辊
rubber covered roller 橡胶滚轮
rubber-covered wire 橡胶线;橡胶线
rubber covering 橡胶涂层;涂胶
rubber-covering machine 包橡筋线机
rubber cracking 橡胶裂化
rubber crawler 橡胶履带
rubber crumb 橡胶粒
rubber cuff 橡胶封套
rubber cup 橡皮碗;皮碗
rubber-curing agent 橡胶熟化剂
rubber cushion 橡胶(气)垫;橡胶减振垫;橡胶缓冲器
rubber cushion assembly 橡胶阻尼器
rubber cushion fender 橡胶垫块护舷
rubber cushion supporter 橡胶垫支承
rubber dam 橡胶坝
rubber dam clamp 橡胶障夹
rubber dam clamp forceps 橡胶障夹夹持钳
rubber dam clamp holding forceps 橡胶障夹夹持钳
rubber damper 橡胶减振器
rubber dam plumb 橡胶障吊锤
rubber dam punch 橡胶障穿孔器
rubber derivate paint 橡胶衍生物涂料
rubber deterioration 橡胶劣化
rubber diaphragm 橡胶膜
rubber diaphragm separator 橡皮隔膜
rubber diaphrgam pump 橡胶隔膜泵
rubber die 橡胶模
rubber die press 橡胶模压力机
rubber dingey 橡皮艇
rubber disk shock absorber 橡胶盘减振器
rubber disk shock strut 橡胶盘减振支柱
rubber distance piece 定距片;橡胶垫片
rubber door 橡胶门
rubber door leaf 橡胶门扇
rubber door stopper 门上橡胶防碰垫
rubber dough 橡胶泥
rubber dowel 橡胶定位销
rubber down 打磨;磨石
rubber draft gear 橡胶缓冲器
rubber drain 橡胶排水管
rubber drum 软桶
rubber drying rolls 橡胶覆面挤干辊
rubber ear cushion 橡胶耳垫
rubber effect 橡胶效应
rubber elasticity 橡胶弹性
rubber element 橡胶元件
rubber elephant trunk 橡胶溜管;橡胶制(混凝土)输送管
rubber elevator bucket 提升机橡胶吊桶;提升机橡胶戽斗
rubber emulsion 胶乳;橡胶乳液
rubber-emulsion paint 橡胶乳液涂料;乳胶漆;橡胶乳漆
rubber engine mounting 发动机橡胶垫
rubber eraser 橡胶擦
rubber estate 橡胶园
rubber expansion joint 橡胶膨胀节
rubber extension modulus 橡胶拉伸模量
rubber fabric 涂胶带
rubber-faced 带橡胶面的;橡胶贴面的
rubber-faced pad 橡胶面炉垫;橡胶面衬垫
rubber-faced steel plate 橡胶面钢板
rubber-faced tile 橡胶涂面瓦;橡胶面铺地砖
rubber-faced track 胶面履带
rubber face for valve 底阀橡胶圈
rubber facing 橡胶覆盖层;橡胶面层
rubber factory 橡胶厂
rubber felt 橡胶毡
rubber fender 橡胶护舷木;橡胶防撞垫;橡胶挡泥板;橡胶碰垫;橡胶护舷;橡胶防撞设施;橡胶防冲装置
rubber fender for dock corner 码头包角型护舷
rubber fender for tug 拖轮护舷
rubber fender with facing panel 带面板的橡胶护舷
rubber filler 橡胶填充料
rubber film 橡胶薄膜
rubber film aerator 橡胶膜片曝气器
rubber filter 滤胶机

rubber finger-cot 橡皮指套
rubber-fingered conveyer 橡胶爪式输送器
rubber finger-stall 橡胶指套
rubber flange 橡胶凸缘
rubber flap 橡胶刮板
rubber-flight elevator 橡胶刮板式升运器
rubber floor covering 橡胶(涂层)地板；橡胶地板覆盖层；橡胶楼板覆盖层
rubber flooring 橡胶铺面地板；橡胶地面
rubber flooring tile 橡胶铺地砖
rubber foam 橡胶海绵；海绵橡胶
rubber foil 橡胶箔
rubber follower for balance piston 平衡活塞橡胶环
rubber for lubricating oils 润滑油用橡胶
rubber friction draft gear 橡胶摩擦缓冲器
rubber friction drive 橡胶摩擦传动
rubber friction tape 橡胶摩擦带
rubber gaiter 橡胶防护罩
rubber gasket 橡胶垫密片；橡胶垫圈；橡皮垫(片)；橡胶密封垫；橡胶垫圈；橡胶垫片；橡胶垫(环)；止水胶垫
rubber gasket connection 水力压接
rubber-gasketed joint 橡皮衬垫接缝(混凝土路面)；橡皮圈连接
rubber gasket joint 橡胶填实接缝；橡皮垫圈接口
rubber gasket of feed valve 供给阀胶垫
rubber gasket protected by lead angle 导角保护橡胶垫圈
rubber gate stick 橡胶直浇口
rubber ga(u)ge 橡胶测厚计
rubber glazing channel 窗玻璃装配用橡胶槽道；橡胶玻璃密封条
rubber gloves 橡胶手套
rubber glue 橡胶
rubber goods making machinery 橡胶用品制造机
rubber grab 橡胶刮板
rubber grade carbon black 橡胶用炭黑
rubber-grade talc 橡胶级滑石
rubber green 橡胶绿
rubber grip 橡胶手柄
rubber grommet 橡胶垫圈；橡胶密封圈；橡胶环
rubber hammer 橡胶锤
rubber hammer with handle 带柄橡胶锤
rubber hand grip 橡胶把手；橡胶手柄
rubber hand pump 橡胶手揿泵
rubber hardness (degree) 橡胶硬度
rubber hardness tester 橡胶硬度测试器
rubber headed hammer 橡胶头锤
rubber hod 橡胶灰浆桶；橡胶灰浆桶
rubber hold-down strip 橡胶压条
rubber hollow weather strip 挡雨空心橡胶带
rubber hose 水龙带；橡胶(软)管
rubber hose braided with cotton wire 棉编胶管
rubber hose braided with steel wire 钢编胶管
rubber hose clamp 橡胶管卡圈
rubber hose connection 橡皮软管连接
rubber ice 松冰(团)
rubber industry 橡胶工业
rubber ingredients 橡胶配合剂
rubber inner tube 橡胶内胎
rubber insert(ion) 橡胶嵌入片；橡胶垫
rubber-insulated 橡皮绝缘的
rubber-insulated body mount 橡胶绝缘体减振架
rubber-insulated cable 橡胶绝缘电缆
rubber-insulated copper wire 铜芯橡胶绝缘电线
rubber-insulated wire 橡胶绝缘(电)线；皮线
rubber insulating board 橡胶绝缘板
rubber insulating tape 橡胶包带
rubber insulation 合成橡胶绝缘材料；橡胶绝缘
rubber insulation wire 橡胶绝缘电线；绝缘电线
rubber-insulator 橡胶绝缘子
rubber-internal mixer 密闭式混胶机
rubber-isolated cable 外包橡胶电缆
rubber isolator 橡胶隔振器
rubber isomer 橡胶异构体
rubber item 橡胶制品
rubberized asphalt 橡胶(地)沥青；加橡胶沥青
rubberized asphalt sealing strip 橡胶沥青密封条；橡胶地沥青嵌缝板
rubberized breaker cord 橡胶帘布缓冲层
rubberized conveyor belts 橡胶传动带
rubberized cord plies 涂胶帘布层
rubberized cushion 涂橡胶垫层

rubberized fabric 涂橡胶织物；橡胶布；胶布(指上胶的布)
rubberized-fabric envelope 橡胶布外罩
rubberized linoleum 涂橡胶的油地毡
rubberized paint 用橡胶处理的涂料
rubberized sand pump 橡胶内衬沙泵
rubberized tape 胶(布)带
rubberized-tar paving mixture 橡胶化焦油铺路合剂
rubberized ventilation pipe 橡胶风筒
rubberized waterproof seal 橡胶防水封口
rubberizing 涂胶；贴胶
rubberizing machine 注液胶机
rubber jacket 橡胶套
rubber jacket cable 橡胶套电缆
rubber jar (蓄电池的)橡胶壳
rubber joint 橡胶接缝；橡胶接头；橡胶垫
rubber joint filler 橡胶嵌缝材料
rubber knob 橡胶按钮
rubber-lagged 橡胶护面的
rubber landing craft 橡胶登陆艇
rubber landing ship 橡胶登陆艇
rubber latex 橡胶乳汁；橡胶乳液；橡胶胶乳；橡胶浆
rubber latex cement 橡胶乳液黏合剂；橡胶乳液水泥；橡胶浆黏接剂；橡胶乳液黏接剂
rubber latex cement mortar 橡胶浆水泥砂浆
rubber latex floor 橡浆地板；橡胶乳液地面
rubber layer 橡胶层
rubber-like 仿橡胶；橡胶样
rubber-like behaviour 似橡胶特性
rubber-like liquid 似橡胶液体
rubber line 同心橡胶管
rubber-lined 衬有橡胶的；橡胶衬里的
rubber-lined ball mill 橡胶衬球磨机
rubber-lined bearing 衬胶轴承
rubber-lined claw 橡胶夹手
rubber-lined fire hose 橡皮内里水龙带；橡皮内里消防软管
rubber-lined hose 衬橡胶软管；橡胶夹衬软管
rubber-lined pipe 橡胶衬里管；衬橡胶管子；橡胶涂面输送管；橡胶涂面管道；橡胶涂面导管
rubber-lined rake 衬胶的齿耙；橡胶衬里的齿耙
rubber-lined steel pipe 衬胶钢管
rubber-lined steel sheel 衬橡胶钢制壳体
rubber-lined valve 衬橡胶阀门
rubber liner 橡胶衬垫
rubber lining 橡胶衬里
rubber lining of mill 磨机橡胶衬里
rubber lining plate 橡胶衬板
rubber lubricant 橡胶用润滑剂
rubber lump 橡胶凝块
rubber machine belting 橡胶机器带
rubber machinery 橡胶工业机械
rubber mallet 橡胶锤
rubber mallet wood(en) handle 木柄橡胶锤
rubber mass 橡胶基质；胶料
rubber mastic 橡胶油灰；橡胶胶
rubber mat 橡胶垫；胶地席
rubber matrix 橡胶基质；胶料
rubber membrane 橡胶薄膜法；橡胶(薄)膜；橡胶膜
rubber membrane correction 橡胶膜校正
rubber mercury bougie 橡胶水银探条
rubber-metal assembly 橡胶金属结合件
rubber-metal buffer 侧挡缓冲器
rubbermeter 橡胶硬度计；橡胶计
rubber mill 橡胶磨；橡胶加工厂
rubber mill wastewater 橡胶厂废水
rubber mixing machine 炼胶机
rubber-modified asphalt felt 橡胶改性沥青油毡
rubber-modified bitumen 掺橡胶的改性沥青；橡胶改性沥青
rubber-modified plastics 橡胶改性塑料
rubber mo(u)ld 橡皮模
rubber mount 橡胶垫架
rubber-mounted 装汽车轮胎的；装在橡胶软垫上的
rubber-mounted crane 轮胎式起重机
rubber-mounted crusher 轮胎式轧碎机
rubber-mounted derricking jib crane 轮胎式摇臂起重机
rubber-mounted ditcher 轮胎式移动挖沟机
rubber-mounted mobile crane 轮胎式移动起重机
rubber-mounted mobile excavator 轮胎式移动挖土机
rubber-mounted mono-tower crane 轮胎式单塔起重机

rubber-mounted shovel crane 轮胎式挖土起重两用机
rubber-mounted tractor 橡胶轮拖拉机；轮胎式拖拉机
rubber-mounted tractor-drawn scraper 轮胎式拖拉机牵引铲运机
rubber mounting 橡胶垫；橡胶隔振垫层
rubber mudguard 橡胶挡泥板
rubberneck bus 游览车
rubberneck car 游览车
rubberneck wagon 游览车
rubber number stamp 橡皮号码印
rubberoid jet-proof joint sealing (混凝土路面的)防喷气橡胶填缝料
rubber oil seal 橡胶油封
rubber packed coupling 橡胶套弹性联轴节
rubber packing 橡胶衬垫；橡胶(衬)垫；橡胶填料
rubber pad 橡皮垫块；橡胶防振制品；橡胶垫
rubber pad bearing 橡胶垫片支座(桥梁)；橡胶垫片支座；板式橡胶支座
rubber pad blanking 橡胶冲裁
rubber paddle 橡胶叶片
rubber pad drawing 橡胶拉深
rubber pad forming 橡胶成型；橡胶凹模成型
rubber pad forming process 橡胶成型法；橡胶板压形法
rubber pad hydraulic press of tunnel type 圆筒式橡胶囊液压机
rubber pad press 橡胶垫压力机
rubber pad process 橡胶垫式成形法
rubber pad shoe 橡胶垫履带板
rubber paint 橡胶漆
rubber panel 橡胶板
rubber paste 胶浆
rubber patten of grind mill 研磨机橡胶挡板
rubber paving 橡胶铺面；橡胶材料
rubber paving block 橡胶铺块
rubber penetration effect 橡胶嵌入效应
rubber-phobe 疏橡胶性
rubber piece 防擦材
rubber pinch tip 橡胶喷嘴
rubber pipe 橡胶管
rubber piston 橡胶活塞
rubber plant 橡胶厂
rubber plantation 橡胶(种植)园
rubber-plastic waterproofing sheet 橡塑防水片材
rubber plasticizing agent 橡胶增塑剂
rubber plate 橡胶版；橡胶板
rubber plate ink 橡胶版油墨
rubber plating metal 金属镀胶
rubber plinth 橡胶支座(桥梁)
rubber plug 橡胶插塞
rubber poisons 橡胶毒物
rubber polish cup 橡胶抛光杯
rubber polishing wheel 磨光橡皮轮
rubber pot-head 橡胶端套
rubber powder 橡胶粉(末)
rubber pressing 橡胶模成型；橡胶压制；橡胶模成型
rubber processing 橡胶加工
rubber processing oil 橡胶加工用油
rubber processing plant wastewater 橡胶厂废水
rubber producing plant 橡胶植物
rubber profile 橡胶型材
rubber pump valve 橡胶泵阀
rubber quick-joint 橡胶快速接头
rubber reinforcement 橡胶增强材料
rubber resin 橡胶树脂
rubber-resin adhesive 橡胶树脂黏合剂
rubber-resin alloy 橡胶树脂并用胶；耐冲击塑料
rubber-resin blend 橡胶树脂并用胶
rubber rim 橡胶轮圈
rubber ring 橡胶环；橡胶(垫)圈；橡胶垫片
rubber-ring joints for pipes 管道接缝一橡胶圈接缝
rubber-ring packing 橡胶填密圈；密封橡胶圈；填密橡胶圈
rubber-ring seal 橡胶垫圈密封
rubber-ring shock absorber 橡胶圈减振器
rubber-ring-type buffer idler 橡胶圈式缓冲托辊
rubber-ring type gate valve 胶环闸阀
rubber roll 橡胶辊
rubber roller 橡胶辊
rubber roller for hulling rice 龙壳胶辊
rubber rope 防撞钢丝绳
rubber rotary hose 橡胶软管；橡胶水龙带

rubber safety tread 橡胶安全踏板
rubber salvage pontoon 橡胶打捞浮筒
rubber sandwich spring 橡胶夹心弹簧
rubber screen plate 橡胶过滤板;橡胶筛板
rubber seal 橡胶密封;橡胶垫圈;橡胶止水(条)
rubber sealed 橡胶密封的
rubber sealed flexible joint 橡胶密封的柔性接头
rubber sealed valve 橡胶密封阀
rubber sealing boot 橡胶密封套
rubber sealing gasket 橡胶密封垫圈
rubber sealing ring 橡胶密封圈;橡胶密封环
rubber sealing strip 橡胶密封条
rubber seat 橡胶垫
rubber seat holder 橡胶压垫圈
rubber seat retainer 橡胶垫座
rubber-seed oil 橡胶树子油
rubber separator 橡胶隔板
rubber serum cap 橡胶盖
rubber set 假凝
rubber-set jam brush 橡胶座扫灰刷
rubber sett paving 橡胶石块路面
rubber shaft bearing 橡胶轴承
rubber sheath 橡胶套(筒)
rubber-sheathed 橡胶包的
rubber sheathed cable 橡胶铠装电缆;橡胶(外包)电缆
rubber sheathed wire 橡胶外包线
rubber sheet 橡胶布;橡胶垫;地板橡胶
rubber sheet cutter 橡胶切片机
rubber sheeter 压片机
rubber sheet flooring 橡胶地面
rubber sheet(ing) 橡胶垫布;橡胶板
rubber sheeting various color 各色平胶板
rubber sheet lining 橡胶衬里
rubber shim 橡胶薄垫片
rubber shock absorber 橡胶减振装置;橡胶减振器
rubber shock absorber strut 橡胶减振柱
rubber shoe 橡胶闸瓦
rubber side seal 橡胶边缘密封;橡胶边侧密封
rubber silencer 橡胶消声器;橡胶碰门垫
rubber skin 橡胶表层;橡胶膜
rubber slat mud conveyer 橡胶板式泥浆输送器
rubber sleeve 橡胶套(筒);橡胶管套
rubber-sleeve core barrel 带橡胶套的双层岩芯管
rubber-sleeve stabilizer 橡胶箍稳定器
rubber solution 橡胶漆;胶浆
rubber solution strainer 橡胶胶水滤器
rubber solvent 橡胶溶剂
rubber spacer 橡胶间隔片
rubber sponge 橡胶海绵
rubber spreading machine 涂胶机
rubber spring 橡胶防振制品;橡胶弹簧
rubber spring chain 橡胶弹性链
rubber spring shackle 橡胶弹簧钩环
rubber squeegee 橡胶刮(水)板;橡胶滚子;橡胶刮刀
rubber stamp 橡胶图章
rubber stamp decoration 橡胶图章印花装饰法
rubber stock 橡胶混合物
rubberstone 磨石
rubber stop 橡胶挡块;橡胶止水;橡胶水封;橡胶塞
rubber stopper 橡胶塞
rubber strainer 滤胶机
rubber strip 橡胶防水条;橡胶防水带
rubber structural gasket 橡胶结构垫圈
rubber substitute 橡胶代用品
rubber-supported screen 橡胶减振筛
rubber-supporting bearing 橡胶轴承
rubber-surfaced 橡胶包面的
rubber-surfaced paving block 橡胶面铺砌块
rubber-surfaced roll 胶滚
rubber suspension spring (座位的)橡胶悬挂弹簧
rubber swab 橡胶拭子;橡胶拖把
rubber swelling test 橡胶膨润试验
rubber-swelling test fluids 橡胶泡涨试验液
rubber swivel-joint cardan drive 橡胶万向联轴节驱动
rubber-synthetic resin mix(ture) 橡胶合成树脂混合物
rubber tape 橡皮带;橡胶带
rubber tapping knife 橡胶刀
rubber test plug 橡胶试塞(试验管的密实性)
rubber thread 橡胶线
rubber tile 橡胶片(铺地面用);橡胶(面层)砖;橡胶铺地砖
rubber tile flooring 橡胶面砖地面
rubber tiling 橡胶贴面
rubber tire 橡胶轮胎
rubber-tired and inner tube industry 胶车胎和内胎工业
rubber-tired 橡胶轮胎式的;轮胎式
rubber-tired carrier 橡胶轮胎的搬运汽车
rubber-tired compactor 气胎(路)碾
rubber-tired container grantry crane 轮胎式集装箱门式起重机
rubber-tired crane 轮胎吊;轮胎式起重机
rubber-tired ditcher 轮胎式挖沟机
rubber-tired drill 轮胎汽车钻
rubber-tired excavator 轮胎式挖掘机
rubber-tired fender 橡胶轮胎护舷;胶式轮胎缓冲装置
rubber-tired gantry crane 轮胎门式起重机;轮胎式龙门吊
rubber-tired haulage 轮胎车拖运;胶轮运输;无轨运输
rubber-tired loader 装有橡胶轮胎的装卸机
rubber-tired mobile crane 胶轮式起重机;轮胎起重机
rubber-tired port vehicle 轮胎式港口运载工具
rubber tired power blader 轮胎式自动平地机
rubber-tired roller 橡胶轮胎(压路)机;橡胶轮胎辗压机;轮胎式碾压机;轮胎式路碾;气胎(路)碾
rubber-tired shovelloader 轮胎式挖掘装载机
rubber-tired side dumping car 轮胎式侧卸车
rubber-tired tractor 胶轮轮拖拉机
rubber-tired transtainer 轮胎式集装箱门式起重机
rubber-tired truck-wheel fender 橡胶车轮护舷;橡胶车圈护舷
rubber-tired vehicle 轮胎式运输工具;胶轮车辆
rubber tire fender 橡胶轮胎护舷
rubber tire mounted 装有橡胶轮胎的
rubber tire roller 轮胎压路机
rubber tissue 橡胶片;橡胶布
rubber tooling 橡胶加工设备
rubber torsional bearing 橡胶扭力轴承
rubber track 橡胶履带;橡胶跑道
rubber-tracked vehicla 橡胶履带车辆
rubber trackshoe 橡胶履带板;橡胶履带片
rubber traffic cone 橡锥形交通路标
rubber transition 橡胶态转变
rubber tree 橡胶树
rubber tremie tube 橡胶导管
rubber trim 橡胶镶框;橡胶装修
rubber trough 橡胶槽
rubber tube 橡胶管;胶皮管
rubber tube level 橡胶管式水准器;橡胶管式水准仪
rubber tube wiring 胶皮线
rubber-type crane 轮胎式起重机;胶轮式起重机
rubber-type drill jumbo 软胎式钻车
rubber-type retarder 橡胶型缓行器
rubbeer-type vibration isolator 橡胶隔振体
rubber underlay 橡胶垫底层
rubber union 橡胶接头
rubber unit 橡胶部件
rubber universal joint 柔性万向节
rubber vacuum cup 橡皮真空吸盘
rubber valve 胶阀
rubber valve seat 橡胶阀座
rubber varnish 橡胶清漆
rubber velvet 防水棉绒
rubber vibration insulator 橡胶隔振器
rubber vibroinsulator 橡胶防振装置
rubber vulcanizing machine 橡胶热补机
rubber warmer 橡胶热炼机
rubber washer 橡胶垫圈;橡胶垫片
rubber waste 橡胶废料
rubber wastewater 橡胶废水
rubber waterbar 橡胶挡水条;橡胶挡水埂;橡胶止水带
rubber water pipe 橡胶水管
rubber water piping 橡胶水管
rubber waterproofing sheet 橡胶防水片材
rubber waterstop 橡胶止水(带);橡胶止水条;橡胶止水;橡胶水封;橡胶截水条
rubber (water) stop strip 橡胶止水带
rubber waterstop strip with sheet metal 带金属片的橡胶止水带
rubber waterstop with steel flanges 钢边橡胶止水带
rubber water tube 橡胶水管
rubber water tubing 橡胶水管
rubber weather-strip 橡胶垫条
rubber wheel 橡胶轮;橡胶结合砂轮
rubber-wheeled hydraulic excavator 轮胎式液压挖土机
rubber wind seal 橡胶防风条
rubber windshield cleaner 橡胶风挡清洁器
rubber wing 橡胶挡泥板
rubbery 橡胶样的
rubbery copolymer 橡胶状共聚物
rubbery flow 橡胶状流动
rubbery flow zone 橡胶态流动区
rubber yielding plant 橡胶植物
rubbery plateau zone 橡胶态高弹区
rubbery polymer 橡胶状聚合物
rubbery shear modulus 橡胶态剪切模量
rubber(y) soil 橡胶土
rubber(y) state 橡胶态
rubbery substance 橡胶性物质
rubbery tensile modulus 橡胶态拉伸模量
rubbing 研磨;抛光;磨退;摩擦
rubbing-abrading action 磨蚀作用
rubbing area 摩擦面积
rubbing bar 防擦棒材
rubbing bed 圬工磨床;磨床
rubbing-bed hand 磨石工
rubbing block 磨块;摩擦块
rubbing board 刮板;抹子;镘板
rubbing brick 金刚砂砖;磨砖;磨光用砖
rubbing compound 磨料;抛光膏
rubbing contact 摩擦触点
rubbing crack 摩擦裂纹
rubbing down 磨平;磨亮;抛磨
rubbing-down of varnish 清漆磨退
rubbing drum 搓磨滚筒
rubbing edition 拓本
rubbing effect 摩擦作用
rubbing factor 摩擦因数
rubbing fastness 耐摩擦程度
rubbing flat 磨平
rubbing friction 滑动摩擦
rubbing from stone or metal imscription 拓本
rubbing index 摩擦指数
rubbing keel 防擦龙骨;保护龙骨
rubbing leather 搓条皮圈
rubbing machine 搓擦机
rubbing oil 擦油;打磨用油
rubbing packing cup 滑动密封碗
rubbing paste 抛光浆;抛光膏
rubbing paunch 防磨席
rubbing piece 防擦材
rubbing plate 防擦材
rubbing property 打磨性
rubbing speed 摩擦速率;摩擦速度
rubbing stone 磨(面)石;磨光石(料)
rubbing strake 防擦材
rubbing strip 护舷材;防冲桩;防擦材
rubbing stroke 护舷材
rubbing surface 摩擦面
rubbing test 打磨试验
rubbing varnish 无光清漆;可打磨清漆;耐磨清漆;磨光清漆;打光清漆
rubbing velocity 摩擦速度
rubbing wear 摩擦损耗
rubbing work 摩擦功
rubbish 垃圾;碎屑;废物;废石
rubbish barge 垃圾驳(船)
rubbish bin 垃圾桶
rubbish box 垃圾箱;废料箱
rubbish can 垃圾箱
rubbish cartage 废物运费;垃圾运费
rubbish chute 废物斜溜槽;拉紧槽;垃圾滑道;垃圾滑槽
rubbish collecting chamber 废物箱;废物收集室
rubbish collection 废物收集
rubbish container 废物储存器;垃圾箱
rubbish dam 垃圾堤坝
rubbish destructor 废物焚化炉
rubbish disposal 垃圾处理;废物处理
rubbish dump 废料卸下;垃圾倒卸;垃圾卸下;废料倒卸
rubbish grinder 垃圾磨碎机;垃圾碾磨机

rubbish heap 废物堆;垃圾堆
rubbish incieration 废物焚化
rubbish incinerator 垃圾焚化炉
rubbish incinerator plant 垃圾焚化厂
rubbish load 毛石荷载
rubbish lock 废物阻塞
rubbish pit 垃圾坑
rubbish pollution 垃圾污染
rubbish pulley 垃圾滑车
rubbish removal 清除垃圾
rubbish salt 层积盐;碎盐;废盐
rubbish shoot 垃圾滑槽
rubbish storage chamber 垃圾储藏室;垃圾箱
rubbish structure 碎块构造
rubbish tip 垃圾堆场
rubbish utilization 垃圾利用
rubble 荒料;转石;建筑碎料;毛石;蛮石;粗石;齿轮贴纸测绘法;冰砾
rubble aggregate 毛石集料;粗石集料
rubble arch 乱石拱;毛石拱;粗石拱
rubble ashlar 琢石面毛石墙;毛方石;粗料石;方块毛石
rubble ashlar masonry 粗毛石圬工;粗方石圬工
rubble ashlar masonry work 方块毛石圬工
rubble ashlar wall 方块毛石墙
rubble backfill 毛石回填料
rubble backing 毛石底层;片石底层
rubble base 毛石基床;抛石基床
rubble beach 角砾滩
rubble bed 粗毛石基床;粗方石基床;毛石基床;毛方基床;抛石基床;碎石路;碎石底
rubble bedding foundation 明基床
rubble bedding under bottom 暗基床
rubble between return walls 转角墙间填石
rubble breakwater 毛石堆抛防波堤;抛石防波堤
rubble catchwater channel 填石集水沟
rubble catchwater drain 毛石截水沟
rubble chute 溜槽
rubble concrete 旧砖(石)混凝土;蛮石混凝土;块石混凝土;毛石混凝土;埋石混凝土;片石混凝土
rubble concrete wall 毛石混凝土墙
rubble cushion(ing) 毛石垫层
rubble dam 块石坝;毛石坝;堆石坝
rubbled finish 磨光饰面
rubble discharging 抛石
rubble disposition 抛石
rubble drain 乱石盲沟;块石排水沟;毛石排水沟;毛石(盲)沟;盲沟;透水石渠;填石排水沟;填石盲沟;石砌排水沟;砌石沟渠
rubble drift 乱石冰碛;巨砾泥;碎石堆
rubble fill 块石堆筑;毛石填筑
rubble fill foundation bed in trench 暗基床
rubble filling 抛石(防坡堤)
rubble fireplace 毛石壁炉
rubble flow 乱石流;泥石流
rubble footing 块石基础;毛石基础
rubble for toe protection 镇脚石
rubble foundation 块石基础;片石基础
rubble from ruins 墙堆中的乱石
rubble-guiding jetty at brook-outlet 溪口导堤
rubble hewn stone masonry (work) 粗削石圬工
rubble-intercepting dam in brook 溪沟拦河坝
rubble land 砂石田
rubble load 毛石荷载
rubble mason 毛石工;砌石工
rubble masonry 砌毛石圬工;乱砌圬工;块石砌体;毛石圬工;片石圬工
rubble-masonry check dam 毛石砌谷坊;碎石砌谷坊
rubble masonry dam 毛石砌体坝;毛石圬工坝;蛮石圬工坝
rubble masonry work 毛石圬工工程
rubble mound 乱石护坡;毛石护坡;毛石堆;棱体;堆石斜坡堤;粗石堆石堤;抛石堆
rubble-mound breakwater 抛堆毛石防波堤;斜坡式防波堤;块石防波堤;毛石堆防波堤;抛石防坡堤;堆石防波堤
rubble-mound for foot protection 护脚堆石;护脚抛石棱体;抛石棱体
rubble-mound foundation 抛石基床;乱石基础;毛石基础
rubble-mound prism 抛石棱体
rubble-mound seawall 抛石防潮海堤
rubble-mound with a loam core 土砂芯堤

rubble-mounted foundation 抛石基础
rubble outerskin 乱石墙皮
rubble pavement 块石路面
rubble paving 毛石铺砌;块石铺砌;粗石铺砌;粗石铺面;毛石铺面
rubble pile 毛石堆;乱石堆;石堆
rubble pitching 块石护坡;毛石护坡;片石护坡;抛石护坡
rubble pitching work 块石护坡工程
rubble protection 块石护岸
rubble removal 清除碎砖;清除乱石
rubble retaining wall 毛石挡土墙
rubble rock 角砾岩
rubble run rockfill 毛石填筑
rubble-slope (protection) 毛石(护)坡;毛砌坡;乱石护坡;块石斜坡
rubble soling 乱石基底
rubble stone 卵石;块石(料);毛石;二片石;粗石
rubble stone arch bridge 乱石拱桥
rubble stone footing 毛石基础;片石基础
rubble stone masonry bridge 乱石圬工拱桥
rubble stone subbase 毛石基础
rubble stone vault 毛石筒拱;毛石穹顶
rubble stone wall 块石墙
rubble subbase 毛石基础
rubble tipping 抛石
rubble vault 毛石穹顶
rubble wall 毛石墙
rubble wave absorber 消浪乱石堆;毛石消波设备
rubble with binding material 浆砌毛石;浆砌块石
rubble work 乱石工程;乱砌毛石;毛石砌体;毛石工(程);砌石工程;毛石砌筑
rubbly 瓦砾状的;碎石状的;块石的;毛石的;毛石砌成的;多碎石的
rubbly reef 碎裂矿脉
rubbly soil 砾质土
rub brick 磨面砖;磨光用砖;磨平块(混凝土);磨光块
rub check 防摩擦
rub down 磨平;擦干净
rubellan 红云母
rubellite 红电气石
rubene 丁苯
ruberal 废地杂草
rubercrete 铺有多层橡胶面的沥青碎石路面
ruberoid 橡胶沥青层面材料;屋面用油毛毡
rubescence 变红
rub fastness 耐磨程度
rubicelle 橙尖晶石;橙红尖晶石
rubidium antimonide photocathode 锑化铷光电阴极
rubidium deposit 铷矿床
rubidium frequency standard 铷频率标准
rubidium halide 卤化铷
rubidium hydroxide 氢氧化铷
rubidium manganate 锰酸铷
rubidium ore 铷矿(石)
rubidium oxide 氧化铷
rubidium perchlorate 高氯酸铷
rubidium selenate 硒酸铷
rubidium-strontium age 铷—锶年龄
rubidium-strontium age method 铷—锶年代测定法
rubidium-strontium dating 铷—锶年代测定法
rubidium-strontium dating method 铷—锶测年法
rubidium-strontium isochrones 铷—锶等时线
rubidium-strontium method 铷—锶法
rubidium-strontium model 铷—锶模式
rubidium thorium tungsten oxide 钨针铷氧化物
rubidium vapo(u)r magnetometer 铷蒸气磁力仪
rubiesite 杂硫碲铋矿
rubiginous 赤褐色的;锈色的
rubigo 锈斑病;铁锈;铁丹;过氧化铁
rubimentogeosyncline 雏地槽
rubin 品红
rubine 玉红色
rubine red 玉红
rubing grain 玉红磨料
rubinglimmer 针铁矿
rub iron 磨光(用)铁
Rubislaw 灰蓝色花岗岩(英国产)
rub mark 擦痕
rub-on coating 擦涂
rub-out 擦掉;删去;擦去

rub-out character 删去字符
rub-out method 调刀混合法
rub plate 搓板面
rub proofness 耐磨程度
rub resistance 耐磨性;耐摩擦性
rubrax 矿质橡胶
rubric 红字标题;赤铁矿
rubricated account 专款账户;标红字的专款账户
rubrication 用红色刷背景;用红色涂写
rubrozems 腐殖质红壤土
rubsen oil 菜油
rubstone 砂石;磨刀石;磨石
rub test 摩擦试验
rub to attack 打磨到(漆膜)损坏
rub to dull 用溶剂打磨到无光
rub up 擦亮或擦暗
ruby 玉红色;红宝石(的);浅红晶石
ruby alumina 红色刚玉;红宝石色氧化铝
ruby blende 红闪锌矿;红色闪锌矿
ruby boule 梨状深红宝石
ruby continuous laser 红宝石连续波激光器
ruby crystal 红宝石晶体
ruby crystal laser 红宝石晶体激光器
ruby disk laser 红宝石圆盘激光器
ruby glass 玉红玻璃;红(色)玻璃;红宝石玻璃;宝石红玻璃
ruby internal reflection mode 红宝石内反射振荡模
ruby laser 红宝石激光器
ruby laser microscopic radiation meter 红宝石激光显微辐射仪
ruby laser range finder 红宝石激光测距仪
ruby laser scattering apparatus 红宝石激光散射仪
ruby luster 红石红光泽彩
ruby maser 红宝石微波激射器
ruby mica 针铁矿;红宝石云母
ruby optic(al) master 红宝石激光器
ruby pin 红宝石圆盘钉
ruby ranging laser 红宝石激光测距仪
ruby red 宝红
ruby red glaze 宝石红釉
ruby-sapphire deposit 红蓝宝石矿床
ruby silver 红银矿
ruby spinel 红(宝)尖晶石
ruby stain 红色彩料;宝石红着色剂
rubytine 宝石叉瓦
ruby wood 红色装饰硬木(印度产)
ruck 皱褶;废物
ruckbildung 软化现象
Rucker plan 拉克计划
rucklidgeite 碲铅铋矿
rucksack 帆布背包;背囊
ructation 嗳气
rudaceous 砾状的
rudaceous rock 砾状岩;砾屑岩;粗碎屑岩
rudaceous sediment 碎砾沉积(物)
rudder 艄;舵;垂直舵;方向舵(偏角)
rudder action 舵的作用
rudder actuating cylinder 方向舵操纵动作筒
rudder actuator 方向舵传动器
rudder adjustment 舵角调整
rudder angle 方向舵偏角
rudder angle indicator 舵角指示器
rudder angle stopper 舵角限制器
rudder area 舵面积
rudder arm 舵箍(筋);舵臂
rudder balance 平衡舵
rudder band 舵箍(筋);舵臂
rudder bar 方向舵脚蹬
rudder bearer 舵托
rudder blade 舵叶;舵板
rudder braces 舵钮
rudder brake 尾甲板上止舵煞;舵闸
rudder brake pedal 舵闸踏板
rudder carrier 舵托;舵承
rudder case 舵杆管
rudder casing 舵杆管
rudder chain 应急舵链;转舵链
rudder chock 尾甲板上止舵楔
rudder coat 舵头帆布罩;舵套
rudder contactor trolley 自动舵接触器触轮;接触器触轮
rudder control 用舵操纵;航线控制
rudder control cable 方向舵传动索
rudder control unit 舵控制装置

rudder coupling 舵杆联结;舵杆接头
rudder crosshead 横舵柄;舵十字头
rudder deck stop 甲板止舵楔;尾甲板上止舵楔
rudder driver 方向舵传动装置
rudder effect 舵效
rudder eye 舵眼圈
rudder flange 舵杆头凸缘
rudder force 舵压力
rudder frame 舵框
rudder gear 操舵传动装置
rudder gland 舵杆转动填料函压盖
rudder gudgeon 舵枢
rudder hanger 挂舵钩
rudder head 舵杆头
rudder heel 舵踵;舵板根
rudder heel bearing 舵踵承;舵板根承
rudder hole 舵杆孔
rudder horn 应急舵链挂环;方向舵杆
rudder house 舵室;操舵室
rudder indicator 舵角指示器
rudder irons 舵钮
rudder keeper 舵杆接头螺帽防松器
rudder lever 舵杆
rudder lock(ing) 锁紧舵栓
rudder lug 舵承钮
rudder main piece 舵杆
rudder mass 方向舵杆
rudder moment 转舵力矩
rudder order 舵令;操舵口令
rudder order indicator 舵令指示器
rudder paddle 舵叶;舵板
rudder pedal adjuster 舵板调节器
rudder pendant 应急舵链索
rudder pintle 舵针;舵栓
rudder pit 卸舵钩;舵穴;舵坑;舵槽
rudder plate 舵叶;舵板
rudder port 舵轴简;舵杆孔
rudder post 舵柱;方向舵轴;方向舵梁;承舵柱
rudder post stopper 舵柱止舵楔
rudder potentiometer 自动舵分压器
rudder-propeller 推进舵
rudder quadrant 方向舵扇形体
rudder rake 舵后沿形状
rudder ratio adjustment 操舵灵敏度调整
rudder riser 舵钮空挡
rudder roll stabilization 横摇稳定舵
rudder score 舵钮空挡
rudder-screw 推进舵
rudder setter 自动舵管制器
rudder stay 舵箍(筋);舵臂
rudder stock 上舵杆;舵杆
rudder stock trunk 舵杆管
rudder stop(per) 止舵器;止舵楔
rudder stuffing box 舵杆转动填料函
rudder tackle 应急舵缓冲绞辘;缓冲绞辘
rudder telltale 舵角指示器
rudder tiller 舵柄
rudder torque 转舵矩;舵转矩
rudder trim tab 方向舵平衡调整片
rudder trunk 舵杆管
rudder tube 舵杆管
rudder vang 应急舵链
rudder well 卸舵钩;舵穴
rudder wheel 舵轮
rudder with bulb 梨形舵
rudder yoke 横舵柄;舵柄横木
rudding 清除废石
ruddle 赭色;红土;染红
ruddy 红的;浅红色的
ruddy shelduck 赤麻鸭
rude 未加工的;粗糙的
rude cotton 原棉
rude drawing 草图
rude edge-arrissing 粗磨边
rudenture 卷绳饰
rude ore 原矿
rude processing 粗处理
ruderal 宅旁杂草;杂草的
ruderation 卵石砂浆路面;浆砌石路面
rude respiration 吸粗
rude times 原始时代
rudiment 遗迹
rudimentary 退化的;残遗的
rudimentary algorithm 基本(的)算法

rudimentary compound microscope 基本组合显微镜
rudimentary system 初步系统
rudimentogeosyncline stage 雏地槽阶段
rudimentoplatform 雏地台
rudimentoplatform stage 雏地台阶段
rudimentum 遗迹
rudistids 厚壳蛤类
rudite 砾状岩;砾质岩;砾岩;砾屑岩;碎砾岩
Rudolphine table 鲁道夫星表
rudstone 砾屑碳酸盐;砾灰岩
rudyte 砾状岩;砾质岩;砾岩;碎砾岩
Rueping 吕平法;空细胞防腐(法)
Rufereen 鲁福林(一种坚韧的防水毡)
ruff 轴环
ruffed border 皱褶缘
ruffing 亚麻除籽工艺
Ruffini cylinder 罗芬尼圆柱体
Ruffini end organ 热感受器
Ruffini's organs 罗芬尼器
ruffite 泥岩
ruffle 褶边;皱纹;褶边
ruffled membrane 波状膜
Rufflette 鲁福莱特(一种金属窗的窗帘)
ruff sawn 锯纹装饰镶板
rufosity 赤褐色
rufous 带红的;赤褐色的
rufu 乳状
rug 小(块)地毯;厚地毯;毯;地毯;粗地毯
rugby 橄榄球
rugby ball 橄榄球
rugby field 橄榄球场
rugby football 橄榄球
rugged 坚固的;不平坦的;最硬的;崎岖的;粗糙的
rugged area 崎岖地
rugged catalyst 稳定催化剂
rugged coast 高低不等的海岸;起伏海岸
rugged country 崎岖地带
rugged face 粗糙面
rugged facing 硬质粗面贴面
ruggedised instrument 抗振仪表
ruggedization 强化
ruggedized computer 耐振计算机
ruggedized construction 耐振结构
ruggedized fiber optic(al) cable 高强度光缆
ruggedized instrument 抗振仪表
ruggedized machine 耐振计算机;耐振机器
ruggedized packaging 加固组装
ruggedized tube 耐振管
ruggedness 坚固性;耐用性;崎岖不平;凹凸不平度
rugged relief 崎岖地形
rugged shore 崎岖海滨
rugged terrain 崎岖地区;崎岖地带
rugged topography 起伏地形;崎岖地形;断绝地;不平地形
rugged work 繁重的工作
rugosa rose 玫瑰
rugose 皱的
rugose coral extinction 四射珊瑚绝灭
rugose leaf 皱纹叶
rugosimeter 糙度计
rugosity 皱褶;皱曲;折叠状;脉缩;粗糙度;糙率;糙度;凹凸不平(度)
rugosity coefficient 糙率系数;糙度系数
rugous 皱的
rug-overtufting machine 地毯簇绒机
rug weaving frame 地毯织机
rug wheel 抛光轮
ruim marble 块状大理岩
ruin 旧址;废墟
ruined 倒塌的;毁坏的
ruined by fire 毁于火灾
ruin garden 古迹园;废墟园
ruinous building 毁坏的房屋
ruinous earthquake 毁灭性地震
ruinous exploitation 滥伐;毁灭性砍伐
ruinous price 破产价格
ruinous shock 毁灭性地震
ruin probability 破产概率
ruins 遗址
ruins hill 废墟山
ruizite 水硅锰钙石
rulcanello 子火山
rule 规章;规则;规程;条例;定则;法则;守则

rule against perpetuities 反永久所有权规则
rule and lead cutter 裁铅线铅条机
rule and regulation for filling and stowing of container 集装箱货物装载规则
rule base 规则库
Rule brass 铅黄铜
rule connection graph 规则连接图
rule curve 规程曲线;调度图;调车线;操作规则表
ruled counting plate 分格计数板
rule depth ga(u)ge 深度规
ruled grating 刻划光栅
ruled length 规定长度
ruled paper 格子纸;格纸;划线报告纸
ruled surface 直纹曲面
ruled-surface map 直纹曲面图
rule for determining load 确定荷载标准;确定荷载规则
rule for distribution 分配原则;分配的规则
rule for implementation 实施细则
rule for rounding of numbers 凑整规则
rule for the preparation of financial statements 财务报表编制标准
rule joint 折尺铰接;肘形铰接;肘节形接头
rule of classification 分组规则
rule of combination 组合规则
rule of competition 竞争规则
Rule of Conciliation and Arbitration 调解及仲裁规则
rule of conversion 换位规则;变换规律
rule of correspondence 对应规则
rule of diffusion with dilution 稀释扩散规律;稀释作用的扩散规律
rule of direct consignment 直接运输规则
rule of dispatching 调度工作规程
rule of dispatching communication 调度通信规程
rule of elimination 消元规则
rule of existential generalization 存在推广规则
rule of existential specification 存在规定规则
rule of financial accounting 财会制度
rule of footway 人行道规则
rule of induction 归纳法规则
rule of inference 推断规则
rule of logic 逻辑规则
rule of mixtures 混合定律
Rule of Navigation 海上避碰规章;海上避碰规则
rule of nodal point numbering 节点编号规则
rule of Ogino-Knaus 奥一瑙二氏规律
Rule of Practice of Association of Average Adjustions 海损理算师协会理算规则
rule of precedent 先例约束力的原则
rule of road 行车规则
rule of road operation 道路营运规则
rule of safety operation 安全操作规程
rule of sign 正负号规则;符号律
rule of similarity 相似规则
rule of sixty 六十法则
rule of solubility product 溶度积规则
rule of superposition 重叠规律
rule of surface 表面规则
rule of tangential line 切线规则
rule of the road 交通规则;海上避碰规章
rule of the road at sea 海上避碰规章;海上避碰规则
rule of track maintenance 轨道维修规则
rule of transactions 交易规则
rule of three 比例法;比例的运算规则
rule-of-thumb 根据经验;概略估计;无科学根据;经验方法;经验法则;近似计算法;凭经验总结的大致作法;粗略公式;手工业方式;经验公式
rule-of-thumb design 凭经验进行设计
rule-of-thumb method 估计法;凭经验的方法;经验估计法;经验工作法;目力估测法
rule-of-thumb in forecasting of fog 实地预测雾情方法
rule of universal generalization 全称推广规则
rule of universal specification 全称规定规则
rule on orientation 定位规律
rule price 市价
ruler 直规;直尺;划线板;尺子
ruler inversor 直尺控制器
ruler line 刻度线
rules and regulations 制度;规章制度;条例;规定制度
rules and regulations in force 现行规章制度

rules and standards and instructions 规则、标准与说明
rules for adopting international standards 采用国际标准原则
rules for arbitration 仲裁规则
rules for construction of policy 保险单解释规则
rules for drafting symbol and code standards 符号、代号标准编写规定
rules for implementation 施行细则
Rules for Implementation of the Fire Regulations of the People's Republic of China 中华人民共和国消防条例实施细则
rules for quality evaluation 质量评定规则
rules for railway technology management 铁路技术管理规程
rules on administration of utilizing and saving water in city 城市节约用水管理规定
rules relating to rail-water through goods 铁水货物联运规则
rules relating to railway goods tariff 铁路货物运价规则
rule to number of ties 轨枕布置根数
rule with holder 带柄尺
ruling 支配；裁决；裁定
ruling axle 划线轴
ruling condition 控制条件
ruling down-gradient 限制下坡坡度
ruling engine 刻线机；划线机
ruling film 网线板
ruling grade 限制坡度；限制坡道；最大纵坡(度)；控制坡度
ruling gradient 限制坡度；控制坡度
ruling grating 刻线光栅
ruling language 主导语言
ruling machine 刻线机
ruling of a ruled surface 直纹面的母线
ruling of screen 网目线数
ruling pen 直线笔；绘图笔
ruling point 控制点
ruling price 现价
ruling rate of interest 主利率；基准利率
ruling section 等圆断面
ruling span 刻线间距
ruling stem 支配木
ruling up-gradient 限制上坡坡度
rulley 四轮载货车
rull roll 轻度凸面轧辊
rumble 噪声；滚筒；低频音
rumble filter 滤声器
rumble level 转盘噪声电平；盘噪声电平
rumbler 滚筒；清理滚筒
rumble spectrum 振动频谱
rumble strip (道路的)齿纹车道
rumble stripe surface 齿纹铺面
rumbling 转筒混合；滚筒涂漆
rumbling cleaning barrel 转筒清理滚筒
rumbling compound 抛光剂
rumbling mill 清砂滚筒
rumbling noise 车轮噪声
rumbling room 滚光间
rumbow line 再生绳
Rumex acetosa 酸模
Rumford fireplace 伦福德壁炉
Rumford photometer 伦福德光度计
rummage 海关检查；搜查
rummage sale 清仓拍卖
rummel 渗水坑(苏格兰语)；干井
rummer 大酒杯
rump 隆起部(分)
rumpfite 淡绿泥石
rumpometer 水压式漆膜抗张强度计
rumpus 地下室娱乐室
run 印数；开动；经营；挤提；回次【岩】；驱动；梯段水平距离；踏步宽度；导管尾；船尾端部；浮动
run aboard a ship 船相撞
runabout 小马达船；轻便小汽车；轻便货车
runabout crane (小型的)移动式起重机；流动式起重机；轻便起重机
run a business at a loss 赔本
run a cooperative 办合作社
run a curve 作特性曲线；绘出曲线(发动机特性曲线)
run aground 搁浅
run a level 水准测量队

run a line 拉线；铺设管道；定线；敷设管线
run along the surface 从土表流走
run along the ventral surface 沿腹面的外表伸展
run a mo(u)lding 拉线脚；拉出线脚
run and sag 流动下垂；流垂
run an engine 开动发动机
run an experiment 做试验
run a rabbit 通入检验器
runaround 临时线路；井底绕道；二层平台
run-around ramp 回返坡台
run-around track 回转线【铁】
runaround way 绕道
run a screw 钻进一个立轴长度
run a shop aground 冲滩
run ashore 搁浅；使船搁浅
run a survey 进行测量；测量
run at angle 转角走向
run a traverse 布设导线【测】
runaway catch siding 避难线【铁】
runaway corrosion 剧增腐蚀；脱落腐蚀(氧化膜)
runaway device 跑车装置
runaway economy 经济失控
runaway effect 失控效应
runaway girder 行车大梁；吊车梁
runaway governor 过速调节器；失控调节器；超速调节器
runaway hoist 移动式电葫芦
runaway inflation 无法控制的通货膨胀
runaway locomotive or car 机车车辆溜逸
runaway market 脱离控制的市场
runaway of bit 跑钻
runaway protecting device 防溜装置
runaway rate 失控的工资率
runaway reaction 失控反应
runaway reactor 失控堆
runaway shop 逃亡工厂
runaway speed 极限转速；空载转速；失去控制的速率；失控速度；飞逸转速
runaway switch 转辙器；防跑车道岔；安全道岔
runaway task 超越任务
runaway temperature 剧增温度；失控温度
runaway velocity 极限速度；空载速度
runback 回流管；反流
run back into the hole 下钻
run barefoot 裸眼完井
run before the wind 顺风航行
run board (楼梯)踏脚板
run book 运行资料；运行手册；操作说明书
run button 快动按钮
run by gravity 重力自流
run by the local people and subsidized by the state 民办公助
run chart 运行表
run coal 软烟煤
run command 运行命令
run continuous samples 进行连续取样
run copal 软化库巴树脂；熔凝库巴树脂
Runcorn planning 兰考恩规划(1967年英国利物浦新城规划)
run curve 车辆调度曲线
run days 日班工作
run derrick 竖立钻塔
run diagram 运行图
rundle 绞盘头；绞车筒；梯级；绳梯横杆
rundlet 小桶
run down 停止；扫描一周；波形下降；慢停堆留时
run-down a coast 沿着海岸航行
run-down box 观察器
run-down circuit 扫描电路
run-down equipment 磨损了的设备
run-down idle 空转
run-down legs 垂直的溢流管
run-down mill 中间机座
run-down of drill string 跑钻
run-down pipe 下放的管子
run-down pipe run-down piping 溢流管
run-down tank (storage) 馏出油接受罐；倾入槽
run-down test 自由下落试验
run-down tube 下放的管子；溢流管
run-down tubing 溢流管
run dry 在干燥的情况下操作；无滑润运转
run duration 运转周期；运动时间
run error (测微器的)行差
run error of micrometer 测微器行差

run faultlessly 运转正常
run free 顺风航行；自由活动
rung 梯级棍；梯的横木；舵轮幅
Runge-Kutta method 龙格—库塔法
Runge-Kutta numerical integration 龙格—库塔数值积分
Runge vector 龙格矢量
rung ladder 高凳；直爬梯；轻便梯；攀梯
run goods 走私货物；私货
run gravel 冲积砾石
run harshly 运转不规律(发动机)
run heavily 不稳定运转
run high 物价上涨；海浪汹涌
runholder 持证者
run hot 运转中发热；高炉热行
run hydraulic test 进行水力试验
runic character 象形字母(古代北欧文字的字母)
Runic cross 凯尔特十字架
runic texture 象形文结构
run idle 窝工；空转
run in 收进；插入；磨合运转；跑合试车
run indicator 运行指示器
run-in high 高速钻进
run-in stand 磨合试验台
run-in synchronism 同步运转
runin table 输入辊道
run-in test 空车试验
run-in time 磨合期
run into 撞上；碰上
run into debt 开始负债；借债
run into port 进港
run-in wear 磨合期磨损；跑合期磨损
R unit R 单位(德国)
runite 文象花岗岩
run-length 游程长度；进尺深度；扫描宽度
run-length coder 运行长度编码器
run-length coding 扫描宽度编码
run length limited code 游程长度受限码
run-length prediction 扫描宽度预测
runlet 小溪；小河
run light 空转
run line 拉线；画线
run location 运行位置；运行操作位置
run marks (铸造)滑移线
run mode 运行方式；执行方式
run motor 运转电动机
run mo(u)lding 灰线(脚)；灰泥线(脚)
runnel 小溪；小河口；潮沟
runner 运转者；运转体；叶轮；压碎机；转轮；工作轮；流道；流槽；临时水手；滑行装置；滑橇；滑道；横浇道；爬犁原；偷越国境的船；动单绞辘；导滑车；垂直木板桩；传递手；传辊；奔子
runner and tackle 动单绞辘串联单双复绞辘
runner and whip 动单绞辘串联定单复绞辘
runner angle 承辊角钢
runner-at tached boot 装有滑脚的开沟器
runner back 磨床托盘；磨床平卡盘
runner band seal 转轮下止漏环；转轮下密封环
runner bar 磨掌；磨盘肋
runner basin 浇口杯；池形外浇口
runner blade 叶片；转轮叶片；主动舵叶；轮叶；桨叶
runner blade control mechanism 协联机构；桨叶控制机构
runner blade opening 叶片开口；叶片开度；桨叶开度
runner blade servomotor 桨叶接力器
runner blade trunnion 叶片轴颈；桨叶轴颈；桨叶枢轴
runner boss 泄水锥；转轮中心体；转轮轮毂
runner box 浇注箱；浇口箱；冒口保温箱；分叉浇口
runner boxing 分流浇注
runner brick 阳道砖；流钢砖
runner bush 浇口圈；浇口杯
runner cap 转轮顶盖
runner casing 转轮外壳
runner chain 爬犁脚链
runner chamber 转轮室
runner collar 转环；镜板
runner cone 泄水锥；转轮锥
runner crew 传缆手(靠码头)
runner crown 转轮轮冠；上冠
runner crown seal 转轮上止漏环；转轮上冠密封
runner cup 漏斗形浇口；浇口杯
runner cut 磨料伤痕；抛光盘刻痕

runner disc 转轮轮盘
runner end cover 外顶盖;顶盖
runner end lid 外顶盖;顶盖
runner extension 横浇口延伸端
runner eye diameter 转轮入口直径;转轮进口直径
runner fittings (滑动门的)滚动零件
runner gate 横浇口
runner head 废铸物
runner housing 转轮室;转轮壳
runner hub 转轮毂;动轮毂
runner inlet 转轮进水口
runner inner lid 转轮支持盖;转轮内顶盖
runner level 转轮中心高程;转轮安装高程
runner opening 叶片开口;叶片开度;桨叶开度
runner outline 转轮轮廓线;转轮草图
runner overflow 浇口溢流
runner pin 直浇口棒;浇口棒
runner pipe 铸道;流通管
runner piston pump 转子活塞泵;回转活塞泵
runner pitch circle 转轮节圆
runner plate 转环;镜板
runner refractory 填沟耐火材料
runner removal access 转轮检修通道;转轮拆卸通道
runner riser 直接冒口;补缩横浇口
runner scrap 回炉浇冒口
runner seal ring 转轮止水环;转轮密封环
runner series 转轮型谱;转轮系列
runner shirt 转轮下环
runner shoe 仿形滑脚
runner shroud 转轮下环
runner stick 浇口棒
runner stone 上磨盘;动石辊
runner system 浇注系统;冒口系统
runner tackle 定单绞辘串联单双复绞辘
runner tooth adapter 活动齿座
runner trough 出渣口流槽;出钢口流槽
runner tube 流通管
runner vane (涡轮机的)工作轮叶片;转轮叶片;桨叶
runner vane servomotor 工作轮叶片伺服发动机
runner wagon 游车
runner wheel 碾轮
running 转轮;运行;流通;流挂;拉模;浇注;继续进行;跑模
running ability 运转能力;运行能力;营运性能
running a blank 做空白试验
running account 流水账;来往账户
running accumulator 连续累加器;后进先出存储器
running accuracy 旋转精度
running adjuster 运转调整器
running against current of traffic 反向行车【铁】
running against pressure 顶压下管
running against the seas 逆流航行
running agreement 航次延伸协会
running aground 搁浅
running alignment 跑合对中
running a mo(u)lding 用样板拉线脚
running and feeding system 浇冒口系统
running a position 经营部位
running at constant speed 等速运行
running at full capacity 满负荷运转
running at increased speed 加速运行
running at reduced speed 减速运行
running attention 运转维护;运行维护
running average 移动式平均
running away 不黏着(涂料)
running axis 工作轴
running axle stand 车轴座
running balance 动(态)平衡
running balance indicating machine 动平衡机
running balance statement of account 账户滚存余额结单
running balk 纵向顶梁
running bar 纵向顶梁
running batch 正常配合料
running before the seas 顺浪航行
running below normal velocity for construction 施工慢行
running block 动滑车;传动滑车
running board (车辆的)踏脚板;布线板
running-board antenna 汽车底架天线
running-board lamp 踏脚板灯

running board shield 踏脚板护罩
running bond 侧砖叠合;满条砌法;顺砖(压缝)砌合
running book 运行说明书
running bowline 活套
running brake test 制动机运行试验
running broker 流动经纪人
running casing 下套管
running casting 浇铸铸型
running center 活顶尖
running channel 浇道
running characteristic 运行特性;工作特征;工作特性
running characteristics of vehicle 车辆运行性能
running characteristics of an electrode 焊条使用特性;焊条使用说明书
running charges 行车费(用);营运费(用);营运成本
running chart 运行图
running check 运行检查;经常检查
running class 运行等级
running clearance (轴和轴承间的)转动间隙;齿侧隙
running commentary 实况广播
running condition 运行条件;工作状态;运行状态;运行情况;运行工况;工作条件
running control 日常检查
running coring 取岩芯
running cost 行车费(用);运转费(用);运营费(用);运营成本;运行费(用);运行成本(用);营运成本;连续成本;经常费(用);日常费用;维持费(用);使用费;生产成本;操作成本
running current 运行电流;正常工作电流
running curve 运转曲线;工作曲线
running curve simulator 工作曲线模拟计算机
running cycle 运行周期
running days 连续天数(通常用于租赁条款);连续(装卸)日(风雨和假日均计在内的装卸期);连续日
running depot 机车运用段
running differential mechanism 运转差动机构
running direction 行车方向
running direction indicator 转向指示器
running direction interlock 转向联锁装置
running dog 回转卷饰;波形涡卷饰;波形卷涡饰;波浪形雕饰;波状涡纹饰
running down 全熔化
running down clause 碰撞(责任)条款
running down of battery 电池耗尽
running dry 干钻;无润滑运转
running edge 压延边缘;轧制边缘
running efficiency 运转效率;机械效率
running empty 空转
running end (钢绳、链子的)活端
running expenses 行车支出;行车费(用);营运费(用);营运成本;经常开支;经常费(用);日常费用
running experience 运转经验
running fit 转(动)配合;间隙配合;松配(合);松动配合;动配合
running fix 移线定位;移动定位(航空或航海)
running flange 移动法兰
running flow 运行流动
running flux 运行通量
running foot 延尺;延英尺;按每英尺计
running form 报告格式
running form of profit and loss statement 连续式损益表
running free 空转;空运转;顺风驶帆
running fret 连续回绞(装)饰
running gate 直浇口;滑动门
running ga(u)ge 定程挡块
running gear 行走装置;行走系统;行动装置;工作部件;运转装置;运行装置;运行机构;运动机构;移动装置;走行装置;走行部分;轮轴;传动装置【铁】;传动系统;传动机构
running girder 连续大梁
running ground 流动地基;流沙地基;流动围岩;流动土(层);流动地层
running-ground timbering 松散岩层支护
running hitch 滑结
running hours 运转时数;连续时数(通常用于租赁条款);运营小时;运行小时;工作小时
running hours indicator 运转小时指示器
running hours sheet 工作小时报表
running ice 移动冰

running-in 扭击;发动机的空转;试转;下钻;磨合;跑合;试运转
running-in ability 磨合性
running-in characteristic 磨合性能;跑合性能
running-in compound 磨盘钻探用的复合润滑油
running index 运转指数
running indicator 运转指示器
running-in machine 磨合装置;试运转装置
running-in of assembly unit 总成的磨合
running-in of automobile 汽车的磨合
running-in of engine 发动机试转
running-in of the drill pipe 钻管钻入;钻管运转
running-in oil 磨合用润滑油
running-in parallel pumps 并联水泵
running-in period 走合时期;试运转时期;磨合期;跑合期;投产期
running-in reverse direction in double track district 复线反方向行车
running-in schedule 磨合时间表
running-in speed 跑合速度;试谐速度
running interval 行车间隔
running-in test 试运行试验;空转试验;空转调试
running-in time 磨合时间
running into tank 装入桶内
running joint 施工缝;伸缩(接)缝
running knot 活结;滑结;圈套结;套绳
running landing numbers 搬运顺序
running lane 行车车道
running large 顺风驶帆
running laydays 装卸货时间;连续停泊时间
running legs 钻塔主腿
running life 运转寿命
running lift 小型凿井用泵
running light 舷灯;航行灯;航向指示灯;动行指示灯;空载运行;空转
running-light current 空载电流
running-light test 空载实验
running line 行车线;接引绳;通行线(指铁路)
running load 工作负荷;流动荷载;活动荷载
running maintenance 巡回小修;例行保养;经常性维修;日常维修;日常保养
running margin 流动差额(英国)
running mark 流痕
running masonry bond 顺砖压缝坍工砌合
running material 运转材料;运转(所需的)物料
running mean 相继平均值;相继平滑值;移动式平均;移动平均(数)
running measure 纵长量度;长度
running mechanism 行走机构;走行机构;走向机构
running meet 不停车会车【铁】
running meter 延米;纵长米;直线米;按每米计
running moor 前进抛双锚;倒退锚一字锚
running mo(u)ld 抹灰滑动橡胶样板;灰线模子;连续花饰;模饰样板;模筛样板;灰线模具
running mo(u)lding 制备石膏线脚(在原来位置)
running nip 运转辊距
running nipple 外丝接头;短螺纹接套
running noise 运转噪声;运转噪声
running no-load 无载运行
running numbers 接连号码
running of a line of position 船位线转移
running of cold oil 输送冷石油
running of extra or withdraw of trains 加开或停运列车
running of extra train 列车加开
running off 流失;线脚终饰
running off of a portfolio 未了责任到期满为止
running off water 径流水;流出水;地表径流水
running of gear 齿轮的跑合
running of glaze 釉层流动
running of grout 串浆
running-on 套罗立
running on beam sea 横浪航行
running open 开放运行
running operation 运转操作
running order 运转次序;维护细则;动作次序
running ornament 连续花饰;流水形(装)饰件;连续花饰
running out 漏炉;超差
running-out machine 分批挤压机
running-out pit 流出槽;接玻璃液池
running-out system 自然到期制

running over core 套取脱落岩芯
running over the gradient by momentum 动能闯坡
running parameter 工作参数;变化参数
running part 生产备件;动端
running pea ballast 流动的砾石碴;流动的豆石碴
running performance 运行性能;运行特性
running period 运转周期,运转时间;运营周期;运行周期
running piping 输送管(道)
running plank (移动涵洞模架的)滑行板;跳板;车道板
running policy 长期有效保险单
running position 运转位置;运行位置;工作位置
running power 运转功率
running power spectrum 行进功率谱
running program(me) 运转程序;运行图表;运算程序;操作程序
running program(me) language 运行程序语言
running protective system 运行安全系统
running pulley 滚子
running pump time 开泵时间
running quality 运转质量能,运转的质量
running rabbit 跑动干扰信号;窜动干扰信号
running rail 运行钢轨;连续轨;导轨;行车(钢)轨;走行轨
running rate 运转率
running recorder 运转记录器
running repair 小修;巡回小修;现场修理;验修;临时修理;经常修理;日常修理;日常维修;日常检验;日常检修
running requirement 运转要求
running resistance 行走阻力;行驶阻力;运转阻力;运行阻力
running rigging 活动索具;动索
running roller 导辊
running rope 动索;动绳
running royalties 连续提成
running rule 靠尺;抹灰准尺;灰线靠尺
running sample 标准样品;平均样品;平均样本;平均试样;抽取样本;抽取试样
running sand 流沙(也称砂);松砂
running sand filtration 流沙过滤
running sand foundation 流沙地基
running saturation test 运行饱和度试验
running screed 冲筋;抹灰准带;抹灰准条;定墙灰厚度用的准条
running service 日常维修;日常勤务
running sheave 滑动轮;滑车
running shed 车辆修理厂;运用库;车辆保养厂
running shoe 架模靴;(灰线模上的)滑靴
running signal 行车信号(机)
running slag 流动性熔渣
running slope 纵坡(车辆行驶方向的坡度)
running spare 生产备件
running speed 行驶速度;行驶车速;行车速度;下钻速度;运转速度;行车速度;运行速度;车速;工作速度;跑速
running spiral design 涡纹
running stability 运行稳定性
running state 运行状态
running state of a process 进程的运行状态
running status 运行状态
running stock 流动库存;经常库存
running structure 运行结构
running subscript 游动下标
running support 试车架
running surface 轮轨接触面;路面;跑合面;(道路的)车行道表面;波状表面
running surface of wheeltrack 轮轨接触面
running survey 勘测;航行测量;踏勘
running system 行走系统
running system of locomotive 机车运转制
running temperature 运转温度,运行温度
running test 运转试验;运行试验;正常负荷试验;试探性试验;额定负荷试验
running test of whole tank vehicle/trailer 整车运行试验
running text 正文
running the anchors 起锚;抛锚
running the books 经理账簿
running the kiln till cold 冷窑运转;冷窑运行
running the tools into the well 将钻具下入孔内
running the tubing string into well 下油管

running through curve 通过曲线
running time 行驶时间;行车时刻;行车时间;运行时间;运转时间;执行时间;航行时间;实际工作时间;操作时间
running time counter 运行时间计数器
running time meter 运行计时器
running time of dredge(r) 挖泥船运转时间
running to coke 蒸馏到焦炭
running to cylinder stock 蒸馏到汽缸油料
running token 行车凭证
running tool 送人工具
running to putty 调成腻子;调成油灰
running torque 转矩;额定转速转矩
running torque converter 运转变扭器
running track 通行轨道(指铁路);行车线;通行线(指铁路)
running track resurfacer 流挂(痕迹)修饰剂
running trap 双弯存水弯;存水弯;U形存水弯
running trial 运转试验;航行试车
running tunnel 区间隧道
running turbine components 透平旋转部件
running turk's head 活打花篮
running under braking 制动运行
running under power 牵引运行
running under pressure 顶压下管
running unloaded 卸荷运转
running-up accident 试车事故(发动机);启动事故;碰撞事故
running-up accident analysis 碰撞事故分析
running-up test 起动试验
running variable 游动变量
running velocity 运行速度
running vine 编织使用的枝条(泥浆)
running viscosity 运转黏度(泥浆)
running voltage 运转电压;运行电压;工作运行电压;工作电压
running warm 运转发热
running warn 运转预报
running water 自来水;流水;流动的水;活水
running water ecosystem 流水生态系统
running wave 行进波;行波
running ways 滑橇
running weight type dynamometer 移动配重式测力计
running wheel 工作轮;机车主动轮
running wire 动钢索
running working days 连续工作日
running yard 码长(度)
running yield 流动的收益率;连续收益率;当前收益率
runny paste 软膏
run of a furnace 开炉
run-of-bank gravel 原案砾石;河卵石;石膏原矿;岸边砾石;未(过)筛砾石;河岸石堆
run-of-crusher stone 轧石机筛余屑;破碎时拣出的废石;轧石场弃石;未筛机碎石料
runoff 泄漏;雨水排放量;流量;满期为止;出轨;曲线缓和长度;流量率;经历时限标准
runoff amount 径流量
runoff analysis 径流分析
runoff and seepage 径流和渗漏
runoff and soil loss 径流和土壤流失
runoff area 泄水面积;径流区;径流面积
runoff box 排放箱(筛余渣);溢流箱
runoff calculation 径流计算
runoff capacity 径流量
runoff characteristic 径流特征
runoff coefficient 径流系数
runoff coefficient method 径流系数法
runoff component 径流组成部分
runoff condition 径流条件
runoff condition of groundwater 地下水径流条件
runoff conduit 泄水管(道)
runoff contaminant 径流污染物
runoff corner 切角
runoff curve 径流曲线
runoff cycle 径流循环
runoff data 径流资料
runoff depth 径流深
runoff desert 径流沙漠
runoff detention 径流滞留量;径流累积量
runoff distribution 径流分布;径流分配
runoff distribution curve 径流分布曲线;径流分配曲线

runoff dry season 枯水径流
runoff dynamics 径流动力学
runoff erosion 径流侵蚀
runoff event 有径流时候
runoff experiment(al) station 径流实验站
runoff factor 径流因数;径流系数;径流因素
runoff feeder 排渣冒口
runoff field of groundwater 地下水径流场
runoff flow 径流
runoff forecast 径流预报 runoff formula 径流(量)计算公式
runoff-frequency data 径流频率资料
runoff-frequency information 径流频率资料
runoff from fee-lots 从施用场所的流失
runoff from glacier 冰川径流
runoff from snow 融雪径流
runoff-generating model 径流形成模型
runoff-generating precipitation 产流降水
runoff-generating rain 产流降雨
runoff generation 径流形成
runoff hydrograph 流量过程线;径流过程线
runoff impact erosion 径流冲失
runoff in depth 渗流
runoff in drought season 枯水期径流量
runoff in flood season 汛期径流量
runoff inflow volume 径流流入量
runoff intensity 径流强度
runoff intercepting terrace 截流梯田;截流阶地;截流地埂
runoff in yield 产流量
runoff losses from slope 坡地径流损失
runoff low 枯水径流
runoff map 径流分布图;径流等值线图
runoff mass curve 径流累积曲线;径流积分曲线
runoff measurement 径流测定;径流测量
runoff-measuring flume 径流反应
runoff mine coal 毛煤
runoff model 径流模型
runoff modulus 径流模数
runoff modulus map of aquifer 含水层径流模数图
runoff modulus of aquifer 含水层径流模数
runoff modulus of groundwater 地下水径流模数
runoff number 径流数
runoff of dry season 枯季径流
runoff of low water 枯水径流;枯季径流
runoff of melting snow 融雪径流
runoff of water 径流
runoff percentage 径流百分数
runoff phenomenon 径流现象
runoff pipe 泄水管(道)
runoff pit 溢流池
runoff plant 径流式(发)电站
runoff plat 径流场;水平衡场
runoff plate 引出板;助焊板;溶焊助焊板
runoff plot 径流试验区;径流区域
runoff point 离开处
runoff pollution 径流污染
runoff pollution control 径流污染控制
runoff pollution load 径流污染负荷
runoff precipitation 降雨径流
runoff process 径流过程
runoff producing precipitation 产流降水
runoff producing rain 产流雨;产流降雨
runoff pump 径流泵
runoff quality 径流水质
runoff-quality model 径流水质模型
runoff-quarry stone 采石场原石
run-off rails 脱轨
runoff rate 径流强度;径流率
runoff regulation 径流调节
runoff retardance coefficient 径流减速系数
runoff retardation 径流阻滞量
runoff riser 排渣冒口
runoff-river hydroelectric(al) power plant 径流式水电站
runoff-river plant 径流式水电站;径流式(发)电站
runoff-sediment discharge curve 径流输沙率曲线
runoff-sediment relationship 径流泥沙关系
runoff-sediment relationship curve 径流泥沙关系曲线
runoff separation 过程线分析;过程线分割;径流分割
runoff slag 溢出渣
runoff source 补给水源;径流来源

runoff station 径流实验站
runoff-stream plant 径流式水电站
runoff super-elevation slope 超高顺坡【铁】
runoff surface 地面径流；径流面
runoff suspended-sediment record 径流悬移质记录
runoff tab 引出板
runoff the straight 发生倾斜
runoff type 匡流型
runoff utilization factor 径流利用系数
runoff volume 径流(总)量；径流体积
runoff water 活水；地表径流(量)；流水
runoff yield 产流量
runoff zone of groundwater 地下水径流带
run of ice 流冰；漂凌；漂冰；淌凌
run-of-kiln lime 原石灰；未筛分石灰；未分选的石灰
run of micrometer 测微器行差
run-of-mill 选矿厂接收处理的矿石
run-of-mill ore 去掉废石的原矿
run-of-mine 未筛分的；原矿
run-of-mine bin 原矿仓
run-of-mine coal 原煤
run-of-mine coal grade total 毛煤总计
run-of-mine gypsum 石膏原矿
run-of-mine-milling 无介质磨矿
run-of-mine type ore 原矿(石)
run-of-mine shaker 原矿摇动筛
run of pipe 排管
run-of-pit 原坑石料；原(采)石料
run-of-pit-quarry 原石
run-of-pit stone 未筛(分)石料
run-of-quarry 原石；未筛分石料
run-of-quarry material 毛料；未筛(分)石料
run-of-quarry stone 采石场(统货)石料；未筛(分)石料
run of rays 光路
run-of-river plant 径流式河床发电站；径流式水电站
run-of-river plant with open bay 具有敞开式河湾的径流电站
run-of-river plant without pondage 无调节池的径流电站
run-of-river plant with pondage 有调节池的径流电站
run-of-river power plant 径流式(发)电站
run-of-river scheme 径流式水力发电站规划
run of steel 钻杆长度
run-of-the bank filling 岸土流失
run-of-the coast 岸线走向
run-of-the ice 冰的漂程
run-of-the-mill 一般质量
run-of-the mine coal 原煤
run-of-the wind 风程
run-of-water scheme 低水头电站
run-on 接排材料
run on a bank 挤兑
run-on chapters 章回接排
run on gold against U.S. dollars 美元挤兑黄金的风潮
run on grassland 在草地上放牧
run-on plate 引弧板
run-on point 转动点
run on rocks 坐礁；触礁
run on/run off 滑上滑下
run on shore 搁岸
run-on tab 引弧板；起弧板
run-on time 运转时间；连续(运转)时间
run-out 散热器到立管的横向连接口；旁送口；溢流；振摆；漏料；期满；跑封；跑钢；铸跑火；给水量；漏炉；流域总(径)流量；偏心率；偏斜；退刀纹；铁水穿漏
run out a contract 合同期满
run-out contouring 外送运输机
run-out conveyer 外送运输机
run-out distance 着陆减速滑跑距离
run-out field of bubble 磁泡形成场
run-out groove 引出纹槽
run-out key 输出键
run-out of cash 现金短缺
run-out of controls 失去操纵
run-out of gas 燃料耗尽
run-out of thread 螺纹尾(部)；部退刀纹
run-out of timing gear 正时齿轮的摆动
run-out of work 窝工；停工
run-out table 输出辊道

run-out time 停车时间
run-out tolerance 跳动公差；摆差
run-out to sea 出海
run over 沸腾翻出
run over in the hole 落孔内
run over the facing points 通过对向道岔
run over the trailing points 通过顺向道岔
run phase 运行阶段
run pipe 流通管；卷取导管；盘条卷取导管
run plank 跑道板
run position 运转状态；运转位置；运行位置
run program(me) file 运行程序文件
run rampant 泛滥成灾
run rider 跟车工
run risk 冒风险
run rockfill 岩崩
run rough 运转不平稳(发动机)
run-round plough 单向犁
run scheduling subsystem 运行调度子系统
runstitching 初缝
run stream 运行流
run test 游程检验；运行试验
run-through 浏览【计】
run-through operation 连续运转
run-through ruling 不中断的划线
run-through train 通过列车
runtime 运转时间；运行时间
runtime accumulator 运行时累加器
runtime administration routine 运行时管理程序
runtime data area 运行时数据区
runtime environment 运行时环境
run time library 运行时间库
runtime stack top 运行时栈顶
run-time statistics 运行时间统计
runtime storage allocation 运行存储分配
runtime subroutine 运行子程序
runtime value 运行值
run timing 运行定时
run-to-completion 从运行到完成的工作方式
run-to-completion technique 从运行到完成的技术
run to putty 筛浆料；拌成腻子；拌成油灰
run to run 两次运转对比
run-to-waste drilling 不取心钻进
run true 无跳动运转
run tube 流通管
run under the lee of the shore 在背风岸边航行；在岸下风航行
run unit 执行单位
run untrue 跳动运转
run-up 加快速度；爬高；上坡；上爬；涌起；迅速增长；上涌；涌上；上装
run-up area 起飞前试车区
run-up elevation 上涌高程
run-up factor 上冲系数
run-up height 爬高速度
run-up noise suppression 加速噪声抑制
run-up of engine 发动机转数增加
run-up on dam face 抬高坝面
run-up pad 待飞坪(飞机)
run-up time 起动时间
run-up to maximum power 加速运转到最大功率
run-up track 试车轨道
runway 悬索道(料运石)；流水道；机场跑道；滑道；河沟；跑道；吊车道；单轨架空道；飞机跑道
runway access taxiway 进入跑道的滑行道
runway alignment indication light 对准跑道指示灯(飞机)
runway arrangement 跑道配置
runway arrester 跑道拦阻装置
runway arrester net 跑道拦阻网
runway arresting gear 跑道拦阻装置
runway axis 跑道轴线
runway barrier 跑道拦阻装置
runway base 跑道基础
runway base course 跑道基层
runway beam 行车梁
runway bearing capacity 跑道承载能力
runway board 施工(台)跳板；(机场的)跑道板
runway boundary 跑道界线
runway bracket (机场的)跑道托架
runway capacity 跑道容量；跑道承载能力
runway caution lights 跑道头警告灯
runway centerline 跑道中心线
runway check 跑道检查

runway cleaner 清轨器；跑道清洁器
runway clearance 清扫跑道
runway condition 跑道状况
runway condition report 跑道状况报告
runway configuration 跑道构型
runway control 跑道管理
runway control van 跑道边指挥车
runway distance marker 跑道距离标志
runway ecnterline lights system 跑道中心线信号灯系统
runway edge light 跑道边线灯；跑道边界灯
runway edge lighting 跑道边界照明
runway ejection 跑道上弹射
runway ejection seat 跑道弹射座椅
runway end identification lights 跑道端识别灯光
runway end marking 跑道头标志
runway end safety area 端安全道
runway environment 跑道环境
runway escape 跑道上离机
runway extension 跑道延长段
runway facility 跑道设备
runway floodlight 跑道泛光灯
runway gear 跑道设备
runway girder 天车大梁；行车大梁
runway grooving 跑道浅槽(防止降雨时飞机滑移危险)
runway guidance 跑道引导
runway heading 跑道方向
runway hoist 移动式电动葫芦
runway landing 跑道上降落
runway layout 跑道布置方案
runway light 跑道标志灯；跑道照明灯；跑道指示灯；跑道
runway lighting (机场的)跑道照明
runway lighting equipment 跑道照明设备
runway lighting installation 跑道照明设备
runway localizer 着陆航向信标台
runway localizing beacon 跑道定位信标；飞机场跑道标灯
runway locating system 跑道位置指示系统
runway marking 跑道标志
runway observation 跑道气象观测
runway obstruction 跑道障碍
runway of the landing strip 简易着陆机场跑道
runway overrun area 跑道头保险道
runway pattern 跑道分布图
runway pavement 跑道路面；跑道面；跑道铺砌层
runway portable tower 起飞线塔台车
runway roll-out 沿跑道滑跑
runway roll-out guidance 跑道上着陆滑跑操纵
runway safety shoulder 跑道的侧安全道
runway shoulder 跑道道肩
runway strengthening 跑道加固
runway strip 跑道
runway surface ice detector 跑道道面结冰探测器
runway surface icing detector 跑道道面结冰探测器
runway surface lights 跑道照明灯
runway surfacing 跑道铺面；跑道道面
runway sweeper 跑道清扫机
runway sweeping 清扫跑道
runway temperature 跑道温度
runway threshold (机场的)跑道界限；跑道进口
runway threshold light 跑道界限指示灯；跑道入口灯
runway threshold marking 跑道入口标志
runway visibility 跑道能见度
runway visual 跑道视距
runway visual marker 机场标志板
runway visual range 跑道视程
runway visual range equipment 跑道视距测定装置
runway weight bearing capacity 跑道承载能力
run with tide 乘潮前进
Ruolz alloy 鲁尔茨合金
rupes recta 直壁
rupideserta 石质荒漠群落
rupture 折断；遮断；破裂；破坏；断裂
rupture appearance 破碎现象
rupture area 破裂范围
rupture bending angle 断裂时弯曲角
rupture bending moment 破坏弯矩
rupture circle 破裂圆
rupture compressive strength 断裂抗压强度
rupture condition 断裂条件
ruptured 破裂的

ruptured grain 破裂木纹
rupture diagram 破裂圆；破裂图
rupture disc 破裂片；断裂盘；破裂盘；防爆膜；爆破膜；安全隔膜
rupture disk device 防爆膜装置；爆破膜装置
ruptured zone 龟裂区；破裂区；石层风化带
rupture elongation 断裂伸长(率)
rupture envelope 破裂包线
rupture grain 破裂木纹
rupture in bending 弯曲损坏
rupture in buckling 失稳断裂；屈曲断裂
rupture length 破裂长度
rupture lief 持久强度
rupture life 破裂寿命
rupture life of die 模具寿命
rupture limit 破坏极限
rupture line 破裂线；破坏线；断裂线
rupture load 破坏荷载
rupture load in bending 抗折破坏荷载
rupture member 破裂构件
rupture membrane 防爆膜
rupture metamorphism 压碎变质
rupture modulus 破裂模量；折断模量；挠折模量；断裂模数；断裂模量
rupture moment 断裂时力矩
rupture motion 破裂运动
rupture of felt 油毡断裂
rupture of modulus 断裂模量
rupture of oil film 油膜断裂
rupture of rail 钢轨折断
rupture of round window membrane 圆窗膜破裂
rupture plane 破坏面
rupture point 断裂点
rupture pressure 破坏压力
rupture process 破裂过程
rupture proof security 防裂保护
rupture propagation 破裂传播
rupture range 破裂范围
rupture rate 破裂速率
rupture shearing strength 断裂时剪切应力；断裂时剪切强度
rupture speed 破裂速度
rupture stage 破损阶段
rupture strain 破裂应变；破坏应变；断裂应变
rupture strength 抗折强度；破裂强度；断裂强度
rupture stress 致裂应力；破裂应力；断裂应力
rupture stress and strain 破裂应力与应变
rupture stress condition 断裂时应力状态
rupture surface 破坏(表)面；断裂面
rupture tensile stress 断裂时张应力；断裂时拉应力
rupture test 破裂试验；破坏(性)试验；断裂试验；韧度试验
rupture velocity 破裂速度
rupture zone 破坏区域；龟裂区；破裂带；断裂带
rupturing capacity 遮断容量；遮断功率
rupturing capacity test 遮断容量试验；遮断功率试验
rupturing current 切断电流
rupturing duty 灭弧工作；切断功率
rupturing elongation 断裂时伸长
rupturing operation 遮断操作；切断操作
rural 农村的
rural activity 农业活动
rural agricultural industry 乡村农用工业
rural and small town enterprises 乡镇企业
rural architecture 乡村建筑；农村建筑
rural area 乡村地区；乡村地带；农业区；农村地区
rural building 农村房屋
rural church 乡村教堂
rural community 乡村社区；乡镇；乡村乡镇
rural conglomeration 农村集合村
rural construction 农村建筑；农村建设
rural country 乡村国家
rural development 乡村开发；农村开发
rural disease 农村病
rural distribution wire 乡村配电线；乡村电话线；郊区配电线；郊区电话线
rural district 农村；乡区；乡村地区
rural domestic waste(water) 农村生活污水；农村生活废水
rural domestic water 农村生活用水
rural drinking water 农村饮用水
rural drinking water pump station 农村饮用水泵站
rural economy 农村经济

rural electrification 农村电气化
rural enterprise 乡村企业；乡镇企业
rural environment 农村环境
rural environmental planning 农村环境规划
rural erosion rate 农田水土流失率
rural exodus 农村人口外流；迁离农村
rural highway 乡村公路；农村公路
rural home 农村房屋
rural hospital 乡村医院
Rural Housing Alliance 乡村住房联合会
rural housing loan 乡村住房贷款；农村建筑贷款
rural housing site loan 郊区房屋场地贷款；乡村住房地基贷款
rural hygiene 农村卫生
rural income 农村收入
rural industry 乡镇工业；乡村工业
rural intersection 乡村道路交叉口
rurality 田园风光；农村景色；农村特征
rural job creation 农村就业机会
rural land use 农村用地
rural land use strategy 农村土地利用对策
rural location 郊外地区
rural mail delivery route 乡村邮政路线
rural motorway 乡间公路；乡间汽车道
rural penetration 农村小路
rural planning 乡村规划；农村规划
rural planning object 农村规划目标
rural population 乡村人口；农村人口
rural resettlement 农村移民安置
rural resettler 农村移民
rural residence 农村住宅
rural road 乡间土路；乡间道路；乡村公路；郊区道路
rural route 乡间道道
rural section 郊区；乡村地区；乡村地
rural settlement 村落
rural sewage 乡村污水系统；农村排水工程
rural sewage disposal 乡村污水处理；农村污水处置
rural sewage disposal farm 乡村污水灌溉田
rural sewage farm 农村污水灌溉田
rural sewage system 农村污水系统
rural sewage treatment 农村污水处理
rural sewerage 农村污水；农村排污系统；农村排污工程
rural station 乡村车站；村镇电台
rural stmosphere 农村大气环境
rural stormwater pollution 乡村暴雨污染；农村暴雨水污染
rural structure 农村结构
rural township 乡镇
rural traffic survey 乡村运量观测；乡村交通调查
rural-urban fringe 城乡边缘地带
rural urbanization 农村都市化；农村城市化
rural-urban migration 乡村人口移入城市；农村向城市移民
rural water supply 乡村供水；乡村给水；农村供水；农村给水
rural wire broadcasting network 农村有线广播网
rural zone 农村地带
rurban 乡镇接壤地带；城镇郊区(城乡中间区)
rurp 绳钩(爬山攀冰用)
rusacovite 水磷钒铁矿
rusakovite 水磷钒铁矿
Ruselite 耐蚀压铸铝合金
rush 琴片
rushabar 黑风(土耳其克尔迪斯坦一带的干冷东北风)
rush agate 灯草玛瑙
rush current 冲击电流
rush discharge 电晕放电
rush engineering order 紧急工程订货
rushes 蠕动波
rush for goods 抢购物资
rush-hour direction 高峰小时行车方向
rush-hour overcrowding 上下班交通拥挤
rush hours 交通高峰时间；拥挤时间；高峰时间；上下班时间
rush-hour service 上下班时间服务
rush-hour station 高峰时间停车站
rushing flow 急流
rushing-out water 涌出的水；渗出水
rush marsh 灯心草沼泽
rush mat 苇席；草席；草垫
rush mat bale 草席包
rush of current 电流激增

rush of water 水冲击
rush order 紧急订货
rush period 高峰时间；农忙期
rush repair 抢修
rush repair job 紧急修理
rush-repair work 抢修工程
rush season 忙季
rush to deal with an emergency and provide disaster relief 抢险救灾
Rushton schist 鲁施顿片岩
rush work 突击工作；突击工程
rush work for emergency 抢险
Russar 鲁西萨尔超宽角照相物镜
russell cord 罗素棱纹呢
russellite 钨铋矿
Russell's angle 罗素角(光通量测量用角)；球体平均光度角
Russell's diagram 罗素图
Russell's effect 罗素效应
Russell's flask 罗素烧瓶
Russell's mixture 罗素混合物
Russell's model 罗素模型
Russell's test 罗素试验
Russell's traction 罗素牵引
russet 赤褐色
russety 带黄褐色的
Russian architecture 俄罗斯建筑
Russian blue 蓝灰色；浅蓝色
Russian chimney pipe 俄罗斯式壁炉管
Russian platform 俄罗斯地台
Russian rion 发蓝薄钢板
Russia old land 俄罗斯古陆
Russ pavement 俄罗斯式铺面
rust 锈(色)；铁锈；生锈；锈蚀
rust blister 锈疱；锈浮包
rust bloom 锈霜
rust bowl 生锈地区
rust cement 铁屑水泥；铁管接合油泥；防锈胶；防锈膏
rust cleaning 除锈
rust colo(u)r 铁锈色
rust colo(u)red glaze 铁锈花釉；铁透花釉
rust-converting primer 防锈底漆
rust creep 锈徐变；锈蠕变；锈迹
rust-eaten 锈蚀
rusted surface 已锈表面
rustenburgite 等轴锡铂矿
rust film 锈膜
rust formation 生锈
rust forming 生锈
rust grade scale 锈蚀等级
rust grease 防锈油脂；防锈润滑脂
rust growth test 长锈试验
rustic 粗制的；粗琢的；粗面的；结构粗糙；粗凿石；粗面石饰
rustic arch 粗琢石拱；粗毛石拱；乱石拱
rusticated 粗犷；粗琢的
rusticated ashlar 凸角石；粗琢方石
rusticated basement 粗石面的地下室
rusticated column 粗面柱；分块柱；箍形柱
rusticated dressing 粗琢面
rusticated joint 明显缝；凸形勾缝；粗接缝；粗糙缝；粗琢缝
rusticated masonry work 粗石琢面圬工
rusticated quoin 凸出角石
rusticating 粗琢(的)；表面粗化
rustication 凹槽(混凝土或砖墙装饰用)；粗面石工
rustication strip 凹槽板条；刻槽条
rustic bird house 田间禽舍
rustic brick 饰面砖；毛面砖；粗面砖
rustic cabin 粗整修的小屋；粗整修的客舱
rustic chamfered work 粗面切角工程；粗面切角工作
rustic column 无饰柱；无装饰的柱
rustic cottage 村舍
rustic dressing 粗琢面
rustic finish 水洗石饰面；乡趣园；水刷石
rustic home 农村房屋；农村房舍；田野木屋；乡村房屋；田野木舍
rusticity 朴素；乡村式；乡村特点
rustic joint 深凹灰缝；粗面石墙缝；石面凹缝；粗接合；凹入粗缝；凹槽接缝；粗琢缝
rustic masonry 粗石工；粗(琢)圬工；粗面圬工；粗面砌体

rustic or washed finish 水刷石饰面
rustic plaster 粗琢石面饰
rustic quoin 粗琢转角石
rustic road 农村道路
rustics 粗面砖
rustic siding 互搭板;披叠板;粗面壁板
rustic slate 粗面石板;粗砌石板
rustic stone 粗石;毛石
rustic terrazzo 粗面水磨石;水刷石
rustic tile 乡村朴实瓦片
rustic ware 田园风光陶器;粗面器皿
rustic washed finish 水刷石饰面
rustic woodwork 原木装饰品;原木制品;乡野家具;粗木作(业);粗木工
rustic work 简陋房屋(用树枝树皮等造成的);粗琢石作;粗石工;粗面石堆
rustification 粗琢石作
rusting 铁锈;生锈;锈蚀
rusting corrosion 锈蚀
rusting degree 锈蚀度
rusting grade 锈蚀等级
rusting-inhibiting 防锈的
rusting mechanism 生锈机理
rusting-preventing 防锈的
rusting process 锈蚀过程
rusting rate 锈蚀速率
rust-inhibiting additive 防锈添加剂
rust-inhibiting lubricant 防锈润滑脂
rust-inhibiting paint 防锈漆
rust-inhibiting primer 防锈底漆
rust inhibition 防锈作用
rust-inhibitive 防锈的
rust-inhibitive agent 防锈剂
rust-inhibitive pigment 防锈颜料
rust-inhibitive wash 防锈洗涤剂;防锈洗涤底漆
rust inhibitor 阻锈剂;抗腐蚀附加剂;防腐蚀抑制剂;防锈(剂);防腐添加剂
rust joint 锈接;锈结;天沟水密接头
rust layer 锈层
rustless 不锈的
rustless iron 不锈铁
rustless metal 不锈金属
rustless process 不锈钢冶炼法
rustless property 不锈性
rustless steel 不锈钢
rust mark 锈印;锈斑
rust preventative 防锈剂
rust preventative treatment 防锈处理
rust preventer 防锈剂
rust preventing 防锈的
rust preventing agent 防锈剂
rust preventing coat 防锈层
rust preventing compound 防锈混合剂
rust preventing device 防锈装置
rust preventing grease 防锈油脂;抗锈膏;防锈酯
rust preventing hard gloss paint 防锈硬性有光泽的油漆
rust preventing oil 防锈油
rust preventing paint 防锈漆;防锈涂料
rust preventing pigment 防锈颜料
rust-preventing pump 防蚀泵
rust preventing wrap 防锈包裹层
rust prevention 防锈
rust prevention during manufacture 工序间防锈
rust prevention in interstore 中间库防锈
rust prevention test 防锈试验
rust prevention treatment 防锈处理
rust preventive 防锈剂;防腐蚀剂

rust-preventive action 防锈作用
rust-preventive enamel 防锈搪瓷;防锈瓷漆;防锈硬质有光漆
rust-preventive material 防锈材料
rust-preventive oil 防锈油
rust-preventive paint 防锈(油)漆
rust-preventive pigment 防锈颜料
rust-proof coating 防蚀涂料;防锈涂料
rust-proof(ing) 防锈;不生锈;不锈的;防锈的;防锈处理;抗锈
rust-proofing agent 防锈剂
rust-proofing primer 防锈底漆
rust proofing treatment 防锈处理
rust-proof life 防锈期
rust-proof material 防锈材料
rust-proof oil 防腐蚀油;防锈油
rust-proof paint 防锈漆
rust-proof paper 防锈纸
rust-proof pigment 防锈颜料
rust-proof solution 防锈溶液
rust-proof steel 不锈钢
rust protection 防锈
rust protection coating 防锈涂层;防锈漆
rust protection device 防锈装置
rust protection paint 防锈漆
rust protection treatment 防锈处理
rust quality 抗锈蚀性
rust red 锈红色;铁锈红色
rust removal 除锈
rust removal of metal 金属除锈
rust remover 去锈药水;除莠剂;除锈剂;除锈机;防锈器
rust-removing preparation 除锈制剂
rust-removing procedure 防锈程序
rust resistance 耐锈力;耐锈性;防锈性
rust-resisting 防锈的
rust-resisting material 防锈材料
rust-resisting oil 防腐蚀油
rust-resisting paint 防锈涂料;除锈漆
rust-resisting property 防锈性能;抗锈蚀性
rust resistive 防锈性能
Rust's azimuth diagram 拉斯特方位图
rust solvent 溶锈剂
rust spot 黄锈渍;锈斑;锈点
rust stain 锈渍;锈斑
rust staining 污点;锈斑
rust streak 锈纹
rust test for engine oil 内燃机润滑油防锈试验
rust through 锈蚀透;全部生锈
rustumite 鲁硅钙石
rusty 生锈的
rusty-brown 锈棕色的
rust yellow 浅锈黄;铁锈黄色
rusty gold 锈金;难泥汞的金
rusty-leaved 赭叶杜鹃花
rusty part 生锈部分
rusty pot 锈斑
rusty quality 锈蚀质量
rusty rail 生锈钢轨
rusty scale 锈层
rusty spot 锈斑
rusty stain 锈迹
rusty state 锈蚀状态
rut 车辙;轮距;波涛声;凹槽;岸边激浪声
rut depth 车辙深度
rut formation test 车辙成型试验
Rutger's process 木材压力防腐工艺;拉特格工艺
ruthenarsenite 砷钌矿

rutheniridosmine 钌铱锇矿
ruthenium 自然钌
ruthenium halide 卤化钌
ruthenium hydrochloride 氯钌酸
ruthenium ore 钌矿
ruthenium red 钌红
ruthenium tetroxide 四氧化钌
ruthenosmiridium 等轴钌锇铱矿
rutherford 卢瑟福(放射性单位)
rutherfordine 纤碳铀矿;菱铀矿
Rutherford scattering 卢瑟福散射
Rutherford's model 拉瑟福德模型
ruth excavator 多斗式挖掘机
Ruths accumulator 路斯型蓄气器
rutilated quartz 金红石英
rutile 金红石(矿)
rutile-calcium composite 钛钙白
rutile ceramics 金红石陶瓷
rutile content 金红石含量
rutile mixed phase pigment 金红石型混相颜料
rutile modification 金红石型
rutile ore 金红石矿石
rutile pigment 金红石颜料;金红石染料
rutile powder 金红石砂
rutile prism 金红石棱镜
rutile resonator 金红石谐振器
rutile structure 金红石型结构
rutile titanium dioxide 金红石型钛白;金红石型二氧化钛
rutile type electrode 钛型焊条
rutilization nuclei 金红石型晶种
rutilize 金红石化
rutted 形成车辙的;有车辙的
rutter 航线海图;刨
rutterite 微斜钠长石
ruttier 航线海图
rutting 路面形成车辙
rutting in the wheelpaths 轮迹车辙形成
rutting of subgrade 路基上形成车辙
rutty 多车辙的;有车辙的
ruturning materials to storeroom 退料入库
ruyi brand 如意牌
R-value 热阻值
R-wire 塞环引线
Rya 里亚毯
rybat(e) 门窗侧壁砖面;门窗侧壁石面;门窗侧壁木面;孔口边顺砌石块
rydberg 凯塞
Rydberg 里德伯(光谱学测量单位)
R-year flood R 年一遇洪水
R-year wave R 年一遇波浪
R-year wind R 年一遇大风
rye starch 黑麦淀粉
rye straw 黑麦秆
rye terms sound delivery 裸麦完好交货条款
Ryhage separator 赖哈格分离器
rymer 铰刀;铰床
rynersonite 钽钙矿
ryotron 薄膜感应超导电装置
Ryukoin Temple kyoto 大德寺龙光院(京都)
Ryukyu geosyncline 琉球地槽
Ryukyu trench 琉球海沟
Ryzner index 赖兹纳指数
Ryzer stability index 赖兹纳稳定指数;稳定指数
Ryzner index 稳定指数
Rzeppa constant velocity joint 分杆式等速万向联轴器;球笼万向节
Rz-powder 喷雾法铁粉

S

Saalian basin 萨勒盆地
Saalian glacial epoch 萨勒冰期
Saalian glacial stage 萨勒冰期
Saalian orogeny 萨勒运动
Saalic orogeny 萨勒造山运动
saba 沙巴织物;
sabach 钙积层
Sabathe cycle 萨巴蒂循环
sabatierite 硒铊铜矿
Sabattier effect 萨巴蒂尔效应
saber saw 军刀锯;电动手锯;战刀形电动手锯
sabicu 萨比丘木(古巴一种质地坚硬的贵重木材)
Sabina chinensis 圆柏
sabinaite 碳钛锆钠石
Sabin(e) 赛宾(吸声单位,相当于一平方英尺的全吸声面)(1 英尺=0.3048 米)
Sabine absorption 赛宾吸声量
Sabine absorption coefficient 赛宾吸声系数
Sabine coefficient 赛宾系数
Sabine formula 赛宾公式
Sabine reverberation formula 赛宾混响公式
Sabinite 萨拜因隔声灰泥
Sabin process 赛宾防锈法
Sabkha 萨布哈
Sabkha deposit 萨布哈沉积
Sabkha facies 萨布哈相
Sabkha metallogenic model 萨布哈成矿模式
Sabkha sulphate 萨布哈硫酸盐
sable 黑的;深黑色的
sable brush 黑貂毫笔
sable pencil 黑貂毛漆刷;黑貂毫画笔
sable's hair pencil 黑貂毛漆刷;黑貂毫笔
sable writer 毛笔;黑貂毫笔
sabot 桩靴;木鞋;锉刀垫板
sabotage 破坏行动;破坏活动;破坏的行为;怠工
Sabouraud's pastille 萨布罗纸碟法
Sabouraud's agar 萨布罗琼脂
sabre = saber
sabre butt 曲根材
sabstitute money 替代货币
sabtractive primary 相减合成基色
sabugalite 铝钙铀云母
sabulite 烈性炸药
sabulous 含砂的;砂质的;多砂的
sabulous clay 砂质黏土;亚黏土
sabulous loam 砂壤土
sabulous soil 砂质土(壤);砂土
sac 袋状凹湾
saccharide 糖类
saccharin 邻磺酰苯甲酰亚胺;糖精
saccharinity 含糖量
saccharite 砂糖石
saccharoidal 砂糖状
saccharoidal marble 砂糖状大理岩
saccharoidal texture 砂糖状结构
saccular process 成囊过程
sacculation 成囊
sacellum (露天的)小型罗马祭坛
Sach's method 萨赫(残余应力测定)法
sack 袋;粗布袋;布袋
sack and bale machine 打包机
sack application rate 插袋速度
sack barrow 两轮手推车;运袋车;手推车
sack borer 小型钻机
sack borer stem 小型钻机钻杆
sack cargo 袋装货物;袋物
sack-cloth 粗布袋;麻布;麻袋布
sack-cloud 囊状云
sack conveyer 袋装货运输带
sack counter 袋数计数器
sack dam 麻袋堰
sack deflector 水泥包转向器
sack-drying method 袋装干燥法
sack-duster 袋式除尘器
sacked cement 装袋水泥
sacked concrete 袋装混凝土
sacked concrete revetment 袋装混凝土护坡;袋装(水下)混凝土护岸

sacked concrete revetment work 袋装混凝土护岸工程
sacked material 袋装料
sacked riprap 袋装石块
sacker 装袋器
sacker auger 装袋螺旋推运器
sacker elevator 装袋升运器
sack erosion check dam 土包防冲谷坊
sacker weigher 装袋秤
sack filler 包装机
sack filling machine 装袋机
sack finish 袋擦饰面
sack gabion 袋形铅丝石笼
sack grip 麻袋夹持器
sack handling 麻袋装卸
sack hoist 麻袋起重机
sack holder 麻袋夹持器;夹袋器;袋夹
sacking 袋擦饰面;装袋;麻袋布;粗麻布
sacking cement 装袋水泥;水泥装袋
sacking filler 灌包机
sacking filling machine 灌包机
sacking operation 装袋作业;灌包作业
sacking plant 灌包机
sacking spout 装袋口
sacking-weighing machine 过秤灌包机
sacking-weighing plant 过秤灌包设备
sack joint 袋擦平缝
sack-like body 囊状矿体
sack-like residuum 囊状风化壳
sack-like structure 囊状构造
sack loader 装粮装机;袋包装船机
sack mill 十字型四辊万能轧机
sack of cement 袋装水泥;成袋水泥
sack of coals 囊状云
sack packer 装袋机
sack paper package 纸袋包装
sack piler 堆袋机
sack rub 袋擦(用麻布袋擦平混凝土表面);麻布抹面
sack rubbed finish 布纹分析
sack-rubbed surface 麻布(袋)拉平表面
sack scrubbing 布袋抹刷纹
sack sewing machine 缝袋机
sack-shaped cave 袋状溶洞
sack stacker 粮袋码垛机;堆袋机
Sack's theorem 萨克定理
sack transporter 运袋器
sack truck 运袋(卡)车
sack-type filter 袋式过滤器
Sacramento basin 萨克拉门托盆地
sacrarium 神龛;圣堂
sacred alter 神圣祭坛
sacred bamboo 南天竹
sacred building 宗教建筑物;奉献给上帝建筑物
sacred cows 圣牛角【给】
sacred grove of trees 神圣的小树林
sacred place 圣地
sacred precinct 神圣的地区;神圣的境界
Sacred rock 神圣的岩石
sacred street 神圣的街道
sacrifice 亏本出售
sacrifice of ship 牺牲船舶
sacrifice property 牺牲财产
sacrifice sale 亏本出售
sacrifice-ware colo(u)r 法华(指陶器)
sacrifice-ware decoration 法华(指陶器)
sacrifice-ware green 法绿(指陶器);法翠
sacrifice-ware tricolo(u)r 法华三彩
sacrifice-ware yellow 法黄
sacrificial altar 祭坛
sacrificial anode 牺牲阳极;电镀阳极;防蚀消耗阳极
sacrificial anode protection 牺牲阳极(保护)法
sacrificial corrosion 牺牲阳极腐蚀;牺牲腐蚀保护法;阴极保护腐蚀
sacrificial metal 牺牲金属;防蚀消耗金属;保护性金属
sacrificial metal coating 牺牲金属保护层
sacrificial pit 牺牲墓穴;献身墓穴
sacrificial protection 电镀保护法;牺牲涂层式防护

sacrificial system 防蚀消耗系统
sacrisy 教堂餐具室;圣器室
sacrofanite 水钾钙霞石
sadanaguite 砂川闪石
sad colo(u)r 深暗颜色
saddening 加深;暗淡处理;小变形锻造;轻轧
saddening-down 油漆面变暗
saddle 圆枕木锅炉座;座块;支管架;枕木垫;岭鞍;(装窑用)耐火鞍形棒;马鞍;滑动座架;滑鞍;起重小车;坯架;托座;三角鞍形棒;大拖板;床鞍;承架;鞍形物;版鞍;凹座;凹谷;鞍座;山鞍;鞍状构造;鞍形构造【地】;鞍形低压;鞍接;鞍部;陶瓷耐火土托座(窑具)
saddle arch 炉条碹;鞍形拱
saddle axis 背斜轴;鞍轴
saddle-back 鞍形山(脊);鞍形状脊柱;两坡压顶;屋脊状支架;鞍状物;鞍状无云区;鞍状峰;鞍形屋顶;哑口【道】;顶板圆弯;分煤鞍板;鞍背;鞍板(装煤舱导板)
saddle-back board 衬底板;门槛;鞍形板;门槛嵌板
saddle-back brick coping 鞍形砖盖顶
saddle-back capping 双坡盖顶
saddle-back cope 鞍形顶盖
saddle-back coping 双坡盖顶;鞍形压顶;鞍(背)形盖顶;鞍形砖盖顶;两坡压顶;鞍形顶盖
saddle-backed 鞍形的
saddle-backed coping 鞍形压顶
saddle-backed girder 鞍(背)形桁梁;鞍(背)形大梁
saddle-backed harrow 垄作;鞍形耙
saddle-backed horse 鞍状马
saddle-back gutter 鞍形天沟
saddle-back harrow 鞍形耙
saddle-backing 凸起
saddle-back joint 咬口接头;鞍形接头;鞍形接合
saddle-back nose 鞍鼻
saddle-back roof 双坡屋顶;双跨双坡屋面;鞍形屋面;鞍形屋顶
saddle-back shell 鞍形薄壳(结构)
saddle-back temperature curve 鞍状温度曲线
saddle bag 鞍囊
saddle-bag groin 鞍袋形石笼丁坝;鞍袋形石笼沉排丁坝
saddle-bag groyne 鞍袋形石笼沉排丁坝
saddle bar 撑条;窗格鞍形嵌条;撑棍;彩画玻璃嵌条;玻璃嵌条
saddle bead 马鞍形窗间条;玻璃压条
saddle bearing 鞍状轴承;鞍状支座;鞍形支承;鞍承
saddle bend 鞍形弯头;鞍顶
saddle blanket 鞍毡
saddle block (铲土机中的)吊杆旋转块
saddle board 屋脊板;鞍形板;斜屋顶屋脊板
saddle boiler 马鞍形锅炉;鞍形锅炉
saddle boom monotower crane 鞍形悬臂单塔式起重机
saddle-bottomed car 凹底车;鞍形底盘车
saddle-bottomed wagon 鞍形地自卸车;两边底开门自卸车;凹底车;鞍形底盘车
saddle-bow 鞍头
saddle brick 鞍形砖
saddle cam 溜板凸轮;滑板凸轮
saddle check chair 高背靠椅
saddle clamp 床鞍夹紧;鞍座夹
saddle clamping lever 鞍架夹杆
saddle clip 分道管;管卡;扒钉;撑棍;鞍形夹;鞍夹
saddle-cloth 鞍褥
saddle coil 卷边偏转线圈;产生均匀磁场的马鞍型线圈;鞍形(偏转)线圈
saddle coil magnet 鞍形线圈磁体
saddle coil storage 鞍形卷座
saddle control 刀架控制
saddle coping 鞍形压顶;鞍形顶盖
saddle cramp 鞍形夹(钳)
saddle cylinder base 鞍形汽缸座
saddle decollement 鞍状虚脱
saddle deflecting yoke 鞍形偏转线圈
saddle distribution 马鞍形分布
saddle duster 驮载式喷粉机
saddled with debt 负债

saddle embolus 马鞍栓塞;鞍骑性栓子
saddle feather 鞍羽
saddle feed screw 鞍架进给螺杆
saddle-field 鞍形场透镜
saddle fitting 鞍形配件;鞍形管件
saddle flange 鞍形凸缘;鞍形法兰;弧形法兰
saddle flashing 鞍形泛水;鞍形防雨水
saddle fold 鞍状褶皱;鞍形褶皱【地】
saddle forging 马鞍形扩孔锻造
saddle friction-plate 索鞍摩阻板
saddle function 鞍形函数;鞍式函数
saddle-fused joint 鞍形管件熔接接头
saddle gall 鞍伤
saddle gang stitcher 长臂订书机
saddle gear 鞍座机构
saddle glacier 鞍状冰川
saddle grafting 鞍接
saddle grate 鞍形炉栅
saddle guidance 刀架导座;鞍架导槽
saddle hatch(way) 鞍形舱口(装煤分舱导口)
saddle hill 马鞍峰
saddle horse 鞍马
saddle hub 凹形管座
saddle inclusion angle 鞍座包角
saddle jack 有座千斤顶;夹头起重器
saddle jib monotower crane 鞍形转臂单塔式起重机
saddle joint 阶形接榫;咬口接头;咬接头;鞍状关节;鞍形接头;鞍(形)接合
saddle key 鞍形键
saddle leather 鞍皮
saddle mark 鞍印
saddle nose 鞍(状)鼻
saddle of carriage 拖板鞍架
saddle orebody 鞍状矿体
saddle packing 马鞍形填料;鞍形填料
saddle pebble 马鞍石
saddle piece 台架;(屋脊相交处的)鞍形座板;鞍头块;斜沟小屋顶;鞍形垫块
saddle pillar 鞍座柱;鞍座支杆
saddle-pin 鞍栓
saddle pivot pin 鞍座支销;鞍座系紧螺栓
saddle planting 畦植
saddle plate 鞍形板;(装煤分舱导板的)鞍板
saddle point 马鞍点;鞍形区;鞍点
saddle-point azeotrope 马鞍点共沸体
saddle-point equation 鞍点方程
saddle-point game 有鞍点对策;鞍点对策;鞍点的矩形特性
saddle-point method 最速下降法;鞍点法
saddle-point problem 鞍点问题;鞍点定理
saddle-point theory 鞍点理论
saddler 鞍工
saddle radiator 鞍形散热器
saddle rail 鞍形轨
saddle reef 鞍状矿脉
saddle repair clamp 鞍形修理管夹
saddle ring 鞍环
saddle roof 双跨双坡屋面;鞍形屋面;鞍形屋顶
saddlery 马具店;鞍具
saddle scaffold 跨立式脚手架;鞍形脚手架;屋脊鞍形脚手架
saddle seat 鞍座(指自行车类);鞍形座(椅)
saddle-shaped 鞍形的
saddle-shaped arch 鞍形拱
saddle-shaped coping of a wall 鞍形墙盖
saddle-shaped curvature 鞍形曲度
saddle-shaped form 鞍形模子
saddle-shaped lathe 鞍形车床
saddle-shaped shell 鞍形壳体
saddle-shaped tower packing 马鞍形塔填充物
saddle-shaped washer 鞍形垫圈
saddle shell 马鞍(形)壳;鞍形壳
saddle siphon 鞍形虹吸管
saddle sore 鞍伤
saddle spillway 凹形(鞍形)溢洪道
saddle spring 鞍座弹簧
saddle stair(case) 鞍形楼梯
saddle stich 鞍形针迹
saddle stone 鞍形拱顶石;房脊石;屋脊石;山墙顶石;(山墙的)顶石;鞍形石
saddle strap 摇动吊簧;托架;托板
saddle stroke 大刀架行程;床鞍行程
saddle structure 鞍状构造【地】

saddle support 马架;鞍形支座;鞍式支座
saddle supports for horizontal vessels 卧式容器的鞍形支座
saddle surface 鞍形表面
saddle tank 鞍形油箱
saddle tank locomotive 鞍形水柜机车
saddle tank machine 驮箱式喷雾机
saddle tank sprayer 驮载液箱式喷雾机
saddle tee (固定管子的)鞍形柱;马鞍形三通
saddle tie 鞍形系扣;鞍形扣件
saddle tractor 载拖式牵引车
saddle traverse screw 鞍形管件进给螺杆
saddle tree 鞍架;(美洲)百合木
saddle-turret lathe 鞍座式转塔车床
saddle type 鞍形
saddle type cantilever beam 鞍形悬臂梁
saddle type conveyer 鞍式运输机
saddle type lantern-light 鞍形提灯
saddle type milling head 滑鞍型铣头
saddle type monitor 鞍式监测器
saddle type monitor roof 鞍形通风屋顶;鞍形光屋顶
saddle type pin boss 鞍形销座
saddle-type skylight 鞍形天窗
saddle-type turret lathe 鞍(座)式转塔车床
saddle valley 鞍形谷;鞍状谷
saddle valve 鞍形阀
saddle vein 鞍状矿脉
saddle wear of rail end 钢轨接头马鞍形磨损
saddle welding 鞍形焊(接)
saddle yoke 鞍座辊;鞍座叉杆
saddling 座板;鞍锻
Sadgrove maturity figure 塞特洛夫混凝土强度增量特征值
safe 稳妥的;安全的
safe access 船内安全通道
safe accommodation of pedestrian movements 行人活动的安全措施
safe against buckling 防止压曲;免受弯曲
safe against overturning 防止倾倒;免受倾倒
safe aground 安全搁浅
safe allowable load 许用荷载;安全允许荷载;安全允许应力;安全容许荷载;安全容许负载
safe allowable stress 安全允许应力;安全容许应力
safe anchorage 有掩护的锚(泊)地;有掩蔽的锚(泊)地;安全锚(泊)地
safe and economical bridge 安全和经济的桥梁
safe and punctual running of a train 安全正点
safe approach speed 安全趋近车速
safe area generator 画框发生器
safe arrival 安全到达
safe aspect 安全问题
safe atmosphere 安全空气
safe bearing capacity 容许承载力;安全承载力
safe bearing load 安全容许荷载;容许荷载;许用承载能力;轴承容许荷载;容许荷载;安全荷载;安全承载
safe bearing power 许用承载能力;轴承容许荷载
safe belt 安全带
safe berth 安全船席;安全泊位
safe bonus 安全奖金
safe box 保险柜
safe car 安全车
safe carrying capacity 许用承载能力;容许负载量;容许负荷量;安全载流量;安全过水能力;安全承载(能)力
safe carrying capacity power 许用承载能力
safe carrying capacity property 许用承载能力
safe channel capacity 河槽安全行洪能力;河槽安全泄(水)量;安全容量;安全输水能力
safe channel discharge 河槽安全泄(水)量
safe class 安全等级
safe clearance 安全高度;安全余裕;安全余量;安全余度;安全距离;安全净空
safe clearance zone 安全区
safe coefficient 安全系数
safe coefficient of oversaturation 过饱和安全系数
safe coefficient of supersaturation 过饱和安全系数
safe concentration 安全浓度
safe concentration of pollutant 污染物的安全浓度
safe condition 安全状态;安全条件;安全工况
safe conduct 安全通行证;(交通)通行许可;安全进行
safe container 安全集装箱

safe crop period 安全收割期;安全施药间隔期
safe curb 安全路缘(石)
safe current 容许电流
safe current carrying capacity 安全载流能力;安全载流量
safe current velocity 安全流速
safe curve speed (汽车的)弯道安全速度
safe custody box 寄存保险箱
safe custody charges 保管费
safe cut 稳定的路堑边坡;开挖的稳定边坡
safe deposit 保险仓库;信托仓库
safe deposit box 保险箱;保管箱;安全信托柜
safe deposit company 保管公司
safe deposit vault 保管库;保险库;安全储存室;安全库
safe device 安全设备
safe device against over-speed 超速保护装置
safe device against rope break 断绳保护装置【机】
safe device against slack rope 松绳保护装置【机】
safe discharge 安全下泄量
safe discharge capacity 安全下泄量
safe distance 安全距离
safe distance of approach 会遇安全距离
safe dose 安全剂量
safe dose of environment(al) carcinogen 环境致癌物的安全剂量
safe draft 安全吃水
safe drinking water 安全饮用水
Safe Drinking Water Act 安全饮用水法令(美国)
safe drinking water source 安全饮用水源
safe drinking water standard 安全饮用水标准
safe drip pan 集油盘;承油盘
safe driving 安全行车;(车的)安全驾驶
safe driving glasses 安全驾驶镜
safe edge (电梯门的)带形安全探测器
safe edge file 安全锉
safe edge heart 带护边心形平墁刀
safe exposure level 安全曝光水平
safe expression 安全表达式
safe fast road 安全快速道路
safe feeding 安全饲养
safe fire control line 安全防火线
safe firing current 准爆电流;安全放炮电流
safe following distance 安全行车间距
safe food 安全食物
safe format 安全格式化
safe formula 可靠公式;安全公式
safe fuse 安全引信
safe gap 安全空隙
safe gate 安全闸(门)
safe geometry 安全几何条件
safe gird 安全护栏
safe glass 安全玻璃
safe grounding 安全接地
safe ground wire 安全地线
safeguard 防护物;防护设备;保险器;保险板;保护措施;保障措施;安全装置;安全防护;安全保障;安全保护
safeguard analysis 安全保障分析
safeguard clause 保障条款;保留条款;保护条款
safeguard construction 安全结构
safeguard door chain 防盗门链
safeguard door lock 防盗门锁
safeguard glasses 防护眼镜
safeguard industries 保护工业
safeguarding 安全设备;安全防护;保护
safeguarding device 安全防护装置
safeguarding duty 保护性关税
safeguarding mechanism 安全机构
safeguarding structure 防护构筑物
safeguarding system 防护系统
safeguard material control system 安全监督材料控制系统
safeguard practice 保护措施
safeguard practices 保护性措施
safeguard preventions 安全保障措施
safeguard procedure 防护措施
safeguard provision 安全保障条款
safeguard requirement 安全条款
safeguards for the environment 保障环境措施
safeguard station 救生电台
safeguard structure 防护结构

safeguard valve 安全防爆阀
safe habitat 安全生境
safe handling 安全运转;安全运行;安全运输;安全操作
safe harbo(u)r rules 避风港规则
safe hasp 安全搭扣
safe hatch 安全舱口
safe haven 容许幅度标准
safe headway 安全车头时距
safe heat transfer 安全传热
safe indicator 安全指示器
safe inoperation 安全操作;安全运行
safe interclear spacing 安全净间距
safe intermodal transport 安全交替运输
safe international multimodal transport 安全国际多式运输
safe interval 安全间隔;安全车距
safe island 安全岛
safe joint 安全接头
safe keep 保管
safe-keeping fee 保管费
safe lamp 安全灯
safe landfill 安全土地填埋
safe landing 安全着陆
safe leg load (脚手架的)立柱安全荷载
safe length 安全长度
safe level 安全水位;安全辐射强度;安全程度
safe level earthquake 安全级地震
safe levitation 安全悬浮
safe light 安全光;安全灯
safe light filter 安全灯滤片
safe light screen 安全灯屏
safe limit 安全极限
safe load 容许荷载;容许负载;容许负荷;安全载重;安全(使用)荷载;安全(使用)负载;安全(使用)负荷
safe-load alarm 安全载重警报器;安全荷载警报器
safe load capacity 安全载重量
safe load carrying capacity 许用承载能力;安全负载;安全负荷;安全承载能力
safe load carrying power 安全承重能力
safe load coefficient 安全荷载系数
safe load factor 安全荷载系数
safe load indicator 安全负荷指示器
safe loading 安全载重;安全荷载
safe load table 安全荷载表
safely horizontal distance 安全水平距离
safe margin 安全极限
safe mark 安全标志
safe mass 安全质量
safe minimum standard 安全最低标准
safe net 安全网
safe offing 为安全而驶出海面
safe offside overtaking 安全外档超车
safe operating temperature 容许运行温度
safe operation 安全作业;安全运作;安全运行;安全工作
safe overhead clearance 安全净空高度
safe pan 涵洞承盘
safe parapet wall 矮围墙
safe passage of train 列车安全通过
safe passing distance 安全超车距离
safe passing sight distance 最小超车视距;安全超车视距
safe performance 安全运转
safe period 安全期
safe piloting (船、飞机等)安全驾驶
safe pipe 安全管
safe pollutant concentration 安全污染物浓度
safe port 安全港(口)
safe practice 安全技术;安全措施;安全操作方法
safe precaution 安全预防措施
safe pressure 容许压强;容许压力;安全压力
safe productivity 安全生产率;安全产水率
safe prosecution of voyage 安全续航
safe protector for puncher 冲床安全保护器
safe pull 安全拉火
safe pump output 安全抽水量
safe quantity of explosives 安全装药量
safe railway operation 铁路安全运营
safe range 安全范围
safe range of stress 应力安全范围;疲劳极限
safe rate 安全投资收益率
safe recovery area (失控车辆的)安全恢复区

safe region 安全区域;安全区
safe regulation 安全规章;安全规则;安全规程
safe riding surface 安全行车路面
Saferite 沙菲莱特玻璃;两面磨光嵌网玻璃
safe route 安全疏散路线
safe-running of vehicles 安全行车
safe sailing 安全航行
safe service life 安全使用期限;安全使用年限
safe service period 安全使用期限;安全使用年限
safe shutdown 安全停机
safe shutdown earthquake 安全停堆地震
safe sieve 安全筛
safe sight distance 可靠视距;安全视距
safe sign colo(u)r 安全标志色
safe slipping wire 安全溜索
safe speed 安全航速
safe speed of rotation 安全转速
safe spin 安全螺旋
safe starting position 安全起动位置
safe stopping distance 安全制动距离;安全停车距离
safe stopping sight distance 安全停车视距
safe storage 安全储藏
safe stress 许用应力;安全应力
safe system 安全系统
safe technical specification 安全技术规程
safe temperature 安全温度
safeticurb 安全路缘(石)
safe tolerance 安全允许公差
safe to operate 安全操作
safe tracer dosage 安全示踪剂量
safe traffic 安全行驶
safe stock 安全储备
safe tube 安全管
safe turnaround loop 安全回车道
safety 保险;安全性;安全
safety accessory 安全设备
safety accommodation 安全设备
safety action 屏蔽效应;保证动作;保险装置;保护作用;安全措施
safety administration regulation of hazardous chemicals 化学危险品安全管理条例
safety administration system of hazardous chemicals 化学危险品安全管理制度
safety afloat 安全漂
safety against buckling 抗弯安全;抗曲安全度
safety against bulging 防止膨胀安全措施
safety against cracking 防止开裂;免受开裂;抗裂安全;开裂安全度
safety against explosion accident 防爆安全性
safety against failure 抗破裂的安全措施
safety against foundation water pressure 抗基础水压力的安全措施
safety against overturning 抗倾覆安全度
safety against rupture 防止断裂;免受断裂;抗破坏安全
safety against shear failure 抗剪切破坏的安全措施
safety against sliding 防止活动的安全措施;抗滑动安全度;抗滑安全
safety against uplift 防止升起的安全措施;抗浮稳定
safety aground 安全搁浅
safety aids 安全设备
safety alarm 安全警报(器);安全报警器;安定警报
safety alarm device 安全警报装置;安全报警装置
safety allowance 安全宽限;安全补偿
safety analysis 安全(性)分析
safety analysis logic trees 安全分析逻辑树
safety and arming device 保险执行机构
safety and health 安全和健康
safety and health control system 安全卫生管理体系
safety and health regulation for construction 施工中安全和健康调节
safety and industrial gloves 工作手套;劳保手套;保护手套
safety and protection system 安全防护系统
safety and relief valves 安全泄压阀
safety and sanitation management system 安全卫生管理体系
safety angle 安全角
safety apparatus 救护器械;安全装置;安全设施;安全设备
safety appliance 保险设备;保安装置;安全用具;安全设备;安全器具

safety appliance standard 安全器具标准
safety approval plate (集装箱的)安全合格牌照
safety arch 辅助拱;分载拱;安全拱
safety area for radioactive goods accident point 放射性货物事故地点安全区
safety arrestment 安全制动
safety aspect 安全方面;安全问题
safety assessment 安全性评价;安全鉴定书
safety assessment for chemicals 化学品安全评价
safety assurance system 安全保证系统
safety at sea 海上安全
safety audit 安全审计
safety band 安全(保险)带;安全保护带
safety bar 保护杆;安全棒
safety barrel 定筒
safety barrier 停机拦截网;防护栅;安全栅栏;安全栅
safety basin 避风港池;安全停泊水域;安全泊地
safety bearing capacity 安全承载力
safety bearing load 安全承载力
safety bearing power 安全承载力
safety belt 保险带;保安带;安全带
safety belt for the seat of vehicle 车座安全带
safety berm 安全平台
safety berth 安全泊位
safety blasting 安全爆破
safety bleeder valve 安全泄放阀
safety block 保险块
safety board 安全台
safety bolt 保险螺栓;安全螺栓
safety bolt catch 保险销机
safety bonus 安全奖金
safety boot 安全长筒靴
safety boss 安全员;安全工程师
safety box 安全箱
safety brace 安全支撑;安全支柱
safety brake 安全制动器;保险闸;安全闸
safety builders fittings 安全施工配件
safety builders furniture 安全施工设备
safety bulkhead 安全隔墙;安全堵墙;安全舱壁
safety buoy 救生圈
safety buoyancy 预备浮力;储备浮力;保留浮力
safety button 安全纽扣
safety butts 安全铰链
safety by-pass 安全旁通管;安全旁路
safety cable 安全索;安全电缆
safety cage 安全车栅;安全网罩;安全室(油罐操作用);安全罐笼
safety cam 安全凸轮
safety can 安全密封罐;安全(灭火)罐
safety cap 安全罩;安全盖
safety cap lamp 安全帽灯
safety car 救生吊车;安全(矿)车
safety carrying capacity 安全承载量;安全装载量;安全载(重)量
safety cartridge 安全炸药包
safety castnet 安全撒网
safety catch 保险扣;安全制动装置;安全挡(块);安全掣子
safety catches cam 断绳防坠器抓爪
safety catches jackshaft 防坠器撑轴
safety certificate 安全证书
safety certification 安全认证
safety chain 防坠链;保险链;保安链;安全链
safety chain eye 安全链眼
safety chain hook 安全链钩
safety chain of operation 安全运行链【铁】
safety channel 安全航道
safety chassis 安全底盘
safety check 安全逆止阀;安全检测器;安全检查
safety chuck 安全钻夹
safety cinder notch stopper 安全渣口塞
safety circuit 保安电路;安全电路
safety clamp 安全管卡
safety clamp jaw 安全卡瓦
safety classes of structure 结构安全等级
safety clause 安全条款
safety clearance 许用间隙;允许间隙;容许间隙;安全界限;安全净空;安全界限;安全间隙
safety climbing device 安全爬升装置
safety clip 保险夹
safety clothing 安全服;防护衣(服)
safety clutch 保险制动器;安全离合器
safety cock 安全旋塞;安全栓

safety code 安全码;安全规范;安全法则;安全法典
safety coefficiency 安全效率
safety coefficient 储备系数;保险系数;安全系数;安全率
safety coefficient of steel wire 钢丝绳安全系数
safety collar 安全环
safety colo(u)r 安全标志色;安全色
safety communication 安全通信
safety communication equipment 安全通信设备
safety compensation 安全补偿
safety concentration 安全浓度
safety condenser 安全电容器
safety condition 安全状态;安全条件;安全工况
safety conscious 遵守安全规程
safety construction certificate 船舶安全结构证书;安全构造证书
safety contact (开闭器的)安全接点
safety control 事故防护;保安控制;保安措施;安全控制;安全监督;安全保障;防护装置
safety control device 安全控制装置
safety controller 安全控制器
safety control system 安全控制系统
safety copy 第二底图;二底图;备用原图
safety cork 安全管塞
safety cotter pin 保险开尾销
safety coupling 保险联结器;安全联轴器;安全联轴节;安全离合器;安全接头
safety coupling chain 安全联结链
safety cover 防护罩;保护罩;安全罩
safety crank 安全曲柄
safety criterion 安全准则
safety crutch 安全拐杖
safety curb 井栏;安全路边石;安全缘石;安全路缘(石);安全侧石
safety current 安全电流
safety curtain 防火幕(帘);防火帘;防火窗帘;安全屏障;安全幕;安全卷席
safety cushion 安全垫
safety cushion head 安全垫上盖
safety cut 安全切断;安全切除
safety cutoff 保安开关;安全开关;安全断路器;安全断流器;安全切断
safety cutoff device 安全切断装置
safety cutout 熔断开关;安全切断器;安全开关;安全断路器;安全断流器;熔丝断路器;熔断器;保安器;保安断路器
safety cylinder lock 圆筒形保险锁;圆筒形安全锁
safety deck 安全甲板
safety deforestation 安全伐林
safety-deposit box 保管箱
safety design 安全设计
safety despatching 安全调度
safety detaching hook 安全分离钩
safety detector 安全检测器
safety detent 保险卡销;安全卡爪
safety device 过载安全装置;紧急停车装置;保险装置;保护装置;安全装置;安全设施;安全设备;安全防护装置
safety device against wind 防风装置
safety dike 防洪大堤
safety disc 安全片
safety discharge 安全泄量
safety discharge capacity 安全过水能力
safety distance 安全间距;安全(行车)距离
safety distance of explosion 爆破安全距离
safety dog 保险挡;安全轧头;安全箍;安全挡块
safety domain 安全领域
safety dose 安全(剂)量
safety double glass 安全中空玻璃
safety draft 安全吃水
safety drilling method 安全钻孔法
safety drinking water act 安全饮水法案
safety driver 保险夹;安全轧头
safety driving (车辆的)安全驾驶
safety during construction 施工安全
safety during demolition 拆毁建筑物时的安全措施
safety earthing 安全接地
safety edge 安全边
safety education 安全教育
safety element 事故棒(原子堆);安全元件
safety enclosed switch 金属盒开关;密封式开关;密封式闸刀开关;盒保险开关
safety engineer 安全工程师
safety engineering 保安工程学;安全技术;安全工程

safety equipment 设备安全
safety equipment certificate 设备安全证书
safety equipment certificate of vessels 船舶设备安全证书
safety equipment requirement 设备安全要求
safety error 安全误差
safety evacuation instruction 安全疏散指令
safety evaluation 安全(性)评价
safety evaluation of chemicals 化学物安全(性)评价
safety evaluation of traffic 运输安全评估
safety exhaust 安全排气
safety exhaust valve 安全排气阀
safety exit 安全出口
safety explosion 安全爆破
safety explosive 安全炸药
safety extinguishing fire 安全扑火
safety facility 安全设施;安全设备
safety factor 可靠系数;储备系数;保险系数;安全因子;安全因素;安全系数;安全率;安全(度)因数;信贷安全系数
safety factor against shear failure 抗剪破坏安全系数;剪切破坏安全系数
safety factor against sliding 抗滑安全系数
safety factor for cracking load 开裂荷载安全系数
safety factor for ultimate load 极限荷载安全系数
safety factor of shear(ing)stress 剪应力安全系数
safety factor of sliding resistance 抗滑安全系数
safety factor of strength 强度安全系数
safety fairway 安全航道
safety feature 安全装置
safety fence 防护栅(栏);安全栅(栏)
safety film 不燃性胶片;保护膜;安全胶片
safety filter element 安全过滤元件
safety finger 叉头钉
safety firebreak 安全防火林
safety fire line 安全防火沟
safety first 安全第一
safety-first engineering 技术保安;安全技术
safety fitting 安全配件;安全装置
safety flange 安全轮缘;安全法兰
safety flap 安全调节板;安全瓣
safety flare 安全照明灯
safety flask 安全烧瓶
safety flow 安全流(程)
safety flow velocity 安全流速
safety forest administration 安全森林管理
safety fork 保险叉;安全夹
safety free-board 安全超高
safety friction clutch 保险摩擦离合器
safety fuel 防爆燃料;(闪点不高的)安全燃料
safety fund 安全基金
Safety Fund Act 安全基金法
safety fund system 安全基金制度
safety funnel 安全漏斗
safety funnel tube 安全漏斗管
safety furniture 安全设备
safety fuse 保险丝;安全引信;安全信号引线;安全熔断器;安全导火线
safety fusible plug 易熔安全塞
safety gap 安全(空)隙;安全间隙;安全放电器
safety gate 安全闸(门);安全门
safety ga(u)ge 安全计量器
safety gear 安全装置
safety glass 防护玻璃;不碎玻璃;保险玻璃;保护玻璃;安全玻璃
safety glass door 安全玻璃门
safety glasses 防护眼镜;护目镜;安全眼镜
safety goggles 护目镜;防护眼镜;安全护目镜
safety governor 限速器;危急安全阀;安全调压器;安全调速器;安全调节器
safety gripper 安全夹
safety gripping gear 安全夹
safety groove 保险槽;安全带
safety ground 安全接地
safety guard 护栏;保险板;安全罩;安全板
safety guarding 安全防护
safety guard rail 安全护栏
safety handle 保险手柄;安全手柄
safety hanger bracket 安全吊座
safety hardware 安全五金
safety harness 安全设备;安全吊带;安全背带
safety hasp 安全搭扣
safety hat 安全帽

safety hatch 安全舱口
safety hat covering 安全帽帽套
safety hat with broad edge 大边安全帽
safety haulm pulverizer 安全茎叶切碎机
safety hazard 危害安全
safety head 安全(头)盖
safety head lamp 安全头灯
safety height 安全高度
safety helmet 安全(头)盔;安全帽
safety hoist 安全起重机
safety hood 安全罩
safety hook 安全钩;安全吊货钩;安全吊钩;保险钩
safety hub 安全轮毂
safety illumination 安全照明
safety indication 安全指示
safety ink 防涂改油墨;安全(油)墨
safety in operation 运行可靠性;运行安全性;工作可靠性;操作安全
safety in production 安全生产
safety in service 安全服务
safety inspection 安全检查
safety inspection of traffic 运输安全检查
safety inspector 安全检查员
safety installation 保险装置;安全装置;安全设备
safety interlock 安全联锁装置
safety interlocking 安全联锁;安全闭锁
safety interlock(ing)device 安全联锁装置
safety inventories 安全存货
safety island 安全地带;安全岛;中心安全岛;交通(安全)岛
safety isle 中心安全岛;安全岛
safety jack 安全支柱
safety jaw 安全钳口;安全吊
safety joint 安全接头
safety joint with left-hand release 左扣可卸式安全接头
safety kerb 井栏;安全路缘石;安全路边石
safety kick-off 安全断路;安全止动
safety knob 安全按钮
safety ladder 安全爬梯;太平梯;疏散爬梯;安全梯
safety lamp 安全灯
safety lamp gasoline 安全灯用汽油
safety lane 安全航道
safety latch 安全栓;安全插销
safety latch(cargo)hook 自锁安全吊货钩
safety level 安全水平;安全电平;安全程度
safety level of electromagnetic radiation 电磁辐射安全标准
safety lever 安全操纵杆;安全杠杆;保险片;保险杆;安全阀杆
safety licence system of civil nuclear device 民用核设施安全许可制度
safety light 安全照明灯;安全信号灯;安全光;安全灯
safety lighting 安全照明
safety lighting fitting 安全照明装置
safety limit 安全限度;安全界(限);安全极限;安全范围
safety limit controller 安全极限控制器
safety limit switch 保险总开关
safety limit value 安全限度值
safety line 安全绳索
safety-line street marker (路面)安全线路标
safety lining 安全衬
safety link 安全吊环
safety lintel 辅助过梁;安全过梁【建】;眉梁
safety liquid relief 安全泄液
safety load-carrying capacity 安全承载能力
safety load factor 安全荷载因数;安全荷载系数
safety load hook 安全载重钩
safety load(ing) 安全荷载;安全负载;安全负荷;保险费安全系数附加费;容许荷载
safety lock 保险锁;保险机构;安全锁
safety lockage 应急闸门
safety lock gate 保险闸门;应急备用闸门
safety locking bolt 安全锁紧螺栓
safety locking pin 安全锁销
safety locking screw 安全止动螺钉
safety locking wire 安全锁线
safety lock magnet 安全锁闭电磁铁
safety lock plunger 保险活销
safety lockwire 保险锁线
safety log hoist 安全原木起运装置
safety log-wagon 安全运材车
safety loop 保险圈;安全圈

safety lug 保险凸耳
safety lumber milling machine 木工安全刨床
safety machinist's hammer 安全钳工锤
safety magnet valve 安全电磁阀
safety management 安全管理
safety management of traffic 运输安全管理
safety manual 安全手册
safety margin 安全裕度;安全余裕;安全余量;安全余度;安全限度;安全线;安全系数;安全界(限);安全范围;安全储备;安全边缘;安全边际量
safety margin width 安全富裕宽度
safety mark 安全标志
safety match 安全火柴
safety measures 防护措施;安全措施
safety measure of soil slope 土坡安全稳定措施
safety measures to fish catch 安全捕鱼措施
safety mechanism 安全机构
safety meeting 两船安全会遇
safety member (原子堆的)事故棒;安全棒
safety mesh 安全网
safety message 安全信文
safety method 安全技术;安全措施
safety monitoring 安全监测
safety needs 安全需要
safety net 保护网;安全网
safety netting 安全网
safety nosing 楼梯踏步上的防滑条;防滑条
safety number 安全数
safety nut 保险螺母;保险螺帽;安全螺母;安全螺帽;锁螺帽
safety of blasting 爆破安全
safety of ecologic(al) environment 生态环境安全
safety of effluent use 出水使用安全性
safety officer 安全员
safety of guaranteed main dike 确保大堤安全性
safety of life at sea 海上人命安全
safety of railway traffic 铁路运输安全
safety of traffic 行车安全;运输安全;交通安全
safety of use 应用安全性
safety-operated switch 保险闸;保险盒;安全开关
safety operation 安全作业;安全操作
safety operation area 安全工作区
safety operation regulation 安全操作规程
safety operation rule 安全操作规程
safety operation specification 安全操作规程
safety overall garment 安全潜水衣
safety pad 保险垫;安全垫
safety padlock 安全挂锁
safety paper 保险纸
safety pathway 安全通路;安全通道
safety patrol 安全巡查
safety pawl 安全掣子;安全制轮爪;安全凸爪
safety pedal 安全踏板
safety performance 安全性能
safety period 安全周期
safety piece 安全零件
safety pillar 安全支柱
safety pin 保险销;保险丝;安全销机;安全销(钉);安全别针
safety pin coupling 安全销联轴器
safety pinion 保险齿轴
safety pipe 安全管
safety pipette 安全吸管
safety plate glass 安全平板玻璃
safety platform 安全工作台
safety plug 熔线塞;熔丝塞(子);保险塞;安全塞;安全插头
safety plug socket 熔线塞座;安全塞座
safety pole 安全杆
safety post 路边护柱;安全柱;安全标
safety practice 安全操作方法
safety precaution 安全(保障)措施
safety pressure 安全压力
safety pressure margin 富裕压力
safety problem 安全问题
safety program(me) 安全计划;安全规划
safety prop 安全柱
safety protection 安全防护;安全措施
safety protection building 安全防护建设
safety protection installation 安全防护装置
safety protective coverall 安全保护服
safety protective system 安全防护系统
safety provision 安全设施;安全措施
safety quantity 安全存量

safety radio telegraphy certificate 无线电设备安全证书
safety rafting 安全放排
safety rail 护轨;安全栏杆
safety railing 安全栏杆
safety range 安全范围
safety rate 安全率
safety razor 保安刀架;安全剃刀
safety recess 避车洞
safety recording instrument 安全记录仪
safety records 安全记录
safety reflective glass 热发射安全玻璃
safety region 安全区域
safety regulation 安全规程;保安规章;保安规程;安全守则;安全法规
safety regulations for salvage operation 救捞安全操作规程
safety regulations in coal mine 煤矿安全规程
safety relay 安全继电器
safety release 脱开安全器
safety release spring 安全放松弹簧
safety relief and pressure relief valves 安全泄放和泄压阀
safety relief device 安全泄压装置
safety relief valve 安全泄压阀
safety remote control 安全遥控
safety remote control equipment 安全遥控设备
safety requirement 安全要求;安全规格;安全必需品
safety reset valve 安全回复重调阀
safety responsibility 安全责任条例
safety responsibility act 安全责任制
safety responsibility regulation 安全责任条例
safety return line 安全回流管路
safety reverse lever 安全回动杆
safety reverse lever spring 安全回动杆簧
safety ring 保险环;安全圈;安全环
safety rod 保险杆;安全杆;安全棒
safety roller 保险圆盘
safety rope 安全索;安全绳
safety rule 安全规章;安全规则;安全规程;安全条例
safety rules for lifting appliances 起重机械安全规程
safety rules in compressed air 压缩空气安全规则
safety running 行车安全;安全运转
safety scaffold(ing) 安全脚手架
safety screen 安全遮板;保险挡板;安全网;安全筛;安全屏(蔽);安全隔层;安全挡板
safety screw 保险螺栓;安全螺栓;安全螺旋
safety screw chuck 安全螺栓卡盘
safety seal 安全密封
safety service 安全业务
safety set collar 安全固定轴环
safety set screw 安全(止动)螺钉;安全螺栓
safety shaft 安全竖井
safety shear pin 保险切断销
safety shield 安全挡板
safety shim 保险垫片
safety shoe 安全靴;安全鞋;带形安全探测器
safety shut-down 安全制动;安全关闭
safety shutoff device 安全光气装置;安全关气装置
safety shutoff valve (煤气的)安全关闭阀;安全中止阀
safety siding 安全限度;安全线[铁];安全避车道
safety siding turnout 安全让车道[铁]
safety sieve 安全筛
safety sight glass with double layered glasses 双层玻璃安全视镜
safety sign 安全提示牌;安全标志
safety signal 安全信号
safety signal light 安全信号灯
safety sinking door 凿井吊盘安全门
safety siphon 安全虹吸管
safety slip clutch 安全滑动离合器
safety snap clutch 滑转式安全离合器
safety solution 安全解
safety spark gap 安全火花隙
safety specification 安全技术条件
safety spectacles 护目镜;防护眼镜
safety speed 安全速度
safety spring bushing 保险弹簧套
safety stake 安全标杆
safety standard 安全标准
safety standards for buildings 建筑安全标准
safety standard specification 安全标准规范

safety starting crank 安全起动曲柄
safety statistics 安全统计
safety status indication 安全显示
safety stay 安全牵条
safety stiich 安全缝
safety stirrup 安全卡盘
safety stitch 安全线迹
safety stock 保险库存量;安全库存(量);安全存量
safety stock concept 安全存量概念
safety stop 阻车器;紧急制动装置;保险限位装置;安全停止(器);安全断路器;安全断流器
safety strap 安全(吊)带
safety stress 安全应力;容许应力
safety strip (道路中的)安全分隔带;安全地带
safety super-critical reactor 安全超临界反应堆
safety supervision 安全监督;安全监察
Safety Supervision Administration Regulations of Civil Nuclear Device of the People's Republic of China 中华人民共和国民用核设施安全监督管理条例
safety supervision of traffic 运输安全监察
safety supervision regulation of pressure vessel 压力容器安全监察规程
safety supervisor 安全监督员;安全技术主管
safety supplement 安全补充
safety supplies 人身安全防护用品
safety support vessel 安全支持船
safety suspension device 保险悬挂装置
safety switch 安全开关;快速停堆按钮;紧急开关;保险开关;安全按钮
safety system 安全系统;安全设备
safety system engineering of traffic 运输安全系统工程
safety tackle (有增强措施的)安全绞辘
safety tank 安全水舱
safety tap 安全旋塞
safety target 安全指标
safety technical measure 安全技术措施
safety technics 安全技术;安全技术
safety technique 保险技术;保安技术;安全技术
safety techniques for construction 建筑安全技术
safety temperature 安全温度
safety tension 安全张力;安全拉力
safety test 安全试验;安全检测;安全测试
safety theory 安全理论
safety thermal relief valve 超温安全阀
safety thermostat 安全恒温器
safety timber-floating channel 安全流材道
safety tongs 安全钳
safety toolbox talk 工地安全座谈
safety track 尽头线[铁];安全航道
safety traffic 安全运行;安全行驶
safety tread 安全踏步;防滑踏步;安全梯级;踏板;安全轮胎
safety trip 安全脱断装置;安全释放机构;安全断路
safety trip beam 带安全器的犁辕
safety trip device 安全脱开装置
safety trip level 事故防护程度
safety trip lever 紧急触摆杆
safety trocar 安全套管针
safety tube 安全漏斗;安全管
safety tubing joint 油管安全接头
safety turnout 安全让车道[铁];安全避车道
safety-type switch 保险开关;安全开关
safety valve 卸压阀;保险阀;安全活门;安全阀
safety valve escape 安全阀泄漏
safety valve lifting lever 安全阀提升杆
safety valve of boiler 锅炉安全阀
safety valve operation test 安全阀动作试验
safety valve pressing sleeve 安全阀压套
safety valve seat 安全阀座;安全阀门座
safety valve spring 安全阀簧
safety valve turret 安全阀座
safety valve weight 安全阀锤重
safety velocity 容许流速;安全速度
safety vent 安全气孔;安全孔;安全放空
safety vest 安全背包
safety vestibule 安全通廊
safety voltage 安全电压
safety walkway 安全通路;安全通道
safety wall 保护墙;安全墙;安全堤
safety warning system 安全警报系统
safety washer 锁紧垫圈;安全垫圈
safety water 安全给水

safety weight 许可重量;安全重量;安全荷载;容许荷载
safety width 安全宽度
safety window glass 安全玻璃窗
safety wire 保险丝;保险缆
safety work 技术保安;安全作业;安全操作
safety working boots 安全工作靴
safety working depth 安全开采深度
safety working layer 安全开采层
safety working load 安全负载(能力);安全负荷
safety working power 安全使用(马力)
safety working pressure 安全操作压力
safety working shoes 安全工作鞋
safety yield 安全出水率;安全出水量;安全产水量
safety zone 安全区(域);安全地带
safe valve 安全阀
safe valve spring 安全阀弹簧
safe vault 保险库
safe velocity 容许流速;安全流速;许可流速;安全速度
safe velocity of canal 渠道安全流速
safe waste 无害废水
safe water 安全给水;安全水域
safe water level 安全水位
safe waters mark 安全水域标
safe water stage 安全水位
safe water supply 安全供水
safe weight 容许荷载
safe working 安全操作;安全开采;安全工作
safe working area 安全工作区
safe working capacity 安全工作荷载
safe working load(ing) 许用荷载;允许工作负荷;容许工作荷载;额定起重量;安全工作使用荷载;安全使用负荷;安全操作荷载;安全工作荷载;安全工作负载;安全工作负荷
safe working pressure 允许工作压力;容许工作压力;安全工作压力;安全操作压力
safe working strength 安全工作强度;安全操作强度
safe working stress 许用应力;安全使用应力;安全工作应力;容许应力
safe working temperature 安全工作温度
safe working voltage to ground 对地安全工作电压
safe yield 可靠产水量;保证水量;保证出产量;安全出水率;安全出水量;安全产水量;安全产量
safe zone 安全区域
safflorite 斜方砷钴矿
safflower oil 红花油
saffron 番红花;藏红花柱头;藏红色
saffron crocus 藏红花
saffron yellow 橘黄(色)
safing 栅栏装置;上保险;挡火用不燃材料
safranine 沙黄;番红;藏红;碱性藏红
safranine test 番红试验
safrene 黄樟烯
safrol 黄樟油精
safrole 黄樟脑素
sag 向下风面漂移(船);下陷;压向下风;塌芯;上箱下沉;垂(度);垂弛;弛度;凹形变坡点;鞍状山口;下垂;流挂;凹陷
sag adjustment 挠度调节;垂度调节
sag and swell 起伏
sagapenum 波斯阿魏
sagathy 萨加塞毛织物
sag bar 吊杆;防垂(吊)杆
sag bend 垂弯曲
sag bolt 顶板下沉测量锚杆;防垂螺栓
sag calculation 垂度计算
sag control agent 流挂控制剂;防流挂剂
sag correction (测量带尺的)中沉校准;垂直改正;垂度校正
sag crossing 跨越洼地
sag cross-section 凹形(纵)断面
sag curve 垂度曲线;挠度曲线;垂直曲线;弛垂弯曲;凹形曲线
sage green 灰绿色
sage green paint 灰绿色漆
sagenite 网金红石
sag-flow test 下垂流淌试验
saggar 匣钵;耐火匣体;耐火泥土;烧盆
saggar body 匣钵坯体
saggar cement 匣钵修补用耐火泥料
saggar clay 火泥;烧钵体;烧粘土;泥箱式;制钵用黏土;火泥箱土
saggar grog 匣钵渣;碎匣钵熟料

saggar house 烧釉窑
saggar placing 装匣钵
saggar press 匣钵压机
saggar setting 装匣钵
sag ga(u)ge 垂度计;弛度计
sagged 下垂的
sagged bevel 凹面倒边
sagged weld 下垂焊缝
sagger 匣钵;耐火匣体;烧盆
sagger clay 泥箱土;火泥箱土;烧钵土
sagging 自重烘弯;中段下垂;中垂;(喷射混凝土立面或顶面上的)挂留;结构下垂;漆面流坠;松垂;船舶中垂;槽沉;表面陷坑;下垂;流挂;凹陷
sagging axle 下挠轴
sagging beam method 下沉梁法
sagging bending moment 正弯矩
sagging cable 下垂索
sagging condition 中垂状态
sagging curve 下垂曲线;垂度曲线
sagging effect 垂度影响
sagging kiln 槽沉炉
sagging moment 下垂力矩;中垂力矩;正弯矩
sagging of arch 桥拱下沉
sagging of ceiling 天花下垂
sagging of paint 漆面流坠
sagging of support 支承点的下陷
sagging plate 中间管板
sagging point 下垂点;软化点
sagging soil 融陷土;融沉土
sagging stress 中垂应力
sagging tendency 下垂趋向
sag indicator 垂度计
sag in grade 坡度凹陷
sagitta 下垂的指针;挠曲的指针;石拱的锁石;拱锁石;弓形高;矢耳石;矢【数】
sagittal 前后向的
sagittal diameter 矢状径
sagittal focal line 径向焦线;弧矢焦线
sagittal focus 弧矢焦点;次级焦点
sagittal groove 矢状沟
sagittal image 矢形图像
sagittal axis circular current 竖轴环流
sagittal plane 径向平面;矢状平面
sagittal section 矢状平面
sagittal surface 矢状纵断面;弧矢面
sagittal suture 矢状缝
sagitta of arc 弧矢
sagittate leaf 箭头形叶片
sag length 下垂长度
sag-length ratio 垂长比
Sagnac's interferometer 萨尼亚克干涉仪
Sagnac's experiment 萨尼亚克实验
sag of bucket chain 斗链下锤度
sag of cable 悬索的下垂;缆索垂度
sag of line 管线垂度;电线垂度
sag of protecting coating 保护层脱落
sag of rope 绳索的下垂
sag of span 跨中(垂)度;跨度下垂度
sag of tape 卷尺垂曲
sag of the grade line 坡度凹变点
sag of the span 挡中弛度
sag of towing rope 拖缆垂度
sag of wire 电线垂度
sago palm 西谷椰子树属
sag pipe 倒虹吸管
sag point 槽沉温度
sag pond 断陷湖;断陷槽;沉陷洼地
sag prevention 防流挂法
sag profile 垂度纵剖面;凹形纵断面;凹形轮廓图
sag ratio 垂跨比;垂度率;垂度跨度比
sag ratio of bridge cable 桥索垂跨比
sag resistance 抗下垂(作用);抗流挂(作用);拖索中垂阻力
sag rod 系杆;吊杆;防垂(吊)杆
sags and crests 凹凸不平
sag subduction 沉陷俯冲作用
sag tension 垂度拉力;垂弛张力
sag test 下垂试验;垂度试验
sag tester 流挂试验机
sag tie 吊杆;防垂(吊)杆
sag to leeward 向下风漂流
sag vertical curve 悬垂直曲线;凹形竖曲线
sah 伊斯兰寺院天井
Sahalin meridinal tectonic belt 库页岛经向构造带

sahamalite 碳铈镁石
Sahara 撒哈拉沙漠
Saha's equation 萨哈方程
sahel 萨赫尔风
Sahel observatory 萨赫尔观象台
sahlite 次透辉石
said by shipper to contain 据发货人申报的内容
said to be 据报(提单批语)
said to weigh 据称重量
sail 开航;海上航行;帆
sailaba 泛滥平原;泛滥灌溉区
sail against current 逆流航行
sail angle 帆角
sail area 帆面积
sail arm 风车翼板;风车的翼板
sail-assisted ship 风帆助航船
sail-assisted unmanned robot ship 无人驾驶帆船
sail before the mast 普通海员
sail board 小型客车帆船;帆板
sail broad 顺风驶帆
sail canvas 船帆布
sailcloth 家用粗布;帆布
sail diving plane 指挥台水平升降舵
sail duck 船用帆布
sailed 已开航
sailed distance 航段距离
sail flying 滑翔飞行
sail halyard bend 扬帆结
sail hook 缝帆钩
sail in 入港
sailing 开船;航行计算方法;航行;航海
sailing area 航区
sailing barge 带帆驳船
sailing board 开航公告牌
sailing boat 帆船
sailing chart 航行图
sailing course 带形砌筑层;腰线;航线【航海】;航道;挑出砖层
sailing craft 小帆船
sailing date 开航日期;启航日;船期
sailing directions 航行指南;航路指南;航海通告
sailing dispatch 航行调度
sailing draft 出港吃水
sailing free 顺风驶帆
sailing ice 稀疏浮冰;漂凌
sailing in ballast 压载航行
sailing instruction 开航通知;航海指南;航海说明书
sailing intervals 起航密度
sailing large 顺风驶帆
sailing line 航行路线;航线;航迹【航海】
sailing list 开航日期表;船舶开航日期表
sailing master 航行舰长
sailing method 航行法
sailing notice 开航通知单
sailing on her ear 满帆倾斜行驶
sailing on her own bottom 私营商船
sailing order 开航通知;开航命令;起航命令
sailing path 航迹(带)【航海】
sailing permit 开航许可证;出航证
sailing plan 航行计划
sailing preparation 航行准备
sailing raft 帆筏
sailing schedule 预定航行计划;航行时间表;配船计划;船期表;船舶班期;发船计划
sailing ship 帆船
sailing ship fitted with auxiliary motor 机帆船
sailing speed 自由航速;航行速度【船】
sailing telegram 开航电报
sailing term 航海术语;驶帆术语
sailing though 航行通过
sailing time 开航时间
sailing track observation 航迹观测
sailing trim 适航纵倾
sailing velocity 航行速度【船】
sailing vessel 帆船
sailing vessel with auxiliary engine 机帆船
sailing wire 帆绳
sailing zenith telescope 导航天顶仪
sail into wind 逆风航行
sail loft 帆布品制造车间
sailmaker's eye splice 顺纹眼环插接
sailmaker's needle 缝帆针
sailmaker's palm 缝帆顶针
sailmaker's seaming 制帆缝法

sailmaker's splice 顺纹插接
sailmaker's tool 制帆工具;缝帆工具
sailmaker's whipping 帆工式扎绳头
sailmaking 制帆术
sail-making-knot 平结
sail needle 缝帆针
sail on a bowline 抢风航驶
sail on one's bottom 独立经营
sailor 海员;水手
sailor-chest 水手柜
sailor home 海员之家
sailorizing 船艺实习
sailorman 水手
sailor pocket knife 水手刀
sailor's knot 方结
sailor's shore pass 船员登陆证
sailor whipping 扎绳头
sailor-working suit 水手工作服
sail out 出航
sail-over 挑台;挑楼;挑出
sail plane 滑翔机;高性能滑翔机
sail reaper 摇臂收割机
sail room space 帆缆舱;帆库
sail-shaped roof 帆形屋顶
sail training vessel 驶帆训练船
sail trawler 拖网帆船
sail twine 帆线
sail varnish 帆布清漆
sail vault 帆船拱顶
sail with the tide 乘潮
sail yard 帆桁
sainfeldite 水砷钙石
Sainflou's formula 桑弗罗公式
Saint Andrew's cross bond 圣安德鲁交叉砌合(英国式交叉砌合)
Saint Elmo's fire 桅上电灯;圣爱尔摩火
Saint Lawrence river 圣劳轮斯河
saint's church 圣人礼拜堂
saint tomb 圣人墓
Saint Venant equation 圣维南方程
Saint Venant model 圣维南模型
Saint Venant principle 圣维南原理(即等力载原理)
Saishike basin 赛什克盆地
sajong soil 砂姜土
Sakata model 板田模型
Sakawa orogeny 佐川造山运动
Sakawa orogeny cycle 佐川造山旋回
sakhaite 碳硼钙镁石
Sakhalin 萨哈林岛
sakharovaite 脆硫铋铅矿
Sakishima Shoto uplift region 先岛群岛隆起地带
Sakmarian(stage) 萨克马尔阶【地】
sakt marsh plain 盐性草沼平原
sakuraiite 铟黄锡矿
sal 沙尔硬木(印度产);硅铝带;硅铝层
sala 大型客厅
salable coal sample 商品煤样品
salable expenditures of product 成品推销费
salable expenses by unit product 单位成品推销费
salableness 销路
salable profit 销售利润
salacetos 柠檬盐
salad 沙拉
Salado formation 沙拉多建造
Salair cycle 萨拉伊尔构造旋回【地】
Salair orogeny 萨拉伊尔构造作用幕【地】
Salamander 萨拉门特炉
salamander 高炉炉底铁块;炉瘤;炉底结块;烤火钢盆;耐火保险箱;筒形烘炉
salamander grill 烤盘;烤架
salamander stove 烘箱
salamander's wool 石棉
sal amarum 苦盐
sal ammoniac 氯化铵;铁屑灰泥;腻子;卤砂;油灰
sal-ammoniac cell 氯化铵电池
salaries, allowance and emoluments 薪金、津贴和酬金
salaries and allowances 薪金和津贴;酬劳薪金及津贴
salaries bat 薪金税
salaries expenses 薪金费用
salaries or wages on time basis 记时薪金或工资
salaries payable 应付薪金
salaries tax 薪金税;薪俸税

salary accounting 薪金计算
salary administration 薪金管理
salary classes 薪金等级
salary club 薪金俱乐部
salary distribution sheet 薪金分配表
salary increase 加薪
salary increment 薪金自动增长
salary insurance 薪资保险;薪金保险
salary plus bonus 薪金加奖金制
salary progression curve 薪金增长曲线图
salary rate 薪金率
salary roll 薪金清册
salary scales 薪金表
salary schedule 薪金表
salary standardization 薪金标准化
salary structure 薪金结构
salcing effect 放大效应
saleable height 商品高度
sale and leaseback 出售后返租;返租;销后租回
sale and lease back of real estate 不动产的卖租方式
sale-and-repurchase 销后买回
sale as seen 凭检视买卖
sale at an exchange 在交易所出售
sale at arm's length 彼此独立的买卖
sale at a valuation 评价买卖
sale at negotiated price 议销价格
sale by bulk 批发
sale by credit 信用买卖
sale by description 凭文字说明买卖;凭说明书销售
sale by grade 凭规格销售
sale by inspection 验货销售
sale by real cash 现销;现款销售
sale by sample 看样成交;凭样(品)销售;凭样(品)买卖
sale by specification 凭说明书买卖;凭规格售货
sale by standard 凭标准销售;凭标准买卖;按标准销售
sale by trade mark 凭商标售货
sale by type 凭型号售货
salecite 镁磷铀云母
sale commission 销售佣金
sale confirmation 售货确认书
sale contract 销售合同
saleeite 镁铀云母
sale finance company 销售金融公司
sale for cash 现金交易
sale in bond 保税货买卖
sale incomes of fixed assets 固定资产变现(清理)收入
sale-leaseback 租回;出售后返租;出售租回
sale-leaseback-buyback 出售—回租—回购
sale literature 推销刊物
sale lot 销售堆
Salem limestone 萨勒姆石灰岩
sale of business 出盘
sale of fixed assets 固定资产变卖
Sale of Goods Act 货物买卖法
sale of goods for charity 义卖
sale of personal residence by elderly 老人出售个人住宅(美国55岁以上的老人出售个人住宅可以免去部分所得税)
sale of property 财产的变卖
sale of securities 出售证券
sale on account 赊销
sale on description 凭说明销售
sale on good merchantable quality 凭适销品质买卖
sale or transfer clause 出卖或转让条款
sale price 廉价
sale resistance 销路呆滞
sale retail sales department 门市部
saleroom 销货处;售货场
sales 销售额
sales account 销货账
sales accounting 销售会计
sales agency 分销处
sales agent 销售代理商
sales agreement 销售合同
sales amount 销售额
sales analysis 列清单
sales and marketing motivation study 销售激励研究
sales and profit comparisons 销售额与利润的比例

sales area 售货区;销售区
sales area test 销售区试验
sales as per specification 凭规模买卖
sales-assessment ratio 估值与售价之比
sales associate 销售经纪人助理
sales at state fixed prices 平价销售
sales audit 销售核查
sales book 销货账簿
sales brochure 推广小册子
sales budget 销货预算
sales by description 货物分类出售
sales by public auction 公开拍卖
sales by tender 招标出售
sales catalogue 销售目录
sales chain 销售链系
sales commission 销售佣金;销售代理业务;回扣
sales company 销售公司
sales comparison approach 销售比较法
sales confirmation 销售确认书
sales contest 销售竞争;售货竞争
sales contract 销售合同;销货合同;买卖合同;售货合同
sales contract receivable 销售合同应收账款
sales control research 销售控制研究
sales convention 销货会谈
sales cost 总成本费用
sales counter 售货柜台
sales daybook 销货日记账
sales department 营业部;销售部门;门市部
sales depot 商店
sales depth test 销售深度检查
sales determined by products 以产定销
sales discount 销货折扣
sales dollar 销货现金
sales engineer 销售工程师;商业工程师;推销工程师
sales engineering 销售工程
sales escrow instructions 由第三者执行的让售委托书
sales estimate 销售估计
sales exhibition 展销
sales expenses 销售费用;销货费
sales finance company 销售贷款公司
sales force 推销力量
sales-force-composite method 销售人员综合市场预测法;推销人员意见综合法
sales forecast 销售额预测
sales in hard cash 现款销售
sales-inventory ratio 销售—库存比率
sales invoice 销售发票;销货发票
salesite 羟碘酮矿
sales journal 销货日记账;销货日记账
sales kit 样品箱
sales knowledge 销货知识
sales ledger 销售分类账
sales made in advance of delivery 超过供应量的销售
salesman 推销员
sales management 销售管理
sales manager 销售经理;营业主任;营业经理
salesman room 售货员室
salesmanship 推销术
sales margin 销售纯利率
sales maximization hypothesis 销售额最大化假设
sales-maximization principle 销售极大化原理
salesmen's traveling and entertainment expenses 推销员差旅及交际费
sales mix 销售组合;销售构成
sales network 销售网
sales note 售货单
sales objective matrix 销售目标矩阵表
sales of commission 委托销售
sales office 商行;销售处;营业部
sales of investment securities 出售投资证券
sales of merchandise from stock 库存商品销售
sales of property 财产出售
sales on approval 试销
sales on draft terms 凭汇票付款条件
sales opportunities 销售机会
sales order 销售订单;销货订单
sales order processing 销售订单处理
sales orientation 销售导向
sales-oriented 与销路有关的
sales package 销售包装
salesperson 推销员

sales plan 销售计划
sales policy 销售方针
sales potential 销售力量;服务潜力
sales price 售价
sales profit 销售利润
sales profit and loss 销货损益
sales promotion 销售促进;扩大销售;促销
sales promotion cost 推销成本
sales promotion expenses 推销费用
sales promotion network 推销网
sales promotion planning 推销计划
sales quota 销售份额;销售定额;销售限额
sales rebate 销货回扣
sales rebate and allowance 销售回扣及折让;销货回扣及折让;售货回扣
sales record 销售记录
sales region contribution statement 分区销售贡献表;分区销售分配表
sales register 销货登记簿
sales representative 销售代表
sales resistance 拒绝购买
sales return and allowance journal 退货及销货折让日记账
sales return ratio 退货比率
sales returns 销货退回;退货
sales returns book 退货记录簿
salesroom 门市部;拍卖场;货售处;售货场;展销室
sales salaries 销售人员薪金
sales service organization 销售服务组织
sales status 销售情况
sales stimulation factor 销售刺激因素
sales strategy 推销战略
sales structure 销售结构
sales system 销货法
sales territory 销售区域
sales ticket 销货发票
sales value 销售额
sales value method 销售价值法
sales volume 销售量;销售规模
sales volume quota 销售量责任额
sales voucher 销货凭证
sales warrant 销售保证
sale tar 页岩焦油
sale tax 销售税
sale technique 销售技术
salethyl 水杨酸乙酯
sale value 销售价
sale volume 销售额
sale wedging 盐劈作用
salfemic 硅铝铁镁质
Sal-ferricite 沙费里塞(一种防水材料)
salford magnetic sorting bridge 电磁鉴别仪
salgso 秘鲁赤潮
salic 积盐层
salicyl alcohol 水杨醇
salicyl aldehyde 水杨醛
salicyl aldehyde production wastewater 水杨醛生产废水
salicyl aldehyde spectrophotometry 水杨醛分光光度法
salicylamide 水杨酰胺
salicylanilide 水杨酰替苯胺
salicylate 水杨酸盐
salicylate amide 水杨酸胺
salicylic acid 柳酸;邻羟基苯甲酸;水杨酸
salicylic acid phenyl ester 水杨酸苯基酯
salicylic amide 水杨酰胺
salicylic anhydride 水杨酐
salicylide 水杨酸内酯
salience 显极性;凸出(部)
saliency = salience
salien-pole generator 凸极(式)发电机;凸板发电机
salient 突起地;突出部;凸起(的);凸出的
salient angle 凸墙外角;凸角;房屋的外角
salient corner 凸(出)角
salient date 主要数据
salient feature 特征
salient instrument 凸装型仪表
salient junction 外角;凸出连接;凸出的角
salient point 折点;凸点
salient pole 显极;凸极
salient pole alternator 凸极交流发电机;凸板交流发电机
salient pole armature 凸极电枢

salient pole machine 显极电机;凸极式发电机;凸极电机
salient pole synchronous generator 凸极式同步发电机
salient pole synchronous-induction motor 凸极式同步感应电动机
salient pole type 凸极式
salient pole type rotor 显极式转子;凸极式转子
salierous 产盐的
saliferous thermal water 盐化热水
salification 盐化(作用);积盐(作用);成盐作用
salify 成盐
saligenol 水杨醇
salimeter 盐液比重计;盐度计;盐度表(指仪表)
salimetry 盐分析法
salina 盐沼地;盐田;盐滩;盐碱滩;盐碱地;盐场;高盐湖;海水蒸发槽
Salina formation 全盐组【地】
salinate field 盐渍地;盐碱地
salination 用盐水处理
saline 盐沼地;盐碱滩;含盐的
saline-alkali condition 盐碱情况;盐碱条件
saline-alkaline damage 盐碱害
saline-alkaline plant 耐盐碱植物
saline-alkaline soil 盐碱土
saline-alkaline water 盐碱水
saline-alkali soil 盐碱土
saline-alkali soil forestation 盐碱地造林
saline and brine water overflow zone 咸卤水溢出带
saline area 盐清地区
saline bog 咸水沼泽;盐沼(泽)
saline bottle 盐水瓶
saline concentration 盐浓度;盐分;含盐(浓)度;含盐量
saline contamination 盐(渍)污染;盐沾污;盐水污染
saline current 咸水流;盐水流
saline density current 盐水异重流
saline deposit 盐类矿床;盐(类)沉积;盐分沉淀物
saline dome outer edge bar 岩穹穿丘外缘坝
saline dropper 盐水滴管
saline dye wastewater 含盐染料废水
saline encroachment 咸水入侵;盐水侵蚀;海水入侵
saline facies 盐相
saline flo(a)tation method 盐水漂浮法
saline form 盐形式
saline formation 盐生植物群系;含盐层
saline groundwater 地下咸水
saline harbo(u)r 咸水港口
saline infusion needle 盐水针
saline irrigation 含盐灌溉(水)
saline lagoon facies 咸化泻湖相
saline lake 咸水湖;盐湖
saline lake and interior sea water 咸湖内海水
saline land 盐土;盐碱地;盐地
saline lick 盐类沉积
salinelle 盐泥火山
saline marsh 盐沼(泽);盐性草沼
saline matter 盐分
saline meadow 盐生草甸
saline method 盐稀释法
saline mineral 盐类矿物
saline mud-flat 盐泥滩
salineness 含盐度
saline ores 盐矿
saline pollution 盐(水)污染
saline port 咸水港口
saline rock 盐渍化岩石;盐岩
saline sea 咸海
saline sediment 盐沉积
saline sediments 盐类沉积物
saline sewage 含盐污水
saline-sodic soil 盐渍苏打土;盐渍钠质土;盐碱土
saline-sodic soil amelioration 盐碱土改良
saline soil 盐渍土(壤);盐土;盐磺土;盐碱土;含盐土(壤)
saline soil of grassy marshland 草甸盐土
saline solution 盐(水)溶液
saline spring 咸泉;盐水泉;盐泉
saline stratum 含盐地层;含盐层
saline taste 咸味;盐味
saline thermal water 含盐热水
saline wastewater 含盐污水;含盐废水

saline water 咸水;盐水;含盐水;海水
saline water conversion 咸水淡化;盐水转化
saline water conversion plant 盐水转化厂
saline water demineralization 海水淡化
saline water descending 咸水下移
saline water ecosystem 咸水生态系统
saline water intrusion 盐水入侵
saline water reclamation 咸水淡化;盐水淡化
saline weathering crust 含盐风化壳
saline wedge 咸水楔;盐水异重流;盐水楔
Salinian formation 全盐组【地】
Salinia terrane 沙林尼亚地体
saliniferous 含盐的
salinification 盐渍化;盐碱化;盐化(作用)
salinimeter 盐液比重计;盐量计
salinity 咸度;盐浓度;盐度;含盐度
salinity after intrusion 入侵后的矿化度
salinity and alkalinity of groundwater 地下水盐碱度
salinity anomaly 盐度异常;盐度反常
salinity before intrusion 入侵前的矿化度
salinity bridge 电桥式盐变仪;测盐计
salinity content 盐分;含盐量
salinity control 盐度控制;防盐水入侵
salinity convection 盐度混合作用;盐度对流
salinity current 因盐度差异引起的海流;盐度流
salinity front 盐分前锋
salinity gradient 盐分梯度;盐度梯度;含盐量梯度
salinity gradient energy 浓度差能
salinity identification by the naked eye 肉眼鉴定盐渍度
salinity indicator 盐度指示器;盐度计;测盐计
salinity indicator system 盐度指示器系统
salinity intrusion 咸水入侵;盐水入侵;盐分入侵
salinity logging 含盐测量记录井
salinity measurement 盐度测定
salinity meter 盐度计;含盐量测定计
salinity of irrigation water 灌溉水含盐度;灌溉水的盐度
salinity of lagoon 泻湖水盐度
salinity of sea-ice 海冰盐度
salinity of seawater 海水盐度
salinity of water 水盐度
salinity profile 含盐示意图
salinity range 盐度范围
salinity stratification 盐度分层
salinity survey 盐渍化调查
salinity table 盐度表
salinity-temperature depth monitor system 温盐深度测量装置
salinity temperature depth recorder 海水盐度温度深度记录器;温盐深度自记仪
salinity temperature depth recording system 温盐深度测量装置
salinity temperature depth system 温盐深度系统
salinity test 盐分测定法;盐度测定
salinity tolerance 耐盐性
salinity variation 盐度变化
salinization 盐渍化(作用);盐(碱)化(作用)
salinization of soil 土壤盐碱化
salinization range along ditch 渠旁盐渍区
salinization range along reservoir 库岸盐渍区
salinized alkaline soil 盐化碱土
salinized chestnut soil 盐化栗土
salinized soil 盐渍土(壤);盐渍土
salinized zone in desert and desert steppe 荒漠及荒漠草原盐渍
salinized zone in grassland 草原盐渍区
salinized zone in overflowed plain 泛滥平原盐渍区
salinized zone in seashore 海滨盐渍区
salinocline 盐度跃(变)层
salinograph 盐量图
salinometer 盐液密度计;盐液比重计;盐量计;盐度计;盐度表(指仪表);含盐量测定计;测盐计
salinometry 盐量测定(法)
salinoness 含盐度
salinoscope 盐量仪
salinows 盐的
salite 次透辉石
salithion 水杨硫磷
salitrite-annite jacupirangite 铁云霞霓辉岩
Salix babylonica 垂柳
Salix cathayana 中华柳
Salix cavaleriei 云南柳

Salix gracilistyla 细柱柳
Salix purpurea 红柳木
Salix variegata 秋华柳
Salix viminalis 青刚柳
Salix xerophila 崖柳
salle (法庭的)大厅
sal limonis 柠檬盐
Sal log 流压测程计
sallow 柳条;山毛榉;阔叶柳;灰黄色的
sally 钝角;橡头;凸出部
sallying the ship 摇摆船身脱浅
sally port 堡垒地道;冲出口;突破口;太平门;暗门
salmiac 卤砂
Salmian(stage) 萨尔姆阶【地】
sal mirabile 芒硝
salmon 赭色;橙红色的
salmon brick 欠火(低质)砖;未烧透砖;低质红砖
salmon colo(u)r 橙红色
salmon factory ship 捕鲑船队
Salmon gum 萨蒙胶
salmon ladder 过鱼道;鲑鱼梯;鱼梯
salmon migration 鲑洄游
salmon orientation 鲑鱼的定向
salmon pink 鲑肉红色;橙红色
salmonsite 黄磷锰铁矿
Salmon wood 萨蒙红褐色硬木(印度产)
salol 萨罗【化】
salometer 盐液比重计
salometry 盐浓度测定法;盐量测定(法)
salomonica 螺旋形柱;麻花形柱;绞绳形柱(西班牙巴罗建筑特征)
salo(o)n 沙龙;客厅
salo(o)n bar 酒吧(单)间
salo(o)n cabin 头等舱
salo(o)n car 轿车;头等列车;头等客车
salo(o)n dancing 跳舞厅
salo(o)n deck 客舱甲板;一等客舱甲板;头舱甲板;餐厅甲板(船)
salo(o)n passenger 一等舱旅客
salo(o)n stores 船上小卖部
Salopian(series) 萨洛普统【地】
salpausselka 后退冰碛
Salpeter process 沙尔皮特过程
salse 泥火山
sal sedatirum 硼砂
sal sedative 硼酸
salsima 硅铝硅镁层
sal soda 洗涤碱;天然苏打
salt accumulation 盐分蓄积;盐分积累;盐分积聚
salt accumulation zone 盐分堆积带
salt action 盐侵蚀
salt air 有盐分的空气;充盐的空气
salt-air corrosion 含盐空气腐蚀
salt-air heat exchanger 盐空气热交换器
salt and pepper sand 黑白混合砂;杂灰色沙
salt and sand spreader 岩砂撒布器
saltant 突变型
salt anticlinal trap 盐背斜圈闭
salt anticline 盐背斜
salt application 撒盐(路面)
sal tartari 碳酸钾
saltating grain 跃移颗粒
saltating particle 跃移颗粒
saltating soil particle 跃移土粒
saltating soil velocity (泥沙的)跃移速度
saltating transport 跃移质输移;跃移质输送
saltation 河底滚沙
saltation height of sediment 泥沙颗粒的跃移高度
saltation layer 河底滚沙层;(水中的)突变层
saltation load (泥沙等)跃移质;跃移泥沙;河底滚沙负荷;跳动负荷;调冀负载
saltation load discharge 跃移质输移量
saltation point 沉积点
saltation step of sediment 泥沙颗粒的跃移长度
saltation transport 跃移质输移;跃移质输送;跳跃搬运
saltation velocity (泥沙的)跃移速度
saltation zone 突变区
salt atmosphere 充盐大气;盐雾
saltatory 跳跃演化
saltatory conduction 跳跃传导
salt-avoiding species 耐盐(物)种
salt balance 盐量平衡;盐分平衡
salt balance of groundwater 地下水盐均衡

salt barrier pillar 阻盐岩石(柱)
salt base 盐基
salt basin 含盐盆地
salt bath 盐浴池;盐浴(槽)
salt-bath brazing 盐浴钎焊;盐槽焊接
salt-bath chromizing 盐浴渗铬
salt-bath descaling 盐浴除鳞
salt-bath electric(al)furnace 盐浴电炉
salt-bath furnace 盐浴炉
salt-bath hardening furnace 盐浴淬火炉
salt-bath heat treatment 盐浴热处理
salt-bath patenting 盐浴索氏体化处理
salt-bath quenching 盐浴淬火
salt-bath sintering 盐浴烧结
salt-bath tempering furnace 盐浴回火炉
salt-bearing 含盐的
salt-bearing formation 含盐建造
salt-bearing rock 含盐类岩石
salt bed 盐床;盐层
salt bed storage 盐层库
salt block 盐砖;盐块
salt block barrier 盐体遮挡
salt bog 盐性酸沼
salt bond 盐键
saltbox 斜盖盐箱形房屋;盐盒式建筑
salt box type 两坡不对称的硬山顶住宅(美国)
salt breeze 咸风
salt brick 盐砖
salt bridge 盐桥
salt brine 盐汁;盐水
salt brine wedge 咸水楔
salt bubble 盐泡
salt burden 盐分负荷
saltbush 含盐灌木
salt cake 芒硝;盐饼
salt cake glass 硫酸盐玻璃
salt cake pan 芒硝锅
salt cake roaster 芒硝炉
salt cake rotary screen 芒硝回转筛
salt carrier 运盐船
salt cast 盐斑
salt ceders 圣柳丛
salt cementation 盐类胶结
salt chuck 盐水原木池
salt clay 盐土
salt concentration 盐(分)浓度;含盐(浓)度;含盐量
salt-containing 含盐的
salt-containing organic wastewater 含盐有机废水
salt-containing wastewater 含盐废水
salt contamination 盐污染
salt content 盐含量;盐分;含盐量;含盐度
salt content hydrology 盐浓度
salt content of crude oil 原油含盐量
salt correction 盐效校正
salt corrosion 盐腐蚀
salt crust 盐壳;盐结皮;盐结度
salt-crusted soil 盐壳土
salt crystal 盐晶体
salt crystal cement 盐晶胶结
salt crystallization 起霜
salt cutting 盐侵
salt damage 盐害
salt damaged 受盐损害的
salt dehydration 岩盐脱水法
salt deposit 盐矿床;盐层储[贮]存;盐层
salt desert 盐漠
salt diapir 盐丘;盐刺穿;生底脐;刺点盐丘
salt diapirism 盐底脐(作用)
salt-dilution method 盐液淡化法;盐水稀释法;盐液稀释法;盐稀释法
salt dip 砖的薄涂层
salt dissolving tank 化盐桶
salt distributor 撒盐机;盐粒撒布机
salt dome 盐穹;盐丘
salt-dome cap rock lens trap 盐帽透镜体圈闭
salt-dome drilling 盐丘钻进
salt-dome oil-gas field 盐丘油气田
salt-dome oil pool 盐丘油藏
salt door 泥渣孔
salt dune 盐丘
salt dust 盐尘
salted 用盐处理的
salted and dried plum 白梅
salted condition 含盐情况

salted soil 盐渍土(壤)
salted weapon 加料武器
salt effect 盐效应
salt efflorescence 霜华;盐华;盐粉化;盐霜
salt elimination 除盐
salt endurance 耐盐性;耐盐度
salter 加盐机
saltern 砖的上釉;盐田;碱土;盐场
saltern brick 瓷砖
saltern ware 陶瓷制品
salt error 盐误差
saltery 盐场
salt exclasion 除盐
salt exclusion of soil 土壤脱盐作用
salt expansion 可溶性盐吸湿膨胀
salt export 食盐出口
salt extract 盐渍提液
salt-extracted 盐萃取的
saltfield 盐田
salt filled exhaust valve 盐冷排气阀
salt filter 滤盐器
salt finger structure 盐指构造
salt flat 盐碱滩;盐田
salt flat deposit 盐坪沉积
salt fog 盐雾
salt fog corrosion rate 盐雾腐蚀速率
salt fog resistance 耐盐雾性
salt fog spraying cabinet 盐雾箱
salt fog test 盐雾试验;盐水喷雾试验
salt fossil 岩盐
salt-free 无盐的
salt-free process 无盐过程
salt furnace 煮盐锅
salt garden 盐池;盐场
salt ga(u)ge 盐水比重计;盐浮计
salt geology 石盐地质学
salt glacier 盐川
salt glaze 盐釉
salt-glazed 盐釉的
salt-glazed brick 盐釉(面)砖;盐釉瓷砖
salt-glazed earthenware pipe 盐釉陶瓷管
salt-glazed finish 盐釉涂面
salt-glazed stoneware 盐釉炻器
salt-glazed structural facing unit 盐釉建筑贴面块
salt-glazed tile 盐釉面砖;盐釉瓦
salt-glazed ware 陶土制品
salt glazing 盐釉
salt graben 碱土
salt grainer 成盐蒸发器
salt ground 撒盐地
salt ground water 含盐地下水
salt growth 熔盐急冷
salt gypsum 海盐石膏
salt hardening 盐浴淬火
salt haze 盐霾;含盐的薄雾
salt hopper 盐漏斗;盐斗
salt hydrate 水合盐
salt ice 盐冰
saltier cross bars 十字拉杆;交叉拉杆
saltierra 盐土
salt incrustation 盐壳
salt index 盐分指标;含盐指数
saltiness 含盐性
salting 潮水淹没的土地;加盐;撒盐
salting in 盐溶;盐助溶
salting-in effect 盐溶效应
salting liquid 盐溶液
salting of soil 土壤盐碱化
salting-out 加盐分离;盐析
salting-out agent 盐析剂
salting-out chromatography 盐析色谱法
salting-out distillation 盐析蒸馏
salting-out effect 盐析效应
salting-out electrolyte 盐析电解质
salting-out elution chromatography 盐析洗脱色谱
salting-out evapo(u)rator 析盐蒸发器
salting-out extraction 盐析萃取法
salting-out method 盐析法
salting-out of solvent 溶剂的盐析
salting-out paper chromatography 盐析纸色谱
salting-out point 盐析浓度
salting-out strength 盐析强度
salting-out tank 脱盐槽
saltings 盐滩岸

salting stain 盐斑
salt injury 盐害
Saltire cross 圣安德鲁十字架
saltish ground water 微咸地下水
salt karst 岩盐盐溶
salt laden air 充满盐的空气
salt laden atmosphere 含盐大气
salt lake 咸水湖；盐湖；盐池；含盐湖
salt lake basin 盐湖区；盐湖地域
salt lake facies 盐湖相
salt lake karst 盐湖岩溶
salt-lake-type boron deposit 盐湖型硼矿床
salt-lake-type lithium deposit 盐湖型锂矿床
salt leaching zone 盐分溶滤带
saltless process 无盐过程
salt lick 盐渍地；盐碱地
salt lime 硫酸钙；石膏
salt lime calcium sulphate 石膏
salt liquor 盐水
salt load 含盐量；盐负；盐分负荷
salt loving bacteria 嗜盐细菌
salt manufacture 制盐
salt marsh 盐碱地；盐沼(泽)；盐性草沼；盐土沼泽；盐水沼泽；盐碱滩
salt marsh channel 盐性草沼水道
salt marsh hay 盐土沼泽地干草
salt marsh plain 盐沼平原
salt meadow 盐渍草地；盐土草原；盐生草甸
salt metabolism 盐代谢
salt meter 盐液密度计；盐浮计
salt method 盐液测流法
salt mine 盐矿
salt mine disposal 盐矿处置
salt mining 采盐矿
salt mist 盐雾
salt mix in mud 泥浆盐浸
salt mud survey 含盐泥浆测井
saltness 盐浓度
salt nuclei 盐核
salt of lead 铅盐
salt of lemery 硫酸钾
salt of organic acid 有机酸盐
salt of oxyacid 含氧酸盐
salt of phosphorus 磷酸氢铵钠
salt of tartar 碳酸钾
salt oven 煮盐锅
salt pan 盐田；盐锅
saltpeter 硝石；硝酸钾；钾硝
saltpeter cave 硝石洞
saltpeter earth 硝石土
saltpetering 起碱；泛碱(砖或混凝土表面上泛起的白色盐基类粉末)；凝酸
saitpeter rot 湿墙上的白斑
saltpeter salt 硝石盐
saltpetre = saltpeter
salt pillar 盐柱；岩盐柱
salt pillow 盐枕
salt pit 采盐场
salt plant 盐土植物
salt playa 干盐湖
salt plug 盐柱；盐栓
salt plug pool 盐栓油藏
salt plug trap 盐栓圈闭
salt pollution 盐污染
salt pollution of groundwater 地下水的盐污染
salt pond 盐池
salt prairie 盐碱草原
salt provisions 咸货
salt rack (家畜的)食盐架
salt rampart 盐障
salt regulation 盐分调节
salt rejecting membrane 脱盐膜
salt-removal 脱盐的
salt residue 残留盐分
salt-resistance 抗盐性
salt resistance of concrete 混凝土抗盐性
salt-resistant flooring 耐盐蚀楼地面
salt-resistant property 耐盐性能
salt-resistant vegetation 耐盐植被
salt resisting 耐盐的
salt-resistive insulator 耐盐绝缘子
salt respiration 盐呼吸
salt return of soil 土壤返盐作用
salt return period 返盐期

salt ridge 盐脊
salt rock 岩盐
salt rust 盐霜
salts 盐类
salt scaling 盐剥落
salt scaling of concrete 盐碱使混凝土剥落
salt screen 荧光屏
salt sensitive 盐敏性的
salt sensitiveness 盐分敏感性
salt setting tank 盐沉清器
salt settler 盐沉清器
salt sick 盐病
salt slug 使溶液的盐分增大
salt slurry 盐泥
salt soda 碳酸钠
salt soil 盐(性)土
salt solution 盐水(溶液)；盐溶液
salt solution breccia 盐溶角砾岩
salt solution method 盐溶液法
salt solution method of ga(u)ging 盐液测流法
salt solutions concentration 盐溶液浓度
salt spray 盐喷
salt spray climax 盐雾演替顶级
salt spray corrosion test 盐喷腐蚀试验
salt spray test 盐雾试验；盐水喷雾(耐蚀)试验；盐喷试验；喷盐(雾)试验(快速耐腐蚀试验)；撒盐试验
salt spray testing instrument 盐雾试验箱
salt spreader 撒盐机；撒盐器；盐粒撒布器
salt spread in switched and on platform 在道岔和站台上撒盐
salt spring 盐(水)泉
salt stabilization 食盐稳定法
salt-stabilized road 食盐稳定道路
salt status 盐分状况
salt steppe 盐渍草原；盐土(干)草原
salt stock 盐株
salt store 盐库；盐店
salt string 盐层套管
salt swamp 盐沼(泽地)；盐性林沼；盐土沼泽
salt swamp beach 盐沼海滩
salt swell 盐隆
salt system 盐系统
salt titration method 盐液滴定法
salt tolerance 耐盐性；耐盐度
salt-tolerant 耐盐性的
salt-tolerant plant 耐盐植物
salt-tolerant species 耐盐(物)种
salt tongue 盐舌
salt transportation zone 盐分搬运带
salt-treated runway 撒盐跑道
salt treatment 撒盐处理
salt-type flocculation 盐形絮凝
salt-type wood preservative 护木盐
salt up well 被岩盐堵塞的井
saltus 跃幅；急变；不连续
saltus of discontinuity 不连续振幅；不连续度
salt-velocity meter 盐液流速计
salt-velocity method 盐液流速法；盐液测流法；盐测流法
salt-velocity process 盐液流速法
salt wall 盐墙
saltwash member 盐洗段
salt waste 盐土沙漠
salt water 硝水；咸水；盐水
salt-water algae 海水藻
salt-water barrier 咸水屏障
salt-water circulating pump 盐水循环泵
salt-water clay 盐水黏土
salt-water cooling tower 盐水冷却塔
salt-water corrosion 盐水侵蚀
salt-water creep 咸水冷流变
salt-water disposal 盐水排除；盐水处理
salt-water drilling fluid 盐水钻井液
salt-water encroachment 咸水入侵；海水入侵；盐水入侵
salt-water environment 咸水环境
salt-water fish 海鱼
salt-water front 咸水前锋
salt-water heater 海水加热器
salt-water ice 咸水冰；海上成的冰块；海冰
salt-water immersion test 盐水浸渍试验
salt-water intrusion 咸水入侵；盐水侵入；海水入侵；盐水入侵

salt-water lagoon 咸水泻湖
salt-water lake 咸水湖
salt-water light 咸水光
salt-water load line 海水吃水线
salt-water lock 挡潮(船)闸；防咸水闸
salt-water mud 盐水泥浆
salt-water pollution 盐水污染；海水污染
salt-water resistance 盐的耐水性；抗盐水性；耐盐水性
salt-water resistant 盐的耐水性
salt-water resistant test 耐盐水试验
salt-water sanitary pump 海水卫生泵
salt-water service pump 盐水泵
salt-water soap 海水皂；海水用肥皂；船用肥皂
salt-water swamp deposit 咸水沼泽沉积
salt-water system 盐水系统(消防灭火用)；海水管系
salt-water underrun 咸水潜流；海水潜流
salt-water wedge 海水楔(体)；咸水楔；盐水楔
salt-water well 盐水井
salt wedge 盐水楔；盐劈
salt-wedge estuary 盐楔湾；盐水楔河口
salt well 盐井
salt wind 海风
salt works 盐场
salty 盐质的；含盐的
salty basin 盐斑
salty groundwater 含盐地下水
salty mud 盐泥
salty soil 盐渍土(壤)；含盐土(壤)
salty taste 咸味
salty water 咸水
salty wind damage 盐风损害；盐风灾害
salty wind protection forest 防盐林；防海风林
Saltzman conversion factor 萨尔茨曼系数
salt zoning 盐分带出
salubrious 增进健康的
salubrity 增进健康
salutarium 疗养地
salutatorium (中世纪教堂置于门廊或圣器室局部的)议事室
salute 放礼炮
salvage 救助；救捞；海上救助费；海上救难；海难救助；抢救；打捞；残余价值；被救船舶
salvageability 可救助性
salvageable 可抢救的；可救助的；可打捞的
salvage account 残废料处理
salvage agreement 救助协议；救助契约；海难救助协定
salvage anchor 救助锚
salvage and salvage charges 救助与救助费
salvage and special charges 共同海损、救助及特别费用
salvage appliance 打捞工具；营救设备；营救用具；抢救设备；抢救用具；救助设备
salvage association 救捞协会；海难救助协会
salvage at sea 海上救助
salvage award 救助报酬
salvage boat 救助船；海滩抢救船；打捞作业船；捞船
salvage bond 救助契约
salvage by agreement 根据协定救助
salvage by contract 按约救助
salvage by injecting plastic foam 喷射泡沫塑料打捞法
salvage by patching and pumping 封舱抽水打捞法
salvage by sealing 封舱压气抽水打捞法
salvage by voluntary action 自愿救助
salvage charges 救助费(用)；海上打捞费；打捞费
salvage clause 救助条款
salvage company 救助公司；救捞公司；海难救援公司；海难救援公司；打捞公司
salvage corps 消防队；救难船队
salvage cost 利旧费；救济费
salvage count 可再利用粒数
salvage crane 救险起重机
salvage crew 打捞队；营救人员；营救人员；救助队；抢救队
salvaged bort 回收的金刚石粉粒
salvage department 废物利用部门；废料利用车间；废料间
salvage disposal 残废料处理
salvage dive 潜水打捞

salvage diver 救生潜水员;救捞潜水员
salvaged lining 废炉衬
salvage dock 打捞用的船坞;打捞浮坞
salvaged pipe 旧管子
salvaged tie 利用轨枕
salvaged water 回收水
salvage expenses 救助费(用);救护费用
salvage fee 救助费(用)
salvage floating dock 救助浮船坞
salvage gear 救助装置
salvage hawser 应急拖缆
salvage lien 救助留置权
salvage lifting vessel 打捞救生船
salvage loss 扣损余全损;救助损失;救护损失;海难抢救损失;海难救助损害;残值下降损失
salvage master 打捞监督
salvage material 废物利用材料
salvage money 救助费(用);救助报酬;救援费;救难费
salvage of a project 工程抢修
salvage of casting 铸件修补
salvage of engineering 工程抢修
salvage of plant and equipment 厂房和设备残值
salvage of waste material 废品回收
salvage of waste paper 废纸回收
salvage operation 救助作业;海难救助作业;打捞作业
salvage point 废品收集处;废品回收站
salvage pontoon 打捞浮筒
salvage present-worth 残值折现
salvage prize 救助报酬
salvage procedure 打捞作业
salvage proceeds 救助收入
salvage pump 应急排水泵;救助泵;施救泵
salvager 救助人员;救援人员;打捞人员
salvage remuneration 救助报酬
salvage repair 海损修理
salvage retailers 废品市场
salvage service 救助场;救助服务机构;船舶救助
salvage sheer-leg vessel 救捞把杆起重机
salvage ship 打捞船(舶);营救船舶;救助船;救捞船;海事救援船;海难救助船
salvage shop 修理(工)厂;修理车间;三废综合利用工厂;废物综合利用工厂
salvage sinking-fund 残值偿债基金
salvage station 废品收购站
salvage store 残余物料
salvage strap 绕匝索套
salvage suction hose 救助泵吸入软管
salvage sump (炼油厂下水道的)石油捕集器;废油捕集器
salvage synthesis 补救合成
salvage tool 打捞器
salvage towing 拖救
salvage trailer 运金属屑用拖车;金属废料运输拖车
salvage tug 救助拖轮;救难拖船;海事救援拖轮;海难救助拖轮;打捞拖轮
salvage value 余值;折旧(价)值;可挽救价值;处理价值;残值(指变卖改用低价部分);残余价值;废物利用率;废品价值
salvage value of fixed assets 固定资产残值
salvage vessel 救助船;海事救援船;海难救助船;打捞作业船;打捞施救船
salvage waist band 救生腰带
salvage water 回收水
salvage water rights 回收水权
salvage weight 打捞重量
salvage winch 打捞用绞盘;救援绞车
salvage with cofferdams 围堤打捞法;围堰(堰)打捞法
salvage work 救护工作;海上救助工作
salvage work boat 海难救助船
salvaging 利用废料;海难救助作业;抢修工程;打捞;船舶打捞;废物处理;废品(废料)处理
salvaging casting 焊补铸件
salvaging department 废料利用车间
Salvarsan 撒尔佛散
salvation 救济;救助
Salvation Army Hostel 救世军旅舍
salve 抢救;救援船舶;打捞;油膏剂;救助
salved value 获救价值(海难救出物的价值)
salver 救助人(员);救难者;金属盘
salviacim pavement 半柔性路面
salving vessel 救援船

salvo 除外例;保留条款
salvo breadth 齐射扇面宽度
salvo dispersion 齐射散布
salvo jettison 齐投弃
salvo jure [拉]其他权利不受损害
sal volatile 挥发盐;碳酸铵
salvo pattern 齐射散布
salvo point 齐射目标
salvor 救助人(员);救援人员;救援船;救难者;难船救助人员;难船救助船
salvo release 齐投
salvo signal 翼次射信号;齐射信号
Salzagitter reverse circulation drilling rig 萨尔兹吉特逆循环钻井机
Salzagitter reverse circulation method 萨尔兹吉特逆循环法
samar 柚木代替材料
samaric hydroxide 氢氧化钐
samarium boride 硼化钐
samarium-neodymium ages 钐—钕年龄
samarium-neodymium isochron 钐—钕等时线
samarium-neodymium method 钐—钕法
samarium-neodymium model 钐—钕模式
samarium nitride 氮化钐
samarium ores 钐矿
samarium oxide 氧化钐
samarium silicate 硅酸钐
samarous chloride 氯化亚钐;二氯化钐
Samarovo glacial stage 萨马洛夫冰期
Samarovo stade 萨马洛夫冰阶【地】
samarskite 铌钇矿含量;铌钇矿
same 相同的;同样
same adjucement curve 同向曲线
same-age tree 同龄树
same body chain 同体链
same body relation 同体关系
same broad soil groups 相间的广土类
same change of magnitude in relative value 相对价值的同样量的变化
same condition 同等状态
same constant error 相同固定误差
same day funds 同日资金
same dipping direction 倾向一致
same direction adjacent curve 同向曲线
same direction and proportion 同方向同比例
same direction but in different proportions 同方向不同比例
same direction intersecting 顺向交叉【铁】
same distribution 相同分布;同分布
same equilibrium condition 相同平衡条件
same frequency 相同频率
same frequency repeater 同频增音机
same labour exercised during equal periods of time 同样时间内的同一劳动
same layer 同层
samel brick 未烧透砖;粗制砖;半烧砖
same loci 相同位点
same manner 同样办法
same order 同阶
same order bias estimators 同阶有偏估计量
same phase 同相(位)
same-phase horizontal antenna 同相水平天线
same-phase vertical antenna 同相垂直天线
same platform transfer 同站台换乘
same position 相同位置
same purpose 同样作用
same purpose as lime 和石灰具有相同作用
same quality sample 同品质样品
same sea and coast 同一海域和海岸
same sea and country 同一海域和国家;同一海域和地区
same-sense curve 同向曲线
same sequence 相同顺序
same-size ratio 一比一;等量比;等比量
same size reproduction 等大复制
same type 同型
samite 金刚砂;碳化硅
sammel crystallization 聚粒结晶
Sammenxia Key Water Control Projects 三门峡水利枢纽工程(中国)
samming machine 均湿机
samoite 多水硅铝石
sampan 舢板
sampan hitcher 舢板搭钩

sample 样值;样本;货样;土样;试样;实例;抽样;标本
Sample A 样品甲
sample activity 试件放射性
sample agreement 抽样协议
sample analysis 样品分析;样本分析
sample analyzer 试样分析器
sample and hold 样本及保存
sample-and-hold amplifier 取样与同步放大器
sample-and-hold bridge 抽样保持桥路
sample-and-hold circuit 取样维持电路;取样保持电路
sample-and-holed technique 取样保持技术
sample area 样地面积;标准地
sample arithmetic(al) mean 样本算术均数
sample autocorrelation function 样本自相关函数
sample average 样品平均值;样本(平)均值;样本平均数
sample averaging 取样平均
sample-averaging iteration method 取样平均迭代法
sample bag 取样袋;土样袋
sample barrel 土样筒
sample-batch granulator 试样批成粒机
sample board 试样板
sample book 样本
sample boring 取样钻探;取样钻进
sample bottle 样品瓶;取样瓶;沙样瓶;采(水)样瓶
sample bottle sterility control 采样瓶无菌控制
sample brick 样品砖
sample bucket 样品桶;(井口泥浆)采样桶
sample bulb 样品瓶
sample can 样品罐;土样盒
sample canonical correlation coefficient 样本标准相关系数
sample capacity 样本(容)量
sample car 陈列车
sample card 样品卡;样本卡片
sample carrier 试样容器
sample catcher 样品收集器
sample cell 吸收样品池;吸收杯池;样品试池;样品管
sample census 样本普查
sample central moments 样本中心矩
sample certainty 样本确定性;抽样确定性
sample chamber 样品槽
sample change compaction 标样变更压密
sample change compression 样本变更压缩
sample changer 样本转换装置;样品交换器
sample change unit 样品转换装置
sample channel 样品槽钢
sample channels ratio 样品道比
sample characteristic value 样本特征值;样本征值
sample circuit 取样电路;采样电路;幅度—脉冲变换电路
sample circumstances 样品情况
sample cock 取样龙头
sample code 标本编号
sample code in laboratory 样品实验编号
sample coefficient 样本系数
sample collection 取样;试样收集;采样
sample collection tube 样品收集管
sample collector 取样器;试样收集器
sample colo(u)r 样品颜色
sample community 样品集体
sample concentration 样品浓度;试样浓缩
sample concentrator 试样浓缩器
sample condensation 试样冷凝
sample conditioning time 试样调理时间
sample conductivity 样品电导率
sample connection 取样口
sample container 样品袋;样品存储器;样品储存器;水样容器;试样容器;砂样瓶;沙样瓶
sample contamination 样品污染;试样污染
sample content 样品容积;样本含量
sample contract 样品契约
sample control system 采样控制系统
sample copy 样图
sample core 岩芯样品
sample correlation coefficient 样品相关系数;样本相关系数;试样相关系数
sample correlation matrix 样本相关矩阵
sample covariance 样本协方差;试样协变性
sample covariance coefficient 样本协方差系数

sample covariance function 样本协方差函数
sample covariance matrix 样本协方差矩阵
sample crusher 试样破碎机
sample cutter 取样爪；劈样器
sampled analog computer 抽样模拟计算机
sampled analog(ue)data 时间量化连续信息；抽样模拟数据
sample data 样本数据；抽样数据
sampled beam 脉冲调制射束；抽样射束
sampled current polarography 采流极谱法
sampled data 采样数据；取样数据；抽样数据
sampled-data computer 采样数据计算机；抽样数据计算机
sampled-data control 抽样数据控制
sampled-data control system 采样数据控制系统；取样数据控制系统；数据采集控制系统；抽样数据控制系统
sampled-data filter 取样数据滤波器
sampled-data measurements 样本数据量度
sampled-data mode 抽样数据方式
sampled-data system 取样数据系统；采样数据系统；数据取样系统；抽样数据系统
sampled date 取样日期
sampled depth 取样深度；采样水深
sampled device 采样设备
sample decile 样本十分位数
sample degradation 试样降解
sample delay 抽样延迟
sample depth 取样深度
sample depth of soil 土样深度
sample design 样品设计；样本设计
sample deviation 样本偏差
sampled grade 试样品位
sampled imagery 采样图像
sample dimension 样品规格
sample discount 样品折扣
sample dispersion 样品离差；样本离差；试样分散性
sample distance 取样距离
sample distribution 样品分布；样本分布；采样分布
sample distribution of mean 样本均值分布
sample divider 分土器
sample divider pass 试样缩分
sampled offer 附样品的发盘；附样报盘
sample dot 取样点
sampled phase-locked loop 取样锁相环
sampled population 抽样总体
sample drawing 取样；采样；样品图；抽样器；抽样
sampled region 被抽中区域
sample drilling 取样钻探；取样钻进
sampled signal 取样信号；抽样信号；采样信号
sampled threshold 取样阀值
sampled value 样本值
sampled well 采样井
sample edge 取样边
sample EH value 样品氧化还原电
sample enrichment 样品富集
sample error 样本误差；抽样误差
sample evapo(u)ration 试样汽化
sample examination 样品检验
sample expenses 样品费用
sample export 样品输出；样品出口
sample extruder 退岩芯装置；推样器
sample fair 样品展览会
sample fair ship 展览船
sample flip-flap 抽样触发器
sample fluctuation 样本波动
sample for chemical analysis 化学分析样品
sample for determination identification 鉴定样品
sample for electron-probe analysis 电子探针分析样品
sample for fabric analysis 岩组分析样品
sample for fossil identification 化石鉴定样品
sample for full analysis 全分析样品
sample for gas analysis 气体分析样品
sample for granularity analysis 粒度分析样品
sample for heavy mineral sands determination 重砂鉴定样品
sample for isotopic age determination 同位素年龄测定样品
sample for isotopic analysis 同位素分析样品
sample for mineral analysis 矿物分析样品
sample for mineral determination 矿物鉴定样品
sample for multielement analysis 多元素分析样品
sample for natural heavy mineral 自然重砂样品

sample for ore analysis 矿石分析样品
sample for ore determination 矿石鉴定样品
sample for pal(a)eomagnetic measurement 古地磁样品
sample for person heavy mineral 人工重砂样品
sample for reference 参照样本
sample for rock analysis 岩石分析样品
sample for rock determination 岩石鉴定样品
sample for simple analysis 简项分析样品
sample for soil analysis 土壤分析样品
sample for spectrometric analysis 光谱分析样品
sample for sporopollen identification 孢粉鉴定样品
sample for unit weight of ore 体重样品
sample for water analysis 水分析样品
sample fractile 样本分位数
sample freezing 试样冰冻(试验)
sample function 样本函数；抽样函数
sample gear 抽样齿轮
sample generator 样本产生符
sample grabber 取样器；岩芯提取器；取样员；试样抓取器
sample grinder 样品研磨机
sample handling 样品运送
sample-handling system 处理样品系统
sample hand reducing 手工缩样
sample-hold 样本保持；取样保持
sample-hold circuit 抽样维持电路；采样保持电路
sample-holder 试样支持器
sample-hold switch 抽样保持开关
sample hole 取样孔
sample horizon 取样层位
sample household budgets 样本住户收支
sample hunter 样品商
sample identification 样品鉴定
sample indexing 试样存取
sample injection 试样注射
sample injection port 试样注射口
sample injector 进样器
sample input 样本(值)输入；抽样输入
sample inquiry 样本调查
sample inspection 试件检验
sample intensity 样品强度
sample interval 取样(时间)间隔；抽样间隔；采样周期；采样间距；采样间隔
sample introduction 进样
sample introduction under pressure 压力下进样
sample invoice 样品发票
sampleite 氯磷钠铜矿
sample kit 采样箱
sample label 土样标签
sample length 样品长度
sample level of significance 样本显著性水平
sample light beam 试样光束
sample likelihood function 样本似然率函数
sample location sample point 采样点
sample log 钻井采样记录；取样剖面(记录)；取样记录；测井取样剖面
sample logger 样品记录器
sample loop 样品回路管；样品环管；进样环管
sample loss 试样损耗；试样混合
sample making machine 打样机
sample mark(ing) 样品标志；试样标记
sample material 样品种类
sample mean 样品平均值；样品平均数；样本(平)均值；样本平均数；取样平均值；试样平均值；采样平均(值)
sample mean iteration method 取样平均迭代法
sample mean square contingency 样本均方列联；样本均方或然性
sample median 样品中位数；样本中位数；样品中位值；抽样中位值；抽样中位数
sample merchant 样品商
sample merging 样品合并
sample mode 样本众数；抽样状态
sample mold 样品模；试样模
sample moment 样本矩
sample moment of order r 样本 R 阶矩
sample monitoring 样本监督
sample mo(u)ld 实验用模型
sample multiple correlation coefficient 样本多重相关系数
sample name 试样名称；标本名称
sample nozzle 样品喷嘴；取样喷嘴；采样喷嘴

sample number 样品(编)号；样本数；抽样号码；采样号
sample number in the field 样品的野外编号
sample number in the laboratory 样品的室内编号
sample observations 样品观测值
sample of cement 水泥试样
sample of colo(u)r 颜色样品
sample of erratic high-grade 特高品价样品
sample of existing goods 现货样品
sample offer 附样品报盘
sample of material 材料样品
sample of matrix 基质样品
sample of no value 免费货样
sample of pavement 路面试样
sample of signal 信号取样
sample operating system 样本操作系统
sample order 样品订单；试购
sample original number 原始样号
sample-out count 设备内无样测量；本底测量
sample pan 样品盘
sample paper of chemical analysis 化学分析送样单
sample paper of fossil 化石送样单
sample paper of rock and mineral analysis 岩矿分析送样单
sample partial correlation coefficient 样本偏相关系数
sample passers 样品检查员
sample path 取样路径
sample path length 溶样厚度
sample percentage 样本百分比
sample percentiles 样本百分位数
sample period 取样周期；取样时间间隔；采样周期
sample pH value 样品酸碱度
sample piece 样品；试件；样件；试块；标本
sample pipe 取样管
sample plan 样本方案
sample plot(ting) 样图；样地；取样地；试验地；标准地
sample point 采样点；样本点；取样点；抽样点
sample point number 采样点编号
sample polishing machine 试件抛光机
sample pollution 样品污染
sample port 取样口
sample preparation 样品制备；试样制备
sample preparation method 试样制备方法
sample pretreatment 样本预处理；试样预处理
sample probability measure space 样本概率测度空间
sample probe 试样探针；试样探棒
sample process 采样过程
sample processing 样品加工；水样处理
sample processing circuit 样品加工流程
sample processing formula 样品加工公式
sample production 样本生产
sample profile 取样剖面
sample program(me) 取样计划；取样程序；抽样程序
sample proportion 样本比例
sample proportion of successes 样本成功比例
sample pulse phase 取样脉冲相位
sample quantity 样本数量；试样数量
sample quartering 四分缩样法
sample quartile 样本四分位数
sampler 选样器；样品检查员；样本检验员；快速转换器；进样器；采样员；取样器；取样机；抽样机；采样员；采样器
sample range 样本范围；样本极差；抽样范围；采样范围
sample rate 样本率；取样率；抽样率；采样速率；样速度
sample ratio 取样比；抽样率
sampler barrel 取样筒
sampler conduit 取样管
sampler cutting edge 采样机刀口
sampler drill 取样钻机
sample record 样本记录；试样编录
sample record in laboratory 样品实验室记录
sample recovery 试样回收
sample reducing 缩样
sample reduction 试样粉碎
sample regression 样品回归
sample regression function 样本回归函数
sampler for dredging 挖泥采样器

sampler for undisturbed samples 原状土取样器
sampler head 取土器端部
sampler jerker 取样员
sampler liner 取土器内套
sampler log 采样器记录
sample rock container 月球岩样储[贮]存箱
sample rod 取样杆
sample room 样品间;样品(陈列)室
sampler pipe 取样管
sampler room 试样陈列室
sampler space 取样空间
sampler tube 采样管;取样管;取土器容纳管;取土器衬套
sample sack 样品袋
sample sagger 样品烧箱
sample section 取样断面;取样地;典型截面;典型断面
sample segment 取样部分
sample(selection) 选样
sample selection rule 样本抽选规则
sample sensusing 采样调查
sample significance level 样本显著性水平
sample site 采样点
sample size 样品数量;样品(容)量;样品大小;样品尺寸;样本(容)量;样本规格;样本大小;样本尺寸;进样量;试样量
sample size code letter 样本量字码
sample size in decision problem 决策问题中的样本容量
sample size letter 试样尺寸码
sample size loss 试样量损耗
sample size requirements 样本容量要求
sample skein winder 缕纱测长机
sample skewness coefficient 样本偏斜系数
sample smear 试样涂污
samples measuring method 标本法
samples of geochemical gas survey 气体测量的样品
sample solution 试样溶液
sample space 样本空间;取样空间;采样空间
sample spacing 取样距离
sample specification 样品规格;样品说明书
sample spectral function 样本谱函数
sample spinner 取样旋转器
sample splitter 取样器;样品分离器;纵开取样器;取土分样器;试样劈裂器;试样分取器;试样分流器;分样器
sample splitting 样品缩分;样品分离
sample splitting device 试样劈裂器;试样分取器
sample spoon 取样筒;取样勺
sample spot 试样斑
sample stabilization 样本稳定化
sample stand 取样柱
sample standard deviation 样品标准偏差;样本均方差;样本标准差
sample standards of corrosion test 溶蚀试验试样规格
sample station 取样站
sample statistics 样本统计量
sample storage 样品保存
sample storage time 样品保存时间;样品存储[贮]时间
sample streaker 试样划痕器
sample strength 样品强度
sample strip 样带;带状标地;采样带
sample survey 样品鉴定;样本调查;抽样检验;抽样检查;抽样调查
sample system miniaturization 取样系统小型化
sample taken at random 任选样品;随机选样
sample taker 取样工;取样器;采样器
sample taking and preparation of coal 煤的采样和制样
sample taking of aggregate 骨料取样
sample temperature 试样温度
sample test 样品试验
sample testing 抽样测试;样品测试
sample testing physical property 物性测试样品
sample testing technical property 技术测试样品
sample theorem 取样原则
sample thief 取样器;样本器
sample throughput 试样探针;试样处理量
sample time 抽样时间;采样时间
sample time aperture 抽样时间截口
sample time interval 采样时间间隔

sample time point 样本时点
sample tool 取样工具;采样器械
sample total 样本总值
sample traffic survey 典型交通调查;抽样交通检查
sample transfer operation 试样转移操作
sample transport 样品运送;试样输送
sample trap 岩粉收集器
sample treatment 样品加工;样品处理
sample tree 样木;样本树;标准木
sample trial 样品试验
sample trough (有隔板的)岩粉沉淀槽
sample type 取样类型;试样种类;标本类型
sample type of corrosion test 溶蚀试验试样类别
sample uncertainty 不定性取样;不定性抽样
sample unit 样本单位
sample value 取样值;抽样值
sample valve 进样阀;采样阀门
sample variance 样本方差;采样离散;采样方差
sample variance-covariance matrix 样本方差—协方差矩阵
sample variance matrix 样本方差矩阵
sample variation 样本变差
sample volume 样品体积;样本容量
sample wafer 样片
sample waveform 样本波形
sample weight 样品重量
sample window 陈列窗
sample work piece 样体;样件
sampling 样本化;进样;取土样;试件采样;标本采集;变为脉冲信号
sampling action 选样;抽样;脉冲作用;取样(动作);作用;采样(动作)
sampling amount 采样数量
sampling amplitudes 采样振幅
sampling analysis 样品分析;抽样分析
sampling and analysis program(me) 取样分析程序
sampling aperture 取样孔;抽样孔
sampling appliance 取样设备;取样器
sampling around the plant 绕植株采样
sampling assembly 取样组件
sampling at a definite depth 规定深度采样
sampling at a definite horizon 规定层位采样
sampling at a definite size fraction 规定粒度采样
sampling at list time 按表列时间采样
sampling at the shadowy side of plant 植物阴面采样
sampling at the sunny side of plant 植物阳面采样
sampling bag 取样袋
sampling barrel 取样筒
sampling bias 取样偏差;采样偏移
sampling boat 取样舟;采样船
sampling boat technique 采样船技术
sampling bore hole 取样钻孔
sampling bottle 取样瓶;采样瓶
sampling bucket (井口泥浆)采样桶
sampling by box 箱式取样器取样
sampling by centrifugation separation 离心分离法采样
sampling by classification 分类抽样
sampling by dredge(r) 挖泥船取样;拖网取样
sampling by electrostatic precipitation 静电集尘法采样;静电沉淀法采样
sampling by filtration 过滤采样
sampling by free fall coring 自落式取芯器取样
sampling by free fall grabber 自落式抓斗取样
sampling by grabber 抓斗取样
sampling by gravity-sampler 重力取样器取样
sampling by impaction 撞击取样;撞击采样
sampling by piston coring 活塞取芯器取样
sampling by pumping 抽水取样;抽气采样
sampling by thermal precipitation 热沉淀法取样
sampling by vibro-sampler 振动式取样器取样
sampling by water displacement 水置换法取样
sampling camera 取样摄像机
sampling cell 采样格子
sampling check 抽样检验
sampling circuit 取样电路;抽样电路;幅度—脉冲变换电路
sampling clock 取样钟
sampling cock 取样旋塞
sampling collection 样品取集
sampling condition 进样条件;取样条件
sampling configuration 采样模式

sampling connect 取样连接管
sampling connection 取样连接管
sampling connector 取样连接器
sampling control 取样控制;采样控制
sampling controller 抽样控制器;采样控制器
sampling control system 采样控制系统
sampling cooler 取样冷却器
sampling core 中心取样
sampling correlator 抽样相关器
sampling counter clear 取样计数器复位
sampling counter maximum 取样限数计数器
sampling cross-section 采样断面
sampling cup 取样杯
sampling cycle 抽样周期
sampling data 采样数据
sampling data system 抽样数据系统
sampling date 取样日期;采样日期
sampling date of water 水样采取的时间
sampling density 采样密度
sampling depth 取样深度;采样井深
sampling depth of water 水样采取深度
sampling design 取样设计;采样设计
sampling detector 取样传感器
sampling device 进样装置;取样装置;取样设备;取样器;采样设备
sampling directioned current meter 取样定向流速仪
sampling distortion 取样畸变
sampling distribution 取样分布;抽样分布;标本分布
sampling distribution of sample mean 样本平均值的抽样分布
sampling distribution of the statistic 统计量的抽样分布
sampling distribution plan of ore deposits 矿床取样平面图
sampling disturbance 取样扰动;采样扰动
sampling documentation 取样编录
sampling drill 取样钻机
sampling drilling 取样钻探
sampling efficiency 采样效率
sampling electrode 取样电极
sampling equipment 取样设备
sampling equipment to be used 采用的取土设备
sampling error 取样误差;抽样误差;采样误差
sampling factor 扩大系数
sampling fo a stack 烟囱口取样
sampling for chemical analysis 化学分析采样
sampling for dredging 挖泥取样
sampling for fossil identification 古生物鉴定采样
sampling for mineralogic(al) and petrologic(al) determination 岩矿鉴定采样
sampling for physical and technical determination 技术采样
sampling for technologic(al) determination 加工技术采样
sampling fraction 样本分数;取样率;抽样率;抽样分数;抽样比;采样比
sampling frequency 量化频率;取样频率;采样频率;采样率
sampling function 采样函数;抽样函数
sampling gate 取样门
sampling grid 采样网格
sampling gun 取样枪
sampling head 取样头;抽样头;采样抽头
sampling hole 采样钻孔
sampling horizon 取样层位;采样层位
sampling image 取样图像
sampling in duct 管道采样
sampling inspection 取样检验;取样检查;抽样检验;抽样检查;采样检查
sampling inspection by attribute 计量取样检验
sampling inspection by variable 计量取样检验
sampling inspection for continuous production 连续生产型抽样检验
sampling inspection plan 抽样检查方案
sampling inspection with adjustment 调整型抽样检验
sampling inspection with screening 挑选型抽样检验
sampling installation 取样设备;采样装置
sampling instrument 取样仪器;采样仪表
sampling interval 抽样区间;抽样间隔;采样间距;采样间隔

sampling interval of wavelet 子波采样间隔
sampling in underground working 地下工程采样
sampling investigation 采样调查
sampling kit 取样全套工具
sampling label(l)ing 土样标记
sampling layer of water 水样采取层位
sampling layout 采样布局
sampling line 取样线路;取样管道;采样线路
sampling liner 取样器衬套
sampling location 取样位置;取样点;采样现场;采样位置;采样点
sampling loop 采样回路管
sampling loss 取样损失;采样损失;采样错漏
sampling machine 采样机
sampling manhole 采样井
sampling manifold 采样总管;采样歧管
sampling material 试样
sampling mean 样本平均值;样本平均数
sampling mechanism 观测机构;取样机构
sampling mesh 采样网格
sampling method 取样方法;抽样方法;采样方法
sampling method and tool 取样方法和工具
sampling method of soil 土壤取样法
sampling method of soil gas 土壤气取样法
sampling method of underground water 地下水取样法
sampling method of water 水样采取方法
sampling mill 碎样机;采样磨机
sampling mo(u)ld 取样模
sampling network 信号取样网络;采样站网;采样网(络)
sampling normal distribution 样品正态分布;取样正态分布;抽样正态分布
sampling normal distribution curve 样品正态分布曲线
sampling nozzle 取样喷嘴;取样管;采样喷嘴
sampling number 抽样数
sampling number of each trace 每道样点数
sampling observation 取样观测
sampling of aeolian sediment 风积物采样
sampling of aerosol 气溶胶采样
sampling of aggregate 骨料采样
sampling of alluvial sediments 冲积物采样
sampling of attributes 属性抽样
sampling of auto exhaust 汽车排气取样;汽车排气采样
sampling of cones 锥体取样
sampling of effluent 污水采样
sampling of eluvial sediment 残坡积物采样
sampling of glacial sediment 冰川沉积物采样
sampling of lake sediment 湖泊沉积物采样
sampling of lake water 湖水采样
sampling of marine sediment 海洋沉积物采样
sampling of particulate matter 颗粒物(质)采样
sampling of river water 河水采样
sampling of rock and mineral identification 岩矿鉴定取样
sampling of seawater 海水采样
sampling of soil values 土壤值采样
sampling of spring water 泉水采样
sampling of stream sediment 水系沉积物采样
sampling of stream water 溪水采样
sampling of surface sediments 表层沉积物取样
sampling of volcanic sediment 火山沉积物采样
sampling of wastewater 废水取样
sampling of water 采水样
sampling of well water 井水采样
sampling organizer 采样单位
sampling oscilloscope 取样示波器;抽样示波器;采样示波器
sampling outlet 取样出口;采样出口
sampling pair 样本对(儿)
sampling parametric(al) computation 抽样参数计算;采样参数计算
sampling parametric computation 取样参数计算
sampling parent 样品介质
sampling pattern 取样型式;取样模型;典型模式;采样模式
sampling period 取样周期;抽样周期;采样周期
sampling phase 脉冲调制相位
sampling piece 采样品
sampling pipe 取样管(采样点);采样管
sampling plan 取样方案;抽样检验方法;抽样方案
sampling plan by attributes 品质抽样方案

sampling plan of coal seam 煤层取样平面图
sampling plan of mining level 中段取样平面图
sampling plan of orebody 矿体取样平面图
sampling plan of ore district 矿区取样平面图
sampling plotter 取样标绘器
sampling point 样点;取样位置;取样点;采样点
sampling point type of water 取样点类型
sampling polynomial 抽样多项式
sampling position 取样位置;采样位置;采样地点
sampling precision 采样精度
sampling probe 取样针;取样头;采样探针;采样探头
sampling procedure 取样程序;取样步骤;抽样法
sampling process 取值过程;取样过程;抽样过程;采样过程;采样方法
sampling profile 采样剖面
sampling program(me) 取样程序
sampling property 抽样性质
sampling pulse 取样脉冲;抽样脉冲
sampling pulse generator 取样脉冲发生器;抽样脉冲发生器
sampling pulse width 取样脉冲宽度
sampling pump 取样泵
sampling purpose 采样目的
sampling quality 取样质量
sampling quality supervision 抽查型质量监督
sampling quantity of water 采样水数量
sampling range 采样范围
sampling rate 取值速率;取样(速)率;取样频率;抽样(速)率;采样速度;采样(速)率
sampling ratio 取样比;抽样比;采样比
sampling record 取样记录
sampling reduction 缩样
sampling regression distribution 抽样回归分配
sampling reliability 样本可信度
sampling report 取样报告
sampling requirement 采样要求
sampling rig 取样机
sampling rock unit 采样岩石地层单位
sampling room 取样室
sampling schedule 取样时间表
sampling scheme without replacement 不放回抽样方式
sampling scope 取样示波器
sampling sediment 泥沙样品
sampling sensusing 采样调查
sampling servomechanism 采样伺服机构;抽样伺服机构
sampling servosystem 脉冲随动系统
sampling signal 采样信号
sampling site 取样点;采样点位;采样点
sampling soil 取土
sampling spacing 取样间隔
sampling spark chamber 取样火花室
sampling spoon 取样斗;取样筒;取样(环)勺;取样环刀
sampling stand 采样架
sampling station 监测站;采样站
sampling statistics 样本统计量;抽样统计量
sampling survey 抽样检验;抽样检查;抽样调查
sampling survey method 抽样调查法
sampling switch 取样转换器;取样交换机;取样变换器;抽样变换器
sampling synchronization pulse 取样同步脉冲
sampling system 进样系统;进样法;取样系统;抽样系统;采样系统
sampling taking 取样
sampling tap 取样阀
sampling target 取样靶
sampling technique 取样技术;抽样技术;采样技术
sampling tessellation 取样图案
sampling test 样品试验;样本试验;取样试验;抽样试验
sampling theorem 取样原理;取样定理;抽样定理;采样定理
sampling theory 抽样理论
sampling thickness 采样厚度
sampling thief 取样器
sampling time 取样时间;抽样时间;采样时间
sampling time interval 取样时段;采样时间间隔
sampling tin 样品罐
sampling tip 取土钻头部
sampling tolerance 抽样公差
sampling tool 取样工具;采样工具

sampling tool and equipment 采样用品
sampling train 取样装置列
sampling tray 试样盘
sampling tube 样品管;取样管;取土器;采样管
sampling type 采样种类
sampling undisturbed soil 取原状土样
sampling unit 取样装置;取样单元;取样单位;抽样单位;采样装置
sampling valve 取样阀(门);采样阀(门)
sampling valve indicator 取样指示计
sampling variance 抽样(分布)方差
sampling vertical(line) 取样垂线
sampling voltage 取样电压;分层取样电压
sampling voltmeter 采样伏特计
sampling volume 采样量
sampling wave form 抽样波形曲线
sampling well 取样井
sampling with a pail 提桶取样
sampling without ordering 无序抽样
sampling without replacement 无退还抽样;不重复抽样;不回置抽样
sampling with repetition 重复取样
sampling with replacement 有退还抽样;退还抽样;放回抽样
sampling with replacements 回置抽样
sampling with unequal probability 不等概率抽样
samploscope 取样示波器;抽样示波器
sampson 希腊神话中的大力士
sampson anchor 将军锚
sam(p)son knee 缆桩倒肘
sam(p)son post 吊杆柱;吊杆支柱(俗称将军柱)
sam(p)son post brace 吊杆柱系杆;吊杆柱支撑
samsonite 硫锑锰银矿
samson knee 缆桩倒肘
samson post 吊杆柱;起重柱
samuelsonite 羟磷铝铁钙石
San Andreas fault 圣安德列斯断层(美国)
San Andress transform fault 圣安德列斯转换断层
San Andress-type structure 圣安德列斯人字型构造
sanatorium 休养所;疗养院
sanbornite 硅钡石
Sanborn map 山傍地图(一种显示土地价格的地图)
sanction 承认;批准;许可
sanction for test 试验许可证
sanction of custom 按惯例的制裁
sanctuary 动物保护;避难所;庇护所;野生动物保护区;圣殿
sanctuary game refuge 禁猎区
sanctuary of Appolo 阿波罗禁猎区;阿波罗圣地
sanctuary of Asclepius 阿斯克莱禁猎区;阿斯克莱圣地
sanctuary of Zeus 宙斯保护区;宙斯禁猎区;宙斯圣地
sanctum 内殿;密室;书房
sanctus bell 教堂的钟;圣钟;圣铃
sand 砂(子);模砂;
sand abrasion test 耐砂磨损试验
sand accretion 淤沙(地);泥沙淤积;泥沙积聚;砂涨地;沙滩地
sand addition 加砂;型砂加入剂
sand addition device 加砂设备
sand-aerating apparatus 吹砂装置
sand aerator 破砂机
sand aggregate 砂集料;砂骨料
sand aggregate ratio 砂石比;砂骨比
sandal 檀香(木)
sandal brick 等外砖;次品砖;未烧透的砖
sandalwood 宽果苦槛蓝;檀香(木);白檀
sandalwood oil 檀香(木)油
sand analogy 砂模拟
sand anchor 沙地锚
sand and coarse aggregate ratio 砂石比;砂与粗集料比;砂与粗骨料比
sand and gravel 沙砾(石)
Sand and Gravel Association of Great British 英国沙砾协会
sand and gravel bedding 砂石垫层;沙砾(石)垫层
sand and gravel bed load 砂与砾石推移质
sand and gravel concrete 砂砾石混凝土
sand and gravel cushion 砂石垫层
sand and gravel deposit 砂石矿床;沙砾矿床
sand and gravel dredge(r) 砂石挖掘船
sand and gravel exploitation 砂石料开采
sand and gravel extraction 砂石精选

sand and gravel foreshore 沙砾混合滩;沙砾混合滩
sand and gravel mining 砂石料开采
sand and gravel overburden 沙砾覆盖层
sand and gravel overlay(er) 沙砾覆盖层;沙砾覆盖层
sand and gravel overlaying layer 沙砾覆盖层
sand and gravel plant 沙砾厂
sand and gravel trap 沉沙段;沉沙井;沉沙(砾)池;沉沙槽
sand and gravel wash 沙砾洗涤;沙砾(石)冲积层
sand and gravel washer 沙砾冲洗机;沙砾冲洗器
sand and gravel washing plant 沙砾冲洗设备
sand and mud foreshore 泥沙混合滩
sand and slag inclusions 夹砂与含渣
sand and spreader 砂撒布机;砂盐撒布器
sand and storm 暴风沙
sand and total aggregate ratio 砂与集料总量比;砂与骨料总量比
sand apron 沿礁砂;砂裙;砂封板
sandarac(h) 山达脂;香松树胶;桧树胶
sandarac(h) gum 香松树胶;山达(树)脂
sandaracolic acid 山达酸
sand argillaceous texture 砂泥质结构
sand arresting hedge 拦沙栅;防沙篱
sand asphalt 沥青胶砂;沙子沥青混合料;砂—沥青;地沥青砂沥青砂;地沥青砂
sand asphalt cushion 沥青砂垫层
sand asphalt facing 砂沥青护面
sand asphalt mixture 地沥青砂浆;地沥青砂混合料
sand-asphalt pavement 地沥青砂路面
sand-asphalt protection[protective] coat 地沥青砂保护层
sand-asphalt surface 地沥青砂路面
sand auger 尘旋
sand avalanche 砂崩
sand backfill 砂(土)回填;回填砂
sand-bag 砂袋;砂包;堆砂袋
sand-bag bunker 砂袋掩体
sand-bag cofferdam 砂袋围堰;草袋围堰
sand-bag-core dam 砂袋填心坝
sand-bag damming 砂袋筑堤(法);砂袋筑坝;砂袋修筑隔墙;砂袋封堵
sand-bag embankment 砂袋填筑堤
sand-bag mattress 砂袋垫层
sand-bag revetment 砂袋护坡;砂袋护岸
sand-bag stopping 砂袋隔墙
sand-bag wall 砂袋护墙;砂袋挡墙
sand-bag walling 砂袋筑墙;砂袋墙设置;砂袋挡墙
sand bailer (打井用)提砂筒
sand ballast 压舱砂;砂石渣;砂道砟;砂压载
sand ballast type roll 砂石渣碾压机;砂石渣压路机
sand bank 沙海滩;沙洲;沙滩;沙坝;沙岸
sand bar 河口沙洲;沙洲;沙滩;沙礁;沙坝;拦门沙;拦江沙
sand barge 砂驳
sand barren 沙漠
sand barrier 河口三角洲;河口潜洲;沙坝
sand barrier of band shape 行带状沙障
sand barrier of grid shape 格状沙障
sand basin 滤沙池;沉沙井;沉沙池;沉沙槽
sand-bath 喷砂清洗;砂浴
sand bath tempering 砂浴回火
sand beach 砂质海滩;沙滩(海岸)
sand-beach community 沙滩群落
sand-bearing 携带泥沙的
sand-bearing argillaceous texture 含砂泥质结构
sand-bearing dolomite 含砂质白云岩
sand-bearing limestone 含砂质灰岩
sand-bearing method 砂载法;试验砂箱法(试验管压力)
sand-bearing method for testing clay pipe 陶管砂床试验法
sand-bearing mudstone 含砂质泥岩
sand-bearing river 携带泥沙的河流
sand-bearing siltstone 含砂粉砂岩
sand-bearing test 砂承重试验;砂承载力试验;砂承试验
sand-bearing wind 含沙风;带沙风;风沙流
sand-bedded river meander 沙底河弯
sand bed(ding) 砂(垫)层;铺砂层;砂床;砂质河床
sand bedding course 砂垫层
sand-bed filter 砂床过滤器
sand-bed river 砂床河流
sand-bed stream 砂床河川

sand belt 磨带;砂布带;抛光带;砂带
sand-bentonite slurry 砂浆混凝土
Sandberg process 碳钢局部索氏体化处理
sand bin 砂箱;储(贮)砂斗
sand binder 型砂黏合剂;固沙植物
sand-binding grasses 固沙草
sand binding plant 固沙植物
sand bit 采沙坑
sand bitumen 沥青砂(土);砂沥青
sand bitumen core wall 沥青砂土心墙
sand blanket 过滤层;砂盖层;砂垫层;砂滤层;砂毡
sand-blast 喷砂器;喷砂;喷砂清除法
sand-blast apparatus 喷砂设备;喷砂器;喷砂机;喷砂装置
sand-blast barrel mill 喷砂滚筒
sand-blast cleaner 喷砂清理器
sand-blast cleaning 喷砂除锈
sand-blast cleaning room 喷砂清理室
sand-blast decorative finish 喷砂装饰装饰
sand-blast device 喷砂设备
sand-blasted 喷砂的
sand-blasted concrete 喷砂饰面混凝土
sand-blasted finish 喷砂磨光;喷砂饰面
sand-blasted glass 喷砂玻璃
sand-blasted panel 喷砂处理过的板
sand-blast equipment 喷砂(磨光)设备
sand-blaster 喷砂器;喷砂(浆)机
sand-blaster for rust removing 除锈喷砂机
sand-blast finish 喷砂修整;喷砂饰面
sand-blast hose 喷砂胶管;喷砂软管
sand-blasting 喷砂(使表面粗糙);喷砂处理;喷砂除污;喷砂除锈;喷砂冲刷;砂冲;吹砂磨蚀作用
sand-blasting abrasive 喷砂磨料
sand-blasting box 喷砂箱
sand-blasting drum 喷砂筒
sand-blasting finish 喷砂蚀面
sand-blasting gun 喷砂枪
sand-blasting hose 喷砂软管
sand-blasting machine 喷砂(清理)机
sand-blasting method 喷砂清理(法)
sand-blasting nozzle 喷砂口;喷砂嘴
sand-blasting test 喷砂测试
sand-blasting unit 喷砂装置
sand-blast mower 喷砂灭草机
sand-blast nozzle 喷砂嘴
sand-blast obscuring 喷砂蚀面
sand-blast resistant mask 防喷砂护罩
sand-blast sand 喷砂用的沙子;喷用砂;喷砂沙子
sand-blast shop 喷砂车间
sand-blast test 喷砂试验
sand blender 混砂设备;混砂车;松砂机
sand blister 轮胎起泡;砂眼;砂泡
sand block 砂箱墩
sand blow 喷砂
sand blower 喷砂装置;喷砂器;喷砂机
sand blowing 喷砂打光
sand blowing nozzle 喷砂嘴
sand body 砂层
sand boil 涌沙;冒水翻沙;喷水冒沙;喷沙冒水;沙涌
sand boiling 喷砂
sand bond 型砂强度
sand borrow area 取砂区
sand borrow areas for reclamation 填海用取砂区
sand bottom 砂底
sand-bottomed 砂底的;沉沙井
sand box 砂箱(千斤顶);沉沙井
sand box model 砂槽模型;砂箱模型
sand box strainer 砂箱滤器
sand brake 捞砂卷筒制动
sand break 防沙林带;田间沙地
sand break forest 防沙林
sand brick (擦甲板转角用的)小砂砖
sand bridge 砂桥
sand bucket 铲砂斗;活塞式抽筒;砂桶
sand buckle 铸件起皮
sand bulking 砂的体胀;砂的湿胀
sand burning 烧结砂
sand burrowing 掘沙的
sand burying 沙埋
sand by-pass 旁通输沙道
sand by-passing 人工输沙;排沙设施;(绕道的)输沙设备
sand cake 泥芯块
sand calamity 沙害

sand-calcite 砂质方解石
sand cap 防沙罩
sand carpet 砂磨耗层;砂垫层
sand carpeting 铺砂
sand carrier 携沙液;运沙船;载沙液
sand carrier with grab bucket 带抓斗运沙船
sand-carrying capacity 挟沙量;挟沙能力;含砂量
sand-carrying capacity of lime 石灰含砂量
sand casing of well points 井点砂套管
sand cast 砂铸;砂型铸件
sand cast finish 砂型铸件处理
sand casting 砂型铸造;砂模铸造;翻砂
sand-casting process 砂型铸造法
sand-cast pig 砂型铸生铁;砂模生铁
sand-cast pig iron 翻砂铸铁
sand-cast pipe 砂模铸(铁)管
sand-cast tube 砂模铸管;砂型铸铁管
sand catcher 捕砂器;集沙器(又称集砂器);泥沙采集器;水中集砂器;砂粒器;沉砂器;捕砂室;除砂器
sand catching basin 集砂池;集砂井
sand cay 小沙洲;(海上低潮露出的)沙洲;小沙坝;小沙岛
sand cement 砂粉水泥;掺有砂粉混合料的水泥;掺砂水泥;水泥(稳定)土;砂土水泥
sand-cement coating support 喷射砂浆支护
sand-cement grout 砂—水泥砂浆;水泥砂浆
sand-cement grout mixer 水泥砂浆搅拌器
sand-cement mixture 砂—水泥混合浆
sand-cement mortar 水泥砂浆
sand cement ratio 砂灰比
sand-cement slurry 砂土水泥浆
sand classifier 沙子分选机;沙子分级器
sand classifying equipment 沙石分级设备
sand classifying plant 砂子分级设备
sand clay 含黏土砂(约含10%的砂);砂土;砂(拌)黏土
sand clay base 砂(质)黏土地基;砂泥土地基
sand clay liquefaction 砂土液化
sand clay loam 砂质黏壤土
sand-clay matrix 砂土结合料
sand-clay mix(ture) 砂黏土混合物
sand clay pavement 砂黏土路面
sand clay road 砂黏土路;砂土路
sand clay road surface 泥沙路面;黄土路面;砂泥路面
sand-clay soil 砂黏土土壤
sand-clay surface(ing) 砂土面层;砂土路;砂泥路面
sand cleaner 喷砂(式)清洁器
sand-cleaner jig 洗砂跳汰机
sand cleaning by hand 人工洗砂
sand cleaning door 除砂门
sand cloth 砂纸;砂布;金刚砂布
sand-coarse aggregate ratio 细粗骨料比;砂石比;砂率;砂骨比
sand coated wall 砂面墙
sand coating 型砂覆膜
sand collar 防尘垫圈;防尘挡圈
sand collector 砂收集器;沉沙池
sand-colo(u)red 沙色的
sand column 砂桩;沙柱
sand compaction 型砂紧实度
sand compaction method 砂桩加固法
sand compaction pile 挤密砂桩
sand compaction pile method 压实砂桩法;砂桩加固法;砂井加固法
sand compaction ratio 砂的挤实比
sand concentration 砂集聚
sand concrete 砂混凝土
sand conditioner 型砂配置机
sand conditioning 型砂制备;型砂处理
sand-conditioning machine 旧砂处理机
sand-conditioning plant 配砂工段
sand-conditioning system 砂处理系统
sand cone method 灌砂法;砂锥法
sand consolidation 固沙
sand container 砂箱
sand containing fine grains 含细粒砂土
sand containing lightly fine grains 微含细粒砂土
sand content 用砂量;含砂量;砂粒含量;沙粒含量
sand content ga(u)ge 泥浆含砂量计
sand content measuring 测含沙量
sand content of drilling fluid 砂浆含砂量

sand content of mud 泥浆含砂量
sand content of river 河流含沙量
sand content of wash liquid 冲洗液含沙量
sand content of well water 井水含沙量
sand control 型砂质量调整；型砂控制；治沙；沙子管理；砂型控制；沙盘调度；沙漠控制；防沙
sand control dam 防沙堤
sand control technique 防沙技术
sand conveyance 型砂运送
sand cooler 旧砂冷却器；砂冷却器
sand cooling 砂冷
sand cooling equipment 旧砂冷却装置
sand cooling plant 砂冷却设备
sand core 泥芯；喷砂芯；型型芯；砂芯
sand-cored water jacket passage 砂芯铸空的水套通道
sand count 砂层总厚的确定
sand course 沙地试验路线
sand cover 砂盖层；沙埋
sand cover of slice shape 片状沙埋
sand cover of tongue shape 舌状沙埋
sand crack 砖的砂裂（砖面出现的裂缝）
sand cracking 沙子裂解
sand crusher 研砂机；砂块破碎机
sand crystal 砂晶
sand culture 砂耕；沙基培养
sand curing （混凝土用）砂养护
sand cushion(base) 砂垫层
sand cut 磨；冲砂
sand cutter 混砂机；移动式联合混砂机；碎砂机
sand cutting 松砂（工作）；砂蚀；砂侵；拌砂
sand cutting and screening machinery 调砂及筛砂机
sand cutting machine 拌砂机
sand cutting-over 和砂；拌砂
sand cylinder 研磨筒；砂筒；砂磨筒
sand dam 砂坝
sand defense forest 防沙林
sand dehydrater 砂脱水机
sand dehydrator screw 砂脱水机螺杆
sand delta 砂质三角洲
sand density 型砂密度；型砂比重
sand deposit 砂矿床；采砂坑；沉积沙
sand depot 砂点；沙仓
sand desert 沙漠
sand devil 砂暴；沙暴；尘旋
sand dewaterer 砂脱水器
sand dewatering screw 砂脱水器螺杆
sand diameter 泥沙直径；砂粒直径
sand dike 砂岩墙；沙堤
sand disc 磨盘；研磨圆盘；抛光圆盘；砂轮
sand discharge 输沙率
sand discharge facility 输沙设施
sand displacement method 换砂法
sand distributor 撒砂机；铺砂机；砂分配器
sand dome 砂室
sand down 去光（旧漆面）；清面
sand drag 运砂机；刮砂机；砂耙
sand drain 排水砂井；砂井（渗水暗沟）；砂井排水
sand drain method 砂井排水法
sand drain vacuum method 砂井真空排水法
sand dredge(r) 挖砂机；挖砂船；采砂船
sand dredging 捞砂
sand dressing 铺砂
sand dressing machine 撒砂设备；铺砂设备；撒砂机；铺砂机
sand drier 烘砂器；烘砂炉；烘砂滚筒
sand drift 小沙丘；流沙（也称流沙）；漂沙；沙丘；沙流；沙堆
sand drift fence 漂沙栏
sand-driving wind 挟沙风；风沙流
sand drop 掉砂
sand dry 脱砂干燥；不黏砂干；（油漆的）砂干阶段
sand-drying bed 污泥干化场；砂干化床
sand-drying stove 烘砂炉
sand-drying surface 砂干表面；（指油漆面干到不黏砂）
sand dump 砂场；砂堆
sand dune 沙岗；流动沙丘；沙丘；沙垄
sand dune and sand ridge 沙丘和沙梁；沙丘和沙脊
sand dune area 沙荒
sand dune coast 沙丘海岸
sand dune fixation 沙丘固定
sand dune stabilization 沙丘稳定；沙丘固定

sand dune terrain 沙丘地带
sand dust 砂粒级配
sanded 磨光的；铺砂的；多砂的；撒砂的；打磨过的
sanded bitumen felt 砂面油毡；粘砂沥青油毡；砂面沥青油毡
sanded board 砂光板
sanded brick 砂制砖；砂砖
sanded cement grout 水泥砂浆
sanded-face shingle 磨光墙面板
sand edge 型砂挡肋
sanded grout 薄砂浆；加水水泥浆
sanded in 被砂堵塞的
sanded packed snow 撒砂的压实积雪
sanded plaster 掺砂石膏浆
sanded rail 撒砂钢轨
sanded runway 撒盐跑道；撒砂的跑道
sanded sheet 在玻璃上吹砂
sanded siding 砂闭路；堆砂禁行路
sanded up 堵砂钻孔
sanded wheel 砂轮
sand ejector 喷砂器；喷砂机；砂喷射器
sand elimination 除砂
sand eliminator 除砂设备；除砂器；除砂设备
sand embankment 砂路堤；沙堤
sandenol 合成檀香
sand enveloped with cement 灰裹砂；水泥裹砂；砂壳混凝土
sand enveloped with cement concrete 水泥混凝土裹砂；包砂混凝土
sand equivalent 砂当量；含砂当量
sand equivalent method 含砂当量法
sand equivalent shaker 砂当量振动器
sand equivalent test 含砂当量试验；砂当量试验
sander 砂轮磨光机；抛光机；喷砂装置；喷砂器；喷砂机；砂光机；撒砂器；打磨器；打磨机
sander aqueoglacial deposit 冰水沉积（物）
sanderite 二水泻盐
sand erosion 砂冲蚀
Sander's formula 桑德公式
Sanderson drying time meter 桑德森干燥时间测量仪
sanders pipe 撒砂管
sand eruption 喷砂
sand escape 泄砂闸（门）；排砂闸（门）；排沙道
sand etch 喷砂蚀面
sand expansion 型砂膨胀；沙子膨胀；砂层膨胀
sand expansion ratio 砂膨胀比
sand explosion 砂眼
sand extraction 砂提取；砂石开采
sand fabrication 人工制砂
sand face 井底面
sand-faced 砂面的
sand-faced brick 砂面砖
sand-faced clay brick 砂面黏土砖
sand face injection pressure 井底注水压力
sand facing 撒砂（罩）面
sand factor （混凝土中）含砂成分
sandfall 滑落面
sand fallout 沙的沉降
sand feed 砂填；淤砂；填砂
sand feeder 给砂机；加砂器
sand feeding device 加砂设备
sand feeding funnel 加砂漏斗
sand feeding rate 加砂率
sand fence 漂沙栏；砂栅；防沙栏栅
sand fill 填砂；砂填充；水砂充填；吹填砂
sand-filled 填（满）的
sand-filled caisson 内填砂沉箱
sand-filled cushion 填砂垫层
sand-filled drainage well 填砂渗水井；填砂排水井；砂井
sand-filled epoxy 环氧树脂砂浆
sand-filled polyester 聚酯砂浆
sand filler 砂质填缝料
sand filling 回填砂；填砂；砂填料
sand filling method 水砂充填法
sand fill reclamation 填砂围垦；吹填围垦
sand filter 砂滤池；砂滤层；砂（过）滤器
sand filter bed 砂滤床；砂滤层
sand filter blanket 砂滤层
sand filtering 砂路
sand filter layer 砂滤层；砂过滤层
sand filter trench 滤沙沟
sand filter trenches 砂滤槽

sand filtration 砂过滤；砂路；砂滤（法）
sand filtration ditch 砂滤沟
sand fineness 沙子细度；沙子（平均）粒度
sand finish 用砂修整；灰泥墙面粗糙化处理；抛光饰面；砂饰面；砂面
sand-finished brick 砂面砖
sand fixation forest 固沙林
sand flag 细粒薄层砂岩；层状砂岩
sand flash valve 排沙阀
sand flat 沙滩；沙坪
sand flat deposit 砂坪沉积
sand flo(a)tation 砂浮法
sand flo(a)tation process 尾砂浮选法
sand-float finish 粗抹面；浮砂罩面；干黏砂饰面
sand flood 沙洪；沙暴
sand flow 压砂；沙流；出砂；风沙流
sand flowability 型砂流动性
sand flow filtration 流沙过滤
sand flow method （沉管隧道用）压砂法
sand fluke 瘦尖形鲆
sand flushing 冲沙
sand-flushing canal 冲沙道；冲沙渠
sand-flush installation 冲沙设施
sand flush of bore hole 钻孔涌砂
sand flush valve 排砂阀
sand forehore 沙滩
sand for foundry 铸型用砂
sand for glass 玻璃用砂
sand for mortar 灰浆用砂
sand formulation 型砂配方
sand for raising sites 填高场地用砂；填筑用砂
sand for sett paving 小方石铺路砂
sand foundation 砂基；砂地基；砂质基础
sand foundry 铸造厂；翻砂厂
sand fraction 砂粒级配；沙子部分；砂粒组；砂粒粒组；砂粒分级；砂分
sand frame 箱圈
sand-free gravel 砂卵石
sand funnel 灌砂漏斗
sand fusion 黏砂
sand gall 沙管
sand gate 排沙闸（门）
sand ga(u)ge 量砂器
sand ga(u)ge box 量砂箱
sand glass 计时沙漏；砂玻璃
sandglass structure 沙钟构造
sand grab(bing bucket) 挖砂机抓斗
sand grading 沙子粒度分级；砂粒级配；砂粒分级
sand grading chart 沙子颗粒组成图
sand grain 泥沙颗粒；砂粒
sand grain distribution 沙子粒度分布
sand grain grade 砂粒组
sand grain meter 含砂（量）计；水中集砂器；砂粒器；砂粒计
sand-grain volume 砂粒体积
sand-grass 沙草
sand gravel 沙砾
sand-gravel aggregate 沙砾（石）集料；沙砾（石）骨料
sand-gravel-bituminous mixture 沥青沙砾混合料
sand-gravel concrete 沙砾混凝土
sand-gravel cushion 砂石垫层
sand-gravel filter 沙砾过滤机
sand-gravelly soil 沙砾类土
sand-gravel pile 沙砾桩
sand-gravel road 沙砾路
sand-gravel shoal 沙砾石浅滩
sand-gravel surface 砂石路面；沙砾面层；沙砾路面
sand grind 砂磨
sand grinder 砂轮机；砂轮磨床；砂磨机
sand grinding 砂磨机研磨
sand grinding mill 研砂机；磨砂机；碎砂机
sand grinding stone 砂磨石
sand grip 砂箱内壁凸条；持砂条
sand grit 沙砾；粗砂岩
sand gritter 铺砂机
sand-gritting 黏砂
sand grout 薄砂浆；砂浆
sand grouting 砂内灌浆
sand guard 防砂罩
sand guiding and regulating device 输沙措施装置
sand gushing in mine 矿井涌沙
sand-gypsum plaster 砂石膏灰浆
sand handling equipment 沙子输送设备

sand hardness 砂型硬度
sand-haydite concrete 陶粒砂混凝土;砂陶粒混凝土
sand hazard 沙害
sand heap analogy 堆砂模拟(法)
sand hedge 沙丘围护
sand hill 沙岗;沙丘
sand-hill region 沙丘地区
sand hitch 加砂打捞工具
sand hog 捞砂器;挖砂工人;挖沟工人;水下隧道工人;地下工程工人;沉箱工;隧道工
sand holder 砂箱
sand hole 坑点;砂眼;砂孔;沙坑
sand hole and inclusions 砂眼及夹杂物
sand hopper 漏砂斗;砂斗
sand horn 砂岬;沙嘴
sand house 储[贮]砂间
sand humidity 型砂水分
sandification 沙化
San Dimas flume 桑迪马斯计算槽
sand inclusion 夹砂;砂土包裹体
sandiness 多砂;带红的黄色
sanding 挟沙;用纸打磨;用砂打磨;(混凝土表面的)起砂;铺砂;输砂;沙子打毛;砂纸打磨;砂(纸打)光;撒砂;打砂(纸);打砂磨光;打磨;掺砂
sanding agent 打磨剂;研磨剂
sanding and polishing machine 砂磨与抛光机
sanding apparatus 用砂打磨机具;砂打机具
sanding belt 砂磨带
sanding block 打砂纸用垫块
sanding brick 刚砂石
sanding cloth 砂布
sanding disk 研磨砂纸盘;喷砂盘;砂轮;磨轮
sanding equipment 撒砂设备
sanding formation 出砂地层
sanding gear 铺砂装置;铺砂器;喷砂装置;喷砂器
sanding lacquer 磨光漆
sanding machine 铺砂机;抛光机;磨光机;砂纸打磨机;砂纸打光机;带式抛光机
sanding property 可打磨性
sanding sealer 掺打磨剂涂料;可打磨封闭底漆;嵌腻子;掺砂涂料;掺砂火漆;用密封剂;打底腻子
sanding service 撒砂服务;铺砂服务
sanding sheet 薄砂层
sanding skip 打磨遗漏
sanding substation 送电变电站
sanding surfacer 可打磨二道浆
sanding-up 填砂;铺砂;淤积
sanding vehicle 撒砂车
sand in teeter 悬浮沙
sand interceptor 截沙室;截沙器;截沙井;截沙池;拦砂设施
sand inundator (冲水的)量砂斗
sand ionization formula 沙堆模型
sandish 砂质的
sand island 沙岛
sand island method 沙岛法
sand island of river mouth 河口沙岛
sandiver 玻璃沫;玻璃浮渣
sand jack 砂箱千斤顶;砂箱墩;砂箱顶座
sand jet 喷砂处理;喷砂(器)
sand jet method 水力喷砂法
sand-jetting (method) 喷砂法
sandkey 小沙洲;小沙岛;小砂坝
sand kicker 行走式砂处理机
sand kneader 叶片式混砂机
sand laboratory 型砂试验室
sand laden mud 含砂泥浆
sand launder 砂槽
sand layer 砂层
sand layer under a sett paving 小方石路面下的砂垫层
sand leach 砂路
sand lead 撒缆头
sand ledge 砂箱内壁凸条
sand lens 扁豆状砂层
sand lens trap 砂岩透镜体圈闭
sand lift 扬砂管
sand-lime brick 砂灰砖;灰砂砖
sand-lime brick lintel 灰砂砖过梁
sand-lime brick machine 制灰砂砖机
sand-lime facing brick 灰砂面砖
sand-lime mix(ture) 灰砂混合料
sand limestone 砂灰碎块;沙质灰岩
sand limestone concrete 黄砂—石灰石混凝土

sand line 捞砂绳;砂层线;输砂管
sand-lined mo(u)ld 砂模;砂型
sand-line drum 抽筒绳卷筒
sand-line pulley 捞砂绳滑轮
sand-line sheave 抽筒吊绳滑轮
sand-line spool 捞砂绳卷筒
sand lining 砂衬
sand liquefaction (of soil) 砂土液化
sand load 含沙量;砂配重;砂负载
sand loader 装砂机
sand loam 亚砂土;砂质垆姆;砂质壤土
sand lobe 朵状砂体
sand lock 拦砂闸(门)
sand loess 砂黄土
sandlot 沙地游戏场
sand lump 砂块
sand lump breaking 砂块破碎
sand lump crushing 砂块破碎
sand manufacture 轧砂;砂的制备
sand map 砂地图
sand mass 砂堆
sand mastic 沥青砂胶
sand mastic asphalt 沥青砂胶
sand mastic method 沥青砂胶法;砂胶灌缝施工法
sand match 砂胎模
sand mat of subgrade 排水砂垫层
sand mat(tress) (混凝土养护的)砂层;砂垫层
sand maturing 型砂熟化;型砂回性;型砂调匀
sand measuring box 量砂箱
sand measuring container 量砂箱
sand method 砂层积储[贮]藏法
sand migration 沙丘移动
sand mill 混砂机;碾砂机;砂磨机
sand mist 沙雾
sand mix 型砂
sand mixer 混砂机
sand mixing machine 混砂机
sand model test 泥沙模型试验
sand moisture 沙子湿度
sand moisture content 型砂水分
sand mortar 砂浆
sand mo(u)ld 砂型;砂模
sand mo(u)ld casting 砂型铸件
sand mo(u)lded brick 砂模砖
sand mo(u)lding 砂造型;砂模
sand mo(u)ld pig iron 砂型生铁
sand mound 砂堆
sand muller 摆轮式混砂机
sand mulling 混砂
sand nozzle 砂喷嘴
sand of storm 风沙
sand-oil blender (油层水力压裂时用)砂搅拌机
San Domingo mahogany 圣多明戈红木
sand outlet 排砂口;泄砂孔
sandow 填砂支柱
sand pack 砂袋
sand packing 砂层夯实
sand paper 砂纸
sandpapering 用砂纸打磨;砂纸打磨;砂纸打光;打砂纸;砂纸摩擦
sandpapering machine 砂纸打磨机;砂纸打光机
sandpapering method 砂纸打磨法
sandpapering with water 水磨砂纸打光
sandpaper marks 砂纸磨痕
sandpaper surface 粗糙路面;砂纸状饰面
sandpaper surfacing 粗糙面层;砂皮状饰面;粗糙路面
sand particle 砂粒
sand particle bituminous concrete 砂粒式沥青混凝土
sand particle size 砂粒形状
sand parting 砂隔层
sand patch 砂斑
sand patch method 铺砂法
sand patch test 铺砂(法)试验;砂斑试验
sand path 铺砂小路
sand pattern 砂制模型
sand pear 沙梨
sand-pebble protection layer 砂卵石覆盖层
sand-pebble riverbed 砂卵石河床
sand penetration 黏砂;机械黏砂
sand percentage 细骨料含量;含砂率;砂率
sand permeability 型砂透气性;砂粒渗透性
sand pile 砂堆;排水砂桩;砂桩

sand pile method 砂桩加固法
sand piler 堆砂机
sand piling 砂桩
sand piling barge 砂打桩船
sand pillar 砂柱;沙卷
sand pin 铸型用销钉
sand pine 沙松
sand pipe 撒砂管;砂管
sand pit 沙坑;采砂场
sand placement 砂的铺筑
sand plain 砂原
sand plant 烘砂车间;烘砂设备;轧砂设备;砂场;采砂场
sand plaster 灰砂粉刷;细砂粉刷;水砂粉饰
sand plate (夯实土用的)夯板;砂钢板
sand plow 刮砂板
sand plug 砂堵
sand pocket 混凝土中缺乏水泥的砂团;砂窝;砂囊(混凝土中缺乏水泥的砂团)
sand point 砂点
sand polished thin section method 砂光薄片法
sand polishing 打砂磨光
sand pond 磨料粉砂沉积池
sand pool 沉砂池
sand poor in clay 少砂黏土
sand porosity 型砂空隙度
sand precoating 型砂覆膜
sand preparation 型砂制备
sand preparation plant 砂处理装置
sand preparing machine 砂处理设备
sand preparing system 砂处理系统
sand pressure 砂层压力
sand pressure control 砂层压力控制
sand prevention 防砂
sand processing machine 铺砂机;沙子加工机
sand processing plant 沙子加工车间;沙子加工设备
sand producer 出砂油井
sand producing formation 出砂地层
sand producing machine 沙子制备机
sand product 砂质产品
sand production 砂料开采;出砂
sand projection machine 抛砂机
sand proportioning box 配砂箱
sand protecting dam 防砂坝
sand protecting dike 防沙坝
sand protecting plantation 防沙(植)林
sand protection facility 防砂设施
sand protection plantation 防沙林
sand pulp 砂浆
sand pump 吸砂泵;扬砂泵;砂水泵;泥沙泵;泥泵;泥粒级泵;砂浆泵;砂泵;砂粒级泵;出砂筒;抽砂泵
sand pump bucket dredge(r) 链吹挖泥船
sand pump container 砂泵容器
sand pump dredge(r) 吸扬(式)挖泥船;装有吸砂泵的挖泥船;采砂船;泵挖泥船;砂泵采掘船;装有泥泵的挖泥船;砂泵挖泥船;抽砂挖泥船
sand pump grab 捞砂筒打捞爪
sand pumping 捞砂【岩】
sand pumpings 砂泵岩样
sand pump pulley 抽砂泵滑轮;扬砂泵滑轮
sand pump sampler 砂样采取器;砂泵取样器
sand pump valve 砂泵阀
sand putty 含砂油灰
sand quarry 采砂场
sand race track 铺砂跑道
sand rammer 气动砂冲子;气动冲砂器;抛砂机型砂捣击锤;捣固机
sand ratio 砂率
sand recirculation filter 循环砂滤池
sand reclaimer 旧砂再生装置
sand reclaiming machine 沙子回收机
sand reclaiming plant 砂回收设备
sand reclaiming screw 沙子回收螺旋装置
sand reclamation 型砂再生
sand reconditioning 旧砂处理
sand recovery 沙子回收
sand recovery machine 沙子回收机
sand recovery plant 砂回收设备
sand recovery screw 沙子回收螺旋装置
sand recycle system 沙子循环系统
sand reef 拦江沙;沙埂;沙坝;砂礁;砂脊
sand reel 泥泵绳轮;轮
sand-reel reach 抽筒卷筒摩擦离合式控制杆
sand removal 排沙;除沙;冲沙

sand removal structure 排沙构筑物
sand remover 除砂器
sand replacement method 灌砂法;换砂法
sand resources 沙资源
sand ribbon 砂带
sand riddle 砂筛
sand ridge 砂脊;沙丘线;沙脊;沙埂;沙波
sand ripple 砂波纹;沙波痕;沙面波纹;沙波
sand ripple motion 沙波运动
sand ripple movement 沙波运动
sand river 沙河
sand road 沙路
sand rock 泡砂岩;泡砂石;砂岩;砂石
sand roll 软面轧辊
sand roller mill 碾轮式混砂机
sand rope 捞砂绳
sand rubbed 磨光(面);砂磨(面)
sand-rubbed finish 砂磨饰面;砂磨修整
sand rubbing 砂磨;屋面油毡铺装
sand running-down 砂土流失
sand sagebrush 沙蒿
sand sample 砂样(品)
sand sampler 取砂器;采砂(样)器
sand sausage 沙枕
sand-sawed finish 颗粒纹理石面;砂锯石面
sand scale 砂垢
sand scouring 砂向炮刷
sand scouring flow 走沙水流
sand scraper 刮砂板;混砂刮板
sandscratcher 水手
sand screen 筛砂器;筛分器;砂筛
sand screening 脱砂
sand screening machine 筛砂机
sand screw 送砂螺旋(机)
sand scrubber 旧砂干法清洗装置
sand sea 纯砂沙漠
sand seal 砂封(层)
sand sealing 砂填
sand seal plate 砂封板
sand seam 砂层
sand sediment pipe 沉砂管
sand sediment trap 沉砂井;沉砂槽
sand sediment tube 沉砂管
sand separation 砂的筛分
sand separator 型砂分选装置;砂石分选机;砂粒分离器;分砂器
sand settling tank 砂沉淀池;沉石箱;沉沙罐
sand shadow 沙影;背风堆沙区
sand shaker 筛砂器;筛砂机;筛分器;筛砂
sand-shale ratio 砂页岩比(率);砂页比
sand shallow 砂影堆积
sand shallows and holms 浅滩和沙洲
sand sheave 捞砂绳滑轮
sand sheet 小砂原;砂席;薄砂层;小砂层
sand shell 贝壳砂
sand shifter 摆动筛
sand shifting 砂移动
sand-shifting control forest 固沙林
sand shoal 浅沙滩;砂质浅滩
sand shop 砂处理工段
sand shovel 砂铲
sand shows 砂岩油苗
sand sieve 砂筛
sand sieving machine 筛砂机
sand sifter 筛砂器;筛砂机;筛分器;砂筛
sand-sifting 选砂
sand silo 储[贮]砂筒仓
sand silting 砂泥淤积
sand-silt-mud 砂—粉砂—泥
sand sink 砂沉法
sand size 砂粒大小
sand-size analyzer 泥沙颗粒分析器
sand skeleton 砂土骨架
sand skin 砂皮
sand slime separator 砂泥分离器
sand slinger 抛砂造型机;抛砂机;投砂器
sand slinging machine 投砂器
sand slope 砂坡
sand sluice 泄沙闸(门);排沙闸(门);冲沙闸(门)
sand sluicing 冲沙
sand sluicing canal 冲沙道
sand sluicing gate 泄沙闸(门);冲沙闸(门)
sand slurry 砂浆;砂泥浆

sand smoothing 打砂磨光
sand snow 砂性雪
sand soil 含砂土(壤);砂质土(壤);砂土
sand soil foundation 砂土地基
sand-soil-peat matures 砂质泥炭土
sand source 砂源
sand specimen 砂样
sand spit 沙洲;沙嘴;砂岬
sand splay 粗砂决口扇
sand spout 砂喷注;沙柱;尘卷风
sand sprayer 喷砂机;撒砂器
sand spraying 撒砂(法)
sand spraying and water oozing 喷砂冒水
sand spraying device 喷砂器;喷砂机;撒砂设备
sand spraying gear 喷砂装置
sand spraying machine 撒砂机
sand-spreader 撒砂机
sand spreading 铺砂;撒砂法
sand spreading machine 铺砂机;撒砂机
sand-spun casting 砂型离心浇注
sand-spun pipe 砂模旋制管
sand-spun process 砂型离心铸造法
sand-spun tube 砂模旋制管
sand stabilization forest 固沙林
sand stabilizer 型砂稳定剂
sand stem 砂炮泥
sand stemming 砂质炮泥;(炸药之间的)夹砂层
sand-stemming between the charge of explosives 炸药之间的砂夹层
sand sticking 砂疤
sandstone 砂岩;砂石
sandstone ashlar 砂岩方琢石
sandstone ball 砂岩球
sandstone band 砂岩夹层
sandstone concrete 砂石混凝土
sandstone cube 砂岩方块
sandstone curb 砂岩路边石;砂岩路缘石
sandstone curtain(wall) 砂岩幕墙
sandstone deposit 砂岩矿床
sandstone dike 砂岩(岩)墙
sandstone disc 砂轮
sandstone facies 砂岩相
sandstone facing 砂岩覆盖层
sandstone for brick-tile 砖瓦用砂岩
sandstone for cement burden 水泥配料用砂岩
sandstone for glass 玻璃用砂岩
sandstone for metallurgic(al)use 冶金用砂岩
sandstone grit 粗角粒砂岩
sandstone inclusion 砂岩包裹体
sandstone kerb 砂岩路边石;砂岩路缘石
sandstone meal 砂岩粉
sandstone(paving)sett(铺路用)砂岩小方石
sandstone pipe 砂岩筒
sandstone plum 砂岩碎石;(加于毛石混凝土中用)砂岩石
sandstone portal 砂岩隧道门;砂岩桥门
sandstone powder 砂岩粉末
sandstone quarry 砂岩采石场
sandstone reservoir 砂岩储集层
sandstone rock asphalt 砂岩(地)沥青
sandstone rubble 砂岩毛石;砂岩乱石
sandstone rubble curtain wall 砂岩乱石幕墙
sandstone slab 砂岩板
sandstone slab floor cover(ing) 砂岩地面板
sandstone soil 砂岩土壤
sandstone streak 露砂条形麻面
sandstone texture 砂岩结构
sand storage 砂库
sand storage bin 储[贮]砂斗
sand storage case 砂箱
sand storage hopper 储砂斗
sand storage of water 砂地储水
sand storage pit 储沙坑
sand storm 尘暴;风沙;大沙暴;沙漠风暴
sand strainer core 滤渣砂芯
sand stratum 砂岩层;砂层
sand streak 砂脊;砂条;砂岩夹层;(混凝土表面的)起砂;(混凝土中的)露砂条形麻面;露砂;砂纹(混凝土跑浆造成的纹痕)
sand streaking 砂岩夹层;砂纹;起砂
sand stream 砂流
sand stream abrasion method 砂流磨耗法
sand strength 型砂强度;造型材
sand strength testing machine 型砂强度试验机

sand strip 砂带;挡砂条
sand-struck brick 脱坯砖
sand sucker 吸泥(式)挖泥船;吸泥机;(吸水质土用的)吸泥(砂)机;砂泵;采砂船
sand suction dredge(r) 吸扬采砂船
sand sugar 砂糖
sand surface asphalt felt 砂面油毡
sand-surfaced 表面用砂磨光的
sand surfacing 磨砂表面
sand suspension 砂悬浮体
sand sweeping 回砂
sand sweeping equipment 回砂机
sand swell 沙隆
sand system scrubber 砂系洗涤器
sand table 矿砂摇床;沙盘;除沙盘
sand table exercise 沙盘作业
sand-tamped well 砂填实的钻孔;砂填实的井
sand tank 沉砂池
sand test 型砂试验
sand tester 含砂量测定器
sand test specimen 型砂试样
sand textured building coating 砂壁状建筑涂料
sand thickener 沙子沉降槽;沙子稠化器
sand thickness 含油层厚度;砂层厚度
sand thrower 抛砂机
sand throwing machine 抛砂机
sand tornado 沙龙卷
sand-total aggregate ratio 含砂率;砂与集料总量比;砂率
sand toughness number 沙子韧性读数
sand tower 砂滤塔
sand track 上砂线;砂塔;堆砂轨道
sand transport 沙子采运;沙子运输
sand transportation truck 运砂车
sand transport meter 输砂计
sand transport pipe 输砂管
sand trap 拦砂坑;拦沙阱;拦沙井;拦沙设施;截沙井;沉沙池;集沙器;集沙坑;砂槽;沙阱;沙槽;撒砂器拦沙井;除沙盘;沉沙坑;沉沙井;沉沙池;防沙存水弯
sand trap dam 防沙堤
sand trap grate 截沙栅
sand trap model 砂槽模型
sand trap scour gate 沙阱(式)冲沙闸(门)
sand underlay 砂铺地基层;砂铺底;砂垫层
sand up 被泥沙堵塞;用砂堵塞
sand-up well 砂堵钻孔
sandur 冰水沉积平原
sandur deposit 冰水平原沉积
sand valve 撒砂阀
sand valve connection 砂阀接头
sand valve stem 砂阀杆
sand vegetation 砂地植被
sandveld 沙草原
sand vent 喷砂口
sand volcano 砂火山
sand wash 砂型涂料;砂堆积;地表沙砾层;冲砂;砂堆
sand washer 洗砂设备;洗砂机
sand washer plant 洗砂场
sand-wash filter 连续洗砂过滤池
sand washing 冲砂;喷砂清洗;喷砂清理(法)
sand-washing equipment 洗砂设备
sand-washing machine 洗砂机
sand-washing plant 洗砂场;洗砂厂
sand washover 冲砂
sand-watch structure 沙钟构造;沙漏构造
sand-water separator 砂水分离器
sand wave 沙浪;沙波
sand wave formation 沙波的形成
sand wave movement 沙波运动
sandwave system 沙波系
sand wedge 沙楔
sand weighing barrel 黄沙称筒
sand weir 砂袋围堰
sand well 出砂油井
sand well foundation 砂井基础
sand wheel 泥砂轮;抽砂筒砂轮
sand wheel band 加宽轮缘
sandwich 结构夹芯板;夹芯(式);夹芯结构;夹层的;层状(结构);层状构造;防水夹层
sandwich arrangement 交错重叠布置;砂料堆集场;堆砂场;层叠布置
sandwich beam 组合板梁;夹芯结构梁;夹合梁

夹层梁;合板梁;多层叠合梁;叠层梁;层结梁
sandwich belt conveyer 夹皮带输送机
sandwich belt elevator 夹皮带提升机
sandwich belt unloader 夹皮带卸船机
sandwich blender 夹层混合机
sandwich board 夹芯板;夹层板;胶合板
sandwich braze 夹芯钎接;夹芯焊接;夹层钎焊;夹层焊接
sandwich cable 叠层钢索
sandwich channel 夹层槽
sandwich coat 复层涂层;夹芯涂层
sandwich compound 夹层化合物;夹芯化合物
sandwich concrete 夹层混凝土;分层混凝土
sandwich construction 夹层建造法【道】;蜂窝夹芯胶接结构;层状结构;夹层板结构;夹芯结构;夹合结构;夹层构造
sandwich construction method 夹层建筑法
sandwich construction panel 夹层结构板
sandwich corrosion 夹层腐蚀
sandwich course 夹芯层
sandwich covering 夹层蒙皮
sandwich cylinder 夹层柱壳
sandwich damp course 复合防潮层
sandwich digit 中间数位;中间数(字)
sandwiched girder 组合大梁;木材钢板夹层梁
sandwich film 层状胶片
sandwich fire-wall 夹层防火壁
sandwich foam layer 夹芯泡沫层
sandwich frame 双构架
sandwich girder 夹合梁;夹层大梁;夹合梁(由方木与金属板组成)
sandwich glass 层压玻璃;夹层玻璃
sandwich heating 双面加热
sandwich hologram 重叠全息照相法;层状全息图
sandwich hologram interferometry 层状全息干涉测量术
sandwich inclusion compound 夹层包合物
sandwiching 夹层铺料
sandwich injection 夹层注射;夹层灌入
sandwich insulant panel 夹芯层隔热板
sandwich irradiation 夹层辐照
sandwich kiln 夹层窑;扁形宽截面窑
sandwich layer 夹层;芯层;夹芯
sandwich lease 转租
sandwich like compact 夹芯状坯块
sandwich manifold 积层式油路板
sandwich material 芯材;夹合材料;夹层材料
sandwich member 夹芯构件
sandwich method 夹层(铺筑)法
sandwich of spun glass 玻璃纤维夹层
sandwich panel 夹层预制板;夹合板;夹层板
sandwich plate 多层(垫)板;夹合板;夹层板;夹芯板
sandwich-plate method 夹层板方法;多层板法
sandwich pocket 内贴袋
sandwich process macadam 碎石路面的夹层铺筑法
sandwich ribbon fibre 层状长条纤维
sandwich rolling 夹芯轧制;夹层碾压;夹层滚轧;多层板轧制法;包层轧制
sandwich rolling process 复合轧制法
sandwich roof(ing)slab 夹层屋面板
sandwich shell 夹层薄壳
sandwich skin 夹层蒙皮
sandwich steel 夹芯钢
sandwich structure 夹芯构造;夹层构造;多层叠合结构
sandwich structure honeycomb 蜂窝夹层结构
sandwich system (泥结碎石路面等的)夹层铺筑体系;夹层铺筑法
sandwich technique 夹层技术
sandwich type 交错多层形;夹层式;夹层的
sandwich type cell 多层型电池
sandwich type cement-bound macadam 夹层式水泥结碎石路
sandwich type concrete road 夹层式(水泥)混凝土路
sandwich type construction 夹层结构
sandwich type element 多层元件;多层结构元件
sandwich type panel 夹层式镶板;夹芯板
sandwich type premoulded joint 夹层式预缩缝
sandwich type rubber bearing 夹层式橡胶支座
sandwich type shell 夹层壳
sandwich type wall panel 夹芯墙板
sandwich unit 夹层单元

sandwich wall 夹层墙;夹芯墙
sandwich wall panel 夹层墙板
sandwich welding 夹紧冷焊
sandwich winding 交错(多层)绕组;盘形绕组;盘式绕组;分层绕组
sandwich wound coil 盘式线圈
sand wick 袋(装)砂井
sand winning 采砂
sand with clay 砂混黏土
sand worm 砂栖蠕虫
sandy 含砂的;砂质的;多砂的
sandy aggregate 砂质集料;砂质骨料
sand yard 砂料堆集场;堆砂场
sandy beach 砂质海滩;沙滩
sandy bed 砂质河床
sandy bentonite 长石斑脱岩
sandy blanket 冲积砂层;砂罩面
sandy borrow 取沙坑
sandy bottom 砂质底;砂底
sandy braided-stream deposit 砂质辫状河沉积
sandy breccia 砂质角砾岩
sandy casting cement 型砂水泥
sandy cement 砂质水泥
sandy chert 砂质燧石
sandy chimney 砂质排水竖井
sandy clast 砂屑
sandy clastic texture 砂屑结构
sandy clay 亚黏土;含砂黏土;砂(质)黏土
sandy clay impervious core 砂质泥土防渗墙
sandy clay loam 砂质亚黏土;砂质黏壤土
sandy clay with a trace of gravel 混少量砾石的砂质黏土
sandy clinker 砂质熟料
sandy coast 多砂海岸;砂质海岸
sandy conglomerate 砂质砾岩
sandy culture 砂培(法)
sandy cushion 砂垫层
sandy deposit 砂质沉积(物);砂质沉淀物
sandy desert 沙漠;砂质荒漠
sandy desert soil 砂质漠壤土
sandy-earth tunnel 砂土地层隧道
sandy filling 砂质充填
sandy filter 砂质过滤层
sandy formation 砂质层系
sandy foundation 砂质基础
sandy gravel 含砂砾土;含砂砾石;砂质砾(石);砂砾石
sandy gravel aggregate 含砂砾石集料;含砂砾石骨料
sandy gravel cushion 砂砾(石)垫层
sandy gravel stratum 砂砾层
sandy ground 砂地(基);砂土;砂砾地
sandy harbo(u)r 砂滨港
sandy kaolin deposit 砂质高岭土矿床
sandy land 沙地;沙田
sandy land fill 沙填土地
sandy limestone 砂质石灰岩;砂质石灰石;砂质灰岩
sandy limestone chip(ping)s 砂质石灰石屑
sandy loam 轻亚粘土;砂(质)壤土
sandy loess 砂质黄土
sandy marl 砂质泥灰岩
sandy mobile bed 砂质动床;砂粒流动的河床
sandy mud 砂质(淤)泥
sandy ore 砂质矿石
sandy phosphorite 砂状磷块岩
sandy plain 砂原
sandy pocket 砂窝
sandy sediment 砂质宙积物;砂质沉积物;沙质沉积物
sandy shale 砂质页岩
sandy shoal 砂质浅滩
sandy shoal blasting 砂质浅滩爆破
sandy shore 砂质海滨;砂质海岸;砂质滨
sandy silt 砂质淤泥;砂质粉土
sandy size 砂粒粒径
sandy-skeletal 粗骨砂
sandy slate 砂质板岩
sandy soil 含砂土;砂质土(壤);砂性土;砂土;砂类土
sandy soil vegetation 砂土植物群
sandy subgrade 砂质路基;砂质地基
sandy waste(land) 沙荒
sandy well 出砂油井
sand zone 砂区

saneroite 杉硅钠锰石
Sanford-Day car 桑福特—迪型底卸式矿车;底卸式矿车
sanforizer 预缩整理机
sanforizing 防缩处理;防松处理
sanforizing agent 防缩剂
Sangamon 散加蒙间冰期【地】
sanga(r) (原始的)木桥
sangging 粗修表面
Sangtean gneiss 桑干片麻岩(太古界)
sanguine 红润的;血红色;赤铁矿
sanguisorba 地榆
sanidaster 板星
sanidine 透长石;玻璃长石
sanidine feldspar 玻璃长石;透长石
sanidine nephelinite 透长石霞岩
sanidine trachyte 透长石粗面岩
sanidinite 透长岩
sanidinite facies 透长岩相
sanies 腐液
Sanio's beam 萨列阿横梁
Sanio's law 萨列阿法则
sanitarial district 疗养区
sanitarian 公共卫生学家;卫生学家;保健及公共卫生工作者
sanitarium 疗养所;疗养院
sanitary 正体的卫生单间;卫生的
sanitary accommodation 卫生设备
sanitary analysis 卫生分析
sanitary and anti-epidemic affair 卫生防疫工作
sanitary and medical equipment 卫生医疗设备
sanitary apparatus 卫生器具
sanitary appliance 卫生设备
sanitary articles 卫生物件;卫生物品
sanitary authority 检疫部门
sanitary bacteriology 卫生细菌学
sanitary base 瓷砖转角的凹弧形线脚
sanitary board 卫生管理部门
sanitary building 卫生所
sanitary building block module 卫生模数;公共厕所结构单元
sanitary building sewer 房屋生活污水管
sanitary chemical analysis 卫生化学分析
sanitary chemical analysis of sewage 污水的卫生化学分析
sanitary-chemical feature 卫生化学特征
sanitary chemistry 卫生化学
sanitary china 卫生瓷
sanitary clearance 环境卫生清理
sanitary condition 卫生条件
sanitary conveniences 卫生设备;厕所
sanitary corner (为保持清洁而做成的)光面墙角
sanitary corner cast(ing) 壁角式卫生器铸件
sanitary corner pressing 卫生设备角隅边条压制
sanitary cove component 踢脚板元件
sanitary cove heating 踢脚板采暖
sanitary cove member 踢脚板部件
sanitary cove radiator 踢脚板暖气片
sanitary cove unit 踢脚板单元
sanitary criterion 卫生标准
sanitary cross 卫生四通;三承四通;三承十字管
Sanitary Custom-house Regulations 海关卫生条例
sanitary device 卫生器具
sanitary disposal 卫生处理
sanitary distance 卫生距离
sanitary door 光面门
sanitary drainage 卫生排污工程;生活污水排放系统;卫生排污系统;生活污水排泄系统;生活排污
sanitary drainage system 卫生污水排泄系统
sanitary drainage works 卫生排污工程
sanitary drinking fountain 卫生饮水(喷)泉
sanitary drinking water 卫生饮用水
sanitary earthenware 卫生陶器
sanitary engineer 环境卫生工程师
sanitary engineering 环境卫生工程;卫生工程(学)
sanitary equipment 卫生设备;卫生洁具
sanitary facility 卫生用品;卫生设施
sanitary faience 彩瓷卫生设备
sanitary fill 垃圾填埋;垃圾堆(积)场;垃圾堆
sanitary fireclay ware 卫生洁具
sanitary fitments 卫生配件;卫生用具
sanitary fittings 卫生设备(配件)
sanitary fixture 卫生设施;卫生器皿;卫生器具;卫

生活具配件;卫生设备(配件)
sanitary incinerator 卫生焚化炉
sanitary inspection 卫生检查;卫生监督
sanitary inspection certificate 卫生检验书
sanitary installation 卫生装置;卫生设备
sanitary landfill 垃圾填土;垃圾填坑;垃圾堆(积)场;卫生(土地)填埋;卫生填池;卫生垃圾堆
sanitary landfill density 卫生掩埋场
sanitary landfilling 卫生填土
sanitary landfill liner 卫生掩埋场不透水层
sanitary landfill method 卫生填埋法
sanitary line 卫生管道
sanitary measure 卫生措施
sanitary microbiology 卫生微生物学
sanitary napkin 卫生棉
sanitary napkin disposal 卫生纸废弃场
sanitary of water source 水源卫生
sanitary pipe 卫生水管
sanitary pitting 倾倒垃圾
sanitary plumbing 卫生工程管理;卫生设备;卫生管道;污水管道
sanitary plumbing equipment 卫生上下水道设备;卫生水工设备
sanitary plumbing fixtures 卫生装置零件(陶瓷)
sanitary porcelain 卫生陶瓷;卫生瓷
sanitary pottery 卫生陶瓷;卫生洁具陶瓷
sanitary pottery fittings 卫生(洁具)陶瓷配件
sanitary principle 卫生原则
sanitary privy 卫生厕所
sanitary protection of water source 水源(环境)卫生防护
sanitary protection zone 环境卫生防护区;卫生防护带;卫生保护区
sanitary protective range 卫生防护区
sanitary protective zone of groundwater 地下水卫生保护区;地下水卫生保护带
sanitary provision 卫生设备
sanitary pump 海水泵;卫生泵;生活污水泵
sanitary quarantine 卫生检疫
sanitary refuse collection 垃圾收集
sanitary refuse collector 垃圾收集器
sanitary regulation 卫生条例;卫生法规;卫生规则
sanitary requirement of illumination 照明卫生要求
sanitary requirement of waste residue disposal 废渣处理卫生要求
sanitary rule 卫生规则
sanitary science 公共卫生学;环境卫生学
sanitary sewage 生活污水;厕所污水
sanitary sewage pipe 卫生工程的污水管子
sanitary sewer 下水道;污水管(道);卫生污水(管)道;生活污水管(道)
sanitary sewer design 污水管道设计
sanitary sewer manhole 下水道检查井;污水管道窨井;生活污水窨井;生活污水窨井
sanitary sewer network 卫生工程污水管网
sanitary sewer system 生活污水(管道)系统
sanitary shoe 凹形线脚
sanitary sources of pollution 生活污染源
sanitary standard 卫生标准
sanitary standard for drinking water 饮用水卫生标准;生活饮用水卫生标准
sanitary standard of warming 采暖卫生要求
sanitary stoneware 卫生陶器;卫生炻器
sanitary stop 卫生止门件
sanitary supervision 卫生监督
sanitary survey 公共卫生情况调查;环境卫生调查;卫生调查;水污染调查
sanitary system 卫生管系
sanitary tank 盥洗水柜;卫生水柜
sanitary tee 卫生三通;双承丁字管
sanitary tipping 倾倒垃圾
sanitary treatment 卫生处理
sanitary tub 浴槽
sanitary unit 卫生设备
sanitary unitized unit 卫生间建筑模数;卫生组合单元
sanitary value of soil 土壤卫生值
sanitary ventilation 卫生通风
sanitary wares 卫生陶瓷;卫生清洁用具;卫生器皿;卫生器具;卫生洁具;卫生陶器
sanitary waste 卫生设备排出的废物;生活废物
sanitary waste water 生活废水;卫生(设备)废水;厕所污水
sanitary water 卫生的水

sanitary well protection 水井卫生保护
sanitary works 卫生工程
sanitary zone for water source 水源环境卫生区
sanitation 下水道设备;盥洗设备;环境卫生;清洁;卫生
sanitation and anti-epidemic station 卫生防疫站
sanitation cutting 卫生伐
sanitation engineer 公共卫生工程师
sanitation engineering 公共卫生工程(学)
sanitation equipment 卫生设备
sanitation facility 卫生设备;卫生设施
sanitation felling 合理地砍伐;卫生伐
sanitationman (维护城市环境卫生的)清洁工
sanitation method 卫生方法
sanitation of bath-room 浴室卫生
sanitation of buildings 房屋卫生设备;建筑卫生
sanitation of spring water 泉水卫生
sanitation of swimming pool 游泳池卫生
sanitation of well water 井水卫生
sanitation service fee 清洁卫生费
sanitation store 防腐库
sanitation system 下水道(设备)系统;卫生系统
sanitation use of water 卫生用水
sanitation worker 清洁工(人)
Saniter process 萨尼特脱硫法
sanitization 卫生处理
sanitize 卫生处理
sanitizer 卫生消毒剂;卫生洗涤剂
sanitizing agent 卫生消毒剂;卫生洗涤剂
sanitizing compound 杀菌剂
sanit sanitation 卫生设备;环境卫生
San Juan 圣胡安淡褐色硬木(产于中美洲)
sanjuanite 水磷铝矾
sanmartinite 钨锌矿
sannaite 霞闪正煌岩;棕闪正煌岩
Sanodar 萨诺达油毛毡
Sanoisian(stage) 散诺阶【地】
sanosuire 盐质泻湖
sansan 云冠
sansar 桑萨风
Sanskrit script 梵文
Sanson Flamsteed projection 桑森弗兰斯蒂投影法
Sanson projection 桑森投影法
sansoulres 湿冲积土
sans serif 灯芯体
Santa Ana 桑塔阿那风
santaclaraite 羟蔷薇辉石
santafeite 针钒钠锰矿
Santa George basin 圣乔治盆地
santal 檀香木
Santa Lawrence geosyncline 圣劳伦斯地槽
santalol 白檀油烯醇
Santalum album 檀香(木)
Santa Maria 圣玛丽亚深条纹的红色硬木(产于中美洲)
santanaite 黄铬铅矿
Santa Rosa storm 圣罗沙风暴
santite 水硼钾石
santobrite 五氯酚钠(抗真菌化学溶液)
santocel 硅酸瓷
Santonian 桑托阶(晚白垩世纪)【地】
Santopour 散陶普尔(一种降凝添加剂)
Santorin ash 圣多伦火山灰
Santorin cement 圣多伦水泥
Santorin earth 火山灰土;圣多伦土
Santorini's duct 山多利尼管
Santos constant 桑托斯常数
Sanu sea 萨怒海
sap 风化岩石
sapanthracite 腐泥无烟煤
sapanthracon 腐泥烟煤
sapanthracon stone 含油页岩
sap displacement 树液置换处理
sapele 红木;萨佩莱木
sapele mahogany 香味桃花心木
sap green 浅橄榄绿;浓黄绿色;树汁色;暗绿色
sap gum 白木质橡胶树
sap-head 坑道头
saplier 打包粗布
sapling 小树;树苗
sapling forest 幼龄林
Sapodilla 萨波蒂拉装饰用硬木(产于洪都拉斯)
saponated cresol 煤酚皂
saponated cresol solution 煤酚皂溶液

saponated petroleum 皂化石油
saponide 合成洗涤剂
saponifiability 皂化性
saponifiable 可皂化的
saponifiable content 可皂化物含量
saponifiable matter 可皂化物
saponifiable oil 皂化油
saponifiable substance 可皂化物
saponification 皂化(作用)
saponification agent 皂化剂
saponification column 皂化塔
saponification degree 皂化率;皂化程度
saponification equivalent 皂化当量
saponification flask 皂化瓶
saponification number 皂化值;皂化系数;皂化值;皂化价
saponification of fats 油脂皂化
saponification rate 皂化(速)率
saponification ratio 皂化率
saponification resistance 皂化阻碍;耐皂化性
saponification risk 皂化危险
saponification test 皂化试验
saponification value 皂化值;皂化价
saponified acetate 皂化醋酸纤维
saponified emulsifier 皂化乳化剂
saponified oil 皂化油
saponified oleine 皂化的甘油三油酸酯
saponified vinsol resin 松香皂热塑性树脂
saponified wood resin 皂化木质树脂
saponifier 皂化器;皂化剂
saponify(ing) 皂化
saponifying agent 皂化剂
saponin wastewater 皂素废水
saponite 皂石;滑石粉
Sapote 萨波脱各种花色硬木(产于中美洲)
Sapoznikov's penetrometer test 胶质层指数测定
sappare 蓝晶石
sapparire 蓝晶石
sap peel 边材剥皮
sapper 挖掘器;挖掘者;坑道工兵
sapphire 蔚蓝色的;蓝宝石;青玉
sapphire bearing 蓝宝石轴承
sapphire cap 蓝宝石帽
sapphire deposit 蓝宝石矿床
sapphire filament strengthener 蓝宝石纤维增强剂
sapphire jacket 蓝宝石套
sapphire point 蓝宝石轴承
sapphire quartz 深蓝色石英
sapphire spinel 蓝宝石尖晶石
sapphire substrate 蓝宝石基片;蓝宝石衬底
sapphire transmittancy 蓝宝石透射比
sapphirine 青玉色的;像青玉色的;像蓝宝石的;假蓝宝石
sapphire quartz 蓝石英
sapping 基蚀
sap pressure 液压
sapregen 腐生生物
saprist 高分解有机土
saprium 腐生植物群落
saprobe 腐生菌;污水生物;腐生生物
saprobia 污水生物
saprobic animal 污水动物
saprobic classification 污水生物分类(法)
saprobic index 污水生物指数
saprobic system 污水生物系统
saprobic zone 污水生物带;污染带;腐生生物带
saprobien classification 污水生物分类(法)
saprobien system 污水生物系统
saprobiont 腐生生物
saprobiotic 腐生的
saprocol 硬腐泥;灰质腐泥
saprocollite 胶泥煤;灰质腐泥煤岩
saprodil 腐泥软煤
saprodite 腐泥亚烟煤;腐泥褐煤
saprofication stage 腐泥化阶段
saprofization 腐泥化作用
saprogen 死木寄生菌
saprogenous ooze 腐生软泥
saprolite 残余物;腐泥土
saprolith 残余土
sap-roller 坑道滚筒
sapromixite 藻煤;含藻煤
sapromixtite 微腐泥混合煤
sapropel 腐殖质(淤泥);腐(殖)泥
sapropel-clay 腐泥黏土

sapropelic 腐泥的
sapropelic anthracite 腐泥无烟煤
sapropelic bituminous coal 腐泥烟煤
sapropelic clay 腐泥黏土
sapropelic coal 腐泥煤;腐泥煤
sapropelic groudmassinite 腐泥基质体
sapropelic-humic type 腐泥腐殖型
sapropelic type 腐泥型
sapropelic type source rock 腐泥型源岩
sapropelinite group 腐泥化组
sapropelite 泥煤;残余土;腐泥煤
sapropel-peat 泥炭质腐泥
sapropel putrid ooze 腐殖淤泥(沼池)
sapropetitie 腐泥岩
saprophage 腐蚀;腐生物
saprophile 适腐动物
saprophyte 腐生植物;典型腐生菌;腐生植物;腐生菌
saprophyte chain 腐生链
saprophytic 腐生的
saprophytic bacteria 腐生(细)菌
saprophytic form 腐生型
saprophytic fungus 腐生真菌
saprophytic nutrition 腐生营养
saprophytic phase 腐生阶段
saprophytic plant 腐生植物
saprophytic transitional infection 腐生过渡侵染
saprophytism 腐生现象;腐生(生活)
saproplankton 污水浮游生物
sapropsammite 砂质腐泥
sap rot 干枯(指木材腐朽);外部腐朽;腐木;边材腐朽
saprovitrinite 腐泥镜质体
saprovitrite 微腐泥镜煤
saproxylobios 死木生物
saprozoite 腐生动物
sap stain 边材变色;木材变色
sap vesicle 液泡
sap wood 边材;边料;白木质;液材
sapwood forest 边材树
sapwood rot 边材腐朽
sapwood sleeper 边材木枕
sapwood tree 边树木;边材树(木)
sarabauite 硫氧锑钙石
Saracenic arch 马蹄形拱;萨拉森式拱
Saracenic architecture 萨拉森式建筑
saran 赛纶树脂;偏氯纶树脂
Sarcina acidificans 酸化八叠球菌
sarco 石棺
sarcocyte 外中层
sarcolite 肉色柱石
sarcolysin 苯基丙氨酸氮芥
sarcophagus 石棺;雕刻精美的石棺;希腊石灰石
sard 肉红玉髓
sardachate 肉红玉髓
sardine 肉红玉髓;沙丁鱼
Sardinian orogeny 撒丁运动
sardonyx 缠绕玛瑙
Sargasso Sea 马尼藻海
Sargent 萨金特钢板(一种木底衬的钢板)
Sargent cycle 萨金特循环
sarin 沙林
sark 衬垫物
sarking 屋面衬板;衬垫层;衬垫材料;薄镶板;油毡衬垫;望板
sarking board 望板;屋面衬板;屋面望板;衬垫板
sarking felt 油毡衬;油毛毡衬(层);屋面衬毡
sarking felt underlining felt 屋面衬毡
sarkinite 红砷锰矿
Sarma method 萨尔玛法
Sarmatian(stage) 萨尔玛特阶【地】
sarmientite 砷铁矾
sarn 石阶;石路面
saropide 斜磷锰铁矿
saros 萨罗斯周期(日食和月食出现周期)
sarrasine 防御性铁栅吊门;防御性木栅吊门
sarsen 砂岩漂砾;砂岩怪石;残留岩体;砂岩遗址(史前建筑物遗址)
sarsen stone 残留岩体
Sartan glacial stage 萨尔坦冰期【地】
Sartan stade 萨尔坦冰阶【地】
sartorite 脆锑砷铅矿
Sartorius sedimentation balance 电笔记录沉积秤
saryarkite 磷硅铝钇钙石
sasaite 多水硫磷铝石

sash 锯框;推拉窗扇;窗扇(框架);窗框(钢)
sash adjuster 窗扇调整器
sash and frame 框格窗;窗柜及框格
sash angle 窗扇角钢;窗框角铁
sash balance 吊窗复位弹簧;吊窗的平衡锤;吊窗平衡锤;吊窗衡重
sash-balance hardware 吊窗平衡重小五金
sash bar 钢窗料;梃子;门窗框条;窗芯子;窗梃;窗框条
sash bar window 有中梃的窗
sash bead 中旋窗扇压条;窗扇压条
sash bit 深孔木钻
sash block 吊窗滑轮
sash bolt 窗插销
sash brace 排架横撑;水平加劲支撑
sash bracket 窗框支架
sash casing 窗锤箱
sash center 窗扇心轴;窗扇转动轴;翻窗合页
sash chain 提拉窗吊链;吊窗链
sash chisel 窗框凿;窗梃凿;窗梃鏨
sash cord 窗框吊带;吊窗索;开关拉索;提拉窗吊索;提拉吊索;升降窗扇的皮带;吊窗绳
sash-cord iron 双悬窗系吊带的铁件
sash counterweight 吊窗平衡锤;吊窗平衡重;提拉窗平衡重
sash cramp 窗扇钳夹
sash door 镶玻璃门;装有升降玻璃的车门;半截玻璃门
sash fabrication 窗扇装配
sash fast 窗撑杆;窗扇闩扣
sash fastener 窗扇闩扣;吊窗定位销;窗风钩
sash fillister 窗扇槽刨;窗扇榫口;门窗框槽刨;窗框槽刨
sash frame 窗框;窗架
sash gang 一套窗框
sash gang saw 毛方排锯
sash handle 窗拉手
sash hardware 窗扇小五金;窗扇附件
sash holder 窗支杆;窗提手;窗风钩
sash hook 吊窗钩
sash iron 窗框钢
sash knife 窗框槽刨
sash knob 窗捏手;门框上球形把手
sash latch 窗锁闩
sashless window 无框格窗;无窗格窗
sash lift 窗扇提升装置;吊窗提手;提拉窗扣手;吊窗拉手;窗框提手
sash lift and hook 带锁钩的吊窗拉手
sash lift knob 提拉窗扇的吊窗拉手
sash line 窗框吊带;提拉窗索;提拉窗吊链;吊窗绳
sash lock 窗锁;吊窗定位销;窗风钩
sash lock rack 窗止铁
sash mortise chisel 软木榫凿;窗框榫凿子
sash operator 框格升降器
sash pin 门窗框合角加固棱栓;窗插销
sash pivot 窗枢轴;车窗枢销
sash plane 窗框刨;起线刨
sash plate 窗扇转轴器
sash pocket 吊窗锤箱;窗锤箱;吊窗平衡重切口;窗斗
sash pole 推窗杆
sash pole hook 推窗杆钩
sash pole socket (高窗上加窗用的)钩杆开关
sash pull 吊窗拉手;窗拉手
sash pulley 提拉窗滑轮;吊窗滑轮;窗用滑轮
sash putty 镶玻璃油灰
sash rail 窗框横档;窗口窗框;窗扇横冒头;窗扇横冒头
sash ribbon 吊窗钢带
sash roller 窗扇滑轮;窗框滑轮
sash rope 吊窗绳
sash run 双悬窗平衡锤滑框;提拉窗滑车槽;悬窗滑轮窗梃
sash saw 框锯;窗扇开榫锯;弓锯
sash sill 窗框底盘;窗台
sash spring bolt 窗扇弹簧插销
sash staff 门、窗扇材料
sash stile 窗边梃
sash stop 窗止条;护条;吊窗挡条;窗闩
sash stuff 窗扇材料;窗扇用规格材
sash timber 窗框木材
sash tool 焦油刷;窗用漆刷;圆形漆刷;窗扇漆刷
S. A. Shukarlev classification S. A. 舒卡列夫分类法
sash weight 提拉窗平衡重;吊锤;吊窗锤

sash window (上下拉的)直拉窗;框格窗;上下推拉窗;上落窗
sassafras 黄樟
sassafras oil 黄樟油
Sassanian architecture 萨桑(王朝)建筑
sasse 通航河道船闸;水闸
sassolite 天然硼酸;天然硼砂
sastruga 雪面波纹;雪波;波状砂层
sastrugi 砂波;波状砂层
satchel charge 炸药包
Satco alloy 萨特科铅合金
satcom 卫星通讯社;卫星通信
sate-conduct 安全条例
sateen 棉缎
sateen-weave fabric 棉缎织物
satellite 卫星距离差定位;卫星
satellite aberration 卫星光行差
satellite-acoustics integrated positioning system 卫星声学组合定位系统
satellite aerial 卫星天线
satellite-aided navigation 卫星导航
satellite airfield 辅助机场;卫星机场
satellite altimetry 卫星测高
satellite altitude 卫星高度
satellite altitude determination 卫星测高
satellite and space probe measurements 卫星及空间探测器测量
satellite and space tracking system 卫星和空间目标跟踪系统
satellite antenna 人造卫星天线;卫星天线
satellite antenna radiation pattern 卫星天线辐射方向图
satellite asset 卫星装置
satellite astronomy 人造地球卫星天文学;卫星天文学
satellite attack warning system 卫星攻击警报系统
satellite attitude 卫星姿态
satellite automatic monitoring system 自动卫星监视系统;卫星自动监控系统
satellite automatic tracking antenna 人造卫星自动跟踪天线
satellite automation system 卫星自动控制系统
satellite auxiliary power unit 卫星辅助动力装置
satellite availability 卫星可用率
satellite azimuth 卫星方位角
satellite balloon 卫星气球
satellite band 卫星(频)带;附属普带
satellite base 附属基地;辅助基地
satellite-based system for coastal oceans monitoring 沿岸海洋卫星监视系统
satellite battery 人造卫星用电池
satellite boat 附属观测艇
satellite-borne 卫星装载的
satellite-borne remote sensing system 卫星装置遥感系统;卫星装置遥测系统
satellite-borne sensing system 卫星(上的)传感系统
satellite-borne sensor 卫星上的传感器;卫星传感器;卫星装置传感器
satellite broadcasting 卫星广播
satellite broadcasting service 卫星广播业务
satellite building 附属建筑;卫星候机楼
satellite buoy communication 卫星浮标通信
satellite camera 人造照相机
satellite capture and retrieval 卫星捕捉与收回
satellite cartography 卫星制图学
satellite center 卫星中心
satellite channel 卫星信道
satellite charges 卫星费用
satellite circuit 卫星线路
satellite circuit noise 卫星线路噪声
satellite city (大城市的)卫星城镇;卫星城(市)
satellite cloud photograph 卫星云图
satellite cloud picture 卫星云图
satellite coating 人造卫星壳体涂层
satellite collection of meteorological observation 卫星收集的气象观测资料
satellite communication 卫星通信
satellite communication agency 卫星通信局
satellite communication center 卫星通信中心
satellite communication control facility 卫星通信控制设施
satellite communication controller 卫星通信控制器
satellite communication control office 卫星通信

控制室
satellite communication link 卫星通信线路
satellite communication ship 卫星通信船
satellite communication system 卫星通信系统
satellite communication technology 卫星通信技术
satellite communication terminal 卫星通信终端(设备)
satellite computer 卫星计算机
satellite configuration 卫星配置
satellite constellation 卫星座
satellite control center 卫星控制中心
satellite control facility 卫星控制设备
satellite control machine 从属联锁机;分联锁机;分集中联锁机
satellite control network 卫星控制网(络)
satellite control system 卫星控制系统
satellite cooler 多筒冷却器;多筒冷却机
satellite coronagraph 人卫日冕照相术
satellite coupling 卫星(轨道)接合
satellite coverage 卫星覆盖范围
satellite coverage area 卫星覆盖区
satellite data 卫星数据
satellite data communication system 卫星数据通信系统
satellite data modulator 卫星数据调制器
satellite data processing system 卫星数据处理系统
satellite data recorder 卫星数据记录器
satellite data system 卫星资料系统;卫星数据系统
satellite depot 地面卫星站
satellite designation 卫星名称
satellite detection line 卫星探测线
satellite differential 行星齿轮差速器
satellite digital and analog(ue) display 卫星数字模拟显示
satellite Doppler navigation system 人卫多普勒导航系统;卫星多普勒导航系统
satellite Doppler positioning 卫星多普勒定位(法)
satellite Doppler sonar integrated navigation system 卫星多普勒声呐组合导航系统
satellite drift 卫星偏移
satellite dynamics 卫星动力学
satellite early warning system 卫星预警系统
satellite earth antenna for household use 家用卫星接收天线
satellite earth coverage 卫星地球覆盖区
satellite earth station 卫星地面站
satellite educational and informational television 卫星教育和新闻电视
satellite electron 卫星电子
satellite electronic countermeasures system 卫星电子对抗系统
satellite elevation 卫星仰角
satellite emergency position indicating radio beacon 卫星紧急示位无线电信标
satellite ephemeris error 卫星星历误差
satellite equator 卫星赤道
satellite equipment 卫星装置
satellite exchange 电话支局
satellite experiment 卫星实验(重点实验之外的相关实验)
satellite fairing 人造卫星整流罩
satellite fault 附属断层
satellite field 卫星机场
satellite file 外围资料档
satellite fix 卫星定位;卫星船位;卫导船位
satellite fixing by range difference measurement 卫星测距差定位
satellite fixing by range measurement 卫星测距定位
satellite flight 卫星飞行
satellite for gravimetric experiment 重力试验卫星
satellite for lunar probes 月球探测卫星
satellite frequency error 卫星频率误差
satellite geodesy 人卫测地;卫星大地测量学
satellite geodetic network 卫星大地网
satellite global monitoring 卫星全球监测
satellite graphics system 卫星图形系统
satellite gravimetrical survey 卫星重力测量
satellite gravimetry 卫星重力测量
satellite ground station 人造卫星(地面)站;卫星(通信)地面站
satellite hole 辅助钻孔;辅助炮眼
satellite identification number 卫星识别编号

satellite image 卫星图像;卫星像片
satellite image analysis 卫星像片分析
satellite image map 卫星像片图
satellite imagery 卫星图;卫星成像
satellite imaging 卫星成像
satellite industry 卫星工业
satellite inertial guidance integrated positioning system 卫星惯导组合定位系统
satellite information 附属信息
satellite information center 卫星情报中心;卫星信息中心
satellite information processor 卫星信息处理机
satellite information processor operational program(me) 卫星信息处理机操作程序
satellite infrared radiation spectrometer 卫星红外辐射光谱仪
satellite infrared spectrometer 卫星红外光谱仪;卫星红外分光计
satellite inspection network 卫星检测网
satellite lake 卫星湖
satellite laser ranger 卫星激光测距仪
satellite laser ranging 人卫激光测距;卫星激光测距
satellite launching 卫星发射
satellite launching facility 卫星发射装置
satellite launching rocket 卫星发射火箭
satellite launching ship 卫星发射船
satellite launching vehicle 卫星运载火箭
satellite library information network 卫星图书馆情报网络
satellite line 卫线;(摄谱线的)伴线;伴线
satellite link 卫星线路
satellite mapping radar 卫星测绘雷达
satellite measurements 利用卫星测量
satellite measurement station 卫星测站
satellite meteorology 卫星气象学
satellite-missile observation system 卫星观测系统
satellite mobile communication system 卫星移动通信系统
satellite monitor 卫星监视器
satellite monitoring 卫星监控;卫星监测
satellite multiple-access technique 卫星多址技术
satellite navigation computer 卫星导航计算机
satellite navigation earth station 卫星导航地面站
satellite navigation equipment for user 卫星导航用户设备
satellite navigation map 卫星导航地图
satellite navigation system 卫星导航系统
satellite navigator 卫星导航定位仪
satellite network 卫星网络
satellite observation 利用卫星测量;卫星观测
satellite observation ship 卫星观测船
satellite observation station 卫星观测站
satellite oceanography 卫星海洋学
satellite ocean surveillance evaluation center 卫星海洋监视鉴定中心
satellite ocean surveillance system 卫星海洋监视系统
satellite-Omega integrated navigation system 卫星欧米伽组合导航系统
satellite optic(al) surveillance system 卫星光学监视系统
satellite orbit 卫星轨道
satellite orbital period 卫星轨道周期
satellite orbital tracking 卫星轨道跟踪
satellite orbit control 卫星轨道控制
satellite orbit data 卫生轨道运行数据
satellite orbit determination 卫星轨道的测定
satellite orbit parameters 卫星轨道参数
satellite orbit shadow time 卫星轨道背日时间
satellite orientation 卫星定向
satellite orientation device 卫星定向系统
satellite orientation system 卫星定向系统
satellite ozone data 卫星测得的臭氧资料
satellite-pair 伴偶
satellite pass 卫星通过
satellite path 卫星轨迹
satellite peak 卫星峰;伴峰
satellite period of revolution 卫星运行周期
satellite perturbation 卫星摄动
satellite perturbation motion 卫星摄动运动
satellite photo 卫星照片
satellite photogrammetry 卫星摄影测量
satellite photograph 卫星照片;卫星像片
satellite photographic(al) study 卫星照片判读

satellite photography 卫星摄影(术)
satellite photo map 卫星像片图
satellite picture 卫星像片
satellite plant 辅助加工设备
satellite platform 辅助平台
satellite point 偏心测站;卫点
satellite port 卫星港
satellite position 卫星位置
satellite position determination 卫星位置的确定
satellite positioning 卫星定位
satellite position prediction and display 卫星位置预测与显示
satellite power supply 卫星电源
satellite probe 人造卫星探测器
satellite processor 卫星计算机;卫星处理机
satellite radio communication 卫星无线电通信
satellite radio line 卫星无线电线路
satellite radio station 卫星电台
satellite rangefinder 卫星测距仪
satellite range measurement 卫星距离测量
satellite receive centre 卫星接收中心
satellite receiver subsystem 卫星接收机辅助设备
satellite recovery 卫星回收
satellite recovery system 卫星回收系统
satellite relay 卫星中继器
satellite relay of data 卫星转播资料;数据卫星转播
satellite relay station 卫星中继站
satellite remote sensing 卫星遥感;卫星遥感测量
satellite remote sensing calibration site 卫星遥感校正站
satellite remote sensing system 卫星遥感系统
satellite rendezvous system 卫星回合系统
satellite repeater 卫星转发器
satellite report 辅助报告
satellite resonance effect 卫星共振效应
satellite retroreflector 卫星逆向反射器;卫星反射器
satellite road test 环道试验【道】
satellite rocket 卫星火箭
satellite scanning and imagery system 卫星扫描与成像系统
satellite searching radar 卫星搜索雷达
satellite-sensed event 卫星可测事件
satellite series 卫星系列
satellite's ground path 卫星轨道的地面投影
satellite ship earth station 海事卫星船站
satellite simulator 卫星模拟器
satellite sounding 卫星探测
satellite space geodetic network 空间卫星大地网
satellite spacing 卫星间隔
satellite speed 卫星速率
satellite station 星际站;卫星(台)站;卫星电台;附属测站
satellite station-keeping 卫星位置保持
satellite status 卫星状态
satellite studio 小型中继站
satellite surveillance 卫星监视
satellite switching 卫星转换
satellite system 卫星附属系统;卫星(辅助)系统;候机楼卫星系统
Satellite Telecommunication Coordinating Committee 卫星通信合作委员会
satellite telecommunication system 卫星电信系统
satellite telecommunication with automatic routine 自动导航卫星通信
satellite telemetering 卫星遥测
satellite telemetry automatic reduction system 卫星遥测自动数据简化系统;卫星遥测数据自动处理系统
satellite telemetry station 卫星遥测站
satellite television 卫星电视
satellite television distribution service 卫星电视分配业务
satellite terminal 卫星(车)站
satellite test center 卫星试验中心
satellite test range 人造卫星试验场
satellite-to-satellite tracking method 卫星跟踪卫星法
satellite town (大城市的)卫星城镇;卫星城(市)
satellite tracking 人卫跟踪;卫星跟踪
satellite tracking antenna 卫星跟踪天线
satellite tracking camera 人卫跟踪相机
satellite tracking center 卫星跟踪中心
satellite tracking data 卫星跟踪数据
satellite tracking facility 卫星跟踪设备

satellite tracking installation 卫星跟踪设施
satellite tracking laser 卫星跟踪激光仪;卫星跟踪激光器
satellite tracking laser system 卫星跟踪激光系统
satellite tracking observation 卫星跟踪观测
satellite tracking ship 卫星跟踪船
satellite tracking station 卫星跟踪站
satellite tracking telescope 卫星跟踪望远镜
satellite trail 卫星轨迹
satellite trajectory 卫星轨道
satellite trajectory ground track 卫星轨道的地面投影
satellite transmission 卫星发送
satellite transmitter 卫星发射机;辅助发射机
satellite transponder 卫星转频器;卫星转发器
satellite treatment system of wastewater 卫星污水处理系统
satellite triangle and altitude measuring system 卫星三角测高系统
satellite triangulation 人卫三角测量;卫星三角测量
satellite triangulation station 卫星三角测量站
satellite trilateration 卫星三边测量
satellite visibility 卫星能见度
satellite-warning system 卫星警报系统
satellite wastewater plant treatment 卫星污水厂式处理
satellite X-mass tree 卫星井采油树
satellitosis 卫星状态
satellization authority 卫星机构
satellocentric system 卫星中心坐标系
satelloid 准卫星;载人飞行器;太空船;带动力装置的人造卫星
satelloon 气球卫星
satellorb 模拟人造卫星的观察与研究气球
satholith 残留土
satiation 充分满足;饱满
satiety 饱满感
satiety price 市场饱和价格;饱和价格
satimolite 水氯硼铝石
satin 加光泽;磨褪;缎光;半光
satin alcyonne 表里双色缎
satin black 缎纹黑
satin bright alumin(i)um finish 缎光铝饰面
satin bronze finish 缎光青铜饰面
satin-embossed finish 缎光轧花涂装法
satin etch 细毛面酸蚀;丝光面酸蚀
satin etching of sandblasted surface 细毛面酸蚀法
satin fabric 缎纹织物
satin finish 无光饰面;细毛面饰;毛面光洁度;丝光酸蚀面;刷光;缎光处理;蛋壳光面漆;半光面漆;研光
satin finish glass 珍珠光玻璃;细毛玻璃;丝光玻璃
satin finishing 擦亮;抛光;磨褪;研光
satin finish lacquer 段子般的喷漆
satin finish varnish 毛面清漆粉刷;无光清漆粉饰
satin-glazed tile 光亮釉面瓦;有光泽的釉面砖;有光泽的釉面瓦
satin gloss black 缎面光泽黑色
satin ice 纤维状冰
satinizing 磨褪处理;打磨处理
satin nickel 光泽镍
satin paint 缎光涂料
satin paper 蜡光纸
satin-polished 缎光抛亮的
satin sheen 光泽似缎的(漆膜)
satin spar 纤维石(膏)
satin stone 纤维石
satin-vellum glaze 缎光釉;半无光釉
satin walnut 美国枫香
satin weave 缎纹(组织)
satin-weave frabic 缎纹织物
satin white 白色;缎光白
satinwood 金红色椴木;椴木
satinwood finish 半光漆
satisfaction 履约;满足;满意;清偿;赔偿;归还
satisfaction for a debt 偿还债款
satisfaction of lien 解除留置权
satisfaction of mortgage 抵押贷款清还证书
satisfaction of wants 满足需要
satisfaction piece 还款收据
satisfaction proof 满意证明
satisfaction to mortgage 解除抵押
satisfactory 符合要求
satisfactory performance 性能优良

satisfactory profit 令人满意的利润
satisfactory quality 满意质量
satisfactory result 满意效果
satisfactory settlement of a claim 索赔圆满解决
satisfactory solution 满意解
satisfiability 可满足性
satisfiability problem 满意性问题
satisfied 满意的
satisfied hydrocarbon 饱和烃
satisfy debt 偿清债务
satisfying and adaptively rational decision 满足化与适应性的合理决策
satisfying criterion 满足标准
satisfying model 满足的模型
satisfying planning 符合要求计划法
satisfy the demand 满足要求
satisfy the liabilities 清偿负债
satisfy the needs 满足需要
satisfy the prescriptive period 满足有效期
satisfy wants 满足需要
sat linkage 盐键
satpaevite 黄水钒铝矿
Sat Stream 卫星传输业务系统
Sattahip Port 梭挑邑港(泰国)
satterlyite 三方羟磷铁石
saturability 饱和度;饱和能力;饱和额
saturable 可浸透的;可饱和的
saturable absorber 可饱和吸收器
saturable absorption 饱和吸收
saturable absorption Q-switching 可饱和吸收 Q 开关
saturable autotransformer 饱和自耦变压器
saturable choke 饱和抗流圈
saturable core 饱和铁芯
saturable-core generator 饱和铁芯发电机
saturable-core magnetometer 可饱和铁芯磁强计
saturable-core reactor 饱和电抗器
saturable dye Q-switch 可饱和染料 Q 开关
saturable inductor 饱和线圈;饱和感应线圈
saturable magnetic circuit 饱和磁路
saturable organic dye 可饱和有机染料
saturable reactor 饱和扼流圈;饱和电抗器
saturable reactor A C arc welder 饱和电抗器型交流弧焊机
saturable resonator 可饱和共振腔
saturant 浸渍料;浸渍剂;饱和剂
saturate 浸透
saturate colo(u)r 鲜纯色
saturated 浸透的;饱和的
saturated absorption 饱和吸收
saturated absorption frequency stabilization 饱和吸收稳频
saturated absorption peak 饱和吸收峰
saturated acid 饱和酸
saturated activated carbon 饱和活性炭
saturated activity 饱和放射性
saturated acyclic(al) hydrocarbon 饱和无环烃;饱和链烃
saturated-adiabatic lapse rate 饱和绝热直减率
saturated adsorbent 饱和吸附剂
saturated adsorption capacity of rocks 岩石饱和吸附容量
saturated adsorptive capacity 饱和吸附容量
saturated air 饱和(空)气
saturated air mixture 饱和的空气混合物
saturated aliphatic compound 饱和脂肪族化合物
saturated aliphatic hydrocarbon 饱和脂肪烃
saturated alloying 饱和合金
saturated ammonia 饱和氨
saturated and coated bitumen amount of malthoid 油毡浸涂总量
saturated and coated roofing felt 沥青油毡
saturated aqueous solution 饱和水溶液
saturated asbestos felt 浸油石棉毡
saturated asphalt rag felt 浸沥青的破布油毡
saturated bitumen asbestos felt 浸沥青(石)棉毡;石棉纸油毡
saturated bitumen felt 纸胎油毡;低脂油毡
saturated bitumen glass-cloth felt 玻璃布油毡
saturated bitumen mineral wool felt 矿棉纸油毡
saturated brine 饱和盐水;饱和盐溶液
saturated calomel electrode 饱和甘汞电极
saturated carbon ring 饱和碳环

saturated characteristic 饱和特性
saturated circuit 饱和电路
saturated clay 饱和黏土
saturated clay deposit 饱和黏土沉积物
saturated clay sample 饱和黏土样
saturated clay sediment 饱和黏土沉积物
saturated clay stratum 饱和黏土层
saturated coefficient 饱和系数
saturated cohesive soil 饱和黏性土
saturated cohesive soil for normal consolidation 正常固结饱和黏性土
saturated colo(u)r 鲜纯色;彰色;纯色;饱和色
saturated compound 饱和化合物
saturated concentration 饱和浓度
saturated condition 饱和状态
saturated conductivity 饱和导水率;饱和土壤传导度;饱和传导率
saturated converted coefficient 饱和层换算系数
saturated core 浸透的岩芯;饱和铁芯
saturated core reactor 饱和扼流圈
saturated density 饱和密度
saturated dihalide 二卤代烷
saturated diode 饱和二极管
saturated dissolved oxygen 饱和溶解氧(量)
saturated dissolved oxygen concentration 饱和溶解氧浓度
saturated dissolved oxygen level 饱和溶解氧水平
saturated energy 饱和能量
saturated erase 饱和抹音法
saturated erasure 饱和消磁法
saturated expanded cork brick 饱和的膨胀软木砖
saturated factor 饱和系数
saturated fatty acid 饱和脂肪酸
saturated fatty alcohol dibasic acid ester 饱和脂肪醇二羧酸酯
saturated felt 浸透沥青的油毛毡;油毛毡;沥青油纸;浸油毡;浸沥青毡
saturated flow 饱和水流;饱和流(量)
saturated free-face flow 饱和自由面水流
saturated glaze 饱和釉
saturated gradient 饱和梯度
saturated gravel 饱和砾石
saturated ground 含水饱和土;水涝地;饱和土
saturated gun 饱和电子枪
saturated humidity 饱和湿度
saturated humidity mixing ratio 饱和湿度混合比
saturated hydrocarbon 饱和烃;饱和碳氢化合物
saturated hydrocarbon/aromatic hydrocarbon 饱和烃/芳烃
saturated hydrocarbon of petroleum 石油的饱和烃
saturated ice frozen soil 饱和冰冻土
saturated index 饱和指数
saturated inductance 饱和电感
saturated level 饱和水平
saturated line 浸润线;饱和线
saturated liquid 饱和液(体)
saturated liquid of mud-sand 泥沙饱和的液体
saturated logic circuit 饱和(型)逻辑电路
saturated magnetization 磁饱和
saturated market 饱和的市场
saturated material 饱和材料
saturated media 饱和介质
saturated methylene diphenyl-4 饱和亚甲基二苯基二异氰酸酯
saturated mineral 饱和矿物
saturated mixture 饱和混合物
saturated moisture 饱和湿度
saturated noise 饱和噪声
saturated normal mineral 饱和标准矿物
saturated oil 饱和油;饱和气石油
saturated paper 饱和纸
saturated permafrost 饱和永久冻土
saturated permeability 饱和透水性
saturated polybasic acid 饱和多元酸
saturated polyester resin 饱和聚酯树脂
saturated pore medium 饱和孔隙介质
saturated potassium chlorate solution 饱和氯酸钾溶液
saturated pressure 饱和气压
saturated reactance method 饱和电抗法
saturated recombination rate 饱和复合率
saturated rock 渗透岩石;饱水岩石;饱和岩(石)
saturated rock mass 饱和岩体
saturated roofing felt 浸沥青屋面油毡;浸沥青毡

saturated saline flo(a)tation method 饱和盐水浮集法
saturated salt mud 饱和盐水泥浆
saturated salt-water drilling fluid 饱和盐水冲洗液【岩】
saturated sample 饱和样品;饱和试样
saturated sand 饱和砂
saturated sands and gravels 饱和砂和砾石
saturated sediment content 饱和含沙量
saturated silt 饱和淤泥
saturated silt charge 饱和含沙量
saturated slope 饱和坡
saturated soft clay 饱和软黏土
saturated soil 饱和土壤;饱和土
saturated soil zone 饱和土壤带
saturated solid solution 饱和固溶体
saturated solute diffusion method 饱和溶质扩散法
saturated solution 饱和溶液
saturated specimen 饱和试样
saturated state 饱和状态
saturated steam 饱和蒸汽;饱和水蒸气
saturated steam cylinder oil 饱和蒸汽汽缸机油
saturated steam locomotive 饱和蒸汽机车
saturated steam-water mixture 饱和汽水混合物
saturated steel 共析钢
saturated strength 饱和强度
saturated strength of soil 土的饱和强度
saturated surface 地下水位;地下水面;饱和表面;潜水面(指地下水);饱和面
saturated-surface-dried 面干饱和的
saturated-surface-dried condition 表面干燥内部饱和状态;饱和面干状态
saturated-surface-dried sample 饱和面干试样
saturated surface dry 饱和面干(状态)
saturated-surface-dry condition 饱和面干状态;面干饱和状态
saturated-surface dry weight 饱和面干重集料;饱和面干重骨料
saturated-surface-dry weight 面干饱和重量
saturated temperature 饱和温度
saturated thickness 饱和厚度
saturated thickness of aquifer 含水层饱水厚度
saturated time 饱和时间
saturated two-phase condition 饱和二相土状态
saturated two-phase soil 饱和二相土
saturated unconfined flow 饱和非承压流
saturated undisturbed sample 饱和原状土样
saturated undisturbed soil 饱和原状土
saturated unit weight 饱和重度;饱和容重;饱和容量;饱和单位重(量)
saturated unit weight of soil 土的饱和容重
saturated vapo(u)r 饱和蒸汽
saturated vapo(u)r pressure 饱和蒸气压(力)
saturated velocity 饱和速度
saturated virtual temperature 饱和虚温度
saturated water 饱和水
saturated water-absorptivity 饱水率
saturated water vapo(u)r pressure 饱和水气压
saturated weight 饱和重量
saturated zone 饱水带;饱和区;饱和带;饱和层
saturate surface dry 饱和表面干燥
saturate with water 水饱和
saturating 浸透的;饱和的
saturating agent 饱和剂
saturating asphalt 饱和沥青
saturating capacity 饱和量;饱和含水量
saturating capacity of paper 纸的饱和性
saturating characteristic 饱和特征
saturating composition 饱和组分;浸透组分
saturating material 饱和材料;浸渍材料
saturating of capillary 毛细饱和法
saturating of intrusion water 浸水饱和法
saturating paper 油纸
saturating phase 饱和相
saturating plant 浸渍车间;饱和车间
saturating signal 饱和信号
saturating supply 磁饱和电源
saturating temperature 饱和温度
saturation 因子函数;章度;浸透;温度饱和;色饱和度;(作)用
saturation-absorption spectrum 饱和吸收光谱
saturation activation 饱和激励
saturation activity 饱和活性;饱和放射性(强度)
saturation adiabat 饱和绝热线

saturation-adiabatic lapse rate 饱和绝热倾率;湿绝热直减率;饱和热直减率
saturation-adiabatic process 饱和绝热过程;湿绝热过程
saturation-adiabatic rate 饱和绝热率
saturation air temperature 饱和空气温度
saturation amplification 饱和放大
saturation analysis 饱和分析法
saturation area 饱和区;饱和量
saturation boiling 饱和沸腾
saturation bombing 饱和轰炸
saturation boundary 饱和界限;饱和界线
saturation capacity 饱和(容)量;饱和能力
saturation capillary head 饱和毛细管水头
saturation characteristic 饱和特性
saturation characteristic curve 饱和特征曲线;饱和特性曲线
saturation characteristics 饱和特性曲线
saturation charge 饱和电荷
saturation clay 饱水黏土
saturation coefficient 饱和度;饱水系数;饱和系数
saturation collection 饱和收集
saturation colo(u)r 饱和色
saturation compression strength 饱和抗压强度
saturation concentration 饱和浓缩;饱和浓度
saturation condition 饱和状态;饱和条件
saturation constant 饱和常数
saturation constraint 饱和限制
saturation content 饱和含量
saturation control 饱和控制;饱和度调整;饱和度调节
saturation core regulator 铁磁饱和稳压器;饱和铁芯稳压器
saturation correction 饱和度校正
saturation current 饱和电流
saturation current density 饱和电流密度
saturation curve 饱和曲线
saturation curve equation of phreatic water 潜水浸润曲线方程
saturation deep diving 大深度饱和潜水
saturation deficiency 饱和差
saturation-deficient 饱和不足
saturation deficit 饱和差
saturation degree 饱和度
saturation discrimination 彩度鉴别本领
saturation dissolved oxygen 饱和溶解氧(量)
saturation dive 饱和潜水
saturation diver 饱和潜水员
saturation diving 饱和潜水
saturation diving apparatus 饱和潜水装置
saturation diving platform 饱和潜水平台
saturation diving record 饱和潜水记录
saturation diving system 饱和潜水系统
saturation effect 饱和效应
saturation efficiency 饱和效率
saturation energy 饱和能量
saturation extract 饱和提取(液);饱和浸出液
saturation factor 饱和因子;饱和因数
saturation field 饱和场(强)
saturation flo(a)tation 饱和气浮(法)
saturation flow 饱和流(量)
saturation flux 饱和通量;饱和磁通
saturation flux density 饱和通量密度;饱和磁通密度;饱和磁感应强度
saturation gradient 浸润线梯度;地下水位梯度;饱和梯度
saturation ground 饱和土
saturation humidity 饱和湿度
saturation in back-iron LEM 带有饱和实心铁轭的直线电机
saturation index 饱和(度)指数
saturation index of calcite 方解石饱和指数
saturation index of dolomite 白云石饱和指数
saturation index of mineral 矿物饱和指数
saturation induction 饱和感应;饱和磁感应强度
saturation induction density 饱和磁感应强度
saturation intensity 饱和(强)度;饱和磁化强度
saturation irradiation 饱和辐射
saturation isomerism 饱和异构现象
saturation isothermal remanent magnetization 饱和等温剩余磁化强度
saturation level 饱和面;饱和度;饱和电平
saturation lime 饱和石灰
saturation limit 饱和限度

saturation limiter 饱和限幅器
saturation limiting 饱和限幅
saturation line 地下水位线;浸润线;饱和线
saturation logic 饱和逻辑
saturation loss 饱和损耗
saturation lqiuid 饱和液(体)
saturation magnetiation 饱和磁化强度
saturation magnetic flux density 饱和磁通密度
saturation magnetic moment 饱和磁矩
saturation magnetic recording 饱和磁记录
saturation magnetism 饱和磁化
saturation magnetometer 饱和磁强计
saturation method 饱和(方)法
saturation mixing ratio 饱和混合比
saturation moisture content 饱和含水率;饱和含水量
saturation moments 饱和力矩
saturation noise 饱和噪声
saturation of concrete 混凝土饱和水
saturation of dissolved oxygen 溶解氧饱和度
saturation of forces 力的饱和
saturation of peak hours 高峰饱和度
saturation of silt flow 泥沙流饱和度
saturation of the air 空气饱和
saturation output 饱和产量
saturation overland flow 饱和坡面漫流
saturation paper 油纸
saturation parameter 饱和参数;饱和参量
saturation percentage 饱和百分率;饱和百分数
saturation period 饱和周期
saturation photocurrent 饱和光电流
saturation plane 饱和面
saturation point 饱和点
saturation-point pressure 饱和点压力
saturation polarization 饱和极化
saturation pressure 饱和压力
saturation profile 纵剖面饱和度分布图
saturation prospecting 饱和勘察
saturation range 饱和区
saturation rate 饱和速率
saturation ratio 饱和率;饱和比
saturation reactance 饱和电抗
saturation recording 饱和记录
saturation regime (多孔介质中的)饱和(动)态
saturation region 饱和区
saturation remanent flux 饱和剩磁
saturation remanent level 饱和剩余磁平
saturation resistance 饱和电阻
saturation resume method 饱和恢复法
saturation signal 饱和信号
saturation signalling system 饱和传信系统
saturation soil 饱水土(壤);饱和土
saturation soil paste 饱和土浆
saturation solubility 饱和溶度
saturation solution 饱和溶液
saturation specific humidity 饱和比湿度
saturation spectroscopy 饱和光谱学
saturation speed 饱和速度
saturation spreader 饱和扩张器
saturation state 饱和(状)态
saturation steam pressure 汽化压力;饱和蒸汽压(力)
saturation storage time 饱和储存时间
saturation swelling stress 饱和湿胀应力;饱和膨胀应力
saturation tank 浸油池;浸润池
saturation temperature 饱和温度;饱和点
saturation tension 饱和张力
saturation test 饱和程度试验
saturation testing 饱和试验;饱和测试
saturation thickness 饱和厚度
saturation time 饱和时间
saturation time constant 饱和时间常数
saturation tower 饱和塔
saturation traffic flow 饱和交通量
saturation transfer 饱和转移
saturation transformer 饱和变压器
saturation two-phase soil 饱和二相土
saturation type circuit 饱和型电路
saturation type magnetometer 饱和式磁力仪
saturation value 饱和值
saturation vapo(u)r 饱和水汽;饱和水气
saturation vapo(u)r pressure 饱和蒸气压(力);饱和水汽压;饱和气压

saturation vapo(u)r pressure method 饱和蒸气压法
saturation vapo(u)r tension 饱和蒸气张力
saturation velocity 饱和速度
saturation voltage 饱和电压
saturation voltage drop 饱和压降
saturation volume 饱和体积;饱和流量
saturation water 饱和水
saturation water vapo(u)r pressure 饱和水蒸气压力
saturation zone 饱和区;饱和带;饱和层
saturator 浸渍机;饱和塔;饱和器(操作员);饱和剂
Saturdays clause 星期六条款
saturer 饱和器
saturite 饱和溶液沉积物
saturnic 中铅毒的
saturnic red 铅母
saturnigraphic(al) coordinates 土面坐标
saturnigraphy 土面学
saturnine red 铅丹
saturnism 铅中毒
sauce pan 平底锅;蒸锅;釜
sauce pot 平底锅
saucer 桅座;托盘;舵针承座;垫盘;茶托;茶碟
saucer bosh 高炉炉腰;炉腰
saucer crater 碟形环形山
saucer dome 蝶形穹顶
saucer lake 浸没湖;浅盆湖
saucer-like flow mark 碟状流痕
saucer shape 凹形
saucer-shaped depression of vitreous body 玻璃体碟状凹
saucer-shaped dome 碟形屋顶;碟形穹顶
saucer-shaped pit 碟状坑
saucer-shaped wheel 碟形砂轮
saucer spring 盘状弹簧
saucer-type disc sealer 碟状密封圆盘
saucer washer 凹面弹簧垫圈
saucer wheel 盘型砂轮;碟形砂轮
saucisse 沉笼;沉辊;大型长梢捆;柴把滚笼
saucisson 大型长梢捆;柴把滚笼;护河网辊
sauconite 锌皂石
Saul's formula 索尔公式(用于测定混凝土成熟度)
sauna 蒸汽浴室;桑拿浴室
sauna bath 桑拿浴;蒸汽浴(器)
sauna bath installation 蒸汽浴设施
sauna installation 蒸汽浴装置
sauna stove 蒸汽浴用的火炉
Saunderson correction 桑德森修正
Saunderson-Milner zeta space colo(u)r difference equation 桑德森—米尔纳空间色差方程
sausage antenna 笼形天线;圆柱形天线
sausage balloon 圆柱形系留气球
sausage construction 铅丝(网)石笼建筑
sausage dam 铅丝(网)石笼坝;石肠坝
sausage engineer 铅丝网石笼工程
saussurite 槽化石
saussuritization 槽化作用【地】;钠黝帘石化
Savage principle 塞韦奇原则
savane armee 多刺高灌丛
savane epineuse 多刺高灌丛
savanna climate 稀树草原气候
savanna(h) 无树大草原;热带大草原;草原;热带草原
savanna(h) climate 热带草原气候;稀树草原气候
savanna(h) woodland 稀树草原疏林;热带草原疏林
Savart plate 萨伐特片
Savart polariscope 萨伐特偏光镜
save 营救;节省;挽救
save-all 节省器;节约装置;裙衬;挡雾罩;防溅器;白水回收装置;节约装置;滴油盘;承尿盘;承水盘
save-all net 安全网
save and divert funds to 腾出资金
save area 保存区
save area table 保存区表
saved area description entry 保存区描述体
saved system 已保存系统;保存系统
save energy 节能
save file 副本文件
save fuel 节约燃料
save instruction 保存指令
save memory 内存保留区
save money 省钱
save on transport cost 节约运(输)费(用)

save-our-sea movement 挽救海洋运动
save our ship 呼救信号;船舶呼救信号
saver 节约器;回收器
save register 保存寄存器
save-rinse tank 冲洗回收池;冲洗回收槽
save routine 保存(例行)程序
save statement 保存语句
save value 保存值
Savian orogeny 萨夫造山运动
Savic orogeny 萨夫造山运动【地】
savin(e) 新疆圆柏
saving 节约的;储蓄
saving account 储蓄账户
saving and investment plan 储蓄投资计划
saving arch 安全拱;辅助拱
saving bank 储蓄银行
saving clause 附加条款;例外条款;但书
saving coefficient 储蓄系数
saving deposit 储备积蓄
saving grafting 修补嫁接
saving in energy 节省能量
saving in investment cost 投资费用的节省
saving in labo(u)r 节省劳动力
saving in money compensation 节约现金补助
saving in space 节省空间
saving-investment model 储蓄投资模式
saving in weight 节省重量;减轻重量
saving lock 节水船闸;省水船闸
saving of labo(u)r 节省劳动力;省工
saving of travel(l)ing time 旅行时间节约
saving on labo(u)r time 劳动时间的节省
savings account 存款账户
Savings Association Insurance Fund 储蓄联合会保险基金
savings at call 通知储蓄存款
savings bank 储蓄银行;储蓄所
savings clause 保留条款
savings deposit 储蓄存款
saving sequence 保存序列
savings in steel 节约钢材
saving social labor economy of social labor 节约人工的投资
Saving the Ozone Conference 保护臭氧层会议
saving time 节约时间
saving water 节约用水
savite 钠沸石
Savonius rotor 萨沃纽斯转子
Savonius rotor current meter 萨沃纽斯转子流速计
Savonius rotor sea current meter 萨沃纽斯转子海流计
Savonius windmill 萨沃纽斯风车
savory 圆顶顶棚区段;圆顶顶棚一跨
saw 锯(子);锯开
Sawakin deep 萨瓦金海渊
sawara cedar 日本花柏
saw arbor 圆锯座架的轴;铣槽刀轴;锯轴;锯齿铣槽刀轴;花键铣刀刀杆
sawback 锯齿山脊
saw band 锯带;锯条
saw belt 锯带
saw bench 圆锯台架;锯台;锯木台
saw bit 锯片;齿状钻头
saw blade 锯条;锯片
saw block 锯木垫块;锯木架;木马
saw bow 锯弓;锯架;锯框
saw brazing clamp 锯条焊(夹持)器
saw buck 锯台;锯木架
saw carriage 锯座
saw chop (锉锯时)固定锯片的工具
saw cut 温度缝;锯截;锯痕;陡峻峡谷
saw-cut joint (混凝土路面)锯成缝
saw cutting 锯割;锯切
saw-cut veneer 锯切饰面表皮;锯切饰面表层
saw cylinder 锯齿式滚筒
sawder 焊剂;焊料;焊药;焊锡
saw diamond abrasive 锯片用金刚石磨料
saw doctor 锯齿修整器;修锯工;磨锯齿机;锉锯(齿)机
saw dust 锯屑;锯末;木屑
sawdust aggregate 木屑集料;木屑骨料;锯屑骨料;锯屑集料
saw-dust-and-clay stem 木屑黏土炮泥
sawdust and peak 锯末和草碳
sawdust brick 锯末砖;锯屑砖

sawdust cement 锯屑水泥;锯末水泥
sawdust chip 木屑片
sawdust cleat 木屑片
sawdust collector 锯屑收集器
sawdust concrete 锯屑混凝土;锯末混凝土;木屑混凝土
sawdust firing 埋烧
sawdust joint filler 锯屑填缝材料
sawdust mortar 锯屑灰浆;锯末砂浆
sawdust scrubber 锯屑清洗机
sawdust tar 锯屑焦油
sawdust tire 锯屑轮胎
sawed 锯成的;锯开的
sawed alumin(i)um plate 锯割铝板
sawed back 有锯痕的背面
sawed edge 锯后形成的边
sawed engineered timber 工程设计制定的锯材
sawed face 锯过的面
sawed finish 锯成的石面
sawed finish of stone 锯成石面
sawed groove (混凝土路面)锯成缝槽
sawed joint 锯缝
sawed-off 锯断(的)
sawed shingle 锯成的木(片)瓦
sawed stone 锯石
sawed timber 锯材
sawed veneer 锯切单板;锯制层板
sawer 锯木工人
sawer's trestle 锯木架
saw file 细齿锉;整修锉;磨锯锉
saw-fly 锯绳
saw for two men 二人锯
saw frame 锯弓;锯框;锯架;锯齿形框架
saw gang 锯木组
sawgate 锯架
saw grinder 磨锯机;锯齿磨床
saw grinding machine 锯齿修磨机;锯齿刃磨机;锉锯(齿)机
saw gullet 圆锯齿间间隙
saw gumming 锯快圆锯齿;磨锯齿
saw gumming wheel 磨锯齿的砂轮
saw head 锯削头
saw horse 锯木凳;木马;锯木(支)架;锯架
sawing 锯开;锯切
sawing concrete 锯切混凝土
sawing disk 圆盘锯
sawing lumber 制材
sawing machine 锯床;锯机;电锯
sawing of rail 锯轨
sawing of sleeper 锯枕木
sawing pattern 锯剖图
sawing procedure 下锯法
sawing through the pith 破心下料
saw jack 锯木架
saw jointer 连锯器
saw jumper 锯齿器
saw kerf 锯口;锯截口;锯缝;锯劈痕
saw line 锯线;锯路
saw log 可锯木;可锯大圆木;锯架;锯材原木
saw lumber 锯成的材
sawmill 锯木厂;大型锯机;制材车间;制材厂;锯机;锯板厂
saw mill refuse 制材厂废屑
saw mill residue 制材厂废料
saw mill waste 制材厂废料
sawn alumin(i)um plate 锯割铝板
sawn back 有锯痕的背面
sawn edge 锯后形成的边
sawn face 锯成(面);锯开石面
sawn finish 锯过的面;锯成的石面;锯开石面
sawn finish of stone 锯成石面
sawn full 锯切毛尺寸的
sawn goods 成材
sawn joint 锯缝;(混凝土路面)锯成缝
sawn log 锯开的圆木;锯材
sawn-off 锯断(的)
saw nose dozer 锯树机;锯头推土机
saw notch 锯痕
sawn plank 锯制板
sawn shingle 锯齿形屋顶
sawn sized lumber 锯成规定尺寸材
sawn specimen 锯齿试件
sawn square 方材
sawn stone 锯石;锯开石料;锯成石料

sawn timber 锯成木材;锯材;成型木材;成材;锯成木料
sawn timber falsework 方木脚手架
sawn veneer 锯切单板;锯制单板;锯制(薄木)板;锯开薄板
sawn wood 锯成的木材;成材;制材
saw pit 锯木坑;锯坑
saw-pit horse 锯木坑木架
saw powder 锯屑;锯末
saw roof 锯齿状屋顶
saw sample process 锯样法
saw set 锯齿修整器;料度调整器;(锯的)料度;整锯器
saw set pliers 整锯钳
saw setting 锯齿错齿;锯齿修整器;整锯器
saw setting pliers 整锯钳
saw shaper 开锯齿机
saw sharpener 磨锯机;锉锯(齿)机
saw sharpener's bench 整锯台
saw sharpening machine 锯齿修磨机;锯齿刃磨机;磨锯(齿)机;锉锯(齿)机
saw-sized lumber 锯制规格材
saw table 锯台
saw teeth wave form 锯齿波形
saw tex(tured)锯纹
saw timber 锯木;锯材(原木料);锯材林木
saw tooth 锯齿(形)
saw-tooth arrangement 锯齿状排列
saw-tooth barrel 齿状捞抓筒
saw-tooth barrel shell roof 锯齿形桶状薄壳屋顶
saw-tooth bit 锯齿状钻头;锯齿形钻头
saw-tooth ceiling 锯齿形顶棚
saw-tooth chipping 锯齿形裂纹
saw-tooth cotton gin 锯齿轧花机
saw-tooth crown 锯齿状钻头
saw-tooth crusher 锯齿破碎机
saw-tooth current 锯齿形电流
saw-tooth current-wave generator 锯齿电流波发生器
saw-tooth cutter 锯齿形切割器
saw-tooth cylinder cleaner 锯齿滚筒式清洁器
saw-tooth cylinder segment roof 锯齿形圆柱弓形屋顶
saw-tooth cylindrical shell 锯齿形圆柱形薄壳
saw-tooth drive 锯齿状激励
saw-toothed 锯齿形的
saw-toothed blade 锯齿形刀片
saw-toothed joint 锯齿形接合
saw-toothed skylight 锯齿形天窗屋顶
saw-tooth finger 候机楼锯齿形廊式(平面)
saw-tooth former 锯齿形成器
saw-tooth hob 锯齿滚刀
saw-tooth hypar shell frame(d) building 锯齿形双曲抛物面构架建筑
saw-tooth joint 锯齿(形)接合;锯齿状结合;锯齿状接合
saw-tooth-like eaves plate 齿形檐板
saw-tooth linter 锯齿剥绒机
saw-tooth milling cutter 锯齿铣刀
saw-tooth modulated jamming 锯齿形调制干扰
saw-tooth mo(u)lding 饰锯齿形的线脚
saw-tooth output 锯齿波输出
saw-tooth parking arrangement 斜列停车;锯齿形的停车排列
saw-tooth pattern 锯齿型
saw-tooth phase grating modulator 锯齿形相栅调制器
saw-tooth platform 锯齿形站台;锯齿形月台
saw-tooth projection 锯齿形凸起
saw-tooth pulse 锯齿形脉冲
saw-tooth rainwater gutter 锯齿形雨水边槽;锯齿形雨水边沟
saw-tooth retrace 锯齿形信号回程
saw-tooth roof 锯齿形屋顶
saw-tooth roof structure 锯齿形屋面结构
saw-tooth roof truss 锯齿形屋架
saw-tooth-shaped hydrograph 锯齿形过程线;多峰过程线
saw-tooth shaped time series 锯齿型时间数列
saw-tooth shell 锯齿形薄壳
saw-tooth shell roof 锯齿形薄壳屋顶
saw-tooth signal 锯齿形信号
saw-tooth skylight 锯齿形天窗;锯齿式天窗
saw-tooth slot 锯齿形缝

saw-tooth stoping 锯齿形回采法
saw-tooth sweep 锯齿形扫描
saw-tooth system 锯齿状排列
saw-tooth thin shell 锯齿形薄壳
saw-tooth thin shell reinforced concrete structure 锯齿形钢筋混凝土薄壳结构
saw-tooth thread 锯齿形螺纹
saw-tooth truss 锯齿形桁架;锯齿式屋顶
saw-tooth type cutter 锯齿铣刀
saw-tooth voltage 锯齿波电压
saw-tooth wave 锯齿波
saw-tooth wave-form 锯齿波形
saw-tooth wave generator 锯齿波发生器
saw-tooth wharf 锯齿式码头
saw truss 锯齿状桁架
saw-type freight platform 锯齿状站台
saw type platform 锯齿形月台
Sawuer orogeny 萨乌尔运动
saw vice 整锯虎钳;锯机虎钳;磨锯虎钳
saw-wave form 锯齿波形
sawway 锯缝;锯道线
saw web 锯条;锯片
saw wrest 锯齿修整器;扭锯器
saw yard 锯木厂;锯木场
sawyer 锯木者;锯工;锯木工(人)
sawyer's trestle 锯工台架;锯木架
sax 修整石板用凿刀;修琢石板的尖锤;石(板)斧
saxboard 最高一列舷板;舢板
saxe blue 浅蓝色;水蓝
saxifragant 熔石的
Saxifrage 裂石草(虎耳草属)
saxol(ine)液体石蜡油
Saxon 撒克逊建筑(古英国)
Saxon architecture 撒克逊式建筑
Saxon facade 苏格兰的房屋立面;撒克逊式(建筑)饰面
Saxonian orogeny 萨克森造山运动
Saxonian(stage)萨克森阶
saxonite 方辉橄榄岩
Saxon masonry work 英格兰砌筑;撒克逊式砖石工程
Saxon shake 撒克逊木瓦;撒克逊木板屋面
Saxon slab 撒克逊玻璃板
Saxon style 撒克逊式
Saxon tower 撒克逊式塔楼
Saxon window 三角楣窗;撒克逊式窗
saxophone 顶馈线性天线阵
Saxophone gate 萨克管状浇口
saxoul wood land 梭梭林
saxsophone 低压管容器
Sayan Mountains 萨彦岭
Saybolt 赛氏重油粘度计
Saybolt centistoke 赛波特黏度单位
Saybolt colo(u)r 赛波特色
Saybolt colo(u)rimeter 赛波特比色计(又称赛氏比色计)
Saybolt-Furol viscosimeter 赛波特—富洛重油黏度计
Saybolt-Furol viscosity 赛波特—富洛(重油)黏度;赛波特重油黏(滞)性;赛波特重油黏(滞)度
Saybolt ring number of kerosene 煤油的赛波特环值
Saybolt scale 赛波特黏度级
Saybolt seconds 赛波特黏度秒;赛波特秒数
Saybolt seconds universal 赛波特通用秒
Saybolt thermometer 赛波特温度计
Saybolt thermoviscosimeter 赛波特热黏度计
Saybolt universal seconds 赛波特通用黏度计秒数
Saybolt universal viscosimeter 赛波特通用黏度计
Saybolt universal viscosity 赛氏通用黏度;赛波特通用黏度
Saybolt visco(si)meter 赛波特黏度计
Saybolt viscosity 赛波特黏度
say-cast 尾部粗毛
say ladle 取样勺
sayrite 水铅铀矿
Say's Law 萨伊定律
sazhinite 硅铈钠石
Sb-bearing lead ore 锑铅矿石
S-bend S形弯管
Sb-imgreite 锑伊碲镍矿
sborgite 水硼钠石
SBS modified asphalt 苯乙烯丁二烯苯乙烯橡胶改性沥青

scab 盐泡;铸件表面黏砂;钢锭结疤;起夹子;拼接板
scabbed 有疤的;拼接
scabbed casting 糙铸
scabbed stone 粗琢石
scabbing (混凝土表面)成疤;(混凝土表面)疤痕;(表面处治等的)局部露骨;碎甲
scabbing hammer 尖头锤;粗琢锤
scabble 粗琢
scabbled dressing 粗琢面;粗琢
scabbled rubble 粗琢石;粗琢毛石;荒料石
scabbled rubble work 粗琢石工程;粗琢石工作
scabbler 石工
scabbling 粗琢(石作业);粗琢石面
scabbling hammer 尖头锤;粗琢锤
scabbling pick 粗琢镐
scabblings 石片;石屑
scabellum 高的独立的台座
Scabious 山萝卜属
scab land 崎岖地
scabland 劣地;崎岖地;剥蚀熔岩高原
scabland topography 崎岖地形
scabrock 岩石劣地
scabrous 不平滑的;粗糙的
scacchite 钙镁橄石;氯锰石
scaena (古剧院舞台后的)临时建筑
scaena ductilis (古剧院作演出背景的)活动隔屏
scaena fronts 正面(装饰华丽的舞台后房屋)
scafco 钢脚手架
Scaffixer (一种用于捆脚手架的)链条
scaffold 脚台架;架子;棚料;脚手架
scaffold board 脚手架;脚手板
scaffold bridge 栈桥;脚手架踏板
scaffold cradle 悬挂(式)脚手架
scaffolder 架子工
scaffold erector 架子工
scaffold for maintenance work 维修工程用的脚手架
scaffold frame 脚手(架)框架
scaffold-guard 脚手架护网
scaffold height 步架高度;脚手架步距
scaffold-high 需用脚手架的工程高度
scaffolding 搭脚手架;拆卸式脚手架;脚手架;脚手板
scaffolding bearer 脚手架托梁
scaffolding board 脚手架板
scaffolding chain 脚手架链条
scaffolding coupler 脚手架连接件
scaffolding erection 立脚手架;搭脚手架
scaffolding erector 搭脚手架的工人
scaffolding fittings 脚手架配件
scaffolding frame 脚手架构架
scaffolding man 架子工
scaffolding of pipeline 管架
scaffolding plank 脚手架木板
scaffolding platform 脚手架平台
scaffolding pole 脚手木杆;脚手架用杆
scaffolding rope 脚手架绳
scaffolding tube 架设脚手架的管子
scaffolding winch 脚手架绞车
scaffolding work 脚手架工程
scaffold knot 脚手架结点
scaffold lashing 扎捆脚手架;绑扎脚手架
scaffold ledger 脚手架大横杆;顺水杆
scaffold limb 大主枝
scaffold nail 骑马钉;双帽钉;双头钉;扒锔子;扒钉
scaffold on bracket 悬臂式脚手架
scaffold pipe 脚手管;搭脚手架用管
scaffold plank 脚手板
scaffold pole 脚手架支柱;脚手(架)木;脚手架立柱;脚手杆;顶撑
scaffold service stand 工作台
scaffold standard 脚手架支柱;脚手架立柱;架立杆
scaffold suspended 悬挂式脚手架
scaffold tower 脚手井架;塔式脚手架
scaffold tower hoist 井架卷扬机;塔式脚手架卷扬机
scaffold trestle 脚手架排架;脚手架台;脚手架柱;脚手架车
scaffold tube 脚手架管子;脚手管;脚手架用管
scaffold-type shoring 脚手架式模板支撑
Scaglia 斯卡belle雷红色石灰石
scagliola 假大理石;人造大理石;仿云石
scagliola marble 人造大理石
scagliola slab 仿云石板
scalability 可量测性;可扩缩性;可称量性

scalable 可攀登的
scalable architecture 可扩展体系结构
scalage 衡量;估量
scala media 中阶
scalar 无向量的;无矢量;梯状的;纯量;标量
scalar access control 标量存取控制
scalar-array operation 标量数组运算
scalar assignment 标量赋值
scalar attribute 标量属性
scalar axis 无向量轴;标量轴
scalar calculation 标量计算
scalar computation 标量计算
scalar computer 标量计算机
scalar constant 标量常数
scalar coupling 标量耦合
scalar covariance matrix 纯量协方差矩阵
scalar density 标密度;标量密度
scalar diffraction theory 标量衍射理论;标记衍射理论
scalar electrical conductivity 标量电导率
scalar equation 无向量方程
scalar expansion 标量扩张
scalar expression 标量表达式
scalar field 无向量场;标量场
scalar flux 无向量通量;无向量流;标量通量
scalar function 无向量函数;纯量函数;标量函数
scalar horn 梯形喇叭;标量喇叭
scalariform 梯子状的;(阶)(梯)状的;梯状(的)
scalariform conjugation 梯形结合;梯形接合
scalariform duct 梯纹导管
scalariform marking 梯纹
scalariform perforation 梯状穿孔
scalariform perforation plate 梯状穿孔板
scalariform pitting 梯纹孔式;梯纹加厚
scalariform vessel 梯状穿孔导管;梯纹导管
scalar impedance 无向量阻抗
scalar instruction 标量指令
scalar intensity 标量强度
scalar interaction 标量相互作用
scalar irradiance 标幅照度
scalar item 纯量项;标量项
scalarization 标量化
scalar lepton pairs 标量轻子对
scalar Liapunov function 标量的李亚普诺夫函数
scalar magnetic potential 标量磁位
scalar magnetometer 标量磁力仪
scalar magnetotelluric method 标量大地电磁法
scalar matrix 纯量矩阵;标量矩阵
scalar mean 代数平均流
scalar meson 标量介子
scalar meson field 标量介子场
scalar meson theory 标量介子理论
scalar metabolism 无向量代谢
scalar multiplication 纯(标)量乘法;标量乘法
scalar-multiplier coefficient unit 常系数装置(指输入输出之比为常数)
scalar nature 标量性质
scalar network analyzer 标量网络分析器
scalar neutrino 标量中微子
scalar operator 标量算符
scalar pair theory 标量偶理论
scalar parameter 标量参数
scalar particle 标量粒子
scalar photon 标量光子
scalar pipeline 标量流水线
scalar potential 标位;标势;标量(位)势;标位
scalar potential field 标量位场;标量势场
scalar pressure 标量压强;标量压力
scalar processing 标量处理
scalar processing unit 标量处理单元
scalar processor 标量处理机
scalar processor unit 标量处理部件
scalar product 内积;无向积;数(量)积;纯量(乘)积;标量(乘)积;标积
scalar product cycle 无向量积循环;标量积循环
scalar product matrix 纯量积矩阵
scalar property 无向性;无向量;数量特征
scalar quantity 无向量;纯量;标量
scalar random variable 标量随机变量
scalar register 标量寄存器
scalar result 标量结果
scalar-return register 标量返回寄存器
scalar-save register 标量保存寄存器
scalar singlet 标量单态

scalar speed 标量速度
scalar stream 标量流
scalar sum 数和
scalar system 分级体制
scalar-tensor effect 标量张量效应
scalar-tensor theory 标量张量理论
scalar triple product 纯量三重积;成标三重积
scalar type 标量类型
scalar value 纯量值
scalar variable 标量变量
scalar-vector effect 标量矢量效应
scalar wave 标量波
scalar wave equation 标量波(动)方程
scalar wave theory 标量波(动)理论
scalary 有阶段的;如梯的
scala tympani 鼓阶
scald 烫伤
scalding 脱釉
scale 比例(尺);比尺;规模;尺度;结垢;锈皮;锈(氧化皮);硬壳;鳞屑;鳞片;鳞皮;进位制;检尺;计数法;级别;品级;攀登;污垢沉积物;天平;水碱;水垢;度盘;除锈;除水垢;分划;标度;磅秤
scale accuracy 比例尺精度
scale-adjusted 按比例调整的
scale-adjustment 按比例校正
scale advantage 规模损失;规模收益
scale along central median 中央纬线比例尺;中央经线比例尺
scale analysis 尺度分析
scale and corrosion inhibitor 阻垢缓蚀剂
scale arc 标度弧
scale arm 天平(秤)杆;秤杆
scale at mid-point 中点比例尺
scale base 刻度单位;标度
scale batcher 称量配料器
scale beam 天平秤杆;天平梁;秤杆
scale board 胶合板;极薄木板;薄木板;玻璃框镜背板;玻璃镜框背板
scale-borer 磁力除垢器;除垢器
scale box 称料箱
scale breaker 氧化皮清除器;水垢清楚器;氧化皮清理机;锅垢清除器;除鳞机
scale-breaking term 无标度性破坏项
scale bucket 称量斗;储鳞箱
scale buoy 除垢电浮子
scale car 秤量车;称量车;磅秤车
scale card 刻度盘
scale change 尺寸变化;标度改变;标度变化;比例尺变换
scale changer 比例尺变换器
scale characteristics 刻度特征
scale check 比例尺检定
scale cinder 氧化皮
scale circle 刻度圈
scale class 比例尺种类
scale classification 比例尺分类
scale coated surface 水垢覆盖面
scale coefficient 污垢系数;水垢热阻系数;比例因子;比例系数
scale compression 标度压缩;比例尺缩小
scale computation 比例尺计算
scale condition 缩尺条件;比尺条件
scale control 结垢控制;尺度控制;防结垢
scale conversion 标度变换;比例尺换算
scale conversion table 尺度换算表
scale converter 标度变换器
scale crust 锈块;垢层;水垢层
scale cutting 鳞片插
scaled 鳞片状;有刻度的;屋瓦排列成鳞状的;成比例的
scaled decimal arithmetic 带比例的十进算术运算
scaled distance 折合距离;换算距离;投影距离
scaled down 缩小比例的
scaled down version 缩尺表示
scaled drawing 按比例作出的图
scaled dune 鱼鳞状沙丘
scaled energy 折合能量;换算能量
scale denominator 比例尺分母
scale departure 比例尺偏差
scale deposit 积垢;污垢沉积物;水垢沉积(物);水垢
scale deposition 积垢
scale descaler 氧化皮清除器
scale design 尺度设计;标度设计
scale designation 标度标记

scale deterioration 比例尺误差
scale determination 比例尺确定
scale dial 度盘秤
scale difference 比例尺误差
scale discharge 按比例卸货
scale disk 标度盘
scale distance 标距
scale distance of air base 依比例尺空中基线距离
scale distortion 缩尺变态;尺度变形;比例尺变形;比例变形
scale distribution 比例分配
scale division 刻度;分刻度;分度(间距)
scaled model test 模型试验;缩尺模型试验
scale down 递减;按比例缩小;按比例缩减;按比例减小;比例缩小
scale down loose rock 剥落疏松岩石
scale down model 按比例缩小的模型
scale down price 降价
scale down test 模型试验;模拟试验
scaled plating 鳞片状镀层
scaled pulse 定标脉冲
scale drawing 缩小图;缩尺图;比例图(样);用比例尺画的图
scaled rule 量尺
scaled sample 封样
scaled-up 比例增大
scaled value 折合值;换算值
scale economics 按比例的经济
scale economy 规模经济
scale effect 比例作用;刻度效应;结垢影响;缩尺效应;水垢影响;尺度效应;超范围误差;放缩效应;放大效益;标度效应;比例效应;比例(尺)影响;比尺效应
scale effect problem 尺度效应问题
scale equation 标度方程
scale error 刻度误差;定标误差;尺度误差;长度变形;标度误差;比例(尺)误差
scale error card 刻度误差表
scale error of gaussian projection 高斯投影长度变形
scale exaggeration 比例尺放大
scale expansion 刻度扩展;标尺扩展;比例扩展
scale experimental model 数量试验模型
scale extension 尺头
scale eyepiece 带标目镜
scale factor 换算系数;污垢系数;缩尺因数;缩尺比例;定标因素;定标因数;尺度因子;尺度因数;标度因子;标度因数;标度系数;比例因子;比例系数;比例(尺)因子
scale factor check 换算系数校验;比例因子校验
scale factor designator 比例因子指示符
scale factoring 标度选配;比例尺选择
scale factor method 比例因数法
scale factor potentiometer 比例系数电位器
scale feeder 称量式给矿机;称量给料机
scale for load vehicles 载重车磅秤
scale formation 生垢;氧化皮的形成;结成水垢;结成锅垢;生成水垢
scale forming 结成锅垢;生成水垢
scale for moments 力矩标度
scale for sea 风浪等级
scale for-swell 涌浪等级
scale for varying latitude 不同纬线的比例尺
scale free 无锈皮的;无氧化皮的
scale free heating 无氧化加热
scale-free heating furnace 无氧化加热炉
scale gutter tile 鳞状沟瓦
scale-handling 清除氧化皮;清除水垢
scale-handling system 氧化皮排除系统
scale height 刻度高度;均质大气高度;标尺高
scale hook 秤量吊钩
scale hopper 计量(装)料斗;称量斗;计量漏斗
scale incrustation 锅垢;结垢;水锈;水垢
scale index error 分划指示误差
scale indication 刻度指示
scale indicator 刻度指示器
scale infrared theodolite 标尺红外经纬仪
scale inhibition 阻垢
scale inhibition and dispersion 阻垢分散
scale inhibition and dispersion agent 阻垢分散剂
scale inhibition performance 阻垢性能
scale inhibition substance 阻垢剂
scale inhibitor 阻垢剂;结垢抑制剂
scale in parallel 平行圈尺度

scale input 比例输入
scale insect 介壳虫
scale interval 刻度间距;刻度间隔;分划间隔
scale invariance 标度不变性
scale-invariant theory 标度不变理论
scale jacket 氧化皮壳
scale label 标尺刻度
scale lamp 指示灯;度盘照明灯
scale lamp dimmer control 刻度照明亮度控制
scale law 比例法则;比例定律
scale layer 锈皮层
scale leaf 鳞片叶
scale length 分划长;标度(尺)长度;标尺长度
scale length of faults 断层规模
scale-less of free heating 少无氧化加热
scale level 刻度水准器
scale lever 秤杆
scalelike 鳞状
scale-like glaze 鳞片状釉
scale limitation 仪器刻度范围;比例尺范围
scale line 直线比例尺;标度线
scale load 标准荷载;标准负荷
scale loading 比例加荷
scale lock 度盘锁档
scale lofting 比例放样
scale loss 氧化皮损耗
scale map 有比例地图;标尺图;比例(尺)图
scale mark 氧化麻点;刻度(线);尺度分划;分划标记;分度;氧化铁皮印痕
scale master 计量师
scale matching analysis technique 比例尺匹配分析法
scale merit 规模价值;规模指标;规模效益
scale micrometer 千分尺;分划尺测微器
scale microscope 刻度显微镜;读图显微镜;分划尺显微镜
scale mill 除锈滚筒
scale model 按比例缩小的模型;相似模型;几何相似模型;缩制模型;缩尺模型;标度模型;比例模型;比尺模型
scale model flow 模型流动
scale model investigation 模型研究;缩尺模型研究
scale model law 比例模型律
scale model layout 比例模型设置
scale model study 缩尺模型研究
scale model study of obstacles 障碍缩尺模型研究
scale model test 比例模型试验
scale model turbine 相似模型涡轮机
scale model vehicle 比例模型车
scale modulus 尺度模数;标度模数
scale multiplier 标度乘法器
scalene spheric(al) triangle 任意球面三角形;不规则球面三角形
scalene triangle 不规则三角形;不等边三角形
scalene triangle network method 不规则三角网法
scalenohedron 偏三角面体
scale numbering 刻度数码;标度尺数字
scale of abscissa 横坐标轴比例尺;横坐标比例
scale of a drawing 图纸比例
scale of aerial photograph 航空照片比例尺
scale of annual fee applicable 每年的费用标准
scale of assessment 分摊比例额;分摊比例表;分摊比额表
scale of bursting water 突水规模
scale of cave collapse 洞穴塌陷规模
scale of charges 收费比例;费用率
scale of chart 海图比例尺
scale of chord 锐角弦长的缩尺
scale of coefficient of kinematic viscosity 运动黏(滞)度系数比例
scale of construction 建设规模
scale of depth 深度比例(尺)
scale of depth axis 深度轴比例尺
scale of distance 距离刻度
scale of dwelling unit 居住单元规模
scale of earthquake intensity 地震强度分级;地震烈度表
scale off 刮掉(锅垢、铁锈等);片落;脱皮;鳞落
scale of fees 费用表
scale of feet 英制比例尺
scale of fissility 劈度表
scale of flow resistance factor 水流阻力系数比例
scale of flow shear force 水流剪力比例
scale of fold 褶皱规模
scale of fusibility 熔度标
scale of geologic(al) surveying 地质调查比例尺
scale of gradations 分等运价表
scale of hardness 硬度结垢;硬度等级;硬度表;硬度标准
scale of height 高度比例尺;竖比例
scale of horizontal equivalent 坡度尺
scale of intensity 烈度表;强度等级;强度表
scale of intensity axis 强度轴比例尺
scale of investigation 调查比例
scale of investigation work 已有调查工作比例
scale of just temperament 自然音阶
scale of latitude 纬度比例尺
scale of length 长度比例(尺)
scale of longitude 经度比例尺
scale of magnification 放大比例(尺)
scale of map 地图比例尺
scale of meter 仪表刻度;公制比例尺;米制比例尺
scale of mine 矿山规模
scale of notation 记数法;记数单位
scale of operation 定标操作;经营规模;操作规模
scale of outcrop 露头规模
scale of photography 摄影比例尺
scale of points 鉴定标准
scale of pollution 污染分级标准;污染标度
scale of professional charges 业务收费标准
scale of project 工程规模
scale of proportions 比例尺度;比例大小;分配等级
scale of rates 运费表;费率表
scale of ratios 协调比例;相称比例;比值标准
scale of reduction 缩小比例尺;缩微率;缩尺
scale of regional geologic(al) survey 区域地质调查比例尺
scale of relative values 相对价值的比例表
scale of roughness 粗糙(度)标度;糙度等级
scale of saturated sediment content 饱和含沙量比例
scale of sediment content of density current 异重流含沙量比尺
scale of sediment discharge per unit width 单宽输沙量比例
scale of sediment of specific weight 泥沙容重比例
scale of sediment transport 输沙率比例
scale of seismic intensity 地震强度计;地震强度分级;地震强度表;地震烈度表
scale of shear force of sediment initiation motion 泥沙起动剪切力比尺
scale of sixteen 十六分标
scale of slope 坡度(标)尺
scale of state of sea and swell 风浪和涌浪状况等级
scale of station 车站规模
scale of surf 破波级;碎波级
scale of survey 测图比例尺
scale of swell 涌浪等级
scale of tax 税率表
scale of tectonic movement 构造运动规模
scale-of-ten 十分标
scale-of-ten circuit 十分标电路;十进制换算电路;十进位定标器
scale of the photograph 照片比例(尺)
scale of thermometer 温度标;温标
scale of tolerance 耐性等级
scale of topographic(al) map 地形图比例尺
scale of turbulence 流紊度;紊流尺度;紊流标度;湍流尺度;湍流标度
scale-of-two 二进位;二分标;二定标
scale-of-two counter 二进位计数管;二分标度计数装置
scale of visibility 能见度(等级);视级
scale of wind-force 风级(表)
scale of wind waves 风浪等级
scale of works 工程规模
scale operation 换算操作;标度操作
scale order 分段订单
scale out 超过刻度范围;超过尺寸范围;超出尺寸范围
scale pair 标度星对
scale pan 天平盘;秤盘
scale pan arrester 天平盘托
scale paper 坐标纸;计算纸;方格纸;比例纸
scale paraffin 片状石蜡
scale parameter 尺度参数;标度参数;标尺参数
scale pattern 铁皮痕
scale picoammeter 对数标度微微安培计
scale picture 鳞纹;缩尺图形
scale pit 铁皮坑;水垢槽;地磅槽
scale plate 刻度盘;刻度板;标尺;标度盘
scale pocket 称量筐
scale point 标度点
scale preservation 比例尺不变性
scale preventive 水垢防止剂;防垢剂;防垢
scale-production water 生垢水
scale projector 刻度投影器;标度投影器;标度放大器
scale quantity 标量
scaler 检尺员;脉冲计数器;脉冲分频器;换算装置;换算器;换算(电路);去垢剂;水垢净化器;定标器;除锈器;除铁鳞器;除壳器;铲拷工具
scale range 刻度量程;刻度范围;比例范围
scale ratio 尺度比;标度比;比例尺比率;比例;比尺
scale-ratio mosaic 拼接镶嵌图
scale ratio of roughness 糙率比尺
scale reading 刻度(盘)读数;标度读数
scale reading projection 标度投影
scale-reading technique 读数法
scale reciprocal 比例尺倒数
scale relation 缩尺关系;比例关系;比尺关系
scale relationship 比例关系
scale removal 去除氧化皮;清除氧化皮;除鳞
scale removal system 氧化皮排除系统
scale representation 比例表示法
scale resistance 水垢热阻
scale-resistant concrete 抗剥落混凝土
scale ring 刻度环;分度环
scaler invariance breaking 标度不变性破坏
scaler-printer 印刷换算装置;自记脉冲计数器
scaler-type indicator 定标式指示器
scale rule 比例尺;刻度尺
scales 体重秤
scales and diagrams for geochronology and isotope geology 地质年代学及同位素地质学图表
scale scar 鳞痕
scale series 比例系列
scale setting 刻度调整
scale settling tank 氧化皮沉淀池
scale-seven wind 七级风
scale shape 刻度形状
scale-shaped plain tile 鳞状平瓦
scale slate 刮板;刻度板
scale slovent 水垢溶剂
scale sluice system 氧化皮沟道排除系统
scale sluicing 冲洗氧化皮
scale solvent 溶垢剂;除垢剂
scale spacing 刻度间距
scale span 刻度间隔
scale-stable 比例尺不变的
scale stick 检尺杖
scale stone 锂云母
scale sump 氧化皮沉淀池
scale survey 按规定比例尺测图
scale system 度量制
scale tank 计量槽
scale test 比例模型试验
scale tile 鳞片状瓦
scale ton 体积吨;比例吨(50立方英尺=1比例吨)
scale trace 鳞迹
scale track 称重轨道;轨道衡器
scale transfer 比例变换
scale transfer point 比例变换点
scale transfer prism 刻度传递棱镜
scale transformation 尺度变换(法);标度变换;比例变换
scale transformation point 比例变换点
scale trap 氧化皮收集槽;铁屑槽;固体沉降器;固体沉降槽
scale truck 称量车
scale-type center drill 尺型中心钻
scale unit 格值;刻度单位;换算器;分划值;标度单位;比例尺单位
scale unit of log curve 测井曲线横向比例尺单位
scale-up 递增;放大;比例放大;按比例增加;按比例加大;按比例放大
scale-up effect 比例放大效应;比例放大效果
scale-up factor 按比例增加系数
scale-up model 放大模型

scale-up of thickener 浓缩池扩大
scale-up problems of technical process 工艺过程放大问题
scale value 标值;标度值
scale-value of gravimeter 重力仪标值
scale value of level 水准器角值;水准器分划值
scale value of radiometer 射气仪格值;辐射仪格值
scale variable 尺度变数;标量变数;标度变数
scale variation 比例尺变形
scale wax 粗石蜡;鳞状蜡;片状蜡
scale weight 称砣码
scale wood 木片板;层板
scale-work 鳞状排列
scale zero 刻度零点
scale zone 刻度区域
scaliness 起鳞程度
scaling 起氧化皮;消除危石;岩面清理;找顶;凿平;刮治术;结垢;去锈;起皮;起鳞;片落;缩放;定标;定比例;除锈;层状剥落;剥釉;剥落;标度无关;标定
scaling action 剥落作用
scaling analysis 标度无关分析
scaling assumption 标度假设
scaling bar 敲锈棒;除锈棒
scaling bottom 船底刮铲
scaling brush 除锈钢丝刷
scaling chip 剥落碎屑
scaling chipper 风凿;风镐
scaling chisel 除锈凿(子);铲锈凿
scaling circuit 校准电路;脉冲计数电路;定标器
scaling coating 剥落的窑皮
scaling constant 度量常数;标定常数
scaling data table 换算数据表
scaling-down 按比例缩小;按比例缩减;比例缩小
scaling-down process 降低计数比率过程;缩小比例尺度过程;分频过程
scaling drawing 设计图
scaling effect 形成氧化皮
scaling factor 计数递减率;换算因子;缩尺因数;缩尺比;定标因素;定标因子;定标因数;标度系数
scaling function 标度函数
scaling hammer 炉垢锤;琢面锤;錾平锤;锅锈锤;敲锈锤;敲水垢锤;除锈锤
scaling index 腐蚀指数
scaling inhibition ratio 阻垢率
scaling inhibition test 阻垢试验
scaling inhibitor 阻垢剂
scaling ladder 消防梯;云梯
scaling law 缩尺定律;比例规律;比例定律
scaling loss 结垢损失;烧损
scaling machine 去锈机;去垢机;除锈机;剥皮机床
scaling measurement 比例法
scaling method 密封法;缩尺法;定标法;标度方法;比例法
scaling model 调整模型比例尺
scaling of concrete 混凝土剥落
scaling off 去除水垢;脱落
scaling of logs 原木检尺;检尺
scaling of model 模型缩放
scaling of multiplicative utility function 积性效用函数的度量
scaling of variables 变量的比例
scaling on map 图上量测
scaling point 定标点
scaling position 小数点位置
scaling powder 助焊剂
scaling property 标度性
scaling rate 结垢速率;水垢增率
scaling ratio 定标因数;标度比
scaling relation 尺度关系;比例关系
scaling resistance 抗片落性;抗剥落性(能);耐起皮性
scaling rig 带升降平台的钻车
scaling rule 定标法则;比例法则
scaling system 比例换算系统;设计图;计算图(表);换算电路;定标系统
scaling the variable 按比例换算变量
scaling tool 去垢工具;清垢工具
scaling translator 比例尺转换器
scaling unit 换算电路
scaling-up 按比例加大;比例加大;按比例放大
scaling value 标定值
scaling violation 标度无关(性)破坏

scall 松岩;松软岩石
scallage 运尸体门
scallop 卵形洼地;刷帮;扇形花纹;扇贝;垂直像素畸变;贝壳形花纹;贝壳饰;屋檐板扇形花纹;粗糙不平的毛边;切成扇形
scallop boat 采扇贝船
scallop culture 扇贝养殖
scallop dredger 采扇贝船
scalloped 有圆齿的;褶形花边的;薄片材料边缘剪成曲线的
scalloped arch 扇形皱褶拱
scalloped bestpleat 扇形餐馆帘褶
scalloped bevel 半圆贝壳形斜边
scalloped capital 扇形柱头;贝壳柱头;贝壳饰柱头
scalloped leaf 扇形叶片
scalloped mo(u)lding 贝壳饰线脚
scalloped orbit 扇弧形轨道
scalloped rows of dots 扇贝形点线
scalloped strap butt 波形边搭板平接
scalloped surface 扇形表面;贝壳面
scalloper 采扇贝船
scallop hole 扇形孔
scalloping 开扇形孔;航向信号摆动;切成扇形
scallops 扇形凹口;贝壳型
scalp 刮平道路;突山顶;海岸岩石
scalped anticline 削峰背斜;裂口背斜
scalped extrusion ingot 修整的挤压锭
scalper 粗筛;筛石机;振动平筛机;护辊粗网;全面修整机;筛分机;锤劈石机;初级碎石机;分筛机
scalper screen 头道筛;粗筛
scalping 筛屑;石屑;筛出粗块;去表面层;投机买卖;筛除粗物料;筛除粗粒;倒卖;铲除植被;铲除草皮
scalping grizzly 粗格筛
scalping machine 清选机
scalping method 剔除法
scalping No. 通用筛号
scalping number 通用筛号
scalping screen 检查筛;去皮筛;头道筛;粗格筛;初筛机
scalping shoe 上筛架
scalp rock 筛余废石料
scaly 鳞状的
scaly coating 鳞片状涂层
scaly exfoliation 鳞状剥落;鳞剥作用;磷剥作用
scaly fabric 鳞状组织
scaly figure 鱼鳞纹;鳞状纹
scaly flute-like mo(u)ld 鳞形似槽模
scaly leg 石灰脚病
scaly part 鳞片
scaly structure 鳞状构造;鳞片构造
scaly talc 鳞片状滑石
scaly texture 鳞状结构
scaly zone 鳞片带
scamilli (抬高立柱或雕像用的)底座
scamillus 第二道底座;柱脚下的第二阶方石座
Scamozzi 斯卡摩齐柱式
scamp work and stint material 偷工减料
scan 扫视
scan access 扫描存取
scan address 扫描地址
scan-agraver 制版用光电自动装置
scan amplifier 扫描放大仪
scan amplitude 扫描振幅
scan area 扫描面积
scanatron 扫描管
scan axis 扫描轴
scan backward 反向扫描
scan band 扫描频带
scan control unit 扫描(控制)装置
scan-control voltage 扫描控制电压
scan conversion 扫描变换
scan conversion device 扫描变换器
scan conversion equipment 扫描变换设备
scan converter 扫描转换器;扫描变换器
scan converter tube 扫描转换管;扫描转换存储管
scan counter 扫描计数器
scan current waveform 扫描电流波形
scandent shrub 攀缘灌木
scan diagonal 扫描对角线
Scandia roofing tile 斯的亚屋面砖
scan-digitizing 扫描数字化
Scandinavian architecture 北欧建筑;斯堪的纳维亚建筑

Scandinavian plaster 斯堪的纳维亚石膏粉饰;斯堪的纳维亚灰浆
scan display system 扫描显示系统
scan distribution 扫描分布
scandium boride 硼化钪
scandium chloride 氯化钪
scandium deposit 钪矿床
scandium halide 卤化钪
scandium nitride 氮化钪
scandium ore 钪矿
scandium oxide 氧化钪
scandium silicate 硅酸钪
scan efficiency 扫描效率
Scanex 回转窑筒体表面温度监测系统
scan flyback interval 回描时间;扫描回程时间
scan forward 正向扫描
scan frequency 扫描频率
scan frequency racon 扫描频率雷康
scan gate number 扫描门数
scan image 扫描图像
scan-in 扫入;扫描输入
scan in bus 扫描输入总线
scan interlace ratio 扫描隔行比
scanister 扫描器
scan laser 扫描激光器
scan length 扫描长度
scan limit 扫描界限
scan line 精测线;扫描线
scan linearity 扫描线性
scan line drawing 扫描线绘制
scan line pixel 扫描线像素
scan line segment 扫描线段
scan magnification 扫描放大
scan method 扫描方法
scannable travelling wave phototube 扫描行波光电管
scanned aperture 扫描孔径
scanned area 扫描范围
scanned beam holography 扫描射束全息摄影
scanned detector hologram 探测器扫描全息图
scanned holography 扫描全息术
scanned IR interferometry 扫描红外干涉测量法
scanned-laser 扫描激光器
scanned laser infrared microscope 扫描激光红外显微镜
scanned-laser microscope 扫描激光显微镜
scanned picture 扫描图像
scanned point 扫描点
scanned spot 扫描点
scanner 雷达操纵员;扫视程序;扫描装置;扫描设备;扫描器;扫掠机构
scanner detector 扫描检测器
scanner display window 扫描器显示窗口
scanner input 扫描输入
scanner interrupt 扫描器中断
scanner-printer 扫描打印机
scanner recorder 扫描记录器;多点记录器;图像记录器
scanner selection 扫描器选择
scanner selector 扫描选择器
scanner simulator 扫描模拟器
scanner stabilization 扫描器稳定
scanner tube 扫描式发送管
scanner unit 雷达天线装置
scanner with infrared detector 红外(线)探测器扫描装置
scanner with photomultiplier detector 光电倍增管探测器扫描器
scanning 搜索目标;扫描
scanning accuracy 扫描精度
scanning acoustic(al) microscope 扫描声学显微镜
scanning aerial 扫描天线
scanning agent 扫描剂(同位素)
scanning algorithm 扫描算法
scanning amplifier 扫描放大器
scanning amplitude 扫描幅度
scanning angle 扫描角
scanning antenna 扫掠天线
scanning aperture 扫描孔(径)
scanning apparatus 扫描器
scanning area 扫描区域;扫描范围
scanning band 扫描带
scan(ning) beam 扫描射束;扫描电子束;扫描波束
scanning-beam current 扫描束电流

scanning-beam guidance 扫描波束制导
scanning-beam noise 扫描射束噪声
scanning calorimetry 扫描量热法
scanning camera 扫描相机
scanning circle 扫描圈
scanning circuit 扫描电路
scanning constant 扫描常数
scanning control 扫描控制
scanning control register 扫描控制寄存器；扫描控制记录器
scanning conversion 扫描转换
scanning converter unit 扫描变换装置
scanning copy 扫描稿
scanning current 扫描电流
scanning densitometer 扫描密度计
scanning density 扫描密度
scanning device 扫描装置；扫描器；巡回检测装置
scanning diaphragm 扫描膜片；扫描光栏
scanning digitizer 扫描数字化仪；扫描数字化器
scanning direction 扫描方向
scanning disc 扫描盘
scanning distortion of video signal 扫描引起的图像失真
scanning electrochemical microscope 电化学扫描显微镜
scanning electronic microscope 扫描电子显微镜
scanning electron micrograph 扫描电子显微照相；扫描电子显微照片
scanning electron microscope 扫描电子显微器；透射电子显微镜
scanning electron microscope image 扫描电镜图像
scanning electron microscopy 扫描电子显微（镜）术
scanning electron photomicrograph 扫描电子显微照片
scanning element 扫描元件；扫描像素
scanning equipment 扫描设备
scanning factor 扫描因数
scanning field 扫描场
scanning field probe method 扫描场探针法
scanning fluorimeter 扫描荧光计
scanning frame 扫描帧
scanning frequency 扫描频率；行速
scanning frequency racon 扫描频率雷康
scanning frequency radar responder 扫描频率雷达应答器
scanning function 扫描函数
scanning gap 扫描间隙
scanning gate 扫描门；扫描脉冲
scanning generator 扫描发生器
scanning genlock 扫描强制同步
scanning harmonic optical microscope 扫描谐波光学显微镜
scanning heating 渐进式感应加热；扫描热；操作过程加热
scanning high energy electron diffraction 扫描高能电子衍射
scanning hole 扫描孔
scanning hole distortion 扫描孔失真
scanning image dissector 扫描析像管
scanning impulse generator 扫描脉冲源
scanning in darkness 暗（靶）扫描
scanning infrared inspection system 扫描红外检查系统
scanning infrared spectrophotometer 扫描红外分光光度计
scanning instrument 扫描仪器
scanning integrity check 扫描完整性检验
scanning interval 扫描间隔
scanning in transmission 图像分解扫描
scanning ion microscope 扫描离子显微镜
scanning laser beam 扫描激光束
scanning laser radar 扫描激光雷达
scanning laser radiation 扫描型激光辐射
scanning lens 扫描透镜；扫掠透镜；瞄准透镜
scanning light beam 扫描光束
scanning light spot 扫描光斑
scanning line 扫描行；扫描线；分解行
scanning linearity 扫描线性
scanning linearization 扫描线性化
scanning line density 扫描线密度
scanning-line displacement 扫描线移动
scanning line frequency 扫描线频率；行速

scanning line length 扫描线长度
scanning line raster 扫描行光栅
scanning line spacing 扫描行间距
scanning line structure visibility 扫描线结构可见度
scanning line width 扫描线宽度
scanning loss 扫描损耗
scanning machine 扫描机
scanning mechanism 扫描机理
scanning method 扫描法
scanning microdensitometer 扫描微密度计
scanning micrograph 扫描显微照片
scanning microscope 扫描显微镜
scanning microscopy 扫描显微术
scanning microspectrophotometer 扫描显微分光光度计
scanning microwave spectrometer 扫描微波光谱仪
scanning mode 扫描方式
scanning monitor 扫描控制器；扫描监控器
scanning monochromator 扫描单色仪
scanning monochrometer 扫描单色器
scanning mosaic system 扫描镶嵌系统
scanning motion 扫描移动
scanning motor 扫描电动机
scanning multichannel microwave radiometer 多通道微波扫描辐射计
scanning object beam 扫描物光束
scanning objective 扫描物镜
scanning ocean bottom sonar 海底扫描声呐
scanning optical microscope 扫描光学显微镜
scanning optics 扫描光学系统
scanning output stage 扫描输出级
scanning parallax 扫描器视差
scanning parameter 扫描参数
scanning pattern 扫描图形
scanning pattern image 扫描图像
scanning pencil of light 扫描光束
scanning periscope 扫描潜望镜
scanning photograph 扫描摄影
scanning photography 扫描摄影术
scanning pitch 扫描行距；扫描节距
scanning plane 扫描平面
scanning plotter 扫描绘图机
scanning point 扫描点
scanning prism 扫描棱镜
scanning process 扫描过程
scanning proton microprobe 质子探针
scanning radar 扫描雷达
scanning radiometer 扫描辐射仪；扫描辐射计
scanning radiometry 扫描辐射测量
scanning raster 扫描光栅
scanning rate 扫描速率；扫描速度；扫描频率
scanning ray 扫描光线
scanning record 扫描记录
scanning recorder 扫描记录仪；带材边缘定位仪
scanning refractometer 扫描折射计
scanning refractometer 扫描折射计
scanning reticle 扫描调制器
scanning reversal 扫描倒相
scanning room 扫描室
scanning search 扫描搜索；扫描查找
scanning section coil 扫描线圈
scanning sensor 扫描传感器
scanning sequence 扫描序列；扫描顺序；扫描次序
scanning signal 扫描信号
scanning-size adjustment 扫描尺寸调整
scanning slide 扫描头
scanning slit 扫描（狭）缝
scanning slit plane 扫描狭缝平面
scanning sonar 扫描声呐
scanning sonar system 扫描声呐系统
scanning source 扫描光源
scanning spectrometer 扫描光谱仪；扫描分光仪
scanning speed 扫描速率；扫描速度
scanning spherical-mirror interferometer 扫描球面镜干涉仪
scanning spot 扫描点
scanning spot control 扫描点控制
scanning spot focusing control 扫描点聚焦控制
scanning stabilization 扫描稳定
scanning standard 扫描标准
scanning stereoscope 扫描立体镜
scanning stereoviewer （胶片可转动的）航空立体观察镜

scanning strip 扫描(天)线；扫描带
scanning stroke 扫描行程
scanning survey 扫描测量
scanning system 扫描系统
scanning technique 扫描技术
scanning telemeter 扫描遥测计
scanning telescope 搜索目标望远镜
scanning thermography 扫描温度记录仪
scanning time 扫描时间
scanning time base 扫描基线
scanning track 扫描方向
scanning transmission electronic image 扫描透射电子像
scanning transmission electron microscope 扫描透射式电子显微镜
scanning traverse 扫描横移
scanning velocity 扫描速度
scanning-velocity modulation 扫描速度调制
scanning watch 搜索守所
scanning waveform 扫描波形
scanning without gap 无间隙扫描
scanning X-ray micro-analyzer 扫描 X 射线微型分析器
scanning yoke 扫描线圈
scannogram 扫描记录
scan non-linearity 扫描非线性
Scano 斯康诺（一种预制木屋）
scan-off beam 消抹射束；描绘射束；扫描射束
scanography 扫描照相术
scan oscillator 扫描振子
scan-out 扫描输出；扫出
scan out bus 扫描输出总线
scan output 扫描输出
scan path 扫视途径
scan pattern 扫描模式
scan period 扫描周期
scan pointer 扫描指针
scan polarography 扫描极谱学
scan positioned distortion 扫描位置变形
scan prism 扫描棱镜
scan protection 扫描防护
scan pulse 扫描脉冲
scan rate 扫描率
scan registration 色度信号配准
scan resolution 扫描分辨率
scan retrace 扫描逆程
scan roof construction 预制铁片屋面建筑
scan-round 循环扫描
scan scanatron 扫描管
scan sensitivity 扫描灵敏度
scansion 图像分解
scan-spot 扫描光点
scan survey 扫测
scant 尺寸不够；克扣
scan table 扫描表
scan task 扫描任务
scant demand 需求不足
scanting 材料尺寸
scantle 测石板瓦用的量规；材料尺寸样板
scantle slating 铺砌小石板
scantling 小木方；小木料；小特厚材；小块石料；小块木料；小枋；样品草图；船体(构件)尺寸；草图
scantling draft 结构吃水；船材尺度吃水
scantling ga(u)ge 样板
scantling lumber 小方木
scantling number(al) 船材尺度定数
scantling of the structure 结构尺寸
scantling plan 总结构图
scant size 不留后备尺寸的
scant supply of materials 材料供应不足
scant supply of water 供水不足
scant wind 微风
scanty 不够尺寸的
scanty cover 薄覆盖层
scan type 扫描类型
scan-type graphics system 扫描型图形系统
scanty runoff 枯水径流
scanty water discharge 枯水流量
scanty water level 枯水位
scanty water runoff 枯水流量
scan visual 扫描信号发生器
scan width 扫描宽度
scape 凹线形脚；柱座凹线；柱身；柱底放大部分；柄节
scaphander 潜水器具

scaphe 仰仪
scapling screen 粗粒筛
scapolite 方柱石
scapolite hornfels 方柱石角岩
scapolite marble 方柱石大理岩
scapolite sharn 方柱石矽卡岩
scapolitization 方柱石化(作用)
scapple 石的粗琢；削平
scappled stone 粗形石块
scappling 外表粗的砖石建筑；粗形石面；粗形石块
scapus 羽轴；柱形物；柱身
scar 孤岩；炉疤；海滨平台；片痕；伤痕；断崖；陡岩坡；陡崖；低潮时露出的礁石；创痕
scarbroite 羟碳铝矿
scarce 稀有的
scarce currency 硬通货
scarcely noticeable 勉强可感地震
scarcely populated country 人口稀少国家
scarce means of production 紧缺生产资料
scarcement 砖墙收分；堤岸凹凸部；墙凹凸部；墙凹台；壁阶；踏脚处；墙上凹处；坑道梯架；公路路肩；堤岸平架
scarce metal 稀有金属
scarce raw materials 稀有原料
scarcity 稀缺性；缺货；短缺
scarcity of budgetary resources 预算经费短缺
scarcity price 缺货市场；缺货价格
scarcity value 缺货价值；稀缺价值；缺货价格
scare currency 稀有通货
scared bamboo 南天竹
scared strap (装配工人或钢铁工人的)安全带
scarf 斜嵌槽；斜面剂接；切口；嵌接件一端；嵌接；修切边缘；斜切面；斜面加工；斜接；凹线
scarf box 接箍
scarfbutt joint 斜对接接头
scarf cloud 积云云帽
scarf connection 斜嵌连接；嵌接
scarfed 斜接的
scarfed joint 木材搭接；斜面粘接；斜面胶合
scarfer 钢坯烧剥器；火焰清理机；嵌接片
scarfing 嵌接；斜接；斜嵌槽；火焰(表面)清理；气割清理
scarfing bed 火焰清理台
scarfing cinder 清渣
scarfing dock 烧剥室
scarfing half and half 对半嵌接
scarfing joint 斜搭接；斜口接合；嵌接
scarfing machine 火焰清理机；嵌接机
scarfing roll 焊管坯斜边成型轧辊
scarfing yard 火焰清理工段
scarf joint(ing) 斜接(接口)；斜口接合；楔面接头；嵌接(接头)
scarf planer 斜口刨床
scarf splice 镶嵌拼接
scarf tack 嵌钉
scarf together 斜接；嵌接；嵌接合
scarf weld(ing) 楔口焊接；斜面焊接；嵌焊；两端搭接焊
scarification 磨蚀处理；划痕法；松土皮；翻松
scarification method 刻划法
scarification test 划痕试验
scarificator 划痕器
scarifier 划痕器；耙路机；松土耙；松土机；翻土机；翻路机
scarifier attachment 松土设备；翻路设备
scarifier plough 松土犁；翻路犁
scarifier plow 松土犁
scarifier-scraper 松土平整机；松土平地机
scarifier shank 松土机手柄
scarifier tine 松土(机)齿杆
scarifier tooth 松土机耙齿
scarifier tyne 松土(机)齿杆
scarify(ing) 翻松；划痕；刨毛
scarifying attachment 翻土设备；松土设备
scarifying depth 翻松深度
scarifying tooth 松土齿
scariose 干膜质的
scarlatina maligna 恶性猩红热
scarlet 绯红；猩红；鲜红(色)；深红色
scarlet acid 猩红酸
scarlet chrome 铬猩红
scarlet fever 猩红热
scarlet lake 酸性红色淀
scarlet phosphorus 猩红磷

scarlet red 猩红色；深红色
scarlet vermilion 深朱(红)
scarlet vermilion pigment 鲜红银朱颜料
scarp 悬崖；险坡；险波；直切；马头丘；沙坎；陡崖；陡坎
scarped plain 梯级平原
scarped ridge 单斜山；单面山
scarper puhser 铲运推土机
scarp face 悬崖坡
scarph 嵌接
scarp heap 废料堆
scarping 崖底侵蚀
scarp-iron 碎铁
scarplet 滑坡；小崖【地】
scarp line 悬崖线
scarp side 悬崖壁
scarp slope 悬崖坡；陡坡
scarp steel 碎钢
scarp wall 陡坡墙
scarred pebble 凹痕卵石
scars 冻结物
SC asphalt 慢凝沥青
scatter 分散器；散射
scatter absorption coefficient 散射吸收系数
scatter admittance 泄漏电导
scatterance 散射比(率)；散布
scatterance meter 散射比计
scatter baffles 散射障板
scatter band 分散频带；分散带
scatter blue E-4R 分散艳蓝 E-4R
scatter channel simulator 散射信道模拟器
scatter coating method 分散涂覆法
scatter coefficent 散布系数
scatter communication 散射通信
scatter communication system 散射通信系统
scatter communication transmitter 散射通信发射机
scatter detection and ranging 散射探测与测距
scatter diagram 试验点散布图；散列图解；散点图；散布图；散布分析；点子散布图；点图；分散图；分布图
scatter diagram of size parameter 粒度参数离散图解
scatter dispersion 分散
scatter echo 散射体回波
scattered 杂散天空
scattered allocation 分散布局
scattered band 散射频带宽度；散布范围
scattered clouds 散云
scattered covering 散布覆盖
scattered distribution 分散布局
scattered dose 散射剂量
scattered drilling (不按勘探网的)零散钻探
scattered echo 散射回波
scattered electron 散射电子
scattered feeding system 分散供电方式
scattered field 散射场
scattered freight storehouse 零担仓库
scattered funds 闲散资金
scattered gamma-ray log 散射伽马测井
scattered ice 稀疏浮冰
scattered ion 散射离子
scattered light 光散射；散射光；分散光
scattered light wave 散射光波
scattered load 分散负荷；分散荷载
scattered mixed forest 散生的混交林
scattered motif 散乱图案
scattered observation 分散观察
scattered-out beam 散射束
scattered outgoing wave 散射行波
scattered planting 散植
scattered plots of unutilized land 闲散土地
scattered power 散射率；散射功率
scattered quantum 散射量子
scattered radiation 散射辐射
scattered ray 散射线
scattered reflection 扩散反射；漫(反)射；散反射
scattered rocks 散开的礁石
scattered sheaf 散射层
scattered shoal 散滩；散乱浅滩
scattered shower 大面积降雨；大面积降水；分散稀疏阵雨
scattered signal 散射信号
scattered stone 散点石

scattered stray light 散射杂光
scattered striation 散射条纹
scattered trap 散射阱格栅
scattered wave 散射波
scatterer 扩散器；散射体；散射器
scatter format 分散格式
scatter-fringe interferometer 散射条纹干涉仪
scatter gather operation 分散集中操作
scattergram 分布图
scattergram method 点聚图法
scattergraph 分布图
scattering 扩散；漫射；漫散；撒布；播散；散射
scattering absorption 散射吸收
scattering amplitude 散射振幅；散射幅度
scattering angle 漫射角；散射角
scattering angular distribution 散射角分布
scattering area 散射面积
scattering area coefficient 散射面积系数
scattering area ratio 散射面积比
scattering atmosphere 散射大气
scattering attenuation coefficient 散射衰减系数
scatteringbeam intensity 散射束流强度
scattering beater 撒布轮
scattering board (撒播机的)撒播板
scattering body 散射体
scattering by crust inhomogeneity 地壳不均匀性散射
scattering by irregular interface 不规则界面散射
scattering cell 散射池
scattering center 散射中心
scattering coefficient 漏损系数；散射系数
scattering coefficient meter 散射系数表
scattering contrast 散射反差
scattering converter unit 扫描变换装置
scattering corrected value 散射修正值
scattering cross-section 散射(横)截面
scattering detector 散射探测器
scattering diagram 散射图
scattering differential 散射微差
scattering diffusion tensor 散射扩散张量
scattering effect 散射影响；散射效应
scattering efficiency 散射效率
scattering factor 散射因数
scattering field 散布范围
scattering flight 撒料勺；撒料板
scattering formula 散射公式
scattering frequency 散射频率
scattering function 散射函数
scattering geometry 散射几何条件
scattering-in 内部散射
scattering intensity 散射强度
scattering interference 散射干涉
scattering inverse method 散射求逆方法
scattering law 扩散定律；散射定律
scattering layer 散射层
scattering length 散射长度
scattering loss 散射损失；散射损耗
scattering loss in optic(al) thin-film 光学薄膜的散射损耗
scattering machine 散射机
scattering matrix 散射矩阵
scattering mean free path 散射平均自由程
scattering measurement 散射测量
scattering mechanism 抛撒装置；散射机理
scattering medium 散射介质；分散介质
scattering meter 散射仪
scattering microcomputer network 分布式微机网络
scattering model 散射模型
scattering nature 散射性质
scattering noise 散射噪声
scattering of light 光的散射
scattering of particle 粒子散射
scattering of radiant energy 辐射能的散射
scattering of radiation 辐射散射
scattering of sound 声散射
scattering of the points 点的分散(度)
scattering of wave 波散射；波浪散射
scattering of wave packet 波包散射
scattering operator 散射算子
scattering parameter 散射参数
scattering pattern 散射图样
scattering phase 散射相
scattering phase shift 散射相移

scattering polar diagram 散射极坐标图
scattering potential 散射势垒
scattering power 散射长度;散射本领;散射(能)力
scattering probability 散射几率
scattering process 散射作用
scattering propagation 散射传播
scattering property 散射特性
scattering reaction 散射反应
scattering-read 分散读取;分散读出
scattering reflection 漫反射
scattering region 散射区域
scattering scoop 撒料勺
scattering state 散射状态
scattering stone 飞石
scattering strength 散射强度
scattering surface 散射面
scattering system 散射体系
scattering tail 散射尾部
scattering target 散射靶
scattering theory 散射理论
scattering threshold 散射阈
scattering tinting strength 散射着色力
scattering-type pressure ga(u)ge 散射式压力计
scattering volume 散射体积;分散容积
scattering wave 散射波
scattering wave communication 散射波通信
scatter interferometer 散射干涉仪
scatter load(ing) 分散装入程序;分散装入(法);分散装料
scatter load method 分散装入法
scatter matrix 散布矩阵
scatter of points 点据分散;点的分布
scatter of value 数值离散性
scatterometer 散射仪;散射计;散射测定计
scatterometry 散射测量
scatter operation 分散操作
scatter plate 散射板
scatter plotter 分布图
scatter range 分散范围
scatter read 分散读入;分散读取;分散读出
scatter read and write 分散读写
scatter reading 分散读入
scatter read operation 分散读入操作
scatter rug 小块地毯
scatter seed 播种
scatter sheave crown 五滑轮冕形滑车
scatter site housing 零散住房;分散住宅
scatter sounding 散射探测
scatter storage 散布存储;分散存储
scatter storage technique 分散存储法
scatter table 分散表
scatter unsharpness 散射所致模糊
scatter warning system 散射报知通信系统
scatter write operation 分散写操作
scaur 陡崖
scavenge area 扫气孔面积
scavenge blower 扫气鼓风机
scavenge delivery 油泵输出管
scavenge gears 油泵回油器
scavenge gradient 扫气梯度
scavenge line 扫气管道
scavenge manifold 扫气箱
scavenge oil 轴承回油;废油
scavenge oil disposal 废油处置
scavenge oil pump 换油泵
scavenge oil pump body 换油泵体
scavenge phase 扫气期
scavenge pipe (内燃机的)回油管;清洗管;排泄管(道)
scavenge port 换气口;排泄口
scavenge pressure 除垢压力;换气压力
scavenge pump 扫气泵
scavenger 纯化剂;选油;净化剂;清扫垃圾者;清洁工(人);清道工(人);清除器;清除机;清除剂;扫街机;尘垢清除器
scavenger circuit 清洗回路
scavenger concentrate 扫选精矿
scavenger effect 清除剂效应
scavenger return pipe (内燃机的)回油管(
scavenger fan 换气风扇;排气风扇
scavenger floatation circuit 扫选循环
scavenger pipe 排出管;扫线管
scavenger precipitation 清除剂沉淀
scavenger pump 回油泵;换气泵;清除泵

scavenger's cart 扫街车
scavenger system 清洗系统;吹气系统
scavenger tube 排出管
scavenger valve 部分回油阀
scavenge space 扫气腔
scavenge sump 回油油箱
scavenge system 清除系统;扫气系统
scavenge trunk 换气管;扫气通道
scavenge tube 清除管
scavenge unit 清除装置
scavenge valve 扫气阀
scavenge valve chest 扫气阀箱
scavenge water seal 清洗水封
scavenging 换气;清洗;清除的;扫选;扫气;纯化;除气法
scavenging action 清除作用
scavenging action of iron hydroxide 铁氧化物的清除作用
scavenging action of manganese hydroxide 锰氧化物的清除作用
scavenging agent 净化剂;清除剂
scavenging air 清洗(积垢)空气;清净空气;清洁空气;清除空气;反吹风
scavenging air belt 扫气道
scavenging air box 扫气箱
scavenging air damper 扫气调节风门
scavenging air non-return valve 扫气空气止回阀
scavenging air pipe 扫气管
scavenging air pump 换气泵
scavenging air receiver 扫气箱
scavenging air trunk 扫气总管
scavenging air valve 换气阀;扫气阀;打气阀;吹洗空气阀;反吹风阀
scavenging arrangement 扫气装置
scavenging blower 鼓风清扫器;换气鼓风机;清除鼓风机
scavenging duct 换气管;扫气导管
scavenging efficiency 扫气效率
scavenging ejector 清洗喷射泵
scavenging engine 清除废气发动机;带有扫气泵的发动机
scavenging machine 清扫机;清除机
scavenging material 净化材料
scavenging period 扫气期;吹气期
scavenging piston 换气活塞
scavenging port 换气口
scavenging pressure 扫气压力
scavenging process 清除过程
scavenging pump 回油泵;清洗泵;清除泵
scavenging ratio 换气比;扫气比
scavenging scheme 规划
scavenging service 打扫服务;清除服务
scavenging stroke 清除冲程;扫气冲程
scavenging tower 净化塔
scavenging valve 扫气阀
scavenging with gas 气体清除
scaw 陡崖
scawtite 片柱钙石;碳硅钙石;水碳硅钙石
SCD system 预加应力混凝土设计体系
scedasticity 离中趋势;方差性
scena 古代剧场建筑
scenario 概要;格式;方案
scenario analysis 情况分析
scenario comparison method 方案比较法
scenario method 描述法
scence paint 布景漆
scend 上涌(船在波浪作用下的一种运动);(船体的)上颠;船的上升运动;船被波浪抬高
scend of sea 波浪抬升作用
scene 现场;一幕;一场;镜头;景象;舞台;出事地点;场面
scene analysis 景象分析;景物分析
scene brightness 景物清晰度
scene-building 舞台建筑
scene chart 景物测试卡
scene contrast 景物对比度
scene-corrected image 单张校正像
scene-dependent 随景物而变化的
scene dock 舞台布景存放处;道器布景储存室
scene matching 景像匹配
scene of accident 事故现场
scene of fire 火灾现场
scene of operation 作业地区

scenery 景物;景色;舞台面;舞台布景
scenery and scenic elements 风景与景观构成的单元
scenery of humanities 人文景观
scenery setting 布景
scenery store 布景库
scenery wagon 布景车;带轮布景台;推拉台;车台
scenes room 大型道具室
scene storage 布景储存室
scenic area 风景区
scenic beauty forest 风景林
scenic design 布景设计;风景设计
scenic easement 路界外景观保护权;路界外风景保护权;景点建筑
scenic effect 舞台效果
scenic focal point 对景
scenic forest 风景林
scenic highway 风景(好的)公路;游览公路
scenic lighting 舞台照明
scenic lines 风景线
scenicology 风景学
scenic overlooks 观景
scenic-overlooks intrusion 景观干扰
scenic railway 风景区小铁路
scenic reserves 风景保护区
scenic road 景色宜人的道路
scenic route 游览路线
scenic spot 名胜(地);风景区;风景名胜
scenic spot and historic resort 名胜古迹区
scenic spot reserves 风景保护区
scenic spots and historic(al)sites 名胜古迹
scenic tourist city 风景旅游城市
scenic view 风景观赏
scenic vintage point 景色眺望点
scenic zone 风景区
scenioscope 景象管;超光电摄像管
scenioscope machine 布景机
scenograph 透视图
scenographic design 透视图设计
scenography 写景法;透视图法;透视画法
scented guarea 呈灰白桃花心木颜色的硬木(产于尼日利亚)
scented mahogany 香味桃花心木
scented poplar 大叶钻天杨
scented ware 芳香陶瓷
scentometer 气味计
scent plant 芳香植物
scent spray 全景雷达
scent test 嗅试法(测挥发性物质密封程度);闻味探测(下水管找漏用);嗅味试验
scent test method 嗅试法
sceptron 声频滤波器
sceuophylacium 圣器室
S-C foliations S-C 面理
schachnerite 六方汞银矿
schadeite 水磷铝铅矿
Schaeffer's diagram 雪费尔焊缝组织图
schafarzikite 红锑铁矿
Schaffer floor 薛佛地板
Schaffer's acid 薛佛酸
schairerite 凶钠矾;氟氯钠矾
schalfnerite 铜钒铅锌矿
schallerite 砷硅锰矿
schalstein 辉绿凝灰岩
schamot(te) 耐火黏土熟料;烧耐火黏土
schamot(te) brick 黏土质耐火砖
scharnitzer 寒北风
schatzite 天青石
schaum 鳞片状
Schaumann's bodies 球形钙化层状小体
schaumgyps 鳞石膏
schaumkalk 鳞霰石
schaurteite 水锗钙矾
schcarite 硅重晶石
schedugraph 轨道作业计划图
schedulable variable 可调度变量
schedular system of taxation 分类税制
schedule 细目表;一览表;进度(表);进程表;计划表;日程(表);排定;时刻表;时间进度表;时间表;时间安排表;调度表;安排表
schedule air pipe 喷射导管
schedule analysis 进度分析;计划研究
schedule by probability mode 概率调度
schedule card 进度卡

schedule chart 工作
schedule control 预定输出控制;进度控制;进度管理
schedule control chart 日程控制图表
schedule controlled behavio(u)r test 程序控制操作行为试验
scheduled 预定的;列入计划
scheduled cam 程序凸轮
scheduled completion date 预定完工日期;计划完工日期;计划完工期发
scheduled cost 列举成本;编定成本
scheduled course-line 预定航线
scheduled daily yardage 计划日挖掘量;计划日产量
scheduled data 预计数据
scheduled date 指定日期;计划期限
scheduled days per week 计划每周天数
scheduled downtime 计划停工时间
scheduled energy 预计能耗;预计能量;额定能耗;额定能量
scheduled engineering time 预定工程用时间
scheduled flight 定期航班;班机
scheduled for repair 按计划修理的
scheduled frequency off-set 规定频率差度
scheduled hours per shift 计划每班时数
schedule diagram 进度计划表
scheduled interruption 计划断电
scheduled line capacity 有时刻表的线路容量
scheduled line capacity utilization coefficient 有时刻表的线路容量利用系数
scheduled maintenance 预定维修;例行维修;例行维护;进度维修;计划维修;日常维修;定期维修;定期维护;调度维护
scheduled maintenance time 规定维修时间;例行维护时间
scheduled net interchange 规定交换量;计划交换量;计划交换功率
schedule docket 工程(用款)清单
scheduled operating time 规定操作时间
scheduled operation 被调度的操作
scheduled outage 计划停电
scheduled overhaul 定期大修
scheduled passenger-cargo liner 客货班轮
scheduled payment 定期付款
scheduled plan 进度计划
scheduled production 计划产量
scheduled program(me) for allocating resource 分配资源调度程序
scheduled purchasing 计划采购
scheduled radio communication service 定时开放的无线电通信业务
schedule drawing 工序图;工程图;计划图
scheduled repair 定期修理
scheduled repayments on credits 信贷的计划偿还;安排的信贷偿还
scheduled running speed 规定运行速度
scheduled sailing time 预定开航时间
scheduled service 定期船运输
scheduled shifts per days 计划每天班数
scheduled speed 额定速率
scheduled substance 列入清单物质
scheduled target 预计目标
scheduled time 行车时刻;预定时间;规定时间;计划时间;按规定日
scheduled time standard 预定时间标准
scheduled time standard method 预定时间标准法
scheduled train number 核心车次
scheduled works 预定工程
schedule engineering time 排定的工程时间;安排工程时间
schedule file 调度文件
schedule for repayment of principal 偿还本金的期限
schedule income tax 分类所得税
schedule into production 列入生产计划
schedule job 调度作业
schedule list 日程表
schedule method 列表法;图表法;表格法
schedule number 管子长号;表示管壁厚度系列的号码;表号
schedule of accommodation 供应商时间表;设备清单
schedule of accounts payablere(ceivable) 应收账明细表;应付账明细表
schedule of administrative fee 行政费表
schedule of amortization 摊销明细表

schedule of apportionment 分配明细表
schedule of bidding price 投标报价表
schedule of charges 费率表
schedule of collection 收款清单
schedule of commission charges 手续费率一览表
schedule of construction 工程进度表;建筑一览表;计划施工进度;施工(进度)计划;施工进度表
schedule of construction operation 施工进度表
schedule of contract rates 合同单价表
schedule of contribution 分摊数额表
schedule of control 工程管理;进度控制
schedule of depreciation 折旧明细表
schedule of design 设计进度(表)
schedule of dilapidations 房屋承租人的修房清单;损毁修理费清单;住房失修项目一览表
schedule of disbursement 支出清单
schedule of doors and windows 门窗一览表
schedule of doors and windows and ironmongery 门窗及五金表
schedule of earth-works 土方工程进度表
schedule of erection work 安装工程进度表
schedule of financial statements 财务报表附属明细表
schedule of fixed charges 固定费用明细表
schedule of fixed overhead charges 固定制造费用明细表
schedule of freight rates 运价表;运费率一览表
schedule of increase and decrease of property 财产增减表
schedule of inspection and testing 检验与考核表
schedule of insurance 保险清单
schedule of manufacturing overhead 制造费用明细表
schedule of materials 材料表
schedule of materials on order 已定购材料表
schedule of operation 营运计划;作业计划;操作程序;施工进度表
schedule of payments 付款清单;付款表;付款计划
schedule of periodic(al) tests 定期测试时间表
schedule of prices 估价表;价目表;价格清单;价格表
schedule of progress 进度表
schedule of quantities 工程数量表;数量清单;数量(明细)表
schedule of rate(s) contract 单价(表)合同
schedule of rates 单价表;保险费率表
schedule of rates and prices 单价及价目表
schedule of requirement 投标要求表
schedule of room finishes 房间做法表
schedule of salvage operation 打捞施工进度表
schedule of sizes 尺寸清单
schedule of structural steel 钢材一览表
schedule of supplementary information 补充资料表
schedule of tangible fixed assets 有形固定资产明细表
schedule of terms and conditions 收费率表
schedule of transfer 转拨清单
schedule of turnover rate of circulating capital 流动资金周转率计算
schedule of usual tare weights 常用皮重清单
schedule of values 付款说明书;(建筑合同的);工程分项价值表;分项列价表价值一览表;价格一览表;价格表
schedule of work in process 在制品明细表
schedule of works 工程进度表
schedule of work to be performed during a week 一周工作计划
schedule panel 程序转换盘
schedule performance evaluation and review technique 调度性能评估技术
schedule planning 日程计划
schedule policy 分类保险单
schedule programming 日程计划
scheduler 生产计划员;调度程序;调度部件;程序机
scheduler module 调度(程序)模块
scheduler program(me) 调度程序
scheduler task 调度任务
scheduler work area 调度程序工作区
scheduler work area data set 调度工作区数据组;调度程序工作区数据集
schedule sailing time 计划开航时间
schedule sheet 工作日程表;日程表

schedule speed 规定速率;旅行速度【铁】;表定速度
schedule statistics 作业进度统计
schedule step 计划步骤
schedule time 预定时间
schedule weight 额定重(量)
schedule work 按进度表工作;按计划工作
scheduling 制时间表;工期安排;进度控制;进度(计划)安排;时间安排;调度;程序安排;编制图表;编制目录;编制计划;编排进度
scheduling algorithm 调度算法
scheduling control automation 自动程序控制
scheduling control automation by network system 网络系统自动程序控制
scheduling discipline 调度规则
scheduling intent 调度意向;调度意图
scheduling management display 调度管理显示
scheduling model 调度模型
scheduling monitor computer 调度监控计算机
scheduling operation 编制操作
scheduling pattern 日程安排形式
scheduling priority 调度优先(权);调度优先(级)
scheduling problem 时间问题;调度问题
scheduling process 调度过程
scheduling processor 调度处理机
scheduling program(me) 日程安排;排日程计划;调度程序
scheduling queue 调度排队
scheduling system 调度(程序)系统
scheduling theory 调度理论
scheduling user task 调度用户任务
scheduling work 安排工作
Scheele's green 席勒绿;亚砷酸氢铜
scheelite 重石;白钨砂;白钨矿
scheelite ore 白钨矿矿石
scheelite skarn deposit 矽卡岩白钨矿矿床
scheelitine 钨铅矿
scheererite 板晶蜡
schefferite 锰锥辉石;锰透辉石
Scheibel-York extractor 赛贝尔萃取器
Scheiber oil 赛帕油
Scheider kiln 施奈德型立窑
schema 模式
schema category 模式范畴
schema chart 模式图
schema data description language 模式数据描述语言
schema description language 模式描述语言
schema entry 模式项(目)
schema name 模式名
schematic 概略的;图解的
schematic arrangement of rock bolt 锚杆布置示意图
schematic arrangement of wire mesh 钢筋网示意图
schematic circuit 原理图;草图;原理电路
schematic circuit diagram 电路图
schematic-cross-section 示意剖面图
schematic design 纲要性设计;初步设计;草图设计;方案设计
schematic design documents 初步设计文件;方案设计文件
schematic design drawing 纲要性设计图;初步设计图
schematic design phase 初步设计阶段;方案设计阶段
schematic diagram 原理图;框图;简图;模式图;示意图
schematic diagram of constant head permeameter 恒定水头渗透仪示意图
schematic diagram of control panel 控制原理图
schematic diagram of falling-head permeameter 变水头渗透仪图示
schematic diagram of local lighting 局部照明示意图
schematic drawing 简图;示意图;初步设计图;方案图
schematic drawing of magnetic compensation 磁补偿结果示意图
schematic drawing of work program(me) 工序图
schematic eye 简化眼;模型眼
schematic graph of volcanic development stages 火山活动阶段示意图
schematic illustration 简图;示意图
schematic manner 图示法

schematic map 概图;概略地图;草图
schematic model 图解模型;图表模型
schematic outline 方框图
schematic piping diagram 管道系统示意图
schematic plan 平面示意图;平面方案;平面草图
schematic presentation construction 结构略图
schematic representation 示意图;简图;略图
schematic section 示意断面图
schematic two-phase soil 图示二相土
schematic way 图解法
schemation diagram 示意图
schematism 系统性组合
schematograph 视野轮廓测定器
scheme 流程图;图像;设计(图);大纲;草图;方案
scheme arch 扇形拱;缺圆拱;弧形拱;平弧拱
scheme comparison 方案比选;方案比较
scheme competition 方案竞赛
scheme design 草图设计
scheme drawing 计划图;方案图
scheme for irrigation and drainage 排水计划
scheme of a plan 计划方案
scheme of classification 分类表
scheme of colo(u)r 配色方案;着色法
scheme of concentration 选矿方案
scheme of connection 连接方案
scheme of distribution 分红表
scheme of erection 装配简图
scheme of fire 射击计划
scheme of flood prevention 防汛方案;防洪规划
scheme of framework 框架工程方案
scheme of haul 运土计划
scheme of installation 安装示意图
scheme of loading by stations 站别装车计划
scheme of operation 作业计划
scheme of parallactic distance measurement 视差环节
scheme of scantling 船体构件明细表
scheme of unloading by stations 站别卸车计划
scheme of wiring 线路图;配线图
schemer 计划者
scheme system 优先进入线路的信号联动系统
Schenick machine 谢尼克金属疲劳试验机
Schenkel doubler 申克尔电压倍增器
Schenk type filter 申克型过滤器
scherbinaite 钒石
schertelite 磷镁铵石
Scherzer roller bascule bridge 谢撤尔滚动式竖旋桥
scheteligite 钨锑贝塔石;锑钨烧绿石
schiefer spar 层方解石
schieffelinite 水硫碲铅石
Schiff base 席夫碱;希夫碱;甲亚胺
Schiff's reagent 品红试剂
schileren mirror 纹影设备镜
schiller 色彩闪变
schillerization 闪光现象;色闪动现象;呈现光彩
schillerization Aduralia 闪光冰长石
schiller spar 绢石
schiller structure 闪光构造
schilling rudder 先令舵
schindylesis 夹合缝
schirmerite 块辉铋铅银矿
schist 片岩
schist-carbonate formation of greenschist facies 绿片岩相型片岩—碳酸盐建造
schist formation of zonal andalusite/cyanite type 带状红柱石/蓝晶石型片岩建造
schistic 片状
schistoid 似片岩的
schistoid limestone 片岩石灰石
schist oil 页岩油
schistose 片状的
schistose clay 片岩状黏土;片岩质黏土;板状黏土
schistose cleavage 片状劈理
schistose crystalline rock 片状结晶岩
schistose gabbro 片状辉长岩
schistose granular 片粒状
schistose mica 云母片岩
schistose rock 片理化岩;层状岩体;层状岩石
schistose structure 片状构造;层状组织
schistose subbase 片岩基础
schistose texture 片状结构
schistosity 片理
schistosity belts 片理带
schistosity cleavage 片状劈理
schistosity plane 片理面
schistosity sheet 片岩板
schistosity zone 片理化带
Schistosoma 血吸虫属
schistosome 血吸虫
schistosome dermatitis 血吸虫皮炎
schistosomiasis 血吸虫病
Schistosomum 血吸虫属
schistous amphibolite 片状闪石
schistous sandstone 片状砂岩
schizokinesis 反应分裂
schlagblech 压纱板
Schlage lock 谢拉格锁(一种筒状锁)
schlicht function 单叶函数
schlieren 析离体;异离体;纹影;条纹照相;条纹法摄影
schlieren apparatus 纹影仪
schlieren device 纹影仪
schlieren interferometer 纹影干涉仪
schlieren method 条纹照相法
schlieren-method of photography 施利伦照相法
schlieren optic(al) screen 纹影屏
Schlieren optical system 舒利莱恩光学系统
schlieren photograph 条纹摄影
schlieren photography 纹影摄影术
Schlieren projection lens 施利仑投影透镜
Schlieren projection system 施利仑投影系统
schlieren set-up 纹影法摄影装置
Schlieren stop 施利仑光阑
Schlieren technique 纹影术
schloss 庄园建筑;德国城堡
Schlumberger apparatus 施卢姆贝格尔热长度试验仪
Schlumberger array 施卢姆贝格尔台阵
Schlumberger comb sorter 施卢姆贝格尔热梳片式长度试验仪
Schlumberger dipmeter 施卢姆贝格尔倾角测量仪
schlumberger sounding 对称四极测深
schlumberger sounding method 对称四极测深法
Schmidt aspheric(al) corrector plate 施密特非球面校正板
Schmidt camera 施密特照相机;施密特投影箱;施密特摄影机
Schmidt correction plate 施密特校正板
Schmidt corrector 施密特校正板
Schmidt discriminator 施密特甄别器
Schmidt hammer 回弹仪;施密特锤
Schmidt law 施密特定律
Schmidt lens 施密特透镜
Schmidt lens system 施密特透镜系统
Schmidt limit 施密特极限(核磁矩的上下限)
Schmidt line 施密特线
Schmidt mirror system 施密特反射镜系统
Schmidt motor 施密特活塞液压电动机
Schmidt net 施密特网
Schmidt optical system 施密特光学系统
Schmidt optics 施密特学光系统
Schmidt orthogonalization 施密特正交化法
Schmidt plot 施密特图
Schmidt prism 施密特棱镜
Schmidt projection 施密特投影
Schmidt projection optics 施密特光学投影系统
Schmidt reaction 施密特反应
Schmidt rebound hammer 施密特回弹锤
Schmidt system 施密特光学系统
Schmidt telescope 施密特望远镜
Schmidt television projector 施密特投影电视机
Schmidt theodolite 施密特经纬仪
Schmidt toggle circuit 施密特触发电路
Schmidt torque meter 施密特转矩表
Schmidt-type optical system 施密特反射望远镜系统
schmiederite 羟硒铜铅矿
schmitterite 碲铀矿
Schmitt projector 施密特投影机
schmutzdecke 滤膜层
schneebergite 硝铵炸药;铁锑钙石
Schneider furnace 谢利利得高频感电炉
schneiderhoehnite 陕砷铁矿
Schneider kiln 普通立窑
Schneider propeller 谢利得推进器
schnorkel 潜艇通气管;潜水罩;潜水呼吸管
schoderite 水磷钒铝石
Schoenflies-international symbol 圣佛利斯—国际符号
schoenfliesite 羟锡镁石
Schoenflies symbol 圣佛利斯符号
Schoenherr-Hes-sberger process 电弧固氮法
schoepite 柱油矿
schohartite 重晶石
schola (古希腊体育场的)休息室;(罗马浴场中设浴缸的)凹室
scholar 学者
scholarship 学问;奖学金
scholastic achievement 学术成就
scholastic attainment 学术成就
scholasticism 烦项哲学
schollensis 堆积浮冰群;浮冰丘
scholzite 磷钙锌矿
Schonrock autocollimating eyepiece 施恩罗克自准直目镜
Schonrock halfshade 施恩罗克偏光半荫仪
Schonrock prism 施恩罗克棱镜
school 学派
school architecture 学校建筑(学)
school assembly hall 学校会堂
school atlas 教学地图集
school attendance sphere 走读半径
school auditorium 学校大礼堂
school block 学校街坊
school building 校舍
school building program(me) 校舍建筑规划
school bus 学校用公共汽车
school caravan 临时活动教室
school cartography 教学地图制图学
school complex 学校综合建筑
school construction 学校建筑
school crossing sign 学童过街标志
school edition 教学版
school enrollment 学校注册人数
school environmental education 学校环境教育
school for deaf-mutes 聋哑学校
school for feeble minded 弱智者学校
school garden 学校园;校园
school gymnasium 学校体育馆
school health 学校卫生
schoolhouse 校舍
school hygiene 学校卫生学
schooling functions 教育函数
school library 学校图书馆
school map 教学挂图
school microscope 教学显微镜
school nursery 托儿所;幼儿园
school of applied art 实用美术学校
school of architecture 建筑学院
school of business administration 商业管理学院
school of Delft 德尔夫脱学派(1955年前荷兰占优势的战后建筑风格)
school of general instruction 职业学校;专科学校;正规学校;普通学校
school of industrial administration 工业管理学院
school of management 管理学院
school of nurses 护士学校
school of technology 工业学校
school plant 校办工厂
school playground 学校操场
school room 课堂
school run by the factory 厂办学校
school ship 教学船;教练舰;教练船;实习船
school site 校址
school structure 学校建筑
school table 课桌
school term 学期
school town 学园城市
school village 学校村
school wall map 教学挂图
school with honeycomb layout 具有蜂窝状布局的学校
school yard 校园
schooner 纵帆船
schooner guy 中稳索
schoonerite 磷铁锰锌石
schooner stay 两吊货杆的联动绞辘;双吊杆联动绞辘
Schoop process 金属丝喷涂法;金属喷镀;斯库普法喷镀

Schopper's bursting strength tester 肖伯织物顶破强力机
Schopper's cloth abrasion tester 肖伯织物耐磨试验仪
Schopper folding tester 肖伯耐折度测定仪
Schopper's tensile strength machine 肖伯抗张强度试验机
Schopper's tensile tester 肖伯抗拉试验仪
Schopper's torsion meter 肖伯扭力仪
Schori 肖瑞抗腐雾
Schori process 肖瑞金属喷涂法
schorl 黑电气石
schorl blanc 白榴石
schorl noir 辉石
schorlomite 钛榴石
schorl rock 石英黑电气岩
schorl rouge 金红石
schorl-schist 黑电气片岩
schorre 盐草潮滩
schort 环线;短线
Schottky barrier 肖特基势垒
Schottky defect 肖特基缺陷;空位缺陷
Schottky disorder 肖特基位错
Schottky emission 肖特基放射
Schottky noise 散粒噪声
schradan 八甲磷
Schrage motor 施拉盖电动机
Schreiber's fictitious observation equation 史赖伯虚拟观测方程
schreibersite 陨磷铁镍石;陨磷铁矿
Schreiber's method in all combination 史赖伯全组合测角法
Schreiner finish 缎光整理;电光整理
schreinerize 缎光整理
schreyerite 钒钛矿
schriftgranite 文象花岗岩
Schrodinger representation 位置表象
schroeckingerite 板菱铀矿
schroetterite 蛋白铝英石
schrund line 冰斗基线
schubnelite 三斜水钒铁矿
Schuermann cupola 肖尔曼热风冲天炉
Schuermann furnace 肖尔曼热风冲天炉
schuettite 汞矾
schuilingite 铜铅镱石;碳铜铅钙石
Schuldschein loan 借款凭证
schulenbergite 羟碳锌铜矾
Schuler pendulum 舒勒摆
Schullar process 滚筒法
Schulta-Charlton reaction 变白反应
schultenite 透砷铅矿;透砷铅矿
Schulz alloy 舒尔茨锌轴承合金
Schulz formula 舒尔茨公式
schumacherite 钒铋石
Schumann metal (lic) aggregate mortar cover (ing) 舒曼金属骨料砂浆覆盖层
Schumann's ceiling 舒曼顶棚
schungite 次石墨;不纯石墨;半石墨
schuppen structure 叠瓦构造
Schuster ga(u)ge 磁性测厚计
Schuster mechanism 舒斯特机制
Schuster problem 舒斯特问题
schwartzembergite 氯碘铅矿;羟碘铅石
Schwarz's inequality 施瓦尔兹不等式
Schwarzschild coordinates 施瓦茨希尔坐标
Schwedler's dome 施魏德勒穹隆
Schwedler's girder 施魏德勒大梁
Schwedoff model 什卫道夫模型
Schweinfurt(h)green 施温福特绿颜料;亚砷酸铜复盐;巴黎绿
Schweizerische National Bank 瑞士国家银行
Schweizerischer Bankverein 瑞士银行公司
Schweizerischer Handels-und Industrie-Verein 瑞士工商联合会
schweizerite 叶蛇纹石
Schwertmannite 施氏矿物
sciagraph 房屋纵断面图;投影图
sciagraphy 房屋纵断面图;X射线照相术;投影法
sciametry 日月食理论
scid 锡特
Scidmore eyepiece 赛得莫尔目镜
science abstracts 科学文摘
science and technology 科学技术
Science and Technology Commission 科学技术委员会
science and technology development fund 科技发展基金
science and technology development strategy 科学技术发展战略
science and technology forecast 科学技术预测
Science and Technology for Environmental Protection 科学和技术促进环境保护方案
science and technology input 科技投入
science block 科学街(坊)
science branches 科学分支
science building 科学馆;科学建筑
science citation index 科学文献索引
science congress 科学会议;科学大会
science fabrics 化学纤维织物
science house 科学楼
science museum 科学博物馆
science of architecture 建筑科学
science of building 营造学
science of engineering rock and soil 工程岩土学
science of environment(al)law 环境法学
science of fortification 防御工事的科学
science of ground soil 土质学
science of light reflection 反射光学
science of nuclear power 核动力科学
science of personnel 人才学
science of personnel management 人事管理学
science of science 科学学
science of strength for materials 材料强度科学
science of the sea 海洋科学
science of tide 潮汐学
science of total environment 总环境科学
science-oriented 根据科学成果制成的;采用科学成果的
science park 科学园区;科学馆
science policy 科学政策
science satellite 科学卫星
science-technology industrial park 科技工业园
scienology 科学学
scientific administration 科学管理
scientific advance 科学进展
Scientific Advisory Committee 科学咨询委员会
scientific agriculture 科学耕作
scientifically designed 科学设计的
scientifically oriented individuals 科技人员
scientific and educational diving 科教潜水
scientific and technical achievement 科学技术成就
scientific and technical advance 科学技术发展
scientific and technical development program (me) 科学技术发展规划
scientific and technical forecast 科学技术预测
scientific and technical information 科技信息;科技情报(资料)
scientific and technical literature 科学技术文献;科技文献
scientific and technical payoffs 科技成果
scientific and technical results 科技成果
scientific and technical terminology 科技术语
scientific and technologic(al) achievements 科技成果
scientific and technologic(al)circle 科技界
scientific and technologic(al) information 科技情报
scientific and technologic(al) information system 科技情报系统
scientific and technologic(al)personnel 科技人员
scientific and technologic(al) progress 科学技术进步
scientific apparatus 科学仪器
Scientific Apparatus Makers Association 科学仪器制造协会
scientific approach 科学方法
scientific balloon 科学气球
scientific ballooning 科学气球探测
scientific calculation 科学计算
scientific calculator 科学计算器
scientific classification 科学分类;精确分类
Scientific Committee on Antarctic Research 南极研究科学委员会
Scientific Committee on Ocean Research 海洋研究科学委员会
Scientific Committee on Problems of the Environment 环境问题科学委员会;国际科联环境问题科学委员会
Scientific Committee on Water Research 水(问题)研究科学委员会
scientific community 科学团体;科学界
scientific computer 科学(用)计算机
scientific computing program(me) library 科学计算程序库
scientific consultation 科学咨询;科学顾问
scientific data 科学数据
scientific data processing 科学数据处理
scientific diving 科研潜水
scientific documents 科学文献
scientific drilling 科学钻探
scientific effort 科研计划;科研工作
scientific emerald 合成绿柱石
scientific expedition 科学考察
scientific experiment 科学实验
scientific forecasting 科学预测
Scientific Hydrologic(al) Association 科学水文学会
scientific hydrology 理论水文学;基础水文学;水文科学
scientific information 科学情报
scientific information system 科学信息系统
scientific institution 科学协会
scientific instruction set 科学指令系统;科学指令集
scientific instrument 科学仪器
scientific investigation diving 科学调查潜水
scientific landscape theory 科学景观论
scientific language 科学计算语言
scientific literature 科学文献
scientific lunar exploration 月球科学研究
scientific management 科学管理
Scientific Manpower Commission 科学人力委员会
scientific method 科学方法
scientific method of amortization 科学推销法
scientific name 学名
scientific notation 科学记数法;科学表示法
scientific paper 学术论文;科学著作
scientific papers 科学论文集
scientific park 科学公园
scientific pay-off 科研成果
Scientific Personnel Development Bank 科技人才开发银行
scientific potential 科学潜力
scientific prediction 科学预见;科学预测
scientific procedure 科学程序
scientific proceedings 科学论文集
scientific program(me) 科学计算用程序
scientific progress 科学进展
scientific rainmaking 科学人工造雨
scientific reasoning 科学推理
scientific research 科学研究
scientific research cost 科研费用成本
scientific research institution 科研机构
scientific research item 科研项目
scientific research personnel 科研人员
scientific reserves 科研保护区
scientific satellite 科研卫星
scientific sports 体育科学化
scientific station 科学站
scientific study 科学研究
scientific symposium 科学专题论文集;科学讨论会
scientific system 科学用系统;科学计算系统
scientific tariff 科学(性)关税
scientific-technical information 科学技术情报
scientific terrn 科学名词
scientific terminology 科学术语;科学名词;科学词汇;术语
scientific theory 科学理论
scientific verification 科学论证
scientific worker 科学工作者
scientist 科学家;科学工作者
scientize 科学化
scimmer scoop 浅挖掘机
scinticamera 闪烁照相机
scinticounting 闪烁计数
scintigram 闪烁(扫描)图;闪烁曲线
scintigraph 闪烁扫描器
scintigraphy 闪烁扫描术;闪烁法
scintilla 闪烁
scintillant 闪烁体;闪烁材料
scintillating disk 盘状闪烁体
scintillating light 特急闪光
scintillating liquid 闪烁液体

scintillating material 闪烁物质;闪烁材料
scintillating medium 闪烁介质
scintillating phosphor 闪烁体
scintillating solution 闪烁溶液
scintillating target 闪烁目标
scintillation 火花;闪烁现象
scintillation analysis 闪烁(光)分析
scintillation camera 闪烁照相机
scintillation chamber 闪烁室
scintillation chamber method 闪烁室法
scintillation chamber type emanation apparatus 闪烁型射气仪
scintillation conversion efficiency 闪烁转换效率
scintillation converter 闪烁转换器
scintillation counter 闪烁计数器
scintillation counter crystal 闪烁计数器晶体
scintillation counter head 闪烁计数器探头
scintillation counter monitor 闪烁计数监测仪
scintillationcounter telescope 闪烁(计数器)望远镜
scintillation counting 闪烁计数
scintillation-counting crystal 闪烁计数晶体
scintillation crystal 晶体闪烁体;闪烁晶体
scintillation detector 闪烁探头;闪烁探测器;闪烁检测器;闪烁计数器
scintillation effect 闪烁效应
scintillation error 振幅起伏误差
scintillation fading 起伏衰落;调制载频衰落
scintillation fission counter 闪烁裂变计数器
scintillation glass 闪烁玻璃
scintillation index 闪烁指数
scintillation jamming 起伏干扰
scintillation jamming spectrum 起伏干扰频谱
scintillation method 闪烁法
scintillation noise 闪烁噪声
scintillation phosphor 闪烁磷光体
scintillation probe 闪烁探针;闪烁探测器
scintillation process 闪烁过程
scintillation pulse 闪烁脉冲
scintillation pulse survey radon meter 闪烁脉冲测氡仪
scintillation scanner 闪烁扫描器
scintillation screen 闪烁屏
scintillation solution 闪烁液
scintillation spectrometer 闪烁光谱仪;闪烁分光计
scintillation spectroscope 闪烁光谱仪
scintillation spectrum 闪烁光谱
scintillation statistics 闪烁统计学
scintillation technique 闪烁技术
scintillation telescope 闪烁计数望远镜
scintillation vial 闪烁管
scintillation well counter 井式闪烁计数器
scintillator 闪烁仪;闪烁体;闪烁器;闪烁剂
scintillator conversion efficiency 闪烁体转换效率
scintillator fast neutron fluxmeter 闪烁体快中子通量计
scintillator liquid 闪烁液
scintillator material 闪烁器材料
scintillator-photomultiplier probe 闪烁体光电倍增管探头
scintillator prospecting radiation meter 探矿闪烁辐射计
scintillator slow neutron detector 闪烁体慢中子探测器
scintillography 闪烁照相法
scintillometer 闪烁计(数器);闪烁计测量
scintilloscope 闪烁镜;闪烁仪
scintilogger 闪烁测井计数管
scintiphoto 闪烁照相
scintiphotograph 闪烁照相
scintiphotography 闪烁照相术;闪烁照相法
scintiscan diagram 闪烁扫描图
scintiscanner 闪烁扫描器
scintiscanning 闪烁扫描
scinti-tomogram 闪烁断层扫描图
scintitomography 闪烁断层照相
scintled 有对角线通风的砖坯
scintled brickwork 墙面凹凸不齐的砌砖方式
scintling 砖坯留孔排放;砖坯留孔通风干燥;砖坯呆空干燥
sciograph 投影图;房屋纵断面图;房屋剖面图
sciography X 射线术;房屋纵断面图;投影法
scion 插扦
scion grafting 枝接
scion grafting on root 根枝接

sciophyte 阴地植物
scission 裂变
scission reaction 裂解反应
scissor 枝剪;剪技
scissoring 修剪;剪取移动
scissoring vibration 剪式振动
scissors 剪形夹;剪刀
scissors bonding 剪刀焊
scissors bridge 剪式桥
scissors concrete stair(case) 剪刀式混凝土楼梯
scissors crossing 交叉跨越;交叉渡线;剪刀式交叉口;双渡线
scissors crossover 剪刀式渡线;交叉渡线;双渡线
scissors cut 剪纸
scissors escape stair 剪刀式疏散梯
scissors fault 剪形断层;旋转断层;剪式断层;剪断层
scissors gait 剪刀步态
scissors inversor 菱形控制器;剪形控制器
scissors jack 剪式千斤顶
scissors junction 剪式连接;剪刀式交叉;锐角交叉
scissors lift 杠杆式升降机;剪式升降机
scissors lifter 杠杆式起重机
scissors-like flow 剪刀水
scissors-paper-stone game 剪刀—纸—石头对策
scissors stair(case) 剪(刀)式楼梯
scissors track 线路双交叉
scissors truss 斜撑桁架;剪式桁架
scissors type stair(case) 剪刀式楼梯
scissor type 剪刀式布置
scissor type spring 剪形弹簧
Sclattie 灰花岗岩(细颗粒的,产于英国阿伯丁)
sclerite 骨片
sclerobase 硬化基部
sclerocrystallin 硬黄质结晶
sclerodermite 硬皮
sclerogen 硬化原
sclerolac 紫胶树脂
sclerometer 肖氏硬度计;硬度计;硬度分析仪;回跳硬度计;回弹硬度计;回弹仪;测硬器;反跳硬度计
sclerometric hardness 硬度计测定的硬度;冲撞硬度
Scleron 司克莱龙(铝基)合金
scleronomic constraint 稳态约束
sclerophyll 硬叶
sclerophyllous forest 硬叶常绿林
sclerophyllous plant 硬叶植物
sclerophyllous shrub 硬叶常绿灌木
sclerophyllous woodland 硬叶树木
sclerophyte 硬叶植物
scleroscope 金属硬度计;肖氏硬度计;硬度显示器;硬度分析仪;落球回弹硬度计;回跳硬度计;测硬器;验硬器
scleroscope hardness 肖氏硬度;回跳硬度;验器示硬度
scleroscope hardness test 肖氏回跳硬度试验;回弹硬度试验;反跳硬度试验
scleroscopic hardness 回跳硬度
sclerostenosis 硬缩
sclerotic tylosis 硬化侵填体;硬化浸填体
sclerotinite 菌类体
sclerotite 微菌类煤
Sclerotium hydrophilum 喜水小核菌
sclerotization 硬化
sclerous reaction 硬化反应
scmutzdecke 污泥覆盖物
scobinate 表面粗糙的
scobs 锉屑;锯屑;锯末
scoen 薄板砖
scoinson 门窗内侧呈八字形斜面
scoinson arch 半入墙拱;墙拱;对角斜拱;三角拱
scolecite 钙沸石
scolecodont 虫牙
scolerite 蛭沸石
scolite 虫牙
scollop 扇形图案;扇贝形图案
scolloped 有圆齿的;裙形花边的;薄片材料边缘剪切成曲线的
S-component 缓变分量
sconce (钉在墙上的)灯台;遮蔽;烟囱座;壁上烛台;防护套;堡垒
sconcheon 门窗内侧呈八字形斜壁;门窗框槽沟;门窗洞侧墙面;(木梁的)嵌槽口
sconcheon arch 开有槽沟的拱;门窗洞测墙面的拱;嵌槽拱;背拱

scondary inductance 副线圈电感
scone 长向半砖;锭剂
scone brick 平砖;扁平砖(修补用)
scontion 框内屋角石;门窗框内屋角石
scoop 落料导槽;花铲子;戽斗;挖泥斗;椭球体泛光灯;水瓢;勺;凹处
scoop box 料斗
scoop bucket 铲斗
scoop capacity 铲斗容量
scoop channel 斗式(水)槽
scoop condenser 自流式冷凝器
scoop controlled 可控式填充量的
scoop crowding 铲斗推铲装料
scoop digger 斗式挖掘机
scoop dredge(r) 斗式挖泥船;勺式挖泥船;斗式挖土机;斗式挖泥机;铲斗(式)挖泥船;勺式挖泥机
scooper 铲斗;戽斗;勺子;斗式升运机
scooper loader 铲运机
scooper type harvester 贝耙型采捕机
scoop excavator 斗式挖掘机
scoop feeded 斗式装载料机
scoop feeder 进料斗;勺式进料器;斗式进料器;斗式给料器;翻斗加料器
scoopfish 走航采样(戽)斗;在航采样戽;在航采样斗
scoop flight conveyer 戽斗提升机
scoop fork 勺形叉
scoop for nursery planting 苗圃移植铲
scoopful 一满斗
scooping machine 戽水机;斗式挖土机;铲斗挖土机
scoop light 杓状聚光灯
scoop loader 勺斗式装载机;斗式装载机;斗式装运机;斗式装料机;斗式卸载机;铲斗装载机;铲斗装岩机;铲斗车
scoop machine 斗式挖土机
scoop pan self-loading scraper 斗式自动装载刮土机
scoop pole (增加伸缩式钻塔高度的)延伸杆
scoop screen 箕式清沟器
scoop shovel 勺(形)铲;杓铲;斗式挖土机;单斗电铲;铲斗(挖土机)
scoop side-dump(ing)car 侧卸式斗车
scoop tail rope 拖斗尾绳
scoop tube 铲管
scoop-type 戽斗式
scoop-type suction dredge(r) 耙吸式挖泥机;耙吸式挖泥船
scoop-type turbine 水斗式水轮机;冲击式水轮机
scoop vane 挖掘叶
scoop wheel 斗式挖泥车轮;戽斗水车;戽斗轮;斗式挖掘车轮;斗式水车;水轮;扬水轮;疏浚轮;疏浚车轮
scoop-wheel distributor 勺轮式排出装置
scoop wheel feeder 轮式翻斗加料器
scoop wheel pump 斗轮泵
scoop with fluid drive 液压传动斗
scoot 美国灰分次品硬木级别的方法
scooter 小型摩托车;小型开沟犁;注射器;窄式开沟铲;喷水炮
scooter deflector 拖斗转向装置
scope 园地;规模;观测设备;雷达显示器;视测仪;范围
scope analyser 示波分析仪
scope and depth of exchange 交换的广度和深度
scope and range 范围
scope and rework 废品和返工
scope attribute 作用域的属性
scope calibration 示波器校正
scope cover 责任范围
scope face 显示器荧光屏;显示屏;指示盘
scope horizontal sweep 示波器水平扫描
scope of agreement 协议范围
scope of application 应用范围;适用范围
scope of assessment 课税范围
scope of authority 权限范围;委托范围;授权范围
scope of a variable 变量(作用)域
scope of business 业务范围;经营范围
scope of cable 出链范围
scope of command facility 命令设备域;命令设施范围
scope of construction item 施工规模
scope of construction project 建设项目规模
scope of construction work 施工范围
scope of control 控制范围
scope of cover 责任范围;保险范围

scope of declaration 说明作用域
scope of design 设计范围
scope of distribution 分配全域;分配范围
scope of evaluate 评价的视野
scope of examination 检查内容;审查范围
scope of execution 执行范围
scope of experiment 实验范围
scope of experimentation 实验范围
scope of identifier 标识符作用域
scope of invention 发明的范围
scope of operation 经营内容;经营范围;活动范围
scope of project 项目范围
scope of repair 检修范围
scope of repairing course 修理范围
scope of service 服务范围
scope of standard 标准范围
scope of statistics 统计范围
scope of supply 供应范围;供货范围
scope of survey 测图范围;测量范围
scope of survey for tunnel 隧道测量范围
scope of symbolic name 符号名作用域
scope of taxation 课税范围
scope of tender design 供投标用设计的范围
scope of the name contents 注记负载
scope of use 使用范围
scope of works 工作范围;工程内容;工程范围
scope of works and supply by contractor 承包人供货和工程范围
scope of works and supply by owner 业主供货和工程范围
scope range of the fluctuation of voltage 电压波动范围
scope response 示波器响应
scope return 显示器的上回波信号
scope setting 装定光学瞄装器
scope-sight reticule 光学瞄准具十字线
scope sweep 示波器扫描
scope unit 显示器
scoping 对问题范围的研究
scopolite rock 方柱石岩
scopometer 视测浊度计
scopometry 视测浊度(测定)法
scorbutus nauticus 船员坏血病
scorch 烧焦
scorched earth 焦土
scorched ingot 粗晶铸锭
scorched pencil 烧画笔
scorched rubber 早期硫化橡胶
scorched steel 柱状晶钢;粗晶钢
scorcher 纸型干燥机
scorching 灼伤;橡胶过早硫化;疾驰;横晶;电极跳动;穿晶粒
scorching hot 灼热
scorching of wood 木材烧焦
scorch-pencil 炭笔
scorch retarder 防焦剂
score 割痕;刻分;记分;毛刺沟痕;得分;带槽;冰川擦痕
score biotic index 计分制生物指数
scoreboard 记录板;记分牌;记分板
score cutter 刻痕крист岩
score cutting 刻痕面砖
scored 拉毛的
scored card 缺口卡片;成型卡片;分数卡
scored crank pin 刮伤曲柄销
scored cutter 刻痕面砖
scored cylinder 刮伤汽缸
scored-finish tile 划纹装饰面砖;带槽面砖
scored pulley 槽轮
scored surface 粗糙面;拉毛面;刻痕面
scored tile 槽面砖;有槽空心砖
scored variable 得分变量
score line 路面线纹
score mark 划痕;划出标记
score-pencil 炭笔
score rail 槽形钢轨
scoresheet 战绩标志
score sum system 得分总值系统
score system 评分法
scoria 金属渣;火山(岩)渣;熔渣;渣堆;矿渣;[复] scoriae
scoria agglomerate 火山渣集块岩
scoriaceous 矿渣的;熔渣的;渣状;火山渣的
scoriaceous basalt 火山渣玄武岩

scoriaceous lava 泡沫熔岩;渣状熔岩
scoriaceous lava concrete 泡沫熔岩混凝土
scoriaceous lava concrete wall slab 泡沫熔岩混凝土墙板
scoriaceous structure 熔渣构造
scoria cinder 岩渣
scoria pavement 熔渣路面
scoria surface 熔渣面层;熔渣路面
scoria tuff 火山渣凝灰岩
scorification 析取;渣化法;烧熔(试金法);成渣
scorifier 渣化皿;试金坩埚
scoriform 渣状;熔渣状的
scorify 渣化;造渣
scoring 沟纹;(凹)槽;刻槽;刻痕工作;记分;划线;划痕;擦痕;冰川擦痕
scoring equipment 划线设备;切割设备
scoring hardness tester 划痕硬度试验仪;划痕硬度计
scoring knife 刻痕刀
scoring machine 刻痕机;切割机
scoring of brick 砖面沟纹
scoring of tiles 砖面沟纹
scoring resistance 刮伤表面阻力
scoring speed 切割速度
scoring stage 录音室
scoring susceptibility (表面摩擦时的)刮伤趋势
scoring test 划痕试验
scoring wheel 切割刀轮
scorodite 臭葱石
scorpionism 蝎螫中毒
scorpion sting 蝎蜇伤
scorpion sting poisoning 蝎蜇中毒
scorpion venom 蝎毒
scorpion venom poisoning 蝎毒液中毒
scorzalite 多铁天蓝石
scosystem characters of stabilization pond 稳定塘生态系统特征
scot blower 吹灰机
scotch 制动楔;止转棒;加刻痕;切口;圬工小锤;停车器;挡车块;擦痕;粉碎
scotch block 线路遮断器;制转楔;制动块;止轮器;止车楔
Scotch boiler 苏格兰(式)锅炉;烟管锅炉
Scotch bond 苏格兰砌合
Scotch bracketing 苏格兰支架(一种粉刷用支架);苏格兰式檐口隔撑;苏格兰式檐口托座;平顶斜角板条
scotch cleaner 清齿刮铲
scotch club cleaner 弧形底脚修形笔
Scotch crank 苏格兰曲柄
scotch derrick crane 刚性支腿桅杆起重机;德立克吊
Scotch dressing 司各契上浆工序
Scotch fir 银松
Scotch glue 透明胶
Scotch glue in pearl form 球状苏格兰胶
scotching 琢石;打出麻面
scotching of wheel tread 踏面擦伤
scotch light 反射光线
scotchlite 反射标志(道路上用的一种反光玻璃材料)
scotchman 固定支索的防擦板
scotch marine boiler 苏格兰式船用锅炉
Scotch pine 北欧赤松
scotchprint 半透明覆盖层
Scotch stone 蛇纹石;苏格兰石;去痕石
Scotch tape 苏格兰胶带;卡带;透明胶带
scotch wall 半分隔挡墙
Scotch yoke 苏格兰曲柄;止转棒轭;停车器轭;挡车轭
scotia 柱基凹弧边饰;凹圆线
scotia mo(u)lding 凹圆线
Scotia plate 斯科舍板块
scotia scaper 凹圆形柱身
scotlandite 苏格兰石
scotogram 暗室显影片
scotograph 暗室显影片;X射线照片
scotographic 暗室显影的
scotography 暗室显影术;暗室显影法
scotoma 盲点;暗点
scotomagraph 暗点描记器
scotometer 暗点计
scotometry 暗点测量法
scotomization 暗点发生
scotophor 暗迹粉;暗光磷光体;暗光磷光粉
scotopic 暗视的

scotopic curve 适暗曲线
scotopic eye 适暗眼
scotopic luminosity curve 暗光光度曲线
scotopic svectral luminous efficiency 暗视觉光谱光视效率
scotopic vision 暗视(力);微光视觉;微观视觉;暗视觉
scotoscope 增辉望远镜
Scots fir 苏格兰冷杉
Scots pine 苏格兰松;欧洲赤松
Scott cement 透明石膏水泥(生石灰加5%石膏);斯科特水泥
Scott connection 斯科特接线法(把三相电变为二相电)
Scott connexion 斯科特接线法
Scott-Darcy process 斯科特—达西法
Scott former 斯科特真空圆网成型装置
Scottish topaz 黄水晶
Scott jockey 斯科特型杆式连接装置
Scott's cement 斯科特水泥
Scottsconizing 斯科迪斯康法(一种不锈钢表面硬化法)
Scott's connection 斯科特接线
Scott's effect 斯科特效应
Scott's evapo(u)rator 斯科特蒸发器
Scott's jockey 斯科特型杆式连接装置
Scott's tester 斯科特试验机
Scott's viscosimeter 斯科特黏度计
Scott system serigraph 斯科特式纱线强力仪
Scott tensile tester 斯科特拉伸强力试验仪
Scott volumeter 斯科特容量计
Scottwood 斯科特预制住宅;斯科特预制住房
scour 洗刷;挖底;脱壳;淘蚀
scourability 洗涤性能
scourage 洗余的水;冲洗水
scour and deposition 冲积
scour and fill 冲积;边冲边淤;冲淤(作用)
scour and fill changes of shoal 浅滩的冲淤变化
scour-and-fill structure 冲淤构造
scour and silting tendency 冲淤趋势
scour at bridge crossing 桥渡冲刷
scour basin 冲刷盆地
scour below dam 坝下淘刷;坝下冲刷
scour cast 冲刷铸型
scour channel 冲刷(河)槽;冲蚀槽
scourclean (脱粒机的)清选筒
scour criterion 冲刷标准
scour culvert 冲砂涵洞;冲沙涵洞
scour cycle 冲淤周期
scour-deposition cycle 冲淤周期
scour depression 冲刷源槽;冲刷沟;冲淤深槽;冲沟
scour depth 淘蚀深度
scoure 冲刷器
scoured base 冲刷底面
scoured hole 冲刷孔;冲刷坑
scourer 洗刷器;洗净剂;打光机;擦洗者;擦洗剂;擦净剂
scourer cylinder 谷粒脱壳滚筒;碾谷滚筒
scour-fill tendency 冲淤趋势
scour finger 指状冲刷痕
scour forecast 冲刷预防;冲刷预测
scour gallery 冲刷槽;冲砂廊道;冲沙槽
scour gate 冲沙闸(门)
scour hole 冲刷孔;冲刷坑;冲蚀穴;冲沙孔;冲沙坑
scouring 洗净;洗涤;木抹搓平;河床冲刷;素烧坯滚磨抛光;冲刷(作用);冲蚀;擦磨;擦净;冰川刨蚀
scouring abrasion 冲擦
scouring action 扫清作用;冲刷作用
scouring agent 洗涤剂;擦洗剂;擦净剂
scouring and bleaching 练漂
scouring and silting 冲淤;冲刷及淤积
scouring at dike root 坝根淘刷
scouring away the silt 冲沙
scouring basin 蓄潮冲游塘;水闸蓄水池;疏浚水道蓄水池;冲淤盆地;冲沙池
scouring bit 清孔钻头
scouring bowl 洗毛槽
scouring box 磨光卷筒
scouring by contraction works 束水攻沙
scouring capability 侵蚀能力;冲刷能力
scouring cinder 侵蚀性熔渣
scouring cloth 抹布
scouring depth 冲刷深度
scouring effect 冲刷作用;冲刷效应

scouring equipment 冲淤设备;冲刷设备;冲砂设备
scouring escape 冲刷水排出管
scouring force 冲刷力
scouring gallery 冲沙廊道
scouring liquor 擦洗液;擦净液
scouring machine 洗涤机;水洗机;冲浇机
scouring mill 脱壳机
scouring of foundation 基础冲刷
scouring pad 擦洗用的钢丝圈
scouring pinch out 冲刷变薄尖灭
scouring pipe 冲洗管;排淤管
scouring pit near dike head 坝头冲刷坑
scouring plain 侵蚀平原
scouring powder 去污粉;擦净粉
scouring sand 冲洗砂
scouring-silting stage 冲淤水位
scouring slope 刷坡坡面
scouring sluice 冲刷闸门;排沙闸(门);冲刷闸门;冲砂孔;冲砂阀;冲淤闸(门);冲沙道
scouring solution 擦洗溶液
scouring speed 冲刷速度
scouring stand 不锈钢带抛光装置
scouring table 修坯台
scouring test 冲刷试验
scouring velocity 冲洗速度;冲刷流速;冲刷速度
scouring water 洗涤水
scouring wheel and buffing wheel 打磨轮与抛光轮
scour lag 冲刷滞积物
scour-loading 冲刷荷载
scour mark 冲(刷)痕;冲蚀痕
scour outlet 泄水口;冲砂孔
scour outlet channel 冲淤(渠)道;冲砂泄水道;冲沙泄水(渠)道
scour pad 防冲衬垫
scour pattern 冲刷型
scour pipe 排淤管;冲砂管道;冲沙管
scour pit 冲刷坑
scour preventer 防冲沙装置
scour prevention 冲刷防护;防(止)冲刷;防冲(设备)
scour protection 防止冲刷;淘蚀保护;防冲刷面层;防冲刷保护;防冲
scour protection apron 防冲护坦
scour protection measure 防冲刷措施
scour-provoking current 冲刷水流
scour rate 冲刷速率;冲刷率
scour-remnant ridge 冲刷残留脊
scour resistance 抗冲刷性
scour sedimentation equilibrium 泥沙冲淤平衡;冲淤平衡
scour sluiceway 冲砂泄水道
scour the ditch 冲刷水沟
scour-toilet 抽水马桶
scour tube 冲沙管
scour tunnel 冲砂隧洞;冲沙隧洞;冲沙隧道
scour under a pier 桥墩下的冲刷
scour under pier 墩柱下沟蚀
scour valve 排淤阀;冲洗阀;冲淤阀;冲沙道
scour vent 冲沙孔
scourway 冲刷路径;冲刷通路;冲刷沟;冲沙道
scout 侦察员;侦察舰;侦察机
scout boring 试钻;地质探眼
scout car 巡逻车
scout cruiser 侦察巡航艇
scout drilling 钻探
scout hole 侦探孔;勘探孔;普查孔
scouting 初步勘探;初步勘察
scouting course 搜索航线
scouting distance 搜索距离
scouting line 搜索线
scouting speed 搜索速度
scouting test 探索性试验
scouting-type calculation 试探性计算
scout-observation seaplane 侦察观测水上飞机
scout prospecting 勘探
scout sheet listing 勘探记录一览表;调查记录;查勘记录
scout ship 侦察船
Scove kiln 斯寇夫窑;泥封窑
Scove kiln-burnt brick 泥封窑烧的砖
scow 甲板驳;平底方驳;平底船;吊货盘;大型平底船;敞舱驳(船);方驳
scow banker 平底驳船管理人

scow bow 平头前倾船首
scow house 带房的平底驳船
scowing 防锚咬底法
scowman 平底驳船员
scow pontoon 方驳趸船
scow sling 吊货盘
scow sucker 平底吸泥船
scrab 活动滑车
scrabbled 粗面的(指方工)
scrabbled rubble 碎石圬工
scrab scrub resistance 耐洗刷性
S-crack S形裂纹
scraffeto 括花粉刷
scragging 冲击荷载试验
scraggy 凹凸不平的;参差不齐的
scrag mill 双平行圆锯制材厂
scram 快速停堆;紧急停堆;紧急制动;车急停
scramble 紧急起飞;扰频的
scramble competition 资源均匀;分摊竞争
scrambled image 杂乱影像;失真影像;失真图像
scrambled merchandising 跨行业销售
scramble net 登录用网
scramble network 杂混网络;不规则网络
scramble non-return-to-zero change on one 加扰逢1变化不归零制
scramble pattern 不规则格式;不规则模式
scrambler 置乱器;量化器;密码器;脉冲量化器;扰频器;扰码器;倒频器;保密器
scrambler circuit 保密倒频电路
scramble rule 杂混规则
scramble time 零星时间;编码时间
scrambling 置乱;扰频
scrambling and unscrambling network 播散与收集网络
scrambling matrix 密码矩阵
scram button 应急按钮;快速停堆按钮
scramjet 超声速燃烧冲压式发动机;超声速冲压式喷气发动机
scramming mechanism 快速停止机构;快速停堆机构
scram position 关断位置
scram protection 快速故障防护
scram rod 快速停堆棒;安全事故棒;安全棒
scram system 快速停堆系统
scrap 拆碎;炸碎;零头;回炉料;破片;生产废料;残余;废钢铁;边角料;船舶报废;报废
scrap accumulation 废钢积累
scrap a contract 撕毁合同
scrap aluminium alloy 废铝合金
scrapalurgy 废钢利用
scrap a machine 报废机器
scrap and coke process 废钢焦炭炼钢法
scrap and hot metal practice 废钢铁水炼钢法
scrap a ship 船舶报废;拆船
scrap baling 切屑堆
scrap baling maching 废钢打捆机
scrap baling press 废料压实打包;废铁压块机
scrap baller 切边卷取机;废铁压块压力机
scrap balling 废钢打包
scrap balling press 废铁压块压力机;废料压块压力机
scrap bit inserts 碎片硬合金镶嵌物
scrap body material 废坯料
scrap box sling 装卸废钢铁的吊货工具
scrap brass 废黄铜;碎黄铜
scrap breaker 废钢铁破碎机
scrap bridge 搭边
scrap briquetting 切屑堆
scrap briquetting press 切屑压块机
scrap build 设备改装;设备改造
scrap bundler 废钢打包机
scrap bunker 废料仓
scrap bushing 废旧漏板
scrap capacity 铲运机铲斗容量
scrap celluloid 赛璐珞废料
scrap charging box handling crane 废料装箱起重机
scrap charging crane 料箱起重机
scrap chopper 废料切碎机;碎边剪切机
scrap chute hanger 料箱吊架
scrap coke 下脚焦炭
scrap collection system 铁屑收集装置
scrap copper 废铜
scrap crepe 碎屑皱片

scrap crushing plant 废料破碎设备
scrap cutter 切废料装置;碎边剪;废料刀断装置
scrap dealer 废品商
scrap diamonds 碎块状废金刚石
scrap drop crane 废料堆放起重机;废料吊放起重机
scrape along 擦及
scrape boring bar 刮孔器
scrape chiller 刮刀结晶器
scraped board 刮光板
scraped film evapo(u)rator 刮膜式蒸发器
scraped finish 刮磨(面);刮面;刮毛饰面;扒拉石;扒柱石
scrap edge 切边
scraped heat exchanger 刮削式热交换器
scrape dozer 铲运推土机;铲运推土机
scraped plot 已经刮草的地块
scraped rendering 刮磨抹灰;刮光的拉毛粉刷
scraped straight edge 刮准直边
scraped stucco 刮光的拉毛粉刷
scraped surface 刮研表面;刮光面
scraped-surface exchanger 刮板式换热器;刮面热交换器;刮刀式套管结晶机
scraped surface heat exchanger 刮面式换热器;刮刀式换热器
scrape off 擦去
scrape paint 刮漆
scrap equipment 报废设备
scraper 刮(削)器;刮土机;刮泥器;刮料装置;刮料机;刮具;刮管器;刮棒;滤饼刮刀;漆铲;平土机;耙的刮刀;扒矿机;铲运机;铲土机;擦子
scraper bar 刮板(机)板
scraper bit 刮刀头
scraper blade 刮具;刮板(机)板;平土铲;铲运机铲刀
scraper boss 刮板轮毂
scraper bowl 铲运机(土)斗
scraper box 箱式铲斗
scraper bucket 刮土斗;索曳刮斗;刮土斗;铲斗
scraper bucket cleaner 铲斗式清除(泥土)机
scraper-bucket excavator 铲斗挖土机
scraper cable 铲运机钢丝绳;刮具缆索
scraper capacity 电耙耙斗容量
scraper chain conveyer 链板刮土运送机;链式土输送机;链式刮板运送机;链刮式输送机;链式刮板输送机;刮板运输机;刮板式输送机
scraper cleaner 清扫刮板
scraper compression ring 刮油压缩环
scraper conveyer 链板运输机;刮板运输机;刮式运送机;刮板式输送机;刮板式输送器
scraper dewater 刮板脱水机
scraper-dozer 铲运推土机
scraper dredge(r) 刮板式挖泥船;钢耙挖泥船;耙泥船;铲式挖泥船;刮刀式挖泥船
scraper elevator 刮板式提升机;刮板升渣器
scraper excavator 铲斗挖掘机
scraper-fed concrete mixer 抓斗进料式混凝土搅拌机
scraper feeder 刮板喂料机;刮板式给煤机;刮板给料机
scraper fleet 刮土机组;铲运机组
scraper flight 刮土机耙板;刮板
scraper flight conveyer 刮板(式)输送机;刮板运输机;链式刮板输送机
scraper for land leveling 平地铲运机
scraper grab 刮土机抓斗
scraper haulage 铲运机拖运
scraper hitch 铲运机挂钩
scraper hoist 耙斗提升机;扒矿绞车
scraper hone 刮刀磨石
scraper loader 耙斗运载机;刮斗(式)装载机;铲铲上料机;刮板式装填器;耙斗装载机;耙斗装料机;电耙;铲土装运机
scraper loading 铲运机装载
scraper machine 耙式分级机
scraper pan 刮土耙斗;铲斗
scraper plane 刮泥刨刀;刮刨
scraper plant 铲运机设备
scraper pump 铲运机泵;刮板泵
scraper pusher 铲运推进机
scraper ring 刮油垫圈;油圈;刮油环
scraper roller 清洁辊;抄钢丝辊
scraper rope 铲运机钢丝绳
scraper seal 防尘圈
scraper tail gate 耙斗后壁

scraper teeth cutter 刮刀
scraper tractor 铲运拖拉机
scraper trailer 铲运机拖车
scraper transporter 刮运机;刮板(式)运输机;链式刮板运送机
scraper trap 刮屑收集器;刮刀套
scraper trough conveyor 槽形链板运输机;槽式刮板输送机
scraper type bucket elevator 铲斗式升降机;铲斗式提升机
scraper type clarifier 刮泥澄清池
scraper type flocculator 刮泥絮凝池
scraper type hand pump 刮板手摇泵
scraper type loader 耙斗式装岩机
scraper type rock loader 耙式装岩机
scraper type settling tank 刮泥式沉淀池
scraper winch 电耙运输车;扒矿绞车
scrap factor 废品系数
scrap glass 碎玻璃
scrap grapple 废料抓钩;废料抓斗
scrap handling 废料处理
scrap heap 碎铁堆;废物堆
scrap heat 废钢炼钢法
scrap hopper 废料仓
scraping 铲刮;刮削;撤除;扒矿;铲土;刮研;切除坯体余泥;拆船;船舶划拨;划痕
scraping and cutting 机械搅刮法
scraping and grinding method 刮研磨法
scraping and scuttling 机械搅刮法(维特深航槽法)
scraping apparatus 刮管器;刮刮器械
scraping arm 刮刀臂
scraping bar 防擦护条
scraping belt 刮链;链板;带刮板运输带
scraping blade 刮墨刀(片)
scraping blade conveyor 刮板运输机
scraping chain 刮链
scraping conveyer 电溜子链刮式运输机
scraping cut 刮除很细切屑
scraping cutter 刮刀
scraping dredge(r) 刮板式铲泥机
scraping edge 刮油棱缘
scraping effect 刮垢效果;擦尘效果
scraping effectiveness 刮油效率
scraping equipment 刮泥设备
scraping grader 刮土平地机;刮板式平路机;铲土平地机
scraping iron 刮研产生的铁屑;刮秽
scraping knife 刮刀
scraping land 刮油边
scraping machine 铲土机;刮土机;扒矿机
scraping of a filter 滤池洗砂;滤池刮砂
scraping of filter 滤池刮砂
scraping operation 刮除
scraping out cutter 拉刀
scraping plate 刮
scraping plate type eliminator 刮板式除渣机
scraping rake 取料耙
scraping ring 刮油环
scraping stirrer 铲刮搅拌器;刮刮搅拌机
scraping straight edge 刮尺
scraping test 铲刮试验
scraping-throwing machine 刮抛机
scraping tool 刮运机;刮土机;刮刀;铲刮工具
scraping transporter 刮板(式)运输机
scraping work 铲刮工作
scrap iron 碎铁;铁屑;废铁
scrap iron inner electrolysis 铁屑内电解法
scrap iron pretreatment 铁屑预处理
scrap iron rails 废铁栏杆
scrap iron shearing machine 废铁剪机
scrap lead 铅废料;废铅
scrap leather 碎皮;废皮
scrap log book 航海日志草本
scrap machines 报废机器
scrap mark 废边印痕
scrap material 废料
scrap metal 金属废料;废金属
scrap metal processor 金属废料处理设备
scrap mica 云母屑
scrap out a hole 清理炮眼
scrappage 废品;报废率;废物
scrapped parts 残品;报废零件
scrapped project 报废工程

scrap pipe 废弃管道
scrapple 粗琢石;石的粗琢
scrappling 石料的初琢
scrap plywood 胶合板废料
scrap practice 废钢炼钢法
scrap preheater 废钢预热装置
scrap press 废料压块压力机;废料打包压力机
scrap process 废钢法
scrap processing 废料加工
scrappy 碎料的
scrap rail 废钢轨
scrap rate 切屑量;废品率
scrap reel 废料卷取机
scrap report 废料报告单
scrap return 废料回炉;废钢(铁)回收
scrap ribbon 条状废料
scrap rubber 废橡胶(料)
scraps 成型余泥;废泥料
scrap seed 废种子
scraps from metal processing 金属加工碎屑
scrap shearing machine 废钢剪切机
scrap shears 废钢剪切机;废钢剪床
scrap sleeper 废轨枕
scrap steel 废钢
scrap structural iron 建筑铁废料
scrap structural steel 建筑钢废料
scrap tire 废轮胎
scrap tyre rubber 废轮胎橡胶
scrap value 残值;残余价值;废品残值;废料价值;报废时价值
scrap value of fixed assets 固定资产残值
scrap view 局部视图
scrap wire 废线
scrap yard 废品场;废料场
scratch 小刻线刀;刻线;划痕;撤销;擦伤;擦痕;标线;刮痕;划痕
scratchability 刻痕度;易刻性;柔软度
scratch a file 撤销一个文件
scratch awl 画针;通针;钻子;尖锥;尖铲
scratch board 刮板;刮画板
scratch brush 刮毛刷;钢丝刷
scratch-brushed finish 刷光饰面
scratch brushing 钢丝刷刷光;刷光
scratch brush wheel 钢丝磨刷轮
scratch coat 打底砂浆层;划痕抹灰层;抹灰底层;底层抹灰;刮毛底层;水泥砂浆底涂层;底子灰;打底灰泥层
scratch course 抹灰打底层;刮糙
scratch data cartridge 临时数据盒式盘
scratch device 刻针装置
scratch drawing test 针划涂层黏结力试验
scratched figure 划杠的数字
scratched pebble 擦痕卵石
scratcher 粉刷工具;(泥水工用的)拉毛工具;削皮器;刮刀;抹灰用划痕器;划痕器
scratcher chain saw 直齿链锯
scratcher electrode log 滑动接触法测井
scratcher electrode log curve 滑动接触法测井曲线
scratch file 废文件;暂时停用文件;过期文件;撤销文件;草稿文件
scratch hardness 刻痕硬度;刻划硬度;刻痕硬度;擦硬度
scratch hardness test 刮痕硬度试验;刻痕硬度试验;擦痕硬度试验
scratch in 刮腻子;填缝
scratching 刻削;抹灰面划痕;划毛;底灰划毛;刮痕;划痕
scratchning brush 金属丝刷子
scratchning coat 划痕涂层;划痕打底
scratchning finish of stucco 划痕饰面
scratchning hardness 划痕硬度
scratchning hardness number 划痕硬度值
scratchning hardness test 划痕硬度试验;划痕测硬法
scratchning hardness tester 划痕硬度试验仪;划痕硬度计
scratching machine 刮擦试验机
scratching of pantograph 刮弓
scratchning recorder 划痕记录器
scratch lathe 磨光车床;擦光机
scratch mark 抓痕;划痕
scratch off 刮去
scratch-pad 高速暂存;便笺式暂存
scratch-pad area 高速存储区;临时存储区

scratch-pad memory 工作存储器;高速暂存储器
scratch-pad register 中间结果寄存器;便笺式寄存器
scratch-pad storage 高速暂存存储器;便笺式存储器
scratch paper 便条纸
scratchproof 耐擦;耐刮;耐划
scratch removal 去除擦痕;去除擦伤
scratch resistance 耐刮伤性;抗刮伤(性);抗刮性;抗擦伤性;耐擦伤性
scratch-resistant coating 抗磨涂层
scratch resistant hardened glass 防划痕硬化玻璃
scratch start 摩擦起弧;起弧;擦弧
scratch system 磨粉系统
scratch tape 暂存带;草稿带;废带
scratch template 针刮样板;钉刮样板
scratch test 刻痕硬度试验;硬度划痕试验;抓破试验;刻痕试验;划磁试验;划痕试验;擦痕试验
scratch test for hardness 刻痕法硬度试验
scratch test of paint film 漆膜刮痕试验
scratch tool 刮糙工具;划痕工具;抹灰的拉毛工具;刮刀;刻线工具;划痕器
scratch volume 废卷宗
scratch wire brush 钢丝刷
scratch work 刮毛工;乱涂
scrawl 乱涂
SCR brick 多孔空心砖
screaming 发动机啸声
scree 悬崖下的屑堆;岩屑堆;碎石堆;石屑堆;山麓碎石;崩落
screeching 发动机啸声
screeching halt 紧急制动
scree-collector pipeline 网式收集管线
scree cone 岩锥;碎石堆
screed (压实混凝土路面的)整平板;裂片;冲筋(标志抹灰厚度的窄条)
screed accelerating agent 找平层促凝剂;砂浆层促凝剂
screed admix(ture) 砂浆层外加剂;砂浆层掺和料
screed agent 砂浆层外加剂
screed base 砂浆层底
screed bay 冲筋间距
screed board 修平板;刮尺;刮板;平泥板;样板
screed board vibrator 振动样板;样板式振动器;平板式振动器;平板式振捣器
screed chair 刮板支座;刮板支架
screed-coat 刮平面层;找平层
screed cracking 砂浆找平层开裂
screeded bed 刮铺垫层
screeder 找平器
screed finish 找平(修整);饰面找平
screed finisher 刮铺(混凝土路面)修整机;刮平修整机
screed floor 整块浇成的地面;修平地面
screed floor cover(ing) 地面砂浆找平层
screed guide 找平样板;靠尺
screed improver 找平砂浆改进器
screeding 找平;刮铺(法);刮平;抹平;冲筋工作
screeding beam 样板
screeding board 刮平样板;样板(刮平混凝土用)
screeding compound 抹平料
screed joint 找平砂浆接缝
screed modifier 找平砂浆改良剂
screed mortar 装配砂浆
screed of coarse stuff 粗料整平
screed of plaster of Paris 熟石膏浆找平层
screed plate 刮板;样板
screed rail 抹灰样板;找平样板;找平靠尺;抹灰靠尺;杠尺
screed seal(ing) 找平密封层
screed strip 刮尺;抹灰靠尺;冲筋(标志抹灰厚度的窄条);抹灰冲筋
screed surface 找平砂浆表面
screed vibrator 样板式振动器;振动样板;
screed wire 整平直线绳;标准地线;基准钢丝;整平直线器;样板绳;接地导线
screed work 刮平工作
screefing 除草皮
screen 格网;遮护;遮蔽;过滤网;帘;幕;屏障;屏风【建】;网眼;网线胶片;网屏;网板;丝印;丝网;粗筛;保护屏
screenability 筛选性能
screenable matter 可筛选物
screen absorptance 屏幕吸收系数;屏幕吸收比;屏蔽吸收系数
screen actinic efficiency 屏的光化效率

screen address 屏幕地址
screen afterglow 荧光屏余辉
screenage 影像
screen analysis 砂子粒度分析;筛析;筛分分析;筛分试验;筛分(分析)
screen analysis chart 筛析曲线图;筛分分析图
screen analysis curve 筛析曲线
screen angle 网目角度
screen aperture 筛径;筛眼;筛孔
screen area 滤网有效面积;筛子有效面积;屏幕面积;筛网面积
screen assembly 滤网组件
screen-back hardboard 背面网纹硬质纤维板;一面光硬质纤维板
screen banks 筛组
screen bar 栅条;屏条;算条
screen basin 筛选盆
screen basket spinner 带掩蔽罩旋涂器;带掩蔽罩的旋转器
screen bed base 筛网架
screen belt loader 筛的带式装载机
screen blinding 筛子阻塞;筛子堵塞;筛孔堵塞
screen board 遮板;隔板;屏板;筛板
screen body 筛箱
screen bottom 筛底
screen bottom heating 筛底加热
screen bow 弧形筛
screen-bowl 筛选离心机
screen box 筛箱;筛身
screen brush 清筛刷
screen building 屏蔽房屋;筛分车间
screen bulkhead 轻型舱壁;屏壁;非水密隔舱;非水密舱壁;遮蔽舱壁
screen bumper 清筛用撞击器
screen burning 荧光屏烧毁
screen camera 荧光屏照相机
screen cap 滤筛盖
screen capacity 筛子生产能力;筛效率;筛容量
screen capacity of factors 筛分能力系数
screen capture 屏幕捕获
screen carriage 筛选架
screen centrifuge 筛网离心机
screen chamber 格滤室;滤网室;拦污栅室;筛室
screen change 网目变换
screen changer 屏幕转换器
screen circular feeder 筛式圆盘喂料机
screen clarifier 过滤式沉降器;筛式沉降器
screen classification 筛分
screen classifier 筛分机
screen cleaner 清筛器;筛式滤选机;拦污栅清污器;拦污栅清污机;筛网滤清器;筛式清选机
screen cleaning machine 拦污栅清污机
screen cloth 滤网;金属网布;网纱;窗纱;筛网;筛布
screen cloth opening 筛网孔径
screen coil motor 屏蔽线圈电动机
screen collector 幕式集尘器;水幕除尘器
screen colo(u)r 荧光屏颜色
screen-colo(u)r characteristic 屏色谱特性
screen condensation temperature 外壳冷凝温度
screen connection 屏蔽连接
screen conveyer 筛式输送机
screen coordinates 屏幕坐标
screen copy 挂网片
screen count 出入越阻线的交通量计数
screen cupboard 纱橱
screen curve 屏曲线
screen cylinder 筛鼓
screen decanter 筛式沉降器
screen deck 筛底;筛面;筛层
screen deck area 筛面面积
screen deck heating 筛底加热
screen density 网线数
screen device 筛选设备
screen diagonal 屏幕对角
screen dike[dyke] 隔堤
screen discharge 筛分出料;筛选出料
screen discharge ball mill 筛分球磨机;周边卸料带筛球磨机
screen discharge mill 筛网排料式磨矿机
screen display 屏幕显示
screen dissipation 帘栅极损耗;帘栅耗散
screen distance 网屏距离;网目距离
screen door 屏门;网格门;铁丝网门;纱门
screen-door latch 纱门锁止器;纱门(门)闩

screen dot 网点;扫描点
screen drier 百叶窗式分离器
screen drum 筛式滚筒
screen earth 屏蔽接地
screened 筛过的
screened aerial 屏蔽天线
screened aggregate 过筛集料;过筛骨料
screened antenna 屏蔽天线
screened area 加网区
screened base 挂网阴片
screened cable 屏蔽电缆
screened camera 屏蔽照相机
screened casement window 带纱扇窗
screened coal 筛过的煤
screened coaxial cable 屏蔽同轴电缆
screened coil 屏蔽线圈
screened concrete aggregate 筛过的混凝土集料;筛过的混凝土骨料
screened diapositive 挂网阳片
screened feeder 屏蔽馈线
screened film 屏蔽膜
screened fluorescent tube 屏蔽日光灯
screened glass 网眼玻璃
screened glass powder 筛选的玻璃粉
screened gravel 过筛砾石
screened gravel mix(ture) 过筛的砾石混合料
screened host 遮挡式主机
screened indicator 遮蔽指示剂;筛选指示剂
screened insect proof house 防虫纱室
screened intake 有拦污栅的进水口
screened light 限照灯光
screened loess soil 过筛黄土
screened material 过筛的物料;过筛材料;筛分材料
screened negative 挂网阴片
screened oil accumulation 隔离石油层
screened ore 筛选矿石
screened pentode 屏蔽五极管
screened pipe 过滤管;衬管
screened positive 加网阳片
screened product 过筛产品
screened refuse 筛出废料
screened room 屏蔽室
screened sand 过筛砂
screened slag 筛分渣
screened sparking plug 屏蔽火花塞
screened top surface 顶面整平
screened well 过滤器井;过滤井
screen effect 遮帘作用;屏障效果
screen effective diameter 荧光屏有效直径
screen effect of pile 桩的遮前作用
screen efficiency 筛选效率
screen element 网目点;扫描点
screen equivalent 网线数
screener 筛选工;筛分机;筛分工;筛;分级器;分级机
screenerator 筛砂松砂机
screen fabric 筛布
screen facade 屏隔式立面
screen-factor 屏蔽系数
screen failure 筛蔽损坏
screen feed 筛料;筛选进料
screen-feed bin 筛分机给料仓
screen feeder 筛式给料机
screen feeding 筛分机给料
screen fence 掩蔽篱笆;网格式围篱
screen filter 网式滤油器;筛滤器;筛滤板
screen filtering 筛滤截留法
screen filter pressure 滤器屏压
screen filtration media 砂滤介质
screen filtration method 筛滤法
screen fineness 筛分细度
screen flange gasket 滤网凸缘衬垫
screen flap 小窗扇(纱窗上的,供人手伸出)
screen floating 筛余物
screen floor 筛子安装平面
screen for side light 舷灯遮光板
screen for two screening directions 双向筛选筛
screen fraction 筛分部分
screen-frame filter 筛框式过滤器
screen frame(unit) 筛架;滤网框架;银幕架;纱窗框;银幕框架;筛体
screen gear 网屏齿轮
screen grading 筛级
screen-grid 屏栅;帘栅板【电】;帘栅板【电】

screen grid circuit 屏栅电路
screen-grid dissipation 屏栅耗散
screen-grid modulation 屏栅调制
screen-grid tube 屏栅真空管;帘栅管【电】
screen grillage well 拦污滤网井
screen halftone 网目半色调
screen head 加入筛中的料
screen height 屏幕高度
screen holder 滤网架;筛孔板支架
screen hole 筛眼;筛孔
screen house 筛分车间
screen illumination 屏幕照度
screen image 显示图像
screen inclination 银幕斜度;屏幕斜度;滤网斜度;筛子斜度
screening 掩蔽;遮护;过筛;过滤;普查检;屏蔽;筛选制版;筛选;筛检;筛分;筛除法;电屏蔽
screening action 筛分作用
screening agent 掩蔽剂
screening aggregate 筛分集料;筛分骨料
screening analysis 筛选分析
screening and landscaping 屏蔽及景观设计
screening and storage 筛选和储存区
screening and washing plant 筛分与冲洗设备
screening angle 屏蔽角;筛面倾角
screening area 网眼面积;筛孔面积
screening arrangement 筛分装置
screening audiometer 筛式听力计
screening bale press 筛渣填料压机
screening bioassay 筛选生物检定
screening bottom 筛底
screening box 屏蔽箱;屏蔽盒
screening building 筛选厂
screening cage 屏蔽笼
screening can 屏蔽罩;屏蔽外壳
screening capacity 筛分能量;筛分能力;筛选能力;遮蔽能力
screening cell 筛分箱
screening chain 分选输送链
screening characteristic 屏蔽特性;筛选特性;筛分特性
screening classification 颗粒分级
screening cloth 筛布
screening coefficient 屏蔽系数
screening constant 屏蔽常数
screening curve 筛选曲线;筛分曲线
screening data 筛分数据
screening dedusting 过筛除尘(法)
screening design 筛选设计
screening device 滤网设施;滤食结构筛分机;筛选设备;筛机;筛分设备
screening dewatering 筛分脱水(法)
screening disintegrator 网目分离器
screening drum 筛分转筒
screening effect 屏蔽效应;遮蔽效应
screening efficiency 筛分效率
screening equipment 筛分设备;筛选装置
screening examination 过筛检查
screening experiment 筛选试验
screening fence 屏蔽栅栏
screening flume 筛选水槽
screening force 筛选力
screening for light 限光遮板
screening glass 护目镜片
screening grader 筛选机
screening hopper 筛分料斗
screening hypothesis 筛选假设
screening imaginary 挂网图像
screening inspection 筛选检查
screening installation 筛分设备;筛选设备;分选设备
screening into wide size ranges 筛分成大粒级范围
screening level 筛选级
screening machine 筛选机;筛机;筛石机;筛分机
screening machinery 筛分设备
screening material 屏蔽材料
screening method 筛选法
screening number 屏蔽数
screening nursery 甄别苗圃
screening of bids 投标审核;审定投标
screening of import commodities 筛进口商品
screening of nucleus 核屏蔽
screening opening 筛眼
screening operation 筛选操作;筛选工序

screening orthophoto 挂网正射相片
screening out 隔开;挡住;筛分;筛出
screening plane 筛面
screening plant 筛选装置;筛选厂;筛石厂;筛分装置;筛分设备;筛分工场;筛分车间;筛分厂
screening portion 筛选部分
screening practice 筛选工艺;筛选方法
screening procedure 筛选程序;筛选步骤
screening process 筛选程序;筛选方法;筛选过程
screening protector 线路拖流圈;屏蔽保护器
screening quality 筛选质量;筛分质量
screening rate 筛选率
screening refuse 筛集粗料;筛选残渣
screenings 筛选过的物质;筛屑;筛除物
screening sample 筛选样品
screenings crushing 筛下产品破碎
screenings dewatering 筛渣脱水
screenings disintegrator 筛渣破碎机;筛余物破碎机;筛下物粉碎机
screening seal coat 石屑封层
screenings grinder 筛渣粉碎机
screenings gutter 筛渣槽
screening shredder 筛渣撕碎机
screening simulation method 筛选模拟法
screenings incineration 筛渣焚化
screenings incinerator 筛渣焚化炉
screening slide 遮护滑板
screening smoke 烟幕
screenings passage 屏风过道(中世纪客厅布置形式)
screening specification 筛选规格
screenings press 筛渣压机;筛余物压滤机
screenings pulp 木节纸浆
screenings refiner 筛渣再磨机
screenings texture 筛余物结构
screenings triturator 筛渣捣碎机
screening surface 筛子(有效)面积
screenings wrapping 木节包装纸
screening system 滤水系统;筛分系统
screening technique 筛选法
screening test 预试验;预备测试;筛选试验;筛选实验;筛选检查;筛分析;分选试验;甄别试验
screening the bids 审核投标
screening tower 滤网塔架
screening trommel 滚筒筛
screening unit 筛分装置
screening validity 筛检真实性
screening varnish 遮盖漆;罩漆;丝漏清漆
screening washer 筛选器;筛洗器;筛式洗矿机
screening wire 屏蔽线
screening without breakage 无损耗的筛分
screening works 筛分工场
screen input/output 屏幕输入输出
screen intensity 荧光屏亮度;屏幕亮度
screen kneader 筛式捏合机
screen kneader with two shafts 双轴筛式捏合机
screenless printing 无网印刷
screen line 交通越阻线;分属查核线
screen-line count 网线计数
screen liner 滤管
screen loading rate 筛子加料率
screen lock 筛子定位器
screen loss 筛分损失
screen luminance 屏幕亮度
screen man 筛分工
screen memory 屏障记忆
screen menu 屏幕菜单
screen mesh 网眼;筛眼;筛网;筛孔;筛号(效);筛号纲目
screen mesh opening 筛孔
screen mesh size 筛孔尺寸
screen method 挂网法;屏蔽法;筛法
screen mill 筛磨
screen mode 屏幕方式
screen modulation 帘栅调制
screen modulus 筛分模数
screen motion 筛分运转
screen mo(u)ld 筛框线脚;边框饰条
screen mo(u)lding 纱窗压条;纱窗线脚
screen nest 成套筛子
screen noise 荧光屏噪声
screen number 筛孔数目;筛号
screen of bars 筛条
screen off 筛去

screen of sidelight 舷灯遮板
screen open area 筛选工作面积;筛面工作面积
screen opening 筛眼;网目孔;筛孔
screen out 筛出;筛去;筛分
screen overflow 筛余粗料;筛出废料;筛上溢流;筛上物
screen oversize 筛余粗料;筛出废料;筛上物
screen painting 丝网涂漆
screen pan 网屏蒸发器
screen partition 屏风隔断
screen perforation 筛孔
screen persistence 荧光屏余辉保留时间
screen persistence 荧光屏余辉
screen phosphor 屏荧光粉
screen photography 网目摄影
screen picture 屏幕图像
screen pipe 穿孔管;衬管;过滤管;滤水管;井管滤网;花管(进水管有滤网的一段);筛滤管
screen pipe openings 滤管孔眼
screen plane 承影面
screen planting 障景种植;遮蔽栽植;屏障栽植
screen planting strip 树篱带
screen plate 滤板;网目板;遮蔽板;筛板;宿栅极板板
screen point 网点
screen porch 纱窗阳台
screen potential 屏蔽电位
screen print 加网印刷品;加网
screen-printed wallpaper 丝网印刷的墙纸
screen printing 丝网印刷;筛网印花
screen printing glaze 印刷釉
screen process 银幕合成;彩屏法;背景放映
screen process printing 丝网印刷
screen-protected 用阿屏护的
screen protection 屏蔽保护
screen protection range 屏蔽保护范围
screen rack 挡泥板;拦污栅
screen rack well 拦污栅格架井孔
screen rake 拦污栅清理耙;拦污清理机;筛网倾斜度;筛刮;清筛器
screen reflector 金属网反射器
screen reinforcement 钢丝网;网状钢筋
screen reject 筛余粗料;筛出废料
screen residue 筛渣;筛余物;筛余(料)
screen roll 清筛辊
screen ruling finder 检线表
screen sash 纱窗窗扇
screen scale 筛网尺度
screen scarifier 清筛机
screen separation 屏风分隔;筛分
screen series 筛网成套;筛网成组;筛网系列
screen server 屏幕保护程序
screen set 成套筛
screen set bar 筛板拉杆
screen settler 过滤式沉降器
screen settling 荧光物质沉积法
screen shaker 振动筛;筛子撞击器;筛子抖动器
screen sheet 投影片
screen shoe 筛架
screen shredder 筛网式撕碎机
screen side 纤维板有网纹的背面
screen sieve 分离筛
screen size 屏幕尺寸;筛号;网眼尺寸;网线大小;筛孔尺寸;筛分粒度
screen sized material 筛分分级材料
screen size gradation 筛序;筛孔大小序次;筛号分级;筛分级配
screen size television 大屏幕电视
screen sizing 用筛分级;筛分分级
screen sizing test 过筛分析试验
screen slope adjusting shaft 筛网倾斜调节轴
screen slot size 筛网网眼尺寸
screen sludge 筛选泥浆
screen sludge crusher 筛余污泥压碎机;筛泥压碎机
screens passage 屏风过道
screen speed 响应速度
screen spline 屏幕安装条;装纱网压条(门窗)
screen status area 屏幕状态区
screen stencil 筛网
screen stone 屏石
screen storage time 荧光屏余辉时间
screen structure 网状结构;筛架
screen substructure 网架;网状下部结构
screen sunshade 挡板式遮阳
screen surface 筛面

screen surface load 筛面荷载
screen surface support frame 筛框;筛面支承框
screen system 滤网系统
screen table 风力摇床;风力淘汰盘
screen tailings 筛渣;筛余料;筛屑
screen temperature 百叶箱温度
screen temse 筛子
screen test(ing) 筛分试验;甄别试验;试镜头
screen thermometer 百叶箱温度计;百叶箱温度表
screen thickness 屏厚度
screen-throughs 筛屑;过筛料
screen tint 网目色调
screen tone 透明坐标纸
screen tower 筛选塔;塔式筛
screen transmittance 屏幕透射比
screen tray 筛盘
screen tube 捕渣管
screen-type centrifuge 筛筒式离心机
screen-type drier 筛式干燥器
screen-type film 增感型胶片
screen-type mill 周边排矿式磨碎机;筛型磨碎机;筛网排料磨机
screen-type separator 筛式清选器
screen underflow 筛下物;筛下范围;筛下底流
screen undersize 筛下产品
screen unit 筛架
screen used in chemical processing 化工用筛
screen value 网目分量
screen vibrator 振筛器
screen wall 屏蔽墙壁;漏花墙;挡墙;矮墙;影壁;照壁;花格墙;护墙;屏挡墙
screen wall counter 幕壁计数管;屏蔽计数管
screen washing analysis 筛洗分析实验
screen well 滤网井
screen well of pierced concrete blocks 穿孔混凝土砌块的护墙
screen well point 滤网式井点;滤网管井点
screen window 纱窗
screen wiper 刮水器
screen wire 筛网(金属)丝
screen wire cloth 纱窗用金属丝布;铁丝纱
screen with twin crank-shaft drive 双曲轴驱动筛
screen with two counterbalanced frames 有两个平衡架的筛子
screen working area 筛面工作面积
scree slope 坡积层
screw 螺旋;螺纹;螺钉
screw action 捻旋作用
screw adjusting double caliper 螺钉调整内外卡尺
screw adjusting inside caliper 螺钉调整内卡尺
screw adjusting outside caliper 螺钉调整外卡尺
screw adjustment 螺钉调整
screw agitator 螺旋搅拌器;螺旋搅拌机
screw air compressor 螺杆式空气压缩机
screw alley 轴隧道
screw anchor 螺旋固定锚;螺旋形锚;螺旋锚(杆);锚定螺栓
screw anchoring system 螺旋锚碇系统;螺旋锚碇装置
screw anchor pile 螺旋锚着桩;螺旋锚碇桩
screw and double nut 螺钉和双螺母
screw and mushroom sinker 螺旋底托瓣式浮标锚碇杆
screw-and-nut mechanism 螺杆螺母传动机构
screw-and-nut steering gear 螺杆螺母式转向器;螺杆螺母转向装置
screw and post anchorage device 螺丝端杆锚具
screw and toggle press 螺旋肘杆式压力机
screw and worm gate lifting device 螺旋蜗杆式闸门启闭机
screw antenna 螺旋形天线
screw aperture 螺旋桨框穴;推进器框穴
screw arbor 螺杆
screw auger 螺旋推进器;螺旋钻;麻花钻;推运螺旋
screw axis 螺旋轴;螺旋对称轴
screw bakelite lamp holder 胶木螺口灯头
screw barrel 螺旋桶
screw base 螺旋管底;螺口灯座;螺口插座;螺钉座
screw base fuse 螺旋式熔断器
screw batch charger 螺旋投料机
screw batten lamp holder 螺口平灯泡
screw bead 螺钉垫圈
screw bell 钻杆打捞器;打捞母锥
screw belt 螺钉带

screw belt fastener 螺钉带扣
screw blade 螺旋式叶片;螺旋桨(浆)叶;螺旋刀片
screw blank 螺栓毛坯;螺钉坯件
screw block 螺旋顶升器;螺旋式千斤顶;螺母块体;千斤顶
screw body 螺钉杆
screw bolt 方头螺钉;螺栓;螺杆
screw bolt and nut 螺栓和螺母
screw bolt boring machine 螺栓钻孔机
screw boss 桨毂;车叶毂
screw brake 螺旋制动(器);螺杆制动(器);螺杆闸
screwbrake with band wheel 带手轮螺旋制动
screw brake with crank handle 带摇把螺旋制动
screw bulb 螺口灯泡
screw bushing 螺旋轴衬
screw cap 螺旋帽;螺旋盖;螺钉(管)帽;螺母;螺口盖;螺口灯头
screw capping 拧盖
screw case 螺杆箱
screw case box 螺杆箱机座
screw case suppressing metal 螺杆箱紧固件
screw casing anchor packer 可卸锚尾式封隔器
screw centrifugal pump 螺旋式离心泵
screw centrifuge 螺旋卸料离心机
screw charger 螺旋加料器;螺旋加料机
screw chaser 修正丝用丝锥;螺纹梳刀;螺纹扳手
screw chases 螺纹版框
screw chasing 螺纹切削
screw chasing machine 制螺旋机;螺纹切削机;螺丝车床
screw chasing tool 螺纹梳刀
screw chisel 旋凿
screw chuck 螺旋夹盘;螺杆卡紧卡盘
screw clamp 手螺旋夹;螺丝夹具;螺旋夹;螺丝夹(钳)
screw clamp for brake-shoe lining 装制动摩擦夹片
screw clamp for lubricating leaf spring 钢板弹簧上油卡子
screw classifier 螺旋(式)分选机;螺旋(式)分类器;螺旋分级机
screw clip 管钳
screw coal conveyer 螺旋式输煤机
screw coal unloader 螺旋卸煤机
screw collar 螺旋环
screw comb 螺纹梳刀
screw compasses 弹簧圆规;螺纹罗盘仪
screw compressor 螺旋压气机;螺旋(式)压缩机
screw compressor clamp 螺旋压缩机管夹
screw cone 螺旋锥体
screw connection 螺纹连接;螺纹管节;螺钉连接
screw connector 螺旋连接器
screw contact 螺丝触点
screw control 压下控制
screw controller 螺旋链杆掣
screw conveyer 螺旋出土器;螺旋(式)输送机;螺旋(式)输送器;绞刀输送机
screw conveyer flight 螺旋刮板
screw conveyer for bulk cement 散装水泥螺旋输送机;水泥绞刀;散装水泥用裸轮输送机
screw conveyor for bulk cement 散装水泥螺旋输送机
screw correction 螺旋形失真校正
screw coupling 螺旋连接器;螺旋联结器;螺纹配合;螺纹连接器
screw coupling lever 螺旋联轴节手柄
screw cramp 螺旋夹钳
screwcrete 螺旋泵送混凝土
screwcrete pile 螺旋形钻孔灌注桩
screw crusher 螺旋式破碎机
screw cup 螺钉垫圈;木螺钉垫圈
screw current 螺旋浆流
screw current meter 螺旋测流计;旋转流速计;旋浆式流速仪;螺旋(浆)式流速仪
screw cutter 螺纹刀具;螺纹铣刀
screw cutting 螺纹切削;切削螺纹;切螺纹;车螺纹
screw-cutting dies 板牙
screw-cutting lathe 螺纹车床;车螺纹车床
screw-cutting machine 螺纹切削机;螺纹切削机(床)
screw cylinder 螺旋管柱
screw decanter 螺旋卸料离心机;螺旋澄清器;螺旋沉降器;螺旋倾析器
screw depth control 螺杆式深度调节器

screw depth regulator 螺旋式入土深度调节器
screw diad 二重螺旋轴
screw diameter 螺旋直径
screw die 螺丝(钢)板;螺纹模;螺纹铰板;螺纹板(牙);平纹板
screw discharge 螺旋卸料;螺旋卸载
screw discharger 螺旋卸料器
screw discharge sedimentation centrifuge 沉降式螺旋卸料离心机
screw dislocation 螺旋位错;螺形位错;螺线位错
screw displacement 螺旋位移
screw distributor 螺旋分配仪
screw-dolly 螺丝铆钉顶撑;螺旋铆顶
screw-down 下拧;压下机构
screw-down bibcock 螺旋式弯嘴管水龙头
screw-down cap (螺钉上的)旋压盖
screw-down cock 螺旋式(水)龙头
screw-down control 螺旋下降式调整
screw-down conveyer 螺旋式输送机
screw-down gear 压下装置;轧辊压下装置;螺旋压下机构
screw-down hydrant 螺旋式消防栓;螺旋式给水栓
screw-down mechanism 螺旋压下机构
screw-down motor 轧钢机用电动机
screw-down non-return valve 截止止回阀
screw-down pattern draining tap 旋压式放水龙头
screw-down stop valve 螺杆向下制动阀
screw down structure 压下装置
screw-down tap 螺距下调螺塞;螺距下调开关;螺距下调龙头
screw-down valve 旋(开)式闸阀;旋压阀;螺压下阀;闸阀;螺旋式阀(门)
screw dredge(r) 螺旋式挖泥船
screw drill 螺纹钻
screw drive 螺旋推动;螺杆传动
screw-driven 丝杆驱动的;螺旋传动
screw-driven planer 螺杆传动刨床;丝杠式刨床
screw-driven planing machine 螺杆传动刨床
screw-driven shaper 螺杆传动牛头刨;丝杠式牛头刨床
screw-driven slotter 丝杠式插床
screwdriver 旋凿;改锥;螺丝刀
screwdriver bit 螺丝刀;手钻尖头
screwdriver bit holder 螺丝刀夹具;手钻尖头握柄
screwdriver for cross-head screw 十字螺钉头的螺丝刀
screwdriver handle 螺丝刀的握柄
screwdriver set 成套螺丝刀
screwdriver set with wood(en) handle 木柄套装螺丝刀
screwdriver slot 螺丝起子槽
screwdriver socket 螺旋套筒;螺旋插座
screwdriver socket for cross-head screw 十字螺帽的螺旋套筒;十字螺帽的螺旋插座
screwdriver socket for Phillips screw 带十字槽头菲利普螺钉的螺旋套筒;带十字槽头菲利普螺钉的螺旋插座
screwdriver socket for Pozidriv screw 波芝特利夫螺钉的旋凿套筒
screwdriver socket for Torx screw 托克思螺丝刀的旋凿套筒
screwdriver with black plastic handle 黑塑柄螺丝刀
screwdriver with isolated blade 独立刀身的旋凿
screwdriver with through tang 穿心螺丝刀
screwdriver with through tang wood handle (en)木柄穿心螺丝刀
screwdriver with voltage tester 试电笔
screwdriver with wood(en) handle 木柄螺丝刀
screwdriver with wooden scales 夹柄螺丝刀
screw-driving machine 螺杆传动机床
screw drum 螺旋转鼓
screw dryer 螺旋干燥机
screwed bonnet 螺纹阀盖
screwed centrifugal nozzle 螺旋离心式喷嘴
screwed column 螺旋形柱
screwed conduit 螺纹管
screwed conduit coupling sleeve 两端有螺纹的螺纹管耦合套
screwed connection 滤网连接;螺纹连接;螺纹套管接头;丝扣连接
screwed core plug 泥芯孔螺纹堵塞
screwed coupling 拧紧的接头;拧紧的耦联器

screwed coupling pipe 拧紧的连接管子
screwed cutout 保险螺丝;螺旋保险器
screwed dowel 螺纹木塞
screwed end 螺纹端
screwed fitting 螺纹接头
screwed flange 螺旋法兰;旋上凸缘;螺旋凸缘;螺口法兰盘;螺纹法兰
screwed flange of pipes 管子上螺旋法兰;管子旋上凸缘;管口栓接法兰
screwed flange of tubes 螺旋栓接法兰
screwed gas pipe 螺纹煤气管
screwed-gland joint 攻螺丝套筒
screwed joint 螺栓接头;螺纹套管接头;螺旋接合;螺丝接合
screwed jointing 螺旋接合
screwed nipple 螺丝接头
screwed-on hinge 拧上的铰链
screwed-on lock 拧上的锁
screwed-on pipe 拧上的管子;螺纹连接的管子
screwed pack 多层挤压水
screwed pile 螺旋柱
screwed pipe 螺纹管
screwed pipe coupling 螺纹管连接
screwed pipe joint 螺纹(套)管接头;螺管接头;螺纹套管插口;螺(纹)管连接
screwed pipe support 螺纹管支架
screwed plug 螺纹堵;丝堵;螺旋塞
screwed reinforcement bar 螺纹钢筋
screwed rivet 螺纹铆钉
screwed shank tool 螺旋手柄工具
screwed sleeve 螺纹套管
screwed socket 螺丝承插口
screwed socket adapter 螺口插座
screwed socket for union 螺旋承窝
screwed socket joint 螺旋套管接合;螺纹(套)管接头
screwed stay 螺旋撑条
screwed staybolt 螺头撑杆
screwed steel conduit 螺头钢导管;带螺纹的钢导线管
screwed tooth 螺旋齿
screwed tube 螺纹管
screwed type 螺纹式(连接)
screwed valve 螺旋阀
screwed work 螺旋形切削(木料)
screw effect 螺旋桨效应
screw elevation 螺旋提升
screw elevator 螺旋(式)提升机;螺旋(式)升降机;螺旋升运器
screwer 螺纹刀
screw extractor 断(头)螺丝取出器;起螺丝器
screw extruder 螺旋压出机;螺旋挤压机;单螺杆挤出机
screw extrusion machine 蜗杆挤压机
screw extrusion press 螺旋挤压机;螺旋挤出机;螺杆压榨机;螺杆挤压机
screw eye 羊眼螺丝;带眼螺栓;羊眼圈;羊眼螺眼;螺丝眼
screw fan 轴流式扇风机;螺旋扇风机
screw fastener 螺钉扣
screw fastening 螺丝接合;螺钉连接
screw feed 螺旋推进(器);螺旋式送料(器);螺旋桨送料器;螺旋桨送料
screw-feed centrifugal pump 螺旋形进料离心泵
screw feed diamond drill 螺旋给进式金刚石钻机
screw feed diamond rig 螺旋给进式金刚石钻机
screw feeder 螺旋输送机;螺旋喂料机;螺旋式进料器;螺旋(式)给料机;螺旋加料器;螺旋加料机;螺旋供料器;螺杆给料机;喂料绞刀
screw feed gear 螺旋给进器
screw feed grease cup 滑脂杯
screw feed head 螺旋给进器
screw feed machine 螺旋给进式钻机
screw feed mechanism 螺旋给进机构
screw feed pump 螺旋喂送泵;卧式螺旋泵
screw feed rig 螺旋给进式钻机
screw feed swivel head 螺旋给进钻机回转器
screw feed system 螺旋给进系统
screw feed type bucket elevator loader 螺旋进料斗式提升装载机
screw filtering press 螺旋压榨机;螺旋压滤机
screw fitting 螺纹管节
screw fixing 螺丝固定
screw flange 螺栓连接法兰;栓接法兰;螺纹法兰

screw flange coupling 螺旋法兰联轴节;螺旋法兰连接;螺钉凸缘联轴节
screw flange joint 螺旋法兰连接
screw flange jointing 螺旋法兰连接
screw flowmeter 螺旋翼轮表;螺旋流量计
screw for bearing wheel of oscillating weight 自动锤轴承螺钉
screw for framework of automatic device 自动框架螺钉
screw for indicator guard 日历指示器护板螺钉
screw for jumper spring 跳针簧螺钉
screw for motor starter 发动机起动器螺钉
screw for setting lever spring 压簧螺钉
screw for starter fixing clamp 起动器固定夹螺钉
screw friction press 螺旋摩擦压力机
screw gate jack 螺旋闸门千斤顶
screw gate lifting device 螺旋式闸门升降机;螺旋式闸门启闭机
screw ga(u)ge 螺钉号;螺纹样板;螺钉规格;螺旋规尺;螺纹量规;螺纹规
screw ga(u)ge for taper thread 锥形螺纹量规
screw gear 螺旋联动装置;螺旋齿轮;螺纹传动装置;压下装置
screw gearing 螺旋齿轮传动装置;蜗轮蜗杆传动装置
screw gill drawing frame 螺旋杆式并条机
screw grab 打捞公锥
screw grab guide 打捞公锥导向器
screw grader 螺旋分级机
screw guard 螺旋桨式护管;推进器护材
screw gun 螺旋式油枪
screw head 螺旋帽;螺钉头;螺丝头;螺旋头
screw head connector 螺旋索头联结器
screw head file 螺钉头锉
screw height control 螺杆式高度调节器
screw hoist 螺丝起重器;螺旋起重机
screw holder 螺丝夹持器;螺丝灯头;螺口灯头
screw holding forceps 持螺丝钳;螺钉固持钳
screw hole 螺丝孔;螺孔;螺旋眼
screw home 拧紧;拧到头
screw hook 带螺旋的钩;螺丝钩;带钩螺栓
screw hoop 环首螺栓
screw horsepower 螺旋桨功率;推进器功率
screw hour 工组小时
screw housing clearance 推运螺旋与壳间间隙
screw ice 旋冰堆;小冰堆;挤压冰;碎冰;冰屑
screw-impelled compactor 螺旋推力压缩机
screw-impelled pump 螺旋叶轮泵
screw impeller 螺旋(式)叶轮;螺旋式搅拌叶轮
screw impeller pump 螺旋桨式泵
screw in 拧入,旋入
screw-in compression 止动螺钉
screw-in depth 旋入深度
screw-in filter cell 拧入式滤光片环
screwing 旋入螺钉;旋紧螺钉;上螺钉;车削螺纹
screwing and unscrewing machine 瓶阀式启闭机
screwing brass 螺丝黄铜
screwing chuck 螺丝灯头;螺丝钢板盘
screwing die 外螺纹板牙;板牙
screwing interval 螺钉间距
screwing lighter 安放螺旋式锚作业船
screwing machine 螺丝加工机
screwing on 螺旋连接的;拧上螺钉的
screwing pack 旋转浮冰群
screwing stock 板牙扳手
screwing tap 螺丝龙头;螺丝锥;螺丝攻
screw injection machine 螺杆注塑机
screw-in socket 螺旋插座
screw-in type thermometer 旋入型温度计
screw jack 螺旋(式)千斤顶;螺旋起重器;螺旋起重机;螺旋顶重器;螺旋升降器;千斤顶;螺丝千斤顶;螺杆千斤顶
screw jack prop 螺旋起重机支柱
screw jaw with lug 带舌螺扣接杆
screw jaw with movable lug 带活舌螺扣接杆
screw joint 螺纹连接;螺钉连接;螺旋接头;螺旋接合;螺纹套(管)接头;螺管接头
screw joint casing 丝扣连接套管
screw-joint coupling 螺丝套帽接头
screw key 螺栓扳钳;螺旋键;螺丝扳手
screw-kneader 螺旋捏和机
screw lead 螺距
screwless adapter 无螺纹接器套筒
screwless extruder 无螺杆挤出机
screwless knob 无螺钉门把手
screwless rose 无螺纹淋浴喷头;无螺纹连蓬头
screw lid 螺旋盖
screw lift 螺旋提升机
screw lifting jack 螺旋起重器
screw lifting mechanism 螺旋提升机构
screw line 螺旋线
screw-locking device 螺钉锁紧装置
screw lower bridge of automatic device 自动装置下夹板螺钉
screw luffing 螺杆变幅
screw machine 螺杆传动机械;自动车床;制螺钉机;螺杆压出机
screw making machine 制螺钉机
screw mandrel 螺旋心轴;螺旋立轴;丝杆
screw micrometer 螺旋测微器;螺杆测微计
screw micrometer cal(l)iper 螺旋千分卡尺
screw microscope 螺旋测微镜
screw mixer 螺旋(轴)搅拌机;螺旋搅拌器;螺旋拌和机
screw mooring 螺旋形锚碇块(体);螺旋式锚系
screw motor 螺杆马达
screw mount 螺旋框
screw nail 螺纹钉;螺丝钉;螺钉;木螺钉;木螺丝
screw nail of wood 木螺钉
screw nicking machine 螺钉头开槽机
screw nipple 螺纹接管;丝对
screw nut 螺帽;螺母
screw off 拧下(螺丝);拧松;拧开(钻头或钻杆);旋出
screw of tangential cutter head 切向进刀架丝杠
screw oil pump 螺杆油泵
screw on 拧上;拧紧;旋上
screw-on bead 玻璃定位条
screw on cold 冷拧上
screw-on cutter 螺旋铣刀
screw out 拧出
screw packer 螺杆封隔器
screw paddle conveyer 桨叶式螺旋输送机
screw pair 螺旋副;丝杠螺母副
screw pelletizer 螺杆造粒机
screw pile 螺旋桩;螺纹桩
screw pile foundation 螺旋桩基(础)
screw pile mooring dolphin 螺旋桩系船墩
screw pile quay 螺旋桩码头
screw pile shaft 螺旋桩桩身
screw pile wharf 螺旋桩码头
screw pile with two turns 双向螺旋桩
screw pin 螺旋销;埋头螺丝扣
screw pinch 螺旋箍缩
screw pinch-cock 螺旋箍活嘴夹
screw pip coupling 螺丝管连接
screw pipe coupling 螺纹管节
screw pipe joint 螺纹管节
screw-piston type injection mo(u)lding machine 螺杆活塞式注塑成形机
screw pitch 螺纹节距;螺距;螺距
screw-pitch deviation 螺距偏差
screw-pitch ga(u)ge 螺纹规;螺距规
screw plasticating injection mo(u)lding 螺杆塑化式注塑
screw plate 螺丝模;螺丝(钢)板;搓丝机;搓丝板;车丝板;板牙
screw plate loading test 螺旋板荷载试验
screw plate test 螺旋荷载板试验
screw plate tester 螺旋荷载板试验仪
screw plug (用于排水试验的)螺旋水栓;螺旋塞;螺旋接线柱;螺丝堵头;丝扣知堵
screw-plug cut-out 螺塞保险
screw-plug fuse 旋入式熔断丝
screw plug ga(u)ge 螺纹塞规
screw plug header 端部带螺旋塞的管
screw point bit 螺旋锥形钻头
screw post 螺旋接钉;推进器尾柱;车叶柱
screw press 螺旋压滤机;螺杆压滤机;螺旋压砖机;螺旋压(力)机;螺旋冲压机
screw prop 千斤顶
screw propelled tug boat 螺旋桨拖船
screw propeller 螺旋推进器;螺旋桨
screw propeller shaft 螺旋桨轴;尾轴
screw propellor 螺旋桨
screw propellor for boat 船用螺旋桨推进器
screw propulsion 螺旋桨推进
screw protector 护丝
screw pump 螺旋泵;螺杆泵;蜗杆泵
screw punch 螺旋冲压机
screw pusher 螺旋式推车机
screw race 螺旋桨流
screw rack 斜齿条;螺旋齿条
screw rail bender 螺杆弯轨机
screw red lamp bulb 螺口红灯泡
screw release 螺旋式限时解锁器
screw remover 去螺钉器
screw retarder 螺旋式缓行器
screw retention 螺旋防松装置;防松螺栓
screw reverse gear 螺旋回动装置
screw rewasher sand tank 砂料螺旋复洗机
screw ring 环首螺钉
screw ring ga(u)ge 螺纹环规
screw rivet 螺旋铆钉
screw riveting machine 螺旋铆机
screw rod 螺旋轴;螺(旋)杆
screw rolling machine 滚丝机
screw-rudder 转向螺浆
screw rule 螺旋定则
screws 潜水员病
screw scanning type sonar 螺状扫描声呐
screw shackle 张力套筒;螺旋套管;螺旋销卸扣;螺旋销卡环;螺旋扣;螺纹连接环;花篮螺丝
screw shaft 螺旋轴;螺杆轴;尾轴(推进器轴);推进器轴
screw-shaft bushing 螺旋轴套
screw-shaft cone 轴圆锥段
screwshaft liner 轴衬套
screw-shaft tube 尾轴套;尾轴管
screw-shaft tunnel 地轴轴隧;尾轴隧
screw shank 螺钉杆
screw shaped 螺旋状的
screw shaving machine 制螺母及螺栓头机
screw sheel 螺旋套筒;灯头螺口
screw shell 灯头的螺口;螺旋套筒;螺旋套管;螺灯泡座;螺纹套筒
screw ship 螺旋桨船
screw ship unloader 螺旋式卸船机
screw shoe 螺丝箍
screw sleeve 螺纹套筒;螺纹套管
screw-slot milling machine 螺杆头沟槽铣床
screw snap ga(u)ge 螺纹卡规
screw snow plough 螺旋扫雪机
screw socket 螺纹承插口;螺灯头;螺丝承插口;螺口灯座;螺口插座;螺旋套筒;螺纹套筒;螺丝套筒;螺口灯头
screw socket joint 螺纹套管插口
screw socket jointing 螺纹套管插口
screw spanner 螺丝活络扳手
screw spike 木螺钉;螺旋道钉;螺纹杠杆;螺纹道钉;螺钉
screw spike driver 螺纹道钉拧入器
screw spindle 螺纹轴;螺旋轴;螺旋柄
screw spreader 螺旋分布器;螺旋分布机;螺旋传送器
screw stair(case) 螺旋(楼)梯;转梯;盘旋梯;盘旋式楼梯
screw stake 螺旋桩
screw stay 螺旋拉条
screw steel 螺钉钢;螺丝钢
screw steering gear 螺杆滑块操舵装置
screw stem 螺杆
screw stock 板牙架;螺纹板牙;丝锥扳手;螺旋扳手;螺丝柄;螺钉条料
screw stopper 螺旋塞;螺旋锚链犁
screw stud 柱螺栓;双头螺栓柱螺栓
screw support for adjusting regulator 调整快慢针的螺钉座
screw surface 螺旋面
screw swinging mechanism 螺旋摆动机构
screw take-up 螺杆拉紧装置
screw tap 螺丝口(龙头;螺丝攻(丝锥);丝锥
screw templet 螺丝样板
screw terminal 螺旋式接线柱;螺丝接线端
screw testing interferometer 螺杆检验干涉仪;检验螺杆干涉仪
screw thread 螺线;螺纹;丝扣
screw-thread angle 螺纹角
screw-thread bush 螺纹衬套
screw-thread cutting head 螺纹切削头
screw-threaded rod 螺纹板
screw-threaded steel 螺纹钢

screw-thread fit 螺纹配合
screw-thread form 螺纹扣形
screw-thread ga(u)ge 螺纹(卡)规
screw-threading tool 螺纹刀具
screw-thread micrometer 螺旋千分尺
screw-thread milling 铣螺纹
screw-thread milling machine 螺纹铣床
screw-thread rolling machine 滚丝机;螺纹滚轧机
screw-thread steel 螺纹钢筋
screw-thread system 螺纹制
screw-thread tool ga(u)ge 螺纹梳刀样板
screw tightener 压紧螺钉
screw tightening device 螺旋式系紧装置
screw tight fitting 紧配螺旋
screw tip 蜗杆梢
screw together 拧接
screw-together fitting 整体式接头
screw tool 螺纹刀具
screw-to-rod adapter (螺旋立轴接与钻杆连接的)异径接头
screw tow bar 螺旋式牵引杆
screw track spike 螺纹道钉
screw triad 三重螺旋
screw trough 螺纹槽
screw tube 螺纹管
screw tube conveyer 螺旋管式输送机
screw tube conveyer feeder 螺旋管式输送喂料机
screw tube-type conveyer 管式绞刀
screw-type anchor 螺旋式锚
screw-type centrifugal nozzle 螺旋离心式喷嘴
screw-type compressor 螺旋式压缩机;螺旋式制冷机;螺杆式压缩机
screw-type concrete spreader 螺旋式混凝土撒布机;螺旋式混凝土布料机
screw-type cooler 螺旋式冷却器
screw-type cooling bed 螺旋式冷床
screw-type core pusher 螺旋式芯退出器
screw-type coupling 螺旋式连接器
screw-type feeder 螺旋给料机
screw-type flowmeter 螺旋桨式流量计
screw-type kickoff 螺旋推出机;螺旋拨料机
screw-type lens mount 螺旋形透镜框
screw-type lubricator 螺旋式注油器
screw-type mixer 螺旋式混合器
screw-type pump 螺旋泵
screw-type rheostat 螺旋式变阻器
screw-type rotary pump 螺旋形旋转泵;螺旋式回转泵
screw-type sand classifier 螺旋式砂筛分机;螺旋式砂分级器
screw-type ship unloader 螺旋式卸船机
screw-type sinking method 旋形凿井法
screw-type stop 螺旋限制器
screw-type valve 螺旋式阀(门);螺丝式阀门
screw-type ventilator 螺旋桨式通风机
screw-type washer 螺旋垫圈;螺旋式洗砂机
screw unit 组合机床的动力头
screw up 拧紧;旋紧
screw value 螺钉周值
screw valve 螺旋球阀
screw vault 斜拱顶
screw vent 螺杆通孔
screw ventilation 螺旋通风器;螺旋式风机
screw vessel 螺旋桨船
screw vice 螺旋虎钳
screw viscosity pump 螺旋式黏度泵
screw washer 螺帽垫圈;螺旋洗砂器;螺旋洗矿机;螺旋垫圈
screw wedge 丝杆调节楔
screw wheel 螺旋齿轮
screw wiper 螺旋式擦拭器
screw wire 螺钉用螺丝
screw with diamond knurls 金刚石滚花头螺钉
screw with flat head 平头螺钉
screw with knurled head 滚花头螺钉
screw without nut 无帽螺钉
screw with straight-line 直线滚花头螺钉
screw with straight-line knurls 直线滚花头螺钉
screw wrench 活络扳手;螺丝扳钳;螺旋(活络)扳手;螺旋扳钳;螺丝扳手;螺钉扳手
screwy 螺旋形的
scribbing-block 拍纸簿
scribbled 凸口对接缝和层面修琢过的
scribe 合缝

scribe and strip testing 划线剥落试验
scribe awl 划线计;划线器
scribe-base material 刻图片基
scribe board 线形划划板
scribe coat 刻图膜
scribe coat film 刻图膜片
scribe-coating 刻线
scribe-coating protection 刻图膜保护层
scribed adjustment 刻图修改
scribed circle method 圆形刻网格法
scribed-coated base 刻图膜片
scribed joint 对接;合缝接头
scribed key 刻绘底图
scribed map 刻图原图
scribed origin 刻绘原图
scribed surface 刻图膜面
scribe-head 刻图头
scribe projection 划线投影
scriber 刻图仪;划划刀;画线器;划针;划线器;划片器
scriber point 划针侧块
scriber type 划针型号
scribe sheet 刻图版
scribe tool 刻图工具
scribing 刻图法;划线;合缝;对缝;雕合;放样划线
scribing aids 刻圆仪
scribing and breaking 划线折断法
scribing block 画线架;划针盘;划线架
scribing cal(l)ipers 内外卡钳
scribing coating 刻图膜
scribing compass 划线规
scribing diamond 刻图钻石
scribing dimension 缮写尺寸;划线尺寸
scribing gouge 竖槽凿
scribing grid 划线栅格;划线网
scribing guide 刻图底图图形
scribing head 刻图头
scribing instrument 刻图工具
scribing iron 刻划刀
scribing knife 括缝刀;划线笔刀;放样间划线笔刀
scribing material 刻图材料
scribing needle 刻图针
scribing on glass 玻璃板刻图
scribing on plastic 塑料片刻图
scribing plate 画线板
scribing plotting display 划线绘图显示器;机电笔绘显示器
scribing point 刻针
scribing sheet 刻图版
scribing system 划片系统
scribing technologic(al)scheme 刻图工艺
Scribner log rule 斯克莱布诺材积表
scrim 粗织平纹布;粗织网眼织物;稀松窗帘用布;麻棉布(窗帘用);网服布;纤维织品
scrim back 网衬;稀松织物衬
scrimble 用织物纤维增强抹灰层
scrim cloth 网格布
scrim for reinforcing films 增强薄膜用网布
scrim for reinforcing grinding wheel 增强砂轮网布
scrimming 接缝处铺设增强布
scrim or scrimp 墙板接合处的覆盖层;粗织物纤维灰浆层;粗帆布灰浆层
scrimp 粗织平纹布
scrimshaw (鲸骨或者鲸牙制的)雕刻工艺品;贝壳雕刻工艺品
scrip dividend 以暂时股票付红利;期票股利
scrip holder 临时股票持有者
script 原本;正本;脚本;手写体;笔迹
script graphics tablet 手迹图形输入板
scriptorium 缮写室;(修道院的)书房
scripture hall 经堂
scrod 幼鱼
scroll 漩涡形;旋涡饰;旋涡花样;迂回扇;卷状物;卷滚;卷动;卷成卷轴形;上卷;船头装饰;翻卷
scroll arrow 卷动箭头
scroll bar 螺旋形沙坝;蜗旋沙坝;涡形沙坝
scrollbar 卷动条
scroll box 卷动盒
scroll case 螺旋形箱;蜗壳;涡壳(水轮机)
scroll-case access 蜗壳通道;蜗壳进入孔;蜗壳进口;涡壳型通道
scroll-case connection 蜗壳接头
scroll-cased drain pipe 蜗壳排水管
scroll-case drain 涡壳排水管

scroll-case vent 蜗壳通气管
scroll centrifuge 涡壳离心机
scroll chuck 三爪自动定心卡盘
scroll cover 螺旋套
scroll directrix 涡卷准线
scroll discharge centrifuge 螺旋卸料离心机
scroll drum 圆锥形滚筒;(起重机的)圆锥鼓轮
scrolled 卷形的;涡旋形的
scrolled gable 旋涡形山墙
scroll expander 蜗杆式开幅器
scroll file 滚动文件
scroll gear (直径逐渐变化的)蜗形齿轮
scroll head 船头装饰
scroll holder 画筒
scrolling 卷动
scroll lathe 盘丝车床
scroll-lock key 卷动锁定键
scroll meander 内侧堆积曲流;环流河湾;环流河曲
scroll mill 松粉机
scroll milling 铣螺旋线
scroll mo(u)lding 凸圆带形线脚;漩涡形线脚;旋涡形线脚
scroll opener 蜗杆式开幅器
scroll ornament 旋涡饰
scroll picture rack 挂图架
scroll pivoter snips 圆口剪
scroll saw 窄条锯;竖锯;线锯;钢丝锯;锯曲线用的带形锯;曲线锯;云形截锯;带形锯
scroll sawing machine 线锯床
scroll shaft 涡形轴
scroll-shaped 旋涡形的
scroll-shaped capital 旋涡形柱顶
scroll-shaped gable 旋涡形山墙
scroll shear 涡形剪切机;角剪;涡形剪床;曲线剪床
scrolls rack 卷轴架
scroll steps 盘旋梯阶;卷形踏步;盘旋梯级;盘梯级;盘旋阶梯
scroll tank 盘香管灯;盘香管容器
scroll-wheel 涡轮
scroll work 漩涡(形)饰;涡卷装饰;以卷形为主题的装饰;涡形装饰
scroll work design 涡形花纹
scropionism 蝎蚕中毒
scroth wall 凸出墙
scrotiform 囊状
scrpa metal 碎金属
scrub 洗刷;密灌丛;气体洗涤;刷子;洗涤气;低矮丛林;桉树灌木丛
scrubbability 刷洗性;擦净性
scrubbable 可擦净的;可刷洗的
scrubbed 粗面的
scrubbed concrete 拉毛混凝土;刷毛混凝土
scrubbed concrete facing 拉毛混凝土面层;刷毛混凝土面层
scrubbed concrete flower trough 拉毛混凝土花槽;刷毛混凝土花槽
scrubbed concrete slab 拉毛混凝土板;刷毛混凝土板
scrubbed concrete surface 洗干净的混凝土表面
scrubbed concrete tile 擦洗过的混凝土瓦;擦洗过的混凝土砖
scrubbed extract 洗涤过的萃取液
scrubbed finish 混凝土表面用稀盐酸擦洗处理;外露骨料饰面;水刷石饰面;擦洗处理
scrubbed gas 纯气体;纯煤气
scrubbed granolithic finish 水刷石
scrubbed plaster 擦洗过的粉刷层
scrubber 洗涤机;洗涤塔;洗涤设备;洗涤(除尘)器;旧砂再生装置;减震器;湿式滤清器;湿式除尘器;湿法除尘器;涤气器;擦洗机;擦洗粉;擦布;板刷
scrubber brush roll 洗涤器刷辊
scrubber collector 洗气收集器;洗涤收集器;洗涤集尘器
scrubber condenser 洗涤冷凝器;涤气冷凝器
scrubber condensor 涤气冷凝器
scrubber filter 洗涤式滤尘器
scrubber liquid 洗气器液体
scrubber liquor 洗气器液体
scrubber plate 洗涤器板
scrubber sludge 洗涤器淤渣
scrubber tank 洗气罐;洗涤罐
scrubber(wash)tower 洗涤塔;湿式洗气塔;洗气塔;洗气罐

scrubbing 洗涤;涤气;擦洗
scrubbing-and-drying unit 洗涤干燥机;刷洗干燥机
scrubbing brush 擦洗刷;硬毛刷;板刷;擦洗用硬毛刷
scrubbing effect 洗涤效果
scrubbing filter 涤气过滤器
scrubbing finish 擦洗处理
scrubbing head 擦洗机顶部
scrubbing machine 擦洗机;清刷洗涤机
scrubbing oil 洗油
scrubbing raffinate 洗涤废液;洗余液
scrubbing resistance 耐刷洗性
scrubbing resistance test 耐刷洗试验;耐擦洗试验
scrubbing solution 洗气溶液;洗涤液
scrubbing test 洗涤试验;擦洗试验
scrubbing tower 气体洗涤塔
scrubbing unit 刷洗机
scrubbing water 洗涤水
scrub board 洗衣板;护墙板;踢脚板
scrub board component 踢脚板构件
scrub board heater 踢脚板采暖器
scrub board heating 踢脚板采暖
scrub board member 踢脚板构件
scrub board radiator 踢脚板散热器
scrub board unit 踢脚板单元
scrub brush 擦洗刷
scrub clearer 灌木切除机;灌木清除机
scrub-clearing machine 灌木铲除机
scrub column 洗涤塔
scrub cutter 灌木切除机
scrub forest 灌木林
scrub growth 灌木林
scrub land 长满灌木之地
scrub mark 刷状痕
scrub mark on glaze 釉面擦伤
scrub plane 粗刨
scrub pulverizer 灌木切碎机
scrub rake 灌木搂集机
scrub resistance 耐擦洗性
scrub sink 洗手盆(外科手术用)
scrub slasher 灌木铲除机
scrub solution 洗提剂
scrub-up room 外科洗手室
scrub washing test 擦洗试验
scruff 锡渣;炉渣
scruff(y) plate 镀锡薄板;锡污板
scruple 吩(英美药衡单位,1吩=1.295克)
scrupulosity 顾虑过度
scrutator 观察者
scrutineer 监票人
scruting of vouchers 凭证的审核
scrutinizing program(me) 检验程序
scrutiny 详细研究;细看;仔细研究;仔细检查;评核;审查
scuba 水下呼吸器
scuba diver 水下自持呼吸器的潜水员
Scuba diving 斯库巴潜水
scud 碎雨云;顺风行驶
scudding 刮面
scudding knife 切纸刀
scuff 磨损处
scuffie knife 单面平切铲
scuffing 咬接现象;刮伤;毛刺砂光;划痕
scuffing grind 砂粒表层滚磨
scuffing mark 咬合痕
scuffing of cylinder bore 拉缸;汽缸工作面划伤;汽缸镜面擦伤
scuffled 未经加工的粗糙锯面
scuffle hoe 推式锄;推拉锄
scuffler 摔土率
scuff mark 刷状痕
scuff preventative treatment 防擦伤处理
scuff-proof 抗磨损
scuff resistance 耐揉搓性;耐磨损性;耐磨强度;耐擦性
scuff-resistant coating 耐磨镀层;耐磨面层
scuff wear 熔着磨损
scull 摇橹;结壳;浇包渣壳;浇包结壳;凝壳;划小桨;短桨;单人双桨赛艇
scull boat 摇橹划子
sculler 双桨轻划艇
scullery 炊具储藏处;洗碗槽;碗碟洗涤装置;碗碟洗涤室;餐具洗涤室
scullery basin 洗碗槽

scullery table 碗碟洗涤台
sculling notch 橹槽
sculp 片裂板岩
sculping 劈裂板岩
sculpting tool 雕刻工具
sculptor 雕刻师;雕刻家
sculptress 女雕塑家;女雕刻家
sculpt technology 造型技术
sculptural 雕塑品;雕刻的
sculptural arts 雕刻艺术;雕塑艺术
sculptural block 雕刻装饰部件
sculptural decoration 雕刻装饰
sculptural decorative feature 雕刻装饰特色
sculptural detail 雕刻细部
sculptural enrichment 雕刻美化装饰
sculptural ornamental feature 雕刻装饰特征
sculptural ornamental finish 雕刻装潢修饰
sculptural pattern 雕刻花样
sculptural style 雕刻风格;雕刻形式
sculptural type block 雕刻形式装饰部件
sculpture 刻触;捏塑;雕塑;雕刻术;雕刻(品)
sculpture and architecture 雕塑与建筑
sculpture blasting 光面爆破
sculptured 浮雕型的;仿浮雕的
sculptured acoustic(al)tile 浮雕(花纹)天花板
sculptured bust 胸像
sculptured carpet 凹凸花纹地毯;浮雕绒头地毯;浮雕花纹剪绒地毯
sculptured concrete 雕塑用混凝土
sculptured effigy 雕像
sculptured foliage 雕刻叶饰
sculptured frieze 雕刻饰带
sculptured glass 雕刻玻璃
sculptured ornament 雕刻装饰
sculptured pattern 浮雕花纹
sculptured pillar 雕花柱
sculptured plaque 雕塑饰板
sculptured wallcovering 表面有凹凸花纹的墙纸
sculpture gallery 雕塑馆
sculpture garden 雕塑公园
sculpture in ivory 象牙雕刻
sculpture in stone 石雕
sculpture in wood 木雕
sculpture park 雕塑公园
sculpture porcelain 瓷雕
sculpture room 雕塑室
scum 沟管积垢;釉面无光;渣滓;鳞屑;撇渣;泡沫;浮垢;浮渣
scum area 浮渣区面积
scum baffle 浮渣挡板
scum barrier 拦渣栅;除渣栅
scumble 涂不明色;涂暗色;套色漆;薄涂涂料
scumble for fair-faced concrete 薄层刷刮使混凝土表面平整
scumble glaze 晕色用罩光漆(上光剂);半透明薄涂层;暗釉
scumble stain 不透明着色剂;暗色着色剂;晕色用着色剂
scumbling 去色;薄涂暗色;晕色;除色;暗色施工法
scumbling technique 使色彩柔和的技术;涂暗色技术
scum board 刮渣板;拦渣板;浮渣隔板;浮渣(挡)板
scum box 浮渣槽
scum box and baffle 浮渣室和挡板
scum breaker 浮渣破碎器;浮渣破碎机
scum chamber 浮渣室;浮渣井
scum cock 锅炉排污旋塞;排污开关;排沫旋塞;放泡沫旋塞
scum collector 沉淀池浮渣收集器;浮渣收集装置;浮渣收集器
scum concentration and incineration 浮渣浓缩和焚化
scum crusher 浮渣破碎器;浮渣破碎机
scum-forming bacteria 成泡沫细菌
scum gutter 排沫沟
scum incineration 浮渣焚化
scum layer 浮渣层
scum line 泡沫线;浮垢条纹
scummer 除渣勺;撇(浮)渣机;排渣杓;除渣器
scumming 肭版;光泽减弱;浅色斑点;起脏;撇渣;泛霜;浮渣
scum nozzle 除渣喷嘴
scum pipe 浮渣导管
scum pit 浮渣坑

scum plate 拦渣板
scum pump 吸泥泵;泥浆泵
scum removal 除泡沫
scum remover 除泡沫器
scum riser 集渣冒口;集渣包;除渣冒口
scum rod 挡渣棒
scum rubber 泡沫橡胶
scum skimmer 除沫器;除沫剂
scum space 浮渣空间
scum spot 渣孔
scum tank 除渣池
scum trough 浮渣排除槽;浮渣(排水)槽
scum weir 浮渣堰;浮渣排除堰
scum well 浮渣井
scun 细矿脉
scuncheon 门窗洞侧墙面;门框槽沟;窗框槽沟
scuncheon arch 开有槽沟的拱
scuntion 门窗内侧呈八字形斜壁
scupper (排除平屋顶或楼板积水在墙上开的)排泄口;水沟;泄水口;甲板泄水口
scupper drain 平屋顶墙角排水口
scupper grating 排水口格子盖
scupper hose 排水软管【船】;排水软管
scupper nail 铅钉;大头钉
scupper pipe 排水管(道)
scupper shoot 排水筒
scupper valve 排水口阀门
S-curve 单位线相应的径流积分同线;反向曲线;S形曲线;回头曲线;发针形曲线;反转曲线
S-curve hydrograph S形水文曲线
scurvy 坏血病
scutate 盾形的
scutch 刨锛(转起工具);刨锤;瓦工砍刀;圬工小锤;石划;石工小锤
scutcheon 穹隆顶棚中央的浮雕(饰);饰有纹章的盾;小盾片;钥匙孔盖
scutcher 砍石锤;开幅机;打麻机;瓦工小锤
scutcher waste 梳理下脚
scutching 锤琢;打麻
scutching cylinder 打麻滚筒
scutching drum 打麻滚筒
scutching machine 打麻机
scutching rotor 打麻滚筒
scutellum 盾片
scutellum node 盾片节
scuttle 小舱口;煤桶;煤斗;气窗;天窗;雕窗
scuttle a ship 船底凿洞使其沉没
scuttle blind 舷窗盖
scuttle bulkhead 前舱壁
scuttle butt 饮水喷头;饮水柜台;饮用水桶;饮水龙头;饮水缸;淡水桶
scuttle cover 小舱口盖;舷窗盖
scuttle down 紧紧关上舷窗
scuttle for ships 船用舷窗
scuttle frame 舷窗框
scuttle hatch(cover) 舷窗铁盖;升降口盖;煤舱口
scuttle hole (通向屋顶的)人孔;天窗(窗洞)
scuttle key 舷窗水密盖;舷窗扳手;甲板开口水密盖扳手
scuttle lid 舷窗水密盖;甲板开口水密盖
scuttle tank 浮箱式闸门
scuttle ventilation control 天窗通风控制
scuttle washer 舷窗(橡胶)垫圈
scuttling 故意弃船
scyelite 闪云橄榄岩
scyphoid 杯状的
scythe 大镰刀;长柄大镰刀;草地割草机
Scythian stage 赛特阶
sea, air and land 海陆空
sea abeam 舷侧波
sea access 海洋航道
sea acorn 海蛎子;藤壶
sea advance 海进
sea agitation 海面波动
sea air 海洋性气候;海洋空气;海上空气;海岸性空气
sea-air nuclear precession magnetometer 海空核子旋进磁力仪
sea-air temperature difference correction 海水空气温度差改正;海水和空气温度差改正量
sea algae 海藻
sea algae vegetation 海藻植被
sea allowance 航行津贴
sea anchor 流锚;海锚;浮锚;帆布海锚;捕鲸索浮标
sea and air navigation tables 海空航用表

sea and land wind 海陆风
sea and swell chart 波浪及涌浪分级表
sea and swell scale 风浪和涌浪等级；波浪及涌浪等级；波级
sea-anemone 海葵
sea approach 近海的；通海水道；登陆地板
sea arch 海蚀桥；海蚀拱；海穹
sea area 海域；海洋区域
sea area boundary line 海区界线
sea area environmental impact assessment 海域环境影响评价
sea area pollution 海域污染
sea area situation chart 海区形势图
sea arm 海湾
sea atlas 海图集
sea atmosphere corrosion 海洋大气腐蚀
sea backthern 沙棘
sea bacteriology 海洋细菌学
sea bag 水手袋；大行李袋
sea bank 海岸；防波堤；海塘；海堤
sea barge 海驳
sea barge carrier 载驳(货)船
sea barge clipper 载驳快船；海上载驳快船；海蜂型子母船
sea basin 海水池；海盆
sea bathing 海水浴
sea-bathing area 海水浴区
sea beach 海涂；海滩；海滨(沙滩)；沙滩
sea-beach placer 海滩砂矿；海滨砂矿
sea-bear 白熊
sea beat(en) 海浪冲击；海浪拍击
seabed 海底；海床
seabed area 海底区(域)；海床区域
seabed condition 海底条件
seabed configuration 海底形态
sea-bed contour 等深线
seabed disposal 海底处置法；海床处置；海床处理
seabed drifter 支柱式海底开挖机；海底漂流瓶
seabed echo 海底回声
seabed elevation 海底高程
seabed exploration 海底开发
seabed insertion 埋藏海底泥土中
seabed leveller 海底平整器
seabed material 海底物质
seabed mineral resources 海底矿物资源
seabed mining 海底采矿
seabed pipe intake method 海底管式取水法
seabed resources 海底资源
seabed scour(ing) 海床冲刷；海底冲刷
seabed sonar survey system 海底声呐探测装置
seabed sweep(ning) 扫海(测量)
seabed testing system 海底试验系统
seabed thermal water 海底热水
seabed treatment 海床处理
seabee 蜂窝块(一种防波堤异形块体)
sea-bee barge 西比驳
sea-bee carrier 海上载驳货船；西比驳(海蜂式载驳货船)；海蜂式载驳(货)船
sea-bee lighter 西比驳
sea-bee ship 海蜂型子母船；海蜂式载驳(货)船；海蜂船
sea berth 开敞泊位；海上停泊处；海上泊位；外海码头；外海泊位
sea bloom 浪花；海浪开花
sea blubber 海蜇
sea blue 海蓝色
sea board 海滨；海岸(线)；沿海地区
seaboard coastline 海岸线
sea board line 海岸线
sea boat 耐波浪的船；海船
sea-book 航海图
sea boots 高统橡皮靴；橡胶长靴
sea bore 涌潮；特高海浪
sea boring offshore drilling 海上钻探；海上钻井
sea-borne 海生的；海产的；海运的
sea-borne aircraft 舰载飞机
sea-borne articles 海外进口货
sea-borne bottom gravimeter 海底重力仪
sea-borne dredging 海上疏浚
sea-borne goods traffic 海上货物运输
sea-borne gravity measurement 海上重力测量
sea-borne invasion 海上入侵
sea-borne lift 海上疏运
sea-borne machinery 船用设备；船用机械

sea-borne magnetic survey 海上磁力测量
sea-borne only 限海上范围；仅限海上(指保险)
sea-borne supply 海上补给
sea-borne tracking and ranging station 海上跟踪与测距站
sea-borne trade 海外贸易；海运(贸易)
sea-borne trade sea trade 海上贸易
sea-borne traffic 海运运量；海上交通
sea bottom 海底
sea-bottom basement 海底基岩
sea-bottom contour 海底地形；海底等深线
sea-bottom crust texture 洋底地壳结构
sea-bottom cultivation 海底养殖
sea-bottom deposit 海底淀积(物)；海底沉积(物)
sea-bottom drift 海底漂移
sea-bottom embedment 海底埋设
sea-bottom exploration 海底勘探
sea-bottom flora 海底植物群；海底植物区系
sea-bottom geology 海底地质(学)
sea-bottom gravimeter 海底重力仪
sea-bottom gravity meter 海底重力计
sea-bottom mining 海底采矿
sea-bottom multiple 海底多次波
sea-bottom percolation 海底渗漏
sea-bottom petroleum 海底石油
sea-bottom plain 海底平原
sea-bottom relief 海底地形
sea(-bottom) seismograph 海底地震仪
sea bottom survey 海底勘测
sea-bottom tension 海底张力
sea-bottom topographical map 海底地形图
seabound 四面环海的；被海水包围的
sea boundary delimitation 海上划界
sea bow 船首虹影
sea breach 海堤决口
sea breeze 海风
sea breeze front surface 海风锋面
sea breeze of the second kind 冷锋状海风
sea brief 中立船舶证(明)书
sea brine 盐海水
sea buoy 海口浮标；海界浮标；海港口门浮标
sea burial 海洋埋藏
sea calming oil 镇浪油
sea canal 海运河
sea cannal 通海运河
sea cap 浪花
sea captain 船长
sea cargo 海运货物
sea carriage 海洋运输
sea casualty 海难
sea cave 海蚀(岩)穴；海蚀洞
sea cell 海水电池
sea central point 海控点
sea-chalk 海白垩
sea channel 海峡；海沟；海底峡谷；海底峡沟
sea chart 航海地图；海图
sea chart zero 海图零点
sea chasm 海蚀崖龛；海蚀深穴
sea chest 通海吸水箱；水手柜
sea clam 海蛤
sea clam wave energy generator 海蚌式波能发电机
sea claw anchor 海钩锚(小型长掌锚)
sea-clay 海相黏土；海成黏土
sea cliff 海(蚀)崖
sea climate 海洋气候
sea clutter 海面(杂乱)回波；海面干扰；海面反射干扰波；海浪杂波；海浪回波；海浪干扰
sea clutter simulation 海面反射模拟
sea clutter suppresser 海面反射干扰消除装置
sea clutter suppression 海浪干扰抑制
sea-coast 海滨；海岸
sea coastal protection 海岸防护
sea-coast defence 海岸保护；海岸防护
sea coast harbour 海岸港
sea cock 海水阀；海底阀；船底阀
sea colo(u)r 海色
sea colo(u)r index 海色指数
sea compass 船用罗盘；船用罗经
sea condition 海洋水文；海况；波浪等级
sea conditional sign 海洋惯用符号
sea condition going tug 海上拖轮
sea condition going vessel 海船

sea condition 海面状况；海面情况；海面风浪状态
sea condition type 水文分类
sea connection 通海接头
sea construction 海洋建筑物；海工建筑(物)
sea container 海运集装箱
sea container chassis 海运集装箱底盘车
sea conveyance 海上运输
sea cook 船上厨师
sea crab 海蟹
sea craft 小型出海船舶；海船
seacrete fishing boat 钢丝水泥渔船
sea-crust 船体水下部分附生物
sea cucumber 海参
sea culture 海水养殖
sea current 海流
sea current direction 海流方向
sea current dynamics 海流动力学
sea current energy 海流能
sea current meter 海流计
sea current strength 海流强度
sea current velocity 海流速度
sea-cut 浪蚀的
sea-cut notch 海蚀龛
sea damage 海损；海水损害条件
sea damaged goods 海损货；湿损货
sea damaged terms 水损条款
sea-damage for seller's account 海损由卖方承担
sea damage terms 海损条件；海水损害条件
sea day 平太阳日
sea defence 海堤
sea defence works 海岸堤防工程；海岸防护设施；海岸防护工程
sea defense 海岸堤防
sea delivery 压载舱通海水管道
sea demaged 海水损害
sea deposit 海积
sea depth 海深
sea dike 海岸防护堤；海岸防波堤；海塘；海堤；海岸堤防
sea dike construction 海堤建设；海堤工程
sea dike works 海堤工程
sea direction 波浪方向
sea disaster 海上灾难
sea disposal 海洋处置；倾弃于海洋；投海处置
sea dog 海豚
sea dragon 海龙
sea dredged aggregate 海底集料；海产集料；海产骨料
sea dredge(r) 出海挖泥船
sea dredging 海滨挖泥；海上挖泥
sea drift 海上漂散木
sea drilling platform 海上钻探平台；海上钻井平台
seadrome 海面(飞)机场；水上(飞)机场；水面(飞)机场
seadrome contact lights 水上机场降落灯
sea dumping 海洋投弃；海洋倾废
sea duty 海上钻探工作
sea dye 指示色标
sea earth 海底电缆接地
sea echelon 海上梯队
sea echo 海水反射回波；海区反射信号；海面回波；海面反射信号；海浪杂波；海浪回波
sea echo scattering 海面反射波散射
sea effect overwater 海洋效应
sea electric(al) power cable 海底电力电缆
sea elevation 海拔高程
sea embankment 海岸堤防；海堤；防潮堤
sea endurance 续航力
sea energy 海洋能；波能
sea environmental data acquisition system 海洋环境数据采集系统
sea-eroded platform 海蚀台地
sea-eroded terrace 海蚀阶地
sea erosion 海蚀(作用)
sea expedition 海洋考察
sea expedition station 海洋考察站
sea experience 航海经验
sea exploration vessel 海洋勘探船
sea facies 海相
sea fan 海(底)扇；海底沉积扇
seafarer 航海者；海员；水手；船员
Seafarer's International Union 海员国际联盟；海员国际联合会

seafaring 航海的
seafaring life 航海生活
sea faring man 海员;水手
sea farm 海上养殖场
sea farming 海洋养殖;海水养殖
sea fern 石帆
sea ferry 海上轮渡
sea fertility 海洋肥力
sea fire 海上荧光
sea fisheries 海上渔业
Sea Fisheries Committee 海洋渔业委员会
sea fishery 海洋渔业
sea-fishing 海洋渔区;海洋捕鱼
Sea-fix system 西菲克斯定位系统
sea floe 海上浮冰;浮冰
sea floe group 浮冰群
sea floor 海底;海床
sea-floor basalt 洋底玄武岩
sea-floor brine 海底热卤水
sea-floor coring 海底取岩心
sea-floor deposit 海底矿床;海底淀积(物);海底沉积(物)
seafloor earthquake 海底地震
sea-floor earthquake measurement system 海底地震测量系统
sea-floor exploration 海底勘探
sea-floor exploration by explosives 爆破海底探测
sea-floor Fe-Mn nodules 海底铁锰结核
sea-floor foundation 海底基础
sea-floor fueling station 液体燃烧水下储存柜
sea-floor geomorphy 海底地貌
sea-floor gravimeter 海底重力仪
sea-floor gravity survey 海底重力调查
sea-floor height 海底高地
sea-floor high 海底高地
sea-floor manganese nodule 海底锰结核
sea-floor mapping systems 海床绘图系统
sea-floor material 海底物质
sea-floor polymetallic ooze 海底多金属软泥
sea-floor profile survey 海底地层剖面测量
sea-floor relief 海底地形
sea-floor relief map 海底地形图
sea-floor reveval 海底更新
sea-floor sampler 海底取样器;海底采样器
sea-floor sampling 海底(质)采样
sea-floor scan survey 扫海测量
sea-floor soil 海底土
sea-floor soil sampling 海底土质取样;海底底质取样
sea-floor spreading 海底延伸;海底扩张
sea-floor spreading center 海底扩张中心
sea-floor spreading hypothesis 海底扩张说
sea-floor spreading theory 海底扩张说
sea-floor stability 海底稳定性
sea-floor stability survey 海底稳定性调查
sea-floor station 海底测站
sea-floor surficial sampling 海底表层取样
sea-floor surveillance system 海底监视系统
sea-floor topographic(al)survey 河底测量
sea-floor trench 海沟
sea-floor valley 海底谷
sea fluctuation 海面波动;海面变动
sea fog 海雾;海上蒸汽雾;海面蒸汽雾
sea fog prediction 海雾预报
seafolk 海员
sea food 海产食物;海产食品;海味
seafood factory 海产食品加工厂
seafood industry 海产食品工业
seafood processing plant 海味加工厂
sea for cooling 冷却用海水
sea freight 海运货物;海运费(率);海上运输
sea freight liner 海上定期直达列车
sea freight rate 海运费率
sea fret 海涌雾(苏格兰东北沿海)
sea from aft 尾浪
sea front 海滨人行道;城市临海地段;面临海的场所;海边;海岸;临海地区;临海地段;滨海区
sea-front industry 临海工业
sea gate 海闸;海峡;闸门;海口;通海闸门;通海地点;出海口
sea ga(u)ge 海水测深计;海水测深杆;海上验潮仪;吃水
sea gets up 起浪
sea-girt 四面环海的

sea glow 海面辉光
sea goes down 浪静下去
sea going 从事航海事业的;海上的;向海航行;航海的
sea-going ability' seagoing capacity 适航性【船】
sea-going barge 海驳;出海驳船
sea-going capability 航海性能
sea-going catamaran 双体海船
sea-going destroyer 远洋驱逐舰
sea-going draught 出海吃水
sea-going dredge(r)出海挖泥船;海上挖泥船;自航式挖泥船
sea-going drilling island 海上钻井平台
sea-going drilling platform 海上钻井平台
sea-going fishery 远洋渔业
sea-going floating crane 航海浮式起重机
sea-going launch 出海汽艇
sea-going launcher 舰上发射装置;海上发射装置
sea-going lighter 海驳;出海驳船
sea-going pipeline 海驳油管;海底管道
sea-going platform for acoustic(al)research 海上声学研究平台
sea-going powered barge 机动海驳;出海机动驳船
sea-going push-tow unit 出海顶推船队
sea-going quality 航海性能
sea-going ship 海船;远洋舰艇
sea-going shipping 海洋航行
sea-going submersible 海上潜水器
sea-going taxi 海上游艇
sea-going traffic 航海交通
sea-going transport 海上运输
sea-going trawler 远洋拖网渔船
sea-going tug 远洋拖轮;海上拖轮;海上拖船;出海拖轮
sea-going velocity 航海速度
sea-going vessel 航海船;海船;海轮;远洋轮(船)
sea-going vessel waterway 海轮航道
sea grass 海韭菜;海草
sea-grass bed 海草海床
sea-grass building mat 海草建筑垫席
sea-grass ecosystem 海草生态系统
sea-grass insulation grade strip 海草绝缘等级带
sea gravel 海卵石;海砾石
sea gravimeter 海洋重力仪
sea green 海绿色
sea green celadon 海绿色青瓷
sea gully 海底谷
sea harbo(u)r 海港
sea harbo(u)r accommodation 海港设备
sea harbour entrance 海港入口
sea hazard 海险
sea-hedgehog 海胆
sea-high tableland 海底高地
sea horizon 海平线;海平面;海地平(线);水天线
sea horse 海马
sea ice 海上浮冰;海冰
sea ice-atmospheric interaction 海冰—大气相互作用
sea-ice climatic cycle 海冰气候周期
sea-ice coding 海冰数据编码
sea-ice dynamics 海冰动力学
sea-ice forecast(ing) 海冰预报
sea-ice margin 海—冰分界面
sea-ice observation 海冰观测
sea-ice platform 焊接冰平台
sea-ice prediction 海冰预报
sea-ice shelf 海冰架;浮冰架
sea-ice variability 海冰可变性
sea information investigation 海区资料调查
sea inlet 海港;海水吸入口;海水进口;小海湾;海湾;通海阀
sea inlet cock 通海吸入阀
sea insurance 海运保险
sea island 海岛
sea islet 海岛
sea jeep 水陆两用吉普车;水陆两用装甲车
sea-jet-one 海空联运
seakeeping 经得起海上风浪的;耐波性;适航力
seakeeping ability 抗风浪性能;耐波性;航海性能;适航性【船】
seakeeping characteristics 适航性能
seakeeping quality 抗风浪性能;航海性能
seakeeping trial 适航性试验
sea kindliness 海上航行平稳性

seakindly ship 航海性能良好的船
sea knife 水手刀
seaknoll 海丘;海底(圆)丘(从海底升起不到1000米)
seal 印章;止水;盖章;盖印;绝缘装置;密封体;火漆;焊封;图章;戳记;封印;封条;封铅;封接;封焊;封层;封闭
sealability 可密封性;密封能力
sealable equipment 密封装置;密封器械
sealable tank 密封罐
sealable transparent cover 可密封的透明覆盖物;可密封的透明面层
sea-lab(oratory) 海底实验室
sea ladder 海上用梯子
seal-air pipe 密封空气管
seal-air pressure 气密封压力
Sealalloy 西尔艾洛伊铋合金
seal a mountain pass 封山
seal and block with grouting and pressure heightening 注浆升压封堵
sea-land method 海陆法
sealand service incorporation 海陆运输公司
sea-land system 海陆公司方式
sea lane 航路【航海】;海洋航线;海洋航路;海上航线;海上航路
seal angle(iron) 止水角铁
sealant 止水材料;密实剂;密封腻子;密封胶;密封剂;密封膏;密封层;密封(材)料;嵌缝膏;封面料;封口料;封缝材料;封闭剂;封闭层;防渗漏剂
sealant backing 密封膏背衬材料
sealant compound 嵌缝材料
sealant penetration 封面料透入度
sealant profile 密封面
sealant strip 密封条
sealant tape 密封带;封缝带
seal apron 闸门挡板
seal assembly 密封组件
seal bellows 密封皱纹管;密封波纹管
seal block 盖砖
seal boot 封闭罩
seal box 密封盒
seal broken, repaired 封条弄破后重修补
seal by arrow-like ring 箭形圈密封
seal by bimetallic O-ring with teflon 外包聚四氟乙烯双金属
seal by direct contact of conical faces 单锥无垫密封
seal by double conical frustum 双截顶圆锥密封
seal by double knife-edges double knife-edge seal 双刀口密封
seal by duplex conical gasket 双锥垫密封
seal by elastomeric delta gasket 橡胶三角垫密封
seal by flat gasket 螺纹平垫密封
seal by metallic delta gasket 金属三角垫密封
seal by metallic oval-ring 金属椭圆环密封
seal by plug 堵塞密封
seal by precision fit 研合密封
seal by screwed union 螺纹连接密封
seal by simplex conical gasket 单锥垫密封
seal by sleeve 套管密封
seal by special wedge-like gasket 特种楔形密封
seal by stainless conical ring 不锈钢锥环密封
seal by wedge-like gasket 楔形垫密封
seal by wedg-like gasket and O-ring 组合式楔密封
seal by welding 焊接连接密封
seal cage 隔离环;水封环
seal cap 加封盖
seal casing 密封壳体
seal cavity 密封腔
seal cavity pressure rise 密封腔压力上升
seal cement 固形接合剂
seal clutter suppression 海浪杂波抑制
seal coat 止水层;密闭层;封层;封闭(底)漆;封闭衣;封闭(涂)层
seal-coat concrete 密封涂层混凝土
seal-coat emulsion 密封涂层乳液;密封涂层乳胶
seal coating 密封涂层;密封敷层;保护膜
seal coat work 铺筑封层(工作)
seal construction rope 带芯股钢索
seal contact face 密封面
seal course 密封层;密闭层;封层
seal cover 密封罩;密封压盖
seal cracked, resealed 封条破裂后重加封
seal cup 封口圈;密封环;溢流盘

seal documents 密封的文件
seal drain leakage 密封耗损泄漏
seal drain port 密封排放口
seal drive sleeve 密封(动环)传动套;动环传动套
seal driving pin 动环传动销
sea leather 海产皮革
sealed against dust seepage 防尘密封的
sealed aggregate 封闭表面集料;封闭表面骨料(低吸水率)
sealed area 封闭区(域)
sealed basin 闭ައ␣流域
sealed beam lamp 聚光灯;密封式前大灯;密封式光束灯泡
sealed bearing 密封轴承
sealed bearing of a bit 钻头密封轴承
sealed bid 密封的投标书;密封投标;密封出价;秘密投标;非公开招标;密封递价
sealed bid pricing 投标定价法
sealed-bid tender 密封标单投标
sealed book 天书
sealed cabin 严封坚固舱;增压舱;密封(坚固)舱;密闭座舱
sealed cabinet 密封培养
sealed cap 密封帽;密封盖
sealed capsule 密封管
sealed cavity 密封空腔谐振器
sealed cell sponge rubber 闭孔海绵胶
sealed centrifuge 密封式离心机
sealed chamber 气密室
sealed charging linkage 密封进料联动装置
sealed chart-drive motor 密封式记录纸驱动电动机
sealed coil relay 线圈密封式继电器
sealed compartment 密封舱
sealed concrete 密封混凝土;嵌缝混凝土;填缝混凝土
sealed conduit 封闭管
sealed contact type 密封接触型
sealed container 密封容器;密闭集装箱;气密封罐
sealed contract 已盖章合同;正式契约;正式合同
sealed cooling 闭式冷却(法)
sealed cover 密封盖
sealed cowling 密封罩
sealed double glazed unit 密封双层玻璃单元
sealed double-glazing unit 密封双层玻璃窗
sealed double glazing window 密封双层玻璃窗
sealed dust-proof van 密封防尘车
sealed earth 填封土
sealed end 焊头
sealed environment 密封环境
sealed expansion vessel 密封抗压容器
sealed factor of trap 圈闭的封闭因素
sealed fault 闭合断层【地】
sealed flue 密封烟道
sealed fluid coupling 闭式液力耦合器
sealed-for-life 长期密封的;永久性密封
sealed-for-life bearing 闭式式一次润滑轴承;全期油封轴承
sealed-for-life vane pump 永久密封叶片泵
sealed ga(u)ge milled bit 密封保径铣齿钻头
sealed ga(u)ge sliding milled bit 密封保径滑动铣齿钻头
sealed ga(u)ge sliding tungsten carbide insert bit 密封保径滑动镶齿钻头
sealed ga(u)ge tungsten carbide insert bit 密封保径镶齿钻头
seal edge 封边
sealed glass unit 中空(隔热)玻璃
sealed glazing unit 密封上釉单元
sealed groove 封闭槽
sealed head 密封磁头
sealed housing evaporative emission determination test 密封外壳燃油蒸发排出物确定试验
sealed hydraulic coupling 闭式耦合器
sealed-in 封闭的;密封的
sealed instrument 已签名盖章的文据
sealed insulating glass 密封保温玻璃;密封隔热玻璃
sealed insulating glass unit 密封隔热玻璃制品
sealed jet milled bit 铣齿密封喷射钻头
sealed jet tungsten carbide insert bit 镶齿密封喷射钻头
sealed jet tungsten carbide insert sliding bit 镶齿密封喷射滑动钻头
sealed joint 填实缝;密封缝;止水缝;封闭链;密封缝

sealed junction 密封结
sealed lead-acid battery 密封铅酸蓄电池
sealed lithium battery 密封锂电池
sealed lubricator 密封润滑器
sealed metal container 密封金属容器
sealed mica ring 密封云母环
sealed-off 脱焊的;封离的
sealed-off counter 封离计数器;非密闭计数器
sealed-off light valve 密封光阀
sealed-off vacuum system 封离真空系统
sealed outer housing 密封外壳
sealed pack-age 密封包装
sealed pad 密封包件
sealed pipe 密封管;封闭管
sealed plug 屏蔽式电热塞
sealed plug sheathed plug 密封式电热塞
sealed polygon 封闭的多边形
sealed pore 封闭气孔
sealed porosity 封闭式多孔性;封闭气孔;封闭孔隙率;闭气孔率
sealed pressure balance 内封补偿
sealed proposals 密封递价;投标
sealed pump 密封泵
sealed pumped vacuum system 封离抽气系统
sealed pumpless vacuum system 封离的无泵的真空系统
sealed quench furnace 封闭式淬火炉
sealed rectifier 密封式整流阀
sealed refrigeration compressor 密封式制冷压缩机;密封冷冻压缩机
sealed relay 密封(式)继电器
sealed roller bearing 密封式滚柱轴承
sealed roller milled bit 密封滚动铣齿钻头
sealed runway 接缝密封的跑道
sealed sample 密封样品
sealed shoulder 加固的路肩
sealed sintering 密封烧结
sealed sliding milled bit 密封滑动铣齿钻头
sealed soil sample 密封土样
sealed source 密封源;密闭源;封闭放射源
sealed spark gap 密封放电器
sealed spring loaded safety valve 封闭式弹簧安全阀
sealed stirrer 封闭搅拌器
sealed system 密封系统;封闭系统
sealed tender 密封投标;密封递价;秘密投标
sealed tube 密封管;封闭管
sealed unit 中空玻璃
sealed upright silo 气密储存[贮]塔
sealed(water)system 闭式系统(指水循环)
sealed window 密封窗
Seale filler wire rope 西尔式带填充丝钢丝绳
sea legs 不晕船
seal end plate 静封压盖;密封压盖
seal energized by medium pressure 自紧密封
sealer 压盖填缝料;盖盖人;盖印人;密封物;密封器(垫);密封机;海豹猎船;嵌缝材料;封闭剂;封闭底漆;表面保护漆
Seale rope 带芯股钢丝绳
sea letter 中立船舶证
sea levee 海岸堤防;海堤
sea level 海平面;海拔高程;潮位
sea level canal 海平面运河
sea level change 海平面变化
sea level chart 海面天气图
sea level correction 海平面校正;海平面改正;海面高程校正(值)
sea level departure from normal 潮位偏差;潮位离差
sea level elevation 海面高度;海面高程;海拔高度;海拔高程
sea level height 海面高度;海面高程
sea level interface 海陆界面
sea level observation 验潮;海平面观测
sea level pressure 海(平)面压力;海(平)面气压
sea level reduction 海平面归算
sea level rise 海(平)面升高;海平面上升
sea level surface 海平面
sea level variation 海平面变化
Seale Warrington wire rope 西尔瓦灵顿混合式钢丝绳

seal face 压紧面
seal fan 密封风机
seal fitting 密封配件;密封接头
seal fluid connection 密封液接口
seal fluid outlet 密封液出口
seal flush 密封冲洗
seal flush liquid 密封冲洗液
seal flush port 密封冲洗口
seal flush pressure 密封冲洗压力
seal for pipe joints 管道密封
seal for pressure vessel 压力容器密封
seal friction 密封摩擦
seal gland cartridge 密封套
seal groove 密封企口
seal grouting 密封灌浆
seal gull 海鸥
seal gum 密封胶
seal hanger 密封悬架
seal head 动环
seal housing 密封壳体
sea life 海洋生物
sea lift 海上补给
sealift replenishment at sea 海上补给
sealift ship 海上补给船
sea lily 海百合
seal in 密封;封入
seal-in air cushion 密封气垫
seal in contact 自保持接点
sea line 压辊条纹;海岸线;水天线;水平线;海上交通线
sealing 加封;密封(处理);嵌缝;铺筑封层;涂封(底)漆;封闭;封口
sealing ability 密封能力;密闭能力
sealing agent 密封剂;封缝涂料;封闭剂
sealing and blocking the point of bursting water 封堵突水点
sealing and ca(u)lking 密封和填缝
sealing and marking of bids 投标的密封和标志
sealing area 密封面积
sealing asphalt 封缝沥青;封缝(层)地沥青
sealing bar 熔接棒
sealing bare wood 木胎钻生;白坯封闭;白茬封闭
sealing bead 密封焊接;密封焊道;缝焊
sealing box 密封箱;电缆盒
sealing bucket 密封戽斗;(自动加压管的)封水戽斗
sealing bush 密封填料;垫料
sealing by fusing 熔封
sealing by hydration 水合封闭处理
sealing cap of metal 封口用金属帽
sealing catalytic decomposed method 密封催化消解法
sealing cement 胶结材料;密封油灰;密封黏结剂;密封油膏;密封蜡
sealing cement channeling 封串
sealing coat 密封层;隔离层;止水深层;封面层;封闭底涂层;封闭(涂)层
sealing coating 封缝涂料
sealing coat of paint 涂料密封层;油漆密封层
sealing coil 密封盘管
sealing collar 止水圈
sealing component 密封件
sealing composition 封缝涂料
sealing compound 止水材料;密封油膏;密料;密封接合;密封剂;堵漏腻子;电缆膏;封塞料;封口料;封口胶;封口膏;封缝涂料;封缝胶;封缝混合料;封口料
sealing concentrate 密封浓缩物
sealing concrete 封混凝土
sealing cone 密封锥面
sealing constriction 密封压缩;密封缩颈
sealing course 密封层;封闭层
sealing cover 密封盖
sealing cup 密封碗;密封皮碗
sealing cup body 密封皮碗套
sealing device 止水设备;密封装置
sealing door 密闭门
sealing driving rod 密封拨杆
sealing edge 密封唇(口)
sealing element 密封件
sealing end 封端
sealing engineering 密封技术;密封工程
sealing face 密封面
sealing face leakage 密封面泄漏
sealing failure 密封失效

sealing felt(ed fabric) 密封毡
sealing filler 密封填充物;封闭填孔剂
sealing fillet 密封窄条
sealing film 密封薄膜
sealing fin 密封鳍状物
sealing flange 密封凸缘;密封法兰
sealing fluid 密封液(体);封闭液(体)
sealing fluid inlet 密封液入口
sealing foil 密封箔;密封薄膜
sealing forceload 密封力
sealing for pressure vessels 压力容器密封
sealing for tubing joints 管道连接密封
sealing gasket 止水衬垫;密封衬片;密封垫;防漏垫
sealing gasket joint 密封衬垫接缝
sealing gland 密封装置;密封(压)盖;气封装置
sealing glass 封接玻璃
sealing glue 封口胶
sealing groove 密封槽;止水槽
sealing gum 密封胶
sealing hold 封舱
sealing-in 焊接;封接
sealing-in burner 密封用气体喷烧器
sealing installation 止水设施
sealing installation at tail of shield 盾尾密封装置;盾构尾部密封装置
sealing interval 封孔区间
sealing jacket 密封套
sealing joint 密封接头;密封(接)缝;止水缝
sealing joint strip 密封胶条
sealing layer 封闭层;密封层;防水层
sealing leak 密封泄漏
sealing leakage 堵漏
sealing life 密封寿命
sealing lip 密封唇(口);密封圈
sealing liquid 隔离液;密封液(体);封液;封闭液(体)
sealing-liquid duct 水封入口
sealing load 密封压力;密封紧力
sealing machine 密封机;熔接机;封口机;封接机
sealing mass 密封物质;密封块
sealing mastic 密封玛琋脂
sealing mastic glue 黏结胶
sealing material 嵌缝材料;密封材料;封填材料;封塞料;封面材料;封缝材料;止水材料
sealing medium 密封用物质;密封介质
sealing member 密封件
sealing method 密封方法
sealing mix(ture) 密封混合料
sealing mortar 封口砂浆;密封砂浆;填缝砂浆
sealing nipple 密封螺纹管接头
sealing of a casting 铸件浸补
sealing of a mo(u)ld 铸型密封
sealing of boreholes 钻孔的封堵
sealing of cracks 裂缝密封;封缝;填缝;封闭裂缝
sealing-off 封口;开焊;脱焊;堵塞;拆焊
sealing off inrushes of water 防水密封;堵塞涌水
sealing(off)pressure 封闭压力(钻井或钻孔灌浆后)
sealing of hole 封孔
sealing of soil 土壤(表面)板结
sealing of tunnel lining 隧道衬砌填料
sealing of vessel 容器密封
sealing on 焊上
sealing paint 封闭油漆;密封涂料
sealing paint finish 密封油漆面饰
sealing paint for electroplating 电镀封闭漆
sealing paper 高级包装纸
sealing parts factory 橡胶密封件厂
sealing parts of piston 活塞不漏气部分
sealing paste 封口软膏
sealing plate 密封垫;密封板;嵌缝片
sealing pliers 封印钳
sealing plug 封口插座
sealing pot 隔离液罐;封液包
sealing pressure 密封压力
sealing property 密封性能
sealing putty 封口腻子;封口油灰
sealing quality 封孔质量
sealing reliability 密封可靠性
sealing resin used on colo(u)r TV set 彩电用灌封树脂
sealing rib 密封肋

sealing ring 密封圈;止水环;密封环;挡风圈;封闭圈
sealing ring cap 密封环盖
sealing ring of gyrosphere 陀螺球密封环
sealing rock 封面石
sealing rod 止水条;密封条
sealing rope 密封绳索;密封条;密闭条;嵌缝绳
sealing rubber 密封橡胶
sealing rubber ring 密封橡皮圈
sealing run 焊缝;封底焊
sealing screw 密封螺钉
sealing sheeting 密封薄片
sealing stamp 封口印章
sealing steam box 气封蒸汽室
sealing steam desuperheater 气封蒸汽冷却器
sealing stopper 密封塞子
sealing strand 封闭索
sealings trength test 石蜡密封度试验
sealing strip 止水条;密封条;密封片;密封带;密封边;密闭条;汽封片;填缝片
sealing surface 密封面;封接面
sealing system 密封系统
sealing tape 黏结胶带;密封条;密封带;密闭带
sealing test 封闭性试验
sealing the caisson 封闭沉箱
sealing the leakage of stern tube 尾轴管堵漏
sealing thread 封闭螺纹
sealing treatment 密封处理
sealing up 防漏;止水;封舱
sealing varnish 封闭清漆
sealing voltage 闭合电压
sealing washer 密封(垫)圈
sealing water 密封水
sealing water in borehole 井孔止水
sealing water pipe 水封管
sealing water pipe connection 水封管接头
sealing wax 火漆;封蜡;封口蜡;封闭蜡
sealing-wax wood 制取封口蜡的木材;制取火漆的木材
sealing welding 封焊
sealing wire 密封引线;焊接线
sealing with fines 细粒止水;细粒填缝
sealing without gaskets 无垫密封
sealing works 密缝工事;止水工程;填缝工作;封闭工程
sea lion 海狮
seal jacket 密封套
seal joint 密封接头
seal lac 封口漆
seal lead 铅封;封铅
seal leakage 气封
seal leg 闸式阀门;料封管;气动提升器;垂直管式闸门
sealless pump 无密封环的泵
seal lock 孔堵;塞
seal loop 水封环
seal material 密封材料
seal mortar 密封砂浆
seal nipple 密封接头;封闭接头
sea loading pipeline 海上装油管路
sea lock 海上通海船闸;堵孔塞(海)(船)闸;通海闸门;通海船闸
Sealoclean 西罗克净(一种混凝土表面清洁剂)
Sealocrete 西罗防水剂
seal of bearing 轴承密封
seal off 封住;烫开;拆封;封口脱开;封堵
seal-off mercury-arc rectifier 封闭汞弧整流器
seal-off stain 底色封闭;封闭底色
seal of inspection societies 检验合格徽
seal of tube 管密封
seal of vessel 容器密封
seal oil 密封用油
seal oil control equipment 油密封控制装置
seal oil system 油密封系统
Sealontone 西隆酮(一种水泥液体硬化剂)
seal packing 密封填料;密封垫圈
seal packing ring 密封圈
seal pipe 密闭管
seal plate 止水板;密封板
seal point 液封点
seal pot 密封罐
seal pouring 密封注胶
seal press 压印
seal profile 成型密封带
seal replacer 密封盖拆卸工具

seal retainer 密封定位架;密封座;密封护圈
seal ring 密封圈;垫圈;止水圈;封密环
seal ring for pressure cooker 高压锅圈
seal ring shank 橡胶密封环钎尼
seal rope 填缝绳索
seal running dry 密封干运转
seals and sealing 密封装置与封接
seal screw 密封螺钉
seal seat 密封静环
seal seat retainer 密封静环护圈
seal set 密封组件
seal sleeve 气封套筒
seal spring compression ring 密封弹簧压圈
seal stationary ring 静环
seal steam 气封蒸汽
seal style 密封形式
seal supply pressure 密封供应压力
seal tightness test 致密性试验
seal treatment 封层处治;封闭处理
seal tube 密闭管
seal up 查封
seal up time 密封时间
Sealvar 一种铁镍钴合金
seal water 水封
seal water pipe 水封水管
seal water pump 轴封水泵
seal water tank 水封水箱
seal water valve 水封水阀
seal weld 致密焊缝;密封焊缝;密封焊(道);填焊
seal welding 密封焊接
seal welding device 封焊设备
seal with annular groove 油封槽
seal with arrow-rings 箭形圈密封
seal with axis deviation 轴线偏斜密封
seal with cup packing 杯形垫密封
seal with double cone 双锥密封
seal with elastomertic washer 橡胶圈密封
seal with flat gasket 平垫片密封
seal with metal cone ring 锥形毛密封
seal with metal lens ring 透镜垫密封
seal without gasket 无垫密封
seal with screwed joint 螺纹连接密封
seal with semi-circular annular groove 半圆形油封槽
seal with single-cone 单锥密封
seal with stainless cone-ring 不锈钢锥环密封
seam 金属板咬口;节理;贴边;缝合;缝;薄层
sea magnetic gradiometer 海洋磁力梯度仪
sea mail 海运信件;海邮
sea main thermocline 海洋主温跃层
seaman 海员;水手;船员
seaman apprentice 实习水手
seamanite 鳞硼锰石
seamanlike 像水手的;像海员的
sea manners 海员习惯;海上礼让;海上惯例
seaman's 水手的
seaman's chest 水手柜
seamanship 航海技能;航海术;船艺
sea map 海图
sea margin 加算马力【船】
sea mark 助航标志;航海标(志);海上航标;海上航路标志(设在浅水中);海面标志;最高潮位线
sea marker 海上标志
sea marsh 海边沼泽地;滨海沼泽
sea mat 海生苔藓虫群体
seam blast(ing) 缝隙爆破;缝隙装药爆破;缝合处爆破
seam bonding 缝粘合
seam check 模缝裂纹
seam closing 接缝咬口(薄板)
seam composition 捻缝胶
seam cross-section 接缝截面
seam detector 线缝探测器
seam doubling 煤层加倍变厚
seam dustiness 煤层含尘量
seamed 有缝的
sea medical evacuation 海上医疗后送
seamed pipe 有缝管
seamed tube 有缝管
seamen's certificate 海员证
seamen's club 海员俱乐部
seamen's competency certificate 海员执照
seamen's discharge book 海员解雇书
seamen's disgrace 锚链绞缠

seamen's effects 海员私人财物
seamen's head 水手厕所
seamen's hostel 海员招待所;海员宿舍
seamen's lien 海员留置权
seamen's passport 船员护照
seamen's quarters 水手舱
seamen's records 海员记录簿
seamen's register 海员注册
seamen's union 海员工会
seamer 卷边钳;卷边接缝手钳;卷边封口器
sea mew 海鸥
seam face 自然矿层形的石墙面
seam-faced 细纹凿面的
seam finish type 接缝形式
seam-free product 无缝制品
seam hammer 弯板锤;矿层锤
sea mile 海里
seaminess 裂纹现象
seaming 接缝;卷边接缝;咬口(接缝);咬口接合;卷边接合;缝拢
seaming die 搭接模
seaming machine 折边机;卷边接合机;合片机
seaming needle 帆针
seaming pliers 卷边钳;卷边接缝手钳;卷边封口器
seaming twine 帆线
seaming welder 缝焊机
seam inundation 充水岩层
sea mist 海雾;海上蒸汽雾
seam lap 两块连接网板的重叠部分
seamless 整压;无缝(的)
seamless alumin(i)um 无缝铝管
seamless article 无缝制品
seamless bloom 无缝钢管坯
seamless boat 无缝舢舨
seamless bottom 无缝底
seamless casing 无缝钢套管;无缝钢管的套管
seamless cathode 无缝阴极
seamless cold 冷拔无缝
seamless cold drawn steel pipe 冷拉无缝钢管
seamless condenser tube 无缝冷凝器管
seamless conduit 无缝管
seamless copper pipe 无缝铜管
seamless copper tube 无缝铜水管;无缝紫铜管
seamless copper water tube 无缝紫铜水管
seamless crown 无缝冠
seamless crown machine 无缝冠冲压机
seamless door 无缝门
seamless drawn steel pipe 无缝控制钢管
seamless drawn steel tube 无缝控制钢管
seamless drawn tube 拉制无缝(钢)管
seamless drill pipe 无缝钻管
seamless floor cover(ing) 无缝地板(面层)
seamless floor finish 无缝楼面修饰
seamless floor finishing 无缝地面处理;无缝楼面处理
seamless floor(ing) 无缝地面;无缝地板;整块地面
seamless flooring coating 无缝地板涂层
seamless forming of bowl 盆的无缝成型法
seamless gold crown 无缝金壳冠
seamless ground 无缝地面
seamless head 无缝封头
seamless hollow ball bearing 无缝空心球轴承
seamless hot 热拔无缝
seamless hot-drawn steel pipe 热拉无缝钢管
seamless mild-steel pipe 无缝软钢管
seamless mug with cover 带盖杯
seamless pipe 无缝(钢)管
seamless pipe mill 无缝钢管厂
seamless pipe rolling mill 无缝管轧机
seamless piping 无缝管
seamless ring rolling 无缝环轧制
seamless rolled pipe 轧制无缝管
seamless rolled steel pipe 轧制无缝钢管
seamless rolled steel tube 轧制无缝钢管
seamless rolled tube 轧制无缝管;无缝轧制钢管
seamless roof skin 无缝屋面表层
seamless shell 无缝筒体;无缝壳体
seamless shell course 无缝锅炉筒
seamless stainless steel pipe 无缝不锈钢管
seamless steel pipe 无(焊)缝钢管
seamless steel tube 无(焊)缝钢管;无缝套筒
seamless steel tube for drilling 钻探用无缝钢管
seamless steel tubing 无缝套筒
seamless steel tubing plant 无缝钢管厂

seamless tube 无缝管
seamless tubing 无缝管
seamless tubing fabric 无缝管状织物
seam line 中心线纹;拉制条纹;模缝线;接合线
seam machine 接缝机
seam mark 模缝痕
sea moat 海壕;海底环状洼地
seam of silt 淤泥夹层
seam orientation 石层裂缝方位
seamount 海山;海峰;海底(锥形)山(从海底升起1000米以上)
seamount chain 海山链;海底列山
seamount group 海山群;海底山群
seamount method 试探法
seamount range 海岭;海(底)山脉
seamount terrace 海底山阶地;海山阶地
seam penetration 穿透缝
seam pitch 矿层倾斜
seam placket 贴边开口缝
seam roll 钳卷材;卷边接缝的加固条
seam roller 接缝压辊;压缝滚子(裱糊墙纸用);叠缝滚压机
seam roof cladding 焊接的屋面覆盖层
seam roofing 有焊缝的屋面
seam roof sheathing 焊接的屋顶望板
seam saw 缝锯
seam-slag cleaner 焊缝清理机
seam smoother 焊缝清理机
seam spacing 焊缝间隙;接缝间距
seams plan 地层图
seam strap 对接贴板;边接贴板
seam strength 接缝强度
seam strip 边接贴板
seam structure map 矿层构造图
seam tube 有缝管
sea mud 海泥;海滨淤泥
sea mule 机动方驳;顶推船;方形小机动船
seam welder 线焊机;滚焊机
seam weld(ing) 缝焊(接);线焊接;滚焊
seam-welding electrode 滚极
seam-welding machine 缝焊机
seamy rock 有缝岩石;裂缝岩石
sea noise 海洋噪声
sea of transgression 侵陆海
sea ooze 海泥
sea otter 海獭
sea-out 海蚀的
sea outfall 河道出口入海处;海洋排泄口;入海口;污水排海口;出海管
sea outfall hydraulics 污水排海口水力学
sea painter 艇首缆
sea pass 中立船舶证(明)书
sea passage 海上航道
sea patrol 海上巡逻
seapeak 海(底)峰;海底锥形山(从海底升起1000米以上)
seapeak barge 海峰驳船
sea peat 海藻泥炭;海成泥炭;海底泥炭
seaphone 海上电话
sea piece 海景画
sea pilot 海上引航员
sea pink 海石竹
sea placer 海岸砂矿
seaplane 水上飞机
seaplane alighting area 水上飞机降落场
seaplane alights lands on the water 水上飞机降落在水面
seaplane anchorage 水上飞机停泊场
seaplane base 水上飞机基地
seaplane basin 水上飞机起落区域
seaplane berth 水上飞机停泊处
seaplane boat 水上飞机牵引艇
seaplane carrier 航空母舰
seaplane depot ship 水上飞机供应舰
seaplane dolly 水上飞机地面拖车
seaplane fighter 水上战斗机
seaplane float 水上飞机浮筒
seaplane hull 水上飞机船身
seaplane landing area 水上飞机登陆处
seaplane mother ship 水上飞机母舰
seaplane takes off from the water 水上飞机离水面起飞
seaplane tank 水上飞机试验水槽
seaplane tender 水上飞机勤务支援船

sea plant 海草;海洋植物
sea plateau 海底高原
sea plug 通海旋塞
sea pollution 海洋污染;海上污染
sea pollution prevention act 海洋污染防止法规
sea port 港埠;海港;港口;海口
seaport entrance 海港入口
seaport transshipment 海港转运
sea-pouce 下层逆流漩涡
sea power 制海权;海上力量;海上兵力
sea prawn 对虾
sea pressure 海上压力
sea products 海产(物)
sea protest 海事声明;海难抗辩书
sea proton magnetic gradiometer 海洋质子磁力梯度仪
sea puss 沿岸急流;离岸急流;退潮流;下层逆流漩涡
seaquake 海震;海啸(地震);海底地震;地震海啸
seaquake intensity 海震强度
sea quelling oil 镇浪油
sear 烙印;解脱杆
sea rainbow 海洋虹;海上虹
sea ranch 海洋养殖场;海洋牧场
sea range 海底山脉
search 检索;探索;探查;调查;查找;搜索
searchable information 可供查找信息
search and clear 搜索与扫荡
search-and-find 搜寻与营救
search and fire control radar 搜索与火控雷达
search and height finder radar 搜索与测高雷达
search and kill method 搜索删除法
search and mapping radar 搜索与测绘雷达
search and navigation radar 搜索与导航雷达
search and ranging radar 搜索与测距雷达
search and rescue 搜寻与救助;搜索与救助;搜索和营救;搜救
search and rescue aid 搜索与救援
search and rescue and homing 搜索救援与自动导引无线电信标
search and rescue area 搜索救援区
search and rescue beacon equipment 搜索和救援信标设备
search and rescue boat 搜寻与救助艇;搜索与救助艇
search and rescue center 搜寻与救助中心;搜索与救助中心
search and rescue chart 搜寻援救飞行图;搜寻与救助用位置图;搜索援救飞行图;搜索与救助用位置图;搜索营救用海图
search and rescue communication 搜寻与救助通信;搜索与救助通信
search and rescue control center 搜寻与救助控制中心;搜索与救助控制中心
search and rescue craft 搜寻与救助艇;搜索与救助艇
search and rescue exercise 搜寻与救助演习;搜索与救助演习;搜索与救援演习
search and rescue expenses 搜寻与救助费;搜索与救助费;搜索营救费用
search and rescue mission coordinator 搜救任务协调员
search and rescue net 搜索救援网
search and rescue operation 搜寻与救助行动;搜索与救助行动
search and rescue region 搜寻与救援区;搜救区
search and rescue satellite 搜寻与救助卫星;搜索与救助卫星
search and rescue satellite-aided tracking 搜救卫星辅助跟踪
search and rescue service 搜救业务
search and rescue spacecraft 搜索救援航天器
search and rescue telephone 搜索与救援电话
search and rescue vessel 搜寻与救助船;搜索与救助船
search and target acquisition radar 搜索与目标捕捉雷达
search and track radar 搜索跟踪雷达
search angle 探索角
search-angle control 探测角控制
search antenna 探测天线;搜索天线
search area 搜寻范围
search band 搜索范围
search capability 搜索能力
search circle 搜索圆圈区

search clue 查找线索
search coil 指示器线圈;搜寻线圈;测试线圈
search coil magnetometer 探索线圈磁强计
search coil method 试探线圈法
search coil test 探测线圈测试法
search coil transducer 探测线圈传感器
search cost 求职成本
search coverage 探测区域;搜索范围
search craft 搜索艇
search cycle 检索周期;搜索周期;查找周期
search data 搜索数据
search datum 搜寻基点
search dial 搜索度盘
search direction 查找方向
search directory system call 查找目录系统调用
search disconnect 搜索切断
search distillation 蒸馏式检索
search engine 网上查询器
search equipment 搜索设备
searcher 寻觅器;海关检查人员;探查仪;搜索艇;搜索装置;调查者
search for a moving target 对移动目标的搜索
search for an immobile hider 对不动隐藏者的搜索
search for existing information 搜集已有的资料
search for experience 摸索经验
search for information 搜集资料
search for mineral deposits 矿产普查
search for missing boat 搜寻失踪船只
search frequency generator 搜索频率发生器
search gas 探索气体;示踪气体
search gate 搜索选通脉
search graph 搜索图
search helicopter 搜索直升机
search indicator 搜索显示器
search information 搜索情报
searching 搜索;搜寻;查找
searching area 搜索区
searching control 搜索控制
searching current 探察电流;探测电流
searching ephemeris 搜寻星历表
searching fluid 高流动性液体
searching lamp 探照灯
searching light 探照灯
searching lighting 水平扫描
searching method 搜索(波)法
searching sector 搜索区
searching storage 内容定址存储器
searching surface 探测表面
searching the site 地址调查(工程)
searching tube 探管
search jammer 搜索式干扰机
search key 检索关键词;查找码;查找键;查找关键字
search length 探索长度;查找长度
search light 探照灯
searchlight and sound locator 探照灯与声波定位器
searchlight antenna 探照灯式天线
searchlight beacon 探照灯标
searchlight beam 探照灯光束
searchlight belt 探照灯照射区
searchlight canopy 探照灯照射的天空
searchlight car 探照灯车
searchlight co-operation 探照灯的协同
searchlight flat mounting base 探照灯安装座
searchlight generator 探照灯发电机
searchlighting 束光搜索;探照(灯搜索)
searchlighting antenna 搜索天线
searchlight intersection 探照灯光交叉
searchlight lamp 探照灯灯管
searchlight lens 探照灯透镜
searchlight mirror 探照灯反射镜
searchlight mounting 探照灯固定支架
searchlight platoon 探照灯排
searchlight radar 探照灯雷达
searchlight reflector 探照灯反射器
searchlight screen 探照灯滤光镜
searchlight sector 探照灯防区
searchlight shelter 探照灯掩蔽部
searchlight signal 探照式信号;探照式色灯信号机;探照式信号(机)
searchlight site 探照灯阵地
searchlight sonar 探照灯式声呐
searchlight support 探照灯灯光保障
searchlight tandem axle trailer 探照灯双轴挂车
searchlight truck 探照灯汽车

searchlight unit 探照灯部队
searchlight vehicle 探照灯车
search line center 搜索航线中心
search-lock-on radar 搜索和自动跟踪雷达
search means 检索工具
search memory 内容定址存储器
search method 寻优法
search mission 搜索飞行
search mode 搜索(工作)方式
search officer 海关检查人员
search operation 查找操作
search party 搜索队
search patrol 搜索巡逻
search pattern 搜寻方式
search pattern of documents 文献检索标志
search periscope 搜索用的潜望镜
search phase 搜索阶段
search plan 探索优先法;探索布局;探索计划
search plane 搜索飞机
search plotter 搜索情况标图板
search problem 探索问题(优选法)
search problem in the presence of false target 出现伪靶时的搜索问题
search procedure 查找过程
search process 觅数过程;搜索过程
search program(me) 查找程序
search pulse 搜索脉冲
search radar 搜索雷达
search-radar installation 搜索雷达站
search radar set 搜索雷达
search radar terminal 搜索雷达终端
search range 搜索距离
search rate 搜索速率
search-read function 寻读函数;阅览功能
search receiver 搜索接收器;搜索接收机
search record 咨询记录
search reflector 搜索反射器
search result register 查找结果寄存器
search scope 搜索雷达显示器
search sector 搜索扇形区
search sequence 探索次序
search sky 搜索天空
search sonar 搜索声呐
search speed 搜寻速度
search statement 查找语句
search stopper 搜索抑制器
search-stopping mechanism 搜索停止机构
search strategy 搜索策略
search surveillance 搜索监视
search sweep 搜索扫描
search syndicate 探矿辛迪加
search system 搜索系统
search technique 搜寻技术;搜索技术;搜索方法
search theory 搜索理论
search time 检索时间;查找时间
search track 搜寻线路;搜索跟踪
search tree 搜索树;查找树
search tuning 搜索调谐
search turn 搜索圈
search unit 探寻装置;探测装置
search work 探索工作
seare 击发阻铁
sea reach 近海河段;通海直水道
sea reclamation 填海扩地;海洋围垦;海岸围垦;填海造陆;填海造地
sea regime 海面状况
sea remote intension 海洋遥感
sea rescue 海上救助
sea return 海面(杂乱)回波;海面反射信号;海面反射干扰波;海浪杂波;海浪干扰
sea ridge 海脊
sea rights 领海权
sea risks 海难;海险
sea-river interface 海—河界面
searlesite 水硅硼钠石
Searle visco(si)meter 西尔(旋转圆筒式)黏度计
sea road 海路
sea room 回旋余地【船】;海上活动域;船舶回旋余地;宽广航行海面
sea route 海洋航线;海洋航路;海上交通线;海上航线;海路
sea rover 海盗船
Sears Building 希尔斯大厦
sea-run 海行

sea runway 水上跑道
seas 海域
sea salt 海盐
sea salt nuclei 海盐核
sea-salt nucleus 海盐核
sea sampler 海水采样器
sea sand 海沙(也称海砂);砂质海滩;沙质海滩
sea sat(ellite) 海洋(探测)卫星
sea scale 浪级(表);风浪等级;风波等级;波级
seascan marine radar 船用环扫雷达
seascape 海洋景色;海上景观;海景
sea scarp 海(蚀)崖;海底陡岸;海岸(陡岸)
sea scurvy 船员坏血病
sea search 入海顺河段
sea seismic method 海底地震勘探法
sea service 海上资历;海上服务
sea shell 海(贝)壳
seashells 海洋贝类
seashine 海面闪光
sea ship 海船
sea shock 海震
seashore 沿岸;海滨;海岸
seashore beach 滨海盐土
seashore drilling 海域采矿
seashore freeway 沿海岸的高速公路
seashore gravel 海滨砾石
seashore industrial reservations 沿海工业地带;临海工业地带
seashore lake 海滨湖
seashore new land 海滨新生地
seashore reach outside estuary 口外海滨段
seashore saline soil 滨海盐土
seashore terrace 海滨阶地
sea-sick 晕船的
sea sickness 晕船
seaside 海滨(的);海边;海岸
seaside city 海滨城市
seaside development 沿海开发;沿岸开发
seaside development area 沿海开发区
seaside gallery 码头前沿廊道
seaside hotel 海滨旅馆
seaside-orientated industries 沿海工业
seaside park 海滨公园
seaside pine 海边松
seaside promenade 海滨游戏场
seaside resort 海滨避暑地;海滨浴场;海滨胜地;海滨疗养地
seaside sanatorium 海滨疗养院
sea silt 黏质土
sea-skimming equipment 水面浮油刮集装置
sea slick 海面层
sea slide 海底坍塌;海底滑坡;海底地滑
sea slope 向海陆坡
sea smoke 蒸气雾;海烟;海雾;海上蒸汽雾
sea snake venoms 海蛇毒液
sea soil condition 海底土质
sea soil condition survey 海底土质调查
season 季(节);陈化
seasonable 适合时机
seasonal adjustment 季节(性)调整
seasonal aids 季节性航标
seasonal air conditioning 季节性空调
seasonal area 季节区域(干舷规定)
seasonal aspect 季相(变化)
seasonal balancing 季节平衡
seasonal brackish waters 季节性微咸水域
seasonal cargo 季节性货物
seasonal change 季节(性)变化;季变
seasonal channel 季节性航道
seasonal character 季节特征;季节特性
seasonal circulation 季节性环流
seasonal colo(u)ration 季节色泽
seasonal coefficient 季节系数
seasonal control station 季节性控制站
seasonal correction of mean sea level 平均海面季节改正
seasonal cracking 季节性破裂;季节性裂缝
seasonal credit accommodation 季节性贷款
seasonal creep 季节性蠕动
seasonal current 季节性海流;季节性流
seasonal cycle 季节循环
seasonal demand 季节性需求
seasonal depletion 季节亏耗
seasonal deviation 季节性偏差

seasonal discharge permit 季节排污许可证
seasonal discount 季节性折扣
seasonal disease 季节病
seasonal duty of water 季节性用水量;季节性需水量;季节灌溉水量
seasonal effect 季节影响;季节性效应
seasonal effect in time series 时间序列的季节效应
seasonal evaporation rate 季节性蒸发率
seasonal factor 季节(性)因子;季节(性)因素;季节系数
seasonal flood distribution 季节洪水分布
seasonal flow 季节性水流;季节流
seasonal fluctuation 季节性涨水;季节性升降;季节性起伏;季节性波动;季节变化;季节变动
seasonal fluctuation zone 季节交替带
seasonal forecasting 季节预报
seasonal forest 季相林
seasonal frequency 季节频率
seasonal frost region 季节性冻土区
seasonal frozen ground 季节性冻土
seasonal goods 季节性货物
seasonal grazing 季节(性)放牧
seasonal heat(ing) load 季节性热负荷
seasonal housing 季节性住房
seasonal incidence 季节影响
seasonal index 季节性指数
seasonal inventory 季节性库存量
seasonal inversion 季节性转换;季节性转化;季节性对流;季节性变换
seasonality 季节性
seasonal jobs 季节性工作
seasonal labo(u)rer 季节短工
seasonal lag 季节性滞后
seasonal lake 季节湖;时令湖
seasonal light 季节灯
seasonal load 季节性荷载;季节性负荷
seasonal load curve 季节性负荷曲线
seasonal loadline area 季节性载重线区
seasonal loan 季节性放款
seasonal loans for circulating funds 季节性流动资金
seasonally adjusted annual rate 季节变动调整年率
seasonally adjusted data 调整季节变动后的数据
seasonally adjusted figure 调整季节变动后的数字
seasonally banded clay 季节性条带形黏土
seasonally changed gloomy zone 季节变化的弱光带
seasonally dry forest zone 季节性干旱林区
seasonally frozen ground 季节性冰冻地带
seasonally frozen soil 季节(性)冻土
seasonally navigable channel 季节通航航道
seasonally navigable waterway 季节通航航道
seasonally protected plot 季节性围栏保护区
seasonal maximum 季节性最高量
seasonal mean value 季均值
seasonal migration 季节性迁移;季节性迁徙;季节性洄游
seasonal minimum 季节性最低量
seasonal movement 季节性涨落;季节性移动
seasonal navigation 季节通航
seasonal nitratenitrogen content 季节性硝态氮含量
seasonal output 季节性出力
seasonal peak load 季节性高峰负荷
seasonal peak on inventory 季节性库存高峰
seasonal period 季节期
seasonal phenomenon 季相景观
seasonal plan 季度计划
seasonal polymorphism 季节多态
seasonal port 季节性港口
seasonal prevalence 季节分布
seasonal price 季节性价格
seasonal price difference 季节差价
seasonal rainfall 季节性降雨
seasonal rain forest and rain forest region 季雨林和热带雨林带
seasonal rates 按季节计算的费率
seasonal rate schedule 季节价目制
seasonal recharge 季节补给
seasonal recovery 季节性回升
seasonal regulation 季节调节
seasonal reserves 季节性储备
seasonal reservoir 季节性调节水库
seasonal restriction 季节性限制

seasonal rhythm 季节性韵律
seasonal ring 年轮(树木)
seasonal rise 季节性升高
seasonal rise in level 季节性涨水
seasonal river 季节性流水;季节性河流;季节河
seasonal runoff 季节性径流;季节流
seasonal sampling 季节抽样
seasonal schedule 季度时间表
seasonal spring 季节泉
seasonal storage 季节性蓄水;季节储[贮]水;季节性储水量;季节性库容;季节调节库容
seasonal storage reservoir 季节性蓄水库
seasonal stream 间歇河(流);季节性流水;季节性河流;季节河;时令河
seasonal stream flow 季节性水流
seasonal structure 季节性构筑物
seasonal succession 季节性演替
seasonal tariff 季节性税率
seasonal thaw layer 季节性融化层;季节活性层
seasonal thermocline 季节性温跃层
seasonal trade 季节性贸易
seasonal traffic density 季节平均日交通量
seasonal traffic pattern 交通量季节变化图;季节(性)交通量变化图
seasonal tropical area 季节热带区域
seasonal unemployment 季节性失业
seasonal variation 季节(性)变化;季节(性)变动
seasonal variation factor 交通量季度系数
seasonal variation in earth's temperature 地温季节变化
seasonal variation in flow 流量季节变化
seasonal variation in precipitation 降雨季节变化;降水季节变化
seasonal variations in wastewater flow 废水流量季节变化
seasonal volume 季节性水量
seasonal water 季节性水
seasonal waterway 季节性水航道
seasonal weather 季节性天气
seasonal wind 季节风
seasonal work 季节性工作
seasonal worker 临时工;季节工
season character 季节性
season check 风干裂缝;干裂;干裂
season colo(u)r 应时色
season crack 内应力裂缝;干裂;干枯裂纹;天然时效裂纹;风干裂缝
season cracking 应力腐蚀裂缝;干燥开裂;老化开裂;季节性开裂;季节裂缝;时效破裂;陈化开裂;风干裂缝
season dynamics 季节性动态
seasoned 有龄期的;风干的
seasoned lumber 晾干木材;风干木材
seasoned mortgage 长年抵押
seasoned timber 干(木)材;晾干木材
seasoned wood 晾干木材;风干木材
season for planting 植树季节
seasoning 储[贮]放;干燥;晾干;老炼;调味剂;天然时效;时效处理;时效;陈化;风干
seasoning by water immersion (木材的)浸水干燥
seasoning chamber 空气干燥室
seasoning check (木材的)干燥裂缝;干裂;干燥裂隙;干燥裂纹
seasoning of timber 木材干燥(法)
seasoning of wood 木材干燥(法)
seasoning timber 晾干木材
season lag 季节性迟后
season long grazing 全期放牧
season of flax retting 沤麻期
season of growth 生长季节
season recovery 潜水面季节变化
season sale 季节性减价
season-ticket 月票
season timber 晾干木材;风干木材
sea spectrum 波谱
sea speed 营运航速;航海(营救)速度;常用航速【船】;服务航速
sea stack 海蚀柱;海边岩柱
sea state 海洋状况;海面状况;海况;波浪等级
sea state prediction 海况预报
sea state sensor 海况测量装置
sea state spectrum 海况谱;波浪谱
sea steps 舷梯
sea stock 海上供应品;船用物料及食品

sea stores 海上供应品
sea strait 海峡
seastrand 海滨;海岸
sea structure 海洋建筑物;海工建筑(物)
sea suction 海水吸入口
sea supremacy 制海权
sea surface 海面
sea-surface air temperature 海面气温
sea-surface ambient noise 海面环境噪声
sea-surface buoy system 海面浮标系统
sea surface chlorophyll concentration 海面叶绿素浓集
sea surface evapo(u)ration 海面蒸发
sea surface noise 海面噪声
sea-surface observation 海面观测
sea-surface patch 海面声径
sea-surface pressure 海面气压
sea-surface probe 海面测温仪
sea-surface radar model 海面雷达模型
sea-surface reverberation 海面混响
sea surface roughness 海面粗糙度
sea-surface scattering 海面散射
sea-surface scattering coefficient 海面散射系数
sea-surface scattering strength 海面散射强度
sea-surface search 海面搜索
sea surface temperature 海水表层水温
sea-surface temperature 海面温度;海面水温
sea-surface temperature anomaly 海面温度异常
sea-surface wind 海面风
sea-surface wind speed 海面风速
sea surveillance 海域监视
sea swell 海浪
sea-swell wave recording system 涌浪记录装置
seat 座椅;座(位);装填;炉盘;垫子;场所
seat adjuster 座位调节装置
seat adjustment 凳调整;椅调整
seatainer 船用集装箱
seat and chute entanglement 伞椅缠绕
seat and chute involvement 伞椅缠绕
sea tangle 墨角藻
seat angle 垫座角钢;座角钢;支座角钢;钢板弹簧座角;阀座倾斜角
seat arm 座椅臂
seat back 座椅靠背
seat back cushion 座椅靠背软垫
seat backrest 座位靠背
seat base 座椅底座
seat bath 坐浴
seat bath tub 坐式浴缸
seat beam (桥的)座梁;卧梁
seat belt 座舱安全带;(车座位上的)安全带
seat belt assembly 座位安全带总成
seat board 座板
seat bottom channel 座底槽板
seat carriage vertical release 座椅滑架垂直移动机构
seat catapult 座椅弹射器
seat clay 火泥;耐火黏土;底黏土(层)
seat connection 座接(法)
seat connection of beams to columns 梁柱托座(连)接头;梁柱托座(连)接法
seat connection of beam to columns 梁柱座接法
seat cover 座罩;座套
seat cover for vehicle 车辆用座套
seat cushion 座垫;弹性座垫
seat cut 支柱嵌切;椽子檐端截口
seat cutter 阀座修整刀具
seat diameter 阀座口径
seat division 座椅隔臂
seat earth 根土;煤层底板;底黏土;底层
seated armature 入位衔铁
seated colossal statue 坐式庞大塑像
seated colossus 坐式巨像
seated connection 托座连接
seated directional valve 座式换向阀
seated figure 坐式人像
seated frame 支座框架;座椅框架
seated gas generator 固定式燃气发生器
seated L-square 连座角尺
seated number 座位数
seated property 粗制基座
seated statue 坐式塑像
seated valve 座阀
seated valve pump 座阀配流泵;止回阀调节的泵

单向阀配流泵
seat-ejection initiator 座椅弹射起爆器
sea temperature chart 海洋等温线图;海水温度图
seat end wall 座位端墙
sea term 航海用语;航海术语
sea terminal 海运码头;海上装卸站;深水码头
sea terrace 沿岸台地;海成阶地
sea test 海上试验
seat filler material 座垫填充材料
seat fire 座椅弹射
seat for auxiliary machinery 辅机基座
seat for heating tube support 炉管支承架底座
seat frame 座位支架;座架
seat framework 座席构架
seat grinder 座口铰削机
seat grinding 研磨
seat guide 座椅调节导轨
seat guide rail 座椅导轨
seat harness 座椅安全带
seat heater 座椅取暖器
seat horizontal release 座椅水平移动板
seating 座位量;座位安排;支座
seating accommodation 专用坐车;乘坐室;座舱
seating and rowing test 乘座与划桨试验
seating and sealing unit (泵的)座封装置
seating block 锅炉基座
seating brick 座砖
seating capacity 座位数;座位(容)量;座位定额;载客定员;客车定员;可容纳人数;座席容量
seating configuration 座位配置
seating diagram 汽车车身内的座位布置图
seating directional valve 座式换向阀
seating drive 预备贯入;预打;预贯入击数;座式贯入装置
seating face 座表面
seating force (气门等的)座合压力
seating layout 座位配置方案
seating load 固定荷载
seating of boiler 锅炉座
seating of engine 发动机座
seating plane 底座面
seating rake 座位坡度
seating ring 座环
seating room 座位的空间
seating section 坐席区
seating shoe 支持管靴;垫座;底座;柱脚
seating slab 坐板
seating space 坐席区;座椅布置间隔
seating surface 支持面
seating the pump 泵的密合度
seating time 磨合时间;跑合时间
seating time of piston rings 活塞环的磨合时间
seating valve 座阀
seating washer 座垫
seat initiator 座椅弹射起爆器
seat-kilometre cost 客座千米成本
seat layout 座位布置
seat leakage test pressure 阀座密封性试验压力
seat leak test 阀座漏泄(密封性)试验
seat leg 座椅腿;座脚
seat level(l)ing valve 支座调平阀
seat lid (便桶上的)座盖
seat lift 座椅升降机
seat material 阀座材料
seat mile 客运里程(以一人一英里为单位);人英里
seat occupancy rate 客座占用率
seat of a fire 火的影响范围
seat of a rafter 椽脚
seat of business 营业处所
seat of explosion 爆炸中心
seat of fatigue 疲劳部位
seat of plane curve 平曲线位置
seat of settlement 沉陷区;沉陷中心;沉陷影响范围;沉陷影响范围
seat of the pants management style 席不暇暖式的管理
seat of vertical curve 竖曲线位置
seat of war 战场
seat on the exchange 交易所席位
seat-pack parachute 座包式降落伞
seat pad 座垫;垫块
seat pillar 座柱
seat pin 座销
seat pitch 座椅间隔

seat pull 座椅拉手
seat radio 车箱收音机
seat rail 坐凳栏杆;座椅横档;座椅导轨
sea train 出海火车渡轮;海上火车轮渡
sea train line 海上列车运输公司
sea transport(ation) 海上运输;海运;海洋运输
sea transport fleet 海运船队
seat regulator 座椅调整机构
seat reservation 座位预约
seat reservation system 座席预定系统;座位预定系统
seat retainer seat ring 座圈
sea trial 航行试车;海上试航;试航【航海】
sea triangulation 海上三角测量
seat riser 座椅升降机构
seat rock 基岩;底层岩石
seat row hole 辅助炮眼
seat screw 定位螺钉
seat seal 底层封闭
seat separation equipment 座椅分离机构
seat slide 座椅滑轨
seat spacing 座椅间距;座位间距
seat spring 座椅弹簧
seat stand 椅子架;座椅看台;座台;座桥看台
seat stay 座撑
seat support 座支架;座椅座
seat surface 座面
seat switch 定位开关
seat tapping plate 座椅安装铁
seat test 阀座试验
seattle cargo hook 具有旋转环的货钩
seattle pattern cargo hook 升降式吊货钩
Seattle Port 西雅图港(美国)
Seattle-Vancouver/Japan route 西雅图—温哥华—日本航线
seat track 座椅调节导轨
sea tug 海上拖轮
sea tug boat 海上拖轮
sea turn 海雾风
seat warmer 座椅加温器
seat warmer control 座椅加温控制
seatway 座间通道
sea-urchin 海胆
sea valley 海(底)谷
sea valve 通海阀
sea van 海运集装箱;海上运货车
sea van chassis 海运集装箱底盘车
seawall 海堤;护岸石堤;海墙;海塘;防浪堤
seawall works 海塘工程
seaward 向海方向;向海的;向海上;朝海方向;朝海(的)
seaward basin 向海港池
seaward boundary 向海疆界
seaward breakwater 向海防堤坝
seaward channel 通海航道
seaward deltaic progradational mode 向海三角洲前积模式
seaward migration 向海洄游;降海洄游
seaward navigation canal 入海航道
seaward navigation channel 入海航道
seawards 向海上
seaward slope 向海坡面;朝海坡度
sea-ware 被浪冲上岸的海草
sea-washed cliff 海蚀岩
seawater 海水
seawater acetic acid test 海水乙酸测定
seawater analysis 海水分析
seawater analysis chemistry 海水分析化学
seawater separation aquiculture area 海水养殖面积
seawater attack 海水侵蚀
sea-water bath 海水浴(场)
seawater battery 海水(活化)电池
seawater breeding area 海水养殖面积
seawater cell 海水电池
seawater chemical corrosion 海水化学腐蚀
seawater chemistry 海水化学
seawater circulation 海水环流
seawater coarse filter 海水粗滤器
seawater colo(u)r 海水水色
seawater comprehension utilization 海水综合利用
seawater concrete 海水混凝土
seawater condenser 海水冷凝器
seawater conversion 海水转化;海水淡化
seawater conversion plant 海水淡化设备

seawater conversion ship 海水淡化船
seawater cooling pump 海水冷却泵
seawater corrosion 海水侵蚀;海水腐蚀
seawater corrosion test 海水腐蚀试验
seawater culture 海水养殖;海面养殖
seawater damage 水损(货物受水渍);海上水损
seawater damaged 海水损害
seawater demineralizer 海水淡化器
seawater densitometer 海水密度计
seawater density 海水密度
seawater desalinating unit 海水淡化装置
seawater desalination 海水脱盐(作用)
seawater desalinization 海水淡化
seawater desalting 海水脱盐
seawater desalting equipment 海水淡化设备
seawater desulfurization 海水脱硫
seawater distillation 海水蒸馏
seawater distillatory 海水淡化装置
seawater electric(al)conductivity 海水电导率
seawater encroachment 海水入侵;海水浸蚀(地);海水倒灌
seawater evapo(u)rator 海水蒸发器
seawater extract 海水提出物
seawater geothermal system 海水地热系统
seawater gravimeter 海水密度计
seawater immersion test 海水浸渍试验
seawater inlet 海水进口
seawater inlet tunnel 海水进水渠
seawater intake facility 取水设备;海水取水设备
seawater intrusion 海水入侵;海水侵蚀
seawater intrusion pollution 海水入侵污染
seawater isotopic composition 海水同位素成分
seawater level 海平面
seawater lift pump 海水提升泵
seawater magnesia 海水氧化镁;海水镁砂
seawater magnesia clinker 海水镁砂熔块
seawater manure 海水肥料
seawater mercurometer 海水测汞仪
seawater mixing 海水混合
seawater molecular physics 海水分子物理学
seawater osmotic pressure energy 海水渗透压能
seawater physics 海水物理学
seawater pipe 海水管
seawater piping 海水管系;海水管道
seawater pollution 海水污染
seawater pollution by oil 海水油污染;海水石油污染
seawaterproof 防海水的
seawaterproof cable winding 防海水电缆圈
seawater pump 海水泵
seawater pump station 海水泵站
seawater purging 海水净化
seawater purification 海水净化作用
seawater purifying 海水净化
seawater quality standard 海水水质标准
seawater resistance 耐海水性
seawater-resistance low alloyed steel 耐海水低合金钢
seawater-resistant 耐海水的;防海水的
seawater resource 海水资源
seawater reverse osmosis 海水反渗透
seawater salinity 海水含盐量;海水盐度
seawater sampling 海水采样
seawater sampling bottle 海水取样金属瓶
seawater science 海水科学
seawater stratification 海水分层
seawater temperature 海水温度
seawater transparency 海水透明度
seawater type 海水类型
seawater wave swimming pool 海浪游泳池
sea wave 海浪
seawave analyzing system 涌浪记录装置
sea wave electric(al)power system 海浪发电系统;波能发电站;波力发电站
sea wave observation 海浪观测
sea wave prediction 海浪预报
sea wax 海蜡
seaway 航路【航海】;航进;航道;海上航线;海道
sea waybill 海运提(货)单;海上运输单;海上货运单
sea(way) canal 通海运河
sea wear 海洋服
seaweed 海苔;海藻;海草
seaweed biosorbent 海藻生物吸附
seaweed fiber 海藻纤维

seaweed gel 海藻胶
seaweed glue 海藻胶
seaweed green 海藻绿
seaweed meal 海藻干粉;海草粉
seaweed theory 海藻成油学说
seaweed wax 海藻蜡
sea wind 海风
sea wind wave 海洋表面波;风浪;海洋风浪
sea works 海岸维护工程;海岸堤防工程;海事工程;港湾工程
Sea World Inc. 海洋世界公司(美国)
sea worm 蛀木虫;海中蛀船虫
seaworn 海蚀的
seaworthiness 耐风浪性;耐航力;耐波性;水上适航性;适航性(能)【船】;适航(能)力
seaworthiness admitted clause 适航性认可条款
seaworthiness certificate 适航证书【船】
seaworthiness ship 航海性能良好的船
seaworthy 耐风浪的;适航状态;适航的
seaworthy packing 海运包装;耐航包装;适合海上运输包装;适航包装
sea wrack 被浪冲上岸的海草;海草
sea wreck 失事船的漂浮物
sea zone 海丘
sebacate 癸二酸盐(或酯)
sebaceous salt of copper 铜的脂肪盐
sebacic acid 癸二酸
sebacic dinitrile 癸二腈
sebacylic acid 癸二酸
sebastian salt 无水芒硝
sebate 癸二酸盐(或酯)
sebesite 透闪石
sebkha 盐沼(泽);碱滩
seborrhea cerea 蜡样皮脂溢
seca 塞卡风
secant 正切;正割
secant bulk modulus 割线体积弹性模数
secant circle 正割圆
secant cone projection 割圆锥投影
secant conic(al) chart 割圆锥投影地图;双标准纬线等积圆锥海图
secant conic(al) projection 圆锥正割投影;双标准纬线等积圆锥投影
secant conic projection 双标准纬线圆锥投影
secant curve 正割曲线
secant damping 割线阻尼
secant formula 正割公式
secant galvanometer 正割检流计
secant integral function 正割积分函数
secant law 正割定律
secant line 割线
secant meridian 割经线
secant method 正割法;割线法
secant modulus 正模数;正割模量;割线(变形)模量;视模数
secant modulus of elasticity 正割弹性模量
secant ogive 割面拱形体;割面尖拱
secant piles 割切桩
secant plane 切断平面
secant stiffness 割线刚度
secant theory 割线理论
secant yield stress 正割屈服应力
Secar cement 塞卡水泥
secateurs 整枝剪;剪枝剪
secavalent chrome 六价铬
sec-butyl propionate 丙酸仲丁酯
Seccam 赛康姆型胶轮驱动运输系统
Secchi's classification 塞齐分类法
secchi's depth 透明度极限
Secchi's disc 透明度板;塞齐试验圆板
secco (抹灰面上的)水彩壁画
secession 直线派;直线式
secessionism 直线式;分离式
secessionist 直线派建筑家
sechard 塞查德风
seclude the country from the outside world 闭关锁国
seclusion 闭塞(过程);闭关锁国;偏僻处;隔离;隔绝
sec-n-amyl alcohol 二乙基甲醇
seco 预制装配式房屋;闭联
secobarbital 司可巴比妥
secodont 切齿型
secohm 国际亨利;秒欧
secohmmeter 电感表

Secomastic 瑟科玛琦脂(一种衬垫金属的接合剂)
second 次要的
second address 第二地址
second adsorbate 二次吸附质
second anode 第二阳极
second approach section 第二接近区段
second approximation 第二近似值
secondary 第二级;第二次的;从属的;次要的;次生的;次级绕组;次级的;次等的;辅助的;二级的;二次的
secondary aberration 第二像差;第二级像差
secondary accelerator 助促进剂
secondary acetate 次级醋酯纤维;二代乙酸盐
secondary action 次(生)作用;二次作用
secondary adaptation 次级适应
secondary addition time 二次加法时间
secondary address vector table 辅助地址向量表
secondary adsorbate 二次吸附质
secondary adsorption 二次吸附
secondary advance 二次掘进
secondary after shock 第二期余震
secondary air 助燃空气;次级空气;二次空气;二次(进)风;补充空气
secondary air channel 补风管
secondary air classification 二次空气分级
secondary air duct 二次风管
secondary air inlet pipe 二次空气入口管
secondary air pollutant 次生空气污染物;二次空气污染物
secondary air register 二次风门
secondary air return 二次回风
secondary alcohol 仲醇
secondary alkane-sulfonate 二级烷基磺酸盐
secondary alkyl sulfonate 仲烷基磺酸盐
secondary alkyl sulphate 仲烷基磺酸酯
secondary allocation 增量分配;二次分配
secondary alloy 再生合金;再熔合金
secondary altering system 辅助警报系统
secondary alumin(i)um 二级铝;次铝;再生铝
secondary amine 仲胺
secondary amine base 仲胺碱
secondary amine value 仲胺值
secondary amyl acetate 乙酸仲戊酯
secondary annealing 二次退火
secondary annulus 第二内齿轮
secondary anomalous inclusion 次生异常包裹体
secondary anomaly 次生异常
secondary answering jack 副应答塞孔
secondary application 次(站)应用;二次应用;二次应用
secondary application block 次应用块;辅助应用块;二次应用块
secondary application program(me) 二次应用程序;次端应用程序
secondary approach 二次近似法
secondary approximation 二次近似;二次逼近
secondary arc 支弧;次生弧
secondary arch 二级弓
secondary area 副区
secondary armo(u)r rock 辅助护面块石
secondary arsenate 二代砷酸盐
secondary assignment 二次动作元件分配
secondary assimilation 二次同化
secondary assistant engineer 二管轮【船】
secondary association 次级联合
secondary attack rate 续发率
secondary attempt connection 二次接续
secondary attribute 次级表征
secondary axis 次轴;副(光)轴
secondary backing 第二层背衬;(地毯的)加劲背衬层
secondary balance shaft 第二平衡轴
secondary ball mill 二次球磨机
secondary bare area 次生裸地
secondary battery 蓄电池组;二次电池
secondary beam 次梁;次级射束;次级电子注;副梁
secondary bearing structure 次要的承重结构
secondary bedding 次生层理
secondary belt 次生带
secondary bending moment 附加弯矩
secondary bending stress 附加的弯曲应力
secondary benefit 间接收益;次要利益;次要利润;次等收益
secondary biological treatment of wastewater 废水的二级生物处理
secondary biological treatment method 二级生物处理法
secondary blasting 二次爆炸;岩石小爆破;二次破碎;二次爆破
secondary blasting borehole 二次爆破钻孔
secondary blasting drill hole 二次爆破钻孔
secondary blasting drilling 二次爆破钻进
secondary block 二级群
secondary blower 二次风鼓风机
secondary blowhole 二次气孔
secondary body wave S 波
secondary bond 次键;副键
secondary bonding 后胶结
secondary bonding force 次键力
secondary bottom 二级河床;二级底面
secondary brake shoe 从动制动蹄片
secondary branch 辅助支管;二级(排水)支管
secondary break 二次断头
secondary breakage 二次断裂
secondary breaker 二级破碎机;二级粉碎机;二次破碎机
secondary breaking 二段破碎;二次破碎
secondary breccia 次生角砾岩
secondary brush 副电刷
secondary brush coat 第二道粉刷
secondary buffering 继电缓冲
secondary building 次要建筑
secondary bulbil 次生株芽
secondary butanol 仲丁醇
secondary butyl acetate 乙酸仲丁酯
secondary butyl alcohol 仲丁醇
secondary butyl benzene 第二丁苯
secondary butyl cyclohexane 第二丁基环己烷
secondary cabin 二等舱
secondary cable 副绕组电线;辅助索
secondary calcium phosphate 磷酸氢钙;二代磷酸钙
secondary cambium 次生形成层
secondary canal 支渠;灌溉支渠;二次渠道
secondary capacitor 低压电容器;次级电容器
secondary carbon atom 仲碳原子
secondary carrier's freight 二程运费
secondary cash reserves 准现金储备
secondary category gassy mine 二级瓦斯矿
secondary category of commodities 二类商品
secondary category of goods 二类物资
secondary category vessel 二类容器
secondary cathode 次级阴极;二次阴极
secondary cell 蓄电池;再选浮选槽;副电池;二次电池
secondary cellulose acetate 二乙酸纤维素
secondary cementite 二次渗碳体
secondary center 次级中心
secondary change 次生变化
secondary channel 支汊;副河槽;辅助通道
secondary chromatic aberration 二级色像差;二次色像差
secondary circle 次大圆;副圈;副大圆
secondary circuit 次要电路;次级电路;二次回路;二次回流;二次电路
secondary circuit schematic diagram 二次接线图
secondary circulating system 第二循环管系
secondary circulating water system 第二循环水管系统
secondary circulation 第二循环;次级环流;副环流;二级环流;二次循环
secondary circulation pipe 二次水循环管路
secondary circulation piping 二次水循环管路
secondary clarification 二次澄清
secondary clarification tank 二次澄清池
secondary clarifier 次级澄清器;二次澄清池
secondary class 二级;二等的
secondary class fish-plated bolt 二级鱼尾螺栓
secondary class fish-plated nut 二级鱼尾螺母
secondary class highway 二级公路
secondary classification 二级分类
secondary class ore 二级矿石
secondary clay 次生黏土;冲积黏土;二次黏土
secondary cleaner 清滤器
secondary cleaning 再次精选
secondary clearance 副后角

secondary cleat 次生割理
secondary cleavage 次生劈理
secondary clock 子钟
secondary cluster 二次簇
secondary coal recovery 煤二次开采
secondary coast 壮年海岸;次生海岸
secondary coastline 次生海岸线
secondary coat 第二道粉刷;第二道抹灰;二度漆;二次涂敷
secondary coating 二次被覆层
secondary coating optic(al) fiber 二次被覆光纤
secondary coat of plaster 二道抹灰
secondary coil 次级线圈;副线圈;二次线圈
secondary cold front 次冷峰;副冷锋【气】
secondary collecting duct 次级集合管
secondary colony 次生菌落
secondary colo(u)r 间色;合成色;调和色;次生颜色;次级色;复色
secondary combustion 再燃烧;二次燃烧;补充燃烧
secondary combustion zone 延燃区;辅燃区
secondary coming pressure 二次来压
secondary commninution 二次粉碎
secondary compensating lateral 二次补偿支管
secondary compensation 二次补偿
secondary compensator 次级补偿器
secondary compilation 辅助编绘原图
secondary component 次成分
secondary compression 次压条;次固结;二次压缩;二次压固
secondary compression coefficient 次固结系数
secondary compression index 次固结指数
secondary computer 副计算机;副机;辅助计算机
secondary computer room 辅助计算机房
secondary concretion 次生结核
secondary condenser 次级冷凝器;二次冷凝器
secondary consolidation 次生固结;次固结(作用);二次固结
secondary consolidation coefficient 次固结系数
secondary consolidation index 次固结指数
secondary consolidation settlement 次固结沉降量;次固结沉降
secondary constant 二次常数
secondary constriction 次缢痕
secondary consumer 二次小费者
secondary containment 二次安全壳
secondary contaminant 二次污染物
secondary contradiction 次要矛盾
secondary control 大地测量二级控制;辅助操纵;二等控制
secondary controller 次级控制器;副调节器;二级(信号)控制器
secondary control point 次控制点
secondary control unit 辅助控制装置
secondary conversion 再加工
secondary coolant 二次冷却剂
secondary cooler 二次冷却器
secondary cool(ing) 二次冷却
secondary cooling system 二次循环冷却方式;二次冷却系统
secondary copper 再生铜
secondary copper loss 次级铜耗
secondary copper removal 二次除铜
secondary core 副核心
secondary cortex 次生皮层;次级皮质
secondary cosmic radiation 次宇宙辐射
secondary cosmic ray 次宇宙射线
secondary cost 间接费(用)
secondary coupling 次级管箍
secondary course 二道锉纹
secondary cover layer 副保护层;八护面层
secondary cracking 次生裂缝
secondary creep 后期蠕变;蠕变恒速区;第二阶段蠕变;次蠕变;附加徐变;附加蠕变;二级蠕变
secondary creep stage 稳定蠕变阶段
secondary criterion 二次判据
secondary criterion for decision 决策的辅助判据
secondary critical speed 二阶临界转速
secondary crop 次要作物
secondary crushed 二次轧碎的
secondary crusher 中碎机;次碎机;次级轧石机;二级粉碎机;二级轧碎机;二次破碎机
secondary crushing 二段破碎;二次破碎
secondary crystallization 后期结晶;二次结晶
secondary culvert 支涵

secondary curing 二级发酵
secondary current 次级线圈电流;次级(电路)电流;副流;二次电流
secondary curvature 次要弯曲;二次弯曲
secondary curve 二次曲线
secondary cut 二道纹
secondary cut file 二道纹锉
secondary cycle 后期循环;次(要)循环;副循环
secondary cyclone 次生气旋;副气旋
secondary cyst 次生囊
secondary dam 副坝
secondary damage 连带损坏;次生灾害
secondary data 辅助资料;辅助数据
secondary data set group 次(级)数据集组;辅助数据集组
secondary data users station 二级资料使用站
secondary dealings 二级交易
secondary decomposition 再度分解
secondary deflection 次挠曲
secondary deformation 二次变形
secondary degeneration 次级退变
secondary degree surface 二次曲面
secondary deposit 次生矿床;次生沉积;二次矿床
secondary depression 次生低压;副低(气)压【气】
secondary derivative 二阶导数;二次导数
secondary derivative action 二次导数作用
secondary derivative map 二阶导数图
secondary derivative spectroscopy 二阶导数谱学
secondary destination 次目的地;辅助目的地
secondary destruction 次生破坏
secondary device 次设备
secondary dewatering 二次脱水
secondary diaphragm 推迟膜片
secondary difference 二阶差分
secondary digestion 二次蒸煮;二次消化
secondary digestion chamber 再消化室;再腐化室;二次蒸煮室
secondary digestion tank 二次消化池
secondary dike 次要堤防
secondary dilution 二次稀释
secondary dilution method 二次稀释法
secondary dilution solvent 二次稀释溶剂
secondary discharge 次级放电
secondary discontinuity 次生结构面
secondary dispersion 次生分散;二次色散
secondary dispersion halo 次生分散晕
secondary dispersion pattern 次生分散模式
secondary display 二次显示
secondary distribution 二级分配
secondary distribution network 低压配电网;二次侧配电网
secondary division 附尺
secondary division of gas cleaning 精洗煤气设备
secondary documents 二次文献
secondary door 边门;侧门
secondary drain 辅泄水管
secondary drain(age) system 辅助泄水系统
secondary drilling 二次凿岩
secondary drinking-water standard 二级饮用水标准
secondary drying chamber 第二干燥室
secondary dump system 辅助转储系统
secondary dust 次生灰尘
secondary earth 次级线圈接地;二次线圈接地
secondary earthquake 次(地)震
secondary echo 二次回波
secondary editing 复校
secondary effect 次效应;次级效应;次步效应;副效应;二次影响;二次效应
secondary effect evapo(u)rator 二效蒸发器
secondary effluent 二级(污水处理)出水;二次排出物;二次出水;二沉池出水
secondary effluent standard 二次出水标准
secondary electrolysis 次级电解
secondary electron 次级电子
secondary electron conduction 次级电子传导
secondary electron conduction target vidicon 次级电子导电视像管
secondary electron counter 次级电子计数器
secondary electron emission 次级电子发射;二次电子发射
secondary electron image 二次电子图像
secondary electron multipactor limiter 次级电子倍增限幅器
secondary electron multiplication 次级电子倍增
secondary-electron multiplier 次级电子倍增器
secondary electron spectrum 二次电子光谱
secondary element 次要构件;次生要素
secondary elements in seawater 海水次要元素
secondary elevator 二级升降器
secondary emission 次级发射;二次排放;二次发射
secondary emission ceramics 二次电子发射陶瓷
secondary emission coefficient 次级发射系数;二次发射系数
secondary emission multiplication 二次电子发射倍增
secondary emission multiplier 次级发射倍增器
secondary emission photo cell 二次发射光电管;次级发射光电管
secondary emission quantometer 次级发射光量计
secondary emission ratio 次级发射系数
secondary emission stage 次级发射级
secondary emission tube 次级发射管;二次发射管
secondary emitter 次级电子发射体
secondary employment 附带就业
secondary energy 二次能量;二次能源;次级能源;次级能量;不定时电能
secondary energy source 二次能源
secondary enlargement 次生扩大;次生加大
secondary enrichment 次生富集
secondary enrichment of sulphide deposit 硫化矿床次生富集作用
secondary enrichment zone 次生富集带
secondary entrance 次要入口
secondary entry point 次入口点;辅助入口点;二次入口点
secondary environment 第二环境;次生环境
secondary epicyclic train 二级周转轮系
secondary equation 辅助方程
Secondary era 中生代
secondary error 次要误差;二次误差
secondary excavation 二次开挖
secondary excitation 二次激励
secondary expansion 二次膨胀(使用时膨胀)
secondary extinction 次级消光
secondary extinction coefficient 次级消光系数
secondary extra reflection 次级额外反射
secondary failure 从属失效;二次失效;二次故障;波及故障
secondary fairway 支航道;副航道
secondary fan 二次风机
secondary fault 次生断层;二次故障
secondary feather fractures 次级羽状裂隙
secondary feature 副景
secondary feeder 备用馈路;备用馈电线
secondary felling 受光伐
secondary fermentation 二级发酵
secondary fiber pulping 二次纤维浆化
secondary fiber 次级纤维
secondary field 次级场;副磁场;二次场
secondary field amplitude 二次场振幅
secondary field phase 二次场相位
secondary filament 二次纤维
secondary file 辅助文件;二次文件
secondary filling-in network 二等补充网
secondary filter 精滤器;次色滤光镜;二次滤片;二次滤池;二次过滤器
secondary filtration 二次过滤;精密过滤
secondary filtration after filtration 二次过滤
secondary filtration tank 二级滤池
secondary finder 二次寻线机
secondary fire regenerator 辅助加热用蓄热室
secondary fissure 次裂
secondary fixing 二次安装
secondary flo(a)tation 再浮选;二次浮选
secondary flo(a)tation column 二级气浮塔
secondary float feed 二次浮选给料
secondary flood channel 高水位支汊;高水位河槽
secondary floor 二楼
secondary floor plan 第二层平面图(美国);二层平面(美国)
secondary flow 次生水流;副流;二次流
secondary flow-and-return pipes 二次水循环管路
secondary flow pattern 二次流型;二次流谱
secondary flow pipe 副回流管;副流管
secondary flux of atmospheric γ-ray 大气伽马射线次级通量

secondary focal point 次焦距
secondary foci 次焦点
secondary focusing 二次聚焦
secondary fold 次生褶皱
secondary force 次力;辅助力;二次力
secondary forest 再生林;次生林
secondary forest growth 再生林
secondary forming 二次成型
secondary fractionator 二级分馏塔
secondary fragment 二次破片
secondary fragmentation breaking 二次破碎
secondary free lime 二次游离石灰
secondary frequency 第二选用频率
secondary front 副锋【气】
secondary fuel filter 二次燃料油滤器
secondary fuel oil filter 燃油精滤器
secondary function 次功能;辅助功能
secondary fundamental chain 二等基本锁
secondary gamma ray spectrum log 次生伽马能谱测井
secondary gamma-ray spectrum log curve 次生伽马能谱测井曲线
secondary gas 精洗煤气
secondary gas cap 次生气气顶气
secondary gear 二挡齿轮
secondary gear shaft 中间转动齿轮轴
secondary gear torque 二档齿轮转矩
secondary generation 第二代;二次组成;二次发生
secondary geologic(al) environment 次生地质环境
secondary geosyncline 次生地槽
secondary girder 次梁
secondary glass 加工玻璃
secondary gliding surface 次级滑面
secondary glue line 二次胶层
secondary gluing 二次胶合
secondary gonocyte 次级生殖母细胞
secondary Gothic 第二期哥特建筑放射式
secondary grade of minerogenetic prospect 二级成矿远景区
secondary grade of oil-gas prospect 二级油气远景区
secondary grade pf coal-forming prospect 二级成煤远景区
secondary grade structure 二级构造
secondary graphite 次生石墨
secondary graphitization 二次石墨化
secondary graphitizing 二次石墨化
secondary great circle 次大圆;副大圆
secondary grid emission 次级栅极发射
secondary grinding 二次研磨;二次磨(碎);终磨;二级磨;二次粉磨
secondary grinding circuit 二次粉磨系统
secondary grinding compartment 细磨仓;二次粉磨仓
secondary grinding mill 细磨机;二次磨机
secondary grout hole 二期灌浆孔
secondary grouting 辅助灌浆;二次注浆
secondary growth 次(生)生长;二次扩大;次生加大
secondary gyratory crusher 中碎用圆锥式破碎机;二次破碎用圆锥破碎机;二次回转破碎机
secondary gyroscope 阻尼陀螺仪;进动陀螺仪
secondary half-session 次对话端;辅助单边会话
secondary hand equipment 二手设备
secondary hand goods 二手货
secondary hand ship 二手船
secondary harbo(u)r 次等港(口)
secondary hardeness 二次硬度
secondary hardening 二次硬化
secondary harmonic 二次谐波
secondary harmonic generation 二次谐波发生
secondary harmonic oscillation 二次谐波振荡
secondary haulage 转运运输;次要运输
secondary hazard 次生灾害;次生危险
secondary heading 次标题;次航向;次导坑;次导洞
secondary heat exchanger 二次热交换器
secondary heat pump 二次热泵
secondary heavy medium cyclone circuit 再选重介质旋流器系统
secondary hexyl alcohol 仲己醇
secondary high explosive 次级猛炸药
secondary highway 支线公路;次要公路
secondary humic acid 次生腐殖酸
secondary humus tank 二次沉淀池

secondary hydrocyclone 再选水力旋流器
secondary idle-hole 怠速副喷口
secondary image 次像;重像;二次图像
secondary imaging 二次成像
secondary impedance 次级阻抗
secondary impeller 二次破碎用转子
secondary inclusion 次生包裹体;二次加热时形成的夹杂物
secondary income 非营业收入
secondary index 次索引;辅助索引;辅索引
secondary index data base 次级索引数据库
secondary index entry 辅助索引项
secondary indexing 副索引
secondary indicating instrument 二次指示仪表
secondary indicator 辅助指示符
secondary indicator of maturity 成熟度的辅助指标
secondary inductance 次级线圈的电感
secondary induction coil 副感应线圈
secondary industry 第二产业;二级产业;二次工业
secondary inertia force 二次惯性力
secondary infection 再次侵染;第二次侵染
secondary influence 次影响
secondary information 二次情报
secondary ingredient 辅助配料
secondary insect 次生害虫
secondary installation 附属设施;附属设备
secondary instrument 次级仪表;辅助仪器;辅助仪表;二次仪表
secondary intercalated clay layer 次生夹泥
secondary interference 副干扰
secondary internal force 二次内力
secondary interstice 后生间隙;次生裂隙;次生间隙
secondary invader 第二次侵害者
secondary ion 二次离子
secondary ion emission microanalyzer 二次离子发射显微分析仪
secondary ion exchanger 二次离子变换器
secondary ionization 次级电离;二级电离
secondary ionization constant 二级电离常数
secondary ion mass spectroanalyzer 二次离子质谱分析器
secondary ion mass spectrometer 次级离子质谱仪;次级离子质谱计;二次离子质谱仪
secondary ion mass spectrometry 二次离子质谱法
secondary ion mass spectroscopy 次级离子质谱(法)
secondary ion microprobe mass spectrometry 次级离子探针质谱法
secondary isotope effect 二级同位素效应
secondary iteration 第二迭代
secondary jaw crusher 二次颚式轧碎机;细碎用颚式破碎机
secondary joint 次生节理
secondary joist 次龙骨
secondary junctional fold 次级接头褶
secondary key 次关键字;辅助键;辅助关键字
secondary key encrypting key 辅助加密键
secondary key retrieval 辅助关键字检索
secondary kitchen 附属厨房
secondary lateral process 次级侧突
secondary laterite 次生砖红壤
secondary lead 再生铅
secondary leaf fall 第二次落叶
secondary leakage 二次泄漏
secondary leather cup 副皮碗
secondary lens fibers 续发晶状体纤维
secondary letter of credit 附属信用证
secondary levee 次堤;次要防乡;副堤;辅堤
secondary level address 二级地址【计】;二次地址
secondary level addressing 二级寻址
secondary level calibration 二级刻度
secondary level definition 二级定义
secondary level directory 第二级目录
secondary level of prevention 第二级预防
secondary lift 二级提升
secondary light source 第二光源;辅助光源;二次光源
secondary limit 次级限值;次级极限
secondary line 支线铁路;地方铁路;次要线【铁】;次要路(线)
secondary line merchandise 二等品
secondary line switch 第二级寻线机
secondary lining 永久衬砌;次衬砌;辅助衬砌;二次支护;二次衬砌

secondary link controller 辅助链路控制器
secondary link station 次(级)链路站;辅助网站;辅助链路站
secondary lithium battery 二次锂电池
secondary load(ing) 附加荷载;次要荷载;次要负荷;二次负载;二次负荷;二次荷载
secondary lobe 旁波瓣;旁瓣;副瓣
secondary lobule 次级小叶
secondary location 第二备选位置;非最佳位置
secondary loess 再积黄土;再生黄土;次生黄土
secondary logical unit 次(级)逻辑单元;辅助逻辑部件
secondary logical unit key 次逻辑单元键
secondary loop 次级回路
secondary loss 二次侧损耗
secondary low 次生低压;副气旋;副低(气)压【气】
secondary lubricating pump 第二滑油泵;从动滑油泵
secondary maceral 次生显微组分
secondary machine 辅机
secondary machine and electric(al) products 二类机电产品
secondary magma 次生岩浆
secondary magnetic field 二次磁场
secondary main 二级总管
secondary main road 次干道(厂内)
secondary mark 副标
secondary market 次要市场
secondary mass selection 继续群选
secondary master clock system 二级母钟系统
secondary mate 二副
secondary material 在用材料;辅助材料
secondary maximum 次级最大
secondary maximum contaminant level 最大二次污染容许量;二次最大污染水平
secondary measuring system 辅助测量系统;二次测量系统
secondary member 次要构件;次要杆件;副构件
secondary member of equation 方程的右边
secondary memory 辅助存储器;二级存储器
secondary meristem 次生分生组织
secondary metabolism 次级代谢;二级代谢
secondary metal 再生金属;回炉金属;次生金属;重熔金属;再用金属
secondary metamorphosis 次级变态
secondary meter 分户(水)表;分户电表;二次仪表
secondary migration 次生运移;二次运移
secondary mill 二次磨碎机
secondary mineral 次生矿物;二次矿物
secondary mineral deposit 次生矿床
secondary mineral mater 次生矿物质
secondary mineral or elements 次要矿物或元素
secondary mineral reserves 次要矿物储量
secondary minimum coagulation 第二极小凝聚
secondary mining 二次回采
secondary mirror 次镜(反射式望远镜的);副镜
secondary model 次级模型
secondary modem 次级调制解调器
secondary modulation 二次调制
secondary modulus of elasticity 第二弹性模数
secondary moment 二次(力)矩;次弯矩;附件弯矩;次(要)力矩;副力矩;附加力矩;二阶矩
secondary moment area method 二次矩面积法
secondary mortgage 二次抵押(权)
secondary mortgage market 二次抵押(贷款)市场
secondary mother-tree 次生化母树
secondary motion 副运动;附加运动
secondary mountains 次生山脉
secondary mullite 二次莫来石
secondary mullite formation 二次莫来石形成
secondary multiplexing 二次复用
secondary municipal effluent 二次城市污水处理后的出水
secondary mycelium 次生菌丝体
secondary negative 二底图
secondary network 二次侧电力网
secondary network control program(me) 次级网络控制程序
secondary neutral grid 二次中心点接地网
secondary nitro-compounds 仲硝基化合物
secondary nodule 次级小结;二级小结
secondary noise 次声;二次噪声
secondary nondisjunction 次级不离开

secondary nonferrous metal 再生有色金属
secondary nuclear fuel 次生核燃料
secondary nutrient 中量元素肥料
secondary offering 间接出售;二级分配
secondary officer 二副
secondary of the same tenor and date unpaid 副本尚未兑付
secondary oil recovery 二次采油
secondary opening 次生孔口;次生间隙
secondary operation 二次加工
secondary operation work 二道操作
secondary operator control station 辅助操作员控制站
secondary optic(al) axis 副光轴
secondary optic(al) vesicle 次级视泡
secondary optimization 次生最优化
secondary option menu 二级选择表
secondary order 二阶;二级的
secondary order day number 二阶日数
secondary order determinant 二阶行列式
secondary order difference 二阶差分
secondary order filter 二阶滤波器
secondary order formula 二阶公式
secondary order functional value 二阶函数值
secondary order linear differential equation 二阶微分线性方程
secondary order phase transition 二级相变
secondary order spectrum 二级光谱
secondary order subroutine 二级子程序
secondary order transition 二级转变
secondary order traverse 二等导线
secondary order viscoelastic theory 二阶黏弹理论
secondary ore 次生矿石
secondary orebody 次要矿体
secondary organ 附属机构
secondary orientation 从属定向
secondary orogeny 次生造山运动;次级造山运动
secondary oscillator 次级振荡器
secondary output 次级输出
secondary paging device 辅助分页设备;辅助调页设备;二级分页装置
secondary parameter 副参数;二次参数
secondary parasite 次寄生
secondary part 从动部分
secondary particle 次级粒子;二次颗粒物
secondary particle of star 星体的次级粒子
secondary partition 二级分区
secondary path 次要途径;次要路径
secondary peak 第二波峰;次峰;副峰
secondary periderm 次生周皮
secondary phase 二次相
secondary phase factor 二次相位因子
secondary photochemical reaction 二次光化学反应
secondary photocurrent 次级光电流
secondary photoelectric(al) effect 次级光电效应
secondary piezoelectric(al) effect 二级压电效应;二次压电效应
secondary pilot 副驾驶员
secondary pipe 二次缩孔
secondary pipe line 次级管道
secondary piping 二次侧配管;二次侧路管
secondary planet pinion 第二行星小齿轮
secondary plant 附属设备
secondary plasticizer 助增塑剂;次级增塑剂;辅助增塑剂
secondary ploughing 复耕
secondary podzolized soil 次生灰化土
secondary point 二等点
secondary pollutant 继发性污染物;次生污染物;次级污染物;二级污染物;次级污染物
secondary pollution 次生污染;次级污染;二级污染;二次污染
secondary pollution effect 次生污染影响
secondary pollution of water body 水体二次污染
secondary pore system 次生孔隙
secondary porosity 次生孔隙(度);次级孔隙
secondary port 次级港;次等港(口);潮汐次要观测站;副港;副潮港;辅助潮汐站;辅助潮位站
secondary power 次等出力;附加功率;二级电力;二次幂;二次功率;敷余电力
secondary power distribution 二次配电
secondary prevention 二级预防

secondary principal stress 次主应力
secondary problem 次要问题
secondary process 二次处理;次级突起
secondary product 次要产品;次生产物;副产品;二次产品;二步产物;次级产品
secondary production 次级生产
secondary productivity 次级生产力
secondary program(me) 次(级)程序
secondary program(me) status vector 次程序状态向量;辅助程序状态向量
secondary programming 二次规划(法)
secondary propulsion 辅助推进
secondary propulsion device 辅助推进装置
secondary propulsion system 辅助推进系统
secondary propulsion unit 辅助推进装置
secondary protection tube 热电偶护管
secondary protective barriers 二次防护层
secondary protective coating 二次防护涂层
secondary pump 副泵;二次泵
secondary pumping station 二级泵站
secondary pylon 附加塔柱
secondary quality product 二等品
secondary quantization 二次量子化
secondary quartzite 次生石英岩(矿)
secondary quenching 二次淬火
secondary question 次要问题
secondary radar 应答器;应答雷达;次级雷达;二次雷达
secondary radiation 次级辐射;二次辐射
secondary radiation effect 次生辐射效应
secondary radiator 二级辐射器;二次辐射器
secondary railroad 次要线[铁];支线铁路;铁路支线
secondary railway 铁路支线;支线铁路
secondary rain bow 副虹
secondary range 次等草场
secondary road 乡间土路;乡间道路
secondary rate 二等的
secondary raw material 次生原料;二次原料
secondary raw materials cost 辅助原材料费用
secondary raw mill 二次原料磨
secondary ray 次级射线
secondary reaction 次级反应;副反应;二级反应;二次反应
secondary re-bar 补充的钢筋
secondary receiver 差转接收机;二次接收机
secondary reconstruction source 次级再现光源
secondary recovery 二次开采;二次回收;二次回采
secondary recovery method 二次回收法;二次开采法
secondary recrystallization 次生重结晶;二次重结晶;二次再结晶
secondary reduction 二次还原(法);二次减速
secondary reduction pin 二次减速销
secondary reference fuel 副标准燃料
secondary reference fuel secondary standard fuel 第二参比燃料
secondary refining 二次精炼
secondary reflecting image 二次反射像
secondary reflection 人工爆炸;二次反射
secondary reflex 二级反射
secondary reformer 二段转化炉
secondary reformer tubes 二段转化炉管
secondary refrigerant 二级制冷剂;载冷剂;冷媒
secondary register set 次级寄存器组;辅助寄存器组
secondary regulating pond 副调节池
secondary regulation 二次调节
secondary regulator 副调整器;副调节器
secondary reinforcement 次筋;辅助钢筋;次级钢筋
secondary reject elevator 排放中间产品提升机
secondary relay 二次继电器
secondary relief 副后隙面
secondary remanent magnetization 次生剩余磁化强度
secondary remote subroutine 二级远程子程序
secondary remove subroutine 二级子程序
secondary request 次级请求;辅助请求
secondary rescue facility 辅助营救设备
secondary reserves 二级准备金;二次采油储量
secondary reservoir 次级水库
secondary reservoir and connecting aqueduct 辅助水库及连接水道;连接水道的调节水库
secondary resinite 次生树脂体
secondary resistance 次级线圈电阻;二次侧电阻
secondary resistance starter 二次电阻起动器

secondary resistance starting 转子外接电阻起动
secondary resonance 次级线圈谐振
secondary response 再次应答
secondary retarder 副缓行器
secondary retarder control 副缓行器控制
secondary retarder position 副缓行器位置
secondary return 二次回路;二次回流
secondary return air 二次回风
secondary ridge 次分水岭
secondary risks 投资上的一般风险
secondary river 后生(顺向)河;次要河流;次顺向河
secondary road 次干路;支路;公路支线;次要道路;二级公路;二级道路
secondary road network 次级道路网
secondary rock 次生岩(石)
secondary rolling 二次碾压
secondary root 次生根
secondary routes 次要路由
secondary row 辅助行
secondary runway 副跑道;辅助跑道
secondary rural road 郊区次道路;次要乡村道路;次要郊区道路
secondary saddle 副座板
secondary safety mechanism 辅助安全机构
secondary safety valve 二级安全阀
secondary safety zone 二级安全区
secondary saline soil 次生盐土
secondary salinity 次生盐度
secondary salinization 次生盐(渍)化
secondary salinization of soil 土壤次生盐渍化;土壤次生盐碱化
secondary salinized soil 次生盐渍土
secondary saloon passenger 二等舱旅客
secondary salt 仲盐;副盐
secondary salt effect 副盐效应
secondary sample 次级样本;二级样本;二次取样
secondary sampling 二次抽样
secondary sampling unit 二次抽样单位
secondary school 高级中学;中学;中等学校
secondary school buildings 中等学校建筑;中学建筑
secondary screen 次级筛;二道筛
secondary seal 二级封闭;二次密封
secondary seaman 二台
secondary sector 加工制造业;第二产业;二级部门
secondary security 附属抵押品
secondary sedimentation 二次沉淀
secondary sedimentation basin 二次沉淀池
secondary sedimentation tank 二次沉淀池
secondary seed 二次气泡
secondary seeding temperature of glass marbles 玻璃球二次气泡温度
secondary seepage cave 次生渗流洞
secondary segment 次段;辅助段;辅段
secondary separation 二次分离
secondary separator 第二次分离装置;二次分离器
secondary service 次要业务
secondary service area 第二工作区
secondary set rejection 二次排斥
secondary settle calcined gypsum 二次煅烧石膏
secondary settlement 二次沉降
secondary settling tank 二次沉淀池
secondary settling tank effluent 二次沉淀池出水
secondary sewage effluent 二级污水处理后出水
secondary sewage sludge 二次下水污泥;二次处理污水的污泥
secondary sewage treatment 二级污水处理;二次污水处理
secondary sewer 二次下水道;二次污水管
secondary sex 次级性别
secondary shaft 第二轴;次要矿井;传动轴;副轴
secondary shearing 二次剪切
secondary sheave 辅助滑轮
secondary sheet 副本
secondary ship wave 次生船行波
secondary shoot 次生茎
secondary shooting 二次爆破
secondary shore 次生海岸
secondary shoreline 次生海岸线
secondary shrinkage 二次收缩
secondary shrink hole 二次缩孔
secondary side 原边
secondary signal 副信号;二次信号
secondary silica 次生二氧化硅

secondary sizing operation 二次定型加工
secondary slip band 次级滑移带
secondary sludge 二次污泥;二沉污泥
secondary sludge digestion 二次污泥消化
secondary sludge sewage sludge 二级污水处理后的污泥
secondary sludge stream 二次污泥流
secondary sludge streamline 二次污泥流线
secondary smelting 二次冶炼
secondary soil 次生土(壤)
secondary solid solution 次生固溶体
secondary solid state battery 二次固体电池
secondary solvent 辅助溶剂;二级溶剂
secondary solvent effect 二级溶剂影响
secondary sorting siding 辅助编组线
secondary source 二次光源
secondary space allocation 辅助空间分配;二次空间分配
secondary spar 辅助翼梁
secondary specialized 中专生
secondary specialized school 中等专业学校
secondary species 次级产物
secondary spectroscopic(al) standard 二级光谱标准
secondary spectrum 次级光谱;二级光谱
secondary spectrum doublet 次级光谱的双线
secondary speed 涡轮转速;从动轮转速;二挡速度
secondary speed control 二次调速
secondary spheric(al) wave 次级球面波
secondary spore 次生孢子
secondary spring 辅助弹簧
secondary spudding date 二开日期
secondary stabilizer 二级稳定器
secondary stage 二级(梯级、平台等)
secondary stage annealing 第二阶段退火
secondary stage burner 二级燃烧器
secondary stage lining 二次衬砌
secondary stage of creep 二段蠕变
secondary stage oxidation treatment 二级氧化处理
secondary stage sludge 二级污泥
secondary stage treating 第二阶段处理
secondary stage treatment 第二阶段处理
secondary stair(case) 辅助楼梯
secondary standard 次级标准;副基准;副标准;二级标准
secondary standard material 二级标准物质
secondary standard sample 二级标准样
secondary standard substance 副基准物
secondary star 伴星
secondary state of stress 次应力状态
secondary station 从站;次站;副台;附属电台;辅助台;二级站;二等(测)站;二等(测量)点
secondary station algorithm 次站算法
secondary stator (油泵的)副定位子
secondary status 次站状态
secondary steam 二次蒸汽
secondary stone crusher 二次石料轧碎机
secondary storage 辅助存储器;二级存储器
secondary store 辅存储器;二级存储器
secondary stratification 次生层理;次级层理
secondary stratigraphic trap 地层圈闭
secondary stream 后生(顺向)河;次要河流;次顺向河;副流
secondary street 次要街道;次级街道;二级街道
secondary stress 次应力;重分布应力;二次应力
secondary stress condition 二次应力状态
secondary stress state 围岩二次应力状态
secondary structural elements 再次构造
secondary structure 次要结构;次生结构;次生构造;二级结构
secondary subroutine 二次子程序
secondary subscript 副下标
secondary succession 次生演替;次级演替
secondary sulfide zone 次生硫化物带;次生硫化带
secondary sun wheel 第二太阳齿轮
secondary supply water resources 辅助水源
secondary surface 副表面
secondary surveillance radar 二次监视雷达
secondary suspension 第二系悬挂;二次悬挂
secondary switching center 辅助交换中心;二级交换中心
secondary system 次要系统
secondary system control facility 辅助系统控制设备
secondary system error 次系统误差
secondary tap 二锥;二次中间抽头
secondary target 次要射击目标
secondary task 辅助任务;辅任务
secondary taste 余味
secondary technical school 中等专科学校
secondary tectonic stress field 次级构造应力场
secondary tectonite 次生构造岩
secondary tensile stress 副拉应力
secondary tension 副拉力
secondary tensioning 二次张力
secondary terrace 二级阶地
secondary tertiary alcohol 仲叔醇
secondary test board 副测试台
secondary textile effluent 二级纺织废水处理后出水
secondary thermal system 二次热力系统;二次回路;二次回流
secondary thickening 次生加厚
secondary tidal station 临时测潮站
secondary tidal wave 次生潮汐波;副潮波
secondary tide 次生潮汐;副潮
secondary tide station 次要潮期观察站;次要潮期观测站;辅助验潮站;辅助潮汐站;辅助潮位站;辅助测潮站;二等水位站
secondary tillage 再耕;复耕
secondary timbers 次要木材
secondary time effect 次时间效应;次固结作用
secondary tint 次色调
secondary tissue 次生组织
secondary titanium cooler 二段钛材冷却器
secondary tools 二次工具
secondary tooth 恒齿
secondary-to-primary flow 辅助数据流到主数据流
secondary-to-primary-turn ratio 变压系数;次级对初级的匝数比
secondary trajectory cable 辅助正交索
secondary transition 次级转变
secondary transmission 二级传动
secondary transverse current 副横流
secondary treated sewage effluent 二级处理过的污水出水
secondary treated wastewater 二级处理过的废水
secondary treatment 二级处理;二次处理
secondary treatment for sewage 污水二级处理
secondary treatment of chlorination 二级污水氯化处理
secondary treatment of no chlorination 二级污水非氯化处理
secondary treatment of sewage 污水二级处理;二级污水处理
secondary treatment of wastewater 废水二次处理;废水的二级处理;二级污水处理
secondary treatment plant 二级污水处理装置
secondary treatment process 二级污水处理工艺;二级处理法
secondary treatment standards of wastewater 废水二次处理标准
secondary trend 二级趋势
secondary trial 二次接续
secondary triangulation 二等三角测量
secondary triangulation net 二等三角测量网
secondary triangulation station 二等三角测量站;二等三角测量点
secondary troostite 回火屈氏体
secondary trunk 次干道
secondary trunk road 次要干线;次干路(城市);次干道(厂内)
secondary trunk road of port 港口次干道
secondary trunk route 次干路(城市)
secondary truss 次要桁架;次桁架
secondary truss member 次要桁架杆件;屋架辅助杆件;次要桁架构件
secondary tubing 二次侧配管;二次侧路管
secondary turbine 后置式汽轮机
secondary twinning 次生双晶;二次孪生;变形双晶
secondary twister for glass fibre 玻璃纤维复捻机
secondary tympanic membrane 次鼓膜
secondary undulation 次生波动;副振动
secondary unit 次级处理设备
secondary unpaid 副本尚未兑付
secondary use 次生利用
secondary utilization 次生利用
secondary valence 次化合价
secondary variation in depth 次生深度变化
secondary vector 第二向量;辅向量
secondary vegetation 次生植被
secondary ventilation 辅助通风
secondary venting 复管通风
secondary vent stack 副通气立管
secondary vertical system 次生直立枝法
secondary vibration 非缓冲质量的振动
secondary vibration in rear suspension 非缓冲后悬置质量的振动
secondary vitreous 续发玻璃状体
secondary voltage 次级电压;二次电压
secondary wall 次生壁;附加墙
secondary washbox 再选跳汰机;再洗跳汰机
secondary waste heat boiler 第二废热锅炉
secondary wastewater effluent 二级污水处理后出水
secondary wastewater makeup 二次废水补给
secondary wastewater treatment 污水二级处理;废水二次处理
secondary wastewater treatment system 二级污水处理系统
secondary water 外来水;次生水;二次水
secondary water pollution 次生水污染;二次水污染
secondary waterway 次要水道
secondary wave 地震横波;次(生)波;次级波;副波;二级波;二次波
secondary wavelet 次级子波
secondary wave zone 次波带
secondary weir 副堰;辅助堰
secondary white washing 第二道粉刷
secondary winding 副线圈;原边线圈;次级绕组;副线组;二次绕组
secondary wiring stretched-out view 二次结线展开图
secondary woody growth 再生林
secondary work 辅助工作
secondary working 复耕
secondary X-ray 次级 X 射线
secondary yard 第二车场;辅助编组场
secondary yield(ing) 辅助供水量;次级发射产额
secondary zone 次生带
second beat 秒拍
second beater (逐稿器的)第二逐稿轮
second best 次的;次优
second bottom 一级阶地;潮漫滩上的台地
second boundary problem 第二类边值问题
second bower 左首锚
second capital 陪都
second carriers freight 第二船运费
second channel 第二信道
second channel interference 镜像干扰;图像干扰
second chronograph 秒表
second class 第二流的;二级的;二等
second-class car 二等车(国外铁路)
second-class conductor 第二种导体;第二类导体
second-class electrode 第二类电极
second-class maintenance 二级维护
second-class road 二级道路
second-class wood pile 二级木桩
second clearance angle 第二后隙角
second cleat 次级劈理;次生解理
second cleavage 次级劈理
second clock 秒钟
second coat 第二漆;二道涂层;二道抹灰
second coat of plaster 二道抹灰
second commodities 次要矿产
second component 第二分量
second contribution 二次分布
second control 秒控制
second controller 秒控制器
second coordination sphere 外配位层
second copy original 第二底图
second counter 秒数计数器;秒表
second critical micellar concentration 第二临界胶束浓度
second critical speed 第二临界转速
second crop 第二次收成;第二茬作物
second cubic(al) foot 秒立方英尺
second cubic(al) foot day 秒立方英尺/天
second curvature 第二曲率
second curvature radius 第二曲率半径
second cut 二道锉齿纹
second cut file 二道纹锉;中细锉;中(号)锉

second cut flat wood rasp 中扁木锉
second data multiplexer 第二数据多路转接器
second deck 第二层甲板
second-degree burn 二度烧伤
second-degree parabola 二次抛物线
second delay blasting cap 秒延迟雷管；秒差雷管
second delay detonator 秒延迟雷管；秒差雷管
second delay electric(al) detonator 秒延迟电雷管；秒差电雷管
second departure section 第二离去区段
second derivation control 按二次导数调节
second derivative gravity anomaly 重力二次导数异常
second derivative of gravity potential 重力位二阶导数
second derivative test 二阶导数检验法
second development 第二次显影
second dike 副堤
second display 秒显示
second ditch-cut 第二次挖沟
second division segregation 第二次分裂分离
second DO 第二个循环
second doctor 船医助手
second drive belt 带中间皮带的传送装置
second eccentricity 第二偏心率
second edition 再版
second electron lens 第二电子透镜
Second Empire architecture 第二帝国建筑形式
second environment 第二环境；次生环境
second equation of Maxwell 麦克斯韦第二方程
second equivalence point 二次等量点
second estate 第二产业
second evapo(u)rator 蒸发器中的第二效次；二次蒸发器
second explosion 余爆(炸)
second failure surface 第二破裂面
second-feet 秒·英尺
second-feet-day 秒·英尺·日
second fixing 第二道装修；抹灰以后安装的部件；二次安装
second floor 第二层楼(美国)；第三层楼(英国)
second floor plan 第三层平面图(英国)；第二层平面图(美国)
second flow 秒流
second-foot 秒·英尺
second-foot-day 秒·英尺·日
second-forbidden spectrum 二次禁带谱
second fundamental form of a surface 曲面的第二基本形式
second gear 第二速度齿轮
second generation 改进型；第二代
second generation computer 第二代计算机
second generation container 第二代集装箱
second generation container terminal 第二代集装箱码头
second generation synthetic fibre 第二代合成纤维
second grade gravity base station for exploration geophysics 物探重力二级基点
second-grade of national discharge standard 国家二级排放标准
second-grade timber 二级木材
second growth 次生材；次生林
second-growth isotope 次生同位素
second-growth timber 粗纹木(材)；次生木材；小尺寸木材
second hammer 秒回零杆锤头
second hammer damping stop 秒锤杆制动器
second hammer maintaining washer 秒锤杆定位垫圈
second hammer spring foot 秒锤杆簧位钉
second-hand 旧的；用过的；秒针；第二手的
second-hand block 旧砖瓦；旧砌块
second-hand bookstore 旧书店
second-hand brick 旧砖
second-hand car 旧车
second-hand data 第二手资料
second-hand dealer 旧货家
second-hand machine 二手货机器；旧机器
second-hand masonry materials 用过的砌筑材料
second-hand packing 旧包装
second-hand rail 再用轨
second-hand relaying rail 再用的旧轨
second-hand roof(ing) tile 旧的屋面瓦
second-hand ship 旧船

second-hand shop 旧货店；停秒
second-hand tap 中丝锥
second-hand tie 旧轨枕
second harmonic 第二谐波
second harmonic generation 倍频效应
second heart 秒桃轮
second heart spring 秒桃轮簧
second highest crest 次最高波峰
second horizontal derivative 水平二阶导数
second house 别墅
second intention 第二期愈合
second interface (层状体系的)第二层界面
second intermediate rolls 第二层中间辊
second involute of the circle 圆的第二渐伸线
second ionization constant 次级电离常数
second jumper wheel 跳秒轮
second jumper wheel driver 跳秒轮驱动器
second jumper whip star 跳秒杆星轮
second junction 第二中心结
second kind 第二类
second kind collision 第二类碰撞
second kind electrode 第二类电极
second land 第二槽脊
second law of motion 第二运动定律
second law of thermodynamics 热力学第二定律
second leaf width and length 第二片叶宽度及长度
second leaky system 第二越流系统
second-level 二级的
second-level address 二级地址【计】
second-level addressing 二级编址
second-level directory 次级目录；二级目录
second-level index block 二级索引块
second-level interrupt 二级中断
second-level message 第二级信息；二级信息
second-level message member 第二级信息成员；二级信息成员
second-level of packaging 二级组装
second-level storage 次级存储器；二级存储器
second line 第二线
second line recovery 第二线中修
second log 第二段原木
second lower punch 第二下冲杆
second lowest trough 次最低波谷
second mean-value theorem 第二中值定理
second member 右端
second membrane 次生膜
second messenger 第二信使
second mode 第二振型
second moment criterion 二阶矩准则
second moment of area 截面二次矩；面积的二次矩；断面惯性矩
second moment of mass 惯性极矩
second month of autumn 仲秋
second month of spring 仲春
second month of summer 仲夏
second month of winter 仲冬
second mortgage 第二抵押
second-motion drive 间接传动；减速器传动
second-motion engine 第二运动发动机
second-motion shaft 第二运动轴；中间转动轴
second nodal point 第二节点
second of arc 角秒
second operand 第二运算数；第二操作数
second operation machine 第二操作机
second operation work 第二道工序
second order 第二级；第二次；第二的
second-order aberration 次级像差
second-order aberration coefficient 二级像差系数
second-order and high-order transformation 二次变换与高次变换
second-order autoregressive model 二阶段自回归模型
second-order autoregressive scheme 二阶自回归形式
second-order average 二维平均值
second-order bench mark 二级水准点；二等水准(标)点
second-order branching 第二级分支
second-order calculation 二级近似计算
second-order central moment 二阶中心矩
second-order change 第二级转变
second-order climatological station 二级气候站
second-order compressibility rule 二阶可压缩性法则

second-order condition 二阶条件
second-order correction 布隆尼改正；二级修正
second-order correlation function 二阶相关函数；二次相关函数
second-order derivative of gravity potential 重力位二阶导数
second-order design 二阶设计
second-order deviation 二阶离差
second-order difference transformation 二阶差分变换
second-order differential equation 二阶微分方程；二级微分方程
second-order differential system 二阶微分方程组
second-order diffraction spot 二阶衍射斑点
second-order dynamic(al) system 二阶动力系统
second-order effect 二阶效应；二级效应
second-order equation 二阶方程
second-order error 二次误差
second-order estimate 二阶估计
second-order factor 二次因子
second-order factor analysis 二阶因子分析
second-order fault 次级断层
second-order feedback system 二阶反馈系统
second-order fluorescence 二次荧光
second-order folds 次级褶皱
second-order goal 二阶目的
second-order homogeneous non-stationary process 二阶齐次非平稳过程
second-order increment 二阶增量
second-order infinitesimal 二阶无穷小
second-order kinetic law of Lagergren 拉格尔格伦二级动力学方程
second-order level(l)ing 二等水准测量
second-order linear system 二阶线性系统
second-order logic 二阶逻辑
second-order Markov process 二阶马尔柯夫过程
second-order maximum condition 二次极大条件
second-order moment around mean 二阶中心矩
second-order necessary condition 二阶必要条件
second-order parabola 二次抛物线
second-order perturbation 二阶摄动
second-order phase transition 二阶相变
second-order polynomial 平方多项式；二阶多项式
second-order radiation 二级辐射
second-order reaction 二阶反应；二级反应；二级反射
second-order servo 二阶伺服系统
second-order spatial coherence 二级空间相干性
second-order spin patterns 二级自旋图样
second-order stable system 二阶稳定体系
second-order station 二级气象站
second-order stationary 二阶平稳
second-order stationary process 二阶平稳过程
second-order subroutine 第二级子例行程序
second-order sufficient condition 二阶充分条件
second-order system 二阶系统
second-order term 次级效应项；二阶微量项
second-order theory 二阶理论
second order transition temperature 第二次转变温度
second-order triangulation 二等三角测量
second-order triangulation net 二等三角测量网
second-order triangulation point 二等三角点
second-order triangulation station 二等三角测量站；二等三角测量点
second-order wave force 二阶波浪力
second outlet 第二条出口
second-pair back 三楼后房(英国)
second-pair front 三楼前房(英国)
second parts 半旧零件
second party 乙方
second party audit 第二方审核
second party inspection 乙方检验
second pawl wheel 秒棘爪轮
second pendulum 秒摆
second pendulum clock 秒摆时针
second per revolution 每一转所需秒数
second-phase 第二相
second phase design 第二次设计
second-phase dispersion 第二相弥散
second pinion 秒齿轴
second pitch of coil 线圈第二节距
second plane of rupture 第二破裂面；第二滑动面；第二断裂面

second plate 第二板极
second power 平方
second pressing 再压
second principal stress 第二主应力
second probability distribution 第二概率分布
second process machine 光制机
second-quality material 次等材料
second-rate 二等品;二级品;第二流的
second reduction machine 精碎机
second reduction pinion 第二级减速小齿轮
second-remove subroutine 第二级子例行程序
second reset device 秒复位装置
second-ring groove 第二活塞环槽
second ripping 挑顶
second rolling 第二次碾压;次压
second roots 次生根
seconds 等外品;次品材料;次品;次等材料;二级品;二级面砖;二等品;二级码
second Saybolt furol 赛波特重油黏度秒数
second Saybolt universal 赛波特通用黏度秒数
second scion 第二接穗
seconds counter 计秒器
second seasoning 第二次陈化;第二次风干(木材)
second septum 第二隔
second setting deck watch 测天用秒表
second side 后加工面
second signal 第二信号
second signal system 第二信号系统
second sound 第二声
second source 第二资源;第二电源
second spatial moment 第二空间矩
second species 第二类
second speed 第二速度;二挡速度
second speed gear 第二速度齿轮
second-stage 第二级
second-stage annealing 第二阶段退火
second-stage biochemical oxygen demand 第二阶段生化需氧量
second-stage biological aerated filter 第二级曝气生物滤池
second-stage cofferdam 第二期(工程的)围堰
second-stage cooler 第二级冷却器
second-stage development 二期开发
second-stage ejector 第二级抽气器
second-stage excavation 二期开挖
second-stage grafting 第二阶段芽接
second-stage graphite 第二阶段石墨
second-stage graphitization 第二阶段石墨化
second-stage of primary design 初设二期
second-stage of reheat 第二阶段中间加热
second-stage pump 二级泵
second-stage pumping station 二级泵站
second-stage rotating biological contactor 第二级生物转盘
second steel sheet 精整薄钢板
second stop marker 第二停车区段
second stop watch 停表
second-stor(e)y 第二层楼的
second strike 第二次打击
second stud gear 第二变速齿轮
second surface mirror 后表面反光镜
second surplus reinsurance 第二益额再保险
second tap 中丝锥
second thermocline 次(生)温跃层
second-time-around echo 第二次反射回波
second time at right angles to the first 第二次与第一次呈直角
second to none 第一流的
second totalizer 累计计秒器
second trace echo 第二次扫描回波;二次扫描回波
second track 第二线
second transit 第二次中天
second trial 中关
second-trip echo 第二次反射回波
second undercoat 第二道底漆
second vertical derivative 垂向二阶导数
second vertical derivative calculation of gravity anomaly 重力异常垂向二阶导数换算
second vertical derivative of gravity 重力垂直二阶导数
second vertical gravity derivative map 重力垂向二阶导数图
second vibration mode 第二振型

second virial coefficient 第二维里系数
second viscosity 第二黏度
second viscosity coefficient 第二黏性系数
second wave 地震横波;S 波
second wheel 秒轮
second wheel additional lock 秒轮附加锁板
second wheel bridge 秒轮夹板
second wheel driving pinion 秒轮传动齿轴
second wheel friction spring 秒轮摩擦簧
second wheel lock 秒轮锁杆
second wheel lock spring 秒轮锁杆簧
Second World Climate Conference 第二次世界气候会议
second year face 第二年割面
second year ice 隔年冰;二年冰
secos 庙宇内殿;特权者使用的房间;古希腊神坛
secrecy 机密;保密
secrecy system 保密通信制;保密体制
secret abutment 隐桥台;暗桥台
secretarial staff 秘书人员
secretariat 秘书处
secretary 写字台;秘书
secretary general 秘书长
secretary office 秘书室
secret block 隐轮滑车
secret cable 隐蔽的电缆
secret code 密码
secret colo(u)r porcelain 秘色瓷
secret colo(u)r ware 秘色瓷
secret communication 保密通信
secret cover sheet 暗盖板
secret decoration 暗花装饰
secret documents 保密文件
secret door 裱糊的门;密门;隐蔽门
secret dovetail 暗鸠尾接头;暗楔(形)榫;暗燕尾榫;暗马牙榫
secret dovetailing 暗鸠尾榫接头;暗燕尾榫接;暗楔榫;暗榫接合
secret dovetail joint 暗燕尾榫接;暗接接头;暗鸠尾榫接头
secrete door 隐蔽门
secrete project 保密工程
secret fixing 暗装法;隐蔽式装配;隐蔽式固定;隐蔽式安装;暗装配
secret fountain 隐头喷泉
secret gate latch 暗门插销;暗门锁
secret gutter 暗檐沟;屋檐漏水槽;暗檐槽;暗(天)沟;暗藏式天沟
secret heading joint 暗榫结合;暗榫接合
secret hinge 暗铰链;暗合页
secret ink 隐显油墨
secret installation 隐蔽装置
secretion 藏匿;分泌液;分泌物;分泌体
secretion sclerotinite 分泌菌类体
secretion vitrinite 分泌镜质体
secret joggle 暗企口接;暗啮合扣接
secret joint 暗楔接合;暗接(头)
secret ledger 秘密分类账;暗账
secret message 秘密情报
secret miter 暗斜接(头)
secret mitre joint 暗榫斜接
secret nail 隐钉;暗钉
secret-nailed 暗钉的
secret nailing 暗钉;埋头钉;暗钉连接
secretory granule 泌颗粒
secret partner 秘密合伙人
secret passage 暗道
secret patent 机密专利
secret pipe 暗管
secret profits 秘密利润
secret purchasing 秘密采购
secret reserves 秘密准备(金)
secret screw 暗螺钉
secret screwing 暗螺钉连接;暗螺钉接合
secret screw joint 暗螺钉接合
secret signalling 保密通信系统
secret spring 秘簧
secret switch 密封式开关;锁盖开关
secret tack 暗平头钉;暗焊片固定
secret telephone 保密电话(机)
secret telephony 保密通话
secret tenon 暗榫头
secret trust 秘密信托
secret valley 屋顶排水暗沟;暗榫头

secret valve 暗阀
secret wedging 暗楔榫连接;暗楔固;暗楔法
secret wiring 暗线
sectile opus 可分立墙;花砖地面;拼花地面
sectility 可切削性
section 截面;工区;截点;节;磨片;区域;区间【铁】;切面;切断面;剖面;断片;断面;段;地区;部门;部分
section 1 区段一
section 2 区段二
sectionable photographic equipment 分段摄片器
section after the departure 离去区段
sectional 截面的;区域的;剖面的;片式连接的;断面的;地区的;分节的;分级的;分段的
sectional aerated concrete panel 分段加气的混凝土板
sectional aeration 分区充气
sectional aeronautical chart 区域航空图
sectional alidade 横断面仪;断面照准仪
sectional area 截面(面)积;剖面面积;断面面积
sectional area of concrete 混凝土截面积
sectional area of flow 水流横截面
sectional area of reinforcement 钢筋截面积
sectional area of sample 样品断面面积
sectional arrangement drawing 纵剖面图;装配剖视图
sectional assembling 分段拼装(法)
sectional assembly 局部汇编;分段汇编程序
sectional automation 分段自动化
sectional back-stuffing with clay and cement 水泥黏土分段回填
sectional bailer 多节捞砂筒
sectional balance sheet 分部资本负债表
sectional balancing 分组试算;分组结平;分部试算;分部结平
sectional balancing ledger 分部结平分类账
sectional bar 型材
sectional barge 分节驳船
sectional bench 分段排列长凳(会场内)
sectional blasting 分段爆破
sectional boiler 组合式锅炉;片式锅炉;对片锅炉;分节锅炉;分段(拼接)锅炉
sectional boom 分节起重臂;分段起重臂;分段吊杆
sectional box dock 分节(式)浮坞;分段厢式船坞;分段式箱形浮船坞
sectional breaker 分段开关
sectional building 具有地方特色的房屋;可拆卸房屋;地方特色房屋;预制部件构成房屋
sectional casing 组合壳
sectional cast-iron boiler 片式铸铁锅炉;组合式铸铁锅炉
sectional center 地区总局;地区中心;二级控制中心
sectional check block system 区间照查闭塞【铁】
sectional coefficient 剖面系数
sectional coil 分段线圈
sectional constant 断面常数
sectional construction 预制构件拼装结构;分段结构;预制部分构造;预制结构
sectional-continuous function 分段连续函数
sectional conveyer 分段(式)输送机
sectional core 拼合型芯
sectional core barrel 加长岩芯管;多节岩芯管;分段芯管
sectional core type radiator 分段式散热器
sectional curvature 截面曲率
sectional die 镶块式模;组合式模具
sectional dimension 截面尺寸
sectional dock 分段(箱)式船坞;分节(式)浮坞
sectional draw 分段穿综法
sectional drawing 截面图;剖面图;断面图
sectional drawing of building 房屋剖面图
sectional drawing of condensate gas pool 凝析气藏剖面图
sectional drawing of deposit 沉积断面图
sectional drawing of gas pool 气藏剖面图
sectional drawing of oil and gas pool 油气藏剖面图
sectional drawing of oil pool 油藏剖面图
sectional drilling 迂回钻进;分段钻进
sectional drill rod 可卸式钻杆;可卸式钎头
sectional drive 分段驱动
sectional dryer 分段烘干机
sectional earth pulling process 分段拉土法

sectional elevation 切面图;立视图;立剖面;立面(剖视)图;截视立面图;截面立视图;横剖视图;剖视图;竖直剖面
sectional enlarging 局部放大
sectional erection 分段安装
sectional fare 分段票制
sectional feeder 区间馈电线;分格轮(式)喂料机
sectional feeder cabinet 区间馈线柜
sectional fixture 安装夹具
sectional flask 分格砂箱
sectional flight 分段扬料板
sectional flight conveyer 分组螺旋叶片式输送器
sectional floating dock 分节(式)浮坞;分段(式)浮船坞
sectional flywheel 组合飞轮
sectional form 断面形状;分段式
sectional formwork 预制模板
sectional furniture 组合式家具;装配式家具
sectional glue-jointed panel 分段胶接的板
sectional groove 型钢轧槽
sectional header boiler 分联箱式锅炉
sectional hub 组合轮毂
sectional inspection 分段巡检
sectional insulation 组合式绝热材料;预制绝缘材料;预制绝热材料;预先成型的绝缘材料;预先成型的绝缘材料
sectional insulator 分段绝缘子;分段绝缘器
sectional iron 型铁
sectional isolator 分段绝缘子;分段绝缘器
sectionalization 分组;分节
sectionalization test 分段测试
sectionalize 分段
sectionalized 分区的;分段的
sectionalized antenna 分节天线
sectionalized bus bar 分段母线
sectionalized casing 分段式外壳
sectionalized coil 分段绕制线圈
sectionalized construction 分段施工
sectionalized continuous brick laying method 分段连续砌砖法
sectionalized house 装配式住房
sectionalized jacket 分段式导管架
sectionalized line 分段线路
sectionalized machine manufacture 通用机械制造;分段组装机械制造
sectionalized steel form 组合式金属模板
sectionalized tractor 组合式拖拉机
sectionalized transformer 分节变压器
sectionalized vertical antenna 分段竖直天线;分段垂直天线
sectionalized winding 分组绕组
sectionalizing breaker 分段断路器
sectionalizing post 分区亭
sectionalizing switch 分段开关
sectionalizing valve 分段(截流)阀;主管断流阀;分区截流阀
sectional ladder 组合式梯
sectional ledger 分组分类账;分部分类账
sectional lifter 分格扬料器
sectionally linear region 分段线性区域
sectionally smooth 分段平滑的
sectional management 行业管理
sectional material 型材
sectional method 断面法
sectional mining 分区开采
sectional mixing method 全断面混合法
sectional model 拼合模;分段模型
sectional modulus 截面模数;截面模量;断面模数;断面模量
sectional mo(u)ld 镶合砂型;拼合型
sectional mounting 片式串联连接
sectional multi-stage centrifugal fresh water pump 分段式多级离心清水泵
sectional nave 可拆卸衬套
sectional number 断面号数
sectional observation 断面观测
sectional overhead door 卷帘门;门上摇头窗
sectional pattern 组合模样;分块组合模
sectional perspective 剖面透视图
sectional plan 平面剖视图;水平剖面(图);分段平面图;分区平面图;分段剖视图
sectional plane 水平剖面(图);截面;剖视面;剖切平面
sectional plating 分段添纱

sectional ploughing method 分段刨煤法
sectional pontoon dock 底筒分节式浮坞;分节式浮船坞;分节(式)浮坞;方驳式分段浮船坞;连尼氏船坞
sectional position 截面位置
sectional profile 剖面图
sectional property 断面性质
sectional pushed tow 分节顶推驳船队
sectional pushed-tow fleet 分节顶推船队
sectional push-tow train 分节顶推驳船队
sectional quadrant 舵扇
sectional radiator 可分式散热器
sectional radiography 断面X射线照相术
sectional rate 分段运价
sectional rate of investment 部门投资率
sectional remote signal(l)ing 区段遥信
sectional retaining wall 分段挡土墙
sectional ring rubbers(for packers)(封隔器的)多段橡胶环
sectional rod 组合钻杆;拧接式钻杆
sectional roll 分组镇压器
sectional roller 分节(式)压路机;分节路碾
sectional route release 进路分段解锁
sectional sensitivity 铸件壁厚的敏感度
sectional shaft 组合轴
sectional side elevation 侧剖图
sectional sketch 断面草测
sectional smooth 分段光滑
sectional sofa 组合沙发
sectional staff ga(u)ge 分段立式水尺
sectional stamping 分段冲压
sectional steel 型钢;组合钻杆;可拆卸钢钎
sectional steel drilling 接杆钻进;接杆凿岩
sectional structure 分段结构
sectional survey 分段测量;断面测量
sectional switch 分段开关;分段区域开关
sectional system 断面系统
sectional tamping rod 加长捣实炮棍(深孔装药用)
sectional tank 拼装油箱;拼装水箱;组装油箱;组装水箱;型钢水池
sectional tap 组合丝锥
sectional tiller 组合舵柄;舵柄弧
sectional tows 分节驳船队;分节顶推船队
sectional track 移动式轨道;轻便轨道
sectional traffic remote control 区段遥控
sectional train 区段列车
sectional trash racks 分节式拦污栅
sectional travel(l)ing speed 区间车速
sectional turbodrill 多级涡轮钻具
sectional type 分块式;分段式
sectional type bituminous mixing plant 分段式沥青搅拌设备
sectional view 截面图;截视形;剖视图;剖面图;断面图
sectional view of bars 钢筋断面图
sectional view of steel shell 钢外壳混合柱剖面视图
sectional warper 分段整经机
sectional warping 分段整经
sectional warping machine 分段整经机
sectional wheel 扇形轮
sectional wind(ing) 分段绕组
section ama 焊缝喉部截面积(指焊缝长度与喉厚之积)
section area 分段区
section area of collars 钻铤截面积
section area of drill pipe 钻杆截面积
section at crown of arch 拱顶处理断面
section at key 拱心断面;中央悬臂断面(拱坝);中央断面;拱坝拱冠断面
section bar 型材;型钢
section base 分条定位
section-bending machine 型材弯曲机
section between block posts 所间区间
section between stations 站间区间
section bit 区段位
section blocked 区间闭塞
section boss 养路班长;路段养路班长;分段工长
section boundary 采区边界
section box 电缆交接箱;分电箱
section break contact 分断接头
section building 可拆卸房屋
section by section method 分段法
section center 分区中心

section chief 工段长;工长
section chief construction site 工段
section cleared 区间空闲
section cleared up 区间封锁
section closed due to construction 因施工段封闭段
section closed up 区间封锁
section coefficient 翼型系数
section configuration 断面形状
section congestion 区间阻塞
section constant 截面常数;断面常数
section construction 预制部分集合构造;预制构件拼装结构;拼装构造;组合(式)结构;拼装式结构;分段施工
section control 断面控制
section controller 区间值班员
section corner 区域边界角
section coupler panel 区间联络开关屏
section crack 剖面裂缝;断面裂纹
section crew 养路班;路段养路班
section curve 断面曲线
section cutter 切片机
section-cutting 切片法
section cut vertical map 由垂直图切水平面图
section deformation 截面变形
section depth 截面深度
section design speed 路段设计速度
section diagram 剖面图
section discharge determination in rivers and canals 河渠断面测流法
section distribution 分条排列
section drawing 截面图
section economy 截面的经济性
sectioned bar diagram 分段条形图;分段条(线)图
sectioned lens 分割透镜
sectioned polynomial 分段多项式
sectioned specimen 焊接试样
sectioned Y-deflection plate oscilloscope tube 垂直偏转板分段式示波管
section engineer 工地主任;工程主任;工务分段长;工段工程师;路段工程师;分段工程师
sectiones isthmi 峡切面
section fabrication 分段造船法
section factor 截面系数;断面系数
section factor for critical flow 临界水流断面因数
section factor for uniform flow 均匀水流断面因数
section factor of channel 航道断面系数
section feeder panel with bypass disconnector 带旁路短路的区间馈线柜
section forces 维修养护队
section foreman 养路班长;工务段工区领工员;路段养路班长;分段工长
section for ship 船舶车辆及民用建筑型材
section frame 型钢框架
section gang 道班;养路班;路段养路班
section gap 重叠间隔;分段间隙
section girder 拼装大梁
section gradient 落差
section graph 部分图;分段图
section groove 异形孔型
section grouting downwards 分段下行注浆
section grouting upwards 分段上行注浆
section hand 路段工人;道班工人;铁路工段道工
section head 节头
section header 节头;节标题;段头;段标题
section house 工务段办公室;道班房;养路工区房屋;工区房屋
section indicator 分区指示灯
sectioning 换段【铁】;切片法;电分段;分割;分段装置
sectioning examination 分段检验
sectioning method 分段法
sectioning of welded joints 焊接连接的切片
sectioning point 电分段装置
sectioning search 分割搜索;分段搜索
sectioning valve 分段阀;放泄阀
section insulator 隔离开关
section iron 型铁;型钢
section iron chassis 型铁架
section label 段标
section lamp 分组指示灯;分区指示灯
section leader 小组负责人
section lease 分条分绞
section length 区间长度

section level(l)ing 断面水准测量
section light 区间照明灯
section lighting distribution box 区间照明配电箱
section lighting piping 区间照明配管
section lighting wiring 区间照明放线
section line 管界线；截面线；区划线；断面线；段管线；分区线；切面线；剖面线
section-line profile 断面面积
section line spacing 断面线间距
section lining 剖面线法
section locking 区间锁闭；区段锁闭
section man 线路工区工人；路段养路工人；养路工人；区段工长(道路)
section marks 分条痕
section material 型材
section mill 型钢轧机
section modulus 截面模量；截面模数；断面模量；剖面系数；剖面模数；断面系数
section modulus of wall 墙的截面积模量
section moment 截面(力)矩
section mo(u)ld 成型模；截面形状和尺寸样板；石膏制品的模具；型模；型煤
section name 段名
section number 节号
section number of screen well installation 滤水管安装degnotes 管安装段数
section occupancy indication 区间占用表示；区段占用表示
section occupied 区间占用
section of a fuzzy relation 模糊关系的截口
section of alternating pit 分段跳槽
section of a recurrent structure 四端网络节
section of a valley 屋顶排水沟的截面
section of bosh 炉腰剖面
section of building 建筑物剖面图
section of colo(u)rs 彩色设计
section of column 柱断面
section of construction project 工段
section of curve 曲线段
section of drill hole 钻孔剖面图
section of earthwork 工区土石方量图；土方剖面；土方截面；土方断面
section of easy grade 缓坡地段
section of excavation 掘进断面；开挖断面
section of gentle grade 缓坡地段
section of heading 导坑断面
section of insufficient grade 非紧坡地段【铁】
section of insufficient gradient 非紧坡地段【铁】
section of land 一平方英里(等于640英亩)
section of level(l)ing 区段；水准区段
section of line 线段
section of main 直管段
section of multiple 复式塞孔部分
section of nozzle 喷嘴截面积
section of pipeline 直管段
section of reservoir 地层单元；地层单位
section of road 路段
section of screen(ing) deck 筛底部分
section of shear 剪切面(积)
section of slope 缓坡地段
section of soil 土地断面
section of structure 结构剖面图
section of sufficient grade 紧坡地段【铁】
section of the belt 皮带截面
section of the rail 钢轨剖面
section of track 轨排
section of unsufficient grade 非紧坡地段【铁】
section of valley 河谷横断面
section-out 分段
section overhead 段开销
section passengers per hour per direction 每小时单方向断面客流量
section paper 方格纸
section party 养路班；路段养路班
section pillar 分段支柱
section plan 剖面平面图；平剖面
section plane 剖面；剖面平视图
section post 管界标；分区所；分段标
section preparation 切片标本
section projector 截面投影仪
section property 截面性质
section pump 分组泵
section reinforcement 型钢(配筋)
section ribbon 切片带

section roll 型材轧辊
section rolling mill 型材轧机
sections 站线；船体横剖线
section scanner 断层扫描器
section selecting for highway or railway 道路选线
section selecting for piping 管道选线
section sensitivity 断口敏感性
section shape 平面形状
section shaped 定形的
section shape of boudin 石香肠断面形态
section sheet 方格纸
section sign 管界标
section signal 区间信号
section size 剖面尺寸
section size of tunnel 坑道断面尺寸
section staining 切片染色
section steel 型钢
section steel concrete structure 型钢混凝土结构
section steel rolling mill 分类轧钢机
section straightener 型钢矫直机
section straightening 矫形机
section-straightening machine 型材矫直机
section strip 异形带钢；锯条带钢
section surveying 断面调查
section switch 分段开关；区域开关；分段断路器
section symbol 剖面符号；断面符号
section team 养路班
section termination 段终端
section term of estimation variance 面估计方差
section thickness 截面厚度；断面厚度
section through rock 岩石断面
section tool house 道班工具房
section topography 纵剖法；剖面分析法
section track 区间轨道
section-type mounting 片式串联连接
section valve 普通阀；隔离阀
section velocity 断面流速
section view 截视形；剖视(图)
section weight of frame 框架分段重量
section wire 型材钢丝；异形钢丝
section wire method 断面索法
section without canalization 未渠化河段
section wood 型材
section working machine 定型材料工作机
sector 两脚规；区段；片区；四分仪；视界弧；扇(形)物；扇(形)面；扇区；扇段；区段；地段；齿弧
sector accelerator 扇形加速器
sector address 扇面地址
sector address system 区段地址方式
sectoral assessment 部门评价
sector alignment indicator 信号对准指示符
sector area 扇形面积
sector arm 扇形臂
sector barrage 区域拦阻射击
sector bevel gear 扇形圆锥齿轮
sector boss 地段队长
sector boundary 扇(形)边界
sector brick 扇形砖
sector buffer 扇形缓冲器
sector cable 扇形芯电缆；扇心电缆
sector chaining 区段链接
sector characteristic curve 扇形区域特性曲线
sector chart 扇区图
sector conductor 扇形引出线；扇形芯线
sector configuration 扇形结构
sector correction factor 扇形校正因数
sector dam 扇形坝；扇形堰
sector diagram 扇形图
sector disk 扇形盘
sector display 扇形显示(器)；扇形扫描；扇形扫掠
sector-display indicator 扇形显示器
sector distortion 扇形失真
sector distribution of investment 按部门的投资分配
sectored beam 扇形光束
sectored light 具有不同色光弧的灯标；扇形光；分区灯
sector electromagnet 扇形磁铁
sector feeder 扇形加料器
sector field 扇形场
sector-field accelerator 扇形场加速器
sector-field cyclotron 扇形场回旋加速器
sector-field direction focusing 扇形场方向聚焦
sector-field spectroscope 扇形场分光镜

sector-focused isochronous cyclotron 扇形聚焦等时性回旋加速器
sector format 区段格式
sector gate 弧形(闸)门；扇形(闸)门；扇形船闸
sector gate weir 扇形闸门堰
sector gear 扇形齿轮；齿弧
sector gear bearing 扇形齿轮轴承
sector graticule method 扇形量板法
sector hole 扇区孔；段孔
sectorial 扇形的；分横脉
sectorial area 扇形面积
sectorial chimera 区分嵌合体；扇形嵌合体
sectorial coordinate 扇形坐标
sectorial harmonic function 扇谐函数
sectorial harmonics 扇形谐和函数；扇谐函数
sectorial horn 扇形喇叭(筒)
sectorial magnetic field 扇形磁场
sectorial spheric(al) harmonic function 扇球谐函数；扇形球面调和函数
sectorial tides 扇形潮
sectorial watershed 扇形流域
sectorial wave 扇形波
sector imperfection 扇形场畸变
sectoring device 扇形转向装置
sector instrument 扇形仪表装置；扇形仪表；扇形场仪器
sectorization 功能分区(按不同使用功能将城市分为若干区)
sectorization of functions 功能分区
sector length 扇面长度
sector light 扇形照明；扇形光
sector lock 扇形船闸
sector magnet 扇形磁铁
sector mark 段标；区段标记；分段标志
sector marker 区段标志；区段标识符
sector mark-information 扇区标记信息
sector method 扇形测量角法；分区法
sector mirror 扇形镜
sector mode 区段方式
sector model 扇形模型
sector number 区段数
sector of breakdown 分类的部分
sector of demand 需求部门
sector of fire 射击区域
sector of light 光弧；能见光弧
sector of search 搜索地区
sector of the economy 经济成分
sector pattern instruments 扇形仪表
sector photometer 扇形光度计
sector pivot gate 弧形旋转式闸门；扇形旋转式闸门
sector planning 局部规划；小区规划
sector plan position indicator 扇形平面位置指示器
sector plate 扇形板
sector queuing 区段排队
sector rack 扇形齿条；扇形齿板
sector radio marker 扇形无线电指向标
sector rainer 扇形喷灌器
sector register 区段寄存器
sector regulator 滚筒式空心节制闸门；滚筒式空心节水围堰；扇形闸门
sector resonance 扇形共振
sector-rim type rotor 扇形叠片磁轭转子
sectors 分脉
sector scan indicator 扇形扫描显示器
sector scan(ning) 扇形扫描；区域扫描；区域扫掠；扇形搜索
sector scan(ning) sonar 扇形扫描声呐；扇形扫测声呐
sector screen 弧形遮光板
sector screens of sidelight 红绿灯板框
sector search 分区搜索
sector search pattern 扇形搜寻方式
sector selection 分区选择；分段选择
sector sequence 区段序列；区段顺序
sector shaft 扇形齿轮轴；扇形板轴；扇形齿轮轴
sector shaft end play 扇形齿轮轴轴向间隙
sector shaft module 扇形轮组件
sector shaft pilot 扇形齿轮轴导向端
sector shaft thrust adjustment 扇形轴限制器调整
sector shaped coil 扇形线圈
sector-shaped conductor 扇形导线
sector shutter 扇形快门
sector sluice 扇形溜槽
sector sluice valve 扇形泄水阀

sector sprinkler 扇形喷灌机
sector structure 扇形结构
sector switch 扇面开关
sector teeth 扇形齿轮的齿
sector theory 扇形理论
sector-type 扇形
sector-type lock gate 扇形船闸闸门
sector type relay 扇形继电器
sector velocity 扇形速度[点相对于给定中心的速度];扇形速度
sector weir 弧形堰;扇形堰
sector wheel 扇形轮
sector wind 扇形风
sectorwood 扇形胶合板制材法
sector worm 扇形蜗杆
Sectra system 色克特拉式体系建筑
sectrix curve 等分角线
sectroid 穹棱间曲面
sectrometer 真空管滴定计
secular 长期的
secular aberration 长期光行差
secular acceleration 长期加速度
secular acceleration and retardation of the moon 长期月球速度的增减
secular architecture 非宗教性建筑
secular balance 长期平衡
secular basilica 长方形会堂(非宗教性的)
secular change 缓慢变化;时效变化;多年变化;长期而缓慢的变化;长期变异;长缓变化;长缓变化
secular coefficient 长期项系数
secular cycle 多年周期;长期周期;长年周期;百年周期
secular depression 缓慢沉降
secular distortion 时效变形
secular drift 缓慢漂移
secular effect 长期效应
secular equation 特征方程;长期方程
secular equilibrium 永恒平衡;长期平衡
secular error 永久仪差;长期误差;长期变化引起的误差【测】
secular factor 长期因素
secular Gothic structure 非宗教哥特式结构
secular Gothic style 非宗教哥特式风格
secular magnetic change 长期磁变
secular monument 非宗教纪念物;非宗教纪念碑
secular movement 长期(缓漫)运动;长缓运动
secular order 长期订货
secular parallax 长期视差
secular period 长期周期
secular perturbation 长期摄动
secular polar motion 长期极移
secular precession 长期岁差
secular rise 长期上升;长期缓升
secular rise of zero-reading 零点长期上升;冰点长期上升
secular settlement 长期缓慢沉降
secular sinking 长期缓沉
secular stability 长期稳定性;长期稳定度
secular structure 民用结构
secular subsidence 长期缓漫沉降;长期缓沉
secular term 长期项
secular time effect 长期时效
secular trench 长远趋势
secular trend 长期趋势
secular trend in climate 气候长期趋势
secular upheaval 长期缓升
secular variable 长期变量
secular variation 世纪变化;时效变化;多年变化;长期缓慢变化;长期而缓慢的变化;长期变异;长期变化
secular variation of geomagnetic field 古地磁场的长期变化
secular variation period 长期变化周期
secunda oil 页岩太阳油
secure 使安全;缚;安全的
secure a bid 取得递盘
secure an adequate supply 保障供应
secure a painter 系牢艇首缆
secure by means of screw 螺钉固定
secure claims 担保债权
secure communication 保密通信
secure communication satellite 保密通信卫星
secure condition of readiness 三等战备状态
secured bond 担保债务;保证债券;具抵押品的债券;抵押债券
secured creditor 有担保的债权人;具抵押权人
secured creditors' right in bankruptcy proceedings 破产清理中的保证债权
secured deposit 保证存款
secured loan 抵押贷款;具抵押贷款
secure electronic payment protocol 安全电子支付协议
secure for sea 做好出海准备
secure handfill 安全填库
secure hyper-text transport protocol 安全超文本传输协议
secure job 可靠的工作
secure line 保密线路;安全线(路)
secure oneself against accidents 投保人身意外险
secure printer 保密打字机
secure rock 稳固岩石
secure rotationally 利用控制保持转动
secure sockets layer 安全套层
secure speech system 保密话音系统
secure storage area 安全储[贮]存区
secure system 可靠系统
secure telegraph equipment 保密电报设备
secure telephone 保密电话机
secure telephone system 保密电话系统
secure the diving stations 取消下潜状态
secure to double buoys 系双浮
secure to single buoy 系单浮
secure transactions technology 安全交易技术
secure video conferencing interface equipment 保密电视会议接口设备
secure visual communications 保密视觉通信
secure voice 保密语音
secure voice unit 话密装置
Securex 瑟库勒斯连接(一种铜管连接)
Securicor 安全服务社(一种私营的安全服务机构)
securing 固紧
securing attachments 栓固元件
securing device 固定装置
securing loop 固定匝
securing nut 扣紧螺母
securing of the profit 取得利润
securing of the wires 钢丝固定
securing pin 安全销(钉)
securing plant 防护植物
securing rod 锁定杆
securing screw 固定螺栓;固定螺钉
securing strap 紧固索带;安全带
securing strip 安全带
securing to a buoy 系浮筒
securing to buoys fore and aft 双浮筒系泊;双浮标系泊
securing wire of adjustment screw 调整螺钉的固定钢丝
securite explosive 安全炸药
securities business 保证金买卖
securities employment 具保谋职
securities exchange 证券交易所
securities for fulfillment 履约抵押品;履约担保;履约保证
securities in trust 信托证券
securities issued 已发证券;已发行证券
securities issued by affiliated companies 联合公司发行的证券
securities portfolio 所持有的有价证券;证券组合
securities proceeds 出售证券收入
security 证券;警戒;稳固;抵押器;抵押品;担保品;防御措施;保卫;保密(性);保密措施;安全(性)
security alarm system 安全警报系统
security against 防御
security agreement 债务协定
security arrangement 安全措施
security assessor 可靠性鉴定器
security axiom 安全性公理
security bond 保证书
security capital 安全资本
security certificate 安全验证
security class 机密安全等级;安全类
security classification 机密等级;保密级别;保密等级;保留等级
security classification of map 图件密级
security classification of report 报告密级
security code 密级代码;保密码;安全码
security consideration 安全性考虑
security constraint 安全性约束
security contract 担保合同
security control 保安措施;安全控制;安全技术
security control center 安全控制中心
security cylinder lock 安全圆柱形锁
security date set 安全数据集
security department 保卫科
security deposit 担保存款;押金
security device 安全装置
security dispatching 安全调度
security door 防盗门
security dump 安全转储(程序);安全卸出
security equipment 保密设备
security exchange 债券交易所
security explosive 塑性安全炸药
security film 安全制片
security filter 安全(过)滤器
security fire door 安全防火门
security flaw 安全性缺陷
security for a loan 借债抵押品
security for claim 诉讼假扣押
security for cost 讼费保证金
security for debt 债务担保;债务保证
security for fulfillment 履约保证
security for interest 物权担保
security for loan 借款担保
security for payment of contribution 共同海损分摊保证
security glass 防盗玻璃;防弹玻璃;保险玻璃;安全玻璃;夹层玻璃
security glazing 镶安全玻璃
security grille 防护格栅
security hinge 固定销铰链;保险铰链
security in container yard 堆场上的安全作业
security information 密级信息
security in litigation 诉讼担保
security installment land contract 担保分期付款土地合同(美国加州的一种抵押贷款方式)
security instrument 安全证明
security interest 抵押权益;物权担保
security kernel 安全核心
security keylock 保密键锁
security level 保密级别;保密等级;安全水准
security maintenance 保密维护;安全性维护
security manual 安全手册
security market 证券市场
security measure 安全措施
security mechanism 安全机构
security mirror 监视镜;单向透明玻璃镜;安全反射镜;安全发射镜
security monitor microprocessor 安全性监视微处理机
security number 保密代号;安全号码
security of data 数据保密
security of loan 贷款担保
security of service 安全运行;安全行驶
security of tax payment 纳税保证金
security optical fibre 安全光纤
security policy 安全策略
security price 证券价格
security printing 保密印刷
security regulations 保密条例
security requirement 安全性要求
security reserves 保证准备
security restore 安全再存程序
security screen 安全屏;安全栅栏;防护屏
security section 保卫科
security siding 安全线【铁】
security status 可靠现状
security system 安全系统
security valve 安全阀
security vault 证券金库;安全库
security window 防盗窗;保险窗;安全窗;防盗钢栅窗
security zone of water source 水源保护区
sedan 小客(轿)车
sedan coupe 二门轿车
sedarenite 沉积岩屑砂屑岩
sedative 镇静剂
sedentary 定居的;残积的
sedentary benthos 定居底栖生物
sedentary deposit 原地淤积;原地沉积
sedentary fold(ing) 敷挂褶皱;原地褶皱
sedentary herd 定居群

sedentary mineral 原生矿物
sedentary population 定居人口
sedentary product 残积物
sedentary soil 原生土(壤);原地土(壤);定积土;残积土
sedentary trade 静坐作业
sederholmite 六方硒镍矿
sedge 芦苇
sedge moor 苔草沼泽
sedge peat 泥炭
Sedgwick-Rafter filter 塞奇威克—拉夫特滤器
Sedgwick-Rafter method 塞奇威克—拉夫特法
sedilia 祭司席
sediment 渣滓;泥沙(沉淀物);沉降物;沉淀(物)
sediment accumulation 泥沙淤积;泥沙积聚
sediment accumulation rate 淤积速率
sediment aggregation 沉积团粒
sediment allowance 沉降容许高度
sediment ammonia demand 底泥需氨量
sediment analysis 泥沙分析;底泥分析;沉淀分析
sediment and water 水杂淀积
sediment and water test 水和不溶物含量的测定
sedimentary 沉积的
sedimentary aggregate 沉积岩集料;沉积岩骨料
sedimentary analysis 沉积分析
sedimentary apron 沉积裙
sedimentary area 沉积面积
sedimentary association 沉积组合
sedimentary band 沉积带
sedimentary base level 沉积基准面
sedimentary basin 沉积盆地
sedimentary basin model 沉积盆地模式
sedimentary basin type 沉积盆地型
sedimentary bauxite deposit 沉积铝土矿床
sedimentary bed 沉积岩层;沉积土层;沉积河床
sedimentary bed in different levels 沉积层的不同深浅
sedimentary bottom population 沉积海底种群
sedimentary boundary 沉积界限
sedimentary breccia 沉积角砾岩
sedimentary chlorophyll degradation products 沉淀叶绿素降解产物
sedimentary clay 沉积黏土;沉淀黏土
sedimentary coast 沉积海岸
sedimentary column 沉积柱
sedimentary complex 沉积杂岩层
sedimentary condition 沉积条件
sedimentary constant 沉淀常数;沉淀常量
sedimentary contact 沉积接触
sedimentary-contact coastline 沉积接触海岸线
sedimentary-contact coastline shoreline 沉积接触海岸线
sedimentary context 沉积相;沉积条件;沉积环境
sedimentary control 沉积物控制
sedimentary control agent 沉积物控制剂
sedimentary cover 沉积(覆)盖层
sedimentary cycle 堆积轮回;沉积循环;沉积旋回
sedimentary data 泥沙资料
sedimentary debris 沉积碎屑
sedimentary deposit 成层沉积;沉积矿床;沉积层
sedimentary deposit by colloidal chemistry 胶体化学沉积矿床
sedimentary deposit of lake facies 湖相沉积矿床
sedimentary derivation 沉积来源
sedimentary diastem 沉积暂停期
sedimentary differentiation 沉积分异作用
sedimentary dike 水成岩墙;沉积岩墙;沉积岩脉
sedimentary discontinuity 沉积结构面
sedimentary drift 沉积物迁移
sedimentary enclave 沉积岩包体
sedimentary environment 沉积环境
sedimentary exploration 沉积勘探
sedimentary fabric 沉积组构
sedimentary facies 沉积相(组)
sedimentary facies and litho-facies 沉积相与岩性相
sedimentary fault 沉积断层
sedimentary floor 沉积底板
sedimentary force 沉积力
sedimentary formation 沉积岩层;沉积建造
sedimentary framework 沉积建造;沉积格局;沉积格架
sedimentary gap 沉积缺失
sedimentary ga(u)ge 沉积计

sedimentary geochemistry 沉积地质化学
sedimentary gradient 沉积坡度
sedimentary groin 沉积丁坝
sedimentary injection 沉积贯入(作用)
sedimentary intercalation 沉积夹层
sedimentary interface 沉积界面;沉积间面
sedimentary intergranular porosity reservoir 沉积型孔隙热储
sedimentary intrusion 沉积侵入(作用)
sedimentary iron deposit 沉积铁矿床
sedimentary isopach map 沉积物来源图;沉积等厚线图
sedimentary kaolin 次生高岭土
sedimentary lag 沉积滞后
sedimentary laminae 沉积层纹
sedimentary layer 沉积层
sedimentary manganese deposit 沉积锰矿床
sedimentary mantle 沉积(覆)盖层
sedimentary markers 沉积标志
sedimentary material 沉淀物;沉积物
sedimentary material in suspension 悬浮中的沉淀物
sedimentary mechanics 沉积力学
sedimentary mechanism 沉积机制;沉积机理
sedimentary melange 沉积混杂体
sedimentary metallization 沉积成矿作用
sedimentary-metamorphosed pyrite deposit 沉积变质硫铁矿床
sedimentary movement 沉积运动
sedimentary ore-forming process 沉积成矿作用
sedimentary ore-forming process of colloid(al) chemistry 胶体化学沉积成矿作用
sedimentary organic material 沉积有机物
sedimentary organic matter 沉积有机质
sedimentary origin 沉积源;沉积成因
sedimentary outer arc 沉积外弧
sedimentary overlap 海侵超覆;沉积超覆
sedimentary parent materials 沉积母质
sedimentary particles in suspension 悬浮质中沉淀颗粒;悬浮物中的沉淀颗粒
sedimentary particle suspension 沉淀悬浮物
sedimentary peat 沉积泥炭
sedimentary petrography 沉积岩岩相类学;沉积岩相学
sedimentary petrologic province 沉积岩区
sedimentary petrologic provincial alternation 沉积岩交互出现区
sedimentary petrology 沉积(岩)岩石学
sedimentary phosphorus 沉积磷
sedimentary porosity 沉积孔隙度
sedimentary prism 沉积柱
sedimentary process 沉积作用;沉积过程
sedimentary product 风化产物
sedimentary profile 沉积剖面
sedimentary pyrite deposit in coal series 煤系沉积硫铁矿床
sedimentary quartzite 沉积石英岩
sedimentary rate 沉积率
sedimentary region 沉积区
sedimentary reservoir 沉积岩热储
sedimentary rhythm 沉积韵律
sedimentary ripple 沉积波纹;波痕
sedimentary rock 水成岩;堆积岩;沉积岩石;沉积岩
sedimentary rocks 沉积岩系;沉积岩类
sedimentary sampling 沉积物取样
sedimentary section 沉积剖面
sedimentary sequence 沉积序列;沉积顺序;沉积层序
sedimentary series 沉积层系
sedimentary setting 沉积背景
sedimentary soil 沉积土壤;沉积土
sedimentary specific weight 泥沙重率
sedimentary stratiform copper deposit 沉积层状铜矿床
sedimentary stratiform lead-zinc deposit 沉积层状铅锌矿床
sedimentary stratigraphic(al) map 沉积地层图
sedimentary stratigraphy 沉积地层学
sedimentary stratum 沉积层
sedimentary structure 沉积构造;沉积床结构
sedimentary succession 沉积层序
sedimentary system 沉积体系
sedimentary tank 沉积池;沉积槽
sedimentary tectonics 区域沉积构造;沉积构造作用;沉积构造学
sedimentary tester 沉积试验器
sedimentary texture 沉积结构
sedimentary time 沉积时间
sedimentary transport 沉积物运移;沉积物迁移;沉积物搬运
sedimentary transport curve 沉积物迁移曲线
sedimentary transverse transport 沉积物横向移动;沉积物横向迁移
sedimentary trap 沉积物捕留区;沉积圈闭
sedimentary trough 沉积盆地;沉淀池;沉淀槽
sedimentary tube 沉淀管
sedimentary tuff 沉积凝灰岩
sedimentary type 沉积类型
sedimentary veneer 覆盖层
sedimentary volcanism 沉积火山作用
sedimentary water 沉积水
sediment assessment standard 底泥评价标准
sedimentated dust 沉降灰尘
sedimentation 淤积;离心沉淀法;堆积作用;淀积;沉降;沉积作用;沉积(法);沉淀(作用);沉底
sedimentation accumulation rate 淤积速率
sedimentation agent 沉降剂;沉淀剂
sedimentational syngenesis 沉积同生作用
sedimentation analysis 沉降分析(法);沉降法粒度分析;沉积分析;沉淀分析研究;沉淀分析(法)
sedimentation apparatus 沉降器
sedimentation at dam vicinity 坝区泥沙淤积
sedimentation balance 沉降天平;沉积天平;沉积平衡
sedimentation balance method 沉降天平法
sedimentation basin 澄清池;沉沙池;沉积盆地;沉积池;沉淀池
sedimentation binding 沉淀物胶结
sedimentation boundary 沉降界面
sedimentation centre 沉积中心
sedimentation centrifuge 离心沉降器;离心沉降机
sedimentation chamber 降尘室;沉降室;沉淀室
sedimentation coefficient 沉降系数;沉淀系数
sedimentation compartment 沉淀室;沉淀室
sedimentation constant 沉降常数
sedimentation curve 沉淀曲线;沉淀曲线
sedimentation cycle 沉积循环;沉积旋回
sedimentation diameter 沉降粒径;沉积直径
sedimentation effect 沉积效应
sedimentation engineering 泥沙工程学
sedimentation equilibrium 沉降平衡;沉积平衡;沉淀平衡
sedimentation equilibrium method 沉降平衡法
sedimentation-equivalent particle 沉积等效颗粒
sedimentation experiment 沉淀实验
sedimentation factor 沉淀因数
sedimentation feature 沉积作用特征
sedimentation field flow fractionation 沉降场流动分级
sedimentation/flow-balancing tank 流量均衡—沉淀池
sedimentation force 沉淀力
sedimentation indicator 沉积指示器
sedimentation in discharge pipeline 排泥管内沉淀
sedimentation intensity 回淤强度
sedimentation mechanics 沉积力学
sedimentation mechanism 沉积机制;沉积机理;沉淀机理
sedimentation method 沉积法;沉降法
sedimentation migration 沉积作用迁移
sedimentation mode 沉积方式
sedimentation model carbonate slope 碳酸盐斜坡沉积模式
sedimentation model of epeiric sea 陆表海碳酸盐沉积模式
sedimentation model of organic reef 生物礁沉积模式
sedimentation model of pericontinental sea carbonate 陆缘海碳酸盐沉积模式
sedimentation model of shallow-sea carbonate 浅海碳酸盐沉积模式
sedimentation model of turbidity current 浊流沉积模式
sedimentation model of Wilson's carbonate standard facies belts 威尔逊碳酸盐标准相带沉积模式
sedimentation monitor 泥沙沉淀监测
sedimentation number 沉降值

sedimentation of reservoirs 水库沉积;水库淤积
sedimentation of suspension 悬浮物沉积
sedimentation-oxidation process 沉淀氧化法
sedimentation particle 沉降颗粒
sedimentation pattern 沉降式样;沉降谱
sedimentation period 沉淀时间
sedimentation pit 沉淀坑;沉降池
sedimentation plant 沉淀池;沉淀设备
sedimentation pollution control 底泥污染控制
Sedimentation Pollution Control Act 底泥污染控制法令
sedimentation pond 沉淀池
sedimentation potential 沉降势;沉积电位
sedimentation preventer 防沉剂
sedimentation process 沉降过程;沉淀过程;沉淀(池)处理
sedimentation rate 回淤强度;淤积速率;沉降(速)率;沉降率;沉积(速)率;沉淀速率
sedimentation rate test 沉降速率测试
sedimentation reaction 沉降反应
sedimentation reservoir 沉沙水库
sedimentation section 沉淀部分
sedimentation size analysis 沉积量分析
sedimentation sizer 沉降法测粒度
sedimentation sizing method 沉降分级法
sedimentation soil 冲积土(壤);沉积土壤
sedimentation stability 沉降稳定性
sedimentation/storm-sewage tank 雨水污水沉淀池
sedimentation survey 淤积调查;淤积测量
sedimentation tank 澄清池;沉沙池;沉降槽,沉积池;沉淀池;沉淀槽
sedimentation techniques 沉降技术
sedimentation test 静沉试验;沉淀试验
sedimentation testing 沉淀试验
sedimentation theory 沉淀理论
sedimentation time 沉降时间;沉淀时间
sedimentation trap 沉淀井
sedimentation treatment 沉淀处理
sedimentation treatment of wastewater 废水沉降处理
sedimentation trough 沉淀海槽
sedimentation-type centrifuge 沉降式离心机
sedimentation unit 沉积单元
sedimentation velocity 泥沙沉速;泥沙沉降速度;沉降速度;沉淀速度;沉淀速度
sedimentation velocity method 沉降速度法
sedimentation volume 沉降容积
sedimentation water 沉积水
sedimentation with coagulation 混凝沉淀作用
sedimentation zone 沉积带;沉淀区
sedimentator 分离器;沉淀器
sediment balance 砂量平衡
sediment bar 河漫滩;边滩
sediment barrage 沉积堰
sediment barrier 拦沙坝;拦沙设施;沙挡;防沙堤
sediment basin 淤沙坑;沉沙池;沉淀池
sediment-bearing river 夹沙河流
sediment-bearing stream 挟沙河流;夹沙河流
sediment bed 泥沙河床
sediment bed load 沉积泥沙
sediment binder 胶结沉积物的生物
sediment blanket 泥沙(覆)盖层
sediment-borne 泥沙携带的
sediment bottom sluice 排沙底孔
sediment boundary layer 沉积物边界层
sediment bowl 澄清池;沉淀器;沉淀池;沉淀杯
sediment box 沉淀箱
sediment bulb 取沙袋
sediment bulk 泥沙采样器;沉淀器
sediment bypassing 泥沙输移
sediment-carrying capacity 挟沙量;泥沙挟带能力;挟带泥沙能力;挟沙能力;夹沙能力;泥沙携带能力
sediment-carrying capacity of flow 水流挟沙能力
sediment-carrying capacity of river 河流挟沙能力
sediment-carrying capacity per unit width 单宽挟沙能力
sediment-carrying estuary 夹砂河口;含泥沙河口;挟沙河口;夹沙河口
sediment carrying power 挟带(泥沙)能力
sediment-carrying river 挟沙河流;夹沙河流
sediment-carrying stream 夹沙河流
sediment catcher 泥沙取样器;泥沙拦截装置;沉淀物取样器
sediment channel 集沙渠(道)
sediment charge 沉积物负荷量;泥沙含量;含沙量
sediment charge scale 含沙量范围
sediment classification 冲积物分类
sediment coagulation 泥沙絮凝(作用)
sediment-collecting canal 集沙渠(道)
sediment collector 沉积收集器
sediment community oxygen consumption 底泥群耗氧量
sediment compaction 沉积物压实
sediment composition 底泥成分;沉积物组分
sediment concentration 泥沙浓度;泥沙含量;含沙量;底泥浓度;沉积物浓度
sediment concentration at a measuring point 测点含沙量
sediment concentration hydrograph 含沙量过程线
sediment concentration of pollutant 污染底泥浓度
sediment concentration study 含沙量研究
sediment consistency 沉积物稠度
sediment-contaminant transport model 底泥—污染物输移模型
sediment contamination 底泥污染
sediment content 含沙量;泥沙含量;沉淀物含量
sediment content scale 含沙量范围
sediment contributing area 产沙区
sediment control 泥沙控制;沙防;防沙
sediment control dam 拦沙坝;防沙坝
sediment control measure 减淤措施
sediment control structure 泥沙控制设施;防沙工程
sediment control works 泥沙控制设施
sediment core 沉积岩芯
sediment corer 海底沉积物取芯管
sediment delivery percentage 排沙率
sediment delivery process 泥沙输移过程;输沙过程
sediment delivery rate 排沙率;输沙率
sediment delivery ratio 输沙比
sediment density 泥沙密度
sediment deposit 泥沙沉积;泥沙淤积
sediment-depositing side of bend 河曲淤积面;河曲凸岸;河曲的淤积侧
sediment deposition 泥沙沉积;沉积泥沙
sediment detention basin 沉沙池
sediment determination 泥沙测定
sediment diffusion 泥沙扩散;沉沙扩散;沉积物扩散;沉积物的扩散作用
sediment discharge 固体径流;泥沙流量;土沙流量;输沙量
sediment discharge capacity 排沙能力
sediment discharge concentration 含沙量
sediment discharge curve 水位输沙量关系曲线;输沙量曲线;泥沙—流量曲线
sediment discharge facility 输沙设施
sediment discharge intensity 输沙强度
sediment discharge measurement 输沙量测定
sediment discharge of river 河流输沙量;河道输沙量
sediment discharge of stream 河流输沙量
sediment discharge per unit width 单宽输沙量
sediment discharge rate 输沙率
sediment-discharge rating 输沙量—流量关系
sediment-discharge rating curve 输沙量—流量关系曲线
sediment dislodging force 泥沙推移力
sediment distribution 泥沙分布
sediment distribution in reservoir 水库沉积(物)
sediment distributor 产沙源;产沙区
sediment diversion 导沙;分沙
sediment diversion ratio 分沙比
sediment diverter 泥沙转移设施;泥沙分流设施
sediment division 沉积物分区
sediment drift 泥沙漂移
sedimented particle 沉降颗粒
sediment ejection 排沙(又称排沙);放淤
sediment ejector 截沙槽(又称截砂槽);排沙装置;排沙设备
sediment encroachment 泥沙侵入;泥沙堆积
sediment engineering 泥沙工程学
sediment enrichment factor 底泥富集系数
sediment-entraining 携带泥沙的
sediment entrainment 携带泥沙
sediment equivalent 含沙当量
sediment equivalent method 含沙当量法
sediment erosion 泥沙侵蚀
sediment escape 冲沙泄水道
sediment extractable phosphorus 底泥可萃取磷
sediment fabric 沉积物结构
sediment-filled 沉积物充填的
sediment flocculation 沉积絮凝(作用)
sediment flushing at low-water level 低水位泥沙冲刷
sediment flushing bay 冲沙段
sediment flushing by contraction works 束水攻沙
sediment flushing efficiency 冲沙效率
sediment flushing outlet 排沙孔
sediment flux 输沙率;输沙量
sediment flux rate 输沙率
sediment-free river 清水河(流);无泥沙河流
sediment-free stream 清水河(流);无砂河流;无泥沙河流
sediment-free water 无沉积物水
sediment grade size 沉积物分级尺寸
sediment grading 泥沙颗粒级配
sediment grading curve 泥沙级配曲线
sediment grain 泥沙颗粒
sediment grain analysis 泥沙颗粒分析
sediment grain analyzer 泥沙颗粒分析器
sediment grain size 泥沙粒径
sediment granulometry 泥沙颗粒分析
sediment gravity flows 沉积物重力流
sediment humus 沉积腐殖质
sediment hydraulics 沉淀水力学;泥沙水力学
sediment hydrograph 泥沙流量过程线;水位含沙量关系过程线;输沙量过程线;泥沙流速仪
sediment incipient motion 泥沙起动
sediment incrustation 积垢;沉积水垢
sediment index 泥沙指数
sediment-induced 泥沙引起的
sediment induced deformation 沉积物诱发变形作用
sediment inflow 来沙
sediment in laden flow 夹沙水流
sediment in suspension 悬移质输沙;悬移质泥沙;悬浮质泥沙
sediment intrusion 泥沙侵入
sediment-laden 含沙量大的;挟带大量泥沙的
sediment-laden flow 夹沙水流;夹沙水流;挟沙水流
sediment-laden observation 含沙量观测
sediment-laden observation station 含沙量观测站
sediment-laden river 挟沙河流;夹沙水流;多沙河流;多泥河流
sediment-laden stream 挟沙港道;挟沙河流;夹沙河流;多沙河流
sediment-laden tidal flow 挟沙潮流
sediment-laden water 挟沙水;夹砂水;夹沙水
sediment load 固体径流;泥沙负荷;河流泥沙;推移质;底质;沉积负荷
sediment-load sample 泥沙试样
sediment-load sampling 泥沙取样
sediment longitudinal transport 沉积物纵向移动;沉积物纵向迁移
sediment measurement 泥沙测验
sediment mixture 泥沙混合物
sediment mobilization 沉积物移动
sediment model 泥沙模型
sediment model test 泥沙模型试验
sediment monitoring 底泥监测
sediment motion 泥沙运动
sediment motion model 泥沙运动模型
sediment movement 泥沙运动
sediment movement model 泥沙运动模型
sedimento-eustacy 沉积性海面升降运动
sedimento-eustatism 沉积性海面升降运动;沉积海面变动
sediment of coastal zone 海岸带沉积(物)
sedimentogeneous rock 沉积岩
sedimentogeneous rocks 沉积岩类
sedimentogenesis 沉积物形成作用
sedimentograph 粒度分布测定仪;沉降测定仪;沉降测定器;沉淀仪
sedimentography 沉积岩相学;沉积(岩)岩石学
sedimentology 泥沙学;沉积学;沉积(岩)岩石学
sedimento-metamorphic deposit 沉积变质矿床
sedimento-metamorphic manganese deposit 沉变质锰矿床
sedimentometer 沉沙仪;沉降速度计;沉积计;沉积测定仪
sedimentometry 沉淀法;沉淀学

sediment-oriented 与泥沙有关的
sediment outflow 排沙（又称排砂）；输沙量；出沙；出库泥沙量
sediment overlaying water 底泥上覆水
sediment oxygen demand rate 底泥需氧率
sediment pan 沉淀桶；沉积盘
sediment particle 泥沙颗粒
sediment particle analysis 泥沙颗粒分析
sediment particle analyzer 泥沙颗粒分析器
sediment particle-diameter 沉降粒径
sediment-phosphorus-nitrogen model 底泥—磷—氮模型
sediment pipe 钻屑收集筒；钻泥管；取粉管；沉沙管
sediment pollution 底泥污染
sediment pollution control 底泥污染控制
sediment pollution index 底泥污染指数
sediment pool 沉沙池；淤沙库容
sediment producing 产生泥沙的
sediment-production rate 单位面积输沙率；产沙率
sediment profile 泥沙淤积（纵）剖面图；泥沙的垂直分布
sediment quality 底泥质量
sediment quality-based waste load allocation 基于废负荷分配的底泥质量
sediment quality criterion 底泥质量标准
sediment quality guideline 底泥质量标准
sediment quantity 沉积物量
sediment range 淤积范围；泥沙分布范围；测泥沙断面；测含沙量断面
sediment rate 输沙率；沉积率
sediment-rating curve 泥沙率定曲线；输沙量曲线；输沙—流量关系曲线
sediment removal 排沙清淤
sediment resuspension 底泥再悬浮
sediment retention works 拦沙设施；拦沙建筑物
sediment ripple 沙纹；沙痕；沙波；小沙波
sediment routing 泥沙演进计算；泥沙推算
sediment runoff 固体径流；泥沙径流；输沙量；沉积物流量
sediment runoff curve 固体径流曲线；水位输沙量关系曲线
sediment runoff modulus 输沙模量
sediments 沉渣
sediment sample 泥沙试样；沙样；底泥样本
sediment sampler 泥沙取样器；泥沙采样器；底泥采样器；沉积物采样器；采泥器
sediment sampling 泥沙取样；底泥采样
sediment sampling equipment 泥沙采样设备；泥沙采样器
sediment separator 沉渣器；沉淀分离机
sediment siltation basin 沉沙池
sediment silting basin 沉沙池
sediment size 泥沙粒度
sediment size analysis 泥沙颗粒分析
sediment size analysis equipment 泥沙颗粒分析设备
sediment size analyzer 泥沙颗粒分析器
sediment size fraction 泥沙粒级
sediment slide 沉积物场滑；沉积物滑动
sediment sorting mechanism 泥沙分选机理
sediment source 泥沙来源
sediment-source area 泥沙源区
sediment space 沉淀空间
sediment spatial concentration 单位水体中含沙量
sediment specimen 沙样
sediment sphericity 泥沙球面比
sediment station 泥沙站；泥沙观测站
sediments thickness contour map 沉积物厚度等值线图
sediment stirred up by wind wave 风浪掀沙
sediment stopping velocity 止动流速
sediment storage basin 蓄沙池
sediment storage capacity 储沙能力
sediment storage capacity of reservoir 水库储沙能力
sediment storage dam 拦沙坝
sediment strength 沉积物强度
sediment supply 泥沙来源；泥沙补给
sediment survey 泥沙测量
sediment suspension 泥沙悬移
sediment tank 沉淀器
sediment thickness 沉积层厚度
sediment toxicity 底泥毒性
sediment tracer 示踪沙

sediment transfer 泥沙转移
sediment transfer coefficient 泥沙输送系数
sediment transmission 泥沙输送；输沙
sediment transport 沉积物迁移；泥沙移动；泥沙输移；输沙；底沙运移；沉积物运移
sediment transportation 泥沙移动；泥沙输移
sediment transport balance study 输沙破坏研究
sediment transport budget 泥沙输移量计算
sediment transport capacity 挟运能力；泥沙输移能力；输沙能力
sediment transport capacity of flow 水流挟沙能力
sediment transport competence 挟运能力；泥沙输移能力；输沙能力
sediment transport competency 挟运能力；泥沙输移能力；输沙能力；底泥运移能力
sediment transport concentration 输沙率；输沙浓度；沉渣载运浓度
sediment transport curve 泥沙输移曲线
sediment transporter 沉积物输送者
sediment transport formula 输沙公式
sediment transport process 泥沙输移过程；输沙过程
sediment transport rate 泥沙输移率；输沙率
sediment transport through pipeline 管道输沙
sediment trap 沉沙弯管；沉沙坑；拦沙坑；拦泥坑；泥沙捕集装置；除沙弯管；沉沙池；沉积阱；沉淀物捕集器；沉淀阱
sediment trapping 拦沙（又称拦砂）
sediment tube 钻屑收集筒；钻泥管；取粉管；沉沙管
sediment tuff 沉积盆凝灰岩
sediment velocity 沉积物的波速
sediment viscosity 沉积物黏性
sediment-water interface 底泥—水界面；沉积物—水界面
sediment-water mixture with hyperconcentration 高浓度浑水
sediment-water system 沉积物水系统
sediment yield 固体径流量；泥沙输出量；泥沙量；河流来沙；产沙量
sediment yield of river basin 流域产沙量
sedimetry 沉降滴定
SE diwa region 东南地洼区
sedovite 铀钼矿
sed-pyroclastic agglomerate 沉火山集块岩
sed-pyroclastic agglomerate breccia 沉火山集块角砾岩
sed-pyroclastic breccia 沉火山角砾岩
sed-pyroclastic breccia tuff 沉火山角砾凝灰岩
sed-pyroclastic rock 沉火山碎屑岩
se-duct 斯式预制暖气道
Seduct heater 斯式加热器（一种安装在墙上的预制暖气道）
Seebeck coefficient 塞贝克系数
Seebeck effect 塞贝克效应
Seebeck thermoelectric(al) effect 塞贝克效应
seebenite 董长角石
seed 小气泡；釉中残余颗粒；种子；加入晶种或晶核；凝聚粗颗粒；返粗颗粒
seedage 播种法
seed attachment 播草籽附加装置
seed bag 亚麻填料
seedbed 苗床
seedbed frame 苗床保护框
seedbed leveler 平苗床器
seedbed preparation 播前整地
seedbed stakes 苗床支柱
seedbed weeder 苗床除草机
seed blast 喷射清理机
seed blasting 软粒喷砂
seed boot 播种开沟器
seed broadcaster 撒播机
seed capital 原始资本
seed cell 槽眼
seed collecting 采种
seed collector 草籽收集器
seed colter 插种开沟器
seed core 强化堆芯
seed core reactor 强化堆芯反应堆
seed cotton 原棉
seed crystal 籽晶；种晶；晶种
seed dresser 拌种机
seed-dressing 拌种
seed-dressing agent 拌种剂
seed drill 条播机

seed drill plough 条播犁
seeded pasture 人工牧场；人工草原
seeded reverse osmosis 植草反渗透
seeded slope 植草边坡
seeded strip（道路的）植草带；绿化分隔带
seeded surface 植草面；播种面积
seed emulsion 种子乳液
seed extraction 脱粒
seed fat 植物油
seed-free time 气泡消失的时间
seed gathering machine 采种机
seed harvester 草籽收获机
seed huller 去皮机
seed husking establishment 脱粒装置
seediness 起粒；起颗粒；返粗
seeding 幼苗；引晶技术；植草；接种；加晶种；畦植；播种造林；播种（法）
seeding action 放入晶种
seeding age 苗龄
seeding and adult 苗株和成株
seeding bed 秧田；苗床
seeding by aircraft 飞机播种
seeding crop 实生林
seeding cutting 初次采伐
seeding growth 野生苗；天然幼树
seeding in nutritive cup 营养钵育苗
seeding lawn 播种草坪
seeding machine 播种机
seeding material 种料
seeding method 晶种法
seeding nursery 苗圃
seeding of embankment 路堤植草
seeding on unfallowing field 茬地播种
seeding polymerization 接种聚合（作用）
seeding potential 发泡趋势
seeding rate 播种率
seeding requirement 播种要求条件
seeding time 播种时间；播种期
seed lac 粒胶；粗紫胶；粗（制）虫胶
seed level 播深
seedling 籽苗；苗木
seedling clearing 实生树植区
seedling coppice 实生矮林
seedling inarching 幼苗靠接
seedling nursery 苗圃
seedling pasture 青苗牧地
seedling-plant 实生树
seedling planting machine 播苗机
seedlings and small trees 树苗和小树
seedling seed orchard 实生苗种子园
seedlings of cereal crops 禾苗
seedling strip 实生树林带
seedling stump 实生树桩
seedling tree 实生树
seedling with soil 带土小苗
seed machine 拌种机
seed mixer 拌种器
seed money 项目实施费用；种子基金；原始资金
seed money loans 项目实施贷款；种子基金贷款
seed orchard 采种园
seed pan 谷粒盘
seed pedicel 种柄
seed picking machine 采种机
seed placing device 定点播种装置
seed plot 苗圃；苗床
seed press wheel 埋种压土轮
seed production area 采种区
seed quantity 播种量
seed shedding 下种；天然下种造林法
seed sieve 筛粉筛
seed size 晶种粒度
seeds of rotary table 转盘挡数
seed spacing 株距；点线播种；播种间距
seed spout 输种管
seed stalk 种柄
seed station 蚕种场
seed storage 种子库
seed stripper 草籽采收机
seed treatment 拌种
seedy 脏刷子引起漆膜斑点；起砂；起粒；有颗粒的
seed yarn 粗节花线
seedy glass 有小气泡玻璃；灰泡玻璃
seedy wool 草籽毛
see figure 见图

see-in-dark rangefinder 夜视测距仪
seeing 星象宁静度;增导注入;视宁度;视力
seeing disc 视影圆面;视宁圆面
seeing distance 视距;视线距离
seeing disturbance 视宁度扰动
seeing image 视影;视宁像
seek 寻找;自导引;探寻;查找
seek area 查找区域
seekay wax 氯化萘蜡状物(房屋建筑中防火橡胶用的)
seek command 查找命令
seeker 自动导引的头部;探寻器
seeker error 寻的系统误差
seek error 查找误差
seeking 搜寻
seeking range 自导引系统作用距离
seek key 检索区;查找键
seek redress 要求赔偿
seek time 查找时间
seelandian(stage) 西兰阶
seelandite 水镁铝矾
Seelastic (一种嵌填框架和连接建筑材料的)塑性混合物
seeligerite 氯碘铅石
seep 水陆两用轻型汽车;渗(出)
seepage 渗水量;渗滤量;渗流;渗出(量)
seepage action of groundwater 地下水渗流
seepage analog(ue) resistance network 渗流模拟电阻网络
seepage anomaly 渗湿异常
seepage apron 防渗护坦;防渗护层
seepage area 渗漏区;渗流区;渗流面积;(井的)渗水面积
seepage area of well 井的渗流面积
seepage basin 渗透池
seepage bed 滤床;渗漏床;渗滤床;渗流河床
seepage belt 渗水带
seepage biological bed 渗滤生物床
seepage boil 渗漏涌土;渗漏涌气
seepage calculation 渗透计算
seepage calculation model 渗流计算模型
seepage coefficient 渗透系数;渗流系数
seepage continuity equation 渗流连续性方程
seepage control 渗流控制
seepage corrosion 渗流腐蚀
seepage course 渗漏层
seepage deformation 渗透变形
seepage deformation of soil mass 土体渗透变形
seepage discharge 渗透流量;渗漏流量;渗流(流)量;渗出量
seepage discharge formula in river and canal 河渠渗漏量公式
seepage distance 渗透距离
seepage ditch 渗(水)沟
seepage erosion 渗流侵蚀
seepage exit 渗透出口;渗流出口
seepage face 浸润面;渗漏面;渗出面
seepage failure 渗透破坏;渗流破坏;渗流破坏
seepage flow 潜水径流;渗入流量;渗流流量;渗流
seepage flow through dike body 坝体渗流
seepage force 渗透(压)力;渗流(压)力
seepage force per unit volume 动水力
seepage from reservoir 水库渗漏(量)
seepage ga(u)ge 渗流计
seepage gutter 渗水沟
seepage halo 渗湿晕;渗出晕
seepage head 渗透水头;渗流水头
seepage hole 渗水孔
seepage interception 截渗;渗流截断;渗截
seepage investigation 渗漏调查
seepage lake 渗漏湖
seepage law 渗透定律
seepage length 渗透长度
seepage line 地下升水线;浸润线;渗润线;渗漏线;渗流线
seepage liquefaction 渗透液化
seepage loss 渗透性漏失;渗流损失;渗漏损失;渗出损失
seepage loss water 渗流损失
seepage material 透水材料
seepage measurement 渗流测定
seepage meter 渗流计;渗流仪;渗透计;测渗仪;测渗计
seepage of stream 河川渗漏;河川渗流

seepage of tank 水池渗漏
seepage of water 水的渗入
seepage of well 井的渗流
seepage outflow 渗漏出流量
seepage parabola 渗流抛物线
seepage path 渗透途径;渗流路径;渗流通道;渗径
seepage pipe 渗流管
seepage pit 渗水坑;渗水井;渗流坑;渗坑
seepage potential 渗透势
seepage pressure 渗透压(力);渗透水压;渗滤压力;渗流压力
seepage prevention 防渗
seepage prevention of river and ditch 河渠防渗
seepage proof curtain 防渗帷幕
seepage quantity 渗流量
seepage rate 渗透率;渗水速率;渗流(速)率
seepage rating 渗透率
seepage recharge of stream 河水入渗灌注
seepage-reducing method 减少渗水量方法
seepage reflection 渗透回流;渗漏回流
seepage reflux(ion) 渗透回流;渗漏回流
seepage resistance strength of soil 土的抗渗强度
seepage river 亏水河;渗漏河
seepage scour 渗流淘刷;渗流冲刷
seepage speed 渗透速度
seepage spring 渗水泉;渗流泉
seepage stream 亏水河;渗漏河
seepage stress 渗流应力
seepage structure 渗水建筑物
seepage surface 浸润面;渗流面
seepage tank 渗漏槽
seepage test 渗透试验;渗流试验
seepage theory 渗流理论
seepage through valley flanks 谷侧渗流
seepage tracking 渗径
seepage trench 渗流沟;渗水管沟
seepage trench disposal 渗出沟处置法
seepage tube 渗流管
seepage tunnel 引水隧洞;引水隧道
seepage uplift 渗透扬压力;渗透上压力;渗透浮力
seepage velocity 渗透速度;渗水速度;渗流速度;渗流流速
seepage volume 渗透水量
seepage warning device 渗漏警告装置
seepage water 渗流水;渗漏水;渗流水;渗出水
seepage water checking point 渗流水抽点检验
seepage water quality 渗滤水水质
seepage well 渗(透)井
seepage zone 渗透区;渗水带;渗漏区;渗流区
see page… 见……页
seep hole 冒水孔;渗水孔
seep in 漏入;渗入
seeping 渗入作用
seeping discharge 渗出水
seeping soil roadbed 渗水土路基
seep oil 渗出石油
seepproof screen 止水墙
seep water 渗流水
seepy 透水的;漏水的
seepy land 渗漏水地
seepy material 透水材料
seepy meepy 透水物料
seersucker 泡泡纱;绉条纹薄织物
seesaw 跷跷板;前后运动
see-saw amplifier 反相放大器;单位增益电压倒相放大器
see-saw circuit 反相放大电路;反相电路;板极输出器电路
see-saw motion 摇摆运动;上下运动
seethe 煮沸
see this drawing 见本图
see through 透明的
see-through building 透明的建筑物
see-through mask 彩色掩模
see-through mirror 窥视镜;监视镜;单向透明(玻璃)镜;安全发射镜
see-through visibility 明察的可见度;明察的视野
see-through window 透明的窗
segelerite 水磷铁钙镁石
Seger cone 塞格尔(测温)锥
Seger formula 塞格尔公式
Seger porcelain 塞格尔瓷
Seger's green 塞格尔绿(耐1050℃的陶瓷彩料)
Seger's rule 塞格尔规则

seggar 火泥;退火箱
seggar clay 匣钵土;耐火黏土
segistration 重叠
segment 整流子片;管片;弓形;节(片);环节;核算单位;片段;扇形体;段;程序段;部分;分节
segment accounting 责任单位会计;分部会计
segment acknowledgement 段确认
segment address 段地址
segment addressing 段编址
segment adjuster jack 管片调整器千斤顶
segmental 节段性;分节(性)的
segmental apparatus 节段装置;环节器
segmental arc 弧(形)段
segmental arch 弓(形)拱;弧形券;缺圆拱;平圆拱;扇形拱
segmental arch culvert 弧形拱涵洞;弓形拱涵;弓形拱洞
segmental arch girder 弧形拱大梁
segmental arch girder roof 弧形拱大梁屋顶
segmental arch system 梁间拱架系统;梁间架拱的梁式楼板系统
segmental arch window 弧形拱窗
segmental averaging 分段平均
segmental baffle 弓形折流板;缺圆挡板
segmental barrel vault 弧形筒穹顶
segmental beam 弓形梁
segmental bearing 弓形轴承;分片瓦轴承;轴瓦块轴承
segmental billet 弓形棒嵌饰;(圆棍段相隔交错的)花饰线脚
segmental bowstring girder 多边形弓弦梁
segmental box girder bridge 分段浇筑箱梁桥;分段拼装箱梁桥;节段箱梁桥
segmental bridge 分段拼装式桥
segmental ceiling 弓形顶棚
segmental chip 节状切屑;挤裂切屑
segmental column 节柱
segmental conductor 弓形截面导线;扇形导线
segmental construction 分割成弓形构造;分段构造;预制(节)段拼装施工法;分段施工
segmental construction method 节段施工法
segmental construction of precast concrete sections 分段预制(拼装)法
segmental coping of a wall 分段的墙压顶
segmental core disk 扇形片铁芯
segmental curve 弓形曲线
segmental cylindrical shell 多边圆柱形薄壳
segmental data 零碎的辅助材料
segmental die 可拆模;拼合式模具
segmental distribution 叶段分布
segmental file 分段文件
segmental fire bar 弓形炉条
segmental gate 扇形闸门
segmental gear 扇形齿轮
segmental girder 弓形大梁
segmental Gothic arch 哥特式弧形拱
segmental grinding ring 扇形磨盘
segmental gully tile with fancy rim 花边瓦
segmental iron 分节铁件
segmental jet 分段喷射
segmental lamination 扇形迭片
segmentally welded (monolayered) cylinder 瓦片式圆筒
segmental member 组合构架;拼装构件;组合构件;预应力拼装构件
segmental mo(u)ld 组合模(具)
segmental nozzle group control 喷嘴组调节
segmental orifice plate 缺圆孔板;扇形孔板
segmental pad (轴承的)扇形瓦
segmental pattern 节段型
segmental pattern plate 组合模板
segmental pediment 弓形的三角山墙;弓形的三角楣饰
segmental pile 拼接桩;分节桩
segmental pipe bend 弓形管弯头
segmental plate 扇形板;弓形板;体节板
segmental plethysmography 节段体积描记法
segmental pointed arch 弓形尖拱
segmental post-tensioned bridge 分节后张法预应力桥
segmental post-tensioned concrete bridge 节段式后张法预应力桥
segmental precast pile 分段预制桩
segmental punching 扇形冲片

segmental rack 扇形齿轮
segmental reflex 区域反射;分节反射
segmental resection 分段切除术
segmental-rim rotor 扇形片磁轭转子(水轮发电机)
segmental roof truss 折线形屋架
segmental rotor 扇形转子
segmental saw 镶齿圆锯
segmental shell 弓形壳
segmental shutter 扇形快门
segmental slab 弓形板
segmental sluice gate 弧形冲刷闸门;弧形泄水闸门;扇形闸门
segmental stamping 扇形冲片
segmental structure 分段结构
segmental support 弓形架
segmental tile 圆弧瓦;弓形瓦;版瓦
segmental timber arch 短木拱
segmental timbering 装配式支架;框内加强支撑
segmental truss 弓形桁架
segmental truss bridge 弓形桁架桥
segmental type 组合式;分装式
segmental type cellular cofferdam 弧形格形围堰
segmental valve 扇形阀(门)
segmental vault 弓形穹隆
segmental venter 腹面
segmental wheel 拼合轮;扇形砂轮
segment analysis 分部分析
segmentary 分节(性)的
segmentation 分裂;整流子片;节段法;区段;段式调度;段式;程序分段;分节(现象);分割;分段(法)
segmentation caving method 分段崩落法
segmentation error trap 分段误差收集
segmentation hardware 段式硬件
segmentation mechanism 分段机构
segmentation movement 分节运动
segmentation nucleus 分裂核
segmentation of fixed items principle 固定项目划分原则
segmentation of sources and utilization of capital 划分资金来源用途
segmentation overlay 分段重叠;分段覆盖
segmentation paging system 段页式
segmentation register 分段寄存器
segmentation register table 分段寄存器表
segmentation sphere 分裂球
segment attribute 段属性;图块属性;区段特征
segment attribute field 区段特征字段
segment base address 区段基地址;段基地址
segment bearing 多片瓦轴承;轴瓦块轴承
segment blasting 分段爆破
segment blasting method 分段爆破法
segment block 弧形砌块;铸管模
segment call 段调入
segment contribution 分部贡献额
segment conveyer 分段输送机
segment core 扇形铁芯;分段铁芯
segment cover 扇形盖
segment data 段数据;分段数据
segment data buffer 数据段缓冲器
segment decoder 分段译码器
segment definition card 段定义卡
segment delivery 分批交货
segment density of pumping test 抽水试验段密度
segment descriptor 段描述符
segment descriptor word 段描述符字
segment die 组合模(具);拼合模
segment display 节段显示;段式显示
segment distance 段距
segment driver 节段驱动;节段激励器;段驱动器
segmentectomy 分段切除术
segmented capillaries 断开的毛细孔
segmented capillary 分裂的毛细管
segmented circular groove cracking test 分块环形槽热裂纹试验;分割环形槽热裂试验
segmented column 节柱
segmented-core single-mode fiber 分层芯单模光纤
segmented curve 分段曲线
segmented cylindrical rod 分段圆棒
segmented dictionary 分段词典
segmented distributor 分段分配器
segmented encoding law 折线编码律
segmented glass laser 分节玻璃激光器

segmented injector 分段式喷嘴
segmented insert 分段式衬套
segmented machine 区段式计算机
segmented mirror 分节镜
segmented mirror telescope 拼合镜面望远镜
segmented mode 区段方式;段方式
segmented pattern 多模模板
segmented plate 分段板
segmented processor 区段式处理机
segmented program(me) 分段程序
segmented pumping 分段取水
segmented pumping test 分段抽水试验
segmented-rod 扇形棒
segmented roll 分割式碾路机
segmented roller 分段滚压机
segmented rotor brake 扇块动子制动器
segmented sampling 分段采样
segmented sampling method 分段采样法
segmented secondary reluctance motor 次级分段的磁阻电动机
segmented spring ring 扇形弹簧圈
segmented steel-wheel roller 分割式钢轮压路机;分段式钢轮路碾
segmented thermocouple 扇形热电偶
segmented transport 分段运输
segmented wheel compactor 链式压实机
segment electrode 段电极
segment electrode transistor 分段电极晶体管
segment encode 段编码
segment entry 区段入口
segment erection 分段架设;分段安装
segment erector 砌块举重点;分段架设机;分段安装机;管片拼装架;管片安装机
segment erector guide roller 管片安装机导向辊
segment evaluation 分部业绩评价
segment fault 段失败;段故障;分段差错
segment-feed loop 分段馈电环
segment gate 弧形(闸)门;弧形堰;扇形闸门;弓形闸门
segment gear luffing 扇形齿轮变幅
segment grinding wheel 镶片磨轮
segment guide bearing 分块瓦导轴承
segment hanger 管片吊钩
segment hauler 分段运输机
segment head 圆弧段拱的门;圆弧形门楣
segment-headed counterfort 拱顶扶垛
segment identifier 线段标识符
segment information 分部信息
segmenting of basic information unit 基本信息单位的分段
segment joint 管片接头
segment length 段长(度)
segment length violation flag 区段长度违章标志
segment lifter 管片起重机
segment limit 区段界限;段界;段范围
segment lining 管片衬砌
segment-long spacing crystallites 长节段晶粒
segment management 段式管理
segment management policy 区段管理策略
segment map table 段变换表
segment margin 分部毛利
segment mark 区段标记;段标记;段标
segment market 细分市场
segment mica 整流子片用云母片
segment mold 分块模;组合模
segment name 段名
segment notching press 整流子片切槽专用压力机
segment nozzle 组合式喷管
segment number 线段编号;区段号;段(编)号
segment number of pump tube 泵管节数
segment number of segmentation 分割段数
segment occurrence 段具体值;段出现
segment of a curve 弧段
segment of a cylinder 柱面母线
segment of a parabola 抛物线的一段
segment of a sphere 球截体
segment of blading 透平叶片
segment of circle 弓形
segment of counterweight 平衡重的一部分
segment of cylinder 圆柱体部分
segment of flight strip 航线段
segment offset address 区段偏移量地址
segment of level(l)ing 水准测段
segment of survey 测段

segment output 节段输出;段输出
segment pitch 整流子片距;片距
segment placer 分段摊铺工;分段摊铺机;分段浇筑工;分段浇筑机
segment prefabrication plant 管片预制厂
segment priority 图段优先级
segment production line 片断生产线
segment program(me) 程序段
segment protection 区段保护
segment pushing jack 推进千斤顶
segment register 段寄存器
segment reporting 分部报告
segment results 分部成果
segment ring 分节环;管片环
segment roller 分隔环式镇压器
segment saw 弧形锯;节锯;扇锯
segments bound area 线段轮廓面积
segment search argument 段搜索自变量;段检索变量
segment selection 段选择
segment set 线段组
segment sharing 段共享
segment shutter 弓形快门
segment size 段长度
segment store 分段存储
segment switch 分频开关
segment synchronization 段同步
segment sync set 段同步置定
segment table 段表
segment table entry 段表入口;段表项
segment table origin 段表起始地址
segment the market 区隔市场
segment tool 圆角切刀
segment transformation 图段变换;片段变换
segment translation 段转换
segment translation exception 段转换异常
segment type 段(类)型
segment type code 段类型码
segment-type water-cooled cupola 分段水冷冲天炉
segmentum basale posterius 后基底段
segmentum laterale 外侧段
segment valve 弓形阀;弧形阀门
segment wagon 管片台车
segment weir 弓形堰;弧形堰
segment width 管片宽度
segment without contact 无切线段
segregability (混凝土粗颗粒的)分离能力;离析性
segregant 分离子
segregate 分凝;分开
segregate ballast tank 分隔压载水舱
segregated aggregate stockpiling 离析出的集料混凝土(指混凝土中)
segregated ballast tank 专用压载舱
segregated ballast water tank system 专用压载水舱系统
segregated collection system 分流集水系统
segregated concrete 离析混凝土
segregated frozen ground 岛状冻土
segregated goods 分隔商品
segregated ice 离析冰;分凝冰
segregated oil 析得干性油
segregated pool 离析池
segregated use 分区专用
segregated vein 分凝脉;分结脉
segregating 离析;划分;分开
segregating concrete 离析混凝土
segregating liner plate 分级衬板
segregating unit 隔离器;分离装置;分离机;分类设备
segregation 隔票;隔分;离析(现)象;熔析;偏析;分凝(作用);分结作用;分隔
segregational load 分离负荷
segregation banding 分凝条带;分结条带
segregation distorter 分离畸变
segregation distribution 分隔分布
segregation foliation 分凝叶理
segregation-free 无偏析
segregation gas chromatography 分离气相色谱(法)
segregation index 分离指数
segregation in magma 岩浆分凝(作用)
segregation junction 分凝结
segregation kink bands 分凝膝折带

segregation mutant 分离突变体
segregation of concrete 混凝土离析
segregation of ice 冰的分凝;冰析
segregation of items 项目划分
segregation of particles 分聚
segregation of traffic 交通分隔行驶
segregation of turning traffic 分道转弯
segregation of vehicles and pedestrians 车辆交通与行人交通分隔
segregation of waste 废液分离
segregation of wastewater 污水分流;废水离析
segregation phenomenon 分离现象
segregation pool 离析池;分离池
segregation process 离析法
segregation ratio distortion 分离比偏差
segregation schlieren structure 析离体构造
segregation system 分流制
segregation zone 离析区
segregative 离析的;分凝的
segregator 隔离器;离析器;分凝器;分离器;分类器;分禾器
segregent-assisted growth 分凝辅助生长
segresome 分解颗粒
Seheele's green 谢勒绿颜料;谢勒绿色
Sehofield-Scat model 斯菲尔德—斯科特模型
Seibert Stinnes hollow beam floor 塞伯特斯廷空心梁地板
seibt rectifier 低压真空管全波整流器
Seicento 塞斯图建筑(指意大利17世纪建筑)
seiche 驻波;静振波;假潮;湖震;湖面波动
seiche current 假潮流
seiche denivellation 湖面波动;湖面变化
seiche duration 静振历时;假潮历时
seiche oscillation 假潮波动
Seidel aberration 塞德耳像差
Seidel coefficient of aberrations 塞德耳像差系数
Seidel eikonal 塞德耳程函
Seidel-Glaser dioptrics 塞德耳—格拉泽屈光学
Seidel iteration method 塞德耳迭代法
Seidel method 赛德耳(方)法
Seidel optics 塞德耳光学
Seidel's scotoma (sign) 赛德尔氏暗点
Seidel theory 塞德耳(像差)理论;塞德耳理论
Seidel third order theory of aberration 塞德耳三级像差理论
Seidel variable 塞德耳变量
Seidlitz powder 赛德利茨粉;沸腾散
seidozerite 氟钠钛锆石
seif(dune) 纵向沙丘;蛇形砂丘;赛夫沙丘
Seifert tube 赛弗特X射线管
seignette-electrics 铁电体
Seignette salt 赛格涅特盐
seinajokite 斜方锑铁矿
seine 拖底大围网
seine boat 围网渔轮;围网船
seine float 救生素浮子;围网浮子
seine net 大围网
seiner 围网渔船
seiospore 散落孢子
seisin 依法占有土地;所有物
seism 地震
seismal load 地震荷载
seismetric 地震的
seismic 地震的
seismic acceleration 振动加速度;抗震惯性;地震加速度
seismic acceleration indicator 地震加速指示计
seismic acceleration spectral density 地震加速度谱密度
seismic action 地震作用
seismic activity 地震活动(性)
seismical 地震的
seismically active area 地震活动区
seismically active belt 地震活动带
seismically active coefficient 地震活动性系数
seismically quiet period 地震平静期
seismic amplifier 地震放大器
seismic analysis 地震分析
seismic anisotropy 地震各向异性
seismic anomaly 地震异常
seismic apparatus 地震仪(器);地震计
seismic area 震区;地震区(域);地震带
seismic area map 震区图
seismic array 地震台阵

seismic arrival time 地震波到时
seismic basic intensity 地震基本烈度
seismic bedrock 地震基岩
seismic behavio(u)r 地震机理
seismic belt 地震(地)带
seismic bending moment 地震弯矩
seismic blasting 地震爆破
seismic boulder 地震滚石
seismic building code 建筑物抗震法规
seismic bulge 地震鼓包
seismic calamity 地震震害
seismic center 震中;地震中心;地震源
seismic channel 地震道;地震波信道
seismic cloud 地震云
seismic cluster 地震群
seismic code 抗震规范
seismic coefficient 地震系数
seismic coefficient method 地震系数法
seismic coefficient of effect 地震影响系数
seismic collapse 地震崩塌
seismic compiler 地震剖面编绘器;地震编译器
seismic concept design of buildings 建筑抗震概念设计
seismic condition 地震状况;地震条件
seismic-conscious 有地震感知的;有地震意识的
seismic constant 地震常数
seismic constraint 地震约束
seismic construction 抗震建筑
seismic converted wave method 地震转换波法
seismic core phase 核震相
seismic country 地震区
seismic creep 地震蠕动;地震蠕变
seismic crew 地震队
seismic cross-section 地震剖面(图)
seismic damage 地震破坏;地震损害
seismic data 震源勘测资料;地震资料;地震数据
seismic data holography 地震数据全息图
seismic data interpretation 地震资料解释
seismic data processing 地震数据处理
seismic data ratio correction 地震数据比例因子校正
seismic decoupling 地震解耦
seismic degree 震度
seismic degree for design 设计地震烈度
seismic design 抗震设计;结构抗震设计;地震设计
seismic design intensity 地震设计烈度
seismic design parameter 抗震设计参数
seismic destructiveness 地震破坏性
seismic detecting equipment 测震设备
seismic detection 地震探测
seismic detector 地震仪;地震探测器;地震检波器
seismic detector of displacement 位移式地震检波器
seismic detector of the velocity type 速度式地震检波器
seismic digital amplifier 地震数字放大器
seismic direction 地震方向
seismic discontinuity 地震界面;地震间断面
seismic discrimination 地震鉴别
seismic dislocation 地震错距
seismic disturbance 地震扰动
seismic disturbance pulse 地震扰动脉冲
seismic drill 地震孔用钻机
seismic drilling crew 地震钻井队
seismic dynamic(al) load 地震动力荷载
seismic earth pressure 地震动土压力
seismic effect 地震效应
seismic effect field 地震效应场
seismic effect of explosion 爆炸的地震效应
seismic efficacy 地震功效
seismic efficiency 地震效率
seismic electric(al) effect 震电效应
seismic element 感振元件
seismic element method 地震元法
seismic energy 地震能量
seismic energy release 地震能量释放
seismic engineering 地震工程
seismic environment 地震环境
seismic equipment 测震设备
seismic event 地震事件
seismic excitation 地震激发
seismic expansion 地震伸缩缝
seismic experiment 地震试验
seismic exploration 震波勘测;地震勘查;地震探测;地震勘探;地层勘探

seismic exploration method 地震勘探法
seismic exploration vessel 震波勘测船
seismic explosion 地震爆破
seismic explosive 地震(探查)炸药
seismic explosive vessel 震波勘测船
seismic exposure 地震威胁
seismic facies quantitative analysis 地震相定量分析
seismic failure 地震破坏
seismic fatigue intensity 地震疲劳强度
seismic fault 地震断裂
seismic fault belt 地震断裂带
seismic field working method 地震野外工作方法
seismic figure 地震图
seismic filter 地震滤波器
seismic filtering 地震滤波
seismic filtering location 震源位置
seismic floor joint cover 地面抗震缝盖板
seismic focus 震源;地震源
seismic focus location 震源位置
seismic force 地震荷载;地震力
seismic forward mathematical model 地震正演数字模型
seismic fracture 地震裂缝
seismic fracture zone 地震断裂带
seismic frequency 地震频度
seismic frequency spectrum 地震频谱
seismic gap 地震空(白)区;地震活动空白地带
seismic geophone 震波检测仪;地震检波器
seismic geophysical method 地震物探法;地球物理地震法
seismic geophysical survey 地震物探法
seismic geyser 地震间歇泉
seismic ground motion parameter zonation map 地震动参数区划图
seismic hazard 地震危害(性)
seismic head wave 地震首波
seismic history 地震史
seismic holographic exploration 地震全息勘探
seismic holography 全息地震
seismic horizon 地震地层
seismic impedance 地震阻抗
seismic impulse 地震脉冲
seismic impulse method 脉冲地震法
seismic inactivity 地震不活动性
seismic inertial force 地震惯性力
seismic influence factor 地震影响系数
seismic information 地震信息;地震情报
seismic instrument 地震仪
seismic instrument car 地震仪器车
seismic intensity 地震强度;地震烈度;烈度
seismic intensity expectancy map 预期地震烈度图
seismic intensity field 地震烈度场
seismic intensity increment 地震烈度增量
seismic intensity microregionalization 地震烈度小区域划分
seismic intensity scale 地震强度计;地震烈度表
seismic intensity zoning 地震烈度区划
seismic interpretation 地震解释
seismic inverse mathematical model 地震反演数字模型
seismic investigation 地震研究;地震探测;地震调查
seismicity 震级;受震程度;地震活动度;地震活动(性)
seismicity anomaly 地震活动性异常
seismicity gap 地震空(白)区;地震活动空白地带
seismicity map 地震活动分布图
seismic joint 防震缝
seismic lake 地震湖泊
seismic land subsidence 地震地面沉降
seismic level 地震能级
seismic like event 似地震事件
seismic line profile of three-component 三分量地震测线剖面
seismic lithory model 地震岩性模型
seismic load analysis 地震荷载分析
seismic load(ing) 地震荷载;震动荷载
seismic location 地震定位
seismic log 地震参考测井
seismic log curve 地震参考测井曲线
seismic magnetic integrator 地震磁力积分仪
seismic magnitude 地震(震)级
seismic map 地震(地)图

seismic marker bed 地震标准层
seismic marker horizon 地震标准层
seismic measurement 地震观测;地震测验;地震测量
seismic method 地震探测法;地震(勘探)法
seismic method of exploration 震波勘探法;地震法勘探
seismic method of prospecting 地震法勘探
seismic microzonation 地震小区(域)划(分);地震微区划
seismic microzoning 地震小区(域)划(分);地震烈度小区域划分
seismic migration 地震迁移
seismic model 地震模型
seismic model experiment 地震模型试验
seismic modeling 地震模拟
seismic moment 地震(力)矩
seismic moment tensor 地震张量矩
seismic monitoring 地震监测
seismic monitoring system 地震监测系统
seismic motion 地震运动
seismic mudflow 地震泥石流
seismic network 地震台网
seismic noise 地震噪声;地震干扰
seismic observation 地震观测
seismic observatory 地震观测站
seismic origin 震源;地震成因
seismic oscillation 地震的摆动
seismic overturning moment 地震的倾覆力矩
seismic parameter 地震参数
seismic pattern 地震形式
seismic pattern recognition 地震图形识别
seismic pendulum 地震摆
seismic performance 地震性能
seismic period 地震周期
seismic phase 震相
seismic phenomenon 地震现象
seismic pickup 地震检波器
seismic planning 城市抗震规划;抗震规划
seismic planning of city 城市抗震规划
seismic play-back apparatus 地震回收装置
seismic pore water pressure 地震动孔隙水压力
seismic precursor 地震前兆
seismic prediction 地震预测;地震预报
seismic probability estimation 地震概率估计
seismic processing 震震图分析
seismic profile 地震剖面(图)
seismic profiler 震波水下地形仪;地震剖面仪
seismic profiling 地震剖面绘制
seismic program(me) 地震探测计划;地震勘测计划
seismic property 地震性质
seismic prospect 震波勘探
seismic prospecting 地下爆炸测震;地震勘探(法);地震找矿;地震(波)探查;震震勘探
seismic prospecting recording system 地震勘探记录系统
seismic prospecting system 地震探查装置
seismic prospector 地震预报仪
seismic pulse 地震脉冲
seismic pulse intensity 地震脉冲强度
seismic qualification 地震鉴定
seismic range 地震范围
seismic ray 地震线;地震射线
seismic reciprocity 震时互易原理
seismic recognition of unconformities 不整合面的地震识别
seismic record 地震记录;地震波记录
seismic recording 地震记录
seismic record viewer 地震记录观测仪
seismic reflection 地震反射
seismic reflection amplifier 地震反射放大器
seismic reflection horizon 地震反射水平面
seismic reflection method 反射法地震勘探;地震(波)反射法;反射地震波法
seismic reflection mode 地震反射模式
seismic reflection profiling 地震反射剖面测量
seismic reflection survey 地震反射法勘测;地震反射法调查
seismic refraction 震波折射;地震折射
seismic refraction method 地震折射(方)法;地震波折射法
seismic refraction profile 地震折射剖面
seismic refraction profiling 地震折射剖面测量
seismic refraction survey 地震折射探测;折射法

调查
seismic regime 震情;地震范围
seismic region 震区;地震区(域);地震地域
seismic regionalization 地震区域划分;地震区划(分);地震分区
seismic region map 震区图
seismic reliability analysis 地震可靠度分析
seismic report 地震报告
seismic resistance 抗震
seismic resistance design 抗震设计
seismic resistant capacity 抗震能力
seismic resistant design 抗震设计
seismic response 地震感应;地震响应;地震反应
seismic response analysis 地震反应分析
seismic response characteristics 地震反应特性
seismic response spectrum 地震反应谱
seismic restraint 抗震;耐震
seismic restraint of equipment 设备抗震
seismic rigidity of ground 土体抗震刚度
seismic risk analysis 地震危险性分析
seismic risk zoning 地震危险区划
seismicrophone 地震接收器;地震传声器
seismic rupture 地震破裂(面)
seismics 地震测量学
seismic safety evaluation 地震安全度评价
seismic safety net 地震安全网
seismic safety of building 建筑物抗震安全性
seismic scale 地震烈度表
seismic scattering 地震散射
seismic sea-streamer (装有一系列的水中地震检波器的)地震法海底测量绳
seismic sea wave 海啸;地震津波;地震海啸;地震海浪
seismic sea wave apparatus 海啸仪
seismic sea wave warning system 地震海啸警报系统
seismic section 地震剖面(图)
seismic section plotter 地震剖面(绘图)仪
seismic seiche 地震假潮
seismic sensor 地震传感系统;地震敏感元件;地震传感器
seismic shear 地震剪力
seismic shear wave 地震剪力波
seismic shock 由地震产生的振动;震波冲击;地震振动;地震冲击
seismic shooting 地震喷涌;地震爆破
seismic signal 地震信号
seismic signal attenuation 地震信号衰减
seismic signal processor 地震信息处理机
seismic signature parameter processing 地震特征参数处理
seismic sinking 地震陷落
seismic site intensity 地震场地烈度
seismic sounding 地震测深(法)
seismic sounding equipment 地震测探设备
seismic source 震源
seismic source function 震源函数
seismic source model 震源模型
seismic spectrum 地震(波)谱
seismic spread 地震扩散;地震传播
seismic stability 抗震稳定性;地震稳定性
seismic station 地震台站;地震测站
seismic station network 地震台网
seismic status 地震状态
seismic stimulus 地震激发
seismic stratigraphy 地震地层学
seismic strength 地震强度
seismic stress 地震应力
seismic structural wall 抗震墙
seismic structure 孕震构造;发震构造
seismic study 抗震研究
seismic surface wave 地震表面波
seismic surges 地震海啸
seismic surveillance 地震监视
seismic survey(ing) 地震勘测;地震探测;震波勘测;地震探查;地震勘探;地震调查;地震测验;地震测量;地震勘绘;发射法勘探
seismic survey vessel 震波勘测船
seismic system 地震系统
seismic system model 地震系统模型
seismic technique 地震探测技术;地震勘测技术
seismic tectono-geochemistry 地震构造地球化学
seismic telemetry 地震遥测术
seismic test 地震试验

seismic timer 小型地震仪;地下探查仪
seismic tomography method 地震层析成像法
seismic trace 地震记录迹线
seismic transient 地震瞬时脉冲
seismic transmission 地震传播
seismic travel time 地震(波)传播时间;地震波走时
seismic trigger 激振仪;激振器;地震触发器
seismic type mud-stone flow 地震型泥石流
seismic underwater explorer 地震水下探测器
seismic velocity 弹性波速(度);地震速度
seismic velocity logging 地震速度测井
seismic velocity model 地震速度模型
seismic velocity ratio 地震波速比
seismic vertical 震源至震中连线;地震垂线
seismic vessel 地震探测船
seismic vibration 地震的振动
seismic wave 震波;地震波
seismic wave attenuation 地震波衰减
seismic wave energy 地震波能量
seismic wave field 地震波场
seismic wave form 地震波形
seismic wavelet 地震子波;地震示波
seismic wave path 地震波路径
seismic wave reflection 地震波反射
seismic wave refraction 地震波折射
seismic wave theory 地震波理论
seismic-wave velocity 震波速度
seismic wave velocity change 地震波速变化
seismic world map 世界地震图
seismic zone 地震区;地震(地)带
seismic zone factor 地震区系数
seismic zone map 震区图
seismic zoning 地震区域划分;地震区划(分);地震分区
seismism 震动现象;地震作用;地震现象
seismitron 岩层稳定测试仪
seismo-acoustic activity 地震声波活动性
seismo-acoustic(al) reflection survey 震声反射测量
seismo-acoustic method 地震声法法
seismoastronomy 地震天文学
seismochronograph 地震时间记录仪;地震计时器;地震计时法
seismogenesis 地震成因
seismogenic fault 发震断裂;发震断层
seismogenic structure 孕震构造
seismo-geological map 地震地质图
seismo-geology 地震地质学
seismogram 地震波曲线(图);震波图;震波曲线;地震记录(图);地震(波)图
seismogram interpretation 震波图解释
seismogram of horizontal weight drop 水平撞击记录
seismogram of vertical weight drop 垂直撞击记录
seismogram trace 地震图记录线
seismograph 震波图;地震(记录)仪
seismograph amplifier 地震仪放大器
seismograph array 地震台阵
seismograph calibration 地震仪标定
seismograph drill 地震孔物探钻机
seismographic(al) 地震学的
seismographic(al) disturbance 地震扰动;地震干扰
seismographic(al) map 地震(地)图;地震记录图
seismographic(al) observation 地震观测
seismographic(al) observatory 地震观测站
seismographic(al) prospecting method 地震勘探法
seismographic(al) record 地震波记录
seismographic(al) recorder 地震记录器
seismograph instrument 地震仪
seismograph monitoring network 地震仪监测台网
seismograph network 地震仪台网
seismograph rod 地震孔用钻杆
seismograph station 地震台站;地震观测站
seismograph trace 地震记录迹线
seismograph with electric(al) sensitive paper and electric(al) arc lamp 电敏纸电弧灯地震仪
seismography 地震学;地震记法法;地震测验法;地震测定法
seismologic(al) array 地震台阵
seismologic(al) atlas 地震地图集
seismologic(al) center 地震中心
seismologic(al) consideration 震情会商
seismologic(al) engineering 地震工程(学)

seismologic(al) evidence 地震实迹;地震证据
seismologic(al) history 地震史
seismologic(al) method 地震学方法
seismologic(al) network 地震台网
seismologic(al) observation 地震观测
seismologic(al) observatory 地震观测站
seismologic(al) parameter 地震参数
seismologic(al) pulse 地震脉冲
seismologic(al) record 地震记录
seismologic(al) station 地震台站
seismologic(al) zoning 地震区划(分)
seismologist 地震学家
seismology 地震学
seismology model 地震模型
seismomagnetic effect 震磁效应;地震地磁效应
seismometer 地震(记录)仪;地震检波器;地震计;地震表;测震表
seismometer galvanometer 地震检波器检流计
seismometer pier 拾振器墩子
seismometer station 地震测站;测震台(站)
seismometric observation 地震观测
seismometry 地震测量(术);测震学;测震术;测震法
seismonastic movement 感振运动
seismonastic turgor movement 倾振膨压运动
seismonasty 感震性
seismophysics 地震物理学
seismos 地震
seismoscope 验振器;简单地震仪;地震示波仪;地震记录仪;地震波显示仪
seismoscope data 振震器数据
seismoscope date 地震计数据
seismoscope record 验震器记录;地震计记录
seismoscope response 验振器响应
seismostation 地震台站;地震台
seismostatistic(al) data 地震统计资料
seismostatistic(al) method 地震统计方法
seismostructure survey 地震构造调查
seismotectonic 地震构造的
seismotectonic line 地震构造线
seismotectonic map 地震构造图
seismotectonics 地震构造
seismotectonic zone 地震构造带
seismotectonic zone of continental rift system 大陆裂谷系地震构造带
seismotectonmagmatic belt 地震构造岩浆带
seismotropism 向振性
seismovac 真空压力活塞式振源
seistan 十二旬风
seisviewer 声井下电视
Seitz filter 赛氏滤器
seize 咬住;扣押财产;卡住;缚;捕获
seized door 箱门卡住
seized goods 海关扣押的货物
seized piston 卡住活塞
seize markets 争夺市场
seizen 依法占有土地
seizer 扣押者
seize smuggled goods 没收走私货物
seizin 依法占有的财产;所有物
seizing 擦伤;咬住;捆绑;黏附;捕捉;集;绑扎
seizing acknowledgement signal 占机证实信号
seizing end 捆头(钢丝绳)
seizing line 永久性缠扎用油麻细绳
seizing mark 黏附伤痕;抱轴伤痕
seizing of mo(u)ld 模型卡紧
seizing of piston rings 活塞环烧蚀
seizing signal 约束信号;占机信号;捕获信号
seizing strand 卷缆钢缆用的镀锌细钢丝
seizing wire 卷缆钢缆用的镀锌细钢丝
seizure 依法占有;咬粘;抓住;滞塞;扣押(物);扣留;捕取
seizure and capture 捕捉和俘获
seizure of disease 疾病发作
seizure of goods 扣押货物
seizure of plunger 柱塞卡住
seizure of property 扣押财产;财产没收
seizure of smuggled goods 没收走私货物
seizure resistance 抗咬合性
seizuring load 胶住荷载
seizuring pressure 胶住压力
Sejourne formula 塞氏公式(用于设计桥台翼墙平均厚度)
Sejournet process 玻璃润滑剂高速挤压法
sekaninaite 铁堇青石

sekos 庙宇内殿;特权者使用的房屋;古希腊神坛
sekurit glass 安全玻璃;钢化玻璃
Sekur machine 斯库尔携带式硬度计
selacholeic acid 鲨油酸
selagite 黑云粗面岩
Selatan 印度尼西亚的强干南风;塞拉坦风
selbite 杂菱银矿
Seldovian(stage) 赛多维阶(植物阶)
select 1 in 2 二取一
select 1 in 3 三取一
select 1 in 4 四取一
select 2 in 3 三取二
selectable 可选择的
selectable provision 可选择采用的规定
selectable secondary distress alerting network 可选择第二遇险报警网络
selectable unit 可选部件
select address and contrast operate 选择地址及对比操作
selectance 选择系数;选择度
select bit 选择位
select chain 选链
select command 选择命令
selected 被选的;选择的;精选的
selected address 被选择的地址;被选地址
selected alternative 选定的方案
selected area electron diffraction 选定区电子衍射
selected bid 选定的投标商;合格标商
selected bidder 中选的投标者
selected bidding 选择招标;有限招标;邀请招标
selected calling 选择呼叫
selected calling system 选择呼叫系统
selected cell 寻址单元;选址单元;被选单元
selected closure scheme 选定截流方案
selected data 选用(的)资料
selected depth 选定深度;规定深度
selected design alternative 选定的设计方案
selected diffraction 选区衍射
selected economic indicators 主要经济指标
selected fill 选择填料
selected filling 精选的填料;挑选的填充物
selected function groups 选定的函数群
selected glazing quality 优质窗玻璃
selected goods 精选货(品)
selected heading 选定航向;选定的航向
selected individual 选择的个体
selected information 选用(的)资料
selected ion electrode method 离子选择电极法
sleected ion monitoring gas chromatography-mass spectrometry 离子选择检测气相色谱法—质谱法联用
selected ion monitoring method 离子选择检测法
selected list of bidders 投标人选择名单;合格标商名单
selected lump lime 石灰粉;粉状石灰
selected marker 选择(性)标记
selected master file tape 被选的主文件带
selected material 选定材料
selected mating 选配
selected merchantable 优质木材;优质商品;选定的商品
selected mine 选定水雷
selected ordinate method 选择纵坐标法;选择波长法;选定波长法
selected organic compound 选择性有机化合物
selected plan 选定的计划
selected-point method 选点法
selected program(me) 选定的计划;选定的方案
selected project alternative 选定的(设计)方案;推荐项目;推荐工程;推荐方案
selected pulse 选定脉冲
selected quality 精选品;上选品质
selected quicklime 精选生石灰
selected range indicator 被选范围显示器
selected reference 选定目标;选定基准
selected rockfill 精选堆石(料)
selected scenes 选场
selected scheme 选定的方案
selected segment 选择节段
selected soils 选择性土
selected stars 选用恒星
selected temperature 选定温度
selected time 选定时间
selected track 选定的航迹

select finger 选择指针
select good persons 选择人才
select-hold 带保持的选择
selecting address technique 选址术
selecting card 选针纹板
selecting circuit 选择电路;选取电路
selecting cutting 择伐
selecting data 选择数据
selecting lever 操纵系统杠杆
selecting magnet 选择磁铁
selecting mechanism 选择机构
selecting operation 选择操作
selecting perennial vegetable 多年生蔬菜选择
selecting point method 定点(方)法
selecting priority 选择优先权
selecting sequence 选择顺序
selecting single plant 进行单株选择
selecting transfer cam 转移三角
selecting unit 选择器
selection 选择(作用);精选
selection algorithm 选择算法
selection among multiple carriers 多载频的选用
selection approach 选择手段
selectionary tariff 选择关税
selection bias 选择偏差
selection by sifting 筛选
selection call 选择呼叫
selection check 选择校验;选择检验
selection circuit 选择电路
selection coefficient 选择系数
selection control 选择控制;选线控制;拨号控制
selection coppice system 择伐矮林作业
selection core matrix 选择磁芯矩阵
selection cutting 择伐式间伐;择伐
selection differential 选择差别
selection dump 选择性转储[贮]
selection element 选择元件
selection felling 择伐
selection forest 择伐林
selection function 选择函数;选择功能
selection index 选择指数
selection index method 选择指数法
selection in segregating generations 分离世代的选择
selection instruction 选择指令
selection intensity 选择强度
selectionism 选择学说
selection key 选择(电)键
selection level meter 选频电平表
selection lever 选择杆
selection limit 选择极限
selection matrix 选择矩阵
selection mechanism 选择机构
selection menu 选择菜单
selection method 择伐作业;选择法
selection of activated sludge reactor type 活性污泥器选型
selection of aggregate quarry 料场选择
selection of a high yielding variety 选择高产品种
selection of average 平均数的选择
selection of bid 选标
selection of bridge site 桥位选择
selection of capacity 装机容量选择
selection of cargo flow 水运货流选择
selection of centrifugal fan flower 离心通风机选择
selection of channel 运道选销通路
selection of community 群落选择
selection of container 集装箱选择
selection of dam site 坝址选择
selection of design 设计评选
selection of diamond tools 金刚石钻具的选择
selection of dredge(r)s 挖泥船选择
selection of factory site 厂址选择
selection of feed 进给选择
selection of firing position 选择射击阵地
selection of geologic(al) variable 地质变量的选择
selection of materials 材料的选择
selection of mix(ture) 混合料的选择
selection of plant location 选址
selection of plant site 站址选择
selection of port and harbo(u)r site 港址选择
selection of port site 港址选择
selection of risks 风险的取舍
selection of route 线路选择;选线

selection of single traits 单独性状选择
selection of stars 选星
selection of station 车站选址
selection of tender 选标
selection of the names 地名注记选择
selection of the second best course of action 选择次优方案
selection of the second best scheme 选择次优方案
selection of urban land 城市用地选择
selection of water source 水源选择
selection overlap 透图
selection parameter 选择参数
selection path 选择通路
selection piece 选针片
selection plugboard 选择插件板
selection pressure 选择压力
selection principle 选择原则
selection pulse 选择脉冲;脉冲选取
selection ratio 选择比
selection replacement technique 选择替换技术
selection response 选择反应
selection revision 部分修订
selection route 选择进路
selection rule 选择定则
selection rules for atomic spectra 原子光谱选择定则
selection scheme 选择方案
selection schemer 选择路线
selections from introductions 从引种中选择出来
selection signal 选择信号
selection signal(l)ing installation 选择信号装置
selection slit 选择狭缝
selection standard 选取标准
selection statement 选择语句
selection statistic 选择统计量
selection switch 选择开关
selection system 择伐作业
selection system of natural regeneration 择伐更新法
selection terminal 指定终端站;指定终点站
selection theory 选择学说
selection thinning 择伐式疏伐
selection unit 选择单位
selection winding 选择绕组
selection wire 选择线
selection with water 水选
select issue rates 选择产次率
selective 选择性的
selective absorbability 选择吸收性能
selective absorbent 选择性滤光片
selective absorber 选择性滤光片;选择吸收器
selective absorption 选择(性)吸收;选择性吸光;选择吸收作用;选择吸附;优先吸附
selective absorption of light 光的选择吸收
selective absorption of sound 选择吸音
selective-access display 随机显示
selective acidizing 选择性酸化;选择性酸处理
selective action 选择(性)作用
selective addressing 选择寻址;选择定址
selective addressing memory 选择寻址存储器
selective administration 选择参入
selective adsorbent 选择吸附剂
selective adsorption-catalytic process 选择性吸附—催化法
selective advance 选择增效
selective advantage 选择有利性;选择优势
selective affinity 选择(性)亲和力
selective aggregation 选择性聚集
selective amplifier 选择(性)放大器
selective analysis 重点分析
selective annealing 选择退火;局部退火
selective anodic process 选择性阳极过程
selective assembly 选择装配
selective assembly manufacturing 选择装配式制造
selective attachment 选择性黏附
selective audio amplifier 选择声频放大器
selective automatic radar identification equipment 选择性自动雷达识别装置
selective availability 选择可用性
selective azeotropic distillation 用选择溶剂的共沸蒸馏
selective bacteriostasis 选择制菌作用;选择性抑菌作用

selective beacon radar 选择信标雷达
selective bid(ding) 邀请招标;选择性招标;选择性投标
selective bid method 选择投标方法
selective bioaugmentation 选择性生物强化
selective block sequence 选定多层焊次序;顺序分段焊接法
selective bottom-up 自底向上选择
selective breaking 选择性破碎
selective breeding 选择育种
selective broadcast receiving station 选择性广播接收台
selective broadcast sending station 选择广播发射台
selective bullet gun perforator 选择性子弹射孔器
selective call 选择呼号
selective call equipment 选择呼叫设备
selective call line telephone 选呼电话
selective call rejection 选择呼叫拒绝
selective call ringing 选择呼叫振铃
selective capture 选择俘获
selective carburizing 局部渗碳
selective catalytic conversion 选择催化转化
selective catalytic cracking 选择催化裂化
selective catalytic reduction process 选择催化还原法
selective category 选择范畴
selective channel 选择性信道
selective check 选择检验
selective chelating resin 选择性螯合树脂
selective chloridination 选择氯化(法)
selective chopper radiometer 选择斩波辐射计
selective circuit 选择(性)电路
selective classification of solid waste 固体废物分选
selective code 选择代码
selective coefficient 选择性系数
selective collection 选择性收集
selective combustion 选择性燃烧
selective combustion method 选择性燃烧法
selective consonance 选择谐振
selective consumption 选择性消费
selective contact 选择性接触
selective control 选择(性)控制;选线控制;局部控制
selective control system 选择性调节系统;选择控制系统
selective conversion 选择性转化
selective corrosion 选择性腐蚀;局部腐蚀作用
selective course 选修课
selective cracking 选择裂化
selective cracking process 选择裂化过程
selective credit control 选择性信用管理;选择性信贷控制;信用质量控制
selective crude 特定原油
selective crushing 选择性破碎;优先破碎
selective culture medium 选择性培养基
selective cupping 伐前选采法
selective cutting 选择砍筏;选择砍伐
selective damage 选择性破坏
selective demodulation 选择性反调制
selective deposition 选择性沉积
selective depressant 选择性抑制剂
selective detection 选择性检测
selective differential relay 选择性差动继电器
selective diffuser 选择(性)扩散器
selective diffusion 选择扩散
selective digging 选择挖掘(按照不同土类进行挖掘);分类挖土
selective digit emitter 选择数字发送器
selective dimerization 选择性二聚合
selective displacement 选择性顶替
selective display 选择显示
selective dissemination of information 选择性信息提供;信息选择传布;信息的选择性传播;定题情报提供
selective dissemination of information on-line 联机定题情报服务
selective distortion 选择性扭曲
selective distribution 选择分配
selective distribution of fund 选择性分配资金
selective drainage method 选择排流法
selective dump 选择转储
selective dyeing process 选择性染色翻版法
selective electrode 选择电极

selective electrode method 选择性电极法
selective elimination 选择性消灭
selective elution 选择性洗脱
selective emission 选择放射
selective employment tax 特定就业税
selective enlargement 选择放大
selective enrichment 选择(性)富集
selective entraining agent 选择性携带剂
selective epitaxy 选择外延
selective erase 选择消磁;局部擦除
selective erasing 选择(性)擦除;随机擦除
selective erasure 选择性消除
selective etching 选择刻蚀
selective evapo(u)ration 选择性蒸发
selective examination 抽查
selective excitation mechanism 选择激发机理
selective excitation method 选择激发法
selective exclusion sign 车种限制标志
selective extraction 选择(性)提取;分别抽出
selective extraction technique 选择提取技术
selective fading 选择性衰落
selective feeder 选择进料器
selective felling 选伐
selective fermentation 选择发酵
selective filling 选别充填
selective filter 选择性滤光器;选择性滤光镜;选择滤波器
selective finishing 选择精制
selective fit 选择配合
selective flo(a)tation 选择性浮选;优先浮选
selective flocculation 选择性氯凝
selective flocculator 选择性絮凝剂
selective flooding routing 选择扩散式路径确定
selective focus 选择聚焦
selective fracturing 选择性压裂
selective freezing 选择结晶
selective frequency control 选择频率控制
selective fusion 选择熔化
selective gamma-gamma log 选择伽马—伽马测井
selective gamma-gamma log curve 选择伽马—伽马测井曲
selective gapper 选择式间苗机
selective gas precipitation 通气选择性沉淀
selective gate 选通电路
selective gear 配换齿轮;变速箱轮;变速机构
selective gear drive 选择齿轮传动
selective gear shift 选择性变速
selective gear shifting 选择变速
selective gear transmission 选择齿轮式变速器
selective getter 选择性吸气剂
selective grants 特定补贴
selective grazing 选择(性)放牧
selective grinding 选择性研磨;选择性磨矿;选磨
selective growth 选择性生长
selective hardening 选择硬化;局部硬化
selective heading machine 部分断面掘进机
selective headstock 床头箱;变速箱
selective heating 选择加热;局部加热;差别加热(法)
selective herbicide 选择性除草剂
selective high pruning 选木打枝
selective hydration 选择性水化
selective hydrolysis 优选水解;优先水解
selective identification 选择识别
selective identification feature 选择鉴别特征
selective illumination 选择照明
selective impulse 选通脉冲
selective indexing 选择性标引法
selective indicator dilution curve determination 选择性指示剂稀释曲线测定
selective information 选择(性)信息
selective information retrieval 定题情报检索
selective instability 选择不稳定度
selective interchangeability 选择互换性
selective interference 选择性干扰;窄带干扰
selective inventory control 选择存货管理;分级存货管理
selective inverse feedback 选择性负反馈
selective ion exchange 选择性离子交换;离子选择交换
selective ionization ga(u)ge 选择性的电离压力计
selective isotope absorption 选择性同位素吸收
selective jammer 选择干扰机
selective jamming 选择性干扰

selective jettison 选择投弃
selective key 选择键
selective leaching 选择浸出;局部浸出
selective level measuring equipment 选择性电平测量设备
selective listing 选择列表
selective loading 选择装载
selective locale 布点
selective localization 选择性定位
selective logging 选择砍筏;选择砍伐;选伐
selective loop transferring 选择移圈
selective mark insertion 选择标记插入
selective maturation 选择成熟
selective mean 选择平均值
selective measuring system 选择测量系统
selective medium 选择性培养基
selective membrane 选择性膜
selective membrane of ions 离子选择性膜
selective metasomatism 选择交代(作用);分布交代【地】
selective method 选择方法
selective migration 选择迁移
selective mining 选择性开采(法)
selective mixed-phase cracking 选择混合相裂化
selective mutant 选择性突变性
selective mutation 选择性突变
selective network 选择性网络
selective nitriding 局部渗氮
selective noise-detection circuit 选择噪声检波电路
selective optical lock-on 选择性光学自动跟踪
selective osmosis 选择性渗透作用
selective oxidation 选择氧化;局部氧化;分别氧化
selective oxidation process 选择氧化工艺
selective oxide flo(a)tation 氧化物优先浮选
selective pairing 选配
selective permeability 选择透过性;选择通透性;选择渗透性;选择穿透性
selective permeation 选择性渗透
selective pesticide 选择性农药
selective photoelectric(al) effect 选择性光电效应
selective photoeletric emission 选择性光电子发射
selective plating 选择镀
selective plugging 选择性层位封堵
selective poison 选择毒物
selective polymer 选择聚合物
selective polymerization 选择聚合
selective polymerization process 选择聚合过程
selective power 选择能力
selective precipitation 选择(性)沉淀;优先沉淀
selective preference 优选
selective preferential adsorption 选择优先吸附
selective-preferential type 优择方式
selective pressure 选择压力
selective process 选择过程
selective protection 选择性保护
selective protection method 选择性保护法
selective pulse 选择脉冲
selective purchase 选购商品
selective quenching 选择淬火
selective quota 选择性配额
selective radiation 选择辐射
selective radiator 选择辐射体
selective radiometer 选择(性)辐计
selective reabsorption 选择性重吸收;选择性再吸收
selective reaction 选择(性)反应
selective reactivation property 选择再活化性
selective reagent 选择(性)试剂
selective receiver 选择接收机
selective reception 选择接收
selective recrystallization 选择性再结晶;优先再结晶
selective rectification 选择精馏;选择分馏
selective reduction 选择性还原
selective reflection 选择反射
selective reflectivity 选择反射能力
selective refraction 选择折射
selective reject 拒选
selective relay 选择性继电器
selective replacement 选择交代(作用);分布交代【地】
selective repressuring 选择性恢复压力
selective resonance 选择共振;选择共鸣
selective retention 选择性保留
selective revision 部分修订

selective ringing 选择振铃;选择呼叫
selective rodenticide 选择性杀鼠剂
selective rule 选择规则
selective sales taxes 特别销售税
selective sampler 选择采样器
selective sampling 选择性采样;选择抽样;抽样
selective scattering 选择(性)散射
selective scattering of light 光的选择性散射
selective screen 选择滤色片
selective screwfeed 选择性螺旋给进
selective searching 选择搜索
selective sensitivity 选择性灵敏度
selective separation 选择性分离
selective sequence calculator 选择程序计算器
selective sequence computer 选择程序计算机
selective sequential 选择顺序
selective signal 选择信号
selective sliding gear 滑移配速轮;滑动变速齿轮
selective solubility 选择性溶解
selective solubility diffusion 选择性溶解扩散
selective solvent 选择(性)溶剂;有选择性溶剂
selective solvent extracted oil 选择溶剂抽提油
selective solvent extraction 选择性溶剂抽提;选择溶剂提取
selective solvent extraction method 选择性溶剂提取法;选择性溶剂萃取法
selective solvent method 溶剂选择法
selective solvent process 选择性溶剂过程
selective-solvent-refined oil 选择溶剂精制油
selective sorbent 选择性吸着剂
selective sorting 优先拣选
selective species 选择种
selective speed gear 配速齿轮
selective spheric(al) agglomeration 选择性球状团聚
selective stacker 选择接卡箱
selective staining 选择染色
selective standard 选择标准
selective stripping 选择性反萃取
selective structure 选择性结构
selective switch 选择开关
selective switchgear 选择转换开关
selective system 选择系统
selective system of gear changing 变速箱的选择机构
selective system of preference 选择性优惠关税制度
selective tempering 局部回火
selective tender(ing) 邀请招标;选择性招标;选择性投标
selective test 选择性试验;分层测试
selective thinning 树丛剪修;(树木的)适当稀疏;选择砍筏稀疏;选择砍伐稀疏
selective tint correction 选择色调校正
selective top-down 自顶向下选择
selective toxicity 选择毒性
selective trace 选择跟踪
selective tracing routine 选择追踪程序
selective track 选择道
selective transducer 选择性换能器
selective transmission 选择(式)变速器;变速传动
selective transportation 选择搬运
selective transposition 选择交叉
selective treatment 选择性处理
selective tripping 选择断路
selective tripping system 选择断路系统
selective tucking 选针集圈
selective type herbicidal oil 选择性除莠油
selective updating 选择性信息更新
selective value 选择值
selective volatilization 选择挥发
selective water source 选择水源
selective weathering 选择风化;选择风化作用
selective weeding 选择性除莠
selective wetting 选择性溶湿润;优先润湿
selective zone anchor 选层锚
selectivity 选择率;选择性
selectivity characteristic 选择特性;时限特性
selectivity clause 选择条款
selectivity clear accumulator 选择性清除累加器
selectivity coefficient 选择(性)系数
selectivity constant 选择常数
selectivity control 选择性控制;选择性调整
selectivity curve 选择性曲线

selectivity diagram 选择性图
selectivity factor 选择因数
selectivity function 选择性函数
selectivity index 选择(性)指数
selectivity of reagent 试剂选择性
selectivity ratio 选择比
selectivity scale 选择标度
select life table 选择生命表
select line 选择线
select list 选择列表
select lumber 优选木材
select magnet 选择磁铁
select material 选料
select merchantable 优质商品板
select normalization transformation 选择规格化变换
selectophore 选择基
selector 选择者;分离机;选择器;干扰雷达用的金属带;换挡器;分选炉;选线器
selector bar 转换杆
selector bit 选针片
selector blade 分选器叶片
selector button 选择(器)按钮;变速按钮
selector carrying capacity 选择器负荷容量
selector channel 选择通道
selector circuit 选择(器)电路
selector code 选择码
select order 选择指令
selector dial 选择器标度盘
selector disc 选择圆盘
selector fork 换挡叉;变速叉
selector installation 选择器装置
selector lever 变速杆;选拉手柄;换位选择杆
selector marker 选择指示器;选择指点标
selector mechanism 选择机构
selector panel 选择器架
selector pen 选择笔
selector pen attention 光笔注意信号
selector plug 选择器塞头
selector-pulse generator 选择脉冲发生器
selector pushbutton 选择器按钮
selector range 选择器量程
selector register 选择器的寄存器
selector relay 选择(器)继电器
selector-repeater 选择器增音机;增音器;区别机;分区断接器
selector rod 选择器杆;变速叉杆
selector rod housing 变速杆罩
selector scale 选择器刻度
selector shaft guide 选择器轴的导槽
selector shelf 选择器架
selector stepping magnet 选择器步进磁铁
selector switch 选路开关;选择(器)开关;选线开关
selector valve 选择(器)阀;选择活门;手控阀;分配阀
selector valve body 分配阀壳体
selectron 超电子
selectro screen 可变偏心筛
select set 选择组
select structural wood 优质结构(木)材
select suitable material 选材
select tape 选择带
select target 选择目标
select the best alternative 选择最优方案
select the most favorable alternatives 挑选最佳方案
select tight knot 节结致密的优质(木)材
select transmit frequency 选择发射频率
select type and design 形式选择及设计
select verify 选择检验
Selek 施乐克(一种连接金属管的粉末)
selemium oxyfluoride 氟氧化硒
selenalyte 玻璃(6种不同颜色的)
selenate 硒酸盐
selenates tellurates 硒酸盐碲酸盐
selene 月形物
selenic acid 硒酸
selenic photoelectric(al) cell 硒光电管;硒光电池
selenide 硒化物
selenio-siegenite 硒硫镍钴矿
selenious 二价硒的
selenite 亚硒酸盐;(透(明)石膏
selenite cement 石膏水泥
selenite plate 石膏试板

selenites tellurites 亚碲酸盐亚硒酸盐
selenitic cement 含5%熟石膏的水泥；透明石膏作促凝剂的水泥；透明石膏质水泥；掺透石膏作促凝剂的水泥
selenitic lime 石膏水泥；掺石膏快硬石灰；含5%熟石膏的石灰
selenitic lime cement 含5%~10%熟石膏的石灰水泥
selenitum 玄精石
selenium 硒
selenium bromide 一溴化硒
selenium cell 硒质光电管；硒整流片；硒光电管；硒光电池
selenium-cell relay 硒光电池继电器
selenium compound 硒化合物
selenium conductive cell 硒光敏电阻
selenium contamination 硒污染
selenium copper 硒铜
selenium deposit 硒矿床
selenium dialkyl 二烷基硒
selenium dipropyl 二丙硒
selenium disulfide 二硫化硒
selenium halides 卤化硒
selenium hexafluoride 六氟化硒
selenium intoxication 硒中毒
selenium-krypton method 硒—氪法
selenium mixture 硒占剂
selenium nitride 氮化硒
selenium ore 硒矿
selenium oxide 二氧化硒
selenium oxychloride 二氯氧化硒
selenium oxyfluoride 二氟氧化硒
selenium photometer 硒光度计
selenium pile 硒堆
selenium poisoning 硒中毒
selenium pollution 硒污染
selenium propyl 二丙硒
selenium rectifier 硒整流器
selenium rectifier for battery charging 电池充电用硒整流器
selenium rectifier welder 硒整流焊机
selenium red 硒镉红颜料
selenium ruby glass 硒(宝石)红玻璃
selenium soil 硒土
selenium stack 硒堆
selenium stack type rectifier 叠片型硒整流器
selenium stainless steel 硒不锈钢
selenium steel 加硒钢
selenium sulfide 硫化硒
selenium toner 硒调色剂
selenizing 硒化
selenocentric coordinate 月心坐标
selenocentric coordinate system 月心坐标系
selenocentric trajectory 月心轨迹
selenocosalite 硒铜辉铅铋矿
selenocystathionine 丙氨酸丁氨酸硒醚
selenodesy 月面测量(学)
selenodetic control 月面测量控制
selenodetic network 月面网
selenofault 月面断层
selenograph 月面图
selenographic(al) coordinate 月面坐标
selenographic latitude 月面纬度
selenographic longitude 月面经度
selenographic projection 月面投影
selenography 月面物理学
selenolite 氧硒矿；石膏岩
selenology 月球学
selenomorphology 月球地貌学
selenophone 照相录声机
selenophysical measurement 月球物理测量
selenophysical phenomenon 月球物理现象
selenophysics 月球物理学
selenosis 硒中毒
selenotectonics 月球构造学
selenothermy 月热
selenous 二价硒的
selen-tellurjum 硒碲
seleroscope 反跳硬度计
self-absorption 自吸(收)
self-absorption and self-reversal 自吸和自蚀
self-accelerating decomposition 自加速分解
self-accelerating decomposition temperature 自加速分解温度

self-acceleration 自身加速度；自加速(度)
self-act 自动
self-acting 自动式；自动的
self-acting air bearing 动压空气轴承
self-acting boring machine 自动镗床
self-acting clutch 自动离合器
self-acting control 自动控制
self-acting discharge valve 自动泄水阀
self-acting door 自动开关(双扇)风门
self-acting drilling machine 自动钻床
self-acting ejector 自动抽逐器
self-acting feed 自动送料；自动进料；自动进给；自动进刀
self-acting feeder 自动送料器
self-acting gravity incline 自重运输斜坡道
self-acting grinding machine 自动磨床
self-acting idler 自调托滚
self-acting incline 轮子坡；自动(运输)斜道道；重力运输
self-acting injector 自动喷射器
self-acting intermittent brake 自动间歇制动器
self-acting lathe 自动车床
self-acting lubricator 自动润滑器
self-acting milling machine 自动铣床
self-acting moveable flood dam 自动泄洪活动坝
self-acting mule 自动机
self-acting piano link regulator 天平杆自动均匀装置
self-acting pump 自行启动泵；自动泵
self-acting pump station 自动抽水站；自动泵站
self-acting reco-rding meter 自动记录计
self-acting rope haulage 自重运输
self-acting scale 自动秤
self-acting seal 自紧密封
self-acting slope 自动斜坡
self-acting thermos 自动恒温器
self-acting thermostat 自动恒温器；自调恒温器
self-acting trailer coupling 自动拖挂装置；自动拖车联结器
self-acting travel 自动行程
self-acting valve 自动活门；自动阀
self-action 自作用；自动
self-action control 自动控制
self-action valve 自动阀
self-activated switch 自激放电器
self-activating 自激活动
self-activating brake 自加力制动器
self-activating brake band 自加力制动带
self-activating orange smoke signal 自然橙色烟
self-activation 自激活
self-actor 自动机
self-act travel 自动行程
self-actuated 自行的
self-actuated control 自行控制
self-actuated controller 自动控制器；直接调节器
self-actuated regulator 测量元件控制调节器
self-actuating 自动的
self-actuating brake 自动制动
self-adaptation 自适应
self-adaptation autopilot 自适应自动驾驶仪
self-adaptation command receiving technique 自适应指令接收技术
self-adaptation control 自适应控制
self-adaptation system 自适应系统
self-adapting 自适应(的)
self-adapting filter 自适应滤波器
self-adapting system 自适应系统
self-adaption 自适应调节
self-adaptive-adjusting autopilot 自适应式自动驾驶仪
self-adaptive autopilot 自适应式自动驾驶仪
self-adaptive control 自适应控制
self-adaptive controller 自适应控制器
self-adaptive control system 自适应控制系统
self-adaptive-correcting autopilot 自适应式自动驾驶仪
self-adaptive damping 自适应(性)阻尼
self-adaptive system 自适应系统
self-adaptivity 自适应性
self-adhering film 伸缩薄膜
self-adhering polymer modified bituminous sheet 自黏聚合物改性沥青油毡
self-adhesion 自黏作用；自黏合
self-adhesion method 自黏法

self-adhesive 自黏性；自黏的
self-adhesive film 自黏的薄膜
self-adhesive insulating tape 自黏性绝缘胶带
self-adhesive laminate 自黏层压板
self-adhesive material 自黏(合)材料
self-adhesive tape 自粘胶带；自粘带
self-adhesive tile 自粘的面砖
self-adjoint 自伴的
self-adjoint boundary value problem 自伴边值问题
self-adjoint differential equation 自伴微分方程
self-adjoint differential operator 自伴微分算子
self-adjoint elliptic differential operator 自伴椭圆型微分算子
self-adjoint elliptic operator 自伴椭圆型算子
self-adjoint equation 自伴方程式
self-adjoint extension 自伴扩张
self-adjoint linear differential equation 自伴线性微分方程
self-adjointness 自共轭性
self-adjoint operator 自伴算子
self-adjoint partial differential operators 自伴偏微分算子
self-adjoint spectral measure 自伴谱测度
self-adjoint system 自伴随系统
self-adjoint transformation 自伴变换
self-adjustable 自动调整的；自调的；可自动调节的
self-adjustable bearing 自动调整轴承；自动调节轴承座
self-adjustable drill chuck 自校钻夹；自调式钻卡
self-adjustable shock absorber 自动调整减震器；自调式减振器
self-adjustable sliding bearing 自调整滑动轴承
self-adjusting 自动调整(的)
self-adjusting arc welding 自动调节电弧焊；自调(式)电弧焊
self-adjusting automaton 自动调节自动机
self-adjusting balance 自调平衡
self-adjusting bearing 自动调整式轴承
self-adjusting brake 自动调整(的)制动器
self-adjusting brake head 自动调整闸瓦托
self-adjusting control 自调整控制
self-adjusting device 自调装置
self-adjusting draghead 自调耙吸头
self-adjusting expansion chamber 自动调节膨胀室
self-adjusting jaw vice 自调口虎钳
self-adjusting level 自动调整的水平仪；自动安平水准仪；自调水平仪
self-adjusting line of sight 自调(整)照准线
self-adjusting load-proportional brake 按承载自动调节的制动机
self-adjusting magazine 自调胶卷盒
self-adjusting mechanism 自动调整机理
self-adjusting model 自调整模型
self-adjusting seal 自紧密封
self-adjusting stop 自动调整限动器
self-adjusting system 自整定系统；自调整系统
self-adjusting tappet 自动间隙挺杆；自调式挺杆
self-adjusting type nozzle 自动调整型喷管
self-adjusting weigher 自动称量器
self-adjusting weir 自动调整堰；自动调节堰
self-adjustment 自动调整；自动调节；自调(整)
self-adjustment of river 河流自动调整
self-admittance 自导纳；固有导纳
self-advancing chock 自进式楔垫块
self-advancing prop system 自进式支柱系统
self-advancing support 自移式支架；自进式支架
self-aerated flow 自动曝气水流
self-aging 自然老化
self-aid navigation 自备式导航
self-aiming 自导引
self-aliging porous bearing 自调多孔轴承
self-align 自动调整
self-aligned 自对准的
self-aligning 自动对准(的)；自调整
self-aligning ball-bearing 自位滚珠轴承；自动调心滚珠轴承
self-aligning bearing 自位轴承；自动调心轴承；球面轴承；调心轴承
self-aligning carrying idler 自调位承载托滚
self-aligning centring roll 自动对心辊
self-aligning control 跟随控制
self-aligning control system 自校正控制系统
self-aligning coupling 自动调整联轴节

self-aligning double row ball bearing 自位对排滚珠轴承;双列自动调心球面轴承
self-aligning double row bearing 双列自动调心轴承
self-aligning effect 自定位效应
self-aligning gate 自控栅
self-aligning guide idler 自位导向支承滚轴
self-aligning idler 自调位托辊
self-aligning level 自动安平水准仪
self-aligning return idler 自动调位回转段托滚
self-aligning roller 自调位托辊
self-aligning roller bearing 自(调)位滚柱轴承
self-aligning seat washer 调心座圈
self-aligning spheric(al) rolling bearing 自动调心球面滚子轴承
self-aligning support 自(调)位轴承支座;自调支座
self-aligning system 自对准系统
self-aligning torque 自位转矩;自回正力矩
self-aligning wedge grip 自动调整楔形夹
self-alignment 自(定)位;自照准
self-align type 自调整式
self-amelioration 自身改良土壤
self-amortizing mortgage loan 分期偿还抵押贷款
self-anchorage 自锚(定)
self-anchored 自锚式;自锚碇的
self-anchored cable-stayed bridge 自锚式斜拉桥
self-anchored post tensioning prestressing 后张自锚法预应力
self-anchored stay cable 自锚拉索
self-anchored suspension bridge 自锚式悬索桥;自锚式吊桥
self-anchored suspension system 自锚式悬索体系
self-anchored system 自锚体系
self-anchoring 自锚固;自锚式;自锚定的
self-angling discs 自动板
self-annealing 自行退火;(锻后的)自退火;自身退火
self-aspirating turbine 自吸涡轮
self-assembling polar nano-composite 自组合纳米复合材料
self-assembly 自动装配
self-backwashing filter 自冲滤池
self-bailing 自动排水
self-bailing boat 自动泄水救生艇
self-bailing cockpit 自动泄水尾阱
self-baking 自焙
self-balanced 自动平衡
self-balance protection 自(动)平衡保护
self-balancing 自行平衡的;自平衡
self-balancing amplifier 自平衡放大器;零电平自动稳定放大器
self-balancing bolometer 自动平衡式辐射热计
self-balancing compensation measurement 自平衡补偿测量
self-balancing extractor 自动平衡脱水机
self-balancing force 自平衡力
self-balancing hot wire anemometer 自平衡热线风速计
self-balancing instrument 自(动)平衡仪表
self-balancing magnetic servo amplifier 自平衡磁伺服放大器
self-balancing measuring equipment 自平衡测量仪器
self-balancing phase inverter 自动平衡倒相器
self-balancing potentiometer 自平衡式电位计;自动平衡电位计
self-balancing receive-recorder controller 自平衡接收记录控制器
self-balancing recorder 自动平衡记录器
self-balancing strain ga(u)ge 自平衡应变仪
self-balancing system 自平衡系统
self-balancing type recorder 自动平衡记录器
self-baling(life)boat 自排水救生艇
self-ballasted lamp 自镇流灯
self-bearing capacity of surrounding rock 围岩自承能力
self-bearing wall panel 自承重墙板
self-bias 自(生)偏压
self-bias cut-off effect 自偏(压)截止效应
self-biased amplifier 自偏放大器
self-biased off 自偏压截止;自动偏压截止
self-biased tube 自偏管
self-biasing effect 自偏压效应
self-biasing impedance 自偏阻抗
self-biasing resistor 自偏电阻器

self-binder 自动割捆机;自动打捆机
self-blast circuit-breaker 自遁风断路器
self-bleeding 自动排气の
self-blocking 自动中断;自动联锁;自闭塞;自动阻塞;自动关闭
self-blowing drum filter 自吹鼓式过滤机
self-bonding 自粘法
self-bonding material 自结合材料
self-boring pressuremeter 自钻式旁压仪;自钻式横压仪
self-boring type torsion shear apparatus 自钻式扭剪仪
self-brake 自行制动
self-braking 自(动)制动的;自动停止
self-braking device 自动制动装置
self-brazing 自焊的
self-breakover 自转折
self-breakup unit 自毁装置
self-bucking hammer 自顶式铆枪
self-builder 自己动手施工者;自己动手建造者
self-build housing society 自建房屋协会
self-buoyant 自行浮起的
self-buoyant unit 自浮装置
self-bursting 自爆的
self-bursting fuse 自爆熔断器
self-burying anchor 自埋锚
self-bypassing filter 自动旁通滤油器
self-calibrating block adjustment 自检校区域网平差;自检法区域网平差
self-calibrating oxygen system 自检氧系统
self-calibrating system 自动校准系统
self-calibration 自动校准
self-calibration method 自检校法
self-cancelling 自相抵消(的)
self-cancelling direction indicator 自动消除信号方向指示器
self-cancelling trafficator 自动消除信号方向指示器;息灭式转向指示灯
self-capacitance 固有电容
self-capacity 本身电容;固有电容;本身容量
self-catalyzed reaction 自催化反应
self-centered 静止的
self-centering 自动归心;自动对中;自动定心
self-centering action 自动定心作用
self-centering apparatus 自动定心装置
self-centering brake 自行调心制动器
self-centering capability 自动定心能力
self-centering chuck 自(动)定心卡盘;(立轴的)自(动)定心夹盘
self-centering entry guide 自动定心的进口导板
self-centering formwork 自对中模板;自动定心模板
self-centering idler 调心托辊
self-centering internal measuring instrument 自定心内表面测量仪
self-centering lath 自立式钢板网
self-centering mandrel 自动定中心轴
self-centering punch 自定心冲孔器
self-centering screen 自动定心式筛分机
self-centering seal 自动定心密封
self-centering steering 自动回正转向机构
self-centering tripod 自动对中三脚架
self-centering trunnion 自定中心耳轴
self-centering unit 自动回中装置;自动定心机构
self-centering vibrating screen 自定中心振动筛
self-centering vibratory screen 自定中心振动筛
self-centering wheel 自动定心轮
self-centreing column seating 自动定位的柱座
self-certification 自行认证
self-chambering 自装
self-changing gear 自动变速齿轮
self-changing gearbox 自动变速箱
self-charge 自具电荷
self-charging 自充电
self-check 自检查;发票人支票
self-checking 自检;自动检验;自动检核
self-checking code 自校验码;自检验代码;自检编码
self-checking device 自动校准装置
self-checking digit 自校验数位
self-checking number 自动校验号码;自检(验)数
self-checking number feature 自校验数控装置
self-checking numeral 自校验数
self-checking system 自校验方式
self-checking unit 自检装置
self-check key 自校准开关

self-check procedures 自检程序
self-check program(me) 自检程序
self-choking chute 自卡卡式溜口
self-clarification system 自清系统;自净化系统
self-cleaning 自动清除;自洁;自净;自消洗;自然净化作用;自动清洗
self-cleaning ability 自净化能力
self-cleaning air filter 自净化空气过滤器
self-cleaning bit 自洁式钻头;自洁式钻头
self-cleaning bucket 自清铲斗
self-cleaning capacity 自净容量;自净能力;自动净化能力
self-cleaning car 自动卸料车
self-cleaning centrifuge 自动清洗离心机
self-cleaning cloth screen dust collector 自行清灰袋式集尘器
self-cleaning contact 滑动接点;擦拭触点
self-cleaning drilling 水冲钻探;仰孔钻进;自动除粉钻进;水冲钻孔
self-cleaning effect 自行清洁作用
self-cleaning electrode 自净电极
self-cleaning enamel 自洁陶瓷
self-cleaning enamel coating 自洁搪瓷
self-cleaning experiment of soil 土壤自净率实验
self-cleaning filter 自净(化)滤池;自净(化)过滤器
self-cleaning flexible pipe filter 自净软管过滤器
self-cleaning(flow)velocity 自净流速
self-cleaning grade 自净坡度
self-cleaning gradient 自净坡度
self-cleaning grizzl(e)y 自清式格筛
self-cleaning rotating screen filter 自净转筛过滤器
self-cleaning rotation screen filter 自动清洗转筛过滤器
self-cleaning screen 自净拦污栅;自动清污栅(网)
self-cleaning screw 自洁螺杆
self-cleaning separator 自清洗式分离器
self-cleaning spark plug 自净式火花塞;自动清洁的火花塞
self-cleaning strainer 自清洗式滤器
self-cleaning system 自净化系统
self-cleaning tank 自动清洁式油箱
self-cleaning test of water quality in field 野外水质自净试验
self-cleaning test of water quality in laboratory 室内水质自净试验
self-cleaning twin-screw reactor 自动清洗式双螺旋反应器
self-cleaning type centrifugal separator 自动卸料离心分离机
self-cleaning velocity 自清速度;自净流速;自净速度;自净化流速
self-cleaning viscous filter 自净黏液过滤器
self-clean oven 自洁式烤箱
self-cleansing gradient 自净梯度
self-cleansing velocity 自清速度;自净流速
self-cleansing velocity of sewer 管道自清流速
self-clearing 自动清洗;自动清除;自动净化的
self-clearing car 自动卸料车
self-climbing steel-form 自升式钢模
self-climbing tower crane 自升(式)塔式起重机
self-clinching fastener 自紧夹头
self-clocking 自动计时
self-clocking recording method 自定时记录法
self-closing 自接通;自动关闭;自闭合
self-closing air-door 自闭合空气门
self-closing butterfly valve 自动关闭蝴蝶阀;自动蝶形阀
self-closing device 防火门自动开关装置;自动关门装置;自闭装置
self-closing door 自动关闭门;自关风门;自闭水密门;自闭门
self-closing electric(al) gate 自闭式电动闸门
self-closing faucet 自关龙头;自闭旋塞;自闭(水)龙头
self-closing fire assembly 自闭防火窗配件;自闭防火窗五金
self-closing fire door 自关闭防火门;自动(关闭)防火门;自闭(式)防火门
self-closing gate 自闭闸门
self-closing ga(u)ge seal 自动关闭式量油口盖
self-closing grab 自合式抓斗
self-closing hinge 自动关闭铰链
self-closing link 自动关闭节

self-closing paper valve sack 自封口纸袋
self-closing stop valve 自动断流阀
self-closing valve 事故自动切断阀；自关闭阀；自闭(式)阀
self-cocking action 自动起爆
self-collapse 自动倒塌
self-collapsing gate 翻板闸门；自动翻倒式闸门
self-colo(u)r 一色的；本色；自然色
self-colo(u)red 单色的；本色的
self-colo(u)red material 自着色材料
self-combustible 自燃的
self-combustion 自燃作用
self-combustion firing period 自燃发火周期
self-combustion reason 自燃的原因
self-combustion tendency class 自燃倾向等级
self-combustion tendency index 自燃倾向指标
self-commutated inverter 自动换向逆变器
self-compacting 自动挤紧
self-compacting concrete 自密实混凝土；流态混凝土
self-compaction 自密实；自挤压
self-comparison method 自比较法(短寿命核测量方法)
self-compensated machine 自补偿电机
self-compensated motor 自补偿电动机
self-compensating 自动补偿的
self-compensating clutch 自动补偿离合器
self-compensating code 自补代码
self-compensating device 自动补偿装置
self-compensating hair-spring 自动补偿游丝
self-compensating independent pad type thrust bearing 单块瓦自补偿推力轴承
self-compensating magnetic balance 自补偿磁平衡
self-compensating pressure system 压力自动补偿系统
self-compensating system 自补偿系统
self-compensating take-up 自动补偿拉紧装置
self-compensating underground water detector 自动补偿地下水位电测仪
self-compensating unit 自动补偿装置
self-compensation 自补偿
self-complementary antenna 自互补天线
self-complementary code 自补码
self-complementary counter 自补计数器
self-complementing code 自补码
self-complementing counter 自补计数器
self-compound excitation 自复激励
self-compression interruption unit 自加压断路器单元
self-condensation 自缩合作用；自冷凝(作用)
self-conditioned power plant 自动调节动力装置
self-conductance 自(电)导
self-congruent 自相一致的
self-conjugate conic 自共轭二次曲线
self-conjugate element 自共轭元素
self-conjugate element of a polarity 配极的自共轭元素
self-conjugate graph 自共轭图
self-conjugate nuclei 自共轭核
self-conjugate operator 自共轭算子
self-conjugate particle 自共轭粒子
self-conjugate partition 自共轭分类
self-conjugate quadric 自共轭二次曲面
self-conjugate subgroup 正规子群
self-conjugate tetrahedron 自配极四面形
self-conjugate triangle 自配极三角形
self-conjugate vector space 自共轭向量空间
self-consistency 自洽性
self-consistent 独立的
self-consistent family of curves 自相一致的曲线族
self-consistent field 自洽力场
self-consistent field method 哈特里方法
self-consistent field scattered wave technique 散射波自洽场法
self-consistent units 自相一致的测量单位
self-consolidation 自重固结；自动固结
self-constricted beam 自收缩束
self-consumption 本单位消费
self-contact 自力接触
self-contained 配套的；整体的；自备的；整装固定的；机内；设备完善的；设备齐全的；独立的；不需电缆的；闭关自守的
self-contained air cleaner 送风机内装式空气净化设备
self-contained air conditioner 组合式空(气)调(节)器；整体式空气调节器；柜式空调器；自动空调器
self-contained air conditioning 整装式空气调节
self-contained air conditioning unit 带冷凝器的空气调节装置；组合式空(气)调(节)器；整体式空气调节机组；成套空调设备；成套空调机组
self-contained airstair(case) 随机客梯
self-contained apartment 成套公寓
self-contained automatic navigation system 自备式自动导航系统
self-contained-base range finder 内基线测距仪
self-contained bathing tub 带足浴盆
self-contained battery 自备电池；固定电池
self-contained bearing 不可分离型轴承
self-contained blade grader 自行式平路机
self-contained boiler 整装锅炉
self-contained breathing apparatus 自携式呼吸器；自我呼吸装置；自带潜水呼吸器
self-contained breathing system 自带呼吸系统
self-contained cardan joint 球窝万向节
self-contained complete 配套的
self-contained controller 独立式调节器
self-contained cooling unit 自给冷却机组；冷风机组；独立冷风机组；成套冷却器；成套供冷器
self-contained crash tender 自备式抢救车
self-contained data base 独立语言数据库
self-contained digital recording data acquisition system 自容式数字记录资料收集装置
self-contained dock 设备齐全的船坞
self-contained drainage system 配套齐全的排水系统；独立的排水系统
self-contained drill 自备动力钻机；独立式钻机；带发动机的钻机
self-contained drilling platform 独立式钻井平台
self-contained drinking water cooler 整体式饮水冷却器
self-contained dwelling 设备齐全的住房；独立住户
self-contained ecosystem 自给生态系统
self-contained engine 独立发动机
self-contained engine drive 单独发动机传动
self-contained equipment 独立设备；成套设备
self-contained fire detector unit 配套的消防探测装置
self-contained flat 设备齐全的居住单元
self-contained graphic(al) recording profiler 自容式图表记录剖面仪
self-contained hopper 自备泥舱(挖泥船)
self-contained instrument 整装仪表；不需辅助设备的仪表
self-contained language system 独立语言系统
self-contained lubrication 压циркуля循环润滑法；强制循环润滑
self-contained mobile home 设备齐全的活动房屋；独立的可移动的住房
self-contained motor drive 单独电动机驱动；单独电动机传动
self-contained navigation aid 自足式助航设备；自载导航设备
self-contained navigator 自备式导航仪
self-contained ornament 完整装饰品
self-contained oxygen generator 独立式氧气发生器
self-contained penetrant inspection unit 成套式渗透探伤仪
self-contained plant 自给企业；自动装置；自备动力设备
self-contained platform 自给(式)平台；自备动力工作平台；独立式平台；独立工作的钻探平台；独立工作的钻井平台
self-contained portable electric(al) lamp 自给式携带电灯
self-contained power plant 独立动力厂
self-contained power supply equipment 自备供电设备
self-contained press 带有液压泵的压制机
self-contained pump 独立泵
self-contained radar 自导引系统雷达
self-contained range finder 自动控制测距仪；直观式测距仪；光学测距仪
self-contained recording system 独立记录系统
self-contained sanitation 配套的卫生设备
self-contained slurry machine 设备齐全的泥浆(拌和)机
self-contained start 自持起动
self-contained supply 自保持电源
self-contained system 自律系统；自控生态系统；整体装置；配套系统；独立系统
self-contained thermostat 自动定温器
self-contained underwater breathing apparatus 自携式水下呼吸器；自带潜水呼吸器；自带呼吸系统
self-contained unit 自给系统；整装装置；整套设备；独立设备；成套设备
self-contained vehicle 具有卫生设备的车辆；设备齐全车辆
self-contained vertical drilling machine 独立立式钻床
self-containing unit 自律系统
self-contradiction 自相矛盾
self-control 自动操作；自动调整；自动控制
self-control enemator 自动控制气压灌肠器
self-controlled 自控式的
self-controlled oscillator 自控振荡器
self-controller 自动控制器
self-conveyer feed 螺旋加料器；螺旋加料机
self-cooking place 自炊处
self-cooled 自冷的
self-cooled machine 自冷式电机
self-cooled nozzle 自冷式喷管
self-cooled transformer 自冷式变压器
self-cooling 自行冷却的；自冷式；自动冷却
self-cooling vane 自冷式叶片
self-correcting 自动调整；自校；自动校正
self-correcting automatic navigation 自修正自动导航；自校自动导航
self-correcting automatic navigator 自校正自动领航仪
self-correcting capability 自校正能力
self-correcting code 自校编码
self-correcting homeostasis 自校稳态
self-correcting memory 自校正记忆
self-correcting system 自校正系统
self-correction 自校正
self-correction code 自校编码
self-correlation 自相关
self-correlation function 自相关函数
self-correlator 自相关器
self-corresponding 自对应的
self-corrosion 自蚀
self-cost 成本
self-coupling 自动联结器；自动连接
self-coupling device 自动联结装置
self-coupling hitch 自动联结器
self-curing (混凝土的)自动养护；自动硫化
self-curing adhesive 自固黏结剂；自固黏合剂
self-curing binder 自硬黏结剂
self-curing cement 自动硫化胶
self-curing sand mo(u)ld 自硬砂型
self-curing sealant 自熟化嵌缝膏
self-damping 自阻尼
self-detaching hook 自分离钩
self-debugger 自调式程序
self-decomposition 自分解
self-defining 自定义的
self-defining data 自定义数据
self-defining file 自动定义文件
self-defining term 自定义项
self-defining value 自定义值
self-degreasing oil filter 自去油过滤器
self-demagnetization 自去磁；自动消磁
self-demagnetization factor 自去磁因数
self-demagnetization force 自去磁力
self-demagnetization loss 自去磁损耗
self-demagnetizing factor 自退磁系数
self-demarcating code 自分界码
self-demarking code 自解码；自检码
self-deploying space station 自展式航天站
self-description 自描述
self-desiccation 自身除湿；自脱水；自(行)干燥
self-destroying 自毁
self-destroying device 自毁装置
self-destroying fuse 自毁引信
self-destruct charge 自毁装药
self-destruction 自毁作用
self-destruction device 自毁装置
self-destruction equipment 自毁装置

self-destruction feature 自毁装置
self-destruction system 自爆系统
self-destruction-type fuze 自毁引信
self-destructor 自毁器
self-destruct sensor 自炸装置传感器
self-detaching 自分离
self-detaching hook 自动脱绳钩
self-detecting control voltage 自检控制电压
self-detecting signal frequency 自检信号频率
self-diagnosing 自诊断的
self-diagnostic 自诊断的
self-diagnostics 自诊断
self-differentiation 本身分化；自主分化
self-diffusion 自扩散(作用)；自行扩散
self-diffusion coefficient 自扩散系数
self-digestion 自溶；自体溶解
self-directing 自动定向的
self-direction 自动引导
self-discharge 局部放电；自身放电；自放电；自动卸载
self-discharging 自动输电；自动放电；自卸(的)；自动卸载；自动卸货；自动泄水；自动排料
self-discharging barge 自卸驳(船)
self-discharging body 自卸式车厢
self-discharging bucket 自卸式吊桶
self-discharging bulk carrier 自卸散装货轮
self-discharging car 自卸车
self-discharging collier 自卸式运煤船
self-discharging purifier 自动排渣净油机
self-discharging ship 自卸船
self-discharging system 自卸系统
self-discharging trailer 自卸拖车
self-discharging truck 自卸载货汽车
self-discharging wagon 自卸(汽)车；自翻车；自动料车；自动卸货车
self-disengaging 自动脱开
self-dispersing 自分散
self-distributivity 自分配性
self-division 自身分裂
self-docking 自行入坞
self-docking dock 分节(式)浮坞；自坞修船坞；自承式分段浮坞；可分段拆卸的浮船坞；分段接合浮坞
self-docking type 自行入坞式
self-dowelling (混凝土路面的)自动准合；(混凝土路面的)自动榫合
self-drain 自排水
self-drainage 自排水工程
self-draining 自疏水；自(动)排水
self-draining arrangement 自动排水装置
self-draining cockpit 自动泄水尾阱
self-draining hydrant 自动防水消防栓
self-draining lifeboat 自排水救生艇
self-draining pipe 自动排水管
self-draught beam 自动索引木干
self-drawbar 自动索引木干
self-dredging 自然刷深
self-drilling screw 自攻螺钉；自钻孔攻丝螺钉
self-drilling tapping screw 自钻孔自攻丝螺钉
self-drive 自动推进
self-drive motor 自起动电动机
self-driven 自驱动；自动推进的；自行驱动的
self-driven ingot buggy 自动送锭车
self-driven instrument 自驱动仪表
self-driven line-scanning circuit 自驱行扫描电路
self-dual function 自对偶函数
self-dual groups 自对偶群
self-dual identities 自对偶恒等式
self-dual linear space 自对偶线性空间
self-dual matroid 自对偶拟阵
self-dual system 自对偶系
self-dumping 自动卸载；自动卸料；自动倾卸(的)
self-dumping barge 自卸驳(船)
self-dumping cage 自卸罐笼；翻转罐笼
self-dumping car 自卸矿车
self-dumping grab 自卸抓斗
self-dumping hopper suction dredge(r) 装舱自卸扬挖泥船
self-dumping log carrier 自卸式运木船
self-dumping scroll 自动卸载曲轨
self-dumping skip 自卸吊斗
self-dumping truck 自卸汽车；自卸(货)车
self-editing data 自编数据
self-editor 自编辑程序

self-electromotive force 自然电动势
self-electrooptic(al)effect unit 自光电效应器件
self-elevated platform 自升式平台
self-elevating 自升的
self-elevating barge 自升式平底船；自顶升驳船
self-elevating drilling platform 自升式钻探平台；自升式钻井平台
self-elevating drilling ship 自升式钻探船；自升式钻井船
self-elevating platform 自升式水上(作业)平台；自升式平台；自升式驳船；自升降平台；自顶升平台；海上作业平台
self-elevating pontoon 自升式驳船
self-elevating shutting 自升式模板；滑动模板
self-elevating working platform 自升工作平台
self-elevation 自升式
self-elevation offshore platform 自升式海上钻探平台；自升式海上钻井平台
self-embedding 自嵌入
self-embedding property 自嵌入特性
self-emission electrode 自发射电极
self-emitting light source 自发光源
self-employed and private business 私营经济
self-employed and private businesses 个人经济和私营经济
self-employed individual 个体户
self-employed worker 个体劳动者
self-emptying 自动卸载；自动放空；自动出料
self-emptying cockpit 自动泄水尾阱
self-emptying hopper 自卸式料仓
self-emptying trailer 自动卸货拖车
self-emulsifiable 可自乳化的
self-emulsifying 自乳化
self-energized brake 自行加力制动器
self-energized seal 自紧式密封
self-energizing 自激励的
self-energizing action 自助力作用
self-energizing brake 自增(力)式制动器；自动制动闸；自动增力制动器
self-energizing brake band 自加力制动带；自动制动带
self-energizing resilient metal gaskets 自紧式弹性金属垫片
self-energy 自身能量；自具能(量)；固有能(量)；内禀能量；本征能量
self-equalibrating 自动平衡
self-equalizing 自动平衡
self-equalizing preamplifier 自(动)均衡前置放大器
self-equilbrating 自平衡
self-erecting 自动架设的；自行装配；自动装配；自动竖立；自动安装
self-erecting batch type mixer 自动投配式搅拌机；自动投配式拌和机
self-erecting crane 自动运转起重机
self-erecting tower crane 自行架设塔式起重机
self-established tide 自振潮；独立潮
self-etch primer 磷化底漆；自蚀底漆；洗涤底漆
self-evapo(u)rative quality 自蒸发能力
self-evident 不言而喻的；不证自明的
self-excitation 自(激)励；自激(发)
self-excitation low-frequency generator 自激低频发电机
self-excitation method regeneration generator 自激式再生发电机
self-excitation oscillation 自激振荡
self-excitation phenomenon 自激现象
self-excitation transductor 自激磁放大器
self-excitation type 自激型
self-excitation winding 自励(磁)绕组
self-excited 自励的；自激的
self-excited alternative current generator 自激交流发电机
self-excited alternator 自激交流发电机
self-excited amplifier 自激放大器
self-excited booster 自激式升压机
self-excited circuit 自励电路；自激电路
self-excited dynamo 自励发电机；自激发电机
self-excited generator 自激发电机
self-excited grounded grid oscillator 栅地自激振荡器
self-excited machine 自激电机
self-excited magnetohydrodynamic generator 自激式磁流体发电机
self-excited motor 自励电动机

self-excited multivibrator 自激多谐振荡器
self-excited oscillation 自振动；自励振荡
self-excited phase advancer 自励式进相机
self-excited regenerative braking 自励再生制动方式
self-excited series winding 自励串励绕组
self-excited shunt field 自励并激磁场
self-excited system 自激励系统
self-excited transductor 自激磁放大器
self-excited transmitter 自激发射机
self-excited vibration 自激振动
self-excited vibration frequency of bridge span 桥梁自振频率
self-exciter 自励发电机
self-exciting 自激的；自励的
self-exciting condition 自激条件
self-exciting dynamo 自励磁发电机
self-exciting modulating amplifier 自激式调制放大器
self-exciting vibration densimeter 自激式振动密度计
self-exciting winding 自激绕组
self-expanding cement 自膨胀水泥；自应力水泥
self-expanding piston ring 自动伸缩活塞环
self-explanatory 不证自明的
self-extinction of arc 自消弧
self-extinguish(ing) 自动灭火；自熄灭性；自熄性材料
self-extinguishing circuit-break 自灭弧断路
self-extinguishing resin 自熄性树脂
self-faced 无须加工的石面；自然石面的；未修整的(石板)表面；天然表面
self-faced slab 无须加工石板
self-faced stone 天然表面石材；无须加工的石板；未雕凿石材
self-feed 自进给；自动供料；自动给料；能自给的
self-feedback 自回受；自反馈
self-feedback amplifier 自反馈放大器；内回授放大器
self-feeder 自给器；自动喂料机；自动投料器；自动投料机；自动送料器；自动进料器；自动加料器；自动供料器；自动给料器；自动给料机
self-feeding 自动送料；自动加料；自动给料的
self-feeding carburetor 自吸式气化器
self-feeding portable conveyor 自动上料移动式运输机
self-field 固有场
self-filling drinking bowl 自动添水式饮水器
self-filling windscreen washer 自充水式风窗玻璃洗涤装置
self-finance 自筹资金；内部筹款；企业内部筹款
self-financed 自筹的
self-financing 自负盈亏；自筹资金
self-financing ratio 自筹资金比率
self-finish 清水面
self-finished bitumen asbestos felt 沥青石棉卷材(英国)
self-finished bitumen felt 沥青毡(英国)
self-finished roofing felt 自身饰面的屋面毡；沥青浸渍屋面毡；自饰面屋面油毡
self-fission 自裂变
self-flaring fitting 自动扩口接头
self-flashing 自闪光
self-floating pipeline 自浮管线
self-flocculation 自动絮凝(作用)
self-flooding 能自行灌水的
self-flowing grout 自流灌浆
self-flowing well 自流井
self-flow reducer 自流式(排种)节流活门
self-flush tank 自动冲洗(水)箱
self-fluxing 自生熔剂
self-fluxing alloy 自助熔合金
self-fluxing brazing alloy 自带熔剂的硬钎料
self-fluxing nature 自熔性
self-fluxing ore 自熔性矿石；自熔矿
self-fluxing sintered ore 自熔性烧结矿
self-focused laser beam 自聚焦激光束
self-focusing 自聚焦
self-focusing apparatus 自动对光装置；自动对光设备；自动调焦装置
self-focusing beam 自聚束
self-focusing fiber 自聚焦纤维
self-focusing laser 自聚焦激光器
self-focusing mechanism 自聚焦机制

self-focusing optic(al) fiber 自聚焦光(学)纤维
self-focusing rectifying apparatus 自动调焦纠正仪
self-focusing stream 自聚集束流
self-folding strip chart 自折叠条形记录纸带
self-force 自作用力；本身力
self-formed 自形成的
self-formed model 自然模型
self-forming 自成模板
self-forming bio-dynamic membrane 自生生物动态膜
self-forming dynamic membrane 自生动态膜
self-forming dynamic membrane bioreactor 自生动态膜生物反应器
self-fractionating oil pump 自分馏油泵
self-fractionating pump 净化泵
self-free 不需辅助设备的
self-frequency 自频；固有频率
self-fulfilment management style 自我目标管理形
self-furring 自行垫高料；自垫高板条
self-furring metal lath 自垫高金属拉网；自垫高金属条板
self-furring nail 自垫高板钉；自垫高条钉
self-fusible 自熔的
self-gating 自穿透
self-generated graphics 自然生图像
self-generated pressure 自生压力
self-generating method (齿轮的)滚铣法
self-generating transducer 自生换能器
self-generation 自然发生
self-geosyncline 自然地槽
self-glazing 自动上釉；自行施釉；自动研光
self-governing 自治的
self-governing city 自治城市
self-grading 自动分级
self-granulated sludge 自粒化污泥
self-gravitation 自吸引
self-grinding action 自磨作用
self-grinding machine 自动磨床
self-gripping general purpose pliers 自闭多功能扁嘴钳
self-gripping jaws 自动夹爪
self-grown 自生的
self-grown river 自生河
self-grown stream 自生河
self-guarded frog 自护式辙叉
self-guidance 自导引
self-guided 自动导向的；自导式的
self-guided weight-loaded accumulator 自导向的重锤式蓄能器
self-hardening 自硬性；自(行)硬化
self-hardening of air 空气硬化
self-hardening property 自硬性
self-hardening sand(mixture) 自硬砂
self-hardening slurry cut-off wall 自凝灰浆截水墙；自凝灰浆防渗墙
self-hardening steel 气硬钢；自(淬)硬钢；空气硬化钢；空冷淬硬钢；气冷硬化钢
self-hauling drill 自行式钻机
self-heal 夏枯草属
self-healing 裂缝的自动闭合；自愈性；自愈合；自复的；自动强化；自动合拢；(裂缝的)自动闭合
self-healing ability 自愈能力
self-healing ability of concrete 混凝土自愈能力
self-healing action 回自作用；自复作用
self-healing coating 自愈合土层
self-healing effect 自复原效应
self-healing hybrid ring 自愈混合环
self-healing network 自愈网
self-healing soil 自愈性土(料)
self-healing tendency 自动愈合趋势
self-heated thermocouple 自热式热电偶
self-heat(ing) 自(动)加热；自热
self-heating thermionic tube 自热式热离子管
self-heaving 自绞
self-help 自行处置
self-help house 自助的房子
self-help housing 自助(公建)住宅；自助住宅；自建住房
self-help housing program(me) 自助住宅方案
self-help laundry 自助洗衣间
self-help program(me) 自助计划
self-heterodyne 自拍；自差
self-hoisting 自升的
self-hoisting tower crane 自升(式)塔式起重机

self-hold capital 自有资金
self-hold(ing) 自动夹紧(的)；自锁
self-holding contact 自保持触点
self-holding screwdriver 弹力持骨螺钉镟
self-holding taper 自动夹紧拔销
self-hooped penstock 自箍式压力管
self-hunting 自动寻线
self-identification 自识别
self-igniting 自燃
self-igniting bipropellant 自燃双组分推进剂
self-igniting light 救生灯浮；烟火信号
self-ignition 自燃(着火)；自发火；自动点火
self-ignition engine 自发火发动机
self-ignition index 自燃指数
self-ignition method 自行点燃法
self-ignition point 自燃温度；自燃点
self-ignition point temperature 自燃着火点
self-ignition temperature 自燃(着火)温度；自发火温度
self-illuminated dot 光点
self-illuminated sign 自亮标志
self-illuminating sight 自身照明瞄标
self-impedance 固有阻抗
self-impedance function 自阻抗函数
self-impinging injector 自冲击射流喷嘴
self-indexing 自动分度(法)
self-indicating 自动指示
self-indicating scale 指示秤
self-induced air burner 自动雾化喷嘴；自动进气燃烧器
self-induced drift 自感漂移
self-induced magnetic flux 自感磁通
self-induced oscillation 自激振荡
self-induced scrubber 自激式除尘器
self-induced transparency 自感应透明
self-induced vibration 自感振动；自激振动
self-inductance 自感现象；自感量；自感(应)系数
self-inductance coefficient 自感系数
self-induction 自感(应)
self-induction coefficient 自感系数
self-induction coil 自感线圈
self-induction current 自感应电流
self-induction electromotive force 自感电动势
self-induction voltage 自感电压
self-inductor 自感线圈
self-inductor coil 自感线圈
self-infection 自干扰
self-inflammable 可自燃的
self-inflammable material 自燃物质
self-inflammable mixture 自燃混合物
self-inflating 自行充气
self-inflating lifeboat 自动气胀式救生筏
self-inflation surface target 自充水面目标
self-inflection 自屈曲
self-information 自信息
self-inhibition 自抑制
self-initialize 自初始化
self-initializing test sequence 自引导测试序列
self-injection 自喷射
self-inspection 自检
self-instructed carry 自动进位
self-insurance 自保
self-interaction 自身相互反应
self-interest in economic management 经济经营中的本身利益
self-interference 自干扰
self-interrupter 自行断续器；自动断续器
self-interrupting circuit 自行断续电路
self-interval timer 自动间隔计时器
self-investment 自筹投资
self-ionization 自电离(作用)
self-irradiation 自辐照
self-irradiation damage 自辐照损伤
self-irradiation effect 自辐照效应
selfix 加速电机(用于公共房屋的热水系统)
self-jacking offshore drilling vessel 自升式海上钻探平台；自升式海上钻井平台
self-jamming 自锁
self-jetting well point 自射式井点
self-killing 钢液自然脱氧
self-latching bolt 自动插销
self-laying track-type industrial tractor 自动导向履带式工业拖拉机
self-learning system 自学系统

self-level(l)ing 自动取平的；自调平
self-level(l)ing device 自动调平装置
self-level(l)ing floor coating 自流平地面涂料
self-level(l)ing floor finish 自动找平饰面剂；自动定平面层
self-level(l)ing floor plaster 自流平地面石膏
self-level(l)ing instrument 自动找平水准仪；自动定平水准仪；自动安平水准仪
self-level(l)ing laser 自动调平激光器
self-level(l)ing level 自平水准仪；自动测水准仪；自动安平水准仪；自校准水平仪
self-level(l)ing sealant 自流平密封膏；自动找平密封胶
self-level(l)ing shock absorber 自动调平式减震器
self-level(l)ing shoe 自动调平式筛架
self-level(l)ing spreader 水平自动调节吊具；自动水平调节器具；自动水平调节吊具
self-level(l)ing suspension 自动调平悬架
self-level(l)ing unit for removing oil pollutants 清除水面油污的自动平衡刮集装置
self-level(l)ing unit for removing pollution 油污自动平衡刮集装置；水面油污自动平衡刮集装置
self-lift 自动提升器；自动提升机；自动起落器
self-lifting forms 自提升模板
self-lifting resistance 抗咬起性能
self-lifting system 自动提升系统
self-lift plough 自动升降犁
self-lift plow 自动起落犁
self-limiting reactor 自调节反应堆
self-linking coefficient 自环绕系数
self-liquidating investment 自行还本生利的投资；能迅速还本生利的投资
self-liquidating loan 自动清偿贷款
self-load 自重；自输入
self-loaded 自动装料的
self-loading 自行铲装式；自装入；自装的；自动装填
self-loading and unloading ship 自动装卸船
self-loading autoweighing blender 自加料自动称重掺和机
self-loading device 自动装料器
self-loading elevator 自动装置升降机
self-loading excavator 自装载挖土机
self-loading hauler 自装载运输机
self-loading hauling unit 自装运输机组
self-loading mechanism 自动装填机构
self-loading mixer 自装搅拌车
self-loading motorized buggy 自动装载机动手推车
self-loading nut 自锁螺帽
self-loading program(me) 自装入程序
self-loading scraper 自装载铲运机；自装刮土机；自装铲运机；自动铲运机
self-loading ship 自动装载船
self-loading skidder 自装低轮集材车
self-loading skip 自装箕斗
self-loading tractor bucket-elevator 自动装载牵引斗式提升机
self-loading trailer 自动装载拖车
self-loading vehicle 自动装货式汽车；自动装货车
self-loading wheeled scraper 自动装轮式铲运机
self-lock(ing) 自锁；自动制动
self-locking action 自锁作用
self-locking angle cock 自锁角旋塞
self-locking catch 自动闭锁器
self-locking circuit 自锁电路
self-locking differential(gear) 自锁(式)差速器
self-locking fastener 自锁紧固件
self-locking gear 自动闭锁装置；自动闭塞装置
self-locking nut 自锁螺母；自锁螺帽
self-locking steering gear 自锁转向装置；不可逆转向装置；不可逆转向器
self-lock mechanism 自锁原理
self-lock pin 自锁销
self-loop 自循环；自回路
self-lube bearing block 自润滑轴承箱组
self-lube cartridge unit 自润滑环形轴承箱组
self-lube flange cartridge unit 自润滑圆形法兰式轴承箱组
self-lube four-bolt flange unit 自润滑四螺栓方形法式轴承箱组
self-lube hanger unit 自润滑悬挂式轴承箱组
self-lube pillow block unit 自润滑枕式轴承箱组
self-lube take-up unit 自润滑滑块式轴承箱组
self-lube two-bolt flange unit 自润滑双螺栓菱形法兰式轴承箱组

self-lubricate 自动润滑
self-lubricated 自动润滑的
self-lubricating 自动润滑(的)
self-lubricating bearing 自(动)润滑轴承
self-lubricating block 自润滑车
self-lubricating brush 自润滑电刷
self-lubricating bush 自润滑轴衬
self-lubricating gear notch 自润滑传动滑槽
self-lubricating journal 自润滑轴颈
self-lubricating lower track roller 自动润滑式链轨支重轮
self-lubricating material 自润滑材料
self-lubricating property 自润滑性能
self-lubricating pump 自给润滑泵
self-lubricating sheave 自润滑轮
self-lubrication 自身润滑；自动润滑
self-lubricity 自身润滑性
self-luminescent lamp 自发光灯
self-luminescent material 自发光物质；自发光材料
self-luminosity 自发光度
self-luminous 自发光的
self-luminous body 自发光体
self-luminous colo(u)r 自发光色
self-luminous compound 自发光化合物
self-luminous dial 自发光度盘
self-luminous paint 自发光漆；夜光涂料；自发光涂料
self-luminous(perceived) colo(u)r 自发光(知觉)色
self-luminous pigment 自发光颜料
self-made 自制的
self-made quicksand 自成悬浮体
self-magnetic 自磁(感)的
self-magnetic energy 自磁能
self-magnetic field 自磁场
self-maintained circuit 自激振荡电路
self-maintained discharge 自持放电
self-maintained discharge detector 自持放电检测器
self-maintaining 自然维护；自然冲淤
self-maintaining discharge 自持放电
self-maintaining nuclear chain reaction 自持链式核反应
self-maintaining reactor 自持堆
self-maintaining system 自我维持系统
self-maintenance 自动维护
self-maintenance and regulation 自行维持和调节
self-management 自主管理权
self-manufactured semi-finished products in stock 库存自制半成品
self-marker hypothesis 自身标记假说
self-measuring device 自动计量装置
self-measuring tank 自动计量水箱
self-mixing 自动混合
self-mixing trailer 自动搅拌混合饲料车
self-mobile 自动的
self-modeling 自模化
self-mode-locking 自锁模；模同步
self-modulated oscillator 自调制振荡器
self-modulating oscillator 自调制振荡器
self-modulation 自调制
self-monitored autopilot 自监控式自动驾驶仪
self-monitoring 自行监控的
self-monitoring device 自监控装置
self-monitoring system 自监控系统
self-mounting 自动安装(的)
self-mounting tyre chain 自装式轮胎防滑链
self-movement 独立运动
self-mover 自动推动器
self-moving 自动推进的
self-mulching soil 自覆盖土壤
self-mutilation 自毁
self-mutual inductance 自互感
self-natural frequency 固有频率
self-navigation 自动导航
self-neutralizing frequency 自中和频率
self-neutralizing valve 中间自位阀
self-noise 内噪声；自噪声
selfoc fibre 自聚焦纤维
self-oil 自动润滑
self-oiler 自动加油器
self-oil feeder 自动给油器
self-oiling 自动上油的；自动加油的
self-oiling bearing 自(动)润滑轴承

self-oiling block 自润滑车
self-oiling guide bushing 自润滑导套
self-open 自然裂隙
self-opening 自动开启的
self-opening die head 自动开合螺丝板牙切头
self-opening gate 自动快门
self-opening purifier 自动排渣净油机
self-opening reamer 自张式扩孔器
self-operated control 自行控制；自力控制
self-operated controller 自动控制器
self-operated measuring unit 自动测量装置
self-operated thermostatic controller 自动调温器
self-operating fountain 自动喷泉
self-optimalizing 自寻最佳点
self-optimizing control 最优自动控制；最佳自动控制；自优化控制
self-optimizing control system 自寻最佳化控制系统
self-optimizing decision system 自寻最佳决策系统
self-optimizing filter 自寻最佳化滤波器
self-optimizing system 自寻最佳系统；自动最优化系统
self-organization 自组织
self-organizing 自组织
self-organizing control 自组织控制
self-organizing controller 自组织控制器
self-organizing equipment 自组织设备
self-organizing machine 自编机
self-organizing program(me) 自组织程序
self-organizing system 自组织系统；自编系统
self-orientating 自动定向的
self-orientation 自动定向
self-orthogonal 自成正交
self-orthogonal block code 自正交块码
self-orthogonal convolutional code 自正交卷积码
self-oscillating 自激振荡
self-oscillating linear induction motor 自振荡直线感应电机
self-oscillating regime 自振状态
self-oscillating rubidium vapo(u)r magnetometer 自振铷气磁强计
self-oscillating system 自振荡系统
self-oscillation 自(生)振荡；自激振荡
self-oscillation elimination 自振荡消除
self-oscillation method 自动振荡法；固有振动法
self-oscillator 自激振荡器
self-oscillator radio telegraphy transmitter 自激振荡器式发报机
self-oscillatory 自激振荡的
self-oscillatory transducer 自激振荡式换能器
self-osculation 自密切
self-packing 自行包装；自行填实；自动填实
self-parallel curves 自平行曲线
self-parking 停车自理；自理停车场
self-parking wiper 自动开关刮水器
self-perpetuating 自保持
self-perpetuating bureaucracy 永久机构
self-phase conjugation 自相位共轭
self-phase modulation 自相位调制
self-plating solution concentration 自镀液浓度
self-plating solution heating time 自镀液加热时间
self-plating solution temperature 自镀液温度
self-plugging rivet 抽芯铆钉
self-poise 自平衡力；自动平衡
self-poking arrangement 自动搅拌装置
self-polar 自配级的
self-polar curve 自配极曲线
self-polarizing relay 自极化继电器
self-polar surface 自配极曲面
self-polar tetrahedron 自配极四面形
self-polar topology 自配极拓扑
self-polar triangle 自配极三角形
self-polishing wax 有光蜡
self-polymerization 自聚合(作用)
self-positioning 自动对位；自动定位
self-potential 自(然)电位；自电势；自位
self-potential curve 自电位曲线；自电势曲线
self-potential difference 自电位差
self-potential log 自然电位录井曲线；自然(电位)测井曲线；自电位测井
self-potential method 自然电位法
self-potential prospecting 自电位勘探；自电势勘探
self-potential survey 自电位勘探；自电势勘探
self-powered 自行驱动的；自己供电的；自动的；自备能源的

self-powered cargo vessel 自航驳(船)
self-powered destructor 自供电自炸装置
self-powered detector 自供电探测器
self-powered platform 自升落式工作平台
self-powered scraper 自行式铲运机
self-powered vehicle detector 自电源车辆检测器
self-power neutron detector 自给能中子探测器
self-pressurize 自动加压(供油)
self-pressurized system 自加压系统
self-prestressed cement 自预应力水泥
self-prime 自动运转；自动启动；自灌；自吸
self-prime pump 自吸式水泵；自启动水泵
self-primer 自动起爆装置；自动起爆器
self-priming 自动点火的(火箭发动机)；自动注油启动的；自动充满的；自吸的；自灌；自动起注油；底面两用漆
self-priming centrifugal pump 自吸(式)离心泵
self-priming device 自吸设备
self-priming pump 自启动水泵；自引泵；自吸式水泵；自吸式离心泵；自吸泵；自动引水泵
self-priming pumping house 自灌式泵房
self-priming recharge 自灌充
self-priming tank 自动调整液面水箱；自行上水水箱；自动调节液位箱；自调液面水箱
self-priming top coat 底面两用面漆；底面合一面漆
self-programming 自动编制程序；程序自动化
self-programming computer 程序控制计算机
self-programming software 自编程序软件
self-programming system 自动编程序系统
self-propagating 自动传布的；自动扩展的；自动传播
self-propagating high-temperature synthesis 自蔓延高温合成法
self-propelled 自行式的；自(动)推进的；自航(驳)式的
self-propelled assault bridge 自行式架桥车
self-propelled barge 自航驳(船)；机动驳(船)
self-propelled barrack ship 自航式兵营船
self-propelled bathyscaphe 自航式深海探测艇
self-propelled blader 自动平路机；自动平地机；自行式平路机
self-propelled bottom dump barge(r) 自航开底泥驳
self-propelled bucket dredge(r) 自航链斗挖泥船
self-propelled car 自动车
self-propelled carrier 自动底盘
self-propelled chain-type side-rake 自走链式侧向搂草机
self-propelled chassis 自动(推进式)底盘
self-propelled combine harvester 自动联合收割机
self-propelled concrete car 自动行驶式混凝土车
self-propelled concrete pump 自动混凝土泵
self-propelled concreting plant 自拌式混凝土厂；自行式混凝土浇筑装置；自行式混凝土浇筑设备；自行式混凝土浇筑机；自行式混凝土浇灌装置；自行式混凝土浇灌设备；自行式混凝土浇灌机
self-propelled crane 移动式起重机；自走式起重机；自行式起重机
self-propelled crawler mounted crane 自动推进爬行起重机
self-propelled crushing plant 自动轧石机；自动轧石厂；自动破碎设备
self-propelled crust breaker 自动推进式打壳机
self-propelled cutter suction dredge(r) 自航绞吸式挖泥船
self-propelled diesel railcar 自动推进式柴油机有轨车
self-propelled ditch digger 自走式挖掘机
self-propelled dredge(r) 自航进挖泥船；自航式挖泥船
self-propelled drill(er) 自行式钻机
self-propelled drilling platform 自航式钻井平台
self-propelled dump rake 自走式横向搂草机
self-propelled electric(al) locomotive 自备电源式机车
self-propelled excavator 自动挖土机
self-propelled floating crane 自航浮式起重机
self-propelled floating revolving crane 自航旋转浮吊；自航回旋浮吊
self-propelled frame 自行式底盘
self-propelled gantry manure crane 自走式厩肥高架起重机
self-propelled grab dredge(r) 自行抓斗挖泥船

self-propelled grab hopper dredge(r) 自航装舱抓斗挖泥船
self-propelled grader 自动平地机
self-propelled grass mower 自动推进的除草机
self-propelled gritter 自动铺砂机
self-propelled hopper barge 自行开体泥驳
self-propelled hopper dredge(r) 自航式料仓采砂船；自航式料仓采砂船
self-propelled hopper suction dredge(r) 自航耙吸式挖泥船
self-propelled hydraulic rotary auger face drill 工作面自行式液压回转螺旋钻机
self-propelled ingot buggy 自动送锭车
self-propelled jumbo 自行式凿岩台车
self-propelled lifeboat 自航救生艇
self-propelled lighting equipment 自行式照明设备
self-propelled loader 自行式装载机
self-propelled main machine 自行主机
self-propelled mine-collecting machine 自航集矿机
self-propelled mobile reactor 自行式堆
self-propelled model 带发动机的模型
self-propelled motor atomizer 自走式动力喷雾机
self-propelled mount(ing) 自行架；自动架
self-propelled mower 自走式割草机
self-propelled mud barge 自航泥驳
self-propelled pickup baler 自走式检拾压捆机
self-propelled platform 自航式平台
self-propelled pneumatic roller 自行式气胎压路机；自行式气胎碾
self-propelled power frame 自行式底盘
self-propelled power mower 自走式动力割草机
self-propelled puller winch 自行牵引绞车
self-propelled rig 自动推进式钻井机
self-propelled road roller 自行式压路机
self-propelled rock drill(er) 自行式凿岩机
self-propelled roller 自行式压路机；自动压路机；自动路碾
self-propelled rubber-mounted twin batch paver 自动轮胎式间隙式铺路机
self-propelled sand dredge(r) 自航挖砂船
self-propelled sand spreader 自行式撒砂机
self-propelled scraper 自行式铲运机；自推式铲运机；自动刮土机；自行式刮土机
self-propelled sheep-foot roller 自行式羊足碾
self-propelled ship 自航船
self-propelled ship model 自航船模
self-propelled side drag hopper dredge(r) 自航边耙挖泥船
self-propelled single-row harvester-thrasher 自走式单行收割脱粒机
self-propelled single-row harvestor 自动单行收割机
self-propelled skip 自推进式翻斗车
self-propelled split barge 自行开体泥驳
self-propelled spoil barge 自航泥驳
self-propelled sprayer 自走式喷雾器
self-propelled sprayer for hand 自行式手喷雾器
self-propelled sprinkler 自走式喷灌机
self-propelled sprinkler lorry 自推进式洒水车
self-propelled street sprinkler 自动推进式街道洒水车
self-propelled submersible 自航式潜水器
self-propelled tar distributor 自动推进式煤沥青喷布机
self-propelled test 自航试验
self-propelled tool carrier 自动底盘
self-propelled tool chassis 自动底盘
self-propelled tower 自动推进式塔架
self-propelled tracked loader 自行履带式运载装填车
self-propelled transfer car 自行式运输车
self-propelled travel(l)ing 自行
self-propelled travel(l)ing speed 自行速度
self-propelled truck loader 履带式自动装载机
self-propelled underwater research vehicle 自动水下考察船
self-propelled unit 自航式平台
self-propelled vehicle 机动车辆；自动推进式车辆
self-propelled vessel 自航船；机动船
self-propelled vibratory tandem roller 自动振动式双轮压路机
self-propelled vibratory trench roller 自动振动式沟槽压路机

self-propelled vibro-plate 自行式振动板
self-propelled wagon drill 自行式钻车
self-propelled water barge 自行水驳
self-propelled wheeled crane 自动推进的轮胎式起重机
self-propelling 自行的；自航(驳)式的
self-propelling barge 自航驳(船)
self-propelling dredge(r) 自航式挖泥船；自动推进挖泥机
self-propelling ferry 自动推进式渡轮
self-propelling hopper dredge(r) 自行式装料斗采砂船
self-propelling landing gear 自行推进式起落架
self-propelling machine 自动推进机
self-propelling tripper 自动推进倾料器
self-proportioning 自动配料的；自动投配的
self-propulsed 自航(驳)式的
self-propulsed barge 自航驳(船)
self-propulsion 自航
self-propulsion model ship 自航船模
self-propulsion point 自驱点
self-propulsion point of model 船模自航点
self-protected transformer 自保护变压器
self-protecting 自行保护；自保护的
self-protecting type pump 自护型泵
self-protection 自行保护
self-protective 自保护的
self-protective transformer 自保护式变压器
self-provided foreign exchange 自备外汇
self-provided investment 自筹投资
self-provided or leased wagon of enterprise 企业自备车及租用车
self-pruning 天然脱枝
self-pulsed modulation 脉冲自调制
self-pulsed oscillator 自调脉冲振荡器；脉冲自调制振荡器
self-pulsed system 自脉冲系统
self-pulsed transmitter 自动脉冲调制发射机；脉冲自(动)调制发射机
self-pulsing 自生脉冲
self-purging 自清；自排气的；自动净化
self-purging cell 自动净化池
self-purging trap 自动净化槽
self-purification 自清作用；自净(化)作用；自净(化)
self-purification ability 自清能力；自净(化)能力
self-purification ability of stream 河流自净能力
self-purification activity 自净能力；自净活性
self-purification amount 自净量
self-purification capability 自净能力
self-purification capacity 自(动)净(化)能力
self-purification characteristic 自净性能；自净特征；自净特性
self-purification constant 自净(作用)常数
self-purification electronde 自净电极
self-purification grade 自净坡度
self-purification of river 河流自净(作用)
self-purification of seawater 海水自净(作用)
self-purification of stream 河流自净(作用)
self-purification of water body 自净化水体；水体自净(化作用)
self-purification of waters 水体自净(化作用)；水体净化
self-purification parameter 自净参数
self-purification process 自净过程
self-purification velocity 自净速度
self-purifying 自净
self-purifying ability 自净化能力
self-purifying capacity 自净(化)能力
self-purifying effect 自净效应
self-purifying of aquifer 含水层自净作用
self-purifying system 自我净化系统
self-quenching 自消灭；自淬灭
self-quenching counter 自淬灭计数器
self-quenching counter tube 自灭式计数管
self-quenching system 自灭系统
self-radiation 自身辐射
self-radiation impedance 自辐射阻抗
self-radiolysis 自辐射分解
self-raising flour 自发粉
self-raising forms 自升高模板；自升式模板
self-raising fund 自筹资金
self-raising tower crane 自升(式)塔式起重机
self-rake reaper 摇臂收割机

self-reactance 固有电抗
self-reacting 自反应的；自动补偿的
self-reacting device 自动化仪器
self-reading 自读；自动读出的
self-reading device 自读仪
self-reading dosimeter 自读剂量计
self-reading instrument 自动读数仪器
self-reading level(l)ing rod 自读式水(准标)尺；自读数水准尺
self-reading level(l)ing-staff 自读型水准仪
self-reading level rod 自读水准杆；自读水准尺
self-reading pocket dosimeter 自读式袖珍剂量计
self-reading quartz fiber dosimeter 自读式石英丝剂量计
self-reading rod 自读水准杆；自读标尺
self-reading staff 自读式水(准标)尺；自读标尺
self-reciprocal function 自反函数
self-reclosing breaker 自复断路器
self-recognition 自身识别
self-record discharge meter 自记流量计
self-recorder 自记仪；自(动)记录器
self-recorder log 自记仪器测井
self-recording 自记的；自动记录(的)
self-recording accelerator 自动记录加速器
self-recording accelerometer 自记式加速计
self-recording anemometer 自记风速计；自记风速表；自记风力计；自记风力表；风力计
self-recording aneriodograph 自记无液气压计
self-recording aneroid barometer 自记无液气压计
self-recording apparatus 自动记录设备
self-recording audiometer 自动记录测听计
self-recording avapo(u)rimeter 自记蒸发器；自记蒸发计
self-recording barometer 自录气压计；自记气压计；自记气压表
self-recording bottom-mounted wave recorder and analyzer 海底波浪自记分析仪
self-recording compensator 自记补偿器
self-recording current meter 自记流速仪；自记流速表；自记海流计；海流自记仪
self-recording device 自动记录仪；自动记录器
self-recording displacement seismograph 自动记录位移地震仪
self-recording flow meter 自记流速仪；自记流速表；自记流量计；自记流量表
self-recording gold-leaf electroscope 自记式金箔验电器
self-recording high-speed instrument 自记录高速仪表
self-recording hygrometer 自记湿度计；自记湿度表；自动记录湿度计
self-recording instrument 自动记录仪(器)
self-recording interferometric gas analyzer 自记干涉式气体分析仪
self-recording machenism 自动记录机构
self-recording magnetic balance 自记磁天平
self-recording meteorograph 自记气象记录仪
self-recording meter 自记仪表
self-recording micrometer 自记测微器；自记测微仪
self-recording rain-ga(u)ge 自记雨量计
self-recording serial colo(u)rimeter 自录式连续比色计
self-recording snow ga(u)ge 自记量雪器；自记量雪计
self-recording spectrometer 自记分光仪；自记分光计
self-recording spectrophotometer 自记分光光度计
self-recording tachogram 自记转速图
self-recording tachograph 自记转速仪
self-recording thermo-barograph 自记温度气压记录器
self-recording thermometer 自记温度计；自记温度表；温度记录器
self-recording tide ga(u)ge 自记验潮仪；自记潮位计
self-recording titration 自记滴定
self-recording unit 自记仪；自动记录装置；自动记录器
self-recording water level ga(u)ge 自记水位仪；自记水位计
self-recording water level log 自记水位测井
self-recording wattmeter 自动记录瓦特计
self-recording wave meter 自记测波仪
self-recovery 自动复原；自动返回

self-recovery capability 自恢复能力
self-recovery winch 自救绞车;自动拉出用绞车
self-rectified transmitter 自整流发射机
self-rectifier type x-ray apparatus 自整流型 X 射线装置
self-rectifying alternator 自整流式交流发电机
self-rectifying X-ray tube 自整流 X 射线管
self-reducing tach(e)ometer 自动归算速测仪;自动归算视距仪;自动归算测速仪;自计视距仪
self-reducing tachymeter 自动归算速测仪;自动归算视距仪
self-reduction 自吸收
self-refrigerating container 冷藏集装箱
self-refrigeration 自动冷凝(作用)
self-regenerative process 自再生过程
self-register 自动记录器
self-registered technology 自对准工艺
self-registering 自记;自记录的
self-registering anemometer 自记风速计;自记风速表;自记风力计;自记风力表
self-registering apparatus 自动记录仪
self-registering barometer 自记气压计;自记气压表
self-registering ga(u)ge 自动记录计;自记水位仪;自记水位计
self-registering instrument 自动记录(式)仪(器)
self-registering micrometer 自记测微器
self-registering thermometer 自记温度计;自记温度表
self-registering tide ga(u)ge 自记潮位计
self-registration 自配准
self-regulated machine 自调节电机
self-regulated vaporization cooling system 自调蒸发冷却系统
self-regulating 自动调节的;自调(节)的
self-regulating alternator 自调节交流发电机
self-regulating arc welding transformer 自调节弧焊变压器
self-regulating barrage 自动调节堰坝;自动调节拦河坝
self-regulating characteristics 自调特性
self-regulating feedback control 自动调节反馈控制
self-regulating fuel pump 自动调节燃料泵
self-regulating generator 自动调节发电机
self-regulating mass transfer system 自调质量转移系统
self-regulating power transformer 自调整功率变换器
self-regulating process 自(动)调节过程
self-regulating reactor 自调节堆
self-regulating system 自调系统
self-regulation 自平衡性;自动调整;自动调节;自调
self-regulation dam 自动调节坝
self-regulation device 自动调节装置
self-regulation rate 自调整率
self-reinforced composite material 自增强复合材料
self-reinforcing 自增强
self-relative address 自相对地址
self-relative addressing 自相对寻址
self-relative computer 自相关计算机
self-relative subroutine 自相对子程序
self-release coupling 自释联轴节
self-releasing 自然缓解;自动卸开的
self-releasing grappling device 自动放开的夹具
self-releasing gripper 自动开闭式夹纱器
self-releasing hanger 自动脱钩吊具
self-releasing hook 自动脱钩;自动解缆钩
self-releasing slag 自动脱落渣
self-reliance 自力更生
self-relieving chain harrow 自适应链耙;自动适应地形的链耙
self-relocating program(me) 自浮动程序
self-relocation 自浮动
self-rendering towing winch 自动收放拖缆绞车
self-rendering winch 自动收紧绞车
self-renewal resources 自(然)更新资源
self-repair 自修复
self-repairing capability 自修复能力
self-repairing circuit 自恢复电路
self-repairing computer 自修复计算机
self-repairing system 自修复系统
self-replenishing contact 自补偿触点
self-reproducing automaton 自再生自动机
self-reproduction 自增殖;自复制
self-reproductive system 自繁殖系统

self-rescue apparatus 自救呼吸器
self-rescuer 自我援救装置;自救器
self-reset 自动还原;自动复位
self-reset manual release control 自动复原手动释放控制
self-resetting loop 自重置循环;自恢复循环;自复位循环;自复位回线
self-resetting timer 自回复定时器
self-resistance 自电阻;固有电阻
self-restorability 自动复原性
self-restoration 天然更壮;自然更新;天然更新
self-restoring 自动恢复的
self-restoring capacity of nature 自然的自我恢复能力
self-restoring coherer 自恢复式粉末检波器
self-restoring drop 自复吊牌
self-restoring indicator 自动归算指示器
self-restoring insulation 自复绝缘
self-restoring relay 自复归继电器
self-restraint agreement 自愿限制协定;自限协定
self-restraint of boundary stress 边缘应力的自限性
self-retailing shop 自助零售商店
self-retarding cement 缓凝水泥
self-retention 自制动;自行制动;自锁
self-return 自动回位
self-reversal of remanent magnetization 剩余磁性的自反转
self-reversal of thermoremanent magnetism 热剩余磁性的自反转
self-reversing diesel engine 直接逆转式柴油机
self-right effect 自动回正作用
self-righting 自整流(的)
self-righting boat 自动扶正艇
self-righting effect 自动回正作用
self-righting lifeboat 自动扶正救生艇
self-roasting 自焙
self-rotated stopper 自转停止器;自动旋转向上式凿岩机;自动旋转伸缩式凿岩机
self-running sweep 自激扫描
self-sagging temperature 槽沉温度
self-sale 自销
self-same capital 同一个资本
self-saturated magnetic amplifier 自饱和磁放
self-saturated reactor 自饱和电抗器
self-saturation 自饱和
self-scan 自然扫描
self-scan barograph 自扫描条图
self-scan imaging display panel 自扫描图像显示板
self-scanned array 自扫描阵列
self-scanned image sensor 自扫描图像传感器;自扫描成像器件
self-scanned linear array 自扫描线性阵列
self-scanned photosensing array 自扫描光敏série
self-scanned solid image sensor 自扫描固态像传感器
self-scanned solid-state image sensor 自扫描固态成像传感器
self-scattering 自散射
self-scattering effect 自散射效应
self-scouring 自然冲淤;自然冲刷
self-screening jamming 自屏蔽干扰
self-screening property 自蔽性能
self-screening range 自屏蔽范围
self-seal by pressure coupling 无垫压力偶密封
self-sealed bearing 自密封轴承
self-sealed geothermal system 自封闭地热系统
self-sealing 自封闭;自密封;自封式
self-sealing coating 自密封涂层
self-sealing container 自封闭容器
self-sealing coupling 自封联轴节;自封连接器
self-sealing crack 自愈合裂缝(混凝土)
self-sealing fuel tank 自封燃料箱;安全油箱
self-sealing geothermal field 自封闭地热田
self-sealing inner chamber 自身密封式内腔
self-sealing inner tube 自封式内胎
self-sealing oil seal 自紧油封
self-sealing oil system 自封油系统
self-sealing packing 自紧密封
self-sealing paint 自封(闭)漆;自封闭涂料
self-sealing plastics 自动封闭塑胶混合剂
self-sealing pneumatic tube system 自密封气压管路系统
self-sealing ring 自紧式密封环
self-sealing rubber septum 自密封橡胶隔片

self-sealing soil 自愈性土(料)
self-sealing strip 自密封条
self-sealing tank 自封箱;自动封闭桶
self-sealing tyre 自封轮胎
self-sealing wrapping 自封包装
self-seeding 自给籽晶
self-serve station 自动加油站
self-service 自助
self-service bookstall 无人售书处
self-service department(al)store 超级市场;顾客自选的百货商店
self-service display case 自取式陈列柜
self-service elevator 自动电梯;自开电梯
self-service facility 自我服务设备;自我服务设施
self-service grocery 自选副食品
self-service installation 自我服务装置
self-service instrument 自动工作仪表
self-service laundry 自理洗衣店;自助洗衣店;顾客自洗洗衣店
self-service market 自选市场
self-service refrigerator 自助冰箱;自助冷藏箱
self-service restaurant 自助餐馆
self-service shop 自动售货商店;无人售货商店
self-service snack bar 自助快餐店;自助快餐馆
self-service store 自助商店;自选商店
self-servicing battery 自理充电蓄电池
self-servo action 自增力作用
self-set binder 自硬化黏结剂
self-setting 自凝(固)的;自硬性;自动调整
self-setting bearing 多向调整轴承
self-setting bentonite slurry 自凝膨润土泥浆
self-setting device 自动安平装置
self-setting resin sand 自固化树脂砂
self-setting silicate process 水玻璃自硬砂
self-setting unit 自动安平装置
self-sharpening 自锐性;自动磨锐
self-sharpening bit 自锐式钻头;自磨钻头
self-sharpening blade 自磨刃式刀片;自磨刃式锄铲
self-sharpening device 自动磨刀片装置
self-sharpening effect 自磨效果
self-sharpening share 自磨刃犁桦
self-shield 自保护
self-shield coil 自屏蔽线圈
self-shielded arc welding 自保护电弧焊
self-shielded poison 自屏毒物
self-shielded welding wire 自保护焊丝
self-shielding coil 自屏蔽线圈
self-shielding effect 自屏蔽效应
self-shielding factor 自屏蔽因子;自屏蔽系数
self-shielding function 自屏函数
self-shifting 自动转换;自动换向
self-shifting synchronizing clutch 自动移位同步离合器
self-shifting transmission 自动转换开关变速箱;自动变速(箱)
self-shooter 自动闸门
self-shrinkage cylinder 自增强圆筒
self-shrinkage high-pressure cylinder 自增强高压圆筒
self-shunt excitation 自并励
self-shutting hinge 自闭铰链;自关铰链
self-signal(l)ing electronic separation 自发信号电子选矿
self-similar antenna 自相似天线
self-similar elliptical crack 自垂椭圆裂隙
self-similarity 自相似
self-similar solution 自相似解
self-simulating equation 自模拟方程
self-siphonage 自动虹吸作用
self-skimming 自动除渣
self-skinned foam 自结皮泡沫
self-slagging impurity 自生熔渣杂质
self-slip of green body 湿坯表面光洁层
self-slipping transportation 自溜输送
self-solidifying 自凝的;自动凝固
self-solidifying liquid 自凝液体
self-sown flora 自然播种植物;自播植物
self-spillway dam 无闸门溢流坝;自由溢流坝;过水坝
self-spreading bucket 自撒料斗
self-stability 自动稳定性
self-stabilization 自稳定
self-stabilization capacity of surround rock 围岩自稳能力

self-stabilization time of tunnel 坑道自稳时间
self-stabilizing effect in braking 汽车制动自稳效应
self-stabilizing steering 自动稳定转向系
self-stable 不需拉杆的；本身稳定的
self-standing system 自固定系统
self-start 自动起动；自起动
self-starter 自(起)动器；自动起动装置；机械起动器
self-starter pinion 自动起动小齿轮
self-starter ring gear 自动起动器齿环；自动起动机环形齿轮
self-starting 自行启动；自动起动的；自起动
self-starting centrifugal pump 自起动离心泵
self-starting injector 自动吸入喷射器
self-starting method 自动起动法；直接起动法
self-starting motor 自起动电动机
self-starting pump 自起动泵
self-starting rotary converter 自起动旋转变流机
self-starting siphon 自启动虹吸管
self-starting synchronous motor 自起动同步电动机
self-start motor 自起动电动机
self-start synchronous motor 自(动)起动同步电动机
self-start-up capacity 自起动容量；自起动能力
self-start-up load 自起动负荷
self-steering 自动驾驶
self-steering mechanism 自转向机构
self-steering microwave array 自控微波天线阵
self-sterilizer 自动消毒
self-stick alumin(i)um roll roofing 自黏铝箔屋面卷材
self-sticking coefficient 固有黏附系数
self-stiffing 固有刚度
self-stiffness 固有刚度
self-stimulated emission 自受激发射
self-stopping gear 自停机构；自动制动传动机构
self-stowing gate 自行充填开采巷道
self-stowing mooring winch 自动排缆绞车
self-stratifying coating 自分层涂料
self-strengthening 自补强
self-stress 自应力
self-stressed concrete 自应力混凝土；膨胀混凝土
self-stressed prestressed concrete 自应力的预应力混凝土
self-stressing 自应力
self-stressing cement 膨胀水泥；自应力水泥
self-stressing concrete 自应力混凝土
self-stressing concrete pipe 自应力混凝土管
self-stressing grout 自应力水泥浆
self-stressing mortar 自应力砂浆
self-stress recovery 自应力恢复
self-sucking pump 自吸泵
self-sufficiency 自给自足
self-sufficient navigation system 自导向系统
self-superposability 自叠加性
self-supervision 自我监督
self-supplied manure 农家肥(料)
self-supplying 自给的
self-supply power plant 自备发电厂；自备动力厂
self-supported opening 天然支护采矿法
self-supported stack 独立式烟囱
self-supporting 自立的；自持的；自承重的；自承式
self-supporting accounting system 独立核算制
self-supporting aerial cable 自支持架空电缆
self-supporting antenna mast 自持天线杆
self-supporting cement container 自承水泥罐
self-supporting coil 自立式线圈；自持线圈
self-supporting grating 自持光栅
self-supporting insulation 自承重隔热
self-supporting mast 自立式电杆
self-supporting pack 自支式充填带
self-supporting pantile 自承式波形瓦
self-supporting partition 自支撑隔墙；自承隔墙；自承重隔断
self-supporting partition wall 自支承的分隔墙
self-supporting property 自身支撑能力
self-supporting rock 自支护岩石
self-supporting scaffold 自支撑脚手架；自支脚手架；自撑式脚手架；支架式脚手架；独立脚手架
self-supporting stairway 自承式楼梯
self-supporting step 自支承踏步
self-supporting structure 自立式钢结构；自承重结构
self-supporting style 自撑式
self-supporting tendency 自撑能力
self-supporting tower 自立式铁塔
self-supporting wall 非承重墙；自承(重)墙
self-support system 自支援系统
self-sustained 自持的
self-sustained emission cold cathode 自持发射冷阴极
self-sustained oscillation 等幅摆动；非阻尼振荡；自激振动；自激振荡；自动振荡；自动摆动
self-sustained underwater breathing apparatus 自持水下呼吸器
self-sustained vessel 自有装卸机械的船舶
self-sustaining 自承的；自持的
self-sustaining chain reaction 自持链式反应
self-sustaining gear 自锁传动；自动制动机构
self-sustaining medium 自持介质
self-sustaining process 自持过程
self-sustaining reaction 自持反应
self-sustaining reactor 自持堆
self-sustaining rock 稳固岩石
self-sustaining structure 自承结构物
self-sustaining vessel 自有装卸机械的船舶
self-sustaining wall 自承重墙
self-synchronism 自动同步
self-synchronized control 自动同步控制
self-synchronizing 自(动)同步；自动同步的
self-synchronizing clutch 自动同步离合器
self-synchronizing code 自同步码
self-synchronizing system 自动整步系统；自动同期系统；自动同步系统
self-synchronizing transmitter 自动同步传感器
self-synchronous 自同步的
self-synchronous device 自整角机；自同步机；自同步装置；同步机
self-synchronous motor 自整角电机；自同步(电动)机
self-synchronous repeater 同步机
self-synchronous system 自动同步系统
self-synchronous transmission 自动同步传输
self-synchronous transmission system 自动同步传输系统
self-synchronous transmitter 自动同步发射机
self-tapping 自动攻丝
self-tapping saddle tee 自开孔鞍形三通
self-tapping screw 自攻丝螺钉；自攻螺丝；自动攻丝螺杆；自钻孔攻丝螺钉
self-tapping thread 自攻螺纹
self-taught and self-made person 自学成才
self-taught expert 土专家
self-temperature compensated strain ga(u)ge 温度自补偿应变片
self-tempering 自身回火；自热回火
self-tensioning winch 自调拉力绞车
self-tension winch 恒张力绞车
self-test and failure warning 自检故障与告警
self-test capability 自检能力
self-test(ing) 自(行)测试；自检测试
self-testing-and-repairing computer 自测试自修复计算机
self-testing factor 自检系数；自测试因数
self-test relay 自测试继电器
self-texturing yarn 自变形丝
self-thermostat 自控恒温器
self-threading screw 自攻丝螺钉；自攻螺针
self-tightening lever clip 自紧式夹杆
self-tightening locknut 自紧防松螺母
self-tightening nut 自紧螺母
self-tightening oil seal 自紧油封
self-tightening seal 自紧密封
self-tilting concentrator 自倾式洗矿槽
self-timed 自计时的
self-timer 延时器；自动(记)秒表
self-timer with infrared remote control 红外(线)遥控自动定时器
self-timing anemometer 自动计时风速计
self-timing gear 自动调时齿轮
self-tipping 自动倾倒式
self-tipping barge 自倾卸驳(船)
self-tipping lorry 自行翻斗车；自动倾卸汽车
self-tipping skip 翻转式箕斗
self-tolerance 自身耐受性
self-tone 原色调
self-topping 自封顶
self-torque 自转矩
self-towing device 自行装置
self-towing equipment 自行式钻机
self-tracking action 自动跟踪作用
self-tracking system 自动跟踪系统
self-traction 自行式
self-traping effect 光束自收效应
self-trapped optic(al) beam 自聚焦光束
self-trapping effect 自陷效应
self-traveling 自行式的
self-trigger 自动触发器
self-triggering 自动触发
self-trimming 自然平舱的；不需进行平舱作业的
self-trimming collier 自动平舱运煤机
self-trimming hatchway 自动平舱舱口
self-trimming ship 自动平舱船(舶)
self-trimming tank 自动平衡(水)柜
self-trimming vessel 自动平舱船(舶)
self-tripping car 自动倾卸车
self-troughing idler 自动成槽托滚
self-tuning engine 自调发动机
self-tuning factor 自调整因子
self-tuning loop 自调谐线圈检测器
self-tuning short-wave transmitter 自动调谐短波发射机
self-turning mechanism 自动旋转机构
self-turning regulator 自校正调节器
self-turn-over plate gate 自动翻板门
self-tying 自动捆扎
self-tying baler 自动捆绑式捡拾压捆机
self-tying mechanism 自动打捆机构
self-tying pick-up baler 自动结捆式拣拾压捆机
selfube iron 多孔铜铁合金
self-unloader 自动卸料机；自卸船
self-unloading 自动卸货
self-unloading barge 自动卸货驳船
self-unloading barge unit 自动驳船队
self-unloading forage box 自卸式饲料分送车
self-unloading hopper 自卸料仓
self-unloading pump 自动卸载泵；自动卸料泵
self-unloading sand barge 自卸砂驳
self-unloading ship 自卸式货船；自卸船；自动卸货船
self-unloading tug-barge unit 自卸拖驳船队
self-unloading vehicle 自动卸料车
self-unloading vessel 自卸船舶；自动卸载船；自动卸货船
self-unloading wagon 自卸拖车；自卸(汽)车
self-ventilated chamber 自动通风室
self-ventilated machine 自通风电机
self-ventilated motor 自然通风(式)电动机
self-ventilation 自(行)通风；自行换气
self-verifying 自动检验(的)
self-vulcanizing 自固化；自硫化
self-wallowing soil 天然波状土
self-washing filter 自洗式过滤器
self-washing wall 自冲洗的墙壁
self-waterer 自动饮水器；自动饮水机
self-watering device 自动饮水装置；自动饮水器
self-weight 自重
self-weight collapse 自重塌陷
self-weight collapse loess 自重塌陷黄土；自重湿陷性黄土
self-weight collapse settlement 自重湿陷量
self-weight collapsibility 自重湿陷性
self-weight collapsible loess 自重湿陷性黄土
self-weight moment 自重弯矩
self-weight non-collapse loess 非自重湿陷性黄土
self-weight stress 自重应力
self-weight vector 自重向量
self-widening finisher 自行加宽(路面)修整机
self-wind 自动上发条
self-winding 自卷的
self-winding clock 自上发条钟
self-winding watch 自上发条表；自动手表
self-wiping twin-screw extruder 自擦拭双螺杆挤出机
seligmannite 硫砷铅铜矿；砷车轮矿
Selinene 桉叶烯
sellaite 氟镁石
sell-and-leaseback agreement 售出租回合约
sellary 大客厅
sell at a loss 蚀本出售
sell at a profit 赚钱卖出

sell at half price 半价出售
sell at loss 亏本出售
sell at market prices 明码售货
sella turcica 蝶鞍
sell briskly 旺销
sell by lots 分批出售
sell by wholesale 整批出售;批发售出;批发
seller 技术输出方;卖主;卖方
seller financing 售方赊贷
Seller's coupling 塞勒(锥形)联轴节
seller's credit 卖方信贷
Seller's drive 塞勒传动
seller's failure to perform 卖方不履行合同
seller's hedging 卖家套头保值;卖方期货
seller's lien 卖主押留权;卖方留置权
seller's market 卖主市场;卖方市场
seller's monopoly 卖方垄断;卖方独占
seller's option 卖主选择权;卖方选择(权);卖方决定
sellers over 卖主过多;卖者过多
seller's price 卖主控制价格;卖价;卖方价格
seller's rate of exchange 卖价
seller's sample 卖方样品
Seller's screw thread 赛勒螺纹
seller's surplus 卖主剩余;卖方盈余
Seller's taper 塞勒锥度
seller's usance 卖方惯例
seller's usance credit 卖方远期信用证
seller's usual packing 卖方习惯包装
sell goods at a high figure 高价销售
sell goods at high price 高价销售;高价出售
selling agent 销售代理商
selling and administrative cost 销售及管理成本
selling and administrative expenses 销售及管理费用;销售及管理成本;销货与管理费用
selling and administrative expense analysis sheet 销售及管理费用分析表
selling and administrative expense charged to cost 计入成本的销售及管理费用
selling area 营业厅
selling at exposition 展销
selling at less than cost 亏本销售;蚀本销售
selling at premium 高价出卖
selling by grade 分级销售
selling capacity 销售能力
selling charges 销售费用
selling commission 销售佣金
selling concept 销售概念
selling contract 卖汇票预约
selling contract slip 卖汇票预约申请书
selling cost 销售费用;销售成本;推销用;推销成本
selling cost budget 销售成本预算
selling exchange 卖外汇
selling expenses 销售费用;销货费用;推销费用
selling expenses variance 销售费用差异
selling goods at reduced prices 削价处理商品
selling group 销售集团;推销集团
selling group pot 销售集团承销部分
selling hedge 卖期保值;抛售的对冲处置
selling offer 销货要约;卖方发盘
selling off goods at reduced prices 拍卖
selling of ready-built house 商品房屋出卖
selling-out 卖出
selling party 卖方
selling platform 销售纲领
selling price 销售价格;卖价;卖方报价;售价
selling price inventory method 按销售价格盘存法
selling rate 销售率;卖期汇率;卖价;卖出汇率;卖出(汇)价
selling retail department 门市部
selling rice variance 销售价格差异
selling sample 推销样品
selling short 卖空;卖短
selling syndicate 推销集团
selling transportation insurance 出售运输保险
selling wrong ticket 误售
sell large quantities of goods 抛售物资;抛售货品
Sellmeier equation 塞耳迈尔方程
sell off 处理存货;卖清存货
sell off date 处理存货日期
sell on the spot 现货销售
Sellotape 胶纸带
sell out 卖清
sell retail over the counter 门市

sell short 卖空
sell up 卖光
sell wholesale 批发出售;整批出售
Selma chalk 塞尔马白垩层
Selove receiver 塞洛夫接收机
sel-spillway dam 自由溢洪坝
selsyn 自整角机;自(动)同步机
selsyn control 自动同步机控制
selsyn-data system 自(动)同步机数据系统;数据传送的自动同步系统
selsyn device 自动同步设备
selsyn differential 自整角机差动装置
selsyn disc 自动同步机盘
selsyn drive 自动同步机传动
selsyn-drive electric(al) motor 自整角机驱动电动机
selsyn drive gear 自动同步机主动齿轮;自动同步机传动机构
selsyn driver 自动同步传动装置
selsyn generator 自整角发电机;自同步发射机;自动同步发电机
selsyn instrument 自动同步仪表
selsyn model 自动同步模型
selsyn motor 自同步电动机;自动同步电动机
selsyn receiver 自动同步接收机;自整角接收机
selsyn system 自动同步机系统
selsyn train 自动同步传动装置
selsyn transformer 自同步接收机
selsyn transmitter 自动同步发射机
selsyn-type electric(al) machine 自整角电机
selsyn-type electrical motor 自整角电动机
selsyn-type roll-opening indicator 自整角机式轧辊开度指示器
selsyn-type synchronous system 自整角机式同步系统
selt anti-dual function 非自对偶函数
seltrap 半导体二极管变阻器
selva 热带雨林
selvage 镶边;锁孔板;断层泥;布边
selvage edge 锁边,网边(金属);织边
selvage eye 绕匝索环
selvage formation 布边成形
selvage joint 压边接头
selvage strop 绕匝索套
selvage wire rope 编织钢丝绳
selvedge 镶边;滚边;糊墙纸卷的两个边;锁孔板;织边;饰边
selvedge decurler 剥布边器
selvedge guider 导布器
selvedge stamping machine 印边机
selvedge uncurler 剥布边器
Selwyn granularity coefficient 塞耳温颗粒度系数
selwynite 铬铝石
semantide 信息载体
semaphore 信号机;旗语(通信);旗号;探照式色灯信号机;动臂信号机
semaphore arm 信号臂
semaphore code 旗语信号
semaphore flag 手旗
semaphore poel changer 臂板转极器
semaphore signal 杆上信号;塔上信号;手旗信号;手旗通信;臂板信号机
semaphore singal post 臂板信号柱
semar technique 涂抹法
Semastic 瑟玛琦脂(一种砖形铺地材料)
sematic colo(u)rs 保护色
semecarpus vernicifera 台湾漆(大漆)
sem(e)iology 符号学
semen arecae 槟榔
Semendur 塞默杜尔钴铁簧片合金
semenovite 硅铍稀土矿
semester ring 半年轮
semi-able-bodied labour force 半劳动力
semi-able-bodied worker 半劳动力
semi-absolute volt 似绝对伏特
semi-absorbent treatment 半吸声处理
semi-accessible trench 半通行地沟
semi-acid refractory 半酸性耐火材料
semi-active 半活性的
semi-active homing guidance 半主动寻的制导
semi-active laser system 半主动激光系统
semi-active pulse 半主动脉冲
semi-active radar 半有源雷达

semi-active stabilized platform 半有源稳定平台
semi-active tracking system 半主动跟踪系统
semi-actuated controller 半式交通控制器;半活动控制器;半感应控制器
semi-actuated signal 半感应信号
semi-actuated traffic controller 半感应式交通信号控制器
semi-adaptive plan 半适应方案
semi-additive method 半添加法
semi-adherent 半着生的
semi-adiabatic hydration calorimeter 半绝热水化量热计
semi-aerobic landful 半好氧土地填埋
semi-aft 半尾机
semi-aldehyde 半醛
semi-amplitude 半振幅
semi-anaerobic condition 半厌氧状态
semi-analysis solution 半分析解
semi-analytic(al) aerotriangulation 半解析空中三角测量
semi-analytic(al) block aerotriangulation 半解析区域空中三角测量
semi-analytic(al) method 半解析法
semi-analytic solution 半分析解
semi-anarchic state 半无政府状态
semi-anechoic room 半消声室
semi-angle 半角
semi-angle of beam convergence 波束半会聚角
semi-angular field 半视场角
semi-annual 一年两次的;半年一次的;半年的
semi-annual account 半年账单
semi-annual bonus 半年分红
semi-annual component 半年分潮
semi-annual constituent 半年分潮
semi-annual installment 半年一次的分期付款
semi-annual period 半年期
semi-annual repair 半年修理
semi-annual report 半年度报告
semi-annual tide 半年周期潮;半年巡潮
semi-annual variation 半年变化
semi-anthracite 半无烟煤
semi-anthropomorphic atmospheric diving suit 半拟形常压潜水服
semi-aperture 半孔径
semi-apex angle 半顶角
semi-apochromat 近复消色差透镜;半消色差接目镜;半复消色差物镜
semi-apochromatic objective 半消色差物镜;半消色差接目镜
semi-aquatic 半水生的
semi-aquatic plant 半水生植物;半水栖生物
semi-aquatic snail 两栖蜗牛
semi-arboreous desert 半乔木荒漠
semi-arch 半拱
semi-arid 半荒芜;半干旱的;半干草
semi-arid area 半干旱地区
semi-arid climate 半干旱气候;草原气候
semi-arid fan 半旱地扇(半干旱地区冲积扇);平直河
semi-arid land 半干旱地
semi-arid land ecosystem 半干旱地区生态系统
semi-arid mudflow 半干状泥流
semi-arid region 半干旱(地)区
semi-arid soil 半干旱土壤
semi-arid subtropical zone of southern china type 南方型半干旱亚热带
semi-arid to arid subtropical zone 半干旱一干旱亚热带
semi-arid to semi-humid subtropical to tropical subzone 半干旱一半潮湿亚热带一热带亚带
semi-arid tropic 半干旱热带
semi-arid zone 半干燥地带;半干旱地区;半干旱地带
semi-artesian 半自流的
semi-artesian well 半自流井
semi-articulated landing gear 半摇臂式起落架
semi-artificial harbo(u)r 半人工港
semi-artificial port 半人工港
semi-asphaltic 半地沥青的;半沥青的
semi-asphaltic base petroleum 半地沥青基石油
semi-asphaltic flux 沥青助熔剂;石油沥青;半沥青膏
semi-asphaltic oil 半沥青油
semi-asphaltic petroleum 半地沥青石油
semi-auto 半自动的

semi-auto component inserter 半自动插件机
semi-auto folding umbrella 半自动伸缩伞
semi-auto gear hobbing machine 半自动齿轮滚齿机
semi-autogenous 半自磨
semi-autogenous grinding 半自生磨矿(法)
semi-automat 半自动式
semi-automated 半自动的
semi-automated lettering 半自动排字
semi-automated method 半自动(化)方法
semi-automatic 半自动的
semi-automatic action 半自动作用;半自动动作
semi-automatic advance 半自动提前
semi-automatic ampule spray-washing machine 半自动安瓿喷射洗涤机
semi-automatic appliance 半自动设备;半自动用具
semi-automatic arc spot welder 半自动电弧点焊机
semi-automatic arc weld 半自动电弧焊
semi-automatic arc welder 半自动(电弧)焊机
semi-automatic arc welding 半自动弧焊;半自动电弧焊接
semi-automatic arc welding machine 半自动(电)弧焊机
semi-automatic argon arc welder 半自动氩气弧焊机
semi-automatic audiometer 半自动测听计器;半自动测听计
semi-automatic batcher 半自动分批搅拌机;半自动化拌和楼;半自动化拌和斗;半自动配料器
semi-automatic bearing grinder 半自动轴承磨床
semi-automatic bevel gear lapper 半自动锥齿轮研齿机
semi-automatic block 半自动闭塞
semi-automatic block machine 半自动闭塞机
semi-automatic block system 半自动(化)闭塞系统
semi-automatic burner 半自动燃烧器
semi-automatic check 半自动检验
semi-automatic circuit 半自动电路
semi-automatic circuit performance monitor 半自动电路性能监察器
semi-automatic circular sawing machine 半自动圆锯床
semi-automatic circulation 半自动循环
semi-automatic clutch 半自动离合器
semi-automatic coding 半自动编码
semi-automatic command to line of sight 瞄准线半自动控制
semi-automatic computer directed layout machine 半自动计算机控制式划线机
semi-automatic concrete block machine 半自动混凝土砌块制造机
semi-automatic control 半自动控制
semi-automatic controller 半自动控制装置;半自动控制器
semi-automatic controlling machine 半自动机床
semi-automatic copying lathe 半自动仿形车床
semi-automatic copying machine 半自动仿形车床
semi-automatic corking unit 半自动瓶塞机
semi-automatic coupling 半自动车钩
semi-auto matic cutting the double groves 半自动气割双面坡口
semi-automatic cycle 半自动(化)循环
semi-automatic cylindrical grinder 半自动圆磨床
semi-automatic doffer 半自动落纱机
semi-automatic engraver 半自动刻图仪
semi-automatic exchange 半自动化电话局
semi-automatic exposure control 半自动曝光控制器
semi-automatic extraction of radar information 雷达信息半自动提取
semi-automatic feed 半自动送料
semi-automatic flight inspection 半自动飞行检验
semi-automatic focusing 半自动调焦
semi-automatic folding umbrella 半自动折叠伞
semi-automatic forming 半自动成形
semi-automatic gas cutting machine 半自动气割机
semi-automatic gate 半自动闸门
semi-automatic gear hobbing machine 半自动滚齿机
semi-automatic gravity feeding 半自动重力送料
semi-automatic ground environment 半自动化地面防空体系;半自动地面系统
semi-automatic ground environment system 半自动地面防空系统
semi-automatic gun 半自动焊枪
semi-automatic hack sawing machine 半自动弓锯床
semi-automatic hob grinding machine 半自动滚刀磨床
semi-automatic honing machine 半自动搪磨机
semi-automatic hose machine 半自动织袜机
semi-automatic hump 半自动化驼峰【铁】
semi-automatic hump yard 半自动驼峰编组场
semi-automatic hydro-electric(al) station 半自动化水电站
semi-automatic keying circuits 半自动键控电路
semi-automatic lathe 半自动车床
semi-automatic low pressure die-casting machine 半自动低压铸造机
semi-automatic machine 半自动车床
semi-automatic machine for disc insulator forming 高压悬式绝缘子半自动成型机;盘式绝缘子半自动成型机
semi-automatic machine tool 半自动化机床
semi-automatic master fixture 半自动标准夹具
semi-automatic message switching center 半自动报文交换中心;半自动信息交换中心
semi-automatic metal arc welding 半自动金属电弧焊
semi-automatic miller 半自动铣床
semi-automatic milling machine 半自动铣床
semi-automatic mo(u)ld 半自动(塑)模
semi-automatic multicutter lathe 多刀半自动车床
semi-automatic multicut vertical lathe 多刀半自动立式车床
semi-automatic oiler 半自动注油器
semi-automatic operation 半自动运行;半自动化操作;半自动操纵作业;进路操作作业
semi-automatic optical Brinell's hardness tester 半自动光学布氏硬度计
semi-automatic packing and pill-making machine 半自动包衣造粒机
semi-automatic plant 半自动装置
semi-automatic plating unit 电镀半自动装置
semi-automatic position fixing 半自动定位
semi-automatic precision vertical gear hobber 半自动精密立式滚齿机
semi-automatic premilling machine 半自动预磨机
semi-automatic press 半自动压机
semi-automatic pressing 半自动压制
semi-automatic private branch exchange 半自动专用(小)交换机
semi-automatic processing 半自动加工
semi-automatic processor 半自动信息处理机
semi-automatic programmer 半自动程序设计器
semi-automatic programming 半自动程序设计
semi-automatic ratchet screw driver 半自动棘轮式螺丝刀
semi-automatic rectifier 半自动整流器;半自动(刻图)纠正仪;半自动检波器
semi-automatic sand mixer 半自动混砂机
semi-automatic scale 半自动秤
semi-automatic score cutter 半自动切纸机
semi-automatic screwdriver 半自动螺丝刀
semi-automatic seal 半自紧密封
semi-automatic selection 半自动选择;半自动拨号
semi-automatic signal 半自动信号机
semi-automatic slideprojector 半自动幻灯机
semi-automatic sprayer 半自动式喷雾机
semi-automatic sprinkler system 半自动(化)喷水系统;半自动(化)喷灌系统;半自动(化)洒水系统
semi-automatic S/R machine 半自动有轨堆垛起重机
semi-automatic starter 半自动起动器
semi-automatic steam pile hammer 半自动蒸汽打桩机
semi-automatic strapping machine 半自动捆扎机
semi-automatic submerged arc welding 半自动埋弧焊接
semi-automatic substation 半自动变电站
semi-automatic system 半自动制;半自动(化)系统
semi-automatic tape relay 半自动穿孔带中继
semi-automatic telephone 半自动电话
semi-automatic telephone exchange 半自动电话交换机
semi-automatic telephone switchboard 半自动式电话交换机
semi-automatic telephone system 半自动电话系统
semi-automatic toll dialling 长途半自动接续
semi-automatic tracking 半自动跟踪
semi-automatic tracking digitizer 半自动跟踪数字化仪
semi-automatic tracking mechanism 半自动跟踪装置
semi-automatic traffic information and control system 半自动今天信息和控制系统
semi-automatic transmission 半自动变速箱;半自动变速器
semi-automatic trunk board 半自动式中继台
semi-automatic turret lathe 半自动转塔车床
semi-automatic valve 半自动阀
semi-automatic washer-dryer 半自动双缸洗衣机
semi-automatic washing machine 半自动洗衣机
semi-automatic watch 半自动表
semi-automatic weigh batcher 半自动重量称量器;半自动称料机
semi-automatic weighing machine 半自动称量机
semi-automatic weight batcher 半自动称量架
semi-automatic weld 半自动电焊
semi-automatic welder 半自动电焊机
semi-automatic welding 半自动焊接;半自动电焊
semi-automatic welding machine 半自动焊接机
semi-automatic work(ing) 半自动工作;半自动作业
semi-automatic yard 半自动编组场
semi-automation 半自动化
semi-automation of cargo-handling 装卸作业半自动化;装卸工作半自动化
semi-automation operation 半自动化操作
semi-automation routine 半自动顺序法
semi-autonomous 半自治的
semi-auto precision plain grinder 半自动精密平面磨床
semi-autotensioned catenary equipment 半补偿链形悬挂【电】
semi-autotrophic 半自养的
semi-average 半平均(数)
semi-average method 对半平均法;半(期)平均法
semi-awned head 半有芒穗
semi-axial flow pump 半轴流式水泵;半轴流泵
semi-axial flow turbine 半轴流式水轮机
semi-axial hung type 半轴悬挂式
semi-axial thrust 半轴向推力
semi-axis 半轴
semi-axis of an ellipse 椭圆半轴
semi-axis of the central ellipse 中心椭圆半轴
semi-axle 半轴
semi-balance 半平衡
semi-balanced rudder 半平衡舵
semi-balloon tyre 半低压轮胎
semi-band width 半带宽
semi-bankruptcy 半破产(状态)
semi-barge carrier system 半载驳船系统
semi-base load power station 半基荷电站
semi-basement 半地下室
semi-batch chemical reactor 半连续化学反应器
semi-batch process 半分批法
semi-batch reactor 半间歇反应器;半间歇式反应堆
semibeam 悬臂梁
semi-benzene 对二烷基亚甲基己二烯;半苯
semi-bituminous 半沥青的
semi-bituminous coal 半烟煤;半沥青煤;半褐煤
semi-blind shield 半挤压盾构
semi-blown asphalt 轻度氧化沥青;半氧化沥青;半吹制地沥青
semi-bodiless ware 半脱胎器
semi-boggy soil 半沼泽土
semi-boiled process 半沸煮法
semi-boiling process 半煮法;半沸煮法
semi-bolson 半沙漠盆地;半干湖盆地;半干旱封闭洼地;半封闭洼地
semi-bordered pit 半具缘纹孔
semi-box beam 半箱形梁
semi-brackish marsh 半微咸水沼泽
semi-breadth 半宽(度)
semi-bridge 半桥的
semi-bridge system 半桥式
semi-bright coal 半亮煤
semi-bright nut 半光制螺母
semi-brittle solid 半脆性固体

semi-bulk bag 半散装袋
semi-bulk cargo 半散装货
semi-bulk cargo ship 半散装货船
semi-bulk container 集装袋
semi-bungalow 假二层房屋;附有阁楼的平房;附有阁楼的平房
semi-buried cellar 半埋的地窖
semi-butterfly pass 半蝶式孔型
semi-butterfly ring 半蝴蝶式环
semi-canal 半管
semi-cannel 半烛煤
semi-cantilever 半悬臂
semi-cantilever spring 半悬臂弹簧
semi-carbazide 氨基脲
semi-carbazide hydrochloride 氨基脲盐酸盐;氨基脲化氢氯
semi-carbon paper 单面炭纸
semi-cardinal points 隅点
semi-cargo boat 客货轮;客货船
semi-cellulose 半纤维素
semi-cemented 半胶结的
semi-centennial 五十周年的
semi-central system 半集中式系统
semi-centrifugal casting 半离心铸造
semi-centrifugal clutch 半离心式离合器
semi-centrifugal type 半离心式
semi-chemical bleaching 半化学漂白
semi-chemical liquor 半化学液体
semi-chemical mill 半化学工厂
semi-chemical paper-making wastewater 半化学造纸废水
semi-chemical process 半化学方法
semi-chemical pulp 半化学纸浆
semi-chemical pulping 半化学制浆
semi-chemical pulp mill wastewater 半化学纸浆厂污水;半化学纸浆厂废水
semi-chemical pulp sewage 半化学纸浆污水
semi-chemical pulp wastewater 半化学纸浆废水
semi-chemical wood pulp 半化学纸浆
semi-chilled mo(u)ld casting 半冷模铸造;半金属型铸造
semi-chilled roll 半激冷轧辊
semi-china 低温烧成的瓷器
semi-chord 弧弦的半长
semi-circle 半圆形;半圆(的)
semi-circle cylinder 半圆列缸
semi-circle junction terminal 半环形枢纽
semi-circle orifice 半圆孔
semi-circuit canal 半圆形管渠
semi-circular 半圆形的;半圆形的
semi-circular alcove 半圆形凹室
semi-circular apse 半圆房间;半圆壁龛
semi-circular apsis 半圆形房间
semi-circular arch 半圆券;半圆穹顶;半圆形砖;半圆(形)拱
semi-circular arch bridge 半圆拱桥
semi-circular arched window 半圆拱窗
semi-circular arch section 半圆拱形断面
semi-circular arch support 半圆拱支架
semi-circular barrel vault 半圆筒形穹顶
semi-circular basin 半圆形港池
semi-circular baston 半圆形年地工事;半圆形堡垒
semi-circular bay window 半圆形凸窗
semi-circular breakwater 半圆形防波堤
semi-circular canal 半规管
semi-circular channel 半圆形(断面)渠道
semi-circular cofferdam 半圆形围堰
semi-circular column 半圆柱
semi-circular conduit 半圆形管道
semi-circular control panel 半圆形控制盘;半圆形控制板
semi-circular corrector 半圆自差校正器
semi-circular cross-section 半圆横截面
semi-circular cross-vault 半圆形交叉拱顶
semi-circular cylindrical roof 半圆柱形屋顶
semi-circular deviation 半圆自差;半圆偏差
semi-circular dial 半圆形刻度盘
semi-circular dome 半圆穹顶
semi-circular dormer window 半圆形老虎窗;不需要顶窗
semi-circular drain 半圆形排水管;半圆形排水沟
semi-circular drainage pipe 半圆形排水管
semi-circular drain pipe 半圆形排水管
semi-circular electromagnet 半圆形电磁铁

semi-circular exedra 半圆房间;半圆形露天椅;半圆形前廊
semi-circular groove 半圆槽
semi-circular gutter 半圆(形)天沟
semi-circular hook 半圆形弯钩
semi-circular impact pits 圆弧形撞击坑
semi-circular in shape 半圆形;呈半圆形
semi-circular iron 半圆形铁器
semi-circular junction terminal 半环形枢纽
semi-circular key 月牙键
semi-circular laser resonator 半圆形激光谐振腔
semi-circular line 半环线
semi-circular method 半圆法
semi-circular mo(u)lding 半圆线脚
semi-circular niche 半圆形壁龛;半圆壁龛
semi-circular notch 半圆形缺口
semi-circular oriel window 半圆形凸窗
semi-circular pipe 半圆形管道
semi-circular plate capacitor 半圆板电容器
semi-circular profile 半圆形轮廓
semi-circular profile bit 唇面半圆形钻头
semi-circular protractor 半圆仪;半圆量角器
semi-circular rainwater gutter 半圆形雨水槽
semi-circular rib 半圆形肋
semi-circular roof 半圆屋面;半圆屋顶
semi-circular room 半圆形室
semi-circular section 半圆(形)截面
semi-circular sheet(steel)flume 半圆形钢板渡槽
semi-circular space 半圆形场地;半圆形壁龛
semi-circular steel 半圆形钢材
semicircular stilted arch 超半圆拱
semi-circular strether 半圆形拉伸机;半圆形顺砖;半圆形横条
semi-circular structure 半圆形结构
semi-circular surface crack 半圆形表面裂纹
semi-circular surface flaw 半圆形表面缺陷
semi-circular termination 半圆形终点(站);半圆形终端
semi-circular tower 半圆形塔
semi-circular trash-rack 半圆形拦污栅
semi-circular tube 半圆形管道
semi-circular tunnel vault 半圆形隧道拱顶
semi-circular(type D)fender 半圆形(D形)橡胶护舷
semi-circular vault 半圆形穹;半圆筒拱
semi-circular weir 半圆形堰
semi-circular window 半圆形窗
semi-circular wire 半圆形钢丝
semi-circular wood screw 半圆形木螺钉
semi-circumference 半圆周
semi-circumferential flow 半圆周绕流
semi-classical theory 半经典理论
semi-closed 半封闭式
semi-closed basin 半封闭盆地
semi-closed body of water 半封闭水域;半封闭水体
semi-closed circuit deduster 半闭路收尘器
semi-closed circuit diving device 半闭回路潜水装置
semi-closed circuit diving equipment 半闭回路潜水装置
semi-closed circuit diving gear 半闭回路潜水装置
semi-closed circuit diving unit 半闭回路潜水装置
semi-closed circuit underwater breathing apparatus 半闭式潜水呼吸器
semi-closed cycle 半封闭循环;半闭(式)循环
semi-closed-cycle gas turbine 半闭式循环燃气轮机;半封闭循环燃气轮机
semi-closed interval 半(封)闭区间
semi-closed recirculating diving system 半闭式再循环潜水系统
semi-closed sea 半封闭海(域);半闭(塞)海
semi-closed slot 半闭口槽
semi-closed splash screen 半闭挡泥屏
semi-closed system 半密闭系统;半封闭系统
semi-closed top mo(u)ld 半封顶锭模;瓶口锭模
semi-closed tube diffusion 半闭管扩散
semi-closed-type impeller 半闭式叶轮;半开式叶轮
semi-closed water body 半封闭水域;半封闭水体
semi-closure 半闭(合)
semi-coal gas 半煤气
semi-coaxial 半同轴的
semi-coercive function 半强迫函数
semi-coke 半焦(炭);半焦化(作用);低温炼焦
semi-collinite 半镜质体
semi-colloid 半胶体

semi-colloidal 半胶体的
semicolon 分号
semicolon count 分号计算
semi-colonial 半殖民地
semi-column 半附墙柱;平圆附柱;半(圆)柱;半露柱;半附墙柱
semi-commercial 半商业性的
semi-commercial plant 中间试验厂;半工业化工厂
semi-commercial production 中间试制;中间试剂;中间工厂规模生产
semi-commercial scale 半生产规模
semi-commercial scale test 半生产规模试验
semi-commercial unit 半工业装置
semi-compiled 半编译的
semi-compiler 半编译(的)程序
semi-compound grain 半复粒
semi-compreg 半胶压木材
semi-computable predicate 半可计算谓词
semi-computer 准计算机
semi-concealed closer 半露式门器;半隐蔽关门器
semi-concealed coalfield 半出露煤田;半暴露式煤田
semi-concealed fencing 半露式拦沙障
semi-conducting alloy 半导体合金
semi-conducting ceramics 半导体陶瓷
semi-conducting coating 半导电涂层;半导电敷层
semi-conducting compound 半导体化合物
semi-conducting crystal 半导体晶体
semi-conducting diamond 半导体金刚石
semi-conducting glass 半导体玻璃;半导电玻璃
semi-conducting glass fibre 半导体玻璃纤维
semi-conducting glaze 半导体釉
semi-conducting material 半导体材料
semi-conducting metal 半导体金属
semi-conducting paper 半导体纸
semiconducting polymer 高分子半导体
semi-conduction 半导电性
semi-conductive 半导体的;半导电
semi-conductive ceramics 半导体陶瓷
semi-conductive loading tube 半导体装药管
semi-conductive mineral 半导体矿物
semi-conductive thin-film detector 半导体薄膜检测器
semi-conductor 半导体
semi-conductor amplifier 半导体放大器
semi-conductor bolometer 半导体辐射热测量器
semi-conductor capacitor 半导体电容器
semi-conductor cartridge 半导体拾音头
semi-conductor ceramics 半导体陶瓷
semi-conductor chemistry 半导体化学
semiconductor chip 半导体芯片
semi-conductor colo(u)rization 半导体着色
semi-conductor component 半导体元件
semi-conductor crystal 半导体晶体
semi-conductor detector 半导体检测器
semi-conductor device 半导体器件;半导体装置
semi-conductor diode 半导体二极管;晶体二极管
semi-conductor diode amplifier 半导体二极管放大器
semi-conductor diode laser 半导体二极管激光器
semi-conductor diode parametric amplifier 半导体二极管参量放大器
semi-conductor dynamic(al) memory 半导体动态存储器
semi-conductor electrochemistry 半导体电化学
semi-conductor electron multiplier 半导体电子倍增器
semi-conductor element 半导体元件
semi-conductor encapsulation 半导体包胶
semi-conductor freezer 半导体冰箱
semi-conductor ga(u)ge 半导体应变片
semi-conductor glass 半导体玻璃
semi-conductor heterostructure 半导体异质结构
semi-conductor injection laser 半导体注入式激光器
semi-conductor intrinsic(al) property 半导体的本征性质
semi-conductor laser 半导体激光(器)
semi-conductor laser amplifier 半导体激光放大器
semi-conductor laser demodulator 半导体激光解调器
semi-conductor laser detector 半导体激光检测器
semi-conductor laser diode 半导体激光二极管
semi-conductor laser range finder 半导体激光测

距机
semi-conductor laser transmitter 半导体激光发射器
semi-conductor light emitter 半导体光发射器
semi-conductor luminescent diode 半导体荧光二极管
semi-conductor material 半导体材料
semi-conductor memory 半导体存储器
semi-conductor memory test system 半导体存储器测试系统
semi-conductor noise eliminator 半导体噪声消除器
semi-conductor optical amplifier 半导体光放大器
semi-conductor optical shutter 半导体光学快门
semi-conductor optoelectronic display 半导体光电子显示器
semi-conductor oscillistor 半导体振荡器
semi-conductor parametron 半导体参量器;半导体参量激励子
semi-conductor photocatalysis 半导体光催化
semi-conductor photocatalyst 半导体光催化剂
semi-conductor photo diode 半导体光电二极管
semi-conductor photoelectronic device 半导体光电子器件
semi-conductor physics 半导体物理学
semi-conductor point thermometer 半导体点温度计
semi-conductor processing 半导体加工
semi-conductor quantum well laser 半导体量子阱激光器
semi-conductor radiation detector 半导体辐射探测器
semi-conductor rectifier 半导体整流器
semi-conductor rectifier cell 半导体整流元件
semi-conductor rectifier tube 半导体整流管
semi-conductor rectifier welder 半导体整流焊机
semi-conductor refrigerator 半导体制冷机
semi-conductor store 半导体存储器
semi-conductor strain ga(u)ge 半导体应变片;半导体应变计
semi-conductor superlattice 半导体超晶格
semi-conductor technology 半导体工艺
semi-conductor thermocouple 半导体温差电偶
semi-conductor thermometer 半导体温度计
semi-conductor thermosensitive resistor 半导体热敏电阻
semi-conductor time-lag relay 半导体时延继电器
semi-conductor trap 陷阱
semi-conductor triode 半导体三极管
semi-conductor varnish 半导体漆
semi-confined aquifer 漏水含水层;半承压含水层
semi-confining bed 半阻水层;半隔水层
semi-conic 半圆锥体的
semi-conscious 半意识的
semi-consciousness 半意识
semi-conservative replication 半保留复制
semi-consolidated sandstone 弱胶结的砂岩
semi-construction 半整体式结构;半整体式建筑
semi-contacted mosaic 半控制镶嵌图
semi-container 半大集装箱
semi-container ship 部分集装箱船;半集装箱船;半货柜船
semi-container vessel 半集装箱船
semi-continuity 半连续性
semi-continuous 半连续的
semi-continuous activated sludge 半连续性活性污泥
semi-continuous activated sludge test 半连续性活性污泥试验
semi-continuous automatic centrifuge 半连续式自动离心机
semi-continuous casting 半连续铸造
semi-continuous casting machine 半连续铸造机
semi-continuous casting melting 半连续铸造熔化
semi-continuous channel 半连续信道;半连续通路
semi-continuous culture 半连续培养
semi-continuous distillation 半连续(性)蒸馏
semi-continuous electrolytic process 半连续电解法
semi-continuous function 半连续函数
semi-continuous furnace 半连续式炉
semi-continuous hotstrip mill 半连续式带材热轧机
semi-continuous kiln 半连续(式)窑
semi-continuous leaching 半连续(性)浸取

semi-continuous mapping 半连续映射
semi-continuous mill 半连续轧钢机
semi-continuous mining technology 半连续开采工艺
semi-continuous process 半连续(性)过程
semi-continuous rolling mill 半连续式轧机
semi-continuous rolling process 半连续压延法
semi-continuous running 半连续运转
semi-continuous stoking 半连续推料
semi-continuous type oven 半连续式烘炉
semi-continuous wide strip mill 半连续式宽带钢轧机
semi-continuous wirerod mill 半连续式线材轧机
semi-convection 部分对流;半对流型
semi-convection type pipe still 半对流型管式炉
semi-convergent 半收敛的
semi-convergent series 半收敛级数
semi-coordinate paper 半对数坐标纸
semi-coursed 大致成层的
semi-court 半开敞的庭院
semi-covered coalfield 单掩盖式煤田
semi-crystal 半水晶;半晶体
semi-crystal glass 半晶体玻璃
semi-crystalline 半晶状的;半晶质(的);半结晶的
semi-crystalline calcium silicate hydrate 半晶质水化硅酸钙;半晶态水化硅酸钙
semi-crystalline-porphyritic 半晶斑状
semi-cubical parabola 半立方抛物线
semi-cultivated variety 半栽培品种
semi-cupola 半圆顶
semi-cure 半硫化;半固化
semi-curtain wall 半幕墙
semi-curtain wall between windows 窗间半幕墙
semi-cyclic 半环的
semi-cyclic link 半环键
semi-cyclic linkage 半环键
semi-cylinder 半柱面;半圆柱体
semi-cylindric(al) die 半圆柱体模
semi-cylindric(al) reflector 半圆柱形反射器
semi-cylindric(al) tile 筒瓦
semi-cylindric(al) urinal 半圆筒形小便槽
semi-daily 半日一次的
semi-daily tide 半日潮(汐)
semi-decidability 半决定性
semi-decidable 半可判定的
semi-deck trailer 半平板拖车
semi-decussation 半交叉
semi-deep sea 半深海
semi-deep water 半深水
semi-definite 半定
semi-definite form 半定型
semi-definite matrix 半定矩阵
semi-definite operator 半定算子
semi-definite quadratic form 半定二次型
semi-definite system 半正定系统
semi-definitive time 半确定时
semi-deoxidized steel 半脱氧钢
semi-desert 半沙漠【地】;半荒漠
semi-desert basin 半沙漠盆地
semi-desert region 半荒漠地区
semi-desert steppe soil 半漠境草原土
semi-destructive examination 局部破坏性检验
semi-detached 半独立的;房屋一侧与它屋相连的;半分离的
semi-detached building 双连式房屋;半独立(式)房屋
semi-detached dwelling 联立式住宅;双户连体住宅;半分立式住宅;半独立式住宅
semi-detached home 半独立式住宅
semi-detached house 双连式住宅;半独立式住宅;半独立(式)房屋;有公隔墙的两毗连的房屋
semi-developed nations 半发达国家
semi-diagonal 半对角线
semi-diameter 半径
semi-diameter correction 半径修正;半径校正;半径改正(量)
semi-diameter ga(u)ge 半径规
semi-diaphanous 半透明的
semi-diesel 烧球式柴油机
semi-diesel engine 柴油机;热球式发动机
semi-diffusion 半扩散
semi-digital aero-triangulation 半解析空中三角测量
semi-digital block aero-triangulation 半解析区域

空中三角测量
semidine 半联胺;重苯胺;苯氨基苯胺
semi-dine rearrangement 半联胺重排作用
semi-diode 半导体二极管
semi-direct connection 半直接连接
semi-direct connection ramp 半直接连接匝道
semi-direct expenses 半直接费用
semi-direct furnace 半煤气窑
semi-direct heating 半间接暖气装置
semi-direct illumination 半直接照明
semi-directional 半定向
semi-directional interchange 半定向式立交
semi-direct lighting 半直接(采)光;半直接照明
semi-direct measurement 半直接测量
semi-direct ramp 半直接式匝道
semi-discretization 半离散化
semi-distributed coil 半分布线圈
semi-distributed hardware system 半分布式硬件系统
semi-distributed winding 半分布绕组
semi-diurnal 半天的;半日的;半潮汐日的
semi-diurnal age of tide 半日潮龄
semi-diurnal barometric(al) variation 半周日气压变化
semi-diurnal component 半日分潮
semi-diurnal constituent 半日分潮
semi-diurnal current 半日周潮流
semi-diurnal force 半日生潮力
semi-diurnal tidal current 半日潮流
semi-diurnal tidal dock 半日潮汐船坞
semi-diurnal tidal harbo(u)r 半日潮港
semi-diurnal tide 半日潮(汐)
semi-diurnal tide-producing force 半日生潮力;日引力潮
semi-diurnal variation 半日变化
semi-diurnal wave 半日波
semi-dome 半圆屋顶;半穹顶;半圆形穹顶;半圆(穹隆)
semi-dominance 半显性
semi-dominant diagonal 半主对角线
semi-dominant gene 半显性基因
semi-dormancy 半休眠
semi-double 半重瓣的
semi-double strength 中厚玻璃
semi-drop center rim 半凹槽轮缘
semi-dry developing process 半干显影法
semi-drying oil 半干性油
semi-dry pressing 半干压法;半干压成型;半干法成型
semi-dry press process 半干压砖法;半干压制法
semi-dry process 半干压工艺;半干法(工艺);半干处理
semidry process for making mineral wool slab 矿棉板半干法制板工艺
semi-dry state 半干状态
semi-dull 近无光的;近暗的
semi-dull coal 半暗煤
semi-dull silk 半无光丝
semi-dull yarn 半无光纱
semi-duplex operation 半双工操作
semi-duplex process 半双联法
semi-durable adhesive 半耐久性黏合剂
semi-durables 半耐用商品
semi-ebonite 半硬橡胶
semi-ebonite hose 半硬胶管
semi-eggshell ware 半薄胎器
semi-elastic 半弹性的
semi-elastic condition 半弹性状态
semi-elastic deformation 半弹性变形;半弹性变形
semi-elastic foundation 半弹性地基
semi-elastic impact 半弹性碰撞;半弹性冲击
semi-elastic subgrade 半弹性路基;半弹性地基
semi-electronic switching system 半电子交换系统
semi-electronic telephone exchange 半电子式电话交换机
semi-electronic typesetter 半电子式照相排字机
semi-ellipse 半椭圆
semi-ellipse arch 半椭圆拱
semi-ellipsoid 半椭圆体;半椭球体
semi-ellipsoidal mirror 半椭球面镜
semi-elliptic(al) 半椭圆形的
semi-elliptic(al) arch 半椭圆(形)拱
semi-elliptic(al) arch bridge 半椭圆形拱桥
semi-elliptic(al) flaw 半椭圆形缺陷

semi-elliptic(al) head 半椭圆封头
semi-elliptic(al) leaf-spring 半椭圆形片弹簧
semi-elliptic(al) scroll spring 半椭圆形涡簧;双弯头半椭圆形钢板弹簧
semi-elliptic(al) section 半椭圆截面
semi-elliptic(al) sewer 半椭圆形(断面)下水道;半椭圆形下水管;半椭圆形污水管
semi-elliptic(al) spring 半椭圆(形)弹簧;弓形弹簧
semi-elliptic(al) surface crack 半椭圆形表面裂纹
semi-empiric(al) 半经验
semi-empiric(al) calculation 半经验计算
semi-empiric(al) equation 半经验方程(式)
semi-empiric(al) formula 半经验公式
semi-empiric(al) method 半经验法
semi-empiric(al) modification 半经验修正
semi-empiric(al) procedure 半经验程序
semi-empiric(al) relation 半经验关系式
semi-empiric(al) relationship 半经验关系;半经验公式
semi-empiric(al) rule 半经验规则;半经验公式
semi-employed 半失业
semi-enclosed 半开的;半封闭的
semi-enclosed cut-out 半闭式保险器
semi-enclosed impeller 半封闭式叶轮
semi-enclosed machine 半封闭式机器
semi-enclosed motor 半封闭式电动机
semi-enclosed place of outdoor assembly 户外集会的半围蔽场地
semi-enclosed sea 半封闭海(域)
semi-enclosed slot 半封闭(式)槽;半闭口槽
semi-enclosed space 半围蔽处所
semi-enclosed type 半封闭式
semi-enclosed type machine 半封闭式电机
semi-enclosed type motor 半封闭式电动机
semi-enclosed water body 半封闭水域;半封闭水体
semi-engineering brick 半工程砖
semi-evergreen plant 半常绿植物
semi-exciting type regenerative generator 半激磁式再生发电机
semi-exclusive pedestrian-vehicle phase 行人—行车半专用信号显示
semi-expendable 半消耗的
semi-experiential formula of interfering well 干扰井半经验公式
semi-explicit 半显化
semi-exposed 半敞露的
semi-exposed coalfield 半暴露式煤田
semi-exposed framing glass curtain wall 半隐框玻璃幕墙
semi-factor 半因子
semi-faience 半瓷釉
semifast transport 中速行驶运输工具
semi-feathering propeller 半活叶螺旋桨
semi-fibrated 半纤维化的
semi-field 半视场
semi-filled steel 半钢
semi-filling 半充填
semi-film coating 半薄膜衣
semi-fined paraffin wax white 半精炼白石蜡
semi-fine screw 半细螺钉
semi-finished 半制成的;半完成的;半(精)加工的;半光制
semi-finished blank 半精制坯件
semi-finished bolt 半光制螺栓
semi-finished countersunk head bolt 半光埋头螺栓
semi-finished hexagon head bolt 半光六角螺栓
semi-finished hexagon nut 半光六角螺母
semi-finished inventory 半成品储备
semi-finished material 半产品材料
semi-finished metal 半成品扎材
semi-finished nut 半光制螺母
semi-finished parts 半产品部件
semi-finished product 毛坯;半制成品
semi-finished product standard 半成品标准
semi-finished round pin 半光圆销
semi-finished screw 半光制螺钉
semi-finished steel 半制钢
semi-finished working 半细加工
semi-finishing 半精加工
semi-finishing tooth 半精切齿
semi-finishing tooth of broacher 拉刀的半精切齿
semi-fireproof 半耐火的;半防火的
semi-fireproof construction 半防火建筑

semi-fixed 半固定的
semi-fixed bridge 半固定桥
semi-fixed cost 半固定成本
semi-fixed dune 半固定沙丘
semi-fixed girder 半固定梁
semi-fixed length record 半固定长度记录;半定长记录
semi-fixed penstock 半固定式压力水管
semi-fixed sand 半固定沙
semi-fixed support 半固定支点
semi-fixed variable resistor 半固定可变电阻器
semi-flash condition 半闪光条件
semi-flat 表面稍粗糙的;半光泽的;半光
semi-flat band method 半带法
semi-flat rim 半平底轮辋
semi-flexible 半柔性的
semi-flexible board 半软性印刷线路板
semi-flexible coupling 半挠性联轴器
semi-flexible joint 半柔性结点;半柔性节点;半柔性接头;半柔性接合;半挠性连接
semi-flexible pavement 半柔性路面
semi-flexible polymer 半韧性聚合物
semi-flexible screen 半活动筛
semi-flexible wire rope 半硬钢丝绳
semi-flexion 半屈
semi-floating 半浮动
semi-floating axle 半浮(式)轴
semi-floating axle shaft 半浮式半轴
semi-floating piston pin 半浮式活塞销
semi-floating rear axle 半浮式后轴
semi-fluctuating 半波动的
semi-fluid 半流质(的);半流体;半流动的
semi-fluid friction 半液动摩擦
semi-fluid grease 半流润滑脂
semi-fluidized bed 半流化床
semi-fluid mass 半流动体
semi-fluid solution 半流体溶液
semi-flush coupling casing 半内平连接套管(套管末端加厚,薄壁按箍与套管外径相同)
semi-flush inlet 半埋入式进气口
semi-focal chord 半焦弦
semi-focusing 半聚焦
semi-formula 半经验公式
semi-friable alumina 半脆性氧化铝
semi-full budgeting 编制半期全面预算
semi-fusain 半丝煤
semi-fused 半熔化
semi-fusinite 半丝质体;半丝炭煤素质
semi-fusinized circleinite 半丝炭化浑圆体
semi-fusinized groudmassinite 半丝炭化基质体
semi-fusinized sclerotinite 半丝炭化菌类体
semi-fusinoid group 半丝质组;半丝炭化组
semi-fusite 微半丝煤
semi-gantry crane 半龙门起重机;半龙门吊
semi-gas firing 半煤气烧成;半煤气加热
semi-gel 半凝胶(炸药)
semi-gelatin 半凝胶
semi-gelatin dynamite 半胶质硝甘炸药
semi-gelatine(type of) explosive 半凝胶炸药
semi-gelatins 硝铵基半凝胶炸药类
semi-gelation 半凝胶(炸药)
semi-gelified groundmassinite 半凝胶化基质体
semi-gelified sclerotinite 半凝胶化菌类体
semi-gelifusinite-collinite 半胶丝质无结构体
semi-gelifusinite-posttelinite 半胶丝质次结构体
semi-gelifusinite-precollinite 半胶丝质似无结构体
semi-gelifusinites 半胶丝质类
semi-gelifusinite-telinite 半胶丝质结构体
semi-gelinite-collinite 半胶质无结构体
semi-gelinite group 半凝胶化组
semi-gelinite-posttelinite 半胶质次结构体
semi-gelinite-precollinite 半胶质似无结构体
semi-gelinites 半凝胶类
semi-gelinite-telinite 半胶质结构体
semi-geometrical transformation 半几何变换
semi-girder 悬臂梁
semi-glassy 半光的
semi-glassy paper 半光像纸
semi-glaze 半釉
semi-gloss 半光涂层;半光泽的;近有光
semi-gloss enamel 半光瓷漆
semi-gloss epoxide-polyester powder coating 半光环氧聚酯粉末涂料
semi-gloss lacquer 半光(喷)漆

semi-gloss latex enamel 半光乳胶瓷漆
semi-gloss oil paint 半光泽油漆
semi-gloss paint 半光(泽)漆;半光涂料
semi-gloss quick-dry enamel 半光快干瓷漆
semi-glossy paper 半光面相纸
semi-Gothic arch 半哥特式拱;半尖拱
semi-graben 半地堑
semi-granular 半颗粒状态
semi-graphic(al) 半图解的
semi-graphic(al) annunciator 半图解式信号器
semi-graphic(al) control panel 半图解仪表盘
semi-graphic(al) keyboard 半图解键盘
semi-graphic(al) method 半图解法
semi-graphic(al) panel 半图式面板;半图解仪盘表
semi-graphic(al) resection 半图解后方交会法
semi-graphite 半石墨
semi-gravel 半砾石
semi-gravity type abutment 半重力式桥台
semi-gravity type retaining wall 半重力式挡土墙
semi-gravity wall 半重力式墙
semi-gross weight 半毛重
semi-group 半组;半群(组)
semi-group homomorphism 半群同态
semi-group method 半群法
semi-grouser 半稳定桩(拖拉机履带片上的低抓板)
semi-grouting 半泥灌浆的;半灌浆的
semi-guarded machine 半防护式电机
semi-hammerhead boom 半锤头式吊杆
semi-hand 半手工(的)
semi-hard 半硬的
semi-hardboard 半硬质合成板;半硬质纤维板
semi-hard-drawn aluminum wire 半硬铝线
semi-hard drying (涂层的)半硬化干燥
semi-hardening 半硬化
semi-hard magnetic alloy 半硬磁合金
semi-hard magnetic material 半永磁材料
semi-hard packing 半硬填料
semi-hard rock 半坚硬岩石
semi-hard rocks 半坚硬岩石类
semi-hard rubber 半硬质橡胶
semi-hard steel 半硬钢
semi-hard-type balancing machine 半硬支承平衡机
semi-head rock 中硬岩
semi-healthy stream 轻度污染河流
semi-heavy 半稠的;半重的
semi-heavy loading 半重负荷
semi-hermetically sealed condensing unit 半封闭式压缩冷凝机组
semi-hermetic compressor 半封闭式压缩机
semi-hermetic refrigerant compressor 半封闭式制冷剂压缩机
semi-hermetic refrigeration compressor 半封闭制冷压缩机
semi-high level bridge 半浸水桥
semi-hollow drawn shape 半空心拔制型材;半空心拉制型材
semi-hollow extruded shape 半空心挤压型材
semi-hollow shape 半空心型材
semi-homogeneous liquid 半均相液体
semi-homogeneous strain 半均匀应变
semi-housed 半露天的
semi-housed stair(case) 靠墙(楼)梯
semi-housed stringer 楼梯斜梁贴面板
semi-humid 半湿性;半湿润;半湿度;半(潮)湿的
semi-humid region 半湿润(地)区;半潮湿地区
semi-humid soil 半湿土(壤)
semi-hyaline 半透明的
semi-hydrate 半水化(合)物;半水合物
semi-hydrate gypsum 半水石膏
semi-hydraulic earth dam 半游填土坝
semi-hydraulic excavator 半水力挖土机;半液压挖掘机
semi-hydraulic fill dam 半水力冲填坝;半冲填坝
semi-hydraulic fill earth dam 半水力冲填式土坝
semi-hydraulic fill method 半水力冲填法
semi-hydraulic lift 半液压提升机
semi-hydraulic lime 半水硬(性)石灰
semi-immersed liquid-quenched fuse 半浸液淬灭熔断丝
semi-immobile dune 半流动沙丘
semi-impervious 半透水的
semi-impervious glaze 半透水釉面
semi-improved air-field 粗加工道面机场

semi-inclusive 半包含
semi-increment 半增量
semi-independent 半独立的
semi-independent territory 半独立地区
semi-indirect fitting 半间接附件
semi-indirect illumination 半间接照明
semi-indirect lighting 半间接照明
semi-indurated 半固结的
semi-industrial installation 半工业生产装置
semi-industrial scale 半工业规模
semi-industry scale 中间工业生产规模
semi-inertinite 半惰性煤素质
semi-infinite 半无限(的)
semi-infinite aquifer 半无限含水层
semi-infinite bar 半无限长杆件
semi-infinite beam 半无限长梁
semi-infinite body 半无限体;半空间体
semi-infinite cylinder 半无限圆柱;半无限筒柱
semi-infinite elastic solid 半无限弹性(固)体
semi-infinite elastic subgrade 半无限弹性地基
semi-infinite half-space 半无限空间
semi-infinite homogeneous radiator 半无限均匀辐射体
semi-infinite ingot 半无限长锭料
semi-infinite layer 半无限层
semi-infinite length 半无限长度
semi-infinite medium 半无限介质
semi-infinite motor 半无限长电动机
semi-infinite plate 半无限平板
semi-infinite reservoir 半无限水库
semi-infinite soil mass 半无限土体
semi-infinite solid 半无限固体
semi-infinite stratified medium 半无限层状介质
semi-infinite tape 半无穷带
semi-inflation 准通货膨胀;部分膨胀;半通货膨胀
semi-inner product 半内积
semi-installment 准分期付款
semi-instrumental survey 半仪器测量
semi-insulated 半绝缘的
semi-insulating 半绝缘的
semi-insulating polycrystalline silicon 半绝缘多晶硅
semi-insulator 半绝缘体
semi-integral 半整体的;半悬挂式的
semi-integral plow 半悬挂式犁
semi-integrated tow 半组合式驳船队;半分节驳船队
semi-integrated type 半结合式
semi-intermediate section 半中型部件
semi-intrados 半拱腹;半拱凹面
semi-intuitive skill 半直觉技术
semi-invariant 半不变式;半不变量
semi-inverse method 半逆解法;半反算法
semi-invisible flame operation 浅暗火操作
semi-irreducible graph 半不可约图
semi-isolated route 半封闭线路
semi-iterative method 半迭代法
semi-iterative process 半迭代过程
semi-jib boring machine 半臂镗床
semi-jig boring 半坐标镗削
semi-Kaplan turbine 半卡普兰式水轮机
semi-killed ingot 半镇静钢锭
semi-killed steel 半镇静钢
semi-knockdown 半组合式商品;半拆散式输出;半散件商品;半拆散式销售
semi-knockdown export 半拆散式输出
semi-kraft paper 单面牛皮纸
semi-laminar flow 半层流
semi-lattice 半格(点)
semi-latus rectum 半通径
semi-leaky type 半漏泄型
semi-lean solution pump 半贫液泵
semi-lethal 半致死的
semi-lethal mutant 半致死突变体
semi-lift bascule bridge 半竖升开启桥;半升启仰开桥
semi like-grained concrete 半似晶质混凝土;半均质混凝土
semi-linear 半线性(的)
semi-linear partial differential equation 半线性偏微分方程
semi-liquid 半液体(的);半流体;黏稠液状的
semi-liquid flushing 半液体冲洗
semi-liquid lubrication 半液体润滑
semi-liquid phase process 半液相过程

semi-liquid stage 半液体阶段
semi-liquid state 半液体状态
semi-liquid waste 半液体废物
semi-live skid 半活动小架车
semi-locked coil wire rope 半密封钢丝绳
semi-locking differential 半闭锁分速器
semi-log 半对数(的)
semi-logarithmic 半对数曲线;半对数的
semi-logarithmic abscissa 半对数横坐标
semi-logarithmic chart 半对数表
semi-logarithmic coordinate paper 半对数坐标纸
semi-logarithmic curve of drawdown-distance 降深—距离半对数曲线
semi-logarithmic curve of drawdown-time 降深—时间半对数曲线
semi-logarithmic curve of drawdown-time-distance 降深—时间距离半对数曲线
semi-logarithmic graph 半对数图
semi-logarithmic graphy 半对数图解
semi-logarithmic linear chart 半对数线图
semi-logarithmic model 半对数模型
semi-logarithmic paper 半对数纸
semi-logarithmic scale 半对数标尺;半对数标度
semi-log diagram 半对数图
semi-loggia 半凉廊
semi-log graph paper 半对数坐标纸
semi-log model 半对数模型
semi-log paper 半对数(图)纸
semi-log plot 半对数(坐标)图
semi-log transformation 半对数变换
semi-longitudinal fault 偏斜走向断层;斜断层
semi-low head machine 立弯式连铸机
semi-low-loader 半低装载机
semi-lozenge riveting 半阶式铆接
semi-lucent 半透明的
semi-lunar 半月的
semi-lunar body 半月形体
semi-lunar deformed bar 月牙纹变形钢筋
semi-lunar fold 半月皱襞
semi-lunar nucleus 半月核
semi-lunar period 半月周期
semi-lunar space 半月状隙
semi-lunation 半太阴月(用于潮汐)
semi-lunation tide 半太阴月潮汐
semi-lune 半匝线圈;半月形
semi-lustrous 半光泽的
semi-luxation 半脱位
semi-luxury 半奢侈品
semi-machine 部分机械加工;半精加工
semi-macrinite 半粗粒体
semi-macrinoid group 半粗粒体组
semi-magnetic controller 半磁(性)控制器
semi-major axis 长半轴;长半径;半长轴;半长径
semi-major axis of beam cross-section 束流截面的长半轴
semi-major axis of satellite orbit 卫星轨道长半轴
semi-mall 半步行街
semi-malleable cast iron 半可锻铸铁
semi-manufactured goods 半制品
semi-manufactured material 半制成品
semi-manufactured product 半成品
semi-Markov process 半马尔科夫过程
semi-mastic 半黏稠状防潮材料
semi-matrix 半矩阵
semi-mat(te) 半暗淡的;半无光;半暗色
semi-mat(te) glaze 半无光釉
semi-mat(te) glaze tile 半光泽釉面砖
semi-mat(te) lacquer 半无光漆
semi-mat(te) paper 半光像纸
semi-mat(te) photo 半无光相片
semi-mat(te) picture 半无光相片
semi-mat(te) point 半无光像点
semi-mat(te) surface 半粗糙表面
semi-mat(te) varnish 半无光清漆
semi-mature soil 半熟土
semi-mean axes 半中轴
semi-measurement data 半计量资料
semi-mechanical 半机械化
semi-mechanical installation 半机械化设备
semi-mechanization 半机械化
semi-mechanization of cargo-handling 装卸工作半机械化
semi-mechanized 半机械化的
semi-mechanized operation 半机械化操作

semi-mechanized shield 半机械化盾构
semi-member 半构件
semi-membrane type cargo tank 半膜式液货舱
semi-menstrual inequality 半月的日潮不等
semi-meridian 半子午线
semi-metal 半金属
semi-metallic 半金属的
semi-metallic coach 半金属客车
semi-metallic gasket 半金属填密片;半金属垫片
semi-metallic luster 半金属光泽
semi-metallic packing 半金属填料
semi-metope 半块雕饰板
semi-metric 半度量
semi-metro 半地下
semi-micelle 半胶束
semi-micro 半微量的
semi-microanalysis 半微量分析
semi-micro analytical balance 半微量(分析)天平
semi-micro aniline point test 半微量苯胺点试验
semi-micro calorimeter 半微量量热计
semi-micro-chemistry 半微量化学
semi-micro-fractionation 半微量分馏
semi-micro hydrogenation apparatus 半微量加氢装置
semi-micro method 半微量法
semi-micro organic synthesis 半微量有机合成
semi-microscopic organism 半微型生物
semi-minor axis 短半轴;半短轴;半短径
semi-mobile 半移动式;半移动(的);半机动式
semi-mobile batching plant 半移动式拌和设备
semi-mobile crusher 半移动破碎机
semi-mobile ground 半流沙地
semi-molten 半熔化
semi-monocoque 半硬壳式机身
semi-monocoque construction 半硬壳结构
semi-monocoque structure 半硬壳式结构
semi-monolithic 半整体
semi-monolithic construction 半整体式结构;半整体式建筑;半整体式构造
semi-monthly 半月刊
semi-moon hoe 半月形锄
semi-moon shaped 半月形的
semi-motor 往复旋转液压油缸
semi-mo(u)lded 半模制
semi-mo(u)ldering 半模作用
semi-mounted 半悬挂式的;半安装式的
semi-mounted disk plow 半悬挂式圆盘犁
semi-mounted moldboard plow 半悬挂式铧式犁
semi-mounted mower 半悬挂式割草机
semi-mounted nine-row grain drill 半悬挂九行谷物播种机
semi-mounted plough 半悬挂式犁
semi-mounted rear loader 半悬挂式后装载机
semi-mounted rotary grass cutter 半悬挂式旋转割草机
semi-mounted seed drill 半悬挂播种机
semi-mounted sprayer 半悬挂式喷雾器
semi-mounted tool 半挂农具
semi-mounted tractor drill 半悬挂机引条播机
semi-mounted type construction 鹅颈式建筑;半悬挂式建筑
semi-mounted wagon 半挂车;鹅颈式货车
semi-movable 半移动式
semi-muffle furnace 半套炉;半马弗炉;半隔焰炉
semi-muffle kiln 半马弗窑;半隔焰窑
semi-muffle lehr 半隔焰退火炉
semi-muffle type furnace 半马弗式炉
seminar 学术研究会;学术讨论会;研讨会;研究班;专题研究会;专题研讨班;专题讨论(会);专题报告;讨论教室;讨论会(学术方面)
seminar-cum-study tour 研讨会暨考察
seminar on surveying and mapping 测绘学术讨论会
seminary 苗圃;苗床;神学院
semi-natural 半自然的;半野生
semi-natural economy 半自然经济
semi-natural harbour 半天然港(口)
semi-natural port 半天然港(口)
semi-natural vegetation 半自然植被;半野生植被;半天然植被
semi-non-combustible material 半非可燃物;准不燃材料
semi-norm 半模;半范数
semi-normal 半当量浓度的;半当量

semi-numerical symmetrical method 半数字对称法
semi-octave 半倍频程
semi-official 半官方的
Semionotus sandstone 半椎鱼属砂岩
semiopal 普通蛋白石;半蛋白石
semi-opaque 半透明的
semi-open 半开式的;半开敞的
semi-open-air installation 半露天装置
semi-open anchorage 半开敞锚地
semi-open cutter 半开式铰刀
semi-open impeller 半开敞式叶轮
semi-open slot 半开口槽
semi-open texture 稀疏纹理;半开结构
semi-open type 半开式;半开敞式;半敞开式
semi-open type electric(al) equipment 半开敞式电气设备
semi-orbicular 半环形的
semi-order 偏序
semi-organic chemistry 半有机化学
semi-oriented specimen 半定向标本
semi-outdoor 半户外的;半露天的
semi-outdoor power plant 半露天式电厂
semi-outdoor-type power plant 半户外式发电设备
semi-overt 半公开的
semi-owner peasant 半自耕农
semi-oxamazide 氨基草酰肼
semi-packed 半包装
semi-palmate 半掌状的
semi-parabolic 半抛物线的
semi-parabolic girder 半抛物线大梁
semi-parabolic girder with sloping end posts 有倾斜柱的半抛物线大梁
semi-parabolic vibrating screed 半抛物线形振动整平机
semi-parasite 半寄生物
semi-passenger ship 客货轮;客货船
semi-paste paint 半厚(油)涂料;半厚(油)漆;半膏状原漆
semi-patented rod 半铅淬火盘条
semi-path 半路径
semi-pavement 半刚性路面
semi-pearl polymerization 半旋浮聚合;半悬浮聚合
semi-pegmatitic texture 半文象结构
semi-pellet 中间球
semi-penetration macadam 半贯入式碎石路(面)
semi-penetration treatment 半灌沥青处理
semi-perched groundwater 半栖留地下水
semi-perched groundwater table 半滞水地下水位;半栖留地下水位
semi-perched water 半栖滞水;半栖留水;半静止地下水
semi-perched water table 半滞水地下水位;半栖滞潜水面
semi-perimeter 半周长
semi-period 半周期
semi-perishable goods 半易腐烂货物
semi-permanent 半永久性的
semi-permanent anti-cyclone 半永久性反气旋
semi-permanent bridge 半永久性桥(梁)
semi-permanent building 半永久性建筑
semi-permanent connection 半永久性连接;半永久性连续
semi-permanent current 半恒流
semi-permanent data 半永久数据
semi-permanent depression 半永久性低压
semi-permanent drawbar 半永久牵引杆
semi-permanent fluctuation 半永久性波动
semi-permanent high 半永久性高压
semi-permanent mo(u)ld 半永久性模;半永久型;半金属型
semi-permanent pasture 半永久性草地
semi-permanent pigment 半永固型颜料
semi-permanent piling 半永久性桩
semi-permanent snow line 半永久性雪线
semi-permanent store 半固定存储器
semi-permanent structure 半永久性结构
semi-permeability 半透(气)性;半透过性;半渗透性
semi-permeable 半透性的;半透性(的);半可透的
semi-permeable material 半透水材料
semi-permeable membrane 半透性膜;半透(明)膜;半透薄膜;半渗透(薄)膜;半渗膜
semi-permeable surface 半渗透曲面

semi-persistent virus 半持久性病毒
semi-pervious 半渗透的;半透水的
semi-pervious material 半透水材料
semi-pervious riverbed 半透水河床
semi-pervious streambed 半透水河床
semi-pervious zone 半透水区
semi-physical simulation 半物理仿真
semi-piercing 半冲孔
semi-planar electro-optic modulator 半平面电光调制器
semi-plane 半平面
semi-plant 试验工厂;中间试验(工)厂;中间工厂
semi-plant equipment 半工厂设备
semi-plant scale equipment 中间试验设备
semi-plant test 中间试验
semi-plastic 半塑性的
semi-plastic concrete 低塑性混凝土
semi-plastic explosive 半塑性炸药
semi-plastic making 半塑性成型法;半可塑成型法
semi-plastic process 半塑性成型法;半可塑成型法
semi-plastic stage 半塑性阶段
semi-plastic state 半塑性状态
semi-pneumatic tire 半充气轮胎
semi-pneumatic tyre 半实心轮胎;半充气轮胎
semi-polar 半极性的;半极化
semi-polar bond 半极性键
semi-polar capacitor 半极化电容器
semi-polar double bond 半极性双键
semi-polarity 中极性;中级性;半极性
semi-porcelain 炻器;半瓷器
semi-porous 半疏松的;半多孔的
semi-portable 半移动式;半移动
semi-portable agricultural sprinkling system 半轻便式农田喷水设备
semi-portable batching and mixing plant 半移动配料搅拌设备
semi-portable crane 半门式起重机(双门柱不等长)
semi-portable engine 半移动式原动机
semi-portable plant 半移动式设备
semi-portal bridge crane 半龙门桥式起重机
semi-portal crane 半门式吊车;单柱高架起重机;单脚高架起重机;半门式悬臂起重机;半门式座重机
semi-portal gantry 半门式起重机(双门柱不等长)
semi-portal jib crane 半(龙)门座臂架起重机
semi-portal slewing crane 半龙门座回转起重机
semi-portal type of pedestral crane 半门座台架式起重机
semi-positive mo(u)ld 半溢式塑模;半溢式模具;半密闭式模具
semi-potentiometer 半电位计;光电势计
semi-precious metal 半贵金属
semi-precious stone 次宝石;半宝石
semi-precision measuring tool 半精密量测工具
semi-predicate function 半谓词函数;半确定函数
semi-premix injector 半预混式喷头
semi-preparative 半制备的
semi-preparative column 半制备柱
semi-preparative separation 半制备规模分离
semi-probabilistic design 半几率设计
semi-probabilistic statistics 半概率性统计
semi-processed unit 中间产品
semi-product 半成品
semi-production 中间生产
semi-production equipment 中间生产设备
semi-production line 半生产流水线
semi-professional 半职业性的
semi-pronation 半旋前
semi-prone 半俯卧位的;半陡(斜)的
semi-prone position 半俯卧位
semi-proportional control 半比例控制
semi-protected harbo(u)r 半防护港
semi-protected motor 半防护型电动机
semi-protic solvent 半质子性溶剂
semi-public building 半公共建筑;半公共房屋
semi-public pool 半公共游泳池
semi-public space 半公共空间
semi-purse net 轻型围网
semi-pyritic smelting process 半自热熔炼法
semi-qualitative approach 半定性近似法
semi-quantitative 半定量
semi-quantitative analysis 半定量分析
semi-quantitative assessment 半定量估值

semi-quantitative determination 半定量测定
semi-quantitative estimate 半定量估计
semi-quantitative spectroanalysis 光谱半定量分析
semi-quantitative spectrochemical analysis 光谱半定量分析
semi-radial drilling machine 半万能摇臂钻床
semi-radial reciprocating compressor 扇形往复式压缩机;星形往复式压缩机;半圆辐射型往复压缩机
semi-ranching 半牧半耕的
semi-random 半随机的
semi-random access 半随机存取
semi-random access memory 半随机访问存储器;半随机存取存储器
semi-random access storage 半随机访问存储器
semi-random model 半随机模型
semi-random telegraph signal 半随机电报信号
semi-range of the tide at port on a day of mean equinoctial springs 港口潮高系数
semi-range of tide 半潮差;半潮幅
semi-real time 半实时
semi-real time processing 半实时处理
semi-recessed village 半隐蔽村落
semi-reclining position 半卧位
semi-recumbent 半卧的
semi-refined cottonseed oil 半精炼棉籽油
semi-refined oil 半精制油料
semi-refined paraffin wax 半精炼石蜡
semi-refined wax 半精制蜡
semi-reflecting mirror 半反射镜
semi-reflector 半反射体
semi-refractory 半耐熔的;半耐火的
semi-refractory brick 半耐高温的砖
semi-regular 半规则状
semi-regular directed graph 半自则有向图
semi-regular survey 半正规测量
semi-regular variables 半规则变星
semi-reinforcing agent 半促进剂;半补强剂
semi-reinforcing furnace black 半补强炉法炭黑
semi-relieved type 半卸载式
semi-remote 半远距
semi-remote cut-off 半遥截止
semi-remote handling 半远距离操纵;半遥控
semi-remote manipulation 半远距离操作
semi-reservoir 半储热层
semi-reverberation 半混响
semi-reversibility 半可逆性
semi-reversible sorption isotherm 半可逆吸附等温线
semi-reversible steering 半可逆转向
semi-revolution 半周
semi-rigid asbestos-bitumen sheet 半刚性石棉沥青板;石棉沥青板
semi-rigid base (course) 半刚性基层
semi-rigid board 半硬板
semi-rigid board of glass wool 玻璃棉半硬板
semi-rigid cable 半硬性电缆
semi-rigid connection 半刚性(框架)连接;半刚性结合;半刚性接合
semi-rigid connector 半刚性接头
semi-rigid container 半硬容器
semi-rigid cradle 半刚性支船架;半刚性船台
semi-rigid foamed plastics 半硬质泡沫塑料
semi-rigid frame 半刚架;半刚性框架
semi-rigid frame construction 半刚架结构
semi-rigid framing 半刚性框架;半刚性框架(做)法
semi-rigid insulation board 半刚性隔热板
semi-rigid joint 半刚性连接;半刚性结合;半刚性结点;半刚性节点;半刚性接头;半刚性接合
semi-rigid lining 半刚性衬砌
semi-rigid model 局部动床模型;半刚性模型
semi-rigid pavement 半刚性路面
semi-rigid penstock 半刚性压(力)水管
semi-rigid plastics 半刚性塑料;半硬质塑料
semi-rigid polyvinyl chloride sheet 半硬质聚氯乙烯片材
semi-rigid receptacle 半刚性容器
semi-rigid structure 半刚性结构
semi-rigid suspension 半刚性悬架
semi-rigid tubing 半刚性管
semi-rigid type base 半刚性基层
semi-rimming steel 半镇静钢
semi-ring 半环
semi-ring porous wood 半环孔材

semi-rope trolley 钢丝绳半牵引小车;半索牵式起重小车
semi-rotary 摆的
semi-rotary actuator 摆动马达
semi-rotary davit 半旋转式吊艇柱
semi-rotary hand pump 手动摆动泵
semi-rotary melting furnace 可倾式熔炉
semi-rotary motion 半转动
semi-rotary pump 半转轮泵;半回转泵
semi-rotary valve 半旋转阀
semi-rotative machine 半旋转机
semi-rough 半粗制的;半粗糙的
semi-round 半圆的
semi-round bar 半圆钢棒
semi-round nose bit 半圆形冠部钻头
semi-rubbed 半磨光的
semi-rubbed finish 半磨光面
semi-rule 半经验规则
semi-rural environment 半野外环境;半乡村环境
semi-saprophyte 半腐生植物
semi-saprophytic parasite 半腐生寄生物
semi-saturation 半饱和
semi-scale production 半工业化生产
semischist 次片岩
semi-scroll case 半蜗壳
semi-sealing paint 单向透气涂料
semi-seal plug-type fuse 半封闭插入式熔断器
semi-section 半剖面
semi-sectional view 半剖面图
semi-selective ringing 半选择振铃
semi-self-fluxing ore 半自熔性矿石
semi-self-fusible ore 半自熔性矿石
semi-selfpropelled combine 半自走式联合收获机
semi-self-propelled machine 牵引式机具;半自走式机具
semi-self-service 半自助
semi-senior accountant 中级会计师
semi-separated rail fastenings 半分开式扣件
semi-separbility 半可分离性
semi-serialized manufacture 半连续性生产
semi-shrouded impeller 半开式叶轮
semi-shrubby desert 半灌木荒漠
semi-silica brick 半硅砖
semi-silica fireclay brick 半硅质耐火砖
semi-silica refractory 半硅质耐火材料
semi-siliceous refractory 半硅质耐火材料
semi-simple matrix 半单矩阵
semi-simple ring 半单环
semi-sintered condition 半烧结状态
semi-streamlined form 半流线型
semi-sintering 半烧结的
semi-skeleton shoe 半骨架式履带板
semi-skilled 半熟练的
semi-skilled labo(u)r 半熟练工人;半熟练劳动(力)
semi-skilled man 半熟练工
semi-skilled manpower 半熟练劳动力
semi-skilled worker 半熟练工人;半技能工人
semi-slewing rampway 半旋转式跳板
semi-sloping quay 半斜坡式码头
semi-sloping wharf 半斜坡式码头
semi-slotted upper roll 半槽式上辊
semi-slow cement 中凝水泥
semi-soft 半软的
semi-soft clay 半软质黏土
semi-soft clay ore 半软质黏土矿石
semi-soft packing 半软包衬
semi-soft steel 半软钢
semi-solid 半固体
semi-solid asphalt 半固体沥青;半固态沥青
semi-solid bituminous material 半固体沥青材料;半固体沥青物质
semi-solid(core)door 半实心门
semi-solid culture medium 半固体培养基
semi-solid dispersion 半固体分散体
semi-solid floor 半实心地板
semi-solid flush door 半实心夹板门
semi-solid grease 半凝润滑脂
semi-solid laminated partition 半实心叠层隔墙
semi-solid lubricant 半固体润滑剂;塑性润滑剂
semi-solid soil 半固态土
semi-solid stage 半固体阶段
semi-solid state 半固态
semi-solid waste 半固体废物

semi-solid weir 半实体堰
semi-space 半空间
semi-spade rudder 半悬挂舵;半铲型舵
semi-special trade 半专门贸易
semi-speculum 半窥镜
semi-sphere 半球形;半球
semi-spheric(al) 半球形(的);半球的
semi-spheric(al)bit 半球形钻头
semi-spheric(al)bowl 半球形钵体
semi-spheric(al)combustion chamber 半球形燃烧室
semi-spheric(al)dome 半圆体屋顶;半球形圆顶;半圆形穹顶
semi-spheric(al)emittance 半球发射率
semi-spheric(al)lamp 半球形灯
semi-spheric(al)tip 半球形电极头
semi-spiral case 不完全蜗壳
semi-splint coal 半暗硬煤
semi-split 半锯开(木)材
semi-stability 半稳定
semi-stabilized dolomite 半稳定白云石
semi-stable 半稳定
semi-stable bituminous dispersion 中裂乳化沥青
semi-stable dispersion 半稳定沥青分散液;半稳定分散体
semi-stable dolomite brick 半稳定性白云石砖
semi-stable dolomite clinker 半稳定性白云石熔块
semi-stable dolomite refractory 半稳定性白云石质耐火材料
semi-stable emulsion 半稳定乳液
semi-stable energy level 半稳定能级
semi-stable state 亚稳态
semi-stainless steel 半不锈钢
semi-stall 半失速;局部滞止;局部分离
semi-stamping 半模锻;半冲压
semi-static 半静态的
semi-static position 半静态位置
semi-stationary 半固定的
semi-stationary process 半平稳过程
semi-stationary wave 半驻波
semi-steel 低碳钢;高级铸铁;钢性铸铁;半钢
semi-steel casting 半钢性铸件;低碳铸件
semi-steel pipe 高强度铸铁管
semi-steep 半陡峭的;缓倾斜的
semi-stiff extrusion process 半硬挤出工艺
semi-stiff fireclay 半硬质耐火黏土
semi-stiff mud process 半硬泥制坯法
semi-stop 半截止头
semi-streamlined body 半流线型车体
semi-streamlined form 半流线型
semi-submerged 半淹没的;半浸水的
semi-submerged air-cushion vehicle 半浸式气垫船
semi-submerged barge 半潜驳
semi-submerged catamaran 半潜式双体船
semi-submerged offshore nuclear generating station 海上半潜式核发电站
semi-submerged platform 半潜(式)平台
semi-submerged rig 半潜式钻机
semi-submerged ship 半潜船
semi-submerged shipway 半浸式船台
semi-submersible 半潜式
semi-submersible crane vessel 半潜式起重船
semi-submersible drilling platform 半潜式钻探平台;半潜式钻井平台
semi-submersible drilling unit 半潜式钻机
semi-submersible drilling vessel 半潜式钻井船
semi-submersible drill rig 半潜式钻探设备;半潜式海上钻探设备
semi-submersible offshore(drill)rig 半潜式海上钻探设备
semi-submersible pipe-laying barge 半潜式管线铺设驳船
semi-submersible platform 半潜式(钻井)平台
semi-submersible production platform 半潜式生产平台
semi-submersible rig 半潜式钻机
semi-submersible tug boat 半潜式驳船
semi-subsistance low-income sector 半自给自足的低收入阶层
semi-subsistence 半自给
semi-successive brazing furnace 半连续钎焊炉
semi-sunk switch 半埋装式开关
semi-supination 半仰卧位;半旋后
semi-supine 半仰卧位的

semi-surface 半曲面
semi-symmetrical 半对称的
semi-symmetry 半对称
semi-synthesis 半合成
semi-synthetic 半合成品
semi-synthetic(al)fiber 半合成纤维
semi-synthetic(al)sand 半合成砂
semi-synthetic(al)sand mix 半合成型砂
semi-tandem rolling 半连轧
semi-tanned skin 半硝皮
semi-taper pin boss 半倾斜式销座
semi-technological factor 半技术因素
semitelinite 结构半镜质体
semi-tempered glass 轻度强化玻璃;轻度钢化玻璃;半钢化玻璃
semi-Thue system 半图厄系统
semi-tidal basin 半感潮港池
semi-tidal dock 半感潮港池;半潮汐船坞;半潮差船坞
semi-tight 半渗透的;半密封;半不渗透
semi-tight borehole 资料有限的钻孔
semi-tight well 资料有限的钻孔
semi-tone 中间色调;半色调;半音
semi-topping 半顶面铣
semi-topping cutter 顶部倒棱剃齿刀
semi-total station distance theodolite 半站型测距经纬仪
semi-tracked tractor 半履带式拖拉机
semi-tracked vehicle 半履带式车辆
semi-trailer 单轴拖车;半拖车;半挂车;双轮拖车
semi-trailer axle 半拖车轴
semi-trailer bogie 半拖车双后轴
semi-trailer combination 半挂车组合体;半拖连接车
semi-trailer dump wagon 半挂式翻斗车
semi-trailer for car 运输汽车用半拖车
semi-trailer for construction equipment 装卸施工设备用的半拖车
semi-trailer for transport of excavating machines 装运挖土机用的半拖车
semi-trailer system 半拖车方式
semi-trailer tractor 半拖车牵引车;半挂车牵引车
semi-trailer type 半挂(车)式
semi-trailer type construction 半挂(车)式构造
semi-trailer type mastic asphalt boiler 半挂(车)式沥青玛琋脂熔制锅
semi-trailer type supply tank 半挂(车)式给水箱
semi-trailer wagon 半拖车货车
semi-trailer wheel 半挂车车轮
semi-trailer wheel-base 半拖车轴距
semi-transless 半无变压器式
semi-translucent 半透明的
semi-transparent 半暗的;半透射的;半透明的;半透明体
semi-transparent cathode 半透明阴极
semi-transparent colo(u)r 半透明色(料)
semi-transparent film 半透明膜
semi-transparent mirror 半透明镜;半透明反射镜
semi-transparent mirror splitter 半透明镜分波器;半透明反射镜式分束器
semi-transparent photocathode 半透明光电阴极
semi-transparent pigment 半透明原料
semi-transparent sintered 半透明烧结的
semi-transparent stain 半透明着色涂层
semi-transparent thin-film 半透明薄膜
semi-transverse axis 半贯轴
semi-transverse fault 斜断层;偏斜倾向断层
semi-transverse ventilation 半横向通风
semi-transverse ventilation system 半横向通风系统
semi-triangular form 半三角形
semitropics 亚热带
semi-truss 半桁架
semi-tubular 半管形的
semi-tubular rivet 半管形铆钉
semi-tubular rivette 半空心铆钉
semi-tunnel 半隧洞;半隧道
semi-turbulence 半动荡
semi-turn-key 交钥匙半包
semi-turnkey contract 半启钥匙总合同
semi-umbrella type generator 半伞式发电机
semi-unattended 半自动的
semi-underground bin 半地下仓

semi-underground garage 半地下式车库
semi-underground powerhouse 半地下式(水)电厂;半地下式厂房
semi-underground storage 半地下油库
semi-underground storage tank 半地下储罐
semi-uniform 半均匀的
semi-uniform concrete 半匀质混凝土
semi-universal radial drill 半万能摇臂钻床
semi-universal surface 半万有曲面
semi-urban 半城市的
semi-valence[valency] 半价
semi-van 双轮拖车
semi-variable capacitor 半可变电容器
semi-variable condenser 半可变电容器
semi-variable cost 半可变成本;半变动成本
semi-variable expenses 半变动费用
semi-variable overhead 半变动管理费(制造费用)
semi-variation 半变差
semi-vault 半穹(隆)顶
semi-vertical breakwater 半直立式防波堤
semi-vertical-face quay 半直立式码头
semi-vertical-face wharf 半直立式码头
semi-vibration 半振动
semi-vitreous 半吸水性的;半透明的;半熔融态的;半瓷化的;半玻化的
semi vitreous china 半透明瓷;半玻化瓷
semi vitreous earthenware 半透明瓷;半玻化瓷
semi vitreous whiteware 半玻化白坯器皿
semi vitrified 半瓷化的;半玻化的
semi-vitrified wheel 硅酸盐砂轮
semi-vitrinite 半镜质组
semi-vocoder 半声码器
semi-volatile organic carbon 半挥发有机碳
semi-volatile organic compound 半挥发有机化合物
semi-volatile organics 半挥发有机物
semi-vulcanization 半硫化(作用)
semi-wall 半墙
semi-water gas 半水煤气
semi-wave 半波
semi-weathering bedrock horizon 半风化层
semi-weekly 半周刊;每周两次的
semi-wet cement manufacturing 半湿法水泥生产
semi-wet frame 半湿分纺细纱机
semi-wet process 半湿法
semi-white glass 半白玻璃
semi-wild 半野生
semi-wildcat 半野外锚孔
semi-wildcat well 半盲目开掘的油井
semi-wild palm 半野生油棕
semi-window 半窗
semi-window-semi-wall 半窗半墙
semi-winterness 半冬性
semi-work 试验工厂;实验工厂;中试车间;中间(试验)工厂;中间车间
semi-work scale plant 半工厂装置
semi-works production 中间试制;中间试剂;中间工厂规模试验;中间工厂规模生产
semi-xylenol orange 半二甲酚橙
semi-xylinite 半木煤体
semi-xylotinite 半木质镜煤体
semi-xylovitrofusite 微半木质镜煤一丝炭
Semmentum 赛蒙顿水泥(一种防水水泥)
sempervirine 常绿树;常绿钩吻碱
Semple plunger 离心保险销
semprax 门上附件;不锈钢配件
sems 螺钉组;带垫圈的机螺钉
semseyite 板硫锑铅矿;板辉锑铅矿
semtex 防滑无缝地板
senaite 铅锰钛铁石;铅锰钛铁矿
Senamont polarimeter 偏振仪
Senarmont compensator 塞拿蒙补偿器
senarmontite 方锑矿
Senarmont polarimeter 塞拿蒙偏振计
Senarmont prism 塞拿蒙棱镜
senary 六进的
send 上颠(船体);船被波浪抬高
send-back system 回送系统
send down price 压价
sender 引向器;记发器;送波器;发信机;发送人;发送机;发射机;发码器;发送人
sender transmitting station 信号发送台
sender unit 信号发送器;发送装置
sender unit cover 传感器壳盖

sender unit insulated terminal 传感器绝缘接柱
send filter 发送滤波器
send in account 报销
sending allowance 传输衰耗
sending antenna 发射天线
sending apparatus 发送装置
sending area 发送区
sending channel 发送信道;发报信道
sending distributor 发送分配器
sending dotter signal 发点信号
sending-end 送电端;输电端;发信端
sending-end crosstalk 发送端串话
sending-end impedance 输送端阻抗;输入端阻抗;发送端输入阻抗;送电端阻抗
sending-end voltage 送电端电压
sending-file readout 发送文件读出【计】
sending filter 发送端滤波器
sending item 发送项
sending machine 发射机
sending modem 发送调制解调器
sending of Chinese specialist abroad 派遣中国专家
sending power 发送功率
sending set 发送设备;发射机
sending status 发送状态
sending substation 送电发电站
send jack 发送塞孔
sendout 送出量;发货量
send-out goods 发货
sendout pressure 供气压力
send pacing 发送(调)步
send-receive-ground-switch 收发接地转换开关
send-receiver 收发报机
send-receive switch 收发转换开关
send signal element timing circuit 发信信号单元定时电路
send through others 托带
Sendust 铁硅铝磁合金;山达斯特合金;铝硅铁粉
sendust core 铝硅铁粉磁芯
send window 发送窗口
Sendzimir coating process 森吉米尔镀锌法;分解氨热镀锌法
sendzimir galvanizing process 分解氨热镀锌法
Sendzimir mill 森吉米尔式极薄钢板多辊轧机;森吉米尔式轧机;20辊冷轧机
Sendzimir planetary mill 森吉米尔行星轧机
Senecio oryzetorum 田野千里光
Senegal gum 塞内加尔树胶
senegalite 水磷铝石
senescence 老年期;衰老
senescent lake 老年湖;衰老湖
senescent ratio 衰老比
sengierite 水钒铜铀矿;钒铜铀矿
Sengler viscosity 恩氏黏度
senhouse slip 脱钩链段
senhouse slip shot 滑钩链段
senile change 衰老变化
senile lake 老年湖
senile river 老年河;衰老河
senile soil 老年土;衰竭土
senile stream 老年河;衰老河
senility 老年期;侵蚀终期;衰老
senior 高级的
senior accountant 高级会计师
senior architect 高级建筑师
senior citizens housing 敬老院;老年人住房
senior common room 教员休息室
senior creditor 先受偿债权人;优先债权人
senior design engineer 高级设计工程师
senior engineer 高级工程师
senior high school 高级中学
senior inspector 高级检查员
senior issue 高级证券
seniority 高位数;前辈
seniority-based wage 按年资晋升的工资
seniority flag 官方旗
seniority order wage system 年资工资制
senior member 正会员
senior middle school 高级中学
senior partner 主要合伙人;大股东
senior programmer 高级程序员
senior securities 一级证券
senior shares 优先股
senior system designer 高级系统设计员

sennet 扁编绳
Senni beds 僧尼层【地】
sennit 扁编绳
sennit line 鳝鱼绳(旗索用);编索
Senn process 森方法
Senonian 森诺统
Senperm 森泊姆恒合金
sensation 感觉
sensation area 听觉范围
sensation level 知觉水平;听觉阈值;阈上水平;阈上级;感觉水平
sensation of acceleration 加速度感觉
sensation of dryness 干湿感
sensation of hearing 听觉
sensation quantity of sound 声感觉量
sensation scale 感觉尺度
sensation threshold 感觉阈值;感觉阀
sensation unit 声感单位
sense 信号方向;断定;定边
sense antenna 定边天线;测向的定边天线;辨向天线
sense byte 读出字节
sense-class 指向类
sense coil 读出线圈
sense command 检测命令;读出指令;读出命令
sense console 读出控制台
sense data 读出数据
sense determination 定边
sense-digit line 位读出线
sense finder 指向测定器;正负向测定器;探向(指)器;辨向器
sense finding 定边;单值定向
sense-finding switch 定向开关
sense head 读出头
sense indicator 读出指示器;方向指示仪;方向指示器
sense information 读出信息
sense light 传感指示灯
sense line 读出线
sense loop 读出环
sense of business 经营意识
sense of cold 冷觉
sense of compression 紧压感
sense of continuity 连续感
sense of current 电流方向;电流的方向
sense of harmony 和谐感
sense of hotness 热感
sense of line 线的指向
sense of movement 动向
sense of oppression 闷压感
sense of orientation 定向
sense of participation 参与感
sense of rotation 旋转感;旋转方向;自转方向
sense of worth 价值观
sense organ 感官
senser 敏感装置;探测设备
sense refresh amplifier 读出更新放大器
sense response 读出响应
sense reversing 逆向(的)
sense signal 读出信号;单向性信号
sense station 读出机构
sense strand 有意义股(链)
sense strobe 读出选通脉冲
sense switch 更变开关;感测开关;检测开关;读出开关
sense wire 读出线
sensibility 感应性;感觉力;敏感性
sensibility analysis 灵敏分析;敏感分析
sensibility control 灵敏度控制
sensibility of leading marks 导标灵敏度
sensibility of level 水准器灵敏度
sensibility of shield 盾构灵敏度
sensibility of voltage regulator 稳压器敏感性
sensibility reciprocal 灵敏度倒数
sensibilization 增感(作用);敏化作用;敏化
sensibilizer 介体;敏化剂
sensible 敏感的
sensible air current 感觉气流
sensible atmosphere 阻尼大气;可感大气
sensible clay 敏感黏土
sensible cooling 等湿冷却
sensible cooling load 显冷负荷
sensible heat 显热;余热;感热;感觉热;可感热;湿热
sensible heat cooler 显热冷却器
sensible heat cooling effect 显热冷却效果

sensible heat factor 空调的显热与总热荷之比;湿热因子;湿热比
sensible heat flow 显热流;可感热流
sensible heat flux 显热通量
sensible heating 等湿加热
sensible heat load 显热负荷;湿热负荷
sensible heat quantity 显热量
sensible heat ratio 显热比
sensible horizon 天文地平;感觉地平圈;视地平圈;视地平;地面真地平;测者真地平
sensible plan 切合实际的计划
sensible shock 有感地震;可感地震
sensible temperature 感觉温度;可感温度
sensicon 光导摄像管
sensigenous 产生感觉的
sensilized decomposition 敏化分解
sensillometer 感光计
sensillum ampullaceum 坛形感器
sensillum campaniformium 钟形感器
sensillum sagittiforme 矢形感器
sensillum styloconicum 栓锥感器
sensing 感觉;感测;偏航指示;定边;传感;测向器件;方向指示;辨向
sensing and switching device 读出转换装置
sensing antenna 辨向天线
sensing cable 传感电缆
sensing circuit 读出电路;传感电路
sensing coil 传感器线圈
sensing component number 敏感元件数
sensing crystal 敏感晶体
sensing detector 传感探测器
sensing device 辨热器;感应装置;感察器;敏感装置;敏感元件;敏感器件;读出装置;传感装置;传感器
sensing element 传感器;感应元件;敏感元件;感应元件
sensing equipment 指向设备;传感设备;测向设备;测定设备
sensing filament 敏感热丝
sensing head 灵敏头;敏感头;读出头;传感元件;传感头
sensing hole 读出孔
sensing instrument 灵敏仪器;灵敏仪表;敏感仪表
sensing marker 读出指示器
sensing material 传感材料
sensing mechanism 传感机构
sensing pin 读出针
sensing probe 敏感探头;传感探测器
sensing reading 传感读出
sensing station 感测站;读出站;传感站
sensing surface 敏感表面
sensing switch 传感器开关
sensing system 感觉;传感系统;测读系统
sensing system for motion compensator 运动补偿器的敏感系统
sensing tape 读数带
sensing transducer 传感器
sensing unit 敏感元件;敏感部件;传感装置;传感器
sensing zone technique 敏感区技术
sensistor 正温度系数热敏电阻;硅电阻;热敏电阻
sensistor compensation 热敏电阻补偿
sensitimeter 感光器
sensitised door mat 感应门踏板
sensitiser 增光剂
sensitive 敏感的
sensitive adjustment 精确调整
sensitive altimeter 灵敏高度计;气压高程表
sensitive analysis 敏感性分析
sensitive area 敏感面;敏感(地)区
sensitive axis 灵敏轴;量测轴(线)
sensitive balance 灵敏(度高的)天平
sensitive bench drill 高速手压台钻
sensitive bioassay 敏感生物测定
sensitive bolometer 灵敏电阻测辐射热器;灵敏测辐射热计
sensitive bridge 灵敏电桥
sensitive clay 灵敏黏土;敏感黏土
sensitive coating 感光膜
sensitive colo(u)r 敏感色
sensitive colo(u)r plate 灵敏色板
sensitive composition 敏感成分
sensitive continuous recording equipment 灵敏度连续记录仪
sensitive detector 灵敏探测器

sensitive determination 感光测定
sensitive device 敏感器件
sensitive drill 高速平压钻机
sensitive drilling machine 手动进给钻床;手压钻床
sensitive drill press 灵敏钻床
sensitive edge 敏感边缘
sensitive element 感受元件;灵敏元件;灵敏部分;敏感元件;传感器
sensitive emulsion 感光乳剂
sensitive explosive 敏感(性)炸药
sensitive fault program(me) 故障敏感程序
sensitive film 感光胶片;感光胶卷
sensitive form 敏感型;敏感形式
sensitive formation 敏感性地层
sensitive galvanometer 灵敏电流计
sensitive goods 敏感商品
sensitive goods price 敏感商品价格
sensitive hue 灵敏色调
sensitive indicator 灵敏指示物
sensitive infrared telescope 灵敏红外望远镜
sensitive ink 支票印墨;防涂改油墨;安全油墨
sensitive items 敏感项目
sensitive junction 灵敏结
sensitive layer 感光膜;敏感层
sensitive lever 敏觉杆
sensitive line 灵敏线
sensitive local materials 敏感性当地材料
sensitive market 敏感(性)市场;敏感(性)商品
sensitive material 感光材料;灵敏材料;敏感材料
sensitive metal 敏感金属
sensitive microbiologic(al) detector 灵敏微生物探测器
sensitive monitor 敏感检测器;敏感监测器
sensitiveness 感应性;灵敏(度);敏感度
sensitiveness of governor 调速器灵敏度
sensitiveness of light 光敏感性
sensitiveness test 灵敏度试验;敏感试验
sensitiveness to impact 炸药冲击感度
sensitiveness to light 感光性
sensitiveness to shock 震动灵敏度
sensitive paper 感光纸
sensitive part 灵敏部分
sensitive percussion fuze 瞬发引信
sensitive period 感应期;灵敏周期
sensitive photographic(al) material 照相感光材料
sensitive pick-up 灵敏元件;敏感元件;传感器
sensitive plant 敏感(性)植物
sensitive plate 感光干板;感光板;灵敏板
sensitive point 灵敏点
sensitive price 敏感性价格;敏感价格
sensitive priming 敏感起爆
sensitive product 敏感性产品;敏感产品
sensitive property 敏感性能
sensitive question 敏感性问题
sensitive radiation thermocouple 灵敏辐射温差电偶;灵敏辐射热电偶
sensitive region 灵敏区
sensitive regulation 灵敏调整
sensitive relay 敏感继电器
sensitive resistor 敏感性自动调整电阻器;敏感电阻器
sensitive segment 敏感段
sensitive shale 敏感性页岩
sensitive species 敏感种;敏感性物种
sensitive spot 灵敏点;敏感点;反应点
sensitive stock 热敏原料
sensitive switch 快动开关;传感开关
sensitive tapping machine 灵敏攻丝机
sensitive test 敏感性试验;敏感性检验
sensitive thermometer 敏感温度计
sensitive time 感应时间;感光时间;灵敏时间
sensitive tint plate 灵敏色辉片
sensitive to contamination 对污染敏感的
sensitive to friction 对摩擦敏感的
sensitive to heat 热敏材料;热敏(材料)
sensitive to light 感光;对光敏感的
sensitive to pollution 对污染敏感的
sensitive to smog 对烟雾敏感的
sensitive upright drilling machine 立式灵敏钻床
sensitive vibration detector 灵敏检振器
sensitive volume 灵敏体积;灵敏区;灵敏范围;敏感体积;热敏区
sensitive volume of a detector 探测器的灵敏体积
sensitive water body 敏感水体

sensitive waters 敏感水体
sensitivity 感光度;感度;灵敏性;灵敏(度);敏感性;敏感度
sensitivity analysis 灵敏度分析;敏感性分析;敏感度分析
sensitivity analysis of water quality model 水质模型灵敏度分析
sensitivity bolometer 灵敏测辐射热计
sensitivity calibration 灵敏度校准
sensitivity centre 敏感中心
sensitivity characteristic 敏感特性
sensitivity coefficient 灵敏(度)系数;敏感系数
sensitivity coefficient of price 价格敏感系数
sensitivity compensator 灵敏补偿器
sensitivity constant 灵敏度常数
sensitivity control 灵敏度调节
sensitivity correction 灵敏度校正
sensitivity curve 灵敏度曲线
sensitivity decay 感光度衰减
sensitivity decrease 灵敏度下降
sensitivity distribution 灵敏度分布
sensitivity drift 灵敏度漂移
sensitivity equation 灵敏度方程
sensitivity factor 灵敏度因数
sensitivity figure 感光度数字
sensitivity function 灵敏度函数
sensitivity guide 感光指南
sensitivity increment 感光度增高
sensitivity index 感光度指数;灵敏度指数;灵敏度指标
sensitivity level 敏感度;灵敏度;响应级
sensitivity margin 感光度范围
sensitivity of aero-survey system 航测系统灵敏度
sensitivity of a photoresistive cell 光敏电阻管灵敏度
sensitivity of a range 叠标敏感度
sensitivity of cohesive soil 黏性土的灵敏度
sensitivity of control system 控制系统的灵敏度
sensitivity of detector 探测器灵敏度
sensitivity of follow-up system 随动系统灵敏度
sensitivity of governor 调速器灵敏度;调速器的灵敏度
sensitivity of gravimeter 重力仪灵敏度
sensitivity of instrument 仪表灵敏度
sensitivity of K-channel 钾道灵敏度
sensitivity of propagation 传爆感度
sensitivity of range marks 导标灵敏度
sensitivity of regulation 调节灵敏度
sensitivity of screening 筛检灵敏度
sensitivity of soil 土壤灵敏度;土的灵敏度
sensitivity of survey channel 测量道的灵敏度
sensitivity of tester 探伤仪灵敏度
sensitivity of Th-channel 钍道灵敏度
sensitivity of the level 气泡灵敏度
sensitivity of tracer detection 示踪剂检出灵敏度
sensitivity of U-channel 铀道灵敏度
sensitivity points method 敏感点法
sensitivity range 感光(度)范围;灵敏度范围
sensitivity rate 感光速率
sensitivity ratio 灵敏率;灵敏度比
sensitivity shift 灵敏度变化
sensitivity speck 敏化中心
sensitivity spectrogram 灵敏度谱图
sensitivity test(ing) 感光度试验;敏感度试验;敏感性试验;敏感性检验;灵敏(度)试验
sensitivity testing method 灵敏度测试法
sensitivity threshold 灵敏阈值
sensitivity-time control 灵敏度时间控制;灵敏度时间调整
sensitivity to contamination 对污染的敏感性
sensitivity to initiation 起爆感度
sensitivity to light 感光灵敏度
sensitivity to pollution 对污染的敏感性
sensitivity to toxic substances 对有毒物质的敏感性
sensitivity to variation in temperature 对温度变化的敏感性
sensitivity value 感光值
sensitivity velocity 感光速度
sensitivity volume 灵敏区
sensitization 致敏作用;感受作用;激活;敏化;敏感;促爆;变应化作用
sensitization corrosion 敏化腐蚀
sensitization luminescence 敏化发光

sensitization test 敏感试验
sensitization time 敏化时间
sensitizatlon 敏化作用
sensitized 激活的;敏化了的
sensitized cathode 敏化阴极
sensitized cube 敏化立方
sensitized door mat 感应门踏板
sensitized etch material 光刻材料
sensitized fabric 敏化织物
sensitized fluorescence 敏化荧光
sensitized luminescence 敏化发光
sensitized material 敏化材料;敏感物质;敏感材料
sensitized medium 感光媒质
sensitized paper 感光纸;敏化纸
sensitized phosphorescence 敏化磷光
sensitized photocell 敏化光电管
sensitized photodecomposition 敏化光解
sensitized screen 敏化筛网
sensitized surface 敏化面
sensitized zone 敏化区
sensitizer 增光剂;增感剂;感光剂;敏光剂;敏化剂
sensitizer ion 敏化剂离子
sensitizing 敏化过程;敏化法;敏感过程
sensitizing action 增感(作用)
sensitizing agent 增感剂;敏化剂
sensitizing center 敏化中心
sensitizing compound 敏化物
sensitizing dye 增感染料
sensitizing dyestuff 敏化染料
sensitizing effect 敏化效应
sensitizing intensity 敏化强度
sensitizing layer 敏化层
seocnd potential 敏化潜力
sensitizing pulse 照明脉冲
sensitizing range 敏感区
sensitizing solution 敏化液
sensitizing treatment 敏化处理
sensitlzing effect 敏感效应
sensitogram 感光(度)图
sensitometer 感光仪;感光计;感光度测定计;曝光表
sensitometric control 传感控制
sensitometric curve 感光(深浅)曲线;感光度测定曲线
sensitometric data 感光测定数据
sensitometric measurement 感光度测量
sensitometric refraction 感光折射
sensitometric test 感光度测定
sensitometric wedge 感光度测定楔
sensitometry 光敏学;感光学;感光度测量法;感光度测定法;感光测量;感光测定(术)
sensor 传感装置;感应元件;传感器;感察器;激光流量传感器;敏感装置;敏感元件;探测设备;受感器;读出器;传感仪;测定仪
sensor alarm 传感报警器
sensor amplifier 读出放大器;传感放大器
sensor analysis combination system 传感分析联合系统
sensor aperture opening ratio 传感器开口率
sensor array 阵列传感器;传感阵列
sensor-based 以传感器为基础的设备;传感设备;传感器化的
sensor-based computer 传感(器用)计算机
sensor-based system 基于传感器的系统;传感器(用)系统;采用传感系统
sensor board 传感插板
sensor capability 传感器性能
sensor coil 敏感线圈;传感(器)线圈
sensor detector 传感探测器
sensor element 传感元件
sensor fiber 传感光纤
sensor glove 传感式手套
sensor group 传感器组
sensor information 传感信息
sensor interface adaptor 传感器接口适配器
sensor material 传感器材料
sensor matrix 读出矩阵
sensor of remote sensing 遥感传感器
sensor parameter 传感器参数
sensor peen 传感尖端
sensor response 传感器响应
sensor sensitivity 探头灵敏度
sensor signal 传感器信号
sensor simulation experiment 传感器模拟试验
sensor simulation material 传感器模拟资料

sensor simulation system 传感器模拟系统
sensor switch 读出开关
sensor system 敏感元件系统;传感器系统
sensor transmitter 敏感变送器
sensor wiring 传感器布线
sensory club 感觉棒
sensory control 传感控制
sensory information 敏感元件信息
sensory measurement 感觉测定
sensory organ 感官
sensory package 传感组件
sensory plate 感觉板
sensory spot 感觉点
sensory system 感觉系统;敏感度系统
sensuous index 感官性指标
sensuous pollution index 感官污染指数
sent down the hole 下入孔内
sentence 判决
sentence of bankruptcy 破产判决
sentential calculus 句子演算
sentimental damage 推定损害(保险)
sentinel 压链锤;航行灯熄灭警报器;始终标记;传送器;标志;标识(符号)
sentinel code 标识码
sentinel pyrometer 高温计;传送式高温计;标志高温计
sentinel relief valve 超压信号阀;操纵阀
sentinel valve 报警阀
sent request circuit 发送请求电路
sentry box 岗亭
sentus 毛刺
senulate shoreline 锯齿形海岸线
seocnd intermediate host 第二中间宿主
seondary tap 二道螺丝攻
Seoul old land 汉城古陆
sepaloid 萼片状的
separability 可分性;可分离性
separable 可分离;可分的
separable attachment plug 连接插头
separable benefit 可分效益
separable blade 可拆卸螺旋桨叶
separable characteristic 可分特性;可分离性
separable code 可分码
separable collection 分类收集
separable contract 可分合同
separable cost 可分费用;可分成本;可以独立计算的成本;单独成本
separable cost-remining benefits method 可分费用剩余效益法
separable coupling 可拆卸联轴节;可拆式联轴器;可拆式联轴节
separable directed-graph 可分有向图
separable electric(al) connector 分离式电连接器
separable extension 可分扩张
separable fixed cost 可分的固定成本
separable function 可分(离)函数
separable graph 可分图
separable kernel 可分离核
separable library 可分隔程序库
separableness 可分性
separable obligation 可分(的)债务
separable optic(al) coupler 可分离光学耦合器
separable planning 可分规划
separable polynomial 可分多项式
separable programming 可分(离)规划
separable rim 可分开轮辋
separable slide fastener 分开拉链
separable space 可分空间
separable zipper 分离式拉链
separant 隔离剂
separant coating 涂隔离剂
separate 隔离(的);独立的;单独的;分离(的);分开;分隔
separate accounts 专账;独立核算
separate adjustment 分部平差;分别调整
separate air conditioner system 分隔空调方式
separate air friction clutch and brake units 分离式空气摩擦离合制动装置
separate amplifier 分立式放大器
separate application 分别敷用;分开涂胶
separate-application adhesive 分涂型胶粘剂;分施胶黏剂
separate assembly 分别汇编
separate base 单独底座

separate blade 分离叶片
separate block 间隔铁(护轮机)
separate board 分隔板
separate built-in cooking unit 分立式内装灶具
separate burner 分隔式燃烧器
separate bus 分离总线
separate cargo tank 分立液货舱
separate carrier 分离载波
separate carry adder 分离进位加法器
separate carry representation 分离进位表示法
separate cast test bar 单铸试件
separate chamber 隔腔
separate charging 单独加料;分别加料
separate checks 分列式账单
separate code 分离码
separate collection 分类收集;分别收集
separate collection of refuse 垃圾分类收集
separate collector 分离式集热器
separate combustion chamber 炉外燃烧室;分燃室
separate compensation 特别报酬;单独补偿
separate compilation 个别编译;分离编译
separate conditioning 单独调节
separate connection 单独联结;单独连接
separate contract 分类合同;分(包)合同;单项合同;分项(招标)合同;分项承包(合同)
separate contractor 单项承包商;分项承包商
separate controlled integrator 单独控制积分器
separate curb 分式路缘
separate cut-off valve 独立停气阀
separate cylinder 分铸汽缸;分段汽缸
separated accumulator 隔离式蓄能器
separated addressing 分隔导址
separate aggregate 分级集料;分级骨料
separated angle of offset beams 差波束分离角
separated application adhesive 分辨涂敷黏结剂;分敷黏合剂
separate data 分散数据
separate data processing 分散数据处理
separated beam system 分束系统
separated broadcaster 离心式撒播机
separated casting 分割铸造(法)
separated charge 分离电荷
separated clock 分离时钟
separated data 分离数据
separated drafting 分版清绘
separated drainage system 分流制排水系统
separate demodulation 分开解调
separate device 分离装置
separated excitation 分激
separated excitation direct current motor 分激直流电机
separated filter 分隔滤波器
separated fissile fuel 分离裂变燃料
separated fissile material 分离可裂变物
separated flame 分离(式)火焰
separated flow 分流;分流(水)流
separate digestion tank 分开消化池;单独消化池;分离消化池
separated inferior aquifer 隔离劣质水层
separated isotopes 分离同位素
separate dispersion 局部分散
separate dispersion matrix 分离差量矩阵
separated layer oil production rate 分层产油量
separated layer water cut 分层含水量
separated left-turn lane 分离式右转车道
separated lift 分离升力
separated lining 离壁式衬砌
separated manager 脱离群众的经理
separated mineral sample 单矿物样品
separated multiclad laser 分离多包层激光器
separated oil 析得干性油
separated oil skins 防油污衣裤
separated production and united sales 分产合销
separated pulse 分离脉冲
separated rail fastening 分开式扣件【铁】
separate drainage system 分流制排水系统
separate drawing 分色清绘
separated regenerator 分隔式蓄热室
separated regions 可分区域
separated resonance 分离共振
separate drive 单独驱动
separated set 分离的集合
separated sewage membrane bioreactor 分体式污水膜生物反应器

separated sewerage overflow 分流式下水道溢流
separated sludge 离析污泥
separated space 分离空间
separated storm drainage system 分流式暴雨排水系统
separated subgrade 分离式路基
separated synchronizing pulse 分流同步脉冲
separated thin lense 分离薄透镜
separated turning lane 分隔式转弯车道
separated turn isochronous cyclotron 分离圈等时性回旋加速器
separated type inner tube 分离式锭胆
separated type of dam and lock 闸坝分离式
separated type spindle 分离式锭子
separate efficiency 分部效率
separate electric(al) motor 单独电动机
separate elements tread 分格纹
separate end bracket 独立端肘板
separate entity system 纳税单位分列计税制
separate excitation 他激(电)
separate excited 分激
separate-excited block oscillator 他激式间歇振荡器
separate excited motor 分激电动机
separate exciter 他励励磁机;独立励磁机
separate export department 分设出口部
separate face mask 眼鼻面罩
separate facility 分隔设施
separate families 可分类
separate flat 独立单元(住房);成套住房
separate flow and return system 热水、回水分流管系统
separate flow system 分流管系
separate flushing 旁侧给水冲洗
separate function 分流函数
separate galvanometer 分装式检流计
separate game 分离对策
separate gang 单独工区
separate gas holder 单独储气器
separate gate 分离栅
separate gate type 分离栅极型
separate grade 分离式立体交叉
separate grade crossing 分离式立(体)交(叉);立体交叉路(口);立体交叉道
separate gravity structure 独立式重力结构;分离式重力结构
separate grinding 单独粉磨;分别粉磨
separate head 分开头
separate heater valve 旁热式真空管
separate heterodyne 他激外差法;分激外差法
separate heterodyne reception method 分激外差式接收法
separate house 独立式房屋
separate injection well 分层注水井
separate intensifier system 分离式增压器
separate layer 隔离层
separate layer fluid production rate 分层产液量
separate layer quantitative oil production 分层定量采油
separate lead-sheathed cable 分铅包型电缆
separate lifting forces 分项浮力
separate-line calciner 离线分解炉
separate lime calciner kiln 双列预分解炉窑
separate lining 离壁衬砌
separate loading 分装式
separate local oscillator 独立本地振荡器
separate lubrication 局部润滑;分别润滑法
separate lubricator 个别润滑器
separate luminance camera 分离亮度摄像机
separately cast test bar 单铸试棒
separately charged traffic 分段计费的运费
separately compute and pay tax on each category 分别计算纳税
separately controlled 单独操纵的
separately cooled machine 外冷式电机;他冷式电机
separately deduct expenses 分别减除费用
separately driven circuit 他激励电路;分激电路
separately excited circuit 分激电路
separately excited dynamo 他励发电机;分激发电机
separately excited generator 他励发电机;分激发电机
separately excited induction generator 分激感应式发电机

separately excited inverter 分激式变流器
separately excited motor 他励电动机
separately excited oscillator 他激振荡器;分激振荡器
separately excited phase advancer 分激式进相机
separately extinction of arc 外力消弧
separately fired supercharger 分烧式过热器
separately fired superheater 独立燃烧式过热器
separately heated cathode 旁热阴极
separately leaded cable 铅皮分包电缆
separately-lead sheathed cable [SL cable] 分析铅套电缆
separately mounted fuse 单独装置熔断器
separately stowed 隔离码垛
separately ventilated machine 外通风式电机;他力通风式电机
separately ventilated type 外力通风式
separate machine 分割机
separate management 分散经营
separate manuscript 分色原图
separate mesh 分离网
separate mesh electrode 分离电极
separate mesh vidicon 分离网式视像管
separate mesh vidicon pickup tube 分离网光导摄像管
separate mode 分离模式
separate modelling 分离式模型
separate motor 独立式电动机
separate nozzle gas cutter 串列式喷嘴气割机
separate of sewerage 分流下水道系统
separate optic(al) element 分离光学元件
separate out 析出
separate oven 独立式烤箱
separate ownership 分户单独所有权(土地)
separate payment 单独支付
separate phase 独立相;分离相
separate phase operating relay 分相动作继电器
separate pipe transmission 分管输送
separate platform 侧式站台
separate printing 分色印刷
separate processor mode 分隔处理方式
separater 离析器
separate ratio estimate 分别比率估计值
separate reactor 分离反应器
separate regression estimate 分别回归估计值
separate regulation 单行条例;单行规定;单行法
separate ring 分离型套圈
separate roller 分离辊
separate room 单间(餐馆中与大餐厅分开的小间)
separate room heating 单室采暖
separate sanitary sewer 分流污水管(道)
separate-scheme data 分离横式数据
separate school 非公立学校
separate sea lane 分隔航道
separate sewage system 分流制排水系统;污染分流系统;分流制污水系统;分流污水管系统
separate sewer 单用下水道;单用污水管;分流制下水道;分流沟道;分流污水道
separate sewerage 分流式排水;分流排水系统
separate sewerage system 分流制(污水系统);单用污水系统;分区污水系统;分流式下水道系统;分流污水系统
separate sewerage treatment system 分区污水处理系统
separate sewer system 分流式下水道系统;分流排污系统;分流制下水道系统
separate sheet 分幅
separate signal phase (交通灯的)分隔信号相
separate slit image 分离的狭缝像
separate sludge digestion 单独污泥消化;分离式污泥净化(法);分池污泥消化
separate sludge digestion tank 单独污泥消化槽;分隔式污泥消化池;分池污泥消化池
separate sluiceway structure 单独冲淤(道)结构;单独泄水(道)结构
separate sources 不同来源
separate spillway 单独溢洪道
separate sprays 析枝纹(陶瓷纹饰之一)
separate statement 附加声明
separate Stirling cycle refrigerator 分置式斯特林循环制冷机
separate storage rings 分离存储环
separate storm sewer 分流制雨水管;分流制雨水沟渠

separate stratum control 分层管理
separate switchboard 分立的交换台
separate system 分流排水系统;雨污水分流制;雨(水)污水分流系统;离散系统;分流系统;分流制
separate system of sewage 污染分流系统
separate table drive 转盘单独驱动
separate tank 分离池
separate thrust collar 斜挡圈
separate tracing 分色原图
separate trade contractor system 分行业承包制
separate turning lane 分流式转弯车道;分隔式转弯车道
separate twin box-girder 分流式双箱大梁
separate twin box-girder bridge 分流式双箱桥梁
separate type room air conditioner 分体房间空调器
separate type scanning field alternative gradient accelerator 分离型扫描场交变陡度加速器
separate unit 分立单元
separate valve box type piston pump 独立阀箱式活塞泵
separate ventilation 外通风;单通风;分区通风
separate vent pipe 单独透气管;单独通气管;单独通风管
separate VM cycle refrigerator 分置式威勒米尔循环制冷机
separate wastewater treatment system 分区污水处理系统
separate water supply 分散式给水
separate water supply system 分区供水系统;分区给水(管)系统
separate water volume 析水量
separate weighing 分别称量
separate winding 分绕绕组;分离线圈
separate wood fibre 分隔木纤维
separate zone 分隔带
separate-zone producer 分层采油井
separate zone production 分层开采
separate zone waterflooding 分层注水
separating 独立的;离析的;分隔的;分选的;分离的;分开的
separating action 分选作用
separating agent 分离剂
separating air 选粉气流
separating and flattening of upgoing and downgoing events 上行波和下行波分离并拉直
separating ball 间隔球
separating bath 重介质分选箱;重介质分选槽
separating beater 分离逐稿轮
separating belt 分离带
separating bowl 分离篮
separating by dilution 稀释分离
separating calorimeter 分水量热器;分离式热量计;分离式干度测量器
separating capacity 区分能力;分离能力
separating cell 分离槽
separating centrifuge 离心分离机;分离式离心机;分离机
separating chamber 沉降室;选分仓;分离室
separating character 舍入字符;分离(字)符
separating coal seam thickness 煤分层厚度
separating coefficient 分离系数
separating column 分离塔
separating compound 隔离材料;脱模剂
separating conduit 分流管
separating cone 圆锥选矿机;圆锥分选机;分选锥
separating control character 分离控制符号
separating course 分离横列
separating cylinder 圆筒分选筛;分选滚筒
separating dam 分水坝
separating degree 分离度
separating density 分选密度
separating different plot on soil fertility 分成地力不同的地块
separating disk 分离盘
separating draw thread 分离纱
separating drum 分选鼓;分选筒;分离汽锅
separating efficiency 选粉效率;分离(总)效率
separating element 滚动体隔离体;分离单元;分隔器
separating equipment for purification 净化分离设备
separating excavating process 分层开采法
separating factor 分离因子;分离因数

separating filter 分色滤光镜
separating flash 分离瓶
separating floor 分隔楼层;分隔楼面;分隔楼板
separating funnel 分液漏斗;分离漏斗
separating grade 分选品位
separating grid 分离筛
separating guide 分离导槽
separating hyperplane 分离超平面
separating keyword 分离(式)关键字
separating knife 分离刀
separating layer 底衬层;分离层
separating levee 分流堤
separating line 分像线;分隔缝
separating lug 定距耳
separating machine 分离机
separating material 隔离材料;脱模剂
separating mechanism 脱粒机构;分离设备
separating medium 分选介质
separating medium density 分选介质密度
separating medium specific gravity 分选介质比重
separating of grease 润滑脂分出
separating of oil 油的分离
separating operation 选别作业
separating pipe 分流管
separating plant 分隔设备;分选设备;分选厂
separating plate 分离板
separating powder 脱模粉
separating power 分离能力
separating prism 分像棱镜
separating process 分选作业;分选过程
separating property 分离性能
separating rate 分离速度
separating resonance 分离共振
separating ridge 分水岭
separating roller conveyor 分片辊道
separating sample 分别取样
separating saw 分牙锯
separating screen 选分筛;选别筛;筛分用筛;分(离)筛;分极筛;分级筛
separating settling tank 分流沉砂池
separating sewer 分流下水管;分流污水管(道);分流式下水道
separating shield 分隔屏蔽
separating sieve 分选筛
separating size 分离尺寸
separating solution 分离剂
separating strip 车道分隔带;车道间的分隔段;分车带
separating system 分流(式)系统
separating tank 沉砂池;分离池;澄清槽;分离槽
separating the infrastructure from operation 网运分离
separating train yard 分解列车车场
separating trammel 滚筒筛
separating tube 分流管
separating unit 分离装置;分离单元
separating upgoing wave downgoing wave and side wave 分离上行波下行波和侧面波
separating valve 隔离阀;分流阀;分隔阀
separating wall 隔(仓)板;挡板;分户墙;隔墙
separating weir 分流堰;分流堤;分流坝;雨污水分流堰;下水道管道截流堰
separating yarn 分离纱
separating zone 分离区
separation 选矿;析出;分离;隔离;分析;分距;间隔;膜分离;区分;提取法;分选,分离(作用);分隔(带)
separation air conditioner 分离式空调(器)
separation allowance 分居津贴
separation and identification 离析与检定
separation and purification 分离和提纯
separation angle 夹角;释放角
separation area 隔离区;水流分流区;气流分流区
separation axiom 分离公理
separation barrel 分离滚筒
separation board 隔墙木板
separation-break-away connector 分离脱落式连接器
separation bridge 立交桥
separation bubble 气流分离;分离气流
separation by crystallization 结晶析出
separation by deep refrigeration 深冷分离法
separation by development 置换分离
separation by displacement 置换分离

separation by electrolysis 电解分离
separation by precipitation 沉淀分离法
separation by sedimentation 沉降分离法
separation by shape 形状选矿(法)
separation by sorption 吸着分离法
separation center 军队复员中心
separation chamber 展开槽;分选室;分离槽;分隔燃烧室
separation characteristic 分选特征
separation circuit 分离电路;分隔电路
separation cloth 隔票布
separation coal 洗选煤;精选煤
separation code 分隔码
separation coefficient 分凝系数;分离系数
separation command 分离指令
separation condition 可分条件
separation conduit 分水管
separation connector 脱落插座;分离脱落式接插件
separation constant 分离常数
separation control 分离控制
separation control character 分隔控制符号
separation corona assembly 分离电晕器
separation corona discharge 分离电晕放电
separation corona wire 分离电晕丝
separation coupling 可拆联轴节
separation course 分离层;分隔层
separation curve 分离曲线
separation density 分选密度;分选槽
separation device 分隔设备
separation displacement 置换置置
separation distance 防火间距;间距
separation drafting 分色清绘
separation drag 分离阻力
separation eddy 分离涡流
separation efficiency 选粉效率;分选效率;分离效率
separation energy 分离能(量)
separation equipment 分选设备
separation faction 分离装置
separation factor 分选系数;分离因子;分离因数;分离系数;分解因素
separation fence 隔离栅
separation filter 频带分离滤波器;分离滤波器;分隔滤波器
separation fracture 张裂隙;构造裂隙;分离剥离;剥离破裂
separation gel 分离胶
separation grade 分选品位
separation groin 折流丁坝;导水丁坝;分水丁坝
separation groyne 分水丁坝;导水丁坝
separation indicator 分离指示器
separation installation 隔离设施
separation in streaming current 水流分选
separation instrumentation 脱离设备
separation intensity factor 分离强度因数
separation kerb 隔离围墙
separation latent image 分色潜像
separation layer 分离层;隔层;离层;分隔层
separation levee 隔堤;横隔堤;分水堤;分流堤;分离堤
separation line 分离线;分隔线
separation loss 间隙损失;分离损失;分离损耗
separation master 分色复印片
separation mat 隔票席
separation mechanism 分离机理;分离机构
separation mechanism of substance 物质分离机理
separation medium 隔离介质;(混凝土下的塑料薄膜垫底)
separation membrane 防潮层(基础用);隔离膜;分隔膜
separation negative 分色度片;分色底片
separation net 隔票网
separation number 分离数
separation of active fault 活断层断距
separation of aircrafts 飞机间隔
separation of boundary layer 边界层分离(现象)
separation of butane pentane carbon isotope value 丁戊烷碳同位素值间距
separation of carbon stable isotope value 碳稳定同位素值间距
separation of contour line 等高线版间距
separation of density and velocity 速度与密度分解
separation of emulsions 乳(化)液分离
separation of ethane propane carbon isotope value 乙丙烷碳同位素值间距

separation of events by angle of incidence 按入射角分离同相轴
separation of fault 离距
separation of flow 气流分离
separation of fragments 碎片飞散
separation of gases 气体分离
separation of groundwater flow 分割地下径流;地下(水)径流分割
separation of groundwater runoff 地下水径流分割
separation of hydrograph 过程线分割;水文过程线分析;水文过程线分割;按补给分割过程线
separation of hydrographic(al) components 水文因素分离
separation of intersecting points 交叉点分散
separation of isotopes 同位素分离
separation of kerogen 干酪根分离
separation of layers 分层
separation of levels 分层
separation of losses 损失区分;损耗分离
separation of oil sample 油样分离
separation of parallel runways 平行跑道间距
separation of principal part 主部分离
separation of propane butane ethane carbon isotope value 丙丁烷碳同位素值间距
separation of roadway 道路分离;道路分隔;车行道的分隔
separation of roots 隔根法
separation of signals 信号分离
separation of soluble organic matter 可溶有机质分离
separation of soluble salt 可熔盐的分离作用
separation of spectra 谱线间距
separation of tread 胎面剥落
separation of two frequencies 两频率间的间隔;双频间隔
separation of variables 分离变数法;分离变量法;变数分离(法);变量分离(法)
separation of water layer 漫流水层的分离
separation of waybill from shipment 票货分离
separation operation 分选作业
separation orifice 分离孔板
separation panel 分流排
separation parameter 分离条件;分离参数
separation partition 隔墙;分隔墙
separation pay 解雇费;遣散薪金
separation phenomenon 分离现象
separation picture 分幅像
separation pier 墩墩;分水墩
separation plane 分离面
separation plant 分离设备;分离工厂
separation plate 分色板
separation point 脱钩点【铁】;分离点
separation point distance 分离点距离
separation polynomial 分离多项式
separation potential 分离势
separation principle 分选原理;分离原理
separation process 分选作业;分选过程
separation property 分离性
separation rate 解雇率;分离率
separation region 分离区
separation resistance 分层阻力
separation ring 分离环
separation roller 分离辊
separation screen 分级筛
separation sensor 分离传感器
separation sewer 分流式下水道
separation sewer system 分流式下水道系统
separation sharpness 分选效率
separation size 分离度;分离粒径;分离粒度;分级粒度
separation size resolution 分离度
separation stage 分离阶段
separation standard 间隔标准
separation strip 分隔带;分车带
separation structure 高架桥;分隔结构
separation surface 分离面
separation system 分离系统
separation tank 隔油池
separation temperature 分离温度
separation test facility 脱离试验设备
separation test for greases 润滑脂分油试验
separation theorem 分离理论;分离定理
separation value 分离值
separation wall 隔墙;分隔墙

separation weir 分流堰
separation with heavy liquid 重液分离法
separation works 分水工程;分流设施;分离功
separation zone 分离带;分离段
separative duty 分离本领
separative element 分离单元
separative power 分离(功)率
separative sign 分隔符
separative signal 分离信号
separative work content 分离功量值
separator 脱模剂;分车设备;定距隔片;定距隔块;(两梁之间的)分隔板;除尘器;分析器;液滴分离器;滑剂层;区分标记;汽包;脱粒器;除沫室;分液器;分牙器;分选机;分梳辊;分离器;分离挡板;分隔器;分隔件;分隔符;分隔带;捕捉器
separator accumulator 隔离式蓄能器
separator blade 选粉机叶片
separator block 间隔垫块
separator capacity 分离器处理量
separator chamber 选分仓
separator circuit 分离线路
separator clutch 脱粒装置离合器
separator clutch control 脱粒装置离合器操纵手柄
separator column 分离柱
separator cone 分液漏斗;圆锥选矿机;分选圆锥;分离器圆锥体
separator conveyer 分选输送机;筛分运输机;筛分输送机
separator diaphragm 隔板隔膜
separator disk 分离器圆盘;分离盘
separator drum 鼓形选矿机;分选鼓;分选筒
separator efficiency 选粉机效率
separator-filter 分离过滤器
separator fines 选粉机细粉
separator gas sampling 取分离器气样
separator inducer 分流式导流轮
separator in series 串流式选粉机
separator inserter 分隔
separator load 选粉机负荷
separator mill 带选粉机的磨机;分离破碎机
separator of sample 样品分离装置
separator oil sampling 取分离器油样
separator operated by circulating air 循环空气(式)选粉机
separator page 分离页面
separator piston 隔离活塞
separator-press 分离压榨机
separator pressure 分离器压力
separator riddle 逐稿器
separator roller damper 分离辊制动器
separator roller gear 分离辊传动机构
separator roller sprocket 分离辊链轮
separator scrubber 分离涤气器;分离涤气机;分离洗涤机
separator sediment 分离器沉积物
separator tailings 选粉机粗粉
separator temperature 分离器温度
separator thruput 分离器处理量
separator tube 缓冲管;分离管
separator valve 缓冲管
separator water sampling 取分离器水样
separatory funnel 分液漏斗;分料漏斗
separatory wall 隔板
separatrix 分界线;分界面;分隔号
separometer 分离仪
sepatation etching 分离腐蚀
sepcified limits 规定极限
seperated-orbits cyclotron 分离轨道回旋加速器
Sephadex 塞法戴克斯
sepia 乌贼颜料;深棕色照片;棕褐色;黑褐色;乌贼墨(色);深棕色图纸
sepia print 深棕色蓝图;深褐色蓝图
sepia toning 深棕色调色法
sep index fibre 阶梯形折射率光纤
sepiolite 海泡石
sepiolite adsorbent 海泡石吸附剂
sepiolite clay 滑石黏土;海泡石黏土
sepiolite clay deposit 海泡石黏土矿床
sepiolite deposit 海泡石矿床
seprobia 污水生物
sepsometer 空气有机物质测定计
septage 化粪池污泥
septal area 隔区
septal cusp 隔尖瓣

septal defect 间隔缺损
septal folding 隔壁褶皱
septal fossula 内沟
septal neck 隔颈
septal ostium 隔壁孔
septal penetration rate 间隔穿透率
septangle 七边形;七角形
septangular 七角的
Septanix 塞普坦尼克斯数学板
septarian boulder 龟背石
septarian clay 龟背石黏土
septarian nodule 龟背石
septarium 龟背石
septate coaxial cavity 隔片同轴空腔
septate mode 旁模
septate waveguide 隔片波导
septation 分隔
septavalency 七价
September equinox 秋分(点)
septenary 七进制的;七个的
septenary foil 七分反射板(高空测风用)
septenary number 七进数
septenary system 七进制
septendecimal 十七进制的
septfoil 七叶形饰物
septic 腐败性的
septic condition 腐败状态;腐化状态
septic disposal system 家庭污水处理系统
septic gas 腐化气体
septicine 腐鱼尸碱
septicization 腐化作用
septic sewage 腐化污水;腐败性污水;腐败的污水
septic sludge 腐化污泥
septic system 化粪池系统
septic tank 厌氧处理槽;腐化槽;化粪池;污水池;腐化池
septic tank capacity 化粪池容量
septic tank design 化粪池设计
septic tank effluent 化粪池污水;腐化池污水
septic tank effluent pump 化粪池污水抽吸;腐化池污水抽吸
septic tank sludge 腐化池污泥;化粪池污泥
septic treatment plant 腐化处理设备;腐败物处理厂
septic wastewater 腐化污水;腐败废水
septic zone 腐败带
septifragal 轴开裂的
septilateral 七边形
septimal 七的
septinary number 七进制数
Septobasidium carbonaceum 煤状隔担耳
septum 格板;隔墙;隔壁;隔膜;隔片;[复]septa
septum buffer 隔膜缓冲器
septum electrode 隔膜电机
septum interatriale 房间隔
septum magnet 切隔磁铁
septum pectiniforme 梳状中隔
septum pellucidum 透明隔
septum primum 第一隔
septum secundum 第二隔;次隔
septum transversum 原始横隔;横中隔
septum-type 切割板型
septum-type separator 切割板型分离器
septuple space 七维空间
sepulcher 墓地;坑工墓;石墓;坟墓
sepulchral 附属于坟墓的
sepulchral architecture 坟墓建筑
sepulchral chamber 墓穴;墓室
sepulchral chapel 坟堂;坟墓建筑;骨灰堂
sepulchral cross 坟墓的十字架
sepulchral effigy 墓像
sepulchral monument 墓碑
sepulchral mound 冢
sepulchral pit 墓穴
sepulchral pottery 殉葬陶瓷器
sepulchral slab 坟墓石板
sepulchral stone 墓石;墓碑
sepulchral temple 墓地寺庙
sepulchral voice 低沉的声音
sepulchre 坟墓;墓穴
Sepulchre kiln 塞普克窑
sepulcrum of a mensa 墓桌
sepulture 坟墓;墓穴;埋葬
Sequanian(stage) 赛宰安亚阶
sequenator 顺序分析仪

sequence 序列;排列次序;天气集合;顺序(机);数贯;时序;定序;次序;层序组
sequence alternator 序列交流发电机;序列改变器
sequence analysis 序列分析
sequence analyzer 顺序分析仪
sequence automatics 顺序自动装置
sequence axiom 正合序列公理
sequence blasting 顺序爆破
sequence break 序列断点
sequence call 顺序呼叫;顺序调用
sequence calling 序列调用;顺序调用
sequence camera 连续照相机
sequence casting 连包连铸;全连铸
sequence chart 序列图表;(开关的)转接顺序图;时序图;程序图表
sequence check 序列检验;序列检验;序理检测;顺序检验;时序检验
sequence checking 序列检查;顺序检验;次序检验
sequence checking routine 序列校验程序;序列检验程序;顺序检验程序
sequence check mode in action range 作用范围顺序校核方式
sequence circuit 时序电路
sequence code 顺序(代)码;时序码
sequence contact 程序接点
sequence control 序列控制;顺序控制;顺序操作;程序控制
sequence control calculator 序列控制计算器
sequence control counter 顺序控制计数器;时序控制计数器;程序控制计数器
sequence control element 顺序控制元件
sequence-controlled 程序控制的
sequence-controlled computer 程控计算机;程序控制计算机
sequence-controlled contacts 程序控制触点
sequence controller 顺序控制器;顺序监控器;程序控制器
sequence-controlling advance 顺序控制移动
sequence control register 时序控制寄存器
sequence control system 顺序控制系统;顺序检验系统
sequence control tape 序列控制带;程序带
sequence counter 指令计数器
sequence criterion 序列判据
sequenced copolymer 顺序共聚物
sequenced crossover 迎面平行航向双机编队通过
sequence decay rate 连续衰减率
sequenced ejection system 顺序弹射系统
sequenced flashing approach light 顺序闪光进近灯
sequenced program(me) 次序程序
sequence ejection 依次弹射
sequence entropy 序列熵
sequence error 序列误差;顺序误差;顺序错误
sequence etching 顺序浸蚀
sequence filter 序列过滤器
sequence flash lights 顺序闪光灯
sequence generator 序码发生器;序列生成程序;程序发生器
sequence information data 序列信息数据
sequence interlock 顺序联结
sequence in time 时序
sequence length 序列长度
sequence length distribution 序列长度分布
sequence matching 顺序匹配
sequence monitor 序列监视器;顺序监控器;监督程序;顺序监测器
sequence number 序列数;顺序号
sequence number of stacked trace 叠加道序号
sequence number readout 程序号读出
sequence of Bernoulli's trials 伯务利试验序列
sequence of biologic(al) absorption 生物吸附序列
sequence of cartography 制图工序
sequence of columns 支柱系列;柱子系列
sequence of construction 建筑系列;施工顺序
sequence of control 控制顺序
sequence of crank 曲柄顺序
sequence of crystallization 结晶顺序
sequence of current 潮流序列;潮流顺序
sequence of cyclic(al) stress 循环应力序列
sequence of development 开发程序
sequence of earthquakes 地震序列
sequence of element migration 元素迁移序列
sequence of erection 架设安装顺序

sequence of events 事件序列
sequence of excavation 开挖顺序
sequence of flood diversion operation 分洪运用程序
sequence of function 函数序列
sequence of getting 开采顺序;获取顺序
sequence of grain sizes 颗粒尺寸系列
sequence of information 信息的顺序
sequence of injection 注浆顺序
sequence of instructions 指令序列
sequence of layers 层序【地】
sequence of linear functionals 线性泛函序列
sequence of lining placement 衬砌安装顺序
sequence of mapping 映射序列
sequence of marine sediment 海洋沉积物层序
sequence of measurement 测定次序
sequence of migration of elements 元素迁移系列
sequence of number 数列
sequence of operations 作用顺序;作业顺序;工序;动作顺序;操作序列;操作顺序;操作程序
sequence of operators 算子序列
sequence of partial sums 部分和序列
sequence of photograph 图片顺序
sequence of points 点序列
sequence of positive numbers 正数序列
sequence of potentials 位势序列
sequence of prestressing 预加应力程序
sequence of processes 操作顺序;工作顺序
sequence of quotient group 商群序列
sequence of random variables 随机变量序列
sequence of real-valued functions 实值函数序列
sequence of recording 记录顺序
sequence of repeated trials 重复试验序列
sequence of rock 岩序
sequence of sedimention 沉积层序
sequence of statement 语句序列
sequence of strata 地层序列;地层层序;地层层次;层序【地】
sequence of tense 时式序列
sequence of the information 信号顺序
sequence of tide 潮汐序列;潮汐顺序
sequence of trades 工序;施工顺序
sequence of volatility 挥发顺序
sequence of welding 焊接顺序
sequence of well 炮井编号
sequence operation 顺序运算
sequence photography 连续摄影;顺序照相
sequence program(me) control method 顺序程序控制方法
sequence program(me) machine 序列程序机
sequence programmer 顺序编制程序员
sequencer 序列发生器;定序器;定序程序;定序带;程序装置;程序控制器;程控装置
sequence recognition 顺序识别
sequence register 顺序寄存器;顺次序寄存器;定序寄存器;程序计算器
sequence relay 顺序动作继电器
sequence representation of hydrographic(al) data 水文资料系列代表性
sequence representation of hydrologic(al) data 水文资料系列代表性
sequencer program(me) 序列程序
sequence rule 优先法则
sequence search 顺序检索
sequence separator 顺序分隔符
sequence signal 序列信号;时序信号
sequence space 序列空间
sequence starting 按顺序启动
sequence start system 连续启动方式
sequence step nubmer 定序步数
sequence-stressing loss 序列应力损失;系列应力损失;构件受压变形引起的预应力损失;依次张拉的应力损失
sequence switch 序轮机;程序开关
sequence symbol 顺序符号
sequence table 序列表;程序表
sequence test 顺序试验;程序试验
sequence timer 序列时间调节器;顺序时间调节器;时序脉冲发生器
sequence token 顺序记号
sequence transformation 序列变换
sequence unit 程序装置
sequence valve 顺序阀;程序阀门
sequence welding 分段焊接

sequence weld timer 序列焊接时间调节器
sequencing 排序;定序;程序化
sequencing batch activated sludge process 序批式活性污泥法;间歇式活性污泥法
sequencing batch activated sludge reactor 序批式活性污泥反应池;间歇式活性污泥反应池;间歇式活性污泥反应池
sequencing batch airlift reactor 序批式气提反应器
sequencing batch anaerobic bioreactor 序批式厌氧生物反应器
sequencing batch biofilm reactor 序批式生物膜反应器
sequencing batch membrane bioreactor 序批式膜生物反应器
sequencing batch membrane reactor 序批式膜反应器
sequencing batch moving bed biofilm reactor 序批式(移)动床生物膜反应器
sequencing batch reactor 周期反应式活性污泥法;顺序间歇式反应器;序批式反应器
sequencing batch reactor activated sludge process 序批式反应器活性污泥法
sequencing by merging 归并排序;合并排序
sequencing circuit 顺序回路
sequencing computation 顺序计算
sequencing concrete batching plant 间歇式水泥混凝土搅拌站;间歇式水泥混凝土搅拌厂
sequencing equipment 定序设备
sequencing key 定序键
sequencing model 排序模式
sequencing problem 排序问题
sequencing routine 排序程序
sequencing theory 顺序理论
sequencing unit 命令发生器
sequencing valve 程控阀
sequency call 预约电话
sequent 相继的
sequentail comparison index 连续比较指数
sequent depth 共轭水深
sequent geosyncline 后继地槽
sequential 序的;顺次的;时序的
sequential absorption 连续吸收;连续吸附
sequential access 顺序作业;顺序存取;按序存取
sequential access display 顺序存取显示;顺序扫描显示
sequential access file 顺序存取文件
sequential access medule 按序存取模块
sequential access memory 按序(存取)存储器
sequential access method 顺序存取方法
sequential access storage 按序存取存储器
sequential addressing 顺序寻址;按序编址
sequential aerobic sludge blanket reactor 序批式好氧污泥层反应器
sequential alarm 顺序报警器
sequential algorithm 按序算法
sequential allocation 顺序分配
sequential allocation of stack 堆栈的顺序分配
sequential allocation of tables 表格的顺序分配
sequential analysis 时序分析;序列分析;序贯分析;逐次分析;顺序分析;按序分析
sequential analyzer 序列分析仪;序列分析器
sequential and numerically controlled machine 程序及数字控制机器
sequential attribute 顺序属性
sequential automatic control 顺序自动控制
sequential automation 序列自动化
sequential backwashing filter 序批式反冲洗滤池
sequential basin 序列池
sequential calculation 顺序计算
sequential carry 逐次进位;顺序进位;按序进位
sequential chlorination disinfection technology 序批式氯化消毒工艺
sequential circuit 顺序电路;时序线路;时序电路
sequential classification algorithm 顺序分类算法
sequential coding 序列编码;序贯编码;连续编码;时序编码
sequential collating 按序整理;按序排序
sequential collation 按序整理
sequential collation of range 距离顺序整理系统;不间断校对距离
sequential colo(u)r difference signal 顺序色差信号
sequential compactness 列紧性
sequential comparison 逐次比较

sequential comparison index 序贯对比指数;顺序比较指数;程序比较指数
sequential compounding of games 对策的序贯复合
sequential computation 循序计算;顺序计算;时序计算
sequential computer 顺序(操作)计算机;时序计算机
sequential constructed wetland 序批式人工湿地
sequential construction 序列施工法
sequential control 序列控制;连续控制;顺序控制;时序控制;程序控制
sequential control algorithm 时序控制算法
sequential controller 时序控制器
sequential control mechanism 时序控制机构
sequential control system 顺序控制系统
sequential correcting element 时序校正元件
sequential correlation 序列相关;系列相关
sequential correlation coefficient 序列相关系数;系列相关系数
sequential cost system 工程项目综合成本制
sequential counter 时序计数器
sequential data bit 顺序数据位
sequential data change 顺序数据变换
sequential data set 顺序数据集
sequential data storage 顺序数据存储
sequential data structure 顺序数据结构
sequential decision 序贯决策
sequential decision-making problem 连续性决策问题
sequential decision problem 连续决策问题;顺序判定问题
sequential decision procedure 顺序判定方式;顺序决定方法
sequential decision process 序贯决策过程;顺序决策过程;次序决策过程
sequential decoding 序列解码;序贯译码;序贯解码器
sequential decomposition 顺序分解
sequential dependent segment 时序相关段
sequential design 序贯设计
sequential detection 序列检测;序贯检测
sequential determination 顺序测定
sequential digital servomechanism 顺序数字伺服机构
sequential discrimination 序贯判别
sequential earth pulling process 顺序拉土法
sequential effect 连续作用
sequential element 时序元件
sequential encoding 序列编码;顺序编码
sequential error 序列误差;顺序误差
sequential estimation 序列估计;序贯估计;顺序寻优
sequential excavating process 顺铲开挖法
sequential excavation 分步开挖
sequential exposure 顺序曝光
sequential extraction method 循序提取方法
sequential fault diagnosis 顺序故障诊断
sequential feature decision mechanism 顺序特征判别机制
sequential feedback interactive retrieval system 顺序反馈人机对话检索系统
sequential file 顺序文件
sequential file access 顺序文件存取
sequential file structure 顺序文件结构
sequential filtering 序列滤波
sequential firing 顺序起动
sequential flasher circuit 顺序闪光回路
sequential fuel injection system 顺序喷油系统
sequential function 时序函数
sequential geosyncline 后继地槽
sequential harbo(u)r basin 序列港池
sequential hazard 时序险态
sequential hierarchy 时序层次
sequential hierarchy segment 时序层次片段
sequential hypothesis testing 序贯假设检验
sequential information 连续信息
sequential investigation observation 序贯调查观测值
sequential landform 继承地形;后继地形
sequential leave 轮休
sequential light amplifier 序贯光放大器
sequential lines 相继行
sequential link 时序链环

sequential lobing 顺序波束定向;时序波瓣法;波瓣顺序扫掠法
sequential lockage 按序过闸
sequential logic 顺序逻辑;时序逻辑
sequential logic element 时序逻辑元件
sequential logic equation 时序逻辑方程
sequential logic network 时序逻辑网络
sequential logic system 时序逻辑系统
sequentially complete space 序列完备空间
sequentially lobed radar 顺序直线扫探雷达
sequentially prolonged junction terminal 顺序延伸式枢纽
sequential machine 时序机
sequential marked graph 时序标记图
sequential maximization 时序极大化;时序放大
sequential measurement 顺序检测
sequential memory 顺序存储器
sequential memory location 顺序存储单元
sequential method 序贯法;逐次逼近法;顺序(逼近)法
sequential method of data 数据时间序列法
sequential mode 按序方式
sequential monitoring 顺序监视;顺序监控
sequential multielement analyzer 多元素顺序分析仪
sequential network 时序网络
sequential notation 顺序表示法
sequential observation 序贯观测
sequential observer 序贯观测者
sequential operating of switches 道岔顺序转换
sequential operation 顺序运算;顺序操作;时序操作
sequential operator 顺序运算符;顺序算子;顺序操作符
sequential optimization 顺序优化
sequential organization 顺序结构;顺序编排;时序组织
sequential organization file 顺序编排文件
sequential oxidation 连续氧化
sequential parameters 序列参数
sequential parsing scheme 顺序剖析方案
sequential picture coding 顺序图像编码
sequential point 连续点
sequential prime implicant form 时序质隐含式
sequential probability ratio 序列概率比
sequential probability-ratio test 序贯概率比检验;序列概率比试验
sequential procedure 序贯程序
sequential process 顺次过程
sequential processing 连续处理;顺序处理;按序处理
sequential program(me) 顺序规划;顺序程序
sequential programming 顺序程序设计
sequential pulse 顺序脉冲
sequential queue 顺序队列;按序排队
sequential range gating technique 作用距离连续接通技术
sequential raster 顺序(扫描)光栅
sequential reaction 接续反应
sequential reception 序列接收
sequential recognition 序贯识别
sequential record 顺序记录
sequential relationship 前后演变序列;时序关系
sequential relationship matrix 时序关系矩阵
sequential relay 顺序继电器;程序继电器
sequential response 序贯反应
sequential rest 轮休
sequential retrieval 顺序检索
sequential risk function 序贯风险函数
sequential rule 顺序法则
sequential sample 序贯样本;序贯试样
sequential sampling 序贯抽样;序贯采样;序贯抽样;顺序取样;顺序抽样;顺序采样(法)
sequential sampling inspection 序列抽样检验;依次取样检验;逐次抽样检查
sequential sampling inspection plan 序贯抽样检查方案
sequential sampling method 按序抽样法;按序采样法
sequential sampling plan 序列抽样法;逐次验收抽样法
sequential scan(ning) 顺序扫描;顺次扫描;连续扫描
sequential scan system 顺序扫描系统
sequential scheduler 顺序调度程序

sequential scheduling 顺序调度;时序调度
sequential scheduling system 顺序调度系统;顺次调度系统;时序调度系统;按序调度系统
sequential search 序列搜索;序贯寻优(法);序贯选优法;顺序寻优法;顺序检索
sequential selection 序贯选择;顺序选择
sequential shot-firer 连续爆破工;程序引爆器
sequential signal 顺序信号
sequential significance test 显著性序贯检验
sequential similarity detection algorithm 时序相似性检测算法
sequential simplex method 序贯单形法
sequential stacked job control 时序堆栈作业控制
sequential starting of switches 道岔顺序启动
sequential steps 顺序阶梯
sequential stochastic programming 序贯随机规划;顺序随机规划
sequential storage 序排存储[贮]器
sequential storage device 顺序存储装置
sequential strategy 序列战略;序贯策略
sequential structure 序列结构;顺序结构
sequential switcher 顺序换景器
sequential switching 按序转接
sequential switching network 时序开关网络
sequential switching system 顺序开关系统
sequential system 序列系统;依次传送制;顺序制;时序系统
sequential tank 序列槽
sequential test 序列检验;序贯检定;逐次测定
sequential testing 顺序测试;按序检测
sequential test of statistic(al) hypotheses 统计假设序贯检验
sequential three-dimensional angle intersection 顺序空间测角交会
sequential time delay 序时延迟;顺序时延
sequential timer 顺序定时(起爆)器
sequential transducer 顺序转换器;时序变换器
sequential transmission 顺序传输
sequential trial 序贯试验
sequential trial design 序贯试验设计
sequential tripping 顺序脱扣
sequential trough 序列槽
sequential turbocharging system 连续涡轮增压系统
sequential unconstrained minimization technique 序列无约束极小化方法
sequential value 序列值
sequential welding 连续焊(接);程序焊接
sequential working 时序工作
sequester 扣押;多价螯合剂;查封
sequestering activity 螯合活性
sequestering agent 隐蔽剂;掩蔽剂;多价螯合剂;螯合剂
sequestrant 多价螯合剂;螯合作用;螯合剂
sequestrate 扣押
sequestration 隐蔽作用;扣押;多价整合作用;分配破产人的财产
sequestrator 扣押令执行者
S-equivalence S 等价
sequoia 美洲杉;红杉
Serabriansk polarity chronzone 塞莱布里昂斯克极性时间带
serac 冰河的冰流;冰雪柱;冰塔;冰峰
seraglio 宫殿(土耳其);(伊斯兰教建筑的)闺房
serai 客栈
serandite 针钠锰石
Serapeum 建筑物或建筑群;祭祀埃及下界之神塞拉皮斯的地方
seration 交错群落
seraya 马来红柳桉
Serbian architecture 塞尔维亚建筑
serch 木材体积单位(等于 0.71 立方米)
serdab 古埃及雕像室;起居室(近东地区住宅地下室中)
sere 演替系列;植生系列
Serebriansk polarity chron 塞莱布里昂斯克极性时
Serebriansk polarity zone 塞莱布里昂斯克极性带
serein 晴空雨
serendibite 盐硅硼钙石;钙镁非石
sergeevite 水碳镁钙石
serger 拷边机
serging 锁边(地毯)
seriah 马来红柳桉
serial 顺次的;串行的;串联的;分期付款的

serial access 顺序存取;串行访问;串行存取
serial access memory 串行存取存储器
serial access storage 串行存取存储器
serial access system 串行存取系统
serial accumulator 串行(式)累加器
serial adder 串行加法器
serial adder and subtracter 串行加减电路
serial addition 串行加法
serial aerial camera 连续航空摄影机
serial air survey 连续航线摄影测量
serial air survey camera 连续航空测绘(照)相机;连续航测摄影
serial analysis 串行分析
serial arithmetic 串行运算
serial arithmetic operation 串行算术运算
serial arithmetic unit 串行运算部件;串行运算器;串行运算装置;串行四则运算
serial batch system 串行(成)批处理系统
serial binary adder 串联二进制加法器
serial binary computer 串行二进制计算机
serial bit 位串行
serial bit string 串行位串
serial bond 系列债券;分期分批偿还的债券;序列债券;定期偿还的债务;分批还本债券;分批偿还债券
serial-by-bit processing 按位串行处理
serial-by-character 按字符串行处理
serial camera 自动摄影机
serial camera with wind motor 风动连续摄影机
serial capacitor 串联电容器
serial carry 串行进位
serial clause 串行子句
serial coding 连续编码
serial communication 串行通信
serial communication controller 串行通信控制器
serial comparator 串行比较程序
serial compensation 串联补偿
serial computer 串行计算机
serial construction 连续施工(法)
serial contactor 串动接触器
serial contract 系列合同
serial control processor 串行控制处理机
serial correlation 序列相关;系列相关;连续相关
serial correlation coefficient 序列相关系数;系列相关系数
serial correlation model 序列相关模型
serial correlation test 序列相关检验
serial counter 串行计数器
serial data 序贯数据;串行数据
serial data analyzer 串行数据分析器
serial data controller 串行数据控制器
serial data input 串行数据输入
serial data output 串行数据输出
serial data set 串行数据集
serial data transfer 串行数据传送
serial data transmission 串行数据传送
serial data transmitter 串行数据发送机接收机
serial decomposition 串行分解
serial design 序列设计
serial design method for cost function 费用函数系列设计法
serial device 串行设备
serial digital adder 串行数字加法器
serial digital computer 串行数字计算机
serial digit operation 串位数字运算
serial dilution 连续稀释
serial distribution 连续分配;串联式布置;(渗水系统的)串联布设;序列式分布
serial engineering model 系列工程模型
serial entry 串行输入
serial extinction 连续消光
serial fabrication 连续生产
serial feed 串行馈送;串联馈电
serial feeding 串行输送
serial file 串行文件
serial file organization 串行文件结构
serial flow 串(行)流
serial format 串行格式
serial full adder 串行全加器
serial full subtractor 串行全减器
serial geologic(al) mapping based on remote sensing 遥感系列地质图件编制
serial-gram 系列照片;连续照片
serial gravity concentration 连续重力选

serial half adder 串行半加器
serial half subtractor 串行半减器
serial hand tap 成套手用丝锥
serial homology 连续同源
serial independence 序列独立
serial input 串行输入
serial input-output 串行输入输出
serial input-output address 串行输入输出地址
serial input-output buffer 串行输入输出缓冲器
serial input-output command 串行输入输出命令
serial input-output interface 串行输入输出接口
serial input-output interrupt 串行输入输出中断
serial intelligent cable 串行智能电缆
serial interface 串行接口;串连接口
serialization 序列生产过程;系列化;连续化;串行（化）;编序
serialization products 系列化产品
serialize 串行化
serialized manufacture 流水生产;连续生产
serializer 串行器;并串行转换器
serializer-deserializer 串行器—解串器
serial kymography 连续式记波摄影
serial lag correlation 序列滞后相关
serial learning 系列学习
serial line Internet protocol 串行线网际协议
serial logic 串行逻辑
serial loop 串行环
serially connected dredge pumps 串联泥泵
serially connected track circuit 串联式轨道电路
serially correlated error process 序列相关误差过程
serially linked reservoirs 串联水库;梯级水库
serially produced 系列生产的
serially reusable 可连续重用(的);串行可重复使用的;可再用的
serially reusable attribute 可连续重用属性
serially reusable load module 可用入装入模块;串行可重复加载模块
serially reusable module 可连续重用模块
serially reusable program(me) 可用程序;可连续重用程序
serially reusable resources 串行可重用资源;可连续重用资源
serially reusable routine 可连续重用例行程序;可连续重用例程
serially shared resource 串行共享资源
serially transmitted data 串行发送数据;串行传输数据
serially uncorrelated disturbance 序列不相关扰动
serial machine screw tap 成套机用丝锥
serial maturities 分次到期
serial mean 序列平均值;系列平均值
serial memory 串行存储器
serial mode 串行方式;串联方式
serial model No. 系列型号
serial mosaic 像片略图
serial multiplier 串行乘法器
serial number 顺序数(编号);序列号;系列号;流水号;串行数;串联数;成批号码;产品编号;编号
serial number control 串行数控制
serial numbering 串行编号
serial number of container 集装箱箱号
serial number of project 工程序号
serial observation 定点分层观测
serialograph 系列照相装置;连续拍摄照片;连续 X 线照相装置;连续 X 线照相器
serialography 系列照相术;连续 X 线照相术
serial operation 串行操作;串联操作
serial operation(al) word 串行操作字
serial output 串行输出
serial output-input interface 串行输入输出接口
serial output register 串行输出寄存器
serial parallel 串并行(的)
serial parallel addition 串并行加法
serial parallel analog(ue)/digital converter 串并行模/数转换器
serial parallel control 串并行控制
serial parallel conversion 串并行转换;串并行变换
serial parallel converter 串行(行)转换器
serial parallel multiplier 串并行乘法器
serial parallel mutiplication 串并行乘法
serial parallel operation 串并行操作
serial parallel register 串并(行)寄存器
serial parallel switch 串并联开关
serial parallel transmission 串并行传输

serial photogrammetric camera 连续测量摄影机
serial photograph 连续相片
serial port 串行(接)口
serial port connection 串联接口
serial processing 串行加工;串行处理
serial process(ing) system 串行处理系统
serial processor 串行处理机
serial production 序列生产;系列生产;成批生产
serial production line 连续生产线
serial program(me) 串行程序
serial programing 串行程序设计;串行编程序
serial programming 串编程序
serial punch 串行穿孔
serial radiographic apparatus 连续 X 线照相装置
serial radiography 系列射线照相术
serial rapids 滩群
serial rapids of shoals 滩群
serial read 串行读出
serial read punch 串行读出穿孔机
serial redemption 分期还本债券
serial register 串行寄存器;串接寄存器
serial regulator 串联调节器
serial representation 数的串行表示;串行表示
serial sampling 连续取样;连续采样
serial sampling inspection 序列抽样检验
serial scan 串行扫描
serial scheduling 串行调度
serial section 连续切片;连续剖面;顺序剖面
serial shift 串行移位
serial shift register 串行移位寄存器
serial ship model 系列船模
serial shoals 滩群
serial sort 串行排序;串行分类
serial station 定点观测站
serial storage 串行存储器;串行储存器
serial store 串行存储器
serial system 串行系统
serial taps 成套丝锥
serial task 顺序任务
serial terminal 串行终端设备
serial time-sharing system 串行分时系统
serial to-parallel converter 串并转换器;串并行转换器;串并变换器
serial transfer 串行转送;串行传送
serial transmission 连续传输;串行传送;串行传递
serial tube 标准电子管
serial type adder 串联式加法器
serial type of task 串行任务类型
serial variance 序列方差;系列离散
serial variation 序列变差
serial vein 序脉
serial word operation 串行字操作
serial work flow 串行工作流程
seriate 顺次排列;不等粒
seriate blastic texture 不等粒变晶结构
seriate poikilitic texture 不等粒嵌晶结构
seriate texture 连续不等粒结构;不等粒结构
seriation 系列化
seriation method 配列方法
sericite 绢云母
sericite-gneiss 绢云母片麻岩
sericite in powder 绢云母粉
sericite lumps 绢云母块
sericite mica 丝云母
sericite phyllite 绢云母千枚岩
sericite quartzite 绢云母石英岩
sericite-quartz schist 绢云母石英片岩
sericite sandstone 绢云(母)砂岩
sericite-schist 绢云石片
sericitization 绢云母化
sericolite 纤维石膏
series 序系;系列;系【地】;级数;群列;统【地】
series-aiding connection 相助串联
series anchor 串联锚
series anchor mooring 串联锚系碇;串联锚泊
series and-shunt tee 串联和并联的 T 形接头;串并联三通;串联并联 T 形接头
series and tail series 整数系列与零头系列
series arc furnace 串联电弧炉
series architecture 串行体系结构
series arc lamp 串联弧光灯
series arc regulator 串联电弧调节器
series arc welding 串联电弧焊;串联电弧焊接
series arm 串联支路;串联臂

series arrangement 顺序排列;串联装置
series assignment algorithm 串行分配算法
series battery 串联电池组
series board 串列交换台;串联复式交换机
series bond 系列债券
series booster 串联增压机;串联升压机
series booster regulator 串联升压机调节器
series bottle sampler 串联瓶式采样器
series branch 串联分支
series break switch 串联断点开关
series built ship 成批建造的船舶
series camera 连续摄影机
series capacitance 串联电容
series capacitor 串联电容器
series carry 串列进位
series cascade action 串行级联作用
series characteristic 串联特性;串激特性
series characteristic motor 串激特性电动机;反速电动机
series choke 串联扼流圈
series circuit 串联管路;串联电路
series circuit design 串联电路设计
series circuit firing 串联电路起爆
series classification 连续分级;顺序分类
series coil 串联线圈
series combination 串联组合
series commutator motor 串励换向器电动机
series compensated amplifier 串联补偿放大器
series compensation 级联补偿
series computer 串行计算机
series condenser 串联电容器
series conduction motor 交流串励换向器电动机;单相串励换向器电动机
series conductor 串联导体
series connected 串联的
series connection 串行连接;串联(连接)
series connection meter 串联水表
series connector 直接插板
series construction 串联结构;串联施工
series contactor 串联接触器
series control 连续控制;串联调整
series convergence 级数收敛
series convey 串传送
series copper refining 串联铜精炼
series decay 链式衰变;放射性系
series decomposition 串行分解
series designation of maps 地图系列图幅编号
series development 展成级数;级数展开
series digital computer 串行数字计算机
series docks 串联式港池
series drive 组合传动
series dynamo 串励发电机;串励电动机;串激发电机;串联发电机
series dynamotor 串励电动发电机(合并的电动机及发电机)
series electronic self-balancing instrument 连续式电子自动平衡仪
series element 串联元件
series energizing type 串联激励式
series equalizer 串联均衡器;串联截断电路
series excitation 串励;串激
series excitation coil 串励线圈
series excitation electromotor 串联电动机
series excited generator 串励发电机同步信号
series excited machine 串励电机
series excited radiator 串励天线;串励辐射器
series expansion 级数展开(式)
series fault 串联故障
series fed 串联供电
series feed 串联馈电
series feedback 连续回授;连续反馈;串联回授;串联反馈
series feedback amplifier 串联反馈(运算)放大器
series feed oscillator 串馈振荡器
series feed vertical antenna 串馈馈电竖直天线
series field 串励(磁)场
series field coil 串联磁场线圈
series filament 串联灯丝
series filtration 连续过滤
series fire 连续引爆
series firing 连续点火;连续爆破;连续引爆
series flow 串流
series flow basin 串流池
series flow turbine 单轴汽轮机;单轴多缸涡轮机;

串流式涡轮机
series forecasting methods 系列预测法
series gap 串联放电器
series gap condenser 串联空气隙电容器;串联间隙电容器
series gasoline meter pump 自动加油计量泵
series generator 串联发电机;串励发电机;串激发电机
series half shift register 串行平分移位寄放器
series heater 串联灯丝
series hole connection 炮孔串联
series hydraulic circuit 串联液压回路
series impedance 串联阻抗
series impulse 串脉冲
series in closed form 闭式级数
series inductance 串联电感
series in-series out register 串入串出寄存器
series interface board 串行接口板
series lighting 串联照明
series limit 线系极限
series limiter 串联限幅器
series line servomotor 串联伺服马达
series load 串联负荷
series loading 串联加载
series loop 串联回路
series loop system 单管回路系统;串联回路系统
series machine 串行计算机;串励电机
series magnet coil 串联磁化线圈;串励线圈
series manufacture 系列生产;成批生产
series manufacturing 连续生产
series maps 地图系列
series matching resistor 串联匹配电阻器
series modulation 串联调制
series motor 串联电动机;串励电动机
series multiple 多重串联
series multiple arrangement 串联反复配置法
series multiple board 串列式交换机
series multiple connection 串并联;混联
series negative feedback 串联负反馈
series negative impedance 串联负阻抗
series network 串联网络
series number 序列号;顺序号;编号
series observation at fixed point 定点连续观测
series observation in place 定点连续
series of cells 电池组
series of compounds 化合物系
series of concept 相关概念系列
series of conduits 水管串联
series of curves 一组曲线;曲线组;曲线族;曲线系列;曲线簇
series of dams 连续堤坝;堤坝系列
series of decreasing powers 降幂级数
series of fixed wheels 固定轮系列
series of f-number 光圈数系列
series of holes 孔列;钻孔系列
series of increasing powers 升幂级数
series of lines 谱线系
series of locks 梯级船闸
series of master-keyed locks 有万能钥匙的系列
series of mean 均值数列
series of measurement 观测列
series of natural numbers 自然数据级数
series of observations 观察组;观测组;观测(系)列;连续观测
series of options 系列期权
series of payment or receipt 分期分批付款或收款
series of payments 分期连续付款
series of peaks 峰丛
series of pipes 水管串联
series of points 直线点组;点集;点簇
series of positive terms 正项级数
series of potentials 排优次序;电动次序
series of powers 幂级数
series of power station 梯级电站;电站群
series of resales 一系列的转卖
series of river engineering model 系列河工模型
series of rocks 岩系
series of runner 转轮型谱;转轮系列
series of sedimentary formation 沉积建造系列
series of ship model 系列船模
series of sieves 梯级筛子
series of similar experiments 相似试验序列
series of spectral lines 光谱线系
series of standards 标准系列;准则系列

series of strata 地层系;岩系
series of test 试验顺序
series of tolerance 基本公差
series of tubes 水管串联;列管束
series of turbine 水轮机系列
series of turntables 转车盘组
series of variable terms 变项级数;变量项级数
series of variates 变量数列
series of voyage 系列航次
series operation 系列运行
series-opposing connection 反向串联
series over-current relay 串联过电流继电器
series parallel 混联;串并的
series parallel battery control 串并联电池控制
series parallel circuit 串并联电路
series parallel compensation 串并联补偿
series parallel connecting switch 串并联联络开关
series parallel connection 串并联
series parallel control 串并联控制
series parallel controller 串并行控制器;串并联控制器
series parallel field control 串并联励磁绕组控制
series parallel firing 混合联电路起爆;串并联点火
series parallel motor 串并联电动机;串并励电动机
series parallel network 串并联网络
series parallel operation 串并联操作
series parallel pipe circuit 串并联管路
series parallel requisition 串并联条件
series parallel starter 串并联换接起动器
series parallel starting 串并联起动
series parallel switch 串并联(转换)开关
series parallel system 串并联系统
series parallel winding 串并联绕组
series peaking 串联峰化
series phase resonance 串联相位谐振
series pipe 串联管路
series pipe still 管组蒸馏釜;管组蒸发炉
series planting 列植
series position 串联位置
series printer 串行式印刷机
series process 串联法
series production 系列产品;连续生产;大量生产;成批生产
series projection welding 单边多极凸焊
series pyrolysis 系列热解
series radio tap 串联无线电分支
series reactor 串联扼流圈;串联电抗器;串联电感器
series regulator 串联调整器;串联调节器
series relay network 继电器串联网络
series repeater 串联增音器
series repulsion motor 串联推斥电动机;串励推斥电动机
series requisition 串联条件
series resistance 串联电阻
series resistance lamp 串联电阻灯
series resistor 串联电阻器
series resonance 电压谐振;电压共振;串联谐振;串联共振
series resonance circuit 串联共振电路
series resonance frequency 串联谐振频率
series resonant circuit 串联谐振电路
series ringer 串接振铃机;串联感应器
series ring winding 串联环式绕组
series running 串联运行
series seam welding 单面多极滚焊;单边双电极焊缝;单边多极滚焊;串联缝焊
series sequence 串联顺序;串联次序
series shot-firing 串联爆破;串联起爆
series shots 连续爆破
series shunt 串励分流电阻
series source resistance 串联源电阻
series spark gap 串联火花隙
series spot welding 单面双极点焊;连续点焊;单面(多点)点焊;单边多极点焊;串联点焊;串点焊
series stabilization 串馈稳定
series stage 串联级
series storage battery 串联蓄电池组
series submerged arc welding 横列双丝串联埋弧焊
series system 串联制;串联系统
series to-parallel converter 串至并行转换器
series track circuit 串式轨道电路
series transductor 串联饱和电抗器
series transformer 级联变压器;串联变压器;串联变换器;串接变压器
series transistor regulator 串联晶体管稳压器
series trickling-filter and activated-sludge process 串联生物滤池活性污泥法
series triggering 串联触发
series trip coil 串联跳闸线圈;串联解扣线圈
series tripping 串联跳闸
series tuned circuit 串联调谐电路
series tuning 串联调谐
series turbocharging 串联式涡轮增压
series turns 串联线匝
series type adder 串联加法器
series type limiter 串联限幅器;串联式限幅器
series type phase advancer 串联式相位超前补偿机;串联式相位补偿器
series ventilation 单路串列通风;串联通风
series voltage 串联电压
series voltage regulator 串联稳压器
series welding 系列焊(接);单边多极焊接;串(联)焊
series winding 串联绕组;串激绕法
series wire terminal 串联线接头
series wiring 串联接线
series wound arc lamp 串联弧光灯
series wound dynamo 串联磁电机
series wound generator 串激发电机
series wound motor 串绕电动机;串励电动机
serif 衬线
serimeter 验丝计;验丝机
seriograph 系列照相装置
seriography 系列照相术;连续射线摄影术
serioparallel 串并行的;串并联的
serioscopy 连续实体照片投影检查
serious accident 严重事故;大事故
serious breach of contract 严重违约;严重违背合同
serious burn 严重烧伤
serious causality 严重事故
serious consequence 严重后果
serious damage 严重破损
serious damage severe damage 严重破坏
serious degradation ecosystem 严重退化生态系统
serious disease 重病
serious environmental health risk 严重环境健康风险
serious fault 严重故障
serious gas 重瓦斯
serious incident report 严重事故报告
serious-injury road accident 严重伤人的道路交通事故
serious loss 严重全漏失
serious mud pumping 冒浆严重
serious pollution 严重污染
serious water pollution problem 严重水污染问题
serir 砾漠
serir deposit 戈壁沉积;卵石沙漠沉积
seriscission 线切术
serite 钛铁矿
serivice function 服务函数
Serlian motif 瑟类奥拱门
serlite 膨胀陶粒;膨胀黏土集料;膨胀黏土集料
serotinous 晚成的
serous effusion 浆液性渗漏液;浆液性渗出物
serous exudate 浆液性渗出物
serous infiltration 浆液浸润
serous material 浆液材料
serous material concentration 浆液浓度
serous material flow 浆液流量
serous secretion 浆液分泌
serous type 浆液类型
Serpa harp screen 塞尔白栅筛
serpeggiante 木纹黄石
serpenitite 蛇纹岩
serpentarium 蛇类展览馆
serpentine 盘管;螺旋管;蛇形管;S形曲线;蛇纹岩;蛇形线;蛇纹(石)
serpentine asbestos 温石棉;蛇纹石棉
serpentine brick 蛇纹石砖
serpentine-chlorite schist 蛇纹绿泥片岩
serpentine coil 蛇形盘管
serpentine conduit 蛇形管
serpentine contact chamber 蛇管接触室
serpentine cooler 舌形冷却管;阶式冷却器;盘式冷却器;蛇形冷却器;蛇(形)管冷却器
serpentine domain 螺旋形磁畴

serpentine duct 蛇形导管
serpentine folding door 弯曲形折(叠)门
serpentine heater 蛇管加热器
serpentine highway 蜿蜒公路;盘旋公路
serpentine-jade 蛇纹玉
serpentine jigging tenter 盘旋形摆动拉幅机;蛇形摆动拉幅机
serpentine kame 蛇形丘
serpentine layer 蛇状压条
serpentine layerage 波状压条法
serpentine locomotion 蛇形运动
serpentine marble 蛇纹石大理岩;蛇纹大理石
serpentine path 蛇线轨迹
serpentine pipe 蜿蜒管;盘管;螺旋管;蛇形管
serpentine plate 蛇纹石板
serpentine rock 蛇纹岩
serpentine rock slab 蛇纹石板
serpentine schist 蛇纹石片岩
serpentine soil 波状地
serpentine spit 蛇状沙嘴
serpentine-talc 蛇纹滑石
serpentine tube 蛇形管
serpentine turnings 蜿蜒;曲折
serpentine twill 人字形斜纹
serpentine wall 蛇形墙;蛇纹石砖(石)墙;蛇纹形石或砖墙;蛇纹石墙
serpentining 蛇形排管;蛇形盘管
serpentinite 蛇纹岩
serpentinite for chemical firtilizer 化肥用蛇纹岩
serpentinite for flux 熔剂用蛇纹岩
serpentinization 蛇纹岩化;蛇纹石化
serpentinous marble (暗绿色的)蛇纹大理石
serpent kame 蛇纹石丘
serpentuator 蜿蜒管;蛇形管
serpex 塞佩克斯碱性耐火材料
serpierite 锌(钙)铜矾
serpiginous hollow 盘旋砂眼;盘旋气泡
Serpollet boiler 快速发汽锅炉(蒸汽机车用);色波里特锅炉
serpophite 胶蛇纹石
serpulid reef 龙介礁
serpulid reef coast 龙介虫礁海岸
Serra geral polarity chron 塞拉杰拉尔极性时
Serra geral polarity chronzone 塞拉杰拉尔极性时间带
Serra geral polarity zone 塞拉杰拉尔极性带
serrasoid 锯齿波
serrate 锯齿;齿状的;齿形的
serrated 锯齿形的
serrated base 锯齿状基底;齿状底座
serrated belt 齿形三角带
serrated blade 锯齿形刀片
serrated camshaft 凸轮花键盘
serrated casing 锯齿状套管
serrated casing shoe 锯齿状套管头
serrated change 锯齿形改变
serrated die 锯齿状板牙
serrated disk 缺口圆盘
serrated edge trowel 齿边泥刀
serrated end (管子的)收缩端
serrated face 锯齿面
serrated fin 锯齿形翅片
serrated form 齿形
serrated grating 锯齿形炉栅
serrated guard 刻齿割刀护刃器
serrate discontinuity 锯齿状结构面
serrated knife 刻齿动刀片;齿刀
serrated ledger plate 刻齿定刀片
serrated lock washer 锯齿状垫圈
serrated nut 细齿螺母
serrated pick 锯齿状手镐
serrated profile 锯齿形(纵)断面;锯齿形剖面
serrated pulse 锯齿脉冲;缺口脉冲;槽脉冲
serrated ring spanner 细齿环形扳手
serrated roller 齿形辊;槽纹棍;槽纹滚筒
serrated root 锯齿形根
serrated rotary slip 齿形旋转卡瓦
serrated saddle 锯齿形鞍具;齿形坯架
serrated section 刻齿动刀片
serrated section sickle 刻齿刀片割刀
serrated shaft 细齿轴
serrated shoe 齿状管头
serrated slips 牙板;卡瓦
serrated spatula 锯齿形泥刀;齿形刮刀

serrated strip 锯齿形条;(道路的)齿纹车道
serrated-surface 锯齿形表面
serrated suture 锯状缝
serrated topography 锯齿形地形;齿形山脊
serrated trowel 锯齿形泥刀
serrated vertical synchronizing signal 交错垂直同步信号
serrated weir 齿形堰
serrated wheel 锯齿轮
serrated wheel hub 锯齿轮毂
serrate granoblastic texture 齿形粒状变晶结构
serrate profile 细齿形;锯齿形轮廓
serrate ridge 齿形脊
serrate topography 齿状地形
serrating 刻齿纹
serration 细齿;锯齿(形);碾磙;三角花键
serration between points 冠叉
serration broach 细齿拉刀
serration ga(u)ge 花键廓量规
serration hob 细齿滚铣刀
serration hobber 细齿滚刀
serration joint 细齿连接
serration plate 锯齿(形)板
serration plug ga(u)ge 细齿塞规
serration ring ga(u)ge 细齿环规
serration shaft 细齿轴;锯齿形花键轴;三角形(齿)花键轴
Serret-Frenet formulas 弗雷涅—塞雷公式
serrodyne 线性调频转发器
serum 橡浆清
serum albumin 乳清白朊;血清蛋白
serum albumin glue 乳清白朊胶
serum preserving bottle 储血瓶
serum reference bank 血清保存中心
serval 浸透沥青的石棉(纤维)
servant 雇员
servant brake 随动闸
servant key 服务员开启房间的钥匙
servant's bath 仆人浴盆;仆人浴室
servant space 服务性空间
servants quarters 仆人居住部分
servant's room 仆人室
servant's security agreement 雇员保险合同
serve 卷缠
serve-operated control 非直接调节
server 备餐间;服务者;服务器
server-based 专用式
server layer network 服务层网
server operator 服务器操作员
server push 服务器推技术
servery 备餐室
servette 小折叠桌
service 业务;劳务;检修;技术维护;机关;维护;送达;船舶动态业务;常用;服务;保养
service abbreviation 业务缩语;业务略语
serviceability 工作能力;性能最佳性;有用性;可维修性;可维护性;可服务性;耐用性;适用性;使用性能;使用能力;使用可靠性;操作性能;操作上的可靠性;服务能力
serviceability criterion 适用性准则
serviceability index 功用指数(路面行车);耐用性指数
serviceability level indicator processing 可服务性级别指示处理;服务性级指示器处理
serviceability life 可使用年限;使用年限
serviceability limitation 使用性能范围
serviceability limit state 正常使用极限状态;使用性能极限状态
serviceability rating 使用性能级别(路面)
serviceability ratio 可服务时间比;耐用率;使用可靠性系数
serviceability requirement 使用要求
serviceability temperature 使用温度
serviceability test 耐用性试验
serviceable 能操作的;合用的
serviceable car 可用车辆
serviceable chest 多用柜
serviceable condition 可使用状态;完好状态
serviceable length 有效长(度);使用长度
serviceable life 可用年限;可使用年限;耐用年限;使用寿命;使用期(限);使用年限;服务年限
serviceable range 使用范围
serviceable temperature 使用温度
serviceable time 可服务时间

serviceable ton-kilometers per wagon per day 货车日产量
serviceable vehicle 实用车辆
serviceable work-done per car day 货车日产量
serviceable work done per car per day 运用车日产量
service accessibility 服务可得到性
service access point 接入点
service access point address 服务存取点地址
service access point identifier 服务接入点识别符
service account 劳务账;劳力项目
service activities 服务活动
service action 维修;服务
service action drawing 维修图
service action log 维修记录
service action parts list 维修零件一览表
service adapter 服务连接器
service advantages 操作优点
service after sale 售后服务
service age 使用年限;使用期
service agreement 服务协议
service aid 服务工具
service and maintenance manual 使用保养手册;使用保养说明书
service and repair 业务和维修
service and support 服务和配套
service and test restrictions 使用限制及试验的限制
service apartment 服务性公寓
service application position 常用制动位
service application zone 常用制动区
service area 供应区;有效作用区;有效工作区;业务区;供电区;工作区;可收听区域;辅助面积;服务区(域);服务面积
service at cost 按成本定价
service band 业务波段;公务波段;服务频带
service bar 酒吧服务柜台
service basin 工作水池;给水池;配水池;储水池
service basis 计提折旧的使用情况基础
service bay 装配间;装配段
service before and after the production process 产前产后服务
service behavio(u)r 运转情况;工作性能;工作情况;使用状况
service behavio(u)r record 工作状况记录
service below street level 街道地下(公共设施)管道
service bend 接户弯头
service bin 作业料仓
service bit 辅助位;服务位
service block 服务区
service boat 工作艇;工作船;港务船
service box 进宅(电)接线盒;接护居室;厕所水箱;引入线箱;供电线路箱;进线箱;接户阀箱;接户阀室
service brake 行驶制动器;行车制动器;工作制动;脚踏间;常用制动器;常用制动
service braking 常用制动
service braking deceleration 常用制动减速度
service branch 入户支管;用户支管
service bridge 施工便桥;临时便桥;专用桥(梁);工作(便)桥;辅助桥
service building 服务大楼;办公大楼
service-building elevator 建筑用提升机
service buildings for port 港口生活辅助建筑物
service by publication 公布送达
service cable 供电电缆;入户电缆;用户电缆;引入电缆;服务电缆
service call 业务通话
service capacity 运行出力;整备能力【铁】;工作容量;服务容量;服务能力
service car 公务车;维修服务车;服务车
service carriageway 服务性道路
service cat ladder 检修爬梯
service ceiling 使用升限;实用升限
service center 修理站;检修站;服务中心;服务站
service center accounting 服务中心会计
service channel 公务信道;公务波道;勤务信道;服务通道
service channel in series 串联服务通道
service charges 劳务费;接线费;手续费;服务收费;服务费(用)
service charges for public water 公共水用户收费
service chart 维修图表;使用图表
service chute 服务性滑槽;楼内勤务溜槽

service circuit 业务通信电路
service clamp 水龙头接管夹具
service class 服务级别
service cleaner 接户管清扫器
service cleaning 设备清洗
service clip 固定用户管的管夹;用户管管夹
service cock 用户引水龙头
service code 业务代码
service command 服务命令
service company 劳动服务公司;服务公司
service compartment 辅助格间
service completion time 服务完成时间
service compressor 工作压缩机
service condition 工作条件;使用状态;使用状况;使用条件;操作条件;业务情况;工作情况;航行状态
service conductor 引入线;供电导线;接户线
service conduit 供水管
service connection 用户连接管;引入连接;房屋连接管
service contract 业务合同;劳务合同;服务合同
service controller 常压控制器
service core 建筑(物的)服务内筒
service core floor 服务层
service core wall 服务设施井筒的墙
service corridor 操作廊;服务走廊
service cost 劳务成本;服务成本
service cost centre 服务成本中心
service counter 服务(柜)台
service counter door 服务台窗口
service counter terminal 服务台终端
service couplings and fittings 用户管接头及配件
service court 杂务院;后院
service craft 杂用船;工作艇;工作船;基地勤务船
service crane 维护用起重机;备用起重机
service credit 服务信用
service criterion 服务准则
service curve 服务曲线
service cut 公用事业掘路壕
service cut of public utilities 公用事业掘路堑
service cycle 工作周期;服务周期
service danger 运输危险
service data 运输数据;运行数据;业务资料;技术资料;技术数据;维护数据;使用数据;服务数据
service data unit 服务数据块
serviced ceiling 装有水电等线路的顶棚
service dead load 工作恒(荷)载;静使用荷载;计算死荷载;计算恒载
service debt 偿还债务
service delay 服务点的延时
service delivery 提供服务
service demand factor 需求系数
service department 辅助部门;服务部门
service department allocated cost variance 辅助生产部门费用分配差异;辅助部门费用摊配差异
service department budget 服务部门预算
service department cost 辅助生产部门成本;辅助部门费用;辅助部门成本;服务(或辅助)部门成本
service department expenses 辅助生产部门费用
service depot 服务中心
service desk 服务台
service deterioration 使用损耗
service diagram 行车表;运行图;服务线路图
service discharge 用水量;供用量;供水量
service discipline 设备专业;服务训练
service disconnecting means 供电断开装置
service distribution 服务分布
service district 供煤气区;供电区;供水区;服务区
serviced land 有设施用地
service door 工作者的门;作业用门;车门;服务入口;服务门
serviced period of daylight 天然光利用时间
service drain (由户内通到街道污水管的)排水管;下水管户线;用户排水管;工作排水道;工作面排水(沟)
service drainage (用户通至街道污水管的)排水管;用户排水管
service drawing 设备图(纸)
service drop 架空引入线;架空服务线
service drop conductor 用户引入线;进户线
serviced town-district 检修城区
service duct 供应通道;管路户线通道
service durability 耐用性
service dynamo 照明发电机

service economy 服务经济
service efficiency 使用效率
service elbow 带螺纹的弯管接头;接户线弯头
service elevator 辅助电梯;工作电梯;服务梯;服务电梯;人货两用电梯
service ell (卫生管道中的)肘形弯管;接户线管子;接户线弯头;L形螺纹管接头;L形检修(用)延长部分
service engine 维修发动机
service engineer 维护工程师;维修工程师
service engineering 服务(性)工程
service entrance 供电引入段;进户线;进线口;进户导线;通勤口;服务入口
service entrance conductors 引入线;进户线
service entrance conductors-overhead system 进户线—架空线系统
service entrance equipment 分户(配电)设备
service entrance switch 供应控制板;入宅电源开关;电源总开关;输电总开关
service entry 供应入口;用户管穿管
service equipment 供电设备;分户配电设备;修理工具;养路设备;间接生产设备;检修设备;维修设备;维护设备;辅助设备;服务设施;服务设备
service establishments 服务性企业
service expenses 劳务费;服务费(用)
service experience 运行试验;维修经验
service extraction 接出用户线
service facility 服务机构;维修设备;保养设备;生活服务设施;服务设施;服务手段条件
service factor 过载安全系数;负载系数;持续过载系数;运行率;工作因数;工作(条件)系数;开工因数;使用因素
service failure 供电中断;使用(中)损坏;使用(过程中)损坏
service fee 手续费
service feeder 供电馈线;入户导线
service fee norm 取费标准
service firm 服务性企业
service firm business 服务企业
service fittings 入户配件;供给配件;辅助配件
service flat 为居住者备餐的楼房;服务层;共用厨房的住房;供膳食、服务的成套公寓房间
service floor area 辅助面积
service form 适用形式
service-free 无需保养的;不需维修的
service-free life 无保养使用期
service function 服务职能
service gallery 公共管线廊道;执勤廊;维修廊道;辅助坑道
service gangway 施工便道;(脚手架的)工作走道
service garage 服务站
service gate 检修闸门
service gate slot 检修门槽
service generator 营运发电机;日用发电机
service generator room 厂用发电机室
service girder 支模板梁;脚手架梁
service governor 用户调压器;常压控制器
service ground 工作接地;接地线
service hangar 维修飞机库
service hardboard 辅助硬质纤维板
service hatch 领菜窗口;送膳窗口;送菜窗口;膳窗
service head 服务水头
service header 分配管;分户管;接户管集管
service hoist 副井绞车
service horsepower 工作马力;常用马力
service hours 运行时数;上班时间
service index 营运指标
service industry 第三产业;服务(性)工业;服务行业;服务产业
service in random order 随机服务
service instruction 业务规章;业务规程;工作细节;使用细则;使用维修规程
service integrated ceiling 声、光、空调等设施组合的吊顶
service integration 业务复用
service interrupt(ion) 服务中断;线路不通;业务中断;停电
service inventory 业务细目;服务项目
service island 汽车维护区(运输企业中划分出来专供汽车维护用的区域)
service kit 维修箱;维修包
service ladder 工作梯
service lamp 轻便灯
service lane 后巷

service lateral 供电(分)支线;用户(给水)旁管;用户(给水)支线
service launch 工作艇
service layer 服务设施安装工
service layout 管线布置;屋内排水管;屋内管线布置;室内管线布置
service lease 提供维修服务的租赁;服务性租赁
service level 供货水平;工作水位;工作水平;服务水准;服务水平
service level cost curve 供货水平费用曲线
service level of line network 线网服务水平
service level update 服务级更新
service library 服务程序库
service life 运行寿命;工作期限寿命;可用年限;耐用年限;使用寿命;使用期限;使用年限;服役寿命;服务期限;服务年限
service-life method 使用年限法
service life of die 模具寿命
service life of well 水井使用年限
service lifetime 维修寿命;使用寿命
service lift 工作电梯;送货吊机;送货电梯;送菜提升器
service limits 使用限度
service line 业务线;引入线;作业线;整备线;供电线(路);接户线;动力管道;支管;用户管;入户线;入户管
service line material 接户管材料
service linkage 服务联系
service live load 活使用荷载;工作活荷载;实际工作活荷载
service load 有效荷载;资用荷载;工作(用)载荷;工作(用)荷载;工作(用)负荷;工作(用)负载;实有荷载;实用荷载;服务荷载
service load spectrum 使用负荷频谱
service locomotive 路用机车
service machine 工作机
service macro 服务宏指令
service main 给水总管;配水干管;用户干线;支干线;接户干管;分干线;服务干线
service maintenance shop 维修车间
serviceman 修理工;维修人员
service management system 业务管理系统
service manual 工作手册;检修说明书;维修手册;维修规程;维修手册;使用细则;使用手册;操作手册;保养细则
service mark 服务商标;服务标记
service market 服务市场
service mechanism 服务机构
service message 服务消息
service meter 服务计数器;用户水表;接户水表
service mode 预检方式
service module 服务舱
service monthly ticket 公用月票
service network 服务网(络)
service node 业务节点
service node interface 业务节点接口
service observation 业务检查
service observing board 服务观测板
service of locomotive 机车整备
service of notice 送达通知书
service oil tank 日用油箱;日用油柜
service opening 维修孔
service operation console 维护操作控制台
service order 服务顺序
service order management 服务订单管理
service order program(me) 服务指令程序
service order table 服务顺序表;服务命令表;服务次序表
service oscillator 射频发生器;测试振荡器
service output 运行输出功率
service output depreciation method 生产数量折旧法
service outstanding debt 支付未偿还债务
service panel 配电盘
service pantry 小卖部;食品备用室
service park 汽车修配厂;汽车修理厂
service parts 修理用部件;配件;备用零件;修理用零件
service parts catalogue 备件目录
service pass 公用乘车证
service payments 偿还额
service percentage 服务百分比
service performance 运转情况;工作性能;使用性能;使用情况;操作性能

service performance of road 道路的使用品质
service period 使用时期;维修时期;服务时期;运行期;自然采光时间
service period after completion 使用年限
service pipe 室内燃气管道;进宅导线管;用户管;入户管;进宅支管(水管、燃气管);引入管;供水户线;供水管;工作管道;给水管(道);进给管;接户管
service pipeline 入户(给水)管线
service pipeline gallery 公用管线廊道
service pipeline subway 公用管线廊道
service plant 维修工厂
service plate 路牌
service platform 工作(平)台;检修平台;操作平台
service point 服务点;维修点;进户点;检修测试点
service pole 架有公用线的柱杆
service population 服务人口
service port 检修窗孔;服务港
service port function 业务端口功能
service position 开车位置
service potential 服务潜力
service power 伺服功率;使用功率;厂用动(力);厂用电源;厂用电(力)
service pressure 正常工作压力;工作压力;使用压力;服务压力
service pressure head 工作压力水头
service probability 业务概率;服务概率
service processor 服务处理机
service program(me) 使用程序;服务程序
service proofing cycle 运转检验周期
service property 使用质量;使用性能
service protocol 业务协议
service pulling boat 交通用划桨小船
service pump 值勤泵;杂用泵;日用(水)泵;伺服泵;常用泵;辅助泵;备用泵
service qualified rate 服务合格率
service quality 有效性;作用质量;功能质量;使用质量;服务质量
service quarter 服务(用)房
service queue 服务队列
servicer 业务人员;燃料加注车;服务车
service raceway 进线口硬套管
service rack 洗车架;洗车台;检修台
service radius 服务半径
service radius of station 车站服务半径
service ramp 出入坡道
service rank 服务级别
service rate 服务速度;服务率;额定工作压力
service rate of utilities 公用事业收费率
service rating 运转功率;使用等级;额定工作压力
service record 运转记录;运行记录;服务记录
service refrigerator 冷冻商品柜
service regulation 工作规则;维修规则;使用规则;使用规划;工作规程;维护规程;服务条例
service regulator 用户调压器
service representative privilege class 服务代表特权级;服务表示特许级
service request 服务请求
service request bit 服务请求位
service request block 服务请求块
service request output 服务请求输出
service reservoir 常用油箱;工作油箱;储水池;备用水库;小型水库;小型水池;配水库;配水池;地区水库
service reservoir for farming 提供农田用水的水库
service restoring relay 恢复运行继电器;重合闸继电器
service restriction 业务限制;使用限制
service rig 修井机
service riser 室内立管;立管
service road 便道;沿街面道路;工地便道;临街道路;通道;厂用道路;副路;服务性道路
service road index 辅助道路指数
service road way 便道;临街道路;前沿道路;一般在户外);维修阀;辅助阀
service room 服务性工作间;服务人员房间;服务间;服务室
service routine 例行服务工作;使用程序;辅助程序;服务(例行)程序
service-routine address 服务程序地址
service saddle 用户管管夹;接户管鞍
service saddle and connector 用户管鞍座和接头
service screen 滤网;筛门
service sector 服务行业;服务部门
service-seeking 查找服务;服务查询

service-seeking pause 查找服务暂停;服务查询暂停
service set 厂用机组
services for business negotiations 洽商服务
services for trade talks 洽商服务
service shaft 辅助竖井;供应设施用竖井;副井;辅助矿井
service-shock resistance 实用上的耐冲击性
service shop 修理间;检修厂;维修车间
service sign 服务设施标志
service signal 业务质量
service-simulated condition 运转服务条件;模拟使用条件
services in a group of associated enterprises 联营企业集团内部提供的劳务
service sleeve 维修用套袖
service slide 供餐用的推拉窗
service space 工作区;服务区(域);服务面积
service speed 运行速(度);营运速度;工作速度;满载正常航速;常用速度;常用航速【船】;操作速度;服务航速
service stair(case) 后楼梯;服务楼梯;杂务梯;旁门楼梯
service stairway 服务楼梯;房后楼梯
service stand 工作架
service standardization 服务标准化
service state 使用状态;服务状态
service station 修理站;加油站;汽车加油修理服务站
service station of inflatable life rafts 气胀式救生服务站
service stock 维修储用器材
service stop 用户管总阀(一般在户外)
service street 沿街面道路;辅助街道;服务性街道
service stress 使用应力
service stuff 缠填材料
service subprogram(me) 服务子程序
service system 接户系统;服务系统
service table 配餐台
service tank 用户水箱;用户水塔;小型水池;工作油罐;日用油箱;日用油柜;日用水舱;常用油箱;辅助油箱;备用槽
service tee T 形管接头;接用户三通;接户管三通
service telephone plant 内部电话装置
service temperature 工作温度;操作温度
service temperature range 使用温度范围
service test 性能试验;运转试验;运行试验;工作试验;使用试验;出力试验
service-test machine 使用试验机
service test model 性能试验模型
service time 保养时间;业务时间;使用时间;服务时间
service time distribution 服务时间分布
service time of investment 投资使用年限
service tolerance 保养允许公差
service tool 维修工具
service tower (建筑的)垂直运输塔;工作塔
service track 整备线;通行线(指铁路)
service trade 服务贸易;服务(性)行业;服务(行)业
service train 公务列车;通勤列车
service tree 花椒树
service trench 公共设施沟槽
service trial 运行试验;使用试验
service truck 服务车
service tube 供水管
service tunnel 辅助坑道;副隧道;辅助隧道;服务隧道;工作隧洞
service-type business 服务型行业
service-type labor 服务性劳动
service-type test 移交试验;使用状态试验
service unit 扩房群;服务装置;服务单元;服务单位
service utility 服务效用
service value 使用价值;服务价值
service valve 工作阀(门);检修阀(门);备用阀;用户管总阀(一般在户外);维修阀;辅助阀
service vehicle 服务车
service vehicle garage 服务车库
service velocity 运行速度;标准初速
service vessel 工作船
service voltage 供给电压;供电电压;使用电压
service volume 服务流量;可行交通量
service walking 工作通道
service walk vehicle 执勤车
service walkway 服务性人行道;工作便道
service water 生活用水;杂用水;工业用水;工厂用

水;家用水
service water heater 供水加热器;生活用水加热器
service water pump 杂用水泵
service water reservoir 生活用水水库
service water supply 用户供水
service water system 工厂用水系统;厂用水系统
service wear 工作磨损;使用性磨损;使用(性)磨耗;使用损耗
service wear index 实用耐磨指数
service week 工作周;服务周
service weight 使用重量;工作重(量)
service weldability 使用可焊性;使用焊接性
service well 辅助井;补给井
service wholesaler 提供服务的批发商
service wire 用户进户线;引入线;进户电线;入户线
service yard 杂作场;服务院;服务场地
service-yield basis 计提折旧以产出量为基础
service zone 服务区
servicing 检修技术维修;维修;服务;保养
servicing account 维持账户
servicing agreement 押款协议书
servicing center 加油站;服务站(汽车);检修站;汽车加油修理服务站;维修站;服务中心
servicing crew 技术维护班
servicing depot 检修站;维修站
servicing facility 维修设备
servicing factor 服务因素
servicing hints 修理须知
servicing installation 维修设备
servicing machine 维修机械
servicing material 维修材料
servicing of locomotives 机车整备作业
servicing period 运用期
servicing position 整备台位
servicing standard 服务标准
servicing system 服务系统
servicing terminal 折返站
servicing time 预检时间;检修时间;维修时间;维护时间;服务时间
servicing track at depot 机车整备线
servicing unit 修井设备;辅助设备
servient estate 让出有限用地权权力一方
servient tenement 别人有权通行的地产
serving 电缆外包层;被覆物
serving board 卷缠木槌
serving hatch 传递窗;上菜窗口
serving jute 黄麻被覆(电缆)
serving kitchen 配膳厨房
serving line 整备线
serving of cable 电缆外皮
serving room 配膳间
serving stuff 卷缠用的小绳
serving table 备餐桌
serving time 配膳时间
serving track 整备线
servitude 地役权;可处分外一人财产的权利
servo 伺服电动机;执行机构;继动;伺服(机)
servoaccelerometer 伺服(式)加速度计
servoaction 继电作用;伺服作用
servo-actuated control 从动控制
servo-actuated regulating system 随动调节系统
servo-actuator 伺服助力器;伺服电机;伺服执行机构;伺服拖动装置;伺服拖动机构
servo-amp(lifier) 伺服放大器
servo-analog(ue) computer 伺服模拟计算机
servo-analog(ue) device 伺服模拟装置
servo-analyzer 伺服分析器;随动系统分析器
servoand automatic gain control detector 伺服及自动增益控制检波器
servo-assisted brake 继动闸
servo-assisted steering 继动转向
servobalance potentiometer 随机电位计
servobearing transmission 伺服方位发送
servoboard 伺服机构试验台
servobrake 加力制动器;助力刹车;伺服制动(器)
servocamera 伺服摄影机
servochannel 伺服信号电路;伺服通道
servoclutch 伺服离合器
servo-control 继动控制;伺服控制;伺服操纵;随动控制;随动操纵
servo-control for linear motion 直线运动的伺服控制
servo-controlled aperture setting 伺服控制光圈调节

servo-controlled circuit 伺服控制电路
servo-controlled drive 伺服控制传动
servo-controlled motor 伺服控制马达
servo-controlled scan(ning)interferometer 随动控制扫描干涉仪;伺服控制扫描干涉仪
servo-controlled theory 伺服控制论
servo-controlled transport device 伺服控制传送装置
servo-controlled transport equipment 伺服控制传送装置
servo-controlled transport unit 伺服控制传送装置
servo-controller 伺服控制器
servo-control mechanism 伺服控制机构
servo-control mirror 伺服控制镜
servo control of lensiris 透镜光阑的伺服控制
servo-control receiver 伺服控制接收机
servo-control system 伺服控制系统
servocontrol unit 随动操纵装置
servo-curve plotter 伺服绘图器
servo-cutout switch panel 伺服断路开关盘
servocylinder 伺服唧筒;伺服缸
servo-drive 伺服传动器;助力传动;随动传动;伺服拖动;伺服驱动(装置)
servo-drive motor 伺服传动电机
servo-driven 伺服系统驱动
servo-driven plotting table 伺服驱动绘图机
servo-driver 伺服传动装置;伺服传动器
servo-drive system 伺服系统
servo-dynamics 伺服机构动力学
servodyne 伺服系统的动力传动(装置)
servoelement 随动系统元件;伺服元件
Servofrax 塞沃弗雷克斯(三硫化二砷透红外材料)
servo-function generator 伺服函数发生器
servo-gear 伺服机构
servo-generator 伺服发电机
servo-governor 伺服调速器
servo-head 伺服磁头
servo-head wheel 伺服磁鼓
servo hydraulic actuator 液压伺服执行机构;伺服液压激振器
servohydraulic cylinder 液压助力缸
servohydraulic rotary motor 液压伺服转子马达
servohydraulics 伺服液压系统
servo-increment 伺服增量
servo-integrator 伺服积分器
servojack 伺服千斤顶
servo-link 伺服传动装置
servo-loop 单环伺服机构;伺服系统;伺服回路
servo-loop control 闭环随从自动调整系统
servo-loop model 伺服回路模型
servo-loop system 伺服回路系统
servo-lubrication 中央分布润滑法
servo-magnet 伺服电磁铁
servo-manipulator 伺服机械手
servo-manometer 伺服压力计
servo-mechanical system 机械伺服系统
servo-mechanism 助力机构;伺服装置;伺服机制;伺服机构;伺服传动(系统)
servo-mechanism tester 随动系统调谐操纵台;随动系统测试仪;伺服系统测试仪
servo-mechanism types 伺服机构类型
servo-modulation 伺服调制
servo-modulator 伺服调制器
servomotor 接力器;继动器;继电器;伺服(控制)马达;伺服机构能源;伺服电动机;辅助电动机
servomotor capacity 接力器容量;调速功;伺服电动机容量
servomotor cylinder 接力器缸
servomotor cylinder cap 接力器缸盖
servomotor response time 接力器响应时间;接力器不动作时间
servomotor valve 发动机助力阀
servo-multiplier 随动乘法器
servo-noise 伺服系统噪声;伺服干扰
servo-operated control 利用伺服机构控制
servo-operated mechanism 伺服传动机构
servo-operation circuit 伺服运算电路
servo-parameter 随动系统参数;伺服系统参数
servo-piston 活塞式随动传动器;伺服活塞
servo-positioning 伺服定位
servo-potentiometer 伺服分压器;伺服电位计;伺服电势计
servo-programming 伺服程序设计
servo-pump 伺服泵

servo-recorder 伺服记录器
servo-relief valve 继动安全阀;伺服安全阀
servo-resolver 伺服解算器;伺服分解器
servo-response rate 伺服响应率
servo-scanning 伺服扫描
servo-scribe 伺服扫描
servo-scribe recorder 伺服(扫描)记录器;伺服笔绘记录仪
servo-selsyn system 随动自动同步机系统
servo-sensor 伺服式传感器
servo-simulator 模拟伺服机构;伺服模拟装置;伺服模拟机;伺服仿真机
servo-stabilization 人工稳定;伺服稳定(性)
servo swap 交替换带
servo-system 继动系统;从动系统;随机系统;随动系统;伺服系统;伺服设备;伺服机构
servo-system drive 伺服传动装置
servo-tab 伺服补偿机
servo-theory 伺服理论
servo-transducer 伺服式传感器
servo-tuning 伺服调谐
servo-unit 助力器;助力机构;伺服机械;伺服机构;伺服单元;伺服单位
servo-valve 从动阀;伺服阀;伺服操纵阀
sesame oil 香油;芝麻油
sesame spot 麻点
sesame spot glaze 麻点釉
sesquibasic 倍半碱价的
sesquicarbonate 倍半碳酸盐
sesquicentennial 150周年纪念
sesquioxide 三氧化二物;倍半氧化物
sesquioxidic clay 铁铝性黏土
sesquiplicate 二分之三次方的
sesquiquadrate 一个半象限差
sesquiquinone 倍半醌
sesquisalt 倍半盐
sesquisideband transmission 倍半边带传输
sesquisilicate 倍半硅酸盐
sesquisulfide 倍半硫化物
sesquiterpene 倍半萜烯
sesquiterpenoids 倍半萜类
sessible drop method 躺滴法
sessile 固着的
sessile benthos 无柄水底植物
sessile bubble method 固定泡法;静泡法
sessile diatom 固着硅藻
sessile dislocation 不(滑)动位错
sessile drop method 悬滴法;液滴法(测定表面张力);固着液滴法;固定滴法;静滴法
sessile flora 固着植物区系
sessile form 固着种类;固着型
sessile organism 固着生物;底栖固着生物;附着生物
session 会议;会期
session commitment unit 管理工作间隔的约会块
session deactivation 对话撤销
session deactivation request 对话撤销请求
session layer 会晤层;会话层
session parameter 对话参数
session protocol data unit 会晤协议数据单元
session quarantine unit 不可分的约会部件
session seed 对话初始值
session termination 会话终止
session termination request 对话中止请求
seston 浮游物
set 置位;整锯齿;整齿;固着;抹灰罩面层;批;对准;定势;长插条;放置;安置;凝定
setable count 可置位计数
seta brake in operating position 将制动机置于动作位
set-accelerating admixture 促凝剂
set-accelerating agent 促凝剂
set accelerator 速凝剂;促凝剂
set a distance 划定距
set a file 集合文件
setaflash closed tester 闭杯闪点测定器
set after break 扯断永久形变
set algebra 集合代数
setal map 毛序图
setamine colo(u)rs 毛胺染料
set analyzer 通信机试验器;接收机检验器;测定器
set and locking screw 止动及锁紧螺钉
set and try method 试验法
set a new record 创纪录
set a precedent 开先例

set a saw 拨锯路
set aside 闲置;拨出
set assembling 总成装配
set-associative organization 成组相连结构
set back 向后运动;移后;(安放钻杆的)立根盒;红线;建筑后退;逆转;回置距离;退进;退步;台阶式;缩进;收进(砌墙等工作)
setback building 阶梯形建筑物;退进房屋;台阶式建筑;由红线后退进去的房屋
setback buttress 从墙角收进的扶垛;角扶壁;收进式扶垛
setback corner joint 内收转角缝
setback distance of outer wall 外墙退后距离
setback force 后退力
setback frame 阶梯形框架
setback front axle 后移前轴
setback levee 套堤
setback line 房屋后退线;(道路两侧的)建筑限制线;建筑收进线;退入界线;收进线
setback motion 后退运动
setback of building line 建筑线的缩进;建筑线的后退
setback of piping 靠在指梁上的管子
setback of tubing 靠在指梁上的管子
setback ordinance 后退条款;缩进规定
setback pin 惯性销
setback scale 零调整刻度盘;零点调整刻度盘
setback type 台阶式
set bar 盖板;联动销;连接板;锁栓
set behavio(u)r 凝结性能;凝结状态
set bit 嵌镶钻头
set block 整齿铁钻
set block paving 石块铺砌
set bolt 固定螺栓;定位螺栓;冲头销;防松螺栓
set breakpoint 设置断点
set bubble 使气泡居中
set casing 放置套管;用水泥固定套管;固定壳
set casing shoe 镶嵌套管鞋;套管靴钻头
set cement 凝结水泥
set chisel 尖凿;斧凿
set clause 集子句
set clock 置位时钟
set coil 置位线圈
set collar 压紧环;固定轴承环;固定环;棚顶梁;定位轴环;定位环;顶梁
set command 置位指令
set composite 双信回路;复合回路;报话复合制组件
set concrete 凝固的混凝土
set control 凝结控制;置位控制
set control agent 凝结时间调节剂
set-controller 调理剂;调凝剂
set-controlling admixture 调稠剂;控凝(掺合)剂;调凝剂
set-controlling agent 调理剂;调凝剂
set copper 凹铜
set course 定航向
set covering 集合覆盖
set covering problem 集合覆盖问题
set-day 座/天
set deformation 永久变形
set-depth (柱的)埋入深度
set description entry 系描述项
set design 布景设计
set diameter (金刚石钻头的)镶嵌直径
set diaphragm 调整光圈
set die 铆钉模
set difference 集合差
set-down 减水【水文】
set drill 安装钻机
set edge 支边
set enable 可置位
setenosemicarbazibe 氨基硒脲
set entry 系项
set equality 集合相等
set equations 方程组
set fair 墁平的泥灰面;天气稳定晴朗
set feeler 调整触点;定位触点
set fire-to 放火;点燃
set flashing 墙挡板;密封板;挡板
set flush 顶格排版
set formula 固定公式
set forth 阐述
set forward 移前
set forward device 提前量计算器

set forward point 预期前方点
set forward position 提前位置
set forward rule 提前量计算尺
set forward scale 提前量分划盘
set frame 制条机;整条机
set free 解除;释放;游离(状态)
set function 集合函数
set gate 置位门;随模浇口
set ga(u)ge 定位规
set girt(h) 支架横撑
set going 开动
set grazing 固定放牧;定牧
set hammer 堵缝锤;陷型锤;压印锤;击平锤;扁锤
set-hands dial 调整指针刻度盘
set hard 结硬;凝固
set head 镦头;钝头
set-head rivet 片头铆钉;镦头铆钉;型头铆钉;钝头铆钉
set heat 凝结热
set-hold circuit 置位保持电路
set hoop 紧定环箍;定位环
set in 铺砌;砌体层收分
set in bond 砌合
set in bookcase 壁橱式书架;附在墙上的书橱
set in broken stones 用碎石嵌实;用碎石铺砌
set inclusion 集合蕴涵
set index instruction 设定变址指令
set inhibit 禁置位
set in meiyu 入霉
set in motion 开动
set in operation 运转;运行
set in order 检修
set inside diameter (金刚石钻头的)镶嵌内径
set interference 机内干扰;本身干扰
set intersection 集合交
set in use 开机率
set-iron 金属样板
set key 柱螺栓键
set level 支平;放平
set lever 摇尺杆;锁紧手柄
set light 照明设备;照明装置;成型照明灯光
set line-and-grade stakes 设置平面与坡度桩
set location mode 系存放方式;置位定位方式
set member 系成员;集合成员;棚架构件;从记录;成员组
set-meter 沉降仪
set method 永久变形测定法;测回法
set mining method 支柱采矿法
set mode 给定状态;给定方式
set-modifying agent 调理剂;调凝剂
set name 集名;集合名
set noise 机内噪声;本身固有噪声;本机杂音
set nut 定位螺母;制动螺母;制动螺帽;紧定螺母;调整螺母;调整螺帽
set occurrence 集具体值
set of arch 拱架揣数
set of axes 轴系
set of bands 谱带组;带组
set of basin 盆地组合
set of bills 成套单据
set of bills of loading 整套提单
set of boundary condition 边界条件系统
set of bracket 撑
set of brushes 电刷组
set of car 车组
set of cement 水泥凝结;水泥凝固
set of characteristics 特征线系
set of claims 理赔
set of colo(u)r plate 分色板组
set of colo(u)rs 三原色
set of compasses 全套罗盘仪
set of complex numbers 复数集
set of conduits 管组
set of conventional signs 图像式符号表;图式符号系统
set of coordinates 坐标系(统)
set of cross-current 横流方向
set of current 流向
set of current meter 流速仪装置
set of curves 一组曲线;曲线组;曲线族
set of degeneracy 退化点集
set of departure sidings 出发线群
set of diagrams 图组
set of directions 方向组

set of drawing instruments 绘图仪器
set of drills 一套钻机;钎子组
set of eddy flow 涡流方向
set of electrodes 电极组
set of elementary events 基本事件集(合)
set of engines 发动机组
set of equations 联立方程(式);方程组
set of equipment 整套设备
set of exchange 联单汇票
set off 衬托;墙的突出物;墙的凸出部分;墙壁的凸出部;突出部;凸缘;断流;迂回管;划开;拔碌
set of faults 断层带
set of feeder cables 馈电电缆组
set-off expenses 补偿开支
set-off of ink 油墨粘脏;纸背面蹭脏
set of fork spanners 成套叉形扳手
set of formation siding 编组线群
set of function 函数集合
set of furniture 一套家具
set of hachures 晕线【测】
set of holes 炮眼组
set of homologous lines 均称谱线组
set of inequalities 不等式组
set of inequalities of probability 概率的不等式组
set of instructions 指令组
set of knives 动刀片组
set of lenses 透镜组
set of machines 管列
set of maps 图件统一编号
set of matrices 矩阵集合
set of node 节点集
set of non-negative integer 非负整数集
set of operating condition 操作条件组
set of particles 质点集
set of patterns 模式集
set of penetration 贯入度
set of pile per blow 每锤桩的贯入量;每锤桩的贯入度
set of pipes 管组
set of plotting instruction 绘图指令组
set of points 点集
set of posts 接线柱组
set of reception siding 到达线群
set of residuals 残差集
set of rolls 全套轧辊
set of rotor vanes 转子叶栅组
set of rules 规则组
set of screens 套筛
set of shunting siding 调车线群
set of sieves 套筛;筛组
set of sorting siding 编组线群
set of spare parts 成套备件
set of spare units 成套备用零部件;备用组件
set of splitting-up siding 解体线群
set of stream 流向;水流方向
set of structural equations 结构方程组
set of symbols 图像式符号表;符号组
set of teeth 锯齿偏侧度
set of terminals 端子对
set of the current 潮水流向
set of the rolls 辊筒组
set of tide 潮流;潮水方向
set of tools 成套工具;全套工具
set of tracks 线群
set of transmission gears 传动齿轮组
set of tubes 管组;成套管子
set of variables 变数组
set of vectors 矢量集
set of waves 波列
set of wheels 轮系;齿轮系;车轮组合
set of wires 梳针定位
setoglaucine 毛鹥蓝;若杜林蓝
set om centre 安装对中
set-on 安置;挂线
set on center 安装对中
seton wound preventing plate 线伤痕防护板
set operation 集合运算;集合操作
set option 集合选择
set ordering criterion 系定序准则
set out 起程;装饰;修饰;砌出一些;测定
set out line 定线
set-out of mold rains 出梅
set-out of plum rains 出梅
set outside diameter (金刚石钻头的)镶嵌外径

set-out track for wagons forbidden from loose shunting 禁溜线【铁】
setover 让给;偏置;偏距
setover method 跨距法;偏置法
setover screw 偏距螺钉
set owner 集的所有者
set pair 集合对;集对
set partition 集划分
set partitioning problem 集合划分问题
set per blow 每次冲击下陷深度;每次冲击贯入深度;每锤桩的贯入尺寸
set piece 调整块;定位块
set pin 销钉
set pin for scraper blade 刮板固定销
set pipe 固定套管柱
set piston 调整活塞
set plaster 硬化烧石膏
set plate 压板
set plumb 挂直
set point 规定值;给定值;给定点;检查点;调定点;设定值;设定点;沉淀点;凝结点;选点;置位点;凝结点
set point adjuster 设定点调节器;给定值调节器;定值调节器
set point adjustment 设定值调整
set point computer control 计算机整定点控制
set point computer control system 给定值控制系统
set point control 设定值控制;设定点控制
set point of temperature 规定温度
set-point scale 定点标度
set-point value 选定点值;给定值
set prescription 成方
set pressure 规定压力;容许压力;调定压力;整定压力
set pressure for safety valve 安全阀的调定压力
set prevention 防止凝结
set price 固定价格
set pulse 置位脉冲
set ram 人工夯具;手夯
set rate 凝结速度
set reaction 凝结反应
set reaming shell 镶嵌金刚石扩孔器
set recording 沉降记录
set regulator 调理剂;调凝剂
set retarder 缓凝剂
set-reset 置位复位
set-reset bistable 置位复位双稳
set-reset control 置位复位控制
set-reset flip-flop 置位复位触发器
set-reset pulse 置位复位脉冲
set-reset word 置位复位字
set-retarding 缓凝(作用)
set-retarding admixture 缓凝(外加)剂;外加缓凝剂
set-retarding agent 缓凝剂
set right 矫正
set round 矫圆
set rule 定位尺
set run 装定的航程
sets 机组
set sail 张帆;开航;启航
sets and lagging 支架和横撑
set scene 立体布景
set screw 止动螺钉;定位螺钉;调整螺钉;止动螺丝;紧定螺钉;调节螺钉;定位螺旋;制动螺钉
set-screw flange 带固定螺钉的法兰
set-screw piston pin 固定螺钉式活塞销
set-screw spanner 止动螺钉扳手;螺丝钳
set-screw wrench 止动螺钉扳手;固位螺钉扳手;固定螺钉扳手
set section 系节
set selection 系值选择;集选择
set selection criterion 集合选择准则
set shackle 反装卸扣
set shoe 注水泥管鞋
set shrinkage 凝结收缩
set shutter 拨好快门
set sill 支架底梁
sets of joint 节理分组
set solid 结硬;满排
set speed 规定速度;给定速度;额定速度
set spring 离合杆簧
set square 检验角尺;斜角规;三角尺;三角板
set squares for chart work 海图作业三角板

sets room 大型道具室
set stock 定牧畜群
set-stock grazing 定区放牧法
set stocking 固定放牧;定牧
set strength 凝结强度
set strength of cement 水泥强度
set subentry 系子登记项
set support 支护
set swinging 支架安装
set symbol 集合符;设定符
set system algebra 集系代数
sett (打桩时用的)送桩;方格图案;锻工用切割工具;小方石;镶板夹杆;铺路小方石;车组;长方形木块
settable orifice 可变节流口
set tap 手用丝锥
sett burning 铺路石块加热
settee 小型沙发椅;沙发;长靠椅
setter 型锤;装窑工;装定器;固定装置;给定装置;码砖机;棚板;调整工
setter cell 给定单元
setter out 放样技工;放样(木)工
setter slab 装窑垫板
set test 凝结试验;凝结时间测定
set testing apparatus 凝结试验仪
sett feeder 填缝器(用于石块路)
sett frame 成条机
sett grease 铺路小方石润滑脂
sett-handling bucket 石块输送吊斗
set the azimuth 装定方位角
set-theoretic(al) language 集合理论语言
set theoretic(al) definition 集合论定义
set theory 集(合)论
set the standard 控制标准
set the tone 定调子
set the watch 布置值班
set thrust plate 推力挡板
set tile 盖瓦
set time 凝固时间;凝结时间;指定时间;规定时间;设定时间;变定时间
set time counter 置位时间计数器
setting 摇尺;座果;作业点;置位;整定;静置;(液体变为固体的)凝结;调试;动力集材区;衬垫;沉实(谷类散货);变定;背景;安置;凝定
setting a bridge plug 下桥塞
setting accelerator 加速剂;加速沉淀器;加速沉淀池;机械搅拌沉淀池;凝固加速剂;凝固催速剂;促凝剂
setting accelerator additive 促凝剂
setting accelerator admix(ture) 促凝外加剂
setting accuracy 整定精确性;调整精度;定位精度;安平精度
setting accuracy on work 加工坐标精度
setting agent 凝结剂;凝固剂;固定剂
setting anchor 抛锚
setting and hardening 凝结硬化
setting and lustring 定形及上光
setting angle 装置角;安装角
setting an object 测定物标方位
setting appliance 调节设备
setting area 布景区
setting back 退缩;复原
setting back of bank 河岸退进;河岸后缩;河岸凹入;岸线后退
setting bar 圬工撬棒
setting base for counter weight 配重块底座
setting basin 沉沙池;沉淀池
setting beacons 设置航标
setting bed 砂浆底层;(水磨石下层面的)砂浆垫层;铺贴基层;底层(水磨石墙)
setting behaviour 凝固特性
setting belt 凝结皮带
setting bin 沉淀仓
setting bit 钎子墩套
setting block 装配夹具;镶嵌夹具;塞块;锯锉;固定架;调整器;调整垫块;镶嵌垫块;整锯器;定位块
setting bolt spring 压簧
setting box 沉淀箱
setting cap 垫帽
setting capacity 沉降量;沉降力;装窑容量
setting casing depth 套管下入深度
setting casing diameter 套管下入直径
setting casing time 下套管时间
setting cement 慢凝水泥

setting charges 重镶金刚石钻头费用
setting chisel 调整钎子;安装钎子
setting circle 寻星(度)盘;定位圆;定位度盘
setting clay brick 定位黏土砖
setting cloth 铺放绒布;垫布
setting coat 第三道油漆;罩面(层);精修饰涂层;末道粉刷层;抹灰(罩)面层;上涂(石灰膏涂层)
setting coat plaster 第三道油漆;边涂层;上涂;抹灰罩面层;石灰膏涂层
setting cold adhesion 冷凝黏合
setting command 整定指令;整定命令
setting compound 定位剂
setting condition 凝固状况
setting course 结合层
setting cradle for shield 盾构装配支架
setting current 整定电流
setting curve 沉降曲线;凝固曲线
setting cylinder 起动汽缸;调节体
setting data 安置数据
setting density 装窑密度
setting depth 埋深
setting depth of admixer 混合器下置深度
setting depth of surface casing 外套管的设置深度
setting depth of water pipe 水管下置深度
setting depth of wind pipe 风管下置深度
setting designing 布景设计;舞台设计
setting device 调节设备;自动记录器;装定装置;整定装置;整定元件;校准装置;调整装置;调整机构;调用装置;调节装置;定位装置
setting device for sludge sample 岩粉沉淀器
setting dial 仪器标尺;调节盘
setting die 可调冲模;校正模
setting disturbance 凝结扰动
setting-down period 采取措施期
setting drawing 安装图;装置图(样);装配图(样)
setting effect 凝固效应
setting element 定位元件;安装构件
setting elevator 铲车
setting energy 凝结能量
setting error 装定误差;安置误差
setting expansion 凝结膨胀
setting finger 定位指
setting forward of bank 河岸前伸
setting forward of the bank 河岸凸出
setting free the crown 树冠疏枝
setting ga(u)ge 镶嵌规(金刚石钻头);定位规;标准量规
setting glass 安装玻璃;装置玻璃
setting hammer 型锤
setting heat 硬化热;凝固热;水化热
setting height 装置高度;安装高度
setting horizontal point of portal 峒口投点
setting-in 嵌装;弯铅皮用的硬木棒;嵌入物
setting in a kiln 码窑
setting-in brick 嵌入棒
setting-in coincidence 符合对准
setting-in motion 起动
setting-in stick 弯铅皮用的硬木板;嵌入棒
setting-in watering 定植浇水
setting knob 调整旋钮;调定旋钮;安置钮
setting leak 砌体漏泄
setting lever 移动杆;倒换杠杆;操纵杆
setting limit 落限
setting line 安置线
setting load 初撑力;安装荷载
setting machine 码坯机;定形机
setting mallet 捻缝木槌
setting mark 定位符号;参考标志
setting matter 沉淀物
setting mechanism 凝结机理;调节机构;整定机构;固定装置
setting medium 凝固介质
setting monument 埋石
setting movement 定位运动
setting nut 调整螺母;调节螺帽
setting of bridge location 桥位测设
setting of concrete 混凝土凝结;混凝土凝固
setting of diaphragm 光阑调节
setting off 关闭;断流
setting of focal length 焦距调整
setting of ground 地基沉降;地面下沉;地面沉降;地基下沉;地基沉陷
setting of headlights 装定前灯
setting of ink 油墨的固着

setting of moon 月落
setting of mortar 灰浆凝固;灰浆凝结;砂浆的凝结
setting of particles 颗粒沉降
setting of supports 支座下沉
setting of the focal length 焦距调节
setting of valve 阀的装配
setting of wall tiles 面砖铺贴
setting on points of control 按控制点定位
setting orientation element 安置定向构件
setting-out 出发;标志位置;放线;启程;定线;放样;压水法;画线
setting-out a foundation 基础放样;基础放线;基础放线
setting-out angle 放样角
setting-out by intersecting angle coordinate 角度交会法放点
setting-out correct position 正确位置测设
setting-out curve 定位曲线
setting-out data 放样数据
setting-out detail 定线详图;放样详图;放线详图
setting-out detail points of curve 曲线细部点测设
setting-out diagram 放样草图
setting-out element 定位构件
setting-out for tunnel(l)ing 隧道掘进定位
setting-out level 定位高程
setting-out line 定位线;放样线;划定的线;测定的线
setting-out machine 除屑机
setting-out map 放样图
setting-out of building 建筑物放样;建筑放线
setting-out of building axles 建筑物轴线放样
setting-out of building center line 建筑轴线测量
setting-out of center line of tunnel 隧道中线标定
setting-out of centre peg 中桩测设
setting-out of circular curves 圆曲线测设
setting-out of cross line through shaft center 井筒十字中线标定
setting-out of cross-point 交叉点放样
setting-out of curve 曲线测设;曲线测量
setting-out of cut 路堑放样
setting-out of embankment 路堤放样
setting-out of footing foundation peripheral points 填筑轮廓点测量
setting-out of foundation 基础放样;基础放线
setting-out of main axis 主轴线测设
setting-out of route 放线
setting-out of side plumb-bob 边垂线标定
setting-out of technologic(al) edge 技术境界标定
setting-out of trench 基槽放线
setting-out of tunnel 隧道放样;隧道坡度标定
setting-out peg 定位桩;放线桩;定线桩
setting-out plan 放样平面图
setting-out point 放样点
setting-out rod 定线杆;定线杆
setting-out shaft center [英] setting-out shuft centre 井筒中心标定
setting-out sketch 放样草图
setting-out stake 小木桩
setting-out survey 放样测量
setting-out the intersection point by alignment 穿线轿车点定线
setting-out using the perpendicular coordinate 直角坐标法放点
setting-out value 定位值
setting-out work 定线工作;放样工作;放样定线;放线工作
setting pattern of diamond (钻头切削表面的)金刚石分布图
setting period 硬化期;凝结期
setting pit 沉淀池
setting plan 装配平面图;平面装配图
setting plate 垫板
setting plug (金刚石钻头内出刀的)检查塞规
setting point 下落点;下放点;凝固点;凝(定)点;调整点;凝结点
setting-point test 凝固点试验
setting power 沉淀能力;凝结力
setting pressure 给定压力;定压;预撑力;设定压力;初撑力;安装压力
setting procedure 安装程序;凝固过程
setting process 沉淀法;凝结过程;硬化过程
setting promoter 沉降促进剂;沉淀催化剂
setting property 沉降性质;凝固性质
setting pulse 置位脉冲
setting punch 铆头模;铆钉用具

setting quality 下沉性;沉降性;凝固性
setting range 整定置位;整定范围;调整范围
setting rate 硬化速度;凝结速度;凝结率;凝固速度
setting ratio 成果率
setting reaction 凝结反应
setting retarder 缓凝剂
setting retarding agent 缓凝剂
setting ring 调节环;引信固定环;(金刚石钻头内出刃的)检查环规
setting rod 放置钻杆;定线杆
setting room of diamond bit 金刚石钻头镶嵌车间
setting scenery 布景
setting screw 定位螺钉;定位螺旋
setting shack 安装钻棚
setting shrinkage 干缩;凝结收缩;凝固收缩
setting shrinkage crack 凝结收缩裂缝
setting space 安装空间;调整空间
setting speed 沉降速度
setting stake 定线桩
setting strength 硬化强度;凝结强度;凝固强度
setting strip 配合条
setting stuff 修饰材料;抹灰材料;罩面层
setting-sun sign 落日征
setting switch 转接开关;置位开关
setting system 码坯系统
setting system for drilling fluid 钻井液的沉砂系统
setting tank 沉沙池;固体沉降槽;沉淀池
setting temperature 固化温度;变定温度;凝结温度;凝固温度
setting test 凝结实验;凝固试验
setting tester 凝结试验机;凝结时间测定仪
setting testing apparatus 凝固试验仪(器)
setting the pace for heavy industry 突出重工业
setting the packer 座封封隔器
setting threshold 置位阈
setting time 置位时间;整定时间;固化时间;建立时间;凝结时间;凝固时间;凝定时间
setting-time apparatus 凝结时间测定仪;凝固时间测定仪;凝结时间试验器
setting-time of concrete 混凝土凝结时间
setting-time retarder 缓凝剂
setting-time test 凝结试验;凝结时间试验
setting tool 调整用工具;镶嵌工具;切肩用具
setting to touch 指触干
setting true to perpendicular 置平(测)
setting-type deflectometer 置放式挠度仪
setting-type horizontal declinometer 置放式水平测斜仪
setting up 增稠;调定;安装
setting-up agent 硬化剂;凝固剂;固定剂
setting-up business 开业
setting-up cost 准备成本;生产调整准备成本
setting-up expenses 准备费用;开业准备费用;生产调整准备费用
setting-up of beam 束流调定
setting-up piece 固定垫片
setting-up pressure 顶锻压力
setting-up procedure 调整顺序
setting-up screw 固定螺钉;调距螺旋;调距螺栓
setting-up sequence 装配顺序;调整顺序
setting-up shop 装配车间
setting-up temperature 顶锻温度
setting-up time 接续时间;调定时间;(混凝土的)凝硬时间;凝结时间;安装时间
setting-up underground protecting pollution curtain 设置地下防污帷幕
setting-up well rows for protecting pollution 设置截污井排
setting value 调整值;凝结率;给定值;设定值;整定值
setting velocity 沉降速度
setting velocity of suspended particles 悬浮颗粒沉降速度
setting water 凝结水
setting winch 安装卷扬机;安装绞车
setting work 装配工作
setting worm 调节蜗杆
sett jointer 填缝器(用于石块路)
sett joint filler 块石缝填充料
sett joint sand 块石填缝砂
sett joint sealing 石块接缝填料
settle 解决;结清;澄清;长靠椅
settle a bargain 订约;订(立)合同
settleability 沉降性

settleability test 沉淀性试验
settle a bill 付账
settleable matter 可沉降物;可沉淀物(质)
settleable solid 可沉淀的颗粒;易沉降固体;可沉淀固体
settleable solids test 可沉固体试验
settleable solid volume fraction 可沉降固体量比
settle accelerator 凝集剂
settle account 结账;结算;清结
settle accounts in US dollar 用美元结算
settle a claim 解决索赔;清算债务;付账
settle a debt 结算债务;结清债务
settle a debt by payment 付清债券
settle a lawsuit 结案
settle an account 结算
settle bill 结账;付账
settle chair 扶手长凳;高背长椅
settle claims 处理索赔;理赔;清理债权
settle contour 沉降等值线
settled account 已结清账目;已结清账款
settle date 结算日期
settled bed 澄清床
settled dust 沉积的灰尘
settled grease 成层滑脂
settled ground 沉实地层
settled organism 附着生物
settled price 落定价
settled production 固定产量
settled return 固定收入
settled sewage 澄清的污水;沉淀污水;沉淀后的污水
settled sludge 沉积淤泥;沉积污泥;沉淀污泥
settled sludge volume loading rate 沉积污泥量负荷率
settled solution 澄清溶液
settled tar 沉积焦油
settled wastewater 净化后的废水;沉淀后的污水;沉淀废水
settle mark 收缩标志
settlement 新住宅区;移民点;租借地;解决;结算;清;确定;清偿;定居点;澄清;沉陷;沉降量;沉降
settlement account 清偿账款
settlement action 沉降作用
settlement after construction 工后沉降
settlement allowance 沉降容许高度;预留沉降量;沉降限差
settlement analysis 沉陷分析;沉降分析(法);沉淀分析
settlement area 居民区;沉降面积
settlement block 沉降区
settlement book 住房抵押贷款手册
settlement by agreement 协议解决;协商解决;协商和解
settlement by blasting 用爆炸填土法
settlement by soaking 湿陷
settlement calculation 沉降计算
settlement calculation depth 沉降计算深度
settlement carter 沉降漏斗;沉降坑
settlement cell 沉降盒;沉淀盆
settlement change 下沉变化
settlement chart of footing 基础沉降图
settlement chart of foundation 基础沉降图
settlement check 清算支票
settlement coefficient 沉降系数
settlement computation 沉积计算;沉降计算
settlement contour 沉积等高线;沉积分布图;等沉降线;沉降等值线
settlement correction coefficient 沉降经验修正系数
settlement correction factor 沉降计算经验系数
settlement cost 结算款;成交费用
settlement crack 沉陷裂缝;沉降引起的裂缝;裂缝
settlement cracks of walling 墙身沉陷裂缝
settlement crater 沉降坑;沉降漏斗
settlement currency 结算货币
settlement curve 沉陷曲线(图);沉降曲线
settlement date 清算日
settlement datum 沉降基准面
settlement day 结算日;交割日;计息日;清算日
settlement deflection ratio 弯沉比
settlement device 沉陷测定装置;沉淀仪
settlement difference 沉降差
settlement discount 结算折扣
settlement due to blasting 爆炸引起的沉陷

settlement due to compression 压缩沉降
settlement due to consolidation 固结沉降
settlement due to excavation 开挖引起的沉陷;开挖引起的沉降
settlement due to shear failure 剪切破坏引起的沉陷;剪切破坏引起的沉降
settlement effects chart 相邻影响曲线图【岩】
settlement environment 聚落环境
settlement equation 沉降方程
settlement estate duty 遗产税
settlement expectancy 预计沉降量
settlement experiment 沉淀实验
settlement factor 沉降系数
settlement failure 沉陷破坏
settlement forecast 沉积预报;沉降预测;沉降预报
settlement friction 沉降摩擦力
settlement fund 结算资金;清算资金
settlement funnel 沉降漏斗
settlement ga(u)ge 沉陷计;沉降仪;沉降计;分层沉降管
settlement ga(u)ge for each layer 分层沉降标
settlement house 公社中心;社区中心建筑
settlement isoline 等沉降线;沉降等值线
settlement joint 沉降缝
settlement loan 结算贷款
settlement management 结算管理
settlement measurement 沉降观测;沉降测量
settlement meter 沉陷计;沉降(测量)计
settlement method of transportation income 运输进款清算办法
settlement modulus 沉降模量
settlement movement 沉降移动
settlement observation 垂直位移测量【岩】;沉降观测;沉降观测
settlement observation platform 沉降观测平台
settlement observation point 沉降观测点
settlement of accident 事故处理
settlement of account 账款清单;结账
settlement of account by letter of credit 信用证结算
settlement of account for the latter half period 后期决算
settlement of action 诉讼和解
settlement of aggregate (新浇混凝土或砂浆在初凝前的)骨料下沉;集料下沉
settlement of arch crown 拱顶下沉
settlement of a structure 结构的沉降
settlement of balance 结清余额;结清差额
settlement of ballast 渣床沉降
settlement of ballast bed 道床下沉
settlement of building 建筑物沉降
settlement of claims 理赔部;理赔
settlement of collecting consignment 委托收款结算
settlement of concrete slabs 混凝土路面沉降
settlement of debt 清偿债务
settlement of disputes 纠纷处理;争议的解决
settlement of embankment 路堤沉陷;路堤沉降
settlement of exchange 结汇
settlement of fill 填土沉陷;填方沉降
settlement of foundation 地基沉陷(量);地基沉降(量)
settlement of ground 地基沉降;地基下沉;地基沉陷
settlement of income from self-adminstrated district 管内现收清算
settlement of income from through transportation 直通作业清算
settlement of individual pile 单桩沉降
settlement of influence factor 沉降影响系数
settlement of loss 赔付损失
settlement of pile foundation 桩基沉降
settlement of roadbed 道床沉降
settlement of services supplied to each other 互补劳动相互清算
settlement of sewage 污水沉淀
settlement of soil 土壤沉陷;土的沉陷
settlement of structure 结构物沉降;建筑物沉降
settlement of supports 支撑下沉;支架下沉;支座下沉;支柱下沉;支座沉降
settlement of tax 税款清算
settlement of tripod 三脚架下沉
settlement of underpinning pile 托换桩下沉
settlement of walling 墙身沉陷
settlement of wharf 码头沉降量

settlement on account 记账结算
settlement option 结算方式选择权;清算方式选择权;赔款选择保险
settlement out of court 法院外和解
settlement period 清算期间
settlement planning 居住地规划;村落规划
settlement plate 下沉板;沉载板;沉降观测平板;沉降板
settlement platform 沉降(观测)平台
settlement plug 沉降观测点
settlement point 沉陷测点;沉降(观测)点
settlement prediction 沉陷预测;沉降预测;沉降预报
settlement prediction technique 沉降预估方法
settlement-pressure curve 沉降量—压力曲线
settlement-pressure plot 沉降量—压力图
settlement prevention 防沉措施
settlement price 结算价格;清算价格
settlement probe 沉降探测头
settlement problem 沉降问题
settlement process 沉降过程
settlement range 沉降(观测)范围
settlement rate 沉陷率;沉降(速)率
settlement ratio 沉降比
settlement record 沉降记录
settlement risk 清算风险
settlement separate 沉降分离
settlement shrinkage 沉降收缩
settlement similarity 沉降相似
settlement space 沉降空间
settlement statement 房地产销售说明
settlement strain 沉降引起的应变
settlement stress 沉陷应力;沉降应力;沉降引起的应力
settlement tank 沉淀池;沉降池
settlement terms 决算期;清算条件
settlement test 沉降试验;沉淀试验
settlement test for emulsified asphalt 乳化沥青沉淀试验
settlement through accounts 非现金结算
settlement-time curve 沉降时间曲线
settlement total 总沉降
settlement variation 清算变量
settlement volume 沉降量
settlement within quota 限额结算
settlement with retiring partner 退伙人股份清偿
settle other's account 垫账
settle out 稳定下来;降落下来;沉淀出来;沉析
settle out of the court 庭外和解
settle-out time 进入终态时间
settler 增稠剂;沉淀器;澄清槽;沉淀器;沉降槽;沉淀器;财产授予人
settler clarifier 斜板沉淀池
settle sand tank 沉淀沙池
settle tank 澄清槽;沉淀槽
settle the dispute 解决争议
settle time 稳定时间
settle up 付清
settle with creditors 与债权人清算账目
settling 沉积;沉淀;沉底
settling accelerator 加速沉淀池;沉淀加速器;沉淀促凝剂
settling accelerator head tank 助沉剂高位槽
settling account 结算
settling agent 索赔代理人;助沉剂;理赔代理人
settling aids agent 助沉淀剂
settling allowance 安置补助费
settling analysis 沉速分析;沉降分析(法)
settling annulus 环形沉降段
settling apparatus 沉淀装置;沉淀器
settling area 沉降面积;沉降截面
settling bank 结算银行
settling basin 沉砂井;滤水池;沉降池;沉积盆地;沉淀(沙)池
settling bath 沉淀槽
settling bowl 沉淀池
settling box 沉淀箱;沉淀槽
settling box dust eliminator 沉降除尘器
settling capability 沉降性能
settling capacity 沉淀能力
settling centrifugal pump 沉降式离心泵
settling centrifuge 沉降式离心机
settling chamber 沉淀井;沉淀池;降尘室;沉淀室;沉积室;沉淀室

settling characteristic 沉降特征;沉淀特性
settling characteristics of particles 颗粒沉降特性
settling characteristics of sludge 污泥沉降特性
settling classifier 沉降分级机
settling coefficient 沉降系数
settling column 沉淀筒
settling compartment 沉淀隔间;分隔沉淀室
settling cone 锥形沉淀管;圆锥沉淀槽;沉降锥体;沉淀锥(体)
settling contour 沉降等值线
settling convection 沉降对流作用
settling curve 沉淀曲线
settling date 结算日期
settling day 结算日;交割日;清算日
settling down 沉降;沉入
settling drum 沉淀鼓
settling efficiency 沉降效率
settling equipment 沉淀装置
settling experiment 沉淀实验
settling fee 理赔费(用)
settling glass 沉淀玻璃锥(测可沉固体用);方玻璃片
settling height 沉降高度
settling in flowing water 动水沉降
settling joint 沉降缝
settling lag 沉降滞后
settling lagoon 沉淀湖
settling matter 沉降物质;沉降物;沉积物
settling method 沉降法;沉淀法;沉淀法
settling of particles 颗粒沉降
settling of sediment 泥沙沉降
settling of supports 支座下沉
settling out 沉(淀)出
settling out amount 析出量
settling particle 沉淀颗粒
settling performance 沉降性能
settling period 沉淀时间;沉积时间;沉降时间
settling pipe 沉降测管;沉淀管
settling pit 沉淀坑;沉淀池
settling plant 沉淀装置;澄清处理厂
settling pocket 沉降室
settling pond 沉积池;澄清池;沉淀池
settling position 稳定位置
settling process 沉降处理;沉淀处理;沉积法;沉淀法;沉降法
settling property 沉降性能;沉降特性
settling rate 沉降速度;沉淀速度;下沉速度;沉降速率;沉淀速率
settling ratio 下沉比;沉降系数;沉降比
settling regime 沉淀方式
settling reservoir 沉沙水库;沉淀水池
settling section 澄清段;沉降段
settling solid 可沉淀的颗粒;可沉淀固体;沉降性固体;沉淀物;沉淀固体
settling speed 沉降速度
settling system of mud 泥浆沉淀系统
settling tank 澄清箱;澄清桶;澄清器;沉沙箱;沉清槽;沉淀池;沉积槽;沉淀柜;沉淀池;沉淀槽
settling term 决算期
settling test 金属液脱氧度试验;沉降试验;沉淀试验
settling the account 结清账目
settling time 校正时间;建立稳定时间;稳定时间;沉降时间;沉淀时间
settling trap 沉淀井
settling tube 沉淀筒
settling up cost 清偿成本
settling vat 沉降桶;沉淀缸;沉淀桶
settling velocity 沉速;沉降速度;沉淀速度
settling velocity formula 沉速公式
settling velocity of sediment 泥沙沉速;泥沙沉降速度
settling velocity of sedimentation tank 沉淀池沉降速度
settling vessel 沉降器
settling vessel of mud 泥浆沉淀池
settling volume 沉淀体积
settling well 沉淀井;沉降井
settling zone 沉降区;沉淀区
settlor 财产授予;信托财产授予人;财产授予人
sett machine 小方石块机
sett-making 小方石块制作
sett-making machine 小方石制造机
set to exact size 调整到正确尺寸

sett of timber 一车组材
set-top box 机上盒;机顶盒
set-to-touch time 指触干时间
set to zero 置零;零调整;调到零(值)
sett-paved 铺小方石块的
sett-paved bank 铺石块的河岸;铺石块的边坡
sett-paved gutter 铺石块的流槽;铺石块的明沟
sett-paved road 小方石路;石块路;石砌道路
sett-paved strip 铺石块的简易机场
sett pavement 铺石路(面)
sett paving 小方石铺砌;条石路面;方块铺面层
sett paving joint pouring compound 灌注石块路面缝的混合物
sett paving mastic joint sealer 石块路面用玛碲脂填缝料
sett paving rammer 石块铺面夯实机
sett paving tamper 石块铺面夯实机
sett paving vibrator 石块铺面振动器
sett paving work 石块铺面工程;石块铺面工作
sett piling 桩钉
sett roughening machine 石块打毛机
sett tester 石块试验机
set tub 洗衣盆
set type 类型;集合类型
setulf 似槽痕
set union 集合并
set-up 机构;体型;体系;歇后增长;组装;装配调整;增水【水文】;建立;机构;编排;安置;安设;装置妥(当);开办;配套;调试;设立;变定
set-up a business 开业
set-up aids to navigation 航标设置
set-up a machine 调整机器;使用机器
set-up amplitude 顶托幅度
set-up box 装配箱
set-up circuitry 调定电路
set-up cost 调整费用;生产准备成本;设置成本
set-up derrick 起井架
set-up diagram 配置图;准备工作(框)图;设置图
set-up error 调定误差;安全误差
set-up in business 开始营业
set-up instrument 调定仪表;仪器布置
set-up man 调整机器设备的工人
set-up menu 安装菜单
set-up of drill 安装钻机
set-up of instrument 仪器安装;安置仪器
set-up of project 项目组成
set-upper limit 规定的上限
set-up procedure 准备程序;准备过程;装配程序;建立程序;调整步骤
set-up production 正规生产
set-up register 设定寄存器
set-up scale 无零标度
set-up scale instrument 无零点的仪表(刻度不是从零开始的仪表)
set-up scale meter 无零点刻度仪表
set-up sheet 装配图表
set-up spring 调整弹簧;预压弹簧;起压弹簧
set-up stress 建立应力
set-up the finishing mill 调整精整机
set-up the tool joint 固定接头连接
set-up time 安装时间;准备时间;装工具时间;展开时间;生产准备时间;设置时间
set-up unit 装定机构
set-up wheel 研磨整形砂轮
set value 规定值;假定值
set-value adjuster 设定值调整器
set-valued mapping 集值映射
set value of controlled variable 受控变量给定值
set water 凝结水
set weight (钻头上金刚石的)镶嵌重量
set with hard alloy 硬合金补强
set-work 摇尺机构;板条墙双层上抹灰;板条上抹灰
set wrench 止动螺钉扳手
set yielding 残余形变;残余变形
seven barks 八仙花
seven-bar pattern 日字形七划图形
seven-day strength 七天强度
seven-domed pilgrimage church 七圆顶朝圣教堂
seven-eighth rule 八分之七定则
seven-moment equation 七弯矩方程
seven place logarithms 七位对数
seven-ply door (心板两面为夹芯板的)七层门
seven points moving average 七点移动平均

seven-purlin(e)beam 七架梁
seven-roll Abramsen miachine 艾布拉姆森七辊矫直机
seven seas 全世界各大洋
seven-step ped pyramid 七台阶方锥体;七台阶金字塔
seven-tenth rule 十分之七定则
seven-third rule 三分之七定则
seven-unit code 七单元码
seven-wire 七股钢丝
seven-wire strand 七丝钢绞线;七股钢绞索
Seven Wonders of Antiquity 古代世界七大奇迹
seven wreaths of short turret 七小塔花饰山墙(式样)
sever 分隔;切断
severability clause 合同终止条款
severable contract 可分开合同
several complex variables 多复变数
several fold 几倍
several-for-one 多对一
several generations 几代
several hundred kilometers 数百公里
several-layer solenoid 多层螺线管
several liability 各自责任;负连带责任的债务;分担的债;分别债务
severally liability 单独的责任;分别负责
several phenotypes 一些表现型
several searchers 多个搜索者
severalty 个人产业;(土地的)单独所有;个人保有权
severance and conservation taxes 开采和保护税
severance felling 离伐
severance pay 离职金;解雇金;解雇费;退职金
severance tax 采掘税
severe atmosphere 侵蚀性气氛;恶劣环境
severe attack 严重侵蚀
severe avalanche 强烈崩坍
severe climate 严寒气候
severe cold 严寒
severe competition 激烈竞争
severe contamination 严重污染;重污染
severe damage 严重破坏
severe destruction 严重毁坏
severe drought 严重干旱
severed spur channel 截岭河道
severe duty 重(负)载;繁重的工作
severe duty drive 重载传动
severe duty test 重载试验;在恶劣条件下试验
severe earthquake 强烈地震
severe environment computer 抗恶劣环境计算机
severe error 严重错误
severe eutrophication condition 严重富营养化状况
severe eutrophic water body 严重富营养化水体
severe flood 特大洪水
severe frost 严霜
severe heat 酷暑
severe heating 急剧加热
severe intrusion 严重入侵
severe jolt 强烈振动
severe knock 强烈地震;强烈爆震
severely affected plot 严重受影响的小区
severely classical architecture 纯古典建筑
severely errored cell block ratio 严重出错信元块比
severely errored second 严重误码秒
severely errored second ratio 严重差错秒比
severely humid 很湿的
severely impair 严重危害
severely restricted chemical 受到严格限制的化学品
severely swelling-shrinkage foundation 较严重胀缩地基
severe radiation belt 强辐射带
severe rugged environment 苛刻环境
severe scaling 严重剥落
severe service 繁重任务;繁重的业务
severe service condition 困难的运行条件
severe shortage 严重缺乏
severe storm 狂烈风暴;飓风;猛烈风暴;强烈风暴;大暴雨
severe-storm observation 强风暴观测
severe storm warning 猛烈风暴警报;烈风警报
severe strain 危险应变
severe stress 危险应力
severe style 紧凑的风格;简洁式样
severe test 严格试验;苛刻工作条件下试验;(建筑上的)危险考验

severe thunderstorm 猛烈雷暴
severe tolerance 严格容限
severe tropic(al)storm 强热带风暴;猛烈热带风暴
severe wear 严重磨损
severe weather 严酷天气;恶劣天气
severe winter 严冬
severing concrete 切开的混凝土
severity 严重性;严酷性;严重性;强烈程度
severity code 严重出错码;严格代码
severity factor 严重程度因子;强度系数
severity index 烈度指数;事故严重度指数;车祸严重度指数
severity of braking 制动强度
severity of fire 火灾严重性
severity of grind 研磨强度
severity of injuries 事故严重程度
severity of quenching 激冷程度
severity of sediment contamination 底泥污染强度
severity rate 严重事故率
severization 严格化
sever supply 切断供给
severy 穹顶分块;筒拱顶隔块;一开间;祭坛华盖;哥特式建筑穹顶的分隔间
severy bay 穹顶的分隔间(哥特式建筑);(哥特式建筑穹顶的)分隔间;壁跨
sevice 卷绳用的油麻绳
seving mallet 卷缠木槌
Sevres blue 塞夫勒蓝
sewage 污水;阴沟水
sewage aeration 污水曝气
sewage amount 污水量
sewage analysis 下水分析;污水分析
sewage bacteria 污水细菌
sewage batch 污水的一次混合量
sewage biology 污水生物学
sewage biolysis 污水生物分解
sewage boat 垃圾船;排污船
sewage capacity 污水容量
sewage characteristic 污水特征;污水特性
sewage charges 污水费
sewage chlorination 污水氯化(作用);污水加氯(消毒)
sewage clarification 污水净化;污水澄清;污染澄清
sewage clarifier 污水净化池
sewage collection 污水汇集;污水聚集;污水收集
sewage collection system 污水汇集系统
sewage composition 污水成分
sewage condition 污水条件;污水状况
sewage conduit 污水总管;污水管道
sewage conduit system 污水管道系统
sewage construction 污水建筑物
sewage contamination 污水污染
sewage control well 污水检查井
sewage decolouration 污水脱色
sewage decomposition 污水分解
sewage detention tank 污水滞留沉淀池
sewage digestion 污水消化(作用)
sewage digestion tower 污水消化塔
sewage discharge 下水排水;污水排放;污水流量
sewage discharge in harbo(u)r 污水海湾排放
sewage discharge in river central-line 污水江心排放
sewage discharge pipe 污水排放管;污水排出管
sewage discharge standard 污水排放标准;废水排放标准
sewage discharge tube 污水排放管;污水排出管
sewage discharging diffusion pipe 污水排放扩散管
sewage discharging station 污水排泄站
sewage disinfection 污水消毒
sewage disposal 污水处理;污水处理;污水排除
sewage disposal area 污水处理区
sewage disposal facility 污水处理设施
sewage disposal plant 污水处理场;污水处理厂
sewage disposal process 污水处置过程;污水处理过程;污水处理工艺
sewage disposal pump 污水泵
sewage disposal standard 废水排放标准
sewage disposal system 污水处置系统;污水处理系统
sewage disposal system by soil 污水土地处理系统
sewage disposal vessel 污水处理船
sewage disposal works 污水处理工程
sewage distribution 污水系统分布

sewage distributor 污水喷洒机
sewage district 污水区
sewage dose 污水的一次混合量
sewage drain 下水道;排污管;排污沟
sewage drainage standard 排污标准;污水排放标准
sewage ecology 污水生态学
sewage effluent 污水溢出物;净化污水;污水出流;污水处理后出水;污水处理厂出水
sewage effluent quality 污水处理后出水水质
sewage effluent standard 污水排放标准
sewage ejector 排污喷射器;污水射流泵;污水喷射器;污水排出器
sewage ejector submersible pump 喷射式潜入污水泵
sewage electrolytic treatment 污水电解处理法
sewage engineer 污水工程师
sewage engineering 污水工程
sewage examination 污水检验;污水检测
sewage extraction method 污水萃取法
sewage facility 污水设施;污水装置
sewage farm 用污水灌溉的农田;污水田;污水灌溉(农)田;污水地段;污水处理场
sewage farming 灌溉田【给】;污水养殖;污水灌溉
sewage farming of land treatment 污水灌溉法;污水灌溉处理
sewage field 污水区
sewage film 污水膜
sewage filter 污水滤器;污水滤池
sewage filtration 污水(过)滤池
sewage final settling basin 污水最终沉降池;污水最终沉淀池
sewage final settling tank 污水最终沉降池;污水最终沉降槽;污水最终沉淀池
sewage fish culture 污水养鱼
sewage fish pool 污水池;污水养鱼塘
sewage flooding irrigation 污水漫灌
sewage flow 污水流(量);污水量
sewage flow meter 污水流量计
sewage flow rate 污水流量
sewage flow volume 污水流率
sewage flushing 污水冲洗
sewage fly 污水蝇;生物滤池灰绳
sewage force main 压力污水管
sewage from water closet 水厕污水
sewage fungus 污水真菌;污水菌
sewage gallery 污水坑道
sewage gas 沼气;垃圾沼气;污水(分解)气体;污泥气
sewage grease 污水油脂
sewage grit 下水淤渣;下水道淤渣;污水淤渣
sewage grit chamber 污水沉沙室;污水沉沙池
sewage handling pump 污水泵
sewage hoisting pumping station 排水提升泵站
sewage hose 污水软管
sewage inflow 污水流入量
sewage intake basin 污水引水池
sewage interceptor 污水截留井
sewage investigation 污水调查
sewage irrigation 污水灌溉;污灌
sewage irrigation by sprinkler 污水人工降雨灌溉
sewage irrigation norm 污水灌溉定额
sewage irrigation region 污灌区
sewage irrigation system 污水灌溉系统
sewage joining well 污水连接井
sewage lagoon 污水氧化塘;污水塘;污水池
sewage land treatment 污水土地处理
sewage law 污染法
sewage layer 污水层
sewage legislation 污水立法;污水法规
sewage lifting pump station 污水提升泵站
sewage lift station 污水提升站
sewage line 下水道;污水管道
sewage load(ing) 污水负荷量;污水浓度;废负荷
sewage lorry 污水运输车
sewage main collecting pipe 污水截流干管
sewage management 污水管理
sewage management district 污水管理区
sewage manhole 污水人孔
sewage mechanical treatment 污水机械(法)处理
sewage of strong concentration 高浓度污水
sewage outfall 排污口;污水流出口
sewage outfall tunnel 排污口隧洞
sewage outlet 排污口;污水排放口
sewage outlet area 污水泄水区

sewage outlet water 污水厂出水
sewage oxidation 污水氧化
sewage oxidation basin 污水氧化塘
sewage oxidation pond 污水氧化塘;污水氧化池
sewage particle 污水颗粒
sewage particulate 污水中的粒子;污水颗粒
sewage per capita per day 每人每日污水量
sewage per head per day 每人每日污水量
sewage pipe 污水管
sewage pipeline 污水管线
sewage pipe trench 污水管沟槽
sewage piping 污水管
sewage pit 污水坑
sewage plant 污水处理厂
sewage plant effluent 污水(处理厂)出水
Sewage Plant Manufacturing Association 污水处理厂厂主协会
sewage plume 污水卷流
sewage pollution 污水污染
sewage pollution value 污水负荷;污水污染值
sewage pond 污水池;污水塘
sewage practice 污水实践;污水实施
sewage preliminary basin 污水预沉池;污水初次沉淀池
sewage preliminary settling tank 污水初次沉淀池;污水初次沉淀槽
sewage preliminary tank 污水预沉池
sewage pretreatment 污水预处理
sewage pump 排泄泵;排污泵;污水泵
sewage pump house 污水泵房
sewage pumping plant 污水泵站
sewage pumping room 污水泵房
sewage pumping station 排污泵站;污水泵站
sewage pump plant 污水泵房
sewage purification 污水处理;污水净化
sewage purification plant 污水处理厂;污水净化厂
sewage purifier 污水净化池;污水净化设备
sewage quality 污水水质
sewage quantity 污水量
sewage raising plant 污水提升装置
sewage rate 污水流率;污水流量;污水费率
sewage recharge 污水补给
sewage recirculating pump 污水循环泵
sewage relift station 污水提升站
sewage reservoir 污水池
sewage reuse 废水回用
sewage ridge and furrow irrigation 污水垄沟灌溉
sewage river sediment 排污河道底泥
sewage sample 污水水样
sewage sampling method 污水采样方法
sewage scheme 污水方案;污水规划
sewage screen 污水隔筛;污水网筛;污水筛网;污水滤网
sewage shore discharge 污水岸边排放
sewage sink 污水渗坑;污水盆
sewage sludge 污水处理厂的污泥;下水道污泥;污水污泥
sewage sludge application 污水污泥施加负荷
sewage sludge culture 污水污泥培养
sewage sludge dewatering 污水机械脱水
sewage sludge digestion 污水污泥消化
sewage sludge disposal 污水污泥处置;污水污泥处理;污泥排放
sewage sludge drying 下水污泥干化;污水污泥干化;水污泥干化
sewage sludge drying bed 污水污泥干燥床;污水污泥干化床
sewage sludge dumping 污水污泥清除
sewage sludge gas 污水污泥气体
sewage sludge incineration 污水污泥焚烧;污泥焚化;污泥焚烧;污泥焚化
sewage sludge treatment 污水污泥处理
sewage sludge washing 污水污泥洗涤;污水污泥淘洗;污泥冲洗
sewage sluice valve 污水闸阀
sewage solid 污水沉淀物
sewage solid extractor 泥渣抽出器;污水渣滓抽出器;污水渣抽出器
sewage specimen 污水试样;污水采样
sewage sprinkling 污水喷洒
sewage sterilization 污水消毒
sewage stock pond 污水储[贮]存池
sewage stream 污水水系;污水流
sewage strength 污水浓度

sewage sump 污水坑
sewage system 下水道系统;污水(管道)系统
sewage system overflow 下水道系统溢流
sewage tank 污水沉淀池;化污池;化粪柜;化粪池;排污池;污水柜;污水池;污水沉淀柜;污水槽
sewage treated wastewater 污水处理过的废水
sewage treatment 污水处理(技术)
sewage treatment equipment 污水处理设备
sewage treatment level 污水处理水平
sewage treatment method 污水处理法
sewage treatment optimization model 污水处理优化模型
sewage treatment plant 污水处理场;污水处理装置;污水处理厂
sewage treatment plant design 污水处理厂设计
sewage treatment plant discharge 污水处理厂排放
sewage treatment plant effluent 污水处理厂出水
sewage treatment plant simulation model 污水处理厂模拟模型
sewage treatment plant sludge 污水处理厂污泥
sewage treatment process 污水处理工艺;污水处理过程
sewage treatment residue 污水处理余渣
sewage treatment station 污水处理站
sewage treatment structure 污水处理结构;污水处理建筑物;污水处理构筑物
sewage treatment system 污水处理系统
sewage treatment tank 污水处理池
sewage treatment technological process 污水处理工艺流程
sewage treatment works 污水处理厂;污水处理工程
sewage tunnel 污水隧道;污水坑道
sewage ultra-filtration 污水超滤
sewage utilization 污水利用
sewage utilization act 污水利用条例;污水利用法
sewage ventilation 污水通风
sewage vessel 垃圾船;排污船
sewage water 污水
sewage water pump 污水泵
sewage water recycling 污水重复利用
sewage water valve 排水阀;排污阀
sewage works 下水道工程;污水(处理)厂;污水工程
sewage works effluent 污水处理厂出水
sewed up 退潮时船搁浅
sewer 排水管渠;下水管;下水道;排污水;污水道;污水管道
sewerage 下水道工程;排水工程
sewerage analysis 污水分析
sewerage barge 污水驳
sewerage clarifier 污水净化池
sewerage conduit 污水管道
sewerage culvert 排水隧洞;排水隧道
sewerage design practice 排水工程设计实践
sawerage district 下水道区域
sewerage drawing 污水图
sewerage dredge(r) 挖槽机;挖泥机;挖沟机
sewerage ejector 排污喷射器
sewerage engineering 下水道工程学;排水工程(学);污水工程
sewerage farm 污水灌溉
sewerage filter 污水(过)滤器;污水(过)滤池
sewerage force 排水压力
sewerage force main 排水压力干管
sewerage gas 下水道气体
sewerage grease 污水油脂
sewerage grit 下水道淤渣
sewerage implementation project 污水治理工程
sewerage installations 排水设施
sewerage lagoon 污水池
sewerage law 污水工程法;敷设下水道法规
sewerage load(ing) 污水负荷量
sewerage of separate system 分流制下水道
sewerage overflow 下水道溢流
sewerage pollution 污水污染
sewerage pollution value 污水污染值
sewerage pumping house 排水泵房
sewerage station 废水排放场
sewerage system 下水管道系统;下水道(管道)系统;下水道工程系统;排水系统;排水体制;污水(管道)系统;污水工程系统
sewerage system layout 排水系统布置
sewerage system of radial patterns 辐射式下水道系统

sewerage system overfall 下水道系统溢流
sewerage system overflow 下水道溢流
sewerage territory 下水道区
sewerage treatment works 污水处理厂
sewerage tunnel 排水隧洞;排水隧道
sewerage utilization 废水利用
sewerage ventilation 下水道通风
sewerage works 污水工程;污水(处理)厂
sewer alignment 管道方向
sewer alignment design 管道定线设计
sewer appurtenances 污水系统附属设备;管道附属设备;污水道附属物;下水道附属物
sewer arch 下水道拱顶;污水道拱顶
sewer article 下水道制品
Sewer Authority 污水管理局
sewer bottom 下水道基础;污水管基础
sewer branch 污水支管
sewer brick 下水道(用)砖;阴沟砖
sewer capacity 下水(管)道排水能力;下水道容量;污水管容量;地下水道排水能力
sewer catch basin 截泥井;集污面积;下水道窨井;下水道沉泥井;阴沟沉泥井;污泥井;沉泥井
sewer chimney 污水竖管;跌水井竖管
sewer cleaner 清沟机
sewer cleaning 下水道清洗;下水道清除;污水管清洗
sewer cleaning device 下水道清洗设备
sewer cleaning sludge 从下水道清出的污泥
sewer connection (公路下的)私人下水管道;下水道支管;管道连接
sewer construction 下水道建筑物;污水管施工
sewer corrosion 管道腐蚀
sewer cradle 管道护脚;沟管基座
sewer crossing 污水过渡段
sewer culvert 污水涵管
sewer cutoff 阴沟隔断
sewer design 下水道设计
sewer distribution line 污水线网
sewer district 下水道区(域)
sewer downstream 下游管道
sewer duct 污水输送管
sewer evacuator 污水管(道)排泄设备;污水道排泄设备
sewer facility 污水管设施
sewer fitting 污水管件
sewer flow 污水流
sewer flusher 下水道冲洗器
sewer flushing 管道冲洗;污水管冲洗
sewer for combined foul and surface water 污水及雨水两用排水管;污水及雨水混合排水管
sewer foundation 管道基础
sewer gallery 污水坑道
sewer gas 沟道气;污水管沼气;管道气体;污水道(秽)气
sewer goods 排水管制品
sewer grade 下水道坡度
sewer hog 挖沟机
sewer inlet 阴沟进口;污水管进口
sewer inspection 污水管检查
sewer interchange 下水道互通式交叉
sewer invert 下水道内底;下水道倒拱
sewer jointing compound 下水道接头油灰
sewer junction 管道交汇
sewer layout plan 阴沟分布图
sewer leakage 管道渗漏
sewer line 污水管道;下水管道;下水管线;下水管;污水管线
sewer line connection 排水管道连接
sewer line construction 排水管道施工;排水管道建筑
sewer line trench 排水管道沟槽
sewer main 总排水管;主下水管道;污水总管;污水干管
sewer maintenance 污水管维修;污水管保养
sewer man 下水道工人
sewer manhole 污水管检查井;污水窨井;进人孔;下水道进人孔;下水道检查口;下水道检查井;下水道人孔;下水管检查孔;污水道检查井
sewer manway 污水检查井;污水窨井
sewer map 下水道图
sewer net(work) 污水管网
sewer ordinance 污水条例;污水管条例
sewer orifice 污水管出口

sewer outfall 污水管出口;污水道排出口;下水道排水口
sewer outflow 管道出流
sewer outlet 污水排泄口;污水管出口
sewer patrolling 巡查下水道
sewer pill 通沟木牛;通沟木球;污水管清管球
sewer pipe(drain) 污水排泄管;沟管;排水管;下水管;下水道管;下水道;污水管
sewer pipe press 污水管压制机
sewer plank 阴沟方木
sewer product 排水管制品
sewer rat 下水道清理工
sewer rate 污水管流率
sewer-rehabilitation alternatives 污水管修建方案
sewer rod 通阴沟条;通沟条
sewer saddle 污水管鞍座
sewer scouring 污水管冲刷
sewer section 下水道截面
sewer sediment 污水管道沉积物
sewer septicity 沟渠腐化力
sewershed boundary 排污管域边界
sewershed hydrology 排污管域水文学
sewershed level 排污管域范围
sewer size 下水道尺寸
sewer slime 污水管黏垢
sewer slope 下水道坡度
sewer sludge 下水道污泥;通沟污泥
sewer sluice valve 污水管闸阀;污水管冲洗阀
sewer stoneware 陶器下水道;下水道用陶器
sewer syndrome 下水道综合征
sewer system 排水系统;下水道系统;下水道(管)道系统;排水制(度);污水管(道)系统;污水道系统
sewer territory 下水道区
sewer tile 污水(瓦)管
sewer trap 下水道防臭弯管;污水管存水弯
sewer trench 污水沟槽;壕沟【给】;污水沟
sewer trunk 污水干线
sewer tunnel 污水管沟;污水(管)隧道;下水隧洞;污下水道隧洞;污水隧洞
sewer upstream 上游管道
sewer utility 污水公用事业;污染公用事业
sewer ventilation 管道通风
sewer zone 下水道地区;下水道范围
sewgee 苛性硷水
sewing 串线订
sewing needle 缝合针
sewing pole 缝杆
sewing-press 锁线装订机
sewing room 缝纫室
sewing thread 缝纫线
sewing thread on cone 宝塔线
sew-knitting machine 缝编机
sewn bottom 缝合底
sewn building mat 缝制建筑保温毡
sewn building quilt 缝制建筑保温毡
sewn kraft paper sack 机缝牛皮纸袋
sexadecimal digit 十六进制数位
sexadentate 六配位体
sexagesimal 六十进制的
sexagesimal circle 六十分制度盘;六十度制度盘
sexagesimal counting table 六十进制计数表
sexagesimal cycle 六十干支周
sexagesimal division 六十分制分度
sexagesimal graduation 六十分制刻度
sexagesimal measure 六十进制度量
sexagesimal measure of angle 六十进制角度度量
sexagesimal system 六十分制
sexangle 六边形;六角形
sexangular 六角(的);六边形的
sexcentenary 六百年
sex dimorphism 两性异形
sexfoil 六叶形(装饰图案);六叶饰
sexfoil window 六叶形窗;六叶(式)窗;六瓣叶窗
sexiphenyl 联六苯
sexless connector 等同接合连接器
sexpartite 分成六部分的
sexpartite 六肋拱穹
sexpartite rib(bed)vault 六肋拱顶
sexpartite vault 六肋拱穹顶
sex-separation for bedrooms 大子女分室
sex structure of urban population 城市人口性别构成
sextant 六分仪(星座);六分周;纪限仪

sextant adjustment 六分仪校正;六分仪调准
sextant altitude 六分仪高度
sextant altitude correction 六分仪高度修正;六分仪高度校正
sextant angle 六分仪测角定位
sextant chart 六分仪后方交会计算图
sextant error 六分仪误差
sextant imperfection 六分仪缺陷
sextant rangefinder 六分仪式测距仪;六分仪式测距器
sextant reading 六分仪读数
sextet 六重线
sextile 六十度角距
sextile aspect 六分(相互)方位
sexto 六开本
sextodecimo 十六开本
sextuple 六重;六倍
sextuple denotation counter 六进制计数器
sextuple effect evapo(u)rator 六效蒸发器
sextuple-flow turbine 六排汽口汽轮机
sextuple space 六维空间
sextuple star 六合星
sextuple tracks 六线铁路
sextupole 六极
sexuality 性别
sexual propagation 有性繁殖
sexual reproduction 有性繁殖
Seyfert rectifier 赛弗特整流机
Seyler's classification 赛勒分类法
Seyssel 克拉里奇地沥青(产于法国图罗山)
Sezession 赛泽恣(奥地利新艺术的变异)
Sezession style 分离派(建筑)
S factor 修正率
sferics 远程雷电;天气测定法;天电(测定法);风暴电子探测器;大气干扰
sferics fix 天电定位
sferix 低频天电
SF suspension preheater SF 型悬浮预热器
sgabello 矮木椅
sgare chain 面积尺
S glass S 玻璃
sgraffi(a)to 釉雕(陶瓷);透雕型粉刷;彩色露地拉毛粉饰;刮花;五彩拉毛陶瓷;五彩拉毛粉刷;仿雕刻装修
sgraffi(a)to material 釉雕材料
sgraffi(a)to tile 彩色毛面瓷砖
shabynite 水氯硼镁石
shack 木造小房;茅屋;棚子;棚屋;窝棚
shack area 棚户区
shack-dwellers 棚户
shack hole 沉陷灰岩坑
shackle 卸扣;钩键;钩环;链节;联结环;扣环;卡环;束槽;带销U形环;带销U形钩;穿心绝缘子
shackle bar 车钩;拔钉器
shackle block 卸扣滑车;带钩环滑车
shackle bolt 联结环上的螺栓;卸扣栓;钩环螺栓;连杆螺栓
shackle bracket 钩环架
shackle crow 撬前扣栓的棍
shackle for pontoon slings 浮筒索套卡环
shackle hook 大钩;联结环上的钩
shackle insulator 茶台绝缘子
shackle insulator 锁扣绝缘子;穿心绝缘子;茶托绝缘子
shackle joint 钩环接头;扣环联结
shackle marks 锚链节标志
shackle meter 链节示数器
shackle of cable 链条卸扣;链节
shackle of coupling 联轴器钩环;连接器钩环
shackle pin 钩环销;链销
shackle rod 连接杆
shackle stud 钩环柱螺栓
shackle swivel hook 提引水龙头
shack up 租房暂住
shadbolt 扇形窗拉杆;楣窗拉杆
shaddock 柚(子)
shade 阴影;阴处;遮罩;遮阴棚;遮篷;遮光物;遮光板;遮光(窗)帘;遮蔽;色荫;色灰度;色光调;彩色深浅
shade and shadow 光影;阴与影;投影画
shade area 雷达盲区
shadeband 遮阳带
shade-bearer 阴性树
shade-bearing tree 阴性树

shade bears 耐阴性
shade card 色卡
shade cloth 窗帘布
shade colo(u)r 暗色
shade curve 浓淡曲线
shaded area 阴影面积;普染面积
shaded computer graphics 浓淡计算机图形
shade deck 遮阳板
shade deck vessel 遮阳甲板船
shaded effect 阴影色调效果;遮阳效果;遮光效果
shade demanding 喜阴的
shade density 郁闭度;遮阴度
shade design 浓淡设计
shaded glaze 晕色釉
shaded lamp 装罩灯
shaded picture 浓淡图;渲面
shaded pole 罩极;屏蔽极
shaded pole motor 罩极(式)电动机
shaded region 增长密区
shaded relief 晕渲地貌;地貌晕渲
shaded-relief edition 地貌晕渲版
shaded-relief map 晕渲地形图;地貌晕渲图
shaded sandblast 着色喷砂
shaded transducer 束控换能器
shaded walk 遮阳步行道
shaded windscreen 着色挡风玻璃
shaded windshield 着色挡风玻璃
shade endurance 耐阴性
shade enduring plant 庭荫树
shade error 阴影误差;色镜误差
shade factor 增长因数
shade glass 遮光玻璃;色镜
shade holder 灯罩夹;灯罩座
shade line 遮线;影线;阴影线
shade lining 地貌晕渲表示法
shade-loving plant 喜阴植物
shade of gray[grey] 灰色深浅度;灰度
shade paint 遮光漆
shade pine 甜松
shade plant 阴生植物;阴性植物
shade planting 绿荫栽植
shade roof 遮阳棚
shades 遮光帘
shades and shadow 阴影学;投影画
shade screen 遮阳帘;遮阳篷
shade serum 阴地植物
shade-shed 凉棚;凉棚
shades of gray 灰度梯度
shade species 阴性树种
shade tolerance 耐阴性
shade-tolerant plant 耐荫植物
shade tolerant tree 浓荫树;庭荫树
shade tree 成荫树木;行道树;阴性树;绿荫树
shade tree section 绿荫区
shade tube 遮光筒
shading 少量的折扣;细微差别;荫影;遮阳;遮光;遮蔽;明暗法;描影法;描影;调整色光;束控
shading adjustment 噪声电平调整
shading amplifier 寄生信号放大器;黑点放大器
shading and blending 明暗融合
shading coefficient 遮光系数;遮光率;遮阳系数
shading coil 荫蔽环;罩蔽线圈;校正线圈;屏蔽线圈
shading coil starting method 短路环起动法
shading correcting 荫影校正
shading correction 黑点校正;图像斑点调整
shading cure 遮盖养护
shading data 晕渲数据
shading error 图像暗影
shading of colo(u)r 色调
shading pattern 阴影图形
shading-pole 罩极;磁极屏蔽
shading-pole type relay 短路环继电器
shading program(me) 晕渲程序
shading ring 荫蔽环;罩极环;屏蔽环
shading signal 黑斑信号
shading soil 遮被土壤
shading system 地貌晕线法;地貌晕翁法
shading wedge 屏蔽楔
shading with oblique lighting 斜照晕宣法
shading with zenithal lighting 直照晕渲法
shading yellow 暗黄
shadlunite 硫铁铅矿
shadoof 吊杆
shadow 影子;阴影;屏蔽保护范围;暗影

shadowal block 有凹槽混凝土块
shadow angle 影锥角;荫影角;阴影角
shadow area 阴影区;投影面积
shadow attenuation 阴影衰减;阴影区衰减
shadow baffle wall 挡火墙
shadow bands 影带
shadow bar 投影杆
shadow blue glaze porcelain 青白瓷
shadow boundary 阴影边界
shadow box 玻璃盖匣
shadow building 未经精密设计的造船
shadow casting 投影电子显微术
shadow casting technique 投影技术
shadow channel 影道
shadow cone 影锥;阴影锥
shadow corporation 影子公司
shadow correction 阴影校正
shadow cylinder 阴影圆柱
shadow details 暗调层次
shadow director 幕后董事
shadow disc 影子盘
shadow dot 暗点
shadow effect 影子效应;荫影效应;阴影效应;屏蔽效应
shadow exchange rate 影子汇率
shadow factor 影蔽因数;荫蔽因数;阴影因子;阴影系数
shadow factory 伪装的军事工厂;隐蔽工厂
shadow filter 阴影滤波器
shadow foreign exchange rate 影子外汇率
shadow fringe 阴影条纹
shadow fringe test 阴影条纹试验
shadow gradation 暗调层次
shadowgraph 阴影图;逆光摄影;影子照片;影像图;阴影照相法;投影画;X 光照片
shadowgraph camera 阴影照相机
shadow graphing 阴影图解
shadowgraph scales 投射标度天平
shadowgraph system 阴影照相系统
shadowgraphy 影相术
shadow grid 阴影栅;镜像栅
shadow ground 荫蔽地区
shadow image 阴影像
shadowiness 朦胧状态;有阴影状态
shadowing 荫屏(作用);阴影;遮蔽
shadowing factor 阴影系数
shadowing method 阴影法
shadowing of crossing traffic 交叉车辆的避车安全
shadow interest rate 影子利率
shadowless illumination 无影照明
shadowless lamp 无影灯
shadowless lighting 无影照明
shadow line 光影划线
shadow line label 阴影线标
shadow loss 阴影损耗;屏蔽损耗
shadow-mark 透影
shadow mask 显像管荫罩;影孔板;阴影掩模;遮罩;遮蔽屏;障板;低调蒙片;彩色显像管阴罩
shadow-mask screen 影孔板式荧光屏;荫罩屏;阴影掩模屏
shadow mask tricolo(u)r tube 荫蔽三色管
shadow-mask tube 荫罩管;影孔板彩色管
shadow measurement 影测量法
shadow-meter 阴影仪
shadow microdiffraction 阴影微绕射
shadow microscope 影显微镜;阴影显微镜
shadow microscopy 阴影显微术
shadow minimum 影子极小
shadow-Moire method 影移法
shadow-Moire test 阴影莫阿试验
shadow of worm 虫体阴影
shadow page table 阴影页表
shadow pattern 阴影图
shadow pattern analysis 阴影模式分析
shadow peak 暗影最大值
shadow photometer 影像光度计;比影光度计
shadow picture 阴影照片
shadow price 虚价;预兆价格;影子价格;潜在价格
shadow projection microscope 阴影投影显微镜
shadow project method 影子工程法
shadow projector 投影检验器
shadow pyramid 阴影锥形
shadow RAM 影子内存
shadow region 影区;阴影区(域);静区

shadow relief tile 凹凸影瓷砖
shadow resource 阴影资源
shadow scattering 阴影散射
shadow section 阴影段
shadow shield 阴影屏蔽;遮(阳)光板;遮阳布棚
shadow stop 暗调光圈
shadow stripe 阴阳条织物
shadow-sunlight time 阴影日照时间
shadow surface 阴影面
shadow system 阴影仪
shadow table 影像表
shadow test 检影法;暗影试验
shadow tone 阴影色调
shadow-tree zone 遮阴影区
shadow-tuning indicator 阴影式调谐指示器
shadow value vector 影子价值向量
shadow wage 影子工资
shadow wage rate 影子工资率
shadow wall 断垣残壁;掩体墙
shadow weathering 影区风化
shadowy blue glaze porcelain 影青瓷
shadowy blue ware 影青瓷
shadow zone 影区;影带;阴影区;阴影地带;震波微弱带;屏蔽区
shady migmatite 阴影混合岩
shady slope 阴坡
shady structure 阴影状构造
shady walk 成荫的小路
Shaff 水柱;旋塞;水龙头
Shafferd cellar control gate 沙菲尔德防喷器
Shafferd flask 沙菲尔德采样瓶
Shaffer stroke function 舍菲尔环节函数
shaffle 桅杆上承座
shafranovite 沙水硅锰钠石
shaft 旋钮;炉胸;矿井;井筒;升降井
shaft adit 作业井施工法
shaft alignment 轴的对准;轴校正;轴对正
shaft alignment ga(u)ge 轴定心卡规;找正测微仪
shaft alley 轴隧;隧道(又称隧洞);地轴轴隧
shaft allowance 凿井加工留量
shaft and bell dimension 井身和大头尺寸
shaft and tunnel drainage 井巷排水
shaft and tunnel drainage method 井巷排水方法
shaft and tunnel spillway 喇叭形溢洪道;落底式溢水口;堕式溢水口
shaft angle 轴角
shaft arm 柄臂
shaft arm pin 轴臂销
shaft balancing 轴的平衡
shaft bar 钻杆;钎杆
shaft base 竖井基础;轴承底座
shaft-base system 基制制
shaft bearing 轴承;尾轴轴承
shaft bearing of upper tumbler 上导轮轴承
shaft bearing plate 轴承板
shaft bearing replacer 轴承拆卸器
shaft blank 轴坯
shaft-blind slope development 竖井—盲井联合开拓
shaft borer 竖井钻机
shaft boring 钻凿竖井;钻井;竖井钻进
shaft boring machine 钻井机;竖井钻进机
shaft boring method 竖井钻进法
shaft bossing 轴包套;轴包架;尾轴毂
shaft bottom 竖井底板;井底车场;竖坑底
shaft bottom map 井底车场平面图
shaft box 井筒附近的储[贮]仓
shaft box conveyor 轴箱传送带
shaft bracket 传动杆架;操作杆架
shaft bucket 井吊桶
shaft building 矿井建设;建井
shaft building survey 建井测量
shaft bushing 轴衬
shaft cable (输送工人及物品的)竖井钢索;竖坑缆
shaft cage 竖井罐笼
shaft cap 轴盖
shaft carrier 竖井运输机
shaft casing 轴套;轴壳
shaft center flange 辊轴法兰
shaft chimney 竖井通道
shaft cladding 井壁材料
shaft clevis 连接叉
shaft clinker cooler 立式熟料冷却器;立式熟料冷却机

shaft clip 轴籍
shaft collar 轴颈;轴环;井筒锁口盘;凸缘
shaft column 竖井导管;轴承支柱
shaft compartment 井筒隔间
shaft concentricity 轴同心度
shaft connection survey 竖井联系测量
shaft construction 竖井施工;竖井构造;竖井结构
shaft construction survey 立井施工测量;竖井施工测量
shaft contact 轴接点
shaft cooler 立式冷却器;立式冷却机
shaft counter 轴转速表;轴转数计
shaft coupling 联轴器;联轴节
shaft coupling side 联轴器侧
shaft-cover 竖管罩
shaft cover for water catching 回水竖管罩
shaft cross-section 竖井截面
shaft crucible furnace 竖筒坩埚炉
shaft crusher 井底破碎机
shaft-cup 轴盖
shaft-cup spring 轴盖簧
shaft current 轴电流;轴承电流
shaft current(circulating)protection 轴电流保护(装置)
shaft deepening 竖井延深
shaft-deepening survey 井筒延伸测量
shaft depth 采井深度
shaft depth measurement 竖井深度测量
shaft diameter 轴径
shaft digging method 沉井法
shaft displacement 轴的摆度
shaft drainage 竖井排水
shaft drilling 钻粒钻机钻井;钻井;竖井钻进
shaft drill jumbo 钻井机;凿井机;凿井钻车
shaft drive 轴转动;轴(的)传动装置;轴传动
shaft-drive motorcycle 轴动机器脚踏车
shaft-driven 轴驱动的
shaft-driven rotary 万向轴传动转盘
shaft-driven rotary table 轴传动回转台
shaft-driven travel(l)ing crane 轴动移动起重机;轴传动移动起重机
shaft during sinking 正在下沉的井筒
shaft eccentricity 轴偏心度
shafted impost 柱丛拱墩
shaft enclosure 竖井封闭
shaft encoder 转轴编码器;轴端编码盘;计数鼓
shaft end 轴端
shaft-end clearance 轴承端部间隙
shaft-end output 轴端出力
shaft excavation 开挖竖井;井筒掘进;竖井挖掘机;竖井开挖;竖井掘进;大竖井
shaft extension 轴外伸部;轴伸出
shaft extension type tubular turbine 轴伸贯流式水轮机
shaft fan 竖井通风机
shaft filler washer 轴垫圈
shaft filter 轴式过滤器
shaft fish lock 竖井式鱼闸
shaft fit 轴配合
shaft flange 轴法兰
shaft floor 竖井底板
shaft for cover close 机盖压紧轴;筒体紧固轴
shaft formwork 烟囱模板;竖井模板
shaft for retarding the flow 缓流井
shaft for risers 提升机井
shaft for stilling basins 缓流井
shaft foundation 管柱基础;井筒基础
shaft furnace 直井炉;高炉;竖窑;竖(式)炉;井式炉
shaft furnace smelting 鼓风炉熔炼
shaft gear 轴齿轮
shaft generator 轴(向)发电机;轴(向)发动机;轴传动发电机
shaft generator system 轴带发动机系统
shaft gland 轴密封盖;轴封;尾轴筒压盖
shaft governor 轴调速器
shaft grab 竖井抓岩机
shaft grave 竖穴坟墓
shaft guide 轴导承
shaft hammer 杠杆锤;冲击器
shaft hanger 轴架;挂轴器
shaft headframe 井塔;井架
shaft helical spring 轴螺旋弹簧
shaft hoist 竖井提升机;矿井提升机
shaft hole 尾轴孔(单推进器船)

shaft hopper 轴料料斗
shaft horse 轴向马力;轴向功率
shaft horsepower 轴马力;轴功率
shaft house 升降机架(矿井);竖井顶机房
shaft identification 钻进记号
shafting (中世纪建筑的)柱丛构造方式;轴系
shafting hanger 传动轴用吊架
shafting lathe 制轴车床
shafting of ship 船舶轴系
shafting oil 轴油;传动油
shaft inland lock 竖井闸
shaft inspection 检查井筒
shaft installing 轴套(密封用)
shaft installing sleeve 密封用轴套
shaft intake 竖井进水口
shaft job 开凿井筒
shaft-joint 联轴节
shaft jumbo 竖井钻架;竖井钻车;竖井挖掘机;竖井掘进机;竖井井架;盾构掘进机
shaft key 轴键
shaft kiln 立窑;立式烘炉;竖(式)窑
shaft kiln cement 竖窑水泥;立窑水泥
shaft kiln discharge grate 立窑卸料篦子
shaft kiln installation 立窑设备
shaft kiln lime 立窑石灰;立窑石灰
shaft kiln plant 立窑设备;立窑车间
shaft kiln stone 竖窑原料;立窑原料
shaft labyrinth 轴迷宫密封
shaft lagging 孔壁背板;井壁背板
shaft lathe 轴车床
shaft leakage 轴端漏泄
shaft lever 轴杆
shaft lime kiln 竖式石灰窑;立式石灰窑
shaft line 轴线
shaft liner 轴衬;炉身内衬
shaft lining 炉身里衬;井筒支护;井壁;竖井衬砌
shaft loading 轴荷载;竖井装载
shaft locating 轴定位
shaft lock 井式船闸;竖井式船闸
shaft locking 轴锁紧
shaft machine 轴类加工自动机
shaft masonry(work) 竖井圬工
shaft mastaba (古埃及一种长方形平顶斜坡的)竖穴坟墓
shaft material 轴材料
shaft mining 竖井采矿;竖井开采
shaft mixer 轴式混合机;轴式混合器;竖式混合器
shaft-mounted pinion 轴装小齿轮
shaft mucker 凿井装岩机
shaft mucking 凿井装岩
shaft navigation lock 竖井式船闸
shaft neck 轴颈
shaft nut 主轴螺帽
shaft of a hammer 锤柄
shaft of an axe 斧柄
shaft of chelicera 螯肢杆
shaft of column 柱身
shaft of rivet 铆钉体;铆钉杆
shaft ore box 井筒附近的矿仓
shaft orientation survey 竖井定向测量
shaft packing 轴封填料
shaft-packing leakage 轴封漏气
shaft partition wall 竖井隔板
shaft passage 地轴轴隧
shaft pillar 矿井柱
shaft pin 轴销
shaft plate 轴包板
shaft platform 竖井平台
shaft plumb bob (机械的)轴承垂球
shaft plumbing 竖井铅垂定位;投点;竖井垂线定向法
shaft plumbing tools 竖井垂线井下定向工具
shaft plumbing wire 竖井铅垂定位金属丝;竖井铅直线;竖井吊锤线;预应力高强度钢丝
shaft plumb line 矿井垂线
shaft plummet 竖井铅锤
shaft pocket 井底矿仓
shaft position digitizer 轴角模拟数字转换器;轴角模数转换器
shaft position encoder 轴角编码器
shaft position indicator 轴位移指示器
shaft position mechanism 调轴机构
shaft position-to-digital converter 轴位数字变换器
shaft potential 轴电位

shaft power 轴(输出)功率
shaft power station 竖井式电站
shaft preheater 立筒预热器
shaft preheater kiln 立筒预热器窑
shaft propeller 轴推进器
shaft pump 轴流泵;矿井泵
shaft pumping station 竖井(式)抽水站
shaft raising 向上凿井;向上打井;竖井反掘
shaft revolution 轴转速
shaft revolution indicator 转速指示器;轴转数计
shaft ring 柱环饰;柱身条形圆箍线脚
shaft ring machine 轴环机
shaft run-out 轴的径向跳动;轴的摆度
shaft screen 转筒筛;筒筛
shaft seal 轴的密封装置
shaft seal grease system 轴封润滑系统
shaft seal housing 轴封体
shaft sealing 竖井密封;轴封
shaft seal pump 轴封泵
shaft seat 竖井基座
shaft service 凿井作业
shaft shoulder 轴肩
shaft shuttering 竖井模板
shaft signaling 竖井信号;矿井提升信号
shaft signaling system 竖井信号系统
shaft sinker 凿井工
shaft sinking 沉井;向下打井;凿竖井;挖竖井;竖井沉陷;打直井;打竖井
shaft-sinking blasting 竖井开挖爆破
shaft-sinking by borehole 钻孔法凿井
shaft-sinking by caisson method 沉箱法凿井
shaft-sinking by drilling 用钻孔法沉井
shaft-sinking company 沉井公司
shaft-sinking drill 凿井(用)大钻机
shaft-sinking machine 井筒开凿机
shaft-sinking method 沉井法
shaft-sinking pump 沉井用泵;凿井吊桶;竖井抽水泵
shaft-sinking stage 沉井平台
shaft skip 竖井箕斗
shaft sleeve 大轴轴面
shaft spillway 直井式溢洪道;喇叭口式溢洪道;竖井(式)溢洪道
shaft spline 轴键
shaft stair(case) 竖井阶梯
shaft steering system 轴传动操舵系统
shaft straightener 轴整直器;轴矫直机
shaft straightening machine 轴整直机
shaft submergence 淹水
shaft support 竖井衬砌;竖井支撑;轴支架;轴的支架
shaft supporting column 炉身支柱
shaft surge tank 竖井调压水槽;竖井调压水箱
shaft survey 竖井测量
shaft survey instrument 竖井测量仪器
shaft tank furnace 井式池窑
shaft test 井探
shaft timbering 竖井支撑;坑井支撑
shaft tip 轴梢;轴端
shaft top 炉顶
shaft top arrangement 井口设施;井孔布置
shaft top works 井口构筑物
shaft torque 轴转矩
shaft tower 井塔
shaft transfer survey 矿井联系测量
shaft trunk 竖井;地轴轴隧
shaft tube 鞘管;尾轴管
shaft tunnel 地轴轴隧
shaft tunnel alley 轴隧;船尾轴隧
shaft turning gear 盘动装置
shaft type 轴式
shaft-type governor 轴调节器
shaft-type mixer 轴式搅拌机
shaft ventilation 竖井通风
shaft ventilation method 矿井通风方法
shaft vibration 轴振动
shaft wall 竖井墙壁;通风墙;炉身;孔壁;井壁
shaft wall system 通风墙系统;竖井墙系统
shaft washer 止推轴承外圈;紧圈
shaft way 竖井;井道(升降机);竖或斜过道
shaft well 竖井;井窄
shaft well digger 竖井挖掘机;竖井掘进机
shaft winding 竖井提升
shaft with blades 带叶片的轴

shaft with installed services 设有公用设施的竖井
shaft zoning ventilation 矿井分区通风
shag 粗毛;长绒
Shagara deep 沙加拉海渊
shag carpet 长绒地毯
shaggy 表面粗糙的
shaggy pattern 表面粗糙型
shag mat 绳毛垫
shagreen 粗粒草;鲨革;绿革;粗粒革
shagreen paper 糙面纸
shag rug 长绒地毯
shakable 可震动的;可摇动的
shake 振动;振荡;振摇;轮裂;木瓦;抖动;动摇
shake allowance 松动量
shake apparatus 振荡仪
shake-bol 摇动螺栓
shake-cabin 棚屋
shake culture 摇动培养法;震荡培养
shakedown 新工艺试验;安定状态
shakedown cruise 新船的试验航行;试航【航海】
shakedown load 振荡破坏荷载;安定状态荷载
shakedown temperary 临时的
shakedown test 调试
shakedown theory 安定(性)理论
shake flask culture 摇瓶培养
shake flask test 摇瓶试验
shake flat ladder step 防扭踏板
shake-hole 落水洞
shake in timber 木材环裂
shakeless deckle 稳定框
shake mill 摆动筛架
shake-motion wave 蛇行波
shaken bed 摇动床层
shaken helicoids 摇动螺旋面;振动的螺旋体
shake off 摆脱
shake off the other ship 摆脱他船
shakeout 实力薄弱的开发商被淘汰掉(经济危机中);落砂(设备);去砂;铸件振动脱砂;打箱
shakeout equipment 振动落砂机;落砂设备;振动装置
shakeout grate 落砂栅;落砂格子板
shakeout grid 落砂栅;落砂格子板;脱砂箱格子
shakeout machine 落砂机;摇动机(测定石油中沉淀物与水分含量的离心机);振动落砂机
shakeout operation 落砂工序
shakeout sand 落箱砂
shakeout temperature 打箱温度
shakeproof 抗振;耐振的;防振的;防振
shakeproof corrugated paper board 防振瓦楞纸板
shakeproof lock washer 减振锁紧垫圈
shaker 摇动者;摇动器;摇床;振动器;振子;振动(试验)器;振动机;振荡器;振打器;混合器;筛
shaker apparatus 摇筛机
shaker arm 振动臂
shaker bed 分离筛
shaker belting 振动输送带
shaker chute 摇动溜槽;簸动槽
shaker conveyer 摇动输送机;振动输送机;摆动式输送机
shaker cooling 振荡冷却;晃动冷却
shaker crank 逐稿器曲杆
shaker drier 振动式干燥机
shaker feeder 振动给料器;振动给料机
shaker gear 振动器齿轮
shaker grate 振动式炉算
shaker hearth electric(al) furnace 摇床式电炉
shaker hearth furnace 振底式炉
shaker mechanism 摇动机构
shake rot 裂纹腐蚀
shaker pan 振动筛盘;平格振动筛;平板振动筛;盘形振动筛
shaker pitch 摇动筛倾角
shaker rack 分离筛
shaker screen 摇动筛;振动筛;大筛;分离筛
shaker separator 振动分离筛
shaker share 振动犁铧
shaker shoe pitman 振动筛连杆
shaker-type conveyer 振动式运输机
shaker type(flat) chip screen 振动式木片平筛
shaker washer 振动式清洗机
shake table 振动台
shake test 振动试验
shake up 摇荡

shake willey 振荡式除杂机
shakhovite 锑汞矿
shaking apparatus 摇动装置;振动式混合机;摇动器
shaking appliance 摇动装置
shaking bar grizzly 振动筛
shaking-bed sequencing batch reactor 摇床序批间歇式反应器
shaking-bed sluicing device 摇床流槽选矿装置
shaking-bottle incubating test method 摇瓶培养试验法
shaking channel 摇动槽
shaking chute 摇动漏斗;振动(斜)槽;振动(式)溜槽
shaking conveyer 振动输送机;摆动(式)运输机
shaking conveyer ballframe carriage 振动式运输机滚珠支架
shaking conveyer carriage 振动式运输机支架
shaking conveyer pan 振动式运输机槽
shaking conveyer rod 振动式运输机连杆
shaking conveyer screw jack 振动式运输机螺旋顶杆
shaking conveyer supply truck 振动式运输机给料车
shaking conveyer swivel 振动式运输机回转座
shaking culture 摇动培养法
shaking device 振动器
shaking equipment 振动试验设备
shaking feeder 摇动给料机;振动喂料器;振动喂料机;振动式供料器;振动(式)给料器;振动式矿机;振动给料机;振动加料器
shaking flask 振动烧瓶
shaking grate 摇动式筛分机;振动炉箅
shaking grate cooler 振动箅式冷却器
shaking grate stoker 摇动炉箅
shaking grate-type cooler 摇动栅式冷却器
shaking grid 振动落砂机
shaking grizzly 摇动(式)格筛
shaking intensity and soil condition 振动强度和土质条件
shaking ladle 摇动包
shaking launder 振动溜槽
shaking loader 振动装载机
shaking machine 振动器
shaking mixer 摇动混合器;振动混合机
shaking of steering 转向操纵机构振动
shaking-pan conveyer 振动输送机
shaking pan loader 振动槽式装载机
shaking period 摆动期
shaking pick breaker 摇动针碎机
shaking picker 簸动筛
shaking picking table 摇动式拣矸台
shaking plate 摇动式混汞板
shaking platform 振动平台
shaking screen 摇(动)筛;振荡筛;摆动栅;摆动筛;振动筛
shaking-screen suction table 摇动筛纤维吸出装置
shaking-screen washer 振动式洗矿筛;震动式洗矿筛
shaking shale 黏土页岩
shaking sieve 摇筛;振动筛;跳桌
shaking-sieve digger 振动筛式挖掘机
shaking sleeve 摇动套筒
shaking table 摇床;振动(工作)台
shaking table method 振动台法
shaking test 摇振试验;筛振试验;摇动试验;振动试验;手撼试验;抖动试验
shaking test apparatus 振动试验设备
shaking the needle 摇针
shaking trough 摇动斜槽;振动溜槽
shaking up 摇混
shaky steering 转向机构摇动
shale 页岩
shale ash 页岩灰;页岩渣
shale band 页岩夹层
shale break 页岩夹层
shale brick 页岩砖
shale cement 页岩水泥
shale ceramisite 页岩陶粒
shale clay 页岩黏土
shale clay brick 页岩黏土砖
shale concrete 膨胀页岩轻集料混凝土;膨胀页岩轻集料混凝土
shale concrete plank 页岩混凝土板
shale content 泥质含量

shale-control mud 页岩控制泥浆
shale cutter 页岩刨削机;刨削机
shale deposit 页岩矿床
shale diapir 页岩底辟
shale for brick-tile 砖瓦用页岩
shale for cement burden 水泥配料页岩
shale fragment 页岩碎片
shale gasoline 页岩汽油
shale grit 页岩质粗砂岩
shale heap 废物堆
shale laden mud 泥基泥浆
shale lime 页岩石灰
shale line 基准线;基线
shale oil 页岩油
shale oven 干馏炉
shale paraffin(e) 油页岩石蜡
shale-pebble conglomerate 页岩砾砾岩
shale pit 页岩槽;页岩坑;废渣坑
shale planer 页岩刨削机;刨削机;刨土机
shale reservoir 页岩储集层
shale retorting 页岩干馏
shale rock 页岩
shale screen 泥浆振动筛
shale shaker 泥板页岩振动器;泥浆振动筛
shale stringer 页岩脉理
shale tar 页岩沥青
shale tar pitch 页岩硬焦油沥青脂;页岩沥青
shale thickness in coal formation 夹矸的厚度
shalification 页岩化(作用)
shall not be applied to 不应用于
shallow 页海的;浅的;浅薄
shallow aeration activated sludge process 浅层曝气活性污泥法
shallow air saturation diving 浅深度空气饱和潜水
shallow and narrow channel 窄浅水道
shallow apse 浅的教堂后半圆室
shallow arch 浅拱
shallow arch bridge 坦拱桥
shallow article 盘类制品
shallow basin 浅水池
shallow bay 浅海湾
shallow beam 矮梁;浅梁;低截面梁
shallow bed 浅床
shallow-bed groundwater 浅层地下水
shallow bed technique 浅床法
shallow bending 浅弯
shallow blanket 薄地板;薄底板
shallow blasting 浅孔爆破
shallow bore hole 浅钻孔
shallow borehole explosion 浅孔爆破
shallow bottom sediment 浅底沉积物
shallow bowl 浅钵;浅凹形(地面);浅浮槽
shallow bowl toilet 蹲式马桶;浅钵形便池
shallow box 浅箱
shallow breaking waves 浅水破碎波
shallow burialism 浅埋作用
shallow bushing 浅槽漏板
shallow caisson 浅海沉箱
shallow canal 浅水运河
shallow cast 浅水测深
shallow cellular deck 浅空心板
shallow circulation 浅层环流
shallow coastal zone 岸边浅水区
shallow coefficient 浅化系数
shallow compaction 浅层压实
shallow continental shelf 浅大陆架
shallow cooling pond 浅水冷冷却池
shallow cut 浅挖土;浅挖方;浅挖;浅切削;浅切口;浅路堑;浅开挖
shallow cutting 浅挖
shallow dam 隔板;挡板
shallow deposit 浅水沉积(物)
shallow-depression 浅低压
shallow depth 浅啮合;浅埋
shallow depth of field 浅景深
shallow-depth sedimentation 浅层沉淀
shallow-depth tunnel 浅埋隧道
shallow diffusion 浅结扩散
shallow digging 浅挖
shallow dipping structure 缓倾构造;微倾构造
shallow ditch 浅沟;小沟
shallow diving 浅深度潜水
shallow dock 浅坞
shallow draft 轻微压下;吃水浅的

shallow-draft bottom-dumping barge 浅吃水开底泥驳
shallow-draft craft 浅吃水船
shallow-draft lock 浅水船闸
shallow-draft navigation 浅水航行
shallow-draft quay 浅水码头
shallow-draft ship 浅吃水船
shallow-draft split barge 浅吃水开底泥驳
shallow-draft vessel 浅(吃)水船
shallow-draft waterway 浅水航道
shallow-draft wharf 浅水码头
shallow draught 小量镦粗
shallow-draught boat 浅水船
shallow dredge 浅挖(机);浅部挖掘船
shallow dredging 浅挖;浅滩疏浚;浅滩开挖;浅水疏浚;浅疏浚
shallow drill hole 浅钻孔
shallow earthquake 浅震;浅(层)地震
shallow earthquake focus 震源浅
shallow embedment 浅埋
shallow energy level 浅能级
shallow estuary 浅水河口
shallow event 浅层事件
shallow excavation 浅(开)挖;浅坑
shallow exploratory boring 浅孔钻探
shallow explosion 浅层爆炸
shallow factor 浅化系数
shallow fault 浅断层
shallow faulting 浅断层作用
shallow fill 低填土;低路堑;低路基;浅填
shallow filter 浅滤池
shallow fixture 浅式设备
shallow flat ravine 坳沟
shallow floor 浅肋板
shallow flow system 浅部水流系统
shallow focus earthquake 浅震;浅源地震
shallow fog 浅雾;低雾
shallow fold and thrust belts 表层褶皱和冲断层带
shallow fold belt 浅层褶皱带
shallow folding 浅折
shallow footing 浅基脚;浅基(础)
shallow fording 浅水涉渡
shallow foundation 浅基(础)
shallow foundation of bridge 桥梁浅基(础)
shallow foundation structure 浅基础结构
shallow frame 浅肋骨
shallow freezing 表面冻结;表层结冰;表层冻结
shallow gas 浅层气
shallow gas in coal field 煤田浅层气
shallow gas zone 浅含气带
shallow glass pattern 玻璃浅花纹
shallow gravel 浅部砂矿床
shallow grit chamber 浅沉砂池
shallow groove of loessland 黄土区浅沟
shallow ground 浅滩
shallow groundwater 浅层地下水;浅部地下水
shallow grouting 浅层灌浆
shallow harbo(u)r 浅水港
shallow hardening 表面硬化
shallow-hardening steel 浅淬硬钢;低淬透性钢
shallow hexagonal nut 六角形浅螺帽
shallow hoisting 浅井提升
shallow hole 浅炮眼
shallow hole drilling 浅孔钻进(深度30米以内)
shallow hole gamma ray survey 浅孔伽马测量
shallow holes (道路上的)浅坑
shallow hole spectrum survey 浅孔能谱测量
shallow hole temperature 浅孔温度
shallowing 变浅
shallow inland sea 浅内陆海
shallow jet pump 浅井喷射水泵
shallow junction 浅结
shallow karst 浅层岩溶;浅层喀斯特
shallow lake 浅水湖泊;浅湖
shallow lake deposit 浅湖沉积
shallow lake facies 浅湖相
shallow land burial 陆地浅坑掩埋
shallow landslide 浅层滑坡
shallow layer 浅层;薄涂层;薄层
shallow layer landslide 浅层滑坡
shallow layer of soil 表土层
shallow layer profile hydrophone 浅地层剖面仪
shallow layer profile method 浅地层剖面法
shallow lens 薄透镜

shallow level 浅能级;浅层次
shallow lift 薄层(剥落)
shallow lift of concrete 混凝土薄层剥落
shallow light fitting 浅照明装置
shallow low 浅低压
shallow luminaire(fixture) 浅照明装置
shallow-lying 浅埋藏的
shallow manhole 检查井;检查坑;浅人孔
shallow marginal sea 浅陆缘海;浅边缘海
shallow marine clastic deposit 浅海碎屑沉积
shallow marine conglomerate 滨海砾岩
shallow marine type 浅海型
shallow muck 浅游泥层;薄淤泥;薄层淤泥
shallow nut 浅螺帽
shallow of lift 薄层剥落(混凝土)
shallow oil zone 浅含油带
shallow open-cut surface mining 浅部露天开采
shallow pan 浅池;浅盘;浅盆
shallow pan closet 浅盘式便池;浅盘式便桶
shallow pan filter 浅盘过滤器
shallow pass 空轧孔型;空轧道次
shallow patch 浅滩
shallow pattern 浅花纹
shallow penetration 浅近突破
shallow percolation 浅层渗透
shallow phreatic cave 浅潜流洞
shallow pit 浅坑
shallow pitch 低坡度
shallow placer 浅成砂矿
shallow plate water extraction method 浅盘水淬法
shallow ploughing 浅犁;浅沟;浅耕
shallow ploughing and careless cultivature 浅耕粗作
shallow plutonic body 浅位侵入体
shallow-pocket classifier 浅槽分级机
shallow-pocket free settling classifier 浅袋型自由沉降分级机
shallow pond 浅水池
shallow pool 浅水池;浅熔池
shallow pycnocline 浅密度跃层
shallow reach 浅(滩)段;浅水河段
shallow recessing 压凹;浅成型
shallow reef 浅礁
shallow region of lake 湖泊浅水区
shallow reservoir 浅层热储
shallow ridge slot 浅槽
shallow-rise cupola 低矮圆屋顶
shallow-rise dome 低矮穹隆
shallow riverbed 浅水河床
shallow-rooted 浅根的
shallow rooted plant 浅根植物
shallows 浅滩;浅水处
shallow sampling with a soil auger 麻花钻浅层取样
shallow sands 浅砂层
shallow saturation diving 浅深度饱和潜水
shallow scattering layer 浅水散射层
shallow scraping 浅层刮削
shallow scraping of foam 浅撇泡沫
shallow sea 浅海波;浅海(区)
shallow sea clastic sedimentation model 浅海碎屑沉积模式
shallow-sea deposit 浅海沉积(物)
shallow sea environment 浅海环境
shallow sea facies 浅海相
shallow-sea sedimentary soil 浅海沉积土
shallow sea water 浅海水区
shallow-sea wave 浅海波
shallow sediment 浅水沉积(物)
shallow sedimentation 浅水沉积作用
shallow seismic prospecting 浅层地震勘探
shallow seismic reflection technique 浅层地震反射技术
shallow seismic refraction 浅层地震折射
shallow shaft 浅竖管
shallow shaft exploration 井探
shallow shaft hoist 浅井提升机
shallow shell 浅球壳;平扁壳;扁壳体;扁壳
shallow-shell theory 片扁壳理论
shallow shoal 浅滩
shallow shock 浅震
shallow slide 浅层滑动
shallow slip in rock mass 岩体内浅层滑动

shallow soil 浅层土;薄层土壤
shallow soil seeded rice 旱稻
shallow sounder 浅水测深仪
shallow sounding 浅水测深
shallow sounding apparatus 浅水回声测深仪
shallow sounding rod 浅水测深杆
shallow source 浅源
shallow source earthquake 浅源震
shallow-spheric(al) shell 扁球壳
shallow spot 浅点
shallow spread foundation 明挖基础;浅挖基础;浅平基(础);浅扩基础
shallow structure 浅部构造
shallow submarine hydrothermal eruption 浅海水下水热喷发
shallow submerged hydrofoil 浅浸式水翼
shallow sump 积水坑
shallow tank 浅池
shallow teeth (牙轮的)短齿
shallow teleseism 浅源远震
shallow temperature field 浅层温度场
shallow thrust belt 浅层冲断层
shallow tillage 浅耕
shallow tray basin 浅盘式池
shallow tray settling basin 浅盘沉淀池
shallow trial pit 浅井试坑
shallow tube well 浅(层)管井
shallow-turbidite trap 浅海浊积岩圈闭
shallow tunnel 浅埋隧道
shallow underground burial 浅层埋藏
shallow underground burst 浅层地下爆炸
shallow underground disposal 浅地层处置
shallow understand disposal 浅地层处置
shallow underwater burst 浅水爆炸
shallow-vaulted roof 囤顶
shallow vein zone deposit 浅成脉状矿床
shallow waisted船 浅腰船
shallow water 浅水
shallow water blackout 浅水黑视
shallow water cable 浅海电缆
shallow water clapotis 浅水驻波
shallow water component 浅水分潮
shallow water constituent 浅水分潮
shallow water correction 浅水订正
shallow water correction factor 浅水修正系数
shallow water delta 浅水三角洲
shallow water deposit 浅水沉积(物)
shallow water dive 浅水潜水
shallow water effect 浅水影响;浅水效应
shallow water fauna 浅水动物区系
shallow water hydrography 浅海水文学
shallow water lunar 1/4 diurnal tidal component 太阴浅水四分之一日分潮
shallow water lunar 1/6 diurnal tidal component 太阴浅水六分之一日分潮
shallow water luni-solar 1/4 diurnal tidal component 太阴—太阳浅水四分之一
shallow water maneuvering 浅水机动
shallow water mud pump 浅水泥浆泵
shallow water quay 浅水码头
shallow waters 浅水域
shallow water scattering layer 浅水散射层
shallow water sediment 浅水沉积(物);浅海沉积(物)
shallow water species 浅水种
shallow water splash 试车浅水池
shallow water stands for a long time 浅水长期淤积
shallow water survey 浅水区测量
shallow water tide 浅水潮汐;浅海潮汐
shallow water turbidity deposit 浅水浊流沉积
shallow water vessel 浅水船
shallow water wave 浅水波
shallow waterway 浅水航道
shallow water wharf 浅水码头
shallow water zone 浅水区;浅水带
shallow weld 浅焊
shallow well 浅(水)井
shallow well injection 浅井注入;浅井灌注
shallow well jet pump 浅井喷射式水泵
shallow well oil well cement 浅井油井水泥
shallow well pump 浅井泵
shallow-winnowed-crestal trap 浅海簸脊圈闭
shallow work 浅耕;浅部开采工作

shallow workings 浅部巷道
shallow zone 浅区
shallow zone of lake 浅水湖带
shallow zone of sea 浅海带
Shaluk 沙卢克风
shaly 页岩(状)的
shaly bedding 页状岩层
shaly blaes 页岩状黏土;页岩状砂岩
shaly clay 页岩(质)黏土
shaly-graywacke formation 页岩—硬砂岩建造
shaly limestone 页岩质石灰岩
shaly sand 页岩砂
shaly structure 页岩状构造
shaly texture 页岩状结构
sham 假货
shamah-duplex (钢结构和塑性墙板组成的)预制房屋
sham beam 假梁
shambles 屠宰场
sham door 假门
shammy 油鞣革;麂皮
shampoo bowl 洗发盆
shampoo spray 洗涤剂喷射
shampoo spray arch 洗涤剂喷射拱门
shamrock plate 三角眼铁;三角眼板
sham ruin 伪装废墟
Shamvaian cycle 沙姆维造山旋回
sham wood 模造材
shan 多瘤节的坏木材
shandite 硫铅镍矿
shank 旋柄;镜筒;镜身;抬包;手柄
shank angle 刀柄角
shank base 支撑面
shank chisel 固定座凿;带柄的刻刀
shank clamp 铲柄卡夹
shank collar 钻头挡;钎肩
shank diameter 柄直径
shanked 带柄的
shanked drill rod 有肩钎杆
shank end 钎尾
shank end mill 带柄端铣刀
shank gear cutter 带柄齿轮切刀具
shank ladle 手转铁水桶;手转铁水包;手浇包;端包
shank length 钢体长度;尾部长度
shankless die 无柄模具
shank milling cutter 柄式铣刀
shank of a hook 钩身
shank of bit 钻杆
shank of bolt 螺栓杆
shank of connecting rod 联杆体;连杆体
shank of drill 钻头柄
shank of rivet 铆钉杆
shank of screw 螺钉杆
shank of tool 刀具柄
shank pinion cutter 带柄插齿刀
shank plug 方头塞
shank reamer 带柄铰刀
shanks 玻璃安装工用钳
shank safety hook 轴柄安全型吊钩
shank size 钎尾尺寸
shank taper 钎尾锥度;尾部螺纹
shank tools 带柄刀具
shank type cutter 柄式铣刀
shank type fraise 带柄铣刀
shank type milling cutter 柄式铣刀
shank type reamer 柄式铰刀
shank with driving slot 带卸刀槽的刀柄
shank with driving square 带驱动方头的刀柄
shank with groove 沟槽刀柄
shank with half tang 半尾刀柄
shank with square tang 方尾刀柄
shank with tang and keyway 扁尾键槽刀柄
shannonite 钙镁橄榄石
Shannon projector 山农投影机
Shannon's theorem 山农定理
Shannon-Weaver's diversity index 山农—韦弗指数
Shansep approach 应力历史与归一化土工参数法
Shan State massif 掸邦地块
Shantung soil 非石灰性棕壤
shanty 钻机房;路旁小房;临时小房(屋);临时工房;窝棚
shanty boat 水上棚屋
shanty man 伐木工

shanty method 手工制作
shanty town 陋屋区;陋居区;贫民窟;棚户区;水上棚屋
shaoling effect 浅滩效应
shape 形状;形态;号型
shapeability 可成型性;随模成型性
shapeable 可塑造
shape above water 水面上形状
shape a course 定出航向
shape analysis 形状分析;形态分析法
shape anisotropy 各向异性形状
shape beam 成型波束
shape broach 成型拉刀
shape casting 成型铸造
shape change 形变
shape check device 形状检验装置
shape coding 形状编码
shape coefficient 形状系数
shape concept 形状概念
shape constant 形状常数;梁常数
shape control 形状调整
shape cutting 成型切割;仿型切割
shape-cutting machine 成型切割机;定形氧割机
shaped 形似的;喇叭形的
shaped bar 小尺寸的异形钢材;成型杆;异形钢材
shaped beam 特形光束;成型光束
shaped-beam antenna 成型波束天线;赋形波束天线
shaped bevel 异型边
shaped blasting 聚能爆破
shaped bloom （轧制工字梁用的）异形钢坯;异形初轧坯
shaped brick 异形砖;型砖
shaped casting 成型铸造;成型铸件
shaped ceiling 造型顶棚
shaped ceramic fiber product 陶瓷纤维异型制品
shaped-chamber manometer 异型腔流体压力计
shaped channel 定型河道;定形河道
shaped charge 成型填药;锥孔装药;聚能装药;定向爆破装药
shaped charge drilling 聚能炸药爆破钻进
shaped charge perforation 聚能炸药穿孔
shaped charge perforator 聚能炸药穿孔器
shaped conductor 异形导线
shaped-conductor cable 特性导线电缆;特形铁芯电缆
shaped demolition charge 锥形爆破装药;锥形爆破炸药
shaped dish reflector 变形碟式反射器
shaped dual reflector 异形双反射面
shaped earth road 整形土路
shape definition 形定义
shaped electrode 成型电极
shaped engraving 异形刻花
shape description 形状描述
shape detection 形状检定
shaped fabric 异形织物
shaped field stop 成型的场光阑
shaped form 曲线形
shaped gable 齿形山墙
shaped glass grinding machine 异形玻璃研磨机
shaped goggles 电焊工安全眼镜
shaped goods 定型制件;定形制件
shaped grinding wheel 定形磨轮
shaped groove 曲形槽
shaped hole 成型钻孔组
shaped insulating product 定形隔热制品
shaped insulation 异形绝热件
shaped insulator 异形绝缘子
shape(d)iron 型铁;异形钢
shape distortion 形状畸变;变形
shape distortion ratio 变形比
shape distribution 形状分配
shaped joist 异型格栅
shaped like a pear 梨形
shaped natural stone 定型天然石料
shaped nozzle 特形喷管
shaped orifice 型线孔口
shaped piece 定形配件;定型配件
shaped plate 形板
shaped plate stem 弯板首柱
shaped plywood 成型胶合板
shaped pressure squeeze board 异形压实板
shaped profile milling cutter 成型铣刀

shaped pulse 整形脉冲;成型脉冲
shaped reflector antenna 成型反射器天线
shaped refractory 定形耐火材料
shaped roll 型钢轧辊
shaped sea wall 斜面海堤;斜面防浪墙;斜面海塘;定型海堤壁
shaped size information 形状尺寸信息
shaped steel 异形钢材;型钢
shaped steel plate 型钢板
shaped stone 料石;人工磨圆金刚石
shaped stonework 成型石工
shaped susceptor 成型的衬底器
shaped to pattern 按模型制成的
shaped tree 整形树
shaped wire 异形钢丝;成型线
shaped work 定型工作;造型细木工艺;曲线木作
shape-elastic scattering 形状弹性散射
shape fabric 形状构架
shape fabricator 成型装配工
shape factor 建筑物体型系数;形状因子;形状因素;形状因数;颗粒形状系数;成型系数;波形因素;波形因数
shape factor of ballast 道砟颗粒形状系数
shape factor of cross-section 断面形状系数
shape factor of vehicle 车辆的形状系数
shape factor test of aggregate 集料形状系数试验
shape feature of discontinuity 结构面形态特征
shape fill 状态填充
shape for interior work 室内装潢型材
shape function 形状函数;形态函数;形变函数
shape function matrix 形状函数矩阵
shape index 形状指数
shape information 波形信息
shapelessness 不均匀性;不定形性
shape library 图形库
shape maker 模型;成型器
shape measure 形状测量
shape memory ceramics 形状记忆陶瓷
shape memory effect 形状记忆效应
shape memory polymer material 形状记忆聚合物材料
shape model 形模;形状模型
shape of bar 沙坝形态
shape of basin 流域形状
shape of beach 海滩形态
shape of bearing plate 承压板形状
shape of block 块体形状
shape of bottom roughness 底部糙率型
shape of car ferry 汽车渡船型
shape of circular features 环形体形态
shape of cleavage domain 劈理域形态
shape of coastline 海岸线形态
shape of collapse depression 塌陷洼地形状
shape of cross-section 断面形状
shape of cup 杯状
shape of delta 三角洲形态
shape of diapir core 底辟核形状
shape of horse 断片形状
shape of hydrograph 过程线形状
shape of inclusion 包裹体形态
shape of intrusion 岩体形态
shape of landslide crack 滑坡面形状;滑坡裂隙形状
shape of line 线划形状
shape of lineament 线性体形态
shape of nodule 结核形状
shape of orebody 矿体形状
shape of particle 粒子形状;颗粒形状
shape of response spectrum 反应谱形状
shape of saddle 马鞍形
shape of strip 板形
shape of texture body 结构体形态类型
shape of the earth 地球形状
shape of the ground 地貌特征
shape of the moment diagram 力矩图形状
shape of the root cap 根冠形状
shape of the sea floor 海底地形
shape of tool 刀具形状
shape of tooth 齿形
shape of window 窗户形状
shape oscillation 形状振荡
shape parameter 形状参数
shape piece 定型配件
shape pruning 成形修剪
shaper 木成型机;木铣床;整形器;整形机;牛头刨床;成型机;成形器

shaper amplifier 脉冲整形放大器;脉冲形成放大器
shaper and planer tool 刨刀
shaper chuck 牛头刨夹具;牛头刨虎钳
shaper-divider 整形分频器
shape recognition 形状识别
shape recovery 形状恢复
shape resistance 形状阻力;体型阻力
shape retention 形状保持性
shape-retentiveness 定形性;形状保持性
shape-righting 矫形
shaper part 牛头刨零件
shaper rail 成型轨
shaper tool 刨刀
shapes and bars 碳素钢板和型材
shapes of magnetic bodies 磁性体形状
shape sorting for diamond 金刚石选形
shape sorting vibrating table for diamond 金刚石振动选形机
shape stability 形状稳定性
shape steel 型钢
shape steel beam bridge 型钢梁桥
shape steel rolled stock 异形型钢
shape straightener 型钢矫直机
shape syringe pear 梨形洗涤球
shape up 码头临时工的挑选;挑选临时职工
shape wave 浪形
shape welding 有型焊接;异形焊;成型零件堆焊
shape wood 型材
Shap Fell 红褐色花岗石（产于英国威斯特摩兰郡）
shaping 修整;整型;整平;造型;立铣;成型;刨销;刨光
shaping amplifier 整形放大器
shaping and development of educational economics 教育经济学形成与发展
shaping by handcraft work 雕镶成型
shaping by stock removal 切削成型
shaping chuck 牛头刨夹具;牛头刨虎钳
shaping circuit 成型电路
shaping die 成型模
shaping filter 整形滤波器
shaping filtering 整形滤波
shaping for diamond 金刚石整形
shaping groove 成型孔型
shaping lathe 成型车床;刨削车床
shaping machine 牛头刨床;成型机
shaping method 成型法
shaping mill 冷弯机;成型机
shaping network 整形网络;成型网络
shaping of metal powders 金属粉末成型
shaping of mo(u)lding 线脚成型
shaping of plastics 塑料成型
shaping of teeth 齿的成型
shaping operation 整形作用;成型操作;刨削操作
shaping pass 粗轧孔形
shaping plate 型板
shaping property 成型性能
shaping stand 粗轧成型机座
shaping technique 整形技术
shaping technique for light pulse 光脉冲整形技术
shaping time 成型时间
shaping tool 样制刀;成型切刀
shaping unit 信号形成器;信号形成部件
shaping without pressure 无压成型
shaping without stock removal 无屑成型;非切削成型
Shapley-Snow procedure for matrix game 矩阵对策的夏普莱—斯诺方法
shapometer 形状计;形状测定器
shapr change of direction 急剧地改变方向
shapr-cutting edge 锋利的切削刃
shard 碎片;玻屑
share 共用;均分;锄铲;份额;份;分担量;分担（保险）
shareable image 可共享映像
shareable program(me) 可共享的程序
share alike 平均负担租;平均负担税
share apex 共角顶
share assets of divisions 分厂的分享资产
share beak 锋尖
share bonus 以股票作红利
share broker 股票经纪人
share capital 股金;股份资本
share ceded 分出的成分
share certificate 股票

share consolidation 股份合并
share control takeover 投资控股
share count 共享数目;共享计算
share cover 铧式覆土器
share crop 交给地主的租谷
share cropper 佃户
share-cropping 分成(制)
shared access 共享存取
shared address 共享地址
shared air-shed 共有空气域
shared appreciation mortgage 分享增值抵押贷款
shared assembly program(me) 共享汇编程序
shared batch area 共用成批程序区域;共享批处理区
shared bay 共有海湾
shared bus 共用总线;共享总线
shared cell 共享单元
shared channel 共用信道;同频信道
shared class 共享分类
shared communication resource 共享通信资源
shared control 共用控制;共享控制
shared control unit 共用控制器;共享控制器;共同控制器;分配型控制器
shared data base 共享数据库
shared data set 共享数据集
shared-dead-time method 均分死时间法
shared device 共用设备;共享设备;公用设备
shared direct access storage device 共用直接访问存储器
shared direct access storage device option 共享直接存取存储器设备选择
shared dwelling unit 合住单元
shared environment 共享环境
shared experience 分享经验
shared file 共享文件;共用文件;公用文件
shared file system 分时用外存储器系统;共享文件系统
shared flue 共同烟道
shared hardware 公用硬件
shared index database 共享索引数据库;共用索引数据库
shared input-output device 共用输入输出设备
share directories 共享目录
shared kiln 搭烧窑
shared lane 合用车道
shared limit 共享限量
shared lock 共享型锁;共享锁;共享封闭
shared logic 共享逻辑
shared-logic-system 逻辑共享系统
shared main storage 共享主存储器
shared main storage multiprocessing 共享主存多重处理;共享主存储器多道处理
shared management knowledge 共用管理知识
shared management of catchment points 共享流域测站管理
shared memory 共享存储(区)
shared-memory system 共用存储器系统
shared natural resources 共有的自然资源
shared occupancy 合住
shared operating system 共享操作系统【计】
shared page table 共享页表
shared path 共用路径
shared peripheral 共用外围设备
shared port 共享入口
shared processor 共用处理装置;共享处理机
shared protection ring 共用保护环
shared resources 共享资源;共有资源
shared revenue 分成收入
shared ride 合乘
shared routine 共用(例行)程序;共用例程;共享程序
shared row assignment 共用行分配
shared segment 共享段
shared source multichannel processing 共用资源多道处理
shared space 共享空间
shared spooling 共享伪脱机;共享假脱机
shared storage 共用
shared storage manager 共享存储管理程序
shared structure 共享结构
shared subchannel 共用分通道;共享子通道;共分通道
shared subroutine 共用子程序;子程序
shared system 共享系统
shared task set 共享任务组

shared tax 分享的税收;分成税收
shared tenant services 共享租赁
shared track 共用磁道
shared variable 共享变量
shared variable set 共享变量集
shared virtual area 共用虚拟区;共享虚存区
shared visual memory 共享虚拟存储
shared waters 共有水域
shared watershed 共用(的)流域
shared waters management 共享水域管理
share edge 共棱
share experience 交流经验
shareholder 股东;股票持有人
shareholder's equity 股东拥有的资产价值
share holding 股份持有
shareholding equity 总股本
share in profits 分享利润;分红
share investment 股份投资
share limit 共用限制
share losses expenses 分担损失费
share milker 分成牧工
share of faster-disbursing projects 拨款较快的项目比例
share of industrial siding 专用线共用
share of market 市场分配
share of trade 贸易份额
share operating system 共用操作系统
share out equally 平均分摊
share plane 共面
share plough 铧式犁;铧式犁
share point 铧尖
share pole 轻便井架加固杆
share premium account 实缴股本
share profit 分红
sharer 共用者;共用管理程序;共享者;共享程序
share ratio 分担率
share request 共享请求;请求共用
share responsibility for 分担
share-ride 分享乘坐
shares 股票;股份
share shaft 铲柄
shares outstanding 净发股票
share splitup 分股
share supervisor 共用管理程序
shares worth 股票值
share the expenses 分担费用
share the expenses equally 均摊费用
share the price difference 分担差价
share-the-work 分享工作
share time 分时
share transfer 股份转让
share-type lifter 铧式挖掘铲
shareware 共享软件
share warrant 股份证书
sharing 共用;同层;分享;分摊;分担
sharing circuit 分配电路
sharing database 共用数据库
sharing data set 共用数据集
sharing extra profit 超额利润分成
sharing multiprocessing 共用多重处理;共享多重处理;分担多重处理
sharing of losses 亏损分摊
sharing of procedure segments 过程段的共享
sharing part of load 荷载分量
sharing ration 分摊比率
sharing resources 共享资源
sharing subchannel 共享子通道
sharing system 共用系统;共享系统;划分制
sharing the entire profit 全额利润分成
sharing the total profit 全额利润分成
shark-fin hydrophone 敛形水听器
sharki 考斯风
shark liver oil 鲨肝油
sharks' mouth 天幕通椸孔;天幕叉口
sharnaid 类矽卡岩
sharp 尖锐;急剧的;明显的;明确的;锐利的;锐的;陡的
sharpable bit 磨锐式钻头
sharp aggregate 多角集料
sharp angle 锐角;锐棱的
sharp-angled 成锐角的
sharp backed angle 内锐角钢
sharp bank 大坡度转弯
sharp beam 窄光束;尖波束;锐光束

sharp bend 急弯(管);急弯段;锐弯头;锐弯(管);锐角弯曲;陡弯;突跃弯头
sharp-bend-induced traffic-hazard 急弯险滩
sharp bend reach 急弯河段
sharp bilge 尖舭
sharp bit 木工凿;锐利钻头
sharp bottom boat 尖底船
sharp boundary model 锐边界模型
sharp bow 尖形船首
sharp-bowed 尖头的
sharp braking 突然制动
sharp cash 即付现金
sharp central curve 中心清晰度
sharp chisel 木工凿
sharp coarse beach sand 多角粗海砂
sharp coat 铅白油涂料;薄腻子涂层;白铅油涂层
sharp contact 明显接触
sharp corner 小半径转角;尖棱角隅;尖角;急弯;锐角转角
sharp-cornered entrance 锐角入口
sharp-cornered well 锐角势阱
sharp crest 尖峰;急剧转折(岩层);锐顶
sharp-crested 尖口;锐缘的
sharp-crested measuring weir 锐顶量水堰
sharp-crested weir 锐口堰;锐缘堰;锐顶堰;薄壁堰
sharp curve 小半径曲线;急弯曲线;锐曲线
sharp-cut filter 锐截止式滤波器
sharp cut-off 锐截止
sharp cut-off characteristics 锐截止特性
sharp cut-off filter 锐截止滤光片
sharp cut-off model 锐截断模型
sharp-cutoff tube 锐截止管
sharp decline 急剧下降
sharp definition 清晰度
sharp dissection 锐器解剖法
sharp dock 簇花酸模
sharp draft 强制通风;强力通风
sharp drill 尖钻头
sharp drop 急剧下降
sharp ear hook 锐耳用钩
sharped dental curet 锐匙口牙刮匙
sharp edge 锐边;清晰边沿;陡沿;飞边
sharp-edged 规整边;锐缘的;边缘清晰的
sharp-edged circular orifice 锐缘圆孔口
sharp-edged cones 锐缘锥体
sharp-edged crest 锐缘堰顶
sharp-edged ga(u)ging weir 锐缘量水堰
sharp-edged gust 突发阵风
sharp-edged intake 锐缘进气口
sharp-edged measuring weir 锐顶量水堰;锐缘测堰
sharp-edged orifice 锐孔流量计;锐缘孔口
sharp-edged orifice meter 锐孔流速计
sharp-edged orifice plate 锐边孔板
sharp-edged slate 尖边石板瓦
sharp-edged stem 带锐边前柱
sharp-edged timber 锯成的方边木;方木(材);锯成方缘木
sharp-edged weir 锐顶堰;锐缘堰
sharpen 磨刀;锐化
sharpened 磨利的;磨尖的
sharpened pier 破冰墩
sharpened spur 削尖山嘴
sharpened steel 锐钢钎
sharpener 削刀;磨刀器;锐化电路;刃磨器;刀具磨床;锉锯(齿)机
sharpener amplifier 锐化放大器
sharpening 削尖;磨尖;磨快;磨锐;磨刀;锐化
sharpening angle 研磨角
sharpening guide 刃磨导柱
sharpening machine 磨床;修磨机;磨钻机
sharpening of a saw 锯齿磨锐
sharpening saw 锯齿尖打磨
sharpening stone 磨石
sharpening tool 磨刀器;磨锐刀具
sharp entrance 尖进流段
sharpest focus 最精确聚焦
Sharpey's fibre 贯穿纤维;夏贝纤维
sharp fall 激跌
sharp filter 锐截止滤波器
sharp fluting 锐棱柱槽
sharp focus 锐聚焦
sharp focusing 准确调出图像;精确对光;锐聚焦
sharp-focus lens 锐聚透镜
sharp freezer 快速冷冻间;强力制冷器;低温冻结间

sharp freezing 快速冰冻;低温冻结
sharp-freezing room 速冻间
sharp front 陡锋
sharp front wave 陡锋波
sharp-grained 尖锐颗粒的
sharp grit 角砾石
sharp ground 坚硬岩层
sharp height 锐高
sharp horizontal and vertical curvatures 小半径的平曲线和竖曲线
sharp image 清晰影像;清晰图像
sharping dye 阻光染料
sharping jig 绘点器
sharp instrument 准确仪表;灵敏仪器;锐利工具
sharp instrument injury 锐器损伤
sharp interface 明显的分界面
sharp iron 捻缝尖刀;板端捻缝凿
sharpite 水碳铀矿;水菱铀矿
sharp knife 快刀
sharp layer 特薄片;特薄层
Sharples type ultracentrifuge 沙晋尔斯超速离心机
sharp line 尖锐线;锐谱线
sharp line spectrum 锐线光谱
sharp-lip inlet 锐缘进气口
sharply defined image 清晰像
sharply focused 精确对光的
sharply focused image 精确聚焦像;明显聚焦图像;锐聚图像;清晰图像
sharp melting point 明确熔点
sharp mouth 锐齿
sharpness 尖锐性;明锐性;明显度;敏锐度;锐度;清晰度
sharpness angle 研磨角
sharpness definition 临界清晰度;清晰度锐度
sharpness factor 清晰度系数
sharpness fall-off 清晰度下降
sharpness index 敏锐指数
sharpness of creasing 折皱锐度
sharpness of curve 曲线锐度
sharpness of definition 可辨清晰度;清晰度;分辨力锐度
sharpness of detail 碎部清晰度
sharpness of dots 网点识别;网点清晰度
sharpness of fractionation 分馏精确度
sharpness of monocular 单眼视觉清晰度
sharpness of regulation 调节精密度
sharpness of resonance 谐振锐度
sharpness of selection 选择锐度
sharpness of separation 分选精度;分离陡度
sharpness of sight 视觉清晰度
sharpness of vision 视觉清晰度
sharpness screen 高分辨率显示屏
sharpness turn 最急转弯
sharp-nosed 尖头的
sharp-nosed pier 尖鼻墩
sharp nose pliers 尖嘴钳
sharp notch 尖锐缺口
sharp odo(u)r 剧烈气味
sharp paint 快干涂料;快干漆
sharp peak 最高峰(值);陡峰
sharp point 图像边界点
sharp-pointed 尖锐的;急尖的;削尖的
sharp-pointed nose 尖锥形头部
sharp pounding 剧烈振动;强烈振动;强烈冲击
sharp practice 不正当行为
sharp prismatic crystal 尖棱结晶
sharp processing of imagery 图像清晰化处理
sharp product 精确积
sharp projecting 尖锐突出物
sharp pulse 尖脉冲
sharp quenching 急冷
sharp radius 小半径
sharp-radius curve 小半径曲线
sharp-radius curved bridge 小半径弯桥
sharp resonance 锐共振
sharp rise 激涨;猛涨
sharp sand 纯砂粒;磨刀砂;有棱角的砂;精硅砂;尖砂;尖角砂;多角砂;纯砂
sharp sand grains 多角砂粒
sharp screw 锐角螺纹
sharp separation 清晰分离
sharp series 锐线系光谱;锐线系
sharp shock 强激波
sharp-shooter 窄方头铲

sharp spoon 锐匙
sharp starter cracks 尖锐起始裂纹
sharp stern 尖削船尾;尖尾(船型);尖船尾
sharpstone 锐石;尖角岩屑;尖角石
sharp stone chips 多角石屑
sharpstone conglomerate 沉积角砾岩
sharp tackle 锐刀
sharp teeth profile milling cutter 尖齿铣刀
sharp thread 光螺纹;光扣;锐角螺纹
sharp tool 锐利工具
sharp-toothed 光齿的
sharp tower 尖塔
sharp trace 明显痕迹;清晰的扫迹
sharp transition 突变段;突变过渡
sharp tuning 锐调谐
sharp turn 急(转)弯;陡弯
sharp turning 微调
sharp twist 急扭
sharp V thread 锐角螺纹;非截顶三角螺纹
sharp wave 狭频波;尖波;锐波
sharp wedge formation 锐楔形队
sharp wire start 尖丝引弧
sharp work 快速作业;紧张的工作;突击工作
sharp yield 明显屈服
sharp yield point 明显屈服点
Shasta red fir 红松【植】
Shatsky rise 沙茨基海隆
shatter 击碎;破碎;碎裂片
shatter belt 震裂带;断裂带
shatter breccia 震裂角砾岩
shatter coal 碎裂煤
shatter cone 震裂锥
shatter crack 震裂裂缝;破碎裂缝;微裂纹;碎裂(纹);白点
shatter cut 龟裂掏槽(隧道爆破)
shattered belt 碎裂带
shattered clay 碎黏土
shattered fault zone 断层破碎带
shattered rock 震烈岩石;破碎岩石
shattered zone 碎裂带
shatter index 震裂指数;震裂系数;破碎指数;破碎率;粉碎系数
shattering 震裂
shattering 粉碎理论;破碎;粉刷
shattering action 震裂作用
shattering effect 破碎效应
shattering explosive 烈性炸药
shattering force 破碎力
shattering of nodules 料球炸裂
shattering of pellets 料球炸裂
shattering power 粉碎能力
shattering process 破碎过程;爆破(方)法
shatter mark 震裂痕;发裂痕
shatter point 破碎点
shatter-proof 耐震的;防震裂的;不碎的
shatter-proof glass 耐震玻璃;防震玻璃;防碎玻璃;不碎玻璃
shatter-proof glass door 不碎玻璃门;防震玻璃门;防震玻璃门;耐震玻璃门
shatter-proof insulating glass 不碎隔热玻璃;防碎隔热玻璃;防震隔热玻璃;耐震隔热玻璃
shatter-proof plate glass 不碎平板玻璃;防碎平板玻璃;防震平板玻璃;耐震平板玻璃
shatter-proof sheet glass 不碎玻璃片;防碎玻璃片;防震玻璃片;耐震玻璃片
shatter-proof window glass 不碎窗玻璃;防碎窗玻璃;防震窗玻璃;耐震窗玻璃
shatter-resistant 抗碎裂
shatter strength 震烈强度
shatter test 硬度坠落试验;震裂试验
shatter zone 震裂带;碎裂带;断裂带
shattuckite 羟硅铜矿
shave 刮削;刨
shave hook 铅锉;刮刀;钩形刮刀;镰刀钩
shaver 刨刀;刮刀
shaver's disease 铝尘肺
shave tooth 划齿
shaving 修整表面;修整;刮削;刮面;票据高升水
shaving allowance 修边余量;整修余量
shaving arbor 剃齿心轴
shaving board 偏錾板;刨花板
shaving board and attaching veneer 刨花板及贴面
shaving die 整修模;刮边模;精整冲裁模;切边模
shaving horse 刨工台

shaving machine 削匀机;削片机;刮里机;剃齿机;刨花机;刨板机
shaving plane 刨板机
shaving press 整修压力机
shavings 切屑;剃齿屑;刨花
shaving saloon 理发店
shavings exhauster 废屑排除器
shavings separator 切屑分离器
shaving stock 剃齿留量
shaving stone 刮面石
shaving tool 精车刀(具)
Shaw guard 肖氏防护罩
Shaw hardness 肖氏硬度
Shaw kiln 肖氏窑
Shaw process 肖氏铸造法;肖氏精密铸造法;陶瓷型铸造;陶瓷型法
shcherbakovite 硅铌钛碱石
shcherbinaite 钒赭石
shcott 浅盐湖
shea butter 牛油树脂
sheaf 捆;束;草捆
sheaf binder 割捆机
sheaf blower 禾捆吹送器
sheaf carrier 禾捆积运器
sheaf conveyer 禾捆输送器
sheaf cutter 禾捆切割机
sheaf discharger 卸捆器;抛捆器
sheaf elevator 禾捆升运器
sheaflike structure 束状结构
sheaf of coordinate axes 坐标轴束
sheaf type intersection 米字交汇
sheal(ing) 牧羊小屋
sheal or shealing 庇护所;简陋小屋
shear 剪切;剪力;垂直掏槽
shear along edge 边缘剪刀
shear angle 剪切角
shear apparatus 剪切仪;剪力仪
shear apparatus equipped with strain control 应变控制剪力仪
shear approach table 剪机给料辊道
shear area 剪切面积;剪切区;剪力面积
shear bar 抗剪钢筋;剪切钢筋;剪切刀片;底刀片
shear beam 剪切梁
shear-bearing type 剪切型
shear-bearing wall 抗剪墙
shear belt 剪力带
shear blade 剪切刀片;剪刀片;底刃
shear blade grinding machine 剪切刀片磨床
shear block 剪切块;滑车组
shear board 挡泥板;防塌护板;护坡坚板
shear bolt 保险螺栓;受剪螺栓
shear bolt shank 带剪销的支柱
shear-bond failure 剪切—黏合破坏
shear boundary 剪切边界
shear-bow 剪切弯曲
shear box 剪切盒;剪力盒;剪切匣
shear box clearance 剪切盒开缝
shear box test 剪力盒试验
shear breakdown 剪切破坏
shear buckling 剪切造成材料压弯;剪切屈曲
shear building 剪切型建筑;剪切形建筑
shear-building type 剪切型建筑物
shear bulk 人字起重架船
shear cake 炉门
shear capacity of fender 护舷的抗剪能力
shear castellation 堞形抗剪销
shear cell 剪力盒
shear center 剪心;剪切中心;剪力中心
shear cleavage 剪劈理
shear coefficient 切变系数
shear coefficient of viscosity 剪切黏滞系数;剪切黏性系数
shear collar 剪力环
shear-compression failure 剪压破坏
shear compression strength 剪压强度
shear compression type of failure 剪压型破坏
shear condition 剪切条件
shear cone 剪切锥
shear-cone mixer 剪切锥混合机
shear connector 抗剪联结件;抗剪结合器;剪力接头;剪力(接合)器;受剪接合部件
shear crack 剪切裂隙;剪切裂缝;剪力裂缝;海冰受剪产生的裂缝;切变裂缝
shear crack zone 剪切裂缝区

shear crush zone 剪碎带
shear cut 剪切
shear deep fracture 剪性深断裂
shear deformation 剪切形变;切变形
shear deformation effect 剪切变形效应
shear deformation of beam 梁的剪切变形
shear deformation parameter 剪切变形参数
shear degradation 切变降解
shear delay 剪切滞后
shear device 剪切仪
shear diagram 剪切图;剪力图
shear diaphragm 抗剪隔板
shear dislocation 剪切位错
shear-displacement curve 剪力—位移曲线
shear dissected plane 横向切割面
shear distortion 剪切形;切向交错层理
shear distribution 剪力分布;剪切分布;剪力分配
shear drag 剪切阻力
sheared circle diameter covering a tube bundle 管束包切圆直径
sheared edge 剪切边(缘);剪切口;剪切端面;剪断毛边
sheared hedge 剪篱
sheared hysteresis 切变磁滞
sheared length 剪切长度
sheared plate 切边钢板;切边板材
sheared plate mill 切边中厚板轧机
sheared specimen 整形独立样;整形标本树
sheared strip 剪切的条料
shear elasticity 剪切弹性模量
shear elasticity modulus 抗剪弹性模量;切变模量
shear elastic wave 剪切弹性波
shear equation 剪力方程式
shearer 滚齿刨机;剪断机;齿滚刨煤机
shearer gearhead 采煤机机头
Shearer plastometer 席勒塑性仪(测泥浆流动性用)
shear failure 剪破裂;剪力破坏;地基破坏;剪力破坏
shear failure calculation 剪力破坏计算
shear failure formula 剪力破坏公式
shear failure load 剪力破坏荷载
shear failure risk 剪力破坏危险
shear fatigue test 剪切疲劳试验
shear fault 剪切(式)断层;剪(力)断层
shear field 切变场
shear fillet 受剪贴角焊缝
shear fissure 剪切裂纹
shear-flocculation 切变絮凝
shear flow 剪切流(动);剪流;剪力流(动);黏性流
shear flow-induce aggregation 同向凝聚
shear flow theory 剪切流理论;切变流理论
shear flow turbulence 剪切流紊动;切变流紊动
shear fluctuation 剪力絮凝
shear fold 剪褶皱;剪切褶曲
shear folding 剪切褶皱
shear force 剪切力;抗剪力;切力;切变(应)力
shear force factor 剪力系数;抗剪力系数
shear forming machine 剪切成型机
shear for punching hole 剪力冲孔
shear fracture 剪切破裂;斜面断口;剪力破坏;剪切裂隙;剪切断裂;剪切断口;剪切裂;剪裂(面)
shear fracture percentage 塑性断口百分率
shear fracture transition temperature 剪切断裂转变温度
shear friction 剪切摩擦(力);摩擦剪力
shear-friction factor 剪摩擦系数
shear-friction factor of safety 剪摩擦安全系数
shear front 剪切面
shear gate 剪(刀)式闸门
shear ga(u)ge 定尺机
shear-ga(u)ge length indicator 定尺指示器
sheargraph 剪应力记录仪
shear gravity 纯重力
shear gravity wave 切变重力波
shear head 剪切器头;抗剪柱头
shear head lashing 人字起重架绑绳
shear height 倾斜切刃两端高度差
shear hinge 剪切铰
shear hulk 起重机船
sheariness 表面失去光泽;切变性
shearing 直立截槽(法);剪切(作用)
shearing action 剪切作用
shearing and punching machine 冲剪两用机
shearing area 剪切面积;剪切面;受剪面积

shearing beam 剪切梁
shearing belt 剪切带
shearing block 抗剪块体
shearing box 剪切盒
shearing bushing 凿环
shearing center 剪心;剪切中心
shearing check 剪切检验
shearing clutch 剪销式超载安全离合器
shearing coefficient of seismic force 地震荷载剪切系数
shearing compression 剪切压缩
shearing compressive plane 扭压性结构面
shearing connection 剪切接合
shearing crack 剪切裂缝
shearing crocodile machine 杠杆式剪断机
shearing current 切变流
shearing curve 剪切曲线
shearing cut 剪断切削
shearing deformation 剪切变形
shearing deformation angle 剪切变形角
shearing design 抗剪设计
shearing diagram 剪力图
shearing die 剪切模;剪断模
shearing difference 剪力差
shearing dislocation 剪切位错
shearing distortion 剪切变形
shearing edge 剪切刃口
shearing effect 剪切效应
shearing elasticity 剪切弹性
shearing equipment 剪切设备
shearing face 受剪面
shearing factor 剪切系数
shearing failure 剪切破坏;剪断断裂;切变断裂
shearing failure calculation 剪切破坏计算
shearing failure formula 剪切破坏公式
shearing failure load 剪切破坏荷载
shearing fault 剪切断层
shearing field 剪切场
shearing figure 剪切图
shearing flow 切变流(动)
shearing force 剪(切)力
shearing force diagram 剪力图
shearing force transmission 剪力传递
shearing force value 剪力值
shearing instability 剪切不稳定性;切变不稳定性
shearing intensity 剪切强度
shearing interferometer 切变干涉仪;错位干涉仪
shearing length 剪断长度
shearing limit 抗剪极限
shearing limit value 抗剪极限值
shearing load 剪切荷载;剪切负荷
shearing machine 地毯剪平机;剪切机(械);剪(断)机;剪床
shearing machine table 剪床工作台
shearing modulus 切变模量;剪切模数;剪切模量
shearing modulus of elasticity 抗剪弹性模量;弹性剪切模量
shearing moment 剪切力矩
shearing motion 剪切运动;切变运动
shearing movement 剪切运动;切变运动
shearing nut 抗剪螺栓
shearing-off 剪切;切断
shearing pin 剪断销;安全销(钉)
shearing plane 剪断面;剪切面
shearing pressure 剪压力
shearing principle 切变原理
shearing punch 切料冲头
shearing range 剪切范围
shearing rate 剪切率
shearing resistance 抗剪(阻)力;抗剪强度;剪切抗力
shearing rigidity 抗剪刚度;剪切刚度
shearing ring 剪力环
shearing section 剪切断面;剪切截面;承剪断面
shearing spectrum 剪切谱
shearing stiffness 抗剪刚度;剪切刚度
shearing strain 剪切变形;剪(切)应变;切应变
shearing strain energy criterion 剪应变能强度理论
shearing strain rate 剪应变速率
shearing stream homogenizing silo 切变流均化库
shearing strength 剪切强度;抗剪强度;切剪强度;抗剪(断)强度
shearing strength of serous material 浆液抗剪强度

shearing strength of soil 土壤抗剪强度
shearing strength of wood along the grain 顺纹抗剪强度
shearing strength parallel to the grain 顺纹剪切强度;顺纹抗剪强度
shearing strength test 抗剪强度试验
shearing stress 剪(切)应力;切应力
shearing stress component 剪应力分量
shearing stress diagram 剪应力图
shearing stress distribution 剪应力分布
shearing stress flow 剪应力流
shearing stress gradient 剪应力梯度
shearing stress time history 剪应力时程
shearing stress trajectory 剪应力轨迹
shearing stress velocity 剪应力速度
shearing surface 剪切面
shearing tensile plane 扭张性结构面
shearing test 抗剪(强度)试验;剪切试验;剪力试验
shearing test of soil 土的抗剪试验
shearing thickness 切断厚度
shearing tool 剪切刀具
shearing transmission 剪切传递
shearing type shredding 剪断破碎
shearing wave 切变波
shearing work 剪切工作;剪工工作
shearing zone 剪切区
shear instability 剪切失稳
shear joint 剪力连接;剪力节点;剪(切)节理;受剪节点
shear key 抗剪榫;抗剪键;剪力键
shear kink bands 剪切膝折带
shear knife 剪切机刀片;剪切刀
shear lag 剪滞;剪力滞后;剪力后滞
shear layer 剪切层
shear-leg crane 人字(吊臂)起重机;三脚(式)起重机
shear-leg derrick 人字吊臂起重机;三脚(式)起重机
shear-leg derrick crane 合撑式起重机;双肢起重杆;双脚式起重机
shear legs 起重机挺杆;起重机三脚架;人字起重机;人字(起重)架;三脚(吊)架;三脚架;动臂起重架;剪形起重机
shear line 剪切线;剪力线;切变线
shear lip 切变裂痕
shear load 剪切荷载;剪力荷载
shear log pole 起重木(杆)
shear machine 剪切机
shear mark 剪刀印
shear measuring gear 剪切机活动定尺挡板
shear member 受剪构件
shear mill 剪切磨机
shear mixing 剪切混合
shear modulus 抗剪弹性系数;剪切弹性模数;剪(弹性)模量;剪变模量;切变模量
shear modulus of elasticity 剪切弹性模量;剪切弹性模量
shear moment 剪切力矩
shear moment equation 剪切力矩方程(式)
shear motor 剪切机电动机
shear mouth 剪状齿
shear multibond fender 多层结合的剪切型护舷
shear of lines of forces 磁力线的剪切
shear of magnetic field lines 磁力线的剪切
shearometer of mud 泥浆切力计
shear on girder edge 梁缘剪力
shear-out 切开
shear pattern 剪切型;剪切形式;剪切模式
shear pin 剪力销;剪切销;剪切锁销;受剪销钉
shear pin bush 安全销衬
shear pin type coupling 安全销钉连接器
shear pin type relief valve 自断式安全阀
shear pin type safety joint 安全销钉连接
shear plan 船体线型图
shear plane 剪切(平)面;剪力面;切应面;切变面
shear plate 抗剪加劲板;剪切板;剪力板;切边的中厚板
shear-plate connector 抗剪连接板;剪盘连接件
shear-plate connector joint 剪盘连接
shear-plate nozzle 涡流片式喷嘴
shear pole 轻便人字井架加固杆
shear pole derrick 两足人字起重机
shear-pole draw bridge 支柱吊桥
shear position shift model 剪切位错模型
shear problem 剪力问题

shear property 剪切性能
shear property of soil 土的抗剪性
shear rams blowout preventer 剪切闸板防喷器
shear range 剪切范围
shear rate 剪切速率;剪切程度;切应变速率;切变速率
shear-rate hardening 切速硬化
shear-rate softening 切速软化
shear-rate thickening 切速稠化
shear-rate thinning 切速稀化
shear rate thixotropy 切速触变性
shear ratio 剪切比
shear redistribution 剪切重新分布
shear reinforcement 抗剪配筋;抗剪钢筋
shear reinforcing bar 抗剪钢筋
shear release 安全塞
shear resistance 抗剪阻力;抗剪性;抗剪强度;剪切阻力
shear-resisting 抗剪阻力
shear-resisting element 抗剪构件
shear response 剪力反应
shear rigidity 抗剪刚度
shears 剪刀;人字起重架;起重三脚架;升降托架;大剪刀
shear screw 安全螺钉
shear-seal type valve 平板遮盖阀
shear section 剪切断面
shear-sensitive pigment 易分散颜料;对剪切灵敏颜料
shear shoulder 桨叶附根
shear shredder 剪切式碎渣机
shear single-unit 单组抗剪装置
shear slide 土崩;剪切滑坡;剪切滑坍;剪切滑动;剪力滑坡
shear slip 剪切滑移
shear slump 剪切型坍落
shear softening 剪切稀化;剪切软化
shear span 剪切跨度;剪跨
shear span length 剪跨长度
shear span to depth ratio 剪跨比
shear span to ratio 剪跨比
shear speckle interferometry 剪切散斑干涉测量
shear spectrum 剪切谱
shear spinning 剪力旋压
shear splice 抗剪拼接
shear spring rate 剪延伸率
shear stability 抗剪稳定性;剪切安定性
shear stability test 剪节稳定性试验
shear stabilization 剪切致稳
shear steel 高速切削钢;刃钢;刀具钢;刀剪钢;剪力钢筋;剪(切)钢
shear stiffness 抗剪劲度;剪切劲度
shear strain 剪(力)应变;切向应变
shear strain curve 剪应变曲线
shear strain of given direction 给定方位的剪应变
shear strength 抗剪强度;剪切强度
shear strength envelope 抗剪强度包(络)线
shear strength of cohesionless soil 无黏性土抗剪强度
shear strength of cohesive soil 黏性土的抗剪强度
shear strength of fissured clay 裂缝黏土剪切强度
shear strength of footings 基础的抗剪强度
shear strength of soil 土的抗剪强度
shear strength parameter 剪切强度参数
shear strength test 抗剪强度试验
shear strength test of froze soil 冻土抗剪强度试验
shear stress 剪(切)应力;黏性摩擦应力;切向应力;切变应力
shear stretch 切变拉伸
shear-strip indicator 剪力板指示器
shear structural plane 扭性结构面
shear structural system 扭动构造体系
shear structure 剪力结构;剪切构造;剪力构造;切变结构
shear stud 抗剪无纹栓杆;抗剪短杆
shear surface 剪切面
shear-susceptible 易被剪切的
shear tectonic system 扭动构造体系
shear test 抗剪试验;剪切试验
shear test block 抗剪试件;剪切试件;剪力试件
shear testing machine 剪切试验机
shear test of concrete blocks on rock surface 混凝土与岩石接触带的抗剪试验
shear test of discontinuities 软弱结构面的剪切试验;不连续面的剪切试验
shear test of rock mass 岩体剪切试验方法
shear test of soil 土壤抗剪试验;土的剪切试验
shear test of weak plane 软弱结构面的剪切试验
shear tests for soil 土壤剪切试验
shear theory (材料破坏的)剪切理论;切变理论
shear thickening 胀流性;胀流型流动;剪切增稠;剪切稠化;剪切变稠;剪力增长;触稠性
shear thickening effect 剪切稠化效应
shear thickening fluid 剪切增稠流体
shear thinning 剪切软化;剪切稀化;剪切变稀;剪力减小
shear thinning effect 剪切稀化效应
shear thinning fluid 剪切变稀流体
shear thixotropy 剪切触变性
shear through-the-thickness 穿透厚度剪切
shear thrust 剪切推力;剪冲断层;切变推力
shear tie 剪力拉杆
shear to moment ratio 剪弯比
shear tool 剪刨刀
shear transfer 剪切传递
shear transformation 切变(相变)
shear-type boudin 剪裂型石香肠
shear-type cutting off die 带条型冲孔
shear vane 土壤抗剪强度环形测定仪
shear vector 切变向量;切变矢量
shear velocity 剪切速度;剪切流速;切变速度
shear viscosity 剪切黏性;切变黏度
shear-viscosity coefficient 剪切黏滞系数;剪切黏性系数;切变黏滞系数
shear-viscosity function 剪切黏度函数
shear wall 抗剪力墙;剪力墙;结构墙
shear wall core 剪力墙筒体;剪力墙筒体
shear wall-frame system 具有剪力墙的框架系统
shear wall structure 剪力墙结构
shear wall with opening 开洞剪力墙
shearwater (拦木材的)浮式拦河埂;(防止冰块冲击桥墩的)护栏木设施
shear wave S—波;剪切波;剪力波;横浪;切变波
shear wave geophone 横波检波器
shear wave velocity 剪切波速
shear web 抗剪腹板
shear wedge failure 剪切楔形破坏
shear welder 剪切—焊接机组
shear work 剪切功
sheary 切变性
shear yield stress 剪切屈服应力
shear zone 剪碎带;剪切区;剪切带;剪力区;剪力带;切变区;断层区
sheath 钢筋套管;钢筋鞘管;护套;铅包的;外皮覆盖物;外层;套(管);挡板
sheath coat 钢筋稍管涂层(预应力混凝土)
sheath column 鞘柱
sheath current 表皮涡流;表皮电流
sheathe 覆盖;套;插入销
sheathed 装有护套的
sheathed cable 铠装电缆;有甲套电缆;包皮电缆
sheathed compound beam 钉板组合梁
sheath eddies 外皮涡流
sheath eddy current loss 铠皮涡流损耗
sheathed electrode 敷皮焊条;敷有药皮焊条
sheathed explosive 铠装炸药
sheathed flame 屏蔽火焰
sheathed plug 屏蔽式电热塞
sheathed ship 包板船
sheathed wall 夹层墙
sheathed wire 铠装线;金属护皮电缆;金属包皮导线
sheathed wire rope 密封钢丝绳
sheath effects 外皮效应
sheath extrusion 护套挤压
sheath fold 剑鞘褶皱;鞘褶皱
sheath for magnet coils 电磁铁护罩
sheath gas 屏蔽气
sheathing 夹衬板;木望板;模壳板;模板;韧性橡皮护皮;望板;外白板;船帮材;覆套;复材;保护套管;包皮;护套
sheathing board 挡土板;(屋顶瓦下的)夹衬板;模板;护墙板;屋面板;衬板
sheathing brick 望砖
sheathing bronze 镶饰青铜
sheathing compound 涂装用混料
sheathing felt 覆盖油毡;沥青浸渍油毡;衬底油毡
sheathing material 外壳材料
sheathing nail 壁板钉
sheathing paper 防潮纸;隔热纸;绝热纸;屋顶用油毛毡;衬纸;柏油纸
sheathing plate 盖板;模板;衬板
sheathing tape 覆盖钢带
sheathing tile 望砖
sheathing wire 铠装线;被覆线
sheath knife 皮鞘刀
sheath-like structure 鞘状结构
sheath loss 包皮损耗
sheath of hydration 水化膜
sheath of sting 刺鞘
sheath of tendon 预应力筋套管
sheath pile 套桩;板桩
sheath post 板柱
sheath-reshaping converter 外壳整形变换器
sheath restrictive coating 铠装
sheath rolling 包壳轧制
sheath-type curtain wall 全玻璃悬墙;壳状悬墙;挡板式幕墙
sheave 井口滑轮;绞缆轮;滑轮;滑车轮;皮带轮
sheave beams 滑轮梁;(支承电梯的)钢梁
sheave block 滑车组
sheave bolt 模板螺栓
sheave boss 滑轮毂
sheave drive 槽轮传动
sheave fork 绞缆(轮)叉
sheave groove 滑轮槽
sheave of friction 摩擦轮
sheave pulley 滑车(滑轮);滑车
sheave roof truss 单坡屋顶桁架
sheaves (绞盘的)滑轮组
sheave shaft 滑轮轴
sheave to 绞辘拉到头
sheave-to-wire ratio 摩擦轮和钢丝直径比;槽轮和钢丝直径比
sheave transmission 滑车传动
sheave wheel 滑轮;槽轮
sheave wheel horizon 天轮水平
sheave wheel platform 天轮平台
shebang 酒店;赌场;陋室
shebeen 无执照酒店;下等酒店
sheblack rubber insertion sheet 夹布胶片
she bolt 阴螺栓拉杆;凹孔螺栓系杆
shectrock 石膏夹心纸板
shed 中空多面体(一种防波堤块体);栈;机库;货栈;棚子;窝棚;脱落;舍;散发;洒;存货入库
shed area 货棚面积
shed car 棚车
shed cover 库棚顶
sheddage 堆存费
shedder 剥离器;卸件装置;卸货装置;抛料机;推料机;抽出机
shedding 泄出;开口
shedding frequency 散发频率
shedding mechanism 卸载装置;卸料机构
shedding tappet 凸轮推杆随动件
shed dormer 棚顶窗;三角形天窗;单坡老虎窗;屋顶窗;天窗;牛眼窗
shed drying stock 棚干材
shed for cement 水泥仓库
shed forms 梭口形式
shed for periodic(al) mantenance 定修库
shed for shunting locomotive 调车机车库
shedkeeper 仓库管理员;仓库管理人
shed line 分水线;分界线
shed roof 棚顶;单坡屋顶;单面坡屋顶
shed roof truss 单坡屋顶桁架;单坡桁架
shed storage test 储[贮]存稳定性试验
shed-type clinker storage 棚式熟料库
sheelite 自钨矿
sheen(gloss) 糙面光泽;平光光泽;光泽
sheen gloss meter 平光光泽计;糙面光泽计
sheepback rock 羊背石【地】
sheep country 养羊地区
sheep dipping 羊毛浸渍
sheepfold 羊圈
sheep-foot roller 羊蹄路碾;羊蹄压路机;羊滚筒;羊足压路机;羊足碾;羊脚碾
sheep-foot roller compaction 羊足碾压实
sheep-foot tamper 羊脚捣路机;羊脚夯击机;羊捣路机;羊蹄夯击机;羊足夯;羊足碾;羊脚碾
sheep-horn double barrel roller 双筒羊角碾
sheep pen 羊圈
sheep rock 羊背石【地】

sheep run 养羊场地
sheep('s)foot roller 羊蹄压路机;羊蹄滚筒;羊足压路机;羊足碾;羊脚碾
sheepsfoot roller 羊足碾(压机)
sheep shank 缩结
sheep shank knot 长绳临时缩结
sheep shearing unit 剪毛装置
sheep silver 云母
sheepskin 羊皮(纸)
sheepwalk 牧羊场
sheepweed 盐吸草
sheer 舷弧;崖面
sheer aft 船尾舷弧
sheer alongside 斜驶靠拢
sheer at stem 船首舷弧
sheer at stern post 船尾舷弧
sheer cliff 悬岸;峭壁
sheer curve 舷弧线
sheer draft 船体型线图
sheer draught 船体型线图
sheer drawing 船体型线图
sheer earning power 绝对的获利权
sheer forward 船首舷弧
sheer gravity 绝对重力
sheer head lashing 人字起重架绑绳
sheering batten 舷弧样板条
sheer lawns 稀薄细麻布
sheer leg derrick 两脚临时吊杆
sheer legs 起重三脚架;人字起重机;起重机支架;三脚桅座
sheer legs for righting a capsized vessel 倾覆船扳正扒杆
sheer line 舷弧线
sheer mast 人字桅
sheer mo(u)ld 舷弧模
sheer mo(u)lding 舷线护缘
sheer off 驶离
sheer of gunwale 舷弧
sheer plan 船舶侧面图;船体型线图;船体侧面图;侧视图
sheer plank 舷侧厚板
sheer plate 舷侧厚板
sheer pole 桅梯第一级横杆
sheer precipice 绝壁
sheer precipice and overhanging rock 悬岩绝壁
sheer rail 舷线护缘
sheers 人字起重机;起重柱;三脚桅座;人字起重架;起重三脚架
sheer sand 纯砂
sheers derrick 人字起重架
sheer strake 舷侧厚板;舷顶列板
sheer up 斜撑起来
sheer weight 净重;净力
sheet 应力线图;岩席;张(薄钢板);漫洪;片材;图幅;床单;帆脚索;薄板;板片;板(材)
sheet action 圆盘作用;薄板作用
sheet adhesive 片状胶黏剂
sheet aluminium 铝板
sheet alumin(i)um core 薄片铝芯
sheet anchor 备用主锚;备用锚;备用船首大锚;船首船锚
sheet anchorage 船着副锚
sheet and coil grinding machine 薄板和带材磨床
sheet and half cap 小四开图纸
sheet-and-half foolscap 四开图纸
sheet and kieselguhr filter 板框硅藻土过滤机
sheet and rill erosion 薄层及细沟型冲刷
sheet and third foolscap 小四开图纸
sheet-annealing base 薄钢板退火用固定式炉底
sheet antenna 平顶天线
sheet area 图幅面积
sheet asbestos 片状石棉;石棉片;石棉板
sheet asphalt 片状沥青;片地沥青(混合料);砂质地沥青(混凝土);薄沥青面层
sheet asphalt pavement 片地沥青路面;沥青砂混合路面;片沥青面层
sheet assembly 接图表;图幅接合表
sheet backing 背板;护背
sheet backing coat 衬垫板层
sheet bar 板料;铁条片;薄钢片;薄板坯
sheet-bar and billet mill 薄板坯小坯轧机
sheet-bar mill 薄板坯车间;薄板坯轧机
sheet-bar shears 薄板坯剪切机
sheet bar stacker 薄板坯垛放机

sheet billet 钢板坯
sheet bitt 小挽缆桩;缭绳挽桩
sheet block 缭绳滑车
sheet blowing 薄膜吹制
sheet bonding adhesive 片状胶黏剂
sheet border 图廓
sheet brass 黄铜薄板;黄铜片;黄铜皮
sheet breaker 平板纸压光机
sheet caisson 金属薄板沉箱
sheet calendar 平板纸压光机
sheet capacitance 箔电容
sheet cassette 金属薄板沉箱
sheet casting 薄片注浆法
sheet cavitation 平面空蚀;片状汽蚀;片状空蚀;表面汽蚀
sheet cementing agent 片状胶黏剂
sheet classifier 薄钢板分选机
sheet clippings 钢板的边废料
sheet cloud 层状云
sheet-coater 板材涂漆机
sheet coffer 金属薄板围堰;金属薄板沉箱
sheet composting 层状堆肥
sheet conductance 薄层电导
sheet control 纸张控制器
sheet copper 紫铜板;紫铜片;红铜片;铜片;铜皮;铜皮;薄铜皮
sheet core 薄片铁芯
sheet corner 图廓点
sheet corner abscissa 图廓点横坐标
sheet corner coordinates 图廓点坐标;图角坐标
sheet corner latitude 图廓点纬度
sheet corner longitude 图廓点经度
sheet corner ordinate 图廓点纵坐标
sheet corner value 图廓点坐标值
sheet coverage 图幅面积;图幅范围
sheet cover(ing) 金属薄板覆盖层
sheet cows 片牛角
sheet cullet 碎平板玻璃
sheet current 表流
sheet cutting table 切片台
sheet dam 薄板坝
sheet demolition charge 薄片爆破装药
sheet deposit 席状矿床
sheet designation 图名;图幅编号
sheet detector 供纸检测器;记录纸检测器
sheet diaphragm 薄铁振动片
sheet dielectric 片状电介质
sheet diked swarm 席状岩墙群
sheet dimension 图幅尺寸
sheet division 图幅分幅
sheet-doubling machine 薄板折叠机
sheet drainage 排水板法
sheet drawing process 平板玻璃引上法
sheet drift 工艺系统图;程序表;流程表
sheet drift water 溢流水
sheet dryer 纸页干燥器
sheeted 席状
sheeted complex 席状杂岩
sheeted excavation 打板桩开挖
sheet edge 图廓
sheet edge cutting device 薄膜切边装置
sheet editing instruction 图幅编辑计划
sheeted lorry 有篷布遮盖的运料车
sheeted ore body 席状矿体
sheeted sand body 席状砂体
sheeted structure 板状结构
sheeted vein 席状脉
sheeted vein deposit 席状矿床
sheeted veins 席状脉群
sheeted wagon 全钢货车
sheeted zone 席状矿带
sheeted-zone veins 薄膜状矿脉
sheeter 席桩;薄钢板竖撑板;护墙板;铁皮屋面工;薄板工;压片机;轧板机;卷筒纸切纸机;轻型插板
sheet erosion 层状冲蚀;面状侵蚀;片状侵蚀;片流侵蚀;薄层冲刷;表面侵蚀
sheet explosive 片状炸药
sheet extrusion 板的挤压
sheet facing 片材饰面
sheet-fed cutter 对开切纸机
sheet-fed offset perfector 单张纸双面胶印机
sheet-fed offset press 单张纸胶印机
sheet-fed offset rotary press 单张纸轮转胶印机
sheet-fed press 手续纸印刷机;单张纸印刷机

sheet-fed rotary letterpress 单张纸凸版轮转印刷机
sheet-fed rotogravure 单张纸轮转凹印机
sheet-fed single colo(u)r perfector 单张纸双面单色胶印机
sheet-fed stop cylinder press 单张停转式凸版平台印刷机
sheet feed 板料送进
sheet feeder 供纸器
sheet feeding device 板料送进机构
sheet field 飞靶射击场
sheet film 散张胶片;散页片;单张胶片
sheet filter 板式过滤机;板框(式)压滤机
sheet flood 漫流;漫洪;片蚀;片流;薄层地表水;表流
sheet-flood deposit 漫流沉积
sheet floor 薄地板
sheet floor cover(ing) 地面覆盖片材
sheet flooring 卷材地面
sheet flow 漫流;坡面漫流;片(状)流;层流;薄层水流;薄层地表水;坡面径流
sheet-folding machine 折页机
sheet format 表格式
sheet former 纸页成型器
sheet forming 纸幅成型;片材成型;板料成型
sheet-forming machine 板材成型机
sheet frost 片霜
sheet furnace 薄板加热炉
sheet-galvanizing equipment 薄板镀锌设备
sheet-galvanizing line 薄板镀锌作业线
sheet gasket 密封垫圈;密封垫片;填密片
sheet ga(u)ge 板规;厚薄规;板料规;薄板厚度
sheet ga(u)ge system 板规体系
sheet glass 普通窗玻璃;平板玻璃;片玻璃;玻璃片;玻璃板
sheet glass drawing 平板玻璃拉制
sheet glass drawing machine 平板玻璃拉制机
sheet glass drawing process 平板玻璃拉制法
sheet glass furnace 平板玻璃池窑
sheet glass process 平板玻璃制造法
sheet glass tank 平板玻璃池窑
sheet grab 板状物件抓具
sheet gum 生板胶
sheet handling machine 码纸机
sheet history 图历表
sheet-holder 薄板定位销
sheet ice 冰层;片(状)冰;水面薄冰层;浮冰;薄冰层
sheet index 接图表
sheeting 席理;页状剥落;护墙板;片状剥离;挖方支撑;挡板;粗布;薄片材料
sheeting backing 背板
sheeting board 护堤板
sheeting board facing 护墙板饰面
sheeting board lining 护墙板衬垫
sheeting cap 板桩帽
sheeting clip 板夹
sheeting cofferdam 板桩围堰
sheeting cover(ing) 护墙板覆面
sheeting dryer 网式干燥器
sheeting driver 板桩帽;打板桩机
sheeting floor(ing)finish 楼面薄板终饰
sheeting for excavation 挖方支撑
sheeting jack 支撑沟槽挡板的千斤顶;挡土板花篮撑杆;板桩千斤顶
sheeting line 薄板精整作业线
sheeting loom 被单织机
sheeting of dam 护堤板;坝面护板
sheeting of trench 槽沟挡板
sheeting pile 板桩
sheeting pile cofferdam 板桩围堰
sheeting piling 板桩组
sheeting plank 板桩;护堤板;桩板;挡土板
sheeting planks of dam 护堤板;护坝板桩
sheeting planks of dike[dyke] 护堤板桩
sheeting plastics 塑料护板
sheeting rail 横档;板轨
sheeting roll 薄板轧辊
sheeting rubber 橡胶护板
sheet(ing)structure 片状构造
sheeting twill 二上二下斜纹
sheeting wall 挡板墙
sheet iron 薄钢坯;铁片;钢皮;铁器;铁皮;铁板;薄铁皮;白铁皮
sheet-iron chute 铁板溜槽
sheet-iron crane 装卸铁板起重机;铁板起重机
sheet-iron encasing 铁(皮)镶板

sheet-iron ga(u)ge 钢板厚度计
sheet-iron panel 铁皮
sheet-iron plate 铁板
sheet-iron shears 钢皮切断机
sheet-iron-siding 薄铁板壁
sheet-iron strip 铁带
sheet-iron tube 薄钢管;薄钢焊接管;焊接管
sheet jelly 粉皮
sheet joint 席状节理;页状节理
sheet jointing 页状剥落
sheet lamination 板的层积
sheet landslide 浅层滑坡
sheet lath 金属拉网;薄板型灰板条
sheet layout 图廓整饰样图
sheet lead 铅皮;薄铅板
sheet letter 图幅编号字母
sheet leveller 薄板校平机;薄板矫平机
sheet lifter 薄板升降台
sheet lightning 片状闪电
sheet-like 薄片状的;席状;板状的
sheet-like pore 片状孔
sheet line 图纸中线;图廓线;地图分幅线;船壳板边线
sheet-line system 地图分幅系统
sheet lino(leum) 油地毯;漆布片
sheet making 地图制作
sheet margin 图边
sheet mastic 沥青玛琋脂板
sheet material 卷材;板材
sheet memoir 地图说明
sheet metal 薄钢板;金属片;金属皮;金属薄板;薄金属板
sheet-metal casing 钢板蜗壳;金属蜗壳
sheet-metal chaplet 箱式泥心撑;金属片芯撑;箱式金属片芯撑
sheet-metal closure 金属板外墙;金属板围墙
sheet-metal cornice flashing 白铁皮挑檐泛水
sheet-metal cover(ing) 金属片材盖板
sheet-metal door 薄钢板门;金属薄板门;空心金属门
sheet-metal drawing 钣金工图
sheet-metal duct 板状管道;金属薄板管道;薄壁管道;薄板管道
sheet-metal fabrication 钣金加工制造
sheet-metal fitter 薄金属板工;钣金工
sheet-metal flashing 金属片材泛水
sheet-metal flat roof 金属板平屋顶
sheet-metal four edge folding cell 板材四面折边单元
sheet-metal free from oxides 酸洗薄钢板
sheet-metal gate 金属板院门
sheet-metal ga(u)ge 钣金标准圆度规;板规;厚薄规
sheet-metal grader 薄板分选工
sheet-metal grinder 板料磨床
sheet-metal lagging 金属薄板外套
sheet-metal masonry wall flashing 砖石墙金属片材泛水
sheet-metal panel 金属薄板板材
sheet-metal parts 钣金零件
sheet-metal perforating machine 金属薄板冲孔机
sheet-metal press working 板料冲压
sheet-metal processing 钣金加工
sheet-metal roller 金属板滚筒
sheet-metal roof cladding 金属片材屋面盖板
sheet-metal roofing 金属(板)屋面
sheet-metal screw 固定金属板的螺钉;钣金螺钉;自攻丝螺钉;自攻螺钉;金属板螺钉
sheet-metal shears 剪板机
sheet-metal sheath 薄钢板望板;包薄钢板
sheet-metal sheathing 金属片材望板;金属片材钢板
sheet-metal shop 钣金车间;白铁车间
sheet-metal shrinker 板收缩机
sheet-metal smoothing roll(er) 金属板平轧机;薄板矫正辊
sheet-metal stringer 带筋的金属板
sheet-metal strip 金属带材
sheet-metal valley gutter 金属片材屋顶排水沟槽
sheet-metal water jacket 金属板水套
sheet-metal water stop 薄钢板止水条
sheet-metal work 金属薄板工程;金属板制品;金属板工件;钣金工程;钣金工工作;铁皮制品;钣金作业

sheet-metal worker 金属薄板工;冷作工(人);钣金工
sheet-metal working 钣金加工
sheet mica 云母片;片状云母
sheet microfilm 缩微平片
sheet mill 薄板轧机;薄板延压机;薄板压延机
sheet-mill shears 薄板剪切机
sheet-mill stand 薄板轧机机座
sheet mineral 页状矿物
sheet motion 片状运动
sheet mo(u)lding compound 片状模塑料
sheet mo(u)lding compound skin 片状模塑料蒙皮
sheet movement 片状运动;层移运动
sheet name 图幅名称
sheet neoprene 氯丁橡胶板;氯丁橡胶片材
sheet number 图号;图幅编号;图表编号
sheet numbering 地形图编号
sheet numbering system 地图分幅编号系统
sheet of armor 装甲板
sheet of curved surface 曲面的叶
sheet of felt(ed fabric) 油毡织物片材
sheet of hyperboloid 双曲面的叶
sheet of reclaimed cellulose for warpping 包装用再生纤维纸
sheet of regional geologic(al) reconnaissance 区调图幅
sheet of the bit 牙轮钻头抓
sheet of water 水层;一片水;一层水;水面
sheet of waters 水域
sheet oil 雨布清油;雨布漆
sheet pack 钢板叠;薄板叠
sheet packing 密封填料;填密片;垫片;封密片
sheet pack separator 叠钢板分离机
sheet pack turnover 板叠翻转机
sheet pallet 薄板托盘
sheet panel 护墙板
sheet paper 硬纸板;衬纸
sheet pasting machine 包装粘糊机
sheet pavement 片地沥青路面;整片式路面;连续路面;片形铺砌
sheet peat 席状泥炭
sheet pick-up 薄板分送机
sheet pile 板桩;组合式板桩
sheet-pile abutment 板桩式桥台
sheet-pile anchorage 板桩锚(固);板桩锚碇装置;板桩锚碇(结构)
sheet-pile anchor tie 板桩拉杆
sheet-pile anchor wall 锚碇板桩墙
sheet-pile breakwater 板桩防波堤
sheet-pile bulkhead 板桩码头;板桩堤岸;板桩岸墙
sheet-pile bulkhead quaywall 板桩岸壁
sheet-pile bulkhead with anchor-wall support 锚碇墙支撑板桩堤岸;有锚碇支撑的板桩堤岸
sheet-pile bulkhead with batter-pile support 斜桩支撑板桩堤岸;有斜桩支撑的板桩堤岸
sheet-pile cell 板桩格笼;板桩格体
sheet-pile-cell structure 格形板桩结构
sheet-pile cell type wharf 板桩格笼式码头
sheet-pile cofferdam 板桩围堰
sheet-pile connector mechanism 钢板桩连接器机械装置
sheet-pile curtain 板桩帷幕;板桩排(用于河岸或松软土壤);板桩截水墙
sheet-pile cut-off 板桩截水墙
sheet-pile cut-off wall 板桩截水墙
sheet-pile dike 板桩坝
sheet-pile dock wall 板桩坞墙
sheet-pile driving 打板桩
sheet-piled wharf 板桩码头
sheet-pile enclosure 板桩围堰
sheet-pile extracting 板桩拔除
sheet-pile extractor 板桩抽出机;板桩拔出机
sheet-pile foundation 板桩基础
sheet-pile hammer 板桩锤
sheet-pile head 钢板桩帽
sheet-pile interlock 板桩锁口
sheet-pile levee 板桩堤
sheet-pile method 板桩法
sheet-pile of tongue and groove 板桩的企口接缝
sheet-pile penetration 板桩入土深度;板桩贯入度
sheet-pile placed by boring 钻孔钢板桩
sheet-pile puller 板桩拔出机
sheet-pile quay(wall) 板桩码头;板桩驳岸;板桩岸墙

sheet piler 垛板机
sheet-pile retaining wall 桩板式挡土墙;板桩(式)挡土墙
sheet-pile screen 板桩帷幕;板桩截水墙;板桩防护栏;板桩遮板;板桩围幕
sheet-pile seawall 板桩驳岸
sheet-pile structure 板桩锚碇结构
sheet-pile structure in anchor 板桩锚碇结构
sheet-pile supported by batter piles 斜拉桩板板桩(结构)
sheet-pile technology 板桩工艺
sheet-pile thickness 钢板桩厚度
sheet-pile type lock wall 板桩式闸墙
sheet-pile type quay 板桩式码头
sheet-pile type retaining wall 板桩式挡土墙
sheet-pile type structure 板桩式结构
sheet-pile type wharf 板桩式码头
sheet-pile wall 板桩墙;板桩挡墙
sheet-pile walling 板桩横栏
sheet-pile wharf 板桩码头
sheet piling 板桩墙;板桩岸壁;一排板桩;打板桩;板桩支撑;板桩(组);板桩作业
sheet piling cell 钢板桩单元
sheet piling cofferdam 板桩围堰
sheet piling enclosure 打板桩围堰
sheet piling machine 打板桩机
sheet piling wall 板桩墙
sheet piling walls for landslide protection 防治滑坡板桩墙
sheet plastic 塑料板
sheet plate 铜板
sheet polarizer 偏振层板
sheet polymer 片状聚合物
sheet pump 泥浆泵;沙泵
sheet-punching machine 冲片机
sheet quarry 水平节理花岗岩采石场
sheet rectifier 薄板矫直机
sheet-reinforced ceramics 金属板加固陶瓷
sheet release agent 脱模用膜材
sheet residuum 面状风化壳
sheet resistance 薄层电阻;表面电阻
sheet resistivity 薄片电阻率
sheetrock 纸面石膏板;石膏板;石膏纤维板
sheet-roller chain 金属板链环滚子链
sheet rolling mill 薄板车间;钢板轧机;钢板厂;薄板轧机
sheet roofing 屋面板
sheet-roofing felt 屋面油毛毡卷材
sheet-roofing nail 薄板屋面用钉
sheet rubber 橡胶板;橡胶片;生胶片
sheets 划艇的首尾部分
sheet sand 席状砂;平伏砂层;砂席;层状砂层
sheet sand deposit of delta front 三角洲前缘席状砂沉积
sheet sandstone 席状砂岩
sheet separation 板(间)间隔
sheet separator 分纸器;板料分离装置
sheet series 图组
Sheet's formula 希氏公式(一种设计混凝土路面的古典公式)
sheet shears 薄板剪(刀)
sheet silicate 页硅酸盐
sheet-skin pass mill 薄板平整机
sheet slab 大板坯
sheet slide 层滑
sheet stackers 垛板机
sheet stamping 薄板冲压
sheet steel 钢片;钢皮;薄钢板
sheet steel casing 薄钢板外壳
sheet steel enamel 钢片搪瓷;钢板搪瓷
sheet steel flume 钢板渡槽
sheet steel form 钢模(板);钢轨模板;钢板模板;薄钢模板
sheet steel lining 薄钢板衬垫
sheet steel mill 薄钢板厂
sheet steel pipe 钢钢管
sheet steel rolling mill 薄钢板压延机
sheet steel straightening unit 薄钢板矫平机
sheet steel(sur)facing 薄钢板饰面
sheet stock 薄板品种;薄钢板
sheet-straightening machine 薄板矫直机
sheet stretcher 薄板矫直机;钢板延伸机;板材拉伸矫平机

sheet strip 薄板条
sheet strip copper 带材铜
sheet structure 片状结构;席状构造
sheet surface flow 层面流
sheet system of topographic(al)map 地形图图幅
sheet-temper mill 薄板平整机
sheet texture 片状结构
sheet thickness calibrator 薄板测厚仪
sheet thickness ga(u)ge 薄板厚度卡规
sheet thrust 席冲断层
sheet tin 锡板;马口铁(皮)
sheet title 图名
sheet train 挤片生产线
sheet vein 席状脉
sheet vinyl floor finish 乙烯地板面层
sheet waffle 格纹板
sheet wall with relieving platform 有效卸荷平台的板墙
sheet wash 洗刷作用;漫洪;洪水泛滥;片状冲刷;片状冲蚀;片蚀;片流
sheet washing 片状冲蚀;薄层冲刷
sheet water-proofing 板材屋面防水;片材屋面防水
sheet wheel 薄片砂轮
sheet winding 扁线绕组
sheet wrench mark 薄板折印
sheet zinc 镀锌铁皮;薄锌板
sheet zinc roof cladding 盖屋面白铁皮
sheet zinc roof cover(ing) 盖屋面白铁皮
Sheffield plate 包银铜片
Sheffield paddles of aeration tank 曝气池浆板风曝气
Sheffield process 舍菲尔德厂炉法
sheilding can-type centrifuge 屏蔽型离心机
sheilding can-type pump 屏蔽型泵
Shelby soil sampler 谢尔贝取土器;薄壁取土(样)器
Shelby tube sampler 谢尔贝薄壁取土器;不扰动土样薄壁取土器
sheletal-frame wagon 钢骨架货车
shelf 搁架;陆棚;陆架;架(子);货架;棚板;底岩;冲积层下岩石
shelf ag(e)ing 搁置老化
shelf angle 锚固砖石砌体用角钢;支座角钢;支撑并锚固砖石砌体用角钢;座角钢;支承角钢
shelf area 陆架区
shelf basin 陆架盆地
shelf bed 薄矿层
shelf board 搁板
shelf bracket 搁板托座;搁板支架
shelf break 陆架坡折;陆架间断;大陆架外缘陡坎
shelf burner 搁架炉;盘架炉
shelf burning 盘架燃烧
shelf car 棚架车
shelf channel 陆架水道;陆架谷
shelf characteristic 储[贮]存性能
shelf cleat 搁板托木
shelf continental slope 大陆架斜坡
shelf corrosion 闲置腐蚀;搁置腐蚀
shelf deposit 陆架沉积(物);大陆架矿床
shelf deposition 陆架沉积作用
shelf depreciation 闲置折旧;跑电
shelf dryer 盘架干燥器;盘架干燥炉;柜式干燥机
shelf drying 盘架干燥
shelf dynamics project 陆架动力学计划
shelf edge 陆架外缘;大陆架边缘
shelf-edge earthquake 陆棚边缘地震;陆架边缘地震
shelf-edge sand trap 陆棚边棱砂体圈闭
shelf energy 架能
shelf facies 陆棚相;陆架相
shelf front basin 陆架前缘盆地
shelf furnace 搁架炉;盘架炉
shelf glacier 陆棚冰川
shelf goods 小罐包装漆
shelf herbaceous facies 陆棚草本相
shelf hydraulic regime 陆架水力状态
shelf ice 大陆边缘冰层;陆架冰;浅海冰层;冰架
shelf interaction 陆棚相互作用
shelf kiln 搁架炉;盘架炉
shelf label 架子标签
shelf life 闲置寿命;搁置时间;搁置寿命;适用期;库存寿命;适用寿命;存放期(限);储存期(限);储藏寿命;储期限;保存期(限)
shelf location 架上安装
shelf margin deposit 陆架边缘沉积
shelf nog (支搁板的)木桩;搁板的木砖;搁板木

桩;搁板木砖
shelf oven 搁架炉;盘架炉
shelf pin 搁板支撑;搁板支架;搁板座;搁板销架
shelf rest 搁板托架;搁(板)座
shelf retaining wall (具有卸载平台的)钢筋混凝土挡土墙;搁架式挡土墙;搁板(扶壁)式挡土墙;衡重式挡土墙
shelf sea 陆棚海;陆架海;浅海
shelf sediment 陆架沉积(物);大陆架沉积(物)
shelf stand 搁座
shelf storage 搁置储存;备用存储器
shelf strip 搁板托木
shelf support 搁板活动支架;搁板座;搁板托架;搁板钉
shelf test 闲置试验;储[贮]存试验;存放试验
shelf thermometer 百叶箱温度计;百叶箱温度表
shelf tile 壁龛方砖
shelf-time 储[贮]存时间;存放期限
shelf trapped wave 陆架拦获波
shelf-type core oven 架式烘芯炉
shelf-type cylinder head 架式汽缸盖
shelf-type retaining wall 搁板扶壁式挡土墙
shelf water 陆架水
shelfy 多石头的
shell 型壳;壳(层);介壳;机壳;毛坯管;滑车壳;壳体;汽锅身;汽缸;坯材疤疙;外套;外壳;外壁;筒体;塔外壳;斗殴;电铸版壳;薄壳
shellac 紫胶(片);胶(漆皮);假漆;漆干片;虫胶漆;虫胶(片)
shellac and white lead 虫胶和铅白粉
shellac and white-lead mortar 虫胶铅白胶泥
shellac bond 虫胶黏结剂;虫胶黏结剂
shellac colo(u)r 虫胶色素
shell account 外壳账户
shellac ester 虫胶酯
shellac insulating paint 虫胶绝缘漆
shellack 涂假漆;涂充漆;虫胶漆
shellac plastics 紫胶塑料;虫胶塑料
shellac resin 虫胶树脂
shellac spirit varnish 虫胶清漆
shell action 薄壳作用
shellac varnish 紫胶清漆;胶漆;虫胶(清)漆
shellac wax 紫胶蜡;虫胶蜡
shellac wheel 虫胶(黏结)砂轮;虫胶结合(剂)砂轮
Shell ag(e)ing test 壳牌(公司)老化试验;薄膜老化试验
shell aggregate 贝壳(类)集料;贝壳(类)骨料
shell analogy 类似壳体
shell analysis 薄壳分析
shell-and-auger bore 砂螺旋钻钻探
shell-and-auger boring (套管螺旋钻的)钻土取样;冲击与螺旋钻探;砂泵螺旋钻探(法)
shell and coil ammonia condenser 壳盘管式氨冷凝器
shell and coil condenser 壳式蛇管冷凝器;壳(盘)管式冷凝器
shell and coil evapo(u)rator 壳盘管式蒸发器
shell and milling cutter 空心端铣刀
shell and plate heat exchanger 板壳式换热器
shell-and-tube 壳管式
shell-and-tube condenser 管壳式冷凝器;列管式冷凝器;壳管式冷凝器
shell-and-tube cooler 管壳式冷却器;管壳式冷凝器
shell-and-tube evapourator 管壳式蒸发器;列管式蒸发器;壳管式蒸发器
shell-and-tube exchanger 管壳式热交换器;管壳式换热器
shell-and-tube fermentor 管壳式发酵器
shell-and-tube heat exchanger 列管式换热器;壳管式热交换器;管壳式换热器;管壳式热交换器;管壳换热器
shell-and-tube preheater 壳管式预热器
shell-and-tube reactor 列管式反应器
shell-and-tube-type floating head heat exchanger 管壳式浮头换热器
shell-and-tube water condenser 壳管式水冷凝器
shell aperture 壳口
shell arch 薄壳拱
shell arch bridge 薄壳拱桥
shell arched dam 薄拱坝
shell area 壳体面积
shell auger 合抱钻;壳钻(又称清孔钻);清孔钻
shell axis 壳体轴线
shell bank 贝壳沙滩;贝壳堤;贝壳岸

shell beach 贝壳沙滩
shell bearing 轴承壳
shell bearing boring machine 轴承壳车孔机
shell-bearing clay 含贝壳黏土
shell-bearing system 壳体承载体系
shell bed 贝壳层
shell bedding 砌空隙垫层;双灰条铺砌
shell belt 壳体圈带
shell bending theory 壳体弯曲理论
shell bevel 贝壳状斜边
shell bit 壳形钻;筒形钻;筒式钻头;木工打孔钻头
shell boiler 火管锅炉;筒式锅炉
shellboring organism 钻贝生物
shell bossing 轴包架;尾轴毂
shell boundary 壳体边界
shell-boundary stress resultant 壳体边界
shell broach 套式拉刀
shell buckling 壳体压曲
shell calculation 薄壳计算
shell capacity 油槽车容量;槽车容量;薄壳容器
shell carving 贝雕
shell carving picture 贝雕画
shell-cast concrete chimney 现浇薄壳混凝土烟囱
shell casting 壳形铸件
shell center line 薄壳中线
shell clam dredge(r) 抓斗式挖泥船
shell clearance 扩孔器与钻头直径之差;套管间隙
shell coated sand 复模砂
shell coefficient 薄壳系数
shell concrete 贝壳混凝土
shell concrete construction 贝壳混凝土建筑
shell concrete pile 外壳混凝土桩
shell concrete roof 贝壳混凝土屋顶
shell configuration 壳体外形
shell constant 薄壳常数
shell construction 壳体结构;壳体建筑;薄壳结构;薄壳建筑;薄壳构造
shell construction method 薄壳施工方法
shell cooling 炉身冷却
shell core 壳芯;薄壳芯
shell core blower 壳芯吹制机
shell core drill 筒形扩孔钻
shell core process 壳芯法
shell cover 壳体封头;端盖
shell cross-section 壳体横断面
shell crown 壳体顶点;壳体拱度;壳冠
shell cupola 薄壳圆顶
shell curvature 壳体曲度
shell curved in two directions 双向弯曲薄壳
shell cutter 外壳切割;套式铣刀
shell dam 薄壳坝;硬壳坝
shell dead load 壳体静载
shell debris 贝壳碎屑
shell deflexion 薄壳挠曲;薄壳挠度
shell design 薄壳设计
shell displacement 船壳排水量
shell dome 薄壳穹顶
shell dredge(r) 贝壳挖泥船
shell drill 壳型钻;筒形钻;套装扩孔钻;套式扩孔钻
shell dungeon 壳体城堡主楼;壳体城堡主塔
shelled concrete pile 有套管混凝土桩;带套管的钻孔灌注桩;带壳混凝土桩;包壳混凝土桩;套管混凝土桩
shelled corn cleaner 玉米清选机
shelled cylinder 脱粒滚筒
shell edge 壳体边缘
shell effect 壳体效应;外壳效应
shell element 薄壳构件
shell end mill 圆筒形端铣刀;空心端铣刀;套装立铣刀;套式立铣刀
shell end milling cutter 套装立铣刀
shell equation 薄壳方程
sheller 去皮机;脱粒机
sheller cylinder 搓擦滚筒
shell erection 壳板装配
sheller unit 脱粒装置
shelless 挖孔灌注桩
shell expansion 外板展开图
shell expansion indicator 汽缸膨胀指示器
shell expansion plan 外板展开图
shell feed plate 铁托板
shell fiber 贝壳本质
shell fine tooth face milling cutter 套式细齿面铣刀
shell firm 空壳公司

shellfish 水生有壳类动物；水生贝壳类动物；贝壳鱼类
shellfish contamination 贝(壳鱼)类污染
shellfish environment 贝类环境
shellfishery 贝类养殖业
shellfish farming 贝类养殖场；贝壳养殖
shellfishing 贝类养殖；贝类捕捞
shellfish reserves 贝类保护地
shellfish water 贝类养殖水
shellfish waters 富于贝类的水域
shell flange 壳体法兰
shell flange of channel end 管箱侧壳体法兰
shell flange of cover end 盖侧壳体法兰
shell flower 贝壳花
Shell fluid catalytic cracking 壳牌流体催化裂化
shell force 壳体力
shell forces and moments 薄壳的受力和弯矩
shell form 壳体形状
shell formation 蛋壳形成
shell forming organism 成壳生物
shell formula 薄壳公式
shell formwork 饰面模板；面板模板(钢筋混凝土)；镶面模板(钢筋混凝土)
shell foundation 壳体基础；薄壳基础
shell frame 舷肋骨
shell gate 薄壳闸门
shell gimlet 套式手钻
shell gold 金色；油金；金饰
shell gravel 贝壳砾石
shell head 壳体封头
shell hole 弹坑
shell ice 壳冰；薄冰盖；冰壳
shelliness 溅麻面
shelling 网状裂纹；龟裂；去皮；剥裂；剥离
shelling border 脱边；掉边
shelling drum 搓擦滚筒
shelling of rail 钢轨表面剥落
shelling of refractory 耐火材料剥落
shelling-out 涂刷
shelling seed husking 脱粒
shell innage 容器充满部分；槽车剩余量
shell instability 薄壳不稳定性
shell insulator 马鞍形天线绝缘子
shell joint 壳接
shell keep 壳体(城堡)高楼
shell key 壳体顶部
shell knocker 敲击器
shell landings 外板纵搭接
shell lease 空房租约
shell-less pile 无套管钻孔灌注桩；无套管混凝土灌注桩；无壳桩
shell-less type of pile 无套桩
shell-like flow mark 贝壳状流痕
shell lime 贝壳(石)灰
shell limestone 贝壳石灰石；贝壳(石)灰岩；介壳灰岩
shell liner 筒体衬板
shell load 壳体荷载
shell longitudinal 船壳纵骨
shell manhole 壳体人孔
shell manway 容器壁孔上人孔；壁侧人孔
shell marl 介壳泥灰岩；贝壳泥灰岩
shell material 壳体材料
shell membrane 壳体(薄)膜；壳膜
shell model 薄壳模型；壳模型；壳层模型
shell moisture content 外部含水量
shell mo(u)ld 原版
shell mo(u)lded casting 壳形铸件
shell mo(u)lding 壳形造型；壳模法
shell mo(u)lding casting 薄壳铸造
shell mo(u)lding pattern 壳型铸造用模型
shell mo(u)ld machine 壳形造形机
shell-mound 贝冢
shell normal 薄壳标准
shell of aggregate 集料外壳
shell of boiler 锅炉外罩；锅炉(外)壳；锅炉套
shell of column 柱段
shell of double curvature 双曲薄壳
shell of hyperbolic paraboloid 马鞍形壳体；双曲抛物面壳体；双曲抛物面薄壳
shell of negative curvature 负曲率薄壳
shell of pile 桩套管；桩壳
shell of pipe 管壳(壁)；管壁
shell of positive curvature 正曲率薄壳

shell of radiation chamber 辐射段外壳
shell of rectifier 整流器外壳
shell of revolution 旋转(面)壳；旋转面壳体
shell of rotational symmetry 对称旋转壳体
shell of single curvature 单曲率薄壳
shell of spark plug 火花塞壳
shell of tank 储[贮]罐壳体；罐体；容器外壳
shell of tile 空心砖外壁
shell of tube 管壁
shell of uniform strength 等强度壳
shell of well 井壁
shellolic acid 紫胶酸；虫胶醇酸
shell oscillation 壳体振荡；壳体摆动
shell outage 罐空残留容积；容器未充满部分；槽车减耗量
shell pass 壳程
shell pavement 贝壳铺面；贝壳路面
shell-perm process 盾流处理；乳化沥青防渗灌浆法；包壳防渗法
shell pile 带壳桩；包壳桩
shell pile Aba-Lorenz 现浇钢筋混凝土桩；阿巴洛伦茨薄壳桩
shell plate 壳板；锅炉板；外壳板
shell plate shop 壳板加工车间
shell plating 外板；船壳板
shell point 壳体顶点
shell powder 贝壳粉
shell power feed 自动推进器导向滑板(凿岩机)
shell-pressure pile 有外壳的压力桩
shell problem 壳体问题
shell process 包壳料球法
shell-proof structure 避弹结构；防弹结构
shell pump 泥砂泵；壳(式)泵；砂泵；出泥筒
shell puncture 外壳击穿
shell radial keyway cutter 套式弧形键槽铣刀
shell radius 壳体半径
shell reamer 空心绞刀；套式绞刀；筒形铰刀；套式铰刀
shell receiver 压气爆破筒
shell reinforcement 壳体加固；壳体钢筋
shell research 壳体研究
shell ridge 贝壳堤
shell ring 筒夹；环夹；容器筒节；筒节
shell riveter 外壳铆接机
shell road 贝壳(铺)路
shell rock 贝壳岩
shell roof 薄壳屋顶；薄壳穹顶；壳顶
shell roofed 薄壳屋顶的
shell sand 壳形砂
shell sandstone 贝壳砂石；贝壳砂岩
shell sanitizer 消毒药
shell scanner 筒体扫描仪
shell sculpture 贝雕
shell sections and heads 筒节和封头
shell sector 壳体扇形面
shell segment 壳体断片
shell sequence 壳层序列
shell shake 壳裂；环裂；皮裂
shell shape 壳体形状
shell skin plate 面板
shell-slab 薄壳板
shell slope 壳体斜度
shell slotting cutter 套式插齿刀
shell source 壳源
shell source model 壳源模型
shell still 简单壳式蒸馏釜；简单蒸馏壶
shell strake 船壳列板
shell stress pattern 薄壳应力模式
shell structural system 壳体结构体系
shell structure 壳式建筑物；壳体结构；壳层结构；薄壳结构
shell structure roof 壳体屋顶
shell support 薄壳支座
shell supporting system 壳体支承体系
shell surface 贝壳路面
shell system 壳体体系
shell tap 筒形丝锥
shell test apparatus 筒体椭圆度测量仪
shell theory 薄壳理论；壳体理论
shell theory with moments 有力矩理论
shell theory without moments 无力矩理论
shell thermocouple 壳式热电偶
shell thickness 薄壳厚度；壳体厚度；蛋壳厚度
shell top 壳体顶部

shell-to-shell distance 壳间距离；设备外壳间距
shell-tube heat exchanger 壳管式换热器
shell type 密闭式；贝壳形；铁壳式
shell-type baffle 迷宫式障板；同心环形障板
shell type boiler 管壳式锅炉
shell-type construction 壳体结构
shell-type edge seamer 贝壳式缝边机
shell-type motor 封闭型电动机
shell-type surface attemperator 管壳式表面减温器
shell-type transformer 铁壳变压器
shell-vault 薄壳拱筒；薄壳穹顶
shell vertex 壳体拱度；壳体顶点
shell vibration 壳体振动
shell wall 壳壁；外壳壁
shell weight-carrying system 壳体承重体系
shell width 壳体宽度
shell with edge vertical beams 有垂直边梁的壳体
shell with ribs 加肋薄壳；带肋薄壳
shell with valve 有瓣膜的壳体
shell work 贝壳工艺品；壳体作业；壳体工程；壳体施工；薄壳作业
shelly 壳状；壳裂的；多贝壳的；贝壳(状)的
shelly agglomerate 贝壳状集块岩
shelly crack 黑点龟裂(轨条)
shelly facies 壳相；贝壳相
shelly formation 破裂岩层
shelly ground 破裂地层
shelly limestone 介壳灰岩；贝壳(石)灰岩；贝壳灰石；含贝壳石灰石
shelly pahoehoe 有壳绳状熔岩
shelly rock 贝壳石灰岩
shelly sand 含贝壳砂
shelly sandstone 贝壳砂岩
shelly structure 珠光结构；珍珠结构
shelly texture 贝壳状结构
Shelly tube sampler 薄壁管取土器
shelter 隐蔽地；隐蔽处；掩护；遮蔽(物)；屏障；防风雨罩；庇护；保护地；百叶箱
shelter belt 护田林带；海防林；防护林带；防风林带；屏蔽带
shelter belt nets 防护林网
shelter breakwater 避风防波堤
shelter cave 有顶洞
shelter-clad 金属铠装；金属防护罩
shelter cost 储[贮]藏费用；保管费
shelter deck 遮蔽甲板
shelter decker 遮蔽甲板船
shelter deck sheer-strake 遮蔽甲板空间
shelter deck ship 遮蔽甲板船
shelter deck space 遮蔽甲板空间
shelter deck vessel 遮蔽甲板船
sheltered anchorage 有遮蔽的锚地；有掩护的锚(泊)地；有掩蔽的锚(泊)地；有屏障锚地；避风锚地
sheltered area 有屏障水域；隐蔽水域
sheltered area of waters 有掩护的水域；有掩蔽的水域
sheltered arm of the sea 有掩护港口；有掩护海湾；隐蔽海湾
sheltered basin 有屏障港池
sheltered bay 避风海湾；避风河湾
sheltered bedrock 隐蔽基岩
sheltered berth 有屏障泊位
sheltered care home 护理所
sheltered car place 车库；有棚停车处
sheltered coastline 有屏障海岸线；隐蔽海岸线
sheltered dock 有屏障港池
sheltered ground 遮蔽地；防护地区；保护地
sheltered harbo(u)r 有掩护港口；有掩护海湾；有屏障港湾；隐蔽港湾；屏蔽港
sheltered harbo(u)r area 有掩护的港域；有掩蔽的港域
sheltered platform 罩棚月台
sheltered position 遮蔽阵地
sheltered refuge 有棚站台
sheltered shallow water area 防波堤内的浅水区
sheltered shoreline 有掩护海岸线；隐蔽海岸线
sheltered site 隐蔽场所
sheltered situation 隐蔽场所
sheltered station 有遮蔽的观测站
sheltered trade 保护贸易
sheltered water basin 有掩护的锚(泊)地；有掩蔽的锚(泊)地
sheltered waters 有屏障水域；有掩护水域；隐蔽水

域;遮蔽水域
sheltered works 隐蔽工程
shelter facility 容留所
shelter forest 护田林;防护林
shelter forest belt 防护林带;防风林带
shelter forest on farmland 农田防护林
shelter ground 防护地
sheltering 提供住所;提供庇护所
sheltering coefficient 遮蔽系数
sheltering effect 隐蔽作用;庇护效应
sheltering factor 遮蔽系数
sheltering harbo(u)r 避风港
sheltering post 收容所
shelter pore 遮蔽孔隙
shelter rent 居住租金
shelter tent 两人用帐篷
shelter thermometer 百叶箱温度计;百叶箱温度表
shelter tunnel 防空隧道
shelter wall 防护墙;护墙;挡浪墙
shelterwood 庇护木
shelter-wood coppice 择伐式矮林;渐伐矮林
shelter-wood cutting 渐伐
shelter-wood felling 伞伐
shelter-wood group system 划法作业;群状伞伐作业
shelter-wood system 渐伐作业法
shelter-wood uniform system 全林伞伐作业
shelter-wood wedge system 楔状伞伐作业
sheltery 防护
shelve 徐徐倾斜;渐次倾斜
shelves and shoals 暗礁与浅滩
shelving 装入架子;搁板材料;慢慢倾斜;缓斜坡;缓倾面;倾斜;品物架;水平熔灶;搭架子
shelving beach 缓斜海岸;浅滩
shelving bottom 缓倾坡海底;缓倾海底;缓坡河底
shelving bracket 架子托座
shelving coast 缓斜坡海岸
shelving section 搁板部分
shelving shore 缓斜坡海岸;倾斜岸滩
shelvy 多浅滩的;多暗礁的
Shepherd-corrected estimator 谢泼德校正估计量
Shepherd pentration-fracture test 谢泼德断口淬透试验
shepherd test 高碳工具钢的淬火性试验;高碳工具钢淬火试验
shepwood 上釉砌块;上釉砖
sherardise 钢的镀锌;粉末镀锌;镀锌防锈;锌粉热镀
sherardising 镀锌粉;扩散镀锌法
sherardize 钢的镀锌;粉末镀锌;镀(防锈);锌粉热镀
sherardizing 喷镀;镀锌防锈法;(粉(末)镀锌
sherardizing galvanizing 粉末镀锌
sherardizing process 粉末镀锌防锈法
sherd 陶瓷碎片
shergottite 辉熔长石无球粒陨石
sheridanite 透绿泥石
sheriff 舍利夹拉杆(一种开闭扇形窗或天窗的拉杆)
sheriff's deed 官契(由县法院出具的所有权证书)
sheriff's sale 官卖(由法院组织的拍卖)
Sheringham valve 谢林哈式铰链调节瓣通风装置
sherrite process 连续薄板电镀锌法
sherry 干白小岛
Sherwin electromagnetic vibrator 舍温电磁振动器
sherwoodite 柱水钒钙矿
Sherzer bascule bridge 滚动式仰开桥;滚动式竖旋桥;滚动式开合桥
shet 塌落(的)顶板
shetterwood strip system 带状渐伐作业法
shew brick 拱脚砖
Shewhart chart 谢瓦尔特图
Shewhart control chart 谢瓦尔特控制图
Shewhart control model 谢瓦尔特控制模型
shibuol 柿汁丹宁;柿涩酚
Shida's number 志田数
shide 灰板条;木瓦;屋面板
shielced carbon-arc welding 气保护碳弧焊
shield 掩护支架;遮挡板;遮蔽;铠装;铠甲;溶残席;屏蔽物;盾形饰;盾构;地层;防护物;防盾;保护物;保护
shield accessories 盾构附件
shield-arc welding 封闭电弧焊;掩护电弧焊接
shield area 防护区
shield assembly shaft 盾构拼装井
shield basalt 玄武岩盾

shield bearing 防尘轴承
shield block 防护块;屏蔽块;盾块
shield board slat 灰板条
shield body 盾构支承环
shield budding 盾状芽接
shield cable 屏蔽电缆
shield-carrying cylinder 盾构装载的轧辊闸门;有护罩的汽缸;屏蔽圆辊闸门
shield chamber 盾构工作室;盾构洞室
shield chamber with prefabricated shaft 盾构拼装竖井
shield chamber with shaft 盾构拼装竖井
shield coefficient 屏厚系数
shield concrete 防辐射混凝土
shield cone 盾状火山锥
shield construction 盾构构造;盾构施工法;盾构法施工
shield construction survey 盾构法施工测量
shield-cooling duct 屏蔽冷却管道
shield disassembly shaft 盾构拆卸井
shield dismantling chamber 盾构拆卸室
shield door 防辐射门;屏蔽门
shield-driven 盾构推进的
shield-driven bore 盾构掘进
shield-driven from working shaft 工作井推出的盾构
shield-driven method 盾构掘进法
shield-driven tunnel 盾构(开挖)隧道;盾构掘进的隧道
shield-driven tunnel(l)ing 用盾构掘进法开挖隧道;盾构隧洞掘进法;铠架隧洞掘进法
shield driving 盾构掘进
shield driving method 盾构(推进)法
shielded 铠装的;有屏蔽的;装有屏蔽的;隔离的;防护的
shielded angle 屏蔽角
shielded antenna 屏蔽天线
shielded arc welding 保护电弧焊;气体保护电弧焊;有保护的电弧焊
shielded assembly 屏蔽装置
shielded box 屏蔽箱
shielded bridge welding 有保护电弧焊
shielded cable 装甲电缆;铠装电缆;屏蔽电缆
shielded carbon arc welding 碳弧保护焊
shielded carbon dioxide arc welding 二氧化碳气体保护电弧焊
shielded case 屏蔽罐;屏蔽箱
shielded cathode 屏蔽阴极
shielded coil 屏蔽线圈
shielded conductor 屏蔽导线
shielded container 屏蔽容器
shielded cooling system 屏蔽冷却系统
shielded cost 屏蔽造价
shielded counter 屏障计数器;屏蔽计数器
shielded cutting 封闭切割;保护式切割
shielded door 屏蔽门
shielded effectiveness 屏蔽效率
shielded engineering 屏蔽工程
shielded facility 屏蔽设施
shielded factor 遮蔽系数;屏蔽系数
shielded flame 屏蔽火焰
shielded flask 屏蔽容器
shielded flexible cable 屏蔽软电缆
shielded fluorescent tube 带罩的日光灯
shielded from interference 防止干扰
shielded galvanometer 屏蔽式检流计
shielded inert gas metal arc welding 保护式惰性气体金属电弧焊
shielded joint 屏蔽接头
shielded line 屏蔽线路;屏蔽线
shielded loop 屏蔽环形天线
shielded macro camera 屏蔽宏观照相机
shielded magneto 隔电磁电机
shielded measuring instrument 屏蔽型测量仪表
shielded metal arc 气金属电弧焊
shielded metal arc welding 敷层金属保护电弧焊;掩护电弧焊接;自动保护金属极电弧焊;熔化极自动保护电弧焊;气(体)保护金属极电弧焊;保护式电弧焊
shielded metal manual electric arc welding 手弧焊
shielded nucleus 屏蔽核
shielded opening 屏蔽孔
shielded oscillator 屏蔽式振荡器
shielded pair 屏蔽线对

shielded pair cable 屏蔽线对电缆
shielded pit 屏蔽坑
shielded plate tube 屏蔽管
shielded plug 屏蔽(火花)塞
shielded pole instrument 屏蔽极仪表
shielded pole motor 屏蔽极电动机
shielded pond 屏蔽池
shielded probe 屏蔽探针
shielded pump 屏蔽泵
shielded relay 屏蔽计数器
shielded ring 屏蔽环
shielded room 屏蔽室
shielded stud welding 气体保护电栓焊
shielded test apparatus 屏蔽试验装置
shielded test facility 屏蔽试验装置
shielded test reactor 屏蔽试验堆
shielded thermocouple 屏蔽式热电偶
shielded transformer 屏蔽变压器
shielded twisted pair cable 屏蔽双绞线
shielded water 屏蔽水
shielded welding 保护焊
shielded window 屏蔽窗
shielded wire 屏蔽线
shield-erecting chamber 盾构拼装室
shield erection 封闭安装;盾构安装
shield excavation station 盾构车站
shield faced with grid 网络式盾构
shield frame 盾构车架
shield gas 保护气体
shield gas flow rate 保护气体流量
shield glass 驾驶室风窗玻璃;屏蔽玻璃
shield-grid thyratron 屏栅闸流管
shield-hood (盾构的)切口环;(盾构的)前檐;盾构前檐
shield inert gas metal arc welding 惰性气体保护金属极弧焊
shielding 掩盖;屏蔽(物);电屏蔽;保护屏
shielding action 屏蔽作用;屏蔽效应
shielding angle 遮光角;逆光角;屏蔽角
shielding arc welding 盾构电弧焊
shielding atmosphere 保护气体;保护气氛
shielding-barrier equipment 屏蔽设备
shielding block 防护砌块;屏蔽砌块
shielding box 屏蔽箱;屏蔽盒
shielding building 防护建筑;屏蔽房屋
shielding can 隔离罩;屏蔽罩
shielding case 屏蔽罩;屏蔽(壳)箱;屏蔽盒
shielding castle 屏蔽容器
shielding coil 屏蔽线圈
shielding concrete 屏蔽混凝土;防护(用)混凝土
shielding configuration 屏蔽布置
shielding constant 屏蔽常数
shielding construction material 防辐射构造材料
shielding container 屏蔽容器;防护容器
shielding cooling system 屏蔽冷却系统
shielding design 防护设计;屏蔽设计
shielding device 屏蔽装置;屏蔽设备
shielding door 防护门;屏蔽门
shielding door for radioactive rays 放射线防护门
shielding effect 屏蔽作用;屏蔽效应;防护作用
shielding efficiency 屏蔽效率
shielding facility 屏蔽装置
shielding factor 屏蔽系数;屏蔽因子;屏蔽因数
shielding flask 屏蔽容器
shielding gas 保护气(体)
shielding gas atmosphere 保护气氛
shielding geometry 屏蔽几何形状
shielding glass 防护玻璃;屏蔽玻璃
shielding harness 屏蔽系统
shielding heat 隔热;热屏
shielding helmet 屏蔽头盔
shielding layer 屏蔽层
shielding length 屏蔽长度
shielding line 屏蔽线
shielding material 屏蔽材料;防护材料
shielding of nuclear charge 核电荷屏蔽
shielding of nucleus 核屏蔽
shielding parameter 屏蔽参数
shielding plug 屏蔽塞
shielding pond 屏蔽水池
shielding practice 钢流保护法
shielding property 屏蔽性能
shielding shaft 盾构井
shielding slab 屏蔽块;屏蔽板

shielding spectacles 屏蔽眼镜
shielding surface 屏蔽面积
shielding system 隔离系统；屏蔽系统
shielding tank 屏蔽罐
shielding tile 防护砖；屏蔽砖
shielding wall 防护墙；屏蔽墙
shielding window 屏蔽窗；防护窗
shield jack 盾构千斤顶
shield machine 泥水加压盾构；盾构机
shield method 盾构施工法；盾构法
shield of rain ga(u)ge 雨量器防护罩；雨量器挡风罩
shield plate 挡风板；防焰板；防火板
shield propulsion 盾构推进
shield retaining 隔离器
shield ring 挡圈
shield room 隔离室
shield-row tube 屏蔽排管
shield sealer 盾构密封剂
shield-shaped 盾形的
shield shell 盾壳；盾构外壳
shield skin 盾构外壳；盾构机壳体
shield sleeve 护套
shield starting shaft 盾构拼装井
shield steering 盾构纠偏；盾构驾驶
shield support 盾构支护
shield tail 盾尾；盾构尾部
shield tail void 盾尾空隙
shield-test facility 防护试验装置
shield three pair multi-cable 屏蔽六芯电缆
shield through cofferdam 过沉井盾构
shield tube 保护管（铺设电缆用）
shield tunnel 盾构隧道
shield tunnel(l)ing 盾构法掘进隧道；盾构法开挖隧洞；盾构挖进隧道；盾构开挖隧道
shield tunnel(l)ing machine 盾构（挖掘）机；掘进机
shield tunnel(l)ing(method) 盾构法
shield unit of the high leveled platform pier for berthing 高桩码头靠船构件
shield volcano 火山岩穹；火山(丘)；盾状火山；盾形火山
shield wall 防护墙
shield with air pressure 气压盾构
shield with compressed air 全气压盾构
shield with partial compressed air 局部气压盾构；半气压盾构
shield work 盾构工程
shieling（牧场附近的）小屋
shift 移项；移距；移动；中心偏移；值班；工(作)班；轮换；轮班；换挡；换班；偏移；台班；调挡；错箱；班次；漂移
shiftable bulkhead 可拆换的舱壁
shiftable haulage line 移动线路
shiftable mo(u)ld-board 可拆换的刀片
shift aft 挪后桩（船靠码头）
shift anchorage 移锚位
shift and group business accounting 班组经济核算
shift and rotate 移位及轮转
shift and rotate instructions 移位及轮转指令
shift bar 开关柄；换挡杆；变速杆
shift bar arm 换挡杆
shift bed 砂床；动床
shift-boss 当班工长；值班工长
shift chain 移位链
shift changing rule 交接班制【铁】
shift channel 游移河槽
shift character 移位符号
shift code 移位码
shift collar 结合环；离合器活动环箍
shift comb 变位梳
shift control 移位控制；偏移控制；偏移调整；位移调整；变位调整
shift conversion of carbon monoxide 一氧化碳变换
shift conversion unit 变换单元
shift converter 相移变换器；移相变换器；移频变换器；变换装置；变换炉；变换反应器
shift count 移位计数值；移位次数
shift cultivation 轮种；变化种植
shift current 游移水流
shift defect 移位缺陷
shift device 移动装置

shift divisor 移位除数
shift down 下移；调低速挡；接低挡
shift driven shaft 变速被动轴
shift driver 移位激励器
shift driving shaft 变速主动轴
shifted channel 改道河槽
shifted diagonal test 对角线移动测试
shifted divisor 移位除数
shifted gradient 移动定差
shifted indicator diagram 移相示功图
shifted matrix 移位矩阵
shifted penalty function 移位罚函数
shifted signal 漂移信号；偏移信号；时移信号
shifted single-mode fiber 分层心色散位移单模光纤
shifted structure 错列结构
shifted tax 已转嫁的税收
shifted traffic 转变交通
shift effect hypothesis 移动效应假设
shift end 移位终止
shift engineer 值班工程师
shift equilibrium 变换平稳
shifter 移位器装置；移头器；移动装置；转换机构；工班长；换相器；切换装置；切换机构；倒相器
shifter arm 换挡轴
shifter cylinder 关闭汽缸
shifter fork 换挡叉；齿轮拨叉
shifter fork groove 换挡叉轮槽
shifter gate 移动门；开关门
shifter housing 变速杆罩
shifter hub 拨叉凹口
shifter plate 开关板
shifter pole 移动皮带杆
shifter shaft 换挡轴
shifter shaft lock ball spring 移动轴锁球弹簧
shifter shaft support 移动轴支架
shift factor 平移因子
shift fault 位移断层；变位断层
shift filtering 移位滤波
shift foreman 值班工长
shift fork 拨叉
shift forward 挪前桩（船靠码头）
shift frequency 移频
shift gate 移位门
shift gear 调挡齿轮
shift good traffic plan 货运工作计划
shift governor 换挡调速器
shift-group accounting 班组核算
shift helm 反舵
shift-in 移入；移进
shift inboard 向内侧移动
shift-in character 向下转移符号；换入符
shift in demand curve 需求曲线的移动
shift in demand supply 需求供给的转变
shift indicator light 换挡指示灯
shifting 换挡；转移；改组目录；海面起伏；倒垛；拨道；变更分类；移位；移动；漂移
shifting accumulator 移位累加器
shifting agriculture 轮垦农业
shifting amplitude off channel 河槽摆幅
shifting anchor 移锚
shifting anchor without engine stop 移锚不停车【疏】
shifting arm 变速臂
shifting axle 移动轴
shifting ballast 可移动压载物
shifting banks 变迁河岸
shifting bar 游荡性沙洲；移动杆；流动沙洲；迁徙性洲滩；变速杆
shifting base 可变基数
shifting base period 移动基期；变动基期
shifting beam 舱口活动梁
shifting beam carrier 舱口活动梁座
shifting bearing 活动支座
shifting bed 可移动的底部；游荡性河床；不定河床
shifting block 拨头
shifting board 止移板；防动板
shifting board stanchion 防动板支柱
shifting bolt 移动螺栓；延伸圈
shifting bonds 移动键
shifting bottom 不稳定河床；不稳定河底；可移动的底部
shifting cam 换挡凸轮
shifting center 稳心
shifting channel 游荡性河槽；河槽变动河床；不稳定水道；不稳定河槽；不固定河槽；不定河床

shifting charges 移泊费；码头搬运装卸费；搬移费
shifting chock 铰链式艇座
shifting chute 移动槽
shifting coil voltage regulator 移圈式调压器
shifting control 位移控制
shifting control section 移动性控制断面
shifting counter 移位计数器
shifting coupling 变速联轴节
shifting cultivation 轮垦；轮换种植；林农轮作；毁林轮垦
shifting current 不定流；变量流
shifting cylinder 千斤顶
shifting dune 移动沙丘
shifting eye 移动环；吊环；延伸圈
shifting field 移动场
shifting finger bushing 变换指衬套
shifting fork 齿轮换挡叉
shifting fork shaft 移动叉轴
shifting function 移位函数；移位操作
shifting ga(u)ge 划线规
shifting gear 变速
shifting guy system 联杆吊货法
shifting in both ratios 二级传动比调挡
shifting interference fringe 位移干涉条纹
shifting iron 防动板钩扣
shifting lever 变速杆
shifting lever ball 变速(杠)杆球端
shifting lever boot 变速杆罩
shifting lever for exhaust air 排风切换杆
shifting lever spring 移动杆弹簧
shifting links 移动键
shifting method 移袋法（船舶脱浅法）
shifting of a divide 分水岭转移
shifting of brushes 电刷(的)移位
shifting of cargo 货物移动
shifting of channel 改槽
shifting of container 集装箱捣箱；集装箱倒箱
shifting of divide 分水岭移动；分水界移动；分水界迁移
shifting of earth 泥沙游移；泥土移动；土体位移；地表移动
shifting of house tax 房屋税的转嫁
shifting of navigable channel 航道变化
shifting of river 河流改道；河道迁移；河道变迁
shifting of soil 土体位移
shifting of tax 赋税转嫁
shifting of taxation 赋税的转嫁
shifting of the burden of proof 举证责任迁移
shifting of waterway 水道变迁
shifting of wind 风向改变
shifting operator 移位算子；移位算符
shifting order 移位指令
shifting oscillator 漂移振荡器
shifting permit 移泊许可证
shifting plank 止移板；防动板
shifting point 动点
shifting process 移位过程
shifting pump 转货泵
shifting ram 换挡顶杆
shifting reserves 移动预备林
shifting responser 换挡反应器
shifting responser transmission shaft 换挡反应器传动轴
shifting ring 移动环；调整圈；调整环
shifting riverbed 不稳定河床
shifting rock arm bushing 变速摇臂衬套
shifting sand 移动的砂；流动的砂；沙沙(也称流沙)；浮沙
shifting sand control 流沙防治
shifting-sand control forest 防沙林
shifting sand-current 风沙流
shifting schedule 倒班安排
shifting sequence 移动次序
shifting sledge 移动滑轨
shifting sleeve 爪形联轴节；凸轮联轴器
shifting slide gear 滑移齿轮
shifting soil 不固定土壤
shifting spanner 活扳子；活动扳手；活络扳手
shifting stock anchor 活杆锚；活动横杆锚
shifting system of railway enterprise 铁路工作班制
shifting tank 压载水柜
shifting the base 变动基期

shifting theorem 移相定理;移位定理;平移定理
shifting T-square 活动丁字尺
shifting type entry guide 移动式导口
shifting unit 移动装置
shifting up 换挡加速
shifting wind 变向风
shifting winding reel 浮动式卷取机
shifting yoke 拨叉
shifting yoke pin 移动轭销
shift in investment 转移投资
shift instruction 移位指令
shift interval of time window 时窗移动间隔
shift invariant 移位不变性
shift joint 移动接头;错缝
shift key 换位键
shift keying 移位键控
shift knob 开关按钮;翻转开关
shift left 左移(位)
shift length 工班时间
shift lever bracket 变速杆托架
shift lever interlock plunger 变速杆联锁柱塞
shift lever shaft 变速杆轴
shift lock 移位锁定
shift logic(al)instruction 移位逻辑指令
shift matrix 移位矩阵
shift motion 移位动作;移动行程
shift multibyte 移位多数元组
shift network 移位网络;移数网络
shift of butts 端接避距【船】
shift of focus 焦点位移
shift of origin 原点转移;原点移动
shift of pressure center 压力中心移动
shift of spectral line 谱线频移
shift operating plan 班计划
shift operation 移位运算;移位计算;转移操作
shift operation shifting 移位操作
shift out 移出
shift-out character 向上转移符号;移出符号
shift out pulse 移出脉冲
shift parameter 位移参数
shift pedal 控制踏板
shift plunger 移动插塞
shift position 移位位置;移动位置
shift pulse 移位脉冲;偏移脉冲;位移脉冲
shift rail 可移动的轨条;拨叉导轨
shift reaction 转移反应;变换反应
shift reactor 变换反应器
shift reagent 位移试剂
shift register 移位寄存器
shift register code 移位记发码
shift right 右移
shift right logic 逻辑右移
shift right open 向右移位开启
shift roster 轮值表
shift sand 流动沙
shift-ship 工作艘班
shift sleeve 连接套筒
shifts operated per week 每周工作班数
shifts system 轮班制度;移位系统;轮班
shift theorem 移位定理
shift to the right 右移
shift traffic working plan 运输工作班计划
shift transmission 控制机械装置;变速器;逆转换;转换装置
shift-type decade counter 移位型十进位记数器
shift unit 移位器装置;变换装置
shift up 上移
shift-up mode 上移位方式
shift-up operation 上移位操作
shift valve 导阀;控制阀;换挡油阀(自动换挡用)
shift variant 移变
shift velocity 漂移速度
shift winding 移位绕组
shift work 倒班工作制;轮班工作(制)
shift-working 轮班作业
shifty 不稳定的
shift zone 变位带
shikimol 黄樟素
shiki silk 垫布
shikkara 印度寺院的高塔
Shikoku basin 四国海盆
shilf 破损的石板
shilkinite 水硅铝钾石
shilling 先令(英国银币)

shilling-stroke 斜划(线)
shim 楔形填隙片;夹铁;嵌条;偏薄板;垫隙片;垫片;刀垫;分隔片;补偿垫片
Shimadzu sedimentograph 岛津式自记沉降仪
shim and feather 砍石肋条
shim block 垫片
shim bolt 固定衬垫螺栓
shim control 粗稳定调节
Shimer process 氰化浴中快速渗碳法
shim for rail joint 钢轨接头垫片
shim liner 铁夹;填隙垫衬
shimmed 调整垫片
shimmer 微光;闪晃
shimmer error 闪烁误差
shimming 填垫作用;垫补法
shimming cell 单边刮沫充气式浮选槽
shimming plate 填隙板
shimmy 摆动
shimmy damper 转向轮减振装置;减摆器;摆振阻尼器
shimmy damping device 减摇装置;减摆装置;前轮摆动阻尼器
shimmy detector 前轮摆动测定器
shimmy die 侧边打孔模
shimmy-free landing gear 无摆振起落架
shimmy of front wheels 前轮摆动
shimmy-proof 不摆动的
shimmy stop 转向轮抗摆头装置
shim piece 填隙片
shim plate 填隙片
shim rod 粗调整棒;补偿棒
shim rod actuator 补偿棒传动机构
shim washer 填隙垫圈
shiner 发光体;大面;磨边残留;毛石墙边扁石;无光漆发花现象
shine skin 光板皮
shine skin papaya 光皮木瓜
shine test 光泽度试验
Shineton Shales 闪顿页岩
shingle 遮板;盖(屋)板;卵石;挤渣;木瓦(板);木片瓦;平板瓦;屋顶板;屋面盖板;大瓦板;粗砾石;薄屋面板;薄木片;扁砾石
shingle applicator 铺设工;屋顶木板;屋顶瓦板
shingle backer 盖板垫板;大瓦背衬板
shingle bank 卵石滩;砾石滩;碛坝;粗砾岸
shingle barchan 新月形砾滩
shingle barrier 粗砾滩
shingle beach 砾(石)滩;砾石岸滩;石子滩;砂砾海滨
shingle beach ridge 砾石堤
shingle board 削薄木板
shingle bolt 制瓦木段;木瓦短圆材
shingle butt 面板暴露边;盖板突出边
shingle carrying river 输沙河道;输砾河流
shingle construction 套筒式;木瓦结构
shingle cover(ing)屋顶木板;屋顶木瓦
shingle cutter 屋顶木瓦切断机;木瓦切断机;瓦切割机
shingle drift 砾石漂移
shingle flat 砾石滩;碛坝
shingle hanging 屋顶木瓦吊挂;木瓦板吊挂
shingle hatchet 屋面工人使用的木斧
shingle lap 单向搭接;木瓦搭接;瓦片(搭)接
shingle-lap tile 搭接瓦
shingle layer 粗砾层;屋顶木瓦层;屋顶板层;扁砾石层
shingle lining 木板衬垫
shingle nail 屋面板钉;墙面板钉;大瓦钉;木瓦(用)钉
shingle panel 墙面板;嵌板;屋面板;镶板
shingler 挤渣压力机;挤压操作工
shingle ridge finish 波斯顿屋脊;大瓦屋脊铺砌法;波纹脊饰
shingle roll 屋面卷材
shingle roof 木片屋顶;大瓦板屋面
shingle roof cladding 木瓦覆盖屋顶
shingle roof covering 木板屋面;木板铺屋顶
shingle roofing 木板铺屋面;用木板铺屋面;木(瓦)板屋面
shingle saw 瓦片锯
shingle shell construction 瓦片式炉壳结构
shingle siding 鱼鳞板;木板墙面
shingle stain 盖板色斑;盖板疵点;屋顶墙面板涂料;大瓦涂料;板墙油色;板墙涂料
shingle-stone 粗砾石;扁砾石

shingle structure 叠瓦构造
shingle tile 木板屋面瓦;屋面板瓦;木瓦;大平(黏土)瓦
shingle trap 沉砂井;拦沙墙;拦沙坎
shingle underlayment 油毡瓦垫层
shingle valley 木板瓦天沟
shingle wall 木板墙
shingle wall lining 木板墙衬垫
Shinglewood 西部红杉;大瓦木材
shingle work 大片瓦工作
shingling 采用木瓦盖屋面;挤渣锻铁法;叠瓦作用
shingling hammer 压紧锤;锻(炼)锤
shingling hatchet 铺大瓦用短柄斧;起钉斧
shingling press 挤渣压力机
shingling roll 挤渣轧辊
shingly 多砾石的
shingly shore 粗石子石岸
shining intensity of phosphor 磷光体发光强度
shining soot 发亮炭黑;发亮烟煤灰
shining through 透光
Shinmel emulsion 铵力散混剂
Shinto shrine 神道神殿(日本)
shiny clay 发光的黏土;伊利石
shiny-leaved yellow horn 文冠果
shiny surface 光滑面
shioze colo(u)r checked 彩格纺
shioze colo(u)r striped 彩条纺
ship 船舶;倒卸车
ship abandoner 弃船人
ship-and-galley tile 防滑(铺)地砖
ship and skid system 吊滑饼用法
ship and warship 舰船
ship apparent time 船视时
ship arms 船用武器
ship a sea 冒着波浪;波浪打上甲板
ship assembly yard 船舶装配厂
ship astern 在后面的船
ship auger 船用钻
ship axe 船用宽口斧
ship ballast 压舱水;船上压舱水
ship barnacle 茗荷儿
ship-based aircraft 舰载飞机
ship-based hydraulic excavator 船舶式液压挖掘机
ship beacon 船舶导航用信标
ship beam 船宽;船幅
ship bell 船上报时铜钟
ship berth 船泊位;停船处;船台
shipboard 舷侧;在船内;靠近(他船)舷侧;船舷;船上;船侧
shipboard, off shore cable 船舶及海上平台电缆
shipboard cargo crane 船上载货吊车
shipboard cargo-handling equipment 船上装卸设备
shipboard cargo-handling plant 船上装卸设备
shipboard communication equipment test station 船舶通信设备试验站
shipboard coring technique 船上钻探取样技术;船上岩芯钻取技术
shipboard crane 甲板起重机;船上起重机
shipboard domestic waste disposal 船上生活废物处置
shipboard domestic waste disposal system 船用生活污水处理系统
shipboard double crane 船用双钩起重机
shipboard earth station 船载地面站
shipboard gantry crane 船上桥式吊(运)车
shipboard gravimeter 船用重力仪
shipboard gravity survey 船载ома重力调查
shipboard handling equipment 船上装卸设备
shipboard handling plant 船上装卸设备
shipboard ice navigation support system 船载冰区航行支持系统
shipboard integrated maintenance program(me)船上综合维修计划
shipboard integrated man-machine system 船用综合人—机系统
shipboard measurement 船上测定
shipboard oceanographic(al)observation system 船用海洋观测系统
shipboard oceanographic(al)survey system 船载海洋调查系统;船用海洋调查系统
ship-board radar 船载雷达
ship-board radar transponder 船载雷达应答器
shipboard satellite communication system 船用

卫星通信系统
shipboard sonar 船载声呐;船用声呐
shipboard synoptic(al) system 船用电子观测和记录发送系统
shipboard technique 船用取心技术
shipboard-type motor 船用电动机
shipboard ultrasonic wave height sensor 船用超声波高计;船用超声波高度传感器;船用超声波测波仪
shipboard waste 船上废物;船舶废弃物;船舶垃圾
shipboard waste treatment system 船舶废物处理系统
shipboard wave meter 船舷测波仪
shipboard wave recorder 船用自记测波仪;船用波浪计
ship body 船体
ship body design 船体设计
ship body offsets 船体尺度平衡表
ship body stress 船体应力
ship borer 蛀木虫;蛀船虫;凿船虫;船蛆
shipborne 船载
shipborne barge 子驳;载驳货船的子驳
ship borne expendable bathythermograph 海水测温仪
shipborne expendable bathythermograph 船用抛弃式温深仪
shipborne magnetic survey 船磁测量
shipborne plane 舰载飞机
shipborne radar 船用雷达
shipborne service 船舶运输业务
shipborne sonar 舰载声呐
shipborne weather instrument 船舶气象仪
ship bottom anti-corrosion paint 船底防锈漆
ship bottom anti-fouling paint 船底防污漆
ship bottom composition 船底油漆
ship bottom fouling 船底污垢;船底生物结污
ship bottom paint 船底(油)漆
ship bottom paint for steel ship 钢壳船底漆
ship bottom varnish 船底清漆
ship-breaker 拆船业者;废船包拆人
ship breaking 拆(废)船;废船拆卸业
ship-breaking yard 拆船厂
ship breasting force 船舶挤靠力
ship broker 水险掮客;船务经纪人;船舶买卖经纪人;船舶经纪人
shipbuilder 造船者;造船公司;造船工作者
shipbuilder draught 船体线型图
shipbuilding 造船(学);船舶建造
shipbuilding basin 造船场
shipbuilding berth 造船台;造船船台
shipbuilding berth with greased launching way 涂油滑道【船】
shipbuilding circles 造船业者;造船界
shipbuilding corporation 造船公司
shipbuilding crane 造船起重机
shipbuilding dock 造船(船)坞
shipbuilding engineer 造船工程师
shipbuilding facility 造船设备
shipbuilding gantry crane 造船门式起重机
shipbuilding hanger 船体车间
shipbuilding industry 造船(工)业;造船业者
shipbuilding lien 造船留置权
shipbuilding material 造船材
shipbuilding rules of classification society 船级社造船规则
shipbuilding section 船用型钢;船用钢板
shipbuilding slip 造船滑道
shipbuilding tower crane 船台塔式起重机
shipbuilding trade 造船业者
shipbuilding way 造船滑道
shipbuilding works 造船厂
shipbuilding yard 造船场;造船所;造船厂;船坞
ship bulk-load pump 装船气力输送泵
ship by rail 铁路运送
ship caisson 船坞闸门;船形沉箱;船形沉箱;浮箱式坞门;浮门;浮式沉箱
ship caisson gate 船形浮坞门
ship caisson with constant draught 吃水深度固定的浮式沉箱
ship caisson with variable draught 吃水深度变化的浮式沉箱
ship calculation 船舶计算;造船设计
ship canal 航运河道;海轮航道;通航河道;深水运河;深水河道;海运河道

ship-canal route 船用渠线
ship cargo 船货
ship carpenter 船木工
ship carrying chamber 承船厢
ship-carrying chassis 承船(底盘)车
ship ceiling 船舶铺仓
ship chamber 升船厢
ship chamber driving device 承船厢驱动装置
ship chandler 船用商品店;船具商;船舶供应商
ship chandlery 船用品
ship channel 航行间距;船用通道;通航水道
ship characteristic 船舶特征
Ship Characterization Board 船舶性能研究委员会
ship cleaning 洗舱
ship clearance 两船之间的横向安全距离;航迹间富裕宽度;船身净宽
ship coal 船用煤(炭)
ship coaling plant 船上装煤设备
ship coating 船用油漆
ship code 船舶代号
ship collision force 船舶撞击力
ship construction 造船学;造船;船体结构;船的构造
ship constructor 造船师
ship control centre 船舶控制中心
ship craft 船舶(尤指小型船舶)
ship crane 船用吊架;船用起重机;造船起重机
ship curve 船用曲线;造船曲线板
ship data 船舶资料
ship decking 船甲板木
ship demurrage 船舶延期费
ship design 造船设计
ship despatching 船舶调度
ship detained 船被扣押
ship-discharging operation 卸船作业
ship dispatching 船舶调度
ship dock 载驳货船码头
ship draft indicating system 船舶吃水指示系统
ship dressing 船舶旗布
ship drift 船只漂移测流法
ship dry 船用干燥剂
ship dunnage 货舱垫料
ship earth station 船舶地球站
ship echo 船舶回波
ship effect station control mode 船站控制方式
ship effect station data test level 船站数据测试电平
ship effect station identification 船站识别
ship elevating plant 升降机上墩下水设备
ship elevator 升船机
ship engineering 船舶工程(学)
ship entry 申报进口
ship equipment 船舶设备
ship fitter 船体装配工;造船装配工;造船划线工
ship flow 船流
ship foreman 装卸组长
shipform 船式;舰船形状;船体线型
ship formation 船舶队形
ship fouling 船体污垢;船体生物结污
ship fouling control 船舶污垢控制;船舶结污控制
ship fuel 船用燃料
ship gammaradiac 船用伽马辐射级仪
ship garbage 船舶垃圾
ship gate 船形浮坞门
ship gear 船上起重设备;船上吊装机械;船上吊货设备
ship girder 船体承受总纵弯曲的构件
ship goods insurance policy 海上货物保险单
ship goods without permit 私自下货
ship handling 船舶操纵
ship handling behavio(u)r 船舶操纵性能
ship handling model 船舶操纵模型
ship handling simulator 船舶操纵模拟装置;船舶操纵模拟器
ship hauling trolley track 引船小车轨道
ship heading mark 船首标志
ship heading marker 船航向标志
ship holder 船东;船舶所有人
ship-hour rate 船时量
ship hours 装卸工作时数;船时
ship-house 造船棚
ship hull 船体;船壳
ship hull paint 船壳漆
ship ice log 船上观测的冰情日记
ship impact force 船只冲撞力;船舶撞击力

ship in American bottoms legislation 以美国轮船载运的立法
ship in ballast 压载航行船;空放船舶
ship incline 斜面升船机
ship incline with turntable bogie truck 转盘式斜面升船机
ship incline with two-level wheels 高低轮式斜面升船机
ship in commission 现役船
ship in disaster 船舶遇难
ship in distress 遇险船只;遇难船只;船舶遇难
ship induced magnetism 感应船磁
ship inertial navigational system 船舶惯性导航系统
ship inspection certificate 验船证书;船舶检验证书
ship insurance 船舶保险
ship jack 船起重器;升船起重机
shipkeeper 看船人
ship lane 船行道;航道
shiplap 错口(接合);搭叠
shiplap board(ing) 搭接板;鱼鳞板;搭叠壁板
shiplap flooring 错口地板
shiplap joint 搭接(接头);错口接合;截口
shiplap lagging 拔叠;搭叠(壁板)
shiplapped joint 搭接接头
shiplapped lumber 搭接材;搭叠材
shiplap sheet(ing) piling 搭接板桩;搭叠板桩
shiplap siding 搭叠壁板;鱼鳞板
shiplap timber-sheet-pile 搭接缝木板桩
shiplap wooden siding 鱼鳞板;搭叠木墙板
ship latest position 船舶最新位置
ship launched underwater transponder 船舶发射的水下应答标
ship launching 船下水
ship launching ramp 船舶下水滑道
shiplift 升船机;船用升船机
shiplift platform design 升船机平台设计
shiplift with counterweight 平衡重式升船机;平衡重举船机
shiplift with floating balancing beams 浮筒平衡梁(式)升船机
shiplift with pontoon 浮筒式升船机
ship light 船灯
shiplike 船样;船形;船式
shipline 船体型线
shipline correction 航线的修正【航海】
shipline survey 航线测量【航海】
ship list 船籍;船的倾侧
ship load 船舶载重;载船量;船货;船的载货量;船舶装货量;船舶荷载
shiploader 船用装料机;装船机
shiploader bridge 装船机(伸)臂
ship loading 船舶装载;装船
shiploading bridge 装船机(伸)臂
shiploading capacity 装船能力
shiploading machinery 装船机械
shiploading operation 装船作业
shiploading plant 装船设备
shiploading system 装船工艺系统
shiploading tower 装船塔架;塔式装船机
shipload lot 整船货物
ship lock 船闸
ship lock headwork 船闸闸首工程
shiplofting 船体型放样
ship log 航海日志;航海日记
ship lost or not lost 不论船舶损失与否
ship magnet induction 船磁感应
ship main antenna 船舶主用天线
shipman 上船工作的装卸工人;船员
ship maneuverability 船舶操纵性能
ship maneuvering 船舶操纵
ship maneuvering model 船舶操纵模型
ship maneuvering simulator 船舶操纵模拟装置;船舶操纵模拟器;操船模拟器
ship manifest 船舶装货清单
ship mass 船舶质量
shipmaster 船机长;船长
ship mate 驾驶员(船);同船船员
shipment 装运;装船;载货;货载;托运;船载运量;发运;发送的货物;发货
shipment advice 装船通知
shipment and delivery cost 运送成本
shipment as soon as possible 尽速装船;尽快装船
shipment by available steamer 有船即装

shipment by cheapest route 以船运费最廉的航线装运
shipment by first opportunity 优先装运；一遇机会即装运；一有机会就装船
shipment by installments 分批装船；分批装运
shipment by steamer 以船舶装运
shipment certificate textile 纺织品装船证明
shipment control system 运输控制系统
shipment documents 发运单据
shipment drawing 装运图
shipment in bulk 散装装运
shipment index 交货指数
shipment instruction 装运须知
shipment method 装运方法
shipment not in accord with the sample 来货与样品不符
shipment of crude 原油船舶输送
shipment of general cargo 件杂货装运
shipment of hazardous waste 危险废物的装运
shipment order 发运通知单
shipment outward 寄销
shipment price 交运价格；发送价格
shipment request 装货申请（书）
shipments billed above cost 计价高于成本之发货
shipments billed at cost 发货按成本计价
shipment store 发料材料厂
shipment term 出货期限
shipment to branch 发往分店商品
shipment wreck 船舶失事
ship meteorological observation 船舶气象观测
ship model 船模
ship model basin 船模（型）试验池
ship model experiment 船模试验
ship model guide method 船模导航法
ship model pilot 船模驾驶员
ship model resistance test 船模阻力试验
ship model self-propulsion test 船模自航试验
ship model tank 船模（型）试验池
ship model test 船模试验；船模航行试验
ship model test towing tank 船模拖曳试验水池
ship mooring 系船设备
ship-mooring bollard 系船柱
ship mortgage 轮船抵押；船舶抵押权
Ship Mortgage Act 船舶抵押法
ship motion 跳动；船体运动；船舶运动
ship motion analyzing computer system 船舶运动分析计算系统
ship motion sensor 船舶运动传感器
shipmotor 船用电动机
ship movement 船舶运动
ship movement service 船舶动态业务
ship name board 船名牌
ship navigation 航海导航；船舶导航
ship navigation computer 船舶导航计算机
ship noise 舰船噪声；船只噪声；船舶噪声
ship observation 船舶观测
ship of destination 收报船
ship off to 用船运开
ship of opportunity 辅助观测船
ship of war 战舰
ship operating expenses 船舶营运费用
ship-operation(al)time 船舶作业时间
ship operator 船舶营运者
ship ordnance 船用武器
ship out of commission 退役船
ship oven 船式炉
shipowner 船方；船主；船东；船舶所有人
shipowner furnished equipment 船东提供设备
shipowner identification 船东证明
Shipowner's Association 船东协会；船东航业公会；航运同业公会；船舶所有人协会
Shipowner's Clubs 船东互保协会；航运同业公会
shipowner's husband 船东代理人
shipowner's liability 船东责任；船舶所有人的义务
shipowner's lien on cargo 船东的货物留置权
shipowner's limitation of liability 船东的有限责任
Shipowner's Mutual Association 船东互助协会
Shipowner's Mutual Protection, Indemnity Association 船东互相保护与保险协会
shipowner's responsibility 船东责任条款
shipowner's risks 船舶所有人承担风险
shipowner's statutory power 船东法定权限
shipowner's trial 船东自行试航
ship-owning company 轮船公司

ship paint 船舶油漆
ship painting 船舶油漆
ship pass 过船道；通航水道；通航渠道；通海水道；过船道
ship-passing structure 过船建筑物
ship-passing through facility 过船设施
ship-passing works 过船设施
ship pattern 船型
ship pays dues 船方负担税金
shipped bill of lading 已装船提单；装船提单
shipped in good condition 装载良好
shipped in the hold 装在船舱内
shipped on board 已装船
shipped on deck 装在甲板
shipped on deck at shipper's risk 货主负担甲板货风险；甲板货托运人自负风险
shipped quality terms 装船量条款
shipped weight 装运重量
shippen 牛棚
shipper 畸形砖；货主；委托者；托运人；托运方；船长；发货人
shipper arm 铲斗臂
shipper bearing 鞍式轴承
ship performance analyser 船舶性能分析器
ship performance monitor 船舶性能监控器
ship permanent magnetism 永久船磁性
shipper owned container 企业自备集装箱
shipper owned sheet 货主自备篷布
shipper owned wagon 自备货车
shipper-packed unit 成组件件
shipper-receiver 发货—收货人
shipper/receiver difference 发方/收方计量差
shipper rod 移动杆
shipper's guide for proper stowage 发货人正确装载手册
shipper shaft 铲斗传动轴
shipper's letter of instructions 空运货物委托书
shipper's load 货主装载
shipper's load and count 发货人自装自理
shipper's load and tally 发货重量和理货；发货人自装自理
shipper's packed container 货主装箱
shipper's papers 货运文件
shipper style 批发式
shipper's usance 货主运期汇票；货主汇票习惯付款期限
shipper's weight 货方重量；拖运重量；发货人报的重量
shipping 运输业务；发运；装货；起运；船舶运输；发货
shipping, forwarding agent 船舶运输代理行
shipping accident 运输事故；航运事故
shipping accumulation number 装船累积数；发货累积数
shipping administration 航务管理机构
shipping advice 装运通知；装船通知书
shipping afloat 航行中的船只
shipping agent 运输（代办）行；运输代理商；运货代理商；航运代理商；船舶运输代理人；船舶业务代理行；船舶业务代理人
shipping amount 装运量
shipping and loading bridge 船货装卸桥
shipping and storage container 海运和储[贮]存的集装箱
shipping and traffic 水陆运输
shipping application 货运舱位申请书；托运申请书；托运申请单；订舱单
shipping articles 船员名单；船员名册
shipping berth 装船泊位
shipping bill 装货清单；货物装船准单（海关）；装运单据；船舶清单；出口清单；舱单
shipping board 航运局；航务局
shipping box 包装箱；装运箱
shipping bracket 装运支座；装运托架
shipping broker 运输经纪人；货运经纪人；航运经纪人
shipping bulletin 航务公告
shipping business 航运业（务）
shipping by chartering 租船运输
shipping by rail(way) 铁路运输
shipping cage 邮寄笼
shipping canal 运河
shipping capacity 装货容量
shipping case 运输箱；装运箱

shipping cask 运输容器
shipping casualties 船舶海事
shipping channel 航道；通航孔
shipping charges 装船费(用)
shipping claim 装运索赔
shipping clearance 航运净空
shipping clerk 运输业务办事员；货运业务员；航运业务员
Shipping Commission 航运委员会
shipping commissioner 海员契约监护官
shipping communities 航运界
shipping company 航运公司；海洋运输公司
shipping concentration 船舶密集
shipping condition 航运状态
shipping conference 运价协议会；运价公会；航业协会
shipping congress 航运会议
shipping container 运输用集装箱
shipping contract 船员雇用合同
shipping control 船运管理
shipping control procedures 运输控制程序
shipping conveyer 装船输送机
shipping cost 装运费用
shipping cycle 运输周期
shipping damage 运输损伤
shipping data 船运数据
shipping date 装运日期；装船日期；开船日期
shipping day 开船日期
shipping department 航运部；成品库
shipping designator 海运标号
shipping destination 收点
shipping development 航运开发
shipping development project 航运开发工程
shipping device 装运装置
shipping dimension 装运尺寸；航运尺寸；海运尺寸；发送体积
shipping dock 装卸码头
shipping documents 装船单据；装运单据；货运文件；货运单据；航运文件；运货文件
shipping documents for international goods through transport 国际联运货运票据
shipping-dry 干运的
shipping enterprise 航运企业
shipping exchange 航运交易所
shipping expenses 装运费用
shipping expenses of container ship 集装箱船营运费用
shipping federation 航业工会
shipping fever 航运热；船运热
shipping formality 装运手续
shipping gallery 装船廊道
shipping house 轮船公司
shipping in coastal waters 近海航运
shipping industrier 航运业者
shipping industry 造船工业；航运业；海运业；航运界
shipping inspection 发货检查
shipping instruction 装运指示；装运指令；装船指示；装船须知
shipping insurance 航运保险
shipping insurance company 航运保险公司
shipping interest 航运业者；海运事业；航运界
shipping invoice 装货发票；货物运装单
shipping lane 大洋航线；船行道；航线；航道
shipping law 海运法
shipping line 航运线；航运公司；航线【航海】；航海线；海洋航线；海洋航路
shipping-line map 水路交通图
shipping list 装箱单；装货单；货单；发货清单
shipping load 起运载重
shipping manifest 装船舱单
shipping mark 运输标志；装运标志；装船标志；货物标志；发货标记
shipping master 海员契约监护官
shipping memo 装货通知书
shipping memorandum 装运备忘录
shipping note 装货通知书；装船通知单
shipping notice 装运通知单
shipping number 装货票号；装船序号；船运编号
shipping office 船员管理处；运输事务所；货运处；货运办事处
shipping operator 航商
shipping order 装运单；装货通知单；装货清单；装货单；装船通知单；交运单；发运单
shipping ore 一级矿石；直接外运矿石

shipping origin 发点
shipping package 外包装
shipping papers 货运文件
shipping parcel 小件货物
shipping passage 航行通道
shipping permit 准运单;准许装货单;装运许可证;装运通知单;装运许可证;装船通知单
shipping plan 运输计划
shipping plug 装运堵塞
shipping point 装运站;装运点;装船地点;航运点
shipping pollutant 航运污染物
shipping pollution 航运污染;船舶航运污染;船舶航行污染
shipping pool 航业联营
shipping port 装运港;装船港口
shipping process 装运手续
shipping protection 航运保护(主义)
shipping quality 装船品质
shipping receipt 装货收据
shipping register 船舶登记主管人;船舶登记
shipping resistance 航运阻力
shipping returns 航运利润
shipping room 码头(上的)发货仓库;货运室
shipping route 航线【航海】
shipping sack 集装袋
shipping sailing pollution 船舶航行污染
shipping sample 装船货样
shipping schedule 船期表
shipping shock 装运碰撞
shipping space 舱位
shipping space broker 舱位经纪人
shipping specification 航运规程
shipping station 航运站(海运)
shipping subsidy 航运补贴
shipping tag 货运标签;发料标签;发货标签
shipping tare 卖方皮重
shipping terminal 靠船码头;装货码头;终点港;海洋转运站;船运终点
shipping terms 装运条件
shipping ticket 运货单;装运单
shipping time 运送时间;(船的)运输时间;货运时间
shipping ton 装载吨(等于40立方英尺);载货吨;计费吨;货运吨;体积吨
shipping tonnage 船舶吨位;船舶吨数
shipping trade 航运业
shipping traffic 航运
shipping traffic control signal station 航行指挥信号站;船舶交通指挥信号台
shipping transportation cost 发送运费
shipping tribunal 航运法庭
shipping unit 装运装置
shipping value 货运价值(包括货物价值、转运费和保险费)
shipping weight 运输重量;装运重量;装货重量;装运重量;离岸重量;航运重量;起运重量;起运载重;启运重量;出运重量
shipping weight/quality 离岸重量或质量
shipping weight terms 装货重量条款
shipping yard 成品发运工段
ship-plant 造船厂
ship plate 造船(用)钢板;船用钢板
shippon 牛舍;牛棚
ship position buoy 船位浮标
ship position report 船位报告
ship power cable 船用电力电缆
ship products 船舶产品
ship protest 海损异议书
ship pulpit 船操纵台
ship quality 船舶质量
ship queuing time 船舶排队时间
ship radar 船用雷达
ship radio aids 船舶无线电助航设备
ship radio authorization 船舶无线电台试验准许
ship radio service 船舶无线电业务
ship radio station 船用无线电台
ship railway 斜面升船机;船排;移船轨道
ship receipt 大副收据
ship reception chamber 承船厢
ship recycling 拆船;废船解体
ship registered length 船舶登记长度
ship repair basin 修船港池
ship repairer's liability 修船方责任
ship repair industry 修船工业
ship-repair(ing) 船舶修理

ship repairing and building yard 修造船厂
ship-repairing charges 船舶修理费
ship-repairing depot 船舶修理站
ship-repairing station 船舶修理站
ship-repair pier 修船码头
ship-repair quay 修船码头
ship-repair work 船舶修理工程
ship-repair yard 修船区;修船厂
ship report 船舶报告
ship reserve antenna 船舶备用天线
ship resistance 航行阻力;船舶航行阻力
ship return 船舶反射信号;船舶回波
ship rigger 索具工人
ship routing 船舶定线制
ship's accessories 船舶属具
ship's acoustic(al) navigation system 船用声呐导航系统
ship safety fairway 安全航道
ships afloat and building 现有的和建造中的船舶
ship's age 船龄
ship's agency service 船舶代理业务
ship's agent 船舶(运输)代理人
ship sailing resistance 船舶航行阻力
ship's apparel 船具
ship's bill of lading 随船提单;船长提单
ship's boat 船载小艇
ship's borer 海中蛀船虫
ship's bread 压缩饼干(救生艇用)
ship's breadth 船宽
ship's business 船舶管理业务
ship's cable 船舶缆索
ship scaffold 悬吊式脚手架
ship's call 船舶呼救
ship's call sign 船用呼号
ship's call signal 船用呼号
ship's carpenter 船舶木工
ship's class 船级;船舶等级
ship's classification 船级;船舶等级
ship's clerk 装卸交货总管理员;船舶货运员
ship's clock 船钟;船用时钟
ship's computer system 船用电子计算机系统
ship's crane 船用起重机;船上起重机
ship screw 船用螺旋桨
ship's crew 船员
ship's curves 造船曲线板
ship's days 装卸天数;船舶作业时间
ship's departure time 船舶离港时间
ship's depth 船深
ship's direction-finder 船舶测向仪
ship's distress signal 船舶呼救信号
ship's doctor 船上医务人员;船上医生;船舶医生
ship's documents 船舶文件
ship self-contained navigation system 船载导航系统
ships entered harbo(u)r 进港船舶
ship's entry 船进港;船舶入港申请书
ship sewage 船舶污水
ship's expenses per operation(al) day 船舶营运艘天费用
ship's expenses per operation(al) ton-day 船舶营运吨天费用
ship's facility 船舶设备
ship's field error 船的场误差
ship's fittings 船舶装备;船舶属具
ship's gear 船上吊装机械;船上传动装置
ship's grade 船舶等级
ship's handling 船舶作业
ship-shaped caisson gate 船形浮坞门
ship-shaped floating dock 船形浮船坞
ship's head(ing) 船首向;船的航向
ship's heading marker 船首标志
ship sheathed-armoured cable 船用护套铠装电缆
ship sheathed flexible cable 船用护套软缆
ship's height from light loadline to tip untitled mast 固定建筑物高度【船】
ship's hold 船舱
ship's holiday 船上假日
ship-shore safety check list 船岸安全核查表
ship-shore triangular method 船岸三角交会法
ship-shore triangulation 船岸三角测量;船岸间三角测量
ship's hull 船身
ship's husband 船舶代理商;船舶代理人

shipside 码头
ship-side and wharf apron track 码头前沿铁路线
shipside boring 船侧钻孔;船侧钻探
shipside delivery 船边提货
shipside drilling 船侧钻孔;船侧钻探
shipside operation 船边作业
ship-side paint I 一号船边漆
ship-side paint II 二号船边漆
ship's inertial navigation system 船用惯性导航系统;船舰导航惯性系统;船舶惯性导航系统
ship's island 甲板室【船】;船上的上层建筑
ship's journal 航行日志;船舶日志
ship's length 船长
ship's line 船舶型线;船体型线(图)
ship slipway 滑船道;修船架滑道
ship's load cargo 整船货物
ship's loading and unloading time 船舶装卸作业时间
ship's log book 航海日志
ship's luffing crane 船上旋臂起重机
Ship's Machinery Committee 船舶机械委员会
ship's magnetism 船铁磁性;船磁
ship's main engine 船用主机
ship's maneuverability 船舶驾驶性能
ship's manifest 轮船载货清单
ship's mean time 船舶所在地时间
ship's medical officer 船上医务人员
ship smoke generator 舰船烟幕施放器
ship's mooring 船舶锚泊
ship's mo(u)ld line 船体型线
ship's name 船名
ship's nationality 船籍;船泊国籍
ship's number 通信呼号;船名号码旗
ship's operation 船舶作业
ship's option 承运船的选择权
ship source oil pollution damage 船舶油污染灾害
ship's own tackle 船上自备索具
ship's papers 船舶文件;船舶证件及报表
ship's particulars 船舶详细资料
ship speed 航速【船】
ship-speed correction 航速校正
ship-speed log 航速仪【船】
ship's personnel on duty 船方值班人员
ship spike 方形长钉
ship's plan 船舶平面图
ship's plant 船舶设备
ship's port time 船舶在港时间;船舶停港时间
ship's position 船位;船舶位置
ship's position finding 船位测定
ship's principal dimensions 船舶主要尺度
ship's protest 海事声明;海事报告(书);海难抗议书
ship's radio equipment 船上无线电设备
ship's radio station 船电台
ship's ramp 船备斜坡道
ship's register 船名录;船舶登记(局)
ship's registry 船籍
ship's release 提货单
ship's routing 船舶定线制
ship's self-contained navigation system 船舶自主式导航系统
ship's signal 船用呼号
ship's signal letters 船名呼号
ship's size 船的尺寸
ship's space 舱位
ship's speed in still water 船舶静水航速
ship's speedometer 船舶速度仪
ship's stagnation at a rapid 吊滩
ship's steel side scuttle 钢质舷窗
ship's steel weather tight door 船用钢质风雨密门
ship's stopping distance 船舶制动距离
ship's stores 船用品
ship's stores and equipment 船舶物料与设备
ship's superstructure 船舶上层建筑
ship's surgeon 船上医务人员
ship's sweat 船舱汗湿
ship's sweat damage 潮腐险
ship's swing 船舶摇摆
ship stabilizer 船舶稳定器
ship-stabilizing gear 船舶减摇装置
ship's tackle 船舶(吊货)索具
ship's tackling 装备船舶索具
ship stadimeter 舰船舶视距仪
ship stainer 船上集装装卸机
ship station 船舶(无线)电台

ship station carrying on continuous day and night service 昼夜工作的船舶电台
ship steel 船用钢材
ship's test readiness evaluation procedure 船舶试验准备情况鉴定程序
ship's time 船用时
ship's time in harbo(u)r 船舶在港时间
ship's time in port 船舶在港时间
ship stopper 止滑器
ship stopping area 制动水域【船】
ship stores 船用物料及食品;船用物料;储备品(船上)
ship stowage 船舶积载
ships turnover rate 平均航次周转次
ship subpermanent magnetism 半永久船磁性
ship superstructure 船上的上层建筑
ship surveyor 验船师;海事检查人
ship's wake 追迹流;船舶伴流
ship's weather instrument 船舶气象仪
ship's weather report 船舶天气报告
ships without nationality 无国籍船舶
ship synoptic(al) report 船舶天气报告
ship synoptic code 船舶天气电码
ship tackle 船用滑轮组
shiptainer 船用集装箱装卸桥
ship tank 油船
ship-tanker 运油船
ship telescope 舰船望远镜
ship-terminal 靠船码头
ship-terminal utilities 码头公用设施;航运终点设施
ship the oars 放桨
ship timber 造船木材;船用木材
ship time 船时
ship-to-ship communication 船与船的通信联系
ship to shore 船到岸
ship-to-shore call 船对岸呼叫
ship-to-shore communication 船对岸通信
ship-to-shore pipe line 舶岸输油管
ship-to-shore radio 海对陆电台;船岸通信无线电台
ship-to-shore radio station 船对岸电台
ship towed 被拖船
ship track 航迹【航海】
ship trade 航运业
ship trial 试航【航海】
ship turnaround 船舶周转
ship turning area 船舶掉头区
ship-type caisson 船式沉箱
ship type tractor 船式拖拉机
ship unloader 卸船机
ship unloader for piece cargo 件货卸船机
ship-unloading operation 卸船作业
ship-unloading plant 卸船设备
ship-unloading technology system 卸船工艺系统
ship varnish 船用清漆
ship watering hydrant 船舶给水栓
ship wave 船(行)波;船起伏;航行波;船波动
shipway 造船台;造船滑道;过船道;滑船道;船台;船架;滑道(船坞)
ship weather report 船舶天气报告
ship wet dock 造船湿坞
ship with corrugated side 有波形舷部的船
shipworker 装卸组长
ship worm 蛀木虫;蛀船虫;海蛆;船蛆
shipwreck 遇险船;遇难船;失事船(舶);船舶失事
shipwrecked 失事
shipwrecked crew 失事船船员
shipwrecks sunken vessel 沉船
shipwright 造船工(人);船体装配工;造船木工;船木工
shipyard 造船厂;船坞;船厂;船舶修造场;船舶修造厂
shipyard cableway 船坞索道
shipyard facility 船厂设备
shipyard floating crane 造船用浮式起重机
shipyard machinery 造船机械
shipyard machine tool 造船机械
shipyard period 船厂周期
shipyard tug 船厂拖轮
shirt 高炉炉衬;裙板
shirt band 衬布
shirt board (皮带运输机的)边缘裙板
shirt dissipator 裙板消能工
shirting 本色细布;衬衣布料
shirting block 门贴脸座

shirting duct 踢脚板内管道
shirting heater 踢脚板内暖气管
shirting tile 踢脚砖
shirt plate 挡板
shirt-supported pressure vessel 裙式支承压力容器
shirt-tail packer 用麻布作为填充物封隔套管环状间隙
shirt-type core lifter 带导向管的环状卡簧岩芯提断器
shish-kebab 串型多晶结构
shist 页岩
shit blasting shop 喷丸车间
shive 切削;切成薄片;未充分分离纤维的木料;屋面油毛毡的裂片;亚麻杆碎片
shiver 页岩;蓝色板岩;碎块;颤动;发抖
shivering 碎裂;碎花;瓷釉剥落
shives 含木质下脚
shiving 切片(分层进行挖掘)
shoad 漂流矿石
shoad stone 矿脉露头的脱散块;矿砾;漂流矿石
shoal 浅滩;砂洲;沙洲;暗滩
shoal area 淤积区;浅(水)区
shoal area of sand 沙洲浅水区
shoal at a tributary mouth 支流河口浅滩
shoal at estuary 潮汐河口浅滩
shoal at river bend 弯道浅滩
shoal breccia 浅滩角砾岩;潮滩角砾岩
shoal canal 浅水运河
shoal coefficient 浅化系数;变浅系数
shoal crossing 过渡段浅滩
shoal deposit 淤积物
shoaler 沿海船(员)
shoal evolution 浅滩演变
shoal group 滩群
shoal head 沙嘴;暗礁
shoal in a bend 弯道浅滩
shoal in a branching channel 汊道浅滩
shoal indicator 浅滩标志
shoaling 淤浅;回淤;浅水作用;变浅
shoaling breaking waves 浅水破碎波
shoaling channel 浅水槽
shoaling coefficient 变线系数;浅水系数(波浪)
shoaling effect 淤浅作用;浅滩影响;滩浅影响
shoaling factor 浅水系数(波浪);浅化系数
shoaling habitat 成群栖所
shoaling rate 回淤率
shoaling water 浅水
shoaling waters 浅水区
shoaling wave 浅化波浪;浅水波
shoal in lake area 湖区浅滩
shoal inside estuary 口门内浅滩
shoal inside outfall 口门内浅滩
shoal mark 浅滩标志
shoal materials 淤积物;浅滩组成
shoal near a tributary mouth 支流河口浅滩
shoal near river bed 滨河床浅滩
shoal of tidal estuary 潮汐河口浅滩
shoal of tidal funnel 潮汐河口浅滩
shoal patch 点滩
shoal point 浅嘴;浅点
shoal process 浅滩演变过程
shoal reach 浅滩河段;浅滩段
shoal reef 浅礁
shoal regulation 浅滩整治
shoal regulation stage 浅滩整治水位
shoal retreat massif 浅滩后退砂体;浅滩残余砂体
shoal rock 浅礁
shoals and rapids 滩险
shoal section 浅滩河段
shoal sounding 浅水
shoal spit 浅嘴
shoal stretch hard to be scoured 难冲刷的浅滩段
shoal variation 浅滩演变
shoal water coral 浅水珊瑚
shoal water influence 水深变浅的影响
shoal waters 浅水区
shoal water zone 浅水带;浅海带
shoaly 多浅滩的
shoaly land 湖田;浅滩地
shoaly reach 多浅滩河段
shock 震扰;振荡;振动;跳弓;冲撞;冲击;爆音
shock absorber 消震器;消振器;阻尼器;减震器;减振器;缓冲器;冲击吸收器;避震装置
shock absorber arm 减振器臂

shock absorber ball joint 减振器球节
shock absorber ball stud 减振器球头柱螺栓
shock absorber base 防振架
shock absorber body 减振器体
shock absorber bracket 减振器托架
shock absorber combustion head 减振燃烧缸头
shock absorber compression valve 减振器压缩阀
shock absorber conical rubber 减振器锥形橡皮
shock absorber cylinder 减振器筒
shock absorber end cap 减振器端盖
shock absorber end cap gasket 减振器盖衬
shock absorber expander 水封护罩
shock absorber eye bolt sleeve 减振器有眼螺栓衬套
shock absorber filler plug 减振器注入塞
shock absorber filler plug gasket 减振器注油塞垫密片
shock absorber filler screw 减振器注入塞
shock absorber fluid 减振器油
shock absorber for relays 继电器防振器
shock absorber friction disc 减振器摩擦盘
shock absorber gland nut 减振器压盖螺母
shock absorber intake valve 减振器进油阀
shock absorber link 减振器连杆
shock absorber link bushing 减振器连杆补套
shock absorber link pin 减振器连节销
shock absorber link rod 减振器拉杆
shock absorber link rod pin bronze bushing 减振器拉杆销铜套
shock absorber link rod pin steel bushing 减振器拉杆销钢套
shock absorber oil 消振油
shock absorber packing ring 减振器密封环
shock absorber piston 减振器活塞
shock absorber recoil spring 减振器反冲阀弹簧
shock absorber recoil valve 减振器回跳阀
shock absorber relief valve 减振器旁通阀
shock absorber relief valve gasket 减振器保险阀垫密片
shock absorber rod guide 减振器杆导管
shock absorber rubber 减振器橡皮
shock absorber shaft 减振器轴
shock absorber strap 减振器环
shock absorber strut 减振器支柱
shock absorber support 减振器支架
shock absorber type drill 减振式钻机;减振式凿岩机
shock absorber wiper seal 减振器水封
shock-absorbing bushing 减振衬套
shock-absorbing capacity 缓冲性能;防冲能力
shock-absorbing circuit 压力缓冲回路
shock-absorbing cradle 减振架
shock-absorbing fluid 消振液
shock-absorbing guiding handle 减振的导向手柄
shock-absorbing mount(ing) 减振架
shock-absorbing seat 减振座
shock-absorbing spring 减振弹簧
shock-absorbing spring can 减振弹簧筒
shock-absorbing spring for vehicle 车辆用缓冲弹簧
shock-absorbing stud 减振支柱
shock-absorbing stut 减振柱
shock-absorbing suspension 减振弹簧支承
shock-absorbing wagon 带缓冲器车辆
shock absorption 缓冲;消振;减震(法);减振;冲击吸收
shock accelerator 冲击负荷加速表
shock accelerometer 冲击加速计;冲击加速度表
shock action 突击动作;冲击作用
shock ag(e)ing 快速蒸化
shock angle 激波倾斜角
shock arrester[arrestor] 振动吸收装置
shock-associated noise 冲击室噪声
shock-associated screech 冲击室尖叫声
shock at entry 入口冲
shock attenuation 减振
shock attenuation device 减振装置;缓冲设备
shock bending 冲击弯曲
shock bending test 振动弯曲试验
shock blaster 空气炮
shock blasting 振动爆破
shock blow stress 冲击应力
shock bottle 瓶状激波系
shock bottom furnace 振底炉

shock brigade 突击作业班;突击工作班
shock bump 冲击突出;冲击地压
shock chamber 骤冷室;激冷室
shock chilling 骤冷
shock chilling medium 激冷介质
shock coagulation 冲击凝固
shock coefficient 冲击系数
shock component 冲击分力
shock compression curve 冲击压缩曲线
shock concrete 振捣混凝土
shock cooling 骤冷;激冷
shock cord 减压绳;减振绳
shock cord takeoff 橡筋绳牵引起飞
shock crack 冲击裂缝
shock curve 激波曲线
shock damper 阻尼器
shock deformational fabric 冲击变形组构
shock detector 爆炸波检测仪
shock diamond 钻石型激波
shock diffuser 振动滤器
shock discharge 撞击排料;振动排放
shock discontinuity 激波间断
shock driver 冲击式螺丝刀
shock dynamics 冲击动力学
shocked flow 激波流;冲波气流
shocked sheaves 禾捆堆
shock effect 振动作用;冲击影响;冲击效应;冲击效果
shock elasticity 冲击弹性
shock eliminator 消振器;减振器;除振器;避振器
shock energy 振动能;冲击能(量)
shocker 振动器;集草车;冲击器
shock excitation 振激;激振;碰撞激励;冲击励磁;冲击激励;冲击激发
shock excite 振激;冲击激励
shock excited oscillator 振激振荡器;冲击激励振荡器
shock exciter 强行励磁机;冲击励磁机
shock factor 冲击系数
shock failure 冲击破坏
shock feed 冲击加料
shock feeder 冲击加料器
shock fitting 激波拟合
shock flexure test 振动弯曲试验
shock flow nozzle 激流喷嘴
shock formation 冲击形成
shock front 激震前沿;激震波面;激波前沿;激波前;冲波阵面;冲波面
shock-generating body 激波发生体
shock hanger insert 防振悬架垫
shock hazard 触电危险性;突然灾难;振动危险;电击(事故);冲击危险性
shock heat(ing) 冲击波加热;激波加热
shock induced boundary layer 冲击诱发边界层;激波诱导附面层
shocking house 晒全株玉米用的木架
shocking jack hammer 冲击式凿岩机
shocking rammer 冲击夯
shocking rotary jack hammer 冲击旋转式凿岩机
shock insulation 隔振
shock insulation behavior 隔振性能
shock insulator 隔垫
shock interaction 激波相交
shock isolation 振动隔离
shock isolation system 隔振系统
shock isolator 隔振体;隔振器;冲击隔离器
shockless jarring machine 无振冲击机
shockless jolt 弹簧微振振实
shockless jolt mo(u)lding machine 微振造型机
shock line 激波线
shock liquid 防振液
shock lithification 冲击石化作用
shock loading 振动荷载;振动负载;突加荷载;突加负载;冲击荷载;冲击负载;冲淡负荷
shock loss 激波损失
shock machine 冲击机
shock Mach number 冲击马赫数
shock measurement 冲击测量
shock metamorphic effect 冲击变质效应
shock metamorphic facies 冲击变质相
shock metamorphism 振动变质作用;冲击变质作用
shock mirror 形象失真的镜子;形象扭曲的镜子;哈哈镜
shock-mitigating material 缓冲材料

shock mitigation 缓冲
shock mitigation system 减振系统
shock motion 激震运动;激振运动;冲击运动;冲击波传播;振动
shock mount 隔振器;隔振垫;减振架;减振器;防振座;减振器
shock-mounted 减振的
shock mounting 减振支架
shock noise 冲击噪声
shock oscillation 激波振荡;激波脉动
shock oscillator 冲击振荡器;冲击振荡器
shock pad 减振垫
shock phenomenon 冲激现象
shock ply 冲击层
shock polar 激波极线
shock polaric diagram 冲击波极线图
shock polarization 激振极化
shock power 突击力
shock pressure 激波压力;冲击压力
shock pressure absorber 水击吸收器;冲击压力吸收器
shock pressure experiments 冲压实验
shock processing 快速工艺
shock produced deformation 冲击变形
shock-proof 抗振的;防振的;防冲击;耐振的;耐冲击
shock-proof cable terminal 带防振装置的电缆端子
shock-proof coating 防振涂层
shock-proof electric(al) apparatus 耐振电气设备;耐振电气设备
shock-proof glass electrode 防振玻璃电极
shock-proof lamp 耐振灯泡
shock-proof level 耐振水准仪
shock-proof material 防振材料
shock-proof mount(ing) 减振架;耐振台座;防振座
shockproofness 振动强度
shock-proof steering 无反冲转向;防振转向
shock-proof structure 防振结构
shock-proof switch 防电击开关
shock-proof test 抗振性试验
shock-proof watch 防振表
shock pulse 振动脉冲;冲击脉冲
shock recovery 激波恢复
shock reducer 缓冲器;减振器
shock reflection 冲击波反射
shock region 激波区域;激波区;冲击区(域)
shock relief vent 减振孔
shock resistance 抗冲强度;抗振物;抗冲(能)力;抗冲击性能;抗冲击(性);耐振(强度);冲击阻力;防振;耐冲击性
shock resistance rack 防振架
shock-resistance test 耐碰撞试验
shock-resistant 抗振的;防振的
shock resistant sealant 防振密封剂
shock resisting 抗振的
shock resonance 冲击共振
shock response spectrum 冲击响应谱
shock ring 减振环
shocks 缓冲装置
shock sampling 加密采样
shock seeding 刺激起晶
shock seismograph 冲击地震仪
shock-sensitive 不耐振的;对冲击灵敏的
shock sensitivity 冲击灵敏度;振动灵敏度
shock sequence 振动序列
shock severity 冲击强度;冲击程度
shock shoe (保护犁铧用的)缓冲滑脚;防冲滑脚
shock sink 冲击逸除;冲击壑
shock smearing 激波涂抹
shock spark 冲击火花
shock spectrum 冲击谱;激波谱;冲击谱
shock speed 冲击波速率
shocks received during transit 运输途中受震
shock stall 滞止激波;激波失速;激波分离
shock stall effect 激波离体效应
shock stop pin 减振销
shock strength 抗振强度;冲击(击)强度;冲击强度
shock strength of road metal 道砟抗撞击强度
shock stress 冲(击)应力;冲胁强
shock strut 减振柱
shock sub 减振接头;减振装置
shock suppresser [suppressor] 吸振器;防冲击器;消振器
shock surface 冲击面;冲击表面
shock synthesis 冲击合成

shock table 振动台;跳桌;冲击台;冲击试验台
shock test 振动试验;抗振试验;动载试验;冲击试验
shock tester 振动试验器;抗振试验机;冲击试验机
shock testing machine 冲击试验机
shock test machine 冲击试验机
shock test of extreme-pressure lubricants 特压润滑剂的振动试验
shock test of grease 润滑脂振击试验
shock transition region 冲击波过渡区
shock transmissibility 冲击传递率
shock tube 振动管;激波管;冲击(激励)管
shock tunnel 激波风洞
shock turbulence interaction 激波与紊流相互作用
shock value 冲击值
shock velocity 激波波速;冲击值;冲击速度;冲击波速率;冲击波速
shock vibrator 摇动式振捣器;颠振振捣器;冲击振捣器;振摇式振捣器
shock voltage 脉冲电压;电压脉冲
shock wave 振波;空气冲击波;激(振)波;人字形激波;冲击波
shock-wave angle 激波倾斜角
shock-wave boundary layer interaction 激波与边界层相互作用
shock wave discontinuity 冲击波不连续性
shock wave drilling 振激波凿岩;冲击(波)钻进
shock wave front 冲击波前沿;冲击波面
shock wave gas ionization 冲击(波)气体电离
shock wave heating 冲击波加热
shock wave high-pressure technique 冲击波高压技术
shock wave in open channel 明槽冲击波
shock wave ionization 冲击波电离;激波电离
shock wave luminescence 激波发光;冲击(波)发光
shock wave method 冲击波法
shock wave noise 爆声
shock wave polymerization 冲击波聚合作用
shock wave propagation 冲击波传播
shock wave-pumped laser 冲击波抽运激光器;冲击波泵浦激光器
shock wave reducing room using gravel 砾石消波室
shock wave reflection 冲击(波)反射
shock wave refraction 激波折射
shock wave shadow 冲击波后的涡痕;激波阴影照片;冲击波后的涡ది;激波痕
shock wave solidification 冲击波固化
shock wave technique 冲击波技术
shock wave tests 冲击波试验
shock wave thickness 激波厚度
shock worker 突击手;突击工作者
shock zone 激振区域
shod 装金属箍头的;滑锚
shoddy 劣等的;劣等材料;冒充的好货
shoddy product 冒牌货;次品
shoddy work 劣等结构;劣等机件;劣等工程;小方石;低等花岗石
shoddy work and inferior material 偷工减料
shode 矿砾;漂流矿石
shoe 桩靴;管头;履带片;履带板;开口坩埚;90度承插弯管;集结板;基础板;起动导轨;瓦状物;地脚板轨座;导向板;触靴;承座;插入式支座
shoe an anchor 锚尖加锚(防滑措施)
shoe area 履带面积
shoe attachment 桩靴附件
shoe base 插入式支座
shoe block 靴形滑车;制动(滑)块
shoe brake 制动瓦;闸瓦制动(器);块状闸;蹄式制动器;蹄式闸;制动块
shoe-button cell 鞋扣形电池
shoe colter 靴式开沟器
shoe connection 箍松接合
shoe contact pressure 制动蹄片接触压力
shoe counter 测射计数器
shoe elbow 90度底弯;弯底弯;底弯肘管
shoe guide 引鞋
shoe joint 管柱底部
shoeless track 无瓦履带
shoe mo(u)lding 踢脚板底缝压条;踢脚板压条
shoe mounted ball mill 滑履支承球磨机
shoe nog plate 固定底板;马蹄铁形磨盘
shoe-nose shell 带有切削刃的管靴(硬黏土钻进)
shoe of pile 桩靴
shoe of timber pile 木桩靴

shoe piece 尼框底部
shoe pitman 清粮器传动连杆
shoe plate 支撑板;闸瓦;龙骨端包板;底(座)板;座板
shoe polisher 擦鞋机
shoe pressure force 制动闸块压力
shoe process 沉井凿井法
shoe rail 楼梯栏杆基脚
shoe reef 裙礁
shoe retaining clips 闸瓦固定夹
shoe retracing spring 制动片复位弹簧
shoe runner 靴式开沟器
shoe-shaped drag head (挖泥船的)靴状吸嘴;鞋形耙头
shoe shell 钻泥提取器;泥浆泵;抽筒【岩】
shoe shop 鞋店
shoe spring 制动蹄复位弹簧
shoe squeeze tool 注水泥设备(加压下)
shoe store 鞋店
shoestring 小额资本;电线;带状体【地】
shoestring-form 鞋带状
shoestring gully 带状冲沟
shoestring sand 窄条饱和油砂层;透镜状沉积砂;鞋状砂矿层;鞋状砂层;带状砂层
shoestring sand pool trend 鞋带状油气藏趋向带
shoestring trading 小本经营
shoestring washing 细流冲刷;细沟冲刷;细沟冲蚀;纹沟冲刷;纹沟冲蚀
shoe-type furrow opener 靴式开沟器
shoe wheel brake 蹄式车轮制动器
shoftboard 软板
shogaol 姜烯粉
shogged stitch 波纹组织
shoji screen 障子;屏风
sholaing habitat 集群栖所
shole 撑柱垫板
Shone ejector 肖恩喷射器
shonkinite 等色岩
shoofly 临时道路;临时线路;临时轨道(指铁路)
shook 现成的成套待装配板料;桶板料;成捆桶板材
S-hooks and locks 连接管道的金属紧固设备
shoot 卸槽;枝条;根出条;流下;坑;加速;急流水路;(滑运木材);滑槽;人工爆炸;抛出;拍摄;投掷;摄影;富矿体
shootability 可喷性
shoot a string of tools 爆炸法处理卡钻
shoot at right angles 直射
shoot capacity 注射量
shooter 清管发送器;炮工;射手;放炮工;发射机;爆破工
shoot for oil 石油地震勘探
shooting 立射俑
shooting aperture 摄影光圈
shooting block 削台;刨木用导板
shooting board 修边夹具;刨木导板
shooting board fixture 刨光板夹具
shooting box 猎屋;狩猎小屋
shooting brake 电视车
shooting concrete 喷射混凝土
shooting distance 喷射距离
shooting flow 射流;超临界流
shooting gallery 打靶场;射击场;靶场
shooting into the solid 过深炮
shooting lodge 猎屋
shooting method 追赶法
shooting of oil wells 油井爆破
shooting on-the-free 多自由面爆破
shooting plane 拼缝刨;修边刨;大刨;长刨
shooting position 摄影位置
shooting preserve 禁猎区
shooting ramming 射ામ紧实
shooting range 射击场;打靶场;靶场
shooting sheet shutter 棱片快门
shooting valve 喷射阀
shoot knot 渔网结
shoot loose 散卸
shoot mutation 梢变
shoot off 刨去;刨光
shoot-out 起弧
shoot pipe 管子射孔
shoot-root ratio 条根比率
shoot squeeze molding machine 射压造型机
shoot system 茎轴系统
shoot the well 井底爆炸;井底爆破

shoot up 飞涨
shop 工场;商号;店铺
shop administration 企业管理
shop air 工厂用压缩空气;车间气源
shop allowance 工厂容差;商店容差
shop assembling 工厂装配;车间装配
shop assembly 工厂装配;车间装配;厂内组装
shop assistant 店员
shop bill 工厂材料单
shop block 店面房屋;车间房屋
shop box 厂箱
shop building 车间建筑
shop-built forms 工厂预制模板;厂制模板
shop burden 车间管理费;分部制造费(用)
shop calibration 工厂检定
shop car 厂内运货车
shop card 车间工作卡(片)
shop characteristic 工艺性能
shop coat 厂内油漆涂层;底漆;出厂前油漆涂层;车间预涂层
shop coating 车间预涂
shop condition 车间(生产)条件
shop control 车间监控
shop crane 车间用起重机;厂用起重机;厂房起重机
shop cutting panel 等外品板
shop detail drawing 工厂施工详图;加工详图
shop drawing 装配图;车间生产用图;制作图;制造图;工场加工图;加工图;施工图;车间图纸;车间工作图;安装图
shop driven rivet 工厂铆合的铆钉;厂合铆钉;厂铆铆钉
shop effluent 工厂排水
shop equipment 工厂设备
shop erection 厂内安装
shop expenses 车间经费
shop-fabricated 车间预制的;厂制(的);预制的;工厂装配的;工厂预制的
shop-fabricated boiler 快装式锅炉
shop-fabricated component 车间预制件
shop-fabricated joint 工厂预制接头
shop-fabricated member 工厂预制的构件;预制件装配
shop-fabricated parts 车间制作组装件
shop fabrication 工厂装配;车间预制;预制
shop-field erection of tank 工厂内油罐的焊接装置
shopfitter 厂内装配工;厂内钳工
shop fitting 装配车间
shop floor 工厂地面;车间地面
shop foreman 车间工长
shop front 店铺门面;门面;铺面;店面
shop front decorating 店面装饰
shop front finish 店面装饰
shop front profile 店铺门面的外形轮廓
shop front section 店面剖面图
shop ga(u)ge 工作量规;车间量规
shop hoist 车间起重机
shophouse 商店住宅并用
shop illuminator 工厂照明器
shop instruction 工厂工作细则;厂制说明书;出厂说明书;车间守则
shop joint 工厂预制接头
shopkeeper's strike 罢市
shop laboratory 车间化验室
shop line 车间风管
shop lumber 半成材;门窗装修木材;工厂定级木材
shop-made 工厂制造的;定做的
shop management 工场管理;工厂管理;车间管理
shop manual 工厂手册;工厂规程
shop manufacture 厂内制造;车间制造
shop material report 工厂材料报告单
shop miter 车间内斜接
shop office expenses 工厂事务费用
shop of tradition Chinese medicines 中药店
shop order 工作单
shopowner 财东
shop painted 出厂前油漆的;厂内油漆的
shop paint finish 工厂上油漆
shop painting 上底漆;车间预涂;车间油漆
shop-perforated pipe[tube] 厂制带孔管;厂制穿孔管
shopping and business area 购物和商业区
shopping and living complex 商业和生活综合体
shopping arcade 商业拱廊
shopping area 商业区

shopping bag 采购用纸袋
shopping block 商业街坊;商业街区
shopping building 商业建筑
shopping center area 商业中心区
shopping center district 购物中心区
shopping center (围绕大型停车场的)市郊商店区;购物中心;商业中心
shopping complex 商业综合体;商店群
shopping district 商店区
shopping goods 选购(产)品;挑选购买品
shopping hall 商业大厅;营业厅
shopping hours 商店营业时间
shopping lane 商业小弄;商业小巷
shopping level 商业水平
shopping mall 购物商场;(禁止车辆通行的)商业区;只限行人活动的商业区;商业区林荫路;商业街(道);步行商业街
shopping oscillator 间歇作用发生器
shopping passage(way) 商业通道
shopping precinct 购物区;商业步行街;商店区
shopping promenade 购物散布广场
shopping rush 抢购
shopping square 商业广场
shopping street 商业区街道;商业街(道)
shopping street for pedestrians only 步行购物街
shopping terminal 商业区交通终点站
shopping trip 购物旅行;采购出行
shop preassembly 厂内预安装
shop premise 车间办公室
shop primed 车间上底漆;商品;成品
shop primer 预涂底漆;工厂底漆;车间底漆;保养底漆
shop priming 在车间涂底漆;车间预涂底漆
shop railway 升船滑道
shop repair 厂内修理;厂修
shop rivet 预铆铆钉;工厂(铆合的)铆钉;厂合铆钉;厂铆铆钉
shop-riveted 车间铆接的
shop riveting 工厂铆接;车间铆接;厂铆
shop rule 工厂规章
shop safety 车间安全
shop sign 牌号
shop splice 工厂拼接;厂内拼接
shop stall 商业摊位;商业档位
shop stock 工厂存料
shop stor(e)y 商店楼面
shop supplies 车间备品
shop sweeping 车间废料
shop talk 有关本行的谈话;行话
shop temperature 操作间温度
shop term 生产术语
shop test(ing) 工厂试验;车间试验;厂内试验
shop test specification 工厂试验规范
shop track 工厂专用线;段管线
shop traveler 天车;厂内桥式起重机
shop trial 出厂试验;出厂试车;车间试车;厂内试运转
shop truck 修理车;工程修理车;工程车;流动修理车
shop weld 厂焊
shop welded 厂焊的
shop welded connection 厂焊接头
shop welding 工厂焊接;车间(中)焊接
shop window 橱窗;铺面橱窗;商店橱窗
shop window glass 橱窗玻璃
shop-window lighting 橱窗照明;店铺橱窗照明
shopwork 厂造(法);工厂作业;工厂制造;厂加工产品
shoran 精确近程导航系统;近程导航
shoran 肖兰精密导航装置(一种短程导航系统)
shoran chart 短程导航图
shoran control 肖兰控制
shoran controlled photography 肖兰控制摄影
shoran-line crossing 肖兰穿线飞行测距法
shoran method 肖兰法
shoran net 肖兰控制网
shoran network 肖兰控制网
shoran position fixing 肖兰定位
shoran problem 肖兰问题
shoran range 肖兰作用距离
shoran reconnaissance 近程雷达测位侦察
shoran reduction 肖兰距离校正
shoran set 近距离助航仪
shoran straight line indicator 肖兰直线指示器
shoran triangulation 肖兰三角测量

shoran trilateration 肖兰三边测量
shoran wave path 肖兰波路径
shore 海岸;顶柱;顶撑杆;撑柱;岸
shore accommodation 陆域设施;陆域设备
shore accomodation 陆战设施
shore accretion 海岸增长
shore ambient noise 海岸环境噪声
shore anchor 里挡锚;内锚;岸侧锚
shore and terminal operating costs 港口费(用)
shore approach 跳板
shore arm 支柱臂;撑臂
shore bank 沿岸沙洲;沿岸浅滩
shore base 岸上基地
shore-based 岸基地
shore-based radar chain 岸基雷达链
shore-based radar system 岸上雷达系统
shore-based radar unit 海岸雷达装置
shore-based recorder of low-frequency ocean waves 岸用低频自记测波仪
shore-based station 岸台
shore-based warping station 岸上施绞站
shore beacon 岸上航标;岸标
shore bearing 岸边方位标
shore bird 岸禽类
shore board 跳板;登陆跳板;上岸跳板
shore boat 摆渡艇
shore boulevard 岸边林荫道
shore-bound locomotive 纤船的机车
shore bridge 岸边栈桥;栈桥(码头);横栈桥;顺岸栈桥
shore building 海岸形成
shore change 海岸演变
shore clause 陆运条款;陆上条款
shore clearance 沿岸冰间水道;沿岸冰间航路
shore clearing 沿岸冰间水道;沿岸冰间航路
shore cleat 岸上系缆柱;岸上挽缆柱
shore cliff 滨崖
shore-connected breakwater 突堤;半岛式防波堤
shore connecting plant for pipeline 排泥管接岸装置
shore connecting structure (码头的)接结构
shore connection 岸上连接点
shore connection box 岸电接线箱
shore container crane 岸边集装箱装卸桥;岸壁集装箱起重机
shore control system 岸控制系统
shore cover 岸到岸全程保险
shore crane 岸吊
shore current 沿岸流;海岸流;滨流;岸(边)流
shore cutting 海滨切蚀;海滨冲刷;岸边冲刷
shore defence work 海岸保护工事
shore delta 海滨三角洲;滨海三角洲
shore deposit 湖滨沉积(物);海滨矿床;海滨沉积物;岸边淤积;沿岸沉积
shore development 海滨开发
shore direction-finding station 岸边测向台
shore discharge pipeline 岸上排泥管线
shored landslide 推动式滑坡
shore drift 沿滨沉积物;沿岸漂积物;海滨漂积物;海滨堆移物;海岸漂沙;海岸漂积物;滨漂沙;滨海漂积物;海边漂积物
shore dune 滨岸沙丘;岸滨沙丘
shored-up 用支柱撑住的
Shore durometer 肖氏硬度计
shore effect 海岸线效应
shore-end cable 岸电电缆;浅海电缆
shore-end of breakwater 堤根;防波堤堤根
shore end of submarine cable 海底电缆接岸的一端
shore end optical cable 岸边光缆
shore erosion 沿岸侵蚀;沿岸冲刷;海岸侵蚀;滨侵蚀;岸坡侵蚀
shore-face 海岸;近岸;滨面(低潮以下的拍岸浪区);水下浅滩;水下岸坡
shoreface beach 临滨滩
shoreface deposit 滨面沉积
shoreface storm deposit 临滨风暴沉积
shoreface terrace 近岸台地;滨前阶地;临滨阶地;水下阶地;岸面阶地
shore facility 陆域设备;岸上设备
shore fast 岸缆
shore floe 沿岸固定冰;贴岸冰;岸冰;岸边浮冰
shore formation 海岸沉积层
shore-front 岸坡前的

shore-front-trials 陆上试验
shore gravel 海岸砾石;海滨砾石
Shore hardness 肖氏硬度
Shore hardness number 回弹硬度(数);肖氏硬度(数)
Shore hardness scale 肖氏硬度级
Shore hardness test 肖氏硬度试验;肖氏硬度实验
Shore hardness tester 肖氏硬度试验仪;肖氏硬度实验仪;肖氏硬度计;回跳硬度计
Shore hardness testing method 肖氏硬度试验方法
shore head 支柱头;(支柱的)顶部横撑
shore ice 贴岸冰;滨冰;岸冰;岸边冰
shore ice belt 滨冰带;岸冰带
shore ice edge 岸冰界
shore identification code 岸站识别码
shore installation hose 输油软管
shore intake 河岸进水口;河岸进水建筑物;河岸进水构筑物;海岸进水建筑物;岸边进水构筑物
shore investigation 海岸调查
shore karst 滨海岩溶
shore lagoon 海岸礁湖
shoreland 沿岸土地;滨地;岸滨地
shore lead 沿岸通道;沿岸冰间水道;沿岸冰间航路;海岸冰沟;冰岸水道
shore leave 登陆假
shoreline 海岸线;海滨线;滨线;岸上管线;岸(边)线
shoreline bypass 岸边排放岔道
shoreline cycle 岸线循环
shoreline denudation 海岸剥蚀
shoreline deposit 海滨沉积;海滨矿床;滨线沉积
shoreline effect 海岸线效应
shoreline erosion 河岸冲刷;海岸(线)侵蚀
shoreline feature 海岸线特征;岸线形状;岸线特征
shoreline geometry 岸线几何形状
shoreline harbo(u)r 海岸港
shoreline of deposition 沉降滨线
shoreline of depression 下沉岸线;沉降海岸线;沉降滨线
shoreline of elevation 上升海岸线;上升滨线
shoreline of emergence 上升海岸线;下降海滨线;上升滨线;上升岸线
shoreline of retrogradation 蚀退滨线
shoreline of submergence 下沉(海)岸线;下沉滨线
shoreline port 海岸港
shoreline profile 岸线纵断面
shoreline reconnaissance 海岸踏勘;海岸线踏勘
shorelines clastic deposit 滨岸碎屑沉积
shore line survey 岸线测量
shoreline tsunami 海岸海啸
shoreline unsurveyed 未精测海岸线;未精测滨线
shoreline vegetation 岸线植被
shore man 陆上工作人员【港】
shoremark 岸标
shore material 海岸物质
shore mat(tress) 护岸柴排;护岸垫层
shore meadow 海岸草地
shore pier 岸墩
shore pipe 岸边管(疏浚);吹泥管;岸管
shore pipe connection 岸管连接
shore pipeline 岸上管线
shore plain type 滨海平原型
shore platform 浪蚀台地;浪蚀平台;海滨平台
shore pollution 海岸污染
shore polynya 沿岸冰穴;沿岸冰间湖;岸边冰穴;岸边冰湖
shore power 岸电(源)
shore process 海滩演变过程
shore profile 海岸轮廓
shore progress 海岸进展;海岸进展
shore protection 护岸;海岸防护;海岸保护
shore protection embankment 海岸防护堤;海岸堤防
shore protection facility 海岸防护设施
shore protection forest 海岸防护林
shore protection structure 护岸结构(物);护岸建筑物;海岸防护建筑物
shore protection works 海岸防护设施;海岸防护工程;护岸工程
shore radar 海岸雷达
shore radar chain 海岸雷达链
shore radar installation 海岸雷达装置
shore radar system 海岸雷达系统
shore radar television 沿岸雷达电视;海岸雷达电视
shore ramp 岸上坡道

shore recession 海岸后退
shore reef 裙礁;滨礁;滨海礁;边礁;岸礁;岸边礁石;岸边暗礁
shore rights 停泊权
shore sand 岸边沙
shore sand bar deposit 海滨沙坝沉积
Shore scleroscope 肖氏回跳硬度器;肖氏硬度计;肖氏硬度试验仪;回弹硬度计
Shore scleroscope hardness test 肖氏回跳硬度试验
shore sediment 湖滨沉积(物)
shoreside 海岸边;岸边(的)
shoreside accretion 岸边淤积
shoreside collapse 岸边坍塌
shoreside deposit 岸边淤积
shoreside melting ice 岸边融冰
shoreside operator 岸上操作员
shoreside receiving depot 岸边接收站
shoreside receiving terminal 岸边接收站码头
shoreside sediment 岸边淤积物
shoreside storage 岸上库场
shoreside tank 岸上油罐
shoresman 陆上勤务人员
shore sonar 岸用声呐
shore span 边跨(桥、闸等);近岸跨(度);(指桥梁的)近岸跨;接岸跨
shore spur 码头舷侧撑柱;边撑
shore stabilization 海滨稳定;海岸稳定
shore staging 顺岸式栈桥码头
shore station 海岸电台;岸台;岸上信号台
shore structure 岸边结构;护岸建筑物
shore subsidence 海岸坍陷
shore tackle 岸吊
shore tank 岸上储[贮]罐
shore terminal 沿岸码头
shore terrace 沿岸阶地;海成阶地;海滨阶地;海岸阶地
shore-to-ship 岸到船
shore-to-ship call 岸对船呼叫
shore-to-ship communication 陆对海通信;岸对船通信
shore-to-shore clause 陆上条款;陆运条款;岸到岸全程保险
shore trial 陆上试验;(船用机械的)试验台试验
shore type 岸边式
shore up 撑住;撑起;有支柱撑住
shore wall 海滨驳岸;冰成冰堤;岸墙;岸壁;护岸墙壁
shoreward 向岸的;朝岸(的)
shoreward mass transport 向岸的大量运输;向岸水团运动
shoreward thrust 向岸推力
shore wave recorder 岸用波浪自记仪
shore weir 岸上拦欲堰
shore wharf 滨海码头
shore wind 岸风
shore zone 海滨带
shore-zone sequence 滨岸带层序
shoring 斜支撑法;斜撑系统;支撑(工程);临时撑;临时撑(柱);加固(撑);横支梁;顶撑工程
shoring and lashing 撑垫捆扎
shoring beam 支撑梁
shoring cage 支撑框架
shoring column 支柱;支撑
shoring equipment 临时支撑设备;临时加固设备
shoring layout 支撑设计;支架布置;支撑布置(图)
shoring of foundation 基础支撑
shoring of trench 沟槽支撑;基坑支撑;基槽支撑
shoring plan 支撑平面图
shoring procedure 支撑方法
shoring sheeting 临时木板支撑;打支撑板桩;撑板
shoring system 支撑体系
shoring to walls 墙身顶撑
shoring up 用柱子支撑
shoring work 支撑工程;顶撑工程
shorl 黑电气石
shorl rock 黑电气石岩
short 零头料;空头交易;空头;短锯材;短的;脆性
short-access 快速存取
short account 空头账户;卖空账户
short address 短地址
short afterglow 短余辉
shortage 缺少;缺量;缺乏;缺额;短缺(额);不足额

shortage cost 缺货成本
shortage/damage list 货损货差报表
shortage/damage report 残短单
shortage in weight 重量不足;缺重(量);缺量
shortage of capital 资金短少
shortage of drinking water 饮用水不足;缺乏饮用水
shortage of funds 缺乏资金
shortage of heat 热量不足
shortage of housing 住房短缺
shortage of manpower 人力短缺;人力不足
shortage of money 资金短缺
shortage of oil 油不足
shortage of soil moisture 土壤水分不足
shortage of space 舱位不足
shortage of water 缺水
shortage penalty 缺额罚款
shortage penalty cost 缺货费;短缺惩罚费
short age strength 短龄期强度
short and intermediate term credit 中短期贷款
short annealing 快速退火
short anneal(ing) furnace 快速退火炉
short arc 短弧;低振幅
short arc lamp 短弧灯
short arc metal transfer 短弧过渡
short arc method 短弧法
short arc method of satellite orbit 卫星轨道短弧法
short arc reduction method 短弧归算法
short arc welding 短弧焊
short arm 短臂
short armed 短臂的
short-auger drill 短螺旋钻机
short axis 短轴
short bar 短路杆
short base 短轴距
short baseline acoustic(al) positioning system 短基线水声定位系统
short baseline sonar 短基线声呐
short-baseline system 短基线系统
short-bath tub 小浴盆
short beam 短跨梁;短(距)梁
short beam shear 短梁剪切
short beam shear strength 短梁剪切强度
short bearing of tooth 轮齿窄短承载区
short belt drive 短皮带传动
short bench-cut method 短台阶法
short bill 短期期票
short blast 短声
short block 短块;短记录块
short board 短程拖(帆船)
short-bodied sardine 沙丁鱼
short-body cast-iron fittings 铸铁短配件
short body reducing tee 缩径短三通
short bole 矮干材
short bond 短路棒
short bone 短骨
short boom 短悬臂;短吊杆
short boom drag 短悬臂索斗铲
short-boom loader 短转臂式装载机
short-bored piling 短钻孔灌注桩
short borer 开孔钻头
short bort 劣等金刚石
short brace 摇柄木钻
short branch 短转移
short branchlet 短枝
short bridge 短桥楼(长度不大于船长的15%)
short-brittle 热脆(的)
short-brittleness 热脆性
short burst 短期暴;短脉冲
short cane pruning 短梢修剪
short card 短卡片
short carriage knitting machine 短三角座滑架横机
short center belt drive 短中心距皮带传动
short centre vee-rope drive 短径三角皮带驱动
short chain(age) 短链
shortchanging on work and materials 偷工减料
short channel 短沟道
short-chord winding 短弦绕组;短节距绕组
short-circuit 短路的;短接
short-circuit admittance 短路导纳
short-circuit admittance matrix 短路导纳矩阵
short-circuit analysis 短路分析
short-circuit between conductors 线间短路
short-circuit between plates 极板间短路
short-circuit brake 短路制动器

short-circuit braking 短路闸
short-circuit calculation 短路计算
short-circuit calculator 短路计算器
short-circuit characteristic curve 短路特征曲线;短路特性曲线
short-circuit condition 短路状态
short-circuit contact 短路触点
short-circuit control 短路控制
short-circuit current 短路电流
short-circuit current gain 短路电流增益
short-circuit current of feeding section 供电臂短路电流
short-circuit current rush 短路电流冲击
short-circuit detector 短路探测器
short-circuit device 短路装置
short-circuited 短接;短路
short-circuited armature 鼠笼电枢;短路电枢
short-circuited armature coil 短路电转子线圈
short-circuited brush 短路电刷
short-circuited coil 短路线圈
short-circuited compensating winding 短路补偿绕组
short-circuited loop 短路回线
short-circuited resistance 短路电阻
short-circuited turn 短路线匝
short-circuited winding 短路绕组
short-circuit end 短路侧
short-circuiter 短路器
short-circuit fault current 短路故障电流
short-circuit ground 对地短路
short-circuit impedance 短路阻抗
short-circuiting 短流
short-circuiting arc 短路电弧
short-circuiting arc welding 短路过渡电弧焊
short-circuiting bar 短路棒
short-circuiting circulation 短路循环
short-circuiting current 短路电流
short-circuiting device 短路装置;短路器
short-circuiting jumper 短路金属片
short-circuiting piston 短路活塞
short-circuiting ring 短路环
short-circuiting switch 短路开关
short-circuiting switch blade 短路器
short-circuiting transfer 短路过渡;短路传送
short-circuit input admittance 短路输入导纳
short-circuit input capacitance 短路输入电容
short-circuit key 短路键
short-circuit loss 短路损失;短路损耗
short-circuit magnetomotive force 短路磁通势
short-circuit of terminals 端子短路
short-circuit output capacitance 短路输出电容
short-circuit period 短路持续时间
short-circuit plug 短接插头
short-circuit power 短路功率
short-circuit proof 耐短路的
short-circuit-proof output 防短路输出
short-circuit protection 短路电流保护;短路保护(装置)
short-circuit protection switch 短路保护开关
short-circuit ratio 短路比
short-circuit resistance 短路电阻
short-circuit ring 短路环
short-circuit spark 短路火花
short-circuit spring 短路弹簧
short-circuit stability 短路稳定性
short-circuit strength 短路强度
short-circuit subtransient time constant 短路初期瞬变时间常数
short-circuit switch 短路开关
short-circuit terminals 短路接头
short-circuit test 短路试验;短路事故检查
short-circuit torque 短路转矩
short-circuit transfer admittance 短路转移导纳
short-circuit transition 分路换接过程
short-circuit voltage 短路电压
short-circuit wire 短路线
short circulation sticking 短路循环卡钻
short clay 低塑性黏土
short clip 短路线夹
short code 短码
short-cold brittle 冷脆的
short column 短柱;瓜柱
short-columnar habit 短柱状习性
short column formula 短柱公式

shortcoming 缺点
short-contact switch 短触点开关
short-continued load 短暂连续荷载
short-continued static load 短暂连续静载
short contract 空头交易合同;卖空合同
short-cooling fuel 短期冷却燃料
short cord 短考
short corrugated panel 短波纹板;短瓦楞板
short count 短期积日制度
short-count station 短时交通量观测站;短时计数站
short-count traffic survey 短时交通调查
short-coupled 短背的
short covering 补空;补进(短缺)
short credit 短期赊欠
short-crested wave 短脊波;短峰波
short crop 低产作物;低产量
short cross garnet 短丁字形折页;短丁字形合页;短丁字形铰链
short cultivation 短期培育
short culvert 短廊道
short culvert conveyance 短廊道输水
short cut 捷径;捷水道;捷路;快捷方式;简捷
short-cut calculation 简便算法;简便计算;简化计算
short-cut channel 短捷航道;捷径航道
short-cut design 快速设计;简略设计
short-cut division 简捷除法
short-cut line 捷径线
short-cut method 捷算法;简捷(算)法;简捷(方)法;简便方法
short-cut multiplication 简捷乘法
short-cut nitrification-denitrification process 短程硝化—反硝化法
short-cut process 简捷(方)法
short-cut road 捷径道路
shortcutting 抄近路(走);裁弯取直(河道);简化
short cutting works 裁弯工程
short-cut turn 近路转弯;捷径转弯
short cycle 短周期
short-cycle annealing 短时退火
short-cycle malleable iron 快速退火可锻铸铁
short-cycle variation 短周期变化
short cylinder 短圆筒
short cylindrical roller bearing 短圆柱滚子轴承
short cylindrical shell 短圆柱形壳体
short damping electric(al) meter 短阻尼电表
short dashed line 虚线
short day 短日照
short-day crop 短日照作物
short-day-long night plant 短日照长夜植物
short-day plant 短日照植物
short-day plant management 短日植物管理
short-day regions 短日照地区
short deep beam 短的深梁
short-delay blasting 微差爆破;短时延迟爆破;微差延发爆炸
short delay blasting method 瞬时延发起爆法;短时延迟起爆法
short delay detonator 瞬时延发雷管;短时延迟雷管
short delay fuse 瞬时延发引信;瞬时延迟雷管;短时延迟引信;短时延迟雷管
short delay fusing 短时定时引线;短期引线
short delay line 短延迟线
short delivery 交货短缺;货物短缺;短交货(物)
short demy 矩形图纸
short diagonal of indentation 压痕的短对角线
short distance 近程
short-distance beam 近光
short-distance carrier telephone 短距离载波电话
short-distance communication 短距离通信
short-distance navigation 短距离航行;短程航行
short-distance navigation aid 近程导航设备
short-distance navigational aid 短距离导航设备
short-distance of migration 短距离运移
short-distance photogrammetry 近景摄影测量
short-distance reception 近距离接收
short-distance reconnaissance 近距离空中侦察
short-distance scatter 近距散射
short-distance sender 近距离发送器
short-distance traffic 短程交通;地方性交通
short-distance transport(ation) 短途运输
short-distant sprinkler system 密集喷灌系统
short division 简便除法
short dot 短路点
short drafts 短绵

short dry process kiln 干法短窑
short dung 发酵过的粪肥
short duration 短时期；短时间；短期间；短历时
short duration air dive 短时间空气潜水
short duration dive 短时间潜水
short duration diving 短时间潜水
short duration impulse 短持久脉冲
short duration load 瞬时荷载
short duration load striker 瞬时荷重冲击仪
short duration pulse 短脉冲
short duration satellite 短期运转卫星
short duration sustainer 短期工作的主发动机
short duration trace 短持续时间的扫迹
short duration warning 短期警报
shorted factor of wave length 缩波系数
shorted out 短路的
shorted turn indicator 短路线圈指示器
shorted-turn tester 短路线匝探测器
short element 短单元
short end 小头；细端
short end link 链节末端短链环
shortened blade 削短叶片
shortened code 截短码
shortened connection at ends of tracks 缩短式股道终端连接
shortened crossover 缩短渡线
shortened ladders 缩短梯线
shortened McLeod ga(u)ge 缩短型麦氏真空计
shortened nozzle 缩短型喷管
shortened process 缩短的工艺流程
shortened ratio 裁弯取直率
shortened river section 裁直河段
shortened section of river 裁直河段；裁弯河段
shortened title 简称标题
shortener 缩短器
shortening 收缩；缩短
shortening coefficient 缩短系数
shortening hook 缩短项吊钩
shortening-in 缩短锚链
shortening of basement 基底缩短作用
shortening of channel 河道缩短
shortening of pitch 缩短节距
shorten range 近瞄准距；近程
shorten sail 缩帆
shorten the heat 缩短冶炼时间
shorter haul 短程航运；航程较短的
shorterizing 火焰表面硬化处理
shorter process 火焰表面硬化处理
shortest distance 最短距离
shortest focusing distance 最小调焦距离
shortest geodesic 最短测地线
shortest path 最短路径
shortest path of percolation 最短渗流路线
shortest photo-baseline 最短摄影基线长度
shortest route 最短路径
shortest route problem 最短路径问题
short excursion dive 短时间巡回潜水
short extended 列入细数
short extension 用以伸长的渗水件
short face 短面板
shortfall 短缺；不足额
shortfall freight 亏舱运费
shortfall in output 欠产
short fiber 短纤维
short fiber reinforced thermoplastic 短纤维增强热塑性塑料
short fibred 短绒的；短纤维的
short finish 消光；灰斑；抛光不足；粗磨面
short fire 氧化焰
short flame 短火焰
short flame coal 短焰煤
short flamed 短火焰的
short flame gas burner 短焰燃烧嘴
short-flaming 短焰
short flaming coal 锅炉煤；短焰煤
short-flashing buoy 闪光浮标
short flashing light 短闪光(灯)；短程闪光灯
short flute spiral pointed tap 短槽螺尖丝锥
short focal length high aperture lens 短焦距强光力物镜
short-focus camera 短焦距照相机；短焦距摄影机
short-focus lens 短焦距透镜；短焦距镜头
short-focus objective 短焦距物镜
short-focus telescope 短焦距望远镜

short forecastle (不足船长的20%)短船首楼
short form 简体
short form appraisal report 产业估价简表
short-format instruction 短格式指令
short form bill of lading 略式提单；简式提单
short form documents 简要文件
short form report 核数简报
short forward rate 一个月内的期货汇率
short fuse 短导火线引爆
short-gate gain 短时间选通增益
short glass 快凝玻璃；短性玻璃
short glaze 欠釉；施釉不足
short grain 短纹；短晶粒；带斜纹的木材；脆纹；短纤维
short-grained 小颗粒的
short grass 短草
short grassland 矮草型草地；矮草草甸
short-grass plain 矮草大草原
short grass type 矮草型
short gravity wave 短重力波
short groin revetment 短丁坝护岸
short ground 易脆顶板；脆性岩层
short ground return 短线接地
short guide drill wagon 短导轨钻车
short half paper 短半张
shorthand 简写形式；速记(法)
short-handed 缺乏人手；人手不够的
short-handle float 短柄镘板
shorthand notation 简化符号
short haul 短拖；短途运输；短途的
short-haul call 短距离通话
short-haul connection 近程接续
short-haul convoy 短程护航队
short-haul grading 短运距土方整平
short-haul modem 基带调制解调器
short-haul plane 近程飞机
short-haul traffic 短距离运输；短距离交通
short-haul train 短途列车
short-haul tropospheric scatter radio system 近程对流层散射通信设备
short-head cone 短头圆锥
short-head cone crusher 短颈锥式破碎机；短头锥式破碎机；短头圆锥破碎机
short head type cone crusher 短头型圆锥破碎机
short head type Symon's cone crusher 短头型西蒙圆锥破碎机
short heavy swell 短狂涌
short hedge 空头对冲；卖空套头
short high swell 短狂涌
short-hold mode 短暂保持方式
short hole 浅炮眼；浅孔
short-hole blasting 浅孔爆破
short-hole blasting method 浅孔爆破法
short-hole drill 浅孔钻机
short-hole drilling 浅眼凿岩；浅孔钻进(深度30米以内)
short-hole grouting 浅孔注水泥；浅孔灌浆
short-hole method 浅孔注水泥法；浅孔爆破法
short-hole work 浅孔钻进(深度30米以内)
short hopper closet 短漏斗形大便器
short-hour motor 短时工作电动机
short hundredweight 一百磅；短担(等于100磅)
short immediate instruction 短立即指令
short in dispute 短缺待查
short infrared 短波红外线
shorting bar 短路棒
shorting clip 短路夹
shorting contact switch 短路(接触)开关
shorting-out engine 短路发动机
shorting stub 短接线
short ink 高颜料分不稠油墨；短丝油墨
short instruction 短指令
short interest 卖空者
short international voyage 短程国际航行
short internode trait 短节间性状
short interruption rate 瞬断率
short iron 脆(性)铁
short irradiation 短时间照射
shortite 碳(酸)钠钙石
short jack 短千斤顶；矮千斤顶
short jib 短悬臂；短吊杆
short jump 短转移
short laid 硬搓绳
short-landed 短卸

short-landed account 缺账户
short-landed and damaged 短缺及损坏报告
short-landed and over landed cargo list 货物溢短残清单
short-landed coal 短卸煤
short-landed goods 短卸货物
short-landed memo 货差证明；短卸单
short landing 货差
short lane 短车道
short lateral curve 短梯度测井曲线
short lateral device 短梯度电极测深装置
short-leaf pine 短叶松
short lease 短期租赁
shortleaved pine 短叶松
short leg 挽头；短引线
short legged 活动半径小的；短航程的
short-length 短料
short-length of casing 短套管
short length rule 短长度规
short lengths of double track 双线插入段
short life 短期破坏的；不持久；短效；短寿命
short-life 短寿命的；不耐用的
short-life isotope 短寿命同位素
short-life radioactive goods 短寿命的放射性货物
short-life radioisotope 短寿命放射性同位素
short lift build 短程圆式纱筒成形
short line 短途运输线
short linearity controller 近距线性调整器
short-line railroad 短距离铁路
short-line seeking 短行回车
short link 短链环
short link chain 核桃链；短环链；短环节链
short-lived 较短寿命的；短暂的；短寿命的；不耐用的
short-lived activity 短期活度
short-lived burst 短寿命暴
short-lived crack 短期裂缝
short-lived grass 短木生草
short-lived material 不耐用材料；短命材料；非耐用物料
short-lived pesticide 短效农药；非残留农药
short-lived phenomenon 短暂现象
short-lived radioisotope 短寿命放射性同位素
short-lived seed 短命种子
short-lived tracer 短暂示踪物
short-lived transient 短寿命瞬态
short-lived waste 短寿废物
short loaded 减载
short loan 短期借款
short locomotive runs 短交路
short-log bolter 纵剖小圆锯机
short-long-day plant 长日短日植物
short-long flashing light 短长闪光(灯)；断续短长闪光
short loop detector 短环框式检测器
short loop drying and heat setting range 短环烘燥定型机
short loop resin finishing range 短环树脂整理机
short low swell 短轻涌
shortly attenuate 短渐狭的
short magnetic lens 短磁透镜
short material 边缘材料
short maturity 近期到期(外汇)
short measure 短尺寸；尺寸不足
short-memory radiation detector 短记忆辐射探测器
short method 简捷方法；捷征法
short moderate swell 短中涌
short-necked bottle 短颈瓶
short-necked flask 短颈烧瓶
shortness 松脆；脆性
short nipple 短螺纹套接；短内接头；短接管
short normal curve 短电位曲线
short normal device 短电位测井装置
short notice 临时通知
short nut tap 短柄螺母丝锥
short oak 短橡木
short of bunker 燃料不足
short of cash 缺少现款；缺乏现金；短缺现金
short offset 短位移
short offset addressing 短偏移量寻址
short of weight 短量险
short oil 聚合程度不大的油；短油(度)；短油漆(油漆中油脂较少)
short-oil alkyd 短油(度)醇酸树脂；低油度醇酸树脂

short-oil alkyd resin 短油醇酸树脂
short-oil linseed oil alkyd 短油亚麻籽油醇酸树脂
short-oil type phthalic resin varnish 短油性邻苯二甲酸树脂清漆
short-oil varnish 硬油清漆;聚合度小的清漆;少油清漆;镀金黏料;短油(度)清漆(含油量少的清漆)
short operation 短操作
short out 使短路;使变短;短路;捷径
short paid 欠资的
short-pan apron conveyer 短平板裙式输送机
short pass 短路
short pass filter 窄带滤光器;窄带滤光片;窄带滤波器;窄带滤波片
short-path distillation 极窄留分蒸馏
short peak 近锚泊(锚链长约水深1.5倍与海底大约成45°)
short peak-period 短时高峰期
short peck line 短划(虚)线
short-period 短周期的;短寿命的;短期
short-period array 短周期台阵
short-period circuit 短期使用电路
short-period comet 短周期彗星
short-period constituent 短周期分潮
short-period cost 短期成本
short-period data 短期资料
short-period fading 短周期衰落
short-period fluctuation 短(周)期波动;短期变动
short-period forecast(ing) 短期预报
short-period ground motion 短周期地面运动
short-periodic(al) nutation 短周期章动
short-period microseism 短周期微震
short-period mode of motion 短周期运动
short-period motion 短周期运动
short-period motor 短期工作发动机
short-period oscillation 短周期振动
short-period perturbation 短周期摄动
short-period price 短期价格
short-period sea wave 短周期海波
short-period seiche 短周期假潮
short-period seismograph 短周期地震(记录)仪;短周期地震计
short-period seismometer 短周期地震计
short-period structure 短周期结构
short-period term 短周期项
short-period test 短周期试验
short-period variable 短周期变量
short-period variation 短周期变化
short-period wave 短周期波
short pile 短桩
short pile method 簇绒法(地毯织法);短绒法
short pipe 短管
short pitch corrugated sheet 小波形板;小波纹板
short-pitch factor 短节距因数;短节距系数
short-pitch winding 短节距绕组;分支绕组;分节距法
short-pitman mower 短连杆式割草机
short-plate apron conveyer 短平板裙式输送机
short poop 短尾楼
short position 空头(部位);卖出的期货合同;短缺头寸;短缺期货;超卖;不足部位(外汇)
short precision 短精度
short primary line engine 短初级直线电机
short-prismactic 短柱状
short production run 小批(量)生产
short-proof ring 引弧环
short pulse 短脉冲
short-pulse blaster 短脉冲放炮器
short-pulse laser 短脉冲激光器
short-pulse train 短脉冲序列
short quotation 短期汇率
short racemes 短总状花序
short-radius curve 短半径曲线;小半径曲线
short rail 短轨
short railway 地方铁路
short ramp 短匝道;短坡道
short-range 近距离;近程;短量程;距离的;短程
short-range aggregate 均匀集料;均匀骨料
short-range aggregate aerated concrete 短级配集料加气混凝土;短级配集料加气混凝土
short-range aggregate mortar 短级配集料砂浆;短级配集料砂浆
short-range aids 近程导航设备;短程导航设备;短程导航设备

short-range aids-to-navigation system 近程航标系统
short-range air-defense complex 近程防空综合设施
short-range autograph 近景自动测图仪
short-range budget 短期预算
short-range characteristics 近距特性
short-range chart 短程导航图
short-range communication 近距离通信
short-range coverage acquisition 近距截获
short-range decision 短期决策
short-range design 近期规划
short-range direction finding 近距离探向
short-range electronic distance meter 短距离电磁波测距仪
short-range fading 近距离衰落
short-range field 短程场
short-range firing 近距离射击
short-range force 近程作用力;短程力
short-range forecast(ing) 当天天气预报;短期(天气)预报
short-range ground position indicator 近距地面位置指示器
short-range guidance 近程导航
short-range indicator element 近程指示元素
short-range interaction 近程相互作用;短程相互作用
short-range investment 短期投资
short-range modem 短距离调制解调器
short-range navaid 近距导航设备
short-range navigation 高精度兰昌导航系统;近程导航;短程导航
short-range navigation(al) aids 短程无线电导航设备;近程导航设备
short-range navigation(al) chart 短程导航图
short-range navigation(al) system 短程无线电导航系统;精确近程导航系统
short-range navigator 精确近程导航系统
short-range order 近程有序;短程有序;短序
short-range order structure 近程有序结构;短程有序结构
short-range perturbation theory 短程微扰理论
short range photogrammetry 近景摄影测量
short-range pick-up television relay equipment 近距离电视中继装置
short-range plane 近程飞机
short-range positioning system 近程定位系统
short-range product 窄粒级产品;粒级均齐的产品
short-range radar 短程雷达
short-range radio beacon 近距离无线电信标
short-range radio location system 短程无线电定位系统
short-range requirement 短期需求量
short-range scanning 近距离扫描
short-range ship 短程船
short-range sprinkler 短射程喷灌机
short-range survey operation 短程测距作业
short-range transport 短程运输机
short-range waviness 短程波纹
short-range weather forecast 短期天气预报
short-rate 对不可靠保险单的保险费(计算);短期保险率
short red brittle 热脆的
short relay blasting 瞬时迟发爆破
short relay detonator 瞬发雷管
short residence time 短(暂)停留时间
short residue 减压渣油
short residuum 减压渣油;浓缩残油;脆残油
short rib 假肋
short rod 端水准尺
short rope connecting gear 短缆系结装置
short rotation 短期轮作
short round nose pliers 短圆嘴钳
short routing 短交路
short run 小批量;短期运转;短回次
short run average cost 短航程平均价格
short-run cost 短期成本
short-run die 小批生产用模具;简易模
short-run forecast(ing) 短期预报
short run marginal cost 短程限界价格;短程极限价格
short-run maximization 近期最大利润
short-run planning 短期计划
short-run power 短期功率

short-run production 小批(量)生产;短期生产;小量生产
short runs 短行程;钻具稍提
shorts 筛网粗粒;短于6英尺
short sale 卖空(交易)
short scan 快扫描
short scarf 直齿对搭接
short screw 短螺钉
short sea 碎浪海面;短促浪;三角浪;近海运输(租船合同用语)
short-sea trade 海上短程运输;近海贸易
short-sea traffic 短途海上运输
short seller 卖空者
short selling 卖空行为;卖空交易
short series 短系列
short-shaft borer 短轴镗床
short-shaft suspended-spindle crusher 短悬轴式旋回破碎机
short shank anchor 短杆锚
short shank auger 短柄木工钻;短柄麻花钻
short shank drill 短柄钻
short shank stockless anchor 短杆无杆锚
short shell 短壳
short shift controller 近距位移调整器
short shipment 装载不足;短装货运;短缺载货
short shipped 退关
short-shipped cargo 短装货物
short shoot 短枝
short shot 中景;欠注射;喷丸不足
short shunt 短分流器
short-shunt compound 短并复绕
short-shunt compound winding 短并激式复绕组;短分流复励绕组
short side round edge tile 短边圆角釉面砖
short sight 近视;短视线
short sight distance 短视距
short sign-on 简化签到;简化开式;短签到;短开式
short slab 短板
short slag 短渣
short sleeve 短外接头
short slope 短坡
short-slot coupler 短槽耦合器
short-sonar 近程声呐
short spaced neutron log 短源距中子测井
short spaced sonde 短电极系
short spacing 短间距
short spacing curve 短极曲线
short spacing neutron-neutron log curve 短源距中子—中子测井曲线
short span 小跨度
short-span barrel vault 短跨筒形拱顶
short-span roof 小跨度屋盖;短跨屋顶
short-span slabs for precast gypsum 预制短跨石膏板
short-span structure 小跨度结构;短跨结构
short spar 短杆;尺寸不够的圆材
short splice 短拼接;短接;短插接
short split 短小开裂
short spur needle 弯针
short spur protection 短丁坝护岸
short square core 悬臂芯
short squeeze 空头轧平;期货卖方抢购
short stack 浅栈;短排气管;短堆栈
short-stage airline 短途直飞航线
short staker 为享受而打短工的人(俚语)
short-stalked cyprion golden oak 短柄黄栎
short-stalked glandbearing oak 短柄抱frames
short-stalked plant 矮秆作物
short-staple cotton 短绒棉
short stapled 矮纤维的
short-staple upland cotton 短绒陆地棉
short stay 近锚泊(锚链长约水深1.5倍与海底大约成45°角);短链抛锚;抛近锚
short steam 低压蒸汽
short steep wave 短陡浪
short stick 短码尺
shortstop 停显液
short stopped chain 断链
shortstopper 速止剂
short stopping 急速中止;速止
short stopping agent 速止剂
short stopping distance 最短停止距离
short stopping of chain 链的中止;链的速止
shortstop pump 速止泵

short stowage 填隙货
short stowage cargo 填隙货物
short-straw sieve 碎稿筛
short stress path rolling mill 短应力线轧机
short stroke 短划(荡桨)
short-stroke engine 短冲程发动机
short-stroke magnetic shoe brake 短行程电磁闸瓦制动器
short-stroke press 短程压缩
short-stroke solenoid 短冲程螺线管
short-stroke steam engine 短冲程蒸汽机
short-stroke valve 短行程阀
short sunlight photoperiod 短期日光照
short support jack 矮柱千斤顶
short sweep 短弯头
short swell 短涌
short system 快速系统
short table screen printing machine 短台板筛网印花机
short tack 短程掉抢(帆船)
short tail 短尾巴
short takeoff and landing 短距起落
short takeoff and landing aircraft 短距起落飞机
short taper ga(u)ge 短锥度规
short tax year 不满一年的纳税年度
Shortt clock 肖特钟
short tee hinge 短T形铰接
short telescope 短筒望远镜
short-term 近期;短期
short-term aeration 短期曝气法
short-term airfield 临时机场
short-term allowable bearing capacity 短期允许承载力
short-term allowable stress 短期许应力
short-term arrangement 短期安排
short-term autostabilization 短周期运动的自动稳定
short-term behavio(u)r 短期性能
short-term bond 短期债券
short-term borrowing 短时借贷
short-term buffering 短暂缓冲
short-term capital 短期资本
short-term capital gain 短期资金收益
short-term change 短期变化
short-term climatic variation 短期气候变化
short-term concentration 短期浓度
short-term concentration of contaminant 污染物短期浓度;短期污染物浓度
short-term contract 短期合同
short-term cost 短期费用
short-term credit 短期信贷
short-term data 短期资料
short-term debt 短期债券
short-term decision 短期决策
short-term development 短期研制
short-term duty 短时运行方式;短时负荷
short-term earthquake prediction 短期地震预报
short-term effect 近期效应;短时效应;短期作用;短期影响;短期效应
short-term emergency limit 短期应急限值
short-term environmental effect 短期环境效应
short-term exposure 短期接触;短期暴露
short-term factor 暂时因素
short-term fading 快衰落
short-term film 短期胶片
short-term financing 短期集资;短期资金筹措
short-term fire danger 短期火险性
short-term flood prediction 短期洪水预报
short-term fluctuation 瞬间变化
short-term forecast(ing) 短期预报;短期预测
short-term foreign capital 短期外资
short-term frequency stability 短期频率稳定度
short-term fund 短期资金
short-term gains 短期收益
short-term goal 近期目标;短期目标
short-term information 短时信息
short-term instability 短周期不稳定性
short-term insurance 短期保险
short-term investment 短期投资
short-term lease 短期租赁
short-term lending 短期信贷
short-term liability 短期负债
short-term load audit 短期借款审计
short-term load(ing) 短时负载;短时荷载;短时负荷;短期荷载;短暂荷载;

short-term loss 短期损利
short-term margin cost 短期边际成本
short-term material plan 短期物资计划
short-term memory 短时记忆;短期储存
short-term money rate 短期利率
short-term nutrient absorption 短期养分吸收测定法
short-term objective 近期目标
short-term operation 短时工作;短期运行
short-term operation decision-making 短期经营决策
short-term oxygen demand 短期需氧量
short-term parking 短期停车
short-term plan 短期规划
short-term planning 短期计划
short-term prediction 短期预测
short-term premium rates 短期保险费率
short-term pre-screening study 短期初筛试验
short-term pulse 短暂脉冲
short-term rating 短时定额;短期功率
short-term record 短期观测记录
short-term report 短期报告
short-term requirement 短期需求量
short-term road network 近期线网
short-term sampling 短时采样
short-term schedule 短程调度;短期进度
short-term schedule and medium-term schedule 短程调度与中程调度;短期与中期进度
short-term stability 短时间稳定性;短期稳定性;短期稳定度
short-term stabilization 短周期运动的稳定
short-term static load 短暂静载
short-term station 短期测站
short-term stiffness 短期刚度
short-term storage 短期仓库;短期储[贮]存;短期存贮(器);短期储存
short-term storage reservoir 短期调蓄水库
short-term strength 短期强度
short-term test 短时实验;短期试验
short-term testing 短期测试
short-term tide station 短期验潮站
short-term toxicity test 短期毒性试验
short-term trend 短期趋势
short-term variation 短期变化
short-term waviness 短程波纹
short-term weather forecast 短期天气预报
short-term yield 短期出水量
short terne plate 镀铅锡合金的薄钢板材
short test 简易试验
short T hinge 短T形铰接
short thread casing 短扣套管
short-thread milling machine 短螺纹铣床
short tie 短枕
short tillage 短期耕作
short time approach 短期预报方法
short-time climatic variation 短期气候变化
short-time constant 短时间常数
short-time current 短时间电流
short-time cycle 短周期
short-time dead load 短期静载
short-time diffusion 短时间扩散
short-time duty 短时运行方式
short-time fading 短时衰落
short-time homomorphic signal analysis 快速同态信号分析
short-time interval 短时段
short-time load condition 短时负荷状态
short-time load(ing) 短时负载;短期荷载;短时荷载
short-time overload test 短时过载试验
short-time rating 短期功率;瞬时定额;短时功率;短时定额;短时出力
short-time running 短时间运转
short-time solution 短时间的解
short-time static load 短期静载
short-time statistics 短时统计特性
short-time storage reservoir 短期蓄水水库;短期调蓄水库
short-time switching-off 短时切断
short-time test 快速试验;(土的)快剪试验;短时;短期试验
short-time thermal rating 短时发热容许电流
short-time variation 短期变动
short title 简明标题(指文献的);文献代号
short ton 小吨;美吨(1美吨=2000磅);每吨(等

于2000磅);短吨(美国)
short traffic ocean carrier 近海运输船舶;短程海运船舶
short transverse 高度方向(横截面中的短边)
short trip 提起钻具;短程旅行
short trouble 短路故障
short-tube evapo(u)rator 短管蒸发器(浓缩器);短管浓缩器
short-tube vertical evapo(u)rator 短管竖式蒸发器;短管立式蒸发器
short type water swivel 短式水龙头(坑道钻用)
short valve 短路阀
short varnish 短油清漆(含油量少的清漆)
short volume ratio 低容积比
shortwall 短壁(工作面)
shortwall coal cutter 短壁截煤机
shortwall coal mining method 短壁采煤法
shortwall loader 窄工作面装煤机
shortwall mining 短壁开采法
shortwall undercutter 短壁工作面用截煤机
short-wave 短波
short-wave adapter 短波转换器
short-wave antenna 短波天线
short-wave band 短波波段
short-wave beam 短波束
short-wave broadcasting 短波广播
short-wave broadcast transmitter 短波广播发射机
short-wave capacitor 短波电容器
short-wave choke 短波扼流圈
short-wave choke coil 短波扼流圈
short-wave communication 微波通信;短波通信
short-wave converter 短波变频器
short-wave cut-off filter 短波截止滤波器
short-wave deflecting spur wall 短波反射突堤
short-wave directional antenna 短波定向天线
short-wave direction finder 短波探向器;短波测向器
short-wave fadeout 短波消逝;短波衰退
short-wave frequency 短波频率
short-wave frequency band 短波频带
short-wavelength 短波长
short-wave length infrared vidicon 短波红外视像管
short-wavelength irregularity 短波长不平顺
short-wave length limit 短波限
short-wavelength radiation 短波辐射
short-wave limit 短波极限
short-wave-pass filter 短波通滤波器
short-wave propagation 短波传播
short-wave radiation 短波辐射
short-wave radio telephone terminal 短波电话终端机
short-wave radio transmitter 短波无线电发射机
short-wave receiver 短波接收机
short-wave receiving antenna 短波接收天线
short-wave receiving station 短波收信台
short-wave reception 短波接收
short-wave reflecting spur wall (短的)波浪消能丁坝
short-wave sender 短波发射机
short-wave signal 短波信号
short-wave SSB transmitter 短波单边带发射机
short-wave telegraph telephone transmitter 短波报话发射机
short-wave tower 短波塔
short-wave transceiver 短波收发信机
short-wave transmitter 短波发射机
short-wave transmitting station 短波发信台
short-wave ultraviolet radiation 短波紫外(线)辐射
short-wave undulation wear 短波浪磨耗
short week 非完整工作周
short weight 重量不足;缺重(量);缺量;欠重;份量不足;分量不足
short working plaster 和易性差的灰浆;劣质灰浆
shot 硬粒;注塑量;注料量;渣球;钢丸;镜头;金属中硬夹杂;摄影;弹丸;放炮;玻璃球;爆破
shot and grit 铁丸和铁砂
shot backing 硬丸衬垫;填丸加固
shot bag impact test 落袋法冲击强度试验
shot-bag test 毂弹袋试验
shot bit 冲击钻;钻粒钻头;冲击式钻头
shot-blast 喷净法;喷丸除锈;喷丸加工;喷砂加工
shot-blast cabinet 喷砂间;喷丸室
shot-blast chamber 喷丸(清理)室

shot-blast cleaning barrel 抛丸清理滚筒
shot-blast debris 喷丸残渣
shot-blasted roll 喷丸加工轧辊
shot-blaster 喷丸除锈机
shot blasting 喷丸清面;喷净;喷丸清理;喷丸处理;喷丸(除锈);喷丸清理;喷砂法处理;喷净法
shot-blasting cabinet 喷丸清理室
shot-blasting machine 喷丸除锈机;抛丸机
shot-blasting surface 喷丸处理表面
shot-blast processing 喷丸处理
shot-blast room 喷丸室
shot-blast unit 喷丸装置
shot borehole 钻粒钻进炮孔
shot boring 钻粒钻进;钢粒钻孔
shot-boring drill 钻粒钻机
shot bort 金刚砂
shot box 摄影箱
shot break 爆炸信号;爆炸时刻;爆炸时间标志
shot capacity 注塑量
shot cardigan 交替畦编组织
shot casing 炮孔套管
shot casting 金属丸铸造
shot chamber 压射室;爆破室
shot cleaner 钢珠除灰器
shot cleaning unit 喷丸清砂装置
shot cold hardening 喷丸硬化处理
shot concrete lining 喷射混凝土衬砌
shot content 渣球含量
shot copper 铜粒
shot core drill 钻粒岩芯钻机;冲击取芯钻
shot core drilling 钻粒岩芯钻进
shot-coring machine 冲击取芯机;冲击取芯钻;旋转式钻(孔)机
shotcrete 喷射水泥(砂)浆;喷射混凝土;喷浆混凝土;喷混凝土
shotcrete admixture 喷射混凝土外加剂
shotcrete and bolt lining 喷锚衬砌
shotcrete and rock bolt 锚喷
shotcrete and rockbolts protection 喷锚支护;喷射混凝土锚杆支护
shotcrete and rock bolts support 喷锚支护;喷射混凝土锚杆支护
shotcrete application 喷混凝土应用
shotcrete-bolt construction method 喷锚构筑法
shotcrete ceiling 喷浆吊顶
shotcrete dust 喷射混凝土粉尘
shotcrete equipment 混凝土喷枪;喷射混凝土设备
shotcrete facing 喷射混凝土饰面
shotcrete gun 混凝土喷枪;水泥砂浆喷枪
shotcrete hose 喷射混凝土软管
shotcrete-lined shaft 喷射混凝土支护的井筒
shotcrete lining 喷射混凝土衬砌;喷浆混凝土衬砌
shotcrete machine 混凝土喷射机;混凝土喷布机;喷射机;喷浆机;水泥砂浆喷射机
shotcrete method 喷混凝土方法
shotcrete operation 喷射混凝土操作
shotcrete quality control 喷射混凝土质量控制
shotcrete rebound 喷混凝土回弹
shotcrete shell 喷混凝土壳体
shotcrete sprayer 喷射机;混凝土喷射机
shotcrete support 喷射混凝土支护
shotcrete support design 喷射混凝土衬砌设计
shotcrete surfacing 喷射混凝土饰面
shotcrete system 喷射混凝土支护法;喷浆法
shotcrete technical parameter 喷射混凝土工艺参数
shotcrete usage 喷混凝土应用
shotcreting 混凝土喷射;喷射混凝土;喷浆;喷混凝土
shotcreting gun 混凝土喷枪;水泥砂浆喷枪
shotcreting machine 混凝土喷射机;混凝土喷布机
shotcreting system 喷射混凝土体系
shot cycle 压射周期
shot depth 装药深度;爆炸深度;爆破深度
shot drill 钢珠钻机;冲击钻(机);蔓状取芯钻
shot-drilled shaft 钻粒钻进的井筒
shot drill hole 钻粒钻进钻孔;冲击钻孔;钻粒钻进的炮孔
shot drilling 钻粒钻进;钢砂钻井;钢砂钻进;冲击钻探法;钻粒钻眼法;钢珠钻探
shot drilling bit 钻粒钻头
shot-drilling machine 钻粒钻孔机
shot drilling method 钢珠钻孔法
shot drill method 钻粒钻进法;钢珠钻孔法;冲击钻探法
shot easer 辅助炮眼

shot effect 散射效应;散粒噪声;散粒效应
shot elevation 装药高度
shot exploder 爆炸装置
shot faced tile 粗粒面面砖
shot feed 钻粒给进装置;钢粒供给器
shot feeder 钻粒供给器
shot feeding 钻粒供给;供给钻粒
shot feeding per run 投砂量
shotfirer 熟练炮工;装药引爆工人;装药放炮工人;放炮器
shot firing 引爆;放炮;爆破作业;炮眼爆破
shot-firing apparatus 放炮器
shot-firing battery 放炮器;起爆器;起爆电池
shot-firing cable 导火线;引燃电缆;爆破母线;引火线
shot-firing circuit 引爆电路;引燃电路
shot-firing cord 放炮电线
shot-firing in rounds 炮孔成组起爆
shot gate 铁丸通道
shot glass 小酒杯
shot grain 霰弹粒
shot group 射击散布图
shotgun approach 突击销售法
shotgun pattern 散乱点
shot hanger blast 悬链式连续抛丸清理机;连续喷丸清砂
shothole 蛀虫孔;弹痕;爆破洞;爆破孔;炮眼
shothole blasting method 炮眼爆破法
shothole casing 炮眼套管
shothole consumption per cubic entity rock 每立方米原岩炮眼消耗量
shothole consumption per cycle 每循环炮眼消耗量
shothole consumption per meter 每米坑道炮眼消耗量
shothole disturbance 显示到地表的孔内爆炸作用
shothole drill 炮孔凿岩机;钻粒钻孔机;炮眼钻
shothole nomenclature 炮眼名称
shothole parameter 炮眼参数
shothole rig 爆破孔钻机;钻炮眼设备
shothole weevil 穿孔象甲
shot hopper 钻粒供给漏斗
shot house 雷管库
shot-induced earthquake 爆炸诱发的地震
shot iron 铁丸
shot lead 爆炸引线
shot-like particle 粒状物
shot lubrication 油枪润滑;注射润滑
shot lubricator 滑脂枪
shotman 爆破工
shot material 爆破石料
shot metal 铁豆
shot method 压浆方式
shot noise 起伏噪声;射击噪声;散射噪声;散粒噪声
shot noise current 散粒噪声电流
shot noise reduction factor 散粒噪声降低因数
shot-noise voltage 散粒噪声电压
shot of a cable 一节链
shot pattern 射击散布图
shot-peened surface 喷丸处理表面
shot peening 喷丸硬化(处理);喷射硬化;喷丸加工;喷丸处理
shot peening coverage 喷丸范围
shot peening forming 喷丸成型
shot peening strengthening 喷丸强化
shotpin 制动销(钉);制动箱;止销;止动销
shot-plastering 喷涂抹灰
shot point 爆破孔;爆破点;爆炸点;炮点
shot point elevation 炮点高程
shot point increment 炮点增量值
shot point seismometer 井口检波器
shot point site 炮点位置
shot popper 逐点爆炸法
shot primer 长效型预涂底漆
shot-proof 防弹的
shot protection system (硬盘)防碰保护系统
shot rock 崩落岩石;爆碎石料;爆破(的)岩石
shot rope 入水绳
shot rust removing 喷射法除锈
shot sawn 表面喷丸加工
shot-sawn finish 粗石面
shot separator 铁丸砂分离器
shot stroke 压射冲程
shot tank 粒化槽;弹丸槽

shotted 切槽
shotted fused alloy 粒化合金
shot time 爆破时间
shotting 细流法制造钻粒;制铁丸;制粒;金属粒化;倾注造球
shot to 猛涨
shot tower 工地上盥洗室;工地厕所
shot tumblast 抛丸清理滚筒
shot weld(ing) 瞬间点焊;脉冲焊接;点焊
should-be cost 应计成本
shoulder 拱肩;割面肩;谷肩;炉腹;肩形突出物;瓶肩;突出部;凸肩;凸出部分;台肩;钝边;挡肩;池窑的收缩部分
shoulder angle 凹模台肩角度
shoulder architrave 肩托过梁
shoulder bag 背包
shoulder bed effect 围岩影响
shoulder block 带肩滑车
shoulder bracing test 撑肩试验
shoulder carrying mower 背负式割草机
shoulder contrast 路肩分色
shoulder curb 路平缘石(即平式缘石)
shoulder ditch 路肩沟;肩沟;坡肩截流沟;坡肩截水沟
shoulder dovetail halved joint 腰缘鸠尾嵌接
shoulder drag 路肩拖刮器
shoulder drive multi-use bit 肩座式活钎头
shouldered arch 肩拱;并肩形拱;肩形拱;托臂支承门楣;肩托拱梁
shouldered architrave 门顶肩托过梁;肩过梁;出肩筒子板;肩托过梁
shouldered face housing 明榫纳接
shouldered housed joint 套入式接头
shouldered joint 肩接
shouldered nut 带肩的螺帽
shouldered shaft 带肩轴
shouldered sole plate 肩托衬垫;异形板;钩形底板;肩衬垫
shouldered tap bolt 带肩螺栓
shouldered tenon 肩榫
shoulder-elbow height 肩肘高
shoulder elevation 路肩标高
shoulder eyebolt 轴肩有眼螺栓
shoulder face 轴肩端面
shoulder finisher 路肩整修机;路肩修整机
shoulder grader 路肩整平机;路肩(用)平地机;平肩机
shoulder grafting 肩接
shoulder grinding machine 台肩磨床;挡边磨床
shoulder guard 防护板
shoulder gun 轻便抛绳枪
shoulder harness 肩安全带;(车座位上的)安全带
shoulder heads 不缩格的小标题
shoulder height 台阶高度
shoulder-high 齐肩高的
shouldering hole 阶梯孔;齐肩高炮眼
shouldering 垫灰条;铺垫肩砂浆;切角平搭瓦;修筑坡肩
shouldering half-torching 小三角形瓦块;屋面板下半嵌灰泥
shoulder length 肩长
shoulder line 路肩线;路基边缘
shoulder line stripping 路肩线标划
shoulder maintenance 路肩养护
shoulder material 路肩材料
shoulder nipple 螺纹口套管;螺纹接管;肩形螺纹接套
shoulder of ballast 渣肩
shoulder of ballast bed 道床碴肩
shoulder of base-plate 垫板肩
shoulder of box 母接头台肩
shoulder of embankment 堤肩
shoulder of hole 孔内台阶
shoulder of mountain 山肩
shoulder of road 路肩
shoulder of slope 坡肩
shoulder of spindle 轴肩
shoulder of subgrade 路肩
shoulder of the box 锁接头突肩
shoulder of the joints 钻杆接头端面
shoulder of weld 焊缝根部钝边
shoulder pad 垫肩
shoulder padding 垫肩
shoulder peak 肩峰

shoulder pilot 带台阶的导正销
shoulder pitch 路肩斜度
shoulder plane 切肩刨刀；榫槽刨
shoulder pole 扁担
shoulder punch 台肩式凸模
shoulder rest 肩靠
shoulder ring 轴肩挡圈
shoulder safety belt 肩安全带
shoulder screw 有肩螺钉；带肩螺钉
shoulder segment 肩部砌块
shoulder slide plate 带肩式滑床板
shoulder sling 带肩吊环
shoulder slope 路肩(横)坡度
shoulder sole plate 肩托衬垫
shoulder stabilization 路肩稳定
shoulders thickness 台肩厚度
shoulder stone 钻头端部外缘金刚石；外边刃金刚石
shoulder strap 肩安全带
shoulder stud 带肩螺栓
shoulder tap 处理机间通信技术；分支肩
shoulder tire 肋胎
shoulder tool 割肩刀具；肩刀具
shoulder to square up 钻平孔底
shoulder trimmer 路肩整平器
shoulder turnaround system of locomotive running 肩回运转制【铁】
shoulder turning 轴肩车削
shoulder turning lathe 轴肩车床
shoulder turnout 路肩让车道
shoulder-type bit 可卸式钻头；可卸式钎头
shoulder tyre 肋胎
shoulder up 拱起；合扣
shoulder wear 钻头端部外缘磨损
shoulder width 路肩宽度
shoulder work ahead sign 前面路肩施工标志
shoutouts 退关货物
shove 装炉；挤浆；顶进
shoved dredge(r) 单斗挖土机
shoved joint 灰浆挤缝；挤浆缝(用挤浆法砌砖形成的砖缝)；挤浆砖缝；挤灰竖缝
shoved moraine 推碛
shove fault 走向滑(移)断层；逆掩断层
shove joint 挤接
shove joint brickwork 挤浆砌(砖)法
shovel 挖掘机，铁锹；勺子；铲子；铲
shovel access 挖土机工作半径；挖掘机工作半径
shovel arm 挖土机臂
shovel arm sleeve 挖土机臂套管
shovel attachment 挖掘属具
shovel axle 挖土机轴杆
shovel bit 宽头车刀
shovel blade 铲刀
shovel boom 挖土机悬臂；挖土机转臂
shovel boom winch 挖土机转臂曲柄
shovel bow 匙形船首
shovel box 铲箱
shovel box cover 铲箱盖
shovel box coverlatch 铲箱盖闩
shovel box coversupport 铲箱盖托
shovel bucket 挖土机铲斗；挖掘机铲斗
shovel bucket attachment 挖土机挖斗附件
shovel bucket boom 挖土机挖斗悬臂
shovel bucket capacity 挖土机铲斗容量
shovel bucket digging angle 挖土机挖斗挖掘角度
shovel bucket digging width 挖土机挖斗挖掘宽度
shovel cable 挖土机钢丝绳；挖土机钢索
shovel car 铲车
shovel car unloader 推铲卸车机
shovel chain 挖土机链条
shovel control station 挖土机控制站
shovel-crane 挖掘起重两用机
shovel crawler 履带挖掘机；履带式单斗挖土机；履带式单斗电铲
shovel-crowding engine 机铲推压机构
shovel dipper 挖土机铲斗；挖掘机铲斗
shovel dipper arm 挖土机挖斗臂
shovel dipper arm sleeve 挖土机挖斗臂套管
shovel dipper capacity 挖土机挖斗容量
shovel dipper door 挖土机挖斗门
shovel dipper handle sleeve 挖土机挖斗曲板套管
shovel dipper ram 挖土机挖斗顶杆
shovel dipper slide 挖土机挖斗闸板
shovel dipper stick 挖土机挖斗臂
shovel dipper stick sleeve 挖土机挖斗臂套管

shovel dipper strip 挖土机挖斗倾翻器
shovel dipper tooth 挖土机挖斗刃瓣；挖土机挖斗齿
shovel door 挖土机挖斗门
shovel dozer 铲式推土机
shovel dozer attachment 刮铲推土机附件
shovel dredge(r) 单斗挖泥机
shovel engine 挖土机发动机；挖土机引擎
shovel excavator 单斗挖土机；铲形挖土机；铲斗式推土机
shovel front 机铲工作装置
shovelful 一满斗
shovel gear 煤铲轮齿装置
shovel handle 挖土机曲柄
shovel handle sleeve 挖土机曲柄套管
shovel-hoisting engine 机铲提升机构
shovel jib 挖土机旋臂
shovel jib winch 挖土机旋臂曲柄
shovel(l)er 挖土机驾驶员；挖掘工；翻扬机；挖土工；铲土工
shovel(l)ing blade 铲刀片
shovel(l)ing machine 挖土机；铲土机；机铲；挖掘机
shovel lip 挖土机铲刃；挖掘机铲斗刀口；铲刃
shovel loader 装载机；支臂式装车机；铲装机；铲斗式装载机；铲斗车
shovelman 挖土机驾驶员；机铲司机；挖土工；铲工；挖掘机司机
shovel method 勺取法
shovel mining 机铲开采
shovel motor 挖土机电动机
shovel mucker 铲式软土挖运机
shovel-nose tool 平口车刀；铲刀
shovel oiler 挖掘机加油器；挖掘机加油工
shovel outreach 挖土机工作半径
shovel plough 铲犁
shovel ram 挖掘机顶杆
shovel reach 挖土机工作半径；挖掘机工作半径
shovel-run 铲挖的；铲程
shovel runner 挖土机驾驶员；挖掘机驾驶员
shovel sampling 用铲取样
shovel shaft 铲柄
shovel shaft bearing casing 铲柄套
shovel stick 挖土机臂
shovel stick sleeve 挖土机臂套管
shovel stoker 铲加煤机
shovel supporting mat 挖土机支承垫
shovel sweep 箭形中耕铲
shovel swing engine 机铲回转机构
shovel teeth 挖土机(铲)斗齿；铲齿；挖土机齿
shovel-trench-hoe unit 反铲挖土机
shovel trip 挖土机倾翻器
shovel truck 机铲汽车；汽车挖掘机
shovel type cover 铲式覆土器
shovel type crane 铲式起重机
shovel type dozer 铲式推土机
shovel type ejector 铲式出件装置
shovel type excavator 铲式挖土机；铲式挖掘机；单斗挖土机
shovel type load 斗式自动装载机
shovel type loader 铲式装载机；铲式装载机；铲斗车；斗式自动装载机；单斗式装载机
shovel type mucking machine 铲式软土挖运机
shovel use 铲式挖泥船附件；铲式挖掘机附件
shovel with round mouth 圆口铲式
shove moraine 走向滑移冰碛层
shoving 推动；推挤；挤浆
shoving jack 推力千斤顶
shoving method of bricklaying 挤浆砌法
shoving of shield 盾构推进
shoving rake 推料耙
show 显示；展览会
Showalter stability index 肖沃特稳定指数
show bill 招贴
show bottle 陈列瓶
show business 展示会上
show cabinet 陈列室
show card 货样纸板；广告牌；样品卡
show card colo(u)r 蛋黄胶水调颜料；广告涂料
show case 橱窗；陈列窗；玻璃柜台；试片电影院；商店橱窗；陈列柜
show drawings 展览图(片)
shower 宇宙射线簇射；骤雨；阵雨；阵性降水；淋浴装置；淋浴设备；淋浴器；淋浴；淋洒器；空气簇射；喷淋管；喷灌；簇射
shower bath 淋浴室；淋浴间；淋浴(装置)

shower bath drain 淋浴排水口；淋浴间地漏
shower bath enclosure 淋浴小间
shower bath partition(wall) 淋浴隔墙
shower bath room 淋浴浴室；洗澡间
shower bath tray 淋浴槽
shower chamber 淋浴间
shower cloud 阵雨云
shower compartment 淋浴间；淋浴隔间
shower control 淋浴调节
shower cooler 淋水冷却器；喷射冷却器；喷淋冷却器
shower cooling 淋水冷却；喷射冷却；淋浴式冷却
shower counter 簇射探测器
shower cubicle 淋浴(小)间；淋浴隔间
shower curtain 淋浴室帘；淋浴挡完杆；淋浴帐帘；淋浴帘
shower curtain rod 淋浴挂帘杆
shower drain 暴雨排水管
shower fittings 淋浴装置
showerflow restrictor 淋浴水限流器
shower furnace 悬浮炉
shower head 喷射头；莲蓬头；莲蓬式喷嘴
shower head injector 莲蓬头式喷头；莲蓬式喷嘴
shower head with rubber hose 带皮管淋浴莲蓬头
shower hose 淋浴软管
shower installation 淋浴设备；淋浴装置
shower meteors 陨石雨
shower mixer 淋浴水混合器
shower mixer valve 淋浴器混合阀
shower nozzle 淋浴喷头；莲蓬头
shower of ash 尘阵
shower of sparks 闪电阵雨
shower of volcanic dust 火山尘雨
shower pan 淋浴盆；淋浴盘
shower particle 簇射粒子
shower partition 淋浴墙；淋浴隔断；淋浴间隔墙
shower pipe 喷水管
shower-proof 防雨的
shower-proof cloth 防雨布
shower proofing 防雨布
shower radiance 簇射辐射
shower receptor 淋浴盆；淋浴缸
shower recess 淋浴墙凹进处
shower roasting 漂悬焙烧；闪速焙烧
shower room 浴室；淋浴室
shower room without cubicle 未隔成小间的淋浴室
shower rose 淋浴喷头；淋浴喷嘴
shower screen 淋浴喷头滤网；淋浴管式滤网
shower scrubber 喷淋除尘器
shower set 淋浴装置
shower stall 淋浴室；淋浴间；淋浴隔间
shower stall door 淋浴间门
shower tray 淋浴集水喷；淋浴缸；淋浴槽；淋浴浅池
shower tub 淋浴缸
shower unit 淋浴装置；簇射单位路程
shower valve 淋浴阀(门)
shower wall 淋浴墙壁
shower water 喷淋水
showery rain 阵雨
showery snow 阵雪
show ground 展览会场
show house 剧院；花房；花草陈列馆
show-how 技术示范
showily 过分华丽地
showing hole 推料孔
showing of oil and gas 油气显示
showing pan 推料盘
showing through 不盖底
show metal sheet 陈列的金属(薄)板
show methods of coal analysis result 煤质分析结果的表示方法
show piece 展品；范例
show place 展出地；供参观的场所；名胜地；展览场所
show rafter 明椽；外露椽
show repeated deficits 亏累
show-ring 展览会场；评比会场
show room 展览室；展销室；陈列室
show shop 展销店
show slight breaks 略现瑕疵；显示轻微开裂
show stand 展销台
show-through 透字；透印；底层瑕疵透过面层显露
show type 展览型
show-up time rule 上班即计时惯例
show window 陈列窗；商店橱窗；橱窗
show window lamp bulb 橱窗灯泡

show window lamp shade 橱窗灯罩
show-window lighting 橱窗照明
showy 华丽的
shraff 陶瓷废料
shread head 山墙尖(呈斜坡的);山墙尖;半山头【建】
shread-head roof 歇山屋顶
shred 裂片;碎片
shredded cellophane 堵漏用赛璐芬屑
shredded chips 击碎木片
shredded cullet 水淬法碎玻璃
shredded fodder 铡碎粗料
shredded refuse 破碎(的)垃圾
shredded wire 钢回丝;细碎的金属丝
shredded wire underhread tire 细碎金属丝底线轮胎
shredded wood fibre board 碎木胶合板
shredder 切碎机
shredder bar 切碎机定刀;切碎刀
shredder compactor 垃圾撕碎压实机
shredder machine 撕碎机
shredder-mower 割草切碎机
shredder truck 装有撕碎机的垃圾收集汽车
shredding 锯板条;击碎;切碎
shredding cylinder 切碎滚筒
shredding drum 切碎滚筒
shredding equipment 破碎装置
shredding machine 切碎机
shredding mechanism 切碎装置
shredding roll 撕碎辊
shredmaster 撕碎机
shred scrap 碎废钢
SH reflection profile SH 波反射剖面
shreddded cullet 淬火过的碎玻璃片
shrikes 伯劳(一种鸟)
shrimp boat 捕虾船
shrimp canning waste 虾罐头加工废物
shrimp pink 深红色
shrimp trawler 拖网捕虾船
shrine 庙;圣地;神龛;神殿
shrink 留矿(切割槽);拉回
shrink able film 收缩薄膜
shrinkage 皱缩;留矿法;起皱;缩短;损耗;收缩(性)
shrinkage across machine direction 横向皱缩
shrinkage action 收缩作用
shrinkage aggregate 收缩性集料;收缩性骨料
shrinkage allowance 允许收缩量;预留收缩量;预留收缩长度;容许收缩量;收缩允许量;收缩余量;收缩容许值;收缩容许量;收缩容差;收缩留量
shrinkage anchoring 收缩(性)锚固
shrinkage and temperature reinforcement 收缩与温度钢筋
shrinkage and temperature steel 收缩和温度钢筋
shrinkage apparatus 缩性试验器;收缩仪;收缩性试验器
shrinkage area 留矿法采区
shrinkage bar 抗收缩钢筋;收缩钢筋;伸缩缝接头钢筋
shrinkage behaviou(u)r 收缩性能
shrinkage cavity 缩孔(一种漆病)
shrinkage character 收缩特性
shrinkage clay 收缩黏土
shrinkage coefficient 干缩系数;收缩系数
shrinkage compensated concrete 补偿收缩混凝土
shrinkage compensates cement 收缩补偿水泥;无收缩水泥
shrinkage-compensating 收缩补偿
shrinkage-compensating cement 收缩补偿水泥;补偿收缩水泥
shrinkage-compensating concrete 收缩补偿混凝土;无收缩混凝土;补偿收缩混凝土
shrinkage control 留矿法放矿控制;收缩度控制
shrinkage correcting action 防缩作用
shrinkage correction factor 收缩修正系数
shrinkage crack(ing) 缩裂;热裂;暗裂;干缩裂缝;泥裂;缩缝;收缩开裂;收缩裂隙;收缩裂纹;收缩裂缝;收缩发纹
shrinkage-creep relationship 收缩蠕变关系
shrinkage crimping 收缩卷曲
shrinkage cross the grain 横向皱缩
shrinkage curve 收缩曲线
shrinkage deformation 收缩变形
shrinkage degree 收缩度;收缩程度

shrinkage depression 缩注
shrinkage difference 收缩差别;收缩微差
shrinkage distortion 收缩变形
shrinkage down-stream deep 萎缩的下深槽
shrinkage effect 收缩效应
shrinkage factor 收缩率;压缩因子;压缩因素;压缩因数;收缩因素;收缩系数
shrinkage fit(ting) 热压配合;红套(冷缩)装配;冷缩(热胀)配合;热装;热套;收缩配合
shrinkage force 收缩力
shrinkage fracture distress 收缩断裂破损
shrinkage-free 无收缩的
shrinkage front 收缩线
shrinkage gap 收缩裂隙
shrinkage ga(u)ge 铸造缩尺
shrinkage glaze 收缩釉
shrinkage gradient 收缩(因数)增减率;收缩梯度
shrinkage groove 缝两边的)的收缩槽缝
shrinkage head 收缩头;气冒
shrinkage hole 缩孔;收缩孔
shrinkage in casting 浇铸收缩
shrinkage index 收缩指数
shrinkage index of soil 土的收缩指数
shrinkage-induced cracking 收缩开裂
shrinkage in length 径向收缩
shrinkage joint 伸缩缝;收缩节理;收缩接缝;收缩缝
shrinkage level 收缩程度
shrinkage limit 土壤极限;缩性界限;缩性极限;缩限(含水量);收缩限度;收缩极限
shrinkage limit apparatus 收缩限量仪
shrinkage limit of soil 土壤缩限;土壤收缩限
shrinkage limit of swelling soil 膨胀土缩限
shrinkage line 收缩线
shrinkage loss 干缩量;收缩损失
shrinkage mark 收缩皱纹;收缩标志
shrinkage mass 收缩量
shrinkage meter 收缩测定仪
shrinkage method 留矿法
shrinkage mining 留矿法开采
shrinkage modulus 收缩模数;收缩模量
shrinkage of a casting 铸体冷缩
shrinkage of area 面积收缩
shrinkage of line 线收缩
shrinkage of mo(u)lding compound 模塑料收缩性
shrinkage of petroleum 石油的收缩率
shrinkage of soil mass 土体收缩
shrinkage of swelling soil 膨胀土收缩
shrinkage of vitreous 玻璃体皱缩
shrinkage of volume 体积收缩(率)
shrinkage of wood 木材干缩(性)
shrinkage on drying 干缩
shrinkage porosity 缩松;松心;收缩疏松;收缩孔隙
shrinkage prevention 防止收缩(措施)
shrinkage property 收缩特性
shrinkage rate 收缩率;收缩速率
shrinkage ratio 收缩率;收缩比
shrinkage reinforcement 抗收缩钢筋;收缩钢筋
shrinkage rib 收缩肋
shrinkage rod 收缩钢筋
shrinkage rule 收缩比例尺
shrinkage scale 缩尺
shrinkage section 收缩断面
shrinkage segregation 收缩偏析
shrinkage settlement 收缩沉降(量)
shrinkage shortening 收缩缩短
shrinkage stope 收缩的工作面;留矿采矿工作面;留矿采矿场
shrinkage stope face 留矿开采法工作面
shrinkage stoping 收缩回采工作
shrinkage stoping method 仓储[贮]采煤法;留矿采矿法
shrinkage strain 收缩应变;收缩形变;收缩变形
shrinkage stress 收缩应力
shrinkage temperature 收缩温度
shrinkage tendency 收缩趋势
shrinkage tensile stress 收缩拉应力
shrinkage tension 收缩拉力
shrinkage test 干缩试验;收缩试验
shrinkage test apparatus 收缩试验仪(器)
shrinkage theory 冷缩说
shrinkage value 收缩值
shrinkage volume 收缩体积
shrinkage vug 收缩穴

shrinkage water 收缩水
shrinkage wave 收缩波
shrink bob 补缩包
shrink-control backing material 控控片基
shrink-control backing sheet 控控衬板
shrink-on fit 冷缩配合
shrinker 金属板波纹压机;收缩机
shrink film 塑料薄膜护罩
shrink-film package 缩膜包装
shrink fit 冷装配;红套配合;热套配合;热套;垫压合座;烧嵌;热压配合
shrink fit pressure 热套配合压力;热套表面压力
shrink-fit vessel 套合式容器;缩合式容器
shrink flange 收缩法兰
shrink flanging 收缩摺缘;收缩翻边
shrink-flanging die 收缩翻边模
shrink-forming 冷缩成型
shrink graduations 收缩分度
shrink grip tool joint 收缩夹紧头
shrinkhead 冒口
shrink hole 缩孔;收缩孔
shrinking 收缩;皱缩
shrinking and thinning 收缩和细化
shrinking away from the shell 离壳
shrinking head 冒口
shrinking mapping 收缩映射
shrinking measure 收缩度
shrinking of raster 光栅变小
shrinking of the earth 地球的收缩
shrinking of tool joint 接头冷缩(烘装时)
shrinking on 冷缩配合;热装配合;冷缩装配
shrinking percentage 收缩百分比
shrinking processes 预缩工序
shrinking raster 收缩光栅
shrinking-raster method 图像收缩法
shrinking splitting 退缩式分岔
shrinking tolerance 收缩公差
shrinking transformation 收缩变换
shrinking wave 减幅波;收缩波;阻尼波
shrinking zone 收缩区
shrink leather 皱纹革
shrinkless concrete 不收缩混凝土;无收缩混凝土
shrink link 系紧夹(指在热状态安上的)
shrink mark 收缩痕;缩沉;收缩痕迹
shrink member 冷缩配合构件;热套构件
shrink-mixed concrete 搅拌混凝土
shrink-mixing 部分混合
shrink-mixing of concrete 混凝土缩拌
shrink-proof 防收缩的;不缩的;防缩的;不收缩的;抗缩的
shrink-proof finishing 防缩加工
shrink range 过盈量
shrink resistance 防缩性
shrink resistant 不缩水
shrink ring (凹模的)预应力圈;热压围环;热套圈;热套环;套环;缩紧环;紧缩环;紧圈;收缩圈
shrink-ring commutator 套环换向器
shrink rule 收缩定律
shrink stress 收缩应力
shrink-swell potential 潜在膨胀势;潜在膨胀能力;缩胀势
shrink welding 预应变焊接
shrink wrap 热缩塑料包
shrink wrapping 热塑收缩包装;收缩包装
shrivel finish 皱纹面漆
shriveling 皱纹纹;干皱的
shrivel varnish 皱纹清漆;波纹清漆
shroud 销轮的销座;掩蔽;轮盖;笼罩;屏幕;屏板;桅杆侧支索;桅的左右支索;外套;覆盖(物);覆板
shrouded gear 带有端面凸缘的齿轮
shrouded impeller 闭式叶轮
shrouded pinion 凸缘小齿轮
shrouded propeller 导流管式螺旋桨
shrouded screw conveyer 闭封式螺旋输送机;密封式螺旋输送机;密封绞刀
shrouded turbine 闭式涡轮机
shrouded wheel 凸缘小齿轮
shrouding 遮板;罩盖;降落伞吊索;覆盖
shrouding disc 盖盘
shrouding wire 拉筋;屏蔽线;包箍线
shroud knot 结绳结
shroud laid 单绳芯四股绳
shroud laid rope 四股正搓绳
shroud line 吊伞索

shroud ring 罩环;箍环;护环;尼轴止环;套环;包箍
shrub 灌木(的);灌丛
shrub bed 灌木花坛;灌木园
shrub belt 灌木林带
shrubbery 灌木林;灌木丛
shrub border 灌木花境
shrubby 灌木丛生的;多灌木的
shrubby desert 灌木荒漠
shrub-coppice dune 灌丛沙丘
shrub forest 灌木林;灌丛林
shrub garden 灌木园
shrub-grassland 灌木型草地;灌草丛
shrub head 灌木丛喷洒头
shrub land 灌木地;灌丛带;疏灌丛
shrub sand mound 灌丛沙滩
shrub savanna 灌木干草原
shrub species 灌木树种
shrub zone 灌木列
shrunk-and-peened flange 皱合法兰
shrunk-and-rolled flange 皱卷法兰
shrunk dolomite 硬烧白云石
shrunken raster 皱缩光栅
shrunk fit 热嵌配合;烧嵌
shrunk in 压入式
shrunk-in shaft 红装轴;热套轴
shrunk joint 冷缩管接头;热装接头
shrunk-on cylinder 热套式圆筒
shrunk-on disc 套装叶轮
shrunk-on joints 接头热压合
shrunk-on multilayered cylinder 热套式多层圆筒
shrunk-on-pipe joint 热套管接头
shrunk-on ring 缩圈;加热条件下嵌环
shrunk-on ring carrier 冷缩配合环座;热装环座
shrunk-on sleeve 热配套管
shrunk ring 紧圈;热压轮圈;烧嵌环
shrunk rubber collar shank 橡胶垫钎肩式钎尼
shubnikovite 水砷钙铜石
shub tuner 短截线调谐器
shucker 剥壳器;剥壳机
shucking machine 剥壳机
shudder 发抖
shuff 爆裂砖;炸纹废砖
shuffle 正移
shuffle bar 活动杆
shuffle box 换气箱
shuffle valve 梭阀
shuffling 慢慢移动
shuffling time 贴压时间
shuffs 等外品
shuga 糊状冰;白松冰团
Shuiquliu clay 水曲柳黏土
shullhouse (墓葬挖掘出的)尸骨停放房
shumac 盐肤木;漆树属植物的木材;苏模鞣料
shunpike 支路(避车或疏散车辆)
shunpiker 在支路驾车者
shunt 转流木;旁通管;旁路;分路器;分路;分流术;分道叉;分叉
shunt admittance 并联导纳
shunt-back 矿车自动折返装置
shunt bond 分路导接线【铁】
shunt box 分流器箱
shunt cam 分路凸轮
shunt capacitance 寄生电容;分路电容;并联电容
shunt capacity 旁路电容
shunt character 并激特性;并激特性
shunt characteristic 并激特性(曲线)
shunt chopper 并联逆变器
shunt circuit 旁路;分流电路;并联回路;并联电路
shunt closing release 分离分闸释放
shunt coil 分路线圈;分流线圈;并绕线圈;并联线圈
shunt-compensated video amplifier 并联补偿视频放大器
shunt compensation 并联补偿
shunt compensator 并联补偿器
shunt conductance 分流电导;并联电导
shunt-conduction motor 并激整流式交流电动机
shunt connection 旁路连接
shunt current 分路电流
shunt dynamo 分激发电机;并励发电机;并激发电机
shunted 分路的
shunted field control 励磁分路调速
shunted instrument 带分流器仪表
shunted state of track circuit 轨道电路分路状态
shunter 扳道员;转辙器;调车机车;调车员

shunter's cabin at hump crest 峰顶调车员室
shunt excitation 并励;并激
shunt excitation loss 并激损失
shunt-excited 并励的;并激的
shunt-excited antenna 并联馈电天线;并馈天线
shunt-excited machine 并励电机
shunt excited magnetizer 并励励磁机
shunt-fed motor 并联供电电机
shunt feed 并联馈电
shunt-feed antenna 并联馈电天线;并馈天线
shunt feedback 并联反馈
shunt feed back amplifier 并联反馈放大器
shunt feed high-voltage switch cabinet 并联馈电高压开关柜
shunt-feed oscillator 并联馈电振荡器
shunt-feed vertical antenna 并馈竖直天线
shunt field 分激磁场;并励磁场
shunt field coil 分路线圈;并激线圈
shunt field relay 磁分路继电器;并励场电路继电路;分激继电器
shunt field rheostat 分路变阻器;并励场变阻器
shunt filter 支管过滤器;并联过滤器
shunt flow 分水流
shunt generator 分激发电机;并励发电机
shunt impedance 分路阻抗
shunt inductance 并联电感
shunt inductor 分路感应线圈
shunting 调车【铁】;分接;分导流
shunting access route 调车进路
shunting action 分流作用
shunting capacitor 并联电容器;分路电容器
shunting condenser 分接电容器
shunting contact 分路触点
shunting control device 磁削装置
shunting controller 调车员
shunting control position 调车控制台
shunting effect 分路效应;分流作用
shunting engine 调车机车
shunting equipment 分流(调度)设备
shunting impedance 分流阻抗
shunting indicator 调车表示器;分路指示器;分路标识
shunting indicator circuit 调车表示器电路
shunting line 调车线【铁】
shunting locomobile[locomotive]调车机车
shunting neck 牵出线【铁】;调车梯度
shunting on shunting neck 牵出线调车
shunting operating cost per hour 调车作业小时成本
shunting operating plan 车站作业计划
shunting operation 调车作业
shunting operation plan 调车作业计划
shunting place 分路处;分车处
shunting resistance 分流电阻
shunting resistor 分路电阻器
shunting sensitivity 分路灵敏度
shunting set of siding 调车线群
shunting siding 调车侧线
shunting sign 调车号;分路信号
shunting signal 调车信号;调车号志;分路信号
shunting signal for prohibitive humping line 去禁溜线信号
shunting speed 调车速度
shunting spring 并接簧片
shunting station 调度站【铁】
shunting track 调车线【铁】;岔道
shunting tractor 调车牵引车
shunting traffic 调度交通
shunting trip 调车钩;调车程
shunting vehicle 调度车辆
shunting work 调车工作
shunting yard 调车场;编组场【铁】
shunting zone 调车场分区
shunt leads 分流器引线
shunt limiter 并联限幅器
shunt loaded line 并联负载线路
shunt loading 分路负载;并联加感
shunt loss 分路损耗
shunt meter 并联电流计
shunt motor 分流电动机;并绕电动机;并励电动机
shunt neutralization 电感中和;分路中和
shunt of mobile contacts 活动触头分流器
shunt-opposed limiter 并联反接式限制器
shunt peaking 分路升高法;并联建峰;并联峰化

shunt phaseadvancer 并励进相机
shunt position 分流位置
shunt ratio 分路系数
shunt reactor 分路电抗器;分流扼流圈;联电抗器
shunt-regulated amplifier 并联调节放大器
shunt regulator 分流器调整;分路调节器;并联稳压器;并联调整器
shunt regulator tube 分流稳压管;并联稳压管
shunt release 分离释放
shunt repeater 并联增音器
shunt resistance 分路电阻;分流电阻
shunt resonance 分路谐振;并联谐振
shunt-resonant circuit 并联谐振回路
shunt rheostat 分路变阻器
shunt ringer 分流振铃机;分流感应器
shunt running 滞缓;潜动;漂移
shunt sequence 并联顺序
shunt switch 分路开关;并接开关
shunt track 编组轨道;调车轨道
shunt transformer 并联变压器
shunt transition 分路换接过程
shunt trap 并联陷波电路
shunt trip 并联跳闸装置
shunt-trip attachment 分路跳闸机构
shunt trip coil 并联脱扣线圈
shunt tripping 并联跳闸
shunt valve 旁路管;分水龙头;分水阀;分流电子管
shunt voltage regulation 分路电压调节
shunt voltage regulator 并联电压调节器
shunt volume 分流量
shunt winding 分路线圈;分路绕阻;分流线圈;并励绕组;并激绕组
shunt winding parallel winding 并联绕法
shunt wound 并绕的;分绕的
shunt-wound arc lamp 分绕弧光灯;分路弧光灯;并绕式弧光灯
shunt-wound exciter 并激励磁机
shunt-wound generator 并励发电机
shunt-wound motor 并励电动机;分励电动机
shut 锁住
shut-down 关闭;停止运转;停风;停产;扯下;关机;断路;停机;停工;停堆;停车
shut-down amplifier 停堆放大器
shut-down breakdown 停工检修
shut-down channel 停堆通道
shut-down circuit 关闭线路;停止电路
shut-down condition 停运状态
shut-down cooler 停堆冷却器
shut-down cooling pump 停堆冷却泵
shut-down cost 停工成本;停产成本
shut-down day 关闭日;停工日
shut-down device 止动设备;止动装置;停止装置
shut-down expenses 停工费用
shut-down feature 断流装置;停止装置
shut-down handle 停车手柄
shut-down heat 停堆余热
shut-down hydraulic pilot valve 停车液压操纵阀
shut-down inspection 停工检查
shut-down in steps 阶梯式停堆
shut-down kiln 停窑
shut-down maintenance 停工维修;停工检修
shut-down margin 停堆裕度
shut-down mechanism 切断机构;停止机构
shut-down monitor 事故停机监察器
shut-down of service track 封锁线路
shut-down operation 停止操作
shut-down part of seismic fault 地震断裂闭锁段
shut-down period 封冻期;停运期;停业时期;停炉期;停机时期;停工(时)期;停工时间
shut-down phase 停运期
shut-down plunger 停车柱塞
shut-down point 亏损点;停止生产点
shut-down power 停堆功率
shut-down procedure 停机程序;停堆程序
shut-down reactivity 停堆反应性
shut-down relay 断路继电器;停车继电器
shut-down rod 安全瓣;停堆棒
shut-down schedule 停工期工作计划;停堆日程
shut-down servomotor 停车伺服马达
shut-down signal 关闭信号;停堆信号;停车信号
shut-down slide valve 自动停车滑阀
shut-down stage 停运期
shut-down switch 断路开关;停止开关;停堆按钮;停车开关

shut-down system 关闭系统;停机系统;停车系统
shut-down time 停运期
shut-down valve 停机阀;停车阀
shut-down well 暂时停钻的井
shut-down work 停堆操作
shut height 闭合高度;停止高度
shut height indicator 闭合高度指示仪
shut in a well 关(闭)井
shut in bottom hole pressure 闭井井底压力
shuting 檐沟
shuting-off device 关闭装置
shuting stile 装锁门梃
shut-in pressure 闭井压力;关井压力;关闭(井)压力;承压水压力;闭合压力
shut in production 最高产量与允许产量之差
shut in storage 闭井储油
shut in the land 被陆地遮住
shut in time 闭井时间
shut in well 关闭井
shut-off 关闭;切断;挡砖;关断
shut-off block 关闭部件;开关组
shut-off circuit 闭合电路
shut-off cock 用户管阀;切流旋塞;关断旋塞;截止旋塞;截止塞门;切断旋塞;切流龙头;切断龙头
shut-off contact 断路接触器
shut-off cylinder 停车缸
shut-off damper 截止挡板
shut-off device 关闭器;切断装置
shut-off electromagnet 电磁关闭器
shut-off gate 截止门;阻断闸板
shut-off ga(u)ge 闭井压力表
shut-off head 闭井压头;关闭压头;全闭压头;无流时水头
shut-off horsepower 空转功率
shut-off mechanism 断路接触器
shut-off member 停堆控制元件
shut-off of loss 堵漏
shut-off of water 止水
shut-off period 停滞时间
shut-off plug 断路塞
shut-off pressure 关闭压力
shut-off rotary valve 旋转截止阀
shut-off sequence 切断序列
shut-off signal 关机信号;切断信号;停堆信号
shut-off solenoid 切断螺线管;关闭螺线管;切断网络管;关闭网络管
shut-off spindle 停车轴
shut-off stroke 关闭行程
shut-off time 停滞时间
shut-off unit 切断装置;关闭装置
shut-off valve 闭井阀;截止阀;断流阀;闸阀;关闭阀;截流阀
shut-off wave 阻断波
shut out 脱钩;退关;闭厂;退装(航运)
shut-out cargo 退装货物;退关货物
shut-out goods 退装货物
shut-out memo 退装货物清单;退关单
shutter 旋开式快泄板;闸门;闸板;光阑;关闭者;快门;快开堰顶闸;开闭器;节流板;桨门板;门窗;屏蔽闸;调节板;天窗;断路器;百叶窗;安装模板
shutter adjuster 进风门调节器
shutter-aperture interlock 快门光圈联锁装置
shutter axis 快门轴
shutter bar 百叶窗调节杆;百叶门
shutter blade 百叶窗叶片;快门叶片;快门薄片
shutter-blade actuating cam 快门叶片推动凸轮
shutter-blade arrester 快门叶片抑制器
shutter blind 快门卷帘;百叶帘
shutter-board 模板
shutter box 卷帘板壁龛
shutter butt 百叶窗铰链;轻型合页
shutter cocking 快门待发装置
shutter condenser 百叶窗式冷凝器
shutter contact duration 闪光接触持续时间
shutter control 百叶窗开关控制
shutter cream 脱模(油)剂
shutter curtain 百叶窗帘
shutter cutoff 快门关闭
shutter dam 闸门坝;活板坝;翻板坝
shutter disc 快门片
shutter dog 遮光板支撑;遮光窗窗钩
shutter door 卷帘门;
shutter door aperture 百叶门门洞
shutter door guide 百叶门导向杆

shutter door height 百叶门高度
shutter door leaf 百叶门页片
shutter door lintel 百叶门楣;百叶门过梁
shutter door operation 百叶门操作
shutter door operator 百叶门操纵器
shutter doorway 百叶门门道
shutter dozer attachment 辊式推土机附件
shutter drum 百页筒
shuttered fuse 带隔离机构的引信
shuttered socket 三角插座;安全插座
shutter efficiency 快门光学有效系数
shutter engine 压路机发动机;压路机引擎
shutter flap 遮光窗铰链;遮光窗活络板
shutter floor 可开关的地板
shutter for large buildings 大型建筑的百叶窗
shutter gate 翻板(节制)闸门;百叶式闸门
shutter grate 百叶炉篦;可调炉篦;水篦;尘格
shutter guard 百叶门保险
shutter handle 卷轴手柄
shutter handle sleeve 压路机曲柄轴套
shutter hinge 闸门铰链;长翼铰链
shuttering 立模(板);模板支撑;壳子板;模壳;门子板
shuttering agent 脱模剂
shuttering aid 模板支撑件
shuttering assembly work 模板装配工作
shuttering board 模板用木板;制(作)模板的木板;模板的木板
shuttering board cleaning machine 模板洗涤机;模板清洁器
shuttering element 模板部件
shuttering erector 模板安装器;模板安装工
shuttering form(work) 模板
shuttering for shell 薄壳模板
shuttering handler 模板管理人;模板装卸装置
shutteringless 不用模板
shuttering lining 模壳衬垫;模板衬垫
shuttering lube 脱模剂
shuttering mark 模板痕(迹)
shuttering oil 脱模剂
shuttering paint 模壳涂料;模板涂料
shuttering panel 制模壳的板;模板
shuttering paste 脱模膏
shuttering plate 制模壳的板
shuttering removal 拆模
shuttering sealer 模壳密封料;模板密封料
shuttering sheet 制模壳的薄板
shuttering tie 模壳横档;模壳栏杆;模板横档;模板拉杆
shuttering wax 模壳蜡;模板蜡
shuttering works 模板工程
shutter jib 压路机旋臂
shutter jib winch 旋臂曲柄
shutter lath 模壳板条;模板板条
shutter leaf 百叶门页片;百叶门页板;快门
shutterless camera 无快门照相机
shutter liner 模板垫条;模板垫衬
shutter lock 快门闭锁
shutter mat 压路机垫
shutter mechanism 快门机构
shutter motor 压路机发动机
shutter mucker 辊式软土挖运机
shutter oil 脱模油
shutter oiler 压路机加油器
shutter opening angle 快门开角
shutter operator 百叶窗操纵杆;百叶窗开关
shutter-priority mode 速度优先模式
shutter pulse 快门脉冲
shutter rail 卷帘轨
shutter-ramming technique 版筑技术
shutter rate 快门速率
shutter release 快门开关
shutter ridge 错断山脊
shutter setter 快门装置
shutter slat 模壳板条;模板板条
shutter-slit 快门缝隙
shutter slit width 快门缝隙宽度
shutter speed 快门速度
shutter speed dial 快门速度刻度盘
shutter speed setting ring 快门速度调节圈
shutter speed testing 快门速度检验
shutter stay 遮光窗的支撑
shutter sweep correction 快门扫描校正
shutter theodolite 快门经纬仪

shutter time 曝光时间
shutter transit 快门经纬仪
shutter tube 快门管
shutter type ventilator 百叶窗式通风筒
shutter velocity 快门速度
shutter vibration 闸门振动;模板振动
shutter vibrator 模壳振捣器;模板振捣器;外部振捣器;附着式振捣器
shutter weir 活动堰;活板堰;活板坝;横轴闸门堰;翻板堰
shutter worker 百叶窗操纵杆;百叶窗开关
shutting clack 闭断瓣
shutting-down device 关断装置;停车装置
shutting flap 关断瓣
shutting light method 分段曝光法;部分遮光晒印法
shutting-off device 关断装置
shutting off water 封闭水层
shutting post 门门杠;关门木桩;关门横木;关闭横木
shutting shoe 门门座;门栓座
shutting stile 装锁的门梃;门的锁梃;装铰链的门梃;装铰门梃
shuttle 振荡输送机;列车;空间渡船;货叉;前后移动;气压传送装置;梭子;穿梭运动
shuttle-action scraper 梭式刮板输送器
shuttle-adjusting machine 梭子校正器
shuttle air-pressing machine 往复式空气压缩机
shuttle apron conveyer 梭式板形运输机
shuttle armature 梭形电枢
shuttle armature type 梭形电枢型号
shuttle belt 振动带
shuttle belt conveyer 往复式皮带输送机;梭式皮带运输机
shuttle block pump 梭挡转子式(水)泵;梭挡转子水泵
shuttle board 抖动板
shuttle bombing 穿梭轰炸
shuttle boom 梭动式悬臂
shuttle box 梭箱
shuttle bus 小型公共汽车;联络车;交通车;区间公共汽车;区间车;穿梭公共汽车
shuttle-bus speed 区间车速
shuttle cableway 往复移动式缆索道
shuttle car 梭(行矿)车;往复来回料车;往复式料车;梭车;短距离运行车辆;穿梭式机动矿车
shuttle car driver 梭式矿车驾驶员
shuttle car gathering 梭式矿车集矿
shuttle carriage 穿梭式小车
shuttle checking 梭子制动;制梭
shuttle chuck 梭动夹头
shuttle chute 落梭斜槽
shuttle-clamp type continuous extrusion machine 梭夹型连续挤出机
shuttle-cock 羽毛球;梭芯
shuttle conveyer 自走式输送机;可逆式输送机;梭动运输机;穿梭式运输机;梭式运输机;往返式皮带运输机;游动输送机;梭运输器;穿梭输送机
shuttle conveyer belt 游动输送带
shuttle cream 脱模剂
shuttled block train 循环直达列车
shuttle detector 探梭器
shuttle die 杆式移送多工位模具
shuttle drive house 梭动机操作间;梭动驾驶间
shuttle dumper 往复来回卸料车;往复来回翻斗车
shuttle dump truck 往复式自卸车
shuttle extension/retraction limiter 货叉伸缩行程限位器
shuttle eye 梭眼
shuttle-eye cutter 梭眼剪
shuttle feeder 梭式送料装置;梭式给料机
shuttle feeler 探梭指
shuttle freight train 循环货物列车
shuttle haul(age) 往复来回运输;短距离运输
shuttle kiln 台车窑;梭式窑
shuttle loader 梭式送料器
shuttle loading 往复形装载
shuttle magazine 梭库
shuttle marks 梭箱油渍
shuttle mechanism 货叉伸缩机构;往返机制;穿梭机制
shuttle mine car 梭式矿车
shuttle over-travel stops 货叉超行程停止器;货叉超行程挡块
shuttle pallet 梭动随行工作台
shuttle path 梭道

shuttle peg 梭芯
shuttle point 梭尖
shuttle propeller 全回转螺旋桨
shuttle pulse 梭动式脉冲
shuttle race 梭床
shuttle receiver box 落梭箱
shuttle ropeway 往复式架空索道
shuttle saw 滑座锯
shuttle service 往复行车；短途(往返)疏运；短距离的区间车；短程往返交通；穿梭交通
shuttle-shaped column 梭柱
shuttle ship 短途疏运船；短途集散船；定期往返班船；定期班船
shuttle short-haul block train 固定车底的短途循环列车
shuttle stabilizer 梭子稳定装置
shuttle system 交替运输制
shuttle table 穿梭台
shuttle tanker 输油船；穿梭运输油船
shuttle through block train 固定车底的直达循环列车
shuttle tongue 梭芯舌头
shuttle traffic 来回交通；往复行车；穿梭交通
shuttle train 定点定线列车；穿梭式运载列车
shuttle train speed 区间车速
shuttle tram 梭车
shuttle-type 梭式；摆动式的
shuttle-type conveyer 梭式输送机
shuttle-type feed 梭动式进料；摆动式供料
shuttle valve 两用活门；换向阀；滑动活门；往复活门；往复阀；梭形滑阀；梭动阀
shuttle winding machine 往复式绕线机；往复式卷扬机
shuttle working 梭行作业
shuttle yard 梭形车场
shuttling 往复行车
shuttling motion 摆动；往返运动
shut up 冷塞
shut up a business 停止营业
shut well 停产井
Shüller's sliver machine 棒法毛纱机
SH-wave 水平横波
S-hydrograph S 过程线
shylark resource rocket 云雀资源火箭
Siacci method 赛西法
Si activation log 硅活化测井
sial 硅铝带(层)
sialic 硅铝质的
sialic crust 硅铝壳
sialic mineral 硅铝矿物
sialite 黏土矿物
siallite 硅铝土
siallitic 硅铝的
sialma 硅铝镁带
Sialon 赛龙陶瓷(含钇的硅铝氧氮陶瓷)
Sialon ceramics 硅铝氧氮陶瓷；赛龙陶瓷
sialsima 硅铝镁带
sialsphere 硅铝层
sialspheric fracture 硅铝层断裂
Siamese bore design 萨米迟缸体设计
siamese connection 二重连接；管道的二重连接
siamess joint (管道的)二重接头
siatom ooze 硅藻泥
Siberian fir 西伯利亚冷杉
Siberian floral region 西伯利亚植物地理大区
Siberian foraminigeral region 西伯利亚有孔虫地理大区
Siberian high 西伯利亚高压
Siberian larch 西伯利亚落叶松
Siberian nautiloid region 西伯利亚鹦鹉螺地理大区
Siberian plate 西伯利亚板块
Siberian platform 西伯利亚地台
Siberian platform carbonate bank trilobite region 西伯利亚地台碳酸盐岩滩三叶虫地理大区
Siberian spruce 西伯利亚云杉
Siberia paleoclimate epoch 西伯利亚古气候期
Siberia paleocontinent 西伯利亚古陆
Siberia polarity hyperchron 西伯利亚极性巨时
Siberia polarity hyperchronzone 西伯利亚极性巨时间带
Siberia polarity hyperzone 西伯利亚极性巨带
siberite 西伯利电气石
sibirskite 西硼钙石
Sibley alloy 铝锌合金

sibling species 相似种；姐妹品种
siccation 干燥作用
siccative 干燥剂；干燥的；干料；催干剂；促干剂
siccative oil 干性油
siccative varnish 快干清漆；快干凡立水
siccolabile 不耐干燥的
Sichromal 硅铬铝耐酸钢
sichromal 铝铬硅
sichromal steel 铬铝合金钢；铝铬硅合金钢
Sicilian marble 西西里大理石(意大利)
Sicilian rose 西西里蔷薇
Sicilian sumac 西西里漆树
sick absence statistics 病伤缺勤统计
sick activated sludge 不正常活性污泥
sick aircraft 有故障飞机
sick bay 军舰卫生所；船上医务室
sickbed 病床
sick benefit 病假津贴
sick berth attendant 卫生员
sick-building syndrome 设计不良建筑物综合征
sick flag 检疫旗
sickle 镰刀(形物)；切割器
sick leave 每年可拿工资的病假天数；病假
sickle bar 镰刀杆
sickle beach 新月形沙滩；镰刀形沙滩
sickle grinder 磨刀石；切割刀片磨石
sickle guard 切割器护刃器
sickle knife 切割器动刀片
sickle-leaf acacia 镰叶相思树
sickle pump 活翼式泵；叶轮泵；镰式泵
sickle register 刀片对心
sicklerite 磷锂锰矿；褐磷锂矿
sickle scaler 镰状刮器；镰形刮器
sickle section 切割器的动刀片
sickle-shaped 镰刀形的；镰刀状
sickle-shaped arch 镰刀形拱；新月形拱
sickle-shaped oblique rib 镰刀形斜肋
sickle-shaped oil passage 镰刀形油道
sickle-shaped roof 镰刀形屋顶
sickle-shaped trussed arched girder 镰刀形桁结构拱梁
sickle spanner 镰刀式(薄)扳手
sickly and off-type plants 去劣去杂
sickness 病假
sick officer's quarters 军官病房
sickout 托病旷工
sick pay 病假工资
sick-room 病室；病房
sick soil 耗竭土(壤)；病土
sicomat 光电跟迹切割机
SiC particle reinforced alumin(i)um alloy composite 碳化硅颗粒增强铝合金复合材料
sicromal 铝铬硅耐热钢
SiC varistor 碳化硅可变电阻
sidac 硅对称二端开关元件；双向开关元件
siddhanta 历数书
side 作业组；作业面；搁在一旁；侧面；侧部；侧(边)；刨平木材的侧面
side abutment pressure 边缘拱座压力；边墩压力；边帮压力
side acceleration 侧向加速度
side adjustment 横向调节；边线调整；边平差
side air inlet louver 进气用侧百叶窗
side aisle 边侧通道；侧廊；侧道
side aisle bay 教堂侧翼
side aisle gallery 侧廊楼层
side aisleless 边上无通道
side aisleless church 边上无通道教堂
side aisle passage 边上侧廊通道
side aisle pier 边上侧廊支柱
side aisle roof 边上侧廊屋顶
side aisle vault 边上侧廊拱顶
side aisle wall 边上侧廊墙壁
side aisle wall window 边上侧廊墙窗
side altar 侧圣坛
side anchor 边锚
side anchoring winch 边锚绞车
side and-face cutter 侧平两用铣刀
side angle 侧角
side angle measurement network 测边测角网
side angle press 角压机
side application 次要应用
side apposite an angle 角的对边
side arch 侧面拱砖；侧拱；边拱

side arch brick 侧拱砖
side arch span 侧拱跨
side arm 前架；(气窗两侧的)推拉杆；单面线担；单面横臂；侧臂
side armature 边衔铁
side armature relay 边衔铁继电器
side arm electrode 侧臂电极
side arm water heater 循环热水器
side attached 侧悬挂式的
side attached mower 侧悬挂式割草机
side awning 舷遮；边篷
side axe 单面劈斧
side ballast tank 边压载舱；压载边舱
sideband 旁瓣；旁频带；侧带；边频带；边带(容量)
sideband amplifier 边带放大器
sideband amplitude 边带振幅
sideband analyzer 边带分析器；边带分析仪
sideband attenuation 边频带抑制；边带衰减
sideband capacity 边带容量
sideband characteristics 边带特性
sideband clipping 边频带限制；边带削波
sideband effect 旁带效应
sideband energy 边带能量
sideband filter 边带滤波器
sideband frequency 边带频率
sideband image 边带像
sideband interference 边带干扰
sideband inversion 边带倒转
sideband modulated signal 边带调制信号
sideband power 边带功率
sideband spectrum 侧带谱
sideband suppression 边带抑制
sideband system 边带制
sideband telephone 边带电话
sideband transmission 边带发送；边带传输
side banks 河流两岸；边坡；岸坡
side bar 侧沙坝；边滩；侧杆
side bar keel 帮板龙骨
side bar upstream of crossing 上边滩
side bay 次房；明间旁的一间；次间(古建筑)
side bay arch 侧跨拱
side beam 侧梁
side bearer 侧支承；侧架
side bearing 旁承；侧轴承
side bearing casing 侧轴承匣
side bearing foundation 侧向受力基础
side bearing oak block 侧轴承橡木垫块
side bearing plate 侧承滑块
side bench 舷侧坐板
side bend 侧弯
side-bend specimen 侧弯试样；侧弯试件
side-bend test (焊件的)侧弯试验
side-bend test specimen 边弯曲试件
side bilge 污水沟
side bitt 舷侧带缆柱
side blade 侧刀片
side blister 侧旁瞭望窗；侧瞭望窗
side block 侧盘木；舣墩(木)
side blow 旁击；侧向鼓风；侧面风；侧击；侧冲(击)
side blown 侧吹的
side blown basic Bessemer steel 侧吹碱性转炉钢
side blown converter 侧吹转炉；侧吹炉
side board 侧面挡板；侧(挡)板；侧壁；侧切板；餐具架；餐具柜；边材板；边板；供应桌；栏板；橱柜
side boarding 餐具柜
side board system 舷侧装卸方式；利用舷门的装卸方式
sideboard-table 备餐桌
side bollard 舷侧带缆柱
side bolting 侧向锚杆支护
side boom 起重机侧臂；侧置起重臂
side-boom attachment 侧向吊杆附件
side-boom cat 侧臂吊管机；侧吊杆式履带拖拉机
side-boom crawler tractor 侧向吊杆的履带式起重机
side-boom discharge pipe 边抛管
side-boom dredge(r) 边抛挖泥船
side-boom ladder 边抛管架
side-boom pipe 边抛管
side-boom pipe-laying attachment 侧向吊杆的管道安装附件
side-boom tractor 侧臂吊管拖拉机；边带杆式拖拉机；边吊管式拖拉机；侧臂吊管机
side borrow 路旁借土；取土坑

side borrow operation 路旁取土工作;路旁借土工作
sideborrow work 路旁借土工作
side bottom 旁角;旁脚;塔下部引出物
side box 边厢(剧院);剧院边厢
side bracing 侧拉条
side bracket 边肘板
side brake 侧制动器
side branch 侧向支管;侧枝
side buffer 侧缓冲器
side building 建筑物的配楼;建筑物的侧翼
side bulkhead 舷舱纵向隔舱壁
side bumper 边挡
side bumper guard 侧保险杆
side bunker 舷侧燃料舱;边燃料舱;边煤舱
side buoyancy tank 边浮力舱
side-by-side 并列式布置;并排;(两船的)并靠;并肩
side-by-side bicomponent fibre 并列型双组分纤维
side-by-side(bull)dozer 双机并联推土机
side-by-side(bull)dozing 两台推土机并列推土
side-by-side combination refrigerator 对开门式冰箱
side-by-side connecting rod 并列连杆
side-by-side display 并列显示器
side-by-side dual control 并座复式操纵
side-by-side listing 并排列表;并排编目
side-by-side mounting 两侧悬挂
side-by-side test 平行试验
side-by-side turbine 两轮式水轮机
side-by-side valve 并列阀
side-by-side warp beams 并列经轴
side camber 侧边弯曲
side canal 支运河;边渠
side cap 旁盖;犁侧板后踵
side cap bolt 侧帽螺栓
side-car 边车
side car body 边车身
side car buffer lever 边车缓冲杆
side car connecting arm 边车连接臂
side car frame angle plate 边车架角板
side car frame brace 边车架撑
side car frame brace clamp 边车底架拉条夹
side car handle bar 侧车手柄杆
side car leaf spring 边车叠片弹簧
side car machine 跨斗式摩托车
side car mudguard 边车挡泥板
side carrier 侧向叉车
side car wheel axle bearing 边车轮轴轴承
side car wheel axle bearing bushing 边车轮轴轴承衬套
side cast distance 边抛距离
side caster 侧喷吸扬式挖泥船;侧抛挖泥船
side cast(ing) 边抛(疏浚);坑边堆土;侧抛;侧堆;路边弃土
sidecasting arrangement 边抛装置
sidecasting device 边抛装置
sidecasting dredge(r) 边抛挖泥船
sidecasting dredging 边抛式施挖
sidecasting dredging technique 边抛疏浚技术
sidecasting hopper dredge(r) 边抛耙吸(式)挖泥船
sidecasting method 边抛法
side-casting suction dredge(r) 侧喷吸扬式挖泥船
sidecasting trailing suction dredge(r) 边抛耙吸(式)挖泥船
sidecasting valve 边抛阀
side ceiling 货舱舷侧护条
side chain 侧链支链;侧链;边链
side chain bromination 侧链溴代作用
side chain carbon 侧链碳(原子)
side chain compound 侧链化合物
side chain isomer 侧链异构体
side chain isomerism 侧链异构作用
side chain liquid crystal polymer 侧链液晶聚合物
side chain motion 侧链运动
side chain nitrogen 侧链氮(原子)
side chain radical 侧链基
side chain structure 支链结构
side chain theory 侧链学说;侧链理论
side chair 无拱手的单椅;边椅
side chamber 边室;旁室;侧室
side chamfer 边斜切
side-channel 侧(沟);边沟;横浇铸道;旁侧水道;侧流槽;侧(流)槽;边渠
side channel interference 邻频道干扰

side channel spillway 侧流渠式溢洪道;(侧)流槽式溢洪道;侧梁式溢流道
side channel type spillway 侧渠式溢水道;侧渠式溢洪道
side chapel 附属小教堂
side charged furnace 侧装料式炉
side charging 侧部加料
side cheek (颚式破碎机的)颊板
side chemical reaction 支反应
side-chute lagging 溜槽护壁
side circuit 侧电路;半面电路
side circuit loading coil 侧电路负载线圈;半面电路负载线圈
side circuit repeating coil 半面电路中继线圈
side clamp 侧向夹合器
side clause of charter party 租船附则
side clearance 侧隙;侧面间隙
side clearance angle 旁锋余隙角
side cleat 侧面防滑条
side coaming 舱口边围板
side coefficient 岸坡系数
side column 边柱
side combustion chamber 侧燃烧室
side combustion stove 侧烧式热风炉
side compartment 边舱
side compression 侧压缩;侧压(力)
side condenser 船底冷凝器
side condition 边条件;副条件
side conductor rail 旁置接触轨
side connection 侧边接合
side construction tile 侧砌砖;边结构面砖;横孔砖;侧面铺砌的瓷砖;侧端结构瓦
side contact spike 旁插棒;侧插棒
side container crane 集装箱侧面吊运机
side contraction 侧(边)收缩;边收缩
side contraction coefficient 侧收缩系数
side control 侧面控制
side core 边芯
side corridor 侧面走廊;外廊
side counter timber 尾端加木
side coupling 侧面耦合
side cover 侧盖;边盖
side crank 轴端曲柄
side crank arrangement 轴端曲柄装置
side crank shaft 轴端曲柄轴
side creel 落地纱架
side culvert 支渠;侧涵;侧边涵洞;边涵(洞)
side curtain stop 侧帘式光栏
side cut 切边;无髓心板;侧部掏槽;弦切;椽子端部斜切口;避心下锯法;侧面切削
side-cut brick 切边砖;削边砖
side cut distillate 侧线馏出物;侧取馏出物;侧留分
side cut mower 侧悬挂式割草机
side cut shears 剪边机
side cut stripper 侧线馏分气提塔
side cutter 侧切钳;推板式切坯机;斗侧齿;侧置式切割器;侧牙轮;侧面刃铣刀;侧面切坯机,侧刃;侧边切坯机;外围牙轮;三面牙铣刀
side cutterbar 侧置式切割器
side cutter ga(u)ging 侧刀定位
side cutting 路旁借土;路堤旁取土;山边开挖;堤旁取土;堤旁借土;侧方截根;边坡挖土;侧面切削
side cutting combine 侧割型联合收机
side cutting edge 斜切削刃;副切削刃
side cutting edge angle 旁锋缘角
side cutting flail 侧切式甩刀
side cutting pliers 边剪钳;旁剪钳
side cutting reamer 侧铣铰刀
side cutting tooth 侧切齿
side daylight illumination 侧面日光照明
sided bead 轮脂缘
side deep 侧渊;边缘海渊
side delivering crop loader 侧送式作物装载机
side delivering manure spreader 侧送式撒肥车
side delivery 侧方喂送;侧方输送
side delivery attachment 侧送附加装置
side delivery chute 侧送式卸料槽
side delivery complete beet harvester 侧卸式甜菜联合收获机
side delivery conveyer 侧送式卸载输送机
side delivery elevator 侧卸升降器
side delivery feed 侧卸式排肥器
side delivery rake 侧耙
side delivery reaper 侧向铺放收割机

side delivery spreader 侧送式撒布机
side delivery straw chopper 侧喂式茎稿切碎机
side digging 侧向挖掘
side direction 侧向
side discharge 侧面卸料;侧面泄水;侧面卸货;侧面排水;舷侧卸货;侧向卸料;侧边卸料
side discharge assemblage 边抛装置
side discharged furnace 侧出料式加热炉
side discharge hopper wagon 侧卸式斗车
side discharge truck paver 侧卸铺料机;侧卸铺路机
side discharging 侧卸;侧向卸载
side discharging car 侧卸汽车;侧卸车
side discharging trailer 侧卸挂车
side displacement 侧向位移
side ditch 路沟;路边水沟;路边明沟;侧(部)沟;沟沟
side ditch cover slab 侧沟盖板
side dog 端末弯垫仪
side door 侧(开)门;便门;边门
sidedoor basket 旁通打捞筒
sidedoor choke 旁通气门
sidedoor container 侧门式货柜;侧开门集装箱
sidedoor elevator 旁开式提升器;旁开式吊车
sidedoor elevator with slips 卡瓦旁开式提升器;卡瓦旁开式吊车
sidedoor hopper barge 侧倾抛石驳;侧开式料斗船;侧开驳;侧开料斗船
side downcomer 侧面降液管
side dozer 斜角推土机;斜铲(式)推土机;万能推土机;侧向推土机;侧铲推土机
side draft 牵引侧向分力;牵引侧向阻力
side draft carburetor 侧吸式化油器
side draft on the drawbar 作用在挂钩上的横力
side drag 侧面阻力;(自航耙吸挖泥船的)边耙
side drag arm hopper dredge(r) 侧吸耙挖泥船;边耙挖泥船
side drain 路边阴沟;边沟;侧面排水;边排水
side drain age 路边排水
side drain of sample 试样壁排水条件
side draught 侧吸
side draw 侧线;侧取馏出物;侧留分;侧拉力;边部(平衡)拉引
side draw control 侧取控制
side draw pin 侧抽销
side draw plate 侧线抽出塔板【化】
side draw tray 侧线抽出塔板【化】;侧取塔板【化】
side dressing 旁施;侧施肥料;侧面修整
side dressing method 旁施法
side drift 边侧导坑;侧壁导坑
side-drift method 侧导坑法;侧壁导坑法;边侧导坑法(隧道开挖用)
side drive 侧驱动
side drive motor mower 侧驱式动力割草机
side-dump 旁卸的;侧卸的;侧卸式;侧倾
side-dump ballast car 侧倾石渣车
side-dump ballast wagon 侧倾石渣车
side-dump car 侧卸式矿车
side-dump discharge 侧卸卸
side-dumper 侧卸汽车;侧卸车;侧向倾卸车;侧边式倾车
side-dumping 侧卸
side-dumping angle 侧卸角
side-dumping barge 侧卸抛石驳
side-dumping body 侧卸式车身;侧卸车身
side-dumping bucket 侧卸戽斗;侧卸(式)料斗;侧卸铲斗
side-dumping bucket loader 侧卸斗式装料机
side-dumping car 侧卸车;侧向倾卸车;侧面卸货车
side-dumping cylinder 侧卸液压缸
side-dumping equipment 侧卸装置
side-dumping hopper 侧卸斗
side-dumping loader 侧卸装载车;侧卸式装载机;侧卸式装碴机;侧向装载机;侧向装载车
side-dumping scow 侧卸抛石驳;侧倾抛石驳
side-dumping skip 侧卸式箕斗
side-dumping trailer 侧卸(式)拖车;侧卸挂车
side-dumping truck 侧卸式自动卸货汽车;侧卸式运货汽车;侧卸汽车;侧卸货车;侧向倾卸运货汽车
side-dumping type loader 侧卸式装载机;侧卸式装碴机
side-dumping wagon 侧卸运输车;侧卸拖车;侧卸车;侧倾式翻斗车
side-dump lorry 侧向倾卸货车
side-dump mine car 倾卸矿山车

side-dump scow 倾卸甲板驳
side-dump semi-trailer 侧卸半挂车
side-dump skip bucket 爬式翻斗加料桶
side-dump type 旁卸式
side dunnage 货舱舷侧垫衬
side echo 假回波;旁瓣回波;侧向回声;侧(向)回波;边缘回波
side effect 次要的影响;次生效应;副作业;副作用;边界效应
side elbow 支流弯管;侧接弯头
side elevation 侧视图;侧面图;侧立面
side elevation drawing 侧视图
side elevator harvester 侧置升运器式收获机
side emission 旁放射;侧向发射
side engined chassis 侧发动机底盘
side entering type agitator 旁入式搅拌器;侧伸式搅拌器
side entrance 侧入口;侧门;侧进口
side-entrance gulley 侧入口排水沟
side entrance manhole 侧入(口)人孔;旁进式入孔;旁侧进口人孔
side entry blade 侧装叶片
side entry combustion chamber 侧面进气的燃烧室
side entry nozzle 侧向进液喷嘴
side entry serration root 侧装式锯齿型叶根
side equation 边方程
side equation test 边方程检验
side erase 侧向清洗;侧边抹去
side erosion 侧向侵蚀
side error 定镜差(六分仪);侧误差;边差(六分仪)
side etch 侧面腐蚀
side exhaustion 侧方排气
side explosion 单边爆破
side face 条面;侧面
side faced brick 侧边面砖
side face seam 侧面接缝
side facing 路面
side factor 岸坡系数
side-factor equation 河宽因素方程
side fascia 边梁外饰带;外墙板
side fastening bolt 侧紧螺栓
side fed water wheel 侧射式水轮
side feed baler 侧喂式捡拾压捆机
side feeding 侧向送风
side feet (电机的)两侧底脚
side fence 翼墙;翼篱
side fender 侧面护舷
side fiducial mark 边框标
side fillet weld 侧面贴角焊;侧面填角焊(缝);侧面角焊缝
side fillet welding 侧面角焊
side fill(ing) 侧面加料;侧向堆填
side filtration 旁滤
side-fired furnace 侧焰炉;单侧加热炉
side-fired kiln 侧烧窑;横焰窑
side-fired recirculating still 炉侧加热与气体循环管式炉
side-fired tank furnace 横火焰池窑
side fire helical antenna 侧向辐射式螺旋形天线
side firing 侧烧;侧面燃烧
sideflash 侧面放电;侧击雷
side flat 边滩
side flat of shingle and sand 沙砾石变滩
sideflat-overtopping discharge 平滩流量
sideflat-overtopping stage 平滩水位
side flight 边梯(双回楼梯的返回梯段);侧梯;边梯
side flow 侧流
side flow weir 侧流堰
side flue 侧烟道
side force (侧)(向)力;横向力
side force coefficient 侧力系数
side fork 侧叉架
side forklift 侧面式叉车
side fork-lift truck 侧向叉式装卸车;侧叉式运货车
side form 侧模(板);边模(板)
side form paving 两面立模浇灌混凝土施工法
side form spacers 侧模垫片;侧模间隔物
side formwork 侧边模板【道】
side fraction 侧分数
side frame 舷肋骨;侧梁;侧框梁;侧构架;侧壁框架【建】;侧肋骨;边框【建】;边架
side frame plough 侧架犁
side frame plow 侧架犁

side frame type bogie 侧架式转向架
side framing 舷边构架
side frequency 旁频(率);边频(率)
side-frequency current 边频电流
side friction 横向摩擦力;侧面摩擦(力)
side frictional force 侧向摩擦阻力
side friction block 侧面摩擦块
side friction factor 横向摩擦系数
side friction resistance 侧向摩擦阻力
side furring 侧墙木梁;侧壁罩面板条
side gallery 分支巷道;侧面走廊
side gangplank 边跳板
side gate 旁侧控制极;大门旁侧的小门;侧浇口;边浇道
side gatherer 侧扶导器;缩流装置
side ga(u)ge bar 侧定位杆
side gear 侧面齿轮;侧齿轮;边齿轮
side girder 旁纵桁;侧桁材
side glance 暗示;斜视;侧视
side glass 定镜(六分仪)
side grade 侧面坡度
side grafting 侧接法
side grain 平行于木纹的板面;纵纹;平切面;侧面纹理
side grinding 侧磨
side groove 侧槽
side group 侧群;侧基
side guard 侧导板
side guard manipulator 翻侧推床
side guard plate 侧护板
side guard-type repeater 侧导式围盘
side guidance 侧导槽
side guide 导向角钢;侧压板;(指闸门的)侧(向)导轨;侧(面)导板
side guide bar 侧导向杆
side guide wheel 侧向导轮;导导轮
side gun 机身外侧机枪
side gutter 侧向天沟;边沟;泛水
side guy 侧面稳索
side hall 配殿
side hammer 边锤
side hanger 侧吊杆
side hardness 侧面硬度
side haulage dock 横向滑道
side haul marine railway 横向滑道
side-haul railway dry dock 横牵式轨道干船坞
side head 侧冒口
side heading 侧导坑;侧壁导坑(法)
side heading method 侧壁导洞法
side heel 管外扩孔器
side hill 山侧;越过作业线的边坡;山坡;边坡
sidehill canal 傍山沟;旁山沟
sidehill cut 山坡开挖;山坡开挖;半填挖
sidehill ditch 傍山沟;旁山沟
sidehill excavation 山坡开挖
sidehill fill 半挖填土;半挖半填;半路堑
sidehill lug 坡地抓;地爪
sidehill section 山坡断面
sidehill stream 沿山河道
sidehill tunnel 傍山隧道
side hinge 侧面铰链
side hinged window 侧铰(链)窗
side hole 爆破钻孔;周边眼;旁孔;侧孔;边缘炮眼;边眼
side hole bit 侧孔式钎头;侧孔(式)钻头
side hood 侧吸(风)罩;侧方吸气罩
side hook 台卡(使工件不移动的设施);(木工工作台上的)边夹
side hopper barge 侧卸泥驳;侧倾抛石驳;侧开泥驳;侧开料斗式船
side horse 鞍马
side-hung (窗的)平开式
side-hung casement 推开窗扉;推关窗扇
side-hung casement window opening inwards 向内开启的侧悬窗
side-hung casement window opening outwards 向外开启的侧悬窗
side-hung door 平开门
side-hung opening window 侧面悬挂开启式窗
side-hung sash window 侧悬窗
side-hung window 侧悬窗;平开窗
side idler 侧面托辊
side illumination 侧面照明
side index plate 侧分度盘

side inflow 旁侧入流;旁侧来水
side inlet 侧面进水口;路边进水口
side intake 旁侧进水口
side intercostal keelson 间断边内龙骨
side intercostal stringer 间断边纵骨
side interlocking 侧边咬接
side intersection 侧向交会;侧方交会(定位法)【测】
side intersection by plane-table 平板仪侧方交会
side intersection method 边交会(测量)法
side-inverted 两侧倒置
side island 边侧分车岛
side issue 次要问题
side-jacking test 侧向顶压试验(顶压坑侧木板的试验)
side jamb 门柱;门窗边框;边框;门梃;门边框
side jar socket 侧振击式打捞筒
side joint 侧面接缝;竖缝;侧接
side jointing 用油石修整锯齿侧边
side joint of timber members 木构件侧接
side keel 边内龙骨
side keelson 旁内龙骨
side knob screw 球形把手止位螺栓
side knock 横向冲击
side ladder 舷梯
side lagging 巷道两侧背板
sidelamp 侧灯
side land 空地;边地
side lane 靠边车道
side lantern 侧灯
side lap 旁侧重接;侧边搭接;侧向重叠
sidelap of corrugations 互楞边搭;波形瓦侧边搭接
side lapping mill 侧面精研机
sidelap sealer 边部搭接密封剂
sidelap weld(ing) 边搭接焊接;侧面搭接焊;侧搭焊;弯边搭接焊缝;搭边焊接;侧边搭接焊
sidelap width 旁向重叠宽度;侧边搭接宽度
side launching 横向下水
side lay 侧面定位装置
side lay adjuster 侧面定位调整器;侧安排调整器
side lay edge 侧边导规
side layer 侧边料
side layer material 侧面铺料;边铺料
side leakage 侧面漏泄
side leg 曲拐
side length 边长
side length in plane 平面边长
sideless 无端
side lever 侧杆
side lever engine 边杆发动机
side lever press 杠杆压机
side lift 侧置提升器;侧吊作业;侧吊机
side lifting 边抬打捞
side lift truck 侧面起重车
side light 舞台的侧光;舷灯;舷窗;侧射光;侧灯;侧窗;耳光;侧光;边灯;边窗
side lighting 侧向照明;侧方采光;侧窗采光;侧光照明;侧面照明;侧向采光
side light plug 舷灯插头;舷窗外盖
side light screen 舷灯遮光板
side light tower 舷灯架
side line 横线;旁线;边线;侧道;副业;边线
sideline of dredge-cut 挖槽边线
sideline product 副线产品;副业产品
sideline production 副业(生产)
sideline production resources 副业资源
sidelines of science 科学的边缘区域
side lining 内墙板;侧牵引;侧壁内衬板;内衬板;侧内衬;边内衬
side linkage 侧悬挂装置
side load 侧向负荷;侧壁负荷;边载重;边载(荷);边荷载
side loader 侧面装卸机;侧向装卸机;侧向叉车
side loader system 侧向装载机方式
side loading 侧向荷载;边缘荷载;侧面装载;侧面加载;侧加料
side loading container 侧装货柜
side loading forklift truck 侧向叉车;侧面叉式卸车;侧向装载叉车
side loading fork truck 侧装载自动装卸车
side loading paver 从旁装料的摊铺机
side loading ramp 舷侧卸货跳板
side loading refuse collector 侧装载废品运输车

side loading vehicle 侧装载车
side lobe 旁波瓣;侧瓣;副瓣;旁瓣
sidelobe blanking 旁瓣消隐
sidelobe cancellation 旁瓣对消
sidelobe echo 旁瓣回波
sidelobe level 旁瓣电平
sidelobe subtraction 减旁瓣法
side lobe suppression 旁瓣抑制
side lock 侧锁定
sidelocking sonar 旁视声呐
side-lock tile 侧销定瓦
sidelong ground 斜地;山边
side longitudinal 舷侧纵材;船侧纵骨
side looking 侧视
side-looking aerial radar 机载侧视雷达;航空侧视雷达
side-looking aerial radar image 航空侧视雷达图像
side-looking airborne radar 侧视雷达;机载侧视雷达;航空侧视雷达
side-looking airborne radar imagery 机载侧视雷达成像
side-looking airborne radar system 航空侧视雷达系统
side-looking detector 侧视(辐射)探测器;边缘光电效应探测器
side-looking radar 侧视雷达;旁视雷达
side-looking radar display 侧视雷达显示器
side-looking radar survey 侧视雷达测量
side-looking sonar 侧视声呐
side loudspeaker 侧边扬声器
side louver 侧遮光板
side lurch 侧倾
side mark 边标
side masking 侧台遮挡(指边幕、景片)
side member (机车的)纵梁;(机车的)大梁;侧梁
side member extension 车架纵梁伸出部分
side mill 侧铣刀
side milling 侧铣;边铣
side milling cutter 侧面(刃)铣刀
side mirror (工具经纬仪望远镜的)侧镜
side mixing nozzle 侧向混合式喷嘴
side mode suppression ratio 边模抑制比
side mount 侧装
side-mounted 装在侧面的;侧悬挂(式)的
side-mounted condenser 侧装式凝汽器
side-mounted rod 侧插棒
side mounting 侧悬挂
side movement 侧向移动
side nailing strip 侧钉木条;侧部地板梁
side natural lighting 侧面自然采光
side neck 侧筒;侧管嘴
side net 舷边安全网;货网
sidenote 旁注
side number plate 侧路牌
side oblique air photograph 空中倾斜摄影
side observation 旁点观测
side-occupation 副业
side of an anticline 背斜的边;背斜翼
side of an excavation 挖方的边坡
side of hull 船体侧面
side-of-pavement line 路面边缘线
side of piston ring 活塞环端
side of ship 船舷
side of the road 路肩;路边
side on collision 侧面冲突
side on photomultiplier 侧窗式光电倍增管
side opening 侧面孔洞;船侧开口;侧孔
side opening baler 侧喂式捡拾压捆机
sideopening hatchcover 旁侧开式舱盖
side opening stall 侧出式挤奶台
side opening tipping bucket 侧卸翻斗
side opening two-row tandem stalls 侧出式双行串列挤奶台
side open mud barge 侧开式泥驳
side opposite the angle 对角边
side or end lap 边接或端接距离
side ornament 端部装饰品;侧面装饰品
side orthogonality 侧边垂直度
side oscillation 侧摆动
side outlet 侧管;侧向出口;侧孔
side outlet reducing tee 侧口异径三通
side outlet T 支ícios三通(管)
side outlet tee 支íios三通(管)
side outlet tee reducing 缩径侧流三通

side over-flow 侧向溢流
side over-flow pipe 侧面溢流管
side overflow weir 横向越流堰
side overlap 旁向重叠
side overturned car 横翻车
side paddle dyeing machine 边浆式染色机
side paddle tug 边浆式拖船
side panel 门侧玻璃;测板;(集装箱的)侧面;侧板
side paneling 侧边镶面板
side-parking 路旁停车;路边驻车
side-parking lamp 路旁停车信号灯
side part 侧面部件;侧面机件
side part forming machine 侧边成形机
side passby 越行线
side path 侧道;便道
side patio 侧面庭院
side pavement 人行道
side payment 边支付;侧向支付
side penetration 侧面熔透;侧面熔深
side piece 车尼机架;测板;侧板
side pier 束水堤
sidepiercing cam actuated die 侧楔作用侧向冲孔模
side pile 边缘桩
side pile displacement 位移边桩
side piling 打边桩;弃土堆;边桩
side pillar 边柱
side-pilot method 侧壁导坑法
side pipe 侧管
side plane 边刨
side planer 侧刨床;边缘刨床
side planing 侧刨
side planing machine 边刨机
side plate 旁板;船侧外板;测板;侧衬板;侧板;侧边
side plate angle 侧片角
side plate of cage 保持架边框
side plate rudder 复板舵
side platform 月台;侧向收割台;侧式站台
side plat-form combine 侧置割台式联合收割机
side plating 旁板;船侧外板
side play 轴向间隙;轴端余隙;侧隙;侧面误差;侧面间隙
sideplay amount 偏移量
side plot line 旁侧(土地)区划线
side plough 侧刮板;侧刨煤机;侧开沟机
side pocket 侧柱插座;清除口;隧道避车洞
side point 旁交点
side pole 侧杆;沿线式电车杆
side pond 节水池;船闸旁(水)池;船闸储水池;侧塘(船闸补水用)
side pontoon 侧面浮箱;边浮箱
side pool 侧塘(船闸补水用)
side port 舷门;舷侧装货门;载货舷门;侧向孔口;侧舷门;侧喷火口;侧孔
side portal 侧面入口;侧门;侧门架
side port(er) 驳门
side porter system 舷门装卸方式
side port filling system 侧向孔口充水系统(船闸)
side port loading-unloading 侧舷门装卸
side port mouth 侧面喷火口
side port nipple 旁通管节
side port system 舷侧装卸方式
side-port tank furnace 横火焰池窑
side port vessel 侧舷门船
side post 侧壁柱;边柱;(屋架上的)侧柱
side power takeoff 侧置式动力输出轴
side pressed out 片帮[岩]
side pressure 侧压下;侧压(力);侧面压力
side pressure test 旁压试验;侧压试验
side product 副产品
side projection 侧向投影
side property line 旁侧地界线
side protecting plate 侧护板
side pull 风压差;侧向拉力;滑移
side pull lay 侧拉纸规
side rabbet plane 侧榫刨;边槽刨
side rack 侧护板
side radiation 侧面辐射
side radiation tube 侧辐射管
side radiator 侧散热器
side rail 舷梯扶手;路缘轨模;护栏;护轨;铁路护轨
side rake 侧向搂草机;侧前角
side rake angle 横向前角;副前角
side raking 侧向搂集

sideramine 高铁胺
side ramp 舷门斜跳板;边跳板;侧撞杆
side ramp collecting ballast and reshaping machine 边坡收砟机
side ram press 角式水压泵
side rasp 打捞钩
side ratio 边长比;边比
siderazot 氮铁胺
siderazote 二氮化五铁
side reach 横偏;侧向伸出长度
side reaction 支反应;副反应
side reaction coefficient 副反应系数
sidereal astronomy 恒星天文学
sidereal chronometer 恒星时计
sidereal clock 恒星(时)钟;恒时钟
sidereal day 恒星日
sidereal focus 恒星焦点
sidereal hour 恒星小时;恒星时
sidereal hour angle 恒星时角
sidereal midnight 恒星时子夜
sidereal month 恒星月
sidereal noon 恒星时正午
sidereal period 恒星周期
sidereal revolution 恒星周
sidereal table 恒星表
sidereal time 恒星时间;恒星时;天文时
sidereal unit 恒星日单位
sidereal watch 天文表
sidereal year 恒星年
side rebate plane 边槽刨
side recess 闸门边槽;闸门边木;侧面壁龛;(闸门的)边槽
side reducing tee 侧接缩径三通
side reflected 侧向反射的
side reflector 侧反光板
side reflux 线侧回流;侧回流
side relative to track 线路边侧
side relief 刀背;侧面放气;副后角
side relief angle 侧锋后让角
sidereostat 定星镜
side rib 横肋;侧肋
side rim 可卸轮辋
side ring 边环
side ring of rim 可卸轮辋锁环
side riser 侧冒口;边冒口
siderite 陨铁;菱铁矿;蓝石英
siderite cement 菱铁矿胶结物
siderite concretion 菱铁矿结核
siderite mudstone 菱铁矿泥岩
siderite ore 菱铁矿矿石
siderite rock 菱铁矿岩
side road 支路;旁路;岔路
side rod 拉紧螺栓;机车动轮连杆;边连杆
side rod bearing 边杆轴承
side rod crank pin 边杆曲柄销
side rod drive 连杆传动
sideroferrite 自然铁
siderograph 恒星仪;恒星时计时器
siderography 钢版雕刻术;钢刀板;磷铁矿表面研究
siderolite 石铁陨石
side roller 衔接轮;舷窗滚筒;闸门侧滚轮;侧向滚轮;侧面托辊;侧面滚轮;侧附滚轴;侧边滚轮
side rolling hatchcover 侧滚式舱盖
side rolling ladder 侧面滚动梯
siderology 冶铁学
sideromagnetic 顺磁的
sideromelane 铁镁矿物;碎云玻璃
sideromycosis 铁质沉着性真菌病
sideronatrite 纤钠铁矾
sideronitic texture 海绵陨铁结构
side roof rail 顶部桁材
sideropenia 铁质缺乏
siderophile 亲铁的
siderophile element 亲铁元素
siderophore 铁载体
siderophyllite 针叶云母
siderophyry 古铜鳞英铁镍陨石;古铜辉石鳞石英陨石
sideroplesite 镁菱铁矿
sideropyrite 黄铁矿
sideroscope 铁(屑)检查器
siderosilicosis 铁矽末沉着病;铁矽末沉着;铁矽尘肺;铁硅尘肺
siderosis 铁质沉着;铁肺;铁尘肺

siderosphere 铁帽
siderostat 定星镜
siderotil 纤铁矾;铁矾
side roughing turning tool 侧粗车刀
side rudder 边舵
side run-off 塔侧抽出物;侧线馏分;侧向径流
sides 开挖侧帮
side saddle 女用鞍
side sampling 堤旁取土
side scan equipment 侧向扫描设备
side scanner 侧扫描器
side scanning 侧扫描
side scanning sonar 侧视声呐;侧扫描声呐
side scan sonar 侧向扫描声呐;侧(扫)声呐
side scan sonar photo 声呐扫描仪
side scan sonar prospecting 旁侧(扫描)声呐勘探;旁侧(扫描)声呐勘测
side scan sonar sounding 旁侧(扫描)声呐探测
side scan sonar survey 旁测声呐测量
side scan sub-bottom profiler 侧视河底浅层剖面仪;侧视海底浅层剖面仪
side scattered waves 侧面散射波
side scene 侧面布景
side scrap 切边
side scraper 边坡平整器;刮沟机;边坡平整机;边沟刮铲机;壁沟铲刮机
side scraper reclaimer 侧面刮板取料机
side screen 遮阳幔
side screw 侧螺钉
side-scroll 侧面涡卷装饰
side scuttle 舷孔;舷窗
side scuttle blind 舷窗内盖
side scuttle plug 舷窗外盖
side seal(ing) 侧水封;边缘止水;侧边止水条;侧面填缝;侧(面)密封;侧封
side sealing of tin bath 锡槽边封
side seam 水平焊缝;长缝;边缝
side seamer 缝边机;边封机
side seat 边座
side section 侧构件;侧剖面
side securing fitting 侧向连接件
side select signal 面选择信号
side service spillway 边侧辅助溢洪道
side set 错齿;斜边差;侧面倾斜;边接
side shaft 侧轴
side-shake digger 抖动侧送式挖掘机
side-shake mill 横振筛架
side-shake screen 侧振动筛
side shake vanner 侧捣淘矿机
side share 单面平切铲
side-shattering effect 侧面爆破作用
side sheathing 外侧墙板
side sheet 侧壁(板);侧板;边板
sideshift 侧移
side-shiftable 可向侧面移动的
side shifter 侧移器;侧移叉
side shift fork 侧移式货叉
side shift system 侧移方式
side shiplift incline 横向斜直升船机
side shipping berth 横向滑道
side shoal 边滩
side shoe 夹紧装置;夹爪
side shore 横撑木;船边撑木;侧撑;边撑
side shoring 侧面斜撑;侧撑;边撑
side shot 旁点观测;旁测点;侧向爆炸
side shower 侧簇射
side shuttering (道路的)边模板;侧模(板)
side sight 旁点观测
side sill 侧梁
side skew 侧面拱脚砖
side skid 横向滑动;横滑;侧向滑动;侧向滑板;侧面滑动;侧滑
side skidding 侧滑
side skirt of the fender 保护板侧板
side slab 侧面支承板
side slacker system 侧向提箱机方式(指装卸)
side slat 侧条
side slide 侧滑块
side sling 侧吊索
side slip landing 侧滑着陆
side slip of embankment 路堤滑坡
side slip oscillation 侧滑振荡
side slip(ping) 侧面滑动;横向滑动;侧向滑动;侧滑(移);横滑

side-slipping slipway 横滑滑船台;横滑修船台
side-slipway 横滑滑道
side slipway and wedged chassis 横向斜架滑道
side slipway with high and low rail tracks 横向高低轨滑道
side slipway with launching cradle of two level wheels 横向高低腿下水车滑道
side slipway with rail tracks 横向船排滑道
side slope 斜坡;边坡
side slope cutting 开挖边坡
side slope erosion 边坡侵蚀
side slope factor of channel 航道边坡系数
side slope of bank 河岸边坡
side slope of dredge-cut 挖槽边坡
side slope of sedimentation tank 沉淀池边坡
side slope packer 边坡夯拍机
side slope seepage ditch 边坡渗沟
side slopes of canal 渠道边坡
side slope stability 边坡稳定(性)
side snap 侧方摘取
sides of excavation 开挖侧帮
sides of the original track and with the tails of frogs facing each other 异向岔尾相对道岔
sides out of parallel 端面平行差
sides out of square 端面不方正
side spacing 旁侧间距
side span 旁跨;边跨
side sparring 货舱舷侧护条
side spectrum 二次光谱
side spillway 旁侧溢洪道
side-spillway channel 侧槽式泄水槽
side-spillway dam 旁溢洪道式坝
sidespin 侧旋
side spined 侧旁有刺的
side spit fuse 侧面点燃导火线
side splash 邻近频道干扰
side splatter 邻信道干扰
side spray 侧向飞散破片
side squeeze 侧面压制
side stable relay 侧稳继电器;边稳定极化继电器
side stage 侧台
side stage loading door 侧台大门
side stake 边桩【测】
side stay 侧撑
side stay-bolt 火箱侧面螺撑
side steel facing 侧面金属饰面
side steering 变向机构
side step 侧面梯级;侧面台阶;侧梯;旁梯梯级
side-step ladder 有侧面梯级的梯;有旁侧踏步的梯
sidestick 紧版楔
side stiffener 侧面加强板;侧壁扶强材
side strake 舷侧外板;旁列板;船舷侧板
side stream 支流;旁流(水);侧线(馏出物);侧流
sidestream elevated pool aeration 河边高架池曝气
sidestream filter 旁流过滤器
sidestream filtration 旁流过滤
sidestream heat exchanger 侧流换热器
side streaming log 舷边计程仪
sidestream process 旁流法
sidestream stripper 侧流汽提器
sidestream treatment 旁流水处理
sidestream withdrawal 侧流抽出口
side street 横街;小巷;交叉街道;旁街;次要街道
side street yard 临旁侧街道的庭院
side striking 侧击
side stringer 船he纵骨
side strip 道路路边分隔带;路边分隔带
side stripper 侧线汽提塔;侧流汽提器
side stroke 侧击
side stroke rake 侧向搂草机
side stroke reel 侧向搂草轮;侧向搂草机
side substitution 侧链取代
side suction 犁体水平间隙
side surface 端面;侧面
side sway 侧倾;侧摆;侧移
side-sway eliminator 减侧摆器
side swing hydraulic cylinder 侧摆液压汽缸
side swing speed 横移速度
side swipe 擦边撞击
side swipe accident 侧撞事故
sideswipe reflection 侧向岩面多次反射
side T 侧接通
side table 条桌;茶几
side tank 翼柜;边水柜

side tee 侧接通;侧接三通
side telescope 辅助望远镜;侧副镜
side tenon 边榫
side terminal 侧面端子
side terrace 边侧阳台
side tester 侧壁取样器
side thrust 侧压;推冲器;侧向推力;侧推(力)
side thruster 侧向推进器;侧推(力)器
side thrust of piston 活塞侧推力
side thrust unit 侧推(力)装置
side tilt 旁向倾斜
side tilt car 侧卸式送锭车
side timber 横檩条;侧支柱;侧支木
side timber purlin(e) 檩条
side-tip bucket 侧装式铲斗
side-tip car 侧卸车
side-tip dumper 侧向自卸车;倾向自卸车
side-tip lorry 侧卸式四轮车
side tipper 侧卸转翻笼;侧边式倾车
side-tipping 侧卸;侧倾;侧翻式
side-tipping body 侧倾卸式车身;侧翻式车身
side-tipping bucket 侧倾式铲斗;侧倾式铲车
side-tipping car 侧面翻斗车
side-tipping dump car 侧卸车
side-tipping loading bucket 侧倾料斗
side-tipping muck-car 侧卸式软土运车
side-tipping semi-trailer 侧卸式半挂车
side-tipping skip 侧卸式翻斗车
side-tipping truck 侧卸式(载货)汽车;侧卸货车
side-tipping wagon 侧倾卸拖车;侧卸货车;侧翻式翻斗车;侧向自卸车;侧卸式翻斗车;侧倾车
side tippler 侧向装卸装置;侧向翻倾装置
side-tip truck 侧卸式自动卸货车;侧卸货车;侧向倾卸车
side-tip wagon 倾卸车;侧卸车
side title 说明字幕
side-to-main span ratio 边跨与主跨比
side tone 侧音
sidetone level 侧音电平
sidetone ranging 侧音测距
side tongs 边口钳
side tongue graft 舌状腹接
side tongue-grafting 舌状腹接
side tool 偏刀;侧刀
side tool box 侧刀架
side tool head 侧刀架
side-to-side adjustment 横向调节
side-to-side anastomosis 侧侧吻合
side-to-side baffles 单缺圆折流板
side-to-side floating 横向自动定位
side-to-side movement 左右运动
side-to-side pendulum-type distributor 横摆式撒布机
side-to-side setup 横向排列
side-to-side vibrations 左右振动
side tower 阙楼;侧塔
side track 铁路侧线;越行线;铁路支线;待避线;岔线【铁】;侧线;侧轨;支线
side track borehole 另钻新孔(在孔内)
side-track drilling 侧钻
side-tracking 造斜【岩】;偏斜钻法;侧钻新眼;侧钻
side-tracking bit 钻侧孔眼钻头
side-tracking coring 人工偏斜补取岩芯
side-tracking skate 侧向滑动装置
side transeptal portal 旁侧建筑翼部入口;旁侧教堂十字形耳堂门
side transverse 船侧竖桁材
side trawler 舷拖(网)渔船;侧拖网渔船
side tree (巷道中的)侧支木
side trench 边沟
side trimmer 侧边修边机
side trimming line 侧边纵剪作业线
side trip 路缘带
side trip margin verge 路边分隔带
side tube 侧管;支管
side-tube flask 支管烧瓶
side-tunnel fiber 边隧道光纤
side tying 侧栓系
side type jockey 侧面连接抓叉
side underflow 侧面地下水流
side valency 副价
side valve 旁路阀;侧阀;测阀;侧阀;分流活门
side valve engine 侧置气门发动机
side valve head 侧置气门汽缸盖

side vane 侧叶
side veneer-grafting 镶合腹接
side vent 旁透气管;旁边透气管;侧通气口
side verge (道路的)边缘
side view 侧影;侧视[图];侧面形状;侧面图
side view drawing 侧视图
side-volute 侧面涡旋形饰
sidewalk 边道;侧道;人行道;便道
sidewalk arm 边门拦木;人行道拦木
sidewalk bracket 人行道托梁;人行道托架
sidewalk café 街边咖啡座;人行道咖啡座;路边饮食铺
sidewalk cantilever bracket 人行道悬臂牛腿;人行道悬臂托架
sidewalk concrete flag(stone) 人行道混凝土石板
sidewalk door 人行道门
sidewalk edging 人行道边缘
sidewalk elevator 侧边升降机;人行道电梯;人行道升降平台
sidewalk encroachment 占用人行道
sidewalk expansion joint 人行道温度缝;人行道伸缩缝
sidewalk flag(stone) 人行道石板
sidewalk for the blind 盲道
sidewalk gate 人行道拦木
sidewalk glass 人行道玻璃砖砌块
sidewalk greening 人行道绿化
sidewalk joist 人行道格栅
sidewalk kiosk 人行道公共书报亭;人行道公共电话亭
sidewalk loading 人行道荷载
sidewalk of bridge floor 桥面人行道
sidewalk pavement 人行道铺装;人行道铺面
sidewalk paver 人行道铺路工;人行道铺路机
sidewalk paving 人行道铺装;人行道铺面
sidewalk roller 人行道路碾;人行道碾压机
sidewalk shed 人行道棚
sidewalk slab 人行道板
sidewalk space 人行道范围
sidewalks provided on both sides 双侧人行道
sidewalk teller 人行道记数员
sidewalk tree 行道树
sidewall 轮胎侧壁;轮胎壁;边墙;岸墙;孔壁;井壁;池壁;侧墙;侧壁
sidewall air cushion craft 侧壁式气垫艇
sidewall air cushion ship 侧壁式气垫船
sidewall air-cushion vehicle 侧壁式气垫艇
sidewall air supply 侧送风
sidewall block 池壁砖
sidewall body brace 侧墙车体斜撑
sidewall bulging 边墙膨胀
sidewall ceiling support 侧墙顶板支座
sidewall concrete 边墙混凝土
sidewall construction 侧壁式结构
sidewall core 孔壁岩芯;井壁岩芯
sidewall core device 孔壁取样器
sidewall core sample 孔壁岩芯样品;井壁岩芯样品
sidewall coring 孔壁取芯;井壁取芯
sidewall coring gun 孔壁射击式取芯器;井壁射击式取芯器
sidewall coring method 孔壁取芯方法;井壁取芯方法
sidewall coring tool 井壁取芯工具;孔壁取样器;井壁取样器
sidewall craft 侧壁气垫船
sidewall culvert with side port system 侧墙廊道侧向孔口输水系统
sidewall diagonal 侧墙对角斜撑
sidewall drift 侧壁导坑;侧墙的平洞
sidewall drill(er) 侧壁钻机;水平孔钻机
sidewall echo 侧壁回波
sidewaller 侧壁气垫船
sidewall filter 侧壁过滤器
sidewall flashing 侧墙泛水
sidewall formwork 侧墙模板
sidewall hovercraft 侧壁气垫船
sidewall installation 侧墙安装
sidewall left 左端墙
sidewall neutron log 井壁中子测井
sidewall neutron log curve 井壁中子测井曲线
sidewall of crucible 坩埚壁
sidewall of dock 坞墙
sidewall of pass 孔型侧壁
sidewall paneling 侧墙镶面板

sidewall port filling and empty system 船闸侧壁充水、放水系统
sidewall port system 侧墙孔口输水系统(闸门)
sidewall protection 侧墙保护
sidewall register 侧墙出风口
sidewall right 右端墙
sidewall sampler 墙面喷水器(消防用);侧墙采样器;孔壁取样器;井壁取样器
sidewall sample taker 孔壁取样器;井壁取样器
sidewall sampling 孔壁取样;井壁取样
sidewall scaling 片帮【岩】
sidewall segment 边墙梳形木
sidewall slabbing 片帮【岩】
sidewall spalling 片帮【岩】
sidewall stock divider 侧面余量均分装置
sidewall strength 侧壁强度
sidewall strength test 侧壁强度试验
sidewall test 侧壁试验
sidewall type hovercraft 侧壁式气垫船
sideward motion of earthquake 地震水平运动;地震侧向运动
side water course 舷侧污水道
side water depth 周边水深;侧水深;侧边水深
side water hole 侧水眼(钻头)
side water supplying device 旁侧供水装置
side wave 侧面波;旁波;边波
side wave section 侧面波剖面
sideway 斜的;沿边的;侧向的;(道路用地范围内的)人行道;旁路;小路;侧移;侧倾;边道
sideway buckling 侧向屈曲
sideway extrusion 横向挤压
sideway floating 横向自动定位
sideway force 横向力;侧向力
sideway force coefficient 侧向力系数
sideway launching 横向下水
sideway lead 横向通道
sideway moment 旁移力矩;侧倾力矩
sideway restraint 侧向限制
sideway running 侧接运行
sideway feed 侧向馈送
sideway skidding 侧向滑溜
sideway skid resistance 横向抗力;侧向抗滑力
sideways launching 侧面下水
sideways sum 数位叠加和
sideway stiffness 侧斜刚度;侧倾刚度
sideway swivel 侧旋转
side wear 侧磨;侧面磨损;侧面磨耗
side wear of rail head 轨头侧面磨损
side web 侧腹板;边腹板
side weir 侧流溢洪道;侧堰
side weir overflow 侧堰溢流
side weld 侧焊缝;边缝
side welding 边焊接
side wheel 侧轮
side wheeler 边轮式车
side wheel press 侧轮压力机
side wind 横向风;旁风;侧风;侧气流
side winding 侧面绕组
side window 侧窗;边窗
side window counter 带侧窗膜的计数器
side window photomultiplier 侧窗式光电倍增管
side window viewfinder 侧窗取景器
side wing 边翼;侧翼
side wire 边缆
side-wire winch controlling unit 边缆绞车控制装置
sidewise attraction 侧引力
sidewise buckling 侧向压曲;侧向翘曲;侧向扭曲;侧向挠曲
sidewise component 侧向分力
sidewise cut 侧面陶槽
sidewise mismatch 侧向错移
sidewise restraint 侧向约束
sidewise scattered 侧向散射的
sidewise shifting speed 横移速度
sidewise skidding 横向滑动
sidewise skid resistance 横向抗力
side witness 旁证
side wobble 侧面摆动
side work 侧压下
side yard 侧庭;边院庭院;旁院;院;房边空地
side yard adjoining a street 临街侧院
siding 站线;鱼鳞板;股道【铁】;墙板;披叠板;配线【铁】;围护板;铁路支线;索道滑轨;岔线【铁】;防风雨板;壁板;备用线;板壁

siding board 护墙板
siding for car disinfecting 货车消毒线
siding for exchange wagon 交换车停留线
siding for ga(u)ge changing 过轨线
siding for splitting train 列车解体线
siding ga(u)ge 护墙板量规;护墙板卡规
siding length 股道长度
siding machine 边缘修整机;边坡修整机;修坡机
siding material (房屋的)板壁材料
siding retarder 编组线缓行器
sidings 侧线
siding shake (手工劈成的)墙面板
siding shingle 侧墙板;侧墙盖板;护墙板;外墙墙面保护板;外挂板;外墙覆盖板
siding spacing 股道间距
siding steel 护壁钢
siding steel plate 护壁钢板
siding track for girder-erecting 架梁道岔
siding wall 板墙
siding with section 区间岔线
sidorenkite 碳磷锰钠石
Sidot's blende 合成硫化锌
Sieber number 西伯值
Siebert furnace 西伯特三相电弧炉
siegbahn notation 西格巴恩标志
siege 窑底;砖工脚手台;坑工脚手台;炉底;炉床
siege block 窑底大砖
siege joist 炉底托梁
siegenian(stage) 西根阶【地】
siegenite 块硫镍钴矿
Siegwart 西格瓦特式
Siegwart floor 西格瓦特(式)楼板
Siemens alloy 西门子锌基轴承合金
Siemens dynamometer 西门子功率计
Siemens electrodynamometer 西门子式电功率计
Siemens furnace 西门子炉
Siemens-Martin furnace 西门子—马丁平炉
Siemens-Martin plant 平炉车间
Siemens-Martin process 平炉炼钢法
Siemens-Martin steel 平炉钢;西门子—马丁钢
Siemens Method 西门子法
Siemens open hearth process 平炉炼钢法
Siemens producer 西门子煤气发生炉
Siemens setting checker 西门子式格子砖
Siemens static voltmeter 西门子静电伏特计
Siemens steel 西门子平炉钢
Siemens straight packing 西门子格子体
Siemens unit 西门子电阻单位
sienite 正长岩
sienna 黄土颜料;土黄色;赭色颜料;富铁黄土
sienna brown 土黄棕色
sienna deposit 浓黄土矿床
siennese drab 土黄褐色
sierozem 灰漠钙土;灰壤;灰钙土
Sierozem soil 灰钙土;塞罗泽土壤
sierra 锯齿山脊;山脉
Sierra Club 山岳协会
sierra leone copal 塞拉利昂柯巴脂;产自非洲的(树脂)
sieve 过筛分析;筛子;筛
sieve action 筛孔效应
sieve analysis 筛孔分析;颗粒度分析;筛选;筛析;筛分颗粒度分析;筛分颗粒度分析;筛分分析;筛分法;筛分
sieve analysis curve 筛分(析)曲线
sieve analysis method 筛选法;筛析法
sieve analysis of powdery coal 煤粉筛分试验
sieve aperture adjuster 鱼鳞筛筛孔调节器
sieve area 筛域
sieve bend 弧形筛
sieve bottom 筛底
sieve box area 筛箱的工作面积
sieve brush 筛刷
sieve classification 筛析;筛分分级
sieve cleaner 清筛器;筛式清选机
sieve cleaning mechanism 清筛机构
sieve cloth 筛网布;筛网;筛布
sieve compartment 筛分室
sieve correction 筛分析校正
sieve curve 筛分(析)曲线
sieved aggregate 过筛集料;过筛骨料
sieve deposit 渗流沉积
sieve diameter 筛孔直径;筛径
sieve diaphragm 隔膜
sievedigger 筛式挖掘机

sieved material 过筛材料;筛分材料;筛出物
sieve drum drier 筛网圆筒烘干机
sieved sample 筛分样品
sieve effect 筛孔效应;筛分效应
sieve element 筛(管)分子;筛管成分
sieve field 筛场
sieve fineness 筛孔尺度;筛分细度
sieve float-and-sink analysis report table 筛分浮沉试验报告表
sieve fraction 筛分粒级;筛分粒度率;筛分粒度级
sieve ga(u)ge 筛量规
sieve grating 筛箅
sieve hole size 筛孔尺寸
sieve kneader 带筛捏合机;带筛搅拌机
sieve-like distribution 筛状分布
sieve loss(es) 筛选损失
sieve machine 筛选机;筛分析机
sieve membrane 筛膜
sieve mesh 筛孔;筛眼;筛目;标准网目
sieve mesh number 筛网目数;筛孔号
sieve mesh opening 筛孔
sieve mesh size 筛孔大小
sieve method 逐步淘汰法;筛选法;筛分法;筛法
sieve number 网目;筛孔号;筛号
sieve of Eratosthenes 爱拉托逊斯筛法
sieve of molecular 分子筛
sieve opening 筛眼;筛孔
sieve oversize 筛上产品
sieve pan 筛架
sieve part 筛部
sieve particle-diameter 筛分粒径
sieve pitting 筛状纹孔式
sieve plant 筛分厂
sieve plate 栅板;过滤板;筛板
sieve-plate column 筛选塔;筛板柱;筛板塔
sieve-plate(column)tower 筛板塔
sieve-portion 筛部
sieve problem 筛问题
sieve ragging 筛上产品
sieve ratio 筛分比;筛比
sieve region 筛域区
sieve residue 筛渣;筛余(渣);筛余物;筛下料;筛上剩余物料
sieve-residue content 筛余量
sieve residue log 岩屑录井
sieve residue percentage 筛上百分率
sieve retention 筛滞留
sieve scale 节垢;筛制;筛垢;筛比
sieve series 筛序;分级筛
sieve series standard 筛号标准
sieve set 成套筛
sieve shaker 振动筛分机;摇筛机;摇筛器
sieve shoe 筛架
sieve shovel 筛铲
sieve sifter 筛网;筛面
sieve size 筛孔尺寸;筛号;筛眼孔径;筛眼尺寸;筛孔孔径;筛分粒度
sieve size gradation 筛号分级
sieve size range 筛析粒度范围
sieve sizing 筛分分级
sieve sorbent 筛状吸附剂
sieve sorption pump 分子筛吸附泵
sieve tailings 筛余料
sieve test(ing) 筛分试验;过筛试验;筛析试验;筛孔测定
sieve test result diagram 筛分试验曲线
sieve texture 筛状结构;变嵌晶状
sieve tissue 筛管组织
sieve tray 筛盘;筛板
sieve tray column 筛盘塔
sieve tray extraction tower 筛盘式萃取塔
sieve tray tower 筛盘塔;筛板塔
sieve tube 筛管
sieve tube element 筛管分子
sieve tube member 筛管节
sieve tube segment 筛管节
sieve vessel 筛管
sieving 过筛;筛析;筛分
sieving action 筛分作用
sieving area 筛分面积
sieving capacity 筛分能力
sieving classification 筛分级
sieving data 筛分数据
sieving extractor 筛分分离装置;筛分取出装置

sieving machine 筛选机;筛石机;筛分机
sieving machinery 筛分机械
sieving mechanism 筛选装置
sieving of anomaly 异常筛选
sieving rate 过筛率
sieving specification 筛分规范
sieving test 筛分试验
Sifbronze 西夫青铜;钎焊青铜焊料
siffanto 西范托风
sift 挑选;淘汰;筛分
siftage 筛下物
sifted 筛过的
sifted cement 过筛的水泥
sifted material 过筛材料
sifted plaster 筛过的灰泥
sifted sand 过筛沙
sifter 过滤器;精查者;筛子;筛者;筛分机
sifting 详细审查;过筛;精查;筛选
sifting disk 筛盘
sifting machine 筛分机;筛选机
sifting method 筛法
sifting property 筛选性质
siftings 筛屑
sifting screen 细筛;细分筛
sifting sort 筛选分类法
sifting tea 茶片
sift mineral 软矿物
sift-proof bin 防漏式储[贮]存罐
sift-proof packaging 防撒漏包装
sight 瞄准(器);视力;视界
sight adjustment 瞄准具装定
sight alidade 照准仪;全闭式照准仪;视准仪
sight alignment 瞄准线与目标重合
sight amplifier 瞄准放大器
sight angle 视角
sight angles board 视角板
sight axis 瞄准具轴线;瞄准具光轴;视准轴
sight base distance 瞄准基准距离
sight bill 来取即付的汇票;即期汇票
sight blade 准星片
sight board 照准板
sight bracket 瞄准具卡座;视准尺架
sight cap 观察孔盖
sight carrier 尺垫【测】
sight case 瞄准具盒
sight chamber 观测室
sight check 卡片目视检查;目视检验;目视检查;同位穿孔校验法
sight compass 觇板罗盘
sight control 直视控制;观察检验
sight derrick 安装钻塔
sight distance 能见距离;视距;视界
sight distance at intersection 交叉视距
sight distance for stopping 停车肉眼距离
sight distance of intersection 路口时距
sight draft 见票即付票据;见票即付汇票;即期汇票
sight draft bill of lading 即期付款信用证
sight entry 临时查验报关单
sight extension 瞄准具升降器
sight feed 可视进料
sight feed body 可视进料器体
sight feed glass 透进材料玻璃管
sight feed glass base 侧望镜座
sight feed indicator 明给油滴指示器
sight feed lubricator 注油壶;明给润滑器;可视油润滑器
sight feed nozzle 明进给嘴
sight feed oil cup 滴油杯
sight feed oiler 明给加油器;外视注油器
sight feed oil lubricator 可视润油器;可视加油器
sight feed siphon lubricator 明给虹吸润滑器
sight feed tube 可视进料管
sight flow indicator 液流目视指示器;可视流量指示器
sight for sag 垂度计;垂度仪
sight ga(u)ge 观测计;水位计
sight glass 玻璃水位管;观察窗;看窗;观察孔;玻璃水位计;监视孔;观察镜;观测镜;观测玻璃;看火镜;视镜;窥镜
sight glass for powder material 粉状物料视镜
sight glass for vacuum equipment 真空设备视镜
sight glass shield 视镜护罩
sight glass with covering 带罩视镜
sight glass with nozzle 带颈视镜

sight glass without nozzle 不带颈视镜
sight glass with scraping device 带刮板视镜
sight glass with washing device 带冲洗装置视镜
sight graduation 瞄准具刻度;瞄准具分划
sight gravity feed oiler 重力注油示油器
sight gun 瞄准器
sight head 瞄准用准星;瞄准具头部
sight headrest 瞄准具炉额垫
sight hole 人孔;验火孔;观察孔;窥视孔;窥火孔;看火孔;检查孔;瞄准孔
sight indicator 观测指示器;可视指示器;目测指示器
sighting 照准;观察;观测;瞄准
sighting agent 可视指示剂
sighting anchor 起锚检视
sighting angle 瞄准角
sighting bar 瞄准杆
sighting board 觇板【测】;照准板;测视板
sighting centering 照准点归心
sighting colo(u)rs 标志染料
sighting condition 通视条件
sighting conduit 水位管
sighting cylinder 照准圆筒
sighting device 瞄准设备;照准装置;照准设备
sighting diagram for bombing 瞄准图
sighting disc 觇板【测】;圆形觇板
sighting error 照准误差;瞄准误差
sighting for sag 垂度仪
sighting glass 光学照准镜
sighting harmonization 瞄准具校靶
sighting head 照准头
sighting hole 检查孔
sighting hub 侧杆中心
sighting instrument 瞄准装置
sighting line 视域;瞄准线;视线
sighting line method 瞄直法;瞄线法
sighting mark 照准标;瞄准标志;觇标
sighting marker 指示器
sighting mast 瞄准塔;瞄准杆
sighting offset 瞄准提前量
sighting pendant 瞄准锤
sighting piece 目镜
sighting pipe 水位管
sighting platform 瞄准台
sighting point 照准点;瞄准点;视准点
sighting pole 标杆
sighting potential 目视位势
sighting range 瞄准距离
sighting rod 视标;水准尺;标杆
sighting rule 视准仪;瞄准仪;觇尺
sighting slit 窥视缝
sighting slot 觇隙
sighting station 瞄准站
sighting string 瞄准刻线
sighting survey 目测检验
sighting system 瞄准系统
sighting target 照准标(志);观测目标;视标;觇标;测标
sighting telescope 瞄准(望远)镜
sighting tower 瞄准塔;定位塔
sighting tube 水位管;望筒
sighting-tube ring 四游仪
sighting vane 照准器;觇板【测】
sighting wire 照准丝;照准十字丝
sight leaf 表尺板
sight letter of credit 见票即付信用证;即期信用证
sight level 水准仪
sight level ga(u)ge 观测水准仪
sight level glass 视镜
sight liabilities 不定期债务
sight line 视线;照准线;边界线
sight line coverage 视线范围
sight line velocity 视向速度
sight locating 观察定位法
sight mechanism 瞄准镜装置
sight meter 透光率计
sight microscope 瞄准显微镜
sight mirror 观察镜
sight mount 瞄准具支架
sight obstruction 视线障碍物;碍视物
sight oil feed 开式供油
sight oil ga(u)ge 目测油表;外视油表;外观油表
sight oil glass 外观油表
sight oil indicator 目测油标;视油规;示油规
sight payment letter of credit 见票即付信用证

sight peg 龙门桩
sight photometer method 视测光度法
sight pole 标杆
sight proof device 防视装置
sight proof glass 防视玻璃
sight proof louver 遮挡百叶;防视百叶
sight pump 重力泵;手摇观察泵
sight radar range light 瞄准具雷达距离指示灯
sight rail 横向望板;(挖沟槽时的)龙门板;标尺;照准轨;基度横挡(建筑工地);横标杆;视准轨
sight ramp 表尺座
sight reach 瞄准极限
sight-reading chart 杆位指示牌;模拟系统图
sight reduction 视野归算;测天计算;测天归算
sight reduction table 高度方位表;视野归算表
sight reduction tables for selected stars 恒星高度方位表
sight reticle camera 光环摄影机
sight reticule 照准(十字)丝;瞄准具十字线
sight rod 标杆;水准尺;花杆(测);视距标尺;测杆
sight rule 觇尺;照准仪;照准尺
sight scale 瞄准器刻度
sightseeing 观光;游览
sightseeing area 观光区
sightseeing bus 游览公共汽车;游览车
sightseeing city 游览城市
sightseeing flight counter 游览航班服务台
sightseeing harbo(u)r 游览港
sightseeing resort 游览区;观光胜地
sightseeing road 观光道路;游览路;游览道路
sightseeing ship 观光船
sightseeing tower 瞭望塔;观景塔
sightseer 游客
sight sensitivity 瞄准具灵敏度
sight-shot 视觉
sight size (窗的)视域宽度;(窗的)进光的玻璃口宽度;透光尺寸
sight the anchor 起锚检视
sight tracking line 跟踪瞄准点
sight tree 界标树
sight triangle 交叉口视线三角线;视距三角(交叉口)
sight tube 观察管;瞄准镜筒
sight unit 瞄准件
sight vane 瞄准器;觇板;照准板;瞄准孔;透视孔
sight vane alidade 测斜照准仪
sight well 观察井
sight window 观察孔;观察窗
sigillariotelinite 印印木结构镜质体
sigloite 黄磷铝铁矿
Sigma blade kneader 西格玛叶片式捏合机
Sigma blade mixer 西格玛浆式混合机;曲拐形叶片混合机
sigma factor 西格玛因子
sigmalium alloy 西格玛铝基合金
sigma memory 求和存储器
sigma mixer 弓形混合机;曲拐式搅拌机
sigma phase 西格玛相
sigma recording methanometer 丁烷火焰瓦斯检定器
sigma storage 求和存储器
sigmatron 高能X射线仪
sigma welding 西格玛焊接
sigma zero 零度下海水密度
sigmoid C 形反曲线;反曲;S 形
sigmoidal 反曲的;S 形的
sigmoidal vein S 脉型
sigmoid curve S 形曲线
sigmoid curve theory S 形曲线理论
sigmoid distortion S 形失真
sigmoid growth curve S 形生长线
sign 正负号;征兆;宫;签名;符号;标志;标记
sign a bond 具结
sign a certificate 签署证书
sign a contract 签约;签署合同;签订合同;签订合同
signal 信号;可听声信号;号志
signal acquisition 捕捉信号
signal acquisition unit 信号捕捉装置
signal activity 信号活动
signal adapter 信号转换器
signal ahead sign 前面有信号标志
signal alarm 喇叭;警报信号
signal alert 警报信号
signal ambiguity 信号多重性;信号不定性
signal amplification 信号放大
signal amplifier 信号放大器
signal amplitude 信号振幅
signal amplitude distribution 信号幅度分布
signal analyser 信号分析器
signal analysing and measuring instruments 信号分析与测量仪表
signal analysis 信号分析
signal appendant 信号附属品
signal approach control circuit 信号接近控制电路
signal area 信号区域;机场信号区
signal arm 信号臂
signal aspect 信号形态;信号方式
signal at clear 信号开放
signal at stop 信号关闭
signal attenuation 信号衰减
signal attribute 信号属性
signal autocorrelation 信号自相关
signal averager 信号平均器
signal averaging 信号平均法
signal azimuth display 信号方位显示
signal balancing 信号调零
signal ball 信号球
signal balloon 信号气球
signal bandwidth 信号带宽
signal base 信号机底座
signal battalion 通信营
signal beacon 信标
signal bell 信号钟;信号铃;号铃
signal bias 信号偏移;信号偏压
signal block system 信号闭塞装置
signal board 信号牌
signal bond 信号接合
signal book 信号书
signal box 信号箱;信号所;信号室【铁】;信号楼【铁】;信号房;号志房
signal box for electrical interlocking 电气集中信号楼
signal bracket 信号托架
signal bridge 信号台;信号桥
signal bronze 铜锡合金
signal bumping 信号冲撞
signal buoy 信号浮标
signal cabin 信号室;信号楼;号志房【铁】
signal cabin power source changeover box 信号室电源切换箱
signal cable 信号电缆
signal calibration 信号校正
signal capture 捕获信号
signal card 信号卡
signal carrier 载波
signal-carrier FM recording 信号载波调频记录法
signal-carrier frequency 信号载波频率
signal carrying line 信号载线
signal carry-over 信号延长
signal center[centre] 信号中心
signal central control 信号中央控制
signal central(control)room 信号中央控制室
signal channel 信路;信号通路;信号传输通道;信道
signal channel miscellaneous interference 信号通路的各种干扰
signal channel probe 信道探头
signal check 信号检查
signal chest 信号箱;旗箱
signal circuit controller 信号电路控制器
signal classifier 信号分类机
signal clock 信号钟
signal code 信号(代)码;信号符号;信号电码;码组;通信密码
signal code capacity 信号编码容量
signal-code table 信号代码表
signal coding 信号编码
signal coding equipment 信号编码装置
signal colo(u)r 信号颜色
signal colo(u)ration 标志色
signal combining equipment 信号组合设备
signal command 信号命令
signal communication 信号通信;通信联络
signal communication system 信号通信系统
signal-comparison method 信号比较法
signal compensator 信号补偿器
signal condition 信号状态
signal conditioner 信号调节器
signal conditioning 信号加工;信号调整;信号调节;信号波形修整
signal-conditioning unit 信号调整器
signal constraint 信号限制
signal contrast 信号对比度
signal control 信号控制
signal control circuit 信号控制电路
signal-controlled intersection 有交通信号控制的交叉口
signal control relay 信号控制继电器
signal conversion 信号转换;信号变换
signal conversion equipment 信号转换装置;符号转换装置;信号变换设备
signal converter 信号转换器;信号变换器;符号变换器
signal converter storage tube 信号转换存储管;信号变换存储器
signal coordinates 信号坐标
signal coordination points 联动信号点
signal correction 信号校正
signal counter 信号计数器
signal coupling loop 信号耦合环
signal cover 信号覆盖范围
signal-crosstalk ratio 信串比
signal curbing 信号抑制;信号遏制
signal current 信号电流
signal cut 信号切断
signal cycle 信号周期
signal cycle traffic volume 信号周期交通量
signal data converter 信号数据转换器;信号数据处理变换器
signal data processor 信号资料处理机;信号数据处理器
signal data recorder 信号数据记录器
signal data recorder unit 信号数据记录器
signal deconvolution 信号反褶积
signal degradation 信号降低
signal delay 信号延滞;信号延迟
signal delay time 信号滞留时间;信号延迟时间;信号迟延时间
signal demodulator 信号解调器
signal design 信号设计
signal detection 信号检测
signal detection probability 信号检测概率
signal detection theory 信号检测理论
signal detection threshold 信号检测阈
signal detector 信号检测器
signal device 信号装置
signal diode 信号二极管
signal disc 信号圆盘
signal discrimination 信号鉴别
signal display 信号显示
signal distance 信号间距;代间距
signal distortion 信号失真;信号畸变
signal distortion generator 信号失真发生器
signal distortion rate 信号失真率
signal distribution equipment 信号分配装置
signal distribution panel 信号分配盘
signal distribution unit 信号分配装置
signal distributor 信号分配器
signal drift 信号漂移
signal drop 信号吊牌
signal dropout compensator 信号失落补偿器
signal duration 信号持续时间
signal earth 信号用接地;信号通地
signaled train movement 信号通知的列车调动
signal effect 信号效果
signal element 信号码元;信号单元
signal encrypt system 信号加密系统
signal enhancement 信号增强
signal equipment 信号设备;通信设备
signal erection 造标;建立标志;树立觇标
signal excursion 信号偏移
signal extraction 信号提取
signal face 信号(表示)面
signal facility 信号设备
signal fadeout 信号衰落
signal fading 信号衰落
signal field 信号场
signal field period 信号场周期
signal field strength 信号场强度
signal filtering 信号滤波
signal flag 测旗
signal flagman 信号旗手
signal flare 信号灯

signal flasher 信号闪光灯
signal-flow chart 信号流图
signal flow graph 信号流程图;信号流图
signal fluctuation 信号起伏
signal for day and night 昼夜通用信号
signal for help 呼救信号
signal for shunting front and back 双面调车信号(机)
signal frequency 信号频率
signal frequency amplifier 信号频率放大器
signal frequency range 信号频率范围
signal frequency shift 信号频率偏移
signal frequency side band 信号边频带
signal gain 信号增益
signal generated noise 信号引起的噪声
signal generating means 信号发生法
signal generating source 信号发生源
signal generating station 信号产生站
signal generation 信号发生
signal generator 讯号发生器;信号机;信号发生器;交流信号发电机
signal-generator loop 信号发生器环形天线
signal glass 信号玻璃
signal grid 信号栅极;控制栅极
signal ground 信号用接地;信号地;信号场
signal group 信号群
signal gun 信号枪;信号炮
signal halyard 信号绳;旗绳
signal hammer 信号铃
signal-handling apparatus 信号处理装置
signal-handling equipment 信号处理系统
signal head 信号器
signal head assembly 信号柱;信号机头部装置
signal hole 信号孔
signal horn 信号喇叭;号角
signal hypothesis 信号假说
signal identification 信号识别
signal image 信号图像
signal imitation 信号模拟
signal impulse 信号脉冲
signal indication 信号显示;信号表示
signal indicator 信号指示器
signal initiator 信号引发器
signal injection 信号注入
signal injector 信号注入器
signal input 信号输入
signal input channel 信号输入通道
signal input interrupt 信号输入中断
signal input waveform 输入信号波形
signal installation 信号装置
signal integration 信号集合;信号积累
signal intelligence 通信情报
signal intelligence report 通信情报报告
signal intensity 信号强度
signal interface unit 信号接口装置
signal interpretation 信号译码;信号解释
signal in throat section 咽喉信号机
signal inversion 信号极性变换
signal inverter 信号逆变器;信号极性变换器;信号变换器
signalization 信号作用
signalized intersection 信号控制交叉口;有交通信号的交叉口
signalized point 人工标志(点);标志点
signal kit 信号旗组
signal lag 信号滞后
signal lamp 信号灯;号志灯
signal lamp flicker 信号灯闪烁器
signal lamp on chimney 烟囱障碍灯
signal lamps with adjustable potential 可调压信号灯
signal leading edge 信号前沿
signal(l)ed crossing 有信号管理的交叉口
signal(l)ed speed 用信号通知的速度
signal length 信号长度
signal(l)er 信号员
signal(l)er for stub-end track 尽头信号机
signal letters 信号字;通信呼号
signal level 信号(能)级;信号电平
signal level control 信号电平控制
signal level detecter 信号电平检测器
signal level distribution 信号电平分布
signal level fluctuation 信号电平波动
signal level meter 信号电平表

signal level test film 信号电平试验片
signal lever 信号握柄;信号杆;号志柄
signal light 信号(灯)光;信号灯;回光灯;通信闪光灯;标灯
signal lighting circuit 信号机点灯电路
signal lighting power source 信号机点灯电源
signal limiter 信号限值器
signal line 信号线;信号绳
signal(l)ing 发信号;信令;信号传输;直流振铃
signal(l)ing alarm equipment 信号报警设备
signal(l)ing apparatus 信号机具;信号装置
signal(l)ing at stations 车站信号
signal(l)ing attention 发引起注意信号
signal(l)ing by hand flags or arms 手旗或手臂通信
signal(l)ing capacitors 发信电容器
signal(l)ing channel 传信信道;信号通路
signal(l)ing condition 信号状况
signal(l)ing contact 信号触点
signal(l)ing data link 信令数据链路
signal(l)ing distance 信号通信距离
signal(l)ing engineering 信号工程(学)
signal(l)ing equipment 传信设备
signal(l)ing flag 信号旗
signal(l)ing frequency 铃流频率;振铃信号频率
signal(ling) installation 信号装置
signal(l)ing key of pull type 拉引式信号开关
signal(l)ing key of push type 按压式信号开关
signal(l)ing lamp 通信闪光灯
signal(l)ing line 信号线路;信号通道
signal(l)ing link 信令链路
signal(l)ing method 通信方式
signal(l)ing network function 信令网功能
signal(l)ing-nework-management signal 信号网络管理信号
signal(l)ing panel 信号布板
signal(l)ing point 信令点
signal(l)ing procedure 信号发送程序
signal(l)ing rate 信号发送速度
signal(l)ing rocket 烟火信号弹
signal(l)ing route 信令路由
signal(l)ing speed 信号发送速度
signal(l)ing strategy 发信号策略
signal(l)ing system 信令系统;信号系统
signal(l)ing terminal 信令终端
signal(l)ing terminal equipment 信令终端设备
signal(l)ing test 信令传送测试
signal(l)ing time slot 信令时隙
signal(l)ing tone 信令音
signal(l)ing virtual channel 信令虚通路
signal location 信号点
signal lock 号志键
signal locker 信号箱;旗箱
signal make-up 信号组成
signalman 信号员;线路所值班员
signal man's cabin 信号房
signal mark 信号标志
signal mast 信号桅杆;信号桅;信号杆
signal measuring apparatus 信号测量器
signal message 信号消息
signal meter 信号指示器
signal mixer 信号混合器
signal mixer unit 信号混合装置
signal mixing 信号混合
signal modulation 信号调制
signal modulation component 信号调制成分
signal modulator 信号调制器
signal monitor 信号控制器;信号检查器;信号监视器
signal monitoring 信号监视;信号波形监视
signal multiplexing unit 信号多路传输装置
signal multiplication 信号放大
signal multiplier 信号倍增器
signal neon lamp 信号霓虹灯
signal node 信号结
signal noise 信号噪声
signal noise ratio 信噪比;信号噪声比
signal normalization 信号标准化
signal of communication 通信信号
signal of distress 海上遇险信号;遇险信号;遇难信号
signal office 通信处
signal officer 通信官
signal offset 信号时差
signal of giving way 让路信号

signal of prestart 启动预告信号
signal oil 燃灯油;照明油;信号油
signal operation 信号控制;信号操纵;信号操作
signal operation battalion 通信作业营
signal operation instruction 通信作业须知
signal operation room 通信作业室
signal operations center 通信作业中心
signal operator 信号员
signal oscillator 信号振荡器
signal out of use sign 信号无效标
signal output 信号输出
signal output current 信号输出电流
signal output uniformity 信号输出均匀性
signal output waveform 输出信号波形
signal overflow 信号溢出
signal paint 标记漆
signal panel 信号盘;信号板
signal parameter 信号参量
signal passage 信号通路
signal path 信号通道
signal pattern 信号图像
signal pedal 信号踏板
signal pennant 信号旗
signal period 信号周期
signal phase 信号相(位)
signal phase conversion 信号相变换
signal phase fluctuation 信号相位起伏
signal phase shift 信号相移
signal phase split 信号相位分配
signal phasing 信号相(位)
signal pipe 信号管
signal pipe line 信号管线
signal pistol 信号枪
signal plate 信号板;摄像管输出板
signal polarity 信号极性
signal polarity switch 信号极性转换开关
signal post 信号柱;信号杆
signal power 信号功率
signal power source 信号电源
signal preamplifier 前置信号放大器
signal preemption 交通标志优先权
signal printed board 信号印刷电路板
signal probability 信号概率
signal procedure 通信程序
signal process(ing) 信号处理
signal processing antenna 信号处理天线
signal processing system 信号处理系统
signal processing unit 信号处理单元
signal processor 信号处理机
signal production 信号发生
signal projector 信号放映机
signal propagation 信号传输
signal pulse interval 信号脉冲间隙
signal-pulse repetition frequency 信号脉冲重复频率
signal quality detector 信号质量检测
signal quantization 信号量化
signal race 信号追赶
signal range 信号区域
signal ratio 信号比
signal reading 信号读出
signal receiving circuit 信号接收电路
signal-receiving electrode 信号输入电极;信号接收电极
signal recognition 信号识别
signal reconstruction 信号恢复
signal record book 通信日记簿
signal recorder 信号记录器
signal-recording telegraphy 信号记录电报术
signal rectifier 信号整流器
signal redundancy 信号冗余度
signal reflector 信号反射镜
signal reflex 信号反射
signal regeneration 信号再生
signal register power source 信号寄存器电源
signal regulations 信号发射规则
signal relay 信号接力;信号继电器
signal relay point 信号中站
signal repeater 信号中继器;信号复示器
signal replacement training center 通信兵补充人员训练中心
signal reporting codes 信号质量代码
signal representation 信号表示法
signal reshaping 信号整形;信号再生

signal resistance 信号源电阻
signal restoring 信号复原
signal reverberation 信号混响
signal rod 信号杆
signal rotor 信号转子
signal roundtrip time delay 信号往返时间滞后
signal routing 信号通路
signal sampling iteration method 信号取样迭代法
signal sash fastener 高窗扇固定器
signal scan 信号扫描
signal scanner 信号扫描器
signal search light 信号探照灯
signal-seeking receiver 信号搜索接收机
signal selector 信号选择器
signal selector relay 信号选择继电器
signal sensitivity 信号灵敏度
signal separation 信号分离
signal-separation filter 信号分离带通滤波器
signal separator 信号分离器
signal setting 信号配时
signal shaping 信号整形
signal-shaping network 信号成形网络;信号波形校正网络
signal-shield ground 信号屏蔽接地技术
signal shifter 信号转换器
signal shutdown switch 信号停止开关
signal slot 信号选择器;信号选别器
signal source 信号源
signal space 信号空间
signal spacing 信号区间
signal speed 信号速度
signal splitter 信号分配器
signal stabilization amplifier 信号稳定放大器
signal staff 信号杆
signal standardization 信号整形;信号标准化
signal static ratio 信号强度与天电强度之比
signal station 信号站;信号台;信号旗站
signal stay 信号索;横牵索;桅间索
signal stimulus 信号刺激
signal storage 信号存储器
signal street intersection 有信号的道路交叉口
signal strength 信号强度
signal strength adjustment 信号强度调整
signal-strength meter 信号强度计
signal strength scale 信号强度等级
signal strip 信号布条
signal supervisory control 信号管理控制(器)
signal support 信号机头支持装置
signal synchronizing system 信号同步系统
signal system 信号制;信号系统
signal systems control center 通信系统控制中心
signal tail 信号尾部
signal teaching equipment 信号教学设备
signal terminal 信号终端
signal terminal area radar 信号终端区域雷达
sign-alternating 正负号交变;变符
sign-alternating focusing 变符聚焦
signal tester 信号试验器
signal-testing apparatus 信号测试设备
signal theory 信号论
signal threshold 信号阈;信号可辨阀;信号可辨度界限
signal timing 信号同步;信号定时;放行信号有效时间比
signal timing diagram 信号配时图
signal timing dial 信号配时;信号定时键盘
signal-to-deviation ratio 信号偏差比
signal-to-mask ratio 信噪比
signal-to-noise and distortion ratio 信号对噪声和失真比
signal-to-noise characteristic 信噪特征
signal-to-noise for playback 放音信噪比
signal-to-noise improvement factor 噪声改善系数
signal-to-noise performance 信噪比
signal-to-noise ratio 信噪比;信号噪声比;信号干扰比;信杂比
signal-to-noise ratio for record-playback 录放声信噪比
signal-to-noise ratio for record-playback channel 录还信噪比
signal-to-quantizing noise ratio 信号与量化噪声比
signal tower 信号塔;信号楼【铁】;号志塔
signal tower in depot 车辆段的通号楼
signal tracer 信号式线路故障寻找器;信号示踪器;信号故障检寻器
signal tracing 信号追踪
signal-tracing instrument 信号跟踪仪器
signal train 信号序列
signal transducer 信号传感器
signal transfer function 信号传递函数
signal transferring lag 信号传送滞后;信号传送延迟
signal transformation 信号转换;信号变换
signal transformer 信号变压器;信号变换器
signal transition 信号跃迁
signal transit time 信号渡越时间
signal translation 信号变换
signal transmission 信号传送;信号传输
signal transmission form 信号传输方式
signal transmitter 信号发送器
signal truck 通信载重汽车
signal-tuned amplifier 信号调谐放大器
signal value 信号值
signal velocity 信号(传播)速度
signal voltage 信号电压
signal voltage generator 信号电压发生器
signal wave 信号波
signal-wave envelope 信号波包络
signal waveform 信号波形
signal waveform elements 信号波形成分
signal winding 信号绕组;控制绕组
signal zone adapter module 信号模块
sign and currency symbol character 符号和货币符
sign-and-magnitude code 符号幅度码
sign-and-magnitude notation 符号数值表示法
sign and warning indication 符号和警告标志
sign an undertaking 具结
sign area 招牌可见范围
sign a referendum contract 草签合同
sign articles 签字文件;签字条款
signatory 签字人
signatory authorities 签证机关
signatory of a contract 合同签字方
signatron 数字式液晶显示管
signature 印鉴;签字;签署;签名;船磁特性;船磁记录图;标记图
signature analysis 信号特征分析
signature analysis technique 特征分析技术
signature correlation of amplitude attenuation 振幅衰减特点对比
signature letter 贴码
signature line 贴码线
signature loan 无担保贷款
signature log 特性测井
signature mark 贴码
signature method 特征法
signature of pile 桩的特征图像
signature payment 签字费
signature symbol 贴码
sign awning 标志遮缝
sign bit seismograph 符号位地震仪
sign board 路标;标志牌;标牌;招牌
signboard varnish 招牌(用)清漆
sign border 标志镶边
sign change 符号改变
sign changer 正负变器;符号交换器;符号变换器
sign-change test 符号变更检验
sign changing amplifier 符号变换放大器
sign changing unit 符号交换器
sign character 符号字符
sign check 符号校验;符号检定;符号检查
sign check indicator 符号校验指示器;符号检验位;符号检查(指示)器
sign code 符号代码
sign comparator 符号比较器
sign compare flip-flop 符号比较触发器
sign condition 符号状态;符号条件
sign contract 签署合同
sign control flip-flop 符号(控制)触发器
sign-controlled circuit 符号控制电路
sign control symbol 符号控制符
sign convention 符号规定;符号惯例
sign converter 符号交换器
sign digit 符号数(字);符号数位;符号标记
signed 已签署的
signed and sealed 签名密封;签名盖章
signed and sealed contract 签字蜡封合同
signed binary 有符号二进制数
signed binary arithmetic 带符号二进制运算
signed bit 符号位
signed constant 有符号常数;带符号常数
signed declaration 签署声明
signed digit 符号位
signed digraph 符号有向图
signed field 带符号字段;带符号信息组
signed graph 带符号图
signed invoice 签署发票
signed magnitude arithmetic 绝对值运算;加减法运算;符号量值运算;符号化数值运算
signed measure 符号测度
signed multiplicand 带符号的被乘数
signed packed decimal 带符号压缩十进制
signed rank 符号秩
signed rank sum test 符号秩和检验
sign field 符号字段;符号区;符号段
sign flag 符号标记
sign flip-flop 符号触发器
sign forbidding 禁止标志
sign for substitute 代签
sign for uphill traffic lanes 登坡车道标志
sign for water pipe 水道标【给】
sign frame 符号框架
sign frontage 门面长度(商店)
sign gantry (多层平面车道上的)信号架
significance 显著性
significance exception 有效位异常
significance level 显著性水准;显著性水平
significance limit 显著性界限
significance of correlation coefficient 相关系数显著性
significance of difference 差异显著性
significance of slopes 边坡的重要性
significance probability 显著性概率
significance ratio 显著性比率
significance test 显著性试验;显著性检验;显著性测定
significant amount 巨数
significant area 有效面积
significant behavio(u)r 重要性态
significant bit 有效比特
significant condition 有效状态
significant conditions of modulation 调制的有效状态
significant correlation 显著相关
significant damage 严重破坏
significant deficiency 重大缺陷
significant degree 有效程度
significant depth 有效深度
significant difference 显著(性)差异;显著差别
significant digit 有效数字;有效数位
significant digit arithmetic 有效(数)位运算
significant digit code 有效数字码
significant earthquake 重要地震;重大地震
significant error 显著误差;有效位误差
significant figure 有效数字
significant graduation 有效分度
significant height 有效高度
significant hydraulic parameter 特征水力参数
significant instant 有效瞬间
significant interval 有效间隔
significant level 显著(性)水平;有效电平
significant level of correlation 相关显著水平
significant link 有效类目
significant notation 有效标记
significant point 显著点;有效点
significant pollution risk 值得考虑的污染风险
significant shipform for dock and harbo(u)r 设计标准船型
significant space 有效距离
significant surface 有效面
significant test 显著性测验
significant toxic effect 持久毒性效应
significant turn-up 有效爬高值【水文】
significant value 有效值
significant volume 有效体积
significant wave 有效波;主效波
significant wave height 有效波高;主效波高
significant wave length 有效波长
significant wave period 有效波周期;主效波周;指示波浪周期
significant wave prediction chart 有效波浪推算图
significant weather 显著天气;重要天气
signify 符号化

sign illumination system 标志照明系统;标志照明设施
sign in 签收
sign indicator 符号标志
signing 打信号;加符号;加路标;立标牌;签字
signing authority 签字权
signing ceremony 签字仪式
signing signal 蜂鸣信号
sign inspection 符号检查
signinum opus (罗马沟渠内的)灰泥或水泥抹面
sign-journeyman 招牌上装电器的人
signless integer 正整数;无符号(整)数
sign magnitude 符号数值
sign magnitude arithmetic 符号数值运算
sign magnitude code 符号数值码
sign magnitude representation 符号数值表示法
sign-master 招牌上电器设备设计及安装技师
sign mo(u)ld 符号模型
sign of division 除号
sign of equality 等号
sign of erasure 涂改痕迹
sign of evolution 根号
sign off 解聘
sign-off power 审批权;停止权
sign of inequality 不等号
sign of integration 积分号
sign of manufacturer 制造厂名代号
sign of multiplication 乘号
sign of operation 运算记号;运算符号
sign of proportion 比例号;比例符号
sign of ratio 比号
sign of root 根号
sign of rubber stick 卷筒标记
sign of subtraction 减号
sign of summation 和号;累加号
signomial 符号式
signomial geometric programming 符号几何规划
sign-on 开始指令
sign-on verification security 开始证实安全性;符号开始安全校验
sign paddle 手动标志板
sign paint 广告牌漆
sign pattern 符号模式;符号类型
sign pen 信号笔
sign 'per pro' 以代理人资格签署
sign picture character 符号图像字符
sign position 符号位置
sign post 竖标示牌;标(志)杆;标语牌;指向柱;路标;标志柱
signpost and odometer 信号标杆加路码表法
sign posting 立标志
sign pulse 符号脉冲
sign register 信号寄存器;符号寄存器
sign reversal 变号
sign reverser 符号变换器;反演器
signs and symptoms 征象
signs of failure 断裂标志
signs of wear 磨损标志
signs of zodiac 黄道十二宫(符号)
sign statistic 符号统计量
sign structure 标牌支撑构筑物
sign symbol 正负符号
sign test 符号检验;符号检查;符号测试
sign treaty 条约签字
sign truss 标志桁架
signum 正负号函数
signunum opus 沟渠内灰泥灰水泥抹面(古罗马)
sign up 签(合同)
sign variation 标志变动度
signwriting 广告字体;标志字体;标志书写
Sigur brick 硅藻土质隔热砖
sig water 稀碱液
Sikes hydrometer 赛克斯比重计
sikhara 塔状寺顶(印度庙宇建筑中);塔状抹面(印度庙宇建筑中)
sikussak 峡湾里的多年冰(一部分由雪组成)
silafont 硅铝合金
silage 青饲料
silage after math 再生草
silage crop 青饲作物
silage cutter 切草机
silage fermantation 青饲料发酵
Silanca 塞朗克铜锑合金
silandrone 硅雄酮

silane 硅烷
silanediol 硅烷二醇
silanes 硅烷(类)
silanion 硅负离子
silanization 硅烷化
silanizing agent 硅烷化剂
silanol 硅烷醇
silantriol 硅烷三醇
silastic 硅橡胶(密封物)
silastomer 硅塑料
silatranes 硅灭鼠
silazane 硅氮烷;硅氨烷
Silcaz alloy 硅钙钛锌硼合金
silchrome 硅铬合金钢;铬硅耐热钢
silchrome steel 硅铬钢
silcrete 硅质壳层;硅结砾岩
Silcurdur 耐蚀硅铜合金
Silel cast iron 硅铸铁
silence cab 隔声室
silence cabinet 隔声室
silenced camera 低噪声摄影机
silence of fog signal 雾号间歇
silence period 静寂时间
silence pit 消声坑
silencer 静噪器;消声器;静噪器
silencer-boiler 消声废气锅炉;废气锅炉消声器
silencer chamber 消声器腔;减声室
silencer-filter 消声滤气器
silencer-filter type air intake 消声滤气器型空气进口
silencer mounting 消声器
silencer section 消声段
silencer support 消声器支架
silence signal 停机信号
silencing 消声;静噪
silencing device 消声装置
silencing end 消声器出口
silencing equipment 消音设备
silencing of noise 消声;消噪
silencing pot 消声筒
silencing unit 消声装置
silent arc 静弧
silent block 橡胶金属铰节;隔声装置
silent-block mounting 无声铰链架置
silent blower 无噪声通风机
silent blow off valve 无声放气阀
silent chain 无噪声电路;无声链
silent chain conveyer 无声链式运输器
silent chain drive 无声链传动
silent chain sprocket hob 无声链轮滚刀
silent check valve 消声防逆阀
silent circuit 无噪电路
silent cistern 消声储[贮]水器
silent cop 十字路口的交通指挥装置
silent cutter 无噪声绞肉机
silent discharge 无声放电;无声波放电
silent door hook 门后钩
silent earthquake 稳定期地震
silent electric(al) discharge 无声波放电
silent element 完全不活动的元素
silent engagement 无声啮合
silent escapement 无声擒纵机构
silent falling ball tap 消声形龙头
silent fan 无(噪)声风扇
silent feed 消声进给
silent film loop 无声循环式幻灯片
silent-flame gas burner 无声煤气炉;无声煤气灯
silent gear 无声齿轮;塑料齿轮
silent gear change 无声变速
silent map 填充图
silent mesh 无声啮合
silent-mesh gearbox 无声换挡变速器
silent-mesh gearset 无声啮合轮组
silent-mesh transmission 无声啮合变速器
silent motor 无噪声电动机
silent operation 无声运转
silent paint 隔声油漆
silent partner 匿名合作伙伴;隐名合伙人;匿名股东;秘密合伙人
silent pawl 无声掣子
silent period 静止时间;寂静时间
silent pile-driver 无声打桩机
silent pile driving 无声打桩法
silent pile driving system 无声打桩系统;消声打桩设备

silent piling 静压桩(工程)
silent point 零拍点;静点;无感点;特征点
silent preignition 无爆震敲击声的早燃
silent ratchet wheel 无声棘轮
silent riveter 无声铆接机
silent running 减声运行;平稳运行;无声传动;无声运行;无声航行
silent-running engine 无声发动机
silent salesman 无声售货员
silent satellite 无信号卫星
silent service 潜艇部队
silent-sound 超音频的
silent speed 无声工作速度
silent stock support 无声料托
silent stock tube 塑料管
silent switch 静噪开关;去噪开关
silent technicolo(u)r loop 无声彩色循环幻灯片
silent third and fourth gearset 无声三速和四速动齿轮组
silent third gearbox 减声三速变速箱;无声三速传动齿轮箱
silent third speed 无声的三挡速度
silent tire 无声轮胎
silent tuning 无噪调谐
silent zone 影区;静区
silex 燧石;石英粉;石英玻璃
silex glass 石英玻璃
silexite 英石岩;硅石岩;燧石岩
silex lining 硅质内衬
Silflex process 西尔弗莱克斯镀锌法
Silfos 西尔福斯铜银合金
Silfram 铬镍铁耐热合金
Silfrax 碳化硅高级耐火材料
silhouette 廓影;侧面图像;轮廓;侧影(像)
silhouette effect 轮廓线效果;侧影效果;背景效果
silhouette photography 剪影照片
silhouette picture 廓影;廓影;剪影
silhouette target 影像目标
silhouetting underwater detecting system 廓影法水下探测系统
silhouettograph 廓影照片
silhydrite 水硅石
silica 氧化硅;硅质;硅氧;硅石;二氧化硅
silica adjuvants 硅佐剂
silica aerogel 硅土气凝胶
silica-alumina catalyst 硅铝催化剂
silica alumina gel 硅铝凝胶
silica-alumina refractory brick 硅铝耐火砖
silica alumina weathering crust 硅铝型风化壳
silica and alumina content 硅氧铝氧总含量
silica-base catalyst 硅基催化剂
silica-based cation exchanger 硅基极阳离子交换剂
silica-bearing thermal water 含硅热水
silica black 硅酸黑
silica block 硅酸盐砌块;硅砖
silica boat 石英舟
silica breccias 二氧化硅角砾岩
silica brick 硅酸盐砖;硅砖
silica-brick checkers regenerator 硅砖格子蓄热室
silica cement 硅土水泥砂浆;硅水泥;硅石水泥;胶泥;二氧化硅胶结物
silica cladded fibre 石英包层光纤
silica coating 硅石敷层
silica colloid 硅胶体
silica concentration 二氧化硅浓度
silica crucible 石英坩埚
silica crust 硅藻壳
silica deposit 硅酸沉积;二氧化硅沉积
silica dish 石英皿
silica drop 硅滴
silica dust 硅粉;石英粉;硅石粉尘;硅尘;石英粉尘
silica-estimated temperature 二氧化硅法估算的温度
silica-ferric sub-formation 硅铁质亚建造
silica fibre 硅石纤维;硅石纤维;石英光纤
silica-filled epoxy resin 含硅环氧树脂
silica fine sand 石英细砂;硅质细砂
silica fire clay 硅质耐火粘土
silica flour 硅石粉;石英粉
silica flour wash 石英粉涂料
silica fouling 二氧化硅污染
silica fume 硅灰;硅粉
silica fume concrete 硅尘混凝土

silica gel 矽胶;氧化硅胶;硅胶
silica gel column chromatography 硅胶柱
silica-gel desiccant 硅胶干燥剂
silica-gel desiccator 硅胶干燥器
silica-gel drier 硅胶干燥剂
silica-gel filler 硅胶填充剂;硅补强剂;白炭黑
silica-gel grease 硅胶填充润滑脂
silica-gel leak detector 硅胶探漏器
silica-gel particle size 硅胶粒径
silica-gel percolation 硅胶渗透
silica-gel plate 硅胶板
silica-gel sphere 硅胶球
silica-gel system 用硅胶干燥冷却空气法;硅胶冷气系统
silica-gel thin-layer chromatography 硅胶薄层色谱法
silica geothermometer 二氧化硅地热温标
silica glass 石英玻璃
silica glass fiber 二氧化硅玻璃纤维
silica glass-vitreous silica 硅石玻璃
silica graded fibre 石英渐变型光纤
silica hydrated 白炭黑
silica lamp 石英水银灯
silica layer 硅氧层
silica-lime cement 石英石灰水泥
silicalite 硅质岩
silic-alum rate 硅铝率
silicam 二亚氨基硅
silica-manganese sub-formation 硅锰质亚建造
silica material 石英材料;硅质材料
silica microsphere 硅微球体
silica mineral 硅氧矿物
silica modulus 硅氧系数;硅酸比
silica mortar 硅质砂浆
silica nanoparticles stable dispersion solution 硅纳米粒子稳定分散液
silicane 甲硅烷;四烃基硅
silica particle 硅石微粒;二氧化硅微粒
silica pigment 硅补强剂;白炭黑
silica powder 硅石粉
silica ratio 硅酸率;硅酸比;(水泥的)硅率
silica refractory 硅质耐火材料;硅土耐火材料
silica removal 除去二氧化硅;除硅
silicarenite 石英砂岩
silica-rich thermal water 富硅热水
silica-rich zeolite 高硅沸石
silica rubber 硅氧橡胶
silica rundown 硅涕
silica sand 硅酸砂;硅砂;石英砂
silica sand fines 石英细屑
silica sandstone 硅质砂岩
silica screting organism 泌硅生物
silica scum 石英浮渣
silica-sesquioxide ratio 硅铝铁氧化物分子比;硅铝铁率
silica sheet 硅氧片层
silica silicarenite 石英砂岩
silica sinter 硅华
silica soil 硅土
silica sol 硅酸溶胶;硅溶胶
silica sol binder 硅溶胶黏结剂
silica spring 硅泉
silica steel 硅钢
silica-supersaturated thermal water 二氧化硅含量过饱和热水
silicate 矽酸盐;硅酸盐
silicate binder 硅酸盐黏结剂
silicate blast-furnace cement 高炉矿渣硅酸盐水泥
silicate bond 硅土烧结;硅酸酯黏合剂;硅酸盐结合剂
silicate bonded grinding wheel 硅酸酯黏结砂轮
silicate-bond wheel 硅酸盐胶结磨轮
silicate brick 硅酸盐砖
silicate cement 硅酸固粉;硅酸盐水泥;硅酸盐胶
silicate cement filling 硅黏固粉充填
silicate cement plugger 硅粘固粉充填器
silicate cement spatula 硅酸固粉调刀
silicate ceramics 硅酸盐陶瓷
silicate coat 硅酸盐涂层
silicate concrete 硅酸盐混凝土
silicate cotton 熔渣纤维;矿渣棉;矿棉
silicate cotton block 矿棉砌块
silicate cotton covering cord 矿棉包覆绳索
silicate cotton felt 矿棉毡

silicate cotton insulation 矿棉保温;矿棉绝热
silicate crown 硅酸盐冕玻璃
silicated 硅酸盐化的
silicated macadam 硅化碎石(路)
silicated road 硅化碎石路(面);硅化路
silicated road surfacing 硅化道路面层
silicated water-bound surface 硅化水结碎石面层;硅化水结碎石路面
silicate ester 硅酸盐酯
silicate expansive cement 硅酸盐膨胀水泥
silicate facies iron formation 硅酸盐相铁建造【地】
silicate filler 硅酸盐填料
silicate flintglass 硅酸盐燧石玻璃
silicate flux 硅酸盐焊剂
silicate-forming matter 硅酸盐形成物质
silicate glass 硅酸盐玻璃
silicate grinding wheel 硅酸盐胶结砂轮
silicate groundwater 硅酸盐类地下水
silicate hydrate 水化硅酸盐
silicate impregnated asbestos 硅酸盐浸渍石棉
silicate inclusion 硅酸盐夹杂物
silicate industry 硅酸盐工业
silicate material 硅酸盐材料
silicate melt 硅酸盐熔化物
silicate mineral 硅酸盐矿物
silicate mineral spring water 硅酸矿泉水
silicate of alumina 硅酸铝
silicate of alumin(i)um 铝硅酸盐
silicate of carbon 碳化硅
silicate of iron 铁硅酸盐
silicate of lime 硅酸钙
silicate of magnesia 硅酸镁
silicate of magnesium 硅酸镁;镁硅酸盐
silicate of potash 硅酸钾
silicate of soda 硅酸钠
silicate of zinc 硅酸锌
silicate ore 硅酸盐矿石
silicate oxyapatite laser 硅酸氧磷灰石激光器
silicate paint 硅酸盐涂料;硅酸盐漆;硅胶漆;水玻璃涂料
silicate paint coat 硅酸盐涂层
silicate pneumconiosis 矽酸盐肺
silicate process 水玻璃二氧化碳硬化砂法
silicate product 硅酸盐制品
silicate rock 硅酸盐岩(石)
silicates 硅酸盐类
silicate scale 硅垢
silicate slag 硅酸盐矿渣
silicate slag cement 硅酸盐矿渣水泥
silicate solution 硅酸盐溶液
silica test 硅含量测定
silicate system 硅酸盐系统
silicate tetrahedron 硅氧四面体
silicate wheel 硅酸盐砂轮
silicate wool 矿棉
silication 硅酸盐(化)作用;硅化(作用)
silicatization 硅化作用;灌注硅酸钙
silicatization process 硅化法
silicatosis 矽酸盐肺;硅酸盐沉着病;硅酸盐肺
silica tube 石英管
silica valve 石英真空管
silica ware 石英器皿
silica wash 硅涕
silica white 白炭黑
silica window 石英玻璃观察孔
silica wool 石英棉
silica-zirconia catalyst 硅-氧化锆催化剂
silicea 硅剂
siliceous 硅质的;含硅的
siliceous aggregate 硅质集料;硅质骨料;砂岩骨料
siliceous arenite 硅质砂屑岩
siliceous-bearing mudstone 含硅质泥岩
siliceous breccia 硅质角砾岩
siliceous brick 硅质砖;含硅砖
siliceous calamine 含硅的冰锌矿;含硅的菱锌矿
siliceous-calcareous phosphorite ore 硅钙质磷块岩
siliceous cement 硅质水泥;硅质胶结物;硅质胶结料
siliceous clay 硅质黏土;硅质黏粒;硅土
siliceous concrete aggregate 硅质混凝土骨料;硅质混凝土集料
siliceous concretion 硅质结核
siliceous conglomerate 硅质砾岩

siliceous deposit 硅质沉积(物);硅沉积
siliceous dust 硅质粉尘;硅尘
siliceous earth 硅藻土;硅藻土;硅土
siliceous earthenware 半硅质陶器
siliceous environment 硅质环境
siliceous facies 硅质岩相
siliceous fire clay 硅质耐火黏土
siliceous fireclay brick 硅质(黏土)耐火砖;半酸性耐火砖;半硅砖
siliceous glaze 高硅质釉
siliceous hornfels 硅质角岩
siliceous kaolin 硅质高岭土
siliceous laterite 硅质红土
siliceous lime 硅质石灰;含硅石灰
siliceous limestone 含石英的石灰石;硅质石灰岩;硅质灰岩;硅化灰岩
siliceous loam 硅质壤土;硅质垆坶
siliceous magnesian limestone 白云石化硅质石灰石
siliceous manganese steel 硅锰钢
siliceous marl 硅藻土
siliceous material 硅质材料
siliceous materials deposit 硅质原料矿床
siliceous matrix 含硅的胶结料;含硅的杂矿石
siliceous melt 硅酸盐熔化物
siliceous mortar 硅质泥浆
siliceous mud 硅质泥
siliceous mudstone 硅质泥岩
siliceous nodule limestone 硅质结核灰岩
siliceous oolite gravel 含硅的鲕石卵石
siliceous ooze 硅质软泥
siliceous phosphorite ore 硅质磷块岩矿石
siliceous phyllite 硅质千枚岩
siliceous quartz sandstone 硅质石英砂岩
siliceous refractory 硅质耐火材料
siliceous refractory product 硅质耐火材料制品
siliceous reinforcing agent 硅补强剂
siliceous reservoir 硅质储集层
siliceous rock 燧石板岩;硅质岩
siliceous rock deposit 硅质岩矿床
siliceous sand 硅质砂;硅砂;含硅砂;纯石英砂
siliceous sandstone 玻璃砂;硅质砂岩
siliceous schist 硅质片岩
siliceous sediment 硅质沉积(物)
siliceous shale 硅质页岩
siliceous silt mud 硅质淤泥
siliceous sinter 硅渣;硅华
siliceous skeleton 硅结构
siliceous skin 硅皮
siliceous slag 硅土矿渣
siliceous slate 硅质板岩
siliceous soil 硅质土壤
siliceous sponge 硅质海绵;硅藻海绵
siliceous streaked limestone 硅质条带灰岩
siliceous tufa 硅华
siliceous tuffite 硅质沉凝灰岩
siliceous volcanic formation 硅质火山建造
siliceous volcanic rock 硅质火山岩
silichrome steel 硅铬钢
silicic 硅石的
silicic acid 硅酸
silicic acid anhydride 硅酸酐
silicic alteration 硅质蚀变
siliciclastic 硅质碎屑的
silicicole 砂生植物;砂生植物;沙生植物
silicic rock 硅质岩
silicic volcanic rock 硅质火山岩
silicide 硅化物
silicide ceramics 硅化物陶瓷
silicide coating 硅化物涂料;硅化物敷层
silicide dye 含硅染料
silicide of carbon 碳化硅
silicide refractory 硅化物耐火材料
silicide resistor 硅化物电阻器
silicificated surface 硅化路面
silicification 矽化作用;硅化(加固)
silicification grouting 硅化灌浆
silicification method 硅化法
silicification zone 硅化带
silicified 硅化的
silicified carbonate rock 硅化碳酸盐岩
silicified igneous rock 硅化状火成岩
silicified rock 硅化岩(石)
silicified tectonite 硅化构造岩

silicified wood 硅化木
silicify 硅化
silicilith 硅质岩;硅质生物岩
silicious dust 矽尘
silicious manganese ore 高硅锰砂
silicious sand stone 玻璃砂
silicium dust 含矽粉尘
silicium steel 硅钢
silicoacetic acid 硅乙酸
silicoaluminate 硅铝化合物
silicoblast 硅胞
silico butane 丁硅烷
silicocalcium 硅钙合金
silico-calciummagnesium alloy 硅钙镁合金
silicochloroform 硅氯仿
silicochromium 硅铬合金;硅铬
silicoethane 乙硅烷
silicoferrite 硅铁
silicofluoric acid 硅氟酸
silicofluoride 硅氟化物
silicofluoroform 三氟硅烷
silicoformer 硅变压整流器
silicoglaserite 硅钙石
silicohydride 硅烷
silicoide 硅质矿物
silicolites 硅质岩类
silicolous 硅生的
silicomagnesiofluorite 氟硅镁石
silicomanganese 硅锰合金
silicomanganese alloy steel 硅锰合金钢
silico-manganese bronze 硅锰青铜
silicomanganese slag cast stone 硅锰渣铸石
silicomangan(ese)steel 硅锰钢
silicomethane 硅甲烷;甲硅烷
silicomolybdic acid 硅钼酸
silico-molybdoheteropoly acid 硅钼杂多酸
silicon acid 硅酸
silicon activation 硅活化核反应
silicon alkoxide 硅酸烷基酯;烷氧基硅烷
silicon alkoxide coating 硅酸烷基酯涂料;烷氧基硅烷涂料
silicon alkyl 硅烷
silicon-alumin(i)um alloy 硅铝合金
silicon analyzer 硅含量分析仪
silicon avalanche light source 硅雪崩光源
silicon-avalanche photodiode detector 硅雪崩光电二极管检测器
silicon based inoculant 硅基孕育剂
silicon battery array 硅电池方阵
silicon bilateral switch 硅双向开关
silicon blow 脱硅吹炼
silicon blue cell 硅蓝光电池
silicon boride 硼化硅
silicon brass 硅黄铜
silicon brick 硅砖
silicon briquette 硅块
silicon bromide 溴化硅
silicon bronze 硅青铜;坚铜
silicon bronze alloy 硅青铜合金
silicon bronze wire 硅铜线
silicon-calcium alloy 硅钙合金
silicon-calcium compound type polyaluminum ferric chloride 硅钙复合型聚合氯化铝铁
silicon-calcium compound type polyaluminum magnesium zinc chloride 硅钙复合型聚合氯化铝镁锌
silicon capacitor 硅电容器
silicon carbide 金刚砂;碳化硅
silicon-carbide abrasive 碳化硅磨料
silicon-carbide alumin(i)um phosphate binder 磷酸铝碳化硅胶结体
silicon-carbide brick 碳化硅砖
silicon-carbide ceramics 碳化硅陶瓷
silicon-carbide composite 碳化硅合成物
silicon-carbide cutting blade 碳化硅刀片;碳化硅切削片
silicon carbide fiber 碳化硅纤维
silicon carbide furnace 碳化硅电炉
silicon carbide grinding wheel 碳化硅砂轮
silicon-carbide mortar 碳化硅砂浆
silicon carbide reflectivity 碳化硅反射率
silicon carbide refractory 碳化硅耐火材料
silicon carbide refractory product 碳化硅耐火材料制品
silicon carbide rod 硅碳棒;碳化硅棒
silicon carbide slab 碳化硅板
silicon carbide varister 碳化硅变阻器
silicon carbide waterproof paper 碳化硅耐水砂纸
silicon carbide whisker 碳化硅晶须
silicon-carbon rod 硅碳棒
silicon cell 硅电池
silicon charger 硅充电机
silicon chip 硅片;硅基片
silicon chloride 氯化硅;四氯化硅
silicon chrome-steel 硅铬钢
silicon chromium 硅铬铁合金;硅铬
silicon chromium steel 硅铬钢
silicon coating 硅涂层;硅树脂涂层
silicon colloid 硅胶体
silicon-containing dust 含矽粉尘
silicon-containing fertilizer 硅肥
silicon controlled 可控硅
silicon controlled multi-purpose machine tool 可控硅多用机床
silicon controlled rectifier 硅(可)控整流器;可控硅整流器
silicon controlled rectifier locomotive 可控硅整流器机车
silicon controlled switch 硅(可)控开关;可控硅开关
silicon controlled valve 可控硅阀
silicon controller 可控硅装置
silicon copper 硅紫铜;硅铜(合金)
silicon crystal 硅晶体
silicon crystal mixer 硅晶体混频器
silicon damping transistor 硅阻尼晶体管
silicon dead layer 硅失效层
silicon detector 硅检波器
silicon dice 硅片
silicon dichloride 二氯化硅
silicon die 硅片
silicon diiodide 二碘化硅
silicon diode 硅二极管
silicon dioxide 二氧化硅
silicon dioxide powder 二氧化硅粉
silicon disulfide 二硫化硅
silicone 硅酮;聚硅油;聚硅烷;聚硅酮
silicone adhesive agent 硅酮黏合剂
silicone alkyd 有机硅醇酸树脂
silicone anti-foaming agent 高效有机硅消泡剂
silicon earth 硅(质)土
silicone base 硅氧基
silicone bonded 带有机硅填充剂的(绝缘材料)
silicone building sealant 硅酮嵌缝膏
silicone-carbide paper 碳化硅砂纸
silicone-ceramic coating 硅酮陶瓷敷层
silicone-clad silica fibre 硅酮包层石英光纤
silicone coat 有机硅涂层
silicone coated 硅酮包衣的
silicone coated fabric 硅酮涂层织物
silicone coated glass fabric 硅酮涂层玻纤织物
silicone coated polyester fabric 硅酮涂层聚酯织物
silicone cooling 硅酮冷却
silicone coupling 硅黏合
silicone defoamer 有机硅消泡剂
silicone elastomer seal ring 硅化人造橡皮密封环
silicone epitaxial planar transistor 硅外延平面晶体管
silicone-epoxy coating 硅酮环氧敷层
silicone ester 硅氧烷酯
silicone finishing agent 多功能有机硅后整理剂
silicone fluid 硅氧烷流体
silicone foam 硅酮泡沫塑料
silicone foam plastics 硅泡沫塑料
silicone glue 硅氧胶
silicone grease 硅酮润滑脂;硅氧烷润滑脂
silicone hose 聚硅氧烷塑料软管
siliconeisen 低硅铁合金
silicone jelly 硅氧胶
silicone lacquer 有机硅树脂漆
silicone lubricant 硅润滑剂
silicone-modified 改性硅氧烷
silicone oil 硅油;硅酮油
silicone paint 有机硅树脂漆;硅酮涂料;有机硅涂料
silicone parting solution 硅酮脱模液
silicone paste 硅氧烷浆
silicon epitaxy 硅外延
silicone plastics 有机硅塑料
silicone polish 聚硅酮抛光剂
silicone polymer 硅氧烷聚合物
silicone primer 硅氧烷底涂料
silicone proofer 硅氧烷防水剂
silicone protection agent 硅氧烷防护剂
silicone release agent 硅油脱模剂
silicone repellent 硅氧烷防水剂
silicone repellent admixture 硅氧烷防水掺和料
silicone resin 有机硅树脂;硅氧烷树脂;硅酮树脂
silicone resin coating 有机硅树脂涂料
silicone resin emulsion 硅酮树脂乳液
silicone resin solution 硅酮树脂溶液
silicone rotor dies 硅酮滚轴模
silicone rubber 硅橡胶;硅酮橡胶
silicone rubber coating 硅酮橡胶敷层
silicone rubber composite flat membrane 硅橡胶平板复合膜
silicone rubber impression material 硅橡胶印模材料
silicone rubber pad 硅酮橡胶印头
silicone rubber putty 硅橡胶腻子
silicone rubber sealant 硅氧橡胶密封料;硅橡胶密封剂
silicone rubber seal(ing) composition 硅氧橡胶密封混合物
silicone rubber septum 硅橡胶隔片
silicone sealant 硅酮嵌缝膏;硅酮密封剂;硅酮密封层;硅密封胶
silicone ester 硅氧烷酯;硅酸酯
silicone surfactant 硅酮表面活性剂
silicone tape 硅氧烷带
silicone transfer printing 硅酮轻移印花
silicone treatment 硅酮处理
silicone tube 硅胶管
silicone waterproof(ing) agent 硅氧烷防水剂
silicone wax 硅酮蜡
silicon face 硅表面
silicon fluid 硅液;硅流体
silicon fusion transistor 硅熔接晶体管
silicon gate 硅栅极
silicon gate self-aligned technology 硅栅自对准工艺
silicon gel 硅胶
silicon glass cloth laminated sheet 硅玻璃布层压片
silicon grease 硅润滑脂;硅润滑油
silicon halide 卤化硅
silicon hex rod 六棱硅棒
silicon hydride 硅烷
silicon insulation 硅树脂绝缘
silicon integrated circuit 硅集成电路
silicon intensifier vidicon 硅增强视像管
silicon iron 硅铁;硅钢
silicon iron lamination 硅铁叠片
silicon iron pipe 硅铁管
silicon iron sheet 硅钢片
silicon iron tube 硅铁管
siliconit 硅炭棒
siliconize 硅化处理
siliconized plate 硅钢片
siliconized rubber-insulated 硅橡皮绝缘
siliconized synthetic(al)fiber 硅化合成纤维
siliconizing 硅化处理;扩散渗硅;渗硅
silicon lacquer 硅漆
silicon large power switch triode 硅大功率开关三极管
silicon liner 耐火材料;硅砖;耐火衬垫
silicon lubricant 硅润滑剂
silicon-manganese 硅锰
silicon-manganese alloy 硅锰合金
silicon-manganese steel 硅锰钢
silicon metal 金属硅;结晶硅
silicon method 硅法
silicon micro encapsulization 硅微囊
silicon monoxide 一氧化硅
silicon nitride 氮化硅
silicon nitride based cutting tool 氮化硅基切削工具
silicon nitride-bonded brick 氮化硅结合砖
silicon nitride bonded silicon carbide refractory 氮化硅结合碳化硅耐火材料
silicon nitride ceramics 氮化硅陶瓷

silicon nitride film 氮化硅薄膜
silicon nitride passivation 氮化硅钝化
silicon nitride reaction-bonded 氮化硅反应结合
silicon nitride refractory 氮化硅耐火材料
silicon on insulating substrate 绝缘衬底上外延硅
silicon on insulator substrate 绝缘衬底上外延硅
silicon on sapphire 硅蓝宝石工艺;蓝宝石硅(片)
silicon-on-sapphire technique 蓝宝石上外延硅工艺;硅蓝宝石技术
silicon oversaturation 硅过饱和
silicon oxide 氧化硅
silicon-oxygen isotope correlation 硅一氧同位素相关性
silicon-oxygen tetrahedron 氧化硅四面体
silicon oxynitride 氮氧化硅
silicon-oxy tetrahedron 硅氧四面体
silicon pellet 硅片
silicon photocell 硅太阳电池;硅光电管;硅光电池
silicon photodetector 硅光电探测器
silicon photodiode vacuum ultraviolet detector 硅光电二极管真空紫外探测器
silicon photoelectric(al)cell slice 硅光电池片
silicon photoelectric(al)diode 硅光电二极管
silicon photoelectric(al)generator 硅光电发生器
silicon photoelectric(al)transistor 硅光电晶体管
silicon phototransistor 硅光电晶体管
silicon plastic 硅塑料
silicon plate 硅钢薄板
silicon polycrystal 多晶硅
silicon polymer 硅聚合物
silicon power diode 硅功率二极管
silicon rate 硅率
silicon rectifier 硅整流器
silicon rectifier equipment 硅整流设备
silicon rectifier for charging 硅充电整流器
silicon rectifier for electroplating 电镀硅整流器
silicon rectifier for welding 硅焊接整流器
silicon rectifier welder 硅整流焊机
silicon rectifying equipment 硅整流设备
silicon resin 硅(烷)树脂
silicon resin adhesive 硅树脂胶黏剂
silicon resin varnish 硅树脂清漆
silicon resistor 硅电阻(器)
silicon rubber 硅橡胶
silicon rubber adhesive 硅橡胶胶黏剂
silicon rubber glass paint cloth 硅橡胶玻璃漆布
silicon rubber insulated cable 硅橡胶绝缘电缆
silicon rubber insulation 硅橡胶绝缘
silicon rubber insulator 硅橡胶绝缘子
silicon saturation 硅饱和
silicon semiconductor detector 硅半导体探测器
silicon series prescription 硅系配方
silicon sheet 硅钢片;硅钢薄板
silicon sheet polisher 硅片磨光机
silicon single crystal 硅单晶
silicon single crystal rod 单晶硅棒
silicon slice 硅片
silicon slice detergent 硅片清洗剂
silicon slice polishing liquid 硅片抛光液
silicon software 硅软件
silicon solar battery 硅太阳电池
silicon solar cell 硅太阳电池
silicon solar cell array 硅太阳电池方阵
silicon solar cell panel 硅太阳电池板
silicon solid photographic sensor 硅固体摄像传感器
silicon stack 硅堆
silicon steel 矽钢;硅钢
silicon steel plate 硅钢片
silicon steel sheet 硅钢片
silicon steel sheet coating 硅钢片漆
silicon steel sheet varnish 硅钢片清漆
silicon strip 硅钢带
silicon substrate 硅基片;硅衬底
silicon surfactant 硅表面活性剂
silicon switching triode 硅开关三极管
silicon switch transistor 硅开关晶体管
silicon symmetric(al)amplitude limiter 硅对称限幅器
silicon-symmetrical switch 硅对称开关
silicon symmetric(al)twin-diode 硅对称双二极管
silicon target 硅靶
silicon tetrachloride 四氯化硅
silicon tetrafluoride 四氟化硅

silicon tetrafluoride gas 四氟化硅气体
silicon thyratron 硅闸流管
silicon transistor 硅晶体管
silicon unilateral switch 硅单向开关
silicon unsaturation 硅不饱和
Silicon Valley 矽谷
silicon variable capacitor 硅可变电容器
silicon varistor 硅压敏电阻器
silicon varnish 硅树脂清漆;硅基漆
silicon varnish coated glass cloth 硅漆涂玻璃布
silicon voltage-regulative transistor 硅调压晶体管
silicon voltage stable transistor 硅稳压晶体管
silicon wafer 硅圆片;硅片
silicon water repellent 硅树脂增水剂;硅树脂防水剂
silicon water repellent paint 硅树脂憎水涂料;硅树脂抗水涂料
silicon welding rectifier 硅焊整流器
silicon zener voltage regulator 硅齐纳稳压器
silicoorganic compound 有机硅化合物
silico-phosphate cement 硅质磷酸盐水泥;硅磷酸锌黏固粉
silicopropane 丙硅烷
silico-rudite 硅砾岩
silicosiderosis 矽铁肺;铁矽尘肺
silicosis 矽肺病;硅肺病;石末沉着病
silicospiegel 硅镜铁
silico-spiegel iron 硅镜铁
silicotic nodule 硅结节
silico-tungstic acid 硅钨酸
silico-tungstic acid test 硅钨酸试验
silicovanadomolybenum blue 硅钒钼蓝
silification 硅化作用
Silit 硅碳耐火材料;碳硅电阻材料
silk 研光;丝绸;蚕丝;半光面漆
silk agate 蚕丝玛瑙
silkaline 充丝薄棉布
silkatine thread 充丝有光棉线
silk boiling 煮丝
silk bolting cloth 丝筛布
silk cloth 丝绸
silk cotton 木棉;丝光木棉
silk-covered cord 丝包电线
silk-covered wire 丝包线
silk drafts staple diagram 精棉纤维排列图
silk embroidered picture 刺绣画
silken map 丝绸地图
silk fabrics goods 丝织品
silk figure 彩扎
silk filature classification 厂丝等级
silk finish 丝光涂层
silk finish calender 缎光机
silk finishing calender 丝光矸光机
silk gauze 绢网
silk guide roller 导丝辊
silkiness 丝光性
silking (涂层上的)细条纹;漆膜走丝;丝纹;丝条;丝光
silk insulation 丝绝缘
silk insulation tape 丝绝缘带
silk knit goods 丝织品
silk lamp shade 绸灯罩
silk lantern 绸灯
silk-like sheen 丝状光泽;丝光光泽
silk mill 丝厂
silk-mill wastewater 丝厂废水
silk net 绢网;丝网
silk noil yarn 抽丝
silk painting 绢画
silk paper 薄纸
silk reeling frame 络丝机
silk reeling machine 络丝机
silk ribbon 丝带
silk road 丝绸之路
silk screen 丝网;丝屏罩
silk-screened wall panel 丝网印刷墙面板
silk screening 丝网(印刷)法
silk screening process 丝网屏蔽法
silk screen ink 丝网印刷油墨;丝漏(印刷)油墨
silk screen lacquer 丝网印刷清漆
silk screen method 丝网印花法
silk screen paint 丝网印刷涂料;丝网漆
silk screen paste mixer 线网感光胶混合机
silk-screen printing 丝网印刷;绢网印花
silk shavings 丝绸薄片

silk spreader 展棉机
silk spun 绢丝
silk-stocking district 富人区
silk surface 绸绞钡地纸
silk suture 丝缝线
silk tapestry 丝织锦
silk textile 丝织业;丝织品
silk-tree 木棉树;合欢树
silk wall cover(ing)丝绸墙(壁饰面)布
silk wall lining 丝绸墙壁衬垫
silk wall(sur)facing 丝绸墙壁覆面
silk waste 废丝
silk weaving mill 丝织厂
silk white 丝光白
silk winding 络丝
silk winding frame 络丝机
silkworm egg paper 蚕种纸
silkworm-motion wave 蚕行波
silkworm nursery 养蚕室
silkworm(rearing)room 养蚕室
silky 丝绢;丝光漆
silk yarn 绢丝
silky fracture 丝状断裂面;丝状断口;丝光断口
silky lustre 绢丝光泽
silky oak 银华树
silky pig iron 锦生铁
sill 下槛;岩床;基木;门槛;海槛;海底凸槛;侵入岩席;排水舌;坞槛;垫底横木;底木;底梁;底槛;坝底线
sillage 坏架
sill anchor 横木板锚栓;底梁锚固件;下槛锚固螺栓;槛锚;地脚锚栓;地脚螺栓
sill anchorage 地脚螺栓
sill and bolster brace 承梁拉条
sill angle 底槛
sill bar 底杆
sill basin 海槛盆地
sill bead 窗框槛条;门窗口凸圆;下扯窗盖条
sill beam 槛梁;底梁
sill block 洞口混凝土砌块;门槛砌块;窗台砌块;闸首板;界砖
sillboard 窗台板
sill casting 门槛铸件
sill cock 洒水栓;小龙头;软管龙头;水龙带旋塞;水龙头龙头
sill course 虎头砖;窗台砌层;窗台层
sill cover 窗台罩面(板)
sill cross bar 横底杆
sill depth 岩床深度;闸槛水深;槛顶水深;海槛深度;潜坝深度
sill-drip mo(u)ld(ing) 附加窗台;窗台滴水线脚;窗台披水
silled basin 孤立海盆;隔离海盆;受限盆地
sill elevation 基石标高
sillenite 软铋矿
sill for jump control 控制水跃的底槛
sill head 窗槛梁;窗台板
sill head height 窗台(顶面)高度
sill heating 窗台下供暖
sill height dimension 从地板到窗口高尺寸
sill high 门槛高
Silliman bronze 铝铁青铜
sillimanite 矽线石;夕线石;硅线石
sillimanite biotite gneiss 矽线石黑云母麻岩
sillimanite biotite quartz schist 矽线石黑云母石英片岩
sillimanite-biotite schist 矽线石黑云母片岩
sillimanite brick 矽线石砖
sillimanite crucible 硅线石坩埚
sillimanite deposit 矽线石矿床
sillimanite dimicaceous schist 矽线石二云母片岩
sillimanite garnet potash-feldspar and plagioclase gneiss 矽线烯留二长片麻岩
sillimanite garnet potash-feldspar gneiss 矽线石榴钾长片麻岩
sillimanite muscovite gneiss 矽线石白云母麻岩
sillimanite muscovite quartz schist 矽线石白云母石英片岩
sillimanite-muscovite-schist 矽线石白云母片岩
sillimanite plagioclaso orthoclase granulite 矽线石斜长正长麻粒岩
sillimanite-quartz schist 矽线石石英片岩
sillimanite refractory 矽线石耐火材料
sillimanite refractory material 硅线石耐火材料

sillimanite two mica gneiss 矽线石二云片麻岩
sillimanite two mica quartz schist 矽线石二云母石英片岩
sill joint 垫板接合;垫板接缝
sill level 底层;底部水平;底部平基
sill lip 坝坎;坝槛;槛挡;槛槽
sill loss 潜坝;无槛
sill of conditional simulation 条件模拟基台
sill of non-conditional simulation 非条件模拟基台
sill of side-scuttle 舷窗槛
sill of weir 堰槛;堰坎
sill-over signal 信息漏失
sill plate 木骨架的底木条;窗台板;下槛;闸首底板;门槛板;底系定板
sill rail 窗台栏
sill rooted in the bank 岸中伸出的潜坝
sill-sealer insulation 门栏密封绝缘材料
sill sluice 槛下泄水
sill splicing 梁的戳换;拼梁;拼槛
sill step 脚蹬
sill strip 底条
sill tile 窗台砖;槛砖;垫砖
sill timber 基台枕【岩】;底木;槛木
sill value 基台值
sill wall 槛墙;裙墙
sill wall window 槛窗
Silmalec 西尔玛雷克铝硅镁合金
silmanal 银锰铝特种磁性合金
Silmelec 硅铝耐蚀合金
silmet 板状镍银
silmethylene 硅亚甲基;聚亚甲基硅烷
silmethylene bond 硅亚甲基
silmo 硅钼特殊钢
silo 圆筒仓;储[贮]存库;谷仓;料斗;料槽;料仓;筒仓;发射井
silo aeration 库内充气
silo bin 圆筒仓;筒式存仓
silo block 混凝土圆筒仓砌块;筒仓群
silo bottom 筒仓底;库底卸料器;库底
silo bunker 筒仓漏斗
Sil-o-cel brick 硅筒砖(一种硅砖的商品名)
silo cell 圆筒仓室;窖室;仓筒;绝热砖
silo coating plastic 筒仓涂布塑料
silo compartment 格子式料仓室;筒仓分隔间
silo construction 筒仓建筑;谷仓构造
silo content 物料库容量
silo conveying pump 仓式输送泵
silo discharge 筒仓出料
silo-discharge cone 圆库卸料锥体
silo-discharge device 圆库卸料装置
silo-discharge system 圆库卸料系统
silo drawing channel 筒仓出料槽
silo effect 筒仓效应
silo fed by bucket elevator 用斗式提升机加料的筒仓
silo-filler's disease 窖仓装料工病
silo for high moisture silage 高水分青储[贮]料仓
silo formwork 筒仓模板
silo gate 筒仓卸料门
silo heating 筒仓供暖
silo hopper 筒仓装料斗
silo installation 筒仓装置
silo launch 井下发射
silo launcher 井下发射装置
silo launch facility 井下发射设施
silo launch test facility 井下发射试验设备
silo level detector 筒仓液面探测器;筒仓料位探测器;筒仓库存指示器
silo-lift 井内升降机
silo mercerizer 浸碱桶
silometer 料仓料位计;筒仓料位指示器;仓储[贮]指示器
Silon 聚酰氨纤维
silo of concrete blocks acting as lost formwork 无模板的混凝土块材筒仓
silo operating time 井下作业时间
silo outlet 筒仓出口
siloplot 仓位指示器
silo pressure 谷仓压力;筒仓压力
silo program(me) 仓储计划
silo seepage 筒仓渗漏
silo set 库凝
silo side discharge set 库侧卸料器
silo storage 筒仓设备;筒仓储藏

silo tile 筒仓砖
silo-type incinerator 筒仓式焚化炉
silo wagon 粮秣车
silo wall 筒仓墙
siloxane 硅氧烷
siloxane grease 硅氧烷润滑脂
siloxanol 硅氧烷醇;硅醇
siloxen 硅氧烯
siloxene 环己硅氧烷
siloxen indicator 硅氧烯指示剂
siloxicon 硅碳氧;硅碳耐火料;氧碳化硅;硅碳耐火材料
siloxy 甲硅烷氧基
Silsbee effect 西尔斯比效应
Silsbee rule 西尔斯比定律
silt 淤泥堵塞;研磨用砂浆;泥沙;河道沉积泥沙;粉土;粉粒;粉沙
silt accumulation 泥沙沉积;泥沙积聚
silt algae 软泥藻类;沉泥藻类
silt and clay 粉沙和黏土
silt argillaceous texture 粉砂泥质结构
silt arrester 拦沙设备;淤地坝;拦沙坝;拦沙设施;拦沙坝
siltation 淤填;淤泥淤塞;淤泥充塞;淤积
siltation in dike field 坝田淤积
siltation in estuary 河口淤塞
siltation pattern 淤积形式
siltation-relieving measures 减淤措施
siltation volume 淤积量
silt bank 淤泥沙洲;淤积层
silt bar 淤泥沙洲
silt basin 淤泥池;淤泥池;沉沙井;沉泥池;沉泥井
silt-bearing argillaceous texture 含粉砂泥质结构
silt-bearing mudstone 含粉砂泥岩
silt bed 淤泥质河床
silt-bottomed valley 淤积平底谷
silt box 淤泥沉积箱;集粉砂盒;集沙箱;沉沙箱
silt bucket 淤泥铲斗
silt-carrying capacity 泥沙含量;夹沙能力;泥沙携带能力;输沙能力;带沙能力;携沙量;挟沙能力
silt-carrying capacity of flow 水流挟沙能力
silt-carrying capacity of river 河流挟沙能力
silt-carrying flow 挟沙水流;含沙水流
silt-carrying river 挟沙河流;含(泥)沙河流;多沙河流
silt-carrying stream 含沙河流
silt charge 含沙量;含砂量;含沙量;含沙率;泥沙含量
silt clearance 清淤
silt coal 淤泥煤
silt coast 淤泥质海岸
silt collector 淤泥收集器
silt combination 泥沙粒径组成
silt concentration 泥沙浓度;泥沙含量;含沙浓度;含泥量
silt concentration study 含沙量研究
silt container 集沙箱
silt content 泥沙含量;含泥(沙)量;粉粒含量
silt content test 含泥量试验
silt content test for fine aggregate 细骨料粉砂含量试验;细骨料粉土含量试验
silt control 淤泥控制;淤积控制;泥沙控制
silt-covered 淤塞的;淤泥覆盖的
silt curtain 挡淤帘
silt density current 泥沙异重流
silt density index 淤泥密度指数
silt deposit 淤泥处置;泥沙沉积;泥沙沉积物;淤泥淤积;淤泥沉积
silt deposition 粉砂沉积;淤泥沉积
silt detent period 拦沙期
silt-determination measure 泥沙测验
silt discharge 输沙量
silt discharge of creep-movement 蠕移输沙量
silt discharge of jump-movement 跳移输沙量
silt discharge rating 输沙量率定
silt-discharging rating 输沙量—流量关系
silt displacement 泥沙移动;淤泥清除;盾构排泥;隧洞掘进法
silt draghead 淤泥耙头
silt drain 泥沙泄出口
silt dune 淤沙丘
silted land 淤积地
silted model 淤积模型
silt ejector 泥沙喷射泵;排泥器;排泥管

silt elevator 淤泥喷吸器;粉沙提升机;淤泥提升机
silt factor 泥沙因子;泥沙系数;粉沙系数
silt field 磨料粉砂沉积池
silt filter 缝隙滤水管;槽式滤水器
silt flora 软泥植物区(系)
silt flow 泥沙流;含有泥沙的水流
silt flow supersaturation 泥沙流过饱和
silt flow unsaturation 泥沙流不饱和
silt-flushing sluice 冲沙闸(门);放淤闸
silt flux of bed load discharge 推移质泥沙量
silt foundation 粉沙地基
silt fraction 粉土部分;粉粒组;粉粒粒组
silt-free river 无泥沙河流
silt-free stream 无泥沙河流
silt grade 粉沙大小的材料;泥沙粒度
silt grain 粉砂颗粒
silt grain grade 粉粒组
silt gravel 粉砾
silthiane 硅硫烷
silting 沉泥;淤积;泥沙堆积;沉淤
silting coefficient of riverbed 河床淤塞系数
silting index 淤塞系数
silting method 淤填法;淤积法
silting of backwater area 回水区淤积
silting of reservoir 水库淤积
silting of resource of silt 淤积厚度
silting reduction 减淤
silting speed 淤积速度
silting-up 淤高;淤淀;淤塞;淤泥淤塞
silting-up of casing well 管井堵塞
silting up of leak 淤填防渗
silt injection method 粉沙注入法
siltite 泥砂岩;粉砂岩
siltized intercalation 泥化夹层
silt jetty (河口三角洲处粉细砂淤出的)长沙洲;淤泥埂;粉砂质狭长沙洲
silt-laden 夹带大量泥沙的;塞满淤泥的
silt-laden flow 挟沙水流;夹沙水流;含沙水流;含泥水流
silt-laden river 含沙河流;多(泥)沙河流
silt-laden stream 夹沙河川;夹沙河流;含沙河流
silt layer 淤泥层;粉沙层
silt lens 粉砂透镜体
silt load 输沙量;输砂量
silt loam 粉(砂)壤土;粉砂垆姆
silt logging 泥沙沉积
silt lutite 粉砂质泥沙屑岩
silt meter 测沙仪;测淤计
silt monitor 泥沙监测器
silt mud 淤泥;粉砂泥质
silt orifice 排泥口;排沙口
silt pan 粉砂磐
silt particle 粉砂颗粒;粉粒
silt-pelite 粉砂泥岩
silt pollution 淤泥污染;泥沙污染
silt pressure 淤泥压力;泥土压力;泥沙压力
silt range 粉沙地区;粉沙范围
silt-releasing sluice 冲沙闸(门);放淤闸
silt rock 粉砂岩
silt sampler 泥沙取样器;取沙器
silt sand 粉质砂土;粉砂(质)砂
silt sand beach 粉砂海滩
silt sandstone 粉砂质砂岩
silt screen 防泥屏
silt seam 淤泥层;粉土层;粉砂层
silt sediment 粉土沉积物
silt shale 粉砂页岩
silt size 淤沙粒度;粉土粒径;粉土粒度
silt-sludge beach 粉砂—淤泥滩
silt sluicing 淤泥流槽冲洗
silt slurry 淤泥浆;泥浆
silt soil 淤泥土;粉土质土(壤);粉土;粉砂土(壤);粉粒土;粉砂质土(壤)
silt soil containing coarse grains 含粗粒粉质土
silt soil containing lightly 微含粗粒粉质土
silt stable channel 冲淤平衡稳定河槽;冲淤平衡渠道;不淤不冲渠道
silt-stable channel 冲淤平衡的稳定河槽;不冲不淤河道
siltstone 粉土岩;泥砂岩;粉砂岩;粉砂石
siltstone inclusion 粉砂岩包裹体
siltstone-mudstone facies 粉砂—泥岩相
siltstone reservoir 粉砂岩储集层
silt stratification 淤泥层

silt suspension 悬移(泥)沙;泥沙悬浮物(质)
silt texture 粉砂结构
silttil 粉砂冰碛
silt transport 淤泥输运;泥沙搬运;输沙
silt transportation 泥沙输移
silt trap 集泥器;淤泥氹;淤泥沉积地;截沙器;截沙坑
silt-trap dam 拦沙坝
silt trapping principle 拦泥原理
silt trench 积砂沟
silt up 淤积
silt-up channel 淤塞渠槽;淤塞河槽;淤塞航槽;淤塞渠道
silt water 淤泥水
silt-water mixture 泥水混合物
silty 粉土质的;粉砂质的
silty clast 粉砂屑
silty clastic texture 粉屑结构
silty clay 粉质黏土;粉土(质)黏土;粉砂(质)黏土
silty clay loam 粉(土)质黏壤土;粉砂质黏壤土
silty fine sand 粉质细砂
silty fine sand with some clay 粉质细砂混黏土
silty gravel 粉质土砾;粉砂质砾石
silty heavy loam 粉质重亚黏土
silty heavy mild sand 粉质重亚砂土
silty light loam 粉质轻亚黏土
silty light mild sand 粉质轻亚砂土
silty lime 粉砂石灰
silty loam 粉(质)壤土;粉砂质垆姆;粉(砂)壤土
silty loam clay 粉质壤黏土
silty middle loam 粉质中亚黏土
silty mud 粉砂质泥
silty peat 粉土质泥炭土
silty sand 淤泥砂;粉砂质砂
silty sandstone 粉砂质砂岩
silty sandy clay loam 粉质砂黏壤土
silty sandy loam 粉质亚砂土;粉质砂壤土
silty sediments 粉砂质沉积物
silty shale 粉砂质页岩
silty size 粉砂粒径
silty slate 粉砂质板岩
silty soil 粉质土;粉性土;粉土质土(壤);粉砂质土(壤)
silty water 浑水;含沙量大的水
silumin 硅铝明合金;高硅铝合金;铝硅合金
silundum 硅碳刚石
Silurian great regression 志留纪大海退
silurian paper 花纹纸
Silurian period 志留纪
Silurian rock 志留纪岩石
Silurian system 志留系
Siluric 志留系
silva 林木;森林志
silvan 树木多的;森林区的
silve 检查分立方米
silver 有银白光泽的;银色;银白色(的);白银
silver acetate 乙酸银;醋酸银
silver acetylide 乙炔银
silver-activated Geiger counter 银激活兽革计数器
silver alloy 银合金
silver-alloy brazing 银钎焊;银合金焊
silver alloyed contact 银合金接触头
silver amalgam 银汞合金
silver amide 氨基银
silver and foreign currencies 收兑金银外币
silver-antimony alloy 银锑合金
silver arsenate 砷酸银
silver arsphenamine 阿斯凡纳明银盐
silver assay 白银鉴定
silver bar 银条
silver base filler alloy 银基填充合金
silver bath 银浴器;银盐溶液槽
silver-bearing 含银(的)
silver-bearing copper 银铜;含银铜
silver-bearing lead 含银铅
silver belt fish 带鱼
silver bicarbonate 碳酸氢银
silver bichromate 重铬酸银
silver birch 银桦
silver-block bond 花墙砌法
silver boat 银舟
silver brazing 银钎焊;银焊
silver-brazing alloy 银钎料
silver bromide 溴化银
silver bronze 银铜合金

silver bullion 银条
silver bushing 银衬套
silver-cadmium storage battery 银镉蓄电池
silver carbonate 碳酸银
silver certificate 白银证券
silver chlorate 氯酸银
silver chloride 氯化银
silver chloroplatinate 氯铂酸银
silver chromate 铬酸银
silver citrate 柠檬酸银
silver-clad 包银
silver-clad copper 包银铜
silver-coated copper wire 镀银铜线
silver-coated reflector 镀银反射器
silver coating 银镀膜;涂银;镀银层
silver coating film 涂银膜
silver compound 银化合物
silver-constantan thermocouple 银康热电偶
silver contact 银接触;银触点
silver-copper alloy 银铜合金
silver cyanate 氰酸银
silver cyanide 氰化银
silver deposit 银矿床;镀银层
silver diamminohydroxide 氢氧化二氨合银
silver diazo process 金属重氮法
silver dichromate 重铬酸银
silver diethyl dithiocabamate spectrophotometric method 二乙基二硫代氨基甲酸银分光光度法
silver disk pyrheliometer 银盘日射强度计;银盘日射强度表
silver dithionate 连二硫酸银
silvered 镀银的
silvered anodized alumin(i)um 本色氧化铝
silvered-bowl lamp 镀银白炽灯泡
silvered capacitor 镀银层电容器
silvered condenser 镀银层电容器
silvered face 镀银面
silvered film 镀银层
silvered glass 镀银玻璃
silvered mica capacitor 镀银云母电容器
silvered mirror 镀银镜
silvered plated glass 镀银玻璃
silver electrode 银电极
silver electroplating 电镀银
silver-faced 镀银的
silver fahlore 银黝铜矿
silver falsebeech 银山毛榉
silver figure 银光纹理
silver film 银膜
silver finished shade 镜面罩
silver fir 银枞松
silver fluoride 氟化银
silver foil 银箔
silver frost 银光霜
silver fulminate 镭银;雷酸银
silver fuse 银熔丝
silver gasket 银垫圈
silver gauze 银丝网
silver gel 银凝胶
silver glance 辉银矿;硫化银
silver glass 镀银玻璃
silver grain 银光纹理;银光花纹;径截纹;虎斑纹
silver grain wood 辐射纹木材;银光纹木材
silver-graphite 银石墨制品
silver-graphite brush 银石墨电刷
silver gray 银灰色
silver green 银绿色
silver halide 卤化银
silver halide emulsion 卤化银乳胶
silver halide process 银盐复制法
silver halide sensitized photochronic glass 卤化银光致变色玻璃
silver hawthorn 银山楂
silver holographic plate 银盐全息干板
silver hydroxide 氢氧化银
silver hypochlorite 次氯酸银
silver hyponitrite 连二次硝酸银
silver hypophosphate 连二磷酸银
silver image 银像
silver-impregnated zeolite 浸银沸石
Silverine 铜镍耐蚀合金
silvering 银色光泽;镀银;包银
silvering conveyer 带式镀银机
silver ingot 银条

silvering plate glass 制镜用板玻璃
silvering powder 镀银粉
silvering quality 镀银质量
silvering quality of glass 镀银优质玻璃
silvering shop 镀银车间
silvering silver coating 镀银
silver ink 银墨
silver iodate 碘酸银
silver iodide 碘化银
silver iodide collochion process 碘化银胶棉法
silver iodide colloidale 胶体碘化银
silver iodide paper 碘化银相纸
silver ion toxicity 银离子毒性
silver jacketed wire 镀银线;镀银导线
silver jamesonite 银毛矿
silver lacquer 银器喷漆
silver lame 银薄片
silver lead ore 银方铅矿
silver-lead solder alloy 银铅焊条合金
silver leaf 银箔
silverline system 银线系
silver-lock bond 空斗墙砌合;空斗墙砌法;银锁砌合
silver luster 银光泽彩;银光彩料;亮银(陶瓷器边饰)
silver-magnesium-nickel alloys 银镁镍合金
silver maple 银槭
silver market 白银市场
silver membrane filter 银膜过滤器
silver migration 银迁移
silver mirror reaction 银镜反应
silver mirror test 银镜试验
silver-molybdenum 银钼合金
silvern 似银
silver nitrate 硝酸银
silver nitrate spectrophotometry 硝酸银分光光度法
silver nitrate titrimetry 硝酸银滴定法
silver nitride 一氮化三银
silver number 银值
Silveroid 铜银装饰合金;镍银
silver ore 银矿(石)
silver orthoarsenate 原砷酸银
silver orthoarsenite 原亚砷酸银
silver orthophosphate 磷酸银
silver oxalate 草酸银
silver oxide 氧化银
silver oxide battery 氧化银电池
silver oxide cell 氧化银电池
silver paint 银色漆;银粉漆;铝粉漆
silver paper 锡纸;银箔
silver paste 银膏
silver pastebrushing 涂银
silver perchlorate 高氯酸银
silver permanganate 高锰酸银
silver peroxide 过氧化银
silver phosphate 磷酸银
silver phosphate glass 磷酸银玻璃
silver phosphide 二磷化银
silver plate 银极;银板;镀银件
silver-plated 镀银的
silver-plated brass 镀银铜
silver-plated copper 镀银铜
silver-plated copper conductor 镀银铜线
silver-plated glass 镀银玻璃
silver-plated tin alloy 镀银锡合金
silver-plated ware 镀银器具
silver plating 镀银
silver-plating bath 镀银槽
silver-plating katathermometer 高温镀银卡他温度计
silver-plating plant 镀银装置
silver ply steel 不锈覆盖钢
silver point 银熔点;银的凝固点;点银绘画(法)
silver poisoning 银中毒
silver potassium cyanide 氰化银钾
silver potassium fulminate 雷酸银钾
silver powder 银粉
silverprint 银盐感光照片
silver process 银盐工艺
silver process 银盐复制法
silver protein 蛋白银
silver pyrophosphate 焦磷酸银
silver pyrovanadate 焦钒酸银
silver recovery 银(的)回收

silver redactor 银还原剂
silver rink 铝粉油墨
silver salt 银盐
silver salt solution for silvering 镀银用银盐溶液
silver sand 细砂；银砂；白砂
silver sarcophagus 镀银石棺
silver screen 银幕
silver selenide 硒化银
silver sensitized material 银盐感光材料
silver shot 银弹；银球
silver-silver chloride electrode 银—氯化银电极
silver-silver chloride reference electrode 银—氯化银参比电极
silver soap 银光皂
silver sodium chloride 氯化银钠
silver solder 银钎焊料；银钎料；银焊料
silver soldering 银焊
silver spot 银点
silver spruce 银云杉
silver stain 银黄着色
silver staining 银染色法；银黄扩散着色
silver staining test 银镜试验
silver standard 银本位
silverstat regulator 银触型电压调整器；接触式调节器
silverstat relay 银触头继电器
silver steel 银器钢；银亮钢
silver steel wire 银亮钢丝
silver stock 白银储备
silver stonewood 银达理木
silver storm 银光风暴；冰暴
silver suboxide 低氧化银
silver sulfate 硫酸银
silver sulfide 硫化银
silver sulphate 硫酸银
silver tetrahedrite 银黝铜矿
silver thaw 雨凇（下降时呈液态，着地后就冰冻）；银光霜
silver thiocyanate 硫氰酸银
silver thiosulfate 硫代硫酸银
silver-tin amalgam 银锡合金汞齐
silver-tipped 点银的
silver titration coulometer 银滴定电量计
silvertoum testing set 地下电缆故障测定设备
silvertoun 电缆故障寻找器
silver tungsten 银钨合金
silver-tungsten bimetallic band 银钨双金属带
silver-tungsten contact material 银钨电接触器材
silverware 银器；银制品
silverware shop 银楼
silver wattle 银合欢【植】
silver white 纯铅白；银白颜料；苍白色的
silver wire 银线；银丝
silver-work 银饰品
silver yellow 银黄
silvery grey 银灰
silvery grey paint 银灰漆
silvery pig iron 高炉硅铁；高硅生铁
silvery-white 银白色
silver zeolite 银沸石
silver-zinc accumulator 银锌蓄电池
silver-zinc battery 银锌电池
silver-zinc storage battery 银锌蓄电池
silvester zone 湖泊外界
silvialite 硫酸方柱石
silvichemical 林产化工产品
silvicultural rotation 更新轮伐期
silvicultural system 森林作业法
silviculture 造林（学）
Silvola recovery 碱回收法
silylanization 硅烷化
silylation 甲硅烷基化（作用）
silylidyne 次甲硅基
Silzin 硅黄铜
sima 葱形饰；波纹线脚；反曲线线脚；古典式房屋上雨水槽；硅镁带；硅镁层
simaldrate 硅镁铝
Simanal 硅锰铝铁合金
sima recte 上凹下凸反曲线；混枭线脚（古典柱式中）
sima reverse 上凹下凸反曲线；枭混线脚
sima sphere 硅镁圈
simaspheric fracture 硅镁层断裂
Simater sand filter 西马特砂滤池

simatic 硅镁质
simatic crust 硅镁壳
Simcar aerator 西姆卡曝气器
simdig 数字化器模拟设备
simethicone 二甲基硅油
simex 联合法提取
simex process 联合法提取过程
similar 相似的；同样
similar airway 相似风巷
similar argument 相似变量
similar basin 相似流域
similar calibration 相似检定
similar class 相似类目
similar conics 相似二次曲线
similar decimals 同位小数
similar device 类似装置
similar face 相似面
similar figures 相似图形
similar flexure 同向弯曲
similar flow 相似流（动）
similar fold(ing) 相似褶皱
similar geometry 相似几何
similar index 类指数
similarity 相似性；共性；类似（性）；均一性
similarity coefficient 相似系数
similarity consideration 相似条件
similarity criterion 相似（性）准则
similarity criterion of initiation of motion 起动相似准则
similarity criterion of settling velocity 沉降速度相似准则
similarity deviation 相似偏离
similarity discrimination 相似性判别
similarity factor 相似系数
similarity hypothesis 相似性假说
similarity in appearance 形似
similarity index 相似（性）指数
similarity-index analysis 相似指数分析
similarity isomorphic 相同构的
similarity law 相似律；相似定律
similarity level 相似性水平
similarity matrices 相似矩阵
similarity method 相似法
similarity model 相似模型
similarity number 相似数
similarity of competitive environment 竞争环境的趋同
similarity of flow motion 水流运动相似
similarity of flow movement 水流运动相似
similarity of sediment discharge 输沙量沿程变化相似；输沙量相似
similarity of sediment motion 泥沙运动相似
similarity of sediment movement 泥沙运动相似
similarity parameter 相似参数
similarity principle 相似原理
similarity ratio 相似比
similarity reduction 相似性化简
similarity relation(ship) 相似关系
similarity rule 相似法则；相似定律
similarity-rule estimation method 按相似律计算法
similarity ship 相似的船舶
similarity test 相似性试验
similarity theorem 相似（性）定理
similarity theory 相似（性）理论
similarity theory for atmospheric diffusion 大气扩散相似理论
similarity theory of turbulence 紊流的相似理论
similarity transformation 相似变换
similarity variable 相似变量
similar joint action 相似的联合作用
similar kinds of nutrients 相同类型的养分
similar law 相似律；相似法则
similar matrices 相似矩阵
similar model 相似模型
similar operating condition 相似工况
similar orbits 相似轨道
similar ordered aggregates 相似有序集
similar parameter 相似参数
similar permutation 相似排列
similar polygons 相似多边形
similar polyhedrons 相似多面体
similar quadrics 相似二次曲面
similar regions 相似域
similar representation 相似表示

similars 相似导线
similar square matrices 相似方阵
similar system(s) of numbers 相似数系
similar terms 同类项
similar test 相似检验；相似检定
similar to sample 同样品相似
similar tree 相似树
similar triangles 相似三角形
similar turbine 相似水轮机
similar year 相似年份
simili paper 冲沙纸
similitude 相似性；类似物；比拟
similitude criterion 相似（性）准则
similitude effect 相似效应
similitude law 相似律
similitude method 相似特性法
similitude principle 相似原理
similitude theory 相似理论
simlaite 铝海泡石
simmer 徐沸；慢慢煮沸；缓慢沸腾
simmer burner 文火燃烧器
simmer control 文火调节器
Simmer gasket 轴密封垫圈；翻唇垫圈
simmering lamp 预燃灯
Simmer ring 轴密封环；翻唇垫圈
si-mn 硅锰
simo chart 齐动图；同步动作图解
Simon diffusion 西蒙扩散
simonellite 西烃石
Simon-Goodwin charts 西蒙—古德温图表
simonkolleite 水氯羟锌石
Simon poplar 小叶杨
Simon's drilling theory 西蒙钻进理论
Simonsen phenomenon 西蒙森现象
Simonys' formula 西蒙尼公式
simple 简单的；单一的；单纯的
simple absolute expression 简单绝对表达式
simple absorption pipette 单吸移液管
simple aerator 引流竖管曝气器
simple aerial cableway 轻便铁索；轻便索道
simple aggregative index (number) 简单综合指数
simple alternating current 正弦交流
simple alternating function 单交函数
simple alternative 简单备选假设
simple-alternative detection 简单双择检测
simple analysis 简项分析
simple analysis of petroleum 石油简易分析
simple and convenient calculation 简便算法；简便计算
simple and opposite leaves 对生单叶
simple approach lighting system 简单进近灯光系统
simple arbitrage 简单套汇
simple arc 简单弧
simple arithmetic expression 简单算术表达式
simple arithmetic index number 简单算术指数
simple arithmetic mean 简单算术平均数
simple array assignment statement 简单数组赋值语句
simple asphyxiant 单纯性窒息性气体
simple astigmatism 单纯散光
simple automatic electronic computer 简单自动电子计算机
simple automatic frequency control 简单自动频率控制
simple automatism 简单自动作用
simple average 简单平均（值）
simple average loading 简单平均负荷
simple axis of symmetry 简单对称轴
simple balance 简单平衡
simple band brake 简单带式制动器
simple bar 简单棒
simple barrier 简单膜
simple batch distillation 简单间歇蒸馏
simple bath distillation 单程蒸馏
simple bath still 单程蒸馏釜
simple beam 简支梁；简支架（两点支持）；单波束；单靶射束
simple beam equation 简支梁方程
simple bearing 简支座；简支点
simple bending 简支梁弯曲；简单弯曲；单纯弯曲；纯曲率；纯挠
simple blocking game 简单封锁对策
simple block letter 简单方形字码

simple Boolean expression 简单布尔表达式
simple Bouguer gravity value 简单布格重力值
simple boundary 简支边界;简单边界
simple brass 纯黄铜
simple bridge 简支桥
simple buffering 简单缓冲法
simple building 简易房屋
simple bundle 简单丛
simple burst 简单暴
simple cantenary equipment system with single contact wire 单接触线的简单链形悬挂
simple cap 简单压顶块
simple capital structure 简单资本结构
simple carburetor 简单汽化器
simple cascade 简单级联
simple catenary 简单悬挂【电】;单吊线
simple catenary equipment with twin contact wire 双接触线的简单链形悬挂
simple catenary suspension 单式式悬链
simple chain 简单链系;单铰链
simple channeling method 刻线法
simple character 简单特征标
simple check 单裂
simple checkpoint 简易检查点;简单检验点
simple chemical analysis 简易化学分析
simple chlorination 初级氯化处理
simple chopper blower 简易切碎吹送机
simple circuit 简单围道;简单环道
simple circular arch 简单圆拱
simple classifier 普通分粒器
simple cleavage 简单裂解;简单解理
simple closed curve 简单(封)闭曲线
simple column 单柱
simple common subexpression 简单公因子表达式
simple comparative method 简易比较法
simple compiler 简单编译程序
simple compounds 简单化合物
simple compression 单纯压缩;纯压缩
simple compression method 纯压缩法
simple condition 简单条件
simple condition expression 简单条件表达式
simple conic(al) projection 普通圆锥投影;单圆锥投影
simple continued fraction 简单连分数
simple continuous distillation 平衡闪蒸法
simple continuous layout 简单连续观测系统
simple contract 口头合同;简单合同
simple contract creditor 普通债权人
simple control system 简单调节系统
simple cooperative game 简单合作对策
simple coordination 简单协作
simple cornice 简单的挑檐线脚;简单檐的头线脚
simple correlation 简单相关;单项关系
simple correlation coefficient 简单相关系数
simple coulometric method 普通电量滴定法
simple coupled 简单耦联的
simple covered type 单层掩盖型
simple credit 扣账信用证;扣押信用证
simple cubic(al) lattice 简单立方晶格
simple curve 简(单)曲线;单(心)曲线
simple curviline 单曲线
simple cuspate foreland 对称尖岬
simple cycle 简单循环
simple cycle gas turbine engine 简单循环燃气轮机
simple cycle operation 简单循环作业
simple cylindric(al) projection 单圆柱投影
simple data terminal equipment 简易终端
simple debenture 纯信用债券
simple decision rule 简单判定规则
simple declaration 简单说明
simple decrepitation method 简易爆裂法
simple defining 简单定义法
simple deformation 简单形变
simple dichotomy 简单二分支
simple differential 简单差速器
simple diffusion 简单扩散
simple dike 单堤;单岩墙
simple dimension(al) method 单量纲法;单量钢法
simple direct shear test 纯剪试验
simple discontinuity 简单间断性
simple distillation 简单蒸馏作用;简单蒸馏(法)
simple distribution 简单分布
simple domain 单纯域

simple double acting pump 单缸式复动泵
simple-duplex modem 简单双向调制解调器
simple-duplex pump 简单双数泵
simple eaves 简单檐口
simple eigenvalue 简单特征值;简单本征值
simple elongation 简单伸长
simple embankment 简单填筑堤;简单填筑路基
simple engine 单级发动机;单缸发动机
simple entry 简单分录
simple equation 一元方程;一次方程(式)
simple equivalent circuit 简化等效电路
simple event 单纯事件
simple event flood 单峰洪
simple exponential equation 简单指数方程式
simple exponential model 简单指数模型
simple extended array 简单扩充台阵
simple extension 简单拉伸
simple extensional flow 简单拉伸流动
simple eye 单眼
simple fan-in argument 简单扇入变元
simple fault 简单断层
simple fendering 简单木护舷
simple fish-stick 单头鱼叉
simple fixed frame 简单固定框架
simple flexure 纯挠曲
simple flow 简单流
simple fluid 简单流体
simple fluorite ore 单矿物萤石矿石
simple fold 简单褶皱;简单褶皱
simple fold systems with steeply dipping axial plane 具陡倾轴面的简单褶皱系
simple folium 单叶线
simple footing 简单底座;单基础底
simple forecasting model 简单预测模型
simple form 单形;单纯型
simple formal parameter 简单形式参数
simple fraction 简分数
simple frame construction 简单框架结构
simple frame core box 敞开式芯盒
simple frame steel structure 简单框架钢结构
simple (frame)work 简单构架;完整构架;简单刚架
simple free flexure bar 简单双端弯曲棒
simple freezing-in 简单凝入
simple frequency resonance decoupling method 单频率共振去耦法
simple friction 纯摩擦力
simple function 简单函数
simple function city 单功能城市
simple gate 单门
simple ga(u)ge relation 单一(相应)水位关系
simple geared train 单式轮系
simple gearing 简单传动装置;单纯的齿轮传动
simple glass 钾钠玻璃;普通玻璃
simple governer 简易调节器
simple grade separation 分离式立体交叉
simple grain 单粒
simple graphical partition 简单的图划分
simple gravity circulation 简单重力循环;简单重力环流
simple group 单群
simple halo 简单晕
simple hand excavation 人力开挖
simple hand jig 简单手动跳汰机;简单人工跳汰机
simple handle 简单操纵
simple harmonic 简谐的
simple harmonic alternating current 简谐交变电流
simple harmonic current 简谐电流
simple harmonic curve 简谐曲线
simple harmonic electromotive force 简谐电动势
simple harmonic group 简谐波组
simple harmonic law 正弦定律;简谐定律
simple harmonic motion 谐运动;简谐运动;简单谐和运动
simple harmonic motion cam 简谐运动凸轮
simple harmonic oscillator 简谐振荡器
simple harmonic period change 简谐周期变化
simple harmonic period load 简谐周期荷载
simple harmonic progressive wave 简谐前进波
simple harmonic quantity 简谐
simple harmonic sine wave 简谐正弦波
simple harmonic vibration 简谐振动
simple harmonic wave 简谐波
simple harmonic wave motion 简谐波动
simple header boiler 整联箱式锅炉

simple hinged frame 简单铰接框架
simple hoisting machinery 简单起重机械
simple hydrograph 简单过程线;单峰过程线
simple indexing 简单分度法
simple index number 简单指数
simple infrared system 简单红外系统
simple inhibition 简单抑制
simple injector 简单喷射器
simple insertion 简单嵌入;简单插入
simple insert needle 普通插入式针头
simple integer invariant 简单整型不变量
simple integral 简单积分
simple-integral viscoelastic fluid model 简单积分型黏弹性体模型
simple interest 线性增长;单利(息)
simple interest formula 单利公式
simple interferometer 简单干涉仪
simple intersection 简单交叉口
simple knot 反手结
simple Kriging method 简单克立格法
simple labo(u)r 简单劳动
simple lag network 简单相移网络
simple lap 简单搭接
simple lattice 简单点阵
simple lay 单绞
simple leaf 单叶
simple lens 单透镜
simple lever 单杠杆
simple lever-type weighing machine 杠杆式简易秤
simple licence contract 普通许可证合同
simple licensing agreement 普通许可证协议
simple life insurance 简易人身保险
simple light 普通光
simple line 单线
simple linear 单线的
simple linear representation 简单线性表示
simple linear vibration system 简单线性振动体系
simple link 单节
simple liquid 简单液体(切变率与切应力成正比);简单流体;单纯液体
simple list 简单表格
simple list form 简单编目形式
simple list structure 简单表结构
simple lithology type 岩性较单一型
simple live axle 简单传动轴
simple loading system 简单装卸系统
simple loan 普通贷款
simple lock 套闸
simple locomotive 单胀式机车
simple loop of bubble 磁泡简单环
simple loss function 简单损失函数
simple machine 简单机械
simple macro-scheme 简单宏功能方案
simple magnetic anomaly 简单磁异常
simple magnetic network 单一磁网络
simple magnifier 简单放大镜
simple mail transfer protocol 简单邮件传输协议
simple manipulator 简易机械手
simple mapping 简单映像
simple Markovian queue 简单的马尔可夫排队
simple matrix language 单纯矩阵语言
simple mean 简单平均(值)
simple mean value 简单平均值
simple metal 纯金属
simple meteorologic(al) observation 简易气象观测
simple microelastic solid 简单微弹性固体
simple microscope 简易显微镜;简单显微镜;单式显微镜
simple mixed flow 简单混流
simple mixed strategy precedence grammar 简单混合策略优先文法
simple money circulation 简单货币流通
simple monochromator 简易单色仪
simple monoclinic 简单单斜
simple monoclinic structure 简单单斜结构
simple morphologic(al) 形态简单的
simple motion 简单运动
simple moving average 简单移动平均
simple multiple 简单复联
simple multiple reflection 简单多次反射
simple network management protocol 简单网络管理协议
simple number 基数;单值

simple numerical system 简单数字制
simple object 简单对象
simple ocular 单透目镜
simple order 全序
simple oriented path 简单的有向路径
simple orthorhombic 简单正交
simple oscillator 谐振子;简谐振子;简单振子
simple parabola 简单抛物线
simple parallel cable system 简单平行钢索体系
simple parallel winding 简单并联绕组;单并联绕组
simple parameter 简单参数
simple parity 简单奇偶校验
simple parity check code 简单奇偶检验码
simple pass tubular heater 单程列管加热器;单程管式加热器
simple path 简单通路;简单道路
simple payment loan 一次偿清贷款
simple pendulum 单摆
simple perforation 单穿孔
simple perforation plate 单穿孔板
simple periodic(al) motion 简单周期(性)运动;简周期运动;简谐运动
simple phase gas in fluent model 单相气体渗流模型
simple piezoelectric(al) probe 简易压电探针
simple pile frame 简易桩架
simple pinch out 简单变薄尖灭
simple pipe tee reactor 单管三通反应器
simple piston rod 单面活塞杆
simple pit 单纹孔
simple pit pair 单纹孔对
simple plastic flow 简单塑性流
simple plasticizer 低分子增塑剂
simple plastic theory 线塑性理论
simple plate 普通板
simple pointer 单指针
simple polygon 简单多边形
simple pore 单纹孔
simple potentiometer 线性电位计
simple power lathe 单面动力车床
simple precedence relation 简单优先关系
simple precedence technique 简单优先技术
simple press tool 单工序冲模
simple probability 简单几率
simple probability curve 简单概率曲线
simple process 单级过程
simple process factor 单级过程分离系数
simple processor 单处理机
simple product 简单积
simple program(me) control 简易顺序控制
simple programming language 简易程序设计语言
simple progressive system 简单行进式(交通信号)系统
simple proportion 单比例
simple pseudo-manifold 简单伪流形
simple pump 单缸泵
simple pumping test from domestic well 民井简易抽水试验
simple purpose plant 目的单纯的工厂
simple purpose reservoir 专功能水库;单目标水库
simple pursuit game 简单追击对策
simple push 简单推力
simple push-down automaton 简单(的)下推自动机
simple quantity 质数
simple radioreagent method 简单放射性试剂法
simple random sampling 简单随机抽样;单纯随机抽样
simple rate of return 简单收益率
simple reaction 简单反应
simple rectification 简单检波
simple rectifier 简易纠正仪
simple recursion 简单递归
simple reflection goniometer 单反射测角仪
simple reflex 简单反射
simple refrigeration 单式冷藏
simple regression 简单回归;单回归
simple regression analysis 一元回归分析;简单回归分析;单元回归分析
simple repeatable robot 简单重复式机器人
simple replacement 简单取代
simple reproduction of fixed assets 固定资产简单再生产
simple resonance 简单共振

simple retaining 简易支挡
simple ring 简单环形构造;单环
simple road 简易公路
simple rolling process 简易轧制过程
simple roof 单坡屋顶
simple roof cover(ing) 单坡屋顶覆盖层
simple roof truss 简单屋架
simple root 单根【数】
simple rotary sieve 简易回转筛
simple rotary-type cooler 单筒冷却机
simple rubble mound construction 简单斜坡式防波堤结构
simple safety 单一安全
simple salt 简单盐
simple sand spit 单沙嘴
simple scan 单点扫描
simple scanning 简单扫描
simple sedimentation 简单沉降
simple sedimentation tank 简单沉降槽
simple segregation equation 简单偏析方程
simple seiche 简单假潮;简单静振
simple selection 单项选择
simple self-excitation system 简单自励系统
simple self-propelled split-hull hopper barge 简易自航开体泥驳
simple selsyn motor system 简化自动同步机方式
simple series 简单级数
simple servomechanism 简式随动系统
simple settlement 简单沉降
simple settlement tank 简单沉降槽
simple shake 单裂
simple shape 简单断面型钢
simple sharn 简单矽卡岩
simple shear 简单切变;简单剪切;简单剪(力);单剪(力);单纯剪力;纯剪切;纯剪力
simple shear apparatus 单剪仪
simple shearing flow 简单剪切流
simple shearing stress 简单切应力
simple shear strain 简单切应变
simple shear test 单剪试验;纯剪试验
simple sieve plate 单筛板
simple signal 简单信号
simple sinusoidal alternating current 正弦交流电流
simple sinusoidal current 简单正弦电流
simple sinusoidal quantity 简单正弦值
simple site 简单场地
simple slab 简支板
simple slab deck 简支式面板
simple slab deck dam 简支板式坝
simple sliding friction 纯滑动摩擦
simple slip 纯滑移
simple solid 简单固体
simple solution 简单题解;单纯溶液;真溶液
simple sound source 简单声源;点声源
simple span 简支跨(度);计算跨度
simple span bridge 单跨桥
simple species toxicity test 单物种毒性试验
simple square crossing 简单十字形交叉
simples solar shield 简易太阳屏
simple stack 简单叠加
simple-stage suspended sediment sampler 单级悬移质采样器
simple staining 简单染色法
simple stationary gas producer 固定煤气发生炉
simple steam-engine 单级膨胀蒸汽机
simple steel 普通钢;碳钢
simple strain 简单应变;单纯应变;纯应变
simple strand 单股
simple stress 简应力;简单应力;简单受力;单轴(向)应力;单纯应力;纯应力
simple structure 简单结构;静定结构;简单构造
simple strutted frame 简易撑架
simple style in lithologic(al) character and geologic(al) structure 岩性及构造简单型
simple substance 单质
simple substitute 简单置换
simple subtractive colo(u)rant mixture 简单减色法混合
simple summation 简单求和法
simple support 简单支承;简支;简单支座
simple-supported 简单支承的
simple supported beam 简支梁
simple supported beam bridge 简支梁桥

simple supported end 简支端
simple surface 简单曲面
simple surge chamber 简式调压室
simple surge tank 简式调压池;简(单)式调压塔;简式调压井
simple system 简单系统
simple table 简单表
simple tally mark 简单记数记号
simple tarifs 简单的塔立夫材积表
simple tension 单向拉伸;单纯拉力;纯拉伸;纯拉力
simple terracing works 露台工程;台地工程;草坪工程
simple test 简单试验
simple test device for rolling stock 车辆简易试验装置
simple tetragonal 简单四方
simple tetragonal lattice 单一正方晶格
simple texture 简单结构
simple theodolite 简便经纬仪
simple thermodynamic(al) material 简单热力学材料
simple tie 简单绑扎
simple time quenching 普通单一水淬法
simple tissue 简单组织
simple tone 单音;纯音
simple torsion 简单扭转
simple trammel 简单转筒筛
simple translocation 简单易位;单易位
simple transportation 纯输移
simple treadle lathe 单踏式车床
simple treatment 简单处理
simple triclinic 简单三斜
simple trommel 单筒筛
simple truss 静定桁架;简支桁架;简单桁架
simple tungsten carbide cobalt composition 纯碳化钨—钴制品;纯钴钨硬质合金制品
simple turbine 单级涡轮(机);单级汽轮机
simple turbine type mixer 简单涡轮式混合器
simple turnout 单开道岔【铁】
simple twin 简单双晶;简单孪晶
simple twist 简单绞合;简单胶合
simple two-splitting 简单二分岔
simple type 简易式
simple type wave generator 简单式波浪发生器;简式生波器(波浪试验室用)
simple umbel 单伞形花序
simple unbalanced-shaft screen 简单不平衡轴筛
simple unit 简单单位
simple unit cell 简单晶胞
simple variable 简单变量
simple vault (无肋、光面的)简单穹隆;简支穹腹;简单穹顶
simple viscoelastic micropolar material 简单粘弹微极性材料
simple viscous liquid 简单黏性液体
simple volcano 单火山
simple water quality analysis 水质简分析
simple water quality model 简单水质模型
simple wave 简单波
simple wave beam 单波束
simple weighbridge 简便称量桥
simple whip grafting 单舌接
simple wood ceiling 简单木顶棚
simple working stress 简单工作应力
simplex 模拟实验;单向(通信);单路(通信);单工;单缸;单(纯)形;单纯的
simplex air ga(u)ge 单针气压表
simplex algorithm 单纯形算法
simplex ball mill 单式球磨机
simplex burner 单路燃料喷嘴
simplex casement 单平开窗
simplex channel 单向通道;单工信道;单工通道
simplex circuit 单工(通信)线路;单工电路
simplex communication 单工通信
Simplex concrete pile 辛普拉斯钻孔灌注桩;单纯混凝土桩
simplex copying 单面复制
simplex criterion 单体判据;单纯形判据
simplex crusher 单式压碎机
simplex dialing 单工拨号
simplex double acting pump 单缸式双作用泵
simplex/duplex modems 单/双工调制解调器
simplex-duplex pump 单效双泵
simplex flyer frame 单程粗纱机

simplex injector 简单喷射器
simplex iteration 单纯形迭代
Simplex joint 辛普拉斯套管连接
simplex kiln 轮窑
simplex lap winding 单叠绕组
simplex method 单体法;单纯形(方)法
simplex method of linear programming 线性规划中的单纯形法
simplex milling machine 单柱铣床
simplex mode 单工方式
simplex motor 同步感应电动机;单工电动机
simplex multiplier 单纯形因子;单纯(形)乘子;单纯形乘数
simplexometer 简易乳胶比重计
simplex operation (带摩擦铰链的)窗五金;单工操作
Simplex pile 辛普拉斯桩
simplex pressure-jet system 简单机械雾化系统
simplex printing machine 单面印花机
simplex procedure 单纯形程序
simplex pull rod 单拉杆
simplex pump 单缸泵
simplex radio communication 单工无线电通信
simplex reciprocating pump 单缸位复泵;单缸往复(式)泵
simplex repeater 单工帮电机
simplex rock bit 牙轮钻头(钻硬岩石用)
Simplex rudder 辛普拉斯型舵;流线型平衡舵
simplex spiral classifier 单螺旋分级机
simplex station 单工(收发)电台
simplex steam pump 单缸汽动泵
simplex strainer 单层滤器
simplex streamlined rudder 单纯流线型舵
simplex system 单向通信系统;单工制;单工(通信)系统
simplex table 单层摇床;单层淘汰盘
simplex tableau 单纯形表
simplex transmission 单路传输;单工传输
simplex turbine aerator 叶轮曝气设备;叶轮充气器;叶轮鼓风机
simplex unit 单工机
simplex wave 单波
simplex wave wind(ing) 单波绕组
simplex winding 单(重)绕组;单排绕组
simplex working 单工通报
simple zero 单零点
simple zone melting 简单区域熔炼
simplicial approximation 单纯逼近
simplicial complex 单纯复形
simplicial homology 单纯同调
simplicial subdivision 单形分割
simplicity 简单化;单纯
simplicity in treatment 简单处理
simplicity of operator 操作简单
simplification 简易(性);简化;单纯化
simplification of figure 图概括
simplification of international trade procedures 国际贸易往来手续简化法
simplified approach 简化法
simplified basin lock 套闸
simplified bearing strength derivation 单承载力偏差
simplified block diagram 简化方框图
simplified calculation 简化解
simplified case 简化情况
simplified character 简化字
simplified circuit 简化电路
simplified code signaling system 简化的编码信号系统
simplified design of footing 基础简单设计
simplified diagram 简化图
simplified equation 简化方程
simplified flyover roundabout 简化双桥式环形立交
simplified forceps 简易镊
simplified foreign body localizer 简易异物定位器
simplified formula 简化公式
simplified hump 简易驼峰【铁】
simplified hump yard 简易驼峰编组场【铁】
simplified hypothermia unit 简式低温仪
simplified ideal adsorbed solution theory 简易理想吸附液理论
simplified instrument 简易仪器;简仪
simplified jumbo 简易钻孔台车
simplified lake and stream toxic analysis program(me) 简易湖泊和河流毒性分析程序
simplified lake ecosystem 简易湖泊生态系统
simplified land consolidation scheme 土地整理略图
simplified method 简化方法
simplified method of measuring discharge 测流简测法
simplified model 简化模型
simplified numerical automatic programmer 简易数字自动程序设计
simplified optic(al) construction 简化光学结构
simplified overhauling 简单维修
simplified performance prediction 特性简化预计法
simplified periodical 定期小修
simplified photogrammetry 简易摄影测量
simplified probit method 简化概率单位法
simplified profile 简明纵断面图
simplified representation 简化表示
simplified retouching 简化润饰
simplified rule 近似法则
simplified saline test 简易盐水试验
simplified schematic 简图;略图;原理图
simplified solution 近似解;简js解法;简化解
simplified sound meter 简易声级计
simplified surface course 简易面层
simplified survey method 简测法
simplified test procedure 简化的试验程序
simplified topography 简易测图
simplified topper 简易拔顶装置
simplified trial load method 简化的试验荷载法;简化试载法
simplified version 草图;示意图;简化的形式;简化型
simplified water disinfection 水的简易消毒处理
simplified water quality model 简易水质模型
simplified working process 简化工序
simplify 简化
simplifying assumption 简化假设
simplify uranium marked limit 简化铀指标限
simplify working process 简化工序
simplotite 绿水钒钙矿
simply-built highway 简易公路
simply connected 单连通的
simply connected curve 单连通曲线
simply connected region 单连通区域
simply connected space 单连通空间
simply-connected surface 单连通面
simply constructed 构造简单的
simply constructed building 简易建筑
simply curved shell 单曲面薄壳
simply equipped ward 简易病房
simply homomorphic 单同态映射
simply ordered set 全序集
simply periodic(al) function 单周期函数
simply supplied ward 简易病房
simply supported 简单支承的;简支的
simply supported at the edges 端简支;边缘简支;边沿简支
simply supported beam 简支梁;简支架(两点支持)
simply supported dam 简单支撑坝
simply supported edge 简支边
simply supported end 简支端
simply supported on four sides 四边简支
simply supported plate 简支板
simply supported slab 简支板
simply supported span 简支跨度
simply supported truss 简支桁架
simply uniformly convergent 简单一致收敛
simpsonite 羟钽铝石;钽铝石
Simpson mill 碾轮式混砂机
Simpson parabolic rule 辛普森抛物线法则
Simpson's 1/3 formula 辛普森三分之一公式
Simpson's diversity index 辛普森多样性指数
Simpson's formula 辛普森公式
Simpson's index 辛普森指数
Simpson's rule 辛普森法则
Sims position 半俯卧位
simulacrum 模拟物
simulant 模拟装置
simulate 模拟
simulate assembler 模拟汇编程序
simulated agent 模拟的化学战剂
simulated altitude 模拟高度
simulated altitude environment 模拟高度环境
simulated analysis 模拟分析
simulated annealing 模拟重结晶法
simulated approach 模拟进场
simulated atmosphere 模拟气氛
simulated attention 模拟注意信号
simulated attention breaking off 模拟引起注意中断
simulated-azimuth 模拟(的)方位
simulated carrier deck 模拟航空母舰甲板
simulated climatic condition 模拟气候条件
simulated clock 模拟昼夜时钟
simulated clock simulation clock 模拟时钟
simulated cold climate test 低温模拟试验
simulated collision warning 模拟碰撞警告
simulated command module 模拟指令舱
simulated condition 模拟条件;模拟工况
simulated converter 模拟转换器
simulated countermeasure 模拟干扰
simulated data 模拟数据;模块数据
simulated decentralization organization 模拟分权组织
simulated depth 模拟深度
simulated diagram 模拟图
simulated diving 模拟潜水
simulated diving device 模拟潜水装置
simulated drop 模拟投放
simulated earthquake vibration stand 模拟地震振动台
simulated emission 模拟放射;模拟发射
simulated environment 模拟环境;模拟的外界条件;环境仿真
simulated equipment 模拟设备
simulated failure 模拟故障
simulated fall-out 模拟沉降物
simulated feel 模拟感觉
simulated fire ball condition 模拟火球条件
simulated flame out 模拟熄火
simulated flameout landing 模拟停车着陆
simulated flight 模拟飞行
simulated flight plan 模拟飞行计划
simulated flight test facility 模拟飞行试验设备
simulated flow 模拟水流;模拟流动
simulated formation 模拟地层
simulated fuel slug 模拟燃料块
simulated gravity 模拟重力
simulated ground motion 模拟震动
simulated gunnery 模拟射击
simulated hearing loss 模拟听力损失
simulated impact test 模拟冲击试验
simulated impact tester 模拟冲击试验机;模拟冲击实验器
simulated input 模拟输入
simulated input condition 模拟输入条件;仿真输入条件
simulated input operation 模拟输入操作
simulated input sensor 模拟输入传感器
simulated input voltage 模拟输入电压
simulated insert bit 密排深水口站头
simulated in-situ method 模拟现场法
simulated installation fixture 模拟装配夹具
simulated interception 模拟截击
simulated interrupt condition 模拟中断条件
simulated interview 模拟面谈;模拟面试
simulated leakage 模拟泄漏
simulated life test 模拟使用寿命试验
simulated line 模拟线;仿真线
simulated locator 模拟定位机
simulated log-on 模拟注册
simulated machine indexing 模拟机器标引
simulated management operation 模拟管理作业
simulated maneuver 模拟机动
simulated market regulation 模拟市场调节
simulated masonry 仿石(砌体)
simulated missed approach 模拟复飞
simulated motion 模拟运动
simulated movement 模拟运动
simulated operating procedure 模拟运算步骤;模拟操作程序
simulated-operations testing 模拟作业试验
simulated optic(al) range tester 模拟光学距离测试器
simulated output program(me) 模拟输出程序
simulated overlay 简化模拟图
simulated part 模拟部件
simulated performance 模拟性能
simulated photography 模拟摄影

simulated pressure 模拟压力
simulated program(me) 仿真程序;模拟(试验)程序
simulated radar image 模拟雷达图像
simulated rainfall 人工降雨
simulated rain trails 模拟降雨试验
simulated range 模拟距离
simulated re-entry environment 模拟的再入大气层环境
simulated regional freight rate 模拟区域运价
simulated relief 地貌立体表示
simulated road condition 模拟道路条件
simulated road test 模拟道路试验
simulated roadway 模拟道路
simulated rocket 模拟火箭
simulated routine 模拟(例行)程序
simulated sampling 模拟抽样
simulated scene 伪造现场
simulated section 模拟剖面
simulated series 模拟序列
simulated service test 模拟应用试验;模拟使用试验
simulated service testing machine 模拟疲劳试验机
simulated silo 模拟发射井
simulated slot 模拟槽
simulated sonic log 模拟声波测井
simulated space condition 模拟空间环境条件
simulated space environment 模拟的空间环境
simulated spark-ignition engine 模拟火花点火发动机
simulated stereoautograph 模拟式立体测图仪
simulated strain 模拟应变
simulated strike 模拟攻击
simulated system 模拟系统
simulated target 模拟目标
simulated test 模拟试验
simulated test of engineering geology 工程地质模拟试验
simulated time 模拟时间
simulated traffic test 模拟交通试验
simulated training 模拟训练
simulated transit center 模拟转接中心
simulated use (在试验时进行的)模拟使用
simulated value 模拟值
simulated viscosity term 模拟黏性项
simulated voice recognition 仿真声音识别
simulated weightlessness 模拟失重
simulated weightlessness condition 模拟失重条件
simulated wind pressure 模拟风压
simulate thermal equilibrium 模拟热平衡
simulating 模拟
simulating audio system 模拟声频系统
simulating chamber 仿真(模拟)室
simulating control 模拟控制
simulating motor 模拟电动机
simulating rectification 模拟纠正
simulating signal 模拟信号
simulating signal(l)er 模拟信号机
simulating strata movement 模拟岩层移动
simulating system of railway price based on region 铁路运输模拟市场分区定价制度
simulating tractive performance 模拟牵引特性
simulation 制作模型;假装;拟态;模拟;模仿;伪装;仿真
simulation analysis 模拟分析
simulation and dynamic(al) programming 模拟和动态程序编制
simulation architecture 仿真结构
simulation chamber 模拟室;仿真(模拟)室
simulation clock 仿真时钟
simulation command 模拟命令
simulation computer 仿真计算机
simulation context 模拟范围
simulation control algorithm 仿真控制算法
simulation control center 模拟控制中心
simulation data 模拟数据;仿真数据
simulation database 仿真数据库
simulation device 模拟设备;仿真设备
simulation earthquake 模拟地震
simulation earthquake motion 模拟地震运动
simulation environment 模拟环境
simulation equation 模拟方程
simulation equipment 模拟设备
simulation error statistic 模拟误差统计
simulation executive 模拟执行

simulation experiment 模拟试验;模拟实验
simulation expert system 仿真专家系统
simulation game 模拟计划;模拟活动;模拟博弈;仿真博弈
simulation games for business 业务管理模拟训练
simulation-game technique 模拟对策技术
simulation hardware 仿真硬件
simulation imput device 模拟输入设备
simulation in ecology 生态模拟
simulation information library 仿真信息库
simulation input device 仿真输入设备
simulation jamming 模拟式干扰
simulation job 模拟作业
simulation laboratory 模拟实验室
simulation language 模拟语言
simulation manipulation 模拟处理;仿真处理
simulation method 模拟方法;模拟法;仿真法
simulation methodology 模拟方法学;模拟方法论;仿真方法学;仿真方法论
simulation model 模拟模型;模拟模式;仿真模型
simulation modelling 模拟型化
simulation model parameter estimation 模拟模型参数估计
simulation model program(me) 模拟样机程序
simulation monitoring 模拟监控;模拟监督
simulation of climate 气候模拟
simulation of cognitive processes 可识过程模拟
simulation of construction process of exploration engineering 勘探工程施工过程模拟
simulation of deposits 矿床模拟
simulation of microclimates 小气候的模拟
simulation of random events 随机事件模拟
simulation of single sludge process 单一污泥工艺模拟
simulation of stratigraphic(al) section 地层剖面的模拟
simulation of traffic flow 交通流量模拟
simulation of wastewater treatment model 废水处理模拟模型
simulation operation center 模拟活动中心
simulation operations control center 模拟活动控制中心
simulation optimization 模拟优选
simulation package 模拟程序包
simulation parameter 模拟参数
simulation period 模拟期
simulation process 模拟过程
simulation program(me) 仿真程序
simulation programming 仿真程序设计
simulation programming language 仿真程序设计语言
simulation pulse transmission 模拟脉冲传输;仿真脉冲传输
simulation report 模拟研究报告
simulation research design 模拟作业研究设计
simulation result 模拟结果
simulation run 模拟运行;仿真运行
simulation science 模拟科学
simulation series 模拟序列
simulation software 模拟软件;仿真软件
simulation software program(me) 模拟软件程序
simulation study 模拟研究;仿真研究
simulation supervisory program(me) 仿真监督程序;仿真管理程序
simulation symposium 仿真论文集
simulation system 模拟系统
simulation technique 模拟技术;仿真技术
simulation test 模拟试验;模拟测试;仿真试验
simulation test deck 模拟测试组
simulation test method 模拟实验法
simulation theory 仿真理论
simulation training system 模拟训练设备
simulation wellbore hole 模拟井筒
simulative 模拟的
simulative counter 模拟计数器
simulative formability test 模拟成型性试验
simulative generator 模拟振荡器;模拟发送器
simulative method 模拟(电路)法
simulative network 模拟网络
simulative relation 模拟关系
simulative type test 模拟型试验
simulator 模拟装置;模拟计算机;模型设备;模拟设备;模拟器;模拟程序;仿真器
simulator cabin 模拟器座舱

simulator cockpit 模拟驾驶舱
simulator compiler 模拟程序的编译程序
simulator debug 模拟程序调试;仿真程序调试
simulator debug utility 模拟程序的调试应用
simulator generation 仿真程序生成
simulator investigation 模拟研究
simulator memory 仿真程序存储器
simulator model 模拟装置模型
simulator of the digitizer 数字化器模拟设备
simulator package 仿真程序包
simulator program 仿真程序
simulator prototype development 模拟样机研制
simulator rig 模拟装置
simulator software 模拟程序软件
simulator software package 模拟程序软件包
simulator software program(me) 模拟软件程序;仿真软件程序
simulator stand 模拟试验台;仿真试验台
simulator time 模拟装置工作时间
simulator training 仿真器训练
simultaneity 同时性;同时发生
simultaneity factor 同时系数;同时率
simultaneous 联立;同时使用系数;同时发生的
simultaneous access 同时存取;并行存取
simultaneous acquisition 同时采集
simultaneous adaptation 同时适应
simultaneous adjustment 联合平差
simultaneous adsorption 同时吸附
simultaneous altitudes 等高;同时观测的天体高度;同时高度;同时地平纬度
simultaneous approximation 同时逼近
simultaneous assault 同时发起的冲击
simultaneous back filling injection 同步回填注浆
simultaneous biological nitrogen and phosphorus removal 同时生物除磷脱氮
simultaneous blast(ing) 齐发爆破;齐爆;同时爆破
simultaneous broadcasting 联播;同时广播
simultaneous bullet gun perforator 联动子弹射孔器
simultaneous bus operation 同时总线操作
simultaneous call(ing) 同时呼叫
simultaneous carry 同时进位;并行进位
simultaneous census 同时普查
simultaneous collection of two-ion beams 双离子束的同时接收
simultaneous collision 同时碰撞
simultaneous colo(u)r television system 色场同时传送制
simultaneous combined braking 同时复合制动
simultaneous communication 同时通信
simultaneous comparison methods 同时比较法
simultaneous conditioned reflex 同时条件反射
simultaneous connection 同时连接
simultaneous contrast 同时对比
simultaneous control 同时控制
simultaneous cross-folds 同构造交叉褶皱
simultaneous decking 同时装罐
simultaneous denitrification 同时反硝化
simultaneous denitrification-dephosphorization 同时脱氮除磷
simultaneous dependence 同时相关
simultaneous detection 同时检测
simultaneous determination 同时测定
simultaneous determination of two components 二组分同时测定
simultaneous difference equations 联立差分方程
simultaneous differential equation 联立微分方程
simultaneous diffusion 同时扩散
simultaneous displacement 同时位移
simultaneous displacement method 同时替换法
simultaneous display 同时显示
simultaneous distribution 联合分布
simultaneous double line 双转点水准测量路线
simultaneous draw texturing machine 同时拉伸变形机
simultaneous drilling 双井筒钻进
simultaneous drying and grinding 同时烘干和粉磨;粉磨兼烘干
simultaneous earthquake 同时(发生的)地震
simultaneous effect 联立效应;同时效应
simultaneous ejection 同时喷出
simultaneous equations 联立方程组;联立方程(式)
simultaneous equations estimate 联立方程估计

simultaneous equations estimator 联立方程组估计量
simultaneous equations method 联立方程组方法
simultaneous equations system 联立方程组
simultaneous equations techniques 联立方程组方法
simultaneous equilibration 同时平衡
simultaneous error 同时出错
simultaneous estimation method 同时估计法
simultaneous exposure 同步曝光
simultaneous feedback error component 同时反馈误差分量
simultaneous filling 立即充填;同时充填
simultaneous finished process 同时整理法
simultaneous firing 同时起动(多发动机);同时点火;同时爆破;同时爆发
simultaneous frequency and amplitude modulation 同时幅频调制
simultaneous ground fault 同时接地故障
simultaneous grouting 同时回程注浆
simultaneous humping 同时流溜放
simultaneous humping of two trains 双溜放
simultaneous ignition 同时点火
simultaneous independent model aerotriangulation adjustment 独立模型法空中三角测量整体平差
simultaneous induction 同时诱导
simultaneous inequalities 联立不等式
simultaneous information 同步信息
simultaneous initiation 同时起爆
simultaneous input-output 并行输入一输出
simultaneous input pulse 同时输入脉冲
simultaneous instruction system 同时指令系统
simultaneous interpretation 译意风;同声传译
simultaneous interruption 同步中断
simultaneous issue 共同投保
simultaneous iteration 同时迭代
simultaneous landing 同时着陆
simultaneous level line 双转点水准测量路线
simultaneous linear equations 联立线性方程
simultaneous lobing system 同时扫掠制
simultaneously extracted metal 同时萃取金属
simultaneously realizable 可同时实现的
simultaneous measurement 同时测量;同时测定
simultaneous mode 同时处理方式
simultaneous mode of working 同时工作方式
simultaneous motion 同时动作
simultaneous motion cycle chart 同时动作图解;同时动作周期图
simultaneous movements 同相运动
simultaneous multielement analysis 同时多元分析
simultaneous multiframe analytic(al) calibration 同时多片解析改正
simultaneous multiplier 同时乘法器
simultaneous mutation 同时突变
simultaneous nitrification-denitrification 同时硝化一反硝化(脱氮技术)
simultaneous nitrogen and oxygen analysis program(me) 同时氮氧分析程序
simultaneous observation 同时观察;同时观测;同步观测
simultaneous observation method 同时观测法
simultaneous offset 同步时差
simultaneous operation 同时操作
simultaneous operation at throat 咽喉区平行作业
simultaneous ordinary differential equation 联立常微分方程
simultaneous oxic and anoxic bioreactor 同时好氧缺氧生物反应器
simultaneous oxidation 同时氧化
simultaneous partial differential equation 联立偏微分方程
simultaneous phosphorus and nitrogen removal 同时除磷脱氮
simultaneous precipitation 同时降水
simultaneous processing 同时处理
simultaneous reaction 联立反应;同时反应
simultaneous reception 同时接收
simultaneous row iteration 同时行迭代
simultaneous scan 同时扫描
simultaneous scanning 同时检索
simultaneous settlement 同时交易
simultaneous settlements clause 同时付清条款
simultaneous shortcut nitrification and denitrification 同时短程硝化反硝化
simultaneous shotfiring 同时起爆
simultaneous signal 同步信号
simultaneous slip theory 同时滑移理论
simultaneous solution 联立解
simultaneous stage 同时水位
simultaneous statistic(al) inference 同时统计学推论
simultaneous summation 同时性叠加
simultaneous system 同时制;同时系统;同步系统
simultaneous system of equations 联立方程组
simultaneous talking 同时通话
simultaneous techniques 同步技术
simultaneous tension 同时张拉
simultaneous test 对比试验
simultaneous thermogravimetric-differential thermal analyzer 同时热解重量—差热分析仪
simultaneous throughput 同时吞吐量
simultaneous titration 联合滴定
simultaneous tone system 同时单音制
simultaneous traffic signal controlling 同时信号控制法
simultaneous translation 同时翻译
simultaneous transmission 同时传输
simultaneous transmission and reception 同时收发
simultaneous triggering 同时触发
simultaneous tripping 同时解扣
simultaneous twist lock system 同步纽锁系统
simultaneous two-way operation 同时双向操作
simultaneous two-way transmission 双向同时传输
simultaneous type 同时型
simultaneous type radio range 同时制无线电信标
simultaneous variables 联立变数
simultaneous variance ratio test 联合方差比检验
simultaneous velocity 同时流速
simultaneous water and non-aqueous phase liquid flow 水和非水相液体混合流
simultaneous water level 同时水位;同步水位
simultaneous wheels 同时轮
simultaneous work(ing) 平行作业;同时工作
sinaite 正长石
sinap(in)ic acid 芥子酸
sincere lithophysa structure 实心石泡构造
sincipital presentation 高直位
Sinclair pressure visco(si)meter 辛克莱式压力黏度计
sincosite 磷钙钒矿
sindon oleatanae 油棉布剂
sine 正弦
sine bar 正弦曲线板;正弦规;正弦杆
sine capacitor 正弦电容器
sine center 正弦对准仪
sine condition 正弦条件
sine-cosine encoder 正弦余弦编码器
sine-cosine mechanism 正弦余弦机构
sine-cosine potentiometer 正弦余弦电位(计)
sine-cosine product 正弦余弦乘积
sine-cosine resolver 正弦余弦旋转变压器
sine-cosme generator 分析器
sine curve 正弦曲线;正弦弧
sine curve fault 舒缓波状断裂
sine-curve shaped structure 正弦状构造
sine curve tectonic belt 舒缓波状构造带
sine-curve transition 正弦型缓和曲线
sine distribution 正弦(曲线)分布
sine divider 正弦分度器
sine electrometer 正弦静电计
sine equivalent candle diagram 正弦等效烛光图
sine flutter 正弦调变度
sine formula 正弦公式
sine formulas of spheric(al) triangle 球面三角正弦公式
sine function 正弦函数
sine galvanometer 正弦电流计
sine integral 正弦积分
sine law 正弦律;正弦定律
sine method 正弦法
sine motion 正弦运动
sine motor 正弦马达;正弦电动机
sine movement 正弦运动
Sinemurian 锡内穆阶【地】
sine parallax 正弦视差
sine plate 正弦规;正弦板
sine product 正弦积
sine protractor 正弦量角器;正弦规;正弦尺
sine pulse 正弦脉冲
sine pump 正弦泵
sine qua non 必要资格;必要条件
sine relation 正弦关系
sine rule 正弦定理
sine series 正弦级数
sine-shaped 正弦曲线形的
sine-shaped impulse 正弦脉冲
sine signal 正弦信号
sinesoid 正弦曲线
sine spiral 正弦螺线
sine-squared approximation 正弦平方逼近
sine-squared colo(u)r test signal 正弦平方彩色测试信号
sine-squared network 正弦平方网络
sine-squared pulse 正弦平方脉冲
sine-squared pulse and bar signal 正弦平方脉冲与方波信号
sine-squared pulse response 正弦平方脉冲响应
sine-squared shaping network 正弦平方脉冲成型网络
sine-squared testing 正弦平方波测试
sine step 正弦阶跃
sine sweep 正弦扫描
sine table 正弦规;正弦表;正弦台
sine term 正弦项
sine theorem 正弦定理
sine transform 正弦变换
sine typo 无模式标本
sine variohm 正弦式可变电阻器
sine wave 正弦波
sine-wave carrier 正弦载波
sine-wave characteristics 正弦波特性
sine-wave clipper 正弦波限幅器
sine-wave convergence 正弦会聚
sine-wave current 正弦电流
sine-wave envelope 正弦波包络线
sine-wave flux pattern 正弦波通量图
sine-waveform 正弦波形
sine-wave generator 正弦波振荡器
sine-wave input 正弦波输入
sine-wave input power 正弦波输入功率
sine-wave modulation 正弦波调制
sine-wave oscillator 正弦波振荡器
sine-wave power-handling capacity 正波功率使用容量
sine-wave response 正弦波响应;正弦波频率特性
sine-wave shaped meander 正弦波形曲流
sine-wave signal generator 正弦波信号发生器
sine-wave sources 正弦波源
sine-wave switching 正弦波转换
sinewy 纤维状
singal aspect and indication 信号显示
singal fictive 虚设信号
singal flag 标旗
Singapore dollar 新加坡元
Singapore Interbank Offered Rate 新加坡银行同业拆放利率
Singapore International Monetary Exchange 新加坡国际货币交易所
Singapore Port 新加坡港
singe 烧焦
singeing machine 点火机
singel-plybellows 单层波纹管
singing 振鸣(振荡);蜂鸣
singing arc 杜德尔弧
singing effect 振鸣作用
singing gallery 歌唱廊;振鸣廊
singing margin 振鸣安全系数
singing of oscillation 振荡鸣声
singing phenomenon 鸣振现象
singing propeller 谐音推进器;蜂鸣推进器
singing signal 振鸣信号
singing stability 振鸣稳定性
singing stovepipe effect 火炉烟囱振鸣效应
singing-suppressor 振鸣压缩器
singing tolerance 振鸣容限
single 单根【岩】;单个的;单独(的);单次的
single 4-strand casting machine 单机四流式连铸机
single access mechanism 单存取机构
single access satellite 单通路卫星
single accumulator system 单累加器系统;单存储器系统

single-acting 单作用式;单作用(的);单效;单向作用;单动式;单动(的)
single-acting air compressor 单动式空(气压)缩机
single-acting air pump 单作用式空气泵
single-acting altitude control valve 单作用高度控制阀;单动高度控制阀
single-acting atomizer 单动式喷雾机;单动式弥雾机
single-acting centrifugal pump 单动(式)离心泵
single-acting clutch 单作用离合器
single-acting compressed-air pile driver 单动压缩空气打桩机
single-acting compression 单效压制;单面压制;单动压制
single-acting compressor 单(向)作用压缩机;单动(式)压缩机;单动空压机
single-acting crankless press 单动偏心式压力机
single-acting crank press 单动曲柄式压力机
single-acting cross head type engine 单作用十字头式发动机
single-acting cylinder 单作用油缸;单(向)作用(汽)缸
single-acting cylindrical cam 单动圆柱偏心轮
single-acting die 单动模(具)
single-acting dobby 单动式多臂机
single-acting door 单动式门;单向门;单面摇门;单开门
single-acting drawing die 单动拉延模
single-acting engine 单作用(式)发动机;单向作用发动机;单动式原动机;单动(式)发动机;单动机
single-acting floor spring 单面地弹簧
single-acting hammer 单作用锤;单动锤;单动(式)汽锤
single-acting hand pump 单作用手摇泵
single-acting hinge 单向铰链;单动铰链
single-acting horizontal extrusion press 单动卧式挤压机
single-acting hydraulic cylinder 单作用液压缸
single-acting hydraulic shock absorber 单向式减震器
single-acting hydraulic stamping press 单动(薄板)冲压液压机
single-acting line 单向作用线路
single-acting pile hammer 单动(式)桩锤;单动打桩机
single-acting piston pump 单作用式活塞泵;单作用活塞泵
single-acting power unit 单动执行机构
single-acting press 单压头式压力机;单动(式)压力机;单动冲床
single-acting pressing 单向压制
single-acting press with drawing cushion 单拉延垫的单动压力机
single-acting printer 单动式印刷装置
single-acting pump 单作用泵;单动(式)泵
single-acting raising machine 单动式起绒机
single-acting ram 单向夯;单动式桩锤
single-acting relay 单侧作用继动器
single-acting steam engine 单动式蒸汽机
single-acting steam hammer 单动蒸汽锤;单动(式)汽锤
single-acting steam pile driver 单动蒸汽打桩机
single-acting steam pile hammer 单动蒸汽打桩机;单打蒸汽机柱锤
single-acting trunk piston type engine 单作用筒形活塞式发动机
single-acting turbine 单动式汽轮机
single-acting turbocharged two-stroke engine 单作用增压式二冲程发动机
single-acting vane pump 单作用叶片泵
single action 黑漆单弹簧铰链
single action black japanned spring hinge 黑漆单弹簧铰链
single-action door 单动门
single-action pump 单动式泵
single action shutter 自动上舷快门
single address 单(地)址
single address code 一地址码;单地址码
single address computer 单地址计算机
single address instruction 单地址指令;一地址指令
single address order code 单地址指令码
single adjustable switch box 单连活络开关箱
single admission 单向入口;单向进冷
single adsorbate 单吸附质
single adsorbate system 单吸附质系统

single aerial photograph 单张航片
single agency 单方代理
single air gripper feed 单一式压缩空气作用爪式送料
single airhydraulic press 单动气液压力机
single-aisle building 单楹房屋
single alternate system 单交替联动信号系统
single alundum 单晶刚玉
single amplifier 单级放大器
single amplitude 单振幅
single anchor 单锚(泊)
single-anchored sheet pile wharf 单锚板桩码头
single-anchored wall 单锚碇墙
single anchor leg mooring 单柱式单点系泊装置;单柱锚系;单锚腿锚泊装置
single anchor leg storage 单柱锚泊储油装置;单锚腿储[贮]油
single anchor mooring 单锚系泊
single and double action spring hinge 单双弹簧铰链
single and double column milling machine 单双柱铣床
single-and-multistart worm 单头多头蜗杆
single angle cutter 单角铣刀
single angle milling cutter 单角铣刀
single annular preventer 单囊式防喷器
single anode mercury arc rectifier 单阳极水银整流器;单阳极汞弧整流器
single anode rectifier 单阳极整流器
single antenna 单天线
single-apertured core 单孔铁芯
single-aperture lens 针孔透镜;单孔透镜
single-arch beam 单弯梁;单拱梁
single-arch dam 单拱坝
single-arch hammer 单柱锤
single-arc method 单弧方法
single-area method 单独流域法
single argument 单变量
single arm anvil 单臂砧;丁字砧
single armature converter 单枢变流机;单枢变换机;单枢整流机
single arm davit 单臂吊艇柱
single armed common anchor 单爪锚
single armed lever 单臂杆;单臂杠杆
single arm guide rod 单臂导杆
single armored cable 单层铠装电缆
single armored wire 单股铠装电缆
single arm pantograph 单臂受电弓
single arm protractor 单臂分度仪
single arm rig 单臂钻车
single arm spectrometer 单臂谱仪
single arm tiller 单臂舵柄
single assignment language 单赋值语言
single astronomical station datum orientation 大地基准单点定位
single atmospheric diving suit 单人常压潜水服
single-atom detection 单原子探测
single-atom detector 单原子探测器
single attribute assessment 单一品质评估
single attribute transformation 单一属性变换
single auger 单式螺旋
single-auger unloader 单螺旋卸载机
single automatic 单一自动的
single-automatic operation 单按钮自动操纵;单项自动操作;单自动操纵
single-axis control 单轴控制
single axis gyroscope 单轴陀螺仪
single-axis laser gyroscope 单轴激光陀螺仪
single-axis linkage 单轴悬挂装置
single axle 单轴
single-axle bogie 单轴转向架
single-axle compactor 单轴夯具
single-axle drill 单轴式条播机
single-axle drive 单轴传动
single-axle load 单轴加载;单轴荷载
single-axle platform car 单轴平板车
single-axle trailer 单轴拖车;单轴挂车
single-axle trailer excavator 单轴拖曳挖土机
single-axle trailer mixer 单轴拖曳拌和机
single-axle travel(l)ing gear 单轴行走部件;单轴支重台车;单轴起落架;单轴底盘
single-axle undercarriage 单轴行走部件;单轴支重台车;单轴起落架;单轴底盘
single-axle unit 单轴部件;单轴装置

single bag 单层袋
single balance 单盘天平
single-ball mandrel 单球心轴
single band 单波段;单频带
single banked boat 单座划桨
single banked rowing 单座划桨
single-bank engine 单列式发动机
single-bank superheater 单级过热器
single bare busbar 裸体单母线
single bar feed 单杆送料
single barge 单平底船;单驳船;独行驳船
single bar guide 单导板
single barrel carburetor 单筒式气化器;单腔化油器
single barrel electric(al) hoist 单筒电动卷扬机
single barrel engine 单缸发动机
single-barrow hoist 单手车卷扬机
single-bar winding 单条绕组
single base 单基座;单基础
single-base method 单基线法
single-batch extraction 单级分批萃取
single-bath extraction 单釜提取
single-bath method 单浴法
single battery 单电池
single bay 单跨
single-bay gable frame 单跨人字屋架;单跨人字框架
single-bay portal 单跨门架
single-bay portal frame 单跨门式刚架;单跨门架形框架
single-bay rigid frame 单跨刚构;单跨刚架
single-bay two-hinged frame 单跨双铰构架
single-bead lap joint 单道搭接接头
single-bead tyre 单缘轮胎
single beam 单跨梁
single-beam Beckman gas analyzer 束贝克曼气体分析仪
single-beam chargescontrolled tube 单束电荷控制管
single-beam colo(u)rimeter 单光束色度计
single-beam crane 单梁吊车;单梁起重机;单轨起重机
single-beam electric(al) crane 电动单梁起重机
single-beam electric(al) hook crane 电动单梁吊钩起重机
single-beam infrared spectrophotometer 单光束红外线分光光度计
single-beam laser recorder 单光束激光记录仪
single-beam oscillograph 单线示波器;单程示波器
single-beam oscilloscope 单束示波器
single-beam photoelectric(al) colo(u)rimeter 单光束光电比色计
single-beam spectrophotometer 单光束分光光度计
single-beam stair(case) 单梁楼梯
single-beam travel crane 单梁自行式起重机
single-beam ultraviolet spectrophotometer 单光束紫外线分光光度计
single-beam ultraviolet-visible-near infrared spectrophotometer 单光束紫外—可见—近红外分光光度计
single-beam ultraviolet-visible spectrophotometer 单光束紫外—可见分光光度计
single bearing 单轴承的
single-bearing radar prediction 单方位雷达预报
single beater scutcher 单打手成卷机
single beater scutcher with automatic lap end 带自动卷拉装置的单打手成卷机
single-beat escapement 单拍擒纵机构
single-beat fish 单面夹板接合
single-beat main stop valve 单座式主气门
single-beat valve 单层阀
single bed 单人床
single bed fluidized bed reactor 单段式沸腾床设备
single bed guest room 单人床客房
single bedroom 单人卧室
single bedroom with convertible couch-bed 备用两用沙发的单人卧室
single bed rotary press 单托床回转压呢机
single belt 单层皮带;单层带
single belting 单带装置
single bench (采料场的)单层工作面
single bend 单编结
single bend electrode holder 单弯电极管
single-bend valve 单弯气门嘴;单弯阀
single bent 单排架

single berth 单泊位
single berth cabin 单铺位客舱
single berth reinforced concrete jetty 单泊位钢筋混凝土突码头
single berth reinforced concrete trestle 单泊位钢筋混凝土栈桥
single bevel 单斜式
single-bevel butt joint 单面斜对焊接
single-bevel corner joint 单面坡口角接头
single-bevel groove 单边V形坡口；单斜面坡口；单斜槽
single-bevel groove weld 单斜角槽焊
single-bevel joint 单面斜接
single-bevel split ring connector 单斜面裂环连接
single-bevel T-joint 单斜面T形接合
single birth 单产
single bit 一位
single-bit error 单个位错误
single bit felling axe 单刃伐木斧
single bitt 单柱系缆桩；单系船柱
single blackwall hitch 单花钩结
single-blade 单翼；单刃；单桨
single-bladed bit 刮刀钻头
single-blade kneading 单桨捏和
single-blade mixer 单桨混合机；单桨搅拌机
single-blade pocket knife 单刃小刀
single blast earthquake 单发型地震
single blind 带单圈的盲板
single block 整块；单阻塞；单轮滑车；单轮车；单滑轮；单程序段；单饼滑车
single-block brake 单块闸；单块式制动器
single-block braking 单闸瓦制动
single-block footing 独块底座
single-block machine 单卷筒拉丝机；单次拉丝机
single-block type 整体式
single-block wall reinforcing 单层砌块墙加固
single-blow beading 单击镦锻
single blow hammer 单打锤
single body 单粒体
single body damper 单体风阀
single body inside micrometer 单体内径千分尺
single bollard 单柱系缆桩
single bolting 单螺栓接合；单螺栓连接
single bond 单约据；单一合同
single boom 单吊臂
single boom drill rig 单臂凿岩台车
single boom hydraulic drill carriage 单臂液压钻机
single boom jumbo 单臂(架)钻机
single boom rig 单臂(架)钻车
single boom roadheader 单臂掘进机
single boom system 摇杆吊货法；单杆吊货法；单臂架系统
single-border-style memorial plaque 单框式纪念碑
single borehole electromagnet wave method 单孔电磁波法
single-bottom plow 单铧犁
single bottom ship 单底船
single boundary belt 单世界植带
single bowl-apron lever 挡板—铲斗合一控制杆
single-box construction 单个盒子结构
single-box girder gantry crane 箱形单梁龙门起重机
single-box siphon 单箱式虹吸管
single brake shoe 单侧闸瓦
single branch 叉管
single-break 单断(点)；单独中断
single-break circuit breaker 单遮断路器
single-break contacts 单断接点组
single-break switch 单断开关
single breakwater 单道防波堤
single breakwater arm 单突堤式防波堤
single breast gin 单腔轧花机
single brick wall 单砖墙
single-bridge probe 单桥探头
single bridging 地板格栅撑；交叉撑加劲；用交叉撑加劲力；简单跨接；单剪刀撑；单层格栅撑
single bucket elevator 单斗提升机
single bucket excavator 单斗挖掘机；单挖土机
single bucket hoist 单斗卷扬机
single bucket loader 单斗装载机
single budget 统一预算；单一预算
single bull nose 单圆角；单外圆角
single bunk 单人铺位

single buoy mooring 单浮筒系泊；单点系泊
single buoy mooring facility 单点系泊设施
single buoy mooring system 单点系泊系统；单浮筒系泊系统
single buoy storage 单浮筒储油装置
single burner 单喉管燃烧器
single burner torch 单头喷灯
single burring machine 单锡林除草籽机
single burst experiment 单菌释放实验
single bus(bar) 单母线
single bus(bar) connection 单母线结线
single bus(bar) sectioned with tie circuit breaker 单母线分段带母联断路器
single bus(bar) system 单母线制；单母线系统
single bus operation 单总线操作
single-button carbon microphone 单容器炭粒传声器
single butt-strap 单对接搭板
single butt-strap riveting 对接单面盖板铆接
single cable hoist 单索卷扬机
single cable plane 单索面
single-cable system 单电缆制
single cableway 单缆索起重机；单索道
single cage hoist 单笼卷扬机
single call 单呼
single camber 单段曲面
single camshaft 单凸轮轴
single-cantilever 单悬臂
single-cantilever arm beam bridge 单悬臂梁桥
single-cantilever gantry crane 单悬臂龙门起重机；单悬臂起重机
single-cantilever pylon 单悬臂塔柱
single-cantilever shell 单悬臂薄壳
single-cantilever slide gate 单悬臂滑动门
single-canvas binder 单帆布带割捆机
single-canvas front mounted binder 单帆布带前悬挂式割捆机
single-capacity process 单容调节过程
single-capacity system 单容量系统
single capstan 单主动轮；单传动轮
single capstan mode 单轮方式
single card vocoder 单卡式声码器
single carriageway 单车道；单幅式道路；单幅式车行道
single carriageway motorway 单车道公路；单车道快车道
single carriageway road 单幅路；一块板道路；单幅式道路
single carrick bend 单花大缆接结
single carrier per channel 每通路单路载波；单道单载波
single-carrier theory 单载流子理论
single carry 单值进位；单一进位；单位进位
single cascade 单列叶栅
single-casement window 单扇窗；单层窗
single casing 单缸式
single cassette 单卡
single-cast 单铸(的)
single-catenary suspension 单链(式)悬挂
single cavity die 整体压型
single cavity joint 单空腔缝
single cavity linac 单腔线性电子加速器
single cavity maser amplifier 单腔量子放大器
single cavity mo(u)ld 整体铸型；单穴模；单腔模具
single cell 单腔式
single cell box girder 单元箱形梁
single cell for electrolyzation 单室电解槽
single cell furnace 单室炉
single cell porosimeter 单室孔隙度计
single-cell-type cofferdam 单孔式围堰
single centerd compound curve 单中心的复合曲线
single centrifugal pump 单(吸)离心泵
single chain 单链
single-chain bale elevator 单链式草捆升运器
single-chain drive 单链传动
single-chain knot (绳结的)单联结
single-chain-ring discharge grab 单环链卸货抓斗
single-chain suspension grab (bing) crane bucket 单链悬吊式抓斗式起重机抓斗
single-chain-type bench 单链拉拔机
single-chain type multipurpose elevator 单链式多用途升运器
single-chain type vertical elevator 单链立式升运器

single-chain unloading grab bucket 单链卸货抓斗
single-chamber furnace 单室炉
single-chamber incinerator 单室焚化炉
single chamber mill 单仓磨
single-chamber producer 单室煤气发生炉
single-chamber rotary drum vacuum filter 无格滚筒式真空过滤器
single-chamber separator mill 带选粉机的单仓磨
single-chamber unit 单室式机组
single-channel 单信道(的)；单通路；单通道；单频道；单路；单波道；
single-channel γ-spectrometer 单道伽马分光仪
single-channel analyzer 单道分析器
single-channel burner 单通道燃烧器；单通道喷煤管
single-channel carrier 单路载波机
single-channel communication 单路通信
single-channel control 单路控制
single channel cordless telephone 单信道无绳电话机
single-channel count 单道计数
single-channel digital exploration seismograph 单道数字地震仪
single-channel encryption device 单通道加密装置
single-channel energy spectrometer 单道能谱仪
single-channel frequency modulation telemeter 单通道调频遥测仪
single-channel galvanometric recorder 单通道电流测定记录仪
single-channel head 单路磁头
single-channel heated stylus recorder 单道加热式笔录记录器
single-channel image 单道影像
single-channel impeller pump 单路叶轮泵
single-channel instantaneous signal recorder 单通道瞬态信号记录仪
single-channel multiplier 单渠道电子倍增器
single-channel oxidation ditch 单道氧化沟
single-channel radiometer 单信道辐射计
single-channel record 单道记录
single-channel reflection seismic 单道反射地震
single-channel scanning pulse analyzer 单通道扫描脉冲分析仪
single-channel seismic refraction system 单道地震反射系统
single-channel seismic survey 单道地震法调查
single-channel simplex 单信道单工通信；单路单工
single-channel single fiber cable 单通道单纤光缆
single-channel streamer 单道拖缆
single-channel thermal infrared scanner 单通道热红外扫描仪
single-channel tracking receiver 单信道跟踪接收机
single-channel tunable spectrometer 单通道调谐调频仪
single-channel type 单通道型
single character 单字符
single-check snow plough 单向雪犁
single chemical compound explosive 单质炸药
single chip 单片
single chip architecture 单片结构
single chip computer 单片机
single chip processor 单片处理机
single-chisel bit 一字形钻头；一字形钎头；单刃钎头
single chopper 单列缺口重型圆盘耙
single circuit 单回路(的)；单工线路
single circuit line 单回路线
single circular cornering cutter 单圆角铣刀
single circular current 单向环流
single clad board 单面敷箔板
single clam-shell gate 单面扇形闸门
single-class estimation 单类的估计
single class passenger boat 单等级客船
single-cleat ladder 单踏步阶梯
single clip microcomputer 单片机
single clock 单一时钟脉冲
single closed magnetic circuit relay 单闭磁路继电器
single coat 单层
single-coat asphalt pavement 单层式地沥青混凝土路面
single coated paper 薄涂料层纸
single coating technique 表面只涂一层的组合屋面生产工艺
single-coat plaster 单层灰泥；薄涂灰泥；单层抹灰
single-coat stucco 一次抹面

single cog 单面齿
single-coil 单线圈的
single-coil clutch brake 单盘簧离合器制动器
single-coiled moderately slow release relay 单线圈缓放继电器
single-coiled relay 单线圈继电器
single-coil ignition 单线圈点火
single-coil spring 单圈弹簧
single-coil spring washer 单盘簧垫圈
single-coil washer 单卷垫圈
single cold and hot heel-shaping machine 单冷热式后跟定形机
single collar pipe 单盘直管
single-collar thrust bearing 单环(式)推力轴承
single collet winder 单头拉丝机
single-colo(u)ration 单色
single-colo(u)r automatic offset press 单色自动胶印机
single colo(u)red glass 单面有色玻璃
single-colo(u)red marble 单色大理石
single-colo(u)r glaze 单色釉
single-colo(u)r monochrome layer tinting 单色分层设色表
single-colo(u)r offset machine 单色胶印机
single-colo(u)r offset press 单色胶印机
single-colo(u)r press 单色印刷机
single colo(u)r print 单色印样
single-colo(u)r printer 单色印刷机
single-colo(u)r reprint 单色翻印
single-colo(u)r sheet-fed offset press 单色自动胶印机
single-colo(u)r wave 单色波
single column 单柱
single-column bridge 独柱式桥
single-column disconnecting switch 单柱型隔离开关
single-column footing 单柱底座
single-column gas chromatograph 单柱式气相色谱仪
single-column hydraulic drawing press 单柱拉伸液压机
single-column ion chromatography 单柱离子色谱法
single-column jig borer 单柱坐标镗床
single-column jig boring machine 单柱坐标镗床
single-column lathe 单柱立式车床
single-column manometer 单管式压力计
single-column milling machine 单柱铣床
single-column pence 单列十二单位方式
single-column pence coding 单列十二进制编码；每列单孔编码
single-column pile pier 单柱桩墩
single-column planer 单柱刨床
single-column planing machine 单柱刨床
single-column punch 单穿孔
single-column punch machine 单柱冲床
single-column radiator 单管暖气
single-column supported platform 单柱支撑平台
single-column type 单柱式
single-column vertical boring machine 单柱立式镗床
single-column vertical boring and turning lathe 单柱立式车床
single combed 单冠的
single command 单命令
single-compartment jig 单室跳汰机
single-compartment kiln 单室窑
single-compartment mill 单室磨碎机；单仓磨
single-compartment septic tank 单室化粪池
single-compartment type thickener 多尔型增稠器；单室增稠器
single completion 单层完井
single component 单成分；单组分
single-component adhesive 单一成分黏合剂
single-component blending bed 单组分预均化堆场
single-component epoxy adhesive 单组分环氧胶黏剂
single-component magnetic survey in borehole 井中单分量磁测
single-component phosphor 单一荧光粉；单一磷光体
single-component polyurethane enamel 单组分聚氨酯瓷漆
single-component polyurethane sealant 单组分聚氨酯密封胶
single-component polyurethane varnish 单组分聚氨酯清漆
single compound explosive 单体炸药
single-compound primary explosive 单质起爆药
single-computer system 单计算机系统
single-condenser filter 单电容滤波器
single conductor 单向导电
single conductor cable 单芯电缆
single conduit 单跟管道；单孔涵洞
single cone bit 单牙轮钻头
single cone clutch 单锥离合器
single-cone vessel closure 单锥容器密封
single-conic(al) saw 单斜圆锯；单凸面圆锯
single conservative pollution source 单恒定污染源
single consignment 整批交货
single constituent 单组分
single contact 单触点；单接点
single-contact controller 自动控制器；单触点控制器
single contact operation 单一接触操作
single-contact system 单触点制
single contract 单独承包合同；独家承包合同；单一合同；总包合约；总包合同；主要契约
single contraction pattern 自由缩尺的模型
single control 单钮控制；单独控制；单按钮控制
single control receiver 单控接收机
single control refrigerator freezer 单控式冷藏冷冻箱
single control wire 单控制线
single cord 单线塞绳；单软线；单股绳
single-core 单芯的
single-core cable 单芯电缆
single-core cable/PVC jacket 单芯电缆/PVC 套
single-core oil-filled cable 单芯充油电缆
single-core optic(al) connector 单芯光学连接器
single-core plastic wire 单芯塑料电线
single-core PVC cable 单芯 PVC 电缆
single-core tube 单管岩芯取样器
single-core wire 单股线
single corner block 单一角块
single-corridor layout 单走道布局
single cost system 单一成本制
single cotton-covered copper wire 单层纱漆包线
single cotton-covered enamel wire 单层纱漆包线
single cotton-covered wire 单层纱包线；单纱包线
single cotton double-silk covered wire 单纱双丝包线
single counter 单个计数管
single couple model 单力偶模型
single course 单层
single course asphalt pavement 单层式地沥青混凝土路面
single course concrete pavement 单层(水泥)混凝土路面
single course construction 单层筑路；单层结构
single course pavement 单层路面
single coursing 单层铺设法
single-covered butt joint 单盖板对接头
single covering 单股包芯纱
single cover plate riveting 单盖板铆接
single crank 单曲柄
single-crank arch press 单曲柄拱式压力机；单曲柄拱式冲床
single-crank double-action press 单曲柄双动压力机
single-crank press 单曲轴压力机
single crawler unit 单履带牵引装置
single criterion 单准则
single criterion under certainty 确定条件下的单准则
single criterion under risks 风险条件下的单准则
single crop farming 单一经营
single crossing over 单交换
single crossover 单开渡线；单渡线
single crystal 单晶体；单晶的；单晶
single crystal alumina 氧化铝单晶；单晶矾土
single crystal analysis 单晶分析
single crystal casting 单晶铸造
single crystal diffractometer 单晶衍射仪；单晶绕射计
single crystal element 单晶元件
single crystal fiber 单晶光纤；单晶纤维
single crystal filament 单晶丝
single crystal furnace 单晶炉
single crystal fused alumina 单晶电熔刚玉
single crystal glass 单晶玻璃
single crystal graphite 单晶石墨
single crystal ingot 单晶锭块
single crystal particle 单晶粉粒
single crystal photoconductor 单晶光导体
single crystal photoelectric conductor 单晶光电导体
single crystal silicon 单晶硅
single crystal superalloy 单晶超合金
single crystal tungsten wire 单晶钨丝
single crystal volume 单个晶体体积
single crystal X-ray diffractometer 单晶体 X 射线衍射仪
single current 单向电流；单流
single-current method 单(向)电流法；单流法
single-current system 单流制
single-current transmission 单电流传输
single curvature 单曲线
single curvature arch dam 等半径拱坝；曲率拱坝
single curvature shell 单曲率壳体
single curvature shell system 单曲面薄壳体系
single curvature surface 单曲率面
single curved glass 单曲面玻璃
single curved spike grid 单曲面钉格板
single curved surface 单曲率面；单曲面
single curve gear 单曲线齿轮
single curve tooth 单曲线渐开齿
single cut bastard half round file 单纹粗半圆锉
single cut file 斜纹锉；单纹锉(刀)
single cut-length 单切割长度
single cutter 单铣刀；单刀切纸机
single-cutter shearing machine 单刀剪毛机
single-cutting band-saw 单刃带锯
single-cutting bit 单刃钻头
single-cutting drill 单刃钻
single cutting tool 单刃刀(具)
single-cycle 单循环(的)
single-cycle error 单周误差
single-cycle key 单行键；单循环键
single-cycle mountain 单旋回山
single-cycle operation 单循环作业
single cyclone dust collector 单旋风除尘器
single-cylinder 单弹子锁；单缸(的)
single-cylinder blowing engine 单汽缸鼓风机
single-cylinder compressor 单(气)缸压缩机
single-cylinder diesel 单缸柴油机
single-cylinder engine 单缸发动机
single-cylinder hay loader 单滚筒式装干草机
single-cylinder hydraulic sounding machine 单缸液压触探机
single-cylinder loader 单油缸式装载机
single-cylinder manometer 单管压力计
single-cylinder motor 单汽缸发动机
single-cylinder mud pump 单缸泥浆泵
single-cylinder multitubular boiler 单筒多管式锅炉
single-cylinder piston pump 单缸活塞泵
single-cylinder plunger pump 单缸柱塞泵
single-cylinder pump 单缸泵
single-cylinder reciprocating pump 单缸往复(式)泵
single-cylinder steam engine 单汽缸蒸汽机
single-cylinder suction nozzle 单筒式吸咀
single-cylinder test stand 单缸试验装置
single-cylinder turbine 单缸汽轮机；单缸涡轮机
single-cylinder two stage air compressor 单缸双级空压机；单汽缸双级空气压缩机
single-cylinder two-stroke motor 单缸二冲程发动机
single-cylinder washer 单滚筒式洗涤机
single damper 单体阀；单气闸
single data set volume 单一数据集卷宗
single-day-load curve 单纯日负荷曲线
single day production 日产量
single day tide 一日潮；日周潮；单日潮；全日潮
single-deck 单板的；单层的
single-deck bus 单层公共汽车
single-deck classifier 单板分级机
single-decked(flat) pallet 单层托盘
single-decked flat strips pallet 单层平边托盘
single decker 单甲板船
single-deck ferryboat 单层甲板旅客渡轮
single-deck oven 单层烤炉
single deck pallet 单面托盘

single-deck screen 单层筛
single-deck ship 单甲板船
single-deck trolley bus 单层无轨电车
single-deck vessel 单甲板船
single-deck vibrating[vibratory]screen 单层振动筛
single declaration 单个说明
single-degree freedom 单自由度
single-degree freedom gyro(scope) 单自由度陀螺(仪)【船】
single degree of freedom 一次自由度;单一自由度
single degree of freedom system 单自由度系统;单自由度体系
single delivery contract 一批交货合同
single densilog 单源距密度测井
single densilog curve 单源距密度测井曲线
single density 单密度
single density drive 单密度装置
single density encoding 单密度编码
single density format 单密度(记录)格式
single density recording 单密度记录
single-depth room 单人深房间
single detail scribing 单要素刻图
single-detection receiver 单(极)检波接收机
single determinant 单行列式
single determination 一次测定
single diamond knot 握索结
single diaphragm pump 单隔膜式泵
single die 简单模
single-die drawbench 单模拉拔机
single-dielectric(al)fiber 单一介质光纤
single diesel engine 单台柴油机
single diffusion test 单向扩散试验
single-digit 单位数
single-digit adder 一位加法器
single-digit coding 单数码
single digit dialing 单码服务
single dike 单堤
single dimension observation 单一尺寸观测
single diode circuit 单二极管电路
single directional working 单方向运行
single direction automatic block 单向运行自动闭塞【铁】
single direction migration 单向迁移
single direction thrust bearing 单向推力轴承
single direction thrust pier 单向推力墩
single disc 单盘(的);单盘开沟器
single-disc boot 单盘开沟器
single-disc brake 单盘制动;单盘制动装置
single-disc clutch 单盘离合器
single-disc colter 单圆盘开沟器
single-disc drill colter 单圆盘条播开沟器
single-disc electromagnetic brake 单片式电磁制动器
single-disc electromagnetic clutch 单片式电磁离合器
single-disc polyphase meter 单盘式多相计数器
single-disc terrazzo grinder 单盘水磨石机
single-disc winding 单盘式绕组
single dislocation 单个位错
single distilled 一次蒸馏的
single distributary estuary 单叉河口
single distribution liquidation 一次分配结算
single dividing method 简单分齿法
single diving 单次潜水
single diving bottom time 单次潜水水底停留时间
single diving time 单次潜水时间
single domain 单域
single domain particle 单畴颗粒
single door 单扇门;单开门
single-door drop-bottom bucket 单底开式加料斗
single-door electric(al)control device 单门电动控制装置
single-door household refrigerator 单门家用冰箱
single-door leaf 单门扇
single-door pneumatic control device 单门气动控制装置
single-doped laser 单掺杂激光器
single dose 单次量
single-double line layout 单双管式布置
single-double-side offset paper 单胶双胶纸
single-double-swing door 单扇双开弹簧门
single drainage 一面坡排水;单向排水
single draw 单动拉延

single drift flight 单偏航飞行
single drill operation 单机凿岩作业
single drive 单传动;单独传动的;单动传动
single-drive pulley 单速皮带轮
single drive way 单车道
single drug prescription 单方
single-drum 单转筒;单鼓(的)
single-drum boiler 单鼓筒锅炉
single-drum central mixer 单筒集中拌和机;单筒集中搅拌机
single-drum concrete mixer 单筒预填骨料灌浆混凝土拌和机
single-drum dryer 单滚筒干燥器;单鼓干燥器
single-drum dryer with splash-feed 飞溅布料式单滚筒干燥器
single-drum hoist 单筒卷扬机;单筒绞车
single-drum roller 单筒振动压路机;单筒振动(路)碾;手扶单轮压路机
single-drum sheep-foot roller 单筒筒式羊蹄压路机
single-drum sheep's foot roller 单筒羊蹄式压路机;单筒滚筒羊足碾;单滚筒式羊蹄压路机
single-drum tilt rig 单滚筒倾斜钻机
single-drum type mortar mixer 单筒式灰浆搅拌机
single-drum vibratory roller 手扶单轮振动压路机
single-drum winch 单(卷)筒卷扬机;单筒绞车;单卷筒绞车
single-drum winder 单卷筒提升机
single duct air conditioning system 单风管空气调节系统;单风道空调系统
single-duct conduit 单孔管道
single-duct system 单风道系统
single dunnage 货舱单层衬垫
single ear anti-motion washer 单耳止动垫圈
single ear horn lock washer 单耳角形止动垫圈
single ear lock washer 单耳止动垫圈
single earpiece 单耳机
single-eave roof 单檐屋顶
single echo 单回波
single edge adze 单刃木锛
single edge blade 单刃刀片
single-edge crack 单边裂缝
single-edged 一字形的
single-edged cutting tool 单刃刀(具)
single-edged push-pull amplifier circuit 单边推挽放大器电路
single-edged skate 单边铁鞋
single-edge grooving saw 单刃割槽锯
single-edge sweep 单刃锄草铲
single edge veneering machine 单边截夹板机
single effect 单效;单向作用
single-effect distillator 单效蒸馏器
single-effect evapo(u)ration 单效蒸发
single-effect evapo(u)rator 单效蒸发器
single-effect lithium-bromide absorption-type refrigerating machine 单效溴化锂吸收式制冷机
single-effect multistage flash 单效应多级闪蒸法
single electric(al)ear control for mill loading 磨机负荷的单电耳控制
single electrochemical activity coefficient 单电化学活度系数
single electrode potential 单极电位;单电极电势
single-electrode system 单电极系统
single electrolyte solution 纯电解质液
single electronegative atom 单负电性原子
single electronic beam 单电子束
single element 单码元
single-element injector 单组元喷嘴
single-element lens 单元透镜
single element map 单要素元素图
single-element relay 一元继电器;单元件继电器
single enamel(l)ing 单层搪瓷
single enclosure 单层防护
single-ended 单端(的);单头的
single-ended 15 degree angle wrench 单头15°斜口扳手
single-ended adjustable spanner 单头活动扳手
single-ended amplification 单端放大
single-ended amplifier 单端放大器
single-ended boiler 单头(燃烧式)锅炉;单面燃烧式锅炉
single-ended boiler survey 单端锅炉检验
single-ended bottle washing machine 单端洗瓶机
single-ended coating line 单端式镀膜(生产)线
single-ended connector 单端连接器

single-ended control 单端控制
single-ended converter 单端变换器
single-ended deflection 单端偏转;不对称偏转
single-ended discriminator 单端鉴频器;单侧鉴频器
single-ended drive 单边齿轮驱动
single-ended hexagon straight wrench 单头六角直口扳手
single-ended hydro washer 单端喷淋式洗瓶机
single-ended input 单端输入
single-ended list 单终点连接表
single-ended mixer 单端混频器
single-ended open-jawed spanner 单头开口爪扳手
single-ended output 单端输出
single-ended output amplifier 单端输出功率放大器
single-ended push-pull amplifier 单端推挽大器
single-ended push-pull circuit 单端推挽回路;单端推挽电路
single-ended shooting 端点激发
single-ended snap ga(u)ge 单头卡规
single-ended spanner 单端环形扳钳;单头扳手
single-ended square straight wrench 单头方形直口扳手
single-ended straight socket wrench 单头直柄套筒扳手
single-ended straight tenoning machine 单头直榫开榫机
single-ended strength and elongation tester 单纱强力伸长试验机
single-ended synchronization 单端同步
single-ended system 单端系统
single-ended tenoning machine 单头燕尼榫开榫机
single-ended triode 单独三极管
single-ended twisting machine 单纱加捻机
single-ended type 单端引线式
single-ended user 单端用户
single-ended wrench 单端环形扳钳
single-end face mechanical seal 单端面机械密封
single-endwall packer 单头井壁封隔器
single-energy reactor 单能中子反应堆
single engine 单机牵引
single-engined 单发动机的
single-engined monoplane 单发动机单翼机
single-engine locomotive 单发动机机车
single engine wheel type tractor-pulled scraper 单引擎轮式拖拉机牵引铲运机
single entry 单一表目;单项;单式分录;单独输入;单吸式
single entry accounting 单式会计
single-entry centrifugal pump 单吸离心泵
single-entry compressor 单面进气压气机
single-entry impeller 单面进气叶轮
single-entry metering 单车驶入限流
single-entry pump 单吸泵
single-entry volume table 一元材积表
single-person household 单人户
single error 单一误差;单个错误
single error correcting 单个错误校正
single error correcting code 单差校正码
single error correction 单一误差校正
single error model 单一误差类型
single exchange rate 单一汇率;单一互换率
single excited state 单激态
single expanding wall 单页折叠式隔墙;单页折叠式隔断
single exposure 一次照射;一次曝光;一次接触
single exposure maximum allowable concentration 一次最高容许浓度
single extended shaft 单向外伸轴
single extent file 单范围文件
single exterior-corridor layout 单外廊布局
single extraction 简单抽提
single-extraction non-condensing turbine 单抽气非凝汽式汽轮机
single eyeglass 镜片;单眼镜
single eye piston rod 单环活塞杆
single face corrugated fiberboard 单面瓦楞纤维板
single face damping machine 单面给湿机
single-faced cladding 单面覆层
single-face(d)corrugated board 单面波纹纸板;单面瓦楞纸板
single-faced covering 单面外罩
single-faced insulation 单贴面绝热制品
single-faced pallet 单面镶灰板

single faced secondary clock 单面子钟
single faced stack 单面书架
single face mechanical sealing 单端面机械密封
single face planer 平刨;单面刨床
single face planing machine 单面刨床
single face woodworking planer 单面木工刨床
single factor 单因子
single factor evaluation 单一指数评价
single factor experiment 单因素试验
single factor optimization 单因素优选法
single factor trap 单因素圈闭
single factor water quality assessment method 单一指数水质评价方法
single factor water quality index 单因子水质指数
single failure 单故障
single fall and skid method 吊滑饼用法
single family attached dwelling 相连式独户住宅
single family detached dwelling 独立式独户住宅
single family detached house 独立式独户住宅
single family dwelling 单户住宅;独户住宅
single family home 独户住宅
single family home mortgage coinsurance 独户住宅抵押共同保险
single family house 独立住房;独户住宅
single family residence 独户住宅
single family semi-detached dwelling 半独立式独户住宅
single family unit 独户单元
single faucet sham-poo spray 单龙头洗头喷水器
single fault 单断层
single fed repulsion motor 单馈推斥式电动机
single feed 单面进料;单(级)馈电
single feeder 单馈电线
single feed hot-runner nozzle 单供料热流道喷嘴
single feeding and multiple receiving 一送多受
single feeding and multiple receiving track circuit 一送多受轨道电路
single fiber 单光纤
single fiber cable 单芯光缆;单芯电缆;单纤光缆
single fiber cable connector 单芯光缆连接器
single fiber jacket 单光纤套层
single fiber splicer 单光纤接头机
single-filament yarn 单纱;单丝
single-file stream 单列车流
single file volume 单文件卷宗
single fillet lap joint 单面搭焊接
single fillet weld 单边角接焊缝
single fillet welded lap joint 单面搭焊接
single fillet welded T joint 单面T形焊接
single filling duck 单纬帆布
single fire 一次烧成;单烧
single fired boiler 单火锅炉
single fired ware 一次烧成的制品
single firing 一次烧成;一次点燃
single firing enamel 一次烧成搪瓷
single firing process 一次烧成工艺
single fission 单分裂
single flag signal 单旗信号
single-flame blow pipe 单焰焊炬
single-flanged butt joint 单卷边对接接头
single-flanged lap joint 单法兰搭接
single-flanged ring 单边(„) 领
single-flanged wheel 单凸缘轮
single-flange roller 单凸缘滚筒
single-flange track roller 单边支重轮
single-flange track wheel 单凸缘履带轮
single-flange weld(ing) 单法兰焊接
single flank meshingtester 单面啮合检查仪
single-flapper valve 单喷嘴挡板阀
single flash 一次蒸馏;一次闪蒸
single flashing light 单闪光
single-flash pipe still 一次闪蒸管式釜
single-flash system 一次闪蒸系统
single flat yard 单向平面调车场
single Flemish bond 丁顺砖交错砌筑面;同皮丁顺砖交错砌筑;单面荷兰式砌合
single flight 单跑【建】
single-flight conveyer 整条输送机
single-flight cutter head 单翼切削土螺钻
single-flighted screw 单线螺杆;单导程螺杆
single-flight stair(case) 单梯段楼梯;一跑楼梯;单跑(式)楼梯;直筒楼梯;直上楼梯
single-flight type augerhead 单翼螺钻
single float visco(si)meter 单浮子黏度计

single floor 单跨楼板;单层楼板;单楼面
single-floor heating system 单层供热系统
single-floor hot water heating system 单层热水供热系统
single-floor house 单层式房屋
single-floor type auditorium 无楼座观众厅
single-floor warehouse building 单层式仓库建筑
single-flow 单流
single-flow boiler 单流式锅炉
single-flow centrifugal compressor 单流离心压缩机
single-flow pipe 单流管
single-flow pump 单流向泵
single-flow turbine 单流式涡轮机;单流式汽轮机
single-flow type boiler 单流式锅炉
single-flue 单烟道
single-flue boiler 单烟道锅炉
single-flue chimney 单烟道烟囱
single fluid cell 单液电池
single fluid method 单液法(灌浆)
single fluid nozzle 单相喷嘴
single fluid process 单液化学灌浆法;单液法(灌浆)
single fluid system 单流体系统
single fluke anchor 单爪锚
single flush type switch 单连暗开关
single flute bit 单槽钻头
single flute drill 单槽钻;单槽螺旋钻
single-focusing 单焦点的
single-focusing mass spectrometer 单聚焦质谱计
single footing 独立基脚;独立基础;单独底脚;单独基础
single force 集中力
single form 单形;单套模板
single formation 单路基道路
single foundation 柱基(础);墩式基础;独立基础;单独基础
single foundation in pillar 柱下单独基础
single foundation in wall 墙下单独基础
single frame 简单构架;构架单体;单张相片;单层框架
single-framed floor 单层楼板;单架屋顶
single frame hammer 单柱式锤
single-frame power hammer 单柱动力锤
single frequency 单频
single-frequency carbon dioxide laser 单频二氧化碳激光器
single-frequency code 单频码
single-frequency conversion 单频变换
single-frequency duplex 单频双工
single-frequency inductor 单频感应器
single-frequency laser 单频激光器
single-frequency laser diode 单频激光二极管
single-frequency noise 单频噪声
single-frequency polarized light 单频带偏振光
single-frequency signal generator 单频(率)信号发生器
single-frequency signal(l)ing 单频信令;单频发信
single-frequency simplex 单频单工
single-friction-surface clutch 单面摩擦圆盘离合器
single frieze balustrade 单钩栏
single-frogged brick 单凹槽砖
single front wheel tractor 三轮拖拉机
single-fuel system 单燃料系统
single functional group 单官能团
single function control valve 单功能控制阀
single-funnel 单管的
single-furnace boiler 单炉膛锅炉
single-furrow deep digger 单铧开沟犁
single-furrow plough 单铧犁
single-furrow plow 单铧犁
single gable(d)frame 单山墙式构架
single gamma spectrum method with truck 汽车单伽马能谱法
single-gang edger 单轴多片圆锯机
single-gang variable condenser 单轴可变电容器
single gap head 单隙磁头
single gate 挡闸闸(门);单扇闸门
single gate lock 单闸门船闸
single gate mo(u)ld 单浇口模
single ga(u)ge 单口卡规;单轨的;单规
single geared drive 单(级)齿轮传动
single-generation household 无子女户
single generation strain 单代品系
single girder 简支梁

single girder crane 单梁起重机
single girder gantry crane 单主梁桥式起重机
single girder overhead crane 单主梁门式起重机
single glass 单层玻璃
single-glassed window 一层玻璃窗
single-glazed unit 单釉面制品
single glazing 简单上釉;简单装配玻璃
single goal 单一目的
single grade wire 单级别钢丝;单品位钢丝
single-grain comminution 单颗粒粉碎
single-grained arrangement 单粒排列
single-grained coal 单粒级煤
single-grained structure 土的单粒结构;非团粒结构
single-grained texture 单粒结构
single-grain sowing 单粒播种
single-grain structure 单粒构造
single grate cooler 单箅床冷却机
single grip high angle smooth bottom wood bench plane 单握手高角平底刨
single grip low angle smooth bottom wood bench plane 单握手低角平底刨
single gripper feed 单爪式送进
single grip regular smooth bottom wood bench plane 单握手普通平底(木)刨
single groove 单坡口的;单面坡口;单槽的
single-groove gathering shoe 单槽集束器
single group link 单群链路
single grouser shoe 单齿履带板
single guest bedroom 单人客房
single guide 单面导轨
single guide crosshead 单滑板十字头
single-gun tricolo(u)r tube 单束三色显像管
single-gyro pendulous gyrocompass 单转子摆式陀螺罗盘
single half hitch 单半结
single halo 单一晕
single hand drilling 单手打眼;单人凿岩
single-handed working 单人操作(作业)
single hand grip 单机械手夹具;单机械手抓手
single-handled saw 单手锯;手锯
single harmonic distortion 单谐波畸变
single harmonic oscillator 单谐振荡器
single harness 单马挽具
single hatching 平行晕线
single-head bevel gear generator 单刀头锥齿轮刨齿机
single-head box socket set 单头套筒扳手
single-head choke tester 单头塞检验器
single-headed arrow 单头指针
single-headed eagle 单头鹰徽饰【建】
single-headed rail 单头轨
single-headed stud 单头芯撑
single header 单式集管
single-head flat scraper 单头平刮刀
single heading flight 测高术航行
single-head jigger 单刀压坯机
single-head scouring-bleaching machine 单头洗漂机
single-head wire drawer 单头拉丝机
single-head wire drawing machine 单线拉丝机;单头拉丝机
single-head wire stitcher 单头铁丝订书机
single-head wrench 单头扳手
single heat 一次加热
single-helical gear 单斜齿轮;单螺旋齿轮
single-helical heater 单螺旋灯丝
single hill 小山
single hinge 单铰链;单铰
single-hinged arch 单铰拱
single-hinged arch bridge 单铰拱桥
single-hinged arched girder 单铰拱大梁
single-hinged frame 单铰构架
single-hinged rigid frame 单铰刚架
single hit theory 单击学说
single-hoisting 单钩提升
single hoistway 单车提升井
single hole 单孔
single-hole directional coupler 单孔定向耦合器
single-hole method 单孔法
single-hole mixer 单孔拌和机
single-hole nozzle 单孔喷嘴
single hologram 单全息片
single hook 单钩
single hook airless shot blasting machine 单钩

真空抛丸清理机
single-hook piston ring 单钩活塞环
single-hook wire sling 单沟钢丝绳扣
single-hop transmission 单反射传输
single horizontal block 卧式单次拉丝机
single horn anvil 单角砧
single-hose nylon fixing seat 单管尼龙固定座
single household 单身住户
single-hub pipe 单壳管;单承口管
single-hue drawing 单色画
single-hue rendering 单色渲染
single hull 单船体
single hump 单驼峰
single-hump(ed)camel 单峰驼
single humping 单溜放
single humping and single rolling 单推单溜
single hump yard 单向驼峰编组场
single-hung window 单悬窗;单吊窗
single identifier 单标识符
single ignition 单发火
single ignition system 单电源点火系统
single image photography 单像摄影测量学
single-impeller impact breaker 单叶轮冲击破碎机
single impeller impact crusher 单转子反击式破碎机
single-impeller pump 单叶轮泵;单叶泵
single impression die 单槽模
single impression mo(u)ld 单穴模
single impulse 单脉冲
single-impulse turbine 单脉冲式涡轮机
single impulse welder 单脉冲焊机
single impulse welding 单脉冲焊(接)
single inclined differential barometer 单管倾斜压差计
single inclined plane 单倾斜面
single index dobby 单重尼杆多臂机
single index hold 单标志孔
single indirect addressing 单重间接型寻址
single inductive shunt 单线圈感应分流器
single-inductor 单电感线圈
single industrial building 单层工业厂房
single information desk 单个服务台
single-ingot pit 单座式均热炉
single injector 单喷嘴
single inlet fan 单面进风通风机
single-inlet impeller 单吸叶轮;单面进气叶轮
single-inlet pump 单吸泵
single in line 单边引出线
single in-line package 单行排齐封装;单列直插式封装
single-in-line package 单列直插式组件
single-input and multioutput light cable 单进多出光缆
single-input and single-output light cable 单进出光缆
single input controller 单输入控制器
single-input single-output system 单输入单输出系统
single instruction 单指令
single-instruction multiple data 单指令多数据
single instruction stream 单指令流
single interest 单利(息)
single-interlocking tile 单镶口瓦;单咬口瓦
single interrupt request 单中断请求
single intersection 单交(道路)
single-intersection truss 单交叉桁架
single-investor enterprise 独资企业
single ion activity coefficient 单离子活度系数
single ion detector 单离子检测器
single ion monitoring 单离子监测
single ion quantification 单离子量化
single-iris 单膜片;单窗孔
single iron pulley block 单轮铁滑车组
single island arc 单岛弧【地】
single island berth 单岛式泊位
single-item contracting 单项承包
single jack 单手用小锤;单螺旋起重器
single jack bar 单柱式钻架;单腿凿岩机架
single jersey knitting machine 单面平针织物针织机
single jet 单喷口
single-jet carburetor 单嘴气化器
single-jet flapper valve 单喷嘴挡板阀
single-jet nozzle 单孔喷嘴

single job 一次作业;一次操作
single job scheduling 单作业调度
single joint 单缝接头;单缝
single joint cutting 单节插
single-jointed drive shaft 单节主动轴
single journey 单程
single-journey ticket 单程票
single-kernel cumulative-drop planter 单粒排种积累式穴播机
single key 单头扳手
single-kick multivibrator 单冲多谐振荡器
single-knife hydraulic rubber cutting machine 单刀液压立式切胶机
single knife switch 单刀开关
single knot 单节;反手结
single lacing 单缀条
single ladder dredge(r)单梯式挖泥船;单链斗挖泥船
single-lamp luminaire(fixture)单灯照明装置
single lane 单船道;单车道
single-lane bridge 单车道桥(梁)
single-lane capacity 单车道通行能力
single-lane channel 单向航道;单线航道
single-lane highway 单车道公路
single-lane operation 单车道操作
single-lane road 单车道道路;单行车路
single-lane traffic 单线交通
single-lane tunnel 单车道隧道
single lantern-light 单一柱灯照明
single lap 单折叠;搭接
single-lap joint 单搭接接头;单面搭接
single-lap roofing tile 单面搭接屋面瓦
single-lap roof tiling 单搭接瓦屋面
single-lap tile 单(向)搭接瓦;平搭瓦
single-lap tiling 单搭接瓦屋面
single-lap weld 单搭接焊
single lath 单重(灰)板条
single latticed 单格子的
single-layer 单层(的)
single-layer aquifer 单层含水层
single-layer area 单层面积
single-layer bit 单层钻头;表面装嵌钻头
single-layer coil 单层线圈
single-layer course 单层线路;单层道路
single-layer crust model 单层地壳模型
single-layer diaphragm plate 单层隔仓板
single-layer drain 单层排水沟
single-layered 单层的
single-layered fabric for roofing wall 屋顶用单层织物
single-layered monolithic masonry wall 单层整体式圬工墙
single-layered roll roofing 单层卷材屋顶
single-layer grid 单层网络;单层隔墙
single-layer hot press 单层热压机
single-layering 单枝压条
single-layer lining 单层衬里
single-layer mode 单层模式
single-layer model 单层模型
single-layer monolithic wall 单层整体墙
single-layer network 单层网络
single-layer particle board 单层碎料板
single-layer partition 单层隔墙
single-layer potential 单层位
single-layer roofing 单层屋面
single-layer solenoid 单层螺线管
single-layer theory (柔性路面的)单层体理论
single-layer three phase winding 单层三相绕组
single-layer timber piling cofferdam 单层木板桩围堰
single-layer welding 单层焊
single-layer wind(ing)单层线圈;单层绕组
single laying of plank covering 单层盖板
single layout 单面布置
single lead 单管线
single lead cable 单芯电缆;单芯铅包电缆
single-leaf 单一门扉
single-leaf bascule bridge 单翼仰开桥;衡重式单翼仰开桥;单叶竖旋桥
single-leaf door 单扇门
single-leaf equalizer spring 单片均衡弹簧
single-leaf hinged bottom bucket 单底开式加料斗
single-leaf room door 单扇房门
single-leaf shutter door 单扇百叶门

single-leaf sliding door 单扇推拉门
single-leaf sliding shutter door 单扇滑动式百叶门
single-leaf sluice gate 单瓣泄水闸门;单片泄水闸门
single-leaf spring 单片(钢板)弹簧
single-leaf swing bridge 单翼平旋桥;单叶平旋桥
single-leaf type rolling gate 单门叶(式)滚动闸门
single-leaf wall 整体墙;单翼墙
single-leg concrete platform 单腿混凝土平台
single-leg manometer 单管压力计;单管测压计
single-leg platform 单腿式平台
single-leg production platform 单腿生产平台
single-leg propeller strut 独脚艉轴架
single-leg unit 单腿式平台
single length arithmetical 单倍长运算
single-length normalization 单倍长度规格化
single-lens 单镜头;单透镜
single-lens camera 单镜头摄影机
single-lens four-channel multiband camera 单镜头四通多谱段摄影机
single-lens instrument 单镜仪器
single-lens objective 单透镜物镜
single-lens observation 单镜观测
single-lens photograph 单镜头摄影像片
single-lens photographic(al)camera 单镜箱航空摄影仪
single-lens reflex camera 单镜头反光照相机
single-lens wide angle camera 单镜头宽角摄影机
single letter signal 单字母信号
single level 单级;个别能级
single-level address 一级地址;直接地址;单级地址
single-level fiber forming line 单层拉丝作业线
single-level garage 单层车库
single-level geometry 单层布置
single-level indicator method 单级显示方法
single-level machine 单级(存储)机
single-level operation 单层作业
single-level polysilicon 单层多晶硅
single-level resonance formula 单能级光振公式
single-level road junction 平面道路交叉口
single-level steam turbine 单级式汽轮机
single-level system 单级系统
single-level turret 单层转塔
single lever 单杠杆
single-lever brake 单杆闸
single-lever control 单杆控制
single-lever relief valve 单杆安全阀
single-lift compaction 单行程挤压;单行程压缩
single-lift construction 一次浇注施工法(混凝土);一次性浇注施工法;单层施工(法)
single-lift dobby 单动式多臂机
single-lift lock 单级船闸
single ligand 单配位体
single light 单灯
single light lamp 单光灯
single line 单线;单车道;单行(的);单线增音器;单绳
single-line aerial photography 单航线航空摄影
single-line automatic block 单线自动闭塞
single-line block 单线闭塞
single-line bridge 单线桥
single-line clamshell 单绳抓岩机
single-line connection 单线联络线
single-lined belt conveyer 单线皮带运输机
single-lined diagram 单线接线图
single-line diagram 单线图;单线边接图;单结线系统
single-line district 单线区段
single-line drawing 单线图
single-line elevator 单环提引器
single-line fault 单线故障
single-line ground 单线接地
single-line grout curtain 单排灌浆帷幕
single-line highway 单线公路
single-line hook 单索吊钩
single-line inclined-hole cut 单向掏槽
single-line lock 单线船闸
single-line operation 单线工作
single-line piping layout 单单线管道布置
single-line preheater 单系列预热器
single-line pulley 单滑车
single-liner 单列分行机
single-line register 单行移位寄存器
single-line repeater 单线中继器
single-line rivet joint 单行铆接

single-line scanning 单线扫描
single-line section 单线区段
single-line semiautomatic block 单线半自动闭塞
single-line speed 单绳速度
single-line store 专业商店
single line tariff 单一税率
single-line traffic 单车道单向交通
single-line two-way track 单线双向运行线路
single-line working of double track line 双线改为单线行车
single-link line 单节线
single-lip wiper 单唇口防尘圈
single liquid 单一液体
single liquid system injection 单液系统注浆
single liquid type manometer 单液式压力计
single load 单一负荷；单(个)荷载；单独荷载；集中荷载
single-load arched girder 集中荷载拱形梁
single-loaded 单面承载的
single-loaded corridor type 单面走廊式
single loading 集中荷载；单荷载
single lobe 单瓣
single lobe positive displacement mud motor drill 单头螺杆钻具
single lock 单厢船闸；单(级)船闸
single lockage 一次过闸；整队过闸
single-locked bridge 单咬口式双翼桥；单锁双翼开合桥
single-locked cross welt 单咬口十字缝
single-locked flat seam 单咬口平接缝
single-locked seam 单焊接缝
single-locked standing seam 单咬口立缝
single-locked welt 单咬口；单线折边
single logarithm coordinate 单对数坐标
single-longitudinal mode argon ion laser 单纵模氩离子激光器
single-longitudinal mode laser 单纵(向)模激光器
single look direction 单觇标方向
single loop 单循环；单回路
single-loop controller 单回路控制器
single-loop feedback 单闭合环路反馈
single-loop regulator 单回路调节器
single-loop servomechanism 单回路伺服系统；单环伺服机构
single-loop system 单回路系统；单环系统
single-loop temperature controller 单环温度控制器
single-loop winding method 可控表面线圈法
single lubricator 局部润滑器；单独润滑器
single lug 单线夹
single machine 单机
single-machine capacity 单机容量；单机功率
single magnetic component 单一磁性成分
single-main distribution 单主管配水
single-main distribution system 单总管(配水)系统；单配管系统
single-main system 单总管系统
single make contacts 单合接点组
single male stacking fitting 单头式堆装件
single mantle lantern 单头纱罩汽油灯
single-mass system 整体系统；单质点系统；单体系(统)
single-mass vibrator 单块振动器
single mast 单杆(井)架
single mast S/R machine 单立柱型有轨巷道堆垛起重机
single-material batcher 单一材料称料器；单一材料配料器
single material fiber 单材料纤维
single material optic(al) fiber 单材料光纤
single-material scale batcher 单一材料称量进料器；单一材料称量配料器
single material weigh batcher 分批称重计
single measure 单面装饰的
single-measure door 双面包夹板门；光面门；双面光的门
single-measured result 单次测量结果
single measurement 单次测量
single mechanical end face seal 单端面机械密封
single media filter 单滤料滤池
single medium bed 单滤料滤床
single medium filter 单滤料滤池
single medium stratified filter bed 单滤料滤分层滤床

single medium unstratified filter bed 单滤料滤不分层滤床
single-melt ingot 单熔锭
single member 单动构件
single-mesh 单网孔
single-mesh filter 单网孔滤波器
single-message sign 单信息标志
single metallic effect paint 单层金属闪光涂料
single metalics coat 单层金属闪光涂层
single mode 单模态；单模
single mode attenuator 单模衰减器
single-mode behavio(u)r 单模性能
single-mode control 单式控制
single mode coupler 单模耦合器
single mode face pumped laser 单模面抽运激光器
single mode fiber 单模纤维
single mode fiber system 单模光纤系统
single mode filter 单模滤光器
single mode generator 单模发生器
single model 单模型
single mode laser oscillator 单模激光振荡器
single model instrument 单模型仪器
single mode operation 单模工作
single mode optic(al) fiber 单模光纤；单模光导纤维
single mode optic(al) fiber connector 单模光纤连接器
single mode oscillation 单模振荡
single mode oscillator 单模振荡器
single mode silica fiber 二氧化硅单模光纤
single mode system 单模系统
single mode trackside fiber optic(al) cable 单模轨旁光纤电缆
single mode wavelength-division demultiplexer 单模波分解多路复用器
single mode wavelength-division multiplexer 单模波分多路复用器
single module 单模块
single molecule 单分子
single monumental arch 单纪念柱拱门
single-motion turbine 单动式汽轮机
single-motion type 单运动式的
single motor 单发动机；单电动机
single motor car 单电动机车
single motor equipment 单发动机装置
single mo(u)ld 单件模型
single mo(u)ld board plough 单壁犁
single mo(u)ld board plow 单臂犁
single mo(u)ld press 单模压机
single mount 单托；单个安装
single nailing 单钉功
single navigation lock 单船闸
single-needle device 单针器件
singleness 单一
single nip calender 双辊压光机；单一压区压光机
single nodded 单节(驻波)
single node cutting 单节插条
single nodule 单体结构
single non-central load 偏心集中荷载
single normal moveout correction section 单次动校剖面
single notching 单凹口；单凹接头；单面刻槽；单(开)槽
single nozzle 单喷嘴
single oblique junction 单斜接头
single observation 单次观测
single observation forecast 单站观测预报
single observation plot 简单观察小区
single observer forecast(ing) 单站预报
single occupancy 独门独户；单独使用
single-office exchange 单中心交换站；单局交换
single offset 一级起步时差
single open-end spanner 单头开口扳手
single open-end wrench 单头开口扳手
single-operand instruction 单操作数指令
single-operating structure 单孔结构
single operation 单工运用；单操作
single operation cycle 单一作业循环
single operation part 单工序零件
single operator arc welding generator 单站电焊发动机
single operator arc welding machine 单站弧焊机
single operator arc welding set 单人电焊机
single operator set 单焊机

single operator transformer 单站焊接变压器
single operator welder 单站焊机
single operator welding machine 单工使用焊机；单站焊机
single optic(al) fiber 单支光纤
single order 一阶(的)；一位；单一顺序；单位的；单阶
single-order subtracter 一位减法器
single ore 单矿
single organic cation bentonite 单有机阳离子膨润土
single orifice fuel injection valve 单孔燃料注入阀
single orifice plate 单孔板
single out 挑选
single outlet 出水口；入海口
single output switch function 单输出开关函数
single output system 单输出系统
single oven Jun glaze 素炉钧釉
single overhung 单悬式
single oxyacetylene welding goggles 单玻璃焊工护目镜
single-pack 简单包装
single package 一罐装；单组分
single package urethane coating 单组分聚氨酯涂料
single-pack coating 部分涂层；简装涂料
single packer 单一打包机
single packer test 单封隔器测试
single-pack etching primer 简装刻蚀底漆
single packet message 单包消息
single-pack plastic 单组分塑料
single-pack pretreatment primer 简装预先处理的底漆
single-pack primer 简装底漆
single-pack self-etch primer 简装自刻蚀底漆
single-pair problem 单一配对问题
single pan digital analytical balance 单盘光学读数分析天平
single-panel(led) door 单板门
single pantograph working 单受弓工作
single parameter estimation of water quality model 水质模型单参数估值
single parameter method 单参数法
single-parity check 单奇偶校验
single part 单一成分
single particle 单个颗粒
single particle comminution 单一颗粒粉碎
single particle laser velocimeter 单粒子激光测速仪
single partition 单分区
single part production 单件生产
single-party line 同线电话线
single-pass 单行程的；单遍扫描；单通；单程(的)
single-pass boiler 单流锅炉；一次通过锅炉
single-pass condenser 单向电容器；单流式凝汽器；单流式冷凝器；单流程凝汽器；单程冷凝器
single-pass conversion 单程转化
single-pass cooler 空气一次通过的冷却器
single-pass dryer 单向干燥器；单程干燥器
single-pass electrogas process 单程电焊气体保护方法
single-pass exchanger 单程(热)交换器；单程换热器
single-pass gain 单程增益
single-pass grinding 开路研磨；开路磨矿
single-pass mixer 单(行)程拌和机
single-pass mixing 单行程拌和
single-pass operation 单程操作
single-pass prism dosimeter 单通棱镜剂量计
single-pass prism instrument 单通棱镜仪器
single-pass program(me) 单遍程序
single-pass soil stabilizer 单行程土壤稳定机
single-pass stabilizer 单程土壤稳定机
single-pass system 开路系统；单程系统；单通系统
single-pass tubular heater 单程列管加热器；单程管式加热器
single-pass turbine 单流式涡轮机；单流程式汽轮机
single-pass weld 单道焊(缝)
single-pass welding 单道焊接
single-pass yield 单程产量
single-pass zone-melting 一次通过区域熔炼
single path 单路径；单流程
single-path system 单路系统
single pattern 整体模(具)
single payment 一次(性)付款；整(笔支)付；一次偿付

single-payment annuity 一次付清年金
single-payment compound amount factor 一次偿付复利因子；一次偿付复利因素
single-payment factor 一次支付因子；一次支付因素
single-payment formula 一次支付公式
single-payment loan 一次支付贷款；一次偿清贷款
single-payment personal cash loan 一次付清的现金贷款
single-payment present worth computation 一次偿付现值计算法
single-payment present worth factor 现值系数；一次支付现值系数；一次偿付现值因子；一次偿付现值因素
single peak anomaly 单峰异常
single peak(endness)condition 单峰条件
single peak flood wave 单峰洪水波
single peak hydrograph 单峰过程线
single peck line 单虚线
single-pedestal desk 一头沉桌；独柱写字台
single-pedestal table 独柱台
single pendulous gyrocompass 单摆式拖轮罗经
single pendulum 单摆
single-perforated grain 单孔药柱
single perforated plate 单层多孔板
single periodic(al)block 单一定期作业分区
single-period model 单时期模型
single-person household 单身户
single-phase 单相(的)
single-phase aircooled transformer 单相汽冷式变压器
single-phase alloy 单相合金
single-phase alternate current traction 单相交流牵引
single-phase alternative current 单相交流电
single-phase alternative current synchronous generator 单相交流同步发电机
single-phase alternator 单相交流发电机
single-phase arc furnace 单相电弧炉
single-phase auto-transformer 单相自耦变压器
single-phase bridge rectifier 单相桥式整流器
single-phase brushless alternator 单相无刷交流发电机
single-phase brushless generator 单相无刷发电机
single-phase capacitor operation asynchronous motor 单相电容运转异步电动机
single-phase capacitor starting induction motor 单相电容起动感应电动机
single-phase circuit 单相电路
single-phase clutch motor 单相离合器电动机
single-phase commutator induction motor 单相整流子感应电动机
single-phase commutator machine 单相整流子电机
single-phase commutator motor 单相整流子(式)电机；单相换向器电动机
single-phase compound motor 单相复绕电动机
single-phase condenser motor 单相电容电动机
single-phase converter 单反动液力变扭器
single-phase current 单相电流
single-phase current locomotive 单相电流机车
single-phase diesel generating unit 单相柴油发电机组
single-phase diesel power set 单相柴油发电机组
single-phase directional short-circuit relay 单相定向短路继电器
single-phase double-frequency generator 单相双频发电机
single-phase double insulation buffing machine 单相双连绝缘抛光机
single-phase double insulation electric(al)drill 单相双连绝缘电钻
single-phase double insulation hand grinder 单相双连绝缘手提砂轮机
single-phase double-speed motor 单相双速电动机
single-phase driving 单相驱动
single-phase earth fault protection 单相接地保护
single-phase earthing 单相接地
single-phase electric(al)arc furnace 单相电弧炉
single-phase electric(al)blower 单相电动吹风机
single-phase electric(al)furnace 单相电炉
single-phase electric(al)pump 单相电泵
single-phase equilibrium experiment 单相平衡实验
single-phase flow 单相流

single-phase full wave connection 单相全波连接
single-phase full wave rectify circuit 单相全波整流电路
single-phase full wave rectifier 单相全波整流器
single-phase generator 单相发电机；单相电动机
single-phase halfwave rectifier 单相半波整流器
single-phase inclusion 单相包裹体
single-phase induction motor 单相感应电动机
single-phase induction regulator 单相感应调节器
single-phase induction voltage regulator 单相感应电压调节器
single-phase kilowatt-hour meter 单相电度表
single-phase KWH meter 单相电度表
single-phase line oscillator machine 单相直线振荡电动机
single-phase load 单相负荷
single-phase locomotive 单相(电力)机车
single-phase machine 单相电机
single-phase meter 单相功率因数表
single-phase microstructure 单相显微组织
single-phase motor 单相电动机
single-phase nozzle 单相喷嘴
single-phase oil immersed transformer 单相油浸变压器
single-phase overcurrent protective relay 单相过流保护继电器
single-phase parallel inverter 单相并联逆变器
single-phase potential transformer 单相电压互感器
single-phase power 单相功率
single-phase power station 单相发电厂
single-phase protective branch switch 单相保护式分路开关
single-phase railway 单相电力牵引铁路
single-phase reaction 单相反应
single-phase recloser 单相重合闸
single-phase rectifier 单相整流器
single-phase regulating transformer 单相调整变压器
single-phase reservoir 单相热储
single-phase rotor 单相转子
single-phase safety plug with socket 单相保险插头座
single-phase saturable reactor 单相饱和扼流圈
single-phase selsyn motor 单相自(动)同步电动机
single-phase series commutator motor 单相串励换向器电动机；单相串励整流子电动机
single-phase series motor 单相串联电动机；单相串励电动机
single-phase short-circuit 单相短路
single-phase shunt commutator motor 单相并励换向器电动机
single-phase shunt motor 单相并联电动机；单相并励电动机
single-phase single-speed motor 单相单速电动机
single-phase speed regulating motor 单相调速电机
single-phase synchronous machine 单相同步电机
single-phase synchronous motor 单相同步电动机
single-phase system 单相制；单相系统
single-phase three wire transformer 单相三线制变压器
single-phase titanium alloy 单相钛合金
single-phase traction transformer 单相结线牵引变压器
single-phase transformer 单相变压器
single-phase two-speed motor 单相双速电动机
single-phase wattmeter 单相瓦特计
single-phase winding 单相绕组
single-phase with neutral wire single socket outlet 单相带零点三孔插座
single-phasing 单相运行
single photoelectron emission 单光电发射
single-photograph measurement 单像量测
single-photograph measuring 单像量测
single photograph measuring apparatus 单像量测设备
single-photon decay spectrometer 单光子湮没谱仪
single-photo-plotter 单像测图仪
single-photo-plotting 单像测图
single-photo-plotting apparatus 单像测图仪
single-photo-plotting device 单像测图仪
single-pickled sheet 一次酸洗薄钢板
single picture 单张相片

single-piece 整体；整件；整块；整段
single-piece casting 整体浇注
single-piece cast runner 整铸转轮
single-piece disposable type share 单片不修复式犁铧
single-piece frame press 整体机身式压力机
single-piece machining 单件加工
single-piece milling 单件铣
single-piece panel 单块板
single-piece piston 整体活塞
single-piece processing 单件加工
single-piece production 单件生产
single-piece work 单件生产；单件工作；单件加工
single pier 单墩
single-pier draft tube 单墩尾水管
single-pier draught tube 单墩尾水管
single pile 单桩
single pile capacity 单桩承载力
single pile dolphin 单桩系船系统
single pile loading test 单桩荷载试验
single pilot 单先导
single-pin(ned)arched girder 单销铰拱大梁
single-pinned frame 单销铰构架
single-pin track 单销履带
single-pin type spanner 单销式扳手
single-pin type wrench 单销式扳手
single-pipe brake system 单管制动系统
single pipe dropping system 单管向下式系统
single-pipe hanger 单管吊架
single-pipe heating system 单管供热系统
single-pipe hot water supply system 单管热水供暖系统
single-pipe jet pump 单管喷射泵
single-pipe loop circuit heating system 水平单管采暖系统
single-pipe meter 单管仪表
single pipe ring 管子卡环
single pipe system 单管系统
single-pipe system of hot water 单管热水供暖系统
single pipe up-feed system 单管向上式系统；单管下行上给系统
single-piston brake cylinder 单活塞制动缸
single-piston explosive press 单活塞爆炸压力机
single-pitched roof 单坡屋顶；单坡顶
single-pitched roof-light 单坡屋顶采光
single pivot 单枢
single-place shift 一位位移
single plan 单一登记法
single plane 单面
single-plane balance test 单面平衡试验
single-plane balancing 单面平衡
single-plane gear 同平面减速器
single plane multicable harp system 单索面竖琴形密索体系
single-plane swivel joint 单面回转接头
single-planetary gear train 单级行星轮系
single-planetary mill 单辊行星轧机
single plane transverse cable arrangement of cable-stayed bridge 斜拉桥单索面横向布置
single-plane turbine 单排式汽轮机
single plank for bridge floor 单层木桥面板
single planking system 单层板壳船
single-plank track 单跑道；单板道路
single-plank wall 单板墙
single plan of standard cost 标准成本的单一记账法；标准成本的单行记账法
single plan of standard cost accounting 标准成本会计的单一计划
single plant 单株
single planting 单植(庭园)；单株栽植；孤植
single plate 单片板
single plate ball bearing 单垫片滚珠轴承
single plate center keelson 单板中内龙骨
single plate clutch 单片(式)离合器；单盘离合器；单闸片离合器
single plate dark slide 单片遮光滑板；单片暗盒
single plate dry clutch 单片摩擦离合器；单片干式离合器
single plate friction clutch 单片式摩擦离合器
single plate keelson 单板中内龙骨
single plate rudder 平板舵；单板舵
single plough 单铧犁
single plow 单体犁
single plunger lever 单活塞杠杆

single plunger plastic injection mo(u)lding machine 单柱塞塑料注射成形机
single plunger pump 单柱塞泵
single ply 一层(指材料);一片(如胶合板薄层)
single-ply bellows 单层波纹管
single-ply canvas 单织帆布
single-ply film 单层薄膜
single-ply roofing 刚性防水屋面
single-ply roofing asphalt felt 单层屋面油毡
single-ply roof membrane 单层屋面片材;单层屋面膜
single-point 奇点;单点
single-point adjustable suspension scaffold 单点挂式可调脚手架
single-point cutter 单齿铣刀
single-point cutting tool 单刃刀(具)
single-pointed scarifier 单齿松土机;单齿耙路机
single-pointer air ga(u)ge 单针气压计
single-point grounding 单点接地
single-point heater 快速燃烧器;瞬时热水器;单龙头热水器
single-point intercalation 单点插入法
single-point interpolation 单点插入法
single-point lifting 单点起吊法
single-point load 单点荷载;集中荷载;集中负载
single-point measurement 单点量测
single-point mercury settlement ga(u)ge 单点水银沉降计
single-point method 单点法
single-point mooring 单点系泊
single-point mooring berth 单点系泊码头
single-point mooring buoy 单点系泊浮筒
single-point mooring pier 单点系泊码头
single-point mooring system 单点系泊系统
single-point mooring tower 单点系泊塔(架)
single-point press 单点压力机
single-point recorder 一点记录仪;单点记录仪;单点记录器
single points 单式道岔;单开道岔【铁】
single-point sampling 单点取样;单点采样
single-point shovel 单尖松土铲
single-point single-acting press 单点单动压力机
single-point stationary press 单点固定压力机
single-point suspension 单点悬挂装置
single-point suspension spreader 单点吊具;一点吊具
single-point thread tool 单刃螺纹刀具
single-point tool thread cutting machine 单刃刀具螺纹铣床
single-point water heater 单头热水器
single-polarity pulse 单极性脉冲
single-polarity pulse-amplitude modulation 单向脉幅调制
single polarization coupler 单偏振耦合器
single polarization fiber 单偏振光纤
single polarization mode maintaining fiber 单偏振模保持光纤
single polarization single-mode fiber 单偏振单模光纤
single polarizing identification 单偏光镜下鉴定
single pole 单极(的);单杆;单刀
single-pole circuit breaker 单极断路器
single-pole cut-out 单极断路器
single-pole distribution switch 单极分路开关
single-pole double throw 单刀双掷;单刀双投
single-pole double throw switch 单刀双掷开关;单刀双投开关
single-pole enclosed fused switch 单极保护式负荷开关
single-pole intermediate switch 单极中间开关
single-pole(knife) switch 单极(闸刀)开关;单刀(式)开关;单刀式电闸
single-pole machine 单极电机
single-pole manual isolator 单极手动隔离开关
single-pole mast 单腿桅杆
single-pole mercury switch 单极水银开关
single-pole motorized isolator 单极电动隔离开关
single-pole on-off switch 单刀(启闭)开关
single-pole-piece magnetic head 单极(片)磁头
single-pole scaffold 单排(立杆)脚手架;单支柱脚手平台;单(支)柱脚手架
single-pole single-throw 单刀单掷
single-pole single-throw switch 单刀单掷开关;单极单投开关

single-pole three-throw switch 单极三投开关
single-pole travel(l)ing gear 单极移动齿轮
single-pole two-circuit double interruption switch 单极双回路双断开关
single-pole two-circuit single interruption switch 单极双回路单断开关
single-pole two-way switch 单极双路开关
single policy 特定的危险保险单
single pollution source 单点污染源
single population of microorganisms 单一微生物群体
single-port actuating cylinder 单口作动筒
single-ported slide valve 单口滑阀
single-ported valve 单口阀
single position hob 单圈滚刀
single post 单柱
single-post row brush dam 单排桩梢料坝
single pot furnace 单坩埚窑
single-pot system 单坩埚体系
single power set running 单机组运行
single precision 单(倍)精度
single precision integer 单精度整数
single pre-erected grade wire 单根预架设坡度线
single premium 一次付清的保险费
single-pressing 一次压制
single-press process 一次压制法
single-pressure condenser 单压凝气器
single-pressure feed valve 单压进给阀
single-pressure stage 单压级
single-pressure steam turbine 单压式汽轮机
single-pressure vessel 单仓泵
single pressurizator 单增压器
single-prism magnifying stereoscope 单棱镜放大立体镜
single-prism spectrograph 单棱镜摄谱仪
single-prism stereoscope 单棱镜立体镜
single processing machine 单一处理系统
single processor uniprocessor 单一处理机
single process scutcher 单程清棉机
single product plant 单一产品工厂
single programmed repeatable robot 单程序重复式机器人
single program(me) initiator 单一程序启动程序;单程序的启动程序
single program(me) operation 单一程序操作;单道程序操作
single-projection system 单投影系统
single projector method 单投影器定向法
single-prong tester 单叉探测器
single proprietorship 独资
single pulley 单体滑车
single pulley block 单门葫芦;单联滑轮组
single-pulley drive 单皮带轮传动
single pull hatch cover 单拉开式舱盖
single pulse 单脉冲
single pulse decatron 单脉冲十进(计数)管
single pulse device 单脉冲装置
single pulse laser 单脉冲激光器
single pulse mode 单脉冲方式
single pulse radar 单脉冲雷达
single pulse sonar 单脉冲声呐
single pulse voltmeter 单脉冲伏特计
single pump 单吸泵;单流向泵;单级泵
single pumping log 单孔抽水泵井
single-punch press 单冲杆压机
single-punch tablet press 单冲压片机
single purchase 简单起重滑车
single purchase counterweight batten 单式吊杆
single purchase crab winch 单起重装置绞车
single purchase pulley 单滑轮
single purchase winch 单卷筒绞车
single-purpose 专用的
single-purpose building 专用房屋
single-purpose computer 专用计算机
single-purpose construction machine 专用建筑机械;专用施工机械
single-purpose fallout shelter 防放射性降尘专用掩蔽所
single-purpose highway 专用公路
single-purpose local-area network 单用途局部地区网络
single-purpose planning 单项规划
single-purpose project 单目标工程
single-purpose reservoir 专用水库;单目标水库;单用途水库
single-purpose road 专用道路
single-purpose roller 专用轧辊
single-purpose statement 专用报表
single-purpose tadpole plot 专用蝌蚪图
single pushbutton control 单按钮控制
single pusher 单排推钢机
single-push-pull 单次推挽
single-quad cable 单四线组电缆
single quadruple box section 四肋式单箱截面
single quote mark 单引号
single-rabbet frame 单铲口门框
single radial diffusion test 单向辐射状扩散试验
single radiant section furnace 单一辐射段炉
single rail 单轨铁路;单轨(铁道)
single rail circuit 单轨轨道电路
single rail motor hoist 单轨电动起重绞车
single rail transportation 单轨交通
single-rail troll(e)y 单轨小车
single railway 单线铁路;单轨线路
single raincoat 单排雨衣
single-ram booster 单柱塞增压器
single-ram intensifier 单柱塞增压器
single ram preventer 单闸板防喷器
single random variable 简单随机变量;单随机变量
single-range 单波段的
single-range indicating instrument 单量程指示仪表
single-range instrument 单量程仪表
single-range wattmeter 单量程瓦特计
single rapids 单滩
single rate 单一汇率
single read instruction 单读指令
single rear axle 单后轴
single reasonable check 单次合理检验
single receptacle outlet 单孔插座出线口
single recessed wheel 单面槽砂轮
single-recycle counter current flo(a)tation process 单段逆流气浮工艺
single reduction final drive 单级最终传动
single reduction gear 单减速齿轮;单级减速装置;单级减速齿轮(箱)
single reduction geared turbine 单级减速齿轮传动式涡轮机
single reduction helical gears 单级减速斜齿轮
single re-entrant winding 单回路绕组
single-regime speed-concentration model 单段型车速—密度模型
single regulation 单调节
single reheat 一次再热
single reinforcement 单筋;单钢筋
single request 单个请求
single-reset system 单重置系统
single resonance optic(al) parameter oscillation 单共振光参量振荡
single resonator 简单共鸣器;单共振器
single-resonator laser amplifier 单谐振腔激光放大器
single-reversal permanent magnet 单反向永久磁铁
single reversing mill 单相可逆式轧机;单机可逆式轧机
single revolving gate 一字闸门
single rib grinding wheel 单齿磨轮
single rigid frame 单刚架
single rim bearing wheel 单边支重轮
single-ring controller 单环(式信号)控制机
single-ring feed 单环馈电
single-ring infiltrometer 单环渗透计
single ring magnet 单环形磁铁
single riser system 单立管系统
single rivet 单行铆钉
single riveted 单行铆接的
single riveted joint 单行铆接
single-riveted lap joint 单行铆钉搭接
single riveting 单行铆接的;单行铆钉
single robe hook 单爪大衣钩
single rod agitator 单杆翻动器
single rod extensometer 单杆伸缩仪
single-rod piston 单杆活塞
single-rod prestressing jack 单杆预应力千斤顶
single roll 单向滚动
single-roll breaker 单辊轧碎机;单辊压碎机;单辊破碎机

single-roll cake breaker 单辊碎饼机
single-roll cast coater 单辊涂铸机
single-roll crusher 单滚筒破碎机;单辊轧碎机;单辊压碎机;单辊破碎机
single-rolled strip 单张轧制的板材
single roller 单滚子;单辊磨
single-roller applicator 单辊式浸润器
single-roller bearing 单辊轴承
single-roller cam 单滚子凸轮
single-roller catch 单碰珠门扣
single-roller chain coupling 单排链式联轴器
single-roller chock 单滑轮导缆器
single-roller crab 单卷起筒起重机;单滚轮起重绞车
single-roller method 单辊法
single-roller mill 单辊轧粉机
single-roll feed 单边滚式送料;单边轧式送料
single-roll feeder 单辊喂料机;单辊进料机
single-roll hammer crusher 单辊击打式破碎机
single rolling hump 单溜放驼峰
single rolling on double pushing track 双推单溜
single rolling on single pushing track 单推单溜
single-roll stripper 单辊式摘棉铃机
single-roll toilet paper holder 单轴手纸架
single roof 密椽屋顶;普通椽屋顶;简易屋顶;单坡屋顶
single roof cladding 单层屋面板
single room 单人房间
single room apartment 一室户;单室户公寓
single-roomed apartment unit 单室户公寓单元
single-roomed flat 单室套房
single room with convertible couch-bed 设有两用沙发床的单人房间
single rope aerial 单索式架空索道
single-rope grab 单绳抓斗;单索抓斗
single-rope grab for ships 船用单索抓斗
single-rope haulage 单钩运输
single-rope orange-peel grab 单索多颚板抓斗
single-rope scraper 单索铲运机;单索刮土机
single-rope traction 单绳牵引
single ropeway 单索道
single-rope winding 单线线圈;单索卷绕的
single-rotor compressor 单转子压气机
single-rotor gas turbine 单转子燃气轮机
single-rotor hammer crusher 单转子锤式破碎机
single-rotor impact crusher 单转子反击式破碎机
single rotor irreversible hammer crusher 单转子不可逆锤式破碎机
single-rotor mixer 单转子混合器;单转轴搅拌器;单转轴拌和机;单转轴式搅拌机,单转轴式拌和机
single-rotor turbine 单转子涡轮机
single round nose bit 端部半圆形钻头
single-round survey 一次性调查
single route 单路径;单进路
single row 单行;单排(的)
single-row angular contact ball bearing 单列向心推力球轴承
single-row ball bearing 单列滚珠轴承;单列球轴承
single-row bearing 单列轴承
single-row boot 单行开沟器
single-row boring 单排钻
single-row brick-on-edge arch 单排侧桩拱
single-row bushing 单排孔漏板
single-row centripetal ball bearing 单列向心滚珠轴承
single-row centripetal bearing 单列向心轴承
single-row chain 单列链
single-row core 单排孔
single row cylindrical roller bearing 单列短圆柱滚子轴承
single-row deep groove ball bearing 单列深槽滚珠轴承
single-row disk harrow 灭茬圆盘耙;单列圆盘耙
single-row drilling 单排钻进
single-row engine 单列式发动机
single-row governing stage 单列调节级
single-row hog house 单列式猪舍
single-row layout 单行配置;单行排列
single-row machine 单行机具
single-row pig house 单列式猪舍
single-row pile cofferdam 单排桩围堰
single-row radial ball bearing 单列向心球轴承;单列向心滚珠轴承
single-row radial engine 单排星形发动机;单列辐射式发动机

single-row radial thrust ball bearing 单列向心推力球轴承
single-row rivet 单行铆钉
single-row riveting 单行铆钉;单行铆
single-row riveting joint 单行铆接
single-row roller bearing 单列滚柱轴承
single-row self-aligning bearing 单列向心球面球轴承
single-row self-aligning roller bearing 单列自位滚柱轴承
single-row separable ball bearing 分离型向心推力球轴承
single-row shot 单排炮爆破
single-row single butt-strap riveting 单对接搭板单行铆接
single-row snap ring ball bearing with shield 单列有罩开口环滚珠轴承
single-row spheric(al) roller 单列向心球面滚子轴承
single-row tapered roller bearing 单列圆锥滚子轴承
single rubber swab 单橡胶环活塞
single ruling 单向晕线
single-run line 单程水准测量路线
single-runner turbine 单转轮式水轮机
single running 单向运行
single runway system 单跑道系统
single-run welding 单头焊;单道焊(接)
singles 单张轧制的板
single sample 一次抽样品;单一样品
single sampling inspection 一次取样检验;一次抽样检验
single-sashed window 单扇水下推拉窗;单扇窗;单框格窗
single satin ribbon 单面缎带
single saw-toothed skylight 单层钝齿形天窗采光
single scale factor 单一标度系数
single scan 单面扫描
single scattering 单一颗粒上的辐射能散射;单(次)散射
single-scoop loading machine 单斗装载机
single scored block 单沟滑车
single screed 单样板
single screen expanding wall 单页折叠式隔墙;单页折叠式隔断
single screen graphics 单屏图形
single-screw 单螺旋(的);单螺纹
single-screw and single-rudder ship 单车单舵船
single-screw compounding extruder 单螺杆配料挤出机
single-screw extruder 单螺杆挤出机
single-screw extrusion machine 单螺杆挤出机
single-screw oil pump 单螺杆油泵
single-screw open-close machine 单螺杆式启闭机
single-screw plodder 单螺杆压条机
single-screw propulsion ship 单螺旋桨船
single-screw propeller 单螺桨推进器
single-screw pump 单螺杆(式)泵
single-screw recycling extrusion system 单螺杆再循环挤出系统
single-screw ship 单螺旋桨船
single-screw(single screw)pump 单螺杆泵
single-screw steamer 单螺旋桨轮船
single-screw thread 单(头)螺纹
single-screw type vise 单螺旋式夹具
single-screw vessel 单螺旋桨船
single seal (路面的)单层式封层
single seam 单焊缝;单接缝
single seated 单座式
single seated main stop valve 单座式主汽门
single-seat valve 单座阀
single section artificial line 单节仿真线
single section filter 单节滤波器;单级滤波器
single section pulverizer 单节碎土镇压器
single section view 单一剖视
single-seed descent 一粒传法
single-seed drill 单粒排种条播机;单粒播种机
single-seed drilling 单粒条播
single seeded 单粒种子的
single seed of grain 单粒谷物种子
single-seed planter 单粒播种机
single-seed seeder 单粒播种机
single-seed spacing unit 单粒精密排种装置
single segmental baffle 单弓形折流板

single segment structure 单程序段结构
single selection 一次选择
single selvage 单边
single service 单路供电;单个服务
single-service article 一次性用具
single-service items 一次性物品
single shading 单色晕渲法
single shaft 单井道;单轴;单杆的
single-shaft arrangement 单轴布置
single-shaft compulsory mixer 单轴强制式搅拌机;单轴强制式拌和机
single-shaft configuration 单轴结构
single-shaft engine 单轴发动机
single-shaft gas turbine 单轴燃气轮机
single-shaft hammer crusher 单轴锤式破碎机
single-shaft hammer paddle arm mixer 单轴桨叶式搅拌机
single-shaft mixer 单转轴搅拌器;单轴轴拌和机;单轴(式)搅拌机;单轴式拌和机
single-shaft paddle-type mixer 单叶片式混砂机
single-shaft pug mill 单轴拌土机
single-shaft pug mill concrete mixer 单轴混凝土拌和机
single-shaft-screw pump 单轴螺旋泵
single-shaft turbogenerator 单轴型汽轮发电机
single-shaft turbo-jet 单轴涡轮喷气发动机
single-shaft turbo-set 单轴型汽轮机机组
single-shaft vibrating screen 单轴振动筛
single shank beam 单体梁
single shank ripper 单齿松土器
single shank unit 单体构件
single shear 单面剪切;单剪(力)
single-shear bolt 单剪螺栓
single-shear double rivet joint 单剪双行铆接
single-shear joint 单剪接头;单剪连接;单剪接合
single-shear method 单面剪切法
single-shear rivet 单车(面)剪铆钉;单面受剪铆钉
single-shear riveting 单面剪铆
single-shear rivet joint 单剪铆接
single-shear single-rivet joint 单剪单行铆接
single sheathing 单层面板
single-sheaved block 单轮滑车;单滑轮
single-sheaved pulley block 单轮滑车
single sheave iron pulley block 单轮铁滑车
single sheave travel(l)ing block 单滑轮游动滑车
single shed insulator 单裙形绝缘子
single-sheet classification system 薄板分选系统
single sheet cutter 单张切纸机
single-sheet grab 单片爪
single-sheet map 单张地图
single-shell 单层薄壳;单壳式
single-shell casing 单层缸
single-shell coach body 全钢薄壳车体
single-shell cylinder 单层缸
single shell insulator 单裙边绝缘子
single-shell lining 单层衬砌
single-shell rotary drier 单筒烘干机;单壳旋转干燥鼓;单鼓干燥器
single shield bearing 单罩轴承
single shift 一班制;单班制
single-shift work 一班制工作
single-ship operating 单船作业
single ship sailing speed 单船航速
single shoal 单滩
single-shot blocking oscillator 单稳态阻塞振荡器;单稳态间歇振荡器;单程间歇振荡器
single-shot booster 单作用增压器
single-shot camera 单幅摄影机
single-shot circuit 单稳电路
single-shot computer 单步计算机
single-shot directional surveying instrument 单点测斜仪
single-shot exploder 单发式轻便电动爆炸装置
single-shot firing 单发爆破
single-shot flip-flop 单行触发器;单击触发器
single-shot grout(ing) 单液灌浆
single-shot inclinometer 单点测斜仪
single-shot instrument 单点测斜仪
single-shot intensifier 单作用增压器
single-shot lubrication 一次注射润滑法
single-shot mode 单脉冲方式
single-shot multivibrator 单周期多谐振荡器;单稳多谐振荡器;冲息多谐振动器;冲息多谐振荡器
single-shot operation 单拍操作;单步操作

single-shot process 一次性处理；一次性过程
single-shot reclosing 一次式重合阐
single-shot seal coat (路面的)单层式封层
single-shot solution grout 单液灌浆
single-shot survey 单点测斜
single-shot trigger 单脉冲触发电路；单稳触发器
single-shot trigger circuit 单稳态触发电路；单冲触发电路
single shoulder tie plate 单扇垫板；单肩垫板
single-shovel plow 单铧犁
single-shrouded impeller 单侧闭式叶轮
single shrouded wheel 半闭式叶轮
single shunt field relay 单闭磁路继电器
single shutoff hose coupling 单向封闭的软管接头
single side 单面型
single sideband 单边带
single sideband amplifier 单边带放大器
single sideband communication 单边带通信
single sideband converter 单边带变换器
single sideband demodulation 单边带解调
single sideband distortion 单边带失真
single sideband modulation 单边带调制；单边带调节
single sideband modulator 单边带调制器
single sideband radio set 单边带无线电收发机
single sideband receiver 单边带接收机
single sideband reduced carrier 减载波单边带电话
single sideband signal generator 单边带信号发生器
single sideband system 单边带系统
single sideband transmission 单边带发送；单边带传输
single sideband transmitter 单边带发射机
single sideband tuning 单边带调谐
single-side bevel gear cutter 单面斜齿轮刀具
single side braking 单侧制动
single-side cutter 单面刀具
single-sided 单面
single-sided amplifier 单边放大器；单端放大器
single-sided argon arc spot welding machine 单面氩弧点焊机
single-sided bicycle stand 单面自行车架
single-sided copying 单面复制
single-sided corrugator 单面瓦楞纸板
single-sided draft hood 单面烟橱通风罩
single-sided gasket 单面密封垫；单面密封条
single-sided heating panel 单侧供暖辐射板
single-sided hood 单面拔气罩
single-sided(hot) bed 单面冷床
single-sided impeller 单侧吸入式叶轮
single-sided linear induction motor 单边直线感应电动机
single-sided pattern plate 单面型板；单面模板
single-sided plate valve 单面板阀
single-side drying machine 单面烘燥机
single-sided shears 开式剪切机
single-sided sprayer 单边作用喷雾机
single-side heat sink 单面散热片
single-side rack 单面机架；单面齿条
single-side stand 单支架
single-side welding 单面焊(接)
single-side wood planer 单面木工刨机
single-signal 单信号(的)
single signal location 信号单置点；单置信号点
single-signal receiver 单信号接收机
single silk-covered copper wire 单丝包线
single silkcovered wire 单层丝包线
single site network 单地域网络
single-size 均一尺寸
single size aggregate concrete 单粒径集料混凝土；单粒径骨料混凝土
single-sized 均匀尺寸的；单一尺寸的；单一粒径的
single-sized aerated concrete 均匀粒加气混凝土
single-size(d) aggregate 均匀集料；均匀颗粒集料；均匀颗粒骨料；同尺寸骨料；粒度均匀骨料；等径集料；等径骨料；单一粒度集料；均一粒径骨料；同尺寸集料；单(一)粒度集料；同粒径骨料；单径骨料
single-sized aggregate concrete 单(一)粒径集料混凝土；单(一)粒径骨料混凝土
single-sized concrete 均匀粒径混凝土
single-size(d) material 均匀颗粒材料
single-sized mortar 均匀粒径砂浆
single-size fraction 单粒径级；单粒度级

single-size material 均尺寸材料；尺寸均一的材料
single-size material concrete 均匀颗粒材料混凝土
single-size mixture 同粒度混合料
single-skew notch 单斜凹槽
single-skin steel dam 单层钢面板坝
single-skin structure 单荷载结构；单层结构
single-skin timber dam 单层木面板坝
single-skip hoist 单斗提升机；单斗卷扬机
single-slag steel 单渣钢
single sleeve valve 单套阀
single slewing derrick 摆式单斗装置
single sliding door 单扇水平推拉门
single sling 单索式；单索吊具(一头吊环，一头吊钩)
single-slip casing bowl 单卡瓦打捞筒
single slip points 单式交分道岔
single slip switch 单式交分道岔
single slope 单斜坡；一面倾斜的
single-slope roof 单坡(屋)顶
single slope trimmer 单斜坡平整机
single slot 单翼缝
single-slot coil 磁铁气隙线圈
single-slot winding 单槽绕组
single sludge activated sludge process 单污泥活性污泥工艺
single sludge nitrogen removal 单污泥脱氮
single socialist market 统一的社会主义市场
single socket 单承窝
single soil sample 单一土样
single solenoid spring return valve 单螺旋弹簧回位阀；单螺旋弹簧回水阀；单螺旋弹簧回流阀
single-solenoid valve 单螺线管阀
single-solute sorption isotherm 单溶质吸附等温线
single-solute sorption process 单溶质吸附工艺
single-solute transport model 单溶质运移模型
single solvent 单一溶剂
single solvent process 单一溶剂提取法
single-source supply 单一水源供水
single space 单间隔
single span 单跨(的)
single-span beam 单跨梁
single-span beam bridge 单跨简支梁桥
single-span bridge 单跨桥；单孔桥
single-span cantilever bridge 单跨悬臂梁
single-span frame 单跨构架
single-span girder 单跨大梁
single Spanish burton 定单与定单复绞辘
single spanner 单翼梁
single-span shaft 单跨轴
single-span skew slab 单跨斜(交)板
single-span slab 单跨板
single-span trussed roof structure 单跨屋架结构
single-span two-hinged frame 单跨双铰构架
single spar 单杆桅
single-spar construction 单梁结构
single spark gap 单火花隙
single spark ignition 单火花点火
single spark-plug ignition 单火花塞点火
single-spar structure 单梁结构
single special frontier node 单一特殊边界节点
single species ecological risk characterization 单物种生态风险表征
single spectrum 单一谱
single speed 单速的
single-speed axle 单速轴
single-speed camera 单速度摄影机
single-speed door panel (左右移动的)单扇门板
single-speed floating control system 单速无静差控制系统；单速浮点调节系统
single speed motor 单速电动机
single-speed synchro data system 单速同步数据系统
single-spindle 单轴的；单杆的
single-spindle automat 单轴自动机
single-spindle automatic 单轴自动的
single-spindle automatic bar machine 单轴自动杆式车床
single-spindle automatic chucking machine 单轴自动卡盘车床；单轴自动夹头车床
single-spindle automatic lathe 单轴自动车床
single-spindle automatic screw machine 单轴自动螺纹车床
single-spindle boring machine 单轴镗床
single-spindle cold rolling former 单轴冷滚压成型机

single-spindle drilling machine 单轴钻床
single-spindle hammer crusher 单轴锤式破碎机
single-spindle lathe 单轴车床
single-spindle machine 单轴机床
single-spindle milling machine 单轴铣床
single spindle mo(u)lding machine 立式单轴木工铣床
single-spindle roving winder 单锭无捻粗纱络纱机
single-spindle turret automatic lathe 单轴六角自动车床
single spindle turret lathe 单轴转塔车床
single spinner 单圆盘撒布机
single-spiral classifier 单螺旋分级机
single-spiral feeder 螺旋给料机；螺旋给矿机；单螺旋给料机
single spiral turbine 单排量蜗壳式涡轮机；单排量蜗壳式透平
single split 单架劈木机
single split mo(u)ld 组合模(具)
single-spool engine 单转子发动机；单路式发动机
single-spool turboprop 单转子涡轮螺桨发动机
single spot welding 单(行)点焊
single-spout packing machine 单嘴包装机
single spray coat application 单层喷射涂层
single spray gun 单喷嘴喷枪
single spread 单面涂布量
single spreading 单面涂胶；单面涂布
single spring 单泉
single-spring bottom oil can 单簧底油壶
single-spring flexure suspension 单簧挠曲悬挂
single squirrel cage rotor 单鼠笼式转子
single-stable channel 单一稳定河槽
single stack coil annealing 带卷的单垛退火
single-stack furnace 单烟筒式炉
single-stack kiln 单竖井式窑
single-stack plumbing 单一烟囱铅管工程
single-stack system 单(立)管系统；(粪水和污水合一排泄的)混合系统
single-staff method 单标尺法
single-stage 单级(的)；一级
single-stage air compressor 单级(式)空气压缩机
single-stage air cooled turbine 单级气冷式透平
single-stage amplifier 单级放大器
single-stage axial compressor 单级轴流压气机
single-stage axial fan 单级轴流风扇
single-stage blower 单级鼓风机
single-stage cementing 单级注水泥
single-stage centrifugal compressor 单级离心式压缩机
single-stage centrifugal pump 单级离心泵
single-stage clarification 单级澄清
single-stage closed circuit grinding 单级闭路粉磨
single-stage closed circuit system 单级闭路(粉磨)系统
single-stage compression 一级压缩；单级压缩
single-stage compressor 单机压缩机；单级压缩机；单级压气机
single-stage converter 单级液力变矩器
single-stage crushing 一段破碎；单级破碎
single-stage curing 带底板养护；单阶段养护；单级养(预制混凝土的)护
single-stage cyclone separator 单级旋风分离器
single-stage digester [digestor] 单级消化池
single-stage digestion 单级消化
single-stage double-acting centrifugal freshwater pump 单级双动式离心清水泵
single-stage drive 一级驱动
single-stage drying and grinding 单级烘干粉磨
single-stage enrichment 单级浓缩；单级富集
single-stage evapo(u)ration 单级蒸发
single-stage evapo(u)rator 单级蒸发器
single-stage extraction 单级提取
single-stage filter 单级滤池
single-stage filtration 单级过滤
single-stage filtration with effluent recirculation 出水再循环的单级过滤
single-stage flo(a)tation 单段浮选
single-stage floating cover anaerobic digester 单级浮盖厌气消化池
single-stage froth flo(a)tation 单级泡沫浮选
single-stage gear reduction 一级齿轮减速
single-stage granular activated carbon treatment 单级粒状活性炭处理

single-stage grinding 单粉磨
single-stage growth curve 单阶段增长曲线
single-stage harmonic drive 单级谐波传动
single-stage impeller 单级叶轮
single-stage impeller pump 单级叶轮泵
single-stage impulse turbine 单级冲动式涡轮机; 单级冲动式透平
single-stage interconnection network 单级互连网络
single-stage intermittent aeration membrane bioreactor 单级间歇曝气膜生物反应器
single-stage lead ages 单阶段铅年龄
single-stage lock 单级船闸
single-stage mixing plant 单阶式搅拌厂
single-stage nitriding 单级氮化
single-stage nitrification-denitrification process 一段消化—脱氮法; 单级硝化反硝化法
single-stage pneumatic conveyor dryer 单段气流输送干燥器
single-stage precipitator 单区电除尘器
single-stage process 一阶段制管工艺; 单级制备法
single-stage pump 单级泵
single-stage pumping 单级泵送
single-stage quenching 单液淬火; 单级淬火法
single-stage reactor 单级反应器
single-stage recarbonation 单级再碳酸化作用
single-stage recuperator 单级换热器
single-stage regeneration 单级再生
single-stage regulator 单极调节器; 单级式减压器; 单级减压器; 单级调节器
single-stage rocket 单级火箭
single-stage sampler 单级采样器
single-stage seawater reverse osmosis 单级海水反渗透
single-stage self-priming overhang type centrifugal freshwater pump 单级自吸悬臂式离心清水泵
single-stage single element ejector 单级单联喷射器
single-stage single suction volute pump 单级单吸离心泵
single-stage steam turbine 单级汽轮机
single-stage supercharger 单级增压器
single-stage suspension 单级悬挂
single-stage switch 单级开关
single-stage system 单级管网系统; 单级装置
single-stage trickling filter 单级滴滤池
single-stage turbine 单级涡轮(机); 单级水轮机; 单级汽轮机
single-stage turbocharger 单级涡轮增压器
single-stage turbocharging 单级涡轮增压
single-stage twin element ejector 单级双联喷射器
single-stage two speed supercharger 单级双速增压器
single-stage vehicle 单级火箭
single-stalk girder 单腹板大梁
single-stalk standing seam 单条直缝
single-stalk stiff jointed frame 单跨刚架
single stamp mill 单锤碎矿机
single stand 单机座
single standard 单一标准
single-stand mill 单机座轧机
single-stand reversing cold-reduction mill 单机座可逆式冷轧机
single-start thread 单线螺纹; 单头螺纹
single-start worm 单头螺纹蜗杆
single-state recycle 单级再循环
single station analysis 单站分析
single stationary detector system 单固定探测系统
single-station blow mo(u)lding machine 单工位吹瓶机
single-station die 单级模; 单工位模
single-station die attachment machine 单工位置板牙接合机
single-station digitizing table 单机数字化器
single station forecast(ing) 单(测)站预报
single station head 单笔位绘图头
single station method 单站法
single-station range finder 固定基尺视距仪
single-station type thermoforming machine 单工位型热成形机
single-station way type machine 单工位式导轨式机床
single steam chest valve 单汽柜阀

single steel intrauterine device 金属单环
single step 一步; 单步; 单槽齿; 单步
single-step accelerating tube 单间隙加速管
single-step air compressor 单级式空气压缩机
single-step compressor 单机压缩机; 单级(式)压缩机
single-step debug 单步调试
single-step instruction 单步指令
single-step iteration 一步迭代
single-step method 单步法
single-step mode 单步方式
single-step operation 单步操作
single stepping test 单步测试
single-step process 单步法; 单步处理
single step reduction gear unit 单级减速机
single-step run 单步运行; 单步运算
single-step switch 单步开关
single-step Turing machine 单步图灵机
single stereomodel 单立体模型
single sticker 单桅(帆)船
single-stor(e)y 单层的
single-stor(e)y annex(e) 单层附属建筑物
single-stor(e)y building 单层建筑(物); 平房
single-stor(e)y factory 单层厂房
single-stor(e)y heating system 单层拱热系统
single-stor(e)y hot water heating system 单层热水加热系统
single-stor(e)y house 单层房屋; 单层住宅; 平房
single-stor(e)y industrial plant 单层工业厂房
single-stor(e)y rigid frame 单层刚架
single-stor(e)y school building 单层学校建筑
single stor(e)y silo 单层库
single-stor(e)y warehouse building 单层仓库建筑
single-storied 单层的
single-strand cable 单股钢缆; 单股钢丝; 单股钢索
single-strand casting machine 单流连铸机
single-strand conformation polymorphism 单链构象多态性
single-strand copper conductor 单股铜线
single-strand crop and cobble shears 单线式端头废料剪切机
single-strand drag conveyer 单路拖运机; 单线输送机
single-stranded 单股的
single-strand jacking device 单股张拉设备
single-strand mill 单线式线材轧机
single-strand repeater 单线围ம்
single-strand roller chain 单股滚链
single-strand strength tester 单纱强力试验机
single-strand tester 单纱强力试验机
single-strand wire rope 单股钢丝绳; 单股钢(丝)索
single strap 单均压环
single-strap butt joint 单面盖板对接
single-strap lap joint 单盖板搭接接头
single stream preheater 单列预热器
single strength glass 普通强度玻璃
single strength window glass 薄窗玻璃
single-string galvanometer 线电流计
single stroke 单次行程; 单冲程
single-stroke bell 单击电铃
single-stroke deep-well pump 单冲程深井泵
single-stroke header 一次冷镦机; 单击冷镦机
single structure piping 单独结构管系
single-strut trussed beam 单柱上撑式桁架梁; 单柱桁构式梁; 单柱桁架(式)梁
single-stub transformer 单短截线变量器
single-stub tuner 单短截线调谐器
single subsonic jet 单亚音速喷气发动机
single suction 单吸式; 单吸(的)
single-suction centrifugal pump 单吸(式)离心泵; 单进口(式)离心泵
single suction impeller 单吸(入口)叶轮
single suction pipette 单吸移液管
single suction port 单吸口
single suction pump 单吸(式)泵
single suction volute pump 单吸螺旋泵
single supercharger 单增压器
single support 简支座; 简支点
single surface dressing 单层式
single surface glazed packing paper 单面光包装纸
single surfacer 单木刨; 单面木刨
single surface treatment 单层式
single surface wood planer 单面木工刨床
single suspension 简单悬挂【电】

single-sweep 单面扫描
single sweep method 单一扫描法
single sweep tee 三通管
single-sweep tinning unit 单锅式镀锡机组
single-swing blocking oscillator 单程间歇振荡器
single-swing door 单扇摇门; 单面摇门; 单面弹簧门; 单向门
single-swing frame 单摇向门框; 单扇弹簧门框
single swinging boom 单杆
single swinging gate 单开门
single switch 单开道岔【铁】; 单极开关; 单刀开关
single switchboard 单连开关板
single symmetric(al) frame 单跨对称框架
single system 一元系统; 单系统
single system accounting 一贯制会计
single-system machine 单系统针织机
singlet 单一; 单线; 单透镜; 单态; 单谱线; 单峰的; 单电子键
single table press 单工作台压机
single tackle 单绞辘
single tandem 单串式
single-tank breaker 单桶式油断路器
single-tank pneumatic conveying system 单箱风力传输系统; 单箱压缩空气传输系统
single tank wagon 单罐货车
single taper 单锥形
single target 单个目标
single target data read-out 单一目标数据资料读取
single-target level(l)ing rod 单面水准尺
single task 单任务
single task operation 单一任务操作
single task system 单一任务系统
single tax 统一税
single tax in silver 一条鞭法
single tax system 单(一)税制
single teasel raising machine 单向刺果起绒机
single-Tee beam 单 T 形梁
single-tee joint 单边焊缝丁字接头
single template jigger 单刀旋坯机
single tender 单项招标
single tendering 单项工作招标
single-tenon 榫舌
single-terminal 单端
single-terminal accelerator 单电极加速器
single-terminal pair network 二端网络
single test 一次试验; 单个试验
single T groove 丁形槽
single-theodolite observation 单点经纬仪观测
single thickness sheet glass 薄窗玻璃; 单料窗玻璃(1~2毫米厚)
single thread 单螺纹
single-threaded 单头螺纹的
single-threaded screw 单头螺纹螺钉; 单螺纹
single-threaded worm 单头螺纹蜗杆
single-thread hob 单头螺纹滚刀
single threading 单线处理
single-thread milling cutter 单头螺纹铣刀
single-thread processing 单线(索)处理
single-thread worm 单线蜗杆
single throat burner 单喉管燃烧器
single-throw 单投
single-throw contact 单掷接点
single-throw crank shaft 单弯曲柄轴; 单拐曲轴
single-throw knife switch 单向闸刀开关
single-throw pump 单冲程泵
single-throw switch 单投开关; 单掷开关
single thrust bearing 单向推力轴承
single-tier formwork 单层模板
single-tier grillage 单层格床
single tie rod 单系杆
single-tier wall 单列墙
single tilting permanent magnetic chuck 单倾永磁吸盘
single timber floor 单层木地板
single timber roof floor 单层木天花板
single time estimate 单一时间估计法
single-time-lag servo 单时间常数跟踪系统
single-time measurement 一次性测定
single tine ripper 单刃平巷掘进机; 单刃耙路机; 单刃松土机; 单刃平巷掘进机; 单刃耙路机; 单刀松土机; 单齿平巷掘进机; 单齿耙路机; 单齿松土机
singlet interaction 单谱线作用
single tire 单轮胎
single-toe rail fastener 单趾弹簧扣件

single-toggle 单肘(板)
single-toggle jaw crusher 单摆颚式破碎机;单肘(板)颚式破碎机
single-toggle(type jaw)granulator 单肋板颚式碎石机
single tombolo 单连岛沙坝
single-tool 单刀
single-tool axle lathe 单刀切削车轴车床
single-tool holder 单刀架
single-tooth crusher 单齿辊破碎机
single-tooth cutter 单齿插齿刀
single-tooth ripper 单齿松土机;单齿耙路机
single tooth X-ray apparatus 单牙X射线机
single tow 单拖
single-towered facade 单层塔架正面
single trace filtering 单道滤波
single trace seismograph with digital display 数字显示单道地震仪
single track 单线轨道;单线的;单线(车道)【铁】;单声道;单轨线路;单(磁)道
single-track all-relay semi-automatic block system 单线继电半自动闭塞
single-track automatic block system 单线自动闭塞系统
single-track bridge 单(铁路)线桥;单行车道桥
single-track car line 单线有轨电车
single-track circuit 单轨条轨道电路
single-track district 单线区段
single-track gathering shoe 单槽集束器
single-track hump 单驼峰
single-track line 单线铁路
single-track operation 单向运行
single-track plane drawbridge 单轨平吊桥
single-track railway 单线铁路
single-track railway tunnel 单轨铁路隧道
single-track section 单线区间;单线区段
single-track sliding door 单轨推拉门
single-track swing parting 单轨甩车调车场
single-track tunnel 单线隧道
single trait 单一性状
single tramway-type equipment 单接触线简单悬挂
single tramway-type equipment trolley wire 单接触线的简单悬挂接线
single transfer device 单传输设备
single-transistor 单晶体管
single transition time assignment 一次转换时间分配
single transition time decomposition 一次转换时间分解
single trawler 单拖网渔船
single treatment 一次处理
single tree 独立树;横撑杆
single triangulation chain 单三角锁【测】
single trip 单程
single trip charter 单航租船
single trip charter party 单航租船合同
single trip container 不回收装箱
single triplet 三张照片组
single-triplet transition 单态和三重态间跃迁
single trip station wax injection machine 单工位压蜡机
single-trip trigger circuit 单冲触发电路
single-triumphal arch 单门纪念性拱券;单门凯旋门
single-trolley system 单触线制
single T roof slab 单T形屋面板
single-trough floatting machine 单槽浮选机
singlet state 单线态
single tube 单层管
single tube core barrel 单管(式)岩芯管;单管芯钻具;单层岩芯管
single-tube double-track section 单洞双线断面
single-tube grouting method 单管灌浆法
single-tube injector 单管喷射器
single-tube inverter 单管倒相器
single-tube manometer 单管式压力计
single-tube oscillator 单管振荡器
single-tube packing machine 单嘴包装机
single-tube porcelain bushing 单筒瓷套管
single-tube rotary dryer 单筒烘干机
single-tube sprayer 单管喷雾器
single tube top pressure type packer 单管顶入式封隔器
single tube tyre 单管轮胎
single-tub wagon tipper 单矿车翻笼

single-tuned amplifier 单调谐放大器
single-tuned circuit 单调谐电路
single tuned interstage 单调谐级际电路
single-tuned transformer coupled amplifier 单调谐变压器耦合放大器
single tuning 单调谐
single tunnel 单行隧道
single tunnel operation 单隧道操作
single turbocompressor rotor cycle 单轴燃气轮机循环
single turn 单匝
single turn-back track 单折返线【铁】
single-turn coil 单圈绕组
single-turn deflection 单圈引出
single-turn extraction 单圈引出
single-turn lock 单旋转锁
single turnout 单开道岔【铁】;单道岔
single-turn potentiometer 单匝电位计;单旋电位计
single-turn transformer 单匝式电流互感器;单匝变压器
single-twist drill 单曲钻;单麻麻花钻
single-twisting 单丝加捻
single-twist knotter 单结扭结器
single type 单型
single type brake 单闸瓦制动
single type jiyu mill 单式自由粉碎机
single-type marshalling station 单向编组站【铁】
single tyre 单轮胎
single tyre load 单轮荷载
single-U groove 单式U形槽
single U groove with vertical sides 单垂直边的U形槽
single-unit 单(机)组;单机制;单一机组
single-unit diesel locomotive 单节(式)内燃机车
single-unit dwelling 独户单元住宅
single-unit furnace 单体炉
single-unit hydroelectric(al)station 单机组水电站
single-unit machine 单程针织机
single-unit of dryer and pulverizer 粉磨烘干联合机组
single-unit operation 单机运行;单机牵引
single-unit pilot 单联动自动舵
single-unit plant 单机组电站
single-unit system 整体装置
single-unit traction 单级牵引
single-unit triaxial test apparatus 单联三轴试验仪
single-unit truck 载重汽车;无拖车的载重汽车;单辆货车
single-unit type blow-off preventer 单一机组类防喷装置
single up 单绑
single upper floor 格栅楼板;上层楼板
single upper timber floor 木格栅上层楼板
single upright machine 单柱车床
single use bit 一次钻头
single-use goods 一次用完货品
single user access 单用户存取
single-user berth 专用泊位
single-user harbo(u)r 专用港(口)
single use room 专用室
single-user port 专用港(口)
single utility program(me) 独立实用程序
single U-training 单马蹄形整枝
single valence anion 一价阴离子
single valley glacier 单式山谷冰
single value 单(一)值
single-valued 单值的
single-valued function 单值函数
single-valued operation 单值运算
single-valued surface 单值曲面
single valve 单阀;单电子管的
single-valved 单阀的
single-valve knapsack pump 背负式单阀泵
single-valve operation 单阀操作
single-vane actuator 单叶片摆动缸
single-vane attenuator 单片衰减器
single-vane impeller 单翼叶轮
single-vane rotary compressor 单滑片回转式压缩机
single-vane rotary pump 单翼旋转泵;单叶片旋转泵
single vapo(u)rization 一次蒸发
single-varactor parametric amplifier 单变容管参量放大器

single-variable control 单变量调节
single-variable system 单变量系统
single variation method 单变法
single V butt joint 单面V形对ijng(焊)接
single V butt weld 单面V形对(接)焊(缝)
single V corner joint 单面V形角焊(接)
single-vee groove V形坡口
single-vee groove with root face 单纯边的V形坡口
single velocity stage 单速级
single vernier 单游标
single vertical aerial photograph 单张垂直摄影航片
single vertical block 立式单次拉丝机
single vertical key 单行键
single vertical sweep guard 简单立式拨出式安全装置
single vessel 单船
single vessel conveyer 单仓输送泵
single V groove 单式V形槽
single V groove with root face 带纯边V形坡口;单钝边V形坡口
single vibration roller 单振动压路机
single vibration tandem roller 单筒振动串联式路机;单筒振动串联式路碾
single vibratory action pile driver 单振沉桩机
single view 单张像片
single virtual storage system 单独虚存系统
single V joint 单面V形对接
single-voltage rating 单额定电压;单电压额定值
single volume 单卷宗
single volume data set 单卷数据集
single volume file 单卷(宗)文件
single-volute pump 单螺旋泵
single vortex strength 单涡强度
single-V weld 单面V形焊接
single wall 单层墙;单壁;单排墙
single-wall buttress 单壁支墩;单垛支墩
single-wall cofferdam 单墙围堰
single-wall construction 单层墙体结构
single-walled anchored sheet pile bulkhead 锚固的单层板桩挡土墙
single-walled floating dock 靠岸单墙式浮船坞
single-wall knot 单绳端结
single-wall open caisson 单层开口沉箱;单墙沉井
single-wall paper bag 单层纸袋
single-wall partition 单层隔仓板
single-wall sheet piled structure 单板桩结构
single-wall wooden sheet pile cofferdam 单层木板桩围堰
single washstand 单洗面台
single water gun 单水枪
single watershed method 单独流域法
single water source 单一水源
single-wattmeter method 单瓦特计法
single wave 单个波;单波
single wave beam 单波束
single-wave compensator 单波补偿器
single-wavelength infrared analyzer 单波长红外分析仪
single wavelength selector 单波段选择器
single wave of translation 单波传波
single-wave operation 单波运行
single-wave oscillator 单波振荡器
single-wave peak 单波峰
single-wave rectification 半波整流
single-wave rectifier 半波整流器
single-wave wind(ing) 单波绕组
single-way cable duct 单向电缆管道
single-way marshalling yard 单向编组场【铁】
single-way railroad 单线铁路
single-way railway 单线铁路
single-way rectifier 单路整流器
single-way road 单向路;单向交通道路
single-way slab 单向板
single-way slab system 单向板系统
single-way suction 单向吸入;单向吸进
single-way supply 单向供水
single-way switch 单向开关
single-way traffic 单向交通
single-way volume 单向交通量
single web 单腹板
single-web beam 单腹板梁
single-web beam bridge 单肋式梁桥

single-webbed 单腹板的
single-webbed girder 单腹板大梁
single-webbed plate arched girder 单腹板拱形大梁
single-webbed top chord 单腹板上弦杆
single web member 单腹杆
single-web plate girder 单片板梁
single-wedge compensator 单楔补偿器
single weeder 单行除草机
single weight 单件重量
single-weight paper 薄相纸
single-weight thickness 薄纸基厚度
single welded butt joint 单面焊搭接接头
single welded joint 单面焊接头;单面焊接合
single welded lap joint 单面焊搭接接头
single welding 单面焊;单点点焊
single weld seam 单面焊缝
single well 单井
single-well oil production system 单井采油系统
single-well platform 单井平台
single-well pumping test 单孔抽水试验
single welt 单咬口
single wheel 独立轮;单坡轮
single-wheel barrow 独轮手推车
single-wheel brake 单轮制动器
single-wheel bump integrator 单轮撞击积算仪;(测定道路现况的)单轮撞击积算机
single-wheeled assembly 单轮装配
single-wheeled roller 单轮压路机
single-wheeled troll(e)y 单轮手推车
single-wheeled vibrating roller 单轮振动压路机
single-wheel grinder 单头砂轮机
single-wheel hoe 独轮行间中耕器
single-wheel impulse turbine 单级冲动式涡轮机;单级冲动式透平
single wheel load 单轮荷载
single-wheel loading 单轮载荷;单轮荷载
single-wheel pulley 单轮钩式滑车
single-wheel railway-wagon shifter 单轮铁路车辆移动装置
single-wheel reciprocating mower 单轮往复式割草机
single-wheel trailer 单轮拖车
single-wheel turbine 单轮式汽轮机
single whip 定单滑轮组;单滑轮
single whip reversed 动单绞辘
single whip tackle 定滑车;导向滑车
single wide line 单根宽谱线
single-wide shaker 平台式逐稿器
single winding head 单头缠绕
single-winding multispeed motor 单绕多速电动机
single window 单层单(扇)窗
single-window extrusion process control system 单窗口挤出过程控制系统
single-window panel 单窗板
single-window transducer 单窗换能器
single wing Christmas tree 单翼采油树
single-wing dock gate 立轴单扇旋转坞门
single-wing door 单扇门
single-winged building 单翼建筑物
single-wing gate 船坞立轴单扇旋转门
single-wing lock washer 单耳止动垫圈
single-wire 单根钢丝;细钢筋
single-wire anchorage 单索锚固
single-wire armored cable 单层铠装电缆
single-wire armo(u)red cable 单线铠装电缆
single-wire armo(u)red cable 单线铠装电缆;单层铁线铠装电缆
single-wire braided hose 单层钢丝软管
single-wire circuit 单线回路;单线电路
single-wire-drawing machine 单次拉丝机
single-wire feed 单线馈送
single-wire feeder 单线馈送线
single-wire helix 单线螺旋(丝)
single-wire line 单线电路;单线线路
single-wire operated semaphore 单线臂板信号机
single-wire prestressing jack 单索预应力千斤顶
single-wire prestressing system 单索预应力系统
single-wire remote control 单线遥控
single-wire stressing unit 单索预应力构件
single-wire stretching 单线张拉
single-wire system 单线制
single-wire transmission 单线传输
single-wire transmission line 单线传输线

single-withe wall 单堵墙
single word length arithmetic 单字长运算【计】
single-working beam 单臂平衡杆
single-working switch 单动道岔
single worm pusher 单螺杆顶车机
single wrench 单口螺丝钳;单口扳手
single-wythe wall 单堵墙
single yarn 单丝;初捻纱
single yarn of glass 单股玻璃纱
single-yarn strength tester 线抗拉强度试验机
single yarn tester 单纱强力试验仪;单纱强力试验机
single zenith distance 单向天顶距
single-zone embankment 均质土堤
single-zone temperature controller 单区温度控制器
singly curved plane 单曲面
singly curved shell 单曲面薄壳
singly linked 单键连接的
singly-operated drive 单独传动(装置)
singly periodic 单周期的
singly reentrant winding 单回路复绕组
singly repeated 单围盘轧制的
singly shoal 磊石滩
singular additive set function 奇加性集函数
singular angle hammer 独角锤
singular arc 奇异弧线
singular automorphism 奇异自同构
singular boundary condition 奇异边界条件
singular boundary cycle 奇异边缘循环
singular chain 奇异链
singular class 奇异类
singular cohomology group 奇异上同调群
singular collineation 奇直射
singular complex 奇异线丛
singular component 奇异分量
singular conic 奇异二次曲线
singular coordinates 奇坐标
singular correlation 奇对射
singular correspondence 奇异对应
singular corresponding point 对应奇点
singular cover 奇异覆盖
singular cover list 奇异覆盖表
singular cube 奇异立方
singular cycle 连续循环
singular direct product 奇异直积
singular distribution 奇异分布
singular divisor 奇异分母
singular eigenfunction 奇点本征函数
singular element 奇异单元;退化单元
singular equation 奇异方程
singular flexibility matrix 奇异柔度矩阵
singular function 奇异函数
singular generator of a ruled surface 直纹面的奇异母线
singular homology class 奇异下同调类
singular homology theory 奇异下同调论
singular hyperplane 奇异超平面
singular initial value problem 奇异初值问题
singular inner function 奇异内函数
singular integral 奇异积分;奇解;奇积分
singular integral equation 奇异积分方程
singular integral manifold 奇异积分流形
singular integral operator 奇异积分算子
singular interface 奇异面
singularity 奇异性;奇异(点);奇性;奇点
singularity function 奇异函数
singularity line 奇直线
singularity matrix 奇异矩阵
singularity of a curve 曲线的奇性
singularity set 奇点集合
singular kernel 奇异核
singular line 奇轨直线
singular locus 奇轨迹
singular matrix 降秩矩阵;奇异矩阵;退化矩阵
singular measure 奇异测度
singular operator 奇异算子
singular orbit 奇异轨道
singular ordinal 特异序数
singular part 奇异部分
singular periodic solution 奇周期解
singular perturbation 奇摄动
singular perturbation system 奇异摄动系统
singular perturbation theory 奇异摄动理论
singular plane 奇异平面;奇平面

singular point 奇(异)点
singular point method from complex variable function 复变函数求奇点法
singular point of a surface 曲面的奇点
singular process 特别过程
singular proposition 奇异命题;特称命题
singular quadric 奇异二次曲面
singular quadric hypersurface 奇异二次超曲面
singular series 奇异级数
singular set 奇异组;奇异系;奇异络;奇异集;无主系;单系
singular solution 奇异解;奇解;特殊溶液
singular-solution ideal 奇解理想
singular spectrum 奇异谱
singular state 奇(异)态
singular stiffness matrix 奇异刚度矩阵
singular support 奇异支集
singular surface 奇(异)曲面
singular tangent 奇异切线
singular transformation 奇(异)变换
singular value 奇(异)值
singular value decomposition 奇异值分解
singular value decomposition of image matrix 影像矩阵的奇异值分解
singular values of a matrix 矩阵奇异值
singular vertex 奇异顶点
sinhalite 硼铝镁石
Sinian cycle 震旦旋回
Sinian ingression epoch 震旦海侵期
Sinian movement 1 震旦运动一
Sinian period 震旦纪
Sinian system 震旦系
Sinian transgression 震旦海侵
Sinian trend 震旦走向
Sinia rhodoleuca Diels 合柱金莲木
Sinicism 中国风味食物;中国风俗
Sinimax 铁镍磁软合金
sinistral 左的
sinistral dislocation 左旋错动
sinistral displacement 左旋位移
sinistral fault 左旋断层;左行断层;左推断层;左侧断层
sinistral faulting 左旋断层作用
sinistral fold 左褶曲;左褶皱;左侧褶皱
sinistrally tilting swing 逆时针翘倾摆动
sinistral-separation fault 左旋错位断层
sinistral strike-slip fault 左旋走滑断层
sinistral torsion 左行扭动
sinistral transcurrent movement 左旋横推运动
sinistral translation 左行直移
sinistral translation movement 左旋平移运动
sinistral wrench fault 左旋平移断层
sinistrorsal 左转的
sinjarite 水氯钙石
sink 汇;信宿;下沉;下落;洗涤盆;洗涤池;洗涤槽;阴沟;转发器;浸透;浇合;集养区;换能器;频变沉区;污水槽;塌陷;缩沉;水槽;渗坑;沉下;沉没;沉埋;陷坑
sinkage 下陷;沉陷(量);凹进处;陷坑
sinkage depth 下沉深度(浮坞)
sinkage in 凹下
sinkage of valve 阀陷
sinkage to float rate 沉浮比
sinkage trial 下沉试验(浮坞)
sinkaging 陷坑
sink a hole 钻孔
sink-and-float separation 沉浮分离法
sinkankarite 水磷铝锰石
sink a shaft 凿井
sink a well 挖井;打井
sink bib 洗涤盆水龙头;小水龙头;弯嘴龙头;水池龙头(厨房)
sink bowl 洗涤盆
sink drain 水池排水管;洗气盆排水管;洗涤盆排水道;洗涤池排水
sink drainage 排污水系统
sink drain board 污洗池排水挡板
sink effect 吸收效应
sink-efficiency 散热器效率
sinker 向下式凿岩机;下向凿岩机;坠子;掘岩工;轻型风钻;排水孔;受油器;手持式凿岩机;次生吸收根;沉水树种;沉水材;(浮标等的)沉锤;沉块;沉锤;测深锤
sinker bar 压载杆;冲击(式)钻杆

sinker bar guide 冲击钻杆导向器;冲击式钻杆的扶正器
sinker chain 沉锂锚链
sinker deadman 锚碇块
sinker drill 凿井用凿岩机;冲钻;钻孔器
sinker groin 前端沉笼丁坝(防冲沉笼裹头)
sinker leg 凿井用凿岩机气压腿
sinker nail 埋头钉;沉头钉
sinker visco(si)meter 沉锤黏度计
sink evacuator 清除泵;污水泵;污秽泵
sink faucet 洗涤盆龙头
sink-float process 浮沉冲洗法
sink-float separation 重介质分选;浮选法
sink-float separation process 浮沉选矿法;浮沉分选法
sink-float separator 重介质选矿分离器
sink-float technique 沉浮法测密度
sink flow 汇流
sink grating 水斗;水坑格栅;污洗池算子
sink grinder 污水槽研磨机
sinkhead 浇口杯;补缩冒口
sink heater 小型热水器
sinkhole 陷穴;污水阱;污水井;渗坑;阴沟口;沉坑;落水洞;灰岩坑;收缩孔
sinkhole lake 溶陷湖;溶坑湖
sinkhole mode 落水洞水系模式
sinkhole plain 灰岩坑平原
sink in a rivet 埋铆钉头
sinking 下沉;下挖;刻槽;减径拨管;缩陷(漆病);沉落;沉景;凹缩;凹进处
sinking advance 凿井速度;凿井进度
sinking agent 沉淀剂
sinking and control 下沉与控制
sinking and driving engineering 井巷工程
sinking and placing 沉放
sinking a well 凿井
sinking barge 沉排驳船
sinking barrel 凿井吊桶
sinking bit 埋头钻
sinking blast(ing) 凿井爆破
sinking bogie 凿井安全盖板
sinking bucket 凿井吊桶;提升大吊桶
sinking bucket crosshead 吊桶导向架
sinking by boring 钻井法凿井
sinking by jacking 用千斤顶沉放
sinking by jet piling 水冲沉桩(法)
sinking by jetting 水冲沉桩(法);射水沉桩法
sinking by mechanical vibration 振动下沉
sinking by water jet 射水下沉法
sinking caisson 沉箱
sinking coast 下降海岸;下沉(海)岸
sinking compartment 凿井格
sinking creek 地下河;隐没河;伏河
sinking curtain 沉落幕
sinking curve 下沉曲线
sinking cylinder foundation 沉桩基础;沉柱基础
sinking depth of wind pipe 风管沉没深度
sinking derrick 下沉式动臂起重机
sinking drill jumbo 凿井钻车
sinking drum method 沉井方法
sinking equipment 钻孔设备;凿井设备
sinking field 沉陷地
sinking force 下沉力
sinking fund 零存整取基金;集累基金;还本准备金;偿债基金
sinking fund accumulation 偿债基金积存款
sinking fund assets 偿债专款资产
sinking fund balance sheet 偿债基金平衡表
sinking fund bond 偿债基金债券
sinking fund cash 偿债基金现金
sinking fund deposit factor 偿债基金存款因子
sinking fund depreciation 偿债基金折旧
sinking fund expenses 偿债基金费用
sinking fund factor 偿债基金因子;偿债基金因素;偿债基金因数
sinking fund income 偿债基金收益
sinking fund installment 偿债基金按期摊拨款
sinking fund interest rate 偿债基金利率
sinking fund investment 偿债基金投资
sinking fund method of depreciation 折旧偿债基金法;偿债基金折旧法
sinking fund mortgage 偿债基金抵押
sinking fund requirement 偿债基金应提额
sinking fund reserves 偿债基金准备

sinking funds bonds 基金债券
sinking fund surplus 偿债资金盈余
sinking fund system 偿债基金制
sinking fund table 偿债基金表
sinking guide 凿井罐道
sinking head 埋头
sinking headframe 凿井井架
sinking head gear 沉头齿轮
sinking hoist 凿井提升机
sinking in-site 就地下沉
sinking installation 钻孔装置;沉井装置
sinking jacking 顶入
sinking kibble 凿井吊桶
sinking lamp 凿井吊灯
sinking land 沉陷地
sinking lining 凿井井壁
sinking machine 凿井机;井筒开凿机
sinking mattress 沉排;沉褥
sinking method 钻井法;水层下沉去油法;凿井(方)法
sinking mill 减径机
sinking of borehole 钻孔【岩】;凿井
sinking of caisson 下沉箱;沉箱下沉
sinking of caisson by loading 载重沉箱法;加载沉箱法
sinking of cylindric(al) open caisson 井筒下沉
sinking of oil slicks 油膜下沉
sinking of pile by water jet 水力沉桩(法);水冲沉桩(法);射水打桩法
sinking of precast tube 预制管段的沉放
sinking operation 凿井作业;凿井工程
sinking pile by water jet 水冲沉桩(法);射水沉桩法
sinking pipe 沉管
sinking pipe casing 下套管
sinking pipe fill stake 沉管灌注桩
sinking plant 凿井设备
sinking platform 凿井平台;沉井平台;凿井吊盘
sinking platform lighter aboard ship 沉没平台载驳船;浮坞式载驳船
sinking pump 凿井用泵;凿井(用)吊泵;凿井水泵;浸没泵;沉没(式)泵;潜水泵
sinking rate 凿井速度;沉下速率
sinking region 沉降区
sinking resistance 下沉阻力(沉箱)
sinking rider 凿井吊桶的导向架
sinking river 蜿蜒(性)河流;伏河
sinking sea floor core drill 沉入海底岩芯钻机
sinking shift 凿井班
sinking site 钻井现场
sinking skip 凿井箕斗
sinking skylight 下沉式天窗
sinking speed 下沉速度;沉下速率
sinking support 柔性支承
sinking system 凿井方法
sinking technique 凿井法
sinking through 钻透
sinking time 沉下时间;沉降时间
sinking tipple 凿井井架
sinking tube 钻管
sinking-type tunnel 沉埋式隧道
sinking velocity 沉陷速度;沉下速度;沉速;沉降速度;下钻速度;下井筒速度
sinking velocity scale 沉速比例尺
sinking volume 缩沉量
sinking walling scaffold 两用吊盘
sinking well 沉井
sinking winch 凿井绞盘;凿井绞车
sinking winder 凿井绞车
sinking work 钻井工作
sink line 汇线
sink(mark) 凹缩;缩痕
sink mat 污洗池底板
sink matter 沉物
sink matter curve 沉物曲线
sink of graph 图的终点
sink point 汇点
sink pump 潜水泵
sink rate 散热器速度
sink region 振荡中断区
sink resistance 抗下陷性
sink roll 导辊
sink sample 沉没样(品)
sink temperature 冷源温度
sink trap 洗涤盆存水弯;洗涤池存水弯

sink unit 洗涤装置
sink unit with drainer 带泄水的洗涤盆
sink waste 厨房废(污)水
sink water 厨房废(污)水
sink water trap 洗涤盆存水弯
sink well 沉井
sinnerite 辛硫砷铜矿
sinnet 扁编绳
Sino-foreign joint ventures 中外合资企业
sino-foreign trade 中外贸易
sinoite 氧氮硅石;氮氧硅石
Sino-Korean axis paraplatform 中朝准地台
Sino-Korean massif metallogenetic province 中朝地块成矿区
Sino-Korean old land 中朝古陆
Sinopotamon yangtsekiense 长江华溪蟹
Sinotrans 中国对外贸易运输(集团)公司
sinter 氧化铁皮;硅华;熔渣;热压结;泉华;石灰华;烧结物;烧结品;烧结块
sinterability 可烧结性
sinterable powder 易烧结粉末;可烧结粉末
sinter apron 泉华裙
sinter barrier 泉华堤
sinter bed 烧结物床层;烧结料层
sinter blocks 泉华块体
sinter box 烧结箱
sinter breaker 烧结矿破碎机
sinter cake 烧结块;烧结饼
sinter cavity 泉华洞穴
sinter clay concrete 烧结黏土混凝土
sinter cliff 泉华峭壁
sinter coating 烧结涂层
sinter concrete 烧结混凝土
sinter cone 泉华锥
sinter cooler 烧结矿冷却机
sintercorundum 烧结金刚砂
sinter dome 泉华丘
sinter duct 泉华管
sintered 烧结的;热压结的
sintered absorbent 烧结吸收剂
sintered aerated concrete aggregate 烧结加气混凝土陶粒;烧结加气混凝土集料;烧结加气混凝土骨料
sintered aggregate 陶粒集料;陶粒骨料;烧结集料;烧结骨料
sintered alloy 烧结合金;粉末冶金合金
sintered alnico product 烧结铝镍铁制品
sintered alumina 烧结矾土;烧结氧化铝
sintered alumina abrasive tool 烧结刚玉模具
sintered-alumina crucible 烧结氧化铝坩埚
sintered alumina from bauxite 矾土烧结刚玉
sintered alumin(i)um 烧结铝
sintered alumin(i)um powder 烧结铝粉
sintered alumin(i)um product 烧结铝制品
sintered article 烧结制品
sintered ash 烧结灰渣
sintered bar 烧结锭条;烧结棒
sintered bearing 粉末冶金轴承;烧结轴承
sintered bearing metal 烧结轴承合金
sintered beryllia crucible 烧结氧化铍坩埚
sintered binder 烧结结合料
sintered bit 烧结粉末胎体(金刚石)钻头;烧结钻头
sintered blade 粉末冶金的叶片
sintered bloating clay 烧结膨胀性粘土
sintered bloating slag 烧结膨胀性矿渣
sintered body 烧结体
sintered bond 烧结结合剂
sintered brass 烧结黄铜
sintered brass compact 烧结黄铜坯块
sintered brass product 烧结黄铜制品
sintered brick 烧结砖
sintered brick of gangue 烧结煤矸石砖
sintered bronze 烧结青铜
sintered bronze facing 烧结青铜饰面
sintered carbide 硬质合金;烧结碳化物
sintered carbide ball 硬质合金球
sintered-carbide die 硬质合金模具
sintered-carbide tool 硬质合金工具
sintered-carbon steel 烧结碳钢
sintered catalyst 烧结催化剂
sintered cathode 烧结阴极
sintered charge 烧结料
sintered chromatographic plate 烧结色谱板
sintered cinder 烧结煤渣

sintered cinder wall slab 烧结煤渣墙板
sintered clay 烧结黏土
sintered clay cement 烧结黏土水泥
sintered clay concrete 陶粒混凝土
sintered clinker 烧结熟料
sintered clinker wall slab 烧结熔渣墙板
sintered coating 烧结敷层;烧结敷层
sintered compact 烧结的坯块
sintered component 烧结元件
sintered contact material 烧结电接触器材
sintered copper 烧结铜
sintered corundum 矾土陶瓷;烧结金刚砂;烧结刚玉
sintered crucible 烧结坩埚
sintered density 烧结密度
sintered deposit 泉华沉淀物
sintered diamond insert 烧结金刚刀头
sintered dolomite 烧结白云石
sintered electric(al)contact material 烧结电接触器材
sintered electric(al)parts 烧结电气零件
sintered expanded clay 烧结膨胀黏土(陶粒)
sintered expanded slag 烧结膨胀矿渣
sintered faced clutch 烧结材料贴面的离合器
sintered ferrite 烧结铁素体
sintered ferrous product 烧结铁制品
sintered fiber 烧结纤维
sintered filter 烧结过滤器
sintered fly ash 烧结粉煤灰
sintered fly-ash aggregate 粉煤灰陶粒;烧结粉煤灰集料;烧结粉煤灰骨料;粉煤灰膨珠
sintered fly ash brick 烧结粉煤灰砖
sintered fly ash concrete 烧结粉煤灰轻集混凝土
sintered foamed metal 烧结泡沫金属
sintered friction material 烧结摩擦材料
sintered friction parts 烧结摩擦零件
sintered fuel ash 粉煤灰膨珠;焦渣;粉煤灰
sintered gear 烧结齿轮
sintered glass 熔结玻璃;烧结玻璃;多孔玻璃
sintered glass bead 烧结玻璃珠
sintered glass fiber 烧结玻璃纤维
sintered glass filter 细孔漏斗;烧结玻璃(过)滤器;垂熔玻璃滤器;玻璃过滤器
sintered glass filter-bulb 垂熔玻璃滤球
sintered glass filter candle 垂熔玻璃滤棒
sintered glass filtering crucible 烧结玻璃过滤坩埚;玻璃滤坩埚
sintered glass funnel 烧结玻璃(板)漏斗;垂熔玻璃漏斗
sintered glass stem 烧结玻璃心柱
sintered gold 烧结金
sintered gravel 烧结砾石
sintered hard alloy 烧结硬质合金
sintered hard metal material 烧结硬质合金材料
sintered hematite 烧结赤铁矿
sintered high-temperature material 烧结耐高温材料
sintered ingot 烧结锭
sintered interface 烧结界面
sintered iron 烧结铁
sintered iron parts 烧结铁零件
sintered lamella filter 烧结板过滤体集尘器
sintered layer 烧结层
sintered light metal 烧结轻金属
sintered light(weight)concrete 烧结轻质混凝土
sintered liner 烧结衬里
sintered lining 烧结炉衬
sintered magnesia 烧结氧化镁
sintered magnesite 烧结菱苦土;烧结菱镁矿
sintered magnet 烧结磁铁
sintered magnetic alloy 烧结磁性合金
sintered magnetic material 烧结磁性材料
sintered magnetic parts 烧结磁性零件
sintered material 烧结材料;烧结材料
sintered matrix 烧结矩阵;粉末烧结胎体
sintered membrane 烧结膜
sintered metal 金属陶瓷;粉末烧结胎体;烧结金属
sintered metal bearing 烧结含油轴承
sintered-metal bit 烧结硬质合金钻头
sintered metal fiber 烧结金属纤维
sintered metal filter 烧结金属滤油器
sintered metal friction material 烧结金属摩擦材料
sintered metallic core 烧结金属磁芯
sintered metallic filter 烧结金属过滤器
sintered metallic magnet 烧结金属磁铁
sintered metal powder bearing 粉末冶金轴承
sintered metal powder filter 烧结金属滤油器
sintered metal powder oil impregnated alloy 烧结含油轴承合金
sintered metal product 粉末冶金制品
sintered metal ring 烧结网领
sintered mixture 烧结混合物
sintered nickel 烧结镍
sintered nickel cup 烧结镍过滤杯
sintered nickel product 烧结镍制品
sintered nylon 烧结尼龙
sintered ore 烧结矿
sintered oxide 烧结氧化物
sintered oxide-coated cathode 烧结敷氧化物阴极;氧化物烧结阴极
sintered parts 烧结零件
sintered piston ring 烧结活塞环
sintered plate 烧结板
sintered polycrystalline diamond 烧结多晶体金刚钻
sintered porosity 烧坯孔隙度
sintered porous nickel cup 烧结多孔镍滤杯
sintered powder 烧结粉末
sintered powder magnet 烧结粉末磁铁
sintered powder metal 烧结粉末金属
sintered powder metal product 粉末冶金制品
sintered precious metal 烧结贵重金属
sintered product 烧结制品
sintered pulverised-fuel[pulverized-fuel]ash 烧结粉煤灰;结块粉状燃料灰;烧结磨细燃料灰
sintered quartz 烧结石英
sintered refractory metal process 难熔金属烧结法
sintered ring 烧结环;烧结环
sintered rod 烧结棒
sintered silicon nitride 烧结氮化硅
sintered skeleton 烧结骨架
sintered slab 烧结板
sintered slag 烧结矿渣
sintered source 烧结源
sintered specimen 烧结试样
sintered sponge 烧结海绵金属
sintered stainless-steel 烧结不锈钢
sintered steel 烧结钢
sintered stoneware tile 烧结陶瓦
sintered tungsten bar 烧结钨棒
sintered ware 烧结制品
sintered wearresistant material 烧结耐磨材料
sintered woven wire sheet 烧结金属丝网片
sinter eggs 间歇泉华岩;泉华蛋
sinter-encrusted basin 有泉华包壳的泉盆
sinter-encrusted scarplet 有泉华包壳的陡坎
sinter fan 泉华扇
sinter flat 泉华台
sinter forging 烧结锻造(法);煅压烧结;粉末冶金锻造
sinter hearth 烧结炉
sintering 熔结;烧结;焙烧
sintering activity 烧结活性
sintering and palletizing wastewater 烧结球团废水
sintering apparatus 烧结炉;烧结设备
sintering area 烧结面积
sintering atmosphere 烧结气氛
sintering belt 烧结带
sintering body 烧结体
sintering briquette 烧坯
sintering ceramics 烧结陶瓷
sintering character 烧结性能;烧结性能
sintering circuit 烧结工序
sintering coal 炼焦煤;烧结煤
sintering compact 烧坯
sintering condition 烧结条件
sintering control 烧结控制
sintering crack 烧结裂纹;烧结裂纹
sintering current 烧纹电流;烧结电流
sintering cycle 烧结周期
sintering diamond bit 烧结金刚石钻头
sintering diamond crown 烧结金刚石钻头
sintering emulsion 烧结乳剂
sintering furnace 烧结炉
sintering grate 烧结炉栅;烧结炉篦
sintering grow 烧结膨胀
sintering index 熔结指数
sintering in electric(al)field 电场中烧结;电场烧结
sintering in sealed box 密封箱烧结
sintering jar 钟形烧结炉
sintering kinetics 烧结动力学
sintering limit 烧结限度
sintering machine 烧结机
sintering mechanism 烧结机理
sintering metal 金属陶瓷;烧结金属
sintering method 烧结方法;烧结法
sintering mill 烧结车间
sintering mix(ture) 烧结混合料
sintering model 烧结模型
sintering of loose powder 松粉烧结
sintering of monometallic systems 单金属烧结
sintering of polymetallic systems 多金属烧结
sintering of rolled powder 轧制粉坯的烧结
sintering operation 烧结操作
sintering ore 烧结矿
sintering pan 烧结盘
sintering period 烧结周期
sintering phenomenon 烧结现象
sintering plant 稍结工厂;烧结设备;烧结车间;烧结厂
sintering point 软化点;烧结温度;烧结点
sintering process 烧过程;烧结法
sintering process with hot pressing 热压烧结法
sintering product 烧结制品;烧结材料
sintering quality 烧结质量
sintering range 烧结范围
sintering schedule 烧结制度
sintering shape 烧坯
sintering shrinkage 烧结收缩
sintering temperature 软化温度;烧结温度;焙烧温度
sintering temperature range 烧结温度范围
sintering test 烧结性试验;烧结试验
sintering time 焙烧时间
sintering treatment 熔融处理
sintering under pressure 加压烧结
sintering unit 烧结设备
sintering warpage 烧结扭曲;烧结变形
sintering with mixture of $CaCO_3$ and NH_4Cl 碳酸钙—氯化铵烧结
sintering with mixture of ZnO and Na_2CO_3 氧化锌—碳酸钠烧结
sintering with Na_2CO_3 碳酸钠烧结
sintering with Na_2O_2 过氧化钠烧结
sintering zone 烧结区;烧结带
sinter lips 泉华唇
sinter machine 烧结机
sinter meter 烧结点测定仪
sinter mo(u)lding 烧结模塑
sinter mound 泉华冢
sinter plant 烧结厂;烧结设备;熔结工厂
sinter pot 烧结锅
sinter powder metal 烧结粉末冶金
sinter rim 泉华垣
sinter ripple mark 泉华波痕
sinter roasting 烧结焙烧
sinter-roasting 烧结
sinter set 粉末烧结胎体
sinter-set bit 烧结硬质合金钻头
sinter sheet 泉华席
sinter shield 泉华盾
sinter skin 烧结皮
sinter slope 泉华坡
sinter terrace 泉华阶地
sintetics 合成产品
sintex 烧结氧化铝车刀
Sintox 陶瓷车刀;烧结氧化铝车刀
Sintropac 加铜铁粉;铁铜混合粉末
sinuidal 正弦形
sinuosity 曲折(度);蜿蜒(度);弯度
sinuosity ratio 蜿蜒比;弯曲比
sinuous 蛇形的;波状的
sinuous channel 蜿蜒水道;蜿蜒河道;弯曲水道
sinuous coil 盘旋管;蜿蜒盘管;蛇形盘管;蛇形管
sinuous flow 正弦波状流;乱流;涡流;紊流;弯曲水流;湍流;反复曲折流
sinuous header 波形连管器
sinuous movement 曲折运动;蛇形移动;波状移动
sinuous resistance 波阻(力)
sinuous ripple 蜿蜒波浪;弯曲波痕
sinuous section 弯曲河段;弯曲段
sinus 中槽;正弦;港湾;湾

sinus aestuum 暑湾
sinus barotrauma 航空鼻窦炎
sinus iridum 虹湾
sinus median 中央湾
sinusoidal 正弦(式)的；正弦曲线形的
sinusoidal angular modulation 调角
sinusoidal arch 正弦拱
sinusoidal arch bridge 正弦曲线仰开桥；正弦曲线开合桥；正弦曲线吊桥；正弦曲线拱桥
sinusoidal boundary 正弦边界
sinusoidal carrier 正弦载波
sinusoidal current 正弦电流
sinusoidal curve 正弦(缓和)曲线
sinusoidal curve distribution 正弦曲线分布
sinusoidal deformation 正弦式变形
sinusoidal denier variation 正弦曲线状细度变化
sinusoidal distortion 正弦失真
sinusoidal distribution 正弦(式)分布
sinusoidal distribution of conductor 导体的正弦分布
sinusoidal envelope 正弦
sinusoidal excitation 正弦的激励
sinusoidal field 正弦场
sinusoidal flow 正弦流动
sinusoidal frequency divider 正弦分频器
sinusoidal function 正弦函数；类正弦函数
sinusoidal impulse 正弦脉冲
sinusoidal imput source 正弦输入源
sinusoidal input 正弦振荡输入；正弦信号输入
sinusoidal interference 正弦干扰
sinusoidal irregularity 正弦不平顺
sinusoidal jump function 正弦跳跃函数
sinusoidal law 正弦定律
sinusoidal light wave 正弦光波
sinusoidal load(ing) 正弦式荷重；正弦式荷载；正弦(曲线形分布)荷载
sinusoidal meander 正弦波形曲流
sinusoidal modulation 正弦调制
sinusoidal motion 正弦运动
sinusoidal oscillation 正弦(式)振荡
sinusoidal oscillator 正弦波振荡器
sinusoidal output 正弦输出
sinusoidal perturbation 正弦扰动
sinusoidal phase-modulation 正弦相位调制
sinusoidal pressure 正弦压力
sinusoidal projection 正弦(曲线)投影
sinusoidal pulse 正弦波形脉冲
sinusoidal quantity 正弦量；正弦曲线量
sinusoidal reactivity perturbations 正弦式反应性扰动
sinusoidal reactivity variation 正弦式反应性变化
sinusoidal response 正弦特性曲线
sinusoidal rf potential 正弦射频电压
sinusoidal rotating flux field 正弦旋转磁场
sinusoidal sender 正弦发送器
sinusoidal sequence 正弦序列
sinusoidal signal 正弦信号
sinusoidal signal generator 正弦信号发生器
sinusoidal spirals 正弦(类)螺线
sinusoidal stationary distribution 正弦定常分布
sinusoidal-strain dynamic(al) test 正弦应变动力试验
sinusoidal strip 正弦带
sinusoidal structure 正弦状构造
sinusoidal subcarrier output 正弦副载波输出
sinusoidal test 正弦试验
sinusoidal time variation 正弦式时间变化
sinusoidal trace 正弦线轨迹；正弦曲线扫迹；正弦曲线
sinusoidal variable 正弦变量
sinusoidal variation 正弦变化
sinusoidal vibration 正弦振动
sinusoidal voltage 正弦电压
sinusoidal vibrometer 正弦振动计
sinusoidal wave 正弦波
sinusoidal waveform 正弦波形
SiO_2 saturability 二氧化硅饱和度
siphon 管状口器；虹吸器；虹吸道；虹吸
siphon action 虹吸作用
siphonage 虹吸作用；虹吸现象；虹吸能力；虹吸法
siphonal 似弯管
siphon barograph 虹吸(自记)气压计；虹吸气压表
siphon barometer 虹吸(式)气压计；虹吸气压表；弯管气压计

siphon barrel 虹吸管
siphon battery 虹吸组
siphon bend 虹吸弯
siphon breaker 虹吸破坏管；虹吸截断器
siphon brick 挡渣砖
siphon can apparatus 虹吸管仪
siphon-can moisture tester 虹吸湿度计
siphon closet 虹吸气便桶
siphon condenser 虹吸冷凝器
siphon conduit 虹吸管道
siphon cooling system 虹吸冷却装置
siphon culvert 虹吸(式)涵洞；虹吸管涵(洞)；倒虹吸涵洞
siphon cylinder method 虹吸筒法
siphon drainage 虹吸排水
siphon duct 虹吸管道
siphon feed 虹吸式投加器
siphon feeder 虹吸给料器
siphon filter 虹吸滤池
siphon flask method 虹吸比重瓶法
siphon ga(u)ge 虹吸管压力计；虹吸表
siphonic action 虹吸作用
siphonic bowl 虹吸式便桶
siphonic closet 虹吸式便桶；虹吸式便器
siphonic collection device 虹吸集水装置
siphonic intake 虹吸式进水口
siphonic outlet 虹吸式出水口
siphonic pan 虹吸盘
siphonic spillway 虹吸式溢洪道
siphonic spillway crest 虹吸式溢洪道口
siphonic transition 虹吸管渐变段
siphonic water collection 虹吸集水(系统)
siphonic weir 虹吸式堰
siphoning 虹吸
siphoning effect 虹吸(式)效应
siphoning installation 虹吸装置
siphoning type 虹吸式
siphon inlet 虹吸进水口；虹吸管进口
siphon intake 虹吸取水口
siphonium 连管
siphon jet type 虹吸喷射式
siphon jet water 虹吸喷水大便器
siphon jet water closet 虹吸喷射式坐便器
siphon lock 虹吸闸；虹吸式水闸
siphon mouth 虹吸管进口
siphon off the clean solution 虹吸抽出清洁溶液
siphon oiler 虹吸加油器
siphon-operated lock 虹吸充水式船闸；虹吸充水式水闸
siphonostele 管状中柱
siphon outlet 虹吸式出水口
siphonozooid 管状个员
siphon pipe 虹吸管
siphon piping 虹吸管(道)
siphon pit 虹吸坑
siphon pressure ga(u)ge 虹吸压力计
siphon pump 虹吸(水)泵
siphon rainfall recorder 虹吸雨量计
siphon rain ga(u)ge 虹吸管雨量管；虹吸管(式)雨量计；虹吸式雨量器
siphon recorder 虹吸记录仪；虹吸记录器；波纹机
siphon refuelling pump 虹吸加油泵
siphon runner 虹吸浇道；底注浇道
siphon separator 虹吸分级机
siphon sluice 虹吸闸；虹吸式(泄)水闸
siphon spillway 虹吸(式)溢洪道
siphon spring 虹吸泉
siphon suction 虹吸抽水；虹吸抽出
siphon tank 虹吸池
siphon transition 虹吸管渐变段
siphon trap 虹吸封气弯；虹吸式防气弯管；虹吸(式)存水弯；虹吸封气管
siphon tube 虹吸管
siphon tube neck 虹吸管颈
siphon type flume 虹吸式渡槽
siphon-type rain-ga(u)ge 虹吸式雨量计
siphon-type wick oiler 虹吸油芯加油器
siphon-vortex flushing action 旋涡虹吸室冲洗功能；旋涡虹吸式冲洗功能
siphon-vortex water closet 旋涡虹吸式坐便器
siphon washdown type 虹吸冲洗式
siphon water closet 虹吸式便桶；虹吸式便器
siphon water-sampler 虹吸式采样器
siphon weir 虹吸堰

siphon well 虹吸管井
siphon wick-feed oiler 虹吸油芯注油器
sipo 桃花心木
sipo mahogany 红木；桃花心木
Siporex 西波雷克斯材料
Siporex concrete 西波雷克斯轻质绝缘集料混凝土
Siporex slab 西波雷克斯板
Siporex wall slab 西波雷克斯墙板
siporter 舣门
siporter system 舣门装卸方式
SIP process 埋入式吹氧法
sipylite 褐钇铌矿
sirafo bed 同温床
siren 旋笛；警报器；汽笛；气笛；报警器
siris tree 大叶合欢树
Sirius 镍铬钴耐热蚀合金
Sirocco di levante 希腊西南焚风
sirocco fan 多叶通风机；多叶片式风扇；西劳可式扇风机；鼠笼式风扇
S-iron S形铁
Sirte basin 锡尔特盆地
Sirufer(core) 细铁粉磁芯
sisal 剑麻
sisal carpeting 西沙尔地毯
Sisal fiber 西沙尔纤维；西沙尔麻织物
sisal hemp 剑麻；西沙尔麻
sisal hemp rope 西沙尔麻绳
sisal kraft paper 西沙尔牛皮纸
sisal netting bag 麻网袋
sisal reinforced 西沙尔麻加强的
Sisal rope 西沙尔麻绳(索)
SI screw 公制螺钉
sismondinite 镁硬绿泥石片岩
sissing 轻微的涂膜疙瘩；轻微的表面涂布不均；起纹；跑漆；缩边(一种漆病)
Sistan massif 锡斯坦地块
Si-steel 硅钢
Si-steel strip 硅钢带
sister 同级；成对的
sister block 姐妹滑车；双轮滑轮；双轮滑车；双轮；复滑轮；复滑车；双滑车
sister city 友好城市；姐妹城市
sister company 姊妹公司
sister hooks 和合钩；成对钩；安全钩；姐妹钩；双钩；双抱钩；叠钩
sister keelson 副内龙骨
sister-sheave block 双轮滑轮；双轮滑车
sister ship 姐妹船
sister ship clause 姐妹船条款；同船舶所有人条款
sistership clause 姐妹船条款
sitall 微晶玻璃
sitallization 微晶化
sit and wait predator 坐等捕食者
sitaparite 方铁锰矿
site 现场；工程地点；立地；位置；所在地；地址；地盘；地方；地点；场所；场地；部位
site-acceleration-magnitude procedure 场地—加速度—震级方法
site acceptance 现场验收
site accessibility 场地通达度
site access road 现场进路
site accommodation 现场供应；现场设备；现场设施；工地生活设施
site accounting 现场会计
site-acquisition cost 土地征用费
site activity 现场活动
site administration expenses 现场管理费
site agent 现场代理人；驻场员
site altitude 工地高度
site amplification 场地放大(作用)
site amplification characteristic 场地放大作用特性
site analysis 建设场地分析
site analysis services of the architect 建筑师的建设市场分析业务
site and season 场所与季节
site-applied maintenance coating 现场施工维修涂料
site appraisal 工程现场评价；工程地质鉴定；工程地址鉴定；坝址评价
site approach 工地入口；现场入口
site approval 场地批准
site architect 现场建筑师；工地建筑师
site area 厂区占地面积；场地面积
site arrangement 工地安装

site assembly 现场组装;现场装配;就地组装
site audit 就地审计
site bath 坐浴盆
site bin 工地料斗;现场料斗
site-binding phase transfer model 点接合相转移模型
site-binding surface complextion model 点接合表面络合模型
site bolt 固定螺栓;装配螺栓
site bolting 现场螺栓连接;工地栓接;现场栓接
site boundary 场地边线;现场边界
site built 现场制作的;现场修建的
site canteen 工地食堂;现场食堂
site cast concrete 现场浇制混凝土;现场浇筑混凝土;现场浇灌混凝土
site cast concrete pile 现场浇制混凝土桩;现场浇筑桩;现场浇注桩;现场浇灌桩
site casting work 现场浇筑工程;现场浇注工程;现场浇灌工程
site casting yard 工地浇制场;现场浇制场
site cast joint 现浇板缝
site category 场地类别
site checker 现场材料管理员
site class 地位级
site classification 场地类别
site cleaning 场地清理;现场清理;工地清理;清理现场;场地清除工作
site cleaning on completion 竣工时现场清理
site clean-up 工地清理
site clearance 现场清理;场地清理
site-clearance activities 场地清理工作(建筑前)
site cleared of buildings 清除现场建筑物
site clearing 清理场地;现场清理;清场
site clearing work 场地清除工作
site climate 地方气候
site coefficient 场地系数
site complexity 场地复杂程度
site concrete 现场制备混凝土
site condition 工地条件;现场条件;生境条件;场地条件
site connection 现场连接
site construction organization 现场施工组织
site contour 现场地形
site control engineer 现场工程控制工程师
site cost 建设场地成本;场地费用
site coverage 现场范围
site data 现场数据;现场资料
site-dependent 随地变化的
site design 现场设计
site designation memorandum 选址意见书
site development 场地开发;现场设计;建设场地开发;场地开发
site diameter 微区直径
site diary 工地日记
site dimensions 建筑场地尺寸;现场尺寸;场地尺寸
site director 工地主任
site disposal 现场处置
site drainage 工地排水系统;工地排出的水
site drawing 总布置图;位置图;地区图
site effect 场地效应
site enclosure 工地围墙;现场围墙
site engineer 现场工程师;(施工单位的)工地工程师
site engineering 竖向规划
site entrance 现场进入权
site equipment 工地设施;场地设备
site error 仪表误差;位置误差;地物方位误差
site evaluation 地点评审
site expenses 现场经费;现场管理费
site exploration 工地调查;现场勘探;现场勘察;现场查勘;勘探场址;基地勘探;现场探测;场地勘探;坝址勘察;坝址勘查;坝址勘测
site fabricated runner 现场组装式转轮;工地组焊的转轮
site fabrication 现场制作
site facility 现场设施;场地设施;工地设施
site facility plan 工地设施平面图;现场设施平面图
site facility program(me) 现场设施计划
site factor 工地因素;现场因素;环境因素;生境因素;生境因素;场地系数
site factory 工地工厂
site fence 工地围篱;工地围栏
site filtering characteristic 场地滤波特性
site flooding 场地淹没;场地注水
site-foamed insulation 现场发泡隔热材料;现场泡沫保温
site for fumigation 熏蒸气消毒场;熏气消毒场
site for harbo(u)r facility 港口设施用地
site formation 工地平整
site foundation 地基
site frontage decrease 基地临街建筑线后退
site furnishings 基地设施;建筑小品;基地小品设施
site geologic(al) condition 场地地质条件
site grading 场地平整
site grouting 现场灌浆
site handling 工地装卸
site heating 工地供热;现场供热
site height 测站高程
site hut 工地工棚;工棚;工地房屋
site hut heating 现场工棚供热
site hut lamp 现场工棚灯
site-improvement cost 建设场地改善费用
site-independent mode 地点无关方式
site index 地位指数
site influence 场地影响
site inspection 现场踏勘;现场视察;现场勘察;现场监督;现场调查;工地检查;坝址调查
site installation 现场装置;现场设施
site installations plan 工地设施平面图;现场设施平面图
site intensity 场地强度;场地烈度
site investigation 现场踏勘;现场调查;工地勘察;工地勘测;工程地址调查;勘察;就地踏勘;基地勘探;施工现场调查;场地勘察
site investigation for dry dock 干坞的现场调查
site investigation for shiplift 升船机的现场调查
site investigation of a factory 厂址调查
site joint 现场接点;现场连接;现场接合
site laboratory 工地实验室;工地试验室
site labo(u)r 工地劳动力
site lantern 工地灯具;现场灯具
site layout 现场布局;工地布置(图);基地布局
site layout plan 现场布置图
site level(l)ing 平整场地;场地平整
site licencing[licensing] 场地许可证
site lighting 工地照明;现场照明;场地照明
site limitation 场地限制
site location 场地位置;工程地点
site management 现场管理;工地管理
site manager 现场经理
site manufacture 工地制造
site map 现场图;地位图
site map for project 建设项目场地地形图
site mapping 绘制生境示意图
site-marking 现场标志;现场划线
site marking device 位置指示仪
site measuring 工地测量;工地计量;现场测量;现场计量
site mechanic 工地技工;工地机修工;现场技工;现场机修工
site meeting 现场会(议)
site megaphone 场地扩音器
site method 现场法
site-mixed concrete 工地搅拌混凝土;工地拌制混凝土;现场搅拌混凝土;现场拌制混凝土;现场拌混凝土
site mixer 现场搅拌机;工地搅拌机
site mix formulation 现场配料组成;现场配料设计
site mixing 工地拌和
site mixing in place 现场就地混合
site mixing plant 工地搅拌装置;工地搅拌厂;工地拌和厂;工地拌和装置;现场搅拌装置;现场搅拌厂;现场拌和厂;现场拌和装置
site model theory 位置模型理论
site mud 工地淤泥
site noise 地形干扰
site observation 现场观测
site of afforestation 造林地
site of conjugation reaction 结合反应位置
site of construction 建筑工地;施工场地
site of dam 坝址
site of exposure 接触部位
site office 工地办事处;工地办公室
site of generation of waste 废物产生地
site of paracentesis 穿刺部位
site of puncture 穿刺部位
site of special scientific interest 有特殊科学价值的地区
site of station 台址
site of the well 井点位置;井的现场
site of work 工作现场;工地
site operation 现场作业;现场实施;现场施工;现场操作
site operation staff 现场运行人员
site ore supervision 现场矿石质量管理
site organization 现场组织;工地组织
site-painted 工地油漆的
site personnel 工地人员;现场人员
site pile 就地灌注桩;现场灌注桩
site pile casting yard 工地桩的浇捣场
site-placed aerated concrete 现浇加气混凝土
site-placed architectural concrete 现浇建筑混凝土
site-placed concrete 现浇混凝土
site-placed mortar 现浇砂浆
site-placed reinforced concrete 现浇钢筋混凝土
site plan 建筑总平面图;总平面(图);工地平面图;工地布置(图);平面布置图;地盘图;场地平面图;厂址计划;坝址平面图;总设计图;总计划
site planning 总平面设计;现场组织设计;建设场地规划;场地计划;总平面规划;场地平面规划
site planning condition 总平面设计条件(如地形、地质、水文等)
site plan of station 车站线路图
site plant 工地车间;现场设备
site plant plan 现场装置平面图;现场设备平面图
site plate 基地平面
site power plant 工地发电站;工地发电装置;现场发电站;现场发电装置
site practice 工地实践;现场实践
site precasting 现场预制;现场预浇
site prefabricated method 现场预制吊装法
site prefabrication 现场预装配;现场预制;现场预浇
site prefabrication method 工地预制吊装施工法
site preparation 施工现场准备;现场平整;现场准备工作;整地;基地整平;场地准备
site preparation programme 现场准备计划
site prestressing 现场预加应力
site quality 立地生产力
site quality of forest 森林地位级
site rail 工地铁轨
site railway 现场铁路;工地铁路;工地铁道
site ranking 场地等级
site reclamation of harbo(u)r 港区吹填
site reclamation of port 港区吹填
site reconnaissance 现场查勘;踏勘;地址勘测;场地查勘
site records 现场记录;工地记录
site regulations 现场规章
site reinstatement 现场地复原
site renovation 建筑现场用地改造更新
site representative 现场代理人
site requirement 场地要求
site response 场地反应
site risk assessment 点位风险评价
site rivet 工地铆钉
site riveted butt joint 现场铆钉对接接头
site riveting 现场铆合;现场铆接;工地铆接
site road 现场道路;工地道路
site safety officer 现场安全员
site sample 现场实样
sites and services 提供场地与基础设施
site security 现场安全
site selection 选址;基地选择;位置选择;场址选择;场地选择;厂址选择;坝址选择
site selection investigation 选址查勘
site selection of factory 厂址选择
site settlement 工地安排
site shop 现场车间
site silo 工地筒仓;现场筒仓
site size 微区尺寸
site sketch 位置略图
site slaking 工地熟化(石灰);工地消化(石灰)
sites of customs inspection 验关地点
site soil 工地土壤;现场土壤
site space 现场空间
site-specific condition 当地特殊条件
site-specific constraint 因为场地受到影响的限制
site stability 场地稳定性
site stability evaluation 场地稳定性评价
site staff 工地职员;现场工作人员
site stoppage 工地阻塞
site stretching 现场张拉
site stripping 场地表土清除;表层土方平整

site structure interaction coefficient 场地(土)结构相互作用系数
site structure of power-house 发电厂厂址结构
site subsidy 工地补助费
site suited 生境适当的
site supervision 现场质量管理
site supervisor 工地负责人
site supervisor's office 工地负责人办公室
site survey 现场测量;现场勘测;工程地址调查;场地测量;坝址勘察;坝址勘测;坝址测量
site surveying 场地勘测
site surveying and test 现场勘察测试
site surveyor 工地监工(员)
site symmetry 位置对称性
site technical director 工地技术主任
site telephone 工地电话
site tensioning 现场张拉
site test 现场试验;现场测试;实地试验;当场试测
site tiling 现场贴砖;现场铺瓦
site tipper 工地自卸车
site topography 现场地势图;现场地形图;场地地形
site traffic 现场交通(施工时);工地交通
site trailer 工地拖车
site transport 现场运输
site treatment 现场处理
site trial 实地试验
site type 立地类型
site utilization 场所的利用
site value 场地价值
site water-and-electricity supply 工地水电供应
site weigh batcher 工地配料器
site welded 现场焊接的
site weld(ing) 现场焊接;工地焊接;安装焊(接)
site wharf 工地码头
site work 现场工作;场地工程
site work shop 工地车间
siting 建设地点的决定;选址(规划);工程选点;建筑工地选择;建筑地点定位
siting analysis 选址研究;选址分析
siting investigation 选址勘察
siting of dry dock 干坞选址
siting of houses 住宅群选址定点
siting of industry 工农业选址
siting of lock 船闸选址
siting of shiplift 升船机的选址
siting stage 选址阶段
siting study 选址研究
siting survey 选址勘察
sitka 云杉木
sitka spruce 银云杉
sitology 食品科学
sit-on-bottom drilling rig 座底式钻井装置
sitotoxin 谷物毒素
sitotoxism 食物中毒;食品中毒
sitotoxismus 食品中毒
sits bath 坐浴(浴)盆
sitter-up 玻璃熔制辅助工
sitting 平面布置
sitting for marshalling 车列坐编
sitting government 本届政府
sitting height 坐高
sitting length scale 坐高尺
sitting on fire 起火
sitting room 起居室;生活室
sitting W. C. pan 坐便器
situation 概况;局面;情况;情景;弯曲;地点;场所;场合
situation analysis 情况分析
situation analyzer 势态分析器
situation approach to management 情况管理原则
situation board 气候情况测绘板
situation control 势态控制
situation display 状态显示;状况显示
situation display computer 情况显示计算机
situation display console 情况显示台
situation-display tube 情况显示管
situation map 形势图;态势图
situation of project 工程状况
situation of world mineral reserves 世界资源储备状况
situation plan 位置图;地盘图
situation recognition 势态识别
situation record 战况图
situation report 情况报告

situation room 情况室
situation summary 情况综合报告
situation variable 情况变量
situ-cast pile 现浇桩
situs 位置;地点;部位
situ strength 现场强度;实测强度
situ theory 原地生成说
sitz bath 坐浴盆
six-axis planimeter method 六轴求积仪法
six-axle 六轴
six-axle locomotive 六轴机车
six bladed grab 六爪抓岩机;六爪抓具;六瓣式抓岩机;六瓣式抓具
six-break oil breaker 六断点油断路器
six by four 六轮载货车
six-by-six 有六个驱动轮的载货汽车
six colo(u)r infrared photometer 六色红外光度计
six-colo(u)r photoelectric(al) photometry 六色光电测光
six-colo(u)r photometry 六色测光
six-colo(u)r press 六色胶印机
six-colo(u)r recorder 六色记录仪
six-cone bit 六牙轮钻头
six-cut finish 中等粗度凿石面
six-cylinder engine 六缸发动机
six degrees of freedom 六个自由度
six-degree zone 六度带
six digit system 六位制
six-element lens 六片透镜
sixfold 六重;六倍
sixfold axis 六重轴
sixfold degeneracy 六度简并度
six-head matcher 六刀头面刨
six-hole joint bar 六孔夹板
six-hole stone drilling machine 六孔钻岩机
six-inch map 六英寸地图
six-jawed heavy duty grab 六颚板重型抓斗
six-keyway broach 六键拉刀
six-layer quarter-wave stack 六层1/4波长片的多层系统
six-lead tach(e)ometer 六导程测速仪
six-lobe tracery 六瓣窗花格
Six maximum and minimum thermometer 薛氏最高最低温度计
six-membered ring 六元环
six-minute sample count 六分钟取样计算(道路交通技术)
six o'clock hour circle 六点时圈
six panel door 六块镶板门
sixpartite vault 六分区穹顶
six-phase circuit 六相电路
six-phase rectifier 六相整流器
six-phase star connection 六相星形接法
six-phase system 六相制;六相系统
six-phase voltage 六相电压
six-ply tire 六层轮胎
six-pocket plastic foam life jacket 六袋式泡沫塑料救生衣
six-point bit 星形钻(头);六刃钻头;六刃钎头
six-pointed star 六角星
six-pointed star core 六角星形芯
six-pointed star core space 六角星形内腔
six point mooring 六点系泊
six-point steel 六翼钻头
six rayed star 六角星
six-roller bit 六牙轮钻头
six sampling intervals per decade 六分度取样
six-spindle polishing machine 六轴抛光机
Six's thermometer 薛氏最高最低温度计;西科斯温度计
sixteen hour service provided by ship station of the second category 第二类船舶电台工作16小时
sixteenth-diurnal constituent 1/16日分潮
six-tenths factor 6/10的因数
six-tenths rule 十分之六准则
sixth-diurnal constituent 1/6分潮
sixth floor 第六层
sixth generation container 第六代集装箱
sixth generation container terminal 第六代集装箱码头
six times 6倍
sixty degree swan-neck 60°楼梯的鹅颈扶手

sixty-two cm brick wall 两砖半墙
six-vesica piscis tracery 六双尖弧形窗花格;六梭形环花窗花格
six-way plate 六流底盘;六爪底板
six way punch pliers 六孔冲孔钳
six-way valve 六通阀
six-wheel-drive truck 六轮驱动运货车
six-wheel dumper 六轮翻斗车
six-wheeler 六轮载货车
six-wheel freight car 六轮货车
six-wheel locomotive 六轮机车
six-wheel truck 三轴转向架
six-wheel vehicle 六轮汽车;三轴汽车
six-wire 六股钢丝
sizable area 广大区域
sizable lump 适当块状
sizable minority 相当大的少数
size 规模;古物贴金用的黏性清漆液;量值;体量;上胶;上浆;大小;尺寸
size adjusting 粒度调整
size adjustment 尺寸调整
size analysis 颗粒级配曲线;粒径分析;粒度分析;筛分试验
size analysis of ash 灰的分析;灰的筛析
size analysis of submicron particles 亚微细粒的分析;亚微细粒的筛析
size analyzer 粒度分析仪;粒度分析器
size and type code 尺寸和类型代号
size applicator 浸润器;施胶装置
size applicator for filaments 单丝浸润器
size block 规块
size box 胶槽
size bracket 标称尺寸;粒度级别;尺码分档
size category fraction 粒级
size-changing rate in heat-treatment 热处理尺寸变化率
size characteristic 粒度特性;筛分特性;尺寸特性
size characteristic curve 粒度特性曲线
size class 大小级
size classification 粒度分级;分选;按大小分选;按大小分级
size classification of soil particles 土的粒径分类
size classifier 粒度分级器
size clause 长度子句
size coarse aggregate 规格的粗集料;规格的粗骨料
size coat 彩印底漆
size colo(u)r 胶类颜色;水粉颜色
size composition 粒径组成;大小组成;尺寸组成
size consist 粒度组成
size content 浸润剂含量
size contraction 截面收缩
size control 粒度控制;块度控制;大小控制;尺寸控制
size cut-point 分离粒径;分离粒度
sized 按大小排好了的
size data 粒径数据;粒径记录;筛分数据
sized coal 筛选煤
sized coarse aggregate 分级粗集料;分级粗骨料;均匀粒径粗集料;均匀粒径粗骨料
size-decline curves 粒度偏斜曲线
size degradation 磨细;碎解;打小
size delivery pump 齿轮输浆泵
size determination 尺寸确定
sized feed 分级给料
sized glass fibre 被覆浸润剂的玻璃纤维
sized gypsum 磨细石膏灰泥;筛分石膏
size dimension 尺寸大小
size distemper 胶水色粉涂饰;水粉画
size distribution 粒径分配;粒径分布;粒度分布;颗粒级配;颗粒分布;大小分布
size distribution analyzer 尺寸分布分析器
size distribution curve 粒度分布曲线;粒度分配曲线
size distribution diagram 粒径分布图
size distribution factor 粒径分布系数
size distribution mode 粒度分布方式
size distribution of dwellings 户室比
size distribution of firms 厂商的规模分布
size distribution of households 户型比
size division 粒级;粒度级别;尺寸划分
sized lumber 规格木材;净规格料
sized material 筛析品
sized ore 过筛矿石;分级矿石
size down 由到小排列
sized paper 涂胶纸

sized slate 规格石板(瓦);条石
size effect 截面因数;尺寸效应;比例影响;比例效应
size effect of rock mass 岩体尺寸效应
size enlargement 造粒;颗粒放大;颗粒变大;截面扩大
size error 长度误差
size error condition 长度错误条件
size exclusion chromatography 体积排阻色谱法;大小不相容色谱法
size factor 粒度系数;颗粒系数;大小因数;尺度因子;尺度因数;尺寸因子
size field 尺寸域
size finder 尺寸显示装置
size-float-and-sink sample 筛分浮沉大样
size for connection 连接尺寸
size formulation 浸润剂配方
size fraction 粒径分级;粒度级别;粒组;粒径组成;粒级;粒度份额;颗粒组;颗粒级别;颗粒的级分
size fraction analysis 粒度分析
size fraction of sample 样品粒度
size fractions of crushed coal 破碎级
size fractions of raw coal 自然级
size frequency curve 粒度频率曲线;粒度分布曲线
size gradation 粒径级配
size grade 粒径等级
size-graded particle board 渐变结构碎料板
size grader 径选机;按尺寸分级机
size grading 粒径级配;粒径分级;粒度分级;颗粒分级
size grading sampler 粒度分级采样器
size in the clear 净尺寸;内尺寸
size-investment ratio 规模投资比率
size limit 粒径限度;粒径范围;粒度极限;合格粒度
size marking 尺寸标注;尺寸标志
sizematic internal grinder 自动定尺寸内圆磨床
size meter 测尺寸器
size migration 浸润剂迁移
size mixing equipment 浸润剂配制装置
size of a fillet weld 角焊缝尺寸;焊角尺寸
size of anode 阳极规格
size of a sample 样品容量;样品大小
size of a test 检验水准
size of a town 城镇的规模;城镇的大小
size of a wrench 管钳长度
size of bigger rat hole 大鼠洞径
size of Burger's vector 布格矢量大小
size of business 企业规模
size of city 城市规模
size of coal 煤的粒度
size of collapse depression 塌陷洼地大小
size of conduit 管径
size of construction 施工规模
size of dam mass 坝体尺寸
size of design vessel 设计船型尺寸
size of end product 成品尺寸
size of establishment by employment 以就业工人人数而论的工厂规模;就工人人数而论的工厂规模
size of geologic(al) map 地质图规格
size of grain 粒径;颗粒大小
size of hook mouth 钩口直径
size of household 家庭结构;户数
size of image 像幅;形象尺寸
size of interpretation feature 解译特征大小
size of jaw 扳口开度
size of liner 缸套直径
size of load 荷载
size of loan 贷款数量
size of measurement 量值
size of memory 存储器容量
size of mesh 筛子大小;筛眼孔径;筛眼号数;筛眼大小;筛眼尺寸;筛孔尺码;筛孔尺寸
size of mouse hole 小鼠洞径
size of new grain 新颗粒大小
size of node 节点的大小
size of oil and gas field 油气田规模
size of opening 筛孔尺寸
size of orebody 矿体规模
size of ore deposit 矿床规模
size of pipe 管径
size of pipe and tubing 管子公称尺寸;管道标称尺寸;管材尺寸
size of piping 管径
size of plate 板块大小

size of population 总数大小;母体量
size of product 产品规格
size of riser 立管的尺寸
size of seal contact face 密封面尺寸
size of section 截面大小;截面尺寸
size of space 间隔尺寸
size of steel line 钢丝绳直径
size of subgrain 亚颗粒大小
size of symbol 符号大小
size of the(bore)hole 钻孔尺寸
size of the jet 喷管尺寸;喷管直径
size of the well 每昼夜油井产量
size of tube[tubing] 管径
size of under ground observation laboratory 地下观测室尺寸
size of underground opening 洞室尺寸
size of weld 焊缝尺寸;焊区尺寸;焊件的大小
size operand 长度操作数
size pick-up 浸润剂浸取
size preparation 粒度准备
size press coated paper 表面施胶纸
size putty 快干腻子
sizer 分粒机;填料器;上胶器;筛子;分组机;分选机;分粒器;分级机
size range 粒径范围;粒度范围;块度范围;尺寸范围
size range of separation 分选粒级
size ratio 粒径比;径向比;径径比
size recovery pipe 浸润剂回流管
size reduction 轧碎;颗粒变小;减小尺寸;减缩尺寸;磨细;磨碎
size reduction energy 捣碎能量
size reduction equipment 粉碎设备
size reduction machinery 粉碎机械
size reduction mill 粉碎磨
size reduction of coal sample 煤样的破碎
size reduction ratio 破碎比
size reduction sizing 破碎筛分
size reduction work 破碎功;粉磨功
size residue 浸润剂残留量
size return pipe 浸润剂回流管
size rolls 定径辊
size sample 筛分样
size scale 粒度比;分级比
size scale effect 比例影响;比例效应
size segregation 粒度偏集(现象);颗粒偏析;按大小离析(现象)
size select 尺寸选择
size separation 粒析;颗粒选分
size sorting 粒度分选
size stabilizer 浸润剂;稳定剂
size state 糊糊状
size stick 尺杆;材料尺寸样板;刮板定位杆
size test 筛析;筛分试验;粒度分析
size tolerance 尺寸容差;尺寸公差
size up 过筛;测定尺寸;估量
size water 胶水
size-weight ratio 样重比率
sizing 粒径分析;粒度调整;粒度测定;颗粒分级;精整;上胶;上浆;筛分;定尺寸;尺寸定位;测定尺寸;分选;分粒
sizing agent 浸润剂;上浆材料
sizing analysis 粒度分析;颗粒分析;筛分析
sizing and edging machine 裁剪修边机
sizing assay test 矿物粒度分析试验
sizing block 垫铁
sizing box 浆槽
sizing characteristic 筛分特性曲线
sizing coat 胶结涂层
sizing content 浸润剂含量
sizing control 尺寸控制
sizing curve 粒度曲线
sizing device 校准设备;碰停装置;筛分设备
sizing die 整形模;精整模;校正模
sizing drum 筛筒;筛鼓
sizing effect 筛分效应;筛析效应
sizing establishment 确定尺寸大小
sizing finish 上浆整理
sizing ga(u)ge 尺寸(控制)量规
sizing grill 筛分格栅
sizing ingredients 上浆成分
sizing instruction 上浆配方
sizing knockout 精压出坯杆
sizing machine 径选机;胶黏机;浆纱机;上胶机;分级机

sizing machinery 精整设备
sizing material 胶料;浆料
sizing mill 定径机;定经机
sizing of equipment 设备尺寸计算
sizing of filament 单丝涂油;单丝浸润
sizing of mineral 矿物分级
sizing of particles 粒度测定
sizing of under material 垫层尺度
sizing percentage 上浆率
sizing plant 粉degrees装置
sizing plot 筛析图表;筛分曲线;筛分分析图表
sizing press 精整压力机;定径压力机
sizing procedure 精压操作程序
sizing punch 精压模冲
sizing reduction 减径量
sizing roll 定径轧辊
sizing roller 浸润器;涂油辊;上浆辊
sizing rolling mill 定径轧机
sizing screen 分选筛;分粒筛;分级筛
sizing shaker 分级摇动筛
sizing size 分级粒度
sizing specification 尺寸说明
sizing stand 定径机
sizing stop 校准支架
sizing stripper 精压脱模杆;精压出坯杆
sizing test 粒度分析(试验);筛分析;筛分试验
sizing test number 筛分试验编号
sizing test report table 筛分试验报告表
sizing test total sample laboratory test result 筛分试验总样化验结果
sizing tool 校准刀
sizing treatment 上胶
sizing up a job 估计工作量
sizing varnish 胶黏清漆;胶性清漆;贴金漆;上浆清漆;封闭清漆
sizing vat 填料瓮
sizing water main 定给水干管大小
sizy 浆刷的;胶水的;胶质的
sjogrenite 水碳铁镁石;水硅铁石
S-joint S 节理
skag 稳拖链
skaking sieve 摇动筛
Skandinaviska Lackteknikers Forbund 斯堪的纳维亚油漆工艺师联合会
skarn 矽卡岩;夕卡岩
Skarne 斯卡尔恩体系建筑
skarnization 矽卡岩化作用
skarn mineral 矽卡岩矿物
skarn molybdenum deposit 矽卡岩型钼矿床
skarn ore deposit 矽卡岩矿床
skarn orefield structure 矽卡岩矿田构造
skarn-thorianitic deposit 矽卡岩型方钍石矿床
skarn-type copper deposit 矽卡岩型铜矿床
skarn-type pyrite deposit 矽卡岩型硫铁矿床
skarn-type tin deposit 矽卡岩型锡矿床
skarnzation 矽卡岩化
skate 滑动装置;铁鞋
skateboard park 溜冰公园
skate brake 制动块
skate changing room 溜冰鞋更换室
skate clip 制动铁鞋夹
skate conveyer 滑道输送器
skate hire room 溜冰鞋租借室
skate retarder 铁鞋缓行器
skate throw-off device 脱鞋器;脱铁器
skate throw-off switch 脱鞋道岔
skate valve 减速器阀
skate with automatic control 自动控制的制动铁鞋
skating rink 溜冰场;滑冰场;冰上运动场
skedasticity 离中趋势;方差性
skedophyre 勺斑岩
skeeling 搭连屋
skeen arch 平圆拱
skeet 犀斗;水瓢
skeg 尾柱底骨;尾踵;舵底承
skeg band 舢板尾柱护环
skeg framing 尾鳍构架
skein 线团;线球;线绞;一束;一绞;毂衬套;混乱
skein dyeing 绞丝染色;绞纱染色
skeining 成绞
skein winding 分布绕组
skein yarn strength tester 缕纱强力试验机
skeletal code 骨架代码
skeletal coding 程序轮廓编码;程序骨架编码;程

序纲要
skeletal crystal growth 骸晶生长
skeletal detritus 骨质碎屑
skeletal diagram 简图；三相图
skeletal fabric 骨骼组织
skeletal facies 骸骨相
skeletal fragment 骨屑
skeletal grain 骨粒；骨骼颗粒
skeletal limestone 骨粒石灰岩；骨粒灰岩；骸骨石灰岩
skeletal limestone matrix 骨粒石灰岩基质
skeletal material 骨料（又称集料）；骨架材料；骨骼物质；骨材
skeletal phosphoraite 骨屑磷块岩
skeletal query 提纲式询问
skeletal representation 梗概表示
skeletal sand 骨屑砂；粗骨砂
skeletal semi-trailer 骨架式半拖车
skeletal soil 石质土；粗骨土
skeletal structure 骸骨构造
skeletal substance 骨架物质
skeletal system 骨架系统
skeletal texture 骨架状结构；骸晶结构
skeletal trailer 骨架式拖车；无底板牵引车
skeleton 骨架(网)；骨骼；概略；架子；画窗图；骸骨
skeleton abutment 骨架桥台；框架式桥台
skeleton bay (厂房的)预留机组段
skeleton beacon 骨架式立标；笼式立标
skeleton body 骨架体
skeleton building 骨架建筑
skeleton case 笼箱；木架箱；木格箱；条板箱
skeleton chart 航线图；航路图
skeleton code 骨架代码；简要码
skeleton construction 框架构造；骨架结构；骨架构造；钢框架结构；框架(式)结构
skeleton construction building 骨架结构建筑；框架结构建筑
skeleton construction type 骨架结构类型；框架结构类型
skeleton core 框格芯；(空心门的)内架；框架芯
skeleton core flush door 框格芯平面门
skeleton crew 骨干船员
skeleton crystal 骸晶
skeleton curve 骨架曲线
skeleton diagram 简图；总图；方框图；构架图；概略图；轮廓图；单线图；方块图
skeleton dictionary 轮廓目录
skeleton displacement 裸船排水量
skeleton drawing 轮廓图；结构图；简图；原理图；骨架图；示意图
skeleton drum 圆鼓干燥骨架
skeleton facies 骨架相
skeleton floor 组合肋板；构架肋板；空心肋板
skeleton form 略图
skeleton frame 骨架；钢骨架；框架；铁骨构架
skeleton-frame-type wharf 桁架式高桩码头
skeletonizing 绘制草图
skeleton key 万能钥匙
skeleton law 骨架定律
skeleton layout 初步布置；草图
skeleton line 中弧线；骨架线；干线；轮廓线；山脉线
skeleton map 概略图；略图；草测图
skeleton member 骨架构件
skeleton method 简要方法
skeleton of building 房屋骨架
skeleton of panel wall 骨架大板墙
skeleton particle 骨架颗粒
skeleton pattern 骨架模；轮廓模型
skeleton pier 构架式码头
skeleton plan 轮廓计划；草图
skeleton polymer 骨架聚合物
skeleton rock bucket 骨架式岩石铲斗
skeleton sheeting 骨架支撑
skeleton shoe 骨架底板
skeleton sketch 构架图；轮廓草图
skeleton soil 石质土；粗骨土；粗骨粉
skeleton steps 框架楼梯踏步；透空楼梯
skeleton structure 杆系结构；骨架结构；框架结构
skeleton survey 草测
skeleton symbol 结构简式；简式
skeleton system 骨架系统；骨架体系；梁柱承重构架体系
skeleton type bucket 骨架式岩石铲斗
skeleton type construction 骨架式结构；框架结构

skeleton-type motor 开敞型电动机
skeleton type selftightening oil seal 骨架式自紧油封
skeleton type switchboard 骨架式配电盘
skeleton unit 预留机组；缩额单位；辅助安装位置；辅机安装位置
skeleton view 透视图；骨架图
skeleton wagon 骨架式货车
skelp 制造钢板；管材；焊接管坯；焊管铁条
skelp blower 空气喷嘴
skelper 焊接管控制机；焊接管拉接机；焊管成型机
skelp mill 管材轧钢厂；管材轧机；焊接坯轧机
skelp steel 焊接管坯钢
Skempton-Bjerrum's method 斯开普顿—贝仑法
Skempton method 斯开普顿法
skene arch 弧形拱；弧拱；平圆拱；平弧拱
skeo 铸桶；箕斗；大篮子；翻斗车；弧拱棚
skerry 孤岩；残岛；悬崖岛
skerry coast 岛礁岸
skerry-guard 护岸岩礁；低矮小岛；防波岩礁
skerry-type coast 崖岛型海岸
sketch 写生画；工作草图；概略；素描；设计简图；打草图；草稿
sketch blank 异形板坯
sketch book 素描簿；写生簿；略图簿
sketch box 画箱
sketch chart 概略图
sketch drawing 略图；示意图；草图
sketched contour 手勾等高线
sketched geologic(al) profile 素描地质剖面图
sketch for the colluvial soils 崩积土图样
sketch graph 草图
sketching 画草图
sketching board 绘图板
sketching charcoal 绘图炭条
sketching contour 草绘等高线；草绘曲线
sketching easel 画架
sketching paper 绘草样纸；素描纸
sketching pencil 绘图笔
sketching technique 技术草图
sketch in perspective 透视速写；画透视图
sketch input point 草图输入点
sketch input system 草图输入系统
sketch map 略图；速写图；素图；示意(草)图；草图
sketch map of earthquake of the world 世界地震分布简图
sketch map of exploratory trench 探槽素描图
sketch map of shallow mine 浅井素描图
sketch master 草稿地图；照片转绘仪；图片转绘仪；草图转绘仪
sketch of drift 坑道素描图
sketch of elevation 立面草图
sketch of historic(al) opening 老窟素描图
sketch of level(l)ing lines 水准路线图
sketch of pit 浅井素描图
sketch of terrain 地形略图
sketch of trench 探槽素描图
sketch of wall of a well 井壁素描图
sketch or preliminary plat 地段草图；初步地段图
sketch out 拟定
sketchpad 草图；略图
sketch plan 草绘平面；简单平面图；平面草图；示意草图；初步计划；草图
sketch plate 异形板；裁边板
sketch processing 草图处理
sketch profile 素描剖面图
sketch projection 投影转绘
sketch projector 投影转绘仪
sketch recognition 草图识别
sketch rectify 草图校正
sketch survey 草测
sketch survey map 草测图
sketch tools 速写工具
sketchy 粗略的；大体的；草图的
sketchy specification 概略的技术要求；概略的规范
Skevas-Zerfus disease 采海绵潜水员病
skew 斜砌石；偏斜；歪斜失真；歪的；山墙角石；不齐量
skew adjustment 不齐量调整
skew angle 斜交角；斜拱角；相交角；歪扭角
skew angle bearer 碹碴承台
skew angle of wind 风的卸交角
skew antenna 斜向辐射天线
skew arch 斜碹；斜交拱；斜拱；不对称拱

skew arch bridge 斜交拱桥
skew axis 斜轴；相错轴；交错轴
skew axis gear 偏轴齿轮系；双曲面齿轮系；偏斜齿轮系
skewback 拱脚斜块；拱座斜块；拱角斜石块；拱基砌拱砖；起拱石；托砖；斜块拱座；斜角拱座；拱座；弯背手锯
skewback angle 碹碴角钢
skewback hinge 拱angle铰链；枢轴接合
skewback saw 曲背手锯
skew barrel arch 斜(圆)筒形拱
skew barrel vault 斜筒形穹顶
skew bending 斜弯曲
skew bevel 斜伞形；斜交角
skew bevel gear 斜伞齿轮；斜齿锥齿轮；歪伞齿轮
skew bevel gear pair 斜齿锥齿轮付
skew bevel wheel 斜伞齿轮；偏轴斜齿伞齿轮
skew blanket feeder 斜毯式投料机
skew block 山墙端挑六；斜形砌块；堞头【建】
skew branch 斜交叉管
skew brick 碹碴砖；斜削砖；斜砌砖；拱脚砖；拱角砖
skew bridge 斜桥；斜交桥
skew bridge slab 斜交桥板；弯曲的桥板
skew butt 压顶石座；升降铰链；堞头【建】
skew catenary 斜锥形悬挂
skew chisel 弯曲的凿刀；斜凿
skew coil 斜线圈
skew coil winding 斜圈绕组
skew control 菱形失真调整
skew coordinates 斜坐标
skew coordinate system 斜轴坐标系
skew corbel 斜座台；斜座石
skew correction 菱形校正；偏斜校正
skew correlation 偏斜相关
skew crossing 斜交道；斜形交叉；斜向交叉；斜交叉道(路)；斜交
skew culvert 斜交涵洞
skew curve 斜曲线；空间曲线；挠曲线；偏斜曲线；偏态曲线；非正态曲线；不对称(空间)曲线
skew cutter 椭圆形刀头
skew determinant 斜对称行列式
skew distortion 菱形畸变；偏斜失真
skew distribution 偏态分布；歪分布；非对称分布；不对称分布
skew distribution curve 不对称分布曲线
skewed 不垂直的；斜交的；斜的
skewed angie 斜角
skewed barrel arch 斜筒拱
skewed boundary condition 偏斜边界条件
skewed bridge 斜交桥；弯曲的桥
skewed connection 歪斜连接；非直角连接；斜连接
skewed crossing 斜交平交道；斜交叉
skewed culvert 斜交涵洞；斜涵(洞)
skewed density function 偏斜密度函数
skewed distance 斜跨距离
skewed distribution 偏斜分布
skewed error 偏斜误差
skewed frame 斜构架
skewed(gear)tooth 斜齿
skewed helix 斜螺旋线
skewed histogram 偏态直方图
skewed junction of masonry walls 圬工墙的斜交点
skewed normal distribution 非对称正态分布
skewed parallelogram plate 斜平行四边形板
skewed peak 斜峰
skewed plate 斜交板
skewed pole 斜形磁极
skewed pole-shoe 斜极靴
skewed projection 斜轴投影
skewed roll 不平行轧辊
skewed scheme 斜移方案
skewed shell 扭曲薄壳
skewed slab 斜交板
skewed slab bridge 斜板桥
skewed slot 斜沟；斜槽
skewed slot squirrel cage rotor 斜槽鼠笼式转子
skewed storage 斜移存储；错位存储器
skewed surface 斜表面
skewed track 斜向磁迹；倾斜磁迹
skewed trolley 斜动小车
skewer 叉状物
skewer pointer 磨木锭子机
skew error detection 偏斜出错检测

skew factor 斜歪系数;偏态因素;偏态因素;槽扭因数
skew failure 歪斜故障
skew field 反称域
skew fillet 墙帽斜棱脚;斜嵌条;(砖石山墙的)压顶嵌条;墙压顶斜棱脚;披水条
skew fish pass 斜槽式鱼道
skew flashing 斜面防雨板;山墙遮雨板;山头(铅皮)泛水;山墙泛水
skew force 斜拘力
skew form 邻位交叉式
skew frequency curve 偏态频率曲线;不对称频率曲线
skew frequency distribution 偏态频率分布
skew front 斜前沿
skew gear 斜齿轮;交错轴齿轮
skew grid 斜框格
skew grillage girder bridge 斜格排梁桥;斜格排
skew helicoid 螺旋面
skew hinge 上斜铰链;上升铰链;升降合页
skew indicator 偏斜指示器
skewing mechanism 交叉装置
skewing scheme 斜移方案
skewing trolley 斜动小车
skew intersection 斜(向)交叉;斜交(交叉口)
skew involution (in space) 双轴对合
skew joint 斜面接合;斜交接合
skew leakage flux 斜槽漏磁通
skew limiter 偏斜限制器
skew line 斜线;斜叉直线
skew master tape 时差校准带
skew matrix 斜矩阵
skew Mercator projection 斜墨卡托投影
skew mill 斜轧式轧机
skew nail 斜钉
skew nailed material 斜钉材料
skew nailing 斜向钉钉子;斜钉(法)
skewness 斜度;粒度分布曲线的偏度;畸变;偏斜(度);偏态;偏斜(度);歪度;失真;分布不均;反称性
skewness coefficient 偏斜系数;偏态系数
skewness coefficient of direction (al) data 方向数据的偏态系数
skew notch 斜槽口
skew notch on wall 墙面斜凹槽
skew pantograph 斜伸缩仪;斜角缩绘仪
skew penetration of masonry walls 圬工墙的斜砌石贯入度
skew plane 斜刃刨
skew plate 斜板
skew plate settler 斜板沉淀池
skew portal 斜洞门
skew probability curve 偏斜概率曲线
skew product 斜积
skew putt 山墙端挑石;斜座石
skew quadrilateral 挠四边形
skew ray 斜射光;斜光(线);偏斜光线;不沿光轴光线;不交轴光线
skew rebate plane 斜槽刨
skew regression 偏斜回归
skew roller table 斜辊道工作台
skew rolling 斜轧
skew rolling mill 斜轧厂;斜轧机
skew ruled surface 不可展直纹曲面
skew running mark 偏移滚迹
skew scarf joint 斜嵌接(头)
skew shaft 偏轴;歪轴
skew shearing force 斜剪力
skew slab 弯曲的板;斜板
skew slab bridge 斜板桥;弯曲板桥
skew span 斜跨结构;斜交桥跨度;斜跨度
skew square matrix 扭方阵
skew surface 扭曲面;斜面;斜曲面
skew-symmetric (al)反(号)对称的;斜对称的
skew-symmetric (al) matrix 反对称矩阵;斜对称矩阵
skew-symmetric (al) operator 反对称算子
skew-symmetric (al) part 反对称部分
skew-symmetric (al) tensor 斜对称张量
skew-symmetric (al) transformation 斜对称变换
skew symmetry 反对称性
skew T 斜三通
skew table 斜榫接;斜座石;斜交石;斜承石板;山墙盖板(座石)

skew table roll 斜辊;垛板辊道的辊子
skew tee T 形管节;斜(叉)三通
skew three-span slab 斜三跨板
skew tooth 螺旋齿
skew two-span slab 斜两跨板
skew wall 斜歪曲面墙(播音室用,避免连续声反射);斜壁;不平行墙
skew weir 斜堰
skew weld 斜焊(接)缝;斜焊接(缝);斜焊
skew wheel 斜齿轮;偏轴斜齿伞齿轮
skew-wire line 绞合线线路
Skhl steel 镍铬铜低合金钢
ski 在滑板上;滑雪板
skiagram 射线照片;投影图;房屋纵断面图
skiagraph 纵断面图;正影描记器;投影图;射线照片;房屋纵断面图;房屋剖面图
skialith 残影体
skiameter X 线量计
skiascope X 射线透视镜
skiatron 暗powered(示波)管
skid 移送板;炉底滑道;滑材;护舷木;垫货板;打滑;闸瓦
skid bank 冷床
skid bars 横道
skid base 活动基座
skid beam 转盘横梁
skid bed 装料架
skid board 滑板
skid chain 防滑链
skid container 滑动容器
skid-controlled axle 打滑控制轴
Skiddavian 斯基道阶【地】;阿伦尼克阶
skidder 集材拖拉机;集材绞车;滑动垫木
skidder-loader 集材装车联合机
skidder-piler 集材归堆联合机
skid derrick 滑动式起重机
skid detection 打滑检测
skid-differential clutch 滑差离合器
skidding 集材(装置);铺道木;侧行
skidding accident 滑溜事故
skidding distance 滑行距离
skidding friction 滑动摩擦力;滑动摩擦
skidding investigation 抗滑阻力测量
skidding of wheel 轮滑
skidding pan 集材滑盘
skidding resistance 抗滑阻力
skidding stability 抗滑稳定性
skidding test 滑行试验
skidding tire 光面易滑轮胎
skidding trail 滑木道;拖拉道
skidding tyre 光面易滑轮胎
skid-fin antenna 翅形天线
skid fins 鳍龙骨
skid force 溜滑力;滑移力;侧滑力
skid frame 滑行架框架;滑橇机架;滑橇机架
skid-free 防滑的
skid-free road 防滑路
skid-free road surface 防滑路面
skid girder 制动梁
skid hazard 滑溜事故
skid-inducing 有滑溜危险的;有滑溜危险的
skid loader shoe 铲装机支座
skid marks 滑行擦痕(路面上);滑行印迹(路面上)
skid-mounted 滑橇安装的;滑道提升的
skid-mounted box-like spreader 装于滑动底板的箱形喷洒器;装于滑动底板的箱形撒料机
skid-mounted drill 滑橇装钻机;滑轨钻机
skid-mounted pump unit 滑动泵装置
skid pad 试车场
skid pile driver 滑动打桩机
skid pipe 炉底滑管;滑道管
skid platform 载重手推车;滑动平台;矮墙托架
skid potentiometer 防侧滑电位计
skid prevention 防滑(措施)
skid-proof 防滑(的)
skid-proof carpet 防滑垫;防滑毡
skid-proofing 抗滑的;防滑的
skid-proof paint 防滑漆;防滑涂料
skid-proof surface 防滑表面;防滑路面
skid rail 滑轨(移动重物用)
skid resistance 抗滑;防滑(性);溜滑阻力
skid-resistance tester 抗滑试验装置;防滑试验装置;抗滑试验机;防滑试验机
skid-resistant 防滑的

skid resistant carpet 防滑地毯
skid-resistant characteristics of bridge floor 桥面抗滑性能
skid-resistant coating 防滑涂料
skid-resistant course 防滑层
skid-resistant surface 抗滑表面;抗滑路面;防滑路面
skid-resisting 防滑的
skid-resisting capability 抗滑动能力
skid-resisting property 抗滑性能;防滑性能;抗滑性
skid ring 溜滑环
skid road 集材道路;(木材、石料等的)滑运道;滑木道;滑道(采石、伐木用);拖拉道;木材滑道
skid-row 无家游民聚集地区;城镇中破落地区
skids 滑动垫木;楞腿;木块;垫板
skid-senser 打滑感应器
skid shock absorber 橇形减振器;制动减振器
skid shoe 制动铁鞋;制动块;仿形滑脚;闸瓦
skid shovel 滑移式铲土机;正铲挖土机
skid steering loader 速差转向装载机
skid tank 移动式储罐;移动式液化气罐;滑橇装油罐
skid the derrick 拖运钻塔
skid the rig 整体移动钻机;整体搬运钻机
skid trail 拖拉道
skid trial 滑木道
skid-type base frame 滑移式支座框架
skid-type frame 滑橇型底座;滑橇式机架
skidway 横木滑道;拖拉道;垫木楞场
skiff 尖头方尾平底小艇
skiffling 粗凿石头尖角;敲棱角
ski-hoist 雪橇式绞车;雪橇式提升机
ski hut 雪橇棚屋
ski jump 滑雪(跳跃)斜坡;滑雪式跳跃
ski-jump energy dissipater 挑流式消能设备;挑流式消能工
ski-jump energy dissipator 滑雪跳跑道消失器;滑雪式消能工
ski-jump spillway 滑雪跳跑道溢洪道;滑雪式溢洪道;挑流式溢洪道
ski lift 架空索道;滑雪用提升机
ski-lift conveyer 架空无极绳式人车
skill 技艺;技巧;技能;熟练;本领
skill acquisition 技巧获取
skill craft 技术
skilled 熟练的
skilled and semi-skilled labo (u) r 熟练和半熟练劳动力
skilled concrete worker 熟练的混凝土工人
skilled craft 熟练技艺
skilled industry 技术工业
skilled labor-intensiveness of export products 出口商品的熟练劳动密集程度
skilled labo(u)r (er)熟练劳动;熟练工(人)
skilled man 熟练人员;熟练工
skilled manpower 熟练技工
skilled personnel 技术熟练人员
skilled programmear 熟练的程序设计员
skilled staff 熟练人员
skilled work 技术工作;熟练工作
skilled worker 技工;技术工人;熟练技工;熟练工(人)
skillet 铸锅;熔锅
skillet cast steel 坩埚铸钢
skill-intensive 技术密集型的;技术密集的
skill intensive industry 技术密集工业
skill migration 技术人才的迁移
skill of labo(u)r 劳动熟练程度
skill score 技术评分
skill transfer 技术人才转移
skill transfer program(me) 技能转让计划
ski lodge 滑雪场小屋
skim 渣痕;去垢;撇渣;撇去;撇除;扒渣;除渣
skim action 填隙作用;垫补作用;粗略调整;粗略调节
skim adjustment 用垫片调整
skim bar 撇渣杆;挡渣棒
skim block 垫环
skim bob 集渣包
skim bob gate 挡渣浇口
skim coat 撇渣面层;罩面层;薄涂层;薄覆盖层;除表层
skim coat calendar 平板砑光机
skim coat calender 擦胶压光机
skim coat plaster 末道抹灰层
skim colter 除茬刮草铲;草皮铲除器

skim core 撇渣泥芯
skim gate 挡渣浇口;除渣器
skim-grading 刮整表面;刮平
skim liner 楔形垫片
skimmer 撇渣器;撇乳器;撇沫器;撇渣机;撇沫板;泡沫分离器;推土机;捣渣器;除沫器;分液器;分离器;刨铲挖掘机;扒渣杆
skimmer attachment 铲削器附件;挖土机附件;挖掘机附件
skimmer bar 清除熔渣通条;撇渣棒;扒渣棒
skimmer blade 刮刀;刮板
skimmer block 挡砖;清渣口异形砖
skimmer brick 挡渣砖
skimmer equipment 铲削工具;臂斗式挖掘机
skimmer filter 撇沫过滤器
skimmer float 刮板;镘刀
skimmer hoist 挖掘机绞车;挖掘机绞盘;挖土机绞车;挖土机绞盘
skimmer mechanism 撇沫机制
skimmer scoop 挖掘机铲斗;挖土机铲斗
skimmer shovel 刮路机;刮土机;铲土机
skimmer tube 除沫管
skimmer wall 拦污墙
skimmer winch 挖掘机绞车;挖掘机绞车;挖土机绞盘;挖掘机绞盘
skimming 蒸去轻油;回收浮油;撇蒸;撇去(浮质);撇取;撇清;撇沫;填隙;挡渣;粗妥尔油;从水面上撇取浮油;分程取水
skimming baffle 撇液板;分离挡板
skimming centrifuge 乳油分离器
skimming chamber 撇油室
skimming coat 抹灰罩面层;罩面层;末涂层;表层
skimming core 撇渣芯
skimming detritus tank 撇渣沉沙池;撇渣沉砂池
skimming device 撇油器
skimming dish 摩托快艇;撇渣器
skimming door 除渣门
skimming kettle 浮渣槽
skimming ladle 撇渣勺
skimming machine 撇渣机;撇沫器;撇渣机
skimming machine of flo(a)tation pond 气浮池刮渣机
skimming machine of flo(a)tation tank 气浮池刮渣机
skimming of wastewater 废水撇渣;废水撇油
skimming panel 撇渣板
skimming(pipe)line 撇油管线
skimming pit 撇油池
skimming plant 撇蒸厂
skimming plate 撇渣板
skimming pond 撇油池
skimming price 撇奶油式价格
skimming pricing 高额定价法
skimming processing 撇蒸过程
skimming rod 浮渣刮棒;玻璃刮料板
skimming scoop 浮渣槽
skimming tank 撇渣池;撇油池;撇油槽
skimming-the-cream price 最高价订价法
skimming tool 刮料工具;撇渣器
skimming wear 刮削损耗
skimming weir 拦沙撇水堰;撇水堰;撇渣堰
ski mobile 履带式雪上汽车
skimp 克扣
skimping 跳汰选矿法
skim pocket 掏渣池
skim riser 除渣冒口
skim serum 胶清
skim share 铲茬犁铧
skim spacer 衬垫隔片
skim strainer 滤渣网;滤网芯
skim-the-cream pricing 撇奶油式订价法
skin 新冰皮;油墨膜;绝缘墙纸;蒙皮;皮(肤);外皮;表皮;表面护板;表层
skin adhesive 表面黏附剂
skin adhesive cup 胶水杯
skin annealing 表面退火
skin antenna 外壳天线
skin area 表皮面积;表面面积
skin bending stress 表面弯曲应力
skin breakage 表皮破裂;表皮裂纹;(钢锭的)表面破裂
skin clinic 皮肤诊所
skin coat 罩面层;面层;表层
skin conductor 表层导体
skin construction 外皮结构
skin contact temperature 皮肤接触温度
skin covering 表层
skin covering of the surface 表面膜
skin cracking 表皮裂缝
skin current 表面流;表流
skin decarburization 表面脱碳
skin depth 穿透深度;集肤深度;趋肤深度;透入深度
skin dimension 表面尺度
skin disease 皮肤病
skin diver 轻装潜水员
skin diving 轻装潜水
skin dried mo(u)ld 面干铸型
skin dry 表干
skin drying 油漆结皮(油漆疵病);表面烘干;表面干燥
skin drying mo(u)ld 表干型
skin effect 集肤效应;趋肤效应;表皮效应;表面效应
skin-effect attenuation 趋肤效应衰减
skin effect correction 趋肤效应校正
skin effect rotor 深槽式转子
skin effect winding 深槽式绕组
skin electric(al) current 集肤电流
skin eliminator 去皮机
skin emissivity 表面发射率
skin factor 表皮系数
skin flo(a)tation 表层浮选
skin forming 表面形成;表皮成型
skin friction 外表摩擦;表面阻力;表面摩阻;表面摩擦
skin friction along the pile 桩侧摩擦力
skin friction(al) resistance 表面摩擦阻力
skin friction coefficient 表摩系数;表面摩擦系数
skin friction distribution 表面摩擦分布
skin friction drag 表面摩擦阻力
skin friction effect 表面摩擦效应
skin friction force 表面摩擦力
skin friction loss 表面摩擦损失
skin friction resistance 表面摩擦阻力
skin friction temperature 表面摩擦温度
skin friction test 面摩擦试验
skin glue 皮胶
skin grade 间断级配
skin graft spatula 压皮片
skin grouting 表皮喷浆;表面喷浆
skin hard 表皮硬化的
skin hardened 表皮渗碳硬化的
skin hardening 表皮渗碳硬化;表面淬火
skin hardness 表皮硬度
skin heating 表面加热
skin hide 皮革
skin hole 表皮气泡
skin lamination 大片分层;表皮分层;表面分层
skin layer 表层
skinless 去皮的
skin loss 趋肤效应损耗
skin-making machine 造皮机
skin material 罩面材料;蒙皮材料;外皮材料(如叠合板多孔材料等的外罩薄板);表皮材料
skin mill 表皮光轧机
skin miller 表皮铣床
skinned bolt 磨损的螺栓;去齿的螺栓
skin needling 毛刺
skinner 刮皮工具;去皮工具;剥皮段;毛石墙边扁石
Skinner box 斯金纳箱
skinnerite 硫锑铜矿
skinner saw 修边锯
skinning 结皮(现象);去皮;起膜;起皮(油漆筒内表面)
skinning knife 剥皮刀
skinning machine 去皮机
skinning surface 蒙皮曲面
skinning-type sealant 结膜型嵌缝膏
skinny 干瘦
skin of casting 铸件表皮
skin of grout 灰浆外皮;胶泥外皮
skin of mortar 砂浆表皮
skin of paint 油漆层
skin of water 水张力膜
skin packaging 贴体包装
skin packing 贴体包装
skin pass 表皮冷轧;表皮光轧;平整道次;调质轧制
skin pass mill 表面冷轧机;表皮光轧机
skin pass roll 表皮光轧机
skin pass rolling 表皮光轧;平整冷轧
skin patch 对沥青路面层修补;表面修补
skin patching 表面修补
skin plate 复板;挡板;车身里里;罩面板;闸门面板;闸门板面;面板;盾构机壳体;包面板
skin plating 闸门面板;面板;船壳板
skin raft 皮筏
skin recovery 再渗碳
skin-remover 去皮机
skin resistance 表皮阻力;表面阻力;表面摩擦阻力
skin rivet 蒙皮铆钉
skin rolling mill 表面光轧机
skin segregate 表皮偏析
skin soaking test 浸皮试验
skin stress 表皮应力;表面应力;表层应力
skin-stressed structure 承力蒙皮结构
skin suction box 吸皮匣
skin temperature 外壳温度;表皮温度
skin temperature meter 表面温度计
skin temperature recorder 表面温度记录仪
skin thickness 表层厚度
skintle 线脚
skintled 虎皮砖;表面凹凸不平;墙面斜交砖;表面凹凸砌体
skintled brickwork 虎皮砖砌体;线脚;乱插花砌法;错乱突出砖砌工程;凸凹砌筑砖工;夹花砖工;不规则砌筑(法)
skin tone 肤色
skin tracking 雷达跟踪
skin treatment 表面处治;表面处理
skin weld 表面焊
skin welding 表面焊接
skin wool 剥皮羊毛
skiny red 积红
skiny red glaze 积红釉
skiodrome 波面图
skiophyte 嫌阴植物;避阳植物
skiou 准平原面交叨
skip 铸桶;高炉翻斗车;料车;加料机翻斗;箕斗;起重箱;跳转;跳跃单位;吊斗;出砟车
skip area 空区;跳越区
skip back 跳跃式回绕
skip band 越程波段;空白带
skip bar 跳跃杆
skip block welding 跳跃块焊接;分段多层跳焊
skip block welding sequence 分段多层跳焊法
skip bucket 料车车体;箕斗;倾卸斗;翻斗;翻倒铸罐;翻倒铲斗
skip bus input 跳跃总线输入
skip cable 料车钢索;翻斗钢索
skip car 上料车;翻斗车;料车;矿车;倾卸小车;斗翻车;斗车;倒卸车
skip car loader 翻斗车装料斗
skip cast 跳跃铸型;跳模【地】
skip chain 跳转链
skip charger 倾倒提升加料机
skip charging 翻斗装料
skip charging system 料车上料系统
skip cleaning station 翻桶清洁站;翻斗车清洁站;翻斗清洁站
skip code 跳越码;跳跃码;跳步码
skip continuous thumper method 跳跃式连续锤击法(地震勘探)
skip corridor floor 跨廊式楼层
skip crane 吊斗起重机
skip distance 跃距;跳越距离;跳跃距离;死区宽度
skipdozer 小型压路机;小型推土机
skip draft 跳穿
skip draw 跳穿
skip dress 间隔修整
skip effect 跳越效应;超越效应
skip elevator 吊斗提升机;翻斗(式)提升机
skip-fading 跳跃衰落
skip feed 跳跃进给
skip field 空(白)指令部分;空白场;跳越信息组
skip flag 跳跃标记;跳跃标记;跳过标记
skipfloat 修平用长镘刀
skip floor 跃层;跳层(楼面)
skip-floor apartment(house) 跃层式公寓
skip format item 跳跃格式项
skip frame 跳幅
skip-graded 间断级配的
skip-graded material 间断级配材料
skip-graded mix 间断级配混合料

skip grading curve 不连续级配曲线
skip hoist 大吊桶;料斗卸车;料车升降机;料车绞车;吊斗提升机;倒卸式提升机;倒卸式起重机;翻斗(式)提升机
skip hoisting 箕斗提升
skip hoist type charging machine 翻斗爬式加料机
skip image 跳像
skip instruction 空操作指令;跳转指令
skip joist system floor 宽支承端(钢筋混凝土)密肋楼板
skip-keying 脉冲重复频率的分频;跳步键控
skip link 跳跃链路
skip load 料车荷重
skip loader 翻转装料斗;箕斗式装载机;翻斗式装载机;翻斗式装料机
skip-loading chute 翻斗装料(斜)槽
skip logging 跳步记录
skip lorry 翻斗货车
skip mark 跳痕;跳痕
skip mixer 翻斗搅拌机;翻斗混合器;翻斗拌和机
skip option 跳跃任选
skip over 跳过
skip-packaging paper 单面光包装纸
skip page 跳页【计】
skip path 跳跃轨道
skipped 跳跃的
skipped thread 跳纱
skipper 小船船长;跳跃者
skipper arm 挖土机斗柄
skipper road 木杆道
skip philosophy 跳跃方法
skipping 漏涂;跳跃;边缘脱釉
skipping a period 跳段(爆破时)
skipping stitch 跳针
skipping transport 跳跃运输
skip program(me) 跳跃式程序
skippy 急速改变的;不均匀涂层;漏涂
skip relay 跳越继电器
skip release mechanism 翻斗卸料机构;翻斗脱钩机构
skip road 分道;支路
skip scale 自动斗定量秤;库秤
skip-searched chain 跳越检索链
skip sequence welding 跳焊
skip sequential 跳跃顺序存取
skip sequential access 跳跃顺序存取
skip shaker gear 翻斗振动齿轮;料斗振动齿轮
skips-in-planting 漏刨
skip speed 跳越速度;跳跃速度
skip stop insert bar 跳跃杆
skip symbol 空符号;跳跃符
skip test 空白检测指令;空白测试
skip-tooth saw 间断齿锯
skip track 料车轨道;料斗轨道;箕斗罐道;滑轨;翻斗轨道
skip trajectory 跳跃轨迹
skip type charger 翻斗爬式加料机
skip type feeder 斗式给料机;吊斗式给料机
skip vehicle 跳越式航天器
skip vibrator 翻斗振动器;料车振动器
skip weigher 翻斗秤
skip welding 跳焊
skip winder 箕斗提升机
skip winding 箕斗提升
skip zone 静区域;跳越区;跳跃区
skirt 连接裙;裙座;导料槽;缘边;侧板
skirt beam 圈梁;边梁
skirt board 踢脚板;护板;挡板;侧壁;侧板;(皮带运输机的)边缘挡板;壁脚板
skirt clearance 裙部间隙
skirt dipole 套筒偶极子
skirt dipole antenna 套筒偶极天线
skirted fender (汽车车轮的)挡泥板
skirted piston 侧缘活塞
skirted plenum craft 围裙气室气垫艇
skirt end of piston 活塞侧缘
skirt expander 活塞底缘张圈
skirting 踢脚
skirting block 贴脸墩;踢脚墩(堵头);线脚固定块;踢脚板后面的砌块;墙角基脚砌块;门磴座;勒脚石
skirting board 壁脚板;踢脚板
skirting board heater 沿踢脚板铺设的暖气管
skirting board heating device 踢脚板供暖装置

skirting brick 踢脚(板)砖
skirting component 踢脚板构件
skirting duct 沿踢脚板铺设的管道;踢脚边沿管道
skirting heater 沿踢脚板铺设的暖气管
skirting heating 沿踢脚板供热
skirting level 踢脚板标高;踢脚板高程
skirting member 踢脚板构件
skirting plant 饰边植物
skirting plate 屏蔽装甲板
skirting radiator 沿踢脚板设置的散热器;墙脚散热器
skirting tile 踢脚(板)砖
skirting unit 踢脚板构件
skirt of bit 钻头扩大部分
skirt of piston 活塞侧缘
skirt plate 碟形边
skirt relief 裙部凹槽
skirt resolution 边缘分辨力
skirt response 边缘响应
skirt retaining wall 斜面下部挡土墙;周边挡土墙
skirt ring groove 活塞侧缘环槽
skirt-roof 裙状屋顶(墙上挑出的假屋顶);重檐(指两层之间加假屋檐);缠腰屋顶
skirt section 裙部
skirt selectivity 边缘选择性
skirt slot 活塞侧缘开缝
skirt support 裙式支座
skirt support of spheric(al) vessel 球形容器裙式支座
skirt-type joint bar 裙形夹板
ski run 滑雪(坡)道;滑雪坡;滑道
ski slope 滑雪坡
skittle alley 滚木球场
skittle pot 柱形坩埚
skive 磨
skiving 磨削;车坯;车齿
skiving cutter 旋刮刀;车齿刀
skiving tool 切线样板精车刀;定形切向刀
ski with steel ski pole 带钢撑棍的滑雪板
skj(a)ergaard 低矮小岛
Skleron 斯克列隆铝基合金
sklodowskite 硅镁铀矿
skookum house 牢房(俚语)
Skoparon lens 斯柯帕龙镜头
skopometer 浊度计
skotogram 暗室显影片
skotograph 暗室显影片
skotoplankton 深水浮游生物;暗层浮游生物
skototaxis 向暗性
skove kiln 方形无顶烧砖窑
skrag mill 双平行圆锯制材厂
Skrinkle sandstones 斯科林克尔砂岩
skrink on 热套装
skull 结瘤;结壳;凝壳
skull cracker 渣壳破碎机;落锤(式)破碎机;破碎球;拆房重锤
skull cracker demolition 重锤拆除法
skull crucible 凝壳炉
skull furnace 凝壳炉
skull guard 防护帽;头盔;安全帽
skull melting 凝壳熔炼法
skull melting furnace 渣壳熔炼炉
skull melting method 冷坩埚法;壳熔法
skull patch 翻皮(钢锭缺陷)
skull trephine 皇冠钻
skunk oil 臭鼬油
skunks 臭鼬臭石油
skutteradite 方钴矿
sky area 天空部分
sky background 天空背景
sky background noise 天空背景噪声
sky-blue 晴空天蓝色;天蓝(色)
sky blue glaze 天蓝釉
skyborne 机载
sky brightness 天空亮度
sky brightness temperature 天空亮度温度
sky chart 星空图
sky clear 晴天
sky compass 天罗盘;天空罗盘
sky condition 天空状况;天空情况
sky count 高空计数(交通量)
sky cover 天空覆盖
sky crane 大型直升机用起重机;架空索吊运机;空中起重机

sky deck 高楼顶瞭望台
sky diagram 星图;天象图
sky-dome 穹顶;围绕并盖于舞台上的天顶
sky error 电离层误差
skyey 天空的
sky factor 日照系数;天空因数;天空投射系数
sky filter 天空滤光片
sky firing 窑顶加柴煅烧
sky fog 天雾
sky-glare 天空强光
sky glow 天空照亮
sky gray 蓝灰色;天灰(色)
sky green 天绿色
sky grey 天灰(色)
sky hook 架空吊运车;安装锚杆
skyhook balloon 高层等高(空)探测气球
sky-hooker 塔上工人(钻探时)
sky-horn switching 天空喇叭开关
Sky House 天居(日本建筑师菊竹清训1958年设计的住宅)
skylab 空中实验室;天空实验室;太空试验室;实验室
skylab earth terrain camera 天空实验室对地摄影机
sky laboratory 天空实验室;天空实验室;太空实验室
skylab photography 天空实验室摄影
sky light 自然光;天空光;天光;遮光棚
skylight cap 天窗顶盖
skylight coaming 天窗围板
skylight cover 天窗盖
skylight dome 开天窗的圆屋顶
skylight filter 光谱滤光镜;天光滤光片;天光滤光镜;天窗滤光片
skylight frame 天窗架
skylight gear 天窗开关装置
skylight glass 低质量的平板玻璃;天窗玻璃
skylight grating 天窗玻璃护栅
skylight illumination 天窗照度;天窗照度
skylighting 天窗采光
skylight in the train diagram 运行图天窗
skylight lighting 天窗采光
skylight net 天窗网格
skylight purlin(e) 天窗檩条
skylight quadrant 天窗开关弧条
skylight sash 玻璃窗框
skylight window 屋顶窗;天窗
skyline 天边;空中轮廓;天际线;水天线;地平(轮廓)线;承载索
skyline cable logging 架空式索道集材
skyline carriage 架空载运车
skyline crane 架空索道吊运机;起运机
skyline of a town 城市轮廓线;城市建筑轮廓线;城市天际线
skyline yarding 架空索道集材
sky lobby (高层建筑的)电梯起止厅
skylounge 空中客车
sky map 云图亮度图
sky master 巨型客机
sky noise 天电噪声
sky noise temperature 天空噪声温度
sky parking 多层停车场;立体式停车场
sky parlor 屋顶室
sky-port 直升(飞)机场
sky portion 天空部分
sky radiation 慢天空辐射;天空(散射)辐射
sky ray 电离层反射波
sky reference 天空基准
skyrocket 烟火;飞涨
skyrocketing prices 物价飞涨;飞涨的物价
skysail-yarder 天帆横桁式船
sky scanning photometer 天空扫描光度计
skyscape 天空风景;天空景
skyscraper 非常高的烟囱;高层建筑(物);摩天大厦;摩天(大)楼
skyscraper block 摩天大楼公寓
skyscraper city 摩天大楼林立的城市
sky shine shield 间接射线屏蔽;天空回散照射屏蔽
sky sign 高空广告牌;楼顶广告牌
sky spectrum 天光光谱
sky truck 大型运输机;(大型)运输机
sky-vault 高拱顶
skywalk 人行天桥;人行旱桥
sky wave 天空电波;天波;电离层反射波
skywave accuracy pattern 天波误差图

skywave backscatter 天波后向散射
skywave correction 天波校正；天波改正量；天波修正(量)
skywave-delay(curves) 天波时延
skywave effect 天波效应
skywave error 天波误差
skywave interference 天波干扰
skywave operating distance 天波作用距离
skywave pattern 天波辐射图
skywave range 天波使用范围
skywave station error 天波电台误差
skywave synchronized long range navigation 天波同步远程导航
skywave synchronized loran 天波同步劳兰
skywave transmission delay 天空电波传输延迟；天波传输延迟
skywave trouble 夜间效应；天波干扰
sky-way 高架公路；空中航线
skyway bridge 高架快速交通桥；快速路桥
sky-writing 天空文字
slab 锯去背皮；极板；厚片；厚木板；厚块；厚板钢块；切片；平板料；棚板；铁块；铁板；松软的帆；大理石配电板；初轧板坯；冰板；扁锭；板皮料；板皮；板(坯)
slab action 板体作用
slab analogy 拟板法
slab analysis 板体分析
slab-and-beam 平板梁
slab-and-beam bridge T形桥梁；梁板桥
slab-and-beam floor 梁板楼面；轨道梁板基础
slab-and-beam rib 密肋梁
slab-and-beam with non-structural fillers 有非结构填料的平板梁
slab-and-beam without fillers 无填料的平板梁
slab and buttress dam 肋板式支墩坝；平板(式)支墩坝；带肋板式坝；板撑墙坝
slab and buttress type dam 板撑式墙坝
slab and column buttressed dam 板柱式支墩坝
slab and column dam 板柱式支墩坝；平板立柱式坝
slab and edging pass 平轧—立轧孔型
slab and girder 平板梁楼面
slab and girder construction 板梁结构
slab and girder floor 梁板式楼面；梁板式楼盖；板梁式楼板；板梁式地板；板梁楼面
slab and girder structure 板梁结构
slab-and-joist ribbed construction 肋形梁板结构
slab and kerb press 板和侧石压制机
slab-and-stringer bridge 梁板式桥；覆板梁式桥
slab and tile grinder 板和瓷砖研磨机
slab and tile press 板和瓷砖压制机
slab anode 板式阳极
slab arch bridge 板拱桥
slab band 薄板梁
slab band floor 薄板梁地面
slab bar bolster 平板钢筋支架
slab base 板式基座；板式基础
slab beam 薄板梁
slab bearing structure 承载板结构
slabbed construction 平板构造
slabbed log 锯去板皮的原木
slab bending 平板弯曲
slabber 擦边锯机；切块机
slabber-edger 轧边扁坯机
slabbiness 成层状性；成板状性；成片状性
slabbing 阔面铣；泥片粘接成形；切片；起鳞；平面铣削；剥裂；板坯破裂作用；制板；铺板
slabbing action 劈裂作用；板坯的垂直下或侧压下
slabbing cut 分片开挖；分片爆破
slabbing machine 切块机；平面铣床；刷帮截煤机
slabbing mill 扁钢轧厂；阔面铣刀；二辊式万能板坯初轧机；扁钢坯轧机；板坯初轧机
slabbing milling cutter 阔面铣刀；长圆柱铣刀
slabbing of ring 垮圈
slabbing pass 扁坯轧辊孔型
slabbing roll 扁坯轧辊；板坯轧辊
slab blasting 分层爆破
slab block 板形砌块；板式住宅
slab board 边料板；背板
slab bolster 钢筋支垫物；板支架；板横撑；平板钢筋支架
slab bridge 板桥
slab broach 阔面拉刀
slab building 板块建筑
slab butts 板坯的切头

slabby 片状；稠黏的；板状的
slabby coal 层状煤
slabby limestone 板桩石灰岩
slabby shape 片状构造；板状构造
slab calculation 平板计算
slab cantilevering in all directions 各向悬臂板
slab cap(ping) 压顶板；盖板
slab caster 板坯连铸机
slab casting machine 板坯连铸机
slab centering roller 初轧板坯定心辊
slab charge 平面装药
slab charger 板坯装料机
slab chipper 板皮削片机
slab circumference 板的周边
slab cogging mill 扁钢坯轧厂；扁钢坯轧机
slab coil 平线圈；盘形线圈
slab collar 板的钢吊环
slab-column structure 大板—柱结构；板柱结构
slab-column system 板柱体系
slab compressive strength 平板抗压强度
slab concreting 浇制混凝土板
slab construction 平板构造；平板结构
slab cooler 板坯冷却机
slab copper 扁铜锭
slab core 板(状)芯；片型芯
slab core machine 板芯塑造机
slab covering 方块水泥板块
slab cross-section 平板断面
slab crushing strength 平板抗压强度
slab culvert 平板涵洞；盖板涵；平板涵；板涵
slab cutter 硬切刀；平面铣刀
slab dam 平板坝
slab deflection 板的挠度；板的变形；板的挠曲
slab depth 平板厚度
slab dielectric(al) optical waveguide 平片介质光波导
slab dielectric(al) wavequide 平片介质波导
slab displacement 平板位移
slab dissolver 平板式溶解器；天板式溶解器
slab door 平板门；(可装配玻璃的)平面门；厚板门；板门
slab edge 天板边缘
slab effect 平板效应
slab end 平板端部
slab-end-beam on artificial foundation 人工地基上的梁和板
slab entry 刷帮加宽的煤巷
slab extractor 板坯出料机
slab facade 平板正面
slab face 加宽工作面
slab facing 平板饰面
slab failure 平板破坏；板状破坏
slab flexible pier 板式柔性墩
slab floor 石板地板；石板楼板；无梁地板；无梁楼板；钢筋混凝土楼盖；钢筋混凝土楼盖；楼板；石板地面
slab floor cover(ing) 无梁楼板
slab floor(ing) finish 无梁楼板终饰
slab footing 平板基础；平板底脚
slab forming 板的成型
slab forming machine 制板机
slab form(work) 板的模壳；楼板模板(工程)
slab foundation 平板基础；板状基础；板式基础
slab geometry 平板几何条件
slab glass 厚块玻璃；玻璃块；板状光学玻璃
slab grab(bing) equipment 板的夹具
slab grating 木格选别机
slab grinder 磨钢板机；初轧板坯清理机床
slab guide 平面波导；片状波导
slab handling crane 板坯搬运起重机
slab heating 扁坯加热
slab hole 辅助钻孔；辅助炮眼
slab house 毛毡木板房屋
slab ice 厚冰
slab in ballast concrete 道砟混凝土板
slabing round 一组钻孔
slab insulant 热绝缘板材；用绝缘板隔热；绝缘板；板隔声隔热材料
slab insulation 板材保温
slab integrity 路面板的整体性
slab iron 铁板
slab iron sand 铁板砂
slab iron sand cushion 铁板砂垫层
slab jacking 路面板抬高

slab joint 板块接缝；面板接缝；板缝
slab jointing 板状节理
slab-joint reinforcement 板缝配筋
slab keel 龙骨补强板；防擦龙骨；保护龙骨
slab kiln 滑板窑；推板窑
slab knee 焊接肘板
slab laser 板条激光器
slab lattice 板栅
slab-layered joint 板状层状节理
slab lining 板块衬砌
slab load 平板荷载
slab loaded in its own plane 在板平面内受荷的平板
slab machine 板压制机
slab marker 板坯打印机
slab method 板式法；主应力法(形变力计算用)；切块法
slab mill 扁钢轧机；板坯机；板坯初轧机
slab miller 大型平面铣床
slab milling 阔面铣；平面铣削
slab-milling machine 扁钢坯铣床；板坯(剥皮)修整机床
slab moment 平板力矩
slab mo(u)ld 板型模
slab of concrete board sleeper 混凝土轨枕板
slab of constant depth 等厚板
slab of glass wool 玻璃棉板
slab of one bay 单跨板
slab of small format 小规格平板
slab oil 胶块油；白色矿物油
slab-on-girder bridge 梁板(式)桥
slab-on-grade 架空平板；斜铺楼板；踏步式楼板；斜坡(混凝土)板
slab-on-grade construction 架空平板结构；斜坡楼板结构；斜坡地板结构
slab-on-grade floor 架空无梁楼板
slab pahoehoe 板形绳状熔岩
slab paper 纸板
slab paper installer 纸板安装者；纸板安装机
slab partition 板式隔断
slab partition(wall) 预制板隔墙
slab pass 扁平孔型
slab patch 弧形钢板
slab pavement 平板路面；平板铺面；板铺路面
slab paving 石板铺砌；用板铺砌；片石铺砌
slab pile 平板桩；初轧板坯垛
slab piler 垛板机；初轧板坯堆垛机
slab pile shoe 平板桩靴
slab piling 垛放初轧板坯
slab pour(ing) 板块浇注
slab press 平板压制机
slab problem 平板问题
slab projection 板的出挑；板的悬伸
slab prop 板的支撑
slab pumping 混凝土板唧泥(道路)
slab pusher 初轧板坯推进机
slab reactor 平板堆
slab rectifier 平板整流器
slab reinforcement 平板钢筋；面板钢筋
slab resolver 平板分解器
slab rigidity 平板刚度
slab roof 平板屋顶
slab rotation 平板旋转
slab rubber 胶块；胶板；板状橡胶
slab scale 全部累进率
slab share 梯形犁铧
slab shear blade 扁钢剪切机刀片
slab shears 板坯的剪切机；初轧板坯剪切机；扁坯剪切机
slab shuttering 板块模壳
slab skew 平板斜交
slab slide 片滑
slab soaking treatment 板坯均匀热处理
slab soffit 平板底面；预应力薄底板(装配式整体结构用)
slab soffit formwork 楼板底模工程
slab spacer 平板钢筋定位架；钢筋垫块；平板钢筋支架；板内垫块
slab span 平板跨度
slab spanning in two directions 双向平板
slab squeezer 齐边压力机
slab staggering 错台
slab steel 板钢
slab step 板块台阶

slab stiffness 平板刚度
slabstone 片石;石板
slab straightener 初轧板坯齐边压力机
slabs training wall 平板式导水墙
slab strength 平板强度
slab stress 板的应力
slab-stringer bridge 梁板桥
slab strip 跨中板带
slab structure 板结构;平板结构;平板构造;板块构造
slab supported on three sides 三边支承的板
slab supporting medium 平板支承
slab(sur)facing 板块铺面
slab synchro 扇形同步机
slab system 平板体系
slab theory 平板理论
slab thickness 板厚
slab-top culvert 平顶板涵洞
slab track 平板(式)轨道;板式轨道
slab translation 板块移动
slab turning gantry 初轧板坯翻转吊车
slab-type apartment house 板式住宅
slab-type block 板式砌块
slab-type block building 板式砌块建筑
slab-type building 板式建筑
slab-type resolving potentiometer 板式分解电位计
slab-type stairway 板式楼梯
slab unpiler 卸板垛机
slab urinal 板块小便池;通长小便池;挡墙式小便器
slab vibrator 板式振捣器;板式振动器
slab wall 板墙
slab wall tongued and grooved 企口的板墙
slab warping 板(体)翘曲
slab washer 洗板机
slab wax 板状地蜡
slab weight-carrying structure 板块承重结构
slab width 平板宽度
slab winding 盘形绕组
slab with edge beam 有边梁的板
slab with one-way reinforcement 单向加筋板
slab with ribs turned-up 反梁板
slab with slightly curved bottom 微弯板
slab with stiffened edges 加强边缘板
slab with two-way reinforcement 双向加筋板
slabwood 边皮材
slab zinc 扁锌锭
slack 沼泽;静止不动;煤屑;浅谷;碎煤;松绳;松弛;弛缓
slack adjuster 拉紧装置;拉紧螺钉;拉紧接头;空隙调整器;松紧调整器
slack adjuster block 缓冲调整块
slack adjuster worm 减速调整器蜗杆
slack adjuster worm gear 减速调整器蜗轮
slack adjuster worm plunger 减速调整器蜗杆柱塞
slack adjuster worm plunger spring 减速调整器蜗杆柱塞弹簧
slack annealing 不完全退火
slack at high-water 高水位时的静止水位;高潮憩流
slack at low-water 低水位时的静止水位;低潮憩流
slack away 放松
slack barrel 渣屑桶;石蜡桶
slack bilge 平顺底仓
slack block (造桥用的)中心楔块;煤屑砌块
slack busbar 松弛结点
slack cables switch 钢绳松弛开关
slack coal 煤屑
slack current 平流
slacken 松劲;放松
slackening 松弛;减弱
slacken leak 松弛泄漏
slack flow 不满流
slack hardening 不完全淬火
slack hold 未装满舱
slack hours 轻负荷时;业务清淡时间;低峰时间(交通)
slack ice 屑冰;稀疏流冰群;稀疏冰
slacking 风化(作用);消解(石灰等);粉化
slacking basin 滞水洼地
slacking off 起重滑车绳索应变的松弛;下钻杆遇阻
slacking of sludge 污泥消化
slacking sky line 松弛式架索道
slacking test 风化试验
slacking time 松弛时间
slack in the couplings 连接松动
slack in the screw 螺旋隙
slack labo(u)r market 萧条的劳动市场
slack lime 熟石灰;消石灰
slackline 松弛缆道;(供挖运用的)松索缆道;松弛的绳索;松绳式索道
slackline cable drag-scraper 拖铲刮运机
slackline cableway 松弛绳索;松弛缆道;刮斗索道;拉铲索道;电铲索道
slackline cableway excavator 拖铲挖土机;拖铲挖掘机;塔式拉索挖土机
slackline cableway scraper 拖铲挖土机;塔式拉索挖土机
slackline scraper 拖铲刮土机;松绳塔式挖土机
slack loop washer 松式环状洗涤机
slack melt copal 半熔状态的珀翅(树脂)
slackness 松弛
slack off cable 松放锚缆
slack out wire 松放缆索
slack path 松弛路线;非要径
slack piston 大间隙活塞;迟滞活塞
slack puller 拉紧索(防牵引索松垂);转环;旋转轴承;旋转接头;旋转节
slack quench 调质
slack quenching 晶粒细化热处理;热处理;断续淬火;不完全淬火
slack rope 下垂的缆索;拖长的缆索
slack rope stop 绳索松弛停止器
slack-rope switch 松绳开关;钢丝绳松弛自动断电开关
slack-running fit 轻转配合
slack seams 皱缝
slack season 淡季
slack shook 松桶板料
slack side 毛边;皮带松边;皮带从动边;单板松面;单板背面;松弛边
slack side of belt 皮带松边
slack size 薄薄上胶
slack spring 释放簧
slack tank 未装满油舱
slack tide 憩流;憩潮;平流;平潮
slack time 萧条期;富裕时间
slacktip 松坍;崩坍
slack traffic 淡季交通
slack traffic period 淡季交通期;交通缓和时间
slack up a hawser 放松大索
slack variable 余裕变数;松弛变量
slack washing machine 松式绳状洗布机
slack water 静水;海水平湖;憩流;平流;平潮
slack-water area 滞留区;平潮区
slack-water canal 无流速运河
slack-water channel 缓流航道
slack-water deposit 静水沉淀
slack-water harbo(u)r 静水港口;无流港口
slack-water level 落潮憩流;憩流水位
slack-water navigation 静水航运;静水航行;平潮航行;过闸航行
slack-water on the ebb 退潮时的静水位;落潮平流;枯潮平流
slack-water on the flood 涨潮时的静水位;涨潮憩流;涨潮平流;满潮平流
slack-water time 平潮时;平潮期;憩潮期
slack wax 疏松石蜡;软蜡;散蜡
slade 坡道;犁底板;斜路
slag 渣滓;渣状火成碎屑岩;渣化;渣堆;造渣;矿渣;火山岩烬;熔渣;多化石脆性页岩
slag ability 成渣性
slag action 炉渣侵蚀;熔渣侵蚀作用
slag activity index 矿渣活性指数
slag aggregate 炉渣集料;炉渣集料;矿渣集料
slag aggregate block 矿渣集料块材;矿骨料块材
slag-alabaster partition 矿渣石膏隔墙
slag backfill 矿渣回填(料)
slag bank 渣坡
slag-basalt fiber 矿渣玄武岩纤维;矿棉纤维
slag base 矿渣基层
slag based cement 砂渣水泥;矿渣基水泥
slag basin 渣桶;渣罐;渣池
slag bath 渣池
slag-bed 矿渣基座;渣床
slag blanket 渣层
slag blister 渣眼
slag bloating 炉渣起泡
slag blob 渣瘤
slag block 渣块;矿渣砌块;矿渣块;矿渣混凝土块
slag blower 吹灰器
slag blowhole 渣孔
slag-bonding 结渣
slag bot(t) 渣口塞
slag breaker 矿渣破碎机;破渣机;碎渣机
slag brick 炉渣砖;炉碴砖;矿渣砖;煤渣砖;熔渣砖
slag bridge 挡渣器
slag bucket 矿渣戽斗;钢渣铲斗
slag buggy 渣罐车;水渣车
slag built-up roof(ing) 矿渣建成的屋顶
slag cake 渣饼
slag calculation 配渣计算
slag car 渣车
slag catcher 接渣器
slag cement 钢渣水泥;炉渣水泥;矿渣水泥;熔渣水泥
slag cement mortar 熔渣水泥砂浆;矿渣水泥灰浆
slagceram 炉渣陶瓷;矿渣陶瓷
slag chamber 沉渣室
slag charge 渣料
slag chip(ping)s 矿渣屑
slag chute 渣撒
slag clarification 矿渣澄清
slag clouds 云状非金属夹杂
slag coarse aggregate 矿渣粗集料;矿渣粗骨料
slag composition 熔渣成分
slag concrete 炉渣混凝土;炉碴混凝土;矿渣混凝土;熔渣混凝土
slag concrete bedding 矿渣混凝土垫层
slag concrete block 矿渣混凝土(砌)块;矿混凝土砌块
slag conditioning 渣成分调整
slag conveyer 运渣机
slag-conveying machine 运渣机
slag-conveying machinery 运渣装置
slag cooling pit 矿渣冷却坑
slag cork 矿渣板
slag corrosion resistivity 抗渣性
slag corrosion test 抗渣试验
slag coverage 渣保护
slag-covered electrode 熔渣保护型焊条
slag crusher 矿渣破碎机
slag crushing plant 矿渣破碎厂;矿渣破碎装置
slag crust 渣皮;渣壳
slag dam 渣坝
slag dam cement 矿渣大坝水泥
slag deposit 渣沉积
slag desledger separator 除渣器
slag detachability 脱渣能力
slag disposal 废渣处理
slag ditch 灰渣沟
slag dump 卸渣场;渣坑;煤灰堆场;堆渣场;渣场
slag dust 矿渣粉尘
slag enclosure 夹砟
slag entrapment 夹渣
slag expansive cement 矿渣膨胀水泥
slag extractor 排渣机
slag eye 渣孔
slag felt 矿棉毡;矿渣毡
slag fiber 矿渣纤维
slag fill 炉渣(充填)底;矿渣填料;填渣
slag filler 矿渣填料
slag filling 矿渣填料;矿渣填充
slag flour 矿渣粉
slag fluidity 炉渣流动性;熔渣流动性
slag formation 渣化;造渣
slag-formation period 造渣期
slag former 造渣剂
slag-forming 造渣
slag-forming constituents 造渣剂
slag-free 不含熔渣的
slag fuming process 渣烟化法
slag furnace 熔渣炉
slaggability 造渣能力
slagged surface 渣化面;渣化层
slagging 渣化;造渣;结渣;化渣
slagging agent 造渣剂
slagging combustion chamber 排渣式燃烧室
slagging constituent 造渣剂
slagging gasifier 液态排渣发生炉
slagging gas producer 放液渣的气体发生炉
slagging hearth 渣底
slagging incineration 排渣式燃烧
slagging level 渣面

slagging medium 焊剂;助熔剂
slagging of refractory 耐火材料熔渣
slagging process 造渣过程
slagging producer 液态排渣发生炉
slagging test 造渣试验
slag glass-ceramics 矿渣微晶玻璃
slag granulating unit 炉渣水碎装置
slag granulation 熔渣水碎;熔渣成粒
slag-gypsum cement 石膏矿渣水泥
slag hammer 除渣锤
slag handling process 熔渣处理法
slag handling system 熔渣处理体系
slag heap 熔渣堆;渣堆
slag heap conveyer 渣堆输送机
slag hearth 渣炉
slag hole 出渣孔;熔渣孔;出渣口;渣孔
slag inclusion 渣孔;夹渣;焊缝夹碴;含渣;熔渣包体;包渣
slag inclusions in welds 焊接中的夹渣
slag ladle 渣桶;渣包;储渣罐
slag ladle and carriage 渣罐车
slag ladle car 矿渣桶车;渣桶车
slag line 渣线;夹渣线
slag lump 渣块
slagmac 黑色矿渣碎石(一种冷铺沥青拌矿渣混合)
slag macadam 矿渣碎石(路)
slag-magnesia Portland cement 矿渣氧化镁水泥;矿渣镁氧水泥
slag-making 造渣
slag mix 矿渣混合料
slag muck 渣堆;废渣
slag nodule 渣瘤
slag notch 出渣槽
slag notch aperture 渣口
slag notch cooler 渣口冷却套
slag notch stopper 渣口堵塞器
slag occlusion 夹渣
slag-off 造渣;结渣;撇渣;排渣;扒渣;除渣
slag overflow 熔渣溢出
slag pan 渣盘
slag pancake 渣饼
slag particle 渣粒
slag patch 渣块;夹渣
slag paving sett 铺路矿渣块料
slag paving stone 矿渣铺路石;铺路用矿渣碎石
slag penetration 炉渣侵蚀
slag pin hole 渣口;渣孔;砂眼(针孔);针状渣孔
slag pit 渣坑
slag plank 矿渣板
slag plaster 掺渣粒的抹灰灰浆;渣粒灰浆;炉渣浆;掺渣粒灰浆;掺矿渣的抹灰灰浆
slag plaster partition 矿渣灰膏隔墙
slag plate 渣板
slag pocket 渣坑;清渣口;沉渣室
slag pollution 渣污染
slag Portland cement 矿渣硅酸盐水泥;矿渣波特兰水泥;矿渣普通水泥
slag pot 渣桶;渣罐
slag powder 粉末状矿渣;熔渣粉
slag pozzolanic cement 矿渣火山灰水泥
slag print 渣印法;渣灰印迹法
slag processing plant 矿渣处理厂;矿渣处理装置
slag pumice 矿渣浮石
slag ratio 渣比
slag receiver 渣罐;集渣前炉;熔渣罐
slag removal 夹渣的清除;除渣
slag remove 清渣
slag resistance 抗渣性;抗熔渣侵蚀能力
slag road 矿渣路;熔渣路
slag rod 挡渣棒
slag runner 渣沟;渣槽;出渣槽
slag sampling 炉渣采样
slag sand 粒状渣;矿渣砂;熔渣砂;碎矿渣
slag sand block 矿渣砂砌块
slag sand concrete 矿渣砂混凝土
slag sand cored block 矿渣砂芯材砌块
slag sand tile 矿渣砂砖瓦
slag screen 捕渣管筛
slag scummer 撇熔砂勺
slag separation 熔渣分离
slag separator 熔渣分离器;撇渣器
slag sett 矿渣块
slag shielding 渣保护

slag sitall 矿渣微晶玻璃
slag skimmer 撇渣器;撇渣棒
slag skull 渣壳
slag slurry bath 渣水浴
slag slurry concrete 湿碾矿渣混凝土
slag specimen 渣样
slag spot 夹渣
slag spout 流渣槽;出渣槽
slag stockpile 渣堆
slag stone 矿渣(硬)块;矿渣石;矿渣路床
slag streak 渣痕
slag stringer 长条夹渣
slag strip 固渣木条(屋顶);平屋顶阻挡条
slag strontium cement 矿渣锶水泥
slag sulfate cement 硫酸盐矿渣水泥;矿渣硫酸盐水泥
slag-sulphate cement 矿渣硫酸盐水泥;石膏矿渣水泥
slag surface 矿渣路面;熔渣路面
slag system 渣系
slag tap 出渣口
slag tap firing 熔灰燃烧
slag-tap furnace 液态排渣炉(膛);液态出渣炉
slag-tap gas producer 出渣气体发生炉
slag tap(ping)液态排渣;出渣(指炉矿渣)
slag test 造渣试验
slag tip 渣场;渣堆
slag track 出渣轨道
slag trap 集渣包;挡渣板
slag trough 渣槽;流渣槽
slag truck ladle 运渣桶
slag tuyere 渣风口
slag wagon 渣车
slag washing wastewater 冲渣废水
slag water ratio 渣水比
slag water slurry (水力输送的)渣水混合物
slag welding 渣电焊
slag wool 炉渣绒;矿棉;渣绒;渣棉;矿渣棉
slag wool block 矿渣棉砌块
slag wool covering cord 矿渣棉包面绳索
slag wool felt 矿渣棉毛毡
slag wool insulation 矿渣棉保温;矿渣棉绝热
slag wool joint filler 矿棉嵌缝材料
slake 消除;烂泥;化灰
slaked lime 消石灰;水化石灰;熟石灰
slaked lime feeding tank 消石灰加料仓
slake-durability test 抗崩解持久性试验
slaker 消石灰器;松弛式集料机
slake tank 消石灰槽;消和槽
slake the quick-lime 化石灰
slaking 消散;消化;熟化;湿化(崩解);潮解
slaking apparatus 消和器;消和池;崩解仪
slaking behavio(u)r 消化性能;熟化性能
slaking box 消化箱
slaking characteristic 崩解特性
slaking clay 水解黏土;水化黏土;水解黏土
slaking drum 灰化滚筒
slaking heat 消化热;消解热
slaking machine 化灰机
slaking of quick lime 生石灰熟化
slaking of sludge 污泥消化
slaking of soil 崩解性
slaking pan 熟化池;消化池
slaking pit 灰化池
slaking process 水淬方法;水化方法
slaking rate of quick lime 石灰消化速度
slaking resistance 耐水化性
slaking screw 消化石灰螺旋输送器;熟化石灰螺旋输送器
slaking slag 水淬矿渣
slaking tank 消和槽
slaking test 水化试验;湿化试验;(土的)崩解试验
slaking trough 化灰槽
slaking value 水化值
slaking value of clay 黏土湿润时间指数
slaking vessel 化灰容器
slaking water content 湿化含水量
slam 使门关上;碰关
slam hang 碰关声
slamming 猛击;上拍运动;砰击【船】
slamming force coefficient 波浪砰击力系数
slamming stile 装锁的门梃;碰锁门梃
slamming strip 碰锁门梃嵌条;碰锁门梃板条
slam shut valve 快动闭阀

slander of title 对所有权的诋毁
slant 斜向;斜线号;一阵风;倾斜煤巷;倾斜(度)
slant agar 琼脂斜面培养
slant angle 斜度;倾斜角(垂直孔指顶角,水平孔指倾角)
slant angle bearing sensor 倾斜方位角传感器
slant angle sensor 倾斜角传感器
slant angle slaved system 倾斜角伺服系统
slant angle type welding 低角度式焊接
slant bar 弯起钢筋(美国)
slant-bottom scoop 斜底的
slant chute 斜溜井;斜溜槽
slant coherent enhancement 倾角相干加强
slant culture 斜面培养(法)
slant depth 斜线深度
slant distance 斜距;直线距离
slant drilling 斜钻;斜向凿岩
slanted bar 斜条
slanted-bar blade 斜条状叶片
slanted bar reticle 斜条形调制盘
slant edge 斜棱
slanted opaque bar reticle 不透明斜条形调制盘
slanted screen 立筛;斜筛
slanted strut 斜撑
slant-ended fiber 端面倾斜光纤
slant height 斜高
slant hole 斜孔
slant hole boring 斜孔钻进
slant hole drilling 斜孔钻进
slant hole technique 斜孔钻进技术
slanting 斜线
slanting asymptote 斜渐近线
slanting boom type trenching machine 斜吊杆式挖沟机
slanting-bottom bin 斜底料仓;斜底仓
slanting butt seam welding 斜对接缝焊
slanting carburetor 斜吸式化油器
slanting construction 倾斜建筑物;倾斜结构
slanting cut 斜切;斜割法
slanting edge 斜棱角
slanting face 斜工作面
slanting ladder ditcher 斜悬臂多斗挖掘机
slanting leg manometer 斜管微压计;斜管式压力计;斜管式测压计;倾斜式差示压力计
slanting readjustment 倾斜性调整
slanting roof 倾斜屋顶;坡屋顶
slanting set valve 斜置阀;斜座阀
slanting well 斜孔;斜井
slanting windshield 倾斜风挡
slant-legged rigid frame bridge 斜腿刚架桥;斜腿刚构桥
slant legged rigid frame structure 斜腿刚构
slant method 斜法
slant mill 斜轧式轧机
slant nailing 钉斜钉
slant of wind 一阵顺风
slant plane 倾斜面
slant prism 斜棱柱
slant product 斜积
slant pyramid 斜棱锥(体)
slant rail 斜导轨
slant range 斜距;倾斜距离
slant-range-altitude switch 斜距高度转换开关
slant-range converter 斜距变换器
slant range correction 斜距校正
slant range image 斜距图像
slant range marker 斜距指示器
slant range resolution 斜距分辨率
slant range visibility 斜线距离能见度
slant resolution 斜分辨率
slant rolling 斜轧
slant-rolling mill 斜轧机
slant stack 倾斜叠加
slant transformation 斜变换
slant unit 斜轴式机组
slant vein 斜交矿脉
slant visibility 斜能见度;倾斜可见距离
slant-wall duct 斜壁通道
slant well 斜井
slantwise convection 斜向对流
slap 涂刷;敲击声;敲击
slap additional taxes 任意加税
slap dash 粗糙粉刷;粗涂抹面;粗涂
slapped cement 粗涂水泥

slapping 钻杆敲击孔壁;波浪对船体的拍击
slapping percussion 拍叩
slapping press 管端扩口压力机
slap ring 松动活塞环
slash 沼泽低地;螺纹旋压;砍痕;猛砍;切碎;采伐剩余物
slash and burn 砍烧耕作法
slash-and-burn agriculture 刀耕火种的农业
slash chopper 枝梢切碎机
slash-cut 砍痕;伤痕
slash disposal 伐区清理
slasher 截木机;浆纱机;切碎器;多片圆锯机;断木机;单轴多片圆锯机;长把砍刀
slasher cylinder machine 烘筒式浆纱机
slasher saw 圆盘锯
slash grain 弦面纹理;弦面木纹;平锯木纹
slash-grained 原木四开锯法
slashing 废材;残木;废料板
slashing agent 上浆剂
slashing blasting 减振爆破
slashing saw 单轴多锯片圆锯机
slash of hole 扩孔
slash pine 坚硬松木
slash saw 开槽锯
slash sawn 沿木材年轮切向锯开
slat 狭条;窄板条;平板条;条板;衬垫用板条;板条
slatan 印度尼西亚的强干南风
slat angle 板条角度
slat-back chair 条板靠背椅
slat bottom 条壁架体
slat bucket 铲斗框;肋骨挖斗
slatch 一阵好天气;一阵;片阀;绳索松弛部分
slat closure 板条围墙
slat conveyer 翻板式输送机;板式运输机;板式运送机;板带式运输机;链板式输送机;条板(式)输送机
slat crate 条板箱
slat crimper 百页卷折机;百叶卷折机
slat detector 盐分探测器
slat door 百叶门
slate 粘板岩;镜头号码牌;石条;石片;石板瓦;石板(色);板岩;板石(岩)
slate-and-a-half slate 一块半宽的瓦;宽石板瓦
slate axe 劈石斧
slate band 板岩夹层
slate batten 挂瓦条;石板瓦挂瓦条
slate bed 石片滤床
slate belts 板岩带
slate black 石板黑(色)
slate blue 石板蓝(色)
slate board 仿承石板瓦的屋面板;屋顶板瓦(衬板);用石板瓦盖屋顶;盖石板瓦屋面
slate-carbonate formation 板岩—碳酸盐建造
slate cement 页岩水泥;板岩水泥
slate clad 石板贴面;石板面层
slate clay 页岩黏土;黏土板岩;黏板岩
slate cleavage 板岩劈理
slate cleaving machine 板岩劈裂机;石板劈裂机
slate coal 页岩煤;板岩煤
slate concrete 膨胀板石轻集料混凝土;膨胀板石轻骨料混凝土
slate coping 石板压顶
slate-covered 石板覆盖的
slate cramp 石板扒钉;鸠尾形石板扣片;石板瓦双燕尾榫;(石或金属制的)银锭榫
slate cutter 石板切割具
slate deposit 板岩矿床
slated roof 石板屋顶;盖石(板)屋顶
slate dust 板岩粉(尘)
slate fixer 石板固定器
slate fixing nail 石板固定钉
slate floor cover(ing) 石板地面
slate flooring 石板楼地面
slate floor(ing) finish 石板地面终饰
slate flour powder 板岩粉
slate gray 瓦灰色;鼠灰色;石板灰;淡棕灰色
slate green 蓝灰色;石板绿(色)
slate hammer 石板瓦工锤
slate hanging 挂石板瓦;挂墙石板;外挂石板瓦;石板墙面
slate hook 石板瓦挂钩
slate hung 石板垂直面
slate in natural cleft surface 天然开裂面的石板
slate intercalation 板岩夹层

slate knife 石斧
slate lath 挂瓦条
slate lime 页岩石灰
slate listing 横铺石板瓦;平挂石板瓦
slate mine 板岩矿
slate nail 铺屋面用钉;石板瓦钉
slate needle 泥质板岩针
slate olive 石板橄榄绿色
slate panel 石板嵌板
slate-phyllite 板状千枚岩
slate powder 石粉;板岩粉
slate quarry 板岩矿场
slater 石板工人;石板瓦工
slater-and-tiler 瓦工
Slater determinant 斯莱特行列式
Slater-determinant wave function 斯莱特行列式波函数
slate ridge 页岩脊瓦;圆石脊;石板(瓦)屋脊
slate roll 石板屋脊;页岩筒瓦;石板瓦屋脊;圆石脊
slate roof 石板屋顶;板岩屋顶
slate roofing 铺石板屋面;石板瓦屋面
slate roof sheathing 石板屋顶望板
slater's axe 石斧
slater's felt 瓦工油毡
slater's hammer 石板工用锤;石板工锤
slater's iron 铺石板工用的修琢工具
slater's nails 石板屋面瓦用钉;瓦面板用钉
slater's roofing 铺石板面
Slater's rule 斯莱特定则
slater's scaffold 石板工(用)脚手架
slater's work 铺石板工作;铺石板工程
slate rule 量石板间的木尺
slate scudding machine 石刀推挤机
slate shears 石板剪断机
slate siding 石板板壁
slate slab 厚石板
slate slab flooring 石板楼地面
slate slag 板岩矿渣
slate stone powder 板石粉
slate switch panel 石板面配电盘
slate table 石板桌面
slate tan 石板褐色
slate type conveyer 板石式运输机
slate valley 石板屋顶排水沟
slate violet 石板紫色
slate wall panel 石板墙板
slate work 石板瓦工程
Slatex 史莱特斯(一种沥青防水剂)
slat feeder 链板给料机
slat fence 石板栅栏
slat-flecked ice 波纹冰
slat gum 条橡胶
slating 用石板瓦铺屋顶;铺石板(瓦);石板瓦工程
slating-and-tiling batten 挂石板瓦和瓦的木板条
slating battens 挂石板瓦(木板)条
slating ga(u)ge 铺石瓦准尺;铺石板瓦规;石板瓦准尺
slating nail 屋面钉;石板瓦钉
slating work 铺石板
slat lift shutter door 板条升举百叶门
slat of Venetian blind 软百叶片;软百叶窗条板
slat packed column 板条填充柱;板条填充塔
slat packing 条片填充
slat rolling shutter 板条转动百叶窗
slat saw 条板锯;板条锯
slat shutter 板条百叶窗
slat shutter door 条板百叶门
slat-side stock car 条板边马车(畜生车)
slat spring 横弹簧
slatted apron slat conveyer 板条式输送机
slatted belt and flicker type distributor 板条带抛撒轮式撒肥机
slatted belt distributor 板条带式撒肥机
slatted blind 板条百叶屏幕
slatted blind window 板条百叶窗
slatted-chain elevator 链板式升运器
slatted drum 杆条式滚筒
slatted elevator 板条式升运器
slatted floor 石板铺的石面;石板的地板;漏缝地板
slatted roller 板条焊制辊
slatted roller blind 板条滚轴百叶屏幕
slatted roller blind gate 板条滚轴百页门
slatted roller blind housing 百叶窗滚轴箱

slatted roller blind lintel 板条滚轴百页门楣
slatted roller blind section 板条滚轴百页门断面
slatted roller blind shape 板条滚轴屏幕形状
slatted roller blind trim 板条滚轴屏幕装饰
slatted roller blind unit 板条滚轴屏幕构件
slatter 硬灰质板岩;石板瓦工;琢石板工
slatting 铺石板;铺板条
slat-tolerant sludge 耐盐污泥
slat type conveyer 板条式输送机;平板运输机;平输送机
slaty 石板状的;板岩状
slaty clay 页岩质黏土;板岩状黏土;板岩质黏土
slaty cleavage 流劈理;板状劈理;板(岩)劈理
slaty limestone 板状石灰岩
slaty loam 页岩垆坶;页岩壤土
slaty marl 深灰色泥灰;板状泥灰岩
slaty sandstone 板岩状砂岩
slaty structure 板状构造
Slatzmen reagent method 斯拉兹曼试剂法
slaughter establishment 屠宰场
slaughter hall 屠宰场
slaughterhouse 屠宰场
slaughterhouse wastes 屠宰场废物
slaughterhouse wastewater 屠宰场废水
slaughtering rate 出栏率
slaughter wastewater 屠宰废水
slave 下位机;苦工;从属设备;从属(的);从属部件;从设备;从动
slave antenna 从动天线
slave application 从属应用;备用机
slave bistable 他激双稳态
slave camera 伺服摄影机
slave clock 子钟;从钟;副钟
slave clock generator 从动时钟发生器
slave computer 从(属)计算机;从动计算机
slave cylinder 附属油缸;辅助油缸;从动油缸;从动瓶;从动缸
slaved control 从动控制
slaved gyro magnetic compass 从属陀螺磁罗盘
slave display 副显示器
slaved operation 从动作用
slave drive 从动;辅助联动
slaved terminal 从属终端
slave end 从动侧
slave flip-flop 自激多谐振荡器;从动双稳态触发器
slave flipflop 从触发器
slave junction box 副接线箱
slave kit 全套辅助工具
slave laser detector 随动激光探测器
slave manipulator 仿真机械手
slave microcomputer 从(属)微计算机
slave mode 从属(方)式;从方式;从动式
slave motor 随动电(动)机
Slave nucleus 斯莱夫陆核
slave operation 从动运行
slave oscillator 他激振荡器
slave pedestal 副台座
slave pendulum 从动摆;副摆
slave piston 联动活塞
slave processor 从属处理机
slaver 垂涎
slave ram 工作油缸
slave receiver 子接收机;分接收机
slave relay 随动继电器;从动继电器
slavery 奴役现象
slave sensor 从传感器
slave signal 副台信号
slave sludge cone 随动污泥锥
slave state 从属态;从态
slave state of the control processor 控制处理机的从态
slave station 从属站;从(电)台;从站;副台;辅台;被控台
slave storage 从属存储器
slave sweep 等待式扫描;从动扫描;触发(式)扫描
slave sweeper 触发式扫描器
slave terminal 从属终端
slave transmitter 中断发射机;从动发射机
slave tube 从属管
slave unit 辅助设备;从属装置;从属单元;从动环节;从动部件;副台
slave valve 随动阀
slavikite 菱镁铁矾

slaving ring 随动机构
slaving system 从属系统
slaving voltage 驱动电压;从动电压
slawsonite 锶长石
SL-cable 铅包三芯电力电缆
sleak(ing) 冲淡
sleck 粗陶土;红砂岩
sled 换辊小滑车
sled connector 滑车连接器
sled cultivator 拖板式耕耘机
sled drag 橇式刮路器
sledge 滑台车;滑动移车板;大铁锤
sledge car 雪橇汽车;机动雪橇
sledge cross peen hammer 横口大锤
sledge drag 橇式刮路器
sledged stone 锤劈石;砍块石;锤制石;砍石块
sledge hammer 双手轮打的大锤;重磅铁锤;双手使用的大锤;大铁锤;大榔头;大锤
sledge microtome 铲式超薄切片机
sledge mill 多排锤式破碎机;锤磨;槌磨
sledge pin 插销
sledge pin connection 插销连接
sledger 破石工;露天采矿工;锤劈石机
sledge type roll changer 滑座式换辊车
sledge-way 牵引道;木材滑道
sledge with handle 带柄大锤
sledging 二次破碎
sledging roller 单辊轧碎机;单辊压碎机;单辊破碎机
sled kiln 滑板窑
sled runner 滑台的滑行装置;滑板的滑行装置
sleek 修型镘刀;修(光);(修光滑的)曲形光镘刀;丝痕
sleeker 城市居民;异型镘刀;角光子;异型镘刀;磨光器;擦光具;磨光工具
sleek field 油污带
sleeking 弄滑;修光
sleeking glass 磨光玻璃;磨革玻璃块;成形玻璃刮刀
sleeking hammer 抛光锤
sleeking tub 磨光盆
sleekness 光滑
sleeky 柔滑的;光滑的
sleeper 枕木;轨枕;可能畅销品;机座垫;卧车;垫木
sleeper bar 轨枕钢;钢枕
sleeper beam 枕梁;垫梁
sleeper bed 轨枕槽
sleeper bed sieve machine 枕底清筛机
sleeper block 流液洞侧壁砖
sleeper box 轨枕盒
sleeper charge 预装药;预埋药包
sleeper clamping plate 枕木夹紧板
sleeper clip 轨枕垫板;小格栅固定夹
sleeper coach 坐卧两用车
sleeper conveyor truck 轨枕输送机
sleeper-crib tamping machine 枕间捣固机
sleeper drill 枕木钻孔机
sleeper drilling machine 枕木钻孔机;轨枕钻机
sleeper fastening 轨枕扣件
sleeper for crane navy 吊车式掘凿机枕木
sleeper inserter 枕木插入机
sleeper joint 枕木连接;枕木接头
sleeper joint plate 底梁接合板
sleeper joist 地板格栅;小格栅;轨枕梁
sleeper lagging 轨枕背板;枕木衬板
sleeper-layer 轨枕铺设机
sleeper laying machine 枕木铺设机
sleeper mo(u)ld 轨枕盒
sleeper pass 钢枕孔型
sleeper pitch 轨枕间距
sleeper planing machine 枕木削平机
sleeper plate 垫板;支承垫板;轨枕垫板;金属轨枕;地龙墙垫板
sleeper respacing 轨枕正方;方正轨枕
sleeper saw 木枕锯;枕木锯
sleeper scrap 废轨枕
sleepers hire 垫木租费
sleeper slab 接缝下地板;垫板(接缝下)
sleeper spacing 轨枕间距
sleeper specification 轨枕规格
sleeper steel mo(u)ld 轨枕钢模
sleeper stiffness 轨枕刚度
sleeper support 格栅托梁下的横木
sleeper tongs 轨枕钳
sleeper wall 小格栅墙;流液洞侧壁;地龙墙
sleepiness (油漆面的)毛斑;失去光泽
sleeping 浸渍

sleeping account 停止账目
sleeping berth 卧铺(位)
Sleeping Buddha Temple 卧佛寺(北京香山)
sleeping bunk 卧铺;床铺
sleeping cable 惰链
sleeping canal 水面静止的渠道;水面静止的运河
sleeping car 铁路卧车
sleeping caravan 卧拖车;活动住宅
sleeping caravan standing 活动住宅停车位置;卧拖车停车位置
sleeping carriage 卧车;铁路卧车
sleeping company 挂名公司
sleeping fluid 浸渍器
sleeping liquor 浸渍液
sleeping partner 隐名合伙人;隐名股东;匿名股东
sleeping porch 晾台
sleeping rent 固定租金
sleeping room 卧室
sleeping room space 卧室空间
sleeping table 固定淘汰盘
sleeping tank 浸渍池
sleeping-terrace 睡眠平台
sleeping trailer for construction sites 施工现场用的卧拖车
sleep mode 静止方式
sleeporium 旅馆
sleepy gloss 晦光
sleet 雪雨霰;雨凇(下降时呈液态,着地后就冰冻);雨夹雪;冻雨
sleet load 冰雪荷载
sleet-proof 防雹的
sleeve 管衬垫;护套;外接头;体壳;套筒轴;套筒;套环;套管轴;套(垫);插塞套
sleeve actuator 滑阀联杆
sleeve anchorage 锚固套筒
sleeve and wedge coupler 套筒楔块式钢筋连接器
sleeve antenna 同轴偶极天线;同轴管天线;套筒天线
sleeve armature 套筒式电枢
sleeve atomizer 套式喷雾器
sleeve barrel 套筒
sleeve bearing 滑动轴承;套筒轴承
sleeve blocking nut 套筒闭塞螺母
sleeve board 压袖板
sleeve brick 套筒砖
sleeve burner 套筒燃烧器;套筒炉
sleeve cap 套管帽
sleeve catcher 带导向管的环状卡簧岩芯提断器
sleeve chimney 套筒烟囱
sleeve chuck 卡套;卡块
sleeve clutch 刚性离合器;套筒离合器
sleeve compound 连接套管的复合体
sleeve connector 筒式管连接头
sleeve control switchboard 套筒控制交换机
sleeve coupler 套筒连接器
sleeve coupling 套筒接头;套筒联轴节
sleeved backup roll 组合式轧辊
sleeved concrete pile 有套管的混凝土桩
sleeve dimensions 套筒尺寸
sleeved injection tube 有套筒的压注管
sleeve dipole 套管偶极子
sleeve dipole element 同轴管内置偶极振子
sleeved pipe 套管
sleeve drill 套管钻机
sleeved roll 双层轧辊
sleeved roller traction chain 套筒滚子牵引链
sleeved tube 套管
sleeved type punch 套筒式凸模;导套式凸模
sleeve expansion joint 套筒涨缩接合;套筒式伸缩接头;套筒式补偿节
sleeve fence 矮栅栏
sleeve flange 套筒法兰;套接法兰
sleeve flexible joint 套筒式伸缩接头;套筒式挠性接头;套筒式接头;套管式柔性接头;弹性联轴器;弹性联轴节
sleeve force due to centrifugal action 套筒压力
sleeve for taper shank drill 锥孔钻套
sleeve foundation 杯形基础
sleeve gasket 密封套;套垫
sleeve gear 套筒齿轮
sleeve governor 套筒式调速器
sleeve grouting 套管灌浆
sleeve grouting method 套筒灌浆法
sleeve guide 导向套管

sleeve half-bearing 半套筒轴承
sleeve injection pipe 带橡皮管的压入管(注水泥用)
sleeve injection tube 带橡皮管的压入管(注水泥用)
sleeve joint 外解接头;筒接头;套筒连接;套管接头;套筒连接;套管连接;套管接合
sleeve liner 衬套
sleeve method of splicing reinforcing bars 钢筋套筒连接法
sleeve mounted propeller 套筒式螺旋桨
sleeve net 套筒网
sleeve nut 轴套螺母;螺旋帽;连接螺母;套筒螺母
sleeve of ball-bearing 滚珠轴承套
sleeve of jack 塞孔套
sleeve of puppet head 随转尾座套筒
sleeve of tail stock 顶针座套筒
sleeve packing 套筒密封垫
sleeve piece 套件;嵌环;套筒(管间接头)
sleeve pin 套筒销
sleeve pipe 预留管
sleeve-port 套筒口;套阀孔
sleeve puller 套筒拉出器;拔套器
sleeve pump 套筒活塞泵
sleeve roller transmission chain 套筒滚子传动链
sleeve rotor 空心转子
sleeve seal 套筒密封
sleeve shaft 筒形轴
sleeve socket 套筒插座
sleeve spindle 套轴
sleeve spring 套弹簧
sleeve stub 同轴振子;套管短柱;套管短截线
sleeve surface milling cutter with coarse teeth 粗齿套式面铣刀
sleeve tubing connector 套管接头
sleeve turnbuckle 套筒松紧螺旋扣
sleeve-type bearing 套筒式轴承
sleeve-type chain 套筒链
sleeve-type clutch 套筒式离合器
sleeve-type compensator 套管式伸缩器;套筒式调整器
sleeve-type coupling 套筒式接头
sleeve-type engine 套筒式发动机
sleeve-type journal bearing 套筒式颈轴承;套筒式(普通)轴承
sleeve valve 套筒式气门;套筒式滑阀;套(筒)阀
sleeve valve shifting tool 滑套循环阀操作工具
sleeve valve type hydraulic self-sealing coupling 套管滑阀式液压自封接头
sleeve wall protection method 套管护壁法
sleeve weld 套筒焊接
sleeve wire 塞套引线
sleeve wrapper 套封包装机
sleeving 管状织物;插塞套
sleeving of bars 钢筋套管
sleeving valve 套阀
sleigh 雪车;撬木
slender 细长的
slender beam 狭梁;细(长)梁;易变形梁
slender body theory 细长体理论
slender column 细长柱
slender cone 细长锥体
slender configuration 细长体
slender contour 流线型外形
slender flexible structure 高柔性构筑物
slender form 细长形;细长型
slenderness 细长度
slenderness degree 细长度
slenderness effect 细长效应;长细影响
slenderness proportion 长细比例;细长比(例);细长度
slenderness ratio 细长比(例);长细比;长径比;长度直径比
slenderness ratio of compression member 压杆长细比
slender particle 细长粒子;细长颗粒
slender piece 细长工件
slender proportion 细长比(例);长细比
slender purpleosier willow 细紫柳
slender ship 瘦长型船
slender spire 细尖塔
slender stream 细长液柱
slender type 细长形
slender wall 细长墙
slender wedge 细长楔
sleptons 超轻子

slevis handle 插杆手柄
slew 转向;摆动;偏荡
slewability 旋转机动性;回转机动性
slewable 可旋转的
slew a full circle 旋转圈
slewing 回转;驱冰航行;破冰前进
slewing and raking pile driving plant 旋转斜桩打桩装置;旋转斜桩打桩机
slewing angle 旋转角
slewing angle of derrick 吊杆偏角
slewing arc 转向扇形齿板
slewing area 转动面积;转动范围
slewing arm 旋转臂;回转臂
slewing baffle 回转隔板
slewing bearing 回转枢轴轴承
slewing boom 旋转吊杆;回转式悬臂
slewing brake 回转机构制动器(起重机)
slewing bridge type ship loader 旋转桥式(散货)装船机
slewing bucket-ladder excavator 旋转梯式多斗挖掘机
slewing chute 旋转滑槽;旋转斜槽
slewing circle 回转圆
slewing column crane 旋转塔式起重机
slewing construction crane 旋转施工起重机
slewing crane 旋转式起重机;旋转式吊机;旋臂起重机;转动起重机;转斗机;旋转式吊机;回转式起重机;回臂起重机
slewing crane barge 旋转起重船
slewing crown 旋转顶部
slewing derrick 旋转吊货杆
slewing device 旋转装置;旋转机构
slewing device brake 旋转装置制动器
slewing distortion 转换失真
slewing drilling machine 回转钻机
slewing drive 旋转传动装置;旋转驱动装置
slewing excavator 回转式挖掘机
slewing floating crane 旋转起重船
slewing gear 转动装置
slewing gear brake 旋转传动装置制动器;旋转设备制动器;起重机回转机构制动器
slewing gear intermediate shaft 逆转横轴
slewing gear ring 转车台
slewing guy 吊杆中部稳索;吊杆旋转稳索
slewing jib 旋转吊杆
slewing journal 止推销轴承;凸轮止推回转轴承;转向节
slewing limiter 回转限位器
slewing loader 旋转式装船机
slewing lock 回转锁定装置
slewing mechanism 旋转机构;回转机构
slewing motion 旋转运动
slewing motor 旋转电机;回转电动机
slewing pillar crane 旋转塔式起重机;转柱式起重机
slewing point 旋转接点;旋转连接
slewing radius 回转半径
slewing ram 旋转油缸
slewing rampway 旋转跳板【船】
slewing range 变速范围
slewing rate 转换速率
slewing ring 旋转环;回转支承盘
slewing ring for bogie 转向架转盘
slewing ring with fixed column 定柱式回转支承
slewing ring with rotary column 转柱式回转支承
slewing road crane 旋转道路起重机
slewing service 旋转变压器
slewing service line 变速设备线;旋转变压器线
slewing shiploader 旋转式装船机
slewing speed 旋转速度
slewing speed of grab machine 抓斗机旋转半径
slewing stern ramp 旋转式船尾引桥
slewing table 回转台
slewing tower 回转塔
slewing trolley 旋转小车
slewing tubular column 旋转管柱
slewing unit 变速装置;旋转装置
slewing unit brake 变速装置制动器;旋转装置制动器
slewing V-block V 形缺口垫板;旋转三角槽垫块
slew line 旋转吊货杆的牵绳
slew mode 旋转状态
slew post 旋转柱;转臂支柱
slew rate 转换速度
slew relay 停止继电器

slew test 转向测试
slice 下水架用楔;炉钎;泥片;切片;长柄铁火铲;长柄铲;薄切片;薄片
slice analysis 切片分析
slice bar 拨火杆;炉钎;薄片铲
slice circuit 限幅电路
sliced 刨切的
sliced block 斜砌方块
sliced blockwork 叠石筑堤工;斜砌块体;斜砌方块砌体;条块砌筑
sliced blockwork breakwater 斜方块砌体防波堤
sliced crystal 切割晶体
sliced layer 斜砌层
sliced mushroom 蘑菇片
slice drift 分段平巷
sliced time 分割时间
sliced veneer 切片;切成层板;平切单板;刨切单板;刨切薄木;刨平的薄镶板
sliced work 斜砌法;方块层叠砌合
slice height 分层厚度
slice lapping 片研磨
slice latch network 片式闩锁屏蔽网络
slice-lettering 透明注记
slice level 限制电平
slice lip 堰唇
slice mark 堰板痕
slice mask network 片式闩锁屏蔽网络
slice memory 单片式存储器
slice method 条分法;薄片法
slice microcomputer 片式微机
slice minicomputer 单片式小型计算机
slice mining method 分层开采法
slice of photoelastic model 光弹模型切片
slice of three-dimensional data block 三维数据体的切片
slice of transient phase data block 瞬时相位数据体切片
slice opening 堰板缝隙
slice pipeline system 片式流水线系统
slice plywood 薄胶合板
slice position 堰板位置
slice processor 片式处理器
slicer 开片机;检选工;脉冲限制级;切片机;单板切片机;单板刨切机;泥刀
slicer knife 切片刀
slice scale 分片级数;部分累进率
slice setting 堰板调节
slice term of estimation variance 片估计方差
slice tungsten carbide bit 片状硬合金钻头
slice velocity 堰板浆速
slicewood 刨切木材
Slichter method 斯利希特法
slicing 修平;限幅;刮料;切片法;切断;双值限幅;分片;分层开采
slicing and caving 分层崩落法
slicing cutting 切割
slicing knife 切片刀
slicing lathe 修边车床;截切车床
slicing machine 钢锭切分机;切分机
slicing method 分层开采法
slicing saw 切片锯
slick 光滑的;海面油膜;平滑(水)面;平滑器;平滑的;水面上的浮油;带纹海面;穿眼凿
slick-back 滑动拱顶管
slick bit 修光钻头
slick condition (路面上的)滑溜状态
slickens 光滑冲积层;水力冲刷浮土法;粉砂泥
slickenside 滑动面;滑面;擦痕;平滑边缘;断层擦痕;(岩石的)擦痕面
slickensided hair cracks 擦痕面发状裂缝
slicker 刮尺(找平用);刮刀;修型慢刀;光器;修光工具;刮子;磨光器;叠板刮路机
Slicker solder 铅锡软焊料
slick hole (装满炸药的)爆破腿
slicking glass 磨光玻璃
slick joint 光滑接头;滑动接头
slick line 活动浇注管;平滑线;下料管
slick line space 滑线间距
slick mill 斯力克锻轧机
slickolite 擦痕岩面
slick paper proof 清样
slick process 摇摆滚锻工艺
slick road 滑溜道路
slick-rock 平滑岩石

slick rod (深井泵的)磨光轴杆
slick sheet 薄钢板;油光薄板;通用遮板
slick spot 不毛区;盐碱结壳区;光滑点
slick style 光亮式
slick-surface 光滑的表面;平滑面;平滑表面
slick-surfaced 平滑表面的
slick tire 光滑轮胎
slick water 有一层油膜的水
slick wire line equipment 钢丝作业设备
slidable 滑动的
slidable lattice gate 活动的网格门
slidable wedge 滑动楔
slidable window 推拉窗;扯窗
slidac 滑线式调压器;滑线电阻调压器
slide 滑坡;滑移;滑梯;滑动;钻杆导槽;载物片;载物薄片;载玻片;栽玻片;盖玻片;拉筒;滑移;滑梯;滑动;导板;玻片;(显微镜的)玻璃片;崩塌;闸板
slide actuating cylinder 闸板操纵缸
slide adjusting valve 滑动调头阀
slide agglutination 玻片凝集法
slide agglutination test 玻片凝集试验
slide area 滑坡带;滑动面积
slide arm 滑臂;滑板臂
slide assemblies 滑板组合件
slide axis 滑坡主轴
slide bar 滑动顶杆;滑块;滑杆;导杆;长把火铲
slide bar bracket 滑杆托
slide bar system 滑动顶杆系统;滑动顶梁支护
slide base 滑动底座
slide base plate 滑床板
slide bearing 滑动轴承;滑动支承
slide bed 滑轮底;滑动床
slide block 滑块;滑动岩块;滑动断块
slide block friction 滑块摩擦
slide block radial drilling machine 滑座式摇臂钻床
slide block universal radial drilling machine 滑座式万向摇臂钻床
slide bolt 横插销
slide box 载物玻片盒;滑阀箱;分气阀箱;玻片盒
slide bushing 滑套
slide cabinet 载物玻片橱
slide calipers 滑动卡规;游标卡尺
slide cam 滑动凸轮
slide carriage 溜板;滑座
slide cause 滑动原因
slide chair 滑床板
slide cleaner 载物片清洗器
slide cliff 滑坡后壁
slide coast 地滑海岸
slide coil 滑触线圈
slide contact 滑动接触
slide control 均匀调整;滑动调节;平滑控制;平滑调节
slide controller 滑动式控制器
slide conveyer 滑轨运送机;滑槽传送器;滑槽运输机;滑槽输送机;滑板
slide coupler 可调耦合线圈
slide coupling 伸缩接头;滑动连接
slide crack 滑坡裂缝
slide culture 载玻片培养;玻片培养法
slide damper 滑动闸板;滑动挡板;插板阀;闸板
slide deposit 滑坡沉积(物)
slide detector 坍方检测器
slide detector fence 坍方检测栅
slide device 滑动装置
slide direction 滑动方向
slide divider 滑动接点分压器
slide door 水平推拉门;拉门;活动门
slide door groove 水平推拉门滑槽
slide down 下滑
slide drill 滑橇式钻机
slide dropper clamp for twin catenary wire 滑动双承力索吊弦线夹
slide eliminated gear 无滑动齿轮
slide face 滑面;滑阀面
slide-failure surface 滑动破坏表面
slide fastener 拉链;滑扣
slide fault 滑动断层
slide feed screw 滑板进给螺杆
slide fire escape 火警太平滑道
slide fit 滑配合;滑动配合
slide floor 滑动地板
slide for filler valve 充油滑阀

slide form 滑动模板
slide frame 滑动基架
slide free 滑移脱开;滑移至不受节制;自由滑动
slide friction 滑动摩擦
slide gate 推拉门;滑动闸门;滑(动)门;挡板阀门;拉门;扎门
slide ga(u)ge 滑尺;滑动水尺;滑动规;卡尺
slide gear 滑动齿轮
slide generator 打滑发电机
slide gib 导轨镶条
slide glass 载片;幻灯片;(显微镜的)玻璃片
slide guide 滑导承;导轨
slide guide bracket 滑导承托架
slide hook 滑钩
slide-in 抽屉式
slide-in camper 车用野营配具
slide-in chassis 抽屉式部件;插入式抽屉
slide into place 滑移就位
slide in unit 滑入装置
slide joint 滑动接头
slide knot 止滑结
slide lamella 滑动层(地)
slid electrolyte probe 固体电解探测器
slide line 滑动线
slide loader 耙斗装料机
slide mass 滑动体;滑移体;滑体;滑坡体
slide matching 顺序拼木法
slide method 玻片法
slide micrometer 载玻片;滑动千分尺;滑标千分尺
slide multiplier 杠杆式乘法器;滑臂式乘法器
slide noise 滑动噪声
slide nozzle brick 滑动水口砖;滑动喷嘴砖
slide-off transfer 薄膜贴花
slide of papilionaceous flower 蝶形花幻灯片
slide oil 滑油
slide operating handle knuckle 滑板操作柄关节
slide operation 滑动操作
slide-out of log 抽条(成捆木材)
slideover seat 轻便座椅
slide pad 浮动块
slide panel 推拉镶板;扎镶板;扎门;推拉门
slide path 滑移轨迹;坍滑路径
slide plane 滑动平面;滑动面;滑泻面
slide plate 滑床板;滑板
slide plate chair 滑板座
slide prevention 防止滑移;防滑动
slide price 浮动价格
slide probe tuner 滑动探针式调谐器
slide projector 幻灯片放映机;幻灯(放映)机
slide protection 滑行保护;打滑保护
slider 滑行物;滑动物;隔离物;木质滑板;滑座;滑块;滑动件;滑触头;滑板
slide rack 载物片架
slide rail 滑动导轨;滑道;滑轨;滑动围栏;导轨
slider-bed-type moving ramp 滑床式电动斜道
slider-bed-type moving walk 滑动扶梯;滑床式电动走道
slider box 滑块
slider coupling 十字滑块联轴器
slide regulator 滑动调节器
slide resistance 滑动阻力
slide-resistant pile 抗滑桩
slide resistor 滑动电阻器;滑触变电阻;滑臂电阻
slide rest 转盘支架;滑动架;滑动刀架;刀架
slide rest of compound type 复式滑动刀架
slide rheostat 滑线变阻器;滑动变阻器;滑触变电阻
slider locking pin 滑座驻栓
slide rod 连杆;拉杆;滑杆;导杆
slide roll ratio 滑动滚动比
slide roof bar 滑动顶板导杆
slider resistance 滑线电阻
slider shunt 滑动分流器
slide rule 计算尺;滑尺
slide rule computational work 计算尺计算工作
slide sample valve 进样滑门
slide scale 滑动标尺
slide shaft 滑动轴
slide show 幻灯
slide slice 滑片
slide slope 坡道
slide staining machine 载物玻片染色机
slide stern 导杆
slide support 滑动支承
slide surface 滑面

slide switch 拨动开关
slide test 抗滑试验
slide throttle 滑阀
slide thrust method 滑坡推力法
slide tongs 管钳
slide tongue 滑坡舌部
slide tool rest 滑坡刀架
slide top 顶部滑板
slide track 滑轨;滑动面;滑道
slide transformer 滑动式变压器;调感变压器
slide-trap 玻片捕捉器
slide treatment 滑坡处理
slide triangle 滑动楔体;滑动楔块
slide tuner 滑动式调谐器
slide type expansion joint 套筒伸缩器
slide valve 闸板阀;滑(动)阀;挡板阀门;分油活门
slide valve box 滑阀箱
slide valve buckle 滑阀套
slide valve bush 滑阀衬套
slide valve case 滑阀箱
slide valve chamber 滑阀箱
slide valve chest 滑阀箱;滑阀瓣室
slide valve circle 滑阀圆
slide valve control gear 滑阀控制齿轮
slide valve cover 滑阀罩
slide valve ellipse 椭圆阀图
slide valve engine 滑阀发动机
slide valve face 滑阀滑面
slide valve gear 滑阀装置
slide valve hydrant 滑阀消防栓
slide valve inside lap 滑阀内余面
slide valve jacket 滑阀套
slide valve lead 滑阀超前
slide valve lever 滑阀杆
slide valve liner 滑阀内衬
slide valve link 滑阀连杆
slide valve opening 滑阀口
slide valve outside lap 滑阀外余面
slide valve pin 滑阀销
slide valve pump 滑阀泵
slide valve rand 滑阀台肩
slide valve reciprocating pump 滑阀往复泵
slide valve rod 滑阀杆
slide valve seat 滑阀座
slide valve slot 滑阀槽
slide valve spindle 滑阀轴;滑阀杆
slide valve spindle gland 滑阀杆压盖
slide valve spring 滑阀弹簧
slide valve strip groove 滑阀压板槽
slide valve stroke 滑阀冲程
slide valve surface 滑阀面
slide valve thimble 滑阀套管
slide valve travel 滑阀行程
slide valve vacuum pump 滑阀式真空泵
slide valve vane pump 滑阀叶片泵
slide vane compressor 滑片式压缩机
slide vane pump 滑阀叶片泵
slide vane-type rotary blower 滑片式旋转鼓风机
slide vane-type rotary pump 刮板式旋转泵
slide velocity 滑坡速度
slide viewer 观景器
slide warmer 载物片加温器
slide watertight door 滑动式水密门
slideway 滑道;滑斜面;滑路;滑轨;滑槽;坡道
slideway grinder 导轨磨床
slideway grinding machine 导轨磨床
slide wedge 滑动楔体;滑动楔块
slide wedge method 滑动楔体法
slidewheel 滑轮
slide wire 滑线;滑触电阻线
slide wire bridge 滑线电桥
slide wire potentiometer 滑线式电势计;滑线电位器
slide wire resistance 滑线电阻
slide wire rheostat 滑线变阻器
sliding 滑行;滑动(的);使滑动
sliding action 滑移作用;滑动作用
sliding agent 润滑剂
sliding and bi-parting door 两门叶对开滑动型车门
sliding angle 滑动角
sliding armo(u)r door 滑动铠板门
sliding avalanche 滑动崩坍;滑动崩塌
sliding axis 活动轴;滑动轴线
sliding axle 滑动轴
sliding axle box 滑动轴箱

sliding bank 滑岸
sliding bar 滑杆
sliding barrier 滑(动)栅门;防滑器
sliding base 滑动底座;导向轨座
sliding-bat kiln 推板窑;滑底隧道窑;多通道推板隧道窑
sliding bearing 滑动轴承;滑动支座;滑动支承
sliding bearing of neoprene pad surface with Teflon coat 氯丁橡胶垫板上加聚氯乙烯油滑动支座
sliding bearing sleeve 滑动轴承套
sliding bed 滑坡床
sliding bedding 滑动层面
sliding bed lathe 床身可接长的车床
sliding bevel 木工角尺
sliding bilge block 机械舭墩;活动舭墩
sliding blackboard 滑动黑板
sliding blade 滑动叶片
sliding block 滑块
sliding-block guide 滑块导框
sliding block side bearing 滑块式旁承
sliding bolt 插销;滑动螺栓
sliding bookshelf 滑动书架
sliding boom 滑动壁架
sliding boom on tilting track 斜轨上滑行的吊杆
sliding bottom 滑动底座
sliding bottom door 平移式泥门
sliding boundary condition 滑动边界条件
sliding bow 集电弓
sliding box feeder 箱形给料器;箱式定量器
sliding box-shaped caisson 滑动式箱形沉箱
sliding bridge 移车台;转车台;转盘(铁路)
sliding budget 滑动预算;弹性预算;变动预算
sliding bulkhead door 滑动水密门
sliding bush 滑套
sliding caisson 活动式沉箱;滑动式沉箱;推拉式坞门
sliding caisson gate 横拉门[港]
sliding cal(l)ipers 卡尺;滑动卡尺;游标卡尺
sliding cam 滑动凸轮
sliding capacitor 滑动电容器
sliding carriage 滑架;刀架滑板
sliding ceiling 滑动顶棚
sliding center position of up-to-danger 最危险滑动中心位置
sliding chalkboard 滑动黑板
sliding characters 滑动特点
sliding chute 滑槽
sliding coefficient 滑动系数
sliding collar 滑动轴环
sliding colter 直犁刀
sliding condenser 滑动电容器
sliding connection 滑动连接
sliding contact 滑动接触;滑动触点;擦拭触点
sliding contact coil 滑动接点线圈
sliding contact electrode 滑动接触电极
sliding contact gear 滑动接触齿轮
sliding contact starter 滑触起动器
sliding coupling 滑叉式万能接头
sliding cover 滑动盖(板)
sliding cradle 滑动支架;斜架下水车[船]
sliding crown bar 滑动式拱顶纵梁
sliding crown-block 移动式滑车
sliding curve 滑动曲线;滑动曲线
sliding damper 活挡;滑动调节器
sliding deck 可动平台;推拉舱盖
sliding deformation 滑坡变形
sliding device 滑溜装置
sliding diaphragm 滑动光阑
sliding die 可调板牙
sliding displacement mechanism 滑移机理
sliding distance of moving clamps 动钳位移
sliding dock gate 横拉式坞门
sliding door 滑(移)门;滑动门;推拉门
sliding door catch 推拉门锁
sliding door fitting 滑动门装置;滑动门配件
sliding door gear 滑动门齿轮
sliding door gear drive 扎门驱动装置;推拉门驱动装置
sliding door guide 滑门导轨
sliding door hanger 推拉门吊件;推拉门吊架;推拉门吊钩
sliding door hardware 滑动门小五金
sliding door key 推拉门键

sliding door lock 推拉门锁
sliding door pocket 推拉门箱
sliding door rail 推拉门滑轨
sliding door stay 滑动定门器
sliding door stop 推拉门门挡
sliding door track 推拉滑道;推拉滑轨
sliding double bearing 滑动复式轴承
sliding drill bit 滑动扁钻;滑动钻头
sliding dropper clamp 滑动吊弦线夹
sliding dropper clamp for messenger wire 承力索滑动吊弦线夹
sliding dropper clamp for twin catenary wire 双承力索滑动吊弦线夹
sliding eliminated gear 无滑动齿轮
sliding expansion joint 滑动式膨胀接头;滑动式伸缩缝
sliding face 滑(动)面
sliding factor 滑动系数
sliding factor of safety 滑动稳定系数
sliding failure 滑移失效;滑移破坏;滑动失事;滑动破坏;滑坍破坏
sliding fire door 滑动防火门
sliding fit 滑配合;滑合座;滑动配合
sliding flange 滑动管缘;滑动法兰(盘);滑动法兰(板)
sliding floor 移动式调车转辙器;滑动地板;滑动底盘;滑动底板;滑动层面
sliding floor ejection 移动式底板推移;移动式楼板推移
sliding flow 滑移流动;滑动流(动)
sliding flue damper 烟道滑动风门;风道滑动风门
sliding foil 滑动金属箔;滑动滑片
sliding folding door 滑动折叠门;推拉折页门
sliding folding grille 滑动折叠格栅
sliding folding louvred shutter 推拉折叠百叶门窗
sliding folding partition(wall) 滑动折壁;推拉折叠隔断
sliding folding shutter 滑动折叠百叶窗
sliding folding shutter door 滑动式折叠百叶门
sliding folding window 推拉式折叠窗;推拉折页窗
sliding force 滑动力
sliding fork oil seal 滑动叉油封
sliding form construction method 滑升模板施工法
sliding form operation 滑模施工(法)
sliding form paver 滑模铺料机;滑模铺路机
sliding form(work) 滑模;滑动模壳;滑动模板;滑升模板
sliding formwork construction 滑模施工(法)
sliding formwork machine 滑模式混凝土摊铺机
sliding fracture 滑动破裂
sliding frame 滑动框架
sliding frame saw 滑式框架
sliding friction 滑动摩擦
sliding frictional force 滑动摩阻力;滑动摩擦力
sliding friction(al) resistance 滑动摩擦阻力
sliding friction bond strength 滑动摩擦黏结强度
sliding friction pivot 滑动摩擦枢轴
sliding garage door 车库推拉门
sliding gate 闸阀;滑动(闸)门;(连铸中间包的)滑动水口;滑动式闸门;横推门;推拉(式)门;船坞横拉门
sliding gate feed 滑槽送料
sliding gate pump 叶片泵
sliding gate system 滑动闸板机构
sliding gate valve 滑门阀;滑动闸阀门
sliding ga(u)ge 游尺
sliding gear 滑动齿轮
sliding gear drive 滑动齿轮传动(装置)
sliding gear mechanism 滑动齿轮机构
sliding gear transmission 滑动齿轮传动(装置)
sliding gear type gearing mechanism 滑动齿轮式传动机构
sliding gear type transmission 滑动齿轮式传动
sliding glass door 滑动玻璃门
sliding glass-pane 滑动玻璃窗
sliding glass wall 推拉玻璃墙;滑动玻璃墙
sliding guidance 滑导承
sliding guide 罐道滑槽
sliding hanger 推拉吊钩
sliding hatch(way) 滑动下舱孔;滑动闸门
sliding head 滑动刀架
sliding head automatic type lathe 滑动刀架自动车床
sliding head steel bar clamp 木工夹

sliding head stock 滑动主轴箱;滑动床头箱
sliding height 滑动高度
sliding hinge 滑动铰
sliding hook 滑钩
sliding hopper door 滑动泥门;平移滑动泥门【疏】;抽屉式泥门
sliding housing 滑动阀壳
sliding in cut 滑坡;堑道滑坡;挖方滑坡
sliding inner window 滑动内窗;扯内窗
sliding iron 抽动杆扳手
sliding jack 横式起重器;滑座螺旋起重机;滑动千斤顶
sliding jamb 滑动门梃(折叠门上的假门梃)
sliding joint 滑动接缝;伸缩缝;滑移节理;滑动接头;滑动接合;滑动缝
sliding keel 下垂龙骨;滑动龙骨
sliding key 推拉键
sliding ladder 滑行搁梯
sliding landslide 正在滑动滑坡
sliding length after stop 停机滑动长度
sliding level 滑动斜面
sliding lifting door 滑动升降门
sliding lifting door fitting 滑动升降门配件;滑动升降门装置
sliding lifting door hardware 滑动升降门小五金
sliding lifting window 滑动升降窗
sliding line 滑动线;滑动线
sliding lock 推拉锁
sliding locking bolt 滑锁螺栓
sliding mass 滑块;滑坡体;滑动体
sliding-mesh gear-box 滑动齿轮变速器
sliding microtome 滑走切片机;滑动式切片机;推拉切片机
sliding mode crack 滑开型裂纹;滑移型裂纹
sliding moment 滑动力矩
sliding motion 滑移运动;滑动
sliding movement 剪切运动;滑移;滑动
sliding nut 滑行螺母
sliding of foundation 基础滑动
sliding oscillation 滑移振动
sliding pair 滑动副;滑动对偶
sliding panel 滑动板
sliding panel tunnel kiln 滑门隧道窑
sliding panel weir 滑板(式)堰;滑板木堰;插板(式)堰
sliding parallel-plate viscometer 滑动平行板式黏度计
sliding parity 可调整平价
sliding partition 推拉隔断
sliding partition(wall) 推拉水活动隔断
sliding patio door 滑动院门
sliding peg exchange rate 爬行固定汇率
sliding phenomenon 滑移现象;滑动现象
sliding piece rotary compressor 滑片回转式压气机
sliding pintle 滑动调节锥栓
sliding plane 滑移面;滑面;滑动(平)面;滑动层面
sliding plane length 滑动面长度
sliding plane shape 滑动面形状
sliding plate 滑动支座板;滑动板;滑床板
sliding plate bearing 滑板支座
sliding plate microviscometer 滑板式微量黏度计
sliding plate palletizer 滑板式擦包机
sliding plate ring 垫圈
sliding plate valve 滑动式阀门
sliding plate viscometer 滑板(式)黏度计
sliding plunger 滑动(柱)活塞
sliding point 浮放道岔
sliding-point buffer 滑动缓冲剂
sliding poise 游码
sliding pontoon 滑动浮筒;推拉式坞门
sliding price 滑动价格;变动价格
sliding price contract 滑动价格合同
sliding prism 滑动棱体
sliding pulley 滑轮
sliding pulse 滑移脉冲
sliding rail 拖拉滑轨;推拉滑轨;滑轨;滑触线;导轨
sliding reaction 剪切反力;滑动反力
sliding resistance 抗滑力;滑动阻力
sliding resistor 滑线电阻;滑动电阻器;滑触变电阻
sliding rheostat 滑动变阻器
sliding ring 滑环;滑动密封环
sliding roadway arch 缓冲式拱架;让压性拱形支架;可缩性路拱
sliding rod 塔式标尺

sliding roll 滑动辊
sliding roof 活动车顶;滑动顶棚
sliding-roof wagon 滑动顶棚的棚车
sliding rope bridge in mountainous district 山区溜索桥
sliding rubber annulus 滑动橡皮环
sliding rubbing 滑动摩擦
sliding rule 计算尺
sliding saddle 滑动座架
sliding sash 水平推拉窗;滑动窗;推拉窗
sliding sash window 扯窗;推拉窗;横向推拉窗
sliding saw 滑式锯;滑座锯
sliding scaffold 滑动脚手架
sliding scale 累进计算法;计算尺;滑尺;天气滑尺;比例增减制
sliding scale agreement 比例增减协约
sliding scale budget 滑动预算;浮动预算
sliding scale commission 累进计算法手续费;滑准法手续费
sliding scale price 浮动价格
sliding scale royalty 滑动提成率;滑动比例提成
sliding scale wage system 滑动工资制
sliding scarifier 滑动耙路机;滑动翻路机
sliding screw jack 滑座螺旋起重器
sliding seal 滑动式密封
sliding sealing device 滑动式密封装置
sliding seat 滑动座
sliding shaft 活动轴;滑动轴
sliding shear failure 滑动剪切破坏
sliding shears for steel band 钢带滑动剪刀
sliding shoe 滑瓦;滑履;仿形弹脚
sliding shutter 滑动遮阳窗;滑动折叠门;滑阀;推拉百叶窗
sliding shutter Boule 布尔滑动闸门
sliding shutter door 滑动百叶门
sliding shuttering 滑动模板;滑升模板(工程);滑动遮光板;滑动窗板;滑动百叶窗;(水闸的)插板闸门
sliding side saddle 活动侧鞍座
sliding skid 滑轨
sliding slab 滑动板
sliding sleeve 滑套;滑动套筒
sliding sleeve connection 滑套连接
sliding sleeve motor 滑阀配气发动机
sliding slot four bar linkage 滑槽四杆链系
sliding slot linkage 滑槽链系
sliding sluice gate 滑动(泄水)闸门;水闸滑动闸门;水闸滑动泄水门
sliding socket joint 滑动套筒接合
sliding spigot 滑动式插口
sliding spool (方向阀的)滑阀芯;滑动柱塞
sliding spool flow control valve 滑阀式流量控制阀
sliding spool valve 滑阀
sliding spring key 滑动弹簧键
sliding square 三角板
sliding stability 抗滑稳定性;滑动稳定(性)
sliding stack 滑动书架
sliding staff 塔尺【测】
sliding stage 活动舞台
sliding standard window 滑动标准窗
sliding stay 钢窗支子;活动窗撑;滑动窗撑
sliding steel door 滑动钢门
sliding stop 滑动挡板;插片式光圈
sliding strain 滑动应变
sliding strength 抗滑强度
sliding stroke 滑动行程
sliding suction 伸缩式吸入管
sliding support 可缩性支撑;滑动支座;滑动支架;滑动支承
sliding surface 滑(移)面;滑动面(积)
sliding switch 小车式道岔;滑动开关;平移型道岔
sliding table 滑动台;滑台
sliding table press 滑动台式压机
sliding tank base joint 罐底滑动接合板
sliding tariff 浮动关税;非固定关税
sliding test 抗滑试验;滑动试验
sliding timber weir 插板(式)堰
sliding time 滑动时间
sliding tire 滑行轮胎
sliding tongs 带箍锻工钳
sliding tool carriage 滑动刀架
sliding top skylight 滑动天窗
sliding track 滑动轨迹;滑轨;推拉滑轨
sliding triangle 滑动三角木楔

sliding tripod 伸缩三脚架;活腿三脚架;伸缩式三脚架
sliding tube sheet heat exchanger 滑动管板式换热器
sliding type building 可滑动式建筑物
sliding type expansion joint 滑动伸缩式接头;滑动式膨胀接头;滑动式膨胀接口
sliding type mechanical rake 滑式机动耙路机
sliding type rake 滑式耙路机
sliding tyre 滑行轮胎
sliding universal joint 滑叉万向联轴器
sliding-up angle 爬坡角
sliding valve 滑动式阀门;滑动阀;滑阀
sliding-vane air compressor 滑片式空气压缩机
sliding-vane compressor 滑片压缩机;叶片式压缩机;滑片式压缩机;滑动叶片压缩机;滑板式压缩机
sliding-vane meter 滑片式流量计
sliding-vane motor 叶片马达;叶片电动机
sliding-vane pump 刮板泵;滑片泵;滑板泵;滑叶式水泵;滑动叶片泵
sliding-vane rotary air compressor 滑片回转式空压机
sliding-vane rotary blower 滑片式旋转鼓风机
sliding-vane-type rotary pump 滑动叶片式回转泵
sliding-vane vacuum pump 滑板式真空泵
sliding vector 滑移矢量;滑动矢量
sliding-wall wagon 滑墙式货车
sliding way 动滑道;上滑道;导轨
sliding wedge 滑楔;滑楔
sliding wedge block 滑动楔形块
sliding wedge bolt 滑动楔形螺栓;滑楔式锚栓;双楔式锚杆
sliding wedge chuck 滑楔卡盘
sliding wedge method 滑动楔(体)法;滑楔法
sliding wedge rangefinder 滑动光楔测距仪
sliding wedge roof bolt 楔缝锚杆
sliding weight 平衡砝码;微动砝码
sliding wheel 轮子;滑轮
sliding wheel shaft 滑轮轴
sliding window 水平推拉窗;拉扯窗;滑动窗口;滑窗;平移窗;推拉窗
sliding window detection 滑窗检测
sliding window handle 滑动窗柄
sliding window sash 滑动窗框;滑动窗扇
sliding wood patio door 滑动庭院木门
sliding wood terrace door 滑动阳台木门
sliding york 滑动叉;滑叉
sliding zone 滑动区;滑动带
slid platform 滑台
slige 微弱
slight 轻微的;轻浪(海浪二级);道格拉斯二级浪
slight attachment 次要附件
slight attack 轻侵蚀
slight breeze 轻风;微风
slight convex curve 微凸曲线
slight crest 弱峰
slight damage 轻微损坏;轻微破坏
slight disturbed 轻微扰动
slight dive 小角度俯冲
slight drag 小阻力
slight effect 轻度影响
slight enriched fuel 低浓缩燃料
slight error 微小误差
slight flocculation filtration 微絮凝过滤
slight ice 微冰
slighting rough 略微粗糙的
slight-injury road accident 轻微伤害的道路事故
slightly acidic 微酸性的
slightly alkalinity 微碱性
slightly clouded sky 少云天空
slightly curved canal 微弯渠道
slightly curved reach 微弯河段
slightly damaged 轻微受损;轻度破坏
slightly damp 稍湿的
slightly dissolution 稍有溶蚀
slightly-embossed carving 隐雕
slightly expansive cement 微膨胀水泥
slightly fissured 轻微裂隙化
slightly fused granule 微熔颗粒
slightly grey 浅灰色
slightly hard 微硬
slightly humid 稍湿
slightly inclined ramp 微倾斜坡道

slightly mobile potassium 微动性钾
slightly movement 微弱运动
slightly pitched gable roof 小坡度人字屋顶
slightly pitched glazing 小坡度倾斜玻璃窗
slightly pitched roof 小坡度倾斜屋顶
slightly podzolized soil 轻度灰化土
slightly polluted raw water 微污染源水
slightly polluted source water 微污染水源水
slightly salified soil 弱盐渍土
slightly sandy shale 含沙少的黏土页岩
slightly sensitive 轻度敏感的
slightly sinuous canal 微弯渠道
slightly sinuous reach 微弯河段
slightly sloped roof 微小斜坡屋顶
slightly soluble 微溶的
slightly soluble salt test 难溶盐试验
slightly swelling-shrinkage foundation 较轻微胀缩地基
slightly turbid 微浊;微混浊的
slightly viscoelastic fluid 微黏弹性流体
slightly warm ground 低温放热地面
slightly weathered 轻度风化的
slightly weathered layer 微风化层
slightly weathered rock 微风化岩石;弱风化岩石
slightly weathered zone 弱风化带;微风化带
slightly weathering 弱风化;微风化
slightly wet 稍湿
slightly yellow limestone 浅黄色石灰石
slight pressure 低压
slight rain 微雨
slight repair 小修(理)
slight sea 小浪(海浪二级);轻浪(海浪二级);三级浪;二级浪
slight shock 微震
slight shower of rain 小阵雨
slight skewness 轻微偏态
slight thunderstorm 小雪暴;小雷暴
slik 潮泥滩
slikke 潮泥滩
slim 低劣
slime 黏垢;岩粉淤泥;矿泥;黏物质;黏泥(层);黏液;煤泥;软泥;生物黏泥;淀渣;沉泥;沉砟;残渣
slime bacteria 黏细菌
slimeball 黏球
slime bank site 黏泥堆放场
slime bath 河泥浴
slime clarification 矿泥澄清
slime coating 软泥涂抹(层);黏泥涂抹
slime concentrate 细粒精矿;矿泥精矿
slime crusher 沉渣碾碎机
slime dam 泥浆池
slime flo(a)tation 矿泥浮选
slime flocculation 岩粉凝聚
slime flux 黏液;流溢
slime-forming bacteria 黏细菌;黏泥形成细菌
slime fungi 黏土菌;黏泥菌;黏菌
slime growth in waste 废水黏菌生长
slime index 黏泥指数;黏泥指数
slime layer 淤泥层
slime mold 黏土霉菌;黏菌;黏霉菌类
slime peat 浆斑点;泥沼;矿泥煤;泥煤
slime peat soil 泥炭土
slime pit 沥青坑;沥青矿井;矿泥储坑
slime product 泥质产品
slime pump 矿泥泵;泥浆泵;泥泵
slimer 细粒摇床;细粒碎机;矿泥摆床
slimes 黏膜;黏垢
slimes concentrator 矿泥浓缩机
slimes dispersant 黏泥剥离
slime sedimentation 岩粉沉淀
slime separation 泥浆分离;脱泥
slime separator 矿泥分离器;黏泥分离器;泥渣分离器;沉泥池
slime shield 黏泥(浆)盾构
slime sludge 泥渣;泥浆
slime sluice 矿泥溜槽
slime suspension 黏泥悬浮质
slime table 岩粉加工台;矿泥摇床
slime tank 岩粉沉淀箱
slime vanner 矿泥带式流槽
slime washer 矿泥洗选机
slime water 矿泥水;泥泥水
slim file 细锉
slim hole 小直径钻孔;小口径钻孔

slim-hole drilling 小眼井钻进;小口径钻进;小孔径钻进
slim-hole drilling machine 小口径钻机
slim-hole joint 小眼接头
slim-hole rig 小口径钻机
slim-hole system 小直径钻进法
slimicide 黏絮除泥剂;抗石灰化剂;除黏菌剂
sliminess 稀黏程度;黏稠度
sliming 细粒化;黏化;泥浆化
slimleaf fig 狭叶榕
slimline fluorescent lamp 细长荧光灯
slim line lamp 细管荧光灯
slimline lamp fluorescent luminaire fixture 细长荧光灯装置
slimline type 细长形
slimline type fluorescence lamp 细长管式荧光灯
slim nose pliers 薄嘴平口手钳
slim saw file 细锯锉
slim well 小口径钻井
slim wirilda acacia 小树枝状相思树
slimy 泥浆状的;含淤泥的;多泥的
slimy cakes 糊状石蜡的饼块
slimy coating 糊状涂料
slimy ground 泥泞地
slimy milk 黏乳
slimy waste material 泥泞的废物
sling 悬挂带;一关货;一吊货;钩索;环索;起吊装置;抛掷(器);索环;绳回环;吊梁;吊具;吊货环;吊带;吊索
sling and placing bench for car body 车体吊具及安放台
sling band 桅横杆上悬挂吊索的铁箍
sling block 吊货框架;滑车组;吊索滑车
sling bolt 连杆螺栓;活节螺栓
sling canvas 吊货帆布兜
sling cart 重型吊车;吊装车
sling chain 吊链
sling chair 吊椅;躺椅;帆布躺椅
sling cleat 挽索木挂钩
sling conduit 悬吊管
sling container 集装带
sling dog 夹钳;吊钩
slinger 轴油道;摘挂钩工人;挂钩工人;抛掷装置;抛油机;抛砂机;投掷器;吊装工;吊索器;挡油板
slinger head 抛砂头
slinger mix 投射料
slinger mo(u)lding 抛砂造型
slinger ring 吊环
slinger seal 抛油环式密封
slinger sleeve 抛油环套筒
slinger type lubrication 抛油环式润滑
slinger-unloader 抛掷装置减荷器;抛掷装置卸载机
sling for lifting loads 吊具
sling guiding device 吊索导向托座
sling here 此处吊起;此处吊挂索
sling hook 吊索钩
sling hoop 桅横杆上悬挂吊索的铁箍
sling hygrometer 摆动湿度计
slinging 捆吊费;起吊费;抛射;抛砂;吊起
slinging chain 吊链
slinging eye 吊索眼
slinging groove 起吊槽;吊链槽
slinging process 投射成型法(耐火材料)
sling-like suspension rod 吊带形悬置杆
sling mark 吊货索环标志
sling net 货网;网兜
sling of net 网兜
sling one's hook 出发前进(船员俗语)
sling package 集装带成组货件
sling pipe 悬吊管
sling psychrometer 旋转式干湿球湿度计;旋转式干湿球干湿表;旋转式干湿表;摇动湿度计;通风湿度计;手摇湿度计;手摇干湿度计;手摇干湿度表
Slingram method 施铃格拉姆电磁勘探法
sling rope 吊索;吊货绳
slings 桅横杆悬挂吊索部分
sling shot 渔网浮标系绳
slingshot model 引力弹弓模型
sling stay 悬吊牵条
sling stay bolt and washer 吊撑螺栓及垫圈
sling stay tee 吊撑丁字头
sling stowing 抛掷充填
sling tally 按关(即一吊货)计数法
sling thermometer 旋转(式)温度计;旋转(式)温

度表;悬式温度计;挂式温度计
sling tube 悬吊管
sling van 吊斗;吊车
sling wire 钢丝吊索;吊货钢丝缆
slino-anthophyllite 斜直闪石
slinoptilite 斜发沸石
slip 纸条;窄条;造型涂料;螺旋桨滑退率;降低率;减少率;门票(集装箱);码头或土堤上斜滑道;滑泥;滑落;滑距;脱锚;突堤码头间水域;插枝;滑脱
slip a cable 弃链
slip agent 油滑剂
slip agitating tank 泥浆搅拌池
slip angle 滑动角;滑移角;横向偏离角;偏离角
slip area 卡瓦夹住段(钻杆)
slip at the anchorages 锚具滑动;锚头滑动
slip at wall 壁上滑移
slip avalanche 滑移雪崩
slip band 滑移带;滑脱带;滑动带
slip bar 滑杆;滑动式预应力筋;滑动杆;非灌浆预应力钢筋
slip basement 滑坡基座
slip bearing bushing 拼合轴瓦
slip bedding 滑动层理
slip block 滑动岩块
slip bolt 锁门栓;伸缩螺栓;插销
slip boundary 滑动边界
slip bowl 卡瓦引掌;卡盘
slip breccia 滑动角砾岩
slip bushing 可换衬套
slip butt hinge 活销铰链
slip caisson 滑动式沉箱
slip cap 滑动盖子
slip cast 流铸;滑动浇注
slip cast block 流铸件;流铸块
slip cast brick 泥浆浇铸砖
slip cast fused silica 粉浆浇铸熔凝硅石;粉浆浇铸石英玻璃
slip casting 注浆成型;流铸(法);泥浆浇注;粉浆浇铸
slip casting by pressure 压力注浆
slip casting process 注浆成型法;流铸程序
slip castings 粉浆浇铸制件
slip cast Mo crucible 粉浆浇铸的钼坩埚
slip chart 偏离图
slip circle 滑移圆弧;滑弧;滑动圆弧
slip circle analysis 圆弧滑动分析
slip circle method 滑弧法;滑动圆弧法
slip clay 易滑土;易熔黏土;易溶黏土;滑泥土
slip cleavage 滑(移)劈理
slip cliff 滑坡后壁
slip clove hitch 活端丁香结
slip clutch 摩擦式离合器;滑动离合器;滑差离合器;安全摩擦离合器
slip coating 上泥釉;化妆土;涂泥浆层;涂泥釉
slip coating machine 涂泥釉机
slip coefficient 滑移系数;滑动系数
slip connection 匝道
slip consistency 泥浆稠度
slip control for impact crusher 锤式破碎机的滑动控制
slip counter 转差计
slip coupling 活接头;滑动联轴器;滑动联轴节;滑动接筒
slip cover 滑动盖(板);滑顶;家具套
slip crack 压裂;滑裂;滑动裂纹;滑动裂缝
slip curve 转差曲线;滑移曲线;滑动曲线
slip-decorated 泥浆装饰的
slip deformation 滑移变形;滑动变形
slip detector 滑动探测器;打滑指示器
slip direction 劈裂方向;层理方向;滑移方向
slip dislocation 滑移位错
slip dock 斜底船坞;修船湿坞;滑道(船坞);滑船道式船坞;滑船道;升船滑道;半坞式船台
slip dogs (卸钻时把钻头固定在转盘上用的)卡瓦块
slip dovetail 滑动鸠尾榫;滑动鸠尾榫
slip drumlin 滑坡鼓丘
slip element 滑移素;滑移单元
slip energy 转差能量
slip erosion 滑移侵蚀;滑泻冲蚀;滑坡侵蚀
slip exhausting 排浆
slip face 滑落面
slip factor 滑移系数;滑动系数;滑动率;滑差系数
slip fall 错落
slip fault 滑断层

slip feather 滑键;滑楔;嵌缝片;舌榫;塞缝片
slip feeding 进浆
slip fineness test 釉浆细度测定
slip fit 滑动配合
slip-fit connection 套式接合
slip-fit liner 滑动配合衬筒
slip fitting 滑动装置
slip flask 脱箱
slip flask mo(u)ld(ing) 脱箱造型;滑脱砂箱铸型;脱箱铸型
slip flow 黏性流;滑移流;滑流
slip flow phenomenon 滑流现象
slip fluidity 泥浆流动性
slip fold 剪切褶曲;滑褶皱
slip form 滑动模板;滑模;滑升模板;滑动模壳
slipform casting 滑模式浇注法
slipform concrete paver 滑模混凝土摊铺机;滑模式混凝土铺筑机
slipform concreting 滑模浇筑混凝土
slipform construction 滑模施工(法);滑模铺筑法
slipform construction method 滑模施工法;滑升模板施工法
slipform curb machine 滑模路缘石铺筑机
slipformed concrete 滑模混凝土
slipformed curb 用滑模修筑成的路缘;用滑模筑成的边缘
slipformed facing element 滑模面板
slipformed slab 用滑模浇筑的混凝土板
slipformed structure 滑模施工结构
slipform element 滑模构件
slipform equipment 滑升机具;滑开机具
slipformer 滑模施工者;滑模装置;滑模机;滑模工
slipforming 滑动成型;滑模施工(法)
slipform lining 滑模衬砌
slipform machine 滑模机
slipform method 滑模施工法
slipform panel 滑模镶板
slipform paver 滑模式(混凝土)摊铺机;滑模铺料机;滑动铺料机
slipform paving 滑模施工(法)
slipform paving train 滑动模板式联合铺路机
slipform shutter 滑模百叶窗
slipform system 滑模体系
slipform yoke 滑模支架
slip fracture 剪切破裂;滑动破裂
slip frequency 转差频率;差频
slip friction clutch 安全打滑摩擦离合器
slip from an anchor 弃锚出港
slip front 滑动前沿
slip function 滑移函数
slip ga(u)ge 量块;块规;浆料稠度计
slip gear 滑动齿轮
slip gearing 摩擦传动
slip glaze 釉浆;泥釉;泥浆釉
slip gradient 滑移梯度
slip grip 卡瓦打捞爪
slip-hazard 滑移事故
slip hitch 滑结
slip hook 活钩;活扣钩;滑脱环;滑钩;卡钩;滑脱钩
sliphook for connecting rope 系缆活钩
slip house 制浆车间;泥浆工段
slip indicator 滑级指示器
slip-in liner 镶套套管;镶套衬垫
slip interference theory 滑移干扰理论
slip jacket 型套;套箱
slip jaw clutch 波纹齿滑动式离合器
slip joint 滑动接榫;活动接合面;剪切节理;内外键槽连接;活接头;滑配接头;滑动式伸缩缝;滑动接头;滑动接扣;滑动接口;别茬结合;伸缩式连接;伸缩(式)接头;伸缩连接;伸缩结合
slip-joint casing 套接套管(套管一端套入另一套管的一端后加固)
slip-joint conduit 带滑动接头的导线管;滑动接头管道
slip-joint coupling 补偿联轴节
slip-joint pliers 滑动铰点钳;鲤鱼钳
slip-joint safety valve 伸缩节安全阀
slip-joint shaft 滑动接合轴
slip-joint sleeve 伸缩接合套
slip key 滑键;伸缩键
slip kiln 烘浆窑
slip knot 活结;滑结;伸缩结
slip lake 滑坡湖
slip layer 润滑层

slip line 滑移线
slip-line field 滑移线场
slip-line field method 滑移线场解法
slip-line field theory 滑移线场理论
slip lining 滑动衬板;滑动套筒
slip lining method 滑动衬里法
slip load 卡瓦负荷(转盘内)
slip-lock 滑锁
slip loss 转差损耗
slip mass 滑坡体
slip matching 滑动配合;顺序拼木法;顺序配板;滑动配板;滑边拼接
slip membrane 滑动防水薄膜;滑动防潮层
slip meter 转差计;转差测定器;滑差计;测滑表
slip mortise 开口榫;活榫槽;滑动榫槽;滑动沟道;滑接榫眼;滑动榫眼;榫槽
slip motion 滑动
slip mo(u)ld 滑模
slip newel 滑动中柱
slip nipple 滑动连接管;伸缩螺纹接口;滑动连接套
slip nose 嵌入式榫尖
slip of bar 钢筋滑动
slip of fault 滑距【地】
slip-off edge 塌边
slip-off slope 滑走坡;滑溜坡;冲积坡
slip-off slope bank 冲积坡岸
slip of impregnated glass cloth 浸胶玻纤布滑移;胶布滑移
slip of induction motor 异步电动机的转差率
slip of propeller 推进器滑脱
slip on 套上;穿上
slip-on attachment 锥形连接;滑套式连接装置
slip-on bit 移动式钻头;锥形连接的可卸式钻头
slip-on filter 插入式滤波器
slip-on flange 滑动法兰(盘);滑动法兰(板);平焊法兰;套入式法兰
slip-on mount 滑动мount支架
slip-on type 移动式;滑动式;活动式
slip orientation 滑移方向
slip pack coating technique 料浆包渗技术
slippage 滑脱;浪纹;经纬滑动;滑移(量);滑坡;滑程;滑差;打滑(量);屋面流淌;屋面滑移;滑动力
slippage crack 滑移裂缝
slippage cross 滑程
slippage factor 滑动因子;滑动因数
slippage flow 滑流
slippage pump 补油泵
slippage test 偏移量检验
slipped 打滑
slipped area of fault 断层滑动面
slipped bank 顺差触排
slipped multiple 顺差触排
slipped region 滑脱区;切变区
slipped screw 打滑螺钉
slipped shoulder 滑落肩
slipped thread 断缺螺纹
slipped tile 防滑地砖
slipper 滑溜程度(路面);抹灰靠尺;抹灰准尺;(阴沟或排水沟顶部的)接高部分;闸瓦;制动铁鞋;制动块;制动机;滑块导架;滑座;滑块;滑动部分;滑触头;滑板
slipper brake 滑块式制动器;滑动制动
slipper dip 流动浸渍;车身浸浴涂漆
slipper guide 滑块导块
slipperiness 光滑性;平滑性
slipperiness of pavement 路面滑溜
slip permeability 泥浆渗透性;泥层渗透性
slipper path 支承导环
slipper plate 滑动面的薄铁板
slipper seal 滑动密封
slipper tank 可弃油箱
slipper type connecting rod 滑动式连杆
slippery asphalt(ic) pavement 光滑的地沥青铺面;光滑的地沥青路面
Slippery elm 红榆
slippery film 润滑膜
slippery hitch 活端丁香结
slippery reef knot 半缩帆结
slippery sett paving 光滑的块石铺面;光滑的块石路面
slippery when wet sign 滑溜警告标志
slip piece 滑条
slipping 滑移;滑动;打滑

slipping agent 增滑剂
slipping an anchor 弃锚
slipping anchor 滑动锚固
slipping apparatus 滑动仪器
slipping area 滑动面积
slipping bank 滑塌岸
slipping cam 滑轮凸轮
slipping clutch 安全摩擦离合器
slipping coefficient 滑动系数
slipping column 滑动支柱
slipping contact detector 滑触检波器
slipping drive 滑动传动(装置)
slipping eccentric 滑动偏心轮
slipping from a buoy 单浮离泊
slipping from alongside 离码头
slipping from alongside with an anchor down 绞锚离码头
slipping from head-and-astern buoys 双浮离泊
slipping limit 空转限制
slipping of bar 钢筋滑动
slipping of belt 皮带滑动;皮带打滑;传动皮带滑动
slipping of clutch 离合器的滑动;离合器打滑
slipping plane 滑动面
slipping plate 滑动板
slipping plate clutch 滑动摩擦离合器;滑动板离合器
slipping protection device 空转防护装置
slipping resistance surface 滑动阻力面
slipping soil wedge (计算土压力的)滑动楔体
slipping stream 平滑流
slipping surface 滑动(表)面
slipping switch 滑动式开关
slipping the tow line 解拖
slipping wire drawing machine 滑动式拉丝机
slip plane 滑移面;滑面;滑动平面;侧滑面
slip plane position of up-to-danger 最危险滑动面位置
slip points 交分道岔【铁】
slip process 湿法工艺调制泥浆过程
slip proof corrugated board 防滑波纹纸板
slipproofing agent 防滑剂
slip puller 拉出器;(泵缸套的)拉板
slippy road 滑溜道路
slip racking 系缆活结
slip rail 活动栏杆
slip ramp 滑斜面
slip rate 滑移速率;滑行速率
slip ratio 转差率;滑动比率;滑动比;滑差系数
slip ratio of hydraulic drive 液力传动转差率
slip ratio of hydraulic transmission 液力传动转差率
slip recovery 空转回收
slip region 滑移区
slip regulator 转差率调节器
slip relay 转差继电器
slip resistance 滑阻力;滑动电阻
slip-resistance tile 防滑地砖
slip-resistant aggregate 抗滑骨料;抗滑集料
slip-resistant floor 防滑地面
slip-resistant surface 防滑面
slip-resistant tile 防滑铺地砖;防滑瓷砖;防滑瓦
slip-resistant tile floor 防滑地砖地板
slip rheology 泥浆流变学
slip ring 集流环;集电环;汇流环;汇电路;滑环;分离推力环
slip-ring(electric) motor 滑环(式)电动机
slip-ring induction motor 滑环式异步电动机;滑环式感应电动机
slip-ring motor 汇电发动机
slip-ring rotor 滑动环转子
slip road 斜坡路面;支路;岔道
slip roof 滑动屋顶
slips 薄砖;卡瓦;索线头
slip scraper 刮土铲运机;滑行挖土机;滑行刮土机
slip sensing relay 空能继电器
slip sheet 滑动岩席;滑动岩层;滑动薄板;衬薄纸
slip sheet handling system 滑动薄板装卸方式
slip sheet structure 滑席构造
slip shutter 滑升模板
slip sill (门窗的)滑轨;滑入槛;滑槛(门窗);齐口窗台;平开窗台
slip sleeve 滑动套伤
slip socket 滑动管套;滑动套节;卡瓦打捞筒

slip-socket bowl 卡瓦打捞筒导管
slip-socket clip 活夹套;卡瓦打捞筒夹钳
slip speed 滑动速度
slip spider 卡盘
slip stage 移动(式)平台;用轨滑动舞台
slip steep face 滑坡陡壁
slip stick 计算尺;(深井泵的)磨光轴杆
slip-stick phenomenon 滑落黏现象
slip stone 磨刀小油石
slip stopper 活钩链掣;滑道制动器
slip stream 推进器尾流;滑移流;切向流
slip stress 滑动应力
slip supported packer 卡瓦式封隔器
slip surface 滑动面;滑移面;滑行表面;滑溜表面;坍坡面
slip surface of failure 破坏滑动面
slip switch 交分道岔【铁】
slip switch track 渡线【铁】
slip system 滑移系
slip teeth lost 掉卡瓦牙
slip tender 渡船码头ящ缆工人
slip terrace 滑坡台阶
slip test 滑移试验;泵的负载特性试验
slip threshold 滑动临界值
slip tile 填充平瓦;滑动平瓦;烧结陶土片条
slip tongue 滑动榫舌;滑键;滑接榫舌;穿条
slip-tongue joint 穿条接合;两槽填条接合;花键接合;方栓接合
slip topography 滑坡地貌
slip trailer 泥浆挤出器;泥浆挤出机
slip trailing 泥浆涂绘
slip-tube shaft 伸缩管轴
slip turnout 交分道岔【铁】
slip-type core lifter 楔形岩芯提断器
slip-type drawing machine 滑动式拉丝机
slip-type elevator 卡瓦式吊卡
slip-type expansion joint 套筒式涨缩器;套筒式补偿器
slip under the brush 容易刷的涂料
slip universal joint 滑叉万向联轴器
slip vacuum treatment machine 泥浆真空处理机
slip velocity 滑移速度;滑动速度;沉降速度(指岩屑)
slip ware 彩色泥浆装饰的器皿
slip water bottle 带阀门取水样器
slipway 滑船道;修船滑道;下水滑道;造船滑台;造船滑道;缆车滑道;滑台;滑道;船台(滑道);船排
slipway and wedged chassis 斜架滑道
slipway cradle 滑道承船车
slipway crane 船台起重机
slipway for steel roller launching 钢珠滑道
slipway with cradle 摇架式滑道
slipway with rail tracks 船排滑道
slip-wedge method 滑动楔体法
slip-weld hanger 卡瓦焊接套管挂
slip wire bridge 滑线电桥
slipyard portal crane 船台门座起重机
slip zone 滑移区
slit 狭孔;狭缝;隙缝;窄剖面;切条
slit antenna 裂缝天线;槽馈偶极天线;缝隙天线
slit aperture 缝隙孔径
slit bolt 楔缝式锚杆
slit bunker 缝式存仓
slit burner 长口燃烧器;缝式燃烧器
slit camera 交向摄影机
slit cathode 裂缝阴极;分瓣阳极
slit cellulose film 纤维素薄膜条
slit coil 窄带卷
slit cutting 纵切;纵割
slit deal 薄板;薄松板
slit deposit 缝隙沉积物
slit diaphragm 缝隙光阑
slit domain 裂纹域
slit edge 切开边;缝缘
slit fabrics 狭长料子
slit failure 缝隙破坏
slit filter 狭缝滤水器;缝隙滤水管
slit flow 缝隙流
slit hopper 有缝料斗
slit injector 缝隙式喷嘴
slit jaws 裂缝夹子
slit lamp 狭缝灯;缝灯;裂隙灯
slit-lamp microscope 裂隙灯显微镜
slit length 缝隙长度

slit lens 开缝透镜
slitless spectrograph 无缝摄谱仪
slitless spectrum 无缝光谱
slit-mask 缝式遮光片
slit membrane 裂孔膜
slit method 缝隙法
slit nosed boring bar 槽端镗杆
slit nut 开缝螺母;切槽螺母
slit of light 光带
slit orifice 槽口;裂缝;缝隙口型
slit pin 开口销
slit planting 切缝栽植;缝栽
slit program(me) 陕缝程序
slit prong 裂缝插脚
slit recuperator 狭缝式换热器;辐射式换热器
slit-rod-and-wedge-type bolt 楔缝式锚杆
slit scan 栅缝扫描
slit seal 间隙式密封
slit selvage 切边
slit-shaped pore 狭缝形孔隙
slit-shirt piston 开口裙式活塞
slit-side solid sampler 孔壁取样器;井壁取样器
slit source 缝隙源
slit spectroscope 狭缝分光镜
slit stay bolt 槽纹撑螺栓
slit stop 缝隙光圈
slit-tape extrusion system 带缝挤出系统
slitted pipe section 开口管套
slitted wall 长条露空槽;开长槽的剪力墙;开长槽墙
slitter 纵向切片机;纵切机;切纸机;切条机
slitter crack 纵切裂痕
slitter edge 废边
slitter knife 纵剪机刀片
slitter line 纵切线
slitter loss 剪切损失
slitter marks 裁切标记
slitter operator 纵切工
slitter ramp 圆盘剪旁的带卷台
slitter rewinder 纵切复卷机
slitting 纵裂缝;开缝
slitting and coiling line 纵剪和卷取作业线
slitting and re-reeling machine 纵切复卷机
slitting and re-winding machine 纵切复卷机
slitting and shearing line 纵剪和横剪作业线
slitting and trimming line 纵剪和切边作业线
slitting cutter 开缝铣刀;槽铣刀
slitting cutter with coarse teeth 粗齿切口铣刀
slitting disc 锯盘;切割圆盘
slitting disk 切石圆锯
slitting file 菱形锉;开槽锉;螺钉头锉
slitting knife 切条刀
slitting line 多刃圆盘剪纵剪作业线
slitting machine 纵切机组
slitting mechanism 纵切机制
slitting mill 滚切机;滚剪机;开片机;开槽铣刀
slitting pass 切分孔型
slitting roll 纵切圆盘刀心杆;纵切辊
slitting roller 滚剪机
slitting saw 开槽锯;切割锯
slitting shear machine 切条机
slitting shears 纵剪(切)机;纵剪机
slitting unit 纵向切分机组
slitting-up 全切开
slitting up method 全切开法
slitting wheel 锯盘;切割圆盘
slit trench 狭长沟槽;狭壕;散兵坑
slit trough 缝隙槽
slit tube 有缝管;缝管
slit type cracking test 有缝试件抗裂试验
slit type mask 开槽式荫罩
slit ultramicroscope 缝隙超显微镜
slit visco(si)meter 缝隙黏度计
slit-wedge(type) rock bolt 楔缝式锚杆
slit width 狭缝宽度;缝隙宽度
slit window 狭长窗
slitwork 薄板
sliver 岩片;自然银;渣粒;裂片;裂片;纱毛;切条;条子;条片;碎裂物;梳条;分裂物
sliver blending 条子混合
sliver can 条筒
sliver condenser 条子集合器
sliver-doubling plate 条卷机导条板
sliver evenness tester 粗条子均匀度试验机

sliver fill 长条回填;狭长式填土
slivering 辊痕(板材带材表面热轧缺陷)
slivering machine 制条机;成条机
sliver intensifier 银加厚液
sliver lap 条卷
sliver lap machine 条卷机
sliver mixing 条子混和
sliver screen 小节筛;碎料筛
sliver separator 分条器
sliver spreading device 条子展幅装置
sliver tooth 刮毛搓齿
sliver unevenness 条子不匀率
slivery particles of aggregate 集料片状颗粒
slivery wire 折叠线材
sloarization 负感现象
sloat 舞台布景升降机
slob 浓ီ൹泥;泥泞地;海面乱冰;冰泥
slobber-belt 装在带式输送机旁的引带
slobbering 垂涎
slob ice 浓密冰泥
sloe 黑刺李
slogan pylon 标语塔
slogging chisel 截钉凿;切螺栓头的大凿
slogging hammer 平锤
sloid alcali sulphide 固体硫化碱
sloop 护卫舰;单桅纵帆船;单桅(帆)船;大型辅助炮舰
slop 料浆;溅出;废油
slop basin 脏水盆
slop-built 粗制的;建筑粗劣的;不良筑造的
slop chest 船上衣物杂用品箱;船上小卖部
slop cut 不合格馏分
slop-deflection equation 坡度挠度方程
slop disposal barge 废物处理驳
slop disposal ship 废物处理船
slop disposal tank 废油处理油柜
slop feeding 湿喂
slop glaze 稠化的釉浆;备用料浆
slope 斜坡;斜率;斜度;坡度;坡;比降;转角
slope above the water 水面以上坡度
slope air course 斜风道
slope amplification 斜率放大
slope analysis 边坡分析
slope analysis map 坡度分析图
slope angle 斜角(度);角度;坡度角;边坡角
slope angle of abrasion shoal patch 磨蚀浅滩坡角
slope apex drainage 坡顶排水
slope apron 陆坡裙
slope-area discharge measurement 比降面积测流法;坡降面积测流法
slope-area method 坡降面积法;比降面积法
slope at inflection point 拐点处的斜率
slope berth 倾斜船台
slope black top paver 倾斜的沥青路面铺路机;倾斜的黑色面层铺料机
slope-blind shaft development 抖井—盲井联合开拓
slope board 坡度样板;坡板
slope bonder 斜砌合墙
slope bondstone 斜砌合石
slope bottom 斜井井底车场
slope break 坡折点
slope brick 坡面砖
slope cage 斜井罐笼
slope calculation 斜率计算;斜坡计算
slope carriage 斜井提升车架
slope category 坡度等级图
slope caving 斜坡沉陷
slope chaining 斜坡丈量
slope channel 河槽
slope circle 坡面圆;穿过斜坡坡面的滑动弧
slope coefficient 斜率系数
slope collapse-sliding 边坡坍滑
slope compaction 坡面夯实;坡面压实;边坡压实
slope compactor 坡道压实机
slope compensating device 倾斜补偿装置
slope computation 斜率计算;斜坡计算
slope concrete paver 斜坡混凝土铺料机
slope consolidation 边坡加固
slope construction 坡面施工;斜坡施工
slope construction equipment 坡道施工设备;斜坡施工设备
slope control 斜度调整;电流升降调节
slope-controlled arc wilding power source 多特性弧焊电源
slope control timer 陡度调整计时器
slope-conveyance method 比降—输水(率)法
slope conveyer 倾斜输送机
slope correction 斜坡校正;斜坡改正;斜面校正(法);斜度改正;倾斜改正;坡度校正;坡度改正
slope course 斜坡道路
slope-covered terrace 坡下阶地
slope creep 边坡滑动;斜坡蠕动
slope crown 坡道凸面;坡道隆起;斜坡凸面;斜坡隆起
slope culture 斜坡栽培
slope current 倾斜流;坡度流;梯度流
slope curve 斜度曲线
slope cut 开挖边坡;斜坡挖土
slope cutting 坡道挖土;开挖边坡;边坡开挖
slope-cutting for lightening load 削坡减载
sloped 倾斜的
sloped beam 斜梁
sloped building pit 倾斜的建筑基坑
sloped bulkhead plating 斜形舱壁板;斜顶板铺板
sloped bulkhead stiffener 斜顶板纵骨
sloped cable 斜缆
sloped camber 路面坡拱
sloped coping 倾斜压顶
sloped curb 斜路牙子;斜坡式路缘石;斜面路牙;斜路缘
sloped debris flow 山坡型泥石流
sloped edge 斜棱;斜边
slope-deflection coefficient 角变位移系数
slope-deflection equation 转角位移方程;角变位移方程
slope-deflection method 角变位移法;角变位法;倾角挠度法
slope deposit 坡地沉积
slope design 坡道设计;边坡设计
slope detection 斜率检测;斜检测
slope detection method 斜率鉴频法;斜率鉴定法
slope detector 斜率鉴频器;斜率检测器
slope development 斜井开拓;放坡
slope deviation 斜度偏差;斜度变化;倾斜偏差
sloped floor 斜坡式护坦
sloped folding 倾斜折叠
sloped footing 斜形基础;斜坡式底脚
sloped foundation 斜坡基础
slope diagram 坡度图
slope discharge curve 坡降流量曲线;比降流量曲线
slope discharge diagram 坡降流量图;比降流量图
slope distance 斜距
slope distance readout 斜距读数
slope ditch 斜沟
slope downwards to open drain 坡向明沟
sloped pile 斜桩
sloped pipe 斜管
slope drain 山坡排水沟
slope drainage 坡道排水;斜坡排水
slope dredging 坡度浚挖
slope dressing 坡道表面处理;坡道整修;斜坡表面处理;斜坡整修
slope driving 斜井开凿;斜井掘进
sloped roman 斜体罗马字
sloped roof area 斜面屋顶面积
sloped step 斜坡式踏步
sloped substrate 坡基层
sloped tube 斜管
slope edge 坡道边缘;斜坡边缘
slope efficiency 斜坡效率
slope erosion 坡道磨损;坡道冲蚀;斜坡磨损;斜坡冲蚀;边坡冲蚀;坡地冲蚀;坡底冲刷;边坡侵蚀
slope excavation 放坡开挖
slope face 斜井掘进工作面
slope facies 陆坡相
slope facing machine 斜坡铺面机
slope failure 边坡破坏;边坡毁损;滑坡;滑动破坏;坡身不稳;边坡坍塌;边坡破坏
slope feature 坡面特征
slope field tractor 斜坡地用拖拉机
slope filter 斜率滤波器
slope finisher 坡道抹面机;坡道面机;斜坡抹面机;斜坡整面机
slope flattening 坡道整平;斜坡整平;坡度平坦化
slope flow 倾斜流;坡度流
slope for drainage 排水坡度
slope function 斜率函数
slope gain 斜度增益
slope ga(u)ge 斜水尺;斜度规;斜标尺;倾斜式水尺;坡度计
slope genesis 坡地成因
slope grader 整坡机;平坡机;带割坡刀的平地机;边坡平路机
slope gradient 边坡陡度;边坡倾斜度;边坡面坡度
slope grading 边坡坡度
slope grading of rock mound 抛石埋坡
slope greening 边坡绿化
slope gully 海底谷;断续海底谷
slope hachure 示坡线
slope heave 边坡鼓起
slope height 坡高
slope hoisting 斜坡卷扬提升
slope inclination 坡道斜度;坡道倾斜角;边坡角;斜坡斜度;斜坡倾斜角
slope indicator 坡度指示器;测斜仪;测角器
slope indicator tube 斜度管
slope instability 边坡不稳定
slope-intercept form 斜截式
slope in water 水下边坡
slope irrigation 边坡灌溉
slope-keypoint compaction 斜度基点压缩
slope ladder 坡道
slope land 坡地
slope landscape 坡面景观
slope length 坡长
slope level 测斜仪;斜准器;测斜器;测坡度计
slope level(l)ing 坡度测量;斜度测量;边坡测量
slope level(l)ing and compacting machine 坡道整平压实机;斜坡整平压实机
slope limit analysis method 土坡极限分析法
slope line 坡道线;斜线;示波线;比降线
slope lining 边坡的护坡;斜坡的护坡
slope-lining machine 斜坡铺面机;边坡铺面机;边坡衬砌机
slope loess 坡地黄土
slope longitudinal section 斜坡纵断面
slope maintenance 坡道养护;边坡养护;斜坡养护
slope map 地形图;坡度图
slope material 坡积物
slope measuring cross-section 比降观测断面
slope meter 斜度仪;斜率计;量斜(度)仪;跨导计;坡度计
slope method 坡度法;比降法
slope mine 斜井开采矿;斜井
slope model 斜坡模型
slope mo(u)lding 倾斜造型
slope mouth 斜井口
slope mower 边坡割草机;斜坡割草机
slope mowing attachment 边坡割草附件;斜坡割草附件;边坡割草装置;斜坡割草装置
slope net working 编篱护坡工程
slope oar 斜划桨
slope of a cornice 台线坡度;台口坡度
slope of a curve 曲线斜率
slope of adsorption curve 吸收曲线斜率
slope of a masonry wall 圬工墙坡度
slope of an excavation 开挖边坡
slope of an isobaric surface 等压面坡度
slope of a stack 材堆坡度
slope of a wall 墙的斜率
slope of banned cultivation 禁垦坡度
slope of basin 流域坡降;流域坡度;集水区坡降
slope of bed 垫衬坡度;基座坡度
slope of bottom 底坡(度)
slope of canal 运河坡度
slope of cave floor 洞底坡度
slope of curve 曲线斜率
slope of cut 堑坡边坡
slope of cutting 切割坡度;开挖坡度;挖方坡度
slope of embankment 路堤边坡;堤的坡度;堤坝边坡
slope of end wall of portal 洞门端墙面坡
slope of equal resistance 等阻坡度
slope of equilibrium 平衡坡度;平衡比降
slope of fault scarp 断层崖坡度
slope of front 前沿坡度
slope of front delta 三角洲前缘斜坡
slope of grain 木纹坡度;木纹角度;纹理斜率;纹理斜度
slope of ground 地面坡度
slope of hydraulic grade line 水力坡降;水压线坡度

slope of instantaneous velocity 瞬时速度斜率
slope of isochron 等时线斜率
slope of moving walk 自动步道坡度
slope of repose 休止坡度；休止角；安息角
slope of river 河流比降；河道坡度；河道比降
slope of river bank 河岸坡度
slope of river(bed) 河床坡降；河床比降
slope of roof 屋面坡度；屋顶坡度
slope of roof pitch 屋面坡度角
slope of roof slope 屋面坡度角
slope of sea-surface 海面坡度
slope of seepage line 浸润线坡度
slope of sheathing 望板坡度
slope of sodding 边坡种草
slope of stair(case) 楼梯坡度
slope of stowed material 堆置材料坡度；填充材料坡度
slope of strata dip away from excavation 反倾向边坡
slope of strata dip cross excavation 切层边坡
slope of strata dip toward excavation 顺层边坡
slope of stream 河道比降
slope of surge 涌浪陡度
slope of the ground 地面坡度
slope of the ore body 矿体斜度
slope of the slipway 滑道坡度
slope of the stream bed 河床坡降
slope of thread 螺纹斜度
slope of timber grain 木纹斜度
slope of tunnel face 洞脸边坡
slope of underlying bed rock 下覆基岩面坡度
slope of watershed 流域坡降；流域坡度
slope of water surface 水面坡降；水面比降
slope of wave 波浪陡度
slope on water 水上边坡
slope overload 斜率过载
slope overload error 斜率过载误差
slope overload noise 斜度过载噪声
slope pavement 斜坡护面；斜面铺面；护坡；边坡护面
slope paving 砌坡；斜坡铺砌；斜面铺砌；砌护坡；边坡护面
slope paving at dike root 坝根护坡
slope paving joint sealing compound 斜坡路面填缝混合物
slope pillar 斜井矿柱；倾斜矿柱
slope planting 边坡绿化；边坡造林；坡地绿化；坡地造林；斜坡绿化；斜坡造林；护坡绿化
slope portion 坡道区段；坡道部分；边坡区段；边坡部分；斜坡区段；斜坡部分
slope potentiometer 跨导调整电位器
slope protection 护岸；边坡加固；边坡护坡；边坡防护；边坡保护
slope protection at dike root 坝根护坡
sloper 斜掘机；整坡机；铲坡机
slope railless driving 斜坡道无轨掘进
slope ramp 引道；接坡；斜坡道；坡道
slope rank 倾斜分级
slope ratio 斜率比；坡率比；坡度比
slope ratio assay 斜率比测定
slope ratio method 斜率比法
slope readjustment 边坡再调整
slope recession 坡后退
slope rectification 斜度调整；陡度调整
slope regulation 斜率调整；斜调
slope remote measurement device 边坡遥测仪
slope removal 斜坡排除；斜坡移动；斜向排除；斜向移动
slope resistance 坡度阻力；坡道阻力；动态阳极电阻
slope retreat 坡后退
slope revetment 边坡护坡；边坡护墙；斜坡护坡；斜坡护墙；边坡护面
slope roller 斜向轧辊；斜坡压辊
slope runnel 偏压隧道
slope runoff 坡面径流
slope scale 比降比尺
slope scale ratio（天然的和模拟的）表面坡度比；坡降比尺
slope scraper 斜面刮土铲运机
slope section 斜坡区段；斜坡道部分；边坡区段；边坡区段；斜坡部分
slope sensitivity 斜率灵敏度；斜度灵敏度；倾斜灵敏度
slope sett 斜坡铺路石块

slope sett paving 边坡块石铺砌；边坡块石铺面；斜坡块石铺砌；斜坡块石铺面
slope shaft 斜井
slope shape 坡形；边坡形式
slope shaping and compaction machine 边坡修整压实机；边坡成型压实机
slope sign 坡度标
slope signal 斜率信号
slope sinking 斜井开挖；斜井掘进
slope-sliding theory 滑坡理论
slope snow 半融雪
slope sodding 边坡植草坪；斜坡植草坪
slope span 斜跨度
slope spring 坡地泉；边坡泉
slope stability 斜坡稳定性；坡面稳定性；深层活动稳定性；边坡稳定(性)
slope stabilization 边坡加固；边坡稳定
slope stabilization by seeding 用种草来稳定边坡
slope-stage-discharge relation 比降—水位—流量关系
slope stake 坡度标桩；边桩；边坡桩；坡度桩
slope stake location 边坡桩测设
slope staking 桩定坡面点；桩定坡面【测】
slope steepness 边坡陡度
slopes test 斜度测定
slopes tolerance 斜率容限
slope storage 斜库容；动库容
slope stretcher 喇叭形张拉器
slope-supply flow 坡面补给流
slope surface drainage 坡面排水
slope surface protection 坡面防护
slopes wall 斜井井壁
slope switch 偏移开关
slope tamping 边坡夯实
slope tape 斜度卷尺
slope taping 斜距丈量
slope tatus 边坡堆积
slope terracing 坡地梯田化
slope theodolite 坡面经纬仪
slope toe 坡底
slope top 坡顶
slope transit 坡面经纬仪
slope transmission 斜发送
slope trimmer 斜坡修整机；斜坡平整机；整坡机
slope trimming 斜坡修整；边坡修整
slope trimming machine 边坡整修机；斜坡整修机
slope tunnel 偏压隧道
slope unit 坡度单位
slope up 递增
slope-value method 坡度表示法
slope variance 坡度变异；坡度变化
slope velocity formula 比降流速公式；比降测流公式
slope wall 护坡墙
slope wall angle of open pit 露天矿边坡角；露采场边坡角
slope wash 斜坡侵蚀；斜坡冲刷；坡面冲刷；坡积物；坡积土；坡地冲刷；坡底冲刷
slope-wash iron deposit 坡积铁矿床
slope wash soil 坡积土；堆积土
slope water 陆坡水
slope-water gyral 陆坡海水环流
slopeway 坡道
slope weighting 边坡压重
slope weir 斜坡式堰
slope work 边坡施工；斜坡施工
slope zone 斜坡带
sloping 倾斜的；成斜坡
sloping adit 倾斜的坑道入口
sloping apron 斜坡护墙；斜坡护底；斜坡海漫闸；斜坡海漫坝；斜坡(式)护坦；斜(面)护坦
sloping aquifer 倾斜含水层
sloping arch 斜礅；斜拱
sloping area 边坡面积
sloping awning 斜张天幕
sloping baffle 曲线挡板；倾斜挡板
sloping bank 斜岸；坡岸
sloping beach 倾斜海滩
sloping beam 斜梁
sloping bed 斜底；倾斜基床
sloping bench 斜台
sloping berth with cable railway 缆车码头
sloping block 斜块（直立防波堤用）
sloping blockwork 斜方块砌体
sloping bond 斜砌砖工；斜砌筑；斜砌(法)；斜面

铺砌；斜层砌合；方块斜层砌合
sloping bottom 斜底座；斜底；倾斜炉底
sloping-bottom bin 斜底储存器；斜底料斗
sloping bottom tank 斜底罐
sloping breakwater 斜坡式防波堤
sloping bulkhead 斜舱壁；倾斜舱壁
sloping cam plate 斜盘
sloping casting 倾斜铸造
sloping catalyst line 倾斜催化剂管道
sloping concrete 斜坡混凝土
sloping control panel 斜面控制盘
sloping convection 斜坡对流
sloping core 斜心墙
sloping course 倾斜层
sloping crack 斜向开裂
sloping curb 斜式缘石；斜式路缘(石)；斜路缘；斜口侧石
sloping desk 倾斜台；倾斜面板
sloping direction 倾斜方向
sloping double plane transverse cable arrangement of cable-stayed bridge 斜拉桥斜双索面横向布置
sloping edge 斜楔
sloping face 斜面
sloping faced breakwater 斜坡护面防波堤；斜坡式防波堤
sloping faced wharf 斜坡码头
sloping field 坡田；坡地
sloping floor 斜地坪；倾斜基床；倾斜地面
sloping floor parking garage 坡道式车库
sloping folding 倾斜褶皱的
sloping footing 斜坡式基础
sloping foundation 斜坡式基础；斜面地基
sloping front boundary 斜前沿边缘
sloping gallery 扒山廊
sloping ga(u)ge 斜向定位装置；斜坡定位装置；倾斜水尺；斜坡水尺；斜度量规；曲线标尺；倾斜标尺
sloping glazing 倾斜玻璃窗
sloping grain 斜木纹
sloping grizzly 斜置栅
sloping ground 斜面地基；倾斜地面；斜坡(地)
sloping hearth 斜炉底
sloping hinge 斜铰接接头
sloping joint 倾斜接缝
sloping keyboard 斜面键盘
sloping land 斜坡地
sloping layer 倾斜层；斜砌层
sloping leg 斜腿（井架的）
sloping-leg bent 梯形排架
sloping light weight roof 斜面轻型屋顶
sloping lobe 鹿角形柱
sloping loop channel 倾斜式活套槽
sloping louver of the hood 发动机罩斜放气孔
sloping masonry wall 斜面圬工墙
sloping member 斜式构件；倾斜构件
sloping mound breakwater 斜坡式防波堤
sloping newspaper rack 斜面报架
sloping of river bank 河岸边坡放缓
sloping outward frame 外斜肋骨
sloping paleoplain 倾斜古平原
sloping panel 斜底板；斜板；倾斜板
sloping parabolic arch 斜式抛物线拱
sloping pipe settling tank 斜管沉淀池
sloping plank-settling 斜板沉淀
sloping plank-settling tank 斜板沉淀池
sloping plank tube settling tank 斜板管沉淀池
sloping plate 倾斜装甲板
sloping platform 斜台
sloping platform drier 倾斜台式干燥机
sloping portion 倾斜面
sloping quay 斜坡(式)码头
sloping rack 倾斜的支架；倾斜的导轨
sloping radiator 斜散热器
sloping radiator grille 斜散热器护栅；倾斜的散热器护栅
sloping reinforcement 斜(向)钢筋；斜向配筋
sloping roof 斜屋面；斜(坡)屋顶；斜面屋顶；坡屋顶
sloping roof furnace 斜顶炉
sloping roof skylight 斜屋顶天窗；斜层顶天窗
sloping screed 斜面刮板
sloping section 斜削面；斜截面；斜切面
sloping shaft 斜轴；斜井；倾斜轴
sloping-shaft drive 斜轴传动

sloping shelf 斜面书架
sloping site 斜面地基;倾斜地段;斜坡地基;斜地盘
sloping soffit of stair(case) 斜的楼梯背面
sloping steel 斜钢筋
sloping surface 斜面表面;坡面
sloping surge chamber 倾斜式调压室
sloping top chord 斜向上弦杆
sloping topography 倾斜地形
sloping top panel 斜顶板
sloping-trench landfill 斜沟掩埋
sloping trough 斜槽
sloping type 斜体字
sloping valve 斜置阀
sloping wall type breakwater 斜墙式防波堤
sloping wave 缓波
sloping way 斜坡道
sloping wedge 斜劈
sloping wharf 斜坡(式)码头
sloping wharf with cable way 缆车码头
sloping windshield 斜挡风玻璃
sloping with grader 用平路机修筑斜坡
slop line 废线;不合格管路
slop mo(u)lded brick 湿模制砖
slop mo(u)lding 湿模制砖;软泥制坯法;手工塑性制砖法
slop mo(u)lding brick 湿模制砖
slop oil 废油;不合格油料;不合格石油产品
slop pail 污水桶
slopping floor 倾斜地板
slopping name 雁列注记
slopping wave 缓浪
sloppy 湿透的;潮湿的
sloppy concrete 稀混凝土;流质混凝土
sloppy condition 稀薄状态
sloppy heat 冷熔
sloppy mix 稀混凝土
slop room 被服储藏室
slops 泔水
slops chute 污水道
slops collecting ship 粪便污水收集船
slops disposal procedures 污液处理程序
slop shop 廉价成衣商店;现成服装店;买现成服装的商店
slop sink 污水盆;污水槽
slop survey 采场测量
slop tank 污油水舱;污水槽;沉淀油柜;不合格油料储罐
slop wasp facies 坡积相
slop water 洗涤盆污水(厨房);污水
slop wax 粗蜡
slop weight 每ого脱浆料的重量
slosed wrench 梅花双头扳手
sloshing 摇动拍溅;水平晃动
sloshing of liquid 液体溅泼
slosh test 燃料激荡试验
slot 小片;箱位;狭孔;狭槽;隙缝;砖缝;窄口;窄槽;(物体表面的)沟槽;开长口;时间段;存储槽;齿缝;长孔;插槽;槽形地带;槽位;榫槽;槽;立刨;开缝
slot anchor 隙缝式锚固
slot and crank 带连杆的曲柄
slot-and-crank drive 曲柄连杆传动
slot and keyway milling machine 开缝键槽铣床
slot-and-wedge bolt 楔缝式杆柱;楔缝式锚杆
slot-and-wedge fastening 楔缝式锚杆支护
slot-and-wedge type 楔缝式
slot antenna 隙缝天线;裂缝天线;槽缝天线;缝隙天线
slot applicator 槽缝式施加器
slot area 齿缝面积
slot array 隙缝天线阵;裂缝天线阵
slot atomizer 缝缝式喷油嘴
slot bar 槽内导条;槽部线棒
slot base 槽底
slot borer 槽钻
slot bound 插孔限制
slot bunker 槽式卸料口料仓
slot burner 扁焰喷烧器
slot charter 箱位租赁权(集装箱)
slot coil 槽线圈
slot combination 槽配合
slot conduit 开缝导管
slot coordination 槽配合
slot-coupled cavity 槽隙耦合空腔谐振器

slot coupling 槽隙耦合
slot cross-section 槽截面
slot cutter 模片切缝器
slot-cutting machine 切槽机
slot damper 槽隙阻尼器
slot depth 槽深
slot diffuser 条形散流器;条缝形风口
slot dipole 槽馈偶极天线
slot discharge 槽形卸料口卸料;槽部放电
slot distributor 缝隙布料器
slot dozing 开缝推土法
slot drill 槽钻
slot driller 开缝钻头;开缝钻
slote 舞台布景升降机
slot effect 齿槽效应;槽效应
slot exhaust hood 槽边排风罩
slot extrusion 缝口模头挤塑
slot factor 线槽系数;筛孔缝隙系数
slot-fed dipole 槽馈偶极子
slot feeder 槽式给矿机
slot field 槽漏磁场;槽磁场
slot frequency 信道间插入频率
slot gate 长缝浇口;缝隙浇口
slot grinder 槽磨机
slot grit chamber 隙缝沉砂室
slot group 时间段组
slot guide 导向槽架
slot-guide intersector 槽缝式旋转针梳机
slot head 槽形螺钉头;一字槽螺丝帽
slot impedance 槽阻抗
slot in frames 框架中的槽
slot inserter 嵌线器;嵌线机
slot insulating liner 槽绝缘衬
slot insulation 槽(隙)绝缘
slot joint 狭缝(连)接;开口接合
slot leakage 隙缝泄漏;槽漏泄;槽漏磁;槽壁间漏磁
slot leakage flux 槽漏磁通
slot leakage reactance 槽漏电抗
slot length and width of bubble cap 泡罩齿缝长度与宽度
slotless 无槽
slot liner 槽绝缘衬;槽衬
slot lips 砖边
slot machine 硬币自动机;自动售货机;投币式自动售货机
slot magnetron 槽缝(型)磁控管
slot mesh 长眼;长缝孔筛
slot mesh plate 长眼(圆孔)筛板
slot mesh screen 长眼(圆孔)筛板;长腿筛板;长方孔筛
slot meter 预付款煤气表;投币式煤气表;投币式电度表
slot mill 槽铣刀
slot miller 键槽铣刀
slot milling cutter 槽铣刀
slot mortise 滑动榫接;狭槽榫接;榫槽
slot mortising bit 槽钻
slot mortising machine 槽榫机
slot name 空位名;槽名
slot nozzle 缝式喷嘴
slot number 箱位号(集装箱);列号;集装箱船之箱位号;时间段号
slot opening 槽缝开度;槽口宽
slot outlet 窄风道口;条形送风口;条缝形风口
slot part 槽部
slot pattern 隙缝式模型
slot permeance 槽(部)磁导
slot piece 有槽拼模块
slot pin 带槽销钉
slot pitch 线槽节距;槽距;槽节距
slot pulsation 槽隙脉动
slot radiator 隙缝辐射器;缝隙辐射器
slot reactance 槽隙电抗;槽电抗
slot-rectifier 缝隙纠正仪
slot ripple 线槽脉动;槽脉动
slot screw 狭缝装置螺钉
slot screwing 槽眼螺钉连接;螺丝长孔接合;长孔拧接
slot scrubber 百叶窗式除尘器
slots exchange 舱位互换
slot sieve 长缝筛
slot signal(l)ing 复归信号器
slot sir diffuser 隙缝式空气扩散器
slot size 开口尺寸;槽缝尺寸;槽距

slot slope 槽斜度
slot sorting 时间段排序
slot space 槽空间
slot space-factor 线槽占空系数
slot squeegee 槽缝式刮浆板
slotted 开缝
slotted adjustment plate 槽形调整板
slotted aerofoil 开缝机翼
slotted angle 开槽角钢
slotted antenna array 隙缝天线阵;裂缝天线阵
slotted aperture 槽孔
slotted bit 开缝钻头
slotted blade 开缝叶片
slotted-blade axial-flow blower 开缝叶片轴流风机
slotted-blade blower 刮板式鼓风机
slotted breakwater 非密封式防波堤;篱笆式防波堤
slotted bucket 齿槽式挑流鼻坎
slotted bucket lip (溢洪道的)槽式鼻坎
slotted burner 缝式燃烧器
slotted casing 开槽套筒
slotted channel piston ring 槽沟活塞环
slotted circular nut 槽顶螺母;槽顶螺帽
slotted core 有槽铁芯
slotted core cable 骨架式光缆;开槽式芯光缆
slotted crank 槽缝曲柄
slotted diaphragm 带篦缝的隔仓板;缝隙光闸;篦板
slotted disc 带槽圆盘
slotted drum 开槽筒
slotted eccentric bit 开缝偏心式钻头
slotted edge of cap 帽罩齿形边缘
slotted eye 椭圆形孔;长圆形孔眼;长圆孔;槽孔
slotted flap 开缝襟翼
slotted floor 条缝地板
slotted-form winding 分段绕组
slotted gravity dam 宽缝重力坝
slotted guide 罐道滑槽
slotted gypsum acoustic(al) plasterboard 开缝的石膏吸声粉刷板
slotted head 开槽头
slotted headless set screw 带槽无头止动螺钉
slotted head screw 一字(槽)螺钉;槽头螺钉
slotted hexagon head screw 有槽六角头螺钉
slotted hole 长圆孔;长孔;槽孔
slotted hole screen 长方孔筛
slotted injector 缝隙式喷嘴
slotted jaw 开口接平
slotted joint 狭槽接合
slotted joint rod 开口接合杆
slotted lap weld 切匙搭接焊
slotted line 开槽段;开槽测试线
slotted liner 开槽衬套;开槽衬垫;割缝衬管;缝隙过滤管
slotted link 槽孔链节
slotted link motion 槽孔链节运动
slotted mounting screw 槽头固定螺钉
slotted nozzle 开缝喷嘴;开缝喷管;槽式喷嘴
slotted nut 六角形螺帽;有槽螺母;开槽螺母;开槽螺帽
slotted opening 槽孔
slotted outlet 条缝出风口
slotted oval countersunk head screw 带槽半沉头螺钉
slotted partition 井字格盒;带篦缝的隔仓板;篦板
slotted pile 槽口桩
slotted pin 接合销;开缝销;开槽销;接合栓销
slotted pipe 条缝滤水管;隙缝滤水管;有缝钢管
slotted plate 有长槽的板;切口板;带槽板;长眼(圆孔)筛板;长方眼筛板;长槽板
slotted rail 有缝钢轨
slotted recess 槽
slotted ridge guide 开槽脊形波导
slotted roller bucket 差动式鼻坎;槽齿式坎
slotted rotor 开槽转子
slotted round nut 有槽圆头螺母
slotted screen 长圆孔筛;长眼(圆孔)筛;长孔筛;长方孔筛;缝筛
slotted screwdriver with wood(en) handle 木柄一字螺丝刀
slotted screw plug 凹槽螺塞
slotted section 开槽段
slotted shutter 缝隙快门
slotted slab 开槽板
slotted spillway bucket 开槽溢洪道消力戽;开槽溢洪道反弧段;开槽溢洪道挑流鼻坎;齿槽式挑

流鼻坎
slotted spring-finger plunger 带有弹簧销的槽式活塞
slotted stator 开槽定子
slotted steel profile 开槽型材;开槽钢材
slotted strainer 篦式过滤器
slotted template 槽孔模片板
slotted templet 切缝模片
slotted templet cutter 模片切缝器
slotted templet plot 缝隙模片辐射三角测量
slotted templet set 全套切缝模片
slotted tie plate 带槽孔轨枕垫板
slotted top screw 槽顶螺钉
slotted tube 开槽管
slotted tube well 开缝管井
slotted tunnel 开缝风洞
slotted vane 槽片
slotted wall 开长槽墙;穿孔墙
slotted washer 开口垫圈;开缝垫圈
slotted waveguide 开槽段;开槽波导(管)
slotted waveguide array 裂缝波导天线阵
slotted wavescreen 带槽孔挡浪板
slotter 铣槽机;刨床;立刨床;插床
slot tie 槽缝拉杆
slotting 立刨;开缝;刨刨;开槽;打孔;冲孔;插削
slotting attachment 铣槽装置;铣槽附件;插削装置
slotting auger 榫孔钻;长孔钻
slotting cutter 切口铣刀;插齿刀
slotting end mill 槽立铣刀
slotting file 开槽锉
slotting machine 立(式)刨床;刨床;开槽机;插床
slotting machine ram 插床冲头
slotting plane 裁口刨
slotting saw sharpening machine 开槽锯刀磨床
slotting saw with coarse teeth 粗齿锯片铣刀
slotting tool 开槽工具;插刀
slotting wheel 刻花砂轮;开槽用砂轮
slot-type 骨架式
slot-type air diffuser 条缝式散流器
slot-type bed load sampler 槽孔式推移质采样器
slot-type belt feeder 槽形带式喂料机
slot-type furnace 缝式炉
slot-type resonator 槽型谐振器
slot-type sampler 槽孔式采样器
slot value 空位值
slot wall 开缝墙;槽壁
slot wave 槽波
slot way 槽路
slot wedge 槽楔
slot weld 长孔焊;槽塞焊;槽焊;开槽焊
slot welding 开槽搭焊缝;切口焊缝;槽焊缝;槽焊接
slot width 齿缝宽度;槽宽;缝隙宽度
slot winding 槽(部)绕组
slough 滑坍;泥沼;泥坑;磨料制备槽;脱落;坍落物;湿地;倒套
sloughed filling 脱纬
sloughed-off weft 脱纬
sloughing 黏膜脱落;泥化;坍陷;滑坍
sloughing bank 坍塌堤岸;坍岸
sloughing in earth cut 路堑边坡滑坍
sloughing of a cutting 开挖坍方
sloughing off 落下
sloughing-off box 平流 V 形稀分级机
sloughing-off cone 排除细泥的大型圆锥分级机;脱泥圆锥分级机
sloughing-off tank 脱泥槽
sloughing of levee 堤防坍塌
sloughing of sand (钻孔内的)砂土崩落
slough off 脱纬;剥落
slough podzol 沼泽灰壤
sloughy 泥沼的;沼泽化的;泥泞的
sloughy channel 泥沼河道
slow 慢速的;缓慢的
slow accelerating wave 慢加速波
slow access 慢存取
slow access memory 慢速(存取)存储器
slow accumulation 慢积集
slow acting 慢转的;慢动的;延迟动作
slow-acting accelerator 缓动加速器
slow-acting cam 慢动作凸轮
slow-acting insecticide 缓效杀虫剂
slow-acting pesticide 缓效农药
slow-acting press 慢动压机;慢电压机
slow-acting relay 慢动作继电器;缓放继电器;缓动继电器
slow action 缓动
slow addressing 缓慢寻址
slow adjustment 慢速调节
slow aging 慢老化
slow ahead 慢速正车;前进一【船舶】
slow ahead ease her ahead 低速前进
slow angular variable 慢角变量
slow application 慢制动
slow assets 呆滞资产;变现慢的资产
slow astern 低速后退;慢速倒车
slow axis 慢轴;慢速轴
slow axis of a birefringment plate 双折射板的慢轴
slow banding 慢裂的
slow banker 绞车自动减速器
slow banking(method) 慢堆填法;慢速筑堤法;分层填筑(法)
slow bead test 纵向珠焊试验
slow bend test 慢弯试验
slow-blow fuse 缓动式熔断器
slow-breaking 慢裂
slow-breaking emulsified asphalt 慢裂乳化(地)沥青
slow-breaking emulsion 缓慢破坏的乳胶体;慢裂乳液;慢解乳油
slow-break switch 缓动断路器
slow-burning 慢燃的;缓燃性;阻燃的;耐火材的
slow-burning construction 难燃建筑;耐燃建筑(物);耐燃构造;耐火建筑;慢燃结构物;慢燃构造;缓燃构造
slow-burning film 不易燃胶片
slow-burning fuse 缓燃引信;暖燃导火线
slow-burning heavy timber construction 缓燃重型木结构;缓燃大木结构
slow-burning insulation 耐燃绝缘;缓燃隔热材料
slow-burning mill construction 缓燃工厂建筑;耐燃工厂建筑
slow-burning plastic 缓燃塑料
slow-burning powder 慢燃药
slow-burning smokeless powder 慢燃无烟火药
slow-burning structure 耐燃建筑(物);慢燃结构物;慢燃构造
slow-burning timber construction 缓燃木建筑;耐燃木建筑
slow-burning wire 慢燃线
slow-butt welding 电阻压力对焊
slow cement 慢凝水泥;慢干水泥;缓凝水泥
slow change 徐变
slow channel 慢通道;低速通道
slow chemisorption 慢(速)化学吸附
slow chopper 慢选择转子;慢断路器
slow chopper neutron spectrometer 慢转中子谱仪
slow closing faucet 慢闭活门
slow closing valve 慢速隔离阀
slow-closure type valve 缓闭阀
slow coagulation 慢絮凝;缓慢凝结
slow coding 慢电码;缓电码;低速电码
slow combination 慢结合
slow combustion 慢燃烧;缓燃烧
slow combustion stove 缓燃(火)炉
slow component 慢部分
slow condensation 慢凝集
slow cooking 慢速蒸煮
slow cooling 慢冷;缓慢冷却
slow cooling box 缓冷箱
slow cooling method 慢冷法
slow cool section 缓冷区段
slow counter 慢作用计数管;慢(速)计数管
slow counting rate 慢计数率
slow crack growth 裂缝缓慢增长
slow creeping 慢蠕变
slow crustal movement 缓慢运动
slow-curing 慢凝的
slow-curing asphalt 慢凝沥青;缓凝(地)沥青
slow-curing cutback asphalt 慢凝轻制(地)沥青;慢凝稀释地沥青;慢凝控制(度)沥青;慢干稀释沥青
slow-curing liquid asphalt 慢凝液体(地)沥青
slow-curing liquid asphaltic material 缓凝液态(地)沥青材料;慢凝液体(地)沥青材料;慢干液体沥青材料
slow-curing oil 慢干沥青
slow-curing paving binder 慢干的液体道路沥青
slow-curing resin 慢固化树脂;慢干树脂;慢焙固
动继电器
slow cut solenoid 低速燃油切断装置;低速断油电磁阀
slow cutting 慢切射
slow depressing region 缓慢下降区
slow development 慢(速)显影
slow device 慢速器件
slow diffusant 慢扩散源
slow direct shear test 慢剪直剪试验
slow discharge 缓慢放电
slow dispersion 慢弥散
slow down 减慢;减缓;降低速度;慢化
slow-down characteristic 降速特性
slow down signal 慢行信号
slow down the rate of resource degradation 降低资源消耗率
slow draining film 慢排液膜
slow drift 慢漂移
slow drift burst 慢漂爆(发)
slow drive signal 慢行信号
slow drive zone 慢行区域
slow drying 缓慢干燥;干燥缓慢
slow drying ink 慢干油墨
slow earth flow 缓慢泥流
slow earthquake 缓慢地震
slow echo wave 慢回波
slowed sweep oscilloscope 慢扫描示波器
slow-effect fertilizer 迟效肥料
slow electron bombardment 慢电子轰击
slow emulsion 慢感光层;低感光度
slower-down 减速剂
slower-moving track 慢转履带
slower ray 慢光
slow-evapo(u)rating diluent 慢挥发稀释剂
slow-evapo(u)rating solvent 慢蒸发性溶剂;慢挥发溶剂
slow exchange 慢交换
slow execution 慢(速)执行
slow extracted beam 慢引出束
slow extraction 慢引出
slow extraction system 慢引出系统
slow extractor 慢引出装置
slow fading 慢衰落
slow fading margin 慢衰落余量
slow fast reactor 慢快中子双区反应堆
slow feed 慢给进
slow feed gear 慢速给进齿轮
slow filling period 缓慢充盈期
slow filter 慢(过)滤器;慢滤池
slow filtration 慢速过滤法;慢滤(法)
slow fire 慢射
slow fission 慢裂变
slow fission reaction 慢裂变反应
slow flame 蔓延火焰
slow flowage 缓流动
slow-flowing river 缓流河
slow-flowing stream 缓慢水流;缓流河
slow flow rate paper 慢流速纸
slow fluid flow 缓慢流
slow freezing 慢速冻结
slow freight train 普通货物列车
slow gear 慢速齿轮
slow Geiger-Muller tube 慢盖格米勒氏管
slow genlock 慢锁定;慢台从锁相
slow goods train 普通货物列车
slow-grown 窄年轮的;缓和生长
slow-grown timber 密纹木(材)
slow-grown wood 慢长木材
slow hardening 慢硬的;慢速硬化;慢凝;缓硬;缓慢硬化
slow-hardening cement 慢硬水泥;缓硬水泥
slow-hardening concrete 缓凝混凝土;慢硬混凝土
slow-hardening Portland cement 缓硬硅酸盐水泥;缓硬波特兰水泥
slow heating technique 慢加热法
slow hole 慢空穴
slow igniter cord 缓燃导火线
slow ignition 慢速点燃
slow infiltration 慢速渗流
slowing 慢化
slowing down 慢化
slowing-down area 慢化面积
slowing-down brake 减速制动器;缓降器
slowing-down chute 缓降溜槽

slowing-down cross-section 慢化截面
slowing-down density 减速密度;慢化密度
slowing-down energy distribution 慢化能量分布
slowing-down group 慢化组
slowing-down integral equation 慢化积分方程
slowing-down kernel 慢化核
slowing-down length 减速长度;慢化长度
slowing-down model 慢化模型
slowing-down moment 慢化矩
slowing-down nonleakage probability 慢化不泄漏几率
slowing-down power 慢化能力;慢化本领
slowing-down radiation 慢化辐射
slowing-down region 慢化区
slowing-down spectrometer 慢化能谱仪
slowing-down theory 慢化理论
slowing-down time spectrometry 慢化时间测谱学
slowing factor 减慢因子
slowing material 慢化材料
slowing property 慢化性能
slowing time 慢速运动时间
slow interrupter 缓动断续器
slow ion 慢离子;大离子
slow jet 慢速喷嘴
slow land treatment system 慢速渗滤系统
slow lane 慢车道
slow leakage 慢漏失
slow lens 小孔径透镜;慢速透镜
slow line 慢车道;慢车线
slow loading 缓慢加荷
slowly-accumulated dose 慢累积剂量
slowly applied load 慢加荷载
slowly changing field 慢变化场
slowly cooled clinker 慢冷熟料
slowly decreasing function 慢递减函数
slowly developing disasters 正在慢性恶化的灾difficulty
slowly drained soil 潮湿土壤
slowly inclined occurrence 缓倾斜
slowly soak out 缓慢浸出
slowly soluble material 慢溶物质;弱溶解物质
slowly taking cement 慢凝水泥
slowly variable flow 缓变流
slowly varied flow 渐变流;缓变流
slowly varying component 缓变分量
slowly varying function 慢化函数;慢变函数
slowly varying system 慢变系统
slow marks 不灵敏叠片
slow match 慢燃引信头;慢燃导火线;慢凝导火线;缓燃引信;缓燃导火线
slow memory 慢速存储器;低速存储器
slow mode wave 慢模式波;慢波
slow motion 慢镜头;慢动(作);微动
slow-motion camera 慢速摄影机;慢速摄像机
slow-motion clamp 慢动夹
slow-motion control 微动控制
slow-motion dial 慢动度盘
slow-motion effect 慢动作效果
slow-motion hand wheel 精确调整用手轮
slow-motion hoist 慢速卷扬机
slow-motion screw 微动螺旋;微动螺丝
slow-motion starter 慢动作起动器
slow moving depression 慢移低气压
slow moving traffic 慢行交通;慢车行车道
slow moving vehicle 慢行汽车;慢行车辆
slow negative film 慢速负片
slowness 迟缓
slowness component 慢度分量
slowness diagram 缓慢图
slowness gradient 慢度梯度
slowness-time domain 慢度时间域
slowness vector 缓慢矢量
slow neutron 慢中子;热中子
slow neutron chopper 慢中子选择器
slow neutron counter 慢中子计数器
slow neutron detecting chamber 慢中子探测室
slow neutron detector 慢(速)中子探测器
slow neutron exposure chamber 慢中子照射室
slow neutron filter 慢中子(过)滤器
slow neutron fission 慢中子裂变
slow neutron interferometer 慢中子干涉仪
slow operate fast-release relay 慢吸快释继电器
slow-operating 延迟动作;缓动的
slow-operating relay 缓动继电器
slow operation 慢动作

slow order 限速;慢行规定
slow oscillation 慢振荡
slow oxidation 慢性氧化;缓慢氧化
slow pack time 缓装时间
slow passenger train 普通旅客列车
slow penetration test 慢渗入试验
slow permeability 慢渗透(性)
slow permutation 慢排列
slow phase 慢相
slow pickup area 缓行区
slow pickup relay 缓吸继电器
slow pickup time 缓吸时间
slow plate 低灵敏度底片
slow powder 缓燃炸药;火药;缓爆炸药
slow process 慢过程
slow progress 慢进度
slow pulsar 慢脉冲星
slow rake-stirrer 慢速耙式搅拌器
slow-rate infiltration-land treatment system 慢速渗滤土地处理系统
slow rate of curvature 缓和曲率变化率
slow ray 慢光
slow reacting substance 慢反应物质;缓慢反应物质;迟缓反应物质
slow reaction substance 慢反应物质
slow reactive substance 慢反应物质
slow reactor 慢中子反应堆
slow read-out 慢读出
slow regeneration 慢速重放
slow registration 计数偏慢
slow relay 慢速继电器
slow release 慢释;缓慢释放;缓释的(指继电器)
slow release action 缓释作用;缓释动作
slow release formulation 缓释剂型;缓释成型
slow release method 缓释方法
slow-released part 缓释部分
slow release pesticide 缓释农药
slow release pesticide system 缓释农药系统
slow release rate 缓释速率;缓慢释放速度
slow release relay 缓放时间;缓释断电器
slow release technique 缓释技术;缓释方法
slow-releasing relay 缓释断电器
slow relief relay 缓动交替继电器
slow response 慢响应;反应慢
slow response action potential 慢反应动作电位
slow retardation axis 慢延迟轴
slow-roll stabilization 转动稳定(作用)
slow rotator 慢转天体
slow-running 慢怠速;慢速运行;低速转动
slow-running control 空转控制
slow-running cut-out 断油开关
slow-running jet 低速喷口
slow-running jet tube 低速喷管
slow-running mill 低速磨机
slow-running needle 低速调整针
slow-running outlet 慢车出油口
slow sand filter 慢砂滤器;慢(砂)滤池
slow sand filter bed 慢速砂滤床
slow sand filtration 慢(速)砂滤
slow scan(ning)慢扫描
slow scanning rates 慢扫描速度
slow scan sync 慢扫描同步
slow scan sync detection 慢扫描同步检测
slow screen 长余辉荧光屏
slow season 淡季
slow section 慢行地段
slow setback 缓慢逆转
slow-setting 慢凝(的);缓凝
slow-set(ting)cement 缓凝(注)水泥;慢凝水泥
slow-setting concrete 慢凝混凝土
slow-setting emulsified asphalt 慢裂乳化(地)沥青;慢凝乳化沥青
slow-setting emulsion 慢凝乳液;慢裂乳液
slow-setting glass 慢凝玻璃;长性玻璃;慢硬玻璃
slow-setting ink 慢干油墨
slow-setting mortar 慢凝灰浆
slow-setting oil-well cement 缓凝油井水泥
slow-settling particles 慢沉颗粒
slow shear 慢剪
slow shear test 慢剪试验
slow sign 慢行标志;缓行标志
slow slaking lime 慢(消)化石灰;缓消石灰;缓熟石灰
slow solvent 蒸发速率低的溶剂;蒸发慢的溶剂

slows orchard sprinkler 果园用低速喷灌机
slow-speed 低速(的);慢转(的);缓速
slow-speed bridge crane 慢速桥式起重机
slow-speed coal plough 慢速刨煤机
slow-speed device 慢门机
slow-speed diesel 低速柴油机
slow-speed digital system 低速数字系统
slow-speed direct recording tape recorder 慢速直接记录式纸带记录器
slow-speed drive 慢速传动
slow-speed engine 低速发动机
slow-speed governor 低速调节器
slow-speed high-torque converter 低速高转矩转换器
slow-speed high-torque drive 低速高转矩传动;低速高转矩驱动
slow-speed magnet action 电磁铁缓吸作用
slow-speed mixer 低速混合器;慢速度混合器
slow-speed motor 低速电动机
slow-speed of revolution 低转速
slow-speed paper tape reader 低速纸带读出器
slow-speed relay 低速继电器
slow-speed rolls 低速辊式破碎机
slow-speed shaft 低速轴
slow-speed signal 慢行信号
slow-speed tape recorder 慢速纸带记录器
slow-speed test 低速试验
slow-speed towing 低速拖动
slow spin 慢旋转;低速螺旋
slow-spiral drill 平螺旋钻
slow splitting 缓慢分岔
slow spontaneous recovery 缓慢自然恢复
slow start 缓慢起动
slow sticking 慢粘接
slow stirrer 慢速搅拌器
slow stirring arrangement 缓慢搅拌设备
slow stoppage 慢行停车
slow storage 慢速存储器;低速存储器
slow store 慢速存储器
slow surface state 慢态;慢表面态
slow surface wave 慢表面波;缓行表面波
slow sweep 慢速扫描
slow-taking cement 缓凝水泥
slow test 减速试验;慢剪试验
slow time division 低速时间分隔
slow time scale 慢速时间比例;慢时标;慢速标度
slow to absorb water 水分吸收很慢
slow-to-operate relay 慢动作继电器
slow-to-release relay 缓释断电器
slow track 慢行线段
slow traffic 慢速运输;低速交通
slow traffic lane 慢车道
slow train 慢车
slow trapping 慢吸收
slow turning 缓转弯
slow turning control circuit 慢转控制回路
slow type 慢型
slow-type analog computer 低速模拟计算机
slow up 活动减弱;减速
slow uplifting region 缓慢上升区
slow vapo(u)rization 慢挥发
slow vehicle detector 慢车检测器
slow vehicle lane 慢车道;缓速车道
slow vibration direction 慢振方向
slow visual search 慢速图像搜索
slow wave 慢波;缓慢波
slow wave buncher 慢波聚束器
slow wave circuit 慢波线路;低速波电路
slow wave line 慢波线
slow wave pitch 慢波(线)节距
slow wave ratio 慢波比
slow wave structure 慢波结构;波减速结构
slow wave structure to coaxial line coupler 慢波结构同轴线耦合器
slow wave system 慢波系统
slow wave waveguide 慢波导
slow wave wavelength 慢波波长
sloyd knife 木雕刻刀
slub 花式粗节
slucing 泄水
slud 初期冰;薄冰
sludge 淤渣;钻泥;烂泥;矿泥;金属碎渣;泥垢;泥肥;软渣;软泥;污泥;初期海水;腐泥;废泥浆;半溶海水

sludge acclimatization 污泥驯化
sludge accumulation 泥渣堆积;污泥累积;污泥积累
sludge acid 淤渣酸
sludge acid phosphate 淤渣酸磷酸盐
sludge acid separating tank 淤渣流酸分离槽
sludge activation 污泥活化
sludge activation tank 污水活化池
sludge activity 污泥活性
sludge aerobic digestion 污泥好氧消化
sludge anaerobic digestion 污泥厌氧消化;污泥恶氧消化
sludge ash process 污泥灰法
sludge asphalt 淤渣沥青;酸渣(地)沥青
sludge assay 岩粉分析;钻泥分析
sludge band 污泥堆积
sludge bank 淤渣滩;淤泥沉积物;污泥沉积;污泥岸(滩)
sludge barge 运泥驳船;运垃圾驳(船);污泥驳
sludge barged to sea 污泥船运至海(排放)
sludge barging 污泥驳(船装)运
sludge barrel 岩粉管;钻屑收集筒
sludge barrel head 取粉管接头
sludge beach 淤渣海滩
sludge bed 污泥(干燥)床;污泥干燥场;污泥(干化)场
sludge bed filter 污泥床滤池
sludge bed height 污泥床厚度
sludge bed reactor 污泥床反应器
sludge biodegradation 污泥生物降解
sludge bioleaching 污泥生物淋滤
sludge biomass index 污泥生物量指数
sludge blanket 泥渣层;悬浮泥渣层;污泥床;污泥层
sludge blanket clarifier 悬浮澄清池
sludge blanket filter 悬浮澄清池;污泥层滤池
sludge blanket reactor 悬浮泥渣(层)反应器;污泥层反应池
sludge blending 污泥掺合;污泥混合
sludge blending tank 污泥混合池
sludge boat 泥浆船;淤渣船;垃圾船
sludge boil 污泥上涌
sludge bound 岩粉管堵塞的;泥包的
sludge box 岩粉箱
sludge breaker 油泥消除剂
sludge bucket 岩粉管;钻屑收集筒
sludge bulking 污泥膨胀
sludge burial 污泥埋藏(处理)
sludge cake 泥水块;污泥块;污泥饼;松软冰块
sludge cake disposal 污泥泥饼处置
sludge cake production rate 泥饼产率
sludge calorific value 污泥热值
sludge carrier 泥浆运输船
sludge cast 沟槽铸型
sludge-catchment basin 岩粉收集器
sludge centrifuge 淤渣离心脱水机;污渣离心脱水机
sludge chamber 淤渣室;淤渣池;污泥室;污泥池;沉淀室
sludge channel 污泥沟;污泥渠;污泥槽
sludge characteristic curve 污泥特性曲线
sludge characteristics 污泥特性
sludge chemical conditioning 污泥化学调理
sludge chlorine oxidation 污泥加氯氧化
sludge circulation (旋钻技术中的)泥浆循环;污泥循环
sludge circulation clarifier 污泥循环澄清池
sludge circulation pump 污泥循环泵
sludge clarification 泥浆澄清
sludge clogging 污泥堵塞
sludge cock 排污泥龙头
sludge co-composing 污泥混合堆肥
sludge co-incineration 污泥焚烧;污泥合并燃烧
sludge coking plant 淤渣焦化厂
sludge collecting board 集泥板
sludge-collecting machine 刮泥机
sludge-collecting mechanism 刮泥机构
sludge collecting tank 污油柜
sludge collection 岩粉收集
sludge collection pipe 集泥管
sludge collector 刮泥机;集泥器;集泥机;污泥收集器;污泥收集池;污泥沉积室
sludge compartment 污泥(分隔)室
sludge component 污泥成分
sludge composition 污泥组分
sludge compost(ing) 污泥堆肥

sludge composting and fermentation treatment 污泥堆肥发酵处理
sludge concentrated hopper 污泥浓缩斗
sludge concentration 污泥浓缩;污泥浓度;残渣浓度
sludge concentration factor 污泥浓缩系数
sludge concentration tank 污泥浓缩池
sludge concentrator 污泥浓缩器
sludge conditioning 污泥调质;污泥处理;淤渣改质
sludge conditioner 污泥调理剂;污泥调理池
sludge conditioning 污泥改善;污泥调理;污泥调节;污泥处理
sludge conditioning pocket 污泥斗
sludge contact process 污泥接触处理法
sludge content 污泥含量
sludge conveyance 污泥运输
sludge conveyer 污泥输送器
sludge co-pyrolysis 污泥合并高温分解
sludge cultivation 污泥培养
sludge cylinder 污泥量筒
sludge dampler 污泥采样器
sludge decay 污泥衰减
sludge decreasing 污泥压缩
sludge degritting 污泥除粒
sludge density 岩粉密度;污泥密度
sludge density index 污泥浓度指数;污泥密度指数
sludge deposit 淤渣沉积物;污泥沉积;污泥沉淀
sludge deposition 油泥沉积;污泥沉积
sludge determination 淤渣测定;污泥测定
sludge dewatering 污泥脱水
sludge dewatering bed 污泥脱水床
sludge dewatering by centrifuge 污泥离心脱水
sludge dewatering by drying bed 污泥干化床脱水
sludge dewatering by horizontal belt filter 污泥水平带滤机脱水
sludge dewatering by lagooning 污泥池塘堆储[贮]脱水
sludge dewatering by vacuum filter 污泥真空过滤脱水
sludge dewatering equipment 污泥脱水机;脱水机
sludge dewatering machine 污泥脱水机
sludge dewatering potential 污泥脱水势
sludge dewatering system 污泥脱水系统
sludge dewatering system process 污泥脱水系统工艺
sludge digester 污泥消化器;污泥消化池
sludge digestion 污泥消解;污泥消化;污泥菌分解
sludge digestion aerobic 污泥好氧消化
sludge digestion chamber 污泥消化室
sludge digestion compartment 污泥消化间
sludge digestion gas 污泥消化气
sludge digestion method 污泥消化法
sludge digestion tank 污泥消解槽;污泥消化池
sludge diluent 污渣稀释剂
sludge discharge 排污泥
sludge discharge by static water head 静水压力排泥
sludge discharge conduit 污泥排放管道
sludge discharge pipe 淤渣排(泥)管
sludge discharge tube 残渣排出口
sludge discharging tee 排泥三通
sludge disinfection 污泥消毒
sludge disintegration 污泥分解作用
sludge-dispersing agent 淤渣分散剂
sludge disposal 污泥处理;污泥处置
sludge disposal plant 污泥处置厂
sludge distillate separator 淤渣蒸馏
sludge distribution 污泥分布
sludge distributor 污泥分布器
sludge dose 污泥投配量
sludge drain 污泥排放
sludge draw-off conduit 排泥管
sludge draw-off pipe 排泥管
sludge draw-off pump 泥浆排放泵;污泥排放泵
sludge draw-off tube 排泥管
sludge dryer 污泥干燥器
sludge dryer bed 污泥干化场
sludge drying 污泥干燥;污泥干化
sludge drying area 污泥干燥场;淤渣干燥场
sludge drying bed 污泥干燥场;污泥干化床
sludge drying bed dewatering 污泥干化床脱水
sludge drying facility 污泥干化装置
sludge dumping 污泥倾卸
sludge dumping area 污泥堆积地区

sludge dumping at sea 污泥海洋倾卸
sludge dumping ground 污泥堆积场
sludge dumping into ocean 污泥海洋弃置
sludge elutriation 污泥净化;污泥淘洗
sludge emulsion 淤渣乳胶
sludge equilibrium pond 污泥平衡池
sludge equipment 污泥设备
sludge evacuation 排泥
sludge examination 污泥试验;污泥检验
sludge excess 过剩污泥;污泥过剩
sludge expansion 污泥膨胀
sludge extract 污泥分离
sludge extraction pipe 排泥管
sludge extraction tube 排泥管
sludge extractor 净化池;污泥分离池
sludge fermentation 污泥发酵(法)
sludge filamentation bulking 污泥丝状菌膨胀
sludge filter 污泥过滤器
sludge filtrate 污泥滤液
sludge filtration 油泥过滤;污浆过滤;污泥过滤
sludge flash combustion 污泥闪燃
sludge flash dryer 污泥快速干燥器
sludge float 污泥浮游物
sludge flo(a)tation thickening 污泥浮选浓缩
sludge floc 污泥絮凝物
sludge floe 松软大冰块;冰花
sludge flow 污泥流
sludge fluidized-bed incineration 污泥流化床焚烧炉
sludge foaming 污泥起泡;污泥起沫
sludge formation 泥质地层;污泥生成;污泥层
sludge free management 污泥免费管理
sludge freezing process 污泥冷冻法
sludge from primary sedimentation tank 初沉污泥
sludge from secondary sedimentation tank 二沉池污泥
sludge gas 沼气;污泥气(体)
sludge gas dome 污泥气顶盖
sludge gas holder 污泥气体储[贮]存罐;沼气储藏罐
sludge gas meter 污泥气量计
sludge gas pipe 污泥气管
sludge gas recirculation 污泥气回流
sludge gas tank 污泥气体储[贮]柜
sludge gas utilization 沼气利用
sludge gate 污泥闸门
sludge granular 污泥颗粒的
sludge granulation 污泥颗粒化
sludge granule 污泥颗粒
sludge gravity thickening 污泥重力浓缩
sludge grinder 污泥粉碎机
sludge grinding 污泥磨碎
sludge growing index 污泥增长指数
sludge growth index 污泥增殖指数;污泥生长指数
sludge handling 泥浆处理;污泥处理
sludge handling process 污泥处理工艺;污泥处理法
sludge hauling cost 污泥运输成本
sludge head tank 泥浆高位槽
sludge heat drying 污泥加热干燥;污泥加热干化
sludge height 污泥堆积高度
sludge holding tank 污泥接受器
sludge hole 岩粉样(品)钻孔;出泥孔
sludge homogeneous basin 污泥均质池
sludge hopper 集泥斗;泥斗;污泥储[贮]斗
sludge humus 污泥腐殖质
sludge ice 松软冰;初凝冰;冰花;半溶冰
sludge impoundment 污泥储留;污泥池
sludge impoundment basin 污泥储留池
sludge incineration 污泥燃烧;污泥焚化
sludge incineration plant 污泥焚化厂
sludge incinerator 污泥焚烧炉;污泥焚化炉;污泥燃烧炉
sludge index 污泥指数
sludge inhibitor 油渣抑制剂;污泥生成抑制剂
sludge ladle 渣包;掏渣勺;盛渣桶
sludge lagoon 淤渣塘;污泥塘;污泥池
sludge lagooning 污泥池塘处理
sludge land application 污泥土地应用
sludge landfill 污泥填埋
sludge landfilling 污泥填地
sludge layer 污泥层
sludgeless oil 无渣油

sludge level 淤渣面;淤渣含量
sludge level detector 泥位指示器;污泥面检测器
sludge level sounder 污泥面探测器
sludge lifting machine 污泥提升机;淤泥提升机
sludge lime stabilization 污泥石灰稳定;污泥石灰处理
sludge line 污泥管线
sludge liquid 污泥液体
sludge liquor 淤泥液;污泥液
sludge load 污泥荷载
sludge loading 污泥(干)负荷
sludge loading rate 污泥负荷比(率)
sludge loading ratio 污泥负荷比;污泥负荷比率
sludge long-term storage 污泥长期储[贮]存
sludge lump 不规则松冰团
sludge man 岩粉样分选工
sludge management 污泥管理
sludge management program(me) 污泥管理计划
sludge manure 污泥肥料
sludge mass concentration 污泥质量浓度
sludge melting process 淤渣熔化法
sludge minimization 污泥最小化
sludge moisture content 污泥水分含量;污泥含水率;污泥含水量
sludge morphology 污泥形态
sludge moving mechanism 推泥机
sludge multi-hearth dryer 污泥多层床干燥器
sludge multiple-hearth incineration 污泥多层床焚化炉
sludge number 淤渣值;污泥值
sludge occurrence 油渣形成
sludge oil 淤渣油
sludge oil tar 淤渣油焦油
sludge ore 矿泥
sludge outlet 滤渣(排)出口;排泥口;污泥出口
sludge oxidation 污泥氧化
sludge oxidation demand 污泥氧化定值;污泥需氧量
sludge oxidation lagoon 污泥氧化塘
sludge oxidation ratio 污泥氧化比
sludge pan 大口软钢坩埚
sludge part 污泥部分
sludge particle 污泥颗粒;淤渣颗粒
sludge pasteurization 污泥灭菌
sludge pipe 集泥管;泥浆管;污泥管
sludge pipeline 污泥管线
sludge pit 污泥坑;泥浆坑;污泥斗
sludge plug 污泥栓塞;淤泥栓塞
sludge plug cock 污泥堵塞
sludge pond 泥浆坑;污泥池
sludge pool 污泥塘;污泥池
sludge press 污泥压滤机
sludge press filter 污泥压滤机;盐泥压滤机
sludge press filtration 污泥压滤
sludge pressing 污泥压榨;污泥压缩;污泥压干
sludge pressure filtration 污泥压滤
sludge pretreatment 污泥预处理
sludge prevention 防淤渣
sludge preventive 淤渣生成抑制剂
sludge processing 污泥生产;污泥加工
sludge produced in activated sludge process 活性污泥法污泥产量
sludge production 污泥产量
sludge product recovery 污泥回收产物
sludge promoter 淤渣生成促进剂;油渣促进剂
sludge proof 污泥防护(设施)
sludge property 污泥特性
sludge pump 淤泥泵;净油机排水泵;汲泥泵;泥砂泵;泥浆泵;排泥泵;污泥泵;电动泥浆泵
sludge pumpage 污泥抽取量
sludge pump house 污泥泵房
sludge pump, load and transfer system 污泥、抽吸、装卸和输送系统
sludge pumping 污泥泵送
sludge pure oxygen aerobic digestion 污泥纯氧好氧消化
sludge putrefaction 污泥腐化法
sludge pyrosis 污泥高温分解
sludge quality 污泥质量
sludge quality assurance regulation 污泥质量保险细则
sludge quantity 污泥量
sludger 挖泥机;去残渣工具;扬砂泵;炮眼刮杓;污泥泵;挖泥筒;出泥筒

sludge rake 污泥耙
sludge ratio friction 污泥比阻
sludge reaeration 污泥再曝气;污泥复氧
sludge reaeration method 污泥复氧法
sludge recalcination 污泥再煅烧
sludge recirculation 污泥循环
sludge recirculation ratio 污泥回流比
sludge reclamation 污泥回收
sludge recycle 污泥循环
sludge recycle flow 污泥循环流
sludge recycle flow rate 污泥回流率
sludge recycle ratio 污泥回流比
sludge reduction 污泥缩变;污泥减量
sludge removal 淤渣清除;污泥去除;污泥清除;污泥排除
sludge removal barge 垃圾驳(船)
sludge remover 淤渣清除器
sludge residence time 污泥停留时间
sludge resource recovery 污泥资源回收
sludge retention time 污泥停留时间
sludge return 污泥回流
sludge return apparatus 污泥回流装置
sludge return rate 污泥回流速率
sludge return ratio 污泥回流比
sludge return system 污泥回流系统
sludge reuse 污泥回用
sludge ripening 污泥熟化
sludge rising 污泥上浮;污泥上翻
sludge rising concentration 污泥上浮浓度
sludge sample 岩粉样品;岩粉试样;污泥样本
sludge sampler 采泥器;采滤器
sludge sampling 选岩粉样(品);取岩粉样(品);污泥采样
sludge saver 岩粉分选器
sludge scraper 集泥机;刮(污)机;污泥刮泥机
sludge sea disposal 污泥投海
sludge sediment 淤泥沉积物
sludge seeding 污泥接种
sludge separation 污泥分离
sludge separator 污泥分离器
sludge settling 淤泥沉降;淤泥沉淀;游泥沉淀;污泥沉降;污泥沉积;污泥沉淀
sludge settling characteristics 污泥沉降特性
sludge settling ratio 污泥沉降比
sludge shredder 污泥破碎机
sludge silt 污泥渣
sludge siphon 污泥虹吸管
sludge site 污泥场
sludge softening 淤渣软化
sludge-soil mixture 污泥—土壤混合物
sludge solid 固态污泥;污泥固体(物)
sludge solid balance 污泥固体平衡
sludge solid concentration 污泥固体浓度
sludge solidify(ing) 污泥固化
sludge solvent 淤渣溶剂
sludge splitter 岩粉试样分样器
sludge spray dryer 污泥喷雾干燥器
sludge spreading 污泥摊铺
sludge stabilization 污泥稳定(化)
sludge stabilization-chemical process 污泥稳定—化学处理法
sludge stabilization device 污泥稳定装置
sludge stabilization process 污泥稳定化处理
sludge stabilization-thermal process 污泥稳定热处理法
sludge sterilization 污泥消毒;污泥灭菌
sludge stirrer 污泥搅拌器
sludge stirring 污泥搅动
sludge storage 污泥储[贮]量;污泥储[贮]存;污泥淤积量
sludge storage compartment 储[贮]泥室
sludge storage tank 盐泥储[贮]罐;储[贮]泥池;泥浆储[贮]槽
sludge storage zone 存泥区
sludge stratified sedimentation velocity 污泥成层沉淀速度
sludge stripping machine 污泥剥离机
sludge structure 污泥结构
sludge structure curve 污泥结构图
sludge sump 集泥井;污泥槽;沉泥槽
sludge synergist 污泥增活剂
sludge tank 泥浆箱;泥浆罐;污泥柜;污泥池
sludge tank activation 污泥池灌气法
sludge tank car 污泥槽车

sludge tanker 污泥船
sludge tank truck 运泥车
sludge tank wagon 污泥槽车
sludge test 淤渣试验;污泥试验;污泥检验
sludge thermal reduction 污泥热减缩
sludge thickener 污泥增稠器;污泥浓缩机;污泥浓缩池;污泥稠化器
sludge thickening 污泥浓缩;污泥增稠;污泥浓度;污泥稠化
sludge thickening basin 污泥浓缩池
sludge thickening pond 污泥浓缩池
sludge thickness 污泥浓缩;污泥稠度
sludge trap 岩粉捕集器;泥沙捕集装置;灰泥捕集器
sludge treatment 污泥处理
sludge treatment plant 污泥处理装置;污泥处理厂
sludge trenching 污泥沟埋处置
sludge trough 淤泥输送槽;污泥输送槽;污泥槽
sludge tube 泥浆管;污泥管
sludge ultimate disposal 污泥最终处置
sludge utilization 污泥利用
sludge vacuum filtration 污泥真空过滤
sludge value 污泥值
sludge valve 污泥阀
sludge vessel 污泥船
sludge volume 污泥体积
sludge volume index 污泥体积指数;污泥容积指数
sludge volume ratio 污泥体积比
sludge volume reduction 污泥体积缩小;污泥减容
sludge wasting 污泥的废弃
sludge water 污泥水
sludge water return 污泥回流
sludge weight reduction 污泥重量降低
sludge well 污泥井
sludge wet-air oxidation 污泥湿气氧化
sludge wet combustion 污泥湿室焚烧
sludge wet oxidation 污泥湿式氧化法
sludge withdrawal 污泥排放
sludge withdrawal pipe 污泥管
sludge worm 污泥(蠕)虫
sludge yield 污泥产率;污泥产量
sludge zone 污泥区
sludging 塞泥;淤沉;废浆
sludging in oil line 油管内成渣
sludging process 泥浆冲洗法
sludging test 淤渣试验;成渣试验
sludging up 泥包
sludging value 淤渣值
sludgy 淤泥多的;有淤泥的;泥泞的
sludgy deposit 淤渣沉积
sludgy value 淤渣值
slue 偏荡;突然回转
slueing generator 回转发电机
slue line 旋转吊货杆的牵缆
sluff 积雪滑动;软岩脱落;脱落的泥皮(孔壁)
Slug 斯勒格
slug 芯片;渣球;灌浆封闭裂缝防漏;金属小块;金属片状毛坯;毛坯;缓动环;燃料块;铁芯;速度缓慢的船;栓;大嵌齿;波导调配柱
slug bit 镶嵌钻头
slug breaker 废料切断刀
slug can 释热元件外壳
slug clearance hole 冲裁废料漏料孔
slug die 废料切断模
slug discharge 突然排出
slug disposing 冲压废料处理
slug flow 团状流动;慢流;迟滞流
slugged relay 缓动继电器
slugger (辊式轧碎机中的)齿状物
slugger screen 振动筛
slugger type hammer 齿面锤碎机
slugging 泄水冲刷;击压法;埋焊;腾涌;冲裂;成渣;插嵌片;滞流;嵌物焊;脉流;(喷射混凝土流下的)废浆
slugging roll(er) 齿面辊碎机
sluggish 迟滞
sluggish channel 缓流航道
sluggish circulation 缓慢环流
sluggish current 缓慢的海流
sluggish flow 迂缓(水)流;缓慢水流;缓流
sluggish flow channel 缓流航道
sluggish inversion 慢转变;缓慢转化
sluggish layer 缓动层
sluggish lubricant 黏滞润滑油
sluggish market 萧条市面;市场疲软

sluggish metal 冷金属液
sluggishness 停滞；惰性
sluggish reach 缓流河段
sluggish river 徐流河；缓流河
sluggish river reach 缓流河段
sluggish stream 徐流河；迂缓(水)流；滞流河川；缓流河
slug(in)neck 瓶颈过厚
slug pathway 嵌条通路
slug press 压制毛坯
slug slot 出屑槽
slug test 定容积瞬时提水(或注水)试验
slug the pipe 起钻前钻杆泵入一段重泥浆(使卸钻杆时钻杆无泥浆)
slug tuner 动块调谐器
slug tuning 动块调谐
slug-type bit 硬质合金球齿牙轮钻头
slug type reaming shell (镶焊金刚石的)条带形扩孔器
sluice 泄水闸；泄洪闸；溢水道；堰水；闸口；流水槽；流槽；开立闸泄水；节流器；水闸；闸门；水力冲泄；冲渣沟
sluice across river 拦河闸
sluice board 泄水闸板
sluice box 泄水箱；洗选流矿槽；洗矿槽；流矿槽；溜槽
sluice capacity 泄水闸通道量
sluice chamber 泄水闸室
sluice channel 泄水渠(道)
sluice cock 闸门阀；滑板阀
sluice collar 泄水道截水环
sluice control mechanism 水闸操作机械；水闸操作机构；水闸操纵机械；水闸操纵机构
sluice culvert 泄水涵洞
sluice dam 水闸；水沟挡板
sluice damper 插板阀
sluiced fill 水力填方；水力冲填
sluice door 水闸门
sluiced rockfill 冲洗堆石；冲实堆石体
sluice entrance 泄水道进口
sluice equipment 泄水设备
sluice flow 闸下出流
sluice gate 泄水闸门；水闸(闸)门；水闸；冲刷闸门
sluice-gate bridge 闸桥
sluice-gate chamber 泄水道闸门室
sluice-gate lip 泄水闸门下缘
sluice-gate of water prevention 防水闸门
sluice-gate price 闸门价格(欧洲共同市场规定的最低进口价)；开阔价格
sluice gun 冲洗枪
sluice head 闸头；水闸闸头
sluice installation 泄水设备
sluice line 泄水管线；冲淤管线；冲洗管道；冲泥管线
sluice-mounted screen 洗槽振动筛
sluice nozzle 节流喷管
sluice of caisson 沉箱闸门上的小门
sluice of the dock 坞门
sluice opening 闸孔；水闸孔
sluice operating chamber 泄水道操作室
sluice operating gallery 泄水道操作廊道
sluice panel 闸门板
sluice pipe 泄水管(道)；冲淤管；冲沙管；冲泥管
sluice pipeline 冲淤管线
sluice pump 冲淤泵；冲灰泵
sluice scour(ing) 闸下冲刷
sluice section 泄水(孔)坝段
sluice timber 闸门木梁
sluice trough 冲洗槽
sluice tube 泄水管(道)；冲沙管；冲泥管
sluice tubing 泄水管(道)；冲泥管
sluice valve 泄水闸；闸门阀；闸阀；滑动泄水阀；滑板阀；平板式阀
sluice valve with bottom release 底部泄水的闸阀
sluiceway 从水门流出的水；人工水闸渠道；泄水道；洗砂沟；闸口；闸沟；冲渣沟；冲淤道；冲沙道；冲灰沟
sluiceway canal 泄水渠(道)；冲沙渠道
sluiceway channel 泄水渠(道)；冲沙渠道
sluiceway gate 冲沙道闸；泄水道闸
sluice weir 溢流堰；泄水堰；闸堰；活动堰；水沟堰
sluice work 泄水建筑(物)
sluicing 泄水冲刷；放流；流槽选矿法；开闸泄水；水力除土；水力水运；水冲；冲淤；冲泄；冲刷
sluicing and scouring 闸孔放水冲刷
sluicing canal 泄水渠(道)

sluicing capability 冲淤能力
sluicing capacity 泄水闸流量；冲刷能力；冲刷量
sluicing chamber 冲沙室
sluicing channel 放水槽；放水渠；冲洗槽；冲洗渠；冲洗箱；冲淤箱；泄水渠(道)；冲淤(渠)道
sluicing culvert 冲沙涵洞；冲沙涵洞
sluicing gate 冲沙闸(门)
sluicing noz 水灰喷嘴
sluicing of earth dam 冲填土坝
sluicing pipe 洗气管；洗涤管
sluicing pond 水闸蓄水池；疏浚水道蓄水池；冲淤盆地；冲淤池
sluicing sand 冲沙
sluicing-siltation dam 冲填淤积坝
sluicing-siltation method 水坠法；冲填淤积法
sluicy rain 倾盆大雨
sluing arch 斜面拱；八字拱
slum 贫民区；破房；陋巷；含煤耐火黏土；润滑油渣；贫民窟
slum area 陋巷；贫民区；棚户区
slumber-time diversion 夜间改道制
slum clearance 贫民区改造；陋巷改造；贫民窟的清理；棚户区改造；拆除违章建筑
slum clearance area 陋巷清理区
slum-cleared area 陋巷清理区
slum-cleared site 陋巷清理场
slum-cleared zone 贫民区改造地带；破旧房屋拆迁地带
slum coal 煤垩
slum demolition 拆除棚户区
slum district 陋巷区(域)；贫民区
slumdweller 贫民窟居民；棚户
slum improvement 棚户区改善
slumism 贫民窟现象
slum land 贫民区
slumlord 缺德房东
slump 坍落度；下垂度；滑动沉陷；坍塌；衰退；湿陷；边坡坍塌；暴跌；滑塌；滑动
slumpability 流动惰性
slump bedding 滑动层理
slump belt 崩塌地带
slump block 塌陷砌块；崩坍块体；坍落试验堆；坍落堆(混凝土)
slump breccia 滑塌角砾岩；崩滑角砾岩
slump cone 坍落度试验用圆锥体；坍落度筒；坍落度圆锥筒(混凝土)
slump consistency test 坍塌试验；坍落(稠)度实验
slump constant 坍落度；坍落(度)常数；坍落(度)常量
slump deposit 塌坡沉积(物)
slump deposition 滑塌沉积作用
slumped bed 滑动层【地】
slumped mass 崩滑体
slump fault 正常断层；滑塌断层
slumpflation 萧条膨胀
slump fold 滑动褶皱；崩滑褶皱
slump incoherent 松散滑塌
slumping 陷落效应；坍塌；场落；塌落；(地层结构上的)场方；倒塌；崩滑；滑塌；滑动
slumping bank 滑动岸；崩岸
slumping loess 湿陷性黄土
slumping of fill 填土下沉
slumping slide 推移式滑坡
slumping soil 湿陷性土
slumping type settlement 湿陷量；湿陷沉降
slump limitation 坍落度限值
slump loss 坍落度损失
slump loss preventing admixture 防坍落度损失外加剂
slump mass 滑体；崩滑体
slump meter 坍塌度计；坍落度测定仪
slump of concrete 混凝土坍落度
slump of disastrous proportions 严重衰退
slump-puller 拔根孔；拔根机
slum problem 贫民窟问题
slump settlement 塌陷；下陷；下沉
slump sheet 滑塌岩席
slump specimen 坍落度试样
slump structure 滑塌构造；滑动构造；塌滑构造；崩滑构造
slump test 混凝土坍落度试验；坍落度试验；重陷试验
slump test value 坍落度
slump zone 崩坍地带

slum reclamation 棚户区改造
slum site 贫民区
slum upgrading 棚户区改善
slum zone 贫民区
slung-span 悬跨
slung-span construction 悬跨建筑；悬挂结构；悬挂建筑
slung-span continuous beam 悬跨连续梁
slung-span continuous lattice(d) beam 悬跨连续格构梁
slur 黏合；上涂料
slurb(ia) 大都市郊区贫民窟；市郊贫民区
slurgalls 线圈长度变化
slurring 灌浆；上涂料；型芯粘合法
slurry 淤浆；釉浆；料液；料浆；沥青砂浆；浆(料)；泥浆；煤浆；滑泥；生料；薄泥浆；薄浆
slurry agitator 料浆搅拌机；泥浆搅拌器；泥浆搅拌机
slurry agitator pump 泥浆混合泵
slurry auger 浆状物搅拌输送螺旋；泥浆搅拌输送螺旋
slurry basin 料浆池
slurry bed reactor 浆床反应器
slurry blanket 泥渣层过滤；泥渣层
slurry blasting 浆装药爆破
slurry blending 料浆搅拌
slurry-burning process 泥浆燃烧法
slurry-burning process for hydrocarbon waste 碳氢化合物废物泥浆燃烧处理法
slurry cake 料浆饼；泥壳
slurry cake conveyer 料浆饼输送机
slurry calcinator 料浆蒸发炉
slurry carrier 泥浆船
slurry casting 注浆成形；泥浆浇注；粉浆浇注
slurry-casting method 注浆法
slurry cement 水泥浆
slurry chains 链条装置
slurry charge line 泥浆输送管路
slurry clarification 泥浆澄清
slurry coat 泥浆封层
slurry coating 水浆涂料
slurry coat method 泥浆(套)法
slurry compounding machine 配浆机
slurry concentration 泥浆浓度
slurry concentrator 泥渣浓缩器；淤泥浓缩器；薄浆浓缩器；料浆浓缩机；泥渣浓缩池
slurry concrete 稀混凝土
slurry concrete wall 混凝土防渗墙；地下连续墙
slurry consolidation method 泥浆固结法
slurry convey 泥浆液压运输
slurry correction 料浆校正；料浆调整
slurry cutoff 泥浆防渗墙
slurry delivery pipe 供浆管
slurry density control tank 泥水浓度调整槽
slurry desiccator 料浆烘干机
slurry dewatering 料浆脱水；泥渣脱水；泥浆脱水
slurry displacement 换浆
slurry doctor tank 料浆校正仓
slurry drilled pile 泥浆钻孔桩
slurry drilling 泥浆钻孔
slurry drilling method 泥浆钻孔法
slurry drum 料浆滚筒
slurry dryer 料浆蒸发机；料浆烘干机；泥浆干燥机
slurry drying 料浆烘干
slurry emulsion 乳化软膏；乳化薄浆
slurry equipment 拌浆设备
slurry exchanger 淤浆交换器
slurry explosive 黏合型炸药；浆状炸药；浆装药；塑胶炸药；稀浆炸药
slurry-fall fill dam 水坠坝
slurry feed 泥浆喂料；泥浆进料；稀浆进料；稀浆进料；料浆喂入；料浆喂料
slurry feeder 薄浆送料机；泥浆送料机；泥浆馈给器；输浆器；进浆器；料浆喂料机；煤浆进料机
slurry filter 料浆过滤器；泥浆过滤器
slurry filter-cake 淤浆滤饼；料浆滤饼；泥浆滤饼
slurry filter preheater 料浆过滤器预热器；泥浆滤前预热器
slurry filtration 料浆过滤
slurry flocculation 泥浆絮凝
slurry for stability 稳定液
slurry grinding mill 料浆磨；泥浆磨；生料磨
slurry handling 淤浆处理；泥浆处理
slurry holder 泥浆管支架
slurry hole-boring method 泥浆护壁钻孔法

slurry-infiltrate fiber concrete 注浆纤维混凝土
slurry infiltrated fiber reinforced concrete 渗浆纤维混凝土
slurrying 制浆;泥浆制备;涂浆;涂防护漆
slurry inlet 滤浆入口;料浆入口;料浆进口
slurry lifter 料浆提升板
slurry loss 泥浆损失
slurry mill 料浆磨
slurry mineral 矿粉浆
slurry mining 矿泥浆开采法
slurry mixer 泥浆搅拌器;拌浆器;拌浆机
slurry mixing tank 拌浆桶
slurry mole 泥浆掘进机
slurry mo(u)lding 灌浆成型
slurry mo(u)lding process 灌浆成形加工;陶瓷型铸造
slurry-mud jacked pavement 压入泥浆的路面
slurry net cement 纯水泥浆
slurry of neat cement 净水泥浆;水泥净浆
slurry oil 淤浆油
slurry-oil squeeze 挤水泥—油浆隔离油层
slurry outlet 含尘液排出口
slurry overflow 料浆溢流
slurry packed method 泥浆填充法
slurry packing 匀浆填充(法);浆状填充法;泥浆液充填
slurry packing method 泥浆充填法
slurry packing technique 淤浆填充法
slurry penetration 灌浆
slurry phase biological treatment 泥浆相生物处理
slurry-phase reactor 淤浆反应器
slurry pipe 输泥管道;泥浆输送管道
slurry pipeline 泥浆管道;泥浆管线;泥浆管路;泥浆管道;煤浆管道
slurry polymerization 淤浆聚合(法)
slurry polymerization process 淤浆式聚合法
slurry pond 淤浆池;矿石沉淀池;泥浆池
slurry pool 泥浆池
slurry preforming 料浆喷射形成形
slurry preheating 料浆预热
slurry preparation 稀浆制备;料浆制备;泥浆制备
slurry preparation by flo(a)tation 浮选法料浆制备
slurry preparing tank 浆预拌槽
slurry press 淤浆压榨机
slurry pressing 压滤成形
slurry pressure 泥浆压力
slurry process 淤浆法;油浆法;膨润土法施工
slurry process for making soft-mud brick 制作软土砖的泥浆处理(法)
slurry pump 淤浆泵;料浆泵;泥水泵;泥浆泵;煤水泵;排浆泵;砂浆泵
slurry pumping plant 泥浆泵送车间;泥浆泵送装置;泥浆抽吸车间;泥浆抽吸装置
slurry pumping rate 料浆泵送量
slurry reactor 悬浮反应堆;浆料反应器;泥浆反应器
slurry-reinforced overburden 泥浆增强的覆盖层
slurry replenishing hole 补浆孔
slurry return 泥浆挥流
slurry rheopexy 泥浆触变
slurry ring 料浆圈;泥浆圈
slurry sampler 泥浆取样筒;取浆样筒
slurry screen 料浆筛;矿泥筛
slurry seal 密封膏浆;沥青砂浆表面处治;胶浆密封;泥浆封层;灰浆止水;灰浆密封;灰浆封层
slurry seal coat 沥青砂浆封层;灰浆封层;泥浆封层
slurry seal emulsion 乳化软浆;乳化薄浆
slurry seal machine 封沥青浆机;封浆机
slurry seal operation 乳化软膏操作
slurry seal surface treatment 泥浆封面处理
slurry sedimentation 泥浆沉降
slurry separation 泥水分离
slurry service 油浆系统
slurry settler 油浆沉降器
slurry settling 泥浆沉淀
slurry shield 泥水(加压)盾构
slurry shield tunneling machine 泥水盾构掘进机
slurry silo 料浆库;泥浆库
slurry slump 泥浆状滑塌;散乱崩滑物
slurry spillage 回浆
slurry spreader 涂浆机;泥浆铺设器;厩液洒布机;肥肥洒布器
slurry stable wall 泥浆固壁
slurry steeping 淤浆碱化
slurry storage tank 料浆库;泥浆储[贮]存池;泥浆库

slurry support 泥浆支架
slurry system 矿浆液化装卸法
slurry tank 淤浆槽;岩粉沉淀池;料浆罐;泥浆池;泥浆槽;含尘液储[贮]槽
slurry thickener 泥浆增稠剂;泥浆浓缩器;料浆稠化剂;泥浆稠化机
slurry thixotropy 泥浆触变
slurry thinner 料浆稀释剂;泥浆稀释剂
slurry to be filtered 滤浆
slurry transport 浆料输送
slurry trap 岩浆捕集器
slurry treatment 泥水处理;泥浆处理
slurry treatment plant 泥水处理厂
slurry treatment system 泥浆处理系统
slurry trench 泥水沟槽;泥浆坑;泥浆(沟)槽
slurry trench construction 泥浆固壁法施工
slurry trench cutoff 泥浆槽黏土截水墙;截浆槽
slurry-trenched wall 泥浆槽墙;地下连续墙
slurry trench excavation 泥浆挖槽法
slurry trench method 泥浆挖槽法;槽壁法
slurry trench wall 沟槽泥浆壁
slurry-type blasting agent 浆状炸药
slurry-type concrete 稀混凝土
slurry-type explosive 浆状炸药
slurry-type seed mixer 湿式拌种器
slurry viscosity 泥浆黏度
slurry wall 泥浆(槽)墙;灰浆墙;槽壁;地下连续墙
slurry wall method 连续墙法;槽壁法施工
slush 稀泥浆;淤浆;烂泥;软湿冰雪;软泥;冰泥;冰凌;白铅膏(防锈涂料)
slush basket 捞砂筒
slush casting 中空铸造;空心浇注;空心件铸造;空壳铸造;凝壳铸造;薄壳铸件
slush coat 油灰涂层
slush conduit 淤泥管
slush covered runway 烂泥覆盖的跑道
slushed joint 砂浆勾缝;砂浆填缝;挂刀灰砖缝
slushed-up joint 灌浆封缝;砂浆填(砖)缝
slushed-up joint 砂浆填缝
slusher 活动拖铲;铲泥机;扒斗铲泥机
slush fitted pump 泥浆泵
slush fund 行贿资金
slush-grout 泥浆灌浆
slush grouting 灌缝泥浆;刮浆填缝;泥浆灌注;水泥砂浆喷涂;灌浆填缝堵眼
slush horn 帆缆用的油脂盒
slush ice 河底冰;冰泥(浆);雪水;软冰;松软冰;初凝冰;半溶冰
slush ice chute 冰雪泻槽
slush ice run 冰泥下泻;冰泥流
slush ice sluice 泄冰闸
slush icing 湿积冰
slushier 移动式刮土机
slushing 减水(作用);涂油灰;水力填土法;冲刷涂装法;砂浆灌缝;灰泥填塞;出卖下脚料;出卖船上废油
slushing compound 抗蚀油脂;抗蚀润滑剂;防锈油膏;抗蚀剂;防锈膏;半固体防锈剂
slushing giant 填土用水枪;充填用水枪
slushing grease 抗蚀润滑脂
slushing mo(u)lding 冷凝模塑(法)
slushing oil 抗蚀油(脂);抗蚀润滑油;防锈油
slush layer 泥浆层
slush metal 软合金
slush molding 薄壳模塑
slush nozzles 钻头水眼
slush pipe 淤浆管
slush pit 烂泥坑;泥浆坑;泥浆池
slush-pit launder 沉淀池上泥浆槽
slush pond 泥浆池
slush pulp 液体浆
slush pump 淤浆泵;汲泥泵;抽泥泵
slush pump ga(u)ge 泥浆泵表
slush tube 淤浆管
slushy clay 泥浆般的黏土;含烂泥黏土
slush zone 半融雪带
slype 狭窄通道;有顶走廊(由教堂至修道院的);(教堂中的)有覆盖的走廊
smack 小鱼船;小帆船
smack smooth 甲板上空无一物
Smalian's formula 斯马利安公式
Smalian's method 斯马利安法
smalite 高岭石
small aggregate 细集料;细骨料

small aggregate crusher 细碎机
small airport 小型机场
small airport runway indicator 小型机场跑道指示器
small airway resistance determination 小气道阻力测定
small amplitude 小振幅
small amplitude-low gradient magnetic anomaly 低缓磁异常
small amplitude modulation 小幅度调制
small amplitude pulse polarography 小振幅脉冲极谱法
small amplitude wave 小振幅波;微小振幅波
small amplitude wave theory 小振幅波理论
small amplitued shear oscillation 小振幅剪切振荡
small anchor 轻便小锚
small and medium bridge 中小桥梁
small and medium sized business 中小厂商
small and medium sized enterprise 中小企业
small angle boundaries 小角度边界
small angle boundary 小角度晶界;小角边界
small angle camera 低角摄影机
small angle diffraction 小角度衍射
small angle elastic scattering 小角弹性散射
small angle grain boundary 小角晶粒间界
small angle scattering 小角散射(法)
small angle X-ray diffraction 小角度X射线衍射
small angle X-ray scattering 小角度X射线散射
small animal reservoir 小动物饲养场
small anti-gas apparatus 小型防毒器
small anvil 小铁砧
small aperture separator system 小隙缝分离系统
small appliance 小型设备
small appliance branch circuit 小型电器支回路
small application technology satellite 小型应用技术卫星
small arc 小弧
small arcade 小拱廊
small arch 小型拱
small area flicker 小面积闪烁
small area point detector 小范围点检测器
small arms sling 背带
small articles of daily use 小百货
small ashlar 墙面小块石板(装饰用);小块琢石
small astronomical satellite 小天文卫星
small atlas 袖珍地图集
small atmospheric sterilizer 小型常压灭菌器
small axe 斧头;手斧
small ball 小球
small ball-mill 小球磨
small banquet hall 小宴会厅
small batch production 小批量生产
small batch technology 小批成组工艺
small bay 小海湾;沱
small bayonet cap 小卡口灯座
small beak-iron 尖嘴砧
small biologic(al) cycle 生物小循环
small bitt 小挽缆桩;缭绳挽桩
small block 散斗;三才升;小型砌块;小地块
small boat 小艇
small boat landing 小船登岸处;小船码头
small boiler 小锅炉
small bond 小额债券
small bore 小口径;小孔径;小截面;小内管径(半英寸管,用于水泵助力的集中供热系统)
small bore cylinder 小口径缸
small bore deep-hole method 小直径深孔崩矿法;小眼凿深孔法
small bore engine 小径汽缸发动机
small bore heating pipe 小直径暖气管
small bore heating system 小管采暖系统
small bore mobile type pump 小口径机动泵
small bore pipe 小口径(钻)管
small bore piston 小孔径活塞
small bore system 小管采暖系统;小孔径系统;细管热水集中供暖系统;管式采暖锅炉
small bore tubing 小内径管
small breaker 小型破碎机
small breaking machine 小型破碎机
small brokens 小碎粒
small brook 小溪
small bubble spongy body 小泡状海绵体
small bulb 小灯泡
small bundles of spiral vessels 螺纹导管的小维

管束
small burnet 小地榆
small bus-bar 小汇流条
small business 小(型)企业;个体户
small business finance 小企业融资
small business loan 小企业贷款
small business management 小(规模)企业管理
small business under individual ownership 小商业
small cabin 小舱室
small cabin system 小车箱系统
small café 小咖啡馆
small capacitance series air-gun 小容量系列气枪
small capacity cable 小容量电缆;细芯电缆;少芯电缆
small capacity concrete mixer 小容量混凝土搅拌机
small capacity memory 小容量存储器
small capacity mixer 小容量搅拌机
small capacity of hump 小能力驼峰
small car 小型小客车
small cargo boat 小货船
small cash 小桶
small castle 小城堡
small cave 小洞
small cellular cofferdam 小格型围堰
small chamfer 小削角(方桩四角)
small charge 小型装药
small check 小裂缝;小方格;小额支票
small chisel 小凿
small circle 小圆(航路);位置圈
small circle belt 小圆构造带
small circle girdle 小圆环带
small circular shaft 小圆井
small circular window 小圆窗
small city 小城市
small claims court 小额索赔法庭
small clearance space 窄隙
small clear specimen 无疵小试样
small clod 小(碎)土块;小碎石块
small closed curve 小闭曲线
small coal 末煤;碎煤
small coal mine name 小煤窑名称
small coastal transport 小型沿岸运输船
small coasting trade 沿海短程运输
small cock 小龙头
small column 小型柱;小柱
small commodities 小商品
small commodity economy 小商品经济
small commodity producer 小商品生产者
small communications terminal 小型通信终端设备
small compensation space blasting 小补偿空间爆破
small computer 小型计算机
small computer system interface[SCSI] 小型计算机系统接口
small concentration 低浓度
small construction item 小型建设项目
small construction project 小型建设项目
small container 小型集装;小容器
small contrast 影像反差小
small control switch 组合开关
small cooler 小型致冷器
small cordage 细索类
small coring 小型样芯
small correction 小改正
small count 小计
small craft 小艇;小船
small craft anchorage 小船锚地
small craft berth 小船泊位
small craft chart 快艇航海图
small craft harbo(u)r 小船港
small craftsmen and traders 小工商业者
small craft warning 小船警报
small creek 小河流;小港湾
small crop drying equipments 小件作物干燥设备
small cross-section 小断面
small crusher 小型破碎机
small crushing 细级压碎
small crushing machine 小型破碎机
small crystal 小晶体
small crystals of calcium oxalate 草酸钙小晶体
small cup 小杯
small cupola 小冲天炉
small curved meander 小弯度曲流

small damage 小额损失
small damage club 小损失互保协会
small debts court 小债务法庭
small deflection 小挠度;小变位
small diameter borehole 小口径钻孔
small diameter coaxial cable 小同轴电缆;小直径同轴电缆
small diameter drill hole 小直径钻孔
small diameter drilling 小孔径钻进
small diameter gear 小直径齿轮
small diameter inclinometer 小口径测斜仪
small diameter workpiece 小直径工件
small dimension 小尺寸
small dimension wood 小径材
small dining room 小餐厅
small disc harrow 小型圆盘耙
small displacement 小量位移;微小位移
small displacement engine 小容积发动机
small displacement pile 少排土桩
small distributor 小型配电板
small ditch 小明沟
small diurnal range 小日潮差
small domestic appliance 小型家用设备
small dormer 小老虎窗
small double cap 大四开图纸
small drainage basin 小流域
small drift 刨洞(小断面开挖)
small drill 轻型手持凿岩机
small dumper 翻斗车
small duty pipe-still 小负荷管式
small dwelling unit 小型居住单元
small earthquake focus mass 震源体小
small earth terminal 小型地面终端
small eccentric compression 小偏心受压
small echo 小波
small ecosystem 小生态系统
small Edison screw-cap 小型爱迪生螺旋灯头
small egg-plant screw indicating lamp 小螺口茄形指示灯泡
small electric(al) appliance 小电器设备
small electronic couple 小电偶
small end 小头;小端
small end bell reducer 小承大小头
small end bushing 小头衬套
small end face 小端面
small endless rope tractor 小无极绳牵引机
small end of connecting rod 连杆小头
small engine 小型发动机
small enterprise products 小企业产品
small enterprises in commerce 小商业
smaller elliptic semidiurnal component 副太阴椭圆半日分潮
smaller evectional constituent 副出差分潮
smaller evectional semidiurnal component 副太阴出差半日分潮
smaller lunar diurnal component 副太阴椭圆日分潮
smaller stripper 小批货物托运人
small establishment 小基层单位
smallest algebraic group 最小代数群
smallest closed neighbourhood 最小闭合邻域
smallest common multiple 最小公倍数;最上公倍数
smallest curve radius 最小曲线半径
smallest generally felt 最小一般可感
smallest homogeneous ideal 最小齐次理想
smallest ideal 最小理想
smallest limit 最小下限
smallest of largest regret value method 最小最大后悔值法
smallest permissible radius 最小容许半径
smallest positive integral 最小整数
smallest size drill rods 最小尺寸钻杆
smallest sub-A-module 最小子 A 模
smallest subring 最小子环
smallest subspace 最小子空间
smallest transitive set 最小可迁集
small explosion 小型爆破
Smalley process 斯马雷钢铁脱硫法
small face 窄面
small face plate 拨盘
small faceplate 小花盘
small falls 小瀑布
small farm 小农场

small farm tools 小农具
small faucet 小龙头
small fault 小断层
small field radiometer 小视场辐射计
small float-and-sink sample 小浮沉样
small-flower rose 小花蔷薇
small flow sampler 小流量采样器
small forceps 小钳
small forged parts 小的锻造部件
small format 小规格
small format aerial photography 小像幅航空摄影
small format camera 小像幅摄影机
small formation 小编队
small fracture 小冰隙
small frequencies 小频数
small fruit rose 小果蔷薇
small gang plough 小型开沟犁
small gap 小间隙
small garage 小汽车间;小车库
small garden ornaments 园林小品
small gas field 小气田
small gateway 小门道
small ga(u)ge nail 小号钉
small generating set 小型发电机组
small glass 玻璃屑
small goods elevator 小型货物电梯;小型货物升降机
small grain 细粮
small-grained 小颗粒的
small graphite 粉粒石墨
small gravel 小砾(石)
small groove 小槽
small gross 小罗
small guide strap 小抱箍
small hail 小雹
small hammer 小锤
small handicraftsman 小手工业者
small hangar 小飞机库
small harbo(u)r tug 小型港口拖船
small head 小水位差;小水头
small helm 小角度转舵
small holding 自留地
small holding of tree 零星木
small hole 小口径孔;小孔
small hole boring 小口径钻进
small hole darning 小洞织补法
small hole drill 小口径钻机
small hole drilling 小口径钻进
small hole ga(u)ge 小孔规
small hole method 小口径钻孔法
small hole rig 小口径钻机
small house 小屋
small hut 小棚屋
small hydraulic project 小型水利工程
small hydroelectric(al) plant 小型水力发电站
small hydro-power station 小水电站;小型水电站
small ice-field 小冰原
small ice floe 小浮冰;小冰盘
small ice piece 小冰块
small impression 限定版
small impurity 小杂质
small incinerator 小型垃圾焚化炉
small income 微利
small industrialists 小工业者
small infrared drier 小型红外线干燥机
small inside micrometer calliper 小内测千分卡尺
small internal gear 小内齿圈
small ion 轻离子
small irregularities 小不平度
small island 小岛屿
small isolation circle 小隔离圈
small jet 微量射流
small kitchen 小厨房
small knot 小结(直径为0.5~0.75英寸);小(树)节
small laboratory flume 小型试验槽
small land lessor 小土地出租者
small landslide 小型滑坡
small lead block test 小铅柱试验
small leaf abelia 小叶六道木
small library 小型图书馆
small light 小功率灯
small lime 小块石灰;次石灰
small livestock farming 小牧场
small loan company 小额贷款公司

small log 小圆木
small lot 小批量;小批货物;零担散货
small lot cargo 小批量货
small lot manufacture 小规模制造;小规模生产
small lot producer 小型企业
small lot production 小批(量)生产;小量成批生产;小规模生产
small lot storage 小额储[贮]存
small luggage 小件行李
small lunar elliptic diurnal component 小太阴椭圆日分潮
small machine 小电机
small magnitude earthquake 小地震
small main valve piston 主阀小活塞
small manned submersible 小型载人潜水器
small market 小市场
small mean diurnal range 小平均日潮差
small measure expenses 小型措施费
small-medium city 中小城市
small-medium town 中小城市
small mesh sieve 细眼筛
small mine block 小型地雷场
small molecule 小分子
small mosaic 小块锦砖;小块马赛克
small motor 小型发电机
small motor garage 小型汽车库
small navigational radar 小型导航雷达
smallness 微小
small neutron accelerator 小型中子加速器
small noise tube 低噪声管
small nonlinearity 弱非线性
small number 小数
small numeral 微小的数目
small object detector 小目标探测器
small office home office 小型办公室与家庭办公
small oil field 小油田
small oil volume circuit breaker 少油断路器
small open void 小开型孔隙
small orebody 小矿体
small oscillations method 小振荡法
small packets post 小件邮递
small pair boats 小对船
small panel 小镶板
small panel construction 小镶板构造
small parameter stability criterion 小参数稳定准则
small particle 微粒
small particle emission 微粒排放
small particles 小白点
small pendulum 小摆锤
small piece bonds 小债券
small piece of coal 煤渣子
small pile 小桩
small pipeline 小直径管线;小口径管线
small pipe system 细管供暖系统;细管热水集中供暖系统;细管热水集中供暖设备
small pit 小沉淀池
small pitch pocket 小树脂囊
small pit organ 小窝器
small pivoted wheel 小旋转轮
small plant 小型设备
small plate 小板块
small platform scales 小形台秤
small pole 小圆材;小杆材树
small pore and big pore 小孔隙带大孔隙
small porosity gel 细孔凝胶
small power computer 小功率计算机
small power motor 小功率电动机
small power station 小型发电厂
smallpox 天花
small pressure sprayer 小型压力撒雾器;小型压力喷雾器
small production area 低产区
small-production oil-gas bed 低产油气层
small professional or announcement signs 小型招牌
small profile line 小断面线路
small profit 薄利
small project 小型工程;小规模(计划);小型项目
small pulley 小滑轮
small pulley bearing 小型滑车支承架;小型滑轮支承架
small pulley block 小滑轮组
small pulley carrier 小滑轮组支承架
small pulley nest guard 小滑轮组护挡板;小滑轮组限程器

small pull handle 小拉手
small pulse 小脉冲
small punch 小冲子
small punching 初次钻孔(铆钉分两次钻)
small pyramid 小金字塔
small pyramid glass 小型角锥玻璃
small quantity 小量;少量
small quantity pollution sources 少量污染源
small-radius curve 小半径曲线
small rail car 轨道小车
small rain 小雨
small rate gyro 低速率陀螺
small reflective area target 小反射面目标
small region market 小区域市场
small rescue tug 小型救援拖船
small reservoir 塘堰
small rest home 小疗养院;小疗养所
small retail items 小日用品
small reverse body bush 回动体小衬套
small rhythm 小韵律
small rocket lift device 小型火箭升降装置
small rockets launch operation controled center 小型火箭发射活动控制中心
small rolled parts 小件轧制部件
small roller with shields 带护罩的小型压路辊;带护罩的小型压路机
small roof 小屋顶
small round stone 小圆石
small routine 小程序
smalls 小石板;细末;粉末;细料;小件货物;轻货
small sails 副帆
small sample 小样品;小样本;小包货样
small sample assay system 小样品测定系统
small sample chopper 小样品选择器
small sample distribution 小样品分布
small sample method 小样(本)法
small sample scattering 小样品散射
small sample technique 小样本方法
small sample time-series study 小样本时序研究
small sapling 小幼树(高3~10英尺)
small saver's certificate 小额储蓄者存款证
small scale 小尺度;小尺寸的;小比例尺的;小比例(尺)
small-scale building site 小规模施工场地;小规模建筑场地
small-scale channel regulation project 小型航道整治工程
small-scale chart 小比例尺(海)图
small-scale climatology 小区域气候;小尺度气候
small-scale combustion 小规模燃烧
small-scale commercial planting 小规模栽植
small-scale cylinder test method 小比例柱体试验法
small-scale drawing 小比例尺图
small-scale geologic(al)survey 小比例尺地质测量
small-scale hydraulic model 小比例尺水工模型
small-scale hydroelectric(al) station 小水电站
small-scale incinerator 小型焚化炉
small-scale integrated circuit 小规模集成电路
small-scale integration 小规模集成电路
small-scale landform 小地貌
small-scale manufacturing 小批生产
small-scale map 小比例尺(地)图
small-scale model 小尺寸模型
small-scale model test 小比例尺模型试验
small-scale multiple-use equipment 小型多用设备
small-scale of bursting water 小型突水
small-scale pilot model 小比例尺模型
small-scale pilot plant 小规模试验工厂
small-scale plotter 小比例尺测图仪
small-scale problem 小型问题
small-scale production 小生产;小量生产;小规模生产
small-scale regional geologic(al)surveying 小规模区域地质调查
small-scale temporary facility 小型临时设施
small-scale test 小试;小规模试验
small-scale tester 小型试验机
small-scale topographic map 小比例尺地形图
small-scale tunnel data 小尺寸风洞试验数据
small-scale turbulence 小素流度;小尺度素流
small-scale vibration table 小型振动台
small-scale water conservancy project 小型水利工程
small-scale yield 小范围屈服
small scientific research satellite 小型科研卫星
small screw driver 小螺丝刀
small screw jack 小型螺旋起重器
small screw torch 微型焊炬
small section 小剖面
small section mill 小型型钢轧机
small section rolling mill 小型轧机
small section tunnel 小断面隧道
small self-contained iron and steel complex 配套的小型钢铁联合企业
small serial production 小批生产
small set 小集合
small sett 小方石块
small sett paving 小方石路面;小方石铺砌
small sewage tank 污水池
small sewage treatment plant 小型污水处理装置
small sewage treatment works 小型污水处理厂
small sheave 小滑车;小滑轴
small sheave bearing 小滑车支承架
small sheave block 小滑轮组;小滑车组
small sheave carrier 小滑车组支座;小滑车组支架
small sheave nest 小滑轮组
small sheave nest guard 小滑车组挡板;小滑车组限程器
small shipper 小转运商
small shock 小震动
small shops 小商业
small shot 钻粒
small shrub 小灌木
small signal 小信号
small signal analysis 小信号分析
small signal behavio(u)r 小信号性能
small signal characteristic 小信号性能
small signal gain 小信号增益
small signal parameters 小信号参数
small signal performance 小信号性能
small signal power gain 小信号功率增益
small signal theory 小信号理论
small single-phase induction motor 小型单相感应电动机
small single-phase transformer 小型单相变压器
small-size air target 小型空中目标
small-size bearing 小型轴承
small-size bucket elevator loader 小型斗式提升加料机
small size cement plant 小型水泥厂
small-size collapsible container 小型折叠式集装箱
small-size computer 小型计算机
small-sized 小尺寸的;小型的
small-sized coal 小块煤
small-sized flat 小型公寓
small-sized hut 小型棚屋
small-sized diesel electric(al) locomotive 小型柴油电机车
small-sized kitchen 小型厨房
small-sized mosaic 小块锦砖;小块马赛克
small-sized paving sett 小块铺路方块石
small-sized scrap 小块废钢
small-sized sett paving 小方石块路面
small-sized slab 小尺寸平板
small-sized water source 小型水源地
small-size folding container 小型折叠式集装箱
small-size glass ware 小型玻璃器皿
small-size machine 小型机械
small-size motor 小电动机
small-size nuclear power plant 小型核电站
small-size ore deposit 小型矿床
small-size plate 小型板块
small-size production 小批生产
small-size sample 小筛分样
small-size sheet stacker 小片堆垛机
small-size tanker 小型油轮
small-size tractor 小型牵引车;小型拖拉机
small-size travel(l)ing crane and car shed 小型流动机械库
small-size travel(l)ing power generator 小型流动式发电机
small-size vibration table 小型振动台
small sliding microtome 小型滑动切片机
small sludge treatment works 小型污泥处理厂
small slugs 小蚝蝓
small spaces through which water and air may

pass 水和空气可以通过的空隙
small spatula 小抹刀
small specimen 小试件
small sphere 小球
small splint 小夹板
small splint immobilization 小夹板固定法
small square 小广场
small square file 小方锉
smalls roaster 矿末煅烧炉；矿粉焙烧炉
small starlight scope 小型星光镜
small stature form of stock 矮生砧
small steel shape 小型钢材
small stems 小枝条
small stone bit 细粒金刚石钻头
small stone block 小方石
small stowage 填隙货
small strain 小应变；微小应变
small stress 小应力
small strip 小板条
small structure 小型结构物；小梁结构
small stuff 小油膏；小绳；细绳
small sun gear 小太阳轮
small supermarket 小型超级市场
small swivel 小旋转接头
small tear darning 紧缝
small Tekken type cracking test 小铁研式抗裂试验
Small Temple of Abu-Simbel 阿布辛勒小石窟神庙
small tenon 小榫
small test 小型试验；实验室试验
small throat with connecting large pore 粗孔小喉型
small throat with connecting small pore 细孔小喉型
small thrust hoe 小推锄
small tie-beam 撑头；樑枋
small timber 细木工料；细工用材
small tool 抹灰的网工具；小(型)工具；钳工工具；手工具
small tower 小塔(架)
small town 小市镇；小镇；小城镇
small tractor 小型拖拉机；小拖车；小功率拖拉机
small trade 小量贸易
small tradespeople and pedlers 小商小贩
small transfer line 小运输线；支管道；补充管道
small treatment system 小型处理系统
small tree 小树
small tropic range 小回归潮差
small trough 檐溜；明沟；小排水沟
small truck 小型载重汽车
small truss 小桁架
small tugboat 小拖轮
small type goods lift 小型货运电梯
small type motor 小型电动机
small type rectifier 小型纠正仪
small type ski hoist 小型滑雪量提升机
small union nut 小联管节螺母
small union stud 小联管节柱螺栓
small user 小型用户
small valley treatment 小流域治理
small vein 小脉
small vent 小通气孔；小排气孔
small vessel anchorage 小船锚地
small vibration 小振动
small villa 小别墅
small village 小村
small volume production 小量生产
small volume spring 小流量温泉
small ware 小商品
small ware kiln 小器窑
smallwares 带类织品
small wastewater treatment plant 小型废水处理厂
small wastewater treatment works 小型污水处理厂
small water heater 小型热水器
small water power station 小(型)水电站
small watershed 小流域；小集水区
small water supply system 小型供水系统；简易供水系统
small water treatment system 小型水处理系统
small wave 小波
small well 小口径钻井
small white nail 小白铁钉
small whole stone 细粒金刚石
small window 小窗
small wood 小材
smallwood form factor 小材形数
small yard 天井
smalt 钴蓝色料熔块；蓝玻璃；大青色；大青玻璃粉
smalt blue 青蓝色；大青色
smalto 小片彩色珐琅；小片彩色玻璃；玻璃马赛克；做镶嵌细工用的有色玻璃
smaragd 纯绿柱石
smaragdite 绿闪石
smart 灵活设备
smart building 智能办公楼
smart card 一卡通；智能卡
smart ceramics 机敏陶瓷
smart frost 严霜
smart material 机敏材料
smart money 罚金；赔偿金
smart peripheral 灵巧外设
smart structure material 智能结构材料
smart terminal 智能终端(设备)；灵巧终端(设备)；灵活终端设备
smartweed 杠板归
smash 击碎；粉碎
smashing 压平
smashing machine 压平机
smash plate 镇定板
smashup 崩溃
S matrix 散射矩阵
smaze 烟霾
SMB method 斯孟布法
s/m/cm 微秒/米/厘米
smear 瓶颈表面裂纹；污迹；涂片；涂抹作用；涂抹；涂敷；涂
smear camera 扫描照相机
smear-collection 涂油捕收
smeared cracking model 模糊裂缝模型
smeared out crystal on bed plane 层面滑抹晶体
smearer 消冲电路
smear ghost 拖尼重影
smear glaze 淌釉
smearing 釉层颜料涂敷
smearing effect 模糊效应；拖尼效应；涂污效应
smearing function 涂污函数
smear photograph 扫描摄影
smear photography 模糊照相法；扫描摄影术
smear preparation 涂布标本
smear test 涂色检查；涂色斑点试验；擦拭检查
smectic 近晶的；层列的
smectic compound 近晶化合物
smectic crystal 近晶型液晶
smectic mesophase 近晶型中间相态
smectic phase 近晶相；层型相；层列相
smectite 蒙皂石；蒙脱石
smectites 蒙脱石类
smectite structure 蒙脱石结构
smectite zone 蒙脱石带
smee cell 斯米电池
smegmatite 皂石
smell 气味
smelled cargo 有异味货物；有气味的货物
smelling out 气味外泄
smell test(ing) 臭味试验
smelt 熔块料
smelt alumin(i)um waste treatment 炼铝废物处理
smelt alumin(i)um wastewater 炼铝废水
smelt alumin(i)um wastewater treatment 炼铝废水处理
smelted product 熔炼产物
smelter 冶炼工人；冶炼厂；熔炉；熔炼炉；熔块炉
smelter coke 冶金焦炭
smelter drippings 矿滴；熔滴
smelter dust 熔炼炉飞尘
smelter hearth 熔炼炉
smelter injury 烟害
smeltery 冶炼厂；熔炼厂
smeltery slag 冶金废渣
smeltery waste 冶炼厂废水；冶金废料
smelting 冶炼；熔炼
smelting capacity 熔炼能力
smelting crystal 熔炼水晶
smelting facility 冶炼设备
smelting flux 熔炼焊剂
smelting furnace 冶炼炉；熔炉；熔炼炉
smelting furnace hearth 熔炼炉
smelting in suspension 悬浮熔炼
smelting operation 冶炼操作
smelting period 熔炼期
smelting plant 冶炼厂
smelting point 熔结点
smelting pot 熔炼坩埚
smelting process 冶炼过程
smelting quartz 熔炼水晶
smelting raw material commodities 熔炼原料矿产
smelting recovery 冶金回收率
smelting trial 熔炼试验
smelting unit 熔炼设备
smelting waste 冶炼厂废水；熔炼废渣
smelting works 熔炼厂
smelting zone 熔炼区；熔化层
smeraldino 水合氧化铬
s-meter 信号强度测量器
smirch 污点；污斑
smith 钳工；铁匠
Smith admittance chart 史密斯导纳圆图
Smith alloy 史密斯高温电热线合金
Smith anchor 史密斯型无杆锚
smith and founder 锻铸工
smith and founder work 锻铸工作
smith anvil 铁工砧；锻砧
Smith chart 史密斯圆图
Smith correlation 史密斯关联
smithcraft 金属工艺；锻工
Smith curve 史密斯曲线
Smith drum type machine 史密斯滚筒式染色机
smithereens 碎片
Smitherm process 斯密加热法
smithery 锻冶场
Smith fireproof floor 史密斯防火地板
smith floor 耐火地板
smith forging 自由锻造；锻造
smith forging furnace 手锻炉
Smith impedance chart 史密斯阻抗圆图
smithing 锻炼；锤锻；锻造
smithing coal 锻冶煤
smithite 斜硫砷银矿
Smith kominuter 史密斯磨矿机
Smith-Petersen nail 三刃钉
Smith river willow 史密斯河柳
Smith saddle bearing 史密斯枕座轴承
Smith's alloy 史密斯铁基合金
smith's anvil 锻工砧
Smith's approximate formula 史密斯近似式
Smith's cloth abrasion tester 史密斯织物耐磨试验仪
Smith's coupling 史密斯电磁离合器
smith's hammer 锻工锤
smith's hearth 锻工用炉
Smithsonian Agreement 史密森协议
Smithsonian astrophysical observatory star catalogue 史密森星表
Smithsonian standard earth 史密森标准地球
smithsonite 菱锌矿
Smith's refractometer 史密斯折射仪
Smith's stockless anchor 史密斯型无杆锚
Smith's strength tester 史密斯强力试验机
Smith's work 史密斯工作法
Smith triaxial method 史密斯三轴法
smithwelding 锻接；锻焊
Smith willow 史密斯柳
smithy 砧子；锻工车间；锻造车间
smithy coal 冶铁煤
smock frock 长的宽罩衫
S-mod record 生成物理块记录
smog 浓雾；烟雾；烟尘
smog aerosol 烟雾气溶胶
smog alert 烟雾报警信号；烟雾警报
smog chamber 烟雾室
smog control 烟雾控制
smog control device 烟雾控制器
smog forming hydrocarbons 成烟雾的烃类
smog-free 无烟雾
smoggy 烟雾状的
smoggy bowl 经常污雾笼罩的地方
smog horizon 烟雾高度；烟雾层顶
smog index 烟雾指数
smog injury 烟雾危害
smog layer 烟雾层
smog poisoning episode 烟雾中毒事件
smog poisoning incident 烟雾中毒事件

smog reaction 烟雾反应
smog sensitive 对烟雾敏感的
smoke 烟熏;烟气;烟灰色;烟尘;冒烟
smoke abatement 消烟;除烟(法)
smoke abatement device 消烟除雾设备;烟雾排除设备
smoke agent 烟雾剂
smoke alarm 烟雾报警信号;烟气报警器;烟感报警器;防烟警报
smoke alert 烟雾报警信号
smoke aloft 高空烟
smoke and dust 烟尘
smoke and fire door holder 防火防烟门碰头
smoke-and-fire-proof damper 防烟防火阀
smoke and fire vent 屋顶排烟口;排烟口;屋顶排烟窗
smoke and heat detector 烟热报警器
smoke arch 烟拱
smoke baffle plate 隔烟板;挡烟板
smoke band 烟雾地带
smoke bay 防烟分区
smoke belt 烟罩
smoke black 烟黑;煤烟;炭黑
smoke box 烟箱
smoke-box door 烟箱门
smoke-box front sheet 烟箱前板
smoke-box shell 烟箱筒
smoke-box sprinkler 烟箱洒水器
smoke-box superheater 烟箱过热器
smoke-box tube sheet 烟箱管板
smoke burning apparatus 锅炉烧烟器
smoke canal 烟道
smoke can 烟罐
smoke candle 烟烛;烟雾罐
smoke car(riage) 吸烟车厢
smoke chamber 熏制房;降尘室;烟箱;(壁炉与烟道之间的)烟尘室
smoke channel 烟道
smoke chimney 烟囱
smoke cloud 烟云
smoke compartment 烟室;防烟隔离室
smoke composition 发烟剂
smoke concentration 烟浓度
smoke condensate 烟凝结物
smoke condenser 聚烟器
smoke consumer 完全燃烧装置
smoke consuming material 除烟材料;消烟材料
smoke control 排烟;烟的控制;烟雾控制;控烟
smoke control door 控烟门
smoke control sub-area 防烟控制分区
smoke control system 控烟系统
smoke cover 枪杆套
smoke curtain 烟幕;屋顶挡烟隔板;挡烟垂壁
smoke damage 烟熏损坏;烟害
smoke damper 防烟挡板;烟气调节风门;烟气挡板;烟道调节板
smoke defilade 烟光遮蔽高;炮口烟光遮蔽
smoke deflector 烟气导向器(蒸汽机车)
smoke densitometer 烟雾显像密度计
smoke density 烟浓度;烟密度;排烟密度
smoke density indicator 烟浓度指示器;烟密度指示器;烟尘密度计
smoke density meter 烟(尘)浓度计;烟(尘)密度计
smoke deposition 烟灰沉积
smoke detecting arrangement 烟管式火灾报警装置
smoke detecting plant 烟气探测装置
smoke detecting system 烟气探测系统
smoke detection 烟雾探测
smoke detection appliance 烟气探测设备
smoke detection system 烟感系统
smoke detector 烟雾探测器;烟气探测器;烟感探测器;烟感(检测)器;烟尘探测器;烟尘检测器;火灾烟火探测器;火烟监测器
smoke detector actuated 已驱动的火烟监测器
smoke-developed rating 冒烟指数
smoked glass 熏烟玻璃;烟(灰)色玻璃;灰褐色玻璃
smoked glaze 烟熏釉(层)
smoke discharge facility 排烟设施
smoke discharge fan 排烟风机
smoke discharge manifold 烟幕施放喷管
smoke disposal 排烟
smoke door 排烟活门;排烟门;排烟口
smoked paper 熏烟纸;烟熏纸

smoked paper drum recorder 熏烟纸滚筒记录器
smoked paper drum system 熏烟纸滚筒系统
smoke dried 熏干的
smoke-dried lumber 烘干木材;熏干木材
smoke dry(ing) 熏干;烟熏干燥
smoked sheet 熏烟纸;烟胶(熏干的生橡胶片)
smoke duct 排烟管(道);排烟道
smoke dust 烟尘
smoke-eliminating car 排烟消防车
smoke elimination 消烟
smoke-emanating industry 散发烟雾工业
smoke emission 排烟
smoke emitter 发烟器
smoke enduring plant 耐烟树
smoke evacuation 排烟
smoke evacuation system 排烟系统
smoke exhaust 排烟
smoke exhaust damper 排烟阀
smoke exhaust fan 排烟风机
smoke exhaustion compartment 排烟分隔
smoke exhaustion system 排烟系统
smoke exhaust machine room 排烟机房
smoke extract 烟气抽提;排烟门
smoke extract duct 烟气抽提管
smoke extractor 排烟机;抽烟机
smoke extract shaft 烟气抽提管
smoke eye 烟眼
smoke filter 消烟过滤器;烟雾过滤器;烟尘过滤器;滤烟器
smoke fire damper 防烟防火阀
smoke fire detector 感烟火灾探测器
smoke flue 烟道
smoke fog 烟雾
smoke-free exhaust 无烟排气
smoke from burning coal 煤烟
smoke from cigarette 香烟烟雾
smoke from diesel exhaust 柴油机排出烟气
smoke funnel 发烟筒
smoke gas chamber 烟气室
smoke generation 烟熏司
smoke generator 熏烟剂;烟雾发生器;烟气发生器;烟幕施放器;发烟器
smoke glass 烟(色)玻璃
smoke glass window 烟色玻璃窗
smoke hatch 排烟活门;排烟口
smoke haze 烟霾
smoke helmet 消防面具;防烟面罩;防烟面具;防毒面具
smoke hood 排烟罩;烟囱罩
smoke horizon 烟尘层顶;烟层顶
smoke house 鱼肉熏制厂;熏室;熏肉房
smoke house emission 吸烟室排出物
smoke indicator 烟气浓度计;烟管式火灾报警装置
smoke jack 排烟罩;烟囱帽;烟罩;出烟罩
smoke kiln 熏烟干燥窑
smoke-laden 烟雾弥漫的
smoke-laden atmosphere 含烟空气
smokeless 无烟的
smokeless burner 喷灯;无烟燃烧器
smokeless combustion 无烟燃烧
smokeless exhaust 无烟排气
smokeless flare 无烟火炬
smokeless flash powder 无烟闪光粉
smokeless fuel 无烟燃料
smokeless industry 无烟工业
smokeless powder 无烟火药;无烟炸药
smokeless propellant 无烟推进剂;无烟发射药
smokeless rocket 无烟火箭发动机
smokeless zone 无烟区
smoke level 排烟浓度
smoke mask 防烟面罩;防烟面具;防毒面具
smokemeter 测烟仪;烟度计;烟尘计;烟雾指示器;烟雾计;烟雾测量器;烟尘测量计
smoke monitor 烟尘监测仪
smoke nozzle 喷烟管
smoke nuisance 烟害
smoke obscuring screen 烟幕
smoke observation device 烟气观测装置
smoke ordinance 烟雾管理条例;烟气管理条例
smoke outlet 烟气出口;排烟口
smoke outlet duct 排烟出口管
smoke outlet shaft 烟气出口管
smoke pall 原烟层
smoke pan 烟锅

smoke particle 烟尘颗粒
smoke photometer 烟浓度计
smoke pipe 烟管;烟囱(筒);排烟管(道);排烟道
smoke pipe boiler 火管锅炉
smoke pipe bonding 烟囱防雷
smoke pipe fire alarm system 烟管式火灾报警系统
smoke pipe fire detector 烟管式火灾报警装置
smoke plume 烟羽;烟缕;烟流
smoke pocket (舞台防火幕滑槽中的)排烟腔
smoke point 发烟点
smoke pollution 烟(雾)污染;烟气污染;煤烟污染
smoke pot 发烟罐
smoke prevention 烟雾防护;烟雾防治;除烟法
smoke prevention and dust control 消烟除尘
smoke prevention stair(case) 防烟楼梯(间)
smoke prevention stairwell 防烟楼梯间
smoke printing 喷粉复印
smoke-proof 不透烟的;防烟的;不漏烟的
smoke-proof barriers 防烟阻挡物
smoke-proof compartment 防烟分隔
smoke-proof curtain 防烟幕
smoke-proof damper 防烟阀
smoke-proof enclosure 防烟维护结构;防烟隔离室
smoke-proof enclosure vestibule 防烟封闭前室
smoke-proof stair(case) 防烟楼梯(间)
smoke-proof tower 排烟塔;防烟塔(有出口的内部封闭的楼梯间)
smoke-proof vestibule 防烟前室;防烟围护结构;防烟隔离室
smoke protection 消烟防护;烟雾防护;防烟
smoke protection and exhausting ventilation system 防排烟通风系统
smoke protection damper 防烟闸板;防烟挡板
smoke puff 烟迹
smoke quartz 烟晶
smoke receptor 烟感器
smoke recorder 烟尘记录器
smoke removal 排烟
smoke removal efficiency 烟雾滤除效率
smoke rocket 喷烟测漏器
smoke room 吸烟室
smoker's tongue 吸烟舌
smokes 白浓雾
smoke sampler 烟取样器
smokescope 烟度计;烟雾浓度测定器;烟雾浓度测量器;烟尘浓度测定器;烟尘密度测定器;检烟镜
smoke screen 烟幕
smoke scuttles 排烟窗口;排烟舱口;屋顶烟雾排出口
smoke seasoning 熏干
smoke separation wall 挡烟壁体
smoke shade 烟雾色调;大气污染粒子
smoke shaft 排烟竖井
smoke sheets 烟片
smoke shelf 烟挡;导烟板
smoke shell 烟雾
smoke shutter 烟障
smoke signal 烟雾信号;发烟信号
smoke spot 烟斑
smoke spread 烟播散
smokestack 烟囱;大烟囱
smoke stack base 烟筒座
smoke stack cap 烟囱帽
smoke stack emission 烟囱排放
smoke stack guy 烟囱支索
smoke stack head 烟囱头
smoke stack hood 烟囱帽
smoke stack industries 烟囱工业
smoke staining 串烟
smoke stain resistance 抗烟雾污染性
smoke stick 烟雾测风流法
smokestone 烟晶
smoke-stop 隔烟墙
smoke-stop anteroom 消烟前室
smoke-stop door 隔离烟门;挡烟门;防烟门
smoke-stop partition 防烟隔断
smoke stratus 烟层云
smoke substance 烟雾剂
smoke suppressant 消烟剂;防烟剂
smoke suppressant additive 防烟添加剂
smoke suppressor 消烟剂
smoke technique 烟流技术
smoke telegram 烟度计
smoke test 烟雾试验;烟气试验;火焰高度试验;通

烟试验(检查管道方法);送烟试验;发烟试验
smoke tester 烟(浓)度检验计;火焰高度检验计
smoke-tight 不漏烟
smoke tower 排烟塔
smoke tower window 排烟塔窗;烟塔窗
smoke toxicity 烟的毒性
smoke trail 烟迹
smoke train 冒烟流星余迹
smoke treatment equipment 煤烟处理设备
smoke treatment plant 煤烟处理设备
smoke tube 烟筒;烟管;火管
smoke tube boiler 烟管锅炉;火管锅炉
smoke unit 发烟器;生烟强度单位;烟气单位数
smoke uptake 通烟道
smoke valve 烟闸门
smoke vent 排烟口;排烟管(道);排烟道;出烟孔
smoke ventilator 排烟器
smoke vent system 烟管式火灾报警装置;火警探测器
smoke volatility index 烟挥发指数
smoke warning indicator 烟雾警告指示器
smoke washer 烟气洗涤器
smoke wave 烟波
smoke wind tunnel 烟风洞
smoke zone 烟雾带
smoke zoning 防烟分区
smokiness 烟雾性;发烟性
smoking 熏烟;吸烟;窑炉缓慢加热;烟熏
smoking agent 烟雾剂;发烟剂
smoking burner 发烟灯
smoking combustion 冒烟燃烧
smoking load 烟尘负荷
smoking patch 烟斑
smoking pipe 烟囱
smoking room 吸烟室
smoking stand 烟灰台
smoking trace 烟迹
smoking tracer mixture 发烟迹剂
smoking withdrawal clinic 戒烟诊所
smoko 工作休息间
smokometer 烟尘计;烟密度计;测烟仪
smoky chimney 多烟烟囱
smoky flavour 烟熏味
smoky fog 烟状雾;烟雾
smoky pesticide 烟剂农药
smoky quartz 烟色石英;烟水晶;烟晶;墨晶
smoky quartz carving 墨晶石雕
smoky structure 烟灰状构造
smoky topaz 烟色黄玉
smolder(ing) 徐燃;冒烟;阴燃;熏烧;发烟燃烧;闷烧
smoldering fire 闷火
smolianinovite 水砷钴铁石
smooth 光滑的;校平;滑润的;平整的;平稳的;平滑形;平滑的;道格拉斯一级浪;大光钡地纸;修平
smooth acceleration 平稳加速
smooth-acting 平滑动作的
smooth aggregate 光滑集料;光滑骨料
smooth air 平静空气
smooth air-gap 均匀空气隙
smooth and steady 平稳的
smooth approach 平稳进场着陆
smooth ashlar 光细琢石;光面琢石;平整方石
smooth-averaged sound field 平滑平均声场
smooth back 光滑树皮
smooth bar 普通圆钢;平圆钢
smooth-barked 树皮光滑的
smooth bit 磨钝的钻头;磨钝的钎头
smooth blasting 光面爆破;平稳爆破
smooth-bore copper tube 光壁钻孔铜管
smooth boundary 光滑边界
smooth brass pipe 光滑黄铜管
smooth break 平滑断裂
smooth burning 稳定燃烧
smooth by trowelling 慢平抹光
smooth carpet 平滑地毯
smooth cement finish 水泥粉光
smooth choke 平滑滤波抗流圈
smooth circular bend 平移圆形弯道
smooth cleavage 平滑劈理
smooth coast 平直海墙;平直(海)岸;顺直(海)岸
smooth coastline 顺直(海)岸线
smooth combustion 平稳燃烧
smooth concrete 光面混凝土
smooth condenser 平滑电容器

smooth continuous revetment 平顺护岸
smooth contour 光滑的外形
smooth core 光滑铁芯
smooth core armature 平滑电枢;平槽电枢;无槽电枢
smooth-core rotor 光滑铁芯转子
smooth crabgrass 平滑蒲公英
smooth curvature 平滑曲率
smooth curve 光滑曲线;滑顺曲线;平缓曲线;平滑曲线;顺滑曲线
smooth cut 细纹锉
smooth-cut file 细锉;细纹锉
smooth deceleration 平滑减速度
smooth-delineation copy 出版原图
smooth-delineation map 出版原图
smooth descend type 平滑下降型
smooth design spectrum 平滑设计谱
smooth discontinuity 光滑结构面
smooth dowel 光面铁条棒;光面传力杆
smooth drafting 清绘
smooth drilling formation 高钻速完整取芯岩层
smooth drum roller 光滑滚筒碾压机
smooth drum vibratory roller 光筒振动碾
smoothed 光滑的
smoothed-axed stone face 细剁石面
smoothed distribution 平滑化分布
smoothed edge 细磨边
smoothed frequency curve 修匀频数曲线
smooth-edged knife 光刃动刀片
smoothed glass 磨光玻璃
smoothed non-linearity 平滑非线性
smoothed response spectrum 平滑反应谱
smoothed spectral density 平滑谱密度
smoothed surface stone 光面石块
smoothed theoretical dynamic series 修匀理论动态数列
smoothed value 修匀值
smoothened-brick wall 磨砖对缝墙
smooth engagement 平稳接合;平滑接合
smoother 异型曼刀;整平工人;整平工具;展平滤波器;刮路机;路面整平机;校平工具;镘板;平整器;平滑(滤波)器;修光工具
smoother bar 路平板;(路面的)整平板
smoothest approximation 最平滑逼近
smooth face 光滑面
smooth-face clay tile 光面黏土砖;光面黏土瓦
smooth-faced crushing rolls 光面破碎机辊
smooth-faced drum 光滑表面的滚筒
smooth-face structural clay tile 建筑用光面黏土砖
smooth factor 平滑系数
smooth feature 柔性
smooth fibre 光滑纤维
smooth file 细锉
smooth fillet weld 平填角焊缝
smooth finish 光面修整;光洁度;表面光洁度;磨光面
smooth-finished 表面光滑的;修饰平整的
smooth finish tile 光面瓷砖;光面制砖;模制砖
smooth flat wood rasp 细扁木锉
smooth flight 平稳飞行
smooth flow 连续流;平直水流;平稳流动;平滑(车)流;平顺流
smooth fracture 细粒断裂;晶粒断裂
smooth function 平滑函数
smooth-going 平顺
smooth grain 平滑材面
smooth grinding 细磨;精研磨
smooth ground surface 平滑磨削表面
smooth gypsum plaster 光面石膏粉刷
smooth hillside 渐倾坡;平顺山坡
smooth hole 光滑孔
smooth ice 光滑冰
smoothing 修平;修匀;消除数据波动;滤清;滤除;滤波;平滑;使平整
smoothing adjustment 使光滑
smoothing agent 光滑剂
smoothing beam finisher 样条粉光机;准条粉光机;梁式粉光器
smoothing belt 镘光带(混凝土路面)
smoothing board 压deck器;镘板;浆膜台;粉光板
smoothing board finish 用镘板粉光
smoothing by hand 徒手圆滑
smoothing capacitor 滤波电容器;平滑电容器
smoothing cement 油灰;腻子

smoothing chisel 展平凿
smoothing choke(coil) 平滑扼流圈;滤波扼流圈
smoothing circuit 平滑电路
smoothing condenser 平滑电容器
smoothing constant 修匀常数;平滑常数
smoothing current 平滑常数
smoothing device 滤波器;平滑装置
smoothing effect 平滑作用
smoothing filter 平滑滤波器
smoothing filtering 平滑滤波
smoothing formula 修匀公式
smoothing function 平滑功能
smoothing function of a curve 曲线光滑函数
smoothing hammer 敲平锤;平光锤;平锤
smoothing harrow 细平耙;平地耙;无齿(平土)耙
smoothing iron 烙铁;铁镘板;熨斗;抹光烙铁(沥青路面接缝用)
smoothing kernel 平滑核
smoothing kiln 摊平炉(平板玻璃)
smoothing lacquer 平光漆
smoothing machine 磨光机
smoothing method 修匀法
smoothing off peak discharge 削平洪峰
smoothing of irregularities 不规则性修匀
smoothing of series 序列修匀
smoothing operation 光滑操作
smoothing plane 光滑刨;木工细刨;细刨;光刨
smoothing plane machine 细刨床
smoothing planing 细刨
smoothing planing machine 细刨床;精刨床
smoothing planner 细刨床
smoothing procedure 修匀过程
smoothing process 修匀过程;平滑过程
smoothing quelling seas with oil 布油镇浪
smoothing reactor 滤波电抗器;平滑电抗器
smoothing roll 整直轨辊;钢板精轧辊
smoothing saw 细锯
smoothing screed 粉光准条;粉光灰饼;磨光梁;粉光刮板
smoothing screed finisher 准条粉光机
smoothing stone 磨平石
smoothing technique 平整法
smoothing time-series data 修匀时序数据
smoothing tool 光滑加工工具;光刀;精加工工具
smoothing trowel 粉光用泥刀;粉光用抹子;赶光抹子;镘板;粉光抹子;粉光镘
smoothing with float 木抹抹平
smoothing with trowel 铁镘赶光
smooth jaw plate 平面颚板
smooth joint 光滑连接;光滑接点
smooth leather 光面墨辊
smooth lime finish 光面石灰粉刷;光面石灰修整
smooth line 均匀分布参数线路;平滑线(路)
smooth lining 平整井壁
smooth log 誊清的航海日记
smoothly fluidized bed 平稳流化床
smoothly radiused edge 平滑圆角
smoothly sanded 细砂磨光的
smooth machine finish 光面修整;机器修整;机磨光面
smooth manifold 光滑流形;平滑流形
smooth map 光滑映射
smooth millboard 光面(硬)纸板
smooth motion 平稳运动;平滑运动
smoothness 光洁度;光滑(度);平整性;平整度;平稳性;平滑度;平滑(度);表面光滑度
smoothness assumption 平滑性假设
smoothness check 平滑度检查
smoothness of operation 运转平稳性
smoothness of paper 纸张平滑度
smoothness tester 平滑度试验器
smooth nozzle 平滑喷嘴;平滑喷管;平滑管嘴
smooth oblique descend type 平滑斜降型
smoothometer 不平度测定仪
smooth operation 平稳运转
smooth oriental photinia 平滑石楠
smooth outflow 均匀流出;平稳流出;平缓流出
smooth panel container 内柱式集装箱;平面板式集装箱
smooth passage 光滑通道;畅通
smooth pipe 光面管;光滑管
smooth pipe compensator 光面管伸缩器
smooth pipe expansion joint 光面管补偿器
smooth pipe radiator 光面管散热器

smooth piping compressor 光滑管伸缩器
smooth plain packing 烟道式格子体；平流式排列
smooth planer finish 光面修整；机磨光面
smooth planking 平铺(式)船壳板
smooth plaster finish 光滑抹灰面
smooth plastic pipe 平滑塑料管
smooth plate 光滑板；平钢板
smooth power spectral density function 平滑功率谱密度函数
smooth profile 光滑剖面
smooth reaction 平稳反应
smooth regression analysis 修匀回归分析
smooth regulation 平稳调整
smooth relief 平坦地势
smooth-ride 平稳行驶
smooth riding 平稳行车
smooth-riding quality (路面的)行车平稳性
smooth-riding surface 平稳行车路面；平整路面
smooth-riding track 平稳行车轨道
smooth river stretch 平顺河段
smooth roll 平面轧辊；光辊
smooth roll crusher 平面辊式破碎机；平滑辊式破碎机
smooth rolled 光面碾压
smooth roller 光碾压路机；光面压路机；光辊
smooth roofing 光滑屋面卷材
smooth running 平稳转动；平稳运转；平稳行驶
smooth run-off 平稳退绕
smooth sea 小波；细波；平静海面；无浪；微浪；微波海面；二级浪
smooth sea area 平静海面区域
smooth-section sickle 光刀刀片割刀
smooth shelf 平滑陆架
smooth-shell crushing rolls 光滑外壳轧碎滚筒；光轮破碎辊；光轮破碎机
smooth shore 顺直(海)岸
smooth shoreline 顺直(海)岸线
smooth side 光滑侧
smooth-skinned 表皮光滑的
smooth slope 缓坡
smooth soffit 光面拱腹
smooth-stalked meadow grass 草地早熟禾
smooth start 平稳起动
smooth starting 平稳启动
smooth steel pipe 光面钢管
smooth steel tube 光面钢管
smooth surface 平整路面；光面船壳；光洁表面；光滑(表)面；平整(表)面；表面光滑
smooth surface blasting 光面爆破
smooth surface colloid mill 光面胶体磨
smooth surfaced endoplasmic reticulum 光面内质网
smooth surfaced roll roofing 光面油毡
smooth surfaced roofing 光面屋面；光面屋顶
smooth surface finish 光滑表面修整
smooth surface red rubber hose 光面红橡胶管
smooth surface theory 光面理论
smooth terrain 平坦地形；平坦地(带)
smooth texture 平滑性
smooth-texture grease 均匀结构润滑脂
smooth thrust variation 推力缓变
smooth tile 光滑的砖；光滑的瓦
smooth-tires roller 光滑轮胎压路机
smooth tongue 光舌
smooth torque 平滑转矩
smooth transfer 平滑过渡
smooth transition 平稳过渡
smooth tread 平滑胎面；光纹的
smooth-tread tire 光纹轮胎
smooth turning 光车削
smooth-two-side hardboard 两面光硬质纤维板
smooth type 光滑型
smooth-tyres roller 光滑轮胎压路机
smooth visible under-face 内侧可见的光滑(面)
smooth visible underside 内侧可见的光滑(面)
smooth-wall blasting techniques 光面爆破法
smooth-walled 光滑壁面
smooth-walled hole 孔壁平整的钻孔；井壁平整的钻井
smooth-wall polyethylene pipe 光壁聚乙烯管
smooth-wall polyvinyl chloride conduit 光壁聚氯乙烯导管
smooth-wall polyvinyl chloride plastic underdrain system 光壁聚氯乙烯塑料管地下系统

smooth water 静水；微浪
smooth water area 平静海面区域
smooth water resistance 静水阻力
smooth waveguide 均匀波导
smooth wavelet 微波
smooth-wheeled tandem roller 光轮串联式压路机
smooth-wheel roller 光轮压路机；光轮路碾；光轮轮碾压机；平碾；光滑滚筒碾压机；平轮压路机
smooth-wheel roller with internal vibrator 带内振动器光滑轮碾压机
smooth window length 平滑时窗长
smooth working 平稳工作
smooth zone 光滑带
smother 灭火
smothered arc welding 隐弧焊
smothering 封闭灭火
smothering arrangement 窒息灭火装置
smothering line 窒息灭火管系
smothering method 窒息灭火法
smother kiln 熏窑
smo(u)lder 熏烧；烟化；缓慢燃烧；发烟燃烧；阴燃；闷烧
smo(u)ldering 熏；隐燃
smudge 熏烟；油漆残渣混合物；煤污病；黑污点；污渍；污迹
smudge coal 天然焦炭
smudged 涂黑的
smudge oil 花园用燃料油
smudge pot 烟熏点
smudges 影像斑点
smudging 烟熏；污斑；斑点闪烁
smuggle 走私；夹带
smuggled goods 走私货；私货
smuggle goods into a port 私运进口
smuggler 走私者；走私船
smuggling and sale of smuggled goods 走私贩私活动
smuggling ring 走私集团
smuggling ship 走私船
smut 烟炱；劣质煤；煤垩；黑污点；黑粉病煤尘；片状炭黑；土煤；酸洗残渣
smutch 污点
smut fungi 黑粉病菌
smuts 煤尘
smythite 菱硫铁矿
Smz alloy 斯木兹硅锰锆铁合金
snach plate 防止钻杆冲出的工具(遇高承压水时)
snack bar 小吃店；快餐店；快餐餐厅；快餐部
snack counter 快餐柜台；小吃部；快餐部
snack table 单人小餐桌
snag 枝茬；站杆；浸root；清铲；水中隐树；树桩；沉树(将砍下的树锚碇于岸边以护岸防冲)；沉木；打磨
snag boat 清障船；清除水中障碍船；扫床船
snagging 清除障碍物的设备；清除障碍；荒磨；河道清障；清障；清除障碍物(如树枝等)；打磨
snagging boom 清障吊杆 (清浚船吸口)
snagging equipment 掘伐木机具；筏木机；伐木机
snagging grinder 粗磨床
snagging of river 河流障碍
snagging wheel 粗砂轮
snag grinder 荒磨床
snag hook 重型吊钩
snag one 牢固地卡在提引器上
snail 胡桃木上卷曲饰图；平面螺旋；涡形凸轮；涡形轮；腹足类软体动物
snail countersink 螺旋锥口钻
snail hook 螺旋钩
snailish 爬行的
snail-like movement 慢慢爬
snail-paced 爬行的
snail-shaped nebula 蜗牛状星云
snail-show 爬行的
snail's pace 爬行速度
snail type heavy-medium separator 蜗型重力介质选矿机
snail wheel 蜗形轮
snake 拉线钢带；可弯曲芯棒；填卷；蛇形纹；蛇形棒；玻璃带宽窄不一致；白点
snake bend 蛇形弯头；蛇形弯管
snake bite 蛇伤
snake cage resin 蛇笼树脂
snake core 弯曲芯；蛇形芯
snake domain 蛇形畴

snake drill 蛇形钻杆
snake fashion 来回曲折方法(在岩芯箱中放置岩芯)
snake fence 蛇形栅栏
snake formation search 蛇形队搜索
snake hole 爆破孔；蛇穴形孔；蛇炮眼；底部钻孔；底部炮眼；蛇穴式炮眼；蛇穴式炮眼
snakehole shot 蛇穴式炮眼爆炸
snake hole springing 水平炮眼掏壶
snakeholing 底部钻进；打爆破孔
snakeholing method 蛇穴(爆破)法；蛇穴法二次爆破(大块底部打眼爆破)
snake in the tunnel 坑道中的蛇行
snake-like 蛇形的
snake line 滑车尾绳；牵引索；牵引绳
snake mark 蛇形斑点
snake mode 蛇形(飞行)模式
snake motion 蛇行
snake pit 蛇窖；疯人院
snake poison 蛇毒
snaker 弯轧槽；弯曲模
snake riveting 正常交错铆接
snakes 蛇形泥条
snakes in hole 塌孔
snake skin 蛇皮
snakeskin glaze 蛇皮釉
snakeskin green 蛇皮绿
snakeslip 擦板条
snake's sensor 蛇感受器
snakestone 菊石；结核体；蛇石；苏格兰石；蛇纹石
snake type strander 蛇形捻股机
snake venom 蛇毒
snake wire 绳索包皮
snake without tunnel 没有坑道的蛇行
snake wood 淡褐色；蛇根木；蛇纹木材
snaking 填卷；蛇形飞行
snaking chain 拖链
snaking conveyer 弯曲包铁皮的运输带
snaking motion 曲线运动；蛇形运动
snaking river 曲折河；蜿蜒河流；弯曲河流；蛇曲河(流)
snaking stream 曲折河；曲流河；蜿蜒河流；弯曲河流；蛇曲河(流)
snak-quench technique 均热淬火工艺
snaky 蛇形的
snaky edge 蛇形边
snaky track 蛇形导轨；弯曲辊道
snalge shaft 角条
snap 快拍；急速的；急变；突然折断；天气急变；弹响；短冷期；采板
snap-acting thermostat 快速启动恒温器
snap action 快速动作；快动作；瞬间动作；双位调节
snap action contact 突动触点；瞬动触点
snap action switch 快动开关
snap action valve 快速作用阀
snapback 快回；急速返回；快反向；急变返回
snap back forming 快速回吸成形
snapback hook 弹簧钩
snapback method 反复计时法
snap bench 小工作台
snap bolt 自动门闩；自动门栓；自动栓门
snap button 按钮
snap check 现场抽查；突击检查；抽查检验
snap clasp 弹簧拉钩
snap-close 卡锁；锁扣
snap clutch 弹压齿式离合器；弹压齿式离合器
snap coupling 自动联结器；自动连接；快速联轴节
snap die 铆钉头压形；铆钉头压模
snap-down 排出
snap-dragon 金鱼草属
snape 斜截；密配合；接合
snaped 两端为锥形的原木
snap fastener 撤扭；按扣；弹簧搭扣
snap-finger 扳机状指
snap-fitted 扣接的
snap fitting 快接接头；卡口接头
snap flask 扣合式砂箱；可拆式砂箱；脱箱
snap flask band 套箱
snap flask mo(u)ld(ing) 脱箱造型；无箱铸型
snap ga(u)ge 卡规；外径规；卡板
snap hammer 圆边击平锤；铆钉头圆边击平锤；铆锤
snap hand jet 带柄外径气动量规；带柄气动卡规
snap head 圆头铆钉；圆头
snap head bolt 半圆头螺栓；圆头螺栓
snap head die 铆钉模

snap header 假丁砖;半砖;暗丁砖;暗丁石;圆头铆钉;(砌体表面用作丁砖的)半头砖
snap head machine screw 半圆头机器螺钉
snap head rivet 半圆头铆钉;圆头铆钉
snap head screw 半圆头螺钉
snapholder 弹簧柄
snap hook 链索弹簧扣;旗绳钩;弹簧扣;弹簧钩;安全钩
snap in cover 快速压紧盖
snap jet 气动量规;气动测头
snap lever oiler 杠杆带动给油环;弹簧杆润滑器
snap line 弹线
snap link 链索弹簧扣
snap lock 碰锁;弹簧锁;弹簧搭扣
snap-lock switch 弹簧锁开关
snap magnet 速动永磁铁
snap off timing discriminator 快速关断定时甄别器
snap-on 直接固定
snap-on alumin(i)um trim 扣口式铝框
snap-on ammeter 钳式安培计
snap-on bolt 自动门闩
snap-on cover 弹簧扣盖
snap-on joint 搭锁接合
snap-on red filter 快装式红色滤光片
snap-on tongs 抓钳
snap-out 飞丝
snapover 闪弧
snap over mechanism 弹锁机构
snapped head bolt 圆头螺栓
snapped header 半头砖
snapped rivet 圆头铆钉
snapped work 半砌体
snapper 掀钮;抓泥器;海底取样器;掀钮;瞬动咬合器
snapper sampler 抓式采样器;抓式采泥器
snapping finger 扳机状指
snapping line 弹线
snapping of drill pipe 突然回动(卸钻杆时)
snapping the line 弹线;白粉线;弹线工序
snap piston ring 弹簧活塞环
snappy 瞬息作用
snappy return 弹回运动
snappy rubber 优质弹性橡胶;高弹性橡胶
snappy spring cover 自闭盖
snap remover 开口环装卸器;弹性卡环装卸器
snap ring 锁定环;压圈;止动环;扣环;开口环;卡环;弹性挡圈;弹簧锁环
snap-ring bearing 开口环轴承
snap-ring driver 止动环装卸器
snap ring expander 弹性卡环装卸器
snap ring extractor 止动环提取器
snap ring groove 止动环槽
snap-ring internal 锁圈
snap ring pliers 开口环钳
snap-ring seal 扣环密封
snap rivet head 铆钉半圆头;圆头铆钉
snap riveting 型铆
snap sample 抓取样品;抓取试样;瞬时试样;瞬间水样
snap shears 快速飞剪(剪切钢材前端)
snapshot 快照;拍快照;快摄照片;快镜拍摄;抽点打印
snapshot camera 快拍照相机
snapshot debug 抽点查错
snapshot debugging 抽点打印调试
snapshot dump 机动转移信息;抽样转储;抽点转储[贮];抽点打印
snapshot program(me) 抽点打印程序;抽点程序
snapshot routine 抽点打印程序
snap switch 快动开关;弹簧开关;速动开关;瞬动开关;室内开关
snap tempering 快速回火
snap terminal 搬钮接头;弹簧锁;弹簧夹
snap the line 弹线
snap tie 混凝土墙模板系杆;可卡断的拉杆;弹性夹紧联结杆
snap time 阶跃时间
snap tool 铆钉器械;铆钉工具;铆钉用具
snap up 锁键调节式
snap valve 速动阀
snap washer 开口垫圈
snare 勒除器;圈套器
snare loop 绞断器圈套环
snarl 弄乱;打出浮雕花纹(在金属薄片上)
snarl(ing)test 线材反复缠绕试验;打结试验

snarl iron 打凸纹器
snarl knot 死结
snarls 扭结
snarl yarn 辫子花线
snatch block 紧线滑轮;扣绳滑轮;扣缆滑轮;开口滑车
snatching 冲动
snatch-lift 突然吊起
snatch plate 扣链板
snatch pulley 开口滑车
Sn-bearing lead ore 锡铅矿石
S-N diagram 应力次数曲线(疲劳试验中的);疲劳试验曲线图
S-N diagram of stress 应力循环次数图
snead process 直接加热法
sneak 潜行
sneak circuit 寄生电路;超指标电路
sneak-in 潜入
sneak-off 潜出
sneak-on 潜入
sneak(out)current 寄生电流;潜行电流
sneak path 潜通路;偷渡通路
sneck 乱石;填空小石;塞缝小石
snecked 乱石砌筑的;用乱石砌的
snecked masonry wall 乱砌毛石墙
snecked rubble 杂乱毛方石砌体;乱毛石墙;粗毛石;杂乱毛石;乱石
snecked rubble wall 乱砌毛石墙
snecked wall 乱石砌筑的墙
sneck head 窗钩;窗拉手;门拉手
snecking 毛石砌体
snecks 填空小石
sneezing bar 摇臂;冲击梁
Snellen fraction 斯涅伦小数
Snellen notation 斯涅伦表示法
Snell laws of refraction 斯涅耳折射定律
Snell's law 折射定律;斯涅耳定律
snezhura 雪泥;湿雪
SNG 代用天然气
snib 门钩;窗钩;插销;闩
snibill 有枢栓的铰
S. Nichola's church 圣·尼古拉教堂
snick 截痕
sniffer 吸气探针;吸气器;真空检漏器;检漏头;取样器
sniffer probe 吸气(取样)探针
sniff hole 蓄水箱顶部的通气孔
sniffing 吸入空气
sniffing valve 吸气阀
sniffle valve 吸气阀
snift 吸气取样
snifter 自动充气器
snifting valve 溢气阀;吸气阀;排气阀;喷气阀
snip 剪切;剪片;非常划算的交易
snipebill plane 槽刨
snipe nosed pliers 尖嘴钳
sniperscope 夜视瞄准器;瞄准望远镜;红外(线)夜间瞄准镜;红外(线)瞄准镜;夜望镜;红外线瞄准器
snipe's-bill 槽刨;鸟嘴刨
snipped edge 斜边
snippers 手剪
snipper saw 裁切锯
snips 白铁剪;平头剪;铁丝剪;铁皮剪;切金属片机
snitch 机械测井仪表(记录钻进时间、停歇时间和孔深)
snivet 干扰黑线
snivvey 吊索;牵索
sno 埃尔维阵风
snob effect 逆反效应
Snoek effect 斯诺克效应
snook rectifier 旋转回臂高压整流器
snooperscope 夜视镜;夜望镜;夜视仪;夜望器;夜间探测器
snooperscope spectrometer 红外(线)潜望器分光计
snoop leak detector 漏气检查器
snoot 鼻喷嘴
snore-hole 通气孔
snore piece 吸水管端;水泵吸收管头;单向活门(泵的)
snoring condition (泵的)空吸状态
snorkel 潜水送气管;送气管;柴油机通风管工作装置
snorkel diving apparatus 简易轻潜水设备
snort 喷气声

snorter 单套绳
snort wheel 放风阀绳轮
snotter 氧化铈夹杂物
snout 象鼻;进口锥体
snout bow 撞角型船首
snout of glacier 冰川鼻
snow accumulation 雪的累积;积雪
snow algae 雪生藻类
snow avalanche 雪崩
snowball 雪球
snowball (garnet) structure 雪球构造
snowball sampling 雪球式抽样
snowball texture 雪球结构
snow bank 雪堆;雪堤
snowbank digging 雪堤掘进
snowbank glacier 雪蚀冰川
snow barchan(e) 新月形雪丘
snow barrier 挡雪板;防雪栅栏;防雪栏;防雪板
snow belt 雪带
snowberg 覆雪冰山
snow bin 量雪箱;测雪箱
snow blink 雪照云光
snow block 雪片
snow blockade 雪封;积雪堵塞
snow blower 螺旋桨式除雪机;扫雪机;吹雪机;喷气除雪机
snow blowing machine 旋转扫雪犁;旋转扫雪机;喷气除雪机
snow board 屋面冰块挡板;挡雪木板;雪栏;挡雪板;测雪板
snow bound 被雪封住;雪封的
snow box forming hood 雪花式形成室
snowbreak 雪折;雪障;雪挡;冰雪融化
snow breaker 雪栏;除雪机;防雪林
snow break forest 防雪林
snow bridge 雪桥
snow brom 雪暴
snow-broth 雪水;融化雪泥
snow bube 雪柱收集器
snow cap 雪冠
snow-capped 盖着雪的;积雪的
snow caster 旋转清雪机
snow casting device 抛雪装置
snow-cat 雪地履带式车辆
Snowcem 斯诺森(一种防水的彩色水泥涂料)
snow chain 雪链;雪地防滑链;防雪滑链条
snow-clad 盖着雪的;积雪的
snow clearer 铲雪车
snow clearing 扫除积雪;铲雪
snow clearing machine 除雪机
snow clearing machinery 除雪机械
snow climate 雪原气候;极地气候
snow cloud 雪成云
snow condition 降雪条件
snow-control forest 防雪林
snow core 雪样
snow cornice 雪檐挑檐
snow course 降雪量观测线;测雪线;测雪路线
snow cover 雪盖;雪被;积雪覆盖层;积雪度;积雪
snow cover chart 积雪(分布)图
snow cover density 积雪密度
snow cover-density of watershed 流域雪覆盖密度
snow cover distribution 积雪分布
snow-covered 盖着雪的;积雪的
snow-covered area 积雪地区
snow-covered ice 覆雪冰
snow covered runway 覆雪跑道
snow cradling 挡雪板;挡雪架
Snowcrete 斯诺克里特(一种白色或彩色混凝土);凝固雾
snow crust 雪壳;积雪层薄而硬的表面
snow crystal 雪晶
snow damage 雪害
snow defence 防雪
snow density 雪密度
snow deposit 雪层;积雪
snow depth 雪深;积雪深(度);积雪厚度
snow depth on ice 冰上雪深
snow dike 雪堤
snow discharging 雪中装卸
snow drift 雪堆;吹雪(堆);吹雪
snow drift control 控制吹雪措施
snowdrift glacier 吹雪冰川

snowdrift ice 吹雪冰
snow drifting 雪的吹积;雪的漂动;雪害
snow drift prevention forest 防雪林
snowdrop 雪莲花属
snow-dumping shaft 积雪倾卸井
snow dune 雪丘
snow eater 溶雪雾;溶雪风
snowed-up 被大雪阻住
snow emergency 雪灾
snow erosion 雪蚀
snow evapo(u)ration 雪面蒸发
snowfall 下雪;雪量;降雪量;降雪
snowfall rate 降雪强度;降雪率
snow-feed river 雪水补给河;雪水补给的河流
snow fence 防雪围栏;雪栅;挡雪栏;防雪栅栏;防雪栅;防雪栏
snow field 雪原;粒雪原
snow fighter 除雪机
snow flake 雪片;雪花;白点
snow flanger 扫雪器
snow flood 桃汛
snow flurry 雪阵;阵雪
snow gallery 防雪明洞
snow garland 雪花环
snow ga(u)ge 雪量计;雪规;量雪器;量雪计;测雪计
snow gear 雪地起落架
snow geyser 雪喷泉
snow glacier 粒雪冰川
snow grader 推雪机
snow grain 米雪
snow guard 雪栏;挡雪装置;挡雪板;防雪栅(栏);防雪板
snow guard tile 挡雪瓦
snow gutter 排雪沟;排雪槽
snow handling 雪中装卸
snow handling machine 除雪机
snow hazard 雪害
snow hedge 挡雪屏障
snow hook 挡雪圈
snowhouse 雪屋
snow hurdle 防雪栅;防雪栅栏;防雪栏
snow hydrology 雪水文学
snow ice 雪冰;冻雪;冰雪
snow igloo 雪屋
snow interception 降雪截留量
snow landing 雪地着陆
snowland penstemon 雪地钓钟柳
snow level 噪声电平
snow limit 雪限;雪线;雪界
snow line 雪线
snowline height 雪线高度
snow load 雪荷载
snow loader 装雪机;装雪车;扫雪装车机
snow loader with belt conveyer 带式装雪机
snow loader with continuous buckets 连续斗式装雪车
snow loading 雪载;雪荷载
snow load stress 雪载应力;雪荷载应力
snow load value 雪载值
snowmaker 造雪机
snowman 雪人
snow mantle 雪盖;积雪覆盖层;积雪
snow mat 雪席
snow measuring plate 测雪板
snowmelt 雪水;雪融水;雪融化;解冻水;融雪;融雪
snowmelt by convection 对流融雪
snow melter 融雪装置;融雪器;融雪机
snowmelt excess 融雪水产量;超渗融雪水
snowmelt-fed river 融雪补给河(流)
snowmelt-fed stream 融雪补给河(流)
snowmelt flood 融雪洪水;雪洪;桃汛
snowmelt forecast 融雪预报
snowmelt hydrograph 融雪径流过程线
snow-melting oil 融雪油
snowmelting pipe 融雪管
snowmelting rate 融雪率
snowmelting tank 融雪箱
snowmelt intensity 融雪强度
snowmelt period 融雪期
snow-melt runoff 融雪径流
snowmelt season 融雪期
snow metamorphosis 雪的变质;雪的变性;雪的变态;雪的变形
snow mobile 摩托雪橇
snow mo(u)ld 雪腐病
snow niche 雪蚀凹地
snowpack 积雪(量);雪堆
snowpack water equivalent 雪(的)水当量;积雪水当量
snowpack yard 积雪场
snow paling 防雪栏
snow-patch erosion 雪蚀;雪—霜蚀作用
snow pellets 雪丸;软雹(霰)
snow plough 雪犁;排雪机;扫雪机;除雪机
snow-plough model 雪犁模型
snow plow 扫雪机;雪梨;犁雪器;犁雪机
snow plow attachment 除雪装置
snow-plow car 犁雪车
snow plowing automobile 扫雪汽车
snow point 雪点
snow pollution 雪污染
snow precipitation line 降雨线;降雪线
snow primrose 雪报春
snow protection bank 防雪堤
snow protection device 防雪设备
snow protection facility 防雪设施
snow protection hedge 挡雪栅栏;防雪篱
snow protection plantation 防雪(造)林
snow protection system 防雪系统;雪的保护系统
snow pusher 扫雪机
snowquake 雪崩;积雪震落
snow-rain ga(u)ge 雪量雨量计;雨雪计
snow-rain recorder 自记雨雪量器;自记雨雪计
snow reflecting camouflage paint 雪地伪装涂料
snow removal 铲雪
snow removal equipment 除雪设备
snow removal operation 除雪工作
snow removal truck 扫雪车
snow remover 除雪机
snow research 雪的研究
snow residua 残雪量
snow residuum 残雪量;残留雪
snow retention 积雪;挡雪
snow ripening 雪成熟
snow ripple 风成雪波
snow roller 雪卷;雪尺
snow runoff 融雪径流
snow sample 雪样
snow sample cutter 刮雪样器;采雪样器
snow sampler 雪柱收集器;雪试样;积雪采雪器;取(样)雪样器;取雪管;采雪器
snow sampling 取雪样
snow sampling course 测雪取样路线
snow scale 雪(深)尺;量雪(标)尺
snow-scape 雪景
snow scavenging 雪冲刷
snow scraper 刮雪机;铲雪机
snow screen 雪拦;防雪栅栏;防雪栅
snow set 雪地装夹
snow-settlement 积雪沉陷
snow shed 雪水流域;防雪棚;避雪所
snow shelter 避雪所
snow shelter forest 防雪林
snow shield 防雪栅栏;防雪栅;防雪设备;防雪板
snow shifter 除雪机
snow shoes 雪鞋
snow shovel 雪铲
snow-shovel reflector 雪铲形反射器
snow sky 雪照云光
snowslide 雪崩;雪滑道
snow slide alarm 雪崩报警
snowslide defense 雪崩防御
snow slip 雪崩
snow sludge 雪浆;黏粥冰
snow slush 雪泥;湿雪
snow spreader 刮雪机
snow sprocket 雪地履带与驱动链轮;雪链轮;雪地驱动链轮
snow squall 雪飑;暴风雪
snow stage 成雪阶段
snow stake 雪桩;量雪尺;积雪厚度标尺;测雪桩
snow storage 雪蓄水;雪水储量
snow storm 雪暴;暴风雪;雪花形干扰
snow survey(ing) 降雪调查;测雪;积雪测量
snow sweeper 扫雪机;扫雪车
snow tire 雪地用汽车轮胎;防滑轮胎
snow tractor 雪地用拖拉机;雪地牵引车
snow trap 截雪坑
snow tremor 积雪震陷;积雪震落
snow tube 取雪管
snow virga 雪幡
snow wall 挡雪墙;防雪墙
snow-water 雪水;凌汛
snow water equivalent 雪水当量
snow water flood 融雪洪水;桃汛
snow white 雪白;氧化锌
snow wing 雪犁翼
snow work 雪中装卸;雪天作业
snowy 下雪的
snowy district 多雪地区
snowy weather 下雪天
Sn placer ore 砂锡
snub 迅速掣住;紧急制动;紧急制动;加高底部截槽;缓冲;绕绳防滑;冲击吸收
snubbed 摩擦减振器
snubber 减振垫块;缓冲器;掏槽眼;绳链制动器;减振器;减声器
snubber hole 消振炮眼;掏槽炮眼
snubber sprocket 松放链轮
snubber valve 滞流阀
snubbing 高压井中起下钻具;加高掏槽;核对船舶线型图;抛锚掣住船的运动
snubbing cementing 压注水泥
snubbing equipment 强行下钻具
snubbing line 减速索
snubbing piping 强行下管
snubbing post 系缆柱;码头上挽缆桩;强行下管固定辅助钢绳支柱;带缆栓
snubbing shot 掏槽爆破
snubbing tubing 强行下管
snub out 在钻孔内高压提升钻具
snub pulley (胶带运输机的)下托辊;张紧轮;回空托辊;缓冲滚筒
snub roller 缓冲托辊
snub the chain 滞链
snuff bottle 鼻烟盒
snuff-box 鼻烟壶;鼻烟盒
snuffle valve 泄气阀
snug 减帆;舵臂前端突出插舵针的部位;暗礁
snug anchorage 避风锚地;安全锚(泊)地
snug berth 避风泊位;安全泊位
snug bolt 凸头螺栓
snug down 减帆
snug fit 适贴配合;精细组装;精密装配;密配合;滑配合;滑动配合
snuggery 个人用舒适小室;旅馆酒吧间
snug harbo(u)r 良港口;避风港
snug washer 平垫圈
sny 木船首尾部翘起弧度
Snyder crushing method 斯奈德破碎法
Snyder induction furnace 斯奈德电感炉
Snyder life test 斯奈德寿命试验(一种绝缘油老化试验法)
Snyder sampler 斯奈德取样器
snyed plate 弯曲板
SO_2 gas automatic recorder 二氧化硫气体自动记录器
soak 高温保温;浸泡液;积水洼地;均热;均热处理
soakage 吸水量;吸入量;浸透量;浸湿性;均热
soakage pit 浸渍坑;均热炉;渗滤坑
soakaway 渗水坑井;渗水坑;渗滤坑
soak cleaning 浸洗;化学清理
soak degreasing 浸泡脱脂
soaked and mildewed 浸湿且发霉
soaked and mo(u)lded 浸湿且发霉
soaked and spoiled 浸湿且变坏
soaked bobbin 浸卷机
soaked to a putty 渗入油灰
soaked weight 渗入(重)量
soaker 裂化设备;裂化反应室;立墙泛水;浸渍剂;浸洗机;戗脊泛水片;底泛水片;大雨;防漏嵌条
soaker at hip 戗脊泛水片
soak heating 均热加热
soaking 延长熟料在高温带时间;高温保温;浸透;浸湿;浸泡;浸水;浸润;强激励;强充电;湿;均热处理
soaking area 初熔区;保温区
soaking box 保温箱
soaking drum 裂化反应鼓
soaking furnace 均热炉

soaking heat 徐热
soaking-in 吸入;渗入
soaking in water 水浸渍
soaking-out 剩余放电
soaking period 高温保温阶段;均热期;浸水期;浸湿时间;恒温期;浸润期;(混凝土养护的)恒温期
soaking pit 均热炉;均浸坑;浸池;等浸坑
soaking pit clamp 均热炉夹钳
soaking pit crane 均热炉起重机;夹钳起重机
soaking process 浸水过程;水浸法
soaking reactor 吸收电抗器
soaking sample 浸水试样
soaking section 裂化反应段
soaking temperature 均热温度;保温温度
soaking test 浸水试验
soaking time 渗入时间;高温保温时间;裂化反应时间;烧成时间;保温时间
soaking vat 浸水池
soaking zone 均热区;均热带;保温段
soak in lime 浸灰(法)
soak into a soil 渗入土中
soak period 高温保温时间;浸泡时间;湿机阶段;保温时间
soak pit 渗坑
soak test 浸泡试验
soak-up method 吸收法
soap 皂(类);肥皂
soap and detergent industry waste 肥皂和洗涤工业废水
soap ashes 草灰碱
soap bar 皂条
soap bark 皂皮
soap base 皂坯;皂基
soap-based powder 皂坯洗涤粉
soap body 皂体
soap-box race track 含金属矿石铺设的赛车道;临时性跑道
soap brick 皂砖;长条砖
soap bubble 皂膜;皂泡
soap-bubble analogy 薄膜比拟法;皂膜比拟法
soap-bubble flowmeter 皂膜流量计
soap-bubble leak detection 皂泡检漏
soap-bubble method (测定火焰基本速度的)皂泡法
soap-bubble test 皂泡试验
soap chip 皂片
soap chip drier 皂片干燥机
soap chipper 刨皂机
soap chromatography 皂色谱
soap content 皂含量
soap dish 肥皂盘
soap dish tile 皂碟砖
soap dispenser 给皂水器
soap dye 皂模
soap earth 皂石
soaped for leakage 泉水检漏
soap emulsion 皂质乳液
soaper 皂洗机
soap factor 皂因数
soap film 皂膜
soap film analog(y) 薄膜比拟;皂膜模拟;皂膜比拟(法)
soap film flowmeter 皂膜流量计
soap film gas meter 皂膜气量计
soap film meter 皂膜计
soap film method 皂膜法
soap film tensor 皂膜张力
soap flaking rolls 制皂片辊轧机
soap free emulsion 无皂乳液
soap germicide 皂用杀菌剂
soap holder 皂液瓶;肥皂盒;肥皂盘
soap-hydrocarbon gel 皂一碳氢化合物胶体
soap in emulsion 含皂乳液
soapiness 皂滑性
soaping 皂洗;肥皂洗刷
soap-in-oil system 皂油系统
soap kettle 皂化锅
soapless soap 化学洗涤剂;合成洗涤剂
soap-like 类皂的
soap lye 皂液;皂碱液
soap lye glycerine 皂液甘油
soap-making wastewater 制皂废水
soap manufacturing 肥皂制造
soap-mark 皂渍

soap membrane 皂膜
soap mill 研皂机
soap mo(u)ld 皂模
soap oil 皂油
soap-oil dispersion 皂油分散体
soap-oil emulsion 皂质油乳化液
soap paste 皂糊
soap-phase structure 皂相构造
soap powder 皂粉
soap-proof 防皂的
soaprock 皂石
soaps 穿孔砖
soap scum 皂垢
soap shaving machine 刨皂机
soap solution 皂碱液;皂液;皂溶液
soap spot 皂渍
soapstone 皂石;滑石
soap stone powder 滑石粉
soapsuds 起泡皂水;肥皂水
soapsuds method 皂沫法
soapsuds test 皂沫试验
soap tablet 皂丸
soap tank 皂化槽
soap test 皂水试验
soap tray 皂碟
soap tray tile 皂碟砖
soap-treated 皂化处理的
soap treatment 皂化处理
soap-type emulsifier 皂型乳化剂
soap washing 肥皂洗刷
soap water film 皂水膜
soap wrapping machine 包皂机
soapy alcohol 多脂醇
soapy clay 油性黏土;土;肥黏土
soapy feeling 腻感;滑腻感
soapy waste 含皂废水
soapy water 肥皂水
soar 价格猛涨;猛涨;猛增;飞涨
soaring 猛涨
soaring airplane 滑翔机
soaring prices 物价飞涨
sobole 根出条
sobolevite 磷硅钛钙钠石
sobolevskite 六方铋钯矿
sobotkite 镁蒙脱石
SO-cable 三角断面电缆
soccer field 足球场
social accounting 社会核算
social accounting analysis 社会核算分析
social accounting matrix 社会核算矩阵
social adaptation 社会适应
social affect survey 社会影响调查
social and cultural indicator 社会与文化标志
social animal 群居动物
social anthropology 社会人类学
social appraisal 社会评价
social average expenditures standard 社会平均支付水平
social average value 社会平均价值
social benefit 社会效益;社会利益
social benefit of environmental protection 环境保护社会效益
social building program(me) 社群建筑计划
social center 社交中心
social charges 社会福利费
social city 社会城市
social club 社团会所
social compact 社会契约
social contract 社会契约
social cost 公共成本;社会费用;社会成本
social cost accounting 社会成本会计
social cost-benefit 社会成本效益
social cost of congestion 交通拥挤造成的社会费用
social discount rate 社会折现率;社会贴现率
social diseconomy 社会不经济(性)
social disruption 社会瓦解;社会混乱;社会分裂
social dividend 社会股息
social dominance 社群优势
social dumping 社会倾销
social ecology 社会生态学
social economic benefit 社会经济效益
social economic pressure index 社会经济压力指数
social economic survey 社会经济调查
social economy 社会经济

social educational facility 社会教育设施
social effect assessment 社会影响评价
social engineering 社会工程
social environment 社会环境
social environmental education 社会环境教育
social environmental element 社会环境要素
social environmental structure 社会环境结构
social expenses 交际接待费用
social fringe 社会利益
social goal 社会目标
social hall 社交联谊楼;友谊厅;交谊厅
social health science 社会健康科学
social house-building 交谊建筑
social housing 低价住房;社会住房
social impact assessment 社会影响评价
social impact survey 社会影响调查
social increase of population 人口机械(社会)增长;人口的社会增长
social indicator 社会指数
social industrialization 社会工业化
social infrastructure 社会基础设施
social insurance 社会保险
social insurance fund 社会保险基金
social integration 社会一体化
social investigation 社会调查
socialist construction 社会主义建设
socialist emulation 社会主义竞赛
socialist marker-oriented economy 社会主义市场经济
socialist market economy 社会主义市场经济
socialist market mechanism 社会主义市场机制
socialization 社会化
socialization of sports 体育社会化
socialized and profession alized service 服务的社会化
socialized medicine 国家公费医疗;公费医疗制度
socialized network of investment 社会化的投资网络
socialized sector of the economy 社会化经济部门
social-mined 热心社会工作的
social morphology 社会形态学
social national income 社会国民收入
social nature of property relations 财产关系的社会性质
social needs 社会需求
social net output value 社会净产值
social objective 社会目标
social opportunity cost 社会机会费用;社会机会成本
social overhead 社会基本开支
social overhead capital 社会总资本;社会分摊资本
social overhead cost 社会基本设施费用
social overhead infrastructure 社会固定资本
social overhead investment 社会间接投资
social ownership 社会所有制
social parasitism 群居寄生(现象)
social participation 社会参与
social phenomenon 群集现象
social planning 城市社会规划;社会规划
social practice 社会实践
social product 社会产品
social profit 社会利润
social program(me) 社会规划
social purchasing power 社会购买力
social regional environment 社会区域环境
social replacement capital 社会的补偿资本
social reserve funds 社会后备基金
social risk 社会风险
social science 社会科学
social security 社会保障
social security and welfare service 社会保障和福利事业
social security funds 社会保险基金
social security program(me) 社会保险计划
social security tax 社会保险税
social service 社会服务
social service system 社会服务系统
social sphere 社会圈
social structure 社会结构
social surplus rate of return 社会盈利率
social survey 社会调查
social system 社会系统
social system engineering 社会系统工程
social utility 社会福利事业

social value 社会价值
social value function 社会价值函数
social wealth 社会财富
social welfare 社会福利
social welfare decision making 社会福利决策
social welfare facility 社会福利设施
social welfare function 社会福利函数
social welfare policy 社会福利指标
sociation 基群丛
socies 植丛
society 学会;协会;小群落
Society for Advancement of Management 管理促进协会
Society for the Promotion of Natural Reserves 自然资源保护促进会
Society of Architectural Historians 建筑史学会(美国)
Society of Automotive Engineers 自动化工程学会(美国)
Society of Environment(al)Science 环境科学学会
society of high entropy 高熵社会
Society of Industrial and Applied Mathematics 工业与应用数学学会
Society of International Gas Tanker and Terminal Operators Limited 国际气体船及码头经营人协会;国际气体船及码头操作者协会
society of low entropy 低熵社会
Society of Maritime Engineering 水运工程学会
Society of Naval Architect 造船学会
Society of Naval Architects and Marine Engineers 造船与轮机工程师学会
Society of Nuclear Scientists and Engineers 核能科学家和工程师协会
socimetry 社会计量法
socioeconomic 社会经济学的
socioeconomic accounting 社会经济会计
socioeconomic aspects of human settlements 人类住区的社会,经济方面
socioeconomic data 社会经济数据
socioeconomic development 社会经济发展
socioeconomic environmental indicator 社会经济环境指标
socioeconomic factor 社会经济因素
socioeconomic graphics 社会人口统计
socioeconomic impact of biotechnologies 生物技术的社会经济影响
socioeconomic index 社会经济综合指标
socioeconomic level 社会经济水平
socioeconomic mix 社会经济综合指标
socioeconomic performance 社会经济效益
socioeconomic phenomenon 社会经济现象
socioeconomic planning 社会经济计划
socioeconomic pressure index 社会经济压力指数
socioeconomic problem 社会经济问题
socioeconomic revolution 社会经济革命
socioeconomic statistical system 社会经济统计系统
socioeconomic statistics 社会经济统计(学)
socioeconomic status 社会经济状态;社会经济地位
socioeconomic survey 社会经济调查
socioeconomic systems analysis 社会经济系统分析
socioeconomic systems management 社会经济系统管理
socioeconomic trends 社会经济趋势
socioeconomic variable 社会经济变量
socioeconomy 社会经济(学)
sociologic geography 社会地理学
sociology 社会学
sociometric analysis 社会计量分析
sociometric description 社会计量描述
sociometric device 社会经济的工具
sociometrics 社会计量学
sociometric technique 社会经济技术;社会计量法
sociometric test 社会经济检验
sociometry 社会计量
sociotechnical system approach 社会技术系统方法
sock 铧头;软保护套
socker wedge 插孔楔
socket 套节;砂箱定位套;灯座;穿线环;承窝;承口;插头座;插口【电】;插孔
socket adapter 灯座接合器;灯头插业;插座转接器
socket and spigot bend 承插弯头;承插弯管
socket and spigot joint(ing) 窝接;承窝接合;承插(式)接头;承插接合;插口接合;插承接合
socket and spigot taper 承插缩管
socket antenna 插座天线
socket base 插座底
socket bearing 球窝轴承;球窝支座
socket bend 管节弯头;承接弯管;管节弯头;承口弯管
socket bushing 灯头绝缘套
socket cable 插换电缆
socket cap screw 圆柱头内六角螺钉
socket chisel 套柄凿;榫凿
socket chuck 套筒型夹紧夹头
socket-clevis eye 碗以挂板
socket-compatibility 插座兼容性
socket conduit 承口管
socket-connected footing 杯形基础
socket connector 接插件;套筒接头
socket-contact 插压式接点
socket cover 灯头外壳;插座盖;插口罩
socket cup 浇飘
socket drill 窝钻
socketed bar 套接棒
socketed column 窝形柱
socketed member 窝接杆件
socketed pile 嵌岩桩
socketed pipe 套管
socketed rod 套接棒
socketed screw 凹头螺钉
socketed stanchion 套接支柱;套撑;插接支柱;摆动支座
socketed tube 套管;插接管
socket end 大端;窝接口承端;承接端;承端
socket end clamp 杵座鞍子
socket end fitting 内螺纹终端接头
socket end wedge-type clamp 杵座楔型线夹
socket fittings 承插式管件
socket for electric(al)light 电灯插座
socket for hex screw 六角螺丝插座
socket for lead joint 铅塞承口;填铅承接口
socket former 管口扩大器
socket form mo(u)lding 胎膜成型
socket for position rod 活塞杆承插头
socket for self-tapping screw 自动攻丝螺丝座
socket foundation 杯形基础
socket fuse 插头熔线;插入式熔断器
socket-fused joint 套节熔融接合;承口熔接头
socket ga(u)ge 套规
socket grinding 承口磨光
socket head 套筒扳手(头);插座头;凹头
socket head pipe plug 凹头管塞
socket head screw 内六角螺钉;内六角头螺钉;涡头螺钉;凹头螺钉
socket hoop 承口环箍
socketing 纳入接;榫接;嵌接
socketing pile in prebored hole 预钻插桩法
socket inlet 承接式插头
socket joint 球窝结合;球接合;窝接;套筒连接;套接;插座接合
socket mandrel 锥形心轴;锥顶涨杆
socket of jack 塞孔套管;塞孔座
socket of the sweeping table 车板架轴座
socket outlet 煤气接口;电器插座;承窝接合
socket outlet box 插座盖
socket outlet for cleaners 清灰器窝形出口
socket patent eye 钢索端眼具
socket pilot nut 内六角导向螺母
socket pin 导销
socket pipe 套管;活节管;活接口;套节管;套接管子;套柄管(俗称缩节);缩节(即套接管);承口管
socket plug 管塞(子)
socket power 插座电源;插接电源
socket-power unit 插接电源组
socket press 接线匣压机;套节压机;套管压机
socket ratchet wrench 套筒棘轮扳手
socket ring 座环;套环;插座环
socket rosette 插座
socket scoop 长柄浇勺
socket screw 承接螺钉;沉孔螺钉;凹头螺钉
socket screw hexagon wrench 内六角扳手
socket screw key 凹头螺钉键
socket screw wrench 内六角螺丝扳手
socket share 盒形犁铧
socket sleeve 座套
socket slide 环
socket spade 套筒铲
socket spanner 套筒扳手
socket spring 插座弹簧
socket switch 灯头开关;插座式开关
socket tube 活节管;承口管
socket-type chlorinated polyvinyl chloride plastic fitting 承口型氯化聚氯乙烯塑料管件
socket-type chlorinated polyvinyl chloride plastic pipe 承口型氯化聚氯乙烯塑料管
socket-type footing 钟形基础
socket-type polyethylene fitting 承口型聚氯乙烯管件
socket-type polyvinyl chloride plastic pipe 承口型聚氯乙烯塑料管
socket union 凹口管节
socket wedge 套管楔;凹楔
socket weld ends 承插焊接端
socket weld fitting 接线匣焊接配件;套节焊接配件;管节焊接配件
socket welding 承插焊接
socket-welding fitting 焊接式管接头
socket-welding flange 承插焊接法兰
socket with pin 单环钢丝绳插头
socket with shrouded contact 带屏板的插座
socket with switch 带开关的插座
socket wrench 套筒扳手
socket wrench adapter 套筒扳手接头
socket wrench crossbar 套筒扳手扳杆
socket wrench extension bar 套筒扳手加长杆
socket wrench set 全套套筒扳手;成套套筒扳手
socket wrench with bar 带柄套筒扳手
socking in 电容器电荷逐渐增多
socle 勒脚线;座石;管脚;管底;台石
sod 铺草皮;草皮;草地
soda 钠碱;碳酸钠;苏打
soda-acid extinguisher 碳酸钠灭火器;酸碱灭火器;苏打酸灭火器
soda-acid fire extinguisher 碳酸钠灭火器;酸碱灭火器;苏打酸灭火器
soda alum 钠明矾
soda amphibole 钠角闪石
soda amphibole rhyolite 钠质角闪石流纹岩
soda-asbestos 苏打石棉
soda ash 碱粉;钠碱粉;无水苏打;无水碳酸钠;钠碱灰;苏打灰;苏打粉;纯碱
soda ash compounding tank 纯碱配制槽
soda ash glass 钠玻璃
soda ash production 纯碱生产
soda-ash roaster 苏打粉煅烧机
soda-baryta glass 钠钡玻璃
soda-base grease 钠脂
soda brine 苏打卤
soda carbonatite lava 钠质碳酸岩熔岩
soda cell 苏打电池
soda cellulose 碱纤维素
soda chlorine pulp 氯化法纸浆
soda circulating pump 苏打循环泵
sodaclase 钠长石
sodaclase granite 钠长花岗岩
soda crystals 苏打结晶
sodadiabase 钠辉绿岩
soda feldspar 钠长石
soda felsite 钠质霏细岩
soda finishing 碱精制
soda fountain 冷饮摊;冷饮柜
soda-garnet 钠榴石
soda glass 钠玻璃
soda-granite 钠花岗岩
soda grease 钠基脂
soda gypsum 苏打石膏
soda hornblende 钠铁闪石
sod airfield 草地机场
sodaite 中柱石
soda lake 高碱湖;碱湖;苏打湖
soda lime 钠石灰;苏打石灰;碱石灰
soda-lime feldspar 钠钙长石
soda-lime glass 钠钙玻璃
soda-lime mix 苏打石灰混合物
soda-lime process 苏打石灰处理;苏打石灰(软化)法
soda-lime silica glass 钠钙硅玻璃
soda-lime softening process 苏打石灰软化法
soda liparite 钠流纹岩

sodalite 钠沸石;方钠石
sodalite aegirine augite syenite 方钠霓辉正长岩
sodalite analcine cancrinite microsyenite 方钠石方沸石钙霞石微晶正长岩
sodalite Braun tube 方钠石显像管
sodalite syenite 方钠正长岩
sodalite trachyte 方钠粗面岩
soda-lithium-caesium-potassic pegmatite 钠锂铯钾伟晶岩
soda-lithium-potassic pegmatite 钠锂钾伟晶岩
sodalitite 方钠岩
soda loss 碱减量
sodalumite 钠明矾
soda-lye 氢氧化钠(溶液)
soda lye wash 碱水洗涤
soda-margarite 钠珠云母
soda mesotype 钠沸石
soda mica 钠云母
soda microcline 钠微斜长石;歪长石
Sodamide 氨基钠
soda mint 苏打明
sod and clay plow bottom 草皮黏土地犁体
soda nepeline syenite 钠质霞石正长岩
soda-nepheline microsyenite 钠质霞石微晶正长岩
sodanite 赤道热
soda niter 碱石灰;钠硝石
soda orthoclase 钠正长石
soda-pearl ash glass 钠钾玻璃
soda-porphyry microsyenite 钠钾质微晶正长岩
soda-potash glass 钠钾玻璃
soda-potassic feldspar porphyry 钠钾质长石斑岩
soda-potassic-lithium pegmatite 钠钾锂伟晶岩
soda-potassic nepeline syenite 钠钾质霞石正长岩
soda-potassic nepheline microsyenite 钠钾质霞石微晶正长岩
soda-potassic pegmatite 钠钾质伟晶岩
soda process 苛性钠法;苏打法;烧碱法
soda pulp 碱性纸浆;碱法纸浆
soda pulping process 碱法制浆
soda pulping waste 碱法纸浆废液
soda pulp mill 碱法纸浆厂
soda pulp mill wastewater 苏打纸浆废水;碱法纸浆厂废水
soda pyroxene 钠辉石
soda pyroxene rhyolite 钠质辉石流纹岩
sodar 声雷达;声达(声定位和测距仪)
soda rhyolite 钠流纹岩
soda-saline soil 苏打盐土
soda saltpeter 钠硝石
soda sludge 苏打淤渣
soda smelt 苏打熔融物
soda soap 钠皂
soda-soap-base gease 钠皂基润滑脂
soda softening 苏打软化
soda softening method 苏打软化法
soda softening process 苏打软化法
soda soil 苏打土
soda-solonchak 苏打盐土
soda solonetz 苏打盐土
soda soluble glass 钠水玻璃
soda solution 碱液;氢氧化钠溶液;苏打溶液
soda straw pulp black liquor 碱性草浆黑液
soda-syenite 钠正长岩
soda tar 苏打焦油
sodation 碳酸钠去垢(法)
soda-tremolite 钠透闪石
soda triphylite 钠磷锰铁矿
soda wash 苏打洗涤
soda wash solution 碱洗溶液
soda-wash tower 碱洗塔
soda water 苏打水
soda zeolite 钠沸石
sod bed 铺草的花圃;铺草的花坛
sod bottom 草地用犁体
sod-bound 草皮板结
sod breaker bottom 草地破土犁体
sod cover 草皮覆盖;草皮覆被
sod covering 草皮覆盖层
sod crop 草地作物
sod culture 植草法;生草法;草皮栽培;草皮培植
sod curb 铺草皮路缘
sod cutter 割草机;剪草机;切草皮器
sodded area 铺草的区域
sodded check dam 草皮谷坊

sodded runway 草皮跑道
sodded slope 植草坡;草皮边坡;植草边坡;铺草皮坡;草地斜坡
sodded spillway 草皮溢水道
sodded strip 铺草皮带
sodding 铺种草皮;铺草皮;草皮
sodding area 铺草(皮)面积
sodding lawn 铺草皮块草坪
sodding work 铺草皮作业;草皮护坡
sod ditch check 草皮沟槛
soddy 多草皮的
soddyite 硅铀矿
soddy podzolic soils 生草灰化土
soddy podzolic soil test 生草灰化土试验
soddy soil 草皮土;生草土
sode ash 碱灰(苏打灰)
sod farm 草皮农场
sod grass 蔓生草;蔓草;丛草;草皮草
sod harvester 草皮修剪机
sod house 草皮房子;干打垒
sodian adularia 钠冰长石
sodian deposition 钠沉积
sodian microcline 钠微斜长石
sodian orthoclase 钠正长石
sodian sanidine 钠透长石
sodic akenite 钠质正长岩
sodic amphibole granite 钠质角闪石花岗岩
sodic amphibole microsyenite 钠质角闪石微晶正长岩
sodic amphibole miroganite 钠闪微晶花岗岩
sodic amphibole syenite 钠质角闪石正长岩
sodic amphibole trachyte 钠质角闪石粗面岩
sodic amphibole trachyte porphyry 钠质角闪石粗面斑岩
sodic bentonite ore 钠质膨润土矿石
sodic feldspar porphyry 钠质长石斑岩
sodic-lithium-caesium-potassic aplite 钠锂铯钾细晶岩
sodic-lithium-potassite aplite 钠锂钾细晶岩
sodic microsyenite 钠质微晶正长岩
sodic-potassic aplite 钠钾质细晶岩
sodic-potassic-lithium aplite 钠钾锂细晶岩
sodic-potassic syenite 钠钾质正长岩
sodic pyroxene granite 钠质辉石花岗岩
sodic pyroxene microsyenite 钠质辉石微晶正长岩
sodic pyroxene miroganite 钠辉微晶花岗岩
sodic pyroxene syenite 钠质辉石正长岩
sodic pyroxene trachyte 钠质辉石粗面岩
sodic pyroxene trachyte porphyry 钠质辉石粗面斑岩
sodic soil 钠质土
sodic trachyte 钠质粗面岩
sodified water 固化水
sodio-acetoacetic ester 钠代乙酰乙酸酯;钠代乙酰醋酸酯
sodio-alkylmalonic ester 钠代烷基丙二酸酯
sodio-derivative 钠衍生物
sodio-ethylmalonic ester 钠代乙基丙二酸酯
sodio-ketoester 钠代酮酸酯
sodio-malonic ester 钠代丙二酸酯
sodio-methylmalonate 钠代甲基丙二酸盐
sodio-methylmalonic ester 钠代甲基丙二酸酯
sodion 钠离子
sodium absorption ratio 钠吸收比
sodium acetate 乙酸钠;醋酸钠
sodium acetrizoate 醋碘苯酸钠
sodium acetylide 乙炔钠
sodium acid arsenate 酸式砷酸钠
sodium acid fluoride 氢氟化钠
sodium acid phosphate 酸式磷酸钠
sodium acid prophosphate 碱式焦磷酸钠
sodium acid pyrophosphate 酸式焦磷酸钠
sodium acid sulfide 酸式硫化钠
sodium acloholate 烷氧基钠
sodium adsorption ratio 钠吸附比
sodium alcoholate 乙氧基钠;乙醇钠;醇钠;醇基钠
sodium alginate 海藻酸钠;藻酸钠
sodium alkane sulfonate 烷烃磺酸钠;烷基磺酸钠
sodium alkoxide 醇钠
sodium alkylsulfate 烷基硫酸钠
sodium allylsulfonate 丙烯磺酸钠
sodium alphanaphthy-lamine sulfonate 对氨基

萘磺酸钠
sodium alum 钠明矾;钠矾
sodium aluminate 铝酸钠
sodium aluminate solution 铝酸钠溶液
sodium alumin(i)um fluoride 氟铝化钠
sodium alumin(i)um phosphate 磷酸钠铝
sodium alumin(i)um silicate 铝硅酸钠
sodium alumin(i)um silicofluoride 硅氟酸钠铝
sodium aluminosilicate 硅铝酸钠;硅铝酸钠;铝硅酸钠;钠铝硅酸盐
sodium amalgam 钠汞齐;钠汞合金
sodium amalgamation 钠汞齐化
sodium amalgam extraction 钠汞齐
sodium amalgam-oxygen cell 钠汞齐氧电池
sodium amide 氨基钠
sodium amidotrizoate 泛影钠
sodium aminophenylarsonate 阿散酸钠
sodium aminosalicylate 对氨基水杨酸钠
sodium ammonium biphosphate 磷酸氢铵钠
sodium and potassium oxide content 氯化钠氯化钾总含量
sodium anilinearsonate 阿散酸钠
sodium antimonate 锑酸钠
sodium arsanilate 阿散酸钠
sodium arsenate 砷酸钠
sodium arsenide 砷化钠
sodium arsenite 亚砷酸钠
sodium aryl 芳基钠
sodiumautunite 钠铀云母
sodium azide 叠氮化钠
sodium balance 钠平衡
sodium barbital 巴比妥钠
sodium barbiturate 巴比土酸钠
sodium barium niobate ceramics 铌酸钡钠陶瓷
sodium-base grease 钠基脂;钠基润滑脂;钠皂润滑脂
sodium bentonite 钠基膨润土
sodium benzene sulphinate 苯亚磺酸钠
sodium benzoate 苯甲酸钠;安息香酸钠
sodium benzoate technical 工业用苯甲酸钠
sodium benzoylacetone dihydrate 苯酰丙酮钠二水合物
sodium benzyl sulfanilate 苄基对氨基苯磺酸钠
sodium-betpakdalite 砷钼铁钠石
sodium bicarbonate 小苏打;重碳酸钠;碳酸氢钠
sodium bicarbonate concentration 碳酸氢钠浓度
sodium bicarbonate spring 重碳酸钠泉
sodium bicarbonate spring rich in boric acid and ammonia 富硼酸和氨的重碳酸钠泉
sodium bicarbonate water 重碳酸钠型水
sodium bichromate 重铬酸钠
sodium bifluoride 氢氟化钠
sodium bioxalate 草酸氢钠
sodium biphosphate 硝酸钠
sodium bismuthate 铋酸钠
sodium bisulfite 亚硫酸氢钠
sodium bolteoodite 黄硅钠铀矿
sodium borate 硼酸钠;硼酸;硼砂
sodium boroformate 硼甲酸钠
sodium borohydride 氢化硼钠;硼氢化钠
sodium borosilicate 硼硅酸钠
sodium butadiene rubber 丁钠橡胶
sodium butoxide 丁醇钠
sodium butyrate 丁酸钠
sodium carbide 碳化钠
sodium carbolate 苯酚钠
sodium carbonate 碳酸钠;苏打;纯碱
sodium carbonate concentration 碳酸钠浓度
sodium carbonate fusion 碳酸钠熔融法
sodium carbonates' presence 碳酸钠的存在
sodium carboxy methyl cellulose 羧甲基纤维素钠;钠羧甲基纤维
sodium carrier 钠载体
sodium chabazite 钠菱沸石
sodium channel 钠通道;钠道
sodium chlorate 氯酸钠
sodium chloride 亚氯酸钠;氯化钠;食盐
sodium chloride brine 氯化钠盐水
sodium chloride concentration 氯化钠浓度
sodium chloride thermal water 氯化钠型热水
sodium chloride water 氯化钠水
sodium chromate 铬酸钠
sodium citrate 柠檬酸钠;柠檬素钠
sodium citrate technical 工业柠檬酸钠

sodium clay 钠质黏粒
sodium cloud 钠云
sodium cobaltinitrite test 钴亚硝酸钠试法
sodium compound 钠化合物
sodium conductance 钠电导
sodium-conducting ceramic solid electrolyte 钠传导陶瓷固体电解质
sodium contaminants 钠沾染物
sodium content 钠含量;含钠量
sodium-cooled reactor 钠冷反应堆;钠冷堆
sodium-cooled valve 钠冷排气阀;钠冷却气门
sodium cooling 钠冷却
sodium current 钠电流
sodium cyanate 氰酸钠
sodium cyanide 氰化钠
sodium cyanide poisoning 氰化钠中毒
sodium cyanoacetic ester 钠代氰基乙酸酯
sodium decyl sulfate 癸基硫酸钠
sodium deficit 钠缺乏
sodium diacetate 二乙酸钠
sodium dibunate 地布酸钠
sodium dibutyl dithiocarbamate 二丁基氨基硫羧酸钠
sodium dichloroisocyanurate 二氯异三聚氰酸钠
sodium dichromate 重铬酸钠
sodium diethyl dithiocabamate spectrophotometric method 二乙基二硫代氨基甲酸钠分光光度法
sodium dinitroorthocresylate 二硝基邻甲酚钠
sodium dioxide 氢氧化钠
sodium diprotrizoate 二丙泛影钠
sodium discharge lamp 钠气灯;钠灯
sodium disilicate 二硅酸钠
sodium dithionite 连二亚硫酸钠;连二硫酸钠
sodium diuranate 重铀酸钠
sodium D-line 钠 D 线
sodium dodecyl benzene sulfonate 十二烷基苯磺酸钠
sodium doublet lines 钠双线
sodium dross 钠结渣
sodium error 钠误差;钠差
sodium ethylate 乙氧钠;乙醇钠
sodium feldspar 钠长石
sodium ferricyanide 铁氰化钠
sodium ferrocyanide 亚铁氰化钠
sodium-filled exhaust valve 钠冷排气阀
sodium filter 钠滤色镜
sodium fluoaluminate 氟铝酸钠
sodium fluoborate 氟硼酸钠
sodium fluoride 氟化钠
sodium fluoride-arsenate dinitrophenol solution 氟化钠-砷酸二硝基酚溶液
sodium fluoride glycerine paste 氟化钠甘油糊剂
sodium fluoride poisoning 氟化钠中毒
sodium fluoroaluminate 氟铝酸钠
sodium fluorosilicate 氟硅酸钠
sodium fluosilicate poisoning 氟硅酸钠中毒
sodium fusidate 夫西地酸钠
sodium gold cyanide 氰化钠金
sodium grease 钠基润滑脂
sodium halide 卤化钠
sodium hazard 钠质危害
sodium hexametaphosphate 六偏磷酸钠;六聚偏磷酸钠
sodium humate 腐殖酸钠
sodium hydrate 氢氧化钠;烧碱
sodium hydride 氢化钠
sodium hydro-fluoride 氢氟化钠
sodium hydrogen carbonate 小苏打
sodium hydrogen difluoride 二氟化氢钠
sodium hydrogen sulfate 硫酸氢钠
sodium hydrogen sulfite 亚硫酸氢钠
sodium hydroperoxide 氢过氧化钠
sodium hydrosulfite 连二亚硫酸钠;保险粉
sodium hydrosulphide solid fused 固体熔融硫氢化钠
sodium hydrosulphite 次硫酸钠
sodium hydroxide 苛性钠;氢氧化钠;烧碱
sodium hydroxide method 氢氧化钠法
sodium hydroxide treatment 氢氧化钠处理
sodium hypobromite 次溴酸钠
sodium hypochlorite 次氯酸钠
sodium hypochlorite disinfection 次氯酸钠消毒
sodium hypochlorite process 次氯酸钠法

sodium hypochlorite solution 漂白液;次氯酸钠液
sodium hypochlorite solution alkaline 碱性次氯酸钠溶液
sodium hyponitrite 次硝酸钠
sodium hypophosphite 次磷酸钠
sodium hyposulfite 硫化硫酸钠;海波;大苏打
sodium iodate 碘酸钠
sodium iodide 碘化钠
sodium iodide detector 碘化钠探测器
sodium iodide scintillator 碘化钠闪烁体
sodium iodoacetate 碘乙酸钠
sodium iodomethanesulfonate 碘甲基磺酸钠
sodium ion 钠离子
sodium ion conductor 钠离子导体
sodium ion exchange 钠离子交换
sodium ion exchanger 钠离子交换器
sodium ion inflow 钠离子内流
sodium ioth alamate 碘酞钠
sodium itaconate 衣康酸钠
sodium lamp 钠弧灯;钠光灯;钠灯
sodium laurate 月桂酸钠
sodium lauryl sulfate 十二烷基硫酸钠
sodium lead alloy 钠铅合金
sodium light 钠光;钠灯光
sodium lignin(e)sulfonate 木质素磺酸钠
sodium lignosulphonate 木质素磺酸钠
sodium line index 钠线折射率
sodium loop 钠回路
sodium lopodate 碘泊酸钠
sodium maleate 马来酸钠
sodium malonate 丙二酸钠
sodium manganate 锰酸钠
sodium measilicate 硅酸钠钢
sodium meta-aresnite 偏亚砷酸钠
sodium metabisulfite[metabisulphite] 焦亚硫酸钠
sodium metal 金属钠
sodium metallic 金属钠
sodium metaperiodate 高碘酸钠;偏高碘酸钠
sodium metaphosphate 偏磷酸钠
sodium metasilicate 硅酸钠;偏硅酸钠
sodium metasomatism 钠质交代作用
sodium metavanadate 偏钒酸钠
sodium methiodal 碘甲基磺酸钠
sodium methyl-acetylide 丙炔钠
sodium methyl silicate 甲基硅酸钠
sodium mica 钠云母
sodium molybdate 钼酸钠
sodium monofluoracetate 一氟乙酸钠
sodium monothiophosphate 一硫代磷酸钠
sodium monoxide 一氧化钠
sodium naphthionate 对氨基萘磺酸钠
sodium niobate 铌酸钠
sodium niter 硝酸钠
sodium nitrate 硝酸钠;氮酸钠
sodium nitride 氮化钠
sodium nitrite 亚硝酸钠
sodium octyl sulfate 辛基硫酸钠
sodium oleate 油酸钠
sodium oleate gel 油酸钠凝胶
sodium oleate soap 油酸钠皂
sodium oleic acid 钠油酸
sodium orthoarsenite 原亚砷酸钠
sodium orthoclase 钠正长石
sodium ortho-phenylphenate 联苯酚钠;原苯酚钠
sodium ortho-phenylphenolate 联苯酚钠
sodium orthophosphate 一代磷酸钠
sodium orthophosphate dimetallic 二代磷酸钠
sodium orthosilicate 原硅酸钠
sodium oxalate 草酸钠
sodium oxide 氧化钠
sodium p-aminobenzene sulfonat 氨基苯磺酸钠
sodium para-aminosalicylate 对氨基水杨酸钠
sodium para-nitro phenolate 对硝基酚钠
sodium pentachlorophenate 五氯苯酚钠
sodium pentachlorophenolate 五氯苯酚钠盐
sodium perborate 过硼酸钠;高硼酸钠
sodium percentage 含钠百分数;含钠百分率
sodium perchlorate 高氯酸钠
sodium perchromate 过铬酸钠
sodium periodate 过碘酸钠;高碘酸钠
sodium permanganate 高锰酸钠
sodium peroxide 过氧化钠
sodium peroxy-chromate 过铬酸钠
sodium peroxydisulfate 过硫酸钠

sodium persulfate 过硫酸钠;过磷酸钠
sodium persulfide 过硫化钠
sodium phenate 苯酚钠
sodium phenide 苯基钠
sodium phenolate 苯酚钠
sodium phenolsulfonate 苯酚磺酸钠
sodium phenylacetate 苯基乙酸钠
sodium phenyl-arsenite 苯亚胂酸钠
sodium phenyl-arsonate 苯胂酸钠
sodium phenylphosphinate 苯基次膦酸钠
sodium phlogopite 钠金云母
sodium phosphate 磷酸钠
sodium phosphate tribasic 磷酸三元钠
sodium platinate 铂酸钠
sodium plumbate 高铅酸钠
sodium polyacrylate 聚丙烯酸钠
sodium polyacrylonitrile 钠聚丙烯腈
sodium polysulfide 多硫化钠
sodium-potassium 钠钾地热温标
sodium-potassium-calcium 钠钾钙地热温标
sodium-potassium pump 钠—钾泵
sodium potassium silicate 硅酸钾钠
sodium press 钠丝压制器
sodium propionate 丙酸钠
sodium p-styrene sulfonate 对苯乙烯磺酸钠
sodium pump 钠泵
sodium pump failure 钠泵衰竭
sodium pyrophosphate 焦磷酸钠
sodium pyrophosphate flushing 焦磷酸钠洗井
sodium pyrosulfite 焦亚硫酸钠
sodium pyrovanadate 焦钒酸钠
sodium resinate 树脂酸钠
sodium resistant glass 抗钠玻璃
sodium rubber 钠橡胶;丁钠橡胶
sodium salicylate 水杨酸钠
sodium salt 钠盐;二硝基酚钠
sodium sand 钠砂
sodium-saturated clay 钠饱和黏土
sodium selective electrode 钠离子选择电极
sodium selenate 硒酸钠
sodium selenide 硒化钠
sodium selenite 亚硒酸钠
sodium-sensitized thermoionic detector 钠敏化热离子检测器
sodium sesquicarbonate 碳酸氢三钠
sodium sesquisilicate 倍半硅酸钠
sodium-silica glass 钠—硅氧玻璃
sodium silicate 硅酸钠;水玻璃
sodium silicate adhesive 硅酸钠胶料;水玻璃黏结剂
sodium silicate binder 硅酸钠黏结剂;水玻璃黏结剂
sodium silicate bond 硅酸钠结合剂
sodium silicate bonded sand 硅酸钠黏结的砂;水玻璃砂
sodium silicate cement 水玻璃胶料;硅酸钠水泥;硅酸钠胶结料;水玻璃凝材料
sodium silicate concrete 水玻璃混凝土
sodium silicate mastic 水玻璃胶泥
sodium silicate solid 固体硅酸钠
sodium silicate solution 硅酸钠溶液
sodium silicate test 硅酸钠试验
sodium silicate waterproofer 硅酸钠防水剂
sodium silicate waterproofing agent 硅酸钠防水剂
sodium silicoaluminate 铝硅酸钠
sodium silicofluoride 硅氟化钠;氟硅酸钠
sodium-smoke abatement 钠烟减退
sodium soap 钠皂
sodium-soap grease 钠皂润滑脂;钠皂基润滑脂
sodium soil 钠质土
sodium space 钠间隙
sodium stannate 锡酸钠
sodium stearate 硬脂酸钠
sodium succinate 琥珀酸钠;丁二酸钠
sodium sulfanilate 氨基苯磺酸钠
sodium sulfate 硫酸钠
sodium sulfate soundness test 硫酸钠安定度试验
sodium sulfate water 硫酸钠水
sodium sulfide 硫化钠
sodium sulfide reagent 硫化钠试剂
sodium sulfite 亚硫酸钠
sodium sulfite process 亚硫酸钠工艺过程
sodium sulfite spectrophotometry 亚硫酸钠分光光度法
sodium sulfo benzoate 磺基苯甲酸钠

sodium sulfocyanate 硫氰酸钠
sodium sulforicinate 磺基蓖麻油酸钠
sodium sulforicinolate 磺基蓖麻油酸钠
sodium-sulfur[sulphur]battery 钠硫电池
sodium sulphate 硫酸钠;芒硝
sodium sulphite 亚硫酸钠;硫化钠
sodium-sulphur cell 钠硫电池
sodium super-oxide 超氧化钠;过氧化钠
sodium system 钠体系;钠系统
sodium tannate 丹宁酸钠
sodium tellurate 碲酸钠
sodium test 钠试验
sodium tetraborate 硼砂;四硼酸钠
sodium tetraborate decahydrate 焦硼酸钠
sodium tetraphenylborate 四苯硼酸钠
sodium tetraphosphate 四磷酸钠
sodium tetrathionate 连四硫酸钠
sodium thioantimonate 全硫锑酸钠
sodium thioarseniate 全硫砷酸钠
sodium thiocarbonate 全硫碳酸钠
sodium thiocyanate 硫代氰酸钠
sodium thioglycolate 硫代乙醇酸钠
sodium thiolinate 硫麻子油酸钠
sodium thiosulfate 硫代硫酸钠;海波
sodium trimetaphosphate 三偏磷酸钠
sodium tripolyphosphate 三磷酸钠;三聚磷酸钠
sodium tungstate 钨酸钠
sodium tungsten bronze 钠钨青铜
sodium undecylenate 十一碳烯酸钠
sodium uramospinite 钠砷铀云母
sodium vanadate 钒酸钠
sodium vapo(u)r lamp 钠气灯;钠蒸气灯
sodium vedproate 丙戊酸钠
sodium wolframate 钨酸钠
sodium xanthate 黄原酸钠
sodium xanthogenate 黄原酸钠
sodium zeolite 钠沸石
sodium zippeite 钠钾铀矾
sodium α-toluene 苯基乙酸钠
sod lifter 起草皮铲;草铲
sod mulch 盖草法
sod planting 草皮栽培
sod plow 草地犁
sod plug cutter 草皮切取机
sod-podzol soil 生草灰化土;草皮灰化土
sod revetment 草皮铺面;草皮护坡;草皮护岸
sod seeder 草皮地播种机
sod-seeding 草皮播种
sod strip 带状草皮
sod swale 草地
sod-type ground-working equipment 草地型耕作机具
sod waterway 草皮泄水道;草地泄水道
sodwork 草皮护坡
soehngite 羟镓石
sofa 沙发;长沙发
sofa bed 两用沙发;沙发床;坐卧两用沙发
sofar 声发(水中测声器);声波定位与测距(系统)
sofar channel 声发波道
sofar propagation 声发传播
sofar system 声发系统
sofa sleeper 坐卧两用沙发
sofe moist rot 湿腐
soffione 硼酸喷气孔
soffisian knob 冻丘
soffit 下皮;下端;阳台底面;过梁底面;管道内顶;拱圈内面;拱内侧面;拱腹;楼梯底面;楼板底面;梁底面;涵管内顶面;挑檐底面;底模板;底面
soffit block 拱腹面块;平顶面块;(楼板下的)找平砌块
soffit block floor 拱腹形砖块地面
soffit board 出檐天花板;挑檐底板;(不直接暴露在外面的)腹板;飞檐底板;椽檐底板;背面板
soffit bracket 门顶关门器座;门顶插销座
soffit conductor rail 拱腹式导轨
soffit curve 拱腹曲线
soffit curve of girder 梁腹曲线
soffit cusp 拱腹尖饰;拱键;拱尖;拱顶石;锁环
soffit development 拱腹展开
soffit finish 底面饰面;底面抹灰;底面裂抹灰
soffit floor filler block 拱腹形地板填块
soffit form 底面模板
soffit formwork (模板的)底模;拱腹模板
soffit level 下皮标高;拱腹标高;底板高

soffit lined eaves 底衬屋檐
soffit lining 拱底衬砌;底衬
soffit of arch 拱腹
soffit of girder 梁腹
soffit of structural slab 结构板腹
soffit of vault(ing) 拱顶腹面;穹顶底面
soffit scaffolding 拱腹架;弧形拉模(砌拱用);砌拱支架;砌拱模架
soffit shuttering 底面模板
soffit slab 拱腹板;底板
soffit spacer 下部定位垫铁;定位垫座;(决定下模板与钢筋距离的)支承物
soffit spring 拱背起拱(线)点
soffit springing 拱内弧起拱线
soffit surface 拱腹面
soffit tile 拱腹形砖
soffit tile floor 拱腹形砖地板
soffit width 拱腹宽度
soft 软性的;软调;柔软的;不含酒精
soft acid 软酸
soft action 软性作用;软化作用
soft adder 软加法器;负加法器
soft agent 软化剂
soft aggregate 软质集料;软质骨料;软集料;软骨料;松软集料;松软骨料
soft air 湿空气
soft alkyl benzene 软性烷基苯(指易为生物降解的直链烷基苯)
soft allowance 软补贴
soft alluvium loam 软冲积黏土
soft and altered rock 松散变质岩
soft and hard filth water 软硬垢水
soft annealed wire 软金属线
soft antenna 软式天线
soft aperture 软变孔径
soft arbitrage 软套利
soft arc 软电弧
soft argtum 软橡胶
soft artificial light 人工柔光
soft asphalt 软沥青;软地沥青
soft asphaltic bitumen 软石油沥青
softball diamond 垒球场
softball field 垒球场
softball gloves 垒球手套
soft bark 软皮
soft base 软碱;软基地
soft baseplate wax 软底板蜡
soft base-tar 软煤沥青
soft bitumen 软沥青
soft black 软质炭黑
soft blast 软喷砂处理
soft blue clay 软绿黏土
softboard 低密度纤维板;轻质板;软质纤维板
soft bottom 软底
soft bowl 软质轧辊
soft break down 软击穿
soft breeze 和风
soft brick 软砖;软性砖;低温砖;未烧透砖;欠烧砖
soft bronze 软青铜
soft broom 软扫帚
soft budgetary restraint 软预算约束
soft-burned 轻烧的;低温烧制的
soft-burned clinker 轻烧熟料
soft-burned dolomite 轻烧白云石
soft-burned lime 轻烧石灰
soft-burned magnesia 轻烧镁砂
soft burning 轻烧
soft burnt brick 欠火砖;软烧砖;轻烧砖
soft can 软罐头
soft capsule 软胶囊
soft cast iron 软铸铁;软生铁;易切削铸铁
soft catheter 软导管
soft center 软顶尖
soft-center share 软心钢犁铧
soft-center steel 软心钢;夹心钢
soft-center steel mo(u)ld board 软心钢犁壁
soft chair 软椅
soft chemical oxygen demand 软水化学需氧量
soft chemistry technique 软化学技术
soft clamping 软钳位
soft clause 附条件条款
soft clay 软质黏土;软黏土;软弱黏土;低温烧结黏土
soft clay ground 软黏土地
soft clay ore 软质黏土矿石

soft clay stratum 软黏土层
soft clot 软凝块
soft cloth 软布
soft coal 烟煤;软煤
soft coal asphalt 软煤沥青
soft coal-tar pitch 软煤焦油脂
soft coat 软膜
soft coating 软膜;软镀层
soft coating material 软镀膜材料
soft coke 软焦炭;低强焦炭
soft colo(u)r 软彩;柔和色
soft commissure 软连索
soft component 软成分
soft confirmation 软保兑;附条件保兑
soft consistency 柔软结持度
soft contact 软接触
soft contrast 柔和反差
soft control 软控制
soft copper 软铜
soft copper came 软有槽铜条
soft copper compound strand wire 软铜复绞线
soft copy 软拷贝
soft copy photogrammetry 全数字化摄影测量
soft core 软心轴
soft core potential 软心势
soft core spindle 软心轴
soft cost (建筑的)软成本;软开销(指基建中的保险费、利税、设计费等)
soft cost of building 建筑软成本
soft cover 软封面
soft-covered book 软面书
soft credit 软信贷
soft crystal 软晶体
soft currency 软通货;软(货)币
soft cushioning 软式减震
soft deposit 软土层
soft detergent (易生物降解的)软性洗涤剂;软洗涤剂;软去污剂
soft device 软设备
soft distemper 软性水浆涂料;软性刷墙粉
soft docking 软对接
soft dot 软点子
soft drawn 软拉
soft drawn copper wire 软拉铜丝
soft drawn wire 软拉线
soft drink 软饮料
soft-drink bottling plant 软饮料厂;瓶装清凉饮料厂
soft-drink bottling plant waste 瓶装清凉饮料厂废水
soft-drink bottling waste 软饮料厂废物
soft-drink bottling wastewater 软饮料厂废水
soft-drink industry wastewater 清凉饮料厂废水
soft-drink plant 清凉饮料厂
soft driving 软打桩(在预先钻好的孔内打桩)
soft ductile material 软韧性材料
soft earthenware 软烧质陶器
soft-earth gear 软道面用的起落架
soft edge 轮廓模糊;图像边缘模糊
soften 变软;增塑
soft end 开口端
softened lead 软化铅(脱除砷锑锡等杂质的铅)
softened rubber 软化橡胶;塑炼橡胶
softened water 软化水
softened zone 软化区
softener 增塑剂;增韧剂;软水剂;软麻剂;软化机;软化器;软化炉;软化剂
soft energy paths 软能路线
softener paste 软化脂膏
softening 软化的;水的软化;退火;增塑
softening agent 软化剂;柔软剂
softening agent for fiber 纤维软化剂
softening annealing 钢板软化退火;软化退火
softening apparatus 软化装置;软化器
softening application 加权软化剂
softening bilinear system 软化型双曲线系统
softening by lime 石灰软化(法)
softening coefficient 软化系数
softening coefficient of rock 岩石软化系数
softening coefficient of rock mass 岩体软化系数
softening cracking (混凝土的)开裂软化
softening degree 软化程度
softening factor 软化系数
softening filter 软化滤水器;软化过滤器
softening furnace 软化炉;槽沉炉

softening installation 软化装置
softening interval 软化区间
softening kettle 软化锅
softening lead 软化铅(脱除砷、锑、锡等杂质的铅)
softening machine 软麻机
softening material 退火材料;软化材料
softening method 软化法
softening of water 硬水软化;软化硬水;软化水法;水软化;水的软化
softening plant 软化站;软化装置;软化水厂;水软化(工)厂
softening point 软化温度;软化点
softening point drift 软化点漂移
softening point measurement 软化点测定
softening point of bitumen 沥青软化点
softening point of residue 残渣软化点
softening point test 软化点试验
softening point tester 软化点试验仪(环球法)
softening point thermometer 软化点温度计
softening point tube 软化点测定管
softening power 软化能力
softening price 软化价格
softening process 软化法
softening range 软化幅度;软化温度范围;软化区;软化范围
softening rock 软化岩石
softening sludge 软化污泥
softening tank 软化池;软化槽
softening temperature 软化温度;软化点
softening test 软化试验
softening test of rock 岩石软化性试验
softening treatment 软化处理
softening type 软化型
softening unit 软化装置
softening water treatment 软化水处理
softening wood 软木材
soften spectrum 软化谱
soften the terms 放宽条件
softer 软化剂
softest possible terms 最优厚的条件
softest water 极软水
soft expanded natural rubber 软膨胀性天然橡胶
soft extrusion 软泥挤坯;塑性挤出法
soft eye splice 插接软琵琶头
soft face hammer 软面锤
soft faience 软质彩釉精陶
soft feed water 软给水
soft-feel coating 柔软感涂料;手感柔软涂料
soft feeling 软触感
soft ferrite 软铁氧体
soft fiber core (钢丝绳的)软纤维芯
soft fiber 软(质)纤维
soft-field gear 软道面用的起落架
soft filth water 软垢水
soft fine earthenware 软质精陶
soft fine pottery 软质精陶;低温精陶
soft finish 软质面层(建筑);柔软整理(少用或不用浆料)
soft fire 文火
soft fireclay 软质耐火黏土
soft-fired ware 软烧器皿;轻烧器皿
soft flame 软火焰;轻烧;碳化火焰
soft flocks 毛屑
soft floor covering 软性覆面
soft floor finish 软质地面面层
soft flour 软质面粉
soft flux 软稀释剂;软化剂
soft foam filling 软性泡沫填料
soft focus 软焦点;柔焦;软聚焦
soft-focus attachment 柔焦镜;柔光谱
soft-focus filter 软焦滤光片;柔光镜
soft-focus image 软焦像;柔焦像
soft focusing 软聚焦
soft-focus lens 软焦距透镜;软焦点镜头;柔焦镜头;柔光镜头;柔光聚焦镜
soft-focus objective 软焦物镜
soft fold 柔软折叠
soft formation 软岩层;软地层
soft-formation cutter head 三翼岩芯钻头
soft forming tube 软质绕丝筒
soft foundation 软弱地基;软土地基;软基;软地基
soft fusite 软微丝煤
soft gelatin capsule 软胶胶囊
soft gel-like matrix 软拟胶骨架

soft glass 钠玻璃;易熔玻璃;软质玻璃;软化点温度低的玻璃;软质玻璃;普通玻璃
soft glaze 中低温釉
soft goods 软商品;软货物;非耐用货物;不耐用商品
soft grease 软膏
soft grinding stone 软磨石
soft grit 软质颗粒(金属表面喷射加工用)
soft ground 软土地面;软土地基;软土;软路围岩;软弱地基;软地层;软地
soft ground boring tool 软岩石钻进工具
soft ground downhole testing 软弱地层钻井测试
soft ground etching 软底蚀刻
soft ground instrumentation 软弱地层测试仪器
soft ground physical property test 软弱地层物理性质试验
soft ground tunnel 软弱地层隧道
soft ground tunnel(l)ing 松软地层隧道施工
soft gum 软胶
soft hail 雪丸;软雹(霰)
soft hair polishing wheel 软发抛光轮
soft hammer (锤头用软材料做的)软锤
soft hand 手感柔软
soft-hard-acid base concept 软硬酸碱原理
soft-hardness 低硬度
soft hardwood 软硬木
soft-headed hammer 软头锤
soft heart 软心材;软心
soft heat (平炉的)软熔
soft ice 快融解的冰;软冰
soft image 色调不鲜明图像;低反差像;低对比度图像
soft impact 软着陆
soft inclusions 软弱夹杂物
soft ink 稀油墨;低浓度油墨
soft installation 软式设施
soft intercalated bed 软弱夹层
soft interlayer 软弱夹层
soft iron 软铁;熟铁;易切削铸铁
soft iron armature 低碳钢电枢
soft iron ball 软铁球
soft iron bar 软铁杆
soft iron circuit 软铁磁路
soft iron core 软铁芯
soft iron instrument 软铁芯电磁测量仪表
soft iron oscillograph 电磁示波器
soft iron shunt 软铁分路器
soft iron sphere 软铁球
soft iron wire 软铁丝
soft jaw 铁卡爪
soft jaw chuck 软钢卡爪卡盘;铁卡爪卡盘
soft junction 软结
soft kaolin deposit 软质高岭土矿床
soft knee 曲线缓变弯折处
soft label 软标号
soft laid rope 软搓绳
soft-land 软着陆
soft landed instrument station 软着陆仪器站
soft landing 软着陆
soft landscape 软风景
soft-land tractor 软地拖拉机
soft lay 软搓
soft layer 软弱层;软层;柔层
soft layer underneath 软弱下卧层
soft lead 软铅
soft lead pipe 软铅管
soft lead plate 软铅板
soft lead pressure pipe 软铅压力管
soft lead sheet 软铅皮
soft lead wire 软铅丝
soft leather 软革
soft light 柔光;散射光
soft lighting 柔光照明
soft limiter 软限幅器
soft limiting 软极限
soft line 软绳
soft loan 优惠贷款;软贷款;低息贷款
soft-loan window 软贷款窗口(指对发展中国家的优惠贷款业务)
softly cemented 弱胶结的
soft magnet 暂时磁体;软磁铁
soft magnetic ferrite 软磁铁氧体
soft magnetic material 软磁性材料;软磁材料
soft magnetic property 软磁性质
soft magnetism 软磁

soft malm brick 软白垩砖
soft management 柔软经营
soft mantle 软地幔
soft marine mud 软海泥
soft market 市场疲软
soft material 软材料
soft matrix 软胎体
soft mattress laid 软铺排
soft medium carbon steel 软中碳钢
soft medium structural steel 软中碳结构钢
soft metal 轴承用减磨金属;软金属
soft metal jaw 软金属钳口垫片
soft mission 软任务
soft mode 软膜
soft mode theory 软模理论
soft money 纸币;软币
soft mud 软淤泥;淤泥
soft mud brick 软泥砖;湿法黏土砖
soft mud brick machine 软泥造砖机
soft muddy clay 软泥质黏土
soft mud process 软泥工艺;软泥成型法;软泥制砖法;软泥制坯法
soft mud process brick 软泥砖
soft negative 软性底片;软调底片
softness 软性;软化度;软度;柔软性;柔软度;柔韧性;柔度
softness coefficient 柔度系数
softness factor 柔度系数
softness number 软化度
softness of lithosphere 岩石圈软度
softness-stiffness tester 柔软度与刚度测定仪
softness test 柔软性能测定
softness tester 柔软度试验机
softness value 软化值;软化度
soft nitriding 软氮化法
soft nose plate 弯臂形船首柱钢板
soft nose stem 弯臂形船首柱
soft opening 软开业
soft-open rib 柔性开口肋
soft organic clay 淤泥
soft oscillation 软振荡
soft package 软包装
soft-packed stuffing box 软填料箱
soft packing 软填料;柔软填料
soft packing seal 软填料密封
soft paper 软性像纸
soft paper ticket 软票
soft paraffin(wax) 软石蜡;典软石蜡
soft particle 软颗粒;软弱颗粒
soft paste 软坯
soft paste porcelain 软质瓷;低温瓷(器)
soft patch 螺栓补板;软补丁片;柔性搭板
soft peening 轻凿
soft petroleum ointment 矿脂
soft phase 缓变状态
soft phenolic resin 软酚醛树脂
soft photo tube 软光电管
soft picture 弱反差像;软图像;低对比度图像
soft pig iron 软生铁
soft pine 软松木
soft pitch 软沥青;软焦油沥青;软柏油脂
soft pitch concrete tile 软沥青混凝土砖
soft pitch tile 软沥青砖
soft plastic 软塑(的)
soft plastics 软塑料
soft plug 冷型火花塞;低温形火花塞
soft polymetric material 软聚合物材料
soft porcelain 软瓷器;软瓷
soft position 软式阵地
soft radiation 软辐射
soft rags 软石板瓦
soft rain 小雨;细雨
soft ray 软性射线;软射线
soft readjustment 柔性调节
soft recovery 软回收
soft resin 软树脂
soft rib 柔性肋
soft rime 轻雾凇;雾凇
soft roadbed 软土路基
soft rock 软质岩(石);软岩(石);软石;软弱地层
soft-rock excavation 软石开挖;松石开挖
soft roll 软胶辊
soft rope 软绳
soft rot 软腐朽

soft rubber 软橡胶;软质橡胶
soft rubber ball 橡皮塞
soft rubber insert 软橡胶插入物;软橡胶衬垫
soft rubber washer 软橡胶垫圈
soft sand 细级配骨料;软砂;细砂
soft sandy clay 软砂质黏土
soft science 科学指挥学;软科学
soft sculpture 软雕塑
soft seams 磨耗的帆布缝线
soft seat 软底板
soft seated valve 软座圈
soft sector 浮动段
soft sectoring 扇区软划分
soft sediment 松软沉积(物)
soft segment 柔性链段
soft shadow 软性阴影;柔影
soft shale 软页岩
soft shoulder 软路肩
soft shoulder sign 路肩软弱标志
soft silica 无定形二氧化硅;非晶形二氧化硅
soft simulation 软模拟
soft sinter 松烧结块
soft skin 表面脱碳;表面软化
soft slope 缓坡
soft sludge 软污泥;软膏状油垢
soft soap 钾皂;软皂
soft soil 软土;松软性土
soft soil for normal consolidation 正常固结软土
soft soil foundation 软弱地基
soft soil improvement 软土壤处理
soft soil layer 软土层
soft soil mobility 软土机动性
soft soil performance 软土通过性能(车辆)
soft soil treatment 软基加固
soft solder 软钎料;软焊条;软焊料
soft solder flux 助焊剂;软焊剂
soft soldering 锡焊;软钎焊;软焊
soft soldering alloys 软焊接合金
soft spot 模糊点点;弱点;软点;松软路基;松软地点
soft spring 软簧
soft stainless steel 低碳不锈钢;软不锈钢
soft start 软起动
soft starter 柔性起动器;平稳起动器
soft start type 软起动型
soft state 软状态
soft steel 软钢;低碳钢
soft steel pipe 低碳钢管;软钢管
soft steel water-jacket 软钢水套
soft sticky clay 胶黏土
soft still 软静止图像
soft stone 软石
soft stoneware 软炻器
soft storey sway mechanism 柔层摆动机构
soft stratum 软弱地层;软层
soft stream 缓流
soft strip 软带材
soft stuff 软粉刷砂浆
soft stuffing-box seals 软填料密封
soft subsidy 软补贴
soft subsoil 软弱底土;软地层;软底土
soft substratum 软弱下卧层
soft surface road 软地道路
soft surface runway 软道面跑道
soft surface steel hammer 软面钢锤
soft surrounding rock 软弱围岩
soft-switch modulator 软性调制器
soft target 软目标
soft tar pitch 软焦油沥青
soft taxation 软税收
soft tech 软技术
soft technique of decision 决策软技术
soft technology 软技术
soft technology for decision 决策软技术
soft temper 软化回火
soft Teriary sandstone 软的第三纪砂石;软的第三纪砂岩
soft terms 优惠条件
soft-textured chipped wood concrete 软组织木屑混凝土
soft-textured pigment 软质颜料
soft-textured wood 软组织木材
soft-textured wood fiber concrete 软组织木纤维混凝土
soft-textured wood floor boarding 软组织木地板

soft tin 软锡钎料
soft tire 软轮胎
soft-to-medium clay 软至中硬度黏土
soft tone 软色调;软调
soft top 软车顶
soft top container 软顶(式)集装箱
soft touch control 软触控制;软接触控制
soft track 松软的轨道
soft transmission 软传动
soft twisted-cotton string 软棉线绳
soft type 松软类
soft type detergent (易生物降解的)软性洗涤剂;软型去垢剂
soft type surfactant 软型表面活性剂
soft underlying stratum 软弱下卧层
soft vacuum 低真空
soft vector 软矢量
soft vulcanized fiber 软钢纸
soft walked-on finish 软人行道饰面
soft wall 透声壁
software 软设备;软件;程序系统;程序设备
software address 软化地址
software aids 软件工具
software approach 软件方法
software architecture 软件体系结构
software assembler 软件汇编程序
software center 软件中心
software command 软件指令
software command set 软件指令组
software company 软件公司
software compatibility 软件兼容性;程序兼容(性)
software compatible 软件兼容的
software component 软件成分
software component of an operating system 操作系统的软件成分
software consideration 软件设计考虑
software convention 软件约定
software debugging 软件故障排除;软件调试
software debugging aids 软件调试工具;软件调试程序
software design 软件设计
software development 软件开发
software development process 软件开发过程
software development system 软件研制系统;软件开发系统
software device 软设备
software diagnostic instruction 软件诊断指令
software documents 软件文件;软件文档
software documentation 软件文本
software duplex 软件复用
software engineering 软件技术;软件工程
software firm 软件公司
software flexibility 软件灵活性
software for display and edit 显示和编辑软件
software for local operator workstation 就地级工作站软件
software function 软件功能
software house 软件公司;软件服务站
software integration test 软件集中试验
software interface 软件接口
software interruption 软件中断
software kit 软件包
software legal protection 软件法律保护
software library 软件库
software life cycle 软件生命周期
software maintenance 软件维护
software manuals 软件操作程式手册
software mapping 软件映射;软件变换
software memory mapping 软件存储映射;软件存储变换
software modularity 软件模块化;程序模式化
software modulator 软件调制器
software module 软件模块
software monitor 软件监督程序;软件监督;程序监督器
software of automated cartography 自动制图软件
software-oriented 面向软件
software overhead 软件开销
software package 软设备组件;软件标准部件;软件包;套装软体;程序包
software packet 软件包
software paging 软件编页

software performance 软件功能
software polling 软件轮询
software priority 软件优先权
software priority interrupt 程序优先中断
software procedure 软件编制过程
software process 软件过程
software programmable 可程控软件
software program(me) 软件程序
software project 软件计划
software reliability 软件可靠性
software requirements definition 软件需求定义
software resource 软件资源
software science 软件科学
software security 软件安全性
software service 软件装置;软件服务
software setup 软件安装
software simulator test 软件仿真程序测试
software source 软件资源
software stack 软件栈
software structure 软件结构
software support 软件支持
software support program(me) 软件支持程序
software system 软件系统
software technique 软件技术
software technique system 软件技术系统
software test 软件测试
software timer alarm 软件定时报警
software tool 软件工具
software trace mode 软件跟踪方式
software trade 软件交易
software transportability 软件可移植性
software user's manual 软件用户手册
soft waste 软回收
soft water 软水
soft water outlet 软水出口
soft water treatment 软水处理
soft wax 软蜡
soft wax tank 软蜡槽
soft-weld 铜镍焊条合金
soft white clay 软质白陶土;软质白瓷土
soft white gum 软白胶
soft wire 软线;软焊丝
soft wired number control 软连数控
soft wired system 程序布线方式
soft wire rope 软钢索
soft wiring 软连接
softwood 软质木材;软木;软材;针叶树(材);软(性)木材
softwood board 针叶材板材;软材板
softwood charcoal 软材木炭
softwood cutting 软材插条
softwood distillation 软材干馏
softwood flooring 软木材楼地面
softwood forest 针叶树林;软材林
softwood pack 软木木垛
softwood pitch 软质木材焦油沥青
softwood plank 针叶材的厚板
softwood plywood 针叶树胶合板;针叶树胶合板
softwood size 软质木料尺寸
softwood sleeper 软木轨枕
softwood strip 针叶材板条
softwood tar 针叶材焦油;软质木材焦油;松焦油
softwood tar pitch 针叶树材焦油沥青;松焦油沥青;软材焦油沥青
softwood tie 软木轨枕
soft-working developer 软性显影液
soft woven lining 软织物制动衬层
soft zone 软化区
sogasoid 固气溶胶
sogdianite 锆锂大隅石
sogginess 湿润程度
soggy 未烘透的;浸水的;潮湿的
soggy soil 湿润土(壤)
Sohaerotilus bulking 球衣菌膨胀
soie platte 刺绣丝线
soil 泥土;土壤;土地;土
soilability 污染程度
soil absorption 土壤吸收
soil absorption capacity 土壤吸收能力;土的吸收能力
soil absorption field 土壤渗水场
soil absorption system 土壤吸收系统
soil absorption test 土壤吸水性试验
soil abundance 土壤资源丰富

soil acidity 土壤酸性;土壤酸度
soil acidity or alkalinity 土壤酸碱度
soil actimomycetes 土壤放线菌
soil adhesion 土壤黏附;土壤附着力;土的附着力
soil adsorption of herbicides 土壤吸附除锈剂
soil adsorption system 土壤吸附系统
soil aeration 土壤通气(性)
soil aerator 土壤通气器
soil-affected soil 盐渍土(壤)
soil age 土壤年龄
soil aggregate 集料土;土团粒;土壤团粒(结构);土壤集合体;土砂石;骨料土
soil-aggregate mixture 土集料混合物;土骨料混合物
soil-aggregate size 土壤聚合体大小
soil-aggregate stability 干旱土壤团聚体的稳定性
soil-aggregate surface 土砂石路面;碎石土面层;骨料土面层
soil aggressivity 土壤侵蚀性
soil agitator 碎土器
soil air 土壤中空气;土壤空气
soil-air heat pump 地气式热泵
soil algae 土壤藻类
soil alkalinity 土壤碱度
soil ameliorant 土壤改良剂
soil amelioration 土壤改良;土质改良
soil amendment 土壤改良(剂)
soil amplification 土的放大效应
soil analysis 土质分析;土壤分析;土分析
soil analysis error 土分析误差
soil analysis on-the-spot 简易土壤分析
soil anchor 土层锚杆
soil and differential water load 土和剩余水荷载
soil and foundation engineering 土壤基础工程
soil and oil repellent 抗脏;耐脏
soil and rock 岩土体
soil and rock descriptive term 岩土描述术语
soil and rock mechanics 岩土力学
soil and vent pipe 污物排气管
soil and waste pipe 粪便和生活废水管道
soil and water assessment tool 水土评估方法
soil and water conservation 水土保持;水生保持
soil and water conservation dam 水土保持坝
soil and water conservation district 水土保持区
soil and water conservation planning 水土保持规划
soil and water conservation practice 水土保持实践
soil and water conservation program(me) 水土保持实施计划;水土保持规划
soil and water conservation region 水土保持区
soil and water conservation zoning 水土保持区划
soil and water loss 水土流失
soil and water pathogenic factor 水土病因
soil and water relationship 土和水的关系
soil and water source of disease 水土病因源
soil and water utilization 水土利用
soil animal 土壤动物
soil animal population 土居动物群落
soil anomaly 土壤异常
soil appliance 卫生设备
soil arch 分载反拱
soil-asphalt 地沥青土
soil-asphalt mix(ture) 土壤—地沥青混合物
soil-asphalt road 地沥青稳定土路
soil-asphalt stabilization 地沥青土稳定(法)
soil assessment 土壤评价
soil association 土壤组合;土壤结合
soil atmosphere 土壤空气
soil auger 螺旋取土钻;取土样钻;土钻;土壤螺旋钻;土壤采样钻
soil avalanche 土崩
soil background value 土壤背景值
soil bacteria 土壤细菌
soil bacteriology 土壤细菌学
soil balance 土壤平衡
soil bank 土堤;土岸
soil bank program(me) 土壤银行计划
soil beam test 土梁试验
soil bearing capacity 土壤承载(能)力;土的承载力;土承载力
soil bearing ratio 土壤承载比;土承载比
soil bearing test 土的承载力试验;土壤荷载试验;土承载(能力)试验
soil bedding 土壤层理;土的层理

soil-bentonite mix(ture) 土壤—膨润土混合物
soil binder 土结合料;土的黏结键;土的胶结物
soil bin test 土槽试验
soil biochemistry 土壤生物化学
soil biodynamics 土壤生物动力学
soil biology 土壤生物学
soil bitumen 沥青土
soil bitumen base 沥青土基层
soil bitumen survey 土壤沥青测量
soil bituminous road 沥青土路
soil blister 土体冻胀;土胀
soil block 土块
soil block test (木材防腐的)土壤木块试验
soil blow-off 土壤吹失
soil body 土体
soil boring auger 麻花钻;土钻
soil boring cut(ting) 土的钻凿开挖
soil boring source 土壤钻探原始资料
soil boring yardage 土壤钻探深度测量
soil-borne 土壤传播的
soil-borne disease 经土壤传播的疾病;土源疾病;土传病害
soil-borne transmission 经土壤传播
soil-bound 土结的
soil boundary 土壤界线;土壤界
soil box 土壤箱;土槽
soil branch 排粪管;污水排水支管
soil buffer action 土壤缓冲作用
soil buffering 土壤的缓冲作用
soil building 造土
soil building process 土壤构成过程
soil building property 成土特性
soil bulging 土的膨胀
soil bulk density 土壤颗粒密度
soil bulldozing 土壤推移
soil buoyancy 土的浮容重
soil burial test 土埋试验(纺织品耐腐性试验)
soil burning 土的烧灼
soil cap 覆盖土;表土
soil capability 土壤潜力;土壤能力
soil capacity 土壤能力;土壤耐压力
soil capillarity 土壤毛细管作用;土壤毛细管现象;土壤毛
soil capital 土地资本
soil carbonate survey method 土壤碳酸盐测量法
soil carrying capacity 携污容量
soil carrying property 携污性
soil cartography 土壤制图学
soil category 土壤类型
soil category of site 场地土类别
soil catena 土链
soil cavity 土洞
soil cement 掺土水泥;水泥土壤;水泥稳定土(壤);水泥土混合料;水泥土;水泥及土混合料;水泥固结土;水泥加固土
soil cementation 土的води加固法
soil-cement base 掺土水泥基层;水泥土基层;水泥稳定土基层
soil-cement lining 水泥土衬砌
soil-cement mixture 水泥掺土混合料
soil-cement processing 水泥土加固法;土的水泥加固法
soil-cement road 水泥(稳定)土路
soil-cement slurry 掺土水泥浆;水泥土浆
soil-cement surface course 土和水泥表面层
soil-cement treatment 水泥稳定土处治;水泥稳定土处理
soil census 土壤普查
soil change method 换土法
soil characteristic 土质;土体性质;土体特性;土壤特性;土的特性;土壤性质
soil chemical property 土壤化学性质
soil chemistry 土壤化学
soil circle stress method 圆弧应力分析法;按圆弧应力分析法
soil class 土壤类别;土级
soil classification 土壤类别;土质分类;土壤分类;土的分类
soil classification system 土壤分类系统;土分类体系;土分类法
soil classification test 土壤分类试验;土分类试验
soil classification triangle 土壤分类三角图
soil climate 土壤气候
soil-climate relationship 土壤气候相关性

soil climatology 土壤气候学
soil clod 黏土质岩石;土块
soil coagulant 土壤凝结剂
soil coefficient 土壤性质特征
soil coefficient of external friction 土壤外部摩擦系数
soil cohesion 土壤黏聚力;土壤黏结力;土壤黏合性;土壤内聚力;土的内聚力
soil cohesive strength 土的凝聚强度
soil colititre 土壤大肠杆菌群值
soil colloid 土壤胶体;土胶质;土胶体
soil colloidial particle 土壤胶态颗粒;土壤胶体颗粒
soil column 原状土柱试样;土柱
soil compacting machine 土壤夯实机;土壤压实机
soil compaction 土壤固化;土压实;土壤压缩性;土壤压实;土壤挤密加固法;土壤夯实;土的压缩性;土的压实
soil compaction control kit 土压实控制仪
soil compaction index 土壤紧实指标
soil compaction pile method 挤密土桩法
soil compaction resistance 土壤压实阻力
soil compaction roller 土壤压实滚筒;压土机
soil compaction test 土的击实试验;土壤击实试验
soil compactness 土壤密实度
soil compactor 填土夯实机
soil complex 土壤复域
soil composition 土壤组成;土的组成
soil compressibility 土的压缩性
soil compression 土壤压实
soil compression test 土的压缩试验
soil concrete 掺土混凝土
soil concrete conduit 掺土混凝土管
soil concrete pipe 掺土混凝土管
soil condition 土质条件;土体状况;土壤状况;土壤条件
soil conditioner 土壤结构改良剂;土壤调节剂;碎土整地刀板耙
soil conditioning 改良土壤结构;土质改良
soil conductivity 土壤导热率;土壤导电热率
soil cone penetrator 土圆锥仪
soil conservancy 土壤保持
soil conservation 土壤保持
Soil Conservation and Allotment Act 土壤保护和分配法案
soil conservation area 水土保护(地)区
soil conservation district 土壤保持区
soil conservation extension 土壤保持推广
soil consistency[consistence] 土壤结构特性;土壤结度;土壤稠性;土壤稠度
soil consistency[consistence] constant 土壤结构常数
soil consistency test 土的稠度试验
soil consolidation 土壤固结(作用);土的固结(作用);土层加固
soil constant 土体常数;土壤性质特征;土壤常数
soil-constant plant 恒土植物
soil constituent 土壤成分
soil constitution 土壤组分;土壤组成;土壤结构
soil contact 土壤接触面
soil-container interaction 土与容器相互作用
soil containing a high percentage of rotted leaves 含腐叶百分比高的土壤
soil contains organic materials 土壤里含有有机物
soil contamination 土壤污染;土的污染
soil contamination analysis 土壤污染分析
soil content 含泥量
soil core 钻探土样;土样;土芯
soil core budding 带土柱芽接苗
soil core method of transplanting 带土柱定植法
soil core planting 带土柱定植
soil core sample 土柱样品;土芯试样
soil core technique 带土柱定植法
soil coring tube 土芯钻取筒
soil correctives 土壤校正剂
soil corrosion 土壤流蚀;土壤侵蚀;土壤腐蚀;土力侵蚀
soil corrosivity 土壤腐蚀性
soil cover(ing) 土被;地被物;土壤覆盖层;土壤植被;植被护坡;覆(盖)土;表土
soil crack 土壤裂隙
soil cracking 土裂
soil creep 土的蠕动;土的蠕变;土壤蠕动;土壤蠕变;土壤潜动;土壤滑动;土潜滑;土潜动;土爬(滑)(坡);土层蠕动;土体滑坍

soil crumb 土壤小片;土壤碎屑
soil crust 土壤硬表层;土壤结皮;土壤结壳;土结皮
soil crusting 土壳
soil cultivating machine 土壤耕作机械
soil cultivation 土壤耕作
soil culture 土壤培养;土壤
soil cushion 土垫层
soil cut by spade 用铁锹挖土
soil cutting 土壤挖掘
soil data 土质资料
soil decontamination 土壤净化
soil defence 土壤灭菌;土壤防护
soil deformation 土壤变形;土的变形
soil deformation parameters test 土的变形参数试验
soil degeneration 土壤退化;土壤贫瘠化
soil degradation 土壤退化作用;土壤退化
soil delimitation 土壤区划
soil densification 土壤压实;土壤密实化
soil density 土壤密度;土的密度
soil denudation 土壤侵蚀(作用);土壤浸蚀(作用);土壤剥蚀(作用)
soil depletion 土壤消耗;土壤贫瘠化;土壤耗竭
soil deposit 土壤沉积
soil depth 土层厚度
soil-derived fulvic acid 土壤中得到的富里酸
soil desquamation documentation 剥土编录
soil detention storage 土壤滞水量
soil deterioration 土壤破坏作用;土壤破坏;土壤变质
soil development 土壤开辟利用;土壤发育
soil diggable by spade 可用铁锹挖掘的土壤
soil digging 土壤挖掘
soil digging work 土壤挖掘工作
soil dilution plating technique 土壤稀释平板技术
soil discharge 土壤排水
soil discharge of groundwater 土壤地下排出
soil disease 土壤病症;土壤病害
soil disinfectant 土壤消毒剂
soil disinfection 土壤消毒(作用);土壤灭菌
soil dispersion 土壤分散
soil displacement 土壤沉降;土壤位移;换土
soil disposal method 泥土处理方法
soil distribution diagram 土分布图
soil division 土壤区分
soil drain 污水排水管;排污管
soil drainability 土壤排水性
soil drainage 土壤排水
soil drainage map 土壤排水图
soil drifting 土壤风蚀
soil drill 泥土钻;土壤钻杆
soil drilling by vibration 振动钻土;振动钻孔
soil dynamic(al) property 土动力特性
soil dynamics 土壤动力学;土动力学
soil ecological system 土壤生态系统
soil ecology 土壤生态学
soil ecosystem 土壤生态系统
soil ecotype 土壤生态型
soiled cocoon 污染茧
soiled copy 污损本
soiled ends 污丝
soiled lift 污物电梯
soiled linen chute 脏衣洗涤槽
soiled plate 染污薄板
soiled swatch 污布样
soil effective stress 土壤有效应力(垂直于单位面积的颗粒间平均压力)
soil electric(al) resistivity 土壤电阻比棒
soil electrochemistry 土壤电化学
soil electrolyte 土壤电介质
soil element 土单元
soil embankment 填土;路堤;土堤
soil engineer 土壤工程师;土工工程师
soil engineering 土壤工程学;土工(工程)学;土工
soil enrichment method 土壤加富法
soil environment 土壤环境
soil environmental capacity 土壤环境容量
soil environmental quality assessment 土壤环境质量评价
soil environmental quality index 土壤环境质量指数
soil environmental quality map 土壤环境质量图
soil environmental quality standard 土壤环境质量标准

soil equilibrated phosphorus 土壤磷的平衡
soil equilibrium 土壤平衡
soil erosion 土体侵蚀;土蚀;土壤水蚀;土壤溶蚀作用;土壤侵蚀;土壤浸染;土壤冲刷;水土流失
soil erosion control 水土流失控制;土壤冲刷控制
soil erosion datum 土壤侵蚀基准面
soil erosion index 土壤侵蚀指标;土壤浸蚀率
soil erosion map 土壤浸蚀图
soil erosion modulus 土壤侵蚀模数
soil erosion prediction 土壤侵蚀预报
soil erosion process 土壤侵蚀过程
soil evapo(u)ration 土壤蒸发
soil evapo(u)ration pan 土壤蒸发器
soil evapo(u)rimeter 土蒸发计;土壤蒸发器
soil excavation 挖土
soil excavation work 挖土工作
soil exchange works 换土工程
soil exhaustion 土壤疲劳;土壤消耗;土壤耗竭
soil expansion 地基勘察;土膨胀;土壤膨胀
soil expectation value 土地期望值
soil exploitation 土壤垦植
soil exploration 土质勘探;土质调查;土质查勘;土壤探查;土壤调查;地基勘探
soil exposure test 土壤暴露试验
soil extract 土壤提取;土壤浸出液
soil extract agar 土壤浸出琼脂
soil extractor 土块分离机
soil extract's chemical composition 土壤浸提的化学组分
soil fabric 土壤微结构;土的组构
soil factor 土壤因素
soil fail 土壤破坏
soil failure 土壤破坏
soil fall 土崩
soil fauna 土壤动物区系;土居动物群落
soil feature 土壤特性
soil fertility 土壤肥力;土壤肥度
soil fertility exhaustion 地力消耗
soil fertility remedial measures 土壤肥力补救措施
soil-filled bag 填土的岩洞
soil filling 填土
soil filter of odo(u)r 臭气土壤过滤器
soil filtration 土壤渗滤;土壤渗漏
soil fines 土的细粒部分
soil fitments 污物设备
soil fittings 污物设备
soil fixation 土壤加固
soil fixing 土壤固定
soil flora 土壤植物区系
soil flow 流土现象;泥流;土流
soil-flow cover 泥流覆盖
soil flow net 土中流网;土渗流网;土壤渗流网;土壤地下水渗流网
soil-footing contact 土基接触面
soil for filter layer 过滤层土
soil for impervious core 不透水的土坝芯
soil formation 土壤形成(作用);土壤发生;土的生成
soil formation and similar processes 土壤形成及类似过程
soil formation description 土层描述
soil formation of phreatic water 潜水成壤作用
soil formation type 成壤作用类型
soil-forming 成土
soil-forming characteristic 成土特性
soil-forming factor 土的形成因素;土壤形成因素;成土因素
soil-forming process 土壤形成过程;成土过程;成土方法;成土作用;成壤过程
soil-forming rock 成土岩石;成土母岩
soil foundation 土基;土质地基
soil-foundation system 土基础系统;土壤—基础系统;地基—基础体系
soil fraction 土壤部分;土壤粒级;土壤成分
soil-fracturing agent 破土剂
soil freezing 土壤冰冻;土壤冻结
soil friction circle method 圆弧滑动面分析法;按圆弧滑动面分析法
soil fumigant 土壤熏蒸剂
soil fumigation 土壤熏蒸
soil fumigator 土壤熏蒸仪;土壤熏蒸器
soil fungi 土壤真菌
soil fungistasis 土壤抑菌作用
soil fungistasis and nematophagus fungi 土壤抑菌作用和线虫的真菌

soil gas 壤中气;土壤气体
soil gas survey 壤中气测量;土壤气测量
soil genesis 土壤发生学;土壤发生
soil genetic horizon 土壤发生层
soil genetic layer 土壤发生层
soil geography 土壤地理学
soil geology 土壤地质学
soil gradation factor 土级配系数
soil grade 土壤粒级;土壤粒度;土类;土的分类
soil grain 土壤颗粒;土粒
soil grain distribution test 土(壤)粒径分布试验
soil grain size accumulation curve 土壤粒径累计曲线;土粒径累积曲线
soil granulation 土团粒作用;土壤微团粒;土壤团粒(形成)作用
soil granule 土壤团粒
soil great group 土块类
soil grinder 土壤研磨机;土壤分析用研磨机
soil ground 土基
soil group 土壤群;土类
soil grouting 灌土
soil heterogeneity 土壤差异性;土壤不均匀性;土不均一性
soil hole 土洞
soil horizon 土壤层(位);土层
soil humic acid 土壤腐殖酸
soil humic substance 土壤腐殖物质
soil humidity 土壤湿度
soil humification 土壤腐殖化
soil humus 土壤腐殖质;土壤腐殖物质
soil hygiene 土壤卫生;土壤净化
soil identification 土质鉴定;土壤鉴定;土壤鉴别;土的鉴定
soil identification chart 土壤鉴定图
soil impermeability 土壤的不可渗透性;土的不透水性
soil improvement 土质改良;土质处理;土层加固;地基加固;地基处理
soil improvement agent 土壤改良剂
soil improvement hydrogeology 土壤改良水文地质
soil index property 土壤指示特征
soil indicator 土性指标
soil infection 土壤浸染;土壤感染;土壤传染
soil infiltration 土壤下渗;土壤入渗
soil infiltration rate 土壤渗入速率
soil information 土质资料;土壤资料
soiling 撒土;沾污
soiling agent 染污剂(测去污力用)
soiling food 青饲料
soiling formula 污液配方
soiling particulate 污染颗粒
soiling-point technique 沸点技术
soiling procedure 污染程度
soiling solution 污液
soil inhabitant 土壤习居者;土壤习居菌
soil injection 土壤注射;土壤灌浆
soil in place 现场土
soil insect 土壤昆虫;地下害虫
soil insecticide 土壤杀虫剂
soil in-situ 原生土(壤);原地土(壤);原位土;现场土
soil inspection 土壤检验
soil interpretation 土质鉴定
soil invader 土壤寄居菌
soil invertebrate 土壤无脊椎动物
soil investigation 土质勘察;土质勘测;土质调查;土壤研究;土壤调查;地质勘察;地基勘察
soil investigation device 土壤研究设备
soil irrigation 土壤灌溉
soil lab 土工试验室;土工实验室
soil laboratory 土工试验室;土工实验室
soil-laden water 土壤含水
soil lake porosity 土壤缺乏孔隙
soil lathe 修样器;修土样器
soil layer 土层
soil leaching 土壤淋溶作用
soil legend 土的图例
soilless culture 无土栽培
soil-lime 石灰土
soil-lime pozzolan 火山灰石灰土
soil liming map 土壤石灰施用图
soil line 厕所污水管;粪便管
soil lines of equal pore pressure 土内孔隙水压

力等压线;土内等孔隙水压线
soil liquefaction 土的液化
soil liquefaction failure 土壤液化塌垮
soil liquefaction potential 土壤液化潜势
soil litter 枯枝落叶层
soil loading 污垢荷载
soil loss 土壤损失;土壤流失
soil loss tolerance 土壤流失容许值
soil luck 壤良
soil lump 土块
soil lysimeter 土壤液度计;土壤测渗计
soil macrofauna 土壤大动物区系
soil macroorganism 土壤大生物
soil management 土壤管理;土地经营管理;土地经营
soil management system 土壤管理制度
soil mantle 土壤覆盖层;土壤风化层;土被;覆盖土;表土
soil manual 土壤手册
soil map 自然土壤图;土质图;土壤图;土分布图
soil mapping 绘制土壤分布图;土壤制图
soil mass 土体量;土体;土块
soil mass property 土体性质;土体特性
soil mass shock 土体震动
soil mass slide 土体滑动
soil mass swelling 土体胀缩
soil material 土壤物质;成壤物质
soil matrix 土壤基质;土基
soil maturation 土壤熟化
soil maturation process 土壤熟化过程
soil mechanical characteristic 土的力学性质
soil mechanic property 土力学特性
soil mechanics 土壤力学;土力学
soil mechanics and foundation engineering 土力学与基础工程学
soil mechanics investigation 土壤力学研究
soil mechanics laboratory 土壤力学实验室;土力学实验室;土工试验室;土工实验室
soil mechanics principle 土力学原理
soil mechanics vocabulary 土力学词汇
soil metabolism 土壤代谢
soil metallometric survey 土壤金属量测量
soil microbial population 土壤微生物群系
soil microbiology 土壤微生物学
soil microflora 土壤微植物区系;土壤微生物区系
soil microorganism 土壤微生物
soil miller 旋转锄
soil mineral 土壤矿物
soil mineral matter 土壤矿物质
soil mineralogy 土矿物学
soil mixer 土壤搅拌机
soil mixing machine 土壤混合机
soil mix(ture) 土壤混合(物)
soil-mobility characteristic 土壤的通过特性
soil modulus 土的模量
soil moisture 土中水;土壤水分;土壤湿度;土含水量;土的湿度
soil moisture budget 土壤水量平衡
soil moisture constant 土湿度常数;土壤含水量常数
soil moisture content 土壤墒情;土壤含水量;土的含水量;墒情
soil moisture deficiency 土壤湿度不足
soil moisture deficit 土壤湿度水分差;土壤湿度含水差;土壤缺水量;土壤缺水度
soil moisture-density meter 土的湿度密实度仪
soil moisture experiment 土壤水试验
soil moisture film 土壤附着水;土壤薄膜水
soil moisture forecast 土壤水分预报
soil moisture grade 土壤湿度等级
soil moisture gradient 土壤湿度梯度
soil moisture in aeration zone 包气带土壤湿度
soil moisture index 土壤含水指数
soil moisture measurement 土壤水分测定;土壤湿度测定;土壤含水量测定
soil moisture meter 土壤湿度测定仪;土壤湿度计
soil moisture neutron probes 土壤湿度中子测定仪
soil moisture retention 土壤滞水量
soil moisture sampling 土壤潮湿采样
soil moisture storage 土壤水分蓄水量;土壤水分储量
soil moisture stress 土壤水分应力;土的水分应力
soil moisture suction 土壤吸水负压;土壤水吸力;土壤水分抽汲

soil moisture supply 土壤水分供应
soil moisture teller 土的水分速测仪
soil moisture tension 土中水分张力;土壤水分张力
soil moisture tension meter 土壤湿度张力表
soil moisture zone 土壤含水区
soil monitoring 土壤监测
soil monolith 土壤整段标本;大型土样试块
soil morphology 土壤形态学
soil mortar 泥砂浆;泥灰浆
soil-mo(u)lded concrete 土壤成型混凝土;土壤模ócu混凝土
soil mulch 土壤幂;土壤覆盖;土幂;土被
soil nail 土钉
soil nailed wall 土钉墙
soil nailing 土锚法;土钉法
soil nailing method 土锚法
soil nailing wall 土钉墙
soil nailing wall technology 土钉墙技术
soil name 土壤名称;土名
soil names file 土壤制图文件
soil nematode 土壤线虫
soil nitrogen 土壤氮素
soil nitrogenous pollutant 土壤含氮污染物
soil nutrient 土壤养分
soil of glacial origin 冰川生成土
soil of high permeability 高透水性土壤
soil of low permeability 低渗土壤
soil of relatively formation 较近形成的土壤
soil operation 土壤操作
soil organic matter 土壤有机质;土壤有机物
soil organic matter fractions 土壤有机质部分
soil organism 土壤生物
soil overburden influence 土壤盖层影响
soil packing 土壤的压实
soil pan 土壤蒸发器;土壤蒸发皿
soil parabination 土壤复式组合
soil parameter 土壤参数
soil particle 土壤颗粒;土粒
soil particle culture method 土粒培养法
soil particle inoculating method 土粒接种法
soil particle size 土粒大小
soil paste 土浆
soil pat 土饼块
soil path 土路
soil pattern 土样;土型
soil peak strength 土壤最大强度
soil-peat mixture 土壤泥炭混合物
soil pencil 土壤取样器;土壤取样管
soil penetrometer 土力学触探仪;土壤贯入度仪;土壤贯入器;触探仪;触探器
soil percent saturation 土壤水分饱和度
soil percolate 土壤渗液;土壤渗透
soil percolation 土壤渗滤;土壤渗漏
soil perfusion 土壤灌注
soil permeability 土壤透水性;土壤渗透性
soil persistent pesticide 土壤残留性农药
soil pest 土壤害虫
soil pesticide interaction 土壤农药相互作用
soil phase 土相
soil phophorus 土壤磷
soil physical maturity 土壤的物理适耕性
soil physicochemistry 土壤物理化学
soil physics 土物理学;土壤物理学
soil physics investigation 土壤物理学研究
soil physiology 土壤生理学
soil pile 土桩
soil-pile adhesion 土与桩之间的附着力
soil-pile and lime-soil pile 土桩与灰土桩
soil-pile interaction 土与桩之间的相互作用
soil-pile-structure interaction 土桩结构相互作用
soil pipe 下水道;排水铸铁管;污水管;厕所污水管;粪便污水立管;粪(便)管
soil pipe elbow 污水管弯头
soil piping 土壤管涌;土的管涌
soil pit 土穴;大便槽
soil placement 土壤填筑;土壤堆放
soil plague method 泥石法
soil-plant relationship 土壤植物相互关系
soil-plant system 土壤—植物系统
soil plasma 土壤细粒物质
soil plasticity 土壤可塑性
soil plate method 土壤平板法
soil plug 土塞
soil poisoning 土壤中毒;土壤积毒

soil pollutant 土壤污染物
soil pollution 土壤污染
soil pollution analysis 土壤污染分析
soil pollution by heavy metal 土壤重金属污染
soil pollution by organic matter 土壤有机污染
soil pollution chemistry 土壤污染化学
soil pollution control 土壤污染控制
soil pollution investigation 土壤污染调查
soil pollution map 土壤污染图
soil pollution prediction 土壤污染预测
soil pollution source 土壤污染源
soil pore 土壤孔隙;土孔隙;土的孔隙
soil pore space 土壤孔隙
soil porosity 土壤孔隙度;土壤空隙度;土的孔隙率;土的孔隙度
soil porosity ratio 土壤孔隙比;土的孔隙比
soil potassium 土壤钾素
soil preparation 整土;整地;播前耕作
soil pressure 土压(力);土壤压力
soil pressure cell 土压力盒;土压测定盒
soil pressure condition 土壤压力状态
soil pressure diagram 土压力图
soil pressure line 土压力线
soil pressure measurement 土压力量测;土压力测定
soil pressure phenomenon 土壤压力状态
soil primary shear surface (与机具前进方向平行的)土壤主剪切面
soil priming 土壤充水;土的初始吸水量
soil probe 土壤探测杆
soil processing 土壤制备(试验时)
soil productivity 土壤生产力
soil profile 土壤剖面;土壤断面;土剖面;土层剖面图;土层剖面
soil-profile characteristics 土层剖面特性
soil-profile sketch 土体剖面略图
soil-proofing coating 防沾污涂层
soil property 土的性质;土壤性质;土壤特性
soil property of immersion zone 浸没地段岩性
soil prospecting 土质勘探
soil protecting plant 护土植物
soil protection 土壤保护;土地保护
soil protection wood 林地保护树
soil-protective cover 护土覆盖物;土壤保持覆盖
soil pudding 土捣密工作
soil pulverizer 圆盘耙;土(的)粉碎机;碎土机
soil purification 土壤净化
soil pusher 推土器
soil quality 土壤性能;土质;土壤质量
soil quality assessment 土壤质量评价
soil quality classification 土壤质量分类
soil quality index 土壤质量指数
soil quality standard 土壤质量标准
soil quantity determination 土壤数量的确定
soil ramp 土坡道
soil rating 土的分级;土分级
soil reaction 土壤反应
soil reclamation 荒地垦拓;土壤改良
soil reconnaissance 土质查勘
soil regime 土壤状况;土壤情况
soil regionalization 土壤区划
soil regulation 土壤管理
soil reinforcement 土壤的加筋法
soil relief 土壤地形
soil remediation program(me) 土壤矫正计划
soil removability 去污能力
soil removal 去污;土方搬运
soil removal efficiency 去污效率
soil-removing action 去污作用
soil renewal 土壤再生
soil rent 地租
soil replacement method 换土法
soil replacement works 换土工程
soil requirement 土壤需水量
soil requirements of sewage irrigation 污水农灌土壤要求
soil research 土壤研究
soil residue treatment 土壤残留处理
soil resistance 土壤电阻;土壤阻尼;土壤阻力;土壤抗污性
soil resistant fiber 防污纤维
soil resistivity 土壤电阻率
soil resource 土壤资源
soil respiration 土壤呼吸

soil respirator 土壤呼吸仪
soil retaining 挡土的
soil-retaining structure 挡土建筑物
soil retention 土壤持水量
soil rheology 土体流变学
soil ribbon 土条带；土搓条
soil road （无路面的）土路
soil rot 土壤中腐烂
soil salination 土壤盐碱化
soil salinity 土壤盐渍度；土壤含盐量
soil salinization 土壤盐碱化；土地盐渍化
soil salt survey 土壤盐测量
soil sample 土样；土壤样品；土壤取样
soil sample barrel 取土筒
soil sample characteristic 土壤样品特征
soil sample composition 土壤样品粒径组分
soil sample identification with naked eye 土样目测鉴定
soil sample measure 土样法；土样测量
soil sample of known weight 已知重量的土样
soil sampler 取土器；土壤取样筒；土壤取样器；土壤取样管；土壤采样器；采样土钻；采土器
soil sampler of active value 活阀式取土器
soil sampler of air pressure 气压式取土器
soil sampler of double wall 双壁取土器
soil sampler of piston 活塞式取土（样）器
soil sampler of rotary drilling 回转压入式取土器
soil sampler of single wall 单壁取土器
soil sampler of spheric(al) value 球阀式取土器
soil sampler of thin-wall 薄壁取土（样）器
soil sample test 土样试验
soil sampling 取土样；土壤取样；土壤采样；土圈采样
soil sampling borehole 取土孔
soil sampling locality 土壤采样位置
soil sampling tool 土样取样工具
soil sampling tube 取土管
soils and rocks 土石料
soil sanitation 土壤卫生；土壤净化；土壤保洁
soil saturation 土壤饱和
soil saving dam 截水坝；挡土坝；截土坝；保土坝
soil saving dike 拦土堤；截土堤（公路保护挖坡的挡土墙）
soil saviour 土壤保养
soil science 土壤学；土壤科学
soil scientist 土壤学者；土壤学家；土壤科学家
soil sealant 土样密封剂；土壤封闭层
soil sealing 覆土
soil section 土纵断面；土剖面；土地断面；土层剖面
soil seepage 土中渗流；土壤内渗透；土壤内渗流
soil self-purification 土壤自净作用
soil sensitivity 土壤灵敏度
soil separate 土壤成分；土壤粒组；土壤粒级；土壤颗粒分组
soil separator 土壤分选机；土分选机
soil series 土系；土壤系列
soil settlement 土体沉降；土壤沉降
soil sew(er)age 家庭污水；粪便污水；污水
soil sewer 污水管
soil-shaping measures 土面整修
soil shear 土壤剪切力
soil shear strength 土壤抗剪强度；土的抗剪强度
soil shear test 土的剪切试验
soil shifter 推土机；堆土机
soil shoe 犁盘
soil shovel 运土铲斗
soil shredder 土壤破碎机；碎土器；碎土机
soil shrinkage 収缩；土壤下沉
soil sickness 土壤衰竭；土壤疲劳；土壤衰颓
soil silt size 粉土
soils in use for cropland 土壤用作农业耕地
soil site 土场地
soil size 土壤粒径；土的粒径
soil skeleton 土壤骨架；土壤骨骼；土（的）骨架
soil slide 土滑（坡）
soil sliding length 土壤滑移长度
soil sliding path 土壤滑移线
soil slip 土崩；坍方；土滑；滑坡
soil slipperiness 土壤滑溜状态
soil slope stability and earth pressure 土坡稳定性及土压力
soil slump 土壤崩塌；土崩；坍方
soil slurry-sequencing batch reactor activated sludge process 泥浆—序批间歇反应器活性污泥法

soils mechanician 土壤力学家
soil soak 地面施药
soil soaker 湿土器；土壤浸湿器
soil solidification 土壤加固；土壤稳定作用；土壤稳定（法）
soil solidifier 土壤固化剂
soil solid volume 土粒体积
soil solution 土液；土壤溶液
soil sounding 土层探测
soil sounding device 土壤触探仪；土触探装置；土触探仪
soil sourness 土壤酸性
soil specimen 土样
soil sphericity factor 土的球度系数
soil split preparation 土壤切片制备
soil spring constant 土弹簧
soils subject to moderate limination 受到中等限制的土壤
soils subject to moderate limitations in use 在使用时受到中等限制的土壤
soil stability 土壤稳定性；土的稳定
soil stabilization 围岩稳定处理；土质稳定；土壤稳定作用；土壤稳定（法）；土壤加固（法）；土（的）稳定；土的加固；土层加固；地基稳定（作用）
soil stabilization by electroosmosis 电渗法稳定土体
soil stabilization by freezing 冻结法稳定土体
soil stabilization technique 土的稳定技术
soil stabilization with bitumen 沥青稳定土
soil stabilization with lime 石灰稳定土
soil stabilizer 稳定土拌和机；稳定土路修筑机；土质稳定剂；土壤稳定剂；土加固剂；土（的）稳定剂
soil stabilizing machine 土壤稳定用机械；土壤稳定机械
soil stack 土堆；土壤立管；生活污水立管；厕所污水立管；粪便管；放臭气管
soil stack installation 污水管设施
soil standard substance 土壤标准物质
soil-steel interaction 土壤钢material相互作用
soil sterilant 土壤消毒剂；土壤灭菌剂
soil sterilization 土壤消毒；土壤清毒；土壤灭菌
soil-sticked seedling 带土小苗
soil stickiness 土壤黏性
soil sticky limit 土的黏限
soilstone 土壤岩
soil strain 土应变
soil strain-rate behaviour 土壤应变比特性
soil stratification 土壤层理；土的分层；土的层理
soil-stratification unit 土壤地层单位
soil-stratigraphic(al) unit 土壤地层单位
soil stratum 土层
soil stratum thickness 土层厚度
soil strength 土壤强度；土的强度
soil strengthened by sand drains 砂桩加固法
soil strengthening 土的加固
soil strength inspection 土体强度检查
soil stress 土壤应力；土应力
soil stressed zone 地基持力层
soil stress ga(u)ge 土壤应力表；土壤应力仪
soil stripe 土条带
soil strip map 土壤（分布）图
soil-stripping shovel 表土剥离铲（矿山用）
soil structure 土壤结构；土体结构；土壤构造；土的结构；土的构造
soil-structure interaction 土与结构物的相互作用；土结构相互作用
soil-structure interface 土结构（交）界面
soil study 土壤研究；土质研究
soil study device 土壤仪器；土壤设备
soil suborder 土壤亚纲
soil subsidence 土下沉
soil subsystem 土壤亚系统
soil suction 土壤吸力；土（的）吸力；土的负孔隙压力
soil suction potential 土吸力势
soil suite 土套
soil suspension 土壤悬液
soil superelement 土的超单元
soil supported structure 土壤支承结构
soil supporting layer 地基持力层
soil support value 土壤支承力值
soil surcharge 土壤过载；土壤超载
soil surface 土壤面层；土壤表面；土表（层）
soil surface(d) road 土面路；土路面

soil survey 土质调查；土壤普查；土壤调查
soil survey chart 土质调查表
soil surveying 土壤测量
soil survey manual 土壤调查手册
soil survey map 土质调查图
soil survey report 土的调查报告
soil-suspending ability 悬污能力
soil-suspending action 悬污作用
soil-suspending agent 污垢悬浮剂
soil-suspending power 污垢悬浮能力
soil suspension 土悬液
soil swelling 土壤膨胀
soils with a high acid content 含酸量高的土壤
soils with a high alkali content 含碱量高的土壤
soils with fines content up to 50% 细砂含量达一半的土（壤）
soil symbol 土的图例
soil system 土壤系统；土壤体系
soil tank 装土箱；土壤箱
soil tank modelling 土槽模拟法
soil taxonomic classes 土壤分类级别
soil taxonomic unit 土壤分类单元
soil taxonomy 土地分类
soil technics 土工技术
soil technology 土工技术；土质学；土质技术；土壤技术
soil temperature 土温；土壤温度
soil temperature in aeration zone 包气带土壤温度
soil test data sheet 试验成果总表
soil tester 土壤试验设备
soil test(ing) 土壤试验；土质试验；土工试验
soil testing instrument 土工测试仪器
soil test probe 土壤试验取样器；土壤试验探测器
soil test regulation 土工试验规范
soil test specification 土工试验规范
soil test summaries 土壤测试概要
soil textural classification 土结构分类
soil texture 土壤组织；土壤质地；土壤结构；土的组织；土的结构
soil texture classification 土壤质地分类
soil texture map 土壤质地图
soilthermograph 地温记录器
soil thermometer 地温计；土壤温度计；地温表
soil thixotropy 土的触变性
soil thread 土搓线；土搓条
soil thrust 土推力
soil tillage 整地；土壤耕作；土壤耕地
soil tiller 土壤翻拌机；土翻拌机
soil tilth 土壤适耕性；土壤耕性
soil tone 土色
soil trafficability 土在自然状况下可通车能力
soil transport 土壤运输
soil transportation 土方外运
soil treatment 土质处理；土壤处理；地基处理；地层处理
soil trial wedge method 按楔形破裂面分析法；楔形破裂面分析法（用于土壤稳定计算）
soil type 土组；土质分类；土型；土壤类型；土类
soil-ulmin 有机质
soil unit weight 土的重度
soilure 污斑
soil use classification 土壤利用分类
soil utilization 土壤利用
soil utilization map 土壤利用图
soil variation 土壤变异
soil variety 土壤变种
soil vent 污水排气口
soil ventilation pipe 污水通气管
soil venting quality 土壤透气性
soil vibrator 土壤振实器；土壤振动器
soil void 土壤空隙
soil volume 土壤体积
volume of expansion 土的体胀
volume of shrinkage 土的体缩
soil wash(ing) 土壤冲刷；土壤冲蚀；流积土
soil-waste interaction 土与废料相互作用
soil-waste pipe 粪便污水管
soil water 土中水；土壤中的水；土壤水
soil water belt 土中水带；土壤水带；土壤含水带
soil water content 土壤水水量；土壤含水量
soil water diffusivity 土壤水扩散系数
soil water experiment 土壤水试验
soil water heat pump 地水式热泵
soil water in motion 土中动态水

soil water movement 土壤水分运动;土的水分运动
soil water of stress 土壤水分应力
soil water percentage 土壤水分百分率;土壤水百分率
soil water potential 土壤水势
soil waterproofing 土壤防水
soil water relationship 土—水关系
soil water retention 土壤容水性
soil water system 土—水体系
soil water table 土壤潜水面
soil water tension 土的水压力
soil water zone 土壤水带
soil weathering 土层风化
soil wedge 土楔
soil wedge theory 土楔理论
soil with good tilth 有良好耕性的土壤
soil without cave-ins 无坍塌的土壤;无凹陷的土壤
soil with reinforcement strips installed 铺填加固带的土
soil workability 土壤机械适应率
soil working 土壤耕作
soil zonality 土地地带性
soil zone 土壤地带;土带
sol 液胶;溶胶;太阳
solaire 日出风
sol-air temperature 终合温度;太阳作用气温;室外综合温度
solanaceous vegetable 茄类蔬菜
solano 沙拉拿风
solaode 太阳能电池;太阳电池
solar γ-ray burst 太阳伽马射线暴
solar absorber 日照吸收剂;太阳辐射能吸收器
solar absorption index 太阳吸收指数
solar absorptivity 阳光吸收率
solar activity 太阳活动
solar activity cycle 太阳活动周
solar activity effect 太阳活动效应
solar activity prediction 太阳活动预报
solar activity region 太阳活动区
solar air mass 太阳照射的气团;太阳大气光学质量
solar altitude 太阳高度
solar altitude angle 太阳高度角
solar altitude table 太阳高度表
solar angle 日照角;太阳角;阳光入射角
solar annual tide 太阳周期潮;太阳年周潮
solar antapex 太阳背点
solar apex 太阳向点;太阳奔向点
solar array 太阳能电池阵
solar array drive assembly 太阳能电池阵激励组件
solar atlas 太阳图
solar atmosphere 太阳大气
solar atmospheric(al) tide 日大气潮;太阳大气潮
solar attachment 太阳仪;太阳观测附件;太阳附件
solar azimuth 太阳方位角
solar backscattered ultraviolet 太阳后向散射紫外线仪臭氧探测仪
solar barometric(al) variation 日致气压变化
solar battery 太阳能电池;太阳电池
solar battery modules 太阳电池组件
solar bridge 太阳桥
solar building 太阳房;太阳能建筑(物);太阳能房
solar burst 太阳暴
solar burst radio meter 太阳辐射爆发辐射计
solar calendar 阳历;太阳历
solar camera 太阳摄影机
solar cell 阳光电池;(检验弯沉仪的)日光管;太阳能电池;太阳电池;光电管
solar cell array 太阳电池阵列
solar cell modules 太阳电池组件
solar cell paddle 太阳能电池叶片
solar cell roofing material 太阳能电池屋盖材料
solar chimney 太阳能通道
solar chromosphere 太阳色球
solar climate 天文气候
solar collector 太阳能吸收器;太阳能收集器
solar collector roof panels 屋面板太阳能收集器
solar colo(u)rs 太阳染料(酸性偶氮染料)
solar communication 太阳能通信
solar compass 太阳罗盘
solar component 太阳成分
solar concentrator 太阳能集中器
solar constant 太阳能常数;太阳常数;太阳平均辐射率
solar construction 太阳能建筑(物)

solar continuum emission 太阳连续谱辐射
solar control 日照控制
solar control blind 阳光控制(挡)板;遮阳板
solar control film 阳光控制薄膜;(贴在玻璃上的)遮光膜;阳光控制膜;防晒膜
solar control glass 光照控制玻璃;日光控制玻璃;防阳光玻璃
solar controlled venetian(blind) 阳光控制的软百叶窗
solar control shade 控制阳光遮棚
solar cooker 太阳灶
solar cooking 太阳能蒸煮
solar cooling 太阳能制冷
solar corona 日冕;日华
solar coronal hole 日冕洞
solar coronal loop 日冕圈
solar corpuscle 太阳微粒
solar corpuscular beam 太阳微粒束
solar corpuscular burst 太阳微粒爆发
solar corpuscular emission 太阳微粒发射
solar corpuscular flow 太阳微粒流
solar corpuscular radiation 太阳微粒辐射
solar corpuscular stream 太阳微粒流
solar cosmic particle 太阳宇宙粒子
solar cosmic ray 太阳宇宙射线;太阳宇宙线
solar cosmic ray event 太阳宇宙射线事件
solar cosmic ray flare 太阳宇宙线耀斑
solar crown 日冕
solar cycle 太阳周;太阳活动周
solar cycle effect 太阳周效应
solar cycle variation 太阳周变化
solar daily variation 太阳周日变化;太阳日变化
solar day 太阳日;常用日
solar declination 太阳赤纬
solar depression angle 日落角;日降角
solar disc 日盘;日面;日轮;太阳圆面
solar distance 日地距离
solar distillation 日晒蒸馏;太阳能蒸馏
solar distillation basin 太阳能蒸馏池
solar distillation process 太阳蒸馏法;太阳热蒸馏法;太阳能蒸馏法
solar distiller 太阳能蒸馏器
solar disturbance 太阳扰动
solar disturbed daily variation 太阳扰日变化
solar diurnal tide 太阳全日潮
solar diurnal variation 太阳日变化
solar dryer 太阳能干燥器
solar drying 日光干燥
solar eclipse 日食;日蚀
solar eclipse expedition 日食观测队
solar eclipse limit 日食限
solar electric(al) energy generation 太阳能发电
solar electric(al) power generation 太阳能发电
solar electric(al) roof 太阳能发电屋面;太阳能发电屋顶
solar electromagnetic radiation 太阳电磁辐射
solar electron event 太阳电子事件
solar elevation angle 太阳高度角
solar elliptic(al) component 太阳椭圆分潮
solar elliptic(al) constituent 太阳椭圆分潮
solar emission line 太阳发射谱线
solar energy 太阳能
solar energy absorbing coating 太阳能吸收涂层
solar energy collector 太阳能收集器;太阳能聚集器;太阳能集热器
solar energy converter 太阳能转换器
solar energy drying 太阳能干燥
solar energy evapo(u)rator 太阳能蒸发器
solar energy materials 太阳能材料
solar energy panel 太阳能电池极
solar energy prime mover 太阳能发动机
solar energy ship 太阳能船
Solar Energy Society 太阳能学会(美国)
solar energy source 太阳能
solar energy stove 太阳灶
solar energy system 太阳能利用系统
solar energy technology 太阳能技术
solar energy transmission 太阳能穿透率
solar energy utilization 太阳能利用
solar engine 太阳能发动机
solar engineering 太阳能工程
solar ephemeris[复] ephemeredes 太阳历书
solar equation 日缠
solar eruption 太阳爆发;太阳暴

solar escape 脱离太阳引力区
solar evapo(u)ration 日晒蒸发
solar evapo(u)ration pond 太阳能蒸发池
solar-excited laser 太阳光抽运激光器
solar extreme ultraviolet burst 太阳远紫外线暴
solar eyepiece 太阳望远镜;太阳目镜
solar facula 太阳光斑
solar faculae 太阳谱斑
solar farm 太阳能耕作
solar filament 太阳暗条
solar filter 太阳滤光片
solar flare 耀斑;日辉;太阳耀斑;太阳斑
solar flat plate collector 平板太阳能收集器
solar flocculi 太阳谱斑
solar flocculus 太阳谱斑
solar flux 太阳通量;太阳辐射流
solar flux unit 太阳流量单位
solar focus 太阳焦点
solar forecast center[centre] 太阳活动预报中心
solar fraction 太阳能指数
solar furnace 太阳灶;太阳(能)炉
solar gain 日光采暖;太阳能获得量
solar gale 太阳暴风
solar gear 恒星式减速齿轮
solar generator 太阳能发生器;太阳能发电机;太阳电池组
solar geomagnetic tide 太阳地磁潮
solar glass 防光玻璃;反光玻璃
solar grating 阳光栅板
solar greenhouse 太阳能温室
solar grille 日光格子窗
solar halo 日晕
solar heat 太阳热;太阳辐射热
solar heat controlling property 隔热性能;太阳能热温控性能
solar heat enamel collector 太阳能热搪瓷收集器
solar heat exchanger 太阳能热交换器
solar heat exchanger drive 太阳能热交换器推进
solar heat gain 日照得热量(通过窗户和结构材料传入室内的太阳能)
solar heating 日照加热;太阳能取暖;太阳能加热;太阳能供暖;太阳能采暖
solar heating system 太阳能加热系统
solar heat load 太阳热负荷
solar heat radiation 太阳热辐射
solar heat storage 太阳热储存
solar hour 太阳小时
solar house 日光剥离房;阳光玻璃温室;日光玻璃暖房;太阳屋;太阳室;太阳房;采太阳光用于加热的玻璃房
solarigraph 日射总量计
solar illuminated satellite 太阳照射卫星
solarimeter 日照总量计;日射总量表;日射强度计;日射表;太阳能测量计;太阳辐射强度计
solar infrared 太阳红外线
solar infrared radiation 太阳红外辐射
solar input 太阳输入
solar insolation 日照
solar interplanetary magnetic loop 太阳行星际磁圈
solar interplanetary medium 太阳行星标介质
solar irradiance 太阳能辐射照度
solar irradiation 太阳能辐射
solaris 水下物体寻觅及打捞工具
solarisation 曝晒作用;曝晒
solarium 日光治疗室;日光(浴)室;[复] solaria 或 solariums
solarium curved glass 日光浴室用弯玻璃
solarization 阳光辐射作用;日晒作用;日晒;曝晒作用;晒
solarization curve 负感曲线
solarization density 负感密度
solarization image 负感像;反转像
solarization method 日晒法
solarized image 光致负感像
solar light 日光
solar light radiation 日光辐射
solar limb 日面边缘
solar load 日光负荷
solar local radio source 太阳局部射电源
solar longitude 日面经度;太阳经度
solar luminosity 太阳光度
solar magnetic field 太阳磁场
solar magnetism 太阳磁学

solar mass 太阳质量
solar maximum temperature 最高日射温度
solar maximum year 太阳活动峰年
solar method 太阳法
solar microwave burst 太阳微波暴
solar modules 太阳电池组件
solar month 阳历月；太阳月
solar motion 太阳运动
solar nebula 太阳星云
solar network 太阳色球网络
solar neutrino 太阳中微子
solar neutrino flux 太阳中微子流量
solar neutrino unit 太阳中微子单位
solar noise 太阳噪声
solar oblateness 太阳扁率
solar observation 太阳观测
solar observatory 太阳天文台
solar observing and forecasting network 太阳观测与预报网
solar oil 石油蒸馏在煤油后的馏分；页岩油；太阳油
solar orbit 太阳轨道
solar orientation 朝阳性；(房屋的)光线方位；朝阳方位；朝阳
solar outburst 太阳大爆发；太阳爆发
solar output 太阳辐射输出
solar paddle 太阳电池翼板；太阳电池板；太阳电池叶片
solar panel 太阳能电池板；太阳电池板；太阳能集热板
solar parallax 太阳视差
solar partial eclipse 日偏食
solar particle 太阳粒子
solar particle alert network 太阳粒子警戒网
solar particle events 太阳粒子事件
solar patrol 太阳巡视
solar period 太阳周期
solar periscope 太阳潜望镜(航测用)；太阳导航镜
solar perturbation 太阳摄动
solar phenomenon 太阳现象
solar photocatalysis 太阳光催化
solar photocatalytic decolo(u)rization 太阳光催化脱色
solar photocatalytic degradation 太阳光催化降解
solar photocatalytic mineralization 太阳光催化矿化
solar photocatalytic oxid 太阳光催化氧化反应器
solar photons 太阳光子
solar physics 太阳物理学
solar plage 太阳谱斑
solar plasma 太阳等离子体
solar plasma stream 太阳等离子体流
solar plexus 太阳丛
solar polar wind 极区太阳风
solar power 太阳能动力；太阳能
solar-powered 太阳能动力的
solar-powered alternator unit 太阳能交流发电机
solar-powered navigation 太阳能助航系统
solar-powered radio 太阳能无线电设备
solar-powered satellite 太阳能动力卫星
solar-powered space vehicle 太阳能动力航天器
solar power plant 太阳能发电厂
solar power station 太阳能电站
solar probe 太阳探测器
solar prominence 日珥
solar propulsion 太阳能推进
solar protection device 防晒措施
solar proton 太阳质子
solar proton event 太阳质子效应；太阳质子事件
solar proton flare 太阳质子耀斑
solar proton flux 太阳质子流
solar proton monitor 太阳质子监视仪
solar pumping 日光抽运；日光泵浦
solar photovolt cell 太阳能光伏电池
solar quiet daily variation 静日变化；太阳静日变化
solar quiet daily variation field 太阳静日变化场
solar quiet day 太阳静日
solar radar 太阳雷达
solar radiation 日射；日光辐射；太阳辐射
solar radiation energy 太阳辐射能
solar radiation error 太阳辐射误差
solar radiation event 太阳辐射事件
solar radiation intensity 太阳辐射强度
solar radiation measuring satellite 太阳辐射测量卫星

solar radiation measuring set 太阳辐射测量仪
solar radiation observation 日射观测；太阳辐射观测
solar radiation pressure 太阳辐射压
solar radiation satellite 太阳辐射(测量)卫星
solar radiation simulator 太阳辐射模拟器
solar radiation spectrum 太阳辐射谱；太阳辐射波谱
solar radio astronomy 太阳射电天文学
solar radio burst 太阳射电爆发
solar radio dynamic(al) spectrograph 太阳射电运动频谱仪
solar radio emission 太阳射电辐射；太阳射电发射
solar radio noise storm 太阳射电噪暴
solar radio radiation 太阳射电
solar radio waves 太阳射电辐射；太阳射电波
solar radius 太阳半径
solar-ray electron 太阳射线电子
solar reflectance 日光反射比
solar reflecting glass 太阳反光玻璃
solar reflecting surface 反光面(屋顶)；阳光热量反射面；闪光饰带
solar reflection surface 太阳热反射面；太阳热发射板
solar reflective curing membrane 日光反射养护膜
solar refrigeration 太阳能制冷
solar regenerator 太阳能再生器
solar ring 环式日晷
solar rocket 太阳火箭
solar rotation 太阳自转
solar sail 太阳帆
solar salt 太阳盐；晒制盐
solar satellite 太阳卫星
solarscope 日光模拟仪(研究建筑日照用)
solar screen 遮日篷；遮阳花格墙；遮阳板；遮阳屏
solar screening 日光遮蔽；吸热窗纱；遮阳窗纱
solar screen tile 砌遮阳光花墙的(琉璃)瓦；砌遮阳光花墙的瓷砖
solar sea water still 太阳能海水蒸馏器
solar seeing monitor 太阳视宁度监测仪
solar semidiurnal component tide 太阳半日分潮
solar semidiurnal tide 太阳半日潮
solar sensor 太阳传感器
solar service 太阳服务
solar shading 阳光荫蔽
solar shading device 阳光荫蔽设施
solar shield 阳光防护窗
solar silica glass 日光石英玻璃
solar soft X-ray burst 太阳软 X 射线暴
solar space 太阳空间
solar space vehicle 人造太阳行星；太阳研究航天器
solar spectral flux 太阳光谱光通量
solar spectral intensity 太阳光谱强度
solar spectrograph 太阳摄谱仪
solar spectrum 太阳谱；太阳光谱
solar spicule 针状物；太阳针状物
solarspot 聚散光两用照明灯
solar spot index 太阳黑子指数
solar spotlight 太阳聚光灯
solar star 太阳型星
solar still 太阳蒸馏器；太阳蒸馏锅；太阳能蒸发器
solar still process 太阳能蒸馏法
solar storage 太阳能储存介质；太阳能储存器
solar storm 太阳暴
solar stream 太阳流
solar system 太阳系；太阳能应用系统
solar telescope 太阳望远镜
solar temperature 日照温度；日射温度；太阳温度
solar terms 节气
solar-terrestrial environment 日地环境
solar-terrestrial physics 日地物理学
solar-terrestrial relationship 日地关系
solar theory 太阳理论
solar thermometer 辐射温度计
solar tidal component 太阳分潮
solar tidal wave 日潮波
solar tide 日潮；太阳潮
solar time 太阳时
solar-topographic(al) theory 太阳地形说
solar total eclipse 日全食
solar tower 太阳塔
solar transition region 太阳过渡区
solar triangulation 太阳三角测量
solar turboelectric(al) power unit 太阳能涡轮发电装置

solar-type reaction 太阳型反应
solar-type star 太阳型星；太阳型恒星
solar ultraviolet 太阳紫外线
solar ultravioletimaging telescope 太阳紫外成像望远镜
solar ultraviolet radiation 太阳紫外辐射
solar utilization 太阳能利用
solar watch 太阳能表
solar water heater 太阳能热水器
solar water pump 太阳能水泵
solar wave 太阳波
solar-weather relationship 太阳—气候关系
solar wind 海陆风；日风；太阳风
solar wind boundary 太阳风边界
solar wind geomagnetic field boundary 太阳风地磁场边界
solar X-ray 太阳 X 射线
solar X-ray astronomy 太阳 X 射线天文学
solar X-ray flare 太阳 X 射线耀斑
solar X-ray irradiance 太阳 X 射线辐照度
solar year 自然年；民用年；回归年；太阳年；分至年
solar zenith angle 太阳天顶角
solar zone refiner 太阳能区域精炼炉
solasulfone 苯丙砜
solation 胶流；溶胶化(作用)
solatium 慰藉金；赔偿；赔偿金
solbar 硫钡粉
Solc filter 索尔克滤光器
solcheck 平屋顶上保护地沥青的瓦
sol coating(method) 溶胶涂装法
solder 焊接模拟支撑；焊接；焊(于)剂；软(钎)料；软钎焊；钎焊填充金属；钎焊；低温焊料
solderability 易焊接性；可焊接性；钎焊性；可焊性
solderable 可以焊的
solder acid 焊锡水
solder alloy 焊接合金；合金钎料；钎焊合金
solder bit 焊接烙铁
solder bolt 焊接螺栓；电烙铁
solder brazing 硬钎焊
solder club 焊条
solder-covered wire 锡包线；焊锡包线
solder dip(ping) 浸焊
solder dot 焊点
solder drop junction 焊滴接点；焊滴法
soldered 焊接的；锡焊的
soldered ball 焊球
soldered brass nipple 焊接黄铜嘴
soldered connection 钎焊接头
soldered crown 焊接冠
soldered dot 焊料小球滴；焊点
soldered fitting 钎焊接头；锡焊配件；焊接头
soldered joint 钎焊接头；焊接口；电焊接缝；锡焊接；焊接接头；焊接接合；焊接(缝)；软钎焊头；点焊接头
soldered jointing 锡焊接头
soldered lap joint 锡焊搭接
soldered lug 焊连线夹
soldered pipe 锡焊管
soldered seam 焊缝；焊锡缝；软钎焊缝
soldered thermoelectric(al) refrigerating unit 焊接式热电制冷器
soldered up 焊接住的
solder embrittlement 焊料渗入致脆；钎料脆性
solderer 焊工；钎焊机；钎焊工
solderer's helmet 焊工头罩
solder glass 焊料玻璃；焊接用的玻璃；焊接(用)玻璃
soldering 焊接；软钎焊；软焊；低温焊(接)
soldering acid 钎剂；钎焊用酸
soldering and sealing 封焊
soldering apparatus 焊接设备；焊接器
soldering appliance 焊接工具
soldering blow pipe 焊接用吹管
soldering bolt 焊烙铁；电烙铁
soldering borax 焊接硼砂
soldering bushing 锡焊衬套
soldering chemical 焊接用化学品
soldering coal 炼焦煤
soldering copper 紫铜烙铁；钎焊用铜；钎焊烙铁
soldering cup 防尘套
soldering equipment 焊接设备
soldering fluid 锡焊液；焊油；焊(接)液
soldering flux 焊料；合金助焊剂；焊液；(错焊用的)焊药；焊剂；软钎剂；钎焊剂

soldering forceps 焊接镊
soldering gun 焊枪;枪型电烙铁;钎焊枪;烙铁枪;电烙铁
soldering hammer 钎焊焊接器
soldering installation 焊接设备
soldering iron 烙铁;焊(烙)铁;钎焊烙铁;锡焊铁
soldering iron nozzle 烙铁头
soldering joint 焊接缝
soldering kit 焊接工具
soldering ladle 焊锡杓
soldering lamp 锡焊灯;焊接用喷灯;焊(接)灯;钎焊用喷灯;喷灯
soldering lug 焊片;钎焊片
soldering machine 焊机;钎焊机
soldering method 焊接方法
soldering nipple 焊接短管;焊接丝头
soldering outfit 焊接设备
soldering oven 钎焊炉
soldering paste (锡焊)焊药;焊锡膏;焊油;焊(接)膏;钎焊油;钎焊剂;钎焊膏
soldering pencil 钎焊笔
soldering pin 焊接销
soldering point 焊接点;钎焊点
soldering powder 粉状钎焊剂
soldering preparation 焊接制剂
soldering rod 焊条
soldering salt 焊盐
soldering seal 焊封
soldering seam 焊缝(锡);钎缝
soldering sleeve 焊接套管
soldering solution 焊接液
soldering stick 焊条
soldering temperature 焊接温度
soldering tin 焊锡
soldering tin bar 焊锡条
soldering tongs 焊接夹钳
soldering tool 焊接工具
soldering torch 焊枪;焊接气焰;钎焊炬;钎焊炬
soldering turret 焊钳
soldering tweezers 焊捏钳;焊接镊
soldering welding 熔接法
solder iron 锡焊铁;焊铁
solder joint 锡焊接;钎焊缝;锡焊连接;低温焊接连接;钎焊接头
solder jointing 锡焊接头
solderless assembly 无焊剂装配
solderless connection 无焊连接
solderless fitting 无焊接法
solderless joint 扭接;无缝焊接
solderless splice 无焊接合
solderless terminal 无焊接端头
solderless wiring 无焊布线
solderless wrapped connection 绕接
solder metal 金属焊料
solder paint 焊药膏;软焊涂料
solder paste 焊药;焊锡膏;焊油
solder pot 焊罐
solder powder 焊锡粉
solder set 焊接用品;钎焊用具
solder stick 焊锡条;焊锡棒
solder-tail socket 焊引线管座
solder wire 焊条;焊丝;焊料丝
solder wire extruder 焊条挤压机
soldier 支承土压力的竖桩;立砌砖;模板主柱;模板立柱;竖木柱;兵蚁
soldier arch 立砌石拱;立砌平拱;门窗立砌砖过梁;门窗顶上的立砌砖拱;平砌拱;竖砖平拱
soldier beam 立柱;基坑围护立柱;竖柱;挡土板立柱;挖土前打入土内的竖向木桩或宽翼缘 H 形钢桩
soldier block 竖砌大砖
soldier course 墙柱垫;立砌砖;排砖竖砌层;竖砖砌层;竖砌砖层
soldier pile 挡土桩立档;侧柱;支挡桩;模板支撑;支护桩;(基坑围护的)立桩;竖桩;竖枋木
soldier piles and lagging 大立柱横插板支撑
soldier pile with lateral lagging 立柱横板挡土墙
soldier sheeting 立柱板
soldier's waiting room 军人候车室
Soldner coordinate 索德尔坐标
sole 基岭冰;木隔墙下槛;底面;底部;底;窗台板
solea 教堂通道
sole agency 独家经理;唯一代理;独家代理(商)
sole agent 独家代理(商);总代理商;总代理人
sole and unconditional ownership clause 所有权独有条款
sole arbitrator 独任仲裁员;独任仲裁人
sole bar 车架纵梁;底杆;底部边梁;侧梁
sole bar angle 底角铁
sole construction 独立修建
sole corporation 独资公司
sole cutting edge countersink 单刃锥口钻
sole discretion 全权处理;单独裁量权
sole distributor 独家经销商
sole draft 单张汇票
sole fault 基底断层;底基断层
sole flue 炉底烟道
sole-flue port brick 栅形炉箅砖
sole heel reinforced nailing machine 底跟加固打钉机
soleil 直梭纹织物
Soleil compensator 索累补偿器
soleil weave 重缎组织
sole injection 基底贯入
sole leather roller 底革辊筒机
sole licence contract 排他性许可证合同
sole license 排他性许可证
solely foreign funded enterprise 外商独资企业
sole mark 底痕
sole marking 底面印痕作用
sole-material cutter 裁底料机
solemnity of style 庄严的风格
solenia 管系
solenium 管系
solenoid 圆筒形线圈;中空螺旋管;螺线管;电磁线圈
solenoid-actuated 电磁铁操纵的
solenoid actuator 螺线管执行器;螺线管传动器
solenoidal field 螺线管磁场;无散矢场
solenoidal index 力管指数
solenoidal inductor 螺线管电感线圈
solenoidal magnetization 纵向磁化;螺线管磁化
solenoidal vector field 螺线向量场
solenoid ammeter 螺线管安培计
solenoid-and-plunger 插棒式铁芯螺线管
solenoid brake 螺形线圈制动器;螺线管制动器;电磁制动器
solenoid clutch 电磁圈式离合器
solenoid coil 筒管形线圈;电磁铁线圈;螺线管
solenoid-controlled 电磁铁操纵的
solenoid-controlled device 电磁控制装置
solenoid-controlled valve 电磁控制阀;电磁阀
solenoid directional valve 电磁换向阀
solenoid electric(al) valve 电磁控制阀
solenoid-operated 螺管磁铁操纵
solenoid-operated air valve 电磁操纵空气阀
solenoid-operated closing mechanism 螺线管闭合装置;电磁闭合装置
solenoid-operated clutches 电磁控制离合器
solenoid-operated spoiler 电螺管控制的扰流片
solenoid-operated switch 电磁控制开关
solenoid-operated valve 电磁线圈操纵阀;电磁控制阀
solenoid pilot actuated valve 电磁阀
solenoid shut-off 电磁关闭;电磁阀
solenoid starter 螺旋线圈起动器
solenoid switch 电磁开关
solenoid trapper 电磁振打器
solenoid valve 螺线管操纵阀;电磁阀
solenoid valve box 电磁阀盒
solenoid vibrating feeder 电磁振动喂料机
solenostele 管状中柱
sole offer 独家发盘
sole of rudder 舵底垫板
sole of the glacier 冰川台地;冰川的底部
sole or sole plate 墙脚板
sole piece 墙撑柱拱垫;斜撑脚垫;底座;垫块;垫板;钢筋填块;滑道牵板;尾框底部;底块;垫木
sole plate 斜撑的支座;隔墙底槛;底板;窗台板;支承座板;支撑地板;钢轨垫板;基础底座;基础(底)板;滑道牵板;上梁;垫板;履带板;木隔墙下槛
sole proprietary enterprise 独资经营企业
sole proprietorship 独资经营
soler 阁楼;钎焊剂
sole-source equipment 唯一来源设备
Soletanche system 索勒坦奇体系
sole tile 拱形瓦
sole timber 垫木
sole tree 垫木
sole weight 自重
solfatara 硫质喷气孔
solfatara stage 硫质喷气期
solfataric clay 硫质黏土
solferino 品红;鲜紫红色
sol-gel 溶胶一凝胶
sol-gel method 溶胶一凝胶法
sol-gel process 溶胶一凝胶过程
sol-gel technique 溶胶一凝胶技术
sol-gel transformation 溶胶一凝胶转变;溶胶一凝胶变换
solicitation 揽载
solicited 请求的
solicit operation 请求操作
solicitor's act 律师法
solid 固体的;立体;坚固(的);强固的
solid absorbent 固体吸收剂;固体吸附剂
solid absorbent collecting system 固体吸收剂集系统
solid abutment 实心桥墩;实体桥台
solid acid catalyst 固体酸催化剂
solid active waste 稠活性废液
solid adjustable taps 可调整螺丝攻
solid adsorbent collecting system 固体吸附剂收集系统
solid aggregate state 固体凝聚状态
solid aggregation 固体凝聚
solid alcohol 固体酒精
solid alumina 固体氧化铝
solid analysis 固体分析
solid analytic(al) geometry 立体解析几何(学)
solid anchor 整体锚
solid-and-void contrast 虚实对比
solid angle 立体角;立面角
solid anode 实心阳极
solid apron 重型铺板;强固铺板
solid arch 实体拱
solid arch bridge 实腹拱桥
solid area 普染面积;实面积
solid arrow 实线箭头;实箭头
solid asphalt 固体沥青;固态沥青;硬沥青
solid assembly building 刚性装配体(建筑)
solid axes 空间坐标轴
solid axis 实(心)轴
solid axle 重轴;实心轴;实心车轴
solid axle bearing 实心车轴轴承
solid axle box 实心车轴轴箱
solid back transmitter 竖背式送话器
solid ballast 固体压载
solid ballast system 固体压载系统
solid balustrade 实心栏杆
solid bar 实心棒材;实习型钢
solid-barrel arch 实心板拱
solid-barrel arch bridge 实心板拱桥
solid-barrel sampler (取土管不能分开的)整体式取土器
solid base catalyst 固体碱催化剂
solid base slide plate for adjustable rail brace 可调轨撑的实底滑床板
solid base slide plate of rigid rail brace 固定式轨撑的实底滑床板
solid beam 实体梁;实心梁
solid bearing 整体支座;硬支承座;整体轴承;固定轴承
solid bearing liner 单层轴瓦
solid bed 硬基;整体道床;坚硬层;定床
solid bed-rock 岩基床;硬石层;坚实基岩
solid billet 实心坯;实体短木段
solid bit 硬钻头;实心钻头
solid bitumen 固体沥青;固态沥青
solid bituminous material 固体沥青材料
solid black 颗粒炭黑;纯黑色
solid block 支柱;基线三脚架;实心(砌)块;实心混凝土(方)块;实体块
solid blockage 物体堵塞
solid-block machine 整块材制机;实心块材制机
solid-block masonry(work) 整体材砌筑墙;实心块材砌筑墙
solid block model 木雕船模;实心船模
solid block mo(u)ld 整体材模子;实心块材模子;整体结晶器
solid-block tester 整块材试验机;实心块材试验机
solid-blue 深蓝色
solid body 固体;坚硬物体
solid body mechanics 固体力学

solid body rotation 固体自转
solid borer 钻头
solid boring 钻孔
solid boring bar 深孔钻杆
solid boring head 深孔钻刀头
solid borne 固载的
solid-borne noise 固源噪声；固体(载)声；固体源噪声
solid-borne sound 固体传声；固体声；固体载声
solid-borne sound transmission 固体传声
solid borne vibration 固体载振动
solid bossed 实模撞击成型
solid bossing 固体壳罩；锤碾铅皮成器
solid-bottom bucket 整底吊桶
solid-bottom mo(u)ld 带底钢锭模
solid boundary 固体边壁
solid bowl 无孔转鼓
solid bowl centrifuge 无孔转鼓离心机；卧式离心分选机；勃尔特型沉降式离心脱水机
solid bowl type screw decanter 卧式沉降螺旋卸料离心分离机
solid brick 坚硬砖；坚实砖；实心砖
solid brick chimney 实心砖砌烟囱
solid brick masonry 实心砖砌圬工
solid brick pagoda 实心砖塔
solid brick vault 整砖砌拱顶；实心砖砌拱顶
solid brick wall 实心砖墙
solid brickwork 实心砖砌体
solid bridge 坚固的桥
solid bridge deck 实体桥面板
solid bridge floor 实体桥面
solid bridging (小梁间加固的)剪力撑；小梁间加固的剪力撑；格栅横撑
solid broach 整体拉刀
solid bulk (cargo) 干散货
solid bulk container 固体散货集装箱；散货集装箱
solid bulkhead 整体舱壁
solid bulk ship 干散货船
solid buoyancy material 固体浮力材料
solid bus 硬质母线
solid bushing 固体瓷套管
solid buttress 实体支墩
solid cable 实心电缆
solid cadmium 固体镉
solid calcium silicate brick 实心硅酸钙砖
solid cam 立体凸轮
solid car 整体矿车
solid carbide 整体硬质合金刀具
solid carbon 实心炭棒
solid carbon brush 实心炭刷
solid carbon dioxide 固体二氧化碳(俗称干冰)；固态二氧化碳
solid carburization 固体渗碳
solid carburizing 固体渗碳
solid carrier 固体载体
solid case 实心护圈
solid cast door 实体浇铸门
solid casting 实心铸件；实心注浆；实体浇铸
solid casting method 实体浇铸法
solid-cast insulation 浇铸固化绝缘
solid-cast manganese steel frog 高锰钢整铸辙叉
solid-cast transformer 浇注固化式变压器
solid caustic soda 固体烧碱；固碱
solid caustic soda pot 熬碱锅
solid caustic soda section 固碱工段
solid center 整体顶尖
solid center bit 木工钻头
solid center eyed auger 木工钻
solid centre spot 实心亮点
solid centre auger bit with one spur 单翼木工钻(头)
solid centre bit with one spur 单翼木工钻(头)
solid chalk 白垩岩
solid character 固体性质
solid chimney flue checkerwork 烟道式格子体
solid chock 实心木垛
solid cinder tile 实心煤渣砖
solid circuit 固态电路
solid cladding 固体涂层
solid clay brick 实心黏土砖
solid clay unit 实心黏土砌块
solid clinker block 实心熔渣砌块
solid coal 坚实的煤
solid coating 固体涂料

solid collecting chamber 固体存积腔
solid colloidal substance 固体胶体物质
solid colo(u)r 普染色；非金属光泽色
solid colo(u)r enamel 单一色(磁)漆
solid column 实心柱；实体柱
solid commodities 固体矿产
solid complex 固体络合物
solid component 固体分析
solid concave bit 凹形无岩芯钻头
solid concentration 固体浓度；含泥量
solid concrete 实心混凝土
solid concrete beam 实心混凝土梁
solid concrete block 实心混凝土块；实心混凝土(方)块
solid concrete floor(slab) 实心(钢筋)混凝土楼板
solid concrete slab 实心混凝土板
solid condition 固体状态
solid conduction 固体导电
solid conductivity 固体传导性
solid conductor 实心导线
solid cone 整体锥式
solid-cone(type) nozzle 实心锥喷嘴
solid connection 实心连接
solid constituent 固体组分
solid construction 实体式结构
solid contaminant 固体污染物
solid content 固相含量；固体含量；固体分量；含泥量
solid contraction 固体收缩
solid conveying machinery 固体输送机械
solid copper 固体铜
solid copper wire 实心铜线
solidcor 全板门构造
solid core 整体芯板；(胶合板的)实心芯板；实心铁芯
solid-core boring 实心钻深；实心钻孔
solid-core door 实心门
solid-core drill 整体扩孔钻
solid-core flush door 实心平面门
solid-core method 实心法
solid-core packing 实心填充物
solid-core recovery 整体岩芯采取率
solid-core support 实心支承物；实心载体
solid counter 固态计数器
solid coupling 刚性联轴器；刚性联轴节；刚性连接
solid coverage 立体图；空间图
solid crank 整体曲柄
solid crankshaft 实心曲轴
solid cribbing 实心木垛
solid crib timbering 密集井框支架；实心木垛支护
solid cross-section 实心断面
solid cross wall 实心横隔墙
solid crown bit 金刚石无岩芯钻头；实心钻头
solid cryogenic cooler 固体低温制冷器
solid cubic(al) content 实积
solid culture 固体培养法
solid cure 高度硫化
solid curve line 连续曲线；实曲线
solid cutter 整体刀盘
solid cylindrical rotor 实心圆柱形转子
solid dam 实心坝；实体坝；重力坝
solid deck 整片(式)甲板；实体桥面板
solid deck pier 连片式码头
solid dehumidifying 固体除湿
solid delay line 固体延迟线
solid deodorizer 固体除臭剂
solid deposit layer 固体沉积层
solid deposits 固体淀积物；固体沉积物
solid detrital material 固体碎屑物质
solid diameter 固相颗粒直径
solid diamond bit 实心金刚石钻头
solid diaphragm 加劲片
solid die 整体式凹模；整体模(具)；整体板牙
solid die forging machinery 固体模锻机
solid dielectric 固体介质；固体电介质
solid diffusion 固相扩散
solid disc 整个圆片；整个圆盘；固体圆片；固体圆盘
solid disc brake 整体滑轮制动器；整体圆盘制动器
solid discharge 固体径流(量)；输沙率；输沙量
solid dispersion 固体分散体
solid dissolution 固体溶解
solid dissolved 固体溶液化
solid door 实心全板门；实心门；防火门
solid door frame 实心门框
solid dosage forms 固体剂型
solid-drawn 整体拉制；整体拉伸的；拉制的；整体拉制的

solid-drawn pipe 拉制无缝(钢)管
solid-drawn steel pile 无缝钢桩
solid-drawn steel tube 无缝钢管
solid-drawn tube 拉制无缝(钢)管；拉制管；拉制无缝(钢)管
solid drill 整体钻头
solid drill core 整体岩芯
solid drilling 实心钻头凿岩；实心钻法
solid drill steel 整体钎子；整体钻杆；实心钎杆
solid drive 刚性传动
solid drum rotor 整锻鼓形转子
solid dry dock 实体干船坞
solid earth 固体地球；固定接地
solid earth geophysics 固体地球物理学
solid earth surface 陆地表面
solid earth tide 固体地球潮汐
solid eccentric sheave 实体偏心轮
solid elastic body 连续弹性体；完全弹性体；弹性实体；单一弹性体；纯弹性体
solid electrode 固体电极
solid electrolyte 固体
solid electrolyte battery 固体电解(质)电池
solid electrolyte fuel cell 固体电解质燃料电池
solid electrolyte oxygen sensor 固体电解质测氧仪
solid electrolyte probe 固体电解质探测器
solid electrolyte sensor 固体电极探头
solid electrolytic capacitor 固体电解电容器
solid electronic device 固态电子器件
solid electronics 固体电子学
solid element 实心构件
solid emplacement 固态侵位
solid enclosure 不透气外壳
solid-end car 固定式矿车
solid-end mill 整体端铣刀
solid error 固定性错误
solid expanded cinder concrete block 实心膨胀性炉渣混凝土砌块
solid expansion thermometer 固体膨胀温度计
solid explosive 固体炸药；固体喷气燃料
solid extract 固体萃取原料
solid extruded shape 实心挤压型材
solid-faced 整片面层的
solid feeder 干投器
solid feeding 干投
solid feed inlet 固体原料入口；固体加料口
solid fence 实体围篱
solid ferrite 固体铁淦氧
solid fertilizer 固体肥料
solid fiberboard 合成板；硬质纤维板；高压板；密实纤维板
solid figure 立体形；立体图形
solid filled portion 填实部分
solid filling 完全充填；填实
solid film lubricant 固体膜润滑剂
solid film lubrication 固体膜润滑
solid filter-aid 固体助滤剂；固体助滤剂
solid flat metal gasket 实心金属垫片
solid flat plate 定型底板；实心板
solid flat plate floor 实心平板楼板
solid floor 整片式楼板；直接铺设楼面；实体地面；整片(式)面板；平板肋板；实心楼板；实心地板；实体楼板；实肋板
solid-floor heating 实心楼板供热
solid-floor heating by electricity 实心楼板电气供热
solid-floor heating cable 实心楼板供热电线
solid flow 固体流动；固态流
solid flow control valve 固相流量控制阀
solid flow meter 固体流量计
solid fluorimetry 固体荧光法
solid flux 固体通量；固体焊剂
solid flux analysis 固体通量分析
solid fluxing agent 固体熔剂
solid flywheel 整体飞轮
solid foamed slag concrete block 实心高炉泡沫矿渣混凝土砌块
solid fog 浓雾
solid food 固体食物
solid foot 立方英尺
solid forged backup roll 整锻支承辊
solid forging 整体锻造；实锻
solid form 固态
solid forming machine 固体成形机
solid form(work) 整体模板

solid foundation 坚实基础;稳固地基
solid frame 平板肋板;实心门框;实心窗框;实心窗架;整料门窗框
solid frame work 固体骨架
solid friction 干摩擦;固体摩擦(力)
solid front pressure ga(u)ge 整体表面压力计
solid fuel 固体燃料
solid fuel boiler 固体燃料锅炉
solid fuel combustion chamber 固体燃料燃烧室
solid fuel commodities 固体燃料矿产
solid fuel domestic boiler 固体燃料家用锅炉
solid fuel-fired furnace 固体燃料炉
solid fuel-fired room heater 固体燃料室内加热器
solid fuel-fired stove 固体燃料炉
solid fuel-fired water heating 固体燃料的热水供暖
solid fused 氯化钙
solid fused calcium chloride 固体熔融氯化钙
solid fused caustic soda 固体熔融烧碱
solid fused sodium sulphide 固体熔融硫化钠
solid gas circuit breaker 固体气体断路器
solid-gas sol 固气溶胶
solid ga(u)ge 限界量规
solid geometry 立体几何(学)
solid glass block 实心玻璃砖
solid glass door 全玻璃门
solid glass microsphere 实心玻璃微球体
solid gold 纯金;赤金
solid-gold chain 金项链
solid-gold pendant 纯金挂件
solid gravity dam 实心重力坝;实体重力坝
solid gravity wall 实体重力式岸壁
solid groin 实体丁坝
solid ground 硬地;密实土;直接接地;坚实地基;坚固地基
solid groyne 实体丁坝
solid guide 固态波导管
solid gypsum board 实心石膏板
solid gypsum building board 实心石膏建筑板
solid gypsum partition 实心石膏板隔墙
solid gypsum roof(ing)block 实心石膏屋顶砌块
solid hacksaw frame with wood(en)handle 木柄式固定钢锯架
solid handling 固体处理
solid hand packing 手工密实填充
solid harmonics 立体谐(和)函数
solid head 坚实叶球
solid-head buttress dam 大头坝
solid height of spring 弹簧最大压缩高度
solid helium 固体氦(低温);固氦
solid hold pillar 实心舱内支柱
solid household refuse 城市固体生活垃圾
solid house refuse 家庭固体垃圾
solid hygrometer 固体湿度计
solid ice 固体冰;坚固冰;坚冰
solidifiable 可凝固;可充实
solidification 固体化作用;固结作用;固化(作用);凝固(作用)
solidification ages 固结年龄
solidification gradient 凝固梯度
solidification heat 固化热;凝固热
solidification index 固结指数
solidification of lava 熔岩的固化作用;熔岩的凝固
solidification of radioactive waste 放射性废物固化
solidification of radwaste 放射性废物固化
solidification of sand 流沙固化
solidification of wastes 废物固化
solidification point 固化点;凝固点;冻结温度
solidification pressure 固化压力
solidification process 固结过程;固化法
solidification range 固化范围
solidification shrinkage 凝固收缩
solidification structure 固化结构;凝固结构
solidification tank 固化池
solidification temperature 凝固温度;固化温度;固化点
solidification temperature range 固化温度范围
solidification temperature value 固结温度值
solidification treatment 固化处理
solidification value 凝固点
solidified 固(体)化的
solidified agent 固化剂
solidified alcohol 固化酒精
solidified carbon dioxide 固体二氧化碳;干冰
solidified cement 硬结的水泥;硬化的水泥;固化水泥

solidified gasoline 固体汽油;凝固汽油
solidified glaze 凝固釉
solidified ingot 凝固的锭料
solidified kerosene 凝固煤油
solidified matte 凝固锍
solidified moisture film 固化水膜;固化湿(水)膜
solidified oil 固化油
solidified oleoresin 毛松香
solidified paraffin 固化石蜡
solidified petroleum 凝固石油
solidified petroleum product 固化石油产品
solidified slag 凝固渣
solidified slip 固化泥浆;凝固泥浆
solidified soil 坚实土;加固土
solidified water 固结水分
solidifying 固化;凝固;固体化作用
solidifying front 凝固界
solidifying of slag 熔渣固化
solidifying phase 凝固相
solidifying point 固化温度;固化点;凝固点;凝点
solidifying pressure 固化剂
solid image microscope 立体图像显微镜
solid impurity 固体杂质
solid inch 立方英寸
solid inclusion 固体夹杂物;固体包裹体
solid ingredient 固体组分;固体成分
solid injection 高压喷射燃料;无气喷射
solid injection diesel 固体喷射柴油机
solid injection engine 固体喷射引擎;无气喷射引擎;固体喷射发动机;无气喷射发动机
solid injection system 机械喷油系统
solid inorganic carbon 固体无机碳
solid in-situ cast slab 实心现浇板
solid insoluble precipitate 不溶性固体沉淀物
solid in solution 溶液中的固体;不溶解的固体
solid insulator 固体绝缘子;固体绝缘材料;实心绝缘子
solid integration building 刚性组合体;刚性组合建筑
solid interface 固体界面
solid intrusion 固体侵入
solid ion laser 固体离子激光器
solid iron core 实心铁(芯);实铁芯
solid isotherm 固态(面)等温线
solidity 硬度;固体性;固态;坚硬度;坚实性;坚固度;密实性;完整性
solidity coupled shaft 刚性连接轴
solidity of blades 叶片稠度
solidity ratio 硬度比;密实度比;实积比;充实比
solidium 台座体;柱脚底座的墩台;柱脚体
solid jaw 整体爪;固定爪
solid jaw with lug 带长普通接杆
solid jaw with movable lug 带活舌普通接杆
solid jet 连续射流;密实射流
solid-jet injection system 单射流喷射系统
solid jetty 实体突堤
solid jib 实体式吊机臂
solid joint 刚性接合
solid journal bearing 整体轴颈轴承
solid key 固定键
solid laminated partition 实心层积隔墙
solid laser 固体激光器
solid-leg tripod 固定腿三脚架;定腿三脚架
solid level 固体料面
solid level controller 固相料面调节器;固体料面调节器
solid light(weight)brick 轻质实心砖
solid light(weight)clay brick 轻质实心黏土砖
solid lime sand brick 实心石灰砂砖
solid lime sand wall 实心灰砂墙
solid line 实线
solid line curve 实曲线
solid-liquid blender 固液掺合剂
solid-liquid cyclone 固液(水力)旋流器
solid-liquid equilibrium 固液平衡;[复]equilibria
solid-liquid extraction 固液萃取;固液液体提取;浸取
solid-liquid interaction 固液相互作用
solid-liquid interface 固液(相交)界面
solid-liquid interface system 固液液体藕联系统
solid-liquid micromatography 固液色谱(法)
solid-liquid mixing 固液混合
solid-liquid phase 固液相
solid-liquid separation 固液分离

solid-liquid separator 固液分离器
solid-liquid transformation 固液转变
solid load 固体含量;推移质
solid loading 固体负荷;致密装填;满装炸药;固体荷载
solid loading of biologic(al)filter 生物滤池固体负荷
solid loading of sedimentation tank 沉淀池固体负荷
solid loading rate 固体负荷率
solid loading rate capacity 固体负荷率容量
solid logic technology 固态逻辑技术
solid lubricant 固体润滑剂;润滑脂
solid lubrication 固体润滑法
solid lubrication oil 固体润滑油
solid magnetic pole 实心磁极
solid mandrel 整体轴心;整体心轴
solid manganese steel frog 整铸锰钢辙叉
solid manifold 整体油路板
solid marginal wharf 实体式顺岸码头
solid masonry infill wall 实心砖墙填充墙
solid masonry retaining wall 实体圬工挡土墙
solid masonry unit 实体砖石单元;实体圬工单元;实心砌块;实体砌筑单位
solid masonry wall 实体砌体墙;实体墙
solid masonry(work)实心圬工;实心砌体;实体圬工
solid mass 硬块;实体
solid mass type 坚硬类
solid material 固体物料
solid material in suspension 悬移质;悬浮固体物质
solid matrix 固体基质;骨骼物质
solid matte 固体锍
solid matter 固体物质;密排
solid measure 容量测定;容积;体积
solid mechanics 固体力学
solid medium 固体培养基;固体媒质;固体介质
solid melting wax 固体熔化蜡
solid membrane electrode 固体膜电极
solid metal 固体金属
solid metal guide 固体金属波导
solid metal reflector 实心金属反射器
solid microelectrode 固态微电极
solid microscope 立体显微镜;立式显微镜
solid mineral drilling 固体矿产钻探
solid mining wastes 矿物固体废物
solid mixture 固体混合物
solid moment of inertia 刚体转动惯量
solid mopping 满铺沥青;满涂
solid mo(u)ld 整体模型;整体模具;整体铸型
solid mo(u)lding (木材上雕刻的)装饰线条;(车制的)整体装饰线条;整线脚;整体模制;整体模压;整体模塑;固体造型;雕木线脚;整体铸型
solid mullion window 实心竖框窗;实心中框窗
solidness 硬性;坚实性;坚固度
solid newel 旋梯中柱;螺旋梯的(实心)中柱;实砌(楼梯)旋栓
solid newel of stair(case)螺旋梯实心中柱
solid newel post 实心旋梯中柱
solid newel stair 实心中柱旋梯;(中柱支承悬臂踏步)中柱螺旋式楼梯
solid newel stair(case)带实心中柱的螺旋梯;实心中柱式螺旋楼梯
solid non-metallic impurity 固体非金属夹杂物;非金属夹杂物
solid object 实心图形
solid of light distribution 光强分布曲面图
solid of luminous-intensity distribution 光强分布曲面图
solid of revolution 旋转体;回转体
solid of rotation 旋转体;回转体
solidography 实体摄影法
solid oil 固体润滑油
solid organic carbon 固体有机碳
solid outlet 滤渣(排)出口
solid oxide fuel cell 固体氧化物燃料电池
solid oxygen 固态氧
solid oxygen-buffer 固体氧缓冲剂
solid packing 密实充填
solid pagoda 实心塔
solid panel 实心板;厚镶板;实体板
solid panel floor 厚镶板楼板
solid paraffin 固体石蜡
solid parking 紧密停车

solid particle 固体微粒;固体粒子;固体颗粒
solid particle in suspension 悬浮质的固体颗粒;悬浮固体颗粒
solid particulate 固体颗粒
solid partition 实体隔墙
solid pattern 整体模(具);实型
solid pattern mo(u)lding 整体模造型
solid pedestal 整体轴承
solid petroleum 固体石油
solid phase 固相【物】
solid-phase adsorbed concentration 固相吸浓度法
solid-phase adsorbent 固相吸附剂
solid-phase adsorption method 固相吸附法
solid-phase bonding process 固相焊接法
solid-phase composition analysis 包体固相成分分析
solid-phase composition determination 固相成份测定
solid-phase condensation 固相缩合
solid-phase denitrification method 固相脱氮法
solid-phase denitrification process 固相反硝化法
solid-phase diffusion 固相扩散
solid-phase epitaxial 固相外延
solid-phase extraction 固相萃取
solid-phase extraction-high performance liquid chromatography 固相萃取—高效液相色谱法
solid-phase extraction-reversed phase liquid chromatography 固相萃取—反相液相色谱法
solid-phase of soil 土壤固相
solid-phase poly condensation 固相缩聚(作用)
solid-phase polymerization 固相聚合
solid-phase pressure welding 固相压焊
solid-phase reaction 固相反应
solid-phase reaction process 固相反应法
solid-phase sintering 固相烧结
solid-phase surface 固相面
solid-phase synthesis 固相合成
solid-phase welding 固相焊接;压力焊接;固相压焊
solid-phase zone melting 固相区域熔涤;固相区熔
solid phosphoric acid slurry 糊状磷酸催化剂
solid phosphor laser 固体磷光体激光器;固态磷光体激光器
solid pier 实体(桥)墩;实心桥墩;实体突堤;实体式码头;实体码头
solid pier foundation 实体支墩基础
solid pigment preparation 固相加工颜料;固体颜料制备物
solid pile 整体桩;实心桩
solid pillar 实心支柱
solid pintle 整体舵栓
solid pipe 无缝钢管
solid pipe wrench 固定管扳手
solid piston 实心活塞
solid piston rock drill 活塞式凿岩机;活塞式风钻;往复式凿岩机;往复式风钻
solid-planted 密植
solid planting 密条播
solid planting of clone 纯粹无性系植区
solid plasterwork 实体抹灰工程;实体抹灰;实体粉刷工程
solid-plastic system 固体塑性系统
solid plate 实心板;定模板
solid plate frame 整板肋骨
solid point 固化点;凝固点
solid pole 整块磁极;实磁心
solid pollutant 固体污染物
solid polyethylene insulation 固体聚乙烯绝缘
solid polymer 固体聚合物
solid polymer electrolyte 固体聚合电解质
solid polymer ferric sulfate 固体聚
solid polymerization 固相聚合法;固相聚合
solid portion 固体部分
solid post 实心接线柱
solid precipitates 固体沉淀物
solid precipitation 固态降水
solid precipitation quantity 固态降水量
solid propellant sustainer 固体推进剂主发动机
solid propeller 整体螺旋桨
solid pulley brake 整体皮带轮制动器
solid pulverizing machinery 固体粉碎机械
solid pump 杂质泵;固体物料吸泵;固体(输送)泵;固体吹散泵;坚实抽吸泵
solid punch 钉冲;整体冲头;坚硬冲子;实心冲头
solid quay 实体(式)码头

solid quay wall 实体(码头)岸壁
solid radwaste 固体放射性废物
solid reaction phase 固相反应
solid reagent 固体试剂
solid reamer 实体铰刀
solid recovery equipment 固体物质回收设备
solid rectangular cantilevered step 实心长方形悬臂台阶
solid rectangular flyer 实心长方形梯级
solid rectangular step 实心长方形踏步;实心长方形台阶
solid reduction 固体还原
solid refrigerator 固体制冷器
solid refuse 固体垃圾
solid residence time 固体停留时间
solid residual 固体残渣
solid residue 固体残渣;烘干残渣
solid resistance 固体电阻
solid retainer 整体护圈
solid retaining wall 实体挡土墙
solid retangular beam 实体矩形梁
solid retention 固体停留
solid retention period 固体停留期
solid retention time 固体停留时间
solid rib (大型拱鹰架中的)弧形木支架;坚实拱架
solid-rib(bed) arch 实(腹)肋拱
solid ribbon 粗黑线带
solid rim 实心圈
solid rivet 实心铆钉
solid rock 坚硬岩石;坚岩;坚石;基岩
solid rocket 固体火箭
solid rocket fuel 固体推进剂
solid rocket propellant 固体推进剂
solid-rock excavation 坚石开挖
solid rod 实心棒
solid roll 整体轧制;实心卷轴;实心轧辊;实心辊轧
solid rolled center 整轧轮心
solid roller 实心滚子
solid roller bearing 实心滚子轴承
solid roof 实心屋顶
solid rotary conveyor 固定旋转式输送机
solid rotor 实心转子
solid rubber 硬橡皮;固体橡胶
solid-rubber dielectric 硬橡胶介质
solid rubber tire 实心橡皮轮胎
solid rubber-tire roller 整体橡皮轮胎式压路机
solid rudder frame 整体舵架
solids 实地板;硬粒;固体颗粒
solid sample 固体试样
solid sampler 固体取样器
solid sampler with slit inside 带侧面刮刀取土器
solid sampling 固体采样
solid sand-lime brick 实心灰砂砖
solids balance 固体平衡
solids by volume 体积固体分
solids-carrying capacity 携带岩屑能力(泥浆);携带岩粉能力(泥浆)
solid Schmidt telescope 立体施密特望远镜
solid scintillator measurement 固体闪烁测量
solids concentration of sludge 污泥固体含量
solids-contact clarifier 澄清池;固体接触澄清池
solids-contact reaction 固体接触反应
solids-contact softening 固体接触软化
solids-contact unit 固体接触装置
solids content 含固体量
solids content in mud 泥浆中固相含量
solids disposal system 固体处置系统
solid seat 实心座
solid section 实心截面;实心部分
solid separation 固体分离
solid separation facility 固体分离设施
solid set 固定装置;实心集
solid-set sprinkler system 固定喷灌系统
solids face 固相面
solids flow feeder 固粒流体喂料机
solids flow meter 固体粉粒流量计
solids fluidization process 固体流态化工艺
solid shaft 连续轴
solid shaft(ing) 实心轴
solids handling pump 泥浆泵
solids handling technique 固体物质输送技术
solid shape 固相颗粒形状;实心型材
solid share 单层钢犁铧
solid shear wall 实体剪力墙

solid sheet 实心板
solid shell construction 坚肋结构
solid shield 整体挡板
solid shim 整体垫片
solids holdup 固体粒子停留时间
solid shooting 不掏槽爆破
solid shrinkage 固态收缩
solid shuttering 整块模板
solid skeleton 固体骨架
solid skirt piston 导梁实心活塞
solid slab 实心平板;实心板
solid slag 致密熔渣;干渣
solid sleeve 整体套管;整套管;密实套筒
solids level meter 固体料位计
solid sloping way 实体斜坡道
solid slug 实心块
solid-solid contact 固—固相接触
solid-solid mix(ture) 固—固混合物
solid-solid model 固—固模型
solid-solid phase 固—固相
solid sols 固溶胶
solid solubility 固体溶解度;固体可溶性;固态溶解度;固溶性;固溶量;固溶度
solid solubility limit 固体溶性极限;固溶度极限
solid solution 固体溶液;固溶液;固溶体
solid solution alloy 固溶体合金;固溶合金
solid solution cement 固溶体金属陶瓷
solid solution hardening 固溶强化
solid solution heat treatment 固溶热处理
solid solution phase 固溶相
solid solution saturation ratio 固溶饱和率
solid solution with twinning 双晶固溶体
solid solvent 固体溶剂
solid-source mass spectrometer 气体质谱计
solid-spandrel arch 实腹拱;实拱肩拱
solid-spandrel arch bridge 实腹拱桥
solid spar 实心翼梁;实心杆
solid species 固体物种
solid sphere 实心球;实球体
solid spheric(al) harmonics 球体谐函数
solid spindle 无孔心轴
solid sprinkler system 固定式喷灌系统
solids pump 固体喷射器(在循环中形成泡沫用)
solid spur 不透水丁坝
solid square bar 实心方钢筋
solids reduction 固体降低
solid stage 凝固期
solid stair(case) 整体式楼梯
solid standard substance 固体标准物质
solid state 固体状态;固态
solid state acoustic(al) image converter 固态声学图像变换器
solid-state amplifier 固态放大器
solid-state array 固态阵列
solid-state array camera 固态阵列摄像机
solid-state battery 固体电池
solid-state body 固体
solid-state camera 固体摄像机;固态照相机
solid-state cementation process 固态置换方法
solid-state chemistry 固态化学
solid-state cinder 固态炉渣
solid-state circuit 固体电路;固态电路
solid-state component 固态元件;固体元件
solid-state computer 固态计算机
solid-state counter 固体计数器
solid-state data collector 固态数据收集器
solid-state dendrite 固态树枝状晶体
solid-state detector 固态探测器
solid-state device 固态器件;固态元件;固体器件
solid-state diffusion method 固体扩散
solid-state diffusion technique 固态扩散技术
solid-state diode array 固体二极管阵
solid-state disc 固态盘
solid-state display 固体显示;固态显示器
solid-state electrode 固态电极
solid-state electronics 固体电子学
solid-state electronic system 固体电路
solid-state electronic techniques 固体电子技术
solid-state element 固态元件;固体元件
solid-state focusing waveguide 固体聚焦波导
solid-state graphitization 固态石墨化
solid-state hydration (水泥的)固体水化
solid-state hydrology 固态水文学
solid-state ignition system 晶体管点火系统

solid-state image converter 固态变像器;固体变像管
solid-state image sensing device 固体图像传感装置
solid-state imaging detector 固体成像探测器
solid-state imaging method 固态成像法
solid-state imaging system 固态成像系统
solid-state inductance 固态电感
solid-state infrared modulator 固态红外调制器
solid-state injection laser 固体注入式激光器
solid-state ionics 固态离子学
solid-state ionization detector 固态电离探测器
solid-state junction sensor 固态结传感器
solid-state keyboard 固态键盘
solid-state laser 固体激光器
solid-state laser material 固态激光材料;固态激光材料
solid-state laser operation 固态激光器运转
solid-stat electronics 固态电子学
solid-state light detector 固态光探测器
solid-state light emitter 固态光源
solid-state light valve 固态光阀
solid-state liquid crystal digital 数字固态液晶显示仪表
solid-state logic 固体逻辑
solid-state martensite 固态马氏体
solid-state maser 固态微波激射器;固态脉塞;固体微波激射器;固体量子放大器
solid-state maser oscillator 固态微波激射振荡器
solid-state mass spectrometer 固态质谱仪
solid-state mass spectrometry 固态质谱法
solid-state materials 固体材料;固态材料
solid-state memory 固态储存器;固体存储器
solid-state metering relay 固态计数继电器;固体计数继电器
solid-state microwave amplifier 固态微波放大器
solid-state microwave device 固态微波器件
solid-state microwave oscillator 固态微波振荡器
solid-state modulation 固态调制
solid-state modulator 固态调制器
solid-state multiplexer 固体多路调制器
solid-state nucleation 固态成核
solid-state oscillator 固态振荡器
solid-state phase changes 固态相变
solid-state phase transition 固态相位跃迁
solid-state photodiode 固态光二极管
solid-state photorelay 固态光电继电器
solid-state photosensor 固态光敏元件;固态光敏器件
solid-state photovoltaic device 固态光伏器件
solid-state physical electronics 固体物理电子学
solid-state physics 固体物理(学);固态物理学
solid-state plasma 固态等离子体
solid-state polymerization 固相聚合
solid-state power amplifier 固态功率放大器
solid-state pressure sensor 固态压力传感器
solid-state-proportioning controller 固态比例控制器
solid-state pulse 固态脉冲
solid-state quantum electronics 固态量子电子学
solid-state radar 固态雷达
solid-state reaction 固相反应;固态反应(法)
solid-state reference electrode 固态参比电极
solid-state relay 固态继电器;固体继电器
solid-state ring laser 固体环形激光器;固态环形激光器
solid-state scintillation counter 固态闪烁计数器
solid-state sensor 固态传感器
solid-state sintering 固相烧结
solid-state slag 固态炉渣
solid-state software 固态软件
solid-state spectrum 固态光谱
solid-state sweeper 固态扫频器
solid-state sweep laser 固态扫描激光器
solid-state sweep oscillator 固态扫频信号振荡器
solid-state switch 固体开关;固态开关
solid-state table 数字化台
solid-state thyratron 固态闸流管
solid-state track detector 固体径迹检测器;固态径迹检测器
solid-state transformation 固体转变
solid-state tuner 固态调谐器
solid-state unit 固态元件;固体组件
solid-state unit radar 固体组件雷达

solid-state vidicon 固态光导摄像管
solid-state welding 固态焊(接)
solid steel door 实心钢门
solid step 实心踏步
solid step with inclined nosing 有斜前缘的实心踏步
solid step with profiled nosing 有线脚前缘的实心踏步
solid stock guide 固定式导料装置
solid stone step 整块石踏步
solid stop 整体门框碰头;整体门挡;整体制动器;整体挡料器;固定式挡料装置
solid storage engineering test facility 固体放射性废物储[贮]存工程试验设备
solid stowing 密实充填;完全充填
solid-stream nozzle 窄雾锥喷嘴;实心流束喷嘴
solids treatment system 固体处理系统
solid stress 固体应力
solid structural system 实体结构体系
solid structure 实体结构
solid strut 实心支柱
solid strutting 加固的剪刀撑(小梁间);刚性撑;格栅横撑;坚固(的)支撑
solid substance 固体物质
solid substance pollution 固体物质污染
solid sulfate 固体硫化物
solid support 可靠支持
solid supporting roller 实心导辊
solid surface 固态表面;实底面积
solid suspension 固体有机物;固体悬浮物
solid suspension opacifier 固相乳浊剂
solid switch 固体开关
solid tamping 夯实;捣实;捣固
solid tap 整体螺丝攻
solid tap wrench 固定丝锥扳手
solid target 固体靶
solid tide 固体潮
solid tile 整片瓦
solid tire 实心(轮)胎;实体轮胎
solid-tired 装实心轮胎的
solid-to-liquid ratio 固液比
solid tool 整体刀具;刀条
solid tooth 实齿
solid-tooth saws 整体齿锯
solid top block 顶层实面混凝土砌块
solid top(hotplate) 闭式灶面板
solid top internode 顶端节间实心
solid torus 实心环
solid track bed 整体道床;混凝土道床
solid train (直达的)固定编组列车
solid transport 固体物质输务
solid type 固定式钢锯架;实体式
solid type cable 实心电缆
solid type diamond bit 整体V形凿尖凿;整体金刚砂钻头
solid type of dock construction 实体式码头结构;实体式船坞构造
solid type quay 实体式码头
solid ultrasonic delay-line 固态超声延迟线
solid unit 固体组件
solid unit masonry 实心砌体圬工
solid unit weight 实容重
solidus 斜线号;固相(曲)线;固线;[复]solidi
solidus curve 固相线;固相曲线
solidus line 固相线;固液线
solidus-liquidus curve 固液相曲线
solidus sintering 固相线烧结
solidus temperature 固相线温度
solid velocity 固体粒子混合速度
solid viscosity 固体黏度
solid volume 固体体积;固体容积;实体积;实体材积
solid volume index 固体量指数
solid volume percentage 实体积比率
solid vortex separator 固旋流分离器
solid wall 实心墙;实体墙;实体岸壁;实砌墙
solid wallboard partition 实心墙板隔墙
solid wallboard structure 承重墙结构
solid wall bowl centrifuge 实体鼓壁式离心机
solid wall centrifuge 无孔离心机
solid walled panel structure 承重墙板结构
solid walled structure 承重墙结构
solid wall pressure vessel 整体式压力容器;单层式压力容器
solid waste 固体污泥;固体垃圾;固体废物;固体废料;固态废料

solid waste classification 固体废物分类
solid waste compactor 固体废物压实机
solid waste disposal 颗粒废料处理;固体废物处理;固体废物处置;固体废料处置
solid waste disposal area 固体废物处置区
solid waste disposal technique 固体废物处理技术
solid waste engineering transfer system 固体废物技术转移
solid waste environmental study 固体废物环境研究
solid waste handling 固体废物处理
solid waste incinerator 固体废物焚化炉
solid waste management 固体废物管理
solid waste management law 固体废物管理法
solid waste of mine 矿山固体废弃物
solid waste pollution 固体废物污染
solid waste pollution monitoring 固体废物污染监测
solid waste recovery 固体废物回收
solid waste refuge 固体垃圾;固体废渣;固体废物
solid waste resource recovery 固体废物资源回收
solid waste storage area 固体废物储存区
solid waste transport 固体废物运输
solid waste treatment equipment 固体废物处理装置
solid waste treatment facility 固体废物处理设备
solid wastewater 固体废水
solid water 固态水
solid water column 充实水柱
solid water column nozzle 充实水柱喷嘴
solid water jet 充实水柱
solid water reactor 水固体堆
solid wax 硬接蜡
solid web 实心腹板;实体腹板
solid-web arch 实腹拱
solid-web beam 实腹工字梁
solid-web bridge 实腹式桥;实腹板桥;实体腹板桥
solid-web column 实腹(式)柱
solid-web girder 实体腹板梁;实腹梁;实腹梁
solid-web joist 实腹(钢)梁;实腹(钢)格栅
solid-web plate 实体腹板
solid-web plate beam 实腹板梁
solid-web plate bridge 实体腹板桥
solid-web rigid frame 实腹刚架
solid-web steel joist 实腹钢梁;实腹钢格栅
solid-web truss 实腹(板)桁架
solid-wed girder 实腹板式桁架梁
solid weir 整体堰;实心堰
solid welding wire 实心焊丝
solid wharf 实体式顺岸码头;实体(式)码头
solid wheel 整体车身;钢圈;实心车轮
solid wheel rolling mill 整体车轮轧机
solid window arcade 实心窗框;实心窗架
solid window frame 实心(的)窗框;实心窗架
solid wire 普通焊丝;单股线
solid wood 实心木料;坚实的木料
solid wooden chock 实心木垛
solid wood flake door 实心刨花板木门
solid wood floor 实心木楼板;实心木地板
solid wood staved core door 木块胶合实心板;木块拼镶实心板
solid work 开切工作
solid wrench 死扳手;呆扳手
solid yardage 实方
solid yoke 实心磁轭
solifluction 解冻泥流(翻浆现象);泥流作用;融冻泥流(作用);热融滑坍;冻融泥流
solifluction lobe 土溜舌
solifluction loess 泥流黄土
solifluction sediment 泥流沉积(物)
solifluction sheet 泥流席
solifluction terrace 泥流阶地;融冻泥流阶地;土溜阶地
solifluction tongue 土溜舌
solifluxion 解冻泥流(翻浆现象);泥流作用
solifluction mantle 土表层泥流
soligen drier 快干剂(调和油漆用)
soligenous 中位泥炭沼泽
Solignum 索里蓝(一种无臭木材防腐剂)
soling 敷筑石质基层;铺筑石质基层;铺底大块石;石质基础;船壳板叠接部分
solion 电化增益器
soliqueous 固液态的

soliquoid 悬浮液;悬浮体
solitary 分离的
solitary coral 单体珊瑚
solitary internal wave 孤立内波
solitary island 孤立岛;孤岛
solitary obligation 单独债务;单独责任
solitary pore 单管孔;单独管孔
solitary stone 孤石
solitary volcano 单独的火山
solitary water gravity wave 水重力孤波
solitary wave 孤立波;孤波;单波
Solite 膨胀板石轻集料;膨胀板石轻骨料
solite 全板门构造
soliton 孤立子
soliton transmission fiber 光孤立子传输光纤
sollar 梯子之间平台;隧洞竖井;隧道中中纵向隔板;高级住室;(英国住房的)敞亮阁楼间
Sollit's alloy 索利特合金
solochrome black 酸性媒染黑
solo circuit 单独对讲电路
solo cooler 多筒冷却器;多筒冷却机
solod 脱碱土
solodization 脱碱作用;兑碱化作用
solodized solonetz 脱碱化碱土
solod soil 脱碱土
solo fibre 单纯纤维
sologoite 斜氯硼钙石
sololysis 溶剂解
Solomon basin 所罗门海盆
Solomonic column 所罗门柱
Solomon Sea 所罗门海
solonchak 白盐土
solonchak soil 盐土
solonetz 碱土
solonetzic soil 碱化土(壤)
solonetz-solonchak complex 碱土和盐土复合区
solonization 碱化作用
Soloth soil 脱碱土
solo type motor cycle 单座式机器脚踏车
Solpugida 避日目
sol rubber 溶胶体橡胶
solsola nitraria 向轴猪毛菜
sol solution 溶胶溶液
solstice 至点;二至日;二至点(即夏至点和冬至点)
solstitial colure 至点圈;二至圈
solstitial point 至点时;至点;二至点(即夏至点和冬至点)
solstitial tidal current 至点潮流
solstitial tide 太阳潮;夏至潮;至点潮;冬至潮;二至潮
solstitial tide current 二至潮流
solubility 可溶性;可解释性;可解决性;可混性;溶解性;(溶)解(度)度;溶度等温线
solubility analysis 溶度分析
solubility coefficient 可解度系数;溶度系数
solubility curve 溶解度曲线;溶度曲线
solubility effect 溶解效应;溶度效应
solubility exponent 溶解度指数
solubility factor 溶解系数
solubility gap 溶度间隔
solubility index 溶度指数
solubility in water 水中溶解性;水中溶解度
solubility law 溶解质定律;溶解度定律
solubility limit 溶解度极限
solubility-limited dissolution 极限溶度溶解
solubility method 溶解度法
solubility of aromatic hydrocarbon 芳香烃的溶解度
solubility of bitumen 沥青溶解度
solubility of butane 丁烷的溶解度
solubility of cyclopropane 环丙烷的溶解度
solubility of decane 癸烷的溶解度
solubility of ethane 乙烷的溶解度
solubility of gas in oil 天然气溶于石油
solubility of individual pure hydrocarbon 单烃的溶解度
solubility of iso-alkane hydrocarbon 异构烷烃的溶解度
solubility of isobutane 异丁烷溶解度
solubility of n-alkane hydrocarbon 正烷烃的溶解度
solubility of non-hydrocarbon 非烃的溶解度
solubility of octane 辛烷溶解度
solubility of oil in gas 石油溶于天然气

solubility of oxygen in water 水中氧溶解度
solubility of pentane 戊烷的溶解度
solubility of pure hydrocarbon 纯烃的溶解度
solubility of residue 残渣溶解度
solubility parameter 溶解度参数;溶度参数
solubility product 溶解度积;溶度积
solubility product constant 溶度积常数
solubility product principle 溶度积原理
solubility quotient 溶解度商
solubility table 溶解度表
solubility-temperature curve 溶解度温度曲线
solubility test 溶解度试验;溶解度测定
solubility testing 溶解度试验
solubility value 溶解值
solubilization 增溶化;溶液化;增溶(作用)
solubilization chromatography 增溶色谱法
solubilize 溶液化
solubilized excess sludge 溶液化剩余污泥
solubilized sulfur dye 可溶性硫化染料
solubilized vat dye 可溶性还原染料
solubilizer 增溶剂
solubilizing agent 增溶剂
solubilizing group 助溶基团
solubilizing mechanism 增溶机理
solubilizing phase diagram 增溶相图
soluble 可溶于二硫化碳;可溶的
soluble 5-day biochemical oxygen demand 溶性五日生化需氧量
soluble adsorbate concentration 溶性吸附质浓度
soluble alumina 可溶性氧化铝
soluble alumin(i)um 可溶性铝;溶解态铝
soluble anhydrous gypsum 可溶性无水石膏
soluble anion group 易溶阳离子组
soluble ash content 可溶性灰分含量;溶解性灰分量
soluble biochemical oxygen demand 溶性生化需氧量
soluble biological oxygen demand 溶性生物需氧量
soluble bitumen 溶性沥青
soluble blue 溶性蓝
soluble cation 溶性阳离子
soluble cation group 易溶阳离子组
soluble chemical oxygen demand 溶性化学需氧量
soluble chemicals 可溶性化学品
soluble clay 可溶性黏土
soluble cobalt drier 可溶钴催干剂
soluble colo(u)r 可溶性颜料
soluble complex 可溶性复合物;溶性络合物
soluble compound 溶性化合物
soluble concentration 可溶浓度
soluble constituent of petroleum 石油可溶性组分
solublecored oil cooling gallery 可溶性型芯油冷通道
soluble crystal 可溶性晶体
soluble cutting oil 可溶性切削油
soluble developer 可溶性显色剂
soluble dryer 溶性干燥剂;可溶性催干剂
soluble dye(stuff) 可溶染料
soluble effluent biochemical oxygen demand 溶性污水生化需氧量
soluble elements 易溶元素
soluble evapo(u)rated residue 溶性蒸发残渣;可溶性蒸发残渣;溶解性蒸发残渣
soluble glass 硅酸钠;可溶性硅酸钠;可溶(性)玻璃;溶性玻璃;水玻璃
soluble glass coat 水玻璃涂层
soluble glass mastic 水玻璃玛琦脂
soluble glass mix(ture) 水玻璃混合料
soluble glass paste 水玻璃浆
soluble glass putty 水玻璃油灰;水玻璃腻子
soluble glass solution 水玻璃溶液
soluble glaze 含可溶性盐类的釉
soluble gypseous anhydrite 可溶性无水硫酸钙
soluble humus 可溶性腐殖质;溶性腐殖质
soluble hydroxo complex 溶性羟络合物
soluble immediate oxygen demand 溶性即时需氧量
soluble impurities 可溶性杂质
soluble impurity 溶性杂质
soluble inert chemical oxygen demand 溶性惰性化学需氧量
soluble inorganic matter 溶性无机物
soluble inorganic phosphorus 溶性无机磷

soluble in water 可溶于水;可溶入冷水
soluble iron 可溶性铁;可溶铁
soluble Kjeldahl nitrogen 溶性克耶达定氮量
soluble lead 可溶性铅
soluble manganese 可溶性锰
soluble material 可溶物质;溶性物质
soluble matrix type 基料溶解型
soluble matter 可溶性物质;可溶物;溶性无机物
soluble matter content 可溶物含量
soluble microbial metabolic product 溶性微生物代谢产物
soluble mineral 可溶矿物质
solubleness 可溶性;溶解性;溶解度
soluble nitrogen 可溶性氨;溶性氮
soluble oil 可乳化油;可溶油;乳化油;溶性油;调水油;水溶性润滑剂
soluble oil emulsifier 溶解油乳化剂
soluble oil emulsion 溶性油质乳胶
soluble organic carbon 溶性有机碳
soluble organic content 可溶性有机物含量
soluble organic materials containing nitrogen 含氮溶性有机物
soluble organic matter 可溶性有机物;溶性有机物
soluble organic nitrogen 溶性有机氮
soluble organic oxygen demand 溶性有机物需氧量
soluble organic substance 溶性有机剂
soluble phosphate 溶性磷酸盐
soluble phosphate concentration 溶性磷酸盐浓度
soluble phosphoric manure 可溶性磷肥
soluble phosphorus 溶性磷
soluble poison 可溶毒物
soluble polymer 可溶性聚合物
soluble powder 可溶性粉剂
soluble reactive organic anion 溶性活性有机阴离子
soluble reactive phosphorus 溶性活性磷
soluble reactive silica 溶性活性硅氧
soluble reactive silicon 溶性活性硅
soluble residual product 溶性残渣产物
soluble saline filling 可溶盐充填
soluble salt 可溶盐;可溶性盐
soluble salt content 可溶盐含量
soluble salt core 可溶盐芯
soluble salts 可溶性盐类;溶性盐类
soluble silicate 可溶性硅酸盐;溶性硅酸盐
soluble sodium percentage 溶性钠百分率
soluble solid 可溶(性)固体;溶(解)性固体
soluble solids of water 水的溶解性固体
soluble species concentration 溶性物种浓度
soluble starch 可溶性淀粉
soluble substance 可溶(性)物质;溶性物质
soluble sulfide 溶性硫化物
soluble-sulfide precipitation 溶性硫化物析出
soluble tar 轻木焦油
soluble toxic metal 溶性毒性金属
soluble underglaze technique 可溶性盐染色法
soluble unreactive phosphorus 溶性惰性磷
solum 土体;土层风化;土层;风化土(壤);变质土;悬空地板之下的空间;建筑物内地面土层
Soluminium 锡锌铝铜焊料;铝钎料;索勒米尼铝焊剂
solum of soil 土层
soluseptazine 苯丙磺胺苯二磺酸钠
solusphere 溶解圈
solute 溶质;溶解物(质)
solute atmosphere 溶质气团
solute atoms 溶质原子
solute balance 溶质衡算
solute balance equation of aquifer 含水层溶质均衡方程
solute boundary layer 溶质边界层
solute concentration 溶质浓度
solute concentration difference 溶质浓度差
solute concentration vector 溶质浓度矢量
solute degradation rate 溶质降解速率
solute diffusion 溶质扩散
solute dispersion 溶质扩散
solute distribution 溶质分布
solute excitation 溶质激发
solute field 溶质场
solute flux 溶质流
solute flux of entering volume element 进入元内溶质通量
solute flux of flow-out volume element 流出元

内溶质通量
solute gas 溶解气体;溶解气
solute hardening 溶质硬化
solute molecule 溶质分子
solute movement 溶质移动
solute phase 溶质相
solute plume 溶质烟羽
solute potential 溶质势
solute pressure 溶质势
solute redistribution 溶质再分配
solute runoff 溶质径流
solute segregation 溶质偏析;溶质分凝
solute-solute interaction 溶质—溶质相互作用
solute sorption 溶质吸附
solute striation 溶质线条
solute suction 溶质吸力;渗析吸力
solute transfer 溶质运移;溶质传递
solute transport 溶质运移
solute transport equation 溶质运移方程
solute transport model 溶质运移模型
solute trapping 溶质捕获
solute travel 溶质移动
solute travel time 溶质运移时间
solute viscosity factor 溶质黏度因子
solute-volatilization interference 溶质气化干扰;溶质挥发干扰
solute-water interaction 溶质—水相互作用
solute zone 溶质层
solution 解决办法;解法;解答;溶液;溶体;溶(解)液;溶解
solution adhesive 溶液胶黏剂
solution adsorption method 溶液吸附法
solution alkalinity 溶液碱度
solution annealing 固溶退火;溶体化退火
solution assay 溶液分析
solution balance 溶液天平
solution basin 溶解盆地;溶蚀盆地;溶蚀池
solution behavior 溶液变化过程
solution benzene 苯的溶解度
solution breccia 垮塌角砾岩;溶解角砾岩
solution by a triangle 三角形解法
solution by definite integral 定积分解法
solution by electric(al) analogy 电模拟解法
solution by group 分组解法
solution by iterative method 迭代法求解
solution by polynomials 多项式解
solution by successive approximation 逐渐趋近解法;渐近解法
solution calorimeter 溶液量热器;溶液热量计;溶液量热计
solution catcher 溶液收集器
solution cave 溶蚀洞穴;溶洞
solution cavity 溶蚀洞穴;溶洞
solution ceramic 溶解陶瓷
solution ceramic coating 溶液(型)陶瓷涂层
solution ceramics 表面覆盖金属盐溶深层的陶瓷
solution channel 溶蚀槽;溶解裂隙;溶沟;溶道
solution channel spring 溶洞泉
solution check 解答校验
solution chemical deposition 化学溶液沉积;真溶液化学沉积作用
solution chemistry 溶液化学
solution circuit 溶液循环
solution cleavage 溶解理
solution coating 浸液涂漆;溶液型涂料
solution coating method 溶液涂渍法
solution collapse 溶解崩塌
solution complex 溶液络合物
solution composition 溶液组分;溶液组成;溶液成分;溶解组分
solution composition ga(u)ge 溶液成分测量计
solution concentration 溶液浓度
solution concentration detector 溶液浓度检定器
solution condition 溶液条件
solution conductivity detector 溶液电导检测器
solution containing cyanide 含氰化物溶液
solution contains a certain amount of acid 溶液含一定量的酸
solution control valve 溶液控制阀
solution cooler 溶液冷却器
solution crack 溶隙;溶解裂缝
solution crevice 溶洞
solution culture 溶液培养
solution density detector 溶液密度检定器

solution deposits 溶质沉淀(物);溶液沉淀(物)
solution depression 溶液的冰点低;溶液洼地
solution doline 溶蚀落水洞
solution drain hole 溶液排放孔
solution dyeing 原液染色
solution effect 溶解(作用)
solution equilibrate 溶液平衡
solution equilibrium composition 溶液平衡成分
solution ethylbenzene 乙苯的溶解度
solution example 溶液样品
solution existence 有解
solution feed chlorinator 加氯液机
solution feed dosage 溶液进料剂量
solution feeder 加液器;溶液进料器;溶液加料器;投溶液器;溶液加料器
solution feed pump 供液泵
solution film 溶液膜
solution filter pump 溶液过滤泵
solution fissure 溶蚀裂隙;溶蚀裂缝;溶解裂隙
solution flow rate 溶液流速
solution fluorimetry 溶液荧光法
solution funnel 溶蚀漏斗;溶斗
solution gas 溶解气
solution gas drive 内部气驱;溶解气驱动
solution gas drive pond 溶解气驱油田
solution gas drive pool 溶解气驱油藏
solution gas reservoir 溶解气油藏
solution groove 溶槽;溶蚀沟;溶沟
solution grout 溶液灌浆
solution grown crystal 溶液生长晶体
solution growth 溶液生长
solution growth method 溶液生长法
solution hardening 固溶硬化;固溶淬火;溶液硬化
solution heat 溶解热
solution heater 溶液加热器
solution heattreated condition 固溶热处理状态
solution heat treatment 固溶热处理;溶液热处理;溶流热处理
solutioning machine 刮浆机
solution injection 溶液灌浆
solution isopropyl benzene 异丙苯的溶解度
solutionizing 固溶化
solution joint 溶蚀节理
solution lake 喀斯特湖;溶蚀湖;溶解湖
solution landform 溶蚀地貌
solution line 溶液管道;溶线
solution load 溶解负载
solution medium 溶液介质;溶剂
solution method 溶液法
solution method of growth 溶流生长法
solution mine 溶浸法开采矿山
solution mining 溶液采矿;溶解开采;溶解采矿法
solution mixer 溶液混合器
solution mixing tank 溶液混合池;溶液混合槽
solution of a game 对策的解
solution of ammonia 氨溶液
solution of difference equation 差分方程解法
solution of differential equation 微分方程解法
solution of dyestuff 染料溶液
solution of equations 方程组的解
solution of linear equation 线性方程组解法
solution of linear program(me) 线性规划问题解
solution of non-electrolyte 非电解质溶液
solution of normal equations 法向方程组的解
solution of paraxylene 对甲苯的溶解度
solution of triangle 三角形解算
solution opening 溶蚀通道;溶孔;溶洞
solution path 解算路线;解答路径
solution pH 溶液 pH 值
solution phase 溶液相
solution phase equilibrium 溶液相平衡
solution phase relationship 溶液相关系
solution physical development 溶解的物理显影
solution pit 溶蚀坑
solution plane 溶蚀面
solution poison 溶液毒物
solution polycondensation 溶液缩聚
solution polymerization 溶液聚合
solution pool 溶蚀潭
solution porosity 溶解孔隙
solution potholes 溶蚀锅穴
solution power 溶解力;溶解本领
solution precipitates 溶液沉淀物
solution pressure 溶液压力;溶解压强;溶解压力

solution procedure 求解过程
solution propylbenzene 丙苯的溶解度
solution pump sump 溶液抽吸槽
solution purification 溶液净化
solution pyridine 吡啶的溶解度
solution rate 溶解速度
solution reboiler 溶液再沸器
solution reclaiming method 溶液回吸法;溶解回收法
solution regenerator 溶液再生塔
solution regrowth process 溶液再生长过程
solution regrowth technique 溶液再生长技术;溶液再生长法
solution remains 溶液残渣
solution residue technique 溶液残渣技术;溶液残渣法
solution ripple mark 溶解波痕
solution salt 溶液盐
solution scaly peeled off 溶蚀鳞片状剥落
solution scintillator 溶液闪烁体
solution set 解集
solution/solvent viscosity ratio 溶液/溶剂黏度比
solution space 解空间
solution spectrum 溶液光谱
solution state of matter 物质的溶液态
solution storage tank 溶液储[贮]槽
solution strainer 溶液过滤器
solution strength 溶液浓度
solution strengthening 固溶强化
solution stripping 溶液解吸
solution subsidence 溶解塌陷;溶解沉陷
solution subspace 解子空间
solution sump pump 溶液储[贮]槽泵
solution surface 溶液表面
solution tank 溶液箱;溶液池;溶液槽;溶解箱;溶解池
solution technique 溶液技术
solution temperature 溶解温度
solution temperature sensing element 溶流温度传感元件
solution tension 溶液张力;溶解压
solution theory 溶解理论
solution thiophene 噻吩的溶解度
solution to solvent ratio 溶液和溶剂比
solution trace 溶蚀痕
solution transfer 溶移;溶解转移
solution transfer pump 溶液输送泵
solution treating 溶液处理
solution treatment 固溶退火;固溶处理;溶体处理
solution trial 溶液试验
solution type reactor 溶液型(反应)堆
solution type reactor tank 溶液型(反应)堆罐
solution valley 溶谷
solution vector 解向量
solution volume 溶液体积
solutizer 溶解加速剂
solutizer air regenerative process 空气促溶剂再生法
solutizer-tannin process 丹宁促溶剂过程
solutrope 相溶物
solvability 可溶化性;可溶性;溶剂化能力;溶剂化度
solvability of the system of equations 方程组可解性
solvable 可溶的
solvable group 可解组;可溶组;可解群
solvate 媒合物;溶剂化物;溶剂合物
solvated H-ion 水合离子
solvate theory 溶剂化学说
solvation 溶解;溶剂离解(作用);溶剂化(作用);溶化;溶合(作用)
solvation number of ion 离子溶剂化值
solvation of ion 离子溶剂化作用
solvatochromism 溶剂化显色(现象)
solvatochromy 溶剂化显色
solved node 可解节点
solve-labeling procedure 可解标记过程
solvency 有清偿能力;溶解能力;溶解本领;偿债能力;偿付能力
solvency constraint 清偿力约束
solvency determination 偿债确定能力;确定偿债能力
solvency power 溶解力
solvency ratios 偿债能力比率

solvend 可溶物质;溶油浮选
solvent 有溶解力的;有偿付能力;可溶解的;基本组分;溶媒;溶解的;溶剂
solvent absorber 溶剂吸附体
solvent acrylic sealant 溶剂型丙烯酸密封膏
solvent action 溶解作用
solvent-activated adhesive 溶活型胶黏剂;溶剂活性胶;溶剂活化胶黏剂
solvent adhesive 挥发性胶黏剂;溶剂性胶黏剂;溶剂型黏合;溶剂型胶黏剂;溶剂胶
solvent amount 溶剂量
solvent analysis 溶剂分析
solvent anti-felt finishing 溶剂防毡合整理法
solvent architectural coating 溶剂型建筑涂料
solvent balance 溶剂平衡
solvent base 溶剂基
solvent-based adhesive 溶剂基黏合剂
solvent-based coating 溶剂型涂料
solvent base compound 溶剂型封口
solvent-based paint remover 溶剂基涂料清除剂
solvent blank 溶剂空白
solvent bleaching 溶剂漂白
solvent bleeding 溶剂(性)渗色
solvent blend 溶剂中掺和物
solvent bonding 溶剂黏合
solvent borne 含有溶剂的
solvent carried paint 含溶剂的涂料
solvent carrier exchange system 溶剂载体的交换体系
solvent catch tank 溶剂收集槽
solvent cell 溶剂池
solvent cement 溶剂型胶黏剂;溶剂胶结材料;溶剂型胶黏料
solvent cemented joint 溶剂胶结料连接接头
solvent cementing 溶剂黏结
solvent cleaner 含溶剂的清洗器;溶剂洗净剂
solvent cleaning 溶剂清洗
solvent coat 溶剂涂层
solvent coating 溶剂涂层;溶剂性涂布
solvent column 溶剂塔
solvent composition 溶剂组成
solvent condenser 溶剂冷凝器
solvent contamination 溶剂污染
solvent cracking 溶剂分解
solvent crystallization 溶剂结晶法
solvent cut-back type rust preventive 溶剂稀释型防锈剂
solvent deasphalting 溶剂脱沥青
solvent decarbonizing 溶剂脱碳
solvent decarbonizing process 溶剂脱碳过程
solvent degreasing 溶剂脱脂
solvent delivery system 溶剂输送系统
solvent deoiling 溶剂脱油
solvent deresining 溶剂脱树脂(法)
solvent detergent 溶剂洗涤剂
solvent dewaxing 溶剂脱蜡
solvent dewaxing process 溶剂脱蜡法
solvent dispersion 溶剂分散
solvent distillate 溶剂蒸馏液
solvent drum dy(e)ing machine 鼓筒式溶剂染色机
solvent drying 溶剂干燥
solvent dye 溶剂染料
solvent dyeing process 溶剂染色法
solvent effect 溶剂效应
solvent efficiency 溶剂效率
solvent entrainment 溶剂夹带物
solvent epoxy varnish 溶剂型环氧树脂清漆
solvent etching 溶剂腐蚀法
solvent evapo(u)ration 溶剂蒸发
solvent excitation 溶剂激发
solvent exhaustion 溶剂萃取
solvent extract 溶剂提出物;溶剂萃取物;溶剂抽出物
solvent extractable matter 溶剂抽出物
solvent-extractable organics 溶剂可萃取有机物
solvent-extract analysis 溶剂萃取物分析
solvent extracted 溶剂萃取的
solvent extracted oil 溶剂萃取油
solvent extraction 溶剂萃取法;溶液精;溶剂提取;溶剂浸提作用;溶剂萃取法;溶剂萃取;溶剂抽提
solvent extraction column 溶剂萃取柱
solvent extraction contactor 溶剂萃取器
solvent extraction generator 溶剂萃取同位素发生器;溶剂萃取发生器

solvent extraction method 溶剂提取法
solvent extraction monitor 溶剂萃取监测器
solvent extraction plant 溶剂萃取装置
solvent extraction process 溶剂提取法;溶剂萃取过程
solvent extraction separation 溶剂萃取分离
solvent extraction technique 溶液萃取技术
solvent extraction tower 溶剂萃取塔
solvent extractive purification 萃取净化
solvent extract tower 溶剂提取塔
solvent-fast 溶剂稳固的
solvent feed 溶剂料液
solvent feed tank 溶剂储[贮]槽
solvent flo(a)tation 溶剂浮选
solvent fluid 溶液
solvent fractionation 溶剂选择分离;溶剂分馏
solvent-free 无溶剂的;不含溶剂的
solvent front 展开剂前沿;溶剂前沿
solvent hexane 溶剂己烷
solvent hold-up 溶剂滞留
solvent hopper 溶剂料斗
solvent imbalance 溶剂不平衡
solvent immobilization 溶剂停流
solvent inclusion 溶剂包藏作用
solvent induced crazing 溶剂性银纹
solvent-in-pulp 溶剂矿浆萃取;溶剂浆液萃取
solvent isotope effect 溶剂同位素效应
solventized soap 溶剂化皂
solvent laminating 胶黏合薄板;溶剂法制塑料薄片
solvent layer 溶剂层
solventless 少溶剂的;无溶剂的;不溶解的
solventless coating 无溶剂涂料
solventless epoxy varnish 无溶剂环氧清漆
solventless paint 无溶剂漆
solventless polyester varnish 无溶剂聚酯清漆
solventless varnish 无溶剂清漆
solvent/liquid ratio 溶剂与液体比
solvent loading 溶剂负荷
solvent loss 溶剂损失
solvent makeup 溶剂补充
solvent membrane 溶剂膜
solvent method 溶剂法
solvent migration-distance 溶剂移动距离
solvent mixing method 溶剂混溶法
solvent mix(ture) 溶剂混合料;溶剂混合物
solvent molecule 溶剂分子
solvent monitoring 溶剂监测
solvent mo(u)lding 溶剂模塑法;溶剂挥发造型法;溶剂成形
solvent naphtha 溶剂油;溶剂石脑油;溶剂汽油;轻汽油溶剂;松香水
solvent oil 溶剂油
solvent oil ratio 溶剂对油比
solvent paint 溶剂漆;溶媒漆;溶剂型涂料
solvent pairs 溶剂偶
solvent parameter 溶剂参数
solvent partitioning 溶剂分配
solvent paste wax 液态溶剂蜡
solvent pollution 溶剂污染
solvent polymeric membrane 溶剂聚合膜
solvent polymerization 溶解集合
solvent polymerization method 溶剂聚合法
solvent pop 由于溶剂滞留而引起的起泡现象
solvent popping resistance 抗溶剂滞留起泡性
solvent power 溶解能力;溶剂溶解力
solvent pressure 溶媒压
solvent process 溶剂(精)制法;溶解方法;溶剂过程
solvent programmer 溶剂程序变换器
solvent programming analysis 梯度洗脱分析
solvent proof 防溶剂
solvent pump 溶剂泵
solvent purification 溶剂纯化
solvent radiolysis 溶剂辐解
solvent raffinate 溶剂精制产物
solvent reclamation 溶剂回收
solvent recovery 溶剂回收
solvent recovery column 溶剂回收柱;溶剂回收塔
solvent recovery plant 溶剂回收装置;溶剂回收厂
solvent recovery process 溶剂回收法
solvent recuperation 溶剂回收
solvent refined oil 溶剂精制油
solvent refining 溶剂精制
solvent refining agent 溶剂精制剂
solvent refining coal method 溶剂精炼煤法

solvent regeneration 溶剂再生
solvent regeneration process 溶剂再生法
solvent release 溶剂释放性
solvent release agent 溶剂型脱模剂
solvent release butyl sealant 溶剂挥发型丁基橡胶密封膏
solvent removes 溶剂去污剂
solvent reservoir 储溶剂器
solvent resistance 耐溶剂性
solvent resistant coating 耐溶剂涂料
solvent retention 溶剂滞留性;溶剂存留性
solvent re-treated extract 溶剂二次处理抽出物
solvent saturated 饱和的
solvents carrier 溶剂运输船
solvent scouring 溶剂洗涤
solvent seasoning 溶剂干燥
solvent segregation 溶剂分离法
solvent selectivity 溶剂选择性
solvent sensitive adhesive 溶敏性胶黏剂;溶剂敏感性胶黏剂
solvent sensitivity 溶剂过敏
solvent separator with calibrated tank for water 溶剂分离器带水计量槽
solvent shock 溶剂冲击
solvent silicone varnish 溶剂型有机硅清漆
solvent-soluble 溶剂可溶性
solvent soluble dye 可溶于溶剂的染料;溶剂溶解染料
solvent soluble soap 溶剂溶化皂
solvent spot test 溶剂点滴试验
solvent stabilization operator 溶剂稳定化算符
solvent strength 溶剂强度;溶剂浓度
solvent stripping 溶剂反萃取
solvent support ratio 溶剂载体比
solvent susceptible 溶剂敏感的
solvent swell 溶剂溶胀
solvent-thinned 用溶剂稀释的
solvent thinned coating 溶剂稀释型涂料
solvent tolerance 溶剂中的最大溶解度;溶剂稀释极限;溶剂容许溶解度;溶剂容忍度
solvent tower 溶剂塔
solvent treating 溶剂精制
solvent treating plant 溶剂精制厂
solvent treatment 溶剂精制
solvent trough 溶剂槽
solvent type coating 溶剂型涂料
solvent type plasticizer 溶剂型增塑剂
solvent type waterproofing paint 溶剂型防水涂料
solvent uptake 溶剂吸着
solvent vapo(u)r 溶剂蒸气
solvent vapo(u)r cure finishing 溶剂气体焙固整理
solvent vapo(u)r treatment process 溶剂蒸气处理过程
solvent viscosity 溶剂黏度
solvent washer 溶剂洗涤器
solvent washing 溶剂洗提;溶剂洗涤
solvent washing effect 溶剂洗提效应
solvent water 溶剂水分;浓度型脱水分
solvent-welded joint 溶焊缝
solvent weld fitting 溶度焊配件
solvent weld(ing) 溶剂焊接;溶剂粘接
solvent widewidth-continuous plant 溶剂阔幅连续处理设备
solvent winterization 溶剂冬化法
solvent wipe-off method 溶剂擦法
solver 解算装置;解算器;解算机
solving process 解算法;求解过程
solvmanifold 可解流形
solvolysis 溶剂分解(作用)
solvolytic dissociation 溶剂离解(作用)
solvolytic reaction 溶剂化反应
solvophobic 憎溶剂的
solvophobic chromatography 疏溶剂色谱
solvus 固体溶解度曲线;固溶体分解曲线;熔线;溶解度曲线
Somalia abyssal plain 索马里深海平原
Somalia plate 索马里板块
Somali basin 索马里海盆
Somali current 东非沿岸海流
somascope 超声波检查仪
somatic death 整体死亡
somatic effect of radiation 辐射的体质效应
somatic plate 体节板

somber 浅黑
some breeding stations 几处选种站
some elite plots 几个原种圃
some equipment of isotope analysis 同位素分析设备
some major factors in soil formation 土壤形成的几个因素
some observation on 某些观察
some of the introduction 一些引入材料
someriyoshine 日本樱花
Somerville floor 索米威乐地板(一种耐火地板,由空心加混凝土管组成)
some soilplant relationships 某些土壤与作物关系
some soil property 某些土壤性质
Somes Sound granite 缅因州黑云母花岗岩(产于美国缅因州)
sometimes felt 有时可感
some tropical soil 一些热带土壤
somewhat cloudy 少云
somma 外轮山
Somme abyssal plain 索姆深海平原
sommer 地梁;柱顶石;大梁
Sommerfeld radiation condition 索末菲辐射条件
sommering 平砖拱的辐射接合
sommering line 基石线
sonalert 固体音调发生器
sonar 声呐;音响定位器;水下探测仪;水下声波定位器;水下超声波探测系统;水声探测器;水测定位仪;声呐设备;声波定位仪;声波定位和测距
sonaradiography system 声射线摄影系统
sonar amplifier 声呐放大器
sonar and sweep survey 水深及扫海勘查;水深及扫海勘测
sonar anti-submarine detection 反潜声呐定位探测;声呐反潜艇探测
sonar array 声呐阵列
sonar array sounding system 声呐阵测深系统
sonar attack center 声呐攻击指挥中心
sonar background noise 声呐环境噪声;声呐本底噪声
sonar beacon 声呐信标
sonar beam 声呐射束
sonar buoy 声呐浮标
sonar-buoy compartment 声呐浮标舱
sonar calibration station 声呐校准站
sonar capsule 声呐盒;反射高频声波器
sonar certification test 声呐合格试验
sonar chart 声呐图
sonar communication 声呐通信
sonar control indicator 声呐显示器
sonar control room 声呐控制室
sonar control system 声呐控制系统
sonar countermeasures 声呐干扰
sonar countermeasures and deception 声呐干扰和假象;声呐对抗与诱骗
sonar data 声呐数据
sonar data recorder 声呐数据记录器
sonar detecting-ranging set 探测与测距声呐;声呐探测与测距仪
sonar display 声呐显示器
sonar dome 声呐罩;声呐导流罩
sonar dunking 声呐入水
sonar early warning 声呐早期警戒
sonar echo sounder set 声呐回声探
sonar equation 声呐方程
sonar equipment 声呐装置;声呐设备
sonar evaluation station 声呐鉴定站
sonar extensometer 声呐引伸仪
sonar for fisheries 渔用声呐
sonar gear 声呐装置
sonar hoist and transducer group 声呐投吊与换能器组合
sonar hoist set 声呐投吊设备
sonar housing 声呐导流罩
sonar indicator system 声呐指示系统
sonar information center[centre] 声呐情报中心
sonar instrumentation probe 声呐仪探测
sonar listening set 声呐听音仪;声呐收听装置
sonar method 声呐法
sonar nacelle 声呐吊舱(反潜飞机上)
sonar navigation 声呐导航;声呐导航
sonar noise recorder 声呐噪声记录器
sonar operator 声呐操纵员
sonar operator trainer 声呐操作员教练器

sonar parameter 声呐参数
sonar performance computer 声呐性能计算机
sonar pinger 声呐脉冲发射器
sonar pinger system 声呐脉冲测距系统
sonar plot 声呐标图
sonar position plotter 声呐测位器
sonar position sensor 声呐位置传感器
sonar prediction 声呐预报
sonar projector 声呐换能器;声呐发送器;声呐发射器
sonar range 声呐作用距离
sonar receiver 声呐接收机
sonar receiver-scanner 圆周扫描声呐接收器
sonar recorder 声呐记录器
sonar resolver 声呐分解器
sonar scanner 声呐扫描仪;声呐扫描器
sonar scan(ning)set 声呐扫描设备
sonar school ship 声呐训练舰
sonar search arc 声呐搜索扇面
sonar self-noise 声呐自噪声
sonar set 声呐设备
sonar signal 声呐信号
sonar signal processing 声呐信号处理
sonar signal processing computer 声呐信号处理计算机
sonar signal processor 声呐信号处理机
sonar signal simulator 声呐信号模拟器
sonar simulator 声呐模拟器
sonar sounding set 声呐探测仪;声呐测深设备
sonar source level 声呐声源级
sonar speed measuring set 声呐测速仪
sonar SSB communication 声呐单边带通信设备
sonar surveillance system 声搜;声呐监视系统
sonar survey 声呐测量
sonar sweep(ning) 声呐扫海;声呐扫床
sonar system 声呐系统
sonar target 声呐目标;声呐反射物标
sonar technical characteristic 声呐技术性能
sonar technician 声呐技术师
sonar technique 声呐技术
sonar technology 声呐技术
sonar test barge 声呐试验驳
sonar test set 声呐测试装置
sonar thumper unit 声呐声波发生器
sonar towfish plus wing combination 带翼声呐拖体
sonar towing system 声呐拖曳系统
sonar trainer 声呐教练器
sonar transducer 声呐换能器
sonar transducer array 声呐换能器阵
sonar transmission 声呐传输
sonar transmitter 声呐发射器;声呐发射机
sonar transponder system 声呐应答系统
sonar wave recorder 声呐站自记测波仪
sonar window 声呐透窗;声呐窗
sonde 高空探测装置;高空气象探测仪;探针;探头;探空仪;探管;探测器探空仪;探棒;电极系
sonde construction 电极系结构
sonde factor 测井电极系数
sonde-type vibro-pickup 探头型振动传感器
sondo 干热焚风
sondol 回声探测器
sone 宋(响度单位)
so near there to as she can safely get 船舶能安全到达的附近地点
sone scale 响度标度;音响标度
Song typeface 宋体字
sonic 声波的
sonic absorption coefficient 声吸收系数
sonic agglomeration 声聚
sonic agglomerator 低频波聚灰器
sonic alarm system 声波警报系统
sonic altimeter 回音测高计;回声测高计;声学高度计;声测高度计;声波测高计
sonic analyzer 声波探伤仪;声波分析仪
sonic anemometer 声学风速仪;声学风速计;声学风速表;声波风速计
sonic apparatus 声波仪
sonic apparent formation factor 声波视地层因数
sonic applicator 声能应用设备
sonication wet oxidation 声湿氧化
sonication wet oxidation technique 声湿氧化技术
sonicator 近距离声波定位器
sonic bang 轰声;声振

sonic barrier 音障;声障
sonic bearing 音响方位;声(响)方位
sonic bonder 声波接合器
sonic boom 废声;音爆;轰声;声振;声爆;爆音
sonic boom carpet 声振仪;声振层
sonic boom effect 声振音响效应
sonic borehole televiewer plot 声波电视测井成果图
sonic button hole machine 声波锁眼机
sonic carburetion 声速化油
sonic carburetor 声速化油器
sonic chemical analyzer[analyser] 声波化学分析器
sonic chemiluminescence 声化学发光
sonic cleaning 声波清洗;超声波清理
sonic comparator 声波比较仪;声波比较仪
sonic dedusting 声波除灰;声波除尘
sonic delay device 声延迟器
sonic delay-line 声延迟线
sonic depth finder 回声测深仪;回声测深声呐;回声测深器;声波测深(法)
sonic depth-finding instrument 声波测深仪
sonic depth recorder 回声测深记录仪;回声深度记录仪;声波深度记录器
sonic depth sounder 回声测深仪
sonic depth sounding 回声测深
sonic detector 声波探伤仪;声波探伤器;声波探测仪;声波探测器;声波探测定器
sonic device 声音控制装置;声学仪器;声学仪表
sonic discharge carburetor 声速射流化油器
sonic disturbance 声干扰
sonic drill 振动钻孔机;(浅孔用的)声波钻头
sonic dust removal 声波除灰;声波除尘
sonic echo sounder 回声探鱼仪;回声测深仪
sonic energy 声能(量)
sonic fatigue 音响疲劳;声致疲劳;声能疲劳
sonic fatigue measurement 声波疲劳测试
sonic-film memory 声薄膜存储器
sonic-film storage 声薄膜存储器
sonic fineness tester 声波式细度仪
sonic fix 声呐船位;声定位
sonic flaw detection 声波探伤
sonic flocculation 声波絮凝
sonic flow-measuring assembly 声学测流测深装置
sonic flowmeter 音响式流量计;音响流量计;声流量计;声流量计
sonic frequency 声频
sonic frequency band 声频带
sonic gas analyzer 声学气体分析计
sonic gas cooling system 声气冷却系统
sonic ga(u)ge 声波应变仪
sonic generator 发声器
sonic homing 声自导引
sonic homing system 声自导系统
sonic-idle carburetor 声速怠速化油器
sonic inspection 音响检验;听声检验
sonic instrument 声学仪器;声学仪表
sonic interferometer 声干涉仪
sonic layer depth 声学层深度
sonic leak detection 声波探漏
sonic leak detection device 音响式检漏仪
sonic leak tester 声波测漏仪
sonic level indicator 声波料位指示器
sonic line 音速线;声线
sonic line of position 声响位置线;声位置线
sonic liquid-level meter 声波液位计
sonic location method 声波测位法
sonic locator 水声定向器;声探测器;声定向器;声定位仪
sonic log(ging) 声波测井;声(法)测井
sonic logging record 声波测井记录
sonic log plot 声波测井成果图
sonic marine animal 发声海洋动物
sonic method 声波法
sonic mine 声学水雷
sonic modulus 声波模数
sonic navigation 声响导航(法);声导航;声波导航
sonic noise analyzer 噪声分析器
sonic nozzle 音速喷管
sonic-nozzle carburetor 声速喷嘴化油器
sonic nucleation 声波核晶过程
sonic pen 声笔
sonic pile driver 音频振动沉桩机;声波打桩机
sonic pollution 声污染
sonic porosity 声波孔隙度

sonic precipitation 声波集尘;声波除尘
sonic pressure 音速压力
sonic probe 声探头
sonic propagation 声的传播;声波传播
sonic propagation constant 声传播常数
sonic prospecting 声勘探;声波探查(探查海底地层);声波探测
sonic pulse velocity 声脉冲速度
sonic pump 振动泵
sonic ranging 声响测距
sonic record 音响记录
sonic reflection 声的反射;声波反射
sonic rupture velocity 声学破裂速度
sonics 声能学
sonic scattering layer 音波(深海)散射层;声散射层
sonic screening machine 声波筛分机
sonic search 声检索
sonic shock 声激波
sonic sifter 声频筛
sonic sludge level indicator 声速泥面指示器
sonic soldering 声焊
sonic sounder 测高计;声速测深器
sonic sounding 回声测深;声波探测;声波测深(法)
sonic sounding-ger 回声测深仪
sonic source 声源
sonic spark chamber 声火花室
sonic spark locator 声火花定位器
sonic speed 音速;声速(度);声波速度
sonic storage 声存储器
sonic strain ga(u)ge 声发射应变计;声波应变计
sonic survey 声探测;声波测量
sonic technique 声波法
sonic test 音响试验
sonic testing 声波检验
sonic thermometer 声学温度计;声波温度计
sonic threshold 声障
sonic tool 声波测井仪
sonic transducer 声传感器
sonic transmitter 声波发射器
sonic velocity 音速;声速(度);声波速度
sonic vibration 声波振动
sonic vibrator 声振器
sonic washer 超声波振荡水洗机
sonic washing machine 超声波洗涤机
sonic wave 声波
sonic wave ga(u)ge 声波测定表
sonic wave interval transit time 声波时差
sonic wave interval transit time of fluid in pore 孔隙流体声波时差
sonic wave interval transit time of matrix 岩石骨架声波时差
sonic wave interval transit time of shale 泥质声波时差
sonic well logging 声波法测井
soniga(u)ge 超声波探测仪;声波金属厚度测定器;超声波测厚仪;超声波测厚计
sonims 固体非金属夹杂物;非金属夹杂物
soniscope (探测混凝土强度和裂缝深度用的)音响仪;脉冲式超声波探伤仪;脉冲式超声波探伤仪;超声波探伤仪;声测仪;声仪;声波探伤仪;声波探测仪
sonizon 超声波测厚仪
Sonnar lens 松纳镜头
Sonne 桑尼
sonne beacon 桑尼无线电信标;桑尼航系
sonobuoy 音响浮标;无线电声的浮标;无线电导航浮标;水声浮标;声呐浮标;防潜浮标
sonobuoy array 浮标台阵
sonobuoy barrier 声呐浮标屏障
sonobuoy channel 声呐浮标信号接收波道
sonobuoy equipment 声呐浮标设备
sonobuoy indicator equipment 声呐浮标指示设备;声浮标指示设备
sonobuoy parachute 声呐浮标降落伞
sonobuoy pattern 声呐浮标群
sonobuoy receiver set 声呐浮标接收机
sonobuoy reference system 声呐浮标参考系统
sonobuoy refraction system 声呐浮标折射系统
sonobuoys survey 声呐浮标测量
sonochemical degradation 声化学降解
sonochemical destruction 声化学破坏
sonochemical oxidation-sequencing batch reactor activated sludge process 声化学氧化序批间歇式反应器活性污泥法
sonochemical reactor 声化学反应器
sonochemistry 声化学;超声波化学
sono controlled type buoy system 声控释放式浮标系统
sonodivers 潜水噪声记录仪
sonodome 声呐导流罩
sonoelectrochemistry 声电化学
sono-encephalograph 超声波测脑仪
son of node 节点之子
sonogram 语图;正弦图;声图;声波图
sonograph 声探测仪;声谱仪;声谱图;声波仪
sonoholography 声波全息照相术;声波全息学;声全息术
sonolator 声谱显示仪
sonolite 斜硅锰石
Sonolog 回声测井
sonoluminescence 声波发光
sonolysis 超声分解
sonolytic degradation 超声降解
Sonoma orogeny 索诺马造山运动
sonometer 弦音计;振动式频率计;振动频率计;声强计
sonometry 声谱测量学
sonomicroscope 声显微镜
sonophotocatalysis 声光催化
sonophotocatalytic decomposition 声光催化分解
sonophotocatalytic degradation 声光催化降解
sonophotocatalytic destruction 声光催化破坏
sonophotocatalytic reactor 声光催化反应器
sonophotocatalytic technology 声光催化技术
sonophotochemical degradation 声光化学降解
sonophotochemical destruction 声光化学破坏
sonophotochemical reactor 声光化学反应器
sonophotography 声光摄影术
sonoprobe 声探测;声呐测深器;声测深器
sonoprobe line 声探测剖面线
sonoprobe profile 声探测剖面
sonora 索诺拉雷暴
sono-radio buoy 无线电导航浮标;防潜浮标;无线电声的浮标
sonoradiography 声射线照相术;声波辐射照相术;超声波放射照相术
sonoraite 水羟碲铁石
sonority 响度
sonoscope (探测混凝土强度和裂缝深度用的)音响仪
Sonotube 索诺管
soot 烟熏;烟粒;烟灰;烟黑色;煤烟;炭墨
soot and dust 烟尘
soot arrester 滤烟器;集灰器
soot barrier 煤烟隔板
soot blower 吹灰机;烟灰清除机;吹灰装置;吹灰器;除烟垢器;冲灰器
soot blower system 吹灰系统
soot blowing 烟灰吹除;吹灰
soot blowing equipment 烟垢吹净装置
soot cancer 煤烟癌
soot carbon 炭黑
soot catcher 滤烟器;集灰器
soot collector 滤烟器;集灰器
soot concentration 烟灰浓度
soot deposit(ion) 烟灰沉积
soot door 清灰门;出灰门;出灰口
soot door of chimney 烟囱灰门;烟囱出灰门
soot eliminating device 除尘设施
soot emission 烟尘排放
soot fall 烟灰沉降
sootflake 炭黑片
soot formation 煤烟形成
sooting 熏黑
soot laden 含烟灰的
soot lance 吹灰枪;吹灰器
soot luminosity 焰煜发光度
soot lung 煤肺症
soot particle 灰粒
soot pit 出灰坑
soot pocket 集灰坑;烟囱集灰槽;收尘袋
soot stain 煤烟污染
sooty 被烟熏黑的
sooty black 烟黑色
sooty coal 劣质软煤;泥煤;土煤
sooty mold 烟霉病
sooty mold of tea 茶叶煤污病
sooty mo(u)ld 煤污病
sopcheite 碲钯银矿
sophisticated 精致的;牵涉面较广的;成熟的
sophisticated category 精细分类
sophisticated compound industry 尖端复合工业
sophisticated element 复杂(单)元
sophisticated equipment 现代化设备;先进设备
sophisticated heat exchanger 高级热交换器
sophisticated mathematical model 完善的数学模型
sophisticated science and high technology 高科技
sophisticated smoke and soot collection device 高性能集烟灰装置
sophisticated technique 尖端技术
sophisticated vocabulary 精细词汇表;复杂指令集
sophistication 混杂信号;混杂;掺假;完善化;精致化;尖端;采样先进技术
Sophora japonica 槐树
soponification number 皂化系数
sopping wet 湿透(衣物等)
sopraporta 门头装饰;门头花饰
Sopwith staff 游标测尺;自读式水平标尺;塔尺【测】
sorbate 吸着物;吸附物;山梨酸酯
sorbate layer 吸附层
sorbed agent 吸附剂
sorbed film 吸附膜
sorbed gas 吸附气体
sorbed mass 吸附物质
sorbed material 吸附材料
sorbent 吸着剂
sorbent feed 吸着剂原料
sorbent surface skimmer 吸着剂面撇油装置;吸着剂面浮油回收装置
sorbic acid 山梨酸
sorbic acid wastewater 山梨酸废水
sorbicester 山梨酸酯
sorbierite 山梨醇
sorbing 吸着
sorbing agent 吸附剂
sorbing felt 油毡
sorbing layer 吸附层
sorbing material 吸附材料
sorbitan 脱水山梨(糖)醇;山梨聚糖
sorbitan fatty acid ester 脱水山梨(糖)醇脂肪酸酯
sorbite 糖斑铁
sorbite rail 索氏体钢轨
sorbitol 山梨(糖)醇
sorbitolalkyd 山梨(糖)醇醇酸树脂
sorbitol ester 山梨(糖)醇酯
sorbitol stearate 硬脂酸山梨醇酯
sorbitol tallate 妥尔油酸山梨醇酯
sorbol 山梨(糖)醇
Sorbus commixta 混种花楸
sorbyite 索硫锑铅矿
sordawalite 玄武玻璃
sordid 污色的
soredium 藻体锥
Sorel alloy 索雷尔锌合金
Sorel's cement 索雷尔水泥;索雷尔胶结料;菱镁土水泥;镁石水泥
Sorel's composition 索雷尔组成
Sorel's process 索雷尔电炉炼铁法
sorensenite 硅铍锡钠石
sorethytan 复杂聚氧化乙烯醚混合物
Soret prism 索雷特棱镜
sorghum husk 高粱壳
sorghum stalk curtain 高粱秆帘
sorghum stalk products 高粱秆制品
Soro process 索罗铸轧法
sorority house 临时寄宿
sorosis 妇女俱乐部
sorotiite 硫石铁陨石
sorption 吸着(作用);吸收作用
sorption agent 吸附剂
sorption analysis 吸附分析
sorption and desorption 吸着和解吸作用
sorption balance 吸收天平
sorption capacity 吸附容量;吸附能力
sorption characteristic 吸附特征
sorption curve 吸收曲线
sorption-deformation sensor 吸附变化传感器

sorption density 吸附密度
sorption/desorption 吸附/解吸作用
sorption detector 吸着检测器
sorption equilibrium 吸着平衡
sorption exhaust 吸着稀薄化作用
sorption-extraction 吸附萃取法
sorption gradient 吸着梯度
sorption hysteresis 吸着滞后现象
sorption isotherm 吸着等温线
sorption kinetics 吸附动力学
sorption phenomenon 吸附现象
sorption point 吸附点
sorption property 吸着性能
sorption pump 吸附泵
sorption pumping 吸附泵送
sorption rate 吸着速率
sorption reaction 吸附反应
sorption site 吸着点
sorption strength 吸着强度；吸附强度
sorptive 吸着质
sorptive attraction 吸附力
sorptive material 吸附性材料
sorptive medium 吸着介质
sorptive power 吸着能力
sorrel 红褐色的；红褐色
sort 类别；分选
sort algorithm 排序算法
sort and merge generator 分选归并生成程序；分类归并生成程序
sort application 分类应用
sort balance 分类平衡法
sort blocking factor 分类（分）块因子
sort capacity 分类量
sort control item 等级控制项目
sort control key 分类控制键
sort description 分类说明
sorted 分选的；分类的
sorted bedding 分选层理
sorted bedding structure 分选层理构造
sorted bubble 分级碎石
sorted circle 分选环
sorted coefficient 分选系数
sorted crack 分选组合泥裂
sorted factor 分选因子；分选因素
sorted file 已分类文件
sorted index 分选指数
sorted list 有序表
sorted net 分选网
sorted polygon soil 分选多边性土
sorted rock 分选岩
sorted rubble fill 分级碎石填充
sorted set 分类组合
sorted step 分选阶
sorted strip(e) 分选条带
sorter 选抹机；精选机；定序器；分选器；分选机；分粒器；分选装置；分选者；分选人员；分类器；分类机；分类工；分类读数器；分拣员；分拣机；分级机
sorter comparator 分类比较器
sorter interpreter 分类译释器
sorter reader 分类阅读机
sorter receptacle 分选器
sorter room 分拣室
sort feature 分类特性
sort file 分类文件
sort file description entry 分类文件描述项
sort function 不同类分页功能
sort generator 分类（生成）程序
sortie 飞行架次
sortie plot 航摄区域图；航摄标图
sorting 分选作用；分选；分票；分类法；分拣
sorting action 分选作用
sort(ing) algorithm 分类算法
sorting and merging 分类合作；分类合并
sorting and search operation 分类和检索操作
sorting belt 分选输送器；分选输送带；分级输送带
sorting by forks 用叉分选
sorting by screens 精筛
sorting card 分类卡
sorting chain 分选输送链；分级输送链
sorting charges 分类费
sorting chart 分类图
sorting chute 分群栏
sorting circuit 分类电路

sorting classification capacity of a station 车站改编能力
sorting code number 分类编码
sorting coefficient 粒度分配系数；筛分系数；分选系数
sorting column 分级柱；分级室
sorting conveyer 挑选传送器；分选传送带；分级输送器；分选传送带；分选输送器
sorting corral 分群栏
sorting data 整理资料
sorting-departure track 编发线
sorting deposition 分选沉积作用
sorting depot 分类仓库
sorting device 拣选装置；分选装置；分类设备；分类装置
sorting equipment 拣选装置；分选装置；分拣装置
sorting factor 分选因子；分选因素
sorting flax 分级梳成麻
sorting floor 分选台；分类场地
sorting function 分类组合职能
sorting gap 选材口
sorting gridiron 编组梯线
sorting grizzly 选粒筛；拣选格筛；分选筛
sorting hopper 分级料斗
sorting index 分选指数；分选率
sorting jack 选材锭
sorting key 分类关键字
sorting key word 分类关键字
sorting line in marshalling yard 编组场分去向组组线
sorting local wagons siding 管内工作车编组线
sorting machine 分拣机；分级机；分类机；分选机；分类器
sorting memory 分类存储器
sorting merge 分类归并
sorting method 分类法
sorting of contour points 等高线点分类
sorting of household refuse 家庭垃圾的分类
sorting of information 情报分类
sorting of sediment 泥沙分选
sorting of wood 木材分级
sorting operation 分类作业；泥砂水力分选程序；分拣作业
sorting out 拣出；选出
sorting out and consolidating companies 清理整顿公司
sorting parameter 分类参数
sorting phase 分类阶段；分类段
sorting pile 分级桩
sorting pocket 分级袋；分级料框
sorting process 分类过程
sorting program(me) 分类程序
sorting program(me) phase 分类程序段
sorting pulse 分类脉冲
sorting rack 分类图架
sorting roller 分选辊
sorting room 分拣室
sorting scale 分类秤
sorting screen 清选筛
sorting shed 分类仓库
sorting siding 编组线【铁】；车辆分线【铁】
sorting table 选材台；整理表；手选台；分选表
sorting test 分类测验
sorting tolerance 选别公差；分选公差
sorting traces by depth 按深度分选地震道
sorting track 编组线【铁】；分类调车线【铁】
sorting unit 拣选装置；分选装置
sorting yard 拣选场；调车场；存料场；编组站【铁】；编组场【铁】
sort key 分类键；分类关键
sort mechanics 分类力学
sort merge 分类组合；分类合作；分类合并
sort-merge file description 分类合并文件说明
sort-merge file description entry 分类合并文件描述体
sort-merge generator 分类合并生成程序；分类归并生成程序
sort-merge package 分类（及）合并程序包
sort-merge program(me) 分类归并程序；分类（及）合并程序
sort module 分类模块；分类功能模块；分类程序模块
sort of lime 石灰类别；石灰品种
sort operation 分类操作
sort out data 整理资料

sort pass 分类扫描；分类传送
sort program(me) 分类程序
sort program(me) module 分类程序模块
sort selection 分类选择
sort tree structure 分类树形结构
sort utility 分类程序包
sort work file 分类工作文件
sosedkoite 苏钽铝钾石
sosenvironment 宇宙环境
sosin block instrument 双工闭塞机
sosoloid 固溶体
soss hinge 隐蔽式铰链
sotch 灰岩坑
sotto portico 廊下；矮门廊；骑楼下人行道
soucekite 铋车轮矿
souconite 锌蒙脱石
sou'easter 东南大风
soufflet 花饰窗格的四叶饰
sough 坑道；排水平洞；盲沟
souledras 日出风
souledre 日出风
sound 致密的；健全的；稳妥的；完好的；探子；音响
sound-absorbent 吸声体
sound-absorbent backing 吸声背衬
sound-absorbent blanket 吸声毡
sound-absorbent board 吸声板
sound-absorbent brick 吸声砖
sound-absorbent building unit 吸声建筑构件
sound-absorbent cassette 吸声镶板
sound-absorbent ceiling(board) 吸声顶棚
sound-absorbent ceiling paint 吸声顶棚涂料
sound-absorbent ceiling sheet 吸声顶棚薄板
sound-absorbent ceiling system 吸声顶棚体系
sound-absorbent chamber 吸声室
sound-absorbent coffer 吸声平顶镶板
sound-absorbent construction(al) material 吸声建筑材料
sound-absorbent construction method 吸声构造方法
sound-absorbent cover(ing) 吸声覆盖层
sound-absorbent facing 吸声贴面层
sound-absorbent felt(ed fabric) 吸声毡织物
sound-absorbent felted fabric ceiling 吸声毡织物顶棚
sound-absorbent fibre board 吸声纤维板
sound-absorbent foil 吸声箔
sound-absorbent formboard 吸音模板
sound-absorbent glass 泡沫玻璃；吸声玻璃
sound-absorbent hung ceiling 吸声悬挂顶棚
sound-absorbent lining 吸声衬垫
sound-absorbent masonry wall 吸声圬工墙；吸声砖石墙
sound-absorbent material 声学材料；消声材料；隔声材料；吸声材料
sound-absorbent pad 吸声衬垫
sound-absorbent paint 吸声涂料
sound-absorbent pan 吸声盘
sound-absorbent plaster 吸声粉刷；吸声灰泥
sound-absorbent plaster aggregate 吸声粉刷骨料；吸声灰泥骨料
sound-absorbent plaster ceiling 吸声灰泥顶棚
sound-absorbent quilt 吸声毯
sound-absorbent sheet 吸声薄板
sound-absorbent sprayed-on plaster 吸声喷涂粉刷；吸声喷涂灰泥
sound-absorbent surface 吸声表面
sound-absorbent surfacing 吸声表面
sound-absorbent suspended ceiling 吸声悬挂顶棚
sound-absorbent system 吸声体系
sound-absorbent tile 吸声砖
sound-absorbent tile ceiling 吸声砖顶棚
sound-absorbent treatment 吸声处理
sound-absorbent unit 吸声单元；吸声构件
sound-absorbent waffle 华夫饼式吸声构造；吸声蜂窝表面
sound-absorbent wall 吸声墙
sound-absorbent wall block 吸声墙砌块
sound-absorbent wall brick 吸声墙砖
sound-absorbent wallpaper 吸声墙纸
sound-absorbent wood fibre board 木纤维吸声板
sound-absorbent wood-wool board 吸声木丝板
sound absorber 减声材料；消声装置；吸声体；吸声剂；吸声体；吸声器
sound absorbing 吸声的

sound-absorbing and static pressure box 消声静压箱
sound-absorbing blanket 吸声毯
sound-absorbing coefficient 吸声系数
sound-absorbing construction 吸声构造
sound-absorbing cover 吸声罩
sound-absorbing lining 吸声衬砌
sound-absorbing material 消声材料;吸声材料;吸声材料;隔声材料
sound-absorbing paint 吸声油漆;吸声涂料;吸声漆;吸声油漆;吸声涂料;隔声涂料;消声漆
sound-absorbing porous material 多孔吸声材料
sound-absorbing power 吸声力
sound-absorbing quilt 消声被
sound-absorbing structure 吸声结构
sound-absorbing surface 吸声面
sound-absorbing system 吸声系统
sound-absorbing treatment 吸声处理
sound absorption 吸声作用;吸声力;声音吸收;声吸收
sound absorption by film 薄膜吸收
sound absorption by thin plate 薄板吸收
sound absorption capacity 吸声量;吸声能力
sound absorption coefficient 吸声系数;声吸收系数;吸声系数
sound absorption efficiency 吸声效率
sound absorption in air 大气声吸收
sound absorption in cavity 空腔吸声
sound absorption material 吸声材料
sound absorption power 吸声能力
sound absorption property 吸声性质
sound absorption quality 吸声性质
sound absorption unit 吸声单位
sound absorption wedge 吸声尖劈
sound absorptive 消声的;吸声的
sound absorptive plaster 吸声粉刷
sound absorptive treatment 吸声处理
sound absorptivity 声吸收率
sound accompaniment 伴音
sound adsorbing blanket 吸声毡
sound adsorbing foam glass 吸声泡沫玻璃
sound adsorption 吸声
sound adsorption block 吸声砌块
sound adsorption characteristic 吸声特性
sound adsorption coefficient 吸声系数
sound adsorption factor 吸声因素
sound alarm system 声学警报系统
sound alarm unit 警铃;声响警报装置
sound altimeter 声波测高计
sound amplification system 扩音系统
sound amplifying system 扩音系统
sound amplitude transmission coefficient 声波振幅传输系数
sound analysis 声响分析;声分析
sound analysis data 声波分析数据
sound analyzer 声谱分析器;声分析仪;声波频率分析器
sound-analyzing equipment 声波分析仪
sound-and-chroma detector 伴音和彩色信号检波器
sound-and-fire proof 隔声防火
sound and flash ranging 声光测距
sound aperture 声孔径
sound archives 音响资料库
sound area 震声区;声区
sound arrangements 声布置
sound arrester 消声装置;隔声装置;防声装置
sound articulation 声音清晰度
sound attenuating door 消声门
sound attenuation 消声;减声;声音衰减;声衰减
sound attenuation blanket 吸声毡
sound attenuation coefficient 声衰减系数
sound attenuation constant 声衰减常数
sound attenuation duct 声响衰减管
sound attenuation factor 声衰减系数
sound attenuation measurement 声衰减测量
sound attenuation room 消声室
sound attenuator 减音器;减声器;声衰减器;风道消声装置;消声器
sound baffle 阻声屏;声音反射器
sound band power level 频带声功率级
sound band pressure level 频带声压级
sound bandwidth 声频带宽
sound barrier 声障;隔声装置;声屏障;声垒

sound beam 声束
sound bearing 声源方位;声音方位;传声
sound bearing medium 传声媒质;声媒质
sound bearing station 噪声定向站
sound board 共振板;共鸣板;响板
sound boarded floor 隔声楼板
sound boarding 隔声板
sound box 共鸣器;吸声箱
sound branch amplifier 声分路放大器
sound breakdown 声音中断
sound bridge 声桥;传音声桥
sound broadcasting studio 广播电台;广播室
sound buoy 音响浮标;声响浮标
sound cable 通信电缆;传声电缆
sound camera 录音室;声学试验室
sound card 声卡
sound cargo 完好货物
sound carrier 伴音载波
sound carrier frequency 声音载频
sound carrier sweep 伴音载波偏移
sound casting 致密铸件;坚实铸件
sound cell 声学元件
sound cell microphone 声室式传声器
sound cement 优质水泥;安定性良好的水泥
sound chamber 声室
sound channel 声音通路;声道;伴音信道;伴音通道
sound channel in the ocean 大洋声道
sound collateral tube 伴音管
sound colo(u)ration 声染色
sound communication 声响通信
sound component 声元件
sound conduction 声传导
sound conduction function 传音功能
sound conductivity 传声性能
sound conductor 声导体
sound console 调音台
sound control 声控;噪声控制
sound-control booth 声音控制室;音响控制室
sound-control cassette 隔声箱
sound-control console 音响控制台
sound-control glass 隔声玻璃
sound control material 声控材料
sound-control panel 隔声镶板
sound-control room 声控室
sound-control technique 声控技术
sound corrector 声音校正器
sound cutter 伴音剪辑员
sound cutting 声剪辑
sound damper 静噪器;消声器;吸声器;声阻尼器
sound damping 消声的;吸声的;隔声的
sound-damping check valve 消声防逆阀
sound-damping device 消声装置
sound-damping equipment 消声装置
sound-damping material 消声材料;隔声材料
sound-damping unit 消声装置
sound data 声学数据
sound-dead 隔声的
sound-deaden 吸声的;减声的;隔声的
sound-deadener 减声器;隔声器
sound-deadening 消声作用;消声(法);隔声法;隔声(法)
sound-deadening board 消声板;吸声板
sound-deadening capacity 消声量
sound-deadening coat(ing) 吸声涂层
sound-deadening composition 吸声组合物
sound-deadening compound 吸声组合物
sound-deadening material 吸声材料;吸声材料
sound-deadening paint 隔声涂料;消声涂料
sound-deadening plywood 消声胶合板
sound-deadening quilt 消声被
sound decay 声音消衰;声音衰减;声消衰
sound decay curve 声音衰变曲线
sound defect 健全缺陷
sound deflection 声波折射
sound deflection coefficient 声折射系数
sound demodulator 伴音信号解调器
sound density 声密度
sound-detecting and ranging 声响探测仪
sound detection 声检测
sound detector 测声器;检声器;声探测器;声检测器;测音器;伴音信号检波器;伴音检声器
sound device 声学仪器;声学仪表
sound diffraction 声衍射;声绕射
sound diffusion 声扩散

sound diffusivity 声扩散率
sound direction 发声方向
sound discriminator 声频鉴频器
sound dispersion 声频散
sound distortion 声失真
sound distribution 声分布
sound distribution amplifier 声分配放大器
sound distribution measurement 声分布测量
sound disturbance effect 声干扰效应
sound door 消声门
sound driver 声激励器
sound duct 声道;传声筒
sound ecological balance 健全的生态平衡
sound effect 音响效果
sound effect filter 声响效果滤波器
sound elimination 声音消除;声消除
sound eliminator 消声器
sound emission 声发射
sound emission for non-destructive testing 声发射无损探伤法
sound emission indicator 声发射仪
sound emission method 声发射法
sound emission microphone 声发射接收器
sound emitter unit 发声单元
sound energy 声能(量)
sound energy density 声能密度(单位为焦耳/立方米)
sound energy flow density 声能流密度
sound energy flux 声能通量
sound energy flux density 平均声能流密度;声能通量密度
sound energy reflection coefficient 声能反射系数
sound engineering 声学工程
sound environment design 声环境设计
sound equipment 声测设备
sounder 音响器;探空装置;声码器;声波测深仪;声波测深器;测深员;发音器;发声器
sounder operator 测深仪操作员
sounder resonator 音响器共鸣箱
sound event 声源
sound excitation 声激励
sound exclusion 噪声排除
sound exposure level 声辐射级;暴露声级
sound-face cutting 健全材面锯切法
sound fading 声衰落
sound fading device 声衰减装置
sound failure 声致故障
sound field 声场
sound field calibration 声场校准
sound field distribution 声场分布
sound field fluctuation 声场起伏
sound field plotter 声场标绘器
sound film studio 有声电影制片厂
sound filter 滤声器
sound filtration 滤声
sound finance 健全财务
sound financial footing 自负盈亏;可靠的经费基础
sound finding and ranging 声波测位和测距
sound fixing and ranging 声发(水中测声器);声波定位与测距(系统)
sound flanking path 翼侧声径
sound flux 声通量
sound foci 噪声集中区;声响焦点
sound focus(ing) 声聚焦
sound focus test film 声聚焦试验片
sound foil 声箔
sound follower 声跟随器
sound foundation 坚实地基
sound frequency 声响频率;声频
sound frequency analyzer 声频分析器
sound frequency range 声频范围
sound gate 放声口;伴音拾音器
sound gear 水声仪器
sound generation 发声
sound generator 声源;声发生器
sound gobo 消声隔板
sound goods 完好货物
sound grating 声栅
sound-groove 声纹
sound guidance 音响制导
sound-hard boundary 声学刚性边界
sound hole 传声窗
sound homing 声辐射导航;声波寻的;声波导航

sound housing 良好住房
sound IF amplifier tube 声频信号中频放大器
sound image 声像
sound image method (超声检测的)声像法
sound impermeability 隔声
sound imperviousness 隔声
sound impression 声印象
sound impulse 声脉冲;声波脉冲
sound-induced breakdown 声致断裂
sound-induced vibration 声致振动
sound information 声测情报
sounding 水深测量;测量水深;探通术;探空;探测;声探测;声测深度;触探;测(水)深;发声;音响
sounding alignment 测深定线
sounding and probing 测探与触探
sounding apparatus 土壤贯入度仪;土壤透度仪;触探装置;触探仪;触探器;测深装置;测深设备;测深器
sounding balloon 高空气象探测气球;探空气球;探换气球
sounding board 共振板;共鸣板;水恒污水水深登记黑板;测深图板;测深板;音响壁板;音响板
sounding boat 测深船
sounding bob 探测锤;测深(重)锤
sounding book 测深记录簿
sounding borer 探土钻;钻探机;触探钻;测深钻
sounding bottle 海水取样金属瓶;采泥器
sounding buoy 号笛浮标
sounding by echo 回声测深(仪)
sounding by lead 铅锤测深;锤测
sounding by pole 杆测深
sounding chart 测深图
sounding cone 测深锥
sounding contour 等深线
sounding course 测深线
sounding craft 测深艇;测深船
sounding datum 水深基准;海图深度基准面;水深基(准)面;测深基(准)面
sounding datum (level) 深度基准(面)
sounding depth 实测水深
sounding device 声测探器;触探装置;测深设备;发声装置
sounding diagram 油水柜存量图
sounding doubtful 水深不确实
sounding electrode 探测电极
sounding equipment 探测设备;测深设备
sounding figure 测深数据
sounding fix 测深定位
sounding gear 测深装置
sounding history over-lay 测深资料透明片
sounding hole 测水孔
sounding impulse 探测脉冲
sounding in damages 查实损害
sounding installation 测深装置
sounding instrument for cloud and fog physics 云雾物理探测器
sounding interval 测深点时距
sounding launch 测深艇
sounding lead 水砣;测深铅锤;测深锤;测铅
sounding lead line 测绳
sounding level 测时水位
sounding line 测深线;测深索;测深绳;测铅绳
sounding line correction 测深绳偏斜校正
sounding line layout 测深线定位;测深断面定位
sounding log 声速测井
sounding machine 海水测深器;触探机;测水深仪;测水深机;测深仪;测深机
sounding machine check 机械测深用触钩(判断测深锤到海底用)
sounding machine depth scale 机械测深用的水深比例尺
sounding machine feeler 机械测深用触钩(判断测深锤到海底用)
sounding machine lead 机械测深用水砣
sounding mark 深度注记
sounding measurement 测声工作
sounding meteorograph 探空气象仪
sounding method 测深法
sounding needle 触探针
sounding of atmosphere 大气探测
sounding of cement 水泥安定性
sounding of cross section 断面测深
sounding pipe 测深管
sounding pipe for cargo tank 货油舱测量管

sounding platform 水砣测深台;手砣站台
sounding point 测深点
sounding point spacing 测深点间距
sounding pole 探针;探棒;测深竿;测深杆;测深尺
sounding position 测深船位
sounding positioning 测深定位
sounding profile 探测断面;测深断面
sounding protractor 测深量角器
sounding record 测深记录簿;测深记录
sound ingredients 体积稳定成分;体积稳定组分(混凝土)
sounding relay 音响器
sounding rocket 探空火箭;探测火箭
sounding rod 量水尺;测深竿;测深杆
sounding scale 机械测深用的水深比例尺;测深(量)尺
sounding sextant 水文地理测量六分仪;测深六分仪
soundings generalization 一般化水深
sounding stick 测深竿;测深杆;测深尺
sounding system with multibeam 多波束测深系统
sounding table 油水柜存量表
sounding tachygraph 探测速记图;测试转速表
sounding test 水深测试;触探试验
sounding tester 响板;共鸣板;讲坛反声板
sounding thermometer 深海温度计
sounding tube 测深管;测深玻璃管
sounding tube holder 测深玻璃管铜套
sounding unit 探空装置
sounding vehicle (气球带到高空发射的)高空探测火箭
sounding velocity 测深声速
sounding vertical 测深垂线;测深垂丝
sounding vessel 探测船
sounding weight 测深重锤;测深锤
sounding well 探深管(探测压浆混凝土中灰浆高度用)
sounding winch 测深绞车
sounding wire 机械测深用钢丝绳;测深索;测深绳
sound-insulated bulkhead 隔声舱壁
sound-insulated fire door 隔声防火门
sound-insulated material 隔声材料
sound-insulating 声绝缘;隔声(法)
sound-insulating anteroom 隔声前室
sound-insulating concrete 隔声混凝土
sound-insulating construction 隔声构造
sound-insulating door 隔声门
sound-insulating glass 隔声玻璃
sound-insulating grading 隔声等级
sound-insulating layer 声绝缘层;隔声层
sound-insulating paint 隔声涂料
sound-insulating panel 隔声板
sound-insulating room 隔音室
sound-insulating structure 隔声结构
sound-insulating wall 隔声墙
sound-insulating window 隔声窗
sound insulation 消声;声绝音;声绝缘
sound insulation against structure-borne sounds 对结构载声的隔声
sound insulation assessment 隔声评价
sound insulation board 隔声板
sound insulation capability 声绝缘能力;隔声能力
sound insulation character 隔声特性
sound insulation class 隔声等级
sound insulation coefficient 隔声系数
sound insulation equipment 隔声设施
sound insulation floor construction 隔声楼板构造
sound insulation for building 建筑隔声
sound insulation frequency characteristic curve 隔声频率特性曲线
sound insulation glass 隔声玻璃
sound insulation grading 隔声等级
sound insulation index 隔声指数
sound insulation material 隔声材料
sound insulation measurement 隔声测量
sound insulation of floors 楼板隔声
sound insulation of walls 墙壁隔声
sound insulation reference contour 空气声隔声参考曲线
sound insulation room 隔声室
sound insulation standard 隔声标准
sound insulation standards for dwellings 居住建筑隔声标准
sound insulation test 隔声试验
sound insulation value 隔声性能;隔声值

sound insulator 隔声体;隔声材料
sound-in-syncs 同步传声
sound intensifier 增音器;声音放大器
sound intensity 声强(度)
sound intensity decay 声强衰减
sound intensity fluctuation 声强起伏
sound intensity level 声强级
sound intensity measurement 声强测量
sound intensity measuring device 声强测量仪
sound intensity meter 声强仪;声强计
sound intensity reflection coefficient 声强反射系数
sound intensity transmission coefficient 声强透射系数
sound intercarrier 伴音载波差拍
sound interference 声干涉;声干扰;伴音干扰
sound intermediate 声介质;传声介质
sound intermediate frequency 中音频
sound intermedium 声媒质
sound interval 频程;声音音程
sound in the audible range 可听范围内的声音
sound investment 可靠投资
sound irradiator 声照射器
sound isolating requirement 隔声要求
sound isolation 隔声;隔声度;声隔离
sound isolation requirement 隔声要求
sound knot 不活动的木节;硬节;健全节;坚固木节;坚固节(木料);生节
sound lag 声延迟
sound leakage 声漏
sound lens 声透镜
soundless 无声的
sound level 噪声分级;噪声程度;音级;噪声水平;声音电平;声级
sound level calibrator 声级标定仪
sound level control 声级控制
sound level difference 噪声分级;声隔差
sound level measuring device 声级测量设备
sound level meter 声级计;音级仪;音级计
sound level meter attenuator 声级计衰减器
sound level meter data 声级计数据
sound level recorder 噪声分级仪;声级测定仪;声级记录仪
sound level signal 声级信号
sound limit 声极限
sound line 声频通信线中路
sound locater 声波探测仪
sound location 声定位;声波定位;声波测距(法)
sound locator 声呐;声波测定仪;测声器;声源勘定器;声定位仪;声波探测仪;声波定位仪;声波定位器
sound lock 锁音廊;隔声空间;隔声前室;声闸;声锁
sound log(ging) 声波测井;声测井
sound looping 声循环
sound lumber 好木料
sound maintenance 声测量技术
sound market value 完好货物的价格
sound masking 声掩蔽
sound material 良好材料
sound measurement 测声(工作);声波测距(法)
sound-measuring device 测声器
sound message 声音信息
sound metal 无缺陷金属
sound meter 噪声表;测声仪;测声器;测声计
sound mine 声学水雷
sound mirror 声反射板;声镜;回声面
sound mixer 声音信号混合器
sound mixing 声混合
sound mixing console 调音台
sound model(l)ing 声音建模
sound modulation 伴音调制
sound monitor 监听设备;监听器;声监控器
sound moviola 声画编辑机
sound-muffling 消声的
sound narrow 海峡
sound navigation and ranging 水下探测仪;水声测位仪;声呐;声波定位仪;声波定位和测距;测潜艇仪
soundness 致密性;固定性;坚固性;安定性
soundness and fineness test 安定性和质量试验(水泥)
soundness of cement 水泥安定性
soundness test 固定性试验;耐固性试验;气密性试验;气密性实验;安定性试验(水泥)
soundness test by immersion in cold water 冷

水浸渍法(水泥)安定性试验
soundness test pat 安定性试验的试饼(水泥)
sound-off switch 声断路开关
sound on vision 伴音干扰图像
sound oscillation 声音振动;声振荡
sound output 声效率;声输出
sound overshooting 声过调量
sound part 合格部分
sound particle velocity 声波中质点速度
sound path 声迹;声程
sound pattern 声模式
sound perception 声音感觉;听觉
sound pick-up equipment 拾音设备
sound picture 声画
sound pile driver 声波打桩机
sound pile test 桩质声学检验法;混凝土桩声学检验法
sound pipe 声管;探测管
sound pollution 噪声污染;声响污染;声污染
sound power 声力;声功率
sound power absorption coefficient 声功率吸收系数
sound power calculation 声功率计算
sound power density 声功率密度
sound powered communication system 声能通信系统
sound-powered telephone set 声力电话机
sound-powered telephone system 声力电话系统
sound power level 声能级;声功率级
sound power of a source 声源功率
sound power reflection coefficient 声功率反射系数
sound power telephone 无电池电话
sound power transmission coefficient 声功率传输系数
sound power value 声功率值
sound pressure 音压;声压
sound pressure calibration 声压校准
sound pressure compensation method 声压补偿法
sound pressure correction 声压校正
sound pressure equalization 声压均衡
sound pressure field 声压场
sound pressure hydrophone 声压式水听器
sound pressure level 声音强度;声压级
sound-pressure-level measurement 声压级测量
sound pressure meter 声压计
sound pressure reflection coefficient 声压反射系数
sound pressure response 声压响应
sound pressure sensor 声压传感器
sound-print 声复版
sound probe 探声头;探声器;探声管;探声棒;声探头;声频探头;声波探测器
sound probe sonoprobe 探声器
sound production 发声
sound projection 声投射;声的投射
sound-proof 防噪声的;防(响)的;不透声的;隔声的
sound-proof box 隔声箱
sound-proof bulkhead 隔声舱壁
sound-proof cabin 隔声室
sound-proof class 隔声等级
sound-proof coating 防声涂料
sound-proof coefficient 隔声系数
sound-proof concrete 隔声混凝土
sound-proof construction 隔声建筑;隔声构造
sound-proof course 隔声层
sound-proof cover 隔声罩
sound-proof door 隔声门
sound-proof drilling rig 隔声钻探设备
sound-proof enclosure 隔声围墙;噪声封闭器(桩工)
sound-proof fiber board 隔声纤维板
sound-proof film camera 隔声摄影机
sound-proof floor 隔声地面;隔声地板
sound-proofing 声处理;隔声器材;防噪声;隔声法;隔声;消音
sound-proofing lagging 隔声外套
sound-proofing material 隔声材料
sound-proofing of houses 房间的隔声
sound-proofing property 隔声性能
sound-proof machine 隔声器
sound-proof motor 声处理的马达
sound-proofness 隔声

sound-proof planting 隔声栽植
sound-proof room 隔声室
sound-proof structure 隔声结构
sound-proof tile 隔声砖
sound-proof varnish 隔声清漆
sound-proof wall 隔声墙
sound-proof wheel 防噪声车轮
sound-proof window 隔声窗
sound-proof work 隔声工程
sound propagation 声传播;声波传播
sound propagation constant 声传播常数
sound pulse 音响脉冲;声脉冲;声波脉冲
sound-pulse decay 声脉冲衰变
sound pulse photography 声脉冲照相
sound radar 声雷达;声达(声定位和测距仪);声波定位器;声波测距仪
sound radiation 声音辐射;声辐射;声波辐射
sound radiation impedance 声辐射阻抗
sound radiation pressure 声辐射压
sound radiator 声辐射器
sound radiometer 声辐射计
sound ranger 音响测距仪;声波测距仪
sound range recorder 声频记录仪
sound ranging 音响测距;声测距;声波测距(法)
sound ranging adjustment 声测距(离)调整
sound ranging altimeter 回声测高计;音响测高计;声波测高计
sound ranging base 声测距基线;声波测距基线
sound-ranging company 声测连
sound ranging control 声呐测距控制;声测距控制;声波测距控制
sound ranging instrument 声测仪器
sound ranging location 声测距定位;声波测距定位
sound ranging observation post 声测所
sound ranging platoon 声测排
sound ranging plotting board 声测标图板
sound ranging section 声测分排
sound ranging set 声测距设备
sound ranging station 声测距站;声波测距台
sound ranging technique 声测距技术
sound-rated door 消声门
sound ray 声线;声辐射线
sound ray diagram 声线图
sound ray trace 声线轨迹
sound ray tracking plotter 声线轨迹仪
sound ray tracking set 声线轨迹仪
sound receiver 声接收器;声接收机
sound receiving chamber 接收室
sound receiving room 受声室(隔声测量时受声的房间)
sound reception 声接收
sound recognition 声音的识别
sound recording 录音;录声
sound-reduced 减弱的声音
sound reduction 消声;降声;降低音响
sound-reduction factor 减噪因数;声透射损失因数;音响损失系数;隔声量;声响降低因素
sound-reduction index 声减弱指数;隔声量;声衰减指数
sound reflecting 音频反射;声响反射
sound reflecting board 反射声板
sound reflection 声反射
sound reflection board 声反射板
sound reflection coefficiency 声反射率
sound reflection coefficient 声反射系数
sound reflection factor 声反射因数;声音反射系数
sound reflectivity 声反射率
sound reflector 声反射器
sound refraction 声折射
sound regulation room 调音室
sound-reinforcement system 声音增强系统;扩声系统
sound rejection 声抑制;伴音载波抑制
sound remote sensing 声(学)遥感
sound rendering 声音绘制
sound reproducer 扬声器;重发声器
sound reproducing system 放声系统
sound reproduction 声(的)重发;放声
sound resistance 声阻
sound-resistive[resisting]glass 隔声玻璃
sound resonance 共鸣(声);反响声
sound resonant cavity 伴音谐振腔
sound retardance 减声
sound retardancy 减声性

sound retrieval system 声音恢复系统
sound reverberation 回声;声干扰;混响
sound reverberator 混响器
sound rock 坚硬岩(石);坚固岩层;完整岩石
sound roundabout radiant 声绕射
sound rubber 完善胶片
sound scan(ning)device 声扫描设备
sound-scan(ning)drum 声扫描滚筒
sound scatterer 声散射体
sound-scattering object 声散射物
sound screen 隔声屏;声障板;声幕
sound-sensitive detector 声敏探测器
sound-sensitive vehicle detector 声感检车;音响车辆探测器;声感检车器
sound shadow 声影;声影区
sound shadow region 声影区
sound shadow zone 声影区
sound-shield enclosure 试听室
sound ship 好船
sound-sight broadcast 电视广播
sound signal 音响航标;音响信号;雾航音响信号;声音信号;声频信号;声号;伴音信号
sound signal digital 声信号数字
sound signal(l)ing 音响信号装置
sound signal(l)ing appliance 声响信号器材
sound signal(l)ing system 发声信号系统
sound signal shell 音响榴弹
sound signals in restricted visibility 能见度不良时的声号
sound simulation 音响模拟
sound slit 窄缝光门;声缝隙
sound-soft boundary 声学软边界
sound-soft termination 透声界面
sound source 声源
sound source intensity 声源强度
sound source room 声源室
sound spark 声电火花
sound spectrograph 音频频谱仪
sound spectrometer 声频频谱计
sound spectrum 声谱
sound spectrum analysis 声谱分析
sound spectrum analyzer 声谱分析仪;声谱分析器
sound spectrum measurement 声谱测量
sound speed 声速(度)
sound speed profile 声速垂直分布
sound splice 声接头
sound splitter 消声片;消声器
sound splitting 声音散播
sound spotting 声波探索;声波探测
sound sprocket 声链轮
sound stability 声稳定性
sound stage 声基宽;摄影棚
sound stage width 基宽
sound stream absorber 声流吸声器
sound streaming 声流
sound studio 录音室
sound suppression 声抑制
sound synthesizer 声音合成器
sound system 音响系统;声学系统;放声系统;伴音系统
sound take-off 检出声信号
sound talkback 演播室对讲
sound test 锤击检查
sound testing instrument 声学测试仪器
sound tight knot 不传声的节点
soundtightness 不透声性
sound timber 好木料
sound title 合法有些的物权
sound track 声迹;声道
sound track advance 声迹超前
sound track demodulator 声迹解调器
sound track engraving apparatus 声迹刻纹机
sound tracking 声波跟踪
sound track speed 声迹速率
sound transceiver 音响收发装置
sound transducer 声换能器;电声换能器
sound transfer 声迹复版
sound transmission 声传播;声音传播;声透射;传声
sound transmission class 传声等级
sound transmission class rating 传声等级定额
sound transmission coefficient 声(音)透射系数;声音传播系数;透声系数;传声系数
sound transmission loss 隔声量;声透射损失;传声损失

sound transmission qualities 声传输质量
sound transmission reduction 声传播降低
sound transmission system 声传输系统
sound transmitter 发话机；伴音信号发射器；伴音发射机
sound transparent 透声
sound transparent rubber 传声橡胶
sound trap 聚声器；声锁；声频陷波器；声阱；伴音陷波电路；声频信号陷波器
sound trapping 吸声；陷声
sound travel 声通路；声路
sound treatment 声学处理
sound tree 健全树
sound type 音型
sound value 折余价值；好价值；稳妥价值；重置价值
sound variation 声变化
sound velocimeter 声速仪；声速测量仪
sound velocity 声速(度)
sound velocity calibration 声速校准
sound velocity distribution 声速分布
sound velocity error 声速误差
sound velocity gradient 声速梯度
sound velocity meter 声速仪
sound velocity method 声速法
sound vessel 好船
sound vibration 声振动；声振
sound-vision separator 伴音图像分离设备
sound volume 音量；声量
sound volume control 声量控制
sound volume range 声量范围
sound volume velocity 声媒质体积速度
sound warning buoy 鸣响警报浮标；音响警报浮标；声响报警浮标
sound water use 健全的水资源利用
sound wave 声波
sound wave method 声波法
sound wave photography 声波照相；声波摄影学
sound wave radiation 声波辐射
sound wood 好材；良木；健全木；坚硬木(材)；坚实木材；好木料
soup 硝化甘油；照相显影液；很湿的灰浆混合液
soupcrete 流态混凝土
souped-up 提高效率
soupery 食堂；餐厅
soup kitchen 救济厨房；流动厨房(军事)
soup plate 汤盘
soup tureen 有盖汤盘
Souques' phenomenon 伸指现象
sour 酸的
sour bark 酸性树皮
sour bath 酸浴
sourbush cherry 酸灌丛樱
source 信号源；源头；源地；发送端
source accumulation coefficient 生聚系数
source address 源地址
source address field 源地址字段
source address instruction 源地址指令
source admittance 源导纳
source air pollution 空气污染源
source and application of funds statement 资金来源及运用表
source and dispositions of funds 资金来源与运用
source and sink 源头与壑
source and use of funds 资金来源与运用
source area 源地(区)；来源区；起源地；发源地；补给地
source assessment sampling system 污染源评价抽样系统
source bed 水源层；生气岩层；生油岩层；母岩；原生岩；源岩层；矿层；生油层
source bias effect 源偏压效应
source block 源部件
source body 源体
source bomb 震源弹
source box 电源箱
source branch 源支线
source bubble 源磁泡
source bulk resistance 源电阻
source calibration 源标准化
source calibration laboratory 源刻度实验室
source capsule 源管
source case 出处角色
source category 源种类；源类别
source characteristic 震源特性
source circuit 源电路

source code 源代码；源编码
source code instruction 源代码指令
source computer 源计算机；翻译计算机
source condition 源的工作条件
source conductivity 源电导
source contact 源接点
source container 放射源储存器；源容器；放射源箱
source contribution ratio 污染源排污物组分比率；污染源贡献率
source control of wastewater 废水源控制
source core 源芯(放射性)
source coupler loss 光源耦合器损耗
source coupling efficiency 光源耦合效率
source current 源电流
source data 信息源；源数据；初始数据
source-data automation 源数据自动化
source data entry 源数据项
source data item 源数据项
source data structure 源数据结构
source deck 源卡片叠；原始程序卡片组
source density 源密度
source depth 震源深度
source designation instruction 源赋值指令
source destination code 无操作码
source detect 震源检测
source detected interference 源检测干扰
source detector separation 源距(放射性勘探)
source dimension 震源大小
source distance 震源距(离)
source distribution 源分布
source distribution function 源分布函数
source documents 原始资料；原始文件；原始凭证；原始单据
source-drain characteristics 源漏特性
source-drain voltage 源漏(间)电压
source dynamics 震源动力学
source earthquake 源地震
source editor 源编辑程序
source edit utility option 源编辑应用程序
source effect 震源效应
source efficiency 源效率；光源效率
source electrode 源极；源电极
source emittance 源发射度
source encoder 信源编码器
source failure lamp 电源故障指示灯
source-fiber coupling 光源光纤耦合
source file 源文件；原始文件
source file editor 源文件编辑程序
source flow 源流
source follower circuit 源输出电路
source follower probe 源输出器探头
source forcing 源激励
source form 激发方式
source for public water supply 公用水源
source-free flow 无源流动
source function 源点函数；震源函数
source gas 源气体
source grid electrode 源栅电极
source-holder 源座
source host 源主机
source housing 源室
source image 源像；源程序信息
source impedance 源阻抗
source index 资料索引
source information 源信息；原始资料
source intensity 源强(度)；点源强度
source intercalibration 源的相互校准
source interlock 源联锁装置
source item 源项
source jack-socket 电源插头座
source language 源语言
source level 源电平；声源级
source library 源程序库
source library program(me) 源库存程序
source line 源线
source-located 设在原料产地的
source machine 源机器
source macro-definition 源宏定义
source map 资料图
source material 源(始)材料；素材
source material of mapping 制图资料
source mechanism 震源机制
source mechanism array 震源机制台阵
source modulation rate 源调制频率

source module 源模块；源程序模块
source moment 震源矩
source movement 源移动
source multiplication method 源倍增法
source nipple 短管接头；螺纹接口(管)
source node 源节点
source object 始对象
source of air pollutant 空气污染物来源
source of alimentation 补给来源
source of biodegradable contaminant 可生物降解污染物源
source of bursting water 突水水源
source of carbon monoxide 一氧化碳源
source of cash 现金来源
source of chemical contaminant 化学污染物源
source of chemicals 化学物污染源
source of chemical species 化学物种污染源
source of crude 原矿储量
source of current 电源
source of current capital 流动资金来源
source of damage 损坏原因；事故原因
source of data 资料来源
source of deduction 扣除项目来源
source of dust 扬尘点；尘源；风沙来源
source of early warning 预警设备
source of electromagnetic interference 电磁干扰源
source of energy 能源
source of error 误差原因；误差来源
source off 断开电源
source of failure 故障来源
source of feed 原料来源
source of funds and usage statement 资金来源与运用表
source of graph 图的起点
source of heat 热源
source of heat release 散热源
source of hospital wastewater 医院污水污染源
source of hydropower 水力能源
source of illumination 照明光源；光源
source of infection 传染源
source of information 资料来源
source of intelligence 情报来源
source of labo(u)r supply 人力供应来源；人力供给来源
source of light 光源
source of local materials 当地料源
source of noise 噪声源
source of noise pollution 噪声污染源
source of non-point pollutant 非点源
source of nourishment 补给来源
source of odo(u)r 恶臭源
source of oil field water 油田水来源
source of ore-forming solution 成矿溶液来源
source of organic matter 有机质来源
source of oscillation 振动源
source of parallel light 平行光源
source of personnel 人力来源
source of pollutant in water body 水体污染源
source of pollution 污染源
source of potential phosphorus loading 潜在磷负荷源
source of power 电源；能源
source of public water 公用供水源
source of public water supply 公用供水水源
source of radiation 辐射源
source of remote excitation 远处激励源
source of resonant excitation 共振源
source of revenue 税源
source of river 河源
source of runoff 径流补给源；补给水源
source of salt 盐分来源
source of sample 样品来源
source of secondary radiation 二次辐射源
source of sediments 沉积源地；泥沙来源；沉积来源
source of seismic energy 地震能源
source of self-pollution 自污染源
source of supply 供给水源；供应(来)源
source of target information 目标信息源
source of the jet 射流源
source of trouble 故障原因；毛病原因；损坏原因
source of vibration 振动源
source of wastewater 废水源
source of wastewater contaminant 废水污染源
source of wastewater generation 废水产源

source of water 水源
source of water pollution 水污染源
source of water supply 供水水源；水供应来源
source of water troubles 油田出水来源；钻孔侵水原因
source of waves 波源
source of welding current 焊接电源
source out 断开电源
source output efficiency 源输出效率
source oven temperature 源炉温
source packet 源包
source parameter 源参数；震源参数
source plasma 源等离子体
source plate 源板
source point 源点
source pole 源极
source position 源位置
source positioning 源位置控制
source power 源功率
source preparation 源准备
source process 震源过程
source program(me) 源程序
source program(me) character 源程序字符
source program(me) form 源程序形式
source program(me) library 源程序库
source program(me) of confined water 承压水源程序
source program(me) of finite element method 有限单元法源程序
source program(me) of mixed type 混合型源程序
source program(me) of parameter optimization 参数优选源程序
source program(me) of phreatic water 潜水源程序
source program(me) operation 源程序运算
source program(me) optimizer 源程序最佳化程序
source program(me) pointer 源程序指示字
source program(me) symbol 源程序符号
source program(me) tape 源程序带
source program(me) variable 源程序变量
source program(me) word 源程序字
source programming 源程序设计
source protection 水源保护；源保护
source pump station 起点泵站
source range 源区段；震源范围
source rate 信源率
source reactor 中子源反应堆
source-receptor relation 源—受体关系
source recognition 污染源识别
source record 源记录
source recording 原始记录制作；原始记录
source reduction 源头削减；在污染源减少
source region 源区；源地；震源区；起源地；发源地
source register field 源寄存器字段
source release 震源释放
source release history 污染源排放史
source-representation 源表示
source resistance 内电阻；电源电阻
source restoration 复源法
source rock 生油岩层；生油岩层；母岩；原生岩；源岩
source rocks 生油岩层
source room 声源室
source routine 源程序
source sampler 污染源采样器
source sampling 源取样；源采样；污染源采样
source sediments 母岩沉积
source seismology 震源地震学
source-selector disk 选源盘
source series resistance 源电阻
source signal 震源信号
source-sink method 源汇法
source size 源大小
sources of agricultural nonpoint pollution 农业非点污染源
sources of agricultural pollution 农业污染源
sources of air pollution 空气污染源
sources of atmospheric pollution 大气污染源
sources of contamination 污染源
sources of domestic water 生活用水水源
sources of environment(al) pollution 环境污染源
sources of freight traffic 货源
sources of fund 资金来源
sources of goods 货源
sources of groundwater contamination 地下水污染源
sources of groundwater pollution 地下水污染源
sources of income 财源
sources of information 信息来源
sources of irrigation water 灌溉水源
sources of man-made pollutant 人为污染(来)源
sources of manpower 人力来源
sources of particulate matter 颗粒物质来源
sources of pollution 污染源
sources of revenue 财源
sources of river contaminant 河流污染源
sources of runoff 径流来源
sources of sediment contamination of waters 水体底泥污染源
sources of supply 货源
sources of wastewater 污水源
sources of water pollution 水污染源
sources of water quality 水质水源
sources of water supply 给水水源
source spacing 源距
source spectrum 震源谱
source speed 源速度
source station 源站
source strength 源强(度)
source strength measurement 源强度测定
source strength normalization 源强度归一化
source subtraction 减源法
source tape 源带
source tape cross assembler 源带交叉汇编程序
source tape preparation 源带准备
source temperature 源温
source terminal 源引出线；源极端；源端子
source text 源程序正文
source-time function 震源时间函数
source tock potential 生油岩潜量
source-to-fibre loss 光源光纤损耗
source transducer 信源换能器
source transformer 电源变压器
source tray 源盘
source tube 源管
source type 震源类型；污染源类型
source utility 源实用程序
source version 原本
source voltage 源电压；电源电压
source volume 震源体积
source water 水源水
source wave form 震源波形
source zone 源区
sourcing 纯源化
sour condensate 酸性冷凝物
sour corrosion 酸性腐蚀
sour crude 高硫原油
sour crude oil 含硫原油；酸性原油
sourdine 声音抑制器；消声器；噪声抑制器
sour distillate 酸性馏分
sour dock 酸模
sour dry gas 含硫量高的干燥石油气；酸性干气
sour gas 含硫天然气；含硫石油气；含硫气体；全硫化氢天然气；酸气；臭气
sour gasoline 含硫汽油；酸性汽油
sour grass 酸模
souring 加酸
souring house 回火室
souring plant 回火工厂
souring tower 回火塔
sour leek 酸模
sour natural gas 含硫天然气
sourness 酸度
sour odour 酸味
sour oil 含硫原油；含硫轻油；未中和的油；酸性油；酸性池
sour-service trim 防蚀处理
sour soil 酸性土(壤)；酸土
sour taste 酸味
sour trench 冲刷槽
sour water 酸性污水；酸性水
sour water stripper 酸性水洗提剂；酸性水剥离剂
sour well 酸化井
Sousa chinensis 中华白海豚
sou's easter 东南大风
Sous-Palan clause 过驳交货条款；吊钩下交货条款
sou's wester 西南大风；强烈西南风；大雨帽
souterrain 地下建筑物；地下通道
south 南方
south Africa architecture 南非洲建筑
South African jade 绿钙铝榴石
south African rose-wood 南非玫瑰木
South-Africa shipping line 南非航线
south aisle 南侧廊；南走廊
south America-Africa separation 南美大陆与非洲大陆分离
South American architecture 南美洲建筑
South American plate 南美板块
south American platform 南美地台
South-America shipping line 南美航线
South Asia and Far East cable 南亚和远东电缆
South Atlantic bivalve subprovince 南大西洋双壳类地理亚区
South Atlantic current 南方西洋海流；南大西洋海流
South Atlantic deep and bottom water 南大西洋深层和底层水
South Atlantic disjunction 南大西洋间断分布
South Atlantic high (pressure) 南大西洋高压
South Atlantic ocean 南大西洋
South Atlantic-Siberia crustal-wave system 南大西洋—西伯利亚波系
South Australia 南澳洲
south Australian basin 南印度洋海盆；南澳大利亚海盆
south bound 向南航行
southbound node 降交点
south by east 南偏东
south by west 南偏西
south cardinal mark 南界标；南方位标
south Cathaysian floral region 华夏南方植物地理大区
south celestial pole 南天极；天球南极；天南极
south central 中南部
south central asian coral region 中南亚珊瑚地理大区
South China arid subtropical to tropical zone 华南干旱亚热带—热带
South China coral province 华南珊瑚地理区
South China faunal province 华南动物地理区
South China fold system 华南褶皱系
South China geosyncline system 华南地槽系
South China great regression 华南大海退
South China non-marine ostacode province 华南非海相介形类地理区
South China region 华南大区
South China sea 南中国海；南海
South China sea basin 华南海盆
South China sea platform 中国南海地台
South China sea subsiding basin 中国南海沉降海盆
South China sea trough 中国南海海槽
South China section 华南大区
South China semiarid subtropical to tropical subzone 华南半干旱亚热带—热带亚带
South China-Southeast asia plate 华南—东南亚板块
South China transgression 华南海浸
South China upland 华南高地
South Chine basin 南中国海盆
southeast [SE] 东南；南东
southeast by east 东南偏东
southeast by south 东南偏南
southeast China basin region 中国东南盆地区
southeast China epicontinental volcanic basin region 中国东南陆缘火山盆地区
southeast China hills 中国东南丘陵
southeast China intermontane basin region 中国东南山间盆地区
southeast China littoral-shallow sea 中国东南滨浅海
southeast China marine trough 中国东南海槽
southeast China mountains 中国东南山地
southeast China province 中国东南区
southeast China semiarid to semihumid subtropical to tropical subzone 中国东南半干旱—半潮湿亚热—热带亚带
southeast China shallow sea 东南浅海
Southeast coast deep fracture zone 东南沿海深断裂系
Southeast coastland tectonic zone 东南沿海构造带
southeast drift current 东南漂流
southeaster 东南冷风；东南风；东南大风

southeasterly 偏东南的;东南风
southeastern 东南的
South-Eastern coat fold system 东南沿海褶皱系
southeast facing marginal arc 向东南突出边缘弧
southeast flowing 由东南流来的
southeast tarim downwarping region 东南坳陷区
south east trade drift 东南信风海流
southeast trades 东南信风
southeast trade wind 东南信风
southeast wind 东南风
south elevation 南立面
south equatorial current 南赤道海流;地赤道洋流
souther 南大风
southerly 南风
southerly burster 南寒风
southerly-facing flat collector 朝南平面集热器
southerly limit 南图廓
Southern Africa 南部非洲
southern auroral zone 南极光带
southern balsam fir 南方香脂冷杉
southern celestial hemisphere 南半天球
southern chernozem 南方黑钙土
southern China humid subtropical to tropical zone 中国南部潮湿亚热—热带
southern China section 华北大区
Southern daba platform folded belt 南大巴台褶带
southern elongation 南大距
southern falsebeech 南方假山毛榉
southern galaxy 南天星系
southern hemisphere 南半球
Southern Himalayan diwa region 南喜马拉雅地洼区
southern latitude 南纬
southern lights 南极光
southern limit 南界限
southernly burster 南寒风
Southern Marginal deep fracture zone of the Siberian platform 西伯利亚地台南缘深断裂系
southern mid-Atlantic ridge 南大西洋中脊
Southern Mongolian deep fracture zone 南蒙古深断裂系
southern most 最南端的
southern ocean 南大洋
southern oscillation 南方涛动
southern pine 南方松
southern platform depression 南部台坳
Southern polar front 南极幅合带
southern polarity 南极性;南磁极性
southern polar region 南极区
southern portio 南门廊
Southern red oak 赤橡;红橡木;南方赤栎
southern reference star 南天参考星
Southern sea area 南部海域
southern slope 南向坡
southern spring 南半球春季
Southern Star 南星;南天恒星
southern suburbs 南郊
southern transept (教堂的)南交叉甬道
Southern Wan tectonic knot 皖南构造结
southern window 南窗
southern winter 南半球冬季
southern winter seasonal zone 南半球冬季季节区带
southern yellow pine 南方黄松
Southern Zayu area tectonic segment 察南构造段
south-facing balcony 朝南阳台;南向阳台
south-facing wall 朝南墙壁
south-facing window 朝南窗子
south frigid zone 南寒带
south geographic(al) pole 南地(理磁)极;地理南极
south geomagnetic pole 地磁南极
south-going river 南下河流;南流河
south-going stream 南下河流;南流河
south heavy 南重
south Hebrides trench 南赫布里底海沟
south Hemispheric floral realm 南半球植物地理区系
South Himalayan trough system 喜马拉雅南边缘坳陷系
South Indian current 南印度洋海流
South Indian style 南印度风格
Southing 向南航行;南行航程;南进
South Korean coastal plain 韩国沿海平原

South Korean marine trough 韩国海槽
southland beech 南部地区的山毛榉
south latitude 南纬
south-light roof 南方采光的锯齿形屋顶(南半球);南向采光锯齿形天窗
south magnetic 磁南极
south magnetic dip pole 南磁倾角极
south magnetic pole 南磁极;磁南极
south magnetic pole position 南磁极位置
south-north 从南往北
south-north asymmetry 南北不对称性
south-north route 南北航线
south-north water transfer 南水北调
south orientation 朝南;南向
south Orkney trough 南奥克尼海槽
South Pacific basin 南太平洋海盆
South Pacific current 南太平洋海流
South Pacific disjunction 南太平洋间断分布
South Pacific mahogany 南太平洋红木
South Pacific Ocean 南太平洋
South Pacific Regional Environmental Program (me)南太平洋区域环境方案
South Pacific shipping line 南太平洋航线
south pal(a)eo-Mediterranean region 古地中海南部大区
south point 南点
south point of the horizon 地平面南点
south polar circle 南极圈
south polar distance 南极距
South Pole 南极;磁针指南端
south pole exploration ship 南极考察船
south porch 南走廊;南门廊
south sea timber 南洋材;热带材
south Shetland trough 南设德兰海槽
South-South conference 南南会议
South-South cooperation 南南合作
south southeast 南南东;南东东
south southwest 南南西;南西西
South stoa 南拱廊(古希腊式);南柱廊(古希腊式)
south temperate 南温带的
south temperate bivalve realm 南温带双壳类地理区系
south temperature zone 南温带
south Tethyan region 特提斯南缘大区
south Tethyan trough 南特提斯海槽
south tropical disturbance 南热带扰动
southward 向南;南方的
south water to north 南水北调
Southwell's relaxation method 索思韦尔松弛法
southwest 西南
southwest by south 西南偏南
southwest by west 西南偏西
southwest cap 大风雨帽
southwest China 中国西南
southwest China upland 中国西南高地
southwester 西南大风;强烈西南风;大雨帽
southwesterly wind 西南风
southwestern 西南的
southwestern China metallogenetic province 中国西南成矿区
Southwestern-Kraut (hydromotor) jig 流体传动跳汰机
southwest facing marginal arc 向西南突出边缘弧
southwest monsoon 西南季风
southwest pacific ocean 西南太平洋
southwest tarim downwarping region 西南坳陷区
southwestward 西南方向的
southwest wind 西南风
south wind 南风
southwood 可翻开窗扇
souvenir 纪念品
souvenir shop 纪念品商店
sou'wester 长披帽雨衣
sou-wester 西南大风
souzalite 水丝绿铁石
souzan bentonite 多钙膨润土
sovereign remedy 特效药
sovereign state 主权国家
sovereign(ty) 主权;独立国
sovereignty nuisance 隔离障碍
soviet group 黑云碳酸岩类
sovite 黑云碳酸岩
sow 炉底结块;铁水沟;砂床铸铁;大铸型;大型浇池;大锭块

sowback 底板隆起;山脊;冰碛丘
sow channel 主流槽
sowdel 撑棍;彩画玻璃嵌条
sow down to grass 播种草地
sower 播种者;播种机
sowing 播种
sowing area 播种面积
sowing a slope down to grass 坡地植草;绿化坡地
sowing chart 播量表
sowing in bed 床播
sowing in close dill 窄行条播
sowing in furrow 畦播;条播
sowing in furrows 沟播
sowing in line 行播
sowing in narrow dill 窄行条播
sowing in patches 块状播种;簇状播种;丛植播种
sowing in row 行播
sowing machine 播种机
sowing table 播量表
sowing width 播幅
sowing width number 播幅大小
sowing width regulates 播幅调整
sow iron 沟铁(高炉)
sowneck 地颈
sown pasture 人工草地
Soxhlet's extraction method 索格利特提取法
Soxhlet's extractor 索格利特提取器
soya bean adhesive 大豆黏结剂
soya bean alkyd 大豆醇酸
soya bean cake 豆饼
soya bean glue 大豆胶
soya bean oil 大豆油
soya bean oil fatty acid 豆油脂肪酸
soya casein 豆酪素
soya glue 大豆胶
soybean albumin plastic 大豆蛋白质塑料
soybean export promotion 促进大豆出口
soybean for oil 榨油大豆
soybean glue 大豆胶
soybean investment 存麦豆投资
soybean oil 豆油;大豆油
soybean processing wastewater 大豆加工废水
soybean protein glue 大豆蛋白黏合剂
soybean protein plastic 大豆蛋白质塑料
sozoidol 二碘对酚磺酸
sozoiodolic acid 二碘酚磺酸
spa 矿泉;温泉;(兼作消防房的)冷饮店;矿泉疗养地
space 过行;留空格;留间隔;空行;空间;空号(通信);空地;空白(区);距离;间隔;太空;使隔开;场地
space acuity 空间视觉锐度
space advertising 版面广告
space after 后空格
space age 空间时代;太空时代
space allocation routine 空白区分配例行程序
space allotment 分配舱容
space anomaly 空间异常
space arbitrage 空间套利;空间裁定;不同地区套利
space area 空白区
space astronomy 空间天文学
space attribute 空间属性
space average 空间平均值
space averaged value 空间平均值
space axes 空间坐标轴
space background 空间背景
space band 空白段;间隔嵌条
space bands 齐行楔
space bandwidth product 空间带宽积
space bar 隔条;空键;空间杆;空格拉杆;空格键【计】;间隔键
space-based subsystem 以空间为基地的次系统
space bearing structure 空间承载结构;空间承重结构
space between beams 梁(的)间距
space between building 房屋间距
space between dikes 坝的间距
space between girders 大梁间距
space between joists 格栅空当
space between pilasters 壁柱空当;壁柱净距
space between the holes 孔距
space biology 空间生物学
space bit 空位;空白位
spaceborne infrared cloud mapper 空载红外云层测绘仪

spaceborne optic(al) antenna 空载光学天线
spaceborne radar 航天雷达
space-borne remote sensing 从航天器进行的遥感
space-borne repeater station 星载转播站
spaceborne sun pumped laser 空载太阳泵浦激光器
space borne system 空间载运系统；空间运载系统
space box 空铅盒
space buff 隔层抛光轮
space building 太空(发射)建筑
space calculation 空间计算
space camera 空间摄影机
space cartography 空间制图学
space cavern 洞窟空间
space centers 中距
space chamber 空间模拟室；空间工作箱；空间工作间
space character 空格符；空白字符；间隔字符；间隔符号；退格(字)符
space characteristic of object 地物空间特性
space charges 空间电荷；占用场地费用；空间费用
space charge balanced flow 空间电荷平衡流
space charge current 空间电荷电流
space charge debunching 空间电荷散焦
space charge effect 空间电荷效应
space charge grid 空间电荷栅极
space charge layer 耗尽层
space charge limitation 空间电荷限制
space charge mode 空间电荷波模式
space charge model 空间电荷模型
space charge optics 空间电荷光学系统
space charge region 空间电荷区
space charge tetrode 双栅管
space charter 舱位租用；舱容租赁
space charter system 舱位租约方式
space chemistry 空间化学
space city 空中城市；太空城市
space cloth 空格布
space cluster 空间积群法
space code 空间码；间隔码
space coding 间隔编号
space coherent beam 空间相干束
space column 格架式柱；空腹式柱
space combination 空间组合
space communication 宇宙通信；空间通信
space communication technique 空间通信技术
space composition 三维组合；空间构图
space computing complex 综合空间计算设备
space cone 空间锥面
space configuration 空间构图；空间形状；空间构型
space construction 空间结构；空间构造
space-consuming 空间利用；空间使用
space contact 空号接点
space continuity 空间连续性
space cooling 环流冷却
space cooling load 房间冷负荷
space coordinates 立体坐标；空间坐标(系)
space coordinate system 空间坐标系
space correction 空间校正
space cosmic ray 太阳宇宙射线
space cost 空间成本
space cost curve 空间成本曲线
space counter 空码计数器
space counting distribution 空间车速分布
space coupling 空间耦合；分布耦合
spacecraft 太空飞行器；宇宙飞船；空间飞行器
spacecraft launching site 航天器发射场
spacecraft remote sensing system 卫星遥感系统
spacecraft sterilization 宇宙飞船消毒
spacecraft tracking data 卫星跟踪数据
space creating plane 空间形成的平面
space crossing method 空间交会法
space current 空间电流
space cursor 空格指标
space curvature 空间弯曲；空间曲率
space curve 空间曲线
space cycle 空间循环
spaced-aerial direction(al) finder 分集天线测向仪；分级天线侧向器
spaced antenna 间隔天线；等距(离)天线阵；分集(式)天线；分布式天线
spaced armor 屏蔽装甲
spaced array 等距(离)天线阵
spaced boarding 花铺望板；疏铺望板

spaced centers 中距
spaced cleavage 间隔劈理
spaced columns 间隔设柱；空间柱簇；格架柱；格架式柱；离缝组合；空腹式柱
space deformation 空间变形
space derivative of earthquake motion 地震运动的空间衍生
space design 空间设计
space detection and tracking system 空间探测和跟踪系统
spaced frame loop antenna 分集式环形天线
space diagonal 空间对角线
space diagram 立体图；空间图；矢量图
space-dipole array 双分集偶极天线阵
space discharge 间隔卸料
space distortion 空间畸变
space distribution 空间分布
space distribution occupancy rate 空间占用率
space disturbance forecast 空间干扰预报
space diversity 空间分集
space diversity reception 空间分集接收
space division 空间分隔；空间分割；空分
space division multi-channel system 空间分割多路系统
space division multiplex(ing) 空分复用(制)；空间分隔多工制
space division switching 空间分割交换；空分交换
space division switching system 空间分割交换机
space division system 空间分割系统
spaced loading 间隔装药
spaced-loop direction-finder 分集环形天线测向器
spaced name 散列注记
space domain 空间域
space dome 空间圆穹
spaced on centers 排列在中心上的
spaced payment 分期付款
spaced planting 稀植
space drawing 不均匀牵伸
spaced receiver 分集式接收机；分隔接收机
spaced slating 花铺石板瓦；疏铺石板瓦；疏铺(石)板
spaced steel column 格构式钢柱；空腹式钢柱
spaced winding 疏绕线圈；间绕绕组；间绕
space economy 空间经济
space efficiency 库场利用率；舱位利用率
space effort 空间科学计划
space electronics 空间电子学
space-enclosed 内部空间；空间封闭的
space-enclosing structure 空间封闭结构
space engineering geology 宇宙工程地质学
space environment 空间环境
space environmental monitor 空间环境监测仪
space equipment 航天设备
space equivalent 空间当量
space expand key 空格扩展键
space expenses 占地费用
space exploration 空间探索；空间探测
space factor 线圈间隙系数；占空因数；占空系数；空间因素；空间因数；空间系数
space feed 输格
space ferry 空间渡船
space filling puzzle 空间装填迷题
space filtering coordinates 空间滤波坐标
space filtering operation 空间滤波过程
space filtering reticle 空间滤波调制盘
space-fixed coordinate system 固定于空间的坐标系
space fixed reference 空间固定基准
space flight 航天；空间飞行
space flight spectral measurement instrument 航天波谱测量仪器
space flip-flop 空码触发器
space floor area 楼面面积；建筑面积
space flow rate 空间流速
space for balance 推拉窗之平衡锤箱
space for cargo 货物舱位
space force 空间力
space force system 空间力系
space form(work) 立体模壳；空间形式
space forward intersection 空间前方交会
space for working the ship 船舶操作场所
space frame 立体构架；空间框架；空间构架；空间网格结构

space frame cupola 立体构架圆顶
space frame dome 立体构架穹隆
space frame-moment resisting 抗弯的空间构架
space frame of tubular steel section 钢管空间构架
space frame structure 空间框架结构
space frame tower 立体构架塔楼
space frame-vertical load carrying 承受竖载的空间框架
space frame work 空间构架工程；空间框架；空间(构)架
space frequency 空号频率
space frequency migration 空间频率偏移
space function lever 间隔动作杆
space gain 空间获得；空间增大
space geodesy 空间大地测量
space geodetic network 空间大地网
space grating 空间衍射光栅；空间光栅
space grid 网架；空间网格
space group 空间群；晶架群
space group determination 空间群测定
space group extinction 空间群消光
space guidance 宇宙航行导引
space headway 车头间距
space heater 小型采暖炉；房间取暖器；房间加热器；空气加热器；空间(对流)加热器；空间电热器；热风采暖装置；室内采暖器
space heater of burning town gas 城市煤气用于室内采暖
space heater terminal box 防潮加热器接线盒
space heating 室内供暖；室内采暖；空间加热；室内采暖；环流供暖；对流式供暖；场地加热
space heating appliance 室内采暖器；小型采暖燃具
space heating load 供暖热负荷；采暖热负荷
space heating plant 室内采暖设备
space hold 空号同步；空格保持
space homogeneity 空间一致性；空间均一性
space hour 车位小时
space image 立体像
space image mosaic 镶嵌卫片(遥感地质)
space impulse 间隔脉冲
space independent 空间不相关的
space indicate 空间标志
space insertion 空白插入
space instrumentation 航天设备
space integral 空间积分
space integral of force 力的空间积分
space intersection 空间交会
space interval 空间间隔；区间
space interval method 空间间隔法
space interval system 空间间隔制
space invariability condition 空间不变性条件
space invariance 空间不变性
space inversion 反演
space isoline map of magnetic anomaly 磁异常空间等值线图
space junk 太空垃圾
space key 空格键【计】；间隔键
space lab 空间实验室；宇宙实验室
space laboratory 宇宙实验室；空间实验室
space lattice 空间网络；空间网架；空间栅格；空间晶格；空间格子；空间点阵；立体网格；立体格子
space layout 空间设计；空间布置
space length 空间距离；间距
space lifeboat 宇宙救生船
spacelike 类空间；类空
spacelike event 类空事件
spacelike interval 类空间隔
spacelike path 类空路径
spacelike surface 空间曲面；类空面
spacelike vector 类空矢量
space limit 容积限制；空间限制
space limitation 空间限制
space linkage 三向链合；三向连接；空间链合；空间连接
space link-up 空间对接
space loading 空间荷载
spaceman 宇宙飞行员
space manager 存储空间管理员
spacemanship 充分利用空间的技术
space mark 空间测标；间隔符号
space mass spectrometry 空间质谱分析
space mean 空间平均值
space mean speed 空间平均速度；空间平均车速；

区段平均速率
space mechanism 留间隙机构;间隔机构
space medium 空间介质
space model 立体模型;空间模型
space modulation 空间调制;间隔调制;场调制
space modulation technique 空间调制技术
space moisture load 房间湿负荷
space motion 空间运动
space navigation 宇宙导航;宇航
space navigation system 空间导航系统
space network polymer 立体网形聚合物
space nipple 长管接头
space number 间隔数
space occupancy 空间占有率
space occupancy cost 空间占有成本
space occupied 所占舱容;所用舱容;占有空间的
space of detector and sampling section 探头与取样面间距
space of dune 沙丘间距
space of intermittent weld 断续焊缝间距
space of interval 齿宽
space of observation line 观测线间距
space of observation points 观测点距离
space of pole winding 极绕组空间
space of teeth 齿间隔
space optic(al) communication 空间光通信
space optics 空间光学
space optimization 空间优化
space orientation 空间定向
space oriented 面向空间
space oscillator 空间振荡器
space out 加宽行距
space outline 空间轮廓
space out the difference in the prices 拉开差价档次
space panelling pattern 空间嵌板图案
space parallax 空间视差;空间声差
space parasite 空间寄生植物
space partition 空间分隔
space perception 立体感;空间感觉
space permeability 空间磁导率
space perspective 空间透视(法)
space photogrammetry 航天摄影测量
space photograph 空间摄影照片
space photography 空间摄影测量(学);空间摄影
spaceplane 宇航飞机
space planning 空间规划;空间布局;场地规划
space point 空间点
space polar coordinate 空间极坐标
space polygon 空间多边形
space polymer 立体聚合物
spaceport 航天港
space position 空间频谱
space prism 空间棱镜
space probe 宇宙探测器;太空探测器
space problem 空间问题
space process 空间过程
space propulsion 空间推进
space purlin(e) 间隔檩条
space quadrature 空间90度相位差
space qualified 空间适用的
spacer 预压腔;隔离物;隔离片;隔板固定架;间距短管;间隔物;间隔团;间隔确定装置;间隔基;间隔棒;调节块;定位片;定位块;定位件;定距片;撑挡;衬垫;分隔架
space radiation 宇宙辐射;空间辐射
space radio communication 空间无线电通信
space range 空间范围
space rate (沸腾焙烧炉的)直线速度
space-rated 适用于空间
space rate of change 空间变率
space ratio (水泥纯浆的)孔隙比
space ray 空间射线
spacer bar 间布钢筋条;间布钢筋;架立筋;定位钢筋;撑杆
spacer bending system 分隔架弯折系统
spacer block 隔块;间隙环;间隔铁(护轮架);模具定位块;定准砌块;垫块;衬垫片
spacer conduit 间隔管
spacer cutting saw 间隔框切割机
spacer disc 圆隔盘;圆隔板
spacer drill 精密条播机
space reconnaissance 空间侦察;航天勘探
space record 空白记录

space redding 空间红化
space reference 空间基准
space reference radiation level 空间辐射基准
space reinforcement 三向筋
space-related basis of allocation 按作业面积分摊
space remote sensing 航天遥感
space required 需要面积
space requirement 空间需求;空间要求;间隔要求;每头牲畜所用畜舍面积;场地需要
space requirement work sheet 空间需要量工作表
space research 空间研究
space research plan 空间研究计划
space resection 空间后方交会
space resources 空间资源
space restoration value of time depth transform 时深转换空间归位值
space retainer 间隙固位体;间隙保持器
space revenue curve 空间收入曲线
spacer flange 对接法兰;中间法兰;过度法兰
spacer gel 成层胶
space ring 隔圈;分隔圈
spacer leg 隔离支柱
spacer lug 间距定位块;间隔凸缘;定位销钉
spacer medium 隔离介质(混凝土下的塑料薄膜垫底)
spacer nipple 填空短管
space rod 空间导杆
spacer peristyle 圆列柱廊
spacer pin 限位销;隔离销;调整销
spacer pipe 间隔管
spacer plate 隔板;垫板;分离板
spacer ring 隔离环;间隙圈;间距环;定位环
spacer rod 隔离棒
spacer shaft 机间轴
spacer shim 调整垫片;定位(楔形)片;调距隔片
spacer sleeve 间隔套;挡套
spacer spool 调节法兰盘;间距套管
spacer spring 隔离弹簧
spacer step 间插步骤;间步
spacer strip 隔离棒;垫片;(焊接操作时的)支垫板条
spacer tube 间隔管;定距管
spacer type 隔离液类型
space satellite 航天卫星
space-saver dial 省位拨号盘
space-saving 节省地位;空间节约;节约地方;节省空间(的);节省舱(货)位的;少占空间的
space scale 空间标度
space science 空间科学;太空科学
space selling 提供广告版面
space sensor data 空间传感数据
space separation 空间分隔
space shape 立体形式
space sharing 空间共享
space ship 航天船;太空飞船;宇宙飞船
space shuttering 空间模板
space shuttle 空间渡船
space sickness 宇宙病;宇航病
space signal 间隔信号
space simulation facility 空间模拟装置
space simulator 宇宙空间环境模拟器;空中模拟器;太空模拟器
space spanning 空间跨度;空间跨距
space specific and time specific 空间和时间的特性
space spectroscopy 空间光谱学
space-stabilized 空间稳定的
space stage 空间舞台
space station 宇宙空间站;宇宙飞行站;空间站;太空站
space statistics 空间统计量
space-stop rack 间隔架
space stowage 间隔装舱法
space structure 空间结构
space structure in glass-concrete 玻璃—(钢筋)混凝土空间结构
space supporting structure 空间支承结构
space suppression 空格压缩;间隔控制
space surveillance system 空间监视系统
space symbol 空格符号
space system 空间系统;间隔装舱法
space system for search of distress vessels 搜寻遇难船舶空间形态
space system segment 空间系统部分
space target-tracking laser radar 空间目标跟踪激光雷达

space technology 航天技术;航天工程技术
space telecommunication 空间通信
space telescope 空间望远镜
space temperature variation 区域温差
space tensor 空间张量
space time 空间时间关系
space-time algebra 时空代数
space-time brownian motion 时空布朗运动
space-time characteristic of rainfall 降雨空间特性
space-time concept(ion) 时空概念
space-time continuum 时空连续体;时空连续域
space-time coordinates 时空坐标
space-time coordinate system 时空坐标系
space-time correlation function 时空相关函数
space-time correlogram 时空相关图
space-time curvature 时空曲率
space-time curve 时空曲线
space-time diagram 时空图;空间时间图
space-time element 时空元
space-time equation 时空方程式
space-time filter 时空滤波器
space-time functions 时空函数
space-time law 时空律
space-time manifold 时空流形
space-time metric 时空度量
space-time optics 空时光学
space-time pattern 时空类型
space-time point 时空点
space-time processing 时空配合;时空处理;时差定位法
space-time random process 时空随机过程
space-time stochastic process 时空随机过程
space-time structure 时空结构
space-time transformation 时空变换
space-time unit 时空单位
space-time wind field 风区图表
space-time yield 时空产率
space-to-mark transition 空号—标号转换;空格到符号蜕变;间隔—符号变换;间隔到标志转换
space tracking laser beacon 空间激光跟踪指向标
space tracking system 空间跟踪系统
space track laser 空间目标跟踪激光器
space transportation 航天运输
space travel 宇宙飞行
space triangle network method 空间三角形网络法
space truss 空间桁架
space truss analogy 立体桁架模拟法;空间桁架模拟法
space-truss model 空间桁架模型
space truss structure 空间桁架结构
space unit 空间构件
space utilization 空间利用
space utilization factor 空间利用系数
space variable 空间变量
space variance 空间可变性
space variation 空间变化
space vehicle 宇宙飞船;空间飞行器;空间飞船;太空运载器
space velocity 空间速度;空间流速
space velocity of catalyst 催化剂空间速度
space-view cab(in) 全视野机舱
space warning 房间供暖
space washer 间隔垫圈
space wave 空间波;天波
space weapon 空间武器
space welding 有间隙焊接;间隔焊接
space wound coil 间绕线圈
spachtling 灰腻子
spacing 限位;空格;节距;间距;间隔;位距;定距;车头间距
spacing accuracy 螺距准确度;间距准确度;间隔精度
spacing and direction of sounding profiles 测深断面的间距和方向
spacing and population tests 行距和密度试验
spacing bent-up bars 弯起钢筋分布;弯起钢筋间距
spacing between beams 梁距
spacing between holes 炮眼间距
spacing bias 空号偏压;空格偏移;间隔偏移;间隔偏差;间隔拉开
spacing block 隔离片;间隔块
spacing board 间隔板;定位板
spacing braking 间隔制动【铁】
spacing bubble 定距泡

spacing capability 空间位隔能力
spacing chart 空白表格
spacing coefficient 间隔系数
spacing collar 限位套筒;限位圈;隔圈;间隔圈;定距环
spacing column 定距柱
spacing computation 跨距计算
spacing contact 隔离触点
spacing cropping 宽行栽培
spacing current 间隔电流
spacing densified method 加密法
spacing deviation 定距偏差
spacing device 斜线刻绘仪(阴像刻图用)
spacing distance 间隔距离
spacing dowel 间隔连接筋;间隔传力杆
spacing drill 宽垄播种机
spacing effect 空间位隔效应
spacing error 间隔误差
spacing escapement pawl 间隔擒纵机爪
spacing extender 起空间位隔作用的增量剂
spacing factor 间距系数;间距因数;气孔间距因数
spacing-height ratio 距高比;间高比
spacing hole 定位孔
spacing impulse 间隔脉冲
spacing instrument 齿轮距离检测仪
spacing interval (钻孔之间的)间距;间隔时间
spacing in the row 株距
spacing lever 间距调节杆
spacing loosened method 稀空法
spacing loss 空号损耗;无信号损耗
spacing lug 间隔耳
spacing material 衬隔材料
spacing mechanism 间隔机构
spacing mo(u)ld 隔膜
spacing of adjacent strips 相邻航线间隔(航测)
spacing of bars 钢筋分布;钢筋距离;钢筋间距
spacing of beams 梁间距
spacing of boiler tubes 锅炉管距
spacing of buildings 建筑物间距;房屋间距
spacing of carbon atom net 碳原子网间距
spacing of coal seam 煤层间距
spacing of columns 柱距
spacing of drain 排水间距;排水间隔
spacing of electrodes 电极距
spacing of exploratory work 勘探工程间距
spacing of fins 翅片间距
spacing of frame 肋距
spacing of groin 丁坝间距
spacing of joints 接缝间隔;接缝间距
spacing of obstructive vehicles 交通阻塞时的车距
spacing of passing sidings 站间距离
spacing of piles 桩的间距
spacing of rail tracks 滑道间距
spacing of reinforcement 钢筋间距
spacing of ribs 肋间距;肋间隔
spacing of structural planes 结构面间距
spacing of tanks 储[贮]罐间隔;槽间间距
spacing of terminals 编组站间距
spacing of the rivets 铆钉间隔
spacing of trees 树木间隔
spacing of trusses 桁架间距
spacing of twined wires 缠丝间距
spacing of wells 钻孔间距;井距;井的布置
spacing out 隔空齐行
spacing parameter 间距参数
spacing pattern 布井网;(金刚石在钻头上的)摆放样式
spacing piece 定距片
spacing pulse 空号脉冲;间隔脉冲
spacing punch 定距冲头
spacing rail brake 间隔制动缓行器;区间刹车器
spacing ratio 空间比(两灯间的距离对其工作面距离之比);间距比(率)
spacing ring 隔离垫圈;限位环;隔离圈;间隔圈;间隔垫圈;垫环
spacing rule 轨距叉;定距尺;轨距尺【铁】
spacing screw 空号螺钉
spacing shaft 间隔轴
spacing shim 限位填片
spacing sleeve 隔离套筒
spacing sling dip 沿倾向间距
spacing sling strike 沿走向间距
spacing strip 分隔嵌条(抹灰用);间距条;隔条;定距片

spacing support 定距撑
spacing table 限位工作台
spacing timber 隔条;定位木
spacing-to-mounting height ratio 路灯间距与悬高比
spacing track 间隔道;间隔磁道
spacing tube 隔离套筒;定距管
spacing washer 间隔环;间隔圈
spacing wave 空(号)信号;静止信号;间隔波
spacio-temporal characteristic 时空特征
spacio-temporal distribution 时空分布
spacio-temporal prediction 时控预测;时空预测
spacio-time migration 时空迁移
spacious 宽敞的
spaciousness 空灵;宽敞度;宽敞
spacious waled corduroy 粗条灯芯绒
spackle 腻子料;灰腻子;填泥料
spackling 碎木塞;砂浆石子堵塞;填泥;起泡
spackling compound 嵌缝料
spackling compound coat 填配料涂层
spackling compound for brush application 刷敷填配料
spackling compound seal(ing) 填配料嵌缝
Spackman's classification 斯帕克曼分类
spadable sludge 可铲污泥
spadaite 红硅镁石
spaddle 长柄小铲
spade 铁锹;铲
spade bit 铲形钻头
spade bolt 铲状螺栓
spade bug 铲行爬地齿
spaded concrete 锹夯(实)混凝土
spade drill 扁钻;扁平钻
spade end fitting with hook 铲形连接钩头
spade-end wedge 开口式偏斜楔
spade farm 小园地
spade flange 铲形法兰
spade fluke 铲形锚爪
spade half coupling 接轴扁头
spade hand 铲状手;铲形手
spade handle 铲柄
spade lug 平接线片
spader 挖土机;挖土工具;铲具
spade radium hand 铲形夜光针
spade reamer 双刃铰刀;扁钻形铰刀
spade rudder 吊舵;铲型舵
spade sampler 锹式采样器
spade tag 扁形软线接头
spade terminal 扁形接头
spade tip 铲尖端;锹尖端
spade tuning 铲形盘调谐;薄片调谐
spade-type bit 一字形冲击钻头;铲凿;铲形钻头
spade-type joint 铲式接缝
spade-type vibrator 锹式振捣器;片式振捣器
spade-work 铲土作业;铲土工作;铁锹工作
spadiceous 栗色的;浅褐色的
spading 捣固;铁锹捣实;插捣
spading disk 翻土圆盘
spading fork 翻土叉
spading harrow 铲形耙
spadix 佛焰花序
spa garden 疗养地花园;矿泉疗养公园
spaghetti 绝缘线套;绝缘套管;漆布绝缘管
spaghetti tubing 细麻布绝缘套管
spagnolet 旋转螺栓
Spagnum moss 泥炭沼;泥炭地
spa hotel 疗养地旅馆
Spair hollow panel 斯佩尔空心板
spalder 碎石工人
spall 碎石片;剥落物
spallable rock 易碎石
spallation 剥落;散裂
spallation cross section 散裂截面
spallation product 散裂产物
spallation reaction 散裂(反应)
spall drain(age) 碎石排水沟;碎石盲沟
spalled joint 碎裂缝
spall hammer 碎石锤
spalling 自发破裂;胀裂[地];片帮【岩】;碎裂力
spalling crack 剥裂
spalling effect 激冷激热效应;剥落作用
spalling fatigue 疲劳剥落
spalling force 破损力

spalling hammer 碎石锤;大锤
spalling of brick surface 砖面剥落
spalling of refractory 耐火材料散裂
spalling of refractory materials 耐火材料剥落
spalling of rock 落石
spalling resistance 抗散裂强度;抗热剥裂性;抗剥落性(能);抗剥落力;耐热震性;耐剥落性;散裂抵抗力
spalling resistance of alumina brick 高铝砖抗剥落性
spalling test 散裂试验;剥落指数;剥落试验;剥裂试验
spalling test of fireclay brick 耐火砖散裂试验
spalling wedge 凿石楔;碎石楔
spall shield 防碎层
spall wedge 穿空楔
spalt 制瓦碎片
spaly skirting 上缘抹角墙脚板
span 翼展;中索(双杆作业用的);杆间距离;跨径[道];跨越;跨绳;跨(度);取值范围;生成空间;档距
span between two posts 支柱间距
span block 中索滑车;跨绳滑车
span/breadth ratio 跨度/宽度比;跨宽比
span-by-span cast-in-place method 逐孔现浇法
span-by-span construction 逐跨施工(法);逐孔施工(法)
span-by-span method 移动支架逐跨施工法
span center 跨中;跨距中点
span chain 千斤链
span change rigid frame 跨变刚架
span chisel 展平凿;磨平凿
span-chord ratio 展弦比
span clearance 跨度净空;净跨(度)
spancrete 混凝土跨度板;预制预应力混凝土板
spancrete panel 应力空心墙板
span-depth ratio 跨深比;跨高比
spandex fiber 弹性纤维
spandex stretch fabric 氨纶弹力织物
spandex wastewater 氨纶废水
span dogs 木材夹钳;木材抓起机
spandrel 窗心墙
spandrel arch 拱肩拱(主拱背上的小拱);空腹拱;肩拱;腹拱
spandrel arch frame 空腹拱构架
spandrel beam 托梁;窗下(墙的)边托梁;外墙托梁;外墙托梁;窗心墙托梁
spandrel-braced arch 肩拱拱;桁架式拱;腹部支撑拱;空腹拱
spandrel-braced arch bridge 桁架式拱桥;撑腹式拱桥;空腹拱桥
spandrel cantilevered step 拱肩上的悬臂梯级;拱肩上的悬臂踏步;上下窗空间悬臂踏步
spandrel column 拱肩柱;腹拱柱;拱上立柱
spandrel construction 拱桥上部建筑
spandrel cross wall 拱上横墙
spandrel element 窗下墙构件
spandrel-filled arch 实肩拱;填肩拱
spandrel-filled arch bridge 实腹拱桥;填式拱桥;实肩式拱桥
spandrel frame 拱上框架;拱上构架;三角(形)构架
spandrel girder 托梁;窗心墙托梁;窗肚墙(的)托梁;承重梁
spandrel glass 上下层窗间玻璃;裙边玻璃;拱肩玻璃;幕墙(不透明)玻璃;窗心墙玻璃
spandrel hanger 拱肩吊钩;拱吊杆
spandrel of arch 拱肩
spandrel panel 上下层窗空间盖板;拱肩镶板;拱肩托板
spandrel space 拱上空间;拱肩上空洞;拱肩空间
spandrel step 三角形踏步;三角形截面踏步
spandrel structure 拱上结构
spandrel-wall girder-short column system 深梁矩柱系统
spandrel wall(ing) 窗台墙;拱上挡土墙;拱上墙;拱上侧墙;拱肩墙;上下层窗间墙;窗裙墙;窗肚墙;腹拱墙
spandrel waterproofing 窗心墙防水
spandril 窗心墙;模板;空间垫板;拱上墙;拱肩墙;(上下层的)窗间墙
spangolite 氯铜铝矾
spang wire rope knife 钢丝绳切断器
spanipelagic plankton 稀表性浮游生物
Spanish architecture 西班牙式建筑

Spanish blind 卷帘式百叶窗;西班牙式窗帘;西班牙式百叶窗;西班牙式遮帘
Spanish brown 西班牙棕
Spanish burton (有一个定滑轮和两个动滑轮的)滑车组;定单与定单复绞辘
Spanish chestnut 西班牙栗木
Spanish fox 左搓绳股
Spanish garden 西班牙式庭园
spanish gourd 南瓜
Spanish grass fiber 西班牙草纤维
Spanish green 西班牙绿(浅暗绿色)
Spanish high gothic(style) 西班牙高哥特式建筑
spanishing 堵墨;凹纹印刷
Spanish leather 西班牙皮革
Spanish mahogany 巴西红木;西班牙桃花心木
Spanish oak 西班牙栎
Spanish ocher 西班牙赭石;赭石橙色
Spanish oil of turpentine 西班牙松节油
Spanish paint 白色涂料
Spanish red 朱砂;红红红色;铁红
Spanish red oxide 西班牙氧化铁红
Spanish roofing tile 西班牙式屋面瓦;筒瓦
Spanish stonecrop 西班牙景天
Spanish style 西班牙式建筑
Spanish tile 半筒瓦;西班牙式屋面瓦;筒瓦;平搭瓦;(一端稍宽的)弧形屋面瓦;单搭接式瓦
Spanish topaz 褐石英
Spanish type of gothic 西班牙哥特式建筑
Spanish white 西班牙白;染白料;白垩粉;硝酸铋
Spanish windlass 绞绳法
Spanish yellow 西班牙黄
spank 拍打;船在浪中颠簸拍水
span lashing 跨绳
span length 跨径(长);跨距长度;跨距;跨度(距离);跨长
span length of the installed belt 安装皮带的跨度
spanless 不可测量
span line 架空线(路);高架线
span mooring 用双锚固定的系锚
spanned file 生成文件
spanned item 跨项
spanned record 跨越记录;跨区记录
spanner 钩形扳手;螺帽扳手;紧固器;紧固扳手;水平剪力撑;水平剪刀撑;扳子;扳钳;扳紧器
spanner band 束带
spanner double ends 双头扳手
spanner drill chuck 扳手钻夹头
spanner for anchor brake 锚制动器扳手
spanner fork 扳叉
spanner for spark plug 火花塞扳手
spanner for square nut 方形套筒扳手;方形螺帽扳手
spanner for square screw 方螺钉扳手
spanner hole 扳手孔
spanner man 钻机维修人员
spanner nut 旋紧螺帽;扳手螺母
spanner stay 两吊柱之间的横牵缆
spanner wrench 插头扳手;开脚扳手;活络扳手;外卡扳手;套扳手;插销扳手
spanning 拉线;跨越;跨度
spanning forest 生成森林
spanning in one direction 单向跨越
spanning in two direction 双向跨越
spanning member 框架横梁
spanning subgraph 全涉及子图;生成子图
spanning supergraph 生成母图
spanning tree 生成树
spanning value 跨越值
span of arch 拱跨
span of beam 梁跨(度)
span of bridge 桥梁跨度;(以墩分隔的)桥跨
span of bridge and culvert 桥涵孔径
span of control 管理跨度;控制跨度;控制范围;控制变化范围
span of crane 起重机臂伸距
span of jaw 钳口开度
span of life 寿命
span of management 管理范围
span of rod 杆距
span of roof truss 屋架跨度
span of slab 板跨度
span of tow-rope 拖缆的跨距
span of truss 桁架跨度;屋架跨度
span of vault 穹顶跨度

span piece 系杆;檩条;横木;系梁;拉条;横梁木;水平剪力撑;水平剪刀撑
span pipe line 横跨管线
span point 跨绳端点
span pole 拉线电杆
span reinforcement 跨中钢筋;跨中配筋
span rigging 千斤索具
span roof 等斜(坡)屋顶;普通等斜屋顶;双坡屋面
span rope (钻塔的)拉紧杆;(井架的)绑绳
span saw 框锯
span section 跨中截面
span shackle 三角环
span table 宽度表
span valley bridge 跨谷桥
span valley crossing 跨谷测量
span width 跨越宽度
span wire 悬索;横跨索
span-wire fueling rig 跨绳加油设备(海上两船并列航行加油用)
span-wire method 张索输油法
spanwise 翼展方向;宽度方向
span with fixed ends 固端跨
spa park 疗养地公园
spa promenade 疗养地散步场所
spar 小梁;翼梁;组合太阳望远镜;柱木;亮晶;晶石;闪光矿石;椽子
sparagmite 破片砂岩
spar barge 带支腿方程
spar boom 主梁
spar buoy 圆柱浮标;柱形浮标;杆状浮标;杆形浮标;杆形灯桩
spar ceiling 货舱壁护条;护舱板
spar dash 干粘石饰面
spar deck 工作甲板;轻甲板
spar deck beam 轻甲板梁
spar decker 轻甲板船
spar deck ship 轻甲板船
spar dust 晶石粉尘
spare 泥屑;备用的;备件;备份【计】
spare anchor 预备锚;备用锚
spare area 备用区
spare armature 备用电枢
spare bag 备用袋
spare battery 备用蓄电池;备用电池组
spare bed 临时增加的床位(客户里)
spare bedroom 备用卧室
spare bin 备用储仓
spare bit 备用位
spare blade 备用页片
spare block 备用块
spare bow anchor 备用船首锚
spare bower 备用船首锚
spare cable 备用电缆
spare capacity 闲置生产能力;空闲生产能力;储备通引能力;备用通行能力;备用容量
spare capital 闲散资金
spare cash 余额
spare channel 备用信道;备用通路
spare circuit 备用线路
spare coal bunker 备用煤舱
spare code 备用代码;备用编码
spare craft 宇宙飞船
spare details 备件明细表;备用零件
spare disposable blade 备用可换刀片
spared populated area 人口稀少地区
spare equipment 备用设备
spare face 备用回采工作面
spare fender 护舷木
spare film 备用胶卷(片)
spare-fixed assets 备用固定资产
spare flight 宇宙飞行
spare fuse 备用熔断器
spare gate 备用闸门
spare gear 备用机具
spare generator set 备用发电机组
spare hand 预备人员;替班工人
spare labo(u)r force 剩余劳动力
spare lights 备用号灯
spare line 备用线
spare machine 备用机器
spare materials 备用材料
spare materials storehouse 集中备用物资
spare matrix 稀疏矩阵
spare member 备用构件

spare module 备用模件
spare money 余款
spare motor 备用电动机
spare neither labo(u)r nor money 不惜工本
spare nozzle 备用管口
spare nozzle for furnace inspection 炉内检视孔
spare oil bunker 备用油柜
spare package 储备组件;备用组件
spare part compartment 备用舱
spare parts 零件;散件;备用元件;备用零件;备用件;备用部件;备用部分;备用;备品备件;备件;备用品
spare parts and components 备品配件
spare parts case 备品箱;备件箱
spare parts cleaner 备件清洗机
spare parts cleaning machine 备件清洗机
spare parts collection 备件分类
spare parts for finishing thresher 清选脱粒机备件
spare parts for turnout 道岔备件
spare parts inventory 备件清单
spare parts kit 零件箱;备用零件箱;备件箱
spare parts list 备件清单;备件单;备件表
spare parts management 备件管理
spare parts manual 备件手册
spare parts per unit 每台备件
spare parts service 备件检修;备件服务
spare parts store 备件仓库
spare parts supplier 备件供应商
spare path 备用信道
spare piece 备品
spare pipe 厕所喷水管
spare piston 备用活塞
spare plant 备用装置
spare power source 备用电源
spare processor 备用处理机
spare propeller 备用螺旋桨
spare pump 备用泵
spare rail 备用轨
spare rails per kilometer of road 每千米备用钢轨
spare reserves 备用储[贮]量
spare roll 备用轧辊
spare room 客房;备用室;备用房(间)
spares 备用品;备品
spare sack 备用袋
spares allocation 备件分配
spare set 备用机组
spare shaft 备用轴
spare sieve 备用筛
spare space 空位;备用位置
spare spindle 备用轴
spares planning 备件设计
spare spring 备用弹簧
spares provisioning 备件供应
spares requirements 备件需要量
spares service 备件检修;备件服务
spare stand 备用机座
spares vocabulary 备件清单
spare system 附加系统
spare tank 备用池
spare tape 备用带
spare test piece 备用试样
spare time 业余时间
spare time at sea 海上备用时间
spare-time employment 业余职业
spare-time school 业余学校
spare tire 备用轮胎
spare ton 容积吨
spare towing rope 备用拖缆
spare track 备用磁道
spare track link 备用履带节
spare train paths 备用列车运行线
spare transformer 备用变压器
spare tube 备用电子管
spare unit 备用器件;备用(元)件;备用设备;备用机组;备用机件;备用部件
spare varnish 备用清漆
spare wheel 备用轮
spare wheel carrier 备胎架;备轮架
spare wheel carrier latch 备轮架锁键
spare wheel container 备轮壳
spare wire 备用线
spar fabric 亮晶组构
spar fender 护舷木;防舷木
spar finish 光亮面;闪光荧石饰面;阳光热量反射

表面;反光面
spar flour 晶石粉
sparge 喷雾;喷洒
sparge conduit 喷水管
sparged packed-bed biofilm reactor 喷洒填料床生物膜反应器
sparge pipe 穿孔扩散器;厕所喷水管;喷洒器;小便槽冲洗管;喷液管;喷水管;配水管;多孔冲洗管
sparge pipe for urinal 小便槽冲洗管
sparger 气泡喷出器;配电器;散泡器;多孔管分布器;分布器
sparger diffuser 喷气扩散器
sparge ring 喷射环
sparger turbine 喷气器透平
sparge system 喷射系统
sparge tube 喷水管
sparge turbine 喷雾涡轮机
sparge water 冲洗水
sparging 喷射
sparing 舱侧护条
sparing grain 灌浆不好的籽粒
sparingly soluble 微溶的
sparite 亮晶石灰岩;亮晶灰岩;亮晶;胶质方解石
sparite cement 亮晶胶结物
spark 金刚钻;火花;瞬态放电
spark absorber 消弧器;火花吸收器
spark adjustment 火花调整
spark advance 火花提前
spark arrester 烟囱口网络;集烟罩;避电火花器;抑制火花装置;灭火集尘器;火星消灭器;火花制止器;火花抑制器;火花消灭器;火花消除器;火花避雷器
spark at make contact 闭合火花
spark ball 火花放电球
spark blow-out plate 灭火花板
spark breakdown 发生火花放电击穿
spark capacitor 灭火花(用)电容器
spark catcher 灭火集尘器
spark chamber 灭弧腔;火花放电室
spark circuit 火花电路
spark coil 电火花线圈
spark coil leak detector 火花检漏器
spark condenser 灭火花电容器
spark control lever 点火控制杆
spark counter 火花计数器
spark cutting 火花切割;电火花切割
spark detector 火花检波器
spark discharge 火花放电;火花
spark discharge detector 火花放电检测器
spark discharge forming 火花放电成型;电火花放电成型
spark discharge hardening 火花放电硬化法
spark discharge particle generator 火花放电粒子发生器
spark distance 电花距离
spark distributor 火花分配器
spark drilling 电火花钻进
sparked tape 墙板接缝黏带
sparker 火花捕捉器;电火花器
spark eroding 电火花腐蚀
spark erosion 火花烧蚀;火花腐蚀;火花电蚀
spark excitation 火花激励;火花激发
spark explosion method 火花爆炸法
spark extinguisher 灭火器;火花消除器
spark-extinguishing 灭火花(的)
spark-extinguishing circuit 灭火花电路
spark formation 火花形成
spark gap 火花放电隙;火花(间)隙;电花隙;放电器;放电间隙
spark gap adjustment 火花隙调整
spark gap arrester 火花隙式避雷器
spark gap exciter 火花间隙激发器
spark gap generator 火花隙振荡器
spark gap inspection 放电检验
spark gap modulation 火花隙调制
spark gap modulator 火花隙调制器
spark gap transmitter 火花隙式发射机
spark gap type rectifier 火花隙式整流器
spark gap voltmeter 火花隙伏特计;火花间隙电压表
spark gap width 火花隙宽度
spark generator 火花发生器
spark governor 火花调节器
spark hammer drilling 电火花潜孔锤钻进
spark-hand lever 点火杆

spark-hand-lever tube 点火杆管
spark hardening 电火花硬化
spark igniter 火花点火器
spark ignition 火花点火;强制点火;电火花点火
spark ignition engine 火花点火式内燃机;火花点火发动机
spark ignition fuel-injection engine 火花点火喷油式发动机
spark impact transmitter 冲击火花发报机
spark inductor 火花感应线圈
sparking 点火;打火花;发火花
sparking advance 提前点火
sparking alloy 发火合金
sparking coil 点火线圈;电火花感应线圈
sparking commutator 火花整流子
sparking contact 火花触点;发火花的触点
sparking distance 放电器;火花隙;火花间隙;跳火距离
sparking electrode 火花电极
sparking gear 发火装置
sparking limit 火花极限
sparkingnition combustion cycle 奥托循环
sparking of brushes 电刷发火花
spark(ing)plug 火花塞
sparking plug cleaner 火花塞清洁器
sparking plug ignition 火花塞点火
sparking point 发火点
sparking potential 击穿电压;火花电压;闪电电压
sparking voltage 放电电压
sparking voltage resistance 耐火花电压
spark initiation 火花起爆
spark intensifier 火花增强器
spark ionization 火花电离
spark ionization chamber 火花电离室
spark ionization source 火花离子源
spark ion source mass spectrometer 火花离子源质谱计
spark isostatic press sintering 等静压火花烧结;等静力火花烧结
spark jump 发火花
spark killer 消弧器;减弧器;火花断路器;电火花消灭器
spark knock 火花爆击;点火爆震
sparkle 火花
spark lead 点火提前
sparkle metal 起泡金属
spark length 火花长度
sparkless 无火花的
sparkless commutation 无火花整流;无火花换向
sparkless commutator 无火花整流器
sparkless rectification 无火花整流
sparklet 发泡粉
spark lever 点火杆
spark lighter 点火器
spark line 电花谱线
sparkling 腻子料;火花;涂层打底;填料;灰腻子
sparkling glaze 星盏
sparkling pistol 发火手枪
spark linkage 点火联动装置
spark machining 电火花切割
spark machining electrode 电火花加工电极
spark micrometer 火花放电显微计;火花测微计
spark out 抛光到消失火花
spark over 火花放电;火花;跳火;打火花
spark-over test 火花放电试验;击穿试验
spark-over voltage 火花放电电压;跳火电压;飞弧电压
spark pin 引火针
spark plasma 火花等离子体
spark plate 灭火花板
spark plug 发火枪
spark plug adapter 火花塞座
spark plug body 火花塞壳
spark plug cap 火花塞帽
spark plug centre rod 火花塞中心杆
spark plug cleaner 火花塞清洁器
spark plug cooler 火花塞冷却器
spark plug electrode 火花塞电棒
spark plug engine 火花塞点火发动机
spark plug extension 火花塞伸长部
spark plug fouling 火花塞积炭
spark plug gap 火花塞间隙
spark plug gap feeler ga(u)ge 火花塞间隙规
spark plug gap setting tool 火花塞间隙校正器

spark plug gasket 火花塞填密圈
spark plug ga(u)ge 火花塞规
spark plug hole 火花塞孔
spark plug ignition 火花塞点火
spark plug insert 火花塞套座
spark plug insulator 火花塞绝缘体
spark plug pliers 火花塞钳
spark plug post 火花塞柱
spark plug resistance tester 火花塞电阻测试仪
spark plug resistor 火花塞电阻
spark plug shell 火花塞壳
spark plug socket wrench 火花塞套筒扳手
spark plug taps 火花塞螺丝攻
spark plug terminal 火花塞线接头
spark plug terminal nut 火花塞接头螺母
spark plug tester 火花塞试验器
spark plug tire pump 火花塞插头轮胎打气泵
spark plug type indicator 火花塞式指示器
spark plug wire 火花塞线
spark plug wrench 火花塞扳手
spark point 火花电极
spark potential 火花电势
spark producer 火花发生器
sparkproof 防火花的;不起火花的;无火花的
sparkproof concrete 不发火花混凝土;防火花混凝土
sparkproof mortar 防火花砂浆
sparkproof type 防爆型
spark pulse modulator 火花脉冲调制器
spark quadrant 点火杆扇形齿轮
spark quench device 熄灭火花装置
spark quenching 电火花淬火
spark quenching condenser 火花速熄电容器
spark recorder 火花记录器
spark regulation 火花调节器
spark resistance 火花电阻
spark resistant 抗火花的
spark resistant wire wheel brush 火花电阻圆盘钢丝刷
sparks 电工
spark sending 火花发送
spark setting 火花塞装置
spark sintering 火花烧结
spark sintering machine 火花烧结机
spark source 火花源;电火花震源
spark source mass spectrometer 火花源质谱仪;火花源质谱计;固体源质谱计
spark source mass spectrometry 火花源质谱法
spark spacer 点火分电器
spark spectrography 火花光谱分析
spark spectrum 火花光谱;电花光谱
spark stick 绝缘子测试棒
spark suppressor 火花抑制器
spark technique 火花法
spark test (鉴定钢材含碳量的)火花试验;火花检验;电火试验(检查瓷釉的针眼);放电检验
spark tester 火花检验器;火花检测器
spark time lever 发火定时杆
spark timer 火花计时器;电火花计时器
spark timing 火花定时器
spark tube 闪光管
spark voltage 火花电压
spark welding 火花焊
spark wheel 点火轮
spark working 火花加工
spar loading 翼梁荷载
spar miller 翼梁铣床
spar-mooring 杠撑系泊
spar piece 系梁;拉条;交叉支撑
spar platform 桅式平台
spar powder 晶石粉
sparred seat 板条椅
sparring 货舱壁护条;舱壁护条
sparrite 亮方解石
sparrow brace 雀替
sparrow peck 拍打饰面;琢毛;雀啄纹;拉毛;粗琢面;粗刷面
sparrow pecked 点状花纹(砖面)
sparrow starving 梁端间的塞墙;梁端塞墙
sparry 亮晶的;晶石
sparry calcite 亮晶方解石;亮方解石
sparry iron 菱铁矿
sparry limestone 颗粒大理石;晶石大理石;含似晶石的石灰岩
sparry mosaic 亮嵌晶

sparse 稀疏的;稍密
sparse concealment 稀疏隐蔽
sparse data area 资料稀少(的)区域;资料稀少(的)地区;资料稀缺(的)区域;资料稀缺(的)地区;资料短缺(的)区域;资料短缺(的)地区
sparse inverse 稀疏逆矩阵
sparsely or density branches 枝稀疏或密集
sparsely populated area 人口稀少地区
sparse matrix 稀疏矩阵
sparse traffic 稀疏交通
sparse vector 稀疏向量
sparse vegetation 稀疏植被;稀疏的植物
sparse wood 稀疏林地
sparse zone 波疏带
spar ship 竖筒形海洋研究船
spartalite 红锌矿
Spartan environment 恶劣环境
spar-tree 集材架杆
spar varnish 耐水外用清漆;耐水清漆;清光漆;桅杆清漆;桅杆耐水清漆;船舰用漆
spaser 垫背;撑挡;衬垫;炮塞衬垫
spash-proof electric(al) equipment 防溅式电气设备
spasmodic burning 脉动式燃烧;反常燃烧
spasmodic current 间隙流
spasmodic density current 间歇异重流
spasmodic factor 间断因素
spasmodic motion 间歇运动
spasmodic turbidity current 间歇浊流;间歇异重流;特发浊流
spat (门框底的)护板;流线型罩;流线型轮盖;流线型轮盖;机轮减阻罩
spate 暴风雨;涨大水;猛涨;洪水猛涨;洪水;河水猛涨
spate flow 洪流;河道洪水流;地表暴雨径流;河水猛涨
spate irrigation 洪水灌溉
spate runoff 暴雨径流;地表暴雨径流
s-path 单通路
spath 亮晶
spathe 佛焰苞
spa thermae 温泉
spathic iron 菱铁矿
spathic iron ore 菱铁矿石
spathic ore 菱铁矿石
spathization 亮晶化作用
spatial 空间的
spatial ability test 空间能力测验
spatial acuity 空间视觉锐度
spatial analysis 空间分析
spatial and temporal dependence 时空相关性
spatial angle 隅角;立体角
spatial arrangement 空间排列;空间布置;空间布局
spatial art 空间艺术
spatial attenuation 空间衰减
spatial autocorrelation 空间自相关
spatial autocorrelation function 空间自相关函数
spatial bar 间隔钢筋
spatial brightness 空间亮度
spatial carpenter inverter 圆柱控制器
spatial carrier 空间载波
spatial change 空间变化
spatial character 空间特征
spatial characteristics 空间特征
spatial chemistry 立体化学
spatial city 空间发展城市
spatial coefficient 空间系数
spatial coherence 空间相干性
spatial coherence function 空间相干函数
spatial comparison 空间比较
spatial composition 空间构造;空间构图;空间组成
spatial concentration of sediment 水流体积含沙量浓度;单位容积含沙量;水中泥沙单位;水中泥沙集中区;水中泥沙单位水流体积含沙量
spatial configuration 空间构型;立体构型
spatial configuration of basin facies 盆地相的空间配置
spatial configuration of molecules 分子的空间配置
spatial conjugation 立体共轭
spatial construction 空间构造
spatial continuity 空间连续性
spatial coordinates 空间坐标
spatial correlation 空间相关(性)
spatial curve 空间曲线

spatial data 空间坐标数据;三维坐标数据
spatial decision 空间决策
spatial deformation 空间变形
spatial degeneracy 空间简并度
spatial degeneracy effect 空间简并效应
spatial dendrite 立体枝状冰晶
spatial dendritic crystal 立体枝状冰晶
spatial description 空间描述
spatial design 空间设计
spatial dimension 立体尺寸;空间范围;空间尺度
spatial dimensions of economics 经济空间范围
spatial direction 空间方向
spatial dispersion 空间色散
spatial distance 空间间距
spatial distribution 空间分布;空间布局
spatial distribution characteristics of lines 线的空间分布特征
spatial distribution method of magnetic field 磁场空间分布法
spatial distribution model of point 点的空间分布模型
spatial distribution of forces 力的空间分布
spatial distribution of hypocenter 震源空间分布
spatial distribution of population 人口地区分布
spatial distribution of seismic load 地震荷载空间分布
spatial division 空间分隔
spatial domain 空间域
spatial domain filtering 空间域滤波
spatial economic adjustment 空间经济调整
spatial economics 空间经济学
spatial effective resolution element 空间有效分辨元
spatial emission 空间发射
spatial enclosure 空间封闭物
spatial encoding 空间编码
spatial environment 空间环境
spatial environmental planning 空间环境规划
spatial exclusion chromatography 空间排阻色谱法;空间排阻层析
spatial factor 空间因素;空间因数
spatial fidelity 空间保真度
spatial filter 空间频率滤波器;空间滤光片
spatial filtered image 空间滤波图像
spatial filtering 空间滤波
spatial filtering method 空间滤波法
spatial filtering of marine seismic data 海上地震数据的空间滤波
spatial floating mark 空间浮动测标
spatial form 空间形式
spatial frame(d) supporting structure 空间构架支承结构
spatial frame(work) 空间构架(工程);立体构架;空间框架
spatial frequency 空间频率
spatial frequency characteristic 空间频率特性
spatial frequency component 空间频率分量
spatial frequency discrimination 空间频率鉴别
spatial frequency domain 空间频率域
spatial frequency multiplexing 空间频率倍增
spatial frequency pseudocolo(u)r encoding 空间频率假彩色编码
spatial frequency response 空间频率响应
spatial frequency spectrum 空间频谱
spatial frequency vector 空间频率向量
spatial gain 空间增益
spatial grid 空间格网;浮游格网
spatial grid structure 空间网架结构
spatial hole burning mode 空间烧孔模
spatial homogeneity 空间一致性;空间均一性
spatial hydraulic jump 空间水跃
spatial image 立体像
spatial incoherent light 空间非相干光
spatial information enhancement 空间信息增强
spatial integration 空间积分
spatial integration method 空间积分法
spatial intensity distribution 空间强度分布
spatial interpenetration 空间相互钩搭
spatial invariance 空间不变性
spatial irradiance 空间辐照度
spatial isolation 空间隔离;地区隔离;地理隔离
spatial isomerism 立体异构;空间异构
spatialization 空间定位
spatial lake acidity monitoring network 空间湖泊酸度监测网

spatial layout 空间布置;空间设计
spatial light modulator 空间光调制器
spatial limit 空间范围
spatial load bearing structure 空间承重结构
spatial load-carrying structure 空间承重结构
spatial loaded structure 空间承重结构
spatial load(ing) 空间荷载
spatial location 空间位置
spatially aligned bundle 空间定位光纤束
spatially bound function 空间有缚函数
spatially coherent beam 空间相干光束
spatially coherent light 空间相干光
spatially coherent radiation 空间相干辐射
spatially distributed computer control system 空间分布计算机控制系统
spatially integrate water quality data 空间综合水质数据
spatially invariant system 空间不变系统
spatially multiplexed infrared camera 空间多路传输红外照相机
spatially varied flow 沿程变流量;空间变化水流
spatially varied transport rate 沿程变流量输沙率
spatially varied velocity 沿程变流速
spatial margins to profitability 可能获利的空间极限
spatial mode 空间型
spatial model 立体模型;空间模型
spatial model of bridge 桥梁立体模型
spatial modulation 空间调制;空间调节
spatial modulator 空间调制器
spatial moment 空间矩
spatial monopoly 空间垄断
spatial multiplexing 空间多路法
spatial multiplex tomography 空间多路层析 X 射线照相法
spatial noise filter 空间噪声滤波器
spatial noise filtering 空间噪声滤波
spatial non-uniformity 空间不均匀性
spatial offset technology 空间像素错位法
spatial optimization 空间最优化
spatial orientation 空间定向;空间定位
spatial ozone distribution 臭氧的空间分布
spatial panel(l)ing pattern 空间镶板图案
spatial parallel system 空间平行体系
spatial pattern 空间图形;空间格局
spatial penetration 空间贯穿
spatial periodic inversion saturation 空间周期性反转饱和
spatial phase 空间相位
spatial phase shift 空间相移
spatial planning 空间利用规划
spatial point 空间点
spatial position 空间位置
spatial potential 空间位(势)
spatial potential function 空间势函
spatial prestressing 空间预应力
spatial price discrimination 空间价格差别
spatial pricing policy 区域价格政策
spatial problem 空间课题
spatial purlin(e) 空间檩条
spatial-radial flow 空间径向流
spatial random action 空间随机作用
spatial randomness 空间随机性;地区随机性
spatial resection 空间后方交会
spatial resolution 空间分辨率
spatial response spectrum 空间反应谱
spatial rod 空间导杆
spatial scanning correlator 空间扫描相关器
spatial self-focusing 空间自聚焦
spatial separation 空间隔离
spatial sequence 空间序列
spatial series 空间序列
spatial shape 空间形式
spatial similarity transformation 空间相似变换
spatial sink line 空间汇线
spatial sorting 空间分选
spatial statistics 空间统计量
spatial steric stabilization 空间屏蔽稳定作用
spatial stretching 空间拉伸
spatial structural system 空间结构体系
spatial structural variable 空间结构变量
spatial structure 空间结构
spatial summation 空间总和;空间性总和

spatial supporting structure 空间支承结构
spatial symmetric(al) vibration 空间对称振动
spatial symmetry 空间对称性
spatial system 空间体系
spatial temperature field 空间温度场
spatial-temporal electro-optic light modulator 空间时间电光调制器
spatial tensioning 空间张拉
spatial town 占空间城市
spatial transformation 空间变换
spatial triangulation 空中三角测量;空间三角测量;航空三角测量
spatial truss 空间桁架
spatial truss analogy 空间桁架模拟法
spatial unit 空间单元;空间构造
spatial value 空间值
spatial variability 空间可变性
spatial variation 空间变化;区域差异
spatial variation in flow velocity 流速沿程变化
spatial variation in water and sediment 水沙沿程变化;水沙空间变化
spatial variation type 空间变化类型
spatial velocity 空间(传播)速度
spatial vision 立体视觉
spatial wave 空间波;天波
spatial weight-carrying structure 空间承重结构
spatio-dynamic composition 空间动态组成;空间动态结构
spatiotemporal 空间时间的
spatium episclerale 环球间隙
spat of oyster 牡蛎苗
spatter 溅粘;溅射;溅落熔岩;溅;焊接飞溅;喷溅;喷镀;滴落;飞溅(焊接时)
spatter coating 溅彩涂装法
spatter cone 溅落熔岩锥;寄生熔岩锥
spatter dash 粗涂层;溅彩;喷洒;甩灰打底;甩浆搭面
spatterdash 甩浆搭面
spattered tile 珠状饰方砖
spatter finish 溅彩涂装;溅彩涂层;多彩涂层
spatter glass 混合玻璃;混彩玻璃
spatter gun 溅涂喷枪;溅彩喷枪
spattering 溅涂;溅彩;飞溅涂层;飞溅(焊接时)
spatter loss 溅损;飞溅损失
spatter loss coefficient 飞溅率(焊接时)
spatter method 溅彩涂装法
spatter paint 溅彩涂料;多彩涂料
spatter resistance 抗溅性
spatters 溅出物
spatter shield 防溅屏;防溅挡板
spatter ware 斑纹釉艺术陶器
spatting 喷溅麻点
spattle 抹刀;刮刀;刮勺;软膏刀;油漆刀
spatula 涂敷抹刀;油漆刮刀;铸型修理工具;刮勺;刮刀;刮铲;软膏调刀;修刀;土刀;调漆刀;调刀
spatula blade 刮铲刀
spatula rub-out method 刮刀混合法
spatulate 抹刀形的
spatula test 铝粉浆漂浮率试验
spatule 软膏刀;调漆刀;调刀
spaul 剥落;碎石
S. paul's cathedral 圣·保罗大教堂
spave 有斑痕的木材
spavin 粗黏土
spall 碎片石;剥落;碎石
spawning area 产卵区
spawning bed 产卵场
spawning bed for species 物种产卵地
spawning ground 产卵场
spawning ground for fish 鱼类产卵场
spawning habit 产卵习性
spawning location 产卵地
spawning migration 产卵洄游
spawning season 产卵期;产卵季节
spawning site 产卵地
spawning site for fish 鱼的产卵场(地);鱼产卵场所
spay angle 八字角
spayed edging 八字边口
SPC telex exchange 程控用户电报交换机
speader 展着剂
speak-back circuit 工作联络通话电路
speaker 广播员
speaker baffle 扬声器障板
speaker box 扬声器箱
speaker cabinet 扬声器箱

speaker-dependent 声音辨识器
speaker enclosure 扬声器箱
speaker-phone 扬声器电话
speaker's platform 演讲台
speaker stand 讲台
speaker unit 扬声器装置
speaking circuit 通话电路
speaking example 生动实例
speaking-place 演讲厅
speaking rod 自读式水(准标)尺;自读标尺
speaking trumpet 喇叭筒
speaking tube 话筒;通话管;传声筒;传话筒
spear 矛状体;打捞矛;铲尖
spear grass 针茅
spearhead 打捞矛头;矛头
spearhead bit 矛头形钻头
spearhead dieback 矛头状顶死
spearhead shovel 矛形松土铲
spear marker 矛式划行器
spearmint 留兰香
spearmint oil 留兰香油;薄荷油
spear pilot valve 针阀导阀
spear plate 衬板
spearpoint 矛式夹(一种格栅固定夹,固定于混凝土的)
spear-pointed drill 尖嘴钻;矛式钻
spear pointer 箭头指针;矛形指针
spearpoint shovel 矛形松土铲
spear servomotor 针阀接力器
spear with releasing mechanism 自动松机构的捞矛
spec 说明书
spec house 待售房屋
special 专用的;专门的;特种的;特设的;特别的;附加的;额外的
special acceptance 特别承兑
special accessories 专用附件
special accessory 特殊附件
special account 特殊账户;特别账户
special account for traffic revenue 运输收入专户
special acrylic acid 特性丙烯酸
special add 双精度加法
special additive 特种外加剂
special adhesive 特种黏结剂
special adhesive composition 特种黏结剂混合物
special adhesive compound 特种黏结剂混合物
special administration area 行政特区
special administrative region 特区
special advisory engineer 特别咨询工程师
special advisory message 特别咨询电报
special aeronautical chart 特种航空图
special agency 专门机构
special agent 特定代理商;特别代理(人)
special agglutination 特异凝聚
special air cock 特制放气阀
special air operation 特别空中活动
special air pick-up 特种空中钩取
special air region 空中特区
special allocation 专用拨款;特别拨款
special allowance 特别折让
special alloy 特种合金
special alloy steel 特种合金钢;特殊合金钢
special ammunition depot 特种弹药仓库
special ammunition supply area 特种弹药补给区
special ammunition supply point 特种弹药供应站
special anaerobic filter bed 特种厌氧滤床
special analog(ue) system 特殊模拟系统
special analysis data of oil field water 油田水专项分析数据
special anchor 特殊锚
special anchor shackle 特殊锚卸扣
special application bucket 特殊用途的料斗
special appointment contract 指定分包合同
special appointment work 指定承包工程
special appropriation for special use 专款专用
special appropriation(fund) 专用拨款
special approved stocking funds 特准备资金
special area 萧条地区
special area annotation 分区注记
special areas 特别地区
special areas for process(ing) export products 出口加工区
special arm and hand signal 特种手势信号
special array 特种台阵;特殊台阵

special arts and crafts 特种工艺
special asphalt 特种沥青
special asphalt roofing felt 特种沥青屋面卷材
special assembly system 专用组合系统
special assessment 特别派款;(对房地产征收的)公用事业专用税;特种赋税;特别征税;特别税;特别评价;特别捐税
special assessment district 特殊估价地区
special assessment liens 特赋留置权
special assessment roll 特赋清册
special atmosphere furnace 保护气氛退火炉
special atmosphere oven 保护气氛炉
special attachment 专门附件;特殊附件
special attention paid to the release of suspended solids 特别注意悬移质的泄放
special audit 特种审计;特别审计
special automatic modular type band saws 组装式自动专用带锯
special automatic vacuum welding chamber 特种自动真空焊接室
special bank account 特种银行账户
special bank loans 专用银行借款
special bearing 特殊轴承
special bit 特种钻头
special boat 特种船泊(巡逻船工作船等)
special body 特种车身
special boiling point gasoline 特殊沸程汽油
special boiling point spirit 特定沸点酒精
special boiling point spirits 特定沸点的醇类
special bolt 特种螺栓
special bond 特种砌合
special bonding adhesive 特种胶黏剂
special bonding agent 特种胶黏剂
special bonding medium 特种胶黏剂
special bonus 特别奖金;特别红利
special box roof gutter 箱形屋顶雨水沟
special bracket 特别分档
special brick 特殊砖;特种砖
special bridge 特殊桥梁
special broker 特种经纪人
special bronze 特种青铜;特殊青铜
special bucket 特殊料斗
special budget 特殊预算;特别预算
special builder's fitting 特制的建筑工人用具
special builder's furniture 特殊的建筑用具
special building 特殊建筑物
special building material 特种建筑材料
special building method 特种建筑方法
special build-up welding 特殊堆焊
special burden 特殊举证责任
special burnt brick 特制烧透砖
special buyer 特别买方
special cable 专用电缆
special call 专项通报;特别呼叫
special capital 专用资金;专用资本
special car 专用车;特种汽车
special carburettor 特种化油器
special career program(me) 特殊专业计划
special care home 特殊照顾的收容所
special cargo 特种商品;特种货物;特种船货;特殊货物
special cargo container 特种货物集装箱
special cargo list 特种货物清单
special carrier 专用支座;专用夹头;特种夹头;特殊夹头
special carte 特制纸片;电木纸
special case 特殊情况;特殊表壳;特例
special case report 特别检查报告
special casting 特种浇注;特种铸件;特种配件
special cast iron 特种铸铁;特殊铸铁
special cement 特种水泥
special cementing composition 特种胶凝剂
special cementing compound 特种胶凝剂
special cement pipe 特种水泥管
special census 特别普查
special ceramic fiber 特种陶瓷纤维
special ceramics 特种陶瓷;特殊陶瓷
special channel 专用信道;特种信道;特殊航道
special character 专用字符;特种字符;特殊字符
special character set 特殊字符组
special charges 特别费用;特种费用
special charging machine for cadmium-nickel accumulator 镉镍蓄电池充电机
special chart 特种图

special chemical composition cooling valve 特种化合物冷却阀
special chemical cylinder 化工专用钢瓶
special chemical equipment 化工专用设备
special chemical tanker 化工专用槽车
special chuck jaw 专用夹盘爪
special ciphers 特别密码
special circuit 专用电路
special circumstance 特殊情况
special class 特级
special clause 特殊条款;特别条款
special clauses for wood oil in bulk 散装桐油特款
special clearance coupling 留有间隙的特殊联轴器;留有间隙的特殊耦合
special climate application program(me) 特殊气候应用计划
special clothing 专用工作服
special coal tar pitch 特种煤焦油沥青
special coat 特殊涂层
special coating 专用涂料;特种涂料
special code 特种电码
special code selector 特殊编码选择器
special colo(u)r-index 特殊色指数
special colo(u)r rendering index 特殊显色指数
special column 专栏
special columnar journal 特别多栏式
special column section 特殊的柱截面
special comb filter 特种梳状滤波器
special commerce 特种贸易;特殊贸易
special commercial privileges 商业特权
Special Committee on Global Change 全球性变化问题特别委员会
Special Committee on Solar-Terrestrial physics 日地物理学特别委员会
special commodities 特种矿产
special commodity unit train 特种货物固定编组列车
special communications 特别通信
special communications center[centre] 特别通信中心
special communication system 专用通信系统
special compass angle units 专用罗盘测角元件
special competent bed 特别松软岩石
special components 专用附件
special composition 特种混合料
special composition quality of water 水的特殊成分分析
special compound 特种混合物
special concrete 特种混凝土
special concrete product 专用混凝土制品;特种混凝土制品
special condition 特殊情况;专用条件;特殊条件
special conditions of contract 专用合同条款;合同的特殊条件;特殊合同条件
special construction 特殊结构;特殊工程
special construction(al) material 专用建筑材料;特种建筑材料
special construction expenditures 专项工程支出
special construction profile 专用建筑型材;特种建筑型材
special construction shape 专用建筑型材;特种建筑型材
special construction trim 特种建筑装饰件
special construction unit 特种建筑构件;专用建筑构件
special container 特种容器;特种集装箱
special contingency reserves 特别意外准备金
special contribution 一次性特别缴款
special control 专项治理
special coordinates 特殊坐标
special corporation 特种公司
special correcting circuit 特种校正电路;特殊校正电路
special cost 特定成本
special cost studies 特殊成本调查
special coupling 特种管接头
special covering foil 专用覆盖金属箔片;特种覆盖金属箔片
special crane 特殊吊车;特殊起重机
special credit 特定信用证;特别信用证
special credit columns 特别贷方专栏
special credit fund 特别信贷基金
special credit risk 特别信贷风险
special crop 特用作物

special cross-reference entry 特殊参照
special cross-section steel wire 特殊断面的钢丝
special crushed expanded cinder 特种轧制膨胀性熔渣
special crushed foamed iron blast-furnace slag 特种轧制泡沫高炉熔渣
special curing 特种养护
special current account 特别往来存款
special customs invoice 特别海关发票
special cycle 特种循环
special damage 特异性损害;特殊损害
special debit columns 特别借方专栏
special decompression chamber 特种减压仓;特殊降压闸
special decorative block 专用装饰砌块;特种装饰砌块
special decorative tile 专用装饰砖;特种装饰砖
special delivery 限时专递;快递;特别投递
special deposit 专用存款;专项存款;特别存款;特种存款
special depositary 特别存款银行
special design criterion 特殊设计标准
special designing 特殊设计
special detail drawing 足尺大样图;特殊详图;大样详图
special development area 特别开发区
special device 特种器材
special direct current machine 特种直流机
special direction 特别说明书
special disburser 特别支付代理人
special discharge 特殊使用流量
special discount 特别折扣
special distress signal 遇险特殊信号;遇难特殊信号;特殊遇难信号
special district 特区
special dividend 额外股息
special documentary credit 特别押汇信用证
special drawing rights 特别提款权
special drawing rights certificate 特别提款权证明书
special drawing rights certificate account 特别提款权证券账户
special drawing rights certificate of deposit 特别提款权存款单
special drawing rights parity 特别提款权平价
special drawing rights standard 特别提款权本位
special dredging method 特殊挖泥法
special drill 特殊钻头
special drilling practices 特种钻进工艺
special drilling unit 专用钻削头
special drivage method 特殊掘进法(采矿)
special drive 特殊传动
special-duty electric(al) motor 专用电动机
special-duty motor 特用电动机
special-duty wagon 专用货车
special economic zone 经济特区;特区
special education 特种教育
special educational facility 专用教育设施
special effect 特殊效果;特技效果
special effect generator 特技信号发生器
special elected quality 特优等级
special emergency stair(case) 专用太平梯
special emergency vehicle 特殊应急车辆
special emission 特殊排放
special emission standard 特定排污标准
special emulsion 特殊乳胶体;特殊乳状液
special encyclopaedia 专科性百科全书
special endorsement 特别背书
special engineering geology mapping 专门性工程地质测绘
special engineering plastic 特种工程塑料
special engineering structure 特种工程结构;构筑物
special enlarging printer 专用放大机
special entity 特设机构
special environment 特性环境
special environmental variance 专有环境方差
special equipment 特种设备;专用设备;特殊设备
special equipment for chemical production 化工生产专用设备
special equipment for maintenance 专用维修设备
special equipment for making pipe twist 管子钳专用加工设备
special equipment for silo discharge 水泥底卸料专用设备

special equipment package for depot 车辆段专用设备包
special equivalent 特殊等价物
special equivalent point system 特殊等效点系
special erosion 特殊侵蚀作用
special event 特殊事件
special examination fee 特殊试验费
special excavating method 特殊掘进法
special exception use 特定条件的使用
special exemption of reinvested profits 再投资利润的特殊免税
special expenses 特殊费用;特别开支;特别费用
special explanation 特别说明
special export 专门出口
special exporter 特产出口商
special exporter's declaration 特别出口商声明书
special extraneous risks 特别附加险
special family subsidy 特殊家庭补助
special fare 特殊票价
special fastener 专用固定件;专用紧固件
special feature 特殊特征
special fee 特别费用
special fertilizer for flowers 花卉专用长效肥料
special finance 特别融资
special finish 特殊石面;特殊(饰)面
special finishing hydrated lime 专用罩面熟石灰
special finishing tools 事故处理专用工具
special fire hazard area 特殊火灾危险区
special fittings 专用配件;特种配件
special fixture 专用夹具
special flange 特殊法兰
special flood-reporting network 专用报汛站网
special flood-reporting station 专用报汛站
special flower garden 专类花园
special fluorosilicate 特种氟硅酸盐
special fluxed pitch composition roofing 专用稀释硬煤沥青混合物屋面;专用稀释硬煤沥青混合物屋顶
special fluxed pitch ready roofing 专用稀释硬煤沥青预制屋面
special forecast 专门预报;特殊预报
special format 特种规格
special form brick 异型砖
special formed milling cutter 特形铣刀
special form of commodity exchange 特殊商品交换
special form(work) 特殊模板;特殊材料制的模板
special function 特殊函数
special function generator 专用函数产生器
special fund 专用基金;专项资金;专项工程支出;专款;特种基金;特别基金
special fund appropriation account 特种基金拨款账户
special fund deposit 特种基金存款
special fund receipt accounts 特种基金收入账户
special gang 专司工班
special garden plot 专用绿地
special gear box 专门齿轮箱
special gelatin 特种胶
special geologic(al) condition 特殊地质条件
special geologic(al) data base 专用地质数据库
special geology 特殊地质
special geomorphic map 专门地貌图
special geotechnic(al) engineering 特种岩土工程
special glass 特种玻璃
special glazing for heat insulation 隔热特种玻璃
special goods 特种货物;特殊货物;特别商品
special grade 特级
special greases 特种润滑脂
special grout 特殊薄灰浆;特殊灰浆
special grouting 特殊灌浆(法)
special handicraft products 特种工艺
special handling 特殊装卸
special handling instruction 特殊处理指令
special hardware 专用硬件;专用设备
special hazards insurance 意外事故保险;特殊灾害保险;特种灾害保险
special heating furnace 专用加热炉
special heat vapo(u)r 沸点气化热
special heavy duty hammer breaker 特重型锤式破碎机
special heavy duty hammer crusher 特重型锤式破碎机

special heavy duty hammer mill 特重型锤式破碎机
special heavy section switch rail 特种断面尖轨
special high-grade zinc 特种高级锌
special highway 专用公路
special hoisting bar 专用吊架
special holder 专用刀杆
special hollow block 特种空砌块
special hose 特殊水龙带;特殊蛇管;特殊软管
special hospital 专科医院
special hunting license 特计捕猎证
special hydrogeologic(al) survey 专门性水文地质调查
special identification method 特殊识别法
special import 专门进口;特别进口
special import license 特别进口许可证
special import licensing 专案进口签证
Special import measures act 特别进口措施法(加拿大)
special in carbide tipped 专用硬质合金头
special income tax 特别所得税
special index 专题索引
special index analyzer 专用索引分析器
special indoor system 专门室内系统
special industrial building 特种工业建筑物
special industrial services 特别工业业务
special industrial solvent 特种工业溶剂
special information center 专业情报中心
special information retrieval 专业情报检索
special information tone 特别信息音
special injection and distribution device 特种注射和分配装置
special inorganic fiber 特种无机纤维
special inorganic fiber reinforced ceramics 特种无机纤维增强陶瓷
special inorganic fiber reinforced metal 特种无机纤维增强金属
special inorganic fiber reinforced plastics 特种无机纤维增强塑料
special insertion editing 特殊插入编辑
special inspection 特殊检查
special installation 特殊设备
special instruction 专用指令;特殊指令
special insulating brick 特种耐火隔热砖
special insurance 特殊保险
special integral 特殊积分
special interest edition statement 专业版本说明
special intermediate survey 特别中间检验
special investigation 特别调查
special investigation of railway statistics 铁路统计专项调查
special iron 特种铁
specialisation 专门化
specialism 专门化
special isolate idea 专用细目概念
special issue 专刊;特殊问题;特辑
specialist 专家;特种经纪人
specialist advice 专家建议
specialist commission 专家委员会
specialist contractor 分承包者;专业承包者;分承包人;专业承包商;分承包商;专业承包人
specialist firm 专业公司
specialist manufacturer 专业制造厂家
specialist market-making system 专家经营市场制
specialist operator 专门操作者
specialist plant 特殊厂;特殊车间;特殊设备
specialist report 专题报告
specialist sub-group 专家小组
special item 特殊项目
speciality 特制品;专用项目
speciality goods 名产
speciality of screening 筛检特异度
speciality packet 专业组件;专业标准部件;专门全套部件
speciality shop 名牌货商店
specialization 专业化;专门研究;专门化;专化性;特殊化;特化
specialization and cooperation 专业化协作
specialization of commodities and duty rates 商品与税率的详细分类
specialization of exportation 出口专业化
specialization of industrial production 工业生产专业化
specialization of repair 修理专业化

specialized 专业性的;专门的
specialized agency 专门机构
specialized aquatic flora 特产水生植物
specialized capital goods 专用资本货物
specialized cold store 专业冷藏库
specialized concrete 特种混凝土
specialized contractor 专业承包商
specialized data processing 专业化数据处理
specialized facility 专用设备
specialized farming 专业农业
specialized feature 专化特性
specialized fields 专业门类
specialized fleet 专用船队
specialized freight car 专用货车
specialized freight station 专业性货运站
specialized funds 特别资金
specialized goods station 专业性货运站
specialized government agencies 业务领导机关
specialized harbo(u)r 专业性港口
specialized high school 中等专业学校
specialized homing receiver 专用归航讯机
specialized household producer 专业户
specialized information 专业化情报
specialized information center 专业情报中心
specialized information sources 专业情报源
specialized kill 业务专长
specialized knowledge 专业知识;专门知识
specialized marine terminal 专业化码头;专业化码头
specialized person 专业人员
specialized phases of accounting 会计专门化
specialized pig-farm 专业养猪场
specialized port 专业(性)港口;专业港
specialized practice 专业实习
specialized processing 专业化处理
specialized remote sensing satellite 专用遥感卫星
specialized repair system 专业化修制【铁】
specialized school 专业学校
specialized staff personnel 专业人员
specialized standard 专业标准
specialized strain 专门化品系
specialized team 专业施工队(伍)
specialized terminal 专业化码头
specialized thesaurus 专用叙词表
specialized training 专业训练
specialized trust management 特种管理信托
specialized vessel 专用船
specialized vocabulary 专业词汇
specialized wharf 专业性码头
special jaw 特殊卡爪
special jewel 专用钻
special job cost 特种作业费
special job cover map 特别任务完成情况图
special joint 特种接头;典型接头;典型接缝
special joint for seismic movements 抗震运动特殊接缝
special journal 特种日记账
special juristic person 特别法人
special key 特殊扳手
special kind of commodity economy 特殊的商品经济
special kitchen 特餐厨房
special language 非通用语言
special ledger 特种分类账;辅助分类账
special letter of credit 特别信用证
special library 专业图书馆
special license 特殊许可证;特别许可证;特种执照
special lien 专项留置权;特别留置权
special lift 专用升降机;特制升降机
special lighting 特殊照明
special line 专用线;专线
special line for freight train 货运专线;货位专线
special line for passenger train 客运专线【铁】
special liquid red lead 特等红铅调和漆
special liquid white lead 特等白铅调和漆
special list of equipment 专用设备清单;临时设备表
special live load 特种活载
special load 特殊载荷;特殊荷载
special loading and unloading 特殊装卸
special loan 专用借款;特种信贷;特种存款
special loan for producing export goods 出口专项贷款
special local product 特产
special locomotive 专用机床

special logic record 专用逻辑记录
specialloid piston 特制高硅铝合金活塞
special loss 特别损失
special low-ash coal 特低灰煤
special low-phosphorus coal 特低磷煤
special low-strength coal 特低强度煤
special low-sulfur coal 特低硫煤
special lubeoil 特殊润滑油
specially advised letter of credit 特别通知信用证
specially built (按特种设计要求制造的)特种定货
specially constructed plant 特殊建造的工厂
specially constructed quay 特殊建造的码头
specially constructed wharf 特殊建造的码头
specially crossed check 特别划线支票
specially designated major port 特级大港
specially designed terminal 专门设计的码头
specially formulated oil 特殊配制的油
specially good effect 特效
specially made brick 特制砖
specially made gasket 特制垫带
specially made tool for orientation 特制方向架
specially mo(u)lded brick 异型砖
specially negotiating bank letter of credit 特别指定结汇银行的信用证
specially surfaced metal 特殊加工面金属
special machine 专用机(器);特殊电机
special machinery 专用机械
special manager 特别经理
special manhole 特殊覆盖层
special manufacturer 专业厂
special map 专用图;专门地图
special marine grey 特等轮船灰色
special marine protection area 海洋特别保护区
special mark 专用标志
special material 特种材料;特殊材料
special matrix method 特殊矩阵法
special matrix terrazzo 特制水磨石(地面)
special maximum resolution emulsion 特高分辨率乳胶
special metallic gaskets 特种金属垫
special metameric index 特殊异谱同色指数
special method of shaft sinking 特殊凿井法
special methods of computing depreciation 折旧特别计算法
special milling machine 专用设备多头铣床
special model law 特种模型定律
special modulator 专用调制器
special module testing 专用模块测试
special monitoring 特种目的监测;特殊监测
special mortar 特种砂浆
special mortgage 特定抵押
special motor vehicle 特种机动车
special mounted crane 特殊底盘起重机
special municipality under direct central control 特别市
special name 特殊名
special network 专用网(络)
special non-metallic commodities 特种非金属矿产
special notes 专门用语
special nuclear material 特种核材料
special numeric 特殊符号
special nutritive contact resin 特殊养护浸透树脂
special observation 专门观测;特殊观测
special observation borehole 专门观测孔
special offer 廉售报价;特价提供;特别报价
special oil 高级油
special oil-well cement 特种油井水泥
special operation exception 专用操作异常
special optic(al) fibre 特种光纤
special order cost system 特种定货成本制度
special order for goods 特种订货单
special orders 特殊订货
special organ 专门机构
special ornamental block 专用装饰砌块;特种装饰砌块
special ornamental tile 专用装饰砖;特种装饰砖
special outlay 特别支出
special overhaul 大修
special overhead crane 专用桥式起重机
special overlay 特殊贴面
special package 专用程序包
special paint used on vehicle only 车辆专用漆
special park 特种公园
special partner 特别合伙人

special partnership 特别合伙
special pass 一卡通
special pattern 骨架模
special pencil 特种铅笔
special performance 特殊性能
special periodic(al) survey of hull 船体循环特检
special periodic(al) survey of inert gas system 惰性气体装置循环特检;惰性气体系统循环特检
special periodic(al) survey of machinery 机械循环特检
special permission export 特许出口证;特许出口
special permit 特别许可证;特许证书
special perturbation 特殊摄动
special physical examination 专项体检
special piece 异形管件;成型机件
special pile 特殊桩;特种桩
special pipe 异形管;特种(水)管
special pipe cross 特殊十字管接头
special pipe plug 特殊管塞
special pitch 特制硬煤沥青
special placement residence 特殊安排住所
special planning 特殊设计
special plant 专业设备
special planting 特殊种植法
special plaster 特种抹灰;特种灰泥;特种粉刷;特殊粉刷
special plywood 特种胶合板
special point elevation 特殊点高程
special point type 特征点类型
special policy 特殊保险单
special post allowance 特别职位津贴
special pouring compound 专用浇灌混合料;特种浇灌混合料
special power of attorney 委托人授予特别权力的文件
special power pack set machine 专用联动机床
special pozzolan(a) cement 特种火山灰水泥
special prefabricated element 特殊预制构件
special prefabricated hollow block 特殊预制孔空心块体
special preference 特惠
special preferential treatment 特别优惠待遇
special preparation 特殊制备
special prescriptions 特别说明书
special price 特价;特别价格
special primer 特种底漆
special priority 特别优先权
special priviledge monopoly 特权独占
special problem 特殊问题
special process 特种方法
special processing data interpretation 特殊处理资料解释
special process study 特别程序研究
special product 名产;特殊产品;特产
special production machines indexing table type 工作台分度式特种加工机床
special production order 特制品订单
special product of China 中国特产
special product order 特殊产品订货
special profile 专用型材;特制型材
special program(me) 专用程序;特殊程序
special project 特殊工程
special project program(me) order 特殊工程设计程序
special prolongations on forging 锻件取样块
special property 特有物权
special protecting foil 专用防护(金属)薄片;特制防护(金属)薄片
special protective feature 特殊的防渗构造;特殊防渗结构
special provisions 特别条件;专门设施;特殊设备;特殊条款;特别规定
special pump 特种泵
special purchasing 特殊采购
special-purpose 专用;特种目的;特殊用途;专用的;特殊任务
special-purpose anchorage 专用锚地
special-purpose audit program(me) 专用审计日程表
special-purpose box rainwater gutter 专用箱形雨水槽;专用箱形雨水沟
special-purpose buoy 特殊信号浮标;专用浮标;特殊用途浮标
special-purpose camera 专用摄影机

special-purpose canal 专用运河
special-purpose cement 特殊用途水泥
special-purpose chart 专用图;专用海图
special-purpose check 专用支票
special-purpose chuck 专用夹盘
special-purpose computer 专业计算机;专用计算机
special-purpose concrete 特殊用途混凝土
special-purpose detergent 专用洗涤剂
special-purpose equipment 专用设备
special-purpose facility 专用设施
special-purpose financial statement 特定目的财务报表
special-purpose forest 专用林
special-purpose fund 专用基金
special-purpose impeller 特殊叶轮
special-purpose incinerator 特殊用途焚化炉
special-purpose information retrieval system 专用情报检索系统
special-purpose intelligent terminal 专用智能终端
special-purpose interpretation 专业判读
special-purpose item 专用件
special-purpose ladder 特殊用途梯
special-purpose language 专用语言
special-purpose latex 特种胶乳
special-purpose lathe 专用车床
special-purpose loan 专用贷款;专项贷款
special-purpose machine 专用机器
special-purpose map 专用(地)图;特种地图
special-purpose mark 专用标志
special-purpose module board 专用模件板
special-purpose mo(u)ld 特殊用途的模子
special-purpose outlet 专用出线口
special-purpose packet 专用包
special-purpose planning 专项规划
special-purpose port 专用港(口);(专业(性)港口
special-purpose property 特殊用途地产
special-purpose radar set 特殊用途雷达
special-purpose railway 专用铁路;专用铁道
special-purpose range 专用叠标
special-purpose receptacle outlet 专用插座出线口
special purpose rubber 专用橡胶
special-purpose ship 特种用途船舶;专用船
special-purpose statement 专用决算表
special purpose station 专用站
special-purpose step of stair(case) 楼梯的特型踏步
special-purpose system 专用密码
special-purpose table of allowance 专用装备数量表
special-purpose tax 特种目的税
special-purpose telephone 专用电话
special-purpose test equipment 专用试验设备
special-purpose tile 特殊用途石砖
special-purpose tools 专业工具
special-purpose transit 专用叠标
special-purpose transport 特殊用途运输机
special purpose utility program(me) 专用实用程序
special-purpose vehicle 专用汽车;专用车辆
special-purpose vessel 特殊用途船
special-purpose waterway 专用航道
special-purpose wharf 专业化码头;专业性码头
special-purpose window 专用窗
special putty 特殊腻子
special-quality block 优质砌块;特制砌块
special-quality brick 特制砖;特种砖;优质砖
special-quality fire-brick 特种耐火砖
special rail 铁路专用线
special rapid hardening cement 特快硬水泥
special rate 特种运输费率;特惠折扣;特别运费率
special rates for unitized consignment 统一打包货品特别费率
special recording device 特殊记录仪
special reduced rate 特别减低的价目
special referee 特别裁判人
special refractory 特种耐火材料
special refuse 特种垃圾;特殊垃圾
special regulations 特别规定
special reinforcement(steel) 特种钢筋;专用钢筋
special reinforcing steel 特种钢筋
special relativity theory 狭义相对论
special repair and maintenance equipment 专用修理设备
special repair shop 特约维修店

special repair truck 特种修理车
special report 专题报告;特别报告
special reporting grid 特别报告方格坐标
special representation 特殊表示
special requirement 特殊规格
special requisition 特别申请
special research report 专题研究报告
special reserves 特种准备
special reserve fund 特殊备用金;特种准备基金;特别储备基金
special resistance 专用阻抗;特种电阻
special resolution 特别决议
special revenue sharing 特别收入分成
special right-turn lane 右转专用车道
special rigid frame 不规则刚架
special risk 特殊风险
special risk insurance fund 特殊风险保险基金
special road 专用公路
special road asphalt 特种路用沥青
special road asphalt sand 特种路用沥青拌砂
special road oil 特种路用沥青
special rolled profile 专用轧制型材;特种轧制型材
special rolled section 专用轧制型材;特种轧制型材
special rolled shape 专用轧制型材;特种轧制型材
special rolled steel bar 异形钢
special rolled trim 专用装饰构件;特种装饰构件
special rolled unit 专用轧制构件;特种轧制构件
special rolling-mill section 特殊钢材
special room 专用房间
special rotary converter 特种转动换流机
special rounds 特级浑圆形
special routine 专用程序
special rubber 专业橡胶;特种橡胶
special rubber screen deck 特种橡胶筛网
special rules 特殊规则
special rule zone 特殊规则区
special running series of numbers 特定的流水号数
special safety housing 特殊安全轴套;特殊安全壳罩;特殊安全住宅
special sales agency 特约经销处
special sales promotion 特项推销
special scale 特殊比例尺
special school 专门学校
special screen cloth 特殊筛布;特殊网纱
special search 特别搜索
special seat 特殊座椅;特殊座席;专座;特殊座位
special section 异形钢材;特殊型钢
special sectoral source 特别部门资料源
special security measure 特别保密措施
special selected quality 精选质量
special sensibility 特种敏感性
special series 特定的流水号
special service 特殊用途;特别业务
special service final selector 特种终接器
special service pass 特种乘车证
special service practice 特种练习
special service repeater 特种业务中继器
special service ship 专用船
special service valve 专用阀
special service vessel 专用船
Special Settlement Finality Committee 特别清算终结委员会
special shaft-sinking method 特殊凿井法
special shape 专用型材;特种型材;特殊形式
special shape brick 异形砖;特异型砖
special-shaped armo(u)r block (防波堤等的)异形护面块体
special-shaped armo(u)r unit (防波堤等的)异形护面块体
special-shaped block 异形砌块
special-shaped brick 异形砖
special-shaped caisson 异形沉箱
special-shaped coupling 异形接头
special-shaped insulation 绝热异构件
special-shaped parts 异形配件
special-shaped steel 异型钢材;异形钢材
special shape fireclay brick 特异型耐火砖
special shape mo(u)lding box 专用砂箱
special sheaf 特别射向
special shop 特种商号
special shuttering 专用模板;特殊模板
special siding 岔线[铁]
special signal 专用信号

special sinking method 特殊凿井法
special skill 专长
special slag cement 特殊矿渣水泥
special slide rule 专用计算尺
special slurry 专用稀砂浆;特种稀砂浆
special sod land drill 草皮地专用播种机;草地专用播种机
special soft iron 特种软铁
special soft steel 特软钢
special soil 特殊土(壤)
special soil foundation 特殊土地基
special solution 特殊溶液;特解
special sonar system 专用声呐(装置)
special sorts 特种字符
special sounding 特殊水深
special spackling compound 专用填泥料
special spanner 特种扳手
special-special assessment bond 特种特赋债券
special splice plate 异型拼接缝板
special spring safety valve 特种弹簧安全阀
special sprinkler system 特殊喷水灭火系统
special spur line 铁路专用线
special spur railway 特殊装车点
special spur track 铁路专用线
special staining 特殊染色法
special stamping of combination units 组合设备的特殊标志
special stand 专用支架
special station 专用台(站)
special status 特殊状况
special steel 特种钢;特殊钢
special steel for making instruments 特殊仪表用钢
special steel pipe 特种钢管
special steel plant 特种钢厂
special steel plate 特种钢板
special steel tube 特种钢管
special step 专用台阶
special stopper 专用制动器
special stopping 专用隔板式门
special storage 特别舱储[贮](危险品)
special storage rate 特别储藏费率(对贵重货物)
special stowage (贵重品的)特殊装载
special straight run coal tar 特殊的直馏煤焦油硬沥青
special straight run coal tarpitch 特种直馏硬煤沥青
special strike 特别攻击
special structural concrete 特种结构用混凝土
special structural material 专用结构材料;特种结构材料
special structural profile 专用结构型材;特种结构型材
special structural section 专用结构型材;特种结构型材
special structural shape 专用结构型材;特种结构型材
special structural steel 特种结构钢
special structural trim 专用结构装饰件;特种结构装饰件
special structural unit 专用结构构件;特种结构构件
special structure 特殊结构
special structure concrete 特种结构用混凝土
special study 专题研究;专门研究
special subject 专题
special subject map 主题地图;专题地图
special subject service 专题服务
special sum of foreign exchange 专项外汇
special supervision and examination 专项监督检查
special supplies 特殊补给品
special support equipment 特种支援装备
special surface treatment 特种表面处治
special surplus account 特别公积金账户
special survey 特别检验
special Swedish putty 特种瑞典油灰
special symbol 特殊符号
special synchronous machine 特种同步电机
special system of point-group 特殊点集系
special table 专用表
special table of allowance 特种数量表
special tariff 特殊运价;特定运价
special tariff concession 特种关税减让
special tariff indicator 专用资费标志

special tarred composition roofing 特种铺用煤沥青混合物屋顶
special tarred ready roofing 特种煤沥青预制屋面
special tarred roof(ing) felt 特种煤沥青屋面油毡
special tax 特种营业税;特别税
special tax cut 特别减税
special taxes deduction 特别纳税扣除
special tax reduction 特别减税
special technical report 特种技术报告
special techniques 特技
special technological equipment 专用工艺设备
special telephone system for transverse linkage 专用横向电话系统
special temperature sensor in chemical fiber 化纤专用温度传感器
special term 特别条款
special test 特殊试验
special test and handling equipment 特种试验与操纵装备
special test equipment 专用测试设备
special testing machine 特殊试验机
special theory of relativity 狭义相对论
special three-phase transformer 特种三相变压器
special through rate 特种总价目表
specialties 细节;特别商品
special token 专用记号
special toll trunk circuit 专用长途中继电路
special tonnage tax 特别吨位税
special tool and equipment 特种工具与器材
special tools 专用工具
special tool steel 特种工具钢
special topic 专题
special topic figure 专题图
special trade 专门贸易
special traffic 特种运输
special train 专列
special train for luggage and parcel 行包专列
special training 专业训练;特种训练
special training device 特种训练器材
special transfer paper 特种转换纸
special transformation 特殊变换
special transformer 特种变压器
special transmission system 特别传输制
special transport 特殊转运
special transporter 特殊运输器;特殊运输车
special treatment 额外待遇;特殊待遇;特殊处理
special treatment steel 特殊处理钢
special tree 特留树
special triaxial cell 特殊三轴仪(试验盒)
special triaxial device 特殊三轴仪
special triaxial equipment 特殊三轴仪
special triaxial test cell 特效除莠剂
special triaxial test device 特殊三轴试验仪
special triaxial test equipment 特殊三轴试验仪
special trim 专用装饰件;特种装饰件
special trust 特种信托
special tubes and pipes of alumin(i)um alloy 铝合金异形管
special tuning 特别调整
special tunnel 特殊隧道
special turning lane 转弯专用车道
specialty 专业性;专业;专长;加盖火漆印的契约;特制品;特产
specialty chemicals 专用化学品
specialty coated fabric 特种涂层织物
specialty contract 书面要式合同
specialty contractor 分承包人;专门工种承包人
specialty elastomer 特种橡胶
specialty food store 特种食品店
specialty glass fibre 特种玻璃纤维
specialty goods 特种消费品;特种商品;特殊品;特别商品
specialty hospital 专科医院
specialty oil 特种油品
special type 特殊型;特殊类型
special type container 特种形式集装箱
special type flat wagon 长大货物车
special type motor 特殊电动机
special type of equipment 特殊类型的设备
special type of track 特种线路;特便用途线路
special type of construction 特种类型施工;特种类型建筑物
special type subroutine 专用子程序
specialty plywood 特种胶合板

special tire 特种轮胎
specialty restaurant 特殊餐厅;特色饭店;风味餐厅
specialty rubber 特种橡胶
specialty sheet 特种板材
specialty shop 专门用品商店
specialty siding 特殊墙板
specialty store 专门用品商店;特种商店
specialty ward 专科病房
special unemployment assistance 特种失业补助
special unit 专用构件;特种构件
special unit of account 特别记账单位
special unloading berth 特别卸货泊位
special urgent service 特别加急业务
special usage 特殊用途;特殊用法
special use 专用;特殊用途;特殊用法
special use area 特殊用地(城市规划)
special use map 专用地图
special use permit 专用许可证
special using siding 特别用线段【铁】
special valley gutter 专用屋顶排水沟槽;特种屋顶排水沟槽
special valuation method 特别评价法
special varnish 特种清漆;特种漆
special vehicle 特殊效率;特种车辆
special vehicle axle 特殊车轴
special vehicles and devices 特种车辆及装置
special verdict 特别裁定
special version 特别说明
special vessel inspection certificate 特种船检(验)合格证
special vise 专用虎钳;特殊虎钳
special VSP deconvolution 特殊VSP反褶积
special wagon for carriage of gas 煤气罐车
special wagon for carriage of liquids 液体罐车
special waste 特性废物;特殊废物
special waste pipe 特殊废水管
special watch case 专用表壳
special water dispersion epoxy resin 水分散特种环氧树脂
special waveform signal generator 特殊波形信号发生器
special weapons ammunitions supply point 特种武器弹药补给站
special weapons center 特种武器中心
special weapons depot 特种武器库
special weapons operation center 特种武器作战中心
special weapons storage site 特种武器储存地
special wear-resisting rail 特殊耐磨钢轨
special weather report 特殊天气报告
special wedge 专用楔子
special weed chemicals 特效除莠剂
special welfare fund 特种福利基金;特别福利基金
special wharf 专用码头
special wire 特种钢丝
special wire(d) glass 特种钢丝玻璃
special wire rope 特种钢丝绳
special work expenditures 专项工程支出
special works 特种工程
special work time 特殊作业时间
special wrench 特种扳手
special zone 特区
speciation 物种形成
specialty chemical 特殊化学品
specie par 法定平价
specie room 金库;保险室
species 种类;物种;茶剂
species abundance 物种丰富;物种丰度
species' activity 物种活度
species banks 品种资源库
species certificate 品种证明书
species composition 种类组成;物种组成
species' concentration 物种浓度
species difference 种属差异
species distribution 物种分布
species distribution area 物种分布地区
species diversity 种的多样性;物种多样性
species diversity index 物种多样性指数;特种多样性指数;生物种多样性指数
species evenness 物种均匀度
species extinction 物种灭绝
species flow 物种流
species number 分潮类数
species of heavy metal 重金属形态

species of tide 潮汐种类
species pair 姐妹品种
specie's reactivity 物种反应性
species resource 物种资源
species-rich biomes 物种丰富的生物群落
species specificity 种的特征
species trial plots 树种试种小区
species variation 种属差异
specific 专性的;特有的;特殊的;单位的;比率的;比较的;比的(即单位的)
specific abrasion 磨损率;磨蚀率;磨耗率
specific absorbed dose 吸收剂量比值
specific absorption 吸收比;材料吸水率;比吸收;比吸附
specific absorption potential 比吸附势
specific absorption rate 吸收速率比值
specific absorptivity 比吸收系数
specific accelerating force 单位加速力
specific accumulation 累积率;堆积率
specific acid base catalysis 选择性酸碱催化
specific acoustic(al) impedance 特种声阻抗;声比阻抗;单位面积声阻抗;声阻抗率;比声阻抗
specific acoustic(al) reactance 声阻率;声比抗;比声抗
specific acoustic(al) resistance 声阻率;比声阻
specific action 特异作用
specific activity 专性活动;活性比;放射强度率;比活度;比放射性强度;比放射性
specific address 专用地址;绝对地址;具体地址
specific addressed location 绝对地址定位
specific addressing 特殊编址
specific adhesion 化学黏合;特性黏结;特性黏合(力);特性附着;比黏合;比附着;特性黏附;固有黏附;比黏附力
specific adhesion coefficient 比黏附系数
specific adhesion limit 比黏附限度
specific adsorption 吸水率;吸收率;专性附着;特性吸附
specific adsorption capacity 特性吸附容量
specific adsorption of materials 材料吸水率
specific adsorption rate 特性吸附速率
specific adsorptive capacity 比吸附额
specific agglutination 特异凝集
specific agreement 特别协定
specific air consumption 空气消耗(量)
specific airflow resistance 比气流阻率
specific air resistance 单位空气附加阻力
specific alkalinity 比碱度
specific allotment 特定预算分配
specifically adsorbed adsorbate 特性吸附的吸附质
specifically corrugated board 特殊波纹纸板
specifically adsorber species 特性吸附物种
specific antidote 特效解毒剂
specific approval 具体批准
specific area 比面积;比表面积
specification 技术规格书;性能规格;详细说明书;详细说明;一览表;规格明细表;规格;规范(说明书);规定;工序说明;技术要求;技术条件;区分;清单;说明书
specification cement 标准水泥
specification code 规格法则
specification compliance form 规格响应表
specification configuration 区分结构
specification cost 规定成本
specification curve 标准曲线
specification exception 说明异常
specification for applications 应用的详细规范;应用的详细细则;应用的详细说明
specification for construction 施工技术规格书;构造说明书
specification for construction project 建筑工程规格书
specification for delivery 供应规格
specification for design 设计规范
specification for materials 材料规格;材料标准
specification for matter 材料规格
specification for safe use 安全使用说明书
specification language 规范语言
specification limit 规格界限;说明书规范
specification load 额定荷载
specification number 说明书号
specification of a model 模型说明
specification of an item 项目说明书
specification of cartography 制图规范

specification of contents 内容明细表
specification of derivation form 规格差异表
specification of drilling equipment 钻探设备参数
specification of drill tool 钻探钻具规格
specification of econometric model 经济计量模型的设定
specification of econometric model structural form 经济计量模型结构形式的设定
specification of equipment 设备规格
specification of experimental sample 试验样品规格
specification of heat-treatment 热处理规范
specification of heavy compensation equipment 升沉补偿设备规格
specification of imports 进口货物明细表
specification of materials 材料规格
specification of model 模型说明书;模型设定
specification of offshore rig 海上钻井装置规格;海上钻井装置规格
specification of quality 质量规格;质量规范
specification of screen tube 滤水管规格
specification of surveys 测量规范
specification of the mathematical form of model 模型数学形式的设定
specification of the works 工程规范
specification of variables 变量设定
specification of weight 磅码单
specification of well pipe 井管规格
specification of works 工程技术说明书
specification parameter 技术参数
specification part 规格部分;区分部分
specification policies 详细记载保险单
specification pressure 技术要求压力
specification requirements 规范要求;技术规格要求
specifications and characteristics 规格和性能
specifications for cargo 货物技术规格(书)
specifications for civil works and service 土建和服务技术规范
specifications for goods 货物技术规格(书)
specifications for in-situ monitoring measurement 现场监控量测技术要求
specifications for laying 建筑规范;敷设规范;铺筑规范
specifications for stainless steel weld 不锈钢材焊接规范
specifications for workmanship 操作规程
specification sheet 规格单;说明书
specifications of a car 汽车的性能规格
specifications of linear system 线性系统的技术指标
specifications of railway construction 铁路施工规范
specifications of the products 产品规格
specification standard 规格标准
specification statement 区分语句
specification subprogram(me) 说明子程序;分区子程序
specification subroutine 分区子程序;分类子程序
specificator 区分符;分类符
specific audit objective of compliance test 符合性测试的具体审计目标
specific basic resistance 单位基本阻力
specific basic resistance of car 车辆单位基本阻力
specific basic resistance of locomotive 机车单位基本阻力
specific basic resistance of vehicle 车辆单位基本阻力
specific bed load 河床质比率;床沙比率;推移质比率;底质比率
specific bed load transport 河床质输送比(率)
specific binding agent 特效黏结剂;特效黏合剂
specific birefringence 特定双折射
specific bodies of water 特性水域;特性水体;特定水域
specific body of water 特定水体
specific braking force 单位制动力
specific bucketing work 比挖取功
specific burnup 比燃耗
specific calorific intensity 比热强度
specific calorific intensity of combustion chamber 燃气室的比热强度
specific capacitance 单位电容;比电容
specific capacity 单位涌水量;单位容量;比容(量);比抽水率

specific capacity of a well 井的比容量
specific capacity of boiler 锅炉的比容
specific capacity of furnace chamber 炉缸表面比功率
specific capacity of well 井的单位出水量;井的出水比率
specific capillary retention 毛细水容度;毛细持水度
specific capital 特别资本
specific car resistance 车辆单位阻力
specific category 特定范畴
specific character 特性
specific characteristics 特殊性
specific charge 比(电)荷;填充率
specific chart 特种图
specific chemical reaction 比化学反应
specific chemical relation 特有的化学关系
specific coagulation time 半凝结期
specific code 绝对(代)码;具体码;特种码;代真码
specific coding 专用编码;特种编码;特定编码
specific coefficient 比系数
specific collection area 比收尘面积;比集尘面积
specific combination 特定组配
specific combining ability 特殊配合力
specific combustion intensity 燃烧负荷率
specific commodity rate 特种货品费率;特别货物运费率;特定货物运费率
specific compaction 压实率
specific component 特定零件
specific compression 相对缩短;压缩率;单位压缩量
specific concentration 比浓度
specific concept 特定概念
specific conductance 电导系数;电导率;导电率;传导率;比热导;比电导
specific conduction 导电率
specific conductivity 电导率;导水系数;导水率;传导系数;比电导率;比导电率;比传导
specific constant 比速常数
specific consumption 消耗率;单位消耗量;比消耗;比耗
specific container 特殊集装箱
specific contaminant 特性污染物;特定的玷污物
specific contract 特殊合同
specific corona power 比电晕功率
specific corrosion degree 比溶蚀度
specific cost 专用款项;特定成本;单位费用;比价
specific covered property insurance 特定财产保险
specific creep 蠕变率;比徐变;比蠕变
specific criterion 专用准则;特定(判断)标准
specific crossing 特定划线
specific cross reference 特定相关参照
specific crypto-system 专用密码系统
specific current deposit 特定活期存款
specific curve resistance 曲线单位阻力
specific cutting pressure 切削比压
specific cycle 具体循环;特殊周期;特定循环
specific damping 衰减比;比衰减
specific death rate 死亡率;比死亡速率
specific decay rate 比衰减率
specific decelerating force 单位减速力
specific density 相对密度;特定密度;比密度
specific density table 比重表
specific deposit 比沉积
specific depth 特定深度;比深
specific detectivity 比探测率;可检率
specific dielectric(al) strength 比介质强度
specific discharge 流量率;单位流量;比流量
specific discharge of rainfall 降雨流量率;单位面积降雨流量
specific discharge of soil 单位面积土的流量
specific dispersion 比色散度
specific disposal cost 比处理费用
specific drawdown 单位流量降深;比降落
specific dust load 单位含尘量
specific duties 从量税(率)
specific duty 单位生产量;从重税率
specific dynamic(al) action 特种动力作用;特殊动力作用;特别动力作用
specific effective energy 比有效能
specific efficiency 比效率
specific elastic deflection 特定的弹性弯沉
specific electric(al) conductance 比电导率
specific electrical conductance of oil field water 油田水导电率
specific electric(al) conductivity 电导系数;电导

率;比电导率
specific electric(al)loading 单位电负荷;比电负载
specific(electric(al))resistance 比电阻
specific electric(al)resistance of dust 粉尘比电阻
specific element 专名部分
specific-element method 特定元素法
specific elongation 比伸长;相对伸长;延伸率;比伸长度
specific emission 放射率
specific end use 特定用途
specific energy 能量率;单位能量;比水头;比能量;比能;比内能
specific energy consumption 能量比耗;电能比耗;单位能(量消)耗
specific energy curve 比能曲线
specific energy imparted 比授予能量
specific energy loss 能量损耗率
specific energy of cross section 断面单位能量
specific energy of current 水流比能
specific energy of flow 水流比能
specific energy of shock 单位冲击功;冲击比功
specific energy of tension 拉伸比能
specific enthalpy 比焓
specific entropy 比熵
specific entry 特定标目
specific environment 特性环境;特殊环境;特定环境
specific evapo(u)rative capacity 单位蒸发能力
specific event 特殊事件
specific exhaustibility 特异的耗竭性
specific expansion 比膨胀
specific extinction coefficient 吸光系数;比消光系数;比吸光系数
specific field 专用字段;专用域
specific flow 比流量
specific flux 流量比
specific force 专性力;比力
specific force of gravity 比重
specific frictional resistance 比摩阻(力)
specific fuel comsumption 燃烧比耗
specific fuel consumption 耗油率;燃料消耗率;燃料比耗;比油耗;比燃料消耗量
specific function 特殊函数;特定函数
specific function glass 特殊功能玻璃
specific gammaray constant 特征伽马射线常数
specific gamma-ray emission 特征伽马射线照射率
specific gas consumption 单位耗气量
specific gas flow rate 气布比
specific gasoline consumption 汽油消耗率
specific ga(u)ge reading 水尺特定读数(对应某一给定流量)
specific goods 特定货物
specific gradient resistance 坡道单位阻力;单位坡道阻力
specific grain size 比粒度
specific grain surface 颗粒比表面;比颗粒(表)面
specific granule 特异性颗粒
specific granule sludge 特性颗粒污泥
specific gravitation 比重力
specific gravity 比重值;比重
specific gravity analysis 浮沉分析;比重分析
specific gravity at pumping temperature 泵送温度下的液体比重
specific gravity balance 比重天平;比重秤
specific gravity balance method 比重天平法
specific gravity bottle 比重瓶
specific gravity bottle method (测定液体和骨料比重的)比重瓶法
specific gravity bulk-dry 干燥松散状态下的比重
specific gravity cell 比重电池
specific gravity consist 比重组成
specific gravity curve 比重曲线
specific gravity determination 比重测定
specific gravity factor 比重系数;比重因素
specific gravity flask 比重瓶
specific gravity fractions 比重级
specific gravity hydrometer 液体比重计
specific gravity in dry air 气干比重
specific gravity in green 生材比重
specific gravity liquid 比重液
specific gravity of degassed oil 脱气原油的比重
specific gravity of digester gas 消化池气(体)比重
specific gravity of dust 粉尘密度
specific gravity of grit 沉沙比重
specific gravity of oil field water 油田水比重

specific gravity of ores 矿石比重
specific gravity of petroleum 石油的比重
specific gravity of rock 岩石体重
specific gravity of saturated surface dry aggregate 集料饱和面干比重
specific gravity of serous 浆液比重
specific gravity of soil 土壤比重;土块比重;土的比重
specific gravity of solid 固体比重
specific gravity of wash liquid 冲洗液比重
specific gravity plus or minus 0.1 curve 比重正负0.1曲线
specific gravity selection 比重选种
specific gravity separator 比重清选机;比重分选机;比重分选法
specific-gravity stoner 比重去石机
specific gravity test 比重试验
specific ground pressure 单位面积地基压力
specific growth rate 比生长率
specific growth rate of microorganism 微生物比增长速率
specific growth rate of sludge 污泥比生长率
specific guarantee 特殊担保
specific hardening heat 单位时间水化热
specific head 规定水头;特定水头;比压头;比水头
specific head diagram 单位水头曲线;比水头曲线
specific head loss 水头损失率
specific heat 比热
specific heat at constant pressure 定压比热
specific heat at constant volume 定容比热
specific heat by volume 容积比热
specific heat capacity 土壤比热;特殊热容量
specific heat consumption 单位热量消耗;单位热耗;热耗
specific heating surface of checker 格子体受热比面积
specific heat load 散热强度
specific heat of fuel 燃料比热
specific heat of soil 土的比热
specific heat ratio 比热比
specific heat test 比热测定
specific heat test of froze soil 冻土比热试验
specific historical analogy method 特定历史模拟法
specific humidity 绝对湿度;含湿量;含湿度;比湿(度)
specific hydraulic horsepower 比水马力
specific hydraulic slope 水力坡降率;水力比降
specific hydrogeology maps 专门性水文地质图件
specific impedance 比阻抗
specific impulse 比推力;比冲量;比冲
specific indent 特定订单
specific indicator 专一指示剂
specific inductive capacity 介电常数;电容率;比电容
specific inductivity 介电常数;电容率
specific inertia 比惯量
specific injectivity index 单位注入指数(钻孔)
specific installation 专门设备;特殊设备;单独设备
specific insulation resistance 体积电阻系数
specific insurance 特定保险
specific intensity 光强度比
specific intensity of radiation 辐射比强度
specific interaction 特殊相互作用
specific interfacial area 比界面积
specific internal energy 比内能
specific inventory 比投料量
specific investment 专项投资;单位投资
specific ion effect 专性离子效应
specific ion electrode 专性离子电极;特性离子电极;特效电极;特定离子电极
specific ionization 电离比值;单位电离;比电离
specific ionization curve 比电离曲线
specific ionization loss 电离损失比
specific issue market 特别发行市场
specific items 具体项目;特殊条款
specificity 专一性;专性;特异性
specificity of reagent 试剂特性
specific kilowatt-hour consumption 单位电耗
specific latent heat 比潜热
specific leaf area 比叶面积
specific legacy 指定遗赠
specific length 规定长度;特定长度
specific level 特定水位

specific level of test 特定检验水平
specific lien 特定账产留置权(对借契上指明的财产)
specific load 荷载率;单位负荷;负荷率;额定负荷
specific loading 单位荷载
specific loading condition 单位荷载条件
specific locomotive resistance 机车单位阻力
specific locus test 特定座位测验
specific loss 单位损耗;比损耗
specific loss of head 水头的单位损失;单位水头损失
specific luminance 比发光度
specific luminosity 比发光度
specific magnetic loading 平均气隙磁密;单位磁负荷;比磁负载
specific magnetic resistance 磁阻率
specific magnetic rotation 磁致旋光率
specific magnetism 磁化率
specific magnetization coefficient 比磁化系数
specific magnetizing coefficient 比磁化系数
specific mass 质量密度;密度;比质量;比质
specific mass gravity 么重;比重
specific material demand 单位材料需要量
specific medicine 特效药
specific melting efficiency 熔化率;比熔化效率
specific methane production capacity 比甲烷产率
specific methanogenic acitivity 比产甲烷活性
specific modulus 径流模数;比模量
specific moisture 比湿(度)
specific moisture capacity 容水度
specific moisture content 含水率;单位含水量
specific mortality 死亡专率
specific mortality of cause 死因死亡率
specific mortgage 特定抵押
specific motor retardation 特殊运动迟缓
specific mo(u)lding pressure 单位压制压力
specific natural scenes area 特异景观风景区
specific nitrification rate 比硝化速率
specific normal acoustic(al)admittance 允许平均声强度
specific normal acoustic(al)impedance 法向声阻抗率
specific odo(u)r and taste 特殊的臭味
specific optical density 比光学密度
specific order costing 分批成本法
specific order cost system 分批成本制度
specific order production 特定订单生产
specific output 输出率;单位出力;比输出
specific oxygen consumption rate 比耗氧速率
specific oxygen demand 比需氧化量
specific oxygen uptake rate 比摄氧速率
specific oxygen utilization rate 比耗氧速率
specific paper 专用纸张;特种纸张
specific parachor 比等张比容
specific parasite 定主寄生物
specific penetration 指定贯入度
specific penetration resistance 比贯入阻力
specific permanent reserves 单位静储量法
specific permeability 内渗透性;内渗透率;渗透率;渗水比;比渗透率;比磁通密度
specific permeability of water 水渗透率
specific phase 相点;特定相
specific poisonous substance 特定毒物
specific policy 确定保险单;特定保单;船名缺点保险单
specific pollutant 特性污染物;特殊污染物
specific pore volume 比孔容
specific potential power 单位水力资源;单位水力蕴藏量
specific power 功率系数;燃料比功率;单位功率;单位出力;比功率;比动力
specific power consumption 单位电耗
specific power-plant consumption 动力系统消耗率
specific pressure 单位压力;比压
specific pressure drop 比压降
specific problem 具体问题
specific production 单位繁殖率
specific productive index 特定生产指数
specific productivity 单位生产率
specific productivity index 比采油指数
specific product of labor 具体劳动产品
specific products 特定产品
specific program(me) 专用程序
specific propellant consumption 比燃料消耗量
specific property 特性;比性能

specific prophylaxis 特效预防
specific pulse detectivity 比脉冲探测
specific pulse recurrence rate 特殊脉冲重复频率;特定脉冲重复率
specific pulse repetition interval 特殊脉冲重复周期
specific punch diameter 相对凸模直径
specific purpose container 专用集装箱
specific purpose incinerator 特种用途焚烧炉
specific purpose map 专用地图
specific purpose mobile crane 专用流动式起重机
specific purpose station 专用站
specific radioactivity 放射性比;比(度)放射性
specific rate 分类别的比率
specific rate constant 比速率常数
specific rated load 计算荷载;设计荷载;额定荷载
specific rate of flow 流量模数;流量模量;单位面积流量;单位流量;单宽流量;比流量
specific rate of sediment transport 输沙率
specific ratio 比率
specific reaction 特效反应
specific reaction constant 比反应常数;比反应常量
specific reaction rate 反应比速;比反应速率
specific reaction rate constant 比反应速度常数
specific reagent 专一试剂;特性试剂;特效试剂
specific redox environment 特性氧化还原环境
specific refraction 折射系数;折射度;比折射度
specific refractive power 折射率;比折射力;比折射功率
specific refractivity 折射度;比折射率
specific refractory power 折射率
specific reluctance 磁阻率
specific repetition frequency 特殊重复频率
specific repetition rate 特定重复率
specific request 特殊请求
specific requirements 特殊要求
specific resistance 电阻系数;电阻率;单位阻力;比阻
specific resistance coefficient of dust 粉尘比阻力系数
specific resistance of heat production 比导热阻
specific resistance to filtration 过滤单位阻力;过滤比阻
specific resistivity 电阻率;比抗性
specific resistivity method 比电阻法
specific response 特殊响应
specific response plan 特殊响应计划
specific responsibility 专责
specific restraint device 特殊紧固装置
specific retention 岩土单位滞水量;截水率;含水比容量;单位滞水量;单位滞流量;单位持水量;单位持水度;持水率;持水度
specific retention of capillary water 毛细持水度
specific retention of film water 薄膜持水度
specific retention of moisture absorption 吸湿持水度
specific retention volume 比保留体积
specific rolling resistance 车辆单位溜放阻力
specific rolling resistance due to movement on level tangent track 单位滚动基本阻力
specific rotary power 旋光率
specific rotating speed 比转速
specific rotation 旋光率;比旋
specific rotatory power 比旋光度
specific roughness 比糙率;比糙度
specific routine 专用程序
specific runoff 单位径流(量);比径流
specific sample survey 特定样本调查;典型调查
specific search 专用检索;专题检索;全结构检索
specific sedimentation volume 比沉降体积
specific sinking fund 特定偿债基金
specific sludge growth rate 比污泥生长率
specific sludge production 比污泥产量
specific solid discharge 固体材料磨损
specific solubility 比溶解度
specific soluble substance 特性溶质物质;特殊溶性物质;特殊可溶性物质
specific source 特殊污染源;特定污染源
specific spectral detectivity 比光谱探测率
specific speed 有效比速;折算速度;特定速度;比转速;(水轮机的)比速
specific speed constant 比转速常数
specific speed factor 比转速系数
specific speed of pump 水泵比(转)速;水泵比速
specific stain 特殊染色

specific steam consumption 比汽耗
specific stiffness 单位刚度;比刚度
specific storage 释水度;单位释水系数;单位存量;单位储(水)量;比容(量)
specific storativity 体积比储水系数
specific strain 比应变
specific strength 比黏附力;比强(度)
specific substrate degradation rate 比基质降解速率
specific surface 表面系数;比面(积);比表面
specific surface apparatus 比表面试验仪
specific surface area 面积率;比表面积
specific surface area of adsorbent 吸附剂比表面积
specific surface area of dust particle 粉尘颗粒比表面积
specific surface area of mice sheet 云母片比表面积
specific surface area of particle 颗粒比表面积;粉尘地表面积;粉尘比表面积
specific surface energy 表面比能;比表面能
specific surface excess 比剩余表面
specific surface factor 表面系数
specific surface of cement 水泥的比表面积
specific surface of heat exchange 换热比表面
specific surface of porous glass 多孔玻璃比表面
specific surface of porous medium 多孔介质的比面
specific surface of soil 土壤比表面
specific surface on volume basis 体积基准比表面
specific surface power of electric(al) heating element 电热元件的比表面功率
specific surface resistivity 比表面电阻率
specific surface tension 比表面张力
specific surface test 比表面积试验
specific surface tester 比表面试验仪
specific suspended load 悬移质比率;单位悬浮负载;比悬浮负荷
specific temperature rise 比温升
specific tenacity 比强(度)
specific tension 单位拉应力
specific tensor 比张量
specific term 专用名词
specific test 专门试验;特效试验
specific tester 特定试验
specific thermal capacity 比热容
specific thermal conductivity 比导热率
specific thrust 单位牵引力;比推力;比拉力;比冲
specific time 规定的时间
specific torque 比转矩
specific torque coefficient 比转矩系数;比力矩系数
specific toxic effect 特异毒害作用
specific tractive power 单位牵引力
specific train braking force 列车单位制动力
specific train effort 列车单位牵引力
specific train resistance 列车单位阻力;单位列车阻力
specific train resistance due to curves 单位曲线附加阻力
specific train resistance on inclined track 单位坡道附加阻力
specific treatment method 特定处理方法
specific tunnel resistance 隧道单位阻力
specific turbine 比速涡轮
specific ultraviolet light absorbance 紫外光比吸光度
specific unbalance 不平衡率
specific unemployment 特定部门失业
specific unit compaction 单位压实率
specific unit expansion 单位膨胀率
specific uptake rate 比摄取率
specific utilization coefficient 单位利用率
specific value 特定值;比值
specific valve 比流阀
specific vapo(u)r resistance 比蒸气渗透阻力
specific vehicle resistance 车辆单位阻力
specific vent stack 专用通气竖管;专用通气立管
specific viscosity 比滞性;比滞度;比黏(滞)度;比黏
specific visibility 比能见度;比视度
specific voltage 单位电压;比电压
specific volume 容度;体积度;特定卷宗;比体积;比容积;比容(量);比积
specific volume anomaly 比容偏差
specific volume of coal oil mixture 煤油混合燃料比容

specific volume resistance 体积电阻率
specific volume resistivity 比体积电阻(率)
specific wall paper 特种墙纸
specific waste quantity 比废物量
specific water absorption 单位吸水量
specific water adsorption 吸水率
specific water body 特性水域;特性水体
specific water constituent 特殊水组分
specific water power potential 单位水力资源;单位水力蕴藏量
specific water pumpage 比抽水量
specific water requirement 比需水量
specific water retention 单位持水量;单位持水度;单位保水量
specific water yield 给水度;单井出水量
specific wear 单位磨耗
specific wearability 磨损率;比磨损度
specific weight 特殊权数;单位重量;比重
specific weight determine 比重测试
specific weight scale of sediment 泥沙容重比例
specific well capacity 单位出水量
specific well yield 单位出水量;井的产水率
specific year flood 特殊年洪水量
specific yield 给水度;涌水量;单位涌水量;单位生产量;单位流水量;单位给水量;单位产水量;单位产量;产水率;比流量;比产水量
specific yield of gravel stuffing 填砾的给水度
specific yield of well 井的单位流量;井的出水率
specified 规定的;精确确定;额定的;额定
specified altitude assignment 指定高度
specified amount 特约数量;指定金额;规定量
specified by building code 建筑规范所规定的;建筑规范规定
specified by building specification 建筑规范所规定的
specified cargo vessel 专用船
specified carrying cost 固定资产贬值特定费用
specified circuit 指定路
specified compressive strength 规定抗压强度
specified construction time 规定的施工时间
specified contractor 指定承包者;指定承包商;指定承包人
specified course 规定航线
specified criterion 给定技术条件;技术规范;明细规范
specified cross-section 典型横断面
specified current deposit 特定活期存款
specified data 确定数据
specified date 规定日期
specified discharge 额定排量
specified distribution 给定分布
specified equipment inspection code 专用设备检验规范
specified fender 指定护舷
specified flexural strength 规定挠曲强度
specified form 规定格式
specified format 指定格式;规定格式
specified grading 指定级配;规定(的)级配
specified head 设计水头;规定水头
specified-in-detail item 详细规定项目
specified input 确定输入;额定输入
specified lengths 定尺长度
specified life 额定寿命
specified load 规定荷载;规定负荷;计算荷载;设计荷载
specified loading apparatus capacity 加载设备的标定量;加荷设备的标定量
specified loss of head 单位水头损失
specified lubricant 合规格润滑剂
specified material specification 规定的材料规格
specified method of erection 规定的安装方法;规定的架设方法
specified mix 指定配合比;指定配方;规定配合比
specified operating point 规定的运行点
specified operation condition 规定的运行工况
specified output 额定输出
specified over-all limit 规定的总限度
specified penetration 指定贯入度;规定贯入度
specified performance 设计性能;保证性能
specified period 具体期限
specified policy 特约保险单
specified power 指定(的)权限
specified project 按技术规范编制的设计;按技术规范编制的计算;按规范进行设计

specified range 特定范围
specified rate 规定值;给定量;额定量
specified rated load 额定荷载;设计荷载;条件荷载;计算荷载;规定荷载
specified rate of consumption 规定耗料率
specified sensitivity 规定灵敏度
specified size 公称尺寸
specified speed 额定转速
specified standard 规定的标准
specified strength 规定强度
specified stress cycle 特定应力循环
specified temperature 规定温度
specified testing pressure 规定试验压力
specified testing procedure 规定试验步骤
specified time for completion 规定的完成时间
specified time interval 规定时间间隔
specified total head 规定总压头
specified value 给定值;设计值;额定值
specified weight 额定重(量)
specifier 区分符;分类符
specifier extension 说明符扩展部分
specify feature 规定功能部件
specifying machines and equipment 拟订机器和设备技术条件
specify mass discharge 规定排污量
specify the up and down directions of trains 上下行方向的规定
specimen 样品;样本;试样;试件;范例;标本
specimen alignment 试样对准
specimen analysis 试样分析
specimen assay 试样分析
specimen assembly 试样组(件)
specimen automatically embedding equipment 标本自动包埋器
specimen bottle 标本瓶
specimen box 标本箱;标本盒
specimen carrier 试件(模)盒
specimen clip 标本玻片卡夹
specimen coordinate 样品的坐标
specimen copy 样品目录;样本
specimen cup 样品杯
specimen encasement 试件封套;试件包封
specimen fine adjust equipment 标本微动装置
specimen fixator 标本固定器
specimen geometry 试件几何学(高与直径之比)
specimen heater 标本加热器
specimen holder 样品座;样品夹;试样夹具;标本架
specimen-insertion port 样品送入孔
specimen jar 标本缸
specimen lithologic(al) characters 样品岩性
specimen machine 样机
specimen map 样图
specimen material 样品材料
specimen mo(u)ld 试件模
specimen mounting 试样架
specimen number 样品号
specimen of authorized signature 签署样本
specimen of letter of credit 信用证格式
specimen of letter of guarantee 保函格式
specimen ore 特富矿体
specimen orientation 样品取向
specimen page 样张
specimen paper 样张
specimen part 样品零件
specimen planting 孤植
specimen preparation 样品制备;试件准备;试件制备
specimen rammer 舂砂样器
specimen screen 样品图
specimen shape 样品形状
specimen sheet 样张;样图
specimen size 样品尺寸;试样尺寸
specimen stage 样品台
specimen test 样品试验
specimen-transfer mechanism 样品输送机
specimen tree 园景树
speciment signature 签字样本
specimen tube 舂砂试样筒;标本管
specimen viewer attachment lamp 样品指示灯
specimen volume 样品体积
specimen with V-notch V 形切口试件
speck 亮斑;微片;点;未分散的小颗粒
specking 铁点污染
speckle 小点;散斑;斑纹;斑点

speckle camera 斑纹干涉照相机
speckled 有斑点的;玷污的
speckled anti-nuclear factor 斑点抗核因子
speckled band 分散频带
speckled celadon 飞青
speckled coating 斑纹漆
speckled feature 斑状特征
speckled finish 多彩涂装法;斑点花纹涂装法
speckled glaze 斑点釉
speckled green ware 飞青器
speckled signature 斑状特征
speckled surface 斑纹状表面
speckled surface finish 点凿琢纹饰面
speckled texture 斑状结构
speckled tile 斑纹瓷砖
speckled ware 斑纹器
speckle effect 散斑效应;斑纹效应;斑点效应;斑点效果
speckle field 散斑场
speckle finish 斑纹装饰
speckle-free viewing 无斑纹观察
speckle interference 散斑干涉
speckle interferometer 散斑干涉仪
speckle interferometry 散斑干涉测量(术);斑纹干涉测量
speckle metrology 散斑计量术
speckle modulating screen pseudocolo(u)r encoding 散斑调屏假彩色编码
speckle noise 散斑噪声;散斑干扰;斑点噪声
speckle pattern 斑纹图样;散斑图
speckle pattern photography 散斑照相
speckle photography 散斑照相;斑纹照相术
speckle problem 斑纹问题
speckle random phase shifter 散斑随机相移器
speckle-shearing 散斑剪切
speckle-shearing interferometry 斑纹剪切干涉测量法;散斑剪切干涉计量法
speckle technique 斑纹技术
specklogram 斑纹干涉象
speck of colo(u)r 色斑
speckstone 滑石
specky 仿古加皴(吹制)玻璃
speck yarn 彩点纱
specpure 光谱纯净的;光谱纯的;光谱纯
specs 规格
specs for laying 敷设的详细技术条件;敷设的详细规范;敷设的详细规程;敷设的详细规定;布置的详细技术条件;布置的详细规范;布置的详细规程;布置的详细规定
spectacle 景象;场面;展览物;双环
spectacle blind 带双圈的盲板
spectacle bossing 双推进器轴毂
spectacle frame 双孔构架
spectacle glass 眼镜玻璃
spectacle lenses 柔性焦距透镜组
spectacle method 眼镜法
spectacle plate 双孔板;吊柱耳环
spectacles for stereoscopic viewing 立体观察眼镜
spectacle stern frame 双推进器船尾肋骨;双环式船尾框架
spectacle type shaft bracket 双推进器架;双环式船尾轴架
spectacular sign 巨型招牌
spectator's stand 观众看台;观众席
spectra 带谱
spectra at high and low temperatures 高低温光谱
spectracon 光谱摄像管
spectral 光谱的
spectral absorbance 光谱吸收系数
spectral absorbing coefficient 光谱吸收系数
spectral absorptance 光谱吸收比
spectral absorption 光谱吸收
spectral absorption curve 光谱吸收曲线
spectral absorption factor 光谱吸收因数;光谱吸收率
spectral absorption method 光谱吸收法
spectral absorptivity 光谱吸收力
spectral acceleration 谱加速度
spectral acceleration ratio 谱加速度比
spectral adaptive plastic colo(u)rs 按光谱分层设色
spectral albedo 分光反照率
spectral amplification 频谱放大
spectral amplification factor 谱放大系数

spectral amplitude 频谱振幅;频谱幅度
spectral amplitude ratio 谱幅度比
spectral amplitude technique 频谱振幅法
spectral analysis 光谱分析;谱分析;频谱分析
spectral analyzer[analyser] 谱分析仪;频谱分析仪;光谱分析仪
spectral and non-spectral colo(u)rs 谱色与非谱色
spectral arrangement 按光谱排列
spectral background 光谱背景
spectral band 光谱段;光谱带;谱带;频谱带
spectral band absorption 光谱带吸收
spectral bandwidth 光谱带宽;谱带宽度;频带带宽
spectral branch 光谱支
spectral calculation 光谱计算
spectral carbon rod 光谱炭棒
spectral centroid 光谱矩心;光谱的重心
spectral centroid of light 光的光谱重心
spectral channel 光谱通道
spectral character 光谱特性;谱特征
spectral characteristic 光谱特性
spectral characteristic of aerial object 航摄景物光谱特性
spectral chemical analysis 光谱化学分析
spectral chromaticity coordinate 光谱色品坐标
spectral chromatography 光谱色谱法
spectral class 光谱型;光谱类
spectral classification 光谱分类
spectral classification of stationary random process 平稳随机过程谱分类
spectral colo(u)r 光谱颜色;光谱色
spectral comparative pattern recognizer 谱线比较式图像识别器
spectral comparator 光谱比较仪;光谱比较器;光谱比长仪
spectral component 光谱成分
spectral composition 谱成分
spectral concentration 光谱密度
spectral condensation 光谱收缩
spectral condenser 光谱聚光镜
spectral condition 谱条件
spectral confidence internal factor 谱置信区间系数
spectral content 光谱含量
spectral coordinates 谱坐标
spectral corner 谱角
spectral correlation 光谱相关
spectral coverage 光谱范围;谱段
spectral curve 光谱曲线;谱曲线
spectral data output form 波谱数据输出方式
spectral data processing techniques 波谱数据处理技术
spectral data sources 波谱数据来源
spectral data system 光谱数据系统
spectral decay 谱衰减
spectral decomposition 光谱分析;谱分解;频谱分析
spectral density 光谱密度;谱(线)密度;波谱能量分布
spectral density function 谱密度函数;波谱密度函数
spectral density level 谱密度级
spectral density meter 光谱密度计
spectral density model 谱密度模型
spectral detectivity 光谱探测灵敏度;单色探测率
spectral diaphram 光谱光栏
spectral difference 谱差
spectral differential 光谱差
spectral directional reflectance factor 光谱方向反射因子
spectral discrimination 光谱鉴别
spectral discriminator 光谱鉴别器
spectral dispersion power 光谱色散率
spectral displacement 谱位移
spectral distortion 谱失真
spectral distribution 光谱(强度)分布;频谱分布
spectral distribution curve 光谱分布曲线
spectral distribution function 光谱分布函数;谱分布函数;频谱分布函数
spectral distribution graph 光谱分布图;光谱分布曲线
spectral dye density 光谱染色密度
spectral emissive power 光谱发射率
spectral emissivity 光谱辐射率;光谱发射率;波谱发射率
spectral emittance 光谱发射度;光谱发光度;波谱发射度

spectral energy 光谱能量
spectral energy distribution 光谱能量分布；能谱分布；波谱能量分布
spectral energy distribution curve 分光能量分布曲线
spectral estimate 谱估计
spectral evolution 光谱展开
spectral extinction 光谱消色
spectral extrapolation 光谱外推法
spectral fatoring 谱分解
spectral feature enhancement 波谱特征增强
spectral filter 光谱过滤器；滤光器
spectral flash lamp 光谱闪光灯
spectral frequency 光谱频率
spectral function 光谱函数；谱函数
spectral halfwidth 光谱半宽度
spectral hardening 谱硬化
spectral hardness 谱硬度
spectral hue 谱色调
spectral image 光谱像
spectral index 光谱指数；谱指数
spectral induced polarization system 频谱激电仪系统
spectral intensity 光谱强度；谱强度；频谱强度
spectral intercept curve 光谱截取曲线
spectral interference 光谱干扰
spectral interference pattern 光谱干涉图样
spectral interval 光谱间隔
spectral irradiance 光谱辐照度；分光照度
spectral irradiation curve 谱辐照度曲线
spectralite 光学玻璃
spectral lamp 光谱灯
spectral line 光谱线；谱线
spectral line broadening 谱线增宽
spectral line correlation 谱线相关
spectral line emission 谱线发射
spectral line interference 潜line干扰；谱线干扰
spectral line narrowing 谱线变窄
spectral line of emission 发射谱线
spectral line receiver 谱线接收机
spectral line series 光谱线系
spectral line shape 谱线形状
spectral line shift 光谱线的位移
spectral line source 谱线发射源
spectral line wake wave method 谱线尾波法
spectral line width 光谱线宽(度)；谱线宽(度)
spectral line width method 谱线宽度法
spectral locus 光谱轨迹
spectral luminance 谱线亮度
spectral luminance factor 光谱亮度(因数)
spectral luminous efficacy 光谱光视效能
spectral luminous efficiency 光谱光视效率；光谱发光率；发光度函数
spectrally pure colo(u)r 光谱纯色
spectrally pure reagent 光谱纯试剂
spectrally selective nature 光谱选择性
spectrally selective photodetector 光谱选择(性)光电探测器
spectral match 光谱配色；光谱等色
spectral measure 谱测度
spectral measurement 光谱测量
spectral measurement equipment 波谱测量设备
spectral method 光谱(分析)法；频谱法；波谱法
spectral mirror 光谱镜
spectral moment 谱矩
spectral moments of stationary random process 平稳随机过程谱距
spectral multiplicity 光谱复度
spectral norm 谱范数
spectral order 光谱序
spectral output 光谱输出
spectral parameter 谱参数
spectral pattern 光谱图；光谱特性曲线；光谱分布图
spectral peak 光谱灵敏度峰值
spectral photography 分光照相(术)
spectral photometer 分光光度计；分光测光仪
spectral photometry 分光测定
spectral photoresponse 光谱光响应
spectral power 谱功率
spectral power density 光谱功率密度
spectral power distribution curve 光谱能量分布曲线
spectral power spectrum 谱功率谱
spectral presentation 光谱展开

spectral projector 映谱仪；光谱投影仪
spectral property 光谱特性
spectral property of stationary random process 平稳随机过程谱性质
spectral pseudovelocity 拟速度谱
spectral purity 光谱纯度
spectral purity filter 光谱纯滤光片
spectral pyrometer 窄带高温计；光谱高温计
spectral quantitative spectroanalysis 光谱定量分析
spectral quantum radiance 光谱量子辐射率
spectral radiance 光谱辐射强度；光谱辐射率
spectral radiance calibration 光谱辐射率校准
spectral radiance factor 光谱辐射亮度因子；光谱辐亮度因数
spectral radiance increment 光谱辐射率增量
spectral radiance map 光谱亮度图
spectral radiant emissivity 光谱辐射率；光谱发射率
spectral radiant emittance 光谱辐射率；光谱辐射度
spectral radiant energy 光谱辐射能(量)
spectral radiant exitance 光谱辐射出射度
spectral radiant flux 光谱辐射通量
spectral radiant intensity 光谱辐射强度
spectral radiant power 光谱辐射功率
spectral radiometer 光谱辐射计
spectral radiometry 光谱辐射学
spectral radius 谱半径
spectral range 光谱范围；频谱范围
spectral ratio 光谱比值
spectral receiver 光谱接收器
spectral recognition 光谱识别
spectral reflectance 光谱反射率；光谱反射比；分光反射率
spectral reflectance curve 分光反射率曲线
spectral reflectance factor 光谱反射因数
spectral reflected characteristic curve 反射光谱特性曲线
spectral reflection 光谱反射
spectral reflection characteristic 光谱反射特性；反射波频谱特性
spectral reflection factor 光谱反射系数；光谱反射比
spectral reflectivity 光谱反射系数；光谱反射率
spectral region 光谱区(域)
spectral relative distribution 频谱相对分布
spectral representation 谱表示
spectral resolution 光谱分辨率
spectral resolving power 光谱分辨能力
spectral resonance of photo conductivity 光电导的光谱响应
spectral response 光谱响应；光谱敏感性；频谱响应
spectral response characteristic 光谱灵敏特性
spectral response curve 光谱响应曲线；光谱灵敏度曲线
spectral response peak wavelength 光谱响应峰值波长
spectral responsivity 光谱响应率；光谱响应度
spectral scanning method 光谱扫描法
spectral selection 光谱选择性
spectral selectivity 光谱选择性；频谱选择性
spectral sensitive pyrometer 光谱灵敏高温计
spectral sensitivity 光谱灵敏度；光谱感光度；分光敏度
spectral sensitivity characteristic 光谱灵敏度特性；分光感度特性
spectral sensitivity curve 分光感度曲线
spectral sensitivity distribution 光谱灵敏度分布
spectral sensitivity measurement 分光感度测量
spectral sensitivity peak 最大光谱灵敏度；光谱灵敏度峰值
spectral sensitization 光谱增感
spectral sensitometry 分光敏度测量
spectral sequence 光谱序列；光谱序
spectral series 光谱线系；光谱系；谱线系
spectral shape 谱形状；频谱形状
spectral shape function 频谱形状函数
spectral shift 光谱移动；谱移
spectral shift control 谱移控制
spectral shift reactor 谱移堆
spectral signature 光谱特征
spectral signature bank approach 光谱特征库法
spectral sort instrument 波谱类仪器
spectral technique 频谱技术
spectral term 光谱项
spectral test 谱检验(数值解析)
spectral theorem 谱定理

spectral transmission 光谱透射
spectral transmission factor 光谱透射因数；光谱透射系数；光谱透射率
spectral transmission ratio 分光透射率
spectral transmittance 光谱透射系数；光谱透射率；光谱透射比；光谱渗透性
spectral tristimulus value 光谱三色激励值
spectral tuning laser 光谱调谐激光器
spectral type 光谱(类)型
spectral type curve 谱型图
spectral velocity 谱速度
spectral widening 谱展宽
spectral window 频谱窗
spectral zone 光谱区(域)
spectra-spectroheliography 光谱太阳单色光照相术
spectro absorption index 分光吸收指数
spectroactinometer 分光感光计
spectroanalysis 光谱分析
spectroanalysis instrument 光谱分析仪
spectrobologram 分光变阻测热图
spectrobolometer 光谱能量测定仪；分光测热计；分光变阻测热计
spectrochemical air pollution 光化学空气污染
spectrochemical alkylation 光化学烷基化作用
spectrochemical analysis 光谱化学分析；分光化学分析
spectrochemical analysis for gases 气体光谱分析
spectrochemical buffer 光谱化学缓冲剂
spectrochemical carrier 光谱化学载体
spectrochemical cell 光化学电池
spectrochemical dispersion model 光化学分散模型
spectrochemical equilibrium 光化学平衡
spectrochemical equilibrium theory 光化学平衡理论
spectrochemical fog 光化学烟雾
spectrochemical method 光化学测定法
spectrochemical model 光化学模型
spectrochemical ozone equilibrium 光化学臭氧平衡
spectrochemical pollution 光化学污染
spectrochemical reactive 光化学活性的
spectrochemical reactivity 光化学反应性
spectrochemical reagent 光化学试剂
spectrochemical sensitivity 光化学敏感性
spectrochemical series 光化学系列
spectrochemical smog incident 光化学烟雾事件
spectrochemical smog kinetics 光化学烟雾动力学
spectrochemical smog model 光化学烟雾模式
spectrocolo(u)rimeter 光谱色度计
spectrocolo(u)rimetry 光谱色度学；分光比色法
spectrocomparator 光谱比较仪；光谱比较器；光比长仪
spectrodensitometer 光谱密度计；分光光密度计
spectroelectrochemistry 光谱电化学；分光电化学
spectrofilter 分光滤色镜
spectrofluorimeter 光谱荧光计；分光荧光计
spectrofluorimetry 荧光光谱分析测定法；荧光光谱分析；分光荧光法
spectrofluorometric assay 荧光分光光度测定分析
spectrofluorometry 光谱荧光测量
spectrogoniophotometer 光谱测角光度计
spectrograde 光谱级(的)
spectrogram 光谱图；光谱片；频谱图；摄谱图
spectrograph 光谱仪；谱仪；频谱仪；摄谱仪；分光摄像仪
spectrograph camera 摄谱照相机
spectrograph grating 摄谱仪光栅
spectrographic(al) 光谱的
spectrographic(al) analysis 光谱分析法；摄谱分析；分光描记分析法
spectrographic(al) camera 光谱照相机
spectrographic(al) control analysis 摄谱控制分析
spectrographic(al) determination 光谱法测定
spectrographic(al) grade 光谱级
spectrographic(al) identification 光谱鉴定；摄谱鉴定
spectrographic(al) laboratory 光谱实验室
spectrographic(al) orbit 摄谱轨道
spectrographic(al) purity 光谱纯度
spectrograph slit 光谱仪狭缝
spectrograph tube 光谱管
spectrography 光谱学；光谱法；摄谱学；摄谱术；摄谱分析；摄谱法

spectro-grating 分光光栅
spectroheliocinematograph 太阳单色光电影仪
spectroheliogram 光分光光谱图;日光分光光谱图;太阳单色像;太阳单色光照片
spectroheliograph 日光摄谱仪;日光摄谱计;太阳单色光照相仪;太阳单色光谱摄影
spectroheliography 太阳单色光照相术
spectrohelioscope 日光光谱镜;太阳单色光观测镜
spectroirradiator 分光辐照仪
spectrometer 光谱仪;光谱计;谱仪;频谱仪;多道能谱仪;分光(仪)计;分光(光度)计;波谱仪
spectrometer chamber 分光计室
spectrometer cliff 分光计片
spectrometer method 光谱分析法;分光计法
spectrometer model 能谱仪型号
spectrometer type 谱仪型号
spectrometric 光谱测定的
spectrometric analysis 光谱测定分析
spectrometric characteristic 光谱辐射特性曲线
spectrometric data 光谱测量数据
spectrometric photoelectric(al) method 分光计光电法
spectrometric survey 光谱测定
spectrometry 光谱术;光谱光度学;光谱测量学;光谱测定(法);测谱学
spectromicroscope 光谱显微镜
spectromonitor 光谱监测仪
spectrophone 光谱测声器;光谱本底
spectrophosphorimeter 磷光分光光度计
spectrophotoelectric 分光光电作用
spectrophotofluorometer 荧光分光光度计
spectrophotofluorometric detector 荧光分光光谱检测器
spectrophotography 分光照相(术)
spectrophotometer 光谱光度计;分光光度仪;分光光度计;分光测光仪
spectrophotometer method with salicylic acid 水杨酸分工光度法
spectrophotometric[spectrophotometrical] 光谱光度测量的
spectrophotometric(al) analysis 分光光度分析;分光光度测定分析;光谱光度分析
spectrophotometric(al) cell 分光光度吸收池
spectrophotometric(al) colorimetry[colourimetry] 分光光度测色法
spectrophotometric(al) colo match 分光光度法配色;光谱光度法配色
spectrophotometric(al) curve 分光光度测定曲线;光谱光度曲线
spectrophotometric(al) determination 分光光度测定(法)
spectrophotometric(al) evaluation 分光光度推定法
spectrophotometric(al) gradient 分光光度梯度
spectrophotometric(al) hiding 分光光度测定中的遮盖;光谱光度测定中的遮盖
spectrophotometric(al) method 分光光度分析法;光谱光度分析法;光谱分析法
spectrophotometric(al) method with dithizone 双硫腙分光光度法
spectrophotometric(al) method with phenol disulfonic acid 酚二磺酸分光光度法
spectrophotometric(al) study 光谱光度测定研究
spectrophotometric(al) temperature 分光光度温度
spectrophotometric(al) titration 分光光度滴定
spectrophotometric(al) trichromatic colorimeter[colourimeter] 分光光度三色色度仪
spectrophotometry 光谱光度测色法;光谱光度测定(法);分光光度学;光谱光度术;分光光度法;分光光度测量;光谱光度测定(法);分光测光学
spectropolarimeter 旋光分光计;光谱偏光计;偏振频谱仪;光谱旋光计;分光偏振计;分光偏振镜
spectropolarimetry 旋光分光法
spectro-prism 分光三棱镜
spectroprojector 光谱投射器;分光投射器
spectropyrheliometer 太阳分光热量计;分光太阳热量计
spectropyrometer 高温光谱仪;分光高温计
spectroradar 光谱雷达
spectroradiometer 光谱辐射仪;光谱辐射计;分光辐射谱仪;分光辐射通仪;分光辐射计
spectroradiometric curve 分光辐射曲线
spectroradiometric measurement 光谱辐射测量
spectroradiometric property 分光辐射特性
spectroradiometry 光谱辐射度学;分光辐射度学;

分光辐射度量学
spectroreflectometer 光谱反射计
spectroscope 分光仪;分光器;分光镜
spectroscope prism 光谱棱镜
spectroscopic(al) 分光镜的
spectroscopic(al) analysis 分光镜分析;光谱分析
spectroscopic(al) analyzer 光谱分析仪;光谱分析器
spectroscopic(al) assay 光谱检定;光谱测定
spectroscopic(al) behavior[behaviour] 分光性能
spectroscopic(al) binary star 分光双星
spectroscopic(al) buffer 光谱缓冲剂
spectroscopic(al) camera 光谱摄影机
spectroscopic(al) carrier 光谱载体
spectroscopic(al) data 光谱数据
spectroscopic(al) displacement law 光谱位移律
spectroscopic(al) entropy 光谱熵
spectroscopic(al) hygrometer 分光湿度计
spectroscopic(al) illuminator 光谱照明器
spectroscopic(al) instrument 分光仪器
spectroscopic(al) isotope shift 光谱同位素移动
spectroscopic(al) lamp 光谱灯
spectroscopically pure 光谱纯的
spectroscopically pure zinc 光谱纯锌
spectroscopic(al) measurement 波谱测定;分光测定
spectroscopic(al) method 分光镜法
spectroscopic(al) methodology 光谱法
spectroscopic(al) optics 光谱光学
spectroscopic(al) orbit 分光解;分光轨道
spectroscopic(al) parallax 分光视差;分光镜视差
spectroscopic(al) photography 光谱摄影术
spectroscopic(al) plate 光谱板;分光板
spectroscopic(al) prism 光谱棱镜
spectroscopic(al) pure 光谱纯的;光谱纯
spectroscopic(al) purity 光谱纯度
spectroscopic(al) standard air 光谱标准空气
spectroscopic(al) study 分光镜分析
spectroscopic(al) technique 光谱技术;光谱法;频谱法
spectroscopic(al) term 光谱项
spectroscopic(al) test 光谱试验
spectroscopy 光谱学;频谱学;分光学;光谱术;分光镜检查;波谱学
spectrosensitogram 光谱感光图
spectrosensitometer 光谱感光计
spectrosil 最纯的石英;光谱纯石英
spectrozonal film 光谱带底片
spectrum 光谱;光谱线;波谱
spectrum amplitude 谱幅
spectrum analyser 光谱分析器
spectrum analysis 光谱分析;频谱分析;波谱分析
spectrum analysis along layer 沿层频谱分析
spectrum analysis method 谱分析法
spectrum analyzer 光谱分析器
spectrum apparatus 分光仪
spectrum atlas 光谱图册
spectrum binary 光谱双星
spectrum characteristic curve 波谱特征曲线
spectrum character of ground feature 地物波谱特性
spectrum chart 光谱图
spectrum cluster 波谱集群
spectrum colo(u)r 光谱色;谱色
spectrum comparator 光谱比较仪;光谱比较器;光谱比长仪
spectrum degradation 能谱的软化
spectrum density 谱密度
spectrum-density diagram 光谱密度图
spectrum density level 谱密度级
spectrum dip 谱坑
spectrum discrimination 谱识别
spectrum distribution 光谱分布;频谱分布
spectrum distribution function 光谱分布函数
spectrum documentary value 能谱编录值
spectrum effect 波谱效应
spectrum emission 光频放射率
spectrum envelope 谱包络
spectrum equalization 谱均衡
spectrum expander 频谱展宽器
spectrum factor 频谱因数
spectrum feature space 波谱特征空间
spectrum field 谱场
spectrum function 谱函数
spectrum hardness 谱硬度

spectrum image plane 光谱像平面
spectrum index 谱指数;频谱指数
spectrum intensity 谱烈度;频谱强度
spectrum isotope effect 光谱同位素效应
spectrum lamp 光谱灯
spectrum level 光谱(能)级;谱级
spectrum limitation 频谱限制
spectrum line 光谱线;频谱线
spectrum line natural width 谱线自然宽度
spectrum line number 谱线数目
spectrum line position 谱线位置
spectrum line stabilizer 谱线稳定器
spectrum line width 谱线宽(度)
spectrum locus 光谱轨迹;谱线轨迹;谱线轨道
spectrum-luminosity diagram 光谱亮度图;光谱光度图
spectrum matching 光谱匹配
spectrum measuring projector 光谱测量投影仪
spectrum of aurora borealis 北极光光谱
spectrum of carrier 载波的频谱
spectrum of composite seismic signal 合成地震信号谱
spectrum of ductility 延性反应谱
spectrum of solid 固体光谱
spectrum of spike 脉冲频谱
spectrum of turbulence 湍流谱;湍动谱
spectrum of wave height 波高谱
spectrum of wave period 波周期谱
spectrum order sorter 光谱分阶器
spectrum pressure level 谱压级
spectrum prism 光谱棱晶
spectrum projector 光谱投影仪
spectrum quantitative analysis 光谱定量分析
spectrum range 光谱范围
spectrum-rate of variation relation 光谱光变率关系
spectrum renormalization 频谱重正化
spectrum response curve 波谱响应曲线
spectrum sampling converted coefficient 能谱取样换算系数
spectrum scanner 光谱扫描仪
spectrum selectivity 光谱选择性
spectrum selector 光谱选择器;频谱选择器
spectrum-sensitive foil 谱灵敏探测片
spectrum sensitivity 光谱灵敏度
spectrum sequence 光谱序列
spectrum series 光谱线系
spectrum shape 频谱形状
spectrum signature 谱特征
spectrum softing 谱软化
spectrum stabilizer 稳谱器
spectrum stripping 谱线分离技术
spectrum test 光谱试验;频谱试验
spectrum types 光谱类型
spectrum variable 光谱变星
spectrum widening 谱展宽
spectrum width 谱宽(度)
specular 有金属光泽的
specular alabaster 镜雪花石膏;透光蜡石
specular angle 镜面角;反射角
specular cast-iron 镜铁
specular cross section 有效反射面积
specular density 定向(反射光)密度;单向密度
specular enamel finish 强光泽的釉瓷饰面
specular finish 镜面面层;镜面加工;表面平滑加工
specular glass 镜面玻璃
specular gloss 镜面光泽
specular gypsum 镜石膏;透石膏
specular hematite 镜(赤)铁矿
specular iron 镜铁矿;辉赤铁矿
specularite 镜铁矿
specularite ore 镜铁矿矿石
specularity factor 镜面反射系数
specular layer 镜面(反射)层;反射层
specular microdensitometer 反射式微波密度计
specular reflectance 镜面反射率;镜面反射比;反射
specular reflectance excluded 排除镜面反射的反射度测定
specular reflectance slit 镜反射狭缝
specular reflection 正反射;规则反射;镜面反射照明法;镜面反射;单向反射
specular reflection direction 镜反射方向
specular reflectivity 镜(面)反射率;镜面反射比

specular reflector 镜面反射器;单向反射器
specular scattering 镜面散射;反射散射
specular shist 镜铁片岩
specular stone 云母
specular surface 镜面;定向反射面;反射面
specular tranmission density 镜透射密度
specular transmission 镜面传输
specular transmittance 镜透射比;镜面透射率;镜面透射比
speculate 做投机买卖
speculated result figure of mise a-la-masse 充电法推断成果图
speculated result figure of self-potential method 自然电场法推断成果图
speculation 外汇投机;投机交易;投机倒把;投机
speculation and swindling 投机诈骗
speculation arbitrage 投机套利
speculation business 投机事业
speculation company 皮包公司
speculation curve 投机曲线
speculation demand 投机需求
speculation in land 土地投机
speculation market 投机市场
speculation on the rise and fall of the market 市场上的买空卖空
speculative boom 投机性高涨
speculative builder 但风险的建造;投机性营造商
speculative business 投机商业
speculative component 投机性成分
speculative credit 投机信用
speculative demand 投机(性)需求
speculative efficiency hypothesis 投机效率假说
speculative excesses 投机过度行为
speculative execution 预测执行
speculative investment 投机投资
speculative land investment 投机性土地投资
speculative market 投机市场
speculative motive 投机动机
speculative purchasing 投机性采购
speculative reserves 推测储量
speculative resources 推测资源量;假想资源
speculative risk 推理危险性;投机(性)风险
speculative trade 投机买卖
speculative transaction 投机交易
speculative unemployment 投机性失业
speculative value 使用性会改变的土地价值
speculator 纯理论的;投机者
speculator's demand 投机者的需求
speculum 张开器;制镜合金;开张器;金属镜;[复] specula
speculum alloy 镜用合金;反射合金
speculum laser 激光窥器
speculum metal 镜(用)合金;镜(青)铜;铜锡合金
specus 水槽孔道;水渠
Spedex 德银(一种镍铜锌合金)
spee 特准储备物资
speech analysis 语言分析
speech interference level 说话干扰级
speech inverter 倒频器
speech-noise ratio 信噪比;噪声比
speech power 语言功率
speech reinforcement system 语言加强系统
speech scrambler 倒频器
speech understanding 声音理解
speech volume indicator 语言声量指示器
speed 速率;速度
speed ability 速度性能
speed accumulation curve 速度累积曲线
speed adjusting gear 调速机构
speed adjustment 速度调整;速度调节;速率调整
speed adjustment of motor 电动机调速
speed advantage 速度优势
speed agility run 变速跑
speed-altitude trim position 速度高度配平位置
speed amplification 速度放大
speed and altitude supremacy 速度和高度优势
speed-and-delay study 速度及阻滞调查(交通);车速及阻滞调查
speed and drift indicator 速度偏流指示器
speed and slide generator 速度和打滑电机
speed ball 示速悬球
speed barrier 速度极限
speed belt 调整带;调速带;变速皮带
speed boat 高速艇;快艇

speed brake 离心闸;减速装置;减速板
speed brake setting 减速装置调定
speed bulges 高速鼓包
speed calculator 速度计算器
speed calibration 速度校验
speed calling 缩位号码呼叫
speed capacity 速率适应能力
speed car 高速车
speed cement 快硬水泥;快速水泥
speed change 速率变化;变速杆;变速
speed-change area 车辆变速区段;加宽超车区;变速地区
speed-change box 变速箱
speed-change gear 变速齿轮
speed-change indicator 变速标志
speed-change lane 变速车道
speed changer 变速装置
speed-changer motor 同步器电动机
speed-change valve 变速闸门;变速阀门;变速开关;变速阀
speed change with multimotors 双电机组换速
speed change with multipole motor 变极电机换速
speed changing 速度变换
speed changing box 变速箱
speed characteristic 速度特性
speed-characteristic curve 速率特性曲线;速度特性曲线
speed chart 变速曲线图;变速表;速度曲线表;速度表(指图表)
speed checking 速度校验
speed checking apparatus for vehicle 车辆测速器
speed check tape 测速带
speed circle 速度圆;速度圈
speed claim 航速索赔
speed code 快速码;快速代码
speed coding 快速编码
speed coding system 快速编码系统
speed coefficient 速度系数
speed compensated winder 速度支数补偿拉丝机;变速大卷装拉丝机
speed concatenation control 速度关联控制
speed cone 级轮;变速锥
speed constant 速率常数
speed container system 快速集装箱装卸系统;快速集装箱装卸方式
speed control 转速控制;调速;速率控制;速度控制;速度调节
speed control braking 调速制动
speed control device 调速设备;速度调节器;速度调查器
speed control disc 速度控制圆盘
speed control hump 限速驼峰;限速丘
speed controlled separator 转速可控选粉器;调速选粉机
speed controller 转速控制器;调速器;速度控制器;速度调节手柄
speed control mechanism 转速控制机构;换转机构;变速机构
speed control muffler 出口节流阀
speed control rod 调速杆
speed control servo-motor 调速伺服电动机
speed control signal 速度控制信号;控速信号
speed control switch 速率控制开关
speed control system 速率控制系统;速度控制方式
speed control valve 调速阀;速度控制阀
speed correction table 速度误差修正表
speed corrector 速度误差改正器
speed counter 转速表;转数表;计速器;计速表;速率计数器
speed coupling 快速联轴节
speed course 速度测定基线
speed course latitude error 速度航向纬度误差
speed criterion 感光标准
speed data 速度数据
speed dead band 转速死区
speed detector 转速检测装置;速度检测器;测速装置
speed dialing 快速拨号;缩位拨号
speed difference 速率差;速度差
speed difference measuring device 速差测量装置
speed display 速度显示
speed-distance chart 速度—距离图
speed-distance curve 速度—距离曲线
speed-distance diagram 速度—距离曲线图

speed distance transmitter 速度航程发送器(电磁计程仪)
speed distribution function 速率分布函数
speed distribution line 速度分布线
speed down 减低速度
speed drop 速度降(低);速度减慢
speed effect 速度效应
speed electromotive force 速率电动势
speed envelope 速度界限
speeder 增速装置;增速器;加速器;调速装置
speeder rod 调速杆
speed error 速度误差
speed error corrector 速度误差改正器
speed factor 速率因子;速率因素;速度系数
speed-fax 高速传真
speed film 高感光度胶片
speed flash 速闪;电子闪光灯
speed flo(a)tation 快速浮选
speed-flow curve 速度—流量曲线
speed fluctuation 转速波动;速度变动;车速波动
speed fluctuation coefficient 速度波动系数
speed fluid drive 无级变速液压传动
speed follow-up system 调速随动系统
speed for maximum endurance 久航速度
speed gain 速度增益
speedgate 速度选通
speed ga(u)ge 转速表;速率计;速度计;速度表
speed gear 高速齿轮;变速齿轮
speed generator 测速发电机
speed governing 调速;速度调节
speed governing system 调速系统
speed governor 转速调节器;调速器;速度控制器;速度调节器
speed-hauled weight curve 速度牵引重量关系曲线
speed-height ratio 速高比
speed hold 速度保持
speed-holding servo (拖缆机的)速度随动系统
speed increase 增速
speed increase gear 增速齿轮
speed increaser 增速器;升速器
speed increase unit 增速装置
speed increasing gear 增速器;增速齿轮
speed increasing ratio 增速比
speed increment 速度增长
speed indicating generator 测速发电机
speed indicator 转速表;慢行牌;速度指示器;速度仪;示速计
speeding ticket 超速罚款条
speeding-up gear 增速齿轮;加速齿轮
speeding-up pulley block 增速滑轮组
speed input 速度信号输入
speed in reverse 回速;倒挡
speed insensibility 速度不灵敏度
speed instability 速度不稳定性
speedite casing 端部墩粗的无接头梯形丝扣套管
Speedivac rotary oil pump 快速回转油真空泵
speed lamp 电子闪光灯
speed landing 高速降落(飞机)
speed lathe 高速车床
speed length ratio 速长比
speed lens 快速透镜;大相对孔径物镜
speed letter 快信
speed lever 转速调节杆;速度调节杆
speed lifting 提升速度;升举速度
speed light 电子闪光(灯);闪光管
speed limit 限速运行;速率限度;速率极限;速度限度;车速限制
speed limitation 限速运行;速度限制
speed limitation board 限速牌;限速标
speed limitator 限速器
speed limit device 速度限制器;限速设备
speed-limited sign 限速标志
speed-limited signal 限速信号
speed limiter 限速器
speed limiting brake 限速制动器
speed limiting circuit breaker 限速断路器
speed limiting device 限速设备;限速装置;限速器;受速度限制的设备
speed limiting governor 限速器
speed limiting switch 限速开关
speed limit line 速度限制线
speed limit marking 限速标记
speed limit notice 限速标

speed limit providing total adhesion 总黏着对速度的限制
speed limit road sign 限速路标
speed limit sign 限速标(志)
speed limit zone 限速区
speed line 加速气管;速度线
speedline mixer 快速搅拌机;快速拌和机;快速混合器
speedline spraying outfit 快速喷涂机;快速喷射机
speed load characteristic 转速负荷特性
speed load curve 速度—重量曲线;速度—荷重曲线
speed lock light 速度保持信号灯
speed loss 速度损失;失速
speed made good 已实现的速度;航迹速度;实际速度;实际航迹;对地航速
speed maneuver 速度机动
speed margin 速度储备
speed mark 测速标
speed marker 速度标志
speed match(ing) 速度匹配
speed measurement 感光度测量;速率测量;测速
speed measuring equipment 测速设备
speed measuring motor 测速电机
speed measuring position 测速位置
speed measuring radar 测速雷达
speed measuring radar gun 测速雷达枪
speed measuring range 速度测量范围
speed meter 速率计
speed-mobile carriage 高速行驶的车辆
speed modulation 速度调制
speed monitor 转速监视器
speed muller 快速研磨机;快速混砂机;摆轮式高速混砂机
speed multiplier 倍速器
speed no-load 空载转速;空载速度
speed of action 作用速率
speed of adaptation 自调速率
speed of advance 计划航速;前进速度
speed of agitator 搅拌转速
speed of answer 应答速度
speed-of-approach measuring instrument 两船接近速度测量仪表
speed-of-approach monitoring aid 靠船速度监测设备
speed of a ship 船首速
speed of autorotation 自转速率
speed of burning 燃烧速度
speed of car 车速
speed of combustion 燃烧速度
speed of construction 建设速度
speed of convergence 收敛速度
speed of crankshaft 曲轴转速
speed of cutter head 刀头转速;刀架转速
speed of cutting 切削速度
speed of deformation 变形速度
speed of development 发展速度
speed of drum 鼓转速;滚筒转速
speed of dune movement 沙丘移动速度
speed of engine 发动机转速
speed of evacuation 抽气速度
speed of exhaust 抽气速度
speed of filtration 过滤速度
speed of firing travel 火行速度
speed of formation 形成速度;形成率
speed of free flexural wave 自由弯曲波速
speed of growth 增长速度
speed of impregnation 注入速度;浸渍速度
speed of increase 增加速度
speed of instrument 仪表速率
speed of lens 透镜率
speed of light 光速
speed of light circle 光速圈
speed of light in vacuum 真空光速
speed of loading 装载速度;加载速度;加荷速度
speed of melting 熔炼速度
speed of motor 自(动)机转速
speed of operation 运行速(度);运算速度
speed of paper 记录纸速度
speed of photographic(al) plate 照相片感光速率;照相底片感光速度;摄影干版的感光度
speed of photographic plate 感光板灵敏度
speed of power 作用功率
speed of propagation 传播速度
speed of pump 泵速

speed of pumping 抽气速度
speed of punch advance 冲杆前进速度
speed of railway transportation 铁路运输速度
speed of registration 记录速度
speed of relative movement 相对(运动)速度
speed of response 响应速度;反应速率
speed of return stroke 回程速度
speed of revolution 转速
speed of rolls 轧辊速度
speed of rope 钢索速度;绳索速度
speed of rotation 旋转速率;回转速度;转速
speed of rotation of core barrel 岩芯管钻进速度
speed of rotation of motor 电动机转速
speed of rotation of pump 泵(的)转速
speed of service 运行速度
speed of shear 剪切速度
speed of ship 船舶航速
speed of silting up in harbo(u)r 海湾淤积速率
speed of slide(operation) 运行速度;滑动速度
speed of slip 滑移速度
speed of sound 声速
speed of spindle 轴转速
speed of stress alternation 应力交变速度;应力交替变化速度
speed of suspension 悬浮速度
speed of table 工作台速度
speed of trains on curves 曲线上的列车速度
speed of transmission 传播速度
speed of transverse movement 横移速度
speed of travel 行走速度;轧制速度
speed of tunnel(l)ing 隧道施工速度;掘进速度(隧道)
speed of turn 旋转速度;回转速度;转动速度
speed of vision 视觉速度
speed of welding(travel) 焊接速度
speed of wind 风速
speed of working 工作速度;操作速度
speed of working stroke 工作行程速度
speedomax 电子自动电势计
speedometer 里程碑;转速表;记速仪;速率计;速度仪;速度计;速度表;测速仪;测速计
speedometer base 速度计座
speedometer cable 里程表软线
speedometer cable casing 速度计接线套
speedometer drive gear 速度计主动齿轮
speedometer driven gear 速度计从动齿轮
speedometer flexible shaft 速度计挠性轴
speedometer for measuring rotation 转速测量仪表
speedometer gear 速度计齿轮
speedometer main shaft 速度计主轴
speedometer scale error card 速度计刻度误差表
speedometer set stem 里程计定位杆
speedometer take off shaft 速度计传动轴
speedometer transducer 速度计传感器
speedometer two-speed adapter 速度计双速接头
speedometer with mileometer 带里程计的速度表
speed on grade 坡度上车速;上坡速度
speed on load 带负荷时转速
speed on separation 分离速度
speed oscillation 转速振荡;转速波动
speed overshoot 速度过调量
speed over the bottom 对地航速
speed over the ground 对地航速;地面上空速度
speed panchro lens 快速全色照相镜头
speed per hour 时速
speed pick-up 转速加快;速度检测器;速度传感(器)
speed power coefficient 速度动力系数
speed profile 速度分布图
speed pulley 变速皮带轮;变速滑车
speed range 速率范围;速度范围
speed range data 速度范围数据
speed rate 速率
speed ratio 转速比;速度比(率);速比;传动比
speed ratio control 速率比控制
speed ratio controller 速率比控制器
speed recorder 记速器;速度记录仪;速度记录器
speed reducer 变速齿轮箱;减速机;减速器;减速装置
speed reducing ratio 减速比
speed reduction 减速;速率降低;失速
speed reduction gear 减速齿轮;齿轮减速机器
speed reduction gear box 齿轮减速箱
speed reduction gearing 减速齿轮装置

speed reduction unit 减速装置;减速器
speed-regulated scale 调速秤
speed-regulating arm 调速臂
speed-regulating hand wheel 调速手轮
speed-regulating motor 调速(伺服)电动机
speed-regulating rheostat 调速变阻器
speed-regulating servomotor 调速伺服电动机;调速接力器
speed-regulating system 调速系统
speed-regulating valve 调速阀
speed regulation 调速;速率管制;速率调节;速度管制;速度调整;速度调节
speed regulation characteristic 调速特性曲线
speed regulation electro-pneumatic valve 调速电气阀
speed regulation through eddy current braking device 涡流制动器调速
speed regulator 转速调节器;调速器;速度调节器
speed relay 速度继电器;速动继电器
speed responding 速度反应
speed responser 速度反应器
speed restricted section 限速区段
speed restriction 限速运行;限速;速度限制
speed restriction board 限速牌
speed restrictive section 限制区间
speed ring (水轮机的)速度环
speed rise 转速升高;转速上升;速度提高;速度上升;速度加快
speed road 限制车速的道路
speed sampling pulse 速率取样脉冲
speed selector lever 变速手把
speed selector switch 速率选择开关
speed selector valve 速度选择开关
speed sense 速度感
speed sensing 速度传感;转速显示;转速传感;转速读出;速度显示;速度读出
speed sensing control 速度准确控制
speed sensing device 转速显示器;转速传感器;转速读出装置;速度显示装置;速度读出装置;速度传感器
speed sensing variable timing unit 速度传感可变的定时装置;转速显示可变的定时装置;转速传感可变的定时装置;转速读出可变的定时装置;速度显示可变的定时装置;速度读出可变的定时装置
speed-sensitive release 超速放气阀
speed-sensitive switch 速度控制的开关;高速开关
speed sensitivity characteristics 速度敏感特性
speed sensor 速度传感器
speed servo 速度伺服(机构)
speed servo system 调速随动系统
speed setpoint 转速设定点;转速定点
speed setter 定速器
speed setting 转速整定;速度调节
speed setting cam 转速调节凸轮
speed setting controller 调速控制器
speed setting cylinder 转速调整缸
speed setting gear 转速给定装置
speed setting lever 转速调节杆
speed setting piston 转速调整活塞;配速活塞
speed setting screw 转速调节螺钉
speed setting segment 调速器扇形齿轮
speed setting servomotor 转速调节伺服马达
speed setting unit 转速调节装置
speed setting walking beam 转速调整横梁
speed shank 调速手柄
speed shifting fork shaft 变速叉轴
speed shifting fork spring 变速叉弹簧
speed shifting lever casing 变速杆壳
speed shifting mechanism 变速机构
speed shop 快车部件商店(高速赛车)
speed signal 限速信号;航速号志;速度信号
speed signal for track construction site 线路石膏现场速度信号
speed signal generator 速度信号发生器
speed singal(l)ing 速差制信号
speed skate rink 滑冰比赛场
speed snow plough 轻快雪犁
speed sprayer 高速喷雾机
speed stability 速度稳定性
speed stable 速度稳定的
speed stage 速度级;速度挡位
speed standard 速率标准
speedster 高速快艇;快船;双座高速敞篷汽车;双

人座高度汽车
speed switch 转速开关
speed synchronizer 速度同步器
speed table 航速折算表
speedtainer 快速集装箱
speedtainer system 立体运转方式;快速集装箱装卸系统;快速集装箱装卸方式
speed telegraph 航速车钟
speed teleindicator 遥示速率器
speed test 转速试验;航速试验
speed-testing runway 试车速跑道
speed threshold 速度挡位
speed through the water 对水船速;船速(对水)
speed through turnout 过岔速度
speed through turnout branch 侧向过岔速度
speed through turnout main 直向过岔速度
speed-time curve 速率—时间曲线;速度—时间曲线;时速曲线
speed tolerance 容许转速偏差
speed-to-pressure transducer 转速压力传感器
speed-torque characteristic 转速—转矩特性
speed-torque curve 速度转矩曲线
speed track 快车道
speed-tractive effort curve 牵引力—速度曲线;速度牵引曲线
speed transformation 速度变换;变速
speed transformation device 变速传动装置
speed-transforming transmission 变速传动
speed transition ramp 变速坡道
speed transition zone 速度过渡区;速度变换阶段
speed transmission 变速箱;变速器
speed transmission device 变速箱;变速传动装置
speed transmission unit 变速装置
speed transmitter 速度传感器;传速器
speed trap 汽车速度监视器;速度收集器;速度监视站(汽车)
speed trapezoidal wave 速率梯形波
speed trial 航速试验;速率试验;速度试验;测速
speed trial ground 测速场
speed trial trip 试速航行
speed triangle 速度三角形
speed type 速力型
speed under load 有载速度
speed under power 主动段速度
speed up 加速;提速;增速
speed-up capacitor 加速电容器
speed-up condenser 加速电容器
speed-up effect 增速效应
speed-up function 加速试验功能
speed-up gears 增速齿轮
speed-up press-button 增速按钮
speed-up test 加速试验
speed-up theorem 加速定理
speed-up the process of construction 加快工程进度
speed-up the tempo 加快步伐
speed variation 速度变动
speed variator 无级变速器;变速装置;变速器
speed vehicle sensor 车载速度距离传感器
speed voltage 速度电势
speed voltage generator 测速发电机
speed way 高速公路;高速车道;高速干道;快车道;快行道
speed without load 空载速度
speed wrench 快速扳手
speedy building erection system 建筑快速安装体系
speedy cableway 高速缆道;快速缆道
speedy construction method 快速施工法
speedy design method 快速设计(方)法
speedy drivage 快速掘进
speedy drying 快速干燥
speedy erection system 快速架设系统
speedy moisture meter 含水量快速测定仪
speedy moisture tester 快速含水量测定仪
speed zone 速率限制区段
speed zone ahead sign 速率限制区段预告标志
speed zoning 车速限制区间
speer 固定围屏(古英国)
Spekker absorptiometer 斯佩克吸收测定仪;粉末比表面测定仪
spel(a)eology 洞穴学;成洞学
spelaeo-meteorology 洞穴气象学
speleobiology 洞穴生物学;洞栖生物学

speleocartography 岩洞制图学
speleochronology 洞穴年代学;洞穴年代测定
spell 轮班;短时间中断;短时间间隔;服务时间
spelled fluorimeter 荧光计
spelled pasture 休闲草场
spelled waler 模板横撑
spell of cold weather 一阵寒潮
spell of draught 干旱期
spell of fine weather 一阵好天气;晴朗时期
spell of sick absence 伤病缺勤事例
spell of weather 天气周期
spelly wire 劈裂丝
spelter 锌铜焊剂;锌块;钎料;商品锌;粗锌
spelter bath 镀锌槽
spelter bronze 青铜焊料
spelter coating 锌涂料;锌涂层;涂锌
spelter pot 锌熔液槽
spelter solder 锌铜焊料;硬焊料;锌焊药
spelter soldering alloy 锌铜焊接合金
spence 内室(苏格兰语)
Spence metal 斯潘司金属
spencerite 斜磷锌矿
spencer's solution 钢凹版腐蚀液
spend 花费
spendable average weekly earnings 可供花费的平均周收入
spendable earnings 可花费的收入
spendable income 可支配收入
spendable weekly earnings 可支配周工资收入
spending 消费;开支;经费
spending beach 消能滩;消浪海滩;消波海滩
spending on administration and management 行政管理费
spending pattern 消费方式
spending standard 开支标准
spending unit 消费单位
spending variance 支出差异;开支差异
spend material 废料
spendrel step 三角形截面踏步
spend residue 废物
spend scrub stream 废洗涤液
spendthrift trust 浪费信托
spent acid 用过的酸;废酸
spent bill of lading 已到提单
spent bleaching earth 废漂泊土
spent bleach liquor 废漂白液
spent brine 废盐水
spent catalyst 废催化剂
spent caustic 废烧碱;废碱
spent caustic soda liquor 废烧碱液
spent clay 废土
spent condition 耗尽状态
spent developer 失效显影液
spent etching bath 废蚀刻溶液
spent ferric oxide 废氧化铁
spent fuel 用过的核燃料;废烧过的燃料;烧过的燃料;废燃料;乏燃料
spent fuel pool 乏燃料水池
spent gas 废气
spent gas liquor 废煤气水
spent geothermal field 废地热田
spent grinding sand 研磨废砂
spent iron sponge 失效的海绵铁
spent liquid form wood cooking 木纸浆厂废水;木材蒸馏废液
spent liquor 废液
spent lubricant 废润滑剂
spent lye 废碱液
spent mash 废浆
spent material 消耗物料;使用物料
spent nuclear fuel carrier 废核燃料运输船
spent oil 废油
spent ore 废矿
spent oxide 失效的海绵铁;废氧化物
spent pickle liquor 浸渍废液;酸洗废液
spent pickling solution 浸渍废液;酸洗废液;酸渍废液;废酸浸液
spent plating solution 废电解液
spent process water 过程废水
spent reactivation gas 废再生气
spent reagent 废试剂
spent regenerant 洗脱液
spent residue 废渣
spent shot 闷炮眼

spent soda 废碱
spent solvent 用过的溶剂;废溶剂
spent steam 废汽
spent steel pickle liquor 钢酸洗废液
spent sulfate liquor 废亚硫酸盐液
spent sulfite liquor 亚硫酸盐纸浆废液
spent tan 废鞣料
spent waiting time 消耗的等待时间
spent water 废水
speos 石窟墓室(古埃及);洞穴石坟(古埃及);石窟神庙
spere 门帘;固定围屏
spere-truss (大厅中与服务廊相隔的)带拱木屋架
Spergen limestone 萨勒姆石灰岩
spermaceti wax 鲸脑油;鲸蜡
sperm pump 精泵
sperm whale 巨头鲸;抹香鲸
sperolitic welded tuff 球粒熔结凝灰岩
sperone 黑榴白榴岩;扶垛墙
speroni 扶垛;墙墩
Sperry buddle 斯佩里型圆形淘汰盘;圆形精选台
sperrylite 砷铂矿
Sperry process 斯佩里电解法
Sperry's metal 斯佩里铅基轴承合金
sperry super-sonic reflectoscope 超声波反射探伤仪
spertiniite 斯羟铜矿
sperver 天篷顶框
spessartine 斜煌岩;锰铝(石)榴石
spessartite 斜煌岩;闪斜煌岩
spet cash 现金
S. peter Cathedral 圣·彼得大教堂(公元1506—1626年建于罗马)
spew frost 冻胀
spewing 漆面结膜微粒;污版;渗油
spew relief 溢料空隙
spewy soil 沼泽土(壤)
sphaeristerium 室内球场(古罗马)
sphaerite 球磷铝石
sphaeroberandite 硅铍石
sphagniherbosa 泥炭苔藓草本群落
sphagniopratum 苔藓沼泽
sphagniprata 泥炭藓草本群落
sphagnum 水苔
sphagnum bog 苔藓沼泽;水藓沼泽
sphagnum moor 苔藓沼炭
sphagnum moss 水藓
sphagnum peat 水藓泥炭
sphagnum wetland 苔藓湿地
sphalerite 闪锌矿
sphene 榍石
sphenethmoid 蝶筛骨
sphenion(sphn)蝶点
sphenoconformity 楔形整合
sphenoid 楔状的;楔形的;轴双面;四方四面体;蝶骨的;半面晶形
sphenoid bone 蝶骨
sphenolith 岩楔
spheno occipital suture 蝶枕缝
sphenoparietal index 蝶顶指数
sphenophsida 楔叶类植物
spherangular roller bearing 球锥滚柱轴承;球面滚子轴承
sphercial tank 球形罐
sphere 领域;球状体;球形物;球形;球体;球面;天体仪;地球的圈层
sphere bacteria 球状细菌
sphere can 球形油罐
sphere colloid 球形胶体
sphere data 球面数据
sphere gap 球间隙
sphere-grinding mill 球磨机
sphere illumination 球形照明体
sphere impact 球的撞击
sphere joint 球形关节
sphere node 球节(点)
sphere of action 作用区(域);作用范围
sphere of activities 活动范围
sphere of antagonistic effect 相互作用圈
sphere of attraction 吸引圈
sphere of edoxus 埃多克斯球
sphere of function 作用范围
sphere of gravity 重力范围
sphere of influence 影响圈;影响函数;作用范围;

势力范围
sphere of inversion 反演球
sphere of jurisdiction 管辖范围
sphere of life 生活圈
sphere of powder 粉球
sphere of radius 单位球
sphere packing 球状堆积;球状填充
sphere-packing bound 球包界
sphere photometer 积分球光度计;球状光度计;球面光度计
sphere pole 球状电极
sphere riser 球状冒口
spheres of eudoxus 欧多克斯球
sphere spark-gap 球形火花放电器
spheresthesia 球状感觉
sphere transmission method measurement 球对称透射法测量
sphere valve 球阀
spheric(al) 球的;球状的;球形的;球面的
spheric(al) aberration 球面像差
spheric(al) aberration coefficient 球差系数
spheric(al) accretion 球对称吸积
spheric(al) accumulator 球形蓄压器;球形蓄能器;球形储能器
spheric(al) active carbon 球状活性炭
spheric(al) agglomeration 球团矿
spheric(al) aggregate 球状集合体
spheric(al) albedo 球面反照率
spheric(al) angle 球面角
spheric(al) annulus 球形环状
spheric(al) antenna 球形天线;球面天线
spheric(al) approximation 球近似
spheric(al) armature 球形电枢
spheric(al) array 球面阵
spheric(al) arrester 球形避雷器
spheric(al) astronomy 球面天文学
spheric(al) azimuth 球面方位角
spheric(al) balance spring 球形游丝
spheric(al) ballast 球压载
spheric(al) balloon 球形气球
spheric(al) bearing 球形轴承;球形支座;球面轴承;球面支座;球面方位角
spheric(al) Bessel function 球贝塞尔函数
spheric(al) bevel gear 球形伞齿轮
spheric(al) bifurcated pipe 球形岔管
spheric(al) blast wave 球形爆炸波
spheric(al) body 球体
spheric(al) boiler 球状锅炉;球形蒸煮器
spheric(al) bottom bin 球底料仓;球底料斗
spheric(al) brass cup 球形铜环
spheric(al) bulb 球泡
spheric(al) buoy 球形浮标
spheric(al) bush 球形衬
spheric(al) calotte 球形屋顶;球形帽罩;球面拱顶;圆罩;圆顶
spheric(al) cam 球形凸轮
spheric(al) candlepower 球面烛光;球形烛光
spheric(al) cap 球形顶盖
spheric(al) capsule 球形分离舱
spheric(al) captive balloon 球形系留气球
spheric(al) casing 球形汽缸
spheric(al) catalyst 球形催化剂
spheric(al) cavity 球形腔
spheric(al) cavity resonator 球形空腔谐振器
spheric(al) centre bearing 球心轴承
spheric(al) chain 球面运动链
spheric(al) charge 球状药包
spheric(al) cock 浮球阀;浮球旋塞
spheric(al) cock-shell 球状塞门体
spheric(al) collector 球形收集器
spheric(al) combustion chamber 球形烧燃室;球形燃烧室
spheric(al) compass 球面磁罗经
spheric(al) component 球状子系
spheric(al) condenser-type spectrometer 球形聚光器型能谱仪;球面聚光镜型分光仪
spheric(al) conductor 球形导体
spheric(al) cone 球面锥
spheric(al) conic(al) section 球锥剖面
spheric(al) connection 球头节
spheric(al) console typewriter 球形控制台打字机
spheric(al) container 球形容器
spheric(al) convergence 球面会聚
spheric(al) coordinates 球体坐标;球面坐标

spheric(al) coordinate system 球面坐标系
spheric(al) coupling 球形联轴器
spheric(al) crank chain 球形曲柄链系;球面曲柄链系
spheric(al) crevasses 球状裂纹
spheric(al) crucible 球形坩埚
spheric(al) crucible technique 球形坩埚技术
spheric(al) crystal 球晶
spheric(al) cupola 球形圆顶;球形穹顶
spheric(al) current meter 球形海流计
spheric(al) curvature 球面曲率
spheric(al) cutter 球面刀
spheric(al) dam 球面坝;球形挡水墙
spheric(al) datum surface 基准球面
spheric(al) defect 球面像差
spheric(al) depolished glass globe 球形磨砂玻璃灯泡;球形磨砂玻璃灯罩
spheric(al) detonation front 球状爆炸前峰
spheric(al) diameter 球面直径
spheric(al) diffraction loss 球面浇射损耗
spheric(al) diffraction of ultra-short radio wave 超短波球面绕射
spheric(al) digester 蒸煮球
spheric(al) discrepancy 球面不符值
spheric(al) disk 球面圆盘
spheric(al) distance 球面距离
spheric(al) distribution 球面分布;球分布
spheric(al) divergence correction 球面发散校正
spheric(al) dome 圆穹顶;球形屋顶;球形穹顶;球面穹面
spheric(al) earth 球面接地;球面地
spheric(al) earth attenuation 球形地面衰减
spheric(al) earth factor 球形地面因数
spheric(al) effect 球效应
spheric(al) electrode 球形电极
spheric(al) element 球面元素
spheric(al) emanation source 球形射气源
spheric(al) emitter 球形发射体
spheric(al) end 球形端
spheric(al) error probability 球面概率误差
spheric(al) eutectic grain 球状共晶晶粒
spheric(al) even nuclei 球形偶核
spheric(al) excess 球面角盈;球面角超
spheric(al) explosion 球形爆炸
spheric(al) fabric 球组织;球状粒组构
spheric(al) face 球面
spheric(al) feeder 球形冒口
spheric(al) feed valve 球形进料开关;球形进料阀
spheric(al) float 球状浮子
spheric(al) flying 球面飞行
spheric(al) foam chromatography 球形泡沫色谱法
spheric(al) focused log 球形聚焦测井
spheric(al) focused log curve 球形聚焦测向测井曲线
spheric(al) fuel element 球形燃料元件
spheric(al) function 球面函数;球面波函数;球函数
spheric(al) gas-holder 球形储气罐
spheric(al) gasometer 球形煤气表;球形气量表
spheric(al) Gaussian orbital 球型高斯轨道
spheric(al) geometry 球面几何(学)
spheric(al) glass 球面镜片
spheric(al) grain 球状晶粒
spheric(al) granule 球形颗粒
spheric(al) graphite cast iron 球墨铸铁
spheric(al) grating 球面光栅
spheric(al) grid analyzer 球栅分析器
spheric(al) grid-wall counter 球形栅壁计数管
spheric(al) grinding point 球面研磨点
spheric(al) grips 球窝夹头
spheric(al) ground glass joint 球形磨口玻璃接头
spheric(al) ground state 球形基态
spheric(al) guide 滚珠导轨;球珠导轨
spheric(al) harmoics approximation 球谐函数近似法
spheric(al) harmonic calculation 球谐计算
spheric(al) harmonic coefficient 球谐函数系数
spheric(al) harmonic equation 球谐方程
spheric(al) harmonic expansion 球谐展开;球谐函数展开
spheric(al) harmonic function 球谐函数;球体调和函数
spheric(al) harmonic method 球谐函数法
spheric(al) harmonic potential disturbing function 球谐位扰动函数

spheric(al) harmonics 球谐函数;球函数
spheric(al) harmonics series 球谐函数级数
spheric(al) head 球形头;半圆头
spheric(al) headed niche 上端球面的壁龛
spheric(al) head piston 球形顶式活塞;球顶式活塞
spheric(al) head probe 球形测针
spheric(al) head without folded edge 无折合球形封头
spheric(al) helix 球面螺旋线
spheric(al) hinge 球铰链
spheric(al) hipped end 圆形斜脊端
spheric(al) holder 球形储罐
spheric(al) house 球形房屋
spheric(al) hyperbola 球面双曲线
spheric(al) icosahedrons 球形十二面体
spheric(al) illumination 球面照度
spheric(al) image 球面影像
spheric(al) image amplifier 球面像放大镜
spheric(al) inclusion body 球形包裹体
spheric(al) indenter 球形压头
spheric(al) indicatrix of binormal to a curve 曲线副法线球面指标
spheric(al) indicatrix of scattering 球面散射指示量
spheric(al) indicatrix of tangent to a curve 曲线切线的球面指标
spheric(al) ionization chamber 球形电离室
spheric(al) iron particle 球形铁粒
spheric(al) irradiance meter 球面辐照强度计
spheric(al) joint 球形接头;球形接合;球节(点);球接头
spheric(al) jointing 球状节理【地】
spheric(al) journal 球面轴颈;球面轴颈
spheric(al) lamp 圆球形灯;球形灯
spheric(al) lapping 球面擦准法
spheric(al) lathe 球面车床
spheric(al) lens 球面透镜;球面镜头
spheric(al) lens generating machinery 球面透镜制造设备
spheric(al) lens smoothing machinery 球面透镜精加工设备
spheric(al) level vial 圆水准器
spheric(al) lightning 球状闪电;球状电闪
spheric(al) luminosity of light 球面光照度
spheric(al) luminous intensity 球面光强
spheric(al) lune 球面二角形
spheric(al) map 世界地图
spheric(al) mechanism 球面运动机构
spheric(al) meniscus 球形弯月面;球面凹凸透镜
spheric(al) micelle 球形胶束
spheric(al) microphone 球形传声器
spheric(al) microshell target 球形微壳靶
spheric(al) mirror 球面镜;球面反射器;球面反射镜
spheric(al) mirror cavity 球面镜共振腔
spheric(al) mirror interferometer 球面镜干涉仪
spheric(al) mirror objective 球面反射式物镜
spheric(al) mirror resonator 球面镜揩振腔
spheric(al) model 球状模型
spheric(al) motion 球面运动
spheric(al) motor 球形电动机
spheric(al) movement 球面运动
spheric(al) neighborhood [neighbourhood] 球形邻域
spheric(al) nodule 球状结核
spheric(al) nucleus 球形核
spheric(al) nut 球形螺帽
spheric(al) oil tank 球形油罐
spheric(al) optics 球面光学
spheric(al) oscillation 球状振荡
spheric(al) pair 球面副
spheric(al) parameter 球面参数
spheric(al) particle 球状颗粒;球形颗粒
spheric(al) pendulum 球面摆;球摆
spheric(al) perspective 球面透视
spheric(al) photometer 球状光度计;球形光度计
spheric(al) pitot probe 球形头皮托管
spheric(al) pivot 球形枢轴;球面枢轴
spheric(al) plain bearing 关节轴承
spheric(al) plane 球面
spheric(al) plasmoid 球形等离子粒团
spheric(al) polar 球面极
spheric(al) polar coordinates 球面极坐标;球极坐标
spheric(al) polar coordinate system 球极坐标系
spheric(al) polygon 球面多边形

spheric(al) pore 球体孔
spheric(al) powder 球状粉末；球形粉
spheric(al) pressure container 球形加压容器
spheric(al) pressure tank 球形压力罐
spheric(al) probe 球形探头
spheric(al) profile 球形轮廓
spheric(al) projection 球投影
spheric(al) punch 球状凸模；球底面模
spheric(al) pyranometer 球形总日射表
spheric(al) pyrgeometer 球形地面辐射表
spheric(al) pyradiometer 球形全辐射表
spheric(al) quadrangle 球面四方形；球面四边形
spheric(al) quadrilateral 球面四边形
spheric(al) radiation 球辐射
spheric(al) radiator 全向辐射器；全方位辐射器；球状辐射器；球辐射器
spheric(al) radiometer package 球形辐射计装置
spheric(al) radius 球面半径
spheric(al) radius factor 球半径系数
spheric(al) reactor 球形堆
spheric(al) rectangular coordinates 球面直角坐标
spheric(al) reduction factor 球面折算因数；球面换算系数
spheric(al) reference cavity 球面参考共振腔
spheric(al) reflecting mirror 球面反射器；球面反射镜
spheric(al) reflecting system 球面反射系统
spheric(al) reflection 球面反射
spheric(al) reflectivity 球面反射率
spheric(al) reflector 球面反射器；球面反射镜；球反射面
spheric(al) reflector antenna 球面反射器天线
spheric(al) reflector light 球面反射光
spheric(al) refracting surface 球面折射面
spheric(al) region 球形区域
spheric(al) representation 球面表示
spheric(al) representation of a surface 曲面的球面表示
spheric(al) resolvent 球面预解式
spheric(al) resolver 球坐标分解器
spheric(al) resonator 球形谐振器；球形共振腔
spheric(al) retrodirective array 球面反向天线阵
spheric(al) riser 球形冒口
spheric(al) roller 球面滚子
spheric(al) roller bearing 球面液柱轴承；球面滚子轴承
spheric(al) roller thrust bearing 推力球面滚子轴承
spheric(al) roof 球形屋顶
spheric(al) safety valve 球形安全阀
spheric(al) sailing 球面航迹计算法
spheric(al) scale 球形刻度盘
spheric(al) seat 球形座
spheric(al) seating 球铰座
spheric(al) seating nut 球形支承螺帽
spheric(al) seat nut 球座螺母
spheric(al) section of vessel 容器的球形部分
spheric(al) sector 墙面扇轮；球心角体
spheric(al) sector resonator 球瓣共振腔
spheric(al) segment 截球形；球面弓形；球截形
spheric(al) segment combustion chamber 球形燃烧室
spheric(al) separator 球形分离器
spheric(al) shape 圆球形；球形
spheric(al) shell 球形壳(体)；球形薄壳；球面壳体；球壳
spheric(al) shield 球形屏蔽
spheric(al) shock 球面激波
spheric(al) shutter 球形快门
spheric(al) socket 球形承窝
spheric(al) solid angle 球面立体角
spheric(al) spinner 球形转子
spheric(al) spreading 球面扩展
spheric(al) stalactites 葡萄石
spheric(al) state of stress 各向等应力状态
spheric(al) storage vessel 球形储罐
spheric(al) stress 球应力
spheric(al) strip 球面带
spheric(al) structure 球状构造；球形结构
spheric(al) subsystem 球状次系
spheric(al) surface 球状物表面；球形表面；球面
spheric(al) symmetry 球形对称；球对称
spheric(al) tank 球形柜；球形储罐；球罐
spheric(al) tensor 球张量
spheric(al) thrust bearing 球形推力轴承

spheric(al) top molecule 球形陀螺分子
spheric(al) transformation formula 球面变换公式
spheric(al) triangle 弧三角形；球面三角
spheric(al) triangulation 球面三角形测量
spheric(al) trigonometry 弧三角形；球面三角（学）
spheric(al) tubesheet 球形管板
spheric(al) turning lathe 球面车床
spheric(al) type head 球型头
spheric(al) typewriter 球形打印机
spheric(al) union 球面活接头
spheric(al) union swivel 球面活管接旋转接头
spheric(al) valve 球阀；球形阀
spheric(al) valving surface 球形阀配流阀表面
spheric(al) variance function 球面方差函数
spheric(al) vault 球形穹顶
spheric(al) vessel 球形容器
spheric(al) vortex 球形旋涡；球面旋涡
spheric(al) washer 球面垫圈
spheric(al) wave 球状波；球形波；球面波
spheric(al) wave front 球面波前
spheric(al) wave function 球波函数
spheric(al) wave horn 球面波形喇叭
spheric(al) wave illumination 球面波照明
spheric(al) wave proximity effect 球面波近场效应
spheric(al) wave reciprocity calibration 球面波倒易校准
spheric(al) weathering 球状风化；球形风化
spheric(al) well 球势阱
spheric(al) wheel vehicle 球形车轮车
spheric(al) window photomultiplier 球面窗式光电倍增管
spheric(al) xenon lamp 球形氙灯
spheric(al) zone 球带；地球圈层
sphericity 圆球度；圆度；球状；球形体；球形度；球径率；球度
sphericity factor 球形系数
sphericity interferometer 球面干涉仪
sphericity limit 球度极限值
sphericized lattice cell 晶格等球形晶胞；球形化晶胞
spherics 球面几何(学)；天电学
spherics sounder 远程雷电探测仪
spherochromatic aberration 球面色像差；色球差
spherochromatism 色球差
spheroclast 圆碎屑
spherocobaltite 菱钴矿；球泡酸钴矿
sphero-conic 球面二次曲线
sphero-conic(al) coordinate 球锥坐标
sphero-conic(al) harmonics 球锥函数
sphero-crystal 球晶
sphero cyclic 球面圆点曲线
spherocylinder 球柱体
sphero cylindrical lens 球柱面透镜
sphero-cylindrical lenticular 球柱面冰凸透镜
spheroicizing annealing 球化退火
spheroid 回旋椭圆体；球状容器；球体；水滴形油罐；扁球形；扁球体
spheroidal 球状的；球形的；球体的；椭球状的
spheroidal annealing 球化退火
spheroidal azimuth 椭球面方位角
spheroidal carbide 球状碳化物
spheroidal catalyst 球形催化剂
spheroidal cementite 球状结晶（指钢铁）
spheroidal coordinates 椭球面坐标
spheroidal distance 椭球面距离
spheroidal earth 球状地球；地球椭球
spheroidal excess 球面角盈
spheroidal galaxy 椭球星系
spheroidal graphite 球状石墨
spheroidal graphite cast iron 球状石墨铸钢；球墨铸铁
spheroidal graphite iron ring 球墨铸铁环
spheroidal graphite roll 球墨铸铁轧辊；球墨轧辊
spheroidal group 椭球体群
spheroidal harmonic 球体函数
spheroidal harmonic function 球体调和函数
spheroidal height 椭球面高
spheroidal hypercylinder 球状超柱
spheroidal jointing 球状节理
spheroidal junction 椭球区分线
spheroidal latitude 椭球面纬度
spheroidal meridian 椭球子午线
spheroidal mode 球状模式
spheroidal normal 椭球法线

spheroidal oscillation 球体式振荡
spheroidal particle 球形颗粒
spheroidal pearlite 球状珠光体
spheroidal pulse ionization chamber 球形脉冲电离室
spheroidal recovery 球状复原
spheroidal state 球体状态
spheroidal structure 球状构造
spheroidal vibration 球振荡
spheroidal wave 球体波
spheroidal wave function 球体波函数
spheroidal weathering 球状风化
spheroid geodesy 椭球面大地测量学
spheroidicity 椭球度
spheroiding 球体形成
spheroidisation 球化处理
spheroidised pearlite 粒状珠光体
spheroidite 粒状体
spheroidite cementite 粒状渗碳体
spheroiditic agent 球化剂
spheroidization 球化
spheroidization of graphite 石墨球化
spheroidized carbide 球化碳化物
spheroidized particle 球化颗粒
spheroidized steel 球化处理钢
spheroidizing 粒状化；球状化；球化作用；球化处理
spheroidizing medium 球化剂
spheroidizing of powder 粉末球化
spheroid mass 球体质量
spheroid of revolution 旋转椭球
spheroid of the earth 地球椭球体
spheroid tank 球状油罐
spherolitic kimberlite 金伯利岩球
spherome 球体
spherometer 球径计；球面仪；球面曲率计；球面计；测球仪
spherometer callipers 球径规
spherometry 球径测量术
spherophyre 球粒斑岩
spheroplast 圆球体；球粒体
sphero(potential) 正常地球位
spherosome 圆球体
spherotoric lens 复球面透镜
spherotoric lenticular 复球面双凸透镜
spherule 小球体
spherulite 球状晶体；球粒
spherulite growth rate 球晶生长速率
spherulite radial growth rate 球晶辐向生长速度
spherulite rock 球粒岩
spherulitic 球粒状
spherulitic graphite 球状石墨
spherulitic growth 球晶生长
spherulitic pelitic texture 球粒泥质结构
spherulitic structure 球粒构造
spherulitic texture 球粒结构
sphincter 松紧口装置
sphingometer 光测挠度计；梁和桁架挠度测定器
Sphinx 人面狮身像；斯芬克斯(狮身人面像)
sphinx gate 斯芬克斯门
sphognous swamp 水藓沼泽
spicae tessella 长方形铺路砖
spicatum opus 人字式铺砌
spice 香料
spice berry 香料植物
spice bush seed oil 香叶树子油
spice crop 香料作物
spice curing 香料渍
spice trade 香料贸易
spiciform 穗状（排列）的
spick-and-span 干净整齐的
spicular ice 针状冰
spicule 小疙瘩；针状物；针状体；骨针
spiculite 针锥晶
spider 星轮；中心盘；蜘蛛架；支架；针状盘；横杆；十字叉；多脚架；多脚撑
spider and slips 卡盘
spider arm 星轮臂；车板横臂
spider assembly 星状装配；星状组合；带辐条的轮毂；星形轮
spider band 装有挽缆插栓的桅箍
spider bonding 蛛网形接合；蛛网焊；辐式键合
spider brick 中心流钢砖
spider bushing 卡瓦补心；卡盘衬套；十字叉衬套
spider center 星形轮毂；万向节十字轴；十字轴中心

spider chill 框架式冷铁
spider coil 蛛网形线圈
spider die 异形孔挤压模
spider element 蜘蛛架
spider fox-hole 蛛网形散兵坑
spider frame 蜘蛛架
spider ga(u)ge 细木工规尺
spider gear 星形齿轮;差速轮
spider handle 转向盘
spider heart 蛛心裂(木材缺陷)
spider hoop 装有挽缆插栓的桅箍
spider hub 蜘蛛架中心轴;支架毂
spider iron 装有挽缆插栓的桅箍
spider kit 十字形工具;星形工具
spider legs 蛛纹花纹;云卷花纹;流云花纹
spider leg screw 蜘蛛架导杆螺钉
spider line 交叉瞄准线;辐射线
spider lines 蛛丝
spider mounting 星形机架
spider needle bearing 星形轮的滚针轴承
spider net 蛛网形网眼
spider pin 行星小齿轮轴
spider pipe 卡盘管
spider piping 卡盘管
spider plate 牵引起重机柱顶板
spider reel 十字形卷线轮
spider rim 蜘蛛架外环;支架轮缘
spider slip 十字形滑道
spider spanner 星形(套筒)扳手
spider spring 十字形弹簧
spider's web 蛛网状裂纹(玻璃及搪瓷缺陷)
spider template 蛛网形模片
spider template triangulation 机械模片辐射三角测量
spider templet 蛛网形模片
spider trunnion 万能接头
spider tube 卡盘管
spider tubing 卡盘管
spider Turk's head 花箍结
spider vane 辐射形叶片
spider weave 蛛网组织
spider web 蛛网;蛛网状物
spider-web antenna 蛛网(式)天线
spider-web city 蛛网状城市
spider-web coil 蛛网形线圈;平扁蛛网形线圈
spider-web reflector 蛛网式天线反射器
spider-web type of street system 蛛网形道路系统
spider-web winding 蛛网绕组
spider-wed the rig 内部装有许多拉紧钢绳的钻塔
spider wheel 星形轮
spider-wheel rake 指轮式搂草机
spider wire entanglement 蛛网形铁丝网
spider wrench 星形(套筒)扳手
spiegel(eisen) 低锰合金;铁锰合金;镜铁;低锰铁
spiegel iron 镜铁;低锰铁
spier 挂落;固定围屏
spigot 塞;插口;插端
spigot and faucet joint 窝接
spigot and faucet pipe 承插管
spigot and socket bend with flange 有翼缘的承插式弯管;有翼缘的窝接式弯管
spigot and socket concrete pipe 窝接式混凝土管;承插式混凝土管
spigot and socket connection 承插接合;窝接式接头;套筒接合
spigot and socket joint 承插接合;窝接;承插口接头;承插接
spigot and socket pipe 窝接式接头管;套筒管(俗称缩节);缩节(即套接管);承插接头管;承插式接管;承插管
spigot and socket prestressed concrete pipe 承插式预应力混凝土管
spigot and socket stoneware pipe 承插式陶管;窝接式陶管
spigot and socket tee 承插三通
spigot and socket tube 承插式接管
spigot and spigot reducer 双插大小头
spigot bearing 小荷载轴承;轻载轴承;套筒轴承;导向轴承;导向支承;插口轴承
spigot density test 筛下物密度试验
spigot die 穿透模;插入模
spigot discharge 卸料孔;排料口;分级机排砂口
spigot edge 定心凸出物
spigot end 插口端;插端

spigot end of pipe 套接管插入端
spigot jet nozzle 锥塞喷嘴
spigot joint 承插接合;套筒接合;连接器;插头;接嘴;窝接;套管接口;套管接合;水管插口;插口接头;插管接头
spigot mortar 超口径弹迫击炮
spigot nut 导向螺母;导向螺帽
spigot of conduit 水管插口
spigot of pipe 水管插口
spigot of tube 水管插口
spigot pallet (制造混凝土管子的)插端锤垫;(制造混凝土管子的)插端垫衬
spigot pipe 承插式接头管
spigot shaft 中心轴
spike 高峰值;(电子束焊的)局部深熔;进口整流锥;解缆锥;尖铁;尖峰信号(脉冲);尖峰脉冲;棘波;脉冲尖峰;大模钉;普通道钉(即钩头道钉);耙齿;铁笔;示踪钉;道钉;大(铁)钉;刺针;掺料;峰形;反应堆强化;波峰
spike amplitude 波尖幅度
spike and ferrule installation 钉和箍设置;用大钉和套箍固定的檐沟装置
spike and ware complex 棘慢波综合
spike and ware wave 棘慢波
spike arrester 尖头避雷器
spike burst 尖峰状暴;尖峰暴
spike control 进气道锥体操纵
spiked beater 钉齿式逐稿轮
spiked chain harrow 钉齿网形耙
spiked core 强化堆芯
spiked cylinder 针形滚筒;针刺轮
spiked drum 针形滚筒
spiked-drum type root cutter 钉齿滚筒式块根切碎机
spike deconvolution 脉冲反褶积
spike deflection control 进气道锥体偏转操纵
spike desintegrator 转盘式松砂机
spike disintegrator 棒式松砂机
spiked isotope 示踪同位素
spike diverter 尖脉冲分流器
spiked link harrow 钉齿链耙
spike dog 道钉
spike doublet 脉冲偶极子
spike dowel 钉栓
spike down capacitor 峰值抑制电容器
spike drag 钉板刮路器
spike drawer 道钉撬;拔钉钳;拔道钉机
spike drawing winch 道钉起拔器;道钉拨
spiked reactor 强化反应堆
spike driver 道钉打入机;道钉锤;打道钉机
spiked roller 大型滚轮钻头(井筒钻进用);钉齿辊
spiked sample 加料样品;加标样品
spiked solution 示踪溶液;掺料溶液
spike duration 波尖宽度
spiked-wheel lifter 针刺轮式拔取器
spike eliminator 尖脉冲消除器
spike extractor 起道钉器
spike fastening 斩钉
spike filter 窄带滤波器
spike grid 钉格板(一种木结构接合件);木结构的齿环连接件
spike gun 道钉抢
spike hammer 道钉锤
spike harrow 直齿耙;钉齿耙(路机)
spike heel 钉状柱脚
spike hole 小气孔
spike inlet 带中心锥的进气口
spike iron 平头錾
spike isotope 添加同位素
spike killing 道钉毁损
spike knot 角状节;尖节;长节疤;扁节;斜节;销钉;长形结节;长钉
spike leakage 尖漏
spike leakage energy 波尖漏过能量
spike leakage power 波尖漏过功率
spike maul 道钉锤
spike microphone 钉子传声器
spike mill 针碎机;钉碎机
spike nail 道钉;大钉;长折钉
spikenard oil 甘松油
spike noise 尖峰脉冲噪声;波尖噪声
spike plank 冰区航行瞭望台
spike potential 尖峰电位;峰电位
spike puller 拔钉钳;道钉撬;起道钉器;道钉拔出

器;道钉器
spike pulling 拔棒
spike pulse 窄脉冲;尖(峰)脉冲
spiker 铁道杂工;道钉工(铁路);针齿式圆盘耙
spike rod 道钉型钢
spike roller 销钉辊;钉齿辊
spike rot 轴腐病
spike sensor 尖峰脉冲探测器
spike signal 尖峰信号
spike tongs 道钉钳
spike-tooth concave 钉齿式凹板
spike-tooth cylinder 钉齿式脱粒滚筒
spike-tooth flexible harrow 钉齿活动耙
spike-tooth harrow 直齿耙;钉齿耙
spike-tooth harrow maker 钉齿耙制造厂
spike-tooth roller 钉齿碾
spike-top 焦梢树
spike voltage 峰度电压
spike wave 棘波
spike width 峰值宽度
spiking 尖峰信号;钉道钉;掺加(示踪剂);峰值形成
spiking behavio(u)r 尖峰性能
spiking data 尖峰数据
spiking deconvolution 脉冲反褶积
spiking gang 钉道钉队
spiking hammer 道钉锤
spiking isotope 掺加同位素
spiking method 峰值法
spiking output 最大输出信号
spiky roller 羊角碾
spile 开小孔;打桩子;插管;小栓;小塞(子);木桩;超前板桩;插桩;插板
spile-hole 通气(小)孔(桶的);小气孔
spile worm 船蛆
spiling 超前支护;木桩;样板下料方法;超前伸梁掘进法;超前板桩法掘进;超前板桩法
spiling lath 板桩
spiling river 宽槽河道
spilite 细碧岩
spilite-keratophyre sequence 细碧岩角斑岩系
spilitic association 细碧岩组合
spilitic glass 细碧岩质玻璃
spilitic-keratophyre formation 细碧角斑岩建造
spilitic suite 细碧岩组;细碧岩套
spilitic texture 细碧结构
spilitization 细碧岩化(作用)
spill 泄出;小栓;小塞(子);溢土(量);溢漫;溢出物;溢出(量);漏料;漏出;木片;漫溢;洒;大雨倾盆;放松帆面受风压力
spillage 泄漏量;溢出量;溢出;漏损(量);扫舱货;地脚
spillage collection hopper 收集漏料的漏斗
spillage collector 漏灰回收器
spillage control 溢出控制
spillage hopper 溢料漏斗
spillage loss 溢流损失
spillage of material 材料损耗;材料损害
spillage oil 溢油;溢出油
spillage prevention 防撒漏
spillage return screw 漏灰回收绞刀
spillage solution 溢出溶液;漏出溶液;溅出溶液
spillage tank 溢流池;溢流槽
spill air 溢流空气;放空空气
spill-basin 溢出盆地;溢出水槽
spill beam 寄生束
spill boom 水上溢油刮集臂架
spill box 溢流箱
spill burner 回油式喷嘴
spill channel 溢出管道;溢出沟渠;溢出通道
spill chute 漏料溜子;散落物料溜子
spill cleanup activities 溢漏污染物的清除活动
spill clearance vessel 溢油清除船
spill combating vessel 溢油清除船
spill current 溢流;差电流
spill disposal contractor 溢油处理承包人
spilled oil cleanup kit 漏油清洗设备
spill guard 防溢板
spill hazard 溢漏危险;溢漏风险
spill incident 溢漏事故
spilliness 鳞片;溅麻面
spilling 预支掘进;超前板桩法掘进;滨波破碎;溢出
spilling breaker 卷跃碎波;漫涌碎浪;崩顶破浪;崩顶破波
spilling chamber 溢流室

spilling river 宽槽河道;泛溢河道
spilling spring 溢出泉
spilling surge chamber 溢流式均衡室
spilling water 溢水;溢出水
spilling wave 溢出浪;崩顶破波
spillite-keratophyre 细碧角斑岩
spill loss 溢流损失
spill needle 放泄针
spill off valve 分路式流量阀
spillover 溢出(信号);飞弧放电;息漏失;泄漏放电;溢流
spillover dalta 溢流三角洲
spillover echo 超折射效应回波
spillover effect 超溢效果;漂越效应
spillover efficiency 溢流效率
spillover level 溢出高度
spillover pool 溢流潭
spillover position 溢出位置
spillover power 溢失功率
spillover radiation 漏失辐射
spill overs 额外收益
spillover signals 交会信号
spillover valve 溢流阀
spillover zone 溢出区
spill performance response assessment technique 溢油性能响应评估法
spill pipe 溢流管
spill piping 溢流管
spill pit 溢流池
spill plate 防溢栅;防溢板
spill plate extension 溢流板延伸
spill pore (油泵的)回油孔
spill process 引出过程
spill response product 溢漏事故应急用品
spill response team 漏油应急队伍
spill ring 环形百叶窗;环形格片(照明用)
spill sand 散落砂
spill sand conveyer 卸砂输送机
spill shield 遮光栅;遮光屏;遮光板
spill shield louver lighting 隔栅照明
spill site 溢漏事故现场
spill tag 漏标
spill technology 溢漏事故应急技术
spill-through abutment 穿通式桥台
spill time 出束时间
spill tube 溢流管
spill tubing 溢流管
spill valve 溢油阀;溢液阀;溢流阀;回油阀;调节阀
spill valve type fuel injection pump 溢流阀式喷油泵
spill volume 溢出卷
spill wall 防火墙;防火堤;溢液隔墙
spillway 溢流管;泄水道;水道;溢流道;溢洪道;溢道
spillway apron 溢洪道护坦;溢洪道海漫(即溢洪道防冲铺砌或护祖);防冲护坦
spillway basin 消力池;静水池;溢水池;溢洪道消力池
spillway bay 溢洪道闸孔;溢洪道闸孔;溢洪道墩间跨距;溢洪道墩间跨径;溢洪道闸孔径
spillway bridge 溢洪坝顶桥;溢洪道桥
spillway bucket 滚洞戽斗喷雾器;滚洞戽斗式消能器;溢洪道消力戽
spillway capacity 溢流道排水量;溢洪道泄洪能力;溢洪道排水量
spillway channel 溢流渠(道);溢流道;溢流槽;溢洪水道;溢洪道泄水槽
spillway chute 溢流陡槽;溢洪道陡槽
spillway control device 溢流控制装置;溢流控制设备;溢洪道流量控制装置;溢洪道流量控制设备
spillway crest 溢流顶;溢流道顶(部)
spillway crest gate 溢流堰顶闸门
spillway crest level 溢洪道堰顶高程
spillway culvert 泄洪涵管;溢流涵洞
spillway dam 泄水坝;有水道的坝;溢流坝;溢流坝;滚水坝;溢流堰
spillway dam section 泄洪坝段
spillway design flood 溢洪道设计洪水
spillway discharge 溢流量;溢洪道泄水量
spillway face 溢洪道砌面
spillway flume 溢水口渡槽
spillway for reservoir 水库溢洪道
spillway gallery 溢水集水道
spillway gate 泄洪闸;溢水闸;溢流闸门;溢洪闸;溢洪道闸(门)
spillway lip 溢流堰堰缘;溢流堰缘;溢洪道唇
spillway outlet 溢水口
spillway outlet sill 溢洪道泄水底槛;溢洪道出口底槛
spillway outlet structure 溢水出口结构
spillway overflow 溢洪道溢流
spillway pier 溢洪道闸墩
spillway section 溢洪道截面;溢流段
spillway slab 溢流板
spillway stream 溢出水流
spillway structure 溢洪道构筑物;溢流建筑物;溢洪道建筑物
spillway surface 溢洪道表面
spillway tank 溢流池;溢流槽
spillway training wall 溢洪道导水墙
spillway trough 溢流槽
spillway tunnel 溢流隧洞;溢流隧道
spillway weir 溢洪(道)堰
spillweir 溢水道;溢流堰
spillweir dam 溢洪坝;过水坝;滚水坝
spilly place 未焊透
spilosites 绿点板岩
spilth 垃圾;废物;溢出物
spilt level 夹层
spilt-ring 开口式
spilt test 劈裂试验
spin 旋转;旋压成型;自旋;绕转;拔丝
spina (古罗马运动场的)纵向栅栏
spinal animal 脊椎动物
spinal column 脊柱
spin algebra 自旋代数
spin alignment 自旋排列
spinal lherzolite 尖晶石二辉橄榄岩
spin angular momentum 自旋角动量
spina ventosa 风刺
spin axis 旋转轴(线);自旋轴
spin axis of the gyro 陀螺旋转轴
spin bath 沉降槽
spin burst test 旋转破坏试验
spin chamber 旋流室
spin chuck 旋压车床用夹头
spin chute 反螺旋伞
spin coated 旋转涂敷
spin coating 旋转涂布;旋涂
spin compensation 自旋补偿
spin control 旋转控制
spin correlation parameter 自旋关联系数
spin counter 转速计数器
spin coupling 自旋耦合
spin-decelerating moment 自旋减速力矩
spin decoupling 自旋去耦
spin decoupling method 自旋去耦法
spin decoupling pattern 自旋去耦图
spin degeneracy 自旋简并度
spin delocalization 自旋离域
spin density 自旋密度
spin density wave 自旋磁感应波
spin dependence 自旋有关
spin-dependent force 自旋相关力
spin direction 自旋方向
spindle 心轴;转轴;轴梗;杆状信标;汽车转向节;纺锤体
spindle alignment 主轴对准
spindle angle 接轴倾斜角
spindle arm 主轴臂
spindle arm ga(u)ge 转向臂规
spindle assembly 心轴组件;立轴部件
spindle axis 主轴线
spindle balance 连接轴的平衡
spindle band 锭绳
spindle band stretching machine 锭绳拉伸机
spindle bearing 滚柱轴承
spindle bed 纺锤形圆柱床
spindle bolt 转向节主销;转向关节栓
spindle bore 主轴孔径
spindle bore lathe 大孔径车床
spindle box 主轴箱;轴箱;床头箱
spindle brake 锭子制动器;轴梗制动器;螺杆制动(器)
spindle breaker 回转式碎石机;回转式破碎机
spindle buoy 纺锤形浮标
spindle-cap 轴罩
spindle carrier 主轴托架;轴支持装置;接轴支架
spindle carrier bar 连接轴支架
spindle center 主轴中心
spindle clamp 柄夹
spindle clean oiler 锭子清洗加油器
spindle column 圆柱轴
spindle connecting rod 转向横拉杆
spindle control lever 主轴离合杆
spindle counterweight 接轴平衡锤
spindle cyclide 纺锤形圆纹曲面;纺锤形圈纹曲面
spindle drag pump 农田杆泵
spindle drive 主轴传动机构
spindle drive motor 主轴电动机
spindle drive shaft 摘锭传动轴
spindle driving gear 主轴运动传动机构
spindle drum 主轴鼓轮
spindle extension 延长轴
spindle face 主轴面
spindle feed 锭子进给;立轴进给
spindle feed length 立轴行程;立轴给进长度
spindle feed up and down 立轴上下给进
spindle gearing 床头箱
spindle guide 轴导;杆导
spindle head 主轴箱;主轴头;立轴;床头箱
spindle head feed system 立轴给进系统
spindle head hydraulic system 立轴液压系统
spindle headstock 主轴箱;床头
spindle holder 主轴托架
spindle hole 主轴孔
spindle hole saw 立轴孔锯
spindle inclination 主轴倾斜度
spindle inner diameter 立轴内径
spindle jam nut 轴锁紧螺帽
spindle jaw half coupling 接轴铰接叉头
spindle keyway 转轴上的键槽
spindle-knife shredder 轴刀式茎秆切碎器
spindle level 锭子水平仪
spindle lever linkage 转向臂杠杆式拉杆机构
spindle lift 螺柱升降机
spindle lubricant 锭子润滑油
spindle motor 主轴电动机
spindle mo(u)lder 栏杆造形机;样板式栏杆车床;芯轴造形机;立式成型铣床;样本式栏杆车床
spindle mower 旋翩剪草机
spindle nose 主轴端部;轴头
spindle nut 轴螺母;大螺母
spindle of streamline section 流线剖面轴
spindle oil 主轴润滑油;锭子油
spindle oiler 锭子加油器
spindle operated penstock 主轴控制进水管
spindle outside diameter 立轴外径
spindle press 螺旋压力机
spindle pull maximum 立轴最大提升能力
spindle saddle 主轴滑动座架
spindle seat 心轴座
spindle-shaped 纺锤形的;梭状的
spindle-shaped hysteresis curve 梭形滞变曲线
spindle-shaped solid 纺锤形圆纹曲面
spindle shaped training 纺锤形整枝
spindle shield 摘锭罩
spindle shock absorber 心轴减振器
spindle shoulder 主轴肩
spindle sleeve 主轴套;轴套
spindle slide 主轴滑动座架
spindle socket 轴承窝
spindle stage 旋转针台
spindle stay bolt tap 轴式撑螺丝锥
spindle stock 头架
spindle tape 锭带
spindle traverse 主轴横动
spindle type core drill 立轴式岩芯钻机
spindle type drill 立轴钻机
spindle type drilling machine 立轴式钻机;边坡钻
spindle type valve 纺锤开关;纺锤式阀
spindle type visco(si)meter 纺锤形黏度计
spindle unit 主轴部件
spindle winder 竖锭式卷绕机
spin down 旋转减慢;自旋减慢;缩颈旋压
spin-drier 旋转式脱水机;旋转式干燥器;旋转式干燥机;离心式脱水机;离心式干燥器;离心式干燥机
spindrift 旋转流;浪花
spine 脊柱;脊骨;火山栓
spine and rib line 脊肋线

spine apparatus 棘器
spine beam 脊骨梁;主梁
spin echo 自旋回声
spin echo method 自旋回波法
spin-echo spectroscopy 自旋回波谱法
spin-echo storage 自旋回波存储
spin echo technique 自旋回声技术
spin effect 自旋效应
spin girder bridge 脊骨梁桥
spine line 脊线
spinel(le) 尖晶石
spinel(le) ceramics 尖晶石陶瓷
spinel(le) chromite serpentinite 尖晶石铬铁矿蛇纹岩
spinel(le) deposit 尖晶石矿床
spinel(le) diopside sharn 尖晶石透辉石矽卡岩
spinel(le) law 尖晶石律
spinel(le) pigment 尖晶石型颜料
spinel(le) refractory 尖晶石耐火砖
spinel(le) refractory product 尖晶石耐火材料制品
spinel(le) structure 尖晶石结构
spinel(le) type ceramic colo(u)rs 尖晶石型陶瓷颜料
spinel(le) type ceramics 尖晶石型陶瓷
spinellids 尖晶石类矿物
spinel structure 尖晶石型结构
spin energy 自旋能
spin engine 旋转发动机
spinet pyroxenite 尖晶石辉石岩
spine wall 内纵墙
spin fading 旋转衰荡
spin flip 自转翻转
spin-flip collision 自旋翻转碰撞;自旋反向碰撞
spin-flip laser 自旋反转激光器
spin flipping 自旋变相
spin-flip scattering 自旋反向散射;自旋翻转散射
spin-flop transition 自旋转向转变
spin-forging machine 旋转锻机
spin forming 旋压成型法
spin forming machine 旋压成型机
spinfunction 旋转功能
spin gearing 盘车装置
spin glass 自旋玻璃
spingling 熟铁坯挤压
spin hardening 旋转表面淬火法
spin-hard heat treatment 旋转硬化热处理;旋转表面淬火法
spinifex texture 刺结构
spin in 辊口
spin label 自旋标记物
spin label(l)ing 自旋标记;顺磁标记
spin lattice 自旋点阵
spin-lattice coupling 自旋晶格耦合
spin-lattice coupling constant 自旋晶格耦合常数
spin-lattice interaction 自旋晶格相互作用;自旋点阵相互作用
spin-lattice relaxation 自旋点阵弛豫
spin-lattice relaxation time 自旋—晶格弛豫时间
spin magnetic moment 自旋磁矩
spin magnetic resonance 自旋磁共振
spin magnetism 自旋磁性
spin matrix 旋转矩阵;自旋矩阵
spin microharass method 自旋微扰法
spin model 螺旋试验模型
spin moment 自旋力矩;自旋矩
spin momentum 自旋动量矩
spin-momentum vector 自旋动量矢量
spin moment vector 自旋矩矢量
spin motion 自旋运动
spin motor 旋转发动机
spin multiplicity 自旋多重性
spinnaker 大三角帆
spinnaret 纺器
Spinnbarkeit relaxation 可抽丝松弛
spinner 旋转架;旋涂器;钢丝矫直机;离心头;离心式撒布器;机头罩;抛掷器;纺纱机
spinner basket 离心机吊篮
spinner broadcaster 旋转式撒布机;离心式撒播机
spinner chip spreader 旋转石屑摊铺机;旋动石屑摊铺机
spinner column 旋转塔
spinner disk spreader 旋盘铺砂器;转盘铺砂器
spinner distributor 旋转式撒布机
spinner equipment 旋转离心设备

spinneret 喷丝嘴;喷丝头
spinneret assembly 喷丝板组合件
spinneret draft 喷丝头拉伸
spinneret holder 喷丝板座
spinneret orifice 喷丝板细孔
spinnerette 漏嘴
spinner gritter 旋盘铺砂器
spinner induction heating 离心机感应加热
spinner magnetometer 旋转式磁力仪;旋转磁力仪
spinner magnetometer method 旋转磁力仪法
spinner motor 双转子电动机
spinneron 旋转副翼
spinner runner 旋涡集渣浇口
spinner's chisel 旋压工具
spinner-type distributor for chip(ping)s 旋动石屑摊铺机
spinner-type distributor for stones 旋动石子摊铺机
spinner-type plow 双向犁
spinner-type spreader 旋转式撒布机
spinnery 纺纱厂;纱厂
spinney 树丛
spinning 旋转涂漆法;旋转的;旋压加工;旋压;自旋;卷边铆接;甩丝法;抽丝
spinning apparatus with variable pressure 变压纺丝装置
spinning axis 自转轴
spinning band column 旋带蒸馏柱
spinning bath 凝固浴
spinning blade 旋转刀
spinning block 旋压模
spinning box 纺丝锅
spinning cable 猫头绳
spinning carriage 绕索小车
spinning cathead 旋转转换开关凸轮;旋转套管;旋转吊锚架;旋转锚栓
spinning chain 松扣链条
spinning chuck 旋压模
spinning coater 旋转涂布器
spinning crew 纺织人员;纺丝小组;纺丝工班;纺丝人员;纺丝小组;纺丝工班
spinning die 旋压螺丝钢板
spinning disk 旋流片;涡流盘
spinning-disk distributor 转盘式撒布机
spinning disk humidifier 离心式加湿器
spinning factor 编织因数
spinning former 旋压成型机
spinning-frame 精纺机
spinning gate 离心集渣浇口
spinning gritter 旋压喷砂机
spinning installation 纺织设备;纺丝设备
spinning jet 喷丝头
spinning lathe 旋压机床
spinning line (拧钻杆、套管用的)猫头链;猫头绳
spinning machine (成型混凝土的)成型机;离心式旋制机(水下混凝土);纺纱机;拔丝机;旋压机床;旋压机;旋绕机
spinning magnetometer 旋转式磁力仪
spinning mandrel 旋压芯模
spinning method 离心法(混凝土制品)
spinning mill 纺织厂;纺纱厂;离心式制管厂
spinning mo(u)ld 纺丝模;离心铸型;纺丝模具
spinning mo(u)lding 离心成形(法)
spinning nozzle 喷丝头
spinning number 品质号数
spinning of cable 纺缆
spinning of concrete 混凝土离心法浇制
spinning party 纺织小组;纺丝工班;纺丝小组;纺丝工班
spinning plant 纺织厂;纺丝车间
spinning process 离心法
spinning pump 抽丝泵;纺纱泵
spinning reserves 旋转备用(容量);运转备用
spinning reserve capacity 运转备用容量
spinning reserve content 运转备用容量
spinning riffler 旋转缩分器
spinning riveter 扩口铆机
spinning roller 旋轮(旋压用)
spinning room 纺纱车间
spinning speed 旋转速度
spinning stability 自旋稳定
spinning survey 深井流量计测量
spinning team 纺织小组;纺丝工班;纺丝小组;纺丝工班

spinning top 陀螺
spinning unit 纺织机组;纺丝机组;抽丝装置
spinning wheel 纺轮;绕索轮;纺车
spinning with chip forming 剥皮旋压
spinning with tension 张力旋压
spinning wrench 气动管钳
spin nozzle 切向喷管
spinodal 旋节线;拐点
spinodal curve 旋节溶线;拐点分解曲线
spinodal decomposition 旋节线分解;旋节分解;不稳分解
spinodal line 亚稳分界线
spinodal phenomenon 旋节现象
spin off 附带利益;附属结果;伴随结果
spin off method 脱离法
spin on 旋压
spinor 旋量
spin-orbit coupling 轨旋耦合
spin-orbit coupling diagram method 自旋耦合图解法
spin-orbit interaction 自旋与轨道相互作用
spin-orbit resonance 轨旋共振
spin orbit transition 轨旋跃迁
spin orientation 自旋取向
spinous 刺的
spin over 扩口
spin paramagnetism 自旋顺磁性
spin polarization 自旋极化;自旋的极化
spin precession 自旋轴进动
spin-precession method 自旋进动法
spin-probe technique 自旋取样技术
spin projection 自旋投影
spin projection operator 自旋投影算符
spin quantum number 自旋量子数
spin rate 自旋速率
spin-rate control 自转率调节
spin-rate meter 角速计
spin reference axis 自转基准轴
spin relaxation time 自旋弛豫时间
spin resonance 自旋共振;磁共振
spin-ring meter 钢领丝圈摩擦力测定仪
spin riveter 旋压铆钉机
spin safe 旋转安全
spin scan cloud camera 自旋扫描摄云照相机
spin sequence 自旋顺序
spin space 旋量空间;自旋空间
spin-spin coupling constant 自旋—自旋耦合常数
spin-spin relaxation time 自旋—自旋弛豫时间
spin-spin splitting 自旋裂分
spin splitting 自旋分裂
spin-spreader 旋转式撒布机;旋转式喷洒机;旋转式分布机;自转式洒布机
spin stability 自旋稳定性
spin stabilization 自旋稳定
spin-stabilized geostationary satellite 自旋稳定地球同步卫星
spin-stabilized payload 旋转稳定的有效载重
spin-stabilized satellite 自旋转稳定卫星
spin state 自旋态
spin susceptibility 自旋磁化率
spin temperature 自旋温度
spin tensor 自旋张量
spin-test rig 旋转试验台
spin trapping 自旋捕获
spin turn 原地转向
spin-up 自旋加快
spin-up chain 拧管链条
spin velocity 自转角速度
spin wave 自旋波
spin wave function 自旋波函数
spin-wave resonance 自旋波共振
spin wave spectrum 自旋波谱
spin weeder 旋转除草器
spin welding 旋转熔接;旋转焊接;摩擦焊
spin wind tunnel 螺旋风洞
spiny lobster 多刺龙虾
spiny palm 刺棕
spiomkopite 斯硫铜矿
spipot and socket cast iron pipe 承插式铸铁管
spira 柱脚凸圆线脚;柱子底盘饰
spiracle 排水孔;气门;通气孔;通风口
spiracore 钢带螺旋绕铁芯
spiractor 旋流分离器
spiracular 气门的;气孔的

spirakore 螺旋形钢带制成的铁芯
spiral 螺旋形的;螺旋线,螺旋;螺线;盘旋
spiral agitator 螺旋搅拌器;螺旋搅拌机
spiral air-flow diffusion 气流螺旋扩散
spiral alarm lamp 螺旋报警灯
spiral alarm light 螺旋报警灯
spiral anchor 螺旋锚(杆)
spiral angle 螺旋角;捻角;齿的倾斜角
spiral antenna 螺旋天线
spiral approach 螺旋形进路;螺旋形引桥;螺旋形入口
spiral arm 旋臂
spiral arrangement 螺旋状排列
spiral auger 螺旋钻
spiral axis 螺旋轴线
spiral bacteria 螺旋状细菌
spiral balance 螺旋平衡重;双悬窗的弹簧配重
spiral-band brake 多往复带制动器
spiral bevel axle 螺旋斜轴
spiral bevel gear 斜齿伞齿轮;螺旋锥齿轮;螺旋伞齿轮
spiral bevel gear cutter 螺旋伞齿轮铣刀
spiral bevel gear drive 斜齿伞形齿轮传动;螺旋伞形齿轮传动
spiral bevel gear generating machine 螺旋伞齿轮滚齿机
spiral bevel gear generator 螺旋伞齿轮滚齿机
spiral bevel gearing 螺旋伞齿轮装置
spiral bevel gear pair 螺旋锥齿轮副;螺旋伞齿轮副
spiral bevel gears with offset axes 轴线偏置螺旋伞齿轮
spiral bevel pinion and ring gear 螺旋伞形小齿轮和齿圈
spiral binding 螺旋装订
spiral bit 螺旋钻(头);麻花钻头
spiral blower 叶轮式鼓风机;螺旋鼓风机
spiral books 螺旋簿
spiral bourdon tube 螺旋弹簧管
spiral breaker 螺旋辊揉布机
spiral brush sifter 螺旋刷筛
spiral burner 螺旋形燃烧炉;螺旋形燃烧器
spiral cam 螺旋凸轮;螺线凸轮
spiral car park 以螺旋坡道连接的室内多层停车场
spiral case 螺壳;水轮机蜗壳
spiral case access 蜗壳通道;蜗壳进入孔;蜗壳进口
spiral-cased turbine 蜗壳水轮机
spiral-cased water turbine 蜗壳式水轮机
spiral case floor 蜗壳层
spiral case manhole 蜗壳进入孔
spiral cast 螺旋排绕;反自
spiral-cavity mo(u)ld 螺线腔模
spiral characteristic 螺旋线特性
spiral charge 螺旋形锭料
spiral chute 螺式斜槽;螺旋(式)溜槽;螺旋(式)滑槽;螺旋式(垃圾)溜管
spiral classifier 螺旋形分析器;螺旋(式)分选机;螺旋筛分机;螺旋分级机
spiral cleaner (运输机皮带的)螺旋清洁器;螺旋式选种机
spiral cleavage 旋裂
spiral coal unloader 螺旋选煤机
spiral coil 螺旋(形)线圈;螺旋盘管;螺旋管
spiral coil filter 盘管过滤器
spiral coil filter medium 盘管过滤介质
spiral column 螺旋形蒸馏塔;螺旋筋柱;螺旋柱;螺旋形柱;螺旋塔;螺旋式柱;螺旋筋混凝土柱
spiral concentrator 螺旋选矿机;螺旋精选机
spiral concrete column 螺旋筋混凝土柱
spiral conduit 蛇形渠
spiral cone 锥形螺旋式分离器
spiral cone seed cleaner 螺旋锥式选种机
spiral continuous slab 螺旋形连续板;螺旋形连接板
spiral convergence 螺旋收敛
spiral conveyer 螺旋输送器;螺旋输送机;螺旋输送机;螺旋式运输机
spiral cooling chute 螺旋冷溜槽
spiral core 螺旋状岩芯
spiral counterweight drum 平衡重卷索鼓轮
spiral cross current 螺旋形横向环流;螺旋副流
spiral crusher 螺旋式破碎机
spiral crystal 螺旋形晶体
spiral cupola 螺式小屋顶
spiral current 螺旋(状水)流
spiral curve 螺旋(曲)线;螺旋缓和曲线

spiral-curved tooth 螺旋齿
spiral cut 螺旋掏槽
spiral cut end mill 螺旋槽端铣刀
spiral cut milling cutter 螺旋槽铣刀
spiral cut reamer 螺旋槽铰刀
spiral cutter-head 螺旋形挖掘头
spiral cutting 螺旋切削
spiral delay line 螺线延迟线
spiral densifier 螺旋浓缩机;螺旋稠化机
spiral descent 螺旋下降
spiral deviation 螺旋状弯曲
spiral diagram 螺旋曲线图
spiral diamond reaming shell 螺旋金钢石扩孔器
spiral diposition 螺旋形排列
spiral disc distributor 螺旋撒料盘
spiral discharge system 螺旋卸料系统
spiral disk 旋转分像盘
spiral dislocation 螺旋(形)位错
spiral distortion 螺旋形扭曲;各向异性失真;螺旋形失真;螺旋(形)畸变
spiral distributor 支管;配水管;分流管
spiral ditch 蛇形沟
spiral dowel 定缝螺钉
spiral drill 螺旋钻
spiral duct 螺旋形通道;螺旋形管道;螺旋(式)导管;螺纹导管
spiral easement curve 螺旋缓和曲线
spiral easement vertical curve 螺旋缓和竖曲线
spiraled column 螺旋形(配筋)柱
spiraled transition curve 螺旋缓和曲线
spiral electrode 螺旋形电极
spiral element 螺旋形管件
spiral elevator 螺旋式升运器
spiral end 螺旋形终末
spiral expander 螺旋展平机;螺旋形扩幅装置;螺旋伸张器
spiral feed conveyer 螺旋进料运输机;螺旋进料输送器
spiral feeder 螺旋状送料器;螺旋喂料机;螺旋式给料器
spiral field 螺旋形视野
spiral filament 螺旋灯丝
spiral filament forming machine 绕丝机
spiral flight 盘旋飞行;螺旋刮板;螺旋提升刮片
spiral flow 弯道水流;螺旋流动;螺旋流;涡流
spiral-flow aeration 螺旋流曝气;旋流曝气
spiral-flow aeration system 旋流曝气系统
spiral-flow aeration tank 旋流曝气池
spiral-flow aerator 旋流曝气器
spiral-flow air diffusion 螺旋形空气扩散;螺旋流空气扩散
spiral-flow air diffusion aeration tank 螺旋流空气扩散曝气池
spiral-flow mixer 螺旋流搅拌机;涡流式拌和机
spiral-flow port 旋流风口
spiral-flow tank 旋流曝气箱;旋流加气池;斜流箱;旋流箱;旋流滤箱
spiral-flow washer 旋流式清洗机
spiral flush port 螺旋形水槽
spiral flute 螺旋钩;螺旋槽
spiral fluted 螺旋凹槽的
spiral fluted column 螺旋式凹纹立柱
spiral fluted end mill 螺旋槽端铣刀
spiral fluted expansion hand reamer 螺旋槽可胀式手铰刀
spiral fluted expansion reamer 螺旋槽可胀式铰刀
spiral fluted reamer 螺旋槽铰刀;螺旋槽铰刀
spiral fluted roller 螺槽辊
spiral fluted tap 螺旋槽丝锥
spiral flute hand reamer 螺旋槽手用铰刀
spiral flute hob 螺旋槽滚刀
spiral forming 螺旋形绕管成型
spiral four 简单星绞;四芯扭绞
spiral four cable 星绞四芯软电缆
spiral fracture method 螺旋破裂面法
spiral fringe 螺旋形条纹
spiral function 螺旋函数
spiral gear 斜齿轮;螺旋(形)齿轮
spiral gear drive 螺旋齿轮传动
spiral gearing 螺旋齿轮装置
spiral gear pump 斜齿轮(式)泵
spiral gear teeth 螺旋齿轮齿
spiral gear tooth 螺旋齿

spiral grader 螺旋分级机
spiral grain 螺状纹理;螺旋纹理;(木材的)扭转纹理
spiral grain index 螺纹指数
spiral grain loader 螺旋式装粮机
spiral grating 螺旋形光栅
spiral gravity separator 螺旋重力分离器
spiral gravity shoot 螺旋形重力滑道;螺旋形重力滑槽
spiral groove 螺旋槽
spiral grooved tube 螺旋槽管
spiral-groover roll 螺旋凹槽辊
spiral grooving 形成螺旋状细沟(岩芯上)
spiral growth 螺旋生长;螺线生长
spiral head 螺旋分度头;分度头
spiral header 螺旋管束
spiral heater 旋管加热器;螺旋形加热器
spiral heat exchanger 螺旋板式热交换器
spiral helicine 螺旋状的
spiral highway 螺旋形公路;螺旋形道路
spiral hole 螺旋状弯曲钻孔
spiral hook 螺旋形钩
spiral hoop 螺旋箍
spiral hooping 螺旋箍筋
spiral hoop-iron 螺旋钢箍
spiral hoop reinforcement 螺旋箍筋
spiral hoping 加螺旋钢筋
spiral horn 螺旋状角
spiral house 螺旋形房屋
spiraling 螺旋钻孔
spiraling-beam amplifier 旋束放大管
spiraling-beam oscillator 旋束振荡管
spiraling rate 径向扩张率
spiral inlet 螺旋式入口;蜗管入口
spirality helicity 螺旋形
spiral jacket 螺旋式夹套
spiral jaw clutch 螺旋面牙嵌式离合器;螺旋爪离合器
spiral jaw coupling 螺旋形爪式联轴节
spiral jet 螺旋喷流
spiral launder 螺旋流槽
spiral layer 埃克曼层
spiral lead-in and spiral throw out 盘首纹和盘尾纹
spiral(l)ed knife 螺旋形切刀
spiral(l)ed scale 螺旋形阶;缠绕标度
spiral(l)ed water hose 缠绕输水管
spiral limbus 螺旋缘
spiral line 螺旋线;螺线
spiral(l)ing 螺旋形
spiral(l)ing hole 螺旋形钻孔
spiral lobe(type) compressor 螺杆压缩机;螺杆式压气机
spiral loop 螺旋型环形天线;可调谐环形天线
spirally bound concrete 配螺旋形钢箍的混凝土
spirally coilayered cylinder 螺旋绕带式多层圆筒;绕板式圆筒
spirally concreted column 螺旋筋混凝土柱
spirally grooved roller 螺槽纹辊
spirally grooved tube 螺纹沟槽管
spirally guided holder 斜升式储气罐
spirally reinforced 螺旋环扎筋的;用螺旋钢筋的;加螺旋箍筋的
spirally reinforced column 螺旋形钢筋(混凝土)柱;螺旋箍筋柱;配螺旋钢筋的混凝土柱
spirally welded on wing stabilizer 焊接螺旋翼片式稳定器
spirally wound duct 螺旋卷管(俗称蛇皮管)
spirally wound sheath 皱纹套管
spiral membrane 螺旋膜
spiral method 螺旋破裂法(用于土壤滑坡分析)
spiral micrometer 螺旋测微计
spiral micrometer eyepiece 螺旋式测微目镜
spiral microscope 螺旋显微镜
spiral milling cutter 螺旋铣刀
spiral minaret (伊斯兰教寺院的)螺旋形尖塔;螺壳状的回教寺院的尖塔
spiral mixer 螺旋式混合器
spiral mixing rod 螺旋搅拌棒
spiral mode divergence 螺旋形扩散
spiral motif 螺旋线饰【建】
spiral motion 螺旋运动
spiral mo(u)ld cooling 铸型螺旋孔冷却
spiral mo(u)lding 螺旋形线脚
spiral movement 螺旋运动

spiral newel stair(case) 有中柱的螺旋形梯
spiral of archimedes 阿基米德螺线
spiral oil grooves 螺旋油槽
spiral optic(al) micrometer 螺旋光学测微计
spiral organ 螺旋器
spiral path 螺旋形轨道
spiral pattern 螺旋图样
spiral pattern of aperture 小孔螺旋状排列
spiral picker 螺旋分选器
spiral pier 螺旋形支柱;螺旋形桥墩
spiral pin 螺旋销
spiral pinion 螺旋小齿轮
spiral pipe 螺旋盘管;螺旋管;螺盘管;盘管;蛇管
spiral planetary speed reducer 螺环行星减速器
spiral plate 螺旋板
spiral plate exchanger 旋板换热器
spiral plate heat exchanger 螺旋板换热器;螺旋形片状热交换器
spiral point 螺线极点
spiral pointed tap 螺尖丝锥;枪式丝锥
spiral polishing machine 螺旋抛光机
spiral polymer 螺旋状聚合物
spiral positive-displacement compressor 螺旋形正排量压气机
spiral power spring 发条
spiral pressure ga(u)ge 螺旋管压力计
spiral press working spring 螺旋冲压弹簧
spiral production 轮番生产
spiral prominence 螺旋凸
spiral propeller 螺旋推进器
spiral pump 螺旋泵
spiral quad 星绞四线组;扭绞四芯电缆
spiral quad cable 星绞四芯软电缆
spiral rake thickener 螺旋形挡栅浓缩槽;螺旋形挡栅浓缩机;螺旋形挡栅浓缩器;螺旋形倾斜浓缩槽;螺旋形倾斜浓缩机;螺旋形倾斜浓缩器
spiral ramp 螺旋状坡道;螺旋形滑道;螺旋式溜管;螺旋坡线;螺旋坡道;盘旋式斜坡道
spiral ramp car track 螺旋式坡道停车处
spiral ratchet drill 擦旋钻
spiral ratchet screwdriver 自动旋凿;螺旋棘轮改锥;螺旋槽棘轮旋凿
spiral recording spot 螺旋记录斑点
spiral reel 螺线轴
spiral reinforcement 螺旋状钢筋;螺旋配筋;螺旋箍筋;螺旋钢筋
spiral-ribbed roll 螺旋凸棱辊
spiral ribbon mixer 螺旋带式搅拌机;螺带式搅拌机;螺带混合机
spiral ring 螺旋环
spiral ring structure 旋涡环形构造
spiral-riveted pipe 螺旋铆接管
spiral road 螺旋形道路
spiral rod 螺旋箍筋
spiral roll aerator 螺旋卷式曝气器
spiral-rolled aeration system 螺旋式曝气系统
spiral-rolled drill steel 螺旋扎制钻杆;螺旋扎制钎杆
spiral roller 螺柱滚子;螺旋辊子
spiral rope 螺旋绳;绞绳
spiral roving frame 螺旋杆式粗纺机
spiral rubber hose 缠绕胶管
spiral rule 螺线规
spiral sash balance 螺旋平衡重
spiral scale 螺旋标尺
spiral scan 螺旋式扫描;螺旋扫描
spiral-scan method 螺线扫描法
spiral scanning spot 螺旋扫描点
spiral scraper thickener 螺旋形刮板浓缩机
spiral-screen column 旋筛塔
spiral screw washer 螺旋形螺丝洗선机;螺旋形螺丝分级机;螺旋形螺丝分类机
spiral scroll 旋涡花样;螺旋槽
spiral seam 螺旋缝
spiral seam pipe 螺旋形焊缝管
spiral seam tube 螺旋焊缝管
spiral separator 螺旋选矿机;螺旋分选器;螺旋分离器
spiral separator chute 螺旋选矿机溜槽
spiral septum 螺旋形隔板
spiral shaft 螺旋轴
spiral-shaft dotting machine 旋转刻点仪
spiral-shaped 螺旋形的
spiral-shaped agitator 螺旋形搅动器
spiral-shaped column 螺旋形柱
spiral-shaped rotor 蜗壳形转子

spiral shear connector 螺旋形剪力接合器
spiral sheet metal casing 螺旋形焊缝套管;螺旋金属板套管
spiral shoot 螺状滑道
spiral slide 螺旋式滑(冰)道;螺旋溜板
spiral slit 螺旋缝隙
spiral slot 螺旋形槽
spiral slotted oil control ring 开有螺旋槽的油环
spiral sluice 螺旋溜槽
spiral spectrometer 螺旋形分光仪
spiral spline broach 螺旋花键槽拉刀
spiral spring 螺旋形弹簧;盘簧
spiral spring suspension 盘簧悬置
spiral spur gear 螺旋正齿轮
spiral stair(case) 螺旋(式)楼梯;盘梯;盘旋楼梯;中柱螺旋式楼梯(中柱支承悬臂踏步);旋梯
spiral-staircase coil 毕特线圈
spiral staple chute 暗井螺旋溜管
spiral steel 螺旋钢筋
spiral steel pipe 螺旋钢管
spiral stirrer 螺旋搅拌器;螺旋搅拌机
spiral stirrup column 螺旋箍筋柱
spiral strand 螺旋缆索;螺旋绞线
spiral stringer 螺旋形楼梯斜梁
spiral structure 螺旋形超高缓和段结构;螺旋构造;螺环结构
spiral superheater 螺旋管过热器
spiral surface 螺旋面
spiral table for highway design 公路设计用的螺旋线表
spiral tamping machine 螺旋捣固机
spiral tank reel 螺旋卷盘
spiral tap 螺旋丝锥
spiral taper pipe 锥形螺盘管
spiral taper reamer 螺旋槽式锥铰刀
spiral test 螺旋试验;回转试验
spiral thermometer 螺旋带式温度计
spiral thickening 螺纹加厚
spiral thread of trachead tube 气管螺旋丝
spiral to curve point 缓圆点
spiral to tangent point 缓直点
spiral track 盘山展线
spiral transition curve 螺状过滤曲线;螺旋(形)缓和曲线
spiral traverse 螺旋形排丝器
spiraltron 螺旋线管;径向聚束行波管
spiral tube 螺旋管;盘管
spiral-tube heat exchanger 螺旋管式热交换器
spiral tungsten cable 螺线钨丝
spiral tunnel 螺旋式隧道
spiral turbine 螺旋涡轮机
spiral type heat exchanger 螺旋板换热器
spiral-type hose guard 螺旋形蛇管保护器;螺旋形软管保护器
spiral-type lining 螺旋式装配衬砌
spiral-type whipstock bit 螺旋桨式切口钻头
spiral unloading machine 螺旋卸车机
spiral upflow reactor 螺旋升流式反应器
spiral valve 螺旋瓣
spiral vane disk cutter 螺旋叶盘式截装机
spiral vane flowmeter 螺旋叶片流量计
spiral vessels 螺纹导管
spiral vibrating conveyer 螺旋振动运输机
spiral visual field 螺旋形视野
spiral vortex 螺旋形涡流
spiral washer 螺旋垫圈
spiral water lift 螺旋扬水机;螺旋提水器
spiral waterways 螺旋形水槽
spiral waveguide 螺旋形波导管
spiral weld 螺旋形焊缝
spiral-welded pipe 螺旋焊(接)管
spiral-welded steel pipe 螺旋焊接钢管
spiral welding 螺旋焊
spiral welding steel pipe 螺旋焊钢管
spiral weld-pipe mill 螺旋焊管机
spiral-weld sheet-metal casing 螺旋状焊接套管
spiral wheel 斜螺旋齿轮;斜齿轮;螺旋水轮;螺旋式水轮机;螺旋齿轮
spiral winding 螺旋式盘绕;螺旋绕组;螺旋缠绕;螺线绕法
spiral-winged drill stem 螺旋翼冲击钻杆
spiral wire column 螺线精馏柱
spiral worm 螺旋形杆
spiral-wound electrode 螺旋状焊丝

spiral-wound gasket 缠绕式垫片;螺旋形垫衬
spiral-wound module 螺旋缠绕模件
spiral-wound pipe system 螺旋缠绕管系统
spiral-wound roller 螺旋辊子
spiral wrapped 螺旋形外包装的
spiral yarn 螺旋花线
spirane resin 螺环(烃)树脂
spiraster 旋星体
spire 尖塔;尖顶;塔楼尖顶
spire church tower 教堂尖顶塔
spired 成锥形的
spirelet 小尖塔;小尖顶
spireme 线球
spire roof 尖塔(形)屋顶
spire-steeple 尖塔顶的塔尖;尖顶塔;教堂塔尖
spire tower 尖顶塔
spiring 塔状矗立的;螺旋形上升的
spirit acid 浓乙酸
spirit black 醇溶黑
spirit blue 醇溶青;醇溶蓝
spirit bubble 水平泡
spirit circular level 圆水平仪
spirit colo(u)r 醇溶染料;醇溶苯胺染料
spirit compass 液体罗盘;液体罗经
spirit duplicating 醇溶碳纸复印法
spirit dye 醇溶染料
spirit dyestuff 醇溶染料
spirit engine 酒精发动机
spirit fastness 耐酒精性
spirit flat varnish 挥发性无光漆
spiriting off 擦酒精除油
spirit lacquer 挥发性光漆;挥发性蜡克
spirit lamp 酒精灯
spirit level 酒精水准仪;酒精式水平仪;气泡水准仪;气泡水准器;气泡水准器;平水器;水准仪;水准器
spirit level-boat 酒精船式水平仪
spirit level-bubble 泡形水平仪
spirit level building 梯阶式房屋建筑
spirit level construction 梯阶式房屋建筑
spirit level control 水准测量控制
spirit-leveled benchmark 水准点
spirit-leveled elevation difference 水准测量高差
spirit level(l)ing 几何水准测量;气泡水准测量
spirit-level(l)ing instrument 水准仪
spirit level wind 水准器偏差
spirit matt finish 挥发性无光饰面
spirit of rosin 松脂醇;松香精
spirit of salt 盐酸
spirit of turpentine 松节油
spirit of vitriol 硫酸
spirit of wine 乙醇;火酒
spirit plumb rule 酒精(气泡)垂线规
spirit resistance 抗酒精性
spirit-soluble 能溶于酒精的
spirit-soluble dye(stuff) 醇溶性染料
spirit-soluble resin 醇溶性树脂
spirit solvent 酒精溶剂
spirit stain 酒精着色剂;酒精污斑;醇溶(性)着色剂
spirit tanker 酒精运输船
spirituous 醇的
spirit varnish 达玛树脂清漆;虫胶清漆;醇溶性凡立水;酒精油漆;挥发(性)清漆;挥发漆;醇溶(性)清漆;凡立水
spirocid 醋酰胺肿
spiro-compound 螺环化合物
spirocyclic(al) compound 螺环化合物
spiroffite 啼锰铬石
spiroid 偏轴弧齿近平面齿轮
spiro-jet 螺旋形喷射染色机
Spirorbis 螺旋虫属;石灰虫
spiro ring system 螺环系
spirostane 螺甾烷;豆甾烷
spirotallic gasket 缠绕垫片
spirotheca 旋壁
spirotron 高速粒子减速器
spirt 溅
spirule 螺旋线尺
spiry 盘旋;似尖塔;成锥形的
spissatus 密厚云
spit 小雪;岬;海角;沙嘴
spit and sand bar facies 沙嘴沙坝相
spit cock 安全旋塞
spit cut off works 切嘴工程

spit cutting works 切嘴工程
spite effect 泄愤效应
spit growth 沙嘴增长
spit nail 钉枪钉；弹射钉
spit of land 狭长的浅滩；狭长的暗礁；沙嘴；地角（伸入海中的狭长陆地）；一铲土，一层浅土；岬
spit-out 针眼；喷裂
spit platform 沙嘴台地
spitter 出流管；雨水斗；水落斗
spitting 喷气；喷溅(物)
spitting spray 喷枪喷路颤振；喷路波动
spittle bug 吹沫虫
spittoon 痰盂
spit trap 砂咀圈闭
spitzkasten 角锥形分级机；角锥沉淀地
spivot 尖轴
splanchnoscopy 内窥镜检查
S-plane 岩组平行面
S-plane of flattening 平展S面
S-plane of stratification 层理S面
splash 溅着水前进；溅射；溅出；溅(水)；喷溅；水花飞溅；冲浪
splash apron 挡泥板
splash-back 防溅挡板；防溅墙面
splash baffle 防(喷)溅挡板
splash bar 搅棒
splash basin 溅盘
splash beacon 闪光指向标；闪光信标；闪光灯塔
splash block 散水；挡泥板；挡溅板；水落管下导水砌块；水簸箕；防溅挡板
splash board 外门防风雨板；防雨装饰线条；防溅板；墙面防溅；溅灰板；溅水板；挡泥板；防泥板
splash box 溅水盘
splash brush 洒水刷
splash cooler 飞溅冷却器
splash core 防冲剂芯片
splash cover 挡油护罩
splash dam 临时水闸；临时挡水坝
splashdown 溅水；溅落
splashed glaze 斑点釉
splashed-ink landscape 泼墨山水
splashed lustre 斑点光泽彩
splashed ware 斑驳器(带有颜色条纹的陶器)
splasher 防溅板；挡泥板；折焰器；溅洒器；泼散器
splash erosion 雨淋浸蚀；雨淋冲刷；溅击侵蚀；冲刷
splash feed 喷射送料；飞溅润滑
splash feed system 溅油润滑系统
splash film packing 点滴式薄膜式淋水填料
splash fog 溅水雾花
splash guard 挡泥板(汽车)；切削液挡板；防溅罩；防溅板
splash-guard ring 防溅环
splash head 喷射头
splashing 溅水；泼釉；飞溅；击水音
splashing apron 挡溅板
splashing board 挡溅板
splashing device 喷射设备
splashing plate 挡溅板
splashings 喷溅物；铁豆
splashing sound 震水音；振水音；拍水音；击水音
splash jacket 水套；防溅罩
splash lap 防溅搭接；咬口接头；防溅盖板
splash loading 喷射加料
splash loss (雨量器的)溅水损失
splash lubrication 泼溅润滑；喷射润滑；喷溅润滑；飞溅润滑法；飞溅润滑
splash lubrication system 飞溅润滑系统
splash net 溅落网
splash noise 击冰噪声
splash packing 点滴式淋水填料
splash pan 水簸箕
splash panel 挡水板；遮水板
splash plate 溅水板；溅水板
splash-plate injector 溅板式喷嘴
splash pocket 飞溅油箱
splash-proof 防溅水的；防溅式的；防溅的；防溅
splash-proof enclosure 防溅外壳
splash-proof machine 防溅型电机；防溅式机器
splash-proof motor 防溅式电动机
splash-proof nozzle 防溅喷嘴
splash-proof type induction motor 防溅式感应电动机
splash ring 溅油环；润滑油圈；润滑油环
splash-ring injector 溅环式喷嘴

splash shield 防溅板
splash system 飞溅润滑系统；飞溅润滑法
splash system lubrication 飞溅式润滑
splash trough 溅油润滑油池；飞溅槽
splash type filling 飞沫型填料
splash-wall injector 溅板式喷嘴
splash water 喷溅水
splash zone 溅浪区；浪溅区；溅泼区；激浪带；喷溅区；受溅地带
splash zone corrosion 浪溅腐蚀
splat 椅背的一块竖板；一地板；墙板缝上盖条；椅背中部板条；薄片激冷金属；楣间盖条
splat cooling 急冷；喷涂细片冷却法；喷涂冷却法
splatter 邻信道干扰；溅散
splay 斜削；斜口；池端玻璃带扩展
splay angle 八字角
splay bonder 斜面砌墙石
splay bondstone 斜面系石；斜面束石
splay branch 斜支线
splay brick 斜面砖
splay ceiling 八字形平顶
splay cut 展宽开挖
splay deposit 漫滩冲积物；漫滩冲积泥沙
splayed 抹角的；斜式的；喇叭形的
splayed abutment 翼形拱座；八字形拱台
splayed arch 喇叭拱；八字(形)拱
splayed baseboard 斜顶踢脚板；八字形踢脚
splayed beam 加腋梁
splayed boring tool 锥形钻头
splayed brick 斜面砖
splayed ceiling 八字形平顶；八字顶棚
splayed coping 斜式盖顶；墙压顶斜面；单坡屋顶；薄边式盖顶
splayed curb 斜式路缘(石)
splayed door 八字式门；八字形门；八字门
splayed drill 斜面头钎子
splayed edge 斜削边
splayed fillet 八字抹角
splayed folding 斜式折叠
splayed footing 八字式底脚；八字形基脚；八字形底脚
splayed glass 车边玻璃
splayed ground 斜水托条；斜底层；燕尾形冲筋；燕尾形冲筋；斜边底层
splayed heading (地板的)斜接接缝；八字接
splayed heading joint (地板的)斜接头；(地板的)端头斜接接缝；八字形(端)接头；八字接头
splayed intersection 展宽式交叉(路口)
splayed jambs 八字形门窗帮；斜口侧壁；八字形竖框；八字(形)侧壁；八字侧壁
splayed joint(ing) 斜角连接；斜缝；八字形接合；八字形接头；展宽式接缝；楔形接缝
splayed mullion 八字形直棂
splayed niche 八字形壁龛
splayed reveal 斜口门窗洞
splayed scarf 扇面形嵌接；八字(形)嵌接；八字形嵌接
splayed skirting 上缘抹角踢脚板；斜顶踢脚板；八字形踢脚
splayed spring 喇叭状配置弹簧
splayed track 八字线【铁】
splayed window 八字窗；八字形窗
splay end 斜端；砖的一端
splay fault 分叉断层
splay faults 分叉断层群
splay foot 八字脚
splay header 斜面丁砖
splaying 八字形；斜削；端间隙
splaying arch 斜面拱；喇叭形拱
splaying coping 斜石墙压顶(石)
splaying of window jambs 窗洞八字做法
splay knot 斜节(木材)；尖节；条形节；长节疤；掌状节；条状节(沿木节长度方向切开时露出的节疤)
splay-legged roadway arch 八字形腿的车行路拱；斜腿的车行路拱
splay mo(u)lding 斜式造形
splay of a wall 墙壁斜面
splay piece 鸠尾形键；鸡尾形键；燕尾形键
splay sediment 漫滩冲积物
splay stretcher 斜面顺砖
splay walls 八字形墙
splendid edition 精装本
splice 镶接；固定接头；钢筋接头；接合；铰接；绞合；加板；黏接；拼接；拼接；插管；编接

splice adhesive 拼接胶黏剂
splice angle 连接角钢；铰接角；拼接角钢
splice bar 拼接钢筋；鱼尾板；连接板；连接筋；接头夹板(鱼尾板)【铁】；拼接板
splice block 拼接块
splice board 拼接板
splice bolt 拼接螺栓；用作拼接的螺栓
splice box 电缆分线箱；电缆分线盒；分歧套管
splice cement 拼接黏结剂
splice cleaner 拼接清洁剂
splice cover 拼接盖(板)
spliced 叠接
spliced connection 拼接接合
spliced joint 分层接合
spliced pier 拼装式桥墩
spliced pile 桩接；拼接桩
spliced pole 叠接杆；叠接(电)焊
spliced strut 附加短杆
spliced truss 拼接桁架
splice fished joint 拼连接；拼铰接；鱼尾板接合；夹板连接；拼接缝(木材)
splice grafting 合接；拼接
splice joint 鱼尾板接合；拼接接头；拼合接头
splice length 搭接长度
splice loading 电缆接头加感
splice of reinforcing bars 钢筋拼接；钢筋接合
splice pad 拼合衬垫；拼接垫板；拼接衬垫
splice piece 鱼尾板；连接板；拼接板
splice plate 镶接板；鱼尾板；连接板；接板；接合板；拼接板
splice plate joint 二侧拼接板接合
splicer 光纤熔接机；接头器；接头机；接合器；接带器；铰接器；拼接件；拼缝机；断头接合器
splice tape 拼接带
splice web 直梁接合
splice welding 绑焊
splicing 续接；接尾；接片；铰接；捻接；拼接；插接
splicing adhesive 接头胶水
splicing apparatus 拼接装置
splicing box 电缆分线箱；电缆分线盒
splicing cement 拼接胶；水胶
splicing concrete pile 混凝土接桩；连接混凝土桩
splicing ear 连接端子
splicing fid 插接木笔
splicing fitting 接线夹
splicing glue 拼接胶
splicing length 搭接长度
splicing loss 接头损耗
splicing needle 拼接的指针；拼接的指针；接绳锥
splicing of bars 钢筋搭接
splicing of reinforcement 钢筋接头
splicing of steel structural members 构件拼接
splicing of timber piles 木桩接长
splicing of wire rope 接绳
splicing outfit 接绳工具；穿绳工具
splicing pile 接桩
splicing pole 叠接杆
splicing shackle 链条与绳索连结的卸扣
splicing sleeve 管接；连接(套)管；支承套；管节；连接套筒；接续管；交连套(筒)；交接套(筒)；交叠套筒
splicing tape 接片胶管；拼接带；拼合条；拼合带
splicing tools 接绳工具
splicing wire-line tools 接绳工具
spline 楔条；样条；止转楔键槽；键槽条；曲线规；曲线尺；嵌缝片；穿条；齿槽
spline adapter 键槽接头
spline and keyway miller 花键与键槽铣床
spline and keyway milling machine 花键与键槽铣床
spline approximation 样条函数近似；样条逼近
spline batten 活动曲线规
spline broach 键槽拉刀；花键拉刀；多键形拉刀
spline coupling 链连接；花键联轴节
spline curve 样条曲线
splined 多键的
splined connection 花键连接
splined coupling 齿槽联轴节
splined deck 镶木条甲板
splined feed shaft 多键进给轴
splined flooring 键接地板
splined half shaft 花键半轴；多键半轴
splined hobbing machine 花键滚床
splined hole 花键孔

splined hub 用花键接合的柄;用花键接合的轴;用花键接合的衬套;用花键接合的轮毂;花键套
splined independent axle 用花键接合的独立轴
splined joint 花键连接
splined mandrel 花键心轴
splined nut 带槽螺母
splined shaft 用花键接合的轴;花键轴
splined sleeve 花键套管
splined spanner 花键扳手
splined spindle 花键轴
splined steel hub 用花键接合的钢轴;用花键接合的钢衬套;用花键接合的钢轮毂
splined tire 齿槽型轮带
splined tube 花键管
splined tyre 齿槽型轮带
splined york 滑槽轭
splined york dust seal 滑槽轭防尘圈
spline feed rod 花键进给杠
spline finite strip 样条法
spline fit 花键拟合;花键配合
spline fit approximation 试配逼近法
spline fit curve 样条拟合曲线;试配曲线
spline fit method 试配法
spline fitting 样条拟合;花键座
spline function 样条函数;仿样函数
spline ga(u)ge 花键量规
spline gear 花键齿轮
spline grinder 沟槽磨床
spline grinding machine 花键磨床
spline hob 花键滚刀
spline hobbing machine 花键轴铣床;花键滚齿机
spline housing 花键套
spline hub 花键座
spline interpolation 样条内插;样条插值法;样条插值;仿样内插(法)
spline joint 键槽连接;花键接合;填实缝;穿条接合;销板;键接头;键接合
spline miller 花键铣床
spline milling cutter 花键铣刀
spline milling machine 花键铣床
spline pinion cutter 花键插齿刀
spline plug ga(u)ge 花键塞规
spline push broach 花键推刀
spline ring ga(u)ge 花键环规
splines 花键
spline shaft 多键轴;多槽轴
spline shaft gear 花键轴齿轮
spline shaft grinding machine 花键轴磨床
spline shaft hob 花键滚刀
spline shaft hobbing machine 花键轴铣床
spline shaft milling machine 花键轴铣床
spline shaft yoke 花键轴叉头
spline slip yoke 花键套叉头
spline the points 联结各点
splineway 凹线
splining 镶木条的甲板捻缝法
splining algorithm 样条函数算法
splint 开口销;火柴杆;薄木片
splintage 夹板固定
splint coal 暗硬煤
splint cutting machine 割木片机
splinter 裂片;碎片;劈开
splinter deck 防弹甲板
splintered fault scarp 参差断层崖
splinter forceps 裂片镊(钳)
splintering 削片;分开;裂开;碎裂
splintering fracture 碎片状破裂
splintering off 削成碎片
splinter of stone 碎石片
splinter-proof 防破片的;不碎的
splinter-proof concrete wall 防碎片的混凝土墙;防弹片的混凝土墙
splinter-proof glass 防碎玻璃;防弹玻璃
splinter-proofing 防碎片的;防弹片的
splinter roof 木瓦屋面;木片屋面
splinters 碎玻璃屑
splintery 裂片状
splintery fracture 裂片断口;粗糙断面
splintlet 小夹板
splint wood 边材
split 绿信比;中分面;工字钢 T 形贴角连接;裂口的;夹层煤;拼合口;拼合的;劈;未净股票分割;对开的;对开材;等信号区;穿透裂纹;分裂;分开;分割;分叉的;薄板砖;劈裂

split across 对裂开
split aC unit 分体式空调器
split adapter sleeve 剖分式紧定套
split adjusting collar 开口调整环
split ag(e)ing 两段时效;分级时效
split air conditioner 分体式空调
split air conditioning system 分体式空调系统;分体式空气调节系统
split altitude profile 不连续高度剖面图
split and wedge(type)bolt 开尾螺栓
split anode 双瓣阳极;分瓣阳极
split anode magnetron 分瓣阳极磁控管
split application 分期施用;分期施肥
split applications 分施
split arm flywheel 中分面飞轮;分模面飞轮
split astragal 对开半圆饰;双扇弹簧门缝嵌条
split axle 分轴
split axle box 对开轴承
split back 画幅分割滑板
split-bamboo kiln 劈竹形窑
split barge 拼合泥驳;对开式泥驳
split-bar generator 分离条信号发生器
split barrel 贯入器;可拆式岩芯管;拼合式岩芯管;双开式料筒;对开试样筒;半合管
split-barrel sampler 裂环取样器;带半合管取样器;带半合管取土器
split-barrel sampler with liner 带衬管拼合取土器
split base 拼合底座
split-batch charging 分散配料;分批下料
split-battery system 分组电池装置
split beam 组合梁;裂隙梁;分裂波束
split-beam cathode ray tube 分束阴极射线管
split-beam hologram 分束全息图
split-beam microscope 光束分裂显微(镜)测厚仪
split-beam photometric analyzer 分裂光束光度分析仪
split bearing 开口轴承;可卸轴承;可调轴承;剖分轴承;对开轴承;对开式滑动轴承;并合轴承
split bearing shell 拼合轴衬
split bill of lading 分割提单
split-blip 裂峰信号;双峰
split block 裂纹式砌块;裂面混凝土砌块;劈离砖块;切开的砌块
split block concrete 分块混凝土
split board 层合封面纸板
split bolt 楔缝式锚杆;燕尾螺栓;开口螺栓;开口螺栓;带开尾销螺栓
split bond 半厚砖砌筑;半厚砖砌体
split boot 带分外管的开沟器
split-bottom barge 拼合泥驳
split-bottom dump 有缝的倾倒门;开底倾卸;底拼合的倾卸门
split-bottom hopper 分叉底料斗
split box 拼合式滑动轴承;对开式滑动轴承;分线箱
split brick 劈裂砖;长度方向的半砖;半厚砖
split brush 分块电刷;分瓣电刷
split-bubble level 符合水准仪;符合水准器
split-bucket elevator 分离斗式提升机
split bump 接头噪声
split burner 裂口喷嘴;裂口灯
split bush 剖分轴衬
split bushing 裂口衬套;拼合轴瓦
split bus protection 分析母线保护
split cable 分股电缆
split cameras 分束照相机
split capital 分割资本
split carrier 载波分离
split carrier system 分裂载波制
split casing 中分面式汽缸;对分外壳
split casing pump 裂壳泵;有缝的泵;外壳拼合的泵;剖开式泵
split catalogue 分类目录;分部目录
split cavity 对开模;分裂槽
split central island 中分式中心岛
split charging 分装
split chimney block 对开烟囱砌块
split chlorination 分流氯化(未处理及已处理污水)
split chuck 弹簧卡盘
split chute 裤衩溜子;分流槽
split circlip 开口簧环
split clamp 活动箍;拼合夹具
split clamp crankshaft 夹紧式曲轴
split clamping bearing 拼合夹紧套

split clearance 分开单独清除
split coal 夹层煤;分叉煤层
split coil 分流盘管;分列式盘管
split collar 裂口环;开口环;开口垫圈;拼合环
split collar retainer 拼合环座圈
split collector ring 分瓣式汇流环
split column journal 专栏日记账
split commission 部分佣金
split compressor 分级压气机;二级压缩机
split compressor engine 分级压缩发动机
split concrete block 劈裂面混凝土砌块;劈裂混凝土砌块
split conductor 多芯线;多股绝缘导线
split conductor cable 多股绝缘线电缆;分芯电缆;分股电缆
split conductor protection 分股导线保护
split confounding 裂区混杂
split conic(al)bushing 切口锥形套筒
split construction 分段施工
split construction of ships 分段造船法
split contact 双头接点
split contract 分项承包(合同)
split core 分裂铁芯
split core barrel 拼合式岩芯管
split core box 对分式芯盒
split core rod 拼合芯棒
split core type current transformer 分裂铁芯式变流器
split cotter 开口销;开扁销
split coupling 对开联轴节;开口套管;拼合式联轴器;对开式联轴节;半联轴节
split course 薄砖层;砍薄的砖砌层;半厚砖砌层;削砖砌层;半厚砖砌层
split crack 劈裂缝;纵裂
split crankcase 可拆曲柄箱;对开曲轴箱
split-cube mo(u)ld 对开模
split cup 分瓣感应环
split current 支流
split cutoff 分叉式阻种器
split-cycle control 快速调整
split-cylinder mode 分割同位标磁道组方式
split dead bolt 对开死插销
split decision 非一致性决定
split deck 分离式桥面
split delivery 均分到货;分批到货
split design 分割设计
split developer 双液显影
split diaphragm 拼合隔板
split die 组合模;拼合模;缝口板牙;组合锻模;可拼模;拼合板牙
split-die liner 组合模衬
split-digger blade 开口式挖掘铲
split diopter 层光分割
split disk 分离盘
split-disk film-type bolometer 分裂圆盘薄膜型测辐射热器
split dose irradiation 分剂量照射
split-draft metier 热风分向干式纺丝机
split drum mixer 裂筒式拌和机;分离式滚筒式搅拌机
split dump barge 开体泥驳;拼合泥驳;对开式泥驳
split end 劈头
split fabric 分色多层织物
split face 锯裂面;分块饰面;劈裂面;砍断面
split-face block 裂纹式混凝土砌块(饰面用);机切混凝土块;裂纹式砌块
split-face finish 裂面石材
split-face machine 切石机;剖石机
split fall system 分钩装卸法;分钩装配法
split feed 分路馈给
split-feed control 分路馈给控制
split field 分像视场;分裂场;分离视场;分类信息组;像场分割取景器
split-field analyser 分像分析器
split-field filter 远近镜
split-field finder 场像分割取景器
split-field lens 分场透镜
split-field motor 串励绕组分段式直流电动机;串激绕组分段式直流电动机
split-field photometer 分场光度计
split-field photometry 分场光度学
split field picture 分画面图像;分场图像
split-field polarizer 分场偏振镜
split fire bridge 空冷火桥

split fitting 分支配件；分裂接头件
split fix 四标两角定位；分角定位；分段定位
split flame furnace 分焰炉
split floor tile 片状瓦；裂纹地砖
split flow 分(导)流；分叉流动
split-flow heater 分流加热器
split-flow metier 热风分向干式纺纱机
split-flow pump 分流泵；双联泵
split-flow reactor 分流反应堆
split-flow treatment 分流处理
split-flow wastewater 分流污水
split flywheel 可折卸飞轮
split focus 分裂焦点
split-folded waveguide 分离折叠波导
split fraction 分流分率
split frame 断开式门框
split friction cone 拼合摩擦锥轮
split furnace 开缝炉
split gear 拼合齿轮；双片齿轮
split gland 对开压盖
split gravel 裂变砾石
split guide 剖分式导管
split-half 对开的
split-half correlation 二分法相关
split-half method 半分法
split-half reliability 折半信度
split hammer rotary granulator 旋锤式碎石机；旋锤式破碎机
split heads 脚手板支架；(钢管支撑中的)叉端立柱；轨端裂缝；马凳
split holder 拼合刀架；弹簧夹持器；弹簧刀杆
split holding ring 对卡圈
split hopper discharge 开体泥舱卸泥
split housing 剖分式轴承座；拼合壳；拼合衬套
split hub 可卸毂；拼合毂
split-hub pulley 拼合轮毂式皮带轮
split hull 对开式船体
split hull discharge system 开体卸泥方式
split hull dumping barge 对开式泥驳
split hull hopper dredge(r) 拼合式自航挖泥船
split image 振铃效应图像；双像；分裂图像
split-image field range-finder 裂像测距器
split-image rangefinder 分像切合测距仪
split image wedge 裂像光楔；分像光楔
split indication 等强信号指示
split ingot mo(u)ld 对开钢锭模
split inner-tube core barrel 内管拼合式双层岩芯管
split instruction 分指令
split investment company 分裂投资公司
split jamb 开口侧壁；断开式门框
split-jet injector 缝隙(式)喷嘴
split joint 拼接
split key 开口键；切断电键；切断按钮；开口销
split keyboarding 分开键入
split knob 线路瓷夹
split lagging 分体式滚筒罩壳
split lateral recharge system 横向多路灌水系统
split leaf gate 分扇拼合闸门
split ledger account 专栏分类账；分割分类账户
split leg tripod 伸缩三脚架
split lens 剖开透镜
split-lens interference 对切透镜干涉
split-level 错层式；阁楼层
split-level architecture 错台式建筑
split-level design 错台式设计
split-level dwelling 错层式住宅
split-level house 错台式住宅；错层式住宅
split-level interchange 不同平面交叉
split-level investment trust 两级投资托拉斯
split-level viaduct 多层式旱桥
split lever 拼合式拐臂；对开式拐臂；分裂式操纵杆；分半式拐臂
split lever end 叉式杆端
split line 分模线；输泥管；辅助线；分模线；分离线；分叉绳索
split line of sounding 补空测深线
split link 铆接锚链环；拼合链节；分体式链节
split loading 分开加料
split-load washer 锁紧垫圈；对开垫圈
split lock 开口销
split lock washer 弹簧垫圈；弹簧垫片
split-log drag 木条路刮；木条刮路器
split loop 开口环
split mark 分离测标

split-mode 分裂模
split-mode unstable resonator 分裂模不稳定共振腔
split mo(u)ld 可拆模；组合模(具)；拼合铸模；对开模；分件模
split muff coupling 壳形筒联轴节；开口套筒联轴节
split mullion 拼合竖框；拼合直梃
split nut 拼合螺母；对开螺母
splitnut clasp nut 对开螺母
split off 分裂出来；以股易股；分营
split of ice cover 冰盖层裂缝
split of total freight 运费分摊
split open ended core box 对分式无底芯盒
split order 散户定货；分批订单；分次订单
split outer race bearing 拼合外压圈滚柱轴承
split pair 劈分线对
split paraboloid 裂瓣抛物面
split parking 分岔式停车场
split pattern 球节裂；组合模型；解体模；拼合木模；对分式模型；分体模；分块模拼合木模
split pattern mo(u)lding machine 拼合木模制型机
split phase 分相
split-phase belt winding 分割相带绕阻
split-phase conductor 分裂导线
split-phase control 分相控制
split-phase differential relaying 分相差动继电方式；分析差动继电方式
split-phase enclosed bus 分析封闭母线
split-phase induction motor 分相感应电动机；分析感应电动机
split-phase method 分相法
split-phase motor 分相(电动)机；分析(电动)机
split-phase relay 分析继电器
split-phase relay protection 分相继电保护
split-phase return wire 分裂回流线
split-phase starting 分相起动；分析起动
split-phase starting system induction motor 分相起动式感应电动机
split-phase type 分相式
split-phase type relay 分析式继电器
split-phase winding 分相绕组；分析绕组
split photograph 交向摄影像片
split picture 分裂图像
split pillar 开口支柱
split pin 开尾销；开口销；防松螺钉
split pin extractor 开口销分离机；开口销拔器；起开尾销器；起开口销器
split pin for set piston rod 活塞杆定位销
split pin for set pitman pin 连杆开口销
split pin pliers 开口销钳
split pipe 拼合管；输泥管；对管；分叉管；渠道；剖开管；劈裂管；槽
split piston skirt 沟槽活塞裙
split piston valve 两台肩阀
split plate 裂口垫板；单面模板；分裂板
split plate magnetron 瓣形阳极磁控管
split plate pattern 双面模板
split plot 裂区
split plot analysis of variance 裂区方差分析
split plot arrangement 裂区排列
split plot confounding 裂区混杂
split plot design 裂区设计；裂区试验法；裂区设计；分割设计
split plot experiment 裂区实验；分割试验
split plug 香蕉插头
split point 分流点
split pointer 叉式指针
split pole 分裂磁极
split-pole converter 分裂磁极变流机；分极变流机
split-pole motor 分极电动机；分级电动机
split-pole rotary converter 分极旋转变流机
split pricing 分别订价；非单一价格政策
split product 分裂产物
split projector 分裂式发射器
split pulley 组合皮带轮；拼合皮带轮
split-pump hydraulic system 多泵液压系统
split-pumping 分别给料
split punch pad 拼合式凸模压板
split-rag head 叉形头
split rail web 轨腰裂纹
split range positioner 分段定位器
split ranging control 分程控制
split-rate 分项资本还原率
split ratio 绿信比；分流比
split reactor 分裂电抗器

split-receptacle outlet 分离式插销
split resistance 抗裂性；抗劈强度
split rim 拼合轮辋
split-ring 钢裂环；开口圈；裂环；开口环
split-ring clutch 裂环啮合器；扣环离合器
split-ring commutator 分环整流子
split-ring connection 开环接合；裂环接合；开口环连接
splitring connector 裂环(连)接件；开口环连接件；开环接件
split-ring connector joint 裂环连接件接头
split-ring core catcher 开口式岩芯卡环
split-ring core lifter 开口式岩芯管体断环；开口环提芯器
split-ring lifter 开口环提芯器
split-ring mo(u)ld 组合模(具)；瓣合模
split-ring piston packing 开口式活塞胀圈
split-ring seal 开口毡封圈
split rivet 开尾铆钉；开口铆钉；拼合铆钉
split-rivet wire 开口铆钉用钢丝
split-rod graphite resistor 割口石墨电阻棒
split-rod resistor 割口电阻棒
split roller 分束辊
split roof 木条板屋顶
split room air conditioner 分体式房间空调器
split rotor 分裂式转子
split rotor plate 分瓣转片；分瓣动片
split run 交替通蒸汽(水煤气炉)；交流水蒸气鼓风；分割法
split runner 拼合式转轮；分瓣转轮
split-run test 分割实验
splits 薄砖
split sale 分割销售
split sample method 样品等分法
split sampling 分流进样
split saw 粗齿锯
split screen 屏蔽分割；分区屏幕；分割屏面
split screen display 分区屏幕显示；分屏幕显示
split screen unit 裂屏单元
split seam 分裂煤层
split-second action 紧急行动
split secondary 分节初级线圈
split-second blasting 毫秒爆破；微差爆破
split-second clutch(coupling) 瞬时离合器；快速离合器
split-second control 快速控制
split-second delay cap 瞬时延迟起爆筒
split-second delay detonator 毫秒延时雷管；毫秒延发雷管；毫秒延迟雷管；微差电雷管
split-second-hand 秒分指针
split-second mechanism 双秒机构
split-second pusher 双秒推扭
split-second response 瞬时动作
split seconds chronograph 双针秒表
split-second timer 精密秒表；精密定时器
split-second wheel 双秒轮
split-segment die 组合模(具)；可拆模
split separation of impulses 脉冲分散
split series servomechanism 双绕组伺服电动机
split series servomotor 双绕组串励伺服电动机
split service collar 固定用户铅管的管箍
split shaft 分轴式
split shaft gas turbine 分轴式燃气轮机；分辆式燃气轮机
split shakes 劈裂木板；劈裂木瓦
split share 分割股
split sheet size 分片上浆
split shell 二合器
split shift 交替(分次)轮班；分次上班
split shipment 分批发货
split shooting 中间激发
split shovel 缝口铲
split skirt 沟槽活塞裙；开缝活塞导缘
split skirt piston 裙部开槽活塞
split sleeve 夹板套筒；两片轴；分流套管；对开式套筒；对开式套管
split sluice flap 分料翻板
split sound receiver 伴音分路电视接收机
split-sound system 分离声系统
split spacing 裂缝间隙
split spacing of closure method 裂距前进法
split spline crankshaft 键槽连接式曲轴
split-splitless injector 分流—无分流灌浆器
split spoon 拼合式取土器；对开式取土勺；对开式

取土器
split spoon core sampler 带纵向切口的勺形取土器
split spoon sample 对开式取土样
split spoon sampler 开叉式砖取样器;对开式土壤取样器;对开式取(土)样品
split spray 分离喷涂;不均匀喷雾
split spread 分割加码;分段利差
split sprocket 拼合式链轮
splits system 分立系统
split stator 分瓣定子
split-step fourier migration 分步傅立叶偏移
split stirrup 开口钢箍;分支钢箍
split-stone finish 劈石面;裂面石
split storage 分离存储
split strand 分束丝;分股集中
split stream 分流
split stream sampling 分流进样
split-stream treatment 分流处理
split strength 劈拉强度
split strip packing gland 对开填料压盖
split stuff 劈材
split s-turn 拼合S形转弯
split suit 分开套装
split switch 尖轨式转辙器;对称道岔
split system 分流制;分流系统;分离系统(指地板下的通风道系统)
split system heating 分立系统加热
split t 工字钢T形贴角
split tail 鱼尾锤
split taper cotter 锥形开尾销
split tapered bushing 拼合锥形衬套
split taper pin 锥形开尾销
split tariff 分部税率;分部价格
split-tear 剥撕
splitted level 分裂能级
split tee connection 工字形T形贴角连接
split tensile strength 劈拉强度
splitter 合金块(钻头用);破堰器;拼合管;劈样机;劈木机;劈裂机;泡沫胶截断机;水平锯机;导流板;分支器;分样器;分片工;分流器;分流片;分裂设备;分裂器;分流柱;分离设备;分离器;分离机;分解器;剥片机
splitter blade 分流叶片
splitter box 分样箱;分线箱;分流箱;分流槽;分流沟
splitter chute 分叉漏斗;分流溜槽
splitter/combiner 功分器
splitter damper 分流式气流调节器
splitter for reconstructed stone 再造石材分裂机;人造石材分裂机
splitter pier 分水墩
splitter plate 导流板;分流板;分隔
splitter plate type sound absorber 隔板式消声器
splitter-recombiner 分束—并束器
splitter shield 分叉盾构
splitter tiled roof 拼合的瓦屋顶
splitter-type muffler 分隔式消声器
splitter vane 分水叶片;分流叶片
splitter wall 分界墙;中墩(水闸);隔墙;分水墙
split test 开裂试验;劈裂实验
split test method 裂检验法
split the cost 分担成本
split the layout line 按设计图准确加工
split the price difference on 50% basis 差价对半分
split thimble 拼装套管
split thread ring 开口带线圈
split tile 劈裂砖;劈商砖
split tile brick veneer 割开片砖成为饰面砖
splitting 裂开;开裂;破裂;劈裂;裂解;云母薄片
splitting action 裂解作用;劈裂作用;分水作用;分裂作用
splitting and reassembling technique 分裂与还原技术
splitting angle 分束角
splitting beam 分束
splitting blasting 碎裂爆破
splitting block 切开砌块
splitting box 分流箱;分流槽
splitting by frost 冻裂
splitting chisel 开尾凿;劈石凿;分裂器
splitting comb 分槽束紧器
splitting crack 劈裂裂缝
splitting decomposition 分裂分解;裂解
splitting die 组合模(具);可拆模具;分离模
splitting factor 分离系数

splitting failure 劈裂破坏
splitting fee 分享佣金
splitting forceps 分劈钳
splitting gun 爆破枪
splitting hammer 剖劈锤;劈锤
splitting instrument 分解器
splitting machine 剖层机;水平锯机;分切机
splitting mason 劈石工
splitting maul 锤斧
splitting method of pillar recovery 煤柱分隔回采法
splitting network 分裂网络
splitting of blocks 块的分割
splitting of emulsions 破乳
splitting of energy levels 能级分裂
splitting of horizontal target of cost and expenditures 成本费用横向指标的分解
splitting of vertical target of cost and expenditures 成本费用纵向指标分解
splitting-pad 分裂衰减器
splitting parameter 分裂参数
splitting pinch out 分岔变薄尖灭
splitting plane 分裂面
splitting property 劈裂性能
splitting property of wood 木材的劈裂性能;木材开裂性
splitting pulley 拼合滑轮;拼合滑车
splitting ratio 分束比;分流比
splitting resistance 抗裂强度;开裂阻力
splitting rupture 劈裂破坏
splitting scheme 劈因子格式
splitting shanty 有缝的小屋;分裂的小屋
splitting shot 解炮;碎裂爆破;大块二次爆破
splitting signal(l)ing 选路信号
splitting strength 抗裂强度;劈裂强度
splitting stress 破裂应力
splitting style 分岔形式
splitting technique 分裂技术;分裂技巧;分裂法;分离技术
splitting tensile strength 劈裂抗拉强度
splitting tensile strength test 劈裂抗拉强度试验;劈裂法试验;巴西法试验
splitting tensile test 劈裂抗拉试验;劈拉试验
splitting tension 张裂拉力
splitting tension strength 劈裂抗拉强度
splitting test 开裂试验;抗裂试验;劈裂试验;分裂试验
splitting time 分隔时间
splitting up 劈开
splitting wedge 尖劈;裂楔;劈木楔;劈裂楔
split tolerance 分裂耐受性
split torsion(al) test 扭转断口试验
split tower erection 分解组塔
split trailing suction hopper dredge(r) 对开耙吸挖泥船
split-train drive 拼合齿轮系驱动
split transformation 分裂转变
split transmission 分置式传动装置
split treatment 裂解处理;分流处理;分开处理
split tube 拼合管;对开式钻管(取土样用);对开管口;分叉管;半合管
split-tube core barrel 带半合管的岩芯管
split-tube drive sampler 对开管打入式取土器;对开管打入式取土样器;对开管打入式取(土)样器
split-tube in pumped from inner tube 半合管从内管泵出
split-tube sampler 对开式取(土)样器;带可拆式取土管的取土器;拼合管取土器;对开管式取土器;对开管取样器;带半合管取样器;带半合管取土器
split turbine 分轴式透平
split type 拼合式
split type air conditioner 分离式空调(器)
split type crank case 拼合式曲柄箱
split type heating system 分开式供暖系统
split type housing 分开式后轴壳
split type mixer 裂口出料搅拌机
split type room cooler 分离式房间器
split unit design 裂区试验设计;分割试验设计
split-up 分割;分股;裂开
split-up point allowance 废品限额
split valve guide 剖分式阀导管
split variometer 分挡可变电感器
split ventilation 分流通风;分风式通风

split vertical photography 分行竖直摄影;交向摄影
split washer 开口挡圈
split web near bolt-hole 螺孔周裂
split wedge 拼合楔
split wedge anchor 楔缝式锚
split weld 对接焊缝
split wheel 拼合轮
split winding 抽头绕组;分绕绕组
split winding goniometer 抽头绕组式测向器;分割线圈测向器
split winding transformer 分绕绕组变压器
split winding type synchronous motor 抽头绕组式同步电动机;抽头绕
split wood 劈开木材;裂开木材
split wooden carrying roller 拼合式木支持轮
split-word operation 分字运算;分字段运算
split year 跨年度
splocket 链轮;扣链轮;橡头垫块
SP logging 自发电位测井
splotch 污点
splotches of rust 锈斑
splush carpet 丝绒地毯
splutter 溅射
splutting 金属喷敷
spodic horizon 灰壤淀积层;灰化层
spodiosite 氟磷钙石
spodosol 灰土
spodosol mass foundation 灰土基础
spodumene 锂辉石含量;锂辉石
spodumene-lepidolite-pegmatite deposit 伟晶岩锂矿床
spodumene ore 锂辉石锂矿石
spodzolic horizon 灰壤淀积层
spoil 河底挖出物;奔渣;弃土;弃砟;疏浚土;堵塞窑底烟道用砖;废石料;废石方;废石;废品;损坏
spoilage 损坏(物);酸败;残次品
spoilage and defective work losses 废品损失
spoilage expenses 损毁费用
spoilage in excess of norm 超定额损耗
spoilage random fluctuation 废品不规则波动
spoilage report 废品损失报告书
spoil air-lock 废气气闸
spoil area 倾倒区;弃土场;抛泥场;废渣场;废场;废料场
spoil bank 管沟损坏;弃土场;弃土堆;废土堆
spoil barge 泥驳
spoil bin 碎屑储存器;废料储存器;垃圾储存器
spoil bucket 碎屑斗;废料斗;垃圾斗
spoil chute 滑废土槽;溜泥槽;弃土溜槽;弃土槽;抛土溜槽;废土滑槽
spoil conveyer 弃水输送器;废渣输送器
spoil dam 堤坝
spoil deposit 垃圾堆积;垃圾沉淀;废料堆积;废料沉淀
spoil dike 围埝;吹填区围埝
spoil discharge 排泥
spoil disposal 抛泥;疏浚土处置;疏浚土处理;废渣处理;废物处置
spoil dump 废料坑;垃圾坑;垃圾堆;废料堆;碎屑堆;碎屑坑;碎屑场;垃圾场;弃土场;废料场
spoiled 损坏的
spoiled area 矸石堆
spoiled concrete 腐蚀变质混凝土
spoiled goods 损毁货物
spoiled goods market 废品市场
spoiled material 损毁材料
spoiled material report 坏料报告单;废料报告
spoiled pile 矸石堆
spoiled product 废品;次品
spoiled products management 废品管理
spoiler 扰流板;阻流板;扰流器;方向图变换器
spoiler noise 扰流器噪声
spoil gas 腐败气体
spoil ground 垃圾倾倒区;垃圾倾倒场;倾倒区;弃土场;弃土场;抛泥区;抛泥场;倒泥区;废土堆;废土场
spoil ground buoy 垃圾倾倒区浮标;抛泥区浮标
spoil heap 弃土堆;废石堆;废土堆
spoil hold 泥舱
spoil hopper 泥舱;弃土底斗
spoiling 弃土;抛泥;排泥;出土
spoiling area 抛泥场
spoil island 弃土岛
spoil lime 废石灰

spoil line 输泥管
spoil material 弃土
spoil pile 弃土堆；废石堆
spoil pool 泥浆池；水坑
spoil removing 清渣；清除弃土；清碴；废渣清除
spoil site 弃土区；抛石区
spoil soil and rock 废土石
spoils system 分赃制
spoil stock 尾矿堆
spoil tip 废料场；废土场；废料坑；垃圾坑；垃圾堆；废料堆；碎屑堆；碎屑坑；碎屑场
spoil toe 弃渣坡脚
spoil volume 弃土量
spoil wagon 碎屑车；废料车；垃圾车；装碎石小车
spoil waste 弃土
spoke 梯级；轮辐；轮辐棍；舵轮把柄；辐条；扶梯踏蹬
spoke barrel 辐轮毂
spoke bolt 辐螺栓
spoke centre 轮辐中心
spoke clip for wheel 车轮辐条紧杆
spoked assembly 有辐轮毂；星形轮
spoked disc wheel 冲压辐车轮
spoke-divided flywheel 中分面飞轮；分模面飞轮
spoked key 辐条扳手
spoke dowel 轮辐尖端；辐尖端
spoked wheel 辐条轮
spoked wire 辐条钢丝
spoked wire wheel 钢丝轮辐
spoke head 辐头
spoke modulation 辐式调制
spoke nipple 轮辐钢丝螺母
spoke planing machine 轮辐刨床
spoke reticle 辐条状调制盘
spoke reticle blade 辐条状调制盘叶片
spoke(rib) 轮辐
spoke ring 轮辐环
spokeshave 曲面刨包铁；辐刨包铁；刮刀；曲面刨；滚刨；轴刨；轮辐刀；小圆刨；铁弯刨；铁刨子；辐刨片；辐刨；刨子
spoke shave plane 辐刨；轮刨
spoke-shave with wood(en) handle 木柄弯刨
spokesman role 发言人角色
spoke tread 梯级
spoke-type wheel 辐轮
spokevane current meter 辐叶式流速仪；辐射式流速仪
S-polygon 应力多边形
sponge 海绵状物；海绵动物；海绵；泡沫材料；多孔材料
sponge-backed rubber floor 海绵衬底橡胶地板
sponge ball 海绵铁球；海绵球
sponge blanket 海绵毯
sponge brush 海绵刷
sponge cake 海绵状金属块
sponge carrier 持海绵器
sponge chert 海绵燧石
sponged-diver's disease 采海绵潜水员病
sponged material 海绵状材料
sponge drier 海绵吸水式干燥器
sponge filter 泡沫塑料滤器；海绵(过)滤器
sponge glass 海绵(状)玻璃；泡沫玻璃；毛玻璃
sponge grease 钠皂基润滑脂；海绵状润滑脂
sponge holder 海绵缸；海绵容器；海绵夹
sponge holding forceps 海绵钳；海绵镊；持海绵钳
sponge iron 海绵铁
sponge-like 海绵状；海绵样的
sponge-like structure 海绵状构造
sponge-like texture 海绵状结构
sponge material 海绵材料
sponge method 海绵法
sponge nickel 海绵镍
spongeous 多孔的；海绵状
spongeous structure 海绵状构造
spongeous texture 海绵状结构
sponge plastics 泡沫塑料；多孔塑料；多孔泡沫塑料
sponge platinum 铂绒
sponge product 海绵制品
sponge reef 海绵礁
sponge rubber 橡胶海绵；海绵(状)橡胶；海绵橡胶；多孔(泡沫)橡胶
sponge rubber bucket-type seat 海绵橡皮斗式座位；海绵橡胶斗式座椅
sponge rubber diving suit 海绵橡胶潜水服
sponge rubber model 海绵橡胶模型

sponge rubber quilt 海绵毡
sponge rubber underlay 海绵橡胶底层
sponge sanding 湿磨白坯；湿磨白茬
sponge seat cushion 海绵座垫
sponge silver 海绵状银
sponge soil 橡皮土
sponge structure 多孔性结构
sponge tent 海绵塞条
sponge titanium 海绵状钛
sponge tray 海绵盘
sponge tree 金合欢
sponge-type porosity 海绵状孔隙
sponge underlay 海绵底层
spongework 海绵网格状灰华
spongiform 海绵状的
sponginess 疏松性
sponging 起泡沫
sponginum 海绵剂
spongioid 硅藻海棉；海绵样的
spongiolite 海绵硅灰石
spongiostrome structure 海绵层构造
spongolite 海绵岩
spongy 海绵状的；海绵状；松软；多孔的
spongy bone 海绵状骨
spongy casting 多孔铸件
spongy chromium plating 松孔镀铬
spongy copper 海绵状铜
spongy cure 海绵状硫化
spongy cushion 海绵垫
spongy iron 海绵铁
spongy layer 海绵状层；海绵层
spongy lead 海绵铅
spongy lead paste 海绵状铅泥
spongy lead plate 海绵状铅板
spongy material 海绵状物质；海绵状材料
spongy metal 海绵金属
spongy particle 海绵状粉粒
spongy platinum 海绵状铂；铂绒
spongy porosity 海绵状孔隙
spongy powder 海绵状粉末
spongy rot 海绵状腐朽
spongy rubber 海绵状橡胶
spongy soil 海绵土；海绵田；弹簧土；松软土(壤)；疏松土(壤)；橡皮土
spongy spot 海绵斑
spongy structure 海绵状构造
spongy surface 麻面
spongy texture 海绵状结构
spongy tissue 海绵状组织
spongy top 海绵头
sponsion 私人担保
sponson 舷侧突出部分；舷台；舷侧突出台；船旁保护装置
sponson beam 舷伸梁
sponson mount 车侧突出架
sponsor 主办者；主办人；担保人；出资人；发起人；保人
sponsoring 赞助；主办；发起
sponsoring firm 赞助公司；主办公司；发起公司
sponsor shareholder 发起股东
spontaneous 自生的；自发的；自然的
spontaneous aberration 自发畸变
spontaneous activity 自发活性；自发活动
spontaneous activity emission 自发辐射
spontaneous agglutination 自发凝集(反应)
spontaneous amputation 自然断连
spontaneous annealing 自行退火
spontaneous avalanche-like fission reaction 瞬时突变裂变反应
spontaneous breaking 自发破裂
spontaneous caving 自然崩落
spontaneous change 自然变化
spontaneous coagulation 自发凝固
spontaneous combustible 自燃的
spontaneous combustible substance 自燃物质
spontaneous combustion 自燃(烧)；自发燃烧
spontaneous combustion articles 自燃物品
spontaneous combustion articles and combustion in humidity articles 易燃固体
spontaneous combustion coal 自燃煤
spontaneous combustion of coal dump 煤堆自燃
spontaneous condensation 自发凝聚
spontaneous convection 自发对流
spontaneous corrosion 自然腐蚀

spontaneous cracking 自裂
spontaneous crimp 自发卷曲
spontaneous crystallization 自发结晶
spontaneous curing 自然固化
spontaneous current 自然流
spontaneous decay 自发裂变
spontaneous decomposition 自发分解
spontaneous depolarization 自动除极化
spontaneous detonation 自爆
spontaneous discharge 自发放电
spontaneous dispersion 自发分散
spontaneous doubling 自发加倍
spontaneous elastic recovery 自发弹性恢复
spontaneous electrical current 自然电流
spontaneous electrogravimetry 自发电重量测定法
spontaneous elongation 自发伸长
spontaneous emission 自发发射
spontaneous emission coefficient 自(发)发射系数
spontaneous emission noise 自发射噪声
spontaneous emitter 自发发射体
spontaneous emulsification 自发乳化(作用)
spontaneous emulsion 自发乳状液；稳定乳状液
spontaneous evapo(u)ration 自然蒸发
spontaneous fermentation 自然发酵
spontaneous fire 自然火
spontaneous fire prevention 自燃预防
spontaneous fission 自发裂变；自动裂化
spontaneous fission counting 自发裂变计数
spontaneous fission dating 自然裂变年代测定法
spontaneous fission process 自发裂变过程
spontaneous fission rate 自发裂变率
spontaneous fission source 自发裂变源
spontaneous fission track density 自发裂变径迹密度
spontaneous fluctuation 自然涨落
spontaneous freezing 自然冻结
spontaneous generation 自然发生；无生源说
spontaneous generation theory 自然发生说
spontaneous grain growth 晶粒自发长大
spontaneous growth of the crack 裂缝自发增长
spontaneous hardening 自然硬化
spontaneous heating 自燃热；自然加热
spontaneous ignition 自燃(着火)；自发着火
spontaneous ignition point 自燃点；自然发火温度
spontaneous ignition temperature 自燃着火温度；自燃着火点；自燃温度；自发着火点
spontaneous infection 自发侵染
spontaneous ionization 自然电离
spontaneous laser pulse 自发激光脉冲
spontaneous light 自发光
spontaneous light emitter 自发光发射器
spontaneous liquefaction 液化作用
spontaneously combustible cargo 能自燃货物
spontaneous magnetic domain 自然磁畴
spontaneous magnetism 天然磁性
spontaneous magnetization 自发磁化强度；自发磁化
spontaneous magnetostriction 自发磁致伸缩
spontaneous mutation 自发突变
spontaneous nuclear reaction 自发核反应；天然核反应
spontaneous nuclear transformation 自发转化
spontaneous nucleation 自发核化；自发成核作用；自发成核
spontaneous origin 自然发生
spontaneous oscillation 自发振荡
spontaneous oxidation 自发氧化；自动氧化作用
spontaneous polarization 自然极化；自发极化
spontaneous polarization method 自然极化法；自发极化法
spontaneous potential 自然电位
spontaneous potential gradient log curve 自然电位梯度测井曲线
spontaneous potential log 自然电位测井
spontaneous potential log curve 自然电位测井曲线
spontaneous potential log noise factor 自然电位测井干扰因素
spontaneous potential method 自然电位法
spontaneous potential well logging 自然电位测井
spontaneous process 自发过程
spontaneous radiation 自发辐射
spontaneous reaction 自发反应
spontaneous recombination 自发复合

spontaneous recombination radiation 自发复合辐射
spontaneous recovery 自然恢复
spontaneous recrystallization 自生再结晶;自发再结晶
spontaneous regression 自发消退
spontaneous remanent magnetization 自发剩余磁化
spontaneous remission 自行缓解;自动缓解
spontaneous resistance 自动抑阻
spontaneous restoration 自发复原
spontaneous scattering 自发散射
spontaneous spalling 自发崩裂
spontaneous symmetry breaking 自发对称性破坏
spontaneous transfer reaction 自转移反应
spontaneous transformation 自发转化
spontaneous variation 自发变化
spoofing 电子欺骗
spool 线圈;线管;有边筒子;有边筒管;工字轮;卷轴;胶片卷轴;绕线圈;绕线管;双端凸缘管;双端法兰管;短管;并行联机外部操作;线轴
spool axminster carpet 艾克司明斯特络纱地毯
spool bar 线轴形沙坝
spool bed 线轴形圆柱床
spool box 轴瓦盆;胶卷盒
spool cart 卷筒小车;卷轴小车
spool chamber 阀芯腔
spool control command 假脱机控制命令
spool displacement 阀芯位移
spool donkey 卷轴式牵引机
spooled pseudo-unit 假脱机模拟装置
spool element 芯形件
spool end 阀芯端面
spooler 绕卷轴工人;络纱机;线轴;卷线筒;卷取机构;假脱机程序;筒子车
spool face 阀芯表面
spool flange 卷筒盘
spool gear 长齿轮
spool grizzly 滚轴筛
spoolin 假脱机输入
spooling 假脱机;抛土;打轴;缠卷;并行联机外设操作;绕线圈
spooling equipment 绕线设备;打轴设备;打筒设备
spooling gear 排缆装置
spooling machine 绕线机
spooling program(me) 假脱机程序
spooling system 假脱机系统
spool insulator 线轴式绝缘子;直脚型绝缘子
spool land 阀芯台肩
spool management 假脱机管理
spool-mounted bearing 安装在圆盘耙间距套管上的轴承
spoolout 假脱机输出
spool partition 假脱机分区
spool piece 短管
spool pilot valve 滑阀
spool position 阀芯位置
spool queue 假脱机队列
spool roll 圆盘辊
spool spindle 卷盘轴
spool stand 卷线架;筒管架
spool stroke 阀芯行程
spool tile 筒形砖;筒形瓦
spool travel 阀芯位移
spool traverse 阀芯位移
spool turbojet 筒管式涡轮喷气发动机
spool-type casing head 旋转管头
spool-type control valve 绕线轴式控制阀
spool valve 滑(柱式)阀;柱形阀;短管阀
spool valve pump 滑阀配流泵
spool volume 假脱机容量;假脱机卷宗
spool wheel 大卷轮
spool winding 模型绕组
spoon 勺;匙形(物);匙形刮刀;匙
spoon and square 小镘刀;小墁刀
spoon auger sample 匙形试样
spoon bag 捞泥勺
spoon bit 钻泥抽取器;插筋孔钻;匙形钻;勺(形)钻;匙头钻
spoon-bottom 匙底形的
spoon bow 匙形船首
spoon brake 凹入工作面制动器
spoon dredge(r) 匙式挖掘机;捞泥船;匙式挖泥船;单斗(式)挖泥船

spoon drift 浪花
spoon fashion 叠置舢板
spoonful 一满匙
spoon ga(u)ge 管簧真空规
spoonleaf oak 匙叶栎
spoon-like fault 勺状断层
spoon manometer 管簧压力计
spoon nail 匙状形甲
spoon oar 匙形桨
spoon of the blade 叶片凹边
spoon penetration test 土勺贯入仪
spoon-proof 取料试验
spoon sample 熔池取样
spoon sampler 贯入器(英国)
spoon sampling 环刀取样;杓钻取样
spoon-shaped blade 匙状轮叶
spoon-shaped bottom 凹面底
spoon slicker 修型小勺;匙形刮刀
spoon(soil) sampler 勺形取土器
spoon stern 匙形船尾
spoon test 手勺取样;勺样检验
spophyge 柱端凹线脚【建】
sporadic 不规则的
sporadic building 无计划建筑;无规则建筑;散布式建筑
sporadic development 蔓延式发展;分散发展;零星发展;零散开发
sporadic distribution 不规则分布
sporadic e layer 分散E层
sporadic event 偶合事件
sporadic mutation 自然突变
sporadic permafrost 散见多年冻土;分散多年冻土;分散冻土;散见永冻土;岛状多年冻土;散现永冻层;岛状永冻土;土中零星冻土
sporadic permafrost zone 岛状永冻土带
sporadic reflection 散现反射;不规则反射
sporadic X-ray source 暂现X射线源
sporadie distribution of salinity 斑状盐渍分布
sporangium 孢子囊(体)
spore 壳孢;孢子
spore coats 孢子外膜
spore liptobiolith 孢子残植煤
sporite 微孢子煤
Sporobolus heterolepis 草原鼠尾粟
sporoclarite 微孢子亮煤
sporoclarodurite 微孢子亮暗煤
sporodurite 微孢子暗煤
sporoduroclarite 微孢子暗亮煤
sporomycosis 真菌孢子病
sporopollen analysis 孢子花粉分析
sporopollen identification 孢子花粉鉴定
sporopollen spectrum 孢粉谱
sporopollen statistics 孢子花粉统计学
sporopollen zone 孢粉带
sporo-producing plant 孢子植物
sport area 运动场地;运动场
sport arena 体育馆
sport coupe 双座跑车
sport diving 运动潜水
sport fishery 运动渔业
sport goods 体育用品
sport ground 运动场
sporting driver 赛车运动员
sportman toll 射杀量
sport requisites 体育用品
sports administration 体育管理学
sports advertisement 体育广告
sports architecture 体育建筑
sports building 体育馆;运动房;体育建筑
sports center[centre] 体育中心
sports club 体育俱乐部
sports culture 体育文化
sports deck 运动甲板;遮阳甲板;日光甲板
sports development strategy 体育发展战略
sports economics 体育经济学
sports equipment closet 体育用品壁橱;体育用品室;运动器材室
sports facility 运动设施;体育设施
sports field 运动场;体育场
sportsfield roller 运动场压路机;运动场滚筒
sports forum 体育中心
sports ground 运动场;体育场
sports ground pavilion 运动场帐篷
sports ground type tandem roller 运动场用的双轮压路机;运动场用的串联式压路机
sports hall 运动房;体育馆
sports hygiene 体育卫生
sports market 体育推销
sports palace 体育宫
sports stadium 带看台体育馆;带看台运动场
sports statistics 体育统计
sports structure 体育建筑
sports tourism 体育旅游
sport-tourer 双门敞车
spot 侦察目标;光点;卖方当地交货价格;辉点;污点;点滴;点;地点;疵点;场所;斑点;找正
spot a cementing plug 打水泥塞
spot adjustment 光点调整
spot analysis 点滴分析;斑点分析
spot and butt welding machine 点焊对焊两用机
spot annealing 局部退火
spot application 局部应用;定点施用;点状施用
spot assets 现金资产
spot assuming geometrical optics 光点几何光学
spot beam 点波束
spot beam antenna 区域覆盖天线;点波束天线
spot beam discrimination 点波束鉴别
spot board (抹灰工用的)拌板;现场拌灰板;小拌砂浆板;(搅拌混凝土或砂浆用的)小拌板;灰泥板
spot bond 点状黏合
spot-bonded 局部黏合的;点状黏合的
spot bonding 局部黏合;点状黏合
spot broadcasting 本地广播
spot broker 现货交易掮客
spot burning 点状用火;点状烧除
spot buy 卖现货
spot buyer 现货买家
spot calibration 现场校准
spot canker 斑溃疡
spot cargo 现货
spot cash 现款
spot check 抽检;取样检查;实地检查;滴定法材料检验;抽样检查;抽查;现场抽查
spot checker 成品检验员
spot checking 现场抽查;现场检查
spot chromatography 斑点色谱
spot clearance 少量拆除;部分改良
spot coater 滴涂机
spot colo(u)rimetry 点滴(定)比色法
spot commodity 现货
spot comparism exposure meter 比光点曝光计
spot contact bearing 滚珠轴承;球轴承
spot contract 现货契约
spot control 点控制
spot corrosion 点蚀;点腐蚀
spot cover 现货抛补
spot cure 局部硫化
spot dancing 斑点跳动
spot deal 现货交易
spot definition 光点尺寸;光斑清晰度
spot defocusing 散焦
spot delivery 现货交付;现场交货;当场交货;现场交付
spot depth 深度点
spot desmosome 桥粒点
spot diameter 光斑直径
spot dispatch 现场调度
spot displacement 光点位移
spot distortion 扫描点失真
spot distortion point 光畸变点
spot dredging 扫浅
spot drilling 定心钻
spot elevation 点状标高;点立视图;高程点;某点海拔高度;点高程;标高点
spot elevation error 点高程误差
spot error indicator 斑点式误差指示器
spot estimate 粗略估算;抽查估算
spot exchange 现汇;即期外汇
spot exchange rate 现期汇率;现汇汇率;即期汇率
spot exposure meter 反射光曝光表
spot-exposure microscope 点曝光显微镜
spot face 孔口平面
spot faceplate 刮孔口刀
spot-facing 刮孔口平面
spot film 缩影片;点片
spot finishing 修整漆面污斑;整修花斑
spot fire 飞火
spot-fixed 局部固定的

spot fixing 局部固定
spot flooding 局部注水;分点注水
spot footing 柱式基础
spot foreign exchange 现期外汇
spot foundation 柱式基础
spot frequency 标定频率
spot glare 局部眩光
spot-glued 局部胶合;点胶合
spot glu(e)ing 点状胶结;局部胶合;局部胶结;(木模的)点胶合;点滴胶合法
spot goods 现货
spot gouging 局部表面切割
spot grazing 牧斑
spot grinding 点磨法
spot hardening 局部硬化;局部淬硬;局部淬火
spot heat 单加热;局部加热
spot heating 就地供暖;点状加热
spot height 高程点;地面点高度
spot height reading 高程读数
spot holes 布置炮眼
spot-homogen 铁板铅被覆
spot hover 定点悬停
spot indicator 点滴指示剂
spotiness 斑痕
spot infection 斑点侵染
spot inspection 现场检验;现场检查
spot intensity 光点强度
spot investigation 现场调查
spot jammer 选择性干扰机
spot jamming 局部干扰;特定频率干扰
spot killer 亮点消除器
spot lamp 聚光灯;点形灯
spot level 高程标记;水准点
spot level(l)ing (场地的)分格操平(测量)
spotlight 车头灯;反光灯;聚光灯;投光灯;点光源照明;聚光
spotlight booth 聚光灯间;聚光灯室
spotlight flap 面光
spotlight indication 光点式表示;点光指示
spotlighting 局部照明;定向照明;聚光照明
spotlight transmitter 扫描射线式发射机
spotlight with stand 带支座的聚光灯
spot limiter 脉冲干扰限制器
spot loans 个别申请的独户住宅贷款
spot log (装料的)地点桩
spot map 点图;标点地图
spot market 集市;现货市场;区域市场
spot measurement 光点大小测定;光点尺寸测量;单点测量;斑点测量
spot measurement of flow 单点测流
spot meter 点亮度计
spot method 点滴法
spot-mix plant 现场混合装置
spot net 点网络
spot-next 现货日期的次日
spot of light 光聚区;亮点;(投射到舞台的人物上的)光圈
spot on a landing field 降落机场上的定点
spot option 现货期权
spot patch 零补
spot payment 现付
spot photograph 目标照片
spot photography 定点摄影
spot plate 点滴板;滴试板
spot price 现款价格;现金售价;现金付款的最低价格;现货价(格);即期交割价
spot priming 填补
spot punch 点式穿孔(器);点穿孔(器);单(孔)穿孔机;打点穿孔;补孔器
spot punching 点穿机
spot purchase 现货购进
spot quality test 点滴定性分析
spot quotation 现货时价
spot radiograph 局部射线照片
spot radiographed 局部射线照相
spot rate 即期价格
spot reaction 斑点反应
spot receiver 点波束接收机
spot reconcentration 斑点再浓集现象
spot reconditioning 就地修理
spot reflecting galvanometer 光点反射检流计
spot reflector 点波束反射器
spot removal 斑渍清除;污点清除
spot remover 余辉消除器;光点消除器;亮度消除器;除斑剂
spot repair 现场修理
spot rusting 点蚀
spots 现货
spot sale 现货销售
spot sample 个别取样
spot sampling 现场取样;抽样
spots appear 出现稀疏斑点
spot scan(ning) 光点扫描;点扫描
spot-scan photomultiplier method 光点扫描光电倍增管法
spot segregation 局部偏析;点状偏析
spot seller 现货卖家
spot ship 现期船;现成船舶
spot size 光斑尺寸;点尺寸;斑点大小
spot-size divergence 发射度
spot-size error 光点直径误差;光点尺寸误差
spot-size resolution 光点分辨率;斑点大小分辨率
spot sod method 点铺草皮法
spot softening phenomenon 真空管漏气放电现象
spot speed 点车速;瞬时车速;光点扫描速率;点速(度);地点车速;车道上某一点车速
spot speed survey device 瞬时速率量测设备
spot spray 点射
spot stain test 斑点污染试验
spot surfacing 重点找平
spot survey 现场调查;局限性调查
spot system 定点修理制
spotted 有污点的;玷污的;有斑点的
spotted bamboo sheath 斑竹壳
spotted cane 斑竹
spotted cement texture 斑点状胶结物结构
spotted jade 梅花玉
spotted minutiae 标准点
spotted porphyry 斑点斑岩
spotted sandstone 斑点砂岩
spotted schist 斑点片岩
spotted slate 斑点板岩
spotted state 点状
spotted structure 斑点状构造;斑点构造
spotted survey 定点测量
spotted taraspite 斑点白云岩
spotted type 斑点型
spotted warm anomalies 斑点暖异常
spot temperature 工作地点温度
spotter 现场装卸指挥人员;定位器;打桩机定位横架;中心钻;指挥人(交通);警戒雷达站;除污机;测位仪;测位器
spot test 现场试验;硝酸浸蚀试验法;就地试验;定点抽查;点滴试验;当场试测;抽样检查;抽查试验;斑点试验
spot test analysis 点滴试验分析
spot testing 实地试验;定点抽查
spot test method 点滴试验法
spot test plate 斑点试验板
spot time 工间休息
spottiness 斑点度
spotting 精饰加工;花点;花斑;去油污渍;涂污;定心;点样;飞火蔓延;布点;找正
spotting ability 对位能力;定位能力
spotting board 挖穴器
spotting drill 定位钻
spotting fire 蔓延飞火
spotting gain control 增益校正调整
spotting in 修饰斑点;测定点位
spotting mopping 点粘;点涂
spotting operation 对位作业
spotting out 消除污点
spotting projector 定位投影器
spotting rainstorm pattern 阵雨型
spotting spindle 定位心轴
spotting tool 定中心工具
spot trading 现货交易
spot transaction 现货交易
spot turn 原地转弯
spotty oak 麻栎
spot-type fire detector 点型火灾探测器
spotty rainstorm pattern 阵雨型
spotty steel 白点钢
spotty surface 多斑点表面
spot value date 即期外汇买卖起息日
spot velocity 点流速
spot weld 焊点;点焊缝
spot-weld bonding 胶结点焊
spot-welded joint 点焊接头
spot-welded round bars 点焊接圆钢筋
spot welder 点焊机
spot-weld fixture 点焊接夹具
spot welding 点焊(接)
spot-welding electrode 点焊电极
spot-welding electrode holder 点焊电极夹
spot-welding fixture 点焊夹具
spot-welding gun 点焊钳;点焊枪;点焊枪
spot-welding head 点焊头
spot-welding installation 点焊设备
spot-welding machine 点焊机
spot-welding pattern 焊点分布图;点焊分布图
spot-welding point 点焊点
spot-welding primer 点焊底剂;点焊首涂油;点焊底漆
spot-welding robot 点焊机器人;机器人点焊
spot-welding sealer 点焊用密封材料
spot-welding tongs 点焊钳
spot weld nugget 点焊熔核
spot wind 定点风
spot wobble 光点摇摆
spot zoning 地块规划;个别点区划
spout 斜槽;嘴;龙卷;料料槽;流槽;料槽;喷注;喷出;喷流
spout adjusting lever 滑槽调节杆
spout anchor supporting 喷锚支护
spout block 流槽砖
spout-delivery pump 无压供水泵
spout delivery system (喷粉机的)喷粉压送系统
spouted bed 喷腾层;喷射床;喷动床
spouted bed dryer 喷动床干燥器
spouted bed drying 喷动床干燥
spouted calciner 喷腾式分解炉
spouted fluidized bed 喷动流化床
spouter 喷油井
spout feeder 喷嘴式加料器;喷射式加料器;喷粉式加料器;喂槽
spout guide 出口导槽
spout hole 溢流孔;喷水孔
spouting 喷注;喷射;波峰尖;喷流
spouting explosion 喷窑爆炸
spouting fountain 喷泉
spouting plant 混凝土滑槽运送装置
spouting speed 喷射速度
spouting thickness of layer 一次喷射厚度
spouting time interval of layer 分层间隔时间
spouting velocity 喷射速度;级热降理论速度;喷射出口速度;喷出速度
spout lip 流料口
spout mist sprayer 带软管的弥雾机
spout of stock 喷浆
spout operator 料料槽操作工
spout plane 滑槽刨;斜槽刨
spout plug 注口塞;孔塞;出口塞
spout pouring pot 喷油罐
spout test 喷出试验
spported metal oxide 负载型金属氧化物
sprag 斜撑;制栓木;止轮垫;短支柱;支柱;肋板;拉条
sprag clutch 斜撑式离合器;超越离合器
spragger 车辆下坡运行跟车工
spragging gear 制动齿轮
sprag type one way clutch 斜撑式单向离合器
sprain 扭伤
sprangiospore 孢囊孢子
sprawl 蔓延;爬卧;建成区无规划扩大
sprawled area 市区无规划漫伸地区;市区散漫扩展地区;无规划延伸的地区(市区);市区无规划扩展区
sprawling dike 分布堤
sprawl phenomenon 无规则扩展现象(市区)
spray 枝状物;枝状饰;浪花;日喷;喷雾物;喷射;喷补;喷雾;喷洒;喷淋;喷镀;水雾;射出;洒;分支断层;飞沫
sprayability 可喷化性;可喷洒性;喷雾性;喷涂性;雾化性
sprayable 可喷洒的;可喷射的
spray absorber 喷洒吸收器;喷洒式吸收器;喷淋吸收器
spray acid cleaning 喷雾酸洗
spray aerator 喷雾曝气器;喷洒通气器;喷洒曝气器
spray anaerobic dish 喷雾厌氧培养皿
spray and wipe painting 衍彩法(涂漆)
spray angle 喷雾角(度);喷束角;喷射角(度)

spray apparatus 喷器
spray application 喷射作用;喷射应用;喷洒技术
spray application of mortar 砂浆的喷射使用
spray application technique 应用喷射技术;喷射技术
spray application techniques for crop protection 植物保护的喷射技术
spray applied 喷洒施工;喷射施工
spray applied mineral fiber insulation 喷涂矿物纤维绝热材料
spray arc 喷射电弧
spray arch 弧形喷杆
spray area 喷涂面积
spray arrester 泥浆防喷盒;喷水式避雷器;喷射避雷器
spray atomizer 喷淋雾化器;喷淋雾器
spray attemperator 喷水减温器
spray bar 喷油管;喷嘴架
spray barrow 手推式喷雾车
spray board 防浪板(放在船头);防溅板
spray bonderizing 喷射法磷化处
spray bonding 喷洒黏合法
spray bonding method 喷淋黏膜法
spray booth 喷釉室;喷釉厨;喷涂间;喷涂工作间;喷漆室;喷漆间;喷漆厨;喷柜
spray box 喷箱
spray broom 多嘴喷枪
spray burner 喷烧器;喷灯
spray burning 喷雾燃烧
spray by atomizing 用雾化作用进行的喷洒;用雾化作用进行的喷淋
spray calcining 喷雾灼烧
spray can 喷雾器;喷丸器;喷壶
spray can propellant 喷雾罐中的喷雾剂
spray carburet(t)or 喷雾式气化器
spray casting 喷射凝铸
spray catcher 捕雾器
spray catching 喷雾附着
spray cement coating 水泥喷射涂层
spray chamber 气体喷洗室;喷洗室;喷涂室;喷水段;喷漆室
spray charge 喷射电荷
spray cistern 喷雾槽
spray cleaning 喷淋洗涤
spray coater 喷涂器
spray coating 喷涂
spray colo(u)ration 喷涂着色
spray column 喷涂柱;喷洒塔;喷淋式气体洗涤塔
spray combustion 喷雾燃烧
spray concentrate 浓喷雾液
spray concrete 喷射混凝土
spray condenser 喷雾冷凝器;喷水冷凝器
spray cone 喷雾锥
spray cone angle 喷射锥角
spray configuration 喷射形状
spray congealing 喷雾凝
spray controller 喷涂控制器
spray cooler 喷雾冷却器;喷水冷却器;喷水冷却机
spray cooling 喷雾冷却;喷淋冷却
spray cooling pond 喷水冷却池
spray cooling process 喷雾冷却法
spray cooling tower 喷雾冷却塔
spraycrete 喷射混凝土
spray crystallization 喷雾结晶
spray cup 喷嘴头
spray cup injector 溅板式喷嘴
spray curing 喷水养护
spray curing rig 喷洒养护装置;喷洒养护装备
spray current feedback 喷射电流反馈
spray damper 喷雾润湿器
spray deaerator 喷雾除气器
spray decorated porcelain 吹彩瓷
spray decoration 喷花;喷彩
spray deflector 喷淋折板
spray degreaser 喷雾脱脂剂
spray degreasing 喷淋脱脂
spray deposit 金属层喷涂;喷雾附着
spray desuperheating 喷水降温
spray disc 喷雾盘
spray discharge 高压放电分选
spray disposal 喷水处理
spray disposal method 喷水处理法
spray distributor 喷水机
spray dome 喷水水墩

spray-down 喷涂
spray-down equipment 喷洗设备
spray-down hose 喷洗皮带管;喷洗水龙带;喷洗软管
spray drain 简易排水暗沟;排水暗沟;散水暗沟;柴束暗沟
spray drawing 喷淋拉伸
spray-dried detergent 喷雾干燥洗涤剂
spray dryer 喷雾式干燥机;喷雾烘干机;喷洒干燥器
spray drift 喷淋流程
spray-droplet 喷射油滴
spray drying 喷雾干燥(法);喷射干燥
spray drying chamber 喷雾干燥室
spray drying method 喷雾干燥法
spray drying process 喷雾干燥过程
spray drying process of fume gas desulfurization 喷雾干燥法烟气脱硫
spray drying tower 喷雾干燥塔
spray dust 喷涂粉末;雾尘;喷漆粉尘
spray dust scrubber 喷雾式除尘器
sprayed 喷射的;喷涂的
sprayed acoustic(al) ceiling 喷涂隔声天花板
sprayed acoustic(al) plaster 喷涂吸声抹灰层;喷涂吸声粉刷
sprayed asbestos 喷涂石棉层;喷敷石棉层;喷制的石棉;喷涂石棉隔声材料;喷涂石棉
sprayed asbestos insulation 喷涂石棉的涂层
sprayed atomized cathode 喷涂阴极
sprayed coat 喷涂面层
sprayed coating 喷敷
sprayed concrete 喷(射)混凝土
sprayed cooler 喷淋冷却器
sprayed earth 喷制土
sprayed exterior plastering 喷射的外层粉刷抹灰;喷射的外墙粉刷抹灰
sprayed exterior rendering 喷射的外层粉刷打底;喷射的外墙粉刷打底
sprayed fibrous mix 喷散纤维混合料
sprayed fireproofing 喷涂防火
sprayed fireproofing material 喷涂防火材料
sprayed fire-resistive material 喷涂耐火材料
sprayed insulating concrete lining 喷涂绝缘材料的混凝土衬垫
sprayed insulation 喷涂绝缘;喷淋绝缘
sprayed insulation material 喷涂绝热材料
sprayed lining 喷涂衬垫;喷射衬砌
sprayed material 喷涂材料
sprayed metal 喷镀的金属
sprayed-metal bonding 喷射金属法结合
sprayed metal coating 金属熔融喷镀(涂)层
sprayed mineral insulation 喷涂矿物绝缘层
sprayed mineral wool 喷涂石棉;喷涂矿棉
sprayed mortar 喷涂砂浆;喷砂浆
sprayed on absorbing ceiling 喷涂吸声顶棚
sprayed on acoustical ceiling 喷涂吸声顶棚
sprayed on acoustical material 喷涂吸声材料
sprayed on asbestos 喷涂石棉
sprayed on asbestos external rendering 喷涂石棉外墙粉刷打底
sprayed on asbestos finish 喷涂石棉终饰
sprayed on asbestos insulation 喷涂石棉保温(层);喷涂石棉隔热(层)
sprayed on asbestos plaster 喷涂石棉抹灰;喷涂石棉粉刷
sprayed on asbestos stucco 喷涂石棉拉毛粉刷
sprayed on asphalt 喷涂(地)沥青
sprayed on cement 喷涂水泥
sprayed on cement external rendering 喷涂水泥外墙粉刷打底
sprayed on clear varnish 喷涂透明清漆
sprayed on coat 喷涂涂层
sprayed on compound 喷涂混合物
sprayed on concrete 喷射混凝土
sprayed on cork 喷涂软木(屑)
sprayed on facing 喷涂面层
sprayed on film 喷涂薄膜
sprayed on fireproofing 喷涂耐火材料
sprayed on foam 喷涂泡沫
sprayed on geomembrane 喷涂土薄膜
sprayed on insulation 喷涂保温层;喷涂隔热层
sprayed on lining 喷涂衬垫
sprayed on mass 喷涂物质

sprayed on mastic 喷涂玛琦脂
sprayed on material 喷涂材料
sprayed on membrane 喷涂薄膜
sprayed on metal 喷涂金属
sprayed on metal coating 喷涂金属层
sprayed on method of curing 喷淋养护法
sprayed on mortar 喷涂砂浆
sprayed on paint coat 喷涂涂层
sprayed on plaster 喷浆粉刷
sprayed on plastics 喷涂塑料
sprayed on putty 喷涂油灰
sprayed on rendering 喷涂粉刷打底层
sprayed on roof(ing) membrane 喷涂屋顶薄膜
sprayed on seal(ing) 喷涂封层
sprayed on sound absorbent ceiling 喷涂吸顶棚
sprayed plastic finishing 喷涂塑料饰面
sprayed plastic flooring 喷涂塑料地面
sprayed plastics 喷涂塑料
sprayed polyurethane coating 喷涂聚氨酯涂料
sprayed polyurethane foam 喷涂聚氨酯泡沫
sprayed protective coating 喷涂保护层
sprayed roof 喷淋屋面;喷涂屋面
sprayed rubber 喷制橡胶
sprayed salt mist 喷射盐雾
sprayed tank 喷淋罐
sprayed thermal insulant 喷涂绝热材料
sprayed twice with colo(u)red 喷色灰浆两道
sprayed twice with white wash 喷白灰两道
sprayed vermiculite 喷涂蛭石
sprayed vermiculite plaster 喷涂蛭石砂浆
spray eliminator 喷淋净化器
spray enamelling 喷雾涂搪;喷搪
sprayer 油头;喷雾器;喷丸器;喷涂设备;喷涂器;喷床;喷射装置;喷洒器;喷布机;洒水器;洒布机;播散器
sprayer and duster 喷雾喷粉器
sprayer attachment 喷雾附加装置
sprayer duster 喷雾喷粉机
sprayer-flame 火焰喷射机
sprayer-intermittent 间歇式喷雾器
sprayer jet arc 喷雾器弧形喷管
sprayer plate 雾化板
sprayer simulator 喷射模拟器
sprayer-slide 柱塞泵式喷雾器
spray(er) unit 喷雾装置
sprayer-washer 喷淋洗涤器
spray etcher 喷雾腐蚀机
spray evapo(u)ration 喷雾蒸发
spray evapo(u)rator 喷雾蒸发器
spray fan 喷雾风扇;喷雾风机;喷束
spray fault 分枝断裂;分支断裂
spray feed 喷射送料
spray film 雾液
spray film evapo(u)rator 喷雾薄膜蒸发器
spray film maintenance 喷膜养护
spray finish 喷涂涂饰剂
spray finishing 喷浆整理;喷浆修理
spray finishing system 喷漆光滑法
spray flow 雾状流
spray forming 喷射成形
spray fountain 喷水池
spray freezer 喷淋式冻结装置;喷淋式冻结间
spray freezing 喷淋冻结
spray-freezing method 喷淋式冻结法;喷雾冷凝法
spray fusing 喷熔
spray galvanizing 喷涂涂锌
spray gas scrubber 喷淋式气体吸收器
spray gate 分支内浇口
spray glaze 喷釉;吹釉
spray glazing 喷釉法
spray guard 防喷罩
spray gun 金属喷镀器;泥炮;喷涂枪;喷水枪;喷射枪;喷射器;喷散器;喷枪;喷浆枪;喷镀枪;喷漆枪
spray gun controller 喷枪控制器
spray gun control system 喷枪控制系统
spray-gun mix 喷枪(用)混合料
spray gun process 金属喷镀
spray-gun repair 喷枪修补
spray head 喷嘴头
spray header 喷嘴集管;喷淋头;喷淋水管
spray head sprinkler 喷雾喷灌器
spray helmet 喷雾面罩
spray-hole 喷射孔

spray hood 舢板天幕
spray hose 喷淋皮带管;喷涂皮带管;喷雾皮带管;喷射皮带管;喷淋软管;喷涂软管;喷雾软管;喷射软管;水龙带;喷洒软管
spray humidifier 喷淋加湿器
spray ice 吹沫机
spraying 金属喷敷;喷洗;喷涂;喷水法;飞散;喷雾;喷洒;喷淋;喷镀
spraying agent 喷显剂;喷湿剂
spraying altitude 喷洒高度
spraying apparatus 喷雾器;喷涂设备;雾化装置
spraying arch (喷雾器的)拱形喷杆
spraying atomizer 雾化器
spraying burner 喷油燃烧器;喷射燃烧器
spraying cabin 喷雾匣
spraying car 洒水车
spraying carburet(t)or 喷射式化油器
spraying chamber 喷雾室
spraying cleaner 喷洗器
spraying cock 喷洒龙头;洒水龙头
spraying cone 喷雾锥
spraying curing 洒水养护
spraying device 喷雾设备;喷洒装置;喷洒器械;喷洒器
spraying distributor 喷洒机
spraying dryer 喷雾干燥器
spraying equipment 喷雾设备;喷雾器;喷涂设备;喷射设备
spraying fungicide 喷淋杀菌剂
spraying glazing 喷釉
spraying gun 喷射器;喷枪;喷雾器;雾化器;喷漆枪
spraying head 喷嘴
spraying hose 喷雾胶管
spraying humidification 喷雾增湿
spraying irrigation filter 喷灌滤池
spraying irrigation system 喷灌系统
spraying isolation 喷洒保温层;喷洒隔热层
spraying jet 喷雾射流;喷射流;喷口
spraying lacquer 喷漆;硝基喷漆
spraying lance 手压泵;喷水器;喷枪
spraying machine 喷淋机;喷雾机;喷射机;喷洒机;喷漆机;洒布机
spraying machinery 喷射机械;喷涂机械;喷洒机械;喷雾机械
spraying method 喷洒方法;喷射方法;喷雾法;喷涂法;喷射法
spraying muck 喷射碎屑
spraying nipple 喷雾嘴
spraying nozzle 喷雾嘴;喷嘴;雾化喷嘴;雾化喷头;喷嘴
spraying of asphalt 铺砌沥青
spraying of bitumen 铺砌沥青
spraying on concrete 喷混凝土
spraying on material 喷涂料
spraying overlay 喷镀堆焊
spraying paint 喷漆
spraying painting 喷漆法;喷漆
spraying pipe 喷水管
spraying pistol 喷枪;喷漆枪
spraying plant 喷漆装置;喷漆设备;喷涂设备
spraying-pond cooler 喷水冷却池
spraying pressure 喷雾压力;喷淋压力;喷涂压力;喷射压力
spraying pressure container 喷涂压料罐
spraying process 喷雾法;喷涂法;喷射法
spraying rate 喷射率
spraying reagent 喷显剂;喷湿剂
spraying reagent equipment 喷雾洗涤汽器
spraying rig 喷射凿岩机;喷射凿井机
spraying screen 冲洗筛;冲洗网;喷淋筛;喷淋网;喷水筛
spraying shop 喷漆车间;喷漆间;喷漆场
spraying soap powder 喷皂粉
spraying substance 喷涂物
spraying tanker 喷洒机
spraying test 喷雾试验;喷淋试验;喷涂试验;喷射试验
spraying treatment 喷涂处理
spraying tube 喷水管
spraying unit 喷雾器
spraying varnish 喷(涂)清漆
spraying vehicle 喷洒车
spraying vermiculite 喷涂蛭石
spraying washer 喷洗器
spraying water 冲洗水;喷雾水;喷淋水;喷水

spray injector 射流式喷嘴
spray injury 喷药伤害
spray insecticide 喷洒农药
spray insecticide for control insect pests 撒药治害虫
spray irrigation 喷水灌溉;喷洒灌溉(法);喷灌
spray jet 扩散射流;喷雾嘴;喷雾器;喷雾口;喷丸器;喷水器;喷水口;喷洒器;喷雾器
spray lacquer(ing) 喷漆
spray lance 喷雾器;喷雾杆;喷枪;喷杆
spray lay-up 喷涂积层法
spray lime 喷涂石灰
spray line 喷洒管
spray loss 喷雾损失;喷雾飞逸
spray lubrication 喷射式润滑;喷淋润滑;喷溅润滑
spray machine 喷雾机
spray-making valve 喷溅喷嘴
spray manifold 喷水减温箱;喷水减温器
spray mask 呼吸口罩;喷雾护面罩;喷漆工用面罩
spray-mastic 喷涂玛琋脂
spray metal 喷涂金属
spray metal coating 金属喷涂法;金属喷涂层;金属喷镀法;喷镀金属(法)
spray metal plating 金属喷涂
spraymeter 喷雾计
spray method 喷漆法;喷雾锻造
spray mixed plaster 喷洒混合灰泥;喷射混合灰泥
spray mottle 喷斑
spray needle 喷雾针
spray nozzle 溅水喷嘴;喷雾嘴;喷雾管嘴;雾化喷射器;水雾喷嘴;水雾喷头;喷嘴
spray-nozzle burner 喷嘴燃烧器
spray of electrons 电子流
sprayograph 喷雾质量测定仪;喷雾分散度测定仪
spray oil 喷淋油
spray oil cooling 喷油冷却
spray oiler 喷淋润滑器
sprayometer 喷液比重计
spray on 喷涂
spray on deck 浪花溅到甲板上
spray-on insulation 喷涂绝缘
spray-on insulation material 喷涂绝缘材料
spray-on membrane 喷涂养护剂
spray-on plastic finishing 喷涂塑料饰面
spray-on plastic flooring 喷涂塑料地面
spray-on process 喷镀法
spray-on process spray gun process 喷雾法
spray-on release agent 喷涂脱模剂
spray-on vermiculite plaster 喷涂蛭石砂浆
spray oxidizing process 喷雾氧化法
spray paint 喷漆
spray-painted 喷涂的;喷漆的
spray painter 喷(涂油)漆工
spray painting 喷漆;喷涂;喷涂层;喷染
spray painting equipment 喷涂机
spray-painting plant 喷漆装置
Spraypak packing 金属网交织排列填料;斯普雷帕克填料
spray particle size 喷雾颗粒尺寸
spray pattern 喷雾形状;喷束图形;喷流型;雾化状态
spray penetration 喷雾射程;喷射深度
spray pickling unit 喷溅腐蚀装置;喷淋腐蚀装置;喷射腐蚀装置
spray pipe 喷雾管
spray pipe passage 喷雾管路
spray-pipe system 喷水管道系统
spray pistol 喷射枪;喷漆枪;喷雾枪;喷枪;手枪式喷雾器
spray plastering 喷射粉刷
spray plastic finish 喷洒塑料饰面
spray plate 隔沫板;喷漆挡板
spray-plate injector 溅板式喷嘴
spray plotter 喷墨式绘图仪
spray point 喷电针尖;放电针
spray polymerization 喷雾聚合
spray pond 喷淋池;喷水(冷却)池
spray pond condenser 喷泉式凝结器
spray-pond cooler 喷水冷却池
spray-pond roof 喷水池屋顶;喷淋降温屋顶
spray potential 喷电电势
spray-pressure controller 喷雾压力控制器;喷雾机压力控制器
spray printing 喷印

spray probe 喷雾探头
spray processing 喷雾加工法
spray prominence 喷射日珥
spray proof 防溅
spray pump 喷雾筒唧筒;喷淋
spray pyrolysis 喷雾热分解
spray quenching 喷液淬火;喷水淬火;喷射淬火;雾化淬火
spray reagent 显示剂;显色剂;喷湿试剂
spray refining 喷吹精炼
spray regime 喷射方式
spray region 边缘层
spray release agent 喷涂脱模剂
spray repair 喷补
spray residue 喷药残渣;喷雾沉残渣
spray ridge 吹沫机
spray rig 喷射装置;喷射式凿岩机;喷射式凿井机;喷射装备;喷洒设备
spray ring 喷雾器环
spray rinsing 喷洗
spray roasting 喷雾熔烧
spray robot 喷混凝土机械手
spray roll 喷淋辊
spray rose 莲蓬头;喷淋头
spray sagging 喷釉流痕;喷釉纹
spray scrubber 喷液涤气器;喷雾除尘器;喷射式除尘器;喷淋洗涤器;喷淋洗涤器
spray separator 喷淋分离器
spray sheet 喷流幕
spray shower 喷淋;喷射浇水
spray shower bath 喷射淋浴
spray shower column 喷射浇水塔;喷射浇水柱
spray shower control 喷射浇水控制
spray shower cubicle 喷射浇水控制室;喷射浇水控制台
spray shower curtain 喷淋帷幕
spray shower head 喷射浇水莲蓬头
spray shower hose 喷射浇水水龙带;喷射浇水软管
spray shower installation 喷射浇水装置
spray shower public baths 公共淋浴
spray shower receptor 喷淋水接受器
spray shower recess 喷淋水槽
spray shower room 淋浴室
spray shower rose 喷射浇水喷嘴
spray shower set 喷射浇水设备
spray shower stall 淋浴间
spray shower tray 喷射浇水架
spray shower unit 喷射浇水单元
spray shower valve 喷射浇水阀(门)
spray silvering 喷银
spray sizing 喷雾上浆
spray soldering 喷淋软钎焊;喷镀软钎焊;波峰钎焊
spray solidification 喷淋固化(过程)
spray sprinkler 射流式喷水器
spray starching machine 喷雾上浆机
spray sticker 喷雾黏着剂
spray stucco 喷射拉毛粉刷
spray stuccowork 喷射拉毛粉刷工程
spray supply 喷电电源
spray system 喷雾装置;喷淋装置;喷淋系统;喷笔系统
spray tank 喷水冷却塔;喷水降温器
spray tap 喷射龙头
spray test 喷雾试验;喷淋试验
spray testing 喷雾试验
spray texture 喷涂饰面
spray thrower 喷淋器
spraytight 防喷射的
spray tip 喷头;喷枪喷嘴
spray torch 喷镀枪
spray tower 喷雾塔;喷水冷却塔;喷淋塔
spray tower for absorption 喷雾吸收塔
spray transfer 喷射过渡;射流过渡
spray trap 雾滴分离器
spray tray(type) deaerator 喷雾淋盘式除氧器
spray treatment 喷雾处理;喷涂处理;喷射处理
spraytron 静电喷涂器
spray truck 喷洒车
spray tube 喷管
spray-type air cooler 喷淋式空气冷却器
spray-type air water section 喷嘴密度
spray-type cement 喷用胶浆
spray-type coil heat exchanger 喷淋式蛇管热交换器

spray-type collector 喷洒器式集尘器
spray-type controller 喷水式调节器
spray-type cooler 喷雾式冷却器;喷淋式冷却器
spray-type cooling 喷雾式冷却;喷淋式冷却
spray-type deaerator 喷雾式除氧器
spray-type desuperheater 喷水减温器
spray-type dryer 喷雾式干燥器
spray-type evapo(u)rator 喷淋式蒸发器
spray-type extraction column 喷洒萃取塔;喷淋式抽提塔
spray-type fire extinguisher 喷雾式灭火器;喷雾式冷却器
spray-type fluidized bed dryer 喷雾沸腾干燥器
spray-type heat exchanger 喷雾式换热器;喷淋型热交换器
spray-type humidifier 喷雾增温塔
spray-type recovery furnace 喷射式回收炉
spray-type sulfur burner 喷射式硫黄燃烧炉
spray-up 喷射;喷射成型;喷层;喷雾
spray-up method 喷射成型法
spray-up mo(u)lding 喷涂成型;喷射模塑法
spray-up process 喷射成型
spray valve 喷雾阀;喷淋阀
spray valve cover 喷雾阀盖
spray valve spring 喷雾阀簧
spray-wash 喷洗
spray washer 喷洗机;喷射式清洗机;喷洗器
spray washing 喷射冲洗
spray water 喷淋水
spray water and defoamant system 喷淋水和去染剂系统
spray water head 喷水头
spray water pipe 喷水管
spray water tube 喷水管
spray water unit 喷水器
spray welding 喷雾熔接
spray welding unit 喷雾焊器
spray wire 喷镀用的金属丝
spray wiring 喷涂布线
spray zone 浪花带
spread 展扩;流传;扩展;扩散;涂敷(量);伸展;射流扩散角;锤薄;传播;敷胶量;方帆宽度;不合理组合;蔓延;铺展;涂布
spreadability 可涂覆性;可涂布性;铺展性;覆盖性
spreadable life 活化寿命;适用时间(漆料)
spread advance experience 推广先进经验
spread all over 遍及
spread an awning 张开天幕
spread anchorage 张开式锚碇
spread bar 分布钢筋
spread beam 伸展梁
spread beam sling 带梁的吊索
spread-blade cutter 双面刀盘
spread card record 展开卡记录
spread cargo 分散货
spread city 蔓延城市
spread coater 刮涂机;刮胶机
spread coating 刮涂法;刷涂法
spread concrete 铺混凝土;平铺混凝土
spread control 展宽调节;散布控制
spread device 撒布装置
spread effect 波及效果
spreader 匀流坎;展平锤;刮胶机;沥青喷洒车;集装箱吊具;集装箱吊架;模板撑挡;横柱;铺层器;喷液器;涂敷器;涂胶机;涂布机;涂布机;摊药器;摊铺机;摊铺工人;碎石摊铺机;伸张器具;撒布器;撒布机;撑柱;分流梭;分隔撑板
spreader bar 扩杆;(预制门框的)加固撑杆;撑杆
spreader beam (起吊大型物件的)分荷多支点梁;担梁;撑杆;分布梁;分配梁
spreader block 撒煤砖
spreader box 铺料箱;箱形(石料)摊铺机;布石车厢
spreader car 撒布车
spreader check 分水节制阀;分流节制阀;分布节制阀
spreader cone 扩张锥(体)
spreader device 筑路撒料机;摊铺设备
spreader ditcher 扩张挖沟机;平整刮板机
spreader feeder 抛煤机
spreader-finisher 铺摊修整两用机;铺整两用机;摊铺修整两用机
spreader finisher without hopper 无给料器的沥青路面平整机;无布料器的沥青路面平整机
spreader frame 吊架

spreader hoisting mechanism 集装箱吊具起升机构
spreader hopper 摊铺机料斗
spreader incline device 吊具倾斜装置
spreader level-ling mechanism 集装箱吊具水平保持机构
spreader lorry 平压机推料车;平整机推料车
spreader mark 展痕(薄板表面人字形裂缝)
spreader nozzle 喷洒喷嘴
spreader of crushed stone 碎石撒布机
spreader plate 布水板
spreader process 刮胶法;涂布法
spreader screw 推进加料机;螺旋推进器;搅拌器;布料螺旋;给料螺旋
spreader slewing device 吊具水平回转装置
spreader spool 间距套管
spreader spring 扩张弹簧
spreader-steamer 扩幅汽蒸机
spreader-sticker 黏展剂
spreader stoker 抛煤机炉排
spreader stoker boiler 抛煤机炉
spreader-stoker-fired furnace 抛煤机加煤锅炉
spreader strip 分流防冲带
spreader-trailer 拖车式撒布机
spreader truck 平压机推料车;平整机推料车
spreader yoke 担梁
spread factor 零散因数;宽度系数;分布因数
spread flow 扩散流
spread footing 放宽的底脚;扩展式基脚;扩展基脚;扩展底座;扩大基础;单独基础
spread footing foundation 扩展独立基础
spread footing with circular base 圆底的扩展基础
spread for continuous profiling 连续观测系统
spread foundation 独立底座;独立基础;扩展式基础;扩展基础;扩展底座;扩大基础
spread function 扩展函数;扩散函数;弥漫函数
spread groove 连接纹
spread head 喷头
spreading 滋墨;展宽;接长;加宽焊道;铺平;喷散;摊铺;撒铺;撒布;敷涂;敷;分散风险;变粗;蔓延;铺展;喷敷;涂布
spreading action 扩张作用
spreading agreement 扩大抵押协议
spreading and grading works 场地平整作业
spreading and level(l)ing 找平
spreading and level works 场地平整作业
spreading area 涂布面积;摊铺面积;分水面积;分洪区;分洪面积
spreading ballast 铺渣;摊铺石砟
spreading basin 引渗池;扩散盆地;扇形流域
spreading bead 宽焊道
spreading beam 散光束;摊铺板;扩展板;(起吊大型物件的)分荷多支点梁
spreading belt conveyer 布料用胶带运输机
spreading board 撒布板
spreading box 撒料箱;箱形石料摊铺机
spreading buoy mooring 多浮筒锚泊装置
spreading calender 涂胶压延机;等速矸光机
spreading capacity 平均涂布率;涂布能力
spreading center 扩张中心
spreading chest 压延头
spreading coefficient 展布系数;扩展系数;铺展系数;涂布系数;摊铺系数
spreading concept of seafloor 海底扩张说
spreading conveyer 抛撒输送器
spreading device 涂铺器;摊铺设备
spreading disc 撒布盘
spreading draft pipe 喇叭形尾水管;扩大式尾水管
spreading draft tube 喇叭形尾水管;扩大式尾水管
spreading edge collimator 涂布边缘准直器
spreading edge control 涂布边缘控制
spreading effect 扩散效应
spreading factor 扩散因素
spreading failure 开展性破坏
spreading field of recharging water 补给水扩散范围
spreading fire 火灾蔓延;火力扩展
spreading-flame burner 扁平焰燃烧器
spreading floor hypothesis 海底扩张假说
spreading footing 扩展式基础
spreading form 铺筑型
spreading gear 撒播装置
spreading ground 摊铺底子
spreading hopper 布料斗;摊铺料斗;撒布料斗;铺料斗

spreading in layers 层铺法
spreading knife 涂胶刀;涂布刀
spreading layer 扩散层
spreading lens 发散透镜
spreading machine 铺料机;摊铺机;散布机
spreading machine for adhesive plaster 橡皮膏涂料机
spreading machine without bottom doors 无底门的摊铺机;无底门的铺料机
spreading mixture 摊铺混合料
spreading mooring 多点系泊(法)
spreading of a slick 浮油在水上扩展
spreading of binders 铺路沥青的扩展;铺路沥青的撒布;铺路沥青的涂刮;黏合料的扩展;合料的撒布;黏合料的涂刮
spreading of concrete 铺摊混凝土;摊铺混凝土
spreading of fire 火灾蔓延
spreading of load 荷载分布
spreading of sand 摊铺砂料
spreading of screed 摊铺找平层
spreading of soil 土的撒布;土的摊铺
spreading of spray 喷雾角(度)
spreading of stream 河流扩展;水流扩展
spreading of the rail head 钢轨肥边
spreading pit 渗水坑
spreading plate 撒料盘
spreading power 展布力;涂布能力
spreading process 涂布法
spreading range 扩展范围
spreading rate 涂布面积;摊铺速度;涂面率;涂布率;摊涂量;分布率
spreading rate of column 柱展宽率
spreading reel 抛撒轮
spreading resistance 扩散电阻
spreading roll 扩幅辊
spreading rotor 撒布轮
spreading run 摊铺行程
spreading screw 涂敷螺旋布料器;撒布螺旋布料器;摊铺螺旋布料器
spreading site 摊铺场地;分布场地
spreading stand 宽展机座(用于增大板坯宽度)
spreading the work 分散工作
spreading thickness 摊铺厚度
spreading velocity of recharging water 补给水扩散速度
spreading waveform generator 扩展波形发生器
spreading weld 宽缝焊接;加宽的焊缝
spreading wetting 铺展润湿
spreading width 分布宽度;摊铺宽度
spread in performance 工作特性的变动范围
spread-layer chromatography 涂层色谱(法)
spread length 传播长度
spread lens 发散透镜
spread light 散射光
spreadlight lens 偏光透镜
spreadlight roundel 偏光镜
spread loss reinsurance 分散损失再保险
spread manure 施肥
spread model 推广模型
spread mooring 多浮筒系泊(法)
spread mortar 喷敷砂浆;喷布砂浆
spread of axles 轴距
spread of bearing 方位角摆动范围
spread of distribution 分布曲线的展开范围;分布的分散程度
spread of electrodes 极距(指电解极距)
spread of fire 漫燃;火势蔓延
spread of flame 火焰蔓延
spread of-flame test 火焰蔓延试验
spread of flow 漫流;水面宽(度);水流宽度
spread of holes 孔的宽展
spread of indications 示值分布
spread of points 点的散布
spread of productivity 生产率变动范围
spread of results 试验成果变异
spread of smoke 烟雾扩散
spread of tank 池距
spread of the desert 沙漠扩展
spread of the fork 叉齿距
spread of the variables 变量的离差
spread of viscosity 黏度变化范围
spread of wave function 波函数扩展
spread of wheel 轮距
spread oil cooling 喷油冷却

spreadometer 滋墨测定仪
spread order 套做买卖订单;分散订单
spread out 铺展;摊
spread out straight method 展直线法
spread pesticide 散播农药
spread plate method 涂布培养法
spread ratio 间距比(率)
spread recorder (桥梁试验的)扩张记录仪;(桥梁试验的)扩张记录器
spread reflection 混合反射;发散反射
spread risk 分散风险
spread roll 展毯辊
spread rolling 宽展轧制
spread root 伸展的根
spreads 双向差额套头
spread sheet 棋盘式账目分析表;空白表格软件;棋盘式对照表
spread sheet software 报表软件
spread's length 排列长度
spread span 展宽端跨;伸长端跨
spread spectrum 展布频谱;扩频;扩展频谱
spread spectrum communication 扩频通信
spread spectrum system 扩展频谱系统
spread spectrum technique 展布频谱技术
spread spectrum transmission 展布频谱传输
spread speed 开展速度
spread's type 排列类型
spread superintendent 总施工管理员
spread suspension method 悬滴涂布观察法
spread the sail 展帆
spread velocity 开展速度;散布速度;传播速度
spread velocity of wave 波浪扩散速度
spread voltage 分布电压
Sprengel explosive 斯普伦格尔炸药
Sprengel pump 斯普伦格尔泵
sprig 三角条;扁头钉;黏土浮雕花样;无头钉;钉子;玻璃钉
sprig bit 小锥子;锥钻;打眼钻
sprig bolt 棘螺栓
sprig budding 叉形芽接
sprigged ware 印有浮雕图的器皿;贴有浮雕图样的器皿
sprigging 暂时钉住;插型钉
sprigging operation 幼苗移植工艺;种子移植工艺;绿化
sprig-grass 植草
sprigs 三角条
spring 弹跳;弹力;弹回;弹簧;弹出;水泉;船首迎风现象;反翘;边弯
spring abutment 簧座;弹簧座;弹簧支座
spring accumulation soil 泉水堆积土
spring accumulator 弹簧式蓄能器
spring action 弹簧作用
spring action cock 弹簧旋塞
spring activated firing pin 弹簧推动式击针
spring activity 泉水活动
spring actuated 弹簧启动的;弹簧工作的;弹簧控制的;弹簧调节的;弹簧驱动
spring actuated knockout pawl 弹簧促动式推种爪
spring actuated piston 弹簧作用的活塞
spring additional buffer 弹簧副缓冲块
spring adjuster 弹簧调整器
spring adjusting collar 弹簧调整环
spring adjusting screw 弹簧调整螺钉
spring adjustment 弹簧调整;弹簧调节
spring an arch 砌成拱形;砌拱圈;砌拱洞
spring anchorage bracket 弹簧固定架
spring anchor plug 弹簧固定塞
spring and dashpot model 弹簧和粘壶模型
spring and guide assembly 弹簧导承组合
spring and lever loaded valve 弹簧平衡阀
spring and piston ring elasticity tester 弹簧及活塞环弹性检查器
spring antenna 弹簧天线
spring anti creeper 弹簧防爬器
spring anti-squeak 弹簧减声器
spring application 喷洒施肥
spring arch 窟顶高度;弹簧拱
spring arch clasp 弹簧弓卡环
spring arrangement 弹簧布置
spring assembly 簧片组件;弹簧组
spring-assisted cutter 弹簧式切割器
spring attachment 弹簧压紧装置;弹簧固定
spring-back 回弹;弹簧式书脊;弹性后效;回跳;弹性回跳;弹性回复
spring back cover 弹簧脊封面
spring-backed discharge spout 弹簧返回式闸粮管
spring-backed oil ring 弹簧胀圈油环
spring-backed pilot 弹性导正销
spring-backed quill 弹性套管
spring-backed type indirect pilot 弹簧式间接导正销
spring-back gland 簧压密封
spring-back labyrinth gland 弹簧迷宫气封
spring-back of pile 桩的回跃;桩的回弹;桩回跃
spring balance 弹簧秤;摆动游丝
spring-balanced bell ga(u)ge 弹簧钟形压力计
spring-balanced guard 弹簧平衡的吊挂上板
spring-balance safety valve 弹簧平衡安全阀
spring balance system 弹性平衡系统;弹簧平衡系统
spring balancing sash 弹簧平衡窗框
spring ball joint 弹簧球节
spring band 弹簧箍
spring bar 簧杆;弹簧杆
spring bar connection (边车的)弹簧杆接头
spring barrel 弹簧锁定装置
spring base 弹簧座;弹簧底座
spring basket retainer 弹簧篮式盛土器
spring beam 弹性梁;弹簧杆
spring beard needle 弹簧针
spring bearer plate 弹簧压板
spring bearing 中间轴承;支撑轴承;弹簧轴承
spring bed 弹性托板;弹簧床
springbed mattress 钢丝网床垫
spring beetling machine 弹簧捶布机
spring bellows ga(u)ge 弹簧膜盒压力计
spring bend 起拱点;发条;泉水;弹簧弯曲
spring bender 弹簧折弯机
spring bending 弹簧弯曲
spring bending machine 弯簧机
spring bending tool 弹簧弯曲器
spring-biased valve 弹簧加载的阀
spring-binder 弹簧活页夹
spring bit 打眼钻
springblade knife 弹簧折合刀
spring block 弹簧缓冲块;拱座;拱脚
spring board (体育的)跳板;弹簧板;伐木踏板
spring bolster 弹簧承梁
spring bolt 弹簧螺栓;弹簧插销
spring bolt pressure lubricator 弹簧销油嘴
spring booster 弹簧增力器
spring borehole 钻取样孔
spring bow 弹簧圆规;小圆规
spring bow compasses 弹簧圆规;小圆规;自动规;弹簧小圆规
spring bowl 泉塘
spring box 弹簧箱
spring box for excess pressure head 超压簧箱
spring box for maximum pressure 高压头簧箱
spring box mo(u)ld 弹簧箱铸型
spring bracket 弹簧托架;弹簧支架
spring bracket bushing 弹簧托架衬套
spring brake 弹簧制动器;弹簧闸
spring brake chamber 弹簧制动器室
spring brass 弹簧黄铜
spring break 春季解冻
spring breakage 弹簧断裂
spring break-up 春融开裂;春季翻浆(路面)
spring buckle 骑马螺丝;弹簧扣;弹簧箍
spring buffer 弹簧碰垫;弹簧缓冲器;弹簧护舷(设备)
spring buffer cap 弹簧缓冲器帽
spring buffer chamber 弹簧缓冲室
spring buffer stroke 弹簧缓冲压缩行程
spring bumper 弹簧碰垫;弹簧减振器;弹簧缓冲装置;弹簧缓冲器;弹簧避振器;弹簧避振垫;防撞装置
spring bumper bracket 弹簧减振垫托架
spring bumper cover 弹簧盖板
spring bumper pad 弹簧护舷(设备)
spring bushing 弹簧衬套
spring butt(hinge) 弹簧合页;弹簧铰链(门窗)
spring cable winding drum 弹簧电缆绕线筒;弹簧电缆绕线盘
spring cage 弹簧笼(压机的弹簧压力机构)
spring cal(l)ipers 弹簧规;弹簧卡钳
spring camber 弹簧弯度
spring cap 弹簧座盘;弹簧帽
spring capacity 弹簧容量
spring cap for pile driving 弹簧桩帽
spring cap for piling driving 弹簧桩帽
spring cap nut (喷嘴的)弹簧调节螺塞
spring carrier 弹簧托架
spring carrier arm 弹簧悬架;弹簧挂钩;弹簧吊架
spring catapult (撞车试验用的)弹簧弹射器
spring catch 弹簧(止)挡;弹簧销;弹簧闩
spring catch device 弹簧挡装置;弹簧销装置;弹簧制动装置;弹簧抓取装置
spring cavity 弹簧腔
spring center 弹簧顶尖;弹动中心
spring center bolt 弹簧中心螺栓
spring center clamp 弹簧夹
spring center clutch plate 弹簧心离合器板
spring centered 弹簧定中心;弹簧找中心;弹簧对中心;弹簧对中的
spring-centered direction control valve 弹簧对中的方向控制阀
spring center hump 弹簧心瘤
spring centering 弹簧定中;对心弹簧
spring centering cap 弹簧矫准中心的套;弹簧矫准的盖;弹簧矫准中心的帽;弹簧对准中心的套;弹簧对准中心的盖;弹簧对准中心的帽
spring centrifugal clutch 弹簧延迟离心式离合器
spring chamber 泉水室;泉室
spring chamber parking brake 弹簧室式停车制动器
spring chaplet 螺旋心撑
spring characteristic 弹簧特征;弹出特性
spring chasis bumper seat 弹簧车架缓冲块座
spring chuck 弹簧夹盘
spring circulation 春季环流
spring circulation period 春季循环期
spring clamp 弹簧卡;弹簧夹(具);弹簧卡钳
spring clamp bolt 弹簧夹紧螺栓
spring clamping plate 弹簧夹板
spring cleaner 弹簧清洁器
spring clip 弹性扣板;弹簧卡子;弹簧夹
spring clip bar 弹簧夹条
spring clip bar spacer 弹簧钩簧隔片
spring clipboard 弹簧夹板
spring clip plate 弹簧夹板
spring clip retainer 弹簧保险圈
spring clip spacer 弹簧夹定位架
spring clock 发条钟
spring cloth wind-up motion 弹簧卷布装置
spring clutch 弹簧离合器
spring coil 簧圈;弹簧圈
spring coiling machine 卷簧机
spring coiling machines and apparatus 弹簧圈机和装置
spring coil shank 弹簧圈式铲柄
spring collar 弹性驻环;弹簧定位圈;弹簧挡圈
spring collet 弹簧套筒夹头
spring collet holder 弹簧筒夹架
spring compasses 弹簧圆规
spring compression 弹簧荷载;弹簧压力
spring compression adjustment 弹簧压紧调整装置
spring compression strength 弹簧抗压强度
spring compressor 弹簧帽
spring cone penetrometer 弹簧锥贯入硬度计
spring constant 弹簧(常)数;弹簧常量;弹簧比率
spring contact 簧片接点;弹簧接点;弹簧触点
spring contact arc welding 弹力焊
spring contactor 弹簧接触器
spring contact plug 弹簧接触插头
spring control 弹簧控制;弹簧调整
spring controlled float valve 弹簧操纵浮阀
spring controlled valve 弹簧控制阀
spring core catcher 弹簧式土芯取样器
spring core lifter 环状弹簧式岩芯提断器
spring cotter 弹簧制销;弹簧锁销;弹簧开口销;弹簧插销
spring-coupler damper 弹簧防振器
spring coupling 弹性联轴器
spring coupling plate 弹簧连接板;弹簧箍圈
spring cover 弹簧盖
spring cross shaft 弹簧横轴
spring crust 春季冰壳
spring cultivation 春季中耕
spring cup 弹簧压罩;弹簧盖
spring curve 弹力曲线

spring cushion 弹簧缓冲块;弹簧减振垫;弹簧垫块;弹簧垫层;弹簧垫(子)
spring-cushioned pick-up 弹簧缓冲式捡拾器;弹簧浮动式捡拾器
spring damper 弹簧减振器
spring damper coil 盘簧减振器
spring damper system 弹簧减振系统
spring dart 套管起拔工具
spring decay coefficient method 泉水消耗系数法
spring deflection 弹簧变位
spring depth ga(u)ge 弹簧深度规
spring detent 弹簧爪;弹簧擒纵器
spring device 弹簧装置
spring diaphragm 弹簧膜
spring die 管子丝扣扳钳;管子板牙;可调板牙;弹簧板牙
spring discharge 泉的流量
spring discharge in drought season 泉的枯水期流量
spring discharge in raising season 泉的丰水期流量
spring discharge summation 泉水流量总和法
spring disconnecting mechanism 弹簧分离机构
spring-disk valve 簧盘阀
spring dividers 弹簧量规
spring division 弹簧分规
spring dog 弹簧钩
spring drainage 泉水排泄
spring draught excluder 防风弹簧片
spring draw 弹簧起模钉
spring draw gear 弹簧拖拉装置;弹簧车钩
spring draw nail 弹簧起模钉
spring drive 弹簧传动
spring driven 弹簧驱动的
spring dust cover 弹簧的防尘套
spring dynamometer 弹簧动力计;弹簧测力计
spring element 弹簧元件
spring end 弹簧端部
spring energizer 弹簧增能器
spring equalizing device 弹簧平衡装置
spring equinox 春分点;春分
springer 碹砖;碹脚;拱脚石;拱端石;拱底石;砌拱砖;起拱石;托砖;四瓣弹簧夹
spring exerciser 拉力器
spring exhaustion 泉水枯竭
spring expander 弹簧扩张器
spring expander oil ring 弹簧胀圈油环
spring eye 钢板弹簧卷耳;泉眼
spring eye bolt 弹簧眼螺栓
spring eye bushing 弹簧圆眼衬套
spring-fall range 牧区的春秋羊场
spring fastener 弹簧扣
spring fastening 弹簧支座
spring faucet 弹簧龙头
spring-fed intermittent stream 泉源间歇河
spring-fed pond 泉水注入的池槽
spring fed stream 融雪水补给河;泉水补给河;春季补给河
spring feed 泉水补给;弹簧进给;春季补给
spring feeder 弹簧加料器
spring fender 弹性碰垫;弹性防撞装置;弹簧碰垫;弹簧缓冲设施;弹簧护舷(设备)
spring fender for jetty 突堤式码头的弹性缓冲设施
spring finger 弹簧爪式定位装置
spring-finger pick-up 弹簧指式捡拾器
spring fixed end 弹簧固定端
spring fixed eye 固定簧眼
spring flexibility 弹簧挠性
spring-floated die 弹簧(浮动)模
spring flood 桃汛;春汛;春洪
spring flood forecasting 春汛预报
spring floor(ing) 弹簧地板
spring flowering 春华
spring follower 弹性随动件;随动簧
spring following 弹性随动
spring for air pump 气泵弹簧
spring force 回弹力;弹簧弹力
spring forge 弹簧锻工车间
spring for governor 调节器弹簧
spring for grip chuck 套爪夹头
spring fork 弹簧叉
spring forming machine 弹簧成型机
spring formula factor of safety 振动公式安全系数(提升钢丝绳);弹簧安全系数计算公式

spring fracture 弹簧断裂
spring frame 弹性悬架式车架;弹簧框架
spring frame of seat 弹簧座架
spring free end 弹簧自由端
spring freshet 春汛
spring frog 弹性辙叉
spring frost 晚霜
spring funnels 泉口漏斗
spring fuse 弹簧熔断
spring gas 气лень气
spring gate 自关门弹簧
spring gate stick 弹簧直浇口模棒
spring ga(u)ge 手按弹簧定位装置
spring gauze 春纱
spring gear 回转窑弹性大齿轮;弹簧传动装置
spring-go 弹复
spring gold 弹簧金
spring governor 弹簧控制器;弹簧调速器
spring gravimeter 弹簧(式)重力仪
spring grease cup 弹簧润滑器;弹簧加油杯
spring grinding machine 弹簧磨床
spring group minimum flow analogy method 泉群最小流量比拟法
spring guard 弹簧护板
spring guide 支持环;弹簧导杆;弹簧导承
spring guide collar 弹簧导环
spring guide pin 弹簧导销
spring guide plate 弹压导板
spring guide trunnion 弹簧导杆耳轴
spring hair compasses 弹簧分度规
spring hammer 弹簧锤
spring hammer beetle 弹簧槌布机
spring handle 弹簧手柄
spring hanger 钢板弹簧吊耳;弹簧支承;弹簧挂钩;弹簧吊架
spring hanger bar 弹簧吊架杆
spring hanger bushing replacer 弹簧挂钩衬套换装器
spring hanger gib 弹簧吊架扁栓
spring hanger link 弹簧吊杆
spring hanger pin 弹簧挂钩销
spring harrow 弹簧路耙
spring head 泉源;弹簧头
spring-head roofing nail 圆头屋面螺钉
spring high water 大潮高潮位;大潮高潮面;大潮高潮
spring hinge (旋门的)地板弹簧铰链;弹簧铰链(门窗);弹簧合页
spring hitch 弹簧联结器
spring holddown 弹簧压紧板
spring holder 弹簧座;弹簧支架;弹簧盘;弹簧定位销;弹簧柄
spring holding pressure 弹簧保持压力
springhole 弹性孔
spring hook 旗绳钩;弹簧钩
spring horn 弹簧导架
springhouse (建在冷泉上的)泉上冷藏室;冷藏室
spring housing 弹簧套;弹簧罩;弹簧外壳
spring ice 蜂巢冰
springing 起拱线;起拱(石);起拱点;黏贴浮雕玻璃图样;扩孔爆破;炮眼扩孔;弹簧现象;弹动;拱脚石
springing blast 扩孔底爆破;掏壶爆破;掏爆破
springing blast hole 扩孔爆破眼;扩大炮眼;掏壶爆破眼;掏爆破炮眼
springing block 拱脚;拱基;拱底石;起拱砌块;拱脚石
springing course 起拱(砌)层
springing height 起拱高度
springing line 起拱线;倒缆;斜缆;起圈线;起拱梁
springing needle 托换基础支柱
springing of curve 曲线的起点;起拱点
springing of extrados 拱的起拱线;拱背起拱(线)点
springing of intrados 拱圈内弧起拱线;拱腹线的起拱点;拱腹起拱线
springing of soffit 拱腹起拱点;拱内弧起拱线;拱腹起拱线
springing of the stand 机座的弹跳
springing operation 绿化
springing out 涌出(水、油)
springing shot 药壶爆破
springing wall 墙撑;扶檩
spring injection system 弹簧喷射系统
spring insert 弹簧隔板

spring inside caliper 弹簧内卡钳
spring installation 弹簧装置
spring intercepting structure 阻截泉水的结构
spring investigation 泉水调查
spring jack 弹簧塞孔;触簧开关;侧簧底脚片
spring joint 弹簧接合
spring-joint caliper 弹簧接合卡钳
spring key 弹簧开口销;弹簧键
spring knockout 弹簧式顶件器;弹簧式顶件工具;弹簧顶件器
spring lamination 弹簧板
spring latch 弹簧碰锁;弹簧插销;弹簧(门)闩
spring latch lock 碰锁;弹簧锁
spring layer 锐变层
spring lay rope 麻与金属丝合股绳
spring leaf 钢板弹簧主(叶)片;簧片;片簧;弹力叶片;弹簧片
spring-leaf opener 开簧器
spring-leaf retainer 弹簧座圈
spring leg (独立悬挂的)弹簧支柱
springlet 小泉
spring level 泉水面;起拱高度
spring lever 弹簧杆
spring lifter 开口环提芯器
spring lifter case 岩芯提断器外壳
spring ligament 弹簧韧带
spring line 起拱线;斜系缆线;拱脚线;泉线;泉群分布线
spring link 弹簧连杆
spring-loaded 受载弹簧;弹簧加载的;弹簧支承的;弹簧式;弹簧加荷的;受弹力作用
spring-loaded antenna 弹簧天线
spring-loaded apex seal 弹簧加载径向密封片
spring-loaded apron 弹簧支承的挡板
spring-loaded bayonet thermocouple 弹簧加载插入式热电偶
spring-loaded bolt 弹簧加载的螺栓
spring-loaded brake 弹簧制动(器);弹簧蓄能制动;弹簧加载式制动器
spring-loaded catch 弹簧卡子
spring-loaded characteristic 弹力特性
spring-loaded check valve 弹簧止回阀
spring-loaded clamp lever 弹簧制动杆
spring-loaded clutch 常闭式弹簧离合器
spring-loaded cock 弹簧旋塞;弹簧闭锁开关
spring-loaded crusher 弹簧加压破碎机
spring-loaded cultivator 弹簧加载式中耕机
spring-loaded drill colter 弹簧加力开沟器
spring-loaded floating die 弹簧模
spring-loaded fly governor 弹簧调节器
spring-loaded gear 弹簧加载齿轮(消除啮合间隙的一种齿轮结构)
spring-loaded governor 弹簧调速器;弹簧调节器
spring-loaded grease cup 弹簧滑脂杯
spring-loaded idler 装簧空转轮;弹簧托辊
spring-loaded idling valve 弹簧上怠速阀
spring-loaded leather seal 弹簧压紧皮碗
spring-loaded lip seal 弹簧唇形密封
spring-loaded live centre 弹簧活顶尖
spring-loaded mazier sampler 弹簧美兹取土器
spring-loaded membrane 弹簧承力膜片;承弹力膜片
spring-loaded needle valve 弹簧加载式针阀
spring-loaded nozzle 弹簧加压喷嘴
spring-loaded oil control ring 弹簧油环
spring-loaded oil seal 弹簧油封
spring-loaded overflow valve 弹簧溢流阀
spring-loaded pin 弹簧安全销
spring-loaded piston 弹簧加荷的活塞
spring-loaded plunger 弹簧柱塞
spring-loaded pressure reducing valve 弹簧减压阀
spring-loaded pressure relief valve 弹簧式安全阀;弹簧加载的压力泄放阀
spring-loaded regulator 弹簧加载调压器;弹簧承力调节器
spring-loaded release pin 弹簧释放销
spring-loaded relief valve 弹簧安全阀
spring-loaded roller 弹起轮
spring-loaded safety system 弹簧式安全装置
spring-loaded safety valve 弹簧式安全阀
spring-loaded scissor gear 弹簧加载剪式齿轮
spring-loaded scraper 弹簧加载式土刮器
spring-loaded screw 微动螺旋;微动螺丝;弹簧支

承的螺旋
spring-loaded seal 弹簧加压密封
spring-loaded sheave 弹簧加压皮带轮
spring-loaded shedder 弹簧式顶出装置
spring-loaded spindle carrier 弹簧平衡式轴杆托架
spring-loaded stick 弹簧加力的操纵杆
spring-loaded stop 弹簧加载的挡块;弹簧安全销
spring-loaded thrust meter 弹簧测力计
spring-loaded tine cultivator 弹簧加载齿式中耕机
spring-loaded valve 弹簧阀
spring loaded wedge 楔缝锚头
spring load(ing) 弹簧荷载;弹簧压力
spring load test 弹簧荷载试验
spring lock 撞锁;碰锁;弹簧锁
spring lock device 弹簧锁紧装置
spring locking pin 弹簧锁式销
spring lock washer 弹簧紧固垫圈;弹簧锁紧垫片
spring low-water 大潮平均低潮位;大潮平均低潮面;大潮低潮(面)
spring low-water line 大潮低潮线
spring low-water tide 大潮低潮(面)
spring lubricator 弹簧润滑器
spring manometer 弹簧压力计;弹簧压力表;弹簧式压力表
spring mass 弹性体
spring-mass-damping system 弹簧质量阻尼系统
spring-mass system 弹簧质量系统
spring material 弹簧材料
spring mattress 钢丝垫(子);弹簧垫子
spring maximum 春季高峰
spring mechanism 弹簧机构
spring medium 弹簧介质
spring modulus 弹簧模量
spring motor 发条传动装置
spring-mounted engine 弹簧减震发动机
spring-mounted roller conveyer 弹簧支承滚轴运输机
spring-mounted shank 带弹簧安全器的铲柄
spring mounting 弹簧垫架;弹簧安装
spring needle 弹簧针
spring needle machine 弹簧针针织机
springness form 春种性类型
spring of arch 拱顶高度
spring of carriage 车架弹簧
spring of curve 曲线起点
spring offset 弹簧偏置式(换向阀)
spring offset valve 弹簧移位的阀
spring of intermediate depth 中等深度的泉水
spring of the casing 套管柱拧紧度
spring of the line 钢绳张紧度
spring opener 开簧器
spring opening 泉眼
spring operated 弹簧作用的
spring-operated brake 弹簧闸
spring-operated stripper 弹簧卸料板
spring-opposed air valve 弹簧压力空气阀;弹簧承力空气阀
spring-opposed bellows 弹簧承力波纹管
spring-opposed piston 弹簧承力活塞
spring-opposed valve 弹簧加载的阀
spring over-turn 春翻层;春季对流
spring ozone decline 春季臭氧衰减
spring ozone depletion 春季臭氧层耗竭
spring pack 弹簧组
spring pad 弹簧垫
spring-pad clip 弹簧垫夹
spring-pad clip nut 弹簧垫固定螺帽
spring-pad wear plate 弹簧垫耐磨板
spring pawl 弹簧爪
spring perch 弹簧座;弹簧夹具
spring performance 弹簧特性
spring pilot 弹簧退让式导销;钢板弹簧销
spring pin 弹性销;弹簧销
spring pincers 弹簧镊子;弹簧钳子
spring pinch cock 弹簧夹;弹簧节流夹
spring piot 钢板弹簧销
spring pit 泉坑
spring pivot (装在地板支承摆门的)弹簧铰链;弹簧枢;弹簧销
spring pivot bracket 弹簧销支架;弹簧枢架
spring pivot seat 弹簧支枢座
spring plank 簧板
spring plant 春性植物
spring planting 春植

spring plate 弹簧片;弹簧底板
spring plunger 弹簧柱塞;弹簧锁销;弹簧定位销
spring plunger stripper 弹簧柱塞式卸料板
spring pocket 弹簧腔
spring point 弹簧道岔;弹簧开关;起拱点;弹簧撤叉【铁】
spring pole 弹簧杆;手动冲击钻杆
spring pole drilling 手动冲击钻进;弹弓钻杆钻进
spring pole rig 手动冲击钻架
spring preloading device 弹簧预紧装置
spring pressure 弹簧压力
spring pressure ga(u)ge 弹簧压力表;弹簧(管)压力计
spring pressure mill 弹簧压力磨
spring pressure plate 弹簧压板
spring pressure rod 弹簧加压杆
spring pressure sounding apparatus 弹簧压力探测仪
spring protecting sleeve 弹簧护套
spring pruning 春剪
spring puncture needle 弹簧采血针
spring puttee 弹簧钢板套
spring quadrant 弹簧舵柄弧
spring rail frog 弹簧辙叉【铁】
spring range 弹簧的限程;大潮潮差;平均大潮差
spring of tide 大潮潮差
spring rate 弹簧力;弹簧等级;弹簧应变率;弹簧刚性系数;弹簧刚度;弹簧常数;弹簧比率;大潮流速
spring reaction 弹簧反作用
spring reamer 扩张铰刀;弹簧铰刀
spring rebound clip 弹簧连接夹;弹簧箍圈
spring-rebound plate 弹簧回弹片
spring receiver 弹簧接受器
spring reducing valve 弹簧减压阀安全阀,弹簧减压阀
spring reed 弹性箸
spring reed release motion 弹簧游筘装置
spring regime(n) analysis 泉水动态分析
spring regime(n) analysis method 泉水动态分析法
spring regulator 弹簧调速器
spring reinforced eye 加强簧眼;弹簧加强卷耳
spring release 弹簧式脱开装置;弹簧安全器
spring release hitch 弹簧松脱式联结装置
spring residue 泉渣
spring resonance 弹簧共振
spring-restrained accelerometer 弹簧限制式加速度计
spring retainer 弹簧座圈;弹簧限位器
spring retainer lock 弹簧底座销
spring retaining in position 弹簧座圈
spring retaining plate 弹簧限动板
spring return 弹簧回程;弹力回程
spring-return button 弹簧自动还原按钮
spring-return controller 弹簧复位控制器
spring-return directional valve 弹簧复位的换向阀
spring-return piston 弹簧回程活塞
spring-return rule 弹簧卷尺
spring-return switch 弹簧复位开关
spring reverse motion 弹簧复位装置
spring ride model 弹簧滑块模型
spring rigging 弹簧装置;弹簧配重机构
spring ring 活塞环;铁丝弹簧圈;弹簧圈
spring ring coupling 簧圈联轴器
spring ring dowel 弹性环销
spring rise 大潮升;大潮潮升
spring rod 弹簧杆;弹簧导杆
spring rod bushing 弹簧杆衬套
spring roll balance 轧辊的弹簧式平衡
spring roll crusher 弹簧辊式破碎机
spring roller (轴承的)弹簧滚柱;弹簧辊
spring roll mill 弹力辊磨机;弹簧辊磨机
spring roll(s) mill 弹簧辊碎机
spring rubber bushing 弹簧橡皮衬套
spring run 泉溪
spring runoff 泉径流;春季径流量;春季径流
spring saddle 弹簧鞍座
spring safety hitch 弹簧安全联结装置
spring safety valve 弹簧安全阀
spring sag 弹簧弯度;弹簧垂度
springs and compensating beam 弹簧平衡杆
spring-sapping 泉源侵蚀
spring sash 弹簧窗扇
spring scale 弹簧压力标;弹簧秤
spring scale with stabilized pan 盘式弹簧秤

spring screen 弹簧筛
spring screw 弹簧螺钉
spring season 大潮期;春汛
spring seat angle 弹簧装置角
spring seat bearing 弹簧座轴承
spring seat bearing adjusting nut 弹簧座轴承调整螺母
spring seat bearing cap 弹簧座轴承盖
spring seat center line 弹簧座中心线
spring seat(ing) 弹簧座
spring seating collar 弹簧座挡圈
spring seat lock pin 弹簧座锁销
spring sediment 泉水沉积物
spring seepage 渗流泉
spring seismometer 弹簧(式)地震仪;弹簧地震计
spring separator 弹簧隔离件;弹簧钢板隔片
spring set 片组簧;弹簧组
spring shackle 弹簧挂钩;弹簧钩环;弹簧吊架
spring shackle bar 弹簧钩杆
spring shackle bolt lock 弹簧吊钩保险
spring shackle bracket 弹簧钩托架
spring shackle bushing 弹簧挂钩衬套
spring shackle pin 弹簧钩环销
spring shaft 弹簧轴
spring shank 弹簧铲柄
spring sheet-holder 弹簧定位销
spring sheller 弹压式脱粒机
spring shield 弹簧罩
spring shim 弹簧垫片
spring shock(absorber) strut 弹簧减振柱
spring shock absorber 弹簧减振器
spring shoe 弹簧支架
spring side ga(u)ge cam 弹簧式测边定位楔块
spring single coil 单圈弹簧
spring site 泉点
spring sludge 蜂窝冰;蜂巢冰
spring smithy 弹簧车间
spring snap ring 弹簧开口环
spring snib 弹性(门)窗钩;弹性(门窗)插销
spring snow 春雪
spring snowmelt 春融;春季融雪
spring snubber 弹簧减振器
spring socket 弹簧插座
spring softening system 软弹簧系统;弹簧软系统
spring solid eye 实心簧眼
spring space 弹簧定位器
spring spacer bar 弹簧撑杆
spring spike 弹性道钉
spring spindle 弹簧顶杆
spring spline 弹簧花键
spring split chuck 弹簧卡盘
spring-spoked steering wheel 弹性辐条转向盘
spring-spoked wheel 弹簧辐条轮
spring-spoke steering wheel 弹性辐操向轮
spring spool 弹簧卷筒
spring spreader 弹簧扩张器
spring squeak 弹簧尖叫声
spring starter 弹簧起动器(发动机)
spring starting mechanism 弹簧起动机构
spring stay 桅顶悬索;桅顶横牵索;弹簧撑杆;保险拉索
spring steel 弹簧钢
spring-steel band 弹簧钢带
spring-steel floor member 弹性钢底梁
spring-steel holder 弹簧夹钎器;弹簧持钎器
spring-steel roof member 弹性钢顶梁
spring-steel shutter 弹簧钢卷帘门;弹簧钢活门
spring-steel wire 弹簧钢丝
spring-steel wire screen 弹簧钢丝筛
spring steering wheel 弹性转向盘
spring stiffness 弹簧刚度
spring stock ga(u)ge 弹簧式坯料定位装置
spring stock guide 弹簧式导料板
spring stone 起拱石
spring stop 弹簧限动器;弹簧门闩;弹簧挡板;弹簧插销
spring stop block 弹簧阻车器
spring stop-nut locking fastener 弹簧防松螺母锁紧件
spring strength 弹簧强度
spring strip (门窗的)金属挡风雨条
spring stripper 弹簧卸料板
spring-strip suspension 弹簧片悬挂
spring strut (独立悬挂的)弹簧支柱

spring stud 弹簧柱螺栓
spring-style 草图设计
spring support 簧架；弹性座；弹性支架；弹簧支座；弹簧支架；弹簧支承
spring-supported 弹簧支承的；弹簧支撑的
spring-supported mount 弹簧架
spring-supported screen 弹簧减振筛
spring-supported screen box 弹簧支架筛箱
spring-supported screw 弹簧支承的螺旋
spring-supported vibrating conveyer 弹簧支承的振动输送器
spring surge 弹簧颤动
spring surge frequency 弹簧颤动频率
spring suspended 弹簧悬吊
spring-suspended movable stripper 弹簧式可动卸料板
spring suspension 无弹簧的悬吊；无弹簧的悬挂；弹性悬挂(吊架)；弹簧悬置；弹簧托座
spring suspension damper 弹簧悬挂装置
spring suspension device 弹簧悬挂装置
spring suspension eyebolt 弹簧悬挂有眼螺栓
spring suspension link-pin 弹簧悬挂连接销
spring suspension rod 弹簧悬挂杆
spring swage 弹簧陷型模
spring swench 弹簧冲击扳手
spring switch 弹簧转辙器；弹簧开关；弹簧道岔
spring swivel (冲击钻的)弹簧式回转盘
spring system 泉水系统
spring tab 弹簧调整片
spring tab assembly 弹簧调整片
spring tail 弹尾
spring take-up 弹簧张紧装置
spring tape measure 弹簧卷尺
spring temper 弹性回火；弹簧回火
spring tension 弹簧张力
spring tension adjusting screw 弹簧压力调节螺钉
spring-tensioned oil seal 弹簧自紧油封
spring tensioner 弹簧的调压螺钉；弹簧补偿器
spring tension ga(u)ge 弹簧压力计
spring tensioning 弹簧张紧器
spring tension pawl 弹簧爪
spring tension tester 弹簧弹力试验器
spring tester 弹簧试验仪；弹簧试验器；弹簧检验
spring testing machine 弹簧试验机
spring thaw 春天溶雪
spring thrust 弹簧推力
spring thrust bearing (水轮发电机的)弹簧(支撑式)推力轴承；弹簧推力轴承
spring tidal current 大潮(潮)流
spring(tidal)range 大潮潮差
spring tidal sequence 大潮层序
spring tide 子午潮；望朔潮；朔望潮；大潮；春潮；潮汛
spring tide mark 大潮高水位标记
spring tide time 春潮时间；大潮期
spring tie bolt 弹簧系紧螺栓
spring tiller 弹性舵扇；弹簧舵柄弧
spring timber 早材；簧材
spring time 春天
spring-time harrow 春耙
spring-tine cultivator 弹簧齿中耕机
spring-tine harrow (附于平地机后端的)弹(簧)齿耙
spring-tine shank 弹簧柄
spring-tine weeder 弹簧齿除草耙
spring tip 弹簧端头
spring tire 弹性轮胎
spring to break 断电弹簧
spring tool 弹簧(车)刀；鹅颈刀
spring tool holder 弹簧刀夹
spring-tooth attachment 弹齿附加装置
spring-tooth cylinder 弹齿式滚筒
spring-tooth disk harrow 弹齿圆盘耙
spring-tooth drag (附于平地机后端的)弹齿耙
spring-toothed harrow 弹齿耙
spring-toothed weeder 簧齿除草耙
spring-tooth flexing standard 弹齿挠性铲柄
spring-tooth harrow 簧齿耙；弹齿耙；弹齿耙路机
spring-tooth harrow plough 弹齿耙犁
spring-tooth harrow section 弹齿耙组
spring-tooth rake (附于平地机后端的)弹齿耙
spring to pull off 反拉簧
spring track 弹簧履带
spring travel 弹簧行程
springtree 梯段支架

spring trip 弹簧避障器
spring-trip beam 带弹簧安全器的犁辕
spring-trip hitch 弹簧销拖挂装置
spring-trip shank 带弹回安全器的铲柄
spring-trip shovel 带弹回安全器的锄铲
spring-trip standard 带弹回安全器的铲柄
spring tube bender 弹簧弯管器
spring-tube manometer 弹簧管式压力计；弹性管式压力计
spring tup 弹簧锻模
spring turnover 春季翻转
spring type clutch 弹簧压紧式离合器；常接合式离合器
spring-type forceps 弹簧镊子
spring-type governor 弹簧式调速器
spring-type relay 弹力(式)继电器
spring-type retarder 弹簧式缓行器
spring-type self-starter 弹簧自动起动器
spring-type welding 弹力式焊接
spring u-bolt 弹簧骑马架
spring valve 弹簧开关；弹簧阀门；弹簧活门
spring velocity 大潮流速
spring vent 泉口
spring wall 墙撑；扶垛
spring washer 弹性垫圈；弹簧垫圈；弹簧垫片；簧床垫
spring water 泉水
spring water regime(n) 泉水动态
spring water sample 泉水样
spring water seal 弹簧止水条；弹簧防水圈
spring water stopper 弹簧止水夹
spring wedge plate 弹簧楔片
spring wedge valve 弹簧楔形闸板阀
spring weight 弹簧(承受的)重量
spring-weighted 弹簧加压
spring welding 弹力焊
spring well 扩泉井
spring wheel 弹簧悬架车轮
spring winder 卷簧器
spring wind tape 弹簧卷尺
spring wing rail 弹簧翼轨
spring-wing toggle bolt 簧圈系墙螺栓
spring wire 弹簧钢丝
spring wood 春材；早材；春生木
spring wound clock 发条钟
springy 似弹簧
springy land 湿润土地
springy place 软湿地
sprinkle 喷雾洒水，喷洒，喷撒；微阵雨，洒
sprinkle aggregate 撒布用集料；撒布用骨料
sprinkled plot 人工降雨试验小区
sprinkle fire extinguisher 自动喷水灭火器
sprinkle material 撒布用材料
sprinkle mopping 平屋面涂沥青；平屋顶涂沥青
sprinkler 自动喷水灭火器；人工降雨器；喷水设备；密封垫；喷水器；喷洒水器；喷洒器喷淋头；喷壶；喷灌机；洒水设备；洒水器；洒水机
sprinkler alarm 喷水灭火设备报警器
sprinkler apparatus 喷雾器
sprinkler area 喷洒面积
sprinkler arm 喷洒器支管
sprinkler basin 配水箱
sprinkler box 喷洒机控制箱
sprinkler branch 喷洒器支管
sprinkler chamber 喷水室
sprinkler control 喷灌装置调节器
sprinkler control valve 花洒控制阀；喷淋控制阀
sprinkler coupler 喷灌装置管接头
sprinkler coupling 喷灌装置管接头
sprinkler coverage 喷灌覆盖面
sprinkler curing 喷水养护
sprinkler distribution pipe 喷洒器配水管
sprinklered 自动喷水灭火区；自动灭火区；有自动喷水灭火装置的
sprinkler filter 喷洒滤池；滴滤池
sprinkler filtration 滴滤
sprinkler fire extinguishing system 喷洒灭火系统；洒水灭火装置
sprinkler gun 喷灌头；喷灌枪
sprinkler head 灭火洒水喷头；喷头，喷水头，喷头；洒水(喷)头；撒水喷头；滴水头
sprinkler head system 洒水喷头装置
sprinkler intensity 喷洒器强度
sprinkler irrigation 人工降雨灌溉；喷水灌溉；喷灌

sprinkle riser 喷淋立管
sprinkler jet 喷灌装置的喷管
sprinkler lateral 喷洒器横向侧管；喷洒侧向；喷灌支管
sprinkler lorry 喷洒车
sprinkler main 灭火喷水管；喷水干管(灭火)
sprinkler method 喷洒法
sprinkler nozzle 喷洒嘴；喷洒器嘴；喷洒机喷嘴；喷嘴
sprinkler oscillating mechanism 喷灌装置摆动机构
sprinkler pipe 喷水管
sprinkler pipe system 人工降雨喷管装置；喷管系统
sprinkler plant 喷灌装置
sprinkler plate 喷射板
sprinkler pressure 喷洒(器)压力
sprinkler pump 洒水泵
sprinklerrig 人工降雨装置
sprinkler station 喷洒器站
sprinkler strength 喷洒器强度
sprinkler-sweeper collector 洒水扫集机
sprinkler system 自动喷水消防系统；自动消防系统；喷水系统；喷水灭火装置；喷水灭火系统；洒水灭火系统；喷淋装置；喷淋系统；喷灌系统；洒水装置；洒水灭火系统
sprinkler system-dry pipe 干式消防喷水系统
sprinkler system of lifeboat 救生艇洒水系统
sprinkler system-wet pipe 湿式消防喷水系统
sprinkler tank 喷水灭火水箱；洒水灭火器；撒水灭火箱
sprinkler truck 喷洒车；喷洒车
sprinkler valve 灭火喷头阀门；喷淋器阀
sprinkler wagon 洒水车
sprinkle treatment 喷洒处理
sprinkle treatment of pavement 路面撒集料处治法；路面撒骨料处治法
sprinkle with water 喷水
sprinkling 喷釉；喷水壶；喷洒，喷淋，喷灌，洒水，撒布
sprinkling bar 泼油管；喷水管；洒水管
sprinkling basin 喷水池
sprinkling bed 喷洒花圃；喷洒滴滤床；喷灌场
sprinkling can 喷洒润滑器；喷壶
sprinkling car 洒水车
sprinkling cart 手推洒水车；二轮洒水车
sprinkling chamber 喷雾室；喷淋室
sprinkling device 溅洒设备；喷洒设备
sprinkling fan 喷雾风扇
sprinkling filter 喷水滤器；喷洒滤池，喷滤池；洒滴滤池
sprinkling for frost protection 防冻喷洒
sprinkling hydrant 喷水消防栓
sprinkling installation 喷水装置
sprinkling intensity 喷洒强度
sprinkling irrigation 喷水灌溉；喷灌
sprinkling lawns 喷灌草地
sprinkling liquid chemicals 撒化学药液
sprinkling machine 人工降雨机；喷水器；喷洒装置；喷灌机
sprinkling machine sparger 喷淋器
sprinkling of the wheels (压路机的)轮子洒水
sprinkling outlet 喷水口
sprinkling pipe 洒水管
sprinkling pool 喷水池
sprinkling strength 喷洒强度
sprinkling system 喷淋系统；喷水系统；喷射卸货系统；喷洒系统
sprinkling tank 喷水灭火器
sprinkling truck 洒水车
sprinkling valve package 喷水阀组合
sprinkling wagon 洒水车
sprinkling zone 喷水区域
sprint car 短程泥路赛车
Sprite(signal processing in the element)detector 扫积型探测器
sprit(sail) 斜缸帆
sprocke hole 定位孔
sprocket 星轮；檐椽支撑木；链齿；挑椽接木；带齿卷盘；持链轮；飞檐椽；输片轮
sprocket backlash 带齿卷盘间隙；链轮齿间隙；链轮间隙
sprocket bearing 带齿卷盘轴承；链轮齿轴承
sprocket bit 中导位；空位符号；定位
sprocket chain 链轮上的链条；链轮环链；扣齿链；扁环节链；扁链接链
sprocket channel 链轮道；定位道

sprocket crank 链轮曲柄
sprocket cutter 星形齿轮铣刀；链轮铣刀
sprocket drive 链轮传动；链传动
sprocket drive ratio 链轮传动比
sprocket dynamometer 链轮式拉力表
sprocket(ed)eaves 接椽挑檐
sprocket feed 链轮送纸；链轮传动
sprocket gear 链轮；链传动装置；扣链齿轮；扣齿链轮；输片齿轮
sprocket hob 链轮滚刀
sprocket hole 定位孔；中导孔；扣齿孔；输送孔
sprocket-hole effect 片孔效应
sprocket hub 带齿卷盘衬套；链轮衬套
sprocket piece 短人字木；钉在椽头顶的楔形木块；小椽；飞檐椽
sprocket pulse 中导脉冲；轮齿脉冲；计时脉冲；时钟脉冲；定位脉冲
sprocket pulse generator 时钟脉冲发生器
sprocket rim 带齿卷盘齿圈；链轮齿圈；带齿卷盘边缘；链轮边缘
sprocket shaft 链轮轴
sprocket shaft bearing 链轮轴轴承
sprocket shift claw 链轮换挡爪
sprocket shift claw bushing 链轮换挡爪衬套
sprocket shift claw press ring 链轮换挡爪压圈
sprocket shift fork 链轮换挡叉
sprocket tooth 链轮齿
sprocket track 定位道
sprocket wheel 链轮；锚机持链轮；输片惰性轮；链轮
sprocket winch 带齿卷盘卷扬机；链轮卷扬机；带齿卷盘绞盘；链轮绞盘；链轮绞车
sprout 小支流
sprout fountain (庭园中的)涌泉
sprouting 毛圈突出；发芽
sprout land 萌芽林
Sprowl 史普劳尔(一种脚手架牌号)
spruce 云杉；针枞
spruce fir 红杉
spruce pine 云杉屋面盖板；枞松；山地松
sprue 铸型注入口；铸口；注料口；直浇道；流道；浇铸道；浇口；熔碴；堵块
sprue base 直浇口压痕；直浇口窝；直浇道窝
sprue bush 浇道本
sprue bushing 浇口衬瓦
sprue button 水口钮
sprue cup 漏斗形外浇道；漏斗形浇口；浇铸漏斗；浇口杯
sprue cutter 漏差别型浇口杯；浇口切断机；加工浇口和冒口的铣刀
sprue gate 浇注口
sprue hole 直浇口
sprue mill 浇冒口清理滚筒
sprue pin 直浇口模棒；直浇道模型
sprue plug 浇口塞棒
sprue puller 直浇口拉出器
sprue recess 钳口凹人部分
sprue-runner-gate area ratio 直浇道横浇道内浇道面积之比
sprue wire 铸道针
spruing 打浇口
spruing station 打浇冒口区
sprung 边弯材；翘曲(木材变形)；起拱
sprung arch 拱式炉顶；炉拱；悬拱；弓形拱；工业炉落地拱；起拱点；已砌的拱
sprung axle 簧上轴
sprung borehole 爆炸扩孔
sprung borehole method 爆炸扩孔法
sprung floor 翘曲木地板
sprung floor cover(ing) 悬空地板；弹簧地板
sprung hole 葫芦炮眼
sprung-hub wheel 悬架车轮
sprung mass 悬挂质量；弹簧上部质量
sprung mo(u)lding 内角削斜的线脚；墙顶角饰；弯曲(形)线脚
sprung parts 悬挂部件
sprung planking 弯曲木甲板
sprung roof 悬拱；砌拱屋顶
sprung weight 簧上重量；簧上重力
SPT analyzer data signal 标准贯人试验分析仪数据信号
SPT blow count 标准贯人击数
SPT N-value 标准贯人击数 N 值
spud 支杆；栽植铲；固船柱；钢桩【疏】；井下测站

钉；定位桩【疏】；定位钢桩【疏】；铲削工具；草铲；剥皮铲
spud bit 尖凿；矛式钻头
spud cable 定位桩索；定位桩缆；系船桩索；系船桩缆
spud carriage 钢桩台车【疏】；定位桩台车
spud carriage slideway 定位桩台车滑槽
spud carriage travel(l)ing wheel 定位桩台车行走轮
spud carriage well 钢桩台车挡【疏】
spud carrier well 钢桩台车挡【疏】
spud collar 钢桩抱箍【疏】
spud control wire 钢桩吊缆【疏】
spudded-in hole 钻至基岩的钻孔
spudder 钢丝绳冲击(式)钻机；撤钻
spudding 打定位桩；铲除
spudding bit 亚层土钻进钻头；开眼顿钻钻头
spudding date 开钻日期；开孔日期
spudding drill 冲击式钻具；油井钻探船
spudding drum 旋冲钻的卷扬机
spudding in 钻入
spudding mud 开钻泥浆
spudding pulley 冲击滑轮
spudding ring 张紧环
spudding shoe 钻杆接头；顿鞋；(钢绳冲击钻机的)扯绳套环
spudding through soil 土层钻探
spudding tools 钢绳冲击钻具
spudding up and down 上下活动钻具
spud dredge(r) 定位桩挖泥船
spud drill 钢绳冲击(式)钻机
spud drilling 钢绳冲击钻进
spud fishing 捞起断钢丝和工具(冲击钻进)
spud frame 系船柱架；定位桩架
spud gantry 钢桩架【疏】；钢桩吊架【疏】；定位桩吊架
spud hoist 钢桩起升装置【疏】
spud-hoisting sheaves 定位桩升降滑轮
spud hoist ram cylinder 钢桩顶升液压缸
spud hydraulic ram 钢桩液压顶【疏】
spud in 开始钻进(冲击钻)
spud keeper (挖泥船的)定位桩固定装置
spud lead 导海柱
spud lever 定位桩操作杆
spud line 系船柱索；系船柱线；定位桩线
spud machinery 插入式机械
spud moving delay factor 移定桩延迟系数【疏】
spud pile 定位桩【疏】
spud pole 钢桩【疏】
spud pole keeper 钢桩抱箍【疏】
spud rod 插杆
spud rope 定位桩缆
spud sheave 钢桩滑轮
spud spear 钢(丝)绳打捞钩
spud-tilting installation 定位桩倾放装置
spud-tilting seat 定位桩倾放基座
spud-type suction dredge(r) 系船桩式挖泥船；系船桩式吸泥船；柱固式吸扬挖泥船
spud vibrator 插入式振捣器；插入式振动器；插入式振动棒；振动钻；振动棒
spud vibrator for concrete 插入式混凝土振动器
spud well 挖泥船上的套筒；锚井
spud wire 钢桩吊缆【疏】
spud wrench 长柄扳手
spume 泡沫；起泡沫；浮沫
spumescence 泡沫状的
spun acetate 醋酯短纤维纱
spun-barrel reflector 桶形反射器
spun bearing 离心浇铸轴承
spun bitumen-line pipe 旋制沥青衬里管
spun-blended yarn 短纤维混纺纱
spun-bonded fibre 纺黏纤维
spun-bonded process 黏结法
spun casting 离心旋转法铸造；离心式浇注；离心浇铸；离心浇注
spun casting concrete 旋转浇筑混凝土
spun cast-iron pipe 离心铸铁管
spun cast-iron pressure pipe 离心铸铁压力管
spun cast pipe 离心铸管
spun cement-line pipe 旋制水泥衬里管
spun column 离心柱；离心法制成的立柱
spun concrete 旋制混凝土；离心浇注混凝土；离心法制混凝土；离心成型混凝土
spun concrete conduit 旋制混凝土管
spun concrete drain pipe 离心混凝土排水管

spun concrete mast 离心混凝土(电)杆
spun concrete pile 离心(浇制)混凝土桩
spun concrete pipe 离心浇制混凝土管；旋制混凝土管；离心旋转制造混凝土管；离心浇注混凝土管
spun concrete pipe culvert 旋制混凝土涵管
spun concrete pipe laying plant 旋制混凝土管放置车间；旋制混凝土管铺设车间
spun concrete pole 离心混凝土电杆
spun concrete pressure pipe 离心混凝土压力管
spun concrete process 旋制混凝土法
spun concrete tube 旋制混凝土管
spun cotton 棉纱
spund 履带瓦；履带链板
spun glass 玻璃纤维；玻璃丝
spun glass wall cover(ing) 玻璃纤维墙布
spun glass wall lining 玻璃纤维墙衬
spun glass wall(sur)facing 玻璃纤维墙面
spun-in casting 离心铸造法
spun-iron pipe 离心铸铁管
spun iron pumping main 旋制铁泵泵送总管
spun lead 细铅丝；导向栓
spun lighting column mo(u)ld 旋制电杆模子；旋制灯柱模子
spun line machine 旋转衬里机
spun lining 旋制内衬
spun mast 离心(电)杆
spun pile mo(u)ld 旋制桩模
spun pipe 旋(转)制管；离心制管
spun rayon 人造棉纱
spun-refined iron 离心精炼铁
spun reinforced concrete pressure pipe 离心钢筋混凝土压力管
spun rivet 旋制铆钉
spun rivet connection 旋压铆钉连接
spun roving 定长纤维无捻粗纱
spun tussah cloque 凹凸绉
spunware 旋制制品；薄板圆形模具；旋转制品
spun work 压制工件
spun yard 短纤维
spun yarn 细油麻绳；精纺纱；细纱；细索油麻绳；定长玻璃纤维
spun yarn packing 麻纱软垫；麻丝填料
spur 助抓突角；支撑物；骨刺；陆架山脊；径迹；排出口；排出孔；石嘴；山嘴；垛壁；短木支撑；丁堤；岔坝
spur and helical hobbing machine 正齿轮和斜齿轮滚齿机
spur band 分路频带
spur beam 压墙木；椽子垫木
spur bevel gear 正齿伞齿轮
spur bridge 临时桥
spur canal 运河支线；支渠
spur dam 丁坝
spur differential 正齿轮分速器
spur dike 丁字坝；翼坝；挑水坝；挑流坝；丁坝
spur dike for bank protection 护岸丁坝
spur driving pinion 主动小正齿轮
spur duct 截流沟；截流管
spur friction wheel 筒形摩擦轮
spur gear 直齿圆柱齿轮；正齿轮；直齿轮
spur-gear blank 正齿轮坯料
spur-gear cutter 正齿轮铣刀
spur-gear cutting 正齿轮切削
spur-gear differential 直齿圆柱齿轮差速器
spur-gear drive 正齿轮驱动；正齿轮传动
spur-geared 正齿轮传动的
spur-geared chain block 差动链轮
spur-geared machine 正齿轮传动机械
spur-geared winch 正齿轮绞车
spur-gearing 正齿轮对；正齿轮传动装置
spur-gear lubricant 正齿轮润滑剂
spur-gear lubrication 正齿轮润滑
spur-gear operated valve 正齿轮操纵的阀门
spur-gear pulley 正齿轮滑车
spur-gear pump 正齿轮泵；直齿齿轮泵
spur-gear set 正齿轮传动装置；正齿轮组
spur-gear speed reducer 正齿轮减速机
spur-gear wheel 正齿齿轮
spurge laurel 桂叶芫花
spurging 产生泡沫
spur guard 保险销
spur guide 直导轨
spurious 干扰的
spurious aM rejection 寄生调幅抑制
spurious band 虚假谱带

spurious capacitance 寄生电容
spurious charge 寄生电荷
spurious coincidences 假符合
spurious correlation 伪相关;不纯相关;比例相关
spurious count 乱真计数;假计数
spurious coupling 杂散耦合;寄生耦合
spurious disk 虚圆面;盘状假星象
spurious distortion 假畸变
spurious emission 杂散发射
spurious frequency 寄生频率
spurious image 乱真影像;寄生图像;散杂影像
spurious impedance 寄生阻抗
spurious level 寄生电平
spurious lobe 假波瓣
spurious mode 寄生模式
spurious noise 寄生噪声
spurious oscillation 乱真振荡;寄生振荡
spurious output 乱真输出
spurious pattern 寄生图像
spurious peak 假缝;伪峰
spurious phase modulation 寄生调相
spurious power meter 乱真信号功率测量仪
spurious pulse 寄生脉冲
spurious radiation 杂散辐射;寄生辐射;伪辐射
spurious resolution 乱真清晰度
spurious resonance 寄生振荡
spurious response 信号噪声特性
spurious revenues 虚假收入
spurious signal 虚假信号;乱真信号;寄生信号
spurium 寄生束
spur jetty 支突堤;挑流坝;丁坝
spur level(l)ing line 支水准路线
spur line 支线
spur-line filter 支线滤波器
spur line of level(l)ing 水准支线
spur mark 钉脚痕
spurn water 舷侧导水板;(放在船头的)防浪板
spur of matrix 矩阵的迹
spur pile(for dredging) 支撑桩,水撑桩;斜桩
spur pinion 小正齿波;柱形直齿轮;正小齿轮
spur pipeline 管道支线
spur planetary gearing 行星齿轮传动装置
spur post 斜柱
spur protection 护岸丁坝
spur pruning 短枝修剪
spur rack 直齿齿条;正齿齿条;齿筒
spur revetment 护堤丁坝
spur ring 大齿圈
spurrite 硅方解石;灰硅钙石
spur road 岔路
spur shaper cutter 直齿插齿刀
spur shore 码头撑杆;边撑柱(船坞);边撑;斜撑柱
spur stone 隅石;护角石
spur switch 单开道岔【铁】
spurt 溅射;溅(散);价格突然高涨;喷射
spur teeth 正齿
spur tenon 短凸榫(不出头的暗榫)
spur terrace 短阶地
spurting 溅散;喷射状
spurt in prices 价格猛涨
spurt loss volume 初滤失量
spur track 尽头路;死т胡同;岔线轨道;岔线【铁】
spur-training dike 丁顺坝
spur traverse 支导线【测】
spurt value 瞬时值
spur type 直齿式
spur-type planetary gear 正齿式行星齿轮
spur-type planetary gearing 正齿式行星齿轮装置
spur-type planetary reducer 正齿式行星减速齿轮
spur valley 短屋顶分支排水沟;短支谷
spur vibrator 振动棒
spur wall 齿墙
spur wheel 正齿轮
spur wheel back gear 正齿背齿轮
spur wheel countershaft 正齿轮副轴
spur wheel outrigger gear 正齿轮外运传动装置
spur wheel reversing gear 正齿轮回动装置
sputnik 人造卫星;人造地球卫星
Sputnik program(me) 卫星计划
sputter 溅蚀;溅射(机);溅镀
sputter coating 溅射镀膜;溅射涂膜
sputtered beam 溅射束
sputtered dielectric(al) film 溅射介质薄膜
sputtered film 溅射薄膜;喷镀薄膜

sputtered layer 溅射膜层
sputter etching 溅射蚀刻
sputtering 喷涂;真空镀膜;溅射;阴极真空喷镀;溅散;喷溅涂覆法;溅溅;飞溅
sputtering chamber 溅射室
sputtering equipment 溅涂装置
sputtering hollow cathode discharge 溅射型空心阴极放电
sputtering method 溅射法
sputtering source 溅射源
sputtering system 溅射系统
sputtering technology 溅射工艺
sputtering yield 溅射系数;溅射率;溅镀系数;溅镀率
sputter ion pump 溅射离子泵
sputter pump 溅射泵
spy boat 间谍船
spyglass 小型望远镜;小望远镜
spyhole 窥测孔;窥视孔;观火孔;窥孔;探视孔;观察孔;观测孔
spy-in-the-sky 侦察卫星
spy mirror 监视镜;单向透明玻璃镜
Spyndle 司品ং
spy on 窥视
spy satellite 侦察卫星
squabbing pass 梯形立轧孔型
squabbing roll 阶梯轧辊
squad 小组;班组
squadron 小舰队;一组;中队
squad room 集合厅
squalane 鲨烯;低凝点高级润滑油
squalid 不洁的
squall 狂风;暴风;飑
squall cloud 飑云
squall line 飑线
squall weather 恶劣天气
squally weather 暴风雨气候
squama 鳞屑[复]squamae
squamatization 鳞状化
squame 鳞屑
squamiform cast 鳞形铸型
squamiform mark 鳞形痕
squamosal bone 鳞状骨
squamose structure 鳞片状结构
squamous 鳞状的
squamous suture 鳞缝
squamula 翅基片
squander 浪费
squander fund 滥用经费
square 方形平面;直角尺;正方(形);广场;矩形的;矩尺;锚杆眼;曲尺;平方;四方,方形(广场);方块;二次幂
square account 结账;清算;结清(账目);清理账目
square and flat panel 方形平镶板;方平镶板;方块平镶板
square and rabbet 凸方和凹线圆环线脚
square aperture 方形孔口
square-aperture woven mesh 方孔编制筛网
square-area sprinkler 方面积喷灌器
square array 方阵列
square ashlar 方石板;方石块
square ashlar bond 方石块砌合
square ashlar dressing 方石板饰面;方石块饰面
square ashlar masonry work 方石板砌筑工程;方石块砌筑工程
square ashlar slab 方琢石板
square ashlar structure 方琢石结构
square ashlar vault 方琢石穹窿
square auger 方孔钻
square bands machine 带条剪刀机
square bar 方铁条;方形钢;方筛条;方钢筋;方杆,方棒料
square bar iron 方铁条
square bar spiral spring 方条盘簧
square bar steel 方钢(条)
square base 方底;方基础
square basket 方格形
square basket weave parquetry flooring 正方席纹地板
square basket weave pattern 正方席纹
square bay 方形海湾;方桥跨
square bearing 正接触
square bellmouth transition 方喇叭形渐变段
square bellows 正方形伸缩风箱
square bend 直角弯头

square-bend screw hook 直角螺丝钩;直角弯头螺钉
square bent 直角弯头
square billet 直角错齿饰;方坯料;方坯;方钢短条饰
square bit 十字形钻头;十字钻头
square blade screwdriver 方头螺丝刀
square bloom 方钢坯
square body 船中平行体
square bolt 方螺栓;方杆插销
square-bore ball bearing 方轴孔滚珠轴承
square box socket wrench 方套筒扳手
square box spanner 方套筒扳手
square box wrench 方套筒扳手
square bracket 方括弧;方括号;方格形
square brick 方砖
square bridge 正交桥;直交桥
square budding 方形芽接
square butt joint 平头对接(接头);平口对接;平对接
square butt weld 平头对接焊缝
square butt welding 平头对接焊
square callipers 游标卡尺
square cascade 方块级联
square ceiling 方格天花板
square cell 正方形隔仓(气压沉箱)
square center 方顶尖
square centimeter 平方厘米;平方公分
square chain 平方测链;平方链
square chamber 方形矿房(采矿法)
square channel tile 方槽形瓦管
square chart 方形图
square check irrigation 方格灌溉
square chest 方箱
square chimeny 方烟囱
square chisel 方錾;榫孔机;方凿
square cluster planting method 方形簇种法
square cogging 凸榫接合;齿榫接合
square coil 矩形线圈
square-coil nozzle heater 方形蛇管喷嘴加热器
square collar 方钻链
square column 方柱;方形截面柱
square compensator 方形补偿器
square contingency 平方列联
square control network for construction 施工方格控制
square corner 直角隅角;直角转角;内尖角;方角
square-corner(bath)tub 方形浴缸;方角浴盆
square-corner braun tube 矩形显像管;矩形电子束管
square cornered entrance 方角入口
square-corner halving 方角半开接合;内尖角嵌接
square cotter 方销
square coupling 方头联轴节
square course 矩形起落航线
square cross culvert 方形涵洞
square crossing 正(交)交叉;直角交叉;十字交叉;十字交叉道;垂直交叉
square crossing center line 十字形交叉中心线
square cross-section 方形截面;方形断面;方形断面
square cut 切四边;四边切方;裁方;四面锯切
square-cut glass 方形切割玻璃
square-cut plain tile 四方光面砖;四边切方平瓦
square cut stern 方形船尾
square cut stern ship 方形尾船
square cutting 直角切割;平头切割;垂直切割;横截
squared absolute value 平方模数
squared and parallel element 正方形平行构件;正方形平行组件
squared beam 方木梁
square deal 公平交易
square decimeter 平方公寸;平方分米
square degree 平方度;方角度
square delayed pulse 矩形延迟脉冲
square density of road 道路面积密度
squared error asymptotically efficient estimator 平方误差渐近有效估计量
square designation 方形图幅编号
square detector 方形探测器
square detector composition 方形检波器组合
square deviation 平方(偏)差;平方离差
squared glass plate 方格玻璃板
square diamond mesh 菱形筛眼
square die 方形板牙;方板牙
square digging blade 方头挖掘铲

squared map 方格网地图
squared modulation 方波调制
squared off cargo space 形状规则的舱室
squared off cascade 当量方块级联
square dome 凹圆穹顶;方穹顶;正方底穹隆
square door jamb 门框的方竖框;门框方竖框
square downpipe 方水落管
squared paper 方格纸
squared pitching 方格式护坡;方格式护面
square-dressed 方石块铺筑的
square dressed pavement 方块石(铺砌)路面
square drift 方形冲头
square drop shaft 正方形开口沉箱
squared rubble 方整石墙;毛方石;方块毛石;平毛石墙
squared rubble wall 方块毛石墙
squared splice 方木斜面叠接;互缺搭接
squared stone 琢方石;粒石;粗琢方石
squared stone masonry(work) 正方形石块墙;方石砌体;方石圬工;方块圬工
squared timber 方(正木)材;方木料;方木(材)
square edge 方边;直角边;直角边
square-edged 方棱的;四方形的;方边的
square-edged board 方边板;四边刨方板
square-edged broad crested weir 直角宽顶堰
square-edged flooring 四边琢方石铺地面
square-edged inlet 直角进水口
square-edged lumber 方板料
square-edge door 丁头与面成直角的门
square-edged orifice 方棱孔口;方孔;方缘孔口
square-edged outlet 直角出水口
square-edged timber 方直木料
square-edge floor boarding 铺方边地板
square-edge lumber 方正木材
square edge overpanel 直角边盖板
square-edge timber 方正木料;方正木材
square-edge wood 方正木料
square elbow 直角肘管;直角弯头;直角弯管接头;直角弯管
square end 垂直切头;方端
square-ended 方头的
square-ended shovel 方(头)铲
square ending 方头突出部;方头柄
square endlug 方头柄
square-end nicole prism 方头尼科尔棱镜
square endplate 方形端板
square end-trimmed lumber 截头方材
square engine 等行程发动机;等径程发动机;方形发动机
square error 平方误差;误差平方
square error consistency 平方误差相容性
square error pattern 方形差模型
square exploration grid 正方形勘探网
square eyed auger 方孔木钻
square face 横面
square face weld 无坡口焊缝
square field picket fence reticle 矩形栅状调制盘
square figure 方形
square figure knot 方花结
square file 方锉
square flange 方法兰
square flue 方形烟道
square-foot 平方英尺
square footage per drill hole 每个钻眼均分的工作面积(以平方英尺计)
square foot cost 每平方英尺造价
square footing 方形基脚;方形基础
square foot method 平方英尺比价法
square-foot unit of absorption 赛宾(吸声单位,相当于一平方英尺的全吸声面)
square foundation 方形基础
square frame 船平行中体处(或船的中部的)肋骨;方框
square-framed 方角料框架
square front lamp 方形灯
square ga(u)ge 方规
square grade 平方度
square grid 正方形栅板;正方形网格;正方格网;方网架;方格网
square grid method 方格网法
square-grid pattern 正方形网格排列
square grid sampling 方格网采样
square grit chamber 正方形沉沙池
square groove 平(头)坡口(焊体);无坡口;四方进刀架;方槽;不开坡口
square groove corner joint 平头角接头
square groove weld 无坡口槽焊
square grooving and tonguing 正方形舌槽(接合);正方形企口(接合)
square ground plan 方形底层平面图
square hall 方厅
square harrow 方形耙
square head 干砌围墙;门边梁;四方头;方头(的)
square head bolt 方头螺栓
square-headed 方拱门洞
square-headed loophole 方头孔眼
square-head(ed)screw 方头螺钉
square headed window 方头窗
square head flat point set screw 方头平尖紧定螺钉
square head nut 方头螺母
square head plug 方头旋塞
square head wrench 方扳手
square hewn 四面斧砍的
square hinge flap 方形板片;方形铰链;方形折叶
square hoe 方锄
square hole 方孔
square hole bucket 方孔吊斗
square hole cutting slotting tool 方孔插刀
square hole screen 方孔筛
square hole slotting tool 方孔插刀
square hooping 加方箍(指钢筋混凝土梁柱中的钢箍)
square hysteresis loop material 方形磁滞回线材料
square impact 正冲
square-inch 平方英寸
square indexable carbide milling cutter insert 方形可转位铣刀片
square individual base 正方形单独基底;正方形单独基础
square ingot 方钢锭
square-integrable 平方可积的
square-integrable function 平方可积函数
square iron 方铁;方钢
square iron bar 方铁条
square iron-core coil 矩形铁芯线圈;方形铁芯线圈
square jackscrew 方形起重螺杆
square jamb 方竖框;方角侧墙
square-jaw clutch 矩形凸爪离合器;方齿牙嵌式离合器
square joint 相对式接头【铁】;相对对接【铁】;平头接头;平头接合;平口接头;平接
square kelly 方钻杆
square key 方形销钉;方(形)键
square kilometer 平方千米
square kiosk 方亭
square knot 平结;方结
square land valve 方形台肩阀
square lattice 方点阵
square lattice structure 方格状构造
square-law 平方律
square-law condenser 平方标度电容器
square-law detection 平方律检波
square-law scale 平方标度
square level 方水平仪
square lever 直角形杠杆
square-link chain 方形节链
square lock nut 方头锁紧螺母
square log 大方木料;方原木
square loop 矩形回线
square loop ferrite 矩形回线铁氧体
squarely 成方形
square man 石工;木工
square map grid 地图格网
square map-subdivision 正方形分幅
square matrix 方阵;方矩阵
square matrix building 方阵建筑
square matrix code 方阵码
square measure 面积丈量;面积(单位)
square-measure scale 面积比(率);面积比例尺
square mesh 方筛孔;方网孔;正方格网;方网眼;方孔的
square mesh grating 方孔网格
square mesh screen 方网眼筛;方孔筛网
square mesh screening cloth 方格网筛布
square mesh sieve 方眼筛;方孔筛
square mesh tracking 方格跟踪法
square mesh vibrating screen 振动方孔筛
square mesh wire cloth 方孔金属丝网
square mesh wire net 方眼铁丝网
square meter 平方公尺;平方米
square method 方格法
square mil 平方密耳
square-mile 平方英里
square-mile foot 平方英里英尺;平方哩尺(水力学常数,等于一平方哩面积,一尺深的水量)
square millimeter 平方毫米
square moment 方形力矩
square mosaic 矩形镶嵌;方马赛克
square mo(u)lding 方形线脚
square mounting flange 方形座法兰
square-mouth 方口的
square-mouth navvy shovel 方口单斗挖掘机
square-mouth shovel 方口铲土机
square-mouth tongs 方口钳
square nail 方钉
square neck 方颈
square neck bolt 方头螺栓;方颈螺栓
squareness 正直;直角度;平直度;方正度
squareness factor 矩形系数
squareness ga(u)ge 垂直度测量器
squareness ratio 矩形比
square net 方形网络;方格网
square net emdossing 方形网式透镜栅
square-netting screen 方眼编织的滤网;方形编织筛网
square-nose bit 平面金刚石钻头;平底钻头
square-nosed pliers 方头钳;平口钳
square-nose shovel 方头锹
square-nose tool 方头刀具
square number 平方数
square nut 方形螺母;方螺母;方螺帽
square of crossing 交叉道口广场
square of glass 玻璃广场
square of sum of squares 平方和开方法
square of the hatch 舱口围空间
square oil stone 方油石
square-on reflector 直角初调器
square open caisson 正方形开口沉箱
square opening 方形(开)孔
square opening sieve 方眼筛;方孔筛
square opening single end box wrench 方口单头套筒扳手
square opening wire cloth 方孔金属丝布
square orifice 方形口
squareoval roughing passes 方形—椭圆粗轧孔型系
square pack 覆盖100平方英尺的板材单位
square panel 四方镶板;四方嵌板
square parachute 方形伞
square pass 方形孔型
square pattern 规定模式;方形布置;方格井网
square pavilion 方亭
square paving brick 铺路方砖
square pier 方(头)墩;方形墩
square pierhead 方形墩头;方形堤头
square pile 方(形)桩
square pillar vertical drilling machine 方柱立式钻床
square pin punch 方形尖冲头
square pipe 方管(材)
square pitch 45度屋面坡度;45度斜屋面
square pitch arrangement 正方形排列
square plate 方板;方片;方盘
square-plate washer 方垫圈
square plate with round corner 圆角方形衬板
square plug ga(u)ge 方孔塞规
square point shovel 方头铲
square port 方形舷窗
square position 均衡部位
square pot 方钵
square prism 四角棱柱体
square pulse 矩形脉冲
square punch 方孔冲头
square punching 冲孔
square pyramid 方金字塔;方锥体;正方棱锥
square pyramid space grid 四角锥网架
square quoin (房间的)方角落;方隅石;方角隅石;方突角
squarer 锯木方工人;矩形波形成器;木工;平方器;石工;乘方器
square rafter 方椽

square ram 方形撞杆;方形压头;方形夯锤
square random stone 随意整方的毛石
square ratio 矩形系数
square reactor 方形反应池
square reamer 方铰刀
square rebate 方裁口
square rebated joint 方裁口接头
square rebate plane 方槽刨
square recessed bath tub 置于凹处的方形浴盆
square recessed pointing 方形凹缝
square red clay floor tile 红土方阶砖
square reef knot 平结
square repeater 方形孔型用围盘
square reticle 矩形调制盘
square rib-vault 方肋穹隆
square rig 横帆式
square ring 方环;方形圈;方形密封环
square ring laser 方环形激光器
square ring spanner 方套筒扳手
square riser 方冒口
square rod 角钢;方棒
square roll 方卷饰;方形卷饰
square roller 等长径滚子
square roof 方脊屋顶(两个45度坡直角相交的屋顶);直角脊角屋顶;脊角为90度屋顶;45度斜屋面
square roof-light 方形屋顶天窗;方形屋顶灯
square root 平方根;二次根
square-root deviation 均方差
square root extractor 开方器
square root-floating 平方根—浮点运算
square-rooting algorithm 平方根算法
square root law 平方根(定)律
square root method 平方根法
square root of sum squares method 平方和的平方根法
square root of time fitting method 时间平方根拟合法
square root paper 平方根纸
square root planimeter 平方根面积仪
square-root potentiometer 平方根电位计
square-root rule 平方根法则
square root transformation 平方根变换
square roughing pass 方形粗轧孔型
square rubber gasket 方形橡胶垫片;矩形橡胶垫圈
square rubble 方块毛石
square rubble walling in courses 方块毛石分层砌墙
square rule 直角尺
square ruler 丁字尺
squares 封面伸出书芯的部分
square saddle junction piece 方鞍形连接件
square sail 横帆
square-sawn 锯方整的;四面锯切的
square-sawn lumber 方材;方木
square-sawn timber 锯方木;锯成正方木材;成材
square scraping bottom screen 方形刮底网
square screen opening 方形筛孔
square screw 方纹螺旋
square screw thread 方螺纹
square search 正方形式搜索;展开方形搜索
square section 方截面;方钢;方形截面
square section ring 方截面密封圈
square-section tunnel 四方形隧道
square set 方框支架
square set cap 方框支架横梁
square set flooring 方框支架楼梁
square set hammer 方形压印锤
square set sill 方框支架底梁
square shaft 方钻杆;方轴
square shank 方形柄
square shank bolt 方杆螺栓
square shank drill 方柄钻
square shank ratchet drill 方头钢钻;方柄棘轮钻;方柄扳钻
square-shaped array 矩形天线阵
square-shaped roof 方形屋顶
square shear box 方形剪切盒
square shears 龙门剪床
square shoot 方形木水落管;木水落管;方形(斜)槽
square shoulder 方(形)肩
square shoulder carpenter's pincers 方肩胡桃钳
square-shouldered punch 方形台肩冲头
square shouldered tool joint boxes 方切口锁接头

square signal 方形号志;方信号
square single base 正方形单独基础
square skip 方形翻斗车
square slab 正方形板
square slide 方滑板
square slot bush 带方槽的衬块
square socket 卡瓦提筒;方形打捞器;方形插口
square solid pipe die 整体方形管子板牙
square spade 方铲
square spanner 方头扳手
square spigot 方形头
square splice 方木斜面叠接(屋架拉杆);互缺搭接;方块拼接
square-spray irrigation system 方格式喷罐系统;方格喷浆系统
square spread 方形排列
square spring steel 方弹簧钢
squares sum of between group 段间(离差)方和频谱分析
squares sum of within group 段内(离差)平方和
square staff 护角线条;角缘线;角缘;墙角方线条;方桌线;扩角板条;靠尺板
square stations 站线;船体横剖线
square steel 方钢
square steel bar 方钢条;方钢筋
square stem 直(立)船首柱
square step 实心方石阶步
square sterned boat 方尾艇;方尾船
square sterned ship 方尾船
square stick indication rod 方棒表示杆
square stirrup 方形箍筋
square stock 方料;方钢
square stone 细方石;方琢石;方砖;方石
square stool 方凳
square stroke 等径冲程
square stroke engine 等行程发动机
square structural grid 方形结构格栅
square sum method of product matrix 乘积矩阵平方和法
square sum of residues 残差平方和
square table for eight people 八仙桌
square tank 正方形储[贮]罐;方形储[贮]罐
square tap 方形丝锥
square-tee joint 平头 T 形接合;平头丁字接头
square thread 平顶螺纹;方牙螺纹;方螺纹
square-threaded screw 方螺纹螺钉
square-thread form 方螺纹样
square thread tap 方螺丝攻
square-thread tool 方螺纹车刀
square tie 方枕木;方箍(筋)
square tile 方砖;方形瓷砖
square timber 方木
square tongs 方钳
square toothed plate 方形齿板
square-topped pulse 平顶脉冲
square to roof 北面采光的屋顶
square tower 方塔
square transom stern 方形船尾
square trench 方形沟槽;方基坑;方基槽
square trowel 方头墁刀
square tube 方管(材)
square-tube splicer 方管接头器
square tubing 方形管道
square tuck 方形船尾后端
square tuck stern 方形船尾
square-turned baluster 带有装饰线条的方形栏杆柱;四边线饰栏杆柱
square-turned newel 带有装饰线条的方形栏杆柱;四边线饰螺旋梯中柱
square turning 四方切削
square turret 车床四方转刀架;车床回转刀架
square type level 方形水准仪
square up 找方;修成直角,做成直角;结算清楚;结清
square vase 方瓶
square washer 方形垫圈;方垫圈
square wave 方(形)波;矩形波
square-wave amplifier 矩形波放大器
square-wave analyzer 方形波分析器
square-wave anodic stripping voltammetry 方波阳极溶出伏安法
square-wave carrier 矩形波载波;矩形载波
square-wave contrast transfer function 方波对比传递函数
square-wave electromagnetic flow-meter 方波电磁流量仪
square-wave form 方波形状
square-wave-form oscillator 矩形脉冲发生器;方形波振荡器;矩形波振荡器
square-wave generator 方波发生器
square-wave grating 方波光栅
square waveguide 方形波导管
square-wave modulator 方形波调制器
square-wave oscillator 矩形波发生器;方形波振荡器
square-wave output 方波输出
square-wave pattern 方波图样
square-wave polargraphy 方波极谱(法)
square-wave polar o-graph 方形波极谱仪
square-wave polarograph determination 方波极谱测定
square-wave pulse of light 方波型光脉冲
square-wave response 矩形波响应;方波响应
square-wave response characteristic 矩形波响应特性曲线
square-wave ringer 方波振铃器
square-wave signal 方波信号
square-wave signal generator 方形波信号发生器
square-wave transfer function 矩形波传递函数
square-wave voltage 矩形波电压
square weight 方形坠陀
square weir 方形堰
square wheel 方轮
square window 方窗
square wire 方形金属线;方形金属丝
square wire netting 方形钢丝网
square with the axis 与轴垂直
square wobbler 方辊头
square work 方形工件
square wrench 方头扳手
square yard 横帆桁;平方英码;平方码
square yoke 矩形磁偏转系统
squaring 按面积制作工程量计算书;形成矩形脉冲;矫方;四扭编组;抽方(计算工程量的步骤之一)
squaring amplifier 矩形波形成放大器;方形波成型放大器
squaring circuit 矩形脉冲形成电路
squaring device 求面积器
squaring down 缩绘
squaring head 四方刀架
squaring network 平方网络
squaring of tie 方正轨枕
squaring pusher 对正挡板
squaring shears 平行刃口剪切机;直刃剪床;剪边机;四方剪机
squaring tool 方头刀具
squaring tube 矩形脉冲形成管
squaring up 量方;结算;计算工程量;清算工程量;结算工程用料
squaring up machine 方边刨床
squaring valve 矩形脉冲发生管
squarks 超夸克
squash 压碎
squash court 壁球场
squashed delta 互联三角形连接
squash jacket 扁水套
squash technique 压片技术
squat 下坐【船】;航行尾倾现象;航行附加吃水;尾墩现象;船尾下座;船体下座;船体下垂;船身下沉现象;船舶下坐现象
squat ingot 短钢锭
squatjar 矮罐
squat of ship 船体下座值
squat pier 短胖形墩;蹲墩;短粗型墩;矮墩
squat round tower 矮圆塔
squatter 违章占住者;非法占住者;无权而定居公地者
squatter area 木屋区;棚屋区
squatter house 违章建筑
squatter sectionment 违章建筑
squatter settlement 棚户住区;违章居留地;违章建筑
squatter's rights 占住者权利;土地所有权
squatter village 移民安置村;外来人村落
squatting 航行尾倾;违章建筑
squatting closet 蹲式(大)便器
squatting of floor 楼板吱吱声
squatting pan 蹲式便器;蹲坑
squatting pan with low flush tank 低水箱蹲式大

便器
squatting plate 短而厚的板
squatting pot 蹲坑
squatting speed 航行尾倾临界航速
squatting temperature 下降温度;熔化温度
squatting water closet 蹲式洗大便器
squat value 下蹲值
squat wall 矮抗剪墙
squawker 中音乐扬声器
squeaking of floor 楼板吱吱声
squeal 振鸣声
squealer 声响器;声响(指示)器
squealing 啸声;振鸣声;号叫
squealing on curve 曲线上尖叫声(有轨车)
squeegee 隔离胶;橡胶滚子;橡胶刮板;橡胶地板擦;用橡胶刮板刮拭;平�;刮墨板;刮浆;路刷;污泥刮板;涂刷器;沉淀池耙;浮渣刮板;补缝
squeegee assembly 涂刷装置
squeegee buggy 刮刷车;补缝机;滚压机
squeegee coat 滚光层;抽涂涂层
squeegee gold 网印金膏
squeegee mark 网印标记
squeegee oil 印油
squeegee pump 挤压泵
squeegee roller 挤压轮
squeezability 可压缩性;可压实性
squeezable waveguide 可压缩波导;可压实波导管;软波导(管)
squeeze 用橡皮刮板刮拭;压挤浆;空头轧平;紧排;挤压伤;煤层变薄
squeeze action conveyer 垃圾压榨传送机
squeeze a well 钻井压力灌浆堵漏
squeeze board 压实板
squeeze bottle 塑料挤瓶(挤压时能排出所装之物)
squeeze buggy 滚压车;挤压车
squeeze casting 模压铸造
squeeze cement 挤水泥
squeeze cementing 压力灌浆
squeeze concrete pump 挤压混凝土泵
squeezed borehole 塌落岩石堵塞钻孔
squeezed file 紧排文卷
squeezed fold 挤压褶皱
squeezed joint 挤压接头;黏压结合
squeezed riveting 挤压铆接
squeeze effect 挤压效应
squeeze film bearing 压膜轴承
squeeze head 压头;压块
squeeze-in development 见缝插针开发
squeeze job 挤压作业
squeeze lifter 挤压式挖掘铲
squeeze mortar pump 挤压式灰浆泵
squeeze motion 夹取运动
squeeze mo(u)lding 压实造型
squeeze mo(u)lding machine 压实造型机
squeeze of side wall 边墙鼓胀
squeeze on profit 压缩利润
squeeze out 榨出;挤出(加压时胶料自胶层溢出)
squeeze packer 挤注封隔器;挤水泥封隔器
squeeze piston 压实活塞
squeeze pressure 压入压力
squeeze pumping 压力泵送
squeezer 型砂挤压机;压榨机;压铆器;压紧器;压机;挤干机构;缓行器
squeeze rammer 震压式造型机
squeeze riveter 铆钉机;压铆机
squeeze roll 压水辊;挤压辊;挤干辊
squeeze-roll coater 辊式挤压机
squeeze roller 挤压辊
squeeze(r) pump 挤压泵
squeezer roll 压液辊
squeeze section 可压缩区段;可压缩波导段
squeeze stripper mo(u)lding machine 顶箱压实式造型机
squeeze technique 挤压手法
squeeze test 挤压试验
squeeze time 预压时间(电阻焊时)
squeeze-type concrete pump 挤压式混凝土泵
squeeze-type mo(u)lded seal 压缩式成型密封
squeeze-type pump 挤压泵
squeeze valve 压实阀;压挤阀
squeeze-wheel lifter 挤压轮挖掘铲
squeezing 压榨;挤压;挤法;挤出
squeezing action 挤走;挤压作用

squeezing agent 堵塞剂
squeezing and pressing method 挤压法
squeezing cement 挤水泥
squeezing clear water out of soil 指缝中挤出清水
squeezing cylinder 压实缸
squeezing ground 剪胀地层
squeezing honing machine 挤压珩磨机
squeezing movement 挤压移动
squeezing out of soft soil 软土挤出
squeezing-out pressure 挤出压力
squeezing peat out of soil 指缝中挤出泥炭
squeezing press 挤压机
squeezing rock 剪胀岩石
squeezing safe factor 抗挤安全系数
squeezing screw 挤压螺钉
squegger 自动消失振荡器
squegging oscillation 断续振荡器
squelch circuit 噪声抑制电路
squelch control 静噪控制
squelch system 静噪装置
squelch tube 静噪管
squib 雷管;发火管;电气引爆器;点火管
squibbing 扩孔爆破
squibbing blasting 药壶爆破
squib-initiated igniter 火花塞式点火器
squib shell 爆筒壳
squib shooting 用小炸药包爆破
squilgee 橡胶刮板;橡胶地板擦
squinch 突角拱;对角线拱
squinch arch 跨墙角拱;对角线拱;突角拱
squint 斜倾;斜孔小窗;斜角砖;斜角隅砖;两波束轴间夹角;窥视窗;尖墙角砖;发射偏斜;波束夹角
squint angle 斜射角
squint brick 斜角墙用砖;斜形砖;尖墙角砖;边斜砖;异形砖;斜角隅砖;特制砖
squint-correcting prism 斜视校正棱镜
squint correction 斜视校正
squint error 斜视误差
squint hook 斜视钩
squint pier 钝角转角处墩柱;转角处墩柱;尖角墙柱
squint quoin (建筑物的)突出斜隅角;斜隅石;尖墙角
squint quoin of wall 墙边斜隅石
squint window 窥视窗;斜孔小窗;斜角隅砖
Squire's catheter 分节导管
squirm 绳索的扭曲
squirrel-cage 鼠笼(式)
squirrel-cage antenna 笼形天线
squirrel-cage armature 鼠笼式电枢
squirrel-cage balancing machine 笼型平衡机
squirrel-cage bar 鼠笼条
squirrel-cage blower 鼠笼式鼓风机
squirrel-cage coil 笼形线圈
squirrel-cage disintegrator 笼式打泥机;鼠笼式碎解机;鼠笼式粉碎机
squirrel-cage fan 鼠笼式风机
squirrel-cage grid 鼠笼形栅极
squirrel-cage induction motor 鼠笼式感应电动机
squirrel-cage induction type 鼠笼式感应型
squirrel-cage mill 鼠笼式磨碎机
squirrel-cage motor 鼠笼式电动机;短路式电动机
squirrel-cage repulsion motor 鼠笼式推斥电动机
squirrel-cage rotor 鼠笼式转子
squirrel-cage type fan 多叶片式风扇;鼠笼式风扇
squirrel-cage type filament 鼠笼式灯丝
squirrel-cage winding 鼠笼式绕组;鼠笼绕组
squirrel motor 鼠笼式电动机
squirrel tail 鼠尼形
squirrel-tall pipe jointer 鼠尼形管子连接器
squirt 喷射器;喷出(的水);注射器;喷射口;喷气式飞机
squirt can 喷射油壶;弹性注油器
squirt feeder 喷射送料器
squirt gun 喷射器
squirt hole 喷油孔
squirt hose 喷灌用软管
squirting irrigation 喷灌
squirt pump 注射泵
squirt welding 埋弧斗自动焊;埋弧半自动焊
squish velocity 挤气速度;挤流速度
squitter 间歇振荡器;断续振荡器
squitter pulse 断续脉冲
squoze pack 压缩穿孔卡片叠;压缩程序穿孔卡片叠
squrious response 杂波响应

sraping 隔声材料
sratutory ceiling on pollution 污染的法定最高限度
srcrap diagnosis 废品分析
sreavyite 斯硅钾钍钙石
srew water chiller 螺杆式水冷器
S. R. F. value of rock of mass 岩体的SRF值
srilankite 斯里兰卡石
S/R machine 有轨巷道堆垛起重机
S/R machine stability 有轨巷道堆垛稳定性
SRSS 平方和的平方根法
ss 起泡皂水
S-shaped fold S形褶皱
S. Sophia Cathedral 圣·索菲亚教堂
S-surface 岩组平行面;S面
stab 戳伤;把砖墙凿而糙;凿糙墙面
stab bar 连接筋;凿墙面配筋
stabber 谷物空心联样器;穿索针;扶管工;锥
stabbing 砖墙凿粗纹;铁丝钉套识;对正管接头;砖墙面划粗纹
stabbing board 钻塔的扶管平台(高20~40英尺)
stabbing salve 丝扣密封油
stabil 钢筋定位夹
stabilator 全动平尾;安定面
stabile 固定的雕刻物
stabile against crushing 抗压稳定
stabile against sliding 抗滑稳定
stabile analysis 稳定分析
stabile theory 稳定理论
stabilidyne 高稳定式超外差电路
stabilidyne receiver 高稳定式接收机
stabilimeter 稳定计
stabilised finish 稳定整理;定形整理
stabilised power unit 稳定电源设备
stabilised presentation 稳定显示
stabilised soil 经过处理的土;稳定土
stabilised voltage supply 稳定电源
stabiliser 支脚;锁定附件
stabilising agent 稳定剂
stabilising bid 稳定性出价
stabilising condenser 稳定电容器
stabilising embankment 加固堤堰
stabilist 稳定论者
stability 稳性;稳度;稳定性;稳定度;稳定;安定性
stability after mixing 涂料混合后的安定性
stability against 倾覆及滑走
stability against crushing 抗压稳定
stability against heave 抗管涌稳定性
stability against oscillation 振荡稳定性
stability against overturning 抗倾(覆)稳定性
stability against sliding 抗滑稳定
stability against tilting 抗倾(覆)稳定性
stability against upheaval 抗隆起稳定性
stability against vibration 振动稳定性
stability aggregate 稳定性团聚体
stability alarm 稳定度警报
stability analysis 稳定性分析;稳定分析
stability analysis of cells 格形结构面稳定性分析
stability analysis of soil slope 土坡稳定性分析
stability area 稳定区
stability at the helm circle 回转稳性
stability augmentation 稳定性的增长
stability augmentation system 稳定性加强系统
stability boundary 稳定性范围
stability calculation 稳定性计算;稳定计算
stability category 稳定度分类
stability change 稳定性变化
stability characteristic 稳定特性
stability chart 稳定度图
stability classification 稳定度分类
stability coefficient 稳定性系数;稳定系数;安定性系数
stability coefficient of foundation 基础稳定系数
stability computation 稳定性计算
stability condition 稳定状态;稳定条件
stability condition of region 区域稳定条件
stability conditions of equilibrium 平衡的稳定条件
stability constancy 稳定度
stability constant 稳定(性)常数
stability control 稳定性调节
stability criterion 稳(定)性准则;稳(定)判据
stability criterion of lateral pressure test 旁压试验稳定标准
stability criterion of riverbed 河床稳定性指标
stability curve 稳性曲线图;稳定性曲线;稳定特性

曲线
stability degree 稳定程度
stability degree of landslide 滑坡稳定程度
stability derivative 稳定性导数
stability diagram 稳定图解;稳定图
stability domain 稳定区域
stability drop 稳定性下降
stability effect 稳定作用
stability equation 稳定性方程式
stability exchange principle 稳定性交换原理
stability fabrics 稳定性组构
stability factor 刚性系数;稳定因子;稳定因素;稳定性因子;稳定性系数;稳定系数;稳定度系数
stability factor of safety 稳定安全系数
stability factor of slope 斜坡稳定系数
stability for disturbance 扰动稳定性
stability formula 稳定性公式
stability function 稳定性作用函数
stability gradient of loader 装载机稳定度
stability improvement 稳定性改进
stability in bulk 体积稳定(性)
stability in damaged condition 破舱稳定
stability index 稳定(性)指数;稳定性指标;稳定度指数
stability index of groundwater 地下水稳定指数
stability index of water 水稳定性指数
stability indicator 稳定性指示器
stability in periscopic aimingline 潜望镜瞄准线稳定
stability in storage 储存稳定性
stability in the large 全局稳定性
stability in use 使用稳定性
stability investigation 稳定性研究;稳定性调查
stability isotope 稳定性同位素
stability lagoon 稳定塘;稳定湖
stability lake 稳定湖
stability limit 稳定极限;稳定度极限
stability limit amplitude 稳定极限振幅
stability load 安定性荷载
stability loci 稳定域
stability malfunction 稳定系统故障
stability margin 稳定裕量;稳定裕度;稳定边际
stability matrix 稳定矩阵
stability measurement 稳定性测定
stability meter 稳定性试验仪
stability model 稳定性试验模型
stability moment 稳性力矩;安定力矩
stability number 稳定性数;稳定数
stability of abutment 边墩稳定性;坝座稳定性;坝肩稳定(性)
stability of adsorption 吸收(能力)稳定性
stability of aquifuge 隔水层稳定性
stability of a system 系统稳定性
stability of atmosphere 大气稳定度
stability of ballast slope 道床边坡稳定性
stability of cohesive soil 黏性土的稳定
stability of colloid 胶体稳定性
stability of compression member 压杆稳定性
stability of crystal oscillator of coding clock 编码钟晶振稳定度
stability of dam slope 坝坡稳定性
stability of emulsion 乳液稳定性
stability of excavation face 开挖面稳定
stability of flo(a)tation 漂浮稳定性;浮选稳定性;浮体稳定(性)
stability of floating caisson 沉箱浮游稳定性
stability of foundation 地基稳定
stability of foundation rock mass 地基岩体稳定性
stability of gloss 光泽稳定性
stability of governor 调速器稳定性
stability of model calculation 模型计算结果的稳定性
stability of motion 航向稳定性
stability of offshore platform 海上平台稳定性
stability of rating curve 率定曲线稳定性;水位—流量关系稳定性
stability of regional crust 区域地壳稳定性
stability of regulation 调节稳定性
stability of reservoir slope 水库边坡稳定性
stability of rigid frame 刚架稳定性
stability of rock mass 岩体稳定性
stability of rock slope 岩质边坡稳定性
stability of shape 形状稳定性
stability of sheet pile structure 板桩结构的稳定性

stability of ship 稳性【船】;船舶稳(定)性
stability of slag 矿渣稳定性
stability of sliding 滑移稳定性
stability of slope 斜坡稳定性;边坡稳定(性);边坡失稳
stability of soil slope 土坡稳定(性)
stability of solution 解的稳定性;溶液稳定性
stability of stage-discharge relation 水位—流量关系稳定性
stability of surrounding rock 围岩稳定性
stability of surrounding rock of underground excavation 洞室围岩稳定性
stability of synchronization 同步稳定性
stability of the instrument 仪器的稳定性
stability of tracer 示踪剂稳定性
stability of travel direction 行驶方向稳定性
stability of vibration 振动稳定性
stability of volume 体积稳定性;体积安定性
stability parameter 稳定性参数
stability pond 稳定池
stability power limit 稳定功率极限
stability problem 安定性问题
stability product 稳定积
stability range 稳定区范围;稳定范围
stability reduction 稳定性减小
stability sensor 稳定传感器
stability series 稳定系(列)
stability setting 稳定性定形(指干温热法)
stability simulation 稳定度模拟
stability strength 稳定强度
stability subgroup 稳定子群
stability system 稳定体系
stability test 稳(定)性试验;稳定(度)试验
stability test model 稳定性试验模型
stability test of monitoring control points and deformation point 点位稳定性检验
stability theory 稳定理论
stability theory for large-scale systems 大系统稳定性理论
stability threshold 稳定性阈值
stability to heat 热稳定性
stability to hydrolysis 水解稳定性
stability to light 光稳定性
stability towards chemical reaction 对化学反应的稳定性
stability triangle 稳定性三角形
stability under accident conditions 事故状态稳定性
stability under load 有载时的稳定性;载重稳定性;荷载下的稳定性
stability under own weight 自身稳定(性)
stability under torsion 扭曲稳定性
stability under working conditions 载重稳定性;工作状态稳定性
stability unload 无载稳定性
stability value 稳定度值
stability variation 稳定性变化
stability zone 稳定地带
stabilivolt 稳压
stabilizability 可稳定性;能稳定性
stabilization 坚固;加固作用;加固;稳定作用;稳定化;稳定处理;稳定;定影;安定面
stabilization basin 稳定池
stabilization bonds 安定债券
stabilization by aluminizing 铝化加固
stabilization by compaction 压实稳定;夯实加固
stabilization by consolidation 固结稳定
stabilization by densification 挤密加固
stabilization by replacement of soil 换土加固
stabilization by vibroflo(a)tation 振动加固
stabilization cable 保持稳定的缆索
stabilization circuit 稳压电路
stabilization condition 稳定条件
stabilization efficiency 稳定效率
stabilization factor 稳定因子;稳定因素;稳定系数
stabilization fund 平准基金;稳定基金
stabilization lagoon 稳定塘;稳定湖;稳定池
stabilization lagoon method 稳定塘法
stabilization mixer 稳定土用拌和机
stabilization moment 稳定力矩
stabilization network 稳定网络
stabilization of borehole 钻孔稳定性
stabilization of gasoline 汽油的稳定化
stabilization of level 电平的稳定

stabilization of river bed 固床;河床稳定
stabilization of river flats 固滩
stabilization of sewage 污水稳定(化)
stabilization of soft foundation 软基加固
stabilization of soil 土壤稳定
stabilization of speed 转速稳定
stabilization of water 水质稳定(化)
stabilization operation 稳定运行
stabilization period 稳定期
stabilization plant 稳定装置
stabilization pond 稳水池;稳定塘
stabilization pond method 稳定塘法
stabilization power 稳定功率
stabilization process 稳定过程
stabilization range 稳定范围
stabilization ratio 稳定比
stabilization system 稳定系统
stabilization tank 稳定池
stabilization test 稳定试验
stabilization time 稳定时间
stabilization treatment 稳定处理
stabilization unit 稳定器
stabilization with cement 水泥稳定(法)
stabilization with lime 石灰稳定(土壤)法
stabilization works 稳定设施
stabilizator 稳定剂;稳压器;稳定器
stabilizator rod 平衡杆
stabilize 稳定;安定
stabilize a security 稳定一种证券;维持证券的最低价格
stabilize commodity price 稳定物价
stabilize currency 稳定通货
stabilized accounting 稳定币值会计
stabilized alloy 稳定合金
stabilized amplitude 稳定振幅
stabilized antenna 稳定天线
stabilized approach 稳定进场着陆
stabilized ash 稳定灰渣(石灰或水泥)
stabilized bank slope 稳定岸坡
stabilized bar 稳定沙洲
stabilized base 稳定(性)基层
stabilized beam 稳定束
stabilized blast-furnace slag 稳定高炉矿渣
stabilized bond 指数化债券
stabilized camera mount 航摄仪稳定装置
stabilized channel 稳定水道;稳定渠道;稳定河槽
stabilized chip 稳定石屑
stabilized chlorine dioxide 稳定二氧化氯
stabilized condition 稳定状态
stabilized coupling 稳定连接轴;稳定接头
stabilized course 稳定层
stabilized crude(oil) 稳定原油
stabilized crushed rock soil 稳定碎石土
stabilized current supply 稳流电源
stabilized deflection 稳定偏转
stabilized descent 稳定下降
stabilized diazo compound 稳定的重氮化合物
stabilized dolomite 憎水白云石;稳定白云石
stabilized dolomite brick 稳定性白云石耐火砖;稳定白云石砖
stabilized dolomite clinker 稳定白云石熟料
stabilized dolomite refractory 稳定白云石耐火材料
stabilized dune 稳定的沙堆;固定沙丘;稳定沙丘
stabilized-earth concrete 稳定土壤混凝土
stabilized earth road 改善土路;稳定土路
stabilized ejection 稳定弹射
stabilized elementary transformation 稳定初等变换
stabilized feedback 稳定回授;稳定反馈
stabilized feedback amplifier 稳定反馈放大器
stabilized forsterite 稳定镁橄榄石
stabilized frequency 稳定频率
stabilized front 稳定的正面
stabilized gasoline 稳定汽油
stabilized glass 稳定玻璃
stabilized grade 稳定纵波;稳定坡度;不冲不淤比降
stabilized granular surface 稳定砂石面层;稳定砂石路面
stabilized gravel 稳定砂砾;稳定砾石
stabilized gravel soil 稳定砂砾土
stabilized gravel surface 稳定砾石面层;稳定砾石路面
stabilized image 稳定图像
stabilized insoluble azo dyes 稳定不溶性偶氮染料

stabilized insulator 稳定绝缘子
stabilized landslide 已稳定滑坡
stabilized liquid method (防护钻孔壁面用的)水泥稳定液施工法
stabilized local oscillator 稳定本机振荡器
stabilized macadam 稳定碎石
stabilized mixture 稳定混合物
stabilized mount 稳定装置
stabilized mounting 稳定装置
stabilized natural gasoline 稳定天然汽油
stabilized negative feedback amplifier 稳定负反馈放大器
stabilized non-torque yarn 稳定的无捻回弹力丝
stabilized operating statement 详细预计的经营损益表
stabilized oscillator 稳定振荡器
stabilized photomultiplier 高稳定光电倍增管
stabilized pinch 稳定箍缩
stabilized platform 稳定台地;稳定平台
stabilized power source 稳压电源;稳定电源
stabilized presentation 稳定显示
stabilized prints 稳定复制品
stabilized pulse duration 稳定脉冲持续时间
stabilized rectifier 稳压整流器
stabilized road 稳定道路;稳定处理的道路
stabilized road-mixer 路拌式稳定土搅拌机
stabilized satellite 稳定卫星
stabilized shoulder 稳定(处理)的路肩
stabilized shunt 稳定分路绕组
stabilized shunt motor 稳定并路电动机
stabilized shuntwound generator 稳定并励发电机
stabilized shunt-wound motor 稳定并励电动机
stabilized side slope 稳定(性)边坡
stabilized slope 稳定坡度;稳定边坡
stabilized soil 稳定土
stabilized soil base 稳定土基
stabilized soil base course 稳定土基层
stabilized soil bituminous road 沥青稳定土路;沥青加固土路
stabilized soil-bound surface 泥结稳定土路;泥结稳定面层;泥结稳定路面;稳定泥结面层;稳定泥结路面
stabilized soil mixer 稳定土拌和机
stabilized soil road 改善土路;稳定土路
stabilized speed 稳定转速;稳定速度
stabilized state 稳定状态
stabilized steel 稳定化钢
stabilized steel sheet 稳定化钢板;稳定处理钢板
stabilized strand 稳定处理的预应力钢绞线
stabilized supply voltage 稳定电源电压
stabilized toroidal pinch 稳定环形箍缩
stabilized velocity 稳定速度
stabilized voltage 稳压;稳定电压
stabilized voltage supply 稳压电源
stabilized winding 稳定绕组
stabilized wire 稳定化处理的预应力钢丝
stabilized zirconia 稳定二氧化锆
stabilize market 稳定市场
stabilize price 平抑价格
stabilizer 增衡器;减摇装置;加固器;稳流板;稳定剂;锁定附件;扶正器;安定面;安定剂
stabilizer/unfog additive 稳定/无雾添加剂
stabilizer apparatus 稳定装置;稳定器
stabilizer bar 稳定器杆;稳定拉杆
stabilizer base 稳定剂基;构架基础
stabilizer blade 支脚
stabilizer bottoms 稳定塔底产物
stabilizer bracket 稳定托架
stabilizer cavity 稳定空腔谐振器
stabilizer chord 安定面弦
stabilizer column 稳定塔
stabilizer compensator 稳定补偿器
stabilizer feed pump 稳定塔进油泵;稳定塔进料泵
stabilizer foot 稳定支脚
stabilizer gas 稳定塔气体
stabilizer link assembly 稳定器固定柱
stabilizer of hydrogen peroxide 过氧化氢稳定剂;双氧稳定剂
stabilizer overhead 稳定塔顶排(出)气体
stabilizer plant 稳定装置
stabilizer rod 稳定杆
stabilizer sleeve 稳定接箍(钻杆)
stabilizer tower 稳定塔
stabilizer valve 稳压管

stabilizer vane 稳定叶片
stabilize the market 稳定市场
stabilizing 稳定化;稳定的
stabilizing additive 稳定添加剂
stabilizing adjustment 稳定线路调整;稳定度调整
stabilizing agent 稳弧剂;稳定剂;安定剂
stabilizing aging 稳定化时效
stabilizing amplifier 稳定放大器
stabilizing annealing 稳定退火
stabilizing bag 安定囊
stabilizing bath 稳定液
stabilizing bilge 舷侧防摇水舱
stabilizing by hydraulic jacks 用液压千斤顶稳定
stabilizing circuit 稳定电路
stabilizing colloid 胶体稳定剂
stabilizing connecting bracket 稳定连接托架
stabilizing correction 稳定修正
stabilizing device 稳定装置
stabilizing effect 稳定效应
stabilizing element 稳定性元素(阻止石墨化元素)
stabilizing factor 稳定因子;稳定因素;稳定系数
stabilizing feedback 稳定反馈
stabilizing field 稳定场
stabilizing fin 减摇鳍;稳定翅
stabilizing float 稳定浮筒
stabilizing force 稳定力
stabilizing grout 固结灌浆;加固灌浆
stabilizing guy lines 钢丝绑绳(钻塔、桅杆)
stabilizing gyroscope 稳定陀螺仪
stabilizing inductance 稳流电感
stabilizing influence 稳定影响;稳定化影响
stabilizing ion 稳定离子
stabilizing jack 支撑稳定千斤顶
stabilizing material 稳定材料
stabilizing moment 稳定力矩
stabilizing network 稳定电路
stabilizing plane 安定面
stabilizing potential 稳定电势
stabilizing power 稳定能力
stabilizing process 稳定法
stabilizing reboiler 稳定塔再沸器
stabilizing resistance 稳流电阻
stabilizing resistor 平衡电阻器
stabilizing selection 稳定化选择
stabilizing skirt 稳定裙;安定缘
stabilizing spoiler 稳定襟翼
stabilizing surface 稳定面
stabilizing tank 减摇水舱
stabilizing tester 稳定性试验台
stabilizing the capacity of water supply 稳定供水能力
stabilizing transformer 稳定变压器;稳定变压器
stabilizing treatment 稳定(化)处理
stabilizing turned coil 稳定用调谐线圈
stabilizing unit 稳定器
stabilizing vortex 稳定涡流
stabilizing winding 平衡绕组;稳定绕组
stabilometer 稳定仪;稳定性试验机;稳定度仪
stabilometer design method 稳定度仪设计法
stabilopode 四角稳定体(防波堤用)
stabilotron 厘米波功率振荡管
stabilovolt 稳压管
stabioga(u)ge 稳性测量仪
stabipack 稳压组件
stabit 人字稳定体(一种防波堤异形块体);稳定体(一种防波堤块体)
stab knife 穿刺刀
stable 马棚;马厩;平稳(的);稳定的;安定的
stable acceleration process 稳定加速过程
stable air column 稳定气柱
stable air mass 稳定气团
stable amplitude 稳定振幅
stable arc 稳定(的)电弧
stable area 稳定(地)区
stable area of new tectogenesis 新构造运动停顿地区
stable atmospheric condition 稳定性大气条件;稳定的大气条件
stable auroral red arc 稳压极光红弧
stable bank 稳定河岸
stable base 尺度稳定片基
stable base film 尺寸稳定胶片;片基尺寸稳定的胶片
stable beach 稳定海滩;稳定岸滩

stable bed rock 稳定岩层
stable bend 稳定河湾
stable beta titanium alloy 稳定的贝塔型钛合金
stable bituminous dispersion 稳定沥青分散液(即慢裂乳化沥青)
stable block 稳定地块
stable building 稳定房屋
stable buoyancy equilibrium 稳定浮力平衡
stable burning 稳定燃烧
stable but adjustable par value 稳定但可调整的平价
stable carbon isotope composition of petroleum 石油的碳稳定同位素组成
stable carbon isotopic composition of butane 丁烷碳稳定同位素组成
stable carbon isotopic composition of carbon dioxide 二氧化碳碳稳定同位素组成
stable carbon isotopic composition of ethane 乙烷碳稳定同位素组成
stable carbon isotopic composition of isobutane 异丁烷碳稳定同位素组成
stable carbon isotopic composition of isopentane 异戊烷碳稳定同位素组成
stable carbon isotopic composition of pentane 戊烷碳稳定同位素组成
stable center 稳定中心
stable channel 稳定河槽
stable chemical 稳定化学品
stable chlorine disinfectant 稳定性氯素消毒剂
stable coast 稳定海岸
stable cohomology operation 稳定上同调运算
stable combustion 稳定燃烧
stable complex 稳定络合物
stable component 稳定的环节
stable compound 稳定化合物;安定化合物
stable computation graphic(al) construction 稳定计算图解法
stable condition 稳定状态;稳定条件
stable configuration analysis 稳定构型分析
stable confinement 稳定约束
stable control 稳态控制
stable convergence 稳定收敛
stable court 稳定场地
stable crack 稳定裂缝
stable crack propagation 稳定裂缝扩展
stable creep-resistant titanium alloy 稳定的抗蠕变钛合金
stable criterion of discharge 流量稳定标准
stable currency 稳定货币
stable current 定常流
stabled cows 厩内牛角
stable design 通用设计;标准设计
stable detonation 稳定爆炸
stable diagram 状态图;平衡图;稳定状态图
stable distillate fuel 稳定馏出燃料
stable distribution 稳定分布
stable distribution function 平稳分布函数;稳定分布函数
stable door 马厩门;有小门的大门;荷兰门;双截门;叠开门(上下两半分别开关的门);两截门
stable dredge-cut 稳定挖槽
stable dune 稳定沙丘
stable dung 厩粪
stable ecosystem 稳定的生态系统
stable effluent 稳定污水流;稳定排液;稳定出水
stable electron orbit 稳定电子轨道
stable element 稳定元素;稳定元件;稳定成分
stable empty 无载稳定性
stable emulsion 稳定乳状液;稳定乳(化)液
stable equilibrium 稳定平衡;安全平衡
stable equilibrium phase 稳定平衡相位
stable equilibrium shoreline 稳定平衡海岸线
stable equivalent 稳定等价的
stable equivalent vector bundles 稳定等价向量丛
stable exchange rate 稳定汇率
stable fabric 稳定织物
stable factor 稳定因子
stable film 尺寸稳定胶片
stable filter 稳定滤波器
stable fission products 稳定的裂变产物
stable fixed point 稳定固定点
stable force 静力;平衡力
stable form 稳定型
stable framework 稳定构架;稳定框架

stable frequency signal 稳定频率信号
stable frozen ground 永冻土
stable funds of export commodities 出口商品稳定基金
stable gasoline 稳定汽油
stable gel structure 稳定凝胶结构
stable glass fibre 标准玻璃纤维
stable gravimeter 稳定重力仪
stable grout 稳定浆液
stable hole 硐室;壁龛
stable humus 稳性腐殖质
stable hydraulic section 稳定水力断面
stable hydrogen isotope composition of petroleum 石油的氢稳定同位素组成
stable hydrogen isotopic composition of water 水的氢稳定同位素组成
stable iodine 稳定碘同位素
stable island 稳定岛
stable isobar 稳定同量异位素
stable isotope 稳定同位素
stable isotope analysis 稳定同位素分析
stable isotope dilution mass spectrometry 稳定同位素稀释质谱法
stable isotope dilution technique 稳定同位素稀释技术
stable isotope indicator 稳定同位素指示剂
stable isotope tracer 稳定同位素示踪剂
stable isotope tracing 稳定同位素示踪(法)
stable isotope working standard in china 稳定同位素中国工作标准
stable isotope world-wide standard 稳定同位素国际标准
stable isotopic composition of argon 氩稳定同位素组成
stable isotopic stratigraphy 稳定同位素地层学方法
stable laminar current 稳定片流;稳定层流
stable landmass tectonics 稳定地块构造
stable law 稳定律
stable layer 稳定层
stable limit cycle 稳定极限循环
stable loop 稳定回路
stable manifold 稳定流形
stable manure 动物粪便
stable matrix 稳定矩阵
stable maximum 稳定最高量
stable micro-polluted surface water 稳定性微污染地表水
stable mineral 稳定矿物
stable monetary unit assumption 币值不变假设
stable motion 稳态运动;稳定运动
stable nappe 稳定水舌
stableness 稳定性;稳固
stableness of filter 滤光片的稳定性
stable nuclide 稳定核素
stable ocean platform 稳定式海洋平台
stable operation 稳定运行;稳定操作
stable orbit 平衡轨道;稳定轨道
stable oscillation 稳定振荡
stable oscillator 稳定振荡器
stable oxygen isotopic composition of water 水的氧稳定同位素组成
stable pair 稳定对
stable parallelizable flow 稳定可平行流
stable particle 稳定粒子
stable period 稳定周期;稳定期
stable phase 稳定相;稳定期
stable phase angle 稳定相位
stable phase oscillation 稳相振荡
stable phase parameter 稳相参数
stable phytocoenosium 稳定植物群落;稳定植物群落
stable plasma 稳定等离子体
stable plastics 不变形塑料片;不变形塑料
stable platform 稳定台地;稳定平台;稳定地台
stable point 稳定点
stable polymorphs 稳定多晶型物
stable population 稳定种群;稳定人口
stable potential 稳定电位
stable pour-point 稳定倾点;稳定凝固点
stable price 稳定的价格
stable primary cohomology operation 稳定一阶上同调运算
stable production stage 稳产期

stable propagation 稳定扩展
stable pulse generator 稳定脉冲发生器
stable pulse generator clock 稳定脉冲发生器
stable range 固定距离
stable range rings 固定距离圈;固定距标圈;固定距标
stable reactor period 反应堆稳定周期
stable reduction 稳定约化
stable region 稳定域;稳定区(域);稳定境
stable remanent magnetization 稳定剩余磁性
stable river bed 稳定河床
stable rock 稳定岩石;坚固岩石
stable rock bed 稳定岩基
stable running 稳定转动
stable secondary cohomology operation 稳定二阶上同调运算
stable semi-submersible platform 稳定式半潜平台
stable series 稳定级数
stable servo 稳定伺服系统;稳定伺服机构
stable-shelf facies 稳定陆架相
stable shield 稳定地盾
stable ship 平稳船
stable sliding 蠕动;稳滑;稳定滑动;缓慢滑动
stable-slip behavio(u)r 稳滑方式
stable slope angle 稳定坡角
stable soil road 稳定处理的土路
stable solder glass 稳定焊料玻璃
stable solution (微分方程的)稳态解;稳定解
stable stage of expansion 膨胀稳定期
stable standard of subsidence 沉降稳定标准
stable star 稳定恒星
stable state 稳态;稳定状态;稳定态
stable state of motion 稳态运动;稳定运动状态
stable state of vibration 稳定振动状态
stable state oscillation 稳态振荡
stable static model 稳定统计模型
stable strobe 稳定频闪放电管
stable structure 坚固结构;稳定(的)结构
stable subgroup 稳定子群
stable suspension 稳定悬浮液
stable switch 稳定开关
stable system 稳定系统
stable thickness 厚度稳定
stable time 稳定时间
stable time of subsidence 沉降稳定时间
stable tracer isotope 稳定示踪同位素
stable trajectory 稳定轨迹
stable transverse oscillations 稳定横向振荡
stable trigger circuit 稳态触发电路
stable type 稳定型
stable-type gravimeter 稳定型重力仪
stable uniform flow 稳定等速流(动)
stable valley 稳定谷
stable value 稳定值
stable vector bundle 稳定向量丛
stable vegetation 稳定植被
stable vertical 稳定的竖向准直;垂直陀螺仪
stable water source 稳定型水源地
stable waterway section 稳定河段
stable wave 稳定波
stable wind 稳定风
stable working platform 稳定工作平台
stable zone 稳定区
stabling shed 停车库
stabling siding 停留线;停车线
stably continuous period 稳定延续时间
stably parallelizable 稳定可平行的
stably recovering water level 稳定恢复水位
stably stratified flow 稳定层流
stably trapped radiation zone 稳定捕获辐射带
stab net 深水底刺网
stabproof vest 防刺背心
stab wound 刺伤;戳伤
Stach's microlithotypes of coal 施塔赫分类
stack 烟囱;组叠;栈;归垛;高炉炉身;料垛;立管;茎梗饰【建】;货架;海蚀柱;竖起;石柱;堆栈【计】;堆场;堆;叠片;叠(层);叠加;单层垛;存储栈;层垛;草垛;草堆;戳键;叠式存储器;堆积
stack address 栈地址
stack address control 栈地址控制
stack addressing 栈访问
stack algorithm 堆栈算法
stack alignment 堆栈对齐
stack allocation 堆栈分配

stack arch 烟道碹;烟道拱
stack architecture 栈结构;堆栈结构
stack area 烟道切面积;烟道截面积;堆栈区;堆栈范围
stack assembly 叠片组件;叠层组件
stack automaton 栈自动化装置
stack automator 堆栈自动机
stack base register 栈基址寄存器
stack bit 堆栈数位
stack block 书库楼
stack bond 堆积式砌法;横竖通缝砌法;对缝砌法;通风栅;堆砌法;叠层砌法;通缝砌法
stack brickwork 炉身砌砖;炉身衬砖
stack building 堆垛楼;堆积楼;堆存楼(集装箱)
stack-building equipment 堆垛机具
stack by batches 按批堆放
stack by lots 按批堆放
stack cap 烟囱帽
stack capability 栈能力
stack capacity 竖管流水能力;竖管排烟能力;竖管过水能力
stack card 货垛牌
stack casing 炉身外壳;炉身套壳
stack cast 叠铸
stack casting 叠箱铸型;层叠铸造
stack cell 栈单元;堆栈单元
stack chimney 烟囱
stack condition 烟囱排放参数
stack content 竖管排烟能力;竖管过水能力
stack cooler 炉身冷却板
stack-cooling plate 炉身冷却板
stack crane 堆装起重机
stack cross-section 烟道截面;烟道截面
stack cutting 堆叠切割;叠片切割;成堆切割
stack damper 烟道挡板;烟道闸板;烟囱调节(风)门;烟囱挡板
stack desulfurization 排烟脱硫
stack disposal 烟囱处置
stack door 甬道门;通风管(闸)门
stack downdraft 烟囱倒灌风
stack draft 烟道通风;烟囱气流;烟囱排烟;烟囱抽力
stack draught 烟囱抽力;烟囱抽风
stacked 成堆的;成沓的
stacked anchor 有杆锚
stacked antennas 多层天线
stacked array 多层天线阵;分层天线阵
stacked beam 聚积束;多波束
stacked-beam radar 多层波束雷达
stacked bed 多层床
stacked bond 横竖通缝砌法;对缝砌法;堆砌法;叠层砌法;通缝砌法
stacked bricks 堆放的砖;堆砖
stacked capacitor 叠层电容器
stacked capacity 竖管流水能力
stacked content 层积
stacked cubic feet 层积立方英尺
stacked derrick 折叠式钻塔
stacked-dipole antenna 多层偶极天线
stacked double directional throttle valve 叠加式单向节流阀
stacked filter unit 多层过滤器
stacked-foil mica capacitor 箔式云母电容器
stacked gate structure 叠栅结构
stacked graph 叠式图表
stacked integrated circuit 叠层集成电路;层叠集成电路
stacked job 栈作业;栈式作业;连续作业项;叠式施工;成批处理作业
stacked job control 栈作业控制;暂存工作控制;栈式作业控制
stacked job processing 栈式作业处理;堆积式作业处理;成批作业处理
stacked loops 多层环形天线
stacked memory 栈式存储器;堆栈存储器
stacked on edge 侧边堆放
stacked packing 规则填料
stacked plate 堆积式承载板
stacked ribbon cable 多层带状电缆
stacked section of converted wave 转换波叠加剖面
stacked sheets 成叠薄板
stacked system 叠层方式
stacked tailings 沉积尾矿;堆积尾矿
stacked tank system 叠槽式装置

stacked traces number 叠加道数
stacked unidirectional throttle valve 叠加式单向节流阀
stacked unidirectional valve 叠加式单向阀
stacked volume 层积容积
stacked wafer module 叠式组件;叠片组件
stack effect 堆叠效应;拔风效应;抽吸作用;烟囱效应;烟囱抽吸作用
stack effluent 烟囱排放物
stacker 堆垛工;叉式装载车;货物升降机;堆垛器;自动堆垛机;可升降摄像机台;卡片存储器;接卡箱;集纸箱;集草器;码垛机;码包机;退火窑送料装置;堆码工;堆料机;堆积箱;堆栈机;堆垛机;堆垛工;叠卡子;打垛机;叠式存储器
stacker belt 堆垛机的输送带;堆栈用胶带输送机
stacker conveyer 堆垛机的输送带
stacker crane 叠架起重机;塔式起重机;码垛机;堆垛起重机
stacker for building up fills 坝体堆放机;坝体输送机;路基堆放机;路基输送机;筑堤堆放机;筑堤输送机;填方堆放机;填方输送机
stacker full 接卡箱满
stacker-loader 装载堆垛机
stacker of layers 多层涂敷
stacker-reclaimer 堆取料机;堆垛运输机
stacker truck 叉车;堆垛车;堆料车
stack exit velocity 烟囱喷烟速度;烟囱出气速度
stack factor 烟囱因素
stack fitting 竖管配件
stack flat 叠平;堆平
stack frame 栈构架
stack friction 烟囱阻力
stack funnel 烟囱内的尖塔形通风设备
stack furnace 围窑
stack gas 烟囱气体;废气;烟气;烟道气
stack-gas desulfurization technique 烟道气脱硫技术
stack gate 叠层栅
stack group 立管组合
stack guy 烟囱缆;烟囱拉索
stack handler 栈处理机
stack head 墙上管道出口;堆栈头
stack height 烟囱高度;堆装高度
stack holder 堆叠材料的夹持装置
stack honing 多件孔磨
stack indicator 栈指示器;堆栈指示器;堆栈指示符
stacking 堆积(成垛);码垛;群集度;多层叠加;置;堆料;堆积干燥法;堆垛;分层等待
stacking adapter 双头插销;堆装连接件;堆装连接机;堆码接合装置(集装箱等)
stacking apparatus 堆垛设备
stacking area 堆置场;堆场面积;备用堆置场
stacking bay 堆料区
stacking beam 堆码梁
stacking beam system 堆码梁系统
stacking bed 带收集装置的冷床
stacking capability 堆码能力
stacking capacity of derrick 塔内竖放钻杆容量
stacking chair 旋转椅
stacking conveyer 堆垛输送器
stacking crane 堆垛起重机
stacking density 层积密度
stacking derrick 堆垛起重吊杆
stacking disorder 堆垛无序
stacking elevator 举升堆垛机;堆垛升运器
stacking factor 工作比;绕组占空系数;填充因数
stacking fault 堆积层错;堆垛层错;层积缺陷
stacking fitting 双头插销;堆装件;堆码接合装置(集装箱等)
stacking furniture 叠家具
stacking ground 港区货场
stacking height 堆码高度;堆置高度;堆垛高度
stacking load 堆码负载
stacking machine 码垛机;堆垛机;堆料机;堆高机;堆装机;叉式起重机
stacking machinery 堆垛机械
stacking mechanism 整卡机构;叠卡机构
stacking method 叠加法
stacking of drill pipe 钻杆堆放
stacking of layers 层状堆集
stacking of surges 涌浪叠加
stacking of upgoing and downgoing events 上行波和下行波的叠加
stacking pallet 码垛托盘

stacking park 堆场
stacking sequence 堆积顺序;叠层顺序
stacking straddle carrier 堆装跨运车
stacking strength 栓固强度;堆码强度
stacking table 堆料架
stacking tap density 堆积密度
stacking test 堆码试验
stacking truck 堆垛车
stacking unit 产品堆垛机
stacking velocity 叠加速度
stacking yard 堆场
stack in recursive procedure 递归过程的栈
stack instruction 栈指令
stack interrupt 栈中断;栈溢出中断
stack job 栈式处理
stack job processing 栈式作业处理
stack ladder 架子梯;人字梯(俗称高凳)
stack layers 栈积层
stack leaning 烟囱倾斜
stack level 栈深度;堆栈级
stack limit register 栈界限存储器
stack lining 炉身衬
stack lip to lip 对口装窑
stack load 堆积荷载;堆放荷载
stack loader 堆料装料机
stack-loading freight 堆装货物
stack loss 排烟损失;通风损失(量)
stack machine 堆栈机器
stack manipulation 栈处理;栈操作;堆栈操作
stack marker 栈标志
stack mast 烟囱桅
stack memory 栈存储点;栈存储器
stack method 堆积法
stack migration 叠加偏移
stack mixing 堆积混合
stack mo(u)ld 叠箱铸模
stack mo(u)lding 双面型箱造型;双面型叠箱造型法;分层压制;分层模制
stack mo(u)lding machine 叠箱造型机
stack-mounted 片式连接的
stack nonterminal 栈的非终极符
stack nozzle 烟囱口
stack of bricks 砖垛
stack of conduits 管堆
stack of fuel elements 释热元件组件
stack of ga(u)ge blocks (机械量测的)块规组
stack of laminations 叠片组合件
stack of pipe 管堆
stack of test(ing)sieves 一组试验筛
stack of tube 管堆
stack of upgoing traces 上行波各道叠加
stack of wood 木堆
stack of work 工件垛
stack-on oven 独立式烤箱
stack operation 栈操作;堆栈操作
stack operator 栈算子
stack option 栈选择区
stack order 栈指令
stack outlet 烟囱出烟孔
stack overflow 栈溢出
stack page 栈页面
stack paint 烟囱漆
stack pallet 送料车
stack partition 带直立烟道的隔墙;带直立管道的隔墙
stack pile 同厚度、不同种类硬木材货垛
stack pipe 通风管;立管;竖管
stack pointer 栈指示器;堆栈指针;栈指示字;堆栈指示器
stack pointer alignment 堆栈指示字对界
stack pointer definition 栈指示字定义
stack point register 栈指示寄存器
stack pollutant 烟突污染物
stack pop-up 堆栈向上托;堆栈上弹
stack pouring 叠箱浇注
stack power 竖管排烟能力;竖管过水能力
stack presentation with two-way times aligned 双程时间对齐叠加
stack processing 栈处理
stack processing technique 栈处理技术
stack push-down 堆栈压入;堆栈下推
stack/rail machine for both unit load and order picking 拣选—单元混合型有轨巷道堆垛机
stack rain 烟囱雨

stack reduction 栈归约
stack reference operation 堆栈访问操作
stack register 栈寄存器
stack room 书库
stacks 书库
stack sampling 烟道采样
stack sampling procedure 烟道气采样方法
stack scan 栈扫描
stack segment 栈段
stack silo 堆干草筒仓;堆谷物筒仓
stack solid 烟囱气中的固体
stackstand 堆支撑架
stack stave 炉身冷却壁
stack structure 堆栈结构
stack survey 烟囱检查
stack symbol 栈符号
stack system (书库与阅览室分开设置的)书库叠层方式
stack tape 叠加带
stack temperature 烟囱温度
stack the tools 堆放钻杆;堆放工具
stack top 栈顶
stack top location 栈顶单元
stack transfer device 袋堆传送装置
stack transformation 栈变换
stacktron 陶瓷接收放大管
stack turner 翻垛机
stack underflow 堆栈下溢
stack unwinding 堆栈解退
stack-up 层叠
stack-up plant 高塔式拌和厂
stack up yard 堆积场
stack vector 栈向量
stack velocity 出烟速度
stack vent 屋面排气孔;竖管通风口;下水通气竖管;烟囱排烟道;主管通气管;立管通气孔(口);竖管通气孔;伸顶通气管
stack ventilation system 草垛通风系统
stack venting 立管通气法;竖管通风
stack water 缓流;平潮
stack welding 多层接触点焊
stack yard 港口堆场;堆料场;堆场;堆场
stactometer 滴重计;测滴计
staddle 基础;承架;支撑物;堆垛支架
stad(d)le stone 干草棚柱础;支撑石;堆垛垫石;蘑菇状的矮石柱
stade 冰阶
Stader's splint 斯塔德尔夹
stadhuis 城市住房或市政厅(荷兰)
stadia 古希腊长度单位;视距准尺;视距仪;视距(尺)
stadia additive constant 视距加常数
stadia arc 视距弧
stadia circle 视距弧
stadia computation 视距计算
stadia computer 视距计算器
stadia computing disk 视距计算盘
stadia computor 视距计算机
stadia constant 视距常数
stadia correction diagram 视距校正图表
stadia diagram 视距图表
stadia factor 视距因数;视距常数
stadia formula 视距公式
stadia hairs 经纬仪镜内视觉线;视距线;视距丝;十字丝
stadia hand level 视距手提水准仪
stadia intercept 视距标尺截距
stadia interval 视距间距;视距间隔
stadia level 视距水准仪
stadia level(l)ing 视距水平;视距测平;视距测高
stadia lines 视距丝;准距线;视距线
stadia lines multiplication factor 视距丝乘常数
stadial moraine 后退碛;冰退终碛
stadia measurement 视距测量
stadia method 平板仪测量法;视距法;视距测量法
stadiametric camera 视距测量照相机
stadia metric range finder 视距测距仪
stadiametric range indicator 视距测距指示器
stadia multiplication constant 视距乘常数
stadia plane-table survey 视距平板仪测量
stadia point 碎部点;视距点;视距站
stadia range finder 视距测距仪
stadia reading 视距读数
stadia reduction 视距改正

stadia reduction diagram 视距改正图
stadia reticule 视距丝
stadia rod 水准标杆；视距尺；视距标杆
stadia scale 视距标尺
stadia shot 视距测量
stadia slide ruler 视距计算尺
stadia station 视距测站
stadia survey(ing) 视距测量(法)
stadia table 视距计算表
stadia telescope 视距望远镜
stadia theodolite 视距经纬仪
stadia theodolite survey 经纬仪视距测量
stadia transit 视距经纬仪
stadia transit survey 经纬仪视距测量；视距经纬仪测量
stadia traverse 视距导线
stadia traversing 平面测量
stadia wedge 视距光楔
stadia wire 视距丝
stadia wire ring 视距丝环
stadia work 视距测量工作
stadimeter 小型六分仪；手操测距仪；视距计
stadimetric telescope 视距望远镜
stadiometer 自记经纬仪；曲线长度仪；视距仪
stadion (有看台的)露天大型运动场；赛跑运动场（古希腊的）
stadium 体育场；运动场；[复]stadia
stadium acmes 极期
stadium augment 增进期；进行期
stadium construction 体育馆建筑；体育场施工；体育场建筑
stadium facility 体育馆设施；体育场设施
stadium grandstand 体育场的大看台；体育馆的大看台
stadium increment 增进期；进行期
stadium seat 体育场上的看台座位
stadium type auditorium 阶梯式观众席
staff 棍；雇员；工作人员；杆；平衡轴；导引探子；测尺
staffa 钢筋弯筋器
staff accommodation portion 职工食宿供应区
staff aisle 职工通道
staff amount plan 职工人数计划
staff and cage surmounted buoy 带笼状顶杆的浮标
staff and globe buoy 带球杆形浮标
staff and globe surmounted buoy 球顶浮标
staff and labo(u)r 职工
staff and triangle surmounted buoy 带三角形顶标的浮标
staff and worker 职工
staff and worker's bonus 职工奖励基金
staff angle 护角条；角缘；麻刀灰抹墙角；麻刀灰墙角
staff award fund 职工奖励基金
staff bead 麻刀灰抹圆形线条；做窗格材料；墙外角护角线条
staff block 职工大学；职工生活区
staff building 职工大楼；职工宿舍楼
staff canteen 职工(临时)食堂；职工小卖部
staff car 指挥车
staff catcher 铁路收签机
staff changing room 职工更衣室
staff comparison 标尺比长
staff constant 标尺常数
staff cost 工资成本；工资费用；人员费用；人事费(用)
staff cost control 工资要素费用控制
staff dinning room 职工餐厅
staff division 标尺划分；标尺分划
staffed level crossing 有人看守道口
staff elevator 工作人员电梯
staff engineer 主管工程师
staff entrance 职工入口处
staff error 验潮仪误差；水尺误差；标尺误差
staff executive 幕后参谋主管
staff float 杆式浮子；杆式浮标；长杆浮标
staff function 职能机构
staff gallery 执勤廊
staff ga(u)ge 水位站；量杆；水准标尺；水(位)尺；测杆；标尺
staff graduation 标尺分划
staff holder 标尺手柄
staff house 职工住房；职工住宅
staffing 职员设置；配备人员
staffing function 人事职能

staffing organogram 人员组织机构图
staff interval 标尺分划间隔
staff key container 钥匙路签器
staff level 标尺水准器
staff light 杆桅灯标；灯柱
staff location system 职工寻找系统
staff lounge 职工休息室
staffman 标杆员；立尺员；司尺员；扶尺员
staff managers 辅助工作管理人员
staff member 职员
staff morale 职员纪律
staff objective 参谋部门目标
Staffordshire blues 斯塔弗郡蓝砖
staff organization chart 人员组织图表
staff plate 水准尺(尺)垫；水平尺垫
staff point 标尺点
staff pouch 路签携带器
staff proportion quota 比例定员法
staff punch 平衡轴冲头
staff quarter 职员宿舍；职工宿舍
staff reading 水位尺读数；水准尺读数；水尺读数；测尺读数；标尺读数
staff rental apartment 职工租赁公寓
staff rental dwelling unit 职工租赁住房单元
staff rental flat 职工租赁套房；职工租赁公寓
staff rental living unit 职工租赁起居单元；职工租赁生活单元
staff rod 标尺
staff room 职工室；教员休息室
staff's cafeteria 职工食堂
staff shares 职工股
staff ship 司令舰
staff toilet 职工厕所
staff traffic 职工交通运输
staff training 职工培训
staff tube 执勤隧管
staff vehicle 运人车辆；职员专用车
staff walkway 执勤道
staff warping 标尺扭曲
staff welfare funds 职工福利基金
staff with the key 钥匙路签；钥匙路牌
sta-fit 无基托的玻璃纤维卷材和垫片
stage 舷外作业；戏台；载物台；阶级；阶段；舞台；升级；登台；板架
stage acidizing 分级酸化
stage activated sludge process 分段活性污泥法
stage adapter 级转接器
stage addition 分批加入
stage aeration 阶段曝气；多级曝气；分级通气(指处理活性污泥的一种方法)；分段曝气
stage aeration activated sludge process 分段曝气活性污泥法
stage aeration method 阶段曝气法
stage aeration process 阶段曝气法
stage aeration system 阶段曝气系统
stage aeration tank 分段曝气池
stage after failure stage 破坏后阶段
stage agent addition 分段加药
stage amplitude of variation 水位变幅
stage analysis 阶段分析
stage and progress of migration 运移阶段和过程
stage area 舞台区
stage-area curve method 水位面积曲线法
stage-area method 水位面积法
stage-A resin 甲阶段树脂
stage at dam front 坝前水位
stage average 阶段平均数
stage-based piecework wage 分段计酬计件工资
stage bearer 架板架横撑木；横撑木
stage bent 工作架直柱
stage blasting 分段爆破
stage block 剧场舞台部分
stage board 舷外作业板架；脚手架踏板
stage boat 邮务艇；邮务船
stage box 舞台旁特别包厢；前侧包厢；舞台前部的包厢；舞台口包厢；特别包厢
stage breaking 逐级扎碎；逐级破碎；分级破碎；分段破碎
stage-by-stage 分阶段的
stage-by-stage method 逐级计算法
stage capacity curve 水位库容曲线
stage carriage 驿车；公共马车
stage casing pump 多级套管泵
stage cementing 分级注水泥

stage cementing collar 分期黏结环；分层黏结环
stage coach 公共马车
stage compaction 分级压密
stage compaction method 分级压实法
stage compression 分段压缩
stage concentration 逐段精选法
stage construction 分段施工；多层面构造；分期施工；分期建筑；分期建设；分期建造
stage cooler 阶梯形冷却器
stage cooling 分级冷却；分段冷却
stage coordinate system 车架坐标系【测】
stage correlation curve 水位相关曲线
stage crushing 逐级扎碎；逐级破碎；石料分级粉碎；分级破碎；分级粉碎；分段破碎
stage curtain 舞台幕布
stage cut 分段开挖法
staged 阶梯式；成级；分级；分段的
stage dam heightening 大坝分期加高
staged capacity 分期容量
staged charging system 分段进料系统
staged cost 阶段费用
staged development plan 分期开发计划
staged diversion 分期导流；分阶段引水；分阶段分流；分阶段分洪
stage decompression 阶段减压
stage-depth relation curve 水位水深关系曲线
stage designer 舞台设计师
stage development 阶段开发；分期开发；分期建设；分期发展；分级开发；分层开采法
stage development plan 阶段开发计划；地区开发阶段规划
staged fracturing 分段压裂
stage die 过渡模
stage digestion (用二个或多个池子进行污泥的)逐级消化；分级消化【给】；分段消化
stage digestion of sludge 污泥分级消化
stage dimmer 舞台调光机
stage-discharge curve 水位—流量曲线；水位—流量关系曲线
stage-discharge formula 水位—流量公式
stage-discharge record 水位—流量记录
stage-discharge(relation)curve 水位—流量关系曲线
stage-discharge relation(ship) 水位—流量关系
stage diversion method 分期围堰法
staged laser heating 分级激光加热
staged limited-entry technique 分段限流量法
stage door 后台入口；后台门；舞台后台门
staged operation 分段操作
stage dredging 分期挖泥；分期疏浚
staged research 分段搜索
stage drop 级内压降
stage drying 分段干燥
staged tensioning 阶段张拉
staged transformer 分级变压器
stage duration 水位保持时间
stage-duration curve 水位历时曲线；水位变化曲线
staged water-to-water heat exchanger 分段式水—水换热器
stage effect 阶段收益
stage efficiency 级效率；分段效率
stage entrance 舞台入口
stage equipment 舞台布景设备；舞台设备
stage escape 舞台疏散口
stage evapo(u)ration 分级蒸发；分段蒸发
stage extraction 分段提取；分段萃取
stage filter 分段(过)滤器；分级过滤池；分段过滤器；分层滤器
stage fittings 舞台设备
stage flo(a)tation 连续浮选；阶段浮选；分级浮选；分段浮选
stage fluctuation 水位涨落；水位波动；水位变化
stage fluctuation range 水位波动范围；水位变幅
stage for construction 施工水位
stage forecast(ing) 水位预报
stage for liferaft 救生筏滑架
stage for soil and broken rock 碎落台
stage frequency curve 水位频率曲线
stage gain 级增益；放大级增益
stage gangway 架板
stage ga(u)ge 水位标尺
stage-ga(u)ging station 水位站
stage grafting 分级接枝

stage gridiron 舞台布景升降结构
stage grinding 阶段磨矿;分段磨矿;分段磨矿
stage grouting 阶段灌浆;分期灌浆(法);分级灌浆;分段灌浆法;分层灌浆
stage-hardening 分段淬火
stage hardware 舞台小五金;舞台硬件
stage head 级压头;级间压差;级差
stage heat drop 级的热降
stage heating 逐步加热;分段加热
stage hoist 分段提升机
stage holding dolly 弹ศ支承车
stage horizon 弧形天幕
stagehouse 舞台用房;舞台的台口以内部分的建筑
stage hydrograph 水位图;水位曲线图;水位过程线
stage hydrograph for characteristic year 特征年水位过程线
stage hydrograph of flood 高水位曲线;洪水水位曲线
stage illumination 舞台照明
stage improvement 分期整治;分期改善;分期改建;分级调质
stage indicator 水准指示器
stage in reservoir region 库区水位
stage intercooling 级间冷却
stage invariant 阶段不变量
stage jigging 阶段跳汰
stage lamp 舞台灯
stage lashing 脚手架捆绑法;跳板结
stage left 舞台左侧(以演员面对观众时为准)
stageless transmission 无级传动
stage level 舞台高程;舞台高度
stage lift 升降台;舞台升降机;升降舞台
stage light(ing) 舞台灯光;舞台照明
stage lighting effect 舞台灯光效果;舞台照明效应
stage lighting units 舞台灯光装置
stage liquefaction 分级液化
stage load 级负荷
stage loader 装料传送机;分段装载机
stage loading 分阶段加荷;分级加荷
stage machinery 舞台机械设备
stage matching 级间匹配
stage measurement 水位测量;水位测定
stage measurement cross-section 水位观测断面
stage micrometer 载物台测微计;镜台测微计;镜台测微尺;台式测微计;分级测微计
stage mining 分段开采
stage-mixed concrete 分阶段搅拌的混凝土
stage number 级数
stage observation 水位观测
stage of aggregation 聚合程度;凝聚程度
stage of amplification 放大级
stage of a relative standstill 相对停滞阶段
stage of attack 发作期
stage of bank ruin of reservoir 水库坍岸阶段
stage of comparator 坐标仪照片盘
stage of completion 完工程度
stage of compression 压缩级
stage of construction 施工阶段
stage of constructive map 施工图阶段
stage of detailed investigation 详查阶段
stage of detailed planning 细部设计联合体
stage of development 显影过程;开发阶段;发展阶段
stage of dissolution 溶解期
stage of drying 干燥阶段
stage of earthquake prediction 地震预报阶段
stage of economic development 经济发展阶段
stage of elastic deformation 弹性变形阶段
stage of enriched alumina-silicification 富硅铝化阶段
stage of enriched calcification 富钙化阶段
stage of enriched calorization 富铝化阶段
stage of evaluation 评价阶段
stage of fervescence 发热期
stage of freezing 冻结期;封冻期
stage of gas lift 气举级数
stage of geologic(al) prospecting-exploration 地质勘察工作阶段
stage of introduction 初期
stage of invasion 侵入期
stage of investigation work 已有调查工作阶段
stage of ionization 电离级
stage of liferaft 救生筏滑架
stage of loading 荷载阶段

stage of local shearing 局部剪切阶段;局部剪力阶段
stage of local shearing failure 局部剪切破坏阶段;局部剪力破坏阶段
stage of mechanical fragmentation 机械破碎阶段
stage of melting 解冻期;融雪期
stage of metamorphism 变质期次
stage of mineral exploration 矿产勘查工作阶段
stage of mineralization 矿化阶段
stage of neotectonic movement 新构造运动阶段
stage of planning and selecting site 规划选点阶段
stage of project 工程进程
stage of prospecting and exploration of mineral resources 矿产勘查工作阶段
stage of quaternary system 第四纪分期
stage of reaction 反应期;反应阶段
stage of reconnaissance survey 普查阶段
stage of ripple wave 涟波期
stage of river 河流水位
stage of stability 稳定阶段
stage of stream 河流水位
stage of strength 强度级
stage of tectogenesis 构造分期
stage of tide 潮汐位相;潮汐水位;潮位
stage of turning 回转运动阶段
stage of weathering 风化阶段
stage of works 工程进展
stage of zero flow 流量为零时水位;断流水位
stage one 第一阶段
stage-one bid 第一步投标
stage operating plan 阶段计划
stage peg 粗纹螺栓
stage plank 脚手架踏板;简易码头;架板;跳板
stage planning 分期规划
stage platform 有梯级平台
stage plunger pump 多级滑阀泵;多级柱塞泵
stage pocket 舞台后台电源插销盒
stage projector 舞台聚光灯
stage property 道具
stage pump 多级泵
stage purification 分级净化
stage ratio 水位比率
stage-ratio method 水位比率法
stage reconstruction 分期改建
stage recorder 自记水位站
stage recorder installation 自记水位计台
stage recorder station 自记水位站
stage reduction 逐级扎碎;逐级破碎;分级粉碎;分段粉碎;分度粉碎
stage regulation 分期整治
stage relation 相应水位曲线
stage relation curve 水位关系曲线
stage relation method 水位相关法
stage-relationship curve 水位相关曲线;水位关系曲线
stage rigging 舞台索具
stage right 舞台右侧(以演员面对观众时为准)
stage rope 脚手架支索;架板绳;吊盘钢丝绳
stage rostrum 阶梯讲坛
stage routing 水位演算
stage sampling 阶段抽样
stages and sequence of faulting 断层活动期次
stage scene 舞台布景
stage scenery 舞台布景
stages cyclone preheater 多级旋风预热烧成系统
stage shaft 多级烟囱;多级竖井;多层舞台
stage shop 舞台工场
stages in germination 萌芽期
stage sling 脚手架支索
stage sling slip knot 架板活结
stages of development of diwa region 地洼区发展阶段
stages of economic growth 经济发展阶段
stages of volcanic activity 火山活动期次
stage space 舞台空间
stage speed 级速率
stage spotlight 舞台聚光灯
stage storage curve 水位库容曲线
stage strip light 舞台带形灯
stage teeth 高低齿
stage temperature rise 级温升
stage theater 舞台戏院
stage-time curve 水位过程线
stage-time curve for characteristic year 特征年水位过程线
stage tooth 气封高低齿
stage tower 多级塔楼
stage transmission 分级传递
stage transmitter 水位传送器
stage treatment 分级处理
stage trickling filter 多级生物滤池;分段滴滤池
stage turbine 多级涡轮(机);分式涡轮;分级涡轮(机);分级(式)汽轮机
stage twisting machine 分步法捻线机
stage twisting system 分步法捻线系统
stage two 第二阶段
stage-two bids 第二步投标
stage-two tenders 第二步投标
stage valve 旁路阀
stage variation 阶段变异
stage versus Q/F curve 水位—流量模数曲线;水位径流数曲线
stage wagon 有轮布景
stage wall (舞台的)台口墙
stage wall pocket 舞台墙上电源插板
stage well 多级竖井;多层舞台
stagewise 阶梯
stagewise contactor 多级式接触装置
stagewise correlation procedure 逐段回归(方)法
stagewise operation 分段操作
stagewise process 逐级过程
stagewise regression procedure 逐段回归(方)法
stage with trap 有升降台板的舞台
stage-working 舞台前部
stagflation 通货滞胀;停滞膨胀
stagger 斜罩;拐折;拉出值【电】;交错配置;交错;错开排列;错列
stagger amplifier 参差调谐放大器;参差调谐放大器
stagger angle 交错角
stagger arrangement 错列
stagger blanking 锯齿形冲裁
stagger bond 咬砌
stagger circuit 参差调谐式电路
stagger convertion 交错变换
staggered (空心隔墙中的)错列的墙筋;参差的;交错的;错开的
staggered air heater 拐折空气加热器
staggered apses 参差的半圆室
staggered arrangement 参差排列;参差布置;交错排列;交错布置;错排;错列
staggered arrangement of shotholes 炮眼交错布置
staggered band 错列布置管系
staggered brick 错缝砖
staggered brush arrangement 电刷的交错布置
staggered brushes 参差安置的电刷组
staggered building arrangement 交错组合
staggered channel separator 交错沟槽式集尘器;错列式除尘器
staggered circuit 相互失谐级联电路;参差调谐电路
staggered comparison test 交错比较试验
staggered conformation 参差构象
staggered course 交错层;错列层
staggered cross road 错位交叉路
stagger(ed) cycle 交错周期;交叉周期
staggered cylinders 交错(排列)汽缸
staggered deeps 交错的深槽
staggered difference scheme 交替差分格式
staggered electrode 参差电极
staggered fan 交错叶风扇
staggered feed 不均匀进给
staggered floor 参差的楼层;错层
staggered floor parking garage 错层式车库
staggered form 交叉式;参差式
staggered ga(u)ging 参差调测
staggered gear 交错齿轮;错牙齿轮
staggered header 交错集管
staggered intermittent fillet weld(ing) 交错断续角焊
staggered intermittent weld 交错(式)间断焊;交错断续焊缝
staggered intersection 错位交叉
staggered joint 相互式接头;相错式接头(错接);交错式接缝;错列接头;错列接合;错接接头;错缝接合;错(开接)缝
staggered-joint bond 错缝砌法
staggered junction 错位(式)交叉
staggered line 交错线道

staggered line pattern 交错排列布置
staggered mill 参差轧机之间的通道
staggered multiple rows of tubes 多行管的交错排列；错列多排管
staggered packing 交错棋盘式排列的格子体
staggered parallel runways 参差平行的导轨；参差平行的滑道；参差平行的跑道
staggered parking 交错式停车
staggered partition 错列龙骨隔墙；墙筋交错布置的隔墙
staggered pattern 交错布置；梅花形模式；梅花形布置
staggered perforated plate 错列式多孔板
staggered perspective 交错的透视
staggered piles 交错桩；梅花桩
staggered piling 交错桩；梅花状交错桩；跳打；打梅花桩；错列打桩
staggered pipe arrangement 错列管排
staggered pipes 交错管排；错列管
staggered planting 交错形栽植
staggered plug weld 交错塞焊
staggered pools 交错的深槽
staggered protruding points 交错突嘴
staggered riveting 之字铆接；交错铆接；错列铆钉
staggered rivet joint 交错铆接缝；错列铆接
staggered rolling train 布棋式机座布置
staggered rows of rivets 错列铆钉排
staggered scanning 隔行扫描
staggered seating 错排座席；前后排座位错开的布置形式
staggered section view 阶梯剖视图；阶梯形剖视图
staggered spot-welding 交错点焊
staggered square pitch 正方形错列
staggered stacking 错列堆垛法
staggered stories 交错楼层
staggered-stud partition 错列木柱隔墙；错列筋板条隔墙；双层分离式隔墙；错列龙骨隔墙
staggered stud system 错列立筋体系
staggered toe elevation 锯齿形桩尖标高
staggered tooth 错齿
staggered tooth double-helical gear 交错齿人字齿轮
staggered tooth side cutter 错齿侧铣刀
staggered triangular pitch 正三角形错列
staggered triple tunning 三重参差调谐
staggered tube arrangement 错列管排
staggered tubes 交错管排；拐折管排；错列管
staggered tuning 参差调谐
staggered-wall-beam building 竖向错列的墙或剪力墙建筑
staggered welding 错列焊缝
staggered wells 交错排列布井
staggered working day 交错工作时间
staggered working hours 错开上下班时间；错开工作时间
stagger error 摆动误差
stagger feed 锯齿形送进；交错送料
stagger ga(u)ge 拉出值仪
stagger holidays 轮休
staggering 交错的；交错式排列；参差；错列
staggering advantage 参差效益
staggering joint 分段缝
staggering of upper and lower deeps 上下深槽交错
staggering of upper and lower pools 上下深槽交错
staggering polarity 极性交叉
staggering seats 错列座位
stagger joint 交错接头；错接；错缝
stagger laying 错缝砌筑
stagger moving-target indicator 交错对消式动目标显示器
stagger of stabilizer 稳定器摆动
stagger-peaked 参差峰化
stagger ratio 参差度
staggers 眩晕症
stagger scheme 跳点格式
staggers tubes 交错管排
stagger time 参差时间
stagger-tuned 交错调谐
stagger-tuned filter 参差调谐滤波器
stagger tuning 串联同调；串联调谐
stagger tuning amplifier 参差调谐放大器
stagger tuning circuit 参差调谐式电路

stagger wire 倾角线
stagger-wound coil 笼形线圈；筐形线圈
stag-headed 鹿头状的
staging 构架；工作台；脚手架；级分离；平台架；登台；分期；驿车业；驿车旅行；台架；上演
staging adapter 阶梯转接器；升级适配器
staging area 中继工作场
staging base 飞机中转站
staging drive 阶梯驱动
staging drive group 阶梯驱动组
staging effective data rate 阶梯有效数据传输率；登台数据有效率
staging error 级误差；登台错误
staging field 中间机场
staging hierarchy 分级体系(结构)
staging life 储[贮]存寿命；储[贮]存期限
staging method 架台法
staging of pipeline 管道架
staging pack 登台盘组
staging plank 脚手架板
staging point 中途站
staging pole 脚手杆
staging post 补给站
staging station 中途站
staging system 架台法
staging the pipe 间隙式下钻
staging upright 工作架直柱
stagnancy 停滞
stagnant 停滞的；不流动的
stagnant air 滞止气流；滞(流)空气；停滞空气；闭塞空气
stagnant ambient medium 静止周围介质
stagnant area 死水区
stagnant area corrosion 停滞区腐蚀
stagnant atmospheric condition 气团停滞状态
stagnant basin 死水塘
stagnant bottom water 停滞底水
stagnant catalyst 固定催化剂
stagnant condition 滞止条件
stagnant evolution 停滞进化
stagnant film 滞膜；停滞膜
stagnant-film model 停滞膜模型
stagnant gas 停滞气体
stagnant glacier 静止冰川；停滞冰川
stagnant glass 停滞层玻璃(池窑底部)
stagnant ground water 上层滞水
stagnant inversion 停滞逆温
stagnant liquid film 静止液膜
stagnant mo(u)ld 不动模型
stagnant part 不动部分
stagnant point 停滞点
stagnant pond 死水塘
stagnant pool 滞水池；静止地下水；静水池；积水塘；积水池；死水塘
stagnant sea 带流海
stagnant steam 停滞蒸汽
stagnant time of induced earthquake 诱发地震滞后时间
stagnant vortex 滞留涡
stagnant wake 滞流区；停滞区
stagnant water 滞水；积水塘；停滞水；死水
stagnant zone 滞流带；黏滞区；静止区(域)；难变形区；停滞带
stagnated air mass 停滞气团；停滞空气团
stagnated overpopulation 停滞过剩人口
stagnating margin 停滞边缘
stagnating point 停滞点
stagnating sectors 停滞的部门
stagnation 滞流；停滞作用；停滞性；停滞
stagnation condition 临界点条件；界界点参数；停滞状态
stagnation density 滞止密度
stagnation energy 滞止能
stagnation enthalpy 滞止焓
stagnation flow 滞止流
stagnation groundwater 积滞地下水
stagnation line 滞止线；滞点线；临界点线
stagnation of movement 停驶
stagnation overpressure 临界点剩余压力
stagnation parameter 滞止参数
stagnation period 停滞时间
stagnation point 驻点；滞留点；滞流点；临界点；积滞点；停滞点；停止点
stagnation point movement 临界点位移

stagnation point shift 临界点偏移
stagnation point temperature 滞点温度
stagnation pool 积水池
stagnation pressure 总压；驻点压力；滞止压力；滞点压力；积滞压力；停滞压力
stagnation pressure probe 滞止压力传感器
stagnation pressure reduction 总压力减小；滞止压力减小
stagnation state 滞止状态
stagnation surface 驻面
stagnation temperature 总温；驻点温度；滞止温度；滞流温度；停滞温度
stagnation-temperature control 滞止温度调节
stagnation-temperature probe 滞点温度传感器
stagnation tube 滞压测流管
stagnation value 滞止值
stagnation water 积滞水
stagnation zone 静止区(域)；停滞区
stagnophile 静水生物
stagnoplankton 静水浮游生物
stagnopseudogley 滞止假潜育土
stagnum 滞水体
stagonometer 滴重计；表面张力滴重计；表面张力滴量计
Stahl-eisen slag 斯塔尔—爱森矿渣(炼钢生铁生成的矿渣)
Stahlton 斯塔尔顿式；预应力黏土
Stahlton factory 斯塔尔顿式工厂；预应力黏土工厂
Stahlton floor 斯塔尔顿式砖楼板；预应力黏土砖楼板
Stahlton plank 预应力黏土板；斯塔尔顿式(空心楼)板
Stahlton plank window lintel 斯塔尔顿式砖窗梁；预应力黏土砖窗过梁
Stahlton plant 斯塔尔顿式厂；预应力黏土厂
Stahlton prestressed beam 斯塔尔顿式梁；预应力黏土梁
Stahlton prestressed clay plank 斯塔尔顿式板；预应力黏土砖板
Stahlton prestressed door lintel 斯塔尔顿式门梁；预应力黏土门过梁
Stahlton prestressed floor 斯塔尔顿式地板；预应力黏土地板
Stahlton prestressed lintel 斯塔尔顿式过梁；预应力黏土过梁
Stahlton prestressed roof 斯塔尔顿式屋顶；预应力黏土屋顶
Stahlton prestressed roof slab 斯塔尔顿式屋面板；预应力黏土屋面板
Stahlton prestressed wall panel 斯塔尔顿式墙板；预应力黏土墙板
Stahlton prestressed wall slab 斯塔尔顿式墙板；预应力黏土墙板
Stahlton prestressed window lintel 斯塔尔顿式窗过梁；预应力黏土窗过梁
Stahlton roof 斯塔尔顿式砖屋顶；预应力黏土砖屋顶
Stahlton roof slab 斯塔尔顿式砖屋面板；预应力黏土砖屋面板
Stahlton slab 砖铺底浇灌砂浆的预应力槽形板(瑞士创制的)
Stahlton wall unit 斯塔尔顿式墙体单元；预应力黏土墙体单元
Stahl turbine 斯塔尔汽轮机；双回转反击式汽轮机
staight needle 直针
stain 用调色剂染色；染色；染剂；污损；污染痕迹；污迹；污点；斑点；染污；着色
stain colo(u)rimetric test paper 斑迹比色滤纸
stain control agent 防污剂
stained 着色的；染污
stained by salt water 盐渍
stained clay 呈色黏土
stained flux 着色溶剂
stained glass 有色玻璃；着色玻璃；烧彩玻璃；色玻璃；彩色玻璃；彩色冰屑玻璃；彩绘玻璃；彩画玻璃；彩花玻璃；冰屑玻璃
stained-glass block 彩色玻璃砖
stained-glass brick 彩色玻璃砖
stained-glass wall 彩色玻璃墙
stained-glass window 彩色玻璃窗
stained heartwood 变色心材
stained method 染色法
stained paper 色纸；彩色(糊)墙纸
stained preparation 染色标本
stained rough rolled glass 彩色粗玻璃

stained sapwood 变色边材
stained soil suspension preparation 土壤染色制片法
stained with oil 油污的
stained wood 着色木材;变色木材
stainer 着色液;着色剂;染工;着色工;染色器;调色浆;调色工;色料;筛网;染色剂;滤网;粗滤器
stain etch 染色腐蚀
stain finish 无光光洁度
stainierite 水钴矿
staining 侵蚀;染色(法);刷染法;生锈;腐蚀斑;着色
staining agent 着色剂;染色剂
staining bath 染色槽
staining class 侵蚀等级
staining colo(u)r 沾色
staining compound 着色料
staining dish 染色碟
staining material 染色剂
staining method 着色法;染色法
staining of concrete 混凝土表面污渍
staining plate 染色皿
staining power 着色(能)力;着色强度;染色力;着色本领
staining rack 染色架
staining reaction 染色反应
staining solution 染液;染色溶液
staining stand 染色架;染架
staining technique 染色技术;染色法
staining test 染色试验
staining trough 染色槽
stainless 不锈的
stainless acid resistant steel 不锈耐酸钢
stainless acid-resistant tube 不锈钢耐酸管
stainless acid-resisting steel wire 不锈耐酸钢丝
stainless bar 不锈钢筋
stainless bushing 不锈衬套
stainless-clad plate 镀防锈层(的)板
stainless-clad steel 不锈复合钢;不锈钢钢包层金属;不锈包层钢
stainless-clad steel plate 包不锈钢板
stainless compound steel plate 不锈复合钢板
stainless conveyer belt 不锈钢运输带
stainless eyelet 不锈钢孔眼
stainless foil 不锈箔
stainless gear pump 不锈钢齿轮泵
stainless iron 低碳铁素体不锈钢;不锈铁
stainless oil 不锈油
stainless paint 防污涂料
stainless pin 不锈钢针
stainless rule 不锈钢尺
stainless sheet steel 不锈钢薄板
stainless shower seat 不锈钢淋浴座
stainless steel 不锈钢
stainless steel alcohol stove 不锈钢酒精炉
stainless steel alloy 不锈钢合金
stainless steel angle 不锈钢护角
stainless steel angle rule 不锈钢拐尺
stainless steel ball 不锈钢球
stainless steel ball valve 不锈钢球阀
stainless steel band 不锈钢带
stainless steel bar 不锈钢筋
stainless steel basin and plate 不锈钢盘碟
stainless steel beam 不锈钢梁
stainless steel blank 不锈钢种板
stainless steel bolt 不锈钢螺栓
stainless steel buckle 不锈钢扣
stainless steel burden drum 不锈钢配料桶
stainless steel butterfly value 不锈钢蝶阀
stainless steel cable clamp 不锈钢电缆夹
stainless steel case 不锈钢壳
stainless steel ceiling 不锈钢天花板;不锈钢顶棚
stainless steel clad composite steel sheet 不锈钢复合钢板
stainless steel clamp 不锈钢夹
stainless steel clasp 不锈钢丝卡环
stainless steel cleaner 不锈钢清洗剂
stainless steel cloth 不锈钢丝网
stainless steel condenser 不锈钢冷凝器
stainless steel condiment pot 不锈钢调料罐
stainless steel corrosion-resistant needle valve 不锈钢防腐针型阀
stainless steel crucible 不锈钢坩埚
stainless steel curtain wall 不锈钢幕墙
stainless steel disc 不锈钢盘

stainless steel disc for casting 铸造用不锈钢块
stainless steel door 不锈钢门
stainless steel electrode 不锈钢电极;不锈钢焊条
stainless steel enamel 不锈钢搪瓷
stainless steel external wall lining 不锈钢外墙衬里
stainless steel facing 不锈钢房屋正面;不锈钢饰面
stainless steel fertilizer belt 不锈钢撒肥带
stainless steel fiber 不锈钢纤维
stainless steel filter 不锈钢过滤器
stainless steel flag hinge 不锈钢旗铰链
stainless steel flashing piece 不锈钢泛水(件)
stainless steel flexible hose 不锈钢软管
stainless steel grating 不锈钢滤栅
stainless steel handrail 不锈钢钢扶手
stainless steel heat exchange 不锈钢热交换器
stainless steel heavy duty vernier calipers 不锈钢重型游标卡尺
stainless steel helices 不锈钢填料
stainless steel honeycomb core 不锈钢蜂窝芯子
stainless steel honeycomb panel 不锈钢蜂窝板
stainless steel inside calipers 不锈钢内卡钳
stainless steel kettle 不锈钢锅
stainless steel kitchenware 不锈钢厨具
stainless steel lining 不锈钢隔板;不锈钢衬里
stainless steel mesh 不锈钢筛网
stainless steel mineral spring pump 不锈钢矿泉水泵
stainless steel mother blank 不锈钢母板
stainless steel net-cover of axial fan 不锈钢轴流风机网罩
stainless steel nut 不锈钢螺母
stainless steel nylon hinge 不锈钢尼龙铰链
stainless steel nylon ring square hinge 不锈钢尼龙垫圈方铰链
stainless steel outdoor wall surfacing 不锈钢外墙饰面
stainless steel outside calipers 不锈钢外卡钳
stainless steel overlay 不锈钢铺面;不锈钢表面堆焊
stainless steel paper holder 不锈钢纸托
stainless steel pigment 不锈钢颜料
stainless steel pin 不锈钢枢轴;不锈钢针
stainless steel pipe 不锈钢管
stainless steel plate 不锈钢板
stainless steel plate meeting face 不锈钢板贴接面
stainless steel powder 不锈钢粉
stainless steel products 不锈钢制品
stainless steel pump 不锈钢泵
stainless steel retainer ring 不锈钢卡环
stainless steel retort 不锈钢蒸馏罐
stainless steel rivet 不锈钢铆钉
stainless steel roof cladding 不锈钢屋盖
stainless steel ruler 不锈钢直尺;不锈钢板尺
stainless steel sash window 不锈钢(制)窗扇
stainless steel screen 不锈钢丝网;不锈钢屏
stainless steel seamless thin-walled tube 不锈钢薄壁无缝管
stainless steel sheet 不锈钢片;不锈钢板
stainless steel shelf 不锈钢搁架
stainless steel sink unit 不锈钢水槽;不锈钢洗涤盆
stainless steel square 不锈钢直角尺
stainless steel square hinge 不锈钢方铰
stainless steel submersible pump 不锈钢潜水泵
stainless steel swing door 不锈钢回转门
stainless steel tank electric(al) water heater 不锈钢储水内胆电热水器
stainless steel telescopic joint 不锈钢伸缩节
stainless steel thimble 不锈钢护环
stainless steel try square 不锈钢矩尺
stainless steel tube 不锈钢管
stainless steel vacuum pugmill 不锈钢真空炼泥机
stainless steel valve 不锈钢阀(门)
stainless steel wall panel 不锈钢墙板
stainless steel ware 不锈钢器皿
stainless steel watch 不锈钢表
stainless steel watch band 不锈钢表带
stainless steel watch case 不锈钢表壳
stainless steel watch case blank 不锈钢表壳粗坯
stainless steel wedge-wire screen 楔形不锈钢丝滤网
stainless steel welding electrode 不锈钢电焊条
stainless steel window 不锈钢窗
stainless steel wire 不锈钢丝;不锈钢线
stainless steel wire cable 不锈钢丝绳;不锈钢钢缆

stainless steel wire cloth 不锈钢丝布
stainless steel wire mesh 不锈钢金属网
stainless steel wire rope 不锈钢丝绳
stainless steel wire sieve 不锈钢丝筛网
stainless steel wire suture 不锈钢丝缝合线
stainless steel wire wrapped screen 不锈钢钢丝绕制筛管
stainless steel wrap wire 不锈钢绕丝
stainless strip 不锈钢带材
stainless trim 不锈钢镶边
stainless try square 不锈钢直角尺
stainless valve 不锈钢阀(门)
stain on firing 压花
stain-proof 防污点;防锈
stain-proof paint 防污涂料
stain removal 去污点;去锈
stain resistance 耐污染性;耐蚀性;耐染污性
stain-resistant glass 抗霉玻璃
stain resisting paint 防污涂料
stain soil suspen sion preparation 土壤深色制片术
stain test 污点法;点滴试验;斑痕试验
stain varnish 着色清漆
stair 阶梯的一级楼梯
stair bearer 楼梯段纵梁
stair bolt 楼梯扶手螺栓
stair bracket 楼梯踏板加劲撑板
stair brick 楼梯砖
stair builder's truss 楼梯平台十字梁
stair carpet 楼梯地毯;铺在楼梯上的地毯
stair carriage 楼梯基;楼梯格栅;楼梯纵梁
staircase 梯子;楼梯;楼梯间;阶梯
staircase construction 楼梯施工
staircase core 楼梯井
staircase curve 阶梯形曲线;阶梯曲线
staircase deflection 阶梯式偏移
staircase depth 楼梯间深度
staircase design 阶梯形设计
staircase distribution 阶梯形分布
staircase effective breadth 楼梯有效宽度
staircase effective span 楼梯有效跨度
staircase fault 阶状断层;台阶状断层
staircase function 阶梯函数
staircase generator 阶梯波发生器;阶梯波发生器;台阶形波发生器;台阶式波形发生器
staircase iteration 阶梯迭代
staircase joint 楼梯接头
staircase landing 楼梯休息平台
staircase length 楼梯长度
staircase lock 阶梯式船闸;多级船闸
staircase longitudinal connection 楼梯纵向连接
staircase maser 阶梯式微波激射器;阶梯激射器
staircase moment 楼梯力矩
staircase of bifurcated type 双分楼梯;双分叉式楼梯
staircase of dog-legged type 弯脚式楼梯;狗腿式楼梯;无梯井两跑式楼梯
staircase of half-turn type 180 度式楼梯
staircase of helical type 螺旋式楼梯
staircase of helical type with newel 中柱螺旋式楼梯(中柱支承悬臂踏步)
staircase of open-well type 明梯井式楼梯;带梯井式楼梯
staircase of quarter-turn type 90 度转角式楼梯
staircase of straight flight type 单梯段楼梯;单跑式楼梯;单路式楼梯
staircase of three-quarter turn type 二百七十度转角式楼梯;三跑式楼梯
staircase phenomenon 阶梯现象
staircase post 楼梯柱
staircase railing 楼梯栏杆
staircase shaft 楼梯间;楼梯井
staircase stay 楼梯支撑
staircase stiffener 楼梯加劲(杆件)
staircase stiffening 楼梯加劲(杆件)
staircase stress 阶梯应力
staircase structure 阶梯结构
staircase thickness 阶梯厚度
staircase tower 楼梯塔;楼梯间塔楼;楼梯间上塔楼
staircase transfer 楼梯换乘
staircase transverse joint 楼梯横向连接
staircase turret 楼梯间上塔楼;楼梯间塔楼
staircase voltage generator 阶梯形电压发生器
staircase waveform 阶梯波形
staircase way 梯子

staircase well 楼梯井
staircase width 楼梯宽度
staircase window 楼梯窗
staircase with several flights 多梯段楼梯;多阶段楼梯
staircasing 阶梯效应
stair clearance 楼梯净空;楼梯净空高度
stair clip 固定楼梯地毯条;楼梯地毯条;楼梯地毯夹;固定楼梯地毯条
stair disappearing 隐蔽楼梯;隐蔽阁楼梯
stair exit 楼梯(安全)出口
stair flight 楼梯段;楼梯跑
stair flight form(work) 楼梯梯段模板
stair ga(u)ge 楼梯量规
stair hall 楼梯厅
stair handrail 楼梯扶手
stairhead 楼梯顶口;梯段顶;楼梯顶端
stair headroom 楼梯净空(高度)
stair horse 车架构件;楼梯中线加固梁;楼梯搁架;楼梯踏步梁;扶梯支架;扶梯踏步梁;楼梯(支)架
stair landing 楼梯(休息)平台
stairlift 楼梯升降机
stair light 楼梯灯
stair nosing 楼梯踏步小凸沿;楼梯踏步小突沿;梯级凸沿;踏步前缘
stair platform 楼梯(中间)平台;休息平台
stair post 楼梯柱
stair rail 楼梯扶手;楼梯栏杆
stair rail bent glass 楼梯扶挡用弯玻璃
stair railing 楼梯栏杆;楼梯扶手
stair-rail system 扶手系统
stair raiser 梯级起步
stair reeding 梯级上防滑条
stair resting on arches 拱支楼梯
stair rise 楼梯踏步高度;梯段高度;梯段尺寸
stair riser 楼梯踢板;楼梯踏步竖板;楼梯起步板
stair rod 楼梯地毯压条;楼梯(地)毯棍;楼梯压毯棍
stair run 楼梯段的水平尺寸;楼梯踏板高度;梯段水平距离
stair seat 阶梯形座席
stair shaft 楼梯井
stair shoe 支持栏杆的楼梯梁顶的线脚
stair slab 楼梯段;楼梯休息平台
stair soffit 楼梯(段)背面;楼梯底面;楼梯下端
stair spout 水落管
stair step (楼梯的)梯级;阶梯形的;楼梯踏步;楼梯踏板
stair-step cost 阶梯成本;梯级成本
stair-step signal 阶梯信号;阶级信号;步进信号
stair string(er) 楼梯梁
stair tower 楼梯间(塔楼)
stair tread 楼梯踏(步)板;楼梯踏步平板
stair tread cover(ing) 楼梯踏板面层
stair tread grinder 楼梯踏板研磨机
stair trimmer 楼梯段端部托梁
stair turret 楼梯间塔楼;屋顶楼梯间
stair unit 预制(混凝土)梯段
stairwalking hand truck 能上楼梯的手推小车;能上楼梯的轮椅
stairwall 楼梯承(重)墙;楼梯踢脚;楼梯墙
stair wall string 靠墙楼梯梁
stairway 阶梯;楼梯(间)
stairway enclosure 有围护的楼梯;有护栏楼梯
stairway headroom 楼梯净空高度
stairway of locks 船闸的阶梯
stairway risers and treads 楼梯竖板及踏板
stairway stringer 舷梯侧板
stairway width 楼梯宽度;阶梯宽度
stairwell 楼梯间;楼梯井
stair winder 楼梯斜踏步;楼梯的斜踏步;盘梯;楼梯扇步
stair wire 楼梯地毯棍
staith(e) 转运码头(尤指煤码头);铁路装货台
stake 桩(子);铸型用销钉;柱;拉软床;勘探桩;木橛;栅条;撑住;标桩【测】
stake-and-stone sausage construction 桩及钢丝网石笼护岸建筑
stake boat 航标艇;换拖驳船
stake body 平板车身
stake body trailer 拖车;平板车
stake dam 排桩透水坝;桩坝;排桩滞流透水坝
stake driver 打桩工;打桩机
staked to the underlying soil 把桩打入土内

stake ga(u)ge 桩式水尺;矮桩式水尺
stake hound 得过且过的工人
stake line 定灰线桩;标定线;用桩标出的测线;桩线
stake lorry 可插上边桩的平板车
stake-man 放样工;打桩工;标桩工
stake net 桩围渔网
stake off 标出;立桩;定线;放样
stake of line 定线桩
stake out 用桩定出;定线;放样;以桩支护;立桩划分
stake-out-peg out 标桩定线
stake piles 簇桩
stake pin 测针【测】
stake pole 桩式水尺
stake rack 栅栏
stake-setting 定线;放样;立桩【测】;立标桩
stake staff 桩式水尺
stake truck 带帮载货汽车(装有栅栏的载货汽车);可插上边桩的平板汽车
stake up 用桩围住
staking 立桩定线;立桩;桩式接合;凿缝;立支架;立标桩;嵌缝连接;钉木橛;打标桩;标桩定线
staking a line 用小桩标出的测线;定灰线桩;放线
staking die 凿缝凹模
staking line 定灰线桩
staking machine 刮软机,拉软机;打桩机
staking out 标桩;加桩;立桩【测】;立标;定线;钉龙门桩;打桩放样【测】;放样
staking out a curve (线路测量的)曲线放线;(线路测量的)曲线放样
staking out center line 中桩测设
staking out plan 放样图
staking out the bank line 用桩标出堤线;用桩标出岸线
staking out work 现场定线工作;标定工作
staking pin 定线桩
staking punch 凿缝凸模
staking-setting 立标桩
staking tool 打桩机;冲心凿;冲心錾
stalactic(al) 钟乳石状的
stalactiform 钟乳石状的
stalactite 钟乳状沉积物;钟乳石
stalactite grotto 钟乳石洞
stalactite ornament 钟乳石饰;钟乳石饰
stalactite vault(ing) 钟乳石穹顶;钟乳饰穹顶
stalactite work 钟乳状雕饰;钟乳石饰;钟乳石细工
stalactitic(al) 钟乳石状的;钟乳石质的
stalactitic(al) gypsum 钟乳石状石膏
stalactitic(al) aragonite 钟乳状霰石
stalacto-statagmite 石柱
stalagmetric titration 表面张力滴定法
stalagmite 笋状沉积物;石笋
stalagmitic(al) 生滴石笋的;石笋状的
stalagmometer 滴重计;表面张力滴重计;表面张力滴量计
stalagmometric titration 表面张力滴定法
stalagmometry 表面张力滴重法
Stalanium 斯特拉尼姆镁铝合金
stale air 污浊空气;死空气
stale bath 陈旧液槽
stale bill of lading 过期提单;失效提单;迟期提单
stale check 过期支票;失效支票
stale debt 过时效的债务;失时效的债务
stale flavor 腐败味
stale lime 发潮石灰;已潮的石灰;陈石灰
stalemate 陷于困境;停顿
stale paper 过期相片
stale-proof 不腐的
stale refuse 陈旧垃圾
stale sewage 腐(臭)污水;发臭污水
stale wastewater 陈污水;陈废水;腐化废水;腐臭污水
staling 陈废物质积压
staling product 陈废产物
stalk (钢筋混凝土挡土墙的)直墙;高烟囱;模柄;灯丝杆
stalk cutter 割秆刀
stalked eye 具柄眼
stalk-end transporter 麻杆输送器
stalk-pipe chaplet 管子芯撑;单面芯撑
stalk plate 腹板
stalk portion 柄部
stalk stiffener 腹板加劲肋;腹板加劲件

stall 浴厕隔板;泥窑;麻车;货摊;停转;停住;煅烧炉;车位
stallage 摊租;摊位
stall-barn system 分栏舍饲法
stallboard 货摊搁板;铺面栏板;橱窗下槛;商店货架;橱窗搁板
stallboard light 商店柜台橱窗照明;外墙橱窗下采光窗
stallboard riser 橱窗窗台下的竖壁;橱窗勒脚
stall condition 失速条件;失速状态
stall delay 失速延迟
stall-dive 失速后俯冲
stalled area 失速区
stalled blade 失速叶片
stalled descent 失速下降
stalled flow 失速气流
stalled fluid 失速区内的流体
stalled glide 失速下滑
stalled landing 失速着陆
stalled leakage flow 临界泄漏量
stalled speed 失速转速
stalled torque 电动机制动力矩;反接力矩
stalled traffic 被租塞的交通
stall end 前排席位未满(教堂或戏院)
stall flow 失速流量(涡轮)
stall flutter 失速颤振
stall indicator 失速指示器
stalling 停转;失速
stalling characteristic 分离特性
stalling current 制动电流;停转电流
stalling limit 失速界限
stalling load 停转负荷;失速荷载
stalling mach number 失速马赫数
stalling moment 失速力矩
stalling of engine 发动机熄火
stalling point 极限负荷;极限点;气流分离点;失速点
stalling range 失速范围
stalling region 失速区
stalling relay 失速继电器
stalling run 失速状态试验
stalling torque 制动转矩
stalling trial 失速试验
stalling turn 失速转弯
stalling work capacity 机器的临界工作参数;设备最小工作参数
stall keeper 摊贩
stall limit 失速线(涡轮性能曲线);分离界限
stall line 失速线(涡轮性能曲线)
stall margin 失速边际
stall mode 失速状态
stallometer 临界速度指示仪;气流分离指示器;失速仪;失速信号器
stallout 失速
stalloy 硅钢片;薄钢片
stall performance 失速特性;分离特性
stall point 失速点
stall proof 防失速的
stall propagation 失速传播
stall protection system 失速预防系统
stall recognition 失速判别
stall recovery control unit 失速改出操纵机构
stall recovery failure warning lamp 失速改出系统故障警告灯
stall recovery operating lamp 失速改出系统工作指示灯
stall recovery system 失速改出系统
stall riser 店面栏板下短墙;阶梯形货架背板;阶梯形货架竖板;橱窗勒脚
stall roasting 泥窑焙烧
stall severity 失速的严重程度(发动机)
stall simulation 失速模拟
stall speed (液力变矩器等的)失速速度;失速度
stall spot 失速点
stall start 失速起步
stall torque 极限力矩;失速转矩;失速扭矩
stall torque ratio 反接力矩比;电动机制动力矩比;失速转矩比;失速变矩系数
stall-type shower 分隔式淋浴间
stall-type urinal 落地式小便器;竖式小便器;立式小便器
stall urinal 小便间;小便器;立式小便器
stall velocity 失速度
stall-warning 失速告警的
stall warning device 失速警告器

stall warning indicator 失速警告指示器
stall warning light 失速警告灯
stall-warning system 失速警告系统
stall-warning unit 失速警告装置
stall zone 失速区
Sta-Loc floor system 地板稳定系统
stalodicator (干货船的)货物稳定和配置状况指示器
stamba 纪念柱;(印度建筑中柱顶有大型标记的)独立柱;独立纪念柱
Stamey test 斯塔梅试验
Stamicarbon 斯塔密卡邦公司
Stamicarbon process 斯代米卡邦法
stamina 持久力;支承体
staminate papaya tree 公木瓜树
stammering 重言
stamp 印模;压印;盖章;盖印;敲击棒;图章;捣碎机;捣磨;戳子;戳记
stampable sheet 可冲压成型片材
stamp asphalt 压制(地)沥青
stamp battery 捣锤组;连续重击杵;捣矿机组;捣矿机的锤组;捣矿杵
stamp battery shoe 捣锤组的捣蹄;捣矿杆的捣蹄;连续重击杵的捣蹄
stamp box 捣矿机;捣臼
stamp breaking 捣碎;压碎
stamp clause 特别条款
stamp copper 锤碎铜矿石
stamp crushing 压碎;捣碎
stamp duty 印花税
stamped brick 冲压成型砖
stamped ceramic porous tube 压制陶瓷多孔管
stamped circuit 冲压印刷电路
stamped concrete 压实混凝土;捣固混凝土
stamped dropcenter rim 冲压凹槽轮辋
stampede 抢购风潮
stamped gear 模压齿轮
stamped lino(leum) 印花油漆布;印花油地毯
stamped nail 压制钉
stamped nut 模压螺母
stamped packing ring 冲切式密封环
stamped plate 异形金属板;压制板;成型钢板
stamped policy 已缴印花税的保单
Stamped Pottery Culture 印纹陶文化
stamped sheet 模压板
stamped sieve 冲孔筛
stamped thread 模压螺纹
stamper 圆头锤;压模;模子;模压工;捣碎机;捣矿机;打压机;杵;冲压机;冲压工
stamper box 捣矿机的臼槽
stamp etching 蚀刻图章
Stampfer level 斯坦弗水准仪
stamp for application of code 施用规范印章
stamp forging 型锻;落锻;模锻
stamp gravel 捣碎砾石
stamp hammer 捣锤;冲锤;模锻锤
stamping 压制品;压印花纹;压花;模印;模制配件;模冲;烫字;酸蚀印记;捣碎物;捣固;锤击捣碎;锤击;冲压片;冲压(成形)
stamping article 冲压制品
stamping back 模印
stamping board 托模板;垫模板
stamping breaking 捣碎
stamping copper washer 冲压平铜垫圈
stamping core plate 电枢铁片
stamping design 模锻设计;冲压设计
stamping die 压印模;落模;模锻模;冲压模
stamping form 捣实模板
stamping ground 落脚处;常到的地方
stamping hammer 模锻锤;捣碎锤;冲压锤
stamping ink 压印墨;打印油墨
stamping iron hammer 捣固铁锤
stamping knife 冲刀
stamping machine 压印机;号码机;捣碎机;锤击机;冲压机
stamping manipulation 踩法
stamping-material cost 冲压材料费用
stamping mill 捣矿机
stamping of powder 粉末冲压
stamping press 压箔机;模压机;模压压力机;烫金机;冲压机;压印机
stamping process 模压成型法
stamping room 模锻车间;冲压车间
stamping sheet 冲压片
stamping shop 压模车间

stamping steel ribbon 密封条
stamping tool 冲压工具
stamping trimmer 冲压式切边机
stamping tripod 注记盖章
stamping-welded 冲压焊接的
stamping wheel 成品打印轮
stamp ink 印油
stamp licker 冲洗滚筒
stamp mark 刻记
stamp mass 捣结料;打结料
stamp material 捣碎材料
stamp mill 石碓;捣磨机;捣矿厂;冲击磨机
stamp milling 捣磨
stamp mortar 捣锤研碎机
stamp mortar screen 捣研筛
stamp mo(u)lding 冲压成型
stamp-note 装货执照
stamp-office 印花税务局
stamp of the maker 制造厂的标记
stamp out 模锻
stamp pad 打印台
stamp-pad ink 印刷油墨
stamps 印压;印记
stamp sand 压碎砂
stamp shotting 模具型面刨削
stamp stem 捣杆
stamp tax 印花税
stamp tax act 印花税法
stamp washer 槌式洗呢机
stamp work 压印工作;模锻件
stamp work's crane 碎铁用起重机
stamukha 搁浅堆积冰;搁浅冰山
stamukhi 搁浅堆积冰;搁浅冰山
stanchion 交通立标;牛栓枷;牛栏;竖窗框;窗棂;柱子;支柱;窗间小柱;标柱
stanchion barn 系枷牛舍
stanchion base 柱基;支柱底座
stanchion cap 支柱顶板;柱帽
stanchion casing 钢柱的(混凝土)保护层
stanchion-free 无撑杆的;无支柱的
stanchion sign 移动柱座标志;有移动柱座标志;(可移动的)柱座标志
stanchion socket 支柱插座
stanchion splice 支柱拼接
stand 支持架;个体植物群丛;露天看台;立根【岩】;看台;架子;倾斜试验台;台架
standadized ratio 标准化化率
stand-alone 独立
stand-alone capability 独立能力
stand-alone closed loop temperature controller 可独立应用的闭环温度控制器
stand-alone computer 独立计算机;单计算机
stand-alone data processing system 独立的数据处理系统
stand-alone dump 独立转储
stand-alone network system 独立网络系统
stand-alone operation 独立操作
stand-alone printer 独立打印机
stand-alone processor 单立处理机
stand-alone program(me) 独立程序
stand-alone software 独立软件
stand-alone support program(me) 独立支援程序
stand-alone system 独立系统;单机系统
stand-alone type 独立式
stand-alone utility 独立实用程序;独立应用(程序)
stand-alone word processing equipment 独立的字处理设备
Stand and P I Club 标准保赔协会
standard 原尺;正规的;规范;锯材树;基准;标准器;标准的;准
Standard, Poor's Corporation 标准及普尔公司 (美国)
standard Poor's indices 标准普尔指数
standard absorber 标准吸收
standard absorption costing 标准全部成本法
standard acceleration of free fall 标准自由落体加速度
standard account 标准账户
standard accuracy 标准精度
standard accuracy map 标准精度图
standard acid 标准酸溶液
standard acid test 酸度的标准试验;酸度的标准测定法
standard acoustic(al) signal 标准声信号

standard actinometer 标准曝光表
standard acute toxicity test 标准急性毒性检测
standard addition 标准加入
standard addition method 标准添加法;标准加入法
standard address generator 标准地址发生器
standard adjustable wrench 标准活扳手
standard adjustment 标准调整
standard advanced base units 标准前进基地分队
standard aerobic incubation of membrane filter 标准膜滤器需氧培育
standard agricultural tractor 标准农用拖拉机
standard air 标准空气;标准大气
standard air-actuated brake 标准空气调节刹车
standard air capacitor 标准空气电容器
standard air density 标准空气密度
standard air leak rate 标准空气漏率
standard air munitions package 标准空运军需品集装箱
standard alcohol 标准酒精
standard alkali solution 标准碱溶液
standard ambient condition 标准大气状态
standard ambient temperature 标准环境温度
standard ammeter 标准安培计
standard amounts 标准数额
standard amperemeters 标准电流表
standard amplitude 标准振幅
standard analog circuit 标准模拟电路
standard analysis 标准分析
standard analytical process 标准分析过程;标准分析法
standard analyzer 标准分析程序
standard anionics 标准阴离子表面活性剂
standard annual rate 标准年率
standard annuity 标准年金
standard antenna 标准天线
standard application 标准做法;标准作业
standard approval 标准审批
standard arbitration clause 标准仲裁条款
standard architectural design 建筑标准设计
standard area 标准地区
standard artificial stone 标准人造石
standard artillery atmosphere 标准弹道大气
standard artillery zone 标准弹道带
standard asphalt(ic bitumen) 标准石油地沥青
standard as technical barrier to trade 贸易技术性壁垒标准
standard astrograph 标准体摄影仪
standard atmosphere 标准大气压;常压;标准气压;标准大气
standard atmosphere pressure 标准大气压(力)
standard atmospheric condition 标准大气状态;标准大气条件
standard atmospheric head 标准大气水头
standard atmospheric pressure 正常大气压力;标准大气压(力)
standard atmospheric pressure reading 标准大气压读数
standard attenuator 标准衰减器
standard audit program(me) 标准审计程序表
standard automatic tide ga(u)ge 标准自记验潮仪
standard axial load(ing) 标准轴载
standard axle 标准轴
standard baffle 标准障板
standard balance 标准天平
standard ballistic condition 标准弹道条件
standard bar 检验轴;检验杆;标准条料;标准杆
standard bareboat charter 标准光船租赁合同
standard barometer 标准气压表
standard barometric pressure 标准气压
standard basalt 标准玄武岩
standard base 标准碱
standard base drawing 标准底图
standard baseline 标准基线
standard base-rate structure 工资等级表
standard basic rack tooth profile 原始齿形
standard basic unit 标准基本部件
standard batching (混凝土集料的)标准计量
standard beam approach 标准射束着陆系统;标准波束进场
standard beam approach system 标准波束引导进场系统
standard beam section 标准梁断面
standard bearing 标准型轴承
standard bearing compass 标准方位罗盘;标准方

位罗经
standard bed 标准层
standard behavior 标准性能
standard bell and bell bend 标准双承弯头
standard bell and spigot bend 标准承插弯头
standard berry 直茎浆果
standard bicarbonate 标准碳酸氢盐
standard bight 标准角槽
standard billing rate 标准收费率
standard bill of lading 标准提单;标准格式提单
standard binary 标准二进制
standard binary code 标准二进制码
standard biochemical oxygen demand 标准生化需氧量
standard bit 标准钻头
standard black 标准黑(色料)
standard blackbody source 标准黑体光源
standard blanket policy 标准统保单
standard blasting crater 标准爆破漏斗
standard blasting test 标准爆破试验
standard block 标准砌块;标准单元
standard block ga(u)ge 标准块规
standard block of hardness 标准硬度块
standard blow-off branch 标准排泥分叉管
standard body 标准机构
standard bolt 标准螺栓
standard boom 标准臂
standard branch 标准支路
standard brand 正牌子
standard brass 标准黄铜
standard breaking capacity 标准断路能力
standard bred 标准马
standard brick 标准砖
standard bridge design 桥梁标准设计
standard bright finish (板的)双面光整处理;标准光亮精整
standard briquette 标准(水泥)试块
standard briquette mo(u)ld 标准(水泥)试块模子;标准(水泥)试块模壳
standard briquette strength 标准(水泥)试块强度
standard broadcast band 广播波段;标准广播波段
standard broadcast channel 标准广播信道
standard broadcast(ing) 标准广播
standard budget 标准预算
standard buffer solution 标准缓冲(溶)液
standard building 标准建筑
standard building cost 标准建筑造价
standard bullion 标准银块
standard burden rate 标准负荷率;标准分摊率
standard burden rates of manufacturing 标准间接制造费摊配率
standard burette 标准滴定管
standard buried collector 标准埋层集板
standard burned brick 标准烧成砖
standard bus of public transportation 公共交通标准车
standard cable 标准电缆
standard calculation sheet 标准计量表
standard caliper 标准卡板
standard cal(l)iper ga(u)ge 标准测径规
standard calomel electrode 标准甘汞电极
standard candle(light) 标准烛光;国际烛光
standard capacitance 标准电容
standard capacitance box 标准电容箱
standard capacitor 标准电容器
standard capacity 标准容量
standard car 标准车
standard carbon blade 标准碳素钢锯条
standard card cage 标准卡片箱;标准插件箱
standard cargo container 标准货物集装箱
standard cartographic(al) grid 标准地图格网
standard case conversion factor 标准箱折算系数
standard casement window 标准平开窗
standard casing 标准套管
standard casting 标准铸块
standard cast stone 标准铸石
standard cationics 标准阳离子表面活性剂
standard caustic soda solution 标准氢氧化钠溶液
standard cavity 标准谐振腔
standard cavity circuit 标准空腔谐振电路
standard cell 标准电池;标准单元
standard cellulose 标准纤维素

standard cement 标准水泥;普通水泥
standard cement mortar cube 标准水泥砂浆立方块
standard cement paste 标准水泥浆
standard cement testing sand 水泥试验用标准砂
standard chamber 标准电离室
standard chamotte brick 标准黏土熟料砖;标准黏土耐火砖
standard change gear 标准配换齿轮
standard change notice 标准更改通知
standard channel 标准槽钢
standard channel dimensions 标准航道尺度
standard character 标准字符(串)
Standard Chartered Bank 标准麦加利银行(英国);标准特许银行
standard charter party 标准租船合同
standard charter party form 标准型租船契约格式
standard chemical potential 标准态化学位;标准态化学势;标准化学势
standard chisel-face tungsten carbide bit 标准凿岩碳化钨钻头
standard chi-square test 标准卡方检验
standard chroma subcarrier frequency 标准彩色副载波频率
standard chromatogram 标准色谱图
standard chronometer 标准天文钟
standard cipher 标准密码
standard circle 标准圆
standard civil time 标准民用时
Standard Classification of International Trade Society 国际贸易学会标准分类
standard clause 条文范例;标准条款
standard cleaning bench 标准净化工作台
standard clearance 标准清除率
standard clerical cost 标准办公费用
standard clock 标准钟
standard coal 标准煤
standard coal consumption 标准煤耗
standard coal consumption rate 标准煤耗率
standard coaxial cable 标准同轴电缆
standard code 标准规定
standard code book 标准电码本
standard code for information interchange 信息交换标准代码
standard code for interchange 数据交换用标准码
standard code test signal 标准码试验信号
standard coefficient of permeability 实验室渗透系数
standard coil 标准线圈
standard collar 标准套环
standard collocation point 标准配置点
standard colo(u)r 标准色
standard colo(u)ration 标准色
standard colo(u)r chart 标准色谱
standard colo(u)rimetric observer 标准比色观测仪
standard colo(u)rimetric plate 标准比色板
standard colo(u)rimetric system 标准色度系统;标准比色系统
standard colo(u)rimetric tube 标准比色管
standard colo(u)r-mixture curves 标准配色曲线
standard colo(u)r solution 标准色液;标准比色液
standard colo(u)r system 标准表色系统
standard colo(u)r tone 标准色调
standard column 标准柱
standard column of water 标准水柱
standard commission 标准佣金
standard commodity 标准商品
standard communication link 标准通信链路
standard communication protocol 标准通信协议
standard communication subsystem 标准通信子系统
standard communication unit 标准通信单位
standard compaction curve 标准击实曲线
standard compaction effort 标准压实力
standard compaction test 标准击实试验;标准夯实试验
standard compass 标准罗盘;标准罗经
standard compass course 标准罗经航向
standard compass deviation 标准罗经自差
standard component 标准件
standard composition 标准成分
standard comprehensive profile plane 标准综合剖面图

standard compression strength 标准抗压强度
standard concentration 标准浓度
standard concept 标准概念
standard concrete block 标准混凝土砌块
standard condenser 标准电容器
standard condenser tube alloy 标准冷凝管合金
standard condition 标准状态;标准条件;标准情况;标准工况
standard condition for testing 标准试验条件
standard conductivity 标准电导率
standard cone 标准锥形筒子;标准锥
standard cone crusher 标准圆磨;标准型圆锥破碎机
standard connector 标准接头
standard consistence[consistency] 标准稠度
standard consistency of gypsum 石膏标准稠度
standard consolidated area 标准统一区
standard consolidated statistical areas 标准合并统计区
standard consolidation 标准固结
standard construction 标准结构
standard construction element 通用构件
standard construction method 标准施工方法;标准施工法
standard construction model 标准建造方式
standard content map 标准内容地图
standard contour 基本等高线;首曲线;标准等高线
standard contract 标准合同
standard contract form 标准契约格式;标准合同格式
standard contract provisions 标准合同条款;标准合同规定
standard contrast grade 标准反差等级
standard control 规格控制;标准品对照
standard control system 标准控制制度
standard conversion cost 标准加工费
standard convex program(me) 标准凸规则
standard coordinates 标准坐标
standard coordinate system 标准坐标系
standard copper 标准铜;标准工业铜
standard copper wire 标准铜线
standard core bit 标准岩芯钻头
standard core drill 标准取芯钻井
standard corrugated rolled glass 标准波状滚压玻璃
standard cosmological model 标准宇宙模型
standard cost 标准成本
standard cost accounting 标准成本会计
standard cost accounting system 标准成本会计制度
standard cost calculation 标准成本计算
standard cost center 标准成本中心
standard cost clearing 标准成本结算
standard cost control 标准成本控制法
standard costing 标准成本计算
standard cost sheet 标准成本单
standard cost system 标准成本制度
standard cost tableau 标准成本形态
standard cost variance 标准成本差异
standard counter 标准计数器;标准计数管
standard counting requirement 标准计数要求
standard couple 标准热电偶
standard coupler 标准形车钩
standard coverage policy 标准范围保险
standard cover mixtures 标准混合覆盖
standard cover paper 标准封面纸
standard crab vibrator 标准蟹式可控震源
standard crane pieces 标准起重机部件
standard crate 标箱
standard crop 标准作物
standard cross-section 标准横断面
standard crushing strength 标准破碎强度
standard cube 标准立方体;标准混凝土方试块
standard cube of concrete 混凝土标准立方体(试块)
standard cubic(al) feet 标准立方英尺
standard cubic(al) feet per hour 标准立方英尺/小时
standard cubic(al) feet per second 标准立方英尺/秒
standard cubic(al) meter 标准立方米
standard curing 标准养护
standard curing procedure 标准养护法
standard currency 本位货币

standard current generators 标准电流发生器
standard current meter 标准流速仪
standard curvature of navigation channel 航道标准弯曲半径
standard curvature radius of channel 航道标准弯曲半径
standard curve 首曲线;标准曲线
standard curve method 标准曲线法
standard cut-back asphaltic bitumen 标准稀释石油沥青
standard cutter 标准刀具
standard cylinder 标准圆柱体
standard cylinder for compression testing 标准抗压试验圆柱体
standard cylindric(al) coordinate system 标准柱面坐标系统
standard cylindric(al) ga(u)ge 标准圆柱形规
standard data format 标准数据格式
standard data interchange 标准数据交换
standard data message 标准数据电报
standard data phone 标准数据电话装置
standard datum chart 标准基准面
standard datum elevation 标准基准高程
standard datum level 标准基准面;标准基准高程
standard datum plane 标准基准面
standard day work 标准日工作量
standard d coupler 标准 D 形车钩
standard deadweight press tester 标准静压力机试验机
standard deals 标准厚窄板
standard death rate 标准死亡率
standard declaration 标准说明
standard decomposition 标准分解
standard deduction 标准扣除额
standard deep-sea diving dress 标准重潜水服
standard deep-sea gear 标准重潜水装具
standard definition 标准定义
standard deflection 标准偏转
standard density 标准密度
standard density rate 标准密度级
standard depth 标准水层;标准深度
standard depth dyeing 标准深度染色
standard depth of (navigation) channel 航道标准水深
standard depth-pressure recorder 标准深度压力记录仪
standard depth scale 标准深度分级样卡
standard derivation 标准偏差
standard design 标准设计
standard design flood 标准设计洪水
standard design of bridge 桥梁标准设计
standard design of building 建筑标准设计
standard design plan 标准设计图纸
standard design roadbed 标准设计路基
standard details 标准(设计)详图;通用详图;标准细部图;标准大样
standard developer 标准显影剂
standard deviation 均方误差;平均误差;标准误差;标准偏移;标准偏差;标准离差;标准差
standard deviation ellipse 中误差椭圆;标准偏差椭圆
standard deviation map of K content 钾含量标准偏差图
standard deviation observating 标准差观察
standard deviation of aero-survey 航测标准离差
standard deviation of coal thickness 煤层厚度标准差
standard deviation of difference 差收标准差
standard deviation of mean 平均数标准差
standard deviation of probability distribution 概率分布标准离差
standard deviation rate 标准量
standard die 标准模
standard die set 标准模架
standard diet 标准饮食
standard diluent 标准水
standard dilution method 标准稀释法
standard dimension 标准尺寸
standard dimension ratio 标准尺寸比
standard dimensions of channel 航道标准尺度;航道标准尺寸
standard dimensions of navigation channel 航道标准尺度;航道标准尺寸
standard disk ga(u)ge 标准圆盘规

standard disk plow 标准圆盘犁
standard dispersion 标准散布
standard displacement 标准排水量
standard distance 标准距离
standard distillation test 标准蒸馏试验
standard distributed lag function 标准分布滞后函数
standard distribution 标准分布
standard diving 标准潜水
standard diving dress 标准潜水服
standard diving work 标准潜水作业
standard dobson spectrophotometer 标准多布森分光光度计
standard documents 标准文件
standard dosimetry 标准剂量学
standard draft 标准吃水
standard draft for approval 标准报批稿
standard drawing 标准图(纸);通用图;定型设计图
standard drilling machine 标准钻床
standard drilling method 标准钻探法
standard drill rod 标准钻杆
standard drill tool 标准钻具
standard drive 标准传动装置
standard drop measure 标准量滴器
standard dwelling 标准寓所;标准住房
standard dwelling unit 标准居住单元
standard dyeing time 标准染色时间
standard dynamite 标准硝甘炸药
standard dynamometer 标准测力机
standard earth 标准土;标准地球
standard earth model 标准地球模型
standard earth station 标准地球站;标准地面站
standard economic forecast 标准经济预测
standard economic ordering quantity 标准经济订货;标准经济订货量
standard edge connector 标准印制板边端插头
standard efficiency 标准效率
standard electric(al) log 标准电极测井
standard electrode 标准电极
standard electrode potential 标准电势;标准电极电位;标准电极电势
standard electromotive force of the cell 标准电池电动势
standard electron activity 标准电子活度
standard electronic automatic computer 标准电子自动计算机
standard electroweak model 标准弱电模型
standard element 定型构件;标准构件;标准单元
standard electromotive force 标准电动势
standard elevation 标准高程
standard elevation system 标准高程系统
standard emitter 标准发射体
standard emulsion 标准乳液
standard engineering 标准工程
standard engine rating 标准发动机功率
standard enthalpy 标准焓
standard enthalpy of formation 标准生成焓
standard enthalpy of reaction 反应标准焓
standard enthalpy of volatilization 标准挥发焓
standard entropy 标准熵
standard entropy change 标准熵变
standard environment 标准(系统)环境
standard equation 标准方程(式)
standard equilibrium constant 标准平衡常数
standard equipment 标准设备
standard error 标准误差;标准偏差;标准机误;均方差
standard error ellipse 中误差椭圆;标准误差椭圆;标准误差椭球
standard error of estimate 估值标准误差;估计值标准误差;估计标准误差;标准估计误差
standard error of estimate based on sample 样本估计值的标准误
standard error of mean 均值标准误差
standard error of position of the contours 等高线位置标准差
standard error of proportion 比例的标准误差
standard error of sample mean 样本平均值的标准误
standard error of the mean 平均值的标准偏差;平均数标准误差
standard error of unit weight 单位权误差
standard error test 标准误检验
standard Euro-currency rates 标准欧洲通货利率

standard evaluation procedure 标准评价程序
standard evapo(u)rator 短管立式蒸发器;标准(式)蒸发器
standard examination for approval 标准报批审查
standard export packing 出口标准包装
standard external time code 标准外部时间编码
standard facies belts proposed by wilson 威尔逊标准相带
standard facility 标准装备
standard factory 标准工厂
standard fade-o-meter hour 标准耐晒试验仪小时
standard Fahrenheit thermometer 标准华氏温度计
standard fall diameter 标准沉降直径
standard falling gate 普通卧倒门
standard fall velocity 标准降落速度;标准沉降速度
standard fastener 标准扣件;标准紧固件
Standard Federal snow sampler 联邦标准雪采样器(瑞士)
standard fee 标准费用
standard fiber glass 标准玻璃纤维
standard field generator 标准场强发生器
standard field logging system 标准野外测井系统
standard figure 标准体型
standard file 标准文件
standard file label 标准文件标号
standard files input 标准输入文件
standard files print 标准打印文件
standard film 标准片
standard film projector 标准影片放映机
standard film stock 标准电影片
standard filter 标准速率滤池;标准滤光片
standard fine grain aerial photo paper 标准精细航空照相纸
standard fineness 标准细度
standard fire 标准火焰
standard fire alarm system 标准消防报警系统
standard fireclay brick 标准黏土耐火砖
standard fire extinguisher 标准消防灭火器
standard fire policy 标准火险保单
standard fire rating curve 标准建筑防火等级曲线
standard fire stream 标准消防射流量
standard fire test 标准消防试验
standard fit 标准配合
standard fitting 标准配件
standard fixed cost 标准固定成本
standard fixed length record 标准定长记录
standard fixed-rate mortgage loan 标准固定利率抵押贷款
standard flange 标准法兰(盘)
standard flat wagon 标准平车
standard flip chart 标准航空图
standard flood 标准洪水
standard floor 标准地板;标准楼板;(建筑设计中的)标准层
standard flue gas 标准烟气道
standard fluorescence 标准荧光
standard focal length 标准焦距
standard for building industry 建筑业标准
standard for cargo stored per unit area 单位面积货物堆存量
standard for classified code of information 信息分类编码标准
standard for collecting fee 取费标准
standard for discharge of pollutants 污染物排放标准
standard form 标准式(样);标准形式;标准模板;标准公式
standard format 标准版面;标准开本;标准格式
standard formation 标准建造
standard format of type real 整型标准格式
standard form contract 标准的印就合同
standard form for L/C 标准的印就信用证
standard form for letter of credit 标准的印就信用证
standard form of salvage agreement 救助合同标准格式
standard for potable water 饮用水标准
standard for sound-insulation 隔声标准
standard fossil 标准化石
standard frame 标准框架;标准构架
standard framed block 标准框架建筑
standard framed building 标准框架建筑
standard free energy 标准自由能

standard free energy of formation 标准生成自由能
standard freezing cycle 标准冷冻循环
standard freight car 标准货车
standard freight scale 基准运费率;标准运费率
standard frequency 标准频率
standard frequency and time signal 校准频率和时间信号;频标时号
standard frequency and time signal satellite service 卫星标准频率和时间信号业务
standard frequency and time signal service 标准频率和时间信号业务
standard frequency and time signal station 标准频率和时间信号电台
standard frequency deviation oscillator 标准频偏振荡器
standard frequency generator 标准频率发生器
standard frequency meter 标准频率计
standard frequency service 标准频率业务
standard frequency signal 标准频率信号
standard frequency source 标准频率源
standard frequency spectrum 标准频谱
standard frequency station 标准频率电台
standard frequency system 标准频率制;标准频率系统
standard fresh water 标准淡水
standard freshwater medium 标准淡水溶剂
standard frost penetration 标准冻深
standard full shot 标准全景拍摄
standard full V thread 正三角形标准螺纹
standard function 标准函数
standard funnel 标准漏斗
standard gas 标准气体
standard ga(u)ge 法定度量表;标准水尺;标准量具;标准量规;标准计量;标准计;标准轨距(等于1.435米);标准规;标准隔距
standard ga(u)ge block 标准规块
standard ga(u)ge rail 标准铁轨;标准轨距钢轨;标准轨距的轨道
standard ga(u)ge railway 标准轨(距)铁路
standard Gauss quadrature 标准高斯求积法
standard Gear ratio 标准齿轮速比
standard Generalized markup language 标准通用标识语言
standard Geometric(al) deviation 标准几何偏差
standard Gibbs energy 标准吉布斯能量
standard Gibbs energy of formation 产生的标准吉布斯能量
standard Gibbs free energy 标准吉布斯自由能
standard Gibbs free energy of formation 标准生成吉布斯自由能
standard Gibbs function increase 标准吉布斯函数增加
standard global geochronologic scale 标准世界地质年代表;标准全球地质年代表
standard gold 标金(试金用);标金
standard graball 标准式平差机
standard gradation 标准粒度
standard grade 标准品;标准级
standard grade sand 标准级配砂
standard grading 标准粒度;标准级配
standard grain diameter of aquifer 含水层标准粒径
standard grain size 标准粒度
standard graph 标准图
standard gravitational acceleration 标准重力加速度
standard gravity 标准重力
standard gravity acceleration 标准重力加速度
standard gravity value 标准重力值
standard greetings 标准贺电
standard grey 标准灰色花岗石(挪威产)
standard grey scale 标准灰色分级样卡
standard grid 标准格网
standard ground 标准土
standard ground joint 标准磨口接头
standard group 标准群
standard grouser shoe 标准履带轮爪靴;标准履带齿片
standard guard 标准防护器
standard gugless brick 标准无耳砖
standard guy thimble 标准拉线牵线环
standard half-cell 标准半电池
standard harbo(u)r 标准港(口)
standard hardboard 标准硬质纤维板

standard hardness 标准硬度
standard hardware 标准硬件
standard hardware program(me) 标准硬件程序
standard heat consumption 标准耗热量
standard heat of combustion 标准燃烧热
standard heat of formation 标准生成热
standard heat of reaction 反应标准热
standard heat pipe 标准热管
standard heavy duty engine lathe 标准重型普通车床
standard heavy loading truck 标准载重车
standard height 标准高度
standard hexagonal roof slab 标准六角形屋面板
standard high explosive 标准猛烈炸药
standard high frequency signal generator 标准高频信号发生器
standard high pressure stainless steel tubing 标准高压不锈钢管
standard hook 标准钩(体);标准弯钩
standard hooked bolt 标准型钩螺栓
standard horizon 标准层位
standard horn 标准喇叭
standard horn antenna 标准喇叭天线
standard horsepower 标准马力
standard hour 标准小时;标准工时
standard hour plan 标准时间工资制
standard hours worked 标准工作时数
standard hour system 标准点制度
standard housing unit 标准住房单元
standard humic type 标准腐殖型
standard hundred 长百装件
standard hydraulic cylinder 标准液压缸
standard hydraulic station 标准液压站
standard hydrogen cell 标准氢电池
standard hydrogen electrode 标准氢电极
standard hydrophone 标准水听器
standard identifier 标准标识符
standard illuminant 标准照明体;标准照度;标准光源
standard illumination 标准照度
standard impedance 标准阻抗
standard incandescent lamp 标准白炽灯
standard income statement 标准收益表
standard index number formula 标准指数公式
standard indicator with a unit expressed in kind 标准实物指标
standard inductance 标准电感
standard inductor 标准电感器
standard industrial classification 标准工业分类;标准产业分类
standard infiltration capacity curve 标准下渗容量曲线;标准入渗能力曲线
standard in force 现行标准
standard information work in enterprise 企业标准化情报工作
standard information 标准信息
standard information management 标准情报管理
standard information retrieval 标准情报检索
standard ink 标准油墨
standard ink swatch 标准油墨色样
standard input file 标准输入文件
standard input-output instruction 标准输入输出指令
standard input-output interface 标准输入输出接口
standard input-output request 标准输入输出请求
standard inputs allowed 标准投入量限额
standard inside dimension ratio 标准内径尺寸比
standard inspection procedures 标准检验程序
standard inspection record 标准检验记录
standard instruction 标准指令
standard instruction memonic 标准指令助记符
standard instruction set 标准指令组
standard instrument 标准仪器;标准仪表
standard instrumental variable estimators 标准工具变量估计量
standard instrument approach 标准仪表进近
standard intechangeable ground joint 标准可替换磨砂接头
standard interface 标准接口
Standard international age classifications 标准国际年龄分组
standard international atmosphere 国际标准气压
standard international trade classification 标准国际贸易商品分类

standard interrupt 标准中断
standard inventory 标准储备
standard investment opportunity 标准投资机会
standard involute gear 标准渐开线齿轮
standard involute gear tooth 标准渐开线齿轮齿
standard ion dose 标准离子剂量
standard irradiance 标准辐照度
standard isobaric surface 标准等压面
standardization 规格化;规范化;校准;统一标准;定型化;标准化;标定法;标定
standardization and inspection 标准与检验
standardization base 比长器
standardization construction 建筑标准化
standardization correction 检定校正值;标准校正
standardization equivalent 标准当量
standardization error 检定误差
standardization measures 标准化措施
standardization of accounting 会计标准化
standardization of building 建筑标准化
standardization of freight charges 货运费用标准化
standardization of packing 包装标准化
standardization of parts 零件标准化
standardization of solution 溶液标定
standardization of tape 卷尺检定
standardization of tap(wire) length 基线尺长度检定
standardization of tool 工具标准化
standardization of types 形式规格化;形式典型化;形式标准化
standardization of water pollution 水污染标定
standardization parameter 标准化参数
standardization pollution coefficient 标定水污染系数
standardization pollution loading 标定污染负荷
standardization program(me) 标准化计划;标准化程序
standardization reflectivity 标准反射率
standardization rule 标准化法则
standardization site of tap(wire) length 基线尺长度检定场
standardization temperature 检定温度
standardization transformation 标准化变换
standardization trial 标准试航
standardize 标准化
standardize bankruptcy procedure 规范破产
standardized 标准化的
standardized accounting 标准化会计
standardized administration 规范管理
standardized amplitude 标准化振幅
standardized bioassay method 标准化生物测定法
standardized birth rate 标准化出生率
standardized building 标准化建筑
standardized building elements 标准建筑构件
standardized building unit 定型房屋单元;标准房屋单元
standardized burden rate 标准制造费用率;标准化制造费用率;标准化负荷
standardized classification 标准分项;标准分类
standardized component 定型构件;标准(化)部件
standardized container 标准式集装箱;标准集装箱;标准化集装箱
standardized contract 标准化契约
standardized current assets 定额流动资产
standardized curve 标准化曲线
standardized data 标准化数据
standardized death rate 标准化死亡率
standardized denomination 标准化命名法
standardized design 定型设计
standardized determinant 标准化行列式
standardized deviation 标准化离差
standardized dimension for construction 标准化的建筑尺寸
standardized distribution 标准化分布
standardized element 标准构件;标准零件
standardized fertility rate 标准化的生育率
standardized figure for construction 标准化的建筑外形
standardized financial management 规范化的财政管理
standardized financial statements 标准化财务报表
standardized framing 标准化构架
standardized hangar 标准化飞机棚;标准化飞机库
standardized holes 标准化炮眼组
standardized house 标准住宅

standardized incidence ratio 标准化发病比
standardized index ratio 标准化指标比
standardized instruction 标准化指令
standardized keyboard 标准化键盘
standardized level 标准化水平
standardized liability 定额负债
standardized mass production phases 标准化大批量大生产阶段
standardized method 标准化方法
standardized moments of the range 全局标准化矩
standardized mortality rate 标准化死亡率
standardized mortality ratio 标化死亡比；标准化死亡率比；标准化死亡比
standardized name 标准化名称
standardized normal distribution 标准(化)正态分布
standardized normal variate 标准化正态变量
standardized nuclear unit power system 标准化核机组电站系统
standardized number 标准化数
standardized pallet 标准化托盘
standardized parts 标准部件；标准零件；标准件
standardized products 定型产品；标准制品；标准化产品
standardized program(me) 标准化程序
standardized proportional mortality ratio 标化死亡比例
standardized random variable 标准化随机变量
standardized rate 标准率；标准化率；标化率
standardized rate by direct method 直接法标准化率
standardized rate by indirect method 间接法标准化率
standardized regression coefficient 标准化回归系数
standardized residual 标准化残差
standardized score 标准化比数
standardized shape member 标准截面构件；标准(断面)构件
standardized sign 标准化符号
standardized signal 标准化信号
standardized space launch vehicle 标准化空间发射飞行器
standardized structural element 标准化的结构构件；标准预制构件
standardized symbol 标准化符号
standardized technique 标准化法
standardized test 标准测试
standardized testing procedure 标准试验方法
standardized test solution 标准化试液
standardized trials 逐点加速试航
standardized uniform 标准化制服
standardized value 额定值；标准化数值
standardized variable 标准化变数
standardized work 标准工作
standardize the tax system 规范税制
standardizing 标准化；标定；校正
standardizing body 标准体型；标准化机构
standardizing box 标准化负荷测定机；标准箱；标准负荷测定仪
standardizing load sequence 标准化荷载系列
standardizing lock 规格化锁
standardizing number 规格化数
standardizing of solution 溶液的标定
standardizing order 规格化指令；标准化指令
standardizing reagents 基准和标准试剂
standardizing rheostat 调定变阻器；标准化变阻器
standardizing system 标准化系列
standard jig 标准样板
standard job cost sheet 标准分批成本单
standard key 标准键
standard kiln speed 基准窑速
standard kittel tray 标准克特尔塔板
standard kittel tray with a packed intermediate 板间充填料的克特尔塔板
standard knee 倒装时板(木船上)
standard knot 普通木节；标准树节；标准(木)节
standard label 标准标号
standard laboratory atmosphere 标准试验室气压；标准实验室(大)气压
standard laboratory test 标准试验室试验
standard labo(u)r cost 标准人工成本
standard labo(u)r hour 标准人工工时
standard labo(u)r rate 标准工资率

standard labo(u)r time 标准人工工时
standard ladder 梯架立柱
standard lag line 标准延迟线
standard lamp (支柱可伸缩的)落地灯；普通灯泡；标准(光度)灯；标准电灯泡
standard land excavator 陆用标准挖掘机
standard language 标准语言
standard language symbol 标准(的)语言符号
standard Latin square 标准拉丁方
standard law 标准法
standard lay 标准铺设；标准敷设；标准布置；标准交互捻(绳索)
standard layout 数的标准分布；标准格式
standard lead 标准导程
standard leak 标准泄漏；标准漏孔
standard lean-to 标准坡度
standard learning 标准学
standard least squares regression analysis 标准最小平方回归分析
standard least squares regression coefficient 标准最小平方回归系数
standard legal price 标准法定价格
standard length 标准长度
standard length of rail 钢轨标准长度
standard length rail 标准长度钢轨【铁】
standard letter 标准字体
standard level 标准水准；标准水平
standard lever 调节杆
standard light and medium duty engine lathe 标准轻中型普通车床；标准轻型和中型普通车床
standard light antarctic precipitation SLAP 标准
standard light source 标准光源
standard light weight diving dress 标准轻潜水服
standard limb leads 标准肢体导程
standard lime 标准石灰
standard limestone dust 标准石灰粉尘
standard limit 标准限
standard line 标准线
standard linear elastic design response spectrum 标准线弹性设计反应谱
standard linear model 标准线性模型
standard line weight 标准线划粗度
standard lining ring 标准衬砌环
standard liquid heating system 标准液体供热系统
standard live load for bridge 桥梁标准活荷载
standard load 荷载标准值；标准荷载；标准负荷
standard loaded condition 标准负荷状态
standard load system 标准载重制
standard load test 标准荷载试验
standard load value 标准贯入值
standard log 标准测井
standard logic module 标准逻辑微型组件
standard Loran 罗兰A；标准罗兰
standard lot size 标准批量
standard lumber 标准方木(头)
standard luminance response 标准亮度响应
standard luminosity curve 标准亮度曲线
standard luminosity function 标准发光度函数
standard machine hour 标准机器小时；标准机器工时
standard machinery 普通机械
standard machinery and electric(al) products 标准机电产品
standard machine time 标准机器工时
standard machine tool 标准机床
standard magnet 标准磁铁
standard magnetic amplifier 标准磁放大器
standard magnetic compass 标准磁罗经
standard magnification 标准放大
standard maintenance 标准养护
standard maintenance procedure 标准维护程序
standard maize sheller 标准玉米脱粒机
standard man 标准人
standard mandrel 标准心轴
standard manhole 标准检查井
standard manhole pipe 标准人孔管
standard man-hour 标准工时
standard manual 标准手册
standard manufacturing cost 标准制造成本
standard manufacturing expenses 标准制造费用
standard manufacturing overhead 标准间接制造费
standard map 标准地图
standard mapping film 标准航摄胶片
standard marine navigation vocabulary 标准航海用语

standard mark 标准银符号；(钢筋的)标准代号
standard market merchant 标准市集商人
standard matched 标准拼接的；标准企口砌合的
standard material 标准料；标准材料；标准剂
standard material cost 标准原材料的成本；标准材料成本
standard material price 标准材料价格
standard materials for approval 标准报批材料
standard material usage 标准用料
standard mathematical constant 标准数学常数
standard mathematical function 标准数学函数
standard maturities 标准到期日
standard mean chord 标准平均弦长
standard mean ocean water 标准平均海洋潮位；标准平均大洋水；SMOW 标准
standard mean sea level 基准平均潮位
standard measure 标准量度；标准量
standard measure expenses 标准计量费
standard measurement 标准量测(法)；标准度量；标准尺度；标准测量
standard measure of quantity 标准计量单位
standard measuring signal 标准测试信号
standard member 通用构件；标准构件
standard memory 标准存储器
standard memory interface 标准存储器接口
standard memory location 标准存储单元
standard mercurial barometer 标准水银气压表
standard meridian 标准子午线；标准子午圈；标准经线
standard meridian of map 地图投影标准子午线
standard metal stud 标准钢立筋
standard metal window 标准金属窗
standard meter 标准仪表；标准米；标准计；标准测量仪
standard meter bar 标准米尺
standard metering base 标准计量基准
standard metering orifice 标准测流孔
standard methane production rate 标准甲烷产率
standard method 标准方法；标准措施
standard method for examination of water and wastewater 标准水与废水检测方法
standard method for membrane filtration fecal coliform test 标准滤膜大肠杆菌检验方法
standard method of analysis 标准分析方法；标准分析法
standard method of curing 标准养护方法
standard method of measurement 标准测定方法；额定测算方法；标准测量方法
standard method of measurement of building works 建筑工程标准量测法
standard method of measuring (建筑工程上的)标准计量法
standard method of payment 标准决算方法；标准结算方法
standard method of settlement 结算的标准方法；标准支付方式
standard metre 标准米
standard metre bar 标准米尺
standard metropolitan area 标准大城市市区；标准都会市区；标准大都市地区；标准城市区
standard metropolitan statistic(al) area 城市标准统计面积(一个或数个县,小城市,以5万人口为标准的城市统计面积)；标准(大)都市统计区；标准城市统计面积
standard microphone 标准传声器
standard mile 标准英里
standard military grid 标准军用格网
standard milk 标准产乳记录
standard milled bit 标准铣齿钻头
standard milling machine 标准铣床
standard mineral 标准矿物
standard minimum strength 标准最小强度
standard mix 标准混合料；(混凝土的)标准配比；通用配合比；标准配合比(混凝土)；标准配合；标准混合；标准拌和
standard mixture 标准混合物
standard mode 标准模式；标准方式
standard mode flow table 标准模式流程表
standard mode key 标准状态键
standard model 标准(形)式；标准模型
standard model gibbs free energy of formation 标准吉布斯生成自由能
standard mode of vibration 标准振动模式

standard modular brick 标准模数砖
standard modular system 标准模块系统
standard modular system card 标准模块系统插件
standard modular system of package 程序包的标准模块系统
standard modulation degree 标准调制度
standard module 标准模数
standard moisture regain 标准回潮率
standard molal enthalpy of formation 标准生成焓
standard molal entropy 标准熵
standard molal heat capacity 标准摩尔热容
standard molal heat capacity coefficient 标准摩尔热容系数
standard molal volume 标准摩尔体积
standard molar energy 标准摩尔能量
standard molar energy of reaction 反应的标准摩尔能量
standard molar Gibbs energy of reaction 反应的标准摩尔吉布斯能量
standard money 标准货币;本位货币
standard money unit of account 标准货币记账单位
standard mortality rate 标准死亡率
standard mortality ratio 标准死亡比
standard mortar cube 标准砂浆立方(体)试件
standard mortgage 标准抵押贷款
standard mo(u)ld 标准模型;标准模子
standard mo(u)ld board 标准模板
standard mu 标准亩
standard multimode fiber 标准多模光纤
standard multipoint recorder 标准多点记录器
standard musical pitch 标准音乐音调
standard muzzle velocity 标准腔口速度
standard natural rubber 标准天然胶
standard negative 标准底片
standard net assignable area 标准有效分配面积
standard network access protocol 标准网络接入规程
standard neutron source 标准中子源
standard noise factor 标准噪声因子
standard noise generator 标准噪声发生器
standard noise temperature 标准噪声温度
standard nominal voltage 标准标称电压
standard noon 区时12点;标准时正午
standard normal curve 标准正态曲线
standard normal density function 标准常态密度函数
standard normal deviate 标准正态差
standard normal deviation 标准正态偏离;标准正态离差;标准(化)正态偏差
standard normal distribution 标准正态分布;标准常态分配
standard normalization 标准正规化
standard normal probability 标准正态概率
standard normal probability curve 标准正态概率曲线
standard normal random variable 标准正态随机变量
standard normal value 标准正规值
standard normal variable 标准正态变量
standard normal variate 标准正态变量
standard normal Z distribution 标准正态Z分布
standard normal z transformation 标准正态Z变换
standard notation 标准计数法;标准符号(表示法)
standard numerical attribute 标准数值属性
standard numeric display 标准数字显示器
standard numeric keyboard 标准数字键盘
standard nut 标准螺母
standard observation 定时观测;标准观测
standard observer 标准色度观察者;标准观测员
standard observer data 标准观测数据
standard odor unit 标准气味单位
standard of an executive 行政上的标准
standard of California test 加利福尼亚试验标准
standard of carbon stable isotope 碳稳定同位素标准
standard of care 谨慎程度
standard of classification 分类标准
standard of comparison 比较标准
standard of conduct 行为准则;行为守则
standard of construction 工程标准;施工标准;建筑标准
standard of criterion 判据;判定准则;判定标准
standard of current capital 流动资金定额

standard of deferred payments 延期支付标准
standard of economy 经济上的标准
standard of finishes 装修标准
standard of fisheries water 水产用水标准
standard of fixed number of labo(u)r 定员标准
standard of geotechnical engineering test 岩土工程试验标准
standard of illumination 照度标准
standard of industrial water-quality 工业用水水质标准
standard of length 长度标准
standard of living 生活水平;生活标准
standard of luminous intensity 光强标准
standard of mark 标志标准
standard of materials 材料标准
standard of measurement 计量标准
standard of occupational hygiene 职业卫生标准
standard of perfection 鉴定标准
standard of performance(of pollution) 污染控制标准
standard of potable water 饮用水标准
standard of productive forces 生产力标准
standard of project appraisal 项目鉴定标准
standard of purity 纯度标准
standard of quality 质量标准
standard of recharging water quality 回灌水质标准
standard of scaffolds 脚手架立柱
standard of security 安全标准
standard of stable hydrogen isotope 氢稳定同位素标准
standard of stable isotope 稳定同位素标准
standard of stable isotopic analysis 稳定同位素分析标准
standard of stable sulfur isotope 稳定硫同位素标准
standard of surveying and mapping 测绘标准
standard of swell scale 涌浪分级标准
standard of taxation 课税标准
standard of technical grades 技术等级标准
standard of tolerance 标准公差
standard of track maintenance 线路维修标准;轨道维修标准
standard of water use 用水标准
standard of wavelength 波长标准
standard ohm 标准欧姆
Standard oil company 标准石油公司(美国)
standard oil tank 标准油槽
standard on data to be provided 有关应提供数据的标准
standard opening 标准跨度;标准通道;标准洞;标准窗孔;标准开口
standard opening height 标准开口高度
standard operating instructions 标准操作规程
standard operating procedure 标准作业程序;标准经营活动程序;标准操作过程;标准操作规程;标准操作程序;标准操作步骤
standard operation 标准工作;标准处理过程;标准操作
standard operation clause 标准作业表
standard operation procedure 标准作业程序;标准运行程序
standard operation sheet 标准作业表
standard operation time 标准操作时间
standard operative temperature 标准有效温度
standard operator 标准算符;标准操作符
standard opinion 标准意见
standard optical dynamometer 标准光学测力机
standard orifice 标准孔口
standard orifice flowmeter 标准孔流速计
standard orthogonal basis 标准正交基(底)
standard orthographic projection 标准正交投影
standard oscillator 标准振荡器
standard outfit 标准装备
standard output 标准作业定额;标准输出
standard overcast sky of international commission on illumination 国际照明委员会制定的标准阴天天空
standard overhead cost 标准制造费用
standard overhead rate 标准制造费用分配率
standard oxidation potential 标准氧化势;标准氧化电位
standard oxidation-reduction potential 标准氧化还原电位;标准氧化还原势

standard oxygen requirement 标准需氧量
standard oxygen transfer efficiency 标准转氧(效)率
standard oxygen transfer rate 标准转氧率
standard package 标准组合器;标准程序包
standard packet protocol 标准包协议
standard page 标准页
standard panel 标准(镶)板
standard pantile 标准波形瓦
standard paper 标准纸
standard parallel 标准纬线
standard parameter 标准参数
standard partial regression coefficient 标准偏回归系数
standard parts 标准装件
standard paste 标准净浆;通用水泥砂浆;标准浆
standard patch 临时堵漏用木结构
standard pattern 标准型;标准图案;标准式样;标准模式
standard pattern format 标准图形格式
standard paving brick 标准铺路砖
standard payback period of added investment 标准附加投资偿还年限
standard payback period of investment 标准投资偿还年限
standard payment 标准支付
standard penalty function 标准罚函数
standard penetration blow count 标准贯入击数
standard penetration inspection 标准贯入度检验
standard penetration resistance 标准贯入阻力;标准贯入抗力;葡氏贯入阻力
standard penetration test 标准贯入(度)试验
standard penetration test analyzer 标准贯入试验分析仪
standard penetration test blow count 标准贯入试验(锤)击数
standard penetration test correction 标准贯入试验校正
standard penetrometer 标准贯入仪;标准贯入器
standard percolation test 标准渗透(性)试验
standard percolation test method 标准渗透(性)试验方法
standard perfect white diffuser 标准全白漫射体
standard performance 产量定额;标准业绩;标准性能;标准效能;标准工作成绩
standard period 标准期(外汇或货币市场报价的标准到期日)
standard permeability coefficient 标准渗透(性)系数
standard permeability test 标准渗透(性)试验
standard permeability test method 标准渗透(性)试验方法
standard petroleum densimeter 标准石油密度计
standard phase meter 标准相位表
standard photo comparison method 标准照片对比法
standard photograph 标准相片
standard photopic eye 标准适光眼
standard physical units 标准物量单位
standard picture 标准图像
standard piece time 时间定额(工件)
standard pile 支承板桩的柱;标准桩;标准堆
standard pipe 标准管
standard pipe bend 标准管子弯头
standard pipe bender 标准弯管机
standard pipe prover 标准体积管
standard pipe size 标准规格管;标准管径
standard pipe vise 龙门卡头
standard piston ring 标准活塞环
standard pitch 标准音高;标准螺(纹)距
standard pitch auger 标准螺距螺旋
standard plain metal-slitting 标准普通金属开槽锯床
standard plain milling cutter 标准普通铣刀
standard plan 标准计划
standard plane 标准面
standard plan of amortization 分期摊还标准法
standard plant 标准设备
standard plasma 标准等离子体
standard plate 标准感光板
standard plate count 标准板式计数
standard platinum-rhodium alloy thermocouple 标准铂铑热电偶
standard play 标准放音

standard plug-in unit 标准插件
standard point of photo 相片标准点
standard points 标准道岔
standard polar coordinate system 标准极坐标系统
standard polarity 标准极性
standard pole 标准电杆
standard policy 标准保险单;标准保单
standard policy condition 标准保险单条件
standard population 标准人口
standard port 主潮港;基准检潮站;基准潮位站;参考港;标准港(口)
standard portal frame 标准门式刚架
standard portland cement 普通硅酸盐水泥;标准波特兰水泥
standard positioning service 标准定位业务
standard potential 标准电位;标准电势
standard potential transformer 标准电压互感器
standard power 标准功率
standard power meter 标准功率计
standard practice 标准操作规程
standard practice instructions 标准作法说明书;标准操作规程
standard precipitation coefficient 标准沉降系数
standard preemphasis 标准预加重
standard prelude 标准序部;标准开关
standard preparation 标准制剂
standard preparation hours 标准准备时数
standard pressure 标准压力
standard pressure altitude 标准气压高度
standard pressure ga(u)ge 标准压力表
standard pressure level 标准气压层
standard pressure parts 标准受压件;标准承压部件
standard pressure vessel 标准压力罐
standard price 标准价格
standard price method 标准价格法
standard print 标准相片
standard print file 标准打印文件
standard priority 标准优先权
standard prism 标准光谱;标准棱柱体
standard procedure 标准过程;标准程序;标准操作
standard procedure instructions 标准作业说明
standard procedure manual 标准程序手册
standard procedure program(me) 标准过程程序
standard process cost 标准加工费
standard process cost sheet 标准加工成本单;标准分类成本单
standard processing 标准处理
standard processing mode 标准加工方式;标准处理方式
standard processing technique 标准加工工艺
standard proctor(compaction) test 标准普氏击实试验;标准普罗克托压实试验;标准普罗克托击实试验
standard proctor dry density 标准普罗克托干密度
standard product 标准产品;标准积
standard product cost 标准产品成本
standard production 标准产额
standard production rate 产量定额
standard production volume 标准产量
standard products 标准产品
standard profile 标准型材;标准轮廓;标准外形;标准剖面图;标准剖面
standard profile of gamma spectrum 伽马谱标准剖面图
standard program(me) 标准计划;标准程序
standard program(me) library 标准程序库
standard program(me) procedure 标准程序编制步骤
standard project 标准项目
standard project flood 设计标准洪水;标准设计洪水;标准计划洪水
standard projection 标准投影
standard projection lens 标准投影透镜
standard project storm 标准设计暴雨
standard propagation 标准传播
standard property 标准特点;标准特性;标准性能
standard provisions 标准条款
standard pull tension cable 标准抗拉电缆
standard pulse generator 标准脉冲发生器
standard punched washer 标准冲制垫圈
standard purchase price 标准购入价格;标准采购价格;标准采购成本
standard purify materials 标准纯物质
standard pyranometer 标准辐射强度

standard pyrometric cone 标准测温锥
standard quadrangle 标准图幅
standard quality 标准质量;标准品位
standard quality control 标准质量管理
standard quantity manufacture 标准量制造
standard radiation curve 标准辐射曲线
standard radio aids 标准无线电导航设备
standard radio atmosphere 标准无线电大气层
standard radiosonde 基准探空仪
standard radius 标准半径
standard rail 标准轨
standard railing 标准栏杆
standard rail section 标准钢轨断面
standard railway ga(u)ge 标准铁路轨距(等于1.435米);标准轨距
standard railway live load specified by p. R. China 中华人民共和国铁路标准活载;中一活载
standard rain ga(u)ge 标准雨量计
standard RAM 标准随机存取存储器
standard rammer 夯样机;标准夯机
standard random variable 标准随机变量
standard random walk 标准随机走动
standard rate 规定标准;基准费率;标准费率;标准定额
standard rate digestion 标准负荷消化
standard rated load 标准额定负荷
standard rated output 标准额定出力;标准额定产量
standard rated voltage 标准额定电压
standard rate filter 标准速度滤池;标准负荷滤池
standard rate filtration 标准速率过滤;标准滤率
standard rate of tax 标准税率
standard rate of wages 标准工资率
standard rate single-stage digestion tank 标准负荷单级消化池
standard rate turn 标准转弯
standard rating 标准速率;标准规格;标准额定值;标准定额
standard rating cycle 标准循环
standard rating of refrigeration machine 标准制冷量
standard ratio 标准比率
standard ratio method 标准比较率法
standard reaction force 标准反力(护舷)
standard reactor 标准电抗器
standard recession curve 标准退水曲线
standard recipe 标准配方
standard reconstituted stone 标准铸石
standard recording tape 标准测试带
standard rectifier 标准纠正仪
standard reduction potential 标准还原电位
standard reference material 标准物质;标准参照样;标准参考物质
standard reference position 标准参考位置
standard reference sample 标准参考样品
standard reference state 标准规定状态
standard reference tape 标准带
standard reference temperature 标准参考温度
standard reference water samples 标准参考水样本(美国地质调查局)
standard reference zero 标准听力零级
standard reflux-digestion method 标准回流消解法
standard reflux method 标准回流法
standard refraction 标准折射
standard refractive modulus gradient 标准折射模数梯度
standard regression 标准回归
standard regression coefficient 标准回归系数
standard regression model 标准回归模型
standard reinforced fillet weld 标准加强角焊缝
standard relative visibility 标准相对可见度
standard reliability equation 标准可靠性方程
standard repair method 标准修理法
standard requirement 标准技术要求;标准规格
standard reservation procedure 标准预订程序
standard resistance 标准电阻
standard resistance box 标准电阻箱
standard resistivity of logger 测井仪标准电阻
standard resistor 标准电阻器
standard resolution 标准分解
standard restaurant 标准饭店;标准餐厅
standard reverberation time 标准混响时间
standard rig 标准设备
standard ring ga(u)ge 标准环规
standard risk level 标准危险性水平

standard road tar 标准道路煤沥青;标准道路柏油
standard rod 标准杆
standard rolled section 标准轧制型材;标准异型钢材
standard rolling twist equipment 标准转环假捻装置
standard roof 标准屋面板材
standard rotary rig 标准旋转钻机
standard rotating direction 标准旋转方向
standard round 标准钎子组;标准炮眼组
standard routine 标准程序
standard routine practice 标准操作规程
standard rule 标准规则;标准尺(度)
standard running condition 额定工况
standard run quantity 标准(分)批量
standard salinity 标准盐度
standard sample 标准样品;标准样(本);标准试样;标准货样
standard sample analysis 标样分析
standard sample survey 标准抽样调查
standard sample survey theory 标准抽样调查理论
standard sampling container 标准取样容器;标准采样容器
standard sand 标准砂
standards and methods department 标准与方法部门
standards and performance appraisal 标准与完成评定
standards and specifications for railway design 铁路标准及设计规范
standard saprogenic type 标准腐泥型
standard scale 基准尺;标准刻度;标准尺(度);标准表
standard schedules of development 发育标准表
standard schedules of unit rates 单位估价表
Standards Code 标准规约
standards committee 标准化委员会
standard screen 标准筛
standard screen scale 标准筛制;标准筛序表
standard screw 标准螺纹
standard screw thread form 标准螺纹样式
standard screw thread ga(u)ge 标准螺纹量规;标准螺纹规
standard sea level 基准海平面;标准海平面
standard search 标准检索
standard seawater 标准海水
standard sea-water condition 标准海水条件
standard secant parallel 标准纬线
standard section 标准河段;标准型材;标准剖面;标准截面;标准断面
standard sedimentation diameter 标准泥沙粒径
standard segment 标准块管片;标准块
standard seismic coefficient 标准地震系数
standard seismograph 标准地震仪
standard seismograph station 标准地震台
standard seismometer 标准地震计
standard selected soil parameter 标准化选择土工参数
standard sensitometer 标准感光计
standard sequential file 标准顺序文件;标准顺序排列档
standard serial program(me) 标准串行程序
standard series 标准系列
standard series colo(u)rimetric(al) method 标准系列比色法
standard series method 标准系列法
standard series table 标准系列表
standard service pressure 用气压力
standard setting 标准定位
standard settling velocity 标准降落速度
standard sewage 标准污水
standards for discharge of pollutant 排污标准
standards for wastewater treatment 废水处理标准
standard shaft 基轴
standard shaft system 基轴制;标准轴制
standard shank 标准手柄
standard shape 标准型钢;标准模型;标准形状
standard shape parameter 标准形参数
standard sharp-crested weir 标准锐缘堰
standard shear coefficient 标准剪力系数
standard shear value 标准抗剪(力)值
standard sheer 标准舷弧
standard sheet 标准薄板;标准板
standard sheet metal ga(u)ge 薄钢板标准厚度

standard sheet of geologic(al) map 规整地质图图幅
standard ship 标准船型;标准船(舶)
standard shipment 标准装运
standard ship of public transportation 公共交通标准船
standard ship size 标准船型尺度;标准船型尺寸
standard short-body bend 标准短管弯头
standard short-body tee 标准短三通
standard short cross 标准短四通
standard shortened rail 缩短轨
standard short reducer 标准大小头
standard short tube 标准短管
standard shunting sensitivity 标准分路灵敏度
standard shuttering 标准模板
standard sieve 标准筛
standard sieve number 标准筛号(美国)
standard sieves 标准套筛
standard sieve series 标准筛号
standard sieves of ASTM 美国材料试验标筛;美国材料试验学会的标准筛
standard signal 标准信号
standard signal amplifier 标准信号放大器
standard signal generator 标准信号发生器
standard signal to noise 标准信噪比
standard silicate cement 标准硅酸盐水泥
standard silver 标银
standard single-mode fiber 标准单模光纤
Standards Institute 标准协会;标准学会
standards issued by ministry 部颁标准
standard size 成品尺寸;标准大小;标准尺寸
standard size brick 标准型砖;标准耐火砖
standard size of flange 标法兰尺寸
standard size piston 标准尺寸活塞
standard sizing scale 标准筛序制;标准筛序表
standard sleeper 标准轨枕
standard slop peck 标准泥浆体积
standard socket and spigot taper 标准管端套筒接合
standards of construction 建筑标准
standards of consumption 消费水平
standards of design and construction 设计及建筑标准
standards of feasibility 可行性标准
standards of fisheries water 渔业用水标准
standards of practice 从业标准
standards of professional practice 职业道德规范;同行业标准
standards of transportation planning 交通规划标准
standard software 标准软件
standard software package 标准软设备组件;标准软件包
standard soil 标准土壤
standard soil sample 标准土样
standard solar time 标准太阳时
standard solenoid 标准螺线管;标准螺管线圈
standard solids 标准固体量
standard solution 规定(溶)液;标准(溶)液;标准解法
standard solution addition method 标准液添加法
standard solution concentration 标准液浓度
standards organization 标准组织
standard sort routines 标准分类常规
standard sound level 标准声级
standard sound levels difference 标准声级差
standard source 标准(光)源
standard span 标准窗洞;标准开口;标准跨度;标准跨径
standard span of bridge 桥梁标准跨径
standard special(shape) 标准专用件
standard species 标准种
standard specification 标准技术条件;标准技术规程;标准(技术)规格;标准说明书;标准(技术)规范
standard specification brick 标准规范砖
standard specification for building materials 建筑材料标准规范
standard specification format 标准规范格式
standard specific volume 标准比容
standard specimen 标准样品;标准试样
standard spectroscopic notation 标准光谱符号
standard spectrum 标准光谱
standard speed 标准速度

standard sphere gap 标准球形放电器
standard spheroid 标准椭球
standard split barrel sampler 标准半合管取土器
standard split barrel tube 标准半合管贯入器
standard split spoon sampler 标准半合匙取土器
standard square 标准型砖;标准方块砖
standard staff 标准杆
standard stainless steel window 标准不锈钢窗
standard stair(case) 标准楼梯
standard state 标准状态;标准态
standard state air 标准状态空气
standard state concentration 标准态浓度
standard state condition 标准态状况
standard state environment 标准态环境
standard state fugacity 标准状态逸度
standard state solution 标准态溶液
standard station 主潮港;基准检潮站;基准潮位站;标准站
standard statistic 标准统计量
standard statistical routines 标准统计方法
standard statistical tabulation 标准统计汇总表
standard statistical test 标准统计检验
standard steam pressure 标准蒸汽压
standard steam table 标准蒸汽表
standard steel 标准钢
standard steel construction 标准钢结构
standard steel ingot mo(u)ld 标准钢锭模
standard steel section 型钢标准断面;型钢标准截面;标准型钢
standard steel window 标准钢窗
standard steel window section 标准钢窗断面
standard step method 标准逐步推算法
standard stiffness 标准硬度;标准劲度
standard stock 标准库存;标准材料;标准库存量;标准股票
standard stop 标准光栏
standard storage location 标准存储单元
standard storage tank 标准储[贮]罐
standard store 标准仓库
standard stor(e)y 标准楼层
standard stove 标准炉
standard strain 标准应变;标准品系
standard strain ga(u)ge analyzer 标准应变分析器
standard strength 标准强度
standard stress 标准应力
standard structural system 标准结构系统;标准结构体系
standard structure 标准组织
standard stud 标准双头螺柱
standard subchannel 标准子通道
standard submerged orifice 标准淹没孔口
standard subprogram(me) 标准子程序
standard subprogram(me) interface 标准子程序接口
standard subroutine 标准子程序
standard subroutine interface 标准子程序接口
standard subset 标准子集合
standard substance 基准物;标准物(质)
standard substance selection 标准物质选择
standard sum 标准和
standard support 标准支承
standard surface 标准表面
standard survey(ing) 标准测量
standard swing meter 标准回转表
standard switch 标准开关
standard switching voltage impulse 标准操作冲击电压波
standard swivel 标准水龙头
standard symbol 标准符号
standard system 标准制;标准体系
standard system action 标准系统动作;标准系统操作
standard system dump 标准系统转储
standard system label 标准系统标记
standard system of level 标准水网测量;标准水准系;标准高程系统
standard system program(me) 标准系统程序
standard system scanner 标准系统扫描程序
standard system table 标准体系表
standard system tape 标准系统带
standard table 标准表式
standard table telephone set 标准台式电话机
standard tap 标准丝锥;标准抽头
standard tape 基准带;标准带

standard tape executive program(me) 标准带执行程序
standard tape label 标准带标记
standard taper 正常收缩;标准锥度
standard tapered pipe thread 标准锥形管螺纹
standard taper file 标准锥形锉
standard taper ga(u)ge 标准锥度规
standard taper joint 标准锥形接头;标准锥度接头
standard taper reamer 标准锥形铰刀
standard tapping machine 标准撞击机
standard target 标准目标
standard tar visco(si)meter 标准焦油黏度计
standard tax rate 标准税率
standard tax rate zone 标准税率范围
standard technical specification 标准技术说明书;标准技术规格(书);标准技术规范
standard technique 标准技术
standard tee 标准三通(管)
standard television signal 标准电视信号
standard temperature 标准气温;标准温度
standard temperature and pressure 标准状态;标准温压;标准温度和压力
standard temperature/time curve 标准温度时间曲线
standard tensile-test core 型砂标准拉力试块
standard tensiletest core 标准拉力试块
standard tension 标准张力;标准拉力;标准电压
standard terminal 标准终端
standard terminal arrival route 标准场站到达航线
standard termination 标准负载
standard terminology 标准术语
standard terrazzo 标准水磨石
standard test 标准试验;标准检定;标准测定
standard test block 标准试块
standard test condition 标准试验条件
standard test cube 标准立方试块
standard test for glass viscometer 玻璃黏度计标定试验
standard test frequency 标准测试频率
standard test method 标准试验方法
standard test output 标准实验输出功率
standard test piece 标准试样;标准试件
standard test procedure 标准试验程序
standard test sieve 标准试验筛
standard test specimen 标准试件
standard test tone power 标准测试音调功率;标准音调测试功率;标准试音功率
standard thermocouple 标准热电偶
standard thermodynamic(al) scale 标准温标;标准热力学温标
standard thermometer 标准温度计
standard thermometer glass 标准温度计玻璃
standard thickness 标准厚度
standard thimble 标准套环
standard thread ring ga(u)ge 标准螺纹环规
standard three colo(u)r ink 标准三原色油墨
standard throw blasting charges 标准抛掷药包
standard tide station 标准验潮站
standard tie 标准轨枕
standard tile 标准瓦;标准砖
standard timber 标准方木(头)
standard timber window 标准木窗
standard time 定额时间;额定工时;标准时(间)
standard time interval 标准时距
standard time-temperature curve 标准时间—温度曲折
standard time zone 标准时间区带;标准时区
standard titrimetric substance 标准滴定物
standard tolerance 标准公差
standard tolerance unit 标准公差单位
standard ton 冷吨
standard tone gray scale 标准色调灰阶
standard tool 钢绳冲击钻工具
standard tooth 标准轮齿;标准齿
standard topping 标准面层;标准屋面
standard tower 标准铁塔
standard toxicant 标准毒剂
standard toxicity test 标准毒性检验
standard toxin 标准毒素
standard track 准轨;标准声道;标准轨距的轨道
standard track ga(u)ge 标准轨距
standard track layout at intermediate stations 中间站线路和房屋布置标准图
standard tractor 标准型拖拉机

standard trade terms 标准贸易术语
standard trajectory 标准弹道
standard transit bus 公共交通标准车
standard transition probability 标准转移概率
standard transit ship 公共交通标准船
standard transmitter 标准送话器
standard transportation commodity code 运输货物的标准代号
standard transposition system 标准交叉制
standard transverse test core 标准抗弯试块；型砂标准抗弯试块
standard tray installation method 标准塔板的安装方法
standard treatment 标准处理
standard tree 标准乔木
standard trickling filter 标准滴滤池
standard trim 标准线脚；标准修饰
standard truck 标准货车
standard truck load 标准汽车荷载
standard tube 标准管嘴；标准管
standard tube turbine 轴伸贯流式水轮机
standard turbidity solution 标准浊度液
standard turbodrill 标准涡轮钻进；标准涡轮钻具
standard turnout 标准道岔
standard twist 标准捻度
standard type 标准形式；标准型；标准形态；标准类型
standard type bowl 标准型转筒
standard type farm tread tire 标准型农用胎纹轮胎
standard type mechanical tableting machine 标准机械压片机
standard type mudstone flow 标准型泥石流
standard type purifying table 标准净化工作台
standard type sprue former 标准式铸道形成器
standard type symons cone crusher 标准型西蒙斯圆锥破碎机
standard type wheeled tractor 标准型轮式拖拉机
standard unified electroweak theory 标准统一弱电论
standard unit 标准元件；标准单元；标准单位；标准单件
standard unit cost 标准单位成本
standard unit of accounting 标准记账单位
standard unit of processing 标准处理单位
standard unit of timber 木材标准单位
standard unit of value 标准的价格单位
standard unit rate 标准单位费率
standard units of mean error 均值误差的标准单位
standard unit steel trestle （桥梁结构的）标准组合高架桥；（桥梁结构的）标准组合柱台；（桥梁结构的）标准组合钢栈桥；（桥梁结构的）标准单元高架桥；（桥梁结构的）标准单元架柱台；（桥梁结构的）标准单元钢栈桥
standard unit transportation 标准单位运费
standard-unit type machine 组合机床
standard unloaded condition 标准无负荷状态
standard user label 标准用户标号
standard vacuum 标准真空
standard vacuum ga(u)ge 标准真空计；标准真空表
standard vacuum-meter 标准真空表
standard value 标准值；标准价值
standard value of production 额定成本
standard value table 标准值表
standard variable condenser 标准可变电容器
standard variable costing 标准变动成本计算法
standard variance analysis 标准差异分析
standard variety 标准品种
standard vehicle 标准车辆
standard vehicle loading 标准车辆幅载
standard velocity 标准速度
standard version 标准形式；标准方案
standard vertical tube evapo(u)rator 标准竖管蒸发器；中央循环管蒸发器
standard vessel 标准船（舶）
standard vice 标准平虎钳
standard virtual machine 标准虚拟机
standard visco(si)meter 标准黏度计
standard viscosity liquid 标准黏度液
standard visibility 标准见度
standard vision 标准视力
standard visual range 标准视程
standard voltage 标准电压
standard voltage generators 标准电压发生器
standard voltage oscillator 标准电压振荡器

standard voltmeter 标准伏特计；标准电压表
standard volume 标准体积；标准容积
standard volume indicator 标准音量指示器
standard volume label 标准卷标号
standard wage 标准工资
standard wage and salary 标准工薪
standard wage rate 标准工资率
standard waste 标准污水；标准废物；标准废水
standard wastewater 标准废水
standard water column 标准水柱
standard water depth 标准水深
standard water meter 标准量水计
standard water tank 标准水槽
standard wave 标准电波；标准波
standard waveguide 标准波导
standard wavelength 标准波长
standard wave meter 标准波长计
standard wearing limits for inspection and repairs 检修标准限度
standard weather bureau ga(u)ge 标准雨量计
standard weathering hour 标准耐气小时数
standard weather meter 标准耐气候
standard wedge 标准光楔
standard week 标准工作周
standard weekly hours 标准周工时数
standard weight 标准重量；标准砝码
standard weight car 标准校重车
standard weighted needle 标准重试针（测定凝结时间）
standard weight of cargo 货物标准重量
standard weights and measures 标准度量衡
standard welding symbols 标准焊接符号
standard weston cell 标准电池
standard white 标准白度；标准白
standard white body 标准白体
standard white plaque 标准白瓷板
standard white plate 标准白板
standard white plate of diffuse reflectance 标准漫反射白板
standard white surface 标准白色面
standard width 标准宽度
standard width of channel 航道标准宽度
standard width of navigation channel 航道标准宽度
standard window 标准窗
standard wind pressure 标准风压
standard wire ga(u)ge 标准线规
standard wire rope 标准钢丝绳
standard with cranked head 弓形床架
standard with exterior glue 室外用胶黏剂标准板
standard workday length 标准工作日长度
standard work force 标准工时
standard working-hour 标准工时
standard working hours 标准工作时数
standard working time 标准操作时间
standard working volume 标准工作量
standard working week 标准工作周
standard works 权威著作
standard writer 标准写程序；标准打字
standard yield method 标准收获量法
standard zero 标准零点
standard zircon brick 标准锆英石砖
standard zone 标准带
standar error of quantiles 分位数标准误差
stand away neckline 不贴颈圆
stand bar 支脚；支撑杆；站立进食的酒吧
stand basal area 林分底面积
stand board 森林界线
stand-by 储备的；待用的；备用的；备品；做好准备
stand-by act 救助守候行为
stand-by anchor 准备抛锚
stand-by application 备用应用；备用机工作；备用泵；备用应用；待用
stand-by arrangement 支持性安排；备用协定；备用安排
stand-by assets 备用固定资产
stand-by battery 备用蓄电池；备用电池组
stand-by bin 备用料仓
stand-by block 备用信息块；备用数据块【计】；备用块；备用缓存单元；备用单元
stand-by boiler 备用锅炉
stand-by burning system 备用燃烧系统
stand-by capability 备用能力
stand-by capacity 辅助容量；储存容量；备用生产

能力；备用容量
stand-by car 备用车
stand-by channel 备用信道；备用通路；备用波道
stand-by circuit 备用电路；附加电路
stand-by column 备用柱；备用塔
stand-by commitment 备用承诺；备用承付款项
stand-by compass 备用罗经
stand-by compressor 备用压缩机；备用空气压缩机
stand-by computer 备用计算机
stand-by cost 备用成本
stand-by crane 备用起重机；备用吊车
stand-by credit 救急贷款；备用信用证；备用信贷；备用贷款安排
stand-by crew 扑火外站
stand-by derrick 备用起重机
stand-by diesel generator plant 备用柴油发电机组
stand-by diver 待命潜水员
stand-by drive 备用驱动机构
stand-by electric(al) power 备用电源
stand-by engine 主机准备【船】；常备机车
stand-by equipment 闲置设备；备用设备
stand-by equipment key 备用设备转接电键
stand-by face 备用回采工作面
stand-by facility 备用设备；备用工具；备用贷款
stand-by facility arrangement 备用贷款安排
stand-by fee 贷款保证金；备用金
stand-by firing system 辅助燃烧系统；备用燃烧系统
stand-by fund 预备费
stand-by gas 调峰气
stand-by gas boiler 备用煤气锅炉
stand-by generating set 备用发电机组
stand-by generation system 备用发电系统
stand-by generator 急救发电机；备用发动机；备用发电机；应急发电机
stand-by generator set 备用发电机组
stand-by heat 热备用
stand-by heating 值班采暖
stand-by instrument 备用仪表
stand by kedge anchor 准备小锚
stand by lamp 告警灯
stand-by letter of credit 备用信用证
stand-by letter of guarantee 保证信用证
stand-by lighting 备用照明
stand-by lighting system 备用照明系统
stand-by load 备用荷载
stand-by locomotive 备用机车【铁】
stand-by loss 停工损失；备用损失
stand-by machine 辅助机器；备用机（器）
stand-by mixer 备用混合器；备用搅拌机；备用拌和机
stand-by mode 准备方式
stand-by mode of operation 应急工作状态；备用工作状态
stand-by motor 备用辅助电机
stand-by motor test stand 辅助电机试验台
stand-by or spare wagon 备用车辆
stand-by pay 待用工资
stand-by period 准备期；等待期（间）
stand-by pile 备用桩
stand-by pipe 备用管
stand-by plant 备用电站；工地设备；备用设备
stand-by power 储备功率；备用动力；备用电源；备用电力；后备功率
stand-by power generator 备用发动机；备用发电机
stand-by power plant 备用发电站；备用发电厂；备用电厂
stand-by power source 备用电源；备用能源
stand-by power source system 备用电源系统
stand-by power station 备用发电站
stand-by process 准备处理
stand-by program(me) 备用程序
stand-by pump 应急泵；备用（水）泵
stand-by pump station 备用泵站
stand-by redundancy 备用位；备份冗余；待机准备
stand-by register 备用寄存器
stand-by replacement redundancy 备用替换冗余
stand-by reserves 备用储备【贮】量；药剂固定储备量
stand-by sample 备用试样
stand-by service 备用
stand-by set 备用机组
stand-by set of dies 备用模
stand-by ship 备用船
stand-by siding 备用侧线

stand-by station 备用电站
stand-by still 辅助塔;备用塔
stand-by storage 备用储备
stand-by system 备用系统;备份系统;应急系统;后备系统;等待系统
stand-by tank 备用水箱;备用池
stand-by ticket 剩余票
stand-by time 空闲时间;等待时间;待命时间;待机时间;闲置时间;准备时间;备用时间
stand-by to alongside 准备靠岸
stand-by unattended time 闲置时间;非工作时间
stand-by unit 备用机组
stand-by valve 应急阀(门)
stand-by vessel 救援船
stand chute type drop-manhole 竖槽式跌水井
stand chute type drop-well 竖槽式跌水井
stand clarification 静止脱泥法;静止澄清法
stand clear 不接触
stand cock 竖龙头;竖活塞;(厕所的)竖活栓
stand constitution 林分组成
stand density 林分密度
stand density index 林分密度指数
stand development 静止显影
stand-down 停工
standee 站立的乘客;车上站立的乘客
standee area 站票观座区域
stander 机架;采伐残留木
stander-by 旁观者
stand fire 树干火
stand gathering arm loader 立爪式装载机
stand gear 备用工具
stand height 林分高
stand height diameter 林分高大木直径
stand-in 模拟物
stand in debt 负有债务
standing 直立(的);立射俑;不流动的
standing acoustic(al) waves 声学驻波
standing and parking prohibited sign 禁止停车标志
standing apron 停机坪
standing area 停车场地;堆场;车厢站立面积;停机坪
standing army 常备军
standing axis 竖轴
standing axis error 竖轴误差
standing baffle 固定挡板;固定缓冲板
standing-balance 静态平衡
standing bevel 标准斜角
standing block 定滑轮;定滑车
standing bobby 闷炮眼
standing bolt 固定螺栓;双帽螺栓
standing brake test 规定的制动试验
standing business 继续交易
standing capacity 车场容车能力
standing carry 直通进位
standing charges 固定费(用)
standing cloud 驻云【气】;山帽云【气】
standing column 静止液柱
standing committee 常务委员会;常设委员会
standing cost 经常成本
standing crop 现存量;未收割的作物
standing crop biomass 现存生物量
standing current 驻流;基流;稳定电流
standing derrick 立式起重机;起重桅(杆);起重扒杆;单杆;扒杆
standing director 常务董事
standing eddy 驻涡;竖轴漩涡
standing electromagnetic wave 电磁驻波
standing equilibrium 静态平衡
standing evapo(u)ration loss 储存蒸发损失
standing expenses 经常费(用)
standing expenses order 经常费用单
standing face 稳定面;固定面;静止面
standing factory 停工(的)工厂
standing finish 固定配件;标准装修
standing fisheries committee 常务渔业委员会
standing floe 侧立冰
standing formes 存版
standing gas 滞留瓦斯
standing gear 固定装置;静索
standing geyser 周期性间歇泉
standing grain 储备粮
standing gravity wave 重力驻波;竖向驻波
standing growth 稳定增长

standing gutter 落水槽;立式水槽;竖管
standing guy rope 风缆;固定索;稳定索
standing harbo(u)r 低潮港口(浅水位时使用的小港);临时港口
standing hay 未收割的干草;地面干草
standing heater 独立式供暖器
standing height 立高
standing idle 闲置的
standing instruction 委付款项
standing internal wave 内驻波
standing ladder 直立梯;人字梯;便梯
standing lamp 落地灯
standing lane 短时停车道
standing leaf 固定窗扉;固定门扉;(折叠门的)固定门页
standing level 静水位;静水面;稳定水位;停水位
standing line 固定支索;静索;定索
standing load 持续负载;持续负荷
standing loan 固定贷款
standing lock 立交口
standing locomotive 停留机车
standing loss 储存损失
standing machine 停开的机器;停工的机器
standing marble 标准大理石;直立(墙面)大理石
standing moor 倒退锚一字锚
standing off 冷却阶段
standing off crew 休班船员
standing offer 持续报价
standing-off period 冷却阶段;冷却周期
standing of vehicle 车辆暂停
standing-on-nines carry 高速进位;逢九跳跃进位;逢九进位;标准十进位
standing operating procedure 标准操作规定
standing order 经常性支付授权;经常订单;常规命令;长期订单;现行通令;议事规程
standing panel 立式镶板;立式门板;长镶板
standing part 定端
standing pier 直立桥墩;独立桥墩
standing pipe 竖管
standing population 现存生物量;种群蕴藏量
standing position 立位
standing posture 站式;立位姿势
standing pressure 稳定压力
standing pulley 定滑车
standing rate 定额
standing rating 定额
standing rigging 固定支索;固定索具;静索;定索
standing room (戏院的)站票席位;站立的地位;(剧院的)立看位;(戏院的)站席;站立空间
standing root 向上根
standing rope 固定拉索;稳索
standing rule 规章;办事规则;经营规则
standing sale 立木销售
standing seam 直立咬口;立接缝;直立(接)缝;屋面立缝
standing seam metal roof 立接缝金属屋顶
standing seam plate 直立焊接板
standing seam roof cladding 固定焊缝屋面盖板
standing seam roofing 立接缝屋面
standing seam system 立接缝做法
standing seam tin roof 立缝白铁皮屋面
standing ship way 固定滑道【船】
standing shock wave 驻激波
standing shot 震动爆破;松动爆破
standing siding 停车线
standing signal instructions 通用信号细则
standing space 站票席
standing state 现有状态;定常状态
standing stay 固定支索
standing stone 孤赏石
standing storage 长期储藏
standing support 固定支座;支承结构;常规支护
standing surface wave 静震;假潮;表面驻波;表面立波
standing swell 驻立长波;驻波;立波
standing tackle 定绞辘
standing tank 固定储罐
standing test 静止试验
standing timber 立木;未伐倒的树木
standing time 停歇时间;停留时间;停工时间;存放时间;放置时间
standing tower 标准塔楼
standing traffic 暂停的交通
standing tree 立木

standing tube 竖管
standing turk's head 固定打花箍(绳结)
standing type 立式的
standing type of mooring ring 立式系船环
standing urinal 立式(小)便池
standing value 有效价值
standing valve 固定阀
standing vice 固定虎钳
standing waste 溢水竖管;弃水竖管;污水竖管
standing water 静水;停滞水;停潮;死水;滞水
standing water ecosystem 静水生态系统
standing water level 停水位;常驻水位;静水头;静水位
standing water region 静水区
standing water tide 驻波潮
standing wave 驻波;立波;定波
standing-wave accelerator 驻波加速器
standing-wave amplifier 驻波放大器
standing-wave analysis 驻波分析
standing-wave antenna 驻波天线
standing-wave apparatus 驻波仪器
standing-wave condition 驻波条件
standing-wave detector 驻波检测器
standing-wave field 驻波场
standing-wave flume 驻波槽;水跃挖流槽
standing wave guide 驻波波导
standing wave indicator 驻波指示器;驻波检测器
standing wave length 驻波波长
standing-wave linac 驻波直线加速器
standing-wave line 驻波线
standing-wave linear accelerator 驻波直线加速器
standing-wave loss factor 驻波损耗因数
standing-wave measurement 驻波测量
standing-wave measuring instrument 驻波测量仪
standing-wave memory 驻波存储器
standing-wave meter 驻波指示器;驻波检测器;驻波测定器
standing-wave method 驻波法
standing-wave node 驻波节点
standing-wave operation 驻波运行
standing-wave pattern 驻波图案
standing-wave power ratio 驻波功率比
standing-wave pressure 驻波压力;立波压力
standing-wave producer 驻波激发器;驻波产生器
standing-wave protecting circuit 驻波保护电路
standing-wave ratio (吸声系数用的)驻波比;驻波系数
standing-wave ratio indicator 驻波比指示器
standing-wave ratio measurement 驻波比测量
standing-wave ratio measuring kit 成套驻波比测量器
standing-wave ratio meter 驻波检测器
standing-wave separator 驻波选矿机
standing-wave structure 驻波结构
standing-wave system 驻波系统
standing-wave theory 立波理论
standing-wave tide 驻波潮
standing-wave tube 驻波管
standing-wave tube method 驻波管法
standing-wave type discriminator 驻波式鉴别器
standing-wave voltage ratio 驻波电压比
standing way 下水滑道;固定台架;滑道;底滑道
standing welt 固定盖缝条;固定贴边
stand-in rate 替代汇率
stand insulator 支座绝缘子
stand in the way of 妨碍
stand in to land 驶向海岸
stand leg 机腿
stand linseed oil 亚麻厚油;厚熟亚麻(仁)油;调墨油
stand mat 台式拧螺丝机
stand meal 立餐
stand method 台座法
stand microscope 座架式显微镜
stand moor 低水位沼泽
stand motor 轧机电机
stand of drill pipe 钻管支架;钻杆立根
stand off 驶离;基准距;(钻孔底的)残留岩芯
stand off and on forces 船舶荷载
stand-off basin 中和池
stand-off error 变位误差
stand-off force 离岸力
stand-off glass insulator 高脚玻璃绝缘体
stand-off insulator 支座绝缘子;托脚绝缘子
stand-off operation 间接式爆炸成型

stand-off target acquisition system 机载目标捕获系统
stand-off thread (接头连接后未拧到底的)空扣
stand of rods 钻杆立根
stand of tide 憩流;平流;平潮;停滞潮;停潮
stand oil 抽油;厚油;调墨用熟油;熟油;定油
stand oil Chinese white enamel 亚麻仁油中国白釉瓷
stand oil content 熟油含量;亚麻仁油含量
stand oil gloss paint 熟油光泽涂料
stand oil kettle 熟油壶;亚麻仁油壶
stand oil paint 熟油漆;亚麻仁油漆
stand oil pot 熟油壶;亚麻仁油壶
stand oil synthetic resin enamel 亚麻仁油合成树脂釉瓷
stand oil type varnish 亚麻仁油型清漆
stand oil white enamel 亚麻仁油锌白釉瓷
stand on 直航;航向保持;保持航向和速度不变
stand on end 立置;竖着放
stand on light beacon 照准灯标
stand on vessel 直航船(被让路船);权利船;被让路船
standort 综合环境影响;环境综合影响
stand out 离岸向海上行驶;浮出
stand over 延期;监督
standpipe 消防立管;圆筒形水塔;立管;调压竖管;竖管;疏水器;导向管;测压管
standpipe control valve 立管控制阀
standpipe-first-aid 应急辅助消防水管
standpipe floor 立管台
standpipe for fire-fighting 消防立管
standpipe main 主立管
standpipe manhole 竖管人孔
standpipe of a hydrant 消防栓竖管
standpipe outlet 立管出口
standpipe piezometer 竖管水位计
standpipe pressure 立管压力
standpipe riser 立管
standpipe supply 集中给水站
standpipe suspension load sampler 竖管式悬移质采样器
standpipe system 消防水管系统;立管系统
standpipe tank 竖管式水箱
standpipe-type 立管式
standpipe type drop-manhole 竖管式跌水井
standpipe type drop-well 竖管式跌水井
standpiping 打入法下导向管
standpoint 观点;目标
standpost hydrant 柱式消防栓;柱式消防栓;地面柱式消防栓;地面消火栓;地面消防栓
St. Andrew's cross bond 英式(十字)砌法
stand ring 座环;垫环
stands for slide project 幻灯台
stand sheet 封死窗;不开窗;固定窗扇;竖立板
standstill 搁浅;静止;停滞;停工;停顿;停变期
standstill current 止转电流
standstill locking 止转
standstill policy 停滞政策;停顿政策
stand stop 停止
stand stretch 机架变形
stand structure 林分组成;试车台结构
stand table 林分表
stand test 试验合格
stand to lose 一定失利
stand to sea 向海航行
stand tower 墩标
stand transect 直立式样条
stand type 林分类型
stand under load 受载
stand-up 直立(的);备用
stand-up counter 站立便餐柜台;高登便餐柜台
stand-up-drive 直立式传动装置;安全传动
stand-up fork truck 直立的叉式起重车
stand-up formation 直立地层
stand-up ground 支撑地层
stand upon the course 稳住航向;保持航向不变
stand-up time 自持时间;支承时间;承受时间
stand-up welt 屋面立缝
stand urinal 立式小便器
stand volume table 林分材积表
stand wear and tear 耐磨损
stanfieldite 陨磷钙镁石;磷镁钙石;磷镁钙矿
Stanford joint 斯坦福接头(一种承插管接头,承端设斜口沥青圈使管子在一直线上)

stank 小水坝;塘堰;池塘;堰;水沟
stanleya sop 鸡冠花
stanleyite 斯水氧钒矾
Stanley Kent's fibres 斯坦利一肯特纤维
Stanley mountain terrane 斯坦山地体
Stanley up and down work 斯坦利紧松指示器
stannary 锡矿区;锡矿
stannate 锡酸盐
stannate ceramics 锡酸盐陶瓷
stannekite 煤中树脂状烃
stannic 锡的
stannic acid 锡酸
stannic chloride 氯化锡
stannic disulfide 二硫化锡
stannic hydroxide 氢氧化锡
stannic iodide 碘化锡
stannic oxide 氧化锡
stannic sulfide 硫化锡
stanniferous 含锡
stanniferous faience 锡釉陶器
stanniferous galze 锡釉(陶瓷)
stanniol 高锡耐蚀合金
stanniol steel bimetal sheet 高锡钢双金属板
stannize 渗锡
stannizing 渗锡处理
stannoidite 似黄锡矿
stannomicrolite 锡细晶石
stannopalladinite 锡钯矿
Stannoplus 癸磷锡
Stannoram 癸磷锡
stannosis 锡尘肺;锡尘病
stannous 二价锡的
stannous bromide 溴化亚锡
stanuous chloride 氯化亚锡
stannous chromate 铬酸(亚)锡
stannous fluoboric acid 氟硼酸亚锡
stannous fluoride 氟化亚锡
stannous hydroxide 氢氧化亚锡
stannous iodide 碘化亚锡
stannous maleate 马来酸亚锡
stannous mono-sulphate 硫酸亚锡
stannous oxalate 草酸亚锡
stannous oxide 氧化亚锡
stannous sulfate 硫酸亚锡
stannous sulfide 硫化亚锡
stannous sulphate electrolyte 硫酸盐电解液;硫酸锡电解液
stannum 银铅合金
Stanolind pressure test 斯坦纳林德压力试验
stantionary flow 稳流
stanza 楼中一室;套间;房间
Stapeley volcanic series 斯塔佩利火山岩统
staple 肘钉;卡钉;环钩;骑马钉;V 形钉
staple angle 环套角钢
staple bolt 杆环;卡钉
staple cargo 大宗货
staple conveyer 原料传送带
staple court 商事法院
staple crop 主要作物
staple cutter 短纤维切断机
staple diet 主食
staple fastening 弓形夹系固;钉系固
staple fiber 人造短纤维;定长短纤维;切断纤维;定长纤维
staple fiber cloth 定长纤维布
staple fiber bulk yarn 短纤维膨体纱
staple fibre cutting machine 人造纤维切断机
staple fibre muslin 人造棉布
staple fibre top 人造短纤维条
staple-fibre washing machine 人造短纤维清洗机
staple food 主食品;主食
staple food preparation room 主食加工间
staple glass fiber 短切玻璃纤维;标准玻璃纤维;定长玻璃纤维
staple goods 大宗货
staple gun 射钉枪
staple hammer U 形钉锤;钉书机
staple length 毛丛长度;毛长
staple market 主要市场
staple mat 定长玻璃纤维毡
staple pit 盲井;暗井
staple products 大宗货
stapler 纤维切断机;射钉枪;主要商品批发商;钉书机
staple rayon 人造棉

staple roving 定长纤维粗纱
staple-shaft hoist 暗井提升机
staple sliver 合胶线;定长纤维毛条;粗纺条
staple sorter 纤维长度分析器
staple tissue 定长纤维薄毡
staple vice 立式虎钳;长腿火钳
staple yarn 定长纤维纱
stapling 纤维按长度分级;环套角钢
stapling machine 射钉机;纤维切断机;压钉机;钉书机
stapling tacker 勾钉
star 星状物;星形饰【建】;星形;恒星
star aerial 星形天线
star ager 星形架蒸化机
star algebra 星代数
star alloy 星轴承合金;轴承合金
star altitude curves 恒星等高度线图
star aniseed 八角
star antimony 精制锑;精炼锑;纯锑
star atlas 星图集;星图
star background 星空背景
star batcher 星形(计量)给料器;星形进料器;星形配料器
star bit 梅花钻头;星形钻头
starboard aft 右舷后
starboard aids 右舷航标;右舷标志
starboard anchor 右舷锚;右锚;船首右大锚
starboard anchor winch 右锚绞车
starboard beam 右舷正横;右舷梁
starboard bow 右舷方向;右前方
starboard chamber 右舷仓
starboard crane 右手起重机
starboard easy 向右缓转
starboard forward 右舷前
starboard hand 右舷方
starboard hand buoy 右侧浮标;右岸浮标;右舷浮标;右舷侧浮标
starboard hand mark 右侧标;右岸标志
starboard hand structure 右岸建筑物
starboard inboard 右舷内
starboard light 右舷舷灯;右舷灯
starboard quarter 右后方;船尾右舷方向
starboard quater 右舷仓
starboard(side) 右舷;右舷方
starboard side of channel 航道右侧
starboard side pontoon 右边浮箱
starboard signal 右舷信号
starboard spud 右锚柱
starboard tack 右舷抢风
star body 星形体
star box 星形连接电阻箱
star braid 星带
star burner 星形燃烧器
star burst 星形脉冲组;星爆式
star burst increment 星形脉冲串增量
starburst pattern 星芒图案
star burst test pattern 星形脉冲群试验样图
star burst test tape 星形脉冲试验带
star bus bar 星形母线
star cable-stay bridge 星式斜缆桥;星式斜拉桥
star-cap 星状帽
star catalog 星表
star catalogue 星表;恒星目录
star centralizer 星形扶正器;星形定心器
starch 淀粉
starch adhesive 淀粉胶粘剂
star chain 星形(信标)链;台卡电台链
star chart 星图
star chart of whole sky 全天恒星图
starch bottle 糊料瓶
starch coating 淀粉涂料
starch derivative flocculant 淀粉衍生物絮凝剂
star check 星裂;心裂
star checked knot 星裂节;径裂节
starch factory 淀粉厂
starch flocculant 淀粉絮凝剂
starch glue 淀粉胶;糨糊
starch gum 淀粉胶;糊精
starch gum glue 淀粉树脂胶
starching 上浆
starching machine 上浆机
starch interceptor 淀粉中间收集器;淀粉拦截器
starch iodide 淀粉碘化物
starch iodide indicator 淀粉指示剂

starch iodide reaction 淀粉碘化物反应
starch iodine colo(u)rimetry 淀粉碘比色法
starch laden water 充满淀粉的水
starch mixed wastewater 淀粉混合废水
starch partition chromatography 淀粉分配色谱法
starch paste (淀粉)糨糊
starch powder 淀粉
starch soluble 溶性淀粉
starch solution 淀粉溶液
starch test paper 淀粉试纸
starch waste(water) 含淀粉废水;淀粉废水
star circuit 星形连接电路
star clock 校星钟
star cloud 星云
star cluster 星团
star colo(u)r 星色
star communication network 星式信息交流网络
star compass orientation 星象罗盘定向
star complex 星形复形
star connect 星形连接
star-connected 星形(连接)的
star-connected circuit 星形连接电路
star connection 星形网络;星形连接
star connection method 星形连接法
star connection system 星形接法
star connector 星形接头
star controller 星形控制器
star convergence 星收敛
star conveyer 星轮式输送器
star coordinates 恒星坐标
star coupler 星状耦合器;星形耦合器
star coupling 星形连接器
star crack(ing) 星状裂纹
star crater crack 星形弧坑裂纹
star cross section 星形截面
star cut 星形琢磨花样
star-delta connection 星三角形接法;Y-Δ接法;星形三角接线
star-delta connection method 星形三角接法
star-delta connection wire 星形三角连接线
star-delta control 星三角起动(感应电动机)
star-delta reversing switch 星三角换接开关
star-delta starter 星形三角(形)起动器;星三角启动器
star-delta starting 星形三角起动(感应电动机)
star-delta switch 星形三角形转换开关;星形三角开关
star-delta switching 星形三角转接
star-delta switching starter 星形三角形切换起动器
star diagrams 星状图
star distributed system 星形分布系统
star distribution system 辐射管网
star domain 星形域
star dresser 星形修整工具
star drift 星流
star drill 星形钻(头);小孔钻;梅花钻;十字形钻
stardust 星点;闪光膜(漆病)
star dyeing 星形架染色法
stare 污斑
stared drill 星形凿
stare decisis 依照先例
star elevator 星轮式提升机;星轮升运器
starex 浮油松香
star expansion bolt 膨胀螺栓
star feed 星轮进给
star feeder 星形进料器;星形加料器
star-field mapping 星空测绘
star field tracker 星场跟踪仪
star figure 星芒图
star finder 寻星仪;索星卡
star finder and identifier 索星卡
star finite 星形有限的
star finite complex 星形有限复形
star finite covering 星形有限覆盖
star finite property 星形有限性
star finite space 星形有限空间
starfish 海星
star fixing 测星定位
star flare 星式信号弹
star flashing glaze 星盏釉
star flower 星状花
star flowered 星状的
star fracture 星状裂纹;星形裂纹(平板玻璃)
starframe 星形架

star galaxy 星系
star gate 星形给料闸门
star ga(u)ge 星形测膛规
star gear 行星齿轮;星形(齿)轮
star gem 星光宝石
star glacier 星状冰川
star globe 星球仪;天球仪
star graph 星形图
star grid 星形炉栅
star-grounded 星形图中点接地的
star handle 星形手轮
star hand wheel 星形手轮
star honeycomb 星形蜂巢形
star house 星形平面的塔式住宅
star identification 星的判别
star identification map 索星图
star identifier 寻星仪;索星卡
staring blankly forward 直视
staringite 四方钽锡矿
staring materials 原始物料
star junction 星形交叉
Stark broadening 斯塔克增宽
Stark effect 斯塔克效应
Stark-Einstein law 斯塔克—爱因斯坦定律;爱因斯坦光化当量定律
starkeyite 四水泻盐
Stark induced frequency shift 斯塔克感生频移
star knob 星形捏手
star knob nut 捏手螺母
star knob screw 星形旋钮螺旋
Stark number 斯塔克藩数
Stark shift 斯塔克位移
Stark spectroscopy 斯塔克光谱学
Stark splitting 斯塔克劈裂
star lamp 星灯
star lapping machine 行星研磨机
star lattice 星形格
starlight 星光
starlight illumination 星光照度
starlight medium range observation device 星光中程观察装置
starlight scope 星光镜
starlight telescope 星光望远镜
starlike 星形的
starlike crack 星状裂纹
starlike curve 星状曲线
starlike image 星象
starlike optic(al)object 类星光学天体
starlike ornamentation 星形饰【建】
starling 环绕桥墩的防护桩;桥墩尖端分水桩;桥墩尖端;桥墩护栏;分水桩;杀水桩
starling cope 桥墩沿围
starling coping 墙帽;凸出顶盖;凸出顶层
Starling's law of the heart 斯塔林定律
star longitudinal cable arrangement of cable-stayed bridge 斜拉桥星形纵向索布置
star mark 星形裂纹(搪瓷表面缺陷)
star-mesh transformation 星形网眼变换
star metal 精制锑
star mica 星云母
star 星形线脚
star navigation 恒星导航;天体导航
star net 星形网眼
star network 星形网络;星状网络
star note 星币
star number 星号
star observation 恒星观测
star occultation 掩星法
Star of Africa 非洲之星
star of David 大卫王之星(犹太教的六芒星形标志)
Star of South Africa 南非之星
Star of the South 南星
star ornament 星形装饰(品)
star pair method 星对法
star pattern 星形
star-pattern drilling bit 十字钻头(冲击钻用)
star peak group 星峰岩群
star perforator 铣刀式穿孔器
star place 恒星位置
star plate 星形片
star point 星点;中性点
star polymer 星形高聚物
star program(me) 手编无错程序;无错程序
star prong 星支

star proper motion 恒星自行
star quad 星绞四线组
star quad cable 星绞电缆
star quad construction 星绞结构
star quad formation 星形结构
star quad stranding 星绞
star quad twist 星形四线组扭绞(电缆)
star-quartz 星彩石英
star reamer 星形扩孔器
starred glaze 星状结晶釉;星点釉
starred non-terminal 带星号的非终结符号
starred polygon 星多边形
starred roll 星形卷取
star refinement 星形加细
star representation 星表示
star reticle 星形调制盘
star-ribbed vault 星状肋穹隆;星形肋拱顶
starring 呈星状花纹
star roller 星轮式镇压器
star route 星形相交路线
star ruby 星彩红宝石;星形红宝石
starry sky 星夜;晴夜空
star salesman 首席销货员
star sapphire 星彩蓝宝石
star section 十字(形)截面
star seeker 星球自动定向器
star sensor 星传感器
star shake 星形裂纹;星形环裂;星裂
star-shaped 星形的
star-shaped agitator 星形搅拌机
star-shaped bit 星形钻(头)
star-shaped bruise 星形撞伤
star-shaped building 星形建筑物
star-shaped carbon cell 星状碳极电池
star-shaped diaphragm 星状光圈
star-shaped dispenser 星形隔仓板
star-shaped disposition 星形排列
star-shaped filter element 星形滤油片
star-shaped landing gear 星形着陆装置
star-shaped pattern 星形模型
star-shaped plan 星形平面;星形布置图
star-shaped profile 星状截面
star-shaped set 星形集
star-shaped slug 星形棒
star shell 照明弹
starshine recorder 夜天记录器
star-star connection 星形—星形接线;星形接法;星星形接线
star statics 天体干扰
star steamer 星形架蒸汽箱
star-stone 星彩石
star-streaming 星流
star structure 星形结构
star strut 星形支撑
star system 辐射管网;银河系;恒星系统
start 开始;开工;开动;开端;起始
startability of motor fuel 发动机燃料起动性
start and end shift data 上下班数据
start and stop lever 启闭杆
start a new 从头开始
star target 星标
start at full load 满载起动
start auditing 开始审计
start bit 开始位;起始位;起动位;起动码元;启动位
start building 动工
start business 开业
start button 起动按钮;起步开关
start card 起始(登记)卡
start code 起动电码
start codon 起始密码子
start control 起始控制器
start data collection 开始收集资料
start-data-traffic indicator 启动数据通信指示器
start delay 启动延迟
start dialing signal 起始拨号信号
start distance 起动距离
start distributor 启动分配器
start drill 定位中心站
start element 开始元素;起始码元;起始单元
start element number 开始元素号;起始单元数
start telescope 恒星望远镜;单筒望远镜;测星望远镜
start temple 星形边撑
starter 起辉器;钻孔器;封瓦板;檐口衬板;(檐头屋面板的)底板;开动器;起动器;起动机;启动器

导向钻头;初始值;发动机起动机
starter armature 起动机转子;起动机电枢;起动电动机
starter armature shaft spacer 起动机电转子轴隔片
starter bar (连接用的)露天钢筋;伸出钢筋;搭接钢筋;插接钢筋;露头钢筋
starter barrel 开孔短岩芯管
starter battery 起动器电池
starter board 屋檐铺底板
starter box 起动箱
starter brick 露头砖;起始砖
starter bruch spring 起动机电刷簧
starter brush 起动机电刷
starter brush field coil 起动机磁场线圈电刷
starter button 起动(按)钮;启动按钮
starter button guard 起动机按钮护套
starter cable 起动机电缆
starter cartridge 起爆管
starter casing cover packing 起动器箱盖密垫
starter cathode 启动阴极;触发阴极
starter center bearing bushing 起动机中心轴承衬套
starter center bearing plate 起动机中心轴承板
starter clip 地板卡;石膏板条的起始固定件
starter clutch 起动机离合器
starter clutch spring 起动机离合器弹簧
starter commutator end cover 起动机整流器端盖
starter commutator shield 起动机整流器护罩
starter condenser 起动机电容器
starter contact 起动器触点
starter control 启动机控制;起动机控制
starter controller 起动控制器
starter control lever 起动机控制杆
starter cover band 起动机罩箍带
starter crank 起动摇柄
starter dog 起动棘爪;起动机传动轴棘爪
starter drill 钻机启动器
starter drive 起动机传动
starter drive housing bushing 起动机轴壳衬套
starter drive spring 起动机驱动弹簧
starter driving gear 起动机驱动齿轮
starterdynamo 起动充电发电机
starter engine 起动发动机
starter engine control 启动发动机操作机构
starter field coil 起动机磁场线圈
starter field coil connection 起动机磁场线圈接头
starter field coil equalizer 起动机磁场线圈中间接头
starter formula 初始值公式
starter frame 定位模板
starter frame block 起动机机座
starter frame through bolt 起动机机架穿过长螺栓
starter front cover 起动机前盖
starter gap 起动隙缝;起导间隙
starter gear 起动装置
starter gear housing 起动机齿轮传动壳体
starter gear shifting fork 起动机齿轮拨叉
starter generator 起动(器)发电机
starter ground brush 起动机地线电刷
starter ground strap 起动机接地片
starter handle 起动手柄
starter home 首次置业房子
starter housing 起动机壳
starter interlock 起动机互锁机构
starter interlock circuit 起动机互锁电路
starter joint (靠近套管鞋的)一根套管
starter key 启动机钥匙;起动机钥匙
starter magneto 起动磁电机
starter main shaft bearing 起动机主轴轴承
starter main shaft gear 起动机主轴齿轮
starter mechanism 起动机构
starter motor 起动马达;起动机转机;起动电动机
starter motor commutator 起动电动机整流子
starter motor control 起动电动机控制
starter motor cover 起动机盖
starter motor drive spring 起动马达传动簧
starter motor shift lever 起动机传动臂
starter motor strap 起动电动机固定带
starter motor wire 起动马达导线
starter mounting 起动机架
starter nozzle 起动机喷嘴
starter operation(al) system 初始操作系统
starter paddle 起动踏板
starter pedal 起动踏板
starter pedal return spring 起动器踏板回动弹簧

starter pinion 启动小齿轮;起动机小齿轮
starter pinion and clutch yoke 起动机小齿轮及离合器轭
starter pole shoe 起动机磁极瓦
starter push button 起动器按钮
starter relay 起动继电器
starter relay switch 起动机继电器开关
starter rheostat 起动变阻器
starter ring gear 起动机环齿轮
starter rods 构件外伸钢筋
starter shaft 起动机轴
starter shifting lever 起动机接合杆
starter shift lever 起动机变速杆
starter shift spring 起动机移动叉弹簧
starter size 开孔钻头直径;开孔尺寸
starter solenoid 起动机电磁线圈
starter solenoid coil 起动机电磁线圈
starter solenoid relay cover 起动机电磁继电器盖
starter solenoid switch 起动器电磁线圈开关
starter spring 起动机弹簧
starter strip (檐口屋面下的)防水片条;卷钢卷涂长度计数器;屋檐铺底层
starter stub 接头座
starter switch 起动器开关;起动开关;起动机开关
starter switch contact 起动开关触片;起动机开关接触片
starter switch push rod 起动机开关推杆
starter system 初始(值)系统
starter terminal cover 起动机接线盒盖
starter terminal post 起动机接线端子
starter tile 檐瓦;檐口瓦
starter toggle switch 起动机肘节开关
starter toothed wheel 起动齿轮
starter voltage 起动装置电压
starter winding 起动绕组
star test 星点试验
Startex 史达塔克斯(一种专利纤维板)
start-finish line 起止线
start for 开始钻进直到……深度(钻孔)
start form the very beginning 从头开始
start gain time 起始增益时间
start gate 起动闸门
start ga(u)ge 起始定位装置
start gradient 起始梯度
start horsepower 起动马力
start humping signal 推送信号
start-ignition system 起动点火系统
starting 起动;开端
starting ability 起动性能
starting acceleration 起始加速度;起动加速度;始加速度
starting acceleration of train 列车起动加速度
starting achievement 惊人的成就
starting a cut 开切航道
starting address decoder 起始地址译码器
starting a hole 开孔[岩]
starting aid 起动辅助装置;起动辅助器;起动辅助设备
starting air 起动空气
starting air bottle 起动空气瓶
starting air compressor 起动空气压缩机
starting air consumption 起动用压缩空气消耗量;起动空气耗量
starting air container 起动空气瓶
starting air cylinder 起动空气瓶
starting air distributor 起运空气分配器
starting air distributor drive 起动空气分配器传动装置
starting air indicator 起动空气指示器
starting air line 起动空气管路
starting air master handwheel 起动空气主操纵手轮
starting air master valve 起动空气主阀
starting air pilot valve 起动空气控制阀
starting air piping 起动空气管系
starting air silencer 起动空气消声器
starting air slide valve 起动空气滑阀
starting air stop valve 起动空气截止阀
starting air system 起动空气系统
starting air tank 起动空气瓶
starting air timing valve 起动空气分配阀
starting air valve 起动气阀
starting air valve mechanism 起动空气阀机构
starting air vessel 起动泵的气室;起动空气瓶

starting algorithm 起动算法
starting amortisseur 起动阻尼绕组
starting and acceleration performance 起步
starting and control equipment 起动控制设备
starting and fuel control handle 起动及燃油控制手柄
starting and load-limiting device 起动和负荷限制装置
starting and regulating resistance 起动调节电阻
starting and stopping mechanism 起停装置
starting apparatus 起动装置;起动器
starting arc 引弧;引发电弧;起动电弧
starting autotransformer 起动自耦变压器;起动用自耦变压器
starting a well 开孔[岩]
starting bar 起动柄
starting barrel 起始岩芯管
starting battery 起动电池组;起动用蓄电池
starting bit 开动钻头;起动钻头;开口钻
starting block 起动块;起熔块
starting board 起动盘
starting box 起动用电阻箱
starting bucket 起始相稳定区
starting button 起动(按)钮
starting cable 起动电线
starting cam 起动凸轮
starting capability 起动能力
starting capacity 起动功率
starting carbon dioxide level 进口二氧化碳浓度
starting casing barrel 套管开孔钻头
starting chain 起动链
starting chamber 起动燃烧室
starting characteristic 起动特性
starting circuit 起动电路;触发电路
starting cock 起动旋塞
starting compensator 起动补偿器
starting compressor 开动压缩机;起动(用空气)压缩机;起动压缩机
starting condition 起始条件;起动状况;初始条件
starting condition for oscillation 起振条件
starting contactor 起动接触器
starting control 起始点控制;起动操纵
starting control slide valve 起动控制滑阀
starting core barrel 开孔岩芯管
starting course 檐口屋面板;檐口铺底层;起始层;起动距离
starting crane 汽车起动摇柄
starting crank 起动曲柄
starting crank bearing 起动曲柄轴承
starting crank bracket 起动曲柄托架
starting crank handle 起动摇手柄;起动手摇曲柄;起动曲柄摇手
starting crank jaw 起动曲柄爪
starting crank jaw pin 起动摇把爪销
starting crank nut 起动曲柄螺母
starting crankpin 起动曲柄销
starting crank ratchet 起动曲柄棘轮机构
starting crankshaft 起动摇柄;起动曲柄轴
starting crankshaft spring 起动曲柄弹簧
starting crank socket cap 起动曲柄孔盖
starting crude 起动原油
starting current 起动电流
starting curve 起动曲线
starting cut-out solenoid 起动中止螺线管
starting data address 起始数据地址
starting date 开动日期;开工日期
starting date of a project 开工日期
starting date of contract 合同的生效日期
starting datum mark 起测基点[测]
starting delay 起动延误;起动延迟
starting device 起动装置
starting diameter 开钻直径;开孔直径
starting difficulty 起动困难
starting distributor 起动分配器
starting dog 起动爪;起动用凸块
starting drill 中心孔钻机(头);开动钻机;起动钻机;粗钻头
starting drive spring 起动机驱动弹簧
starting duration 起动时间
starting duty 起动功率;起动负载
starting dynamo 起动电机
starting effort 起动力
starting ejector 起动抽气机
starting electrode 起动电极

starting electro-pneumatic valve 起动电空阀
starting element 起动元件
starting ending points of curve-section 曲线起讫点
starting energy 起动能源
starting engine 起动(发动)机
starting engine operating linkage 起动发动机的操纵连杆
starting engine transmission 起动发动机传动装置
starting equipment 起动装置;起动设备
starting fan 起动风扇;引火用风扇
starting fluid 启动液;开始流动;启动燃料
starting force 起动力;初始力
starting force effort 起动力
starting fraction 起始馏分;初馏分
starting frequency 起始频率
starting friction 起动磨损;起动摩擦力
starting friction loss 起动摩擦损失
starting fuel 起动燃料
starting fuel control 起动燃料供给控制
starting fuel supply 起动燃料供给
starting gear 起动装置;起动齿轮
starting gear shifting bolt 起动齿轮拨叉
starting grade 起动坡度
starting grid voltage 起动栅压
starting grip 起动摇手柄
starting grip voltage 起动着火栅压
starting guide relay 起动用继电器
starting handle 起动摇手柄;起动摇把
starting handle bracket 起动摇手柄支架
starting hand-wheel 起动手轮
starting hip tile 起动也脊瓦
starting hole 切齿起始孔
starting hum 起动交流声
starting impedance-relay 起动阻抗继电器
starting impulse 触发脉冲
starting induction 起始感应
starting inertia 起动惯性
starting ingot 原始铸锭;始锭
starting jaw 起动爪
starting length 开始长;起始带长度
starting lever 起动机操纵杆;起动杆
starting lever hydraulic interlock 起动杆液压联锁
starting lever interlock 起动杆联锁
starting lever reversing interlock servomotor 起动杆换向联锁装置伺服马达
starting lever spring 起始杆簧
starting-lighting-ignition cell 起动点火照明用电池
starting line 起线
starting load 起动负荷
starting load cost 生产准备成本
starting location 起始位置
starting looper 起头器
starting loss 起动损失;启动损失
starting lubrication 起动润滑
starting machine 起动机
starting magneto 起动磁电机
starting manostat 起动稳压器
starting material 起动物料;原材料
starting mechanism 起动机构
starting method 起动方法
starting mix 传火药
starting mixture 起动混合气
starting moment 起始力矩;起动转矩;起动力矩
starting motor 起动马达;起动电(动)机;发动机起动机
starting motor brush 起动机电刷
starting motor flange 起动机连接盘
starting motor pinion 起动机驱动齿轮
starting motor torque 起动机扭矩
starting newel(post) 楼梯栏杆柱;螺旋楼梯中柱;楼梯栏杆端柱;梯级起始拉杆柱
starting nozzle 起动喷嘴
starting number 起步数
starting of construction 开工(尤指土木工程方面)
starting of motor 电动机起动
starting of oscillation 起振
starting open-phase protection 起动时断相保护装置
starting operation 起动运转
starting operation expenditures 始发作业支出
starting operation income 发送作业收入
starting order 起动指令
starting outfit 起动设备

starting paddle 起动踏板
starting panel 起动控制盘;起动操纵盘
starting pay 任职首次薪金
starting period 启动时间;开始时间;起始时间;启动期
starting phase 起始相位;起动相位;启动期
starting pit (隧道工程的)开始井坑
starting place 起始地点;起点
starting place for shipping 起运地点
starting plan 实施计划
starting platform 出发台
starting point 出发点;起始点;起动点;起点
starting point of line 接轨点
starting point of rupture 起始破裂点
starting point of wholesale 批发起点
starting position 起始位置;起动位置;起点位置
starting position of segment 分段起点位置
starting position of survey line 测线起点位置
starting post 起点
starting potential 起动电位
starting powder 原始粉末
starting power 起动功率
starting prechamber 起动预燃室
starting preheater 起动预热器
starting preheater boiler 起动预热锅炉
starting preheater boiler bracket 起动预热锅炉支架
starting preheater burner 起动预热器喷灯
starting preheater fire pan 起动预热火盆
starting preheater lamp nozzle 起动预热灯喷嘴
starting preparation 起动准备
starting pressure 起动压力
starting pressure ga(u)ge 起动压力计
starting probability 起动概率
starting process 起动过程
starting protection contactor 起动保护接触器
starting pulse 起动脉冲;开始脉冲
starting-pulse action 起动脉冲作用
starting ratchet 起动摇把棘爪
starting rate 起动率
starting reactor 起动扼流圈
starting receiver 起动用压缩空气瓶;起动瓶(压缩空气);起动空气罐
starting relay 起动继电器
starting resistance 起动阻力;起动电阻
starting resistor 起动电阻器
starting rheostat 起动变阻器
starting ridge tile 起始脊瓦
starting ring 起动圆盘
starting rod 引弧棒;起动杆
starting rope 起动索
starting run 起飞滑跑
starting sequence 起动程序
starting servomotor 起动伺服马达
starting shaft 起动轴;始发井;出发竖井
starting sheet 起始板;始极片
starting sheet blank 始极片母板
starting sheet cell 种板槽;始极槽
starting shield from shaft 出洞盾构
starting shock 起动震动
starting signal 起动信号;出站信号【铁】;发车信号
starting signaller 出站信号机
starting signal transmitter 起动信号发送器
starting slab 起始(楼)板
starting source 起动源
starting speed 起动速度;初速
starting speed control valve 起动转速控制阀
starting spring 启动簧
starting step 首阶;起始的楼梯踏板
starting strip 屋缘铺层;檐口屋面板;檐口铺层
starting submergence 起动沉没深度
starting switch 起动开关;起动接触器
starting switch terminal 起动机开关接线端子
starting symbol 起动符号
starting system 起动系统
starting system of internal combustion engine 内燃机起动系统
starting tableau 初始表
starting tank 起动燃料箱
starting taper 导锥
starting taper of reamer 铰刀导锥
starting technique 起动技术;起动方法
starting temperature 起始温度
starting test 起动试验
starting the borehole 钻孔开孔;炮眼开孔

starting thrust 起始推力
starting time 起动时间
starting time interval 起动时间间隔
starting time jitter 起动时间抖动
starting time limit 起动时限
starting time of activity 作业开始时间
starting time of fault growth 开始活动时间
starting time of operation 作业开始时间
starting time of volcanic activity 火山活动开始时间
starting to flow a well 诱喷
starting torch 起动火炬;起动吹管
starting torque 起始扭矩;起动转矩;起动扭矩;起动力矩
starting torque of motor 电动机起动转矩
starting track number 起始磁道号
starting traction effort 起动牵引力
starting tractive effort 起动牵引力
starting tractive power 起动牵引力
starting transformer 起动变压器
starting transient 起动瞬变量
starting turbine 起动用涡轮
starting under load 带负荷起动
starting unit 起动部件;起动设备
starting unloader 起动卸载器
starting up 开机
starting-up a screen 运转筛组;运转筛机;开始筛组;开动筛机
starting-up boiler 锅炉开始工作
starting-up course 起头横列
starting-up of speed 起动变加速度
starting-up oil 起动油
starting-up to put into production 投产
starting-up to speed 起动并加速
starting value 初值;起始值;初始值
starting valve 起动阀
starting velocity 起始速度;起动速度;起动流速;初速
starting velocity of sediment 泥沙起动速度
starting vessel 起动器
starting volatility 起动挥发性
starting voltage 阈电压;起动电压
starting vortex 起始涡流;起动涡流
starting waste fiber 手拉粗丝
starting weight 始重;初重
starting winding 起动绕组
starting wind-speed 起沙风速
starting wire clip 起动拉索夹
starting with load 带负荷起动
starting work 起动工作
start input/output instruction 起动输入输出指令;启动输入/输出指令
start interlocking relay 起动联锁继电器
start interval 起动位间隔
start key 起动键;启键;启动键
start lag 起动延滞
start lead 线圈内引线
start light 起动灯
start loss of plant 绿初损失
start map 星图
start margin 起始空白
start mark 起始标记
start mode 起动方式
start node 起始节点
start of a run 行程起点
start-of-chain 链开始;起始链
start-of-format control 格式开始控制;起始格式控制
start of fracture 断裂开始
start of header 首标开始;标示开始符;标题开始
start-of-heading 标题开始
start-of-heading character 标头开始字符;标题开始字符;标题符号的启动
start of message 信息开始
start of pulsing 脉冲发送起始信号
start of record(ing)mark 记录标起始;记标开始
start of switch 道岔起点
start of text 正文开头
start of track control 磁道起始控制器
start of word 起始字
start on journey 起程
start operation 开工
star topology 星形拓扑;星形布局
start option 开始选择;启动任选项
start oscillation 起振

start oscillation condition 起振条件
startover data transfer and processing program (me) 启动数据传送和处理的程序
star town 卫星城(市)
start-pilot device 起动辅助设备
start point 起始点
start point correction 起始点校正
start position 发射位置;起飞位置;起动位置
start pulse 触发脉冲
start pump 启动泵
star tracker 星象跟踪仪;星跟踪式定位器;恒星跟踪器
star tracking 星体的跟踪
star-tracking guidance 星体跟踪天文导航
star-tracking system 星体跟踪系统
star-tracking telescope 星体跟踪望远镜
star trap 舞台地板门(演员从下面上升至舞台)
start readout 起始读出
start relay box 起动继电器箱;起动电阻器
start run cost 试运转费用
start sample 开始卸货时所取样品
start sampling 开始取样
start sensor 起动传感器
start shipment 起运;启运
start signal 起始信号
start signal element 起始信号单元
start spot 起始斑点
start stake number of line 测线起始桩号
start stake number of the first receiver line 首检波点线起始桩号
start state 起始状态
start step synchronism 起始同步
start-stop 起停;开停;起止的;启动终止
start-stop and speed control 起动停止及转速控制
start-stop apparatus 起止装置
start-stop bit 起止位;起停位
start-stop button 开关按钮
start-stop correspondence 起止式通信
start-stop counter 一次计数器;起停式计数器;单程计数器
start-stop distortion 起止失真
start-stop distortion-measuring 起止畸变测试
start-stop double-current channel 启闭式双流通路
start-stop lever 起停操纵杆
start-stop mechanism 起止机构
start-stop modulation 起止调制
start-stop multivibrator 起止多谐振荡器;单稳多谐振荡器
start-stop network 起止式网络
start-stop operation 起止操作
start-stop oscillator 间歇振荡器
start-stop push button 起动和停止按钮
start-stop restitution 起止式解调
start-stop scanning 起止扫描
start-stop self-phasing code 起止式自同步码
start-stop signal 起止信号
start-stop signal distortion tester 起止信号畸变测试器
start-stop signal generator 起止信号发生器
start-stop supervisor 起停管理程序
start-stop switch 启闭转换器
start-stop synchronism 起止同步
start-stop synchronization 启止同步
start-stop system 启闭装置;启闭方式
start-stop teleprinter 起止式电传打字机
start-stop teletypewriter system 起止式电传机系统
start-stop time 起停时间
start-stop transmission 起止传输
start-stop type 起止式
start symbol 起始符号
start symbol probability distribution 起始符的概率分布
start synchronism 起步同步
start the engine 启动发动机
start the holes 标记钻孔网络(在钻探区)
start time 开始时间;起始时间
start-to-leak pressure 起漏压力
start trigger 起动触发器
start-up 开始工作;启动
start-up accident 启动事故
start-up burner 烤窑用燃烧器;点火燃烧器
start-up capability 启动能力
start-up circuit 启动电路
start-up component 启动设备

start-up condition 启动条件
start-up conduit 启动管路
start-up cost 开始成本;开办费用;开办费;起岸成本;试生产费用;筹备费(用)
start-up curve 启动曲线
start-up experience 启动经验
start-up feed pump 启动给水泵
start-up flue 烟道接火;烤窑烟道
start-up from cold 冷起动
start-up function 启动功能;启动操作
start-up in gear 传动装置起动
start-up job stream 启动作业流
start-up main 启动管
start-up man 启动调整工
start-up mode 启动方式
start-up of a project 项目开工
start-up of the furnace 开炉操作
start-up operation 启动操作
start-up period 启动期;试车期
start-up phase 启动相位
start-up pipe 启动管系
start-up pipeline 启动管线
start-up plate 启动板
start-up procedure 启动过程;启动程序
start-up pump 启动泵;启动泵
start-up reducing station 启动降压站
start-up rod 启动棒
start-up source 启动源
start-up stage 启动期
start-up system 启动系统
start-up test 试运转;试车
start-up time 启动时间;投产期
start-up(trial run)procedure 试运转顺序
start-up velocity 启动速度
start-up wear 启动磨损
start value 初值
start vector 启动向量
star twisting 星形绞合
star-type bit 十字形钻(头)
star type cable-stayed bridge 星形斜拉桥
star-type charge 星式装药
star valve 星形阀
starvation 断薪灭火
starvation wages 难以维持温饱生活的工资;不足温饱的工资
star vault 星形肋拱顶;星状肋的拱顶
starved basin 非补偿盆地;不补偿盆地
starved glaze 欠釉;薄釉
starved joint 脆弱接合;缺胶接层;缺胶接层;欠胶接头;(黏结不良的)失效接头
starved ripple mark 不完全波痕
starved surface 不丰满的表面
starving out 漏涂
star voltage 星形接线相电压;相电压;对中性线电压
star washer 星形垫圈
star wheel 星形轮;开关旋轮
star-wheel axle controller 星轮式控轴器(控制矿车运行用)
star-wheeled roller 星轮式镇压器
star-wheel extractor 星轮排渣器
star-wheel mechanism 星轮机构
star-wheel motion 星形轮运动
star-wheel type 星轮式
star-wheel type feeder 星形给料机;回转进料器;星形轮式给料机
star wire rope 梅花股芯钢丝绳
starwise disposition 星形排列
star-zigzag connection 星形曲折接法
stasis 停滞
Stassano furnace 旋转电弧炉;斯塔山诺电炉
Stassfurt deposits 斯塔斯弗特沉积物
stassfurtite 纤硼石
Stassfurt potash salt 斯塔斯弗特钾盐
Stassfurt salt 斯塔斯弗特盐
stat 斯坦脱
statampere 静电安培
statcoulomb 静电库仑
state 指定;情景;陈述
state administration 国家管理
State administration of exchange control 国家外汇管理局
state agency in charge of fishery administration and fishing ports superintendence 国家渔政渔港监督管理机构

state and local income tax 国家和地方所得税
state and property of strain 应变状态和性质
state apartment 大礼堂;(举行隆重仪式的)房间;大厅
state apparatus 国家机器
state appropriation account 国家拨款账户
state assets 国家资产
state assignment 状态指定;状态分配
state atlas 州地图集
state at negative temperature 负温状态
State Audit Bureau of Finland 芬兰国家审计局
State Audit Bureau of Republic of Finland 芬兰国家审计局
state bank 国家银行
state base map 国家基本(地)图
state boundary face 状态边界面
state boundary surface 状态边界面
state budget 国家预算;国家财政预算
state budgeted investment 国家预算内投资
state budget expenditures 国家预算支出
state budget revenue 国家预算收入
state cabin 客轮高等舱位
state capital 法定资本
state chamber 公共集会厅
state classification 状态分类
state constraint 状态约束;状态受限
state contract 国家合同
state control 状态控制;国家监督
state controllability 状态可控性
state controlled goods 统配部管物资
state control register 状态控制寄存器
state convention hall 国家议会大厦
state coordinates 状态坐标
state correspondence error 状态对应误差
state court 高等法院
state credit 国家信贷
stated capital 申报股本;设定资本;法定资本;额定资本
state demesne 国有土地
state-dependent 状态随机变化的
state-dependent service 依赖于状态的服务
state description 状态描述
state diagram 状态图
stated liabilities 账面负债
stated limit 指定范围
state dock 州营码头
stated period 规定期限
stated speed sign 限速标
stated sum contract 包干合同
stated time-limit 规定期限
stated value 设定价值;法定价格
state enable signal 允许状态信号
state enterprise 国有企业
state equation 状态方程
state equivalent 状态等价
state estimation 状态估计
state-estimator 状态判定器
state farm 国有农场;国营农场
state feedback 状态反馈
state fixed price 国家固定价格
state flip-flop 状态触发器
state forces 计日工作;基本劳动力;直接劳动;计日劳动力
state forest 国有森林;国有林
state forest boundary 国有林区界
state function 状态函数;态函数
state fund 国家资金
state government 州政府
state grammar 状态文法
state graph 状态图
state guesthouse 国宾馆
state highway 州道;州辖公路(美国)
State highway department 州公路局(美国)
State highway division 州公路局(美国)
state highway system 州道系统;州道网(美国)
state immunity 国有船舶豁免权
state indicating signal 状态指示信号
state insurance enterprise 国营保险企业
state investment 国家投资
state investment fund 国家投资资金
state land 公有地;公地
state language 地区语
state law 国家法
stateless person 无国籍人

state line 州界
state-line sign 州道路线标志(美国)
statelisted prices 国营牌价
state logic 状态逻辑
stately church 堂皇的教堂
stately gateway 堂皇的门道
stately portal 堂皇的入口
state mandatory production plan 国家指令性生产计划
state matching condition 状态符合条件
state medical service 国家公费医疗
state medicine 国家公费医疗；公费医疗制度
statement 综述；命题；声明书；声明；报告书；报表
statement about transport revenue plan finished 运输收入进款计划完成情况表
statement analysis 决算表分析；报表分析
statement at completion 竣工报表
statement bracket 语句括号
statement form 报告格式；报表格式
statement formula 命题公式
statement from creditor 应付账款清单
statement function 命题函数
statement heading 报表表头
statement identifier 语句标识符
statement label 语句标号
statement label array 语句标号数组
statement label data 语句标号数据
statement law 命题法则
statement logic 命题逻辑
statement number 指令数号
statement of absorption of deficit 亏损处理计算书
statement of account 账目表；账单；客账结单；结算表
statement of actual and estimated expenditures 实际和预计支出对照表
statement of actual and estimated revenue 实际和预计收入对照表
statement of affairs 清算(式)资产负债表；财务情况计算书；财产状况说明书
statement of appropriation of net income 净收益分配表；净收入分配计算表；纯利分配计算书
statement of arrangement and method for construction 施工方案
statement of assets 资产目录
statement of assets and liabilities 资产负债表
statement of auction 拍卖清单
statement of balance 余额清单
statement of budgeted cost of goods manufactured 货物制造预算成本计算表
statement of business operation 业务经营表
statement of cash flow 现金流量表；现金流动表
statement of cash in treasure 库存现金表
statement of cash in vault 库存表
statement of cash receipts and disbursements 现金收支及结存表
statement of changes in financial position 财务状况变动表
statement of changes in working capital 运营资本变动表
statement of charge 起诉书
statement of claim 起诉状；索赔清单；索偿陈述
statement of completion 竣工报表
statement of condition 概况报告
statement of cost and production 成本及产量表
statement of cost of production 生产成本表
statement of current position 流动财务情况表
statement of debt account 负债一览表
statement of earnings 收益表
statement of expenses 费用清单
statement of facts 事实记录
statement of final accounts 会计决算报表；决算说明书；决算报告
statement of financial accounting standards 财务会计标准说明书
statement of financial condition 财务状况计算书
statement of financial income 财务收益表
statement of financial income and expenses 财务收支机费用表；财务收入及费用表
statement of financial operations 财务情况表
statement of financial position 财务状况表
statement of fraction 文字比例尺
statement of fund 资金表；基金表
statement of fund and its application 资金运用表
statement of fund application 资金运用表

statement of funds source and utilization 资金来源与运用表
statement of general average 共同海损理算书
statement of identity 身份说明
statement of income and expenditures 收支表；收益及支出明细表
statement of income and expenses 收入费用表
statement of inspection 检验报告
statement of liquidation 破产清算书
statement of loans approved 已批准的贷款项目表；批准的贷款项目表
statement of loss and gain 损益说明书；损益表
statement of necessity 需要声明书
statement of net quick assets flows 速动资产流转表
statement of objectives 叙述目标
statement of operating expenses 营业费用表
statement of operating revenue 营业收入表
statement of operation 营业报表
statement of order cost 分批成本表
statement of performance 性能说明书
statement of probable construction cost 工程费概算说明；建筑成本概算表
statement of production cost 生产成本表；生产成本报告表
statement of profit and loss 损益计算书
statement of profit calculation 利润计算表
statement of railway transportation cost 铁路运输成本报表
statement of realization and liquidation 变现清算表
statement of record 注册记录
statement of refusal to pay 拒付说明书
statement of requirements 需求报告
statement of resources and liabilities 资金负债表
statement of retained earnings 留存收益表；保留盈余表
statement of service 履历(表)
statement of shorts 短缺货物清单
statement of size 尺寸说明
statement of sources and application of funds 资金来源与运用表
statement of sources and uses of working capital 流动资金来源及运用表
statement of standard accounting practice 标准会计实例说明书
statement of surplus 盈余表
statement of tally account 理货账单
statement of the income and expenditures 收支一览表
statement of variation on profit 利润变动表
statement of work 工作报告书
statement of working capital flow 流动资金流转表
statement on audit standards 审计准则声明
statement scale 说明比例尺
statement separator 语句分隔符
state method 极限状态下的安全检查
state model 状态模型
state monopoly 国营专卖
state monopoly of material supply and marketing 统配包销
state observability 状态可观测性
state observer 状态观测器
state of acknowledge 知识状态
state of activation 激活状态；活化状态
state of affairs 事态
state of aggregate 聚集状态
state of all-round tension 圆周受拉状态；全向受拉状态；两维应力相等
state of a process 进程状态；进程的状态
state of art 艺术水平；目前工艺水平；技术发展水平；工艺状况；技术状况
state-of-art facility 现代化设施
state of capillary equilibrium 毛细管平衡状态
state of charge 充电状态
state of chromatic adaptation 色适应状态
state of compliance 允诺程度
state of compression 压缩状态
state of cooling 冷缺状态
state of correction 校正状态
state of cracking 开裂状态
state of crustal stress 地应力状态
state of cultivation 耕作情况
state of cure 固化程度；硫化程度

state of decompression 消压状态
state of deformation 变形状态
state of delivery 输送情况；交货情况
state of disturbance 扰动状态
state of economy 经济状态
state of elastic equilibrium 弹性平衡状态
state of emergency 紧急状态
state of equalization 均衡状态；均等状态；平衡状态
state of equilibrium 平衡状态
state of failure 破坏情况
state of flow 流态
state of ground 地面状况
state of indoor air 室内空气状况
state of inertia 惯性状态
state of instability 失稳情况；不稳定状态
state of ionization 电离态
state of knowledge 知识水平；科学(发展)水平
state of limit equilibrium 极限平衡状态
state of loading 负荷状态
state of matter 物态
state of motion 运动状态
state of nature 自然状态
state of operation 经营状况
state of oxidation 氧化状态
state of packing 填密状态
state of plane deformation 平面变形状态
state of plane distortion 平面扭曲状态
state of plane stress 近似平面应力状态；平面应力状态
state of plastic equilibrium 塑性平衡状态
state of plasticity 塑性状态
state of polarization 偏振(状)态
state of population 人口静态
state of project 工程状况
state of pure membrane stress 单纯薄膜应力状态
state of rest 休止状态
state of room air 室内空气状况
state of saturation 饱和状态
state of sea 海洋状态；海面状况
state of sewage 污水情况
state of stagnancy 停滞状态
state of starting operating 起动工况
state of stress 应力状态
state of stress and deformation 应力状态与变形
state of super-pressure 超压状态
state of surface 表面状态
state of technology 技术现状
state of temperature 温度状态
state of tension 张力状态；张拉状态；拉力状态；受拉状态
state-of-the art 目前技术水平；目前工艺水平；科学发展动态；技术状态；已发展技术的；技术发展现状；技术发展水平；代表目前技术水平的；非实验性的；发展现状
state-of-the-art facility 现代化设备
state-of-the-art report 发展水平报告；技术现状报告
state-of-the-art technology 最新科学技术
state-of-the-art technology and equipment 现代成熟的技术和设备
state-of-the-art water quality modeling 现代化水质模拟
state of thermal anomaly 热异常状态
state of the sea 波浪等级
state of the sky 天空状况
state of the world environment 世界环境状况
state of trade 贸易情况
state of transit 过境国
state of vector 状态向量
state of vulcanization 硫化程度
state of water body 水体状况
state organization 国家机构
state organs 国家机关
state-owned assets management 国有资产管理
state-owned assets management information system 国有资产管理信息系统
state-owned assets of railway enterprise 铁路国有资产
state-owned assets of railway enterprise management 铁路国有资产管理
state-owned construction company 国有建筑公司
state-owned enterprise 国有企业
state-owned fleet 国有船队
state-owned railway 州有铁路；国家铁路
state-owned shipyard 国营船厂

state ownership 国有制;国家所有制
state ownership of mineral resources 矿产资源的国家所有权
state parameter 状态参数;热力学态函数
state path 状态途径;状态历程
state plan 地区计划
state plane 状态平面
state plane coordinate 国家平面坐标
state plane coordinate grid 州平面坐标
state plane right-angle coordinate system 国家平面直角坐标系统
state planning 地方性计划;省计划;州计划;国家计划
State planning commission 国家计划委员会
state plastic-elastic equilibrium 塑性-弹性平衡状态
state point 状态点
state price 国家价格;牌价
State Price Bureau 国家物价局
state primary highway system 州一级道路系统（美国）
state probability 状态概率
state procedure 状态过程
state-provide credit 国家信贷
state purchase and allocation on a monopoly basis 统一收购统一调拨
state quasi-equilibrium 准平衡（状）态
stater 状态器
state rate 法定利率
state reconstruction 状态重构
state rectangular coordinate grid 州直角坐标网
state rectangular coordinate system 国家平面坐标系统
state reduction 状态简化
state register 状态寄存器
state report 全国性报表
state road 州道
state road system 州路网（美国）
state room 客室;起居室;大厅;轮船卧室;火车卧室;特等舱（室）;特等包厢
state route 州道
state-run 国营的
state-run-enterprise 国有企业
state-run shipyard 国营船厂
State Science and Technology Commission 国家科学技术委员会
state secondary highway system 州二级道路系统（美国）
state selection 状态选择（器）
state selectivity 状态选择性
state-set price 国家规定价格
state simulator 状态模拟程序
state space 状态空间
state-space model 状态空间模型
state space system matrix 状态空间系统矩阵
state space theory 状态空间理论
state-specified standard 国家（规定）标准
state-speed sign 限速标志
State Statistics Bureau 国家统计局
state subsidy 国家补贴
state surface 状态表面
State Surveying and Mapping Bureau 国家测绘总局
state table 状态表
state table analysis 状态表分析
state tape symbol pair 状态带符号对
state tertiary highway system 州三级道路系统（美国）
state-to-state reaction 态—态反应
state trading department 国有商业部门
state transition 状态迁移
state-transition diagram 状态转移图;状态迁移图
state-transition equation 状态转移方程
state-transition function 状态迁移函数
state-transition matrix 状态传递矩阵;状态跃迁矩阵;状态转移矩阵
state-transition method 状态过渡法
state-transition model 状态称模型
state-transition probability 状态转移概率
state treasury 国库
state unified distribution material 国家统配物资
state variable 状态变量;热力学态函数
state-variable equation 状态变量方程式
state variable filter 状态可变滤波器
state variable method 状态变量法

state variable technique 状态变量法
state vector 状态向量;状态矢量
state vector equation 状态矢量方程
state vector space 状态向量空间
state verification 状态测试;国家鉴定
state-wide highway plant 全洲公路设备
state with short coastlines 短岸线国
stathenry 静电亨利
static 静止的;静态的;静力的;静的
static accumulator oil 积聚静电的油
static accuracy 静态精度
static action 静态作用
static adjustable range 静态调节范围
static adjustment workshop 静调库
static adsorption 静态吸附
static adsorption property 静态吸附性能
static advertising 静态广告
static aerobic compost 静态好氧堆肥
static(al) ground 永恒接地
static(al) indetermination 超静定
static(al) investigation 静力研究
static allocation 静态分配
statically admissible 静力容许的
statically admissible field 静容许力场
statically admissible multiplier 静定容许乘数
statically balanced 静平衡的
statically broken 静荷载破坏的
statically determinable 静定的
statically determinate 静定的;静力确定
statically determinate beam 静定梁
statically determinate frame 静定框架
statically determinate framework 静定构架
statically determinate principal system 静定基本系
statically determinate structure 静定结构
statically determinate truss 静定桁架
statically indeterminable 静不定的
statically indeterminable beam 超静定梁
statically indeterminate 静不定;超静定的
statically indeterminate beam 超静定梁
statically indeterminate frame 超静定空间;静不定框架
statically indeterminate framework 超静定构架
statically indeterminate structure 静不定结构;超静定结构
statically indeterminate system 超静定系统;超静定（体）系
statically indeterminate truss 超静定桁架
statically indetermine 静不定的
statically indetermined 超静定的
statically stable atmosphere 静力稳定大气
static(al) method 静力方法
static(al) moment 静力矩
static(al) moment of area 静力面积力矩
static(al) redundant 静力多余约束
statical rupture load 静力破坏荷载
static amplification 静态放大
static analogue test 静态模拟试验
static analysis 静态分析;静力分析
static analysis method 静态分析法
static analysis tools 静态分析工具
static analysis scheme of building 房屋静力计算方案
static and dynamic(al) equilibrium 动静平衡
static attribute 静态属性;静态表征
static axle weight detector 静态轴重检测器
static balance 静力（平）衡;静力均衡
static balancer 静力平衡器;静电平衡器
static balancing 静平衡
static ball indentation test 布氏硬度试验
static bath 静力分选槽
static bearing 静压轴承
static bearing capacity 静力承载能力;静力承载量
static bed 固定床
static behavio(u)r 静力性能
static bending 静弯曲;静力弯曲
static bioassay 静水式生物测验
static blow-out preventer 静防喷器;静力防喷器
static bole 静压测量孔
static bottom hole pressure 孔底静压力
static breeze 对流放电
static budget 固定预算;静态预算
static calculating method 静态计算法
static calculation 静力计算
static calculation method 静力学计算法

static calibration 静态校准;静力校准
static call 静态调用
static capacity 静能力
static cast pipe 静态浇铸铁管
static cell 静态单元
static characteristic 静态特征（曲线）;静态特性（曲线）
static characteristic of arc 电弧静特性
static charge 静电荷
static check 静态校验;静态检验
static checkout unit 静态检验装置
static circuit 静态电路
static coefficient method 静力系数法
static coefficient of friction 静（力）摩擦系数
static compaction 固定压紧
static comparison 静力比较
static computation 静力计算
static condensation 静凝聚;静凝聚
static condenser 静电电容器
static condition 静力条件;静定条件;静态条件
static cone penetration 静力触探
static cone penetration dissipation test 静力触探消散试验
static cone penetration resistance 静力锥形贯入阻力
static Cone Penetration Test 静力触探试验;锥体静力触探;静锥贯入试验;静锥触探试验
static cone penetrometer 锥体静力触探仪;静力锥探仪
static cone test 静力触探试验
static constant 静态常数
static contact force 静止接触力
static contact point （簧片的）静止交点
static control 静态控制;静电控制
static controller 固定控制器
static convergence 静态收敛;静（态）会聚
static conversion 静态变换
static converter 静止换流器;静止变流器
static converter cascade 静止换流器串级
static correction 静校正
static cost （不含施工期利息的）静态投资
static coupling 固定接合
static crusher 固定式破碎机
static data 静态资料
static database 静态数据库
static debugging routine 静态调试程序
static deflection 静态压缩量;静荷挠度;静载挠曲;静载挠度;静变位
static deflection of spring 弹簧静挠度
static deformation 静态变形;静力变形
static delivery head 静扬程;静输送水头
static desorption efficiency 静态脱附率
static determinacy 静定（性）
static determinateness 静定性
static determinate structure 静定结构
static determinate system 静定系统
static deterministic model 静态确定型
static deterministic-random mixed model 静态确定随机混合型
static dial inclinometer 罗盘测斜仪
static discharge head 静泄水水头;排送静压头;排出压头
static discharger 静电放电器
static displacement influence function 静力位移影响函数
static display image 静态显示图像
static downhole temperature 井下静态温度
static draft 静压头;静压（水）头;静力通风;静吃水
static draft head 静吸入水头;静吸水头;静吸出水头
static draft of transducer 换能器静态吃水
static drain test 静力排水试验
static dump 静态（信息）转储
static dust 静电尘
static earth pressure 静止土压力
static economics 静态经济学
static effect 静电影响;静电效应
static effect of conjugation 静态共轭效应
static efficiency 静态效率;静电效率
static elastic behavio(u)r 静力弹性工作性能;静力弹性工作状态
static electricity 静电（学）
static electricity earthing device 静电接地装置
static electricity eliminator 静电消除器
static electricity grounding device 静电接地装置

static electricity removal 除静电
static electricity resisting property 抗静电性能
static electricity spray booth 静电喷釉室
static electrometer 静力静电计
static electron microscope 静电电子显微镜
static electro-regeneration 静态电再生技术
static elevation 位差
static-elevation difference 静压头;静水位差
static eliminator 静噪装置;天电干扰限制器;除静电器
static endurance 静力耐力
static energy 势能
static engine 固定式发动机
static envelope 静态包络线
static equation 静力方程
static equilibrium 静态平衡;静态均衡;静力平衡;静定平衡
static equilibrium condition 静力平衡条件
static equilibrium equation 静力平衡方程
static equilibrium method 静力平衡法
static equilibrium state 静力平衡状态
static equilibrium test 静平衡试验
static equipment 静止设备
static equivalent 静力等效
static equivalent load 当量静负荷
static eraser 静电消除器
static error 静态误差
static excavator 固定挖掘机;不移动挖掘机
static excitation 静止励磁
static exciter 变压器励磁
static exciter silicon rectifier 静止励磁硅整流器
static exhauster 车顶通风口
static expression 静态表达式
static external area 静态外区
static factor of safety 静力安全系数
static failure load 静力破坏荷载
static failure matrix method 静态破坏矩阵法
static father 静态源
static fatigue 静疲劳
static fatigue failure 静疲劳失效
static fault tolerance 静态故障限
static fermentation 静态发酵
static file 静态文件
static flip-flop 静态触发器
static flocculation 静态絮凝
static flow 静流
static fluid 静止流体
static fluid column 静止流体柱
static focus 净焦点
static force 静力
static formula of pile 桩的静力公式
static fracture 静力破裂
static fracture model 静断裂模型
static-free 不受天电干扰
static frequency changer 静止变频器
static friction 静态摩擦;静摩擦(力);静力摩擦
static friction coefficient 静摩擦系数
static fumigation 静式熏气
static function hazard 静函数险态
static governor 静止调节器
static gradient 静力梯度
static gravimeter 静重力仪;静力重力仪
static gravity meter 静重力仪
static gravity survey 静态重力测量
static grizzly 固定条筛;固定格栅;固定格筛;固定篦条筛
static ground 静接地
static grounding device 码头静电接触装置
static grounding rod 接地杆
static groundwater level 静止地下水位
static handling 静态处理
static hardness test 布氏硬度试验
static hazard of fuzzy functions 模糊函数的静态冒险
static head 静压力水头;静压力受感器;静压(水)头;静水头;水头静压
static head of reservoir 水库静水头
static head of water 静水头;静力水压
static head turn indicator 静压转弯指示器
static heeling angle 静横倾角
static hoisting moment 静提升力矩;静卷扬力矩
static hum 天电干扰哼声
statician 统计员
static ice strength 静冰强度

static impedance 静态阻抗
static indentation test 静力球印硬度试验;球印硬度试验;球硬度试验(压痕法)
static indeterminacy 静不定性;静不定;超静定性
static indeterminate structure 超静定结构
static index 静态索引
static induction 静电感应
static inductive effect 静态诱导效应
static ingot 模铸钢锭
static input-output analysis 静态投入产出分析
static input-output model 静态投入产出模型
static input-output system 静态投入产出系统
static instability 静态不稳定;静力失稳
static installation 固定设备
static intensity 静力强度
static interfacial tension 静界面张力
static inverter 静止逆变器;静变流器
static investment (不含施工期利息的)静态投资
staticize 静态化;保存
staticizer 静化器;串并转换器;串并行转换器
static jet 固定喷嘴
static jet pressure 喷流静压(力)
static lake 内陆湖
static level 静止水位;静液面;静力水位;静水头;静水水位;静电级;天电极;天电干扰电平;水柱水位;大气干扰电极
static lift 静水浮力;静吸高度;静上浮力
static limiter 静噪装置;天电干扰限制器
static line 静压操纵管路
static list angle 静横倾角
static live load 静力活荷载;静活载
static load 静载(荷);静向载;静负载;静负荷
static load deflection 静荷载挠度
static load fatigue test 静载疲劳试验
static load(ing) 静力荷载;等效静载;静力加载
static loading motion 静压负向运动
static load(ing)test 静荷载试验;静载试验
static load leveling 静态装载均衡
static load of freight car 货车静载重
static load of wagon 货车静载重
static load stress 静荷载应力
static load test 静载强度试验;静力荷载试验;静载(强度)试验;静负荷(强度)试验
static logic hazard 静逻辑险态
static magnetic storage 磁芯存储器;磁芯储存器
static map 静态图;静态地图
static margin 静余量;静态幅度
static mark 静电标记
static mass spectrometric determination 静态质谱测量
static matrix 静态矩阵
static matte 静止蒙罩法
static maximum rating 最大静态额定功率
static measurement 静态测量;静力量测法
static memory 静态存储装置;静态存储器
static mercury drop electrode 固定型滴汞电极
static metamorphism 静力变质
static meteorology 静力气象学
static method 静态法;静力法
static mixer 静态混合器
static model 静态模型;静力实验模型
static model investigation 模型静力研究
static modulator 静态调制器
static modulus 静(态)模量;静力模量
static modulus of elasticity 静弹性模量
static moment 静力矩
static moment of area 静面积矩;截面矩量
static multifunction pipeline 静态多功能流水线
static noise 天电噪声
static non-accumulator oil 不能积聚静电的油
static object 静态对象
static oceanography 静力海洋学;海洋静力学
staticon 视像管;光导电视摄影机
static opening 静压口;静压测孔
static optimization 静态最优化
static optimization problem 静态最优化问题
static optimum population 静态适度人口;静适度人口
static or equilibrium theory 静态或平衡理论(潮汐理论)
static orifice 静水头量测孔;测静压孔
static oscillograph 静态示波器
static output 静态输出
static overvoltage 静电过电压

static parameter 静态参数
static partition 静态分区
static passes 无振压实遍数
static payoff period 静态返本期
static penetration test 静力贯入(度)试验;静力触探试验;贯入试验
static penetration test(ing) 静态贯入度试验
static penetrometer 静力锥测仪
static picture character 静态图像字符
static pile composting 静态堆肥化
static pile-driving formula 静力打桩公式
static pile press extractor 静力压拔桩机
static pin 固定销
static plate manometer 静压板状压差计
static point resistance 静力触探探头阻力
static Poisson's ratio 静态泊松比;静泊松比
static power down mode 低功耗待命状态
static precipitator 静电集尘器;电气除尘器
static preload 静力预压
static pressing 静压力
static pressing palisade 静力压桩
static press piling 静压力桩
static pressure 静压强;静压力;静压;静压力
static pressure coefficient 静压系数
static-pressure compensation 静压力补偿
static pressure compensator 静压补偿器
static pressure differential 静压差
static pressure for outflow 流出水头
static pressure gradient 静压力梯度
static pressure inlet 测静压孔
static pressure level ga(u)ge 静压液面计
static pressure line 静水压成;静压管;低压输送管
static pressure method 静压法
static pressure of water 静水压强
static pressure probe 静压测针
static pressure ratio 静压比
static pressure survey 测静压
static pressure system 静压系统
static pressure tap 测压孔
static pressure transducer 静压传感器
static pressure vent hole 静压力测孔
static prestress 静力预压
static probabilistic inventory model 概率型静态库存模型
static probe 静压力管;静压力传感器;静力传感器
static problem 静态问题
static process 静态过程
static pulse jet 固定式脉动空气喷气发动机
static quenching 静淬火法
static random access memory 静态存储器
static random model 静态随机型
static ratio 静态比率
static rebound deflection 静载回弹弯沉
static recovery 静态回复
static recrystallization 静态重结晶
static recrystallization way 静态重结晶作用方式
static regain 静压力重新恢复
static regeneration 静态再生
static regulation 静态调节
static relation 静态关系
static remote sensing 静态遥感
static reserve index 静态资源指标
static reservoir 超静态热储
static resolution 静态分辨率
static response 静态响应
static restructuring 静态再构成
static rigidity 静态刚度
static roller 光面压路机;静力压路机;静力式压路机
static routine 静态程序
static routing 静态路由选择
statics 静力学;统计学;天电干扰
static safety factor 静力安全系数
static sampling 定点取样
static sea cock 静压水门;计程仪静压管船底阀
static seal 静液封;静密封;密封垫
static sea valve 静压水门;计程仪静压管船底阀
static segment attribute 静态图块属性
static sensor 静态传感器
static setting 静态设定值
static settling 静止沉淀
statics for structural engineering 结构工程静力学
static similarity 静力相似性
static slot 静压缝口
statics of ocean current 海流静力学

statics of shells 壳体静力学
static sounding 静力探测;静力触探
static sounding rod 静力触探杆
static sounding test 静力触探试验
static space frame analysis 静力空间框架分析
static space truss 静定空间桁架
static split autotacking 静态分离自动跟踪
static split system 静态分流系统
static spontaneous potential 静自然电位;静止自然电位
static spring gravimeter 静力弹簧重力仪
statics rate 静止率
static stability 静稳(定)性;静态稳定(性);静力稳定(度);静力稳定
static stability characteristic 静态安定性
static stability diagram 静稳(定)性曲线图
static stability of atmosphere 大气静力稳定度
static state 静态
static state approach 静态状况研究
static statement 静态财务报表;静态报表
static-state test 定置试验
static state traffic 静态交通
static statistics 静态统计
static status display 静态显示
static stereo photography 静态立体摄影
static stiffness 静(力)刚度
static storage 静态存储器
static store 静态存储
static strain indicator 固定式应变指示器
static strength 静态强度;静力强度
static stress 静(载)应力
static structural analysis 结构静力分析
static subroutine 静态子程序
static subsurface sounding 静力地面下的测探;静力地面下的触探
static suction head 静吸升水头;静吸入水头;静负压水头
static suction lift 静止吸入高度;静吸升高度
static suppressor 天电干扰抑制器
static surface tension 静(态)表面张力
static switching 静态转换
static system 静力系统;静态系统;静定系统
static table 静态表
static tandem roller 静力双滚筒压路机;静力串联式压路机
static target 静止目标
static temperature 静温;静态温度
static temperature of formation 地层静止温度
static tension test 静载张力试验
static test 静载试验;静态试验;静水式试验;静负荷试验
static test for cargo gear 起货设备静负荷试验
static testing pad 静力试验台
static test mode 静态测试方式
static test site 静力试验场
static test stand 静态试验台
static theorem 静力原理
static theory 静力学理论
static theory of tide 潮汐静力理论
static thrust 静推力
static tipping load 静态倾覆荷载
static tipping load rating 静态倾覆荷载额定值
static torque 静力扭距;止转转矩;静扭矩
static torsional test 静载扭转试验
static toxicity test 静水式毒性试验
static tracking error 静态跟踪误差
static track irregularity 静态不平顺
static transformer 静止感应器;静电变压器
static tremor 静止性振颤
static trial 静载试验
static trimming angle 静纵倾角
static tube 静压管
static tube aeration system 固定管曝气系统
static tube aerator 螺旋管曝气器
static tube mixer 固定管混合器
static tube reading 静压管读数
static turtle 静态龟标
static type packer 固定式封隔器
static ultra high pressure and high temperature device 静态超高压高温设备
static ultra high pressure and high temperature process 静态超高压高温法
static unbalance 静不平衡
static undrain test 静力不排水试验

static unifunction pipeline 静态单功能流水线
static unstability 静不稳定性
static vacuum mass spectrometer 静态真空质谱计
static value 静力值
static variable 静态变量
static vent 静压孔
static vibration 静态振动
static voltage detector 静电检电器
static water 静水
static water environment 静水环境
static water in soil 土中静水
static water level 静水水位;静水头;静水水位;静止水位
static water precipitation 静水沉淀
static water pressure 静水压(力)
static water sedimentation 静水沉淀
static water table 静止水位
static water table before back-pumping 回扬前静水位
static wheel balancer 车轮静平衡试验机;静力轮子平衡重
static wick discharger 静电放电刷
static work 静态作业;静力作业
static Young's modulus of elasticity 静力杨氏弹性模数;静力杨氏弹性模量;静力弹性模量
statiflux 静电探伤法
stating the facts net 净额表示
statio equipment 车站设备
station 装配站;岗位;观察站;同生区;台站;电台;车站;场所;测量点
station A A 站
station address 站址
station adjustment 测站平差
station adjustment of fixed angle observation 固定角观测平差
station adjustment of observation in groups 分组观测站平差
station administration rooms 车站管理用房
station agent 值班人
stational distributor 固定布水器
station amplifier 功率放大器
station and lock unit 水闸和发电站
station and yard communication 站场通信
station announce panel 车站预告牌
station anomaly 台站异常;测点异常
station-approach warning sign 进站预告标
stationarity 平稳性
stationarity Gaussian process 平稳性高斯过程
stationarity indices 不动指标
station articles storehouse 车站用品库
stationary 固定的;静止的;平稳,稳定的;常设的;不动的
stationary aggregates plant 固定式混凝土集料厂
stationary air compressor 固定式压缩空气机;固定式空压机;固定式空气压缩机
stationary air pollution sources 固定性空气污染源
stationary animal 定居动物;定层动物
stationary appliance 固定式装置;固定式设备;固定式器具;固定装置;固定设备;固定器具
stationary asphalt cooker 固定式沥青加热炉
stationary assumption 平稳化假设
stationary atmosphere 稳定大气
stationary attenuator 固定式衰减器
stationary axis 固定轴
stationary axle 静止轴
stationary baler 固定式压捆机
stationary barrier 固定支架
stationary bar screen 固定条筛;固定格栅;固定格筛;固定筐条筛
stationary bar type of screen 固定栅条式栅网
stationary base 固定底座
stationary base machinist's vise 固定式机工虎钳
stationary base vise 固定式虎钳
stationary batching and mixing plant 固定配料拌和装置
stationary batch plant 固定式称量装置
stationary battery 固定电池组
stationary bed 固定床
stationary bed flow 定床流
stationary belt conveyer 固定式皮带运输机;固定式带式输送机
stationary binomial random process 平稳二项随机过程
stationary blade 固定叶片;静叶片

stationary blade row 固定叶列
stationary block 固定式卷筒;稳定地块;定滑车
stationary body cylinder 缸体固定的缸
stationary boiler 固定式锅炉;陆用锅炉
stationary breaker contact 固定断电器触点
stationary bucket dredge(r) 非自航链斗挖泥船
stationary bushing 固定衬套
stationary cable crane 固定式缆索起重机
stationary caisson 固定式沉箱
stationary camera 固定式照相机
stationary camera technique 固定摄影观测法
stationary cascade 固定叶栅
stationary cavity 固定空穴;常驻空穴
stationary center 固定顶尖
stationary center of pressure 固定压力中心
stationary channel 平稳信道
stationary cine projector 固定式放映机
stationary clamp 固定夹钳
stationary cloud 驻云【气】;山帽云【气】
stationary collapsible container 固定可折叠容器
stationary combustion chamber 静止燃烧室
stationary combustion source 固定燃烧源;静止火源
stationary communication satellite 静止通信卫星
stationary compactor 固定式压实工具;固定式夯具
stationary compressor 固定式压缩机
stationary concrete pump 固定式混凝土泵
stationary cone classifier 固定式锥形分级机
stationary contact 固定接触
stationary contact member 固定触点部件
stationary contact point 固定触点
stationary continuous brick kiln 固定式连续烧砖窑
stationary continuous extrusion machine 固定连续挤出系统
stationary continuous steaming plant 固定式连续蒸煮设备
stationary cooling bed 非机动冷床
stationary core 固定的型芯
stationary core rod 固定芯棒
stationary counter 固定式计数器
stationary crane 固定(式)起重机
stationary creep 稳定蠕动;稳定蠕变
stationary cross-rail 固定横梁;固定导轨
stationary crusher 固定轧碎机;固定式轧碎机;固定式碎石机
stationary cutter 固定割样器
stationary cutter suction dredge(r) 非自航绞吸式挖泥船
stationary cyclone 滞留气旋;静止气旋
stationary datum 冻结基面
stationary derrick 固定式桅杆吊;固定(式)起重机
stationary derrick crane 固定式桅杆起重机
stationary die 固定压模;固定模子
stationary die half 固定的半型
stationary diesel engine 固定式柴油机
stationary direction finder 固定探向器
stationary distribution 平稳分布;稳定分布
stationary distribution process 平稳分布过程
stationary distributor 固定配水器;固定配电器
stationary disturbance 稳态扰动
stationary dog 固定爪
stationary dredge(r) 固定式挖泥船;固定式采掘船;非自航式挖泥船;固定式挖泥机;固定式采掘机;非自航式挖泥机
stationary drift 固定相移
stationary earthmoving machinery 固定的土方机械
stationary echo 固定目标回波
stationary electric(al) pumping set 固定式电动泵组
stationary electrode 固定电极
stationary electroplating bath 固定式电镀槽
stationary engine 固定式发动机;固定发动机
stationary equipment design 固定设备设计
stationary ergodic noise 平稳偏历噪声
stationary ergodic process 平稳各态历程
stationary evapo(u)ration loss rate 稳定蒸发损失率
stationary extraction cylinder 固定圆筒形抽提器
stationary extrusion 恒位挤压
stationary extrusion pressure 固定挤压压力
stationary face 静止面
stationary fatigue test 静态疲劳试验
stationary fatigue test machinery 静态疲劳试验

机械
stationary field 恒定场
stationary film 固定膜
stationary film camera 固定胶片照相机
stationary fit 固定配合;静配合
stationary flow 平稳流;稳定水流;稳定流;定量流;常定流
stationary flow field 稳定流场
stationary force 平衡力
stationary forehearth 固定式前炉
stationary forge 固定式锻炉
stationary fork 固定式货叉
stationary form 停滞型
stationary formwork 固定式模板
stationary front 滞留锋;静止(前)峰
stationary furnace 固定炉
stationary galvanometer 固定电流计
stationary gas turbine 固定式燃气轮机
stationary ga(u)ging 驻测
stationary Gaussian process 平稳高斯过程;稳态高斯过程
stationary grate 固定隔栅;受料隔栅
stationary greenhouse 固定温室
stationary grid 固定滤线栅
stationary gripper die (平锻机上的)固定头模
stationary grizzly 固定条栅;固定格栅;固定格筛;固定篦条筛
stationary guide shoe 固定导板
stationary guide vane 固定导叶
stationary hoist 永久提升机
stationary hole 不动空穴
stationary hopper 固定式混凝土(储)料斗;固定式料斗;固定式漏斗;储料斗
stationary hopper dredge(r) 定depositing装舱挖泥船
stationary horizontal diesel engines 固定卧式柴油机
stationary hysteresis 稳态滞后现象
stationary independent increment 平稳独立增量
stationary information source 固定信息源
stationary ingot tilting pot 固定式翻锭机
stationary inner-tube core barrel 双动双层岩芯管;双动双层取芯器
stationary interface 静止界面
stationary iteration 定长迭代
stationary jaw 固定爪;固定卡爪;固定颚板
stationary jump 稳态水跃;稳定水跃;停驻水跃
stationary knife 固定刀
stationary knockout grid 固定式落砂栅
stationary ladle 座包
stationary layer 稳定层
stationary lifting device 固定提升设备
stationary line 无位移谱线
stationary liquid 固定液
stationary liquid petroleum gas tank 固定式液化气储罐
stationary liquid phase 液体固定相
stationary loading tower 固定装船机(塔式);固定式装船塔架
stationary lower punch 固定下冲杆
stationary machine 固定式机器;定置机
stationary mast 固定立柱
stationary mastic cooking plant 固定玛琋脂拌制厂
stationary message source 固定信息源
stationary method 静态法
stationary mirror 固定反射镜
stationary mixer 固定式搅拌机;固定式拌和机
stationary mixing plant 固定搅拌装置
stationary mode 固定式
stationary model 平稳模型
stationary monitoring station 固定监测站
stationary monitoring station for atmospheric pollution 大气污染固定监测站
stationary monitoring station for water pollution 水污染固定监测站
stationary mortar mixer 固定式灰浆搅拌机
stationary motion 常定运动
stationary motor pison pump 固定动力活塞泵
stationary noise 平稳噪声;稳态噪声
stationary nozzle 固定喷头
stationary object 固定目标
stationary-opening seed metering device 定孔式排种器

stationary open system 稳定开放系统
stationary operation 定点作业
stationary orbit 静止轨道;同步轨道;定常轨道;不动轨道
stationary orbit system 静止轨道系统
stationary paper carrier 固定式输纸装置
stationary part 固定部分
stationary pattern 静止图像
stationary pawl 固定爪;固定齿轮爪
stationary phase 固定相;平稳期;稳定时期;稳定期
stationary phase approximation 稳相近似法
stationary phase gradient 固定相梯度
stationary phase method 稳相位法
stationary phase shift 固定相移
stationary phenomenon 稳态现象
stationary picture 静止图像;静的形象
stationary piston drive sampler 固定活塞式打入取土器(用于塑性土层)
stationary piston sampler 固定活塞式取样器
stationary piston type sampler 固定活塞式打入取土器(用于塑性土层)
stationary plane 逗留面
stationary plant 陆用设备
stationary plastic flow 柔性静流
stationary plate 固定板;静止板块;不动板(压铸用)
stationary platform 固定平台
stationary platinum microelectrode 固定铂微电极;静铂微电极
stationary point 驻点;驻点;留点;平稳点;不动点
stationary point on a curve 曲线的平稳点
stationary point thermometer 留点温度计
stationary poisson process 平稳泊松过程
stationary pollution source 固定污染源
stationary population 稳定的人口;静止的人口
stationary position 固定位置
stationary potential 稳定势
stationary potential energy 稳定势能
stationary process 平稳过程;稳定过程
stationary pump 固定泵
stationary radiant 固定辐射点
stationary random action 平稳随机作用
stationary random distribution 平稳随机分布
stationary random function 平稳随机函数
stationary random input 平稳随机输入
stationary random process 平稳随机过程
stationary random process in the wide sense 广义平稳随机过程
stationary random sequence 平稳随机序列
stationary raster 稳定光栅
stationary ratchet train 固定式棘轮系
stationary reactor 固定式反应器
stationary refuse compactor 固定的垃圾压紧机
stationary resistance welder 固定式电阻焊机
stationary rivet die 固定的铆钉模
stationary riveting machine 固定铆机
stationary rod 固定杆
stationary rod cylinder 活塞杆固定的缸
stationary sandslinger 固定式抛砂机
stationary satellite 静止卫星
stationary satellite camera 固定人卫照相机
stationary saw mill 固定制材厂
stationary scale 计算尺尺度
stationary screen 固定筛
stationary seal ring 固定止漏环;固定密封环;静环;密封静环
stationary seal ring retainer 密封静环护圈
stationary seat 静环座
stationary set of vanes 固定叶片组
stationary setting 固定装置
stationary shaft 固定轴
stationary shaft factor 固定轴因数
stationary shearing load 恒定剪切荷载
stationary shell 稳定气壳
stationary shock wave 驻激波
stationary simple harmonic motion 稳态简谐振动
stationary single degree of freedom system 平稳单自由度系统
stationary sinter cooler 固定式烧结冷却机
stationary slab 固定板
stationary slewing crane 定柱旋臂起重机
stationary slinger 固定式抛砂机
stationary slip 固定卡瓦
stationary solid bed 固定固体床

stationary solid phase 固定固相
stationary sound 稳定音
stationary source 固定水源;固定源
stationary source of air pollution 固定空气污染源
stationary source of water pollution 固定式水污染源;水污染固定源
stationary sources 静止污染源
stationary spray pump 固定式喷雾泵
stationary sprinkler 固定喷灌器
stationary sprinkler system 固定式喷灌系统
stationary sprinkling system 固定的喷淋系统
stationary squeezer 固定式压铆机
stationary stage 稳定水位
stationary stand grinder 固定式砂轮机
stationary star 稳定恒星
stationary state 固定态;静止状态;能态;稳态;稳定态;定态;定常状态
stationary steady 固定中心架
stationary steel bushing 固定钢套
stationary stepped grate plate 阶梯篦板
stationary stochastic process 平稳随机过程
stationary stochastic series 平稳时间序列
stationary stochastic vibration 平稳随机振动
stationary stop 炉用缓冲器;固定止块;固定挡板
stationary storage barge 不动的储[贮]存驳船
stationary storage battery 固定蓄电池
stationary straw press 固定式茎秆压捆机
stationary stream 稳定流;固定流;静流;平稳流
stationary stripper 固定印料板
stationary suction 固定吸扫【疏】
stationary suction dredge(r) 碇泊式吸泥船
stationary surface wave 表面驻波
stationary swash plate 固定的斜盘
stationary tachograph 固定式转速记录仪
stationary tailstock center 固定尾座顶尖
stationary tank 储液槽;液态气体储[贮]槽;固定式储罐
stationary target 固定目标;固定靶标
stationary temperature 恒定温度
stationary test 静止试验
stationary thresher 固定脱粒机
stationary thrust collar 固定推力环
stationary tidal wave 驻潮波;稳定潮波
stationary time series 固定时间序列;平稳时间序列
stationary tone 恒定音
stationary tower crane 固定塔式起重机
stationary traffic 稳定运输;稳定交通
stationary transition probability 平稳转换概率
stationary transmitter 固定发射机
stationary tripper 固定式倾料器
stationary tube sheet 固定管板
stationary turbine 固定式透平
stationary turbine components 透平静止部件
stationary two-arm resin sand mixing unit 固定式双臂树脂砂混砂装置
stationary type 固定式
stationary type cradle 固定式筐架
stationary tyre 定位轮箍
stationary undulation 固定起伏;固定波动
stationary utility value 平稳效用值
stationary value 平稳值
stationary vane 静止叶片
stationary ventilator 固定式通风器;固定式通风机
stationary vertical diesel engine 固定立式柴油机
stationary vibration 固定振动
stationary vortex 固定漩涡
stationary wave 驻波;立波;定波
stationary wave method 驻波法
stationary wave pressure 驻波压力;立波压力
stationary wave theory 驻波理论;立波理论
stationary wave theory of tide 潮汐驻波理论;潮汐立波理论
stationary wave tube 驻波管
stationary wave tube method 驻波管法
stationary way of operation 定位作业方式
stationary welder 固定式电焊机
stationary welding machine 固定式焊机;固定焊机
stationary window 固定窗
stationary X-ray detection apparatus 固定式X射线探伤机
stationary X-ray equipment 固定式X射线机
station authentication 电台鉴定
station awning 车站雨棚
station-based maintenance 驻站维修

station battery 厂用蓄电池
station benchmark 测站水准点
station bill 应急部署表;应变部署表;岗位表;部署表
station block 站内闭塞
station boiler 电站锅炉
station boundary sign 站界标
station break 电台间断
station building 车站建筑(物);车站房屋
station building site 站房地坪【铁】
station buoy 位置浮标;定位浮标;地点指标;标位浮标
station call letters 电台呼号
station capacity 电站容量;车站容量;发电站容量;发电厂容量
station car capacity 车站容车能力
station centering 测站归心
station center 车站中心
station centring 测站归心
station chain 台链
station circus 车站广场
station classification track 车辆编组线
station cluster 站群集器
station computer 车站计算机
station computer control system 站控微机
station condition 测站条件
station console 车站控制台
station continuity chart 台站连续图
station control 位置控制器;测站控制
station control error 电厂控制误差
station control room 电厂控制室;车站控制室
station coordinates 站坐标;台站坐标
station correction 台站校正;台站改正;测站校正;测站改正;测点校正
station datum 测站基面
station deactivation 站停止
station density 台站密度;测站密度;测点密度
station deposit 车站银行存款
station description 监测站说明;测站描述(说明)
station designator 电台标志
station dial(l)ing 电台拨号
station distribution board 厂用配屏;厂用配电盘
station drainage 车站排水
station drift 测站漂移
station drilling machine 连续自动钻床
station dwelling time 停站时间
station efficiency 电站效率
station elevation 台站海拔高(度)
station entrance 进站口;入站口
station entrance-exit 车站出入口
station-epicenter 台站震中距
station equation 测站方程
station equipment 站内设备
station error 铅垂偏差;台站误差;测站误差;测点误差
stationery 文具
stationery case 文具匣;文具盒
Stationery Office 文书局
stationery room 文具室
stationery shop 文具店
stationery stock room 文具库
stationery store 文具库
station esplanade 车站广场
station exit 出站口
station fence 车站栅栏
station floor level 车站地面层
station for dead-end operation 调头车站
station forecourt 站前广场(火车站)
station for lumped method 堆聚法节点
station for popularizing agricultural technique 农业技术推广站
station for satellite injection 卫星注入站
station front 厂前区
station hall 车站候车厅;车站候车室;车站(候车)大厅
station headwall 车站端墙
station headway 车站车头时距
station hotel 车站旅馆
station house 火车站;消防队;警察派出所;车站
station identification 站址识别;站识别;电台识别
station identification for climate study 气候研究站站址选定
station identification signal 电台识别信号;电台识别呼号
station identifier 局识别符

station in advance of terminal 枢纽前方站【铁】
station index 转位;台站索引;台站号
station index machine 转位加工机床
station index number 台站号
station indicator 车站公告牌;(车站的)火车到发公告牌
stationing 设站;定测站;布员【测】;布点
stationing diagram 布点图
station installation 车站设备
station insulator 电站绝缘子
station interface case 接口箱
station large system 车站大系统
station layout 站场布置;车站布置(图)
station limit 站界
station limit post 站界标
station limit sign 站界标
station line 站间线
station-line facility 台站线路设施
station lines 车站线路
station list 应急部署表;应变部署表
station load factor 发电站负载因数;发电厂负载因数;发电厂负荷率
station location 站址;站位;台址
station location board 车站位置牌
station main building 车站主楼
station map 观测站位图
station mark 站名牌;测站标志;测站标点
station marker 测站标石
station master 站长;车站站长
station master control 车站控制;车站主控制
station master on duty 车站值班站长
station master's room 站长室
station meter 基准仪;基准尺;标准量具
station model 填图格式
station network 局内电力网;厂内电力网;测站网
station nomenclature 站名表
station number 站号;台站号;电台号;车站编号
station occupation time 设站时间
station of full two level design 全双层车站
station of origin 首发台
station open from sunrise to sunset 白天业务电台
station open in night 夜间业务电台
station open to official correspondence exclusively 公务通信专用电台
station operating plan 调车作业计划;车站作业计划
station operating procedure 车站作业技术过程
station operation 站务
station operation controller 车站值班员
station operation log 台站日志
station operation office 站务室
station operator 线路所值班员;车站值班员
station order car 按站顺序零担车
station output 电站输出功率;电厂输出功率
station pair 台对
station place 站前广场
station plan 车站线路图
station plant 车站设备
station plant factor 电厂设备利用率;车站设备利用率;发电站设备利用率
station platform 月台;站台;车站站台
station platform lift 站台升降机
station platform loudspeaker system 站台扩音系统
station platform ramp 站台坡道
station platform roof 站台屋顶
station platform stair(case) 站台楼梯
station platform track 站台轨道
station platform tunnel 站台隧道;站台地道
station plaza 车站前广场;站前广场;车站广场
station point 三角点;测站点;测点
station pointer 双角测定仪;示点器;三角分度仪;三杆分度仪;三杆分度器;三杆定位仪;三臂分度仪
station-pointer fix 三杆定位法(三标两角定位的海图作业)
station-point perspective 定点透视
station pole 标杆
station power load 厂用负荷
station power supply 厂用供电;厂用电源
station pressure 本站气压
station processor 台站数据处理机
station rating curve 测站水位流量关系曲线;测站流率曲线
station restaurant 车站餐厅
station rod 测量标尺;标杆

station roof 站台顶棚;站台屋面;独柱伞式屋顶;车站式顶盖;车站顶棚;站台式屋顶;伞形屋顶;车站顶盖
stations 船体分站(船舶线形图上的竖站)
station scale 车站规模
station selection 选站
station selector 选台旋钮
station-service 厂用
station service branch 厂用支线;厂用电管理部门
station-service distribution switch gear 厂用配电设备
station service electric(al) system 厂用电系统
station-service generator cubicle 厂用发电机开关柜
station service power 厂用动(力);厂用电(力)
station service power consumption rate 厂用电率
station service supply 厂用电源
station service transformer 厂用变压器
station service transformer differential protection 厂用变差动保护
station service turbine 厂自用涡轮机;厂用透平
station shift operating plan 车站班计划
station sidings 车站装配【铁】;车站配线
station sight 测站觇标
station sign 站名牌
station signal 车站信号;测站觇标;测站标志
station site 站坪【铁】;测站
station siting 台站选址;设置测站
station space 测点间距
station spacing 站间距离
station square 站前广场;车站广场;厂前广场
station staff 测量标尺;车站职工
station stage-discharge relationship 测站水位流量关系
station stage operating plan 车站阶段计划
station step-up transformer 电厂升压变压器
station stock level 驻地补给品储存量
station stop zone 停站区
station sub-system of the digital dispatching phone 数字调度电话车站分系统
station sump pump 车站集水坑泵
station supervisor 车站值班员
station surveillance buoy 位置监视浮标
station tanker 驻泊油轮
station technical working diagram 车站技术作业(图)表
station throat 车站咽喉(区)【铁】;车站咽喉
station time 装载结束时刻
station-to-station 站对站
station-to-station calling 距间呼叫
station-to-station distance 台站距
station-to-station transport 站到站运输
station track 火车站轨道;站线【铁】;股道【铁】
station transfer matrix 台站传输矩阵节点
station transformer 厂用变压器
station transit income 车站在途进款
station transmission equipment 车站传输设备
station transmitting notices to mariners 发送航海通告电台
station tube 站台隧道
station tunnel 火车站隧道;车站隧道
station-type 厂用型
station wagon 旅行车;小型客车;厢式车身;沿途零担车;工具车;旅行汽车;客货两用(汽)车
station working statistics 车站工作统计
station yard 站场;货场
station yard post 车站场区标志【铁】
station year method 站年法
statist 统计员;统计学家;统计师
statistic 统计数
statistic(al) absorption coefficient 统计吸收系数
statistic(al) abstract 统计摘要
statistic(al) accelerometer 统计加速度仪
statistic(al) account 统计账户;统计科目
statistic(al) accounting 统计核算
statistic(al) account of environment 环境统计台账
statistic(al) accumulation of apparent effect 表观效应统计累积
statistic(al) accuracy 统计准确度;统计精确度;统计的准确性
statistic(al) agency 统计机构
statistic(al) algorithm 统计算法
statistic(al) analysis 统计分析
statistic(al) analysis and processing of geolog-

ic(al) data 地质数据统计分析处理
statistic(al) analysis in the basic geology 基础地质中的统计分析
statistic(al) analysis method 统计分析(方)法
statistic(al) analysis monitor 统计分析监视器
statistic(al) analysis of direction(al) data 方向数据的统计分析
statistic(al) analysis of experiments 统计实验分析
statistic(al) analysis of lineament 线性体统计分析
statistic(al) analysis of mineral exploration 矿床勘探统计分析
statistic(al) analysis of ore control factors 控矿因素统计分析
statistic(al) analysis of synoptic(al) chart 天气图统计分析
statistic(al) analysis parameter 统计分析参量
statistic(al) analysis technique 统计分析技术
statistic(al) analytic(al) program(me) library 统计分析程序库
statistic(al) and economic analysis 统计与经济分析
statistic(al) annual 统计年鉴
statistic(al) approach to estimation 统计估测方法
statistic(al) arbitration 统计判优法
statistic(al) area 统计区
statistic(al) arrangement 统计排列
statistic(al) association 统计相连；统计关联性
statistic(al) assumption 统计假定
statistic(al) astronomy 统计天文学
statistic(al) authority 统计当局
statistic(al) average 统一平均值；统计平均
statistic(al) averaging method 统计平均法
statistic(al) balance method 统计平衡表法
statistic(al) betatron process 统计回旋加速器过程
statistic(al) bias 统计偏差
statistic(al) bioclimatology 统计生物气候学
statistic(al) body 统计团体；统计机构
statistic(al) book 统计簿
statistic(al) boundary 统计边界
statistic(al) breakdown 统计性分类；统计分析
statistic(al) broadening 统计变宽
statistic(al) bulletin 统计公报
Statistic(al) Bureau 统计局
statistic(al) calculation 统计计算；统计核算
statistic(al) cartogram 统计地图图形
statistic(al) cartography 统计制图法
statistic(al) cell 统计单元
statistic(al) certainty 统计的可靠性
statistic(al) chain 统计链
statistic(al) character 统计特征
statistic(al) characteristic parameter 统计特征参数
statistic(al) characteristics 统计特征值；统计特征
statistic(al) chart 统计图；统计表
statistic(al) circles 统计界
statistic(al) classification 统计分类法
Statistic(al) Commission 统计委员会
statistic(al) communication theory 统计通信理论
statistic(al) compensation 统计补偿
statistic(al) concept 统计概念
statistic(al) confluence analysis 统计合流分析
statistic(al) constant 统计常数
statistic(al) consultation 统计咨询
statistic(al) contamination 统计性沾污
statistic(al) control 统一控制；统计性的控制；统计控制；统计管理
statistic(al) control chart 统计管制图
statistic(al) control chart and variance analysis 统计控制图和差异分析
statistic(al) copy 统计报单
statistic(al) correlation 统计相关性；数理统计相关
statistic(al) correlation method 统计相关法
statistic(al) cost analysis 统计成本分析
statistic(al) criteria tests 统计准则检验
statistic(al) curve 统计曲线
statistic(al) data 统计资料；统计数据
statistic(al) database 统计数据库
statistic(al) data of environment 环境统计资料
statistic(al) data processing 统计数据处理
statistic(al) data recorder 统计数据记录器
statistic(al) decision 统计判定
statistic(al) decision function 统计判决函数；统计决策函数
statistic(al) decision-making 统计决策
statistic(al) decision-making theory 统计决策论

statistic(al) decision method 统计决策法
statistic(al) decision problem 统计判决问题
statistic(al) decision procedure 统计判决过程；统计决策过程
statistic(al) decision rule 统计决策原则
statistic(al) decision theory 统计判决理论；统计判定理论；统计决策论
statistic(al) decomposition 统计分解
statistic(al) demand analysis 统计需求分析
statistic(al) demand function 统计需求函数
statistic(al) department 统计部门
statistic(al) dependence 统计相关；统计上的相关性
statistic(al) descriminant function 统计判别函数
statistic(al) descriminant technique 统计判别方法
statistic(al) description 统计说明
statistic(al) design 统计设计
statistic(al) design of experiments 统计试验设计；统计实验设计
statistic(al) design value 统计设计值
statistic(al) detector 统计检波器
statistic(al) determination 统计测定
statistic(al) diagram 统计图表；统计分析图件
statistic(al) diagram of fabric elements 组构要素统计图
statistic(al) diameter 统计直径
statistic(al) discernibility 统计辨别力
statistic(al) discrepancy 统计误差；统计差异
statistic(al) dispersion 统计离差
statistic(al) distribution 统计学上的分布；统计分布
statistic(al) distribution law 统计分布律
statistic(al) distribution theory 统计分配定律
statistic(al) dynamic(al) prediciton 统计动力预报
statistic(al) ecology 统计生态学
statistic(al) economics 统计经济学
statistic(al) efficiency 统计有效性
statistic(al) energy analysis 统计能量分析
statistic(al) ensemble 统计总体；统计系统
statistic(al) environment 统计环境
statistic(al) equalizer 固定均衡器
statistic(al) equation 统计方程
statistic(al) equilibrium 统计平衡；统计均衡
statistic(al) ergodic theorem 统计遍历定理
statistic(al) error 统计误差
statistic(al) error of aero-survey 航测统计误差
statistic(al) estimate 统计性估计；统计估(计)值；统计估计
statistic(al) estimated value 统计性估计值；统计估(计)值
statistic(al) estimate of error 误差统计估计；统计误差估计值
statistic(al) estimation 统计性估计；统计估计
statistic(al) estimation approach to audit 审计的统计估算法
statistic(al) estimation of parameters 参考统计估值
statistic(al) evaluation 统计学评价；统计评定；统计计算
statistic(al) evidence 统计论据
statistic(al) experiment 统计性试验；统计实验
statistic(al) experiment method 统计试验方法；统计实验法
statistic(al) experiment theory 统计决策理论
statistic(al) extrapolation 统计展延；统计外延；统计外推
statistic(al) fact 统计事实
statistic(al) figures 统计数字
statistic(al) fire 统计内的火灾
statistic(al) fit 统计拟合
statistic(al) fluctuation 统计性涨落；统计性涨落；统计起伏；统计波动
statistic(al) fluid mechanics 统计流体力学
statistic(al) forecast 统计预测；统计预报
statistic(al) forecast methods for minerogenesis by processing remote sensing-geonomy data 遥感地学多数据处理成矿统计预测
statistic(al) forecast of sales 销售额统计预测
statistic(al) framework 统计结构
statistic(al) frequency 统计频率
statistic(al) function 统计函数
statistic(al) game 统计对策；统计博弈
statistic(al) gaps 统计缺口
statistic(al) geography 统计地理学
statistic(al) geology 统计地质学
statistic(al) graph 统计图

statistic(al) graphic(al) analysis 统计图解分析；统计的图解分析
statistic(al) graphic(al) method 统计图法
statistic(al) graphic(al) presentation method 统计图示法
statistic(al) graphic(al) symbols 统计图符号
statistic(al) graphics 统计制图法
statistic(al) graph of program(me) 计划进度统计图
statistic(al) grouping 统计分组
statistic(al) handbook 统计手册
statistic(al) hindcast 统计后报
statistic(al) homogeneity 统计同质性；统计齐性
statistic(al) hot spot factor 统计热点因子
statistic(al) hydrology 统计水文学
statistic(al) hypothesis 统计假设
statistic(al) hypothesis testing 统计假设检验
statistic(al) identifiability 统计可识别性
statistic(al) image 统计象
statistic(al) independence 统计上的独立性；统计的独立(性)
statistic(al) index 统计指数
statistic(al) indicator 统计指标
statistic(al) indicator system 统计指标体系
statistic(al) indices of locomotive drawing load 机车牵引工作量统计指标
statistic(al) indices of locomotive operation efficiency 机车运用效率统计指标
statistic(al) induction 统计归纳
statistic(al) inductive method 统计归纳法
statistic(al) inference 统计推断
statistic(al) inference method 统计推断法
statistic(al) inference research design 统计推论法研究设计
statistic(al) information 统计资料；统计信息
statistic(al) information system 统计信息系统
statistic(al) inquiry 统计调查
statistic(al) institution 统计学会
statistic(al) integration 统计一体化
statistic(al) interpretation 统计解释
statistic(al) interpretive language 统计解释语言
statistic(al) invariance 统计不变性
statistic(al) investigation 统计研究；统计调查
statistic(al) irreversible procedure 统计不可逆方法
statistic(al) item 统计项目
statistic(al) judgement 统计判断
statistic(al) language 统计语言
statistic(al) law 统计规律；统计法则
statistic(al) least square fit 统计最小平方拟合
statistic(al) ledger 统计台账
statistic(al) level analysis 统计电平分析
statistic(al) level distribution 统计电平分析；统计电平分布
statistic(al) linearization 统计线性化
statistic(al) linear model 统计线性模型
statistic(al) literature 统计文献
statistic(al) lithofacies 统计岩相
statistic(al) load 统计荷载
statistic(al) logic 统计逻辑
statistically dependent 统计相关的；统计独立
statistically evident 统计上显著的
statistically independent variable 统计独立变量
statistically non-significant 统计上不显著的
statistically significant 统计上显著的
statistically signification variables 统计有效变数
statistically uniform 按统计均匀的
statistic(al) machine 统计机
statistic(al) macroeconometric model 统计宏观经济计量模型
statistic(al) magnitude 统计数量；统计量
statistic(al) management 统计管理
statistic(al) map 统计图；统计地图
statistic(al) map of structural element 构造要素统计图
statistic(al) map program(me) 统计地图程序
statistic(al) mass 统计总体
statistic(al) material 统计资料
statistic(al) mathematics 统计数学
statistic(al) matrix 统计矩阵
statistic(al) mean 统计平均值
statistic(al) measurement 统计测度
statistic(al) mechanics 统计力学
statistic(al) mechanics of irreversible changes 不可逆变化统计力学

statistic(al) method 平均法;统计方法;统计法
statistic(al) methodology 统计方法论
statistic(al) mode 统计模式
statistic(al) model 统计模型
statistic(al) modelling 统计模拟
statistic(al) models evaluation 统计模型评价
statistic(al) modernization of locomotive operation 机车运用统计现代化
statistic(al) moment 统计势差;统计矩;统计动差
statistic(al) multiplexing 统计复用
statistic(al) noise 统计噪声
statistic(al) norm 统计定额
statistic(al) number 统计数
statistic(al) object 统计对象
statistic(al) oceanography 统计海洋学
statistic(al) ogive 统计学的卵形曲线
statistic(al) orbit determination theory 统计定轨理论
statistic(al) organization 统计组织
statistic(al) overall hot spot factor 统计的总热点因子
statistic(al) package 统计软件包
statistic(al) parallax 统计视差
statistic(al) parameter 统计参数
statistic(al) pattern recognition 统计模式识别
statistic(al) phenomenon 统计事件
statistic(al) physics 统计物理学
statistic(al) pictograph 统计图表
statistic(al) point estimation 统计点估计
statistic(al) point number 统计点数
statistic(al) population 统计总体;统计学群体;统计群体
statistic(al) precision interval 统计精度区间
statistic(al) prediction 统计预测;统计预报
statistic(al) prediction model 统计预测模型
statistic(al) prediction of blind ore 矿区盲矿统计预测
statistic(al) prior information 统计先验信息
statistic(al) probability 统计概率
statistic(al) probability theory 统计概率论
statistic(al) problem 统计问题
statistic(al) procedure 统计程序;统计步骤
statistic(al) process control 统计过程控制
statistic(al) processes 统计过程
statistic(al) processing and analysis 统计处理与分析
statistic(al) processing system 统计处理系统
statistic(al) production function 统计生产函数
statistic(al) program(me) 统计程序
statistic(al) project 统计方案
statistic(al) property 统计性能;统计特征
statistic(al) quality control 统计质量控制;统计品质管理;统计法质量控制;统计法质量管理;质量统计检验
statistic(al) quality control system 统计的质量控制法
statistic(al) quantity 统计量
statistic(al) questionnaire 统计调查
statistic(al) range 统计范围;统计度
statistic(al) reasoning 统计推理
statistic(al) reception 统计接收
statistic(al) record 统计台账;统计记录
statistic(al) regression analysis 统计回归分析
statistic(al) regression model 统计回归模型
statistic(al) regularity 统计规律性
statistic(al) relationships among variables 变量间的统计关系
statistic(al) relative frequency 统计相对频率
statistic(al) reliability 统计可靠性
statistic(al) report 统计报告
statistic(al) report of aids' classification 航标分类统计报告
statistic(al) report on m-of-w facility 铁路工务设备统计报表
statistic(al) requirements 统计要求
statistic(al) result 统计结果
statistic(al) returns 统计报表
statistic(al) risk 统计风险
statistic(al) room 统计室
statistic(al) sample 统计样本
statistic(al) sampled-data control system 统计数据采样控制系统
statistic(al) sample moment 统计样本矩
statistic(al) sampling 统计抽样;统计采样
statistic(al) sampling investigation method 统计抽样调查法
statistic(al) sampling in voucher examination 凭证查检统计抽样法
statistic(al) sampling model 统计抽样模型
statistic(al) scattering 统计散布
statistic(al) segment 统计链段
statistic(al) seismicity 统计地震活动性
statistic(al) seismic response 统计地震反应
statistic(al) seismology 统计地震学
statistic(al) series 统计序列;统计系列;统计数列;统计级数
statistic(al) services 统计工作
statistic(al) significance 统计显著性
statistic(al) simulation 统计模拟
statistic(al) software 统计软件
statistic(al) soil mechanics 统计土力学
statistic(al) sound level 统计声级
statistic(al) sound power absorption coefficient 统计声功率吸收系数
statistic(al) staff 统计工作人员
statistic(al) standard 统计标准
statistic(al) straggling 统计离散
statistic(al) subject 统计主体
statistic(al) summary 统计摘要;统计汇总
statistic(al) surface 统计面
statistic(al) survey 统计调查;统计测定
statistic(al) survey of environment 环境统计调查
statistic(al) system 统计制度
statistic(al) table 统计表
statistic(al) table of actual wagon loadings and tons of goods carried in comparison with those planned 货运计划完成情况统计表
statistic(al) tabulation 编制统计表
statistic(al) technique 统计法;统计技术
statistic(al) test 统计试验;统计实验;统计检验;统计测验
statistic(al) testing method 统计检验法
statistic(al) test method 统计试验方法
statistic(al) test theory 统计试验理论
statistic(al) theory 统计理论
statistic(al) theory for atmospheric diffusion 大气扩散统计理论
statistic(al) theory of extreme value 极值统计理论
statistic(al) thermodynamic(al) method 统计热力学方法
statistic(al) thermodynamics 统计热力学;统计力学
statistic(al) time lag 统计延时
statistic(al) tolerance limit 统计容许极限;统计允许极限
statistic(al) tolerance region 统计容许域
statistic(al) tool 统计工具
statistic(al) treatment 统计整理;统计加工;统计处理
statistic(al) uncertainty 统计误差;统计不准;统计不确定性
statistic(al) uniformity 统计的均匀性
statistic(al) unit 统计单位
statistic(al) validity 统计真实性
statistic(al) variable 统计变数
statistic(al) variable estimation 统计变量估计
statistic(al) variance 统计方差
statistic(al) variation 统计涨落
statistic(al) variation or dispersion 统计的变差或离差;统计变异
statistic(al) weight 统计体重;统计权数;统计权(重)
statistic(al) weight factor 统计权重因数
statistic(al) weighting variable 统计赋权变量
statistic(al) weight theorem 统计权重定理
statistic(al) work 统计工作
statistic(al) yardstick 统计尺度
statistic(al) yearbook 统计年鉴
statistician 统计员;统计学家;统计师;统计人员;统计工作者
statistico-thermodynamic(al) analysis 统计热力学分析
statistics 统计学;统计数字;统计量;统计
statistics act 统计法
statistics by group 统计分组
statistics controller 统计员
statistics department 统计局
statistics files 统计文件
statistics for coal reserves variation 煤炭储量变动统计
statistics for entire group 统计总体
statistics for ore reserves variation 矿产储量变动统计
statistics law 统计法
statistics of aids abnormality 航标失常统计
statistics of aids failure 航标失常统计
statistics of attributes 品质统计
statistics of classification frequency 分组频率统计
statistics of consumption expenditures 消费支出统计
statistics of database 数据库统计
statistics of environment(al) pollution 环境污染统计
statistics of groping the same item 同项归并统计
statistics of locomotive operation 铁路机车运用统计
statistics of mining area 矿区统计
statistics of m-of-w facility 铁路工务设备统计
statistics of natural conservation 自然保护统计
statistics of natural protection 自然保护统计
statistics of ore reserves 矿产储量统计
statistics of population 人口统计
statistics of population change 人口变动统计
statistics of population movement 人口变动统计
statistics of railway line length 铁路线路长度统计
statistics of railway state owned capital 铁路国有资产统计
statistics of railway superstructure 铁路上部建筑物统计
statistics of railway wagon and coach operation 铁路客货车辆运用统计
statistics of safety in mine 矿山安全统计
statistics of social insurance 社会保险统计
statistics of total hours worked 全部工作小时统计
statistics of train operation 铁路列车运行统计
statistics on labo(u)r input 劳动投入统计
statistics on three-dimensional sphere 三维球面上的统计
statistics recording feature 统计记录功能
statistics statements of railway transport 铁路运输统计报表
statitron 充压型静电加速器
statobiolith 固着(力)生物岩
statodyne 定子发电机;单机交流发电机
statokinetic 平衡运动的
statometer 静电荷计
stator 固定盘;静子;碇子;定子;定片
stator ampere-turn 定子安匝数
stator armature 定子电枢
stator blade 固定叶片;静片片;定子叶片;导气叶片
stator case 定子罩;定子外壳
stator casing 定子外壳
stator circuit time constant 定子回路时间
stator coil 定子线圈
stator contactor 定子接触器
stator core 定子芯;定子铁芯
stator current 定子电流
stator current limiter 定子电流限制器
statoreceptor 平衡感受器
stator end plate 定子端板
stator-feed-type poly-phase shunt motor 定子馈式多相并激电动机
stator-feed-type shunt motor 定子馈式并激电动机
stator frame 定子架;定子机座
stator frame turning & boring machine 定子机座车镗床
stator housing 定子外壳
stator impedance 定子阻抗
stator lamination 定子叠片
stator plate 电容器定片
stator-plate assembly 定子叠片组
stator ring 静环
stator-rotor starter 定子转子起动器
stator-rotor unit 定子转子组
stator screw insulator 起动器绝缘螺纹
stator sheet 定子薄钢片
stator slot 定子槽
stator-slot puncher 定子冲槽机
stator tooth 定子齿
stator-to-rotor air gap 定子转子间的空气隙
stator vane 静叶片;定子叶片
stator voltage 定子电压

stator winding 定子绕组
statoscope 高差仪；灵敏气压计；灵敏气压计；灵敏高度表；微动气压器；微动气压计；变压计
statoscope reading 高差仪读数
statoscope record(ing) 高差仪记录
Statoten-siometer 固定荷载纱线张力测定仪
Statron gun 斯塔特朗喷枪
statuary 雕刻家；雕像术；雕塑艺术
statuary biscuit 雕像陶坯
statuary bronze 塑像用青铜；雕像铜；雕像青铜
statuary column 雕像柱
statuary marble 优质大理石（一般指意大利白大理石）；雕像大理石
statuary porcelain 雕塑瓷
statue 用雕像装饰；塑像；雕像
statue chamber 雕像馆
statued avenue 有雕像点缀的大马路
statue marble 雕塑用大理石；雕像大理石
statue niche 雕像龛；铸像龛
statue of a deceased 死者雕像
statue of an apostle 帆船系缆柱上的雕像
Statue of Liberty 自由女神像（美国）
statue of Roland 罗兰雕像
statue pedestal 塑像台座
statuesque 雕像般的；雕饰
statuette 小塑像；小雕像
statumen （古罗马用于铺路的）石灰砂浆
statunit 厘米克秒静电制单位
statury bronze alloys 雕塑青铜合金
status 状况；情况；情景；身份；地位
status alarm unit 工况报警装置
status available 可选用状态
status bit 状态位
status buffer 状态缓冲器
status change 状态改变；暂隐与再现
status change report 状况变化报告
status circuit 状态电路
status code 状态码
status diagram 状态图
status display 现场显示；状态显示
status display area 状态显示区域
status display mode 状态显示方式
status enquiry 状态询问
status filters 状态滤光片
status flag 状态特征位
status format 状态形式
status indicator 状态指示器；状态指示符；状态显示器
status information 状态信息
status input instruction 状态输入指令
status lamp 状态灯
status line 状态行
status modifier 状态修改符
status nascendi 初生态
status note 使用状态说明
status of affairs 事态
status of a legal person 法人身份
status of development 发展状况
status-of-equipment board 装备状况登记板
status-of-equipment report 装备状况报告
status of funds 资金状况
status of taxpayers 纳税人身份
status quo 现状
status quo ante 原状
status read 状态读出
status register 状态寄存器
status report 现况报告；资信报告；情况报告（指工程方面）
status reporting program(me) 库存报表程式
status routine 状态程序
status signal 状态信号
status supervision 状态监控
status switch 状态序列
status switching 状态转换
status switching instruction 状态开关指令
status word error 错误状态字
statutable 法定的
statute 章程；规章；法令；法规
statute-barred 已逾时效的规定
statute labo(u)r 公役服务
statute law 成文法
statute mile 英里；法定英里（1 法定英里 = 1609.3 米）
statute of fraud 防止欺诈法

statute of limitations 法定时效；时效；法令限制
statutes and regulations 法规和条例
statutory 法定（的）
statutory accounting principles 法定会计原则
statutory agent 法定代理人
statutory appropriated retained earnings 法定拨用留存收益
statutory audit engagement 法定审计约定
statutory auditor 法定审计师
statutory audit requirement 法定审计要求
statutory audit responsibilities 法定的审计责任
statutory body 法定团体
statutory bond 法定合同；法定契约
statutory books 法定登记本
statutory capital 法定资本；法定股本
statutory company 法定公司
statutory control of dam 坝的法定管理
statutory copyright 法定版权
statutory deck line 勘定甲板线
statutory declaration 法定声明
statutory dedication 依法献地
statutory depletion 法定折耗
statutory documents 法定文件
statutory duty 法定责任
statutory equipment 规定属具
statutory financial statement 法定财务报表
statutory foreclosure 依法取消抵押品赎回权
statutory formula 法定公式
statutory freeboard 勘定干舷
statutory general meeting 法定股东大会
statutory grouping of gross income 毛收入法定分类
statutory guarantee 法定担保
statutory income 法定所得
statutory inspection 法定检查
statutory interpretation 法令的解释
statutory lien 法定留置权
statutory limitation 法定时效
statutory loading 法定荷载
statutory load line 勘定载重线；法定荷重吃水
statutory meeting 法定会议
statutory merger 法定兼并
statutory minimum 法定最小限度
statutory notice 颁布法令的通告
statutory nuisance 法定公害
statutory order 法定顺序
statutory period 法定期间
statutory period of redemption 法定抵押品赎回期限
statutory procedures 法定手续
statutory profit 法定利润
statutory provision 法规条款
statutory punishment 法定刑罚
statutory qualification 法定资格
statutory rate 法定税率
statutory records 法定记录
statutory report 法定报告
statutory requirement 法定要求
statutory reserves 法定资金储备；法定准备
Statutory Reserve of Deposits 法定存款准备率
statutory right in rem 法定对物权利
statutory right of detention and sale 法定滞留和出售权
statutory rights 法定权利
statutory room cost 法定的居室单价
statutory salary increase 法定加薪
statutory shares of estate 法定遗产继承份额
statutory survey 法定检验
statutory tariff 法定关税
statutory tax rate 法定税率
statutory tenant 法定租户
statutory trust 法定信托
statutory undertaker 法定承包者；法定承包商；法定（公用事业）承办者；法定承担者
statutory water company 法定自来水公司
statvolt 静电伏特
Staub clay lath(ing) 泥镘板条
staubosphere 尘圈；尘层
Stauffer grease 润滑牛油；润滑油脂
Stauffer lubricator 牛油环润滑器
staulog 皮托管水压计程仪
staunch 不透气的；气密的；密的；坚固的；不透空的；不漏水的
staunch and strong 坚实和强固

stauncheon 窗间小柱；支柱
staunching bead 留置垂直缝；止水垫圈
staunching piece 密封件；坝段间垂直止水槽；止水片；止水缝；留置垂直缝（坝工）
staunching plate 密封板
staunching rod 止水条
staunching wall 不漏水墙
stauractine 十字骨针
staurane 十二氢戊搭烯并戊搭烯
staurolite 十字石
staurolite cyanite 十字蓝晶
staurolite-muscovite-schist 十字石白云母片岩
stauros 十字结
stauroscope 十字镜（测定光在晶体中偏振平面方向的仪器）
staurotile 十字石云母片岩
staurotite dimicaceous schist 十字石二云母片岩
Stauwerke gate 斯托沃克闸门
stave 狭板；栅条；桶板（条）；梯级横木；梯级棍；侧板
stave bolt 制桶板用的短木材；桶板短圆材
stave church 木构教堂（斯堪的纳维亚）；板条教堂
stave construction 拼板建筑；环状板条结构（芯盒）
stave core 拼木板芯；细木工板；（木条板拼装成的）桶状物
staved collar drill shank 墩粗轴肩式钎维
stave flume 木渡槽；板条渡槽
stave in 装楼梯踏步；撞破；冲破
stave jointing 拼板接合
stave-jointing machine 拼板接合机
stave ladder 梯级横木
stave off 桶板脱落
stave pipe 木板管；条木管；木质水管；木质板拼成管；木板条水管；板条拼成的桶形管；板条管道
stave sheet 储罐壁板
stave silo 板条木筒仓；桶状青储[贮]塔
stave wood 制桶板材
staving press 扩孔机
stavrte 黑云角闪岩
stay 拉条；门窗撑挡；停留；锁紧片；定位钢筋；窗撑杆；撑条；车立柱
stay adjuster 撑条调整器
Stay alloy 斯特耶铝铜合金
stay bar 门窗横梃；窗风撑；窗撑条；窗风钩；拉条钢筋；窗撑杆；撑条；撑杆；撑窗棍
stay block 拉线桩；锚块
stay bolt 锚栓；拉紧栓；拉杆螺栓；拉撑；锚栓；牵条螺栓；撑螺栓；长螺栓；支撑螺栓
stay bolt body 撑螺体
stay bolt pitch 撑螺距
stay bolt screw 拉杆螺栓螺丝；支撑螺栓螺丝
stay bolt shank 撑螺体
stay bolt tap 撑螺栓丝锥
stay bracket 悬臂轴承架
staybrite 镍铬耐蚀可锻钢
stay cable 拉索；斜张钢缆；斜索；斜缆
stay-chain 系链
stay collar 拉环；牵环
stay cord 支索；拉绳
stayed bridge 拉索桥
stayed-cable bridge 斜张桥；斜拉桥
stayed girder 支承大梁
stayed girder bridge 有系杆的板梁桥
stayed girder construction 有系杆的板梁结构
stayed mast 拉线杆；拉力杆
stayed pole 拉线杆；加撑力柱；牵拉杆
stayed surface 支撑表面
stayed tower 拉线铁塔；拉线式铁塔
stayed truss 斜拉桁架
stayer 支撑物
stay fastener (窗的)撑头；固定拉锁
stay foresail 前支索的三角帆
stay fork 支撑叉
stay-furring tool 清垢器；除垢器
stay green throughout the year 常年总绿的树
stay guy 固定拉索；牵索；拉线
stay hook 拉线钩；撑钩
stay in 链档
staying 加劲；紧固；撑法
staying of pipeline 管架
staying power 耐久性；支撑能力；耐力；持久力
staying quality 持久性
stay in grade 品质稳定
staying wire 固定拉索；拉线
stay-in-place form(work) 永久模板；固定模板

stay knot 合结
stay-lathing 撑条
stay leg 井架大腿
stay light 锚灯
stay line cable 过河缆(测流用);固定测流缆索
stay line method 标线法;缆索测流法;标线测流法
stay log 支撑木;撑架木;支承用大木料
stay nut 撑条螺帽;撑条螺母
stay of arm 臂端撑条
stay of execution 停止执行
stay of lathe 车床撑架
stay of proceedings 停止进行诉讼
stayoverroom 超期用房
staypak 压实材;压定木
stay piece 屋架系杆
stay pile 支承桩;拉索锚桩;拉索锚桩;锚(固)桩;锚碇桩;牵(拉)桩;稳位桩
stay pin 链环档
stay pipe 支持管;支撑管;撑管
stay plate 座板;垫板;系板;缀板;支撑板;撑板
stay-pole 撑杆
stay post 撑柱
stay proceedings 停止手续
stay-put button 锁定按钮
stay ring 穿索环;座圈;座环;顶环
stay ring transition piece 座环过渡段;座环碟形边
stay rod 拉索桩;拉索桩;缀条;终端杆;拉线桩;锁定杆;撑杆
stay rod pin 拉杆销
stay rope 固定拉索;牵索;缆风;拉索
staysail 支索帆;牵帆;三角帆
stays and staybolts 支撑件及拉杆螺栓
Stayset damper 史坦塞牌专利风门(用于机械通风)
stay tacking 钉紧定位
stay tackle 支索绞辘
stay tap 铰孔攻丝复合刀具
stay thimble 电杆拉线的中断环
stay tightener 松紧调节的拉杆;(拉索的)旋转接头;拉线张紧器;拉线紧固体;拉紧收紧器
stay time 停留时间
stay time of container 集装箱堆存期
stay transom 拉条横梁
stay tube 牵管;牵条管;撑管;支撑管;拉管
stay vane 固定导叶
stay vane ring 座环;固定导叶环
stay washer 撑条垫圈
stay wedge 调整楔
stay wire 系留索;系紧线
steadier 底座;支架;支座
steadily accelerated fluid 匀加速流体
steadily convergent series 固敛级数
steadily drop down 平稳下降
steadily lift up 平稳上升
steadily rise 平稳上涨
steadiness 稳定性;常定度
steadiness of commodity 物价稳定
steadiness parameter 稳定参数
steading 农庄;农场内建筑物;小农场
steadite 针硅磷灰石;磷共晶体;磷共晶
Steadman packing 斯特德曼填料
Steadman system 斯特德曼方式
stead state system 稳定状态系统
steady 牢靠的;稳恒的;稳定的
steady acceleration 等加速度
steady allocation method 平稳分配法
steady arm 支持杆;定位器
steady arm heel 定位器管
steady arm with insulating part and offset hook end clamp 绝缘定位器
steady arm with insulating rod and offset hook end clamp 绝缘定位器
steady background 稳定本底
steady baseline 稳定基线
steady beam 稳恒束
steady bearing 稳态方位;方位没有明显变化
steady brace 固定销;稳定用拉条
steady center rest 固定中心架
steady clamp 定位线夹
steady clamp for busbar 汇流排定位线夹
steady climb 稳定上升
steady coefficient of spring 泉的稳定性系数
steady component 稳定部分
steady compression process 稳定压缩试验法
steady condition 稳定条件;常定状态

steady creep 稳态徐变;稳态蠕动;稳定蠕动;稳态蠕变;稳定蠕变;持续徐变;持续蠕变
steady current 稳定电流;定量流;常定流
steady deformation process 稳定变形过程
steady delivery 均衡交货
steady dilution method 稳定稀释法;定浓度法
steady discharge 稳定流量
steady distribution 定常分布
steady dive 稳定俯冲
steady divergence motion 非周期渐强运动
steady drag force 稳定水流阻力
steady draw 稳定放矿
steady drift 稳定偏移
steady energy input 稳定能量输入
steady extrusion 稳定挤压
steady field 规则场;稳定场
steady field demagnetization 稳恒场退磁
steady flame 稳定火焰
steady flow 恒态流;恒流;恒定水流;恒定流;稳流;稳定水流;稳定流量;稳定流动;稳定流变;定流;定量流;常定流(动);常定流
steady flow coefficient 稳定流系数
steady flow combustion 稳流燃烧
steady flow forming 稳定流动成型
steady flow formula of boundary well 边界井的稳定流公式
steady flow model 定常流模式
steady flow non-uniformity 稳定流不均匀性
steady flow nozzle headbox 稳流喷嘴流浆箱
steady flow of heat 稳定热流
steady flow process 稳流过程
steady flow ram jet 稳流冲击式喷气发动机
steady flow reactor 稳定流反应器
steady flow state 稳定流态
steady flow turbine 稳流式涡轮
steady flow viscosity 稳流黏度
steady fluidized bed 平稳流态化床
steady foundation 稳固地基
steady gas dynamics 定常流气体动力学
steady geyser 规则的间歇泉
steady glass 化学稳定玻璃
steady glow 稳定辉光
steady gradient 连续坡(度);均均坡度;均坡;平稳坡度;稳定坡度;持续坡度;不变坡
steady head 稳定压头
steady heat conduction 稳定热传导;稳定导热
steady heat transfer 稳定传热
steady hydraulic jump 定常水跃
steady illumination 稳定照明
steady image 稳定图像
steadying baffle 消浪栅;稳水栅
steadying bar 支撑杆
steadying bracket 稳定夹
steadying effect 旋转质量惯性;稳定效应
steadying line 稳定索
steadying resistance 稳恒电阻
steadying rope 稳定索
steadying sails 稳定帆
steady jump 稳定水跃
steady laminar flow 稳定片流;稳定层流
steady load 静荷载;恒载;恒定荷载;稳定荷载;稳定负载;稳定荷载;稳定负荷;不变荷载
steady load test 稳定负荷试验
steady mark 固定记号;稳恒传号
steady market 市场坚稳
steady motion 匀速运动;恒稳运动;平稳运动;稳定运动;定常运动
steady motion of a fluid 定常流
steady noise 稳态噪声
steady non-uniform flow 稳态非均匀流;稳定非均匀流;稳定不均匀流;稳定变速水流
steady operation 稳定运行;稳定工作;稳定操作
steady persistence 长余辉
steady pin 固定销;锁紧销
steady planar detonation 稳定平面爆震
steady plastic flow 恒定塑流
steady point steel rod 尖端稳定钢杆
steady position deviation 稳定位置偏差
steady potential 稳定电势
steady pressure 稳定压力;定常压力
steady process 稳恒过程
steady progression 稳定进步;稳定演变;稳定发展
steady pull 稳恒拉力
steady pulsation 稳定脉动

steady pumping test 稳定流抽水试验
steady quantity of water bursting 稳定突水量
steady quantity of water injection 稳定注水量
steady rain 连绵雨;淫雨;连续降雨;均衡雨;持续雨;持续降雨
steady rate 稳定速率
steady resistance 稳流电阻;稳定电阻
steady response 定常应答
steady rest 中心架;固定中心架;固定支架;稳定架;辅助支架
steady rise of pressure 压力稳定上升
steady river 稳定河流
steady rolling condition 稳定滚动条件
steady rotation 稳定转动;稳定旋转;定常转动;等角速度转动
steady run 稳定运行
steady running 平稳运转;稳定运转
steady running condition 稳定运转状态
steady seepage 恒定渗流;稳定渗流;等量渗透
steady shear compliance 稳剪切柔量
steady space 稳定车距
steady speed 稳定速度
steady speed deviation 稳定速率偏差
steady speed test (发动机的)稳速试验
steady spin 稳定绕转
steady spring 稳定泉
steady stability 静态稳定(性);静态稳定(度)
steady stage 稳定阶段
steady state 准稳态;恒稳态;恒定状态;平稳状态;稳态;稳恒状态;稳恒态;稳定状态;定态;常定状态
steady-state acceleration 定常加速度
steady-state amplitude 稳态振幅
steady-state analysis 稳态分析
steady-state approximation 稳态近似
steady-state behavio(u)r 稳态性能
steady-state biofilm 稳态生物膜
steady-state boundary condition 稳态边界条件
steady-state boundary layer 稳态边界层;稳恒状态边界层;稳恒附面层
steady-state chain reaction 稳态链式反应
steady-state characteristic 稳态特性;定态特性
steady-state commutation 稳态换向
steady-state compliance 稳态柔量
steady-state concentration 稳态浓度
steady-state concentration of pollutant 稳态污染物浓度
steady-state condition 稳态条件;稳定状态条件
steady-state conduction 稳态热传导
steady-state constant 稳态常数
steady-state creep 稳态徐变;稳态蠕动;稳态蠕变;二级蠕变
steady-state culture 稳态培养
steady-state current 稳态流;稳态电流
steady-state cycle 平稳循环
steady-state damp heat test 恒定湿热试验
steady-state deformation 稳定变形
steady-state density 稳态密度
steady-state deviation 稳态偏差
steady-state diffusion 稳态扩散
steady-state discharge 稳态放电
steady-state distillation 稳态蒸馏
steady-state distribution 稳态分配;稳态分布
steady-state distribution equation 稳态分布方程
steady-state dye plume 稳定染色羽流
steady-state dye study 稳定染色研究
steady-state equation 稳态方程
steady-state error 静差;稳态误差
steady-state estuary 定态河沟;稳态河口
steady-state flow 稳定流;定常流;稳态流
steady-state forced vibration 稳态强迫振动
steady-state frequency control 稳态频率控制
steady-state froth 稳态泡沫
steady-state fuel conversion ratio 稳态燃料转换比
steady-state fusion reactor 稳态聚变堆
steady-state gain 稳态增益
steady-state gradient 稳态梯度
steady-state growth 稳态生长
steady-state heat exchanger 稳态热交换器
steady-state heat transfer 稳态传热;稳定状态的热交换
steady-state incidence 定常迎角
steady-state intensity 稳定强度;稳定密度;稳定亮度

steady-state lateral acceleration 稳态转向离心加速度
steady-state level 稳态级
steady-state lifetime 稳态寿命
steady-state loading 稳态荷载
steady-state measurement 稳态量测;稳态测量
steady-state method 稳态法
steady-state method of test 稳态测试方法
steady-state model 定常状态模型;静态模型;稳态模型
steady-state noise 稳态噪声;稳态噪音
steady-state noise level 稳态噪声级
steady-state noise source 稳态噪声源
steady-state of flow condition 稳定流态;水流条件的稳定状态
steady-state of motion 稳定运动状态
steady-state operation 稳态运行
steady-state optimization 稳态最佳化
steady-state output 稳态输出
steady-state ozone level 稳态臭氧水平
steady-state performance 稳态特性
steady-state period 稳态周期
steady-state permeation 稳态渗透
steady-state plasma 稳态等离子体
steady-state poisoning 稳态中毒
steady-state power coefficient 稳态功率系数
steady-state power limit 稳态极限功
steady-state pressure 稳态压力
steady-state profile 稳态水面线;稳流纵剖面
steady-state reactance 稳态电抗
steady-state reactor 稳态堆
steady-state recombination 稳态复合
steady-state regulation 稳态调节
steady-state residence time 定常状态保持时间;稳态保持时间
steady-state resonance curve 稳态共振曲线
steady-state response 稳态响应;稳态反应;稳定态特性
steady-state response gain 稳态反应增益
steady-state reverberation sound 稳态混响声
steady-state short circuit 稳态短路
steady-state signal 稳态信号
steady-state simulation 定态模拟;稳态模拟
steady-state sinusoidal excitation 稳态正弦波激振
steady-state solution (微分方程的)稳态解
steady-state sound 稳态声
steady-state sound pressure level 稳态声压级
steady-state sound source 稳态声源
steady-state speed 稳态转速
steady-state speed drop 固态转差系数;残留不平衡度
steady-state speed regulation 稳态速度调整率
steady-state speed variation 固态转速变化;稳态转速变化
steady-state stability limit 稳态稳定极限
steady-state stability margin 稳态稳定裕度;稳态稳定系数;稳态稳定储备量
steady-state stream 均衡河流
steady-state system 稳态系统;稳定系统
steady-state tape tension 稳态带张力
steady-state temperature 稳态温度;稳定状况温度
steady-state test 稳态试验
steady-state theory 恒稳态学说;稳态说;稳态理论;稳恒态学说
steady-state thermal anomaly 稳态热异常
steady-state thermal stress 稳态热应力
steady-state transfer 稳态传导
steady-state transistor 稳态晶体管
steady-state transport 稳态运输
steady-state turn 稳态转弯
steady-state type of signal 稳态信号
steady-state uniform flow 稳态均匀流
steady-state variable 稳态变量
steady-state vibration 定常振动;稳(定)态振动
steady-state viscosity 稳态黏度
steady-state wave motion 稳态波动
steady steaming 稳定航行
steady stimulation 稳态激励
steady stimulus 稳态激励
steady strain 稳定应变
steady stream 稳恒流;稳定河流
steady stress 静应力;剩余应力
steady subsonic drag 恒定亚声速曳阻
steady sunspot 稳态黑子

steady temperature 恒定温度;稳定温度
steady temperature damp test 恒温恒湿试验
steady time 稳定时间
steady traffic 均衡运输
steady traffic flow 稳定车流
steady tube 定位管
steady tube for cable support 电缆支架预埋固定管
steady turbulent flow 稳定湍流
steady turning motion 稳定回转运动
steady turning performance 稳定旋转特性
steady turning speed 稳定回转速度
steady uniform flow 恒定均匀流;稳定均匀流;稳定等速水流;稳定等速(动);等速匀流
steady uniform turbulent flow 稳定等速湍流
steady value 稳定值
steady velocity 稳定速度
steady vibration 稳态振动
steady voltage 稳定电压
steady water level 稳定水位
steady water level in aquifer 含水层稳定水位
steady wave 稳定波
steady wind 恒风;稳定风;定常风
steady working condition 稳定工况
steady yield 稳定产量
steak house 牛排餐厅
steal hole 偷进尺
stealing cycle 窃用周期
stealth material 隐形材料;隐身材料
steam 蒸汽;水蒸气
steam accumulator 蓄汽器;储[贮]汽箱;蒸汽收集器;蒸汽储蓄器;储汽箱
steam activation 蒸汽活化
steam admission 进汽
steam admission pipe 蒸汽进气管
steam admission side 蒸汽进气侧;进蒸汽侧;进汽侧
steam-age 蒸汽蒸化
steam ager 蒸汽熟化器
steam-agitated autoclave 蒸汽搅拌高压釜
steam-air activation 蒸汽空气活化
steam-air decoking method 蒸汽空气烧焦法
steam-air die forging hammer 蒸汽空气模锻锤
steam-air forging hammer 蒸汽空气自由锻锤
steam-air gasification 蒸汽空气气化
steam-air heater 蒸汽暖风器;蒸汽加热式空气预热器;蒸汽热风器
steamalloy 铜镍基合金
steam and air mixture 蒸汽空气混合气
steam and gas mixture 蒸汽煤气混合气
steam-and-solvent condenser 蒸汽与溶剂冷凝器
steam-and-water gland 汽水填密胀圈
steam-and-water radiator 蒸汽及热水散热器
steam-and-water separation nozzle 汽液分离喷嘴
steam annihilator 蒸汽灭火装置
steam apparatus 蒸汽养护设备
steam asphalt 蒸汽处理沥青;水蒸气处理沥青
steam-assisted pressure jet burner 蒸汽机械喷油燃烧器
steam atmospheric distillation 常压蒸汽蒸馏
steam atomization 蒸汽雾化
steam atomized flat spray 蒸汽雾化扁平喷嘴
steam atomizer 蒸汽雾化器;蒸汽喷油器;蒸汽雾器
steam atomizing 蒸汽雾化;蒸汽喷雾
steam atomizing burner 蒸汽雾化燃烧
steam atomizing conic(al)jet 蒸汽雾化圆锥形喷嘴
steam atomizing oil burner 蒸汽油雾燃烧炉;蒸汽雾化油燃烧器;蒸汽雾化型油燃烧器
steam attemperation 蒸汽的减温
steam autoclave 蒸汽压力罐
steam automatic hammer 自动蒸汽锤
steam automobile 蒸汽自动车;蒸汽汽车
steam balancing main 蒸汽平衡干管
steam barrier 蒸汽屏层
steam bath 蒸汽浴(器)
steam bell 蒸汽警钟
steam bending 蒸煮弯曲法;蒸汽弯曲法;蒸汽弯曲(法)(木材)
steam blancher 蒸汽烫漂器
steam blanket 蒸汽膜
steam blast 蒸汽喷洗;蒸汽喷净(装置);蒸汽鼓风
steam blast device 蒸汽喷洗装置
steam blow 潮气泡所致的瑕疵
steam blower 蒸汽喷雾器;蒸汽喷雾机;蒸汽喷射

器;蒸汽鼓风机
steam blowing 蒸汽喷吹法;蒸汽吹扫
steam blowing machine 蒸汽吹制机
steam blown 蒸汽吹制
steam-blown asphalt 蒸汽吹制沥青
steam-blown fiber[fibre] 蒸汽喷吹纤维
steam-blown poke hole 蒸汽吹孔
steam-blown process 蒸汽吹制工艺
steam boat 蒸汽机船;汽艇;汽船;轮船
steamboat coal 航运煤
steam boiler 蒸汽锅炉;蒸汽发生器;汽锅
steam boiler for domestic heating 民用蒸汽锅炉
steam boiling 蒸汽熬煮
steam-bottom still 底部蒸汽加热蒸馏釜
steam boundary curve 蒸汽界面曲线
steam box 汽柜;蒸汽器;蒸汽养护室;蒸汽箱
steam brake 蒸汽制动器;汽闸
steam-break accident 蒸汽破裂事故
steam bronze 汽阀青铜
steam bubble 蒸汽包;蒸汽囊;气泡
steam bubbling 蒸汽加热搅拌
steam-bubbling type autoclave 蒸汽鼓泡搅拌式高压釜
steam bypass 蒸汽旁通;蒸汽旁路
steam cabinet 蒸汽室
steam calorifier 蒸汽热水器;蒸汽加热器
steam calorimeter 蒸汽量热器;蒸汽干度测量器
steam can 蒸汽发生器
steam canal 进汽口
steam cans 蒸汽圆筒烘燥机
steam capacity 蒸汽容量;蒸发量
steam capstan 蒸汽绞盘
steam car 蒸汽车
steam cargo winch 蒸汽绞车
steam casting die 蒸汽压迫铸造法
steam catapult 蒸汽弹射器
steam chamber 蒸汽养护室;汽室蒸汽养护室;汽室
steam channel 蒸汽通道
steam channel(l)ing machine 蒸汽挖沟机;蒸汽截煤机
steam chart 蒸汽图
steam chest 进汽室;滑阀箱;集气管;蒸汽箱;蒸汽室
steam circle 作用半径;航行范围
steam circuit 蒸汽循环回路
steam circulating pipe 蒸汽循环管
steam cleaner 蒸汽清洗器;蒸汽清洗装置;蒸汽清洗机
steam cleaning 蒸汽清洗;蒸汽清洁法;蒸汽喷洗
steam cleaning unit 蒸汽清扫装置;蒸汽清除装置;水蒸气清除装置
steam cloud 蒸汽云
steam coach 蒸汽车
steam coal 蒸汽煤;锅炉用煤;锅炉煤;短焰煤;动力煤
steam cock 蒸汽阀门;蒸汽气门;蒸汽旋阀;蒸汽旋塞
steam-coil 蒸汽盘管;蒸汽旋管;蒸汽加热盘管;蛇形蒸汽管
steam-coil heated 蒸汽盘管供热;蒸汽盘管供暖;蒸汽旋管加热的
steam-coil-heater tank car 蒸汽旋管加热的油槽车;蒸汽盘管加热槽车
steam coil heating installation 蒸汽盘管供暖设备
steam coil storage tank 蒸汽盘管式储[贮]水箱
steam collector 蒸汽总管;蒸汽集气管;蒸汽集合器;蒸汽收集器
steam colo(u)r printing 蒸汽印染
steam companions 蒸汽伴热管
steam condensate 蒸汽冷凝液;蒸汽冷凝水
steam condensation 蒸汽凝结
steam condensation number 蒸汽冷凝数
steam condenser 蒸汽冷凝器
steam condesation zone 蒸汽冷凝区
steam condition 蒸汽参数
steam conductivity 蒸汽导电性
steam conduit 蒸汽管(道)
steam cone 蒸汽喷嘴
steam connection 蒸汽接管
steam consumption 蒸汽(消)耗量;耗汽量;汽耗
steam consumption meter 蒸汽消耗计
steam consumption per hour 每小时汽耗
steam content 蒸汽含量;蒸发量
steam control valve 蒸汽调节阀
steam converter valve 减温减压阀
steam convertor 蒸汽发生器

steam cooked 蒸煮的
steam cooker 蒸汽锅
steam cooled slag 蒸汽淬冷渣
steam cooler 蒸汽冷却器
steam cooling 蒸汽冷却
steam cooling coil 蒸汽冷却蛇管
steam cooling system 蒸汽冷却系统
steam corrosion 汽蚀
steam cost 蒸汽成本
steam cracking 蒸汽裂化
steam cracking unit 蒸汽裂化装置
steam crane 蒸汽起重机;蒸汽吊车
steam cultivator 蒸汽机引式中耕机
steam cure 蒸汽熟化;蒸汽硫化;蒸汽固化;蒸汽处治
steam-cured 蒸汽养护的
steam-cured box 蒸汽养护箱
steam-cured concrete 蒸汽养护混凝土;蒸养混凝土
steam-cured expanded concrete 蒸汽养护的膨胀混凝土
steam cured gas(-formed)concrete 蒸汽养护的加气混凝土
steam-cured paste 蒸汽养护水泥净浆
steam-cured unit 蒸汽养护制件;蒸汽养护的(混凝土)制件;蒸汽养护的(混凝土)试件
steam curing 蒸汽养生;蒸汽养护
steam curing at atmospheric pressure (在大气压下进行的)蒸汽养护;常压蒸汽养护
steam curing at high pressure 高压蒸汽养护
steam curing chamber 蒸汽养护室
steam curing cloche 蒸汽养护覆盖物
steam curing cycle 蒸汽养护制度;蒸汽养护周期;蒸汽养护循环
steam curing hood 蒸汽养护罩
steam curing installation 蒸汽养护设施;蒸汽养护装置
steam curing kiln 蒸汽养护窑
steam curing of concrete 蒸汽养护混凝土;混凝土蒸汽养护
steam curing of concrete products 混凝土制品蒸汽养护
steam curing room 蒸汽养护室
steam curtain 蒸汽幕
steam cushion 蒸汽垫;汽垫
steam cushion construction vehicle 汽垫工程车
steam cut off valve 停气阀
steam cutter 小火轮
steam cycle 朗肯循环
steam cylinder 蒸汽缸;汽缸
steam-cylinder assembly 蒸汽缸部件
steam-cylinder gasket 汽缸填密片
steam-cylinder jacket band 汽缸套带
steam-cylinder lagging 汽缸外套
steam-cylinder lubrication 汽缸润滑
steam-cylinder lubricator 汽缸润滑器
steam-cylinder oil 蒸汽缸油
steam-cylinder piston 汽缸活塞
steam-cylinder stock 汽缸油料
steam deaerator 热力除气器
steam degreasing 蒸汽除油
steam density 蒸汽密度
steam-descaling sprayer 蒸汽除鳞喷嘴
steam desorption section 蒸汽解吸段
steam developing 蒸汽显色
steam diagram 蒸汽(压力)图
steam digester 蒸汽蒸缸
steam digestion 蒸煮消化
steam-digestion reclaim 蒸煮(法)回收
steam digger 蒸汽挖掘机;蒸汽单斗挖土机;汽力掘凿机;汽铲
steam direct drive 蒸汽直接驱动
steam discharge 蒸汽排放;蒸汽排出
steam discharge pipe 排(蒸)汽管
steam discharge tube 排(蒸)汽管
steam disengaging surface 蒸发表面
steam dispersion mix 蒸汽喷布拌和工艺
steam dispersion mixer 蒸汽喷布拌和机
steam distillation 蒸汽蒸馏(法);气馏
steam distilled 蒸汽蒸馏的;汽馏的
steam-distilled oil 蒸汽蒸馏松油
steam-distilled wood turpentine 蒸馏的木松节油
steam-distributing pipe 蒸汽分配管
steam-distributing valve 蒸汽分配阀
steam distribution 蒸汽分布;蒸汽配汽
steam distribution system 蒸汽分配系统

steam distributor 蒸汽分配器
steam dome 聚汽室;汽室
steam dredge(r) 蒸汽挖泥船;蒸汽疏浚机;蒸汽挖泥船动力挖泥船
steam drier 蒸汽干燥器;蒸汽干燥机
steam drilling plant 蒸汽钻孔装置
steam drive 蒸汽驱动;蒸汽动力传动
steam drive drill 蒸汽驱动钻机
steam drive drilling rig 蒸汽驱动钻探设备
steam-driven 蒸汽动力的;蒸汽驱动的
steam-driven catapult 蒸汽弹射器
steam-driven digger 蒸汽驱动挖掘机
steam-driven excavator 蒸汽挖掘机;蒸汽驱动挖掘机
steam-driven generator 蒸汽发电机
steam-driven lift 蒸汽升降机
steam-driven pump 蒸汽泵;蒸汽驱动泵
steam-driven pumping system 蒸汽燃气涡轮泵式输送系统;蒸汽燃气涡轮泵式
steam-driven reciprocating pump 蒸汽往复泵
steam-driven slewing crane 蒸汽回转起重机
steam-driven truck 蒸汽机运货车
steam-driven vessel 蒸汽机船
steam driver 蒸汽打桩机
steam drop hammer 蒸汽落锤;锻锤
steam drum 蒸汽锅;蒸汽筒;蒸汽锅筒;蒸汽罐;汽鼓;蒸汽包;汽鼓;汽包;泡包
steamdrum 蒸汽锅筒;上汽包
steam drum winch 蒸汽卷筒绞车
steam dryer 蒸汽烘缸;蒸汽干燥器;蒸汽干燥机
steam drying 蒸汽干燥(法)
steam drying apparatus 蒸汽干燥器
steam-dump system (汽轮机事故时用的)蒸汽排放系统
steam dynamo 直流发电机;蒸汽发电机
steam economy 蒸汽耗用量
steamed concrete 蒸汽养护混凝土
steamed wood 蒸干木材
steam efficiency 蒸汽效率
steam ejection system 蒸汽喷射系统;蒸汽弹射系统
steam ejector 蒸汽喷射器;喷汽机
steam ejector gas freeing 喷汽驱除油气法
steam electric(al)floating crane 蒸汽电力浮式起重机
steam-electric(al)generating station 蒸汽发电厂
steam-electric(al)generation 火力发电
steam electric(al)plant 火力发电厂
steam electric(al)plant cleaning 热电站清洗
steam-electric(al)power generation 蒸汽电力发电
steam electric(al)propulsion 蒸汽电力推进
steam emulsification 蒸汽乳化
steam emulsion number 蒸汽乳化值
steam emulsion test 蒸汽乳化试验
steam end 蒸汽端
steam energy 热能
steam engine 蒸汽(发动)机;蒸汽发电机
steam engineering 蒸汽工程
steam engine governor 蒸汽机调节器
steam engine indicator 蒸汽机示功器
steam engine locomotive 蒸汽机火车
steam engine lubricant 蒸汽机润滑剂
steam engine pile hammer 蒸汽机桩锤
steam engine plow 锅驼机犁
steamer 蒸(汽)锅;蒸汽机船;蒸汽机;蒸汽发生器;轮船;海上拖缆;汽蒸器;汽锅;汽船
steamer chair 折叠躺椅
steamer connection 消防汽车接头
steamer-mixer 蒸煮混合器
steamer pays dues 船方支付费用
steamer traffic 轮船交通
steamer wharf 轮船码头
steam escape valve 放汽阀
steam evapo(u)ration 汽蒸法
steam excavator 蒸汽挖土机
steam exhaust 乏汽;乏回汽
steam exhausting way 排汽道
steam exhaust period 蒸汽排出时间;蒸汽排出周期
steam exhaust pipe 排气管
steam exhaust port 废气排出孔
steam expansion 蒸汽膨胀
steam exploded wood 蒸汽爆裂木材纤维
steam explosion 蒸汽喷发;蒸汽爆炸
steam extraction 抽气

steam extraction cleaning method 蒸汽抽吸清洗法
steam-fan furnace 蒸汽喷嘴炉
steam feed heater 蒸汽式给水预热器
steam feed pump 给料泵;蒸汽泵;汽动给水泵
steam-ferry 蒸汽渡轮;渡轮
steam field 蒸汽田
steam film 蒸汽膜
steam filter 蒸汽过滤器
steam filtration 蒸汽过滤
steam finish 蒸汽光泽
steam fire-extinguishing 蒸汽灭火
steam fire smothering system 蒸汽灭火系统
steam fitter 汽管装配工;蒸汽管道工(人)
steam fitting 蒸汽管道配件;汽管装配;蒸汽锅炉配件;蒸汽管件
steam floating crane 蒸汽浮动吊车;蒸汽浮式起重机;蒸汽浮式起重机
steam flow 蒸汽流;蒸汽流量;气流
steam flow channel 蒸汽流道
steam flow diagram 蒸汽流程图;蒸汽管路系统图
steam flowmeter 蒸汽流量计;蒸汽流量表
steam flow recorder 蒸汽流量记录器
steam flow sealing 蒸汽排气真空封罐
steam flushing 蒸汽清洗
steam fog 蒸汽雾
steam for fire-extinguishing 灭火蒸汽
steam forging hammer 蒸汽锤
steam formation 蒸汽形成;蒸汽产生
steam friction loss 汽阻损耗
steam fumarole 蒸汽喷气孔
steam-gas 高过热气体;过热蒸汽
steam-gas cycle 蒸汽燃气联合循环
steam-gaseous mixture 汽气混合物
steam-gas mixture 蒸汽煤气混合气
steam gate valve 蒸汽闸阀;蒸汽平板阀;蒸汽滑门阀
steam ga(u)ge 蒸汽压力计;蒸汽表;气压计;气压表
steam ga(u)ge cock 气压计旋塞
steam ga(u)ge plate 气压计座
steam ga(u)ge stand 气压计座
steam-generating bank 蒸发管束
steam-generating equipment 蒸汽发生装置
steam-generating furnace 蒸汽锅炉
steam-generating heavy water reactor 蒸汽发生重水反应堆;产生蒸汽的重水反应堆
steam-generating plant 蒸汽发电装置;蒸汽发电厂;蒸汽发生装置
steam-generating reactor 蒸汽发生堆
steam-generating station 蒸汽发电厂
steam-generating unit 蒸汽发生器
steam generation 蒸汽发生;汽化
steam generation plant 蒸汽发生厂;产汽厂
steam generator 蒸汽锅炉;蒸汽发生器
steam generator blowdown system 蒸汽发生器排污系统
steam generator isolation and dump system 蒸汽发生器隔绝与排放系统
steam generator lead 主蒸汽管道
steam generator module 蒸汽发生器组件
steam gland 蒸汽用填密胀圈
steam grate shaker 汽力炉箅摇动机
steam gravity stamp 蒸汽重力捣碎机组
steam grid humidifier 蒸汽管网增湿器;蒸汽管网加湿器
steam hammer 捣锤;汽锤
steam hammer anvil 汽锤砧
steam hammer oil 回火油
steam hammer piling machine 汽锤打桩机
steam harden concrete block 蒸汽养护混凝土砌块
steam haulage engineer 蒸汽机车驾驶员
steam hauler 蒸汽绞车;蒸汽运输机;蒸汽起重机;蒸汽卷扬机;蒸汽吊机;蒸汽吊车
steam hauling machine 蒸汽绞车
steam header 蒸汽微锻机;蒸汽锻造机;(锅炉的)聚汽室;干汽包;蒸汽总管;蒸汽室;蒸汽联箱;蒸汽集器;集汽管;汽包蒸汽联箱;汽包;分汽缸
steam heat 蒸汽热量;(冷凝时放出的)蒸汽热;蒸汽加热
steam heat dryer 蒸汽烘燥机
steam heated 蒸汽加热的
steam-heated calorifier 蒸汽热风机
steam-heated circle 蒸汽加热圈
steam-heated concentrator 蒸汽加热浓缩器
steam-heated dryer 蒸汽加热式干燥机

steam-heated evapo(u)rator 蒸汽加热蒸发器
steam-heated mo(u)ld 蒸汽加热模
steam-heated oven 蒸汽干燥器
steam-heated pipe line 蒸汽加热管道
steam-heated plate dryer 蒸汽加热烘扳机
steam-heated rabble-type hearth drier 蒸汽加热耙膛式干燥机
steam-heated still 蒸汽加热蒸馏釜
steam-heated tempering coil 蒸汽调温蛇管
steam heated water heater 蒸汽热水器
steam heater 蒸汽散热器;蒸汽暖气器;蒸汽加热器;汽热器
steam heater tank car 蒸汽盘管保暖储[贮]油车
steam heating 蒸汽供热;蒸汽供暖;蒸汽采暖;蒸汽加热
steam-heating apparatus 蒸汽供暖装置;蒸汽供暖设备;蒸汽加热设备
steam-heating appliance 蒸汽供暖设备
steam-heating boiler 蒸汽(供暖)锅炉;供暖蒸汽锅炉
steam-heating by gravity 重力式蒸汽供暖
steam-heating chamber 蒸汽加温室
steam-heating coil 蒸汽加热盘管
steam-heating down-feed gravity system 重力上行下给式蒸汽供暖系统
steam-heating down-feed system 上行下给式蒸汽供暖系统
steam-heating of clay 泥料蒸汽加热处理
steam-heating pipe 蒸汽暖气管;暖气管;蒸汽加热管
steam-heating system 蒸汽供暖系统;蒸汽采暖系统;蒸汽加热系统
steam-heating system with dry return 干式回水蒸气供暖系统
steam-heating system with wet return 湿式回水蒸气供暖系统
steam-heating tank 蒸汽加热柜
steam-heating tracer 蒸汽伴随加热管道
steam-heating tube 蒸汽加热管
steam-heating up-feed system 下行上给式蒸汽供暖系统
steam-heating with gravity air removal 重力排气式蒸汽供暖
steam-heating with mechanical circulation 机械循环式蒸汽供暖
steam heat-supply network 蒸汽热网
steam heat-supply system 热水供热系统
steam hoist 蒸汽升降机;蒸汽起重机;蒸汽卷扬机;蒸汽吊机;蒸汽吊车;汽动吊车
steam homo-treatment 蒸汽均匀化处理
steam hose 蒸汽软管;蒸汽胶管;通汽软管;通气软管
steam hose connection 蒸汽软管接头
steam humidification 蒸汽加湿法
steam humidifier 蒸汽加湿器
steam hydraulic power gear 蒸汽液力传动
steam hydraulic press 蒸汽增压式水压机;蒸汽液压机
steam hydraulic shears 蒸汽液压剪切机
steam hydrocarbon reformer 烃蒸气转化炉
steam impulse 蒸汽脉冲
steam-in 蒸汽入口
steam indicator 蒸汽指示器
steam infusion 蒸汽注入
steaming 注入蒸汽;蒸热;蒸汽加工;蒸发作用;汽蒸;汽化;通入蒸汽
steaming and brushing machine 蒸刷机
steaming apparatus 蒸汽发生设备;蒸汽养护设备
steam(ing) chamber 蒸汽室;蒸汽养护室;汽蒸室
steaming coal 熏蒸煤
steaming cycle 蒸汽养护周期
steaming economizer 蒸发式经济器
steaming graded thin-layer chromatography 蒸汽梯度薄层层析法
steaming ground 冒汽地面
steaming ground type field 冒汽地面型热田
steaming ground type geothermal aquifer 冒汽地面型地热含水层
steaming head to sea 顶浪航行
steaming installation 蒸汽设备安装
steaming light 桅灯
steaming masthead light 航行桅灯
steaming of the rods 蒸汽清洗抽油杆
steaming of wood 木材蒸干
steaming operation 蒸汽通入操作
steaming out 吹汽
steaming out tank 吹汽槽
steaming oven 蒸汽锅
steaming pan 蒸锅
steaming part for permeation 高温渗透部
steaming pit 蒸汽坑
steaming plant 蒸汽设备;蒸煮装置;蒸热装置;蒸汽间
steaming point 蒸发点
steaming power 蒸汽生产额
steaming process 蒸汽法;汽蒸法
steaming radius 续航力;续航半径
steaming range 续航力;续航半径
steaming rate 蒸发率
steaming surface 蒸发表面
steaming time 蒸发时间
steaming treatment 蒸汽处理
steaming tube 蒸汽管
steaming with the sea on the bow 偏浪航行
steaming zone 冒汽地带
steam inhalation 蒸汽吸入
steam injection 蒸汽喷射
steam injection equipment 蒸汽喷射装置
steam injector 蒸汽喷射器;汽射器
steam injector type burner 喷汽式燃烧器
steam inlet 蒸汽入口;蒸汽进口;进汽口
steam inlet for snuffing 熄火蒸汽吹入管
steam inlet nipple 进汽管螺纹接套
steam inlet pipe 进气管
steam inlet plug 进气口塞
steam input 蒸汽输入量
steam intensifier 蒸汽增压器
steam iron 喷雾式电熨斗
steam iron generator 蒸汽铁屑生氢器;氢气发生器
steam-jacked kettle 蒸汽夹层锅;带蒸汽夹套煮锅
steam jacket 蒸汽套(管);蒸汽夹套;蒸汽加热套;汽套
steam jacketed 蒸汽套的
steam-jacketed bucket 蒸汽套桶
steam-jacketed mo(u)ld 蒸汽
steam-jacketed pipe 蒸汽套管子
steam jet 蒸汽射流;蒸汽喷嘴;喷汽器;喷汽孔道
steam-jet agitator 蒸汽喷射搅拌器;蒸汽搅动器;蒸汽搅拌器;喷汽搅拌槽
steam-jet air-conditioning system 蒸汽喷射空调系统
steam-jet air ejector 蒸汽喷射式空气抽逐器
steam-jet(air)pump 汽轮泵;蒸汽喷射式空气泵;蒸汽泵
steam-jet atomizer 蒸汽雾化器;气流喷雾
steam-jet attriter 蒸汽喷射磨细器
steam-jet blower 蒸汽喷嘴;喷射蒸汽器;喷汽鼓风机;喷汽送风机
steam-jet blower kiln 蒸汽喷射干燥窑
steam-jet burner 蒸汽雾化式燃烧器
steam-jet chiller 蒸汽喷射式冷冻机
steam-jet cleaner 蒸汽喷气清洁机;蒸汽清洁器
steam-jet cleaning 蒸汽喷气清洗
steam-jet compression 蒸汽喷射压缩
steam-jet cooler 蒸汽喷射冷却器
steam-jet cooling system 蒸汽(喷射)冷却系统;喷汽冷却系统
steam-jet cycle 蒸汽喷射循环
steam-jet drawing process 蒸汽喷射拉伸法
steam-jet ejector 射汽抽气器;射流抽气泵
steam-jet exhauster 排汽器
steam-jet heater 蒸汽喷射加热器
steam-jet hot water heating system 蒸汽喷射热水采暖系统
steam-jet humidifier 蒸汽加湿器
steam-jet injector 蒸汽喷射器
steam-jet kiln 喷蒸干燥窑
steam-jet method 蒸汽喷射法
steam-jet mixer 蒸汽喷射式混合器
steam-jet pump 蒸汽喷射泵;喷气泵
steam-jet refrigerating cycle 蒸汽喷射式制冷循环
steam-jet refrigerating equipment 蒸汽喷射制冷机;蒸汽制冷设备
steam-jet refrigerating machine 蒸汽喷射(式)制冷机
steam-jet refrigerating system 蒸汽喷射式制冷系统
steam-jet refrigeration 蒸汽喷射式制冷
steam-jet refrigerator 蒸汽喷射制冷机
steam-jet sand blast 蒸汽喷砂机
steam-jet siphon 蒸汽喷射管
steam-jet sprayer 蒸汽气流喷射装置;蒸汽喷雾器
steam-jet stoker 蒸汽鼓风层燃炉
steam-jet sucker 蒸汽喷射吸入器
steam-jet texturing 蒸汽喷射变形
steam-jet time 蒸汽注入时间
steam-jet unit 蒸汽喷射装置
steam-jet vacuum pump 蒸汽喷射真空泵
steam-jet water heater 喷汽式热水器
steam jib crane 蒸汽悬臂起重机
steam kettle 蒸汽壶;蒸汽锅
steam kiln 蒸汽养护窑
steam lance 蒸汽喷枪;蒸汽喷管;低压蒸汽除冰器;低压蒸汽除冰机
steam lap 蒸汽余面;排汽余面
steam launch 小火轮;汽艇
steam laundry 蒸汽洗衣房
steam leak-off valve 漏汽阀
steam lift 蒸汽提升器
steam lighter 蒸汽动力驳船
steam-limit curve 汽液界线;蒸汽极限曲线
steam line 蒸汽管路;蒸汽线路;蒸汽管道;供汽管
steam line smothering 蒸汽灭火管系
steam lock 汽封
steam lock(ing) 汽塞
steam loco crane 蒸汽起重机车
steam locomotive 蒸汽机车
steam locomotive crane 蒸汽起重机车
steam locomotive terminal 蒸汽机务段
steam logging winch 蒸汽伐木卷扬机;蒸汽伐木绞车
steam loop 蒸汽环管;蒸汽环道
steam lorry 蒸汽货车
steam loss 蒸汽损耗
steam lubrication 蒸汽润滑
steam main 主蒸汽管道;主蒸汽管;蒸汽总管;蒸汽母管
steam manifold 集汽箱;汽包;分汽缸
steam mass quality 蒸汽干度
steam mechanical wood pulp 蒸汽制木纸浆
steam meter 蒸汽流量计;量汽计
steam meter manometer 蒸汽流量表
steam micronizer 过热蒸汽扁平式气流粉碎机
steam mist 海上蒸汽雾;轻蒸汽雾
steam-mixer 蒸汽混合机;蒸汽搅拌机;蒸煮拌和器
steam moisture 蒸汽湿度
steam molecule 蒸汽分子
steam mooring winch 蒸汽系泊绞车
steam-motivated diamond drill 蒸汽驱动的金刚石钻机
steam motor car 蒸汽车
steam navvy 悬臂长柄挖斗掘土机(每斗0.4~60立方米);蒸汽挖掘机;蒸汽单斗挖土机;汽力掘凿机;汽铲;单斗挖土机
steam nebulization 蒸汽喷雾法
steam nozzle 蒸汽喷嘴
steam off 驶出
steam of thermometer 温度计液柱
steam oil heater 蒸汽式油加热器
steam oil magnetic valve 蒸汽燃油电磁阀
steam-operated 蒸汽驱动的
steam-operated drill 蒸汽驱动钻机
steam operation 蒸汽驱动;蒸汽操纵
steam or hot water inlet 蒸汽或热水入口
steam orifice 蒸汽口
steam out 蒸汽养护;蒸汽吹出;外接蒸汽
steam outlet 蒸汽输出;蒸汽出口
steam outlet valve 蒸汽出口阀;主蒸汽闸阀;主汽阀
steam out line 蒸汽灭火管系
steam output 蒸发量
steam out system 蒸汽灭火管系
steam oven 蒸汽灶;蒸汽锅
steam oxygen gasification 蒸汽氧气气化
steam packing 汽密封
steam parallel slide stop valve 蒸汽滑门阀;蒸汽闸阀
steam parameter 蒸汽参数
steam period 通蒸汽期
steam permeability 蒸汽渗透率
steam phosphating 蒸汽磷化处理法;热蒸汽磷化
steam pile driver 蒸汽打桩机;汽力打桩机;汽锤桩机;蒸汽打管机

steam pile driving plant 蒸汽打桩装置;蒸汽打桩机
steam pile hammer 蒸汽桩锤;蒸汽打桩锤;汽桩锤;打桩汽锤
steam piling winch 蒸汽打桩绞车
steam pipe 蒸汽管;汽缸
steam pipe coil 蒸汽盘(蛇)管;蒸汽旋管
steam pipe driver 蒸汽打管机
steam pipe expansion loop 蒸汽管膨胀圈
steam pipe isolating valve 汽管隔离保安阀
steam pipeline 蒸汽管道;蒸汽管线;加热蒸汽管道
steam pipe oven 蒸汽加热炉
steam pipe plug 汽管塞
steam pipe sleeve 汽管套
steam pipe stud 汽管柱螺栓
steam pipe support 汽管架
steam pipe swivel 汽管旋转接头
steam pipe T connection 丁字形汽管接头
steam pipe tee 汽管丁字管节
steam pipe tee head 蒸汽管丁字头
steam piping 蒸汽管道;蒸汽管
steam piping line 蒸汽汽管线
steam piston 蒸汽(机)活塞
steam piston ring 蒸汽活塞环
steam plant 蒸汽动力厂;蒸汽设备
steam plate 蒸汽板
steam platen press 蒸汽平压
steam plow 蒸汽机引犁
steam plunger pump 蒸汽圆柱式泵;蒸汽滑阀泵;蒸汽柱塞泵;活塞泵
steam pocket 汽囊;蒸汽包
steam point 水汽化点;沸点
steam pollution 蒸汽污染
steam pool 冒汽塘
steam port 汽口
steam power 蒸汽动力
steam powered 蒸汽驱动的
steam powered car 蒸汽动力汽车
steam powered motor car 蒸汽动力汽车
steam powered rotary engine 蒸汽转子发动机
steam power engineering 蒸汽动力工程
steam power machine 蒸汽机械
steam power plant 蒸汽发电厂;蒸汽动力设备;火力发电设备;火力发电厂;热电厂
steam power plant 蒸汽动力装置
steam power station 蒸汽发电站;火力发电站;火电站;火电厂;热电厂
steam press 蒸汽烫衣机
steam pressure 蒸汽压强;蒸汽压(力)
steam pressure automatic regulator 蒸汽压力自动调节器
steam pressure curve 蒸汽压力曲线
steam pressure detector 蒸汽压力指示器
steam pressure device 蒸汽压缩设备
steam pressure ga(u)ge 蒸汽压力表;蒸汽电力表;气压表
steam pressure reducer 蒸汽减压器
steam pressure-reducing valve 蒸汽减压阀
steam pressure test 气压试验
steam pressure type cascade heating system 蒸汽加压(串列)采暖系统
steam-proof 防蒸汽;不透气
steam pulsometer pump 蒸汽双缸泵
steam pump 蒸汽泵;汽泵;气泵
steam purge 蒸汽吹扫
steam purging 蒸汽扫气
steam purification 蒸汽净化
steam purifier 蒸汽清洗装置;蒸汽清洗器;蒸汽净化器
steam purity(value) 蒸汽纯度
steam quality 蒸汽质量;蒸汽品质;蒸汽干燥度;蒸汽参数
steam-rack dryer 蒸汽管架式干燥器
steam radiant heating 蒸汽辐射采暖
steam radiator 蒸汽散热器;蒸汽散热片;汽炉片
steam railcar 蒸汽轨道车
steam raising 蒸汽蒸发;气化蒸发;产生蒸汽
steam raising equipment 蒸汽发生器
steam raising plant 蒸汽发生装置
steam raising tower 蒸汽发生塔
steam raising unit 蒸汽发生器;蒸发器
steam ram 直接作用蒸汽汽缸;汽动撞头;汽动推钢机;汽动锤体
steam rate 蒸汽消耗率;耗汽率;汽耗率
steam rate guarantee 保证汽耗

steam receiver 容汽器;储汽室;储汽器
steam reciprocating engine 蒸汽往复机
steam recovery 蒸汽回收
steam recovery tower 蒸汽回收塔
steam reducing tee 汽管缩径丁字管节
steam reducing valve 减压汽阀;蒸汽减压阀
steam reduction 蒸汽还原(法)
steam-refined 蒸汽精制过的;蒸汽精炼过的;蒸汽精制
steam-refined asphalt 蒸馏沥青;蒸汽精制的沥青;蒸汽法精制(地)沥青
steam-refined cylinder oil 蒸汽提炼汽缸油
steam-refined residuum 蒸汽精制油
steam refining 蒸汽精制;蒸汽精炼;蒸汽蒸馏(法)
steam-reformer 蒸汽转化作用装置
steam-reforming 蒸汽改制;蒸汽转化
steam regulating valve 蒸汽调整阀;蒸汽调压阀
steam regulator 蒸汽调节器
steam reheater 蒸汽再热器
steam release valve 放汽阀
steam releasing surface 蒸发面
steam relieving area 蒸发面积
steam relieving capacity 蒸发量
steam-remedial treatment of lumber 木材蒸汽处理
steam requirement 蒸汽需用量
steam reserves 蒸汽储存;蒸汽备用量
steam reservoir 蒸汽热储;热储系统;储汽筒
steam return line 回汽管;蒸汽回管;蒸汽返回管道;冷凝蒸汽管线;回汽管路
steam reversing gear 蒸汽反向机构
steam rig 蒸汽驱动钻机
steam riser 蒸汽竖管
steam riveter 蒸汽铆钉枪
steam road roller 蒸汽压路机
steam roll 蒸汽加热辊
steam roller 蒸汽压路机;蒸汽路碾;蒸汽滚筒
steam room 蒸汽室;蒸汽浴室
steam run 蒸汽鼓风;蒸汽吹物
steam run gas 蒸汽吹出的气体
steam-sailing ship 张帆汽船
steam sand blaster 蒸汽喷砂机
steam sand blower 蒸汽喷砂器
steam sand ejector 蒸汽喷砂器
steam saturated 蒸汽饱和的
steam saw 汽力锯
steam schooner guy 吊长件货物索具
steam scrubber 蒸汽洗涤器
steam scrubbing 蒸汽清洗
steam scrubbing action 蒸汽清洗作用
steam scrubbing method 蒸汽清洗法
steam seal 蒸汽密封;汽封
steam sealing treatment 蒸汽封闭处理
steam seasoned 蒸汽加热处理
steam seasoning 蒸汽熏干;蒸汽干燥法(木材);蒸汽干燥;蒸汽干材法
steam security valve 蒸汽安全阀
steam seep 气眼
steam separator 蒸汽分离器;隔汽具;截液器;分汽器
steam set 蒸汽凝固;蒸汽固化;湿固化
steam-set ink 蒸汽凝固油墨
steam setwork 汽动摇尺机
steam shielded arc welding 水蒸气保护电弧焊
steam ship 蒸汽机船;汽船;轮船
steamship agent 轮船代理公司
steamship freight conference 运价公会
steam ship line 轮船公司;轮船航线
steam shock 蒸汽激波
steam shop 锅炉间
steam shovel 蒸汽挖土机;蒸汽挖掘机;蒸汽单斗挖土机;蒸汽铲;汽力铲;汽铲
steam shovel excavation 蒸汽铲土机开挖
steam silencer 蒸汽消声器
steam siren 蒸汽警报器;汽笛
steam slewing crane 蒸汽旋臂起重机
steam smothering line 蒸汽灭火管道
steam smothering 蒸汽灭火
steam smothering line 蒸汽灭火管系
steam smothering system 蒸汽灭火装置;蒸汽灭火系统
steam smothering system in machinery space 机舱蒸汽灭火系统
steam soaking 蒸汽清扫;蒸汽除灰

steam space 蒸汽空间
steam spray 蒸汽喷雾
steam sprayer 蒸汽喷嘴;蒸汽雾化器
steam spraying 蒸汽喷涂
steam spraying gun 蒸汽喷枪
steam spraying process method 蒸汽喷涂磷酸盐处理法
steam stamp 蒸汽冲压机
steam stamp battery 蒸汽捣碎机组
steam state 汽态
steam steering engine 蒸汽操舵机
steam steering gear 蒸汽转向装置;蒸汽操舵机
steam sterilization 蒸汽消毒;蒸汽灭菌(法)
steam sterilized soil 蒸汽消毒土壤
steam sterilizer 蒸汽消毒器;蒸汽灭菌器
steam sterilizing 蒸汽灭菌(法)
steam still 蒸汽蒸馏器
steam stirring 蒸汽搅拌
steam stop 蒸汽截止阀
steam stop cock 停汽旋塞
steam stop valve 蒸汽停汽阀;蒸汽节流阀;蒸汽止动阀;蒸汽断流阀;停汽阀
steam strainer 蒸汽滤气器;蒸汽滤净器;蒸汽滤器;滤汽器
steam strainer valve 滤汽阀
steam stream 蒸汽气流
steam stripper 蒸汽脱漆器
steam stripping 蒸汽脱除;汽提
steam-stripping appliance 袖珍式喷蒸汽设备(贴墙纸用)
steam stuffing box 汽缸填密函
steam suction dredge(r) 蒸汽吸浆疏浚机;蒸汽吸浆疏浚船
steam superheater 蒸汽过热器
steam superheating 蒸汽过热
steam supply 蒸汽管道;蒸汽供应;供汽
steam supply and power generation 热电站
steam supply pipe 供汽管
steam supply(pipe)line 供汽管道;蒸汽供应线
steam supply valve 供汽阀
steam switcher 蒸汽调车机车;蒸汽转换开关
steam system 蒸汽系统
steam table 蒸汽表
steam tail piece 进汽部尾管
steam temperature 蒸汽温度
steam temperature rise 蒸汽温度升高
steam tempering 蒸汽回火
steam tensiometer 蒸汽压力仪
steam tension 蒸汽张力;蒸汽压(力)
steam-tent 蒸汽帐
steam test 蒸汽试验
steam thawing 蒸汽解冻
steam throttling 蒸汽节流
steam tight 蒸汽密封;汽密(的);不透蒸汽的
steam tight globe 汽密灯罩
steam tight joint 汽密接合
steam tightness 汽密性
steam tight test 汽密性试验
steam-to-steam heat exchanger 汽汽热交换器
steam trace 蒸汽伴随加热;蒸汽伴热管;加热蒸汽管道
steam traced product 蒸汽保温加热的产品
steam trace heating 蒸汽式伴随加热法
steam tracing 蒸汽保温加热;伴热蒸气管
steam tracing line 蒸汽伴随加热;蒸汽伴热管路
steam traction 蒸汽牵引
steam traction railway 蒸汽牵引铁路
steam transformer 蒸汽交换器;蒸汽发生器
steam transmission 蒸汽渗透
steam trap 冷凝水排除器;回水盒;阻汽器;阻汽排水器;蒸汽疏水器;隔汽具;冷凝罐;凝水排除器;汽水阀;汽阱;疏水器;疏水阀
steam trap valve 滤汽阀
steam trawler 渔轮;蒸汽动力拖网渔船
steam treated timber 蒸汽处理木材
steam treatment of soil 土壤蒸气处理
steam trial 蒸汽动力装置试验
steam tube 蒸汽管
steam-tube bundle 蒸汽管束
steam-tube dryer 蒸汽管干燥器
steam-tube rotary dryer 蒸汽管旋转干燥器
steam tubing 蒸汽管
steam tug(boat) 蒸汽拖轮;蒸汽(机)拖轮;拖轮
steam tumble 汽蒸圆筒

steam tunnel 汽管隧道
steam turbine 蒸汽涡轮(机);蒸汽透平;蒸汽汽轮机;蒸汽轮机;汽轮机
steam turbine automatic remote control system 汽轮机自动遥控系统
steam turbine condenser 蒸汽透平冷凝器;蒸汽透平冷凝器
steam turbine-driven alternator 汽轮发电机
steam turbine-driven windlass 汽轮机驱动起锚机
steam-turbine-electric(al) drive 汽轮机电气驱动
steam-turbine for power plant 电站汽轮机
steam-turbine generating set 蒸汽汽轮机发电机组;蒸汽透平发电机组
steam-turbine generator 汽轮发电机
steam-turbine lead 汽轮机蒸汽管道
steam-turbine lubricating system 汽轮机润滑系统
steam-turbine lubrication 汽轮机润滑
steam-turbine oil 汽轮机油
steam-turbine parts 汽轮机零件
steam-turbine plant 汽轮机装置
steam-turbine power station 蒸汽汽轮机发电厂;蒸汽汽轮机发电站;蒸汽透平发电厂;蒸汽透平发电站
steam-turbine room 汽轮机舱
steam turbo-alternator 汽轮发电机;汽轮交流发电机组
steam turbo-compressor 汽轮压缩机
steam turbo-generator 汽轮发电机
steam turbo-generator set 汽轮发电机组
steam turboset 汽轮发电机组
steam type dust removal 蒸汽除尘
steam under pressure 加压蒸气
steam union 汽管联管节
steam union gasket 汽管дЫ垫密片
steam unit ventilator 蒸汽暖风机组
steam universal excavator 蒸汽万能挖土机
steam vacuum pump 蒸汽真空泵;真空式汽泵
steam vacuum system 真空式蒸汽系统
steam valve 蒸汽阀
steam valve center piece 汽阀中心件
steam valve controller 汽阀控制器
steam valve disk lock nut 汽阀盘防松螺母
steam valve gland 汽阀压盖
steam valve packing 汽阀填密
steam valve packing gland 汽阀填密压盖
steam valve spindle 汽阀轴
steam valve stem 汽阀杆
steam vapo(u)r 水蒸气
steam vapo(u)r curing chamber 蒸汽养护室(混凝土)
steam vapo(u)rizer 蒸汽蒸发器
steam vapo(u)r system 微压式蒸汽系统;低压蒸气系统
steam velocity 蒸汽速度
steam vent 蒸汽排气口
steam vessel 轮船;汽船
steam void 蒸汽空腔;汽穴;汽泡
steam vulcanization 蒸汽硫化
steam vulcanizer 蒸汽硫化器;蒸汽补胎机
steam washer 蒸汽洗涤器;蒸汽清洗设备;蒸汽洗器
steam washing 蒸汽洗舱;蒸汽清洗
steam washing unit 蒸汽清洗设备
steam waste pipe 残汽排出管
steam water heater 蒸汽加热的热水器
steam-water heater exchanger 汽-水换热器
steam-water mixture 汽水混合物
steam-water separation 汽水分离
steam-water separator 汽水分离器
steam-water shock 汽水冲击
steam-water system 蒸汽水系统
steam-water type heat exchanger 汽水换热器
steam-water type mixed heat exchanger 汽水混合式换热器
steam way 蒸汽道;蒸汽槽
steam well 蒸汽井
steam whistle 汽笛
steam winch 蒸汽起货机;蒸汽绞车
steam windlass 蒸汽(起)锚机
steam yacht 汽艇
steamy place 漏汽点
stean 古陶罐
Steane house 斯提奈房屋(一种商品装配式房屋)
steaning 孔壁砌内衬;井壁砌内衬

stearaldehyde 硬脂醛
stearamide 硬脂酰胺
stearate 硬脂酸酯;硬脂酸盐;硬脂酸化物
stearate chromium complex 硬脂酸铬络化物
stearatochromic chloride 硬脂酸氯化铬
stearic acid 硬脂酸;十八酸;二压硬脂酸
stearic acid film 硬脂酸膜
stearic aldehyde 硬脂醛
stearic amide 硬脂酰胺
stearin 硬脂酸;硬脂精;硬脂;三硬脂酸甘油酯
stearin(e)-cored solder wire 硬脂芯焊丝;硬脂芯焊条
stearin(e) oil 硬脂油
stearin(e) pitch 硬脂沥青
stearinery 硬脂制造业
stearo dipalmitin 二棕榈酸硬脂酸甘油酯
stearol 油脂剂
stearolactone 硬脂内酯
stearolic acid 硬脂炔酸
stearone 硬脂酮
stearonitrile 硬脂腈
stearoptene 玫瑰蜡
stearrhea flavescens 黄色皮脂溢
stearyl alcohol 硬脂醇
steatite 块滑石;滑石;冻石
steatite bobbin 块滑石线圈架
steatite ceramics 块滑石陶瓷;滑石瓷
steatite porcelain 块滑石陶瓷;滑石瓷
steatogenous 产生脂肪的
Steckel mill 可逆式炉卷轧机;斯特克尔式轧机
steckel rolling 斯特克尔轧制法
S-tectonite S 岩组构造;S 构造岩
Stedman packing 金属网规则填料;斯特曼填料
steel 钢制(的);钢铁
steel abutment 钢台座
steel accommodation ladder 钢质舷梯
steel adhesive 钢铁黏合胶
steel-alkaline battery 碱性镍铁蓄电池
steel alloy 钢合金;合金钢
steel-alumin(i)um conductor 钢铝线
steel alumin(i)um trolley wire 钢铝电车线
steel analyzer 钢含量分析仪
steel anchor 钢锚件
steel anchor bar 锚固钢筋
steel anchor plate 钢锚板
steel and alumin(i)um trolley wire 钢铝电车线
steel and angle column 钢板角钢组合柱
steel and iron slag cement 钢渣水泥
steel-and-reinforced concrete 钢与钢筋混凝土
steel-and-reinforced concrete structure 钢与钢筋混凝土混合结构
steel angle 角钢
steel angle lintel 角钢过梁
steel angle stanchion 角钢支柱
steel anvil 钢砧
steel apron conveyer 钢的皮带运输机;钢的板式运输机;钢的裙式运输机
steel arch 钢架拱;钢拱(架)
steel arch bridge 钢拱桥
steel arch centering 拱式钢拱架
steel arched girder 钢拱大梁
steel arch(ed) girder bridge 钢拱形板梁桥
steel arched support 拱顶纵梁
steel arched timbering 钢拱支撑
steel arch-gate 弧形钢闸门
steel architecture 钢建筑
steel arch post 钢拱支柱
steel arch support 钢拱支架;钢拱支撑
steel area 钢筋面积;钢筋截面面积
steel area grating 钢窨井算盖
steel area ratio 配筋率
steel armo(u)r 钢铠;包铁皮件;包钢皮件
steel armo(u)red 包钢;铠装的
steel armo(u)red conduit 铠装的管线
steel article 钢制件
steel ash door 出灰铁门
steel back brake shoe 钢背闸瓦
steel-backed 钢背的
steel-backed bearing 钢背轴承
steel-backed bearing alloy 钢背轴承合金
steel-backed bearing metal 钢背轴承合金
steel-backed bronze-faced washer 钢背铜面垫圈
steel backing 钢背
steel backing plate 钢背板

steel baling strap 打包铁皮
steel ball 钢珠;钢球;滚珠
steel ballast 钢平衡器;钢压块
steel ball coal mill 钢球磨煤机
steel ball indent 钢球痕
steel ball lapping machine 钢球研磨机
steel ball lost 掉弹子
steel ball polishing machine 钢球抛光机
steel ball tube mill 钢球磨管机
steel ball-type valve 钢珠活门
steel baluster 钢栏杆小柱
steel band 钢籀带;钢带
steel band butt welder 钢带对焊机
steel-band conveyer 钢带运输机;钢带输送机
steel band tape 钢卷尺
steel bar 钢条;钢筋;钢顶梁;钢材;钢棒;金属条
steel barbed wire 有刺钢丝
steel(bar)bender 钢筋弯曲机
steel bar button head forging 钢筋冷墩
steel butt welding 钢筋冷焊
steel butt welding machine 钢筋对焊机
steel bar cold draw 冷拔钢筋
steel bar cold drawing machine 钢筋冷拉机
steel bar cold-extruding machine 钢筋冷拔机
steel bar cold flattening 冷轧钢筋
steel bar cold-press connecting 钢筋冷压连接
steel bar cold working 钢筋冷加工
steel bar dieing-drawing machine 钢筋冷拔机
steel bar flux press welding 钢筋电渣压力焊
steel bar gas-press welding 钢筋气压焊
steel bar gas-press welding machine 钢筋气压焊机
steel bar header 钢筋镦头机
steel bar heading press machine 钢筋冷镦机
steel bar mechanical connecting 钢筋机械连接
steel bar reinforcement 钢筋
steel bar shearing machine 钢筋切断机
steel bar shears 钢筋切断机
steel bar straightening and shearing machine 钢筋调直切断机;钢筋拉直切断机
steel bar welding 钢筋焊接
steel bascule bridge 钢竖旋开启桥
steel-based plate 钢基印版
steel basement window 地下室钢窗
steel baseplate 钢座板
steel basin 钢盆
steel bath tub 钢浴盆
steel bay 钢材堆场
steel beaker 钢制烧杯
steel beam 钢梁;钢顶梁
steel beam centering 梁式钢拱架
steel beam enclosed in concrete 混凝土包钢梁;外包混凝土钢梁
steel beam for crane 吊车钢轨
steel beam grillage 钢梁格排;钢梁格栅
steel beam trammel 钢杆规
steel bearer 钢支承物
steel bearing 钢支座
steel bearing skeleton 钢支承框架;钢支承骨架;承重钢骨架
steel bearing structure 钢支承结构
steel bed 钢床
steel bed plate 钢基础底板;钢机座
steel beetle 木槌;木夯;大锤;钢锤
steel bell and spigot pipe 钢承插管
steel bellow 钢质波纹管
steel bell ring 钢珠承口环
steel belt 钢带;传动钢带
steel belt conveyer 钢丝绳芯胶带输送机;钢带输送机
steel belted piston 钢带轻合金活塞
steel belt lacing 钢带接头
steel belt pulley 钢皮带(滑)轮
steel bender 弯钢筋工具;钢筋弯具;钢筋工;弯钢筋机
steel bending 钢材弯曲
steel bending machine 弯钢筋机;钢材弯曲机
steel bending yard 钢筋弯曲工场;钢筋弯场;钢筋工场
steel bent bridge 钢排架桥
steel bin 钢料斗
steel binder 钢材黏合剂;钢系杆
steel binding beam 钢系梁
steel binding joist 钢系住托架;钢系住格栅

steel biological shielding wall 钢生物屏蔽墙
steel bit 钢刀片;钢刀刃;钢钻头;空白钻头;未镶钻头
steel blank 钻头钢坯;空白钻头
steel block 钢滑车;钢制滑车;铁滑车
steel bloom 钢锭;大钢坯
steel blue 钢青色;钢蓝色;钢蓝
steel boat 钢船
steel bobwire 钢悬锤线
steel body 钢质车身
steel boiler 钢板锅炉
steel boiler plate 锅炉钢板
steel bomb 钢制反应釜;钢弹
steel bond 铁粉结合剂
steel-bonded carbide 钢结硬质合金
steel-bonded carbide products 钢结硬质合金产品
steel-bonded carbide tool 钢结硬质合金工具
steel-bonded hard alloy 钢结硬质合金
steel book shelf 钢书架
steel book stacks 钢书架
steel bottle 钢瓶
steel bottle screw 伸缩螺丝
steel bottom plate 钢底盘
steel bottom rim 钢底箍
steel bowl 钢辊筒;钢杯
steel box 钢箱
steel box beam bridge 钢箱梁桥
steel box breakwater 钢箱防波堤
steel box caisson 钢箱沉箱
steel box car 钢棚车
steel box girder 箱形钢梁;钢箱形梁
steel box girder bridge 钢箱梁桥
steel box girder bridge with orthotropic deck 正交各向异性板面钢箱梁桥
steel box king pile 钢箱防冲桩
steel box pile 箱式钢板桩;箱形钢桩;箱式钢桩;盒形钢桩
steel bracket 钢牛腿;钢支(承)架;钢托架
steel brad 钢角钉;钢无头钉
steel breakage 钎子折断
steel bridge 钢桥
steel bridge building 钢桥建筑
steel bridge construction 钢桥施工;钢桥工程
steel bridge deck(ing) 钢桥桥面板
steel bridge pier 钢桥墩
steel bridge truss 钢桥桁架
steel bronze 钢性青铜;钢青铜
steel brush 钢刷;钢丝刷
steel building 钢建筑物
steel building block module 钢建筑块体模数
steel building construction 钢房屋建筑
steel building design and construction 钢房屋设计与施工
steel bulkhead 钢闷头;钢隔框;钢隔板;钢板桩驳岸壁;钢板舱壁;钢板驳岸;隔仓;(管的)钢盖头;钢隔壁
steel bur 钢牙钻
steel bushing 钢套
steel butt hinge 钢平接铰链
steel button 钢钮
steel cable 钢索;钢丝绳;钢丝缆;钢绳;钢缆
steel cable baffle ring 钢丝挡圈
steel cable belt 钢丝绳胶皮带
steel cable clamp 钢丝绳夹
steel cable conveyor belt 钢缆输送机皮带
steel cable hoist 钢丝绳式电动葫芦
steel cable roof 钢缆屋顶
steel cable terminal 钢丝绳接头
steel cable yarn 钢丝芯纱
steel cage 钢罩
steel cage construction 钢骨架结构;钢骨架构造
steel caisson 钢船坞闸门;钢藻井;钢沉箱
steel caisson breakwater 钢沉箱防波堤;钢沉箱基础;钢板沉箱防波堤
steel calendar 钢丝压延机
steel can 钢罐
steel cantilever bridge 钢悬臂桥
steel cantilever dam 悬臂式钢坝
steel cantilever truss bridge 钢悬臂桁架桥
steel cap 钢顶梁
steel capacity 炼钢生产能力
steel cap and plastic barrel 钢套塑杆
steel capsule 钢包套
steel casement 钢窗扉;钢窗框;钢窗
steel casement door 钢门

steel casing 钢套管
steel cassette 钢沉箱;钢箱
steel-cast 钢铸的;钢铸的
steel-casting 铸钢;钢铸件;钢铸件
steel-casting foundry 铸钢厂;铸钢车间
steel catch 钢制窗钩;钢制门扣;钢制锁键;钢制轮挡;钢制抓取器;钢制动片
steel ceiling joist 钢顶棚格栅
steel cell 钢壳电解槽
steel cellar window 地下室钢窗
steel cellular bulkhead 钢板桩格型围堰;钢板桩格型岸壁;钢板桩格体围堰;钢板桩格体岸壁;钢板格型岸壁
steel cellular unit 钢分格式单元
steel centering 钢拱架
steel centralizer 钢钎定位器
steel chamber 钢制容器;钢舱室;钢房间
steel change 换杆长度;更换钻头;更换钎头;各种不同长度的钻杆
steel channel 槽钢
steel channel beam 槽钢梁
steel channel column 槽钢柱
steel channel(l)ing 槽钢
steel channel purlin(e) 槽钢檩条
steel channel section 槽钢断面
steel checkered plate cover 钢制格纹管井盖
steel chimney 钢烟囱
steel chimney shaft 钢烟囱通风井;钢烟囱柱身
steel chippings 钢切屑
steel chippings concrete 钢屑混凝土
steel chippings mortar 钢屑水泥砂浆
steel chisel 钢钎;钢扁凿
steel church 钢结构教堂
steel-clad 钢片衬;铁甲;包钢(皮)的;包钢;装甲的
steel-clad brick 钢皮砖;铁皮砖
steel-cladding 钢覆盖;钢覆面
steel-clad rope 钢甲绳
steel-clad switchbox 铠装开关箱
steel-clad wire rope 包钢钢丝绳
steel clamp 钢夹子
steel clamping plate 钢夹板
steel cleat 固定钢楔
steel clip 钢夹
steel cloth sheet 钢丝布
steel coach screw 钢制方头木螺钉
steel coffer 钢沉箱;钢平顶镶板;钢浮船坞;钢围堰
steel cofferdam 钢围堰
steel collar 钎肩
steel collector bar 钢导电棒
steel column 钢柱
steel column cased in concrete 包混凝土钢柱
steel column protection 钢柱保护
steel comb 钢刷;钢梳
steel complex 钢铁联合企业
steel composing stick 钢手托
steel composite construction 钢筋混凝土组合结构;劲性钢筋混凝土结构
steel compression ring 钢压力环
steel compressive stress 钢的压缩应力
steel-concrete 钢筋混凝土
steel-concrete composite beam 钢—混凝土组合梁
steel-concrete composite girder 钢筋混凝土合成梁;钢筋混凝土组合梁;钢材混凝土组合梁
steel-concrete composite slab 钢—混凝土组合板
steel-concrete mark 钢筋混凝土标石
steel conditioning 钢的表面修整;钢表面修整
steel conduit 钢明沟;钢管;钢导管;布线钢管
steel cone concrete column 钢筋混凝土柱
steel construction 钢结构;钢架厂房
steel construction engineering 钢结构工程
steel construction section 钢构件断面
steel construction type 钢结构类型
steel construction work 钢建筑工程
steel consumption 钢材耗用量;钢材消耗
steel container 钢制集装箱
steel contractor's shop 钢结构承包人的工场
steel conveyor belt 钢质运输带
steel cooling coil 钢制冷却蛇管
steel cord conveyor belt 钢丝绳输送带;钢绳运输带
steel cord for tyre 轮胎细钢丝绳
steel cord tire 钢丝织轮胎
steel core 钢芯;钢心
steel core belt 钢丝(绳)芯胶带

steel core column 钢芯混凝土柱
steel core concrete column 钢芯混凝土柱;钢管内填混凝土柱;钢管混凝土柱
steel core concrete pile 钢管内填混凝土桩
steel-cored 钢芯的
steel-cored alumin(i)um 铝线绕钢心电缆
steel-cored alumin(i)um cable 钢芯铝线
steel-cored alumin(i)um strand(wire) 钢芯铝绞线
steel-cored alumin(i)um twisted wire 钢芯铝绞线
steel-cored alumin(i)um wire 钢芯铝线
steel-cored belt conveyer 钢绳芯带式输送机
steel-cored copper conductor 钢芯铜线
steel-cored copper wire 钢芯铜线
steel-cored wire rope 钢芯钢丝绳
steel core rubber belt 钢缆芯胶带
steel-corroding 钢材腐蚀
steel corrosion 钢锈蚀;钢的腐蚀
steel corrosion damage of concrete 混凝土的钢筋锈蚀破坏
steel cotter pin 钢扁销
steel counterbalance weight box 钢平衡重箱
steel creep 钢的徐变
steel crib 钢棚架
steel cross-section 钢筋横截面
steel crucible 钢坩埚
steel culvert 钢涵管
steel cup 钢杯
steel cupola 钢圆顶;钢穹顶
steel current conducting plate 导电钢板
steel curtain wall 钢幕墙
steel cutting 钢材剪切
steel cutting and bending 钢材切断与弯曲;钢筋切断与弯曲
steel cutting curb 沉箱的钢靴;(钢井筒或套管底部的)切割边缘
steel cutting equipment 钢材切割设备
steel cuttings 钢屑
steel cutting shield 钢刃盾构
steel cycle stand 钢自行车架
steel cylinder 钢汽缸;钢筒;钢瓶;钢管柱
steel cylinder pier 钢管桥墩
steel cylindric(al) column 钢管柱
steel cylpebs 钢段
steel dam 钢坝
steel deck bridge 钢上承桥
steel deck girder dam 钢面板式梁坝;钢板梁坝
steel deck(ing) 钢桥梁车行道;钢板面板;钢甲板;钢板屋面;钢板层
steel deck pavement 钢甲板铺面;钢桥面铺面
steel deck plate 钢桥面板
steel deck roof 钢平台式屋顶
steel deck surfacing 钢桥甲板面层;钢桥面面层
steel deck truss bridge 钢上承式桁架桥
steel deck unit 钢平台式单元
steel defect 钢材疵病;钢材缺点;钢材缺陷
steel degree of drill pipe 钻杆钢级
steel derrick 钢制桅杆起重机;钢制人字起重机;钢制塔式起重机;钢制转臂起重机;钢塔架
steel design 钢结构设计
steel diameter 钎杆直径
steel diaphragm (坝内的)钢心墙;钢膜片
steel die 压锻模;压铸模;钢型;钢模具;钢模(板)
steel-die engraving press 钢凹版印刷机
steel disc 钢片
steel disc wheel 钢碟形轮;钢盘轮
steel distributor 钢传墨辊
steel dog 蚂蟥钉;钢索钩;扒钉;闸门钢轧头;钢索钩
steel dolphin 钢系船浮标;钢系船栈;钢护墩桩;钢簇桩
steel dome 钢圆屋顶;钢圆拱顶
steel door 钢门
steel door frame 钢门框
steel double bolt pipe clamp 钢制双螺栓管卡
steel dowel 钢销杆;钢榫;(水泥混凝土路面的)钢传力杆;(钢筋混凝土上的)连接筋;插筋
steel dowel pin 钢榫钉
steel drag 钢板路刮;钢板刮器
steel drain(age) pipe 排水钢管
steel drain water pipe 排水钢管
steel drawings 钢筋图;钢结构图
steel drift 钢冲(机)
steel drill bit 钢钻头
steel drill pipe 钢钻杆
steel drill rod 钢质钻杆

steel drill screw 钻孔钢螺钉;自攻螺钉
steel driving pile 冲击打入的钢桩
steel drop shaft 钢开口沉箱
steel drum 钢桶;铁桶
steel drum curb 沉箱的钢靴;(钢井筒或套管底部的)切割边缘
steel dryer 钢制晾衣架
steel dump car 钢制动(自动)卸料车
steel dump sheet 卸料钢板
steel edged rubber water-stop 钢边橡胶止水带
steel edge processing 钢材边缘加工
steel electrode 钢焊条
steel electrode for low-temperature service 低温钢焊条
steel emery 钢砂
steel-encased beam 劲性钢筋混凝土梁
steel engineering hydraulics 水工钢结构
steel engine parts 发动机中的钢部件
steel engraver 钢板雕刻机
steel engraving 钢雕;钢板雕刻;钢凹板
steel eraser 刮墨水字迹刮刀
steel erecting 钢结构安装
steel erecting crane 钢制装配起重机;钢制装配吊车
steel erection 钢结构架设;钢结构安装
steel erection hoist 钢制装配卷扬机;钢制装配起重机
steel erector 钢结构安装工;钢件安装工;钢架装置机;钢结构安装机
steel erector's labo(u)rer 钢结构安装工人
steel exit gas pipe 煤气出口钢管
steel expansion dowel (水泥混凝土路面)伸缩缝钢膨胀传力杆;钢膨胀销杆;钢膨胀暗榫
steel extension machine 钢筋拉伸机
steel fabric 钢筋网
steel fabricated 钢加工的
steel fabric mat 钢丝网;钢筋网;钢丝网片
steel fabric reinforcement 钢织物加劲件;钢织物加强件;钢丝网(片);网状钢筋
steel fabric sheet 钢丝布;钢筋网;钢丝网片
steel facade 钢房屋正面
steel-faced roller 钢护套滚筒;钢外壳滚筒;固定滚筒;钢外壳的碾压机;固定的碾压机
steel-faced three-wheel roller 钢护套三轮滚筒;钢固定的三轮滚筒;钢固定的三轮碾压机;钢外壳的三轮碾压机
steel facing 钢护套;钢外壳;钢面板;钢棉面板;钢板护面;表面钢化
steel fastener 钢紧固件;钢固定件
steel fastening 钢紧固件
steel feed 钢坯
steel fence picket 围篱钢桩
steel fence post 围篱钢柱
steel fence stake 钢栅栏杆
steel fencing post 钢栅栏柱
steel fiber 钢纤维
steel fiber cement product 钢纤维水泥制品
steel fiber concrete 钢纤维混凝土;钢丝混凝土
steel fiber content 钢纤维含量
steel fiber reinforced concrete 钢纤维束加筋混凝土;钢纤维增强混凝土;钢纤维混凝土;钢筋网混凝土
steel fiber reinforced sulphur impregnated mortar 钢纤维增强浸渍砂浆
steel fiber shotcrete 钢纤维喷射混凝土
steel fiber shotcrete lining 钢纤维喷射混凝土衬砌
steel fibrous shotcrete 钢纤维喷射混凝土
steel file 钢锉
steel file cabinet 钢制卷宗柜
steel fillings 钢锯屑;钢锉屑
steel film selection 钢带式选针
steel filter 钢板过滤器
steel filter pipe 过滤器钢管
steel finger 钢梭子
steel finish 钢表面处理
steel fire-screen door 钢质挡火网门
steel fitting 钢器材;钢套筒;钢接头;钢连接件;钢零部件;钢制接头
steel fitting for windowdoor 钢门窗配件
steel fittings 钢制接头配件
steel fixer 钢筋工;钢筋定位器
steel fixer's pliers 钢筋绑扎钩
steel-fixing 绑扎钢筋作业
steel fixing accessory 钢筋定位附件
steel fixing device 钢装配设备

steel fixing means 钢装配方法
steel flagpole 钢旗杆
steel flange 钢翼缘;钢法兰
steel flat 钢箍铁(条);扁钢
steel flat roof truss 钢屋顶桁架;钢平顶桁架
steel flat slab 扁钢;钢平板
steel flat wire 扁钢丝
steelflex coupling 钢栅挠性联轴节
steel floating dock 钢制浮船坞
steel floor 钢楼板
steel floor cover(ing) 钢楼板覆面;钢楼板覆盖物
steel floor element 钢楼板部件
steel floor girder 钢楼板大梁
steel floor(ing) 钢楼板面层;钢桥面
steel floor(ing) finish 钢地面修饰;钢楼面修饰
steel floor(ing) plate 钢地面板;钢楼面板;楼面用板
steel floor tile 楼面用板
steel foil 钢箔
steel folded plate roof 钢折板屋顶
steel folded slab roof 钢折板屋顶
steel folding door 钢折页门
steel folding rule 钢折尺;钢皮折尺
steel for building construction 房屋建筑用钢
steel for construction 建筑用钢(材)
steel for construction(al) requirement 构造钢筋
steel forging 锻钢
steel forgings 钢锻品
steel fork 钢叉
steel fork head 钢叉头
steel fork with handle 带把钢叉
steel form 钢模(板)
steel form jumbo 钢模
steel form panel 钢制定型模板
steel form mo(u)lding 钢模成型
steel formwork 钢铁模板;钢(质)模板
steel formwork cleaning machine 钢模板清洁机械
steel for special purposes 特殊用钢
steel for sucker rods 制造泵轴的钢材
steel foundation pile 钢基桩
steel foundry 铸钢车间
steel frame 钢门窗框;钢(框)架;钢框格
steel frame concrete 钢架混凝土
steel frame construction 钢框架构造;钢框架结构
steel framed 钢架的
steel frame dam 钢架坝
steel-framed block 钢框架建筑;钢框架房屋
steel-framed building 钢框架建筑;钢框架房屋
steel-framed house 钢框架房屋
steel-framed modular construction 钢框架模数化建筑
steel-framed multi-stor(e)y carpark 钢框架多层车库
steel-framed office building 钢框架办公大楼
steel-framed portal building 钢框架门式建筑
steel-framed reinforced concrete 钢框架钢筋混凝土
steel-framed reinforced concrete column 钢框架钢筋混凝土柱;钢骨钢筋混凝土柱
steel-framed reinforced concrete construction 钢框架钢筋混凝土结构
steel-framed scaffolding 钢构架脚手架
steel-framed structure 钢架结构
steel-framed wall 钢构架墙
steel-framed window 钢构架窗;钢结构窗
steel-framed woodlined silo 钢架木衬青储[贮]塔
steel frame mill building 钢框架厂房;钢架厂房
steel frame mill plant 钢架厂房
steel frame structure 钢框架构造;钢框架结构;钢框格
steel framework 钢框架工程;钢框架;钢构架
steel framework building 钢框架建筑
steel framework weir 钢结构堰
steel framework with riveted joints 铆接节点的钢结构
steel framing 钢构架;钢框格
steel framing plan 钢框架平面图
steel furnace slag cement 钢渣水泥
steel furniture 钢家具
steel gabion 钢丝石笼;铁丝石筐
steel gang 换轨队;铺轨工班
steel garage 钢车库
steel garage door 车间钢门
steel gate 钢制大门;钢闸门
steel girder 钢主梁;钢梁;大梁

steel girder bridge 钢大梁桥
steel girder floor 钢大梁楼板;钢大梁楼层
steel girder joint 钢大梁接头
steel girder span 钢大梁跨度
steel-glass construction 玻璃钢建筑
steel-glass structural system 玻璃钢结构体系
steel glass system 钢玻璃结构体系
steel grade 钢种;钢号;钢的等级牌号;钢的等级
steel grade and type 钢号及钢种
steel grade character 钢级代号
steel grader 钢材分选装置
steel grain 钢粒
steel grating 阴沟盖板;铁格子板
steel gray 钢青色;钢灰色;青灰色;铁灰色;灰青色
steel grey 铁灰色;钢青色;青灰色
steel grid 钢网格;钢筋网格;钢筋格排
steel grid decking 钢格式地板
steel grid floor 钢网格楼面;钢格式地板;钢格楼层
steel grid footing 钢网格基础;钢格排基础;钢格基础
steel grid system 钢栅格系统
steel grillage 钢筋网格
steel grillage footing 钢网格基础;钢格排基础
steel grillage foundation 钢格排基础
steel grille 钢格子窗;钢格栅;钢窗栅
steel grinding rod 钢磨棒
steel grit 硬砂砾;钢丸;钢砂
steel-grit blasting system 钢喷砂处理系统;钢喷砂清除系统
steel grit concrete 钢砂混凝土
steel guard fence 钢保护网
steel guardrail 钢护轨;钢(板)护栏
steel guide 钢罐道;金属罐道
steel guideway 钢制导轨
steel gusset plate 节点钢板
steel guy wire 钢拉线
steel handling 装卸钎杆
steel hangar shutter door 飞机库钢卷帘式门
steel hanger 钢吊架
steel hardening oil 钢材淬火油;钎子淬火油
steel hatch cover 钢质舱口盖
steel hawser 钢索;钢丝缆索;钢缆
steel H bearing pile H型钢承重桩;H型钢桩
steel helix 钢螺旋饰
steel helmet 钢盔
steel-hide picker 钢合皮皮结
steel hinge 钢绞链
steel hinge connection 钢铰链接合
steel hinge purlin(e) 钢铰檩条
steel hipped-plate roof 钢折板屋顶
steel hoist tower 钢制吊装塔架
steel hollow column 钢空心柱
steel hollow frame section 钢空心框架构件
steel hook 钢钩
steel hoop 环箍钢筋;钢箍;环状钢筋;环(形)箍筋;环形钢筋
steel horizontal tie back 钢水平弓弦;钢水平拉杆
steel hose 钢制软管
steel housing 钢住房
steel H-pile H形钢桩
steel hydraulic structure 钢制水利结构
steel I-beam 工字钢梁
steel I-beam bridge with reinforced concrete deck 工字钢梁混凝土桥面
steel I-beam bridge with timber floor 工字钢梁木面桥
steel I column 工字钢柱
steelify 炼钢
steel impression mark 钢印标记
steel industrial rail car 钢制自翻有轨车辆;钢制倾倒有轨车辆
steel industry 钢铁工业
steel industry environment 钢铁工业环境
steel industry waste(water) 钢铁工业废水
steeling 钢化;包钢
steeling inserting die 篷夹冲模
steel ingot 钢锭;钢块
steel ingot scale 钢锭秤
steel ink 钢材画线用墨
steel insert for valve seat 钢阀座圈
steel installation 钢结构安装
steel iron 钢化铁
steel jack 矿用螺旋立柱;闪锌矿
steel jacket 钢质导管架;钢罩;钢套

steel jacket platform 钢套管架平台(近海建筑物用);钢护板平台;钢导管架平台
steel jacks 钢蜘蛛架
steel jamming 卡钻
steel joist 钢托梁;钢龙骨;钢桁条;钢格栅;工字钢
steel joist construction 钢梁构造;钢梁结构
steel jubilee skip 小型钢料车;小型钢制翻斗车
steel jubilee wagon 钢制小型货车
steel knife-edge 钢质刀口
steel ladder 钢梯
steel ladle 钢水包
steel lagging 钢套(筒);钢护壁;钢保护套;钢顶撑;钢挡板;钢背板
steel lamella 薄钢板;薄钢片
steel lamella cupola 薄钢片圆屋顶
steel lamella dome 薄钢片穹顶;钢壳层拱
steel landing mat (楼梯、车站等地方用的)平台钢垫板;铺路钢丝垫
steel lathing 钢板网;钢板条;抹灰(用)钢丝网
steel lattice arch(ed girder) bridge 钢格构拱形板梁桥
steel lattice boom 钢格吊杆
steel lattice bridge 钢格子桥
steel-latticed column 钢格构柱;格构式钢柱
steel-latticed girder 格构钢梁
steel-latticed jib 钢格吊杆
steel-latticed lighting column 钢筋格构照明灯杆;钢筋格构照明柱
steel-latticed lining 钢格构镶衬;钢格构衬板
steel-latticed load-bearing system 钢格构承重系统
steel-latticed loading bucket 钢格构装料勺;钢格构装料罐;钢格构装料斗
steel-latticed mast 格构钢杆
steel lattice rigid frame 钢格构刚架
steel lattice structure 钢格构结构
steel lattice truss 钢格构桁架
steel lattice work 钢格构
steel leaf 钢叶片
steel letters 钢字
steel level 钢水平尺
steel lift bridge 钢升降桥
steel lifting collar 钢提升环
steel lighter 钢驳
steel lighting column 照明钢柱
steel light structure 轻型钢结构
steel lightweight girder 轻型钢梁
steel lightweight member 轻型钢构件
steel lightweight structure 轻型钢结构
steel linear 钢覆面
steel-lined 衬钢的
steel lined concrete pipe 钢板衬砌混凝土管
steel lined plate 钢衬板
steel line drawer 钢线拉线机
steel liner 钢衬(套);钢衬(里);钢衬圈
steel lining 钢内衬;钢板衬砌;钢衬(里);钢板衬砌;钢板衬垫;金属支护
steel link chain 钢链条
steel link conveyer 钢制链板式输送机
steel link harrow 链耙
steel lintel 钢过梁
steel load-bearing skeleton 承重钢构架;承重钢骨架
steel load-bearing structure 承重钢结构
steel log washer 钢制洗矿槽
steel loop 钢扣环;钢环
steel I-plate 铁三角
steel magnet 钢磁体;磁钢
steel maker 炼钢工人;钢铁制造厂
steelmaking 炼钢
steelmaking equipment 炼钢设备
steelmaking furnace 炼钢炉
steelmaking iron ore 炼钢用铁矿石
steelmaking process 炼钢法
steel mall 木槌;木夯;大锤;钢锤
steel manhole cover 检查井钢盖
steel manhole frame 检查井钢框
steel manhole lid 检查井钢盖
steel manufacture 钢铁工业
steel mark 钢(的)标号
steel marking system 钢的标号制
steel marline spike 铁笔
steel mast 钢柱;钢支柱;钢桅杆
steel mat runway 钢垫跑道
steel measuring tape 钢测尺;钢卷尺

steel member 钢构件
steel membrane 钢膜
steel mesh 钢丝网;钢网
steel mesh derrick floor 钢制网格钻塔地板
steel mesh lagging 钢丝网衬板;钢丝网背板
steel mesh(reinforcement) 钢筋网;网状钢筋
steel mesh silo 钢丝网制青储(贮)筒
steel metallurgy 钢冶炼学
steel mill 轧钢厂;钢铁厂;炼钢厂
steel mill bucket 轧钢车间铲斗;轧钢车间料斗
steel mill pollution 炼钢厂污染
steel mill sheet 薄板厂
steel mill shoe 钢轧地脚板;钢轧机座
steel mill slag 钢渣
steel mill waste 钢厂废物;炼钢厂废物;炼钢厂废水
steel mirror 钢镜
steel mitre gate 钢(制)人字门
steel mortar 钢研钵
steel motor lifeboat 钢质机动救生艇
steel mo(u)ld 钢模具;钢模(板);钢锭模
steel mo(u)lder's paint 铸钢涂料
steel mo(u)ld for electric(al) pole 电杆钢模
steel mo(u)lding plate 钢模(板)
steel movable bridge 活动钢桥
steel movable form 钢滑动模板
steel nail 钢钉
steel nailer clip 钢钉板夹
steel needling 蹬梯
steel netting 钢网
steel net forming machine 钢筋网成型机
steel open caisson 钢沉井
steel ore 菱铁矿
steel orthotropic decking 正交异性钢桥面板
steeloscope 钢用光谱仪
steel overhead roadway 钢高架道路
steel pad 钢安全臼
steel pallet 钢集装箱;钢制板台;钢制货架;钢制托盘;钢托盘
steel pan 钢盘
steel panel 钢护板
steel panel radiator 钢板暖气片;钢板散热器
steel pan form 方格顶棚模板
steel pantile 钢波形瓦
steel pantiling 铺设波形钢瓦;波形钢瓦
steel-pan variable speed feeder 变速型钢板给料机
steel-pan wheelbarrow 钢斗木手推车
steel partition 钢隔板
steel partition(wall) 钢隔墙
steel pattern number 钢型号
steel penstock 钢救火龙头;钢闸门;压力钢进水管;压力钢管
steel percentage 含钢筋率
steel perforating rule 打孔用钢尺
steel pick 钢镐
steel pick head 镐头
steel pickling liquor 钢材酸洗液
steel pickling operation 钢(的)酸洗作业
steel pickling waste liquid 钢酸洗废液
steel picks and mattock 钢镐
steel pierced-plank runway 有孔钢板拼装的跑道
steel pig 钢锭;炼钢生铁
steel pig slag 炼钢生铁炉渣
steel pile 钢桩
steel pile anchorage 钢桩锚定
steel pile dolphin 钢桩式系船墩
steel pile type dolphin 钢桩系船墩;钢桩靠船墩
steel pile type mooring dolphin 钢桩式系船墩
steel pile wall 钢桩岸壁
steel piling 钢(板)桩;打钢桩
steel piling frame 钢的打桩架
steel piling wall 钢板桩岸壁
steel pipe 钢管
steel pipe and spike splice for timber pile 钢管和道钉接长木桩
steel pipe arch bridge 钢管拱桥
steel pipe bifurcation 钢叉管
steel pipe butter 钢管端部固定杆
steel pipe clamp 钢制管卡
steel pipe column 钢管柱
steel pipe-column infill 钢管柱内填物
steel pipe construction 钢管结构
steel pipe fittings 钢管零件
steel pipe flange 钢管凸缘;钢管法兰
steel pipe for fire-fighting water 消防水管

steel pipe furniture 钢管家具
steel pipe handrailing 钢管扶手
steel pipe laying plant 公共铺设装置
steel pipe line 钢管线路
steel pipe needle 钢管针
steel pipe penstock 钢救火龙头;钢闸门;压力输送管;压力水管;压力钢管
steel pipe pile 钢管桩;钢板管桩
steel pipe piles filled with concrete 混凝土填心钢管桩
steel pipe pile splicing by butt welding 钢管桩对焊拼接
steel pipe pile splicing by tight steel outside sleeve 用外钢套拼接钢管桩
steel pipe plate 钢管管板;钢管薄板
steel pipe railing 钢管栏杆
steel pipe reinforced concrete structure 钢管混凝土结构
steel pipe scaffolding 钢管脚手架
steel pipe sheeting 钢管桩挡土墙
steel pipe space frame 钢管空间构架
steel pipe string 钢管楼梯斜梁;钢管纵梁
steel pipe structure 钢管结构
steel pipe support 钢管支柱
steel piping 钢管道布置;钢管道铺设;钢管道系统;钢管
steel pits 凹坑(薄板表面缺陷);钢制泥浆罐;(储[贮]存泥浆的)铁箱
steel pivot pin 钢枢销
steel plain fish-plate 平面钢鱼尾板
steel plain splice 钢板平面拼接
steel plain web beam 面面腹板钢梁;实腹钢梁
steel plain web beam bridge 实腹钢梁桥
steel plain web composite girder bridge 实腹组合钢梁桥
steel plain web girder 光面腹板钢大梁
steel plane 钢刨
steel planing 钢板矫平
steel plant 钢铁厂;炼钢厂
steel plant boiler house 钢厂锅炉房
steel plant pollution 炼钢厂污染
steel plant waste 炼钢厂废物;炼钢厂废水
steel plate 钢板
steel plate anchorage wharf 钢板锚固的码头
steel plate beam 钢板梁
steel plate box girder 钢板箱形大梁
steel plate bridge 钢板桥
steel plate catch 钢板抓爪;钢板凸轮挡;钢板制动装置;钢板制动片
steel plate cold rolled drawing sheet 冷轧钢板
steel plate concave 钢板变凹
steel plate convex 钢板凸起
steel plate conveyer 钢板输送器
steel plated 包以钢板的;包钢板(的)
steel plate deformed 钢板变形
steel plate diaphragm 钢隔板
steel plate door 钢板门
steel plate element 钢板构件
steel plate engraving 钢凹版术
steel plate flange 钢板法兰
steel plate floor 钢板楼面;钢板桥面
steel plate galvanized 镀锌钢板
steel plate garage 钢板车库
steel plate girder 钢板梁;钢板大梁
steel plate girder bridge 钢板梁桥
steel plate lagging 钢板挡板;钢板隔板;钢板护壁
steel plate lath(ing) 钢板板条
steel plate lining 钢板衬里;钢板衬层
steel plate member 钢板部件
steel plate paper 钢板雕刻用纸
steel plate radiator 钢制板式散热器
steel plate roadway 钢板车行道;钢板路面;钢板桥面
steel plate rolling 钢材卷圆
steel plates for boiler and pressure vessels 锅炉和压力容器用钢板
steel plates for low-temperature service 低温用钢板
steel plate shop 钢板加工车间
steel plate silo 钢板筒仓
steel plate square washer 方形钢垫板
steel plate stair(case) 钢板楼梯
steel plate stamping 钢板冲压;钢板冲件
steel plate string 钢板楼梯斜梁;钢板纵梁

steel plate structure 钢板结构
steel plate sub-floor 钢板底层地板;钢板毛地板
steel plate system 钢板体系
steel plate thickness meter 钢板测厚仪
steel plate transfer process 钢板转移印花法
steel plate unit 钢板单元
steel plate washer 钢垫片
steel plate window 钢板窗
steel platform 钢站台;钢平台;钢质平台
steel plating 镀钢的
steel-ply form 摺叠式钢模板;折叠式钢模(板)
steel pointed marline spike 铁笔
steel pole 钢杆
steel pontoon 钢浮筒;钢浮船;钢趸船;钢浮码头;钢浮驳
steel portal frame 钢门框
steel post 钢柱
steel powder 钢粉末
steel prestressed concrete 预应力混凝土钢筋
steel prismatic shell roof 钢折板屋顶;钢菱形薄壳屋顶
steel producer 炼钢工人;钢生产者;钢铁生产商;钢铁生产厂
steel production 钢产品;钢铁生产;炼钢
steel products 钢制品;钢材
steel products carrier 钢制品船
steel profile 预应力钢筋烧制;钢材剖面;钢材截面
steel prop 钢撑胶脚;钢支架;钢支撑物;钢柱;钢支柱
steel protection coaming 钢质防围门槛
steel protection fence 钢围护栅栏
steel protection rail 钢保护轨;钢保护栏
steel protractor 钢制分度器;半圆规
steel puller 手摇钻夹具;钎杆提取器;拔钎器
steel punching 钢材制孔
steel purlin(e) 钢檩(条)
steel push barrel 钢筒取样器
steel pylon 钢柱台;钢桥塔;钢标塔;钢支架;钢塔架
steel quality 钢质量
steel rabble 钢渣棒
steel rack 钻杆架;钎子架
steel radiation shielding wall 钢辐射屏蔽墙;钢制辐射防护墙
steel radiator 钢暖气片;钢(制)散热器
steel radiator section 钢暖气片部件;钢散热器部件
steel rail 钢轨
steel rail depreciation apportion and overhaul charges 钢轨折旧摊销及大修费
steel rail grizzly 钢轨棒条筛
steel railing 钢扶手;钢栏杆
steel rake boat 钢耙船
steel rake dredge(r) 钢耙挖泥船
steel raking tool 钢刮刀
steel ratio 钢筋比(率);钢筋百分比(率);含筋率;含钢率;配筋率;配筋百分率
steel ratio of concrete 钢筋配筋比;钢筋混凝土含钢率;钢筋百分比(率);混凝土配筋比
steel reactor 钢制反应釜
steel rectification 钢材矫正
steel reef (海洋钻探平台底座上形成的)钢礁
steel reinforced 加钢筋增强的
steel reinforced alumin(i)um wire 钢芯铝线
steel reinforced concrete 钢筋混凝土;钢筋混凝土
steel reinforced concrete column 钢筋混凝土柱
steel reinforced concrete structure 钢—混凝土结构
steel reinforcement 增强钢筋;钢筋
steel reinforcement bar delivering roller 钢筋输送辊
steel reinforcement bar feeder 钢筋送料机
steel reinforcement bar predrawn platform 钢筋预拉台
steel reinforcement cage 钢筋笼
steel reinforcement laying machine 钢筋敷设机
steel reinforcing bar 钢筋
steel reinforcing rod 钢筋棍
steel relaxation 钢筋的松弛
steel requirement 钢筋需要量;钢材需要量
steel residence 钢住宅
steel residential housing 钢住房
steel retainer 钢挡板;钢格栅;钢定位器;钢止动器;夹钎器
steel retort 钢蒸馏罐
steel ribbed cupola 钢肋圆顶;钢肋穹顶
steel ribbed dome 钢肋穹顶

steel ribbed floor 肋形钢楼盖
steel-ribbed rim with rubber cover(ing) 外包橡胶的钢肋轮缘
steel ribbon 钢条
steel rib support 钢架支护
steel rider 钢辊
steel ridging 钢屋脊
steel rigid foam sandwich element 硬泡沫夹心的钢构件
steel rigid framing 钢刚架
steel rigid salvage pontoon 钢壳打捞浮筒
steel rim 钢轮辋
steel-rimmed wheel 有钢轮缘的轮子
steel ring 钢圈;钢环
steel ring borehole deformation ga(u)ge 钢环式孔径变形计
steel-ring burr cylinder 钢环除杂锡林
steel ring set 钢圈支架
steel rip bit 凿岩钢钻头;凿岩钢钎头
steel rivet 钢铆钉
steel road bridge 钢公路桥
steel roadstud (道路交通用的)道路钢钉
steel rod 钢筋;钢条;钢拉杆;钢棍
steel rod furniture 钢筋家具
steel roll 钢轧辊
steel roller 钢轮压实机;钢轮碾压机;钢轮路碾;钢辊
steel roller bearing 钢滚珠轴承
steel roller shutter 钢卷帘
steel rolling 轧钢;钢材轧制
steel rolling grille 可卷钢格子门窗;可卷钢栅
steel rolling machine 轧钢机
steel rolling shutter 钢制卷帘
steel rood girder 钢屋顶大梁
steel roof cladding 钢屋顶覆盖物
steel roof cover(ing) 钢屋顶覆盖物;钢制屋面层
steel roof deck(ing) 钢屋顶;钢板屋面
steel roof(ing) 钢屋顶
steel roof profile 钢屋顶型材;钢屋顶剖面;钢屋顶断面
steel roof section 钢屋顶构件
steel roof shape 钢屋顶形状
steel roof sheathing 钢屋顶衬板;钢屋顶望板
steel roof system 钢屋顶系统
steel roof tile 钢屋顶瓦
steel roof trim 钢屋顶装饰;钢屋顶贴面
steel roof truss 钢屋(顶桁)架
steel roof truss structure 钢屋架结构
steel roof unit 钢屋顶单元
steel rope 钢丝绳;钢绳;钢缆;钢丝索
steel-rope drive 钢索传动
steel-rope guide 钢丝绳罐道
steel rope roof 钢缆屋顶
steel-rope tracting belt conveyer 钢绳牵引带式输送机
steel round 圆钢材;圆钢筋
steel ruler 钢直尺;钢尺
steel rundle 钢踏步;钢梯级
steel rung 钢梯级;钢踏棍
steel rung ladder 钢制爬梯
steel runway 钢板跑道
steel rust removing 钢材除锈
steel safety fence 钢制护栏
steel safety rail 钢安全扶手;钢保护栏
steel sand drilling 钢砂钻进
steel sash 钢推拉窗扇;钢框格;钢窗
steel sash putty 钢窗油灰
steel sash putty of the exterior glazing type 外装配玻璃型的钢窗油灰
steel sash putty of the interior glazing type 内装配玻璃的钢窗油灰
steel savings 钢材储存;钢的节省量
steel saw 钢锯
steel scaffold(ing) 钢鹰架;钢脚手架
steel scale 钢尺
steel scoria 钢渣
steel scrap 钢渣;钢屑;废钢
steel scraper 刮钢尺
steel screen 钢板网
steel screw 钢螺钉
steel screw mooring 钢螺旋锚系船设备
steel screw pile 钢螺旋柱
steel-seal type cell 铁皮封闭式电池
steel seamless tube 无缝钢管
steel section 型钢断面;型钢;型钢截面

steel section string 楼梯斜钢梁
steel segment 钢制管片;钢管片
steel seizure 钻头卡住;钎卡住
steel set 拱肋;钢支(承)架;矿用钢支架
steels for pressure vessel use 压力容器用钢
steel shaft ring 钢轴环
steel shank 钎子尾;钎尾
steel shape 型钢
steel-shape for tunnel support 隧道支护型钢
steel sharpener 锻钎机
steel sharpening 锻钎
steel shavings 钢切削片;钢屑
steel shear key 钢剪力链
steel shear wall 钢剪刀墙
steel sheath 钢皮
steel-sheathed lead pipe 包钢皮的铅管;包钢铅管
steel sheathing 钢模(板);钢面板;钢棉面板
steel sheave 铁滑轮
steel sheet 压型钢板;薄钢板;钢板
steel sheet corner folded 钢板卷角
steel sheet disk wheel 钢板盘轮
steel sheet edge folded 钢板卷起
steel sheet facing 钢筋覆面(层);钢板贴面
steel sheeting 钢板(桩);钢板栅
steel sheet lining 钢皮衬里
steel sheet panel 钢镶合板
steel sheet pile 钢板桩
steel sheet pile anchorage 钢板桩锚定
steel sheet pile breakwater 钢板桩防波堤
steel sheet pile bulkhead 钢板桩护岸;钢板桩驳岸
steel sheet pile cell (填充混凝土的)钢板桩单体
steel sheet pile cellular cofferdam 钢板桩格形围堰;钢板桩格体围堰
steel sheet pile cellular cofferdam type quaywall 钢板桩格形围堰式岸壁;钢板桩格体围堰式岸壁
steel sheet pile cellular quaywall 钢板桩格形岸壁;钢板桩格体岸壁
steel sheet pile cofferdam 钢板桩防水堰;钢板桩围堰
steel sheet pile cutoff 钢板桩截水墙
steel sheet-pile(d) wall 钢板桩墙
steel sheet pile earth cofferdam 钢板桩填土围堰
steel sheet pile quay 钢板桩码头
steel sheet pile quaywall 钢板桩码头驳岸;钢板桩岸壁
steel sheet pile quaywall with relieving platform 钢板桩承台式(顺岸)码头
steel sheet pile type dolphin 钢板桩靠船墩
steel sheet pile wharf 钢板桩码头
steel sheet piling 钢板桩;钢板支撑;打桩板桩
steel sheet piling cofferdam 钢板桩围堰
steel sheet piling cofferdam with internal bracing frame 用内撑框架的钢板桩围堰
steel sheet piling cut-off 钢板桩截水墙
steel sheet piling type dolphin 钢板桩(式)墩台
steel sheet piling wall 钢板桩墙
steel sheet-piling wharf 钢板桩码头
steel sheet-piling wharf wall 钢板桩码头墙
steel sheet roof cladding 钢皮屋顶
steel sheet roof cover(ing) 钢皮屋顶覆盖物
steel sheet roofing 钢皮屋面
steel sheet roof sheathing 钢皮屋顶覆盖物
steel sheet shears 钢皮剪刀
steel sheet straightener 钢板校直机
steel sheet(sur)facing 钢皮覆面
steel shelf 钢架
steel shell 钢(薄)壳;钢制炉体;钢壳体
steel shell cofferdam 钢壳围堰
steel shelled box type load switch 钢壳负荷开关
steel shelled concrete 钢壳混凝土
steel shelled concrete pile 钢壳混凝土桩
steel shell tube 钢壳沉管
steel shield 钢盾;防护钢壳
steel shielding wall 钢屏蔽墙;钢制防护墙
steel ship 钢船
steel ship bottom paint 钢船船底漆
steel ship hull 钢船体;钢船壳;钢壳船
steel shod 装有钢箍头的;底部包钢皮的
steel shoe 装有钢靴的;钢(制)柱靴;钢挡板
steel shop work 钢结构的车间加工
steel shore 钢圆撑;钢支柱;钢撑
steel shore strut 钢支撑
steel shoring column 钢支柱;钢撑柱

steel short link chain 葡萄链；短环链
steel shot 钢弹；钢丸；钢砂；粒粒
steel shot abrasive 钢砂磨粒
steel shot aggregate 钢球集料
steel shot blasting 钢喷丸器；钢喷丸处理
steel shot blasting system 钢喷丸系统
steel shot boring 钢珠钻探
steel shot concrete 钢屑混凝土；钢砂混凝土
steel shotcrete 钢支撑喷射混凝土
steel shot drill 钢珠钻机
steel shot drill hole 钢珠钻孔
steel shot drilling 钢珠钻探；钢砂钻井；钢粒钻进
steel shovel 钢锹；钢铲
steel shutter door 钢门
steel shutter(ing) 钢百叶窗；钢卷帘；钢模(板)
steel skeleton 钢骨架
steel skeleton block 钢骨架大梁
steel skeleton building 钢骨架建筑(物)；钢骨架房屋；钢骨房屋
steel skeleton car park 钢骨架停车场
steel skeleton compact 钢骨架坯块
steel skeleton construction 钢骨架建筑；钢结构；钢骨架结构；钢骨架构造
steel skeleton house 钢骨架房屋
steel skeleton multi-stor(e)y 钢骨架多层车库
steel skeleton office building 钢骨架办公大楼
steel skeleton structure 钢骨架结构
steel skid 钢滑行架；钢导板；钢导轨；钢滑橇
steel skin 钢皮；钢板衬砌；钢板衬垫
steel skyscraper 钢摩天大厦；钢制摩天大楼
steel slab 钢锭；钢板
steel slab bridge 钢板桥
steel slab prop 钢板支撑；钢板撑
steel slag 钢渣
steel slag cement 钢渣水泥
steel slat 钢条板
steel slat rolling shutter 钢条板卷帘
steel slatted roller blind 钢条板卷帘
steel sledge hammer 钢锤
steel sleeper 钢枕木；钢枕；钢轨枕
steel sleeve 钢套管；钢衬垫
steel sleeve pipe 钢套筒管
steel slide gate 钢制门
steel sliding folding shutter door 钢活动折叠百叶门
steel sliding gate 钢推拉门；钢扯门
steel slip ga(u)ge 钢制块规
steel-smelting 炼钢
steel-smelting furnace 炼钢炉
steel socket 钢套节
steel soldier 钢模板支撑；钢模板立柱
steel sole plate 钢轨垫板；钢基础底板
steel solid-rib arch bridge 钢实肋拱桥
steel solid web beam 实心腹板钢梁
steel solid web girder 实心腹板大梁；实腹大梁
steel sorter 钢鉴别仪
steel space frame structure 钢空间框架结构；空间钢框结构(物)
steel space load-bearing structure 空间承重钢结构
steel spade 钢锹
steel spar 空心铁杆
steel spatula 钢刮刀；钢调刀
steel specification 钢号
steel sphere 钢球；钢粒
steel spigot and socket pipe 承插式接头钢管
steel spigot ring 钢插口环
steel spile 钢插桩
steel spiral casing 钢蜗壳
steel spiral flume (超高压汽轮机的)螺旋形钢水槽
steel spiral stair(case) 钢制盘梯；钢制螺旋梯
steel spirit level 钢水平尺
steel spoke wheel 钢制辐轮
steel spout 出钢槽
steel spiral spring 钢螺簧
steel spring 钢(弹)簧；弹簧钢
steel spring fender 钢弹簧护木；钢弹簧护舷
steel spring isolator 钢弹簧隔离器
steel spring plate 弹簧钢板
steel spring undercarriage 钢簧起落架
steel square 直角钢(尺)；钢曲尺；钢角尺
steel stabilized with titanium 用钛稳定的钢
steel stack 钢(制)烟囱；钢通风管
steel stack shaft 钢烟囱通风井；钢烟囱柱身
steel stair 钢楼梯

steel staircase builder 钢楼梯建造者
steel stanchion 钢(支)柱
steel stand 钢自行车架；钢支架
steel stem 钎杆
steel sticking 夹钎
steel storage 钢油罐
steel storage tank 钢水箱；钢蓄水池
steel storage yard 钢堆场
steel store 钢堆栈
steel strand 钢绳索；钢丝束
steel strand rope 钢索
steel strap 钢条；钢片；钢带
steel strapping equipment 钢带捆扎机
steel street marker 道路钢路标
steel stress 钢应力
steel stretcher 钢筋冷拉机
steel string(er) 钢斜梁楼梯架；钢板楼梯梁；钢纵梁；钢绳弦线
steel string stair(case) 钢斜梁楼梯；钢梁楼梯
steel strip 钢板条；钢带
steel strip die 钢带模
steel structural derrick 钢结构钻塔
steel structural engineering 钢结构工程
steel structural hull 钢结构船体
steel structural plate 钢结构板材
steel structural section 钢结构型材；钢结构剖面(图)
steel structure 钢结构
steel structure design 钢结构设计
steel structure processing technology 钢结构加工工艺
steel strut 钢支柱
steel strutted rigid frame bridge 钢斜腿刚构桥
steel stud 钢龙骨；钢立筋
steel stud anchor 钢门框锚碇夹
steel-studded tyre 钢叠层轮胎
steel subpurlin(e) 钢副檩；钢副椽；钢制副檩条
steel sub-tread 钢踏步底板
steel superstructure 上部钢结构
steel support 钢支座；钢支(承)架；钢圈支架；金属支架
steel supporting hoop 钢支撑箍
steel supporting pipe 支承钢管
steel supporting stock column 钢支柱料柱
steel supporting tube 支承钢管
steel support structure 钢支撑结构
steel surround 钢包裹物；钢围绕物
steel swarf-removing tool 断屑刀片
steel swing bridge 钢平旋开启桥
steel swing door 钢旋转门
steel system 钢体系
steel tackle block 钢轮滑车组
steel tangentially sliding bearing 弧形板活动支座
steel tank 钢罐；钢槽；金属结构罐
steel-tank mercury-arc rectifier 铁壳汞弧整流器
steel tank rectifier 钢壳汞弧整流器
steel tape 钢合尺；钢卷尺
steel tape armo(u)ring 钢带铠装
steel tape distance measurement 钢尺量距
steel tape measure 钢卷尺
steel tape measurement 钢卷尺丈量
steel tape method 钢尺法
steel tape sheath 钢带外皮
steel tape web divider 钢带毛网分条器
steel tapping hole 出钢口
steel tee-plate 铁T角
steel telescoping form 钢嵌套支架
steel tendon 钢筋束；预应力钢筋束；预应力钢筋束；钢丝束；钢筋腱
steel tension bar 钢拉条；钢拉索
steel test piece 钢材取样块
steel thimble 钢套管
steel thread 钢丝
steel tie 钢箍；钢枕；钢系杆；钢轨枕
steel tie bar 钢系条；钢锚筋
steel tie beam 钢拉梁
steel tie-rod 钢拉杆
steel tile 钢瓦
steel tilted slab roof 钢斜板屋顶
steel timber 金属支架
steel timber composite roof truss 钢—木组合屋架
steel timbering 钢支撑
steel tire 钢轮箍
steel tired wheel 钢圈轮

steel to clinker ratio 钢球熟料比
steel toe rubber boots 铁龟头橡胶长统靴
steel tool box 钢壳工具箱
steel tooth 钢齿
steel tooth bit 钢齿牙轮钻头
steel tower 钢制水塔；钢塔架；钢塔；钢标；铁塔；铁架
steel tower for high viaduct 栈桥钢塔
steel track 钢轨道
steel traffic stud (道路交通用的钮状的)钢钉
steel transmission pole 钢质电线杆
steel trap 捕兽夹
steel traveler 钢丝圈
steel tray 钢托架；钢坐垫；钢盘
steel trench sheeting 沟槽钢板桩；钢楼梯踏步
steel trestle 钢高架桥；钢排架桥；钢栈桥
steel trim 钢装饰；钢贴面
steel trowel 镘刀；钢制泥刀；钢抹子；钢镘刀；钢镘(板)
steel troweled 钢抹压光
steel troweled concrete 钢镘抹面混凝土
steel troweling 钢抹镘光；钢抹镘板；镘刀；钢抹子
steel truss 钢桁梁；钢桁架
steel truss bridge 钢桁梁桥；钢桁架桥
steel trussed arched girder 钢桁架反拱大梁
steel trussed girder 钢桁架大梁；钢桁架
steel truss frame 钢桁架式构架
steel truss-span 钢桥桁架；钢桁架跨度
steel tub 钢浴盆
steel tube 钢管
steel tube armo(u)r 钢管铠装
steel tube badly bent 钢管严重碰弯
steel tube benchmark 钢管标
steel tube centering 钢管拱架；钢管膺架
steel tube column 钢管柱
steel tube for fired heater 加热炉用钢管
steel tube foundation 钢管基础
steel tube raker pile 钢管斜桩
steel tube seat 钢管座
steel tubes for heat transfer 热传导用钢管
steel tubing 钢管
steel tubing lattice jib 钢管格构起重机吊杆；钢管格构起重机臂
steel tubing lattice tower 钢管格构式塔架
steel tubing light building crane 钢管轻型建筑起重机；钢管轻型施工吊车
steel tubing plant 钢管厂
steel tubular frame 钢管构架
steel tubular furniture 钢管家具
steel tubular handrailing 钢管扶手
steel tunnel support 钢制支撑；隧道钢支护
steel turnbuckle 花篮螺丝；伸缩螺丝
steel type 钢材型号
steel type of construction 钢建筑类型
steel unit 钢部件
steel unitized unit 钢建筑块体模数
steel upright 钢支柱
steel valve 钢阀门
steel ventilation pipe 金属风筒
steel vessel 钢制容器；钢容器；钢船
steel visual inspection 钢材外观检验
steel volute casing 钢蜗壳；钢开关
steel waffle 钢平顶镶板；钢盒；钢盘
steel waling 钢围图
steel wall 钢墙
steel wall stave 钢墙板条
steel washer 钢垫片；钢垫圈
steel waste pipe 钢污水管
steel water pipe 钢水管
steel wear plate 防蚀增强板
steel web girder 钢腹板梁
steel wedge 钢楔
steel weight box 钢窗平衡锤箱；钢压载箱；钢配重箱
steel weight-carrying skeleton 钢载重骨架
steel weight-carrying structure 钢载重结构
steel weir 钢堰
steel-welding fittings 钢焊接管件
steel weldment 钢焊接件
steel wheel 钢轮
steel wheeled compactor 钢轮压实机
steel wheeled roller 钢轮碾压机；钢轮路碾；钢轮式压路机
steel-wheeled tractor 钢轮拖拉机
steel wheel glass cutter 钢砂轮玻璃刀
steel wind-bracing wall 钢抗风墙

steel window 钢窗	steelwork details 钢结构详图	steeping 浸渍;浸泡
steel window for agricultural use 农用钢窗	steelworker 钢铁工;钢构件及钢筋安装工;炼钢工人	steeping and pressing tank 浸碱压榨机
steel window for industrial building 工业建筑用钢窗	steelwork erection 钢结构安装	steeping brick 浸湿砖
	steelwork erector 钢结构工程架设者	steeping fluid 浸渍液
steel window frame 钢窗框	steel working 钢的加工;钢的生产	steeping preservative 冷浸防腐剂
steel window frame and accessory 钢窗框及配件	steel works 钢制品;钢制件;钢铁工程;钢结构;钢厂;钢铁厂;炼钢厂;钢铁工程;钢件;钢架	steeping press 浸压机
steel window section 钢窗料;窗钢料		steeping process 浸渍法
steel wire 钢线;钢纤维;钢丝;钢绳	steel works equipment 钢厂设备	steeping treatment 冷浸处理
steel wire armo(u)r 钢线铠装	steel workshop 钢材(加工)车间	steeping trough 浸渍槽
steel wire armo(u)red 钢丝铠装的	steel works pollution 炼钢厂污染	steep land 险峻土地;陡坡地
steel wire armo(u)red air rubber hose 钢丝编织空气胶管	steel works waste 炼钢厂废物;炼钢厂废水	steep land soil 陡坡土
	steely 钢制;含钢;似钢的	steep lapse rate 陡递减率
steel wire armo(u)red cable 钢线铠装电缆;钢丝铠装电缆;钢丝铠包电缆	steelyard 钢筋堆场;杆式秤;杆秤;提秤;弹簧秤;吊称;大磅秤;秤	steeple 尖塔;(礼拜堂的)尖阁;尖顶
		steep-lead thread 大导程螺纹
steel wire armo(u)red high-pressure rubber hose 钢丝编织高压胶管	Steelyard moisture meter 斯蒂尔亚德含水量测定仪;吊称湿度计	steeplechase course 障碍赛跑跑道
		steeple-compound stamp 尖柱捣磨
steel wire armo(u)red water rubber hose 钢丝缠绕排水胶管	steel(y) iron 炼钢生铁;炼钢用铁	steeple-compound turbine 叠置式汽轮机
	steel yoke 钢轭铁;钢锁头板;钢叉架;钢叉臂	steeple-crowned 尖塔形的
steel wire belt 钢丝网运输带	steel yoke end 钢叉	steepled 装有尖顶的;有尖顶的;尖塔形的
steel wire braid 钢丝编织层	steely pig 钢性生铁	steeple engine 倒行发动机
steel wire braided high-pressure hose 钢丝编织高压管	steely wool 钢丝毛	steeple-head rivet 尖头铆钉
	steem 加衬	steeplehouse 教堂建筑
steel(wire)brush 钢丝刷	steen 干砌砖石衬壁;古陶罐	steeplejack 尖塔修建工人;烟囱修建工人;高空砌筑作业工
steel wire clamp 钢丝绳卡头	steening 砖衬;石衬;(井或污水池周壁的)砖石或混凝土衬砌;干砌砖石衬壁	
steel wire cloth 钢丝布;铁砂布		steeplejack's mate 高空作业辅助工
steel wire cutter 钢丝刀	Steenrod algebra 斯丁洛特代数	steeple plain roof(ing)tile 教堂尖塔平瓦
steel wire cutter one step quick forming 钢丝刀一次快速成型法	Steenrod square 斯丁洛特平方	steeple tile 塔尖用瓦
	steenstrupine 磷硅稀土矿	steeple top 尖塔状顶部;(教室屋顶上的)尖塔
steel wire dynamometer 钢丝应力测定仪	steep 浸染;峭壁;陡峭的;陡峭;陡的	steeple-window 教堂尖塔上的窗
steel wire embedded water suction rubber hose 埋线工业吸水胶管	steep angle bearing 大锥角滚柱轴承	steep lifting 垂直提升
	steep arch 陡拱	steeply dipping bed 陡倾岩层
steel wire for fences 篱用钢丝	steep ascent 陡升坡	steeply inclined belt conveyer 大倾角胶带输送机
steel wire for prestressed concrete 预应力混凝土用钢丝	steep asphalt 浸渍沥青;陡屋面沥青	steeply inclined conveyer 陡斜的运输机
	steep bank 陡河岸;陡岸	steeply inclined occurrence 陡斜
steel wire for welding electrode 焊条钢丝	steep bank revetment 陡坡护岸	steeply pitched glazing 陡坡屋面装配玻璃
steel wire ga(u)ge 钢线规;钢丝线规	steep bevel 斜边;陡边	steeply pitching seam 急倾斜矿层
steel wire harp-type screen 竖琴式钢丝带	steep channel 陡峭河槽;陡槽	steeply plunging fold 陡斜伏褶皱
steel wire head 钢丝综	steep channel fish pass 陡槽式鱼道	steeply sloped 陡斜坡度的
steel-wire inducer 钢丝导引器	steep coal seam 急倾斜煤层;急倾斜矿层	steeply sloping racks 陡斜的支架;陡斜的导轨
steel wire lashing 钢丝绑扎	steep coast 绝壁海岸;陡海岸;陡岸	steeply sloping roof 陡坡屋顶
steel wire measurement 钢线尺丈量	steep crown 陡路拱	steepness 坡度;尖锐度;陡度
steel(wire)mesh 钢丝网	steep current residual voltage 陡峭电流;冲击残压	steepness of a curve 曲线陡度
steel-wire napper 钢丝起毛机	steep curve 急弯曲线;急弯;锐曲线;陡曲线;陡曲线	steepness of pulse edge 脉冲边缘陡度
steel wire packer 钢丝打		steepness of slope 坡降陡度;边坡陡度
steel wire product 钢丝制品	steep descending gradient 陡下坡	steepness of the terrain 地面坡度
steel wire quarry machine 钢丝绳锯石机	steep descent 陡降坡	steep parabola 陡抛物线
steel wire reinforced belt 钢丝芯加固橡胶带	steep dip 陡倾角	steep parabolic arched girder 陡抛物线拱形大梁
steel wire reinforcement 钢丝网配筋	steep dipping 急倾斜的	steepped face wall 阶梯形面墙
steel wire ring wheel 钢丝圈刷轮	steep down grade 陡下坡	steep pitch 黏质沥青;急倾斜(大于15°);陡坡
steel wire rod 钢线材	steep down gradient 陡下坡	steep-pitched roof 陡坡屋顶
steel wire rope 变形钢筋;钢缆;钢索;钢丝索;钢丝绳;钢丝缆	steep dropping voltampere characteristic 陡降伏安特性	steep pitch of thread 螺纹大节距
	steeped face wall 阶式胸墙	steep-pitch thread 大螺距螺纹
steel wire rope belt conveyer 钢丝绳皮带运输机	steepest ascent 最速上升;最陡上升	steep precipice and cliff 悬岩绝壁
steel wire rope chisel 钢丝绳凿子;钢丝绳铲刀	steepest ascent method 最速上升法;最陡上升法	steep preservative 冷浸防腐剂
steel wire rope clamp 钢丝绳卡	steepest ascent procedure 最速上升;最陡上升	steep pulse 陡前沿脉冲
steel wire rope clip 钢丝绳扎头;钢丝绳卡头;钢丝绳夹子	steepest ascent sampling 最速上升抽样;最陡上升抽样	steep region 陡坡带
		steep ridge 陡峭山脊
steel wire rope graphite-base grease 钢丝绳石墨润滑油	steepest descent 最上下降;最大角度下降;沿最陡轨迹下降	steep rise 急剧上升
		steep river 大比降河流;大比降的河流
steel wire rope grip 钢丝绳执持扎头	steepest descent approximation 最陡下降近似法	steep rock slope 陡峭的岩石斜坡
steel wire rope socket 钢丝绳套环	steepest descent method 最速下降法;最陡下降法;鞍点法	steep roof 陡斜屋顶;陡坡屋顶
steel wire rope stranding-bundling machine 钢丝绳绞束机		steep sea 陡浪
	steepest descent vector 最速下降向量	steep sharply turned mo(u)ldboard 大角度翻转犁壁
steel wire rope thimble 钢丝绳眼心环	steepest gradient 最陡坡度	
steel wire rope type electric(al)block 钢丝绳电动葫芦	steepest gradient method 最陡下降法;最速下降法	steep-sighting prism 测大仰角棱镜
	steep flume 陡槽	steep slope 急坡;陡坡;陡立边坡;大比降
steel wire rope with thimble 带线环钢丝绳	steep folding 过分的褶皱	steep slope belt 陡坡带
steel wire rotary brush 钢丝轮刷	steep-fronted wave 陡峭前缘波;锯齿波	steep slope channel 急流槽
steel wire screen 钢丝筛(网)	steep front wave 陡峭前缘波	steep slope revetment 陡坡支撑;陡坡护坡
steel wire sheath 钢丝外皮	steep gable 陡的山墙;陡坡屋顶	steep slope river 大比降的河流
steel wire stay 钢索拉条	steep grade 陡坡度;陡坡	steep slope stream 大比降河流;比降大的河流
steel wire strand 钢绞线;多股钢丝;多股钢丝	steep grade culvert 陡坡涵洞	steep spectrum 陡谱
steel wire stylet 钢丝针芯	steep grade tunnel(l)ing 陡峭隧洞掘进(法);陡坡隧洞掘进	steep spin 急螺旋
steel wire tow gear 钢丝拖带装置		steep stream 大比降河流;比降大的河流
steel wire tube brush 钢丝管刷	steep gradient 陡坡;大比降	steep taper shank 大锥度刀柄
steel wire wheel brush 钢丝轮刷	steep gradient on road 陡坡道	steep terrain 陡峭地带
steel wiring furniture 钢丝家具	steep gradient section 陡坡段	steep to 陡深(指岸边或浅滩边缘突然变深)
steel wiring reinforced brick wall 钢丝网加劲筋砖墙	steep gravity wave 陡重力波	steep up gradient 陡上坡
	steephead 溶蚀(残)谷;陡头谷	steep wall 陡壁
steel with organic coating 镀有机涂层的钢材	steephouse 教堂建筑	steep-walled canyon 峭壁峡谷;陡壁峡谷
steel-wood furniture 钢木家具	steep incline 陡斜井;陡坡	steep-walled valley 峭壁河谷;陡壁河谷
steel wool 铁惠丝(一团细钢线);钢丝绒;钢丝棉;钢棉	steep-incline belt conveyer 陡斜式输送带	steep wave 陡浪;陡波
	steep increase in prices 物价飞涨	steep wave front 陡峭波阵面
steel wool filter 钢丝绒滤油器		steep working 在陡峭的地层上施工

steer 引导;木垛;驾驶【船】
steerability 可控制性;可驾驶性;可操纵性;驾驶性能
steerable antenna 可控天线
steerable crawler 易操纵的履带拖拉机;可操纵的履带拖拉机
steerable hydrofoil 可操纵水翼
steerable roll(er wheel) 易操纵的滚筒;可操纵的滚筒;易操纵的压路机;可操纵的压路机
steerable walking mechanism 易操纵的步行式机械;可操纵的步行式机械
steerable wheel 导向轮
steer against current 逆流航行
steer against wind 逆风航行
steerage 客船舱舱;舵效;船的转向性能;操舵;客轮上的统舱
steerage fin 导向板
steerage gear 转向装置;操纵齿轮
steerage mess 统舱餐厅
steerage passage 统舱旅客;统舱
steerage passenger 四等舱旅客
steerage way 舵效航速
steer a trick 舵工值班
steer-dampet unit 操纵和缓冲组件
steerer 舵手;操纵器
steering 引导;转向;导引;导向;操纵方向;操纵;操舵
steering actuation cylinder 转向助力缸
steering analysis department 转向系检验工间
steering and brake linkage 转向和制动的联动系统
steering angle 转向角
steering angle indicator 转向角指示器
steering anti-kickback snubber 防反向减振器;防反向缓冲器;转向减震器
steering apparatus 转向装置;舵机;操舵装置
steering arm 转向操纵杆;转向臂
steering arm adjusting screw wrench 转向臂调节螺钉扳手
steering arm ball 转向杆球端
steering arm ball stud 转向臂球头柱螺栓
steering arm block 操纵臂滑轮
steering arm bolt 转向臂螺栓
steering arm nut 转向臂固定螺母
steering arm pin 转向臂销
steering arm shaft 转向臂轴
steering arm shaft felt washer cup 转向垂臂轴毡垫盖
steering arrangement 操舵装置
steering axis 转向节主销轴线
steering axle 转向轴;转向桥;舵轴
steering axle pivot pin 转向轴枢销
steering axle quadrangle 前轴方形转向架
steering ball 转向球
steering ball cap 转向球节盖
steering ball joint 转向球节
steering band 转向闸带
steering bar 操纵杆;导向杆
steering behavio(u)r 转向性能
steering bogey 操舵转向架
steering bogie 操舵转向架
steering booster 转向助力装置
steering booster pump 转向的增压泵;操纵的增压泵
steering box 转向箱;转向机构箱
steering box cover 转向变速器盖
steering box side cover 转向机侧盖
steering brake 转向制动器;转向闸
steering brake bleed nipple 转向制动器放油开关
steering brake cross-shaft 转向制动器横轴
steering brake drum 转向制动鼓
steering brake drum shaft bearing 转向制动鼓轴承
steering brake lever 转向制动杆
steering brake-shoe 转向闸瓦
steering bridge 驾驶室(又称驾驶台)【船】;操舵台;操舵室
steering by brakes 用制动(装置)操纵
steering by rack and pinion 齿条和小齿轮转向
steering cable 操纵钢索;操纵钢缆
steering cam 转向蜗杆凸轮;转向凸轮
steering camshaft 转向凸轮轴
steering chain 舵链
steering circuit 引导电路;控制电路;操纵电路
steering clutch 转向离合器
steering clutch adjusting screw 转向离合器调整螺钉

steering clutch brake 转向离合器制动器
steering clutch control 操纵离合器控制;转向离合器控制
steering clutch control butt 转向离合器杠杆挡块
steering clutch control button 转向离合器控制钮
steering clutch control valve 操纵离合器控制开关;操纵离合器控制阀
steering clutch driven disk 转向离合器从动盘
steering clutch driving drum 转向离合器主动鼓
steering clutch housing 转向离合器壳
steering clutch(lever) 转向离合器;操向杆
steering clutch lining 转向离合器摩擦片
steering clutch linkage 操纵离合器联动装置;操纵离合器连杆
steering clutch plate 转向离合器钢片
steering clutch pressure plate 转向离合器压板
steering clutch push rod 转向离合器推杆
steering clutch push rod arm 转向离合器推杆臂
steering clutch release 转向离合器分离叉
steering clutch release bearing 操纵离合器松脱轴承
steering clutch release crank axle 转向离合器开关曲柄轴
steering clutch release fork 转向离合器分离拨叉
steering clutch release ring 转向离合器分离环
steering clutch release rocking lever 转向离合器分离摇臂
steering clutch release yoke 转向离合器拨叉
steering clutch relief valve 操纵离合器减压阀;操纵离合器安全阀;操纵离合器保险阀
steering clutch shaft 转向离合器轴
steering clutch spring 转向离合器弹簧
steering column 转向柱;转向柱护管;转向盘轴;转向管柱;驾驶杆;舵(工站)台;操纵轮胎
steering column bracket 转向柱管固定架
steering column case 转向柱套管
steering column clamp 转向柱夹
steering column control 转向柱控制;操纵杆控制
steering column gear change 转向柱变速
steering column gear control 转向盘柱变速杆
steering column jacket 转向轴套
steering column jacket oil hole cover 转向轴套油孔盖
steering column lock 转向盘柱锁
steering column mounting 转向柱套管托架
steering column shaft 转向器轴
steering column shaft spline 操纵柱轴齿条;操纵柱轴花键;转向柱轴齿条;转向柱轴花键
steering column support 转向柱支架
steering column tube 转向柱管
steering column universal joint 转向柱万向节
steering column upper bracket 转向盘柱上支架
steering command 操纵信号
steering committee 指导委员会;管理委员会;领导委员会
steering compass 领航罗盘;驾驶发变器;操舵罗经
steering compass course 操舵罗经航向
steering computer 驾驶用计算机;操纵用计算机
steering connection 转向关节
steering console 控制台;操纵台
steering control 转向控制
steering control level 转向控制杆
steering controlling shaft 转向控制轴
steering control rope 操向索
steering control shaft 操纵控制轴
steering control valve 操纵控制阀
steering correction 操纵修正
steering coupling 转向联轴节
steering crank 转向曲柄
steering cross rod 转向横拉杆
steering cross rod jaw 转向横拉杆接头
steering crutch 舵桨索环
steering current 引导气流
steering curve 转向曲线
steering cylinder 转向助力油缸;操纵动作筒
steering cylinder pivot pin 转向圆柱体枢销;转向液压缸枢销;操纵圆柱体枢销;操纵液压缸枢销
steering damper 转向减振器
steering damper assembly 转向减振器组合
steering damper friction assembly 转向减振器摩擦片组合
steering damper rod 转向减振杆
steering damper unit 转向机构中的减振器
steering device 操纵设备;转向设备;转向器;转向装置;操纵装置

steering disc and bevel gear 换向机盘和伞齿轮
steering disk 操向圆盘
steering drag 转向阻力
steering drag link 转向纵拉杆
steering drag link rear end grease fitting 转向纵拉杆后端油嘴
steering drag link rubber boot 转向纵拉杆橡胶套
steering drag rod 转向(纵)拉杆
steering drag rod ball 转向纵拉杆球接头
steering drag rod front end pressure lubricator 转向纵拉杆前端油嘴
steering drag rod spring 转向纵拉杆弹簧
steering drive axle 操纵传动轴;转向传动轴
steering driving axle 转向驱动轴
steering drop arm 转向臂
steering drum 转向离合器闸轮;导向碾轮(压路机)
steering effort 转向力
steering engine 转向(舵)机;舵机;操向机
steering engine room 舵机舱
steering equipment 操舵装置
steering fin 导向滑板
steering flow 引导气流
steering follow-up linkage 操纵随动联杆
steering force 转向操作力;操纵力
steering gate 导流门
steering gear 转向器;转向机构;转向齿轮;舵机;操纵装置;操舵装置
steering gear adjusting nut 转向齿轮调整螺母
steering gear adjusting nut lock 转向齿轮调整螺母保险
steering gear adjustment 转向机构调整
steering gear arm 转向装置臂
steering gear assembly 转向装置总成
steering gear back lash 转向传动间隙
steering gear box 转向齿轮箱
steering gear bracket 转向机支架
steering gear cam bearing 转向机构蜗轮轴承
steering gear case 转向装置箱
steering gear check 转向机构减震器
steering gear checking scale 转向装置检验秤
steering gear clamp 转向齿轮夹箍
steering gear clamp plate 转向齿轮夹板
steering gear column bushing 转向齿轮柱衬套
steering gear column clamp 转向齿轮柱夹紧夹子
steering gear connecting rod 转向联杆
steering gear connecting rod end 转向联杆端
steering gear connection 转向杆系
steering gear flat 舵机平台
steering gear frame bracket 转向柱固定架
steering gear housing 转向器壳
steering gear housing bracket 转向机壳支架
steering gear housing cover 转向器壳盖
steering gear housing cover gasket 转向器壳垫密片
steering gear housing end plate 转向器壳端盖
steering gear housing lower cover 转向机壳底盖
steering gear housing(oil) filler plug 转向装置壳注油塞
steering gear housing plug 转向装置壳塞
steering gear housing side cover 转向器壳侧盖
steering gear housing top cover 转向器上盖
steering gear lock 转向装置保险;转向角限止器
steering gear ratio 转向传动比;转向齿轮速比
steering gear reduction ratio 转向齿轮机构降速比
steering gear room 舵机舱
steering gear shaft 转向齿轮轴;操纵齿轮轴;转向机构轴
steering gear tube 转向柱管
steering gear tube cam 转向装置空间凸轮
steering gear with circulating ball 循环球形转向器
steering gear with rack and pinion 齿条齿轮式转向器
steering gear with worm and nut 蜗杆螺母转向器
steering gear with worm sector 蜗杆齿扇转向器
steering gear worm 转向蜗轮
steering gear worm shaft 转向蜗轮轴
steering gear worm thrust bearing 转向蜗轮止推轴承
steering grommet 舵桨绳叉
steering handle 操纵手柄;舵柄(把);操纵杆;操舵手柄
steering hand wheel 操纵盘;转向轮
steering hardware 转向机构

steering head 转向头
steering head bracket 转向头托架
steering head bushing 转向头衬套
steering head lock 转向头保险
steering head lock key 转向头钥匙
steering horizontal shaft 换向机水平轴
steering house 驾驶室(又称驾驶台)【船】;操舵室
steering hydraulic cylinder 转向油缸
steering hydraulic oil pump 转向液压油泵;操纵液压油泵
steering hydraulic system 转向液压系统;操纵液压系统
steering impulse 控制脉冲
steering index 操纵指数
steering indicator 驾驶指示器;操纵指示器;方向指示标
steering instruction 驾驶说明
steering journal 转向轴颈
steering king pin 转向主销
steering knuckle 转向节;转向关节
steering knuckle arm 转向(关)节臂
steering knuckle bearing grease nipple 润滑转向节轴油嘴
steering knuckle bracket 转向节支架
steering knuckle bush(ing) 转向(关)节衬套
steering knuckle case oil seal retainer 转向(关)节油封座圈
steering knuckle gear rod arm 转向节臂
steering knuckle king pin 转向关节主销
steering knuckle king pin bushing 转向(关)节主销衬套
steering knuckle king pin dust cover 转向(关)节主销盖
steering knuckle king pin dust washer 转向节主销盖
steering knuckle king pin locking pin 转向(关)节主销键钉
steering knuckle king pin lower bearing 转向(关)主销下轴承
steering knuckle king pin lubricator 转向(关)节主销加油器
steering knuckle king pin upper bearing 转向(关)节主销上轴承
steering knuckle lever 转向节臂;转向关节杆
steering knuckle pin 转向万向接头销;操纵关节销;转向关节销
steering knuckle pivot 转向节销;转向关节枢
steering knuckle sleeve 转向节套筒
steering knuckle spindle 转向节上的轮毂轴;转向关节枢
steering knuckle stop screw 转向节的限位螺钉;转向关节止动螺钉
steering knuckle support 转向节支架
steering knuckle thrust bearing 转向主销推力轴承;转向关节主动轴承
steering knuckle thrust washer 转向节止推垫片;转向关节止推垫圈
steering knuckle tie rod 转向节系杆;转向关节系杆
steering knuckle tie rod ball 转向节系杆球端;转向关节系杆球端
steering knuckle tie rod ball joint 转向节横拉杆球节;转向关节横拉杆球节
steering knuckle tie rod ball seat 转向关节系杆球座
steering knuckle tie rod end 转向节系杆端;转向关节系杆端
steering knuckle tie rod end assy 转向节横拉杆接头组件;转向关节横拉杆接头组件
steering knuckle tie rod end pressure lubricator 转向节横拉杆接头油嘴;转向关节横拉杆接头油嘴
steering knuckle tie rod plug 转向节系杆端塞;转向关节系杆端塞
steering knuckle tie rod spring 转向节系杆弹簧;转向关节系杆弹簧
steering knuckle tie rod yoke 转向横拉杆轭
steering knuckle type 转向节形式
steering level 引导高度;操纵杆
steering lever 驾驶杆;转向杆
steering lever arm 转向杆臂
steering lever column 转向杆柱
steering lever handle 转向杆手柄;转向杆柄
steering lever quadrant 转向杆扇形齿轮
steering lever shaft 转向杆轴
steering link 转向连杆;转向杆件;操纵铰接头;转向联动装置;操向拉杆
steering linkage 转向传动杆系;转向连杆系统;转向联动装置;转向连杆系统;转向杆系;导向铰链装置
steering linkage assembly 转向拉杆机构总成
steering linkage reduction ratio 转向杆系降速比
steering lock 转向限角;转向极限;转向保险
steering lock angle 转向保险角
steering locking angle 最大转向角
steering lock lever 转向装置销定杆
steering machinery 转向机械
steering magnetic compass 操舵磁罗经
steering mast 转向柱
steering master controller 操舵主令控制器
steering mechanism 控制机构;操纵机构;驾驶机构;转向机构;操舵机构
steering method 转向方式
steering motor 舵机电动机;电动舵机
steering needle 控制指针
steering nozzle 转向导流管
steering nut 操纵螺母;转向螺母
steering nut gear 控制螺母转向装置
steering oar 舵桨(救生艇);代桨
steering of shield 盾构导向
steering order 控制指令;控制信号;舵令;操纵信号;操舵口令
steering pad 转向试验场
steering pedal 操纵脚蹬;操纵踏板
steering pedestal 驾驶台;舵(工站)台
steering pillar 转向柱
steering pillar shell 转向柱壳
steering pitman 转向装置轴销;转换装置轴销
steering pitman arm 转向舵盘
steering pitman arm shaft felt wahser 转向舵盘轴油封毡垫
steering pivot 转向枢轴;转向装置轴销
steering pivot lever 转向回转杆
steering play 转向余隙
steering plow 操向犁
steering pointer 操向指示器
steering post 转向柱;舵(工站)台
steering post bracket 转向柱座托架
steering post clamp 转向柱夹
steering post clamp insulator 转向柱夹绝缘体
steering post collar 转向柱管轴环
steering post gear shift 转向柱变速
steering post jacket 转向柱套
steering post spacer 转向柱隔片
steering program(me) 导引程序
steering pump 转向助力油泵
steering quadrant 转向扇形齿轮;舵扇
steering quality 转向操纵性能;转向性能
steering quality index 操纵性指数
steering rack 转向齿条
steering rake 转向柱斜度
steering ram 操向油缸
steering range 转向范围;操纵范围;操纵标距
steering ratio 转向机构传动比
steering reduction ratio 转向变速比;转向减速比
steering reference position 转向基准位置
steering relay 控制继电器
steering repeater 驾驶转发器;操舵罗经复示器;操舵分罗经
steering resistance 转向阻力;转变阻力
steering resisting moment 转向阻力矩
steering response 转向路感;可转向性;可操向性
steering reversal 转向回正
steering rim 转向轮辋
steering rod 转向操纵杆;舵链杆
steering roll 导向滚轮;导向滚筒;转向滚轮
steering room 驾驶室(又称驾驶台)【船】;操舵室
steering rope 转舵索
steering routine 执行程序;导引程序;操纵程序
steering sail 掌舵帆
steering screw 转向蜗杆;转向螺杆
steering sector 转向齿弧;操纵齿弧
steering sensitivity 转向灵敏度
steering shaft 操纵杆;舵链杆
steering shaft bearing 转向轴承
steering shaft bearing packing 转向轴承包装
steering shaft grommet 转向轴护环
steering shaft thrust bearing 转向轴止推轴承
steering shaft worm 转向轴蜗杆
steering shock eliminator 转向机构减震器
steering signal 转向信号
steering spindle 转向螺杆
steering spindle bushing 转向轴套
steering sprocket key 链轮键
steering stability 转向稳定性
steering stabilizer 转向稳定器
steering stand 舵(工站)台;操舵台
steering stop 转向停止器;操向限位器;操向机构限位器
steering stress 操纵应力
steering stub 转向节枢;转向关节枢
steering stub axle 转向节短轴
steering surface 转向面
steering swivel 转向旋转面
steering system 控制系统;操纵系统;驾驶系统;转向系统;驾驶设备
steering system vibration 转向系统的振动
steering telegraph 舵令钟
steering telemotor 舵机传动装置
steering test 转向试验;可控性试验;操纵试验
steering tie rod 转向系杆;转向横拉杆
steering tiller 舵柄
steering tire 转向轮轮胎
steering torque indicator 转向力矩测定仪
steering track rod 转向拉杆
steering trapezium 转向梯形
steering tube 转向柱套
steering tube bush 转向轴套管衬套
steering universal joint 转向万向联轴节
steering well 船尾部井型座舱
steering wheel 转向(盘);舵轮;操纵轮;操向轮;操舵轮;方向盘
steering wheel actuator 转向盘操纵器
steering wheel arm 转向轮辐;转向盘辐条
steering wheel clearance checking scale 转向盘自由间隙测量器
steering wheel cover 转向盘套
steering wheel flutter 转向轮摆振
steering wheel for vehicle 车辆方向盘
steering wheel gear change 转向盘下变速杆
steering wheel geometry 转向轮安装几何位置
steering wheel hub 转向(盘)轮毂
steering wheel key 转向轮键;操向轮键
steering wheel lock 转向盘锁
steering wheel movement 转向盘运动
steering wheel nut 转向盘螺母
steering wheel placing 转向盘定位
steering wheel play 转向盘游隙
steering wheel pointer 舵角指示器
steering wheel puller 转向盘拉具
steering wheel rim 转向盘轮缘;转向轮辋
steering wheel(rim) effort 转向盘操纵力
steering wheel shaft 转向盘轴;转向轮轴
steering wheel spider 转向盘轮辐;转向轮十字架
steering wheel spindle 转向盘轴
steering wheel spoke 舵轮把柄
steering wheel steel core 转向盘钢架芯
steering wheel tilting 转向轮隙
steering wheel tire 转向轮专用轮胎
steering wheel torque 转向盘扭矩
steering wheel tube 转向柱套
steering wire 转舵索;舵索
steering worm 转向蜗杆
steering worm bearing 转向蜗杆轴承
steering worm bearing outer ring 转向蜗杆轴承外环;轴向蜗杆轴承外环
steering worm eccentric adjusting sleeve 转向蜗杆偏心调整套
steering worm gear 转向蜗轮
steering worm roller bearing 转向蜗杆滚柱轴承
steering worm sector 转向器扇形轮
steering worm sector bearing 转向器扇形轮轴承
steering worm sector shaft 转向蜗杆齿形齿轮轴
steering worm shaft 转向蜗杆轴
steering worm wheel 转向蜗轮
steering yoke 转向头;转向节叉
steering yoke bolt 转向节枢
steersman 驾驶员【船】;舵手;舵工;撑筏工人
steersmanship 操舵(技)术
steertree 舵柄
steeve 斜桁仰角;吊杆;斜桅仰角;起重桅(杆)
Stefan-Boltzmann constant 斯忒藩—玻耳兹曼常数
Stefan-Boltzmann law 斯忒藩—玻耳兹曼定律

Stefan number 斯忒藩数	stellerite 红辉沸石	stem-operated valve 推杆操纵的阀
Stefan's formula 斯忒藩公式	Stelleroidea 星形纲	stem plate 艏柱板
Steffen's law 斯蒂芬定律	stelliform network 星形网络	stemple 平台托梁;巷道横梁;井筒内横木梁;嵌入梁
Steffen's process 斯蒂芬法	stellite 钨铬钴合金;斯特莱特钨铬钴硬质合金;司太立合金	stem post 船首柱
Steffen's waste 甜菜制糖废液;菜塘厂废水	stellited valve 钨铬钴合金阀	stem pressing 模压
Stegers crazing test 斯特格斯裂试验	stellite faced valve 司太立合金覆面阀	stem pruning 主干修剪
stegma 硅石条	stellite hard facing 硬质合金表面硬化	stem rack 管子架
steigerite 水钒铝矿	Stelmor process 盘条轧制控制冷却法	stem radiation 靶径辐射
Steiner bubble viscometer 斯坦纳气泡黏度计	stem 芯柱;羽轴;转动表;转柄;重型钻杆;逆风航行;艏柱;艉;顶风或顶流前进;导源;词干;船首柱;船首材;船首	stem remover 除梗滚筒
Steinheil lens 斯坦海尔透镜		stem-rot 心材腐朽
steining 衬(井);井内砌筒;竖井衬壁		stem seal 芯轴密封;杆密封
Steinmann pin 斯坦曼钉		stem seal on valves 阀杆防尘盖
Steinmetz's constant 磁滞系数	stem absorption 母体吸收	stem section 梁腹断面;芯柱断面;(T 形梁等的)梁腹截面
Stekelian analysis 司太克分析	stem a hole 钻孔装药封泥	
stela 纪念堂围起面积;石碑	stem anchor 后锚;船首中锚;船首备用锚	stem sintering furnace 芯柱烧结炉
Stelcon concrete raft 司丹康混凝土筏式基础	stem and cage bolt 杆笼螺栓	stem sluice gate 杆式泄水闸
Stelcon industrial floor cover(ing) 司丹康工业地面覆盖物	stem apex 茎尖	stemson 船头护木;副首材
	stem attachment 管材运输设备	stem straightener 钻铤矫直器
Stelcon steel anchor plate 司丹康钢锚板	stem backer 挤压杆支撑垫	stem stream 主流;干流;干河
stele (建筑物或岩石上准备的)刻字平面;中心柱	stem bag 炮泥纸袋	stem sweep 箭形铲
stellar 星状	stem bank 主干堤岸	stem the tide 顶潮航行
stellar aberration 恒星光行差	stem bar 流嘴;墙身钢筋;梁铰钢筋	stem valve coupling 推杆阀快速接头
stellar altitude camera 恒星姿态摄影机	stem bearing 叶柄轴承;导叶轴承;导柄轴承	stem valve self-sealing coupling 推杆阀自封接头
stellar association 星协	stem bit 冲击回转钻头	stem-ware 高脚器皿
stellar astronomy 恒星天文学	stem borer 天牛;茶天牛	stem-ware automatic forming process 高脚杯自动成型法
stellar astrophysics 恒星物理学	stem by step plan 阶段计划	
stellar atmosphere 天体大气	stem calorimeter 支管量热器	stem wave 船首波
stellarator 仿星器	stem cap 船首柱冠板	stem width 支墩宽度
stellarator configuration 仿星器位形	stem casing 杆套	stench 臭气
stellar background 星空背景	stem chaplet 单面芯撑	stench trap 防臭罩;(防臭)存水弯;防臭瓣
stellar bar 梁铰钢筋;墙身钢筋	stem chock 艏导缆孔	stencil 印花板;油印蜡纸;油印;镂花型板;镂花模版;镂花模板;刻蜡版;模绘板;花版;图案;漏印版;模版
stellar camera 恒星摄影机;太阳摄影机	stem class 树干级木	
stellar chronometer 恒星天文钟	stem correction 管茎改正;首水尺修正	
stellar cluster 星团	stem correction table 船首吃水修正表	stencil application 镂花模板涂敷
stellar comparator 天体坐标量测仪	stem cup 高脚杯	stencil bit 特征位
stellar complex 恒星复合体	stem curvature 首柱曲率	stencil brush 镂花刷
stellar crystal 星状晶体	stem date 装舱开始日	stencil colo(u)red print 模版彩色印刷;彩色镂刻印刷
stellar dynamics 恒星动力学	stem dike 干堤	
stellar dynamo 恒星发电机	stem-dish 高脚盘	stencil duplicator 模版复印机
stellar field 星场	stem draft 船首吃水	stenciled decoration 印花装饰
stellar guidance 恒星导航;天文导引	stem effect 支柄影响	stenciler 模板印刷机
stellar inertial guidance 恒星惯性导航;天文惯性导航	stem fender 船首护材	stencil finishing 镂花涂装
	stem flow 茎流	stenciling 模板刷滑;模板喷花;型板印刷;镂花涂装;漏印版;镂版
stellar-inertial guidance system 天文惯性导航系统	stem foot 首柱脚(龙骨前端)	
	stem form factor 干材形数;树干形数	stenciling brush 镂空模板印字刷
stellar-inertial stable platform 天体惯性导航稳定平台	stem gear 连轴齿轮	stencil ink 模版印刷油墨
	stem grain 船头波纹	stencil-like 模板型的
stellar interference 恒星干涉仪	stem grass 分茎草	stencil mark 货物刷印标志
stellarite 星状沥青煤;火星沥青;土沥青	stem guide 柄的导承	stencil master 模版原版;仿形模板
stellar kinematics 恒星运动学	stem head 首柱头;首柱上端	stencil paint 镂空板印花用涂料;镂花涂装用漆
stellar light 背景星光	stem hygrometer 插入式湿度表	stencil paper 钢板蜡纸;蜡纸;油印蜡纸
stellar lightning 星状闪电	stem insulator 棒式绝缘子	stencil pen 铁笔
stellar luminosity 恒星光度	stem-jack 船首旗	stencil plate 蜡板;型板;模板;镂空模板
stellar magnitude 星等;恒星星等	stem knee 船头护木;副首材	stencil printing 模版印刷
stellar map matching 星匹配	stem knob 销钉头	stencil process 镂花涂装法;丝网印刷法
stellar mass 恒星质量	stem lead 底座引线	stencil sheet 蜡纸
stellar motion 恒星运动	stemless funnel 无管漏斗	stencil tissue paper 薄棉纸
stellar orbit 恒星轨道	stemless nozzle 无杆喷嘴	stencil work 用镂花模板复制(图案)
stellar parallax 周年视差;恒星视差;日心视差	stem light 前锚灯;停泊灯	stenhuggarite 砷锑铁钙矿
stellar pattern 星形模型;星形	stem light of push-train 顶推船队船首灯	stenion 横狭点
stellar physics 恒星物理学	stem-like characteristics 茎状特性	stenochromy 彩色一次印刷法
stellar plan 星形布置图	stem machine 芯柱机;模压机	stenode circuit 高选择性超外差接收电路
stellar plate 恒星底板	stemmed 堵炮泥的	stenode radiostet 超外差(式)收音机
stellar population 星族	stemmer 炮棍;塞药棒;除梗机	stenographer 速记员
stellar radiation 恒星辐射	stemmer saw 除梗的刻齿圆盘	stenography 速记(法)
stellar rotation 恒星自转	stemming 装药孔塞柱;炸药孔填塞;炮眼封泥;填塞(物);填炮泥;捣实;泊位申请	stenohaline 狭盐性的
stellar scintillation 天文闪烁		stenohaline algae 狭盐性藻类
stellar spectrometer 恒星光谱仪		stenohaline animal 狭盐性动物
stellar stability 恒星稳定性	stemming bag 封炮泥袋	stenohaline fishes 狭盐性鱼类
stellar statistics 恒星统计学	stemming date 协议装货日期;装舱开始日	stenohaline organism 狭盐分生物
stellar surface 恒星表面	stemming infilling 井内填料	stenohalinity 狭盐性
stellar system 恒星系	stemming list 船舶在装载港等待泊位次序表	stenohygric animal 狭湿性动物
stellar temperature 恒星温度	stemming machine 装填机;捣固机;夯实机;去梗机	stenoky 狭栖性
stellar triangulation 恒星三角测量	stemming material 填塞材料;炮泥材料	stenometer 速记测距仪
stellar vault 带肋穹顶;有星的穹顶;星状肋穹顶;星形肋拱顶;星体装拱顶;星肋穹顶	stemming piece 填高冢;(平屋顶梁上的)垫木;承拱上的承条(壁炉)	stenonite 氟碳铝锶石
		stenopaic-slit scope 测散光镜
stellar vaulting 支肋拱顶;星状拱顶	stemming plug 炮泥	stenophotic animal 狭光性动物
stellated 星状	stemming rod 炮眼封泥棒;填炸药棒;填塞杆	stenotherm 狭温种;狭温性
stellated drainage 星状水系	stem of beam 梁腹	stenothermal animal 狭温性动物
stellate diaphragm 星状隔膜	stem of bridge pier 桥墩墩身	stenothermal organism 狭温生物
stellated ornament 星形装饰品	stem of funnel 漏斗茎管	stenotic 狭窄的
stellated structure 星状构造	stem of mole 堤干(防波堤)	stenotic type 狭窄型
stellated texture 星状结构	stem of stamp 冲杆	stenotopic speices 狭分布种
stellate hair 星状毛	stem of upright wall 直墙堤身;直立堤堤身	stenter 拉幅机
stellate reticulum 星形网	stem on stem(ship) 迎头对遇	

stentering machine 展幅机
stentorphone 强力扩声器；电磁气流扬声器；大功率扩声器
step 节距；阶段；阶步；梯级；台阶；踏步；措施；擦阶；步骤；步位；步进；步调
step acquisitions 分步取得
step action 阶跃作用
step addition polymer 逐步加成聚合物
step-aerated neutralization 分段曝气中和
step aeration 阶段曝气；分段曝气
step aeration method 分段曝气法
step aeration process 分段曝气法
step aeration system 阶段曝气系统
step allocation 阶跃分配
step and hole type image quality indicator 级孔式缺陷指示器
step and-repeat 连晒机；分步重复
step and-repeat camera 分步重复照相机
step and-repeat machine 逐步反复加工机
step and-repeat pattern 分步重复图形
step and-repeat process 分步重复工序；分步重复（法）
step and-repeat system 分步重复系统
step and-repeat technique 分步重复技术
step annealing 逐步冷却退火
stepanovite 氯草酸钠矿；氯草酸钠石
step Aperture 插片式光圈；分步光圈
step area 分段面积；阶梯段面积
step arrangement 阶段配置；阶段布置；梯段布置；台阶布置
step attenuator 阶梯衰减器；分压器；分级衰减器；分段衰减器
step back 逐层后退
step-back method 分段反向焊法；分段反向堆焊法
step-back method of welding 分段反向焊法
step back reclaiming method 往复式取料法
step-back relay 跳返继电器
step-back welding 逆向分段焊接；后向分段焊接；分段退焊法焊接
step bar 阶梯试块
step bearing 立式止推轴承；立式推力轴承；阶式止推轴承；阶式推力轴承
step bearing bracket 立式止推轴承托架
step bit 阶梯式钻头
stepblender 梯形混合机
step block 级形垫铁；多级滑轮；阶梯形垫铁；台阶形垫铁；多组滑轮；挡土坎
step board 踏面；踏步板
step bolt 阶梯形螺栓；半圆头方颈螺栓
step-bore cylinder 异径缸
step bracket （楼梯踏步及梁端部的）牛腿式雕饰；牛腿式装饰
step brass 轴瓦；轴衬衬
step brazing 分层降温铜焊法；层次硬钎焊；层次钎焊；分级钎焊；层次焊接
step budget 分步预算
step budget method 梯级预算法
step button 阶跃按钮；步进按钮
step-by-step 逐步的；逐级；步进的
step-by-step accelerator 级联加速器；步进加速器
step-by-step algorithm 逐步算法
step-by-step approach 逐步逼近法
step-by-step approach to final 逐渐逼近平差法
step-by-step approximation 逐次逼近法；逐步近似法；逐步渐近法
step-by-step automatic system 步进自动制；步进制自动系统
step-by-step automatic telephone system 步进制自动电话交换机
step-by-step braking 阶梯制动
step-by-step calculation 按步计算
step-by-step calibrated procedure 逐次校正法
step-by-step calibration method 逐次校正法
step-by-step carry 逐位进位；按位进位
step-by-step computation 逐位计算；按步计算
step-by-step control 步进控制
step-by-step controller 逐级控制器；步进式控制器
step-by-step conveyer 步进式输送机
step-by-step correction procedure 逐次校正法
step-by-step counter 步进计数器
step-by-step counting 脉冲计数；步进计数
step-by-step decoding 逐步解码
step-by-step design 逐步设计（法）；阶段设计；分段设计

step-by-step drilling 逐步钻法
step-by-step drive 步进传动
step-by-step excitation 逐步励磁
step-by-step feed 逐步进给
step-by-step impulse 步进脉冲
step-by-step instruction 步进指令
step-by-step integration 逐步积分
step-by-step integration procedure 逐步积分法
step-by-step method 循序渐进法；逐次测量法；逐步方法；逐步测量法；顺序渐近法；步进法；渐近法；按步法
step-by-step motor 步进式电动机；步进电机
step-by-step operation 逐位操作；单步操作；步进式操作；步进操作；按位操作；按步操作
step-by-step plan 分期规划
step-by-step procedure 逐步过程；逐步方法；按步法
step-by-step process 逐次逼近求解过程；逐步求解过程
step-by-step reaction 逐步反应
step-by-step recycling 步进循环
step-by-step regulation 逐级调节
step-by-step regulator 阶式调节器
step-by-step relay 步进式继电器；步进继电器
step-by-step roller spot welding 步进式滚点焊
step-by-step saturation method 逐步饱和法
step-by-step seam welding 步进式缝焊
step-by-step selection 步进选择
step-by-step selector 步进式选择器
step-by-step simulation 逐步模拟；步进模拟；步进仿真
step-by-step switch 步进制开关；步进式开关；步进式交换；步进开关
step-by-step switch testing equipment 步进制机键测试仪
step-by-step system 步进制；步进（式）系统；步进方式
step-by-step telephone switching system 步进电话交换机；变进制电话交换机
step-by-step test 逐步试验；阶段试验；步进制试验
step-by-step tracking 步进跟踪
step-by-step transmitter 步进发送器
step-by-step tray 逐步板；逐板
step-by-step type 步进制的
step-by-step type seamless tube rolling mill 无缝钢管轧机；无缝钢管的分段轧制设备
step by step variable gear 多级变速齿轮
step-by-step welding 跨步焊；步进焊缝
step cast 阶梯式地形模型塑造
step change 阶跃变化；阶段变速；级变；单增量变化；步进变化
step-characteristic 阶梯特性
step chromatography 台阶色谱法；台级色谱法
step chuck 阶梯弹簧夹头
step cleaner 梯形开棉机；梯形除杂机
step closure 分段关闭
step component 梯级单元；（楼梯的）梯级
step-cone 宝塔轮
step connection 级式连接；分品连接；分级连接
step control 分级控制；分级管理；分步控制；步控制
step controller 分级控制器
step control system 分级控制系统
step cooler 阶段式冷却器
step core 阶梯形铁芯
step core bit 阶梯式金刚石岩芯钻头
step cost 阶梯成本；梯式成本；梯级成本
step counter 阶跃计数器
step counterbore 阶级式平底扩孔钻
step count set 步进计数装置
step cover 踏步顶盖；轴承盖
step covering 跨盖
step current 阶跃电流
step curve 梯形曲线
step cut 阶梯式开挖；梯形掏削；分段裁切
step cut milling 分段铣削
step cutting 台阶式挖土
step cutting test 分段切削试验
step dams 梯级坝
step delay 分级延迟；步进延迟
step diffuser headbox 方管宽布浆器
step distribution method 阶梯分配法
step dose 分次投药
step down 阶梯式下降；低落；变低；降压

step-down amplifier 降压放大器；降低放大器
step-down ceiling diffuser 下落式顶棚散流器
step-down gear 减速器
step-down lease 递减租金的租赁
step-down method 间接分摊法
step-down ratio 降压比；变压器降压比
step-down schedule 降压表
step-down (sub) station 降压站；降压变压器；降压变电所；降压变电站
step-down substructure 阶梯式基础
step-down transformer 降压器；降压变压器
step-down voltage regulator 降压调节器
step drawdown test 分段抽水试验
step drill 分级钻头
step drill bit 阶梯式钻头
step drilling 分段钻削；分段钻进（深孔）
step duration 步进时间
step error 阶跃误差
step execution 逐步执行
step exposure test 逐级曝光试验
step expression 步长表达式
step face 坎面
step face(design) bit 阶梯状钻头
step faulted belt 阶断带
step faulted region 阶断区
step faulted zone of west philippine islands 菲西断阶带
step faults 阶（梯）状断层
step fault scarp 阶状断层崖
step fault zone 断阶带
step feed 分级进给
step feed activated sludge process 分段进水活性污泥法
step feed drill attachment 分级进给钻削附件
step feeding 分段投料
step filter 阶梯滤光片
step flange 阶式法兰
step flashing 踏步泛水；阶梯形披水板
step flow 台阶流
step fold 阶状褶皱
step footing 梯级式基脚；梯级式大方脚；阶状梯级
step force 台阶力
step forced response 阶跃扰动响应
step fracture 分段裂纹
step freezing 分段冻结
step function 阶跃函数；阶梯函数；阶段函数；阶段功能
step function analysis 阶梯函数分析
step function generator 阶跃函数发生器
step function signal 阶跃函数信号；阶跃信号
step gable 阶梯式山墙；马头山墙；梯形山墙
step gate 阶梯浇口；分层浇口
step ga(u)ge 梯形隔距；台阶形量规
step generator 阶梯信号发生器；级信号发生器
step-graded 间断级配的
step-graded material 间断级配材料
step-graded mix 间断级配混合料
step grate 阶梯式炉算；阶梯炉排；塔式篦子
step grate producer 阶梯式发生炉
step gray scale 灰度等级
step grind 多段研磨
step grinder 分级砂轮机；多级砂轮机；分级磨床
step grind(ing) 分段研磨
step growth 台阶式生长
Stephane 史蒂芬尼头饰
step hanger 踏阶支架；踏阶吊铁
Stephanian 斯蒂范阶
stephanite 脆银矿
step height 阶跃高度；台阶高度；陡度
step height interferometer 阶跃高度干涉仪
Stephen conjack 斯蒂芬集装箱起重装置
Stephenson link 斯蒂芬逊连杆
Stephenson's alloy 斯蒂芬逊减摩轴承合金
step hydroelectric(al) station 梯级水电站
step in 步入
step index 步长指数
step index optic(al) waveguide 阶跃折射率光波导
step index profile 阶跃折射率分布
step index waveguide fibre 阶跃光波导纤维
step induction reguraltor 分接头式感应调压器
step induction regulator 步进感应式电压调节器；步进感应式电压调节器
step injection 阶梯式注入
step input 阶式信号输入；分段进水；阶跃输入；阶

step-in utility cooler 通用装配式冷却器
step iron 附壁爬梯;铁踏步;铁爬梯;踏脚铁条
step iteration[interative] method 逐次迭代法
step jaw 阶梯形卡爪
step joint 齿式接头;阶式接合;台阶状接合;齿式接合;槽齿连接;槽齿接合
step-jointed coupling 梯形丝扣接头
step junction 阶跃结;突变结
step-kiln 自动窑;连续作业窑;连续窑;渐进干燥窑
step labyrinth 阶梯式迷宫密封
step ladder 码头瓜梯;阶梯;活梯;人字梯;梯凳;直爬梯;活梯子
stepladder method 阶梯分配法;间接分摊法;梯式分摊法;梯级分配法
stepladder stage 阶梯状热期
step-laminate 层压薄片
step layout 阶梯式布置
step length 梯级长度;步长
step lens 棱镜;分步透镜
step lens condenser 阶梯形聚光镜
stepless 连续的;无(级)的;平滑;不分级的
stepless acceleration 无级加速
stepless action 连续作用;均匀作用;无分步作用
stepless canal 开敞式运河;无闸通航渠道;无梯级运河
stepless change 连续变速
stepless control 连续控制;连续调整;均匀调整;无级控制
stepless control gearing 无级调节传动
stepless friction transmission 无级摩擦式传动
steplessly variable automatic transmission 无线自动变速器
stepless regulation 平滑调节;无级调节;不分挡调速
stepless sea canal 无闸通海运河
stepless servo-motor 无级调速伺服电动机
stepless speed adjusting gear 无级调速器
stepless speed change 无级变速
stepless speed change device 无级变速装置
stepless speed changing type 无级变速型
stepless speed control 无级调速;无级变速控制
stepless speed regulating 无级调速
stepless speed regulation 无级调速
stepless speed variation 无级变速
stepless transmission 无级变速箱;无级变速器
stepless variable drive 无级变速传动(装置)
stepless variation of torque ratio 变扭比无级变化
stepless voltage regulation 无级调压
stepless voltage regulator 无级调压器
step library 步进程序库
step-like 梯级状
step-like epitaxial growth 台阶状外延生长
step-like function 步进函数
step-like interface 阶梯界面
step-like rates 台阶式运率
step-like surface of denudation 梯形剥蚀面
step-like terrace 梯状阶地
step lining 分级衬板
step load change 分步负载变动
step load(ing) 阶跃荷载;单folio荷载
step logic 步进逻辑
step making machine 梯级制造机
step masonry work 梯级圬工工程;阶梯形圬工建筑
step measuring 步测
step mechanical hand 步进机械手
step member 梯级构件;(楼梯的)梯级
step meter rate 切间电量递减收费制;级差电费制
step method 逐步逼近法;光阶法;分级推算法;分段推算法;分段计算法
step micrometer 阶梯型测微计
step mill cutter 阶梯形端铣刀
step milling 步进式铣削
step mode of control 分段调节法
step motion electric(al) drive 步进电气传动
step-motor 步进电动机
step multiplier 阶梯式乘法器;步进式乘法器
step multiply 步进式乘法
step multiply operation 步进式乘法运算
step name 步骤名
stepney 预备轮胎;备用轮胎
step noise 阶跃噪声
step number 步数
step of coal resources exploration 煤炭资源勘探阶段

step off 量出
step of mine end 矿山结束阶段
step of mining planning 矿山设计阶段
step of production geology 土产地质阶段
step of reaction 反应阶段
step of spiral stair(case) 螺旋梯踏步
step of stair(case) 楼梯踏步
step optic(al) fiber 阶跃光纤
stepout 解列
stepout hole 扩边孔
stepout relay 失调继电器;失步继电器
stepout well 远距离钻井;扩边井
step pattern 阶状纹
steppe 干草原;草原
steppe black earth 草原黑土
steppe climate 草原气候
steppe community 草原群落
stepped 碾磴坡度;阶形的;阶梯形的;分级的
stepped abutment 阶形桥台;阶形拱座
stepped acclimatization 分段式气候驯化
stepped addressing 逐步寻址
stepped aeration 分段曝气
stepped ag(e)ing 阶段时效
stepped annealing 分段退火
stepped appearance 阶梯状
stepped arch 阶(梯)形拱
stepped area 阶段截面积
stepped artificial hill 梯形假山
stepped atomic time 跳跃原子时
stepped austenitizing 分段奥氏体化
stepped baffle 多级跌水坎
stepped beam 阶形梁
stepped bearing 阶式轴承
stepped bed 台阶式渠底;阶梯形渠底;梯底;阶台式渠底;分阶段跌水
stepped bit 阶状钻头
stepped block 阶形方块壁
stepped block wall 阶状方块岸壁
stepped bore 级形孔
stepped bottom 阶形底
stepped bowl centrifuge 多级碗式离心机
stepped breast wall 阶式胸墙
stepped bulkhead 阶形舱壁;槽形舱壁;凹形舱壁
stepped cam 分级凸轮
stepped cap flashing 阶梯形帽盖泛水
stepped cave 阶梯状溶洞
stepped channel 梯级河槽
stepped clamping block 级形垫铁
stepped column 阶形柱
stepped comb 分级伸缩箍
stepped concrete path 有踏步的混凝土通道
stepped cone 级轮;塔轮
stepped cone pulley 锥形柄轮
stepped construction 阶梯式结构;层叠结构
stepped control 步进控制;阶梯控制
stepped cooling 分段冷却;分段冷却
stepped cost 台阶式成本;步序费用
stepped counter flashing 阶梯反向防水板
stepped court 阶梯式庭院
stepped cross-section 阶跃式截面;阶梯形截面
stepped cross-section(al) block wall 阶梯形断面方块码头
stepped cup 阶梯式拉延件
stepped curve 阶梯(形)曲线
stepped curve distance-time protection 阶梯时限特性距离保护装置
stepped cut 分级切断
stepped cut joint 阶形切口对搭接头
stepped cut ring 直切环
stepped cutter 多联齿轮铣刀
stepped development 梯级开发
stepped diameter 等级直径
stepped diameter auger 梯级异径螺旋
stepped discontinuity 台阶状结构面
stepped drain 阶形排水
stepped drive 分级变速传动装置
stepped driven device 多级变速传动装置
stepped driving cone 主动级轮
stepped energy destroying spillway 阶(梯)形消能溢洪道;阶梯式消能溢洪道
stepped extruded shape 阶梯式挤压型材
stepped extrusion 阶段式挤压
stepped face 阶梯形面;台阶式工作面
stepped face sea dike 阶梯面海堤

stepped face sea wall 阶梯面海堤;阶梯面海塘
stepped face wall 阶梯形面墙
stepped factor of safety 分段变动安全系数
stepped fault 阶梯断层
stepped feeding 阶段投料;分段进水
stepped filter 阶梯滤色器;阶梯滤光片
stepped fish pass 梯级鱼道
stepped flashing 阶梯形泛水【建】;踏步形泛水;阶形泛水;阶梯式防漏板(房屋)
stepped floor 阶梯形楼地面;阶形地板
stepped flushing 阶形排水;阶形防雨板
stepped focusing magnet 阶梯型聚焦磁铁
stepped footing 阶梯形底座;阶梯形脚脚;阶(梯)形基础;阶式底座;阶台式脚脚;阶阶式基础
stepped foundation 阶形基础;阶梯形基础;阶形底脚;阶式基础;踏步形基础
stepped funerary temple 阶梯形丧葬用寺庙
stepped gable 阶形山墙;阶式山墙
stepped gable wall 五花山墙
stepped gear 塔轮齿;塔齿轮;分级齿轮
stepped gearing 级齿轮传动装置
stepped gear wheel 台阶齿轮
stepped grate 梯级炉格
stepped grate cooler 阶段箅式冷却器
stepped grate plate 阶梯箅板
stepped grey wedge 阶梯式灰度光楔
stepped grizzly 多级格筛
stepped hardening 分段淬火
stepped head wall 阶梯状端墙(位于船坞内部)
stepped hillside house 坡地住宅
stepped hole 阶梯孔
stepped house 阶梯式房屋;台阶式住宅
stepped increase 台阶式增加
stepped joint 阶形切口对搭接头;阶梯形接合;阶梯接头;挖割的分型面
stepped kiln 阶梯窑
stepped labyrinth gland 高低齿齿迷宫汽封
stepped leader 梯级先导
stepped lens 阶梯形透镜;阶梯式透镜
stepped liner 阶梯形衬板
stepped lining plate 阶梯衬板
stepped load 阶段荷载
stepped loading 分段负载
stepped masonry work 阶梯形砌体工程;阶梯圬工
stepped mastaba(h) 阶梯式长方形斜坡坟墓(古埃及)
stepped merlon 阶梯形(城墙)雉堞
stepped mining 阶梯矿房法
stepped model 阶梯式模型
stepped mortuary temple 阶梯形停尸寺庙
stepped multiple lap 多重阶形切口搭接
stepped-off joint 阶梯式切口接合;阶梯式切口接缝
stepped overflow weir 阶梯式溢流堰
stepped-packing block 阶梯形支柱
stepped piston 级形活塞
stepped plain 阶状平原
stepped plug ga(u)ge 台阶形塞规
stepped-ply belt 厚边输送带
stepped ply conveyer belt 边缘力强运输带
stepped pool 梯级式水池
stepped-pool fish pass 鱼梯;跌塘式鱼道
stepped positive mo(u)ld 阶梯式阳模
stepped post 阶形柱
stepped profile 阶梯形
stepped pulley 级轮;塔轮;阶式滑轮;宝塔式滑轮
stepped pyramid 阶(梯)形金字塔;阶梯式金字塔
stepped quenching 分段淬火
stepped radial interstage seal 级间高低齿式径向密封
stepped ramp 阶形坡道;踏步式坡道
stepped refining 分级精炼;分级调质;分段精炼
stepped resistance 分段电阻
stepped retaining wall 阶形挡土墙;阶式挡土墙
stepped-ring dowel 阶形榫椎;齿形环销
stepped roller 台阶旋轮(旋压用)
stepped roof 阶形屋顶
stepped scarf 梯级斜接
stepped scarf joint 阶梯斜接
stepped screen 分级筛
stepped screw 台阶螺钉
stepped seating 踏步式座位
stepped section 阶形断面;阶梯形断面
stepped shaft of pier 阶梯式墩身
stepped shirting 阶梯形壁脚

stepped side-wall 阶形边墙;阶式边墙
stepped sink hole 阶梯状落水洞
stepped skirting 阶梯形壁脚(板)
stepped slit 阶梯狭缝
stepped soffit 阶梯形拱腹
stepped spillway 阶式溢洪道
stepped spinning 错距旋压
stepped stacking 梯堆垛法
stepped stalls type auditorium 阶梯式观众厅
stepped start(ing) 分级起动
stepped start-stop system 分级起止式;分段起停系统;分步起停方式;步进式起停系统;步进起止系统
stepped stone 阶梯石级
stepped string 锯齿形楼梯斜梁;明楼梯梁;踏步式楼梯梁
stepped stringer 阶梯形楼梁;阶形楼梯斜梁
stepped superstructure 阶梯式上层建筑
stepped surface 级形面
stepped system of dredging 阶式疏浚方式;阶式疏浚方法
stepped taper tube 逐节变直径管
stepped teeth gear 级齿轮传动装置
stepped teeth reamer 阶梯齿形铰刀
stepped temple 阶梯形寺庙
stepped tenon 阶梯式榫
stepped terrace 阶梯形里弄;阶梯形街坊;阶梯形平台
stepped terrace block 阶梯形里弄街坊
stepped terrain base 阶梯地形模型基板
stepped thrust-bearing 踏步式止推轴承
stepped-up 加速的
stepped voltage 步进电压
stepped voussoirs 平顶拱石;阶式拱石;踏步式拱楔块
stepped wall 阶形墙
stepped wall foundation 台阶式墙基础
stepped wave 阶梯式波
stepped waveguide 阶梯形波导(管);阶梯式波导管
stepped weakener 阶梯减光板
stepped wheel 塔轮;塔齿轮
stepped wheel gear 塔轮装置
stepped winding 阶梯形绕组
steppe land 草原
steppe planning 草原规划
steppe plant 草原植物
stepper 分节器;分挡器
stepper motor 步进马达;步进电机;步进电动机
steppe soil 草原土(壤)
steppe sylviculture 草原造林学
steppe type of soil formation 草原成土作用
steppe vegetation 草原植被
steppe zone 草原带
step photometer 步进光度计
steppification 草原化
stepping 踏步木;逐步变化的;进桩;成区;分段开挖法;步进
stepping accuracy 步距精度
stepping angle 步进角
stepping counter 移位计数器;步进计数器;分级计算器
stepping formwork equipment 梯级式模板装置;超前支架设备;(建筑物外伸部分的)模板设备
stepping iron 金属踏步;埋入金属踏步
stepping lightly and heavily alternately 虚实徘徊势
stepping mechanism 阶跃机构;步进机构
stepping method 分段测高法;步进焊法
stepping motor 步进马达;步进电(动)机
stepping motor bank 步进马达群
stepping motor controlled module 步进马达控制模块
stepping motor system 步进电机系统
stepping-off 用角尺分步画线
stepping of foundation 基础踏步
stepping-off place 出发地点
stepping point 嵌接线;榫条的内边线
stepping preselector 步进式预选器
stepping rate 步进速度
stepping reflex 跨步反射
stepping register 步进式寄存器
stepping relay 分级继电器;步进(式)继电器
stepping sampling 分段采样
stepping scanning circuit 分级扫描电路
stepping service girder 超前支架大梁;(建筑物外伸部分的)支架大梁
stepping spud 进位桩;前移桩(指挖泥船)
stepping stone 步石;踏步石;垫脚石
stepping-stone method 踏脚石法(解答运输问题的一种方法)
stepping-stone state 起脚石状态
stepping switch 步进开关;分挡开关;步进继电器
stepping technique 步进技术
stepping technology 步进技术
stepping time 步进时间
stepping-type relay 分段动作继电器
stepping wood 木楼梯的厚踏步板
step piston 阶梯式活塞;级差活塞
step-plank 踏步木板
step potentiometer 分级电位计;步进式电位器
step press 踏步压型机(楼梯)
step printer 步进式印刷机
step profile 阶梯形轮廓;阶梯形断面;阶形断面
step pulse 阶跃脉冲;步进脉冲
step-pulsed welding 步进脉冲焊
step pyramid 阶梯形金字塔
step rail 阶式路轨
step ramp loading chute 斜梯式装车台
step rate 分级费率
step-rate prepayment meter 分级预付电度计
step reaction 阶段式反应;逐步反应
step recovery 阶跃恢复
step recovery effect harmonic generator 阶跃恢复效应谐波发生器
step-recycle counter current flo(a)tation process 分段逆流气浮工艺
step-refraction index 阶跃折射率
step regulator 分级调压器;步进式调节器
step relay 步进继电器
step repeat 分步重复
step reservoirs 梯级水库
step resistance 跳步电阻
step resistance type tide ga(u)ge 分级电阻式验潮仪;分级电阻式潮汐计
step resistance type wave ga(u)ge 分挡电阻式波高仪;分挡电阻式波高计;分级电阻式测波仪
step response 阶跃响应
step response problem 阶跃响应问题
step restart 步式重新启动;步进(式)再启动
step riddle 阶梯式清选筛
step risers 楼级踏板
steps 梯级跌水;一段楼梯;活梯;双折板
step sampling 分层抽样;分层采样
step scale 分级标度;步数标度
step scan 步进扫描
step-scan mode 分段扫描方式
step screen 阶梯式筛
step screw 止杆螺钉;带阶梯螺钉
step seal 阶式密封
step sector 阶梯遮光板
step-sector disk 阶梯扇板
step sequence 步进顺序
step-servo motor 步进伺服电机;步进电机
step shaft 级形轴;塔形轴;变截面轴
step-shaped 阶梯形的
step shaped canal 梯级运河
step-shaped graph 阶梯式曲线图;阶梯形曲线图
step-shaped platform 阶梯形货物站台
step shutdown 分步停运
steps in decision-making 决策步骤
step size 步长
step size change 步长改变
step sizing 逐级筛选;连续筛选;筛分尺寸;分级筛分
step-sizing operation 分段筛分作业;按顺序筛分
step slide switch 分挡滑动开关
step slope 阶梯型边坡
step soldering 分层降温焊接法;逐次焊接;层次软钎焊;分级钎焊
step spectrum method 数阶法
step speed 级速度
step speed change 有级变速
step stanchion 梯柱
steps teller 步数计;步程计;记步器
step stone 阶梯石级;楼梯石级;阶沿石
step-strain phenomenon 阶形应变现象
step street 阶式(人行)街道;踏步式街道
step stress test 逐级加载(应力)试验;级增应力试验;分级加载应力试验;步进应力试验
step strobe marker 逐级选择标记
step structure 台阶形结构
step switch 分挡开关
step-switch converter 步进转换器;步进式变压器;步进式(开关)转换器
step table 桌式梯;踏步桌
step-taper concrete pile 分段渐缩的混凝土桩
step-tapered-cylindric(al) structure 阶梯状锥形结构
step-tapered drainage line 渐细河流线
step-taper pile 阶梯状锥形桩;分段锥形桩;分段截形桩;斜面坡度分段变化桩
step technique 步进法
step terrace 梯状阶地
step test(ing) 阶段试验;阶式试验;台阶试验;分步试验
step-test procedure 分级试验法;逐级试验法;逐步试验法
step the sizes 注明尺寸等级
step tile 阶梯形砖;阶梯形瓦
step timer 步进计时器
step tracking 步进跟踪
step transformer 升降压变压器;升降压变压器;分级变压器
step tread 踏板
step treads and risers 踏步的踏板与踢板
step-tube counter 步进式计数器
step turner 阶台弯板器(用于将铅皮弯成直角);阶形弯板器;阶梯式泛水板弯板
step-type 步进式的
step-type counter 步进式计数器
step-type ladder 阶梯式扶梯
step-type multimode fiber 阶跃型多模纤维
step-type optic(al) fiber 阶跃形光导纤维
step-type recording wave ga(u)ge 分挡电阻式波高仪;分挡电阻式波高计;分级式波高仪;步进分级记录式自记测波仪
step-type seam welder 步进式缝焊机
step-type single mode 阶跃型单模
step-type spot welder 步进式点焊机
step-type support 迈步式支架
step unit 踏步单元;(楼梯的)梯级
step up 升高
step-up-and-down transformer 固定升降变压器
step-up austempering 升温等温淬火
step-up auto-transformer 升压自耦变压器
step-up circuit 升压电路
step-up coil 升压线圈
step-up direct-current transformer 升压直流变压器
step-up drive 增速传动
step-up error 升压误差
step-up exchange 步进制交换机
step-up excitation 升压激励
step-up frequency changer 升频变换器
step-up gear 增速传动装置;加速齿轮
step-up lease 定期升租租契
step-up method 变数增加法
step-up motor 步进电动机
step-up(of voltage) 升压
step-up piston spring 加力活塞弹簧;级差活塞弹簧
step-up piston 加力活塞;级差活塞
step-up pulse transformer 升压脉冲变压器
step-up ratio 升压比;升高系数;变压器升压比
step-up ring 大小连接环
step-up side 高压绕组边;高压侧
step-upstair(case) 上楼梯
step-up substation 升压站;升压变电站;升压变电所
step-up system 增力装置;加浓装置
step-up throat 上升式流液洞
step-up transformer 升压器;升压变压器
step-up transmission 增速传送
step valve 阶式阀;级阀;层式阀
step vane 阶梯式叶片
step-variable capacity cost 梯式变动能量成本
step-variable cost 梯式变动成本
step variation 阶跃变化
step vein 阶状矿脉
step velocity 级速度
step vibrating press 踏步振动压型机(楼梯)
step voltage 阶跃电压;阶迁电压;突跳电压
step voltage regulator 阶跃电压调整器;分级电压调节器;步进式调压器;步进式电压调整器
step wave 阶梯波
step wave form 梯级波形

stepway 踏石小径;梯段
step weakener 级梯减光板
step wedge 楔形梯级;感光谱;阶跃式光楔;灰度梯尺;步进式光楔;梯尺
step wedge sensitometer 阶梯光楔感光计
step-wedge type 级楔式
step weir 阶梯式堰;阶级式堰;梯级堰
step width 步长
step winding 多段绕组
step wire 台阶形线
stepwise 步长;阶梯状布置;逐步的;阶式;步进式
stepwise aeration process 分段曝气过程
stepwise application 逐步应用
stepwise backwards method 逐步向后法
stepwise comparison 逐步比较
stepwise computation 逐级计算(法);逐步计算;分段计算
stepwise construction 分(阶)段施工
stepwise continuous 按步连续
stepwise convergence 步长型收敛
stepwise countercurrent extraction 分段逆流抽提
stepwise decomposition 逐级分解;逐步分解
stepwise depletion procedure 分段枯竭法;分段抽油法
stepwise device 阶梯装置
stepwise discrimination 逐步判别
stepwise distillation 分段蒸馏
stepwise elution 分阶洗脱;分阶段洗脱
stepwise excitation 分步激发
stepwise formation constant 逐级生成常数
stepwise forwards method 逐步向前法
stepwise gradient 阶式梯度
stepwise heterocoagulation 逐步杂凝聚
stepwise hydrolysis 逐级水解
stepwise method 分段逐次计算法;分段依次计算法;分段法
stepwise operated camera 分步摄影机
stepwise polymerization 逐步聚合
stepwise polynomial regression 逐步多项式回归
stepwise regeneration 分步再生
stepwise regression 逐步回归(法);分段回归
stepwise regression analysis 逐步回归分析
stepwise relaxation 逐级松弛;逐次松弛
stepwise relaxation method 逐次松弛法
stepwise solution 阶式解
stepwise stability constant 逐级稳定常数
stepwise synthesis 分步合成
stepwise titration 依次滴定;逐级滴定;分级滴定;分段滴定;分步滴定
stepwise variable optic(al) attenuator 分级可变光衰减器
step with mortar rubble 浆砌片石踏步
step with nosing 带挑口的踏步
stepwork 迎水坡阶梯式护坡工程
step X register X 步进记录器
step Y register Y 步进记录器
steradian 立体弧度;球面角度(立体角单位)
steradiancy 球面发射强度;辐射率
sterance 立体角密度
sterane 甾烷
steranes content 甾烷含量
Sterba antenna 司梯巴天线
Sterba curtain 斯梯巴幛形天线
stercorite 磷钠铵石
stercoroma 积粪
Sterculiaceae 梧桐科
sterculic oil 梧桐籽油
stere 立方米;层积立方米
sterecscope 体视镜
sterene 甾烯
stereo 立体照片;立体的
stereoacuity 立体清晰度;立体量测精度
stereo aero triangulation 空中立体三角测量仪
stereo amplifier 立体声放大器
stereo angle corrected value 立体角校正值
stereo attachment 立体摄影附件
stereoautograph 自动立体测图仪;立体自动测图仪;体视绘图仪
stereoautomatograph 立体自动测图仪
stereobase 立体基准;立体基线;建筑物地下室(无柱底基)
stereobase unit 聚合物的立体异构基本单位
stereobate 土台;建筑物地下室;基座;基础;无柱底基;无柱地基;外露底基;台基

stereobetatron 立体电子感应加速器
stereoblique plotter 倾斜立体测图仪
stereo-block 立体块规;立构规整嵌段
stereoblock adjustment 立体区域网平差
stereobroadcasting 立体声广播
stereocamera 立体照相机;立体摄影机
stereocampimeter 立体视野计
stereocartograph 立体测图仪;体视测图仪
stereo cartridge 立体声拾音头
stereocasting 立体声广播
stereochemical change 立体化学变化
stereochemical effect 立体化学效应
stereochemical formula 立体化学式
stereochemistry 立体化学
stereochrome 有立体感的彩色壁画
stereochromy 立体彩饰法;彩色立体照相
stereo-cinefluorography 立体荧光电影相术
stereocomparagraph 立体坐标测图仪;体视比较仪
stereocomparator 体视比较镜;立体坐标仪;立体坐标量测仪;体视比较仪
stereocompilation 立体测图
stereo crossed microphone 立体声交叉传声器
stereodetection 立体检测
stereodetector 立体检测器
stereo-directed polymer 立体定向聚合物
stereo display 立体显示
stereoeffect 立体声效应
stereo electron microscope 立体电子显微镜
stereo electron microscopy 立体电子显微术
stereoequipment 立体测图仪
stereo-exaggeration 立体放大
stereoexposure 立体摄影曝光
stereo facet plotter 小平面立体测图仪
stereo-fluoroscope 立体荧光屏
stereofluoroscopy 立体荧光术;立体荧光屏透视检查;立体荧光法
stereofunduscope 立体眼底镜
stereo geologic(al) mapping 立体地质填图
stereognomogram 极射心射图
stereognosis 形感觉;立体感觉;实体辨别
stereogoniometer 立体量角仪;体视量角
stereograft polymer 立构规整接枝聚合物
stereogram 立体照相;立体照片;立体图;立体频数;球极投影图;体视图;实体图;三维直方图;赤平图
stereogrammetric plotter 立体测图仪
stereogrammetric plotting procedure 立体测量法
stereogrammetry 立体摄影测量学
stereogram of aquifer structure 含水层立体结构图
stereograph 立体照片;立体像对;立体平面图;体视照片
stereographic(al) 立体平面的;立体图的
stereographic(al) coverage 立体像片重叠
stereographic(al) grid 立体格网;球极投影格网;立体平面格子图
stereographic(al) mapping 球极映射
stereographic(al) net 赤平网;球极投影网
stereographic(al) polar projection 极射赤平投影
stereographic(al) projection 立体照片投影;陆平极射投影;立体投影;球面投影;球极平面投影;球极平面射影;赤平(极射)投影
stereographic(al) (projection) grid 球极平面投影网
stereographic(al) (projection) net 球极平面投影网
stereographic(al) (projection) ruler 球极平面投影尺
stereographic(al) projector 球极投影器
stereographic(al) technique 赤平极射技术
stereography 立体作图法;立体摄影术;立体平画法;立体描图;立体几何(学);立体画法;体视术;体视法
stereograticule 立体视标线
stereo holography 立体全息摄影
stereo-identical points 立体对合点
stereo-inspection 立体镜观测
stereoisomer 立体异构体
stereoisomeric polymer 立体异构聚合物
stereoisomeric specificity 立体异构特异性
stereoisomery 立体异构
stereoisomerism 立体异构现象;立体异构
stereo-lens camera 立体摄影机
stereo lens pair 立体物镜对
stereo line plotter 简易立体绘图仪
stereology 立体测量学;体视学

stereomapping delineation 立体描绘
stereomapping satellite(stereosat) 立体测图卫星
stereo-mate 立体像对;立体配对片
stereome 立体的
stereomeasurement 立体量测
stereomer 立体异构体
stereomeride 立体异构体
stereometer 立体量测仪器;立体测量仪;体视测量仪;体积计;视差测镜
stereometer-type instrument 立体测量仪;立体量测仪器
stereomethod 立体测量法
stereometric(al) camera 立体量测摄影机
stereometric(al) map 立体测量图;立测地形图;实体测量图;立体测绘地形图
stereometric(al) method 立体测量法
stereometrograph 立体测图仪
stereometry 立体几何(学);立体测量学;测体积学;测体积术;测体积法
stereomicrography 立体显微(照像)检验;体视显微术
stereomicrometer 立体测微器;立体测微计;体视测微计
stereomicrophone system 立体声传声系统
stereomicroradiography 立体显微射线照相术
stereomicroscope 立体显微镜;体视显微镜
stereomicroscopy 立体显微镜检查
stereomodel 立体模型
stereomodelling 建立立体模型
stereomodifier 立体结构调节剂
stereomotor 带永磁转子的电动机
stereomutation 立体变更;体积改变
stereonet 赤平投影网
stereonet analysis 赤平分析
stereo-operator 立体测图员
stereooptics 体视光学
stereo-orthophoto 立体正射像片
stereo-orthophoto pair 立体正射像片对
stereo-overlap 立体重叠
stereo-overlapping surface 立体重叠面
stereopair 立体照片对;立体像对
stereopantometer 立体万测仪;立体经纬测角仪
stereoperception 立体感(觉)
stereo perspective 立体透视
stereo phantom channel 立体声仿真通道
stereophone 立体声耳机
stereophonic effect 立体音响效果
stereophonic receiver 立体声收音机
stereophonic sound system 立体声系统
stereophonic system 立体声系统
stereophonic television 立体声电视
stereophony 立体声学;双声道立体声
stereophotogrammetric camera 立体摄影测量照相机;立体摄影测量摄影机
stereophotogrammetric compilation 立体摄影测量测图
stereophotogrammetric method 立体摄影测量法
stereophotogrammetric survey 立体摄影测量
stereophotogrammetry 航测实体制图术;立体摄影测量学;立体摄影测量术;立体摄影测量;地面立体测量(学)
stereophotograph 实体摄影;立体照相;立体照片;立体像片
stereophotography 立体照相术;立体摄影术;立体摄影地形测量学;立体摄影
stereophotomap 立体影像地图
stereophotometer 立体光度计
stereophotomicrograph 立体显微照片
stereophotomicrography 立体显微照相(术);立体显微摄影术
stereo phototopography 实体摄影地形测量;立体摄影地形测量学
stereophysics 立体物理学
stereo pickup 立体声拾声器
stereopicture 立体图像
stereoplanigraph 立体伸缩绘图仪;精密立体测图仪
stereoplotter 立体测图仪;立体制图仪;立体绘图仪;立体测图员
stereoplotting 立体测图
stereoplotting apparatus 立体测图仪
stereoplotting device 立体测图装置
stereoplotting equipment 立体测图设备
stereoplotting instrument 立体测图仪器
stereo power 体视率

stereopricking device 立体刺点装置
stereoprojection 立体投影
stereopsis 立体视觉;立体观测
stereopticon 透射投影仪;投影放大器
stereoptics 立体摄影光学;立体光学
stereo radar 立体雷达
stereo radar interpretation 立体雷达判读
stereoradian 立体弧度
stereo-radiography 立体放射线摄影术;立体X射线摄影术
stereorandom copolymer 石构无规共聚物
stereo range finder 立体测距仪;体视测距仪
stereoreceiver 立体声收音机
stereo reflector 立体反射器
stereoregularity 立构规整性
stereoregular polymer 立构有规聚合物
stereoregular polymerization 立体异构聚合
stereorepeating unit 立体重复链节
stereo-representation 立体表示法
stereorestitution 双像测图
stereo reticle 立体分划板
stereoroentgenograph 立体X线照片
stereoroentgenography 立体X线照相术
stereoroentgenometry 立体X线影像测量法
stereorubber 立体橡胶;立构橡胶
stereosat 立体卫星
stereosatellite 立体卫星
stereoscan 立体扫描
stereoscanner 立体扫描器
stereoscan photograph 扫描电镜照片
stereoscope 立体照相机;立体镜;双眼照相镜
stereoscope picture 立体照片
stereoscopic 立体的;体视的
stereoscopic acuity 立体视觉敏感度;体视锐度
stereoscopic aerial photograph 立体航测照片
stereoscopic agriculture 立体农业
stereoscopic 体视镜
stereoscopically covered surface 立体重叠面
stereoscopic analysis 立体分析
stereoscopic base 立体观测基线
stereoscopic camera 立体照相机;立体摄影机
stereoscopic colo(u)ring 分层设色法
stereoscopic construction 立体影像构成
stereoscopic contact 立体照准;体视接触
stereoscopic correspondence 立体对应
stereoscopic coverage 立体摄影面积
stereoscopic cure oven 立体地图热成型器
stereoscopic depth 立体深度
stereoscopic device 立体观测装置
stereoscopic drawing 立体描绘
stereoscopic dull sight 立体效果差
stereoscopic effect 立体作用;立体效应;立体效果;体视效应
stereoscopic electron microscope 立体电子显微镜
stereoscopic electron microscopy 立体电子显微术
stereoscopic evaluation 立体估测
stereoscopic exaggeration 立体放大
stereoscopic examination 立体镜检查
stereoscopic film 立体影片
stereoscopic fusion 立体凝合
stereoscopic graph 体视图
stereoscopic graticule 立体网
stereoscopic ground control 立测地面控制
stereoscopic harmonics 立体谐(和)函数
stereoscopic heightfinder 体视测距仪
stereoscopic high-speed framing camera 高速分幅立体摄影机
stereoscopic holographic(al) view 立体全息图
stereoscopic image 立体像
stereoscopic impression 立体效应;立体效果
stereoscopic instrument 立体测绘仪(器);体视化仪表
stereoscopic machine 立体观测仪
stereoscopic map 立体图;立体地图;立测地形图
stereoscopic mapping delineation 立体描绘
stereoscopic master terrain model 立体地图原模型
stereoscopic measurement 立体测量
stereoscopic microscope 立体显微镜;体视显微镜;双目立体显微镜
stereoscopic model 立体图样;立体模型;立体感图像
stereoscopic observation 立体像观察(法);体视观测
stereoscopic ocular 立体接目镜

stereoscopic pair observation 立体像对观测
stereoscopic pair photograph 立体像对像片
stereoscopic pantograph 立体刻模机
stereoscopic pantograph router 立体地图缩放刻模机
stereoscopic parallax 立体视差
stereoscopic parallax apparatus 立体视差仪
stereoscopic perception 立体感
stereoscopic photography 立体摄影(术);立体照相;立体摄影术;体视照相
stereoscopic photometer 立体光度计
stereoscopic picture 立体像;体视像片
stereoscopic plastic preform 立体地图模具
stereoscopic plotter 立体测图仪
stereoscopic plotting apparatus 立体测图仪
stereoscopic plotting device 立体测图装置
stereo scopic plotting instrument 立体绘图仪;立体测图仪(器);立体测绘仪(器);立视测绘仪
stereoscopic pointing 立体照准
stereoscopic point transfer device 立体转点仪
stereoscopic power 体视率
stereoscopic presentation 立体显示
stereoscopic print 立体印刷
stereoscopic prism 立体棱镜
stereoscopic profile 立体河床纵剖面
stereoscopic radius of observation 立体观测范围
stereoscopic rangefinder 立体像(式)测距仪
stereoscopic representation 立体表示法
stereoscopic screen 立体屏
stereoscopic sensation 立体感
stereoscopic sense 立体感
stereoscopic slide 立体滑座
stereoscopic strip camera 立体条带照相机
stereoscopic structure 立体结构
stereoscopic synthesizer 立体合成仪
stereoscopic system 立体观察系统
stereoscopic technology 体视技术
stereoscopic telescope 立体望远镜
stereoscopic television 立体电视
stereoscopic terrain model 立体地形模型
stereoscopic training 训练立体镜
stereoscopic transfer 立体转绘;立体转刺
stereoscopic view 立体影像
stereoscopic viewer 立体观测仪;立体观察器
stereo scopic viewfinder 立体检景器
stereoscopic viewing 立体观察
stereoscopic vision 立体观察;立体视觉;立体视法;体视
stereoscopic vision test chart 立体观察能力试验图
stereoscopic wave 立体测波;立体波
stereoscopic wavefront 立体波前
stereoscopic wavefront reconstruction 立体波前再现
stereoscopic wave recorder 立体测波仪
stereoscopic X-ray photograph 立体X射线照片
stereoscopy 立体观察法;立体观测(法);体视学;体视术;体视(法)
stereoscopy stereopsis 立体视法
stereo screen 立体荧光屏
stereoscreen 立体屏
stereoselective reaction 立体选择反应
stereoselectivity 立体选择性
stereoseparation 立体声区分
stereosimplex 简单立体测图仪;简易立体测图仪
stereo slide 立体滑座
stereoslide 立体正片
stereosonic 立体声的
stereospecific adsorbent 立体有择吸附剂
stereospecific catalysis 立体有择催化剂
stereospecificity 立体有择性;立体特异性;立体定向性
stereospecific polymer 立体有择聚合物;立体定向聚合物
stereospecific polymerization 立体有择聚合;立体定向聚合
stereospecific rubber 立体定向橡胶
stereospecific synthesis 立体定向合成
stereosphere 岩石圈;刚性圈
stereospheric(al) chart 球面投影地图
stereo statistic(al) prediction of mineral deposits 矿床立体统计预测
stereo strip camera 立体条带照相机
stereostroboscope 立体动态镜
stereo subcarrier 立体声副载波

stereo symbol 立体符号
stereosystem 立体系统
stereotactic polymerization 定向聚合
stereotaxic apparatus 立体定位器
stereotaxic instrument 立体定位仪
stereotaxic technique 立体定位术
stereotaxis 向实体性;趋触性;定向性
stereotelemeter 立体遥测仪;立体测距仪
stereotelescope 立体望远镜;体视望远镜
stereotelevision 立体电视
stereotelevision receiver 立体电视接收机
stereo-television system 立体电视系统
stereotemplet 立体模片
stereotemplet method 立体模型法;立体模板法
stereotemplet triangulation 立体模板三角测量;立体模片三角测量
stereotheodolite 立体经纬仪;体视经纬仪
stereotome 立体图片
stereotomy 切石法;实体物切割术;石雕刻术
stereotopograph 立体地图测图仪
stereotopographic(al) map 立体地形图
stereotopography 立体地形测量学;实体地形
stereotopometer 立体构型仪;立体地形量测仪
stereo transparency 立体正片
stereotriangulation 立体三角测量
stereo-triplet 三联立体像片
stereotropism 向实体性
stereotype 铅版
stereotype alloy 铅版合金
stereotyped command 标准指令
stereovectograph 偏振立体图
stereovectograph film 立体偏振软片
stereoview ability 立体观察能力
stereoviewer 立体观察器
stereo viewfinder 立体寻像器;立体检景器
stereo view(ing) 立体观察
stereovision 立体视觉;立体观察
stereovision method 立体观察法;立体观测法
stereo-visor 偏光镜
stereo wave 立体波
sterevector 立体矢量
sterhydraulic 固致水力的
steric 立体的
sterically hindered phenol 位阻酚
steric anomaly 比容偏差
steric compatibility 空间相容度
steric conformation 立体构像
steric effect 位阻效应
steric exclusion chromatography 凝胶色谱(法)
steric factor 位阻因素
steric hindrance 立体阻碍;空间位阻;位阻(现象)
steric information 立体信息
steric inhibition of resonance 共振空间阻碍
steric separation 位阻分离(作用)
sterilamp 消毒灯;灭菌灯
sterile 消毒的;无菌
sterile adhesive plaster 无菌橡胶膏
sterile biologic(al) sampler 无菌生物采样器
sterile board 无菌箱
sterile cabinet 无菌操作柜
sterile chamber 无菌容器
sterile culture 无菌培养
sterile deionized water 无菌无离子水;无菌去离子水
sterile distilled water 无菌蒸馏水
sterile elution generator 无菌淋洗发生器
sterile filling 无菌灌装
sterile generator 无菌发生器
sterile latex 无菌胶乳
sterile product 无菌产品
sterile rock 采矿废石
sterile room 无菌室;无菌操作室
sterile shoot 不结果实枝
sterile soil 生荒地
sterile solution 灭菌溶液;无菌溶液
sterile working 无菌操作
steriling test 无菌检验
sterility 不结实性
sterility detector 灭菌指示器
sterility test 无菌试验
sterilization 消毒灭菌;消毒(法);灭菌(法);杀菌
sterilization by chlorine 用氯灭菌;用氯消毒
sterilization by filtration 细菌滤除法;过滤灭菌法;滤过除菌法

sterilization by infrared radiation 红外(线)照射灭菌法
sterilization by oil bath 油浴灭菌法
sterilization filter 除菌滤器
sterilization of greenhouse 温室土壤的灭菌处理
sterilization of sewage 污水杀菌
sterilization of water supply pipe 给水管消毒
sterilization operation 冲销措施
sterilization with dry heat 干热杀菌(法);干热消毒(法)
sterilize 消毒
sterilized absorbent points 消毒纸尖
sterilized beaker 消毒烧杯
sterilized crucible 消毒坩埚
sterilized fennel 消毒漏斗
sterilized filter paper 消毒滤纸
sterilized flask 消毒烧瓶
sterilized fund 封存基金
sterilized ointment 灭菌软膏剂
sterilized pharmaceutics 灭菌药剂
sterilized powder 灭菌散剂;灭菌粉
sterilized powder for injection 注射用灭菌粉剂
sterilized pure water 消毒纯水
sterilized room 杀菌室
sterilized suspension 灭菌混悬液
sterilized talc 灭菌滑石
sterilized test tube 消毒试管
sterilized wastewater 灭菌废水;无菌废水
sterilized water 消毒水;无菌水
sterilizer 消毒器;自毁装置;灭菌器;灭菌机;杀菌器
sterilizer circulator 灭菌液循环泵
sterilizer instrument 消毒设备;灭菌设备
sterilizing 灭菌
sterilizing agent 杀菌剂;消毒剂;灭菌剂
sterilizing apparatus 消毒设备;消毒器
sterilizing chamber 消毒槽;灭菌室
sterilizing effect 杀菌作用;杀菌效应;灭菌作用;灭菌效应;消毒作用;消毒效应
sterilizing equipment 消毒设备
sterilizing facility for staff 工作人员卫生消毒设施
sterilizing forceps 消毒钳
sterilizing installation 灭菌设备
sterilizing mixer 灭菌混合机
sterilizing outfit 消毒器
sterilizing room 消毒室
sterilizing tower 消毒塔
sterilometer 消毒测定器;灭菌计
sterin 硬脂酸精
Sterlin 斯特林铜镍锌合金
sterling 杀水桩;分水桩;桥墩护桩;破冰设备;(桥梁的)冰挡;英国货币;英币
Sterling aluminium solder 斯特林锡锌铝合金焊料
sterling area 英磅区
sterling balance 英镑结存
sterling bonds 英镑证券
sterling constants 英币常数
sterling forward price 英镑远期价格
sterling gold 标准纯金
sterlinghillite 斯砷锰石
sterlingite 氧化锌;红锌矿
sterling picture 英币图像
sterling silver 纯银;标准纯银
sterling white 纯白颜料
sterling worth 真值;真正价值
Sterlite 斯特里特锌白铜
sterluminancy 单位立体弧度的亮度
stern 舰尼;艄;船尼
stern anchor 后锚;尼锚;船尼锚
stern anchoring winch 艉锚泊绞车【疏】
stern approach 船尼引桥
stern bearing 尼轴管轴承;船尼轴承
sternbergite 硫铁银矿
stern board 倒车【船】;船的后退
stern boss(ing) 尼轴毂
stern breast line 船尼横缆
stern breast rope 尼横缆
stern bulkhead 艉舱壁
stern bush 尼轴管衬套;尼轴衬
stern casting 铸钢船尼材
stern chock 艉导缆孔;船尼导缆孔
stern climate 严酷气候
stern compartment 尼舱;船尼舱
stern construction 船尼结构
stern davit 尼锚吊锚杆

stern door 尼门【船】
stern draft 船尼吃水(深)
stern draft correction 尼水尺修正
stern end bulb 船尼球
stern-engine(d)ship 尼机型船
stern fast 艉缆;尼缆;船尼缆
stern-first tow 尼拖
stern frame 尼柱;船尼框(架);船尼构架
stern framing 艉部构架
stern gang-plank 船尼跳板
stern gland 尼轴管压盖;船尼管压盖
stern lantern 船尼灯
stern layer 腹层
stern light 艉灯;尼灯;船尼灯
stern line 系船船尼索;艉缆;尼缆;船尼缆
stern line stopper 尼缆链掣
stern line winch 船尼绞车
stern marker 船尼标志
stern mooring point 尼部系留点
stern mo(u)lding 船尼缘饰
stern-mounted ladder 艉治杊桥架
stern navigating bridge 船尼驾驶桥楼
stern notch 尼槽
stern pipe 船尼导缆孔
stern port 尼门【船】;船尼载货门(装长木材用)
stern post 船尼柱;尼柱
stern ramp 船尼引桥;尼船引道;船尼跳板
stern ramp trawler 艉滑道拖网渔船
stern rampway 尼跳板【船】
stern rope 艉缆;尼缆
stern sea 尼浪;顺浪
stern section 梁腹断面;船尼分段
stern segment 船尼分段
stern settlement 艉部下沉
stern shaft 尼轴
stern sheave 船尼滑车
stern sheets 小艇尾台;敞舱船后面旅客乘坐部分
stern shield 尼部护木
stern signal 后退信号【船】
sternson 船尼龙骨肘材
stern spring 后倒缆
stern superstructure 船尼上层建筑
stern thruster 船尼侧推器
stern to wind when backing 倒车尼找风
stern trawler 尼拖网渔船
stern trim 艉倾;尼倾;船尼纵倾度
stern tube 尼轴管;船尼轴管
stern tube bearing 尼轴衬
stern tube bushing 尼轴衬
stern tube check ring 尼轴管止环
stern tube end plate 尼轴管端板
stern tube gland 船尼管压盖
stern tube packing 尼轴管填料
stern tube(ring)nut 尼轴管螺母
stern tube stuffing box 尼轴管填料涵
stern unit 船尼分段
stern wake 追迹流;船尼伴流
stern wave 船尼浪;船尼波
sternway 倒驶;倒车【船】;船的后退
stern wheel 尼明轮;船尼明轮
stern wheeler 尼明轮船
stern wheel steamer 尼明轮船
sterny 粗粒的
sterone 固酮
Sterro alloy 斯特罗黄铜合金;铜锌铁合金
Sterro metal 铜锌铁合金;特罗黄铜;斯特罗含铁锰黄铜
sterryite 斯硫铅矿
S-test 慢剪试验
stet 不删符号;保留符号
Stetefeldt furnace 斯特德菲尔特炉
stetefeldtite 水锑银矿
stethoscope 金属探伤器;听诊器;导音探测器
steuerspannung 激励电压
stevedorage 装卸费(用);码头搬运费
stevedore 装卸工人;装卸工;码头装卸工(人);码头工人;搬运工(人)(指码头作业工人)
stevedore charges 装卸费(用)
stevedore damage 装卸损坏
stevedore gang 装卸组
stevedore gear 吊货工具
stevedore hook 手钩;搬货用的搭钩
stevedore knot 双八字结
stevedore pallet 货盘

stevedore's gear 装卸工具
stevedores gear and change house 搬运工人用具及更衣室
stevedore's liability 装卸责任
stevedore's pallet 装卸托盘
stevedore's warehouse 装卸工具房;(码头的)搬运工库房
stevedore type pallet 装卸式货盘
stevedoring 装卸作业;装卸(工作);码头装卸工作;船舶装卸
stevedoring barge 装卸货物的驳船;装卸货驳
stevedoring capacity 码头装卸能力
stevedoring company 装卸公司;码头搬运公司
stevedoring crane 装卸起重机
stevedoring firm 装卸公司
stevedoring machine 装卸机械
stevedoring operation 装卸作业
stevedoring operation(al)time 船舶装卸作业时间
stevedoring organization 装制公司;装卸公司
stevedoring pallet 装卸托盘
stevedoring unloader 装卸起重机
Stevens former 斯蒂文思真空圆网纸机
stevensite 斯皂石
Stevenson screen 斯蒂文森百叶箱
Stevin principle 固化原理
stew 鱼塘;噪声;人工养蚝场;热浴(室)
Stewart alloy 斯图尔特铸造铝合金
stewartite 斯图尔特石
Stewart's law 斯图尔特人口分布定律
Stewart's pile 斯图尔特桩(一种现场浇制混凝土桩)
stewing 拉拔时效硬化;静置
stew-pan 炖锅
stew pond 养鱼场
stew-pot 炖壶
steyre (中世纪教堂的)梯级
St. george's cross 圣乔治十字架
sthene 斯乘(力的单位,1 斯乘 = 1000 牛顿)
stiacciato 浅浮雕
stibarsen 砷锑矿
stibialism 锑中毒
stibiated 含锑的
stibiation 锑疗法
stibiconite 黄锑矿;黄锑华
stibine 锑化(三)氢
stibiobetafite 锑贝塔石
stibiocolumbite 铌锑矿
stibiomicrolite 锑细晶石
stibiopalldinite 锑钯矿
stibiotantalite 钽锑矿
stibium 辉锑矿
stibivanite 钒锑矿
stibnide 锑化物
stibnite 辉锑矿
stibnite ore 辉锑矿矿石
stibophen 锑波酚
stibous trifulfide 三硫化二锑
stichobasidium 并列担子
stichtite 菱水碳铬镁石
stick 直木材;棍;杠秤;杆;连续投弹;卡住;木垛;火药柱;棒状炸药;黏附
stickability 胶粘性;黏着性;胶性;黏着能力
stick and-green leaf machine 枝叶清理机
stick and rag work 纤维石膏制品;纤维石膏工作;纤维灰浆抹面工作;石膏幔灰制品
stick antenna 条形天线;棒状天线;棒状天线
stick bit 切槽口
stick bite 割刀
stick block 黏结块
stick-built 木架房屋
stick button 非自复式按钮
stick circuit 自闭电路;保持电路
stick control 手柄控制;手柄操纵
stick displacement 操纵杆角位移
sticked on another 互相黏住
stick electrode 手工焊条
stick-electrode welding 焊条电弧焊
sticker 预埋附固件;透明胶纸;架杆;黏结剂;尖物;黏剂
sticker break 黏结条痕
sticker machine 线脚成型机;装饰木条锯割机;车木机(制木模用)
sticker mark 黏痕
sticker mo(u)lder 线脚成型机

sticker price 标签价格
stick force 驾驶杆力;黏着力;外附力
stick force instrumentation 驾驶杆测力装置
stick force recorder 驾驶杆力自记器
stick-free 脱胶;松开驾驶杆(指飞机、机车等)
stick gate 直浇口
stick ga(u)ge 测量杆
stick grip 驾驶杆
stick harness 塞柱
stickiness 黏性;发黏
sticking 线脚成型;吸持;油墨黏纸;黏结;黏冲;黏着;黏连;黏附;趋稳性;残留图像;板材码垛条隔通风法
sticking agent 黏着剂
sticking board 黏固板
sticking circuit 自保电路
sticking coefficient 黏附系数
sticking drill tools 卡钻
sticking friction 黏着摩擦
sticking index 黏结性指数
sticking label 粘贴标签
sticking mark 黏痕
sticking of a charge 黏料
sticking of contacts 烧结触点
sticking of drilling rod 埋钻
sticking of drill rods 卡钻
sticking on 胶合
sticking patch 贴胶;补片
sticking picture 烧附图像
sticking plaster 橡皮膏
sticking point 黏结点
sticking potential 极限吸电位;饱和电位
sticking relay 吸持继电器
sticking string 卡钻
sticking temperature 黏结温度
sticking together 黏在一起;黏结;黏住
sticking up 黏结
sticking voltage 黏着电压
stick jet gun 杆式聚能射孔器
stick lac 紫梗胶;枝状虫胶;树枝虫胶;原虫胶;原紫胶
stick locking 保留锁定
stick marks 叉杆擦痕
stick micrometer 杠杆式千分
stick motion 驾驶杆运动
stickness 黏性;黏稠性
stick no bills 请勿张贴
stick of dynamite 柱状炸药包
stick of explosive 柱状炸药包;炸药卷;筒装炸药
stick of gelatin(e)dynamite 明胶炸药卷
stick of solder 焊料棒
stick of tools 卡钻;夹钻
stickogram 反射系数图
stick on 贴上;黏上
stickout 突出;外伸长度;(电极或焊丝的)伸出长度
stick-out weld 未熔合点焊接头
stick perforator 金属棒凿孔机;锤击穿孔机
stick phosphorus 棒状磷
stick plot 柱状图;杆状图
stick point 发黏点
stick position 操纵杆位置
stick powder 炸药筒;炸药卷;炸药;筒装炸药
stick pusher 自动推杆器
stick relay 连锁继电器;保持继电器
stick requisition 自闭条件
stick servo 操纵杆随动机构
stick shellac 紫胶棒;棒状虫漆
stick shift 手动变速器
stick signal 限制信号机;保留信号
stick slip 黏滞滑动;黏滑运动
stick-slip behavio(u)r 黏滑方式
stick-slip earthquake 黏滑地震
stick-slip fault 黏滑断裂;黏滑断层
stick-slip genesis of earthquake 黏滑成因
stick-slip mechanics 黏滑力学
stick spectra 不分离光谱
stick spring 自保弹簧
stick system 杆式系统
stick table 带灯独柱桌
stick thinning 丈量疏伐
stick to principle 坚持原则
stickum 黏性物质
stickup (钻孔底的)残留岩芯
stick-up lettering 植字;注记剪贴
stick-up symbol 透视符号

stickwater 滞水
stick wax 黏蜡
sticky 黏性的;黏的
sticky bomb 黏性炸弹
sticky cement 黏性水泥;结块水泥;胶粘水泥
sticky charge 黏性炸弹
sticky clay 胶(粘)黏土
sticky consistency 黏着结持度;黏韧结持度
sticky dust 黏性粉尘
sticky end 黏性末端
sticky ends 发黏纱
sticky formation 黏结性岩层
sticky ga(u)ge 黏滞真空规
sticky grenade 黏性炸弹
sticky limit 黏滞限度;黏限
sticky material 胶结材料;黏着材料
sticky oil 高黏石油
sticky ore 胶粘矿石
sticky point 黏点;脱黏
sticky price 黏性价格
sticky sale 黏泥板岩
sticky soil 黏土
sticky tape 胶带
sticky together 互相黏结在一起
sticky wax 黏蜡
sticky weather 湿热的天气
stick zone 黏着区
stiction 静态阻力;静摩擦(力);接触面阻力
stictolite 斑点混合岩
Stiefel mill 斯蒂费尔式轧机
Stiefel process 自动轧管法;斯蒂费尔自动轧管法
Stieltjes integral 司蒂吉斯积分
Stieltjes transform 司蒂吉斯变换
stiff 刚性的;坚硬的;稠的
stiff arc 刚性弧
stiffback 定位板;强力背板;(模板的)刚性托架
stiff bass brush 硬棕毛刷
stiff batch 稠厚装料
stiff blue clay 硬青黏土;硬绿黏土
stiff-boom crane 固定伸臂起重机
stiff bottom 黏质海底
stiff bottom hole hook up 钢性钻铤
stiff-bristled broom 硬扫帚
stiff broom 硬刷
stiff brush 硬刷
stiff cement grout 稠水泥浆
stiff clay 硬黏土;压密黏土;僵黏土;坚硬黏土;低塑性黏土
stiff-closed rib 劲性封闭肋
stiff concrete 硬稠混凝土;干硬(性)混凝土;干性混凝土;无坍落混凝土;稠(硬)混凝土
stiff connection 加劲连接;加劲节点;刚性接头
stiff consistency 干硬性;硬稠性
stiff consistency concrete 干硬性混凝土
stiff consistency mix 稠拌和
stiff consistency mortar 干硬性砂浆;干硬性灰浆
stiff core 刚性核(心)
stiff coupling 刚度耦合
stiff design 刚性设计
stiff diagram shoeing chemical composition of soil field water 油田水化学组成板式图解
stiff differential equation 僵化微分方程
stiff dowel bar 刚性传力杆;劲性接缝杆;劲性传力杆;刚性接缝条
stiff duty 重税
stiffen 增大船的稳性;使硬化;变硬
stiffened 加固的;加劲的
stiffened arched girder 加劲拱(形)梁
stiffened bracket 加劲托座
stiffened cable suspension bridge 加劲悬索桥
stiffened chain suspension bridge 加劲链索桥
stiffened compression element 加劲承压杆件;加强受压构件
stiffened derrick 固定脚起重机
stiffened edge 加强边
stiffened element 加劲构件
stiffened end 固定端
stiffened expanded metal 加强的金属拉网;加强钢板网
stiffened girder 加固大梁;经加固大梁
stiffened junction 加劲接头;加劲接合
stiffened masonry wall 加劲圬工墙
stiffened mat 加强层(土壤)
stiffened penstock 加劲压力钢管

stiffened plate 加劲板
stiffened roadway 刚性行车道
stiffened seated-beam connection 加强的梁支座连接
stiffened shell 加强壳
stiffened shell fuselage 半硬壳式机身
stiffened shell structure 加强壳结构
stiffened skin 加强蒙皮
stiffened structure 加劲结构
stiffened suspension bridge 加劲悬索桥;加劲悬桥;加劲式吊桥
stiffened tied arch bridge 刚性系杆拱桥
stiffened trussed arch 加劲桁架拱
stiffened type 加劲式
stiffened wire lath 加强钢丝网;加劲钢丝网矢板条;加劲钢丝(灰)网
stiffener 硬化剂;钢圈外包布;刚性元件;加强筋;加强构件;加强杆(件);加强材;加强板;加强肋;加劲(构)件;加劲杆;加劲材;防移袋装货;防挠材;补强件
stiffener angle 加劲角铁;加劲角钢;加筋角钢
stiffener plate 加劲板
stiffener ring 加劲环
stiffening 刚性连接;劲化;加劲;使硬;补强
stiffening against buckling 抗扭曲加劲
stiffening agent 硬化剂
stiffening agent for chemical fibric 化纤布硬化剂
stiffening angle 加劲角铁;加固角钢;加劲角钢;加筋角钢
stiffening angle of web 梁腹加劲角钢;梁腹加筋角铁
stiffening arch 加固拱;护壁拱
stiffening bar 刚性梁;加强筋;加劲筋;防挠材
stiffening bead 肋强波纹;加强梗
stiffening beam 加强梁;加劲梁
stiffening by ballast 加压载增加稳性【船】
stiffening by diagonals 斜杆加劲
stiffening coefficient 劲度系数
stiffening diaphragm 加劲隔板
stiffening effect 劲度效应;加强效果;加劲作用(效应)
stiffening effect of cladding 饰面加强效果
stiffening element 加劲构件
stiffening frame 加劲框架;加固框架
stiffening girder 加强梁;加劲桁材;加劲(大)梁;加固大梁
stiffening grillage 加强格框;加强格排;加劲格框;加劲格排;加劲格栅
stiffening iron 加劲铁件
stiffening member 加强构件;加劲构件
stiffening of rope 绳的韧性
stiffening order 货物全部卸完前装载重或压舱物许可证
stiffening panel 加强板;加劲板
stiffening piece 加劲件;加固件
stiffening pier 加劲的柱;加强的桥墩
stiffening plate 加劲板;加固板;补强板
stiffening purlin(e) 加劲檩条
stiffening purlin(e) brace 加劲檩(条)撑
stiffening rib 加劲肋;加强筋;加劲条;加劲肋(条);防挠肋
stiffening ring 刚性环;加劲圈;加劲环;加固圈;补强圈
stiffening ring attached to vessel 与容器连接的加强环
stiffening ring beam 加强的圈梁;加劲的圈梁
stiffening rubber bearing 加强橡胶支座
stiffening side 舱壁有防挠材的一面
stiffening sleeve 加固轴套
stiffening stile 加固边梃
stiffening temperature 硬化温度
stiffening time 增稠期
stiffening truss 加劲桁架;加强桁架
stiffening truss of suspension bridge 悬索桥加劲桁架
stiffening wall 加劲墙
stiff equation 刚性方程
stiffer 加劲肋(条)
stiffest consistency 最小坍落度;最干硬稠度
stiff extrusion 硬挤坯
stiff extrusion press 硬挤出机
stiff fiber board 硬质纤维板
stiff-finger pick-up 刚性指式捡拾器
stiff fireclay 硬质耐火黏土

stiff fissured clay 裂隙硬黏土;硬裂隙黏土;硬裂纹黏土
stiff frame 刚构架;刚架;坚挺肋骨
stiff frame slab 刚架板
stiff frog 固定辙叉
stiff gantry 刚性台架
stiff gel 硬化凝胶
stiff girder 加劲(大)梁
stiff girder connection 加劲梁节点
stiff glaze 硬性釉;高张力釉
stiffness 硬度
stiffing testing machine 刚性试验机
stiff joint 加劲性接头;刚性接头
stiff-jointed 牢固连接的
stiff-jointed flat frame 刚结平面框架
stiff-jointed frame 刚结的框架;钢结构架
stiff-jointed frame slab 刚结框架板
stiff-jointed planar frame 刚结平面框架
stiff-jointed truss 固结桁架;劲联桁架
stiff layer 刚性层
stiff leaf 密叶饰
stiff-leaf capital 密叶饰柱帽;密叶饰柱顶
stiff leaved plant 硬叶植物
stiffleg 刚性桅杆起重机;刚性支柱;后拉索;后拉杆;背撑;后撑条;复livre河缆;劲性支柱
stiffleg derrick 刚性柱架(起重机);刚性支腿杆吊;刚腿人字起重机;固定支架桅杆式起重机;固定脚起重机
stiffleg derrick crane 定腿式人字起重机
stiffleg derrick travel(l)er 刚性腿移动式转臂起重机
stiffleg mast crane 刚性支腿桅杆吊
stiff life and figure drawing 静物写生画
stiff mill 高刚度轧机
stiff mix 干硬性拌和物;干性混合料;干硬(性)混合物
stiff mixture 干硬性混合料
stiff modulus 刚性模量
stiff-mud 硬泥浆;硬泥
stiff-mud brick 坚硬泥砖;硬泥砖
stiff-mud process 硬泥制砖法;硬泥成型法;可塑法
stiffness 硬性;刚性;刚度;劲性;劲度;僵硬;坚硬;挺度;不倾性
stiffness analysis 刚度分析
stiffness center 刚度中心
stiffness coefficient 硬度系数;刚度系数;劲度系数
stiffness condition 刚度条件;劲度条件
stiffness constant 刚度常数;劲度常数
stiffness control 劲度控制;强度调整
stiffness criterion 刚度准则
stiffness degradation 刚度降低
stiffness-degrading model 刚度退化模型
stiffness distribution 刚度分布
stiffness equation 刚度方程;劲度方程
stiffness factor 刚劲因素;劲度系数
stiffness force 刚度力
stiffness influence coefficient 刚度影响系数;劲度影响系数
stiffness in torsion 抗扭劲度;抗扭刚度
stiffness joint 加劲接头;加劲接合
stiffness loss 劲度损失;刚度损失
stiffness mat foundation 刚性板式基础
stiffness matrix 刚度矩阵;劲度矩阵
stiffness matrix method 刚度矩阵法;劲度矩阵法
stiffness meter 劲度计
stiffness method 刚度法;劲度法
stiffness modulus 刚性模量;劲性模量;劲度模量
stiffness of collars 钻铤刚性
stiffness of coupling 耦合强度
stiffness of cushion block 垫块的劲度
stiffness of drill pipes 钻杆刚度
stiffness of equivalent spring 等效弹簧刚度
stiffness of foundation soil 地基土刚度
stiffness of piles 桩的劲度
stiffness of plate 板的劲度
stiffness of structure 结构刚度;结构劲度
stiffness of subsoil 地基刚度
stiffness ratio 刚度比(例);劲度比
stiffness test 干硬性测定试验
stiffness tester 刚度测量器
stiffness test load 硬度试验荷载;劲度试验荷载
stiffness test of structural member 构件刚度检验
stiffness to density ratio 刚度密度比
stiffness to weight ratio 刚度重量比;比刚度

stiff paint 厚漆
stiff paper form 硬纸模板
stiff paste 浓膏;稠浆
stiff paste paint 干稠性油漆
stiff piston 刚性活塞
stiff-plastic brick-making machine 低水分砖料成型机
stiff-plastic making 硬塑料制造;硬塑料成型法;低水分料成型;半干成型法
stiff-plastic mo(u)lding 硬塑料压制;硬塑料制模;硬塑料模压;硬塑料造型(法)
stiff-plastic process 硬塑压制法;硬塑性成形法;硬塑料成型法
stiff plate 加劲板
stiff pole 刚性杆
stiff reinforcement 劲性钢筋;加劲(钢)筋
stiff rib 刚性肋
stiff safety fence 刚性护栏
stiff scale of index 硬稠性
stiff shaft 刚性轴
stiff-shaft centrifugal compressor 刚性轴离心式压缩机
stiff-shaft internal vibrator 有刚性轴的插入式振动器;有刚性轴的内部振动器
stiff-shaft turbine 刚性轴汽轮机
stiff-shaft vibrator 有刚性轴的振动器
stiff ship 过稳船;稳性过大的船
stiff slope 陡坡
stiff soil 硬土
stiff spring 牵引弹簧
stiff stability 强稳定性
stiff stand 刚性机座
stiff standard 刚性铲柄
stiff structure 刚性结构
stiff test 刚性试验
stiff vessel 过稳船;稳性过大的船
stiff wind 强风(六级风)
stigma 小斑;点斑
stigma of degeneracy 变质特征
stigmata 小斑
Stigmatella 标记菌属
stigmatic 共点的;同心的
stigmatic concave grating 凹面聚焦光栅
stigmatic image 斑点成像
stigmatic imaging 共点成像
stigmatic spectrograph 无像散摄谱仪
stigmatism 折光正常
stigmula 分柱头
stiker alignment 隔条对线
stiker spacing 隔条间距
stiker stain 隔条变色
stilb 提(亮度单位)
stilbene 均二苯代乙烯;二苯乙烯
stilbite 辉沸石
stilb-meter 亮度计
S-tile S 形砖
stile 弩针;立边;门桩;横路栅栏;竖框(梃);窗梃;边梃
stile edging board 门窗贴脸板
stile of door 门轴柱(边梃)
stile of elevator 电梯门框
stile of gate 门轴柱(边梃)
stile of sash 窗梃
stile plate 边梃推板;门推板
Stiles method 斯蒂尔斯法
stilet 管心针
stiletto heel 道钉顶头
still 熙度(亮度单位);蒸馏锅;蒸馏釜;静止的;静片;不起泡的
still adjustment 静像调整
stillage 制坯板;蒸馏釜馏出物;堆装架;垫底的货架;釜馏物
stillage bottoms 釜脚
stillage gas 釜馏气
stillage residue 釜残渣
still air 静止空气
still air range 无风航程
stillcidium (陶立克式建筑的)滴水檐
still column 蒸馏塔
still command 静止指令;静止命令
still-cooled cylinder 自冷式汽缸
still current of gas 静止气流
still dome 釜室
stilleite 方硒锌矿

stiller 消力池
Still-Esslingen swing lift 斯蒂尔—埃斯林根摆动吊运车
still frame 静止帧;静止画面
Stillfried b interstade 斯梯尔弗雷德间冰阶
still gas 蒸馏气(体)
still head 蒸馏头;蒸馏盖
stilliard 放坯架;运坯架;承坯架
still image 静止像;静止图像;静像
stilling 蒸馏;釜馏
stilling apparatus 蒸馏器
stilling baffle 消力坎;消力槛
stilling basin 消力池;静水池;消速池;沉水池
stilling basin in ditch 沟渠中的消力池
stilling box 消速池;消力箱;静水箱;分馏箱
stilling cavern 消力孔洞;消力洞穴
stilling chamber 消涡室;消能室;蒸馏室;镇静室;静水箱
stilling dam 消力坝
stilling device 防波装置
stilling grate 静水栅
stilling grid 消力格栅;静水栅
stillingia oil 梓油;乌桕油
still(ing)pond 消力池;(泄洪道的)消力塘;前池;沉砂砾池
stilling-pond overflow 消力池溢流;消力槽溢流
still(ing)pool 消力池;消力池
stilling test 静水试验
stilling well 消能井;消流井;自记水位井;静水井;调压井
stilling well lag 自记水位井滞后;静水井滞后
still kettle 蒸馏锅;蒸馏釜
still layer 缓动层
still life 静物(画)
still life and figure drawing 静物写生画
still liquor 釜馏液
still optimizable loop 尚可优化循环
still photography 静止摄影
still picture 静止图像
still pond 消力塘
still pond regulation 静水池调节
still pool 消水池
still pot 沉积槽;沉淀釜;沉淀槽
still preheater 蒸馏釜预热器
still pressure 静压力
still process 蒸馏过程
still room 备餐室;饮料储藏室;蒸馏室;茶点整备室
still shell 釜壳
Stillson 管钳;管子扳手
Stillson pipe wrench 可调管子扳钳;可调管子扳手;活动管子扳钳;活动管子扳手
Stillson wrench 管子钳;管子扳手;可调管扳手;斯蒂尔森管子钳;斯蒂尔森扳手;可调管子钳;活动扳手
still speed 静止图像速度;静像速度
still stationary 静止的
still steam 蒸馏釜用蒸汽
still stress 静应力
still switch 静像开关
still tide 平潮;停潮
still tube 蒸馏管
still video 静止图像视频
still video transmission 静止图像发射
still water 静水;静流
still water bending moment 静水弯矩
still water canal 静水运河
still water depth 静水水深;静水深
still water head 流体静压水;静水头
stillwaterite 六方砷钯矿
still water level 静水位;静水(平)面;静力面
still water level(l)ing 静水面高程测量
still water navigation 静水航行;平水航行
still water reach 静水河段
still water resistance 静水阻力
still waters 静水域;静水区
still water surface 静水面
still wax 蜡渣;釜馏蜡
still well 稳水器
stilpnochlorane 鳞绿脱石
stilpnomelane 黑硬绿泥石
stilt (水上房屋的)桩柱;接高的压杆;高跷;装坯用三角支架;支撑物;支材;三臂星窑具;三臂坯托;承坯架
stilted arch 高拱;高跷拱;高矢拱;上心拱

stilted arch bridge 高矢拱桥;上心拱桥
stilted height 支撑架的高度;支撑物的高度;踏上高跷的高度
stilted vault 高矢穹顶;上心穹顶;上心拱;上心拱形穹顶
stilt house 吊脚楼;桩支方屋;高桩房屋
stilting 跷高
stilt marks 坯托印痕
stilt root 支柱根;高跷根
stilt-tractor 高架式拖拉机
Stimsonite 司汀姆森奈塑料
stimulant 兴奋剂;刺激物;刺激剂
stimulated absorption 受激吸收
stimulated absorption transition probability 受激吸收跃迁几率
stimulated action 受激作用
stimulated Brillouin diffusion 受激布里渊漫射
stimulated Brillouin effect 受激布里渊效应
stimulated Brillouin scattering 受激布里渊散射
stimulated compton scattering 受激康普顿散射
stimulated emission 诱发发射;受激发射
stimulated emission detection 受激发射探测
stimulated emission detector 受激发射探测器
stimulated emission of radiation 辐射受激发射
stimulated emission transition probability 受激发射跃迁几率
stimulated fracture 诱发破裂
stimulated geothermal bore 激发地热井
stimulated light emission 受激光发射
stimulated oscillation 受激振荡
stimulated photon 受激光子
stimulated photon liberation 受激光子释放
stimulated radiation 诱发辐射;受激辐射
stimulated Raman effect 受激拉曼效应
stimulated Raman mission 受激拉曼发射
stimulated Raman scattering 受激拉曼散射
stimulated Rayleigh effect 受激瑞利效应
stimulated Rayleigh scattering 受激瑞利光散射
stimulated Rayleigh wing scattering 受激瑞利翼散射
stimulated recharge resources of groundwater 地下水激发补给资源
stimulated recombination radiation 受激复合辐射
stimulated scattering 受激散射
stimulated threshold 受激阈
stimulated transition 受激跃迁
stimulating action 受激作用
stimulating factor 刺激因素
stimulating optical field 受激光场
stimulating phase 发黄期
stimulating substance 刺激性物质
stimulating the economy 搞活经济
stimulation 荧光增强
stimulation earthquake 诱发地震
stimulation fund 鼓励基金
stimulation level 激励电平
stimulation of a well 井的增强
stimulation of geothermal system 地热系统激发
stimulation of precipitation 人工致雨;人工催雨;人工催化降水
stimulation of wells 油井增产措施
stimulator 激励器;激活剂;激动器
stimuli 色质
stimuli generator 激励振荡器
stimulus 激发源;色质
stimulus energy level 刺激能级
stimulus energy value 刺激能量
stimulus-response equation 刺激反应方程
sting 架杆;刺螯
stinger bit 切锉刀片;穿孔刀具;穿孔刀片;切锉刀具
stinger ream 带导向杆的钻头扩孔
sting-out 火焰喷出;喷火
sting-out loss 冒火热损失
sting poison 蜂毒
sting support 张臂式支架
stink 臭气
stink bug 椿象
stink coal 臭煤
stink cupboard 通气橱;通风柜
stinkers 讨厌的工作
stinking fish 发臭鱼
stinking sewage 腐臭污水
stinks cupboard 通风橱
stinkstone 臭石;含沥青石灰;臭灰岩

stink trap 存水弯;阻臭管;水封管;防臭罩;防臭存水弯
stinkwood 臭木
stinky 全景雷达
stint 派定的工作
stipend 津贴;定期生活津贴
stipoverite 超石英
stipple 点状花纹(面砖);点画法;点涂修整法;点刻法;蓓蕾漆糁饰法
stippled 凹凸纹的;点彩饰面的
stippled area 斑点面积
stippled coating 复色拉毛涂层
stippled finish 点彩饰面
stippled pattern 点彩饰面图形
stipple line 点虚线
stipple method 点描法
stipple paint 点彩涂料;复色拉毛漆;蓓蕾漆
stipple pattern 点状符号图形
stipple perforated 点描孔饰面纹理
stipple point 刻点
stippler 点刻刀;扁平漆刷;画人造木纹的工具;点画笔
stippling 点彩;敲点涂法;点涂法;点刻;蓓蕾漆涂装法
stippling brush 点画笔;点彩刷
stippling decoration 点彩
Stippolyte 斯蒂普利特玻璃(一种轧有纹饰的不透明玻璃)
stipulate 写明
stipulated commission 契约规定佣金
stipulated date 预定期限
stipulated depth 规定深度
stipulated price 协定价格;约定价格;规定价格;合同价(格)
stipulated sum 约定的总价
stipulated sum agreement 契约总价协定;付款协议书
stipulated velocity 额定速度
stipulation 契约(条款);规定;合同(条款);条款
stipulation sum 约定的总价
stipulative definition 约定定义
stipulator 约定者;规定者;订约人
stir 搅拌;搅动
stir-bake to brown 炒焦
stir-bake to cracking 炒爆
stir-bake to yellowish 炒黄
stir-in pigment 调入颜料;拌和型颜料
stir-in resin 拌和型树脂
stirling cycle 等温等容循环;回热循环
stirlingite 锌铁橄榄石
Stirling's approximation 斯特林近似法
Stirling's boiler 斯特林锅炉
Stirling's cycle 斯特林循环
Stirling's engine 斯特林发动机;斯特林发动机
Stirling's formula 斯特林公式
Stirling's furnace 斯特林炼锌电弧炉
Stirling's interpolation formula 斯特林内插公式
Stirling's metal 斯特林合金
Stirling's number 斯特林数
Stirling's process 斯特林精炼锌法
stirred ball mill 砂磨机
stirred bed 搅动床(层)
stirred bed agitator conveyer 搅动床搅拌送料器
stirred bed plant 搅动床设备
stirred bed reactor 搅动床反应器
stirred specific volume 搅拌比容积
stirred tank 搅拌池
stirred tank reactor 搅拌池式反应器
stirred vessel 搅拌容器
stirrer 搅拌机;搅拌者;搅动器;搅棒;搅拌器
stirrer bank 搅拌器组
stirrer bar 搅拌棒;拌和杆
stirrer distributor 搅拌布煤器
stirrer paddle 搅拌桨叶
stirrer shaft 搅拌器轴
stirrer with four tilted blades 搅拌棒
stirrer with spiral blades 搅拌棒(旋板式)
stirring 用泵抽送;汲取;搅拌
stirring apparatus 搅动器;搅拌装置
stirring arm 搅拌器叶片;搅拌器臂;搅拌桨叶;搅拌机桨叶
stirring blade 搅拌桨叶
stirring coil 搅动盘管
stirring device 搅炼装置;搅拌装置

stirring dryer 搅拌干燥器
stirring extractor 搅拌萃取器
stirring filter 搅拌式滤池;搅拌过滤器
stirring filtration 搅拌过滤
stirring gear 混合器;翻动器;振动器;搅动器;搅拌装置;搅拌机
stirring hole 搅孔
stirring machine 搅拌器
stirring machinery 搅拌机械
stirring mechanism 搅拌机械;搅炼机械;搅拌机制;搅拌机构
stirring melt 搅动玻璃液
stirring mill 搅拌机
stirring mixer 搅拌混合器
stirring motion 漩涡运动;紊流运动;湍流
stirring paddle 搅拌浆
stirring period 搅拌周期;搅拌时间
stirring rake 搅拌浆
stirring rod 搅棒
stirring screw conveyer 螺旋拌和(搅拌)输送机;螺旋拌和输送机;螺旋搅拌输送机
stirring section 搅动区
stirring shaft 摆动轴;扰动轴;搅动轴
stirring speed 搅拌速度
stirring star 星形摆动装置;星形搅动装置
stirring stat 星形扰动装置
stirring switch 换向开关
stirring tank 搅拌池
stirring test 搅动试验;搅拌试验
stirring time 搅拌时间
stirring-type mixer 搅式拌和机;搅动式拌和机;搅拌式混合器;搅拌式混合机;强制式搅拌机;强制式拌和机;搅炼式拌和机
stirring-type reactor 搅拌式反应设备
stirring unit 搅拌装置
stirring(up) 搅拌
stirring-up rake 翻料耙
stirring-up velocity 扬动流速
stirring velocity 搅拌速度
stirring visco(si)meter 旋转黏度计
stirrup 箍筋;钳具;骑马螺钉;桅横杆上踏脚挂绳;蹬形锓件;蹬索
stirrup bender 钢筋弯箍机;弯箍机
stirrup bending machine 弯箍机
stirrup bolt 钢箍;扒钉;(暖气管上用的)管夹
stirrup feeder 摆门式给料机
stirrup for hoisting 起重箍
stirrup frame 框式机架
stirrup hand pump 便携手压泵;手摇灭火泵
stirrup hanger (梁的)吊箍
stirrup hoop 箍筋弯钩
stirrup link 框式连杆
stirrup method 箍筋法
stirrup of axle-box 轴箱箍筋
stirrup piece (木工用的)蹬形支架
stirrup pump 手摇灭火泵
stirrup ratio 配箍率
stirrup ratio of reinforcement 配筋率
stirrup repair clamp 蹬形夹
stirrup spacing 箍筋间距;钢箍(筋)间距
stirrup strap 悬挂铁条;U形搭接钢带;钢筋箍;卡箍;铁靴;扁钢悬托镫
stirrup stress 钢箍应力
stirrup tensile force 钢箍拉力
stirrup-tie 箍筋
stirrup type pedal 蹬形踏板
stirrup wire 扎钢箍铁丝;钢箍钢筋;箍钢丝;箍钢;钢丝箍
stir speed 搅拌速率
stishovite 斯石英;冲击石英;超石英
stistaite 锑锡矿
stitch 缝线;缝合;绑结
stitch-and-seam weld 点线焊
stitch-and-seam welding 断续缝缝;断续缝焊
stitch bond 针脚式接合
stitch-bonded camping cloth 缝编帐篷布
stitch-bonded carpet 缝编地毯
stitch-bonded monolithic chip 针脚式接合单块片;滚压黏合单块片
stitch bonding 合缝焊;滚焊;自动点焊;针脚式接合法;连续点焊焊接;跳焊
stitch brake lining 制动闸边皮
stitched 缝合的;针绣的
stitched belting 结合皮带;缝合皮带

stitched building mat 缝合的建筑地毯
stitched building quilt 缝合的建筑地褥
stitched canvas 缝合的帆布带
stitched canvas belt 钉接帆布运输带
stitched catenary suspension 弹性链形悬挂
stitched catenary type equipment 弹性链形悬挂
stitched-on pile carpet 簇绒地毯;针织的绒头地毯
stitched simple catenary suspension 弹性简单悬挂
stitched tramway type equipment 弹性简单悬挂
stitched veneer 缝接单板
stitched wire joint 编缝丝结合
stitcher 钉书机;带材缝合机;齿形压辊
stitching 压合;滚压;钉合法
stitching flat wire 扁铁丝
stitching force 缀合力
stitching oil 滚压油
stitch-knitting machine 缝编机
stitch rivet 缀缝铆钉;拼合铆钉;缝合铆钉
stitch riveting 缀合铆钉
stitch scissors 拆线剪
stitch-up 缝补
stitch-varied construction 变化组织结构
stitch washer 构件间垫圈
stitch weld 缝焊;滚焊
stitch welding 自动点焊;针脚点焊法;连续点焊;合缝焊接;跳焊;断续缝焊;叠焊(法);缝合点焊;滚焊;垫缝焊接
stitch wire clamp 接头线夹
stitchwork 刺绣
stithy 打铁铺;铁砧;锻工场
stivane toxilicacid 马来酸
St. Kajetan church 圣卡杰坦教堂
stllwellite 菱硼硅铈矿
St Lucia 圣卢西亚
St. Mark Square 圣·马克广场
stoa (有屋顶的)拜占庭建筑(的)圆柱大厅;(篱笆用的)小立柱;(希腊建筑中的)柱廊或拱廊;拱廊
Stoa of Attalos II. on the Agora at Athens 阿塔洛斯二世的拱顶柱廊(在希腊古雅典广场上的)
Stoa of the Hermae 赫梅的拱顶柱廊(未发掘)
Stoa of Zeus Eleutherios 泽乌斯伊柳塞里喔斯的拱顶柱廊
Stoa Poikile 波伊奇尔拱顶柱廊(未发掘)
stob 长铁楔
Stobie furnace 斯托比电炉
stob-jobber 测量打桩工
stochastic abstracting 随机编制文摘
stochastic accelerator 随机加速器
stochastically independent component 随机独立分量
stochastically seismic clustering model 随机震群模型
stochastic analog 随机模拟
stochastic analysis 随机分析
stochastic approximation 随机近似法
stochastic array 随机阵
stochastic augmented transition network 随机扩展转移网络
stochastic automata inference 随机自动机推断
stochastic automata model 随机自动机模型
stochastic beam 随机束
stochastic calculus 随机计算
stochastic chain rule 随机链规则
stochastic characteristic 随机特征
stochastic continuity 随机连续性
stochastic control 随机控制
stochastic control theory 随机控制理论
stochastic convergence 随机收敛
stochastic data generating system 随机数据生成系统
stochastic decision process 随机判定过程;随机决定过程
stochastic decision tree analysis 随机决策树分析
stochastic deformation model 随机畸变模型
stochastic dependence 随机相依;随机相关
stochastic design earthquake 随机设计地震
stochastic difference equation 随机差分方程
stochastic differentiability 随机可微性
stochastic differential 随机微分
stochastic differential equation 随机微分方程
stochastic dissolved oxygen model 随机溶解氧模型
stochastic distribution 随机分布
stochastic disturbance 随机扰动;随机干扰

stochastic dynamic(al) prediction 随机动态预报;随机动力预测
stochastic dynamic(al) programming 随机动态规划
stochastic dynamics 随机动力学
stochastic effect 随机效应;随机效果
stochastic ejection 随机引出
stochastic element 随机元素;随机成分
stochastic equation 随机方程(式)
stochastic error 随机误差
stochastic error-correction 随机误差修正
stochastic event 随机事件
stochastic extraction 随机引出
stochastic extraction system 随机引出系统
stochastic failure 随机破坏
stochastic faulting 随机断层作用
stochastic fault model 随机断层模型
stochastic finite element 随机有限元
stochastic finite state 随机有限状态
stochastic finite state automata 随机有限状态自动机
stochastic forecast 随机预报
stochastic fracture mechanics 随机断裂力学
stochastic function 随机函数
stochastic functional equation 随机函数方程
stochastic game 随机对策
stochastic heating 随机加热
stochastic hydraulics 随机水力学
stochastic hydrology 统计水文学;随机水文学
stochastic independence 随机独立;随机不相关
stochastic indexed production 随机附标产生式
stochastic input 随机输入
stochastic instability 随机不稳定性
stochastic integral 随机积分
stochastic integral equation 随机积分方程
stochastic inventory 随机性存储
stochastic inverse 随机反演
stochastic kernel 随机核
stochastic linear discrete system 随机线性离散系统
stochastic linear planning 随机线性规划
stochastic linear programming 随机线性规划
stochastic linear system 随机线性系统
stochastic mapping 随机映射
stochastic matrix 随机阵;随机矩阵
stochastic method 随机方法
stochastic model 随机性模型;随机模型
stochastic modelling 随机模拟
stochastic modulation 随机调制
stochastic monotonicity 随机单调性
stochastic motion 随机运动
stochastic movement 随机位移
stochastic network 随机网络
stochastic noise 无规则分布噪声;随机噪声
stochastic non-linear system 随机非线性系统
stochastic observability 随机可观察性;随机可观测性
stochastic occurrence 发震随机过程
stochastic operator 随机算子
stochastic optimal control 随机最佳控制
stochastic parameter 随机参数
stochastic pattern 随机型
stochastic prediction 随机预测
stochastic problem 随机问题
stochastic process 概率分析法;随机过程
stochastic programming 随机规划(法)
stochastic pushdown automata 随机下推自动机
stochastic quantity 随机量
stochastic quantity linear energy 随机量线能量
stochastic queue 随机排队
stochastic queuing systems 随机排队模式
stochastic regime 随机状态
stochastic relation 随机关系
stochastic residual process 剩余随机过程
stochastic response characteristic 随机反应特征;随机反应特性
stochastic retrieval 随机检索
stochastic sampling 随机取样;随机模型;随机抽样;随机采样
stochastic scheme 随机方案
stochastic searching 随机查找
stochastic seismic analysis 随机地震分析
stochastic seismic excitation 随机地震激发
stochastic seismic response analysis 随机地震反应分析
stochastic selection 随机选择
stochastic selection algorithm 随机选择算法
stochastic sequence 随机序列
stochastic sequential model 随机序列模型
stochastic series 随机序列
stochastic service system 随机伺服系统;随机服务系统
stochastic service system theory 随机服务系统理论
stochastic signal 随机信号
stochastic simulation 随机模拟;随机仿真
stochastic source 随机源
stochastic state transition matrix 随机状态转移矩阵
stochastic strain 随机应变
stochastic stress 随机应力
stochastic structural system 随机结构体系
stochastic system 随机系统
stochastic technique 随机技术
stochastic theory 统计理论
stochastic time-series model 随机时间序列模型
stochastic transformation 随机变换
stochastic transition function 随机转变函数
stochastic transition network 随机转移网络
stochastic type of acceleration 随机型加速
stochastic variable 随机变数;随机变量
stochastic variation 随机变化
stochastic vibration 随机振动
stochastic water quality system 随机水质系统
stochastic watershed system 随机流域系统
stock 蓄积;原料;岩株;纸张;纸浆;砧木;轧件;管状矿脉;股票;工具台座;炉架;库存;坯料;盘料坯;存料;粗钢料;船底枕木;储藏;承印物;材料;标准木料;备料;存货
stockable mixture 可存储的混合料
stock account 存货账;股票账本
stock accounting 库存账目
stockade 篱笆;桩打的防波堤;栅栏;围桩;防浪排桩
stockaded village 寨子
stockade groin 挡栅丁坝
stockade groyne 改良的木笼丁坝;栏栅杆;挡栅丁坝
stock adjustment 库存调整
stockage 料库
stockage objective 库存指示
stock allowance 机械加工余量
stock anchor 带销锚栓
stock and dies 螺纹印模的底与模具;螺丝绞牙架;螺丝板牙架;套口设备(钻杆)
stock appreciation 库存增值
stock article 库存物品
stock assessment 资源评价
stock at par 面值股票
stock-at-risk 受威胁群体和资源
stock at valuation 估计存货
stock authorized 额定股本
stock balance 库存余额;库存量
stock barn 牲口棚
stock bin 料仓;配料仓
stock biomass 资源生物量
stock block 垫墩
stock board 砖模底板;规格板材
stock bonus 红利;股息
stock book 存货簿
stockbreeding pollution 牲畜污染
stock brick 普通砖;标准砖
stock-brick work 普通砖工程
stockbridge damper 架空线减振器;储能桥式防振锤
stockbroker 证券经纪人;股票经纪人
stock brokerage 证券经纪业务
stock brush 润砖刷
stock budget 库存量预算
stock cap 风口盖
stock capacity 载畜量
stock car 牲畜车
stock card 存货卡;存货单;材料卡片
stock car double deck 双层牲畜车
stock car single deck 单层牲畜车
stock cattle 存栏牲畜
stock certificate 股票
stock checking 存货盘点
stock chute 牲畜跳板
stock clerf 存货管理员
stock column 料柱

stock commodities 大量生产产品
stock company 股份公司;股份制;存货公司
stock compensation 股票报酬
stock control 库存控制;库存管理;存量管制;存货控制
stock control card 库存控制卡;库存管理卡
stock control level 存量管制基准
stock control system 零件管理方式;库存管理制度;备件管理方式
stock converter 冷料转炉
stock cooperative 股份合作社
stock core 长条泥芯;备用型芯;棒状芯
stock-core machine 挤芯机
stock crane 堆货吊车
stock creep 牲口地下(通)道
stock culture 原种培养;储[贮]备培养;存储培养
stock cum right 附有增权的股票
stock cutter 坯料剪切机;切料机
stock deal 股票交易
stock debit notes 入库凭单
stock debt depletion 储备耗损
stock deposit 证券存款
stock die 管子铰板板牙
stock distributing gear 布料器
stock distribution 物料分配器
stock dividend 红利;股息
stock door 仓库门;机制门
stock door element 仓库门部件
stock dump 储[贮]料堆;料堆
stock-dyed yarn 散纤维染色纱
stock dyeing 散纤维染色
stocked anchor 有杆锚
stocked quadrat 立木样方
stock emulsion 储藏乳剂
stock engine 座式
stocker 加煤机;往复推送器;堆料机;堆料工;储料器
stocker and reclaimer 堆料取料机
stock exchange 股票交易所;交易所;证券交易所
stock exchanger 证券交易所
stock-farm 畜牧场
stock farmer 畜牧业者
stock farming 畜牧业
stock farming district 牧业区
stock feed 条材进给
stock fluctuation 储量变化
stock form 卷纸
stock freezer 冷藏仓库
stock fulling mill 槌打式缩绒机
stock fund 库存基金;材料基金
stock-gang 框锯
stock grazing 放牧
stock ground 料场;堆料场
stock guard 牲畜护栏
stock guide 导料板;板料导向块
stock handling facility 牲畜管理设备
stock head 车床头;车床的头座
stock heap 料堆;矿堆;储料堆
stockholder 持股人;股票持有人;股东
stockholder's equity 股东拥有的资产净值;股东权益
stock holding chest 储[贮]浆槽
stock-holding cost 库存成本
Stockholm black 斯德哥尔摩黑色原料
Stockholm pine tar 斯德哥尔摩焦油
Stockholm tar 斯德哥尔摩松焦油
stock house 料仓;库房;料房
stockhouse set (仓库内的)黏结水泥;库凝;库硬化;储存凝结;在库房内结块的水泥
stockhouse set cement 仓库中的凝结水泥
stock indicator 装料指示器;料线指示器
stock influence 砧木影响
stocking 储[贮]藏;立木度;储量;储存
stocking area 堆料区
stocking capital 储备资金
stocking capital quota 储备资金定额
stocking cutter 预切刀;柄式铣刀
stocking degree 立木度
stocking density 载畜量
stocking filter 袋式集尘器
stocking indicator 装载指示器
stocking limit 放养限度
stocking machine 装机
stocking piling 堆场
stocking quantity of materials 物资储备量

stocking raising 畜牧业
stocking rate 养畜密度;载畜率;载畜量;储占率
stocking tool 粗切刀
stocking yard 料场;堆料场;成品仓库
stock in storage 现存物资
stock in the inventory 存货
stock in the till 备用现金储备
stock in trade 待销存货
stock in transit 在途物资
stock inventory 现有存货;固定资本清单;库存目录;货物库存;存货清单;财产清单
stock investment 股金投资
stock issue 发行股票;股票发行
stock jobber 证券经纪人;场内股票经纪人;股票经纪人;投机者
stock-jobbing 证券经纪业务;股票买卖
stock-keeping 存货保管
stockkeeping unit 库存单位;库存保管单位;存库标识
stock kettle 储水壶
stock key 键
stock layout 排样
stock length 储[贮]料长度
stockless anchor 转爪锚;无杆锚
stockless bower 无杆首锚
stock level 库存水平
stock lifter 板料升降器
stockline 料线
stock line armo(u)r 料线保护板
stock line ga(u)ge 料线指示器
stock line indicator 探料尺装置
stock list 库存单;存货单;材料表
stock locator systems 出货编序系统
stock lock 呆锁;门外装置的锁;门外锁;外装弹簧锁
stock lumber 常用规格木材;库存木材;储存木材;常规木材;储[贮]存木材;锯成木材
stock machine 槌打式缩绒机
stockman 牧场工;存货管理员;仓库管理员
stock map 林型图;林相图;牲畜分布图
stock market 证券市场;牲畜市场
stock material 存料
stock millwork 库存木制品
stock mo(u)ld 砖模;定型模
stock note 木材报价单
stock number 资源量;物料编号
stock of cars 汽车库;汽车总数
stock of drying zone 干燥层储料
stock of equipment 设备储备
stock of fish 鱼群;鱼类资源
stock of housing 住房量
stock of manpower 人力储备
stock of materials 储[贮]料堆;库存材料;物资储备;堆料堆;材料库存;材料库;材料储备
stock of real capital 全部实际资本
stock of urgently needed material 应急用料储备
stock of water 蓄水
stock of wheat 存麦豆投资
stock oiler 座架加油装置
stock on hand 现有存货;现存物资;现存量;库存货;存货
stock on order 定购货物
stock option 职工优先认股权;材料选择
stock order 备货生产
stock out 库存中断;库存耗竭;库存短缺
stockout cost 库存削价成本;存货费用;库存不足费用;缺货成本;脱销费用
stock ownership 股票所有权
stock pass 牲口地下(通)道;牲畜通道
stock pig 存锭
stockpile 料堆;存堆;藏量;材料库;堆存;存货;储[贮]堆;储[贮]备品;货堆;囤积;堆集;堆存;存储;储料堆;储存;储备料堆;存货
stockpile cableway 堆装索道;储[贮]料索道
stock piled material 堆放的材料;储存的材料
stockpile loader 储[贮]料堆装载车;储[贮]料堆装载机
stockpile loading 堆装材料
stockpile measurement 储[贮]料堆测量
stockpile of topsoil 表土(层)的积聚
stockpiler 堆垛工
stockpile(re-)handling 原料储存移动;原料储存输送;原料储存堆放;原料储存移动;原料储备输送;原料储备堆放;材料储存移动;材料储存堆放;材料储备移动;材料储备堆放

送;材料储备堆放;材料储存管理;材料储备管理;原料储存管理;原料储备管理
stockpile ropeway 堆装索道;储[贮]料索道
stockpile tunnel 原料回收坑道
stockpiling 存货;存料;堆装;储[贮]料堆;囤积
stockpiling belt conveyer 堆装胶带机
stockpiling machine 堆装机
stock plots 原种植区;几个原种圃
stock pond 蓄养池;储[贮]水池;牲畜塘;池塘
stock power 股票转让授权书
stock prepartion 备料
stock price 股票价格
stock production 估需生产
stock pump 纸浆泵
stock pusher 装炉推料器;推料机
stock rack 成品放置架
stock rail 尖轨前基本轨长;基本轨;普通铁轨;备用铁轨材料
stock rail end of switch 道岔基本轨端部
stock raising in livestock shed 牛羊圈养
stock ramming 用黏土夯筑围堤;夯土堆筑围堰;黏土夯筑围堰;堆填夯实(法)
stock reclaiming tunnel 原料回收坑道
stock redemption fund 收回股份基金
stock reel 棒料架;杆料架
stock-releasing 抛售储存
stock removal 切削量
stock removal action 切削动作
stock removing 切削
stock-removing efficiency 切削效率
stock-removing machine 切削机床
stock reservoir 储罐
stock right 股权
stock rod 探尺
stock room 商品展览室;仓库;原料库;储[贮]料间;存货室;储料室;储藏屋;储藏室;材料库;库房
stock rotation 存货周期
stock-run material 堆装材料;堆场材料
stocks 过火硬砖;过火面砖;股份;公司股票
stock-sales ratio 库存率
stocksaver 捕浆器
stock saving bank 股份储蓄银行
stock scrip 小额股票
stock shears 台剪(机)
stock sheets 库存板;标准尺寸板
stock size 常备货的尺寸;一般尺寸;资源量;库存量;库存尺寸;存货规格;常规尺寸;标准尺寸;仓库存货数量
stock solution 原液;储[贮]存液;储[贮]备溶液;储液;储备溶液
stocks returned 退回股票
stock stainless steel window 库存不锈钢窗
stock stand 材料台
stock status report 储备情况报告
stock steel window 标准钢窗
stock stop 挡料器;挡料机构
stock subscription 股份认购
stock subway 牲畜地下(通)道
stock support 送料支架;带座支架;材料支架
stock swap 证券交易
stock taking 成果调查;清货;清查存货;盘点;库存盘点;盘货;盘存
stock tank 储油罐;储罐
stock tank barrels 油罐桶数
stock thickness 坯料厚度
stock timber window 库存木窗
stock transfer 股票转让
stock trust certificate 股份信托证券
stock turn 存货周转
stock turnover 库存周转率
stock type 存货类型
stock utilization 材料利用率
stock vice 台式虎钳;台虎钳
stock warrant 认股权证;可转让票据
stock washer 洗浆机
stock water development 畜牧用水开发;畜牧水源开发
stock white 白色浆;白漆浆
stock with par value 面值股票
stockwork 备售制品;网状脉;网脉状矿体;网脉
stock yard 畜生围栏;原料间;储料场;储[贮]矿场;储[贮]料堆;料场;堆料场
stockyard transporter 堆装场运输机;堆料场桥式起重机;堆料场输送器;畜牧场运输机;畜牧场桥

式起重机;畜牧场输送器
stockyard waste 牲畜场废料
Stoclet mansion at brussels 布鲁塞尔的斯托克莱脱大厦
Stoddard solvent 斯托达特溶剂;干洗溶剂汽油
Stoddard tray distributor 斯托达特浅盘布水器
Stodola method 斯托多拉法
stoechiometer 化学计量器
stoep 阳台;游廊
Stohesina wave 斯托克纳波
stoiberite 钒铜矿
stoichiometer 化学计量器
stoichiometric(al) 理想配比的;化学计算的
stoichiometric(al) calculation 化学计算
stoichiometric(al) coefficient 化学计量系数
stoichiometric(al) combustion 化学计量燃烧
stoichiometric(al) composition 化学计量组成
stoichiometric(al) compound 化学计量化合物
stoichiometric(al) concentration 化学计量浓度
stoichiometric(al) constant 化学计算常数;化学计量常数
stoichiometric(al) crystal 化学计量晶体
stoichiometric(al) equation 化学计算方程式
stoichiometric(al) equivalent 化学计算当量;化学计量当量
stoichiometric(al) factor 化学计量因数
stoichiometric(al) mixture 理论混合物
stoichiometric(al) number 化学计量数
stoichiometric(al) point 化学计量点
stoichiometric(al) proportion 化学计量比
stoichiometric(al) ratio 化学计算比;化学计量比
stoichiometric(al) reaction 化学计量反应
stoichiometric(al) relationship 化学计算关系;化学计量关系
stoichiometric(al) softening 化学计量软化
stoichiometric coefficient 理想化合系数
stoichiometric combustion 按化学计量燃烧
stoichiometric compound 定比化合物
stoichiometry 化学计算(法);化学计量学;化学计量(法)
stoichiometry of coagulation 混凝化学计量法
stoke 加煤;添煤;泡(动力黏度单位)
stoke ground 浅棕色巴斯石灰石(萨默塞特产)
stokehold 锅炉舱;汽锅室;烧火间;火舱
stokehold bulkhead 锅炉舱壁
stoke hole 加煤孔;拨火孔
stoker 加煤机;加煤工;喂煤机;添煤器;司炉(工人)
stoker casing 自动加煤机罩
stoker distributor jet 加煤机喷煤器
stoker drive mechanism 机械炉排拖动机构
stoker elevator conveyer 加煤机输煤器
stoker firing 机械加煤焚烧法
stoker rating 炉排燃烧率
stoker's platform 装煤工平台;司炉工平台
stoker surface 炉排表面;抛煤面
stoker timer 加煤定时控制器;加燃料自动控制器
stoker type grate cooler 推动篦式冷却器
Stoke's constant for ellipsoid 斯托克椭球常数
Stoke's diameter 斯托克粒径
Stoke's drift 斯托克漂移
Stoke's equation 斯托克方程
Stoke's flow 斯托克流动
Stoke's function 斯托克函数
Stokesian fluid 斯托克流体
Stoke's integral 斯托克积分
Stoke's integral theorem 斯托克积分定理
stokesite 硅钙锡石;硅钙锡矿
Stoke's law 斯托克定律
Stoke's law of resistance 斯托克阻力定律
Stoke's line 斯托克谱线
Stoke's meter 斯托克参数测量仪
Stoke's number 斯托克数
Stokes number 1 第一斯托克斯数
Stokes number 2 第二斯托克斯数
Stoke's parameter 斯托克参数
Stokes press 斯托克凸轮式压机
Stoke's problem 斯托克问题
Stoke's radiation 斯托克辐射
Stoke's-Raman scattering 斯托克—拉曼散射
Stoke's shift 斯托克移移
Stoke's stream function 斯托克流函数
Stoke's theorem 斯托克定理
Stokes transition 斯托克跃迁
Stoke's viscosity 斯托克黏滞度

Stoke's wave 斯托克波
stoking 连续烧结;烧火;加煤
stoking tool 烧火工具
stolen bill 被窃票据
stolen goods 失窃货物
stolpenite 铁蒙脱石
stolzite 钨铅矿
Stolz stop 斯托耳兹光栏
stoma 气孔【植】
stomach piece 副船首材
stomatal band 气孔带
stomatal index 气孔指数
stomatal movement 气孔运动
stomatal transpiration 叶面散发;气孔蒸腾;气孔散发
stomatic band 气孔带
stomatic cleft 气孔缝
stomatic movement 气孔运动
stomatic resistance 气孔阻力
stomatologic(al) hospital 口腔医院
stomatology 口腔科
Stomer viscosimeter 斯托麦旋转黏度计
stomp clutch in 蹬(脚踏板接)上离合器
stone 石材制品;岩石;磨刀小油石;石头;石料;石块;石(材);成型宝石;材核;碑
stone accompanied by fracture 周围有裂缝的石材
Stone age 石器时代
stone aggregate 碎石集料
stone aggregate concrete 石材集料混凝土
stone altar 石台阶;石祭坛
stone altar piece 石祭坛块件
stone altar reredos 石祭坛屏障
stone altar retable 石祭坛屏障
stone altar screen 石祭坛屏障
stone anchor 石砌体锚固件
stone angle quoin 屋角石块
stone animal 石象生
stone appended to wall 附壁石
stone apron 石护墙(用来保护海塘免遭海水冲刷);石砌护坦;抛石护坦;石围裙
stone arcade 有石拱顶的走廊;石拱廊
stone arch 石法券;石券;砌石拱;石拱
stone arch bridge 石拱桥
stone arching 砌石拱圬工;石拱圬工
stone armo(u)r 块石护面
stone armo(u)red tee-head 块石护面T形盘头
stone arrangement 置石
stone article 石器
stone ashlar 琢石块
stone ashlar arch 琢石拱
stone ashlar bond 琢石砌合
stone ashlar bonder 琢石墙砌块
stone ashlar masonry 琢石砌筑工程
stone ashlar slab 琢石板
stone ashlar structure 琢石结构
stone ashlar vault 琢石拱顶
stone ashlar window 琢石窗
stone asphalt 沥青胶结块石;沥青灌浆块石
stone axe 琢石斧
stone balcony 石楼厅;石阳台
stone ballast 道砟;石渣
stone baluster 石栏杆小柱
stone balustrade 石栏杆;石扶手
stone band 条石
stone bannister 石栏杆小柱
stone bar 石棒
stone barge 石驳
stone barrel vault 石砌圆柱体拱顶;石砌桶形拱顶
stone base 碎石基础;碎石基底
stone basket 石笼
stone batcher 填充料供料定量器;填充料计量箱;填充料秤量器;助熔剂供料定量器;助熔剂计量箱;助熔剂秤量器;石料供料定量器;石料计量箱;石料秤量器
stone beam 石梁
stone beam bridge 石梁桥
stone bedding 石块基层;石垫层;道砟基床
stone bench 石长凳;石实验台;石钳工台;石工作台
stone bin 石料仓;石渣仓
stone block 石砌块;石块
stone block masonry 石砌圬工
stone block pavement 石块路面;石块铺路面;石块铺面
stone block paving 石块铺路面

stone block tie 石枕
stone blockwork 石块制作;石砌块工作;石砌块工程
stone block yield 成荒料率
stone blue 灰蓝色
stone boat 运石平底船;石舫;运石雪橇;运石平底橇
stone bolt 棘头特栓;石榫;石栓;底脚螺钉
stone bond 石砌合;砌石法
stone borer 蛀石(海)虫
stone bottoming 石块铺底
stone bound 石界标
stone boundary wall 石分界(围)墙
stone-bound gravel road 石结砾石路
stone box 沉石箱;废石仓
stone box culvert 石矩形涵洞;石箱形涵洞
stone bracket (从墙上伸出的)石托臂;(从墙上伸出的)石牛腿;(从墙上伸出的)石托架
stone breaker 碎石机
stone breaking plant 碎石设备;碎石工厂
stone breakwater 石防波堤
stone bridge 石桥
stone bridge construction 石桥施工
stone building 石砌房屋;石屋;石砌建筑物
stone building material 建筑石材
stone built building 石料砌成的房屋
stone built groin 堆石丁坝
stone built groyne 石防波堤;石丁坝;石构交叉拱(肋)
stone bull 石料大件
stone cage cofferdam 石笼围堰
stone canal 石渠
stone capping 石料压盖;石料封顶;石压顶
stone capping slab 石顶盖板
stone carrying hopper barge 运石开底驳
stone carved pagoda 石雕塔
stone carving 石雕刻;石雕;石刻
stone-case dam 石笼坝
stone casting 铸石
stone catcher 沉石箱;采石器
stone cavern 石洞
stone china 白色硬质陶瓷
stone chip filler 石屑填充料
stone chip(ping)s 石片;碎石片;石屑
stone chip resistance 抗石击性
stone chisel 石凿
stone church 石砌教堂
stone circle 石环;石圈;分选环
Stone circuit 斯通电路
stone cladding 石料覆盖;石料贴面
stone cladding work 石料镶面工程
stone classification 石料分级
stone clay 泥石
stone-clay core 泥石芯;石黏土芯
stone cleaner 石块分离机;除石机
stone clips 石屑
stone coal 白煤;块状无烟煤;无烟煤;石炭
stone coating test 沥青黏附石料能力试验;石裹覆试验
stone coffin 石棺
stone cold 冷透的
stone collector 捡石机;集石机
stone column 碎石桩;碎石柱;石柱
stone concentration 金刚石浓度
stone concrete 块石混凝土;石子混凝土;砂石混凝土
stone construction 石结构;石建筑
stone construction(al) material 石材建筑材料
stone construction(al) method 石材建筑方法
stone content 金刚石含量;含石量
stone cope 石压顶
stone coping 石遮帽;石盖顶;石压顶
stone coping slab 盖顶石板
stone corbel 石托肩;石牛腿;石梁托;突出墙面石支托;石支托
stone core fascine 梢枕
stone count (钻头上)金刚石颗数
stone counterfort 石撑壁;石拥壁;石扶垛;石砌扶壁;石扶壁;石墩
stone course 石砌层;石铺道路;块石层
stone cradle vault 石支座拱顶
stone cramp 石扣
stone cross 石十字架(坟墓用)
stone crusher 碎石机
stone-crushing 石料破碎
stone crushing installation 石材破碎装置

stone crushing plant 轧石厂
stone crushing strength 石料压碎强度
stone culvert 石砌涵洞;石涵(洞)
stone curb (建筑物上的)石边饰;路缘石;栏石;石路牙;侧石;石井栏
stone curtain wall 石屏障;石幕墙
stone curving 石材弯曲件;石弯道;石弧线
stone cutter 凿石机;截石机;切石机;切割石工;石工;石材切割机
stonecutter's chisel 石工用凿刀;石工凿(刀)
stone cutting 雕石;凿石;切石
stone cutting machine 切石机;石材切割机
stone cutting method 切石法;石材切割法
stone cutting saw 切割石材的锯
stone cylindrical vault 圆柱状石拱顶
stone dam 石坝
stone dead wire 软钢丝;退火钢丝
stone deposit(e) 石矿体
stone desert 戈壁;石质沙漠;石质荒漠;石漠
stone die 石制代型
stone dike 石堤;石砌堤
stone discharge gutter 石砌排水沟;石排水槽
stoned mesh construction 钢丝石笼结构
stone door threshold 石门槛
stone drain 盲沟;砌石排水沟;填石盲沟;石砌排水沟;石砌盲沟;石砌暗沟
stone drainage 填石排水
stone drain(age)gutter 石砌排水沟;石排水槽
stone drawing tool 挖石工具
stone dressed ready for building 琢凿加工好的建筑石材
stone dresser 雕琢工;凿石工;琢石工;石工;石匠
stone dresser's sign 凿石工场的标志;凿石工场的牌号;琢石工场的标志;石工场的牌号
stone dressing 琢石面;琢石装饰面;琢石法;石料琢面;石料饰面;石料砌面;石材修琢法
stone-dressing machine 石雕琢机;石修琢机;精整机;石平整机;石修整机;琢石机
stone-dressing plant 琢石工场;琢石厂
stone-dressing stem 石面修琢凿
stone-dressing work 凿石工作
stone drier 干石机
stone drift 石平巷;岩石坑道
stone drifter 凿岩机
stone drill 钻岩机;凿岩机;石钻
stone dropout 金刚石脱落
stone dumper 抛石船
stone dust 石质粉尘;石屑;石粉
stone dust apparatus 撒金刚石粉器
stone eliminator 石块分离机;除石机
stone embankment 堆石堤;石堤
stone exposure 金刚石出刃
stone extractor 石块分离机;除石机
stone facade 石材(建筑)正面;石材(建筑)立面
stone-faced 砌石护面的;石面的;石料砌面的
stone-faced arch 石饰面拱
stone-faced concrete panel 石材贴面的混凝土板
stone-faced masonry 石饰面圬工;石砌体;石镶面圬工;石面圬工;石面砌体
stone-faced rubble masonry 毛石贴面圬工
stone facies 岩漠相
stone facing 琢石面;砌石护面;石饰面;石料镶面;石料饰面;石料砌面;石材饰面
stone facing slab 石饰面板;石材贴面板
stone fall 石瀑
stone fascine 包石梢捆
stone female figure 石雕女像
stone fence 石砌围墙;石围栏
stone field 冻砾原;采石场;石海
stone-filled 填石的
stone-filled bamboo gabion 填石竹笼
stone-filled crib 填石木笼;石笼
stone-filled crib dam 石笼坝;填石木框坝
stone-filled drain 填(碎)石排水沟;碎石排水沟
stone-filled drainage 碎石排水沟
stone-filled drain trench 填(碎)石的排水沟槽;填碎石的排水沟渠
stone-filled pile revetment 填石排桩护岸
stone-filled sheet asphalt 片地沥青石(即细粒式地沥青混凝土);填石片地沥青(混合料);填石沥青板(粗集料达25%的沥青板)
stone-filled timber crib groin 木笼填石丁坝
stone-filled trench 填石盲沟;填石暗沟

stone-filled work 堆石工程
stone filler 石材垫板;石材垫片;石材填料
stone filling 填石;堆石体
stone filter 碎石滤器
stone finish 石材装修;石磨成精光表面;石饰面;石材饰面
stone fixing 石作安装
stone fixing work 石作安装工程
stone flag 铺路石;石板;板石
stone floor cover(ing) 石材地面覆盖板
stone floor(ing) 石材地板
stone flower bed 石花台
stone font 石材(修筑的)喷水池
stone footing 石底脚;毛石基础;片石基础;石底座;石底基础
stone footpath 石小径
stone footpath paving flag 铺人行道石板
stone forest 溶柱;石林
stone fork 碎石叉;石渣耙
stone for stream work 构筑河流的石块
stone foundation 毛石基础;石基础
stone fragment 碎石片
stone frieze 石柱中楣;石材雕带
stone furrow 磨盘沟纹
stone gall (岩石中的)黏土结核;岩石含的杂质;岩石含的黏土
stone garden 叠石庭园
stone grab 抓石斗
stone grafting 核接
stone granulating 石料轧碎;石面斩凿
stone grapple 攫石器
stone-grid dam 格栅坝
stone grille 石格子宽;石格栅
stone grinder 磨盘;磨石机;磨石工;石磨;石料研磨机
stone grinding machine 石材研磨机
stone groin 钢丝网填石丁坝
stone guard 砂石防护网;除石机
stone gutter 石砌(水)槽
stone hammer 石工锤
stone hand 拼版工人
stone header 露头石;石丁头
stone hedge 石砌围墙
stone henge 巨石柱群围栏(英国原始社会的遗迹);石制镶板;石制面板;石砌围墙
stone hewing work 石材铣磨工作;石材斩劈工作
stone hinge 石头接合;石门枢;石节点;石铰接
stone house 石屋
stone hut 石小屋;石棚屋
stone ice 矿冰;地下冰
stone inclusion 夹杂石块层
stone inscription 石刻
stone jaw crusher 颚式碎石机
stone jetty 石突堤式码头;石防波堤
stone kerb 井口锁口圈;路边石;(建筑物上的)石边饰
stone lantern 石灯笼(常见于日本式庭园中)
stone layer 块石层;(天然的)石层
stone laying 铺石
stone layout 置石
stone levee 石堤
stone lifting bolt 起吊块石螺栓;吊石栓;起吊块石用的螺栓
stone lifting tongs 块石提升夹钳;石块吊升夹具
stone-like 像天然石材
stone-like coal 石煤
stone lime 石灰
stone line 碎石带(土内);石线
stone lined 块石衬砌的
stone lining 石隔板;石衬(砌);块石衬砌
stone lintel 条石梁;石过梁
stone male figure 石雕男像
stoneman 石标;石匠;石工
stone marker 标石
stone mason 砌石工(人);石工;石匠
stone mason mark 石工标记
stone masonry 块石砌体;砌石圬工;砌石(工程);石圬工;石砌体
stone masonry dam 砌石坝
stone masonry gravity dam 砌石重力坝
stone masonry gravity retaining wall 乱石圬工重力式挡土墙
stone masonry lining 料石衬砌;石块衬砌;石工衬砌

stone masonry pier 石砌桥墩
stone masonry retaining wall 砌石挡土墙;石圬工挡土墙
stonemason's hammer 石工锤
stonemason's lodge 石工小屋
stonemason's saw 石工锯
stone mass 石块
stone material 石料;石材
stone matting 石垫层
stone-mesh apron 钢丝网填石护坦;钢丝网填石护底;钢丝网填石海漫;钢丝网填石沉排
stone-mesh construction 石笼结构
stone-mesh groin 钢丝网填石丁坝
stone-mesh groynes 钢丝网填石丁坝
stone-mesh mattress 钢丝网填石护底;钢丝网填石沉排
stone mill 石粉工厂;岩石破碎器;岩石破碎机;琢石机;磨石机;平石磨;碎石机;碎石厂;石粉工场;石粉厂
stone miller 石研磨机
stone milling work 石材铣磨工作;块石成形工程
stone miner 采石工
stone monument 标石
stone monument damaged 标石已损坏
stone monument perfect 标石完好
stone mortar with pestle 带捣杵石臼
stone mosaic 石马赛克
stone mosaix 石拼花地砖
Stone Mountain granite 淡灰色花岗岩(美国佐治亚州产)
stone mulch 石幂
stone mullion window 石质中棂窗;石竖框窗;石直棂窗
stone needle 矼石
stone of apricots 杏仁
stone of one-quarter brick size 四分之一大小的石块
stone oil 石油
stone order 石柱型
stone ornament 石头点缀;点缀石
stone ornamental string course 装饰性石束带层
stone packing 块石基层;填石;填筑块石
stone pagoda 石塔;石宝塔
stone paint 仿石涂料;圬工涂料
stone pavement 铺石路面;石料路面;石块铺面
stone pavilion 石亭
stone paving 块石铺砌;砌石面层;砌石(护面)
stone paving sett 铺路小方块石
stone perfect marker damaged 标石完好标心志已损坏
stone perron 石台阶
stone pick 石镐;石凿
stone picker 选石机;集石机
stone picking machine 集石机
stone pier 石防波堤;石墙墩;石(桥)墩
stone pillar 石桩;碎石墩;石柱;石墩
stone pine 意大利五针松;石松
stone pit 石坑;采石坑;采石场
stone-pitched 石砌的
stone pitched facing 砌石护面
stone-pitched road 石砌道路
stone pitching 干砌石;砌石护坡;砌石护面;砌石护坡
stone pitching with filled joints 有填缝的砌石护坡
stone plaster 涂石墙粉;仿石饰面;斩假石
stone plaster mix(ture) 涂石墙粉混合物
stone plaster stuff 涂石墙粉材料
stone plug 瓷火花塞
stone plums 大块石
stone pockets 蜂窝
stone pockets of concrete 混凝土中蜂窝状气孔;混凝土中的蜂窝;石窝混凝土
stone polisher 磨石工
stone polishing 石材抛光
stone polishing machine 石材磨光机
stone polygon 石质多边形土
stone portal 石洞门;石桥门;石门架;石库门
stone post 界石
stone powder 石粉
stone preservative 石料防腐剂
stone preserving agent 石料防腐剂
stone press 压石机
stone pressure 金刚石上的比压(钻进时)
stone-protected tee-head 块石护面T形盘头

stone pulpit 石布道坛(教堂中);石控制台(教堂中);石讲堂(教堂中)
stone putty 石质油灰
stone putty cementation 石质腻子黏结;石质油灰黏结
stone quality 金刚石质量
stone quarry 采石场
stone quoin 角隅石
stoner 碎石机;除石块机;除石机
stone railing 石围栏;石扶手;石栏杆
stone rake 搂石器;石耙
stone receiving box 石子受料箱
stone releaser 石块排除器
stone remover 除石器
stone retarder 石块分离器
stone revetment 砌石护岸;石砌护岸
stone rib 石肋
stone-ribbed cupola 石肋圆屋顶
stone-ribbed dome 石肋穹隆
stone-ribbed vault 石肋拱顶
stone ring 石环
stone riprap 乱石堆;乱石;抛石体;抛石护坦;抛石护坡;抛石防坡堤;抛石;堆石
stone river 岩石流
stone road 碎石路;碎石路面;石子路
stone roll(er) 石辊;石滚筒;石辗;石碾;石龙
stone roof 石屋顶
stone roof cover(ing) 石屋面;石屋顶覆盖层
stone roof(ing) tile 石屋面瓦
stone roof spire 石屋尖顶;石屋顶尖塔
stone row 石街;巨石纪念碑
stone rubber 磨石工
stone rubbing machine 石料研磨机;石料磨光机
stone rubbish 废矸石
stone rubble masonry(work) 乱石圬工;乱石砌筑;块石圬工;块石砌筑;粗石砌筑;粗石砌筑;碎石圬工;碎石砌筑
stone run 岩石流
stones 碎石
stone sand 轧石砂;碎石砂;石砂;石粉
stone saw 石锯
stone saw blade 切石锯条
stone-sawing machine 锯石机
stone-sawing strand 锯石钢绳
stone screen 石子筛
stone screenings 石屏障;筛出岩粉;轧制石砂;石屑
stone screw 地脚螺钉;棘螺钉
stone scrubber 洗石机
stone sculpture 石雕刻术;石雕刻物
stone seal 倒贯入式沥青路面;厚沥青处治层;掺石封闭层
stone-separating device 除石装置
stone-separating mill 石块分选机
stone separator 石块分离机
stone sett 小方石
stone setter 砌石工(人)
stone-setter's adjustable multiple-point suspension scaffold 砌石工用活动悬吊脚手架
stone setting 块石铺路;块石砌筑
stone-sett pavement 小方石路面
stone sett paving 小方石路面;小方石铺砌
Stonesfield slate 石场板岩
stone shaft 石井
stone shaping work 石材整形工作;石材铣磨成型工作;块石成型工程
stone shingle 石板瓦
stone shoulder 石建筑突额;石建筑凸出部分;石砌路肩;石块路基边缘;石拱座
stone sidewalk 石砌人行道
stone sidewalk paving flag 石砌人行道石板
stone sill 石窗台;石(门)槛
stone sill of door 门槛石;石门槛
stone sill of window 石窗台板
stone size 金刚石粒度;石料尺寸;石块大小
stone sizing plant 石料筛分厂;筛石设备
stone skeleton construction 石骨架结构
stone skip 石料(翻斗)车
stone slab 石板(砌面)
stone slab bridge 石板桥
stone slab cladding 石面板;石板覆层
stone slab correction 中间层校正;布格校正
stone slab facing 石板镶面;石板贴面
stone slab floor(ing) 石板地面;石板路面;石板楼面

stone slab lining 石板护岸
stone slab revetment 石板衬砌;石板护岸
stone slab roofing 石板屋面;石板瓦屋面
stone slate 石板(瓦盖);石板瓦
stone sledge 石工锤
stone slope 碎石边坡
stones per carat 每克拉金刚石粒数
stone spire 石尖顶;石尖塔
stone splitter 石材劈裂机;石材分离机;碎石片
stone spreader 碎石摊铺机;碎石撒布机;石屑撒布机;石料摊铺机;撒石机;布石机
Stone's representation theorem 斯通表示定理
stone stair(case) 石(砌)楼梯;石材楼梯(间);石筑楼梯;石阶梯
stone steps 石级;石阶;石砌台阶
Stone's theorem 斯通定理
stone stopper 除石块机
stone stream 石川
stone strength 石(材)强度
stone string course 石束带层;石线脚
stone stripe 石带
stone structure 石(筑)结构
stone structure soil 石质构造土
stone subdrain 石暗沟
stone supply train 集料火车;石料供应列车
stone support 石材支架
stone surfacer 凿石工具;琢面石;磨面石
stone surfacing 砌石面层;石材镶面;石材饰面;石料贴面;石料护面;石面
stone sweeper 排障器
stone sweep(ing) 石弯道
stone table altar 石台式祭坛
stone table pavilion 碑亭
stone tablet 碑碣;石碑
stone template 石垫;垫石
stone temple 石砌寺庙;石砌教堂
stone threshold 石门槛
stone throne 石王位;石宝座
stone tile 石瓦
stone tomb 石坟墓
stone tongs 夹石钳;吊装夹具(吊装时以夹石块或砖块等);吊石夹钳
stone tool 凿石工具
stone tracery window 石雕花格窗
stone training wall 石顺坝;石导流堤
stone transom 石楣窗;石横档(门窗)
stone trap 沉石箱
stone tunnel vault 石砌圆形拱顶;石砌隧道拱顶
stone turtle 趺
stone two-sided town gateway 城镇两侧石铺道路
stone underlayer 块石垫层
stone vault 石拱顶;石穹
stone-vaulted ceiling 石拱顶棚;石穹屋顶
stone-vaulted roof 石砌拱顶状屋顶
stone veneer(ing) 石料饰面;石料面层;石板贴面;石料护面;石面;石料镶面;石面
stone vessel 石器
stone wagon vault 石砌筒形拱顶
stone wall(ing) 砌石墙;石壁;石围栏;筑石墙
stone wall limit 极限界限
stone wall plaster 石墙粉刷
stone wall tile 石墙面砖
stoneware 缸瓦器;陶瓦器;炻器;石制品;石器;粗陶器;粗陶瓷
stoneware article 陶瓦器件;粗陶制品
stoneware ceiling tile 陶瓦平顶砖;粗陶顶篷砖
stoneware clay 缸瓷土;粗陶土
stoneware discharge gutter 陶瓦排水沟;粗陶排水沟
stoneware discharge pipe 陶瓦排水瓦筒;粗陶排水管
stoneware drain 缸瓦排水管;排水陶管;陶瓦排管
stoneware drainage gutter 陶瓦排水沟
stoneware drainage pipe 陶瓦排水管
stoneware drain pipe 粗陶排水管;缸瓦排水管;陶瓷下水管;石制品下水管
stoneware duct 粗陶瓷管(道)
stoneware facing 陶瓦饰面;粗陶护面
stoneware factory 陶瓦工厂
stoneware filter 陶瓦滤器
stoneware fittings 陶瓦配件
stoneware floor cover(ing) tile 陶瓦地砖
stoneware for industrial purposes 工业用炻器
stoneware for sewer pipes 陶瓦污水管;粗陶污水管

stoneware glaze 炻器釉
stoneware goods 陶瓦制品
stoneware gutter 陶瓦明沟
stoneware half-pipe 陶瓦半管;陶瓦沟槽
stoneware joint(ing) compound 陶瓦水管接头组合物
stoneware junction 陶瓦接合
stoneware lining 陶瓦衬里
stoneware pipe 缸瓦管;陶制管;陶瓷;粗陶土管
stoneware pipe fittings 陶瓦管零配件
stoneware pipe joint cement 陶瓦管接头水泥
stoneware pipeline 缸瓷管道
stoneware plant 陶瓦厂
stoneware product 陶瓦产品
stoneware pump 陶制泵;陶瓷泵
stoneware roof gutter 陶瓦天沟
stoneware sanitary wares 卫生炻器
stoneware sewer joint(ing) compound 陶瓦污水管接头组合物
stoneware small-size mosaic 小尺寸陶瓦锦砖
stoneware spigot and socket pipe 承插式陶瓦管
stoneware(sur)facing 陶瓦修饰;陶瓦贴面
stoneware tank 陶瓦槽;陶瓦桶;陶瓦油箱;陶瓦水箱
stoneware tile 陶板;陶瓦;缸瓷砖;炻器墙地砖
stoneware tile floor cover(ing) 陶瓦地砖
stoneware water pipe 陶瓦废水管
stone washer 洗石机
stone waste 石料废石
stone wastewater 石粉废水
stone watchtower 石阙
stoneway 碎石路;石子路
stone weight 金刚石含量
stone weir 石拦河堰;落石楔;石堰;堆石堰
stone wheel 砂轮
stone window sill 石窗盘;石窗台板
stone with rough-hewn face 粗琢石材
stonework 凿石工程;石作(工程);石圬工;石细工;石砌体;石砌建筑物;石(方)工程;琢石作业;石造物;石结构;砌石结构
stonework box culvert 石砌矩形涵洞;石砌箱形涵洞
stonework box drain 石砌矩形水道;石砌箱形水道
stonework decoration 石板装饰
stonework decorative finish 石材表面装饰加工
stonework enrichment 石材表面装饰加工
stonework explosion 石方爆炸;石方爆破;采石爆炸;采石爆破
stone-working 石料加工
stone-working machine 石材加工机械
stonework ornamental finish 石材表面装饰加工
ston(e)y 铺石的;多石的
Stoney gate 提升式平板闸门;司东尼(式)闸门
Stonhard 石丹哈德(一个厂商牌号,出售防水材料、填充料、硬化剂、防腐剂等建筑产品)
Stoniflex 史丹尼弗莱斯(一种专利屋面防水材料)
stoniness 土壤含石量
stoning 磨加工;磨刀;河床铺石;碎石护岸;碎石过程
Stonington pink gray granite 粉红色与灰色花岗岩(美国缅因州产)
Stonite 史丹尼特(一种专利墙面材料)
stony bottom 多石地犁体
stony clay 坚硬土;多石黏土;含石黏土
stony-coral 石珊瑚
stony desert 多石沙漠
Stony gate 提升平板闸门;模造大理石门
Stony Greek granite 深红色花岗岩(美国康涅狄格州产)
stony ground 多石地
stony-iron meteorite 铁石陨石;石铁陨石
stony-land guard 多石地护刃器
stony-land share 多石地犁铧
stony material 石料
stony meteorite 陨石;石陨石
stony shore 粗石子石岸;石砾石岸
stony soil 硬土;坚硬土;含石土;石质土;多石土壤(碎石土)
stony sprout 石芽
Stoodite 斯图迪特焊条合金
stood off hour 停工小时
stooks 麻架
stool 根株;内窗台;模底板;垫凳;凳子;板凳;小凳;踏脚凳;钢筋支座;便器
stooled sill 窗台端部找平基石
stooling 托架;托板;垫块;窗台端部找平基石;窗

台板;托芯(用托板)
stool layer 根株压条
stool layerage 堆土压条法
stool layering 根株压条
stool pigeon 钻机记录仪
stool plate 垫板
stool sticker 粘底板钢锭
stool tank 台座舱;墩舱
stoop 台阶;月台;门廊;门阶
stoop and-room 柱廊巷道采石;柱廊巷道采矿
stoop and-room mining 房柱式采矿法
stooping 回采矿柱;大气折光现象;回采廊
stoop way 矮巷道
stoothing 钉墙板条;龙骨板条墙;灰板(条)墙
stop 限位器;终止;中止;制动箱;止块;止动销;止挡;光圈;工间休息;检量止点;停顿;停车;停机;定位条
stop a check 停付支票
stop a crack 阻止开裂
stop-adjusting screw 止动挡销
stop adjustment 停止调节
stop a flag 将旗扎上
stop a gap 弥合差距
stop ahead sign 前面(临时)停止通行标志;前面(临时)停止标志
stop alarm test 停止警报试验
stop a leak 堵塞漏洞
stop-all button 故障按钮;全切断按钮
stop-and-check valve 截止阀
stop-and-go-driving 频繁停车操作;停止再行的驾驶(按交通规则)
stop-and-go signal 停车后再行信号;停车后再进信号;停止和通行信号
stop-and-go valve 起停阀;起动阀
stop and license plate light 停车和牌照灯的组合灯
stop and license plate light cable 停车和牌照灯线
stop and release mechanism 停放机构
stop-and reverse rod 止动回动杆
stop-and-start operation 停停走走运行
stop and tail light 停车和尾灯的组合灯
stop an order 停止订货
stop arm 止动杆
stopback 止滴物
stop band 抑制(频)带;抑止(频)带
stopbank 堤岸
stop bar 停止杆;停车挡板
stop bath 停显液
stop bead 吊窗挡轨;门窗压条;钉在门框上的压条
stop bead screw 固定螺丝;止动螺丝
stop belt 玻璃带皮厚
stop bit 终止位;结束位;停止位;停止码元
stop block 限动块;限定块;阻止块;止推块;止轮楔;止轮器;挡块;触止块
stop board 止动板;堵头板;停车牌;挡板
stop bolt 定位销钉
stop bracket 止动块;弓座;挡铁
stop buffer 停车缓冲器;弹性缓冲器
stop buffer track bumper 车挡【铁】
stop bumpers 停车防冲装置
stop bush 止动套管;止动轴衬
stop butt hinge 半开铰链
stop button 制动按钮;止动按钮;停止按钮;停机按钮
stop casing tube 锁口管
stop catch 止动挡
stop chain (阻止防火幕撞击舞台地面的)吊链
stop-chamfer 切角面终止石;切尾凹槽
stop chamfer plane (带调整装置的)倒棱刨
stop claw 止爪
stopclock 停表;小龙头;止水栓;止水阀;止流旋塞;龙头;截水旋塞;停止旋塞;旋塞阀;旋塞;止旋塞;活栓;调节旋塞;水龙头;栓考克
stopcock buret 活塞滴定管
stopcock grease 活栓润滑脂
stopcock plug 活栓塞
stopcock water bottle 活塞式(采)水样器
stop code 终止码;结束符号;停止(代)码;停机码
stop codon 终止密码子
stop collar 止动环;限位环;限动环;限定环
stop conducing 停止导通
stop contrivance 止动设备
stop control 停车管制;停车管理
stop-controlled intersection 有停车管制的交叉口

stop-controlled sequence 停机控制程序
stop-cylinder machine 间歇滚筒
stop-cylinder press 对开平台印刷机;停转式凸版平台印刷机
stop device 停止机构
stop device for towing line 拖缆限位器
stop diameter 光圈直径
stop disc 停止盘
stop distance 停止距离
stop dog 止动块;止动器;碰停块;挡块
stop down 光阑缩小;缩微光圈
stop drill 钻头
stop drum 止动鼓;碰停鼓
stope 回采工作面;梯段形开采面;地下阶梯形开挖;采矿场
stope board 采矿场凿岩台
stope cutting 采区拉底
stope edge 精制边
stope face 采矿工作面
stope filling 采区充填;采矿场充填
stope hoist 采场绞车
stop element 止单元;停止符;停止单元
stop element number 结束符号数
stop end (墙的)平直收尾;临时挡板(浇注垂直施工缝用);(沟槽的)封闭端
stop-end 塞头
stop-end joint 施工缝(混凝土)
stop end piece 端盖
stop-end waterbar 封端挡水条
stop-end waterstop 封端止水器;封端止水剂
stope of coal mines 煤矿回采
stope pillar 采样矿柱;采场矿柱
stoper 矿石回采工;重型气锤
stoper leg 向上式凿岩机支架
Stopes-Heerlen system 斯托普斯—赫尔冷分茨
Stopes's classification 斯托普丝分类
stope stowing 采区充填
stop ferrule 用水供水阀
stop field of view 视场遮光
stop finer 止动指针
stop finger 停车指针
stop firing 歇火
stop flow 截流
stop foot 墙护板
stopgap borrowing 短期债务
stopgap loan 过渡性贷款
stopgap unit 暂时装置
stop gate 快速闸门;截止闸门;速动闸门;事故闸门;断流闸门
stop-gate valve 停止阀;断流阀
stop ga(u)ge 可调整量规;挡料装置
stop gear 停止装置;制动机构;止动机构
stop glazing 玻璃压条
stop-go 交通红绿灯;停停走走
stop-go control 双位控制
stop-go signal 红绿灯交通信号;停车后再行信号;停车后再进信号
stop-height 高程点注记
stop hitch 超载停车联结器
stop imagery 光阑成像
stop impulse 停住脉冲
stop in advance of a signal 机外停车
stop index 光阑指数
stop indicator 停机指示器
stop inertia test 惯性试验
stoping 岩浆侵入;回采;顶蚀作用
stoping area 采区
stoping face 采煤工作面
stoping geologic(al) control 回采地质指导
stoping ground 准备回采区段
stopingsignal 停闭信号
stop instruction 停机指令
stop interval 停车间隔时间
stop iron 挡铁;铁盖板(下水道用)
stop island 停车岛
stop joint 嵌固接头
stop key 阻止钥匙;止动键;停机键
stop knot 防松绳结
stop lamp 停车信号灯;停车灯
stop latch spring 满线跳杆簧
stop-leak compound 密封脂
stop level 自动停止水位;汽车停车场标高
stop lever 制动(操力)杆;止动杆;定位杆;挡杆;制动杠(杆)

stop light 交通指示灯;停车信号灯;停车灯
stop light area 停车区域
stop limit order 限价订单;限价停止订单;限价平仓单
stop limit switch 行程开关
stop line 止动线;停止线;停车线
stop-line delay 停车线延误
stop-line queue 停车线排队
stop list 非用词表
stop location board 停车位置牌
stop log 叠梁闸门;止水板;叠梁;插板
stop-log dam 拦木坝
stop-log gate 叠梁闸门;船坞叠梁门
stop-log gate slot 叠梁闸门槽
stop-log groove 叠梁(门)槽;叠梁闸门槽;插板槽
stop-log guide 叠梁槽;插板槽
stop-log handling equipment 叠梁吊装起重机
stop-log hoisting equipment 叠梁吊装设备
stop-log plate 叠梁闸板
stop-log recess 叠梁槽
stop-log storage site 叠梁存放处
stop-log weir 叠梁闸门堰;叠梁堰;叠梁堰
stop loop 循环停机;中止循环;停机回路
stop-loss 停业损失
stop-loss cover 超额赔款分保合同
stop-loss order 停止损失指令;限价补进或卖出;限亏单;减价平仓单
stop loss reinsurance 超率赔款再保险
stop loss treaty 超额赔款分保合同
stop mechanism 限位机构;停车机构
stopmeter 制动距离测量仪;制动距离记录仪
stop motion 中断动作;停止运动;停止机构;停车动作
stop motion mechanism 止动机构;自动停车机构
stop motion motor 间歇驱动电机
stop-motion switch 停车开关
stop mo(u)lding 收头线脚;不延伸的线脚;线脚止端
stop notch 止动槽
stop number 光圈数;光圈号数
stop nut 止动螺帽;止动螺母;防松螺母;制动螺母
stop-off 中途停留;中途分道;填塞;补砂
stop-off agent 阻流剂(焊接);阻化剂
stop-off core 补砂型芯
stop-off lacquer 涂漆;漆封;电镀隔绝涂料;防镀漆
stop-off material 电镀屏蔽材料
stop-off piece (模型的)不铸出部分;补砂块
stop of frame 框架挡销;框子挡块
stop once sign 一次停车标
stop-on red signal 停车红灯信号
stop operation and undergo shake-up 停产整顿
stop order 限价补进或卖出;中止命令;止付通知;停止委托;停止损失指令
stop-out price 截止价;停止降价
stop-over station 中停站
stoppage 停工;停付;扣除;中止;故障;截止;停机
stoppage at the source 赋税的从源扣缴
stoppage day 停工日;截止日;停机日
stoppage en route 停止运输权
stoppage examination 故障检测
stoppage examination system 故障检测判定
stoppage in transit 运送品中途截留;中止运输(行使停运权);停运权;停运
stoppage joint 新老混凝土接缝;施工缝
stoppage of business 停止营业
stoppage of consignment 停止托运
stoppage of interest 停付利息
stoppage of publication 停止发行
stoppage of work 调整工作;停止工作;停工
stoppage on acceptance of goods 停止承运货物
stoppage period 停工(时)期
stop pawl 制动爪;止动棘子
stop payment 中止支付;止付;停止付款;停止支付
stop payment note 止付通知
stop-payment notice 停止支付通知
stop payment on a check 停止支付支票;停付支票
stop payment order 止付命令
stopped chamfer 脊槽;淤塞的沟槽
stopped check 停止支票
stopped dado 半开榫槽
stopped delay 停车延误
stopped end 封闭端;(墙的)平直收尾;(浇筑垂直施工缝用的)临时挡板;(沟槽的)封闭端
stopped engine landing 停车着陆

stopped-flow method 停流法
stopped flute 凹弧与平肋相间的装饰;肋形柱槽
stop ped mortise 暗榫眼;暗榫孔;堵塞的榫眼
stopped pipe 闭管
stopped process 停止过程
stopped reaction 中止反应;制止反应;停止反应
stopped state 停机状态
stopped status 停机状态
stopped time 停止时间
stop penalty 停车损失
stopper 限动件;闸砖;制止器;制动器;浇口塞;擎缆索;嵌填料;瓶塞;停止器;停机地址;调节砖;塞头砖;塞棒;挡块;挡车器;掣索;闭锁装置
stopper bending tester 密封材料抗弯试验机;填充料抗弯试验机
stopper brick 塞头砖
stopper chain for fastening wire rope 搭索链
stopper coat 腻子涂层;油灰涂层
stopper drill 伸缩式凿岩机;套筒式凿岩机
stoppered bottle 带阀取样器
stoppered graduated measuring glass 带塞刻度量筒
stoppered pouring basin 带塞棒的钢水包
stop(pered)screw 止动螺钉
stoppered test tube 有塞试管
stopper fastener 定塞器
stopper for brush application 涂刷用腻子;涂刷用油灰
stopper gypsum 填嵌石膏
stopper hammer 向上式凿岩机
stopper hammer drill 工地式凿岩机;上凿式凿岩机
stopper handle 舵轮止动销
stopper head 铸罐顶头;柱塞头;塞头砖
stopper head brick 塞头砖
stopper hitch 单掣索结
stopper hole 止裂小孔
stopper knot 防松绳结
stopper ladle 底注(式)浇包
stopper lifting device 开启柱塞装置
stopper mechanism 限动机构
stopper nozzle 注口;浇铸嘴
stopper pin 限动销;止动销;塞头栓
stopper pouring 拔塞浇注
stopper powder 密封粉料;止水粉料;填嵌粉料
stopper ring 止动圈
stopper rod 柱塞杆;塞杆;塞棒铁芯
stopper rod ladle 塞杆式底注浇包
stopper screw 紧定螺钉;止动螺钉
stopper seal(ing) 填嵌物密封
stopper snug 最大舵角限制钮
stop pick-up 止动传感器
stop piece 行程限制器;停止器;停止块
stop piece of lath 齿条行程限制器
stop pin 栓钉;定位销;止动销(钉);挡料销;防转销;制动销(钉)
stopping 向上掘进;阻隔;止动;光阑调节;隔墙;嵌填;嵌填(塞);填补料;风道隔墙
stopping accuracy 停车精度
stopping agent 阻化剂;终止剂
stopping amber 止动黄灯信号
stopping and smoothing down 刮腻子及砂光;填孔及擦平
stopping and waiting time 停待时间
stop ping arrangement 停止装置
stopping at a target point 定点停车
stopping bar 止杆;渣口塞杆
stopping basin 稳定池
stopping bending tester 密封材料抗弯试验机;填充料抗弯试验机
stopping brake 止动闸
stopping braking 停车制动
stopping building area 停建矿区
stopping capacitor 隔直流电容器;隔流电容器
stopping coat 止水涂层;密封涂层
stopping condenser 隔流电容器
stopping criterion 停止准则
stopping cross-section 抑制截面;制动截面
stopping cycle 停车循环
stop ping device 停止装置;停车装置;制动装置;止动装置;停车器
stopping direct current 隔直流
stopping distance 制动距离;停船效率;停船滑行距离;停车效率;停车距离;停车滑行距离;冲程
stopping distance test 停车距离试验

stopping effect 抑制效应
stopping equivalent 阻止当量
stopping for being received 停车待接
stopping for departure 停车待发
stopping knife 断流闸门;刮腻子刀;油灰刀;嵌油灰刀
stopping lever 止杆
stopping limit mark 停车界限标
stopping line 停车横条
stopping machine 回采机械;采矿机
stopping maneuver 制动演习
stopping method 回采方法
stopping mining area 停采矿区
stopping motion 停车装置;自停机构
stopping off 涂保护层;封闭交通;补砂;阻止
stopping of reaction 反应中止
stopping-out 分段腐蚀
stopping-out ink 防蚀油墨
stopping pad 停车坪
stopping payment of check 停止支付支票
stopping performance 停车性能
stopping period 停用期间
stopping place 车站;停止的地方;停车站;停车场
stopping plane 止动薄片
stopping plate 止动薄片
stopping point 停止点;停车点;爆炸点
stopping-position 停车位置
stopping-position indicator 停车位置指示装置
stopping position of survey line 测线终点位置
stopping potential 截止电位
stopping power 阻止本领;阻塞力;制动能力;制动功率
stopping property 密封性能;密闭性
stopping reaction 终止反应;稳定反应
stopping resistance 停止阻力;停车阻力
stopping resistor 制动电阻
stopping rule 停止规则
stopping seal(ing) 填充密封
stopping shadow 停车车影
stopping sight distance 停车视距
stopping sign 停车标志
stopping signal 终止信号;停止信号;停机信号;停车信号
stopping smoothing down 填孔及擦平
stopping switch 停机开关
stopping test 停船试验;冲程试验
stopping the service 停止运行;停止操作
stopping time 制动时间;停歇时间;停车时间;车滑行时间;停止时间
stopping train 慢车
stopping truck heap 阻车堤(厂矿道路)
stopping-up 堵塞
stopping-up of minor side street 封闭支路
stopping valve 停汽阀;断流阀;停气阀
stopping width 回采厚度
stopping window 停车窗
stopping work 建筑物勒停施工
stop plank 止水闸板;止动板;闸板;截流木闸(板);挡空气板;插板
stop plate 制动片;止水闸板;止动片板;止动片;止动板;闸板
stopple plugging machine 堵塞机
stoppling 带气堵管
stop plunger 限位柱塞;止动柱塞
stop point 乘降所
stop position 光阑位置;停止状态;停止位置;停车位置
stop post 停止柱
stop production 停产
stop pulse 停止脉冲
stop push-button 还原按钮
stop ring 制动环;制动箍;止推环;止动环
stop riveting 消除铆接漏缝
stop rod 止动杆;停车杆
stop rod arm 止动杆臂
stop roll 碰停转筒
stop rope 止动索
stop routine work and rectify 停业整改
stop scale 光阑刻度盘
stop scallop's chamfered edge 斜边
stop screw 止动螺钉;止动螺丝;制动螺旋;制动螺丝;制动螺钉;固定螺丝;固定螺钉
stop-second 停秒
stop section 停车区段

stop setting 光阑调整
stop shelter 风雨停车站
stop shift 光阑位移
stop short 空挡(俗称天窗)
stop shoulder 止推肩
stop shutter (混凝土浇捣中的暂设的)止工木条
stop sign 停止信号;停车标志;停车标
stop signal 停车信号;中止信号;停止标志;停车标志
stop signal indication 信号关闭表示
stop siphoning 停止注水;停止虹吸
stop size 光阑大小
stop sleeve 止动套筒;支撑套筒;停止套筒
stop slide 报时开关片
stop spindle 止动杆
stop spring 止动弹簧
stop-start chromatograph 间断式色谱仪
stop-start frequency 停车—起动频率
stop-start switch 停止起动开关
stop-start test 停车起步试验
stop statement 停止语句
stop stone 门槛石;止门石
stop street 停车交通口
stop switch 止动开关;停止(信号)开关
stop system 光阑系统
stop tab 止动挡块
stop tap 止流旋塞
stop time 制动时间;减速时间;停留时间;停机时间;停止时间
stop tube 止动管
stop up a loophole 堵塞漏洞
stop up a water 堵水
stop value 光阑值;光阑值
stop valve 主汽阀;制止阀;制动阀;止水阀;止动阀;截止阀;截门;停汽阀;断流阀;停气阀
stop valve ball 止阀球;截止阀球;停止阀球
stop valve for washing 洗涤水截止阀
stop valve spindle 主汽阀芯子
stop washer 止动垫圈
stop wat 停表
stop watch 秒表;马表;跑表;停表;记秒表
stop watch cycle time 秒表循环时间
stop watch reading 秒表读数
stopwatch time method 码表计时法
stopwatch time study 码表测时法
stopwatch timing 码表测时
stopwater 止水器;止水条;水密填料
stop water loss or entrance 止漏或止水侵
stop wavelength 截止波长
stopway 停机道;止动道;停车道
stopway test 冲程试验
stop work 歇工;制档机件;限紧装置
stop work for the day 收工
stop working 停止办公
stop work order 停工指令;停工命令
storability 可储存性
storable cost 可储存成本
storable property 耐储[贮]性
storage 蓄电;一组服务台;储[贮]运容器;储[贮]积;储[贮]存;储[贮]藏(库);寄存器;集群;堆存;存储;储存器;储存;储藏量;储藏;材料总库;保管费;保管;仓库;储[贮]罐
storage above spillway crest 溢洪道顶以上蓄洪
storage access 存储器存取
storage access conflict 存储器存取冲突
storage access control unit 存储器访问控制部件
storage access width 存储器存取宽度
storage address 存储器地址
storage address display light 存储器地址显示灯
storage administration 存储管理
storage agency 仓库储公司
storage allocation 库容分配;存储器分配;存储分配;存储单元;分配内存【计】
storage allocation algorithm 存储器分配算法
storage allocation location 存储器分配单元
storage allocation plan 存储分配方案
storage allocation routine 存储器分配程序
storage allocation statement 存储器分配语句
storage allocator 存储器分配程序
storage-and-conveyor train 仓式列车
storage-and delivery 储[贮]运
storage and information retrieval system 存储与信息检索系统
storage and retrieval 存储与检索
storage and retrieval system 存储和检索系统

storage and shipping of oil products 油品储[贮]运
storage and transportation 储[贮]运;储运
storage and transportation equipment 储[贮]运设备
storage area 蓄水面积;储[贮]油区;储[贮]木场;库场面积;堆存面积;存储区(域);存储部分【计】;储藏区
storage area definition 存储区限定
storage area operation 库场作业
storage balance 蓄水调节;存储平衡
storage basin 蓄水库;蓄水池
storage battery 蓄电池组;蓄电池;电瓶;二次电池
storage battery car 电瓶车
storage battery locomotive 电瓶机车
storage battery reserves 备用蓄电池组
storage battery room 电瓶间;电池室
storage battery supply 蓄电池供电
storage battery trailer 电瓶拖车
storage bay 蓄水河湾;堆存处
storage bed 储存台
storage bin 储[贮]藏箱;储料箱;储[贮]料斗;储[贮]料仓;储[贮]存器;储[贮]槽;存仓;储料仓;储存箱
storage bin level detector 储存器级位探测器;储存器位面探测器;储存斗位面探测器;储存斗级位探测器;储存器液面探测器;储存斗液面探测器
storage bit 存储位
storage block 存储单元;存储块;存储器部件
storage bottle 储液罐;储液瓶;储气罐;储气瓶
storage box 存储箱
storage bridge crane 仓库桥式起重机
storage buffer 存储器缓冲器;存储缓冲区
storage building 库房;储库;仓库
storage bunker 储煤柜;储[贮]煤柜;储[贮]料仓;存仓;储煤仓;储料仓;储存库;储仓
storage bus in 存储器总线输入
storage bus out 存储器总线输出
storage cabinet 储藏小室
storage calorifier 储存加热器;加热储水罐
storage camera 光电摄像管
storage canal 蓄水渠
storage capability 存储容量
storage capacitance 存储电容
storage capacitor 存储电容器;储能电容器
storage capacity 蓄水容量;蓄水量;储[贮]存量;储[贮]藏能力;储[贮]藏量;库容量;库容;库场容量;库场能力;积聚电容;堆容量;存容量;存储器容量;存储能力;存储量;储水能力;储量;储存器能量;储存能量;存储能力;储藏容量;充电容量;舱容(量)
storage-capacity curve 库容曲线;水库库容曲线;蓄容曲线
storage capacity for deposition 水库淤沙库容
storage capacity for sedimentation 淤沙库容
storage capacity of pond 池塘蓄水量
storage capacity of reservoir 水库蓄水量;水库库容
storage capacity of sedimentation 水库淤沙库容
storage capacity of sewers 管道存储[贮]能力
storage capacity of water shed 汇水流域内的蓄水量
storage capacity per effective unit area 单位有效面积存储量
storage capacity per unit area 单位面积存储量
storage card 货垛牌;货垛
storage carrier 存储载体
storage cathode ray tube 储存式阴极射线管
storage cavern 储存洞穴
storage cell 蓄电池;存储元件;存储(器)单元;存储单元胞
storage cellar 储[贮]藏地窖;储[贮]藏窖
storage center 存储中心;藏书中心
storage chamber 储[贮]藏室
storage change of phreatic water 潜水储存量的变化量
storage channel 存储器通道
storage characteristics 存储特性
storage charges 库房费用;堆装费;堆存费;仓租;仓储费(用);保管费
storage circuit 记忆电路;存储电路;存储电路
storage cistern 储料容器;备用油箱;储[贮]水箱;水箱
storage class 存储类型;存储类
storage class attribute 存储类属性

storage class attribute restriction 存储类属性的限制
storage class attribute with structures 构件的存储类属性
storage closet 储料室;储藏室
storage coefficient 蓄水系数;蓄水率;(水库的)库容系数;储水系数
storage compacting 存储压缩;存储器紧致;存储密化;存储精简;存储紧缩;存储紧缩化【计】;存储紧化
storage compaction 存储器的紧凑
storage compartment 储料室;储藏室
storage constant 蓄水常数
storage constant for aquifer 含水层蓄水常数
storage container 储[贮]存器
storage content 蓄水量;存储内容
storage control 存储器控制;存储控制
storage control condition 存储控制条件
storage controller 存储控制器
storage control unit 存储控制器
storage core 存储磁芯
storage cost 存储成本;保管费(用)
storage counter 累积式计数器
storage curve 库容曲线;容量曲线;储量曲线;储存线;槽蓄曲线
storage curve of spillway 溢洪道蓄水曲线
storage cycle 存储[贮]存周期;蓄水周期;存取周期;存储周期;存储器周期
storage cycle time 存储循环时间;存储周期时间
storage cylinder 储气瓶
storage dam 挡水坝;拦水坝;蓄水坝;储[贮]水堰
storage data 存储数据
storage database 存储数据库
storage data register 存储数据寄存器
storage decoder 存储译码器;存储解码器
storage delay 存储延迟
storage density 存储密度
storage depot 储[贮]料场;储[贮]存仓库;堆栈
storage descriptor 存储器描述符
storage detector 存储探测器
storage device 存储装置;存储设备;存储部件
storage-discharge curve 槽蓄流量曲线;库容-流量曲线
storage disorder 储藏变质
storage display 余辉管;存储显示器
storage display tube 存储显示管
storage distance 存放间距
storage dome 穹顶仓库
storage-draft curve 蓄泄曲线
storage drain 储[贮]池水道;储[贮]池水道
storage drier 储[贮]式干燥机;仓库式干燥机
storage dump 信息转储;存储转储;存储器转储;存储器信息转储;存储器内容打印
storage effect 储[贮]存效应;累积效应;储能效应
storage efficiency 存储效率
storage element 存储元素;存储元件;存储单元;储存元件
storage-elevation curve 高程-库容(关系)曲线;水位-容积曲线
storage energy 蓄能
storage energy power station 蓄能电站
storage equation 蓄水量方程;蓄量方程(式);储量方程;槽蓄方程
storage equipment 仓库设备
storage/excavation ratio 库容挖方比
storage exchange 存储内容交换控制装置;存储内容交换;存储内容互换的控制装置
storage expenses 库房费用;堆存费;仓库费(用);仓库成本;保管费
storage facility 储藏装置;蓄水设备;储[贮]存设施;储[贮]存设备;储[贮]藏室;拦蓄设施;器材库;仓库设备;仓储设施;保管设备;存储设备;储藏设施
storage factor 蓄水率;蓄存系数;(水库的)库容系数;货物入库系数;货物入场系数;存储因数;储能因数;储备系数
storage feed bin 喂料储仓
storage fill 存储填充;存储器填充
storage fire 仓库加热器;仓库火炉
storage flood routing 槽蓄(洪水)演算
storage for dangerous goods 危险品库
storage format 存储格式
storage fragmentation 存储碎片;存储剩余空间;存储器分段存储;存储残片

storage furniture 储[贮]物用家具
storage gain 累积增益
storage gallery 蓄水廊道;集水廊道
storage garage 存车间;储藏库
storage ga(u)ge 蓄水压力计
storage geyser 仓库加热器
storage grid 存储栅极
storage ground 堆货场(地)
storage guard 存储警戒
storage hall 储库
storage handling system 库场装卸工艺系统
storage head 存储磁头
storage heap 储[贮]料堆
storage heater 蓄热(式)供暖器
storage hierarchy 存储器体系;存储器分级体系;存储层次;分级存储器系统
storage holdings 堆储场地
storage hole 存储孔
storage hopper 储[贮]藏斗;储[贮]料斗;储[贮]灰斗;料仓;储料漏斗;储仓
storage image 存储映像;存储图像;存储器映像
storage in abandoned mine 废矿储存
storage in ambient temperature 室温寄存
storage in a reach 河段蓄水(量)
storage-in bus 输入总线
storage in channel 河槽储[贮](水)量;河槽蓄水;河槽储水;槽蓄
storage indicator 存储指示器
storage in fissures 裂缝堵气
storage input 存储器输入
storage input-output 存储器输入输出
storage in reservoir 库蓄;水库蓄水量
storage instruction 存储指令
storage integrator 记忆积分器;存储用积分器;存储积分器
storage interference 存储器冲突;存储干扰
storage in transit 中转保管费
storage in valley 河谷蓄水(量)
storage investigation 水库调查
storage island 岛式库区
storage jug 岩层储油库
storage key 主存储键;存储链;存储关键字
storage keyboard 存储键盘
storage lagoon 储[贮]水塘
storage lake 人工湖
storage level 储存位面;储存级位;储存液面;储量;蓄水高程;储[贮]水位;水库水位;存储[贮]位;存储级别;存储级;储备量
storage level detector 储存位面探测器;储存级位探测器;储存液面探测器;筒仓库存指示器
storage level representation 存储级的表示
storage life 储存有效期;储[贮]存寿命;储[贮]存期限;储[贮]存期;储[贮]存的延续时间;储[贮]藏期限;适用期;存放寿命;储存寿命;储存期(限);储藏寿命;保存期限;保存期
storage life of mo(u)lding compound 模料储存期
storage light 存储器指示灯
storage light amplifier 光储存放大器
storage limit 存储极限;储存限度
storage limit register 存储界限寄存器;存储范围寄存器
storage list 存储表
storage load 蓄热负荷
storage load factor 蓄热负荷系数
storage load module 存储装入模块;存储器加载模块
storage location 存取位置;存取地址;存取场所;存储位置;存储器单元;存储单元;储存单元
storage location selection 存储单元选择
storage loop 库容绳套曲线;槽蓄绳套曲线
storage loss 储[贮]存损失;仓储损耗;储存时的损失
storage losses 储存时的损失
storage macroinstruction 存储器宏指令
storage magazine 储[贮]放库;仓库
storage management 存储器管理;存储管理
storage management strategy 存储管理策略
storage map 存储映像表;存储图
storage mapping 存储映像;存储器映射
storage mapping function 存储映像函数;存储映射函数
storage mark 存储标志
storage matrix 存储矩阵

storage means 存储方法
storage medium 信息存储体;记录材料;存储载体;存储媒体;存储媒介;存储介质;存储材料;保藏培养基
storage mesh 存储栅网;储荷网络
storage meter 仓库计量器
storage method of operation 存储作业法
storage mode 存储模式
storage module 存取模块;存储体;存储器模块;存储模式
storage modulus 储能模量
storage mosaic 存储镶嵌
storage norm per unit of warehouse floor space 仓库单位面积储存量定额
storage number 存储数
storage occupancy 存储器占有率
storage of aquifer 含水层储量
storage of cargo 货物储存
storage of data 资料储存
storage of digester gas 消化池气(体)存储[贮]
storage of energy in aquifer 含水层储能
storage of flood 蓄洪
storage of heat 热量储[贮]存
storage of information 情报存储
storage of inventories 存货保管
storage of lake 湖泊蓄水(量)
storage of liquefied gas 液化气的储[贮]存
storage of luggages and parcels 行李和包裹的保管
storage of material 材料库存量
storage of radioactive solid wastes 放射性固体废物储[贮]存
storage of raw materials 储存原料
storage of samples 样品储[贮]存
storage of storm overflow 暴雨溢流水存储[贮]
storage of surplus water 剩余水量蓄存
storage of valuables 贵重物品库
storage of water 蓄水;缺水
storage of water in aquifer 含水层储水
storage operation 库场作业;水库运作;水库运用;水库管理;水库操作;存取操作;存储器操作;存储操作
storage organization 存储组织;存储结构
storage orthicon 存储正析像管;存储正析摄像管
storage oscillograph 存储示波器
storage oscilloscope 存储示波器
storage-out bus 输出总线
storage-outflow curve 蓄水—泄水曲线;水库泄水曲线;蓄水池出流曲线
storage output 存储器输出
storage over-flow 存储器溢出;蓄水溢流
storage over-lay 存储器覆盖
storage over-lay area 存储器覆盖区;存储覆盖区
storage pad 堆场地坪
storage parity 存储奇偶校验
storage pattern 存储模式;存储模块
storage period 储[贮]存期间;储[贮]存周期;蓄水周期;蓄水时间;储[贮]存期;停放时间;放置时间
storage pile 储料堆
storage pile ropeway 储[贮]料堆索道
storage pipe array 管式储气罐
storage pipe line 仓库管线
storage piping installation 油库管线装设
storage pit 储[贮]料槽;储[贮]料坑;储藏窖
storage place 储[贮]料场;堆栈
storage plan 配置图;堆存计划
storage plant 仓库;蓄水力发电厂;蓄能式发电站;蓄能式发电厂;储[贮]水场;储[贮]藏室;储油场;仓库设备
storage plate 存储板
storage pond 沉淀池;储水池;蓄水池
storage pond for dewatering 脱水沉淀池
storage pool 蓄水池;储[贮]水池;储[贮]存池;存储器的池;存水池;存储池
storage pool on surface 地表蓄水池
storage port 终点港
storage power 储存能力
storage power station 蓄水发电站
storage precipitation 储蓄雨量
storage-precipitation ga(u)ge 储瓶式雨量计
storage premises 库房
storage principle 积累原理;存储原理
storage print 存储内容打印

storage print program(me) 存储器打印程序;存储单元打印程序;存储打印程序
storage procedures 储[贮]存方法
storage processor 存储处理机
storage program(me) computer 程序存储式计算机
storage property 耐储(存)性;存储寿命;保存期限;储[贮]藏性质;储存性
storage protect bit 存储保护位
storage protect feature 存储保护特性
storage protection 存储(器)保护
storage protection block 存储保护区
storage protection key 存储保护键
storage pulverized coal system 仓储[贮]式制粉煤系统
storage pump 蓄水泵;蓄能水泵;蓄能泵
storage queue 存储器队列;存储队列
storage rack 工具架;储料橱架;储[贮]料库橱架;储[贮]料摘架;摘架;带卷垛存台架;存放架;储料架;储存架
storage rain ga(u)ge 蓄存式雨量计;储瓶式雨量计
storage rate 库存量
storage rate of thermal energy 热能储存率
storage ratio 蓄水比;库容挖方比;调节系数(水库)
storage reconfiguration 存储重新配置;存储器再组合功能
storage redundancy 存储冗余
storage reference 存储访问
storage region 存储区(域)
storage register 存储寄存器
storage regulation 保管条例
storage regulator 调节水池
storage relay 存储继电器
storage requirement 需要库容;需要存储量;储[贮]存条件
storage reserve 仓库储备量
storage reservoir 储水池;蓄水库;储[贮]水池;水库;储藏仓库
storage reservoir for wet weather flow 雨天流污水调蓄池
storage resources of groundwater 地下水储[贮]存资源
storage restoration 重新蓄水
storage/retrieval machine 有轨巷道堆垛起重机
storage rings 储存环
storage ripple 存储器重叠
storage room 储料室;储藏室
storage rot 储[贮]藏腐烂
storage routing 库容演算;洪水储蓄路径;调洪计算;水库洪水演算;水库调容演算;水库调洪演算
storage routing in natural channel 河槽调洪演算;天然河槽中洪水储蓄路径追踪;天然河槽中洪水储蓄路径推算
storage rule 存储规则
storage safety 仓储安全
storage scheme 储水方式;蓄水电站;蓄水方案;存水方案
storage screen 储荷屏;存储屏
storage section 存储部件;存储部分【计】
storage-selection circuit 存储器选择电路
storage sewage depression 蓄污洼地
storage shaft 集合竖井;集合井筒;蓄水竖井;蓄水井筒
storage shed 储[贮]料棚;库棚;货棚
storage siding 存车线【铁】
storage signal 存储信号
storage silo 储料筒仓;储[贮]料筒仓;储料罐;储料仓;储仓
storage site 堆存地点;存储位置;储藏场所
storage site assignment 存储位置指定
storage site element 存储位置单元
storage site node 存储位置节点
storage size 存储器容量
storage space 储[贮]藏处;堆场场;堆货场;堆存场地;存储空间;储料场;仓库面积
storage speed 存储速率
storage sphere 球形气瓶
storage stability 蓄水稳定性;储[贮]存稳定性;耐储存性;储存稳定性;储藏稳定性
storage stability test 耐储[贮]性试验
storage stack 存储栈;存储体;存储堆栈
storage stage 存储状态
storage station 蓄能式发电站
storage stress 堆放应力(预制构件)
storage structure 存储器结构;存储结构;储藏建筑物;仓储结构
storage structure definition 存储结构限定
storage structure of anticline 背斜储水构造
storage structure of bed rock 基岩储水构造
storage structure of contact 接触储水构造
storage structure of fault 断裂储水构造
storage structure of monoclinal interlayers 单斜层间储水构造
storage structure of perched 顶托储水构造
storage structure of syncline 向斜储水构造
storage structure of vein 岩脉储水构造
storage structure of weathering fissure 风化裂隙储水构造
storage subsystem 存储子系统
storage sump 蓄水集水坑;储[贮]槽
storage surface 储[贮]荷面;储水面;存储表面
storage switch 存储器开关;存储开关
storage system 蓄能系统;存储系统;存储器系统;存储器
storage system design 存储系统设计
storage tab setting 存储制表置位;存储表设定
storage tank 蓄水箱;蓄水池;油罐;储[贮]存槽;储[贮]藏缸;储[贮]藏罐;水柜;储箱;储桶;储水池;储料罐;储罐;储存箱;储池;储槽
storage tank for wet weather flow 雨天流污水调蓄池
storage tape symbol 存储带符号
storage target 储荷电极
storage technique 存储方法
storage temperature 存放温度;储存温度
storage terminal 港口运转储库
storage theory 存储论
storage thrift lock 节水船闸;省水船闸
storage time 储[贮]存时间;储[贮]存期限;衰减时间;存储周期;存储时间;储备时间;保管时间;保持时间
storage time constant 存储时间常数
storage-to-storage operation 存储器-存储器操作
storage tower 储存塔
storage track 编组线;存车线【铁】
storage-transport tank 储[贮]运罐
storage-transport vessel 储运容器
storage tube 储[贮]能管;记忆管;存储显像管;存储管
storage tube display 存储管显示器;存储管显示
storage-type calorifier 蓄热器;储[贮]存式供暖器
storage-type camera table 存储式摄像管
storage-type camera tube 光电摄像管
storage-type geyser 桶式热水器;存储式热水器
storage-type heater 储[贮]热式热水器
storage-type heat exchanger 储蓄式热交换器
storage-type service water heater 储[贮]藏式供水加热器
storage-type water heater 容积式热交换器
storage under pressure 加压储存
storage under water 水中养护
storage unit 存储器
storage vat 储料桶
storage vault 储存库
storage vessel 蓄水器;储[贮]存容器
storage volatility 存储易失性;存储器易失性
storage volume 蓄水容积;存信息卷;存储体;储量;储存量;储藏量
storage wall 仓库墙
storage warehouse 储备货栈
storage warehouse building 储[贮]存仓库大楼
storage warmer 仓库加热器;仓库取暖器
storage water 储存水
storage water heater 储水(式)热水器
storage water heating 仓库水加热
storage water tank 储水箱
storage word 存储字
storage work 水库工程
storage writing speed 存储写入速度
storage yard 储[贮]料场;储[贮]藏场;港区货场;堆货场(地);堆场;存车场;储料场;材料堆放场
storage yard and ware house 场库
storage yard crane 堆放场起重机;储[贮]料场起重机
storage yard for material 储料场
storage yard of port 港口堆场
storage zone 储[贮]存区域;储气层
storaging time 保存期限
storascope 存储式同步示波器

storativity 给水度；释水系数；储水系数
storativity in early period 早期储水系数
storativity in later period 后期储水系数
storativity of groundwater 地下水储量系数
storatron 存储管
Storch-Morawsk test 斯托希—莫拉夫斯基试验
storde charge transistor 存储电荷晶体管
store 储[贮]存；栈房；入库(存货)；商号；店铺；储存；储藏；仓库储存
store access controller 存储访问控制器
store access cycle 存取周期
store access time 存储器的存取时间；存储访问时间；存储存取时间
store accumulator 存储累加器
store address register 存储地址寄存器
store adjustment 库存调整
store and forward 信息转接；储存和转送；存储转发
store and forward basis 先储存后发送制
store and-forward mode 存储转发方式；存储和向前方式
store and-forward switching 存储转发交换
store and-forward switching center[centre] 存储转发交换中心
store and yard labo(u)r 仓库及堆置场劳务
store area 存放面积；存储面积
store block 存放仓库；存储块
store building 商店房屋；仓库建筑；仓库
store button 存储按钮
store capacity 存储器容量
store card 原材料卡；库存卡
store carrier 供应运输船
store cattle 架子牛
store cell 存储单元
store cellar 存放地窖
store chamber 储[贮]藏室；储藏室
store character 存储字符
store closet 储藏室
store controller 存储控制器
store controller data 存储控制器数据
store controller definition record 存储控制器定义记录
store controller storage 存储控制器存储器
store controller storage save 存储控制器存储器保留；存储控制器存储保护
store control logic 存储器控制逻辑
store crane 仓库起重机
store credit 退料单
store cycle time 存储器的周期时间
stored 存信息的
store data buffer 存储数据缓冲器
stored at room temperature in daylight 储[贮]藏于温室光照下
stored car 备用车
stored carry 存储进位
stored carry multiplication 存储进位乘法
stored charge 存储电荷
stored charge pattern 存储电荷图
stored count 累计计数；累积计数
stored data 内存数据；存储数据
stored data base 存储数据库
stored data definition language 存储数据定义语言
stored data description language 存储数据的描述语言
stored energy 储能
stored-energy curve 蓄能曲线
stored energy function 储能函数
stored energy welding 蓄能焊；储[贮]能焊
store depot 材料库
stored format instruction 存储格式指令
stored gas 压缩气体；罐装气体
stored grain insects 仓库害虫
stored heat 蓄热；存储热量
stored hologram 存储全息图
stored in a stack out of-door 室外堆垛储[贮]存
stored kinetic energy traction 储能牵引
stored locomotive 备用机车【铁】
stored logic 存储逻辑
stored logic computer 存储逻辑计算机
stored logic control 存储逻辑控制
stored logic controller 存储逻辑控制器
stored material 存储材料
stored matrix 存储矩阵
store door latch 仓库门碰锁；仓库门插销
store double precision 双精度存储

stored program(me) 存储程序的；存储程序；内存程序
stored program(me) computer 可存储程序计算机；存储程序计算机；程序存储方式计算机
stored program(me) concept 存储程序概念
stored program(me) control 存储程序控制
stored program(me) controlled exchanger 程控交换机
stored program(me) controller 存储程序控制器
stored-programmed computer 存储程序式计算机
stored-program(me) digital computer 存储程序数字计算机
stored program(me) element 存储程序单元
stored program(me) logic 存储程序逻辑
stored program(me) system 存储程序系统；存储程序方式
stored routine 存入程序；存储程序
stored tank 蓄存槽
stored value 储值
stored water 蓄水
stored-water tank 蓄水池
stored word 存储词
store empty 存储器空
store energy closing 蓄能合闸；储能合闸
store equipment room 仓库设备间
store flood-water for use in a drought 蓄洪防旱
store for samples 样品储[贮]藏间
storefront (商店的)铺面；沿街面；沿街大楼；仓库沿街正面；商店沿街正面；门面；店面
store front profile 店面剖面图
storefront sash 橱窗窗框
store front section 店面剖面图
store front trim 店面修饰
store front unit 店面单元
store fund 材料资金
store goods 囤积货物
store holder 储料器
store house 堆栈；仓库；栈房；储[贮]藏室
store imprest account 仓库预付账；备用材料账
store input 存储器输入
store instruction 存储指令
store issue 仓库发料；发料
storekeeper 材料管理员；仓库管理员；仓库管理人；仓库保管员；保管员
storekeeping 仓库保管；仓库看守
store ledger clerk 材料会计
store ledger sheet 材料登记单
store list 库存单；库存表；物资单
store location 存储单元
store locking 存储器闭锁
store loop 存储回路
store loop driver 存储回路驱动器；存储环路驱动
storeman 仓库管理员；仓库管理人；仓库保管员；保管员；保管人员
store management 存储管理；仓库管理
store memorize 存储
store module 存储体
store moist 湿藏
store negative operation 存储负操作
store of tradition chinese medicines 中药店
store operation 存储操作
store operator 存储操作符
store output 存储器输出
store pests 仓库虫害
store phasing 存储器定相
store protection 存储保护
storer 存储器；仓库保管员
store record card 库存记录卡；仓库记录卡
store requisition 领料单
store returned 退还材料
storeroom 储料室；储藏室；库房；物料间；保管室
storeroom clerk 材料记账员
storeroom expenses 仓库费用；物料库费用
storeroom layout 仓库布置
storeroom receipt 领料单
stores 物品；物料；堆积物；存物品；备用品
stores adjustment 材料调整
stores audit 库存审计；仓库审计
stores balance book 材料余额簿
stores balance report 材料耗存报表
stores coordination 库存调剂
store service 存储服务
stores fund 库存物资基金；材料基金
store shed 储[贮]料棚；库棚；货棚；储料棚；储

棚；材料堆放处
store shelf 商品陈列架
store ship 给养船；粮食补给船；物资供应船；物资船；补给船
store shuffling 存储移动
stores in transit 在途材料
stores inventory report 库存物资盘点报告；库存物资盘存报告；材料盘存报告单
stores inventory sheet 材料盘点单
stores-issued book 发料登记簿
stores-issued material 厂发材料
stores-issuing book 发料登记簿
stores ledger 库存物资分类账；仓库流水账；材料分类账
stores ledger card 库存物资分类户卡
stores ledger clerk 材料记账员
stores ledger sheet 库存物资分类清单
stores ordered ledger 定购材料分类账
store space 储[贮]藏室；店堂间；储[贮]藏场所
stores purchase book 库存物资购入簿；材料进货登记簿
stores received and issued book 库存物资收发簿；材料收发登记簿
stores rejected book 拒收材料登记簿
stores requisition 库存物资清领单
stores reserve 库存物资准备；材料准备
stores returned 退料
stores returned note 退料单
stores ruturned 退回库存物资
stores swap 材料交换
store status 存储状态
store-status save area 存储状态保留区
store support procedure 存储支持程序
store tank 储槽；备用水箱；调节水池
store through 全存储
store through cache 存入高速存储器
store transfer requisition 调拨材料请求单
store transmission bridge 存储传输电桥
store unit 存储器
store(up) 储[贮]藏；储[贮]备；收藏
store up grain 蓄积粮食
store vessel 物料船
store volume 储[贮]藏容积
stor(e)y 楼层；层【建】；林冠层
storey-and-a-half 一层半房屋
storey arch 平拱
store yard 储[贮]藏场；堆场
stor(e)y band 台口线
stor(e)y bar 直钢筋；楼层直通钢筋
stor(e)y beam 楼层直梁
stor(e)y deformation 层间变形
stor(e)y displacement 层间位移
stor(e)y drift 层间位移
stor(e)yed 叠生的
stor(e)yed building 楼阁
stor(e)yed drying machine 多层烘布机
stor(e)yed high forest 分层乔林
stor(e)yed structure 叠生构造
stor(e)y frame 楼层框架
stor(e)y height 层高；楼层高(度)
stor(e)y-height cladding panel 与层高等高贴墙镶板
stor(e)y-height form(work) 与层高等高的模板
stor(e)y-height panel construction 与层高等高的镶板建筑
stor(e)y-height panel wall 大墙板
stor(e)y-height panel wall block 大墙板建筑
stor(e)y-height panel wall building 大墙板建筑
stor(e)y-height panel wall structure 大墙板结构
stor(e)y-height panel wall system 大墙板结构体系
stor(e)y-height shuttering 与层高等高的大模板
stor(e)y landing 楼层平台
stor(e)y level 楼层高程
stor(e)y mechanism 楼层机制
stor(e)y moment 楼层力矩
stor(e)y plan 楼层平面图
stor(e)y pole 量柱杆；木骨架支柱
stor(e)y post 楼梯栏杆木柱；楼层柱
stor(e)y post and beams 楼层的梁与柱
stor(e)y rod 量层杆；楼层标尺；皮数杆
stor(e)y shear 楼层剪力
stor(e)y stiffness 楼层刚度；层间刚度
stor(e)y transfer matrix 楼层转换矩阵

stor(e)y wall 楼层的墙
storied 叠生的
storied building 楼房
storied cambium 叠生形成层
storied cork 叠生木栓
storied element 叠生分子
storied high forest 分层乔林
storied house 楼房
storied structure 叠生构造
storied tapestry 历史画挂毯
storing 储料
storing compartment 存放柜
storing mechanism 存储机构
storing property 储[贮]存性
storing room 储藏室；库房
storm 狂风；十级风；风暴；暴风雨
storm alarm 风暴警报
storm alarm signal 风暴警报信号
storm anchor 风暴锚碇
storm and screen door 有玻璃窗和纱窗的门
storm and tempest insurance 暴风雨保险
storm axis 风暴轴；暴雨轴线
storm beach 后滩；风暴海滩；风暴海滩
storm beach profile 风暴型海滩剖面
storm beat(en) 受暴风雨打击的
storm belt 风暴中心经过的地区；风暴区；风暴地带；暴雨地带；风暴雨地带
storm berm(e) 后滩；风暴浪滩肩
storm board 风暴板
storm boots 风雨靴
storm bound 为暴风雨所阻的
storm building drain 建筑物雨水管
storm building sewer 建筑物雨水下水道
storm burst 噪暴
storm cave 防风地下室；防风地窖
storm cellar 防风雨地下室；防风暴地下室；防风地下室；避风穴；避风窖
storm center 风暴中心；暴雨中心
storm chain 风链(用于门、遮篷等)
storm circle 风暴圈
storm clip 防风雨玻璃窗夹；固定玻璃片的卡钉
storm cloud 风暴云；暴雨云；暴风云
storm collector 暴雨排水设备；雨水管；雨水沟渠
storm-collector system 雨水沟渠系统；雨水管系统
storm cone 锥形风暴信号；极风锥；风暴型号；风暴信号；风暴警报
storm-cover 防浪盖
storm current 风暴流；风暴潮流
storm curve 暴雨曲线
storm cusp 风暴滩角
storm data 暴雨资料
storm delta 越浪堆积；冲溢三角洲；风暴三角洲
storm deposit 风暴沉积
storm detection 风暴探测
storm detector 风暴探测器
storm distribution pattern 暴雨分布型(式)；暴雨分布类型
storm-dominated facies 风暴控制相
storm door 防风雨的外重门；风暴的附加内门；外重门；风雨门；防风门；防风暴门；防风外重门
storm down-pipe 雨水管；风暴雨水落管；暴风水落管
storm drain 雨水排水管；雨水排水道；暴雨下水道；暴雨排水管；暴雨排水沟；雨水排水道；暴雨排水系统
storm drainage 暴雨排泄；暴雨排水(系统)；风暴排水；雨水排除；雨水管出口(工程)
storm drainage orifice 雨水管孔口
storm drainage outlet 雨水管出口
storm drainage pipe 雨水排水管
storm drain area 雨水面积
storm drain combined with sanitary drainage 雨污水合流排水管
storm drain pipe 雨水管
storm drain system 暴雨排水系统
storm drum 风暴型号；风暴雨讯号筒
storm duration 风暴延时；暴雨历时
stormer 暴风中心
Stormer viscometer 斯托默黏度计
Stormer viscosity 斯托默黏性
storm event 一场暴雨；暴雨事件
storm eye 风暴中心
storm flag 风暴信号旗；风暴警报号旗
storm flood 雨洪；暴雨洪流

storm-flow 暴雨流量；雨水量；暴雨径流
storm formation 风暴形成；暴雨形成
storm frequency 暴雨频率
storm gale 强烈风；暴风
storm gate 风暴闸门；防洪闸(门)；防风暴闸门
storm glass 晴雨计；晴雨表；气压表(测大气)；气候预测管
storm grade 特高品位
storm-guyed pole 耐风加固电杆
storm hood 风雨篷
storm hydrograph 雨洪水过程线；暴雨洪水过程线
storm ice foot 由浪花冻结的岸冰
storm ice wind 沿岸风暴冰脚
storminess 风暴度
storm information 暴雨资料；暴雨情报
storm inlet 雨水井；雨水口；雨水进水口；雨水进口
storm in situ 当地暴雨
storm intensity 暴雨强度；暴风强度
storm intensity pattern 风暴强度图；暴雨强度类型；暴雨强度分布形式
storm intercepting well 暴雨截流井
storm investigation 暴雨调查
storm ladder 风暴软梯
storm lag deposit 风暴滞留沉积
storm lamp 汽灯；风雨灯
storm lane 暴雨区；风暴中心经过的地区；风暴路径；风暴轨迹；暴雨路径
storm lantern 汽灯；防风灯
storm level 风暴水位
storm light 抗风灯；风雨灯；风灯
storm loading 暴风负荷
storm loss 暴风损失
storm louver 防雨百叶窗；防暴雨百叶窗
storm match 抗风火柴
storm maximization 暴雨极大化
storm mechanism 风暴机制；暴雨机制
storm microseism 风暴微震
storm model 风暴模式
storm mooring 风暴系船装置
stormograph 气压记录器
storm oil 镇浪油
storm oil bag 布油袋
storm oil tank 镇浪油箱
storm outfall sewer 暴雨排水管
storm overflow 雨水溢水设施；暴雨溢水口；暴雨溢水槽；暴雨溢流
storm-overflow chamber 雨水溢流井
storm overflow gutter 暴雨溢流槽
storm overflow manhole 暴雨溢流井
storm overflow sewer 雨水溢水管；暴雨溢流污水道；暴雨溢流排水管
storm overflow well 暴雨溢流井
storm path 风暴中心经过的地区；风暴路径；暴雨路径
storm pattern 暴雨型；暴雨类型
storm pavement (桥墩式防波堤的)护坡石；护坡铺面；防暴雨护面；波浪消能滩(堆石)
storm period 暴雨期
storm porch 防风门廊；门斗；防风门斗；防风阁
storm precipitation 暴雨降水量
storm profile 风暴示意图
storm prone area 暴雨多发区
storm-proof 防暴风雨的；防风暴的
stormproof ability 抗风暴能力
stormproof building 抗风暴建筑物；防风暴建筑物
stormproof louvres 防风雨百叶
stormproof structure 抗风暴建筑物
stormproof window 防暴风雨窗；风暴窗；防风雨窗
storm protection 防风暴；防暴风雨；暴风雨防护
storm protection breaker 风暴涌浪防护堤
storm pumping house 雨水泵房
storm radar data processor 风暴雷达数据处理机
storm radiation 噪暴辐射
storm rail 风暴扶手；保险索
storm rainfall 暴雨降雨；暴雨流量；暴雨(量)；暴风雨
storm region 风暴区；暴雨区
storm-resistant glazing 耐风压玻璃
storm roller(structure) 风暴卷流(构造)
storm runoff 暴雨径流；雨水量；雨水径流
storm runoff forecasting 暴雨径流预报
storm runoff retention basin 暴雨径流调节池
storm sail 风暴用帆
storm sample 风暴样品

storm sash 外重窗；风暴窗；防风外重窗；防暴风外重窗
storm sea 暴风浪；风暴海况
storm seepage 雨水渗流；壤中流；表层流；暴雨渗流
storm settling tank 雨水沉淀箱；雨水沉淀池
storm sewage 雨水污水；雨水排污；雨水；暴雨污水；暴雨水
storm-sewage diversion weir 雨水污水分流堰
storm-sewage overflow 暴雨污水溢流
storm-sewage system 雨水污水合流系统；雨污水合流系统
storm-sewage tank 雨水污水池
storm sewer 雨水管(道)；雨水沟道；雨水管；暴雨水管；暴雨水沟；暴雨排水沟
storm sewerage 雨水下水道(工程)
storm-sewerage system 雨污水合流系统
storm sewer network 雨水排水管网
storm sewer pipe 暴雨排水管
storm sewer system 雨水排放系统；雨水管网；雨水沟渠系统；雨水沟沟系统；暴雨下水道系统；雨水(管)道系统
storm sheet 防雨板；屋顶卷材
storm signal 暴雨信号；暴风雨信号
storm signal station 暴雨信号站；风暴信号台；风暴信号所
storm stand-by tank 备用雨水池
storm stayed 受暴风停阻的
storm structure 风暴结构
storm surge 风暴涌潮；风暴(大)潮；暴雨巨浪；暴风雨涌潮；暴风潮；巨浪
storm-surge barrier 风暴涌潮防护堤
storm-surge dynamics 冲击波动力学
storm-surge forecast(ing) 风暴潮预报
storm-surge lamination 浪成纹理
storm-surge prediction 风暴潮预报
storm surge protection breakwater 风暴大浪防波堤；风暴防波堤；防暴潮堤
storm surge warning 风暴潮警报
storm survey 暴雨调查
storm system 风暴系统；雨水沟渠系统
storm tank 雨水池
storm terrace 风暴阶地
storm tide 大潮；风暴雨潮；风暴涌潮；风暴潮(位)；风暴波浪；暴风雨洪潮
storm tide gate 挡风暴雨闸门；暴潮门
storm tide model 风暴潮模式
storm-time 暴时分量
storm tossed 受暴风颠簸的
storm track 风暴轴；风暴中心经过的地区；风暴路径
storm transposition 暴雨转移；风暴移置；暴雨移置
storm valve 暴风雨汽门；止回排气阀；节汽阀；排水口止回阀
storm vortex 热带风暴中心
storm wall 防浪(胸)墙
storm warning 风暴警报
storm warnings 风暴警报号型
storm warning signal 风暴警报信号
storm washing 暴雨冲刷面
storm washings 暴雨冲刷物
storm wastewater 雨水废水；雨量废水；暴雨废水
storm/wastewater management model 暴雨与污水管理模型
storm water 表面径流；雨水；暴雨水
storm-water basin 雨水盆地；雨水流域；雨水池
storm-water bed 雨水净化(生物)滤池
storm-water channel 暴雨水排水沟；排雨水渠道；雨水渠
storm-water clarification plant 雨水净化厂；雨水净化装置；雨水澄清厂；雨水澄清装置
storm-water collection 雨水集水管
storm-water control package 暴雨水控制程序包
storm-water detection pond 暴雨水滞留池
storm-water detection pond volume 暴雨水滞留容量
storm-water diversion drain 雨水分流排水沟
storm-water drain 雨水排水管；雨水沟；暴雨(排)水沟；雨水排水道
storm(-water) drainage 雨水排水；雨水沟工程
storm-water filter 雨水净化生物滤池
storm-water flow 暴雨流量；雨水径流
storm-water flume 雨水槽
storm-water inlet 雨水(排水)口
storm-water intercepting sewer 雨水拦截下水道；雨水拦截排水管

storm-water management 暴雨水管理
storm-water management facility 暴雨水管理设施
storm-water management model 暴雨水管理模型
storm-water management model version 3.0 暴雨水管理模型 3.0 版
storm-water network 雨水排水网络;雨水排水管网
storm-water outfall 暴雨溢流
storm-water overflow 雨水溢流;暴雨溢流;暴雨水分流排管
storm-water overflow device 雨水溢流装置
storm-water overflow sewer 暴雨溢流污水道;暴雨溢流排水管
storm-water pollutant 暴雨水污染物
storm-water pollution 暴雨水污染
storm-water pollution control 暴雨水污染控制
storm-water pollution monitoring 暴雨水污染监测
storm-water pumping station 雨水泵站
storm-water quality 暴雨水水质
storm-water quality control 暴雨水水质控制
storm-water quality index 暴雨水水质指数
storm-water quality management measure 暴雨水水质管理措施
storm-water retention and detection pond 暴雨水滞留池
storm-water retention tank 雨水储[贮]水池
storm-water runoff 暴雨水径流
storm-water runoff quality 暴雨径流水质
storm-water runoff quality model 暴雨径流水质模型
storm-water sewer 暴雨排水道;雨水干管;暴雨下水道;雨水沟渠;雨水排水系统
storm-water stage 风暴水位
storm-water standby 辅助雨水池
storm-water stand-by tank 备用雨水池
storm-water system 暴雨排水系统;雨水系统
storm-water tank 暴雨水流储水池;雨水池;暴雨(沉砂)池;雨水管道沉砂井;雨水沉砂井
storm wave 风暴涌浪;风暴大浪;风暴波;风浪
storm weather 暴雨天气
storm wind 暴风
storm window 风暴窗;防风外重窗;防风;防暴风(雨)外重窗
storm window frame 外层防风暴窗的窗框
stormy 暴风雨的
stormy petrels 海燕
stormy weather 风暴天气;暴风天气;暴风雨气候
stormy weather report 风暴天气预报
storm zone 风暴区;暴雨地带;暴雨带
storyboard 故事画板
stoss-and-lee topography 不对称丘;鼻状地形
stoss face 向流面;迎水面;迎风面;顶冲面
stoss side 向流面;向风面;迎风面;迎冰坡;顶冲面
stoss slope 迎风坡
stotic pressure mo(u)lding machine 吹气压实造型机
stottite 羟锗铁石
Stoughton converter 司托顿侧吹转炉
stoup 酒杯;圣水体(教堂入口)
stout carrying handle 结实的搬运把柄
stout heart plyboard 硬芯胶合板
stout post (钢绳冲击钻机的)平衡杆冲击柱
stovarsol 醋酰胺肿
stove 窑;炉子;火炉;烘干炉
stove and range work 炉灶工作
stove black 耐高温沥青漆
stove bolt 小螺栓;炉用螺栓;圆头带槽螺栓
stove bolt tap 炉用螺栓丝锥
stove clay 耐火土
stove coal 大炉级无烟煤
stove coil 炉用盘管
stove connection 火炉装接;火炉连接
stoved alumin(i)um paint 烘焙铝漆
stoved goaf 充填的采空区
stoved resin 烘焙树脂
stove dried 烘干的;炉干的
stove-drying 烘箱烤干
stove enamel 烤瓷;干型瓷漆
stove-enamelled 烘焙搪瓷的
stove enamelling 烤瓷法
stove exhauster 厨房排烟机
stove finish 烘干(的)油漆
stove fuel 火炉燃料油
stove heating 火炉供暖;火炉采暖
stovehouse 温室

stove industrial enamel 烘焙工业搪瓷
stove oil 点炉用的油
stove pipe 外伸的排水管;火炉(烟囱)管;烟囱
stove pipe casing 火炉烟囱罩;大口径铆制套管
stove pipe method 铆接装管法
stove pipe weld 窑接焊接;炉管焊接
stove pipe welding 基地焊接管子
stove pipe well casing 炉管式井壁
stove plant 温室植物
stove plate 厨房灶台板;厨用灶台板
stove polish 擦炉粉
stove room 火炉室;火炉房(间)
stove size 火炉尺寸;火炉大小
stove tile 瓷砖;炉面砖;炉衬砖;搪炉砖;壁炉用砖
stovetop ware 耐热玻璃炊具
stove tube 火炉烟囱管
stovewood varnish 罩面烘漆
stoving 烤干;烘干
stoving colo(u)r 烘烤后的颜色
stoving enamel 烤干瓷漆;烘干型瓷漆
stoving filler 烘填料
stoving finish 热干清漆;烤漆;烘漆;烘干型面漆;烤清漆
stoving gloss 烘烤后的光泽
stoving medium 烘干型漆料
stoving oven 烘箱;烘焙炉
stoving paint 烘漆;烤干料;烘干涂料
stoving quality 烘焙质量
stoving stopper 烘焙制止器;烘焙限制器
stoving temperature 烘烤温度
stoving varnish 烘干面漆涂装;烤清漆;烘干型清漆;烘干涂料
stovis 可移动空气压缩机
stow 码垛;收卷;充填;积载
stowable cargo container 装载式货物集装箱
stow across 横装;横堆
stowage 装载;装货费;装载货物;储[贮]藏物;配载【船】;配货;堆装;储存;积载
stowage certificate 装载合格证
stowage charges 理仓费
stowage coefficient of container 集装箱积载系数
stowage compartment 储[贮]藏舱
stowage control 装载控制
stowage conveyer 充填运输机;充填输送机
stowage factor 装载因子;装载因数;载货系数;积载因素;积载因数;货物积载因数;货物积载系数;堆存系数
stowage factor table 货物积载因数表;货物积载系数表
stowage improper 积载不当
stowage of container 集装箱堆装
stowage of dangerous cargo container 危险货物集装箱装载
stowage of different sized container 不同尺寸集装箱混装
stowage plan 装载图【船】;积载图;货物配载计划图;配载图;船舶积载图
stowage plan for container 集装箱积载图
stowage planning 配载计划
stowage planning center 积载计划中心
stowage rate 积载率
stowage recommendation 推荐装载方法
stowage requirement 配载要求
stowage ship 跨月度船舶
stowage space 有效空间;装载容量;装载空间
stowage survey 装载检查;装载调查;积载检验;积载检查
stowage surveyor 积载检验员;积载检查员
stowage tonnage 装卸吨位
stowage unit 充填设备
stow all over 满装;满堆
stow athwartships 横堆;横装
stow away a boat 整帆待发
stow away from boiler 远离锅炉
stow away from heat 远离热源
stowboard 充填巷道
stow down 装入
stowed and trimmed 装舱及平舱;积载费及平舱费在内的离岸价格
stowed value 复合值
stowed wind 被截风
stower 堆装工人;充填机;充填工
stow fore and aft 直装;直堆
stow in block 分垛堆放

stowing 堆置;充填;装填;堆装
stowing gun 充填箱;充填喷枪
stowing machine 充填机
stowing material 充填材料
stowing of goods in container and on flat 集装箱和托盘货物装载
stowing operation 堆装作业
stowing pipe 充填管
stowing pressure 充填压力
stowing ratio 充填率
stowing tool 推料器
stowing weight 紧实重量
stow pneumatically 气动装载
stow wood 货楔;货垫木;楔木;垫木
St. peteraburg standard 圣·彼得标准(木材标准尺度)
strabismeter 斜视计
strabismus 斜视
strabismus hook 斜视钩
strabismus scissors 斜视剪
strabometer 斜视计
stracekite 斯特拉基石
straddle 跨装;跨立;跨板;期货买卖的期权;同价买卖
straddle alternate current power forklift 插腿式交流电叉车
straddle arm type fork lift truck 插腿式叉车
straddle attachment 跨装附件;跨装法
straddle broach 跨式拉刀
straddle bus 跨运车
straddle carrier 万能装卸车;运木车;跨式运输车;跨运车;跨式搬运车;跨车;长重货物运输车
straddle carrier crane 跨载起重机
straddle carrier lifting lug 跨车起重夹头
straddle carrier system 跨载机法;跨运车方式;集装箱跨运车搬运方式
straddle conveyer 跨式输送机
straddle crane 跨装起重机;跨式起重机
straddle cutter 跨式铣刀;双面铣刀
straddle dump device 跨式倾斜装置
straddle mill 跨式铣刀;双面铣刀
straddle milling 跨铣
straddle mill work 跨铣加工
straddle monorail 跨座式单轨铁路
straddle mounted pinion 跨架置小齿轮
straddle mounted type 跨立式
straddle mounting 跨式架;跨架型
straddle packer 跨式压试机;跨式打包机;跨式装填机;隔离式封隔器;跨隔式封隔器;双用封隔器
straddle packer test 跨隔测试
straddle pole 叉杆;跨杆;撑架梁
straddler 跨车;双体载驳船
straddle scaffold 跨立式脚手架;鞍形脚手架
straddle-shaded mono-railway 跨座式单轨铁路
straddle ship 跨月度船舶
straddle sprocket milling cutter 跨式链轮铣刀
straddle testing 选择性地层试验
straddle-transported crane 龙门起重机
straddle truck 龙门式吊运车;跨运车
straddle type 跨(座)式
straddle-type mill 跨鞍式减径一定径机
straddle wrench 叉形扳手
straddling compaction 跨步式夯实
straddling composition (蛙式打夯机的)踏步式夯实
straetlingite 水铝黄长石
strafe 空中扫射
straflo turbine 全贯流式水轮机
stragetic innovation 策略改进
straggling 离散;歧离;分散
straggling parameter 离散参数;偏差参数
straighening by bending 弯曲法矫形
straight 直道的
straight abutment 直线坝头;一字形桥台;无翼桥台
straight-across-cut 正切
straight across the road 与道路中线直交
straight advancing klystron 直射式速调管;直进式速调管
straight-ahead drilling 持续顺利钻进
straight-ahead method 直前法
straight air brake 直通(式)空气制动器;直通(式)空气制动(机)
straight air brake valve 直通空气制动阀
straight alcohol 纯乙醇;纯酒精
straight alkyd resin 未改性醇酸树脂;纯醇酸树脂

straight alluvial channel 顺直冲积性河槽
straight amplification 直接放大
straight anchor 直端锚固(钢筋)
straight and bellied jaw crusher 平凸颚板式轧碎机;平凸颚板式碎石机;平凸颚板式破碎机
straight angle 锐角;平角
straight arch 平拱
straight arch lintel 平拱过梁
straight argon 纯氩
straight-arm 直桨叶;直臂
straight-arm cutter 开式绞刀
straight-armed pulley 直辐带轮
straight-arm mixer 直叶片拌和机;直线叶式搅拌机;直臂混合机;直臂拌和机
straight-arm stirrer 直臂搅拌机
straight arrival bill 直达汇票
straight artery 直动脉
straight asphalt 直馏沥青;纯石油沥青;纯沥青
straightaway 直线行进;直路;机场跑道
straightaway cut 直线切削
straightaway digging 直线锥挖掘
straightaway mill 串列式轧机;串列式小型钢轧机
straightaway stands 串列式盘条轧机机组
straight axle 直轴(的)
straight-axle landing gear 直轴式起落架
straight back hand saw 直背手锯
straight barrel 平辊身
straight-barrel mission roofing tile 直筒教会式屋面瓦;直筒阴阳屋面瓦
straight baseline 直基线
straight base rim 平底轮辋
straight bead 直线焊缝
straight bead welding 直向迭珠焊接;直线焊接
straight beam 直射束;直梁
straight-beam bracket 桥形机架
straight beam element 直梁单元
straight beam method 直梁法
straight beam pass 梁形直孔型
straight beam probe 直探头
straight bed lathe 普通床身式车床
straight between two curves 曲线间的夹直线
straight bevel gear cutter 直齿伞齿轮刨刀
straight bevel(gear) generator 直齿锥齿轮刨齿机;直齿伞齿轮刨齿机
straight bevel gear pair 直齿锥齿轮付
straight bill of lading 直运提单;直交提单;记名提单;不转让提单
straight bills of lading 直接提单
straight binary 标准二进制
straight binary code 标准二进码
straight bit 一字形钻头;一字形钎头;冲击钻头
straight bituminous filler 纯沥青填(塞)料;直馏沥青填(塞)料
straight blade 直叶浆;直叶片;直线叶片式;直接推土刀;直接推土板;直铲推土刀;直铲推土板;直叶片
straight-blade bulldozer 直式推土机;直刃推土机
straight-blade centrifugal fan 直叶离心式风扇
straight-blade dozer 直铲推土机
straight-blade paddle mixer 直叶片拌和机;直浆叶片拌和机
straight-blade snow plough 直刃雪犁
straight-blade type rail sawing machine 直锯条式锯轨机
straight blow 直吹;直击强风
straight boiler 平顶锅炉
straight boiler tap 直柄锅炉丝锥
straight bond 普通债券
straight bore 直通
straight bottom rail 下侧梁矫直(集装箱)
straight bow 直线型船首;直首柱【船】;直首(柱)【船】
straight box beam 直的箱形梁
straight brace 直拉条
straight-braced frame 直拉条架
straight branch 直支管
straight brass wire cup wheel 直黄铜丝杯刷轮
straight brass wire wheel 直黄铜丝刷轮
straight brick 直型砖;长方形砖
straight brick lintel 直砖过梁
straight bridge 直线桥
straight bulldozer 直式推(土)铲
straight burr 直纹
straight butt weld 无坡口对接接头

straight cable 直线型钢丝束;直缆束
straight cal(l)ipers 内径量规;测孔器
straight cap vault 直的帽盖拱顶
straight cargo 统一种类的货物
straight cement 纯水泥
straight cement concrete 纯水泥混凝土
straight cement mortar 纯水泥砂浆;纯水泥灰浆
straight chain 直链
straight chain aliphatic hydrocarbon 直链脂肪烃
straight chain compound 直链化合物
straight chain hydrocarbon 直链烃
straight-chain paraffin 直链烷烃;直链石蜡
straight chain polymer 直链聚合物
straight chair 直靠背椅子
straight channel 直沟渠;直水道;直槽;顺直河槽;顺直航道
straight-channel light beacon 直道灯标
straight check-valve 直通单向阀
straight chopping bit 击碎式平铲钻头
straight chrome tanning 纯铬鞣
straight-chromium stainless steel 铬不锈钢
straight circuit 直通线路;直通回路;直接式电路
straight clay brick lintel 直的黏土砖过梁
straight-clay slurry 纯黏土浆;纯泥浆
straight clear path 开阔顺直段;开阔的直线段
straight climb 直线上升
straight closure rail 直合拢轨;直导轨
straight coast 直线海岸
straight coastline 平直海岸线
straight cock 直旋塞
straight collecting tubule 直集合管
straight collet 直夹套
straight compression 直接受压
straight condensing turbine 纯凝汽式汽轮机
straight conic(al) draft pipe 直锥形尼水管
straight connecting rod 直连杆
straight connector 直套管;直接管头;直管连接管;直管接头
straight conveyer 直线输送机;直线流水作业
straight cooler (烧结矿的)直列式冷却器
straight corridor 直廊
straight coupling 直通接头;直管箍
straight course 直线船程;上层屋面(木)瓦层
straight course paving 分层块石铺砌
straight crank 直线拐肘
straight crank web 直曲柄板
straight credit 直接信用证(一次使用信用证)
straight crested ripple 直脊波痕
straight crossfall 单一横坡
straight crossing course 直越航线
straight crown block 单滑车
straight cup(grinding) wheel 直角碗形砂轮
straight cup wheel 杯形砂轮
straight curb 直路缘石
straight current 直流
straight curtain hung 链条直线垂挂
straight curved grooving cutter 直内弯旋槽刀
straight cut 纵切的;横切(接头处);纵向切削;直线掏槽
straight cut burr 直纹刻石刀
straight cut gear 正齿轮;直齿圆柱齿轮;直齿齿轮
straight cut off 裁弯取直
straight cut operation 纵向切削操作
straight cutting 直插
straight cycle 直接蒸汽循环
straight cylinder 直列式多缸发动机
straight cylinder type servomotor 直缸接力器
straight cylindric(al) duct 直圆柱孔道
straight dam 直线型坝
straight deck 直纹(木)甲板
straight deposit 纯粹存款
straight die 校平模;平面配合模
straight diesel 直列式柴油机
straight discontinuity 平直结构面
straight distillation 直接蒸馏
straight dive 直线俯冲
straight dovetail 直鸠尾榫
straight down gutter 直通雨水管
straight downward-sloping lines 下斜直线
straight dozer 直铲推土机;直进式推土机
straight-draft furnace 直接通风炉
straight drawn shell 直缘拉制件
straight drill 直钻头;直柄钻头
straight drive 直接传动(的)

straight drop spillway 跃水式溢洪道;直落式溢洪道;落井式溢洪道
straight duct 竖直管道
straight dynamite 硝酸甘油火药;纯炸药
straight edge 平尺;规板;放线尺;直缘;直规;直尺;直边;刮尺;直棱;校正尺
straight-edge beveling machine 直边导棱机
straight-edge boundary 直线晶界
straight-edge cone 直边锥形筒子
straight-edge cutting machine 直线切割机
straight-edged ruler 直边导规
straight-edge glass carboy 直边玻璃酸瓶
straight-edge guide 直边导线
straight-edge level(l)ing 直尺水准测量
straight-edge line 直母线
straight-edge ruler 直线尺;刀口直尺
straight-edge tyre 直边轮胎
straight-edging 用直尺检查平面的平整度;用直尺刮平
straight eight 直排八汽缸;直八式;八汽缸直排式
straight-eight engine 直列八缸发动机
straight electrolytic(tin) plate 双面等厚镀层镀锡薄钢板
straight elevator 直挺
straight embedment of anchorage 钢筋直端锚固
straight embryo 直立胚
straighten 使矫直
straighten corner post 角柱矫正(集装箱)
straight-ended choir 直端唱诗班席位
straight end guide 直端导轨
straight endwall 直端洞口;直端墙
straighten end panels 端梁矫正(集装箱)
straightener 初轧板坯齐边压延机;调直机;校直装置;矫直装置;矫直机;矫正器
straightener for seamless piping 无缝钢管校正器
straightener for seamless tubing 无缝钢管校正器
straightener roll 校平辊;矫直辊;调直辊
straightener stator blade 整流叶片
straight engine 直列式发动机
straightening 裁弯取直;压直;整直;整顿;拉直;截弯取直;校直;矫直;调直;变形矫正
straightening and prestressing bed 校直和预应力装置;校直和预应力台架
straightening anvil 校正砧
straightening apparatus 拉直器
straightening by heating 加热矫直
straightening coat 矫正涂层
straightening cutting machine 矫直切断机
straightening device 整直装置;校直装置;矫直装置;矫直设备
straightening drill tool 纠斜器
straightening force 纠斜力
straightening machine 校直机;矫直机
straightening of bankline 岸线取直
straightening of bars 钢筋调直
straightening of channel 河道裁弯取直;河槽裁弯取直
straightening of hole 矫正孔斜
straightening of the girder lying flat-wise 平放大梁的矫正
straightening operation 矫直操作
straightening plate 矫直板
straightening polyline 拉直折线
straightening press 压直机;校正压力机
straightening press for steel sections 型钢校直压力机
straightening rate 纠斜率
straightening roll 辊式棒料矫直机;矫直辊
straightening roller 校直滚筒
straightening rolls 辊式矫直机
straightening speed 矫正速度
straightening unit 校直装置
straightening vane 整流导叶
straighten out the relationships among taxes, charge, prices 理顺税、费、价三者关系
straighten roof bows 顶梁矫正(集装箱)
straighten steel bar 调直钢筋
straighten top rail 上侧梁矫正(集装箱)
straight excavating and back loading process 正向开挖、后向装土法
straight excavating and side loading process 正向开挖、侧向装土法
straight extinction 直消光;平行消光
straight extrusion die head 顺挤压模

straight faced roughing tool 直面粗车刀
straight face piston ring 圆柱面活塞环
straight fan 直扇状流
straight fender 直碰垫；直靠把
straight fertilizer 单一肥料
straight filing 直锉法
straight fin 平直翅片
straight finger gloves 直手指手套
straight firing 直接加热
straight fitting 直通接头
straight flange 直边；折边
straight flange method 直配孔型轧制法
straight flanked ring 矩形环
straight flash welding 直接闪光焊
straight flight stair(case) 单跑楼梯；直跑楼梯；直行(楼)梯段
straight flow 直流
straight-flow pump 直流(式)泵
straight-flow system 直流系统
straight-flow valve 直流(式)阀；顺流式阀
straight flute 平接瓷砖；平接瓦
straight fluted drill 直槽钻(头)
straight fluted hob 直沟滚刀
straight fluted reamer 直槽绞刀；直槽铰刀
straight fluted tap 直沟丝锥
straight forceps 直镊
straight-forward 简单的；顺向
straight-forward calculation 直接计算
straight-forward circuit 直通线路
straight-forward diagram 直观图
straight-forward fractional distillation 直接分馏
straight-forward junction working 直通汇接呼叫制
straight-forward line 直达线路
straight-forward method 简单法；直接(方)法
straight-forward operation 遥控操作
straight-forward procedure 简单方法
straight-forward surfacing 一次性浇灌面层；流水作业法铺筑路面
straight-forward system 直通制
straight-forward test 直观试验
straight-forward trunking 直通中继
straight fractional distillation 直接分馏
straight frame 直架
straight frame(d)gate 直框式闸门
straight frame(d)ship 直框架结构船；直肋骨船
straight frame semitrailer 直梁半挂车
straight frame trailer 平床式拖车；平板式拖车；平板半挂车
straight freezing 一次冻透
straight gas lift 连续气举；直续气举
straight gasoline 直馏汽油
straight gear(wheel) 直齿轮
straight generator 串列式发电机
straight girder support 直梁支护
straight glide 直线下滑
straight-going traffic 直行交通
straight grade 直坡度；直坡
straight grain 顺纹；(木材的)直行纹理；直纹理；直纹；直木纹；径向纹理
straight grained 直纹理的；直纹的
straight grained board 直纹木板
straight grained wood 直纹(理)木材；直纹(理)木板
straight graph 直图
straight grinder 平面磨床；直线磨具
straight grinding wheel 平行砂轮
straight grip 直把手
straight grizzly bar 直筛条；直棒条筛
straight grooving cutter 直刃铣槽刀
straight grooving iron 直槽刨铁
straight guide(way) 直导轨
straight guiding 直线运动导向器
straight half round 直头半圆
straight half round scraper 直头半圆刮刀
straight halved joint 对合接头；对合接合；对半接合
straight hand lever 直手柄
straight head 平顶的
straight heat dryer 直接浮选
straight hole 直孔，直井；垂直孔
straight hole bit 弯曲钻孔校直钻头
straight hole cut 直眼掏槽
straight hole guide 矿物井内直导向孔
straight hole test 钻孔直线度检查(人工偏斜前)
straight horn 直射式喇叭
straight in 直通
straight-in approach 直线进场
straight-in feed 直进给；横向走刀
straight-in grinding machine 直进外圆磨床
straight-in landing 直线进入着陆
straight-in penetration 直线下降
straight-in run 直线进入目标
straight iris knife 直刃虹膜刀
straight iron powder 纯铁粉
straight jacket 墙加固板；平直套板
straight jaw plate 直线型颚板
straight jet 单回路涡轮喷气发动机
straight job 无拖车的载重汽车
straight joint 直缝拼合(木地板)；(砌体中的)垂直通缝；无分支连接；无分支接头；对接；直线接头；直接口；直缝砌缝；直缝对接；平口接头；无分支连接套管；通缝；暗榫式对接
straight joint bond 直缝砌合
straight joint floor 直拼缝地板；直拼地板
straight joint(ing) 直缝接头
straight joint tile 平接瓷砖；平接瓦
straight kerb 直路缘石
straight knife 直刀
straight land 锋后导缘
straight lead rail 直合轨轨
straight lease 定期定额付租租契
straight-legged roadway arch 直线路面拱
straight-leg tripod 定腿三脚架
straight length of pipe 管子的直线长度
straight letter of credit 一次使用信用证；指定受益人信用证；直接信用证(一次使用信用证)；不能转让信用证
straight lever 直杆
straight ligament 直线吊线
straight line 直线；直排式
straight-line approximation 直线近似
straight-line approximation method 直线近似法
straight-line apron boom 直线护坦栏栅；直线护坦大墙栏；拦河堰
straight-linear 直线状
straight-linear polyethylene 直链聚乙烯
straight-line attack 直线攻击
straight-line basis 直线基线；直基线
straight-line bending 直线弯曲
straight-line block 直线建筑；板式建筑物
straight-line body 直线(型)车身；直线本身
straight-line breaking 直线断裂
straight-line characteristic 直线特性
straight-line code 直线编码；直接式程序；无循环程序
straight-line coding 直线编码；直接式程序编制；无循环程序
straight-line commutation 直线整流
straight-line compressor 直线式压缩机；直立压缩机
straight-line condenser 直线可变电容器
straight-line control 直线控制
straight-line controller 直线控制器
straight-line correlation 直线相关
straight-line course 直线航线
straight-line crane 直线运动起重机；直线移动式起重机
straight-line cross cut saw 直线横切锯
straight-line current 直线电流
straight-line depreciation method 直线折旧法；直接折旧法
straight-line design 直线轮廓的结构
straight-line displacement 直线位移
straight-line distance 直线距离
straight-line distribution 直线分布
straight-line diving attack 直线俯冲攻击
straight-line drive 直线传动
straight-lined vessel 直线型船
straight-line edger 修边锯；修边机；直线磨边机
straight-line engine 直列汽缸发动机
straight-line equation 直线方程(式)
straight-line flight 直线飞行
straight-line flow 直线流动
straight-line-flow method 直线流水作业法
straight-line formula 直线关系(公)式
straight-line frequency 直线频率式
straight-line frequency capacitor 线性频率电容器
straight-line furnace 直线炉
straight-line gale 直击大风
straight-line generator 直母线
straight-line graticule 直角坐标格网
straight-line graver 直线刻绘器
straight-line hammer 直线运动的击锤
straight-line hardening 端部淬火
straight-line harmonic motion 直线和谐运动
straight-line inductive transducer 直线运动感应式传感器
straight-line interpolation 直线内插法
straight-line joint 直缝
straight-line knurling 直纹滚花；直纹滚花
straight-line knurls 直线滚花
straight-line law 直线定律
straight-line leakage path 直线泄漏通路
straight-line lever 直叉锚式摘纵机构
straight-line link motion 直线连杆运动
straight-line lower bound 直线下限
straight-line machine 直线式拉丝机
straight-line mechanism 直线机构
straight-line method of amortization 直线摊提法；直接摊还法
straight-line method of concrete 混凝土应力—应变直线比例法
straight-line method of depreciation 直线折旧法
straight-line motion 直线运动；直线流动
straight-line motion mechanism 直线运动机构
straight-line of infinity 无限远直线
straight-line oscillation 直线振动
straight-line pad sander 往复式压板打磨机
straight-line pass 直线通过
straight-line pattern 直线花型
straight-line pen 直线画笔
straight-line plotting 直线描绘
straight-line pole 直线杆塔
straight-line portion 直线部分
straight-line program(me) 非转移式程序
straight-line rate 按单位平均费率
straight-line relation 线性关系；直线关系
straight-line seam welder 直线缝焊机；直线滚烽机
straight-line section 直线区间
straight-line sintering machine 带式烧结机
straight-line sludge collector 直线污泥吸集器
straight-line theory 应力应变成正比假说；线性分布理论；直线理论；线性理论
straight-line tint 网线普染
straight-line traverse 直伸导线
straight-line tunnel kiln 直线隧道式窑
straight-line type automatic sampler 直线型自动取样器
straight-line type regulator 接点直线运动式调节器
straight-line type unloader 直线运动式取出装置
straight-line vee bottom 直线 V 形底
straight-line wavelength 直线波长式
straight-line wavelength capacitor 波长标度正比电容器
straight-line work 直线加工工作(天然石材)
straight lining 直线衬里
straight link 直线滑车连杆；直线滑机连杆
straight loan 纯信用贷款；信用贷款；完全贷款
straight-lobe compressor 转子压缩机；罗茨压缩机
straight lock 平贴门锁；平销锁
straight magnetic axis 直线形磁轴
straight main motion 直线主运动
straight main movement 直线主运动
straight member 直杆
straight method 直配孔型轧制法
straight method of depreciation 直线折旧法
straight mineral oil 直馏矿物油
straight mortgage 分期付息到期还本的抵押贷款
straight motor 直管式发动机
straight mo(u)ld 直结晶器
straight multiple 直接复接
straight nailing 垂直正面钉入
straight-neck rivet 直颈铆钉；无头铆钉
straightness 直线性(度)；平直(程)度
straightness of weir plate top 溢流堰顶面不平度
straightness tolerance 平直度公差
straight note 到期一次清还贷款的借据
straight numerical system 纯数字系统
straight oar 直叶桨
straight-on starter 直接起动器
straight-pane hammer 平头錾锤
straight panel 直线形墙段
straight part 直部；展长轧辊型缝

straight pass 直孔型
straight path 直线轨迹
straight-path approximation method 直线路近似法
straight-path sampler 直路取样机
straight pattern blasting 直线型爆破
straight pattern shooting 直线型爆破
straight peen 直头;直锤头
straight-peen hammer 直头尖嘴锤;直头尖口锤
straight peen sledge 纵锻锤
straight-penetration method 直接贯入法
straight phenol-formaldehyde resin 纯酚醛树脂
straight piece work system 单纯计件工资制
straight pin 圆柱销;直销;直脚钉;绝缘子直脚钉
straight pin porcelain insulator 直脚瓷瓶
straight pipe 直管;直缝管
straight pipe injector 直管式喷射器
straight pipe-line 直管线
straight pipe tap 直式管用丝锥
straight pipe union 直接接头
straight plate jaw crusher 直板颚式碎石机
straight plowing 顺坡耕
straight-ply belt 中部加强的运输带
straight point scriber 直线画线针
straight polarity 正接;正极性
straight polymer 单一聚合物
straight portion 直部
straight pressing 直接压制(法)
straight product 纯(净)产品
straight profiled bat 直型杆
straight pulsed device 直管状脉冲器件
straight pusher-blade 直的推料叶片;推土机刮铲
straight quay 直线岸壁式码头
straight radial blade 径向直叶子
straight ramp 直坡道;直进航线
straight reach 直线河段;河流直段落;顺直河段;顺直段
straight reaming 直接扩孔;校直扩孔
straight receiver 直接放大式接收机
straight reciprocating motion 直线往复运动
straight reduced asphalt 直馏沥青
straight refinery gasoline 精制纯汽油
straight regeneration 顺流再生
straight regular stream 顺直正常河流
straight reinforcement bar 直配(钢)筋
straight resin 净树脂
straight ring 直线环
straight river 顺直河道
straight river reach 顺直型河段
straight riveted pipe 铆接管
straight roll 平面辊
straight roller 直滚子
straight roller bearing 普通滚柱轴承
straight rolling 等断面轧制
straight rotary crown block 单层天车
straight rotary kiln 直筒形回转窑
straight roughing shaping tool 直粗刨刀
straight rule 划线尺;直尺
straight run 管道直段;直馏产品;直管段
straight-run asphalt 直馏(地)沥青
straight-run asphalt cement 直馏沥青胶泥
straight-run asphaltic bitumen 直馏(地沥青)沥青
straight-run bitumen 直馏(地)沥青
straight-run clear gasoline 直馏无铅汽油
straight-run coal tar 直馏煤沥青;直馏煤焦油
straight-run coal tar dispersion 直馏煤焦硬沥青悬浮液
straight-run coal tar emulsion 直馏煤焦硬沥青乳液
straight-run coal tar pitch 直馏硬煤沥青;直馏煤焦油脂
straight-run coal tar solution 直馏煤焦硬沥青溶液
straight-run diesel distillate 直馏柴油馏分
straight-run diesel oil 直馏柴油
straight-run distillate 直馏馏分
straight-run distillation 直馏
straight-run distillation process 直馏法
straight-run fuel 直馏燃油
straight-run gasoline 直馏汽油
straight-run heavy fraction 直馏重馏分
straight-run landing 直跑休息平台
straight-run light fraction 直馏轻馏分
straight-run motor fuel 直馏发动机燃料
straight-run naphtha 直馏石脑油

straight-run oil 直馏油品;直馏油
straight-run pitch 直馏硬煤焦油脂;直馏煤沥青;直馏(地)沥青
straight-run product 直馏产品
straight-run ramp 直坡道
straight-run residue 直馏渣
straight-run spirit 直馏汽油
straight-run stair 直行楼梯段;直跑楼梯
straight-run stock 直馏油料
straight-run tops 直馏轻汽油
straight-run virgin 直馏分
straight safety valve 直安全阀
straight saw 直锯
straight scale 直尺
straight scarf joint 直嵌接头
straight scissors 直剪
straight seam pipe 直缝管;直缝式管
straight section 直线段;直边段
straight selective flo(a)tation 直接优先浮选;直接选择性浮选
straight sending system 直接发送系统;直接发送式;直接传送系统;直接传送方式
straight serial bond 分期平均偿付债券
straight series 串联接合
straight shackle 直型卸扣
straight shaft 直轴(的)
straight shank 直柄
straight shank arbo(u)r 直柄刀杆
straight shank drill 直柄钻头
straight-shank drill bit 直柄螺钻
straight shank drill holder 直柄钻夹头
straight shank end mill 直柄端铣刀
straight shank milling cutter 直柄铣刀
straight shank punch 直柄冲头
straight shank reamer 直柄铰刀
straight shank rivet 无头铆钉;直颈铆钉
straight shank tager tap 直柄螺母丝锥
straight shank triple grip drill chuck 直柄三牙钻夹头
straight shank twist drill 直柄麻花钻头;无头钢钻
straight sheet pile 扁平板桩;板桨
straight shift transmission 直线变速传动(机动车);前进挡传动(机动车)
straight shoreline 平直海岸线;平直滨线
straight shovel 直铲
straight shunt-wound motor 直并激电动机
straight side 直边
straight-side 4-point double action drawing press 闭式四点双动拉伸压力机
straight-sided axial worm 阿基米德轴向蜗杆
straight-sided bucket 直边料斗
straight-sided choir 直边的唱诗班席位
straight-sided column 直边的柱;直边形柱;棱柱体柱
straight-side(d)core bit 直孔空心钻头;直壁式岩芯钻头
straight-sided flank 直齿面
straight-side double crank press 闭式双点压力机
straight-side flank splines 直边式花键
straight-side flat rim 直边平缘条
straight-side four point press 闭式四点压力机
straight-side frame 闭式机身
straight-side high speed precision press 闭式高速精密压力机
straight-side hydraulic press 液压直边冲压机
straight-side knuckle joint press 闭式双柱肘节式压力机
straight-side length 直边高度
straight-side mechanical power press 机动校直压力机
straight-side panels 侧板矫直(集装箱)
straight side press 双柱压床;闭式压力机;闭式双柱压力机
straight-side press with side slide 闭式侧滑块压力机
straight-side rim 直边轮辋
straight-side single crank press 闭式单曲柄压力机
straight-side single point press 闭式单点压力机
straight-side slot 直线侧缝
straight-side tire 直线轮胎;直边式胎
straight-side transfer press 闭式多工位压力机
straight-side two point press 闭式双点压力机
straight-side tyre 直线轮胎;直边式胎
straight silicon polymer 单一有机硅聚合物;纯有机硅聚合物

straight size 45 degree lateral 同径四十五度分支管
straight size cross 同径四通
straight size elbow 同径肘管
straight size side elbow 同径侧肘管
straight size side outlet tee 同径侧流三通
straight size tee 同径三通
straight skid 直滑
straight sleeve 直套管
straight sliding(shutter)door 直线型推拉(百页)门;垂直滑动百叶门
straight slip surface 平滑面
straight slit 直线裂缝
straight slope 直线坡
straight slope method 直线斜率法
straight slot hand reamer 直槽手用铰刀
straight slot head 一字槽螺丝帽
straight smooth spindle 光滑直钢锭
straight snips 平剪
straight sopke shave 直刃刮刨
straight-split 直劈的
straight split switch 直线尖轨式转辙器
straight spur gear tooth 直线正齿
straight stair 单跑(式)楼梯
straight stair(case) 直通楼梯;直跑楼梯
straight stair(case)with landing 双跑直梯
straight stairway 直跑楼梯
straight steam cycle 纯蒸汽循环
straight steel ruler 钢直尺
straight steel wire cup wheel 直钢丝杯形刷轮
straight steel wire wheel 直钢丝刷轮
straight stem 直线型船首;直首桩【船】;直柱【船】;直首
straight-stem sweep 直尼箭形铲
straight-stern rampway 尾直跳板
straight stock rail 直基本轨
straight strap clamp 直条夹
straight stream 直线型河流;顺直水道;顺直河流;顺直河道
straight streamline 直线型流线
straight stretch 直线距离;顺直河段
straight stretch inserted between two sets of adjoining turnouts 插入直线段
straight strut 直支柱
straight switch rail 直线尖轨
straight system 直接式
straight table rolls 辊道的正置辊
straight tail dog 直尼轧头;直尼夹头
straight tail lathe dog 直尼床子鸡心夹头
straight takeoff run 直线起飞滑跑
straight tamper 平夯
straight taper draft pipe 直锥形尼水管
straight taper draft tube 直锥形尼水管
straight tee 同口径三通;直三通;等径三通(管)
straight tenon 直榫
straight tension rod 直(接)受拉钢筋
straight-term mortgage 分期付息到期一次清还本金的抵押贷款
straight thawing 一次融透;直接解冻
straight thread 圆柱形螺纹;直丝扣;直螺纹
straight throat 平流式流液筒;水平流液洞
straight-through arrangement 直线布置
straight-through cock 直通旋塞
straight-through combustion chamber 冲压式燃烧室
straight-through cracking 直通裂化
straight-through current transformer 穿心式电流互感器
straight-through cyclone 直流旋流分流器;直流式旋风除尘器
straight-through drive 直接传动(的)
straight-through drive condition 直接传动条件
straight-through extrusion 直通压出
straight-through fast acting valve 直通快动阀
straight-through flow 直通流动
straight-through joint 直通连接;直接接头
straight-through labyrinth 直通式迷宫密封
straight-through process 直通过程
straight-through traffic 直行交通
straight-through-type wind tunnel 直流式风洞
straight-through vehicle 直行车辆
straight timber 直纹(理)木材;直纹(理)木板;直木材;直材
straight time 正式工作时间;定时(工资)

straight T-joint 直缝T形接头
straight tongs 直钳
straight tongue 企口榫;直雄榫;直纹凸榫;直榫
straight tonguing and grooving 直形槽与舌;直形企口
straight tool 直头车刀;直锋刀具
straight toolholder 直刀架
straight-tooth 直齿
straight-toothed 直齿的
straight-toothed bevel gear 直齿伞齿轮
straight-toothed conic(al) gear 直齿柱齿轮
straight-toothed cylindric(al) gear 直齿柱齿轮
straight-toothed gear 直齿齿轮
straight-tooth(milling) cutter 直齿铣刀
straight-tooth saw 直齿锯
straight-toothed spur gear 直齿圆柱齿轮
straight top 平顶
straight-topped section 单向横断面
straight track 直向股道;直向道岔
straight triangulation network 直伸三角网
straight-truck buoy 杆式浮筒
straight trunnion end needle roller 直销头滚针
straight tube 直管
straight-tube boiler 直管锅炉
straight tube steam generator 直管式蒸汽发生器
straight tubing valve 直流阀
straight tubular motor 直管式发动机
straight tuner 直接放大式调谐器
straight tunnel vault 直的隧道拱顶
straight turning 直线车削;直车
straight turning between centers 顶尖间直车削
straight-type engine 单排汽缸发动机
straight-type rampway 直跳板
straight-type wedge 开式偏斜楔
straight-type wheel 盘轮
straight union joint 直管活接头
straight upstream pipe 上游直管
straight-vaned fluid flywheel 直叶水力飞轮
straight vault 直的拱顶;平穹顶
straight vein 直脉(型)
straight vibration 线向振动
straight wagon vault 直形拱顶
straight wall 竖墙
straight-wall bit 直壁式岩芯钻头;直壁空心钻头
straight-wall core shell 直壁式校正扩孔器壳体;直壁式扩孔筒
straight-wall coring bit 直壁取芯钻头
straight-wall nozzle 直壁喷管
straight waterway section 顺直河段
straightway 道路的直线段;直线跑道;直线段;直线部分
straightway check-valve 直通单向阀
straightway cock 直通旋塞
straightway drill 直槽钻(头)
straightway(line) valve 直流阀;通直阀
straightway pump 直流泵
straightway setup 串列式布置
straightway travel(l)ing s/R machine 直线运行型有轨巷道堆垛机
straight way valve 直通阀;直流阀
straight web section 直腹式截面
straight-web sheet pile 平腹板型钢板桩;平板型(钢)板桩
straight-web sheet piling cell 腹板式格形板桩
straight-web sheet-piling wall 平腹板桩墙
straight-web steel sheet pile 一字型钢板桩;直腹板型钢板桩;平板型钢板桩
straight-wedge mechanism 直楔机构
straight welded joint 对缝焊接
straight weld(ing) 直缝焊
straight wheel 直轮;平面砂轮;平面磨轮;盘形轮
straight wheel grader 直轮式平路机;直轮式平地机
straight worm 直蜗杆
strain 胁变;族类;应变;族;张紧度;过劳;劳损;拉伤;菌株;菌弱;扭歪;品种;品系;使变形;粗滤
strain acceleration 应变加速度
strain accelerometer 应变式加速度计
strain accumulation 应变积累
strain age 应变时效
strain-aged 应变时效的
strain-aged brittleness 应变时效脆性
strain-aged cracking 应变时效裂纹
strain-age embrittlement 应变时效脆性
strain-aged hardening 应变时效硬化

strain-aged steel 应变时效钢
strain-age-embrittlement test 应变时效脆性试验
strain ag(e)ing 机械时效;应变老化;弥散硬化;应变时效
strain amplifier 应变放大器
strain amplitude 应变幅(度)
strain analysis 应变分析
strain and stress 应变和应力
strain annealing 应变退火
strain around opening 洞室四周应变
strain at failure 破坏应变
strain axis 应变主轴;应变轴
strain bar 应变测定棒
strain birefringence 应变双折射
strain bolt 拉紧螺栓
strain break 应变破裂
strain bridge 测应变电桥;应变电桥
strain bridge accelerometer 应变桥加速度计
strain build-up 应变积累
strain burst 应变爆裂;应变岩石突出
strain calculation 应变计算
strain capacity 应变能力;应变量
strain cell 应变计;应变盒
strain center 应变中心
strain clamp 拉线夹;拉紧夹;耐胀线夹;耐张线夹;耐拉线夹
strain cleat 耐张线夹
strain coefficient 应变系数
strain compatibility 应变相容性;应变协调
strain complementary energy 应变余能
strain complex 复合品系
strain component 应变分量
strain concentration 应变集中
strain concentration factor 应变集中因子;应变集中因数;应变集中系数
strain control 应变控制
strain controlled direct shear apparatus 应变控制式直(接)剪(力)仪;应变控制式直剪设备
strain-controlled load test 应变控制荷载试验
strain controlled test 应变控制试验
strain crack(ing) 应变裂纹;应变裂缝;变形裂缝;应变开裂;变形开裂
strain creep 应变蠕变
strain crystallization 应变结晶
strain curve 应变曲线
strain cycling 应变循环
strain detector 应变探测器;应变检验仪
strain deviation 应变偏移;应变偏量
strain deviator 应变偏量
strain diagram 应变图
strain disk 应变盘
strain-displacement equation 应变位移方程
strain displacement matrix 应变位移矩阵
strain displacement relation 应变位移关系
strain distribution 应变分布
strain due to torsion 扭转应变;扭曲应变
strain dynamometer 应变仪测力计
strained 应变的
strained casting 抬箱铸件;带飞边的铸件
strained condition 应变状态
strained rubber 应变橡胶;弹性疲劳的橡胶
strained zone 应变区
strain ellipse 应变椭圆
strain ellipsoid 应变椭球(体)
strain energy 应变能(量)
strain-energy amplitude 应变能振幅
strain-energy density 应变能密度
strain-energy due to impact 冲击应变能
strain-energy function 应变能函数
strain-energy method 变形能法;应变能法
strain-energy of distortion 畸变应变能;变形能量
strain-energy rate 应变能释放率
strain-energy release rate 应变能释放率
strain-energy storage 应变能存储
strainer 应变器;张紧器;张紧机构;过滤机;滤器;滤漆机;滤板;漏勺;清管用过滤器;松紧螺旋扣;筛滤网;粗滤器;除污器
strainer arch 横向扶拱;拉弦拱;扶拱;扶拱
strainer box 舱底水滤盒;蜂巢箱【给】
strainer cartridge 粗滤片筒
strainer case 滤器箱
strainer chamber 粗滤室
strainer core 滤渣片;滤网状型芯;滤网芯(片);浇口滤片;筛状芯片

strainer core gate 滤网芯浇口
strainer cover 过滤器盖;滤清器盖
strainer cross 滤水格栅
strainer filter 粗滤器
strainer gate 滤网浇口;筛状浇口
strainer-gate core 滤渣芯片
strainer grouting 滤网注浆
strainer head 过滤器水头;滤头
strainer injection 滤网注浆
strainer line 粗滤器品种
strainer mesh 滤网
strainer plate 滤板
strainer plug 滤器塞;粗滤(器)堵塞
strainer rack 细密拦污栅
strainer retainer 滤器卡簧
strainer screen 过滤器滤网;网状过滤器
strainer system 滤头系统;滤头装置
strainer tub 滤渣槽
strainer valve 过滤阀;滤汽阀
strainer well 过滤器管井;过滤井
strain etch method 应变蚀刻法
strain evaluation 劳损评价
strain exponent 应变指数
strain factor 应变灵敏度系数
strain failure 破坏应变;应变破坏;破坏性应变
strain fatigue 应变疲劳
strain field 应变场
strain figure 流线;应变图形;应变图;滑移线
strain finder 应变检查仪
strain foil 应变片
strain force 应变力;变形力
strain-free single crystal 无应力单晶
strain galvanometer 应变检流计;张力检流计
strain-ga(u)ge 应变仪;应变片;应变计;张线式传感器;张力仪;拉力计;电阻应变计
strain-ga(u)ge accelerometer 应变仪式加速度仪计
strain-ga(u)ge amplifier 应变仪放大器
strain-ga(u)ge bridge 应变片桥
strain-ga(u)ge dynamometer 应变仪测力计;应变测力仪
strain-gauge extensimeter 变形测定器
strain-ga(u)ge indicator 应变仪指示器
strain-ga(u)ge installation 应变仪设备
strain-ga(u) ge installed by spraying process 喷涂式应变仪
strain-ga(u)ge instrumentation 应变仪设备
strain-ga(u)ge load cell 应变仪负荷传感器;应变计荷载盒
strain-ga(u)ge logger 应变测定仪
strain-ga(u)ge measurement 应变仪测量
strain-ga(u)ge method 应变仪法;应变计法
strain-ga(u)ge pickup 应变仪传感器
strain-ga(u)ge pressure cell 应变计压力盒
strain-ga(u)ge pressure indicator 应变仪式压力表
strain-ga(u)ge reading 应变仪读数
strain-ga(u)ge recorder 应变仪记录器
strain-ga(u)ge rosette 玫瑰花形应变片;应变花;应变片花;应变片丛
strain-ga(u)ge scanner 应变计扫描器
strain-ga(u)ge scanner recorder 应变计式多点记录器
strain-ga(u)ge sting 应变计针架;应变仪支架
strain-ga(u)ge system 应变仪系统
strain-ga(u)ge technique 应变片技术
strain-ga(u)ge tensiometer 应变张力计
strain-ga(u)ge test 应变试验
strain-ga(u)ge tester 应变片式测试仪
strain-ga(u)ge torquemeter 应变扭转仪;应变片转矩计
strain-ga(u)ge transducer 应变片传感器;应变传感器
strain-ga(u)ge type displacement meter 应变计式位移计
strain-ga(u)ge type pickup 应变仪型传感器
strain-ga(u)ge type sensor 应变型传感器
strain ga(u)ging 应变测量
strain gradient 应变梯度
strain hardenability 应变硬化性
strain-hardened 应变硬化的
strain-hardening 应变强化;应变硬化;冷加工硬化;机械硬化;变形硬化;形变硬化;加工硬化
strain-hardening effect 应变硬化效应
strain-hardening exponent 应变硬化指数

strain-hardening material 应变硬化材料
strain-hardening range 强化阶段
strain-harding steel 应变硬化钢
strain-harding law 应变硬化规律
strain-hardness 应变硬度
strain history 应变历史
strain hoop 应变环
strain increment 应变增量
strain indicating lacquer 应变指示漆;测定应变用的涂料
strain indicator 应变仪;应变指示仪;应变指示器;应变指示计;电阻应变仪
strain induced cracking 应变诱导开裂
strain induced fracture 应变诱导断裂
strain-induced molecular orientation 应变诱导的分子取向
strain-induced precipitation 应变诱导析出
strain influence factor 应变影响因数;应变影响系数
straining 粗滤;应变;下坠;隔滤;拉紧;绷皮操作
straining apparatus 张拉设备
straining arch 扶拱
straining bag 过滤囊
straining beam 横梁;拉梁;跨腰梁;横系梁;缀条支撑;系梁;双柱上撑式桁架拉杆
straining box 筛箱
straining chamber 滤室;滤清室;粗滤室
straining coating 隔滤涂层
straining cord 牵引索;牵引绳
straining device 张拉设备
straining ear 紧定夹持器
straining element 抑制元件
straining equipment 张拉设备
straining filter 渗滤池;粗滤池
straining filtration 渗滤
straining frame 加载架;应变架
straining funnel 粗滤漏斗
straining index 隔滤指数
straining meter 应变仪;应变计
straining piece 水平撑木;拉条;拉杆(双柱上撑式桁架下弦);系杆;跨腰梁;变形部分
straining plant 张拉设备
straining pulley 拉紧轮;张力轮;张拉滑车;拉紧滑轮;牵引滑轮
straining ratchet 棘轮拉紧器
straining ring 紧圈
straining sill 上槛木;双柱桁架的分柱木;双柱桁架撑木;分柱木;二重桁架腰梁;双柱上撑式桁架下弦拉杆
straining standard (钢丝网篱笆的)立柱
straining surface 粗滤面
straining trestle 拉力架
straining tripod 拉力架
straining unit 张拉设备
straining vice 紧线钳
straining weight 张力锤
straining wire 张拉钢丝;拉线;引张线
strain instrument 应变仪
strain insulator 拉紧绝缘子;耐胀绝缘子;耐张绝缘子;耐拉绝缘子
strain intensity 应变强度
strain intensity factor 应变强度因子
strain invariant 应变不变量
strain isolating 菌株分离
strainless ring 无张力环
strainless stereoscopic viewing 无扭曲立体观察
strain level 应变水平;应变大小
strain line 扭曲微细裂纹;应变微细裂纹;应变线
strain magnitude 应变量
strain map 应变图像
strain mark 应变痕
strain markers 应变标志
strain matrix 应变矩阵
strain measurement 应变测量;应变测定
strain measurement technique 应变测量技术
strain measuring apparatus 应变测量设备
strain-measuring bridge 应变测量电桥
strain measuring instrument 应变仪;应变测量仪
strain meter 应变仪;张力仪
strain net method 变形网格法
strain of breed 品系
strain of flexure 挠曲应变
strain-offset 应变截距
strainometer 应变计;伸长计;应变仪;张力计
strain-optic(al) coefficient 应变光学系数

strain-optic(al) equation 应变光性方程
strain-optic(al) relaxation 应变光学松弛
strain pacer 定速应变试验装置
strain path 应变路径
strain pattern 应变图形;应变图像;应变模式
strain percentage 应变百分率
strain pickup 应变传感器;应变拾音器
strain plane 应变面
strain plate 拉线板
strain point 应变点
strain pole 耐张塔杆
strain-pulse 应变脉冲
strain quadric 应变二次式;应变二次曲面
strain quantity 应变量值
strain quenching 应变淬火
strain range 应变幅(度);应变范围
strain rate 应变速率;应变率;变形速度
strain rate effect 应变速率影响
strain rate equation 应变率方程
strain-rate factor 应变速率系数
strain rate tensor 应变率张量
strain rate vector 应变率矢量
strain ratio 应变率
strain ratio method 应变比法
strain reactance 寄生电抗
strain rebound 应变回跳;应变回弹
strain-recording equipment 应变记录仪(器)
strain-recording instrument 应变记录仪(器)
strain recovery 应变恢复
strain recovery characteristics 应变回复特性
strain recovery curve 应变回复曲线
strain recovery ratio 应变恢复率
strain recrystallization 应变再结晶作用
strain regime 应变状态
strain region 应变区
strain relaxation 应变松弛;蠕升
strain release 应变释放;应变分离
strain-release acceleration 应变释放加速度
strain-release point 应变消失点
strain relief 应变消除;应变释放
strain relief clamp 电缆卡子;电缆卡
strain relief method 应变释放法;应变起伏法应;地应变解除法
strain relief tempering 消除应力回火
strain resolution 应变分解;应变分辨率
strain retardation 应变推迟
strain rigidity 应变刚度
strain roll 张力辊
strain rosette 应变片花;应变片丛;应变计玫瑰图;应变花;应变电阻花
strain screw 张力螺钉
strain seismogram 应变地震图
strain seismograph 应变地震仪
strain seismometer 应变地震仪;应变地震计
strain sensibility 应变灵敏度
strain sensibility factor 应变灵敏度系数
strain sensitivity 应变敏感性;应变分辨率
strain sensor 应变(式)传感器
strain shadow 应变影;应变阴影;波动消光
strain sheet 应变图
strain shift 应变偏移
strain size 应变大小
strain-slip 应变滑动
strain slip cleavage 应变裂缝;应变滑劈理;错动劈理
strain slip folding 应变滑动褶皱;滑移褶皱
strain soften form 应变软化型
strain softening 应变软化
strainsometer 伸长计
strain spectrum 应变谱
strain state 应变状态
strain step 应变阶
strain strake 箍条
strain strength 应变强度
strain-strengthening 应变强化
strain-stress curve 应变应力曲线
strain stress diagram 应变应力图解
strain-stress loops 应变应力(环形)曲线
strain-stress relation 应变应力关系
strain system 应变系统
strain tensor 应变张量
strain testing system 品系试验系统
strain theory 应变理论;张力学说
strain thickening 应变稠化

strain thinning 应变稀化
strain threshold 应变阈值
strain tide 应变固体潮
strain-time curve 应变时间曲线;应变历时曲线
strain tolerance 应变公差
strain tower 耐拉铁塔
strain transducer 应变传感器
strain transfer 应变传递
strain triangular element 应变三角形单元体
strain tube 应变管
strain type 应变类型
strain value 应变值
strain vector 应变向量;应变矢量
strain velocity 应变速度
strain viewer 胁变观察器;应力测验仪;应变检查仪;应变测验仪
strain wave 应变波
strain wave velocity 应变波速
strain wire 弦式引伸仪;应变弦;拉线
strain wire ga(u)ge 电阻应变仪
strain work 应变功
strait 海峡;地峡
strait bridge 预制构件窄桥
strait deposit 海峡沉积
straited secondary wall 条纹次生壁
strait flange 窄凸缘
Strait of Dover 多佛尔海峡
Strait of Florida 佛罗里达海峡
Strait of Gibraltar 直布罗陀海峡
Strait of Hormuz 霍尔木兹海峡
Strait of Juan de fuca 胡安德富卡海岭
Strait of Magallan 麦哲伦海峡
Strait of Malacca 马六甲海峡
Strait of Tunis 突尼斯海峡
strait port 海峡港
straits dollar 马来西亚和新加坡元
Straits tin 邦加锡
strait work 狭窄地采掘
strake 洗矿槽;箍条;轮爪;轮箍;列板;手选台;船舷侧板;船侧板;防滑爪链;防滑轮爪;舣外板
strake below sheer 副弦侧厚板
strake below sheer strake 舷侧顶部外板
strake fender 护舷碰垫
Stramit partition(wall) in timber framing 斯特兰米脱木构架隔墙
Stramit slab 斯特兰米脱楼板
stramskiite 兰砷铜锌矿
strand 钢绞线;细条;裸多芯电缆;缆股;绞合线;索状;绳条;湖滨;河边;海荒滩;海滨;绳股;沙岸;导线束;玻璃纤维原丝
strand anneal(ing) 分股退火;带材退火;连续退火
strand cable 股绞电缆;扭绞电缆;成股电缆
strand casting machine 单流连铸机
strand clamp 绞线箍
strand clip 绞线箍
strand core 钢丝绳芯线芯;钢绞线芯
strand cross-section 股绳横截面
strand deflecting device 绞合金属绳挠曲装置;绞合金属绳弯曲装置
strand deposit 海滨沉积物;潮间带沉积(物)
strand diameter 股绳直径
strand dune 海岸沙丘;滨岸沙丘
stranded 绳断股;多芯的;多股的;沉没或失火险
stranded annealed copper cable 退火绞铜电缆
stranded bond 绞成股;绞成索
stranded cable 股绞合电缆;股绞电缆;绞线;绞合电缆;多股绞合电缆
stranded cable superconducting wire 股绞电缆超导线
stranded caisson 开口沉箱;沉箱
stranded conductor 绞线;绞合导线【电】;绞股导线【电】;多股绞合线
stranded copper conductor 多股铜线
stranded electrode 综合焊条
stranded floe ice foot 搁浅冰脚
stranded galvanized steel wire 镀锌钢绞线
stranded goods 海难船漂到岸边的货物
stranded ice 搁浅冰;浅滩冰
stranded ice foot 搁浅冰脚
stranded ignition wire 多股发火电线
stranded mill 多辊型钢轧机
stranded open caisson 开口沉箱
stranded rope 股绞绳;股绞钢绳索;绞绳绳索
stranded ship 搁浅船(舶)

stranded steel wire 绞股钢索
stranded welding wire 绞合电焊丝;股绞电焊条;绞合电焊条;综合焊条
stranded wire 悬缆(线);股绞电缆;钢绞线;多股绞合线;绞合线
stranded wire cable 多股绞合电缆
stranded wire cloth 绞合金属丝布
stranded wreck 部分露出水面的船舶残骸
strand electrode 绞合电焊丝;股绞电焊条;绞合电焊条
strand engaging wire component 钢丝螺旋排线器
strander 制绳机;绞线机;扭绞机;捻股机;绳缆搓绞机
strand extrusion 料条挤出
strand fiber cable 扭绞电缆
strand fishery 沿岸渔业
strand flat 沿海台地;沿海平地;沿岸台地;浪蚀台地;浪蚀平台;海滨边滩;海滨坪;海(滨)滩;海滨浅滩;潮间坪
strand grip 钢索夹具;预应力钢筋束锚具
strand ground 不规则花叶花纹
strand guide 铸坯导架;连铸铸坯导架
stranding 坐礁;搁浅;捻
stranding connection 绞接
stranding factor 铰线系数
stranding grounding 上滩
stranding harbo(u)r 枯潮时干枯的港湾;浅水港
stranding machine 制绳机;铰线机
stranding roll 条钢粗轧机
stranding wire 绳索;多股绞合线
strand insulation 单股导线绝缘
strand-laid rope 多股钢绞绳;绳股捻成的钢丝绳
strand layer 绞合线层
strand line 滨线;岸线;岸边线;海滨线
strand line pool trend 滨线油气藏趋向带
strand losses 涡流附加损失
strand marking 沿岸沉积痕
strand of cable 成股电缆
strand of rolls 粗轧机
strand pelletizer 单丝成粒机
strand pig casting machine 单线铸铁机
strand pitch (钢绳缆的)绳股捻距
strand plain 潮间平原;滨外平原;滨海平原
strand plant 海滨植物
Strand process 斯特兰特方法(轻集料烧结法)
strand pulling machine 穿束机
strand reduction 铸坯轧压
strand rope 绞股绳
strand seven wire 七股钢丝绳
strand shoe 金属尾履带片;多股绞合线终端(套管);索靴
strand sintering plant 多绞合线热压装置;多芯绞线热压结装置
strand socket 钢束索接;钢束承插
strand steel wire 钢绞线
strand templet 绞合线模型;绞合线型板
strand track cable 绞股式轨道索;绞股式承载索
strand vise 钢索钳
strand wire 钢丝绳股;多股绞合;绞合线
strand wire bond 扎线;多股绞线连接
strange 奇异
strangeness conservation 奇异性守恒
strangeness number 奇异数
strangler 阻气塞门;阻气门;节流门
strangler cable 挡风缆索;(汽化器的)阻气门电缆
stransference of patents 出售专利
S-trap 存水弯;S形下水管;防臭水封管弯头;虹吸封气管;双291头水封;反水弯;S形存水弯
strap 蓄电池同极连接片;小舌片;狭条;条板;用带捆扎;终端连接板;固夹(滑车)带;盖板;卡箍;交连带;耦合带;皮带;铁皮带;套上;套箍;索片;对接搭板;吊管带;带片;带箍;搭接片;布带
strap anchor 夹板(对接)锚固;条状锚固铜板
strap and key end 叉端
strap beam 接地片;接地母线;带形梁
strap bevel(l)ing machine 皮带斜切机
strap bolt U形紧固件;蚂蟥螺钉;带形螺栓;带栓;带螺栓;长平头螺栓
strap-bored block 隐轮滑车
strap brake 带闸;带式制动(器)
strap cam 带凸轮
strap clamp 卡板;带夹
strap coil 铜带线圈
strap cover 条状盖板;条状覆盖

strap cutting machine 裁带机
strapdown 捷联式
strapdown inertial navigation equipment 捷联式惯性导航设备
strap footing 连续脚;条形基础;带形底脚
strap fork 皮带移动器
strap frame gridiron 挂轮架
strap hammer 带式落锤
strap hanger 檐槽形吊钩;条形吊杆;吊带;吊挂带
strap head 环箍形连杆头
strap hinge 板式铰链;尖尾铰链;铁皮折页;铁扁担门铰链;带式铰链;长页铰链
strap holder 带夹
strap in 测量所下管子长度
strap iron 扁钢;箍铁(条);铁条;带条;带式轨;带钢;扁铁(条)
strap joint 加箍带的接头;夹板接合
strap laid 扁编股绳
strap lap joint 夹板塔接;夹板结合;夹板接合
strap link 带端链节
strap lock 带结合
strap of burlap 粗麻布带
strap of the brake 闸带
strap out of hole 测量所起拔管子长度
strapped bar line 带杆线
strapped-down inertial navigation equipment 捷联式惯性导航设备;束带式惯性导航设备
strapped-down inertial navigation system 捷联式惯性导航系统
strapped elbow 扁弯头
strapped joint 盖板接头;夹板结合;搭接接合;衬垫接头;夹板接合
strapped magnetron 耦腔式磁控管
strapped wall 木条板墙;板条墙
strapped weld 对接搭板焊接;搭板对接焊缝
strapper 捆扎机;捆包机;包扎机;搬运木材工人
strapping 系板;捆紧;捆带条;带量法;打包;绑扎法;板条
strapping band making machine 打包带生产机
strapping fertilizer colter 带式施肥开沟器
strapping machine 捆扎机;带卷捆扎机
strapping plate 加劲夹板;搭接板
strapping table 计量表(格);容积表
strapping tool 打包工具
strapping wire 双连开关接线法;分开线
strap plate 条铁
strap rail 带式轨
strap rivet 皮带铆钉
strap saw 带锯
strap setting 跳线设置
strap snap 皮带锁扣
strap steel 带钢;扁钢
strap stiffener 加劲窄板;加劲环;加劲箍
strap stub 夹箍形端头
strap tension 闸带张力
strap valve guide 阀导承
strap wire 带状电线
strapwork 交织凸起带饰纹;带箍比条饰
strap-wound winding 带绕绕组;扁绕绕组
strap wrench 带式扳手;带绕扳钳
Strasbourg turpentine 斯特拉斯堡松节油
strashimirite 水砷铜石
strass 有光彩的铅质玻璃;假金刚石
Strassman-type cutter 粗加工铣刀
strata alternation 层的交互变化
strata and rock type investigation 地层岩性调查
strata and rock type that sliding surface locate 滑动面所在地层岩性
strata behavio(u)r 岩层性状;岩层性质;岩层特性
strata-binding ore-forming process 层控成矿作用
strata bolting 岩层锚杆支护
strata-bound 层控
strata-bounded orefield structure 层控矿田构造
strata-bound lead-zinc deposit 层控铅锌矿
strata-bound lead-zinc deposit in carbonate rocks 碳酸盐岩层控铅锌矿床
strata-bound lead-zinc-iron-manganese deposit 层控铅锌铁锰矿床
strata-bound(ore)deposit 层控矿床
strata bridge 岩桥
strata control 岩体控制;顶板控制;顶板管理;地层控制
strata control analysis 矿压分析
strata-control engineer 地层控制工程师;顶板管理工程师

strata cross-section 地层剖面
strata cut through by fault 断裂带穿过地层
strata displacement 岩石移动;岩层移动
strataflow process (沿井筒测液体电阻以确定孔内的)涌水层部位法
strata gap 岩层间隙;地层间断
stratagem 计策
strata group 地层群
strata inclination 地层倾斜度
strata in fold core 核部地层
strata in fold limbs 翼部地层
strata in undisturbed condition 未扰动层理
strata mode 地层模式
strata model 岩层模型
strata movement 地层运动;地层移动
strata of allochthone 外来系统地层
strata of autochthon 原地系统地层
strata of foot wall 下盘的地层
strata of glide system 滑动系统地层
strata of hanging wall 上盘的地层
strata of horse 断片的地层
strata of quaternary period 第四纪地层
strata of rootzone 根带的地层
strata of soft soil 软土层
strata of underlying system 下覆系统地层
strata of volcanics 火山岩地层
strata pinching 地层尖灭;地层变薄
strata profile 岩层剖面;地层剖面图;地层剖面
strata reservoir 层状热储
stratascope 地层仪
strata section 地层断面
strata sequence 地层顺序;地层次序;层序【地】;层系;层理次序;层次
strata spring 矿层泉水;地层泉水
strata stacking method 分层堆料法
strata strength 岩层强度;地层强度
strata strike 地层走向
strata succession 层序【地】
strata thickness above karst 溶洞顶板厚度
strata thickness of downthrow side 下降盘地层厚度
strata thickness of upthrow side 上升盘地层厚度
strata tilting 地层倾倒
strata water 层间水
strategic air base 战略空军基地
Strategic air command 战略空军司令部
strategic alert light 战略警报信号灯
strategic alert sound system 战略警报声响系统
strategically equivalent 策略等价的
strategically equivalent game 策略等价对策
strategically equivalent utility function 策略等价效用函数
strategically value function 策略值函数
strategic point 交通要害点
strategic analysis 战略分析
strategic and critical raw material 战略与稀有重要原料
strategic area 战略地区
strategic arsenal 战略武器库
strategic base 战略基地
strategic city 战略城市
strategic data planning 战略数据规划
strategic environmental assessment 战略环境评价
strategic environmental assessment system 战略环境评价系统
strategic environmental impact assessment 战略环境影响评价
strategic equivalence 策略等价
strategic flexibility 策略灵活性
strategic fortifications 战略性防御工事
strategic grid 战略格网
strategic map 战略地图
strategic materials 战略物资
strategic mineral 战略矿物
strategic objective 战略目标
strategic outline chart 战略目标标示航图
strategic plan(ning) 策略计划;战略计划
strategic planning model 战略计划模型
strategic planning process 策略计划程序
strategic point of view 战略观点
strategic railway 战略铁路
strategic raw material 战略原料
strategic situation 战略

strategic-tactical map 战略与战术地图
strategic target 战略目标
strategic water planning 战略水规划
strategic weapon firing post 战略武器发射控制站
strategy 战略;策略
strategy formation 策略形成
strategy for new product development 新产品经营策略
strategy for urban growth 城市发展战略
strategy objective of environment(al) planning 环境规划战略目标
strategy of economy development 经济发展战略
strategy of ecosystem development 生态系统发展的战略
strategy of environment protection 环境保护战略
strategy of sustainable development 可持续发展战略;持续发展战略
strategy space 策略空间
strategy value 策略值
strath 陆架洼地;宽谷;平底河谷;平底谷
strath terrace 宽谷阶地
staticulate 成薄层的;分层的,薄层的
stratification 成层作用;成层性;成层现象;成层(法);层理【地】;层化(作用);层叠(形成);层叠现象;层次(法);分层现象;分层(法)
stratification line 分层线【地】
stratification mode 层理模式
stratification of atmosphere 大气分层
stratification of concrete 混凝土的分层现象(过湿或过度振捣后)
stratification of hot-gas 热气体分层;气体分层
stratification of sediments 泥沙(分)层;泥沙层理
stratification of snow cover 积雪层理;积雪层结
stratification of the ground 土壤的分层作用;土壤的叠层形成;地面的分层作用;地面的叠层形成
stratification of water 水的分层
stratification plane 层(理)面
stratification ratio 成层比
stratification sampling 分层取样
stratification surface 层面
stratification with layer 次层理
stratified 成层的;分层的
stratified air conditioning 分层空气调节
stratified alluvium 成层冲积土;成层冲积层
stratified analysis 层状分析
stratified atmosphere 成层大气;分层大气
stratified bed 分层基底;分层床
stratified bed ion exchange 双层离子交换
stratified cambium 叠生形成层
stratified charge 分层进气
stratified charge engine 层状燃烧发动机;分层燃烧发动机;分层充气发动机
stratified cluster sampling 分层整群抽样;分层集合体抽样
stratified coal seam sample 分层煤样
stratified combustion 层状燃烧
stratified cone 层状(火山)锥
stratified crown 分层型树冠
stratified current 异重流;成层流;分层流
stratified deposit 成层矿床;成层沉积;层状泥沙;层状沉积物;层状沉积;分层沉积物
stratified development 梯度发展
stratified drift 漂积层;漂积层;漂迹层;成层漂碛;成层漂积物;层状冰碛
stratified earth pulling process 层状拉土法
stratified estuary flow 河口层流;分层河口流
stratified eutrophic lake 成层富营养湖
stratified excavating process 分层开采法
stratified filter bed 分层滤床
stratified flow 多重流;成层流;层流;分层流(动)
stratified fluid 成层流体;层叠流体;分层流体
stratified foundation 成层地基;层状地基
stratified in thick beds 成厚层的
stratified in thin beds 薄层状
stratified lake 层化湖;分层湖;成层湖
stratified material 层状材料;层压材料
stratified media 成层介质
stratified medium 层状介质
stratified mixture 层状混合物;分层混合气流;分层混合料
stratified multistage cluster sampling 分层多级整群抽样
stratified multistage sampling 分层多级抽样

stratified observation 分层观测
stratified ocean 层化海洋
stratified one-stage cluster sampling 分层单级整群抽样
stratified plastics 层压塑料
stratified probability sampling 分层概率抽样法
stratified proportional sampling 分层比例抽样
stratified pumping test 分层抽水试验
stratified pumping up of underground water 分层抽取地下水
stratified purposive sample 分层目的样本
stratified random 分层随机
stratified random sample 分层随机样本
stratified random sampling 分层随机取样(法);分层随机取土样;分层随机抽样(法);分层随机采样;分层任意抽样
stratified random sampling approach 分层随机采样法
stratified random sampling method 分层随机抽样法
stratified randon sampling 分层随机抽样法
stratified ratio estimate 分层比率估计
stratified regression estimate 分层回归估计
stratified relief 层状地貌
stratified reservoir 分部用水水库;分层取水水库
stratified rock 岩层;成层岩(石);沉积岩;层状岩体;层状岩石
stratified rockfill 分层填石;分层抛石
stratified rocks 成层岩类
stratified sample 分层样本
stratified sampling 分层取样;分层抽样(法);分层采样
stratified sampling for proportions 比例的分层抽样
stratified sampling method 分层抽样法;分层采样法
stratified sampling of water 分层取水样
stratified sand 夹砂层;层状砂;层夹砂
stratified sand deposit 成层砂矿床
stratified sediment 层状泥沙;层状沉积物
stratified sedimentary rock 成层沉积岩
stratified sedimentation 层状沉积
stratified segregation 层状偏析
stratified simple random sampling 分层简单随机抽样
stratified single-stage cluster sampling 分层单级整群抽样
stratified soil 成层土(壤);层状土(壤);分层土(壤)
stratified structure 层状结构;层状构造
stratified subgrade 成层路基;成层地基
stratified subsampling method 分层二段抽样法;分层抽子样本法
stratified system 层状体系
stratified systematic sampling 分层等距抽样
stratified test 分层试验
stratified thallus 叠生叶状体
stratified training 层型整枝
stratified volcano 成层火山(岩);层状火山
stratified water 海水跃层;成层水
stratified waterflooding 分层注水
stratiform 层状的;层状
stratiform antimony deposit in carbonate rocks 碳酸盐岩层中层状锑矿床
stratiform cataclastic texture 层状碎裂结构
stratiform cloud 成层云
stratiform copper deposit 层状铜矿床
stratiform deposit 层状矿床
stratiformis 成层状云
stratiform lead-zinc deposit to reef limestone 与礁灰岩有关的层状铅锌矿床
stratiform magnesite deposit 层状菱镁矿床
stratiform mercury deposit in carbonate rocks 碳酸盐岩层中层状汞矿床
stratiform orebody 层状矿体
stratify 阶层化;使层化;使成层;分层
stratify the material bed 材料分层
stratigrapher 地层学家
stratigraphic(al) 地层学的;地层的
stratigraphic(al) age 地层时代
stratigraphic(al) age of water-bearing section 含水段地层时代
stratigraphic(al) analysis 地层分析
stratigraphic(al) barrier 地层遮挡

stratigraphic(al) bore(hole) 构造钻孔;绘制地层图用钻孔
stratigraphic(al) boundary 地层界限
stratigraphic(al) break 地层断缺;缺层;地层缺失;地层间断;层缺
stratigraphic(al) break separation 地层缺失离距
stratigraphic(al) classification 地层分类
stratigraphic(al) column 地层纵剖面(图);地层纵断面(图);地层柱状图
stratigraphic(al) comparative map 地层对比图
stratigraphic(al) condition 地层条件
stratigraphic(al) control 地层控制
stratigraphic(al) correlation 地层相互关系;地层对比图;地层对比
stratigraphic(al) correlation diagram 地层柱状剖面对比图;地层对比图解
stratigraphic(al) correlation of quaternary system 第四纪地层对比
stratigraphic(al) (cross-) section 地层断面;地层剖面;地层横剖面(图);地层横断面(图)
stratigraphic(al) datum 地层基准面
stratigraphic(al) deconvolution 地层反褶积
stratigraphic(al) development division 地层发育程度
stratigraphic(al) discontinuity along strike 走向不连续
stratigraphic(al) division 地层区划;地层区分;地层划分
stratigraphic(al) facies 地层相
stratigraphic(al) facies-map 地层相图
stratigraphic(al) fence section 地层栅栏剖面
stratigraphic(al) fissure 地层裂隙
stratigraphic(al) framework of basin 盆地地层格架
stratigraphic(al) gap 地层缺失;地层间断
stratigraphic(al) geology 地层学;地层地质学
stratigraphic(al) growth sequence 地层生长层序
stratigraphic(al) guide 地层标志
stratigraphic(al) heave 地层平铺;地层平错
stratigraphic(al) hiatus 缺层;地层缺失;地层间断
stratigraphic(al) high resolution dipmeter tool 地层学高分辨率地层倾角测井仪
stratigraphic(al) hole 基准钻孔;构造钻孔;绘制地层图用钻孔
stratigraphic(al) horizon of reservoir 储集层层位
stratigraphic(al) indication 地层学标志
stratigraphic(al) inspection 地层检查
stratigraphic(al) isopach 地层等厚线
stratigraphic(al) map 地层图
stratigraphic(al) model 地层模式
stratigraphic(al) oil field 地层油田;地层油层
stratigraphic(al) oil-gas field 地层型油气田
stratigraphic(al) oil pool 地层油藏;地层油藏
stratigraphic(al) omission 地层缺失
stratigraphic(al) overlap 地层超覆
stratigraphic(al) profile 地层纵剖面(图);地层纵断面(图);地层剖面
stratigraphic(al) provincialism 地层区划;地层区分
stratigraphic(al) range of species 物种地层分带
stratigraphic(al) reef 层礁
stratigraphic(al) reef trap 地层礁圈闭
stratigraphic(al) repetition 地层重复
stratigraphic(al) sampling 分域采样
stratigraphic(al) scale 地质年代表
stratigraphic(al) screened oil accumulation 地层隔离石油层
stratigraphic(al) section 地层剖面图
stratigraphic(al) separation 地层断距;地层落差;地层离距;地层间距
stratigraphic(al) sequence 层序组;地层顺序;地层次序;地层层序;地层层位;地层层次
stratigraphic(al) sequence of the beds 矿层的地层层序;岩层的地层次序;地层的地层次序;地基的地层次序
stratigraphic(al) succession 地层顺序;地层次序
stratigraphic(al) succession of metamorphic rock 变质岩地层层序
stratigraphic(al) terrane 地层地体
stratigraphic(al) test hole 地层试验孔
stratigraphic(al) texture analysis 地层结构分析
stratigraphic(al) thaw 地层间断
stratigraphic(al) thickness of reservoir 储集层地层厚度
stratigraphic(al) throw 地层落差;地层断距
stratigraphic(al) time scale 地层年(代)表;地层

时序表;地层时间表
stratigraphic(al) timetable 地层年(代)表
stratigraphic(al) time unit 地层年代
stratigraphic(al) trap 地层油捕;地层圈闭
stratigraphic(al) unconformity 地层不整合
stratigraphic-unconformity combination trap 地层不整合复合圈闭
stratigraphic(al) unit 地层单元;分层单位;地层分层单位
stratigraphic(al) well 参数井;基准孔;基准井
stratigraphy 地层图;区域地层;地层学
stratigraphy and lithofacies factor 地球及岩相因素
stratit element 钼电阻加热元件
stratobios 底层生物
stratochamber 高空舱
stratocirrus 层卷云;卷层云
stratocumulus 层积云
stratofabric 岩层组构
stratofortress 同温层堡垒
stratofreighter 同温层货运机
stratographic(al) 色层分离的
stratographic(al) analysis 色层分析
stratography 色层(分离)法
stratoid ore body 似层状矿体
stratoid structure 似层状构造
stratoisohypse 地层等厚线
stratomesosphere 平流层中间层
stratometer 土壤硬度计
stratometric survey 地质测量
straton 层子
straton model 层子模型
stratopause 平流层顶
stratoscope 同温层观测镜;地层检查仪
stratose 成层的
stratosphere 最上层;尖端科学领域;平流层(指大气);同温层;似平流层
stratosphere of the ocean 海洋同温层;海洋平流层
stratospheric(al) 平流层的
stratospheric(al) aerosol and gas experiment 平流层气溶胶和气体实验
stratospheric(al) aerosol measurement device 平流层气溶胶测定装置
stratospheric(al) balloon 同温层气球
stratospheric(al) chemical dynamics 平流层化学动力学
stratospheric(al) chemistry 平流层化学
stratospheric(al) chlorine 平流层氯
stratospheric(al) dust 尘埃;平流层微尘
stratospheric(al) dust load 平流层尘埃浓度
stratospheric(al) fallout 平流层回降
stratospheric(al) oscillation 平流层振荡
stratospheric(al) ozone 平流层臭氧
stratospheric(al) ozone loss 平流层臭氧损耗
stratospheric(al) ozone steady state 平流层臭氧稳态
stratospheric(al) pollutant 平流层污染物
stratospheric(al) pollution 平流层污染
stratospheric(al) warming alert message 平流层增温警戒信息
stratotanker 同温层加油机
stratotype 层型
stratovision 卫星电视
stratovolcano 层火山;成层火山(岩)
stratum 地层;层【地】;[复]strata
stratum board 多层硬纸板
stratum compactum 致密层
stratum disjunctum 脱落层
stratum dorsale 背侧层
stratum dynamics 地层动力学
stratum granulosum 颗粒层
stratum longitudinale 纵层
stratum lucidum 透明层
stratum moleculare 分子层
stratum nucleare 核层
stratum of humus 腐殖层;腐蚀层
stratum of invariable temperature 地下恒温层
stratum of regime observation 动态观测层位
stratum of sand 砂层
stratum of water sealing 止水层位
stratum society 层组合
stratum spongiosum 海绵层
stratum spring 夹层泉
stratum thickness 地层厚度
stratum water 地层水;层状水

stratum zonale 带状层
stratus 层云
stratus cloud 层云
Straub curve 斯性脆化曲线;斯特劳伯曲线
Strauss bore(d) pile 斯特劳斯螺旋桩
Strauss test 斯特劳斯试验
straw 稻秆;麻茎;禾秆;稻草
straw and earth cofferdam 草土围堰
straw and hay elevator 禾秆升运机
straw and loam pugging 草土隔声层
straw and willow products 草柳制品
straw automobile cushion 草汽车垫
straw bag 草袋;草包
straw bale 稻草捆;稻草包
straw bale buncher 草捆堆垛机
straw bale conveyer 草捆输送器
straw bale piler 草捆码垛机
straw band 草辫
straw basket 麦秸框;草篮
strawberry 草莓
strawberry tomato 毛酸浆
straw binder 稻秆捆束机;稿草打捆机
straw blower 稿草吹送器
strawboard 马粪纸;硬纸板;稻草板;草纸板;纸板;压制纸板
strawboard factory 草纸板厂
strawboard mill 草纸板厂
strawboard partition(wall) 稻草板隔墙
strawboard waste 草纸板废料
straw boss 助理工长;副领班
straw breaker 铡草机
strawbuncher 集草车
straw carpet 草地毯
straw chopper 铡刀;切草机
straw collector 集草车
straw colo(u)r 稻草色
straw-colo(u)red 稻草色的
straw compressor 稻草压捆机
straw cord 草带
straw covering 草席覆盖
straw cushion 草垫
straw cushion cover 草垫袋
straw cutter 稻草切碎机
straw cutter with feeder 带喂入器的铡草机
straw deflector 集草器
straw fan 草扇
straw fiber 草纤维
straw fibre board 稻草纤维板
straw file 粗齿锉
straw floor mat 草垫
straw from rye 黑麦秸
straw green epoxy polyvinyl butyral enamel 草绿环氧聚乙烯醇丁缩醛瓷漆
straw hat body 草帽胚
straw hat glue 草帽胶
straw hat theater 夏令剧院
straw line 引线
straw-made articles 草制品
straw man 代理人;稻草人;名义上的所有者
straw mat loom 草垫编织机
straw mat shed 草棚
straw mat(tress) 稻草编织品;稻草垫席;草垫;草席
straw meal 稻草粉料
straw mud mortar 草泥灰
straw-mud mortar plaster 草泥灰粉刷
straw mulch 盖草层;草幂;草盖
straw mushroom 草菇
straw oil 锭子油;草黄油
straw panel 麦秸板
straw paper 草纸
straw patchwork 麦秸画
straw pavement 禾秆加强的土路
straw products 草编制品
straw pulp 草纸浆;草浆粕;草浆
straw pulp black-liquor 草浆黑液
straw pulp factory 草纸浆厂
straw pulping wastewater 稻草纸浆废液;草浆浆废水
straw pulp paper-making wastewater 草浆造纸废水
straw-rack louver 逐稿器鱼鳞筛
straw roof(ing) 茅屋顶;稻草屋顶
straw rope 草绳
straw-rope spinning machine 打草绳机

straw-rope twisting machine 打草绳机
straw sack 草包
straw shoe 禾捆梳解器
straw shopping bag 草制购物袋
straw shredder 铡草机
straw stacker 集草车
straw stacking 堆秆
straw suitcase 草提箱
straw tar 稻草焦油
straw thatch(ing) 稻草屋顶
strawtrusser 捆草机
strawwalker 逐稿器曲轴;逐稿器
strawwalker beater 逐稿器逐稿轮
strawwalker crank 逐稿器曲轴
strawwalker feeder beater 第一逐稿轮
straw wall paper 草墙纸
straw wall picture 草墙画
straw ware 草织品
straw-weaved product 草编织品
straw willow 草柳
straw wrapping paper 草浆包装纸
straw-yard benyard 铺草运动场
straw yellow 麦秸黄;稻草黄;淡黄色;草黄色
stray 寄生;迷层
stray air current 离散气流
stray animal 离群动物
stray beam 杂散束
stray capacitance 杂散电容;寄生电容
stray-capacity coupling 杂散电容耦合
stray-capacity effect 杂散电容效应
stray copper loss 杂散铜损
stray coupling 杂散耦合
stray current 杂散电流;漏泄电流;离散电流;迷流
stray current corrosion 杂散电流腐蚀;漏泄电流腐蚀;离散电流腐蚀
stray-current electrolysis 杂散电流电解
stray-current hazard 杂散电流危险
stray current measuring instrument 杂散电流测定仪
stray current protection 防迷流保护
stray current testing cabinet 迷流测试柜
stray earth current 杂散地电流
stray electric(al) current 杂散电流
stray electric(al) field 杂散电场
stray electromagnetic field 杂散电磁场
stray electromotive force 杂散电动势
stray element 杂散元件
stray emission 杂散发射;寄生发射
stray fibre 散丝
stray field 杂散场;漏磁场
stray-field effect 杂散场效应
stray flux 杂散磁通量;杂散磁通
stray heat 漫射热;散失热
stray illumination 杂散光照;寄生照明;散射照明
stray inductance 杂散电感;寄生电感
stray induction 杂散感应;杂散电感
stray light 杂光;杂散光(线);漫射光
stray light cone 漫射光锥
stray light emission 杂散光辐射
stray light error 漫射误差
stray light filter 杂散光滤色
stray lighting 漫射照明;散光照明
stray line 空白线
stray load 杂散荷载
stray load loss 杂散荷载损耗
stray loss 杂散损耗
stray-loss factor 杂散损耗系数
stray magnetic field 杂散磁场
stray magnetic flux 杂散磁通
stray noise 杂音
stray parameter 杂散参数;杂散变量;寄生参数;补充参数
stray pay 杂层
stray pick-up 杂散噪声拾波;感应干扰
stray power 杂散功率
stray power loss 杂散功率损耗
stray radiance 漫射光亮度
stray radiant energy 漫射辐射能
stray radiation 杂散辐射;漫辐射
stray reactance 杂散电抗
stray reflection 杂散反射
stray resistance 杂散电阻
stray resonance 杂散谐振
strays (回声测深仪的)干扰回波;无线电干扰;天电

stray sand 零散油砂层;夹层沙
stray signal 寄生信号
stray transformer 高漏磁变压器;大漏抗变压器
stray value 偏离值
stray voltage 寄生电压
stray waves 杂散波
streak 煤层出露边缘;起条;漆条纹;条纹;条痕(色);色线;色条纹;带色条丝;斑纹
streak camera 快速扫描照相机;条纹摄像机;扫描照相机;超快扫描照相机;超快扫描摄影机;超高速扫描摄影机
streak chlorosis 条纹褪绿病
streak culture 划线培养
streak decoration 条纹装饰
streaked glaze 条纹釉
streaked limestone 条带状石灰岩
streaked necrosis 条纹坏死
streaked pseudofluidal structure 条带状假流纹构造
streak flaw 条状裂痕;条痕
streak image 拖尾图像
streakiness 丝纹;丝光
streaking 图像重影;形成条纹;划线分离;拖影;图像拖尾;斑纹
streaking chart 图像拖尾测试卡
streakless mixing 均匀混合
streak lightning 枝状闪电
streak line 条纹线
streak method 痕色法
streak of colo(u)r 色线
streak photograph 纹影照相;条纹照相;条纹照片
streak plate 划线平面培养法;划线平板法;条痕板
streak reagent 喷显剂;喷湿剂;涂显剂
streak rust 针蚀
streaks of mortar 砂浆条痕
streak test 纹理试验;色线试验;刻痕试验;痕色试验;色条试验
streaky 似条纹的
streaky amphogneiss 条痕状混合片麻岩
streaky coat 条痕状涂层;条痕状漆膜
streaky metal 色波筋
streaky structure 条纹构造;条痕状构造;条带构造
stream 溪流;流路;河流;河沟;射出;潮流
stream abstraction 河流袭夺;河流合流作用;夺流
stream acidification 河水酸化
stream action 河流作用;河流冲刷作用
stream activity 河流船舶动态
stream adjustment 河流调节;河道调整;河道调节
stream aeration 河水复复;河流曝气
stream ager 蒸化箱
stream aggradation 河流淤积
stream allocation procedure 域流分配法
stream analysis method 流路分析法
stream anchor 小锚;中等重量锚;流锚;河川锚;艇锚
stream arm 河流的支流
stream assimilation capacity 河流同化能力;河流同化流量
stream association 河流协会
stream at grade 均衡河流
stream attribute 流属性
stream axis 河流轴线;河流中心线
stream bank 河岸
stream bank cutting 河岸切割
stream bank erosion 河岸侵蚀;河岸冲蚀
stream bank erosion control 河岸侵蚀控制;河岸防冲
stream bank protection 河岸保护
stream bank stability 河岸稳定性
stream bank stabilization 河岸加固
stream bar 河岸;河中潜洲
stream base cation 河水碱性阳离子
stream basin 河流流域;河谷
stream bed 河床
stream bed aggradation 河床淤高;河床抬高
stream bed conductance 河床导水性
stream bed degradation 河槽冲深
stream bed erosion 河床侵蚀;河床冲刷;河床冲蚀
stream bed gradient 河床比降
stream bed intake 河底取水口
stream bed material 河床材料;河床质
stream bed protection 河床保护
stream behavior 河流特性
stream beheading 夺流河;断头河
stream bend 河流弯曲

stream berth 河中泊位
stream betrunking 断尾河;河流干涸
stream bifurcation 河流支流;河流分叉
stream biota 河流生物群;河流生物区
stream bit transmission 数位流传输
stream boat 河船
stream-borne 河水随带的
stream-borne material 河流挟带物;泥沙;河流挟带的泥沙
stream-borne sediment 河流挟沙(也称河流挟砂);河流挟带的泥沙
stream bottom 河底
stream bottom sediment 河底沉积
stream bottom soil 河底土壤
stream boundary 流域界
stream branch 河流岔道;河流叉道
stream breaker 碎流板
stream bridge 河上桥梁
stream built terrace 河成阶地
stream cable 水底电缆;尼锚锚链
stream canalization 河流的渠化工程;河流渠化
stream capacity 河流输沙能力;河流能力
stream capture 河流袭夺;河流截夺;夺流
stream centre line 河流中心线;河流中泓;水流轴线
stream chain 尼锚锚链
stream channel 河流水道;河底;河道沟渠;河道;河床;河槽
stream channel adjustment 河道调整
stream channel degradation 河道刷深;河槽刷深
stream channel deposit 河槽沉积
stream channel erosion 河槽冲刷;河槽冲蚀
stream channel form 河槽形状
stream channel form ratio 河槽形状比
stream channel pattern 河槽航道模式
stream channel shape 河槽形状
stream channel shape ratio 河槽形状比
stream channel slope 河道坡降;河道比降
stream characteristic 河流特性
stream chemistry behavior 河水化学性质
stream chloride concentration 河水含氯浓度
stream classification 河流分类;河道分类
stream coefficient 开工系数
stream condition 水流状态;水流状况
stream conservancy engineer 河流疏浚工程师;水土保持工程师
stream control 河流控制
stream control works 河流整治设施;河流整治工程;河流控制设施
stream corner 河流的拐角
stream course 水道;河道
stream criterion 河流质量判断
stream crossing 跨河桥;渡口;跨越河流;水道交叉
stream crossing bridge 跨河桥
stream crossing in skew angle 斜交水道;斜交渡口
stream crossing site 水道交叉地点;渡口位置
stream cross-section 河流横剖面;河流横断面
stream current 窄急海流;流动很急的窄洋流;河道水流;河川水流;海洋急流
stream curve 河流曲线段
stream cut terrace 河切阶地
stream data specification 数据流说明
stream data transmission 流式数据传输
stream day 连续开工日
stream deflector 折流设施
stream degradation 河底刷深;河道刷深;河床下降;河床刷深;河床加深
stream density 河网密度;河流密度
stream deoxygenation rate 河流脱氧率
stream deposit 河流沉积
stream deposition 河流堆积;河流沉积作用;河川沉积
stream development 河流改建;河流演化;河流开发;河流发育
stream dike 河流防洪堤
stream dike construction 河流防洪堤施工;河流防洪堤结构
stream dimension 流尺度;水流尺度
stream direction 河流流向
stream discharge 河流流量;河道流量
stream discharge record 河道流量记录
stream dissection 河流分支
stream diversion 河流分流;河流改道
stream drainage pattern 河网类型;水系类型
stream dredge(r) 河流挖泥船;河流挖泥机;河流

疏浚船;河流疏浚机
stream drift 河流沉积
stream driven pump 水流驱动泵
stream driving 河道流送
stream dynamics 河流动力学
stream ecology 河流生态学
stream ecosystem 河流生态系统
stream ecosystem process 河流生态系统流程
stream ecosystem restoration 河流生态系统修复
stream ecosystem service 河流生态系统机能
stream ecosystem sustainability 河流生态系统可持续性
streamedlined method 流水作业法
stream elevation 河流高程;河流标高
stream enclosure 截流;河道截流;地下输水道;地下封水道;封堵河流;封闭式输水道
stream encryption 流编码
stream energy 河流能量;水流能量
stream engineering 河流工程
streamer 通栏标题;狭长的纸带;狭长的旗;光幕;流光;极区流光;横幅标语;日冕射线;飘带;拖缆;射束;冲流;测风带;彩色纸带
streamer buoyancy 电缆浮力
stream erosion 流水冲刷;溪流侵蚀;河流侵蚀;河流冲刷作用;河流冲刷;河流冲蚀;河道侵蚀
streamer sensibility 电缆灵敏度
stream factor 运转系数;流水作业因数;水流系数;生产定额
stream fall 河流水位落差;河流落差;水面落差
stream feeder 流水式给纸装置
stream filament 流线;(河,气,潮)流束
stream flat 河滩
stream flood 河泛;槽洪;暴发洪水
stream floor 河底
stream flow 河流;河道水流;河道流量;河川流量;河水流;河川径流
stream-flow control 流量控制;径流控制
stream-flow coupling 内流线(型)接头(冲洗液孔呈锥形)
stream-flow data 河川流量资料;径流资料;河道流量数据
stream-flow data compilation 汇编流量资料
stream-flow depletion 径流消退;水流亏耗;径流亏耗;河川径流亏竭;退水
stream-flow forecasting 流量预报;径流预报
stream-flow ga(u)ging 河川流量测量;河流流量测量;流量测定;水文观测
stream-flow hydrograph 流量过程线;河川流量过程线
stream-flow in open channel 明渠均匀流
stream-flow measurement 水文观测
stream-flow measurement station 河道测流站;流量站;河道观测站;水文观测站
stream-flow measuring structure 测流建筑物
stream-flow pattern 河流模型
stream-flow plant 贯流式电站;径流式(发)电站;河流水电站
stream-flow record 河道流量记录;水文测量记录;测流记录
stream-flow record extension 河道流量记录延长;河川径流资料延长
stream-flow recording 测流仪记录
stream-flow regulation 流量调节;径流调节;河道流量调节;河道径流调节;河川径流调节
stream-flow routing 流量演算;洪水行程;河流洪水演算;河道流量演算;河川水流演算;水文演算
stream-flow sample 河道水样
stream-flow sampling 河水采样
stream-flow separation 河流分叉;水流分离
stream-flow synthesis 流量综合
stream-flow variability 流量变率;河流流量变率
stream-flow water quality model 河道流量水质模型
stream-flow wave 河道流量波
stream fluting 河流槽蚀
stream forecast(ing) 河流水情预报;河道水情预报;河流预测
stream fork 河流分叉口;河流岔道;河流叉道
stream form 河流形状;河道形态
stream frequency 河网密度;河流频数;水流密度频率
stream friction 通行阻滞(交通)
stream function 流量函数;流函数
stream ga(u)ge 流速计;流量计;水位标尺;水标

尺(河、溪流的)
stream-ga(u)ging 河流计量;河道流量测量;河道测(流)量;测流;河道水文测量;水文观测
stream-ga(u)ging flume 流水槽;测流槽
stream-ga(u)ging station 河流水文站;河流水文测量站;河道测站;河道测流站;水文站;测流站;流量站;河道测量站
stream-ga(u)ging(station)network 河流测站网;河道测站网;水文测验站网;测流站网
stream-ga(u)ging weir 测流堰;量水堰
stream gold 河金;砂金
stream gradient 河流降坡;河流坡度;河流比降
stream gradient ratio 河流坡降比;河流比降比(率);比降率
stream gravel 河流砂砾
stream gravel method 河流碎屑法
stream gravel veneer 河流砾石层
stream guidance 导流
stream habitat 河流生境
stream habitat classification 河流生境分类
stream habitat quality 河流生境质量
stream handling 流动式处理;连续供应
stream hardening 喷水淬火
stream head 河源;河流源头;河流发源地
stream health 清洁河流;未被污染的河流
stream heat exchanger 蒸汽换热器
streamhood 河流性质;河流状态
stream hours 连续开工时数;连续工作时数
stream hydraulics 河流水力学;河流水力特性
stream ice 狭长流水区;河冰
stream icing 河流积冰
stream improvement(rectification) 河流整治
streaming 流水存储;拖归费
streaming birefringence 流动双折射
streaming box 水洗槽
streaming current 流动电流
streaming current detector 流动电流检测器
streaming flow 线流;连续水流;缓流;冲流;成股水流
streaming-ga(u)ging cross-section 测流断面
streaming mercury electrode 流汞电极
streaming motion 流注活动
streaming potential 泳动电势;流势;流动电位;水流势能
streaming SIMD extensions 单一指令多数据流扩展
streaming steam sterilization 流动蒸汽灭菌法
streaming suspension 流动悬浮液
streaming tape 汇流带
stream inlet 支流河口;河流引水口
stream input 河流输入
stream in regime 冲淤平衡河流
stream intake 河流进水口;山溪取水口
stream-interacting element 射流相作用型元件
stream invertebrate 水生无脊椎动物
stream islet 河中小岛
stream jam 河道壅塞
stream-laid deposit 河流堆积;河流沉积物;河成沉积(物)
stream landscape 河流景观
stream length 河流长度;河长
stream-length ratio 河长比
streamlet 小溪;小河口;小河;小川;细流
stream level 河流水位
stream level fluctuation 河流水位变化
stream limiter 防溅器
streamline 水流线;通量线;河流线;流线型的;流线;日冕射线
streamline assembly 流水装配
streamline balanced rudder 流线型平衡舵
streamline body 流线型车身;流线(型)体
streamline car 流线型车
streamline casing 流线型外皮
streamline conduit 流线型管
streamline conveyer 流线输送机
streamline coordinate system 流线坐标系
streamlined 流线型的
streamlined aircraft 流线型飞机
streamlined body 流线型体
streamlined car 流线型汽车
streamlined construction of asphalt paving 地沥青路面的整体性结构;地沥青路面的整体施工
streamlined contour 流线型轮廓;流线型外形
streamlined course 速成班
streamlined cross-section 流线型截面;流线型横截面
streamlined gating 流线型浇口
streamlined house 流线型房屋
streamlined housing 流线型机壳
streamline diagram 流线图(谱);流动图
streamlined liner 流线型火车
streamlined method 合理化方法
streamlined method construction 施工流水作业法
streamlined nosing 机头整流罩
streamlined production 流水作业生产
streamlined rudder 流线型舵
streamlined sinker 铅鱼
streamlined specifications 简要施工说明;简要规范;简要规定;合理化规定;施工说明书摘要
streamlined spillweir face 流线型溢流堰面
streamlined strut 流线型支柱
streamlined tool joint 贯眼接头
streamlined unit 流线型机组
streamlined waterway bit 带流线型水槽的钻头
streamlined weight 流线型重锤;铅鱼
streamline field 流线场
streamline filter 流线式(过)滤器
streamline filtration 流线式过滤
streamline flow 线流;流线型流(动);流线流;平滑流动;畅流;层流
streamline form 流线型
streamline hollow shape 流线型空心型材
streamline locomotive 流线型机车
streamline operation 流水线操作
streamline passenger train 流线型旅客列车
streamline pattern 流线型
streamline pipe 流线型管
streamline process 流线过程
streamline production 流水作业
streamline production technique 流水作业生产技术
streamline production technology 流水作业生产工艺
streamliner 流线型物;流线型装置
streamline rod coupling 流线型内接头
streamline section 流线型截面;流线型断面
streamline shape 流线型
streamlines pattern 流线图(谱);流线谱
streamline superstructure 流线型上层建筑
streamline support 流线型支架
streamline theory 一维流动理论;流线理论
streamline tool joint 贯眼钻杆接头;贯眼接头
streamline tube 流线型管
streamline ventilator 连续型通风设备;流线型通风器
streamline vessel 流线型船
streamline wake 势伴流;船尼伴流
streamline wire 流线型张线
streamlining 流线型上层建筑;流线型化
streamlining of way way 水槽流线型化
stream load 河流挟沙(也称河流挟砂);河流挟带物含量;河流输沙量;河流泥沙量;河流荷重;河流负荷;河流的负载;河流含沙量
stream location 河流定位
stream log 放下计程仪
stream meander 河弯;河流弯曲
stream measurement 河道观测;水文观测
stream measurement station 河流水文站;水文站
stream measurer 流速计;水速计
stream mechanics 河流动力学
stream metamorphosis 河流变态
stream meter 流速计
stream mode 流状态;流方式;流水作业;连续方式
stream model 河流模型
stream monitoring 河流监测
stream morphology 河流形态学;河流结构;河貌学
stream mouth 河口
stream net(work) 水系;河网
stream nitrogen and oxygen analysis program (me) 河流氮氧分析程序
stream number 河流数目
stream observation 河道观测
stream obstruction 河道夺流
stream of air 空气流
stream of bit 比特流
stream of business activity 经营活动流程
stream of coordinate 坐标流
stream of cost and benefit 成本和收益流量
stream of electrons 电子流
stream of ice 流冰带;带状流冰;冰流
stream of money 资金流量
stream of people 人流
stream of stress 应力流
stream of traffic 车流
stream of viscousness 黏性流
stream on 在流程中
stream option 流选择
stream order 河流分级;河流等级;水系级别
stream-oriented computation 流式计算
stream-oriented input/output 流式输入输出
stream-oriented transmission 流式传输
stream outfall 河口
stream outlet 河流出口;河口;河道出口
stream oxygen slag 河流氧化性渣
stream pattern 流型;流线谱;河流类型;河道形态;河道类型;河川类型
stream penetrating 喷入射流
stream pier 河流突栈桥;河流防波堤;河流码头;河流桥台
stream piracy 河流袭夺;河流截夺;河道夺流
stream point river point 河流转折点
stream pollution 河流污染;河道污染;河川污染
stream port 河湾;河港
stream power 河流能力;河流功率;水流功率
stream process 河床演变过程;河床形成过程
stream profile 河流纵剖面图;河谷底线;河流纵剖面;河流剖面;河流纵剖面
stream protection 河道保护
stream purification 河流净化作用
stream pusher 内河推轮
stream quality 河流质量
stream reach 河流直段;河流流程;河区
stream reaeration 河流再曝气;河流复氧作用;河流复氧
stream regime(n) 河性;河势;河流状况;河流情态;河流情势;河流情况;河况
stream regulating reservoir 河流调节水库
stream regulation of works 河流整治工程
stream rise 河水上涨;河流涨水
stream robbery 河流袭夺;河流截夺
stream routing 河流洪水演算
stream runoff 河川径流
stream sampling 河流采样
stream sand 河沙
stream sanitation 河川环境卫生
stream scriber 河流刻图仪
stream sea anchor 放下海锚
stream section 河流横剖面;河流横断面;河槽断面
stream sediment 河流沉积物;河流沉积;水系沉积物
stream sedimentation 河流沉积作用
stream sediment sample 水系沉积物样品
stream sediment sampling environment 水系沉积物采样环镜
stream sediment sampling locality 水系沉积物采样位置
stream sediment survey 水系沉积物测量
stream seepage 河川渗漏;河川渗流
stream segment 河流分支;河段
stream self-purification 河流自净机理;河流自净(作用)
stream sheet 流面
stream shingle 扁砾石
streamside 河边
streamside management zone 河边管理带
stream size 河流尺度;河道大小
stream slope 河面坡度;河流坡降;河流坡度;河流陡降;河流比降
stream source area 河源地区;水源地区
stream's oxygen economy 河流中氧的平衡
stream splitter 气流分离器
stream's self-purification 河流自净能力
stream stabilization 河岸稳定
stream stage 河流水位
stream stage ga(u)ging 河道水位测量
stream standard 河川标准
stream station 河流水文站;河道水文站
stream straightening 河道裁弯取直;河道整直
stream stretch 河流伸延;河流伸展;河流伸长
stream structure 河流建筑;河流结构
stream subject to backwater 回水倒灌河
stream subterranean 地下河(流)
stream superstructure 河流的上部结构
stream surface 流面;河流水面

stream surveillance 河流监测
stream survey(ing) 河流调查;河流测量;河道调查;河道测量
stream survey station 河道测量站
stream suspended drift 河流悬浮物质
stream swirl 气流涡流
stream system 河系;河网;河流系统;水系
stream takeoff 跟进起飞
stream temperature 流股温度;气流温度
stream terrace 沿河阶地;河流阶地;河川台地;河成阶地
stream thermal regime 河水温度状况
stream thread 河流方向线;河道主流线
stream tide 河流潮汐
stream time 连续开工时间;连续工作时间
stream tin 砂锡;锡砂;河沙锡
stream-to-stream time 清理与安装时间;上一次停工与下次开工之间的时间
stream toxic analysis 河水毒性分析
stream traction 河流挟运(作用);河流推移(作用)
stream traffic 河流运输;河流交通
stream training 河流导治
stream transmission 流式传输;流传输;数据流传输
stream transmission mode 流传输方式
stream transport(ation) 河流运输;河运;河流搬运(作用)
stream transverse gradient 河流横向坡度
stream trenching 河流下切;河道下切
stream tube 流管
stream tunnel 河流隧道
stream turbidity 河水浊度
stream turbulence 河水紊动;水流紊动
stream underflow 河底潜流
stream unit 流部件
stream upflow 河床渗流
stream valley 河谷
stream variable 川流变量
stream velocity 河流流速;水流速度
stream velocity fluctuation 流速脉动
stream vertical 河道垂线;测流垂线
stream wash 河流冲刷;河流冲积物
stream waste load allocation 河流废负荷分配
stream water 河流水;河水
stream water anomaly 水系水异常
stream water chemistry 河水化学性质
stream water discharge 河水流量
stream water ion chemistry 河水离子化学性质
stream water level 河水水位
stream water purification 河水净化
stream water quality criteria 河水水质基准
stream water quanlity 河水水质
stream water sample 河水样本
stream water sources 河水水源
stream water temperature 河水温度
streamway 河道
streamwise direction 沿水流方向
streamwise flow condition 沿程水流状态;沿程水流条件;沿程水流特性
streamwise flow direction 沿程流向
streamwise friction loss 沿程摩擦损失
streamwise section 沿程水流剖面
streamwise variation 沿程变化
streamwise variation of flow velocity 沿程流速变化
streamwise variation of velocity 流速沿程变化
streamwise water surface slope 沿程水面比降
stream with movable bed 砂质河床流;动床河流
stream with uniform distribution 均匀分布流
stream work 河流工程
streamy 多溪流的
streamy district 多溪区
stream yield 河流产生量
stream zone 河段
street 垫片(钉在履带上以保护路面);街道;马路;大街;非正式交易所
street accident 街道事故
street alignment 街道线路
street appearance 街景
street approach 街道引路
street architectural feature 街头小品
street architecture 沿街建筑;街道建筑
street area 街区
street area ratio 街道面积率;街道面积比
street arrangement 街道布置

street authorization map 街道规划图
street bend 内外接弯头
street bend union 内外接弯头联管节
street block 街坊;街区
street bridge 街道桥梁
street broker 非会员经纪人
street broom 路刷
street building 市区建筑;沿街建筑
street capacity 街道容量
street car 市内电车;电车;有轨电车
street-car crossing 电车道交叉
street-car detector 有轨电车探测器
street-car line[美] 有轨电车线(路);(有轨的)电车线
street-car safety zone 电车安全区
street-car stop 电车站
street-car terminal 电车站点
street-car track 电车轨道
street centerline 街道中(心)线
street cistern 街道水池
street cleaner 清道工(人);扫路车;道路清扫机
street cleaning 街道清扫;清扫街道
street cleaning vehicle 清道车
street cleanser 清洁工(人)
street commercial district 沿街商业区;商业街区
street congestion 街道拥挤
street construction 街道施工;街道建筑
street crossing 街道交叉(口);市区街道路口
street crossing center garden 街心花园
street cross-section 街道横断面
street crown 街道路拱
street crown loading 街道中央乘车
street curb 街道缘石;街道路缘(石);街道侧石;道路侧石;路缘;路牙
street decking 街道路面
street dedication plat 街道地名图
street design 街道设计
street design pattern 街道设计形式
street door 临街门;临街大门
street drain 街道排水沟;街道边沟
street drainage 街道排水
street dust 街道尘埃
street elbow 异径弯头;内外接弯头;弯管接头;异径弯管;带内外螺纹的弯管接头;长臂肘管;异径肘管
street ell 内外接头(管子);L形管接头
street entrance 临街入口
street environment 街道环境
street equipment 道路设备;街道设备
Streeter-Phelps equation 斯特里特—费尔普斯方程
street excavation 街道开挖
street exit 临街出口
street facade 街面;街道外观
street facility 道路公共设施地带
street feeder 街道支线
street floor 街道层
street flush 街道冲洗
street flusher 街道冲洗者;街道冲洗器;洗路机
street flushing demand 浇洒马路用水
street fountain 街道喷水池
street front 沿街建筑立面;沿街屋前空地;街面;街道铺面;屋前空地
street frontage 临街门面
street function 街道功能
street furniture 道路公共设施;街头小品;街道附属设施;街道(建筑)小品
street gas main 街道煤气干管
street gate 街道门
street grade 街道坡度
street grade curb level 街面侧石高程
street greening 街道绿化
street grid 街道方格网;方格式街道网
street gull(e)y 街道排水沟;街道雨水进水口;街道雨水进水井;街道进水口;街道进水井;街道集水沟
street gutter 街沟;路边排水管;路边水沟
street gutter profile 道路边沟纵断面
street hydrant 街道消火栓;街道消防栓
street illumination 街道照明
street improvement 街道改建
street improvement bond 街道维修债券
streeting planting 道路植树
street inlet 街道雨水口;街道进水口;街道侧雨水进口

street intersection 街道交叉(口)
street inventory 街道详录;街道记录
street lamp 路灯;街灯
street lamp stand 路灯柱
street lantern 路灯
street layout 街道网规划
streetlet 小巷(道)
street level 街道水平;街道路面高程;街道路面标高;街道标高
streetlight 路灯;街道照明
street lighting 路灯;街道照明
street lighting column 路灯柱
street lighting luminaire 街道照明设备
street lighting system 路灯系统;街道照明系统
street lighting unit 街道照明单元;街道照明装置
street line 街道边界线;地段线;街道线;道路红线
street litter bin 街道垃圾箱
street loading zone 街上搭车站点(区)
street loans 门市贷款
street main 街道干管;街道干道
street maintenance 街道养护;街道维护
street manhole 街道进入口;街道窨井;街道人孔;街道进入孔;街道检查井
street map 街道图
street marker 道路标志
street market 街道集市
street marking machine 道路标志机械;道路画线机械
street mason 铺路工;道路砖石工
street motorcycle 街道摩托车
street name 街名
street name-plate 街名牌;路名牌
street-name sign 路名牌
street network (指城市中的)道路网;街道网
street noise 街道噪声;街道噪声
street number 门牌号码;街道号数
street observation 街道交通观测
Street of the tombs at pompeii 庞贝坟墓街
street of vortex 漩涡列;漩涡道;涡列;涡道
street opening 堤岸门洞;城墙门洞;街道门洞
street orderly 街道清洁工(英国俗称);清道工(人)
street paper 短期支票
street park 路边公园
street passageway 过街通道
street pattern 街道形式;街道模式
street pavement 街道铺面
street paving 街道铺砌;街道铺面
street picture 街景
street pipeline 街道管线
street pipe system 街道管网;街道管道系(统)
street plan 街道(平面)图
street planning 街道规划
street planting 街道绿化;人行道树种植;道旁栽树
street projection 房屋突出在街道部分
street property line 街道地界线
street railroad 市区铁路;有轨电车道;市区铁路
street railroad track 市区有轨电车道
street railway 有轨电车线(路);有轨电车道;市区铁路;城市铁路
street reducer 单向联轴节的减速器
street refuge 街道安全岛;街道安全带
street refuse 街道垃圾;道路清扫垃圾
street repair 街道整修;街道修理
street roller 压路机;路碾
street running section 街面运行线路段
street run-off 路面径流
streetscape 街景
street scavenging 街道清扫
street scenery from neighbour 街道借景
street sclerosis 街道(停车)困难
street sewer 街道污水渠;街道排水管
street shadow area 街影投射面积
street sign 路牌;路名牌;街道标志
street sprinkler 街道洒水机;街道洒水车;洒水车
street sprinkling 街道洒水;街道洒扫
street surfacing 街道铺面
street survey 街道交通调查;街道交通量观测
street sweeper 清道工(人);扫路机;扫路车;扫街机
street sweeper flusher 街道清洗机
street sweeper with washer 洗路车
street sweeping 街道清扫;街道垃圾
street sweep(ing)machine 街道清扫机;扫街机
street system 街道系统;道路系统

street tee 直管外螺纹三通接头
street track 城市铁路
street traffic 市内交通;街道交通(量)
street traffic control 街道交通控制;街道交通管制
street traffic control light 街道交通管制灯;交通管制灯;交通信号灯
street traffic light 街道交通灯
street traffic marking 街道交通标线;街道交通标志
street traffic sign 街道交通标志
street traffic signal 街道交通信号
street traffic survey 街道交通量观测;市街运量观测
street traffic system 街道交通系统
street tramway 市区电车道
street trap 街道气阀门;街道水阀门;街道陷阱
street travel pattern 道路行驶制度;道路行驶方式
street tree 路树;行道树;街道树
street tunnel 街道隧道;街底隧道
street utilities 街道公共设施;道路公共设施
street value 街道评价
street wall 沿街墙
street wash 街面冲洗;街道污水
street washer 洗街机;街道冲洗车
street washing 冲洗街道用水;街道冲洗物;街道冲洗
street waste 街道污水
street watering 街道洒水
street watering hydrant 街道洒水给水栓
street watering motorcar 街道洒水汽车
street watering standpost 街道取水柱
street widening 街道拓宽;街道扩宽;街道加宽
street width 街道宽度
stregth property 强度性质
strelkinite 钒钠铀矿
stremmatograph 钢轨应力自记仪;道轨受压纵向应力自记仪;道床压力自记仪
strenght at high temperature 高温强度
strengite 红磷铁矿
strength 力量;韧力;强度
strength age relationship 强度—龄期关系
strength analysis 强度分析
strength and elongation 强伸度
strength and sedimentation test 强度和酸度试验
strength anisotropy 强度异向
strength anomaly 强度异常
strength assurance coefficient 强度安全系数
strength asymptote 强度渐近线
strength at pulsating load 脉冲载荷强度
strength beam 宽板横梁;桁架梁;强梁
strength bulkhead 加劲隔壁;耐压舱壁
strength calculation 强度计算
strength capability 强度性能
strength ceiling 强度极限;(轻集料/骨料混凝土的)强度上限值
strength character 强度性能
strength characteristic 强度特征值;强度特性
strength character of rock 岩石的强度特征
strength check(ing) 强度检验;强度校核
strength class 强度分级
strength classification 强度分类
strength classification of concrete 混凝土按强度分等;混凝土按强度分类;混凝土的强度分级
strength component 强度组分
strength concept in soil mechanics 土力学中强度概念
strength condition 强度条件
strength constant 强度常数
strength contour 等强度线
strength contour graph 等强度线图
strength deck 强力甲板;强劲甲板
strength decrease 强度减小;强度减弱
strength-deformation characteristic 强度—变形特性
strength-density ratio 强度—密度比
strength depending on shape 疲劳强度
strength design criterion 强度设计准则
strength design method 强度设计法
strength development process 强度发展过程
strength distribution curve 强度分布曲线
strength draft 强度吃水
strength-duration curve 强度时间曲线
strength efficiency 强力利用率
strength elongation diagram 强力伸长曲线图
strengthen 增强
strengthened beam 加固梁

strengthened blasting crater 加强爆破漏斗
strengthened bulkhead 加强舱壁
strengthened by cold working 冷作强化
strengthened concrete 加筋混凝土
strengthened condition 强化状态
strengthened cone cut 加强锥形
strengthened dispersion alloy 弥散强化合金
strengthened edge 加厚边;加固边缘
strengthened element 加固构件
strengthened glass 增强玻璃
strengthened microelectrolysis 强化微电解法
strengthened section for tunnel 隧道加强长度
strengthened structure 加强结构
strengthened throw blasting charge 加强抛掷药包
strengthened truss 加强桁梁
strengthened wedge-shape cut 加强楔形
strengthen environmental protection 加强环境保护
strengthener 助力器
strengthen glass 强化玻璃
strengthening 增强;加强;加固;强化;补强
strengthening action 补强作用
strengthening activated sludge process 强化活性污泥法
strengthening agent 补强剂
strengthening and toughening 强化与增韧
strengthening beam 加固梁
strengthening building 建筑物加固
strengthening business management 强化企业管理
strengthening corner 加动角;加固角
strengthening course 补强层
strengthening effect 强化效应
strengthening film 增强膜
strengthening grout(ing) 补强灌浆
strengthening layer 强化层;补强层【道】
strengthening measure 加强措施;加固措施
strengthening mechanism 加固机构;强化机理
strengthening method 加固方法
strengthening net 加强网
strengthening of dam 坝的加固
strengthening of dike 坝的加固
strengthening of ebb interval 落潮流间隙
strengthening of foundation 基础加固
strengthening of soil 土壤加固
strengthening of steel bridge 钢桥加固
strengthening of structure 加固结构
strengthening of timber support 补修支撑
strengthening of welded seam 焊缝补强
strengthening plate 加固层;加固板
strengthening procedure 加固方法
strengthening process 强化工艺
strengthening rib 加固肋;加劲条
strengthening ring 加强环
strengthening soil 加固土
strengthening stability 强化稳定性
strengthening stud 加强螺栓
strengthening works 加固工程
strengthenization theorem 强化定理
strengthen of seismic activity 地震活动增强
strengthen pollution control 加大污染治理力度
strengthen the cooperation between the railway and shippers 加强路内外协作
strengthen the greening of the city 加强城市绿化
strength envelope 强度包(络)线
strength factor 强度因子;强度指数;强度系数
strength failure of track 线路松动
strength forecast 强度预测
strength freeboard 强度干舷
strength gain 强度增益;强度增长
strength-gaining time 强度增长时间
strength grade 强度等级;强度标号
strength grade of concrete 混凝土强度等级
strength grading of brick 砖的强度等级
strength grading of clinker 熟料标号
strength grading of mortar 砂浆强度等级
strength grading of stone 石材强度等级
strength group 强度分类
strength imparting material 加固物料
strength improver 强度增进剂
strength in bending 抗弯强度;挠曲强度
strength increase 强度增长
strength-increasing 强度增长
strength-increasing time 强度增长时间

strength in decay 强度衰减
strength in dry state 干燥状态强度
strength in situ 现场混凝土强度
strength inspection of soil 土体强度检查
strength in static bending 抗静弯强度;木材的抗弯极限强度
strength in undisturbed state 未扰动状态的强度
strength investigation 强度研究
strength isotropy 强度同向
strength joint 坚固连接;加顶接头
strength level 强度水平;强度级
strength limit 强度极限
strength loss 强度损失
strength margin 强度极限
strength-maturity factor 强度成熟系数
strength member 重要零件;强力构件
strength-member optic(al) cable 加强构件光缆
strength modulus 强度模数
strength of acid 酸强度
strength of attack 扑火力量
strength of beam 梁的强度
strength of brake current 闸电流强度
strength of brazed joint 钎焊连接强度
strength of breaking 抗震强度
strength of cement 水泥强度;水泥强度等级
strength of cement mortar 水泥砂浆强度
strength of compression 抗压强度
strength of connection 连接强度
strength of current 最大潮流;电流强度;潮流最大速度
strength of dependence 相依强度
strength of diamond 金刚石强度
strength of draft 通风强度
strength of dried body 干坯强度
strength of ebb 最大退潮流(速);最大落潮流(流速);落潮流速
strength of ebb interval 最大落潮流间隙
strength of electric(al) field 电(磁)场强度
strength of explosive 爆力
strength of extension 抗拉强度
strength of field 场强
strength of figure 图形强度;图像强度
strength of flange 法兰强度
strength of flexure 抗弯强度;弯曲强度
strength of flood 最大涨潮流;涨潮流速
strength of flood interval 最大涨潮流间隙
strength of fluid 液体浓度
strength of forklift pockets 叉槽强度
strength of fracture 断裂强度
strength of ice 冰强度
strength of illumination 照明强度;光照强度
strength of joint 连接强度;接头强度
strength of level 能级强度
strength of magnet 磁极强度
strength of magnetic field 磁场强度
strength of materials 连接强度;材料强度(学);材料力学(强度)
strength of moment 稳定力矩
strength of mud and sand flow 海底泥沙流强度
strength of pavement 路面强度
strength of pole 磁极强度
strength of reference 引用的强度
strength of repeated loading 重复荷载强度
strength of resonance 共振强度
strength of rock 岩石强度
strength of rock mass 岩体强度
strength of roof 箱顶强度(集装箱)
strength of rupture 折裂强度;断裂强度
strength of salt return 返扯强度
strength of set(ting) 凝固强度
strength of sewage 污水强度;污水浓度
strength of shearing 抗剪强度
strength of side walls 侧壁强度
strength of soil 土的强度
strength of solution 溶液浓度
strength of sound source 声源强度
strength of structure 结构强度;结构力学
strength of testing 试验强度
strength of the end wall 端壁强度
strength of thermal force 热力强度
strength of thermal force within the combustion chamber 燃烧室空间热力强度
strength of triangulation 三角网强度
strength of virgin glass fibre 新生态玻璃纤维强度

strength of vortex 旋度;涡流强度
strength of wall 墙体强度
strength of wastewater 废水强度
strength of welding seam 焊缝强度
strength of wind 风力
strength of wind-sand current 风沙流强度
strength of wood 木材强度
strength parameter 强度参数
strength per unit area 单位面积强度
strength power 强度功能
strength prediction 强度预计
strength pressure 强度计算压力
strength problem 强度问题
strength progress 强度递增
strength property 强度性质;强度特性
strength quality 强度性质
strength range 强度范围
strength rating 强度定额
strength ratio 强度比
strength recovery 强度恢复
strength reduction 强度削减;强度衰减;强度降低
strength reduction factor 强度降低系数
strength reduction pass 强度削减扫描
strength reduction processor 强度削减处理程序
strength regain 强度恢复
strength requirement 强度要求
strength reserves 强度储备
strength retention 残留强度
strength retention rate 强度保留率
strength retrogression 强度退化;强度衰退;强度递减
strength rivet 强力铆钉
strength riveting 强度铆接
strength safety coefficient 强度安全系数
strength tensioning 张力补偿
strength test 强度试验
strength tester 强度试验器;强度试验机;强度试验仪
strength-testing machine 强度试验机
strength test of compression 抗压强度试验
strength test of structural member 构件强度检验
strength theory 强度理论
strength-to-weight 强度一重量
strength-to-weight ratio 比强(度);强度一重量比
strength under peripheral pressure 围压强度;三向压缩强度;侧限抗压强度
strength under shock 抗震强度;冲击强度
strength under sustained load 长期荷载下强度
strength value 强度值;强度数值
strength-weight 强度一重量
strength-weight characteristics 功率重量特性
strength-weight ratio 强度对重量比;强重比;强度一重量比
strength weld 高强度焊接;强力焊缝;强固焊缝;受载焊缝;传力焊缝;承载焊缝
strength welding 焊接强度
strength-wise 强度方式
strenuous exercise 剧烈运动
strenuous test 强化试验
strenuous vibration 剧烈振动
strephotome 螺钻形刀
Streptococcus foetidus 恶臭球菌
stres issued book 库存物资发出簿
stress 应力;胁强;主要应力;强调;受力(状态)
stress absorbability 应力吸收能力
stress absorbing membrane 应力吸收薄膜
stress accommodation 张装夹具;拉装装置;接紧装置
stress accumulation 应力累积;应力集中;应力积蓄;应力积累;应力积聚
stress activated model 应力活化模型
stress activative flow 应力活化流动
stress adjustment 电压调整;应力调整;应力调节
stress ag(e)ing 消除应力时效;应力时效
stress alleviating membrane 应力缓冲薄膜
stress alleviating membrane interlayer 应力缓冲薄膜中间层
stress alternation 应力交替;应力循环;应力交变;应力变换;交变应力;交变负荷
stress amplitude 应力振幅;应力幅度;应力幅;应力幅幅
stress amplitude ratio 应力循环脉动系数
stress analog(ue) 应力相似
stress analogy 应力模拟
stress analysis 应力分析

stress and strain 应力与应变
stress and strain fields 应力和应变场
stress and strength contrast method of slope foot 坡脚应力与强度对比法
stress anisotropy 应力各向异性
stress annealing 消除应力退火
stress application 加负荷;加力
stress approximation distribution 应力近似分布
stress area 应力面积
stress around opening 孔口应力
stress at-break 断裂应力
stress at transfer 应力传递;传递应力
stress axis 应力轴(线)
stress balancing 应力平衡
stress bearing layer 强度层
stress behavio(u)r 应力形状;应力特性
stress behavio(u)r in natural ground 地层应力状态
stress beyond yield strength 超过屈服强度的应力
stress birefringence 应力双折射;受力双折射
stress block 应力区
stress bolt 拉力螺栓
stress boundary condition 应力边界条件;边缘应力条件
stress-breaker 应力中断器
stress broadening 应力增宽
stress buffer 应力缓冲器
stress build up 应力聚集
stress bulk 应力泡
stress calculation 应力计算
stress case 应力情况
stress change 应力变化
stress check 应力验算;应力校核
stress check calculation 应力检验计算
stress chord 受力弦杆
stress circle 应力圆
stress circulation 应力循环
stress coat 应力涂料;应力涂层;示力涂料;脆性漆涂层
stress coating method 应力涂料法
stress compensation 应力补偿
stress component 应力分量;分应力
stress concentration 应力集中
stress concentration area 应力集中区
stress concentration effect 应力集中效应
stress concentration factor 应力集中因子;应力集中因素;应力集中因数;应力集中系数
stress concentration point 应力集中点
stress condition 应力状态
stress cone 应力锥
stress contour 应力等值线;等主应力线
stress control 应力控制
stress controlled 应力控制的
stress controlled direct shear apparatus 应力控制式直(接)剪(力)仪
stress controlled load test 应力控制荷载试验
stress controller 应力控制器
stress-control winding 可调脉冲电压绕组
stress cord 应力条纹
stress corrosion 冷拔钢丝腐蚀;应力侵蚀;应力腐蚀;金属超应力引起的腐蚀
stress corrosion cracking 应力腐蚀破裂;应力腐蚀裂缝;应力侵蚀开裂;应力腐蚀裂纹;应力腐蚀开裂;应力及腐蚀引起的开裂
stress corrosion cracking test 应力腐蚀裂纹试验
stress corrosion failure 应力腐蚀破坏
stress corrosion fracture 应力腐蚀断裂
stress corrosion test 应力腐蚀试验
stress couple 应力偶
stress crack(ing) 应力裂纹;应力开裂;应力裂缝;应力断裂
stress crazing 应力银纹
stress crystallization 应力结晶
stress-current analogy 应力—电流模拟
stress curve 应力曲线
stress cycle 应力周期;应力循环
stress-cycle diagram 应力循环图;疲劳曲线
stress cycling 应力循环
stress decay 应力松弛;应力衰减
stress decline 应力下降
stress decrease 应力减小
stress deflection chart 垂度—应力曲线
stress deflection diagram 应力—挠度曲线图
stress deformation 应力形变;应力变形

stress-deformation characteristic 应力变形特性
stress-deformation curve 应力形变曲线
stress-deformation diagram 应力—应变图;应力变形图
stress-deformation modulus 形变模量
stress degree of freedom 应力自由度
stress-dependence 应力依赖性;应力依存性;对应力的依存性
stress-dependent 随应力而变的
stress dependent behavio(u)r 应力依从性质
stress dependent moduli 应力影响模量;应力相关模量
stress detector 应力仪;应力计
stress determination 应力的确定
stress deviation 应力偏差
stress deviator 应力偏量
stress diagram 应力图;力线图
stress difference 应力差
stress-difference line 应力差线
stress dilatancy theory 应力切变膨胀理论
stress-director line 应力指示线;应力方向线
stress dispersal beneath footing 基底应力扩散
stress dispersion 应力扩散
stress dispersion capacity 应力扩散能力
stress-displacement relation 应力变形关系
stress displacement vector 应力位移矢量
stress dissipation 应力散失
stress distribution 应力分布
stress distribution approximation 应力分布近似法
stress distribution diagram 应力分布图
stress distribution in earth mass 土中应力分布
stress distribution property 应力分布性质
stress division 应力分担(指钢筋与混凝土在钢筋混凝土中分担应力);应力分配
stress drop 应力(下)降
stress due to driving 打桩时引起的应力
stress due to handling 吊装时引起的应力
stress due to long-term loading 长期荷载应力
stress due to long-time loading 长期荷载应力
stress due to temperature change 温度应力
stress duration test 疲劳试验
stress dyadic 应力并向量;应力并矢式
stressed clay 受应力的黏土
stressed collar 剪力环
stressed concrete design system 预加应力混凝土设计体系
stressed connection 受应力的接头;预应力接合
stressed covering 受力蒙皮
stressed member 受力构件;受力杆件
stressed membrane 受应力的薄膜
stressed plate 应力板
stressed polymer 应力聚合物
stressed sandwich panel 受力夹层板
stressed skin 受力外包层
stressed-skin body 蒙皮车体
stressed-skin construction 应力外皮结构;加载式结构;蒙皮(受力式)结构;薄壳建筑;表层应力结构;承力表层式结构
stressed-skin panel 张紧蒙皮板;外层受力板
stressed-skin roof deck 预应力面层屋面板
stressed-skin structure 承力蒙皮结构
stressed state analysis 应力状态分析
stressed system 应力体系
stressed zone 应力区
stress effect 应力影响;应力效应
stress element 应力单元
stress ellipse 应力椭圆
stress ellipsoid 应力椭圆体;应力椭球(面)
stress-elongation curve 应力伸长曲线
stress-elongation ratio 应力伸长比
stress-endurance curve 疲劳曲线
stress-endurance diagram (材料疲劳试验的)应力反复图;应力包络线图
stress energy 应力能
stress envelope 应力分布图;应力包络线
stress equalizing 应力均匀化
stress equation 应力方程
stress equilibrium 应力平衡
stresser 应激子
stress erosion limiting intensity factor 应力腐蚀界限强度因子
stresses due to friction 摩擦应力
stresses in pile 桩中应力
stresses of traffic 交通压力

stress estimation 应力估计
stress etching 应力腐蚀
stress exponent 应力指数
stress factor 应力系数
stress factor of corner-points 角点应力系数
stress fatigue 应力疲劳
stress field 应力场
stress field in the earth's crust 地应力场
stress flow 应力流
stress flow around an opening 孔口周围的应力流
stress flow around a reinforced opening 加强洞口周围的应力流
stress fluctuation 应力起伏;应力脉动;应力波动;应力变动;交变应力
stress for temporary loading 短期荷载应力
stress-free 无应力
stress-free annealed 退火后无应力
stress-free annealing 无应力退火
stress-free initial strain 应力释放初始应变
stress-free pore 无应力的孔隙;无应力的细孔
stress-free rail temperature 零应力轨温
stress-free strain meter 无应力应变计
stress-free temperature 锁定轨温
stress freezing 应力冻结(光弹试验)
stress-freezing method 应力冻结法;冻结应力法
stress-freezing oven 应力冻结烘箱
stress from volume change 体积变化(引起)的应力
stress frozen method 应力冻结法;冻结应力法
stressful region 应力区
stress function 应力函数
stress function element 应力函数单元
stress galvanometer 应力检流计
stress ga(u)ge 应力仪;应力计
stress generation 应力产生
stress grade 应力等级
stress graded 强度分级材;按强度分级(木材)
stress-graded lumber 应力分级木材
stress-graded timber 按应力分段的木材;按强度分级的木材;应力分级木材
stress grade of lumber 木材应力等级
stress grade of timber 耐应力木材
stress gradient 应力梯度
stress grading 应力分级
stress grading of timber 木材的应力分级
stress guide 应力导
stress hardening 应力硬化
stress heating 应力加热
stress history 应力(历)史;应力变化历程
stress holo-interferometry 应力全息干涉测量
stress in a rock mass 岩体中应力
stress increase 应力增加;应力增长
stress increment 应力增量
stress index method 应力指数法
stress indicating coating 应力指示涂层
stress-induce 应力诱导
stress-induced anisotropy 应力诱导的各向异性;应力引起的各向异性
stress-induced birefringence 应力双折射
stress-induced crystallization 应力诱发结晶
stress-induced growth 应力诱导生长
stress-induced order locking of dislocation 应力诱导有序位错固定
stress-induced orientation 应力诱导取向
stress-induced phase transformation 应力诱导相变
stress-inducing factor (路面板的)应力诱导系数;应力诱导系数
stress in earth crust 地壳应力
stress in earth mass 土中应力
stress in extreme fibre 外缘纤维应力
stress influence factor 应力影响系数
stress influence function 应力影响函数
stressing 施加应力
stressing abutment 锚墩;锚碇桩
stressing area 受应力面积;张拉区
stressing bar 加力杆;预应力钢筋
stressing bed 张拉端
stressing block 加力轮轮;加力滑车
stressing cable 预应力钢缆;预应力钢索;预应力钢丝绳
stressing cement 自应力水泥;自膨胀水泥
stressing device 预应力设备
stressing distance 预加应力长度;预加应力距离

stressing duct 预应力管道
stressing element 预应力构件
stressing end 预加应力端;张拉端
stressing equipment 施加预加应力设备
stressing factory 预应力工厂
stressing force 预加应力的力量;预张力
stressing gang 预应力班组
stressing head 预应力(混凝土)锚具;张拉夹具套;试样夹具套
stressing jack 预应力用千斤顶;张拉千斤顶
stress in glass marble 玻璃球应力
stressing line 预应力工作线
stressing load 预加的应力荷载
stressing moment 预加应力力矩
stressing mo(u)ld 预应力模具;张拉模具
stressing of ground 地基的受压;地基的加压
stressing order 预加应力顺序
stressing process 预加应力的步骤;预加应力的方法;张拉过程
stressing processor 张拉次序
stressing reinforcement 预应力钢筋
stressing rod 预应力钢筋
stressing steel 预应力钢筋
stressing strand 预应力股绳;预应力钢材;预应力钢绞线
stressing tendon 受应力钢筋腱;预应力束;力筋
stressing value 预应力值
stressing wedge 预应力楔片;预应力楔块
stressing wire 预应力钢丝
stressing yard 预应力工场
stressing zone 预应力区
stress in ligaments 管口带中的应力
stress in lining 衬砌应力
stress intensity 单位应力;应力强度;应力大小
stress intensity factor 应力强度因子;应力密度因子;应力密度系数;应力(场)强度系数
stress intensity factor 应力强化因子
stress intensity factor range 应力强度因子范围
stress intensity range under cyclical loading 周期荷载作用下应力强度范围
stress intensity value 应力强度值
stress interval 应力中断期间;应力区间
stress in the earth's crust 地应力
stress in the semi-infinite half-space 半无限半空间中的应力
stress in thin film 膜层应力
stress in three-dimensions 三维应力
stress invariant 应力不变量
stress isobar 等应力线
stress isopleth 应力等值线
stress jump 应力跳动
stress layer 应力层
stressless 无应力的
stress level 应力级位;应力水平;应力级;应力等级
stress liberation 放散温度应力
stress limit 应力极限;极限应力
stress line 应力线
stress loss 应力损失
stress loss due to friction 摩阻应力损失
stress management 压力管理
stress map 应力图形;应力图
stress mark 应力标记;应力斑痕
stress matrix 应力矩阵
stress maximum 应力最大值;应力峰(值)
stress measurement 应力测试;应力测量
stress measuring apparatus 应力测试装置;应力测量装置
stress measuring device 应力测试装置;应力测量装置
stress measuring system 应力检测系统
stress measuring technique 应力测试技术;应力测量技术
stress measuring unit 应力测试装置;应力测量装置
stress member 受力构件;受力杆件
stress meter 应力仪;应力计
stress meter for reinforcement 钢筋计
stress method 应力法
stress-mineral 应力矿物
stress model 应力模型
stress modelling 应力模拟
stress moment 应力力矩
stress-number curve 应力计数曲线;(疲劳试验中的)应力次数曲线
stress-number diagram of stress 应力循环次数图

stress of circumstances 受环境所迫
stress of conductor 导线应力
stress of fluidity 流体的应力;流动性应力
stress of primary rock 原岩应力
stress of sewer pipe 管道应力
stress of soil moisture 土壤水分应力
stress of tangential direction 切向应力
stress of the first kind 第一类应力
stress of weather 气候应力;受恶劣天气所迫;风暴天气
stress of wind 风应力
stressometer 应力仪;应力计;胁强计;机械应力测量器
stress on principal plane 主平面上的应力
stress-optic(al) coefficient 光弹系数
stress-optic(al) constant 光学力学常数;应力光学常数
stress-optic(al) law 应力光学定律
stress overshoot 应力超越
stress path 应力过程;应力途径
stress path method 应力路径法
stress path model 应力路径模型
stress pattern 应力图形;应力形式;应力情况;应力轨迹图;应力分布形式;应力斑
stress peak 应力峰(值)
stress peening 喷丸强化;喷砂强化
stress-penetration curve 应力贯入曲线;荷载—贯入曲线
stress-penetration curve for CBR test CBR 试验的应力贯入曲线
stress plane 应力平面;应力面
stress plate 应力板
stress plotting by computer 计算机应力标绘
stress point 应力点
stress problem 应力问题
stress-producing force 引起应力的外力
stress profile 应力剖面图
stress program(me) 应力程序
stress propagation 应力传播
stress quadric 应力二次式;应力二次曲面
stress raiser 应力上升处;应力集中源;应力集中因数;应力集中区;应力集中处;局部应力集中
stress raising 应力上升
stress range 应力幅度;应力范围
stress rate 应力速度;应力率
stress ratio 应力比;疲劳试验应力比
stress ratio method 应力比法
stress ratio of pile to soil 桩土应力比
stress rearrangement 应力调整
stress recorder 应力(记录)仪;自记应力仪
stress recovery method 应力恢复法;地应力解除法
stress redistribution 应力重分布;应力再分配;应力再分布
stress reduction 应力减少;应力减低
stress reduction factor 应力下降因数;应力折减系数;应力减少系数
stress reflection coefficient 应力反射系数
stress reinforcement 应力筋
stress relaxation 应力松弛;应力缓冲
stress relaxation response 应力松弛反应
stress relaxation under constant load 恒定荷载下应力松弛
stress relaxometer 应力松弛试验机;应力松弛计
stress release 应力释放
stress releasing 应力消除
stress releasing borehole 应力解除钻孔
stress releasing cut 应力解除切口
stress-relief 应力消除;应力解除;应力减除;应力放散(焊接长钢轨)
stress relief annealed 退火的应力释放
stress-relief annealing 消除内残余应力退火;消应回火;弛力退火
stress-relief annealing crack 消除应力处理裂纹
stress-relief blast 卸载爆破;应力解除爆破
stress-relief course 应力消减层(路面)
stress-relief crack(ing) 再热裂纹;应力消除裂纹
stress-relief furnace 退火炉
stress-relief grooves 卸载槽
stress-relief heating 应力消除加热
stress-relief heat treatment 消除应力热处理;消除应力的热处理
stress-relief method 地应力解除法;应力解除法
stress-relief slot 应力解除切槽
stress-relief temperature 消除应力加热温度

stress-relief tempering 消应回火
stress-relief test 应力消除试验
stress-relief treatment 消应处理；应力消除热处理
stress-relieved 应力消失
stress-relieved wire 低温退火钢丝；消除应力钢丝；应力放松钢线
stress-relieving 消除应力的热处理；应力消除热处理；应力消除；应力释放；应力解除；去应力；泄力
stress-relieving annealing furnace 低温退火炉
stress-relieving blasting 预裂爆破法；预裂爆破；解除应力爆破
stress-relieving by beat treatment 热处理消除应力
stress-relieving by local heating 局部加热消除应力
stress-relieving cavity 应力减除穴
stress-relieving furnace 退火炉；消除(残余)应力退火炉；应力解除炉
stress-relieving interlayer 应力消减层(路面)；应力消除层
stress-relieving process 内应力消除过程
stress removal 应力移除；应力消除
stress reorientation 应力重取向
stress repetition 应力反复
stress repletion 应力饱和
stress restoration 应力恢复
stress restoration method 应力恢复法
stress resultant 应力综合；应力合力
stress resulting from anthropogenic sources 来自人类活动的应力
stress reversal 应力反向；应力反复；应力变向
stress ribbon bridge 悬带桥；悬板桥
stress ring 应力环
stress riser 应力梯级；应力梯度
stress rivet 耐力铆钉
stress rupture 应力破裂；应力破断
stress rupture curve 应力断裂曲线
stress rupture strength 应力破裂强度
stress rupture test 应力破坏试验；持久试验
stress rupture testing machine 持久强度试验机
stress-rupture testing of tube 管子的应力破裂试验
stress section 应力检验片
stress sensitive 应力灵敏
stress shear dilatancy theory 应力剪胀理论
stress sheet 应力图；应力计算表；应力表
stress similitude 应力相似
stress singularity 应力奇点
stress-softening 应力软化
stress-softening effect 应力软化效应
stress solid element 应力立体单元
stress space 应力空间
stress spectrum 应力谱
stress stability 应力稳定性
stress state 应力状态；紧张状态；受力状态；受力状况
stress state analysis 受力状态分析
stress step 应力阶跃
stress-strain 应力一应变
stress-strain behavio(u)r 应力一应变状态；应力一应变性状；应力一应变性能
stress-strain characteristics 应力一应变特性
stress-strain characteristics of rock 岩石应力一应变特性
stress-strain circle 应力一应变圆
stress-strain compatibility 应力一应变相容分析
stress-strain condition 应力一应变状态
stress-strain curve 应力一应变图；应力一应变曲线；变形曲线
stress-strain diagram 应力一应变图解；应力一应变图；应力一应变曲线
stress-strain diagram of concrete 混凝土的应力一应变图
stress-strain equation 应力一应变方程
stress-strain function 应力一应变函数
stress-strain ga(u)ge 应力一应变仪
stress-strain-history relation 应力史应变史关系式
stress-strain loop 应力一应变回线
stress-strain matrix 应力一应变矩阵
stress-strain measurement 应力应变测量
stress-strain modulus 形变模量；应力一应变模量
stress-strain plot 应力一应变图
stress-strain property 应力一应变性质；应力一应变特性
stress-strain ratio 应力一应变比
stress-strain relation(ship) 应力一应变关系
stress-strain sample 应力一应变试样
stress-strain state 应力一应变状态

stress-strain state of soil 土的应力一应变状态
stress-strain-temperature space 应力一应变温度空间
stress-strain tester 应力一应变试验
stress-strain time behavio(u)r 应力一应变时间性状
stress-strain time relation 应力一应变时间关系
stress-strain time response 应力一应变时间反应
stress-strain-volume change 应力一应变一体积变化
stress strength 应力强度
stress strength ratio 应力强度比
stress surface 受力面
stress system 应力系统；应力分级法
stress-temperature characteristic 应力温度特性
stress tempering 应力回火；应力调质；加载回火处理
stress tensor 应力张量
stress tensor ga(u)ge 应力张量计
stress test 应力试验
stress testing for pressure vessel 受压容器应力测试
stress theory 应力理论
stress thermo-optic(al)coefficient 应力热光系数
stress thickening 应力加厚
stress threshold 应力阈值
stress time superposition 应力时间叠加作用
stress together 应力产生在一起
stress to rupture 断裂应力
stress to strain ratio 应力一应变比
stress to strength ratio 应力与强度比
stress trajectory 应力图；应力轨迹
stress transfer 应力传递
stress transmission 应力传递
stress transmission coefficient 应力传递系数
stress triangle 应力三角形
stress under compression 压缩应力；压缩应力
stress under impact 冲击应力；冲击应力
stress value 应力值
stress variation 应力变化
stress vector 应力向量；应力矢量
stress-voltage analogy 应力一电压模拟
stress wave 应力波形；应力波；压缩波
stress wave emission 应力波发射
stress wave equation method 应力波方程法
stress-wave theory 应力波理论
stress wave velocity 应力波速度
stress weld 耐力焊接
stress wire 应力筋；应力钢丝
stress work 应力功
stress zone 应力带
stretch 影像拉长；直规，直尺；展宽；路段；拉紧，拉长；加宽；全帆行驶；伸展；伸出；伸长；地段
stretchability 拉伸性
stretch bath 拉伸浴；塑化浴
stretch bender 拉弯机
stretch bending 拉弯；拉伸弯曲
stretch-blow mo(u)lding machine 拉坯吹塑机
stretch breaking machine 牵切机
stretch circuit 脉冲展宽电路
stretch-draw dies 张拉成型模
stretch-draw forming machine 张拉成型机
stretch-draw press 张拉成型压力机
stretched 拉伸的
stretched beam 受拉梁
stretched connection 张拉连接；预应力接头
stretched diaphragm 张紧膜片
stretched fiber 受拉纤维
stretched in pairs 成对张拉的
stretched length 伸直长度
stretched out structure 伸展结构
stretched pebbles 伸长卵石
stretched plate 拉伸板
stretched wire 引线线
stretched-wire joint 张线接合
stretched-wire method 张线法
stretched wire method of aligning 拉线找中法
stretch elongation 拉伸长度
stretcher 引张器；延伸器；走砖；展宽器；展幅机；拉伸器；拉伸机；顺砖；顺砌；伸长器；担架；薄板矫直机
stretcher bar 尖轨连(接)杆；伸张杆；伸缩支承杆；道岔拉杆
stretcher block 顺砌混凝土砌块；顺砌砌块(混凝土)
stretcher bond 满条砌法；条砌；顺砌砌合

stretcher-bond type rowlock paving 顺砖铺砌式竖砖路面
stretcher brick 顺砖；顺砌砖
stretcher course 墙的顺砖层；顺砖(砌)层
stretcher duster 担架式喷粉机
stretcher face 顺面(砖的长侧边)
stretch(er)forming 拉伸成型(法)
stretcher jack (深井泵拉杆的)拉紧器
stretcher leveler 拉伸矫直
stretcher level(l)ing 拉伸矫直
stretcher masonry bond 顺砖坵工砌合
stretcher-mounted sprayer 担架式喷雾器
stretcher of brick 砖侧面；砖的顺边
stretcher paving 顺砖铺地
stretcher store 张拉器储存室
stretcher straightening 拉伸矫直
stretcher strain 拉伸变形；滑移线
stretcher strain marking 拉伸变形纹；(轧材矫直时的)滑移线痕
stretch fault 引伸冲断层；椭曳冲断层
stretch flange 拉伸凸缘
stretch flanging 拉伸摺缘；拉伸翻边
stretch-flanging die 拉伸翻边模
stretch former 延伸压力机；拉伸成型机
stretch forming 延伸造型；张拉成型法
stretch forming machine 拉伸成型机
stretch girder 受拉大梁
stretching 延伸；张拉；锻伸；拉直；拉伸；摊平；伸缩
stretching apparatus 张拉设备；拉直器；拉尺器
stretching area 张拉面积；张拉场
stretching bar 张拉杆
stretching bed 张拉台座；伸展台座；张拉座；张拉台
stretching block 张拉滑轮
stretching bolt 拉紧螺栓
stretching bond 满条砌法；顺砖砌合
stretching cable 张拉缆索；预应力钢绞索
stretching course 顺砌砖层；顺砖层
stretching crack 伸缝
stretching device 张拉设备；张紧装置；拉紧装置；拉紧夹具；紧线器
stretching die 拉伸模
stretching direction 伸长方向
stretching distance 拉伸长度；拉伸距离
stretching distance increment 拉伸长度增量；拉伸距离增量
stretching effect 拉伸效应
stretching element 张拉元件；预应力构件
stretching equipment 张拉设备；拉伸装置
stretching factory 张拉工厂；预应力厂
stretching force 张力；抗张力；抗拉力
stretching hammer 锻扁锤
stretching insulator 耐拉绝缘子
stretching jack 张拉千斤顶
stretching lineation 拉长线理
stretching loss 张拉损失；预应力损失
stretching machine 拉幅机；撑鞋机
stretching method 张拉方法；张拉法
stretching mode crack 拉伸型裂纹
stretching motion 伸缩运动
stretching order 张拉程序
stretching palnt 张拉工厂
stretching piece 支撑杆；连杆；拉条
stretching plant 张拉设备
stretching pole 拉杆
stretching press 拉延压机；薄板伸张压力机
stretching process 张拉过程
stretching pulley 张紧轮；皮带紧轮
stretching reinforcement 张拉钢筋；预应力钢筋
stretching resistance 抗张力；抗拉力
stretching rod 张拉钢筋
stretching roller 拉伸辊
stretching rope 张拉索
stretching screw 张拉螺杆；拉紧螺钉；扩张螺丝；夹紧螺钉；调整螺丝；调整螺杆；伸缩螺栓
stretching slide 张紧滑轨
stretching speed 拉引速度
stretching steel 张拉钢筋
stretching strain 拉应变；拉应变；拉伸变形
stretching strand 张拉钢束
stretching strength 拉伸强度
stretching stress 拉伸应力
stretching system 张拉体系
stretching tendon 预加应力的钢丝束；预应力的钢筋束

stretching unit 张拉设备;张紧装置
stretching value 张拉值
stretching vibration 伸缩振动
stretching wedge 张拉楔片;张拉楔块
stretching wire 预应力混凝土用钢丝;张拉细丝;张拉钢丝;拉索;拉绳
stretching wires in pairs 成对的张拉钢丝
stretching yard 张拉工场
stretching zone 张拉区域
stretch level(l)ing 拉伸校平
stretch of river 河段
stretch-out 用力划;伸展;伸出;拖延工作
stretch-out line 直伸线
stretch-out view 展开图
stretch-proof 抗拉伸;耐拉伸
stretch receptor 牵张感受器
stretch reducer 张力减径机;拉力减径机
stretch reducing (钢管的)张力减径;拉力减径
stretch reflex 牵张反射
stretch roll 张力辊
stretch rolling 拉伸矫直
stretch rotary forming 拉伸回转成型
stretch straightening 张力变形法矫形
stretch thrust 引伸冲断层;拉伸断层;椭曳冲断层
stretch-twister 抽伸捻线机
stretch wrap 拉伸包装
stretch-wrap forming 拉伸卷缠成型
stretchy asphalt roofing felt 弹韧油毡
stretchy nylon 弹性尼龙
Stretton group 斯特雷顿岩群
strewing sand 散砂;散沙
strewn island 散岛
stria baillargeri interna 内粒层纹
stria carpet 条纹地毯
stria(e) 条纹【地】;细擦痕;细沟;柱沟;凸条纹;波筋;擦痕
striae albicantes 白纹
stria longitudinalis lateralis 外侧纵纹
stria mallearis 锤纹
stria projector 条纹仪
striate 条痕的;成纹的;擦痕的
striate body 纹状体
striated 条痕的;擦痕的
striated bedrock 有擦痕的岩基
striated border 纹状缘
striated boulder 擦痕漂砾;擦痕巨砾
striated coal 线理状煤
striated column 细沟装饰柱;凹槽柱
striated crystal 条纹晶体
striated effect 条纹效应
striated finish 条纹饰面
striated finish of concrete 混凝土细条纹饰面
striated glass 有条纹玻璃
striated ground 带状土【地】
striated gypsum 纤维石膏
striated marble 条纹大理岩;条纹大理石
striated mica 分层的云母
striated pattern 线性模式;线性光栅;条纹图形
striated pebble 擦痕卵石;冰川条痕石
striated rock surface (岩石的)擦痕面;擦痕岩面
striated rust pattern 条纹状锈蚀
striated structure 线理状结构
striated surface 条纹表面;(岩石的)擦痕面
stria terminalis 终纹
striation 刻柱身凹槽;线状;疲劳纹;条纹组织;条纹状;条痕;擦痕
striation cast 擦痕铸型;条纹模
striation lineation 擦痕线理
striation on bedding plane 层面擦痕
striation technique 纹影技术
striatum 纹状体
striature 条纹状;条纹【地】
strick 切绢;切绵;绳束;小麻把
stricken area 受灾地区
stricking board 刮型板
strickle 刮型器;刮平器;刮板
strickle board 旋转刮板;造型刮板
strickle board support 造型刮托;车板座
strickle mo(u)lding 刮板造型;车板造型
strickle sweep 旋转刮板
strickling 刮;斗刮
strickling core 刮制芯
strict active constraint 严格有效约束
strict anaerobic autotrophic bacteria 绝对厌氧自养菌
strict bending theory 严格的弯曲理论
strict comparison principle 严格比较原理
strict complementary 严格互补性
strict compliance rule 严格一致原则
strict confidential 绝密件
strict constraint 严格约束
strict control 严格控制
strict imbedding 严格嵌入
strict implication 严格蕴涵
strict inductive limit 严格归纳极限
strict inequality 严格不等式
striction (钢条接近拉断时的)颈缩
strictional self-focusing 紧缩自聚焦
striction strain 颈缩应变
strict liability 严格责任(制);严格赔偿责任
strict local extremum 严格局部极值
strict lower bound 严格下限
strictly classical 正宗古典的
strictly concave function 严格凹函数
strictly confidential 严守秘密;绝密文件;绝密
strictly convex function 严格凸函数
strictly convex game 严格凸对策
strictly convex space 严格凸空间
strictly decreasing 严格递减
strictly decreasing function 严格递减函数
strictly determined matrix game 严格决定矩阵博弈
strictly diagonal dominance 严格对角优势
strictly increasing 严格递增
strictly increasing function 严格递增函数
strictly increasing sequence 严格递增序列
strictly inductive limit 严格归纳极限
strictly initial functor 严格初始函子
strictly linear equation 严格线性方程
strictly linear function 严格线性函数
strictly monotone 严格单调
strictly monotone decreasing 严格单调递减
strictly monotone decreasing function 严格单调递减函数
strictly monotone function 严格单调函数
strictly monotone increasing 严格单调递增
strictly monotone increasing function 严格单调递增函数
strictly positive definite operator 严格正定算子
strictly positive functional 严格正泛函
strictly positive linear mapping 严格正线性映射
strictly positive operator 严格正算子
strictly positive vector 严格正向量
strictly pseudoconvex 严格伪凸
strictly quasiconvex 严格拟凸
strictly same asper attached sample 同附样品严格一致
strictly separate 严格分离
strictly stationary process 严格平稳过程;强平稳过程
strictly stationary random process 强平稳随机过程
strictly transitive operator algebra 严格可递算子代数
strictly triangular matrix 严格三角矩阵
strictly unilateral digraph 严格单侧有向图
strictly unimodal function 严格单调函数
strictly upper triangular matrix 严格上三角矩阵
strictly weak digraph 严格弱有向图
strict nature reserves 完整的自然保护区
strict operating condition 严格操作条件
strict parasites 严格寄生物
strict performance of contract 严格遵守合同条款
strict preference 严格优先;严格偏好
strict rationalism 严格的理性主义
strict root condition 严格开平方条件;严格的开根条件
strict rules 严格的规则
strict-sense stationary series 严格平稳数列
strict sequential order 严格串行顺序
strict simple ordering 严格简单排序
strict standard 严格标准
strict upper bound 严格上限;严格上界
stricture 束缚
strided distance 跨距
stride level 跨式水准器;骑式水准器
stride of crane 起重机起重高度
striding compass 跨乘罗盘
striding level 跨水准器;步测水准器
striding stake 骑马桩【测】
stridor dentium 锉牙
stridulate 摩擦发音
stridulation 摩擦发音
striga 柱槽;柱子上的长条形凹槽;柱身凹槽;[复] strigae
striges 柱身凹槽(饰)
strigil ornament (古罗马建筑)凹槽饰
strigovite 软绿泥石;铁柱绿泥石
strike 岩层走向;走向【地】;撞击;起弧;锁舌片;闪击;地质走向;触击弧坑;触击电镀膜;触发;冲击;拆除;发横财;罢工
strike a balance 结算
strike a bargain 拍板成交;成交
strike a bell 敲钟报时
strike against rock 触礁
strike and dip of sampling strata 采样地层产状
strike arc 打火
strike a reef 触礁
strike backset 撞回距离(撞锁)
strike block 短刨
strike board 木耙;(用以刮平混凝土面的)刮板
strike-breaking 破坏罢工
strike cleat 走向割理
strike cupboard 通风橱
strike delay insurance 罢工延误险
strike-dip survey 产状测量
strike edge 装锁的门窗扇侧边;装锁门窗扇边框
strike expense insurance 罢工追加费用保险
strike exploration line 走向勘探线
strike face 沿走向推进工作面
strike fault 走向断层;断层走向
strike faulting 走向断层作用
strike fold 走向褶皱
strike fund 罢工津贴基金
strike geologic(al) profile 走向地质剖面图
strike-in 纸张对油墨的吸收性
strike insurance 罢工(保)险
strike in sympathy 同情罢工
strike jamb 安锁舌片的窗边框
strike joint 走向节理;节理走向
strike lift fault 走向上升断层
strike line 走向线
strike mast 可放倒式桅杆
strike measure 刮平计量;斗刮量法
strike migration 走向迁移
strike of alignment 线路走向
strike of anomaly 异常走向
strike of anomaly body 异常体走向
strike of bed 地层的走向
strike of breaking zone 破碎带走向
strike off 刮平;抹子;抹去;抹平;勾缝
strike-off an appropriation from the budget 取消拨款;从预算中取消一部分拨款
strike-off an entry 抹去账页
strike of fault line 断层线走向
strike of fault surface 断层面走向
strike off bar 刮料板
strike-off beam (混凝土摊铺机的)整平梁;刮平梁
strike-off blade 整平板;整平刀板
strike-off board 刮平板;刮板
strike-off measure 刮平计量
strike-off method 刮平法
strike-off screed 刮平样板;整平样板;刮平勺泥尺;整平勺泥尺;刮平准条;整平准条
strike of hinge 枢纽走向
strike of indefinite duration 无限期罢工
strike of orebody 矿体走向
strike of ore vein 矿脉走向
strike of polarized body 极化体走向
strike of rock body 岩体的走向
strike of seam 岩层裂缝走向;裂隙走向
strike of the magnetic body 磁性体走向
strike of underground river 暗河的走向
strike on (桩工的)复打
strike on a shoal 触滩
strike out 拟出
strike overlap 走向超覆
strike paralleling 走向平行
strike photography 命中摄影术
strike plate 挡板板;击碎凹板;锁舌片;冲击板;门鼻子
strike plating 触击电镀

strike price 敲定价格
strike projection method 走向投影法
striker 撞针;撞针【机】;击针;斗刮;大铁锤;锤;冲击仪
striker cylinder 冲击唧筒
strike reinforcement 锁舌片加固件
striker fork 操作叉
striker head 击针尖
strike river 走向河
strike rose diagram 走向玫瑰图
striker plate 锁舌片;撞针托板;挡料板
striker spring 击针簧
strikes 罢工、暴动和内乱
strike scale 步测比例尺
strike separation 走向离距;走向间距
strike shift 走向移距;走向变位
strike-shift fault 走向滑动断层
strike-ship movement 平移运动
strike signal 敲击信号
strike slip 走向滑距;走向滑动
strike slip earthquake 走向滑动地震
strike slip fault 走向滑(动)断层;走向滑(移)断层;走滑断裂;平移断层;走向平移断层
strike slip faulting 走向滑断层作用
strike slip focal mechanism 走向滑震源机制
strike slip motion 走向滑运动
strike slip source mechanism 走向滑震源机制
strike stile 安锁舌片的边梃;装锁边梃
strike stream 走向河;后成河
strike surcharge 罢工附加费
strike the bottom 触底
strike-through 透墨;褪色;(胶合板面)透胶的污迹
strike to sand 船底搁沙;船底触沙
strike valley 走向谷
striking 刮制;击法;冲制;拆除支架;爆花(烤花时颜料剥落)
striking and forged tool 锻打工具
striking angle 转角;转动角;入射角;冲击角
striking arm 冲击臂
striking blade 簧舌
striking block 冲头
striking board 刮平板;样板;刮板
striking current 击穿电流;起弧电流
striking distance 击穿距离;起弧距离;放电距离
striking edge 冲击试验机的撞摆
striking end 焊条引弧端
striking end of shank 钻头连接端
striking energy 放电能量;冲击能(量)
striking face 承冲面
striking force 撞击力;冲击力
striking formwork 拆除模板
striking gear 皮带拨杆;打击传动装置;拨动装置
striking hammer 打眼锤;打印锤;重型锤
striking job 拆模作业
striking knife 三棱刮刀
striking lever 操纵杆
strikingly colo(u)r 鲜明的颜色
striking mechanism 打点机构
striking network pattern of white colo(u)r 显著的白色网状图案
striking of an arc 电弧引燃
striking of arch 拆除拱架
striking of centers 拆除模架
striking of cutting 插枝成活
striking off 刮平;勾缝;刮砂
striking off lines 刮平线;嵌缝线;划深痕线;勾画线
striking of shuttering 拆模(板)
striking of the floor form(work) 拆除楼板模板
striking period 拆模时间;拆除脚手架的时间;拆模周期
striking pin 三棱刮刀刃
striking plate 门锁舌片;锁舌片;拆除支架板;(护舷)防冲板
striking point 划弧线的中心;熔断点
striking potential 着火电位;起弧电位;闪击位;闪击电势
striking price 协定价格;定约价
striking stile 装锁边梃
striking surface 电弧触发面;承击面
striking technique 打打手法
striking temperature 玻璃显色温度
striking the arc 引弧(电焊)
striking time 拆模时间
striking tool 撞击工具;修光工具;刮刀

striking-up 刮制
striking velocity 冲击速度
striking voltage 引燃电压;引弧电压;起弧电压;闪击电压
striking wedges 松紧楔(调整高度用的一对木楔);对楔;拆除楔架;拆除架楔;敲掉楔子
striking weight cartridge 打点锤筒
striking winding 起弧线圈
striking winding weight drum 打点绕锤辊
striking work 打时装置;打点机件;报时装置
striking wrench 冲击式扳手
strikle 刮平器;斗刮
string 线道;弦线;细条纹;细脉;一行;一串;钻具组;楼梯斜梁;行;数串;绳
string-a-like assembly 整套电器材合电缆
string assignment 串赋值
string attribute 信息串属性;串属性;串型属性
string bead 直焊道;直的圆凸勾缝;窄焊道
string beading 线状焊;窄焊道;挺杆叠置焊道
string-beads 串球状
string-beads nodule 串珠状结构
string binder 麻绳扎机
string board 楼梯斜梁侧板;楼基盖板;楼梯斜边板
string break 串断点
string built-in function 串行内函数
string bus 柔性母线
string constant 串常数
string control byte 串控制字节
string cornice 楼梯斜梁上线脚饰;楼基楣饰
string course 层拱;带饰;凸砖层;水平箍带;腰线;束带层;带状层
string curve 束束曲线
string data 字符串数据;串数据;串式数据;成串数据
string denotation 串标志
string descriptor 串描述符
string development 带形发展;带状城市建筑
string device mode 字串符设备操作方式
string device state 字串符设备状态
string diagram 线形图;线图;线条图(表)
string drive 弦丝传动
string dryer 带式干燥器;带式干燥机
string dynamometer 弦式测力计
string efficiency 串效率
string electrometer 弦线(式)静电计
stringent assumption 严格假定
stringent cash 银根紧
stringent condition 严峻条件
stringent condition of use 严格的使用条件
stringent control 紧缩控制
stringent regulation 严厉的规则;严格管理
stringent specification 高标准
stringent tolerance 严格容差;精确容差
stringer 小纵梁;小梁;细矿脉脉道;纵梁;纵枕木;楼梯斜梁;架设装置;脉道;牵条;楼梯;穿串装置;沉积薄层
stringer angle 舷侧纵通角钢(在甲板边板上)
stringer bar 舷侧纵通角钢(在甲板边板上)
stringer bead 线状焊道;窄焊条(焊条不横摆);窄焊道
stringer bead welder 窄焊道焊机
stringer bead welding 窄焊道焊接
stringer beam 横撑;横夹木;楼梯梁
stringer bracing 纵梁斜撑;纵梁联结系
stringer bracket 纵梁托架;纵梁拉杆
stringer bridge 纵梁桥
stringer formation 纤维状形成
stringer frame 桁条框架
stringer girder 纵桁架
stringer-head 沉积薄层源头;细矿脉脉道源头
stringer lug 短舷边角钢
stringer plate 纵桁板【船】;甲板边板下纵桁
stringer strake 甲板边板
stringer-type stairway 梁式楼梯
stringer vein 细脉
string(er) wall (支承楼梯踏步的)梯墙
string expression 串表达式
string extensometer 弦式应变计
string file 串文件;成串文件
string filter 绳式过滤机
string form(work) 楼梯斜梁模板
string frame 楼梯斜梁框架;桁条框架;串框架
string function 串函数
string galvanometer 弦线式检流计;弦式电流计;弦式电流计

stringhamite 水硅钙铜石
stringiness 拉丝性(能)
stringing 沿管线敷设;紧线
stringing grip 绳卡
stringing line time 倒钢丝绳时间
stringing mortar 展铺砂浆
stringing pipe 沿线路铺管
stringing test 拉丝试验
stringing the line 倒出钢丝绳(卷筒上)
stringing the wire line 有钢丝绳装备滑车系统
stringing-up 穿线挂车
string item 串项
string knot 带条纹的节瘤
string length 字串长;串长度
string length check 串长度校验
stringless 无弦的
string level 线挂水准器;麻线放样高程;悬挂式水准管;放样高程
string line 放样麻线;标线
string lining 麻线放样;麻线放样;用麻绳放样;用麻绳放样;麻线放样;绳索定线
string lining of curve 绳正法整正曲线
string list 串表
string loading 线状装药;(用小于炮眼直径的)管壳装药
string manipulation 串处理;串操作
string manipulation scheme 串处理方案
string masonry wall 楼梯基坑工墙
string measurement 绳测
string milling 连续铣削
string model 成线束状
string of bathometer 测深索;测深绳
string of cars 车列
string of casing 井管钻具组;套管柱
string of deposits 沉积透镜状排列;沉积带
string of drill pipes 钻杆组
string of drill rods 钻杆柱
string of fathometer 水深仪测绳
string of rods 抽油杆柱
string of tools 钻具
string operand 串操作数
string operator 串运算符;串算符;串操作符
string option 串任选
string oriented symbol language 串式处理符号语言
string oscillograph 线式示波器;弦线式示波器
string over 测量孔深
string packing 线形填料;线形填缝
string parameter 串参数
string pattern 串模式
string pendulum 线摆
string piece 楼梯斜梁;楼梯侧板;楼梯纵梁;码头边板;横木;横梁;纵梁
stringpiece man 码头边装卸工(人)
string pointer 弦控指针;仪表弦控指针;串指针
string polygon 索线多边形;索多边形
string polygon equation 索多边形方程
string potentiometer 弦线电位计
string process(ing)language 串处理语言
string process system 串处理系统
string-pulling behaviour 拉绳行为
string quantity 串量
string quote 符号串的引用符
string reinforced concrete 钢弦混凝土
string replacement rule 串替换规则
string rod 水泵连杆
string-shadow instrument 弦线式仪表
string shot 井下爆破器
string shuttering 楼梯斜梁模板
string sign 线样征
string sort 串分类
string space 行距
string specification operand 串说明操作数
string stair(case) 有楼梯斜梁的楼梯;绳索楼梯
string statement 并串语句
string switch 串开关;成串开关
string symbol 串符号
string the block 滑车装置
string the line 拉紧绳索
string tying 麻绳打捆
string unit 串部件
string up 装备滑车系统
string vacuum filter 绳索式真空过滤器
string variable 字符串变量;串变量

string wire 钢弦;钢丝;预应力钢丝;绑扎钢丝
string-wire concrete 钢丝混凝土;钢弦混凝土
string wreath 索环
stringy knot 筋节
stringy stonecrop 垂盆草
stringy wool 钢丝毛
striograph 影像图
strip 卸去索具;小木板;狭长的土地;细长片;窄条;路带;解装;解吸;磨带;条地;条;梳刷;捣矿机排矿沉淀槽;带;板条
strip aerial triangulation 航带法区域网空中三角测量
strip anchoring system 板条固定系统
strip-and-group system 带状群状渐伐作业法
strip and skelp mill 带材和焊管坯轧机
strip annealing 带材退火
strip aquatic mat 条形水草
strip aquifer 条形含水层
strip area 带形地区
strip a still 开蒸馏器的入孔盖
strip attenuater 条带形衰减器
strip away 冲去
strip bar 窄扁钢
strip board 外层板
strip bonding 带状钢轨导接线
strip borer 露天矿穿孔机;水平钻孔机;水平眼钻车
strip-borer drill 露天矿水平炮眼钻机
strip breakage 带材拉断
strip breakdown 板坯
strip building （平行长排的）简易房屋;条形建筑;带形建筑;长条房屋
strip burner 带状燃烧器
strip burning 带状烧除
strip by strip rectification 逐航线纠正
strip-by-strip transformation 分带纠正
strip calender 条胶压延机
strip camera 航空带状摄影机;条幅式照相机
strip-card reader 磁卡阅读机
strip caster 带坯连铸机
strip casting 漏模造型
strip ceiling 板条顶棚
strip chart 带状记录纸;长图记录纸;长条记录纸;带状图
strip chart autobalancing recorder 长图自动平衡记录仪
strip-chart instrument 记录式仪表
strip-chart recorder 纸带记录器;条形记录仪;条纸记录器;带状图表记录器;长图式记录仪;长条记录纸式记录仪
strip-chart temperature recorder 长条图温度记录仪
strip city 线形城市;条形城市;带形城市
strip clamping device 带钢夹持器
strip cleaning machine 带材清净机
strip coal mining 露天采煤
strip-coated polyester sheet 撕膜聚酯基片
strip coating 可剥涂料;带式镀膜;镶木涂料;条状包覆物
strip coating film 撕膜片
strip coil 带材卷
strip coiler 带材卷取机
strip coiling apparatus 带材卷取装置
strip commercial area 条状商业（地）区
strip conductor 条状导线;条形导线;扁导线
strip contour map 带状等高线地形图
strip copper 条铜;铜带;带状铜板
strip core 细木工板;板条芯胶合板
strip correction 航线的修正【航空】
strip count 刨削材帐
strip cropping 等高条植;等高条播;带状耕作法;带状播种
strip cropping field 等高条植地;等高带状耕植地
strip cultivation 条植;条垄耕作;带状耕作
strip cutting 纵切;带伐
strip cutting machine 带材切割机
strip deformation 航线弯曲度
strip design 航带设计
strip development 条形商业区;条形发展;带料展开
strip diffuser 条形散流器
strip down 剥下;拆散
strip drain 条状排水沟
strip dryer 带材干燥机
stripe 漆条纹;条纹;条板;色条;斑纹
striped amphogneiss 条纹状混合麻岩

striped bed load movement 带状底砂运动;有条纹的推移质运动;有条纹的河床质运动
striped figure 带状木纹;条纹
strip-edge scanning equipment 带材自动定心装置;带材边纹扫描器
striped migmatite 条纹状混合岩;条带状混合岩
striped structure 条纹状构造
striped tile 条纹方砖
striped veneer 条纹单板
strip electric(al) heater 装有（散热）叶片的电热器
strip electrode 框极;极板;带状电极
strip end 成卷带材的端头
strip end engaging 带材端头嵌入
strip enumeration 带状清查
stripe painter 路面标志机;路面画线机
stripe pattern 色条信号图
striper 路面画线机;画条纹刷;漆条纹机;漆条纹工
strip(e) road 纤道;拉船道
stripe screen tube 条屏显像管
stripe signal 条纹信号
stripe veneer 隔色胶合板
stripextraction 反萃取
stripe yarn 长结子花式线
strip feeding device 带料进给装置
strip felt 带形毡
strip figure 条纹图案;带状花纹
strip filling 梳式加料;带式加料
strip film 条状胶片
strip-film camera 条幅式摄影机;条幅式摄像机;自动摄影机
strip filter 带式滤色器
strip fin 条形翅片
strip finder 航线寻景器
strip flipper 板料弹动器
strip floor cover(ing) 木条地板
strip floor(finish) 条纹地板
strip flooring 连续底脚;条形基础;企口木地板;铺狭条地板;木地板;带状地板;镶木条形地板;狭条地板;直条地板;条形地板;条木地板
strip footing 条形基础;带状基础;带状地基;条形底脚;条形地基
strip forms 条形模板;板条模板
strip foundation 条形基础;带状基础;带状地基
strip front end detector 带钢前端检测器
strip fuse 片状保险丝;片状保险器;带状熔丝;保险丝片
strip ga(u)ge 条状应变片
strip-grinding line 带材磨光作业
strip handle 起模手柄
strip handling car 带卷运送车
strip heater 装网格的加热器;片式加热器;电热丝式加热器
strip heating 木条胶接缝加热;条状加热
strip heating pipe 网格式加热管
strip hole framework 条孔骨架
strip index 航ং像片索引图
striping 横线;去胶;断开线;拆开
striping development 带式布置
striping machine 画线机
striping paint 路面标志油漆
strip inspection 解体检查;拆检
strip intrusion 带状入侵
strip iron 铁箍;冷轧钢带;窄带钢;冷轧钢带;带钢
strip irrigation 条形灌溉;带状灌溉
stripite 窄条金属网板;窄板条
strip jasper 缟玉髓
strip-lacquering line 带材涂漆作业线
strip lamp 管状灯;照明带;灯管
strip lath 窄条金属网板;窄板条
strip layer 带卷的圈
strip leaf 烟叶
strip light 登陆点灯标;长条状灯;着陆灯
strip light indication 光带式表示
strip lighting 管状白炽灯照明;灯串照明
strip lighting element 带状照明单元
strip lighting source 带光源
strip-light lamp 天幕灯;顶灯;带形照明器
strip line 条线;电介质条状线;带状线
strip-line circuit 带状线电路
strip-line lighting fixture 条带式照明装置;条带式照明设备
stripling 修剪苗
strip load 条状荷载;条形荷载;带状荷载
strip loading 带状荷载;狭条荷载

strip log 柱状剖面图
strip lumber 木板条
strip machine 拉荒机床;脱模机;抽锭机
strip management 带状管理
strip manipulation 带状处理;符号序列处理
strip map 带状地（形）图
strip mashing process 撕膜法
strip material 撕膜材料
strip matrix 带状矩阵;带形矩阵
strip metal 金属条带
strip method 航线法;带状法
strip mill 薄板轧厂;带钢轧厂;薄板轧机;带钢轧机;带宽轧机
strip mine 露天采石场;露天矿;地面矿湖
strip-mine lake 露天矿坑湖
strip mine reclamation 填筑露天矿
strip mine survey 露天矿测量
strip mining 露天开采作业;露天开采;露天采矿;露天剥采（采矿）
strip mopping 条铺沥青法;条粘;条涂
strip mosaic 像片略图
strip mulching 带状盖草
strip number of investigation 调查条数
strip oak flooring 橡木板条地板
strip of element 元素带
strip off 剥离;剥开
strip of fuses 熔线排
strip of ground 狭长地带;狭长基地
strip of keys 电键排;按钮排
strip of pre-formed filling material 预制填充材料条
strip of successive photography 连续摄影航线
strip of tags 接线条组
strip of tin foil 锡箔条
strip oiler 带钢涂油机
strip overlay welding experiment 板极堆焊试验
strippable 可剥的
strippable coating 可剥性涂料;可剥性涂层;可剥离涂层;脱模涂层
strippable paint 可剥性油漆;可剥(性)漆
strippable paper 撕贴两用墙纸
strippable property 可剥性
strippant 解吸剂
strip park 带形公园
stripped 条带状
stripped atom 剥离原子;剥电子层原子;被剥原子
stripped bedding plane 条带状层面
stripped coating 剥落的涂层
stripped-down style 条纹式样
stripped emulsion 脱基乳胶
stripped gas 已除去汽油烃的干气
stripped glaze 脱落釉层;剥釉
stripped ground 带状土【地】
stripped illite 退化伊利石
stripped joint 裸露的接头;裸头的接头;露出接头
stripped joist 板条格栅
stripped plane 剥露平面
stripped point 露天接头
stripped soil 带状土【地】
stripped solution 贫化液
stripped sour water 溶出酸性水
stripped structural surface 剥露构造面;条带状构造面
stripped surface 清基面;表层剥除的地面
stripped well 低产井
strip peel 条状剥皮
stripper 剥带机;刮漆铲;露天采矿工;汽提塔;贫油井;脱模机;脱模工;涂消除剂;低产井;出坯杆;抽锭机;冲孔模板;抄针器;拆卸器;拆卸板;残油带系;分离装置;剥皮器;剥离器;剥离剂;剥开机;刨土机;板条圆锯
stripper and grinder 磨钢丝车工人
stripper and worker 剥取辊和工作辊
stripper band 钢皮环
stripper bar 清理杆
stripper bush 脱模衬套
stripper cell 种板槽;刮板槽
stripper crane 脱模机;脱模吊车;剥片吊车
stripper cylinder 起模缸
stripper efficiency 解吸塔效应;汽提塔效率
stripper excavator 剥离挖掘机
stripper flash drum 汽提闪蒸罐
stripper knife 电缆刀
stripper loop 清齿环;分隔环

stripper machine 卸料机;脱模机
stripper mechanism 芯棒抽出机构;脱锭机构
stripper pin 起模顶杆;脱模销
stripper pipe 残油管
stripper plant 汽提装置;汽提车间
stripper plate 漏模板;挤压板;脱模板;冲模退料板;分馏柱塔板
stripper-plate mo(u)ld 漏模造型;脱模板塑模
stripper plunger 柱塞式炉用推料机
stripper pump 清舱泵;残油泵
stripper punch 脱模冲头
stripper reboiler 汽提塔再沸器
stripper roller 抄钢丝辊
stripper rolls (自动轧管机的)回送辊
stripper rubber (从提升钻杆上清除泥浆的)橡胶刮管器;密封胶皮
stripper shockless jolt squeeze mo(u)lding machine 顶箱微震压实造型机
stripper shovel 剥离(型)电铲
stripper station 起模工位
stripper tower 汽提塔
stripper unit 摘取装置;梳抓装置
stripper well 衰竭井;低产井
stripper zone 汽提段
strip pickle 带材酸洗装置
strip pilaster 狭条扁柱
stripping 削裂;解吸作用;墨辊脱墨;溶出;清理场地;清基;汽提;挖除表土;脱去旧漆;脱模(混凝土);条纹;扫纹;抽锭;拆箱;反萃取;脱剥;剥离;剥开;剥除表土;剥除
stripping agent 反萃取剂;剥色剂;剥离剂;脱模剂
stripping analysis 溶出分析;汽提溶极谱分析;提溶极谱分析
stripping and ballast pump 残油及压载水泵
stripping apparatus 洗提设备;解吸设备;分离设备
stripping area 剥离区
stripping attachment 退模器;拆卸器
stripping bay 脱模跨
stripping cascade 再生级联;级联分化部分
stripping coefficient 剥离系数;剥采比
stripping column 汽提塔;气提塔;分馏塔;反萃取柱
stripping compound 脱模剂
stripping cracking 汽提裂化
stripping crane 脱模起重机;脱模吊车;脱锭起重机;剥片吊车
stripping crew 露天剥离班组;露天采矿班组
stripping depth 剥离深度
stripping device 脱模机
stripping drum 汽提塔
stripping electrography 纸条电谱法
stripping facility 清舱设备
stripping factor 解吸因子
stripping felt 油毡条
stripping field 带状田
stripping film 可剥(漆)膜;乳胶片;剥离片;剥离胶片
stripping film method 剥裸法
stripping folk 钩形卸料装置
stripping frame 漏模框
stripping gang 露天剥离班组;露天采矿班组
stripping indicator 位移指示器
stripping job (修理深井时)提出抽油杆
stripping knife 铲刀;剥离刀
stripping lacquer 临时涂料;可剥(性)漆
stripping layer 剥离层
stripping line 开沟段;清基线;残油管系
stripping liquid 液吸收塔;解吸液
stripping machine 脱模机;卸料机;画线机;起模机;剥皮机
stripping mask 撕膜片
stripping method 脱模方法;汽提法处理;汽提法;脱锡法
stripping of aggregate 集料剥落;骨料剥落
stripping of coat 拆开;剥落
stripping off 剥分
stripping of falsework 拆除脚手架
stripping of old paint 脱除旧漆
stripping of reservoir 库坝清基;清库工作;水库清基;水库表土清除;水库表土剥离
stripping of seed 铲除植被
stripping of the site 挖除场地表土
stripping operation 拆除作业
stripping order 剥离次序;脱卸次序
stripping overburden 清除覆盖层;清除上部沉积;清除过载
stripping paper 条纹纸
stripping party 露天剥离班组;露天采矿班组
stripping pattern 漏模
stripping piece 便于拆模的斜窄构件
stripping pin 起模顶杆
stripping plate 漏模板;起模板
stripping plate machine 抽模制型机
stripping position 拆卸位置;清除位置;开采位置;剥离位置
stripping post 顶杆
stripping pressure 脱模压力
stripping pump 扫舱泵
stripping rack 剥片架
stripping ratio 采剥比;剥采比
stripping reaction 剥裂反应
stripping resistance 分离阻力;剥离阻力
stripping ring 拆卸环
stripping roller 抄针辊
stripping room 剥片间
strippings 抄针花
stripping sampling 剥层取样
stripping schedule 脱模日程表;脱模时间表;拆模时间表
stripping section 提馏段
stripping shovel 剥离铲;矿山表层剥离机;剥离挖掘机;表土层剥离机铲
stripping soil 剥土
stripping solution 反提取剂
stripping speed 顶模速度
stripping spider 线盘卸料器
stripping star 星形脱模器
stripping steam 解吸用蒸汽
stripping still 汽提蒸馏器
stripping strength 拆模强度;脱模强度;脱锭力;剥离强度
stripping system 残油管系
stripping table 起模台
stripping taper 拔模斜度
stripping team 露天剥离班组;露天采矿班组
stripping technique 分离技术
stripping test 去胶试验;汽提试验;剥落试验;剥离试验
stripping test for aggregate 集料剥落试验
stripping test of bituminous coat 沥青膜剥落试验
stripping the pipe 提出钻杆作业
stripping the rod 提出钻杆作业
stripping the well 起出井管
stripping time 起模时间;脱锭时间;拆模时间
stripping tongs 脱锭钳;剥片钳
stripping topsoil 剥除表土
stripping tower 吹落塔;汽提塔;气提塔
stripping valve 收舱阀
stripping voltammetry 溶出伏安法
stripping wastewater 汽提废水
stripping wheel 切割砂轮
stripping with steam 蒸汽汽提
stripping yard 起卸场所
strip pit 采石场;露天矿;露天石场;露天采掘坑
strip planking 狭条船壳板
strip planting 带状植树法;带状栽植
strip plate 带坯
strip plot 航线图
strippr 抽出机
strip prediction 航线预测;航线预报
strip printer 航带印像机;带层表面涂层印花机;带材表面涂层印花机
strip processing 精整成卷带材
strip-processing mechanism 带材除鳞机
strip processor 带地耕作机组
strip radial plot 航线辐射三角测量
strip radial triangulation 航线辐射三角测量
strip radiant panel 带状辐射板
strip record 条纹记录
strip reeling 带材卷取成卷
strip region 带状区域
strip repeater 窄带材围盘
strip resistance 抗剥离性
strip ridge buckle 带材皱纹
strip road 狭长道路;纤道;拉船路;轨道式道路
strip-rolling 带材轧制
strip rolling mill 带钢轧厂;带钢轧机
strip roughing mill 带钢粗轧机
strip rubble 粗加工石条
strip scan 带状扫描
strip sealing 条缝嵌缝;条状填缝;条状焊接;条状封接
strip seeder 带播机
strip shaped 带状的
strip shaped radiator 条带状辐射体
strip sheet mill 板卷轧机
strip shingle 条形木瓦;带形屋面板;条形屋面板
strip sight glass 条形视镜
strip skimmer 带式撇油器
strip slate 仿板岩橡胶屋顶;沥青瓦
strip slitter 带材纵切圆盘剪
strip slitting machine 带材纵切机组
strip slitting shears 带材纵切圆盘剪
strip soaker 屋面防漏条带
strip steel 带钢
strip steel coiler 带钢卷缠装置;带钢卷缠机
strip steel cutting off machine 带钢切割机
strip steel level(l)er 带钢整平机;带钢校平机
strip steel rolled stock 带材
strip steel section 条带式型钢
strip storage 带材备用量
strip surfacing 带极堆焊
strip survey 航线测量【航空】;带状调查
strip suspension 吊带悬挂
strip-suspension type 轴尖悬置式
Stript 斯特里普特(一种专利除漆剂)
strip taping 盖楼油毡条
strip tensile 狭条抗拉
strip tensile test 布条拉伸试验
strip tension 带钢张力
strip tensioning apparatus 带钢的张力设备
strip tension meter 带钢张力计
strip tension test 狭条拉伸试验
strip terminal 接线条;片接头
strip test 条杆强力试验;长片法
strip the drill pipe 提升钻杆
strip the drill rod 提升钻杆
strip the falsework 拆除脚手架
strip thermocouple pyrometer 带状热电耦高温计
strip the thread of a screw 螺旋螺纹磨损
strip thickness ga(u)ge 带材测厚仪
strip tillage 带状整地
strip timber floor 立板木桥面
strip topographic(al) map 带状地形图
strip transfer method 纸条转移法
strip transmission line 带状传输线
strip tube 条状管
strip tube heater 条管加热器;翅管式加热器
strip tube radiator 条管辐射加热器
strip-type magnetron 耦腔式磁控管
strip type(reaming)shell (镶焊金刚石的)条带形扩孔器
strip uncoiler 带材开卷机
strip water 带状灌溉
strip weeding 带状除草
strip welting 贴边弯下的铁皮条
strip width 航带宽;行宽
strip width meter 带材测宽仪
strip winding 绕带;条带形线组
strip winding machine 绕带机
strip winding vessel 绕带式容器
strip window 条状窗
strip wiper 带式擦拭器;带材擦拭器
stripwise rectification 逐航线纠正
strip with water 水力剥离
strip wooden flooring 木条地板
strip wool 棉条;带状棉
strip-wound armature 铜带绕的电枢
strip zinc 锌带;锌条
strisilane 丙硅烷
strix 柱身突筋;柱射凹槽;柱上凹槽间的隆起线
strobane 氯化松节油
strobe 门电路转换开关;电子闪光
strobe circuit 选通电路
strobe edge 选通脉冲边缘
strobe flash 频闪器闪光
strobe gate 选通脉冲门
strobe input 选通脉冲输入
strobe light 频闪(灯)光
strobe marker 选通脉冲标志
strobe photography 频闪摄影术
strobe pulse 选通脉冲
strobe switch 选通脉冲(转换)开关

strobe unit 选择装置;选通装置
strobing gate 选通门
strobo 闪光仪
strobodynamic balancing machine 闪动平衡机
strobolamp 旋光试验灯
strobolume 高强度闪光灯
strobolux 大型频闪观测器;大功率频闪观测器
strobophonometer 爆震测声计
stroboscope 频闪(观测)仪;频闪观测器;示速器;闪光(测速)仪;闪光测频器
stroboscope photography 频闪观测器照相术
stroboscope tube 闪光管
stroboscopic 频闪的
stroboscopic checking 频闪校验法
stroboscopic direction finder 频闪式测向仪;频闪测向器
stroboscopic disc 频闪观测盘
stroboscopic effect 频闪效应
stroboscopic illumination 频闪照明
stroboscopic indicator 频闪指示器
stroboscopic instrument 频闪式仪表;频闪测速仪
stroboscopic instrumentation 频闪仪式检测仪表
stroboscopic light(ing) 频闪光;高速闪光灯
stroboscopic method 闪光测频法
stroboscopic observation 频闪观察(高速机床试验)
stroboscopic pattern wheel 频闪图案轮
stroboscopic photography 频闪摄影术
stroboscopic polarimeter 频闪仪式偏振计
stroboscopic principle 频闪观测原理
stroboscopic tach(e)ometer 频闪转速仪;频闪仪式转速计
stroboscopic tachometer 闪光转速表
stroboscopic tube 频闪管
stroboscopic xenon lamp 频闪氙灯
stroboscopy 闪光测频法
strobostercoscope 立体动态镜
strobotach 频闪测速器;频闪转速计;频闪测速计
strobotron 频闪管
stroke 行程;桩锤落高;桨程;击打;推摩法;动程;冲量冲击;冲孔;中程;笔画;笔划
stroke adjuster 冲程调整器;冲程调节器
stroke adjusting shaft 行程调整轴
stroke adjustment 行程调整
stroke alteration 行程改变
stroke analysis 笔画分析
stroke-bore ratio 冲程与直径比;冲程内径比;冲程缸径比
stroke capacity 汽缸工作容积;冲程容量;冲程容积
stroke center line 笔画(的)中线
stroke character 短画字符
stroke character generator 笔画字符发生器;笔画符号发生器;线段字符发生器
stroke coefficient 冲程系数
stroke control 行程调节;冲程调节
stroke control knob 行程控制捏手
stroke counter 行程计;压砖计数器
stroke cure 抚熟试验
stroked 凿划石面
stroke density 闪击密度
stroke display 笔画显示
stroke dog 行程挡块
stroke down 下行程;下冲程
stroked work 细琢石料
stroke edge 笔画边缘
stroke edge irregularity 笔画边缘不规则性
stroke end 行程终点
stroke flash 闸门闪光
stroke generator 短画生成器;笔画发生器
stroke-incidence rate 雷击事故率
stroke indicator 冲程指示器
stroke length 行程长度;冲程长度
stroke limiter 冲程限制器
stroke limiting pin 行程限制销
stroke milling 直线走刀曲面仿形铣;行切
stroke mode 笔画方式
stroke multiplier 行程放大器
stroke oar 领桨(手)(左舷最后端的桨手)
stroke oarsman 领桨(手)
stroke of a press 压力机行程
stroke of crane 起重机的起重高度;吊车起重高度
stroke of machine 机器冲程
stroke of piston 活塞冲程
stroke of punch 冲量;冲程
stroke of shear blade 剪片冲程

stroke of slide valve 滑阀冲程
stroke of table 工作台冲程
stroke operation 加横运算
stroke output 每搏输出量
stroke per minute 每分钟冲程次数
stroke rate 冲数
Stroke roller bascule bridge 斯脱劳贝尔滚轴竖旋桥;斯脱劳贝尔滚轴仰开桥
strokes meter 冲程测量仪
strokes of pump 冲数
stroke speed 行速
strokes per minute 每分钟行程;每分钟冲程;冲数
stroke thwart 桨手座板
stroke-to-bore ratio 冲程缸径比
stroke volume 行程排量;工作容积;每搏输出量;冲程容积
stroke width 笔画宽度
stroking mechanism 变量机构
stroll garden 路边花园
stromal lamellae 基质片层
stromatactis 平底晶洞构造
stromatactoid 似平底晶洞状构造
stromatactral structure 层孔构造
stromatite 叠层混合岩;层状混合岩
stromatolite 叠层石;叠层
stromatolith 叠层面;叠层(混合岩)
stromatolithic 叠层的
stromatolithic structure 叠层构造
stromatolitic dolomite 叠层石白云岩
stromatolitic limestone 叠层石灰岩
stromatolitic phosphorite 叠层石磷块岩
stromatolysis 基质溶解
Stromatoporoidea 层孔虫目
stromatoporoid limestone 层孔虫灰岩
Stromberg asymmetrical drift 斯特龙伯格不对称漂移
Stromberg diagram 斯特龙伯格图
Strombolian 斯特龙博利式
strombus 凤螺属
strome reef 层状礁
stromeyerite 硫铜银矿
Stromeyer test 疲劳强度试验
Stromgren four colo(u)r index 斯特龙根四色指数
Stromgren photometry 斯特龙根测光
Stromgren radius 斯特龙根半径
Stromgren sphere 斯特龙根球
strong 烈性的;牢靠的;坚固的;浓的;强固的;深的;放长
strong abrasive formation 强研磨性地层
strong absorption 强吸收
strong acetate 强力醋酯纤维
strong acid 强酸
strong acid acidity 强酸酸度
strong acid anion 强酸阴离子
strong acid anion concentration 强酸阴离子浓度
strong acid anion concentration expected 期望强酸阴离子浓度
strong acid cation 强酸阳离子
strong-acid cation exchange resin 强酸性阳离子交换树脂
strong-acid cure leaching 强酸熟化浸出
strong-acidic cation exchange column 强酸性阳离子交换柱
strong acid ion exchange resin 强酸性离子交换树脂
strong acidity 强酸性;强酸度
strong acid leaching 强酸浸出
strong-acid number 强酸中和值
strong acid resin 强酸性树脂
strong acid-sodium ion exchange 强酸—钠离子交换
strong acid tower 强酸塔
strong acid type ion exchanger 强酸型离子交换剂
strong activity stage 强烈活动阶段
strong adsorption 强吸附
strong after shock 强余震
strong agitation 强力搅拌
strong alkaline water 强碱性水
strong ammonia solution 浓氨溶液
strong and weak points 优缺点
strong anion exchanger 强阴离子交换剂
strong anomaly 强异常
strong aqua 浓氨水
strong axis 强轴

strong back 门门;横杆;人孔撑架;强力背材;十字架;强力背板;(模板的)加强撑架;吊装靠板;吊装垫衬;加劲材;铁弓门门;定位板
strong balanced rapid type 强均衡迅速型
strong balanced slow type 强均衡徐缓型
strong bank 资力雄厚的银行
strong barrier function 强闸函数
strong base 强碱
strong base anion 强碱阴离子
strong-base anion exchange resin 强碱性阴离子交换树脂
strong-base number 强碱中和值
strong base type ion exchanger 强碱型离子交换树脂
strong basicity 强碱性;强碱度
strong basis 强基
strong beam 宽板横梁;桁板梁;强梁
strong bound water bond 强结合水连接
strong box 保险箱;保险柜
strong branch 强分支
strong breeze 强风(六级风)
strong brine 浓盐水
strong broke 完全破产
strong caking coal 强黏结性煤
strong cation-exchange 强阳离子交换
strong cementation 强胶结
strong cement grout 稠水泥浆
strong change 碱析
strong chemical oxidizing agent 强化学氧化剂
strong clay 强黏土;强塑性黏土;不合格泥灰;重黏土
strong coherent background 强相干背景
strong colo(u)r 硬彩
strong column 强柱
strong column and weak beam 强柱弱梁
strong compacting 压的非常紧密
strong compexing agent 强络合剂
strong component 强构件;强分支;强分量
strong compression 强压缩
strong concrete 高强混凝土;大强度混凝土
strong connectedness 强连通性
strong connection 强连接
strong contrast 高反差;强反差的
strong converse theorem 强逆定理
strong correlation 强相关
strong coupling approximation 强耦合近似法
strong coupling model 强耦合模型
strong coupling rotational model 强耦合转动模型
strong coupling theory 强耦合理论
strong currency 硬通货
strong current 强流;强海流;强电流
strong current test panel 大电流测定电源屏
strong current transformer 大电流变压器
strong decrepitation 强爆裂
strong defence 深沟
strong deflagration 急燃;强烈燃烧
strong dielectric(al) glass ceramics 强介电性微晶玻璃
strong differentiability 强可微性
strong digraph 强有向图
strong discontinuity 强间断
strong discontinuity surface 强间断面
strong dispersion 强分散
strong displacement 强震位移
strong dual space 强对偶空间
strong earthquake 强震
strong earthquake belt 强震带
strong earthquake instrument array 强震仪台阵
strong earthquake measurement 强震观测
strong earthquake motion 强地震运动;强地震动
strong earthquake record 强震记录
strong earthquake response spectrum 强震反应谱
strong earthquake sensor 强震传感器
strong earthquake telemetric network 强震遥测台网
strong earthquake wave 强震波
strong earthquake zone 强震带;强地震区
strong eddy 强漩涡
strong electrolyte 强电解质
strong electromagnetic mineral 强电磁性矿物
stronger ammonia water 浓氨水
strongest 极强
strongest oxidant 最强氧化剂
strongest oxidizing agent 最强氧化剂
strong expansive fusion caking 强膨胀熔融黏结

strong explosive 烈性炸药
strong extremum 强性极值
strong fiber[fibre] 强力纤维
strong field 强场
strong-field method 强场法
strong fix 可靠陆标定位
strong focusing 强聚焦
strong focusing accelerator 强聚焦加速器
strong focusing cyclotron 强聚焦回旋加速器
strong focusing lens 强聚焦透镜
strong focusing orbit 强聚焦轨道
strong focusing principle 强聚焦原理
strong focusing synchrotron 强聚焦同步加速器
strong force field 强力场
strong frost heaving 强冻胀
strong fumarole 强烈喷出的喷汽孔
strong function 强函数
strong gale 烈风(九级风)
strong gale warning 烈风警报
strong gas 富煤气
strong girder-weak brace 强梁—弱撑
strong ground motion 强地面震动
strong hold 要塞;堡垒
strong holding device 强力夹具
strong hydraulic jump 强水跃
strong hydraulic lime 强水硬性石灰
strong hydrophilicity 强亲水的
strong intensity 高强度
strong interaction 强相互作用
strong interaction dynamics 强相互作用动力学
strong interaction effect 强相互作用效应
strong interaction model 强相互作用模型
strong-interaction theory 强相互作用理论
strong interference 强干扰
strong interference blocking 强干扰阻塞
strong iodine solution 浓碘溶液
strong jamming 强干扰
strong law of large numbers 加强大数定律;强大数定律
strong light 强光灯标;强光
strong line 粗实线
strong liquor 浓溶液;强碱液;强碱水
strong low-pressure system 强低压系统
strongly absorbing medium 强吸收介质
strongly absorbing region 强吸收区
strongly acid 强酸
strongly acidic cation-exchange 强酸型阳离子交换树脂
strongly acidic solution 强酸性溶液
strongly acidic water 强酸性水
strongly alkaline soil 强碱性土
strongly base 强碱
strongly bound water 强结合水
strongly cemented 强胶结的
strongly competing adsorbate 强竞争吸附质
strongly connected automaton 强连接自动机
strongly connected complex 强连通复形
strongly connected component 强连接成分
strongly connected digraph 强连通有向图
strongly connected direction graph 强连通有向图
strongly connected graph 强连通的图
strongly connected region 强连接区域
strongly connected sequential machine 强连接时序机
strongly connected system 强连接系统
strongly consistent 强相容的
strongly continuous mapping 强连续映射
strongly continuous representation 强连续表示
strongly continuous semigroup 强连续半群
strongly continuous stochastic process 强连续随机过程
strongly contrasting colo(u)rs 色彩对比强烈
strongly convergent 强烈收敛的
strongly corrosion liquid 强腐蚀性液体
strongly coupled system 强耦合系统
strongly curved reach 急弯河段
strongly damped collision 拟裂变
strong lye 强碱液;强碱水
strongly fissured stratum 裂隙发育地层
strongly flattened 极扁平形
strongly flattened and elongate 极扁平和细长
strongly inaccessible ordinal number 强不可达序数
strongly leached 强度淋溶

strongly magnetic material 强磁性物料
strongly measurable function 强可测函数
strongly mixing 强混合性
strongly mixing automorphism 强混合性自同构
strongly monotone 强单调
strongly monotonic function 强单调函数
strongly non-linear differential equation 强非线性微分方程
strongly non-linear system 强非线性系统
strongly overlapping resonance 强重叠共振
strongly oxidizing slag 强氧化炉渣
strongly permeable stratum 强透水层
strongly salified soil 强盐渍土
strongly salined soil 重盐土
strongly soluble rock 易溶性岩石
strongly soluble salt test 易溶盐试验
strongly stationary 强平稳的
strongly weathered 强风化的
strongly weathered granite 强风化花岗岩
strongly weathered layer 强风化层
strongly weathered mudstone 强风化泥岩
strongly weathered sandstone 强风化砂岩
strongly weathered shale 强风化页岩
strongly weathered soil 强风化土
strongly weathered stratum 强风化层
strong magnetic mineral 强磁性矿物
strong market 看涨的行情;看挺的市场;行情看涨
strong measurability 强可测性
strong measures 高压手段
strong melt-settlement 强融沉
strong metal-complexing agent 强金属络合剂
strong metal support interaction 金属载体强相互作用
strong method 强方法
strong mineralized water 强矿化水
strong mixing 强混合
strong mixing hypothesis 强混合假设
strong mixing property 强混合性
strong monotonic property 强单调性
strong mortar 高强度砂浆
strong motion 强震
strong motion accelerogram 强震加速度图
strong motion accelerograph 强震加速度仪;强震加速度计
strong motion displacement 强震位移
strong motion earthquake 强烈震动;强烈地震
strong motion earthquake instrumentation 强震观测
strong-motion information retrieval system 强震信号检索系统
strong motion seismogram 强烈震波图
strong motion seismograph 强震计;强震仪;强动地震仪
strong motion seismology 强震学
strong noise 高噪声
strong non-linearity 强非线性
strong nonlinear patterns 强非线性类型
strong object 强对比度景物
strong odo(u)r 强恶臭
strong oxidation slag 强氧化性渣
strong oxidative free radical 强氧化自由基
strong oxide-interaction 氧化物间强相互作用
strong oxidizing property 强氧化性
strong pillar 坚固支柱
strong point 优点;坚固支撑点;防卫据点
strong post 坚固支柱
strong pulp 高强度纸浆
strong radioactive test 强放射性试验
strong reducing agent 强还原剂
strong reducing environment 强还原环境
strong reductant 强还原剂
strong relative minima 强性相对极小
strong relief 显著起伏;强烈对照
strong relief area 起伏大的地区
strong renormalization effect 强重正化效应
strong rock 硬质岩石
strong room 保险库
strong runoff zone 强径流带
strong salinization 强度盐渍化
strong sand 肥砂;强黏力砂
strong seam 受载焊缝
strong seismogram 强地震图
strong seismology 强地震学
strong sewage 浓污水;强污水;腐蚀性污水

strong shade 强色调
strong shear capacity and weak bending capacity 强剪弱弯
strong shock 强震;强激波
strong shoot 强枝
strong sideband 强单边带
strong signal 强信号
strong signal blocking 强信号阻塞;大信号阻塞
strong-signal theory 强信号理论
strong solution 浓溶液
strong solvent 强溶剂
strong spring 强弹簧
strong stand oil 厚定油;稠热油;稠聚合油
strong stationary stochastic process 强平稳随机过程
strong steel 高强钢;强力钢
strong sulfuric acid-potassium sulfate copper sulfate digestion-distillation-nesslerization 浓硫酸—硫酸钾—硫酸铜消解—蒸馏—等浓比色法
strong supperradiation 强超辐射
strong supporting layer 强有力的支持层
strong surface-coupling nuclear model 强表面耦合的核模型
strong swelling-shrinkage soil 强胀缩性土
strong tidal environment 强潮汐环境
strong tide 强潮
strong tide river mouth 强潮河口
strong tower 强酸塔
strong toxic pollution 极毒性污染
strong tuning point 强调谐点
strong typhoon 强台风
strong unrestrained type 强烈不可制止型
strong upwelling 强上升流
strong variation 强性变分
strong vat 强碱瓮
strong vibration 强振动
strong vibration observation 强振观测
strong viscidity 强黏性
strong-wash slope 强洗刷坡
strong waste 浓污水
strong weathered zone 强风化带
strong weathering zone 强风化带
strong welded agglomerate 强熔结集块岩
strong welded agglomeratic texture 强熔结集块结构
strong welded breccia texture 强熔结角砾结构
strong welded tuff 强熔结凝灰岩
strong welded tuffaceous texture 强熔结凝灰结构
strong welded volcanic breccia 强熔结火山角砾岩
strong well 高产油井;高产井
strong wind 强风(六级风);大风
strong wind fumigation plume 强风下沉型烟羽
strongyle 两头圆针
strongylote 一头圆针
strontianite 菱锶矿;碳锶矿
strontianite ore 菱锶矿矿石
strontioborite 硼钙石
strontiochevkinite 锶硅钛铈铁矿
strontiodresserite 水碳铝锶石
strontioginorite 锶水硼钙石
strontiohilgardite 含锶氯羟硼钙石
strontiojoaquinite 硅钠锶钡钛石
strontio-orthojoaquinite 斜方硅钠锶钡钛石
strontium acetate 乙酸锶
strontium age 锶龄
strontium barium niobate 铌酸锶钡
strontium borate 硼酸锶
strontium boride 硼化锶
strontium cement 锶水泥
strontium chlorate 氯酸锶
strontium chloride 氯化锶
strontium chromate 铬酸锶;锶黄
strontium cyanide 氰化锶
strontium development line 锶同位案增长线
strontium dioxide 过氧化锶
strontium iodide 碘化锶
strontium lanthanum chromate ceramics 铬酸锶镧陶瓷
strontium-lead isotope correlation 锶—铅同位素相关性
strontium method 锶测定法
strontium mineral 含锶矿物
strontium molybdal 钼酸锶

strontium nitride 氮化锶
strontium ore 锶矿(石)
strontium oxalate 草酸锶
strontium oxide 氧化锶
strontium-oxygen isotope correlation 锶—氧同位素相关性
strontium peroxide 过氧化锶
strontium potassium niobate 铌酸钾锶
strontium silicate 硅酸锶
strontium stannate 锡酸锶
strontium sulphate pigment 硫化锶颜料
strontium sulphide 硫化锶
strontium titanate 钛酸锶
strontium unit 锶单位
strontium white 重晶石类白色体质颜料；锶白
strontium yellow 铬酸锶；锶黄
strontium zirconate 锆酸锶
strontium zirconate ceramics 锆酸锶陶瓷
strop 环索；滑车带
strophoid 环索线【数】
stropping block 索带滑车
Strouhal number 斯特劳哈数；风振系数
Strowger selector 史特劳格尔选择器
Strowger system 史特劳格尔制
struck atom 被击原子
struck brick 压制砖
struck capacity (挖掘机的)装载容量；刮平容量；刮斗容量；平装斗容量；平斗容量；平顶容量；挖掘容量
struck core 车制芯
struck joint 勾缝；斜刨缝；斜缝；刮平缝；刮缝；敲接；随砌随手勾缝
struck joint pointing 刮线勾缝
struck joint work 勾斜缝
struck mo(u)lding 斜刮线脚；整线脚；凹入线脚
struck notch 冲凹口
structional drawing of lining waterproofing 衬砌防护构造图
structite 混凝土快速修补剂
structural 构造的；结构的
structural aberration 结构畸变
structural action 结构作用
structural adequacy 结构上合适性
structural adhesive 结构黏剂；结构黏合剂；结构用黏剂；结构胶黏剂
structural adjustment 结构调整
structural adjustment lending 结构调整贷款
structural adjustment program(me) 结构调整计划
structural alloy steel 结构(用)合金钢；合金结构钢
structural alteration work 结构上的改建工程；结构上的更改工作
structural alternation 改建
structural alumin(i)um 结构用铝；铝结构件；结构铝材
structural amplitude 结构因子
structural analog(ue) 结构模拟
structural analysis 构造分析；结构分析；应力分析
structural analysis package 结构分析包
structural analysis problem 结构分析问题
structural analysis program(me) 结构分析程序
structural anisotropy 构造各向异性；构造非均质性
structural application 结构应用
structural approach 结构变革方式
structural approach limit of tunnel 隧道建筑界限；隧道建筑界线
structural approach to pattern recognition 结构化模式识别法
structural arch analysis 结构拱形分析
structural architecture 房屋建筑
structural area 结构面积
structural aspect 结构状态
structural associations 构造组合
structural attachment 结构附件
structural auscultation 结构检查；结构监测
structural axis 结构轴线
structural balance 结构平衡
structural bar 构造筋；构造钢筋
structural basin 构造盆地
structural bay 结构开间
structural bead 结构加强筋
structural bearing 结构支承
structural behavior 结构性状
Structural Behavior Engineering Laboratory 工程结构性能试验室(美国)
structural behavio(u)r 构造品质；结构性能；结构特性
structural birefringence 形状双折射；结构双折射
structural block diagram 构造立体图(解)
structural body of rock mass 结构体
structural bond 结构性砌接；建筑物的砌合；结构黏结；结构型接头
structural bonding adhesive 结构胶粘剂；结构黏结剂
structural bonding agent 结构黏结剂
structural bonding medium 结构黏结材料；结构黏结介质；结构黏结剂
structural boring 构造钻探；构造钻井
structural borne noise 结构噪声
structural borne sound 结构噪声
structural boundary 构造边界
structural break 结构破损；结构破坏；结构断裂
structural breccia 构造角砾岩
structural bulkhead 主构舱壁
structural calculation 结构计算
structural capacity 结构能力；结构力
structural carbon cartridge 结构炭滤芯
structural carbon steel 结构用软碳素钢；结构碳钢
structural carbon steel hard 结构用中碳素钢
structural carbon steel medium 结构用中碳素钢
structural carbon steel soft 结构用软碳素钢
structural carpentry 大木作；木结构制作
structural casting 结构预制；结构浇制；结构浇注；结构铸件；结构浇铸
structural cast-in-place concrete 现浇混凝土结构
structural cation 结构阳离子
structural cave 构造洞
structural cementing agent 结构用胶凝剂
structural ceramic building unit 结构陶瓷建筑单元
structural ceramics 结构陶瓷；建筑陶瓷
structural ceramics articles 结构陶瓷制品；结构陶瓷产品
structural ceramics product 结构陶瓷制品；结构陶瓷产品
structural change 构造改变；结构改变；结构变化
structural channel 结构用槽钢；槽型钢材
structural characteristic 构造特征；结构特征；结构特性
structural chemistry 结构化学
structural chimney 结构烟囱
structural class 结构等级
structural classification of mountain 山地构造分类
structural clay 耐火泥；结构黏土
structural clay article 结构用黏土制品
structural clay block 结构黏土砌块
structural clay block partition(wall) 结构黏土砌块隔墙
structural clay facing block 结构黏土面块；结构黏土面砖
structural clay facing block partition(wall) 结构黏土面砖隔墙
structural clay facing tile 结构黏土地面砖；建筑面砖；黏土结构面砖
structural clay floor tile 黏土结构地面砖；建筑地面瓷砖
structural clay industry 结构用黏土制品工业；结构黏土制品工业
structural clay partition 建筑黏土隔墙
structural clay product 结构(用)黏土制品；建筑用黏土制品
structural clay tile 黏土砖；黏土结构空心砖；空心黏土砖；建筑地板瓷砖；黏土空心砖；贴面瓷砖；结构空心砖；结构瓷砖
structural clay tile partition(wall) 结构黏土砖隔墙；自承重空心砖隔墙
structural clearance 结构间隙
structural clement of rock mass 结构体
structural coefficient 结构系数
structural cohesion 结构黏聚力
structural colo(u)r 结构色
structural compatibility 结构协调性；结构相容性
structural completion 结构上完成
structural component 结构零件；构件；结构部件
structural component connection 构件接合
structural computation 结构计算
structural concrete 结构用混凝土；结构混凝土
structural concrete core 结构混凝土型芯
structural concrete panel 结构混凝土(铺)板
structural concrete plate 结构混凝土板
structural concrete slab 结构混凝土(平)板
structural concrete topping 结构混凝土面层
structural concrete topping slab 结构混凝土顶盖板
structural concrete unit 结构混凝土构件
structural concrete work 结构混凝土工程
structural concreting 浇捣结构混凝土
structural condition 构造条件
structural configuration 结构形式
structural connection 结构连接；结构接头
structural conservation 结构防腐；结构保护
structural constant 结构常数
structural constituent 结构组分
structural constitution 结构成分
structural constraint 结构约束；结构限制
structural construction 结构施工；承重建筑
structural construction material 结构施工材料
structural contact 构造接触
structural contact type 构造接触类型
structural contour map 构造等值线图；构造等高线图
structural contour map of magnetotelluric method 大地电磁法构造等值线图
structural control 构造控制；结构控制
structural control basin 构造控制盆地
structural control measure 结构控制措施
structural conversion 结构转换；结构变更；结构改变
structural conversion work 结构更改工作
structural coordination 结构配位
structural core drilling 确定结构条件的岩芯钻进
structural coupling 结构耦合
structural cracking 内部裂纹；内部裂缝；结构(性)裂缝
structural cross-section 结构断面
structural cross-section map 构造横剖面图
structural crystallography 构造结晶学；结构结晶学
structural cycle 构造旋回
structural damage 结构损坏；结构破损；结构(的)破坏
structural damping 结构阻尼；结构衰减
structural data model 结构化数据模型
structural deck 结构屋面板
structural deficiency 构造缺失【地】
structural deflection 结构挠度
structural deformation 结构变形
structural deformation facies 构造变形相
structural depression 构造洼地；构造凹陷
structural depth 结构高度；建筑高度
structural description 结构描述
structural design 结构设计
structural design assumption 结构设计的假定
structural design criterion 结构设计标准；结构设计准则
structural design curve 结构设计曲线
structural design drawing 结构设计图
structural design equation 结构设计方程
structural design formula 结构设计公式
structural design hypothesis 结构设计假设
structural design language 结构设计语言
structural design of bridge 桥梁结构设计
structural design of building 建筑结构设计
structural design theory 结构设计理论
structural details 结构细部
structural deterioration 结构损坏；结构破坏
structural development 结构改进
structural diagram 结构(简)图；合金状态图；合金平衡图
structural dimension 结构尺寸
structural discordance 构造不整合；结构不和谐
structural disequilibrium 结构性不平衡
structural disorder 构造无序
structural distortion 结构变形
structural distress 结构隐患；结构性损坏；结构事故；结构破损；结构破坏
structural diversity of habitat 生境结果多样性
structural division 结构划分；构造分区；结构隔板
structural dowel 结构传力杆；结构销钉；结构暗榫
structural drag 结构阻力
structural drawing 建筑图样；构造图；构件图；结构图纸；结构图样；结构图；结构简图；结构构体
structural drawing of construction joint waterproofing 嵌缝防水构造图；施工缝防水构造图
structural drawing of portal waterproofing 洞门

防水构造图
structural drill 构造钻机
structural drilling 构造钻探;构造钻井;结构钻探
structural drilling machine 构造钻机
structural ductility 结构延性
structural durable years 结构耐用年限
structural dynamic(al) response 结构动力反应
structural dynamics 结构动力学
structural eccentricity 结构偏心(距)
structural economic model 结构计算经济模型
structural economics 经济结构学
structural effects index 结构影响系数
structural elasticity 结构弹性
structural element 构造要素;构造体;结构元素;结构元件;结构要素;结构体;结构构件;结构单元;结构成分
structural elements of bit 钻头的结构要素
structural elevation 结构标高
structural employment 结构性就业
structural engineer 结构工程师
structural engineering 结构工程(学);建筑工程
structural engineering system solver 结构工程系统计算程序;结构工程系统程序语言;建筑结构分析语言
structural enrichment 结构装饰
structural environment 构造环境
structural equation 结构方程式;结构方程
structural equation model 结构方程模型
structural equation parameter 结构方程参数
structural equilibrium state of glass 玻璃平衡状态
structural erosion 构造侵蚀
structural error 设计误差
structural evaluation 结构评估
structural evolution 结构演变
structural evolution map 构造演化图
structural evolution of basin 盆地构造演化
structural experiment 结构试验
structural exploration 构造勘探
structural extinction 结构消光
structural extruded product 结构挤压制品
structural extruded section 结构挤压型钢
structural extruded shape 结构挤压型钢
structural extruded unit 结构挤压单元
structural fabric 组构
structural-facies map 构造岩相图
structural facing tile 建筑饰面瓷砖
structural facing unit 建筑饰面构件
structural factor 结构因子;结构因数
structural failure 结构(性)破坏;结构误差;结构损坏;结构失效;结构破损
structural fastener 结构扣件;结构紧固件
structural fault 结构缺陷
structural feasibility 结构可行性
structural feature 构造型相;构造特征;结构要点;构造形迹;构造细部;结构细部;结构外貌;结构特征
structural feature researches 构造形迹的研究
structural features of bedding plane 层面构造特征
structural field concrete 现浇结构混凝土
structural figure 结构图;结构简图
structural fill 结构回填(土);人工填筑垫层
structural film adhesive 结构薄膜胶粘剂
structural fire 结构防火措施
structural fire precaution 结构防火
structural fire protection 结构防火
structural flaw 结构裂缝
structural flexibility 结构柔度
structural floor 结构楼板
structural flow (泥浆的)构造流
structural fluid 结构化流体
structural-fluid intersection 结构流体相互作用
structural foam 结构泡沫(塑料)
structural foam machine 结构泡沫成型机
structural foam mo(u)lding 结构泡沫模塑
structural forging 结构锻件
structural form 结构设计;结构格式;结构形式;构造形态
structural form surface map 构造形面图
structural formula 构造式;结构式
structural fracture 结构裂隙
structural frame 结构框架;构架;结构构架;承重框架
structural framework 结构框架工程;结构框架;结构骨架

structural framework system 结构构架体系
structural framing 构架;结构框架;结构构思;结构构架;结构形式构思
structural function 结构函数;结构功能
structural-functional analysis 结构功能分析
structural gasket 结构密封垫;结构填塞材料;结构垫板;加嵌条窗玻璃垫圈
structural gasket joint 结构填实接缝
structural generation 构造世代
structural geologic(al) features 地质构造特征
structural geology 构造地质学
structural geomorphologic(al) map 构造地貌图
structural glass 玻璃砖;墙面玻璃;结构(用)玻璃;建筑玻璃;大块玻璃
structural glass block 承重玻璃砖;建筑用玻璃砖
structural glass panel 钢筋玻璃镶板;结构玻璃板
structural glazing gasket 承重型玻璃密封条;承重型玻璃衬垫;结构镶嵌玻璃密封垫
structural glazing sealant 结构镶嵌玻璃密封胶;结构胶
structural glue 结构胶
structural glued-laminated lumber 结构用胶合层积材
structural gluing 结构胶粘
structural grade 构造级别;结构用的
structural grade of cement 结构用水泥
structural grade steel 结构等级钢
structural granite 建筑用花岗岩
structural grid 结构柱网;结构网格
structural grid elements 组装墙面连接件
structural group 结构基团
structural group analysis 结构族分析
structural grouting 结构灌浆
structural growth history analysis 构造发育史分析
structural hardening 结构淬火
structural height 结构高度
structural high 构造隆起
structural high-alumina-cement concrete 结构用高铝水泥混凝土
structural history 构造演化史
structural hollow section 结构空心型材
structural hollow tile 多孔混凝土砖
structural homogeneity 组织均匀性
structural identification 结构鉴定
structural imbalance 结构性失调
structural impediment 结构阻扰
structural imperfection 结构缺陷;结构不完整性
structural imporosity 结构紧密性
structural-independent operating system 结构独立操纵系统
structural index 结构指数
structural indication 构造标志
structural inflation 结构性通货膨胀
structural influence factor 结构影响系数
structural information 结构信息
structural inheritance 构造继承;结构继承
structural in-situ cast concrete 结构现浇混凝土;现浇结构混凝土
structural in-situ topping 结构现浇顶盖
structural instability 结构不稳定(性)
structural installation diagram 安装结构图
structural installation drawing 安装结构图
structural installation figure 安装结构图
structural insulating board 建筑绝缘板;结构绝热板
structural integrity 构造完整性;结构整体性;结构完整性;结构坚固性
structural interaction 结构相互作用
structural interference belt 结构干涉带
structural intersection 构造交汇
structural intervention 结构性干预
structural inversion technique 构造反演技术
structural iron 结构用铁;结构铁件;结构铁;建筑钢
structuralism 结构主义
structural isomerism 结构同分异构(现象)
structuralist 结构派
structural joint 构造缝;结构连接;结构结点;结构接点;结构缝
structural joint gap 构造轨缝【铁】
structural landform 构造地形;地质构造地貌
structural layer division 构造层划分
structural layer of covering strata 盖层的构造层
structural layout 结构布置
structural length 构造长度
structural lesion 结构损害

structural level 构造层位
structural life-time 结构使用年限
structural lightweight aggregate 结构轻集料
structural light weight aggregate concrete 结构轻骨料混凝土;结构轻集料混凝土
structural light weight concrete 轻(质)结构混凝土;结构轻质混凝土
structural lime 结构石灰
structural limit 结构强度极限
structural line 构造线
structural load 结构负荷
structural-load-carrying capacity 结构的承载能力;结构承载(能)力
structural load(ing) 结构荷载
structural location 构造上位置
structural longitudinal-section map 构造纵剖面图
structural low 构造低洼
structural lumber 结构(用)木材;建筑(用)木材
structurally complete positive sample set 结构完整的正样本集
structurally insulated panel 结构绝热板
structurally-ordered instrumental variables 结构有序工具变量
structurally weak orientation 沿软方向定向(金刚石)
structural management 结构管理
structural map 构造图;构造等高线图
structural map based on remote sensing 遥感构造图
structural markers 构造标志
structural masonry building component 结构圬工建筑组成部分
structural masonry building member 结构圬工建筑构件
structural masonry building unit 结构圬工建筑单元
structural masonry wall 结构圬工建筑墙
structural mass 结构物质;结构质量
structural mast 格子桅;塔式灯架
structural material 建筑材料;结构材料
structural material dealer 结构材料商(人)
structural material delivery 结构材料运输;结构材料输送;建筑材料运输;建筑材料输送
structural material deposit 结构材料储存;建筑材料储存
structural material engineer 结构材料工程师;建筑材料工程师
structural material failure 结构材料的破坏
structural material industry 结构材料工业;建筑材料工业
structural material machine 建筑材料机械
structural material manufacturer 建筑材料制造者;结构材料生产者;结构材料制造者;建筑材料生产者;建材制造商
structural material market 建筑材料市场
structural material practice 建筑材料业务;建筑材料实践
structural material processing 建筑材料处理;建筑材料加工
structural material producer 建筑材料生产者;建筑材料制造者;结构材料生产者;结构材料制造者
structural material production 建筑材料制造;建筑材料生产
structural material quality 建筑材料质量
structural material quality control 建筑材料质量控制
structural material requirement 建筑材料要求
structural material saving 建筑材料节约
structural material scale 建筑材料尺度
structural material show 建筑材料展室
structural materials industry 建材工业
structural material standard 建筑材料标准;建材标准
structural materials testing machine 建筑材料试验装置;建筑材料试验机;建材试验机
structural material storage 建筑材料仓库
structural material store 建筑材料堆栈;建筑材料商店
structural material test 建筑材料试验
structural material testing device 建筑材料试验设备
structural material testing institute 建筑材料试验研究所
structural measure 结构措施

structural mechanics 结构力学
structural mechanics of beam system 杆系结构力学
structural member 构件;结构元件;结构构件;结构杆件
structural member lofting 结构件放样
structural member shop 结构坚固件车间
structural metal 结构金属;金属结构件
structural metallic materials 结构金属结构材料
structural metalwork 承重金属结构
structural mild steel 结构用低碳钢
structural mill 钢梁轧机;钢轨钢梁轧机;结构钢材轧机
structural model 构造型式;构造模型;结构模型;几何相似模型
structural model analysis 结构模型分析
structural model(l)ing system 结构建模系统
structural model(l)ing technique 结构模拟技术
structural model map 构造模型图
structural model shop 结构模型车间
structural model test 结构模型试验
structural modification 结构修正;结构修改
structural module 结构模数
structural mortar topping(slab) 建筑砂浆顶盖
structural motif 结构基元
structural movement 结构位移
structural mutation 结构突变
structural nature of employment 结构性就业
structural nickel steel 结构用镍钢
structural nose 构造鼻;鼻状背斜
structural nose type of hydrodynamic(al) trap 构造鼻型水动力圈闭
structural obstacles 结构性障碍
structural oil-gas field 构造型油气田
structural oil pool 构造油田
structural operator 结构算子
structural optimization 结构最优化;结构优化
structural optimization method 结构优化方法
structural outline map 构造纲要图
structural panel 结构用板
structural parameter 构造参数;结构参数
structural partition(wall) 结构砌块构成的隔墙;结构隔墙
structural parts 结构部分;构件;结构部件;组成部分;结构零件
structural paste adhesive 结构膏状胶粘剂
structural pattern 构造形式;构造模式;结构样式;结构形式
structural pattern diagram 结构模式图
structural pattern of rock mass 岩体构造类型
structural pattern recognition 结构模式识别;结构图形识别
structural performance 结构性能
structural perturbation 结构扰动
structural petrology 构造岩石学
structural phase transformation 结构相变
structural phase transition 结构相变
structural philosophy 结构哲理(学)
structural pile 承重桩
structural pillar 构造柱
structural pipe 结构用管
structural plan 构件图;结构(平面)图;结构平面;结构简图
structural plane 构造面;结构面
structural plane landslide 构造面滑坡
structural plane of compresso-shearing origin 压扭性结构面
structural plane of rock mass 结构面
structural plane of tenso-shearing origin 张扭性结构面【地】
structural plastic profile 结构塑料型材;结构塑料制品;结构塑料构件
structural plastic section 结构塑料型材
structural plastic shape 结构塑料型材
structural plastic trim 结构塑料贴面
structural plastic unit 结构塑料单元
structural platform 构造台地
structural pool 构造油(气)藏
structural position 构造层位
structural position of ore deposit 矿床构造位置
structural potential 结构势能;结构潜力
structural poured-in-place concrete 建筑现浇混凝土
structural precast concrete 预制混凝土构件

structural precast concrete building frame 预制混凝土房屋框架
structural precasting member 结构预制件
structural probability distribution 结构概率分布
structural problem 结构问题
structural profile 建筑剖面(图);构造剖面图;结构型体;结构剖面
structural property 结构特性;构造特征;结构性质
structural protection 建筑物的保护(措施)
structural province 构造区
structural purity 结构纯度
structural quality 结构性质
structural quality plate 规格钢板;结构钢板
structural query language 结构化查询语言
structural quicklime 建筑用生石灰
structural ratio 财务结构比率
structural rationalism 结构合理主义派【建】
structural readjustment 结构再调整
structural re-analysis 结构的重新分析
structural rearrangement 结构重新排列
structural rebate 结构槽口
structural re-calculation 结构的重新计算
structural reference 结构访问
structural reform 结构改革
structural regression relationship 结构回归关系
structural reinforcement 构造筋;结构配筋
structural reinforcement in monolithic line bed 整体道床结构钢筋
structural reinforcement in open cut tunnel 明挖隧道结构钢筋
structural relaxation 结构松弛
structural reliability 结构可靠性
structural reliability of harbo(u)r engineering 港口工程结
structural relief 构造起伏;构造高差;构造地形
structural repair 结构修复
structural repair material 维修用建材;建筑维修材料
structural representation 结构示意;结构表示法
structural requirement 构造要求;结构要求;结构强度规范
structural research 结构研究
structural resolution 结构分解
structural response 结构响应;结构反应
structural response coefficient 结构响应系数
structural return loss 结构回路损耗
structural rib 承重肋;结构肋
structural rigidity 结构劲度;结构刚度
structural risk analysis 结构危险分析
structural riveting 结构铆接
structural rivet steel 结构铆钉钢
structural rock bench 剥蚀阶地;基岩台阶;构造岩石阶地
structural rolled steel 轧制结构钢
structural rubber profile 结构橡胶型材
structural rubber section 结构橡胶型材
structural rubber shape 结构橡胶型材
structural rubber trim 结构橡胶胶粘面
structural rubber unit 结构橡胶单元
structurals 重型刚架(建筑用的)
structural saddle 构造沉降部分
structural safety 结构性安全;结构安全度
structural safety evaluation 结构安全度评定
structural sandwich 结构夹心板
structural scale 构造尺度
structural scheme 结构方案
structural sealant 结构密封膏
structural sealant glazing 结构密封膏镶嵌玻璃
structural section 结构断面;型材构件;构造剖面图
structural seepage 结构渗水
structural sequence 构造序列
structural shape 建筑型材;结构形式;结构型材;结构形状
structural shape factor of wind load 风荷载体型系数
structural shape steel 结构型钢
structural sheet mo(u)lding compound 结构片状模塑料
structural shell 建筑壳体
structural shop 结构坚固件车间
structural silicone 结构硅酮
structural similarity 结构相似性
structural site-placed concrete 建筑现浇混凝土
structural size 建筑尺寸;结构尺寸

structural skeleton 结构框架;骨架;结构骨架;承重骨架
structural skeleton building 结构骨架建筑
structural skeleton member 建筑骨架构件
structural skeleton structure 建筑骨架结构
structural sketch 构造示意图
structural skin 结构蒙皮
structural slab 结构楼板;板件(房屋建筑中的地板、镶板等);结构板
structural slate 建筑板岩;建筑石板;建筑板材
structural slope 构造成因坡
structural solution 结构题解
structural sound insulation 结构隔声
structural space 结构空间
structural spalling 结构散裂;结构剥落
structural spalling test 结构剥落试验
structural specimen 构造标本
structural stability 结构稳定性
structural stabilization 结构稳定
structural-stable 具有稳定结构的;结构稳定的
structural standing seam alumin(i)um roof panel 结构竖立缝铝屋面板
structural standing seam steel roof panel 结构竖立缝钢屋面板
structural statement 结构语句
structural static mechanics 结构静力学
structural steel 钢架;建筑用钢(材);建筑钢;型钢;结构钢(材)
structural steel chassis 结构钢底盘
structural steel coating 钢构筑物涂料
structural steel column 钢柱
structural steel construction 钢筋结构建筑
structural steel contractor 钢结构承包商
structural steel core 结构钢芯材
structural steel derrick 结构钢钻塔
structural steel design 钢结构设计
structural steel electrode 结构钢焊条
structural steel erection 钢结构安装;钢结构吊装
structural steel fastener 钢结构紧固件;钢结构连接件
structural steel frame 钢结构框架
structural steel hollow section 结构钢空心型材
structural steel lintel 结构钢过梁
structural steel material 钢结构材料
structural steel member 结构钢构件;建筑钢件;钢结构杆件
structural steel pile 承重钢桩
structural steel plate 结构用钢板
structural steel profile 结构钢型材
structural steel rolling mill 建筑钢材轧机
structural steel section 结构钢型材
structural steel shape 结构钢型材
structural steel skid mounted 结构钢撬架暗转的
structural steel trim 建筑钢装饰
structural steel unit 结构钢元件
structural steel work 承重钢框架;钢结构(工程)
structural stiffness 结构劲度;结构刚度
structural stone 建筑用石材
structural strain 结构应变
structural-stratigraphic(al) combination trap 构造地层复合圈闭
structural streak 结构条纹
structural strength 结构强度
structural strengthening 结构加固
structural stress 组织应力;结构应力
structural stress field 结构应力场
structural style 构造样式;结构形式;结构式样;结构风格
structural support 支撑结构
structural surface 结构面
structural synthesis 结构综合
structural synthetic(al) design 结构综合设计
structural system 构造体系;结构系统;结构体系
structural system control of ore deposition 构造体系控矿作用
structural system control of ore types 构造体系控矿类型
structural system control of rocks and minerals 构造体系控岩控矿
structural system of equations 结构方程组
structural systems map 构造体系图
structural T T形钢
structural table 结构表
structural tectonite 构造岩

structural tee T形钢
structural terrace 构造阶梯;构造阶地
structural terrace type of hydrodynamic(al) trap 构造阶地型水动力圈闭
structural terra cotta 结构用陶砖
structural test 结构试验
structural test hole 地质构造的勘察钻孔;地层构造探测孔
structural test model 结构强度试验模型
structural theory 结构理论
structural theory for large-scale systems 大系统结构理论
structural tile partition(wall) 结构陶瓷砖隔墙
structural timber 结构木材;建筑用木料
structural timber connector 结构木材连接件;木结构连接件
structural timber grade 结构木材等级
structural tin 结构用白铁;结构用锡
structural tomb 建筑坟墓
structural tomb architecture 结构坟墓建筑学
structural topping concrete 结构现浇混凝土顶板
structural topping mortar 结构现浇砂浆顶板
structural transformation 结构转变;结构变化
structural trap 易积水的地质构造;构造圈闭;圈闭型储油结构
structural trim 建筑装饰
structural trough 构造凹槽
structural tube 结构管材
structural turbulence 结构湍流
structural type 构造形式;构造类型
structural type map 构造形式图
structural type of rock mass 岩体结构类型;土的结构特征
structural type of soil mass 土体结构类型
structural types of volcanoes 火山构造类型
structural unconformity 构造不整合
structural-unconformity combination trap 构造—水动力复合圈闭
structural-unconformity-stratigraphic(al) combination trap 构造—水动力—地层复合圈闭
structural under construction 施工中的建筑物
structural unemployment 结构性失业
structural uniformity 结构均匀性
structural unit 构件;结构元件;结构构件;结构杆件;结构单元
structural unit of network 网络结构单元
structural upheave 构造隆起
structural use of precast concrete 结构用预制混凝土
structural variation index 结构变异指数
structural veneer 制造承重胶合板用的单板;结构单板
structural vibration 结构振动
structural viscosity 构化黏度;结构黏度;结构化黏度
structural viscosity coefficient 结构黏度系数
structural viscosity index 结构黏度指数
structural wall 剪力墙;结构墙;承重墙
structural wall construction 结构墙体建筑;结构墙体施工
structural wall structure 建筑墙体结构
structural wall tile 承重墙(用)空心砖
structural water 构造水;结构水
structural weight 结构重(量)
structural weight per square meter 建筑自重
structural weld 强固焊缝
structural welding 结构焊接
structural wood 结构用木材;建筑用木料
structural woodwork 大木作
structural work 结构工程
structural working drawing 结构施工图
structural wrench 大钳;叉形扳手
structural zinc 结构用锌
structural zonation 构造分带
structure 构筑物;构造物;结构;建筑物;机构;配置
structure-activity correlation 结构—活性相关
structure-activity relationship 构效关系;结构—活性关系
structure adhesive 结构胶黏剂
structure analysis model 构造分析模型
structure analyzer 结构分析器
structure and arrangement of town 城镇结构和布置
structure and texture 结构构造

structure array 结构阵列;结构数组
structure assignment 构件赋值
structure at initial stage 先期结构
structure attribute 构件属性;结构属性
structure beyond repair 无法修理的建筑物
structure/biodegradability relation 结构生物降解性关系
structure bit 管状钻头
structure bore(hole) 构造钻孔
structure boring 构造钻探;构造钻井
structure-borne noise 结构声(载);结构传递噪声
structure-borne sound 撞击声的发生;结构声(载);结构噪声
structure-borne sound insulation 结构载声隔绝
structure-borne sound insulation tile 结构载声绝缘砖块
structure-borne sound intensity 结构载声强度
structure-borne sound material 结构载声绝缘材料
structure-borne sound transmission 结构载声传递;结构传声
structure-borne sound transmission level 结构载声传递程度;结构载声传递水平
structure-borne sound waves 结构传导噪声波
structure breaking 结构破裂
structure-breaking property 结构破裂性质
structure built on the axial principle 按轴向原理建成的建筑物(指长的建筑物)
structure bulkhead 结构舱壁
structure category 结构目录
structure cavitation 结构汽蚀;结构空穴;结构空蚀
structure ceramics 结构陶瓷
structure chart 结构图
structure clash 结构分裂
structure clearance 建筑接近限界
structure closure 构造闭合度
structure component 结构成分
structure composition 结构综合
structure concrete 结构混凝土
structure conduct-performance framework 结构—行为—实绩体制
structure conformity 构造整合
structure constant 结构常数
structure constituent 结构元件;结构组件
structure contour 结构轮廓线
structure-contour map 构造等高线图
structure-contour method 构造等高线法
structure control 结构控制
structure controller 结构控制器
structure controller module 结构控制器模块
structure conversion 体制转换
structure cost 构造造价;建筑物造价
structure crossing over river channel 跨河建筑物
structured analysis 结构化分析
structured analysis and design technique 结构分析和设计技术
structured approach 构造法
structured assembler 结构化汇编程序
structured buffer pool 结构缓冲区
structured coding 结构编码
structured decision 结构化决策
structured design 结构化设计
structured design method 结构化设计方法
structured design process 结构设计过程
structured design specification 结构化设计技术条件
structure declaration and attribute 结构说明和结构属性
structure deformation 结构变形
structure description 结构简述
structure design database 结构设计数据库
structure deterioration 结构恶化
structure determination 结构确定;结构测定
structured graphics 结构化图形
structure diagram 结构框图;构造图;生产结构图
structured interview 结构性采访
structure discipline 结构专业
structure display 结构显示
structure division 结构部分
structured light 结构光
structured microprogram(me) 结构化微程序
structured multiprocessor system 结构式多处理机系统
structured network 结构网络
structured object 结构化事物;结构化对象

structured object representation 结构化对象表示法
structure double refraction 结构双折射
structured packing 规整填料
structured paint 触变性漆
structured programming 结构(化)程序设计
structure drawing 结构图;结构绘图
structure drill 构造钻机
structure drillhole 构造钻孔
structure drilling 构造钻探;构造钻井
structured specification 结构化技术条件
structured system analysis 结构(化)系统分析
structured system design 结构(化)系统设计
structured system implement 结构(化)系统实现
structure ductility 结构延展性
structured value 结构值
structured variable 结构变量
structured walkover 结构分析
structured walk-through 结构预演;结构业务;结构普查;结构化普查
structured wiring 结构化布线
structure economics 结构经济学
structure editor 结构编辑程序
structure element 结构(单元)体
structure engineer 结构工程师
structure engineering 结构工程
structure excavation 结构开挖
structure experiment 结构试验
structure expression 结构表达式
structure factor 构造因素;结构因数
structure failure 结构毁损
structure filtering 构造滤波
structure fissure 构造裂隙
structure fissure water 构造裂隙水
structure-flank trap 构造翼部圈闭
structure flowchart 结构流程图
structure follows strategy 结构随着经营策略而变
structure for berthing 靠船构筑物
structure for cooling water 冷却水构筑物
structure for water purification 净水构筑物;净化水建筑物;净化构筑物
structure framework of basin 盆地构造格架
structure function 结构功能
structure ga(u)ge 建筑(接近)限界
structure ga(u)ging method 建筑物测量法
structure grade 建筑物等级
structure group 结构群
structure height 构造高度;结构高度
structure hole 构造钻孔
structure identifier 结构标识符
structure indeterminate to the nth degree n次超静定结构
structure index 构造指数;结构指标
structure indicator 结构指标;结构阵列;标准层
structure insensitive property 组织钝感性
structure in space 空间桁架;空间结构
structure installation 结构安装
structure length 结构长度
structureless 无定形的;无结构的
structureless soil 无结构土(壤)
structure level number 结构层数
structure line 构造线
structure-lithologic(al)map 构造岩性图
structure load 结构负荷;结构负载
structure low 构造低地
structure management 结构管理
structure map 构造地图
structure matrix 结构矩阵
structure member 结构成员;结构部件
structure memory 结构存储器
structure memory reference 结构存储器访问
structure mineralogy 结构矿物学
structure modify 结构变更
structure name 结构名
structure node 结构节点
structure node memory 结构节点存储器
structure nose 构筑物的突出部分;构造鼻;建筑物突出部分
structure number 组织号数;结构编号
structure of abrasive tool 磨具结构
structure of absorbed water 吸着水结构
structure of a compiler 编译程序结构
structure of a system 系统结构
structure of basement 基底构造

structure of business unit 营业体制
structure of coal 煤的结构
structure of consumer demand 消费构成
structure of consumption 消费构成
structure of contract 合同结构
structure of crust of weathering 风化壳结构
structure of device 设备结构
structure of energy consumption 能源消费结构
structure of engineering 工程建筑物
structure of equipment 设备结构
structure of fault rock 断层岩构造
structure of file 文件结构
structure of freight transport 货物运输结构
structure of glass 玻璃结构
structure of glass block masonry 玻璃砖砌筑建筑物
structure of goods traffic 货运量构成
structure of heating pipeline 供热管线构造
structure of ice 冰结构
structure of igneous rocks 火成岩岩石构造
structure of impact crater 撞击坑的构造
structure of interest rate 利率结构
structure of intrusive body 侵入体构造
structure of investment 投资结构
structure of loan facility 贷款安排结构
structure of melt 熔体结构
structure of metal complex 金属络合物结构
structure of metamorphic rocks 变质岩岩石构造
structure of mineral deposit 矿床构造
structure of mining 采掘业结构
structure of moon 月球结构
structure of offshore rig 海上钻探装置结构;海上钻井装置结构
structure of orefield 矿田构造
structure of ores 矿石构造
structure of passenger train 旅客列车结构
structure of pedosphere 土圈结构
structure of plain web girders 简单实腹板大梁结构
structure of price 价格结构
structure of railway transportation 铁路运输结构
structure of rate 比率结构
structure of reborner 转化炉的结构
structure of rock mass 岩体构造
structure of rocks 岩石构造
structure of sample 样本构成
structure of sedimentary rocks 沉积岩岩石构造
structure of silicate glass 硅酸盐玻璃结构
structure of silicates 硅酸盐结构
structure of size 规模结构
structure of soil 土壤结构;土的结构
structure of solid web girders 实腹大梁结构
structure of standard system 标准体系表的构成
structure of the model 模型的结构
structure of thermocline 温跃层结构
structure of the soil itself 土壤自身结构
structure of thin-film 薄膜结构
structure of threshold function 阈值函数的构造
structure of urban consumption 城市消费结构
structure of urban ecological system 城市生态系统结构
structure of urban production relations 城市生产关系结构
structure of volcanic crater 火山口构造
structure of wagon 货物列车结构
structure of wood 木材结构;木材构造
structure opening 结构跨径
structure operation 结构操作
structure operation request 结构操作请求
structure option 结构选择
structure outline map 构造纲要图
structure page 结构页面
structure pattern 构造模式
structure plan 结构规划;结构图;结构方案
structure plane 结构平面;结构面
structure pointer 结构指示器
structure preservation coat 结构防腐涂层;建筑物防护层
structure preservation emulsion 结构防腐乳液
structure profile 结构剖面
structure program(me) 结构程序;程序结构
structure programming 结构程序设计
structure qualification 结构限定
structure range 结构限界
structure reference 结构引用

structure representation 结构表达
structure representation in memory 存储器的结构表示
structure retrieval 结构检索
structure ring 结构环
structures as passageway for ore fluids 导矿构造体系
structure section 构造剖面;构造断面
structure sensitive property 组织敏感性;结构敏感性
structures for cooling water 冷却水建筑物;冷却水结构
structures for ore housing 储矿构造体系
structure-sharing 结构共享
structure sketch 构造素描图
structure's life 结构使用年限
structures of hierarchical clustering 多级聚合结构
structure soil 构造土;结构土壤
structures on arch ring 拱桥上部建筑
structure spalling 结构剥落
structure specification 结构说明
structure stability 结构稳定性
structure stabilization 结构稳定
structure standard of privately-owned container 私有集装箱结构标准
structure-steel frame 型钢构架;钢框架
structure storage allocation 结构的存储分配
structure strain measurement 构造应变测量
structure stress 结构应力;接合应力
structure surface of rock mass 岩体结构面
structure symbol 结构符号
structure system 结构体系
structure system in bending 受弯结构体系
structure system in single stress condition 在单一条件下的结构体系
structure system in surface stress condition 在表面应力条件下的结构体系
structure system through interpenetration of folded surface 折叠表面互相穿插的结构体系
structure testing machine 结构试验机
structure toe 建筑物基脚
structure to foundation 结构至基础
structure trap 构造圈闭
structure tree 结构树
structure type 建筑物类型;结构类型
structure type bed material sampler 结构型河床质采样器
structure type classification of rock mass 岩体的结构类型分类
structure type mud flow 结构型泥石流;黏性泥石流
structure under construction 施工中的建筑物
structure unit 构造单元;构造单位
structure variable 结构变量
structure water 构造水
structure with perforated face 带开孔面板结构
structure with rubble mound at seaward face 海侧有抛石棱体的结构
structure with solid face 实体墙结构
structure with storage class attribute 构件的存储类属性
structure zoning 结构分区
structuring 结构化
structuring piece 系杆;支撑件;撑木
structurist 结构艺术家
structurization 结构化
structurology 构造学
strueverite 钽金红石
struggle for existence 生存竞争
strum 筛节;防吸滤网;电缆振动;吸入滤网
strum box 舱底水滤盒;蜂巢箱【给】
strum plate 滤板
strunzite 施特伦茨石
strut 斜杆;压杆;支撑(柱);支撑短柱;横撑;水平撑
strut antenna 支杆天线
strut attachment 支柱附件;支承装置
strut beam 支撑梁;支梁;支墩梁;受压梁
strut boom 柱式伸臂;受压伸臂
strut bracing 斜撑;压杆;支柱;支撑;受压联系件
strut frame 压杆;支撑;支撑框架;支撑架;撑架
strut-framed 撑架式;撑架的;下撑式
strut-framed beam 撑架式梁;杆架式梁;撑托梁
strut-framed bridge 撑架式桥;杆架式桥
strut insulator 支杆绝缘子
strut in trench work 明挖挡墙支柱

strut joint 水平撑接头
strut leg 支柱;斜撑;撑杆
strut member 支撑构件
strut of axle guard 车轴护挡支柱
strut of roof truss 屋架斜撑
strut pile 撑桩
strut rail 撑轨;横撑
strut section 支柱剖面
strut sleeve 支撑套;间隔衬套
strut socket 撑杆座
strut support 柱式支架;支柱架
strutted 加固的;支撑的;支撑式
strutted beam bridge 撑架桥;撑梁桥;撑架梁
strutted dolphin 支撑式系船柱
strutted excavation 加撑深开挖
strutted flap gate 后撑式卧倒门【港】
strutted frame 支撑框架;撑杆式框架
strutted purlin(e) roof 支撑檩条屋顶
strutted roof 支撑屋顶;撑杆式屋顶
strutted sheet piling 支撑板桩;有支撑的板桩;支承板桩
strutted timber beam bridge 撑架木梁桥
strut tenon 斜撑杆;撑榫;撑榫;柱撑榫
strutting 支撑物;加柱;支撑
strutting beam 支撑梁
strutting board 支撑板
strutting piece 支撑杆;系杆;双柱上撑式桁架柱间弦杆;拉杆;跨腰梁;支撑件
strutting ring 支撑圈
struvite 鸟粪石
struvite ore 鸟粪磷矿石
Strux 斯特鲁斯高强度钢
Stuart windmill 斯图尔特风车
stub 线头;线墩;接线;中介轴;支脚;截线;接管座;树桩;短支柱;短卡;短截棒;短粗支柱;短粗卡头;导体棒;存根;粗短支柱;残桩;残极;残段;茬;短木柱
stub abutment 直式桥台
stub acme thread 截顶细流
stub aerial 短截线天线
stub angle 短截棒角接头
stub axle 转向轴;转向节;短(心)轴
stub bar 剩余材料;预埋连接钢筋;露头钢筋;料头;搭接钢筋
stubbing 铲除树桩;根除
stubble 残株;残茬
stubble bottom 茬地犁体
stubble cleaner 灭茬机;表土犁
stubble crop 茬地作物
stubble grazing 茬地放牧
stubble mulch(ing) 茬后覆盖物;茬地覆盖层;残茬覆盖
stubble-mulch tillage 残茬覆盖耕作
stubble seeder 茬地播种机
stubble share 熟地犁铧;茬地犁铧
stubble skim plough 灭茬机
Stubbs-Perry system 斯塔布斯—佩里系统
stubby driver 木柄木螺钉刀
stubby screwdriver 短柄螺丝刀
stub cable 连接电缆
stub card 计数卡片;存根卡片
stub column 短柱;短粗支柱;突出短杆
stub culvert 短廊道
stub cut off 防渗矮墙
stub dolly 短替打垫
stub end 原木大头;连杆端(木料大头);焊条裸尾;大头端;残端
stub ended track 尽头线【铁】
stub-end goods station 尽头式货运站
stub-end goods yard 尽头式货场
stub-end passenger station 尽头式客运站
stub-end siding 尽头线【铁】
stub-end station 回程站;终点站;起点站;端点火车站
stub-end taxiway 一头不通的滑行道
stub-end turnbuckle 截头松紧螺旋
stub-end type junction terminal 尽端式枢纽
stub feeder 直接电源馈路
stub frame 短型车架
stubgrafting 桩接接
stub-in 梁头焊上梁腹
stub ingot 残锭
stub key 摇杆插销
stub line 短截线

stub matching 短截线匹配
stub mortise 短粗榫眼；浅榫眼（不穿透木材）；暗榫眼
stub mortise and tenon 暗榫接头
stub nail 短粗钉
stub pile 短粗桩；短桩
stub pinion 短牙小齿轮
stub pipe 短管
stub pole 短粗柱(杆)；主梁
stub post 墩；短粗柱；短粗柱墩
stub puller 树桩挖掘机
stub riser for the access hatch 窨井井身
stub rod 伸出钢筋；预留钢筋
stub screw machine reamer 短型机用铰刀
stubs ga(u)ge 线径规
stub shaft 短轴
stub shaft drill 短轴钻机
stub shaft drilling machine 短轴钻机
stubs of a paying-in book 解款簿存根
stub station 高峰负载发电厂；尽头线火车站
stub street 残段街道
stub-supported coaxial 短截线支承同轴线
stub-supported line 短截线支承传输线
stub-survivor curve 残段生存曲线
stub switch 平端转辙器；钝轨转辙器
stub teeth 短齿
stub tenon 短粗榫舌；粗短榫
stub terminal 尽头火车站；岔段终点站
stub-tooth gear 短齿齿轮
stub track 尽头线【铁】
stub traverse 支导线【测】
stub tube 短管
stub tuning 短线调谐
stub type station 段末车站
stub wall 短墙；矮墙
stub wings 矮翼墙（桥台顶处）
stuc 仿石粉刷；仿石抹灰；人工造石
stucco 拉毛灰泥；拉毛粉饰；拉毛粉刷；毛粉刷；毛粉饰；灰泥(粉饰)；外墙粉刷；粉饰灰泥；半水石膏与生石膏的混合物
stucco architecture 拉毛粉刷建筑
stucco ceiling 拉毛粉刷天棚；拉毛粉刷平顶；抹灰顶棚
stucco cement 粉饰用砂浆；粉饰灰泥
stucco coating 拉毛涂层；拉毛粉饰涂装；粉刷灰泥涂层
stuccocornic ceiling 灰泥粉饰檐
stucco cornice 拉毛粉刷檐饰；拉毛粉饰檐
stucco-embossed surface 粉饰浮雕表面；外墙面上浮雕装饰
stucco-encrusted wall 拉毛粉刷镶饰的墙；灰泥抹面墙
stuccoer's hammer 粉饰工用的锤
stuccoer's larry 粉饰工用的搅浆铲；粉饰工用的搅浆锄
stucco-exterior-ledger-supported system 金属骨架支承的外墙拉毛粉刷
stucco finished building 抹灰饰面建筑
stucco finished wall 混水墙
stucco fluidized bed 沸腾撒砂
stuccoing 撒砂（熔模铸造时）
stuccoist 拉毛粉刷工匠；拉毛粉刷工人
stucco keratosis 灰泥角化症
stuccolustro 大理石拉毛粉刷
stuccolustro rib 大理石拉毛粉刷的肋
stucco mesh 装饰抹灰用的拉展金属网
stucco pattern 拉毛粉刷型；拉毛粉刷式样；毛粉饰型板；灰泥模；拉毛水泥形式
stucco stippling 拉毛粉饰
stucco strap 拉毛粉刷
stucco work 拉毛粉刷工作；拉毛粉刷工程；外粉刷层；拉毛粉刷
stucco worker 拉毛粉刷工人
stuck bit 卡住的钻头
stuck capacity of bucket 铲斗平斗容量
stuck casing 卡住的套管
stuck drill pipe 卡住的钻管
stuck fish 落物卡钻；被卡（在孔内的）落物
stuck jumper 被卡钎杆
stuck mo(u)ld 粘贴线脚；粘贴线条
stuck mo(u)lding 绊住饰条；斜刮线脚；切出线脚；雕刻线脚；凹入线条
stuck on dust 黏附尘埃
stuck on particles 黏附尘埃

stuck pipe 被卡管子
stuck point pointing 泻水勾缝
stuck rods 被卡钻杆
stuck steel 被卡钻具；被卡钎头
stuck tools 被卡（在孔内的）工具
stuck ware 陶瓷废品
stuc mixture 石屑粉刷拌和物；砌石灰浆
stuc stuff 砌石灰浆
stud 栓子；大头钉；芯撑；装饰钉；柱；链环档；链档；立筋；尽端；的；间柱；蘑菇头（用于固定集装箱）；模板支柱；锚钉；嵌钉；调整螺栓；双端螺栓；饰钉；房屋净高度；房间净高度；板墙筋
Studal 斯塔锻造铝基合金
stud anchor 墙筋锚固件
stud and mud 篱笆抹泥墙；墙筋抹泥
stud behind the face frame 面架后的加强壁骨；面架后的加强支柱
stud bolt 销子；短轴；嵌入螺栓；大头螺栓；柱头螺栓；柱螺栓；双头螺栓柱螺栓；双头螺栓；双端螺栓
stud bolt hole 柱螺栓孔
stud bolt thread 双头螺栓螺纹
stud chain 有档链；柱环节链；日字链
stud clamp 间柱夹紧装置
studded armo(u)ring 带柱销的(辊)套
studded-cage beater 钉头笼式打架器
studded connection 双头螺栓连接
studded roller 锁齿式排种轮
studded tile 有钉齿的瓦
studded tyre 装有防滑钉的轮胎；带钉轮胎
studding 框架建筑墙构件；灰板墙筋（材料）；板筋立筋；间柱支撑；间柱（材料）；木甲板室直木柱；房间净高度；栽丝（焊前的）
studding sail boom hitch 扬帆结
studding-sail tack 帆根结
studdle 棚梁中柱；支撑杆；支撑杆；平台支柱；棚间隔梁；井框间支柱
stud dowel 合缝钉
stud driller 老司钻
stud driver 柱螺栓传动轮；击钉器
stud duck 井口把钩工
stud(ended)track 尽头轨道
student assessment 学生评价
student cooperative housing 学生合作住房
student engineer 见习技术员；见习工程师；实习工程师
students and apprentices 学生及实习生
students' dining hall 学生食堂
students' dormitory 学生宿舍
students' dwelling unit 学生居住单元
students' hostel 学生（寄宿）舍；学生招待所
students' quarters 学生寓所
students' reading room 学生阅览室
students' residential hostel (校外的)学生宿舍
student's t distribution 学生 t 分布
student union 学生联合会；学生活动大楼；学生活动大楼
stud extractor 双头螺栓拧出器
stud framed partition 立筋构架隔断
stud framed wall 立筋构架墙；立筋构架隔断
stud gear 柱齿轮；变速轮
stud grade 立筋等级；立柱等级
stud gun 射钉枪；螺钉枪；销钉枪
stud hexagonal nut 柱螺栓六角螺母
stud-horse bolt 主传动轴轴承的固定螺栓
studied precision 原有研究程度
Studies and documentation 国际建筑研究与文献委员会
studies of east asian tectonics and resources 东亚构造和资源研究
studies of the life histories 生活史的研究
studies of the physical processes 物理过程研究
studio 演播室；画室；电影制片厂；播音室
studio absorption 演播室吸声
studio acoustics 演播室声学
studio apartment 小型公寓住宅；画室公寓；一套小型公寓房间（有工作室的公寓）
studio apparatus 演播室设备
studio broadcast 室内广播
studio camera 电影摄影机
studio camera chain 演播室摄像机通路
studio camera dolly 演播室摄像机移动车
studio camera equipment 演播室摄像设备
studio camera pedestal 演播室摄像机三脚架
studio capacity 演播室容量

studio center 演播中心
studio complex 演播中心
studio control booth 播控调整室
studio control console 演播室控制台
studio control room 演播控制室；播控室
studio couch 三用沙发；可以当床用的沙发；沙发床；可作床用的长沙发
studio crane 演播室升降设备
studio decorations 演播室布景
studio display facility 演播室重放设备
studio easel 画架
studio equipment 播控设备
studio facility 演播室设备
studio floor 演播室表演区
studio foldback circuit 演播室返送电路
studio house (绘画、摄影、录音、演奏等用的)作业房屋
studio light boards 演播室灯光板
studio lights 演播室照明
studio mixer 演播室混合台
studio monitor 演播室监视器；广播监听器
studio pottery 艺术陶器
studio scenery 演播室布景
stud kicker 监工员；检查员
stud lathe 柱螺栓车床
studless link 无挡条环
studless metal lath and plaster partition 无墙筋金属板条粉刷隔墙
stud link 有挡链环；高节链节；日字形链节
stud link chain 有挡链；柱环链；日字链；双头环节链
stud-link chair 日字环节链
stud mill 打碎机；冲击磨
stud mount 柱螺栓装置；铰轴安装
stud nut 柱螺栓螺母
stud opening 毛洞口
stud partition 隔墙架；立筋隔断；间柱隔断；木柱隔墙；木筋隔墙；木隔断墙
stud pin 柱螺栓销；柱螺栓；双头螺栓；大头钉
stud pin inserting machine 销钉封接机
stud removal machine 柱螺栓切削机
stud remover 双头螺栓拧出器
stud ring 齿环
stud rivet 螺柱铆钉；螺旋铆钉
stud riveting hammer 螺柱铆钉锤
studs construction type 壁骨构造体系；支柱构造体系
stud screw 螺柱；双头螺栓；柱螺栓螺钉
stud setter 装双端螺栓工具；双头螺栓拧入器；双头螺栓拧出器
stud shaft 螺栓轴；交换齿轮架中间轴；栓轴
stud shooting 螺钉射入；射钉
stud spacer 柱螺栓隔片；栓螺栓隔片
stud spacing 立筋间距
studs set 锚件
stud stave 标桩
studs type of construction 建筑的壁骨类型；建筑的支柱类型；构造的壁骨类型；构造的支柱类型
stud switch 按钮式转接开关
stud terminal 尽端终点站
stud type chain 日字环节链；挡环式链
stud wall 立柱墙；中心立柱墙
stud wall(partition) 立筋隔墙
stud weld(ing) 柱极焊；柱钉焊接；螺柱焊；螺栓焊；电栓焊；螺栓焊接
stud welding gun 螺柱焊枪
stud wheel 柱栓齿轮；柱齿轮
stud wheel plate 柱栓齿轮板
stud work 安装灰板墙工程；安装板墙筋作业；(墙和房屋的)骨架工程；带立筋的砖砌体；安装灰板墙筋工作；有间柱的建筑物
study 研究
study and test fee 研究试验费
study-bedroom 学生宿舍；书室一卧室
study center 研究中心；学习中心
study group 研究组
study house 研究房
studying degree 研究程度
study level of investigation area 调查区研究程度
study method of recent tectonic stress field 现代构造应力场研究方法
study mission 调查任务
study model 研究模型
study of foliation 面理的研究

study of highway 公路的调查研究
study of investment opportunities 投资机会研究
study of local structure 局部构造研究
study of mineral deposit 矿床学
study of multilevel topography 多层地形的研究
study of ore dressability 矿石可选性研究
study of ore-dressing scheme 选矿流程研究
study of regional seismic stratigraphy 区域地震地层学研究
study of regional structure 区域构造研究
study on agrobiology 农业生物学研究
study on controlling of winter injury 冬害防治研究
study on large sample 大样本研究
study on operating efficiency 工作效率研究
study on water management scheme 进一步研究排水计划
study phase 研究阶段
study plot 试验田;试验区;试验地;标准地
study range 研究范围
study reach 研究河段
study room 书房
study scale 研究范围
study tour 科研出差;考察旅行
study year 研究年度
stuetzite 六方碲银矿
stuff 原料;填充料;三股油麻绳;材料
stuff bending machine 人工弯筋机
stuff chest 储[贮]浆池
stuff colo(u)ring 布料着色
stuff cusp configuration 填充会切位形
stuff drawn roller 人工喂料辊
stuff dug shaft 人工挖掘井
stuffed 实心的
stuffed basin 填塞盆地
stuffed mineral 填塞矿物
stuffer yard 衬垫纱线;填充纱线
stufffed animal 剥制动物标本
stuff goods 呢绒;毛织品
stuff grinder 磨浆机
stuffiness 不通气
stuffing 胶合密封;填塞;装箱;货物装箱;填塞料;填密;填充剂;填充
stuffing and stripping capacity 拆装箱能力
stuffing and stripping shed 装箱拆装库(集装箱);集装箱拆装库;拆装箱库
stuffing and unstuffing of container cargo 集装箱货物拆装作业
stuffing box 密封垫;密封圈;密封箱;密封盒;填料箱;填料盒;填料涵盖;填料函;填料槽;封盖盒;防喷管堵头
stuffing box bearing 填料压盖轴承
stuffing box bolt 填料盒螺栓
stuffing box bulkhead 尾尖舱舱壁
stuffing box casing head 盘根盒套管头
stuffing box compensator 填料箱补偿器
stuffing box gland 密封压盖;盘根盒压套;填料压盖;填料涵压盖;填料盖
stuffing box heat exchanger 填料函式换热器
stuffing box model 填料箱型号
stuffing box recess 轴隧尾室
stuffing box thermal compensator 填料函热补偿器
stuffing case 填料函
stuffing chamber 填料压盖室;衬套室
stuffing digit 塞入数字
stuffing gland 密封压盖;填料箱压盖;填料盒压盖;填料函压盖
stuffing rate 塞入率
stuffing ring 填料压盖
stuffy 通风不良的;不通风的
stuffy nose 曲差
stuffy room 不通气房间;不通气的室;不通风的房间;闷热的房间
stugged 粗加工石料;石料粗加工
stugging 轻琢石面;凿石;石料凿毛
stuke 粉刷;拉毛粉刷;抹面;水泥仿石粉刷
Stuko-bak (用于外粉刷背面的)油毛毡
stulken system 翻转吊杆装卸方式
stull 顶梁(巷道);采掘台;支柱;横撑;采空区
stull-divided 横撑分隔的
stulled 横撑支护的(隧洞掘进巷道)
stull floor 横撑支柱假顶
stull piece 加固斜撑木

stull stoping 撑木回采法
stumbling block 绊脚石;障碍物
stump 桩的残根;柱桩;钢凿承托;木桩(障碍物);平茬;树桩;树根桩;树墩;钝钎子;短柱;池窑底部玻璃;残桩;残株;残余部分;残干;伐根
stumpage 伐根年龄;立木山价;立木采伐权;立木;未砍伐的木材;伐木支出
stump auger 挖根螺旋钻
stump clearing digger 树桩挖除机
stumper 掘根机;除根机;伐木机;拔根机
stump fence 树桩围培
stump figure 根株花纹
stumpflite 六方锑铂矿
stump foundation 树桩基础
stump-grubbing 开垦
stump height 伐桩高
stumping off 树根掘除
stump-jump device 跃障复位装置
stump-jump disk plow 跃障式圆盘犁
stump-jumper 带弹簧安全器的犁
stump-jump mechanism 跃障机构
stump mast 无桅帽的短上桅
stump mortise 浅榫眼
stump pasture 森林砍伐后牧地
stump plant 栽植平茬苗;平茬苗
stump puller 挖根机;除根机;拔根机
stump remover 树桩拔除机
stump rooter 拔根机
stump rot 伐根腐朽
stump scale 根际材积
stump tenon 宽脚短凸榫;宽脚榫;宽脚短粗榫
stump topgallant mast 短上桅
stump tracery 德国哥特式花格窗
stump turpentine 明子松节油
stump veneer 树桩饰面板
stumpwood 伐根材
stunner 假支票;假钞票
stunner knot 双花钩结
stunt 矮化
stunt box 限定器;特技匣
stunted brushwood 矮小灌木
stunt end 模数堵头;(浇灌混凝土时用的)堵头板
stunt-head 施工日志进度表;(浇灌混凝土路面接缝的)堵头板;封头板
stunt photography department 特技车间
stupa 印度神龛塔;印度塔;卒塔婆(印度佛塔);佛塔(印度)
stupa base 塔基
stupalox 陶瓷刀(美国商品)
stupa mound 神龛塔墩(印度)
stupa-mound base 神龛塔基(印度)
stupa shrine 神龛塔神座(印度);塔庙
stupa stylobate 塔柱座
stupe 热敷布
stupid 空心砌块制坯机
stupp 粗汞华
sturdy 坚固
sturmanite 硼铁钙矾
Sturmer viscometer 斯特梅尔型黏度计
Sturm-Liouville problem 斯图姆—刘维尔问题
Sturm-Liouville system 斯图姆—刘维系
Sturm motor 斯特姆式叶片液压电动机;斯特姆式叶片液压马达
Sturm's sequence 斯图姆序列
Sturm's theorem 斯图姆定理
S-turn 蜿蜒河段
Sturtevant type air separator 斯笃尔特万特型离心式选粉机
Sturtevant whirlwind classifier 斯笃尔特万特型离心分级机
Sturtevent blower 斯笃尔特万特型鼓风机
sturtite 水硅铁锰矿
Sturt orogeny 斯图尔持运动
Sturtvant type air separator 离心式选粉机
stutite 水丝铀矿
stutter 扫描残迹
stutter shooting 时序截断脉冲发射
St. venant plastic body 圣·维南塑性体
St. venant principle of elasticity 圣·维南弹性原理
St. venant torsion 圣·维南脱扭力
stwining stem 缠绕茎
stybolate 柱基;柱座
styela 海鞘

styela clava 海鞘
stylar 尖的
stylar facade 外正立面(房屋)
style 样式;格式;类型;款式;式样;风格
style-bbok 样本
style character(istic) 风格特性;风格特征
style detail 风格细节
style device 风格特色;风格面貌;(建筑上的)特点
style element 风格元素
style feature 风格面貌;(建筑上的)特点
style herrera 赫雷拉风格
style modern 现代建筑形式
style of architecture 建筑格式
style of basin 盆地形式
style of calligraphy 字体
style of decoration 装饰风格
style of ornamentation 装饰风格
style of the shoe 马掌型;蹄型
style sheet 样图
styles of dressing 修整类型
stylet 通管丝;小尖刀;匕首细探子;管心针;探针
styli density 记录针密度
stylidium 柱体
styliform 笔尖形的
styliform ornament 针状装饰
styling mousse 定型摩丝
stylist 新式样设计师
stylistic apparatus 新款设备
stylistic conception 合时的概念;风格概念
stylistic departure 风格上的变更;风格上的偏离
stylistic development 式样开发;风格形成
stylistic feature 式样特征;风格特征
stylistic forerunner 文体先驱;风格先驱
stylistic form 风格形式
stylistic formula 风格程式;风格定则
stylistic history 风格历史;风格形成过程
stylistic idiom 独特风格;风格习语
stylistic imitation 风格上的模仿
stylistic perfection 风格上的完善
stylistic phase 风格方面
stylistic predecessor 文体先驱;风格先驱
stylistic purity 风格上的纯正
stylistic significance 风格显著特征;风格意义
stylistic tendency 风格趋势
stylistic unity 风格上的统一(协调)
stylization 因袭;风格上的效仿;仿效(风格上的)
stylized 使合于某种风格
stylized symbol 象征符号
stylobate 台基;柱列脚座【建】;柱座
stylode 分枝花柱
styloid 柱状晶体
stylolite (石灰石等中的)小石柱;缝合岩面;缝合面
stylolite line 缝合线
stylolites 细缝
stylolitic 柱状;缝合的
stylolitic cleavage 缝合劈理
stylolitic structure 柱状构造;缝合结构;缝合构造
stylometer 柱斜度量测器;柱身收分测量器;量柱斜度器
stylospore 柄生孢子
stylus 指时针;管心针;刻针;记录针;描画针;铁笔;触笔;笔状突起;笔头(记录仪);笔尖
stylus holder 触针座
stylus pen 铁笔
stylus pin 指销
stylus point 记录针;划线针
stylus pressure 针压
stylus printer 触针打印机;点阵式打印机
stylus type instrument 记录针式仪器
styolite (大理石中的)纵向纹理结构
styphnic acid 收敛酸
styracin 安息香英
styracitole 安息香醇
styrax japonica 野茉莉
stremic 高耐热性苯乙烯树脂
styrenated alkyd 苯乙烯化醇酸;苯乙烯改性醇酸树脂;苯乙烯改性醇酸
styrenated alkyd coating 苯乙烯改性醇酸树脂涂料
styrenated linseed oil 苯乙烯化亚麻油;苯乙烯改性亚麻油
styrenated oil 苯乙烯化油
styrenated phenol 苯乙烯酚
styrenated soyabean oil 苯乙烯改性豆油
styrenated tall oil 苯乙烯改性妥尔油

styrene 苯乙烯
styrene-acrylic emulsion 苯丙共聚乳液
styrene-acrylic exterior wall coating 苯丙外墙涂料
styrene-acrylic interior latex paint 苯丙内墙乳胶漆
styrene-acrylic interior wall matt latex paint 苯丙内墙无光乳胶漆
styrene acrylic latex 苯乙烯丙烯酸乳胶;苯丙乳胶
styrene-acrylic latex coating 硅灰石水性乳胶涂料;苯丙乳胶漆
styrene-acrylic latex primer for metal 苯丙型金属乳胶底漆
styrene-acrylic matt latex paint 苯丙无光乳胶漆
styrene-acrylonitrile 苯乙烯丙烯腈
styrene-acrylonitrile copolymer 苯乙烯丙烯腈共聚物
styrene-acrylonitrile resin 苯乙烯丙烯腈树脂
styrene alloy 苯乙烯合金
styrene-butadiene block copolymer 苯乙烯—丁二烯嵌段共聚物
styrene-butadiene coating 丁苯乙烯涂层;丁苯涂料;涂布丁苯涂料;丁苯涂层
styrene-butadiene compound 苯乙烯—丁二烯化合物
styrene-butadiene copolymer 苯乙烯—丁二烯共聚物
styrene-butadiene latex 苯乙烯—丁二烯乳液;苯乙烯—丁二烯乳胶
styrene-butadiene resin coating 丁苯树脂涂料
styrene-butadiene resin emulsion 丁苯树脂乳液;苯乙烯—丁二烯树脂乳液
styrene-butadiene rubber 充油丁苯橡胶;合成丁二烯橡胶;苯乙烯—丁二烯橡胶
styrene-butadiene rubber modified asphalt 丁苯橡胶改性沥青
styrene-butadiene rubber sealant 丁苯橡胶密封膏
styrene-butadiene-styrene 苯乙烯—丁二烯—苯乙烯橡胶
styrene-butadieric resin coating 苯乙烯—丁二烯树脂涂料
styrene copolymer 丁二烯苯乙烯共聚物
styrene dielectric(al)cable 苯乙烯绝缘电缆
styrene emission 苯乙烯挥发
styrene emulsion 苯乙烯乳剂
styrene foam 泡沫聚苯乙烯
styrene glycol 苯代乙二醇
styrene ion exchange resin 苯乙烯离子交换树脂
styrene isoprene rubber 聚苯乙烯—异戊二烯橡胶
styrene latex 丁二烯苯乙烯乳胶
styrene maleic anhydride copolymer 苯乙烯—马来酐共聚物
styrene-methyl methacrylate 苯乙烯—甲基丙烯酸甲酯
styrene monomer 苯乙烯
styrene oxide 氧化苯乙烯;苯乙烯化氧
styrene plastics 苯乙烯类塑料;苯乙烯塑料
styrene polyester 苯乙烯聚酯
styrene polymer 苯乙烯聚合物
styrene resin 苯乙烯树脂
styrene rubber 苯乙烯橡胶
styrene-rubber fittings 苯乙烯—橡胶管件
styrene-rubber pipe 苯乙烯—橡胶管
styrene-rubber plastics 苯乙烯—橡胶塑料
styrene-solubility test 苯乙烯溶解度测定
styrene sulfonic acid 苯乙烯磺酸
styrene-tall oil copolymer 苯乙烯—妥尔油共聚物
Styrian orogeny 斯提里运动
styroflex 聚苯乙烯
styroflex cable 聚苯乙烯软性绝缘电缆;聚苯乙烯软电缆
styrofoam 泡沫聚苯乙烯
styrofoam insulation 泡沫聚苯乙烯绝缘板
styrofoam insulation material 泡沫聚苯乙烯绝缘材料
styrol alsyd rsein 苯乙烯化醇酸树脂
styrolene 苯代乙撑
styrol reisn 苯乙烯树脂
styropor 膨胀型聚苯乙烯材料
styryl 苯乙烯基
styrylamine 苯乙烯胺
styryl ketone 二苯乙烯酮
styryl methyl ketone 苯乙烯基甲基酮
suaeda glauce 咸蓬
suanite 遂安石
sub 转换接头;代替物;次于;残株

subaccountant 副会计员
subacidity 微酸性
subaclinal 微倾斜的
subacture toxicity test of aquatic organism 水生生物亚急性毒性试验
subacute 稍尖的
subadditive function 次可加函数
subaddress 子地址
subadiabatic layer 微绝热层
subadjoint surface 次拌随曲面
subaeration 底吹法
subaeration flo(a)tation machine 机械搅拌式浮造机
subaeration machine 液下空气吹入式浮选机;底吹式浮选机
subaerator 底吹式浮选机;底吹机
subaerial 近地面的;地表的
subaerial area of delta 三角洲水上部分面积
subaerial delta 水上三角洲
subaerial delta plain 陆上三角洲平原
subaerial denudation 陆上剥蚀;陆地剥蚀
subaerial deposit 陆上沉积物;地表沉积
subaerial-laminated crust 陆上纹理壳
subaerial landslide 陆上滑坡
subaerial levee 陆上堤
subaerial sediment 陆上沉积物
subaerial spring 地表泉
sub-A flo(a)tation cell 丹弗型底吹式浮选机
subagency 分销处;分代理处
subagent 代办助理;代理行分行;付代理人;分代理(人);副代理人
subaggregate 子集
subalgebra 子代数
subalkalic rock 次碱性岩
suballocated file 次分配文件
suballocation 次分配;子分配
suballotment 分拨款;转分配
subalphabet 部分字母
subalpine belt 亚高山带
subalpine flora 亚高山植物区系
subalpine meadow 亚高山草甸
subalpine peat 亚高山泥炭
subalpine region 亚高山(地)区
subalpine rendzina 亚高山黑色石灰土
subalpine zone 亚高山带
subaluminous rock 次铝质岩
subambient chromatography 低温色谱法
subambient gas chromatography 低温气相色谱法
subangle 分角;副角
subangular 略有棱角的;次棱角的;次棱角状
subangular particle 次棱角土粒
Subantarctic intermediate water 亚南极中层水
subantarctic zone 亚南极带
subanthraxylon 半镜煤
subantiparticle 反亚粒子
subapplication 分申请
subaqual landscape 水下景观
subaquatic cable 水底电缆
subaquatic delta 水下三角洲
subaquatic helmet 潜水盔
subaquatic illumination 水下照明
subaquatic plant 水下植物
subaquatic rock breaker 水下碎石机
subaquatic rock excavation 水下开采岩石
subaquatic signal 水下信号
subaquatic spring 水下泉
subaquatic tunnel 水下隧洞;水下隧道
subaquatic work 地面下自流井;半自流井
subaqueous 水下的
subaqueous antifouling coating 水下防腐涂层
subaqueous area of delta 三角洲水下部分面积
subaqueous bar trap 水下坝砂体圈闭
subaqueous blasting gelatin(e) 水下炸胶
subaqueous breccia 水成角砾岩
subaqueous cable 过河电缆;水下电缆;水底电缆
subaqueous cable house 水线房
subaqueous cable marker 水线标志牌
subaqueous concrete 水中混凝土;水下混凝土;水底混凝土
subaqueous concreting 水下浇筑混凝土;水下浇注混凝土;水下灌注混凝土;水下混凝土施工;水下灌筑混凝土;水下灌注混凝土
subaqueous construction 水下施工
subaqueous corrosion 水下侵蚀

subaqueous debris deposit 水下泥石流沉积
subaqueous debris flow 水下泥石流
subaqueous delta plain 水下三角洲平原
subaqueous deposit 水底沉积(物)
subaqueous detonator 水下起爆器;水下雷管
subaqueous dike 潜堰
subaqueous dune 水下丘
subaqueous execution 水下施工
subaqueous foundation 水下基础
subaqueous foundation trench cutting 水下基槽开挖
subaqueous geomorphy 水下地貌
subaqueous glide 水底滑动
subaqueous gravity flow deposit 水下重力流沉积
subaqueous helmet 潜水盔
subaqueous jet 水下射流
subaqueous landslide 水下滑坡
subaqueous levee 水下堤
subaqueous loudspeaker 水下扬声器
subaqueous mat(tress) 水下柴捆垫
subaqueous mining 水下采矿
subaqueous pipe 水下管
subaqueous pipeline 水箱管线;水下管道
subaqueous pump 潜水泵
subaqueous reconnaissance survey 水下勘测
subaqueous rock excavation 水下岩石开挖;水下岩石采掘;水下采掘岩石
subaqueous running survey 水下草测
subaqueous sample 水下土样;水下试样
subaqueous sand ridge 水下沙脊
subaqueous section 水下段
subaqueous shock 水下振动
subaqueous signal 水下信号
subaqueous signal(l)ing 水下信号装置
subaqueous sill 潜槛;水下潜槛
subaqueous slump 水下滑塌
subaqueous slump deposit 水下滑塌沉积
subaqueous slumping 水下滑动
subaqueous sound ranging 水声测距
subaqueous spring 海底泉
subaqueous suit 潜水衣
subaqueous survey 水下测量
subaqueous topographic map 水下地形图
subaqueous tube 水下管
subaqueous tunnel 越江隧道;河底隧道;水下隧洞;水下隧道;水底隧道;暗沟
subaqueous tunnel(l)ing 开掘水下隧道
subaqueous vehicular tunnel 水下交通隧道
subaqueous work 水下施工
subarch 副拱
subarctic 副北极的
subarctic climate 泰加林气候
subarctic region 副北极带
subarctiv zone 副北极带
subarcuation 副拱结构;结构的副拱
subarea 分区
subareal discharge 部分流量(两测流垂线间的)
subarid 亚干旱的;半干旱的
subarkose 亚长石砂岩;次长石砂岩
subarm 辅助臂
subarray 子台阵;分阵列
subarterial 次干路
subarterial railway 铁路辅助干线
subarterial street 辅助干道
subartesian 半自流的
subartesian pressure 半自流水压力
subartesian water 半自流(地下)水
subartesian well 地面下自流井;半自流井
subassemblage 组合部件(建筑框架)
subassembled part 半装配部件
subassembler 局部装配工;部件装配工
subassembles 分段结合体
subassembly 分部(装配);部件(装配);组(合)件;装配组件;局部装配;配件分组装配处;分组合件;分段装置;分段装配
subassembly wrapper 元件盒
subAtlantic epoch 亚大西洋期
subatmospheric(al)discharge 亚大气压放电
subatmospheric(al)heating system 负压供暖系统
subatmospheric(al)hydraulic model 负压水工模型
subatmospheric(al)pressure 大气负压;次大气压力;真空计压力;负压
subatomic particle 次原子粒子

subaudible 次声频的
subaudio 亚音速的;亚音速的
subaudio frequency 声频下频率
subaudio generator 亚音频振荡器
subaudio-grade channel 亚音频级信道
subaudio oscillation 亚音频振荡;亚声波振荡
subaudio oscillator 亚音频振荡器
subaudio wave 亚声波
subauroral latitude 极光下点纬度
subaverage 低于平均值(的)
subballast 底渣;底碴(垫层)
subballast layer 底碴层
subband 次能带;分波段;部分波段
subbase 底基(层);底床;基础下卧层;土基;底盘座;踢脚板;墙裙板;路面底层;基层下层
subbase concrete 铺底混凝土
subbase course 底基层;底垫层;基底层;基础下卧层;基层(下层)
subbase course drain(age) 底基层排水
subbased arch 矮矢拱
subbase deformation 地基变形
subbase deformation modulus 地基变形模数;地基变形模量
subbase drain 基层排水;底基层排水(管)
subbase drainage 地基排水
subbase drainage system 地基排水系统
subbase friction 基层摩擦(力);底基层摩擦
subbase layer 底基层
subbase material 底基层材料
subbasement 副地下室;副地下层;底层下的地下室;下层地下室
subbase mounted 板式安装的
subbase mounting 板式连接
subbase of stone pitching 石砌护坡的底层
subbasin 支流流域;次盆地;分流域
subbeam 副梁
subbed 路基
subbidder 分包投标人
subbids analyzing 分包报价分析
subbill of lading 副提单
subbing 潜水灌溉;地下渗灌;底层涂布
subbiosystem 子生物系统
subbituminous coal 次烟煤
subblock 小组信息;亚区域
subboard 子插件
subboiling still 逊沸蒸馏器
subbolster 辅助模座
subboreal 亚北部的
subbottom 海底浅层;底基
subbottom depth recorder 海深记录仪;海底深度记录仪;海底浅层深度记录仪;海底地质探测仪;海底地层记录仪
subbottom formation 海底地层
subbottom profile 海底地层剖面;水底浅层剖面;浅层剖面
subbottom profile exploration 浅层剖面探测;浅层剖面勘探;水下浅层剖面探测;水下浅层剖面勘探
subbottom profiler 海底浅地层剖面仪;海底地层剖面仪;水底浅层剖面仪;地层剖面仪;浅层剖面仪
subbottom profiling 水底浅层剖面测量
subbottom profiling exploration 水下地层剖面勘探
subbottom profiling system 海底浅地层剖面仪
subbottom structure 海底构造
subboundary 亚晶界
subboundary structure 亚晶界结构
subbox 小格子
subbranch 分店;子分支;分行
subbuck 辅助框架;支承墙面板的框架;未装修的毛门窗
subbuilding drainage system 地下建筑排水系统
subbureau transport profit 分局运输利润
subburst light 扇形气窗
subburst pleats 辐射式褶裥
subcabinet 分线箱
subcaliber equipment 次口径装置
subcapillary interstice 亚毛细孔隙;次毛细间隙
subcapillary openings 亚毛细孔;次毛细孔
subcapillary porosity 亚毛细管孔隙率;次毛细管孔隙率
subcarbonate 碱式碳酸盐
subcarrier 副载波
subcarrier frequency 副载频

subcarrier frequency modulation continuous wave laser ranging 连续波副载波调频激光测距
subcarrier modulator 副载波调制器
subcarrier monitor 副载波监视器
subcarrier multiple access 载波多址接入
subcarrier oscillator 副载波振荡器
subcarrier pulse 副载波脉冲
subcarrier tracking loop 副载波跟踪回路
subcasing 未装修的毛门窗框;毛的门框
subcatchment 支流集水区
subcategory 副类;亚类;子范畴
subcavitating craft 亚空泡水翼船
subcelestial point 星下点
subcell 小格;亚晶胞
subcellar 下层地窖;下层地下室;副地下室;最下层地下室(多层地下室中的最下层)
subcellular structure 亚细胞结构
subcenter 次中心;辅站;主分支点;副中心
subcenter planning 次中心规划
subcentral bay 次间
subcentre planning 次中心规划
subchannel 子通道;支通道;副航道;分通道;分流道
subchannel address 分通道地址
subchannel bus 分通道总线
subcharterer 转租合同承租人
subchartering 转租船舶
subchloride of mercury 氯化亚汞
subchord 短弦;副弦(杆);子弦
subcircuit 支路【电】;辅助电路;分支电路
subcircular 近似圆形;近似环形
subcivic center 城市副中心;副市中心
subcivic center planning 副中心规划
subclass 小分类;亚纲
subclassification 小分类;子分类
subclause 子条款;分条款
subclay 亚黏土
subclient 分委托人
subclimax 亚顶极
subclinical intoxication 潜在型中毒
subcloud layer 云下层
subclutter visibility 副干扰可见度
subcoating 包底衣
subcode 副码
subcommand 子指令
Subcommission on Stability and Load Lines and Fishing Vessel Safety 稳性和载重线及渔船安全分委员会
Subcommission on the Carriage of Dangerous Goods 危险货物运输分委员会
subcommittee 小组;委员会分会;分组委员会;分委员会;小组委员会
Subcommittee on bulk chemicals 散装化学品分委员会
Subcommittee on Containers and Cargoes 集装箱和货物分委员会
subcommittee on fire protection 防火分委员会
Subcommittee on Life-saving Appliances 救生设备分委员会
Subcommittee on Life-saving Search and Rescue 救生和搜救分委员会
Subcommittee on Safety of Navigation 航行安全分委员会
Subcommittee on Ship Design and Equipment 船舶设计和设备分委员会
subcommutation 副换接
subcompact 次紧的;超小型汽车
subcompany 子公司;附属公司
subcompartment 小班
subcomponent 亚成分
sub concrete 底层(混凝土之下)
subconductor 再分导线;次导体
subconfiguration 子格局
subconical 近圆锥形的
subconjunctival capillaries 赤丝
subconnector 辅助连接器
subcon offer 附条件要约
subconsequent river 后生(顺向)河;次顺向河
subconsequent stream 后生(顺向)河;次顺向河
subconsistent 次相容
subconsole 辅助控制台
subconstruction 基础结构;地下建筑
subconsultant 分包咨询者
subcontinent 次大陆
subcontract 分订契约;分订合同;转订合同;转包合同;局部收缩;合同副本;契约副本;分包契约;分包(合同)
subcontract bids 分包工程投标
subcontract change notice 转包合同更改通知
subcontracted work 分包合同工程
subcontract factory 分包工厂
subcontracting 转包;分包(生产)
subcontracting agreement 分承包协议
subcontracting arrangement 转包办法;分包办法
subcontracting cost 转承包(合同)成本;分包成本
subcontracting manufacture 分包生产
subcontracting records 分包付款的核算
subcontracting shift 分包制
subcontracting system 分包制度
subcontract item 转包合同项目
subcontractor 小承包商;二次转包;副承包商;副承包人;分承包人;小包工;转包人;分约人;分包者;分包商;分包人;二包者;二包商;二包人;二包单位
subcontractor bonds 分包人契约
subcontract price 分包价格
subcontrol 副控制
subcontrol panel 分控制板
subcontrol station 分指挥局
subcontrol traverse 采区控制线
subcooled 低温冷却的
subcooled boiling 过冷沸腾
subcooled flake ice 过冷冰片;低温冰片;超冷冰片
subcooled liquid 过冷液体
subcooled state 过冷状态
subcooled water 过冷水
subcooler 再冷却器;过冷却器;过冷器;低温冷却器
subcooling 过冷;局部冷却;低温冷却
subcooling austenite 过冷奥
subcooling coil 过冷盘管
subcooling condensate 过冷凝液
subcooling condenser 再冷凝器;低温冷却器
subcooling heat rejection effect 过冷段排热效能
subcooling liquid 再冷液体
subcountry 城市化的农村地区
subcoupling 异径接头;变径接头
subcovering license 分批输入许可证
subcriber connector 用户活动连接器
subcriber line free 用户线空闲
subcriber loop 用户环路
subcriber network interface 用户网络接口
subcriber's premises network 用户驻地网
subcriber switch 用户开关
subcript 角注
subcripted qualified name 下标限制名
subcritical 低于临界的;亚临界;亚临界的
subcritical annealing 亚临界温度退火
subcritical area of extraction 亚临界开采面积;亚临界开采区;次临界开采面积;次临界开采区
subcritical assembly 次临界装置;次临界反应堆
subcritical austenite 亚临界奥氏体
subcritical centrifuge 亚临界离心机
subcritical condition 亚临界条件
subcritical crack growth 亚临界裂纹增长;亚临界扩展
subcritical filtration 亚临界过滤
subcritical flaw growth 亚临界缺陷增长
subcritical flow 缓流;亚临界流(动);亚临界(水)流;次临界流;平流
subcritical flow installation 缓流设施
subcritical hardening 低温硬化
subcriticality 亚临界度
subcritical mass 亚临界质量;次临界质量
subcritical measurement 亚临界状态的测量;次临界状态的测量
subcritical mineralization 亚临界矿化
subcritical power technology 亚临界发电技术
subcritical pressure 亚临界压力
subcritical pressure boiler 亚临界压力锅炉
subcritical propagation 亚临界扩展
subcritical reactor 次临界反应堆
subcritical rotor 亚临界转子
subcritical speed 亚临界速度
subcritical state 次临界态
subcritical temperature 低于临界温度的温度
subcritical treatment 亚相变处理;亚临界处理
subcritical velocity 亚临界速度;亚临界流速
subcritical volume 亚临界的体积
subcritical water 亚临界水

subcritical water oxidation 亚临界水氧化
subcrop 地下露头
subcrust 位于表层下的地层；次表面层；路面底层；路面底面
subcrustal current 壳下流
subcrustal earthquake 地壳下地震
subcrustal type of gravitative tectonic 壳下型重力构造
subcrystalline 次结晶的
subcrystalline structure 亚晶态结构
subculture 再次培养；分培
subcutaneous 皮下的
subcycle 次要循环；次循环；分周期
subcycle generator 副周波发生器；分周波发生器
subcylindric(al) 接近圆柱形的；近似圆柱形的
subdam 副坝
subdamage 亚损伤
subdata 分数据
subdata base 亚数据库；子数据库
subdealer 支店；分店
subdefinite 子定义
subdelta 子三角洲
subdepartment 分局；支局
subdepot 辅助仓库；附属仓库
subdeterminant 子行列式
subdevice bit 选择位
subdiabasic texture 次辉绿结构
subdiagonal 副斜杆
subdifferential 次微分
subdirectory 子目录
subdiscipline 分支学科；分支科学
subdistribution box for elevator 电梯配电箱
subdistribution box for escalator 扶梯配电箱
subdistribution box for pump 水泵配电箱
subdistribution box for sump pump 集水配电箱
subdistrict office 街道办事处
subdivide 把……分成几分；细分；分水界；支流分水领
subdivided capacitor 分组电容器
subdivided flow 小支流；细水流
subdivided gap 分间隙；分割间隙
subdivided member 再分杆
subdivided panel 镶板小分格；再分节间；复分镶板
subdivided resistance 分段电阻
subdivided shaft 组合轴
subdivided transformer 带分接头的变压器
subdivided truss 再分布桁架；再分(式)桁架
subdivided truss bridge 再分式桁架桥
subdivided Warren truss 再分式华伦桁架
subdividing 再分；分小块
subdividing partition 再分区段；再分隔墙；再分隔断；分块隔板
subdivision 细分类住它小区；细分；子部分；再分；剖分；水密分舱区划；重分；分区；分割；分舱
subdivisional work 分项工程
subdivision control 分段制；分段控制
subdivision controlling 分段控制
subdivision draft 分舱吃水
subdivision for bridge and tunnel 桥隧领工区
subdivision layout 地区分段草图
subdivision length 分舱长度
subdivision load line 分舱载重线
subdivision method 下分法
subdivision of common land 公用小块地
subdivision of degrees of freedom 自由度的重分
subdivision of land 土地重划分；(供出售的)小块土地
subdivision of variance 方差的重分；方差的划分
subdivision of vessel 船舶分舱制
subdivision regulation 土地划分条例；分区管理规则
subdivision survey 重分测量；分区测量
subdomain 子域【数】
subdominant organism 次优势种生物
subdorsal keels 侧背蜡管
subdouble method 二分之一法
subdrain 下层排水道；暗沟
subdrainage 地下排水；暗沟排水
subdrainage structure 地下排水建筑物；地下排水构筑物
subdrain of building 房屋地下排水管
subdrain pipe 暗沟管道
subdrain tile 地下水排水瓦管
subdrain type of surface drainage 地下暗沟型地面排水；地下排泄型地面排水
subdrift 中间平巷；分段平巷
subdrift caving 分段崩落法
subdrift contour 地形
subdrilling 向下钻进；先钻；钻孔加深；底钻；初钻；超钻
subduct assembly 防回流装置
subducted organic gas 俯冲带有机气
subducting plate 消减板块；俯冲板块
subduction 消减作用；潜没；俯冲作用；俯冲
subduction complex 俯冲复合体；俯冲带杂岩
subduction downwarped extracontinental basin 消减型陆外下陷盆地
subduction orogenic belt 俯冲造山带
subduction plate 下降板块；俯冲板块
subducition zone 消减带；削减带；潜没带；俯冲带
subdued colo(u)rs 分层设色表
subdued forms 晚壮年地形
subdued light 柔光
subdued relief 平坦地形
subdued uplift trap 悄平隆起圈闭
subduing 色变暗
subduplicate 平方根的
subeconomic 不经济的
subeconomic reserves 次经济储量
subeconomic resources 次经济资源
subedge connector 片状插座
subelement 子单元；分元件
subelliptical 近似椭圆的
subengineering 超小型工程
subentry address 分入口地址
sub-environment 小环境
subepoch 亚世
subequal spacing structure 近等间距构造
subequator 近岸赤道
subequatorial 亚赤道的
subequatorial belt 副赤道带
suber 软木
suberain 木栓质煤
suberate 辛二酸酯；辛二酸盐
suberect 接近垂直
suberic acid 辛二酸；木栓酸；软木酸
suberic tissue 软木组织
suberification 木栓化
suberin 木栓质；木栓脂；软木脂
suberinisation 木栓化作用
suberinite 木栓质体；木栓质煤素质；木栓质类；木栓质
suberinite-collinite 木栓质无结构体
suberinite-posttelinite 木栓质次结构体
suberinite-precollinite 木栓质似无结构体
suberinite-telinite 木栓质结构体
suberin lamella 木栓质层
suberite 微木栓煤
suberization 栓化作用
suberized 栓化的
suberized wall 木栓化壁
suberone 木栓酮
suberose 木栓状的；软木质的
suberosion 潜蚀作用
suberous layer 木栓层
subexcavated 下挖的
subexcavated section 下挖断面
subexchange 电话分局；电话支局
subexciter 副励磁机
subfabric 亚组构；次组构
subface 底面
subfacies 亚相
subfamily 亚科；分家庭
subfebrile temperature 微热
subfeeder 给水支管；副馈电线；分配电线
subfeldspathic 次长石质的
subfeldspathic lithic arenite 亚长石质岩屑砂屑岩
subfield 子域【数】；分区；分栏
subfigure 子图形
subfile 子文件
subfilling 分申请
subfirst grade structure 次一级构造
subfloor 楼面底板；第一层地板；底层地板；低层地板；下层地板；毛地板；地板垫层；粗地板（用于双层楼板）；桥面底层
subfloor beam 底层地板梁
subfloor filler 毛地板填料
subfloor for lino(leum) 供铺油毡的地砖毛地板；供铺油毡的毛地板
subflooring 下层地板；毛地板(材)料；毛地板；楼面底板；底层地板
subfloor stopper 毛地板填料
subfloor stopping 毛地板腻子
subflow 底流
subfluvial 河底的；水下的
subfluvial tunnel 河底隧道
sub for bit and auger-stem 钻柄和钻头间的连接（宾夕法尼亚索钻）
subforeman 副工长；养路副工长；副领工员
subfoundation 下层基础；基础底层；垫层
subfraction 子分式
subfractional rating 小分数功率
subframe 次框架；副架；副车架；辅助构件；下支架；子帧；子构架；副帧；辅助框架；分帧；支承墙面板的框架；支承门窗框的框架；分隔处的框架
subframing supported panel flooring 架空板材地面（活动地板的常规做法）
subfreezing 冰点以下
subfreezing temperature 冰点下温度
subfreight 搅载的运费
subfrequency 分谐频；分谐波频率
subfrigid zone 亚寒带；副寒带
subfringe 亚条纹
subfringe technique 亚条纹技术
subfunction 子函数
subfuscous 暗淡的；带黑色的
sub-ganger 副领班；副工长
subgelisol 冻下土
subgenus 亚属
subgeostrophic wind 次地转风
subglacial 冰川下的；冰川底部的
subglacial channel 冰下水道
subglacial drainage 冰下水系
subglacial eruption 冰下喷发
subglacial landscape 冰下地形
subglacial moraine 冰河下的冰碛；冰底碛
subglacial river 冰下河流
subglacial stream 冰(川)下河流
subglacial water 冰下水
subglobal metallogenic belt 亚全球性巨成矿带
subglobular 接近球形的
subgoal 附属目标
subgrade 地基(土)；沟底标高；路基(表)面；路基；土基；地基
subgrade accuracy 路基准确度；地基准确度
subgrade bearing capacity 路基承载力；地基承载力
subgrade bearing stratum 地基持力层
subgrade bearing value 路基承载值；地基承载值
subgrade bed 基床
subgrade capacity 路基承载力；地基承载力；路面承载力
subgrade condition 地基情况；路基情况
subgrade construction 路基施工；路基建筑
subgrade cross-section 路基横断面；路基断面
subgrade defect 路基病害
subgrade deformation 地基变形
subgrade deformation modulus 地基变形模数；地基变形模量
subgrade density 路基密度
subgrade design 路基设计
subgrade dilatancy 路基膨胀
subgrade dilation 路基膨胀
subgrade drainage 路基排水
subgrade drilling 超钻
subgrade equipment 路基修筑机械
subgrade erosion 路基冲刷；路基侵蚀
subgrade friction 路基面摩擦；地基摩擦(力)
subgrade frost penetration 地基冰冻深度；路基冰冻深度
subgrade heave 路基隆起
subgrade improved 改良路基
subgrade intrusion 土基浸透；土基浸渗
subgrade irregularity 路基不平整
subgrade lesion 路基病害
subgrade level 路基高程；路基面；路基标高
subgrade machine 路基面平整机
subgrade material 路基材料
subgrade modulus 路基模量；地基模数；地基反力模量；地基变形模数；地基变形模量
subgrade of highway 道路基层
subgrade of special soil 特殊土路基
subgrade paper 路基面(层)防潮纸；地基防水纸

subgrade-pavement interface 路基—路面交界面
subgrade planer 路基整平器;路基面整平机
subgrade preparation 路基(面)准备工作;地基处理
subgrade protection layer 路基保护层
subgrade pumping 路基灌浆
subgrader 路基整平机;路基平整机;路基面(层)修整机
subgrade rating 地基承载力
subgrade reaction 路基反力;地基反力
subgrade reaction modulus 路基反力模量
subgrade reaction theory 床反力理论
subgrade resistance 路基抗力
subgrade restraint 路基约束;基床约束作用;地基的约束;地基的限制
subgrade settlement 路基沉降
subgrade settling 路基下沉
subgrade soil 路基土
subgrade soil type 路基土类型
subgrade stability 路基稳定性
subgrade stabilization 路基稳定技术;路基稳定措施;路基稳定(法)
subgrade stiffness 路基刚度;地基刚度
subgrade strength 路基强度
subgrade strengthening 路基加固
subgrade stress factor 路基应力系数
subgrade support 路基承载力
subgrade surface 路基面
subgrade surface treatment 路基表面处理
subgrade survey 路基测量
subgrade template 路基样板
subgrade test 路基承载试验;地基承载试验
subgrade tester 路基检验器
subgrade traffic factor 路基承载交通系数
subgrade treatment 路基处理
subgrade trimmer 路基修整机
subgrade under the special condition 特殊条件下的路基
subgrade waterproofing 地下防水
subgrade width 路基宽度
subgrade with karst 岩溶型地基
subgrade with soil and rock 岩土交错型地基
subgradient 次梯度;次生梯度;次陡度
subgradient inequality 次生梯度不等式
subgradient wind 次梯度风
subgrading 平整路基;修筑路基
subgrading bed 基床
subgrading machine 路基修筑机;地基修筑机;路基平整机;地基平整机;平整路基
subgrain 亚颗粒;亚晶(粒);副晶粒;二次晶粒;正晶粒
subgrain boundary 亚晶粒边界
subgrain fabric 亚晶粒组织
subgrain structure 亚晶粒结构
subgraph 子图
subgraphite 次石墨;亚石墨;准石墨
subgravity 亚重力;次重力
subgraywacke 亚杂砂岩
subgreen schist 次绿片岩
subgrid scale process 次网格尺度过程
subground condition 地下条件
subgroup 小组;亚群;子群;分群
subgroup A 副族;A族
subgroup assembling 组件装配
subgroup B 副族;B族
subgroup element 副族元素
subgroup movement 侧链运动
subgroup of railway transport statistics 铁路运输统计分组
subharbo(u)r 辅助港
subharmonic 次调和;分谐波的;分频谐波;分次谐波
subharmonic function 次调和函数
subharmonic generator 次谐波发生器;分谐波发生器
subharmonic oscillation 次谐波振荡;次简谐振荡;分谐波振荡
subharmonic oscillator 分谐波振荡器
subharmonic phase modulation 分谐波相位调制
subharmonic radiation 分谐波辐射
subharmonic repsonse 次谐波响应;次简谐响应
subharmonic resonance 次谐波共振;次简谐共振;分谐波共振;次调和共振
subharmonic triggering 分频触发
subharmonic vibration 分谐波振动;分频振动
subharmonic wave 次谐波;次简谐波;分谐波

subhead 小标题
subheader 分联箱;副标题
subheading 小标题;副标题
subhedra grain 半形的晶粒
subhedral 半自形;半形的
subhedron 半自形(晶体)
Subhercynian orogeny 新海西(造山)运动;亚海西运动
subhigh type pavement 次高级路面
subhologram 亚全息图
subhouse drainage system 地下建筑排水系统
subhumic acid 低腐殖酸
subhumid 半潮湿的
subhumid and temperate zones 半湿润温带
subhumid area 半湿润(地)区
subhumid climate 半湿润气候
subhumid region 半潮湿地区
subhumid soil 半湿土(壤);亚湿土
subhumid tropics 湿亚热带;湿副热带
subhumid zone 半湿润地带
subhunter 猎潜艇
subhydrous coal 低氢煤
subhydrous vitrinite 低氢镜质组
subidiomorphic 半自形的【地】
subincline 暗斜井
subinclined shaft 暗斜井
subincrement 子增量
subindex 分指数
subindustry 分支工业
subinertial current 次惯性流
subinertial wave 次惯性波
subinspector 辅助检查员;副检查员
subinternet 万维子网
subinterrupt control block 次中断控制块
subinterval 子区间;分时程
subintradosal block 内拱砌块(坝的承重墙)
subinversion layer 逆温下气层
subinvolution 复旧不全
subirrigation 过滤场污水灌溉;地下渗灌;潜水灌溉;地下灌溉
subirrigation pipe 地下灌溉管
subirrigation system 地下灌溉系统
subisochroneity in evolution 近等时性演化
subitem 分项;子项目
subjacent 下覆的
subjacent bed 下卧层;下层地质
subjacent body 下面的物件;下卧体
subjacent igneous body 下延火成岩体
subjacent karst 下覆岩溶;下覆喀斯特
subjacent support 下面支承
subjacent waters 下层水域
subject 学科;主项;主体;主题;科目;受支配;受验者;题目
subject arrangement 主题排架法
subject authority file 主题标准档
subject bid 可商讨的出价
subject building 估价建筑物
subject classification 主题分类
subject classification table 主题分类表
subject construction cost 主体工程费
subject contrast 拍摄物反差;被摄物对比度
subject coordinate system 主观坐标系统
subject copy 主件副本
subject data 主观数据
subject data base 主题数据库;主库数据库
subjected to foot traffic 从属于步行交通
subject for a motif 母题题材
subject for decoration 装饰的主题
subject frame of reference 主观参考格网
subject guide card 主题导卡
subject heading 主题标目
subject heading list 主题表
subject index 主题索引;标题索引
subject indexing method 主题法
subject investigated 研究对象
subjection 满射
subjective angle of view 主观视角
subjective assessment of noise 噪声主观评价
subjective audiometry 主观测听法
subjective colo(u)r 主观色彩
subjective error 主观误差;人差【测】
subjective evidence 主观证据
subjective expression of worth 价值的主观表现
subjective gloss 主观光泽;对比光泽

subjective information 主观资料
subjective measurement 主观量度
subjective method 主观方法
subjective noise meter 主观噪声计
subjective photometer 主观光度计;目视光度计
subjective photometry 目视测光
subjective probability 主观概率
subjective probability method 主观概率法
subjective probability model 主观概率模型
subjective reverberation time 主观混响时间
subjective sample 主观样本
subjective scale of sensation 感觉的主观尺度
subjective sensation 主观感觉
subjective speckle 主观斑纹
subjective speckle problem 主观斑纹问题
subjective spectrophotometer 目视分光光度计
subjective standard 主观标准
subjective temperature 主观温度
subjective territorial principle 主观的领域原则
subjective thinning 主观疏伐
subjective value 主观价值;商品的主观价值
subjective view of probability 概率的主观观点
subjective weighting variable 主观赋权变量
subjectivity 主观性
subject job 源作业
subject jurisdiction 从属的管辖区
subject key 主题判读样片(遥感摄影)
subject matter 主题;论题;课题;题材;标的物;标的;题目
subject matter of an invention 发明的主题
subject matter of contract 契约标的
subject matter of insurance 保险标的
subject of entry 项目
subject of investment 投资主体
subject of numerous patents 专利权
subject of private right 民事权利主体
subject of standardization 标准化主题
subject of taxation 课税主体
subject open 未落实协议(租船条款)
subject program(me) 源程序
subject property 估价财产
subject retrieval 主题检索
subject ship being free 未落实协议(租船条款)
subject study 课题研究
subject terms 主题词
subject to 受限于
subject to approval 需经批准;须经批准/同意;以批准为条件
subject to approval no risk 确认前须经验证无风险
subject to approval of export license 出口证待领
subject to contract 以签订契据为条件
subject to customs duty 须付关税
subject to immediate reply 立即回答生效
subject to import license 需取得进口许可证的交易
subject to overflow damage 遭受洪水的危害
subject to particular average 仅保单独海损;单独海损担保
subject to prior approval 须经预先核准
subject to rapid pressure fluctuation 承受快速的压力波动
subject to regulation 按……规定
subject to severe limitations 受到严格限制
subject to severe risks of damage 遭受严重破坏
subject to shipping space available 以有舱位为准;以获有舱位为条件
subject to tax 负缴税义务
subject to taxation 课税对象;负缴税义务
subject to third part inspection 接受第三方监督检验
subject wallpaper 主题图案墙纸
subject wave 主体波
subjob 子作业;次作业
subjoin 附加;追加
subjoint 副接头;辅助接头
subjoist 次格栅;次龙骨
subkeel 次龙骨
subke v 亚千电子伏
subkiloton 千吨以下;低于千吨
subkingdom 亚界
sublacustrine channel 湖下水道
sublacustrine hot-spring system 湖下热泉系统
sublacustrine spring 湖底泉
sublaminar layer 层流底层
sublateral 支渠;分支渠;分支管;暗沟分支

sublateral canal 毛渠;斗渠;分支渠
sublateral line 分支管线
sublattice 亚晶格;亚点阵;子格
sublayer 垫层;下卧层;下层;底层;次层
sublayer depth 次层深度
sublayer laminar 次层流
sublayer thickness 次层厚度
sublease 转租;分租;二次租赁
sublessee 三房客;分租承受人;转租承受人
sublessor 分租主;转租人
sublet 转租;分包;转包
sublethal amount 亚致死量
sublethal concentration 亚致死浓度;不致死浓度
sublethal damage 亚致死损伤
sublethal dose 半致死量;亚致死剂量;次致死量
sublethal effect 亚致死效应;不致死效应
sublethal exposure 亚致死照射;亚致死接触
sublethal level 亚致死水平
sublethnal effect test of aquatic organism 水生生物亚急性毒性试验
sublet number 副约号
subletting 分租;分包;转包
sublevel 支能级;次能级;次级;次层;次阶;分段巷道;分段
sublevel bench stoping 分段梯段回采
sublevel caving 分段崩落采矿法
sublevel caving method 分段崩落法
sublevel drift stoping 分段平巷回采
sublevel drive 分段掘进
sublevel fill stoping 分段充填
sublevel limited-entry method 分段限流量法
sublevel-long-hole benching 分段深孔梯段式回采
sublevel stoping 分段回采法
sublevel stoping mining 分段梯段回采
sublevel stopping 分段采矿法
sublevel undercut caving method 分段下部掏槽崩落开采法
sublibrary 子程序库
sublicence 分许可;再次转让许可证;副许可证
sublicense 副许可证;再次转让许可证;分许可证
sublicensing 技术转让
sublight 气窗;腰头窗
sublimate 升汞;纯化;升华(物)
sublimated ice 升华冰
sublimation 理想化;耐蚀铸铁;提净;升华(作用)
sublimation condensation 升华冷凝法
sublimation cooling 升华冷却
sublimation curve 升华曲线;固气平衡线
sublimation drying 升华干燥
sublimation energy 升华能
sublimation from the frozen state 冻结升华
sublimation heat 升华热
sublimation method 升华法
sublimation nucleus 凝华核
sublimation of snow 雪的升华
sublimation point 升华点
sublimation pot 升华皿
sublimation pressure 升华压力
sublimation pump 提净泵;升华泵
sublimation temperature 升华温度
sublimation(transfer)printing 升华印花
sublimation vein 升华矿脉
sublimatogram 升华谱
sublimatography 升华谱法
sublimator 升华器
sublimatory 升华器
sublime 升华的
sublimed iodine 升华碘
sublimed method 升华法
sublimed salicylic acid 工业水杨酸
sublimed sulfur 硫华;升华硫
sublimed white lead 升华铅白;升华白铅
sublimer 升华器
subliminal stimulus 低于限值的刺激
subliming 升华
subliming pot 升华皿
sublimited dose 限下剂量
subline 辅助管(道)
sublinear 亚线性;次线性的
sublinear vibration 亚线性振动
subline differentiation 亚系分化
sublist 子表;副表;分表
sublithwacke 亚岩屑瓦克岩
sublittoral 次大陆架的

sublittoral platform 沿岸台地
sublittoral sheet sandstone 潮下席状砂岩
sublittoral zone 浅海地带;浅海带;低潮下植物带;潮下带
sublocus 亚位点
subloop 次级回路;副回路
sublot 分块土地;小批
sublunar point 月下点
subluxation 半位
submaceral 亚显微组分
submaceral group 显微组分亚组
submachine gun 冲锋枪
submacromethod 超微量法
submain 辅助干管;地下总管;次主管;主干管;辅助干线;分干管;排水次干管;地下干管
submain circuits 分支线路
submain line 副干线
submain sewer 污水次干管;污水次干道
submanager 副经理
submarginal 近缘的;赔钱的
submarginal drainage 近冻缘水系
submarginal granular base material 规格以下的底层粒料
submarginal land 无价值土地;限界以下的耕作土地
submarginal moraine 近缘碛
submarginal resources 边界下资源
submarine 水中的;海底的;潜艇;潜水艇;水下的;水下;水底的;水底
submarine alluvial fan 陆坡冲积扇;海底冲积扇
submarine anchor 潜艇锚;短杆霍尔锚
submarine antenna drive system 潜艇天线驱动系统
submarine area 海区域
submarine armo(u)r 潜水员装备;潜水服
submarine bank 海底洲(滩);海底岸滩;暗沙
submarine bar 海底洲(滩);海底沙洲;海底沙坝
submarine barchan 海底沙丘
submarine bar crest 水下沙脊
submarine barrier 海底阻层
submarine bell 潜水钟;水中雾钟
submarine bell operation 潜水钟作用
submarine biologic(al) noise 水中生物噪声
submarine blasting 海底爆破;水下爆破
submarine block 海底断块
submarine bottom scanner 海底扫描器
submarine bulge 海底凸起;扇形地
submarine(bull)dozer 水下挖泥机;水下推土机
submarine cable 海底送电线;海底电缆;水下电缆;水底电缆
submarine cable area 海底电缆区
submarine cable buoy 海底电缆浮标
submarine cable communication 海底电缆通信
submarine cable equalizer 海底电缆均衡器
submarine cable landing station 海缆登陆站
submarine cable repeater 海缆中继站;海缆中继器;海缆增音机
submarine cable ship 海底电缆铺设船;海底电缆敷设船
submarine cable telegraph system 水底电缆电报系统
submarine camera 海底摄影机;水下摄影机
submarine canyon 海谷;海底峡谷;海底峡沟;海底深谷;海底谷
submarine cave 海底扇
submarine chaser 猎潜艇
submarine city 海底城市
submarine civil engineering 海下土木工程
submarine climate 副海洋性气候
submarine coal field 海底煤田
submarine coaxial cable 海底同轴电缆
submarine communication cable 海底通信电光缆
submarine conduit 海底管道;海底管
submarine cone 海底扇形地;海底冲积锥;深海锥
submarine construction survey 海底施工测量
submarine contour 海底等高线;水底等高线
submarine core drill 海底岩芯钻;水下岩芯钻探
submarine current 下层海流
submarine daylight 海下日光
submarine delta 海底沙洲;海底扇;海底三角洲;扇形地
submarine denudation 海底剥蚀
submarine deposit 海底淀积(物);海底沉积(物)
submarine depot boat 潜艇母舰
submarine depression 海底洼地

submarine-detecting set 潜艇探测器
submarine detecting system 潜艇探测系统
submarine detection 潜艇探测
submarine detection gear 潜艇探测装置
submarine detection satellite 潜艇探测卫星
submarine detector 海底探测器;潜艇探测器
submarine detonator 水下雷管
submarine discharge pipe 水下排泥管
submarine discharge pipeline 水下排泥管线
submarine distributary channel deposit 水下分流河道沉积
submarine domain 海底领地
submarine drilling 海底钻探;海底钻井
submarine drilling engineering 海底钻探工程;海底钻井工程
submarine drilling operation 海底钻探作业;海底钻井作业
submarine dune 海底沙丘
submarine earthquake 海震;海底(地)震
submarine earthquake observation 海底地震观测;海底地震测量
submarine effusion 海底喷发
submarine emergency buoyancy system 潜艇紧急上浮系统
submarine erosion 海底侵蚀
submarine eruption 海下喷发;海底喷发
submarine escape equipment 潜艇脱险设备
submarine escape immersion equipment 潜艇逃生与沉水救援设备
submarine exercise area 潜艇演习区
submarine fan 海底扇(形地);海底(陆源)沉积扇;深海锥;深海扇
submarine fan sequence 水下扇层序
submarine fan valley 海底扇谷
submarine fault 海底断层
submarine fault scarp 海底断崖
submarine feature 海底地形
submarine floater 海底钻探船
submarine fog bell 水中雾号;水下雾钟
submarine fog signal 水中雾号
submarine fresh-water spring 海下淡水泉
submarine frontier 海底边界
submarine furmarole 海底喷气孔
submarine gate 下泻口
submarine geologic(al)structure 海底地质构造
submarine geology 海底地质学;地质海洋学
submarine geomorphologic chart 海底地貌图
submarine geomorphology 海底地貌(学)
submarine geophysics 海底地球物理学
submarine height finding radar 潜艇测高雷达
submarine hot spring vent 海底热泉喷孔
submarine humus 海底腐殖质
submarine hunter 猎潜艇
submarine hydrothermal spring 海底热泉
submarine instrument 水下仪器
submarine integrated sonar system 水下集成声呐系统
submarine isthmus 海底地峡
submarine laboratory 水下实验室
submarine land slide 海底滑坡;海底滑;海底坍塌
submarine laying 海底铺管
submarine lifting pontoon 潜艇打捞浮筒
submarine light(ing) 水下照明;潜水灯
submarine line 海底线路;水底线路
submarine loading boom 水下装油臂
submarine loading line 水下装油管
submarine mapping 河底测绘
submarine meadow 海底草地
submarine mine 海底矿;水下水雷;水底水雷
submarine mine area buoy 水下雷区浮标
submarine mineral deposit 海底矿床
submarine mineralization on the east pacific rise 东太平洋洋隆海底成矿作用
submarine mineralization red sea 红海海底成矿作用
submarine mineral resources 海洋矿产资源;海底矿产资源
submarine mining 海底开采;海底采矿
submarine mining buoy 水雷区浮标
submarine morphology 海底地形学
submarine navigation 水下航行
submarine noise 潜艇噪声
submarine obstacle 海底障碍物
submarine oil 海底石油

submarine oiler 潜水油船
submarine oil field 海底油田
submarine oil formation 海底含油构造;海底含油地层
submarine optic(al) fiber cable 海底光纤电缆;海底光缆
submarine optoelectronic mast 潜艇光电桅杆
submarine ore carrier 水下矿石运输船
submarine oscillator 潜用振荡器;水中振荡器;水下振荡器
submarine passive detection and tracking set 潜艇被动探测与跟踪设备
submarine passive fire control sonar 潜艇被动火力控制声呐
submarine passive sonar 潜艇被动声呐
submarine penetration 海底钻入
submarine peninsula 海底半岛
submarine photography 水下摄影术;水下摄影
submarine photometer 水下光度计
submarine physiography 水下自然地理学
submarine pipe column foundation 海底管柱基础
submarine(pipe)line 海底管线;海底管路;海中管道;海底管道;海底电线;水下管线;水下管路;水下油管;水下管道
submarine pit 海底坑;海底勘坑;海底洞
submarine plain 海底侵蚀平原;海(底)平原
submarine plant 海底植物
submarine plateau 海台;海蚀平原;海底台地
submarine platform 海底(侵蚀)台地
submariner 潜水员
submarine railway 海底铁路
submarine range finder 潜艇测距计
submarine receiving set 潜艇接收装置;水下(超声波)接受器
submarine red mud 海底红泥
submarine reef 暗礁
submarine relief 海底地势;海底地形;海底起伏;海底救援;海底地貌学;水下地形
submarine rescue chamber 潜艇救生舱
submarine rescue ship 潜艇救难舰
submarine rescue vessel 潜艇救生船
submarine resources 海底资源
submarine ridge 海岭;海脊
submarine rift 海底裂谷;海底断裂
submarine rise 海底隆起
submarine river crossing 横贯河底管道
submarine route 横贯河底管道
submariner sampler 海底采样器
submarine safety lanes 潜艇安全航线
submarine safety monitoring system 潜艇安全监控系统
submarine sampling 海底取样;海底采样
submarine sanctuary 潜水艇演习区
submarine sandy soil 海底砂土质;海底沙土质
submarine saw 水下锯机
submarine scarp 海底崖
submarine search 海底搜索
submarine search sonar 潜艇搜索声呐
submarine secure depth sounder 潜艇安全深度探测器
submarine sedimentology 海底沉积学
submarine seismograph 海底地震仪
submarine sentry 水中探礁器;水下拖测器
submarine shock 海底地震
submarine sickness 潜水病
submarine signal 水中信号;水下信号
submarine signal receiving apparatus 水下信号接收装置
submarine sill 海槛;海底岩床;海底山脊
submarine single-mode fiber cable 海底单模光缆
submarine site 水下位置
submarine slide 海底滑动;水下塌方
submarine slump 海底沉陷
submarine slumping 海底滑移;海底崩坍
submarine soil 海底土
submarine soil mechanics 海底土壤力学
submarine soil sampler 海底取土器
submarine sound ranging station 水下音响测距发音站
submarine sound receiver 海底集声器
submarine sound signal 水下音响信号;水底信号
submarine sound velocity 海水中声速
submarine spring 海下泉;海底泉;水下泉
submarine spur 海底石嘴;海底岬

submarine steel pipeline 水下钢管;海底钢管
submarine stratigraphy 海底地层学
submarine structural chart 海底结构图
submarine structure 海底结构物;水下建筑(物)
submarine swell 海底隆起
submarine tectonic geology 海底构造地质学
submarine tectonism 海底构造运动
submarine telegraph 海底电报
submarine telegraph buoy 海底电线浮标;海底管线浮标;海底电线标;海底电缆浮标;海底电话线浮标
submarine telegraphy 水底电报
submarine telephone 海底电话
submarine telephone cable 海底电话电缆
submarine tender 潜艇勤务支援船;潜艇母舰;潜水艇供养船
submarine terrace 海底台地;海底阶地
submarine thermal reactor 潜水艇用热中子反应堆
submarine thermometer 海水温度计;水下温度计
submarine topographic(al) map 海底地形图
submarine topography 海底地形;水下地形
submarine topography surveying 海底地形测量
submarine trench 海沟;海底沟
submarine tunnel 海底隧道;水底隧道;水下隧道
submarine tunnel survey 海底隧道测量
submarine uplift 水下隆起
submarine valley 海谷;海底谷
submarine varnish 海底用清漆(用于海底结构物)
submarine vegetation 水下植被
submarine volcanic activity 海底火山活动
submarine volcanic earthquake 海底火山地震
submarine volcanic pile 海底火山岩锥
submarine volcanism 海底火山作用
submarine volcano 海底火山
submarine weathering 海底风化(作用)
submarine well 海底坑
submarine work 水下建筑(物);水下工程;水底工程;海底钻探作业;海底钻井作业;潜水工作
submaritime conduit 水下管
submaritime pipe 水下管
submaritime tube 水下管
submark 副标志
submaster controller 次控制器;局部主控机
submatrix 子矩阵
submature 早壮年期
submaximal intensity 极限下强度
submaximal oxygen intake 极限下吸氧量
submeander 小曲流;小河弯;河道小曲流;次生河曲
submellite 钙黄长石
submember 副构件
submenu 子菜单;分单
submerge 潜航;放在水中;浸没
submerged 下沉的;水中的;水下的
submerged aerated basin 淹没式曝气池
submerged aerated filter 淹没式曝气滤池
submerged aeration 淹没曝气
submerged aeration system 淹没(式)曝气系统
submerged agitator 淹没式搅拌机
submerged anaerobic membrane bioreactor 淹没式厌氧膜生物反应器
submerged antenna 水下天线
submerged aquatic plant 水下水生植物
submerged aquatics 水下生物
submerged aquatic vegetation 水下水生植被
submerged arc 埋弧
submerged arc furnace 埋弧炉
submerged arc welding 埋弧焊;焊剂层下焊;潜弧焊;水下弧焊;水下电(弧)焊(接)
submerged arc welding machine 水下电弧焊机;埋弧焊机
submerged area 淹没(地)区;浸没地
submerged automatic arc welding 埋弧自动焊
submerged bank 海中滩
submerged bar 沉没沙洲;潜洲;沉溺沙洲;暗沙洲
submerged beach 沉没海滩
submerged bearing 水下支承;水下轴承;水下障碍物
submerged-bed aeration 淹没床曝气
submerged-bed aeration system 淹没床曝气系统
submerged-belt separator 浸入带式磁选机
submerged biofilm 淹没式生物膜
submerged biofilm process 淹没式生物膜法
submerged biofilm reactor 淹没式生物膜反应器
submerged biofilter 淹没式生物接触滤池

submerged biological contactor 淹没式生物接触器
submerged biological filter 淹没式生物接触滤池
submerged body 潜(没)体;水下部分
submerged breakwater 潜堤;水下防波堤;潜防波堤
submerged bucket 潜没式挑流器
submerged bulk weight density of soil 土的水下容重
submerged buoy 潜水浮标;潜没浮标;水下浮标
submerged burner furnace 浸没式烧嘴炉;浸没式燃烧炉
submerged canned pump 液下屏蔽泵
submerged canned pump for chemical industry 化工液下屏蔽泵
submerged cargo vessel 潜水货船
submerged carrier 淹没式填料
submerged cathode cell 沉阴极电池
submerged channel 水下河槽;沉溺河槽
submerged cleaning and maintenance platform 水下清洗和维护平台
submerged coast 下沉海岸;淹没岸;溺岸;沉溺(海)岸;沉没(海)岸
submerged coastal plain 沉没海岸平原
submerged coastline 下沉岸线
submerged coil condenser 沉没式蛇管冷凝器;沉浸盘管式冷凝器
submerged coil evapo(u)rator 沉浸式盘管蒸发器
submerged coil type cooler 沉入盘管式冷却器
submerged combustion 浸没燃烧
submerged-combustion burner 沉没燃烧器;浸没式燃烧器
submerged-combustion evapo(u)ration 潜水燃烧蒸发
submerged-combustion evapo(u)rator 潜水燃烧蒸发器;沉没燃烧蒸发器;液下燃烧式蒸发器
submerged combustion furnace 浸没式烧嘴炉;浸没式燃烧炉
submerged-combustion heater 沉没燃烧加热器
submerged combustion vapo(u)rizer 浸没燃烧式蒸发器
submerged completion 水下完成
submerged concrete 水下混凝土
submerged concreting 水下浇筑混凝土;水下浇注混凝土;水下浇灌混凝土;水下灌筑混凝土;水下灌注混凝土
submerged condenser 浸没式冷凝器;潜没冷凝器;沉浸式冷凝器
submerged condition 浸水状态;水下状态
submerged conduit 淹没式管道
submerged construction 水中结构
submerged contact aeration 淹没(式)接触曝气
submerged controller 淹没式控制器
submerged corrosion 水下侵蚀;水下腐蚀
submerged cover 淹没盖;浸没覆盖
submerged cradle 下潜架
submerged craft 潜水艇
submerged crib 淹没式取水口;淹没式木笼;潜没滤水网框格
submerged cultivation 浸没培养
submerged culture 水下培养;深层培养;沉没培养;沉井培养
submerged cultuse 深层发酵
submerged culvert 壅水涵洞;漫水涵洞;半压力式涵洞
submerged current 潜流
submerged dam 潜坝
submerged damage 水下部分破损
submerged decompression chamber 下潜式减压舱
submerged deflection baffles 淹没式折流板
submerged delta 水下三角洲;沉没三角洲
submerged delta plain 水下三角洲平原
submerged demineralizer system 淹没式水软化装置系统
submerged density 浸没密度
submerged deposit 水下沉积(物)
submerged depth 淹没水深;淹没深度;潜航深度
submerged diffuser 水下扩散器
submerged dike 潜防波堤;淹没式丁坝;潜丁坝;潜堤
submerged discharge 淹没泄流;淹没出流;潜流量
submerged discharge pipeline 水下排泥管线
submerged displacement 排水量
submerged disturbance 水下扰动
submerged draw bar 引砖

submerged dredge pump 潜水泥泵;水下泥泵
submerged drill 沉入式钻机
submerged drilling machine 沉入式钻机
submerged drilling rig 潜孔电钻钻机
submerged earth 漫水土层
submerged efflux 淹没出流;潜伏出流
submerged embankment 浸水路堤;潜堤
submerged estuary 沉溺河口
submerged evaporator 水下蒸发器
submerged exhaust 水下排气
submerged fermentation 深层发酵
submerged filter 淹没式滤池;地下渗滤器
submerged filter method 地下渗滤法
submerged filtration 淹没(式)过滤
submerged float 淹没式浮标;潜没浮标;水下浮筒
submerged flow 淹没水流;淹没流;浸没;暗流
submerged flow diversion works 潜流式分水设施;潜流式分水建筑
submerged foil 浸水式水翼
submerged foundation 潜水基础;水中基础工程
submerged gate 潜没式闸门;水下闸门
submerged generator 沉浸式发生器
submerged grade line 淹水坡度线
submerged grading of foundation 工程基础水中修整;水中基础工程维修;水下基础修整
submerged groin 淹没式丁坝;河底潜堤;潜式丁坝;潜丁坝;水下堤埂
submerged groyne 潜坝;淹没式丁坝
submerged heating 浸渍电加热
submerged hollow fibre membrane reactor 淹没式中空纤维膜反应器
submerged hull 水下部分船体
submerged hydraulic jump 淹没水跃;沉溺水跃
submerged hydro-electric(al)unit 淹没式水轮发电机组
submerged hydrofoil craft 全浸式翼艇
submerged ice 潜冰;水下冰;水内小冰块;水内冰
submerged ice cracking engine 水下破冰机
submerged ice foot 冰锤(冰山或冰丘的水下部分的)
submerged incinerator 液下焚烧器
submerged inlet 淹没式取水口;浸没式进水器;埋入式进气口
submerged intake 淹没式进水口;浸没式进水器;浸没式进水口;埋入式进气口;淹没式取水口
submerged irrigation 淹没灌溉
submerged isothermal current 等温潜流
submerged jet 淹没式射流;淹没射流;淹没喷射流;潜射流;水下射流
submerged jetty 淹没式丁坝;潜突堤;水下突堤
submerged joint 暗缝
submerged jump 淹没式水跃;沉溺水跃
submerged land 淹水地;水淹地(区);水涝地;沉没陆地
submerged length of member 杆件浸入长度
submerged log 水下计程仪
submerged lubrication 浸入式润滑
submerged macrophyte 水底大型植物
submerged margin 淹没边界;水下边界
submerged marine delta 沉没海相三角洲
submerged media anaerobic reactor 淹没式中型厌氧反应器
submerged membrane bioreactor 淹没式膜生物反应器
submerged membrane bioreactor system 淹没式膜生物反应器系统
submerged membrane microfiltration 淹没式膜微滤
submerged membrane process 淹没式膜工艺
submerged membrane separation activated sludge process 淹没式膜分离活性污泥法
submerged meter 淹没式水表
submerged mining ship 水下采矿船
submerged mole 潜突堤;潜堤
submerged multiport diffuser discharge 水下多孔扩散器排放
submerged natural levee deposit 水下天然堤沉积
submerged navigation 潜航
submerged obstacle 水下障碍物
submerged orifice 淹没(式)孔口;浸液隔膜;潜(流)孔口
submerged orifice fish pass 淹没孔口式鱼道
submerged outfall 水下排污口
submerged outlet 淹没式出(水)口;浸没式出水口;埋没式出水口;淹没式出水口;潜流式出水口;水下排污口
submerged paddle aeration 潜浆式曝气
submerged paddle aerator 淹没式曝气器;潜浆式曝气设备;潜浆式曝气器
submerged paleo delta 水下古三角洲
submerged part of upright wall 直墙水下部分
submerged pier 淹没式丁坝;潜突堤;潜水墩;潜丁坝;暗堤;暗桩
submerged piling 沉桩
submerged pipe 淹没式管道;淹没管;水下管;水底管
submerged(pipe)line 海底管线;海底管路;潜水管线;水下管道
submerged plant 沉水植物
submerged plate type membrane 淹没式板式膜
submerged platform 下潜平台
submerged plunger die-casting-machine 柱塞式热压铸机;柱塞式浸注压铸机
submerged portion 水下部分
submerged power plant 水下发电厂
submerged pump 液下泵;浸没泵;潜水泵;沉没泵
submerged reef 下沉礁;沉没礁;暗礁
submerged reef-type breakwater 暗礁式防波堤
submerged repeater 海底中继站
submerged ridge 潜埂
submerged-ridged caused rapids 潜埂型急滩
submerged rig 沉入式钻机
submerged riprap dam 抛石潜坝
submerged roadway 矿坑淹没
submerged rock excavation 水下岩石开挖;水下岩石采掘
submerged roll 沉浸辊
submerged roof 潜水屋面;淹水屋面
submerged root 沉水根
submerged rope-type biofilm reactor 淹没式索型生物膜反应器
submerged rotating polyvinylidene fluoride membrane bioreactor 淹没转盘式聚偏氟乙烯膜生物反应器
submerged rotor aeration 淹没旋转子曝气;淹没转子曝气
submerged sand bar 暗沙滩
submerged screw log 水轮计程仪;水流式计程仪
submerged sediment 水下沉积(物)
submerged sharp-crested weir 尖顶潜堰
submerged shelf 淹没(大)陆架
submerged shoal 潜洲
submerged shoreline 下沉岸线;沉溺海岸线;沉溺滨线;沉没海岸线;沉没滨线
submerged sill 淹没(式)底槛;潜槛;水下底槛
submerged slide 水下滑坡
submerged sluice 潜水闸
submerged soil 深层土(壤)
submerged speed 潜航速度
submerged spillway 淹没式溢洪道;潜溢洪道
submerged spiral type classifier 淹没式螺旋分级机
submerged spring 沉没泉
submerged spur 淹没式丁坝;潜丁坝
submerged spur dike 淹没丁坝
submerged stage 沉水植物阶段
submerged steady bearing 浸没式支撑轴承
submerged storage tank 水下油罐;沉没式油罐
submerged structure 潜水结构;水下建筑(物);水下构筑物
submerged stumps 水下墩;水下柱墩
submerged suction filter 淹没式吸入滤器
submerged switch 埋入式开关
submerged tainter gate 淹没式弧形门
submerged tank 海底油库
submerged telegraph repeater 海缆电报转发器
submerged telephone repeater 海缆电话增音机
submerged tube 淹没式管道;浸没管道;潜管;沉管
submerged-tube boiler 潜管锅炉
submerged tube bridge 漫水管桥
submerged tube condenser 潜管冷凝器
submerged-tube evapo(u)rator 潜管蒸发器;浸没管束蒸发器
submerged-tube F. C. evapo(u)rator 浸没管束蒸发器;浸没管束浓缩器
submerged tunnel 悬浮隧道;水下隧洞;水下隧道;水底隧道;沉埋式隧道;沉管隧道
submerged tunnel(l)ing 沉埋法
submerged turbine 潜水涡轮
submerged turbine aerator 浸没式涡轮曝气机
submerged type 淹没式;沉没式
submerged type bottom vane 淹没式导流板
submerged type construction joint 暗式施工缝;暗式伸缩缝
submerged-type contraction joint 暗式收缩缝
submerged-type desuperheater 内部减热器
submerged unit weight 有效重度;有效容重;潜容重;水下容重;水下浮容重;水下单位重量;单位潜容量;浮重度;浮容重
submerged unit weight of soil mass 土体浮容重
submerged valley 溺谷;潜谷;沉没河谷;沉没谷
submerged vane 潜没式导流板
submerged vegetation 沉水植被
submerged washing machine 潜式洗舱机
submerged waste field 水下废物场
submerged wastewater field 水下废水场
submerged water plant 水中水生植物
submerged water supply 潜水供应
submerged weight 水下重量;浮重
submerged weir 淹没堰;潜堰;潜水堰
submerged well 水下(油)井
submerged work 下潜式工作舱;潜水工作;水下建筑(物);水下构筑物
submerged wreck 水下沉船
submergence 淹没(度);浸入;埋入;沉入;沉没;浸没
submergence coast 下沉岸;溺岸;沉降海岸
submergence coefficient 淹没系数
submergence constant 淹没常数
submergence control 吸上高度调节;浸深调节
submergence factor 淹没系数
submergence head 吸入侧压头
submergence investigation 浸没调查
submergence of ground 地沉;地面下沉
submergence pressure 下潜压力
submergence ratio 淹没比;沉没比
submergence shoreline 下沉滨线
submergent plant 沉水植物
submergible 可潜的;可浸入水中的
submergible bridge 漫水桥
submergible drainage pump 淹没式排水泵
submergible hydro-generator set 淹没式水轮发电机组
submergible power plant 淹没式电站
submergible power station 淹没式电站
submergible pump 可沉浸的水泵;潜水泵
submergible roller dam 淹没式定轮闸门(的)坝;可淹没式圆辊坝
submergible tainter gate 淹没式弧形门
submergible type barge carrier 下潜型载驳货船
submergible vertical lift gate 淹没式平板门;淹没式垂直提升门
submerging coast 沉溺(海)岸;沉没海岸
submerging test 浸水试验
submersed 适于水下使用
submersed wall 池窑液面下隔墙
submersible 可潜的;潜水艇
submersible barge 潜入式驳船
submersible barge platform 全潜式驳船平台
submersible bilge pump 可潜式舱底水泵
submersible boat 可潜小艇
submersible boring rig 潜孔钻机
submersible bridge 漫水桥
submersible core drill 全潜式岩芯钻机
submersible decompression chamber 下潜式减压舱;潜水减压舱
submersible deep well pump 深井潜水泵
submersible drainage pump 潜水式排水泵
submersible dredge(r) 可潜式挖泥器;潜水挖泥船
submersible drilling barge 全潜式钻探驳船;潜水钻探船
submersible drilling platform 座底式钻井平台;全潜式钻探平台
submersible drilling rig 座底式钻机
submersible electric(al)motor 潜入式电动机
submersible electric(al)pump 潜水电泵;沉入式电动泵
submersible electro pumping completion 电动沉没泵完井
submersible fish pump 潜水鱼泵
submersible gate 水下闸门
submersible lifeboat 可潜式救生艇

submersible machine 潜入式电机;水中用电机
submersible motor 孔底动力钻进;可潜(水)发动机;潜水(式)电动机;潜入式发动机;防水发动机
submersible motor pump 潜水电机动泵
submersible observation 潜水器观测
submersible platform 座底式平台;可潜平台;潜水平台
submersible pump 浸没泵;可潜泵;可沉浸水泵;潜水泵;水下泵
submersible quay surface 可淹没式码头面
submersible roller 漫水滚筒
submersible roller dam 可淹没式定轮闸门坝
submersible roller sluice gate 下沉式圆辊闸门
submersible shallow coring rig 海底取样机
submersible shear strength tester 水下剪切强度测试器
submersible tainter gate 水下弧形闸门
submersible temperature and oxygen recording equipment 淹没式温度和氧量记录设备;水下温度与氧记录仪
submersible test unit 可潜试验装置
submersible transducer 潜水换能器;沉没式传感器
submersible turbine pump 潜水涡轮泵
submersible water quality monitor 水下水质测定仪
submersible winch 下潜式绞车
submersion 淹没;浸入;淹没;沉没
submersion area 淹没地区
submersion depth 浸没深度
submersion irrigation 淹没灌溉
submersion irrigation method 淹灌法
submersion-proof 可潜水的;防水的
submersion-proof ignition 防水的点火系统
submersion skimmer 浸没式撇油器;浸没式浮游回收装置
submersion watch 潜水表
submersiprata 水下草本群落
submesothyrid 亚中孔型
submetal 半金属;非金属
submetallic 半金属的
submetallic elements 半金属元素
submetallic luster 半金属光泽
submetallogenetic belt 成矿亚带
submeter(ing) 分表(指仪表);煤气分表
submicro 亚微量
submicro analysis 亚微量分析
submicroearthquake 亚微震
submicrofracture 亚微观断裂
submicrogram 亚微克;次微克
submicrogram quantity 亚微克量
submicro method 亚微量法
submicro-morphological feature 亚微观形态特征
submicron 亚微米;亚微型;亚微细粒;亚微米细粒;亚微粒;次微子;超微型
submicron aerosol 亚微米气溶胶
submicron airborne particle 亚微米空传微粒
submicron film 亚微米薄膜
submicron gate 亚微型栅
submicronic aerosol 亚微粒气溶胶
submicron metal 超细金属粉末;亚微金属粉末
submicron particle 亚微米粒子;超微末微粒
submicron particulate emission 亚微米微粒排放
submicron particulate matter 亚微米微粒物质
submicron-sized particulate 超微细粒
submicro particle 亚微细粒;亚微粒
submicro-particulate emission 亚微颗粒排放
submicro-particulate matter 亚微颗粒物质
submicro powder 亚微粉末
submicrosample 亚微量试样;超微量试样
submicroscopic 亚微观的;超显微
submicroscopic clustering 亚微观束状簇
submicroscopic cracking 亚微裂纹
submicroscopic damage 亚微观损伤
submicroscopic deletion 亚显微缺失
submicroscopic feature 亚微观特征
submicroscopic particle 亚微观粒子;超微观质点
submicroscopic porosity 亚微孔隙
submicroscopic size 亚微观的大小
submicroscopic structure 亚显微结构;亚微观构造;次亚微构造
submicrosecond 亚微秒
submicrosome 亚微粒体
submicrostructure 亚显微结构;亚微观结构
submicrotubule 亚微管

submicrowater filter 超微孔过滤器
submillimeter 亚毫米
submillimeter astronomy 亚毫米波天文学
submillimeter photon counter 亚毫米光子计数器
submillimeter region 亚毫米波段
submillimeter wave 亚微米波;亚毫米波
submin 超小型摄影机;超小型零件
subminiature 超小型元件;超小型的;超小型
subminiature accelerometer 超小型加速表
subminiature camera 超小型摄影机
subminiature capacitor 超小型容器;超小型电容器
subminiature coaxial cable 超小型同轴电缆
subminiature indicator lamp 超微型指示灯
subminiature motor 超小型电动机
subminiature pipe 超小型管
subminiature project 超小型工程
subminiature relay 超小型继电器
subminiature tube 超小型(电子)管
subminiaturization 超小型化;超微型化
subminy timer 超小型记时计;超小型计时器
submission 提交公断
submission date 提交日期
submission for arbitration 提交仲裁
submission of bid 投标
submission of competitive tenders 投标
submission tender 投标
submission to arbitration 提交仲裁
submit 提交
submit a claim 提出赔偿要求
submit a dispute for arbitration 提前仲裁
submit an offer for 认购;报盘
submit a tender 投标;呈送标书
submit a tender for building and outfitting project 建筑安装工程投标
submit bid 投标
submit data 提供资料
submit drawings 提交图纸
submit for approval 提请批准;报请批准
submit state 提交状态
submit tender 投标
submitting the bid 投标
submit to test 提交试验;送请试验
submit to the relevant authority 申报
submodel 分模型;亚模型;子模型;辅助模型
submodel copying 子模型复制
submodular 分模数
submodular size 分模数尺寸
submodulator 副调制器;辅助调制器
submodule 辅助模数;分模数
submolecular plexus 分子下丛
submontane 山麓的
submortgage 再抵押;副抵押权;二次抵押(权)
submountain 山麓的
submountain railway 沿山(脚)铁路
submountain region 山脚
submultiple 因数;分倍数;倍数因数
submultiple resonance 分谐波谐振;分频谐波共振
submultiple unit 分数单位
submultiplex 分路调制
subnanosecond 亚毫微秒;次毫微秒
subnatant 下层清液
subnatant liquid 下层水
subnational boundaries 次国家边界
subnekton 下层自游生物
subnet(work) 粒界网状组织;亚网络;亚晶网;分作业网图;分网图;分网络;部分网络;子网(络)
subnormal 正常以下的;正常下的;低于正常的;低常的;次法线;次法距;不正常的
subnormal depreciation 低度折旧
subnormal integral 次法线积分
subnormal pressure 亚正常水面;低于正常压力;次正常压力;负压
subnormal temperature 正常下温度;低于正常温度;次正常温度;不正常温度
subnormal voltage 低于额定值电压
subnormal voltage operation 低电压运行
subnumber of lot 建筑用地次级号数
subocean fan 海底扇
suboceanic 洋地的
suboceanic region 非远洋海区
suboffice 分理处;分局;分办事处
suboiler 喷油翻土机
suboiling 翻土喷油处理(道路);土壤喷油处理(道路)

suboiling of soil road 土路渗油稳定;土路喷油稳定处理
subopaque 近似不透明
suboperand 子操作数;辅助操作数
suboptic(al) diffraction pattern 亚光频衍射图样
suboptic(al) frequency 亚光频
suboptimal 亚最佳的;最适度下的;次优的
suboptimal control 次最优控制
suboptimal design 次优设计;次优方案
suboptimality 次优性;次优胜
suboptimality bound 次优化界限
suboptimality index 次最优指标
suboptimal option 次优方案
suboptimal solution 次优解
suboptimal stopping rule 次最优停止规则
suboptimal system 次优系统
suboptimization 局部最佳化;次最佳化;次优化;部分最优化
suboptimization basis for decision-making 决策的分级优选基准
suboptimization by conflict 冲突中的次优化
suboptimum 次最适度;次最佳;次优
suborder 亚目;装饰性柱式;从属的建筑风格(由装饰决定的);次要柱式
subordinate 次要的;辅助的
subordinate altar 副圣台
subordinate building 辅助房屋;附属建筑
subordinate business 次要经销商
subordinate class 附类
subordinate complement 从属补体
subordinate crop 补播作物
subordinated 附属付款承诺
subordinated debenture 附属公司债券
subordinated debt 附属债务
subordinate debenture 附属信用公司债
subordinate demandable nations 次要需求国
subordinated ground lease 土地从属租契
subordinated loan 附属贷款
subordinate entrance 次要出入口
subordinate fold 从属褶皱
subordinate function 隶属函数
subordinate industry unit turnover 附业单位资金周转额
subordinate lien 附属留置权
subordinate line 辅线
subordinate load 附加荷载;次要荷载
subordinate mineral 次要矿物
subordinate mines 次要矿山
subordinate mining nations 次要矿产国
subordinate mining regions 次要矿产地区
subordinate productive commerce 次要生产商
subordinate productive nations 次要生产国
subordinate quantum number 次量子数
subordinate station 次要站港;潮汐次要观测站;副港;副潮港;附属站;辅助站;辅助测站;辅助位站
subordinate station of passenger train 车底配属站
subordinate system 辅助系统
subordinate tidal station 辅助验潮站;次要潮汐测站;附属验潮站;辅助观潮站;辅助潮汐站;辅助潮位站
subordinate tide station 辅助观潮站
subordinate unit paying profit 所属上交利润
subordination 隶属关系
subore halo 矿下晕
sub out 分包出
suboutcrop 埋藏露头;隐伏露头;隐蔽露头
subowner 二船东
suboxide 低氧化物;低(价)氧化物;次氧化物
subpackage 分装
subpanel 分面板;安装盘;安装板;副面板
subparagraph 附属条款;小段
subparameter 子参数
subparcel 分区
subparticipation 部分参与
subparticle 亚微粒子
subpartition 辅助分区
subpartner 附股合伙人
subpeninsula 次半岛
subperiodic type 亚周期型
subperiod of sea level relative change 海平面相对变化全球性亚周期
subpermafrost water 永冻层下水;冻土底水;次永久冻土水

subpermanent magnetism 次恒磁
subpermanent set 非永久性变形
subphylum 亚门
subpicosecond 亚微微秒
subpicosecond pulse 亚微微秒脉冲
subplan 辅助方案
subplate 底板
subplate mounted 板式连接的
subplatform （金属楼梯的）起步平台
subplinth 柱础基座;(柱底座下的)柱基
subpoena 传票
subpoint 下点;投影点
subpolar 副极地
subpolar anticyclone 副极地高压;副极地反气旋;亚寒带反气旋
subpolar convergence 亚寒带极限线;副极地辐合带
subpolar culmination of a star 恒星下中天
subpolar glacier 亚极地冰川;副极地冰川
subpolar gyre 副极地环流
subpolar high 副极地高压
subpolar low 亚寒带低压;副极地低压
subpolar low pressure belt 亚寒带低压带;副极地低压带
subpolar region 亚寒带区;亚寒带带;近极地区;副极地区;副极地带
subpolar westerlies 西风带
subpolar westlies 副极地西风带
subpolar zone 亚寒带
subpool 子存区组合
subpool number 子存区组合序号
subport 辅属港
subpost 副柱;辅助支柱;小柱
subpost car frame 底柱式车架
subpower 亚功率
subpower range 低于额定功率范围
subpreputial 包皮下的
subpress 小压机;小压力机
subpressure 次压力
subprincipal 次要承重构件;次承重件;分委托人
subproblem 子问题;部分问题
subprocessor 子处理机
subproduction order 附属生产通知单
subprofessional 专业人员助手
subprogram(me) 子程序;辅助程式;辅程序;部分程序
subprogram(me) library 子程序库
subprogram(me) segment 子程序段
subprogram(me) tape 子程序带
subproportional reducer 次比例尺减薄液
subprovincial administrative region 专区
subpulse 次脉冲
subpunch 先冲;留量冲孔
subpunching 初冲孔;先冲;预冲孔;初次钻孔(铆钉分两次钻)
subpurchase 转买
subpurchaser 转购人;转买人
subpurlin(e) 模板支架;小桁条;小檩条
subpurlin(e) installation 副檩条安装
sub-Qinling mountains 亚秦岭山地
subquadrate 正近方形的
subquality 质量差的
subquality product 不合格产品;次级品
subradar point 雷达下点
subradical 根号下的
subrail (楼梯的)副斜梁;安栏杆的暗楼梯梁
subrail foundation 轨下基础
subrail track bed foundation 轨下基础
subrange 支脉
subraster 小光栅
subraster deflection coil 亚光栅偏转线圈
subraster deflection yoke 小光栅偏转系统
subreach 小河段
subrefraction 次折射;亚折射;标准下折射
subregime 小类;分类
subregion 亚区;分区
subregional harbo(u)r 地方性港口
subregional planning 分区规划
subregional port 地方性港口
subrent 转租人所付租金
subreption 隐瞒事实真相
subresonance 亚共振
subrogate 权利转让
subrogation 代位(行使);权利转移;权利代位;取代债权人

subrogation clause 债权替代条款;代位条款
subrogation form 债权转让书;权利转让书;权利代位书;代位求偿条款
subrogation of compensation 代位赔偿
subrogation of rights 权利转移
subround 接近圆形的
subrounded 次圆状;次圆形的
subrounded particle 次圆形土粒;次浑圆土粒;次浑圆颗粒
subrounded pebbles 圆形砾石
subroutine 子程序;子例程
subroutine address stack 子程序地址栈
subroutine analyzer 子程序用分析程序
subroutine area 子程序区
subroutine call 子程序调用
subroutine call table 子程序调用表
subroutine cell 子程序
subroutine data area 子程序数据区
subroutine entry address 子程序入口地址
subroutine entry and exit 子程序的进口和出口
subroutine for symbols 符号子程序
subroutine instruction 子程序指令
subroutine jump 子程序转移
subroutine level 子程序级
subroutine library 子程序库
subroutine linkage 子程序连接
subroutine name 子程序名字
subroutine nesting 子程序套用;子程序嵌套;例行子程序嵌套
subroutine package 子程序包
subroutine parameter 子程序参数
subroutine pointer 子程序指针
subroutine procedure 子程序过程
subroutine programming 子程序程序设计
subroutine reference 子程序引用
subroutine return 子程序返回
subroutine separation 子程序分离
subroutine status table 子程序状态表
subroutine subprogram(me) 子程序的子程序;子程序的辅程序
subroutine tape 子程序带
subroutine test 程序检验
subroutinization 子程序化
subsalt well 盐水井
subsample 分样品;子样品;子样本;次级样本;重采样
subsamples 附属样本
subsampling 二次取样法;二次抽样
subsampling of unequal clusters 不等群体二级采样
subsand 亚砂土
subsatellite 子卫星
subsatellite point 卫星星下点
subsatral point 星下点
subsaturated erosion 次饱和冲刷
sub saver 辅助螺纹保护器
subscale 内部氧化物;次生氧化皮
subscanner 子扫描程序
subscanning 副扫描
subscience 科学分支
subscrew 副螺杆
subscribe 认缴;认购;签名
subscribed capital 认购资本
subscribed capital stock 已认购股本
subscriber 用户;认股人;认购人;签署者;订购户
subscriber access line 用户入口线
subscriber branch line 用户分线
subscriber-busy signal 用户占线信号
subscriber cable 用户电缆
subscriber circuit 用户电路
subscriber class circuit 用户类别路
subscriber connecting equipment 用户连接设备
subscriber equipment 用户设备
subscriber extension 用户分机
subscriber's drop 电话用户的落线
subscriber line 用户线
subscriber line circuit 用户电路
subscriber loop 用户线
subscriber meter 用户计费表
subscriber multiple 用户复式塞孔
subscriber of direct insertion 直插用户
subscriber plant transmission index 用户设备传输指标
subscribers 捐款人

subscriber's account 用户账单
subscriber's drop 用户吊牌
subscriber selector 用户选择器
subscriber set 电话用户机
subscriber's exchange 用户交换台
subscriber's extension 用户分机
subscriber's extension station 用户分机
subscriber's intercommunication installation 用户内部通话装置
subscriber's jack 用户塞孔
subscriber's line 用户专用线;用户线路
subscriber's line busy signal 用户线占线信号
subscriber's line equipment 用户线路设备
subscriber's line interface circuit 用户线路接口电路
subscriber's loop 用户线路;用户闭路
subscriber's machine 用户机
subscriber's main station 用户总机;用户主机
subscriber's module 用户线模块
subscriber's multiple jack 用户复式塞孔
subscriber's protector 用户保护器
subscriber's register 用户通话计数器
subscriber station 用户站
subscriber's telephonic cables 电话用户电缆
subscriber's uniselector 用户旋转式选择器
subscriber supervisory circuit 用户状态监视电路
subscriber terminal equipment 用户终端设备
subscriber toll dialing 用户长途直拨
subscriber-vision 计时收费制电视
subscribing witness 证人;保证者
subscript 索引;标号;注脚;脚码;脚标;添标
subscript bound 下标界
subscripted variable 下标变量
subscript expression 下标表达式
subscript index 下标
subscription 预约预订;预约;认捐;认购;签署;订购
subscription agreement 认股合同;认购合同
subscription and purchase agreement 认购合同;预定及购买协议
subscription blank 认购单
subscription book 应募公司债簿
subscription call 特约通话
subscription certificate 认购证券
subscription date 认股日期;认购日期
subscription instalment 分期缴款认股登记簿
subscription instalment book 认缴分期支付股款
subscription list 认购清册
subscription period 认可期限
subscription price 认购价格;认缴价值;认购价格
subscription quota 认缴份额;认定份额
subscription realized 已募集的总金额
subscription right 认购权
subscription television 收费电视
subscription to capital stock 认缴股款
subscript list 下标表
subscript position 下标位置
subscript range 下标范围
subscript sign 脚注
subscript variable 下标变量
subscuttle 舱窗风斗
subsea accumulator 水下储能器
subsea apron 海底扇
subsea blowout preventer 海底防喷器
subsea blowout preventer stack 海底防喷器组
subsea cable ship 海底电缆铺设船;海底电缆敷设船
subsea completed well 海底完井
subsea completion 水下完井
subsea completion system 水下完井系统
subsea cone 海底锥;海底冲积锥
subsea control pod 水下控制盒
subsea control system 水下控制系统
subsea current 海底潜流
subsea depth measurement 海底深度测量
subsea drilling equipment 水下钻探设备 水下钻具;水下钻井设备
subsea equipment 海底装置;海底设备;水下器具;水下井口装置
subsea facility 海底设备
subsea hose bundle 水下软管束
subseal 封底;基层封
subsealing 基层防水处理;封底处理;底封;基层处理;封底
subsealing treatment 封底处理
subsea lubricator valve 水下钢丝防喷阀

subsea manifold system 海底管汇系统
subsea oil field reactor 海底油田反应堆
subsea operation 水下作业
subsea outline of dredged areas 挖泥区海底略图
subsea physiographic provinces 海底自然地理区;海底地形区
subsea pipeline 海底管道
subsea production manifold 海底生产管汇
subsea telemanipulation 海底遥控操作
subsea temperature 海底温度
subsea template system 海底基板系统
subsea test tree 海底测试树
subsea topographic survey 海底地形调查
subsea transponder 水下声脉冲转发器
subsea work enclosure 海底工作舱
subsection 细目表;节段;分条;分段
subsection lending 分部门贷款
subseismic case 次地震情况
subsellia 折叠便座(用于客车、剧院等)
subsellium (教堂折椅底的)横挡;椅突板
subsemigroup 子半群
subsequence 子序列;顺次;次序
subsequence unit 子程序装置
subsequent 后续的;后成的;随后的
subsequent acceptance 附条件承兑;付条件承兑
subsequent address message 后续地址信息
subsequent anaeraobic ammonium oxidation 连续厌氧铵氧化
subsequent cross-folds 后成交叉褶皱
subsequent data 子序列数据
subsequent distillate 续馏液
subsequent drainage 后成水系
subsequent events 财务报表编制后发生的事项
subsequent fault 后生断层
subsequent filter 后续过滤器
subsequent filtrate 续滤液
subsequent filtration 后续过滤
subsequent flocculation 后絮凝作用
subsequent fold 后成褶皱
subsequent handling 后续工序
subsequent installation 后装置
subsequent legislation 后续法规
subsequent loading surface (塑性力学的)后继加载面
subsequent lock 定时(关闭)锁
subsequent mineralization 后成矿化作用
subsequent ore deposit 后成矿床
subsequent orientation 连续定向
subsequent payment 分期缴付
subsequent pick-up 继续拾取
subsequent pluton 后成深成岩体
subsequent record 序列记录
subsequent river 后生河;后成河;次成河
subsequent settlement 后续沉降;附加沉降量
subsequent stream 后生河;后成河
subsequent supervision 事后监督
subsequent symplectite 后成合晶
subsequent taxable year 下一纳税年度
subsequent thunderstorm 后续雷雨;后续雷暴
subsequent time period 连续时间;接续时期
subsequent treatment 后续处理
subsequent valley 后成谷
subsequent works 后期工程
subsere 次生演替系列
subseries 子级数;派生系列;次分类
subservience 从属性
subset 子设备;子集;电话用户机;分设备
subsetability 可分解性
subset data 数据子集
subset of a programming language 程序设计语言子集
subshaft 从轴
subsheaf 子层
subshell 亚层;子壳层;支壳层;副壳
subshot 重复放炮
subsidability 湿陷性;下陷性
subsidability of loess 黄土湿陷性
subsidability of loess under its own weight 黄土自重湿陷性
subside 下沉;凹陷
subsided area 沉陷区
subsided coast 沉降海岸
subsided plain 下降平原
subsided surface 下陷面

subsidence 消退;下陷;下沉;平息;坍陷;塌陷;沉陷;沉下;沉降移动;沉降;沉淀;凹陷
subsidence area 塌陷区;沉陷区
subsidence basin 沉降洼地;沉降盆地;沉陷盆地
subsidence belt 沉陷带
subsidence break 地层沉裂
subsidence chamber 沉降室
subsidence coast 沉陷海岸
subsidence curve 沉落曲线;下沉曲线;沉陷曲线(图);沉降曲线
subsidence damage 沉陷破坏;沉陷破坏
subsidence due to lost ground 地层流失导致的沉降
subsidence earthquake 剧烈地震;造成沉陷的地震;沉陷地震
subsidence factor 沉降系数;沉降系数
subsidence factor of dump 排土场下沉系数
subsidence flow 沉降流
subsidence gap 沉降缝
subsidence in fill 填土下沉
subsidence inversion 下沉增温;下沉逆温;沉降性逆转
subsidence inversion aloft 高下沉逆温
subsidence land 沉陷地
subsidence magnitude of bearing plate 承压板下沉量
subsidence magnitude of pile 单桩下沉量
subsidence model 沉降模式
subsidence observation 沉陷观测
subsidence observation content 沉降观测值
subsidence observation time 沉降观测时间
subsidence of a building 建筑物沉陷
subsidence of dike body 坝体沉陷
subsidence of flood wave 洪峰消退
subsidence of ground 地基沉陷;地沉
subsidence of ground surface 地表沉降
subsidence of levee 堤的沉陷
subsidence of reservoir bottom 水库底沉陷
subsidence of track 线路下沉
subsidence range 沉降观测范围
subsidence rate 下陷率;沉陷速度;沉陷率
subsidence settling 泥渣沉积
subsidence site 沉降地点
subsidence slope 沉陷斜度;沉陷坡度
subsidence tank 泥渣沉积槽;沉淀池;沉降池
subsidence temperature inversion 下沉逆温
subsidence theory 下沉说
subsidence trough 下沉盆地;下沉槽池;沉陷槽
subsidence velocity 沉淀速度;沉陷速度;沉降速度
subsidence volume 沉淀容积
subsidence wave 沉降潮(波)
subsidence zone 沉降带
subsidency 沉下
subside particle 沉降粒子
subsider 沉淀池;沉陷槽;沉降槽
subsidiaries 附属公司
subsidiary 子公司;附属机构;附属的;附带;辅助的;补助金;补助的
subsidiary account 辅助账户
subsidiary air supply 辅助空气供给;补充空气供给;补充的空气供应
subsidiary angle 辅助角
subsidiary appropriation ledger 预算拨款明细账
subsidiary attribute 随附属件;附属属性地位
subsidiary beam 辅助梁
subsidiary body 附属机构
subsidiary book 辅助簿记;明细账簿;辅助账簿
subsidiary building 附属建筑;辅助建筑(物)
subsidiary business 副业;附属营业
subsidiary business center 辅助商业中心
subsidiary cable 辅助电缆
subsidiary canal 支渠;辅助渠道
subsidiary channel 支渠;辅助航道
subsidiary civil action 附带民事诉讼
subsidiary class 附类
subsidiary classification 辅助分类
subsidiary commercial transaction 辅助商业交易
subsidiary communication 辅助通信
subsidiary company 附属公司;子公司;辅助公司;分公司
subsidiary condition 附加条款;辅助条款;辅助条件;附加条件
subsidiary conduit 支管道;配线管(道);辅助管(道)

subsidiary constituent 辅助成分
subsidiary conveyer 辅助输送机
subsidiary credit 辅助信用贷款
subsidiary criterion 附则
subsidiary crop 次要林作
subsidiary dam 副坝;辅助坝;辅坝
subsidiary department 辅助部门
subsidiary details 补充碎部
subsidiary development drivage 准备巷道
subsidiary dissipator 辅助消能工
subsidiary documents 附属单据
subsidiary drain 集水暗管;排水支沟;辅助排水沟
subsidiary earthquake effect 次生地震效应
subsidiary equation 辅助方程
subsidiary export company 附属出口公司
subsidiary factory 附属加工厂
subsidiary farming 补充耕作
subsidiary fault 分支断裂;分支断层
subsidiary feed 补充喂料
subsidiary felling 副伐
subsidiary fold 分支褶皱
subsidiary food(stuff) 副食品
subsidiary fracture 张破裂
subsidiary functions research 辅助功能研究
subsidiary grinding 辅助粉磨
subsidiary hospital 附属医院
subsidiary index number 辅助指数
subsidiary insurance 附属保险
subsidiary journal 辅助日记账
subsidiary labo(u)r power 辅助劳动力
subsidiary ledger 明细分类账;辅助总账;辅助账户;辅助分类账(户)
subsidiary ledger account 明细账户(户);辅助分类账(户)
subsidiary ledger for property 财产明细分类账
subsidiary ledger of merchandise in store 库存商品明细账
subsidiary legislation 附属立法
subsidiary letter of credit 附属信用证
subsidiary levee 副堤
subsidiary light 润饰光;辅助灯
subsidiary magnet 辅助电磁铁
subsidiary main 辅助干线
subsidiary main track 副干线
subsidiary material 辅助材料
subsidiary material cost 辅助材料成本
subsidiary maximum 辅峰
subsidiary model test 辅助模型试验
subsidiary occupation 副业
subsidiary operation time 辅助作业时间
subsidiary optic(al)system 辅助光学系统
subsidiary organ 附属机构
subsidiary or vertical method 附属工厂或直式方法
subsidiary patent 附属专利
subsidiary pipe 支管
subsidiary plant 辅助电站
subsidiary point 辅助点
subsidiary producing industries 辅助生产部门
subsidiary production 副业生产;副业;附属生产;辅助生产
subsidiary production plan 辅助生产计划
subsidiary production process 辅助生产过程
subsidiary production specialization 辅助生产专业化
subsidiary production workshop 辅助生产车间
subsidiary products 辅助产品;副产物
subsidiary questionnaire 辅助调查表
subsidiary reaction 副反应
subsidiary record 明细记录;辅助记录;辅助账簿
subsidiary reinforcement 分布钢筋
subsidiary reservoir 辅助水库
subsidiary revenue 辅助收入
subsidiary ring road 次要环路;辅助环路
subsidiary road 支路;辅助道路
subsidiary rod 辅助杆
subsidiary service 附带业务;辅助性服务
subsidiary sewer 支沟
subsidiary signal 辅助信号
subsidiary staff 副签
subsidiary standard oxidation reduction potential 副标准氧化还原电势
subsidiary station 辅助站;辅助电站;分站;补站【测】;补点【测】
subsidiary station peg 辅助潮站木桩

subsidiary stream 支流
subsidiary street 辅助街道
subsidiary stress 附加应力
subsidiary survey 小三角测量;辅助测量
subsidiary test 辅助试验
subsidiary thermal system 辅助热系统
subsidiary trade 辅助职业技能
subsidiary transport 辅助运输
subsidiary traverse 支导线【测】
subsidiary treaty 军援协定
subsidiary triangulation 小三角测量;辅助三角测量
subsidiary tube 支管
subsidiary unit 从属单位
subsidiary valence 副化合价
subsidiary weir 副堰;辅助堰
subsidiary workpoints 补贴工分
subsidiary works 附属工程
subsidiary workshop 附属品车间;辅助车间
subsidies and incentives 津贴和奖金
subsidies for health 保健津贴
subsidies for heating expenses 取暖费补贴
subsiding 下沉
subsiding amplitude of downthrow side 下降盘沉降幅度
subsiding basin 沉淀池;沉淀池
subsiding coast 沉降海岸
subsiding ground 下陷地面;沉降地面
subsiding movement 下降运动;沉降运动
subsiding period 沉降期
subsiding reservoir 沉淀池
subsiding sheet 下沉板片;下沉板
subsiding speed 沉降速度
subsiding stage 下降阶段
subsiding tank 沉淀池
subsiding velocity 下沉速度;降落速度;沉降速度;沉淀速度
subsiding velocity of downthrow side 下降盘沉降速度
subsiding water 沉降水
subsidization 津贴;补助金
subsidization in operation 运营补贴
subsidized apartment 受政府补贴的住房;廉租住房;廉价住房;公助住房
subsidized dwelling 受政府补贴的住房
subsidized export 受补贴的出口
subsidized house-building 受政府补贴的住房
subsidized housing 受补贴住房;有津贴的房屋;补贴住房
subsidized import 受补贴的进口
subsidized industries 政府辅助产业
subsidized line 补助航线
subsidized price 补助价格;补收价格
subsidize trial manufacture of new products 新产品试制费
subsidy 津贴;补贴
subsidy energy 补贴能量;补加能
subsidy for capital expenditures 基建补助金;投资补贴
subsidy for health 保健费
subsieve 微粒的;不能用筛分级的;亚筛度度;亚筛;不能用筛子分级的微粒
subsieve analysis 亚筛分析
subsieve fineness 亚筛细度(不能用筛子分级的微粒)
subsieve fraction 亚筛料末;亚筛份额;亚筛粉(末)
subsieve particle 亚筛粒子;亚筛颗粒
subsieve powder 亚筛料末;亚筛粉(末)
subsieve range 亚筛范围
subsieve size 亚筛粒度
subsieve size apparatus 亚筛粒度分析仪
subsieve size range 亚筛粒度范围(小于325目的粉末)
subsill 窗台披水;窗框托;底槛;副槛;副门槛;副窗台;垫底横木;底梁
subsistence agriculture 自给农业;温饱型农业
subsistence allowance 生活补助费;生活津贴;生活补贴
subsistence farming 自给农业
subsistence level 维持生活水平
subsize 尺寸不足的;过筛的;小尺寸
subsized specimen 小型试件
subslot 槽底沟
subsoil 下卧土;下层土;亚土层;亚层土;地基下层土;地基土;底土;底层土壤
subsoil agricultural drain 底层土壤农用排水
subsoil compacting 下层土压实
subsoil compaction 下层土压实
subsoil condition 土质条件
subsoil drain 农用地下排水沟;地下排水管;下层土排水沟;地下排水沟;底土排水沟
subsoil drainage 基土排水;下层土排水;亚土层排水;亚层土排水;地下排水;底层土(壤)排水
subsoil drainage pipe 地下排水管
subsoiler 喷油翻土机
subsoil erosion 底土侵蚀;管涌;底土冲刷
subsoil exploration 下层土勘探;亚土层勘探;底下查勘;底土查勘
subsoil flow 地下水流
subsoil grouting 底土灌浆
subsoil ice 地下冰;底冰
subsoil improvement 底土改良
subsoiling 翻土喷油处理
subsoil investigation 底土勘察;地基勘察;地基勘探
subsoil irrigation 地下灌溉;底土灌溉
subsoil liquefaction 底土液化
subsoil nature drainage 底层土天然排水
subsoil paper 覆盖在下土层上的纸
subsoil permissible load 天然地基允许荷载
subsoil pipe 下层土排水管
subsoil plough 底土犁
subsoil plow 底土犁
subsoil profile 地基土剖面图
subsoil puddling 基土捣固
subsoil sample 底土试样
subsoil stabilization 地基稳定(作用)
subsoil stratum 底土层
subsoil study 底土调查;底土勘测
subsoil study device 底土勘测设备
subsoil survey 底土勘测
subsoil test 地下试验
subsoil treatment 底土处理;基土处理
subsoil water 潜水;亚土层水;地下水;底土水
subsoil water movement 地下水运动
subsoil waterproofing 底土防水;地下防水
subsoil waterproofing wall 地下防水墙
subsolar point 日下点
subsolar point altitude 日下点高度
subsolar point temperature 日下点温度
subsolidus 亚固线;固相线以下
subsolidus data 亚固线数据
subsolution 海底溶解作用
subsonic 亚音速的;亚声速的;次音速的;次声波的
subsonic aerodynamics 亚声速空气动力学
subsonic area 亚声速区
subsonic boundary layer 亚声速边界层
subsonic compression 亚声速压缩
subsonic flow 亚声速流;亚声速气流
subsonic force coefficient 亚声速空气动力学系数
subsonic frequency 次声频
subsonic jet aircraft 亚声速喷气飞行器
subsonic lift 亚声速升力
subsonic mach number 亚声速马赫数
subsonic motion 亚声速运动
subsonic ramjet 亚音速冲压喷气发动机
subsonic rupture 亚声破裂
subsonic rupture speed 亚声破裂速度
subsonics 次声学
subsonic sonar 次声频声呐
subsonic sound 次声
subsonic speed 亚音速;亚声速
subsonic stability 亚声速稳定性
subsonic velocity 亚音速
subsonic vibrations 亚声频振动
subsonic whistle 亚声哨
subsonic wind tunnel 亚音速风洞
subspace 子空间
subspace iteration 子空间迭代
subspan 部分跨度;子跨度
subspecies 亚种
subspecification water 亚规格水
subspontaneous 半自生的
subs-stabilizing agent 稳泡剂
substable soil 亚稳定土
substage 亚期;亚阶;镜台下部;分台;分期
substage condenser 镜台下聚光镜;台下聚光器
substage condenser iris 聚光器光栏
substage illumination 穿透照明
substage microlamp 显微镜台下灯
substance 物质;物体;实物;单重
substance concentration 物质浓度
substance contour 横向厚度剖面
substance cycle 物质循环
substance liable to cause infection 易于引起传染病的物质
substance of anthropogenic origin 源于人类活动的物质;人为的物质
substance of property 财产实体
substance pumped 泵送物
substance tolerance 厚度公差
substandard 次级标准;低于定额(的);低于标准规格;低于标准的;低定额;低标准;等外;不合格的;不标准的;标准下的
substandard building 不合标准房屋
substandard casing 不标准套管
substandard cement 水泥次品;次品水泥
substandard dwelling 低标准住房
substandard goods 次品;副品
substandard housing 低水平住宅;不合标准住房
substandard instrument 次标准仪表;副标准仪表
substandard lot 不合标准地块
substandard meter 副标准电度表
substandard parts 不合格零件;次等零件
substandard pile 不标准桩
substandard product 等外(产)品;副次产品;等外废品;次品;副品
substandard project 不合标准项目
substandard rate 标准以下的工资率
substandard rim 窄口轮缘
substandard road 等外路
substandard sea water 副标准海水
substandard structure dwelling 低标准住宅
substantia compacta 密质
substantial 基本的;实质的;富裕的
substantial agreement 基本一致
substantial change 重大修改
substantial completion 实际竣工
substantial cost differentials 巨额的成本差异;巨大成本差额
substantial derivative 物质导数
substantial design 基本完工
substantial difference 相当大的差异
substantial increase 显著增长
substantial lane flow 实际车道车流
substantial law of environment 环境实体法
substantially insoluble 基本不溶解
substantial pillar 实体柱
substantial pollution 重污染
substantial results 实效
substantial-unemployment area 严重失业地区
substantiation 证实;证明;具体化;实质化
substantiation of claims 索赔证明
substantiator 立证据者
substantive 直接染色;直接的
substantive agreement 实质协议
substantive convention 实质性公约
substantive dye(stuff) 直接染料
substantive issue 实质性问题
substantive law 实体法
substantive provision 实质性条款
substantive right 实体权利
substantive rules 实体法性质的规定
substanvive input 直接输入
substate 亚态
substating 矿用变电所
substation 变电间;支站;次级站;辅助车站;分站;分台;分局;变压站;变电站;变电所;
substation capacity 变电站容量;变电所容量
substation grounding 变电站接地
substation of district heat supply network 分区供热系统分站
substation of heat supply network 热力站
substation program(me) control testing vehicle 变电站程控测试车
substellar point 星下点
substellar point of satellite 卫星星下点
substended chord 对弦
substep 分步
substitition swap 替代掉换
substituent 取代者;取代基
substitutability 可置换性;可代换性
substitutability analysis 置换分析

substitutable 可代换
substitute 置换物;置换;取代物;替代中心;替代物;代替者;代替;代理人
substitute arbitrator 替代仲裁员
substitute cargo 替代货物
substitute center 替换中心
substitute character 置换符
substitute community 替代群落
substitute consistently 相容置换
substituted benzene 取代苯
substituted cargo 代替货
substituted character 取代字符
substitute defect 等代缺陷
substituted expenses 替代费用;代替费用
substituted hydrocarbon 驱动烃;取代碳氢化合物
substitute discharge 引用流量
substituted phenol 取代酚
substituted security 替代担保
substituted service 替代送达
substituted ship 代替船
substitute fibre 代用纤维
substitute fixed assets 替代固定资产
substitute flag 代用旗
substitute for 替换;替代
substitute frame 替换框架;代用框架
substitute fuel 代用燃料
substitute goods 替代商品;代用品
substitute high speed steel 代用高速钢
substitute influence radius 引用影响半径
substitute material 代用材料;取代材料
substitute mode 转接方式;置换方式;更换方式;替换方式;代替方式;代入方式
substitute natural gas 合成天然气;代用天燃气
substitute of vessel 换船;代替船
substitute parts 代用零件
substitute point 代入点
substitute products 替代商品
substitute radius of well 水井的引用半径
substitute rudder 应急舵;代用舵
substitutes for imported goods 替代进口
substitute signal 代用标志
substitute simply 简单地置换
substitute soldering alloys 代焊接合金
substitute species 替代种
substitute sub 异径接头
substitute timber sleeper 木枕代用物
substitute type 替代群落型
substitute variable 替代变量
substitute vessel 替代船
substitute water head value 引用水头值
substituting effect 取代效应
substituting ticket 代用票
substitution 置换;取代(作用);替换(作用);代用
substitution account 替换账户;替代账户
substitution adsorption 置换吸附
substitutional acceptor 替代式受主
substitutional center 取代中心
substitutional circuit 等效线路
substitutional connection 等效线路
substitutional diffusion 置换扩散;替代扩散
substitutional element 置换元素
substitutional impurity 替代杂质
substitutional material 替换材料
substitutional production function 替代生产函数
substitutional reaction 置换反应
substitutional relation 替代关系
substitutional repulsion 代用品的冲击
substitutional resistance 替代电阻;等值电阻
substitutional solid solution 替代固溶体;置换(型)固溶体;替位固溶体
substitutional type solid solution 置换型固溶体
substitution analysis 替代分析
substitution cipher 代用密码;代用记号;代换密码
substitution clause 调船条款
substitution constant 取代基常数
substitution cover 临时盖
substitution curve 替代曲线
substitution deletion and insertion errors 代换删除和插入错误
substitution detector 置换检测装置
substitution effect 替代效能;替代效应;替代效果
substitution impedance measuring set 置换阻抗测量装置

substitution impurity 置换型杂质
substitution index 取代系数
substitution instance 代换实例
substitution instance of the statement function 命题函数的置换实例
substitution law 替代规律
substitution line 代换系
substitution list 代替表
substitution method 置换法;替代法;代替法;代入法;代换法
substitution method of measurement 置换测量法
substitution method of photometry 光度学代替法
substitution multipliers 替代乘数
substitution of invoice transfer 替换发票的转让
substitution of liability 债务转让;责任代替
substitution of rationing for exchange 配给代替交换
substitution of trustee 变更受信托人的文书
substitution operation 代替运算
substitution principle 替代原理
substitution product 置换产物;取代产物
substitution property 代换性质
substitution rate 替代率;代换率
substitution reaction 取代反应;排代反应
substitution(substit) 调换
substitution theorem 替代原理
substitution value 取代值
substitutive 置换的;取代的
substitutive derivative 取代衍生物
substitutor 替手;代用品
substoichiometric 亚化学计量的;低于化学计量的;不足化学计量的
substoichiometric analysis 亚化学计量分析
substoichiometric compound 亚化学计量化合物
substoichiometric extraction 亚化学计量萃取
substoichiometric isotope dilution 亚化学计量同位素稀释
substoichiometric isotope dilution analysis 亚化学计量同位素稀释分析
substoichiometric separation 亚化学计量分离
substoichiometry 亚化学计量学;亚化学计量法
substope 分段工作面
substore 辅助仓库;备用仓库
substorm 准风暴
substractional solid solution 取代固溶体
substraction chromatography 扣除色谱法
substraction fumigation 减法熏气
substrata formation 下层地层构造
substrate 质地;基片;基件;漆器的胎;底物;底金属;衬底;被涂物
substrate biodegradation 基质生物降解
substrate bonding area 衬底键合面积
substrate concentration 基质浓度
substrate concentration distribution 基质浓度分布
substrate conversion 基质转化
substrate crystal 衬底晶体
substrate cycle 底物循环
substrate decomposition 基质分解
substrate decomposition metabolism 基质分解代谢
substrate destabilization 底质去稳定作用
substrate failure 基层破坏(混凝土墙接缝处)
substrate-fed logic circuit 衬底馈电逻辑电路
substrate feed logic 衬底馈电逻辑
substrate gate 衬底栅
substrate holder 基片座
substrate information collection system 基层信息采集系统
substrate inhibition 底物抑制
substrate level phosphorylation 基质水平磷酸化
substrate magazine 基片暗箱
substrate-mask 基片遮板
substrate material 基底材料
substrate microbial biomass 基质微生物生物质
substrate pollution 基质污染;底质污染
substrate preparation 木纹处理;底材处理;板面处理
substrate reflection polarizer 基片反射偏振器
substrate slice 衬底片
substrate sludge 基质污泥;底质污泥;底泥
substrate stabilization 底质稳定作用
substrate temperature 衬底的温度
substrate transistor 衬底晶体管
substrate transmission 衬底透射率

substrate utilization 基质利用
substrate yield coefficient 底物收率系数
substratosphere 副平流层;亚平流层;副同温层
substratum 基层;土壤下层;下(卧)层;下覆地层;下部地层;胶层;基底;底(土)层;附着层;[复] substrata
substratum drainage 底层天然排水
substratum for waterproofing 防水底层
substring 副行
substruction 路基;基础工事;地下建筑;下层建筑;基体结构;底部结构;底层结构;下部结构;次结构
substructure 下层建筑;下部结构;下部建筑;亚组织;亚结构;子结构;炉底下部结构;基础结构;基础;地下结构(物);地下建筑;底层结构;底部结构;次结构
substructure cost 地下工程造价
substructure for closed drain 地下管道出口建筑物
substructure for electric(al)device 电子元件基板
substructure installation 安装底座
substructure method 子结构法
substructure of bridge 桥墩;桥梁下部结构
substructure work 下部结构工程;基础建筑物;基础工程
substrut 副撑
sub-subcontractor 三包(商);小分包者;小分包商;小分包人;小包(从大承包商中转包部分工程);三包单位
subsubmicron 绝微细粒;次微粒
subsubroutine 次子程序
subsubsystem 子子程序
subsulphate 碱式硫酸盐
subsumed clause 被包含子句
subsumption 归类;类别;包孕
subsuperheavy elements 亚超重元素
subsupplier 供销分店;供销分处
subsurface 地表以下;下层土面;水面下水层;地下;地面下;底面;表液面下的;表面下的
subsurface absorption 地下处置(废水);地下吸收
subsurface air 亚表层空气
subsurface bedrock 地表下基岩
subsurface boiling 地下沸腾
subsurface boring 地下钻探
subsurface canal 地下渠(道)
subsurface circulation 海面下环流;次表层环流
subsurface condition 地下条件;地质情况;次表面条件
subsurface containment of solid wastes 固体废物埋藏地下技术
subsurface contour 构造等高线;水下等高线
subsurface contour map of the key bed 地面下分界层等高线图;地面下标准层等高线图
subsurface correlation 地层对比
subsurface corrosion 表面下腐蚀
subsurface course 路面面层
subsurface coverage 地下覆盖
subsurface coverage section 地下覆盖段
subsurface current 下层流;次表层流
subsurface defect 亚表面缺陷;表层下缺陷
subsurface deposit 地下矿床
subsurface detention 地下滞留;地下截流
subsurface disposal 地下排放;地下处置
subsurface disposal field 地下处置场(废水)
subsurface disposal method 地下处置法
subsurface disposal of refuse 废料地下处理
subsurface divide 地下水界
subsurface drain 地下排水
subsurface drainage 地下排水;底排水
subsurface drainage basin 地下水流域;地下流域
subsurface drainage check 地下排水节制闸
subsurface drainage check gate 暗沟排水节制闸
subsurface drainage check valve 地下排水节制阀
subsurface drainage system 地下排水系统
subsurface draining 地下排水
subsurface drain pipe 地下排水管
subsurface earth exploration 地下土质查勘
subsurface easement 地下使用权
subsurface echo sounding 水下回声探测(法)
subsurface engineer 地下地质工程师
subsurface engineering 地下工程
subsurface erosion 潜蚀;水下冲刷;地下侵蚀;地下潜蚀;地下浸蚀
subsurface excavation 地下开挖
subsurface exploration 地下探查;地下探测;地下

勘探;地下勘察;地下勘测;地下查勘
subsurface exploration drilling 地下勘查钻孔
subsurface explosion 地下爆破
subsurface field 亚表土
subsurface film 次表面膜
subsurface filter 地下(过)滤场;地下滤池;地下过滤器
subsurface filter gallery 地下滤水廊道
subsurface filtration 地下渗透;地下过滤
subsurface filtration field 地下过滤场
subsurface flaw 次表面裂隙
subsurface float 潜没浮标;水下浮子;水下浮筒;水下浮标;深水浮子
subsurface flow 伏流;地下水流;地下径流;次表层流;暴雨渗流;潜流
subsurface flow constructed wetland 潜流人工湿地
subsurface flow constructed wetland treatment system 潜流人工湿地处理系统
subsurface flow system 潜流系统
subsurface flow wetland 潜流湿地
subsurface geochemical exploration 深部化探
subsurface geology 地下水地质学;地下地质学
subsurface highway 低堑道路;下层道路;地下道路
subsurface horizontal water flow 地下水平水流
subsurface hydraulics 地下水力学
subsurface ice 潜冰;地下冰
subsurface illuminator 内反光器
subsurface incineration 地下焚烧
subsurface indicator 地下测验器
subsurface infiltration 地下渗滤
subsurface infiltration treatment system 地下渗滤处理系统
subsurface investigation 地下探查;地下探测;工程地质勘探;地下调查
subsurface investigation drilling 地下勘查钻孔
subsurface irrigation 浸润灌溉;土壤层下灌溉;地下灌溉
subsurface karst 地下岩溶
subsurface land roller 亚表土镇压器
subsurface layer 地下层;次表层
subsurface level 次表层水位
subsurface map 地下地貌图
subsurface mark 地下标志;地下标记;地下标点
subsurface material 下层材料
subsurface mine 地下矿
subsurface mining 地下采矿
subsurface model 地下界面模型
subsurface moisture control 地下水分控制
subsurface opening 地下洞室
subsurface pavement layer 路面下层
subsurface permeameter 地中渗透仪
subsurface picture 地下(构造)情况
subsurface pressure 孔内压力;地下压力
subsurface pressure ga(u)ge 井下压力计
subsurface pump 深井泵
subsurface reconnaissance 地下勘测;地下勘查
subsurface riches 地下财富
subsurface right 地下所有权
subsurface road 下层道路;地下道路
subsurface runoff 亚表土径流;壤中流;潜水径流;地下径流;次表层流
subsurface runoff modulus method 地下径流模数法
subsurface sample 钻孔试样;钻孔中的样品;孔内试样;井下试样
subsurface sampling 钻孔取样;孔内取样;井下取样
subsurface sand filter 地下砂滤系统;地下滤池;地下滤场
subsurface sewage disposal 地下污水处理
subsurface sewage disposal system 地下污水处理系统
subsurface sewerage disposal system 地下污水处置系统;地下污水处理系统
subsurface shops 地下街道
subsurface soil 亚表土;生土;地下土;次表土
subsurface soil injection 地下灌注
subsurface sounding 地层测深;水下触探
subsurface storage 地下储藏
subsurface storage of water 地下储藏水
subsurface storm flow 地下流走的雨水;壤中流;暴雨渗入地下水流;暴雨地下流量
subsurface stratigraphy 地下地层学
subsurface stratum 亚表层

subsurface stress increment 下卧层应力增量
subsurface structure 下层构造;地下结构(物);地下构造
subsurface system 暗管系统
subsurface tectonic map 地下地质构造图
subsurface temperature 井内温度;地下温度;次表层温度
subsurface temperature survey 地下温度测量
subsurface tillage 次表土耕作法
subsurface tramway 地下电车道
subsurface trickle irrigation 地下滴水灌溉
subsurface unconformity 地下不整合
subsurface utility 地下公用设施;地下公共设施
subsurface velocity 水下流速;地下流速
subsurface waste disposal 地下废物处置;地下废料处置
subsurface wastewater disposal 废水地下处置
subsurface wastewater infiltration 地下污水渗滤
subsurface wastewater injection 地下污水注入
subsurface water 潜水;壤中水;地下水;底土水;次表层水
subsurface water basin 地下水盆地
subsurface watering 地下灌水
subsurface water inquiry 地下水的查询
subsurface water measurement 地下水测量
subsurface water pollution 地下水污染
subsurface wave 亚表面波;界面波;内波;水下波
subsurface work acceptance 隐蔽工程验收
subsurface zone 次表层带
subswitch 分机键
subsynchronous 亚同步;低于同步的;次同步的
subsynchronous layer 次同步层
subsynchronous observation 似同步观测
subsynchronous resonance 次同步谐振
subsynchronous satellite 亚同步卫星
subsynchronous speed 次同步速度
subsystem 亚系统;亚系;亚体系;子系统;子体系;次要系统;次要入口;辅助系统;分系统
subsystem equipment of substation integrated automation 变电站综合自动化子系统装备
subsystem index 分系统索引
subsystem information retrieval facility 子系统信息检索设施
subsystem library 子系统库
subsystem of liquidated revenue 清算收入子系统
subsystem of transport revenue 运输收入子系统
subtable 子表
subtabulation 副表
subtangent 次切线;次切距
subtank 水位计平衡器
subtask 子任务;工作分类
subtasking 执行子任务
subtebse method 视距法
subtenancy 转租;转借
subtenant 转租人
subtend 弦对弧;对向;衬托
subtended angle 弧对角
subtense 对边
subtense angle 弦角;视差角
subtense bar 横基尺;横标尺;水平视距尺;对角杆
subtense bar equipment 横基尺设备
subtense(bar)traverse 横基尺视差导线
subtense base 视差基线
subtense instrument 弦线测角仪;横基尺视距仪
subtense method 横基尺视距法;视差角法
subtense method distance measurement 视差法测距
subtense method with horizontal staff 基线横尺视距法;横基尺视距法;横基尺视距法
subtense method with vertical staff 竖基尺视距法
subtense technique 视距法
subtense traverse 视距导线
subterminal eruption 顶生喷发
subterminal outflow 底部出流
subterminal oxidation 次末端氧化
subterrane 地下结构物;洞穴;地下室;下层;下覆基岩;地下的
subterranean 隐蔽的;地下(的)
subterranean aerodrome 地下机场
subterranean algae 土层藻类
subterranean animal 地下动物
subterranean bacteria 地下细菌
subterranean bodies 地下岩体

subterranean brook 地下溪流
subterranean cable 地下光缆
subterranean catchment area 地下集水面积
subterranean chamber 地下室
subterranean deposit 地下矿床
subterranean disposal 地下处置
subterranean drainage 地下排水
subterranean dwelling 地下居所
subterranean heat 地下热
subterranean heat energy 地下热能
subterranean house 地下住宅
subterranean ice 地下冰
subterranean irrigation 地下灌溉
subterranean layer 地下层
subterranean living 地下生活
subterranean outcrop 掩蔽露头
subterranean outflow 地下水渗出
subterranean palace 地下宫殿
subterranean(pipe)line 地下管路;地下管线;地下线路
subterranean(pipe)network 地下管(线)网
subterranean railroad 地下铁道
subterranean railway 地下铁道
subterranean river 地下河(流);暗河
subterranean river course 地下河;地下水道;地下河道
subterranean room 地下室
subterranean shelter 地下掩蔽所
subterranean storage 地下仓库;地下油库
subterranean storehouse 地下仓库
subterranean stream 地下河(流);伏流河;伏流
subterranean stream course 地下河道
subterranean termite 地白蚁;地下蚁
subterranean tunnel 地下隧洞;地下隧道
subterranean water 潜水;地下水;底土水
subterranean water collector 地下水集水管
subterranean water enquire 地下水调查
subterranean water parting 地下水分线;地下水分岭;地下水分界
subterranean watershed 地下分水线;地下分水岭;地下分水界
subterranean work 地下设施;地下工程
subterrane heat 地热
subterraneous cable 地下电缆
subterraneous outcrop 掩盖露头
subterraneous root 地下根
subterraneous works 地下设施
subterrestrial 地下的
subthreshold 低于最低限度;次阈值
subthreshold behavio(u)r 次阈值性能
subthreshold concentration 亚阈浓度
subthreshold dose 阈下剂量
subthreshold resonance 次阈值共振
subthreshold response 次阈值响应
subthreshold spectrum 次阈值光谱
subtidal 潮下的
subtidal community 潮下群落
subtidal deposit 潮滩沉积
subtidal facies 潮下相
subtidal flat deposit 潮下坪沉积
subtidal line 低潮线
subtidal zone 浅海带;潮下带
subtide 潮下
subtie 副拉杆;副系杆
subtile 锐敏的
subtilization 稀薄
subtitle 小标题;副标题
subtle 精during的;锐敏的
subtle anomaly 难捉摸异常
subtle trap 隐秘圈闭
subtopia 市郊;地区一体化(趋势);城乡一体化(趋势);远郊城镇;开发为工业区的乡间地
subtopic 副主题
subtotal 小计;大部的;部分和
subtotaling 中间计算;求部分和;分合计
subtow 分丝束
subtrace 部分追踪
subtract 减去;减
subtract bit circuit 减位电路
subtracter 减数;减法器
subtraction 减去;扣除;减法
subtraction circuit 相减电路;减法电路
subtraction colo(u)r 减色
subtraction operator 减法算符

subtraction sign 减号
subtraction with serial operation 串行减法
subtractive 负号的
subtractive colo(u)r 减色
subtractive colo(u)r compositing 减色合成
subtractive colo(u)r matching 减色法配色
subtractive colo(u)r method 减色法
subtractive colo(u)r mixing 减色法混合
subtractive colo(u)r mixture 减色混色法
subtractive colo(u)r system 减色法系统
subtractive colo(u)r triangle 减色三角形
subtractive combination 差组合
subtractive complementary colo(u)r 减补色;减色法互补色
subtractive direct-positive process 减色法直接正像处理
subtractive mixing of colo(u)rs 减色法混色
subtractive polarity 减极性
subtractive primaries 减色法三原色;减法三原色
subtractive primary colo(u)r 减色法的基色
subtractive process 精制过程;脱除(杂质)过程
subtractive reducer 等量减薄液;等比例减薄液
subtractive synthesis 减色混合法
subtractive theorem 减色原理
subtractor 减法器
subtract pulse 相减脉冲;减法脉冲
subtract statement 减法语句
subtract time 减法运算时间
subtrahend 减数
subtransient 起始瞬态;次瞬态
subtransient current 冲击电流;超瞬变电流
subtransient reactance 次瞬变电抗;起始瞬态电抗;超瞬变电抗
subtranslucent 微透明的;半透明的
subtransmission 中等电压输电;二次输电
subtransmission system 分支输电系统
subtransparent 半透明的
subtraverse 附合导线【测】
subtread 踏步底板
subtreasury lock (耐火保险箱内的)小库锁
subtree 子树
subtriangular 近似三角形的
subtriple 三分之一的
subtriplicate 立方根;开立方
subtroic(al) plant 亚热带植物
subtropic(al) 亚热带的
subtropic(al) animal 亚热带动物
subtropic(al) anti-cyclone 副热带高压;亚热带反气旋
subtropic(al) belt 亚热带;副热带
subtropic(al) calm belt 亚热带无风带
subtropic(al) calms 副热带无风带
subtropic(al) cell 副热带高压单体
subtropic(al) climate 亚热带气候;副热带气候
subtropic(al) conifers 亚热带针叶树
subtropic(al) convergence 副热带辐合带;副热带地区
subtropic(al) crop 亚热带植物
subtropic(al) cyclone 副热带气旋
subtropic(al) district 副热带植物区
subtropic(al) easterlies 热带东风带;副热带东风带
subtropic(al) easterlies index 副热带东风带指数
subtropic(al) ecology 亚热带生态学
subtropic(al) ecosystem 亚热带生态系统
subtropic(al) evergreen forest yellow cinnamon earth zone 亚热带常绿林黄壤带
subtropic(al) exposure test(ing) 亚热带曝晒试验
subtropic(al) forest 亚热带森林;温带雨林
subtropic(al) gardening 副热带园艺
subtropic(al) gyre 副热带环流系
subtropic(al) high 副热带高压;副热带高气压
subtropic(al) high belt 亚热带高压带
subtropic(al) high (pressure) belt 副热带高压带;副热带高压
subtropic(al) high pressure zone 副热带高压区
subtropic(al) jet stream 副热带急流
subtropic(al) karst 亚热带岩溶
subtropic(al) monsoon climate 亚热带季风气候
subtropic(al) monsoon forest red soil zone 亚热带季雨林红壤区
subtropic(al) rain forest 亚热带雨林;副热带雨林
subtropic(al) region 亚热带地区;亚热带;副热带地区
subtropic(al) ridge 副热带高压脊

subtropic(al) soil 亚热带土壤
subtropic(al) subzone 亚热带亚带
subtropic(al) westerlies 副热带西风带;西风带
subtropic(al) zone 亚热带;副热带
subtropics 亚热带
subtroposphere 副对流层
subtruss 脚手架;支承桁架;支撑桁架;次桁架
subturps 松节油代用品
subtype 副型
subtype of soil 土壤亚类
subultramicro method 次超微量法
subultra-microscopic 次超显微镜的
subunguis 爪内片
subunit 原体;亚组;亚基;亚单位;辅助装置
subunit construction 单元结构;部件结构
suburb 郊区;市郊区;市郊
suburban 近郊的;近郊;郊区镇
suburban agricultural pollution 城郊农业污染
suburban agriculture 市郊农业
suburban area 近郊区;市郊;郊区
suburban bridge 市郊桥梁
suburban coach 市郊客车
suburban collector 郊外干管
suburban dispersal 市郊蔓延
suburban district 市郊区;近郊区;郊区
suburban estate 市郊房地产;郊区庄园领地
suburban garden 郊区花园
suburban green area 近郊绿地;市郊绿化区
suburban growth 郊区发展
suburban highway 市郊公路
suburban house 郊区住宅
suburban housing estate 郊区住宅房地产
suburbanite 郊区居民
suburbanization 郊区建造;近郊化;郊区建设;市郊化
suburban line 市郊界线;市郊线路;市郊铁路线;市郊铁路
suburban passenger flow 市郊客流
suburban passenger flow diagram 市郊客流图
suburban passenger train 市郊旅客列车
suburban populated area 市郊居住区
suburban railroad 市郊铁路;近郊铁路
suburban railway 近郊铁路;市郊铁路
suburban rapid transit system 郊区快速交通系统
suburban residential area 市郊住宅区
suburban residential quarter 近郊住宅;近郊住宅区
suburban road 近郊道路;郊区道路;市郊道路;市郊道路
suburban road network 郊区线网
suburban roadway 市郊铁路
suburban run 郊区扩展地段
suburban service 近郊(交通)服务;市郊区服务业
suburban sprawl 郊区蔓延
suburban station 市郊站
suburban track 市郊线路
suburban traffic 郊区交通;市郊交通;近郊交通;市郊运输
suburban train 市郊列车
suburban train station 市郊火车站
suburban transportation 郊区运输
suburban villa 郊区别墅
suburban zone 近郊地带
suburb dispersal 市郊延伸;市郊扩大
suburbia 都市郊区;郊区居民;城市郊区
suburb populated area 市郊居住区
suburb residential area 市郊居住区
suburbscape 市郊景观
suburb sprawl 市郊的扩大
suburb uptown 近郊住宅区
suburdan estate 市郊房地产
subvalue 次值
subvariable spring 亚变化泉水
subvelocity of light 亚光速
subvention 津贴;补助金;补贴
subversion 颠覆
subvertical 垂直支杆;副竖杆;接近垂直
subvertical member (桥梁的)副垂直杆;副竖杆
subvitreous 光泽不如玻璃的;次玻璃的
subvitreous lustre 半玻璃光泽
subvoice-grade channel 亚音频级信道
subvolcanic 潜火山的;次火山的
subvolcanic activity 次火山活动
subvolcanic body 次火山岩体
subvolcanic facies 次火山相

subvolcanic geothermal system 次火山地热系统
subvolcanic-hydrothermal ore deposit 次火山热液矿床
subvolcanic ore-forming process 次火山成矿作用
subvolcanic phase 次火山岩相
subvolcanic rock 次火山岩
subvolcanics 次火山岩
subvolcano 潜火山
subwalk 地下过街人行道;人行隧道
subwatering 地下给水
subwater pipeline 水下管线;水下管路
subwatershed 小集水区;支流(流)域;次分集水区
subwave 部分波
subwaxy phase 次蜡状相
subwaxy soap 次蜡状皂
subway 工作通道(船坞的);地下铁道;地下电缆管道;地下道路;地铁;地道
subway administration 地铁管理
subway bridge 地道桥
subway car 地下铁道车辆
subway crossing 地下人行道;地下交叉口;地下横道
subway environment 地铁环境
subway for pedestrians 行人地道
subway leading to platforms 通站台的地道(车站);站台(间)地道;车站站台间地道
subway line 地下铁路线;地铁线路
subway network 地下铁道网
subway network map 地铁网示意图
subway noise 地铁噪声
subway station 地下(铁)道车站;地铁(车)站
subway train 地下铁道列车
subway transformer 浸没式变压器;地下变压器
subway tube 地下铁道圆形隧道
subway tunnel 地下铁道隧道
subway-type transformer 地道型变压器
subway vehicle 地铁车辆
subway waste 地铁废弃物
subwhole 不完整
subzero 负温度;零度以下;零下(的);低温;低凝固点;负荷温度
subzero coolant 冷处理用冷却剂;冷处理剂;零下冷却剂
subzero cooling 冷处理
subzero equipment 低温处理装置
subzero fractionation 冷冻分离
subzero gear oil 寒区齿轮油
subzero oil 低温润滑油
subzero refrigerator 零下冰箱
subzero temperature 零下气温;零下温度;零度以下气温
subzero treatment 低温处理;零下处理
subzero weather 严寒天气
subzero working 零下温度加工
subzone 小区;附属贸易区;亚带;分区
succah 茅屋;棚舍
succedaneum 替代品
succeeding lift 后继浇筑层
succeeding pump 继动泵;顺序启动泵
succeeding tide 后续的潮汐;后续潮
successful bidder 得标人;中标人;中标者;投标中标人;投标得标人;得标者
successful call 成功呼叫
successful scheduling 成功的安排
successful search 成功查找
successful species 继承种
successful tenderer 中标者
success go-to field 成功的去向区
succession 演替;连续;继续性;继承性;顺序性;顺序
successional habitat 演替环境
successional species 继承种
succession of aspect 季相演替
succession of blows 连续撞击
succession of crop 复种制
succession of generation 世代演替;生成序次
succession of plant 植物演替
succession of quaternary system 第四记地层层序
succession of sedimentary facies 沉积相序列
succession of strata 地层序列;地层层序;地层层位;地层层次;层序【地】
succession of tectonic movement 构造运动继承性
succession tectonics 继承性构造
successive 逐次;连续的;累世的;递次
successive anchors 子母锚;串连锚

successive approximate values 逐次近似值
successive approximation 逐渐逼近;逐次近似法;逐步渐近;逐步逼近法;渐近法
successive approximation analog/digital converter 逐次近似模/数转换器
successive approximation method 逐步近似法;逐步渐近法;分段渐近法
successive approximation type analog (ue)-digital 逐次逼近型模拟数字转换
successive approximation unit 逐步逼近装置
successive bound 逐次跃进
successive carry 逐次进位;顺序进位
successive collapse 连续塌陷
successive comparison 逐步比较
successive component 继承组分
successive contrast 相继对比;逐次对比;连续对比;继时对比
successive co orientation 连续定向
successive course 邻接层
successive cut 连续切削;连续切割
successive cycles 逐次循环
successive decay 逐级衰变;逐次衰变
successive demolition 连续爆破
successive derivative 逐次求导;逐次导数
successive difference 相继差值;逐次差分;递差
successive differential coefficient 逐次微分系数
successive differentiation 逐次微分
successive displacement 逐次替换
successive displacement method 逐次置换法;逐次移位法;逐次位移法;逐次替换法;逐步替换法
successive doubling 顺次加倍
successive elimination 逐次消除(法)
successive elimination method 逐次消元法;连续消去法
successive eruptive events 连续喷发活动
successive failure 逐步破坏;渐进性破坏
successive feedback repression 连续反馈抑制
successive field 相继场
successive fracture treatment 连续压裂作业
successive grain 连续颗粒
successive group overrelaxation 逐次成群超松弛
successive holographic summation 连续全息叠加
successive induction 逐次归纳法
successive integration 逐次积分(法)
successive integration method 连续积分法
successive interchange 连续式道路立体枢纽
successive intervals 逐次区间
successive intervals method 逐次间隔法
successive iteration method 逐次代入法
successive launching method 顶推法
successive layers 相继各层;连续层
successive line 连续谱线
successive linear filtering 逐次线性滤波
successive line overrelaxation 逐次线超松弛
successive losses 连续损失;连续损害
successively worked parallel relay network 继电器并联传递网络
successive minimum points 逐次极小点
successive orientation 连续定向
successive overrelaxation 逐次超松弛
successive overrelaxation method 逐次超松弛法
successive phase of deformation 连续变形幕
successive photograph 连续像片
successive plate 分色板
successive point 连续点
successive point overrelaxation method 逐次点超松弛法
successive point relaxation method 逐点松弛法
successive procession 连续加工
successive pulse groups 连续脉冲组
successive quadrature 逐次求积分
successive range 移动范围;逐次距离;逐次范围
successive reading 连续读数
successive readout 连续读出
successive reciprocating 连续往复
successive reduction 逐次约化;逐次简化;逐步简化
successive river 连贯河流
successive rounds of grader 连续平地机的循环过程
successive route 延续进路
successive row relaxation method 逐次行松弛法
successive sampling method 逐次抽样法;连续抽样法
successive scales 比例尺系列
successive sedimentation 逐次沉淀
successive selection 连续选择
successive slip 连续滑动;渐进性滑动
successive spacial threshold 继时空间阈
successive structure 逐次结构
successive substitution 逐次代法;逐步取代法;逐步代替法;递代法
successive substitution process 逐次代入法
successive summation 相继总和
successive surge 连续涌浪
successive sweep method 逐步淘汰法
successive synthesis 逐次综合法
successive tectonism 连续构造作用
successive term 邻项;逐项
successive transformation 递次变换
successive transport 接运
successive trials 逐次试探(法)
successive undermining 连续地下挖掘
successive value 逐次值;连续值
successive weather maps 逐次天气图
successive withdrawal 逐次撤退
successive years 连年
successor 接班人;继承者;继承人;后继(者)
successor address 后继指令地址
successor and assigner 继承人和受让人
successor block 后继块
successor function 后继函数
successor instruction 后继指令
successor in title 权利继承人
successor list 后继表
successor matrix 后继矩阵
successor set of a set 集合的后继集合
success rate 成功率
succimide 琥珀酰亚胺
succinamide 琥珀酰胺;丁二酰胺
succinanil 琥珀酰苯胺
succinanilic acid 琥珀酰苯胺酸
succinate 琥珀酸酯;琥珀酸盐;丁二酸酯;丁二酸盐
succinate oxidation 琥珀酸氧化
succindialdehyde 琥珀醛;丁二醛
succinic acid 琥珀酸;丁二酸
succinic acid 2 丁二酸—2
succinic acid aqueous solution 琥珀酸水溶液
succinic aldehyde 丁二醛
succinic anhydride 丁二酸酐
succinic anhydrite 琥珀酐
succinic chloride 琥珀酰氯
succinic diamide 琥珀酰胺
succinic monoamide 一酰胺
succinimide 琥珀酰亚胺;丁二酰亚胺
succinite 钙铝榴石;琥珀色;琥珀
succinonitrile 琥珀腈
succinoresinol 琥珀树脂醇
succinyl 琥珀酰;丁二酰
succinyl chloride 琥珀酰氯;丁二酰氯
succinylcholine 丁二酰胆碱
succinyl oxide 琥珀酰化氧
succinylsulfathiazole 丁二酰磺胺塞唑
succory 菊苣属
succulent silage 多叶青储[贮]
succulent thorn-scrub 肉质刺灌丛
succulent wood cutting 软材插条
succulometer 湿度计
succusion-sound 震荡音
sucess-fail value 成败值
suck 吸空气;抽吸;凹形
suck-back device 回吸装置
suck board 稿图架
sucker 吸(引)管;吸头;吸器盘;根出条;进油管
sucker antenna 吸盘天线
sucker frame 吸盘架
suckering root 寄生根
suckering seedling 连根苗
sucker pole 抽油杆;泵杆
sucker rod 吸杆;抽油杆;泵杆
sucker rod elevator 吸光升降机;泵杆吊卡
sucker rod guide 抽油杆导向器
sucker rod hanger 抽油杆悬挂器
sucker rod hook 吸杆钩;抽油杆大钩
sucker rod jar 抽油杆震击器
sucker rod joint 抽油杆接头;抽油杆箍
sucker rod pump 杆式泵;有杆泵
sucker rod socket 泵杆打捞器
sucker rod wax 吸杆蜡
sucker rod wrench 泵杆扳手;吸杆扳手
suck holder 吸气框
sucking 吸釉(窑具吸取油中挥发性成分);吸引;拉细;缩号;抽
sucking action 吸引作用;抽吸作用
sucking and forcing pump 吸压两用泵;抽气压力泵
sucking booster 馈线升压器
sucking coil 吸引线圈
sucking disc 吸盘
sucking disk 吸盘
sucking fit 推入配合
sucking head 吸水头
sucking hole 吸水口
sucking jet pump 吸入喷射泵;吸入射流泵;吸引喷嘴泵
sucking main 吸水干管
sucking mouth parts 刺吸式口器
sucking pest 刺吸式口器害虫
sucking pipe 吸油管;吸水管;吸管
sucking port 吸入口
sucking pump 真空泵;吸抽加压泵;抽吸泵;抽气泵
sucking rate 抽气量
sucking solenoid 吸持螺线圈
sucking tube 吸水管;吸入管;吸气管;吸筒;吸管
sucking water 吸入水
sucking water head 吸水头
sucking(water)well 吸水井
sucking well on shore 岸边吸水井
suckling room 乳儿室
suckout 频带空段
sucrose 蔗糖
sucrose benzoate 苯甲酸蔗糖酯
sucrose octa-acetate 蔗糖八乙酸酯
sucrosic 砂糖状
sucrosic texture 糖粒状结构
suction 吸引术;吸扫[疏];吸入;吸力;中心缸;空吸;抽真空;抽吸器;抽吸;抽力
suction action 吸引作用
suction air 吸入空气;抽吸空气
suction air chamber 吸入空气室;抽吸空气室
suction and compressed air conveying 吸压混合式气力输送
suction and discharge valve 吸水与放水阀簧
suction and discharge valve seat 吸水及放水阀座
suction and exhaust valve 进排气阀
suction and force pump 吸压泵;空气压力泵
suction and-pressure fan 吸压式通风机
suction anemometer 吸管式风速表
suction apparatus 吸引器;吸入装置
suction aspirator tip 吸引器头
suction attachment 吸管接头;吸管;吸附件
suction bailer 吸入式砂泵;汲泥(泥浆或泥沙);抽泥筒
suction basket 水泵进水花管底阀;吸管滤网
suction bay 吸水河湾;吸水池
suction bell 吸入室;吸入(钟)口;吸入喇叭
suction belt 吸附带
suction bend(ing) 吸水管弯头;吸管弯头
suction between ships 船吸
suction blast 逆爆破
suction blower 吸入式通风机;吸入式鼓风机;吸风机
suction boiler 吸入式钻泥提取器
suction bottle 吸引瓶;吸滤瓶
suction bottle frame 吸引器瓶架
suction boundary layer control 吸除式附面层控制
suction box 吸水箱;吸入箱;真空吸水箱;通海吸水箱
suction branch 吸入分叉管;抽气支管
suction bush 吸入套管
suction button 吸引钮
suction cap 排烟罩壳;排气罩壳
suction capacity 吸扬能力;自吸能力
suction casing 进口壳体;入口壳体
suction casting 真空铸造;真空注浆成型;真空吸铸;抽成浇铸
suction catheter 抽吸导管
suction cell 吸槽
suction chamber 吸室;吸入腔;吸气室
suction characteristic 吸水特性
suction check valve 吸管止逆阀;吸入单向阀
suction chute 吸入管
suction clack 吸瓣
suction cleaner 吸式清扫机;吸尘器

suction cleaning 吸气式清洗
suction cock 吸入开关;进水旋塞;抽汽旋塞;抽气旋塞
suction coefficient 抽吸系数
suction compressor 吸除系统抽气机
suction conduit 吸水管;吸入管道;吸管
suction cone 吸锥;吸入锥;吸入喇叭口;吸风锥
suction connection 吸水口;吸入连接管;吸入连接短管;吸入口连接
suction contact 吸入接触
suction conveyer 真空式输送装置;吸引式输送装置;吸力运输机;吸力输送器;真空输送机;抽吸输送机
suction cover 吸入口外壳;吸入端泵盖
suction cowl 吸气盖
suction culvert 吸扬涵洞;吸扬暗渠;吸提暗渠
suction cup 吸盘;吸杯
suction cup tread tyre 吸盘花纹的轮胎
suction curette 吸引刮匙
suction current 吸入水流;吸入流
suction cutter 吸砂头
suction cutter apparatus 吸扬式挖泥沙装置;吸砂机构
suction cutter dredge(r) 旋桨吸泥机;吸扬式挖泥机;吸式挖土机;吸式挖泥船;绞吸(式)挖泥船
suction cutter dredging 旋桨吸泥疏浚
suction cutter head 吸式挖泥头
suction cylinder 吸入缸;吸汽缸
suction damper 吸入口调节器
suction deaerator 真空除氧器;真空除气器
suction depth indicator 吸口深度指示仪
suction device 吸引装置;吸尘装置
suction disc 吸盘;吸板
suction dose 吸入剂量
suction draft 吸气通风;吸引通风
suction drag 吸阻力
suction drainage 吸引流法;吸引导液法
suction draught 吸入式通风;抽引风力
suction dredge(r) 吸扬式挖泥机;吸扬(式)挖泥船;吸式挖土机;吸式挖泥船;吸入式采砂船;吸泥器;吸泥机;吸泥船;抽砂挖泥船
suction dredge(r) for emptying barge 泥驳吸卸泵
suction drilling 吸出式凿岩
suction duct 吸水管;吸入管;吸气管
suction eddy 吸气涡流;吸力漩涡;负压旋涡;负压涡流
suction effect 吸入效应;吸气作用;负压作用
suction efficiency 吸入效率
suction elbow 吸肘管;吸水弯管
suction equipment 吸扬设备
suction extraction 吸出术
suction eye 吸入孔
suction fan 吸气(风)扇;吸风机;吸风风扇;进气通风机;排气风机;抽风机
suction fan gas freeing 抽风驱除油气法
suction feed cup 吸力进给杯
suction feeder 吸气给料装置;真空给纸器;吸气式供纸器
suction feeding 真空吸料法
suction feed type gun 吸入式喷枪
suction filter 吸滤器;真空滤池
suction flange 吸入端法兰;进口法兰
suction flash 吸滤瓶
suction floor valve 吸水底阀
suction flow equalizer (泵的)排量均衡器
suction flue 吸烟道;吸排烟道
suction foot valve 吸水底阀
suction force 负压;吸力
suction forceps 吸引镊
suction funnel 吸入漏斗
suction gas 空吸煤气;吸引瓦斯
suction gas engine 吸入式煤气机
suction gas producer 吸入式煤气发生炉
suction gatherer 真空挑料机
suction ga(u)ge 吸入压力表;吸力计;真空计;低压压力表;负压表
suction gradient 抽吸压力梯度;负压梯度
suction grid 吸水网
suction guide wheel 吸导轮
suction gun 吸枪
suction head 吸(升)水头;吸升力;吸升高度;吸上高度;吸入高度;吸入侧压头;吸泥头;吸泥口;吸料头;吸出高度;虹吸水头;水泵吸入扬程;负压水头

suction header 吸水进水主干;吸水进水头;吸水受水器;吸收器;吸入集管
suction heater 吸入加热器
suction height 吸水高度;吸升高度;吸入高度
suction height controlling 吸水高度调节
suction hole 吸气孔
suction hole condition 吸水口条件
suction hood 吸气罩;吸尘罩
suction hopper dredge(r) 吸扬装舱挖泥船;吸扬式开底挖泥船;自行式吸扬挖泥船
suction hose 吸水软管;吸入软管;吸泥软管;吸力头;抽吸软管;抽吸胶管
suction inlet 吸入口;吸泥口
suction inlet shutoff valve 吸口关闭阀
suction instrument 负压表
suction intake 吸入管
suction intensity 抽吸强度
suction isotherm 吸附等温线
suction jet 抽吸喷射器;抽吸射流器
suction-jet conveying system 射流抽吸式输送系统
suction ladder 吸砂机构
suction lead 吸入导管;抽吸导管
suction leak-proof ring 入口防漏环
suction length 吸水高度
suction lift 吸引高度;吸水高度;吸升扬程;吸升力;吸升高度;吸上高度
suction lift of a pump 泵吸升高度;泵吸吸高度;泵的吸引升力
suction limit 吸升极限
suction line 抽吸管线;吸油系统;吸油管系;吸引管线;吸入管路;吸入总管
suction line filter 吸入管滤油器
suction line refrigerant velocity 制冷吸入速度
suction line screen 吸入管滤网
suction loader 抽吸式装料机
suction loss 吸升损失
suction machine 吸尘机;气吸式捡拾机;抽吸机
suction main 吸水总管;吸水干管;吸入总管
suction manifold 吸水汇管;吸入支管;吸入口底板
suction meter 负压表
suction method 抽吸法
suction method drainage 吸扬式排水
suction method of cleaning 抽吸清洗法
suction mill 吸磨机
suction moisture equivalent 吸收水分当量
suction mo(u)ld 真空铸造;抽吸成型
suction mouth 吸水口;吸入口;吸嘴;吸口
suction mouth level indicator 吸口深度计
suction mouth liner 吸口衬环
suction mouth wearing ring 吸口衬环
suction nozzle 抽吸喷嘴;吸嘴;吸入嘴;进气嘴
suction nozzle palletizer 吸盘式摞包机
suction oiling 吸入加油
suction opening 抽风口
suction operated instrument 吸动仪表
suction overfall 虹吸溢洪道;吸入溢流;虹吸溢水道;虹吸溢流道;虹吸式溢洪道
suction overflow 虹吸溢流
suction pad 吸收垫
suction parameter 入口参数
suction part 抽吸部分
suction passage 吸入孔道
suction phase 吸引相
suction phenomenon 吸引现象
suction pickup 气吸式捡拾器
suction piece 吸入连接管
suction pipe 吸水平管;吸入管;吸(泥)管;空吸管;抽吸管;负压管
suction pipe aboard ship 船内吸泥管
suction pipe bracket 吸管托架
suction pipe gantry 吸泥管吊架
suction pipe joint 吸管接头
suction pipe line 吸水管线;吸入管线;吸泥管线;抽吸管线
suction pipelining 抽吸管线
suction pipe loss 吸水管水头损失;吸管损失
suction pipe strainer 吸水管滤网
suction pipe system with filter system 带滤水器的吸水管
suction pipette 移液管
suction piping 吸水管线;吸水管道;吸水管道系统
suction piping system 吸水管系统
suction pistol 吸枪
suction pit 吸水坑;泥浆池;水源箱

suction plant 吸尘器;吸扬设备;吸尘装置;抽吸泵站
suction plate 抽气晒象框;负压力板;吸板
suction plate apparatus 负压力板仪
suction plate method 测量负压压法【岩】
suction plenum 吸式通风系统
suction pneumatic 气吸的;抽吸空气的
suction pneumatic conveyer 真空式输送装置;吸送式气力输送器;气吸式输送器;负压气力输送泵
suction point 抽吸点
suction pool (泵站的)进水井;(泵站的)吸水井
suction port 吸入口;吸入孔;进口气
suction port of compressor 压缩机入口
suction potential 吸势
suction pouring 真空吸铸
suction power 抽吸能力
suction press 吸水压榨;真空压榨
suction press roll 吸水压榨辊
suction pressure 负压;吸引压(力);吸压力;吸水压(力);吸入压力;吸气压(力);吸力;抽吸压;负压强;负压力
suction pressure ga(u)ge 吸气压力计
suction pressure transfer installation for bulk cement 散装水泥负压输送机设备
suction primer 吸入式起动注水器
suction producer 吸气式发生器
suction property 抽吸性能
suction pulsation 吸管脉冲
suction pump 吸扬泵;吸水泵;吸入泵;吸泵;真空泵;空吸抽机;空吸泵;抽吸泵站;抽吸泵;抽吸机;抽水泵
suction pump reverse circulation drilling 泵吸反循环钻进
suction pyrometer 真空高温计;空吸式高温计;抽吸式高温计
suction quality 抽吸特性
suction range 吸水极限;吸水范围
suction rate 吸水率;吸收率
suction rate of bricks 砖的吸收率
suction ratio 吸附比
suction replenishing valve 吸入补充阀
suction resistance 吸引阻力;吸入阻力
suction reversing valve 吸入转换阀
suction roll 真空辊
suction rose 水泵进水花管底阀
suction runner 吸取转轮
suction sapdisplance meat 抽吸树液置换处理
suction scavenging machine 吸引式输送机
suction scour 波浪退吸冲刷
suction seal 吸入密封;吸力密封
suction shaft 吸引开关
suction side 背风面;抽吸侧;负压侧;吸引端;吸入端;吸入边;吸力侧;真空面;低压面
suction side bearing cover 吸入口轴承压盖
suction side cartridge 吸入侧筒形接头
suction side of calciner 分解炉抽吸侧
suction side of fan 通风机进气边
suction side sleeve 进口轴套
suction skimmer 吸入式撇油器;吸入式浮油回收装置
suction sleeve 吸泥套管;吸泥管套管;吸管橡胶套管
suction-slot spillway 吸水槽式溢洪道;吸水狭长开口槽式溢洪道
suction sorter 吸入式分级机
suction speed 吸入速度
suction spindle 吸针
suction strainer 吸嘴滤网;吸水过滤器;吸入滤网;吸入管过滤器;吸入粗滤器;吸滤器
suction strength 吸力强度
suction strip feed 真空吸力式板料送进装置
suction stroke 吸入冲程;吸气行程;吸气冲程;进气冲程
suction surface 吸力面;真空面
suction sweeper 真空除尘器;吸入式收集器;吸尘器;街道吸尘器
suction system 吸油系统;吸入系统
suction table 吸纸台
suction take-down 吸风式牵拉
suction tank 吸水箱;吸水池;吸水罐
suction temperature 吸入温度;吸气温度;回气温度
suction tension 吸水压(力);抽吸张力
suction trap 液体分离器
suction tube 吸引管;吸筒;吸水管;吸入管;吸泥管;吸管;抽筒【岩】
suction tube automatic winch controller 耙臂绞

suction tube diameter 吸泥管直径
suction tube ladder 吸煤架
suction tube position monitor 耙臂位置监控器
suction tube position monitoring 吸管位置监控
suction tube velocity 吸管速度
suction type 吸扬式
suction-type baghouse 负压袋式收尘器
suction type carburetor 吸入汽化器;吸入型化油器
suction type cloth filter dust collector 真空式布滤集尘器;真空式布滤除尘器;抽吸式布滤集尘器;抽吸式布滤除尘器
suction type draught regulator 吸入式通风调节器
suction type dredge(r) 吸砂机
suction type pneumatic conveyer 吸送式气力输送机
suction type sampler 吸入式取样器
suction type sediment sampler 吸入式泥沙采样器
suction type suspension load sampler 抽吸式悬移质采样器
suction unit 吸扬设备
suction valve 吸水阀;吸入阀;进油阀;进气阀(门);进口阀(门)
suction valve bolt and split pin 吸入阀螺栓及开尾销
suction valve cover 进口阀盖
suction valve disc 吸入阀片
suction valve lifter 吸入阀片提举器
suction valve of pump 水泵吸水管阀
suction valve seat 吸水阀座
suction valve spring 吸入阀簧
suction vane 引流叶片
suction vane gear 引流叶片齿轮
suction velocity 吸入速度
suction velocity at return air inlet 回风口吸气速度
suction velocity at return air intake 回风口吸气速度
suction ventilating fan 吸气(通)风扇;吸气风箱
suction ventilation 吸入(式)通风;吸气通风
suction ventilation mode 吸出式通风方式
suction ventilation system 吸入式通风系统
suction ventilator 吸入通风装置;吸入通风机
suction venting 吸入通风
suction volume 吸入量
suction vortex 吸水漩涡;吸入漩涡
suction washer 吸力洗净器;吸力冲洗器;真空洗涤机
suction washing 吸力冲洗
suction water 吸入水
suction water barrel 抽水桶(凿井用)
suction water column 吸力水柱;负压水柱
suction water head 吸入水头;吸力水头;吸收水头;吸升扬程;吸升扬水
suction water pipe 吸水管
suction wave 稀疏波;吸行波;吸入波;空吸波;负压波
suction well 吸水井
suction well of pumping house 泵房进水井
suction well on shore 岸上吸水井;岸边吸水井
suction zone 负压区;抽吸区
suctorial 吸附的
Sucupira 苏库匹拉硬木(一种产于中美洲光亮深棕色硬木)
Sudan gum arabic 苏丹阿拉伯树胶
sudanophilic 染苏丹的
sudanophilous 染苏丹的
Sudan red 苏丹红
Sudan yellow 苏丹黄
sudation 渗出
sudatorium 热汽浴(室);蒸汽浴(室);热汽浴(复)sudatoria 或 sudatory
sudburite 倍苏玄武岩
Sudbury copper-nickel deposit 肖德贝里铜镍矿床
sudburyite 六方锑钯矿
sudd (阻碍航行的)水面植物漂浮物;大块浮水植被
sudden applied load 突加荷载
sudden catastrophe 突然事故(大的)
sudden change 突跃;突然变化;突变沉积
sudden change of water level in rivers and canals 河渠水位突然变化
sudden change of weather 天气骤变
sudden changing in biology 生物突变
sudden closure 突然关闭

sudden collapse 突然坍塌
sudden commencement 急始
sudden contraction 骤缩;突然收缩
sudden displacement 突然位移
sudden drawdown 骤降(水位);降落;水位突降
sudden drop 急剧下降
sudden drop in pressure 压力突降
sudden drop in temperature 气温突降
sudden enhancement of atmospherics 天电突增;大气干扰突然升高
sudden enlargement 骤扩;突然扩大
sudden failure 突然失效;突发故障
sudden fall 骤退
sudden flooding 突然涌水
sudden frequency deviation 突然频移;突然频率偏差
sudden halt 突然停机
sudden huge profits 暴利
sudden injection method 一次投放法;突然(注入)溶液法(测流)
sudden ionospheric disturbance 电离层突然骚扰;电离层突然扰动
sudden load change 荷载突变;负荷突变
sudden load(ing) 冲击负荷;骤加负荷;骤加荷载;突然荷载
sudden loss of wash liquid 冲洗液突然捕失
sudden lurch 突然倾斜
suddenly applied load 骤加荷载;突加荷载
suddenly-applied short circuit 突然短路
suddenly applied stress 突加应力
suddenly changing interface 突变界面模型
sudden magnetic disturbance 急始磁扰
sudden phase anomaly 相位突然异常
sudden pollution 突发性污染;突变体污染;特发性污染
sudden profuse uterine bleeding 崩证
sudden rupture 突然破裂
sudden rupture of rail 钢轨折断
sudden settlement 突然沉陷;突然沉降
sudden short-wave fade out 短波突然衰退
sudden shower 突降阵雨
sudden stress 骤加应力
sudden subsidence 突然沉陷
sudden thrown 突然丢弃
sudden transition 突变段
sudden turn 急转
sudden variation 突然变化
sudden water pollution 特发性水污染
sudden water release 突然泄水
Sudetian orogeny 苏台德运动【地】
sudial 日晷
sudoite 须藤石;铝绿泥石
suds suppressor 抑染剂
sue and labo(u)r charges 损害防止费用;施救费用
sue and labo(u)r clause 损害防止条款(海上保险船方的救护条款);施救(整理)条款(防止损害条款);救护条款;事故防止条款(海上保险)
suede 绒面革(棕色)
suede coating 新感觉涂料;绒面涂料
sue for a breach of promise 控告违约
suegee 苛性硷水
suessite 硅三铁矿
suevite 冲击凝灰角砾岩
Suez canal 苏伊士运河
Suez graben 苏伊士地堑
suface temperature 地表温度
sufferance 默许;优惠;海关落货许可;容许
sufferance quay 优惠码头;特许码头
sufferance wharf 优惠码头;特许码头
sufficiency 适应度;充分条件
sufficiency condition 充分性条件
sufficiency of contract price 合同价格的充分性
sufficiency of tender 投标书的完备性
sufficiency rating 适应性检定【道】
sufficient approximation 充分近似法
sufficient condition 充分条件
sufficient coupling 足耦合
sufficient estimate 充分估计(量)
sufficient estimator 充分估计式;充分估计量
sufficient lubrication 充分润滑
sufficiently nourished profile 供砂充分的剖面
sufficient reason 充分理由
sufficient receiver 充分接收机
sufficient statistics 充分统计量

sufficient supply 敞开供应
suffix 下标;添标;词尾
suffix experiment 反缀实验
suffix notation 后继表示法;反向波兰表示法
suffix of constituent 分潮指数
suffix signal 词尾信号
suffocating gas 窒息性气体;窒息毒气
suffocating substance 窒息性物质
suffocation 窒息作用;闷死
Suffolk brick 白垩黏土砖
Suffolk co(u)lter 立式开沟器
Suffolk kiln 底烧窑
Suffolk latch 萨福克插销;门窗插销
suffrutescent 半灌木状的;亚灌木
suffruticose 半灌木状
suffuse 弥漫(的)
suffusion 地下淋溶;潜蚀;弥漫
suffusion funnel 潜蚀漏斗
suffusion landform of loessland 黄土潜蚀地貌
sugar apple 番荔枝
sugar basin 糖缸
sugar beet 甜菜
sugar beet drill unit 甜菜条播装置
sugar-beet lifter 甜菜挖堆机
sugar beet pickup loader 甜菜捡拾装载机
sugar beet waste-grown anaerobic sludge 甜菜制糖废物生长厌氧污泥
sugar beet wastewater 甜菜制糖废水
sugar berg 松冰山
sugar bin 糖仓
sugar candy 冰糖
sugarcane 甘蔗
sugarcane board 甘蔗板
sugarcane crusher 甘蔗压榨机
sugarcane fibre 甘蔗纤维
sugarcane field 甘蔗田
sugarcane top 甘蔗叶
sugarcane weigh(ing) platform 称蔗台
sugar content 糖分含量
sugar effluent 制糖废水
sugar factory waste(water) 制糖厂废水
sugar-granular 糖粒状
sugar hardwood maple 糖槭
sugar house 糖厂
sugar industry 制糖工业
sugar industry wastewater 制糖工业废水
sugar loaf 圆锥形高丘;圆锥形丘
sugar-loaf fashion 宝塔糖形;圆锥形式样;塔形;宝塔形;圆锥形
sugar-loaf sea 三角浪
sugar maple 糖槭
sugar mill effluent 制糖厂污水;制糖厂废水
sugar milling machinery 甘蔗压榨机
sugar of iron 铁糖
sugar of lead 铅糖;醋酸铅;乙酸铅
sugar over 美化;粉饰
sugar palm 桄榔
sugar pine 甜松
sugar refinery 制糖厂;糖厂
sugar refinery plant 炼糖厂
sugar snow 浓霜
sugar soap 糖皂
sugar-solubility test 糖溶解度试验法
sugary 砂糖状
sugary quartz 糖状石英
suggest commend 提出建议
suggested amendment 修改意见
suggested common name 建议通用名称
suggested design 推荐方案
suggested no adverse response level 建议无害响应水平
suggested permissible level of noise 噪声建议容许级
suggested specification 建议性规范
suggestibility 易暗示性
suggestion 建议
suggestion box 意见箱
suggestion system 提案制
suggestive selling 提示销售法
sugi gum 日本柳杉脂
sugilite 钠锂大隅石
suglgum 松脂
Suhler-white copper 苏里锌白铜
suicidal stream 自灭河

suicide contactor 自动灭磁开关
suitability 适用性;适配性;适合性
suitability analysis 适宜性分析
suitability for climate condition 对气候条件的适合性
suitability for cycle operation 适合于循环操作的
suitability number 适用性系数
suitable 相配的;合适的;适当的
suitable appropriate transport temperature 运输适温
suitable condition 适宜条件
suitable containerable cargo 适合装箱货
suitable for eating 适于食用的
suitable for limed 适用于石灰性土壤
suitable formations 合适的岩层
suitable habitat 适宜生境
suitable moisture content 适宜含水率
suitable occasion 适当的场合
suitable phase switching 适当相位接入
suitable place 适当场所
suitable port site 合理港址
suitable position 适当位置
suitable rail type 适用轨型
suitable regions of ambient noise of cities 城市环境噪声适用区
suitable running flux 适宜运行通量
suitable technology 适用技术
suitable water 适用的水
suitably designed collection vehicle 适当设计的垃圾收集车辆
suit at law 民事案件
suitcase 手提(皮)箱
suitcase rock 不能再向下钻的岩层
suite 序列;一组;一套房间;岩套;套间;套房;数贯;程序组(合)
suite cabin 特等舱
suite deluxe 高级套房
suite of racks 成套机架
suite of room 整套房间;套房;成套房间
suite room 特等舱;套间房
suite room for patient 套间式病房
suit for contract 合同上诉
suiting local needs and making full use of favo(u)rable condition 因地制宜以而发挥优势
suitor 起诉者
suit pending 案件正在审理中
sujee 苛性碱水
suji-muji 苛性碱水
sukkah 茅舍;棚舍
sulcate 有平行深槽的
sulfabenzamide 苯酰磺胺
sulfacetamide 阿尔布西
sulfacid process 硫黄酸工艺过程
sulfadehyde 硫醛
sulfa drug 磺胺剂
sulfa film 磺胺薄膜
sulfamate 氨基磺酸盐
sulfamation 磺胺化作用
sulfamic 胺磺酰基
sulfamic acid 氨基磺酸
sulfamide 硫酰胺
sulfamine 氨磺酰
sulfamine-benzoic acid 氨磺酰苯甲酸
sulfamoyl 氨磺酰
sulfanilamide 磺胺
sulfanilamide sulfamidyl 氨苯磺胺
sulfanilamide wastewater 磺胺废水
sulfanilamido 磺胺基
sulfanilic acid 对氨基苯磺酸
sulfasuxidine 丁二酰磺胺塞唑
sulfatara 硫质气孔
sulfate 硫酸酯;硫酸盐
sulfate activation 硫酸盐激发作用
sulfateapatite 硫磷灰石
sulfate attack 硫酸盐腐蚀;硫酸盐侵蚀
sulfate bacteria 硫酸盐细菌
sulfate barrier 硫酸盐障
sulfate bearing soil 含硫酸盐土
sulfate blister 硫酸盐气孔
sulfate bloom 硫酸盐华
sulfate bubble 硫酸盐气泡
sulfate carbonate salinized soil 硫酸盐—碳酸盐土
sulfate cement 硫酸盐矿渣水泥

sulfate-chloride salinized soil 硫酸盐—氯化物盐土
sulfate corrosion 硫酸盐腐蚀
sulfate cycle 硫素循环
sulfated oil 硫酸化油
sulfated residue 硫酸化残渣
sulfate expansion 硫酸盐膨胀
sulfate expansion of brickwork 砖砌体因硫酸盐而引起的膨胀
sulfate groundwater 硫酸盐类水
sulfate-H$_2$S sulfur isotope fractionation 硫酸盐二氧化硫同位素分馏
sulfate ion 硫酸离子;硫酸根离子
sulfate-laden soil 硫酸盐土壤
sulfate lignin 硫酸盐木质素
sulfate liquor 硫酸盐溶液;硫酸盐黑液
sulfate method 硫酸盐法
sulfate mineral 硫酸盐矿物
sulfate mineral spring water 硫酸盐矿泉水
sulfate of ammonia 硫酸铵
sulfate of copper 硫酸铜
sulfate of lead 硫酸铅
sulfate of lime 硫酸钙;钙类硫酸盐
sulfate of potash 硫酸钾
sulfate of water 水的硫酸盐
sulfate ore 硫酸盐矿石
sulfate pigment 硫酸盐颜料
sulfate pine oil 硫酸化松油
sulfate pitch 妥尔油(脚);松浆油(脚)
sulfate plaster 石膏浆
sulfate pollution 硫酸盐污染
sulfate process 硫酸盐法
sulfate process titanium dioxide 硫酸法二氧化钛
sulfate pulp 硫酸盐木浆
sulfate pulping 硫酸盐制浆法
sulfate pulp mill wastewater 硫酸盐纸浆厂废水
sulfate radical 硫酸根(自由)基
sulfate reducing bacteria 硫酸盐还原菌
sulfate reducing bacteria degradation 硫酸盐还原菌降解
sulfate reducing condition 硫酸盐还原条件
sulfate reduction 硫酸盐还原法
sulfate reduction zone 硫酸盐还原带
sulfate resin 松浆油
sulfate resistance 抗硫酸盐性;抗硫酸盐腐蚀能力
sulfate-resistant 抗硫酸盐的;耐硫酸盐的;抗硫酸盐剂
sulfate-resistant blast furnace cement 抗硫酸盐矿渣水泥
sulfate-resistant cement 抗硫酸盐水泥
sulfate-resistant classification 抗硫分类
sulfate-resistant oil well cement 抗硫酸盐油井水泥
sulfate-resistant portland cement 抗硫酸盐硅酸盐水泥;抗硫酸盐波特兰水泥
sulfate-resisting 抗硫酸盐的;耐硫酸的
sulfate-resisting cement 抗硫酸盐水泥
sulfate-resisting characteristics 抗硫酸盐特性
sulfate-resisting portland cement 抗硫酸盐硅酸盐水泥;抗硫酸盐波特兰水泥
sulfate rosin 松浆油
sulfate rubber 硫化橡胶
sulfates 硫酸盐类
sulfate salinized soil 硫酸盐盐渍土
sulfate scab 硫酸盐气泡
sulfate seed 硫酸盐气泡
sulfate sludge 硫酸盐污泥
sulfate-sodium type 硫酸钠型
sulfate soundness test 硫酸盐稳定性试验(混凝土试验);抗硫酸盐稳定试验;抗硫酸盐试验
sulfate-sulfide sulfur isotope fractionation 硫酸盐—硫化物同位素分馏
sulfate susceptibility test 对硫酸盐敏感性试验
sulfate test 硫酸盐含量试验
sulfate wood turpentine 木松节油
sulfathiazole ointment 磺胺噻唑油膏
sulfating 硫酸化(作用);硫酸垢
sulfating agent 硫酸化剂
sulfating roasting 硫酸盐化焙烧
sulfation 硫酸盐化(作用);硫酸化(作用);硫化
sulfation rate 硫酸盐化率
sulfbenzamide 苯甲酰磺胺
sulfenamide type accelerators 次磺酰胺类促进剂
sulfenic acid 次磺酸
sulfenone 对氯二苯砜
sulfetrone 苯丙砜

sulfhalides 硫卤化物
sulfhydryl group 硫基
sulfidal 胶态硫
sulfidation corrosion 硫蚀
sulfidation pan 磺化锅
sulfide 硫化物
sulfide accretion 硫化物炉瘤
sulfide-bearing sediment 含硫化物沉淀物
sulfide catalyst 硫化物催化剂
sulfide colo(u)rs 硫化染料
sulfide concentrate 硫精砂
sulfide containing wastewater 含硫化物废水
sulfide dyes 硫化染料
sulfide enrichment 硫化物富集;硫化矿富集(作用)
sulfide-facies iron formation 硫化物相含铁建造
sulfide-H$_2$S sulfur isotope fractionation 硫化物二氧化硫同位素分馏
sulfide inclusion 硫化物夹杂
sulfide method 硫化物法
sulfide mineral 硫化物矿物
sulfide of water 水的硫化物
sulfide ore 硫化(物)矿(石)
sulfide oxidation 硫化物氧化
sulfide pair sulfur isotope fractionation 硫化物对硫同位素分馏
sulfide pollution 硫化物污染
sulfide removal 脱硫化物
sulfide rubber 硫化橡胶;聚硫橡胶
sulfide staining 涂层的硫化物锈蚀;硫化物污染失光;硫化变暗
sulfide toning 硫化物调色法
sulfide treatment 硫化物处理
sulfide zone 硫化物带
sulfidic copper ore 硫化铜矿石
sulfidic lead ore 硫化铅矿石
sulfidic mo ore 硫化钼矿石
sulfidic nickel ore 硫化镍矿石
sulfidic pb-Zn ore 硫化铅锌矿石
sulfidic sb ore 硫化锑矿石
sulfidic zinc ore 硫化锌矿石
sulfiding smelting 硫化熔炼
sulfidion 二价硫离子
sulfidity 硫化度
sulfidizing cementation 硫化置换
sulfindigotate 硫靛酸盐
sulfindigotic acid 硫靛酸
sulfine 硫化物
sulfinuz processing 渗硫处理
sulfitation 亚硫酸化(作用)
sulfitation tower 亚硫酸化塔
sulfite 亚硫酸酯;亚硫酸盐
sulfite cellulose 亚硫酸纤维素;亚硫酸盐纤维素;亚硫酸盐赛璐珞
sulfite cellulose extract 亚硫酸盐纤维素提出物;亚硫酸盐纤维素萃取物
sulfite cellulose liquor 亚硫酸盐纸浆废液
sulfite cooking liquor 亚硫酸盐蒸煮液
sulfite ion 亚硫酸盐离子
sulfite lye 亚硫酸盐碱液;纸浆废液
sulfite lye adhesive 亚硫酸盐木质素;亚硫酸盐碱液胶粘剂
sulfite-mill effluent 亚硫酸盐厂废水
sulfite ore 黄铁矿
sulfite process 亚硫酸盐法
sulfite pulp 亚硫酸盐纸浆
sulfite pulping 亚硫酸盐纸浆化;煮制亚硫酸盐纸浆
sulfite pulping process 亚硫酸盐纸浆化工艺;亚硫酸盐纸浆法
sulfite pulp manufacture industry sewage 亚硫酸盐纸浆制造工业污水
sulfite pulp mill wastewater 亚硫酸盐纸浆厂废水
sulfite-reducing bacteria 亚硫酸盐还原菌
sulfite reduction 亚硫酸盐还原法
sulfite reduction activity 亚硫酸盐还原活性
sulfite solid 亚硫酸盐固体
sulfite spent liquor 亚硫酸盐纸浆废液;亚硫酸盐废液
sulfite spirit 制造亚硫酸盐产生的酒精副产品
sulfite turpentine 亚硫酸盐松节油
sulfite waste liquor 亚硫酸盐废液
sulfite waste liquor by-product 亚硫酸盐废液副产品
sulfite waste lye 亚硫酸盐纸浆废(碱)液

sulfite wood pulp 亚硫酸盐木浆
sulfo 磺基
sulfoacetic acid 磺基乙酸;磺基醋酸
sulfoacid 硫代酸;磺酸
sulfoacylation 磺基乙酰化作用
sulfoaluminate 硫代铝酸盐
sulfoaluminate cement 硫铝酸盐水泥;硫代铝酸盐水泥;磺基铝酸盐水泥;膨胀水泥
sulfoaluminate cement clinker 硫铝酸盐水泥熟料
sulfoaluminate crystal 硫铝酸盐结晶
sulfoaluminate early strength cement 硫铝酸盐早强水泥
sulfoaluminate expansive cement 硫铝酸盐膨胀水泥
sulfoaluminate self-stressing cement 硫铝酸盐自应力水泥
sulfo aluminous cement 硫酸盐矾土水泥
sulfoaminobenzoic acid 磺基氨基苯甲酸
sulfoarsenide 硫砷化物
sulfobenzide 二苯砜
sulfobenzoic acid 磺基苯酸
sulfoborite 硫硼镁石;硼镁矾
sulfocarbolate 酚磺酸盐
sulfocarbolic acid 苯酚磺酸
sulfo chlorinated lubricant 硫氯化润滑剂
sulfocompound 含硫化合物
sulfocyanic acid 硫氰酸
sulfoethanoic acid 磺基醋酸
sulfoether 硫醚
sulfofatty acid salt 磺基脂肪酸盐
sulfofication 硫化(作用)
sulfoform 硫化三苯基;硫仿
sulfogel 硫酸凝胶
sulfohalite 氟盐矾
sulfoid 胶态硫
sulfolane 环丁砜
sulfolane process 环丁砜法
sulfolane process desulfurization 环丁砜法脱硫
sulfoleic acid 磺化油酸
sulfolene 环丁烯砜
sulfonamic 胺磺酰基
sulfonamide 磺胺
sulfonamide formaldehyde resin 磺酰胺甲醛树脂
sulfonamide p 氨苯磺胺
sulfonamide plasticizer 磺酰胺类增塑剂
sulfonaphthol 磺基萘酚
sulfo-naphthyl red 磺基萘红
sulfonaphtine 氨基萘磺酸
sulfonatability 磺化性
sulfonate 磺酸酯;磺酸盐;磺化
sulfonated 磺化的
sulfonated bodies 磺化物
sulfonated castor oil 磺化蓖麻油
sulfonated coal 磺化煤
sulfonated copper phthalocyanine 磺化铜酞菁
sulfonate detergent 磺酸盐洗涤剂
sulfonated hydrocarbon 磺化烃
sulfonated lignite 磺化褐煤
sulfonated oil 磺化油
sulfonated oleicacid 磺化油酸
sulfonated organophosphonates 磺化有机膦酸盐
sulfonated phenol formaldehyde resin 磺化酚醛树脂
sulfonated stearic acid 磺化硬脂酸
sulfonated styrol resin 磺化苯乙烯树脂
sulfonated tall oil 磺化妥尔油;磺化松浆油
sulfonate expansion 磺酸盐膨胀
sulfonate grease 磺化脂
sulfonate surfactant 磺酸盐型表面活性剂
sulfonating 磺化
sulfonation 磺化(作用)
sulfonation index 磺化指数
sulfonation method 磺化法
sulfonation number 磺化值
sulfonation value 磺化值
sulfonator 磺化器
sulfone 砜
sulfone polymer 聚砜
sulfoniazid 苯磺烟肼
sulfonic acid 磺酸
sulfonic acid amide 磺酰胺
sulfonic acid bromide 磺酰溴
sulfonic acid cation exchange 磺酸阳离子交换

sulfonic group 磺酸基;磺基
sulfonic resin 磺酸型树脂
sulfonic acid ion exchange resin 磺酸型阳离子交换树脂
sulfonium compound 硫鎓化物
sulfonium hydroxide 氢氧化三烃基硫
sulfonyl acid halide 磺酰卤
sulfoparaldehyde 三聚乙烯醛
sulfophenylate 苯酚磺酸盐
sulfo-ricinoleic acid 磺基蓖麻酸;磺化蓖麻油酸
sulfosalt 硫盐;磺酸盐
sulfoselenide 硫硒化物
sulfosol 硫酸溶胶
sulfotep 硫特普
sulfovinic acid 烃基硫酸
sulfoxidation 磺化氧化作用
sulfoxide 亚砜
sulfoxylate 次硫酸盐
sulfuonephthalein indicator 磺酞指示剂
sulfur 硫黄
sulfur acid 含硫酸
sulfur acid-resisting binder 硫黄耐酸胶结料;耐硫酸黏结剂
sulfur aggregate 含硫集料
sulfur amino acid 含硫氨基酸
sulfur analysis 硫分析
sulfur and nitrogen free basis 无硫无氮基
sulfur asphalt 含硫沥青
sulfur asphalt mix 硫黄—沥青混合物
sulfurated hydrogen gas 硫化氢气
sulfur attack 硫腐蚀;低温腐蚀
sulfur-autotrophic-denitrification process 硫自养反硝化工艺
sulfur bacteria 硫细菌;硫黄细菌
sulfur ball 硫黄球
sulfur band 硫带(钢材)
sulfur-bearing 含硫的
sulfur-bearing crude(oil) 含硫原油
sulfur-bearing fuel 含硫燃料
sulfur-bearing gas 含硫气体
sulfur-bearing high concentrated organic wastewater 含硫高浓度有机废水
sulfur-bearing oil 含硫石油
sulfur bearing steel 含硫钢
sulfur-bearing wastewater 含硫污水
sulfur bloom 硫霜;硫黄华
sulfur blooming 喷硫(现象)
sulfur blue 硫化蓝
sulfur budget 硫收支
sulfur buffer 硫缓冲
sulfur cement 硫结合剂(用于铁件);硫黄水泥;硫黄胶合剂;含硫结合料
sulfur cement mortar anchorage 硫黄锚固
sulfur coated urea 包膜硫尿素
sulfur compound 硫化合物;含硫化合物
sulfur compound analyzer 硫化合物分析仪
sulfur concrete 硫黄混凝土
sulfur consumption 硫耗;耗硫量
sulfur containing alloy 含硫合金
sulfur-containing fuel 含硫燃料
sulfur-containing substance 含硫物质
sulfur content 硫含量;硫分;含硫量
sulfur content in fuel 燃料含硫量
sulfur content in gas 煤气含硫量
sulfur content reduction program(me) 降低含硫量规划
sulfur crack 硫蚀裂纹;硫裂(焊接缺陷)
sulfur cycle 硫循环
sulfur deposit 硫矿
sulfur dichloride 二氯化硫
sulfur dioxide damage 二氧化硫造成的损害
sulfur dioxide effect 二氧化硫效应
sulfur dioxide emission 二氧化硫排放
sulfur dioxide emission standards 二氧化硫排放标准
sulfur dioxide fumigation 二氧化硫熏蒸法
sulfur dioxide index 二氧化硫指数
sulfur dioxide monitor 二氧化硫监测仪;二氧化硫监测器
sulfur dioxide poisoning 二氧化硫中毒
sulfur dioxide pollutant 二氧化硫污染物
sulfur dioxide pollution 二氧化硫污染
sulfur dioxide reduced chrome 二氧化硫还原的铬鞣液

sulfur dioxide scrubber 二氧化硫洗涤器
sulfur disulfide 二硫化硫
sulfur dust 硫灰尘
sulfureous 含硫的
sulfuret 硫化物
sulfur explosive 硫黄炸药
sulfur family element 硫族元素
sulfur fertilizer 流肥
sulfur flour 粉末硫黄
sulfur-free basis 无硫基
sulfur-free clinker 无硫熟料
sulfur-free fuel 无硫燃料
sulfur-free natural gas 无硫天然气
sulfur gas 硫质气体
sulfuric acid 硫酸
sulfuric acid aerosol 硫酸气溶胶
sulfuric acid anhydride 硫酸酐
sulfuric acid anhydrite 三氧化硫
sulfuric acid bath 硫酸浴
sulfuric acid chamber 硫酸室
sulfuric acid cooler 硫酸冷却器
sulfuric acid industry sewage 硫酸工业废水
sulfuric acid industry wastewater 硫酸工业废水
sulfuric acid mist 硫酸雾
sulfuric acid mist eliminator 硫酸捕沫器
sulfuric acid plant 硫酸工厂
sulfuric acid refining 硫酸精制
sulfuric acid residuum 硫酸渣
sulfuric acid scrubbing 硫酸洗涤
sulfuric acid separator 硫酸分离器
sulfuric acid tower 硫酸干燥塔
sulfuric acid wastewater 含硫酸废水
sulfuric dioxide 二氧化硫
sulfuric material 硫化铁材料
sulfurified lubricant 含硫润滑剂
sulfur impregnated concrete 融硫混凝土
sulfur in ash 灰中硫
sulfuring 吸硫
sulfur insulated switch 硫绝缘开关
sulfur iodide 碘化硫
sulfur ion 硫离子
sulfur isotope geothermometer 硫同位素地质温度计
sulfurization 硫化作用
sulfurized asphalt 硫化(石油)沥青
sulfurized boiled oil 硫化熟油
sulfurized coal tar pitch 硫化煤沥青
sulfurized oil 硫化(切削)油
sulfurizing 渗硫
sulfurless cure 无硫硫化
sulfurless vulcanization 无硫硫化(作用)
sulfurless vulcanizing agent 无硫硫化剂
sulfur limestone denitrification 硫石灰石反硝化
sulfur modified concrete 融硫黄混凝土
sulfur monoxide 一氧化硫
sulfur ore 硫矿石
sulfurous 亚硫酸
sulfurous acid anhydride 亚硫酸酐
sulfurous acid gas 亚硫酸气
sulfurous anhydride 二氧化硫
sulfurous fuel 含硫燃料
sulfurous gas 含硫气体
sulfurous hot spring 含硫热泉
sulfurous london smog 含硫伦敦雾
sulfurous steam 含硫蒸汽
sulfur oxide 氧化硫
sulfur oxide perchlorate 高氯酸硫
sulfur oxidizing bacteria 硫氧化菌
sulfur oxyfluoride 二氟二氧化硫
sulfur oxygen isotope correlation 硫—氧同位素相关性
sulfur pentafluoride 五氟化硫
sulfur plaster 硫黄胶泥
sulfur point 硫沸点
sulfur pollution 硫污染
sulfur pool 硫黄塘
sulfur powder 硫黄粉
sulfur print 硫黄检验
sulfur rain 硫化雨;黄雨
sulfur recovery plant 硫回收车间
sulfur recovery process 硫回收过程
sulfur reducing bacteria 硫还原菌
sulfur reducing bacteria degradation 硫还原菌降解

sulfur removal 脱硫
sulfur removal plant 脱硫车间
sulfur resistant catalyst 抗硫催化剂
sulfur sesquioxide 三氧化二硫
sulfur spring 硫黄泉;含硫矿泉
sulfur subbromide 二溴化二硫
sulfur subiodide 二碘化二硫
sulfur test 硫试验
sulfur tetrafluoride 四氟化硫
sulfur thermal water 含硫热水
sulfur transformation and cycle 硫元素转化与循环
sulfur trioxide 硫酸酐;三氧化硫
sulfur trioxide and sulfuric acid fume poisoning 三氧化硫和硫酸雾中毒
sulfur trioxide mist 三氧化硫烟雾
sulfur water 硫黄泉水;含硫水
sulfuryl 硫酰
sulfuryl chloride 磺酰氯
sulky 轮式集材拱架;双轮运圆木车
sulky derrick 三角吊架
sulky disk plow 乘式圆盘犁
sull 氧化铁薄膜
sullage 熔渣;沉积淤泥;渣滓;淤泥;污水;垃圾;废物
sullage barge 垃圾驳(船)
sullage disposal 洗液废水处置
sullage water 阴沟污水
sullcoating 锈化处理
Sullivan angle compressor 苏利文角形压气机
sulphamate 氨基磺酸盐
sulphamethazine 氨基二甲基嘧啶基苯磺酰胺
sulphamide 磺酰胺
sulphasomidine 氨基邻二甲基嘧啶基苯磺酰胺
sulphate 硫酸酯;硫酸盐
sulphate admixture 硫酸盐外加剂
sulphate aerosol 含硫酸烟雾;含硫酸气溶胶
sulphate attack 硫酸盐侵蚀
sulphate-bearing water 含硫酸盐的水
sulphate conent 硫酸盐含量
sulphate corrosion 硫酸盐侵蚀
sulphate efflorescence 硫酸盐风化起霜
sulphate expansion 硫酸盐膨胀
sulphate expansion of brickwork 砖砌体中的硫酸盐膨胀
sulphate index 硫酸物指数
sulphate ion 硫酸根离子
sulphate-laden water 含多量硫酸盐的水
sulphate lake 硫酸盐湖
sulphate liquor 硫酸盐黑液
sulphate of alumina 铝矾
sulphate of barium 硫酸钡
sulphate of calcium 硫酸钙
sulphate of copper 硫酸铜
sulphate of lime 石膏;硫酸钙
sulphate of patash 硫酸钾
sulphate of potash magnesia 硫酸镁钾
sulphate radical 硫酸根
sulphate resistance 耐硫酸盐;硫酸盐耐力;抗硫酸盐侵蚀性;抗硫酸盐腐蚀能力
sulphate resistance test 抗硫酸盐试验
sulphate-resistant 耐硫酸的
sulphate-resistant cement 抗硫酸盐水泥;耐硫酸盐水泥
sulphate-resisting 耐硫酸的
sulphate-resisting cement 耐硫酸盐水泥
sulphate-resisting portland cement 抗硫酸盐硅酸盐水泥;抗硫酸盐波特兰水泥
sulphate ring 硫碱圈
sulphate-saline soil 硫酸盐渍土
sulphate soundness test 硫酸盐坚固性试验
sulphate sulfur 硫酸盐硫
sulphate thermal water 硫酸盐型热水
sulphate water zone 硫酸水带
sulphating 硫酸处理
sulphating in aggregate 集料硫化物含量
sulphating roasting 硫酸化焙烧
sulphation 硫酸化(作用)
sulphation factor 硫酸化因子
sulphatising 用硫酸酸解;硫酸盐化
sulphenothiazine calcium 吩噻嗪磺钙
sulphetrone 苯丙砜
sulphidation attack 硫化腐蚀
sulphide 硫化物
sulphide colo(u)rs 硫化染料
sulphide disclo(u)r 硫化物变色

sulphide disclo(u)ration 硫化变色
sulphide dyes 硫化染料
sulphide fog 硫化灰雾
sulphide inclusion 硫化物夹杂
sulphide of cadmium 硫化镉
sulphide of copper 硫化铜
sulphide ores 硫化矿
sulphide phosphor 硫化物磷光体
sulphide rectifier 硫化物整流器
sulphide selenides 硫化物硒化物
sulphide staining 硫化物污染失光
sulphide stress corrosion cracking 硫化物应力腐蚀破裂
sulphide sulfur 硫化物硫
sulphide tailings 硫化物尾矿
sulphide thio-ether 硫醚
sulphide toning 硫化调色法
sulphide zone 硫化矿物带
sulphidisation 生硫化物;成硫化物
sulphidity 硫化度
sulphidizing 硫化;磺原酸化
sulphinyl 亚磺酰
sulphite 亚硫酸酯;亚硫酸盐
sulphite cellulose 亚硫酸盐纤维素
sulphite cellulose extract 亚硫酸盐纤维素提出物;亚硫酸盐纤维素萃取物
sulphite cooking liquor 亚硫酸盐蒸煮液
sulphite corrosion 硫化物腐蚀
sulphite ion 亚硫酸盐离子
sulphite liquor 纸浆废液
sulphite lye 亚硫酸盐碱液;黑液
sulphite-mill effluent 亚硫酸盐厂废水
sulphite process 亚法制浆
sulphite pulp 亚硫酸纸浆;亚硫酸盐纸浆
sulphite pulping 亚硫酸盐纸浆浆化
sulphite pulping process 亚硫酸盐纸浆浆化工艺
sulphite pulp mill wastewater 亚硫酸盐纸浆厂废水
sulphite road binder 路用亚硫酸盐结合料;亚硫酸盐路用结合料
sulphite road binder concentrate 路用浓缩亚硫酸盐结合料;浓缩亚硫酸路用结合料
sulphite spent liquor 亚硫酸废液;废亚硫酸盐液
sulphite spirit 制造亚硫酸盐产生的酒精副产品
sulphite waste liquor 亚硫酸废液;废亚硫酸盐液
sulphite waste-liquor byproduct 废亚硫酸盐液副产物
sulpho 磺基
sulpho aluminate cement 硫酸铝水泥
sulpho aluminate early strength cement 快硬硫铝酸盐水泥
sulpho aluminous cement 硫酸铝水泥
sulphoborite 硼镁矾
sulphocarbolic acid 苯酚磺酸
sulphohalite 氟盐矾
sulphonate 磺酸盐
sulphonated carbon 磺化煤
sulphonated oil 磺化油
sulphonating 磺化
sulphonating agent 磺化剂
sulphonating bottle 磺化瓶
sulphonating brown 磺化褐煤
sulphonating metyl brown coal 磺甲基褐煤
sulphonating tannin 苛化丹宁
sulphonation 磺化作用
sulphonator 磺化器
sulphonic acid 磺酸
sulphonic acid ester 磺酸酯
sulphonyl 磺酰
sulphophile element 亲硫元素
sulphopone 锌钙白
sulphotsumite 硫碲铋矿
sulphur = sulfur
sulphur alkali match 硫碱平衡
sulphur amino acid 含硫氨基酸
sulphur anchorage 硫黄锚固【铁】
sulphur asphalt paving 硫黄沥青铺面
sulphurated coal 硫化煤
sulphurator 硫化器;硫黄漂白器
sulphur-bearing wastewater 含硫污水
sulphur black 硫化黑(硫化青)
sulphur box 熏硫箱
sulphur bright green 硫化亮绿
sulphur budget 硫收支;硫平衡
sulphur capacity 含硫量

sulphur cement 硫黄沥青胶结料;硫黄水泥
sulphur cement mortar anchor(age) 硫黄锚固【铁】
sulphur chloride 氯化硫
sulphur concrete 硫黄混凝土
sulphur consumption 硫耗;耗硫量
sulphur containing 含硫
sulphur containing substance 含硫物质
sulphur content 硫含量;含硫量
sulphur corrosion 硫侵蚀
sulphur crack 硫带裂纹
sulphur cycle 硫循环
sulphur deposit 硫矿
sulphur deposition 沉积硫
sulphur dioxide 二氧化硫
sulphur dioxide monitor 二氧化硫监测仪;二氧化硫监测器
sulphur dioxide reduction 减少二氧化硫
sulphur donor complex 硫供体络合物
sulphureous 含硫的
sulphuret 硫化物
sulphur free 无硫的
sulphur free-cutting steel 含硫易切削钢
sulphur free fuel 无硫燃料
sulphur-free natural gas 无硫天然气
sulphur hexafluoride 六氟化硫
sulphuric acid 硫酸
sulphuric acid aerosol 硫酸烟雾;硫酸气溶胶
sulphuric acid anodizing process 硫酸阳极极化法
sulphuric acid carrier 硫酸船
sulphuric acid electrolyte 硫酸电解液
sulphuric acid industry sewage 硫酸工业废水
sulphuric acid industry wastewater 硫酸工业废水
sulphuric acid mist 硫酸雾
sulphuric acid plant 硫酸厂
sulphuric acid plant wastewater 硫酸厂废水
sulphuric anhydride 三氧化硫;硫酸酐
sulphuric cement pile coupling 硫黄胶泥接桩
sulphur(ic)dioxide 二氧化硫
sulphuric oxidation coating 硫酸(阳极)氧化膜
sulphur impregnated concrete 硫(黄)浸渍混凝土;融渍混凝土
sulphur impregnation 硫黄浸渍
sulphur infiltrated concrete 硫黄浸渍混凝土
sulphuring 吸硫;亚硫酸盐处理;烧硫灭菌
sulphur ion 硫离子
sulphurized asphalt 硫化沥青;硫化石油沥青
sulphurized coal-tar pitch 硬化硬煤沥青;硫化煤焦油脂
sulphurized lubricant 含硫润滑剂
sulphurized resin 硫化树脂
sulphurizing 渗硫
sulphur mastic 硫黄胶泥
sulphur mortar pile splice 硫黄胶泥浆锚法接桩
sulphur ore 硫矿石;黄铁矿
sulphurous acid 亚硫酸
sulphurous acid anhydride 亚硫酸酐
sulphurous acrid saline soil 亚硫酸盐渍土
sulphurous wood 含硫木
sulphur pockmark 硫黄麻点
sulphur pool 含硫泉
sulphur powder 硫黄粉
sulphur print 鲍曼硫印
sulphur reducing dyes 硫化及硫化还原染料
sulphur refined 精制硫黄
sulphur removable 脱硫
sulphur spring 硫黄泉;含硫泉
sulphur steel 高硫钢
sulphur tanker 硫黄船
sulphuret toning 硫黄调色法
sulphur treatment kettle 加硫除铜精炼锅
sulphur trioxide 三氧化硫
sulphur trioxide mist 三氧化硫烟雾
sulphur water 含硫水;含硫矿水
sulphury chloride 氯化硫酰
sulrhodite 硫铑矿
sultry 闷热的;热而潮湿的
sultry weather 酷热天气;闷热天气;沉闷天气
sulvanite 硫钒铜矿;等轴硫钒铜矿
Sulzer alloy 苏尔泽锌基轴承合金
Sulzer packing 苏采尔填料
Sulzer twocycle engine 苏尔泽二冲程发动机
sum 小计;总数;总计;总和;总共;总额;金额;合计
sum accumulator 和数累加器

sumac(h) 盐肤木；苏模鞣木；漆树属植物的木材
sumac(h) seed oil 栌木核(仁)油；黄栌核(仁)油；漆树子油
sumacoite 安山碱玄岩
sumac wax 日本蜡
sum aggregate 和集；并集
sum-and-difference net 求和求差网络
sum-and-difference system 和差系统
sum angle 和角
sumaresinol 安息香胶酸
sumatra 苏门答腊风
sumbiont 共生者
sum bit 两位和
sum check 总计检验；求和校验；求和检查
sum check code 检查和代码
sum counter 和计数器
sum difference system of stereo 立体声和差变换系统
sum electrode 积分电极
sum equation 累加方程
Sumerian architecture 苏美尔建筑
Sumet bronze 萨米特轴承青铜
sum formula 求和法
sum frequency 总频率；和频
sum graph 和图
sumi ink 黑墨(汁)
sum insured 保险(金)额
sum in words 大写金额
sum isopleth of distances 等距(离)和曲线
sum lent on a mortgage 已提供押款金额；押款
sumless 不可数的
sum load of current capital 流动资金全额贷款
summability 可和性
summable 可求和的被加数
summable bounded 受限可加
summable function 可积函数；可和函数
summable series 可和级数
summand 被加数
summarization 概括
summarize 总结；概括
summarized information 相加数据；综合情报
summarized principle 总则
summarized table 总结表；总表
summarizing instruction 求和指令
summarizing schedule 汇总表
summary 一览(表)；总结；摘要(总结)；归纳；概要；概述；概略；概括；纪要；汇总表；大略
summary bill enforcement procedure 简化票据生效程序
summary calculation 总计
summary card 总计卡片
summary card punch 总计卡片穿孔机
summary counter 累加计数器
summary gang punch 总计复穿孔机
summary index 总和指数
summary minutes of talk 会谈纪要
summary of cost and expenses classification 成本费用分类表
summary of cost finished parts used 耗用零件成本汇总表
summary of cost of goods sold 销货成本汇总表
summary of cost of manufacturing expenses applied 已分摊制造费用汇总表
summary of cost of returned goods 退货成本汇总表
summary of customer's orders 客户订单汇总表
summary of earnings 收益汇总表
summary of element cost 分项造价汇总
summary of estimated expenditures 费用估计简表
summary of factory payroll taxes 全厂工资税汇总表
summary of factory wages 全厂工资汇总表
summary of housing condition 居住条件摘要
summary of invention 发明摘要
summary of labo(u)r cost 人工成本汇总表
summary of material consumed 耗用材料汇总表
summary of material received 收料汇总表
summary of materials and requirements 材料及设备汇总表
summary of materials returned 退料汇总表
summary of meeting 会议纪要
summary of minutes 纪要
summary of operating revenue and expenses 营业收支汇总表

summary of receipts and expenditures 收支汇总表
summary of reinforcement 钢筋总表；钢筋(明细)表
summary of reinforcing steels 钢筋总材料表
summary of scrap and spoiled materials 材料损耗汇总表；材料耗损汇总表；废损材料汇总表
summary of spoiled work 损毁作业汇总表
summary of trigonometrical points 三角点成果表
summary of volcanic activities 区域火山活动概况
summary on crystallization path 结晶路程一般规律
summary printing 总计打印
summary procedure 简易程序；即决程序
summary punch 总计穿孔；计总穿孔
summary puncher 总计穿孔机；结果凿孔机；输出凿孔机
summary punching 总计穿孔
summary recorder 总计记录器
summary records 纪要
summary register 简略寄存器
summary report 概要报表
summary settlement 汇总会计报表
summary sheet 观察记录表
summary sheet of goods description 货物描述一览表
summary statement of cash receipts and disbursements 现金收支汇总表
summary statement of cost and fee 成本费用总表
summary statement of development credits 开发信贷汇总表
summary statement of loan 贷款汇总表
summary statement of receipts and expenditures 收支汇总表
summary statement of resources and obligations 财力负担汇总表
summary statistics 概统计量
summary table 总表；摘要表
summary tag along sort 沿用标志累加分类
summary technical report 技术报告摘要
summating meter 累计计数计
summating potential 总合电位
summation 累加；累计；求和；总和
summation actinometer 累积曝光表；累积露光计
summation action 相加作用
summational 总和的
summation by parts 部分求和；分部求和法；分步求和
summation check 总计核对；和数校检；合计检查；求和校验；求和检验；求和检查
summation check digit 和数校验位
summation convention 记和常规；求和缩写法
summation curve 累积曲线
summation effect 相加效应
summation flowmeter indicator 总流量指示器
summation formula 求积公式；求和公式
summation graph 总和图
summation hydrograph 总和流量过程线；总和水文过程线
summation index 总指数；求和指数
summation instruction 求和指令
summation instrument 总和仪表；总和器
summation law 累加定律
summation meter 加法器
summation metering 求和测量法；累积计量(法)
summation method 累加法；求和法
summation network 求和网络
summation notation 连加(求和)记号
summation of binomial coefficients 二项式系数求和
summation of forces 力的总和
summation of geometric(al) progression 几何级数求和
summation of losses 损耗求和
summation of parts 分部求和法
summation of percentage 累计百分数
summation of series 级数求和
summation of temperature 温度总和
summation panel 总输出功率记测板；总和配电盘
summation principle 累加原则
summation process 求和法
summation sign 总和符号；连加号
summation tone 和音
summator 加法装置；求和元件

summed ablation 总消蚀(作用)
summed current 总电流
summed-up length 总长(度)
summer 榍；基石；柱顶石；檩条；门窗过梁；墙梁；大梁；加法器；求和器
summer accumulation 夏季堆积
summer air-conditioning 夏季空调
summer air-conditioning system 夏季空调系统
summer annual 夏季植物
summer area 夏季区带
summer beam 盖顶石；门窗过梁；托墙梁；过梁；檩条；柱顶石；大梁
summer berm 夏季海滨阶地
summer black oil 夏用黑色润滑油；回火油
summer buoyancy 夏季浮标装置
summer capacity 夏季冷却容量
summer compound oil 夏季用齿轮油
summer cooling load 夏季的冷负荷
summer cottage area 避署地
summer cutting 夏季插
summer desiel oil 夏季用柴油
summer dike[dyke] 夏堤；子堤
summer draft 夏季载重水线吃水；夏季吃水线
summer draft mark 夏季载重线标志
summer dry region 夏干区
summer electric(al) energy 夏令电能
summer fallow 夏季休闲；夏季休耕
summer fallow field 夏季休闲地
summer flood 夏季洪水；伏汛
summer fog 夏雾
summer freeboard mark 夏季干舷标志
summer freshet 伏汛
summer gasoline 夏季汽油
summer grade 夏季级
summer grade gasoline 夏季级汽油
summer half year 夏半年
summer house 避署别墅；凉亭
summer hydrant 夏令给水龙头；夏令供水龙头
summer irrigation 夏灌
summer kitchen 夏季厨房
summer leaf-drop 夏季落叶
summer levee 夏堤
summer lightning 片状闪电
Summer line 萨母纳位置线
summer loading 夏令载货吃水线
summer load(water)line 夏季装载线；夏季载重线；夏季满载吃水线
summer lubricating oil 夏季用润滑油
summer lubrication 夏季润滑
Summer method 萨母纳天文定位法
summer monsoon 夏季季风
summer monsoon current 夏季季风海流
summer motion picture theater 夏令电影院
summer mo(u)lded draft 夏季型吃水
summer noon 夏令时午中；日光节约正午
summer oil 夏用油；夏季油
summer oil level 夏用润滑液面高度
summer on loadline 夏令载货吃水线
summer palace 夏宫
summer pasturing 夏季放牧
summer patching 夏季修整
summer petrol 夏用汽油
summer piece 挡护板(夏季用)
summer-placed concrete 夏季浇灌的混凝土
summer plankton 夏季浮游生物
summer polder 夏季水边低地
summer port 夏港
summer-pruning 夏季剪修
summer range 夏季牧场
summer refuse 夏季垃圾；夏令垃圾
summer reservoir 夏季蓄水库
summer residence 避署别墅；夏季住宅
summer resort 避署地；避署胜地
summer runoff forecast 夏季径流预报
summer season 夏汛
summer service 夏季运行；夏季服务
summer shelter 凉亭
summer solstice 夏至(日)；夏至点
summer spray 夏季喷药；夏季喷药
summer stagnation 夏季停滞
summer stagnation period 夏季停滞期
summer stone 斜托石；柱帽石；墩帽石
summer tank 夏季油舱
summer theater 活顶剧院；夏令剧场；夏令剧院

summer tide 夏潮
summer timber load line 夏季木材载重线
summer time 夏令时(间);夏季(时间);日光节约时间
summer time stagnation 夏令停滞
summer traffic 夏季交通
summer train working diagram and timetable 夏季运行图和时刻表
summer tree 夏材;木梁;大梁;大(木)材
summer water level 夏季水面;夏季水位
summer wet region 夏湿区
summer white oil 夏季(白)油
summer wood 夏木;夏季木材;夏材;晚材;大木材
summer yellow oil 夏季黄油;粗棉籽油
summer zone 夏季区带
summing amplifier 加法放大器
summing circuit 加法电路;求和电路
summing impact modulator 加法对冲元件
summing integration 加法积分
summing integrator 总和积分器;加法积分器;求和积分器
summing junction 求和点
summing network 求和网络
summing-up 总结;归纳
summit 高峰;丘顶;山峰;顶峰
summit canal 越岭运河;越岭渠道
summit concordance 山顶平齐
summit crater 山顶火山口
summit curve 凸曲线;顶曲线
summit cut off 越岭取直;山脊捷径
summit eruption 山顶火山喷发
summit hydrothermal eruption 顶峰水热喷发
summit level 越岭河段;峰顶面;峰顶高程;分水岭高程
summit level reach 峰上河段;山墙河段
summit level reservoir 峰顶水库
summit line 山脊线
summit metabolism 定点代谢
summit of a mountain 山顶
summit of bend 河弯顶点;弯顶;弯道顶点
summit of height 制高点
summit of hump 驼峰峰顶
summit of the grade line 坡度凸变点
summit of vertical curve 竖曲线顶点
summitor 相加器
summit plane 峰顶面
summit pond 山顶水池;峰顶水塘;峰顶水库;峰顶水池
summit pool 峰顶水塘;峰顶水库;峰顶水池
summit profile 凸形纵断面
summit reach 越岭河段;分水岭上河段
summit reach canal 越岭河段
summit reservoir 峰顶水库
summit tunnel 越岭隧道
summit vertex 顶点
summit yard 驼峰调车场;驼峰编组站;驼峰编组场
sum module 加法组件
summons 传票
Sumner line 塞姆纳线;萨姆纳线
Sumner method 逊纳法;塞姆纳法【航海】
sum of acid anions 酸阴离子总和
sum of a random number of random variables 任意个随机变量和
sum of base cation 碱阳离子总和
sum of-digits method 总和数字法
sum of distribution 分布和
sum of expected life method of depreciation 使用年限折旧法
sum of infinite series 无穷级数的和
sum of maintenance 维护费(用)
sum of money 款项;金额
sum of product 积和;乘积之和
sum of-product canonical form 积之和的正则形式
sum-of-product form 积之和形式
sum of products of mean deviations 离均差积和
sum of series 级数和
sum of square between treatment 处理间平方和
sum of squares 平方和
sum of squares of deviations 离差平方和;差方和
sum of squares of standardized residuals 标准化剩余平方和;标准化残差平方和
sum of squares sum of within group 段内(离差)平方和的和
sum of states 配分函数

sum-of-the-year-digits 年限总和法;数字总和法(美国租赁业一种折旧法)
sum-of-the-year-digits depreciation 快速拆旧(法);总和年数折旧(法)
sum-of-the-year-digits depreciation method 年限总额折旧法;年数总和折旧法
sum of vectors 向量和
sum output 和数输出
sum over states 配分函数
sump 小料浆池;油底壳;储[贮]槽;聚水坑;积垢器;灰渣浆池;曲柄箱;污水坑;污水槽;掏槽;水仓;池;沉淀器;集水坑
sum parameters 综合指标
sump basin 集水池
sump drain 渗坑排水
sump drain tank 集储[贮]罐
sumper 井筒中心炮眼;对角炮眼;底部炮眼;底板炮眼
sump gully 集水沟
sump handling method 集泥舱转送法
sump hole 集水孔;集水井
sumping 下渗;集水坑排水;明排水
supplementary construction equipment 辅助施工设备
sump oil 下脚油;沉积池油
sump pit 集存槽;排液槽;集水坑
sumpplementary benefit 辅助福利
sump pump 潜水泵;深井泵;集水泵;集水坑泵;集水泵;油池泵;井底水窝水泵;浅池泵;排水泵;排除积水泵;水窝泵;水舱泵
sump pumping 集水沟抽水
sump rehandler 抛泥池
sum product output 和积输出
sump shooting 掏槽放炮
sump shot 掏槽爆破
sump strainer 集水坑滤网
sump tank 集油槽;排水池;污水槽;废油罐
sump throat 浸没式流液洞
sumptuous 花钱的;豪华的
sum pulse 集合脉冲
sump(waste oil)pump 污油泵
sump well 排水井
sum rule 求和法则
sum signal 相加信号;积分信号;和信号
sum storage 和数累加器
sum total 总数;总金额;总价
sum-total of element functions 要素作用总和
sum total of physical point 物理点数
sum total of station 测站总数
sum up 总括;总结
sun 太阳
sun altitude 太阳高度
Sunalux glass 透紫外线玻璃
sun-and-planet gear 行星(式)齿轮
sun-and-planet gearing 行星齿轮传动
sun-and-planet motion 太阳行星运动
sun-and-planet strander 筐篮式捻股机
sun-and-planet wheel 行星齿轮
sun angle 日照角;太阳角
sun arc 太阳弧光
sun atmosphere 太阳大气
sunbaked 风干;晒干;日晒的
sunbaked brick 晒干泥砖;晒干砖坯
sunbaked soil 晒干土
sun bath 日光浴
sunbathe 晒日光浴
sunbathing area 日光浴场
sunbathing patio 日光浴内院;日光浴场
sunbathing room 日光浴室
sunbathing terrace 日光浴场
sunbeam 日光束;日光;太阳光(线)
Sunbelt area 阳光地带
Sunblazer 太阳探测器
sun blind 篷盖;遮阳;窗帘;遮帘;百叶窗
sun blind wall 遮光墙
sun blister 晒焦
sun breaker 遮阳板;遮阳装置
sun bronze 钴铜铝合金;钴钴铝合金
sun bronze alloy 太阳青铜合金
sunburn 日晒红斑;晒黑;晒斑
sun burner 太阳灯;太阳光灯;局部过厚
sun burst 太阳状破裂(锆合金的一种腐蚀破裂形式)
sunburst area 日光放射区
sunburst light 扇形气窗;气窗

sunburst pleats 渐宽褶;辐射式褶裥
sunburst processor 辐射式处理机
sun chart 日照图
sun compass 日光罗盘;太阳罗盘
sun compass orientation 太阳罗盘定向
sun control 日光调节器;太阳保护装置
sun control device 日光保护装置
sun control work 日光保护操作
sun crack 干裂;泥裂;日裂;晒裂
sun creen wall 遮光墙
sun cross 十字晕
sun crust 再结雪壳
sun cured 在日光下晒干;晒干的;日晒干燥的
sun cured hay 日晒干草
sun-cured tobacco 晒烟
sun-curing 晒干
sun curtain 遮阳布篷
Sundance series 松丹斯统
Sunday and holidays excepted 星期日、假日除外
Sunday count 假日交通量统计
Sundays excepted in laydays 装卸天数扣除星期日
Sunday holidays excepted 星期日、假日除外
Sundeala 山达板(一种木纤维墙板的牌号)
sun deck 运动甲板;遮阳甲板;日光浴平台;日光浴板;日光甲板;太阳浴层面;日光浴屋面
sun dial 日晷仪;日规
sundiusite 氯铅矾
sun dog 幻日
sun drawing water 太阳吸水
sun-dried 风干的;晒干的
sun-dried brick 晒干砖坯
sun-dried brick construction 泥砖建筑
sun-dried brick masonry wall 泥砖墙
sun-dried mud brick 土坯
sundries 杂件;零星小件;杂项;杂货
sundries for fire-fighting and protection 消防用具
sun-dring 晒干
sundry 杂项
sundry assets 杂项资产
sundry charges 杂项费用;杂费
sundry charges for water transport payable 应付水运运杂费
sundry charges for water transport receivable 应收水运运杂费
sundry charges incurred for water transport 垫付水运运杂费
sundry commission paid 已付杂项手续费
sundry credit 暂时存款
sundry deposit account floating 浮存
sundry goods 杂货
sundry increased prices liquidated revenue 各种加价清算收入
sundry interest paid 已付杂项利息
sundry interest received 已收杂项利息
sundry items 杂项项目;杂项
sundry overhead 杂项管理费
sundry reserves 杂项准备
sundry supplies 杂项耗用品
Sundstrand pump 组合泵
sun-earth observation and climatology satellite 日地观测与气候学卫星
sun-earth relationship 日地关系
sun effect 日得得热量(通过窗户和结构材料传入室内的太阳能);日光效应
sun exposure 日曝;日照
sun filter 太阳滤光片
sun finder 太阳定向仪
sunfish 太阳鱼
sunflower 向日葵
sunflower pattern 辐射状布孔;辐射状布井
sunflower(seed)oil 向日葵油;葵花子油
sun follower 太阳跟随器
sun gear 恒星齿轮;中心齿轮;太阳齿轮
sun-generated electric(al) power 太阳能电源
sun glass 有色玻璃;遮阳玻璃;遮光玻璃
sunglasses 太阳眼镜;太阳镜
sun glint 太阳反辉区
Sunglow 山格洛(一种专利的灯式供暖电热器)
sungulite 蠕蛇线石
sun heat 日光热;太阳热
sun hemp 大麻
sun hour angle 太阳时角
sun-illuminated cloud background 日照云层背景
sun-illuminated cloud brightness 日照云层亮度

sun-illuminated earth background 日照地球背景
sun-illuminated target 日照目标
sun interference 太阳干扰
S-unit meter 信号强度计
sunk 凹下的
sunk basin 集水井;集水盆地
sunk bead 凹圆形线脚;凹入线脚;凹焊缝
sunk bridge 半隐驾驶室【船】
sunk coak 凹榫槽
sunk cost 已投资本;固定成本;无差别成本;沉没成本;不能回收的成本
sunk draft 下陷边(石工);下沉缘琢;沉边琢;石料凹框;房角埋石
sunk drill 埋头钻
sunked garden 下沉式花园
sunken 凹陷的;水底的;凹下的
sunken and raised plating 内外叠板法
sunken arc 潜弧
sunken beam equipment 沉降梁装置
sunken bed 低床
sunken blue 凹陷蓝色彩饰
sunken caldera 陷落火(山)口
sunken danger 水下障碍物
sunken deck 低甲板
sunken evapo(u)ration 埋置式蒸发
sunken evapo(u)ration pan 埋置式蒸发器
sunken eye 凹眼
sunken fascine 沉柴埽;沉柴排;厢埽工
sunken fascine layer 沉柴埽层
sunken fascine mattress 沉排
sunken fence 沉篱
sunken flat 海底台地
sunken flower bed 盆地花坛;凹地花坛
sunken forecastle vessel 低首楼甲板船
sunken garden 沉园;盆地园;沉床园;凹地园
sunken highway 低堑道路;低堑公路
sunken joint 接缝凹陷;拼缝凹陷;暗缝
sunken lane 低堑车道
sunken manhole 底舱人孔
sunken mattress 沉柴排
sunken meadow 淹水草原
sunken oil storage 地下油库
sunken pipe 地下管道
sunken pit 下沉坑;下挖坑
sunken plant 沉水植物
sunken poop 低船尾楼
sunken reef 暗礁
sunken river 隐入河;流失河
sunk(en) road 低堑道路
sunken rock 暗礁
sunken ship 沉船
sunken ship breakwater 沉船防波堤
sunken ship load 沉船荷载
sunken stomata 内陷气孔
sunken strake 内列板
sunken stream 隐入河;流失河
sunken tank 地下储罐
sunken throat 沉降式流液洞
sunken track 下凹铁路线
sunken tree 沉树(将砍下的树锚碇于岸边以护岸防冲)
sunken tube 沉埋管段;沉管
sunken tube tunnel 沉管隧道
sunken water level 下沉水位
sunken waters 下沉水域
sunken well 钻完的井;开口沉箱
sunken wreck 水下沉船;沉船
sunk face 下陷面;下凹面;下凹面
sunk fascine 沉柴排;沉柴埽;沉梢捆
sunk fence 暗墙;矮篱墙
sunk fillet 埋嵌线条;凹线条
sunk fillet mo(u)ld 埋嵌线饰;平条埋嵌线饰;埋嵌线条饰
sunk foundation 埋设基础;埋沉基础;沉埋式基础
sunk gutter 暗天沟;暗檐沟;暗檐槽
sunk handle 沉入把手
sunk heat rivet 埋头铆钉
sunk hydrant 地下消火栓;地下(式)消防栓
sunk key 埋头键;嵌入键;槽键;暗键
sunk mo(u)lding (浅于边框的)凹线脚;凹入形线脚;凹线条;凹饰饰条
sunk or burnt risk 沉没或失火险
sunk panel 埸壁;藻井;埋入式镶板;嵌入式镶板;凹形镶板;凹进镶板

sunk pin 埋头销
sunk relief 凹凸浮雕
sunk rivet 埋头铆钉;沉头铆钉
sunk road 低堑道路;堑路
sunk screw 埋头螺丝;埋头螺钉;沉头螺丝;沉头螺钉
sunk shaft 沉井基础;矿井下沉;竖井下沉;竖井沉基;沉井
sunk shaft foundation 沉井基础
sunk shelf 盆架;盆饰挑出线脚
sunk skylight 下沉式天窗
sunk spot 沉陷处
sunk-stock 沉木
sunk tank 集水箱;集水池
sunk-type switch 埋装式开关
sunk weathered 凹陷的泻水斜面
sunk well 集水井;开口沉箱;沉井
sunk well foundation 沉井基础
sunk wood screw 埋头木螺丝
sun lamp 日光灯;太阳灯
sun light 太阳灯;日光;太阳光
sunlight-driven reaction 阳光引起的化学反应
sunlight-fast 日晒不褪色的
sunlight glare 日光耀目
sunlight greenhouse 日光温室
sunlighting 退休后工作
sunlight lamp 太阳灯
sunlight ordinance 日照标准
sunlight penetration 透光;阳光透入;光线入射
sunlight pressure 阳光压力
sunlight pressure perturbation 太阳光压摄动
sunlight print 日光晒印
sunlight-proof 不透阳光的;耐阳光的
sunlight pumping energy 日光泵浦能
sunlight radiation test 日光辐射试验
sunlight resistant 耐阳光的
sun line 太阳位置线
sunlit 日耀
sunlit aurora 日耀极光
sunlit surface 日耀表面
sun louver 遮阳板;遮光板
sun-loving plant 喜光树
sun-loving species 喜光种
sunmarine rock 暗礁
sunned oils 润滑油
sunny slope 阳坡
sun observation 太阳观测
sun opal 火蛋白石
sun-oriented vehicle 太阳定向飞行器
sun parlo(u)r 阳光充足的房间;向阳间;日光浴室;日光室;向阳客厅;向阳病房
sun path 日光路径;日光光束
sun path diagram 太阳位置图
sun patio 日光浴场
sun pillar 日柱;日光柱;竖光柱
sun pinion 中心小齿轮;太阳小齿轮
sun plant 阳性植物;阳生植物;阳地植物
sun plant community 阳生植物群落
sun porch 日光室
sun power 太阳能
sun printing 日光晒印
sun projector 日光投影仪
sun-proof 防晒(的);防太阳(光)的;耐晒的;不透日光的
sun-proof paint 耐晒涂料;耐晒漆
sun-proof plastics 耐晒塑料
sun protection glass 不透日光玻璃
sun-pump 日光泵浦;日光泵
sun-pumped laser 日光泵激光器;太阳(光)抽运激光器
sun pumping 太阳光抽运;日光泵浦
sun radiation 日光辐射;日射
sun radius 辉球半径(雷达性能监视)
Sunray 日光管(一种专利灯管)
sun reflection 日光反射;欠折射
sun relay 太阳继电器
sun-resistant plastics 耐晒塑料
sunrise 日出
sunrise and sunset transition 日出与日落过渡期
sunrise effect 日出效应
sun-rise industry 朝阳工业
sun-room 日光(浴)室
sun-run-sun fixing 太阳移线定位
sun's altitude angle 太阳高度角
sun satellite 太阳卫星

sun's azimuth 太阳方位角
sun scald 日灼病;晒伤
sun screen 遮阳帘;日光屏;遮阳栅;遮阳百叶板
sun screen drug 遮光剂
sun screener 防晒剂
sun-screening agent 防晒剂
sun's disc 日轮
sun seeker 太阳跟随器;太阳定向器;向日仪
sun-seeking device 太阳定向仪
sun sensor 太阳传感器
sun set 日没;日落
sunset budgeting 日落预算法(系零基预算法的别称)
sunset clause 届满条款
sunset effect 日落效应
sunset-industries 夕阳产业
sunset yellow 日落黄
sun shade 遮阳罩;遮阳物;遮阳设施;遮日罩;遮光罩;遮阳百叶板;遮阳;颜色玻璃;天棚
sunshade grill(e) 遮阳花格
sun shading 遮阳
sun-shading board 遮阳板
sun-shading device 遮阳装置
sun-shading louver 遮阳百叶
sun-shading overhang 遮阳挑檐
sun shadow area 日射阴影面积
sun shadow curve 遮阳曲线;日影曲线
sun shield 遮阳设施;遮阳板;遮光板
sun shielding glass 遮阳玻璃
sun shielding layer 遮光涂层
sunshine 日照;太阳光
sunshine area 日照面积
sunshine carbon arc weathering tester 日光型碳弧(老化试验机)
sunshine collector 阳光收集器
sunshine condition 日照条件
sunshine control 日照控制;日照调节
sunshine deformation survey 日照变形测量
sunshine duration 日照时间;日照时数
sunshine hours 日照小时;日照时数
sunshine integrator 日照累积器
sunshine period 日照期
sunshine record 日照记录
sunshine recorder 日照记录器;日照计
sunshine spacing 日照间距
sunshine time 日照时间
sunshine unit 日照单位;锶单位
sun's input 太阳输入
sun's meridian altitude 太阳中天高度
sun's motion 太阳的运动
sun's output 太阳辐射输出
sunspot 日斑;太阳黑子;太阳灯
sunspot component 太阳黑子分量
sunspot cycle 日斑周期;太阳黑子周(期)
sunspot flare 太阳黑子耀斑
sunspot group 太阳黑子群
sunspot maximum 太阳黑子极大期
sunspot minimum 太阳黑子极小期
sunspot number 太阳黑子数
sunspot periodicity 日斑周期;太阳黑子周(期)
sunspot polarity 太阳黑子极性
sunspot prominence 太阳黑子日珥
sunspot radiation 太阳黑子辐射
sunspot relative number 太阳黑子相对数
sunspot spectrum 太阳黑子光谱
sunspot theory 太阳黑子理论
sunspot zone 太阳黑子带
sun's radiation 太阳辐射
sun steel 聚氯乙烯薄膜色层压花钢板;压花钢板
sunstone 日长石;日长石;太阳石
sun streams 太阳光线
sun strobe 太阳标志
sunstrokde prevention 防暑
sunstroke 日射病
sun's way 太阳路径
sun switch 日光开关
sun-synchronous nearly polar orbit 太阳同步近极轨道
sun-synchronous near polar orbit satellite 太阳同步极地卫星
sun-synchronous orbit 太阳同步轨道
sun-synchronous satellite 太阳同步卫星
sun system 太阳系
sun test 日晒试验

sun time 太阳时
sun time clock 太阳时钟
sun tracker 太阳跟踪仪
sun trap 日晒遮风的露台;日晒遮风的花园;晒太阳的露台
sun valve 日光开关;日光阀
sun-vector sensor 太阳角传感器
sun visor 遮阳棚;遮光板;防太阳玻璃
sun wake 夕阳反照
sun-weather relationship 太阳—天气关系
sun wheel 太阳轮;恒星齿轮
sun wind 海陆风;日风;太阳风
sunwise 日转方向
sun-X film 防热胶片
Sun Yat-sen's mausoleum 中山陵
suolunite 索伦石
suozalite 水磷铝镁石
Supa 苏帕柚楠(一种黄铜色—深棕色圆木,产于印度—马来亚)
supasorb 白色硅藻土载体
sup complete 上半完备的
superabacus 拱墩;拱端托
superabduction 外展过度
superabound 过多
super abrasion furnace black 超耐磨炉黑
super-absorbent resin 高吸水性树脂
super-abundance 过多
super-abundance sludge 剩余污泥
super-abundance water 剩余水
super-abundant observation 超多余观测
super-accelerator 超促进剂
super-achromatic 超消色差的
super-achromatic lens 超消色差透镜
super-acid 超酸;过量酸的
super-acidic catalyst 超强碱催化剂
super-acidity 超酸性;过度酸性
superacidulated 超酸化的
superacidulation 过酸化作用
super-acritical ore fluid 超临界含矿流体
super-actinides 第二锕系元素;超锕系元素
super-activity 超活性
superaddition 添加物;附加物
super-additivity 超加性;超加和性
superadequancy 过分的装修
super-adiabatic 超绝热的;超绝热
super-adiabatic convection 超绝热对流
super-adiabatic lapse rate 超绝热直减率
super-adiabatic state 超绝热状态;超绝热的状态
superaeration 过曝气;过量曝气
super-aerodynamics 超越空气动力学;超高速空气动力学;超高空空气动力学;稀薄气体空气动力学;高空空气动力学
super-ag(e)ing 超老化
super-ag(e)ing mix 超老化混合胶
super-agency 超级管理机构
super-air-filter 高效空气滤净器
super-alethal dose of x-irradiation 超致死剂量照射
super-alimentation 超量营养法
super-alkali 苛性钠
super-alkalinity 超碱性
super-alloy 超透磁合金;超耐热合金;超耐热不锈钢;超级合金;超合金;高温合金;高温高强合金;高级合金;耐高温合金
super-alloy power processing 超合金粉末加工
super-alloy steel 超级耐热合金钢
super-alpha titanium alloy 超阿尔发型钛合金
super-altitude 超高空
super-alumag 超级铝锌镁合金
super-aluminous cement 特快硬高铝水泥
superannuated building 老朽建筑物
superannuated ship 老船;废船(体)
superannuation 退休金
superannuation and retired allowances 养老金及退休金税收优惠
superannuation fund 养老基金;退休金基金
superannuation payment 退休金
super-anthracite 超无烟煤;变无烟煤
super-arc 优弧
super-arch fender 巨型拱形
super-arch rubber fender 超级拱形橡胶护舷
super-ascoloy 超级奥氏体耐热不锈钢
super-atmospheric pressure 表压(力);超大气压力
super-attemperator 超级恒温器

super-audiable 超音频的
super-audiable frequency 超音频
super-audibcy 超听觉频率
super-audible 超声波频的;超声波
super-audio 超音频;超声波频
super-audio frequency 超音频;声频上频率
super-audio telegraph 超音频电信
super-audio telegraphy 超音频电报
superb 高超
super-balloon tire 超低压轮胎
super-bang 超大爆炸
superbar 上横号
super-basic catalyst 超强碱催化剂
super-basin 超级盆地
super-beta transistor 超贝塔晶体管
super-bituminous coal 超烟煤
super-block 大街区;大街坊;超级块段;特殊街坊;超级街区
superblock plan 特殊街坊规划;大街区规划;特大街坊规划
super-bolide 超火流星
super-bomber 超级轰炸机
super-booster 超功率运载火箭
super-bound particle 超束缚粒子
super-brand 超级名牌商品(被世界公认和使用的商品)
super-bridge 特大桥
super-bright 超亮
super-broadening 超展宽
super-bronze 超级青铜;高强度青铜
super-bubble 超级磁泡
super-bulk carrier 超级散货船;巨型散货船
superburning 过烧
super-business 超级公司
super-calender 超级辗光机;高度研光机;高度压光机
super-calendered coverpaper 超级研光封皮纸
super-calendered finish 超级轧光纸
super-calendered paper 超级研光纸;超级压光纸
super-calendered woodfree printing paper 超级研光道林纸
supercalendering 高度压光
super-capacity bucket 大容量斗
super-capacity bucket elevator 大容量斗式提升机
super-capillary 超毛细现象;超毛细管
super-capillary interstices 超毛细孔隙;超毛细空隙;超毛细管间隙
super-capillary percolation 超毛细渗透(作用);超毛细管渗透
super-capillary pore 超毛细管孔隙
super-capillary porosity 超毛细管孔隙率
super-capillary seepage 超毛细管渗透;超毛细管渗流
super-capital 副柱头;拱基
super-carbon steel 超碳钢
super-cardioid microphone 超心形传声器
supercargo 押运员;租船驻船代表;监货员;货运负责人
super-carrier 超级货船;超级航空母舰;巨型货船
super-category gassy mine 超级瓦斯煤矿
super-cavitating craft 超汽蚀水翼船;超汽蚀水翼船;超空蚀水翼船;超空蚀水翼船
super-cavitating flow 超汽蚀水流;超空蚀水流;超空蚀水流;超空泡流
super-cavitating hydrofoils 超汽蚀水翼;超汽泡水翼;超空蚀水翼;超空泡水翼
super-cavitating propeller 超空泡螺旋桨
super-cavitation 超汽蚀;超空蚀;超空泡;超空化
super-cavity flow 超空泡流
super-cell 精制硅藻土助滤剂;超晶胞
super-cell rubber fender 超级鼓型橡胶护舷
super-cement 超级水泥;高级水泥;高标号水泥
supercenter 市郊的大商业中心
supercentral station 区域电厂
super-centrifugation 超速离心法
super-centrifuge 超速离心机;高速离心机
supercharge 超载;超荷;增压
supercharged air 增压空气
supercharged boiler 增压锅炉
supercharged dam 溢水坝;溢流坝
supercharged diesel engine 增压柴油机
supercharged engine 增压式内燃机;增压式发动机;增压器;增压内燃机;增压发动机
supercharged press(ing) 增压压力

supercharged pressure 增压压力
supercharged steam generator 增压锅炉
supercharged turboprop 增压式涡轮螺旋桨发动机
supercharge-exchange interaction 超荷交换相互作用
supercharge load(ing) 超载;过载
supercharge method 增压法
supercharge octane number 增压辛烷值
supercharge pressure 进气压力
supercharger 增压机;超装器;增压器;增加器
supercharger blast gate 增压器节流阀;送风机节流挡板
supercharger control 增压调节
supercharger dam 溢流坝
supercharger impeller 增压器叶轮
supercharger pressure ga(u)ge 增压机压力计
supercharger regulator 增压器调节器
supercharging 增压作用
supercharging blower 增压鼓风机
supercharging boosting 增压
supercharging gas compressor 增压压缩机;增压气机
supercharging device 增压装置
supercharging equipment 增压设备
supercharging pressure 充液压力
super-check 超级检验
superchip 高密度芯片
super-chlorination 过量氯消毒法;过量氯化(法);过量加氯;过氯化(作用)
super-chopper 超速断路器;超高速断续器;特快断路器
super-chromatic lens 超彩色透镜
superciliary 眉纹的
supercilium 雕花过梁;(古罗马建筑中屋檐波状花边上部的)带形线条
super-circulation 超循环;超环量
super-city 超级城市;特大城市
super-civilization 超级文明世界
superclass 超类;总纲;母集合
superclass gas tunnel 超级瓦斯隧道
superclass station 特等站
super-clean 超净(的)
super-clean bench 超净(工作)台
super-cleaned coal 超净煤
super-cleaner 梯形除杂机
supercoat 表面涂层
super-coated 表面涂膜的
supercoated stress 叠加应力
super-cobalt drill 超钻钻头
super-code 超码
super-coil 超外差(式)线圈;超螺旋;超卷曲
super-collier 巨型运煤船
supercolumnar 重列柱式的
super-columniation 叠层柱列;重列柱;重柱式
super-combat gasoline 超级战斗用汽油
super-commutation 超转接;超级分配;超换接;超倍采样
super-compact 超紧
super-compactor 重型压实机;重型压力机
super-compensation 超量补偿
super-complete well 超完善井
super-compressibility 超压缩性
super-compressibility factor 超压缩因子;压缩因数
super-compression 超压缩;过度压缩
super-compression arrangement 超压缩装置
super-compression engine 超压缩发动机
super-compressor 超压缩器;超压缩机
super-computer 超级计算机;超大型计算机;巨型计算机
super-computer numerial control wire cutting machine 电火花数控线切割机
super-concentrated juice 超浓缩果汁
super-concentrated low-sud laundry detergent 超浓缩无泡沫洗衣粉
super-conducting 超导的
super-conducting accelerator 超导加速器
super-conducting alloy 超导合金
super-conducting amplifier 超导放大器
super-conducting antenna 超导天线
super-conducting bolometer 超导体电阻测温计;超导(体)测辐射热计;超导测辐射计
super-conducting bridge tunnel junction 超导桥隧道结
super-conducting cable 超导电缆

super-conducting circuit 超导回路;超导电路
super-conducting coil 超导线圈
super-conducting coil accumulator 超导线圈式蓄电器
super-conducting component 超导元件
super-conducting computer 超导计算机
super-conducting crystal 超导晶体
super-conducting delay line 超导延迟线
super-conducting device 超导装置;低温器件
super-conducting element 超导元件
super-conducting energy storage 超导储能
super-conducting fiber 超导纤维
super-conducting filament 超导丝
super-conducting frequency converter 超导变频器
super-conducting galvanometer 超导电流计
super-conducting generator 超导发电机;超导材料发生器
super-conducting gravimeter 超导重力仪
super-conducting gravity gradiometer 超导重力梯度仪
super-conducting gyro(scope) 低温陀螺(仪);超导陀螺仪;超导磁场陀螺仪
super-conducting helix 超导螺旋线
super-conducting junction 超导结
super-conducting lens 超导透镜
super-conducting line 超导线
super-conducting link 超导链
super-conducting machine 超导电机
super-conducting magnet 超导磁铁;超导磁体
super-conducting magnetic gradiometer 超导磁重力梯度仪
super-conducting magnetic shielding 超导磁屏蔽
super-conducting magnet levitation 超导磁浮
super-conducting magnetometer 超导磁强计;超导磁力仪
super-conducting magnet quenching 超导磁体淬熄
super-conducting material 超导体;超导材料
super-conducting memory 超导存储器
super-conducting memory array 超导存储阵列
super-conducting metal 超导(性)金属
super-conducting microscope 超导显微镜
super-conducting mixer 超导混频器
super-conducting motor 超导马达;超导电机
super-conducting nuclear magnetic resonance meter 超导核磁共振计
super-conducting nuclear particle detector 超导核粒子探测器
super-conducting parametric(al) amplifier 超导参量放大器
super-conducting pulse power supply 超导脉冲电源
super-conducting quantum interference device 超导量子干涉仪
super-conducting quantum interference magnetometer 超导量子干涉磁力仪
super-conducting state 超导(状)态
super-conducting state super cooling 超导态过冷
super-conducting technology 超导电技术
super-conducting temperature 超导温度
super-conducting thin film 超导薄膜
super-conducting thin-film ring 超导薄膜环
super-conducting transition 超导跃迁
super-conducting tunnel effect 超导隧道效应
super-conduction 超导
super-conduction cable 超导电缆
super-conduction phenomenon 超导现象
super-conductive 超导电的;超导的
super-conductive chopper 超导斩波器
super-conductive gravimeter 超导重力仪
super-conductive magnetometer 超导磁力仪
super-conductive state 超导电状态
super-conductive suspension 超导悬浮
super-conductive transition 超导跃迁
super-conductive wire 超导体铌锡合金丝
super-conductivity 超电导率;超导(性);超导技术;超导电性;超导(电)率
super-conductivity bolometer 超导电测辐射热器
super-conductivity memory 超导存储器
super-conductor 超导(电)体
super-conductor alloy 超导体合金
super-conductor bolometer 超导体辐射热测定器
super-conductor fibre 超导体纤维
super-conductor physics 超导体物理学
super-conductor thermometer 超导电温度计
super-coner 高速络筒机
super-consistent 超相容
super-continent 超级大陆;超大陆
super-continent breakup 超大陆解体
super-contracted fibre 超收缩纤维
super-contraction 超收缩
super-control tube 超跨导管;可变互导管;变跨导管;变互导管
super-convergence 超收敛
super-convergent bootstrap 超收敛自举
super-convergent propagator 超收敛传播子
super-converter 超外差(式)变频器;超级转换堆
super-cool 超冷;过冷
supercooled 过冷(却)的
supercooled accelerator 低温加速器
supercooled air 超冷空气;过冷空气
supercooled austenite 过冷奥氏体
supercooled cloud 过冷云
supercooled fog 过冷(却)雾
supercooled liquid 过冷液体
supercooled liquid water 过冷液态水
supercooled liquid water concentration 过冷液态水浓度
supercooled magnet 超冷磁铁
supercooled rain 过冷雨
supercooled soil 过冷土
supercooled state 过冷状态
supercooled steam 过冷蒸汽
supercooled vapo(u)r 过冷蒸汽;过冷水(蒸)气
supercooled water 过冷水;过冷却水
supercooled water droplet 过冷水滴;过冷的水滴
supercooling 过渡冷却;过冷现象;过冷;过冷(的);过度冷却现象
super corona 超日冕;超冕
super corporation 超级公司
super cosmic radiation 超宇宙辐射
super cosmotron 超高能粒子加速器
super couducting film 超导膜
super crevice 超裂缝
super crimp 超卷曲
super criterion 超级(标)准
super critical 超临界(的);上临界的
super critical aqueous solution 超临界水溶液
super critical area of extraction 超临界抽取面积
super critical carbon dioxide extraction 超临界二氧化碳萃取
super critical carbon dioxide technology 超临界二氧化碳技术
super critical chemical reaction 超临界化学反应
super critical condition 超临界条件
super critical cycle 超临界(压力)循环
super critical damping 超临界阻尼
super critical extraction process 超临界抽提过程
super critical flow 超临界水流;超临界气流;超临界流动;超临界流
super critical fluid 超临界流体
super critical fluid chromatograph 超临界流体色谱仪
super critical fluid chromatography 超临界流体色谱法;超临界流体层析
super critical fluid chromatography-Fourier transform infrared spectroscopy 超临界流体色谱法—傅里叶变换红外光谱法联用
super critical fluid chromatography-mass spectrometry 超临界流体色谱法—质谱法联用
super critical fluid extraction 超临界流体萃取
super critical fluid-liquid chromatography 超临界液液相色谱法;超临界液液相层析
super critical fluid-solid chromatography 超临界液固相色谱法;超临界液固相层析
super criticality 超临界状态;超临界性
super critical mass 超临界质量
super critical nuclear chain reaction 超临界链式核反应
super critical nucleus 超临界晶核
super critical oxidation 超临界氧化
super critical phenomenon 超临界现象
super critical plant 超临界参数装置
super critical pressure 超临界压力
super critical pressure boiler 超临界压力锅炉
super critical pressure cycle 超临界压力循环
super critical reactor 超临界反应堆
super critical recovery 超临界回收
super critical river 超临时河流;超临界河流
super critical rotor 超临界转子
super critical speed 超临界速度
super critical state 超临界状态
super critical steam power plant 超临界蒸汽发电厂
super critical steam pressure 超临界蒸汽压力
super critical steam turbine technology 超临界蒸汽涡轮机技术
super critical stream 超临时河流;超临界流速河流;超临界流动;超临界流;超临界河流
super critical system 超临界水系统
super critical velocity 超临界流速
super critical volume 超临界体积
super critical water 超临界水
super critical water oxidation 超临界水氧化
super critical water oxidation process 超临界水氧化工艺
super critical water oxidation reactor 超临界水氧化反应器
super critical water oxidation reactor apparatus 超临界水氧化反应器仪表
super critical water oxidation reactor method 超临界水氧化反应器方法
super critical water oxidation system 超临界水氧化系统
super critical water oxidation technique 超临界水氧化技术
super critical wet oxidation 超临界湿式氧化(法)
super critical wing 超临界机翼
super cross beam 上横梁
supercrust 上层;路面层;表层;顶层
super crustal 上地壳的(指覆盖基底上的岩石)
super crustal erosion 上地壳侵蚀
super crystal lattice material 超晶体材料;超晶格材料
super current 超(导)电流
super current accelerator 超强流加速器
super cut 细纹锉
super-deep drillhole 超深钻孔
super-deep hole 超深钻孔
super-deep hole drilling 超深孔钻进
super-deep soil stabilizer 超深密实土搅拌机
super-deep well oil well cement 超深井油井水泥
super-deformation 超形变
super-deformed states 超形变态
super-delocalizability 超离域性
super-dense lining 高密度炉衬
super-dense star 超密恒星
super-diagonal matrix 超对角矩阵
super-diamagnetic 超抗磁
super-dielectric(al) glass 高介电玻璃
super-dimensioned 超尺寸的
super-dip 超倾磁力仪;超灵敏磁倾仪
super-directive antenna 超锐定向天线
super-directivity 超指向性系数;超锐定向性;超方向性系数
super-dislocation 超位错;超断层
super-disruption 超崩裂
super-distention 膨胀过度
super-dividend 附加的股息
super-divisional organization 超事业部组织
super-dominance 超显性
super-draft 超大牵伸
super-drawing 超拉伸(法);高倍拉伸
super-drill 超深钻
super-dry 过干燥
superduct 上转
super-duper 高超(的)
super-dural 超级硬铝
super-duralumin 超强铝;超(级)硬铝
super-dural umin alloys 超硬铝合金
super-dust 超微粉尘;超微尘(粒)
super-duty 超重型(的);超级(的);超高温的
super-duty brick 高级砖;超(级)耐火砖
super-duty clay brick 特级黏土砖
super-duty fireclay brick 特级黏土耐火砖
super-duty refractory 超级耐火材料;特级耐火材料
super-duty refractory concrete 超级耐火混凝土
super-duty silica brick 超重型硅酸盐砖;高级硅砖;特等硅酸盐砖
super-duty silica refractory 超级硅质耐火材料;特级硅质耐火材料
super-efficiency 超有效性

super-efficient shallow air-floating 超效浅层气浮
super-elastic collision 超弹性碰撞
super-elasticity 超弹性
super-elastic load 超弹性极限负荷
super-elevated 超高的
super-elevated a curve 设置曲线超高
super-elevated and widened curve 超高加宽曲线段;超高和加宽曲线
super-elevated corner 超高转角
super-elevated curve 超高曲线段;超高(的)曲线
super-elevated curve structure 弯道超高结构
super-elevated runoff 超高顺坡率
super-elevated turn 有超高的弯道
super-elevation 超仰角;超高;高角
super-elevation ga(u)ge 超高规
super-elevation in track 外轨垫高
super-elevation marker 超高标记
super-elevation of guide rail 护轮轨轨超高
super-elevation of outer rail 外轨超高
super-elevation of outer rail on curve 曲线外轨超高
super-elevation of the belt 传送带超高
super-elevation of the floor surface of a bridge on a horizontal curve 平曲线桥面的超高
super-elevation on curve 曲线超高
super-elevation runoff 超高缓和段
super-elevation slope 超高顺坡【铁】;超高坡度
super-elevation tag 超高标记
super-elevation templet 超高规
super-emitron 超光电摄像管
super-emitron camera 超光电摄像机
super-energy 超高能
super-energy proton synchrotron 超高能质子同步加速器
super-environment 超级环境
super-equivalent adsorption 超当量吸附
superette 小型自动售货商店;小型超级市场
superevacuation 高度排空
super-exchange 超交换
super-exchange interaction 超交换作用
super-excitation 超激发
superexcition 过激励电机
super-expansibility factor 压缩系数
super-express railway noise 超速火车噪声
super-express train 超级特快列车
superface 顶集
super-facies 超相
super-family 超科
super-fast 超高速的
super-fast loading 快速装料
super-fast plasma bunch 超速等离子体凝块
super-fast polymerization 超速聚合
super-fast screw 快速螺钉
super-fast-setting cement 超快凝水泥;超级快凝水泥
superfatted soap 富脂皂
superfatting agent 富脂剂
super-fault list 超故障表
super-fiber 超强力纤维
super-ferromagnetism 超铁磁性
super-fiche 超微胶片
superficial 浅表的;表面的
superficial air velocity 过滤风速
superficial alternation 表面变化
superficial and rapid pulse 脉浮数
superficial application 表面防腐处理
superficial area 表面积;表面面积
superficial ca(u)lking 表面捻缝
superficial cementation 表面渗碳
superficial change 表面变化
superficial characteristic 表征
superficial charring 木材表面碳化(法);皮焦法;表面炭化(一种木材防腐法);表面烧焦
superficial coat 表层
superficial compaction 表面压实;表层压实;表土夯实
superficial compactor 表面压实机
superficial content 面积
superficial cooling 表面冷却
superficial creep 表面蠕动
superficial current 表面水流;表面气流;表面电流
superficial damage 表面破坏
superficial degradation 表面损坏;表面变质;表层剥蚀

superficial density 表面密度
superficial deposit(ion) 地面沉积;表生矿床;表面沉积(物);表层沉积(物)
superficial deterioration 表面损坏;表面变质;表层剥蚀
superficial dimension 外尺寸;表面尺度
superficial earth characteristic 表土特性
superficial expansion 表面膨胀
superficial floor area 净楼板面积
superficial fold 表层褶曲
superficial foot 平方英尺;方英尺
superficial growth 表面生长
superficial hardness 表面硬度
superficial hardness method 表面硬化度法(无损试验的)
superficial injury 浅表损伤;表面伤害
superficial inspection 外表检查
superficiality 表面性
superficial layer 浅层;表面层
superficial liquid velocity 表面流体流速
superficial loading rate 表面负荷率
superficially porous particle 表面多孔颗粒
superficial mass velocity 表面质量速度
superficial measure 表面量度;表面积计量
superficial microflora 表面微生物区系
superficial microlayer 布面微层;表面细层
superficial moraine 表碛
superficial ooid 薄皮鲕
superficial part 浅部
superficial porosity 表面孔隙率
superficial reflex 浅(层)反射
superficial regolith 风化表层
superficial rockwell hardness tester 轻压力洛氏硬度试验机;表面洛氏硬度试验机;表面洛氏硬度计
superficial root 表土根
superficial sediment 表面沉积(物);表层沉积(物)
superficial sedimentary layer 表面沉积层
superficial speed 表面速度
superficial stratum 表土层;表(面)层;[复]strata
superficial structure 表生构造;表层构造
superficial superficialis 表面的
superficial surface 浅面
superficial treatment 表面处理
superficial treatment of timber 木材表面处理
superficial unit 面积单位
superficial upflow velocity 表面上升流速
superficial vapo(u)r velocity 表面的蒸气速度
superficial variation 表面变化
superficial velocity 表面速度;表面流速;表观速度
superficial water 地面水;地表水;表层水
superficies 表面;外面;外观;外表
super-field 超场
super-filter 超滤器;超滤机
super-filtrol 超滤土
super-fine 超细;特级
super-fine aggregate 极细骨料;极细集料
super-fine aluminium power 超细铝粉
super-fine antimony power 超细锑粉
super-fine calcium carbonate 超微细碳酸钙
super-fine cement 超细水泥
super-fine classifier 超细分级机
super-fine dust 超细粉尘
super-fine fiber 超细纤维;极细填充物
super-fine file 最细锉
super-fine flour 极细粉料
super-fine glass 超细玻璃
super-fine glass fabric 超细玻璃丝布
super-fine glass wool 超细玻璃棉
super-fine grain 超微粒
super-fine grain film 超微粒胶片
super-fine grinding 超细粉碎;超细磨
super-fine ground 超磨的
super-fine powder 极细粉末
super-fine powdered marble 极细大理石粉
super-fine quality 上等货;特上品质
super-fines 超细粉末;超细粉
super-fine sand 超细砂
super-fine structure 超精细结构
super-fine thermoplastic fibre 超细热塑性纤维
super-fine white 超精白
superfine writing paper 高级书写纸
super-finish 超级研磨;超细研磨;超细加工;超精表面;超精加工

super-finisher 超精加工机床
super-finishing 超级研磨;超精加工
super-finishing equipment 超精加工装置
super-finishing machine 超精整机;超精加工机床
super-finishing stone 超精整磨石
super-finish surface 超精加工面
superfische 缩微平片
super-flexible sub-rigid pier 上柔下弯墩
super-flexion 超弯曲现象;屈曲过度
superflood 特大洪水;非常洪水
superfluence 顶流
superfluent 顶流熔岩流;顶流的
superfluent lava flow 顶熔岩流
super-fluid 超流体
super-fluid core 超流体核
super-fluid density 超流体密度
super-fluid gyroscope 超流回转仪
super-fluid helium 超流态氦
super-fluid helium theory 超流氦理论
super-fluidity 超流(体)性;超流动性
super-fluidity of ornamentation 冗饰;多余装饰
super-fluidity theory 超流(体)理论
super-fluid model 超流体模型
superfluity level 富余强度
superfluity of ornamentation 多余装饰
super-fluorescence 超荧光
super-fluorescent 超荧光的
super-fluorescent fiber laser 超荧光光纤激光器
superfluous 过剩的;多余的
superfluous and complicated ornamentation 烦琐装饰
superfluous constraint 多余约束
superfluous information 多余信息
superfluous of ornamentation 多余装饰
superfluous parameter 多余参数;多余参量
superfluous raw material 多余原料
superfluous term 冗余项
superfluous variable 多余变数
superfluous water 过剩水分;多余水
superfluous water vapo(u)r 过剩水气
superflux 泛滥
super-foot true 实片英尺
superfoundation 基础以上的
superfoundation structure 基础以上的建筑物
super-fraction 超精馏
super-fractionation 超精馏
super-fractionator 超精馏器
super-fraction column 超精馏柱
super-frame 超帧
super-frequency 超高频;特高频
super-fuel 高级燃料;超级燃料
superfusion 过熔
super-gain 超增益;超增色
super-gain antenna 超增益天线
super-galactic astronomy 超银河系天文学;超星系天文学
super-galactic coordinates 超星系坐标
super-gasket 耐热垫片
super-gasoline 高抗爆性汽油;超级汽油
super-ga(u)ge 超规范
super-ga(u)ge field theory 超规范场论
supergence 浅成的
supergene 表生的
supergene deposit 表生沉积
supergene enrichment 次生富集;表生富集
supergene enrichment zone 表生富集带
supergene mineral 表生矿物
supergene mobility 表生活动性
super-generator 超级起电机
supergene rock 浅成岩
supergene solution 表生下降溶液
supergene structure 表生构造
supergenetic environment 表成环境
supergene water 下降水
super-geometric(al) distribution 超几何分布
super-geostrophic wind 超地转风
super-giant 超巨型;特大的
super-giant gas field 超级气田
super-giant oil field 超级油田
super-giant star 超巨星
superglacial 冰上的
superglacial river 冰川表面河流
superglacial stream 冰面水流

super-Gleamax 超光泽镀镍法
super-gloss 最光亮;最高级光泽
super-glycerinated oil 超甘油化油
super-goldstone 超哥特斯顿粒子
super-gold tranche 超黄金份额
super-grade 超等级;超标准
super-gradient wind 超梯度风
super-granular cell 超米粒泡
super-granulation 超米粒组织;超粒化
super-graph 超图
super-gravity 超重力;超引力
super-gravity theories 超引力理论
super-grid 超高压输电网;超高压电网;特大功率电网
super-grinding 超精磨
super-ground 极精细研磨
super-group 超群;超级组
super-group allocation 超群分配
super-group assembly 超群组合
super-group connector 超群连接器
super-group demodulator 超群解调器
super-group demultiplexer 超群多路分解器
super-group designation 超群标识
super-group filter 超群滤波器
super-group modem 超群调制解调器
super-group multiplexer 超群多路复用器
super-group translating equipment 超群转换设备
super-group translator 超群译码器;超群变换器
super-grown 超生长
super-grown transister 超生长型晶体管
super-harbo(u)r 超级港口
super-hard 超硬(的);过硬的
super-hard alloy 超硬合金
super-hardboard 特硬木板;超级硬木板;超硬板;超硬质纤维板;超密度级硬质纤维板
super-hard coating 超硬涂层
super-hard material 超硬材料
super-hardness 超硬度;超级硬度
super-hard tool material 超硬工具材料
super-hard X-ray 超强 X 射线
superharmonic function 上调和函数
superheated 过热的
superheated aggregate 过热集料;过热骨料
superheated engine 过热发动机
superheated fumarole 过热态喷汽孔
superheated liquid 过热液体
superheated molten steel 过热钢液
superheated pool 过热沸水塘
superheated region 超热区(域)
superheated steam 超热蒸汽;过热(水)蒸汽
superheated steam boiler 过热蒸汽锅炉
superheated steam cracking 过热水蒸汽裂解
superheated steam current 过热蒸汽流
superheated steam cylinder oil 过热蒸汽缸(润滑)油
superheated steam engine 过热蒸汽发动机
superheated steam locomotive 过热蒸汽机车
superheated steam operation 过热蒸汽操作
superheated vapo(u)r 过热蒸汽;过热蒸气;过热水蒸气
superheated vapo(u)r drier[dryer] 过热蒸气干燥器
superheated water 过热水
superheated water heating 过热水供热;过热水采暖
superheated water heating installation 过热水供暖装置
superheated water main 过热水总管
superheated water network 过热水管网
superheated water radiant panel 过热水散热片
superheated water well 过热水井
superheater 过热器;回热器
superheater elbow 过热器弯头
superheater header 过热器箱
superheater tube 过热器管(子);过热管
superheater unit 过热器
superheat(ing) 过热,过加热;市场狂热
superheating calorimeter 过热量热器;高温测量器;高温测量计
superheating phenomenon of glass 玻璃液过热
superheating surface 过热面
superheating temperature 过热温度
superheat meter 过热计
superheat region 过热区域

superheat resisting alloy 超耐热合金
superheat resisting material 超(级)耐热材料
superheat section 过热段
superheat temperature 过热温度
super-heavy 超重型;超重(的)
super-heavy concrete 超重混凝土
super-heavy-duty winch 超重型绞车
super-heavy element 超重元素
super-heavy element halides 超重元素卤化物
super-heavy hydrogen 超重氢;超重氢
super-heavy ion 超重离子
super-heavy nucleus 超重核
super-heavy particle 超重粒子
super-heavy quantum 超重量子
super-heavy quark 超重夸克
super-heavy vehicle 超重型车辆
super-heavy weight 超重量级
super-helicity 超螺旋性
super-helix 超螺旋;超卷曲
super-helix density 超螺旋密度
super-het 超外差(式)接收机;超外差(式)
super-heterodyne 超外差(式)
super-heterodyne circuit 超外差(式)电路
super-heterodyne converter 超外差(式)变频器
super-heterodyne detection 超外差(式)检波
super-heterodyne interference 超外差(式)干扰
super-heterodyne mixer 超外差(式)混频器
super-heterodyne optics 超外差(式)光学
super-heterodyne radio set 超外差(式)收音机;超外差(式)接收机
super-heterodyne receiver 超外差(式)接收机
super-heterodyne reception 超外差(式)接收(法)
super-heterodyne spectrometer 超外差(式)分光计
super-high compressor 超高压缩机
super-high density 超高密度
super-high density state 超高密态
super-high draft 超大牵伸
super-high early strength cement 超早强水泥
super-high early strength portland cement 超早强硅酸盐水泥;超早强波特兰水泥
super-high energy 超高能
super-high energy cosmic ray 超高能宇宙线
super-high energy particle 超高能粒子
super-high frequency 超高频(率);特高频
super-high frequency band 厘米波段
super-high-frequency radiation 超高频辐射
super-high-frequency radiometry 超高频辐射测量
super-high frequency radio telephone equipment 特高频无线电话设备
super-high frequency transmitter 特高频发射机
super-high frequency wavelength 超高频波段
super-high intensity 超高强度
super-high pressure 超高压
super-high pressure boiler 超高压锅炉
super-high pressure compressor 超高压压缩机
super-high-pressure hose 超高压胶管
super-high pressure hydraulic press 超高压液压机
super-high pressure polyethylene reaction pot 超高压聚乙烯反应釜
super-high pressure sealing 超高压密封
super-high pressure synthetic (al) diamond press 超高压合成金刚石压力机
super-high pressure technique 超高压技术
super-high pressure turbine 超高压汽轮机
super-high pulse 超高脉冲
super-high purity indium 超高纯铟
super-high purity mercury 超高纯汞
super-high reaction force 超高反力(护舷)
super-high refractive index optic(al) beads 超高折射玻璃细珠
super-high-rise 超高层建筑
super-high-rise building 超高层建筑
super-high speed 超高速的
super-high speed alloy 超高速切削合金
super-high speed cutting 超高速切削
super-high speed cutting alloy 超高速切削合金
super-high speed digital intergrated circuit 超高速数字集成电路
super-high speed magnetic levitation sytem 超高速磁浮系统
super-high speed railway 超高速铁路
super-high-speed spinning 超高速纺丝
super-high speed steel 超高速(切削)钢

super-high steel 超高速钢
super-high strength cement 超高强水泥
super-high-strength optic(al)fiber 超高强度光纤
super-high tension 超高电压
super-high tension and wide spacing electrostatic precipitator 超高压宽极距静电除尘器
super-high voltage mercury lamp 超高压水银灯
super-highway 超级公路;高速公路
super-highway-bridge mixer 公路桥梁用大梁混凝土搅拌器
superhighway system 高速公路系统
super-honing 超精磨
super-hopper dredge(r) 巨型自航式挖泥船;超级自航挖泥船
super-huge computer 超大型计算机
super-huge turbogenerator 特大型汽轮发电机
super-huge water source 特大型水源地
super-human scale 超人尺度(指超过人尺度的空间尺度)
super-humid soil 过湿土(壤)
super-hybrid composite material 超混杂复合材料
super-hypabyssal facies 超浅成相
super-iconoscope 超光电摄像机;超光电摄像管;移像式光电摄像管;复式光电摄像管
superimpose 信息叠加;添加;叠上;叠加
superimposed 上叠的;叠加的;叠合的;叠覆的
superimposed anomaly 外加异常
superimposed arches 上叠拱;层叠拱
superimposed back pressure 附加背压
superimposed bed 上覆层
superimposed carving 镶嵌雕刻品
superimposed circuit 叠置电路
superimposed clause 附加条文;附加条款
superimposed coding 重叠编码
superimposed core 预埋芯;春入芯
superimposed crossing 浮放道岔
superimposed current 叠加电流
superimposed dead load 叠加净荷载;附加恒载
superimposed deposit of multiple genesis-multiple stage 多成因多阶段迭生矿床
superimposed drainage 叠置水系;重置水系;重叠排水系统
superimposed earth 加填土
superimposed effect 叠加效应
superimposed fan 叠置扇
superimposed field 叠加场
superimposed fill 加填土
superimposed filter 重叠式过滤器
superimposed flood plain 上叠河漫滩
superimposed flow 叠流
superimposed fold 横舒褶皱;叠加褶皱
superimposed geosyncline 上叠地槽
superimposed glacier 上层迭置冰川
superimposed grid 叠上坐标网;重叠坐标格网
superimposed halo 后叠晕;重叠晕
superimposed hologram 叠置全息图
superimposed ice 叠置冰
superimposed image 重置影像;重叠像;重叠图像
superimposed imagery 叠加图像
superimposed law 叠加定律
superimposed layer 上叠层;叠加层;重叠层
superimposed line method 配线法
superimposed line method for steady flow 稳定流配线法
superimposed load(ing) 超负荷;超负荷载;外加荷载;堆码负重;叠加荷载;附加荷载
superimposed load of building 建筑物附加荷载
superimposed load of dam foundation 坝基岩体附加荷载
superimposed metamorphism 叠加变质作用
superimposed method 叠置片法
superimposed mineral deposit 迭生矿床
superimposed notation 附加批注
superimposed orders 叠柱
superimposed oscillation 叠加振荡
superimposed pairs 重叠像对
superimposed picture 重叠像片
superimposed pluvial fan 叠垒式洪积扇
superimposed point sources 重叠点辐射源
superimposed profile 并列透视断面
superimposed record 重合记录法
superimposed relationship 上叠关系
superimposed ringing 叠加振铃
superimposed river 表生河流

superimposed soil 上覆土
superimposed strains 叠加应变
superimposed stratabound deposit 迭生层控矿床
superimposed stratum 叠加地层
superimposed stream 上置河流;叠置河流;叠置河;表生河流
superimposed stress 应力叠加;附加应力
superimposed tax 特别附加税
superimposed template 叠加模片
superimposed terrace 上叠阶地
superimposed valley 上置谷;叠置谷(地);表生河谷
superimposed wave 叠置波;重叠波
superimposing 叠加
superimposing technique 叠加技术
superimposition 涂布;叠置;叠印;叠加;重接;重叠
superimposition evoked potential 加算诱发电位
superimposition of metamorphism 变质作用叠加
superincumbent bed 悬帮;覆盖层
superincumbent layer 覆盖层
super-individual crystal 超单晶
superinduce 重复诱导
superinduced stream 叠置河
super-inducible mutant 超诱导突变体
super-induction 超诱导;超感应;增加感应;添加
superinfection 过度感染;重复侵染
super-inflate 超膨胀粘胶纤维
super-insulant 超绝缘体;超绝热体
super-insulation 超绝缘(体);超绝热;附加绝缘
super-insulator 超绝缘体
superintegrated memory 高密度集成存储器
superintegrated storage 高密度集成存储器
superintegration 高密度集成
superintendence 管理;监管;监督
superintendency 主管人的职位;监督者的地位
superintendent 主管人;管理(人)员;工务段长;监督员;厂长
superintendent boat 监督艇
superintendent engineer 总轮机长;监理工程师;督察工程师
superintendent of customs 海关监督
superintendent of documents 文件主管人
superintending staff 管理人员
super-Invar 超殷瓦;超因瓦低膨胀系数合金;超恒范钢;超级镍钴钢
superinverse 上逆
super-ionic conductor 超离子导体
super-ionic crystal 超离子晶体
superior 精良;上等的
superior air 高空下沉空气;高层空气
superior alveolar canals 上齿槽管
superior bacteria 高效菌
superior-bore 上位开孔
superior conjunction 上合
superior earnings 超额收益
superior figure 上位(数)字(如指数);上标
superior force 不可抗力
superior function 上函数
superior geodesy 高等大地测量(学)
superior grease 超级润滑脂
superiority 优越性;优势
superior joinery 高级细木作
Superior lake 苏必利尔湖
superior limit 最大限度;上限
superior limit event 上极限事件
superior lobe 上叶
superior lubricant 优质润滑剂
superior mirage 高度晨景【气】;上蜃景
superior mouth 上口鱼
superior passage 上中天
superior planet 外行星
superior quality 品质优等
superior tide 向月潮;上中天潮
superior transit 上中天
superior trochoid 延长摆线
superior wave 主波
superior whorl 上轮
superisocon 分流正摄像管
super-Isoperm 超级恒导磁率铁镍合金;超级导磁钢
superjacent anomaly 上邻异常
superjacent bed 悬帮;上层;覆盖层
superjacent halo 上方晕
superjacent waters 上覆水域;上方水域;上层水域;大陆架水域

super-jet 超音速喷气飞机;超声波速喷气机
super-jet cleaner 气流除杂机
super-key 超码;超关键字
super-large-scale 超大规模
super-large-scale integration 超大规模集成电路
super-large tanker 超级油轮;超级油船
super-laser 高能激光(器)
super-lattice 超晶格;超结晶格子;超格;超点阵;超点群
super-lattice dislocation 超点阵位错
super-lattice high speed switch device 超晶格高速开关器件
super-lattice laser 超点阵结构激光器
super-lattice material 超点阵材料
super-lattice ordering of defects 超晶格有序化
super-lattice quantum well 超晶格量子阱
super-lattice structure 超晶格结构;超点阵结构
super-lattice symmetry 超点阵对称性
super-lattice type magnet 规则点阵型磁铁
super-leak 超漏
super-ledge 超步
super-lethal 超致死量的
super-light insulated sandwich panel 超轻隔热夹芯板
super-light material 特轻材料
super-light particle 超光速粒子
super-linear 超线性
super-linear convergence 超线性收敛
super-linearity 超线性
super-linear variable capacitance transducer 超线性可变电容换能器
super-linear vibration 超线性振动
super-liner 超级邮轮;超级客班轮;超级班轮;超大型客轮
superliner 巨型班轮
super-liquid asset 超速动资产
superload 超载;特定活载;特定荷载
super-lock 超级锁定
super-long dash line 超长虚线
super-long range navigation receiver 超远程导航接收机
super-long range transmission 超长程传输
super-low density sampling 超甚低密度采样
super-low frequency voltmeter 超低频电压表
super-low temperature store element 超低温存储元件
Super-loy 超硬熔敷面用管型焊条
super-luminescene 超发光
super-luminescent diode 超发光二极管
super-luminous star 高光度恒星
super-macroanalysis 超微量分析;超微分析
super-macrocomputer processed moisture tester 超级微机水分测定仪
super-magaluma 超级镁铝(锰)合金;超级铝镁合金板
super-magnet 超强磁铁
super-major bridge 特大桥
super-malloy 超透磁合金;超坡莫合金;超级镍铁磁性合金;导磁合金;镍铁钼超导磁合金
super-many time formalism 超多时格式
supermarket 超级市场;自选市场
supermarket subsystem definition record 超级市场子系统定义记录
supermassive rotator 特大质量旋体
super-master group 超主群
super-mastergroup translating equipment 超主群变换设备
supermatic 高度自动化;完全自动化的
supermatic drive 高全自动化传动;高度自动化驱动;全自动化传动
super-mature 超成熟的;超成熟
super-mechanical 高度机械化的
super-mechanical construction 高度机械化施工
super-mechanized 高度机械化施工的
super-memory gradient method 超记忆梯度法;超存储梯度法
supermetallurgical cement 掺有大量矿渣的矿渣水泥;(掺有大量矿渣的)冶金水泥
super-metal-rich star 特富金属星
super-methylation 超甲基化
super-microcomputer 超微型计算机;超级微(型)计算)机
super-micrometer 超级显微镜;超级测微仪;超级测微计

super-micron mill 超微粉碎机
super-microplanktology 超微浮游生物学
super-microprocessor 超微处理器;超级微处理机
super-microscope 超级显微镜;超级电子显微镜;超倍显微镜;大功率电子显微镜
super-microscopic 超显微镜的
super-miniature 超小型的;超小型
super-miniature capacitor 超小型电容器
super-miniature engineering 超小型工程
super-miniature project 超小型工程
super-miniaturization 超小型化
super-minicomputer 超小型计算机;超微型电脑
super-miser 超级节热器
super-mixer 高速混合器
super-mode competition 超模竞争
super-mode generator 超模振荡器;超模发生器
super-mode laser 超模激光器
super-mode technique 超模技术
super-molecular complex 超分子复合物
super-molecular self assembly 超分子自安装
super-molecular structure 超分子结构
super-molecule 超分子
super-molecule approach 超分子方法
super-monopoly 超级垄断
super-multiplet 超重线;超多重线;超多重态
super-multiplet theory 超多重态理论
super-multiplex structure 超多重结构
super-multiplicity 超多重性
super-mumetal 超导磁合金
super-mumetal alloy 超镍钢合金;超高磁导率镍铁合金
supermutagen 高效诱变剂
supernatant 污泥上层液;上清液;澄液;浮在表面的;浮于上层的
supernatant draw-off pipe 上清排出;上层清液排出管
supernatant flow 上清流量
supernatant fluid 浮面液
supernatant layer 清液层
supernatant liquid 清液层;上层清液;澄液;浮于上层清液;浮液
supernatant liquor 上层清液
supernatant sludge 漂浮泥渣;上层泥渣;澄清液污泥消化
supernate 上层清液
super-national bodies 超国家机构
super-national corporation 超国家公司
super-national organizations 超国家机构
super-neat soap 超液晶皂
super-network 超网络;超级线路网;超级道路网
super-neutrality 超中性
Super-nickel 铜镍耐蚀合金;超级耐蚀铜镍合金
super-nickel alloy 超级镍合金
super-nilvar 超级尼尔瓦合金
super-node 超节点
super-non-expansion steel 超级不膨胀钢
super-normal 超正常;超常的
super-normal conduction 超常传导
super-normal dispersion 超正态离差
super-normal growth 超速成长
super-normal profit 超额利润;超正常利润
super-normal refraction 超正常折射;超常折射
super-normal sign stimulus 超常符号刺激
super-nucleus 超核
supernumerary 超数量的
supernumerary income 额外收益;额外收入
supernumerary pollen tubes 超粉花粉管
supernumerary rainbow 多余虹
supernumerary root 额外根
supernunerary bud 附芽
supernutrition 过量营养
super-octane number fuel 超辛烷值燃料
super-octave 超倍频
super-oilite 超级多孔铁铜合金
super-oilite bearing 多孔铁铜含油轴承
super-oilite machine parts 多孔铁铜机械零件
super-oil tanker 超大型油轮
super-ometer (装在汽车上的)超高测量仪
super-opaque enamel 超乳白搪瓷
super-opaque glaze 超乳白釉
super-optic 超级万能测长机
super-orbital velocity 超轨道速度
super-order 超目;总目
super-ordinate goal 超常目标

superore halo 矿上晕
super-orthicon 超正析像管
super-osculation curve 超密切曲线
super-oxide 超氧化物;过氧化物
super-oxide anion 超氧阴离子
super-oxide ion 超氧化物离子
super-paramagnetic material 超顺磁性材料
super-paramagnetism 超(级)顺磁性
super-parametric(al) elements 超参数元素
superparasite 重寄生物
superparasitic 重寄生的
superparasitism 重寄生(现象)
super-particle 超粒子
super-patent coated board 高级白纸板
super-peak hours coefficient 超高峰小时系数
super-peak hours factor 超高峰小时系数
super-penumbra 超半影
super-performance 超级性能;良好性能
super-period 超周期
super-period effect 超周期效应
superpermafrost 永久冻土之上的;永冻层上的土层
superpermafrost water 永冻土上层水
super-permalloy 超级坡莫合金;超高导磁率合金
super-per mivar 超级坡明伐恒导磁率合金
super-phantom circuit 超幻(像)电路
superphosphate 过磷酸盐
superphosphate fertilizer 过磷酸肥料
super-piezochemistry 超高压化学
super-pipeline 超流水线
super-plastic 超塑性的;超塑料
super-plastic concrete 流态混凝土;超塑性混凝土
super-plastic deformation 超塑性变形
super-plastic flow 超塑性流动
super-plastic forging 超塑性锻造
super-plastic forming 超塑成型
super-plastic forming diffusion bonding 超塑成型扩散焊
super-plastic forming diffusion welding 超塑成型扩散焊
super-plastic forming technology 超塑成形技术
super-plasticity 超塑性
super-plasticity creep 超塑性蠕变
super-plasticized concrete 超塑性混凝土;超塑化混凝土;流态混凝土
super-plasticizer 超增塑剂;超塑化剂;高效塑化剂;高效减水剂
super-plasticizing admix 高增塑剂;高塑化混合物
super-plasticizing admixture 超塑化剂
super-plasticizing concrete 超塑化混凝土
super-plastic material 超塑性材料
super-pneumatic tyre 超压轮胎
super-Poisson distribution 超泊松分布
super-polarizability 超极化率
super-polarization 超极化
super-polishing 超级抛光
super-polyamide 高聚酰胺
super-polyester 高聚酯
super-polymer 超聚合物;超高聚(合)物;高聚(合)物;高分子聚合物
super-poly propylene rope 优质丙纶缆
super-pool in centrifuge 离心机中过水量
super-population 超总体
super-port 超级港口
super portland cement 超级硅酸盐水泥;超级波特兰水泥
superpose 叠置
superposed 位于上部的;叠生的;叠加信号;叠加的;叠覆的
superposed basin 上叠盆地
superposed bud 叠生芽
superposed circuit 重叠电路
superposed consequent system 叠置顺向
superposed dead load 叠加死荷载
superposed drainage 重叠水系;重叠排水系统
superposed fluid 叠加流体
superposed fold 叠加褶皱;重叠褶皱
superposed graph 叠置图
superposed layer 悬浮层
superposed magnetization 叠加磁化;重叠磁化
superposed peak 叠加峰
superposed plant 前置机组
superposed ringing 叠加振铃
superposed ripple mark 叠置波痕
superposed river 后成河

superposed square planting 梅花形栽植
superposed stream 后成河;叠置河
superposed type sFAG accelerator 重叠式扫描场交变陡度加速器
superposition 叠上;重叠;叠置;叠加作用;叠加(信号);叠合原理;重接;重合;被覆
superposition equation 复合方程式
superposition eye 重叠像眼
superposition gravity anomaly 迭重力异常
superposition image 重叠像
superposition integral 叠加积分
superposition law 叠加法则
superposition method 叠置法;叠加法;重叠法
superposition of a storm 暴雨叠置
superposition of flow field 流场叠加
superposition of moments 力矩叠加;弯矩叠加
superposition of potential 势的叠加
superposition of stress 应力叠加
superposition of waves 波叠加;波的叠加
superposition of wave surfaces 波面叠加
superposition principle 叠加原理
superposition rule of potential 势的叠加规则
superposition solution 叠加解
superposition specific heat 叠加比热
superposition structure 叠加构造
superposition theorem 叠加定理
superposition theory 叠加理论
superposition turbine 重叠式汽轮机
super-potential 超势;超电压;超电势
super-power 超功率;超功的;高功率;强功率;特大功率
super-power boiler 特大功率锅炉
super-power electromagnetic 超强电磁吸盘
super-power laser 超(强)功率激光器;高功率激光器
super-power net(work) 特大功率电网
super-power permanent magnetic 超强永磁
super-power plant 超级发电厂
super-power station 超功率电站;超功率电台
super-power system 超级电力系统;超级电力网;特大功率系统
super-power tube 超功率管;强功率管;特大功率管
super-precipitation 超沉淀(作用)
super-premium needle coke 超级针状焦
superpressed plywood 密实胶合板
superpressed wood 密实木材;高压木材
super-pressure 超高压(力);超压(力)
super-pressure boiler 超高压锅炉
super-pressure-high temperature 高压高温
super-pressure lamp 超高压灯
super-pressure plant 超临界压力发电厂;超高压设备
super-pressure pump 超(高)压泵
super-pressure technology 超高压技术
superpress wood 高压胶合板;高密度胶合板
super-profit 超额利润
superproject 特大工程
super-prompt criticality 超瞬间临界性
super-propagator 超传播子
super-proportional 超比例的
super-proportional reducer 超比例衰减器;超比例减薄液
super-protective trade policy 超保护贸易政策
super-proton 超质子;超高能质子
super-pure metal 超纯金属;超纯度金属
super-pure reagent 超净试剂
super-pure silica glass 超纯石英玻璃;高纯石英玻璃
super-pure water 超净水
super-purgative working table 超净化工作台
super-purity 超纯度
super-purity almin(i)um 超纯铝
super-purity metal 超纯金属;超纯度金属
super-push 超推轮
super-quality combination pliers 高级多用钳
super-quantization 超量子化
super-quark 超夸克
super-quench(ing) 超淬火
super-quench oil 高级淬火油;特质淬火油
super-radiance 超发光
super-radiance generator 超辐射发生器
super-radiance loss 超辐射损耗
super-radiant 超辐射的
super-radiant emission 超辐射发射

super-radiant laser 超辐射激光器
super-radiant oscillation 超辐射振荡
super-radiant wiggler interaction 超辐射摇摆互作用
super-radiation 超辐射
super-radiator 超辐射器
super-range rocket 超远程火箭
super-rank gas mine 超级瓦斯矿井
super-rapid hardening cement 高级快硬水泥;超快硬水泥;超高速硬化水泥
super-rate filer 超负荷滤池
super-rate(roughing)biological filter 超负荷(粗滤)生物滤池
super-reaction 超再生
superreducing atmosphere 过度还原气氛
super-refining 超精炼
superrefining process 高度精炼法
super-refraction 超折射;超折光差;过折射
super-refractory 超效耐熔质;超效耐火材料;超级耐火材料
super-regeneration 超再生
super-regeneration reception 超再生接收法
super-regenerative amplifier 超再生放大器
super-regenerative detector 超再生检波器
super-regenerative maser 超再生微波激射器
super-regenerative receiver 超再生式接收机;超再生接收机
super-regenerator 超再生振荡器
superregulator 高灵敏度调节器;极精确调整器
super-relativistic effect 极端相对论性效应
super-repressed mutant 超阻遏变种
super-resolution 超限分辨;超分辨率
super-rocket 超型火箭;超级火箭
super-royal 超王裁
super-rubber 超级橡胶
super-sack 超大袋
super-saline marine 超碱海水
super-saline water 超咸水;强咸水
supersalinity 强咸度
supersaprobic zone 过度污染带;特污带;超腐生带
super-saturated 过饱和的
super-saturated air 过饱和空气
super-saturated deposit 过饱和沉积
super-saturated design 超饱和设计
super-saturated glaze 过度饱和着色氧化物的釉;过饱和釉
super-saturated permafrost 过饱和永冻层
super-saturated solid solution 过饱和固溶体
super-saturated solution 过饱和溶液
super-saturated steam 过饱和蒸气
super-saturated vapo(u)r 过饱和蒸气;过饱和(水)汽
super-saturated water 过饱和水
supersaturation 过饱和(作用);过饱和现象
supersaturation nucleation 过饱和核化
supersaturation ratio 过饱和比;过饱和比(率)
super-scalar technology 超标量技术
super-scanning 超扫描
super-Schmidt camera 超施密特照相机
super-Schmidt meteor camera 超施密特流星摄影系统
super-Schmidt telescope 超施密特望远镜
super-Schmidt telescope system 超施密特望远镜系统
super-scope 超视镜;超宽银幕
superscript 上标
superscript representation 上标表示法
super-scrubber 超级涤气器;超级洗涤机
super-seal repair clamp and coupling 特密封修理管箍和接头
supersecret 绝密
supersecret documents 绝密文件
supersecret file 绝密文件
supersede 免职
superseded edition 废版
superseded map 过时地图
superseded suretyship riders 替代担保附加条款
superseding clause 替代条款
super-selection rule 超选择规则
super-seniority 超年资制度
super-sens 增加感光度;超灵敏度的
super-sensitive 高灵敏度的;超敏感的;超灵敏度的;超灵敏的;增加感光度
super-sensitive clay 过敏性黏土;超敏感黏土

super-sensitive relay 高灵敏度继电器;超灵敏继电器
super-sensitivity 超敏感性;超灵敏度;过敏性
super-sensitization 超色增感;超敏化(作用)
super-sensitizer 超增感剂;超敏化剂
super-sequence 超序列;超层序
superservice station 高级服务站(汽车);高级修车加油站;通用技术保养站
supersession 接替
super-set 超集
super-ship 超级巨轮
super-short base line acoustic(al) positioning system 超短基线音响定位系统;超短基线水声定位系统
super-sicon 超硅靶视像管
super-sifter 超级筛
supersilent motor 高无噪声电动机;绝对无噪声电动机
supersilica brick 高硅质硅砖
super single tire 超宽钢箍;超宽轮胎
super-skyscraper 摩天大楼;特高高耸建筑
super-slurper 超级吸湿材料
supersmooth file 超细纹锉
super-soft 超软牲
super-solenoid 超螺旋管;超螺线管
super-solid 超固体
super-solidification 超凝固现象;超凝固;过凝固(现象)
super-solidus 超固相线的
super-solidus sintering 超固相线烧结
super-solubility 超溶解度;超溶度;过溶度;过饱和现象
super-solubility curve 超溶解度曲线;超溶度曲线
super-solution 超解
supersonic 超音速的;超声波速的;超声波的;超声波;超声
supersonic absorption 超音吸收
supersonic acceleration 超音速加速度
supersonic accretion 超声波速堆积
supersonic aerodynamics 超音速气体动力学;超音速空气动力学;超声波速气体动力学;超声速空气动力学;超声波空气动力学
supersonic aging 超声波时效
supersonic aircraft 超音速飞机
supersonic aircraft engine 超音速飞机发动机
supersonic aircraft oil 超声波航空油
supersonic airfoil 超音速翼型
supersonic airscrew 超音速螺旋桨
supersonic antiship missile 超音速反舰导弹
supersonic atomizer 超声波喷雾器
supersonic axial compressor 超音速轴流压气机
supersonic bias 超声波偏压
supersonic biology 超声波生物学
supersonic boundary surface transmitter 超声波界面变送器
supersonic casting 超声波铸造
supersonic cleaning 超声波清洗
supersonic cleaning tank 超声波清洗箱
supersonic combustion ramjet 超音速燃烧冲压式发动机;超音速燃烧冲压喷气发动机;超音速燃冲击式发动机
supersonic compressor 超音速压缩机;超音速压气机
supersonic control 超音速操纵
supersonic crack detector 超声波裂纹探测器
supersonic dedusting 超声波除尘
supersonic deep drawing 超声波深拉延
supersonic delayline 超声波延迟线
supersonic density transmitter 超声波浓度变送器
supersonic detector 超声波探伤仪;超声波探测器;超声波速探伤器;超声波检测器
supersonic diffuser 超声波速扩压器;超声波漫射器;超声波扩压器
supersonic dispersed oil 超声波分散油
supersonic dust removal 超声波除尘
supersonic echo sounder 超声波回声测深仪
supersonic echo sounding 超声波测深(法)
supersonic echo sounding device 超声波回声测深仪
supersonic engineering 超音速工程;超声工程
supersonic flaw detecting 超声波探伤
supersonic flaw detector 超声波探伤仪
supersonic flaw testing 超音频探伤
supersonic flight 超声波速飞行
supersonic flow 超声气流;超音速流动;超声波速流;超声波流
supersonicflow anemometer 超声波气流风速仪
supersonic flow meter 超声波流量计
supersonic fluorometer 超声波荧光计
supersonic forming 超声波成型
supersonic frequency 超音频;超声波频率
supersonic gas bubble detector 超声波气泡探测仪
supersonic generator 超声波速发生器
supersonic grain refinement 超声波晶粒细化
supersonic hardness tester 超声波硬度测试机
supersonic heat treatment 超声波热处理
supersonic heterodyne receiver 超音频外差式接收机
supersonic impulse transmitter 超声波脉冲传输器
supersonic inlet 超声速进气道
supersonic inspection 超声波探查;超声波探伤;超声波检验
supersonic interferometer 超声波干涉计
supersonic isobar 超声波等压线
supersonic isothermal pachymeter 超声波等温测厚仪
supersonic jet 超音速射流;超声速喷气发动机;超声波射流
supersonic light valve 超声波光阀
supersonic liquid carburizing 超声波液体渗碳
supersonic locator 超声波定位器
supersonic machining 超声波加工
supersonic method 超声波法
supersonic microtome 超声波切片机
supersonic modulator 超声波调制器;超声波调节器
supersonic nitriding 超声波渗氮
supersonic nozzle 超音速喷嘴;超声波速喷管
supersonic oscillations 超声波振荡
supersonic propeller 超音速螺旋桨
supersonic pulse 超声波脉冲
supersonic pulse method (测混凝土强度的)超声波脉冲法
supersonic quenching 超声波淬火
supersonic ramjet 超音速冲压式喷气发动机
supersonic range 超声波频段;超声波段
supersonic reception 超音频接收
supersonic reflection technique 超声波反射技术
supersonic reflectoscope 超声波检视仪;超声波反射探伤仪;超声波反射镜
supersonic relay 超声波继电器
supersonic rupture propagation 超声波破裂传播
supersonics 超声学;超声波速空气动力学
supersonic scanning 超声波扫描
supersonic sound 超声
supersonic sounder 超声波回声测深仪;超声波测深仪;超声波测深器
supersonic sounding 超声波探测(法);超声波测探(法)
supersonicsounding apparatus 超声波检测仪
supersonic sound pulse 超声波脉冲
supersonic source 超声波源
supersonic spectrometry 超声波测谱学
supersonic spectrum 超声波频谱
supersonic speed 超音速;超声波速
supersonic tempering 超声波回火
supersonic test 超声波试验
supersonic testing 超音频测试;超声波检验
supersonic thickness ga(u)ge 超声波测厚计
supersonic thickness meter 超声波测厚仪
supersonic transducer 超声波传感器
supersonic transport 超音速输送;超音速运输
supersonic transporter 超音速运输机
supersonic transport plane 超音速运输机
supersonic turbo jet 超音速透质气发动机
supersonic type detector 超声波检验器;超声波测(水)深计
supersonic velocity 超音速;超声波速
supersonic vibration 超音频振动
supersonic vibrator 超音频振动器
supersonic viscometer 超音速黏度计
supersonic water level measuring equipment 超声波水位仪
supersonic wave 超声波
supersonic wave doppler diagnosing instrument 超声波多普勒诊断仪
supersonic wave drill 超声波钻井
supersonic welding 超声波焊接(法);超声波焊
supersonic welding inspection 超声波焊缝检查
supersonic wind tunnel 超声波风洞
supersonic working 超声波加工
super-sound 超音;超声
supersound projector 大功率扬声器
super-space 超空间
super-spacer 超级胶条
super-span 超跨
super-sparsity 超稀疏
super-species 超种
superspeed 超高速的;超高速
super-speed film 超感光度胶片;高感光度胶片
superspeed filter 高速过滤池
superspeed filtration 超速过滤;高速过滤
superspeed filtration method 高速过滤池法
superspeed filtration process 高速过滤池法
superspeed filtration tank 高速过滤池
super-spiral 超螺旋
super-spiromatic 高级手表游丝自动定长仪
super-stability 超稳定性
super-stabilizer 超稳定器;超稳定剂
super-stable orbit 超稳定轨道
super-stage 超级
super-stainless 超级不锈的
super-stall 超失速
super-standard 超标准(的);超级标准
super-standard kraft wrapping paper 超标牛皮包装纸
super-standard news 超标新闻纸
super-standard propagation 超标准传播
superstation 特大型发电厂;特大功率电台
super-steel 超级钢;超钢
super-steep asphalt 高软化点浸渍沥青
super-stereoscopic effect 超立体感
super-stoichiometric 超化学计量的;过当量的
super-ston 超级斯通合金
super-strain 超应变
super-stratosphere 超平流层
superstratum 覆盖层;上覆层;表面层;[复]superstrata
super-strength 超强度的;超高强度;特大强度
super-strength steel 超高强度钢
super-stress 超应力
super-string 超弦
super-string theory 超弦理论
super-strong carbon fibre 超强力碳纤维
super-strong focusing 超强聚焦
super-strong interaction 超强相互作用
super-strong stimulus 超强刺激物
superstruction 上部建筑;上部结构
super-structural wall 矮挡土墙(桩基承台上的)
superstructure 地上结构;超结构;超型结构;超点阵;轨道线路;浅层构造【地】;上部建筑(物);上部结构;上部建筑;表壳构造
superstructure charges 上部建筑费用
superstructure concrete 上部结构混凝土
superstructure cost 上部建筑成本
superstructure deck 上层建筑甲板;船楼甲板
superstructure facility 上部设施
superstructure freezing 浅层冻结
superstructure freezing soil 浅层冻土
superstructure line 超结构线
superstructure of bridge 桥梁的上部结构
superstructure of track 线路上部建筑
superstructure of upright wall 直墙水上部分
superstructure of vertical breakwater 直立式防波堤上部结构;直立堤上部结构
superstructure paint 船上层涂料
superstructure work 上部设施;上部建筑物
superstructure works 上部结构工程
supersulfate 高硫酸盐
supersulfated cement 高硫酸盐水泥
supersulfated(slag) cement 高硫酸盐(矿渣)水泥;石膏矿渣水泥
supersulphate = supersulfate
super-sulphated cement 富硫酸盐水泥;高硫酸盐水泥
super-sulphated metallurgical cement 富硫酸盐矿渣水泥
super-sulphated slag cement 富硫酸盐矿渣水泥;石膏块材;石膏矿渣水泥
super-sunernova 超超新星
super-super-computer 超巨型计算机
super-suppressor 超抑制因子
super-surfacer 超级刨床

super-symmetric(al)transformations 超对称变换
super-symmetry 超对称性
super-symmetry charge 超对称荷
super-synchronous 超同步的
super-synchronous motor 超同步电动机
super-synchronous speed 超同步速度
super-synthesis aperture 超综合孔径;地球自转综合孔径
super-system 超系(统);超级系统
super-tall building 超高层建筑
super-tanker 超级油轮;超级油船;超级油槽车;超大型油轮;巨型油轮;大型油船
super-tanker wharf 超级油轮码头
supertax 累进所得税;特别附加税;附加税;附加累进(所得)税
super-tension 超限应变;过电压
super-tension network 超高压电力网
superterranean 架空;地上的;地面的
Supertex 超塔克斯(一种专卖的无缝地板)
super-thermal 超热(的)
super-thermal diesel 带中间冷却器的增压柴油机
super-thermal electron tail 超热电子尾
super-therm alloy 超高温合金
super-thermostat 超恒温器
super-thickener 超浓缩机
super-thin asbestos paper 超薄石棉纸
super-thin box fan 超薄型箱式电扇
super-thin speaker 超薄型扬声器
super-threshold 超阈(值)
super-time relay 超级时间继电器
super-trace analysis 超痕量
super-train 超高速火车
super-transuranics 超铀元素
super-truck 巨型载重汽车
super-turbulent flow 超紊流;超湍流
super-turnstile antenna 超绕杆天线;三层绕杆式天线;蝙蝠翼形天线
super-twisted nematic liquid crystal 超扭曲液晶
super-tyre 超级轮胎
super-unification 超统一
super-universal shunt 超万能分流器
superunmerary 额外的
super-varnish 超级清漆;桐油清漆
super-vector computer 超级矢量计算机
super-velocity 超速度;超高速
super-velocity of light 超光速
supervene 附加;并发
super-ventilated foil 超通气水翼
super-video graphic(al)array 增强型图形适配器
superview cab(in) 瞭望机舱;瞭望车厢
super-virulent 超毒力的
super-viscose fibres 超强粘胶纤维
supervised alarm system 监视报警系统
supervised classification 监督分类(法)
supervised image classification 监督图像分类
supervised operation 控制操作
supervised sprinkler system 有监控的喷水灭火系统
supervising architect 监理建筑师;监督建筑师
supervising authority 领导机关;领导单位
supervising(control)system 监控系统;管理系统
supervising device 监视设备
supervising engineer 主管工程师;监理工程师;监督工程师
supervising expenses 监工费
supervising point 观察点
supervising program(me) 管理程序
supervising staff 指导人员
supervising system 管理系统
supervising the duration of fees 费用期限监视
supervision 管理;监视;监理;监督;施工监督
supervision brief 监理大纲
supervision by statistic(al)means 统计监督
supervision computer control 计算机监督控制
supervision contract 监理合同
supervision control 监督控制
supervision cost 监理费
supervision engineer 监理工程师
supervision in advance 事先监督
supervision of construction 施工监督
supervision of open mining 露天开采地质管理
supervision of underground mining 地下开采地质管理
supervision of works 工程监理;工程监督

supervision organization 监理组织
supervision outline 监理大纲
supervision plan 监理规划
supervision program(me) 监控程序
supervision program(me) 监理规划
supervision room for radioactive rays 放射线管理室
supervisor 主管人;管理人(员);管理器;管理程序;领工员;检查员;监视器;监理员;监控制装置;监控者;监控器;监工(员);监督员;监督人(员)
supervisor call 管理程序调用;管理程序调入;请求管理程序;调入管理程序
supervisor call code 管理程序调入码
supervisor call instruction 管理程序调用命令;管理程序调入指令;访管指令
supervisor call interrupt 管理程序调用中断;调入管理程序的中断;进入管理程序中断;监控程序请求中断
supervisor call interruption 管理程序调用中断;管理程序调用命令中断
supervisor call program(me) 管理调用程序
supervisor call routine 管理程序调用命令子程序;访管例行程序
supervisor communication macro 管理通信宏指令
supervisor communication region 管理通信区;管理程序通信区
supervisor console 监视控制台
supervisor control program(me) 管理控制程序
supervisor control system 监视控制系统
supervisor interrupt 管理程序中断
supervisor lock 管理程序锁(定);管理程序封锁
supervisor mode 管态;管理程序状态;管理(程序)方式
supervisor of works 工程现场监工
supervisor operation system 管理操作系统
supervisor overlay 重叠管理程序;管理覆盖程序
supervisor-privileged instruction 管理特许指令
supervisor program(me)test 管理程序测试
supervisor queue area 管理排队区;管理程序队列区
supervisor register 管理寄存器
supervisor request block 管理请求分程序
supervisor resident area 管理程序常驻区域;管理程序驻留区
supervisor routine 管理程序的例行程序
supervisor's desk 监视操作台;监控台
supervisor services 管理服务程序;管理服务
supervisor staff 监督人(员)
supervisor state 管态;管理程序状态
supervisor status 管态;管理程序状态
supervisor test program(me) 管理程序的试验程序
supervisor transient area 管理程序暂驻区
supervisory 监控
supervisory activity 监督活动
supervisory agency 监督管理机构
supervisory and alarm system 监视和警报系统
supervisory authorities 监督管理机构
supervisory authority 管理机构
supervisory capability 管理能力
supervisory circuit 监视电路;监控电路
supervisory computer 管理计算机;监视用计算机
supervisory computer control 计算机监督控制
supervisory computer control system 计算机监督控制系统
supervisory console 管理控制台;监控台
supervisory control 监督控制(系统);监控用的控制系统;监控;监督控制
supervisory control and data acquisition 监控和数据采集
supervisory control and data acquisition system 监控与数据捕获系统
supervisory control computer 监控计算机;监督控制计算机
supervisory control desk 控制台
supervisory control disk 管理控制盘
supervisory controlled robot 监控机器人
supervisory control printer 监控打印机
supervisory control signal 监控信号
supervisory control system 监控系统
supervisory equipment 监视设备
supervisory frame 管理画面
supervisory gear 检测监视装置

supervisory instruction 管理指令
supervisory keyboard 管理键盘;控制键盘;监控键盘
supervisory lamp 监视灯
supervisory level 管理级别
supervisory light 监视灯
supervisory memory 监视存储器
supervisory methods 监督管理方法
supervisory mode 管态
supervisory monitoring 监视性监测
supervisory office of fishery 渔政监督管理机构
supervisory operating system 管理操作系统
supervisory panel 监视信号盘
supervisory personnel 管理人员;监视人(员)
supervisory policy 监督管理政策
supervisory powers 监督管理权力
supervisory process control 管理过程控制
supervisory program(me) 管理程序;监督程序
supervisory program(me) simulation 管理程序模拟;管理程序仿真
supervisory record 监视记录
supervisory relay 管理转接;监视继电器
supervisory role 监督管理作用
supervisory routine 管理例行程序;管理程序
supervisory sequence 管理序列;管理顺序;监督序列
supervisory service 监视设施;管理业务;监听业务;监督服务
supervisory signal 管理信号;监视信号
supervisory staff 管理人员
supervisory sub-routine 管理子程序
supervisory system 管理系统;监视系统;监理系统;监督系统;监管系统
supervisory tone 监控音
supervisory unit 管理设备
super-vital mutant 超活力突变型
super-vital reaction 超生反应
super-voltage 超高(电)压【电】;超电压
supervoltage radiation 高压辐射;高能辐射
super-water proof multi coated lens 超级防水加硬多涂层光学镜片
super-wave flow 超波速流
super-way computer 超大规模计算机
super-weak force 超弱作用力
super-weak interaction 超弱相互作用
super-weak luminescence 超微弱发光
super-welder 超精密焊机;超级焊机
super white cement 超白水泥
superwhole 超完整的;超全的
super-wide-angle 特宽角
super-wide-angle camera 特宽角摄影机;超文角照相机
super-wide-angled object glass 超宽视角物镜
super-wide-angle lens 特宽角镜头
super-wide-angle multiplex 特宽角多倍仪
super-wide-angle photograph 特宽角像片;超广角摄影
super-wide-angle photography 特宽角摄影;超广角摄影术
super-wide-angle projector 特宽角投影器
super-wide array 超宽组合
super-wide band oscilloscope 超宽频带示波器
superzigzag decorative stitch 高级曲折形装饰线迹
super-zone 超区;超带
super-β-transister 超贝塔晶体管
suphate bacteria 硫酸细菌
suphohalite 凶钠石
supine position 背卧位
Supiron 高硅耐酸铁
supper club 高级夜总会;夜总会
supple falls short of demands 供不应求
supple grille 送风格栅
supplement 增补;添加物;附录;附加物;附册;补遗
supplemental 补足的
supplemental agreement 补充协议
supplemental air 储气;补呼气
supplemental appropriations 追加拨款
supplemental area 公寓辅助面积(包括、大厅、会议室、卸货台、游泳池等)
supplemental authorization 补充协议;补充协定
supplemental borehole logging 辅助钻孔测井
supplemental capacity 补供输出功率
supplemental colo(u)r 补色
supplemental condition 补充条件;补充条款

supplemental contract 合同补充书;补充合同
supplemental control point 辅助控制点
supplemental cost 追加成本
supplemental credit 追加信贷;附加信贷
supplemental deed 补充契据
supplemental elevation 过渡高程
supplemental equipment 补充设备
supplemental estimates 补充概算书
supplemental factor 补助因子
supplemental firing 强化燃烧
supplemental grant 补充拨款
supplemental grinding 补充研磨
supplemental illumination 辅助光照
supplemental information 增补信息
supplemental information report 补充资料报告
supplemental irrigation 辅助灌溉;补水灌溉;补充(性)灌溉
supplemental irrigation water 补充灌溉水
supplemental loan insurance for multifamily housing 多户住房的追加贷款保险
supplemental material 辅助物料
supplemental medium 补给性培养基
supplemental pasture 辅助牧草;补充牧场
supplemental pipe 辅助管(道)
supplemental post for survey monuments 指示标桩
supplemental pressure 附加压力
supplemental services 补充服务项目
supplemental survey point 补充测点
supplemental test 补充试验
supplemental trimming filter 辅助微调滤光片
supplemental water 补给水;补充灌溉水
supplemental water-supply sources 辅助供水水源
supplementary 副的;附加的;附带;辅助信息;辅助的;补足的
supplementary account 补充账户;辅助账户
supplementary aerodrome 备降机场
supplementary agreement 补充协议;补充协定
supplementary air 辅助空气;补助空气;补充空气
supplementary air valve 补充气阀
supplementary allowance 补充免税额
supplementary alternate 补充投标;补充比较方案
supplementary altitude 过天顶高度
supplementary angle 辅助角;补角;互补角
supplementary appropriations 追加拨款
supplementary arc 余弧
supplementary balance sheet 补充资产负债表
supplementary bar 辅助钢筋
supplementary bearing 副轴承
supplementary bench mark 辅助水准点;水准补标点;补充水准点
supplementary blasting 二次爆破
supplementary boiler 辅助锅炉
supplementary borehole 补充钻孔
supplementary brake 辅助制动机
supplementary budget 追加预算;补充预算
supplementary budget for annual receipts 追加岁入预算
supplementary budget for expenditures 追加支出预算
supplementary cable 辅助钢缆
supplementary call 增收保险费
supplementary calling frequency 补充呼叫频率
supplementary calls of premium 追加保险费
supplementary calls of remittance 追加股款
supplementary chord 补弦
supplementary civil compensation 附带民事赔偿
supplementary clause 辅助条款
supplementary combination of load 荷载的附加组合
supplementary computer 辅助计算机
supplementary computer room 辅助计算机房
supplementary condition 辅助条款;辅助条件;补充条件;附加条件
supplementary contour 间曲线;辅助(性)等高线;半距等高线
supplementary contract 合同补充书;补充契约;补充合同
supplementary control 辅助控制
supplementary control point 辅助控制点;补充控制点;图根点
supplementary cooling 补充冷却
supplementary cost 附加成本;辅助成本
supplementary course 补充课程

supplementary data 补充数据
supplementary data report 补充数据报告
supplementary design formula 补充设计公式
supplementary diaphragm pump 备用隔膜泵
supplementary direction(al)sign 辅助指路标志
supplementary dividend 追加股息;附加股息
supplementary drawing 补充图
supplementary earning 补充收益
supplementary element 附加成分
supplementary equipment 配套设备;辅助设备
supplementary estimate 追加预算;补充预算;追加概算;补充概算书
supplementary experiment 补充试验
supplementary exploration 补充查勘
supplementary exposure 附加曝光;补充曝光
supplementary factor 补助因子
supplementary fare 附加票价
supplementary feed 补充给水
supplementary fertilizer 补充肥料
supplementary financial information 补充财务信息
supplementary financing 补充资金
supplementary financing facility 补充贷款
supplementary financing to previous credits 补充信贷资金
supplementary-fired steam generator 补充燃烧式蒸汽发生器
supplementary flux 炉外熔剂
supplementary fuel 垃圾燃料;辅助燃料;补充燃料
supplementary gear box 副变速箱
supplementary geologic(al)report 补充地质报告
supplementary group 辅助线组
supplementary grouting 辅助灌浆
supplementary guarantee 补充担保
supplementary heating 补充供暖
supplementary heat source 辅助热源
supplementary hole 辅助眼
supplementary horizontal control 辅助平面控制
supplementary humidification 补充加湿
supplementary income 附加收入
supplementary income tax 追加所得税
supplementary instruction 补充指示
supplementary instrument 辅助仪器;辅助仪表
supplementary insulation 补充绝缘
supplementary irrigation 辅助灌溉;补充灌溉
supplementary item 补充项目
supplementary judgment 补充判决
supplementary labo(u)r 附带劳动
supplementary labo(u)r power 追加的劳动力
supplementary layer of felt 附加油毡层;辅助油毡层
supplementary lens 附加透镜;附加镜头;附加镜
supplementary lighting 辅助照明;附加照明;补充照明
supplementary light source 补充光源
supplementary line 辅助线
supplementary list 补充项目表
supplementary live steam nozzle 辅助进汽嘴
supplementary live steam pipe 辅助进汽管
supplementary loading 附加荷载
supplementary loan 补充贷款
supplementary log book 补充航海日志
supplementary loss 附加损失;附加损耗
supplementary maintenance 附加维修;辅助维修;补充维修
supplementary maintenance time 附加维修时间;辅助维护时间
supplementary mark 辅助标志
supplementary means 辅助手段;辅助工具
supplementary mill 辅助轧机
supplementary note 补充说明
supplementary observation 辅助观测;补充观测
supplementary occupation 辅助职业
supplementary parts 补充配件
supplementary pasture 补充草地
supplementary petroleum duty 补充石油税
supplementary photography 补充摄影
supplementary pipe 备用管
supplementary plate 备用印刷版
supplementary point 补点【测】;补充点
supplementary power 补充供电;补充电力
supplementary power supply set 附加电源装置
supplementary pressure 附加压力
supplementary production 副业生产

supplementary program(me) 辅助程序
supplementary program(me) budget proposal 概算补偿方案
supplementary property 次要性能
supplementary provision 补充条款;补充规定
supplementary pump 辅助泵;备用泵
supplementary radial arm 辅助旋臂
supplementary rate 补充比率
supplementary ratios 补充比率
supplementary recovery method 补充开采法
supplementary regulation 补充规定
supplementary relay 辅助继电器
supplementary reserve 附加准备金
supplementary reservoir 辅助容器;辅助泥浆池
supplementary resource 补充资金
supplementary role 辅助作用
supplementary scheme of triangulation 三角测量补充网;补充三角网
supplementary service 补充业务
supplementary set 补集
supplementary special deposit 补充特别存款;辅助性特别存款
supplementary specification 补充说明
supplementary speed reduction means 辅助减速装置
supplementary spring 附加保险弹簧;辅助弹簧;保险弹簧
supplementary standard colo(u)rimetric system 补充标准色度系统
supplementary statement 辅助报表
supplementary station 辅助站
supplementary steam valve 进汽补充阀
supplementary storage 辅助存储器;辅助存储法;辅助存储
supplementary structure 辅助结构
supplementary subroutine 辅助子程序
supplementary support 附加支护
supplementary survey 补充测量
supplementary survey station 补充测站;辅助测站
supplementary switch 辅助开关
supplementary symbol 补充符号
supplementary table 辅助表;辅助用表;补充表(格)
supplementary tank 补充水柜
supplementary tax 附加税
supplementary tax sensitivity 附加税敏感性
supplementary tax survey 附加税调查表
supplementary tax types 辅助税种
supplementary tide ga(u)ge 补充验潮仪
supplementary tide station 补充验潮站
supplementary treatment 补充处理
supplementary triangulation 填充三角测量;补充三角测量
supplementary tube 备用管
supplementary unemployment benefit 补偿失业福利;辅助失业救济金
supplementary valve 备用阀
supplementary ventilation 辅助通风
supplementary vertical control 辅助高程控制
supplementary volume 补卷;补编
supplementary wage 附加工资
supplementary wall 附加墙
supplementary water 补充水
supplementary water injection 补充注水
supplementary zoning 补充区划
supplementation 补充
supplement cartographic(al)document 补充制图资料
supplement cost 追加成本
supplemented subspace 补充子空间
supplement exploration of mine district 矿区补勘
supplement investigation 补偿勘察
supplement of an angle 补角
supplement of an arc 补弧
supplement report 补充报告
supplements to wages and salaries 附加薪资
supplement supporting graphics 战术目标图
supplement ticket for air conditioner 空调票
supplement to authorized signature 补充印鉴
supplement to sailing directions 航路指南补编
supplement to the draft environment(al) impact statement 环境影响报告书草案附录
supplement transport of goods 货物补送
supplement unemployment benefits 补充失业救

济金
supplementzl equipment 辅助设备要求
supplied air suit 气衣
supplied by the owner 物主提供;业主提供
supplied free issue 赠阅书刊
supplied population 给水人口
supplier 供应者;供应商;供应方;供应厂商;供货者;供货商;供货人;供应者;技术输出方;提供者;承制厂;厂商
supplier benefit 供给者效益
supplier's account 卖主账(户)
supplier's catalogue 供应者目录;厂商目录
supplier's credit 信贷供应者;卖主信贷;卖方信贷
supplier's declaration 供货商声明
supplier's factory inspection report 供货商的工厂检验报告
supplier's invoice 供货商发票;供货人发票;供货方发票
supplier's letter of guarantees 供货人保证信函;供货商保证信函
supplier's technical personnel 供货商的技术人员
supplies 供应品;物品;物料;补给品
supplies compass 供货界限;供货范围
supplies expenses 物料费用
supplies inventory 物料盘存;辅助材料储备
supplies on hand 库存物料
supplies under category 2 二类物资
Supplies under category 1 一类物资
supply 供应;供给;给料;填补;电源;补给
supply a deficiency 补缺
supply air 送风
supply air boot 供气罩
supply air duct 送风管道
supply air duct system 送风管道系统
supply air fan 供气风机
supply air flue 空气供应
supply air grille 送风栅;供气栅
supply air hole 供气孔
supply air line 供气管路
supply air outlet 送风口
supply air pipe 供气管(道)
supply air rate 送风量
supply air system 送风系统
supply air temperature difference 送风温差
supply air tube 供气管(道)
supply and demand equilibrium 供求均衡
supply and disposal services 上下水道设备
supply and erection 供应与装设
supply and installation 供应与安装
supply and maintenance 供应与维修;补给与维修
supply and marketing contract 供销合同
supply and selling accounting 供销会计
supply and transport 供应与运输
supply area 服务区;供应区;补给区
supply authority 电源管理机构;能源管理局
supply barge 供应驳
supply base 补养基地;补给基地
supply basin 给水池
supply bay 电源架
supply bin 供应箱;供应仓
supply boat 供应船
supply bond 供应保证书;供应契约
supply branch 供水支管
supply bridge 馈电电桥
supply buoy 补给浮标
supply cable 供电电缆;输电线
supply cage 供料罐笼
supply canal 给水渠;供水渠
supply capacity 供货能力
supply car 台车
supply cell 供电间隔
supply center 供应中心
supply chain 供应链
supply chain management 供应链管理
supply channel 供水渠
supply channel deposit 补给水道沉积
supply channel irrigation 供水渠灌溉
supply cistern 供水池;给水池
supply coefficient 供水系数
supply colter 栽植开沟器
supply column 供应塔
supply conduit 供水管道;给水管道
supply conduit diversion 改移供水管道
supply conduit supplement 增补供水管道

supply constraint 供给约束条件
supply container 供应容器
supply contract 供应合同
supply control 补给品管理制
supply conveyer 送货带;送货机
supply cord 电源(软)线
supply corporation 供应公司
supply correspondence 供应对应
supply current 供电电流;馈电电流;电源电流;补流;补充电流;补偿气流;补偿海流
supply delay time 供应等待时间;待用时间
supply-demand model 供需模型
supply-demand price 供求价格
supply-demand relations 供求关系
supply department 供应部门
supply depot 供应站;补给仓库
supply detection circuit 供电电流制检测电路
supply district 供电范围
supply ditch 犁沟;畦沟;供水渠
supply duct 供应管道;送风道
supply equipment 供应设备;供水设备
supply exceeds demands 供过于求
supply exhaust ventilation 抽压联合通风
supply factor 供水系数
supply failure 供应间断;供应中断;断电
supply fan 进气风扇;送风机
supply farm reservoir 提供农田用水的水库
supply film cassette 送片暗盒
supply fittings 供水零配件
supply floor 供应楼层
supply flume 供水槽
supply frequency 供电频率;电源频率
supply gate 供料门;补充眼
supply glassing bag 提供玻璃纸袋
supply grill(e) 进气花板板;进气格栅
supply head 馈电线
supply header 料集流管;加料集流管
supply heat 供热
supply hose 供水软管
supply hub 电源插孔
supply hydrogenating solvent method 供氢溶剂法
supply in accordance with special needs 特需供应
supply in full set of 成套供应
supplying branch 供水支管
supplying conduit 供水管
supplying current time 供电时间
supplying pipe 供水管
supplying tube 供水管
supply installation 供应装置
supply irrigation hydrant 提供灌溉龙头
supply irrigation layout plan 提供灌溉系统图
supply jet 供给射流
supply level 供应面;供应面;供水水平
supply line 供水管(线);供应管(道);供电线路;馈电线路;补给线
supply load 供电负载
supply lock arm 供带锁定臂
supply main 供电网;供给干管;供水干管;传输干管;总供水管;总供气管;供水总线;供水总线;供电干线;给水总管;馈电干线
supply main manifold 供应总管
supply main pipe 提供输水管
supply meter 用户电表;电量表;供电电度表;馈路电度表
supply mother seed 提供原种
supply net 供电网(络);给油管网;供水管网;给气管网
supply network 供应网络;供电网(络);馈电网
supply nozzle 供给喷嘴
supply of commodities 商品供应
supply of equipment 设备供应
supply of equipment and materials 设备材料供应
supply of equipment contract 设备供应合同
supply of equipment with erection contract 设备供应和安装合同
supply of fuel 燃料供应
supply of goods 货源
supply of heat 供热
supply of manpower 人力供应
supply of materials 材料供应
supply of means of production 生产资料补给
supply of people 人的供给

supply of personnel 职员的提供;人员的提供
supply of power 电力供应
supply of supply 动力供应
supply of technical information agreement 提供技术情报协议
supply of the market 市场供应
supply of water 给水;供水
supply oscillator 信号源
supply parent material 提供原始材料
supply passage 供应渠道
supply pedigree seed 提供原种
supply perforated sprinkler 提供多孔管喷灌装置
supply pipe 给水管;供应管;给水管道
supply pipeline 供应管道;供水管道
supply piping 供给管线
supply plant 供应厂;供应装置
supply point 供应站;供应处;供应点;补给点
supply pressure 电源电压;供水水压;供应压力;供给压力
supply pressure level 供给压力值
supply pressure null shift 供应压力零漂
supply price 供应价格;供货价格
supply protable sprinkler 提供活动喷灌装置
supply pump 给水水泵;供油泵;供给泵
supply ramp 供货斜面
supply range of complete set 成套供应范围
supply rate 供水率;供气率;补给率
supply reel 供带盘;供带卷轴;绕线车
supply register 调节供气门
supply relay 供电继电器
supply reservoir 供应罐;供风风缸;供电箱
supply-return diffuser 送吸两用风机
supply riser 供水竖管;供水立管;给水竖管;给水立管
supply road 送货道
supply room 供应室
supply route 供应线
supply schedule 供给表
supply section 供电段
supply service base 供应基地
supply service quay 物资码头
supply services 供电线;服务设施
supply service terminal 物资码头
supply service wharf 物资码头
supply shaft 送货竖井
supply ship 供水船;供应船;物资供应船
supply side 供应方面;供电侧;馈电侧;进给端
supply-side subsidy 供方补助
supply socket 电源插座
supply soil sample 提供土样
supply source 供应源
supply spool 供片盒;供卷筒
supply station 供应站;供电站;给水站
supply stock seed 提供原种
supply stor(e)y 供应楼层
supply subsurface water 提供地下水
supply system 送风系统;供给系统;供应系统;供水系统;供气方式;通风系统;补水系统;包干制
supply tank 储[贮]液槽;储[贮]槽;供油箱;供应油罐;供应槽;供水箱;供水柜;给水箱;给水池
supply technical data 提供技术资料
supply tender 供应船
supply terminal 供电点;馈供终端;电源接线端
supply to seal 供给密封
supply to ship 船舶供水;船舶供电
supply transfer unit 补给传递器
supply transformer 供电变压器;馈电变压器;电源变压器
supply tube 供给管
supply tunnel 供水隧洞;供水隧道
supply underground water 提供地下水
supply undertaking 公用事业
supply unit 供电装置;供电设备;电源装置;电源部件;电源部分
supply valve 供应阀;供给阀
supply valve spring 供给阀簧
supply variation 供电波动
supply ventilation 引入式通风;给气式通风
supply voltage 供电电压;电源电压
supply voltage indicator 供压指示器;电源电压指示器
supply voltage line 电源电压输送线
supply voltage variation 电源电压变动
supply water 供水

supply water conduit 给水管
supply water main 供水干管
supply water management scheme 提供排水计划
supply water pipe 给水管
supply-water pump 供水泵
supply water temperature 供水温度
supply water tube 给水管
supply well 供水井
supply wire 供电线
supply without cost[obligation] 无条件供应
supply works 供水工程
supply zone 供应范围
support 支座；支援；支护；支持物；支持(器)；支架；支承；赞助；光学座；托臂；扶助
supportability 支承能力；可支援性
supportable 可支承的；能支承的；可支持的；可援助的；可拥护的
support abutment 支座
support analogy 柱比拟(法)；支柱比拟法
support anchorage 支柱锚固
support and registration from drop tube 隧道吊管上的支援与定位
support arm 支架臂；支臂(架)
support arm shaft 支臂轴
support a single floret 支撑单位
supportation 支承式接头
support bar 支杆；支承杆(件)
support base 柱础；柱基；柱基(础)
support beam 简支梁；支座梁；支撑梁
support bearing 支座
support bent 柱排架
support blade 支撑叶片
support block 支柱
support bolt 支螺栓
support boss 支撑套筒
support bracket 支承吊板；支承牛腿；承托架；支托板
support bracket bolt 支(托)架螺栓
support by rock anchoring and shotcreting 锚喷支护
support cage 支承轮毂
support cage carrier 支承轮毂架
support cap 柱帽；支柱帽
support capacity 支撑承载力
support casing 柱周包围物；柱筒
support center 供应中心
support chip 支撑片
support circuitry 支架电路
support clamp 支承夹子；支承夹具
support coated open tubular column 载体涂层开管柱；涂层载体开关柱
support column 支承管；支承杆柱
support concrete 支承混凝土
support condition 支撑条件
support connection 柱的连接；柱连接
support constraint 支承约束
support contour (履带起重机的)支承轮廓
support core 柱芯
support cost of tunnel 巷道支护费用
support creep 柱的蠕动
support cylinder 支撑油缸
support deed 赡养契约
support density 支柱密度；载体密度
support design 柱设计；支座设计；支柱设计
support device 支护装置；支撑装置
support dimension 柱的尺寸；支柱尺寸
support document 证件
supported adhesive 载体胶粘剂
supported along four sides 四边支承(的)
supported at both ends 两端支承的
supported at circumference 圆周支承
supported at edges 沿边支承
supported at two edges 两边支承
supported bar 支承杆
supported beam 支承梁
supported bedplate 支承底板
supported boom 支承臂
supported by the meteorology development 在气象部门的协助下
supported cable 支撑电缆
supported catalyst 负载型催化剂
supported characteristic 支护特征
supported craft 辅助船舶
supported crane 支承起重机
supported density 支架密度

supported diaphragm 支承隔板
supported end 支承端
supported face 支护工作面
supported film adhesive 胶膜；膜片胶粘剂
supported flange 压紧法兰
supported flange joint 固定突缘接头
supported joint 支承(式)接头；承托接头；支垫式钢轨接头
supported liquid membrane 支撑液膜
supported member 支承构件；支承杆件
supported membrane catalyst 负载型膜催化剂
supported nanometer-sized titanium dioxide photocatalyst 负载型纳米二氧化钛催化剂
supported nanoscale zero-valent iron 负载型纳米零价铁
supported on four sides 四边支承的
supported piston 平衡活塞
supported rail joint 支承式钢轨接头
supported screwed joint 支撑螺丝接头
supported shaker 支架式振动运输机
supported-spindle gyratory crusher 支承轴式旋回破碎机
supported-spindle(type)crusher 支轴式破碎机
supported-spindle(type)gyratory 支轴式旋回破碎机
supported type abutment 轻型桥台；支撑式桥台
support effect 载体效应
support element 支架构件
support end moment 柱端弯矩
support equipment 支护设备；支持设备；地面维护设备；附属设施；附属设备
supporter 支脚；支持者；支撑物；载体；支架
supporter combustion 助燃剂
supporter wages 支护工人工资
support facility 支承设备；地面维护设备；附属设施；附属设备
support facing 柱面装饰；支柱饰面
support fitting 支持金具
support flange 支撑凸缘
support foot 柱脚
support footing 支柱基础
support for conductor-rail cover 导电轨罩支撑
support for gyrosphere 陀螺球座
support forming 柱模
support forms 柱体模壳；柱体模板
support formwork 柱型；柱模
support for soffit conductor rail with damping element 带减震装置的拱腹式导轨支撑
support foundation 支柱基础
support frame(work) 支架；支撑框架
support-free 无支架；无柱
support from workshop roof 车间悬挂支撑
support function 支撑函数
support graph 支座图
support grid 支承栅板
support grid pattern 柱的格式布置
support guard 柱的缘饰
support guide 支架导杆
support hardware 支援硬件；支持硬件
support head 柱头板；柱帽
support height 柱高；支柱高
support hinge 柱上铰链
support hotline 支援热线
support inerval 支柱间隔
supporting 支承
supporting a fence 支撑篱笆
supporting angle iron 支承角钢；支角铁；支撑角铁
supporting angle(steel) 支承角钢
supporting area 支承面积；支承面
supporting arm 托臂；支持臂；撑臂
supporting axis 支承轴
supporting axle 承载轴
supporting back 支背
supporting bar 支承杆；支撑杆
supporting base 衬底
supporting baseplate 支承底板
supporting bead 支承珠；支承垫圈
supporting beam 支承梁
supporting bearing 支座轴承
supporting bedplate 支承底板
supporting berm 压坡棱体平台；支承棱体平台
supporting block 承重砌块；支持凸轮；支承断块；支座；支柱
supporting body 支撑物；承载体

supporting bracket 支承架
supporting brick 承重砖
supporting bridge 承桥
supporting bulkhead 支承舱壁
supporting caisson 支承沉箱
supporting capability 承载能力
supporting capacity 承载能力；支承能力；承重量；承载量；承载力
supporting capacity of a pile 桩的承载能力
supporting capacity of ground 地面承载能力
supporting channel U 形柱；槽式立柱
supporting clay brick cross wall 承重砖隔墙
supporting clip 支持环箍
supporting column 支承柱
supporting cone 支承锥体
supporting construction 支承构筑物；支承结构
supporting course 支承层；持力层；承重层
supporting culture 支撑栽培
supporting cylinder 支持圆柱体；支承圆柱体
supporting device 支承器件(混凝土路面接缝)；支承装置
supporting device for slewing part 旋转部件支承装置
supporting dielectric 支持电介质
supporting disk 支承盘
supporting distance 支援距离
supporting distance of beam 梁的支承间距
supporting documents 证明文件；单据
supporting element 支承构件
supporting energy 支承能
supporting evidence 附属证据
supporting film 支持膜
supporting flange 压紧法兰
supporting floor block 承重地板预制块
supporting floor brick 承重地砖
supporting floor joists 支承地板格栅
supporting force 支承力
supporting frame(work) 支承构架；支架；支承框架；承重构架
supporting fund 配套资金
supporting gas 载气；负载气体
supporting heelplate 支持导向背靠板；支持导架背靠板
supporting housing 支(承)罩
supporting hyperplane 支撑超平面
supporting hyperplane algorithm 支撑超平面算法
supporting in longitudinal direction 纵向支承
supporting instrument 辅助仪器；辅助仪表
supporting in transverse direction 横向支承
supporting iron 支承铁件
supporting jack 支撑千斤顶
supporting joint 支承接头
supporting journal 支承轴颈
supporting lamella 支持板
supporting layer 支持层；支承层；持力层；承重层；承托层
supporting leg 撑脚；支架
supporting length 支柱长度
supporting lever 支承杠杆
supporting line 支撑线
supporting line function 支撑线函数
supporting liquid 支承液体；载液
supporting load 承重量；承载力；承载量
supporting lug 支承凸缘；支撑耳柄
supporting machinery 支承机械
supporting masonry wall 承重坞工墙
supporting masonry(work) 承重砌体
supporting mass 支承体
supporting mast 中间支杆；支架
supporting material 支撑材料
supporting measures 配套措施
supporting mechanism 支承机理；支撑机构
supporting medium 支持媒质；支持介质；支承介质；载体；负载介质
supporting medium sample 支承介质试样(道路建筑)
supporting member 支承构件；支承杆件；承重构件
supporting memorandum 附属单据
supporting mobile structures 机动支承构架
supporting nut 支承螺母
supporting organization 支援组织
supporting out wall 承重外墙
supporting pad 支承垫
supporting pad of open caisson 沉井承垫

supporting part 支承部分;承重部分
supporting partition 承重隔墙
supporting parts 支承部件;受力零件;受力部件;承重部件
supporting performance 支护性能
supporting pier 支墩;支承墩
supporting pile 支承桩
supporting pillar 支持柱;支柱;承重柱
supporting pin 支承销
supporting plane 升力面;支承面;支撑平面
supporting plate 支承板;撑板
supporting platform 底座
supporting plug 支持柱塞
supporting point 支点;支承点
supporting point of drill string 钻具支点
supporting pole 支持杆;支承杆
supporting position 支援阵地
supporting post 支柱;支承柱
supporting power 支承能力;承重量;承载能力
supporting power of a pile 桩的承载能力
supporting pressure 支承压力;支撑压力
supporting process 支持过程;辅助过程
supporting property 支承能力;承载能力
supporting quality 支承能力;承载能力
supporting rack 支持架
supporting rail 支承轨
supporting reaction 支承反力;支座反力;支持反应
supporting record 辅助记录
supporting reflex 支持反射
supporting resistance 支承阻力;支承反力
supporting rib 加强肋;加固肋;支承肋
supporting ring 座环;支承环;支环;垫环;绑环
supporting ring of shield 盾构支承环
supporting rock 基岩
supporting rock sample 基岩试样
supporting rod 支承杆;支杆
supporting roll(er) 支承辊;支承轮;托辊;支承滚柱;随动轮
supporting roof truss 支承屋架
supporting rope 支撑缆索
supporting scaffold(ing) 支承高架;支承脚手架
supporting schedule 补充报表
supporting screw 支承螺钉
supporting seat 支座
supporting section 支承断面
supporting shell 支承壳体;支壳
supporting shoe 支撑块
supporting skeleton 支承骨架;承重骨架
supporting sleeve 支承套
supporting soil 持力(土)层;支持土;基土
supporting soil sample 基土试样
supporting soil stratum 基土层
supporting space structure 支承空间结构;承重空间结构
supporting spring 支承弹簧;承簧
supporting staff 附属人员;辅助人员
supporting stand 支柱;支架
supporting stand hole distance 支座孔距
supporting stand size 支座尺寸
supporting statement of cost and fee 成本费用附表
supporting statements 附表
supporting steel 钢支(承)架
supporting steel plate 支承钢板
supporting strap 支承带
supporting strategies 辅助战略
supporting stratum 地基支承力;承载强度
supporting strength 地基支承力;承载强度
supporting structure 固定架;下部结构;支承结构;受力结构
supporting strut 支柱
supporting stub 支撑短截刃
supporting subgrade 路基基面
supporting surface 支持面;支承面
support(ing) system 支承系统;辅助系统;支架系统;承重体系
supporting table 支撑桌
supporting technology 技术基础;基础技术
supporting three-dimensional structure 空间支承结构;支承三维(空间)结构
supporting trestle 钢塔
supporting truss 托架;支承桁架
supporting tube 支托管;支撑管
supporting type 支撑类型

supporting value 承重量;承载值;承重能力;承载量
supporting value of surface area 表面积承载量
supporting volume 支持篇;附篇
supporting vouchers 凭证;附属票据
supporting wall 承重墙;支承墙
supporting wall construction 承重墙构造
supporting wall structure 承重墙结构
supporting washer 支承垫圈
supporting wheel 支承轮
supporting wire 支撑线;吊线
support in heading 导坑支撑
support instability 柱的不稳定;支柱失稳
support insulator 支柱绝缘子;支持绝缘子;支承绝缘子
support item 配套项目
supportive cushioning 基垫;(地毯的)支承垫
support jaw 支座夹片
support layout 柱的布置
support leg 支脚;支柱
support length 柱长;柱高
support-less 无柱
support lining 支撑衬材;支撑衬里
support load 支承荷载
support material cost 支护材料费
support mechanism 维持机制;承重机构
support microprocessor 辅助微处理器
support moment 支护力矩;支点力矩;支承力矩
support money 补助金
support motion 支座运动
support movement 支点移动;支柱移动;支座运动
support of a game 对策的支撑
support of air duct 风管支架
support of bearing 轴承支架
support of difficult ground 复杂地层支护
support of electric(al) generator 发电机墩
support of embarkation 辅助装货港
support of large vessels for stiffness 大容器支座刚性
support of motor 电动机支座
support of pipeline 管道支架;管道支撑架
support of piping lines 管道支架
support of ridge purlin(e) 脊檩支承
support of table 工作台支架
support of tubelines 管道支架
support of tubing lines 管道支架
support of vessel 容器支承
support oneself or one's family 维持生活
support operation 辅助性操作
support or reinforcing work 支护或加固工程
support package 支持软件包
support pad 支承耳
support pair 对柱;成对支撑
support parameter 支架参数;支护参数
support pattern 支护方式;支护布置;支承方式
support personnel 支援人员;辅助人员
support piece 支架构件;支承块
support pier 支墩
support pillar 承重柱;支柱
support pin 支钉
support pipe 支承管
support plant 支持设备
support plate 支承(座)板;支撑板;支座板
support plate for busbar 汇流排支撑压板
support platform 支承平台;辅助平台
support point motion 支承点运动
support pointwise 逐级点支承的
support power cost 支护动力费
support pressure 支点压力;支座压力;支座反力;支点压力
support price 支持价格;维持性价格
support processor 辅助处理程序
support program(me) 支援程序;支架程序
support radius 柱径
support rail 承轨(铁)
support reaction 支座反力;支护反力;支点反力
support reaction ordinate 支承反力竖标距
support regulation 支护规程
support reinforcement 支比加固
support removal 拆除支撑
support resistance 支承反力;支护反力
support rigidity 柱的刚度;支柱刚度
support ring 护环;支承圈;支撑环;环形支架
support-rod glass 支撑玻璃
support roller 惰轮;支承轮;托轮

supports 撑角
support saddle 鞍式支座
support screw cap 结合螺母
support section 柱的断面;柱的截面;支柱截面
support set 支架集合
support setting 支架安设;架设支架
support settlement 支座沉陷;支架沉陷
support shaft 柱轴;柱体;支架轴
support ship 供应船;火力支援舰;补给运输船
support shuttering 柱体立模;支柱模板
support side 支柱模壳
support signal 支承式信号灯
support signal device 支承式信号装置
support size 柱子尺寸;支柱尺寸
support skid (支承重物的)活动垫木;滑翔起落架
support skirt 支承裙筒;承压支承筒
support slot 支承槽
supports of spheric(al) vessel 球形容器支座
support software 支援软件;支持软件【计】;保障软件
support spacing 柱距;支柱间距;隧道支护间距
support spring 支承弹簧
support stand with bearing 轴承支架
support state 支护状况
support stiffness 柱的劲度;柱的刚度;支柱刚度
support stope 支承回采工作面
supports to prevent distortion 用支撑防止大变形
support strength 支撑力;支撑强度;支柱强度
support stress 柱应力;支柱应力
support stringer 栅篱架设机
support structure 下部结构;支承结构
support stud 支架柱螺栓
support surface 支承表面
support surfacing 柱面装饰
support sustainable urban drainage planning 支持可持续性城市排放规划
support system 支座系统;支援系统;支护系统
support taxiway 辅助滑行道
support-to-floor connection 柱板连接
support-to-footing connection 柱基连接
support tool 后援手段
support-to-slab connection 柱板连接
support-to-support connection 柱柱连接
support trunnion 支轴;支枢
support tube 支承管
support type tool holder 刀架式刀杆
support unit 支架单元;支护装置;支护构件;支承元件;单体支架
support using pulleys 滑轮悬挂支撑结构
support utility 后援实用程序
support value 保值;支承量
support value duty 保价关税
support vector 列向量;列矢量
support vessel 支援船;供应船
support vinyl 人造革(衬垫用织物的乙烯制品)
support volume 支承量
support web 钢柱腹板;支撑板
support width 柱宽
support work 支护工作
support zone 支承范围
supposed fault 推测断层
supposed value 假定值
supposition 假设;假定;推测
suppositional growth 虚拟的增长
suppositional length of railway line 铁路虚拟长度
suppository 栓剂
suppresion current 抑制电流
suppressant additive 抑制添加剂
suppressed 受抑制的
suppressed carrier 抑制载波;载波抑制式
suppressed carrier optic(al) modulation 抑制载频光学调制
suppressed carrier system 抑制载波体制
suppressed carrier transmission 抑制载波传输
suppressed carrier transmitter 抑制载波发射机
suppressed circuit 抑制电路
suppressed contraction 无侧收缩;不完全收缩;不全约束
suppressed demand 被抑制的需求
suppressed field system 抑制场制
suppressed frame 被删的框架
suppressed inflation 被抑制的通货膨胀;被遏制的通货膨胀
suppressed rectangular weir 两端不束的长方形

量水堰
suppressed red response 抑制红色灵敏度
suppressed scale 压缩刻度;压缩度盘
suppressed sideband 抑制边带
suppressed time 抑制时间
suppressed time delay 抑制时间延迟
suppressed tree 被压木
suppressed tree farm 被压木场
suppressed weir 两端不束窄堰;两端不收缩狭堰;无侧收缩堰;长方形量水堰;不收缩堰
suppressed zero 校正零点;零点熄灭
suppressed-zero instrument 无零位仪表;无零点仪器;无零点仪表
suppressed-zero meter 无零点刻度仪表
suppressed-zero pressure ga(u)ge 无零位压力表
suppressed-zero scale 无零位刻度(盘)
suppresser 抑制器
suppresser electrode 抑制极
suppressible point 可抑制点
suppressing antenna 抑制天线
suppressing debiteuse down 压槽子
suppressing of detonation 爆震抑制
suppressing of knocking 爆震抑制
suppress interference 排除干扰
suppression 抑止;压制作用;制止;禁止显示
suppression circuit 抑制电路
suppression condenser 噪声抑制电容器
suppression device 抑制装置
suppression filter 抑制滤波器
suppression hangover time 抑制延迟时间
suppression loss 抑制损耗
suppression of coherent noise 压制相干噪声
suppression of homoeologous pairing 对部分同源配对的压制
suppression of the load 稳压;负抑制
suppression operate time 抑制工作时间
suppression pulse 抑制脉冲
suppression ratio 下降比
suppression recovery 抑制恢复
suppression resistance 抑制电阻
suppression resistor 抑制电阻
suppression symbol 取消符号
suppression test 抑制试验
suppress length indicator 抑制长度指示器;控制长度指示器
suppressor 消除器;抑制栅极;抑制栅;抑制器;抑制剂
suppressor choke 降低干扰扼流圈
suppressor column 抑制柱
suppressor electrode 抑制电极
suppressor factor 抑制因子
suppressor ga(u)ge 抑制型电离真空计
suppressor grid 抑制栅极
suppressor grid amplitude modulation 抑制栅极调幅
suppressor grid ga(u)ge 抑制栅真空计
suppressor grid modulation 抑制栅极调制;抑制栅极调变
suppressor grid modulation circuit 抑制栅调制电路
suppressor grid voltage 抑制栅电压
suppressor holes 抑制栅孔
suppressor impulse 抑制脉冲
suppressor ion ga(u)ge 抑制型电离真空计
suppressor kit 抑制无线电干扰的成套装置
suppressor loss 抑制损耗
suppressor mesh 抑制栅网
suppressor-modulated amplifier 抑制栅调制放大器
suppressor modulation 抑制栅极调制
suppressor nozzle 消声喷管
suppressor overvoltage 抑制过压器
suppressor potential 抑制栅(极)电位
suppressor pulse 抑制脉冲;闭塞脉冲
suppressor resistor 抑制干扰电阻
suppress smuggling 缉私
suppress zero 消零
supra acoustic frequency 超音频;超声波频
supra aqual landscape 水上景观
supraconductivity 超电导率
supraconductor 超电导体
supracrust 外地壳层;上地壳(层)
supraeconomic administrative measures 超经济的行政手段

supraeconomic compulsion 超经济强制
supraeconomic controls 超经济控制
supragelisol 永冻土上层;永冻土活动层
supraglacial debris 冰面岩屑
supralateral tangent arcs 太阳上侧正切弧
supralethal irradiation 超致死剂量照射
supralittoral 潮上的;高潮线上的
supralittoral community 潮上带群落
supralittoral flat 潮上坪
supralittoral sediment 潮上沉积
supralittoral zone 潮上带
suprameatal triangle 麦丘恩氏三角
supramolecular structure 超分子结构
supra-mutagen 超诱变原
supranekton 上层游泳生物
supraneuston 水膜上漂浮生物
supra occlusion 咬合过高
suprapermafrost 永冻线之上的土层
suprapermafrost layer 永冻土活动层
suprapermafrost water 永冻土上的水
supra piston alloy 超强铝活塞合金
supraposition 叠加
suprasegmental structures 上环节组织
suprasellar 蝶鞍上的
suprasterol 超甾醇
suprastructure 浅层构造【地】;上覆构造;表壳构造
suprastructure map 浅层构造图
supratectal 覆盖层上的
suprateneus fold 顶薄褶皱
supratenuous fold 上薄褶皱(同沉积褶皱)
suprathermal electron 超热电子
suprathermal plasma analyzer 超热等离子分析仪
suprathermal proton bremsstrahlung 超热质子韧致辐射
supratidal 潮上的
supratidal community 潮上带群落
supratidal flat 潮上升;潮上坪
supratidal flat deposit 潮上坪沉积
supratidal zone 潮上带
supratopset fan 超电积扇
supraverge 上转
supravergence 上转
supravital staining 超活体染色
suprematism 至上主义
suprematist architecture 至上主义派建筑
supreme 最高的;最优级(木材)
supreme court 最高法院
supremum 最小上界【数】;上确界
suprethmoid 上筛骨
suralimentation 超量营养法
surbase 柱脚花线;柱基座线脚饰;护壁板顶木条;墙裙压顶线;台基上部装饰线脚
surbased arch 矮矢拱(矢拱小于跨度一半);弓弦三心(平圆)拱;扁圆拱;扁拱
surbased dome 扁圆屋顶;扁圆顶
surbased spheric(al)vault 有座圆穹顶;扁球形穹顶
surcharge 要价过多;上置荷载;附加费;附加钌款;边载(荷)
surcharged drain 超埋排水管
surcharged earth 超载土(高于挡土墙顶的土);超载地
surcharge depth (溢流堰顶以上的)超高水深
surcharged steam 过热蒸汽
surcharged wall 超载式挡土墙;超填挡土墙
surcharge effective pressure 堆载有效应力
surcharge level 超蓄水位
surcharge load 过载;超载;附加荷载
surcharge on fuel prices 燃料价格附加费
surcharge period 堆载期
surcharge preloading 超载预压
surcharge preloading consolidation 预压固结
surcharge pressure 超压;超载压力
surcharge pump 增压泵
surcharge sewer 过载污水管
surcharge storage 超高蓄水;超高库容(水库);超额容量;超高水量
surcharge storage capacity 超高蓄水量
surcharge stress 堆载应力
surcharge value 附加金额
surcharging of sewers 管道超负荷
surd 根式;不尽根【数】
surd number 不尽根数
surd root 不尽根【数】

sure event 必然事件
sure-footed safety 可靠的安全;稳妥的保险
surefooting 护脚(岸坡等)
sureness of proprtions 大小的安全性;大小的稳妥性;比例的安全性;比例的稳妥性
sure region 必然区域
sure-seal neoprene sheet membrane 氯丁橡胶薄膜优质密封板
sure-thing principle 确实原则
surety 担保人;保证人;保证
surety bond 保证物;保证金;保证人;担保证书;保函
surety company 保险公司;担保公司
surety obligation 保证债务
suretyship 担保合同
suretyship contract 保证合同
suretyship obligation 保证债务
surety's obligation 担保人的义务
surf 激浪;击岸波;浅浪;破浪;拍岸碎浪;拍岸浪;碎波;岸边碎波
surface 平面切削;外部;浮出水面;表面
surface ablation 表面消融;表面烧蚀
surface abrasion 路面磨损;路面磨耗;表面磨蚀;表面磨耗
surface abrasion course 表面磨耗层
surface abrasion resistance 表面耐磨损性;表面耐磨能力
surface absorber 表面吸收器
surface absorption 表面吸收
surface absorptivity 表面吸收率
surface accident 地面事故
surface accumulation layer 表面增强层;表面积累层
surface accumulation-layer capacitance 表面增强层电容
surface accuracy 表面精度
surface acid-base reaction 表面酸碱性反应
surface acidity reaction 表面酸性反应
surface acoustic(al)wave 声表面波;表面声波
surface acoustic(al)wave amplifier 声表面波放大器
surface acoustic(al)wave bandpass filter 声表面波带通滤波器
surface acoustic(al)wave coder 声表面波编码器
surface acoustic(al)wave convolver 声表面波卷积器
surface acoustic(al)wave correlator 声表面波相关器
surface acoustic(al)wave delay line 表面声波延迟线
surface acoustic(al)wave device 表面声波器件
surface acoustic(al)wave filter 声表面波滤波器;表面声波滤波器
surface acoustic(al)waveguide 声表面波波导
surface acoustic(al)wave oscillator 声表面波振荡器
surface acoustic(al)wave phase shifter 声表面波移相器
surface acoustic(al)wave pressure sensor 声表面波式压力传感器
surface acoustic(al)wave pulse compression filter 声表面脉冲压缩滤波器
surface acoustic(al)wave sensor 声表面波传感器
surface acoustic(al)wave techniques 声表面波技术
surface acoustic(al)wave transducer 声表面波换能器
surface acting fungicide 表面作用的杀菌剂
surface action 集肤效应;表面作用;表面效应
surface-action agent 表面活性剂;表面活化剂
surface activated polymerization 表面敏化聚合作用
surface activating agent 表面活化剂
surface activation 表面激活
surface active 表面活化的
surface active adsorbate 表面活性吸附物
surface active agent 表面活性剂;表面活化剂;表面缓凝剂
surface active agent ad uvants 表面活性剂佐剂
surface active agent micelle 表面活性剂胶团
surface active agent pollution 表面活性剂污染
surface active anion 表面活性阳离子
surface active cation 表面活性阳离子
surface active chemical 表面活性化学剂

surface active composition 表面活性组分
surface active constituent 表面缓凝剂
surface active film 表面活性膜
surface active impurity 表面活性杂质
surface active ion 表面活性离子
surface active load-bearing system 表面活性的承重系统
surface active material 表面活性材料；表面活化物质
surface-active organic compound 表面活性剂有机化合物
surface active organic solute 表面活性剂有机溶质
surface active substance 表面活性物质
surface active weight carrying system 表面活性的承重系统
surface activity 表面活性；表面活度
surface additive materials 表面附着物
surface adhesion 表面附着
surface adhesion separation 黏附选矿法
surface adsorption 表面吸附
surface adsorption film 表面吸附膜
surface aeration 表面曝气
surface aeration basin 表面曝气池
surface aeration equipment 表面曝气装置；表面曝气设备
surface aeration system 表面曝气系统
surface aeration turbine 表面曝气涡轮
surface aeration unit 表面曝气设备
surface aerator 液面曝气器；表面曝气器；表面曝气机
surface aggregate ratio 表面团聚率
surface air 地面空气；地面大气；大气底层
surface air cooler 表面式空气冷却器；表面空气冷却器
surface air leakage 外部漏风
surface air pressure 地面气压
surface air temperature 地面气温；地面空气温度
surface albedo 表面反射率
surface alignment 地面定线（测量）
surface-alloy phototransistor 表面合金光电晶体管
surface alternation 表面变化
surface analysis 故障面分析；表面分析
surface analyzer 表面光洁度检查仪；表面测定器；显微光波干涉仪
surface and boring machine 端面镗床
surface and dimension inspection 表面和尺寸检查
surface and trespass damage 农田破坏；表面及侵入性破坏；农田受损
surface angularity 表面露骨状态；表面凹凸性
surface anomaly 地面异常；地表异常
surface anti-cyclone 地面反气旋
surface appearance 外表；表状况；表面形状
surface application 敷面；表土施肥；表面处理
surface applicator 表土施肥机；表面敷贴器
surface applicats to calcareous soil 石灰性土壤表层施用
surface approximation 曲面逼近
surface arcade 假拱廊；墙面假联拱；墙堵拱；表面连拱
surface area 曲面面积；地面区段；表面面积；表面积
surface area apparatus 表面积仪
surface area coefficient 表面积系数
surface area concentration 表面积浓度
surface area constant method 表面积常法
surface area factor 表面面积系数；表面积系数
surface-area method 表面积法
surface area of aggregate 骨料表面面积；集料表面面积
surface area of dissolved sample 被溶样品表面积
surface area of land 陆地面积
surface area of particle 颗粒表面积
surface area of sea 海洋面积
surface area of soil 土（颗）粒表面积
surface area of terrestrial spheroid 地球椭球面积
surface area per unit capacity of packing 单位填料表面积
surface area test 表面积测定
surface area theory 表面积学说
surface area-volume ratio 表面面积体积比
surface artery 地面干道
surface as forged 黑皮锻件
surface associate 表面缔合物
surface astragal 门碰头（双扇扯门前沿边安装的）
surface availity 表面有效度

surface back-scatter equipment 表面反向散射装置
surface back-scatter method 表面反向散射法
surface bar 测平杆
surface bargaining 敷衍性谈判
surface barrier 表面阻挡层；表面势垒
surface barrier detector 面垒型探测器
surface barrier diode 表面势垒二极管
surface barrier photodiode 表面势垒光电二极管
surface barrier phototransistor 表面阻挡层光电晶体管
surface barrier semiconductor detector 面垒型半导体探测器
surface barrier transistor 表面阻挡层晶体管；表面势垒层晶体管
surface barrier transistor 表面势垒晶体管
surface based observing station 地面观测站
surface based subsystem 地面子系统；地面分系统
surface beacon 地面信标
surface bearing 面支承；滑动轴承；平面轴承；表面支承
surface beat(ing) 水面波动
surface bed 浅床
surface bedrock 地表基岩
surface behavio(u)r 表面特征
surface bit 表面钻头
surface blast-hole drilling 露天爆破孔钻进
surface blasting 露头爆破；露天爆破；地面爆破；表面爆破
surface blasting sand 表面喷砂料
surface blemish 表面瑕疵；表面缺陷；表面疵病
surface blister 表面气泡
surface blistering 表面气泡
surface blowhole 表面气孔
surface-blowing 面吹法
surface-blown converter 侧吹转炉
surface blow off 表面排污
surface boat 水面航行船
surface boiling 表面沸腾
surface bolt 明装插销
surface bond(ing) 表面黏结；表面结合
surface(bore)hole 下表面套管的钻孔；地面钻孔；地表钻孔
surface-borehole mode 地面井中工作方式
surface boundary layer 地面边界层；地表边界层；表面界面层；表面边界层
surface bound by traffic 车结路面
surface box 地面阀闸；地下水小室；窨室；地面操纵器
surface breakage 地表断裂
surface breakdown 表面击穿
surface broach 平面拉刀；扁拉刀
surface broaching 平面拉削；表面拉削
surface broaching machine 平面拉床
surface broaching tool 表面拉刀
surface bucket thermometer 水面提斗式温度计
surface building 地面建筑
surface buoy 水面测流浮标
surface burnishing facet 磨光棱面
surface cable 明敷电线
surface calculate 表面计算
surface calibrate 表面标定
surface capability 表面容量；表面性能
surface capacitance 表面电容
surface capacitor 表面电容器
surface capacity 表面容量；表面性能；表面承载力
surface car 地面车辆
surface carbonizing 木材表面碳化（法）
surface carburetor 表面式油器；表面汽化器
surface carburization 表面渗碳
surface carburizing 表面渗碳
surface car park 地面停车场
surface carrier 地面交通工具
surface casing 地面套筒；地面套管；表面防护罩；表层套管；外壳
surface casing depth 表层套管深
surface catalysis 表面催化作用
surface catalyzed reactions 表面催化的反应
surface cavity 路面凹坑；面穴；地面凹坑；表面凹坑
surface cell 表面电池
surface cement 表面黏合剂
surface channel 表面沟道
surface character 表面特性
surface character(istic) 表面特性
surface characteristic determination of grain 颗粒表面特征鉴定

surface characteristics 表面特性
surface charge 表面电荷；裸露电荷；面电荷
surface charge density 表面电荷密度
surface charge distribution 表面电荷分布
surface charge transistor 表面电荷晶体管
surface charring 表面烧焦
surface chart 海面天气图；曲线图；地面（天气）图
surface chart program(me) 曲面图程序
surface check 表面开裂；面层裂纹；表面裂纹；裂缝；表层裂
surface checking 表面起网状裂纹；表面龟裂
surface chemical analysis 表面化学分析
surface chemical reaction 表面化学反应
surface chemical reactivity 表面化学反应性
surface chemistry 表面化学（性质）
surface chill 外冷铁；表面激冷
surface chromatography 表面色谱法
surface circulation 海面环流；表面环流；表层环流
surface cladding 表面熔涂；表面敷涂
surface classifier 表面分级器
surface clay 地面黏土；地表黏土；表面黏土
surface cleaning 表面清理；表面冲洗扫除
surface clearness 表面清洁度
surface clogging 表面堵塞
surface coarseness profiling instrument 表面粗糙度轮廓仪
surface coat 面层；外涂漆；表面油漆；表层；面层抹灰；表面包层（轻填料）；表面涂层
surface coated 表面涂膜的
surface coated mirror 表面镀膜反射镜
surface coated wall paper 墙纸
surface coating 表面处治；表面包皮（轻填料）；修筑面层；表面涂料；表面覆层；表面镀膜；表面涂层
surface coating composition for concrete 混凝土表面涂料
surface coating industry 电镀业；涂料工业
surface coating(material) 表面涂料
surface coating medium 涂料溶剂
surface coating method for waterproofing 涂刷面层防水法
surface coating operation 面层修筑；表面处治操纵
surface coating treatment 表面涂层处理
surface cock 液面控制旋塞；液面开关
surface coefficient 表面系数
surface coefficient of friction 表面摩擦系数
surface coefficient of heat transfer 放热系数；表面散热系数；表面传热系数
surface collapse 地面塌陷；地表陷
surface collection agent 液面收集剂
surface colling 表面冷却
surface colo(u)r 表面着色；表面色
surface colo(u)rant 表面着色剂
surface colo(u)ration 表面着色
surface colo(u)red eye glass 表面着色眼镜玻璃
surface colo(u)r hologram 面彩色全息图
surface colo(u)ring 表面着色作用；表面上色
surface combustion 表面燃烧
surface combustion burner 辐射板燃烧器；表面燃烧燃器
surface combustion method 表面燃烧法
surface comlexation 表面络合
surface compaction 表土夯实；表面压实；表面夯实
surface compactor 表面捣实器
surface comparator 表面比较仪
surface complex adsorption 表面络合吸附
surface complexes 表面络合物
surface composition 表面组成
surface composition mapping radiometer 表面成分测绘辐射计
surface compressibility 表面压缩性
surface compression 表面压缩
surface concentration 表面浓度
surface concentration excess 表面浓度过剩
surface concrete 室内地面混凝土层
surface concrete membrane 表面混凝土防水膜
surface condensation 表面凝结；表面冷凝
surface condenser 表面式凝汽器；表面冷却器；表面冷凝器
surface condition 水上状态；表面状况；表面条件
surface conditioner 表面调整剂
surface conditioning 表面修整
surface conditioning agent 表面调整剂
surface conductance 表面电导；表面导电

surface conductivity 表面电导率
surface conductivity of glass 玻璃的表面电导率；玻璃的表面导电性
surface conductor 地面导向管
surface configuration 地面构型；地表形态；表面形状
surface connecting system 表面相连管道系统
surface connections 地表设备的管线
surface constant 表面常数
surface constant temperature 地表恒温带深度
surface construction 路面建筑
surface construction in rock 岩石中的地面建筑；岩石面层上的工程
surface contact 表面接触；面接触
surface contact angle 面接触角
surface contact hole 表面接触孔
surface contact pressure 表面接触压力
surface-contact reheater 表面接触式再热器
surface-contact system 面接触系统；表面接触系统
surface containment method 水面浮油封堵法
surface contaminant detector 表面污染物质检测仪
surface contamination 表面污染
surface contamination meter 表面污染测量仪；表面测污仪
surface contamination monitor 表面污染监测仪
surface contamination monitoring 表面污染监测
surface content 表面积
surface continuity 表面连续性
surface continuous row fluorescent fixture 明装荧光灯带
surface contour 地面等高线
surface contour map 地形测量图；等高线地形图
surface contraction 表面收缩
surface contrast 表面分色
surface controlled secondary recrystallization 表面控制二次再结晶
surface contribution 面贡献量
surface cooled 表面冷却的
surface coordinate 曲面坐标
surface cord 表面条纹；表面结瘤
surface corona 表面电晕放电
surface correction 地表校正
surface corrosion 表面腐蚀
surface coupling 表面耦合
surface course 面层；表面层
surface course mix 面层混合料
surface cover 地表被覆
surface coverage 表面覆盖率；表面覆盖度
surface covering 表面覆盖率；表面覆盖度
surface covering material 路面盖料
surface crack 路面裂缝；表面微裂纹；表面裂纹；表面裂缝
surface cracking 表面开裂
surface craft 水面舰艇；水面飞行器；水面船只；水面船舱
surface crazing 表面网纹
surface creep 表土蠕动；表面蠕动；表层滑移；表层蠕动；表层爬动；表面滑动；河床面蠕移；表土塌滑
surface crew 船上人员
surface crew on ship 船上人员
surface crew on vessel 船上人员
surface crust 表层壳层；表壳
surface crystalline glaze 表面结晶釉
surface crystallization 表面微晶化；表面结晶
surface crystallography 表面晶体学
surface cultivation 表土中耕
surface cultivator 表土中耕机
surface culture 表面培养
surface cum seepage drain 地表排水兼渗流排水管
surface current 面流；地表流；表面流动；表面流；表面电流；表层流
surface current classifier 表面分离器；表流分级器；表流式分级机
surface current density of electrode 电极表面电流密度
surface curvature 表面曲率
surface curvature apparatus 路面曲率半径测定仪
surface curve 水面曲线；表面曲线
surface curved in one plane 单曲率曲面
surface curve method 水面曲线法
surface cut 表面切削
surface cutting 表面开挖
surfaced 表面的；刨光(的)
surfaced airfield 有铺筑道面机场

surface damage 表面缺陷；面损伤；表面损伤；表面损坏
surface damage field 表面损伤场
surface damp course 地面防潮层
surface damp-proof membrane 地面防潮层
surface decarburization 表面脱碳
surface decompression 水面减压
surface decontamination 表面净化
surface decoration 表面装修；表面装饰
surface decoration material 表面装饰材料
surface decorative feature 表面装修特征
surface decorative finish 表面装修润饰
surface decorative material 表面装饰材料
surface defect 表面缺陷；表面疵病
surfaced-effective vehicle 气垫车
surface deformation 路面变形；表层变形
surface degradation 表面变质
surface demand diving apparatus 水面充氧潜水装置
surface densifier 表面密实剂
surface density 面层密度；地表密度；表面密度
surface density of site 点表面密度
surface depletion layer 表面耗尽层
surface deposit 表面堆积物；表面沉积(物)；表层沉积
surface depression 凹地；路面沉陷；地面凹陷；面缩陷；表面沉陷；表面凹陷
surface depression storage 洼蓄；洼地蓄水(量)；洼地储积(量)
surface design 表面设计
surface desorption molecular microscopy 表面解吸分子显微镜法
surface detail 表面花纹
surface detection equipment 地面探测设备
surface detention 地面滞流量；地面阻滞；地面滞留(水量)
surface deterioration 表面破碎
surface development 表面显影
surface dewatering 地表疏干
surface diagram 曲面图
surface diffusion 表面扩散
surface diffusion coefficient 表面扩散系数
surface digging 地面平整；地表挖掘
surface digging machine 地面平整机；地表挖掘机
surface dipole 表面偶极子
surface dipole moment 表面偶极矩
surface discharge 表面流排料；表面放电
surface discontinuity 表面不连续性
surface dislocation 表面位移；表面变位
surface dispersion 表面色散
surface disposal 表面处置
surface distress 面层损坏；面层龟裂(混凝土)
surface distribution 面上分布
surface ditch 阳沟；明沟；地表排水；放水沟
surface divergence 表面散度
surface divide 分水线
surfaced lumber 光表面板材；刨削材；刨光木材；刨光木材
surface doming 表面隆起
surfaced on both sides 双面涂层；双面加工
surface double-electric(al)layer 表面双电层
surface downcast pillar 地表陷落柱
surface drag 表面阻力；表面曳力；表面拖曳(力)
surface drain 路面排水；明沟排水；明沟；排涝沟；地面排水(沟)
surface drainage 明沟排水；地表水系；地表排水；地表疏干法；地面排水
surface drainage ditch 地表排水沟
surface drainage divides 地面分水岭；地面分水界
surface dredge(r) 水面漂(浮)物刮集船；表面除油器
surface-dressed macadam 表面处治的碎石路
surface dressing 沥青表面处治；面层；道路表面处治；敷面料；敷料料；表面整修；表面修琢；表面修整；表面修饰
surface dressing chip(ping)s 铺面石屑
surface dressing chip(ping)s distributor 地面石屑撒布机；路面石屑撒布机
surface dressing treatment 敷面法；表面修整处理
surface dressing with tar 表面浇焦油(沥青)
surface-dried condition 表面干燥状态
surface drift 表面漂移
surface drift bottle 漂流瓶；水面浮标瓶
surface drift current 表层漂流

surface drifter 表面漂流物
surface drift net 表层流网
surface drill 撒播机
surface drill hole 地面钻孔；地表孔
surface drilling 地面钻进
surfaced road 有路面的道路；铺面路
surface drop technique 表面涂抹法
surface dry 表干；表面干的；表面干燥
surface-dry aggregate 面干集料；面干集料
surface-dry condition 面干状态(集料的)
surface drying 漆面干燥过快；油漆结皮(油漆疵病)；表面干燥
surfaced timber 刨光木材；光面木(料)
surface duct 地面管道；地面波导；表面波导
surface dump 地面堆放场
surface durability 表面耐久性
surface dust 浮尘
surface dusting 表面起砂
surfaced width 铺有路面的厚度
surfaced with granular cork 粒状软木护面
surface dynamic(al)profilometer 表面动态光洁度检查仪
surface earthing 表面接地
surface easement 地面使用权
surface ecology 浅层生态学
surface eddy 表面漩涡；表面涡流
surface edge curve 表面边缘曲线
surface effect 表面效应
surface effect in test specimens 试件的表面效应
surface effect ship 气垫船；表面效应船
surface effects of tunnel construction 隧道施工所引起的地表影响
surface-effect vehicle 气垫艇；表面效应运载工具
surface electric(al)double layer 表面双电层
surface(electric(al))potential 表面电位
surface electro conductive glass 表面导电层玻璃
surface electrode 表面电极
surface electrostatics 表面静电
surface element 面元(素)；面积元(素)；曲面元素；表面单元
surface element method 面元法
surface element size 面元大小
surface elevation 水面高程；地面高程；地面标高；表面高程
surface elevation change 地表高程变化
surface elongation viscosity 表面伸长黏度
surface emission ion source 表面发射离子源
surface emissivity 表面辐射(系数)；表面发射率
surface emitter 面发射体
surface emitting laser 表面发射激光器
surface-emitting laser diode 表面发射激光二极管；表面发射发光二极管
surface-emitting light emitting diode 面发射发光二极管
surface endurance limit 表面持久极限
surface energy 表面能(量)
surface energy-absorbing layer 表面能吸附层
surface energy level 表面能级
surface energy of fracture 断裂的表面能
surface engraving 表面雕刻
surface enhanced raman spectroscopy 表面强化拉曼散射铺
surface enrichment 表面装修；表面富集
surface entropy 表面熵
surface entry 地面入水口
surface equipment 地面设备
surface erosion 地表冲刷；表面侵蚀；表面冲刷
surface error 表面误差
surface evaluation 路面状况评价
surface evapo(u)ration 水面蒸发；地表蒸发；表面蒸发(作用)
surface evapo(u)ration pan 地面蒸发器
surface evapo(u)rative condenser 表面蒸发凝汽器
surface evenness 表面平整度；路面平整度；面平整度
surface evenness of pavement 道面表面平整度(机场)
surface excavation 地表开挖；地面平整；路面开掘；表面开挖
surface exchanger 表面换热器
surface exclusion 表面排斥
surface expansion 面膨胀；表面膨胀
surface exploration 地表探测

surface exposure 地面露头;地表露头
surface external vibrator 表面式外振动器
surface facility 地面设施
surface factor 细度因子;地形;表面因素;表面系数
surface failure 表面失效;表面破坏
surface fatigue 表面疲劳
surface fault 地表断层
surface faulting 表面断层作用
surface fault trace 表面断层迹
surface feature 场地地貌;地形;地物;地面要素;地面特征;地貌;地表特征;表征;表面形态;表层地形
surface features of grain 颗粒表面特征
surface-fed intermittent river 表源间歇河(流)
surface-fed intermittent stream 表源间歇河(流)
surface feed heater 表面式给水加热器
surface feeding 表面摄食的
surface fermentation 表层发酵
surface filler 表面填料
surface film 表面膜
surface film balance 表面膜平衡
surface film of water 水的表面膜
surface film potential 表面膜势
surface filter 表面型滤器
surface filtration 表面过滤
surface finish 饰面;面层;面层做法;表面装饰;表面修琢;表面抛光;表面加工;表面光制;表面终饰;表面修整;表面修饰
surface-finish analyzer 表面光洁度分析仪
surface finish from formwork 拆模后表面修整(加工)
surface finish indicator 表面光洁度指示仪;表面光洁度检查仪
surface finishing 路面整修;表面精整;表面精加工;表面终饰;表面修整;表面修饰
surface finish ment 表面光洁度
surface finish meter 表面光洁度测量仪
surface finish microscope 表面光洁度显微镜
surface finish of concrete 混凝土表面处理
surface finish quality 表面光洁度
surface finish recorder 表面光洁度记录器
surface finish testing microscope 表面光洁度测量显微镜
surface fire 地表火
surface fire spread 表面火焰扩散
surface fish 表层鱼
surface fitting 曲面拟合
surface fixation test 表面固定试验
surface-fixed hinge 扁担铰链;长脚铰链;汉堡式铰链
surface fixing 放置;面上置放
surface flammability 表面易燃性
surface flash 喷溅;表面溅点
surface flatness 表面光滑度;表面不平度
surface-flatness checker 表面平直度检查器
surface flaw 表面缺陷;表面裂纹;表面裂缝;表面发纹;表面发裂(纹)
surface float 水面浮子;水面浮标;水面测流浮子
surface float coefficient 水面浮标系数
surface floating trawl 表层浮拖网
surface float method 水面浮标(测流)法
surface flood(ing) 地面洪水;地面泛流;地面泛滥;漫灌;地面漫流
surface flooding irrigation 地表漫灌
surface flow 径流;坡面漫流;地面水流;地面径流;地表水流;地表流;地表径流(量);表面水流;表面流动;表流;表层流
surface flow and underground storage water 总水况
surface flow classifier 表面分离器
surface flowing waters sedimentary type 地表流水沉积型
surface flow pattern 面流形态
surface flow ventri scrubber 表流文丘里洗涤器
surface fluorescent fixture 明装荧光灯
surface focus 表面震源
surface folding 表面褶皱;表层褶皱
surface following buoy 水面随波浮标
surface footing 明置基础
surface force 面力;表面力
surface forecast chart 地面预报图
surface for fitting 内插曲面
surface form 面状
surface foundation 明置基础

surface fragmentation of rock 岩石表面破碎
surface free energy 表面自由能
surface freezing 表面冻结
surface freezing index 地表冰冻指数
surface friction 地面摩擦;表面摩阻;表面摩擦
surface friction coefficient 路面摩擦系数;表面摩擦系数
surface friction drag 表面摩擦阻力
surface friction windup device 表面摩擦卷取机
surface front 地面锋【气】
surface frost 地面霜
surface frost heave 路面冻胀
surface functional group 表面官能团
surface ga(u)ge 划线盘;平面规;表面规
surface ga(u)ge type wave recorder 验潮仪式自记测波仪
surface geochemical exploration 地表化探
surface geologic(al) documentation 地表地质编录
surface geologic(al) investigation 地表地质勘察
surface geologic(al) map 表层地质图
surface geologic(al) survey 地面地质测量
surface geology 地表地质(学)
surface geometry 表面几何图形
surface gradient 河面比降;水面比降;地面坡度
surface graft 表面移植物
surface gravity anomaly 地表重力异常
surface gravity wave 表面重力波
surface grinder 磨面机;平面磨床
surface grinding 平面磨削;表面研磨;表面磨削;表面磨工;表面打磨
surface grinding machine 平面磨床;表面磨光机
surface ground fall 地表坍陷
surface-groundwater relationship 地表—地下水关系
surface grouting 接触面灌浆;接触灌浆;表面灌浆
surface hall of metro-station 地铁车站地面站厅
surface hardened 表面硬化的
surface hardened steel 表面硬化钢
surface hardener 表面硬化剂
surface hardening 表面硬化(法);表面淬硬;表面淬火
surface-hardening agent 表面硬化剂
surface-hardening fluosilicate 表面硬化用氟硅酸盐
surface-hardening liquid 表面增硬液
surface hardening rail 顶面淬火钢轨
surface hardening steel 低淬透性钢
surface-harden saline soil 结壳盐土
surface hardness 表面硬度
surface harmonic function 面谐函数
surface harmonics 面谐函数
surface heater 路面加热器;路面加热机;暖面器;热表器;表面式加热器;表面加热器
surface heat exchanger 表面式热交换器;表面式换热器
surface heat flow 地表热流
surface heating 路面受热(量);表面热度;表面加热;表面采暖
surface heat transfer coefficient 表面热转移系数;表面热传递系数;表面换热系数
surface heat-treatment 表面热处理
surface heterogeneity 表面多相性;表面不匀性;表面不均匀性
surface hinge 表面铰链
surface hole drilling 钻下表面套管的钻孔
surface-hole shrinkage method 浅孔留矿法
surface hologram 面全息图
surface horizon 地表土层
surface horizontal displacement 地表水平位移量
surface humidity 地面湿度;表面湿度
surface humus 表面腐殖质
surface hydrant 地面消火栓
surface hydrated cement 表面水化水泥
surface hydraulic control manifold for subsea test tree 海底测试树控制装置
surface ice 水面冰层;水面表层冰;地面冰;表冰
surface ice glazed 表面结冰的
surface ignition 表面点火
surface ignition engine 热点火机;热球柴油机;热球式发动机;表面点火发动机
surface impedance 表面阻抗
surface imperfection 表面缺陷;表面不规则
surface impoundment 地面储存
surface-improved 表层改良的
surface improvement 表面改变;表层改良

surface improver 表面改性剂
surface-improving agent 表面改性剂
surface impurity 表面杂质
surface inactivation 表面钝化
surface incandescent lamp 明装白炽灯
surface inclusion 表面夹杂物
surface in contact 接触表面
surface indentation 路面凹痕;路面凹槽;表面压痕;表面凹陷
surface index 表面指数
surface index of aggregate 集料颗粒表面特性指数
surface indications 地面显示
surface indicator 表面找正器;平面规;表面规
surface induced process 表面诱发过程
surface induction 表面感应
surface inflow 地面来水量;地表渗入;地表入流;地表补给
surface inlet 表面入口
surface inspection 表面探伤
surface inspector 路面检验器
surface instability 表面不稳定性
surface installation 洞外设备;地面设施
surface insulation 表面绝缘
surface integral 面积分;曲面积分;球面积分
surface integration method 面积分法
surface intensity function 表面强度函数
surface internal vibrator 表面式内振动器
surface intersection curve 交面曲线
surface interval 水面间隔时间
surface inversion 地面逆温
surface ionization 表面电离(作用)
surface ions 表面离子
surface irradiation 表面照射;表面辐照;表辐照
surface irregularity 表面不平整;路面不平;地面起伏;表面粗糙;表面不平整度;表面不平度;表面变形;表面凹凸不平
surface irrigation 地面浇灌;地面灌溉;地表灌溉
surface karst 地表岩溶
surface karst form 地表岩溶形态
surface key bed 地表基准层;表层基准层
surface laitance 表面浆沫层(混凝土);水泥浮浆层
surface landing probe 表面着陆探测器
surface lap 表面结疤
surface lapping machine 平面精研机
surface latch 明锁;明插销;明装碰锁
surface law 表面积定律
surface layer 修整层;近地(面)层【气】;加工层;磨削层;地表边界层;覆盖层;表(面)层;保护层
surface layer ni-triding 表面层氮化法
surface layer sampler 表层采样器
surface layer temperature 表层温度
surface layer waters 表层水体
surface lead 表面铅
surface leakage 表面泄漏;表面漏泄
surface leakage conductance 表面漏泄电导
surface leakage current 表面漏泄电流
surface level 水面水平;地面高程;表面能级
surface lifetime 表面寿命
surface lift 地面隆起
surface ligand 表面配位体
surface light source 面光源
surface line 地面likely
surface lining of wells 井表面衬砌
surface liquid cooler 表面液体冷却器
surface load(ing) 表面荷载;表面加载;表面负荷;面荷载
surface loading of sedimentation tank 沉淀池表面负荷
surface loading rate 表面荷载率
surface loosening 路面松散
surface lost coefficient 地面损耗系数
surface low 地面低压
surface lubricant 表面润滑剂
surface lubrication 表面润滑
surface lumber 加工材
surface magnet charge 面磁荷
surface magnetic wave 磁表面波
surface magnetism 表面磁性
surface mail 普通邮件
surface maintenance 路面养护;面层养护;表面养护
surface-man 井上工人;护路工人
surface management 地面整理
surface manganese nodule 表层锰结核
surface manuring 表土施肥

surface map 地面图;地面天气图
surface mapping 地形图绘制;地形填图
surface mark 地面标志
surface marker 表面标志
surface marking 地面花纹;表面印痕
surface mark of aerial photogrammetry 航测地面标志
surface masonry 地面圬工
surface masonry wall 地面砌体墙;地表圬工墙
surface masonry work 地面砌筑工程
surface mat 路面保护层;表面毡;表面污泥法;表面薄毡;土跑道上的网眼钢板
surface material 面层材料
surface matting of roots 根铺表土
surface mean diameter 表面平均直径
surface measure 单面量度
surface measurement 地面测量;表面测量
surface-measuring instrument 表面测量仪
surface medium 表面培养基
surface melting 表面熔解
surface membrane 表面(薄)膜
surface metal raceway 明装的金属电缆管道;面装金属电线管
surface meteorologic(al) observation 地面气象观测
surface meter 表面仪
surface method 表面法
surface microanalyzer[analyser] 表面显微分析仪
surface migration 表面迁移
surface miller 平面铣床
surface milling 辊铣;面铣
surface milling machine 平面铣床
surface mine 露天矿
surface mine car circulation 露天开采车辆运行
surface mine development 露天矿开拓
surface mineral 绿豆砂(屋面保护层);(覆盖屋面沥青的)表层矿料
surface mining 露天开采矿;露天开采法
surface mining methods 地面采矿法分类
surface-mixing method 路拌法
surface mixture 路面混合料
surface mobility 表面迁移率
surface mode container 地面运输用集装箱
surface model 表面模型
surface modeling 曲面造型
surface modification 表层变性;表面修正;表面改性;表面改良
surface modified 表面改性的
surface modified silica 表面改性硅石
surface modified wollastonite 表面改性硅灰石
surface modifier 表面改性剂;表面调节剂
surface modulation 表面调制;表面造型
surface moisture 表面水;表面(自由)水分;表面水量;表面湿度
surface moisture absorber 表面吸湿器
surface moisture capacity 表面吸湿能力
surface moisture content 表面含水率
surface molecular imprinting resin 表面分子印迹树脂
surface molecule 表面分子
surface monitor 表面沾污的检测器
surface monitoring 表面(污染)监测
surface moraine 表碛
surface morphology 表面形态(学)
surface motion 地表运动
surface mount component placement machine 表面安装元件布局机
surface-mounted 明装
surface-mounted astragal (双扇扯门前沿边安装的)门碰头;面装圈带
surface-mounted conduit 地面管道;地面水道
surface-mounted distribution board 地面配水板;地面布水板
surface-mounted hardware 明装小五金
surface-mounted installation 地面管道;地面管线;断面装置
surface-mounted luminaire 安装在表面上的照明体;明装照明设备;吸顶灯具
surface-mounted paper towel dispenser 贴面式纸巾配出器
surface-mounted powdered soap dispenser 贴面式肥皂粉配出器
surface-mounted rod extensometer 表面镶杆式伸缩计

surface-mounted sanitary napkin dispenser 贴面式卫生巾配出器
surface-mounted wire 明线
surface mounting 裱糊面层;镶嵌面层;糊表面层;表面安装
surface mounting device 表面安装装置
surface mounting lampholder 面上安装的灯座
surface movement 地表移动;表层移动
surface movement curve 地表移动曲线
surface movement observation 地表移动观测
surface movement observation station 地表移动观测站
surface mud pressure 孔口泥浆压力
surface mulch 地面腐蚀土层;路面覆盖料;路面腐殖土层;覆盖料(保护地面用的木屑,砾石,纸等)
surface nature 表面性质
surface navigation 水面航行;地表航行
surface network density theory 面网密度理论
surface noise 音纹噪声;地面噪声;表面噪声
surface normal 表面法线;曲面法线
surface nuclear burst 地面核爆炸
surface observation 海面观测;地面观测;地表观测
surface observation station 地面观测站
surface occurrence 地表露头
surface ocean current 表层洋流
surface of adherence 黏附面
surface of adsorbent particles 吸附剂粒子表面
surface of ballast 道砟面;石渣面层
surface of bed 床面
surface of best definition 最清晰表面
surface of bone 骨面
surface of centres 中心距曲面
surface of columnar structure 柱状结构面
surface of compensation 补偿面
surface of concentric(al) shearing 集中剪切面
surface of constant surface 等曲率面
surface of constant total curvature 定值全曲率的曲面
surface of contact 接触(表)面
surface of density discontinuity 密度不连续面
surface of deposition 沉积面
surface of discontinuity 不连续面层;间断面;突变面;不连续面
surface of discontinuity current 水流间断面;水流不连续面
surface of earth 地面
surface of equal density 等密度面
surface of equal dynamic(al) depth 等动力深度面
surface of equal parallax 等视差面
surface of equal path difference 等程差面
surface of equal potentials 等势面
surface of equal specific volume 等比容面
surface of equi-pressure 等压面
surface of esplanade 堤岸面;码头前沿地面;码头面
surface of external wall 外墙面
surface of fatigue break 疲劳破坏表面
surface of filling 填筑面
surface of fracture 破碎表面;破裂面;破坏面;断裂面;断口;折裂面
surface of freeing 分隔面
surface of friction 摩擦面
surface of geoisotherm 等地温面
surface of grade crossing 道口铺面
surface of intensity distribution 光强分布图
surface of internal wall 内墙面
surface of light source 光源面
surface of liquid 液面
surface of low flame spread 低延性面
surface of medium flame spread 中延烧性面
surface of no distortion 无扭面
surface of operation 加工基准面;施工作业参考面;操作参考面;施工参考面
surface of position 位置面
surface of positive curvature 正曲率的曲面
surface of potential 潜伏面
surface of pyramid or cone 方锥体或圆锥体表面(面积)
surface of rapid flame spread 快延烧性面
surface of reference ellipsoid 参考椭球面
surface of revolution 旋转面;转成面;回转面
surface of rolling 滚动曲面
surface of rotation 旋转面
surface of rupture 破裂面
surface of secondary order 二阶曲面

surface of seepage 渗出面
surface of separation 界面;分隔面
surface of settling zone 沉淀层面
surface of shear 剪切面
surface of shear slide 剪切滑动面
surface of sheet type 片式(沥青砂)面层
surface of sliding 滑动面
surface of soil 表土层
surface of specific volume 比容面
surface of spoil bank 弃渣面
surface of stability 稳定面;稳定表面
surface of subsidence 下沉面
surface of translation 平移面
surface of unconformity 不整合面
surface of underground water 地下水位;地下水面
surface of water 水面
surface of weakness 软弱面
surface of weld 焊缝表面
surface of wound 创面
surface of zero velocity 零速度面
surface oil indications 地面石油油苗;地面石油迹象
surface oil pickup 水面油污传感器
surface orientation 地表定向;表面定向
surface ornament 表面装修
surface ornamental feature 表面装修特征
surface ornamental finish 表面装修润饰
surface oscillation 表面振荡
surface outflow 地表溢流
surface overflow 地表漫流
surface overflow rate 表面溢流率
surface overlaying 表面堆焊
surface oxygen decompression 水面吸氧减压
surface ozone 地表臭氧
surface ozone pollution 地表臭氧污染
surface packer 表土镇压器
surface paint 面层油漆;面层涂料;表层漆
surface pan 表土磐
surface parking 地面停车
surface passivation 表面钝化
surface passivation technology 表面钝化工艺
surface patch generation 曲面片生成
surface patching 路面修补;补面;表面修补
surface pattern 砖砌图样;砖砌面式样;砖砌面层;地表水系
surface peening 表面剥落;表面强化处理;表面强化
surface penetration treatment 表层渗入层处理
surface penstock 露天压力钢管;露天压力管道
surface permeability 表面渗透率
surface permeameter 路面透水度测定仪
surface permissible gradient dip of underlying bed rock 地基岩表面容许坡度值
surface personnel 地面人员
surface phase 地面相;表面相
surface phenomenon 表面现象
surface photoelectric(al) effect 表面光电效应
surface photoemission 表面光电发射
surface-piercing(hydro)foil 划水式水翼;半浸水水翼
surface-piercing(hydro)foil craft 划水式水翼艇;浅浸水水翼艇
surface pinholes 表面针孔
surface pipe 地表管
surface pipe irrigation method 地面水管灌溉法
surface pitting (混凝土的)表面点蚀;表面凹孔
surface plan 地面平面图
surface plane 刨面机
surface planeness tolerance 平面度公差
surface planer 木工平刨床;平面刨床;平滑刨床;刨平机;手压刨床;平刨
surface planing machine 面层磨平机;平滑刨床
surface plankton 水面浮游生物;表面浮游生物
surface plant 露天设备;地面设备;地面电站
surface planting 培土浅种法
surface plasmon mode 表面等离子激元模
surface plasmon polariton 表面等离子激元
surface plasmons 表面等离子激元
surface plasmon wave 表面等离子激元波
surface plate 验平板;划线台;平台;平版;平板
surface plot 水面目标标图板;封面图
surface plug receptacle 露出插座
surface pockmark 麻面
surface point 地面点;表面点
surface poise 表面泊
surface poisoning 表面中毒

surface polariton waves 表面极化元波
surface polarization 面极化
surface polishing 表面抛光
surface polishing machine 表面抛光机
surface pollution 表面污染
surface pollution film 表面污染膜
surface pond 地面水源
surface ponding 地面积水
surface pore 地面孔;地面细孔;表面气孔;表面孔隙
surface porosity 表面孔隙率;表面孔隙度
surface portion of highway 公路路面部分
surface potential 表面势;表面电势
surface-potential barrier 表面位垒;表面势垒
surface potential detector 表面(电)位检测器;表面电势检测器
surface potential of double layers 双层表面位
surface potential of simple layer 单层表面位
surface power 表面性能;表面折光力;表面屈光度;表面(能)力
surface power house 地面电站
surface power station 地面电站
surface precipitation isotherm 表面沉积吸附等温线
surface precipitation process 表层沉积过程
surface preparation 表面整备;地表准备;表面准备;表面预加工;表面加工;表面处治;表面处理
surface preparation for adhesive bonding 胶接表面处理
surface preparation technology 表面制备工艺
surface pressing 表面压力
surface pressure 膜压;地面压强;地面压力;表面压力
surface pressure chart 地表气压表;表面压力表
surface pressure coefficient 表面压力系数
surface pressure model 表面压力试验模型
surface pretreatment 表面制备
surface printing 表面印染;凸版印刷
surface printing machine 凸纹辊筒印花机
surface probe 表面传感器
surface process of reductive series 还原潜育系列的表生过程
surface process of reductive sulfides series 还原性硫化系列的表生过程
surface production 表面形成;表面生成
surface profile 水面(曲)线;水面(高程)纵剖面;地面纵断面;表面轮廓
surface projection 路面突起部分
surface proofing liquid 表面防水液
surface property 表面能力;表面性质;表面性能
surface property of glass fibre 玻璃纤维表面性质
surface protectant 表面保护剂
surface protection 表面防护;表面保护
surface protection layer 表面保护层
surface protection work 表面保护操作
surface protective treatment 表面防护处理
surface protuberance 表面凸出物
surface purification 表面净化
surface pyrometer 表面高温计
surface quality 表面性能;表面能力;表面质量
surface quality measurement 表面质量测定
surface quenching 表面淬火
surfacer 压刨(机);整饰涂料;光面器;路面修整机;磨面机;平面刨机;石面刨床;表面修整机;刨压机;刨平机;刨光机
surface raceway 平面布线槽
surface radiation 地面辐射;表面辐射
surface radiation budget climatology project 地面辐射平衡气候学计划
surface radiation coefficient 表面辐射换热系数
surface radiator 表面辐射器
surface railing 路面铁轨;路面护栏
surface railroad 地面铁路;地面铁道
surface railway 路面铁路;地面铁路;地面轨道
surface ranging 地面测距
surface rate equation 表面反应速率方程
surface rating 表面光洁度等级
surface ratio 面积比;表面比
surface reactance 表面电抗
surface reaction 表面反应
surface-reaction control 表面反应控制
surface readout gear 地面读出计
surface reaeration 表面再曝气
surface reboil 表面再生气泡;表面再沸腾
surface receptor 表面受体

surface recombination 表面复合
surface recombination rate 表面复合率
surface recombination velocity 表面复合速度
surface reconnaissance 地面踏勘;地面勘察
surface recording 磁表面记录;表面记录
surface recycling machine 路面翻修机
surface reflection 表面反射
surface-reflection camera 表面反射照相机
surface-reflection method 表面反射法
surface reinstatement 表面填平复元
surface rejuvenation 地面覆生
surface relic 地面残片
surface relief 地形;地势;地面起伏;地表起伏
surface relief method 表面浮雕法
surface removal method 刮面法(一种填缝法)
surface renewal theory 表面更新理论
surface renewing 重修表面
surface repair 路面修补
surface repelling admix(ture) 表面防水剂
surface repelling agent 表面防水剂
surface replica 表面复型
surface resistance 路面阻力;表面阻力;表面摩擦阻力;表面电阻
surface resistance heat transfer 表面换热阻
surface resistivity 表面电阻系数;表面电阻率
surface resistivity measurement 地面电阻率测定
surface restoration 表面恢复
surface retardant 表面阻凝(在模板上刷阻凝剂);缓凝表层;表面阻燃的
surface retarder 表面缓凝剂
surface retarder paper 延时凝结强力纸(作用于表面上的)
surface retarring 表面重涂柏油
surface retention 地面滞留(水量);地面蓄水;地面截流水量;地表滞水;地表滞留;地表流量;地表截流水量;表面滞留
surface reverberation 表面交混回响;表面反射
surface rheometer 表面流变仪
surface rib 穹的露出肋
surface ribbing 路面起搓板
surface rig 地表(钻探)设备
surface rights 地表所有权;地表使用权
surface ripping 表面疏松
surface ripple 表层砂波
surface riser 表面立管;地面立管
surface rising main 表面垂直于干线;地面垂直于干线
surface rock 地表岩石;表层岩石;表层岩
surface roller 水面滚浪;表面镇压器;表面旋滚;表面碾压机
surface rolling 滚压加工;表面冷轧;表面滚压
surface rotational 表面旋翼
surface rotor 表面转子
surface-rough 表面粗糙的
surface roughening 表面粗糙化
surface roughening treatment 表面粗化处理
surface roughness 路面粗糙度;地表粗糙度;表面粗度;表面粗度;表面粗糙率;表面粗度
surface roughness automatic tester 表面粗糙度自动测量分析仪
surface roughness measurement 表面糙率测定;表面粗度测定
surface roughness measuring ga(u)ge 表面粗糙度测量计
surface roughness measuring instrument 粗糙度仪
surface roughness microscope 表面光洁度显微镜
surface roughness of the terrain 地形起伏程度
surface roughness sensor 表面粗糙度传感器
surface roughness specimens 表面粗糙度样块
surface roughness tester 表面粗糙度检查仪
surface running fluid 地表径流(量)
surface runoff 地面流水;地面径流;地表径流(量);表面流出量;地面流量
surface runoff coefficient 地面径流系数;地表径流系数
surface runoff modulus map 地表径流模图
surface runoff of city 城市地表径流
surface runoff peak 地表径流峰值
surface runoff rate 地表径流率
surface runoff water 地表径流水;地表径流水
surface rupture 地面破坏
surface salinity 表层含盐量
surface sample 地面试样;地表样品;表面取样;表层试样

surface sampler 表层采样器;表层采水器;表面取样器;表层取(土)样器
surface sander 磨面机
surface sandsheet 地表砂席
surface sash center 气窗转动轴
surface scaling 金属锈蚀脱皮;道路面层剥落;路面鳞剥;表面掉皮;表面剥皮;表面剥落
surface scarfing 表面嵌接;表面火焰清除
surface scattering 表面散射
surface scour 地面冲刷
surface scouring 表面冲刷
surface scraper 刮土机
surface scratch 表面划痕;表面刮痕
surface screeding 表面刮平;表面抹平
surface scuffing 表面拉毛
surface sea deposit 浅海沉积(物)
surface sealer 表面保护层
surface sealing 表面封闭;表面密封
surface sealing agent 表面密封剂
surface search coordinating vessel 海面搜寻协调船
surface search coordinator 海面搜寻协调船
surface section program(me) 曲面—剖面图程序
surface sediment 表面沉积(物)
surface seep 表面渗漏
surface seepage 地面渗漏;表面渗流
surface segregation 表面偏析
surface seiches 表面湖震
surface self diffusion 表面自扩散
surface set diamond bit 表镶金刚石钻头;表面镶金刚石钻头
surface set diamond reaming shell 表镶金刚石扩孔器
surface-set drill 表面装嵌钻头
surface setting retarder 表面凝结迟缓剂
surface settlement 地面下沉;地面沉降;地基下沉;表面沉陷;表面沉降;隧道地表沉陷;隧道地表沉陷
surface shadow 表面阴影
surface shake 表面裂纹;表面裂缝
surface shape 面形
surface shear viscosity 表面剪切黏性
surface sheet 表面片材;表层板
surface shell 格构外翼;表面薄壳
surface shelter 地上掩蔽体;地上隐蔽所
surface ship 水面舰艇;水面船只
surface-ship gravimeter 船上重力仪
surface-ship gravity measurement 船上重力测量
surface ship gravity metre 船上重力计
surface shrinkage 表面缩陷;表面收缩
surface-shrinkage crack 表面收缩裂缝
surface shrinkage hardened steel 表面收缩硬化钢
surface shut-in pressure 表层关井压力
surface silo 地面筒仓;地面青储[贮]堆
surface site 表面点
surface sites concentration 表面点浓度
surface situation 地势图
surface sizing 上胶
surface sketch 地面草图
surface skimmer 面浮船;面浮车;表面撇渣管
surface skimming 表面撇渣
surface skin 面板;表面皮
surface slide 表层滑动
surface slip agent 表面增滑剂
surface slipperiness 路面滑溜
surface slope 面坡;水面坡降;地面坡度;地表坡度;水面比降
surface slope of water 水面比降
surface smooth 表面光滑的
surface smoothing 表面磨光
surface smoothness 表面平整度;表面平滑度;表面光洁度
surface socket 表面插座
surface soil 地表土;面层土;地面土;表土;表层土(壤)
surface soil creep 表层土蠕动
surface solution 表面溶液
surface source 面光源;地面源;地表水源;地面水源
surface spalling 表面剥离
surface specific gravity of crude oil 地面原油比重
surface speed 表面速度
surface spheric(al) harmonics 球面谐函数
surface spreading 在液面扩展;在表面展开;表面

漫流;表面扩展
surface spread of flame (材料的)表面燃烧速度;面延烧性;火焰表面蔓延;表面延烧性
surface spring 地面泉;地表泉
surface stability 表面稳定性
surface stain 表面变黑
surface staining 表面失去光泽
surface standing wave 表面驻波
surface state 表面光洁度;表面性质;地面情况;表面状态;表面态
surface state effect 表面状态效应
surface step 表面阶梯
surface sterilization 表面消毒
surface storage 地面滞留(水量);地面蓄水;地面截流水量;地表蓄水;地表截流水量
surface storage element 地表储藏要素
surface storm flow 地面暴雨流量;地面暴雨径流
surface strain 表面应变
surface strain ga(u)ge 表面应变仪;表面应变计
surface strain indicator 表面应变仪;表面应变计;表面变形器
surface stratum 表层
surface streamline 表面流线
surface stream pollution 地表河流污染
surface street 地面街道
surface strength 表面强度
surface strength of paper 纸张表面强度
surface stress 表面应力
surface string 表层套管柱;表层管
surface stringer 表面长条夹杂物
surface stripping 地表剥离
surface stripping device 表面脱膜器
surface structure 表面结构;地表构造;表面构造;表层构造
surface structure analysis 表面结构分析
surface structure in building 建筑物的表面构造;建筑物的表面结构
surface structures of lava 熔岩表壳构造
surface subsidence 地面下陷;地面沉陷;地表下沉
surface-supplied diving apparatus 水面供气式潜水装具
surface-supplied diving device 水面供气式潜水装具
surface-supplied diving equipment 水面供气式潜水装具
surface-supplied diving unit 水面供气式潜水装具
surface-support buoy system 海面支撑浮标系统
surface surge 表面涌浪
surface survey 洞外测量;地面测量
surface sweating 表面结露
surface swell 表面涌浪
surface swimmer 表层水游泳动物
surface switch 地面道岔;墙面开关
surface synoptic(al) chart 地面天气图
surface synoptic(al) requirement 地面天气要求
surface synoptic(al) station 地面天气站
surface system 地表微迹系统
surface table 划线台
surface tackiness 表面发黏现象
surface taping 地面测距
surface tarring 表面涂焦油(沥青)
surface technique 堆焊技术;镀面技术;表面修整技术;表面磨削技术;表面加工技术;表面处理技术
surface temperature 地面温度;表面温度;表层水温
surface temperature anomaly 地表温度异常
surface temperature contrast 地表温差
surface temperature effect 表面温度效应
surface-temperature mapping 表面温度测图
surface temperature sensor 表面温度传感器
surface temperature survey 地表温度测量
surface temperature variation 地表温度变化
surface tensile force tester 表面张力试验机
surface tensiometer 表面张力计
surface tension 表面张力
surface tension apparatus 表面张力测定装置;表面张力测定仪
surface tension balance 表面张力天平;表面张力计;表面张力测定天平;表面张力测定杆
surface tension coefficient 表面张力系数
surface tension constant 毛细常数
surface tension method 表面张力法
surface tension modulus of elasticity 表面张力弹性模量
surface tension number 表面张力数

surface tension of moisture films 水薄膜表面张力
surface tension of water 水的表面张力
surface tension reducer 表面张力降低剂
surface tension reducing agent 表面张力降低剂
surface tension test 表面张力测定
surface tension wave 表面张力波
surface tensity 表面张力
surface tester 表面检查仪
surface test tree 地面测试树
surface texture 路面纹理;路面组构;表面状态;表面纹理;表面结构
surface texture of aggregate 集料表面纹理
surface texturing of concrete pavement 混凝土路面的表面扫毛
surface thawing index 地面融化指数
surface thermal flux method 地面热流量法
surface thermal pattern 地表热异常图式
surface thermocouple 表面温差电偶
surface thermometer 海面温度计;贴附温度计;地面温度表;表面温度计;表面测温计
surface thermostat 表面恒温器
surface thrust 地面冲断层;地表冲断层
surface tidemeter 地面潮汐重力仪
surface tillage 表土耕作
surface tillage method 表土耕作法
surface to be exposed to fluid 接触介质的表面
surface to be painted 涂有油漆的表面;涂有涂料的表面
surface to be pressed 拟压面
surface-to-mass ratio 表面与质量之比
surface topography 表面形貌;表面构形;表面地形(学)
surface-to-underwater 舰对水下目标
surface-to-volume ratio 表面积与体积比;比表面积
surface track 明轨
surface traction 表面(牵)引力
surface-transfer type regenerator 表面传热式回热器
surface transformation 表面相变
surface transport 地面运输
surface transportation 地面运输
surface trap 表面陷阱
surface trapping 表面抑制;表面吸收;表面俘获
surface trash 地面上的残碎叶
surface traverse 地面导线
surface traversing 地面导线测量
surface trawl 表层拖网
surface treated road 表面处理的土路;表面处治的道路;表面处理过的路
surface treating machine 表面处理机
surface treatment 表面镀层;表面涂层;路面处治;路面处理;路面表面处治;表面处理
surface treatment of polyolefins 聚烯烃表面处理
surface trowelling 抹面
surface turbulence 地面湍流
surface-type heat exchanger 表面换热器
surface type heating 表面式加热
surface type intercooler 表面式中间冷却器
surface type meter 面板用仪表;面板式仪表
surface-type vibrator 表面振动器;表面(式)振捣器;外附式振捣器
surface unconsistency compensation 表层不平均补偿
surface-underground contrast plan 井上下对照图
surface undulation 地面起伏;表面起伏;表面波荡
surface upheaval 地表隆起
surface vapo(u)r permeance 表面透湿系数
surface varactor 表面变容二极管
surface velocity 面流速;水面流速;表面速度;表面流速;表层流速
surface vent 表面微裂纹;表面通风孔;表面裂纹;表面裂缝
surface vertical displacement 地表垂直位移量
surface vibrating machine 表面振动器;附着式振捣器
surface vibration 表面振动;地表振动;表面振捣
surface vibrator 表面振动器;平面振实器;平板振捣器;附着振捣器;附着式振捣器;表面式振捣器
surface view of epidermis 表皮的外观
surface view plan 地面俯视图
surface viscometer 表面黏度计
surface visco(si)meter 表面黏度计
surface viscosity 表面黏度
surface viscosity of crude oil 地面原油黏度

surface visibility 地面能见度;地面可见度;表面能见度
surface voids 表面孔隙;蜂窝;麻面(混凝土)
surface voltage 表面电压
surface-volume ratio 面积体积比;表面与体积比
surface wall 表面墙
surface wash 地面冲洗;地面冲刷;地面冲积;表面冲积物
surface-washed concrete 表面冲洗过的混凝土
surface wash(ing) 表面冲洗;表面冲刷;地表径流排水
surface washing room 表面冲洗滤池
surface washoff 地表冲刷
surface wash of rain 雨水冲刷
surface waste 地面流失;表土流失;表面流失
surface water 淤水;混合层;地面水;地面流水;地表水;表面水;表层水
surface water acidification 地表水体酸化
surface water balance 表层水平衡
surface water body 地表水体
surface water control 表层水控制
surface water degradation 地表水水质恶化
surface water drain 街沟;路面排水沟;地面水排水道;地面排水沟;地表水排除;地表排水沟;地表排水
surface(water)drainage 地面排水;表面排水
surface water heavy metal pollution 地表水重金属污染
surface water hydrology 地面水水文学;地表水水文学
surface water infiltration 地表水渗入
surface water influence 地表水影响
surface water inlet 地表水入口;地表水进口
surface water intake 地表水取水构筑物
surface water irrigation 地面水灌溉;地表水灌溉
surface water leakage 地表水漏失
surface water management 地表水管理
surface water pollution 地面水污染;地表水污染
surface water pollution map 地表水污染图
surface water-proof agent 表面防水剂
surface water-proofer 表面防水材料;表面防水剂
surface water-proofing 表面防水
surface water-proofing agent 表面防水剂
surface water-proofing emulsion 表面防水乳液
surface water-proofing liquid 表面防水液
surface water protection 地表水保护
surface water quality 地表水质;地表水水质
surface water quality assessment 地表水水质评价
surface water quality model 地表水水质模型
surface water quality sanitary standard 地面水水质卫生标准;地表水水质卫生标准
surface water quality standard 地面水质量标准
surface water repellent admix(ture) 表面排水外加剂;表面斥水外加剂
surface water repellent emulsion 表面排水溶液;表面斥水乳液
surface water-repeller 表面排水剂;表面斥水剂
surface water resources 地面水资源;地表水资源
surface water rice soil 地表水型水稻土
surface water runoff 地表水径流
surface waters 地表水体
surface water sample 地表水样;表层水样
surface water sampler 表层水采样器
surface water sampling 表层水采样;表层采水样
surface water sampling point 表层水采样点
surface water sewer 雨水阴沟;地表水排除管;地面水排水管
surface watershed 地面水流域;地面分水岭;地面分水界;地表水流域
surface watershed line 地表水分界线
surface water source 地面水水源;地表水水资源;地面水源
surface water supply 地下水补给;地面水供应;地面水给水;地表水供应;地表水供给水;地表水补给
surface water survey 地表水调查
surface water system 地表水系
surface water temperature 表面水温
surface water treatment requirement 地表水处理要求
surface water treatment rule 地表水处理规则
surface water treatment system 地表水处理系统
Surface water Yearbook 地表水年鉴
surface wave 面波;海面波;水面波;地面电波;地波;地表电波;船行波;表面波

surface-wave antenna 表面波天线
surface wave hydrophone 面波勘探仪
surface wave instrument 表面波测定仪
surface wave magnitude 面波震级
surface wave measuring 表面波法
surface wave motion 表面波传播
surface wave ness recording instrument 表面波度记录仪
surface wave probe 表面波探测器;表面波辐射测试器
surface wave signal correlator 表面波信号相关器
surface wave testing method 表面波测试法
surface wave transmission line 表面波传输线
surface waviness 条纹;条痕;表面波度
surface weather chart 地面天气图
surface weathering 地表风化
surface weather observation 地面天气观测
surface weather observation network 地面天气观测站网
surface weeds 表土杂草
surface/weight ratio 面积/重量比
surface weir 溢流堰
surface weld 表面焊缝
surface welding 表面焊接
surface weld metal buildup 表面焊缝金属的形成
surface wettability 表面润湿性
surface wetting and adhesion 表面润湿荷黏结性
surface width 路面宽度
surface wind 水面风;海面风;地面风;地表风
surface winding 沿电绕组;表面绕组
surface wind stress 表面风应力
surface wiring 明线布置;明线布线;明线;表面布线
surface wiring switch 明线开关;明装搬把开关
surfacewise 沿着表面
surface with gravel 砾石铺面
surface work function 表面功函数
surface working 露头作业;露头开采场;表面研磨;表面加工
surface works 地面工程;地表工程;地面工作
surfacework trench 地面工程(沟槽);沟槽
surface-wound armature 平滑电枢
surface wrinkling 表面起皱
surface yield 总水量地下水;地表水水量
surface zonal harmonics 球面带谐函数
surfacial spalling 表面剥落
surfacing 轨面找平;路面整修;路面修整;面层材料;铺筑路面;铺面;铺路面;喷焊;撒布料;镀面;表面修整;表面修饰;表面平整;表面磨削;表面加工;表面堆焊;表面处理
surfacing and boring machine 车面镗孔两用机床
surfacing board 面板;表面修饰用板;铺面板
surfacing by fire cracker welding process 躺板极堆焊
surfacing component 表面装修构件;表面修饰用组分
surfacing composition 表面修饰成分
surfacing compound 表面修饰合成物
surfacing concrete slab 表面覆盖的混凝土板;铺面混凝土板
surfacing construction 路面建筑
surfacing course 铺面层
surfacing cut 端面车削
surfacing design 路面设计
surfacing electrode 堆焊焊条
surfacing feed 横向进刀
surfacing foil 表面修饰用箔片
surfacing foundation 路基
surfacing gear 横向进刀机构
surfacing hopper 撒布料漏斗
surfacing lathe 落地车床;端面车床
surfacing laying 路面铺筑
surfacing machine 路面摊铺机;端面切削机床;表面平整机
surfacing mass 表面物质
surfacing mat 路面保护层;表面毡;表面修整用席;表面薄毡
surfacing material 面层材料;装修材料;铺屋面料;铺面材料
surfacing member 表面构件
surfacing mixture 铺面混合料
surfacing of areas 地面铺砌
surfacing of lumber 木料的表面处理;木料面上记号;木料表面处理
surfacing of tunnel lining 衬砌饰面
surfacing panel 表面镶板
surfacing performance 路面(耐用)性能
surfacing power feed 端面机动进给
surfacing quality 路面性能
surfacing restoration 路面修复
surfacing roughness 路面平整度;路面粗糙度
surfacing sheet 表面(修饰用)薄板;铺面板
surfacing sinkage 路面沉陷
surfacing stoneware 表面修饰用石制品;表面修饰用粗陶瓷
surfacing tester 路面试验机
surfacing thickness 路面厚度
surfacing time 出水时间
surfacing train 铺路机械列车
surfacing unit 平面部件
surfacing ware 表面修饰制品
surfacing welding electrode 堆焊焊条
surfacing welding rod 堆焊填充丝;堆焊填充焊棒;堆焊填充棒;堆焊焊丝
surfacing with band-electrode 带极堆焊
surfactant 界面活化剂;表面活性物质;表面活性剂;表面活化剂;表面缓凝剂
surfactant biodegradation 表面活性剂生物降解
surfactant compound 表面活性化合物
surfactant concentration 表面活性剂浓度
surfactant contaminated zone 表面活性剂污染带
surfactant micelle 表面活性剂胶团
surfactant modified montmorillonite clay 表面活性剂改性蒙脱石黏土
surfactant molecular structure 表面活性剂分子结构
surfactant mud 表面活性物质的泥浆
surfactant organic compound 表面活性剂有机化合物
surfactant organic solute 表面活性剂有机溶质
surfactant-producing bacteria 产表面活性剂菌
surfactant property 表面活化性能
surfactant solution 表面活性剂溶液
surfactant wastewater treatment 表面活性剂废水处理
surfaction 表面改质
surfactivity 表面活性;表面活度
surfagauge 表面光洁度测量仪;表面粗糙度测量仪
surfascope 光学表面光洁度测量仪
surf beat 破波拍岸;破波带拍;拍岸浪;岸边碎波拍击
surf-beaten shore 浪击海岸
surf board 冲浪板;防浪板;滑水板
surf boat 碎浪艇
surf boat man 碎浪艇船员
surf breaker 激浪;拍岸浪花
surf cut terrace 浪蚀阶地
surf day 有激浪日;多浪日;风浪日
surfeit 过量
surficial 地表的;地面的
surficial creep 土爬
surficial deposit 地表沉积;表层沉积
surficial dispersion 地表分散
surficial environment 地表环境
surficial geology 地表地质(学)
surficial organism 表面生物
surficial sampler 表层取(土)样器
surficial sampling 表层取样;表层采样
surficial soil 表土
surficial water 地表水
surfing 冲浪运动
surflexion 超弯曲现象
surf line 激浪线;碎浪线;碎波线
surf man 碎浪艇船员
surf port 风浪港
surf ripple 激浪痕
surf similarity parameter 碎波相似性参数
surf ski 冲浪橇
surfused liquid 过冷液体
surf washer 浪式洗矿槽
surfy 有浪花的
surf zone 破浪带;破波带;拍岸浪区;拍岸浪带;碎浪区;碎浪带;碎波区
surge 涌波;纵荡【船】;浪涌;绞盘腰部;急放绳链;急变;冒涨;强风潮;气压波;气象潮(气压变化所引起的潮汐变化);突然放松绳线;上下波动;电涌;电流剧变;颠簸(指船);大浪;冲击波;波涛;波动;冰川涌流
surge absorber 过压吸收器;电涌吸收器;冲击压力吸收器;冲击吸收器;避雷器
surge absorber condenser 过压吸收电容器;电涌吸收电容器
surge admittance 浪涌导纳;特性导纳;电涌导纳
surge amplitude 涌浪振幅
surge anticipator 预调压阀
surge arrester 涌流接地;电涌放电器;防电涌器
surge arrester equipment 涌浪制动设备
surge arrester with parallel gaps 带并联间隙避雷器
surge basin 消能池;消波池;调压池
surge bin 溢涌槽;中间仓;缓冲储[贮]仓;缓冲(存)仓;平衡(料)仓;调压槽;调节槽;调节仓
surge block coil 冲击波阻尼线圈
surge bunker 缓冲仓;调节仓;防止过载斗
surge capacity 浪涌能力;调节能力
surge celerity 涌浪波速
surge chamber 均衡室;均衡风缸;缓冲室;缓冲井;调压室
surge characteristic 过电压特性
surge check valve 充液阀
surge coefficient 涌浪系数
surge column 缓冲柱
surge connector 脉冲接插件
surge control 压力波调压;压力波调节;冲击波调压;冲击波调节
surge crest ammeter 脉冲峰值安培计;电涌峰值安培计;冲击峰值电流表
surge current 涌流;浪涌电流;冲击电流
surge damper 减震器;缓冲器;稳压器
surge distortion 浪涌失真
surge diverter 电涌分流器;避雷针;涌流分流器
surge drum 积储[贮]器;缓冲罐;气液分离器;平衡筒;平衡罐;低压储[贮]液筒
surge effect 涌浪效应
surge excitation 冲击励磁
surge factor 涌浪系数
surge front 涌浪前沿;冲击波前沿
surge generator 脉冲发电器;马克斯发生器;冲击发生器;冲击波发生器
surge generators 冲击发电机
surge guard 振荡闭锁装置;电涌防护
surge header 气液分离器;低压储[贮]液筒
surge height 涌浪高度;涌波高度
surge hopper 缓冲漏斗;调节漏斗
surge impedance 浪涌阻抗;特性阻抗
surge injector 脉冲注入器;脉冲引弧器
surge in pipeline 管(道)中涌波;管道涌波
surge in piping 管中涌波
surge in tube 管中涌波
surgeless relay 无冲动继电器
surge level 冲击电压电平
surge line 涌浪线;浪涌线;海边波涛线;气压波线;风速骤增线;风速突变线
surge load 激增负荷
surge load tolerance 激增负荷容限
surge modifier 电涌吸收器
surge of reciprocating pump 往复泵的脉动(作用)
surgeon 船上医生
surge oscillography 脉冲示波术
surge parameter 波参数
surge pile 储料堆
surge pipe 解涌管;缓冲管;平压竖管;调压竖管;调压管
surge plate 涌浪挡板;防冲板
surge plunger 涌浪缓冲滑阀
surge pocket 缓冲仓;调节料仓
surge point 供压陡降点;起浪点
surge pressure 冲击压力;峰值压力;波动压力
surge prevention 防冲击
surge prominence 冲浪日耳
surge-proof transformer 防电涌变压器
surge protection 过电压保护;高峰电压保护
surge-protect socket 电涌保护插座
surge-protect switch 电涌保护开关
surge pump 冲击泵;涌浪泵;隔膜泵;膜式泵;薄膜式泵
surge relay 冲击继电器;波前陡度继电器
surge sentinel 涌浪标志
surge separator 缓冲分离器
surge shaft 调压(水)柱;浪涌缓冲井;缓冲井;调压井
surge simulator 涌浪模拟器
surge spray 不均匀喷射

surge storage 调节仓库
surge storage pile 电堆
surge strength 冲击强度
surge stress 冲击应力
surge suppression 涌浪抑制
surge suppressor 消减涌浪设备;涌浪抑制器;涌浪消减设备;涌波抑制器;平压装置;电涌抑制器
surge tank 涌浪调整槽;中间罐;均压箱;缓冲罐;气室;平缓罐;平衡箱;平衡罐;稳压罐;调压水箱;调压水塔;调压井;调压池;调浆槽;水电站调压塔
surge test 冲击波试验
surge tester 脉冲试验仪
surge the capstan 松掉绞盘上绳索
surge tower 调压塔
surge valve 间歇作用阀;补偿阀
surge velocity 涌浪速度
surge vessel 平衡容器
surge voltage 浪涌电压;脉冲电压;冲击电压
surge wave 泄放波;涌潮;壅水波;碎波
surgical department 外科
surgical hospital 外科医院
surgical lavatory 医用洗手盆
surgical stitching 缝合
surgical suite 手术间
surgical ward 外科病房
surgidero 锚地
surging (调压塔内的)水位波动;涌波;纵荡【船】;浪涌;冲击电压;冲击;岸坡涌浪;岸边激浪
surging breaker 激散破波;冲涌碎浪
surging current 脉冲流
surging flow 脉动流;脉冲流
surging force 涌浪力;脉冲力;冲击力
surging glacier 间歇冰川
surging lap 反复折叠;波状重皮
surging line 浪涌线
surging pad 摩擦轮垫
surging profile 涌浪波型
surging sea 汹涌海面;涌浪
surging shock 涌浪冲击;液锤;浪涌撞击
surging water 潮水
surging wave 激散破波
surging well 间歇自流井
surinamite 硅镁铝石
Surinam mahogany 苏里南红木
surite 碳硅铝铅石
surmount 饰顶;罩面;超越
surmountable fatigue 可耐受疲劳
surmounted arch 超半圆拱;长脚半圆拱
surmounted dome 超半圆穹隆;超半圆穹顶
surpalite 聚光气
surpass 超越;超过;超额
surperficial alternation 表面温度变化
surperficial hardening 表面淬硬
surplus 盈余;过剩的;净余;顺差;结余;剩余量;剩余的;超量;残料
surplus activated sludge 剩余活性污泥
surplus area 堆积区
surplus assets 剩余资产
surplus at liquidation 清算盈余
surplus at opening(of the) period 期初盈余
surplus available for dividends 可供股息分配的盈余
surplus buoyancy 剩余浮力
surplus cant 余超高
surplus capacity 余量;过剩生产能力;剩余水量;剩余能力;富余容量
surplus capital 过剩资本;剩余资本
surplus charge 冲转盈余;冲转公积的借项
surplus compensation 超额补偿
surplus content 剩余水量
surplus deflection of spring 弹簧挠度余量
surplus deployment 备用机场
surplus disposal plant 余料处理(工)厂
surplus disposal work 余料处理(工)厂
surplus earth 弃土;剩余土方;多余土方
surplus energy 剩余力量;剩余能(量)
surplus factor 剩余因子
surplus flood volume 超额洪量
surplus flow 剩余流量
surplus for current year 本年盈余
surplus for the year 本年盈余
surplus from donation 捐赠差益
surplus from profit 已获盈余

surplus from property reappraisal 财产重估盈余
surplus from revaluation 重估盈余
surplus fund 盈余资金;经费剩余
surplus fund audit 盈余公积金审计
surplus generation 剩余发电量
surplus head of dredge pump 泥泵剩余扬程
surplus heat 余热;过剩热
surplus-heat firing 余热烧成;余热焙烧
surplus heat utilization 余热利用
surplus humidity 余湿
surplus income 额外收入
surplus in preceding financial year 上财政年度结余
surplus in preceding fiscal year 财政年度盈余
surplus in the budget 财政结余
surplus inventories 超储
surplus labo(u)r 剩余劳动力
surplus lining overbreak of tunnel 隧道内超挖
surplus load 剩余荷载;超额荷载
surplus material 剩余材料;剩余物资
surplus materials division 剩余物资处
surplus modulus of groundwater 地下水盈余模数
surplus mortar 过剩砂浆;表面砂浆
surplus of capital 资本过剩
surplus of cash 现金过剩
surplus of national income 国民收入余量
surplus of water 多余的水
surplus oil 剩余油
surplus on appreciation of capital assets 固定资产增值盈余
surplus or deficit of appropriation 经费余绌
surplus or deficit of budget 预算余绌
surplus population 人口过剩
surplus power 过剩电力;剩余功率;剩余动力;剩余电力
surplus pressure 剩余压力
surplus product 剩余产品
surplus profit 剩余收益;结余利润;剩余利润;分余利润;额外利润
surplus receipts 额外盈余;额外收入
surplus reconcilement 盈余调节
surplus reinsurance 溢额再保险
surplus reinsurance treaty 溢额分保合同
surplus reserves 盈余准备
surplus seats and berthes on train 剩余能力【铁】
surplus section 水量过剩河段
surplus share 溢额部分
surplus sludge 剩余污泥
surplus social purchase power 社会节余购买力
surplus soil 剩余土方;弃土
surplus spoil 废石方;弃土
surplus stock 剩余原料;多余原料
surplus superelevation 过超高
surplus torque 剩余扭矩
surplus trade balance 贸易顺差
surplus traffic volume 剩余交通量
surplus transportation 剩余运输工具
surplus value 剩余值;剩余价值
surplus valve 溢流阀
surplus variable 剩余变量
surplus water 溢水;过剩水量;弃水;剩余水
surplus water from river during flood season 河流弃洪
surprint 加印
surprise attack 突然袭击
surprise audit 突袭式审计
surprise check 突袭式核查
surprise checkpoint 突击检查站
surprise count 突袭式清点
surprise factor 意外因素
surrender 解约;交出
surrender bill of lading 交出提单;收回提单
surrender documents 交单
surrenderee 受让者;受让人
surrender for cancellation 交出注销;收回注销
surrender of bills 返还票据
surrender of rights 放弃权利
surrenderor 让与人
surrender value 解约退还的保险费;退保金额;保险单(的)退保值
surrey 双排座四人小客车
surrogate 代用物;代用品;代理
surrogate indicator 代用指标

surrogate parameter 替代参数;代用参数
surrosion 腐蚀增重作用
surround 环绕;围绕;包围
surround curtain 舞台帷幕
surrounding 周围环境;外界;周围的(事物);周围介质
surrounding area 周围地区
surrounding atmosphere 周围大气
surrounding chapels 周围的基督教小礼拜堂
surrounding countryside 近郊;郊区
surrounding dam 围堰
surrounding embankment of a spoil ground 弃土场围堤
surrounding environment 周围环境;生活环境
surrounding fluid pressure 周围液体压力
surrounding line 边线
surrounding loop 外层循环
surrounding masonry wall 围绕的圬工墙
surrounding material 围岩;周围的土(隧洞);周围材料
surrounding region 周围地区
surrounding rock 周围岩石;围岩;峒室围岩
surrounding rock class 围岩类别
surrounding rock pressure 土压或地压
surrounding rock stability 围岩稳定
surrounding rock stress 围岩应力
surrounding rock type 围岩类型;围岩类别
surroundings 环境;外界;生活环境
surroundings investigation 围岩调查
surrounding soil 周围土壤
surrounding soil stratum 周围土层
surrounding strata of diapir 底辟围岩
surrounding structure 包覆组织(指金晶)
surrounding topography 周围地形
surround of a comparison field 视场背景
surround profile 周围剖面;周围轮廓
surround section 座位形状;座位外形
surround slab 镶板地板;镶边楼板
surround trim 周围镶边;围周修饰
surround unit 周围单元
sursassite 锰帘石
sursulfate cement 硫酸盐矿渣水泥
surtax 超额累进所得税;附加税
surtax exemption 附加税减免
surtax survey 附加税调查表
Surtseyan 色特塞式
surveillance 监视
surveillance and control equipment 监控设备
surveillance and evaluation system 监视与计算系统
surveillance approach 监视进场
surveillance calibration satellite 监视校准卫星
surveillance device 警戒装置;监视装置
surveillance mirror 观测镜
surveillance network 观察台网;观测网;监测台网
surveillance of disease 疾病监视
surveillance of environment 环境普查;环境监测;环境监视
surveillance photography 侦察摄影
surveillance radar 监视雷达
surveillance radar station 监视雷达站
surveillance railway statistics 铁路统计监督
surveillance ship 侦察船;监视船
surveillance sonar 监视声呐
surveillance system 观察系统;观测系统;监视系统
survey 测量;观察;概况;勘查;调查;查勘
survey across 直角测器
survey adjustment 测量平差
survey adjustment of observation 测量平差
survey agent 检验代理人
survey aircraft 测量飞机
survey analysis 普查分析
survey and design expenses 勘察设计费
survey and design fee 勘察设计费
survey and drawing 测绘
survey and tender clause 检验与投标条款
survey and test of cargo gear 起货设备检验和试验
survey area 测区面积;测区
survey beacon 测量标志
survey boat 观察船;测量艇
survey borehole 测井孔
survey bridge 测量电桥
survey buoy 测量用浮标
survey bureau 测量局;测绘局

survey by aerial photographs 航空摄影测量;航测
survey by boring 勘探钻进
survey by drilling 勘探钻进
survey by radiation 极坐标测量法
survey by serial photography 连续摄影测量
survey charges 检验费
survey coordinates 测量坐标
survey count rate 测量计数率
survey craft 测量艇
survey crew 勘测人员;勘测队;测量人员;测量班
survey data 勘测资料;测量资料
survey date 检验日期
survey datum 测量基准(面)
survey department 测绘局
survey device 测量装置
survey drawing 测量(制图)
surveyed 已测的
surveyed before shipment 装船前已检验
surveyed coastline 精测岸线
surveyed drawing 实测图
surveyed geologic(al) profile 实测地质剖面图
surveyed line 观测线
survey engineer 测量工程师
survey equipment 勘测设备;测量设备
survey equipment inside boring 孔内探测设备
survey fee 检验费
survey field notes 野外测量手簿
survey flying 测量飞行
survey foot 测量英尺
survey for alternation 改装检验
survey for damage 损坏检验
survey for design stage 勘测设计阶段测量
survey for establishing geologic(al) prospecting network 地质勘探网测量
survey for land grading 平整土地测量
survey for land preparation 平整土地测量
survey for land smoothing 土地平整测量
survey for marking of boundary 标界测量
survey for the finding of ocean(ic) minerals 海洋矿产调查
survey for water table level 地下水位调查
survey gang 测量班
survey grid 控制网【测】;测量格网;测点网
surveying 调查;测量学;测量术
surveying accessory 测量用辅助器材
surveying achievement map 调查成果图件
surveying achievement report 调查成果报告
surveying aircraft 航空勘测飞机
surveying altimeter 空盒气压计;测量用高度计
surveying and laying out the line 测量及划线
surveying and mapping 测绘学;测绘
surveying and mapping instrument 测绘仪;测绘器
surveying aneroid 空盒气压计
surveying area 测量面积;调查区域;测量区域
surveying base point for track laying 铺轨基标
surveying board 测图板
surveying boat 测量船;水道测量艇;测验船
surveying buoy 测量用浮标;测量浮标【水文】
surveying by laser 激光测量
surveying calculation 测量计算
surveying camera 测量摄影机;测量照相机
surveying chain 测链(长66英尺)
surveying classification 调查类别
surveying code 测量规范
surveying compass 测量指向仪;测量罗盘(仪)
surveying computation 测量计算
surveying control network 测量控制网
surveying country 调查国家
surveying craft 测量船
surveying crew 测量队
surveying cruise 调查航次
surveying data 测量资料
surveying dial 测量罗盘(仪)
surveying engineer 测量工程师
surveying error 测量误差
surveying field book 测量手簿
surveying float beacon 测量浮标【水文】
surveying for land level(l)ing 土地平整测量
surveying for site selection 选厂测量
surveying gang 测量队
surveying group 测量组;测量队
surveying guide 调查准则
surveying in parts 碎部测量

surveying institution 调查部门
surveying instrument 测量仪器
surveying level 水准仪;测量水准仪
surveying line 测线
surveying map board 测图板
surveying mark(er) 测量标志;测标;测量标记;测量标
surveying mark on dam surface 坝面测标
surveying mark on rock surface 岩面测标
surveying method 调查方法
surveying of borehole 炮眼斜度测量;钻孔斜度测量
surveying office 测量局
surveying of the ground temperature 地温测量
surveying on devanning container 集装箱货拆箱鉴定
surveying on vanning container 集装箱装箱鉴定
surveying panel 平板仪;测图板;测量平板(仪)
surveying party 勘测队;测量队;勘察队
surveying peg 测桩【测】;测量标桩
surveying picture 调查片
surveying pin 测量针
surveying place 测量区
surveying plane table 地形测量平板仪;平板仪
surveying plate table 测量平板(仪)
surveying point 测点
surveying point elevation 测点海拔高度
surveying pole for differential formation 分层标
surveying practice 测量实践;测量实习
surveying range 测量范围
surveying record 测量记录
surveying rod 水准标尺;测量标尺;测杆;花杆;标杆;标尺
surveying scale 调查比例尺
surveying sextant 测量六分仪;六分仪;水文地理测量六分仪
surveying sheet 测量图;地形图
surveying ship 调查船;测量船
surveying signal 测量觇标
surveying specification 测量规范;测量工作规程
surveying spot 测量地点
surveying squad 测量组
surveying staff 测量准尺
surveying stake 标桩(测量用的);测桩【测】;测量标桩
surveying station 测站
surveying table 测量平板(仪)
surveying target 测量觇标
surveying team 测量组;测量队;勘察队;测量术语
surveying time 调查时间
surveying tool 测量工具
surveying trawl 扫海具
surveying unit 测量组
surveying vessel 测量船;测量艇
surveying work 勘测;测量工作
survey in loading port 装货港检查
survey in reconnaissance and design stage 勘测设计阶段测量
survey in rotation 轮流检验
survey instrument 测量仪器
survey launch 测量艇
survey level 水准仪
survey line 测线;测量中线
survey location 测线位置
survey map 实测地图;测量图
survey mark made of steel tube 钢管标石
survey measurements 测量读数;测量值
survey meter 巡测仪;检测计;探索器;探测仪(器);测量器
survey method 鉴定法
survey method of cost allocation 成本分配的测量法
survey mode 施工方式;调查工作方式
survey monument 测量标石
survey network 测图网
survey network of lower name 图根网
survey not complete 检验不全
survey note 图廓外测图说明;测量记录
survey of a structure 建筑物的勘查;建筑物的测量;结构物的勘查;结构物的测量
survey of car speed 车速调查
survey of cleaning hold 货舱扫舱检验
survey of crane beam location 吊车梁测量

Survey of current business 现代商业概览(美国)
survey of dam breach flood 溃坝洪水调查
survey of details 碎部测图;碎部测量
survey of disasters 灾情调查
survey of elastic deformation of foundation pit 基坑回弹测量
survey of environment(al) statistics 环境统计调查
survey of eolian features 风成地貌调查
survey of existing railway 既有线测量【铁】
survey of glaciated landform 冰川地貌调查
survey of groundwater 地下水调查
survey of igneous rock 火成岩调查
survey of karst morphology 岩溶地貌调查
survey of land 地亩测量
survey of loess landform 黄土地貌调查
survey of metamorphic rock 变质岩调查
survey of morpholithos 岩石地貌调查
survey of morphology 地貌调查
survey of morphostructure 构造地貌调查
survey of mountain morphology 山岳地貌调查
survey of plain morphology 平原地貌调查
survey of plateau morphology 高原地貌调查
survey of pollution sources 污染源调查
survey of quaternary system 第四系地质调查
survey of repair 修理检验
survey of reservoir inundation line 水库淹没界线测量
survey of resource of silt 淤积物来源调查
survey of river morphology 河流地貌调查
survey of sedimentary rock 沉积岩调查
survey of ship 验船;船舶检验
survey of shore morphology 滨岸地貌调查
survey of site 工地测量
survey of stores 供应检查;物料检查
survey of stratigraphy 地层调查
survey of structural geology 构造地质调查
survey of the terrain 地形测量
survey of tow 拖航检验
survey of volcanic cone 火山岩调查
survey on building activities and losses 建筑动态调查
survey on hire 起租检验
survey on scale 按规定比例尺测图
survey on the sea 海上调查
surveyor 验货员;验船师;勘测程序;检验员;检验人;检查员;测量员;测量器
surveyor appointed 指定验船师
surveyor authorized 授权的验船师
surveyor competent 合格验船师
surveyor general's specifications 测绘规范
surveyor of customs 海关检查人员
surveyor of the technical inspection 技术鉴定者
surveyor peg 测量木钉;测量标桩
Surveyor program(me) 月球勘测者计划
surveyor's alidade 平板照准仪;照准仪(平板仪上的)
surveyor's arrow 测钎;测针【测】
surveyor's bob 测量用垂球
surveyor's chain 六十六尺测链;测链(长66英尺)
surveyor's compass 测量指南针;测量用罗盘仪;测量罗盘(仪)
surveyor's cross 测量十字线
surveyor's dial 测量指南针;测量用罗盘仪
surveyor's fee 鉴定费(用)
surveyor's instrument 测量仪器
surveyor's level 水平仪;水准仪;测量水准仪;测量水准器;测量水平仪;测量水平尺
surveyor's pass 检验合格证
surveyor's pole 花杆【测】;测量标杆;测杆;标杆
surveyor's quality certificate 鉴定品质证书
surveyor's report 鉴定证明书;鉴定行报告;鉴定报告
surveyor's rod 曲尺;测杆;水准尺;测量标尺;水平尺;测杆
surveyor's staff 测量标尺
surveyor's stake 测量标桩
surveyor's table 测量平板(仪)
surveyor's target 测量觇标
surveyor's transit 经纬仪;测量经纬仪
surveyor's tripod 测量用三脚架
surveyor's weight certificate 鉴定重量证书
survey parameter of radioactive method 放射性方法测量参数
survey party 测量队;勘察队;勘测队;测量班

survey period 检验周期
survey photograph 航片
survey photography 地形测图摄影
survey photomapping 航片测图
survey pin 测量标针;测钎
survey pipes form the surface 设置在地面上的观测管道
survey plan 测量平面图
survey platform 观测平台;勘探平台
survey plug 水准标桩;测槽
survey post 测标
survey precision 调查精度;测量精(确)度
survey precision of apparent resistivity 视电阻率测量精度
survey questionnaire 情况调查表
survey reading 测量读数
survey report 勘测报告;鉴定报告;检验报告;调查报告;测量报告
survey report centre 测绘资料中心
survey report on quality 品质鉴定证(明)书
survey requested 申请检验
survey results 调查成果
survey rod 标尺
survey sampling 试验性采样
survey satellite 测量卫星
survey scale 测图比例尺;测量比例尺
surveys during placement 拼装阶段的测量
survey service 测量局;测绘局
survey sheet 测量图(表);实测原图;地形图;测图原图
survey sheet comparison 测图比较
survey situation 碎部测量
surveys of attitudes 态度调查
survey stage 普查阶段
survey stake 测量标桩
survey sweep 扫海具
survey table 调查表
survey tape 卷尺
survey task of gravity 重力测量任务
survey team 勘察队;勘测队
survey theodolite 测地经纬仪
survey the terrain 察看地形
survey training 测量训练
survey transit 经纬仪
survey traverse 测图导线;测量导线
survey truck 测量车
survey tunnel 探查坑道
survey umbrella 测伞;测量伞
survey unit 测量装置
survey value of base station 基点观测值
survey vessel 勘探船
survey work during construction 施工期测量
survivability 耐受性;生存性;残存性
survivability criterion 生存性准则
survivable route 保险路线
survival 存活者;残余;残存者
survival bond 残存债券
survival craft 救生艇具;救生艇筏
survival craft station 救生艇筏电台
survival curve 生存曲线;存活曲线
survival equipment 救生设备
survival food 救生食品
survival fraction 存活分数
survival function 剩余函数;残存函数
survival kit 急救装置
survival method 幸存法
survival movement 生存活动
survival of the fittest 适者生存;适应生存
survival percent 成活率
survival period of bacteria 细菌生存时间
survival potential 生存潜能
survival probability 生存概率
survival rate 生存率;存活率
survival ratio 存活比;成活率
survival rations 救生口粮
survival strategy 生存战略
survival suit 保暖救生连衣裤;救生服
survival time 生存时间;存活时间
surviving 残存
surviving company 续存公司
survivor 幸存者
survivor curve 残存曲线
survivor ion 残存离子
survivor's benefit 遗族恤金

survivorship curve 生存曲线
survivorship method 幸存法
Surwell clinograph 瑟韦尔孔斜计
susannite 菱硫碳酸铅矿;三方硫碳铅石
susceptance 电纳
susceptance loop 电纳环
susceptance standard 电纳标准
susceptibility 易感性;灵敏性;敏感性;敏感度;磁化系数;磁化率;磁鼓系数
susceptibility anisotropy value 磁化率各向异性值
susceptibility ellipse 磁化率椭球
susceptibility meter 滋化率仪;磁化率计
susceptibility to change in temperature 感温性
susceptibility to corrosion 腐蚀敏感性
susceptibility to failures 易损性
susceptibility to frost 霜冻耐受性;霜冻感受性
susceptibility to frost action 易冻结性
susceptibility to heat 热的耐受性;热的感受性;热敏性
susceptibility value of sample 标本磁化率值
susceptibility value on outcrops 露头磁化率值
susceptible 易受影响的;易感应的;易感动的;敏感的
susceptible period 易感受期
susceptible to moisture 易受潮的;对湿度敏感的
susceptible to water 对水敏感的
susceptive 敏感的
susceptiveness 灵敏性;敏感性
susceptivity 易感性
susceptivity analysis 敏感性分析
susceptometer 磁化率计
susceptor 感受器;衬托器
susceptor element 感应元件
suspected carcinogen 可疑致癌物
suspected ship 有疫病嫌疑船
suspect terrane 可疑地体
suspend 悬挂;悬浮;中止;暂停;挂起
suspendable 可中止的
suspendable subsystem 可中止子系统
suspend arch 悬索拱(由索和拱组合的结构)
suspend bed sediment 悬浮底质;悬浮床沙质
suspended 悬挂的;悬吊的;被中止的
suspended absorber 悬吊式吸声器;悬挂吸收器;悬挂吸声体;空间吸声体
suspended accounts receivable 未决应收账款
suspended acoustic(al)ceiling 吸声顶棚;吸声吊顶
suspended acoustic(al)plaster ceiling 隔声泥灰吊顶
suspended acoustic(al)tile ceiling 隔声砖吊顶
suspended aggregate removal 悬浮团粒去除
suspended algae 浮游藻类
suspended arch 悬托拱;悬拱;吊毫
suspended array surveillance system 悬挂式基阵监视系统
suspended ash 悬浮灰分
suspended beam 悬吊物件用的梁;吊梁;悬式梁;挂梁
suspended bed reactor 悬浮床反应器
suspended bed sediment 悬浮底质
suspended block 吊砖
suspended body flowmeter 悬浮体流量计
suspended bridge 悬索桥
suspended building 悬挂式建筑
suspended cable 悬索
suspended cable anchor 悬索锚碇(悬索桥缆索各式的锚固)
suspended cable roof 悬索屋顶;悬索屋盖
suspended cable structure 悬索结构
suspended cable system 悬索系统
suspended call 中止的呼叫
suspended cantilever hangar 吊式悬臂飞机库
suspended cantilever roof 吊式悬臂屋顶
suspended capillary water 悬挂毛细水;毛细悬挂水
suspended car 悬空车
suspended carbon 悬浮碳
suspended cargo 悬吊货物
suspended carrier 悬浮载体
suspended carrier biofilm reactor 悬浮载体生物膜反应器
suspended catwalk 悬挂跳板;悬挂脚手架
suspended ceiling 悬吊顶;室内吊顶;吊栅;吊平顶;吊顶棚;吊顶;吊车;悬式天花板
suspended ceiling incorporating services 与水、电等设施组合成的顶棚;与水、电等设施组合成的吊顶

suspended ceiling load 吊顶荷载
suspended ceiling system 吊顶体系
suspended centrifuge 悬挂式离心机;上悬式离心机;上挂式离心机
suspended chain moving aeration technology 悬挂链移动曝气技术
suspended chute 悬挂式溜槽
suspended coil galvanometer 悬圈式电流计;悬挂式电流计
suspended colloid 悬浮胶体
suspended column 雷公柱;枨杆
suspended concrete base 悬浮式混凝土基础;悬浮式混凝土基底
suspended concrete floor 悬浮式混凝土地板;悬式混凝土桥面
suspended concrete(slab)subfloor 悬浮式混凝土毛地板
suspended cone 悬挂锥体
suspended content 悬移质含沙量
suspended core 吊芯
suspended counterweight 悬挂砝码;悬挂平衡重;悬式平衡重
suspended deck 悬桥面;悬空式面板
suspended deck structure 悬空式面板结构
suspended density flow 悬浮异重流
suspended diffuse plate 悬吊式扩散板
suspended diving bell 悬吊潜水钟
suspended-drop current metre 滴漏式流速仪
suspended dust 浮悬尘埃;悬浮尘埃
suspended electromagnet 悬吊电磁铁
suspended expanded metal lath ceiling 钢板网吊顶
suspended fender 吊挂护舷木(码头或船舶用);悬木缓冲器;悬木防护器;悬挂式碰垫;悬挂式护舷;悬挂式防撞装置;悬挂式防撞物;吊悬式护舷(木)
suspended filler 悬浮填料
suspended flight conveyer 悬式刮板输送机
suspended floor 悬空地板;无中间支点的楼板;悬吊式楼板;悬桥面;悬吊面;架空地板
suspended floor beam 悬挂式地板梁
suspended flow velocity 悬浮流速
suspended flyer 悬挂式锭翼
suspended forms 悬吊(式)模板
suspended formwork 悬挂式模板;悬吊模板工程
suspended-frame weir 悬挂式框架堰;悬吊式框架堰;活动式悬架堰
suspended framing 悬吊骨架
suspended furnace 悬挂式炉
suspended girder 悬架大梁;挂梁;吊车梁
suspended glazing 悬挂式玻璃窗
suspended granular sludge bed reactor 悬浮颗粒污泥床反应器
suspended gravity fender 悬挂式重力护舷;悬挂式重力防撞装置
suspended grinder 悬挂砂轮机
suspended ground floor 悬吊式底层地板;悬吊式地面板
suspended-growth biological process 悬浮生长生物法
suspended-growth denitrification process 悬浮生长脱氮法
suspended-growth microorganism 悬浮生长微生物
suspended-growth nitrification process 悬浮生长硝化法
suspended-growth process 悬浮生长处理法
suspended-growth reactor 悬浮生长反应器
suspended-growth sludge 悬浮生长污泥
suspended guide 悬挂式导向体
suspended gutter 悬挂式雨水槽;悬挂式排水沟
suspended highway bridge 悬索公路桥
suspended histogram 悬浮式直方图
suspended hood 吸风罩
suspended hunting area 禁猎区
suspended hydrogenerator 悬挂式水轮发电机
suspended idler 悬挂式托辊
suspended idler structure 悬挂式托辊结构
suspended inorganic matter 悬浮无机物
suspended in water 悬挂在水中
suspended joint 钢轨端悬空接头(接头下面不是以枕木直接垫着的);悬式接头;悬接(式)接头;吊接;浮接
suspended joint of steel rail 钢轨端悬空接头

suspended lamp 吊灯
suspended level 悬式水准仪;悬式水平仪
suspended level visco(si)meter 气承式液柱黏度计
suspended light diffusing ceiling panel 悬吊式平顶漫射照明灯框;悬吊式平顶漫射照明灯盒
suspended light rail system 悬吊式轻轨系统
suspended load 吊载;悬移质沙量;悬移质泥沙;悬荷;悬挂荷载;悬浮质;悬浮物质;悬浮泥沙;悬浮负载
suspended load budget 悬移质平衡
suspended load capacity 总起重量
suspended load discharge measurement 悬移质输沙率测定
suspended load discharge per unit width 单宽悬移质输沙量
suspended load measurement 悬移质测验;悬移质测定
suspended load model test 悬移质泥沙模型试验
suspended load sampler 悬移质取样器;悬移质采样器
suspended load sampling 悬移质采样
suspended load transport 悬移质输送;悬荷搬运;悬浮质搬运
suspended luminaire 悬挂照明体;悬吊式照明器具;悬吊式照明设备
suspended magnet 悬吊磁铁
suspended magnet separator 悬浮磁分离器
suspended mass 悬式质量;悬吊质量
suspended material 悬浮物(质);悬浮泥沙
suspended matter 悬移质;悬浮(物)质;悬浮物
suspended matter concentration 悬浮物浓度
suspended media blanket 悬浮介质床
suspended membrane structure 悬挂膜结构
suspended metal ceiling 悬挂式金属顶棚;悬挂式金属平顶;悬式金属天花板
suspended metal lath 悬式金属条板
suspended microparticle 悬浮微粒
suspended microparticle contaminant 悬浮微粒污染物
suspended microparticle control 悬浮微粒控制
suspended microphone 悬挂式传声器
suspended mineral 悬浮无机物
suspended mixture 悬浮混合物
suspended model 悬挂模型
suspended moisture 悬着水分
suspended monorail railway 高架悬吊单轨铁路
suspended monorail(way) 悬挂式单轨铁路
suspended motor 悬挂式电(动)机
suspended mud 悬浮淤泥
suspended object 悬吊物
suspended oil 悬浮状油;悬浮油
suspended or concealed projects 停建缓建项目
suspended organic matter 悬浮有机物
suspended packed bed 悬浮填料床
suspended particle 悬浮粒子;悬浮粒状物质;悬浮颗粒
suspended particles of dust 悬浮尘埃微粒
suspended particulate 悬浮性粒子;悬浮颗粒;悬浮的粒状物质
suspended particulate matter 悬浮颗粒物(质)
suspended particulate sampling equipment 悬浮颗粒物采样装置
suspended partition(wall) 悬挂式隔墙
suspended pattern 悬垂型
suspended pedal 悬吊式踏板
suspended phase 悬浮相
suspended pipe 悬吊管
suspended pipeline 悬吊管道
suspended plaster ceiling 悬挂式灰墁平顶;悬挂式灰墁顶棚;灰泥吊顶
suspended plastic filler 悬浮塑料填料
suspended platform 悬挂式操作台;悬吊式平台;悬空操作台
suspended pump 吊泵
suspended railroad 悬索铁路;悬吊铁道;悬架铁道;高架铁路;高架铁道
suspended railway 悬索铁路;悬吊铁道;悬架铁道;高架铁路;高架铁道
suspended refractory wall 悬挂耐火墙
suspended reinforced concrete roof 悬挂式钢筋混凝土顶;悬吊式钢筋混凝土屋顶
suspended ribbon bridge 悬带桥
suspended river 悬河

suspended roller mill 悬吊辊式粉碎机
suspended roof 悬挂屋盖;悬挂式屋盖;悬吊式屋顶;悬索屋顶
suspended roof brick 吊顶砖;吊顶砖
suspended sand 悬浮泥沙
suspended sash 平衡窗;吊窗
suspended scaffold(ing) 悬挂(式)脚手架;悬吊式脚手架;挂吊手架;悬式脚手架
suspended scintillator 悬浮闪烁体
suspended screen 悬挂帘;悬挂屏幕;悬挂细栅;悬挂网格;悬挂筛
suspended sediment 悬浮沙;悬移质输沙;悬浮泥沙;悬沙;悬浮泥沙;悬浮沉积(物)悬移质;悬浮沉积(物)
suspended sedimentation 悬浮体渣脚
suspended sediment concentration 悬浮质浓度;悬浮质含沙量;悬浮泥沙浓度;含沙量
suspended sediment discharge 悬移质输沙率
suspended sediment sampler 悬移质采样器;悬浮体渣脚取样器
suspended shadow wall 吊隔墙
suspended-shaft crusher 悬轴式破碎机
suspended shake out device 悬吊式落砂架
suspended shaker 悬吊式簸动运输机
suspended shell 悬挂式薄壳
suspended shuttering 悬挂式脚手架;悬挂模板;间接支模;悬式模板;悬吊模板工程
suspended signal 悬空号志
suspended simple beam 简支梁
suspended slab 悬(式)板;简支板
suspended sloping way 架空斜坡道
suspended sludge 悬浮淤渣
suspended sludge blanket 悬浮污泥层
suspended sludge blanket process 悬浮污泥层法
suspended sludge clarifier 悬浮污泥澄清池
suspended sludge concentration 悬浮污泥浓度
suspended sludge technology 悬浮污泥技术
suspended solids 悬浊固体;悬浮性固体;悬浮(固)体;悬移质
suspended solid analyzer 悬浮物分析仪
suspended solid concentration 悬浮固体浓度
suspended solid contact reaction basin 悬浮固体接触反应池
suspended solid contact reactor 悬浮固体接触反应器;悬浮固体接触反应池
suspended solid content 悬浮固体含量
suspended solid hydrograph 悬浮固体水位图
suspended solid material 悬移质;悬浮固体(物质)
suspended solid matter 悬浮固体物质
suspended solid organic carbon 悬浮固体有机碳
suspended solid particles 悬浮固体颗粒
suspended solid pollutant 悬浮固体污染物
suspended solid removal 悬浮固体颗粒去除;悬浮固体的去除
suspended sonde 悬浮探空仪
suspended sound absorber 悬挂吸声体
suspended span 悬挂结构;悬置跨;悬索桥跨度;悬跨;悬孔;挂孔
suspended-spindle gyratory crusher 悬轴回转碎矿机
suspended-spindle(type) gyratory crusher 悬轴式旋回破碎机
suspended sprayed acoustic(al) ceiling 喷涂隔声吊顶
suspended spring 悬簧;悬挂式弹簧;悬挂泉
suspended s/R machine 悬挂型有轨巷道堆垛机
suspended stacking crane 悬挂桥式堆垛起重机
suspended staging 悬挂式脚手架
suspended stair(case) 悬吊式楼梯;悬挑楼梯
suspended state 悬浮状态;暂行状态;暂停状态;挂起状态
suspended staying 悬挂(式)脚手架
suspended stiffening truss 悬加劲桁架;悬式加劲桁架
suspended structure 悬挑结构;悬挂结构;悬挂建筑;悬吊结构
suspended substance 悬浮物质;悬浮物;悬浮体
suspended sulfate 悬浮硫酸盐
suspended superstructure 悬挂式上部结构
suspended support 悬吊支承
suspended system 悬吊体系
suspended theodolite 悬式经纬仪
suspended tow 悬挂拖运
suspended transformation 悬浮转变作用;悬浮转变

suspended transit 悬式经纬仪
suspended tray conveyer 悬筐式输送设备;悬吊槽式运输机
suspended tray conveyer 吊盘输送机
suspended tray elevator 悬挂盘式提升机
suspended tread floor 悬吊地板
suspended truss 悬置桁架;悬式桁架;悬挂式桁架
suspended turbidity current 悬浮浊流;悬浮异重流
suspended type 悬挂式
suspended(type)generator 悬式发电机
suspended type of furnace 吊挂式炉
suspended type unit heater 悬挂式暖风机;悬挂式单位加热器;悬式散热器(件);悬挂式散热器(件)
suspended valley 悬谷
suspended velocity 悬浮速度
suspended volatile solid 悬浮挥发性固体
suspended wagon 挂篮
suspended wall 隔断板墙;半截板墙;悬墙;吊墙
suspended wall lining 悬吊式壁面铺盖层
suspended water 悬着水;悬浮水分;上层滞水;饱气带水;包气带水
suspended-water zone 悬浮水带
suspended weight 悬挂重量
suspended wire 挂丝
suspended wood(en) lath ceiling 木板条吊顶;板条吊顶
suspended zone melting 悬区熔化
suspender 吊索;悬杆;悬吊物;吊杆;吊材
suspender cable 吊索
suspending agent 悬浮剂;助悬浮剂
suspending a well 暂停钻井
suspending beam scale 悬梁秤
suspending dust 悬浮尘埃;飘尘
suspending liquid 悬浮液体;悬浮液
suspending medium 悬浮介质;悬浮剂
suspending monomer 悬浮单体
suspending operation or curtail production 停工减产
suspending operation or slashing production 停工减产
suspending power 悬浮能力
suspending tool 过滤器中下放的器具
suspending velocity 悬浮速度
suspending wire 挂丝;吊线
suspend level 挂水准器
suspend payment 暂停支付;无力支付
suspend peformance of obligations 中止履行(合同)义务
suspend production 停产
suspend talks 中止谈判
suspend the coaching 截功
suspensate 悬浮质
suspense 悬挂
suspense account 暂记账(户)
suspense credit 暂收款;暂记贷项
suspense debit 暂记借项;暂记款项
suspense debt 暂付款
suspense payment 暂付款
suspense payment account 暂付款账户
suspense receipt 暂收款
suspense receipt account 暂收款账户
suspensible solid of water 水的悬浮固体
suspension 高架铁轨;悬油液;悬置;悬移(作用);悬液;悬丝;悬挂(物);悬浮(现象);悬浮体;悬浮胶体;悬吊(术);中止;暂停;挂起状态;无力偿付;同换映像;停止工作;双角锥
suspension access 悬挂装置检查孔
suspension account 悬账
suspension additive 悬浮添加剂
suspension agent 悬浮剂
suspension anchor cable 悬吊锚碇索
suspension arch 悬拱
suspension arm 悬挂臂;弹簧架
suspension arrangement 悬挂装置;弹簧架吊件
suspension attachment 悬吊工具
suspension attachment points 悬架固定点
suspension-auxiliary components 悬挂辅助件
suspension banker 悬挂式料斗
suspension beam 悬(挂)梁;悬吊梁
suspension bearing 悬置轴承;悬浮轴承;吊轴承;吊悬轴承
suspension bogie 悬挂转向架
suspension bolt 挂环螺栓

suspension boom 吊杆;横江缆索
suspension bow 吊环
suspension box-cofferdam 吊箱围堰;浮箱式围堰
suspension bracket 悬挂(托)架
suspension brick 悬吊砖
suspension bridge 悬(索)桥;吊桥
suspension bridge cable 悬桥索
suspension bridge tower 悬索桥桥塔
suspension bucket scale 料仓秤;悬斗秤
suspension building 悬挂建筑(物)
suspension bunker 悬仓;悬挂料斗;悬吊进漏斗;高悬式储[贮]仓;吊式矿仓;吊仓
suspension cable 悬索;悬缆(线);吊索
suspension cable anchor 悬索锚碇;悬索锚固
suspension cable railway 悬索铁路
suspension cable roof construction 悬索屋顶施工
suspension cable system 悬索系统
suspension cage 吊笼
suspension calciner 悬浮式分解炉
suspension ceiling 悬浮式天顶
suspension cell 悬浮细胞
suspension center 悬点;吊点
suspension chain 悬链;吊链
suspension chord 弦杆;悬杆;吊杆
suspension clamp 悬挂线夹;悬垂线夹;吊架;吊夹
suspension claw 吊爪
suspension clinometer 悬式测斜仪;悬式测斜器
suspension clock sampler 悬钟式采样器
suspension coil 悬吊线圈
suspension colloid 悬胶;悬浮胶体;胶体悬液
suspension combustion 悬浮燃烧
suspension compass 悬式罗盘
suspension construction 悬挂建筑(物);悬挂构造;悬吊构造
suspension copolymerization 悬浮共聚
suspension cord 悬索
suspension crane 吊车
suspension current 悬移流;悬浮流;浊流;混浊流;含沙流
suspension damper 悬挂减震器
suspension deck liquid natural gas tank 吊顶式液化天然气罐
suspension device 悬挂装置
suspension dispersion 悬浮液
suspension-dissolve load 悬胶溶解荷载
suspension dredge(r) 悬挂式挖泥机
suspension dyeing 悬浮染色
suspension effect 悬液效应;悬浮作用;吊悬效果
suspension electrode 悬浮电极
suspension electrodynamometer 悬置式电功率计
suspension equipment 悬挂设备
suspension fertilizer 悬浮肥料
suspension fiber 悬丝
suspension fire furnace 悬燃炉
suspension firing 悬浮燃烧
suspension fixture 悬挂设备
suspension flash preheater 悬浮快速内燃预热器
suspension floor 悬挂式地板
suspension fluid 悬浮液
suspension fork 悬架
suspension fuel 悬浮燃料
suspension furniture 悬挂家具
suspension gasifier 悬浮气化煤气发生炉
suspension gear 悬吊机构
suspension girder 悬置大梁;悬梁;悬(车)梁
suspension grid system 悬挂式格栅体系
suspension grout 悬液浆
suspension grouting 悬液灌浆
suspension gypsum slab 悬挂式石膏板
suspension height 支悬高度
suspension hook 吊钩
suspension hopper scale 悬斗秤;料仓秤
suspension hydraulic generator 悬吊式水轮发电机
suspension insulator 悬式绝缘子;悬挂(式)绝缘子;悬垂绝缘子
suspension insulator string 悬式绝缘子串
suspension joint 悬式接头
suspension kiln 悬吊煅烧窑
suspension ladder 吊梯
suspension laryngoscope 悬吊式喉镜
suspension layer 悬浮层
suspension lever 悬架臂
suspension lever pivot pin 悬架杠杆销轴
suspension light 悬挂式照明;吊灯

suspension light fitting 悬挂式灯具
suspension lighting fixture 悬挂式照明装置
suspension line 吊架线;悬挂线;悬挂索;架空线(路)
suspension link 悬架臂;悬挂链条;悬杆;拉环;吊链;吊钩;吊杆
suspension linkage 悬挂装置
suspension liquid 悬浮液
suspension liquor grinding 悬浮研磨法
suspension load 悬移质;悬浮荷载;悬胶荷载;悬浮负载;吊载
suspension load sampler 悬移质取样器;悬移质采样器
suspension lock device 悬挂锁紧机构
suspension locking handle 悬挂锁紧手柄
suspension log 吊材
suspension loop 挂耳(电解极板上的)
suspension lubrication pipe 悬挂装置滑油管
suspension lubrication pump 悬挂装置滑油泵
suspension lubrication tank 悬挂装置滑油箱
suspension luminaire fixture 悬挂式照明装置
suspension member 吊材;悬索构件;悬索;悬吊部件;竖拉杆;吊索;吊链;吊杆
suspension method 悬浮法
suspension mounting 悬式座架;悬式装置
suspension mounting reinforcement 悬架支承加强板
suspension natural frequency 悬挂系固有频率
suspension of business 休业;停止营业;停业
suspension of contract 中止合同
suspension of driver's license 暂停驾驶执照;扣留驾驶执照
suspension of hire 停租船
suspension of lime 石灰悬浮体
suspension of payment 中止付款;停止支付;停止付款
suspension of prescription 停止时效
suspension of right 停止权利
suspension of shipping/air service 停航
suspension of the dollar's conversion into gold 美元停止兑换黄金
suspension of the plummet 垂球悬挂装置
suspension of the sale of securities 停止出售证券
suspension of works 工程暂停;工程停工;工程停顿
suspension oil 悬浮油
suspension packing 悬浮填充
suspension pantograph 悬式缩放仪
suspension peg 悬挂桩
suspension percentage 悬浊率;悬浮率
suspension period 停息段
suspension piercing 悬吊式火力钻进
suspension-piercing machine 悬吊式火力钻机
suspension pile 悬桩
suspension pin 悬挂轴
suspension pipeline 架空管道
suspension plaster 悬垂石膏
suspension point 悬挂点;悬点
suspension polymerization 悬浮聚合(法)
suspension population 悬浮总体
suspension porcelain 瓷质悬挂隔电子
suspension post 悬挂柱;悬挂杆;吊杆
suspension preheater 悬浮预热器
suspension preheater and flash furnace 悬浮预热器和分解炉
suspension preheater kiln 悬浮预热器窑
suspension preheater without precalciner 不带预分解炉的悬浮预热器
suspension preheater with precalciner 带预分解炉的悬浮预热器
suspension preheater with two strings 双系列悬浮预热器
suspension preheating 悬浮预热
suspension preheating system 悬浮预热系统
suspension process 悬浮法
suspension property 悬浮性
suspension pump linkage 悬挂装置滑油泵联杆
suspension push 悬垂式按钮
suspension pylon 吊塔(架)
suspension rail 悬轨
suspension rail conveying system 悬轨运输系统
suspension rail type travel(l)ing weighbatcher 悬轨式移动称量配料器
suspension railway 高架铁路;高架铁道
suspension rate 悬架刚度

suspension reactor 悬浮液反应堆;悬浮液堆
suspension rheology 悬浮体流变学
suspension ring 吊环
suspension ring of hanging compass 挂式罗盘悬挂环
suspension road bridge 公路吊桥
suspension roast 漂悬焙烧法
suspension roaster 悬浮焙烧炉
suspension roasting 悬浮焙烧;飘悬焙烧
suspension rod 吊索;悬杆;悬吊杆;竖拉杆;吊材
suspension roll center 悬挂滚翻中心
suspension roller mill 悬辊磨
suspension roll geometry effect 悬挂系统侧倾效应
suspension roof 悬浮屋面;悬索屋顶;悬式屋顶
suspension roof support 悬顶支护;杆柱支护;锚杆支护
suspension rope 吊绳;吊索;悬缆(线)
suspension rope equalizer 悬缆平衡器;吊索平衡器
suspension screen 悬浮筛
suspension separation method 悬浮分离法
suspension sequence 同纬映像序列
suspension setting 吊装
suspension shackle 悬架吊耳;悬挂钩环
suspension shaft 悬挂轴;起吊轴
suspension shelving 悬挂式书架
suspension shock absorber for vehicle 车辆用悬挂式缓冲器
suspension similarity 悬浮相似(性)
suspension smelting 悬浮熔炼
suspension span 吊孔
suspension spike 悬挂架销钉
suspension spotlight 悬挂式聚光灯
suspension spring 悬置弹簧;悬架弹簧;悬簧;悬挂弹簧
suspension stability 悬液稳定性;悬浮稳定性
suspension stability control 悬挂稳定性控制
suspension stage 悬挂式舞台
suspension stay 悬挂支撑
suspension stiffness 悬架刚度
suspension stor(e)y 悬挂式楼层
suspension string 悬垂绝缘子串
suspension structural system 悬挂式结构体系
suspension structure 悬挂(式)结构;悬索结构;悬浮结构;悬吊结构
suspension strut 连接吊杆的横向梁
suspension support 悬架支承
suspension support installation 悬式支承装置
suspension switch 拉线开关;悬挂式开关
suspension system 悬挂体系;悬置系统;悬架系统;悬挂装置;悬挂系统;悬浮系统;悬吊系统;摇动法
suspension theodolite 悬式经纬仪;悬挂经纬仪
suspension time 中止时间
suspension tower 吊塔
suspension tower pier 吊塔墩
suspension transit 悬式经纬仪
suspension transmissibility 悬架传递振动的性能
suspension transport 悬浮搬运
suspension truss bridge 桁构吊桥;悬索桁架桥;悬式桁架桥;桁架吊桥
suspension type 悬挂型;悬浮式
suspension type conveyer 悬架式运输机
suspension-type direct desulfurization 悬浮式直接脱硫
suspension type drawing frame 高架式并条机
suspension type fuel 悬浮态燃料
suspension type log haul-up 悬架式拉木机
suspension type oil hydraulic press 悬挂式双动油压机
suspension type preheater 悬式预热器;悬浮式预热器
suspension valve 悬置阀
suspension vertical stiffness 悬挂系统垂直刚度
suspension weight 悬锤
suspension wheelbase 悬架轴距
suspension winch 悬置绞车
suspension wire 悬吊钢丝;挂丝;吊线
suspension yoke 悬吊台
suspensive 悬而未决的;未决定的
suspensoid 溶胶;悬胶液;悬胶(体);胶体悬液
suspensoid process 悬浮体分选法;重介法
suspensoid state 悬胶态
Suspensol 苏斯平索尔(一种多彩石墨防水漆)

suspensor 悬带
suspensory 悬的;吊带
suspensory membrane 悬膜
suspensory wire 悬丝
Sussex bond 苏塞克斯砌合
Sussex garden-wall bond 荷兰式花园墙砌合;三顺一丁砌法;苏塞克斯花园墙砌合
sussexite 硼锰矿;白硼镁锰矿;白锰石
sustainability 可持续性;稳定性;持续性
sustainability of urban water management 可持续性城市水管理
sustainable catch 稳定捕获量;持久捕获量
sustainable development 可持续开发;可持续发展
sustainable development indicator 可持续发展指标
sustainable drainage system 可持续排水系统
sustainable ecosystem 可持续生态系统
sustainable growth 持续增长
sustainable lagoon 稳定槽
sustainable living environment 可持续生活环境
sustainable transport 可持续运输
sustainable urban drainage system 可持续城市排水系统
sustainable use of energy 能源的持久使用
sustainable use of potable water sources 可持续饮用水源利用
sustainable use of the environment 环境的持久作用
sustainable wastewater management 可持续污水管理
sustainable water resources 可持续水资源
sustainable yield 可维持产量;稳定产量
sustained acceleration 持续加速(度)
sustained annual yield working 定期保续收获
sustained arc 持续弧
sustained average low flow 持续平均低流量
sustained average peak flow 持续平均高峰流量
sustained capillary water 支持毛细水
sustained collector voltage 集电极保持电压
sustained combustion 持续燃烧
sustained contraction 持久性收缩
sustained current 持续电流
sustained disinfection 持续消毒
sustained exposure 持续辐照
sustained fault 持续故障
sustained flow 持续流量
sustained grade 持续坡度
sustained growth 持续增长
sustained input 持续输入
sustained load 支承荷载;支持荷载;持续荷载;长周期负荷;长期施加的荷载
sustained load(ing) 持续荷载
sustained modulus of elasticity 持续弹性模量;持久弹性模量
sustained note 持续音
sustained oscillation 等幅振荡;持续振荡;持续摆动
sustained oscillator 持续振荡器
sustained ouput 持续输出
sustained over-load 持续过载;持续超载;长时间超载
sustained over-voltage 持续过电压
sustained peak mass loading 持续高峰质量荷载
sustained pressure test 持续压力试验
sustained production 稳定产量;持续生产量
sustained radiation 等幅波辐射;持续辐射
sustained rate of fire 持续射击速率
sustained runoff 基本径流;持续径流;持久径流
sustained segment 持续片段
sustained short-circuit 持续短路
sustained sound 持续声
sustained speed 持续速度;持久速率
sustained stable growth 持续稳定增长
sustained steep grade 持续陡坡
sustained stress 静应力
sustained surge 持续涌浪
sustained velocity 持续速度
sustained vibration 持续振动
sustained wave 等幅波;持续波
sustained wind 持续风
sustained wind speed 持续风速
sustained wind velocity 持续风速
sustained yield 永续收获;稳定产量;持续收获;持续生产量;持续抽水量;持续产量
sustained-yield management 永续收获经营

sustained yield of water source 水源持续产量
sustainer 支座;支点;支撑;主级发动机;证实者;认可者
sustainer chamber 主级发动机;主发动机燃烧室
sustainer engine 主发动机
sustainer motor 续航发动机
sustainer-pumped laser 持续泵浦激光器
sustainer unit 主发动机组
sustain(ing) 持续
sustaining force 持续力
sustaining layer 持力层
sustaining plane 支持面
sustaining power 支持能力;持续力
sustaining slope 连续斜坡;连续坡(度);顺流持续坡度
sustaining valve 支撑阀
sustaining velocity 悬浮速度;悬浮流速;不淤积速度;不淤(积)流速
sustaining voltage 保持电压
sustaining wall 支承墙;扶墙
sustain loss 蒙受损失
sustain losses in business 赔本
sustanable agriculture 持久农业
sustanable developmentalism 持续发展说
sustanined load 支承荷载
sustentaculum 支持物;支柱
susy transformations 超对称变换
Sutcliffe's sounding apparatus 舍特克利夫(机械)测深仪
sutogamy 自配
Sutro measuring weir 苏特罗式量水堰
Sutro weir 比例式堰(流量与水头成正比);苏特罗式堰;比例式堰
sutruck 单辆货车
suttle 净重(的)
suttle weight 净重
Sutton equation 苏通方程式
Sutton's formula 萨顿扩散公式
Sutton stone 萨顿石
sutural basin 缝合带盆地
suture bobbin 缝线轴
sutured contact 缝合接触
sutured quartz 缝合石英
sutured structure 缝合结构
suture line 地缝合线;缝合线
suture material 缝合材料
suture orogenic belt 缝合造山带
suture scissors 缝线剪
suture scissors titch scissors 拆线剪
suture wire cutter 缝合钢丝切割钳
suture zone 缝合带
suyao 束腰
Suzhou-style decorative painting 苏式彩画
Suzhou-style pattern 苏式彩画
Suzhou traditional garden 苏州园林
suzukiite 苏硅矾钡石
svabite 砷灰石
svanbergite 菱磷铝锶石;磷锶铝矾
Svecofennian orogeny 斯维可芬造山运动【地】
Svenska handelsbanken 瑞典商业银行
Svenska Kullager fabriken 瑞典轴承制作所
Sverdrup-Munk-Bretchneider method 斯孟布法
Sveriges investerings bank 瑞典投资银行
Sveriges Kreditbank 瑞典信贷银行
Sveriges riksbank 瑞典中央银行;瑞典银行
svetlozarite 钾丝光沸石
svitalskite 镁铁云母
SV-wave 垂直横波
swab 拖布;拖把;抽吸作用;敷水笔;抹
swab back into production 抽吸诱导油流
swabber 拖把;清扫工(人);水手;装管工
swabbing 擦洗;揩涂;(模子)上油;刷水;刷色;刷浆;搓涂;抽吸;抽汲;擦涂;抹
swabbing action 抽吸作用
swabbing cloth 抹布
swabbing depth 抽子下入深度
swabbing effect 抽吸作用(深井钻探)
swabbing line 拭子拉杆(管壁清扫)
swabbing pass 送料孔型
swabbing speed 抽提速度
swabbing times 抽汲次数
swab hitch 测锤结
swab holding forceps 持卷棉子钳
swab man 管道清洁工

swab sink 墩布池
swabstick 炮眼清除刷
swadenborgite 锑钠铍矿
swag 倾侧;松垂;垂花饰【建】
swage 铁模;锻模
swage block 横锻砧块;型砧;型模块;花板砧
swage bolt 横锻螺栓
swaged file 梯形锉
swage die 型砧;型模;型锻模
swaged socket 模锻锥套
swage engine 型锻机
swage head 铆钉头;模压锚碇头
swage lock 接头套筒;接头套管;(管子的)接套
swagelok coupling 套管接头
swage machining 旋转模锻
swage nipple 异径接头;大小头短节
swage pliers 轧印钳
swage process 陷型模锻
swager 带锯机;旋锻机;陷型模锻机;压模器;锤锻(机)
swage rivet head 模压铆钉头;模压铆钉
swages 管子接头
swage saw 圆锯
swage set 锯路
swage-setting (圆锯片的)拨鱼尾形锯路;模具拆除装置
swage sharper 整齿器
swage-wedge anchor 型铁楔形锚
swaging 型锻;模压
swaging hammer 模压锤
swaging machine 锻打机;型锻机;锻冶机;锤造器;擦洗机
swaging mandrel 型锻芯轴
swag leaf 悬垂的叶片(装饰);垂花饰叶
swag light 吊灯
swale 放火烧林;沼泽低地;烘底船;浅沟;微洼地;滩槽
swaller 伏流;地下河
swallet 落水洞
swallet hole 陷阱;落水洞;灰岩洞;斗淋
swallet river 伏流
swallet stream 伏流
swallow 绳孔(滑车上)
swallowed up by flood 洪水淹没
swallow hole 落水洞;石灰坑;渗坑
swallowing capacity 临界流量
swallow one's word 取消诺言
swallow tail 燕尾榫
swallow-tailed 燕尾形的
swallow-tailed ironwork 燕尾铁脚
swallow-tailed type outside locking device 燕尾式外锁闭
swallow-tail form 燕尾形
swallow-tail law 燕尾律
swally 向斜层
swamboite 斯铀硅矿
swamp 林沼;木本沼泽;湿地
swamp area 沼泽区
swamp autochthonous organic matter 沼泽原生有机质
swamp boat 沼泽船
swamp buggy 沼泽地用汽车;沼泽车
swamp bulldozer 湿地推土机
swamp community 沼泽群落
swamp deposit 沼泽堆积
swamp ditch 沼泽排水沟;排水渠(道)
swamp drainage 渍水沼泽排水;沼泽(地)排水
swamp ecotype 沼泽生态型;生态型林沼
swamped 泥沼状的
swamped land 湿地
swamp environment 沼泽环境
swamper 驾驶员助手,清洁工;卡车装卸工;矿车跟车工
swamp facies 沼泽相
swamp fever 沼泽热
swamp forest 沼泽林
swamp-forest peat 沼泽森林泥炭
swamp formation 沼泽形成作用;沼泽形成
swamp gum 桉树属
swamping 沼泽化
swamping lagoon facies 沼泽化泻湖相
swamping land 沼地
swamping of soil 土壤沼泽化
swamping product 沼泽化产物

swamping resistance 扩程电阻
swamp lake 沼泽湖
swamp land 沼泽(湿地);低湿地
swamp material sample 沼泽物质样品
swamp meadow 沼泽草甸
swamp muck 沼泽腐殖土;沼泽腐泥土;沼泽丛林
swamp of hat mud 热泥塘
swamp ore 沼铁矿
swamp pine forest 沼泽松林
swamp process 沼泽过程
swamp road 沼地道路
swamp sedimentary deposit 沼泽沉积矿床
swamp shoe 湿地履带板
swamp shooting method 沼泽爆炸法
swamp soil 渍水沼泽土;沼泽土(壤)
swamp stream 沼泽河
swamp sugrudok 沼泽混交栎林
swamp thicket 沼泽灌丛
swamp type 沼泽类型
swamp vegetation 沼泽植被
swamp water 沼泽水
swamp water quality 林沼水水质
swamp woodland 沼泽腐殖土;沼泽丛林
swamp woody facies 沼泽木质相
swampy 沼泽的
swampy caterpillar tractor 沼泽地履带拖拉机;沼泽履带拖拉机
swampy district 沼泽地区
swampy ground 沼泽(地);低湿地
swampy land 沼泽(化)地;湿地
swampy odor 沼泽地气味
swampy soil 沼泽土
swampy tudra soil 治泽化冰沼土
swan base 卡口灯座;卡口接头;插口式灯头
swan-neck 弯曲管;鹅颈形的;鹅颈管;弯头管
swan-neck bearer (转向架的)鹅颈式托架
swan-neck bend(ing) 鹅颈形弯头
swan-neck bent tube 鹅颈弯管
swan-neck boom 鹅颈吊车架;鹅颈起重臂
swan-neck bracket 弯脚
swan-neck chisel 长鹅凿;鹅颈凿
swan-neck downpipe 双弯头水落管
swan-necked davit 鹅颈式吊艇柱;鹅颈式吊艇架;转轴式吊艇柱
swan-necked drainage tool 鹅颈式排水工具
swan-neck handrail 鹅颈扶手;鹅颈形扶手
swan-neck hook 鹅颈钩
swan-neck jib 鹅颈杆;鹅颈臂
swan-neck pipe 鹅颈(式)管
swan-neck shovel 鹅颈铲;鹅颈形铁锹
swan-neck socket 卡口灯座
swan-neck street lamp 马路弯灯
swan-neck tube 鹅颈式管
swan-neck vehicle 鹅颈车
swan-neck ventilator 鹅颈式通风筒
swan reflector 反光玻璃
swansdown twill 三上一下斜纹
swan-shot 大钻粒
swan socket 卡口管座;卡口灯座;插入式插座
swap action valve 交换作用阀
swap agreement 互惠协议;通货互换协定
swap allocation unit 交换分配装置
swap arrangement 汇兑协定
swap body 交换箱体(汽车)
swap credits 互惠信贷
swap data 交换资料
swap data set 交换数据集
swap file 交换文件
swap-in 换入
swap mode 交换方式
swap out 换出
swap(ping) 调动
swapping buffer 交换缓冲区
swap table 交换表
sward 植草;牧草地;生草;草土
Sward hardness 斯瓦德硬度
sward illimerized soil area 生草灰化土区
sward podzolic soil 生草灰化
Sward rocker 斯瓦德摆杆硬度计
Sward-Zeidler rocker 斯瓦德-扎德莱尔摆杆硬度计
swarf 木屑;细铁屑;钢板切边;切屑;砂纸打磨产生的碎屑
swarf box 切屑箱
swarf clearance 切屑余隙

swarf collector for drills 钻屑收集器
swarf extraction 切屑排出
swarf removal 切屑排除
swarf tray 切屑盘
swarm activity 震群活动性;成群活动
swarm earthquake 震群
swarming and migratory 成群迁徙的
swarming of diatoms 硅藻大量密集
swarm of bubbles 气泡群
swarms earthquake 震群型地震
swarm sequence 震群序列
swarm together 蜂聚
swarm-type 震群型
swarm type earthquake 震群型地震
swarth 黑墨
swartzite 碳钙镁铀矿;水钙镁铀矿
swash 上爬;冲刷滩;冲刷口;冲刷坝;冲洗水道;冲溅;波浪上爬;波浪上溅;波浪上冲;波浪冲溅
swash-backwash zone 冲洗回流带
swash bank 海堤上段;波浪爬溅岸
swash bar 冲流沙坝
swash bulkhead 缓冲舱壁;(防止舱内液体晃动的)防晃纵隔舱
swash cam 旋转斜盘
swash channel 冲流水道;冲沟
swash cross bedding structure 冲洗交错层理构造
swash current 上爬水流
swash height 浪溅高度;爬高;冲溅高度;波浪上爬高度;浪爬高
swash height of waves 波浪冲击高度;波浪侵蚀高度
swash letters 花饰大写斜体字
swash limit 波浪上冲界线
swash mark 冲流线;冲流痕;冲痕;潮水痕迹
swash plate 旋转斜盘;斜板凸轮;纵向隔板;缓冲舱壁;防晃纵隔舱板
swash plate aeration 转盘曝气
swash plate axial cylinder pump 斜盘式轴向柱塞泵
swash plate axial piston pump 斜盘式轴向活塞(油)泵
swash plate mechanism 旋转斜盘机构
swash plate motor 斜盘电机
swash plate pump 旋转斜盘泵;摆盘式活塞泵
swash plate type axial plunger pump 斜盘式轴向柱塞泵
swash pool 冲流潭
swash way 狭水道;急流冲刷道;冲流水道
swash width 扫捆行距
swash zone 冲流带
swastika 万字(形)饰;万字廊;纹连续图案
swastika fylfot 万字饰
swatch 样品;样片;样本;装饰单板样片;材料样品
swatch width 扫捆条带宽度
swath(e) (扫测水深的)扫迹道;扫测幅;草条;草铺
swath(e) aerator 草条通风机
swath(e) board 集条板;拨草板
swath(e) distributor 草条撒散机
swath(e) lifter 禾铺捡拾器;草条捡拾器
swath(e) loader 叉式装载机
swather 割晒机;铺条机
swath(e) scatterer 干草铺条撒布器
swath(e) sounding system 条带测深系统;扫测系统(多通道回声测深)
swath(e) stick socket 拨草杆窝座
swath(e) tedder 干草摊晒机
swath(e) turner 草条翻晒机;翻草机
swathing device 铺条装置
swathmaker 铺条机
S wave 地震横波;S波
S-wave velocity 横波波速
sway 摇晃;摇摆;转向;横荡;偏向一边;摆动
swayback 背部凹陷
sway bar 稳定杆;(悬挂装置的)锁定斜杆;摆杆
sway beam 平衡杆;防摇梁
sway brace 抗斜剪刀撑;斜支撑;斜撑块;抗摇摆系杆;抗倾斜的角撑;抗摆系杆;竖向支撑;竖撑
sway braced derrick 斜撑型钻塔
sway bracing 斜向支撑;阻绞支撑系统;抗摇支撑;横拉杆;竖向支撑杆
sway chain 锁定链
sway force 摇摆力
sway(ing) 摇动
swaying motion 船舶晃摇动作

swaying of vehicle 车辆摇摆
sway leg 柔性支退
sway moment 横向弯矩;侧向弯矩
sway of support 摆架共振
sway rod 摆杆;对角斜撑;斜撑杆;抗倾斜杆;防倾杆;斜杆
sway sand screen 摆动筛砂机
sway strut 抗倾斜压杆;斜支撑;竖向(支)撑杆
sweal 烘底船
swealing 手工洗污渍
sweat 结露水;渗豆(铸造用语);热析;水气;烧熔
sweat and heat damage 汗湿及受热损失
sweat-back 热析;出汗
sweat board 手操作混凝土拌和器;舱内防汗湿木板;防潮台板
sweat cargo 货汗
sweat cooling 蒸发制冷;蒸发冷却
sweat damage 汗湿货损;濡损;潮腐蚀
sweated environment 污浊环境
sweated joint 熔焊接头;熔焊接合
sweat equity 改进后净增值
sweat furnace 热析炉
sweat generated in cargo 货物中产生的汗湿
sweating 凝结水;磨损硬币以盗取金银粉末;摩擦引起的漆膜失光;钎焊;出汗;表面凝水;渗出水;熔焊;泌出水;光泽;热析;水气;烧熔
sweating dross 热析浮渣
sweating heat 焊接热;熔化热;熔焊热
sweating out 热析珠
sweating process 发汗工艺
sweating rate 析水率
sweating room 蒸汽室;发汗室
sweating soldering 热熔焊(接)
sweating water 表面凝结水
sweat in hold 舱内凝露;舱内汗湿
sweat iron 焊铁
sweat on container 箱汗(集装箱)
sweat out 出汗(抹灰面);潮坏;热析
sweat phenomenon 结露现象
sweat roll 蒸汽滚筒
sweat soldering 软焊;热熔焊(接)
sweat-soldering oven 软焊炉
Sweco vibrating screen 斯威科振动筛
Sweden mill 瑞典自动轧管机
Sweden's international vehicleregistration letter 瑞典国际车辆注册字
swedge bolt 模锻螺栓
swedged 管端墩粗
swedged nipple 异径接头;锥形体接头
swedged out 矫直管子
Swedish 瑞典梯形钻孔法
Swedish architecture 瑞典建筑
Swedish bit 工字形钻头
Swedish break 分析土坡稳定的瑞典方法
Swedish circle method 瑞典圆弧法
Swedish cone penetration test 瑞典静力触探
Swedish cylindrical surface method 瑞典圆弧法;瑞典滑弧法;瑞典圆柱面法
Swedish employers' confederation 瑞典雇主联合会
Swedish fall-cone method 瑞典落锥法
Swedish Hoganas powder 瑞典塞格纳斯铁粉
Swedish impulse oil firing system 瑞典脉冲燃油系统
Swedish iron 瑞典木炭生铁
Swedish(iron)powder 瑞典海绵铁粉
Swedish method (计算土坡稳定性的)分片计算法;(计算土坡稳定性的)瑞典法
Swedish method of slices 瑞典土坡稳定性分片计算法
Swedish mile 瑞典里
Swedish mill 自动轧管机;瑞典自动轧管机
Swedish modern 瑞典现代派风格(指家具和陈设)
swedish olein 松浆油
swedish pine oil 松浆油
Swedish putty 瑞典油灰
Swedish putty bending tester 瑞典油灰弯曲试验仪
Swedish putty coat 瑞典油灰涂膜
Swedish putty for brush application 涂刷用瑞典油灰
Swedish putty seal(ing) 瑞典油灰密封
swedish resin 松浆油
swedish rosin oil 松浆油
Swedish safestop barrier 瑞典式防撞栏

Swedish sand putty 瑞典砂油灰
Swedish school 瑞典学派
Swedish slip circle method (计算土坡稳定性的)瑞典滑弧计算法
Swedish sounding 瑞典式触探
Swedish trade union confederation 瑞典工会联合会
Swedish window 瑞典式窗(中间装有活动百叶的双层玻璃窗)
swell-and-swale topography 高低相间的地形
sweep 后掠角;推开;扫测线;舵柄弧轨;吹扫;吹除;车板;长桨;反曲;刮除;扫描
sweep account 流动账户
sweep amplifier 扫描放大仪
sweep angle 扫描角
sweep arm 导向杆;车板横臂
sweep arm auger unloader 转臂式螺旋卸载机
sweep auger 刮扫式输送螺旋
sweep away 冲走
sweepback 回扫;后掠
sweepback agitator 后掠形搅拌器
sweepback impeller 后掠式叶轮
sweep balance 扫描平衡
sweep bandwidth 扫描带宽
sweep bar 扫海杆
sweep blade 平铲
sweep blast 清扫级喷射除锈
sweep board 刮型板
sweep boom 扫油板
sweep camera 扫描照相机
sweep capacitor 扫描电容器
sweep check 摇版检验;扫频检验;扫频测试
sweep circuit 扫描电路;扫频电路;扫回电路
sweep coil 偏转线圈
sweep current 扫描电流
sweep current generator 扫描电流发生器
sweep delay 扫描延时
sweep-delay circuit 扫描延迟电路
sweep device 清扫装置;扫描装置
sweep diameter 扫描直径
sweep diffusion 分离扩散
sweep diffusion method 扫描扩散法;分离扩散法
sweep drive 扫描末级前置放大器
sweep driver 扫描激励器
sweep efficiency 清扫效率;波及系数
sweep elbow 肘形弯管;巨肘弯头;肘形弯流;肘形淘弯
sweep electron microscope 扫描式电子显微镜
sweeper 清扫机;清扫车;清理机;清管器;扫频仪;扫描器;扫海具;扫海船;扫除机;地面清扫机
sweeper collector 扫路垃圾车
sweep expander 扫描扩展器
sweep expansion 扫描扩展
sweep finger 刮板指
sweep finish 弯曲加工;盘珠
sweep fitting 弯头;弯曲管件;月弯;弯曲管件
sweep flusher 街道清洗机
sweep frequency 扫描频率
sweep-frequency attenuation measurement technique 扫频衰减测试技术
sweep-frequency audiometer 扫描频率听力计
sweep-frequency generator 扫描频率发生器
sweep-frequency heterodynetype analyser 扫描频率外差式分析器
sweep-frequency modulation 扫描调频;扫描频率调制
sweep frequency radar responder 扫描频率雷达应答器
sweep-frequency receiver 扫频接收机
sweep-frequency reflectometer 扫频反射计
sweep-frequency technique 扫频技术
sweep gas 扫掠气;残气
sweep gate 扫描闸门电路;扫描选通;扫描门电路;扫描脉冲
sweep gate pulse 扫描门脉冲
sweep ga(u)ge 曲线样板
sweep gear 扫雷工具
sweep-generation circuit 锯齿波发生器电路;扫描发生电路
sweep guard 推出式安全装置;拨开式安全装置
sweep hold off 扫描间歇
sweeping 清扫;吹扫;冲淘;车板造型;刮除;扫扫舱工作
sweeping amplifier 扫描放大器

sweeping amplitude 扫描幅度
sweeping anchor 扫海锚;探锚
sweeping appliance 扫海用具
sweeping at definite depth 定深扫海(测量)
sweeping beam 扫描电子束;扫描射束
sweeping board 车板
sweeping car 扫路车;清道车
sweeping coil 扫描线圈
sweeping core 车制砂芯
sweeping curve 连续曲线;急转曲线;曲率不大的曲线;大半径曲线
sweeping depth 扫海深度
sweeping dislocation 扫动位错
sweeping electrode 扫除电极
sweeping equipment 扫描设备
sweeping field 净化场;漂移电场
sweeping gas 吹扫气体
sweeping gas membrane distillation 气扫式膜蒸馏
sweeping gear 扫雷装置
sweeping generator 扫描发生器
sweeping guide 导框
sweeping hold charges 扫舱费
sweeping loam mo(u)ld 刮泥型
sweeping machine 扫街机;清扫机;扫路车
sweeping matrix 淘汰矩阵
sweeping meander 下移弯曲线段;下移河湾
sweeping mechanism 清淤机械
sweeping mo(u)lder's horse 车板架;铁马
sweeping of meander 河弯下移
sweeping of sea 扫海
sweeping of street 街道尘土垃圾
sweeping out method 刮去法
sweeping repetition rate 扫描重复频率
sweeping repetition time 扫描重复周期
sweeping rib 双曲肋
sweepings 扫集物;扫舱货;金属屑;地脚;成堆垃圾
sweeping scheme 扫描方案
sweepings of street 街道尘土垃圾
sweeping sounder 扫床测深仪
sweeping strip 扫描带
sweeping survey 扫海测量;扫测
sweeping system 扫描系统
sweeping table 扫动式淘汰盘
sweeping tackle 车板架
sweeping technique 淘汰技术
sweeping trains 扫海艚
sweeping up 车制砂型
sweeping vehicle 清道车
sweeping width 擦光宽度;清扫宽度
sweep-initiating pulse 扫描起始脉冲
sweep integrator 扫描积分器
sweep intensification 扫描增强
sweep intensifier 扫描增强器;扫描扩大器
sweep intensity 扫描线亮度
sweep intensity control 扫描线亮度控制
sweep interval 扫描时间
sweep jammer 扫频干扰机
sweep length 扫描长度
sweep-length selecting switch 扫描长度选择开关
sweep linearity 扫描线性
sweep lock (安全窗扇间的)紧固插销;弯头锁;安全插销
sweep lock-on jamming 扫频跟踪干扰
sweep long quarter bend 镰刀形长的直角弯管;镰刀形长的直角弯头
sweep magnifier 扫描扩展器
sweep marker generator 扫描标志发生器
sweep mechanism 扫描机构
sweep motor 扫描(用)电动机
sweep mo(u)lding 样板造型;刮板造型;车板造型
sweep net 大拖网
sweep of duct 管道弯头;导管曲线段
sweep of pipe 管道弯头
sweep of tube 管道弯头
sweep operation 清除操作
sweep optic(al) square 平面扫描仪;扫描直角转光器;扫描式光学直角头
sweep oscillator 扫描振荡器
sweep oscillator voltage 扫描振荡器电压
sweep-out 清扫
sweep-out depth 远驱水深
sweep-out effect 扫出效应
sweep-out pattern 清扫面积

sweep-output amplifier 扫描输出放大器
sweep over 掠过;猛烈袭击;扫过
sweep pattern 刮板模型
sweep phase 扫描信号相位
sweep phase of garbage collection 不用单元收集的扫描阶段
sweep phasing 扫描定相
sweep rake 集堆机
sweep ramp 扫描斜坡
sweep range 扫描范围
sweep rate 扫描速率;扫描速度;扫描频率
sweep record 扫描记录
sweep recurrence rate 扫描重复频率
sweep reflectometer 扫描反射计
sweep retrace 扫描回程
sweeps 废屑
sweep saw 弧线锯;弧锯;曲线锯
sweep second hand 长秒针
sweep signal generator 扫频仪;扫描信号发生器
sweep spacing 扫描间隔
sweep speed 扫描速率;扫描速度
sweep speed calibrator 扫速校准器
sweep speed control 扫描速度控制
sweep-spot jamming 扫描瞄准式干扰
sweep stabilization 扫描稳定
sweepstakes route 漂流航线
sweep start 扫描启动器
sweep stone picker 捡石机
sweep-stopping circuit 扫描停止电路
sweep strip 门扇防风刷;(转门上下缘的)挡风条
sweep switching 扫描交换
sweep synchronization 扫描同步
sweep tee 弯曲三通;斜三通;斜 T 形管
sweep template 造型刮板;弧型样板;车板
sweep test 扫描测试
sweep-through 扫频式干扰机
sweep-through jammer 扫掠干扰机
sweep-through jamming 扫掠式干扰
sweep time 扫描期
sweep time range 扫描时间范围
sweep trigger 扫描脉冲;扫描触发器
sweep tube 扫描管
sweep turnoff 扫描停止
sweep-type agitator 扫动式搅拌机
sweep unit 扫描装置;扫描部件
sweep-up pipe 吸尘管
sweep vector 扫描矢量
sweep velocity 扫描速度
sweep voltage 扫描电压;扫描电压
sweep volume 扫描容积;扫过容积
sweep waveform 扫描波形
sweep well 大口(土)井;用吊桶打水的井
sweep width 扫描宽度
sweep wood 沉底木
sweep-work 刮板造型;车板造型
sweet 脱硫的
sweetbark 略带淡玫瑰红色的硬木(澳大利亚昆士兰产)
sweet bay 肉桂
sweet birch 山桦
sweet briar 多花蔷薇
sweet butter 淡黄油
sweet camber 平坦路拱
sweet chestnut 甜栗木
sweet clay 新鲜白土;再生白土
sweet condensate storage tank 除盐凝结水箱
sweet corn 甜玉米
sweet corrosion 无硫腐蚀;非硫腐蚀
sweet crude oil 低硫重油;低硫原油
sweet distillate 脱硫馏分
sweetener 脱硫设备
sweetening of soil 中和土壤
sweetening 脱硫
sweetening process 脱硫过程
sweetening treatment 脱硫处理
sweeter 全景分析器
sweet gas 无硫天然气;无硫气;脱硫气;低硫天然气
sweet gasoline 低硫汽油
sweet glass 易于加工的玻璃;长性玻璃
sweet gum 枫香木;胶皮枫香树(胶);枫香
sweeting 涂焊剂;凝结水
sweetite 四方羟锌石
Sweetland filter 斯维特兰型叶片过滤器
sweet naphtha 低硫石脑油

sweet natural gas 脱硫天然气;低硫天然气
sweet oil 橄榄油;无硫油;无臭油;脱硫油;食用油
sweet orange 甜橙
sweet osmanthus 桂花树
sweet peach 密桃
sweet petroleum product 无硫石油产品;低硫石油产品
sweet reinfery wastewater 糖果加工厂废水
sweet roasting 全脱硫焙烧;死烧
sweet-scented oleander 夹竹桃
sweet-scented osmanthus 木犀【植】
sweet shop 糖果店
sweet smelling 芳香的
sweetsop 番荔枝
sweet spring water 甘泉
sweet water 饮用水;清水;甜水;淡水
sweet-water tank 除盐水箱
sweet white 甜白(一种白釉品种)
swell 增音器;增减音器;增大;高涨;隆起地;隆起;马脊岭;涌;梭箱板;大舒缓穹隆;长浪;涌长;胀箱;胀大
swell abatement 涌浪减退
swellability 可膨胀性
swellable formation 可膨胀地层
swell attenuation 涌浪衰减
swell beach profile 涌浪型海滩剖面
swell-butted 根端膨大的
swell chamfer 波浪式线脚
swell compensator 波浪影响补偿器
swell damage 浪涌损害;浪损
swell decay 涌浪衰减
swell diameter 膨胀直径
swell direction 涌浪方向
swelled chamfer 波形线条;波形线脚
swelled column 鼓形柱
swelled field 膨胀场
swelled frieze 鼓起的雕带;凸出的雕带
swelled ground 膨胀土
sweller 溶胀剂
swell factor 膨胀系数
swell forecast 涌浪预报
swell head 激升水头
swell increment 膨胀量
swell index 膨胀指数
swelling 铸铁长大;肿胀;加粗柱身中部;溶胀;膨胀;膨化(现象);泡胀;温胀;水涌;冻胀;胀箱;胀大
swelling acid 膨胀性酸
swelling agent 溶胀剂;膨胀试剂;膨胀剂;膨松剂;膨润剂;泡胀剂
swelling and shrinkage grade of swelling soil 膨胀土的胀缩性分级
swelling and shrinkage of foundation 地基胀缩
swelling apparatus 膨胀仪
swelling capacity 膨胀能力;膨胀量;膨润度
swelling capacity of soil 土的膨胀能力
swelling capacity test 膨胀量试验
swelling clay 易膨化的黏土;膨胀黏土;膨化黏土
swelling curve 膨胀曲线
swelling factor 膨胀因素
swelling fastness 耐膨胀度
swelling fold 膨胀褶皱
swelling force test 膨胀力试验
swelling formation 膨胀性地层
swelling foundation 膨胀性地基
swelling gel 膨胀胶
swelling grade of foundation soil 地基土的胀缩等级
swelling ground 隆起地基;隆起地层;膨胀性地基;膨胀性地层;膨胀地基;冻胀土;冻胀地
swelling heat 膨胀热
swelling heave 膨胀隆起
swelling index 回弹指数;溶胀性;溶胀度;湿胀指数;膨胀指数
swelling isotherm 等温膨胀性
swelling limit 胀限;膨胀极限
swelling material 膨胀性材料
swelling mode 膨胀方式
swelling number 膨胀指数
swelling of coal 煤膨润
swelling of nucleus 核膨胀
swelling of sea 涌浪
swelling of soil 土壤膨胀;土的膨胀
swelling of the pattern 模型膨胀

swelling of tidal wave 潮波壅高;潮波升高
swelling of wood 木材湿胀性;木材的韧性
swelling phenomenon 膨胀现象
swelling potential 膨胀势力;膨胀潜能
swelling power 胀箱力;溶胀(能)力;溶胀力;膨胀能力;膨胀力
swelling press 膨胀压力机
swelling pressure 膨胀压力;膨胀压;膨胀型土压力;膨润压力
swelling pressure test 膨胀压力试验
swelling process 膨胀法;膨胀过程
swelling property 膨胀性质
swelling rate 膨胀速率;膨胀率
swelling rate of volume 体积膨胀率
swelling ratio 膨胀比(率)
swelling resistance 耐溶胀性;耐膨胀性;抗胀性
swelling rock 膨胀岩石;膨胀性岩石
swelling rubber 膨胀橡胶
swelling salt 膨胀性盐
swelling-shrinkage and disintegration of cohesive soil 黏性土的膨胀性与崩解性
swelling soil 膨胀土
swelling speed 膨胀速度
swelling strain 膨胀变形
swelling strain index 膨胀应变指数
swelling stress 膨胀应力;湿胀应力
swelling tendency 膨胀趋势
swelling test 膨胀试验
swelling test of soil 土的膨胀试验
swelling tide 涨潮
swelling-up 膨胀凸出
swelling value 膨胀值;溶胀值
swelling velocity 膨胀速度
swelling water content 膨胀含水率
swell load 递增荷载
swell measurement 隆起测定
swellmeter 膨胀计
swell-neck pan head rivet 盘头锥颈铆钉
swell of a mo(u)ld 胀沙
swell of pulley 滑轮槽
swell of seafloor 海底隆起
swell pedal 增减音器踏板
swell piece 桨门加强板;桨门承板
swell potential 膨胀势
swell ratio 膨胀比
swell scale 涌浪等级;涌级
swell-shrinkage behavio(u)r 胀缩性
swell-shrink characteristic 胀缩性;胀缩特性
swell-shrinking soil 胀缩土(壤)
swell soil 膨胀土
swell wall 膨胀曲壁
swell wave 长涌浪
sweltering hot weather 闷热气候
Swenson buff antique 淡灰色花岗岩(美国新罕布什尔州产)
Swenson pink 淡红色黑云母花岗岩(美国新罕布什尔州产)
swept and garnished 被打扫修饰一新
swept area 射及区;扫及区;扫海区
sweptback angle 后掠角
sweptback blade 后弯叶片
sweptback vane 后弯式叶片
sweptback wing 后掠翼
swept band 扫掠波段;可变波段
swept brilliance 扫描亮度
swept by wire 用钢丝扫过的海
swept capacity 排量
swept-focus 扫描焦点
swept frequency 摇频;扫掠频率
swept frequency interferometer 扫频干涉仪
swept frequency oscillator 扫频振荡器
swept frequency picture modulation 扫频信号图像调制
swept frequency racon 扫描频率雷康
swept frquency technique 扫频技术
swept gain 扫描增益
swept jamming 扫频干扰
swept-lobe interferometer 扫瓣干涉仪
swept planking 弯曲木甲板
swept pumping 扫掠抽运
swept ripple mark 歪扭波痕
swept signal generator 扫频信号发生器
swept valley 倾斜排水沟;曲线屋面排水沟;屋角斜瓦面

swept volume 排量;置换容积;转容量;工作容积;活塞排量;体积排量
swept volume of cylinder 汽缸工作容积
swept volume of internal combustion engine 内燃机工作容积
swept wing 后掠翼
swerve 折射;偏向
swift 绕线车;锡林滚筒;卷筒;卷盘
swift a capstan 绕好绞盘杆围绳
Swiftcrete 斯夫特克里特(一种快凝水泥)
swifter 绞盘加固束;绞盘杆围绳;舢板舷围绳
swifter a capstan 绕好绞盘杆
swift feed 快速送进
swift flow 奔流
swift growth 快速增长
swiftly moving traffic 快速交通;急行交通
swift race 湍急急流
swift running water 急流水
swifts 雨燕
swiggle 带波浪形铝带的胶条框
swig in 拉紧
swig up 拉高
swill 泔水
swill box 泔脚饲料筒;垃圾箱
swim bow 平头前倾船首
swim end 橇形船头(驳船)
swim fin 脚蹼
swim lane 泳道(池的)
swimmable water quality 可游泳水水质
swimmeret 游泳足
swimming ability 游动能力
swimming and diving pool 游泳和跳水池
swimming area 游泳区
swimming bath 室内游泳池
swimming beach 游泳场;海滨浴场;海滨游泳场
swimming hall 游泳馆
swimming place 游泳场;天然游泳场
swimming pool 游泳池
swimming pool cover(ing) 游泳池盖;游泳池覆盖棚
swimming pool filter 游泳池过滤器
swimming pool hall 游泳池大厅
swimming pool heating system 游泳池加热系统
swimming pool nosing 游泳池池边
swimming pool paint 游泳池涂料;游泳池(用)漆
swimming pool reactor 水冷却原子反应堆;游泳池型反应堆;游泳池式反应堆
swimming pool shelter 游泳池棚
swimming pool water 游泳(池)水
swimming population 游泳人数
swimming sea-mark 浮动航标
swimming vortex 浮涡
swim pool 游泳池
swin arm 摆臂
swindler 诈骗者
swindle sheet 假想岩矿图
swinefordite 锂蒙脱石
swinestone 沥青石灰石;臭石灰;臭灰岩
swing 旋转度;旋转;悬腿架;摇摆;转向;指针最大偏转;机头偏转;回转;秋千;掉头;侧移;峰—峰值;摆幅;摆度;摆动额;摆动;摆程
swing account 自动调整结算;悬欠账户;波动性账户
swing anchor 摆动锚
swing and tilt mechanism 移机机构
swing angle 旋转角;旋角;旋变角度;摇摆角度;回转角(度);回旋角度;横移角度;摆角;摆动角
swing angle of photograph 相片旋角
swing area 回转水面
swing arm 旋臂;摇动臂;摇臂;摆动杆
swing arm discharge pipe 旋转卸料管;回旋卸料管
swing arm mechanical hand 摇臂机械手
swing arm sprinkler 摇臂式喷灌器
swing around(hay)stack 转臂式垛草机
swing around seat 转座
swingback blade 连枷式刀片
swingback release 摆回式安全器
swingback unloading auger 后退转位式螺旋
swing bar 转门门枢;摇闩;摇杆
swing base compass 动座式罗盘
swing basin 调头港池
swing bearing 摇座;绞链支座;回转支承;摆动支座;摆锤支座
swing beater grinder 活动锤片式击碎机
swing-beater pulverizer 摆锤式粉碎机

swing bed 成对的对床
swing belt conveyer 摆动式输送带
swing bend pipe 转动弯管
swing berth 调头泊位
swing block four bar linkage 摆杆四杆链系
swing block linkage 摆块链系
swing boat davit 旋转式艇架
swing bolster 摆动承梁
swing bolster link 摆动承梁连杆
swing bolt 铰接螺栓;铰接支点枢轴;铰接螺栓;活节螺栓;活接头螺栓
swing boom 转向河缆
swing boom crane 悬臂起重机
swing boom system 摆动吊杆装卸方式
swing brake 回转水制器
swing bridge 旋桥;旋开桥;平旋桥;平旋桥;平开桥
swing bridge signal 旋桥信号
swing-bucket conveyer 吊斗式输送机;摆斗式输送机
swing bucket mixer 倾槽拌和机
swing calibration 旋角校准
swing cam 摆动凸轮
swing check valve 平旋止回阀;旋转回止阀;旋启式止回阀;绞链板式止回阀;回旋逆止阀;摆动式止回阀
swing choke 变感扼流圈;摆动扼流圈
swing circle 转盘;摆动圆
swing clause letter of credit 交互计算信用证
swing clear 回转无碍(指单锚泊地的安全性)
swing clear hinge 开足铰链(门窗的)
swing clevis 摆动马桶钩
swing clip 旋转货架
swing clutch 摆动离合器
swing cock 旋转龙头;旋转阀(门)
swing concertina arm 旋转手风琴式挂物架;旋转伸缩悬臂挂物架
swing conduit 吊管
swing conveyer 摆式输送机;摆动式输送机
swing crane 旋枢起重机;摆动式吊机;旋臂起重机;回转式起重机;回旋式起重机
swing credit 信用摆动额;转向信贷;额度高低变化的贷款;摆动信贷
swing cross cut saw 横割摆锯
swing cross cut sawing machine 摆动式锯床
swing curve 摆动曲线
swing cut 旋涡型陶槽
swing cut off saw 切断摆锯
swing cylinder 转臂转动油缸
swing day 倒班日
swing diameter 调头直径
swing diameter of work 工件回旋直径
swing diffuser 旋转喷嘴;旋转扩散体;旋转扩散器;摇臂曝气器;摇摆散布器
swing dipper shovel 摆动式铲斗
swing dolphin 调头系缆墩
swing door 双向摇摆门;双向弹簧门;双开摇摆弹簧门;自关门;弹簧门;双开式弹簧门;双动自止门;摆门
swing door closer 双开弹簧门关闭器
swing door fittings 双开弹簧门附件
swing door furniture 双开弹簧门设备;双开弹簧门装配
swing door grip handle 双开弹簧门手柄;双开弹簧门握手
swing door hardware 双开弹簧门小五金;双开弹簧门硬件
swing door push handle 双开弹簧门推板;双开弹簧门推楗;双开弹簧门推杆
swing door tubular grip handle 双开弹簧门管式手柄;双开弹簧门管式握手
swing drilling machine 回转式钻机
swinger 回转机构;不定位置圆
swing error 旋角误差;摆动误差
swinger spring 回转弹簧
swing excavator 回转式挖掘机;回旋式挖掘机;全回转式挖土机;全回式挖掘机;全旋式挖土机
swing faucet 旋转龙头;旋转阀(门)
swing-forging machine 摆锻开坯机
swing fork lift 摆式叉车装草车;摆动叉起重机
swing frame 旋框
swing frame grinder 悬挂式砂轮机;摇动框架式研磨机;旋架磨床
swing-frame grinding machine 旋架磨床
swing gate 旋转门;回旋式栅门;锁气器式挡板;双向摇摆大门;摆动挡板
swing gaye 旋枢栅门
swing gear 回转大齿轮
swing hammer 摆锤
swing hammer crusher 旋锤式破碎机;转锤式破碎机;锤磨机;摆锤(式)破碎机
swing hammer mill 旋锤式磨机;摆锤式磨机;摆锤磨
swing hammer pulverizer 摆锤式粉碎机;冲击磨
swing hammer rotary granulator 回转摆锤碎石机
swing hanger 摆动吊架;摆动吊挂
swing hinge 旋转铰接
swing hopper 吊斗起重机
swing hose rack 旋转式软管架
swing in 转倒舷内
swing indicator 摆偏转指示器
swinging 吊运
swinging a claim 调整矿区形状
swinging angle 回转角度;回旋角度;摆角
swinging arc 摆弧
swinging arch 调整拱模位置
swinging area 调头区(船舶);回转区域;回船区;回船(港)池
swinging arm 摆动逼;摆臂
swinging arm extractor 摆动臂取件器
swinging axis 摆动轴
swinging basin 调头港池
swinging beam hardness tester 摆杆硬度计
swinging belt conveyer 摆动式皮带输送机
swinging berth 宽敞锚位
swinging block linkage 回转滑块联动机构
swinging block quick return motion 摆杆急回运动
swinging block slider crank mechanism 摆杆滑块曲柄机构
swinging boom 系艇杆;起重机旋动臂;起重机回转臂;吊车旋转杆;撑艇杆
swinging boom stack 转臂式垛草机
swinging boom system 单杆吊货法
swinging bracket 摆动支架;扇形齿轮架
swinging brick 屏蔽砖
swinging-bucket ditch cleaner 摆式清沟机
swinging bucket elevator 回转斗式提升机
swinging buoy 旋回浮标;校正罗经浮筒
swinging choke 变感扼流圈
swinging chute 吊式漏口
swinging circle (挖土机等的)回转底盘;转盘
swinging clevis 摆动 U 形夹
swinging compass 摆动罗盘
swinging conveyer 摇摆输送机;回转式运输机;回转式输送机;摆动式运输机
swinging core line 取土器升降绳;取土器吊绳
swinging crane 摆动起重机;摇臂起重机;摇臂吊车
swinging damper 旋转挡板
swinging delivery tray 摆动转移托架;摆动输送盘
swinging derricks 摆动双杆操作
swinging derrick system 单杆吊货法
swinging die method 摆动辊压法
swinging die press 摆动辊压机
swinging distributor 回转布水器;旋转配水器;旋转配水池
swinging door 双动自止门;转门;双开式弹簧门;摇(扇)门;弹簧门
swinging double sieve 双面摇筛
swinging drawbar 回转拉杆;摆动式联结装置
swinging earth 暂时接地;活接地
swinging erection 转体施工
swinging feeder 摆式输料器;摆动式喂料机;摆动给料器
swinging fluke anchor 转爪锚;无杆锚
swinging fork 摆动叉
swinging fork type bale loader 摆动叉式装草捆机
swinging gate 摆动式闸门
swinging gravity fender 重力摆动缓冲板
swinging grinding machine 摇摆式研磨机
swinging grippers 摆动叼纸牙
swinging ground 自差校正场
swinging hammer breaker 摆锥式轧碎机;旋锥式碎石机;旋锥式破碎机
swinging hammer crusher 旋锥式轧碎机;碎石机;旋锥式破碎机
swinging hammer method 摆锥法
swinging hammer mill 锤式粉碎机;春碎机
swinging hopper 悬吊式漏斗;悬吊式斗;吊斗
swinging-in casement 内开(式)窗扇
swinging jaw 回转卡爪
swinging jib 回转悬臂;回转吊杆
swinging knob 旋角钮
swinging latch bolt 摆动式弹簧锁闩
swinging lever 摇摆式杠杆
swinging load 摆动负荷
swinging machine 打麻机
swinging mark 校正罗经标志;回转标(校正罗经)
swinging mechanical hand 摇摆式机械手;回转式机械手
swinging mirror 摆动反光镜
swinging mooring 单点系泊
swinging mooring buoy 单浮系船浮筒
swinging motion 振动;摇动;摇摆
swinging mounting 摆动支架
swinging off clear 摆开无碍
swinging of meander 河弯侧移;曲流下移
swinging of signal 信号不稳
swinging out casement window 外开窗
swinging-pair hardware 双扇摇门五金件
swinging pillar jib crane 定柱旋臂起重机
swinging plate anemometer 风压板风速仪
swinging plates feeder 摆动板式给料机
swinging platform 回转操作台
swinging post 悬门门柱;铰链门柱;(门的)悬吊柱;摇摆门柱
swinging projecting sign 挑出摇摆式招牌
swinging radius 伸出长度;旋转半径;旋回半径;回转半径;起重机臂的工作半径
swinging radius of buoy 浮标回旋半径
swinging rate 回转速率
swinging reactor 变感扼流圈
swinging room 掉头区
swinging scaffold 悬吊(式)脚手架;挂脚手;悬挂(式)脚手架;摇摆脚手架;吊篮;脚手架
swinging screen 摆动筛;摇筛
swinging shield 摆动板
swinging ship 摆动法罗盘校正
swinging sieve 摆动筛;摇筛
swinging sign 摇摆招牌
swinging sounding 摆动测深
swinging space 调头区(船舶);掉头区
swinging speed 回转速度
swinging spider 吊篮(钢绳冲击钻)
swinging spout 摆式给料器
swinging staff 摇尺测量
swinging stage 吊篮
swinging staging 悬吊式脚手架
swinging strength 摆动强度;摆动力量
swinging the arc 左右摆动六分仪;六分仪弧尺的转动
swinging the ship 旋转船首
swinging valve 平旋阀;旋开阀
swinging-vane oil pump 转叶式油泵
swinging-vane pump 转叶式(水)泵
swinging wedge 摆动楔
swinging weight 回转锤
swinging wet masher 摆动式湿饲料搅碎机
swing-in-mirror 摆动反光镜
swing jack 折叠式千斤顶;横式起重机;脱轨车起重机
swing jaw 活动颚板;动凿板;摆动颚
swing jaw crusher 摆动颚式破碎机
swing jaw crushing plate 摆式颚板
swing jaw shaft 旋摆颚板轴
swing jet perforator 摇杆式射孔器
swing jib 转动臂;摇臂
swing jib crane 旋转挺杆起重机;旋臂(式)起重机;回转臂式起重机
swing joint 悬吊接头;膨胀接头;旋转铰链结;旋转式连接;旋转接头;旋转接合;可接接头
swing knuckle 摆动爪扣
swing leaf 活动门窗扇
swing lever 旋臂
swing (lever) crane 旋臂(式)起重机
swing lift 集装箱装箱车;侧摆式提升机
swing line 吊管
swing link 月牙板;摇杠吊;摆杆
swing link bolster 摆动杆枕梁
swing link carrier 摆动支座
swing link guide 导向摆杆
swing link pin 摆杆销
swing loader 回转式装载机;悬臂式挖掘装载机;

转臂式装载机
swing lock 扭锁
swing lock lever 旋转制动杆
swing man 替班工人;代班人
swing mechanism 摆动机构
swing mechanism brake 回转式制动机构
swing mechanism group 回转摆动机构组
swing meter 摆度仪;摆动仪
swing method 摆动法
swing migration 摆动迁移
swing moor 泥泞沼泽
swing mooring 单点系泊
swing mortgage 借款人以现住房临时作抵得款以购置另一住宅
swing-nose crossing 可动心轨辙叉
swing of climatic regime 气候状况的变化
swing offset 旋ற位移;旋尺支距法;摆支距
swing of meander 河弯下移
swing of photo 像片旋偏角
swing on bridge 悬动桥
swing out 旋出;转倒舷外
swing over-bed 床面以上的摆度
swing over carriage 刀架上最大加工直径;刀架上旋径
swing over carriage 车径
swing over compound rest 刀架滑座以上的摆度;车径
swing parting parameter of slope 斜井甩道参数
swing parting switchyard of slope 斜井甩车调车场
swing pipe 回转管;吊管;摆管
swing place 调头处
swing plate 振动平板;晃板
swing plough 平衡犁;摆杆步犁
swing plow 摆杆步犁
swing plug 旋转插销
swing-position steady rest and ejector 转位中心架和工件自动装卸器
swing post 铰链门梃;(门的)铰链柱
swing press 手动螺旋压机
swing prism 活动棱镜
swing radius 摆动半径
swingram baler 摆动柱塞式捡拾压捆机
swing reactor 摇摆式反应器
swing roller 回转托轮;回转滚子
swing-roof furnace 旋转炉顶电炉
swing room 吸烟室;工人休息室
swing round 调头
swing sample pipe 悬吊取样管
swing sash 旋转式窗扇;弹簧窗扇
swing saw 摆(式)锯
swing scaffold 挂脚手;悬吊式脚手架
swing-scraping 摇摆刮削
swing sheave 摆动滑轮
swing shift 中班(下午四点到半夜)
swing shiftman 中班工人
swing sieve 摇动筛
swing slinger 悬挂式抛砂机
swings of demand 需求波动
swing span 平转跨;平旋跨;摆动架;旋开桥;跳桥
swing speed 旋转速度;摆动速度
swing spout 摆动流嘴;摆动给料器
swing stem 摆杆
swing stop 转动定位器
swing swift 轮班制度
swing-swing method 双旋转法
swing-swing method of relative orientation 单独像对相对定向
swing table 振实台
swing table abrator 抛丸清理转台
swing test 振动试验;停滞试验(检查罗经灵敏度);摆动试验
swing tooth plate 活动齿板
swing to port 向左横移
swing to starboard 向右横移
swing trough 摆动槽
swing tube 吊管
swing type 旋启式
swing type centrifuge 中速离心机
swing type draw bar 摆动式牵引杆
swing type entrance 旋转门入口
swing type loader 旋转式装载机
swing-type plate shears 摆式剪板机
swing-type quickopening handhole 摆柄快开手孔

swing-type quickopening manhole 旋柄快开入孔
swing-type ram 摆动式压实器
swing-type valve 回转阀
swing-up door 上开门;翻门
swing-up garage door 车库翻门
swing up saw 倒摆式截锯
swing valve 摆阀;旋阀;回转阀
swing-vane dredge(r) 旋叶式挖泥船
swing velocity 旋转速度
swing voltage 激励电压;摆动电压
swing wedge 摆动光楔
swing winch 旋转式绞车;横移绞车;摆动绞车
swing winch load indicator 横移绞车负荷指示器
swing winch speed controller 横移绞车速度控制器
swing wire 摆动缆
swing worm drive 摆动蜗杆传动
swining conveyer 回转式运输机;回转式输送机
S-wire 塞套引线
swirl and though cave 旋涡沟槽洞
swirl atomizer 旋涡式喷嘴;旋流(式)雾化器
swirl burner 涡流燃烧室
swirl calciner 涡流分解室
swirl chamber 涡流室
swirl combustion chamber 涡流燃烧室;弯曲燃烧室
swirl defect 漩涡缺陷
swirl diffuser 旋流扩散器;旋转曝气器
swirled amphogneiss 漩涡状混合片麻岩
swirler 旋流器;涡流式喷嘴
swirl error 漩涡误差;涡旋误差
swirl finish 转动修整;弯曲加工;盘绕加工;漩涡形抹灰饰面
swirl flame 旋转火焰
swirl-flow burner 旋流式燃烧器
swirl furnace 涡流分解炉
swirl gate 离心集渣浇口
swirl grain 涡漩纹;涡卷纹;涡卷木纹
swirling action 漩涡作用
swirling clarifier 涡漩澄清池
swirling flow 旋流;回旋流
swirling fluidization type of SF preheater 旋流沸腾式 SF 形预分解炉
swirling jet 涡漩射流
swirling jet-induced cavitation 涡漩射流空化
swirling meter 旋流流量计
swirling nozzle 漩涡喷嘴
swirling type of sF preheater 旋流式 SF 形型预分解炉
swirling vane 旋流叶片
swirl injection 涡漩喷注
swirl jet 旋流式喷嘴
swirl mat 卷曲纤维毡
swirlmeter 漩涡计
swirl nozzle 旋流喷嘴;漩涡式喷嘴;涡流喷嘴
swirl-nozzle type injector 漩涡喷嘴型喷射器;旋流喷嘴型喷射器
swirl plate 涡流片
swirl reducing baffle 减小气流偏扭度的隔板
swirls 涡卷形线脚
swirl speed 涡动速度
swirl sprayer 漩涡喷嘴
swirl-type atomizer 涡流式弥雾机
swirl-type reamer 螺旋状吊槽扩孔器
swirl vanes 回旋叶片
swirtl diffuser 旋流风口
swish pan 快速摇镜头
Swiss air 瑞士航空公司
Swiss architecture 瑞士建筑
Swiss bank corp 瑞士银行
Swiss blue 瑞士蓝
Swiss cottage 瑞士式村舍;瑞士式茅屋
Swiss credit bank 瑞士信贷银行
Swiss federation of trade unions 瑞士工会联合会
Swiss formula 瑞士公式
Swiss franc 瑞士法郎
Swiss gold franc 瑞士金法郎
swiss hammer 回弹仪(测混凝土强度)
Swiss lapis 瑞士青金
Swiss lapis lazuli 瑞士蓝宝石
swiss muslin 薄细布
Swiss parallel gutter tile 瑞士式平行沟瓦
Swiss pattern file 瑞士型锉;瑞士式锉刀
Swiss pine 瑞士松
Swiss pine oil 瑞士松油

Swiss screw-thread 细牙螺纹
Swiss style hammer 瑞士锤
Swiss-type automatic 瑞士型自动机床;单轴自动机床
Swiss welding method 左向焊法
Swiss willow 瑞士柳
switch 转辙器;转换器;开关;机键;换接;切换开关;铁路侧线;电闸;电门;道岔集中操纵楼;分支符号
switchable 可有开关控制
switchable code 可转换代码;交换码
switchable-mode line 可交换方式线路;交换方式线路
switchable network 可转接的网络
switch adjustment 开关调整
switch anchor 转辙器防爬铁撑
switch and-fuse 带保险丝开关
switch and fuse combination 开关与熔断器组合
switch and-lock mechanism 转换锁闭器
switch and lock movement 转辙锁闭器
switch angle 转辙角
switch apparatus 开关设备;开关器具
switch area 道岔区
switch arm 开关转臂;开关臂
switch attendance 道岔维护
switchback 之字线;之字形线路;之字形路线;回头路线;转向线;Z 形线路【铁】;之字形爬山路线(铁路的);Z 形铁道路线【铁】
switchback alignment 之字形定线
switchback curve 回头曲线
switchback development 之字展线
switchback line 人字线
switchback ramp 转弯坡道
switchback station 回头线车站【铁】;人字形车站;之字形路线车站;折返站;车辆调车站;Z 形车站【铁】
switchback track 之字线
switchback turn 之字形回转
switch bank 转换触排
switch bar 转辙杆【铁】
switch base 开关座
switch base plate 大垫板
switch bed plate 道岔垫板
switch bench 开关台
switch blade 辙尖;闸刀开关铜片;闸刀;开关闸刀;道岔尖轨;岔尖
switch block 开关部件
switchboard 开关板;控制板;接线器;总机;交换台;交换机;配电盘
switch board and fault locator 配电盘及故障点标定装置
switchboard cell 交换机塞孔单元
switchboard cord 交换机塞绳
switchboard dial 交换机拨号盘
switchboard fitting 交换机配件
switchboard house 开关室
switchboard panel 配电板
switchboard position 交换台座席
switchboard room 配电室;电话接线室;开关室
switch box 转换器;转换开关匣;转换开关盒;闸盒;开关箱;开关盒;配电箱;电闸盒
switch branch track 道岔侧股轨道
switch building 开关间
switch cabicle control 开关柜控制
switch cabin 电缆配线房
switch cabinet 开关箱;开关间;开关柜
switch cam 超针闸刀
switch cargo 转手货(物)
switch chair 道岔轨座
switch character 开关字符
switch circuit controller 开闭器
switch cleaner's cabin 道岔清扫房
switch clock 开关钟
switch closed up 道岔封锁
switch closure 道岔密贴
switch cock 开关旋塞
switch column 道岔搬杆;转辙杆(道岔就地操纵);道岔扳杆
switch combination 开关组合
switch command 开关命令
switch control circuit 道岔控制电路
switch control computer 转接控制计算机
switch cover 开关罩
switch crossing angle 转辙角

switch cubicle 开关柜
switch cylinder 筒形开关
switch dealer 转手掮客
switch declaration 开关说明
switch design 道岔设计
switch designator 开关命名符
switch desk 控制台;开关台
switch disconnector 负荷开关
switched capcitor network 开关电容网络
switched circuit 开关电路;开关线路
switched circuit automatic network 开关电路自动网络
switched communication network 交换通信网络;交换式通信网
switched connection 交互连接
switched digital video 交换式数字图像视频
switched line 交换线路
switched message network 信息转接网络;信息交换网(络);交换电信网络
switched multimegabit data service 交换式多兆位数据业务
switched network 交换网(络)
switched network backup 交换网络后援;交换网络后备支持
switched operational amplifier 开关运算放大器
switched telecommunication network 交换远程通信网;交换型远程通信网络;交换长途通信网络
switched virtual circuit 交换式虚电路
switcher 转接开关;转换开关;调车机车
switcher's servicing track 调车机车整备线
switch filter 开关滤波器
switch fixtures 转辙器附件
switch for caging 销紧开关(陀螺测斜仪)
switch frame 开关架
switch function 开关函数
switch fuse 开关保险丝;熔断开关;开关熔丝
switch gang 道岔工班
switch gap 开关间隔
switch gear 交换装置;分配装置;操纵机组;开关设备;开关装置;配线装置;配电装置
switchgear and control equipment 配电和控制设备
switchgear bay 开关柜内腔
switchgear cabinet 开关间;开关室;开关柜;开关盒
switchgear in compact 紧凑式气体绝缘开关柜
switchgear pillar 柱形开关装置
switchgear plant 开关厂
switchgear room 开关室;配电间
switch grease 电闸润滑脂
switch group 组合开关
switch guard rail 道岔护轮轨
switch handle 开关把手
switch heater 转辙器的融雪器
switch hook 挂钩开关;钩键;开关钩键;叉簧开关
switch house 控制室;开关控制室;配电室
switch housing 开关器外壳
switch identification 开关标识符
switch identifier 开关标识符
switch impulse 转换脉冲
switch impulse withstand voltage 开关操作冲击耐受电压
switch in 接通;接入
switch indication 道岔表示
switch indicator 开关指示器;开关指示符;交换指示器;道岔标志
switching 转接;转换;开闭;配电
switching action 转换作用
switching algebra 开关代数
switching angle 开关角
switching arc 开关电弧;切换电弧
switching area 交换区;调车场
switch(ing) arrangement 开关装置
switching boundary 开关边界
switching by drop on 浮放道岔调车
switching by float-double track 浮放双轨调车
switching cabinet 开关箱;控制室
switching cable 开关电缆
switching capacity 开断容量
switching center 交换中心;交接中心;交换台
switching channel 转换通道;切换通道
switching characteristic 开关特性
switching charges 调车费
switching circuit 开关电路;转换电路;开关线路
switching coefficient 转换系数

switching column technique 色谱柱切换技术
switching computer 开关计算机
switching configuration 开关组态
switching congestion 转接拥塞;交换拥塞
switching constant 开关时间常数;开关常数
switching contrivance 电门设备
switching control 转换控制;开关控制
switching core 开关磁芯
switching criterion 转换准则;开关准则
switching current 开关电流;合闸电流
switching cycle 转换周期;开关操作循环;开断周期;切换循环
switching delay 开关延迟
switching depot 折返站
switching device 开关装置;开闭装置
switching diode 开关二极管
switching effect 开关效应
switching element 开关元件
switching equation 切换方程
switching equipment 交换机
switching force 合闸力
switching function 开关函数;开闭作用;开闭函数
switching gallery 开关廊道
switching gear 开关装置
switching gradient 溜放坡
switching group 开关组
switching hysteresis 转换滞后;开关滞后
switching impulse residual voltage 操作脉冲残压
switching in 合闸
switching input 开关输入
switching intersecting 调车交叉【铁】
switching interval 转换时间
switching key 开关键
switching kiosk 开关亭
switching knob 圆扭开关
switching lead 牵出线【铁】
switching lead at grade 坡度牵出线【铁】
switching lever 合闸手柄
switching line 开关线
switching loss 开关损耗
switching manipulation 开关操作
switching matrix 开关矩阵;切换矩阵
switching mechanism 转换机构;开关机构;开闭机构;切换机构
switching member 开关构件;切换构件
switching method 调车方法
switching mode 切换方式
switching multiplexer 交换多路复用器
switching network 开关网络
switching node 转接节点;交换节点
switching off 断开
switching office 交换局
switching on 合闸
switching on clock 定时开关
switching operation 调车作业
switching order 合闸顺序
switching oscillator 开关振荡器
switching over 切换
switching over unit 自动转换器
switching over-voltage 操作过电压
switching pad 自动转接衰减器;开关衰减器
switching period 切换时间
switching point 开关点;转折点;转接点
switching position 转换位置;切换位置
switching power 切换功率
switching principle 切换原理
switching procedure 转换程序
switching process 线路动态过程;转换过程;整流过程;翻转过程
switching processor 转接处理机;交换处理机
switching program(me) 开关程序;开闭程序
switching pulse 转换脉冲;控制脉冲
switching pulse generator 切换脉冲发生器
switching push-button 开关按钮
switching railroad [美]联络铁路(线);调车铁路
switching rate 转换速率
switching reactor 开关电抗器
switching regulator 开关式
switching relay 转换继电器;开关操纵继电器;切换继电器
switching rod 开关绝缘棒
switching route 调车进路
switching schedule 切换计划
switching section 转接段;倒接段

switching selector repeater 选局中继器
switching sequence detector 切换顺序检测器
switching service 调车作业
switching signal 转换信号;开关信号;合闸信号
switching speed 开关速度
switching spring 合闸弹簧
switching starter 直接起动器
switching station 控制站;开关站;开闭所
switching substation 开闭所
switching subsystem 交换子系统
switching surface 开关曲面
switching surge 开关涌浪;开关浪涌;开关冲击;操作过电压
switching surge test 操作波试验
switching system 转换系统
switching theory 开关理论
switching time 转换时间;开关时间;切换时间;切变时间;动作时间;磁芯的翻转时间;翻转时间
switching time constant 开关时间常数
switching time of bipolar transistor 双极晶体管开关时间
switching torque (电动机的)换接力矩
switching tower 站调楼【铁】
switching track 车道;调车轨道;错车道调车
switching transient 开关瞬态;操作瞬变现象
switching tube 开关管
switching value 转变值;开关量;开断值
switching variable 开关变量
switching yard 火车调度车站;调车场;编组场【铁】
switch in series 串联开关
switch insertion 开关插入
switch jack 机键塞口;机键插孔
switch jaw contact 簧片接触
switch key 开关
switch knob 开关手柄;开关钮;开关按钮;合闸按钮;切换旋钮;扳钮
switch lamp 道岔标志灯
switch lampholder 带开关的灯头
switch lamp-socket 带开关的灯座
switch layout 道岔配列;道岔布置
switchless section 无岔区段
switch lever 转辙杆【铁】;开关杆;合闸手柄;道岔握柄
switch lever with counter-weight 带锤转辙杆
switch lever with indicator 带指示灯的开关杆;带柄道岔标志
switch lever with reversing counterweight 回动锤转辙杆
switch lighting 开关照明
switch list 开关表
switch lock 转辙船闸;开关锁;开关闭锁;道岔锁闭器;道叉键;岔船闸
switch lock out indication 道岔锁闭表示
switch machine 转辙机
switch machine installation 转辙机安装装置
switch main track 道岔直股轨道
switchman 转辙员;扳道员(铁路的);扳道工(人)
switchman's cabin 扳道工房;扳道房【铁】
switch manual release 道岔人工解锁
switch mode power 开关式电源
switch motion 开关动作
switch normal indication 道岔定位表示
switch off 关灯;关机;开闸;切断;断电
switch off delay 断开延迟
switch of inclined plane 坡面道岔
switch of plane 平面岔道
switch oil 开关油;电闸油
switch on 接通;接入;开灯;接通电源
switch on and off 开关通断
switch on delay 接通延迟
switch on the steering wheel 操向轮开关
switch operated response control 灵敏度调节装置;开关操纵力调节
switch operating rod 道岔操纵杆
switch operation 转手交易
switch order 开关指令
switch out 关闭
switchover 转接;互投;换位机构;切换
switchover capability 转接能力
switchover point 转换点
switchover-travel 开关越程
switch panel 控制屏;开关盘;开关板;配电盘
switch plant 开关设备

switch plate 转辙器垫板；滑床板；拨料板
switch plug 插头；墙上插座；开关插座；电门插座
switch point 辙轨；尖轨尖(端)；道岔点
switch-point derail 尖轨式脱轨器
switch point guard rail 尖轨护轨
switch-point protector 道岔尖轨保护器
switch power supply 开关电源
switch pretravel 开关预行程
switch rack 开关架；机键架
switch rail 转辙轨；辙轨；尖轨
switch rail length 尖轨长度
switch rail of special section 特种断面尖轨转辙器
switch rail tie plate 尖轨垫板
switch rail with block heel joint 间隔铁式尖轨转辙器
switch rail with flexible heel joint 弹性可弯式尖轨转辙器
switch rating 开关额定值
switch register 开关寄存器
switch resistance 道岔阻力
switch resistance loss 道岔阻力能高损耗
switch reverse indication 道岔反位表示
switch rod 尖轨连(接)杆；道岔拉杆
switch roller 尖轨滚轮
switch room 配电室；机键室；转换室；交换室；交换机机房；电压房
switch screen 控制屏
switch section 道岔区段
switch selling 转手销售
switch service 接入服务
switch setting 道岔转换
switch shaft 开关轴
switchsignal 转换信号
switch sleeper 转轨枕木；道岔轨枕；长轨枕；岔枕
switch socket 开关灯头
switch socket outlet 开关电器插座
switch soleplate 转辙器长垫板
switch split 挤岔
switch springs 塞孔簧片
switch stand 转换器；转辙器(座)；扳道机
switch starter 起动开关；开关起动器
switch starting 道岔启动
switch-status condition 开关状态条件
switch steel section 道岔钢材
switch step 切换阶段
switch-stop 开关制动销
switch storage 开关存储器
switch tank 开关油箱
switch target 道岔表示器；道岔标
switch tender 扳道工(人)
switch terminal 开关接线端子
switch the feed 转换进料
switch throttle 节气阀
switch through 接转
switch throw 尖轨动程
switch thrown under moving cars 道岔中途转换
switch tie 转辙器轨枕；道岔轨枕；岔枕
switch time 合闸时间
switch-timing error 开关同步误差
switch tongue 道岔尖轨；辙尖；闸刀开关铜片；岔尖
switch tower 信号塔【铁】
switch track 岔轨
switch trade 转手贸易
switch train 开关列
switch transaction 道岔转换
switch transmission 交换传输
switch tree 开关树
switch tube 开关电子管
switch turnout 道岔【铁】
switch-type limiting integrator 开关式限幅积分器
switch-type modulator 开关式调制器
switch-type voltage regulator 分级电位调整器
switch unit 转换开关
switch valve 转换阀；开关阀
switch valve system 开关阀系
switch wheel 开关轮；机键轮
switch with a movable frog 活动辙叉道岔
switch with follow up movement 带动道岔
switch yard 露天开关站；开关装置；互换机；铁路调车场；铁路编组站；室外配电装置
switzerite 水磷铁锰石
swivel 转环体；旋转体；转环
swivel adjustment 旋转调整
swivel adjustor 旋转调整器

swivel anchor 转环锚
swivel angle plate 转盘角板；倾转台
swivel arm 旋臂；水龙头软管
swivel bail 水龙头提环
swivel bailing 迅速不停地钻进
swivel ball 旋转节中的球体；转球
swivel barrel 旋转式岩芯管
swivel base 转座
swivel bearing 旋转轴承；旋转轴
swivel block 转环滑车
swivel body 水龙头外壳
swivel bolt 活节螺栓
swivel bottom vice 转盘虎钳
swivel bow 转弓
swivel bridge 旋开桥；平转桥；平旋桥；平开桥
swivel bushing 轴承衬套
swivel cantilever bracket in tunnel 隧道腕臂调整底座
swivel cargo hook 旋转吊货钩
swivel carriage 回转刀架
swivel cement head 旋转水泥头
swivel chain 转动链；斗链
swivel chair 转椅
swivel chute 旋转供料槽；回转下料溜子；回转滑槽
swivel clamp 活动钳
swivel clip holder 短支持器；分体支持器
swivel clip holder with hook 带钩支持器
swivel conduit 旋转管
swivel connection 旋转接头；旋转接合
swivel conveyer bucket loader 旋转传送机料斗
swivel couple 转环连接杆
swivel coupling 旋转接头；旋转接合；转环式连接器；转动式车钩；转环联轴节；转动式联轴器；铰链连接；活动接头；活动接合；转环联结器
swivel cowl 旋转式通风帽
swivel crane 回转式起重机
swivel damper 旋转挡板；转动式烟道调节板；转动式风门；转动(式)风挡
swivel davit 转动式吊艇杆
swivel disk colter 自动定位圆犁刀
swivel disk jointer 自位式圆盘小前犁
swivel draw bar 旋臂牵引杆；转动车钩
swivel-elbow type floating pipe 鹅颈管式浮排泥管
swivel eye 水龙头侧轴颈
swivel feed(ing) 回转进给；旋转送料；旋转进给
swivel feeding spout 旋转播料器
swivel flight conveyer 旋转螺旋刮片输送机
swivel frame 摇转竖轴窗框；中旋式(气)窗框
swivel graver 回旋刻度仪
swivel head 旋转头；(立轴式钻机的)回转器
swivel head bevel gear (立轴的)回转器伞齿轮
swivel head chute 转头滑槽
swivel head scriber 回旋刻度仪
swivel hoisting plug 提引塞(钻杆)
swivel hook 旋转钩；转环钩；转钩
swivel hook pin 水龙头大钩销钉
swivel hose 水龙头软管
swivel housing 旋转罩
swiveling pile driver 旋转打桩机
swivel jar 活动振击器
swivel jaw vice 转口虎钳
swivel joint 旋转接头；旋转接合；旋转管接头；转环接合；铰接接头；旋转接合；转节
swivel joint drain line 有旋转接头的排水管
swivel ladder 旋转梯
swivel lever 旋转操纵杆
swivel(l)ing angle 旋转角
swivel(l)ing arm 旋臂；回转臂
swivel(l)ing characteristic 转换特性
swivel(l)ing chute 摆动卸料槽；摆槽；旋转溜槽；回转溜槽；摆动溜槽
swivel(l)ing conduit 旋转管
swivel(l)ing crane 转臂起重机
swivel(l)ing device 回转装置
swivel(l)ing device brake 回转装置制动器
swivel(l)ing discharge chute 旋转式卸料槽
swivel(l)ing drill 旋转式钻机
swivel(l)ing gear 旋转装置；回转齿轮
swivel(l)ing grooving saw 旋转的铁槽锯
swivel(l)ing head 万向接头；旋转头
swivel(l)ing head lamp 旋转式前灯
swivel(l)ing hear brake 旋转装置制动器
swivel(l)ing idler 旋转滚柱
swivel(l)ing jib 旋臂；回转杆；转动起重臂；摆臂

swivel(l)ing lever 回转杆
swivel(l)ing mechanism 旋转机构；旋转装置
swivel(l)ing mechanism brake 旋转装置制动器
swivel(l)ing motion 旋转
swivel(l)ing movement 旋转移动
swivel(l)ing nozzle 旋转喷嘴；旋转喷管
swivel(l)ing pile driver 旋转(式)打桩机
swivel(l)ing pipe 旋转管
swivel(l)ing pressure cloth beam 旋转卷布压辊
swivel(ling) ring 旋转环
swivel(l)ing-roller flanging die 旋转辊凸缘
swivel(l)ing seat 旋转座
swivel(l)ing speed 旋转速率；回转速度
swivel(l)ing spotlight 旋转式聚光灯
swivel(l)ing table 转台；转盘
swivel(l)ing tool-holder 旋转刀架
swivel(l)ing tool post 回旋刀架
swivel(l)ing tube 旋转管
swivel(l)ing turnbuckle 旋转式松紧螺旋扣
swivel(l)ing unit 旋转装置
swivel(l)ing unit brake 旋转装置制动器
swivel(l)ing vane 风标
swivel(l)ing wheel 旋转轮
swivel link 铰节；转环节
swivel loom 挖花织机
swivel luminaire fixture 旋转照明设备
swivel mechanical joint 旋转机械接头
swivel mount 转座；转台
swivel mounted 装转铰的；装转座的
swivel mount pulley 转轴滑轮
swivel neck 水旋转头；旋转轴颈
swivel off 丝扣旋出
swivel outlet head (水龙头的)摆口水头
swivel pen 回转笔
swivel pen drafting 曲线笔绘图
swivel pen point 旋转刻针
swivel piece 转环链段
swivel piler 抛掷装置；回转堆垛装置
swivel pin 旋转枢轴；转向节销
swivel pin angle 主销内倾(角)
swivel pin boss 转向销支撑凸块
swivel pipe 旋转管；立轴
swivel plate 旋转板
swivel plough 双向犁
swivel plow 翻转犁
swivel plug 旋塞
swivel pontoon 转轴浮筒
swivel portion of cutter head 进刀架回转部分
swivel pulley 动滑轮
swivel regulator 旋转调整器
swivel replication 旋转复制
swivel ripper 铰接式耙路机；铰接式松土机
swivel rod 旋转杆
swivel rope socket 活动绳帽；活动绳夹
swivel safety cargo hook 旋转自锁式安全吊货钩
swivel sash 旋转窗扇；转窗
swivel seat 转座
swivels for anchor chain 锚链转环
swivels for cargo block 吊货滑车转环
swivel shackle 旋转卡环；转环卸扣；转环卡环
swivel shaft 回转轴
swivel shanks binge-type ripper 转腿铰接松地机
swivel sheaves pivot shaft 转件枢轴；滑轮枢轴
swivel sheaves support bracket 转轮(件)支座
swivel shell 水龙头外壳
swivel shot 锚端链节
swivel spindle 长插销；活节心轴
swivel stem 水龙头中心管
swivel sub 水龙头异径接头
swivel table 回转工作台；旋转台；旋转工作台
swivel tee fitting 旋转三通
swivel tightener 旋转紧线器
swivel tube 旋转管
swivel tube core barrel 单动双层岩芯管(内管不动)；单动双层取芯器(内管不动)
swivel-type castor 小脚轮；自位轮
swivel type double tube core barrel 单动双管取芯钻具；单动双层岩芯管(内管不动)；单动双层取芯器(内管不动)
swivel union 旋转联管节；转向管接头
swivel valve 旋转阀门(门)；颈形龙头；弯颈龙头
swivel vice 旋转座老虎钳；旋转平口钳；旋转虎钳
swivel water distributor 旋转布水器
swivel weaving 挖花织造

swivel wheel 转轮;自位轮
swivel with eye 旋转单耳
swivel wrench 旋转吊钳
swollen consumption demand 消费需求膨胀
swollen credit 信用膨胀
swollen investment 投资膨胀
swollen soil 冻胀土;膨胀土
SWOPS angle well oil production system 单井采油船
sword 勾缝用的工具(石砌的);泥刀;递钩
sword fingers 箭指
sword mat 防磨绳垫
sword shaped leaf 剑形叶
sworn broker 合法经纪人
sworn commercial invoice 宣誓发票
sworn invoice 宣誓发票
sworn measurer 公证行
sworn measurer and weigher 宣誓度量人
sworn measures 公定计量
sworn measures certificate 衡量单
sworn weigher 公证行
SWR meter 驻波比测量器
Sx board 萨克斯板(一种绝缘墙板)
syalon 赛纶陶瓷(含钇的硅铝氧氮陶瓷)
sybergistic antioxidant 多效防老剂
sycamore 三球悬铃木;小无花果树;大枫树
Sychlophone 旋调管
sychometric room 人工气候室
sychronous homopolar motor 同步单极电动机
sycnhronizing slave unit 同步从动装置
sycomore 大枫树
sycon 双沟型
syenite 黑花岗岩;正长岩
syenite group 正长岩类
syenite porphyry 正长斑岩
syenite sett paving 正长岩小方石铺路
syenitic aplite 正长细晶岩
syenitic granite 正长花岗岩
syenitic pegmatite 正长伟晶岩
syenodiorite 二长岩
syenogabbro 正长辉长岩
syhinite 萨砷氯铅矿
Sykes microphone 塞克斯传声器
sylcine rock 钾盐岩
Sylcum 赛尔卡铝合金
syllabus 提纲
sylphon 皱纹管;涨缩盒;膜盒;波纹管
sylphon bellows 薄壁金属圆筒风箱;波纹管
sylphon seal 波纹管密封
sylva 森林志;森林区
Sylvadure 西瓦度(一种木材防腐剂)
sylvanite 针碲金银矿
Sylvester 铰式拔柱机;手摇链式回柱机
Sylvester process 西尔威斯特工艺
sylvestren(e) 枞油烯;枞萜
sylvestre soil 森林土
sylvic acid 松脂油酸
sylvic oil 松浆油
sylvine 钾石盐
sylvinite 钾石盐岩;钾石盐
sylvite 钾盐;钾石盐
sylvogenic soil 森林土
symap 符号地图
symap program(me) 符号地图程序
symbiont 共生物
symbiont control 共存控制
symbiont manager 共存管理程序
symbiosis 共生关系;共生(现象)
symbiotic algae 共生藻类
symbiotic bacteria 共生细菌
symbiotic ecosystem 共生生态系统
symbiotic nitrogen fixation 共生固氮作用
symbiotic nitrogen fixer 共生固氮生物
symbiotic variable 共生变化
symboints 共存程序
symbol and number of standard 标准代号与编码
symbol by proportional point 依比例符号
symbol colo(u)r association 彩色图式符号
symbol control table 符号控制表
symbol definition 符号定义
symbol definition instruction 符号定义指令
symbol dictionary 符号字典
symbol disc 符号盘
symbol for welding 焊接符号

symbol guide 符号模片
symbolic 象征性的;符号替代;符号的
symbolic address 符号地址
symbolic addressing 符号选址;符号地址法;符号编址
symbolic addressing mode 符号寻址方式
symbolic address program(me) 符号地址程序
symbolic age of neutron 费米年龄
symbolically 用符号表示
symbolic analysis 符号分析
symbolic assembler 符号汇编程序
symbolic assembly 符号汇编
symbolic assembly language 符号汇编语言
symbolic assembly program(me) 符号汇编程序
symbolic check 符号检定
symbolic circuit 简图;框图;作用图;职能图;符号电路
symbolic code 象征码;符号(代)码
symbolic coding 符号编码;符号编号
symbolic coding format 符号编码技术;符号编码形式
symbolic concordance 符号索引程序;符号便览
symbolic control 符号控制
symbolic conversion program(me) 符号转换程序
symbolic conversion routine 符号转换例行程序
symbolic cross assembler 符号交叉汇编程序
symbolic data set 符号数据集
symbolic debugger 符号调试器;符号调试程序
symbolic debugging 符号调试
symbolic deck 符号卡片组
symbolic delivery 象征性交货;象征性交付;推定交货
symbolic device name 符号设备名
symbolic diagram 记号图
symbolic differential equation 符号微分方程
symbolic differentiation 符号微分法
symbolic editing 符号编排;符号编辑
symbolic editor 符号编辑程序
symbolic equivalent 符号等价
symbolic expression 符号表达式
symbolic file 符号文件
symbolic file name 符号文件名
symbolic function 符号函数
symbolic generator 符号发生器
symbolic identifier 符号标识符
symbolic information 符号信息
symbolic input language 符号输入语言
symbolic input/output 符号输入输出
symbolic input/output assignment 符号输入输出分配
symbolic input programme 符号输入程序
symbolic instruction 符号指令
symbolic key 符号键;符号关键字
symbolic label 符号标签;符号标号
symbolic language 符号语言
symbolic layout 符号布局
symbolic library 符号程序库
symbolic logic 数字逻辑;符号逻辑学;符号逻辑
symbolic logical operator 符号逻辑算子
symbolic machine 符号计算机;符号机器
symbolic machine code 符号机器代码
symbolic machine language 符号机器语言
symbolic manipulation 符号操作
symbolic mathematical system 符号数学系统
symbolic mathematics 符号数学
symbolic meaning 象征意义
symbolic memory addressing 符号存储定址
symbolic method 符号法
symbolic microprogram(me) 符号微程序
symbolic model 符号模型
symbolic name 符号命;符号名字;符号名称
symbolic network analysis program(me) 符号网络分析程序
symbolic notation 符号表示法
symbolic number 符号数
symbolic operand 符号运算数;符号操作数
symbolic operating system 符号操作系统
symbolic operation 符号运算;符号操作
symbolic operation code 符号操作码
symbolic operator 符号算子
symbolic optimum assembly program(me) 符号最佳汇编程序
symbolic panel 符号面板
symbolic parameter 符号参数

symbolic parameter initial value 符号参数初值
symbolic parameter standard value 符号参数标准值
symbolic placeholder 符号占位符
symbolic polynomial 符号多项式
symbolic power 形式幂
symbolic programme 符号化程序
symbolic programming 符号程序;符号编码技术;编制符号程序
symbolic programming system 符号程序设计系统
symbolic programming tape 符号程序带
symbolic rate 符号速率
symbolic reference 符号引用
symbolic representation 符号表示法;符号表示
symbolic significance 符号的意义
symbolic solution 符号解法
symbolic source program(me) 符号源程序
symbolic string 符号串
symbolic structure 符号结构
symbolic table control 符号表控制
symbolic table element 符号表元素
symbolic table entry 符号表项目
symbolic table pointer 符号表指针;符号表指示字
symbolic table routine 符号表程序
symbolic terminal name 符号终端名
symbolic test 符号检查
symbolic text editor 符号文本编辑程序
symbolic thought 象征性思维
symbolic tools 符号工具
symbolic unit 符号部件;部件符号
symbolic variable 符号变量
symbolic vector 形式矢量;符号矢量
symbolic virtual address 符号虚拟地址
symbolic weight coefficient 符号权系数
symbol identifier 符号辨识器
symbol input 符号输入
symbolism 象征主义;符号体系;符号化;符号表示
symbolism of forces 力的记号
symbolism of form 形状的记号
symbolization 象征化作用
symbolization of accounts 账户编号;科目编号
symbolize 象征
symbolized output 符号输出
symbolizing system 标记制度
symbol library 符号库
symbol list 图式符号;符号表
symbol manipulation 符号处理;符号操作
symbol map 符号图
symbol mask 符号掩模
symbol of crystal face 晶面符号
symbol of element 元素符号
symbol of geologic(al) time unit 地质年代单位符号
symbol of lithological pattern 岩石花纹符号
symbol of net 面网符号
symbol of numeral 数字符号
symbol of operation 运算符号
symbol of simple form 单形符号
symbol of space group 空间群符号
symbol program(me) 符号程序
symbol programming 符号程序设计
symbol rank 符号秩;符号位置;符号(排)列
symbol request 符号请求
symbols and abbreviations of (nautical) chart(s) 海图图式
symbol sequence 符号序列
symbol sheet 符号模片
symbol sign 符号;标记
symbols of hydraulic system 液压系统图形符号
symbol standardization 符号标准化
symbol subroutine 符号子程序
symbol substitution 符号替换
symbol system of account 结账符号法
symbol table 图例
symbol table algorithm 符号表算法
symbol table control 符号表控制
symbol table entry 符号表项目;符号表登记项
symbol template 符号模片
symbol weight 符号量
symbol word marking 文字标牌(道路交通的)
symb process 淀粉废物微生物处理过程
symcenter 对称中心
sym-dichloroethane 对称二氯乙烷
symmedian point 类似重心

symmetallism 金银混合本位
symmetric(al) 相称的;匀称的;对称的
symmetric(al) achromat lens 对称式消色差透镜
symmetric(al) adjustment 对称调正
symmetric(al) all-dielectric(al) interference filter 全介质对称干涉滤光片
symmetric(al) alternating current 对称交流
symmetric(al) alternating quantity 对称交变量
symmetric(al) alternating voltage 对称交变电压
symmetric(al) anastigmat 对称消像散镜组
symmetric(al) antenna 对称天线
symmetric(al) arch 对称拱;对称式拱
symmetric(al) arch analysis 对称拱分析
symmetric(al) arched dam 对称(式)拱坝
symmetric(al) attenuator 对称衰减器
symmetric(al) avalanche rectifier 对称雪崩整流器
symmetric(al) axis 对称轴线;对称轴
symmetric(al) axis of rotation 旋转对称轴
symmetric(al) balance 对称平衡
symmetric(al) banded structure 对称带状构造
symmetric(al) bandpass filter 对称通带滤波器
symmetric(al) band-reject filter 对称阻带滤波器
symmetric(al) barycenter 对称重心
symmetric(al) bedding 对称层理
symmetric(al) bending vibration 对称弯曲振动
symmetric(al) binary code 对称二元码
symmetric(al) bipolar transistor 对称双极晶体管
symmetric(al) blading 对称叶片
symmetric(al) body 对称体
symmetric(al) cable 对称电缆
symmetric(al) cable carrier communication 对称电缆载波通信
symmetric(al) cable carrier communication system 对称电缆载波通信系统
symmetric(al) cable carrier telephone 对称电缆载波电话
symmetric(al) center 对称中心
symmetric(al) channel 对称信道
symmetric(al) characteristic 对称性
symmetric(al) circuit 对称电路
symmetric(al) circulation 对称环流
symmetric(al) clipper 对称削波器
symmetric(al) closure 对称闭包
symmetric(al) colo(u)r dispersion 对称色散
symmetric(al) component 对称部分;对称分量
symmetric(al) compound 对称化合物
symmetric(al) compressor 对称型压缩机
symmetric(al) configuration 对称图形;对称配置
symmetric(al) construction 对称结构
symmetric(al) cooperative game 对称合作对策
symmetric(al) coordinates 对称坐标
symmetric(al) correspondence 对称对应
symmetric(al) cross-section 对称断面
symmetric(al) crown 对称路拱
symmetric(al) cruciform pier 对称形十字墩子;对称形十字码头
symmetric(al) curve 对称曲线
symmetric(al) cuspate foreland 对称尖岬
symmetric(al) deflection 对称偏移;对称变位
symmetric(al) deformation 对称变形
symmetric(al) determinant 对称行列式
symmetric(al) diagram 对称图解
symmetric(al) difference 对称差(分)
symmetric(al) dipole 对称振子
symmetric(al) directed-graph 对称有向图
symmetric(al) distribution 对称分布
symmetric(al) dom cap 对称顶
symmetric(al) double curve turnout 单式对称道岔【铁】
symmetric(al) double points 对称双开道岔
symmetric(al) double turnouts 复式对称道岔
symmetric(al) dual 对称对偶
symmetric(al) electrometer 对称静电计
symmetric(al) element 对称元素;对称要素
symmetric(al) equilateral arch 对称双心拱
symmetric(al) extinction 对称消光
symmetric(al) eyepiece 对称目镜
symmetric(al) fabric 对称结构
symmetric(al) facade 对称立面
symmetric(al) fiber 对称光纤
symmetric(al) figure 对称形状;对称图形
symmetric(al) filter 对称滤波器
symmetric(al) flooding 对称灌注法
symmetric(al) flow 对称流
symmetric(al) flower 整齐花
symmetric(al) fold 对称褶皱;对称褶曲
symmetric(al) formulation 对称表示法
symmetric(al) frame 对称框架
symmetric(al) fret 对称回纹饰
symmetric(al) function 对称函数
symmetric(al) game 对称对策;对称比赛
symmetric(al) gradient 对称梯度
symmetric(al) grading 对称分品连接
symmetric(al) graph 对称图
symmetric(al) group 对称性群;对称群
symmetric(al) house 对称房屋
symmetric(al) idler arm 对称空转臂
symmetric(al) image 对称像
symmetric(al) indication amplifier 对称式指示放大器
symmetric(al) inductive diaphragm 对称电感窗口片
symmetric(al) input/output unit 对称输入输出装置
symmetric(al) integral form 对称积分形式
symmetric(al) junction 对称联结;对称结
symmetric(al) kernel 对称核
symmetric(al) kink-band 对称曲折带
symmetric(al) laminate 对称层板
symmetric(al) lens 对称透镜;对称镜头
symmetric(al) limiting 对称限幅
symmetric(al) line 对称线
symmetric(al) linear equation 对称线性方程
symmetric(al) linear operator 对称线性算子
symmetric(al) linear programming 对称线性规划
symmetric(al) load(ing) 对称荷载;对称加荷
symmetrically banded vein 对称层脉
symmetrically braced column 对称支撑立柱
symmetrically precompressed precast concrete-pile 对称预压混凝土预制桩
symmetric(al) matrix 对称矩阵
symmetric(al) matrix game 对称矩阵对策
symmetric(al) membrane 对称膜
symmetric(al) method of angle observation 对称测角法
symmetric(al) microthermo analyzer 对称式微热量分析仪
symmetric(al) mirror 对称镜
symmetric(al) mode 对称振型;对称型;对称模式
symmetric(al) model of vibration 对称振型
symmetric(al) molecule 对称分子
symmetric(al) moment 对称力矩
symmetric(al) motion 对称运动
symmetric(al) motor 对称电动机
symmetric(al) mounting 对称装置
symmetric(al) multioscillator 对称多谐振荡器
symmetric(al) multiprocessor 对称多处理机
symmetric(al) multivibrator 对称多谐振荡器;对称多谐荡振器
symmetric(al) network 对称物网络;对称网络
symmetric(al) number 对称数
symmetric(al) o attenuator 对称O形衰减器
symmetric(al) object 对称物
symmetric(al) object glass 对称物镜
symmetric(al) one-legged multistoried frame 对称单柱多层框架
symmetric(al) operation 对称操作
symmetric(al) order for binary tree 二叉树的对称次序
symmetric(al) parabolic loading 对称抛物线荷载
symmetric(al) part 对称部分
symmetric(al) pass 对称孔型
symmetric(al) pattern 对称图案
symmetric(al) -placed 对称放置的
symmetric(al) plane 对称(平)面
symmetric(al) polyphase system 对称多相制
symmetric(al) powering 对称供电
symmetric(al) processor 对称处理机
symmetric(al) profile 对称剖面
symmetric(al) quadratic form 对称二次形式
symmetric(al) quadripole method 对称四极法
symmetric(al) quadripole profiling curve 对称四极剖面曲线
symmetric(al) radiator 对称辐射器
symmetric(al) rank one algorithm 对称秩一算法
symmetric(al) relation 对称关系
symmetric(al) resonator 对称谐振腔
symmetric(al) rhythm 对称韵律
symmetric(al) ring 对称环
symmetric(al) ring synchrocyclotron 对称环形同步回旋加速器
symmetric(al) ripple marks 对称波痕
symmetric(al) sampling gate 平衡取样门
symmetric(al) scattering 对称散射
symmetric(al) section 对称断面
symmetric(al) section steel 对称断面型钢
symmetric(al) short circuit 对称短路
symmetric(al) shunt-type thermoammeter 对称分路型热电偶安培计
symmetric(al) space 对称空间
symmetric(al) spheric(al) triangles 球面对称三角形
symmetric(al) state 对称状态
symmetric(al) stress 对称应力
symmetric(al) stress cycle 对称应力循环
symmetric(al) stress system 对称应力系统
symmetric(al) stretching vibration 对称伸缩振动
symmetric(al) structure 对称结构
symmetric(al) successive overrelaxation 对称逐次超松弛
symmetric(al) surface of revolution 对称旋转面
symmetric(al) syncline 对称向斜
symmetric(al) system 对称系统
symmetric(al) tensor 对称张量
symmetric(al) three-input majority gate 对称三输入端多数决定元件
symmetric(al) three-throw turnout 三开道岔
symmetric(al) top 对称陀螺
symmetric(al) top molecule 对称陀螺分子
symmetric(al) transducer 对称换能器
symmetric(al) transformation 对称变换
symmetric(al) treatment 对称处理
symmetric(al) triangles 对称三角形
symmetric(al) triple diagonal matrix 对称三对角矩阵
symmetric(al) turnout 对称道岔
symmetric(al) two-legged multistoried frame 对称两脚多层构架;对称型两脚多层框架;对称型双柱多层框架
symmetric(al) type 对称型
symmetric(al) vein 对称脉
symmetric(al) vibration 对称振动
symmetric(al) voltage 对称电压
symmetric(al) wave 对称波
symmetric(al) wavefield 对称波浪场
symmetric(al) welding 对称焊
symmetric(al) winding 对称绕阻;对称电枢绕组
symmetrization 对称化(作用)
symmetrization of a game 对策的对称化
symmetroid 对称曲面
symmetry 相称;对称性;对称现象;对称
symmetry axis 旋转轴;对称轴
symmetry axis of rotation 旋转对称轴
symmetry center 对称中心
symmetry concentration 对称浓度
symmetry element 对称元素;对称要素
symmetry element of crystal 晶体对称要素
symmetry element of crystal structure 晶体结构对称要素
symmetry form 对称型
symmetry function 对称变换
symmetry group 对称性群
symmetry law 对称定律
symmetry number 对称数
symmetry of fabric 组构的对称性
symmetry of faces 对称面
symmetry of strategic equivalence 策略等价的对称性
symmetry operation 对称操作
symmetry operator 对称性算子
symmetry plane 对称平面;对称面
symmetry point 对称点
symmetry powder diagram 对称粉末图
symmetry principle 对称原理
symmetry property of wave function 波函数对称性
symmetry transformation 对称变换
symmetry type 对称型
symmict 混沉的
symmictite 混杂陆源沉积岩
symmicton 混杂陆源沉积物
Symons cone crusher 西蒙斯型圆锥破碎机

Symons crusher 西蒙斯型破碎机
sympathetic(al) 共鸣的
sympathetic(al) cracking 感应开裂;对应裂缝
sympathetic(al) damage 同情罢工损失
sympathetic(al) detonation 感应起爆
sympathetic(al) earthquake 感应地震;和应地震
sympathetic(al) ink 隐显油墨;变色油墨
sympathetic(al) radio burst 井振射电暴
sympathetic(al) resonance 共鸣
sympathetic(al) retrogression 连带刷深;同步退潮;同步降低(指上下游水位同时降低)
sympathetic(al) strike 同情罢工
sympathetic(al) vibration 和应振动;谐振;共振
sympathy strike 同情罢工
sympatric species 同域物种
symphysis 膜连
sympiesometer 弯管流体压力计
sympiezometer 双测压管测流(速)计
symplasm 共浆体
symplasmatic state 共质态
symplast 共质体
symplastic state 共质态
symplectite 后成合晶
symplektite 后成合晶
symplesite 砷铁石;砷铁矿
sympodial branch 果枝;合轴枝
sympodially branched 假轴式分枝
symposium 学术讨论会;学术会议;研讨会;座谈会;专题研究会;专题讨论会;讨论会(学术方面)
sym-tetrachlorodifluoroethane 对称四氯二氟乙烷
sym-triazine 均三嗪
synactic foam 综合作用泡沫
synadelphite 辛羟砷锰石;砷铝锰矿
Synadicum marble 彩色大理石(大理石中的一种)
synaesthesia 复合感觉
synagogue 犹太教堂
synantectic river 初生河流
synantectic stream 初生河流
synantexis 岩浆后期蚀变;自变质作用;后期变质
synanthropic 共居的
synaptic stage 联会期
synapticulum 合隔筋
sync amplifier 同步放大器
sync amplitude 同步信号振幅
syncarcinogen 合成致癌物质
syncaryon 合核
sync bit 同步位
sync buffering 同步缓冲
syncburst 同步触发
sync carrier 同步载波
synccharacter 同步字符
sync check 同步检验;同步检查
sync check word 同步检查字
sync circuit 同步电路
sync comparator 同步比较器
sync compression 同步信号压缩
sync control 同步控制
sync direction 同步信号极性;同步方向
sync discriminator 同步鉴别器
sync distribution amplifier 同步分配放大器
sync edge muting 同步边缘静噪
sync equalizing 同步均衡
sync fail 同步失效
sync generator 同步信号发生器
synchorology 群落分布学
synchotransmission 同步传输
synchro 自整角机;自同步机;同步传送
synchroaccelerator 同步加速器
synchroangle 同步角
synchroassembling line 同步装配线
synchroballistic 同步补偿
synchroclock 同步电钟;同步时计
synchrocontrol 同步调整
synchrocontrol receiver 同步控制接收机
synchrocontrol transformer 同步控制变压器
synchrocontrol transmitter 同步控制发射机
synchroconverter 同步换流器
synchro coupling 同步耦合
synchro current direction detector 同步流向仪
synchro cyclotron 稳相加速器;同步回旋加速器
synchrodata 同步数据
synchrodevice 同步机
synchrodifferential generator 差动式自动同步电动机

synchrodifferential motor 差动同步电动机;差动式自动同步电动机
synchrodifferential receiver 差动同步接收机
synchrodifferential transmitter 差动同步发射机
synchro-draw 同步拉伸
synchrodrive 同步驱动机;同步传动
synchrodyne 同步机
synchrodyne circuit 同步电路
synchrodyne mixer 同步混频器
synchro-fazotron 同步相位加速器
synchroflash 同步闪光灯
synchroflash mechanism 闪光同步机构
synchro-fly cutter 双刀辊式切纸机
synchrofollow-up system 同步随动系统
synchrogenerator 自同步发射机;同步机;同步发电机
synchroguide 水平扫描同步制导
synchro-indicator 同步指示器
synchroized operation 同步运行
synchrolift 同步升船装置;同步升船台(装置)
synchrolift system 同步升降系统;同步起升方式
synchrolight 同步指示灯
synchrolock 同步锁
synchromagslip 自同步机
synchro-marketing 同步销售;同步行销
synchro-mating punch 同步配合穿孔机
synchro-measurement 平均海面传递法
synchromesh 同步配合;同步啮合;同步齿轮系
synchromesh change gear 同步变速齿轮
synchromesh clutch gear 同步齿轮离合器
synchromesh gear 机械同步装置;同步啮合齿轮;同步齿轮;同步配合齿轮机构
synchromesh gearbox 同步啮合齿轮副变速箱
synchromesh(type) transmission 同步变速装置;同步啮合齿轮式变速箱;同步啮合变速器;同步配合变速器
synchrometer 同步指示计
synchromotor 同步电机
synchronal 同时的;同步的
synchrone 等时线
synchronia 同步现象
synchroniser 整步器
synchronising(connecting) rod 同步连杆
synchronising torque 同步力矩
synchronism 同时性;同步性
synchronism check relay 同步校验继电器
synchronism deviation 同步偏差
synchronism governor 同步调速器
synchronism indicator 同步指示器
synchronism regulator 同步调速器
synchronization 整步;同步化;同步
synchronization acquisition 同步捕获
synchronization action 同步动作
synchronization adjustment 同步调整
synchronization bay 同步机架
synchronization check 同步校验;同步检验
synchronization check word 同步校验字
synchronization contraint 同步限制
synchronization control 同步控制
synchronization factor 同步因数;同步系数
synchronization frequency 同步频率
synchronization frequency division 同步分频
synchronization generator 同步振荡器
synchronization indicator 同步显示器
synchronization mechanism 同步机构
synchronization of flow 径流同步化
synchronization of food 洪水同步(化)
synchronization of jacking up 同步顶升
synchronization of oscillator 振荡器的同步
synchronization of process 进程的同步
synchronization panel 同步配电板
synchronization regulation 同步调节
synchronization regulator 同步调速器
synchronization sensitivity 同步灵敏度
synchronization signal 同步信号
synchronization stretching 同步展宽
synchronization time 同步时间
synchronization waveform 同步信号波形
synchronization zone 同步区
synchronize 同步化
synchronize and close 并网
synchronized asynchronous motor 同步化的异步电动机
synchronized beacon 联动信号台

synchronized brake 同步作用制动机
synchronized breaker 同步断路器
synchronized changeover contact 同步转换接点
synchronized clamping 同步箝位
synchronized clock 同步电钟
synchronized control 同步控制
synchronized countershaft transmission 同步啮合副轴式变速器
synchronized culture 同步培养
synchronized cycle 同步循环
synchronized division 同步分裂
synchronized driving device 同步传动装置
synchronized grouting system 同步注浆系统
synchronized indicator 同步指示器
synchronized motion 同步运动
synchronized multiple access communication system 同步多址通信系统
synchronized network 同步网(络)
synchronized operation 同步运转
synchronized oscillation 同步振动
synchronized phase modulator 同步相位调制器
synchronized reserves 同步备用
synchronized sampler 同步采样器
synchronized shutter 同步快门
synchronized signal 同步信号
synchronized-signal control 同步信号控制
synchronized-signal generator 同步信号发生器
synchronized sparks 同步火花
synchronized spectrum analyzer 同步光谱分析器
synchronized stroke 同步行程
synchronized sweep 同步扫描
synchronized switch 同步开关
synchronized time schedule 同步进度表
synchronized timing 同步成圈
synchronized-timing oscillator 同步定时振荡器
synchronized transmission 同步传输
synchronized trigger generator 同步触发脉冲发生器
synchronized unit 同步单元
synchronized winch 同步绞车
synchronizer 整步器;同步器;同步机;并车装置
synchronizer cone 同步器摩擦锥轮
synchronizer equipment 同步设备
synchronizer gear 同步齿轮
synchronizer hub 同步器毂
synchronizer ring 同步器闭锁环
synchronizer sleeve 同步器啮合套
synchronizer switch 同步开关
synchronizer trigger 同步触发器
synchronizing amplifier 同步信号放大器
synchronizing angle 同步角
synchronizing apparatus 同步设备
synchronizing arrangement 同步设备
synchronizing at load 负荷同步
synchronizing band 同步区;同步频带
synchronizing beater 同步逐稿轮
synchronizing bus-bar 同步母线
synchronizing capacity 同步能力
synchronizing cash flows 同步现金流转
synchronizing channel 同步信道;同步通路;同步通道
synchronizing characteristics 同步特性
synchronizing circuit 同步回路;同步电路
synchronizing clock 同步时钟
synchronizing clutch 同步离合器
synchronizing cone 同步器摩擦锥轮的锥形工作面
synchronizing controls 同步控制机构;同步调整装置
synchronizing current 同步电流
synchronizing cycle 同步周期
synchronizing delay 同时间阻滞
synchronizing detection 同步探测
synchronizing device 同步器
synchronizing drum 同步转鼓
synchronizing frequency 同步频率
synchronizing function 同步功能
synchronizing gear 同步齿轮
synchronizing generator 同步信号发生器;同步发电机
synchronizing indicator 同步状态指示器
synchronizing information 同步信息
synchronizing lamp 同步指示灯
synchronizing leader 同步引线
synchronizing level 同步信号电平

synchronizing linkage 同步联动装置
synchronizing mechanism 同步机构
synchronizing microprocessor 同步微处理机
synchronizing modulation 同步调制
synchronizing network 同步网络
synchronizing of images 图像同步
synchronizing pattern 同步图形
synchronizing pilot 同步导频
synchronizing potential 同步讯号电平
synchronizing power 同步功率
synchronizing primitive command 同步基本命令;同步的基本命令
synchronizing process 同步过程
synchronizing pulse 整步脉冲
synchronizing pulse generating equipment 同步脉冲发生设备
synchronizing pulse regenerator 同步脉冲再生器
synchronizing pulse-timing error 同步脉冲定时误差
synchronizing quality 同步质量
synchronizing range 同步范围
synchronizing reactor 同步电抗器
synchronizing rectifier circuit 同步整流电路
synchronizing regeneration 同步再生
synchronizing relay 同步继电器
synchronizing selector 同步选择器
synchronizing sensitivity 同步灵敏度
synchronizing separation 同步分离
synchronizing separator circuit 同步分离电路
synchronizing separator tube 同步分离管
synchronizing sequence 同步序列
synchronizing shaft 同步轴
synchronizing signal 同步信号
synchronizing signal amplitude 同步信号幅度
synchronizing signal compression 同步信号压缩
synchronizing signal generator 同步信号发生器
synchronizing signal level 同步信号电平
synchronizing signal separation 同步信号分离
synchronizing signal separator 同步信号分离器
synchronizing signal source 同步信号源
synchronizing sleeve 同步套(齿形离合器)
synchronizing switch 同期开关
synchronizing system 同步系统;同步方式
synchronizing time 同步时间
synchronizing torque 同步转矩
synchronizing voltage 同步电压
synchronizing voltmeter 同步伏特计;整步电压表;零值电位计;同步电压表
synchronizing waveform 同步信号波形
synchronizing wheel 同步轮;同步惯性轮
synchronograph 同步自动电报机
synchronometer 同步指示器;同步流向仪;同步计
synchronoscope 同步指示器
synchronous acceleration 同步加速
synchronous accelerator 同步加速器
synchronous advancer 同步进相机
synchronous alternative generator 交流同步发电机
synchronous alternator 同步交流发电机
synchronous altitude 同步高度
synchronous altitude communications satellite 同步高度通信卫星
synchronous altitude gravity gradient experiment 同步高度重力梯度实验
synchronous altitude spin stabilized experiment 同步高度自旋稳定实验
synchronous angular velocity 同步角速度
synchronous apparatus 同步设备
synchronous apparatus for crane travel(l)ing 起重机移动同步装置
synchronous astrocompass 同步天文罗盘
synchronous asynchronous motor 同步异步电动机
synchronous automaton 同步式自动机
synchronous belt 同步皮带
synchronous booster 同步增压机;同步升压器
synchronous booster converter 同步增压变流机;同步升压变流机
synchronous booster-inverter 同步调压逆变器
synchronous boundary 同期界线
synchronous camera 同步摄影机
synchronous capacitor 同步电容器
synchronous capacity 同步能力
synchronous carrier system 同步载波系统

synchronous circuit 同步电路
synchronous clamping 同步箝位
synchronous clock 同步脉冲;同步时钟
synchronous clock motor 电钟同步电动机
synchronous communication 同步通信
synchronous communication electrodynamic(al)system 同步通信电气动态系统
synchronous communication interface 同步通信接口
synchronous communication satellite 同步通信卫星
synchronous communication system 同步通信系统
synchronous commutator 同步换向器
synchronous compensator 同步调相器;同步调相机;同步补偿器
synchronous computer 同步计算机
synchronous condenser 同步调相器;同步调相机;同步电容器
synchronous condenser operation 同步调相运行
synchronous condition 同步条件
synchronous construction 同步建设
synchronous contactor 同步断续器
synchronous converter 同步换流器;同步换流机;同步变流机
synchronous correction 同步校正
synchronous counter 同步计数器
synchronous culture 同步培养
synchronous data link control 同步数字键路控制;同步数据链控制
synchronous data link control regulations 同步数据链控制规程
synchronous data network 同步数据网(络)
synchronous data transmission 同步数据发送;同步数据传送;同步数据传输
synchronous deposit 同时沉积物;同时沉积;同期沉积;同步沉积物
synchronous detection 同步探测
synchronous detector 互相关器;同步探测器;同步检定器;同步检波器
synchronous developmentalism 同步发展论
synchronous device of hoisting mechanism 起升机构同步装置
synchronous digital cross connect equipment 同步数字交叉跨接设备
synchronous digital hierarchy 同步数字体系;同步数字分级系统
synchronous digital hierarchy management sub-network 同步数字体系管理子网
synchronous digital hierarchy network 同步数字体系管理网
synchronous digital hierarchy physical interface 同步数字体系物理接口
synchronous discharge 同步放电
synchronous disperse address beacon system 同步离散地址信标系统
synchronous distortion 同步失真
synchronous drive 同步传动
synchronous driving 同步传送
synchronous earth observation satellite 同步地球观察卫星
synchronous effect 同步作用
synchronous electric(al)clock 同步电钟
synchronous electric(al)sampler 同步电子开关
synchronous electromotive force 同步电动势
synchronous electronic sampler 同步电子开关
synchronous electronic switch 同步电子开关
synchronous energy 同步能量
synchronous energy gain 同步能量增益
synchronous equatorial orbit 同步赤道轨道
synchronous equipment management function 同步设备管理功能
synchronous equipment timing physical interface 同步设备物理定时物理接口
synchronous equipment timing source 同步设备定时源
synchronous error 同步误差
synchronous executive cycle 同步执行周期
synchronous exit routine 同步出口程序
synchronous exposure 同步曝光
synchronous fading 同步衰减
synchronous frequency 同步频率
synchronous frequency booster 同步频率放大器
synchronous gate 同步闸门;同步门

synchronous generator 同步发电机
synchronous governor 同步调节器
synchronous granite 同期花岗岩
synchronous head 同步头
synchronous homodyne 同步零差
synchronous idle 同步空载
synchronous increment of energy 同步能量增量
synchronous induction motor 同步感应电动机
synchronous initiation 同步起动
synchronous inputs 同步输入
synchronous interface 同步接口
synchronous inverter 同步逆变器;电动机发电机组
synchronous ion 同步离子
synchronous jammer 同步干扰器
synchronous jamming 同步干扰
synchronous leader 同步引导器
synchronous lever 同步杠杆
synchronous lever shaft 同步摇臂轴
synchronous light 同步灯
synchronous linear motor 同步线性电动机
synchronous line control 同步线性控制
synchronous link 同步联动装置
synchronous loop 同步环线
synchronously scanned laser television system 同步扫描激光电视系统
synchronous machine 同步电机
synchronous margin 同步边限
synchronous margin of start-stop apparatus 起止装置同步边界
synchronous mark 同步标记
synchronous meteorological satellite 同步气象卫星
synchronous mode comparator 同步式比较器
synchronous modem 同步调制解调器
synchronous modulation 同步调制
synchronous momentum 同步动量
synchronous motion 同步运动
synchronous motor 同步电机;同步电动机
synchronous motor clock 同步电钟
synchronous motor-generator 同步电动发电机
synchronous motor hunting 同步电动机振荡;同步电动机振荡
synchronous multiplex 同步复用器
synchronous multiplexer 同步多路转接器
synchronous network 同步网络
synchronous nitrification-denitrification 同时硝化—反硝化
synchronous nitrification-denitrification denitrogen technique 同时硝化—反硝化脱氮技术
synchronous oblique air photographs 同步倾斜航摄照片
synchronous observation 同步观测
synchronous operation 同步运行;同步工作;同步操作
synchronous optic(al)network 同步光网络
synchronous orbit 同步轨道
synchronous oscillation 同步振荡
synchronous oscillator 同步振荡器
synchronous output analysis 同步输出分析
synchronous period 同步周期
synchronous phase advancer 同步进相机
synchronous phase converter 同步相位变换器
synchronous phase modifier 同步相位补偿器
synchronous plane 同步平面
synchronous pluton 同期深成岩体
synchronous power 同步功率
synchronous precipitation 同时沉淀
synchronous process(ing) 同步过程;同步处理
synchronous protective 同步保护的
synchronous pull-in torque 同步牵入转矩
synchronous pulse 同步脉冲
synchronous pulse jamming 同步脉冲干扰
synchronously punching 同步穿孔
synchronous radio telegraph channel 同步无线电信道;同步无线电通路
synchronous range 同步范围
synchronous receiver 同步接收器;同步接收机
synchronous receiving motor 同步接收电动机
synchronous recurrent waveform 同步再现波形
synchronous regime 同步状态
synchronous regulator 同步调节器
synchronous relay system 同步中继系统
synchronous reluctance motor 同步磁阻电动机
synchronous remote control 同步遥控

synchronous repair approach 同步修理法
synchronous resonant wave 同步共振波
synchronous rotary switch 同步转动开关
synchronous rotation 同步旋转;同步绕转
synchronous running 同步运转;同步运行
synchronous sampler 同步取样器;同步采样器
synchronous sampling 同步取样
synchronous satellite 同步卫星;地球同步卫星
synchronous satellite communication system 同步卫星通信系统
synchronous satellite orbit 同步卫星轨道
synchronous satellite system 同步卫星系统
synchronous scanning 同步扫瞄
synchronous sediment 同步沉积物
synchronous selector system 同步选择制
synchronous separation 同步分离
synchronous sequential system 同步时序系统
synchronous serial data adapter 同步串行数据适配器
synchronous serial system 同步串联系统
synchronous signal 同步信号
synchronous signal frequency 同步信号频率
synchronous signalling 同步信号法
synchronous spark 同步火花
synchronous spark gap 同步火花隙
synchronous speed 同步转速;同步速度
synchronous speed of rotation 同步转速
synchronous stability 同步稳定性
synchronous start(ing) 同步起动
synchronous station 同步航天站
synchronous storage method 同步存储方法
synchronous supervision mechanism 同步监控装置
synchronous surveying 同步测量法
synchronous switch 同步开关
synchronous switching 同步转接
synchronous system 同步制;同步系统;同步式
synchronous telescope 同步伸缩
synchronous terminal 同步终端
synchronous tie 同步耦合
synchronous time division multiplexer 同步时间分配多路转换器;同步时分多路复用器;同步分时多路换接器
synchronous timer 同步时间继电器;同步秒表;同步计时器
synchronous time selector 同步时间选择器
synchronous timing comparator 同步定时比较器
synchronous torque 同步力矩
synchronous tracking 同步跟踪
synchronous tracking antenna 同步跟踪天线
synchronous tracking computer 同步跟踪计算机
synchronous traffic signal 同步交通信号
synchronous transfer 同步传送
synchronous transfer mode 同步传送模式
synchronous transfer module 同步传送模块
synchronous transfer orbit 同步转移轨道
synchronous transmission 同步传送;同步传输
synchronous transmission device 同步传输装置
synchronous tuning 同步调谐
synchronous type vibrator 同步式振动器
synchronous velocity 胶片同步速度;同步速度
synchronous vibration 谐振;同周期振动;同步振动
synchronous vibrator 同步振子
synchronous voltage 同步电压
synchronous watt 同步瓦;同步功率
synchronous wave 谐振波(浪);同步波
synchronous working 同步工作;同步操作
synchrooscilloscope 同步示波器
synchrophasemeter 同步相位计
synchrophase shifter 同步移相器
synchrophasing 同步定相
synchrophasotron 同步稳相加速器
synchropickoff 同步传感器
synchropickup 同步传感器
synchro-position control 同步位置控制
synchropter 同步交叉式双旋翼直升机
synchropulse 同步脉冲
synchroreader 同步读出器
synchroreceiver 自整角接收机;自同步接收机;同步接收机
synchroresolver 分析器
synchroresonance 同步共振
synchroscope 汽车点火整步器;同步测试仪;同步指示器;同步波仪;同步示波器

synchro-self-shifting clutch 同步自脱离合器
synchro-shear 同步切变
synchro-shifter 同步转换机构;同步移位器
synchro-shutter 同步快门
synchro-speed 同步转速;同步速度
synchro-spiral gearbox 同步斜齿轮变速箱
synchrostat regulator 同步调节器
synchrostep remote control 同步遥控
synchrosystem 同步随动系统;同步系统
synchrotexturing system 同步变形体系
synchrotherm reactor 同步热反应器;同步热反应堆
synchrotie 同步耦合;同步机馈线
synchrotimer 同步定时器;同步计时器;时间同步器
synchroto digitat converter 同步机数字转换器
synchrotrans 同步转换
synchrotransformer 同步控制变压器
synchrotransmitter 同步传感器;同步发送器;自同步发射机;同步系统发送装置
synchrotron 同步加速器
synchrotron damping rate 同步加速阻尼速率
synchrotron instability 同步加速不稳定性
synchrotron light 同步加速器辐射
synchrotron radiation 同步加速器辐射;同步(加速)辐射
synchrotron radiation decay 同步辐射衰减
synchrotron radiation source 同步加速辐射源
synchrotron source 同步加速辐射源
synchrotron spectrum 同步加速谱
synchrotron voltage 同步加速器电压
synchysite-(Nd) 直氟碳钙钕矿
synchysite-(Y) 直氟碳钙钇矿
sync-in pulse 同步输入脉冲
sync input 同步输入
sync insertion 同步信号插入
sync inversion 同步倒相
synclastic curvature 同向曲率
synclastic surface 同向曲面
synclator 同步(信号)振荡器
sync level 同步电平
sync limiter 同步信号限幅器
synclinal axis 向斜轴;顺斜轴
synclinal closure 向斜闭合度
synclinal continental sea 向斜大陆海
synclinal down-warped lake 向斜坳陷湖
synclinal fault 向斜断层
synclinal fold 向斜褶皱
synclinal formation 向斜地层
synclinal hinge 向斜枢
synclinal limb 向斜翼
synclinal mountain 向斜山
synclinal oil-gas field 向斜油气田
synclinal oil pool 向斜油藏
synclinal ridge 向斜山脊
synclinal river 向斜河(流)
synclinal spring 向斜泉
synclinal stratum 向斜地层
synclinal stream 向斜河(流)
synclinal valley 向斜谷
syncline 向斜
syncline fold in rock beds 岩层斜向褶皱
sync line-up 同步信号对齐
synclinore 复向斜
synclinorium 复向斜
sync-lock equipment 同步锁定设备
sync-lock unit 同步锁订单元
sync mark 同步标记
sync motor 同步马达
sync non-linearity 同步信号非线性
syncom 同步通信卫星
Syncore 西科板(一种芯材为黏结碎木的胶合板)
sync out pulse 同步输出脉冲
sync output 同步输出
sync peaks 同步峰值
sync pedestal 同步基座;同步基底电平
sync pip 同步标记
sync process 同步处理
sync-pulse 同步脉冲
sync pulse regenerator 同步脉冲再生器
syncrho control 同步控制
syncrho differential generator 差动自整角发电机
syncrho-switch 同步开关
syncriminator 同步鉴别器
syncrystallization 同结晶作用

sync section 同步部分
sync selector 同步选择器
sync select switch 同步选择开关
sync separating circuit 同步分离电路
sync separator 同步信号分离器
sync separator circuit 同步信号分离电路
sync signal amplitude 同步信号振幅
sync signal compression 同步信号压缩
sync signal generator 同步信号发生器
sync signal insertion 同步信号插入
sync signal level 同步信号电平
sync signal limiter 同步信号限幅器
sync signal purifier 同步信号滤清器
sync signal re-insertion 同步信号重插入
sync source 同步源
sync splits 同步分隔
sync waveform 同步波形
Syncyanosen 共生蓝藻
syndeposit 同沉积作用
syndeposition 同沉积
syndepositional depression 同沉积凹陷
syndepositional fold 同沉积褶皱
syndepositional fracture 同沉积断裂
syndepositional structure 同沉积构造
syndepositional uplift 同沉积隆起
syndet(ergent) 合成洗涤剂
syndiagenesis 早期成岩作用;同(生)成岩作用
syndiagenetic stage 同生成岩期
syndic 理事;经理人
syndicate 银团;企业联合组织
syndicate agreement 辛迪加协议
syndicate banking 辛迪加银行业务
syndicated loan 银团贷款;财团放款
syndicated market research 联合市场研究
syndicated tender 辛迪加投标
syndicate loan 辛迪加贷款
syndicate release 辛迪加解除限制
syndicate restriction 辛迪加限制
syndicate termination 辛迪加限制终止
syndication 辛迪加组织
syndiotacticity 间同立构规正度;间规性;间规度
syndiotactic polymer 间同立构聚合物;间规聚合物
syndiotactic polypropylene 间规聚丙烯
syndiotactic sequence 间同序列
syndiotaxy 间同立构
syndrome 综合征;校验子;特征群;伴随式
syndrome decoding 故障译码
syndrome due to attack of pathogenic dampness 中湿
syndyotaxy 间同立构
syneclise 台向斜
synecology 群体生态学;群落生态学;群聚生态学
synectics 协力创新法
syneklise 陆槽
syneresis 离浆(作用);胶体脱水收缩(作用);凝溢;润滑脂的脱水收缩;脱水收缩(作用)
syneresis crack 脱水收缩裂缝
synergetic effect 协同效应
synergetic index 协和指数
synergetic log 合成测井图
synergetic log plot with computer display 计算机综合显示测井图
synergetic network 协作网络
synergia 增效
synergic curve 最优上升曲线;最佳上升曲线
synergic reflex 合作反射
synergic system 公用系统
synergism 协作作用;协作;协同作用;协同共栖;协同;协力作用;协和作用;相生(现象);增效作用;增效;共同作用;统合(多国性企业);超益互助
synergist 协作剂;协合剂;增效剂;配合剂
synergistic 协同的
synergistic action 协同作用
synergistic agent 增效剂
synergistic agrochemical fungicide 复合农业化学杀菌剂
synergistical action 协和作用
synergistic curve 协和曲线
synergistic degradation 协同降解
synergistic effect 协同效应;协和效应;增强效应;叠加效应
synergistic effect curve 协和效应曲线
synergistic effects of toxic substance 有毒物质

的协同效应
synergistic extraction 协同萃取
synergistic index 协同指数
synergistic toxic effect 协同毒性效应
synergist of pesticide 农药增效剂
synergized action 增效作用
synergizing agent 增效剂
synergy 协和作用;整合效果;增效
synform 向形
synformal 向形的
synformal anticline 向形背斜
synformal basin 向形盆地
synfuel 合成燃料
syngas 合成气
syngenesis 早期成岩作用;同生作用
syngenetic 同生的
syngenetic anomalous inclusion 同生异常包裹体
syngenetic anomaly 同生异常
syngenetic component 同生组分
syngenetic concretion 同生结核
syngenetic deposit 同生矿床
syngenetic dispersion 同生分散
syngenetic dispersion pattern 同生分散模式
syngenetic dolostone 同生白云岩
syngenetic mineral 同生矿物
syngenetic ore-forming process 同生成矿作用
syngenetic ore-forming theory 同生成矿说
syngenetic stratabound deposit 同生层控矿床
syngenetic structure 同生构造
syngenetism 共生作用
syngenic 同生的
syngenite 钾石膏
synhcromotor 同步电动机
syn-hold 保持同步
syniphase 同相
syniphase excitation 同相激励
synkaryon 合核
synkinematic 同构造的
synkinematic granite 同构造期花岗岩
synneusis 聚晶状
synneusis graphite 聚晶石墨
synodal hall 区牧师聚会厅
synodical month 朔望月
synodic(al) motion 会合运动
synodic(al) period 会合(周)期
synodic(al) revolution 会合周
synodic(al) rotation 会合自转周
synodic(al) rotation number 会合自转周数
synodic(al) rotation period 会合自转周期
synodic(al) year 会合年
synodic periods 朔望周期
synoeciosis 片利共生
synonym 类似物;同物异名
synonymy 异名关系
synopsis 一览表;摘要;概要;提要;说明书;大纲
synoptic(al) 天气图的;天气学
synoptic(al) analysis 天气图分析;天气分析
synoptic(al) chart 概况图;天气形势图;天气概要图
synoptic(al) climatology 天气图气候学;天气气候学
synoptic(al) code 天气图电码;天气电码
synoptic(al) diagram 原理图
synoptic(al) disturbance 天气扰动
synoptic(al) experiment 天气试验
synoptic(al) forecast 天气(图)预报
synoptic(al) map 天气图
synoptic(al) map of the sun 太阳综合图
synoptic(al) meteorology 气象学;天气学
synoptic(al) method for weather forecasting 天气图方法
synoptic(al) model 天气图模型;天气模式
synoptic(al) network 天气观测站网
synoptic(al) observation 天气观测
synoptic(al) oceanography 预报海洋学
synoptic(al) perturbation 波浪小扰动
synoptic(al) process 天气过程
synoptic(al) report 天气图报告;天气报告
synoptic(al) situation 天气形势;天气情况
synoptic(al) station 天气观测站
synoptic(al) table 天气分析表;总表
synoptic(al) type 天气型
synoptic(al) wave chart 波浪图
synoptic(al) (weather) chart 气候图;天气图
synoptic(al) weather map 气象图;天气形势图;天气气象图
synoptic correlation 欧拉相关
synoptic oceanography 海况海洋学
synoptic-scale 天气尺度
synoptic-scale circulation system 天气尺度环流系统
synoptic-scale current 天气尺度流
synoptic-scale flow 天气尺度流
synoptic-scale forecast 天气尺度预报
synoptic-scale motion 天气尺度运动
synoptic-scale process 天气尺度过程
synoptic-scale wind 天气尺度风
synoptic terrain photography 地表概要摄影术
synoptic view 衬视
synoptic wave chart 波浪状况图
syn oreforming fracture 成矿期断裂
syn oreforming structure 成矿期构造
synorogenic 同造山期的
synorogenic basin 同造山期盆地
synorogenic coalification 造山期煤化作用;同造山期煤化(作用)
synorogenic period 同造山期
synphitic deposit 同生矿床
synpiontology 古植物群落学
syn-position 顺位
synreflexia 联带反射
synroc 合成岩石
synsedimentary anticline 同沉积背斜
synsedimentary deformational process 同沉积变形作用
synsedimentary fault 同生断层
synsedimentary fold 同沉积褶皱
synsedimentary syncline 同沉积向斜
syntactic(al) check 句法检查
syntactic(al) foam 复合泡沫塑料
syntactic(al) granite 同熔花岗岩
syntax controlling probability 句法控制概率
syntax-directed compiler 面向语言的编译程序
syntaxial growth type 同向生长型结晶纤维
syntaxial overgrowth 衔接复生
syntaxial overgrowth cement texture 共轴生长胶结物结构
syntaxial system 共生体系
syntectic 同熔作用
syntectic magma 同熔岩浆
syntectonic 同构造的
syntectonic crystal fiber 同构造结晶纤维
syntectonic growth vein 同构造生长脉
syntectonic pluton 同构造深成岩体;同构造等深线
syntectonic recrystallization 同构造重结晶
syntenic group 同线群
syntexis 同熔作用
syntexis way 同熔作用方式
synthal 合成橡胶
Synthaprufe 迅达普鲁防潮液(一种由沥青橡胶配制的,用于胶粘木块)
synthermal 同温度的
synthesis 综合作用;综合;合成;拼合;标准工作时间
synthesis compression workshop 合成压缩工场
synthesis converter 合成塔
synthesis cylindric(al) diagram 综合柱状图
synthesis dredge 综合整治
synthesis gas 合成气
synthesis method 综合法;合成方法;合成法
synthesis of ammonia 氨的合成法
synthesis of control system 控制系统综合
synthesis of fuzzy functions 模糊函数的综合
synthesis of hydrograph 过程线综合;过程线合成
synthesis of switching network algorithm 开关网络综合算法
synthesis phase 合成期
synthesis plan of pipelines 综合管道线图
synthesis profile 合成剖面
synthesis radio-telescope 综合孔径射电望远镜
synthesis reaction 合成反应
synthesis reactor 合成反应器
synthesis spectrum 合成谱
synthesis telescope 综合孔径望远镜
synthesis wave 合成波
synthesized beam 综合波束
synthesized hydrocarbon fluid 合成烃润滑液
synthesized mixed crystal 合成混合晶体
synthesized pattern 综合方向图
synthesized seismic record 合成地震记录
synthesized travel desires 计算的交通需要;综合的交通需要
synthesized VSP record 合成VSP记录
synthesizer 综合器;合成装置;合成器
synthesizer mixer 合成混频器
synthesizing tower 合成塔
synthetic(al) 综合的;合成的;综合性的
synthetic(al) action 综合作用;合成作用
synthetic(al) address 生成地址
synthetic(al) adhesive 合成黏结剂;合成黏合剂;合成胶粘剂
synthetic(al) agar 合成琼脂
synthetic(al) aggregate 合成集料;合成骨料;人造集料;人造骨料
synthetic(al) aggregate light concrete 合成骨料轻质混凝土
synthetic(al) aggregate light(weight) concrete 人造轻集料混凝土;人造轻骨料混凝土
synthetic(al) air chart 综合空气状态表
synthetic(al) alcohol 合成醇
synthetic(al) alexandrite 合成变石
synthetic(al) alumina 人造氧化铝
synthetic(al) alumin(i)um base grease 合成铝基润滑脂
synthetic(al) ammonia 合成氨
synthetic(al) ammonia installation 合成氨装置
synthetic(al) ammonia process 合成氨生产
synthetic(al) and many-sided 综合性和多方面的
synthetic(al) anhydrite 合成硬石膏;合成无水石膏
synthetic(al) anhydrite screed(material) 合成无水石膏抹面
synthetic(al) anomaly map of geochemistry 地球化学综合异常图
synthetic(al) anomaly map of geophysics 地球物理综合异常图
synthetic(al) anomaly map of geophysics and geochemistry 物化探综合异常图
synthetic(al) antenna 合成天线
synthetic(al) antenna array 合成天线阵
synthetic(al) aperture 综合孔径;合成孔径
synthetic(al) aperture antenna 合成孔径天线
synthetic(al) aperture radar 合成孔径雷达
synthetic(al) aperture radar antenna 合成孔径雷达天线
synthetic(al) aperture sonar 合成孔径声呐
synthetic(al) aperture technique 合成孔径技术
synthetic(al) apparatus for weld thermal cycle 焊接热循环模拟装置
synthetic(al) asphalt 人造沥青;合成沥青
synthetic(al) assessment of eco-environment 综合生态环境评价
synthetic(al) assessment of geology of mineral deposits 矿床地质综合评价
synthetic(al) assessment of geology of ore district 矿区地质综合评价
synthetic(al) assessment of mineralization point 矿点综合评价
synthetic(al) bag 化纤包
synthetic(al) baking enamel 合成烘漆
synthetic(al) baking enamel white 合成白烘漆
synthetic(al) birefringence 人造双折射;人工双折射
synthetic(al) boat 塑料船
synthetic(al) bonding agent 合成胶粘剂
synthetic(al) borneol 合成冰片
synthetic(al) brake fluid 合成制动液
synthetic(al) bristle brush 人造鬃毛漆刷
synthetic(al) building product 合成建筑制品;合成建材产品
synthetic(al) building syndrome 建筑综合症
synthetic(al) butyl tape 合成丁基胶带
synthetic(al) calcium silicate 合成硅酸钙
synthetic(al) camphor 合成樟脑
synthetic(al) camphor powder 合成樟脑粉
synthetic(al) camphor tablet 合成樟脑片
synthetic(al) casting 合成渣浇铸
synthetic(al) cast iron alloy 合成铸铁合金
synthetic(al) cementing agent 合成胶凝剂
synthetic(al) chart of pipelines 管道综合图
synthetic(al) chemical industry 合成化学工业
synthetic(al) chemistry 合成化学
synthetic(al) chloroprene rubber 人造橡胶
synthetic(al) city 综合都市;综合城市
synthetic(al) climatology 综合气候学

synthetic(al) clinker 合成熟料
synthetic(al) coagulant 合成混凝剂
synthetic(al) coating 合成涂料
synthetic(al) colo(u)r 合成色素;合成色料
synthetic(al) columnar section of strata 综合地层柱状图
synthetic(al) concrete curing agent 合成混凝土养护剂
synthetic(al) concrete curing compound 合成混凝土养护组合物
synthetic(al) concrete with artificial resin admix(ture) 合成树脂混凝土
synthetic(al) construction(al) product 合成建筑制品
synthetic(al) continuous fiber 合成长纤维
synthetic(al) copal 合成酚醛树脂
synthetic(al) core 人造岩芯
synthetic(al) corundum 合成刚玉;人造刚玉
synthetic(al) crude 合成原油;合成石油
synthetic(al) crystal 合成晶体;人造水晶;人工晶体
synthetic(al) curve 综合曲线;合成曲线
synthetic(al) damping 合成阻尼
synthetic(al) data 综合资料;综合性数据;合成数据
synthetic(al) detergent 合成洗涤剂
synthetic(al) detergent industry 合成洗涤剂工业
synthetic(al) diamond 合成金刚石;人造金刚石
synthetic(al) diamond bit 人造金刚石钻头
synthetic(al) display generation 合成显示产生法
synthetic(al) division 综合除法
synthetic(al) drawing method 综合成图法
synthetic(al) drilling fluid 合成钻井泥浆
synthetic(al) drug plant 合成药厂
synthetic(al) drying oil 合成干性油
synthetic(al) drying rust resisting paint 合成快干防锈涂料;合成快干防锈漆
synthetic(al) duration curve method 综合历时曲线法
synthetic(al) dye(stuff) 合成染料
synthetic(al) ecology 综合生态学
synthetic(al) economics 综合经济学
synthetic(al) elastomer 合成弹性体
synthetic(al) emerald 合成祖母绿;合成翡翠绿;人造祖母绿
synthetic(al) emulsifier 合成乳化剂
synthetic(al) emulsion 合成乳剂
synthetic(al) enamel 合成瓷漆;醇酸磁瓷漆
synthetic(al) enamelled copper wire 合成漆包线
synthetic(al) enamels 胶玉瓷漆
synthetic(al) environmental carcinogen 合成的环境致癌物质
synthetic(al) environmental impact assessment for surface water 地表水环境影响综合评价
synthetic(al) expenses 生产综合费用
synthetic(al) extender 合成增量剂;合成填充剂;合成体质颜料
synthetic(al) fabric cleaner 化纤清洗剂
synthetic(al) fabrics 合成纤维织物
synthetic(al) fat 合成脂肪
synthetic(al) fatting acid detergent 合成脂肪酸洗涤剂
synthetic(al) fatty acid 合成脂肪酸
synthetic(al) fault 同向断层;顺向断层
synthetic(al) felt 合成毛毡
synthetic(al) fertilizer 合成肥料;人造肥料
synthetic(al) fiber 合成纤维
synthetic(al) fiber carpet 化纤地毯
synthetic(al) fiber eye splicing 化纤绳眼环插接
synthetic(al) fiber fabric 合成纤维织物
synthetic(al) fiber goods 合成纤维货物
synthetic(al) fiber in dustry 合成纤维工业
synthetic(al) fiber material 合成纤维材料
synthetic(al) fiber paper 合成纤维纸
synthetic(al) fiber plant 合成纤维厂
synthetic(al) fiber rope 化纤绳;人造纤维绳
synthetic(al) fiber rope splice 化纤绳插接
synthetic(al) fiber wastewater 合成纤维废水
synthetic(al) fiber web 合成纤维织物
synthetic(al) fiber wire 合成纤维绳
synthetic(al) filler 合成填塞料
synthetic(al) filter media 合成过滤介质
synthetic(al) filter medium 合成滤料
synthetic(al) fixed nitrogen 合成固定氮
synthetic(al) flocculant 合成絮凝剂
synthetic(al) foam 复合泡沫塑料;合成泡沫塑料

synthetic(al) foam plastics 合成泡沫塑料
synthetic(al) for synthetic(al) ammonia 合成氨原料气
synthetic(al) frequency curve 合成频率曲线
synthetic(al) fuel 合成燃料;人造燃料
synthetic(al) fuel wastewater 合成燃料废水
synthetic(al) furnace 合成炉
synthetic(al) gas 合成气;合成煤气
synthetic(al) gas compressor 合成气压缩机
synthetic(al) gas compressor turbine 合成气压缩机透平
synthetic(al) gasoline 合成汽油
synthetic(al) gem 合成宝石;人造宝石
synthetic(al) glue 合成胶;合成胶结剂
synthetic(al) glue powder 合成胶粉
synthetic(al) glue work 合成粘胶工程
synthetic(al) glycerine 合成甘油
synthetic(al) graphite 合成石墨;人造石墨
synthetic(al) greywater 合成灰水
synthetic(al) grit 人造金刚石粒
synthetic(al) growth index method 综合增长指数法
synthetic(al) gum tragacanth 合成龙胶
synthetic(al) gypsum 合成石膏
synthetic(al) harmfulness coefficient 综合危害系数
synthetic(al) high molecular compound 合成高分子化合物
synthetic(al) high molecular roofing material 合成高分子屋面材料
synthetic(al) hologram 合成全息图
synthetic(al) hospital 综合性医院
synthetic(al) humans 人造腐殖质
synthetic(al) humic acids 合成腐殖酸
synthetic(al) hydrocarbons 合成碳氢化合物
synthetic(al) hydrochloric acid 合成盐酸
synthetic(al) hydrograph 综合水文曲线;综合水文过程线
synthetic(al) hydrology 综合水文学;统计水文学
synthetic(al) indigo 合成靛青;合成靛蓝
synthetic(al) industry 人工合成工业
synthetic(al) ink 合成油墨
synthetic(al) inorganic pigment 合成无机颜料
synthetic(al) instruction 合成指令
synthetic(al) insulating oil 合成绝缘油
synthetic(al) ion-exchange resin 合成离子交换树脂
synthetic(al) iron oxide 合成氧化铁
synthetic(al) lacquer-grade resin 合成蜡克树脂;合成亮级树脂
synthetic(al) language 人工语言
synthetic(al) latex 合成乳胶;合成胶乳
synthetic(al) leather 合成皮革;合成革;人造革
synthetic(al) lifebuoy 塑料救生圈
synthetic(al) lifefloat 塑料救生浮
synthetic(al) lifejacket 塑料救生衣
synthetic(al) light curve 综合光变曲线
synthetic(al) light(weight) aggregate 人造轻集料;人造轻骨料
synthetic(al) lignosulfonate 合成木质素磺酸钠
synthetic(al) lining 合成衬里
synthetic(al) liquid fuel 合成液态燃料
synthetic(al) liquid membrane concrete curing agent 合成膜式混凝土养护剂
synthetic(al) lubricant 合成润滑剂
synthetic(al) lubricant fluid 合成润滑油
synthetically effect coefficient 综合影响系数
synthetically tan wastewater 制革综合废水
synthetic(al) macromolecular compound 合成巨分子化合物
synthetic(al) magnesia clinker 合成镁砂
synthetic(al) magnetite 氧化铁黑;合成磁铁矿
synthetic(al) manure 人造厩肥
synthetic(al) map 合成地图
synthetic(al) map of morphology 地貌综合图
synthetic(al) material 人造材料;化学建材;合成材料
synthetic(al) mechanism 综合机制
synthetic(al) medium 合成培养基;人工培养基
synthetic(al) membrance liner 合成膜村里
synthetic(al) membrane 合成膜
synthetic(al) membrane liner 合成薄膜村层
synthetic(al) metal 烧结金属
synthetic(al) method 综合法;合成法

synthetic(al) mica 合成云母
synthetic(al) mica nacreous pigment 合成云母珠跟颜料
synthetic(al) mineral 合成矿物
synthetic(al) mineral fibre 合成矿物纤维
synthetic(al) motor oil 合成机油
synthetic(al) (mo(u)lding) sand 合成型砂
synthetic(al) mullite 合成莫来石
synthetic(al) natural gas 合成天然煤气;合成天然气
synthetic(al) natural rubber 合成天然橡胶
synthetic(al) neotectonic map 新构造合成图
synthetic(al) observation 内插同步观测
synthetic(al) ochre 合成赭石
synthetic(al) oil 合成原油;合成油;合成石油
synthetic(al) optic(al) cable 综合光缆
synthetic(al) organic aggregate 合成有机集料;合成有机骨料;人造有机集料;人造有机骨料
synthetic(al) organic chemicals 合成有机化学物
synthetic(al) organic compound 合成有机化合物
synthetic(al) organic contaminant 合成有机污染物
synthetic(al) organic insecticide 合成有机杀虫剂;合成有机农药
synthetic(al) organic molecule 合成有机分子
synthetic(al) organic pigment 合成有机颜料
synthetic(al) oriental lacquer 合成大漆
synthetic(al) paint 合成油漆;合成涂料;合成漆
synthetic(al) paper 人造纸
synthetic(al) parameter of cylinder 柱体的综合参数
synthetic(al) parameter of finite sheet 有限板状体的综合参数
synthetic(al) parameter of half-infinite sheet in dipole field 偶极场中半无限板状体的综合参数
synthetic(al) parameter of infinite sheet 无限板状体的综合参数
synthetic(al) parameter of infinite sheet in dipole field 偶极场中无限板状体的综合参数
synthetic(al) parameter of sphere 球体的综合参数
synthetic(al) park 综合公园;多功能综合公园
synthetic(al) pattern generator 复试验振荡器
synthetic(al) perspiration 人造汗液
synthetic(al) pesticide 合成农药
synthetic(al) petroleum 合成石油;人造石油
synthetic(al) pig iron 再制铁;再制生铁
synthetic(al) pigment 合成颜料
synthetic(al) pine oil 合成松油
synthetic(al) plastic coating 合成塑料涂膜;合成塑料涂层
synthetic(al) plastic material 合成塑料材料
synthetic(al) plastics 合成塑料
synthetic(al) plate 合成板
synthetic(al) pollution data 综合污染资料;综合污染数据
synthetic(al) pollution index 综合污染指数
synthetic(al) pollution quota 综合污染指标
synthetic(al) polyamine flocculant 合成聚胺絮凝剂
synthetic(al) polymer 合成聚合物
synthetic(al) polymer coagulant 合成高分子混凝剂
synthetic(al) polymeric flocculant 合成高分子絮凝剂
synthetic(al) polymer membrane 合成高分子膜
synthetic(al) process 合成过程;合成代谢过程
synthetic(al) product 合成产物
synthetic(al) project flood 综合设计洪水
synthetic(al) quartz 合成石英;合成水晶;人造石英;人工水晶
synthetic(al) quartzcrystal 人造石英晶体
synthetic(al) ratio 合成率
synthetic(al) remote control and supervisory device 综合遥控监测装置
synthetic(al) research ship 综合调查船
synthetic(al) resin 合成树脂;人造树脂
synthetic(al) resin adhesive 合成树脂黏结剂;合成树脂黏合剂;合成树脂黏胶剂;合成树脂胶
synthetic(al) resin asbestos board 合成树脂石棉板
synthetic(al) resin backing 合成树脂衬里
synthetic(al) resin base 合成树脂基底
synthetic(al) resin-based bonding adhesive 合成树脂为基材的胶黏剂
synthetic(al) resin based glue 合成树脂胶

synthetic(al) resin bearing 合成树脂轴承
synthetic(al) resin binder 合成树脂黏合剂;合成树脂黏结剂
synthetic(al) resin bond 合成树脂黏结剂;合成树脂黏合剂
synthetic(al) resin bonded 合成树脂黏结的
synthetic(al) resin bonded laminate 合成树脂黏固层压板;合成树脂层压板
synthetic(al) resin bonded paper sheet 层合塑料片;层合塑料板;合成树脂黏合纸板;合成树脂纸板
synthetic(al) resin-bound 合成树脂黏结的
synthetic(al) resin building mastic 合成树脂建筑玛琋脂
synthetic(al) resin cement 合成树脂黏结剂;合成树脂黏合剂;合成树脂胶黏剂;合成树脂胶结剂;合成树脂胶
synthetic(al) resin clear varnish 合成树脂涂膜;合成树脂清漆
synthetic(al) resin coat 合成树脂涂层
synthetic(al) resin coating (material) 合成树脂涂层;合成树脂(防水)涂料
synthetic(al) resin cold water paint 合成树脂冷水溶性涂料;合成树脂冷水溶性漆
synthetic(al) resin concrete 合成树脂混凝土
synthetic(al) resin covering 合成树脂覆盖层;合成树脂涂层
synthetic(al) resin deposition 合成树脂面层;合成树脂覆盖层;合成树脂涂层
synthetic(al) resin dispersion 合成树脂分散体
synthetic(al) resin dispersion base 合成树脂分散基
synthetic(al) resin dispersion finish 合成树脂分散体饰面
synthetic(al) resin emulsion 合成树脂乳液;合成树脂乳胶
synthetic(al) resin emulsion paint 合成树脂乳液涂料;合成树脂乳胶漆
synthetic(al) resin exterior coat 合成树脂(建筑)外部涂层
synthetic(al) resin external coat 合成树脂外墙涂料
synthetic(al) resin filling-packaging machine 合成树脂灌装机
synthetic(al) resin filter 合成树脂过滤器
synthetic(al) resin flat paint 合成树脂无光漆
synthetic(al) resin floor cover(ing) 合成树脂地板覆盖层;合成树脂楼板覆盖层;合成树脂楼面涂层
synthetic(al) resin floor(ing) 合成树脂楼面材料
synthetic(al) resin gasket 合成树脂垫片
synthetic(al) resin glue 合成树脂胶
synthetic(al) resin hydraulic hose 合成树脂液压软管
synthetic(al) resin impregnated 合成树脂浸渍过的
synthetic(al) resin impregnating seal(ing) material 合成树脂浸渍的密封材料
synthetic(al) resin joint 合成树脂结合;合成树脂连接;合成树脂接头
synthetic(al) resin lacquer 合成树脂蜡克;合成树脂喷漆
synthetic(al) resin mastic 合成树脂乳香脂;合成树脂玛琋脂
synthetic(al) resin mastic joint 合成树脂玛琋脂填缝料
synthetic(al) resin material 合成树脂材料
synthetic(al) resin matt (finish) piant 合成树脂无光涂料
synthetic(al) resin-modified 合成改性树脂
synthetic(al) resin mortar 合成树脂灰泥
synthetic(al) resin mo(u)lding compound 塑料粉
synthetic(al) resin outside coat 合成树脂室外涂层
synthetic(al) resin paint 合成树脂漆;合成树脂涂料
synthetic(al) resin paint thinner 合成树脂涂料稀释剂;合成树脂漆稀释剂
synthetic(al) resin penetrating aid 合成树脂渗透辅助剂
synthetic(al) resin penetrating sealer 合成树脂渗透性密封剂
synthetic(al) resin plaster 合成树脂灰浆
synthetic(al) resin prime coat 合成树脂底层漆
synthetic(al) resin primer 合成树脂底漆
synthetic(al) resin product 合成树脂成品;合成树脂产品

synthetic(al) resin putty 合成树脂腻子;合成树脂油灰
synthetic(al) resin ready-mixed paint 合成树脂调和漆
synthetic(al) resin rigid foam 合成树脂硬质泡沫
synthetic(al) resin rubber mix(ture) 合成树脂及橡胶混合物
synthetic(al) resin saturated 饱和合成树脂
synthetic(al) resin seal(ing) ring 合成树脂密封圈;合成树脂垫圈
synthetic(al) resin solution 合成树脂溶液
synthetic(al) resin stand oil enamel 合成树脂熟油搪瓷;合成树脂熟油涂层材料
synthetic(al) resin substrate 合成树脂衬底
synthetic(al) resin surface 合成树脂表面
synthetic(al) resin surface waterproofing agent 合成树脂表面防水剂
synthetic(al) resin surface water-repelling agent 合成树脂表面拒水剂
synthetic(al) resin underlay 合成树脂垫底层
synthetic(al) resin varnish 合成树脂凡立水;合成树脂清漆
synthetic(al) resin water(-carried) paint 合成树脂水溶性涂料
synthetic(al) resin waterproof coating (material) 合成树脂防水涂料
synthetic(al) resin water repellent 合成树脂表面拒水剂
synthetic(al) resin workshop 合成树脂间
synthetic(al) reverberation 人工混响
synthetic(al) rock 人造石
synthetic(al) rock motion 合成基岩运动
synthetic(al) route 合成路线
synthetic(al) rubber 合成橡胶;人造橡胶块;人造橡胶
synthetic(al) rubber accessory 合成橡胶附件
synthetic(al) rubber adhesive 合成橡胶黏结剂;合成橡胶黏合剂
synthetic(al) rubber apron 合成橡胶裙板
synthetic(al) rubber based emulsion 合成橡胶基乳液
synthetic(al) rubber building mastic 建筑用合成橡胶玛琋脂
synthetic(al) rubber contact adhesive 合成橡胶接触黏合剂
synthetic(al) rubber contact solution 合成橡胶催化溶液;合成橡胶接触溶液
synthetic(al) rubber dam 合成橡胶坝
synthetic(al) rubber dispersion 合成橡胶分散体
synthetic(al) rubber emulsion 合成橡胶乳胶;合成橡胶乳剂
synthetic(al) rubber filler 人造橡胶嵌缝料
synthetic(al) rubber latex 合成橡胶乳胶液;人造橡胶乳液
synthetic(al) rubber mastic 合成橡胶玛琋脂
synthetic(al) rubber phenol resin adhesive 合成橡胶苯酚树脂黏结剂
synthetic(al) rubber resin 合成橡胶树脂
synthetic(al) rubber resin adhesive 合成橡胶树脂胶粘剂;合成橡胶树脂黏合剂
synthetic(al) rubber sealant 合成橡胶密封剂;合成橡胶嵌缝料;合成橡胶密封膏
synthetic(al) rubber sealing 合成橡胶密封
synthetic(al) rubber stereo-picture 合成橡胶立体图
synthetic(al) rubber tire 合成橡胶轮胎
synthetic(al) rubber washer 合成橡胶垫圈
synthetic(al) rubber wastewater 合成橡胶废水
synthetic(al) ruby 合成红宝石;人造红宝石
synthetic(al) ruffle 合成金红石;人造金红石
synthetic(al) sand 人工组成砂;人工配砂;组成砂;合成砂
synthetic(al) sandal wood oil 合成檀香油
synthetic(al) sandenol 合成檀香
synthetic(al) sapphire 合成蓝宝石;人造蓝宝石
synthetic(al) satellite system 综合卫星系统
synthetic(al) scheelite 合成白钨
synthetic(al) seawater 人造海水
synthetic(al) secondary wastewater effluent 综合二级武士处理后出水
synthetic(al) seismogram 合成震波图;合成地震图
synthetic(al) seismograph 合成地震记录
synthetic(al) silica 合成硅石

synthetic(al) silica of aerogel and hydrogel 多孔和重量轻的合成二氧化硅
synthetic(al) silica glass 合成石英玻璃;人造石英玻璃
synthetic(al) slag 合成矿渣
synthetic(al) slow release fertilizer 合成缓释肥料
synthetic(al) sludge 合成污泥
synthetic(al) soil 人工合成土
synthetic(al) solvent 合成溶液
synthetic(al) sound 合成声
synthetic(al) spectrum 合成光谱
synthetic(al) spinel 合成尖晶石;人造尖晶石
synthetic(al) stability system 综合稳定系统
synthetic(al) stack trace 合成叠加道
synthetic(al) staple 合成短纤维
synthetic(al) steam cylinder oil 合成汽缸油
synthetic(al) steel 合成钢
synthetic(al) stereo image 合成立体图像
synthetic(al) stone 人造石;铸石
synthetic(al) structural product 合成结构制品;合成建筑用品
synthetic(al) study 综合研究报告;综合研究
synthetic(al) sunshade 综合式遮阳
synthetic(al) surfaceactive substance 合成表面活性剂.
synthetic(al) system 综合系统
synthetic(al) table 综合表
synthetic(al) tannin 合成丹宁
synthetic(al) tanning agent 合成鞣剂
synthetic(al) tanning material 合成鞣料
synthetic(al) tar 合成焦油沥青
synthetic(al) technology 合成技术
synthetic(al) textile fiber 合成纺织纤维
synthetic(al) textile wastewater 合成纺织废水
synthetic(al) thinner 合成稀料;醇酸稀料
synthetic(al) tide 合成潮
synthetic(al) time history 合成时间历程
synthetic(al) ultramarine 合成群青;合成佛青
synthetic(al) unit hydrograph 综合单位水文图;综合单位水文过程线
synthetic(al) utilization 综合利用
synthetic(al) varnish 合成清漆;人造清漆
synthetic(al) varnish resin 合成清漆树脂
synthetic(al) washer 合成洗涤剂
synthetic(al) wastewater 综合废水
synthetic(al) wave 合成潮
synthetic(al) wax 合成蜡;人造蜡
synthetic(al) wetting agent 合成湿润剂
synthetic(al) wood 合成木材
synthetic(al) yellow ocher 合成黄赭石颜料
synthetic(al) zeolite 合成沸石
synthetics 合成物;合成品
synthetics from coal 煤合成物
synthfuel 合成燃料
synthoil 合成石油
synthol 合成燃料
syntholube 合成润滑油
synthon 合成纤维
synthowater heater 合成水加热器
synthronon 教堂长凳
syntonic jar 谐振瓶
syntonization 共振法;谐振(法);同期;同步
syntonizer 谐振器;共振器
syntony 谐振;电流谐振
syntony jar 谐振瓶
syntrophy 互养共栖
syntype 共型;全模式;全模标本
synusia 同型同境群落;层群;层片
synusium 生态群
synvaren 酚醛树脂胶粘剂;酚醛树脂胶合剂
syphomatic filter 虹吸滤池
syphon = siphon
syphonage 虹吸作用
syphon barometer 虹吸气压计;虹吸气压表
syphon cistern 虹吸池
syphon effect 虹吸效应
syphon hood 虹吸罩
syphon inlet 虹吸进水口;虹吸管进口
syphon jet 虹吸喷射
syphon jet type 虹吸喷射式
syphon lubrication 虹吸润滑
syphon lubricator 虹吸润滑器
syphon mouth 虹吸管进口
syphon pipe 虹吸管

syphon rainfall recorder 虹吸式雨量计
syphon recorder 虹吸记录仪;虹吸记录器
syphon runner 底注浇口(内浇口向上倾斜)
syphon spillway 虹吸式溢洪道
syphon trap 水封管
syphon tube 虹吸管
syren 汽笛
Syrian architecture 叙利亚建筑
Syrian asphalt 叙利亚地沥青
Syrian garnet 沙廉榴石
Syrian vault 叙利亚拱顶
Syria type 叙利亚式
Syringa vulgaris 西洋丁香
syringe 注射器;注射管;唧筒;手动活塞泵;冲洗器
syringe boiling sterilizer 注射器煮沸消毒器
syringe buret(te) 注射滴定管
syringe for lubrication 润滑油枪
syringe graduation 注射器刻度
syringe guide 注射导引器
syringe hydrometer 吸管式比重计
syringe nozzle 注射器喷嘴
syringe pipette 注射器吸液管
syringe piston 注射器活塞
syringe reaction 注射器反应
syringe sampling 注射器进样
syringic acid 丁香酸
syrinx 管孔;鸣管;(古埃及石墓中的)隧道
syrup 浆状物;浆
sysop 系统操作员
system 系统,系[地];制度;物系;体制;体系;方式
system acceptance tests 系统验收试验
system accuracy 系统精度
system action command 系统启动命令;系统处理命令
systema digestorium 消化系统
system admittance 系统导纳
system advisory board 系统咨询部
system algorithm 系统算法
system alternatives 系统的选择方法
system analog 系统模拟
system analog(ue) language 系统模拟语言
system analysis 系统分析
system analysis and integration model 系统分析和综合模型
system analysis engineering 系统工程学
system analysis manager 系统分析经理
system analysis model 系统分析模型
system analysis model for pavements 路面的系统分析模式
system analysis of water resources 水资源系统分析
system analyst 系统分析专家;系统分析员
system approach 系统研究;系统途径;系统分析法
system approaches equilibrium 系统近似平衡
system architecture 系统结构;总体结构;体系结构
system area 系统区
system assemble 系统安装
system assembly 综合系统
system assurance program(me) 系统保险程序;系统安全程序
systematic 系统性的;系统的;有系统的
systematical analysis of water resources 水资源系统分析
systematical drainage canal 系统排水渠
systematically laid out pedestrian street 经系统规划的人行道
systematically use 有系统应用
systematic analog network testing approach 系统模拟网络测试方法
systematic analysis 系统分析
systematic approach 系统研究法
systematic arrangement 顺序排列
systematic assessment of water quality of river 河流水质系统评价
systematic botany 植物分类学
systematic building 装配式建筑;系统施工法;装配式施工法
systematic category 系统类目;分类类目
systematic circulation 循环系统
systematic circulatorium 循环系数
systematic code 系统码
systematic convergence error 系统会聚误差
systematic correction 系统修正
systematic damping 体系阻尼

systematic decision model 系统决策模式
systematic design discipline 系统设计规范
systematic design method 系统设计方法
systematic deviation 系统偏差
systematic deviation from linearity 系统性直线离差
system(atic)diagram 系统图
systematic distortion 系统失真;系统畸变;规则性失真;基本畸变
systematic distribution 系统分布
systematic drift rate 系统漂移速率
systematic effect 系统效应
systematic engineering 系统工程
systematic error 系统误差;规律误差;顺序误差
systematic error checking code 系统检错码
systematic evaluation program(me) 系统评价程序
systematic fashion information 系统化信息
systematic graden 植物分类园
systematic grouping 系统分组
systematic identification 系统鉴定
systematic inaccuracy 系统不精确性
systematic inspection 系统检查
systematic integration 系统集成
systematic investigation and study 系统调查研究
systematic isolate idea 系统细目概念
systematic joint 系统节理
systematic joints 节理系
systematic layout planning 系统的布置管理
systematic management 分类控制法;分类管理法
systematic mapping 系列制图;系列成图
systematic mathematical analysis 系统数学分析法
systematic mathematical model 系统数学模型
systematic measure 系统的测度
systematic microbiology 分类微生物学
systematic mutation 系统突变
systematic name 系统名(称);分离名;分类名
systematic negative anomaly 系统负异常
systematicness 系统性
systematic noise 系统噪声
systematic nomenclature 分类命名法
systematic observation 系统观测
systematic order 系统顺序
systematic overrelaxation 系统超松弛
systematic perspective 系统观
systematic point of view 系统观点
systematic procedure 系统过程
systematic production of substitution lines 系统的产生代换品系
systematic random sample 系统随机样本;规律性随机样本
systematic relaxation procedure 系统松弛步骤
systematic requirement criterion 系统要求标准
systematic residuals 系统误差
systematic risk 系统性风险;不可避免的风险
systematic river water quality assessment 河流水质系统评价
systematics 系统学;分类学;分类系统
systematic sample 系统样本;系统抽样样品;系统抽样;程式样品;程式样本
systematic sampling 系统取样;系统化采样;系统抽样方案;系统抽样法;系统抽样;系统采样;等间隔抽样
systematic sampling in two-dimensions 二维系统等距抽样
systematic samplingmethod 系统采样法
systematic sampling scheme 系统抽样模型;系统采样格式
systematic selection 系统选择
systematic self adjustment 系统的自我调节
systematic self organization 系统的自我组织
systematic separation 系统分离
systematic socioeconomic factors 系统的社会经济因素
systematics of science and technology 科学技术体系学
systematic spectral analysis 系统光谱分析
systematic statistic 系统统计量
systematic support 系统化支撑
systematic theory 系统理论
systematic theory method 系统理论法
systematic tree 系统树
systematic uncertainty 系统不确定性;系统不确定度

systematic variation 系统变化
systematist 分类学家
systematization 系统化
systematization of design 设计系统化
systematize 系统化
systematology 系统学;系统论
system availability 系统利用率;系统利用度;系统可用性
system backup 系统后备
system balance 系统补偿
system bandwidth 系统带宽
system behavio(u)r 系统功效
system behavio(u)r model 系统功效模型
system biology 系统生物学
system board 系统插板
system bolting 系统锚固
system boundary 构造边界(地质图);岩层边界
system buffer 系统缓冲区
system buffer element 系统缓冲单元
system building 工业化建筑;系统施工法;体系建筑;预制部件装配式建筑
system building approach 系统建筑方法
system building construction 装配式住房建筑
system building method 装配式建筑方法
system-built concrete structure 预制混凝土结构
system-built facade 预制正面;预制立面
system-built structure 预制结构
system-built wall 预制墙
system bus 系统总线
system bus arbitrator 系统总线仲裁器
system busbar 系统母线
system bus interface 系统总线接口
system bus loading 系统总线负载
system bus width 系统总线宽度
system calibration 系统校准
system call 系统调用;系统调度
system call interrupt 系统调用中断
system capacity 系统容量
system center 系统中心
system centered approach to change 系统导向的变革
system center line 系统中心线
system centralized control 系统集中控制
system centre line 系统中心线
system change 系统变换
system chart 系统框图;系统程序图
system check 系统校验;系统检验
system check module 系统校验模件;系统检验模块
system classification 系统分类
system clock 系统时钟
system clock control 系统时钟控制
system clock control signal 系统时钟控制信号
system cluster 系统聚类
system code 系统码
system command 系统指令;系统命令
system command executive 系统命令执行(语言、程序)
system command interpretation 系统命令解释
system communication 系统通信
system communication processing 系统通信处理
system compatibility 系统相容性;系统协调性;系统兼容性;系统互换性
system completion code 系统完成码
system component 系统元件
system composition 系统组成;系统成分
system concept 系统概念
system condition 系统条件
system configuration 系统配置;系统结构;系统几何尺寸
system connection 系统连接
system console 系统控制台
system constant 系统常数
system constituent 系统组件
system constraint 系统约束;系统制束
system construction 体系工程
system construction method 系统施工法;预制施工法
system control 系统控制
system control centre 系统控制中心
system control interface 系统控制接口
system controller 系统(信号)控制机
system control panel 系统控制面板;系统控制板
system control room 系统调度室
system convention 系统规则

system conversion 系统移交;系统更换
system coordination 信号系统协调;系统的协调
system corrected image 系统校正像片
system damping 系统阻尼
system data acquisition 系统数据收集
system data bus 系统数据总线
system debug 系统调试
system definition 系统定义
system definition information 系统定义信息
system degradation 系统性能降低;系统退化
system delay 系统固有延迟
system description maintenance 系统现状的保存
system design 系统设计;总体设计;体系设计
system design aids 系统设计手段;系统设计工具
system design criterion 系统设计标准
system designer 系统设计员;总体设计员
system design simulator 系统设计模拟器
system detail design 系统细部设计
system determinant 系数行列式;体系行列式
system development 系统开发
system development and analysis program (me) 系统开发和分析程序
system deviation 系统偏差;系统离差
system device 系统设备
system diagnostics 系统诊断程序
system diagram 系统图
system disturbance 系统扰动
system diversity factor 系统不同时率
system dose 系统投配量
system down 系统失灵;系统停机;系统故障
system drawing 系统图
system dump 系统转储
system dynamics 系统动态学;系统动力学
system dynamics forecasting method 系统动态学预测法
system dynamics predictive model 系统动态预示模型
system dynamics simulation model 系统动态学模拟模型
system earth 系统接地
system effectiveness 系统有效性;系统效率
system efficiency 系统效率
system eigenvalue 系统本征值
system electrical power swing 系统电功率摆动
system element 系统元件
system emergency protection 系统应急保护
system encyclopedia 系统百科全书
system engineer 系统工程师
system engineering 系统工程(学)
system engineering evaluation 系统工程学评价
system engineering tools 系统工程工具
system ensamble 综合系统(统计力学)
system entropy 系统熵
system environmental simulation 系统环境仿真
system environment recording 系统现场记录;系统环境记录
system environments 系统环境
system equilibration 系统平衡
system equivalence 系统等价
system error 系统误差
system error recovery 系统错后复原
system error routine 系统查错程序
system evaluation 系统评价
system exerciser 系统试验器
system exercisor 系统试验器
system expansion 系统扩充
system extinction 系统消光
system factor 系统因子;系统因素
system failure 系统故障
system fan 系统风机;装置排风机
system fault 电力网故障
system firmware 系统固件
system flow 系统流程
system(flow)chart 系统流程图;系统框图
system for analysis of soil-structure interaction 土(壤)结构相互作用分析系统
system for digital mapping 数字化制图系统
system for government purchase 政府采购制度
system for information retrieval and storage 信息恢复与存储系统
system for integrated environment(al) and economic accouting 环境与经济综合核算系统
system for non-iventoriable items 不可存储产品系统

system for nuclear auxiliary power 核辅助动力系统;核辅助电力系统
system for quantitative examination on integrated control of urban environment 城市环境综合整治定量考核制度
system for reporting and filing 报告和档案制度
system foulant 系统污垢物
system frequency 系统频率
system frequency excursion 系统频率偏移
system function 系统函数;系统功能
system function design 系统功能设计
system generation 系统形成;系统的生成
system generation time 系统生成时间
system ground 系统接地
system grounding conductor 系统接地导体
system grounding method 系统接地法
system halt 系统停机
system handbook 系统手册
system hardware 系统硬件
system head 系统压头;系统水头
system height 结构长度
system height/encumbrance 结构高度【电】
system hydrology 系统水文学
systemic 系统的;内吸的;全身的
systemic action 内吸作用
systemic circulation 系统循环;体循环;大循环
systemic contamination 全身污染
systemic distortion 基本畸变
systemic fungicide 内吸杀菌剂
systemic gain drift 系统增益漂移
systemic herbicide 内吸收性除草剂
systemic infection 系统侵染
systemic insecticide 内吸杀虫剂
systemics 内吸剂
systemic sclerosis 系统性硬化
systemic selection 系统选择
systemic velocity 质心速度
system identification 系统识别;系统鉴别
system identification theory 系统辨识理论
system impedance 系统阻抗
system impedance ratio 系统阻抗比
system implementation language 系统执行语言
system improvement group 系统改进组
system improvement time 系统改进时间
system information 系统信息;系统情报
system information management 系统情报管理
system initialization 系统预置
system initiation 系统开始
system input device 系统输入装置
system input unit 系统输入装置
system inspection 系统检查
system instability 体系不稳定性
system install 系统安装
system installation 系统安装
system installed capacity 系统装机容量
system integration 系统综合;系统联调
system integration technique 系统集成技术
system integrity 系统完整性
system interconnection 系统互联
system interface design 系统接口设计
system interface module 系统接口模件
system interrogation 系统询问
system interrupt 系统中断
system interrupt action 系统中断动作
system interrupt dispatch routine 系统中断调度程序
system interrupt request 系统中断请求
system in three-dimensional space 三维空间体系
system invariants 系统不变量
system in which all the wall carry loading 全部承重墙系统
system language 系统语言
system layout 系统布置
system library 系统程序库
system life cycle 系统寿命周期
system line 系统线路
system linking 系统耦合
system load 系统负载
system load characteristic 系统负载特性
system loader 系统装入程序
system load factor 系统荷载因数;系统负荷系数
system loading 系统负荷
system loading condition 系统负载状况
system lock 系统封锁;系统封闭

system log 系统运行文件;系统运行记录;系统运行程序;系统记录器;系统记录
system log data set 系统记录数据集
system log function 系统记录功能
system logic 系统逻辑
system loss 系统损耗
system macro instruction 系统宏指令
system main 系统干管
system maintenance 系统维护
system maintenance manual 系统维护手册
system maker 整机制造厂
system malfunction 系统故障;系统功能失灵
system management function 系统管理功能
system manager 系统经理
system manual 系统手册
system map 系统图
system map of open pit bench 露天矿台阶系统图
system map of underground mining pit 地下开采井巷系统图
system mask 系统屏蔽
system matching 系统匹配
system mathematical model 系统数学模型
system matrix 系统矩阵;体系矩阵
system maximum hourly load 系统小时最大负载
system measurement routine 系统测试程序;系统测量程序
system mechanization 系统机械化;综合机械化;全盘机械化
system message data set 系统信息数据集
system mistake 系统错误
system mode 系统模式
system mode instruction 系统状态指令
system model 系统模型
system model(l)ing 系统模拟;系统模造
system-modulation-transfer-acutance 系统的调制传递锐度
system module 系统模件
system monitor 系统监控程序;系统监督程序
system monitor unit 系统监视装置
system multiplex 系统多路转换
system network 系统网络
system network architecture 系统网络体系;系统网络结构
system network structure 系统网络结构
system noise 系统噪声
system noise figure 系统噪声指数
system nomenclature 系统命名法
system of absolute unit 绝对单位制
system of account 合计制度
system of aid to navigation 航标制度
system of air voids 气穴系统;空气孔隙系统
system of all employees signing labo(u)r contracts 全员劳动合同制
system of a measurement 测量制(度)
system of anchoring 锚碇系统
system of assigning crews to designated trains 包乘制【铁】
system of astronomic(al)constants 天文常数系统
system of automatic internal audit 内部自动审计制度
system of axes 坐标系(统);轴系
system of axioms 公理系统
system of axis 坐标轴系
system of binders and joists 托梁系统
system of binders and joists for tower 塔架托梁系统
system of bracing 加肋系统;支撑系统
system of bridges of stage 舞台桥面系统
system of buttresses and flying buttresses 扶壁和飞扶壁系统
system of centered spheric(al)surfaces 同轴球面系统
system of check and acceptance 验收制度
system of civil aviation management 民用航空管理制度
system of civil justice 民事审判体制
system of civil registration 民事登记体系
system of classification 分类法
system of closing hills for forest cultivation 封山育林制订
system of coalition 联合制
system of collecting fees for discharging pollutants 征收排污费制度
system of compensatory financing 补偿贷款制

度;补偿贷款制度
system of compounds 化合物系统
system of comprehensive evaluation of sports 体育综合评价体系
system of conics 二次曲线系
system of conjugate fracture 共轭破裂系
system of contract 合同运费制
system of controls 控制制度
system of coordinates 坐标系(统)
system of crystallization 结晶系
system of crystals 晶系
system of curves 曲线系
system of cutting quota 限额采伐制度
system of decentralization 分权制
system of disjunctive Kriging equations 析取克立格方程组
system of distinct representation 不同表示系列
system of distribution 分配系统
system of dividing taxes 分税制
system of dowelled wood(en)beams 键合木梁系统
system of economic contract 经济合同制
system of economic cooperation 经济合作制度
system of economic responsibility of enterprise 企业经济责任制
system of economizing water use 节约用水制度
system of effluent charges 排污收费制度
system of election of cadres 选任制
system of enterprise operation and management 企业经营管理制度
system of environment(al) impact assessment 环境影响评价制度;环境影响评价体系
system of environment(al) law 环境法体系;环境法律体系
system of environment(al) management 环境管理体制;环境管理体系
system of environment(al) monitoring 环境监测制度
system of environment(al) pollution accident reporting 环境污染事故报告制度
system of environment(al) standards 环境标准体系
system of environmental statistics reporting tables 环境统计报表制度
system of equations 方程组
system of exploratory engineering 勘探工程系统
system of exploring mining and drilling combine engineering 坑钻组合工程系统
system of export credit insurance 出口信贷保险制度
system of export licensing 出口许可制度
system of facing 饰面系统
system of faults 断层系
system of fissure 裂隙组;裂隙系
system of fit 配合制;配合方式
system of fixed ecliptic 固定黄道坐标系
system of fixed quota 定额制
system of flo(a)tation 浮动率制度
system of forces 力系
system of forestry fund 林业基金制度
system of free use of land 土地无偿使用制度
system of groove for roll forging 辊锻型槽系
system of harmonic analysis 调和分析法
system of height 高程系
system of holding adminsitrative director taking full responsibility of environmental protection 环境保护首长责任制
system of holdings 参与制
system of identification 鉴定制
system of import and export license 进出口许可证制(度)
system of incomplete mutual solubility 不完全互溶物系
system of industrial management 工业管理体制
system of inequalities 不等式组
system of in-factory economic accounting 厂内核算制
system of internal check 内部牵制制度
system of inventory control 库存控制系统
system of investment responsibility 投资包干责任制
system of irrigation pipes built under the ground 地下灌溉管道系统
system of job responsibility 责任制(度)

system of jointing 勾缝系统;黏结系统;接合系统;连接接头系统
system of joints 节理系
system of knife edges 刀口系统(天平)
system of Kriging equations 克立格方程组
system of land classification 土类分级法
system of lettering in six direction 六方位注记法
system of licencing 许可证制度
system of linear algebraic equations 线性代数方程组
system of linear equations 线性方程组
system of lines 直线系
system of loads 负载系(统)
system of logarithm 对数制;对数系
system of macroeconomic management 宏观方面的经济管理体制
system of measuring units 测量单位制
system of medical care insurance 医疗保险制度
system of nets 网系;网络系统
system of non-linear equations 非线性方程组
system of notation 记数法;符号系
system of one-head leadership 一长制
system of one-man leadership 一长制
system of on-site inspection 现场检查制度
system of operation 作业系统
system of ownership 所有制
system of parametric(al) line 参数曲线系统
system of partial free supply and partial wages 部分供给
system of particles 质点系
system of periodic(al) maintenance 定期养护制;定期修缮制
system of permits for cutting forest and trees 林木采伐许可证制度
system of personal responsibility 责任制(度);岗位责任制
system of pipes 管道系统;管系
system of plates 电容器片组
system of pollution administration of offshore oil exploration and exploitation 海洋石油勘测开发污染管理制度
system of polymerization 聚合体系
system of power transmission 功率传递系统
system of preventing on pollution damage of coastal construction project 防止海岸工程污染损害制度
system of preventing pollution by land based pollutants 防止陆源污染制度
system of private ownership 私有制
system of processing 加工系统
system of production engineering 生产勘探工程系统
system of production responsibility 生产责任制
system of promotion 晋级制度
system of rafters 屋脊梁系统;椽子系统
system of railway statistic(al) standard 铁路统计标准体系
system of rapid evaluation and sorting of multielement anomalies 多元素异常快速评价分选系统
system of rating 考核制度;定额制度;定额法
system of rays 辐射纹系统
system of recruitment of cadres 考任制
system of rectangular coordinates 直角坐标系
system of rectangular tables 矩形表体系
system of rectangular X and Y coordinates X-Y 直角坐标系
system of reference 参考系(统)
system of regular pay 薪饷制
system of reporting accident of environment (al) pollution and destruction 环境污染与破坏事故报告制度
system of reporting and registering pollutant emission 排污申报登记制度;污染物排放申报登记制度
system of rewarding comprehensive utilization 奖励综合利用制度
system of rewards and penalties 奖惩制度
system of risers 补缩系统
system of sales promotion 推销系统
system of sanitation 环境卫生系统
system of save use of pesticide 农药安全使用制度
system of selection 选择方法
system of semi-private ownership 半私有制

system of sewerage 污水排放系统;下水道系统
system of sewers 阴沟系统;下水道系统
system of share-cropping of net income 净收入分成制
system of shifting down and moving polluting enterprise 关、停、并、转、迁制度
system of signs and signals 信号系统
system of simultaneous equations 联立方程组
system of space axes 空间轴系
system of span pieces 檩条系统;吸杆系统;拉条系统
system of statement 报表制度
system of statistic(al)indicators 统计指标体系
system of statistic(al)report 统计报表制度
system of statistics 统计制度
system of supply 供应制
system of system 超系(统)
system of technical responsibility 技术责任制
system of tectonic migration 大地构造迁移类型
system of three posts 三柱系统
system of tolerance for building(construction) 建筑施工的公差系统;建筑构造的公差系统
system of treating environmental pollution within a prescribed time 限期治理环境污染制度;限期环境污染治理制度
system of uncompensated land use 土地无偿使用制度
system of units 单位制度;单位制
system of universal Kriging equations 泛克立格方程组
system of vectors of reference 参照向量系;参考矢系
system of verification 核查制度
system of vocational assessment 人事考核制度
system of water consumption 用水制度
system of water quality 水质分类法
system of water sprinkler 水喷淋系统
system of water supply 给水系统
system of water supply regulator 给水调节器系统
system of weights and measures 度量衡制
system of wind 风系
system operation 系统操作
system operator 系统管理员;系统操作员
system optimization 系统调优;系统(最)优化;系统最佳化;系统优选
system organization 系统组织;系统结构
system oriented 面向系统的
system oriented computer 面向系统的计算机
system oriented hardware 面向系统的硬件
system oscillation 系统振荡
system outage 系统中断
system overhead 系统总开销
system overshoot 系统过调量
system pack 操作系统程序包
system parameter 系统参数
system partition 系统划分
system peak load 系统尖峰负荷;电网尖峰负荷
system performance 系统性能;系统特性
system performance calibration compound 系统性能校正化合物
system performance effectiveness 系统性能有效度
system performance evaluation 系统性能评价
system performance index 系统性能指标
system performance information 系统性能数据
system performance monitor 系统性能监视器
system phase diagram 体系相图
system planner 系统计划人员
system planning 系统计划;系统规划
system plating 内外叠板法
system point 系统中心
system pool 系统区;系统池
system power source 系统电源
system power supply 系统电源
system preparation routine 系统准备程序
system pressure 系统压力
system probatoire observation de la terrestre 地球观测实验系统
system processing error 系统处理误差
system program(me) 系统程序
system program(me)error 系统程序错误
system program(me)operator 系统程序操作员
system program(me)planning 系统规划程序
system programmer 系统程序员;系统程序设计员;系统程序计划员;系统程序号

system programming 系统程序设计
system programming language 系统程序设计语言
system protection 系统保护
system reaches equilibrium 系统区段平衡
system realm 系统区域
system reconfiguration 系统重组;系统结构改变
system record file 系统记录文件
system recovery 系统恢复
system reform 体制改革
system reliability 系统可靠性;系统可靠度
system research 系统研究
system reserves 系统储[贮]备
system reset 系统重置;系统复原;系统复位
system residence 系统程序常驻区;系统常驻;常驻系统
system residence disk 常驻系统盘
system residence volume 系统驻留容积;常驻系统卷
system resistance 系统阻力;系统电阻
system resistance characteristic 系统阻力特性
system resistance curve 系统阻力曲线
system resolution 系统分辨能力
system resource management 系统资源管理;系统资料管理
system resources 系统资源
system response 系统响应;系统频率特性
system riser 纵向给水主管(建筑物内)
system risk 系统风险
system roll-back 系统重新运行
system running cost 系统运行费用
system's ability for orderly change 序变能力
system safety 系统安全
system safety engineering 系统安全工程
systems age 系统龄
systems alternatives 系统选择方案
system salt permeability 系统透盐率
system sampling 系统抽样
systems analysis 系统分析
systems analysis staff 系统分析人员
systems analyst 系统分析人员
system sand 再生砂;单一砂
systems and procedures 系统和程序;制度与程序
systems approach 系统研究法;系统思路;系统论方法;系统方法
systems approach to management 系统研究法
systems assessment 系统评价
systems audit 系统审计
system save file 系统保存文件
systems behavio(u)r 系统行为
systems building 工业化建筑;预制建造法;体系建筑
system scanner 系统扫描程序
system science 系统科学
systems construction 建筑系统工程
systems design specification 系统设计说明书
systems development 系统发展
systems development corporation 系统发展公司
systems development process 系统发展过程
systems development requirement 系统发展要求
systems ecology 系统生态学
systems engineering 系统工程;总体工程
systems engineering methodology 系统工程方法论
systems engineering of urban ecological system 城市生态系统工程
systems enrichment 系统工程;总体工程
system separation 系统分列
system service control point 系统服务控制点
system service program(me) 系统服务程序;系统处理程序
system service request 系统服务申请
system service transformer 系统供电变压器
systems estimation 系统估计
systems excretion 排泄系统
systems for the identification of numbers 自动识别箱号装置
systems function diagram 系统功能图
systems group 系统小组

system shop 大修厂
system shutdown 系统关闭
systems implementation 系统实施
systems implementation test 系统运行试验
system simulation 系统模拟
system sketches 系统框架
systems life cycle 系统生命周期
systems linking 系统连贯
systems management 系统管理
systems management school 系统管理学派
systems method 系统方法
systems modelling 系统模型化
systems of coordinating traffic signal 联动交通信号系统
systems of first order equation 一阶方程组
system software 系统软件
system software component 系统软件成分
system software package 系统软件包
systems operation 系统运行
systems operational mechanism 系统运行机制
system specification 系统技术说明;系统规范;系统技术规格书
system specific chemical factor 系统特性化学因子
system specific parameter 系统特性参数
system speed 系统速度
system splitting 系统分列
system sponsor 系统赞助人
systems reliability analysis 系统可靠性分析
systems requirements specification 系统要件说明书
systems research 系统研究法
systems revovery management 系统恢复管理
systems school 系统学派
systems selling 配套出售
systems simulation 系统仿真
systems structure 系统式管理组织结构
systems synthesis 系统综合
system stability 系统稳定性
system stability analysis 系统稳定性分析
system stable of first order 一阶稳定体系
system stack 系统堆栈
system standard 系统标准
system standard label 系统标准标号
system start-up 系统起动;系统启动
system state 系统状态
system state table 系统状态表
system statistical analysis 系统统计分析
system statistical information 系统统计信息
system status 系统状态;系统状况
system status interrogation 系统状况询问
systems technical consultant 系统技术顾问
systems technology 系统技术
systems theory 系统理论
system stiffness matrices 体系刚度矩阵
system stop page 系统停止页
system storage 系统存储器
system storage capacity 系统存储量
system storage hierachy 存储器层次结构
system stored energy 系统储能
system's transient response 系统的瞬态响应
system structure 系统结构
system study 系统研究
system subcomponent 系统子成分
system subroutine 系统子程序
system supervisor 系统主管;系统管理程序
system support 系统支援设备;系统支持;系统配套
system support channel 系统后援通道
system support function 系统支援功能
system support program(me) 系统支援程序
system surging 传导系统中的压力脉动
system switching 线路转换;系统转换
system tape 系统带
system task 系统任务
system task partition 系统任务区
system task save area 系统任务保存区
system test complex 系统测试全套设备
system tester 系统试验器;系统测试仪;系统测试程序

system test(ing) 系统试验;系统检验;系统测试
system testing program(me) 系统测试程序
system test mode 系统测试方式
system test set 系统试验装置
system test station 系统试验站
system theory 系统论
system thermal power plant 系统火力发电厂
system timer 系统计时器
system time stamp 系统计时
system trace 系统跟踪
system trace tab 系统跟踪标记
system transfer function 系统传递函数
system transformation 体系转换
system transient output response 系统频率特性;瞬变过程
system transposition 系统交叉
system transter 系统的传递函数
system tree 系统树
system under test 被测系统
system unit 系统处理单位
system user 系统用户
system utility program(me) 系统实用程序
system utility program(me) support 系统应用程序的提供
system utilization 系统利用率
system utilization factor 系统利用率
system utilization logger 系统利用率记录器;系统利用率记录程序
system utilization logging 系统利用率记录
system variable 系统(的)变量
system variable symbol 系统可变符号;系统变量符号
system variation 系统性变化
system virtual address 系统虚拟地址
system voltage 系统电压
system voltage drop 系统电压降
system voltage stabilizer 系统电压稳定器
system volume 系统容积;系统卷宗
system wait(ing) 系统等待
system water balance 系统水平衡
system-wide control 全系统范围的控制
system with alternating columns and piers 柱墩交替系统
system with non-linear characteristics 非线性性质系统
system with one degree of freedom 一自由度系统;一维系统;单自由度系统;单自由度体系
system with ring-type build-up 建立建筑系统
system with several degree of freedom 多自由度系统;多自由度体系;多维系统
system with stacked up three hinged frames 层叠式三铰框架系统
system with thrusting force 推力体系
system with transverse stabilization beams 具有横向稳定梁系统
system with transverse stabilization cables 横向稳定缆索系统
system with variable utilization 可变利用系统
systolic retraction 收缩期回缩
systox 有机磷杀虫剂;地灭通(一种杀虫剂)
systyle 两径间排柱式;双柱径间距柱廊;两柱径式
systylos 窄柱式(间距为柱径的二倍)
sytematize 系统化
sytull 微晶玻璃
syzygial tide 子午潮;朔望潮
syzygy 朔望(位置);大潮期
syzygy tide 朔望潮
szaboite 紫苏辉石
szaibelyite 硼镁石
szaiblyite 硼石
Szechtman cell 塞茨特曼电解池
Szerelmey stonecoat encaustic 萨雷尔迈石液(一种专利材料)
Szerelmey stone liquid 萨雷尔迈防水剂(用于砖、石、水泥墙);萨雷尔迈石用液
szik soil 盐渍土(壤)
szmikite 锰矾
szomolnokite 水铁矾

T

taaffeite 塔菲石;铍美晶石
tab 石板瓦修换;支承块;账目;造册;列表;调整翼片;搭襻;分隔符;补冀;标记
tabacosis 烟尘尘着病
tabacosis pulmonum 烟尘肺
tab assembly 翼片组合;翼片安装
Tabberbbran orogeny 塔伯拉伯造山运动
tabby 地面阀闸;平纹;灰砂;土砂碎石混合料;砂砾土;土质混凝土;波纹绢
tabby cat heart 斑纹心
tabbyite 皱纹沥青;韧沥青
tab-card cutter 制卡刀具
tab code 标记码
tablet(ing) machine 造粒机
tabenacle 礼拜堂
Taber abraser 泰伯尔耐磨性试验仪;泰伯尔耐磨性测定仪
Taber abrasion test 泰伯尔法耐磨试验
tabergite 钠黑蛭石;叶绿泥石
Taber ice 分凝冰
taberna 畜舍(古罗马)
tabernacle 临时住房;临时房屋;甲板上夹桅板;轻便帐篷;桅座;敞舱;舱面桅座;帐篷;壁龛
tabernacle work 神龛顶部装饰(礼拜堂);搭帐篷;墓屋;华盖装饰
tabetisol 不冻地;融冻土层;层间不冻层(在永久冻土之间的)
tabet soil 融解土层;溶冻土壤
tab gate 直角浇口;尖角浇口
tabia 白灰夯实土
tabid 腐烂的
tab labels 连续形式标记
tablature 顶画;刻有铭文的碑;记谱法;壁画
table 一览表;摇床;造册;陆台;列表;平板;顶面;表格;表册
table access 表访问;表存取
table addressing 表寻址;表编址
table address operand 表地址操作数
table antenna 台式天线
table area 工作台面积
table argument 表列自变量
table arm 工作台支架
table assembler 浮动汇编程序
tableau curtain 大幕(由中间向上方两侧提起的);斜拉大幕
tableau format 表格结构;表的格式
table balance 架盘药物天平;托盘天平;台秤
table band saw 平台带锯;小带锯
table base 工作台底座
table base register 表式基址寄存器;表式变址寄存器;表格基地址寄存器
tableberg 桌状冰山
table blast 喷丸清理转台
table block 数据表块;表块
table bushing 转桌轴套
table casting 立式浇筑;台式浇筑
table cast process 辊合间歇压延法
table chair 两用家具(可做桌或椅子使用)
table circuit 摇床流程
table classifier 摇床分级机
table cloth 桌布;台布
table companion microbe 共生微生物;伴生微生物
table concentrate 摇床精矿
table concentration 摇床精选
table concentrator 摇床
table control device 控制台装置
table cooker 台式烤箱灶
table crosswise movement 工作台横向运动
table data 列表数据
table deck 摇床面
table description 表式说明
tabled fish plate 扣榫的接合板;嵌接鱼尼板;带扣榫的鱼尼板;带扣榫的接合板;凹凸接板
tabled fishplate joint 榫槽夹板结合
table diamond 顶面切平的金刚石
table directory 表目录索引
table dispatching method 表上作业法
table display 表显示器

tabled joint 叠层嵌接;上下块之接头
table dog 平台挡
table drive 转盘传动;工作台传动;驱动表
table-drive motor 工作台驱动马达;工作台驱动电动机
table-driven 表格控制的
table-driven compiler 表驱动编译程序;表格控制的编译程序
table driven program(me) 表格驱动程序;表格控制程序
table-driven simulation 表控制模拟;表格驱动模拟
table-driven technique 表驱动技术;表格驱动法
table-driving mechanism 工作台驱动机构
tabled scarf 榫接凹槽;叠嵌接(合)
tabled scarf joint 叠嵌接(合)
tabled steel plate connection 凸片钢板结合
tabled wooden fish plate 带扣榫的木接合板;带扣榫的木鱼尼板
table easel 绘画台
table edger 台式齐边锯
table element 表(格)元素
table elevation 高程表
table entry 表项目
table expense 伙食费
table fan 台扇;台风扇
table feed 工作台进给
table feeder 圆盘喂料机;圆盘给水器;圆盘给料机;平板送料机;平板(式)加料器;平板给料器;平板给料机;盘式喂料器;盘式给料机;盘式送料机
table-feed lever 工作台进给手柄
table-feed screw 工作台进给螺旋
table-feed wheel 工作台进给手轮
table file 表文件
table filter 平面过滤机
table fishplate joint 企口板接合
table flap (折叠式桌面的)折板
table flo(a)tation 摇床浮选;台浮
table foot 工作台脚;台脚
table for laying off curves 曲线测设用表
table for look-up method 查表法
table for look-up procedure 查表法
table for the joint frequency distribution of meteorological data 风向风速稳定度联合频率表
table function 表函数
table glassware 玻璃餐具
table guide 台导承
table handing module 表处理模块
table handling 表(格)处理;表操作
table hand-wheel 工作台手轮
table handwheel knob 工作台手轮捏手
table-hinge 台铰
table iceberg 桌状冰山;平顶冰山;台状冰山
table index 表索引
table indicator 表格指示符
table-inking 调墨板
table interlocker 桌上摇柄;桌上联锁装置
table jack 平台千斤顶
table joint 嵌接
table knoll 海台;平顶山丘;桌状海丘
table lamp 台灯
tableland 台地;高原(台地);高原高地;海(底)台(地)
table language 表格语言
table layout form 表设计形式
table lead screw 工作台丝杠
table level(l)ing screw 绘图桌置平螺旋
table light 台灯
table locking clamp 工作台锁紧夹头
table lockup 表封锁
table longitudinal movement 工作台纵向运动
table look-at 直接查表;查表
table-look-up 一览表
table-lookup instruction 引用表指令;查表指令
table look-up technique 查表技术
table man 摇床工;在工作台边工作的人
table mechanism 工作台机构
table memory 表格存储器
table money 招待费;就餐费;交际接待费用

tablemount 海底平顶山;平顶海山
table mountain 桌状山;平顶山;平顶海山;方山地貌
table-mountain pine 山松;台地松
table-mounted cooker 台式炉灶
table movement 工作台运动
table multi-purpose wood working machine 台式多用木工机床
table napkin 餐巾
Table of Allocation of International Call Sign Series 国际呼号序列划分表
table of allowances 津贴表
table of altitudes 高程表
table of amount of work 工作量一览表
table of binomial coefficients 二项式系数表
table of cases 案件一览表
table of charges 费用表
table of classification 分类表
table of combination 组合表
table of compound interest 复利表
table of contents 目次;内容一览表;目录
table of conventional signs 符号表
table of correction 改正表
table of difference 差分表
table of drawings 图纸目录
table of earthwork 土方表
table of earthwork balance 土方平衡表
table of economic indicators 经济指标表
table of equipment 装备表
table of errors 误差表
table of exponential function 指数函数表
table of exponential random numbers 指数随机数表;按指数分布的随机数表
table of feeder 进料台
table of freight 运费表
table of frequency distribution 频数分配表
table of increase of coordinates 坐标增量表
table of increment 增值表
table of integrals 积分表
table of interest rate 利率表
table of isotopes 同位素表
table of leeway 风压差表
table of limits 限额表;公差表
table of logarithm 对数表
table of measures and weights 度量恒表
table of natural 三角函数表
table of normal air 标准大气表
table of normal atmosphere 标准大气表
table of offset 型值表
table of organization 机构表;编制表
table of organization and equipment 编制与装备表
table of organization augmentation 编制扩充表
table of parallel arcs 平行圆弧长表
table of pipe connection 管口表
table of price 价目表
table of quantities 数量表
table of rainfall 雨量表
Table of random numbers 随机数(字)表
table of rates 税率表;费率表
table of results 成果表
table of scantlings 构件尺度表
table of spectrum lines 光谱线波长表
table of stadia reduction 视距换算表;视距改正表
table of standard air 标准大气表
table of standard atmosphere 标准大气表
table of standard performance 性能表
table of standards 标准表
table of strata 地层表
table of the cumulative normal distribution 累计常态分配函数表
table of tidal information 潮汐表
table of tidal signal 潮信表
table of transfer 换算表
table of trigonometric point 三角点成果表
table of values 树根表
table of weights and measures 度量衡表
table opening 转盘的通孔
table pack 表组合

table planing machine 龙门刨床
table platform 桌状台地
table press 台式压力机
table-rack 桌框
table reef 桌状礁；平顶礁；桌(形)石焦
table reef trap 台礁圈闭
table rest 工作台(支)座
table reverse-lever 工作台回动杆
table reversing dogs 工作台返向轧头
table revolving movement 工作台回转运动
table roll 案辊
table runner 狭长台布(装饰用)
table saddle 工作台滑板；工作台滑鞍；工作台滑座
table salt 食盐
table saw 台锯；圆盘锯
table search 表检索
table segmenting 表(格)分段
table separation 摇床选
table-service restaurant 餐桌服务饭店
table setting 工作台紧固
table settle 高背长靠椅
tables for correcting mean mid latitude 中分纬度改正表
table shake 台式振荡器
table shore 低平海滨；低平海岸
table showing shipping dates for various specific destination 承运日期表
table simulator 算表程序；表模拟程序
table size 表大小；工作台尺寸
table skeleton 表骨架；表表架
table slate 屋面(平)石板
table slide 工作台导承
tables of computed altitude and azimuth 高度方位表(美国)；天体高度方位计算表
tables of equipment 设备单
table spar 硅灰石
table speed indicator 转盘转速指示器
table standard 台上柱形灯
table stone 古桌式墓标；史前巨石墓
table stop 工作台制动器；工作台止动器
table stroke 工作台行程
table structure 表结构
table support 工作台支架
tablet 笠石；小片；料片；片剂；图形输入装置；碑；板状晶体
table tailings 摇床尾矿
table tap 台用插头
tablet-arm chair 扶手椅(课堂用)；扶手课椅
tablet compressing machine 制片机
tablet disintegration tester 片剂崩解测定器
table telephone set 台式电话机
table tennis court 乒乓球比赛场
table tennis hall 乒乓球厅
table tennis room 乒乓球室
tablet hardness 片剂硬度
tablet hardness tester 片剂硬度计
tableting press 制片机
tablet lock 路牌锁塞；路牌锁闭器
tablet mottle 片剂斑点
tabletop 桌面
tabletop conveyer 桌面输送机
table-top glass 玻璃餐具
table-topped 顶上平的
table-top tripod 台式三脚架
table-top unit 台式仪器
table torque 转盘转矩
tablet pattern 输入图形
table transform 表变换
table travel 工作台行程
table traverse 研磨机台的往复运动；工作台横向进给
table-traverse screw 工作台进给螺杆
table tripod 台(式)三脚架
tablets press 造粒机
tabletting 压电
tablet(ting) machine 压片机
tabletting press 压片机
table-tunring mechanism 转台机构
tablet weight variation 片剂重量差异
tablet with the key 钥匙路牌
table-type buffing machine 台式抛光机
table type jack 台座式千斤顶
table type low and medium speed centrifuge 台式低中速离心机
table vertical movement 工作台垂直运动

table vibration 台振
table vibrator 振动台；台式振动器
table vice for wood working 木工台虎钳
table vise 台虎钳
table viewer 台式投影仪
tableware 桌上用具；餐具
tableware and domestic glassware industry 日用玻璃制品工业
tableware detergent 餐具洗涤剂
tableware disinfection 餐具消毒
tableware dryer 餐具干燥器
tableware washing and sterilizing equipment 餐具清洗消毒设备
table worm gearing 工作台涡轮传动装置
tabling 摇床选矿(法)；摇床；制表；嵌合；淘汰盘选矿；餐巾；帆的加强边；造册；压顶；墙帽
tablinum 通道(古罗马建筑正厅)；家谱室(古罗马)
tablite 板石；钠板石
taboleiro 沿海平顶方山
taboo facility 安全装置
taboo probability 禁止概率
tabo(u)ret 圆凳(无扶手和靠背柱形的)；轻便小支架；带脚轮小橱
tabs 加强片；机头布
tab sequential 列表顺序
tab sequential format 分隔符顺序格式；标记顺序格式
tab terminal 焊片引出线
tab test 翼片试验
tabula 横隔
tabular 列表的；平板状(的)；片状；表列；表格式的；扁平的；板状的
tabular alumina 片状氧化铝
tabula rasa 石板；洁白状；光板
tabular bone 扁骨
tabular book 表格(式)账簿；分栏簿
tabular bridge 箱形板梁桥
tabular calculation 表格计算法
tabular calculator 表式计数器
tabular cash book 表格式现金账簿
tabular chart 表列图
tabular compilation 表格汇总；表格汇编
tabular computation 表格计算
tabular cross-bedding 板状交错层理
tabular crystal 片状晶体；片状结晶
tabular data 表列数据；表定诸元
tabular difference 表差
tabular display 表格显示
tabular drawing 表格图；附尺寸表的图纸
tabular fault 平台断层
tabular ferric concretion 板状铁结核
tabular figure 表列数据
tabular form 报表
tabular format 表成格式
tabular freeboard 基本干舷
tabular girder 箱形板梁
tabular iceberg 平顶(的大)冰山；桌状冰山
tabular ice crystal 板状冰晶
tabular information 表列资料
tabular interpolation 表的插值
tabular interpretive program(me) 列表翻译程序；译表程序
tabularium 档案保存所(古罗马)
tabular language 表列语言
tabular ledger 表式分类账
tabular matter 表格排版
tabular ore body 板状矿体
tabular parenchyma 板状薄壁组织
tabular pile 管桩
tabular reef 平台礁
tabular representation 图表；表格表示法
tabular root 板状根
tabular spar 硅灰石
tabular standard 法定指数本位制；物价指数表
tabular structure 板式结构；板状构造
tabular surface 平面
tabular texture 板状结构
tabular type model 表格式模型
tabular type of variable budget 多栏式变动预算
tabular value 表值；表列(数)值
tabulate 制表；列表；平面的；平板状(的)
tabulate coral extinction 床板珊瑚绝灭
tabulated altitude 表列高度
tabulated azimuth 表列方位

tabulated azimuth angle 表列方位角
tabulated data 列入总汇表的数据；表列数据
tabulated form 表格形式
tabulated interpolation 列表插值
tabulated or charted reading 表列或图上读数
tabulated pedigree 附表系谱
tabulated quotation 行情表
tabulated solution 表解
tabulated value 表列(数)值
tabulating card 穿孔卡片
tabulating equipment 制表设备；卡片设备；编表设备
tabulating machine 制表机；列表机
tabulating method 列表法
tabulating stop 制表挡板
tabulation 制表；造册；列表；分类列表；编表
tabulation analysis 列表分析
tabulation method 列表法
tabulation of bill 开账单
tabulation of mixture 混合物配料表；混合料配合表；配料表
tabulation set 制表装置
tabulator 制表人；制表机；列表机构
tabulatum 墙裙；木楼板(古罗马)；顶棚
tabun 塔崩
Tabus massif 塔巴斯地块
T-abutment T 形桥台
tab washer 爪式垫圈；舌片垫圈
Tacan 战术空中导航系统
Tacan antenna 塔康导航天线
tacciometer 表面粘力计
tacharanite 易变硅钙石
tache 斑点；黑点；瑕疵
tache noire 黑斑
tach(e)o-generator 测速传感器；测速发电机；转数表传感器
tacheograph 速度记录器
tacheometer 转速计；测速仪；测距计；速测仪；视距仪
tacheometer shot 视距瞄准
tach(e)ometer singalling device 转速信号装置
tacheometer staff 视距(仪)标尺
tacheometer stop gear 视距仪制动装置
tacheometer theodolite 视距经纬仪
tach(e)ometer with gradient screw 倾斜螺旋视距仪
tacheometric(al) alidade 转速照准仪；视距照准仪
tacheometric(al) book 视距测量手簿
tacheometric(al) level 视距水准仪
tacheometric(al) nomograph 视距诺谟图
tacheometric(al) polygon 视距导线
tacheometric(al) prism attachment 视距光楔
tacheometric(al) protractor 视距仪分度器；视距量角器
tacheometric(al) (record) book 视距记录簿
tacheometric(al) rule 转速计算尺
tacheometric(al) ruler 视距计算尺
tacheometric(al) survey 速度测量；视距测量
tacheometric(al) surveying 视距测量术；视距测量法；视距测量术
tacheometric(al) table 转速计算表；视距计算表
tacheometric(al) theodolite survey 经纬仪视距测量
tacheometric(al) transit survey 经纬仪视距测量
tacheometric(al) traverse 转速导线；视距导线
tacheometry 转速测量法；转速测定法；转数测定法；速测；视距法；视距测量学；视距测量法；视距测量
Tacher 涂抹派(一种抽象画)
tachiol 氟化银
Tachism 行动派；场奇主义；抽象表现派
tachistoscope 视速仪
tacho-alternator 测速同步发电机；测速交流发电机
tachoderived pulse 测速导出脉冲
tachogram 转速记录图；速度记录图
tachograph 旋速记录器；自记转速计；自记速度计；流速图(表)；速度记录图；速度记录器
tachometer 旋速计；转速计；转速表；速度计；测速计；测速机
tachometer adapter 转速表传感器
tachometer chronograph 转速计时器
tachometer connection 转速表接头
tachometer dial 转速表盘
tachometer disc 测速盘

tachometer drive 转速表传动;转数表传动
tachometer drive cable 转速表传动软轴
tachometer drive gear 转数计传动齿轮
tachometer drive quill 转速表空心传动轴
tachometer dynamo 转速计用发电机;转速表用发电机
tachometer electric(al) indicator 转速表电气指示器
tachometer feedback 测速发电机反馈;转速计反馈;测速计反馈
tachometer ga(u)ge 转速计
tachometer generator 转数表传感器;测速发电机
tachometer head 测速磁头
tachometer indicator 转速表指示器
tachometer pulse 测速脉冲
tachometer sender 测速发送机
tachometer sending unit 转速表传动装置
tachometer sight 转速计瞄准器
tachometer signal 磁鼓脉冲信号;测速信号
tachometer stabilized system 转速计稳定系统
tachometer stop gear 转速表止动装置
tachometer testing table 转速表试验台
tachometer torquemeter 转矩转速测量仪
tachometer transmitter 转速计发送器
tachometer unit 转速表组
tachometer wheel 测速轮
tachometric(al) flowmeter sensor 转速表式流量传感器
tachometric(al) method 速度法
tachometric(al) record book 视距测量手簿
tachometric(al) relay 转速继电器
tachometric(al) survey(ing) 转速测量
tachometric electrometer 转速静电计
tachometry 转速测定法
tachometry method by fixed short base bar 短基线测距
tachomotor 测速电动机
tachyaphalite 硅钍锆石
tachyauxesis 加速生长
tachydrite 溢晶石
tachygenesis 快速发生
tachygraphometry 图示测速法
tachylite 玄武黑曜岩;玄武玻璃
tachylite basalt 玻璃玄武岩
tachylyte 玄武玻璃
tachymeter 速测仪;视距仪
tachymeter staff 视距标尺
tachymeter surveying 视距仪测量
tachymeter-transit 视距经纬仪
tachymetric(al) alidade 视距照准仪
tachymetric(al)(record) book 视距记录簿
tachymetric(al) table 视距计算表
tachymetry 速测法;速测;视距测量学;视距测量
tachymetry method by invar subtense bar 横基尺测距
tachyon 超光速子;超光子
tachyseismic instrument 速测地震仪
tachystoscope 速测器
tachytely 快速演化地
tacit acceptance 默示承诺;默认承诺;默契的接受
tacit agreement 默认协议;默契的协议
tacit consent 默许
tacit law 习惯法(规)
tacit renewal 延期或重订;默认合同续订;默契转期;默契的转期
tacitron 噪声闸流管;噪声闸流管
tacit understanding 默契
tack 抢风行驶;抢风;平头钉;图钉;初黏度
tackability 黏着能力
tack and tack 正顶风行驶
tack board 软木布告板;(软木制的)布告板;布告牌
tack bolt 定位螺栓;装配螺栓
tack claw 平头钉拔除器;钉爪;拔钉器
tack coat 沥青黏结层;黏(结)层
tack coater 结合层喷涂器
tack coat(ing) 黏结层
tack course 黏结层
tack driver 平头钉(自动)敲打机
tack dry 黏干状态
tacked spot 定位焊点
tack eliminator 脱黏剂;防黏剂
tacker 定位搭焊工
tack free 指压干;不发黏干;不剥落的
tack-free dry 指触干燥

tack-free time 指触干燥时间(涂层);指压干时间;表干时间
tack hammer 平头锤;平头(钉)锤
tack hole 铆钉孔
tackifier 胶合剂;黏着剂;增黏剂
tackiness 胶粘性(密封剂表面的);黏性
tackiness ability 发黏
tackiness agent 胶粘剂
tackiness paper tape 浸胶纸带;胶带纸
tackiness resistance 抗黏度
tacking 暂时缝上;临时钉住;紧钉;换抢;定位铆;定位焊
tacking dust pollution 粉尘综合治理
tacking mortgage 优先受偿的抵押
tacking rivet 接合铆钉;结合铆钉;定位铆钉
tacking screw 装配螺钉;装配螺栓
tackle 卸货索具;用具;辘轳;绞辘;船具滑轮组;滑车
tackle-block 起重滑车;钢绳滑车;滑车组;复滑车;滑轮组;滑车
tackle burton 复滑车;辘轳;三饼滑车组
tackle-fall 吊索;辘绳;复滑车索
tackle gear 卷扬设备;滑车齿轮;起重装置
tackle hook 提引钩;大钩
tackle-house 船上起重设备间;装有起重滑车设备的房屋
tackle principle 吊钩原则
tackle rigging 滑车设备
tackless strip 防滑压条;地毯卡条
tackling clinker ring 克服后结圈
tackmeter 黏性计
tackness 胶粘性
tack of priming 沥青黏结层
tack point 黏着点
tack producer 增黏剂
tack-producing agent 增黏剂
tack puller 拔图钉的拔钉器
tack rag 揩拭灰尘用的破布;黏性纱布;抹布;漆布;揩布
tack range 黏性期
tack rivet 装配铆钉;临时定位铆钉;紧缝铆钉;平头铆钉
tack room 马具室
tack screw 装配螺栓
tack strength 黏着力
tack system 流水作业
tack temperature 发粘温度
tack weld 断续平头焊接;预焊;临时点焊;定位焊缝;定位焊点;定位点焊
tack welder 定位焊机
tack weld(ing) 平头焊接;定位焊;暂焊;定位焊接;点固焊;点定焊
tacky 发黏;未干透的;胶粘的;发黏的
tacky dry 干燥黏着性;未干的;涂层发黏期;干后黏性
tacky producer 黏附剂
tacky resin 胶粘用树脂;胶粘树脂
tacky state 黏着状态
tacky surface 胶粘面;发黏表面
tac-locus 互切点轨迹
Taconic orogeny 太康造山运动(晚奥陶世)
taconite 铁英岩;铁燧岩;铁燧石
tac-point 自切点;互切点
tacryl 丙烯纤维
tacsatcom 军用通信卫星
tact correlation 事实相关
tactical air control center 战术空中管制中心
tactical air navigation 战术空中导航设备
tactical air navigation system 战术(空中)导航系统
tactical communication satellite 战术通信卫星
tactical communications system 战术通信系统
tactical data communication center 战术数据通信中心
tactical diameter 回转直径;回转初径;回转半径
tactical efficiency 战斗力
tactical locality 战区
tactical map 战术地图
tactical pilotage chart 战术航空图
tactical range 战术航程
tactical scheme 图上战术作业
tactical setting 态势图
tactical situation 态势
tactical situation display 战术情况显示器
tactical symbols 部队番号
tactical target 战术目标
tactical target illustration 战术目标图
tactical target materials program(me) 战术目标

资料图表
tactical trial 回转试验
tacticity 立构规整度;等规性
tactic polymer 有规立构聚合物
tactics 策略
tactile disk 触盘
tactile hairs 触毛
tactile hallucination 触幻觉
tactile keyboard 接触式键盘
tactile lever 触感测量杆
tactile menisci 触盘
tactile sensor 触感传感器
tactile trace instrument 触觉追踪仪;触觉示迹器
tactite 接触岩
tact of production 生产节拍
tactoid 类晶团聚体;触液取向胶
tactosol 有规溶胶
tactron 冷阴极充气管
tact system 流水作业线
tact-system production 节拍装配生产方式
tact time 节拍时间
tact timing 生产节拍时间计算
tactual map 触觉地图
Tacub jig 筛下气室跳汰机
tadem office 汇接局
tadienated fish oil 环戊二烯化鱼油
tadienated soybean oil 环戊二烯化豆油
tadpole 毛圈织物
tadpole plot 蝌蚪图;箭头图
Tadzhik basin 塔吉克盆地
tadzhikite 塔吉克矿
tael 两(重量单位)
taele 冻土
taenia 束带饰
taeniolite 带云母
taenite 镍纹石;天然铁镍合金
taep end detection 带端点检测
Tafel slope 塔费尔斜率
taffarel 船尾上部
taffare log 拖曳式计程仪(船尾舷)
tafferel 船尾上部
taffeta 波纹绸;塔夫绸
taffeta weave 平纹编组织
taffrail 船尾栏杆
taffrail log 拖曳式计程仪(船尾舷);拖曳式航速计;船速仪
taffrail recorder 拖曳式计程仪记录器
tafoni 侵蚀孔;石窝;风化穴
tafrogeny 断裂运动;地裂运动
taft joint 插入焊接头
tag 紧跟在后头;签条;电缆终端接头;标签;标价牌;标记物;标记
tag block 终端板;标签接线板
tag card 特征卡(片);标识卡(片);标签卡(片)
tag card reader 特征卡阅读器
Tag closed (cup) tester 泰格闭杯闪点试验器;泰格闭杯点试验器
tag cloth 标签布
tag converting unit 标签变换器
tag end 末尾
tag end test 末端测试
tageranite 等轴钙锆钛矿
tag field 标记字段;标记域
Tag flash point test 泰格闭杯闪点试验
tag format 标志格式;标记格式
tagged approach 保护的识别原则
tagged atom 示踪原子;标记原子;标记单子
tagged compound 标记化合物
tagged detergent 标记洗涤法
tagged element 标记元素
tagged fish density 标志鱼密度
tagged isotope 标记同位素
tagged molecule 示踪分子;标记分子
tag(ged) word 特征字;标签字
tagger 装箍的人;贴标签人
tagger plate 极薄镀锡钢板(0.18毫米以下)
taggers 薄铁皮;极薄铁皮
taggers tin 镀锡薄钢板
tagging 轧尖;加标签;锻尖;标志放流
tagging method 跟踪法;示踪法;标志放流法;标记法
tagging reader 特征阅读器
tag gun 销钉汽笛
tagilie 假孔雀石

tagliere 浅器皿
tag line 标记线;跟踪线;拉索;带钩绳索;安全索;抓斗稳定索;防漂缆
tag-line method 牵线法
tagline survey 牵索测量
tagma 胶粒
tagman 起重机操作工
tag marker 特征穿孔机
tag mark(ing) 特征记号;标志符号
tag-on-vehicle 车辆贴标签
Tag open-cup tester 泰格开路闪点试验器;泰格敞杯试验器
tag or sticker method 贴标签法
Tag-Robinson colo(u)r 泰格—鲁宾色度
Tag-Robinson colo(u)rimeter 泰格—罗宾逊比色计
tag sale 标签出售
tag slot 标识区;标记存储槽
tag sort 特征分类;标记排序;按特征分类
tagstone 片石
tag storage 特征标记存储器
tag swaging 锻头
tag system 特征系统;标签系统
Tagu glaciation 大姑冰期
tag-value compilation 标识值编组方式
tag wire 终端线
taiga 寒温带针叶林
taiga climate 泰加林气候
Taihu stone 太湖石
tail 限定财产继承权;高沸点溶剂馏分;末尾;脉冲后的尖头信号;滑车尾索;谱带尾;尾状物;尾状条纹;绳尾;色谱技术;盾尾;地脚;岛后沙嘴;岛后沙坝;峰尾
tail anchor 牵引缆索的尾部锚定;尾锚
tail and stop light 尾灯和停车灯
tail arm 挺杆
tail assay 谱带尾色谱技术;尾料分析
tail bay 端跨;下游河段;最后节间;最后开间;后段船闸(水位最低);尾水池;端跨(度);最后间间;尾水河段;下闸首
tail bay wall 尾水池墙;船闸后庭墙
tail beam 梁尾端;端梁;尾(端挑)梁(墙内突出部分)
tail bearing 梁尾端支承;尾支承;尾轴承
tail block 末端滑轮;回空索滑车;后顶尖座;带尾滑车;带索滑车;尾轮
tailboard 跳汰机溢流堰板;后挡板;尾板;车尾板
tailboard gritter 尾板式铺砂机
tailboard scraper 尾板式铲运机
tail bolt (固定石棉屋面板下端的)螺栓
tail bone 尾骨
tail boom 尾桁
tail-box 尾矿槽
tail bracket 尾(轴承)架
tail builder 尾部成型床
tail cable 尾绳
tail car 尾部支点
tail cavity 盾尾空隙
tail centre 尾架
tail chain 尾链;带钩短链
tail channel 出流渠;退水渠;尾水渠
tail clearance of shield 盾尾空隙
tail clipping 后沿斩断;后沿削波
tail clutch 尾离合器;推进器轴离合器
tail collision 尾追冲突
tail conduit 尾管
tail cone 机身末端;尾锥体;尾部整流锥
tail core print 榫尾芯头
tail-coverts 覆尾羽
tail crab 尾绳绞车
tail cup 尾罩;尾帽
tail current 尾电流
tail cut 橡尾锯口
tail damping circuit 尾部阻尼电路
tail-down 机身朝下
tail drive 机尾传动
tail drum 尾绳滚筒
tailed coil 端头板直的带卷
tailed peak 拖尾峰
tail effect 尾端效应
tail end 尾(轴)端;筛尾;末端
tail-end booster 线路终端升压器
tail-end-drive conveyer 后鼓轮驱动的运输带
tail-end process 尾端过程
tail-end treatment 尾端处理

tail-end winch 尾端绞车
taileron 尾部升降副翼
tail fin 尾翼;尾翅;副翼
tail first landing 单点落地(飞机)
tail flap 襟翼
tail fraction 最后馏分;尾部馏分;残余馏分
tail gas 尾气
tail gas absorber 尾气吸收塔
tail-gas analyzer 尾气分析器;尾废气分析器
tail gas buffer 尾气缓冲罐
tail gas condenser 尾气冷凝器
tail gas recovery tower 尾气回收塔
tail gas tower 尾气塔
tail gas treatment 尾气处理;废气处理
tail gate 下游闸门;车尾木板;下(首)闸门;后段船闸闸门;尾(水闸)门;船闸下游闸门
tail-gate 尾板;船闸下游闸门
tail-gate chain 可折后栏板链
tail-gate-chain bracket 可折后栏板链托架
tail-gate-chain catch 可折后栏板链扣
tail-gate hinge-rod 可折后栏板铰链杆
tail-gate ramp 船尾门跳板;滚动装船(斜坡)跳板;汽车轮渡跳板
tail-gate spreader 尾部挡板式石屑撒布器
tail-gate step 可折后栏板踏阶
tail gearbox 尾部减速器
tail-gear spindle 尾轮轴
tail hammer 杵锤
tail head 铆钉镦头;尾水头
tailheaviness 尾部重心
tail-hood 尾盖;尾钩;尾端弯钩
tailhook 尾钩
tailhook spring 尾钩弹簧
tailing 牢固嵌入墙身(悬臂构件);砖石墙挑出块;(砖在墙内的)嵌入部分;谱尾;尾状物;尾部操纵;衰减尾部
tailing brick 尾矿砖
tailing chute 粗粉溜出槽
tailing chute clamp ring 粗粉溜出槽夹紧环
tailing chute flange 粗粉溜出槽法兰
tailing column 尾塔
tailing down 嵌固端;砌入端;固定端
tailing factor 拖尾因子
tailing filling method 尾砂充填法
tailing fraction 尾馏分
tailing hangover 拖尾影踪
tailing in 砌入端;嵌固端;固定端
tailing-in work 收尾工作;收尾工程
tailing iron 悬臂固定钢构件;墙砌型钢;悬臂构件固定铁;墙面挑出铁件
tailing launder 尾矿槽
tailing outlet 粗粉出口
tailing-out rods 拧卸及摆管
tailing peak 拖尾峰
tailing reducer 减尾剂;尾部减缩剂
tailings 尾渣;尾砂;尾馏分;尾料;尾矿(砂);尾粉;尾材;筛余料;粗粉;残渣;残油
tailing sampling 尾矿采样
tailings apron 尾矿排放边
tailings area 尾矿场
tailings assay 尾矿试金
tailings (back)fill 尾矿充填料
tailings bin 尾矿仓
tailings cake 尾矿滤饼
tailings chest 尾浆槽
tailings classifier 尾矿分级机;粗粉分离器
tailings cone 分出粗粉的圆锥体
tailings control 筛余量控制
tailing screw flight 尾部螺纹
tailings dam 尾矿坝;尾矿池;尾矿坝
tailings discharge 尾矿排卸口;粗粉出口
tailings disposal 筛尾处置;尾矿处置;尾矿处理
tailings disposal area 尾矿排放区
tailings elevator 尾矿提升机;尾矿吊斗
tailings fill dam 尾矿坝
tailings filter press 尾矿压滤机
tailings grade 尾矿品位
tailings heap 尾矿堆
tailings impoundment 尾矿库
tailings in flo(a)tation 浮选的尾渣
tailings launder 尾矿排放槽
tailings of separator 选粉机粗粉
tailings ore brick 尾矿砖
tailings outlet port 尾矿排出口

tailings (overflow) weir 尾矿溢流堰
tailings pile 尾矿堆;尾矿场
tailings pond 筛屑坑;尾矿坑;尾矿池
tailings spout 尾喷口
tailings race 尾矿排出沟
tailings recycling 尾矿回收;粗粉循环
tailings reservoir building 尾矿池建设
tailings sample 尾矿试样
tailings screen 尾筛
tailings separator 粗粉分离器
tailings settling tank 尾矿沉淀桶
tailings side 尾矿排出端
tailings zone 尾矿带
tailing wheel 尾轮
tailing zero 尾零
tail into the derrick 拖入井架内
tail jack 机尾起重器;尾部千斤顶
tail jigger 带索滑车
tail joist 托梁支承的格栅;(搁置在短横梁上的)格栅;格栅尾端;挑出的格栅尾端
tail journal 尾轴颈
tail lamp 后灯;尾灯
tail-lamp base 尾灯底座
tailless nail head bonding 无引线钉头式接合
tail light 后灯;尾灯
tail-light socket 尾灯座
taillight switch 尾灯开关
taillight wiring conduit 尾灯导线包皮
tail line 尾绳
tail load 尾部荷载
tail loader 后悬挂装载机;后挂装载机
tail lock 尾水闸门
tailloir 圆柱头顶板
tail nut 尾管螺母
tail of a gale 风暴末期
tail of a hump yard 驼峰编组场尾部
tail of a shield 盾构尾部
tail of a stochastic process 随机过程的截尾
tail of a stream 潮流末期
tail of central bar 洲尾
tail of dog 挡块尾端
tail off 关机;发动机关闭;尾推力终止
tail of hump yard 驼峰调车场尾部
tail of river islet 江心洲尾
tail of shield 隧道护盾尾部
tail of the body 车身尾部
tail of the earth 地尾
tail of the queue 队尾
tail of wave 波尾
tail oil 最后馏分;尾油
tail on 船尾被抬起
tail optic(al) fiber 耦合光纤
tailorability 可修整性
tailored 特制的
tailored version 专用型
tailoring 剪裁
tailoring workshop 缝纫车间
tailor-made 特制的;定做的;定制的;专用的
tailor-made aggregate 特制集料;特制骨料
tailor-made column packing 定做的柱填充物
tailor-made microbe 特制细菌
tailor-made oil 精炼油;特制油
tailor-made tank-truck 非标准型油罐车
tailor's scissors 裁缝剪刀
tailor's shop 成衣铺
tailor's table blanket 裁缝用台毡
tailor's tape 裁缝尺
tail-out period 收尾时期
tailover 筛渣;筛除物
tail packing 盾尾密封材料
tailpiece 尾部;补白图饰;接线头;墙端短梁;尾管;尾端件;执柄;尾水管;加劲格栅
tail(-piece) beam 半端梁
tailpiece of the piston rod 活塞杆端
tail pipe 排气尾管;尾水管;尾喷管;尾管;泵吸管;尾部排气管
tailpipe center-piece 尾管中心件
tailpipe flame 尾管火焰
tailpipe nozzle 尾喷管
tail pit 仰开行桥平衡重竖井
tail plane 横尾翼
tail pond 尾水池
tail post (冲击钻机的)捞砂筒轴柱
tail pulley 尾轮;尾卷筒;尾滑轮;导轮

tail-pulley screw take-up 尾轮螺旋拉紧装置
tail pump 抽油机游梁传动泵;残液泵
tail race 水轮机水槽;尾水沟;出水道;尾水渠;排渣渠(矿渣)
tailrace 悬浮尾矿排出沟;排矿渣渠;尾水渠;尾水沟;尾流
tailrace adit 尾水渠入口
tailrace channel 尾水渠
tailrace dam 尾水坝;尾水池坎
tailrace deck 尾水平台
tailrace gate 尾水闸门
tailrace head 尾水水头
tailrace lining 尾水渠衬砌
tailrace platform 尾水平台
tailrace section 尾水渠段
tailrace slab concrete 尾水渠底板混凝土
tailrace surge basin 尾水调压池
tailrace system 尾水系统
tailrace tunnel 尾水隧洞;尾水隧道
tail race wall 尾水渠墙;下游尾水渠墙
tail radius of crane 后部旋转半径;起重机后部旋转半径
tail rake 杆状延长筛
tail reach 尾水段
tail recession 尾水消退
tail region 尾区
tail resistor 尾电阻器
tail reversing inductance 尾部变号电感
tail ring 挡料圈
tail rod 节制杆;尾杆;导杆
tail rod catcher 尾杆罩
tail roll 从动辊;尾磨
tail rope 拖索;尾索;尾绳
tail-rope system 尾绳运输系统;尾绳运输法
tail rotor 尾桨;尾螺旋桨
tails 尾矿;手指状喷涂图案
tail seal 盾尾密封;盾尾封闭
tail seal greasing 盾尾密封油脂
tail sealing of shield 盾尾密封
tail section 尾部
tail series 零头系列
tail setover 顶尖座跨度
tail shaft 尾轴;尾干
tail shaft governor 尾轴调速器
tail shaft packing 尾轴填料涵;尾轴密封填料
tail shaft passage 尾轴隧
tail-sharpening inductor 后沿锐化线圈
tail sheave 尾绳轮;尾轮
tail sheet 尾筛
tail shock wave 尾激波
tail signal 车尾号志
tail sitter 垂直起落飞机
tail-skid 尾橇
tail skid bar 尾橇杆
tail-skid shoe 尾滑瓦
tail slide 尾冲
tail slip 尾冲
tail soaking test 浸尾试验
tail spar 尾柱木
tail spindle 尾架顶尖套;顶尖轴
tailstock 机尾架;尾架;托柄尾部;床尾顶尖座;尾座;滑轮活轴
tailstock barrel 顶针套;床尾轴
tailstock base 顶尖座板
tailstock center 尾架顶尖
tailstock-center sleeve 尾架顶尖套筒
tailstock clamp-bolt 顶尖座夹紧螺栓
tailstock guide 尾座导轨
tailstock quill 尾架顶尖套筒
tailstock sleeve 顶尖座套筒
tailstock trimmer 靠墙格栅尾座梁;尾座短横梁
tail stripper 带卷直头机
tail strop 带尾滑车带
tail surface 尾面
tail swing 尾部回旋空间;尾摆度
tail symbol 尾部符号
tail tackle 带尾绞辘
tail tag 尾标
tail tank 尾水池
tail tank of condenser 冷凝器的尾槽
tail throat of a hump yard 驼峰调车场尾部
tail-to-tail linking 尾尾连接
tail-to-tail structure 尾接结构
tail to tide 锚泊中船随着潮水涨落而回转

tail tower 尾塔
tail tree 尾木
tail trimmer 贴墙托梁;墙端短托梁;短横梁
tail tube 尾水管;尾管
tail undercarriage 尾部起落架
tail up and down the stream 锚泊中船随着潮水涨落而回转
tail up on a rope 收紧绳索
tail valve 尾阀
tail vane 尾翼;尾舵
tail vat 尾池
tail vehicle 尾车
tail void 盾尾空隙;盾尾净空
tail wall 丁字(形)堤基凸部
tail warning 尾部报警(飞机);尾部警戒
tail warning radar 尾部警戒雷达
tail water 下游水;尾水(从建筑物的下游一端排出的水)
tailwater canal 尾水渠
tailwater channel 尾水渠
tailwater depth 下游水深(坝闸);尾水(水)深
tailwater elevation 下游水位;尾(水)水位;尾水高程
tailwater erosion 尾水冲刷;尾水冲蚀
tailwater gallery 尾水廊道
tailwater gate 尾水闸门
tailwater head 尾水水头
tailwater level 下游水位;尾水水位;尾水高程
tailwater pond 尾水池
tailwater power installation 尾水电站
tailwater power plant 溢流式水电站;尾水电站
tailwater rating curve 尾水水位流量关系曲线
tailwater recovery 尾水回收再用
tailwater recovery irrigation system 尾水回收灌溉系统
tailwater stage 尾(水)水位
tailwater velocity 尾水流速
tail wave 尾波
tail wheel 尾轮
tail-wheel boot 尾轮罩
tail-wheel bumper 尾轮减振器
tail-wheel fork 尾轮叉
tail-wheel landing gear 尾轮起落架
tail-wheel lock 尾轮锁
tail wind 顺风
tailwind approach 顺风进场着陆
tail wire 尾端线;出线
tail with the stream 锚泊中船尾朝着水流方向;锚泊中船随着潮水涨落而回转
tain qing glaze 天青釉
taint damage 沾污损害
tainter 弧形闸门
tainter gate 弧形闸门
tainter-gate movable dam 弧形闸门堰
tainter valve 弧形阀门
Taipingdao-Jindun'ansha depression region 太平岛-金盾暗沙拗陷地带
Taishan group 泰山群【地】
Taishan system 泰山系【地】
tai strip 滑行跑道
Taiwan cypress 台湾杉
Taiwan Strait 台湾海峡
Taj mahal 泰姬玛哈陵
Tajmyr geosyncline 太梅尔地槽
takanelite 塔锰石
take 收获量;成效量;付清
take a bath 投资失败
take a bearing 测方向;测方位
take a chance on wrenching bits out of round 拧管时丝扣拧坏
take a core 取芯;提取岩芯
take a drift 测航偏角
take a fall flier 冒险行事
take a hitch on 用绞车提取
take a lot of work 费工
take altitude 定标高
take-and-pay contract 提货与付款合同
take an offing 驶向海面
take apart 拆装;拆卸
take a photograph 摄影
take a picture 给井内事故打印痕
take a profiteering attitude 投机心理
take a range 测距
take a reading 读数
take a risk 接受保险

take a sample for examination and test 取样检验
take a share in 参与
take a spell 接班
take a stand 表态
take a step 采取措施
take a strain on pipe 夹住钻杆并拉紧
take a stretch on pipe 拉紧钻杆柱测量钻杆拉紧伸长量(以确定卡钻位置)
take a turn 转一圈
take-away belt (conveyer) 引出装置;接取装置
take-away capacity 卸货能力
take back a mortgage 收回抵押权
take back goods 退货
take bearing 定方位
take care of 保管
take charge of 接办;担任;负责人
take delivery 受领
take delivery of goods 提货
take departure 驶出
take-down 支取借款;拆毁;可拆卸部件;拆卸
take down connection 拆线
take-down square 可卸钢角尺
take-down time 拆卸时间;释放时间
take effect 生效;实施;施行
take effect as from this day's date 今天起生效
take elevation 抄平【测】
take full account of 充分估计
take good care of experimental fields 爱护农业试验地
take harbo(u)r 避入港湾
take hold 夹住
take-hold pressure 维持压强
take home engine 应急返航发动机
take-home pay 扣除捐税后的实得工资;扣除税捐等后的实发工资;实际工资;实得工资;实发工资
take in 收捷;工人装窑
take in an aircraft 引导飞机着陆方向
take-in device 夹具
take-in fork 装窑用叉子
take in oxygen 吸进氧气
take in sail 缩帆
take into consideration 加以考虑
take in tow 拖拽
take in water 装水
take it or leave it 自由选择;自由取舍
take legal action 起诉
take level(l)ing 抄平测量【测】
take levels 抄平【测】
take load up 装载
take means 采取手段
take measure of 测定
take measures to prevent losses in advance 事先防止损失
take nut 轴向调整螺母
take-off 材料数量表;揭;取出;起飞;起步;放水沟
take-off a heavy tax 免除重税
take-off airspeed 起飞空速
take-off altitude 起飞海拔高度
take-off and landing deck 起飞降落甲板
take-off angle 偏離角;出射角
take-off area 起飞地带
take-off belt 分选带;传送带
take-off board 起跳板
take-off circuit 供电分支电路
take-off clearance 起飞许可
take-off connection 分支接头
take-off control equipment 起飞控制仪
take-off coupling 分支接头
take-off device 卷取装置
take-off distance 起飞滑行距离;起飞滑(跑)距离
take-off distance available 起飞距离
take-off drift 起飞偏航
take-off end of the runway 跑道起飞端
take-off flight path 起飞航迹
take-off gate 放水闸门
take-off gear 接取装置
take-off ground run distance chart 起飞滑跑距离列线图
take-off heading 起飞航向
take-off horsepower 起飞马力
take-off/landing wheel sets 弹起和降落轮对
take-off line 分支管线
take-off machinery 接取装置
takeoff man 材料估算员;材料表编制者

take-off monitor 起飞指示器
take-off monitoring system 起飞监控系统
take-off noise of aircraft 起飞噪声
take-off path 起飞航迹
take-off pattern 起飞航线
take-off period 经济起飞期
take-off phase 起飞阶段
take-off pipe 取出管;放水管
take-off point 起飞机场;分接点
take-off position indicator lamp 起飞位置指示灯
take-off post 拉杆离合支柱
take-off power 起飞马力
take-off (propeller) shaft 输出传动轴
take-off rate 流出速度
take-off roll 起飞滑跑
take-off run 起飞滑跑距离
take-off run available 起飞滑跑距离
take-off runway 起飞跑道
take-off runway gradient 起飞跑道坡度
take-off safety zone 起飞安全区
take-off setting 起飞位置
take-off shaft 功率输出轴;传动轴
take-off site 起飞场
take-off speed 起飞速度
take-off spot 已测点;测知点
take-off strip 起飞跑道
take-off the-embargo 解除禁运
take-off the-fractions 取出馏分
take-off thrust 起飞推力
take-off tower 提升塔(架);出线塔
take-off weight 起飞重量
take on 担任
take-out 取出;债券买卖的现金盈差;取消银行贷款
take-out an overdraft 透支
take-out channel 排水渠;放水渠
take-out commitment 远期贷款承诺
take-out gate 放水闸(门)
take-out investor 以长期贷款易取短期贷款的投资者
take-out loan 房地产完工后的长期贷款
take-out of service 取出不用
take-out restaurant 外卖餐馆
take-out roller 提升辊
take-out service 外卖饮食店
take over 接替;接管;接办;接管企业
take-over bids 投标购买企业;出价盘进
take over for use 征用(土地)
take over-head 从塔顶取
takeover of land 征地
take over with 洽商
take place 发生
take precautions against earthquakes 防震
take proceeding against 起诉
take profit 取得利润
taker 计件矿工
take readings 取读数(仪表上)
take reel 卷线轮;卷线盘
taker-in 以证券为担保的贷款者
taker-off 工程计量员;读图估料员
take sample 取样
take set 得到残余变形
take shape 成型
takeshapes of closure 封头的成型
take-siding indicator 进入侧线指示器
take sight 抄平(测);测天,观测;瞄准
take something as a pledge 以物为证
take spell 换班
take steps 采取步骤
take stock 盘存;清理存货;清仓查库;盘货
take suction 吸取
take temperature 量温度
take the altitude 确定高程;定高度;测定高程
take the first place 名列第一
take the goods out of bond 提出被海关扣存的货物(完税后)
take the ground 搁浅
take the lead in 发挥带头作用
take the mean 取平均数
take the reading 记录仪表读数
take the sun's altitude 测太阳高度
take the weight 除去荷载
take the whole situation into account 统筹全局
take time 费时
take to piece 卸成部件;拆开

take tour 接班
take transport income as transport revenue 以运输进款作为运输收入
take turns 轮流;轮换;轮班
take turns system 轮换制
takeuchiite 塔硼锰镁矿
take-up 补漏;张紧;松紧装置;收线装置
take-up a bill 赎票
take-up an occupation 就业
take-up assembly 张紧装置
take-up bearing 调整轴承间隙
take-up bill 已付讫票据
take-up block 收线卷筒
take-up bolt 扣紧螺栓;夹紧螺栓
take-up by gravity 重力张紧(装置)
take-up case 卷带盒
take-up counterweight 拉紧装置配重
take-up device 张紧装置;拉紧装置;卷片装置
take-up film cassette 卷片暗盒
take-up gear (stand) 导出装置;卷取装置
take-up magazine 收片盒
take-up mechanism 张紧装置
take-up method (轧材的)收集或收材方法
take-up one's indentures 服务期满
take-up pulley 拉紧轮;拉紧滚筒;收线轮
take-up reel 卷带盘;接收盘;收带盘;引出轴
take-up roll 卷取辊
take-up screw 挠性传动张紧装置
take-up set 夹具;预加应力装置(钢筋)
take-up slack 收紧绳索
take-up spool 卷片轴;卷带轴;收片盒;收带盘;承片轴
take-up stand 收线架(钢丝绳)
take-up (stock) pan 接料盘
take-up strain 张紧应变
take-up tension 卷片拉力;卷带张力;安装张力
take-up the backlash 收紧背隙
take-up the-belt 张紧皮带
take-up the lost motion 收紧游隙
take-up the slack 收紧间隙
take-up torque 卷带转矩
take-up travel 拉紧行程
take-up tumbler 张紧轮
take-up unit 张紧装置
take-up water 吸水
take water 浪上船
taking a bath 发生一笔巨大损失
taking a departure 最后离港定位
taking angle 摄影角度
taking apart for inspection 分解检查
taking a squeeze 取模
taking civil action 起诉
taking disturbance sample 取样扰动样
taking disturbance soil sample 取样扰动土样
taking-down time 玻璃液冷却时间
taking-down work 拆卸作业
taking earth from borrow pits 堤旁取土
taking-in charge 接管
taking-in cock 进给旋塞
taking inventory 盘存
taking lens 取像透镜;拍摄镜头
taking of bearings 方位测定
taking-off 列工程项目(计算工程量的步骤之一);实船丈量;取数;估算工料
taking-off the embargo 取消封港
taking of samples 取样
taking on board 准许上船
taking-out and placing-in of cars 取送调车
taking-out and placing-in of passenger train stock 车底取送
taking out of service 列入报废
taking out turns 换倒铅
taking over an international standard 接受一个国际标准
taking-over certificate 移交证书;工程验收证书
taking-over of sections or parts 区段或部分的移交
taking position 拍摄位置
taking primaries 摄像三原色
taking rate 摄影速度
taking scale 摄影比例尺
taking steel as the key link 以钢为纲
taking substandard products as fine products 以次充好
taking topography 测地形

taking turns in S-shaped motion 轮对蛇行
taking ullage 测量油柜缺量
taking-up equipment 卷取设备
taking-up of play 设置游隙
taking wrong train 误乘;错乘【铁】
takir desert 龟裂漠境
Takla Makan 塔克拉玛干沙漠
takovite 水铝镍石
takspan 瑞典松木屋顶;瑞典木瓦
Takuma's boiler 田熊式锅炉
takyr 泥漠
takyr-like soil 龟裂型土
takyr soil 龟裂土
talasskite 高铁橄榄石
talat 水平冲沟
talayot (巴利阿里群岛史前的)出檐石塔
Talbot bands 塔尔博特带
Talbot duplex process 塔尔博特双联炼钢法
Talbot ingot process 塔尔博特液心钢锭轧制法
Talbot process 塔尔博特平炉炼钢法;塔尔博特方法
talc 滑石(粉)
talc carbonate rock 滑石碳酸盐岩
talc chalk 滑石石笔
talc-chlorite rock 滑石绿泥石岩
talc-chlorite schist 滑石绿泥石片岩
talc-chlorite schist deposit 滑石绿泥片岩矿床
talc deposit 滑石矿床
talc detritus 滑石碎屑
talc flour 滑石粉
talc glaze 滑石釉
talc grain 滑石小粒
talcite 变白云母;滑块石
talc-knebelite 镁锰橄榄石
talcky 含滑石的
talc lump 滑石块
talc lump and grain 滑石碎块
talc-magnesite 滑石菱镁岩
talc-micaschist 滑石云母片岩
talcoid 硅滑石
talcose 滑石的;含滑石的;似滑石的
talcose rock 滑石岩;滑石矿
talc(ose) schist 滑石片岩
talcosis 滑石沉着病
Talcott level 太尔各特水准
Talcott method 太尔各特方法
Talcott method of latitude determination 太尔各特测定纬度法
talc plaque 滑石斑
talc plastics 云母雕刻
talc pneumoconiosis 滑石肺;滑石沉着病
talc powder 爽身粉;滑石粉
talc pressure seal 滑石压力密封
talc rock 滑石岩
talc serpentine schist 滑石蛇纹片岩
talc slate 滑石板岩
talcum 滑石(粉)
talcum powder 滑石粉
tale 白云母薄片
talent bank 人才银行
talent development 人才开发
talented person 人才
tale quale 全数好质量
Tali glaciation 大理冰期【地】
talik (多年冻土上的)融区;融冻土层;(在永久冻土之间的)层间不冻层;不冻地
talipot palm 贝叶棕
talk and listen beacon 声号导航信标
talk-back 站场扩音对讲系统;应答
talk-back circuit 工作联络通话电路;联络通话电路;内线自动电话机;内部对讲电路
talk-back microphone 对讲传声器
talk-back system 回报系统
talk between ships 两船间对话;船舶间通话
talk channel 通话电路
talk content 滑石含量
talker echo 发话人回声
talk(er) key 通话电键
talking 岩层内部破裂声;通话
talking circuit 通话电路
talking radio beacon 音响无线电信标
talk-listen button 通话按钮
talk map 税务地图
talk shop 说行话
talks under contracts 合同谈判

talk to the line [rope] 根据钢绳冲程的特点确定钻头在孔底的工作状态
talktriplite 镁磷锰石
tall alkyd 妥尔油醇酸树脂;松浆油醇酸树脂
tall and slender structure 高耸结构;高柔结构
tallate 妥尔油酸酯;松浆油酸酯
tall block 高层建筑;高大楼房
tall block facade 高层建筑立面
tallboy 烟囱顶(上的瓦)筒管;高脚橱柜;五屉柜
tall building 高耸建筑;高大建筑(物);高层房屋
tall building design 高层建筑设计
tall building facade 高层建筑立面
tall-case clock 落地钟;立地钟
tall chimney 高烟囱
tallest 最高的
tallet 屋顶室;阁楼
tall field layer 高草本层
tall flats 高层公寓(英国)
tall-grass prairie 高草地
tall-grass steppe 高草草原
tall growing 生长高大的
tall guyed tower 高拉索塔;拉索加固的高塔
tall herbaceous layer 高草层
tall industrialized block 工业化建筑(预制)高层(塔架)公寓(英国)
talloel 高油
tall oil 妥尔油;松浆油;高油
tall oil fatty acid 妥尔油脂肪酸;松浆油脂肪酸
tall oil heads 妥尔油的初馏分;松浆油的初馏分
tall oil pitch 妥尔油沥青;松浆油沥青
tall oil rosin 妥尔油松香;松浆油松香
tall oil soap 妥尔油皂;松浆油皂
tallol 妥尔油
talloleic acid 妥尔油酸;松浆油酸
tallot 屋顶室;阁楼
tallow 结晶蜡石;牛脂;黄矿脂;动物脂;脂肪;油脂
tallow amine 牛脂胺
tallow cock 带栓润滑器
tallow compound 调配牛油(固体润滑剂)
tallow oil 牛(脂)油
tallow oleine 牛油软脂
tallowseed oil 乌桕油
tallow wood 红褐色桉树(澳大利亚昆士兰和沃尔斯产);乌桕木
tall pier 高桥墩
tall regular weighing bottle 高型称量瓶
tall slender column 高细长柱
tall stack source 高烟囱污染源
tall structure 高耸建筑;高耸结构
tall timber 偏僻地区
tallus 建筑物任何斜面
tallus wall 土坡挡土墙
tallut 阁楼
tall varnish 妥尔油清漆;松浆油清漆
tally 运算;记账符号;计numberboard板;清点;吻合;手执计数器;点数;标签;打记牌
tally book 理货记录簿;理货簿;计数簿
tally card 盘存计数卡
tally card information 计算卡材料
tally certificate 理货证明书
tally clerk 理货员;点货员
tally cliker 计数器
tally counter 检尺计数器;计数器
tally down 减
tally function 计数功能
tallying fee 理货费
tally items 现货工作项目及内容
tally keeper 理货员
tallyman 拿样品卖货的人;理货员;木材检尺员;点货员;分期付款商
tallyman's association 理货员协会
tally mark 骑缝号
tally on 船尾被抬起
tally order 总结指令;结算指令
tally plan 分期付款(英国)
tally register 计数器
Tallyrondo 椭圆度检查仪
tally room 理货间
tally sale 计算交易额
tally sheet 检验单;计数单;盘存计数单;点数单
tally shop 赊销商店;分期付款方式售货的商店(英国)
tally slate 成材石板块;论数石板瓦;论块出售石板
tally sticks 理货筹码

Tallysurf oxide measurement 泰勒雪夫氧化物测量
tally system 赊销法;赊卖;分期付款赊销法
tally the cargo 点货
tally the trade 赊销
tally trade 赊账贸易;赊销贸易;分期付款赊销法
talmessite 砷镁钙石
Talmi gold 镀金黄铜
talnakhite 硫铜铁矿
talon 息票调换券;息票;爪形条纹;爪饰;附单
Talon fixing 塔能固件(一种专利的镀锌墙板紧固件)
talon mo(u)lding 双曲线形脚;葱形饰线条
talons 爪形条纹;鸽胸雕花
taltalite 电气石
talus 悬岩下的屑堆;岩堆;塌砾;碎屑堆;山麓碎石(堆);山麓堆积;山脚阶积;倒石堆;坝脚抛石
talus accumulation 坡下堆积物
talus apron 坡积裙
talus breccia 山麓角砾岩;倒石堆角砾岩
talus clastics 山麓堆积碎屑物质
talus cone 岩锥;岩屑锥;岩堆;岩堆锥
talus creep 岩屑下滑;岩屑蠕滑;岩屑蠕变;山麓堆积体蠕动
talus deposit 岩锥堆积土;岩屑堆积;崖堆;坡积矿床
talus fan 岩屑扇;坡积裙;冲积扇
talus fine 山麓堆积细粒物质
talus glacier 岩屑流;岩石流;岩石冰川;崩塌冰川
talus material 岩屑堆;坡积物;坡积层;石屑堆
talus ruoble 崩落
talus slide 岩屑滑坡;岩堆滑坡;堆积体滑坡
talus slope 岩屑坡;岩堆坡
talus spring 岩屑泉;山麓堆积泉;堆积泉
talus train 岩屑流
talus wall 挡土斜墙;竖斜面挡土墙;陡壁竖斜面墙;斜面墙;竖斜面墙;收分墙;堆石护岸
talus yard 堆积物
Talwalker-Parmelee plasticity index 塔沃克-珀米利塑性指数;塔—珀二氏塑性指数
talweg 水谷线;主航道中心线;谷底线;河谷线;河道深泓线
Talysurf 粗糙度检查仪;表面光度仪
tam 活塞
tamalin 缅甸郁金香木;缅甸红色黄檀
Tam alloy 塔姆铁钛合金
Tamanori 塔马努里黏合剂(商品名)
tamarack 落叶松(木);北美落叶松
tamarisk 怪柳
tamarugite 斜钠明矾
tambac 沉香
Tambac alloy 顿巴克黄铜合金
tambour 圆形屋顶的柱间墙;鼓形柱座;鼓形柱身
tambour for embroidery machine 刺绣机绷子
tambour-making machine 刺绣品加工机
tame dry 干牧草
tame forage grass 栽培牧草
tame grass mixture 混播牧草
tame pasture 人工草地
taming small rivers 小流域治理
tamis 滤布
Tammann temperature 塔曼温度
Tammann tube 碳粒发热管(石墨管)
tammy 滤汁布;滤布;布筛
tamo 日本槐木
tamp 填实;填塞炮泥;捣棒
tamp a hole 填塞炮眼
tamped backfill 回填(土)夯实;夯实回填土;捣实回填(土)
tamped bottom 打结炉底
tamped concrete 夯筑混凝土;夯实混凝土;捣实混凝土;捣固混凝土
tamped finish 夯实整修面
tamped glass 捣碎的玻璃
tamped joint 混凝土浇件伸缩缝;混凝土浇件压缩缝;填塞接头;浇筑混凝土时的施工缝
tamped liner 捣炉衬
tamped lining 捣实式炉衬
tamped slope of earth dam 路堤边坡夯实;土坝夯土坡
tamped tie 捣实轨枕
tamped volume 装填体积
tamped well 填实孔;填塞炮孔(用炮泥)
tamper 夯实器;夯实机;夯具;夯;填塞者;塌缩隔离器;堵塞物;捣固器;捣固机;打夯机
tamper bar 炮棍;捣棒;振压棒
tamper(beam) finisher 夯板平整机

tamper concrete 夯捣混凝土
tamper-crane rammer 落锤式夯实机
tampered cinder 煤渣拍实
tampered concrete 捣实的混凝土
tamperer 填炮眼工
tamper finisher 混凝土整修机;捣杆平整机
tamper head 捣固机镐头
tampering 收买
tamper-leveller 捣固整平机
tamper-proof 防干扰
tamper-proof spring module 防干弹簧组件
tamper roller 夯实滚;羊足碾;羊脚碾
tamp forging 压锻
Tampico embayment 坦皮科湾盆地
tampico rolls 镀锡薄钢板的除油辊
tampin 开口木楦(铅管);管木楦;黄柳木捣塞
tamping 重锤夯实;夯实;夯打;筑;捣实;捣塞;捣固;打夯
tamping anchor 捣固锚栓
tamping backfill 夯实回填土
tamping bag 填装袋
tamping bar 砸道锤;夯棍;炮棍;填塞杆;塞药杆;捣实锤;捣棍;捣棒
tamping beam 捣杆
tamping beam finisher 混凝土整修机;捣杆平整机
tamping blade 捣固版
tamping bruise 捣固碰伤
tamping car 轨道捣固车
tamping-crane rammer 落锤式夯实机
tamping device 捣实工具
tamping drum 碾压滚筒
tamping foot (碾压滚筒上的)羊足
tamping foot roller 炮脚碾压机
tamping hammer 夯锤;捣固锤;捣锤
tamping in layers 分层夯实
tamping iron 铁撞柱;铁碾;铁夯;捣铁碾;捣板
tamping-level(l)ing finisher 捣固整平机(混凝土);整平捣机;振捣刮平机;混凝土整修机
tamping machine 压型机;砸道机;捣道机;落锤机;夯实机;夯拍机;捣固机;捣打机;打夯机
tamping mass 捣打料
tamping material 填塞材料;捣筑料
tamping of sleepers 夯实轨枕;捣固轨枕
tamping pick 砸道镐;夯镐;捣(固)镐
tamping plant 压型设备;捣筑设备
tamping plate 夯板
tamping plug 炮眼塞;炮泥
tamping pole 炮棍
tamping pole of jointed sections 段接炮棍
tamping rod 夯棍;夯杆;炮眼封泥棒;捣棍;捣固杆;捣棒
tamping roller 夯击压路机;羊足碾;羊脚碾;压土碾;夯捣式压路机;夯击碾
tamping shoe 捣蹄铁
tamping slab 夯板;捣板
tamping speed 打夯速度
tamping stick 装药棒;炮棍;塞药棒
tamping template 夯样板;捣场模板
tamping tool 夯实工具;夯具;捣固机;捣固工具
tamping-type roller 夯实压路机;夯击式压路机;夯击式(路)碾
tamping-vibrating design 振动捣实型
tamping weight 捣锤;夯锤;夯击重量
tampion 炮口塞;塞子
tamp machine 打桩机
tampon 棉塞;软布团(涂漆等用);填塞物
tampon cannula 棉塞套管
tampon forceps 棉球固定镊
tamponing 揩涂;搓擦;擦涂
tamp roller 夯击碾
Tamtam 塔姆锡青铜
tan 黄褐色
Tanagra figure 坦纳格拉无釉陶塑人像
tanalith 塔纳利斯防腐剂
tanalith process 木材防腐工艺
tanalized timber 防腐处理的木材;防腐木材
tanatarite 斜铝石
tanbark 树皮鞣料
tancoite 羟磷铝锂钠石
tan cup 坛盏
tandem 纵列;前后直排;串列布置;串列;串联压路机;串联(的)
tandem accelerator 串列式加速器
tandem ACME tap 阿克姆梯形扣丝锥

tandem arc weld 前后弧焊
tandem arrangement 前后纵列排列
tandem axle 串列(轮)轴
tandem axle drive 串列轴驱动
tandem axle load 串列(轮)轴荷载;双轴荷载;双轴载荷
tandem axles 双轴;串联(传动)轴
tandem bicycle 双人自行车;多人自行车
tandem blade 串列式叶片
tandem bladed inducer 串联诱导叶轮
tandem boost 轴向助推器
tandem-boost configuration 串联配置的助推器
tandem booster 轴向助推器;串联助推器
tandem-bowl scraper 串联斗(式)铲运机;串联铲运机;联斗铲运机;双轴驱动刮刀
tandem bridge 串桥
tandem brush 串联电刷
tandem capacitor 双组定片电容器
tandem center 中立旁通
tandem center valve 中间位置串通的阀
tandem chain drive 串联链条驱动
tandem cold-mill 串列式冷轧机
tandem cold strip mill 串列式带材冷轧机
tandem compensation 级联补偿
tandem compound engine 串排复缸蒸汽机
tandem compound flow turbine 串联双缸式涡轮机;串联双缸式透平;串联复式涡轮机
tandem-compound turbine 单轴多缸涡轮机;串联式涡轮机
tandem compressor 串排复缸压缩机
tandem concrete mixer 联列式混凝土搅拌机
tandem configuration 串列配置
tandem connected turbine 串联式涡轮机;串缸式涡轮机
tandem connection 串联
tandem cylinder 串列(式)汽缸;串联式汽缸
tandem data circuit 串行数据线路;串联数据电路
tandem die 串联模具;重合模;复式拉深模
tandem disc harrow 串联圆盘耙;复式圆盘耙
tandem disk harrow 双列圆盘耙;双行圆盘耙;二纵立式圆盘耙
tandem dock 串联式船坞
tandem double-flow turbine 单轴双排汽轮机
tandem double focusing mass spectrometer 串联双聚焦质谱计
tandem drawing-machine 串列式拉丝机;串列式多次拉拔机
tandem dredge pumps 串联泥泵
tandem drive 前后传动;双轴驱动;串动;串轴驱动;串联(式)驱动;串联传动;前后传动链;双轴驱动链
tandem-drive conveyer 双轴驱动输送机
tandem drive housing 前后传动轴套;双轴驱动轴套
tandem driven 双轴驱动
tandem drive wheels 串联驱动轮
Tandem drying grinding plant 坦登串联式烘干机粉磨设备
tandem dumper 串联双翻车机
tandem duplex bearing 成对串联安装轴承
tandem duplication 连续重复
tandem electrostatic accelerator 串列静电加速器
tandem elevator 双层电梯
tandem engine 串联式发动机
tandem exchange 中继交换机;汇接局
tandem exchanger 汇接交换机
tandem exciter 串列励磁机
tandem fuel pump 串联燃油泵
tandem gas engine 串复缸瓦斯发动机
tandem generator 串联起电机
tandem go-devil 串联清管器
tandem grating pair 串联光栅对
tandem hoist 平行轴双绞筒提升机
tandem hoisting 串联接力提升
tandem inversion 连续倒位
tandem ion analyzer 串联式离子分析仪
tandem knife switch 串(联)闸刀开关
tandem launching 连续下水方法【船】
tandem lift 双层电梯
tandem line 串联辊涂线
tandem mass spectrometer 串联质谱仪;串联质谱计
tandem mass spectrometry 串联质谱法
tandem master cylinder 串联主油缸;串联双制动总泵

tandem method 顺序选择法
tandem mill 连轧机;串列(式)轧机
tandem mixer 联列式(混凝土)搅拌机;复式搅拌机;复式拌和机
tandem moor 长短纵列双锚泊
tandem motor 级联电动机;串列(式)电动机
tandem mounting 片式(串联)安装
tandem office 中继(电话)站
tandem operation 串联工作
tandem piston 串联活塞
tandem plan 接力计划
tandem position 直线排列;串联(式)配置
tandem-powered 串联驱动的;串联式发动机的
tandem-powered scraper 串联驱动式铲运机
tandem processor 串联处理机
tandem propeller 串置螺旋桨;串列螺旋桨;串联推进器
tandem propeller submarine 串联螺旋桨潜艇
tandem propulsion system 串联式推进系统
tandem pull 串列机组牵引架
tandem pulverizer 串联式碎土镇压器
tandem pumps 双联泵
tandem push loading 串联推动装载
tandem queue 串行队列;串联排队;多行排列
tandem reaction sequence 连续反应过程
tandem rear axle 串联后轴
tandem ripping push block 串联松土推块
tandem road roller 串列轮压路机;串列压路机
tandem roller 双轮压路机;双滚筒压路机;串联滚筒;串联式压路机;串联式碾压机;二轮压路机
tandem rolling 连轧;交叉轧制
tandem rope pulleys 串联皮带轮
tandem rudder 串联舵
tandem scraper 串列清管器
tandem sequence 串联多弧焊
tandem sequence submerged arc welding 纵列多丝埋弧焊
tandem set 串联机组
tandem sheet mill 串列式薄板轧机
tandem stabilizer 串联稳定器
tandem static roller 两轮静碾压路机
tandem steam roller 串联蒸汽碾压机
tandem stock option 分前后的股票认购权
tandem stranding machine 串列式捻股机
Tandem support system 串联液压支架推进法
tandem switch 汇接交换机
tandem switching 汇接交换
tandem system 串行系统;串联处理系统;串级系统
tandem tension packer 串联式拉伸封隔器
tandem tin-plate mill 串列式镀锡薄钢板轧机
tandem tractor 串联式拖拉机
tandem trunk 汇接中继线
tandem turbine 纵联涡轮机;级联透平;串联式汽轮机;串联汽轮机;串联复式透平机;串级式透平机
tandem turnout 串列道岔
tandem type 串联式
tandem-type connection 串联接法
tandem type pump 连接泵
tandem type road rollers 两轮压路机
tandem-type rollers 串联式滚柱
tandem unit 单轴机组
tandem vibratory roller 两轮振动压路机;双轮振动压路机
tandem welding 串列多点焊;串联多弧焊
tandem-wheeled spreader 后双轴四轮撒肥机
tandem wheels 双轴车轮;串联车轮
tandem wiper 双刷刮水器
tandem work 依次先后排列施工
tandem working 串联工作
tanden calender 成对砑光机
T and G connection 舌槽接合;企口接合
T and G joint 舌槽接缝;企口接缝
tandour (置于方桌下取暖的)火炉
taneyamalite 塔硅锰铁钠石
tang 插入手柄内部分;锉根;刀根;支索固定器;柄脚;柄根
tangage 纵摇
tangaite 铬磷铝石
tangear (method) 双刀盘切齿法
tangeite 钙钒铜矿;钒钙铜矿
tangency 相切;切触
tangency condition 切向运动条件
tangency point 切点
tangent 正切

tangent adjacement to switch rail 尖轨后插直线
tangent adjacent to toe of frog 辙叉趾端前插直线
tangent angle 正切角
tangent arc 晕切弧【气】;正切弧
tangent bend 双弯曲
tangent between curves 夹直线
tangent chart 正切曲线图
tangent circle 相切圆
tangent circle pattern 旋子彩画
tangent condition 切条件
tangent cone 切锥面
tangent-cone approximation 切线锥近似法
tangent-cone method 截锥法
tangent-cone of a qualric 二次曲面的切线锥面
tangent conic(al) projection 切圆锥投影
tangent curve 正切曲线
tangent cylinder 切柱面;相切柱面
tangent cylindrical projection 切圆柱投影
tangent distance 切线距离【铁】;切距
tangent error 正切误差
tangent galvanometer 正切电流计
tangential 相切;正切;切向(的);切线的
tangential acceleration 切向加速度;切线加速度
tangential actuator 切向调节器
tangential admission 切向进入;切向进汽;切向进给;切向供给
tangential angle 切角
tangential approach principle 切线进场法
tangential approximation (method) 切线逼近法
tangential at bore 与孔相切方向
tangential bar-mechanism 切线棒机构
tangential beam 切向束
tangential bearing 切向支承;弧形板支座
tangential belt 切向传动带
tangential bender 切线折弯机
tangential bending 切向弯曲
tangential bit 切向车刀;切线车刀
tangential bladespacing 切向叶片间距
tangential brush 切向配置电刷
tangential burner 切线式燃烧器;角式喷燃器
tangential casting 切线铸造
tangential channel 切向孔道
tangential chaser 切向螺纹梳刀
tangential circle 切圆
tangential close ray tracing 切向闭合光线跟踪
tangential column supports around equator 赤道正切型柱式支座
tangential combustion 切向燃烧
tangential complex 切线丛
tangential component 切向分量;切向分力;切线分量
tangential cone 切线锥面
tangential connection of curves 曲线共切点连接
tangential controller 切向控制器
tangential coordinates 切线坐标
tangential coupling 切向耦合
tangential cross-bedding 切向交错层理
tangential curvature 切向曲率;切线曲率;切面曲率
tangential cut 切线锯开
tangential cut ring 切向切环
tangential cutter 切向车刀;切线车刀
tangential cutter head 切向进刀架
tangential damping 切线阻尼
tangential deflection 切向偏移;切线挠距;切线挠度
tangential deformation 受剪应变;剪应变;切向变形
tangential derivative 切面导数;切向导数
tangential direction 切线方向
tangential discharge 切向卸料
tangential displacement 切向位移
tangential distance 切线支距;切线距离
tangential distortion 正切畸变;切向交错层理;切向畸变
tangential distortion of lens 透镜切向畸变
tangential division 切线分割法
tangential division 切向分裂
tangential division scale 切线分割规
tangential-entry type cyclone 切线进入式旋流集尘器
tangential equation 切面方程;切线方程
tangential erosion 切向侵蚀
tangential error 切向误差
tangential error of satellite orbit 卫星轨道切向误差
tangential fan 贯流式通风机

tangential fault 切向断层
tangential feed 切线进料;切线进给
tangential feed attachment 切向进给装置
tangential feed mechanism 切向进给机构;切向进刀机构
tangential feed method grinding 切向进给法;切向进给磨法
tangential feed worm hob 切向进给蜗轮滚刀
tangential field 切向场
tangential field curvature 切向场弯曲
tangential filtration 切向过滤
tangential firing 切圆燃烧;切向燃烧
tangential flow 切向水流;切向流动
tangential flow fan 切向流动式通风机
tangential flow hollow fibre ultrafiltration 切向水流空心纤维超滤
tangential flow turbine 切向流动式涡轮机;切向流动式透平;切向流动式汽轮机
tangential focal line 切向焦线
tangential focus 主焦点;正切焦点
tangential force 切向力;切线力
tangential form grinding 切向成型磨削
tangential formula 切线公式
tangential friction force 切线摩擦力;切向摩擦力
tangential frost-heave force (土的)冻胀切力
tangential frost heaving force 切线冻胀力
tangential frost heaving force test 切向冻胀力试验
tangential gate 切线内浇口
tangential grain type (木材层积组合方式中的)环弧型
tangential grinder 切向进给磨床
tangential head cutter 切向进刀架
tangential hob 切向滚刀
tangential hobbing 切向滚削
tangential hole 切向孔
tangential hole swirl atomizer 切线孔漩涡喷雾器
tangential image field 正切像场
tangential incidence 切向入射
tangential inlet 切向进口;切线进口
tangential intercept 切线截距
tangential isotropy 切向同性
tangential jet 切向喷管
tangential joint 切向接合
tangential junction 分线
tangential keys 切向键
tangential law 正切定律
tangential length 切线长(度)
tangential line 切线;正切线
tangential linear space 切线性空间
tangential line load 切向线载荷
tangential load(ing) 切向载荷;切向负荷
tangential maneuver 切向机动
tangential method 切线法;切向法
tangential method with coefficient 带系数切线法
tangential method with multipoint 多点切线法
tangential mirror 切向反光镜
tangential modulus 切向模量;切线模数;切线模量
tangential modulus 正切弹性模数
tangential modulus of deformation 切线变形模量
tangential modulus of elasticity 切线弹性模量
tangential motion 切向运动;切线运动
tangential motion condition 切向运动条件
tangential movement 切向运动;切向移动;水平运动
tangential nozzle 切向喷管;切向接管
tangent(ial) offset 切线偏移;切线垂距
tangent(ial) offset method 切线支距法
tangential partial turbine 切向非整周进水式水轮机
tangential path 直线段;曲线间的直线段;切向直线段
tangential pencil of rays 切向光束
tangential plane 切面;切线面
tangential point 切线点;切(向)点
tangential point load 切向点荷载
tangential point of a cubic 三次曲线的切线割点
tangential polar coordinate 切线极坐标
tangential porous wood 切向孔材
tangential pressure 切向压力;切线压力
tangential property 切线性质
tangential pump 切线增压泵
tangential ray 切向射线;切向光线
tangential ray focus 切向光线聚点

tangential reaction 切向反力;切线反力
tangential resistance 切向阻力
tangential resolution 切向分辨率
tangential resolving power 切向分辨率
tangential road 切线道路
tangential roentgenography 正切X线照相术
tangential roundabout 切线连接式环形交叉
tangential rule 切线法则
tangential run-off 切线延伸段;切线(超高)缓和段
tangential runout 切线缓和段;切线延伸段;切线(超高)延伸段
tangential runway 机场切向跑道;机场直向跑道
tangential sawing 切线锯木法
tangential sawn 弦锯的(木材)
tangential screw 切向螺旋;切向螺纹;微动螺旋;微动螺丝;切线螺旋
tangential section 弦切面;切向切片;切向截面
tangential shadow 切线影
tangential shear 切向剪力
tangential shearing stress 切向剪应力;切线剪应力
tangential shrinkage 切向收缩
tangential slip 切向滑距
tangential slope 切线(的)斜率
tangential slot 切向槽
tangential speed (component) 切向分速度
tangential spiral agitator 切向螺带式搅拌机
tangential spoke 切线轮辐
tangential spoke wheel 切线辐轮
tangential stiffness 切线刚度
tangential stiffness coefficient 切向刚度系数
tangential (storage) silo 切向储库
tangential strain 切向应变
tangential stress 切向应力;剪应力;切(向)应力
tangential stress in flange 法兰(中的)切向应力
tangential stress of the wind 切向风应力
tangential support 切线支座
tangential tensile-strength 切向拉伸强度
tangential thrust 切向推力
tangential tool 切向车刀
tangential tool-holder 切向刀夹
tangential traction 切向引力
tangential tractive force 切向牵引力
tangential turbine 正切透平;切向涡轮机;切向透平;切向水轮机;切向汽轮机;冲击式涡轮机;冲击式水轮机;冲动式涡轮机
tangential vane 切向叶片
tangential vane wheel type flowmeter 切线叶片式流量计
tangential vector 切向量;切线矢量
tangential vector bundle 切向量丛
tangential vector space 切向量空间
tangential velocity 切向速度
tangential viscous resistance 切向黏性阻力
tangential wall 切向壁
tangential (water-) wheel 切向水轮
tangential wave 剪切波;切向波
tangential wave path 切线波路径
tangential wedge 切向楔
tangential wheel 切涡轮机;冲击式水轮机;冲击水轮
tangential wind stress 切向风应力
tangent latitude error 正切纬度误差;纬度差
tangent length 切线长;接线长(度)
tangent meridian 切经线
tangent method 正切法
tangent modulus 切线模量;正切模量
tangent-odd circuit 奇切围道
tangent offset 切线支距
tangent of lead angle 导程角切线
tangent ogive 正切卵形线;切面尖拱
tangent parallel 标准纬线
tangent pile 支护桩;连续桩墙(地下连续墙用)
tangent plane 切平面;切面
tangent-plane projection 切面投影
tangent plane to a surface 曲面的切平面
tangent point 切点
tangent point of drill string 钻具切点
tangent point of the curve 曲线起讫点
tangent point retarder 直点式缓行器
tangent projection 切面投影
tangent-saw 切向锯
tangent-sawn 切线锯木法
tangent scale 正切(标)尺
tangent screw 切线螺旋;正切螺旋;微动螺旋;微

动螺丝;微调螺旋;微调螺丝
tangent section 支线段
tangent sight 正切瞄准具
tangent sphere 切球面
tangent surface 切削曲面;切曲面
tangent system 切面法
tangent tensor 切张量
tangent to curve point 直圆点
tangent to periphery 圆周切线
tangent to spiral 螺旋曲线起点
tangent to spiral point 直缓点
tangent tower 直线塔
tangent track 直向股道;直线线路
tangent track of open route 地面直线段
tangent track tunnel 隧道内直线段
tangent-tube wall 切接管水冷壁
tangent wedge 切向楔
tangent-wedge method 切楔法
tangent welding 多极平行弧焊;多极平衡弧焊
tangerine 橘树;橘红色(的)
T ang G ceiling 企口天花板
T ang G connecting 企口接合
T ang G joint 企口缝
T ang G subfloor 企口地板
tangible advance 有形效益;实际效益
tangible assets 有形资产;可计资产;实物资产
tangible assets turnover rate 有形资产周转率
tangible benefits 显著实惠;有形效益;有形收入;可计利益
tangible capital 有形资本
tangible cost 有形成本
tangible equipment cost 有形设备成本
tangible fixed assets 有形固定资产
tangible fixed assets per regular employee 按正式职工计算的固定资产装备率
tangible goods 有形商品;有形货物
tangible investment 有形投资
tangible losses 可计损失
tangible net worth 有形净值
tangible personal property 有形动产
tangible plant assets 有形工厂设备资产
tangible product 有形产品
tangible proof 可见证据;明确的证据;确切的证据
tangible property 有形财产
tangible returns 有形赢利
tangibles 有形资产
tangible value 有形资产价值;有形价值;实际价值
tangile 红柳桉;多子婆罗双(菲律宾产的桃花心木)
tangiwaite 透蛇纹石
Tang Jun kiln 唐钧窑
Tang kiln 唐窑
tangled shape 缠结形
tangled vegetation 密植度
tangle net 刺网;缠刺网
Tang sancai 唐三彩
Tang style horse 连座坐坡唐马
tanguile 红柳桉;多子婆罗双(菲律宾产的桃花心木);多子红柳安
Tang ware 唐窑
tank 箱池;油柜;液体舱;储[贮]气瓶;储[贮]槽;罐;容箱;水柜;储能电路;储藏容器;池
tank accessories 油罐附件
tankage 箱容量;油箱容量;油柜容量;油罐容量;装罐储[贮]存;装罐储[贮]存;罐容量;容器设备;水柜容量;池容量;槽容量;槽路电容
tankage arrangement 油水舱布置
tankage installation 油罐安装
tank agitator 油罐搅拌器
tank air 油罐中混合气;油罐空气
tank air-mover 油罐空气驱除器
tank anchoring 油罐的固定
tank and container clean 油罐与集装箱清洗
tank band 油罐卡带
tank barge 油罐驳;油槽驳船;油驳;罐柜驳;槽驳
tank base 油罐底
tank base joint 箱座接合
tank batch 油罐顶小门
tank battery 油罐组;油罐场;罐群
tank battery barge 油罐组驳船
tank bay 油箱舱
tank black 饮水舱用无味防腐黑漆
tank bleeding 从油罐放出水及沉积物
tank block 池壁砖
tank block bracing 池壁砖拉条

tank block course 池窑大砖层
tank-body 油槽车身
tank body truck 油槽汽车;油槽罐车
tank boiler 箱式锅炉;火管锅炉
tank bottom 油箱底;罐底(脚);池底;槽底
tank bottom block 储水池底垫块;储[贮]水池底垫块;池底砖
tank bottom chime 油罐底缘
tank bottom connection factor 罐底校正系数
tank bottom replacement 更换罐底
tank breather pipe 油罐通气管
tank breather tube 油罐呼吸管
tank breathing diaphragm 油罐呼吸隔膜;油罐的通气隔膜
tank bulkhead 液舱壁
tank calibration 油罐校正
tank cap 油罐盖
tank capacitance 振荡回路电容;槽路电容
tank capacitor 槽路电容器
tank capacity 水池容量
tank car 油罐车;油槽车;油杯润滑脂;罐车;水槽车;槽车
tank car blower 油槽车吹扫机
tank car bottom shell 油槽车底壳
tank car clearance point 罐车卸油点
tank car dome 油槽车穹室
tank car dome head 油槽车穹顶
tank car for sulphuric acid use 硫酸槽车
tank car freight 罐装货物
tank car head 油槽车顶
tank-car heater 油罐车加热器
tank car ladder 油槽车梯子
tank car lining 油槽衬
tank car loading rack 油槽车装卸台
tank cars controller 罐车调度员
tank car shell sheet 油槽壳皮
tank car steam cleaning depot 罐车蒸洗站
tank car washing 洗罐【铁】
tank car washing shed 洗罐棚;洗罐库
tank cell 槽式电池
tank cession price 油罐转证价
tank circuit 谐振电路;储能电路;槽路;并联谐振回路
tank circuit Q 槽路品质因数
tank circumference 油罐的周围长度
tank classification 油罐分类
tank classifier 槽式分级机
tank cleaning 油罐清洗;油舱清洗
tank-cleaning barge 油轮清洗驳
tank-cleaning equipment 洗罐设备;洗舱设备
tank-cleaning plant 洗舱装置
tank-cleaning pump 洗舱系统泵
tank cleanings 油罐洗出物
tank-cleaning system 油轮清洗系统;洗舱系统
tank-cleaning vessel 油舱清洗船;净舱专用船
tank clean-out opening 油罐舱的清扫口
tank coaming and lid 油轮舱口装置
tank coating 油罐涂料层;油罐涂料;储[贮]罐涂料
tank cock 水箱旋塞;水箱龙头
tank coil 油罐旋管;振荡回路线圈;振荡电路线圈;槽路线圈
tank-coil formula 油罐旋管公式
tank collapsing 油罐压扁
tank coller 油箱冷却器
tank compartment 油罐分隔室;油罐舱
tank concrete pad 油罐混凝土基座
tank condenser 振荡电路电容器;槽路电容器
tank connections 油罐连接管
tank construction 箱式结构
tank container 油缸集装箱;液体集装箱;液(体)货(物)集装箱;罐状集装箱;罐式集装箱;桶状货柜;散液集装箱;槽形货柜
tank container marketing organization 液体货物集装箱交易机构
tank content 水池容量
tank content ga(u)ge 油罐量器
tank contents indicator 油箱油量指示器
tank cooler 油罐冷却器
tank cover 箱盖
tank cross member 油罐横梁;油槽横梁
tank crystallizer 槽式结晶器
tank cup 油罐盖
tank deck 油台;油罐顶
tank deck float 油台浮子;油罐顶浮子
tank design 油罐设计

tank details 油罐结构零件
tank development 罐中显影;槽式显影
tank dilution 油箱润滑油稀释;油箱内稀释;罐内稀释
tank discharge-auger 粮箱卸粮螺旋
tank ditch 油罐周围的沟
tank dome 油罐圆顶;油罐空气包;液舱顶
tank dozer 履带式推土机;坦克推土机
tank draft hole 油箱通风孔
tank drain 油罐放出口
tank drainback 油箱回油
tank draws in 抽入油罐
tank elevation 油罐高度
tank engine 带水柜油机车
tanker 沥青喷灌机;油船;油槽船;液货船;供水车;巨型油轮;救火水车;水槽汽车;石油运输船;槽(式汽)车
tanker aft 油船艉
tanker aircraft 空中加油飞机
tanker ballast water 油轮压舱水
tanker beam 油轮船梁
tanker berth 油轮码头
tanker centre line bulkhead 油槽船中线隔框
tanker cofferdam 油槽船上油舱与锅炉间的隔墙
tanker control system 油轮控制系统
tanker depth 油船深度
tanker discharging 油槽船排出
tanker discharging facility 油轮卸油设备
tanker displacement 油船排水量
tanker draft 油船吃水
tanker erection 油罐建造
tanker erection 油罐安装
tanker facility 油槽船的设备
tanker fleet 油船队;油槽船队
tanker freeboard 油轮干舷
tanker hull 油槽船船身
tanker length 油槽船长度
tanker line 油轮客兑
tanker loading 油槽船装载
tanker loading terminal 油槽船装油码头
tanker longitudinal bulkhead 油船纵向舱壁
tanker lorry 料罐汽车
tanker mooring 油船系泊
tanker oil compartment 油轮船油舱
Tanker Owners Voluntary Agreement Concerning Liability for Oil Pollution 油轮船东关于油污自愿协定
tanker owners voluntary agreement concerning liability for oil pollution 油船主自订油污染责任契约
Tanker Safety and Pollution Prevention 油轮安全与防污条约
tanker ship 油槽船
tanker stokehold 油船锅炉舱
tanker-submarine 潜水油船
tanker terminal 油(船)码头;油轮码头;油船栈桥
tanker time charter 油轮期租
tanker tonnage 油槽船吨位
tanker tonnage committee 油船吨位委员会
tanker transportation 油槽船运输
tanker transverse bulkhead 油槽船横隔框
tanker truck 油罐卡车
tanker turnaround 油槽船周转过程
tanker unloading pipeline 油轮卸油管线
tanker vessel 油轮;散液船
tanker voyage charter 油轮航次租
tanker wagon 槽车加油
tank evapo(u)ration 蒸发池
tank farm 油库;油罐区;油槽场;储[贮]油站;罐区;储罐区
tank farm pipe line 油罐场管线
tank farm stock 油罐场储存油料
tank field 罐区
tank filler 槽注入口;箱注入口;油罐加油口
tank filler cap 油罐加油口盖;油罐加油孔盖帽
tank filling system 油罐装油系统
tank filter 油罐过滤器
tank filter valve 油罐过滤阀
tank finishing 罐修整
tank fire 油罐失火;油罐火灾
tank fire boil-overs 油罐失火时溅出物
tank floating 油罐浮动
tank floor area 池底面积
tank for flushing 冲(洗)水箱;冲水池;冲水槽

tank foundation 油罐基础;液柜座;液舱座
tank framing 液舱舱柜构架
tank freeze-up 池子封冻
tank fuel indicator 油箱油量表
tank full indication 油箱已满指示
tank furnace 玻璃熔炉;玻璃池炉;浴槽炉;池窑;池炉;槽(式)炉
tank furnace brick 池窑砖
tank furnace process 池窑拉丝法
tank gas freeing by steam 用蒸汽驱除油箱气
tank gas-freeing installation 油气驱除设备
tank ga(u)ge 油箱储油量指示器;油罐标尺;油舱标尺;液位计;容器液位计;水柜水尺
tank ga(u)ge transmitter 液面计传感器
tank geometry 槽的几何形状;箱的几何形状
tank glass 池窑熔制的玻璃
tank grounding 油罐接地
tank group 罐群
tank hanger 油罐支架
tank hatch 油罐口
tank head 液舱压头
tank heater 水箱加热器
tank heating coil 油罐加热旋管
tank heating line 油罐加热管线;油舱加热管系
tank heating system 油轮货预热系统
tank heels 罐底脚
tank hood 油罐人孔盖
tank hose 水箱软管
tank-house cell 电解槽
tankies 油罐建造工人
tank inductance 振荡回路电感
tanking 池化防水(地下部分的墙和底板全部做防水层);地下室防水层;地下防水;防水物料;防水层
tanking-down period 玻璃液冷却阶段
tank installation 油罐装置
tank insulation 油罐绝缘;罐绝缘
tank in use 使用中的油罐
tank jack 坦克用千斤顶;坦克用起重器
tank joint 油罐接缝
tank kiln 池窑
tank knee 舱底接角
tank ladder 水柜爬梯
tank leakage 水泄漏量
tank level control 油舱液位控制;油舱液面控制
tank level control system 液位控制系统(油船)
tank level indicator 油箱液面计;柜液面指示器
tank level indicator system 油舱液位指示系统
tank-level sensor 油箱油面传感器
tank lid 水柜盖
tank life 油罐使用期
tank lighter 油罐驳船
tank line 油箱管;槽路传输线
tank lining 油罐衬里;槽衬里
tank liquor 槽液
tank loading facility 油罐装油设备
tank locomotive 水柜(式)机车;带水柜机车
tank lorry 油罐车;油槽汽车;槽车
tank lowering 水箱降低
tank lubrication 油箱润滑
tank lubricator 油箱润滑器
tank maintenance 油罐保养
tank manhole 油罐人孔
tank manifold 油罐支管
tank manifold valves 油罐支管阀
tank metallizing 罐喷镀金属
tank microflora 油罐微生物
tank mileage 油箱英里程
tank miles 油箱英里数
tank mirror 油舱装卸观测反光镜
tank mixed plaster 罐装搅拌好的灰浆
tank model 水箱模型;槽式模型
tank-mounted trailer 油槽拖车
tank on tower 高架水箱;高架水塔;水塔
tank operation 油罐操作
tankoscope 油罐透视灯;透视灯
tank outage 油罐中损耗
tank outlet plugging 塞住油罐出口
tank overflow 水池溢流量
tank pad 油罐座
tank paint 油柜漆
tank park 油罐区
tank piping 油水舱管系布置
tank plate 油罐钢板
tank pocket 油罐底凹处

tank pressure 油箱压力;油罐中压力
tank pressure distributor 沥青压力喷洒机;油罐压力洒油车;压力油灌洒油车;压力喷洒机
tank pressure ga(u)ge 油罐压力计
tank prover 标定器
tank rack 油罐架
tank railing 油罐顶栏杆
tank railway car 铁路油槽车
tank raising 水箱台高
tank reactor 罐式反应堆
tank rectifier 箱式整流器;铁壳整流器
tank refractory 池窑耐火材料
tank regulator 储[贮]液槽调整器;车内液体气体控制设备
tank relief valve 油罐放气阀
tank replacement 更换油罐
tank rim 油罐边缘
tank ring 油罐顶环
tank riser 连接水箱的立管
tank riveting 油罐铆合
tank roof cave 油罐顶棚
tank room 电解间
tank sampler 储油取样器
tank scale 油舱标尺;罐秤;水柜水垢;水柜水尺
tank screw-cap 油箱螺盖
tank seam testing 油罐缝试验
tank selector valve 油罐选择阀
tank semi-trailer 半拖挂式槽车
tank sender 液面信号发送器
tank service 加油车
tank setting 油罐安装
tank sewer 水池排水管
tank-shaped 槽形的
tank sheet 槽用网板
tank sheet-iron 制筒铁皮
tank shell 罐壳;油罐壳体;罐板
tank shell thickness 油罐壳厚度
tank ship 油船;油轮;散液船
tank shop 储罐工场
tankside bracket 双层底翼板;舭肘板
tankside knee 双层底翼板;舭肘板
tank-sinking method 沉箱法
tanks in series 串联池
tanks-in series model 多釜串联模型
tank site 油罐位置
tank sizing 油罐容量计算
tank skin 罐壁
tank slabbing 油罐鞍座垫板
tank sludge 油箱残渣;罐内污油
tank sound pipe 油罐探测管;液舱测量管
tank spacer 油罐隔板
tank spacing 油罐间距
tank sprayer 压力柜喷洒机;压力喷洒机;喷洒车;喷洒机
tank stabilizer 减摇水舱
tank station 油罐站;供水站;火车加水站
tank steaming-out and cleaning system 油舱熏洗系统
tank (storage) capacity 油罐容量
tank strainer 水箱滤器
tank strap 油桶箍条
tank strapping 围测罐周
tank stripping 油舱清底
tank suction 液舱吸口
tank sump 油罐沉积槽
tank support system 油罐支架系统
tank surface reconditioning 油罐表面修整
tank survey 液舱检验;水油柜检查
tank switch 罐区开关
tank system 箱式(热水)系统
tank table 箱形台;油罐校正表
tanktainer 罐式集装箱
tank test 水柜防漏检查;水池试验;水槽试验
tank tester 油罐检查工;油槽车检查工
tank toggle 油桶提闩
tank top 油罐顶盖;内底;货舱底部;双层底顶部;底舱顶板
tank top outfit 罐顶附件
tank-top plate 内底板
tank top replacement 更换罐顶
tank tower 水塔
tank town 小城镇;不重要小镇
tank track 坦克履带
tank trailer 液罐挂车;油(罐)拖车;油拖挂式槽车;水柜拖车;水槽拖车;槽罐挂车
tank trailer (with spray bar) 沥青喷洒拖车(有油嘴管)
tank train 油槽列车
tank train filling rack 油罐列车装油栈桥
tank transformer 油冷(却)变压器
tank treatment 在水池中处理(废水);水池处理;沉淀池处理;槽内处理
tank truck 油罐货车;运油车;运液体汽车;油罐卡车;油箱汽车;液罐汽车;自动洒水车;公路汽车槽车;水槽汽车;汽车槽车
tank-truck circuit-breaker 箱形断路器;槽形断路器
tank-truck for liquefied petroleum gas 液化石油气汽车槽车
tank-truck for liquid ammonia 液氨汽车槽车
tank truck loading rack 油罐车装载架;油槽汽车负载架
tank truck pickup 油罐车运输燃料
tank truck pump 油罐车油泵;油罐汽车泵
tank truck trailer 油槽汽车拖车
tank truing 油箱调整;油罐整形
tank-type hindered-setting classifier 槽式阻带沉降分级机
tank-type hydraulic classifier 槽形水力分级机
tank-type oil circuit breaker 箱型油断路器;铁槽形油断路器
tank-type soap dispenser 罐式肥皂配出器
tank unit 油罐信号发送装置
tank unit terminal 油箱接线头(接地用)
tank unloading 油槽车卸载
tank valve 油罐(呼吸)阀;柜阀;罐壁阀
tank vapour 油罐蒸气
tank vapour recovery 油罐蒸气回收
tank vapour space 油罐蒸气空间
tank vehicle washing line 洗罐线
tank ventilation 油舱换气
tank-ventilation fan 油舱通风机
tank ventilation system 油舱通风系统
tank venting 油舱通气
tank vent pipe 油罐通气管
tank vent valve 油罐放气阀
tank vessel 油船
tank voltage 槽电压
tank volume 储罐容积
tank volume charts 油罐容积图表
tank wagon 油槽车;油杯润滑脂;液罐车;罐(柜)车;汽车槽车;拖罐车;槽车;铁路槽车
tank wagon price 油槽车上交货价格
tank wall 液罐壁;池墙;池壁
tank-wall connector 油箱壁接头
tank washer 洗箱;洗桶
tank-washing system 洗舱系统
tank water riding 油罐浮水移动
tank wear machine 装甲磨损试验机
tank welding 油罐焊接
tank white 储[贮]罐用白漆
tank winch 油罐绞车
tank winding machine 箱式卷线机
tank with floating top 浮板(油)罐
tank with vault top 拱顶罐
tank work 油罐装置
tan liquor 植鞣液;鞣液
tannate 丹宁酸盐
tannbuschite 暗橄榄玄岩
tanned jute 精制黄麻
Tannenberg masonry bond 坦能贝格坊工砌筑法
tannery 制革厂;鞣革厂;皮革厂
tannery effluent 皮革厂流出水;皮革厂废水;皮革厂出水
tannery sewage 皮革厂污水
tannery sludge 皮革厂污泥
tannery waste 皮革厂废料
tannery wastewater 制革工业废水;皮革厂废水
tannic acid 鞣酸;丹宁酸
tannic acid equivalent 丹宁酸当量
tannic acid value 丹宁酸值
tannic substance 丹宁物质
tannin 鞣酸(处理混凝土模板用);丹宁
tannin attack 鞣酸侵蚀
tannin extract 烤胶;栲胶
tanning agent 鞣剂
tanning equipment 制革设备
tanning extract 丹宁提取物
tanning factory 制革厂
tanning liquor 鞣液
tanning material 鞣料;丹宁物质
tanning matter 鞣料
tanning paper 丹宁纸
tanning waste 鞣革废物
tanning wastewater 制革废水
tannin of persimmon 柿汁丹宁
tannin resin 丹宁树脂
tannoy 船上广播机;本地广播网
Tannsteel 坦斯提尔(一种密实高熔点合金钢)
tanny 黄褐色
tant 石板的匠名(石工按尺寸命名石板)
tantalaeschynite-(Y) 钽钇易解石
tantalate 钽酸盐
tantalian-cassiterite 钽锡石
tantalic 含钽的
tantalic acid 钽酸
tantalite 钽铁矿
tantalum carbide 碳化二钽
tantalo hatchettolite 钽铀烧绿石
tantalorutile 钽金红石
tantalous carbide 碳化钽
tantalpyroclore 钽烧绿石
Tantal tool alloy 钽镍铬合金工具钢
tantalum boride 硼化钽
tantalum bromide 溴化钽
tantalum bronze 钽铝青铜
tantalum capacitor 钽电容器
tantalum carbide 碳钽矿;碳化钽
tantalum carbide ceramics 碳化钽陶瓷
tantalum clips applying forceps 钽夹钳
tantalum die 钽模(热压用)
tantalum lamp 钽丝灯
tantalum lining 钽衬
tantalum nitride 一氮化钽;氮化钽
tantalum nitride resistor 氮化钽电阻器;氮化钽薄片型电阻
tantalum ores 钽矿
tantalum pentoxide 五氧化二钽
tantalum rectifier 钽整流器
tantalum rutile 钽金红石
tantalum silicide 硅化钽
tantalum-titanium ion pump 钽钛离子泵
tantamount 等值的;等价(的)
tanteuxenite 钽黑稀金矿
tantiron 高硅(耐热耐酸)铸铁
tantite 钽石
tan yard 制革场
tanyard sludge 制革厂污泥
tanyard waste(water) 制革(厂)废水
tanyard wastewater treatment 制革厂废水处理
tan-ye 黄棕色印画
Tanzania nucleus 坦桑尼亚陆核
Tanzania-Zambia Railway Authority 坦赞铁路管理局
Taoism 道教(中国主要宗教之一)
Taoism temple 道教寺院;道教庙宇
Tao Ji [notes on ceramics] 陶记
Tao Jun 陶钧
tap 旋塞;支管接头;龙头;敲击;排出孔;丝锥;矢锥【岩】;塞子;打孔抽水;出钢量;分接头(电、水);放料;抽头
tapability 成带性
tap a blast furnace 出铁
tapa cloth 塔帕纤维布;树皮布
tap adjuster 分接头调整器
tap a line 管路上钻孔
tap and die 丝锥及扳牙
tap bolt 自攻螺栓;固定螺丝;有头螺栓;螺基;螺钉;龙头螺栓;带头螺栓
tap bore 螺纹底孔
tap borer 锥孔钻;螺纹底孔钻;螺孔钻;开塞子锥
tap bottoming 全扣打捞丝锥
tap box 接线盒
tap certificate of deposit 随开存单
tap-change operation 分接头切换
tap changer 抽头转换开关;抽头变换器;插头变换器;分接头切换装置
tap-change transformer 分接头切换变压器
tap changing 抽头切换;分支变换;分接头切换;接头切换开关
tap changing apparatus 分接头切换装置;分接头切换开关
tap changing device 抽头转换开关

tap changing switch 分头切换开关;分接头切换开关;抽头切换开关
tap chuck 丝锥夹头
tap cinder 高炉下渣;高炉出渣;铁口渣;(搅烧炉渣);出渣(指炉矿渣)
tap cock 旋塞龙头;水管栓
tap conductor 分支导线
tap cooling 流水冷却
tap crystal 分接头晶体
tap cutter 丝锥刀具
tap decode circuit 抽头译码电路
tap density 振实重度;振实密度;堆积密度
tap device 分接装置
tap die holder 丝锥扳牙两用夹头
tap down 抽头降压
tap drill 螺孔钻
tape 线带;狭带;纸带;胶纸带;软尺;条;带(子)
tape access time 读带时间
tape adhesion 胶带附着力
tape air sampler 纸带空气采样器;滤纸式空气取样器
tape alignment correction 卷尺定线校正
tape alternation 带交替;带的交替
tape armo(u)red cable 钢带铠装电缆
tape asperity 带不平度
tape assembler 带型收集器
tape attachment 锥形夹具
tape back spacing 带反绕
tape beginning marker 带始标
tape-bounded Turing machine 带界限图灵机
tape burst signal 带色同步信号
tape butt-seam welding machine 对接缝焊接机;带状对接缝焊机
tape cable 带状电缆
tape cable connector 带式电缆连接器
tape card converter 带卡片转换器
tape cartridge 带盒;穿孔带筒
tape casting 流延成型;带式注浆
tape catridge 穿孔带筒夹
tape cell 带单元
tape certifier 验带器
tape channel 带通道
tape character check 带符号检验
tape chronograph 纸带记时仪
tape clatter 带传动噪声
tape cleaner 带清洁器
tape cleaning machine 带清洗机
tape coating system 黏带涂层法
tape code 纸带码;带码
tape comparator 校带器;带比较器;带比测器
tape condenser 搓条机
tape control 带式控制
tape-controlled carriage 带控制式托架;自动送带机构;自动滚轮
tape-controlled machine 程序带控制机床
tape-controlled milling machine 穿孔带控制铣床
tape-controlled operating 程序控制加工
tape-controlled turret lathe 程序控制转塔车床
tape-controlled turret drilling machine 程序控制的六角钻床
tape-controlled turret punching-machine 程序控制转塔冲床
tape-controlled turret punching-press 程序控制转塔冲压机
tape-controlled winding machine 带控绕丝机
tape controller 带控制器;带锯控制器
tape control unit 带控制器;带控制机构
tape copy 带式副本;穿孔纸带复制本
tape corner tool 墙角敷带工具;封角器
tape correction 钢尺测长修正值;卷尺修正;卷尺校正;卷尺改正
tape counter 带长记录器
tape covering 带状覆盖物;带包绝缘层
tape curvature 带弯曲度
tape cutting machine 条带式裁剪机
tape data process(ing) system 带数据处理系统
tape deck 带组;带运转机械装置
tape depth-ga(u)ge 带尺潮位计;带尺测深计
taped forming and assembling machine 带式成型组装机
taped insulation 带绕绝缘
tape distortion 带失真度
tape divider 分条装置
tape-drive 带驱动(器);带传送机构

tape-drive interrupt routine 带驱动中断程序
tape-drive mechanism 皮带传动装置;皮带传动机构;带驱动机构
tape drum 带鼓
taped slow-wave line 梯形慢波线
tape edit 带编辑
tape editing 带编辑
tape-editing machinery 带剪辑机械
tape elongation 带延伸率
tape entry 带入口
tape error 记错带;带错误
tape facsimile equipment 带式传真机
tape feed 馈带机构;带馈送;带卷盒;送带机构
tape feed motor 卷带电机
tape float liquid-level ga(u)ge 带浮子液位计;自记浮标液位计
tape footage counter 带长计数器
tape forming 流延成型;薄带成型
tape frame pulse 带帧脉冲
tape function 带函数
tape ga(u)ge 自记浮子水位计;链式水(标)尺;活动水尺;带尺潮位计;传送水尺;浮筒式潮位计
tape guide 纸带导轨;带向导装置
tape guide post 带导向柱
tape guide roller 导带辊
tape guide servo 带绕伺服机构
tape guiding drum 导带鼓
tape handler 卷带机;带处理机
tape handling 带处理
tape handling unit 纸带机
tape head 带头
tape helix 螺旋带;条状螺旋线;带绕螺旋线
tape hierarchy 带分级结构
tape holder 带架
tapehole clay 放铁口黏土
tape H pulse 带行脉冲
tape hub 带盘心
tape identification 带标识
tape initialize 带预置
tape injection mold 带注塑模
tape input 带输入
tape input-output control 带输入输出控制
tape joint 纸贴缝;带条黏结缝
tape joint compound 接缝带混合料
tape label 带标签;带标记;带标号
tape labelling system 带加标系统
tape lace 狭条花边
tape layout 带格式设计
tape layout form 带布局格式
tape leader 带首
tapeless jointer 薄板胶合机
tapeless splicer 薄板胶合机
tape library 胶带库;带程序库
tape lifter 挑带柱;带提升器
tape light 带指示灯
tape limited 受带限制
tape line 厂矿联络线;卷尺;带尺;皮尺
tape loading 装带
tape load key 送带键
tape loadpoint 带装入点
tape loop 循环带;带环
tape looseing 带松弛
tapeman 持尺员;测量持尺员;司尺员
tape mark 带记录结束符;带标记
tape marker 带终端标记;带始端标记;带标记
tape mastic 带状玛王帝脂
tape measure 卷尺;带尺丈量;皮尺;带尺
tape measurement 卷尺丈量;尺量法
tape mechanism 卷带机构
tap end 龙头
tape of mo(u)ld 模型锥度
tape operated typewriter 带控制打字机
tape parity 带奇偶校验;带的奇偶校验
tape parity check 带奇偶校验
tape perforator 纸带凿孔机;带打孔机;带穿孔机
tape-placement machine 铺带机
tape printing 穿孔带打印
tape-processing simultaneity 带处理的同时性
tape processing unit 带处理装置
tape puller 拖纸轮
tape punch 穿带机
tape punch(er) 纸带穿孔机
taper 圆锥(体);锥形;锥体;锥度;逐渐渐渐变尖;渐尖形;渐变器;尖锥;尖削;尖梢;平板压

延;退拔扩口管;梯形度;穿孔机;拔斜率;拔梢
taper-and-shim drive fit 锥形垫圈连接;(钻头和钻杆之间的)锥形垫环连接
taper angle 楔角;锥角;磨锐角
taper attachment 锥度附件
taper-bar grizzly 锥形棒条筛
taper base 圆锥底
taper baton hand 锥形棒式针
taper bevel 楔形边
taper bit 锥(形)铰刀;锥形(不取芯)钻头
taper blade 锥形叶片
taper boiler-tap 锥形炉用丝锥(锥度为1/16)
taper-bore cylindrical-roller bearing 锥孔圆柱滚子轴承
taper-bored spindle 锥孔轴
taper-bore mounted 锥形孔安装
taper boring 锥孔镗孔
taper bracket 过渡肘板
taper bridge reamer 锥桥形绞刀
taper calculating 锥度计算
taper claming sleeve 滚动轴承夹套
taper cone pulley 锥形轮
taper crusher 圆锥破碎机
taper cup 锥形杯
taper cup (grinding) wheel 锥杯形砂轮
taper cutter 锥形牙轮;锥形铣刀
taper drill 锥柄麻花钻
taper drum 锥形滚筒
tape-recording machine 带式记录机
tapered 锥形的;锥度的;变断面的
tapered acceleration lane 渐缩型加速车道
tapered acme 锥形锁接式的梯形丝扣
tapered adapter-sleeve 紧定套(筒);锥形接头套筒
tapered aeration 逐减曝气;渐减曝气(法)
tapered aeration activated sludge process 渐减曝气活性污泥法
tapered aeration process 渐减曝气法;渐变曝气法
tapered aeration system 渐变曝气系统
tapered antenna 锥形天线
tapered auger 圆锥形螺旋
tapered bar 楔杆
tapered beam 递变截面梁;楔形梁;加肋梁;变截面梁
tapered bearing 斜面轴承;锥形轴承
tapered bed 锥形床
tapered bend 锥形弯头;锥形弯管
tapered bit 梯形钻头;带锥形接头的可卸式钻杆;带锥形接头的可卸式钎杆;锥形钻头
tapered blade 锥形叶片;变截面叶片
tapered bobbin 锥形筒子
tapered bolt 圆锥销;锥形螺栓
tapered bottom 锥形底;锥底
tapered branch 锥形旁路;锥形支路
tapered brick 大小头砖
tapered bush(ing) 锥形衬套
tapered carriageway section 逐渐收窄的车行道路段;逐渐收窄的车行道断面
tapered casing string 复式套管柱;复合套管柱
tapered cavities 锥形串接腔
tapered center 锥形顶尖
tapered center-column rotameter 锥形芯柱式旋转流量计
tapered channel 锥形孔道;渐缩槽
tapered charge 递减充电;减流充电
tapered chord 不等弦;变截面弦杆
tapered chrome plated bezel 锥形镀铬表玻璃圈
tapered classifying liner plate 楔形分级衬板
tapered collar 锥形轴环
tapered column 楔形柱;递变截面柱;锥形柱;变截面柱
tapered conduit 锥形输水管;锥形管
tapered connection 锥形连接
tapered construction thickness 渐减的构造厚度(横跨行车道宽度)
tapered core bit 锥形岩芯钻头(换径时用);锥形取芯钻
tapered corrugated sheet 锥形波纹铁皮;锥形瓦楞铁皮
taper(ed) cotter 圆锥销;锥形销
taper(ed) coupling 锥形(管)接头
tapered cowl 锥形风帽
tapered cross-section 渐缩断面;变截面
tapered cup grinding wheel 碗形砂轮
tapered cup nut 锥形螺母(装皮碗用)

tapered cut 锥形裁剪
tapered-cylindric(al) structure 斜筒形结构
tapered die 锚锥;锥形模(冷拔钢丝用)
tapered diffuser 锥形扩压器
tapered disk 锥形圆盘
tapered distribution 锥形分布
tapered dowel 锥形暗销;锥形榫;锥形销钉
tapered draft tube 锥形尾水管;锥形弯头;锥形弯管
tapered drilling line 变径钢丝绳(钻井用)
tapered drill string 塔式钻具
tapered edge 楔形边;坡形板边;斜削边;坡口边
tapered-edge plaster board 倒角石膏板
tapered-edge strip 锥形边沿衬条
tapered edge wallboard 斜削边墙板
tapered elbow 锥形弯管
tapered emery-wheel 斜砂轮
tapered-end joint 漏斗形榫接;锥(头)形榫接
tapered enlargement 锥形扩大(管道)
tapered extension 锥形延伸段
tapered extruded shape 锥形挤压型材
tapered extrusion 锥台拉伸;锥形拉伸
tapered file 尖(细)锉;斜面锉
tapered fitting 锥形配件;锥形连接;锥形接合
tapered-flange beam 变截面翼缘钢梁;楔形翼缘梁;楔形渐窄梁;楔形工字钢梁
tapered flocculation 变截絮凝
tapered flow header 锥形行浆管
tapered fluted pile 锥形凹槽桩
tapered fluted pile union 锥形凹槽桩组合
tapered-foot roller 羊足碾;羊脚碾;羊足压路机;羊蹄滚筒
tapered-foot type sheepsfoot roller 锥形羊足碾
tapered foundation 锥状基础
tapered graded index fiber 递变型渐变折射率光纤
tapered grinding ring 环辊磨的锥形磨环
tapered grip 锥形柄
tapered grizzly bar 楔形棒筛条;楔形棒条筛;锥形筛条
tapered-groove valve stem 锥槽阀杆
tapered gutter 断面渐变沟
tapered haunch 拔斜梁腋;斜腋
tapered hole 锥形孔;螺纹孔;收缩孔
tapered-hole disk penetrometer 锥孔盘针入度计
tapered hub 锥形轴套
tapered hull 锥形车体
tapered incrcaser 锥形扩大器
tapered inlet 锥形入口;锥形进口;喇叭形进(水)口
tapered interchangeable ground joint 锥形标准磨口接头
tapered jacket 滑脱套箱
tapered jaw 斜爪
taper(ed) joint 锥形接头;锥形接缝
tapered junction 河流锥形合流处;道路锥形交叉口;锥形连接
tapered-land bearing 斜面轴承
tapered-land thrust bearing 斜面推力轴承
tapered lap 楔形搭接
tapered (ledge) reamer 锥形扩孔器;锥形金刚砂扩孔器
taper(ed) liner 锥形衬条
tapered loading 渐变加感;递变加载;抽尖式加感
tapered mainspring 变截面发条
tapered member 楔形杆件;变截面构形
tapered mold 常锥度的模子
tapered neck rivet 圆锥铆钉
tapered opening 锥形开口
tapered overlap 两斜面接头
tapered packer 锥形封隔器
tapered parapet gutter 锥形天沟;(不等宽的)女儿墙斜天沟;收口匣形天沟(女儿墙后的箱形水槽)
tapered pattern 椭圆形喷涂图案
tapered pavement cross-section 路面渐缩断面
tapered pier 锥形墩;锥形斜面墩
tapered pile 锥形桩
tapered pile standard step 标准分阶锥形桩
taper(ed) pin 锥形销;锥销
tapered pintype jack 锥形栓式千斤顶;锥锚式千斤顶
tapered pipe 异径管接头;锥形管
tapered pipe branch 管子的锥状入口
tapered pipe junction 管子的锥形连接
tapered pipeline 变径管道
tapered piston 锥形活塞
tapered plain clay roof(ing) tile 普通黏土锥状屋瓦

tapered plain tile 拔斜平瓦;楔形平瓦
tapered planking 渐薄船壳板
tapered plate 楔形板;锥形板;变面变面板材
tapered plug 锥状塞子;锥形栓;锥形螺塞
tapered pole 锥形极
tapered portion (加宽车道的)渐减部分;变截面段
tapered potentiometer 递变电阻分压器;抽头式分压器
tapered reducer 锥形减小器
tapered rim 楔形边
tapered ring of tank 油罐的锥形部分
tapered rivet 锥形铆钉;斜削铆钉
tapered rod 锥形连接钎杆;锥形棒
tapered rod string 变径泵杆;复式抽油杆柱
tapered roll 圆锥辊
tapered roller 平板压延辊;锥形滚筒;滚锥
tapered roller and cage 锥形滚柱和保持架
tapered roller bearing 滚锥轴承;锥形滚柱轴承;滚锥形轴承
tapered roller thrust-bearing 锥形滚柱止推轴承
tapered-roll pantile 斜削波形瓦
tapered rope 变断面钢丝绳
tapered screw 锥形螺杆;锥形螺钉;锥度螺纹
tapered screw plug 锥形螺纹堵头;锥形螺旋塞
tapered screw press 锥形螺旋压榨机
tapered section 渐变段;收缩部分
tapered segment 异形管片
taper(ed) shaft 锥形轴;锥形烟囱;锥形柱身;锥形竖井
tapered shank 锥形钎尾
tapered shaped can 锥形罐
tapered shims 楔形填隙片
tapered shoulder 锥形轴环;锥形环
taper(ed) sleeve 锥形套筒
tapered slide valve 锥形滑阀
tapered slip 锥形衬套
tapered slot 斜沟
tapered snorkel 锥形进气管
tapered socket bit 锥形连接钻头;锥形接头钎头
tapered spindle 长锥形摘锭
tapered spline 锥形花键
tapered split-dowel 锥形开口销
tapered steel line 变径钢丝绳
tapered steering wheel hub 锥形转向盘壳
tapered step-core bit 锥形阶梯钻头
tapered string 变径管柱
tapered string of drill pipe 变径钻管(组)
tapered strip 锥形条
tapered strut 锥形支柱
tapered stud 锥形双头螺柱
taper(ed) tab 锥形接头
tapered tap 锥形打捞器;打捞公锥
tapered taper 分层锥削
tapered tenon 斜榫;楔形榫(头);楔削榫;锥形榫
tapered thickness ga(u)ge 锥形厚度规
tapered thimble 锥形套管
taper(ed) thread 楔形螺纹;锥形螺纹;圆锥螺纹
tapered tin 锥形罐
tapered tip 锥形极端
tapered tooth 收缩齿
tapered-tooth spindle 锥齿形摘锭
tapered tower 锥形塔
tapered transformer 锥形变换器
tapered transmission line 锥形波导
tapered tube 锥形管
tapered-tube flow-meter 浮子(式)流量计
tapered-tube rotameter 锥形管柱式旋转流量计
tapered tubing 等压油管柱;变直径油管柱;变径油管柱
tapered type cracking test 锥形抗裂试验
tapered valley (上小下大的)斜天沟
tapered valley clay roof(ing) tile (屋顶排水的)斜沟黏土瓦
tapered wage increase 分级增加工资
tapered wallboard 斜削边墙板
taper(ed) washer 锥形垫圈;斜垫圈
tapered waveguide 锥形波导管;锥形波导;递变截面波导管
tapered wedge 斜楔块
tapered wheel 锥形砂轮
tapered winding 锥形绕组
tapered wire rope 锥头索;(变截面)的钢丝绳;变截面钢丝绳
tapered workpiece 锥形工件

tape reel 带盘;带卷
tape reiative sensitivity 纸带相对灵敏度
tape relay 带转接;带中继
tape relay center 纸条转接中心
taper end miling cutter with coarse teeth 粗齿锥柄立铣刀
tape reperforator 带式自动纸带穿孔机
tape rewind 倒带
tape rewinding 倒带
taper-face compression-ring 锥面活塞压环
taper-face piston ring 锥面活塞环
taper-face ring 锥面活塞环;锥面环
taper female thread 锥形阴螺纹
taper file 斜边锉;斜削锉;圆锥锉
taper fit(ting) 锥度配合
taper fitting bit 锥形连接钻头
taper flask 锥度砂箱;可卸式砂箱
taper foot roller 羊足压路机;羊蹄压路机;羊蹄滚筒;锥足路碾;锥形销齿式镇压器;羊锥足碾
taper freight rate 分等级运费;分等级的运费
taper freight rates 分级运费
taper fuselage 锥形机身
taper gas tap 管用锥形丝锥
taper ga(u)ge 锥度量规;锥度规
taper gib 锥形镶条;调整楔
taper grinding 锥形磨削
taper grip 锥形夹紧装置
taper ground 逐渐磨细的
taper-ground saw 双凸面圆锯
taper guide 分配凸轮机构
taper guide-bar 锥形导杆
taper handle 锥形柄
taper hand tap 锥形手用丝锥
taper hob bing-cutter 锥形(齿轮)滚刀;切向进给滚刀
taper hole 锥形孔
taper honing machine 锥度珩磨机
tapering 锥锥度;渐细;渐尖;变细
tapering aeration 渐减曝气(法)
tapering aeration process 渐减曝气法
tapering arch form 尖形弓
tapering beam 拔斜梁
tapering column 锥形柱;锥形柱;烛状柱
tapering end joint 斜削型端接缝
tapering freight rates 远距离递减的运费率
tapering function 渐变函数
tapering gutter 宽度渐变的沟;斜沟;喇叭口沟;宽端斜沟;端宽沟
tapering machine 车锥度机床
tapering plate 楔形衬环
tapering point 锥形端
tapering rate 随着里程递远递减的运价
tapering scale 锥度尺
tapering shaft 变径杆;锥形柱
tapering shape 锥式
tapering spindle 锥形轴
taper-in-thickness ratio 厚弦比展üksek向变化率
taper joint 管子套口接头
taper key 斜键;楔键
taper knock 圆锥顶销
taper layout 带面格式设计
taper leaf spring for vehicle 车辆锥形钢板弹簧
taper lifter 楔形举扬盘
taper line 锥形塞垫;锥削传输线;过渡线
taper line gratings 锥形线密度镜
taper liner 锥垫;楔形塞垫
taper machine screw tap 锥形机用丝锥
taper male thread 锥形阳螺纹
taper mandrel 锥形心轴
taper matching 锥形匹配
taper mill 楔形扁削轧机;锥形牙轮;锥形铣刀
taper-neck roll 圆锥辊颈轧辊
tape rod 带状卷尺;卷动标尺
taper of face slab 水坝面板斜坡度
taper of groove 轧槽斜度
taper of model 模型锥度
taper of timber pile 木桩锥度
taper out 渐变尖
tape row 带行
taper parallel 调整垫铁;双楔固定平行垫板
taper pile 锥形桩;变截面桩;斜桩
taper pin 锥形钢销
taper pin fixing 锥形销钉紧固
taper pin-hole 锥(形)销孔

taper pin(-hole) reamer 锥销孔铰刀
taper pipe 变径管;锥形管;锥形大小头;楔形管;异形管;大小头管
taper-pipe tap 锥形管子丝锥
taper pipe thread 锥(形)管螺纹
taper-pipe-thread plug-ga(u)ge 锥管螺纹柱形
taper-pipe-thread ring-ga(u)ge 锥管螺纹环规
taper plate 变截面板材;楔形板材
taper plate condenser 递变式平板式电容器
taper plug ga(u)ge 锥形塞规;锥体塞规;锥度塞规
taper plug header 锥塞管头
taper plug mandrel 锥形心轴;粗磨工具
taper-pointed 渐尖的
taper printer 带式打印机
taper ratio 锥度比
taper reamer 锥(形)铰刀
taper reaming shell 锥形扩孔器
taper reducer 锥形异形管
taper reducer sleeve 锥套
taper ring 锥圈;锥形环
taper ring-ga(u)ge 锥度环规
taper rivet 斜梢铆钉;锥形铆钉
taper rod type tie 圆锥棒型联结件;圆锥棒型联结件
taper-roller 锥形滚柱
taper roller bearing 锥形滚柱轴承
taper-roller-bearing hub 锥形滚柱轴承毂
taper roller thrust bearing 锥形滚轴止推轴承
taper rolling 斜坡轧制
taper rolling mill 变断面轧机
taper rope 变径钢丝绳
taper sawed 斜锯(板)
tapersawn 双斜劈面木瓦板;斜锯(板);对斜劈面木瓦板
taper seat 锥形座
taper-seat valve 锥形座阀
taper section 锥形断面
taper sectioning 斜剖面法
taper serew chuck 锥形螺纹夹头
taper serration 锥形键
taper shank 锥形柄;锥柄
taper-shank arbor 锥形塔轴;锥柄轴
taper-shank bridge reamer 锥柄桥工铰刀
taper-shank core-drill 锥柄扩孔钻
taper-shank cutter 锥柄铣刀
taper-shank drill 锥柄钻头
taper-shank drillchuck 锥柄钻夹头
taper-shank drillholden 锥柄钻托
taper-shank end-milling cutter 锥柄立铣刀
taper shank endmill with long edge 长刃锥柄立铣刀
taper-shank reamer 锥柄铰刀
taper-shank self-adjustable drill-chuck 锥柄自调钻夹头
taper-shank triple-grip drill-chuck 锥柄三牙钻夹头
taper shank twist drill 锥柄麻花钻
taper-sheepfoot roller 锥形羊足辗;锥形羊蹄压路机;锥形羊蹄碾
taper sheet pile 尖头板桩
taper shell course 锥形锅炉管
taper sleeve 锥套管;锥套
taper sleeve of ball bearing 锥形滚珠轴承座套
taper slide 斜滑板
taper socket 锥形承窝
taper spindle 锥形主轴
tapersplit 斜坡(板);双斜劈面木瓦板
taper-split shake 斜破板
taper sprocket 锥形链轮
taper steel line 变径钢丝绳
taper-stick 烛台
taper sunk-key 锥形埋头键
taper tap 锥形丝锥;锥形螺丝攻
taper tester 锥度测量器
taper thread 锥螺纹;锥形螺纹
taper throat 倾斜流液洞
taper thrust bearing 锥形止推轴承
taper tools 锥形工具
taper-tread wheel 锥形行轮
taper trimming 修整成斜面
taper trowel 刮平镘刀
taper tube 异形管
taper turning 锥形车削;车锥体
taper-turning attachment 车锥度附件;车圆锥体装置;车圆锥附件
taper-type dropper (向井内下开口楔的)锥形工具

tape rule 皮尺
tape rule in metallic case 钢合尺
tape-running accuracy 带运转精确度
taper-wall bit 内锥面钻头
taper-wall core shell 内锥面扩孔器
taper wedge 调整楔
taper wedge key 锥形楔钥匙
taper winding 锥形卷绕
taper wire and thickness ga(u)ge 锥线厚度组合规
taper work 锥形工件
taper-work grinding 锥形工件磨削
tape sampler (测大气烟尘的)纸带采样器
tape sawing 尖削原木下锯法
tape scratch sort 暂存带分类
tape sealant 粘封带;密封带
tape sealer 胶粘纸封盒机
tape search unit 找带装置
tape set 调梢装置
tape skew 带偏斜;带扭斜
tape skip 越带指令;跳带指令
tape sorting 带的分类
tape speed 带速
tape speed deviation 带速误差
tape speed variation 带速变化
tape splicing 接带
tape standardization 卷尺检定
tape station 带站
tape storage 带存储器
tape storage bin 存带箱
tape stretch 带延伸率
tape stretcher 拉带器
tape string 带行
tape strip 带条
tapestry 挂毯;花挂毯;壁毯
tapestry brick 粗面砖;饰面砖
tapestry-brick wall 缀锦砖墙
tapestry carpet 织绒地毯
tapestry glass 挂毯式玻璃(一种不透明花纹玻璃)
tapestry map 挂毯地图
tapestry portait 绣像
tapestry satin 织锦缎
tapestry wool 刺绣毛纱;彩绒绒
tape-supported adhesive 带式载体黏合剂
tape swap 带交换
tape swapping 换带
tape symbol 带符号
tape synchronizer 带同步装置
tape temperature 卷尺温度
tape tensile strength 带破断强度
tape tension 带拉力
tape tension control 带张力控制
tape test 带测试
tape test set 带测试装置
tape theorem 薄带原理
tape thermometer 卷尺温度计
tape-to-tape converter 带转换器
tape traction 布兜牵引
tape trailer 带尾
tape transmitter 带式发射机
tape transmitter-distributor 带发送分配器
tape transport 走带机构;带传送
tape transport mechanism 带传送机构
tape travel 带运转;带前进
tape unit 纸带机
tape verifier 穿孔带校验机;纸带校对机;穿孔带校核机
tape volume 带卷
tape wap 带卷
tape weave 带弯曲度
tape wheel 传带轮
tape width 带宽度
tape width switch unit 带宽转换机构
tape winder 绕带器
tape winding 带缠绕成型
tape winding machine 布带缠管机
tape wire buck 尺架
tape with selvage 有边带
tapeworm-shaped 条虫状的
tape wound core 带绕铁芯;带绕磁芯
tape-wound resister 带绕电阻器
tape yarn 扁丝
tap-field control 抽头激磁控制
tap-field motor 分段励磁绕组电机

tap financial resources 开辟财源
tap float level ga(u)ge 钢带式浮标液面计
tap for metric thread 公制螺纹丝锥
tap for taper thread 锥形螺纹丝锥
tap for trapezoidal thread 梯形螺纹丝锥
tap funnel 滴液漏斗;分液漏斗
tap gain 分支增益
tap grease 排出油;排出油脂
tap grinder 丝锥磨床
tap-grinding machine 丝锥磨床
tap handle 丝锥手柄;丝锥扳手
tap hob 标准螺丝攻
taphocoenose 埋藏群落
tap holder 攻丝夹头;丝锥柄
tap hole 螺纹孔;出钢口;出渣口;出铁口;放液出口;塞口;放液孔;熔炉出铁口
tap-hole bar 开口钎
tap-hole block 出铁口成型砖
tap hole clay 堵口泥
tap hole drilling machine 炉口钻机
taphole for slag 出渣口
taphole gun 泥炮
tap-hole hammer 放出口塞子;堵口钎子
tap-hole launder 放出口流槽
tap hole of blast furnace 高炉开口机
tap-hole plug 放液口塞
tap-hole plug-stick 渣口塞杆;堵口塞杆
tap-hole rod 钎子;出铁口塞棒
tap-hole stopping-machine 堵铁口机
taphonomy 埋葬学;埋藏学
tap house 小旅馆;酒吧间;酒店;客栈
taphrium 沟渠
taphrogenesis 裂陷作用;断陷作用;断裂运动;分裂运动
taphrogenic 断陷的;地震运动
taphrogenic stage of platform 地台裂陷阶段
taphrogenic trough 裂陷槽
taphrogeny 张陷;断陷作用;断裂运动;地裂运动;分裂运动
taphrogeosyncline 断裂地槽;地堑式地槽
tapia 胶土坯;黏土坯
taping 绝缘绕阻;卷尺度量
taping and sealing 捆轧与铅封
taping arrow 测钎;测针【测】
taping box 分线匣;分线盒
taping buck 尺架
taping by parallactic method 视差法测距
taping by tacheometric(al) method 视距法测距
taping machine 包边机
taping pin 测钎;测针
taping point 踏面基点;穿孔点
taping screw 自动刻螺纹
taping spout 排水口
taping strip 盖缝条;油(毛)毡(条);胶带
tapioca snow 雪珠;雪大
tapioca starch 木薯淀粉
tapis 议事桌;桌毯
tap issue 时价发行
tap joint 抽头接头;分支接头;分接头(电、水)
tap ladle 浇铸桶
tap line 抽头线
tap new resources and economize on expenditures 开源节流
tap-nose grinder 丝锥前端刃磨机
tap off 出铁;出炉;出线;分接;分出
tap-out 敲出;以手指沾墨在纸上抹样
tap-out bar 火棍;通铁棒;开口钎;出铁口钎;出口钢钎
tap-out block 出铁口成型砖;放玻璃孔堵砖;放料砖
tapped 中间抽头的
tapped actuator 偏心轮执行机构
tapped center 中心抽头;抽头中心
tapped circuit 抽头电路;带抽头的电路
tapped coil 多接头线圈;抽头线圈
tapped coil oscillator 电感耦合三端振荡器;带抽头线圈振荡器
tapped-condenser oscillator 电容耦合三端振荡器
tapped control 抽头调节;抽头法调节;分接头调节
tapped down generator 带多抽头线圈振荡器;带抽头线圈振荡器
tapped fitting 螺纹接头
tapped function potentiometer 抽头式函数电位计
tapped heater 抽头式电热器
tapped line 分接头线

tapped nut 螺纹螺母
tapped or flanged connections 螺纹或法兰连接
tapped potentiometer 抽头式电位计
tapped-potentiometer function generator 抽头分压器式函数发生器
tapped resin 放出树脂液
tapped resistance 分接头电阻
tapped resistor 抽头式电阻器
tapped tee 带螺纹的三通接头
tapped transformer 抽头变压器
tapped-tuned circuit 抽头调谐电路
tapped volume 堆积体积
tapped winding 抽头绕组;分组绕组;分组线圈
tapper 螺纹工;打孔工;穿孔机;音响器;攻丝机
tapper tap 螺母丝锥;机用丝锥
tappet 凸轮;挺杆;随行件;随动杆;捣杆
tappet adjusting device 挺杆调节装置
tappet adjusting screw 阀间隙调整螺钉
tappet adjustment 挺杆调整
tappet (arm) 挺杆(臂)
tappet assembly 挺杆组合
tappet clearance 挺杆余隙;挺杆间隙
tappet drum 凸轮鼓
tappet gear 凸轮分配机构;挺杆装置
tappet guide 挺杆导承
tappet-guiding groove 挺杆导槽
tappet head 拧紧头;挺杆头(部)
tappet lock 挺杆锁
tappet loom 踏盘织机
tappet machine 挺杆冲锤式凿岩机
tappet motion 挺杆行程
tappet oil gallery 挺杆润滑油道
tappet plunger 阀门提杆;挺杆柱塞
tappet pull-out guard 挺杆推出式安全装置
tappet rod 挺杆;提杆
tappet rod guide 挺杆导管
tappet roller 止推辊;挺杆滚轮
tappet screw 推杆调整螺钉;挺杆间隙调节螺钉
tappet shaft 凸轮轴;桃子轴
tappet sleeve 挺杆套
tappet spanner 阀挺杆扳手
tappet spring 挺相弹簧
tappet switch 制动开关
tappet valve 凸轮定时阀;凸轮传动阀
tappet wear surface 挺杆磨损面
tappet winding 分段绕组
tappet wrench 挺杆扳手
tapping 引流;攻螺纹;攻螺丝;割漆;割胶;割浆;流出(口);绕组抽头;导出液体;出渣(指炉矿渣);出铁;出钢(量);缠绝缘带;采割松脂;分线;分接;放流;抽头
tapping analysis 钢水分析;出钢时钢水分析
tapping arrangement 排出孔
tapping assembly 井口钻杆密封装置
tapping attachment 攻丝装置;攻丝夹头
tapping bar 敲顶棍;出渣通条手柄
tapping bracket 放液托架
tapping bushing 攻螺丝套套
tapping cable 抽头电缆
tapping capacity 切制螺纹最大孔径
tapping chuck 攻螺丝夹头;攻丝夹头;螺丝攻(夹)头盘;丝攻夹头盘
tapping clamp 钻孔夹具;丝锥夹头
tapping conduit 排放水管
tapping die 螺丝扳(牙)
tapping drill 螺孔(底)孔钻头;螺孔钻头
tapping emission 出料排气
tapping floor 浇铸场
tapping head 螺丝扳(牙)
tap(ping) hole 放流口;螺纹孔
tapping-hole clay 堵口泥
tapping knife 割胶刀
tapping launder 出铁流槽
tapping line 放料管线
tapping machine 测声器;标准撞击器;钻孔机;攻丝机;开口机;开孔机;穿孔机;放液机;放流设备
tapping method 敲击法
tapping of groundwater 潜水集水井;地下水流出口;地下水出流口
tapping of molten steel 出钢
tapping of resources 资金动用
tapping paste 攻丝润滑剂
tapping pin 捅杆
tapping pit 流出槽
tapping plug 出铁口泥塞
tapping point (管道的)引出点;分叉点;(测压管的)取压点;出渣口;出水点(地下下水);抽头点;分接头(电、);出口
tapping pond 放玻璃水池
tapping pressure pipe 在有压管上穿孔
tapping process 流放法
tapping ratio 抽头匝数比
tapping sample 出铁样;出钢样
tapping screw 攻螺丝;自动刻螺纹;自攻丝螺钉;自攻丝螺针;自攻螺丝
tapping side 出炉侧;放渣侧
tapping slag 出渣(指炉矿渣)
tapping sleeve 螺纹套管;穿孔套管;螺纹支管;接头套筒;接品套管;接水套管;旁通套管;分流套管
tapping spout 出铁槽;出钢槽
tapping step 分接级
tap(ping) switch 分接开关;抽头开关
tapping tee 穿孔三通
tapping temperature 出渣温度;出铁温度;出炉温度;出钢温度
tapping the head 击头
tapping tool 螺丝攻
tapping unit 组合攻丝机;多头攻丝机
tapping valve 放液阀
tapping well 吊管井(打在含水层的顶部未穿含水层的井)
tap pipe 泄水管(道)
tap point 穿孔点
tap post 水栓柱
tap rate 时价行情
tap rivet 螺柱铆钉;螺旋铆钉
tap room 酒吧间
taproot 主根
tap root system 主根系
tap sand 界砂;分型砂
tap screw 螺丝攻;螺基;螺钉
tap-size drill 螺孔钻
tapster 酒吧间;酒吧间招待员
tap switch 抽头切换开关;抽头变换开关;分线开关
tap the labo(u)r potential 发掘劳动潜力
tap the natural resources 开发资源
tap the production potential 发掘生产潜力
tap transformer 抽头变压器
tap voltage 抽头电压;分接头电压
tap volume 捣实容积
tap water 饮用水;自来水
tap water degassed with argon 氩除气的自来水
tap web 锥心
tap weight 捣实重量
tap with plain cylindrical pilot 带普通外圆导柱的丝锥
tap wrench 攻丝扳手;螺纹扳手;螺丝扳手;铰杠;丝锥扳手
tar 焦油(沥青);柏油;真振幅恢复
tar acid 焦油酸
tar acid oil 焦酸油
taramellite 纤硅钡高铁石
taramite 绿闪石
taranakite 磷钾铝石
tar and asphalt sprayer 焦油和沥青喷洒机
tar-and-gravel roof cover(ing) 焦油和石屋顶覆盖
tar-and-gravel roofing 柏油碎石屋面料;柏油碎石屋面;柏油豆石屋面;沥青砾石屋面
tarapacaite 黄钾铬石;黄铬钾石
tarare 麻帆布
taras 火山灰;浮石
tar asbestos covering 焦油石棉覆盖层
tarasovite 云母间蒙脱石
tar asphalt 焦油沥青
tar-asphalt composition roofing 焦油—沥青组合屋面材料
tar-asphalt mix(ture) 焦油—沥青混合物
tar-asphalt prepared roofing 焦油—沥青预制屋面
tar-asphalt ready roofing 焦油—沥青预制屋面
tar-asphalt roof(ing) felt 焦油—沥青屋面油毡
taraspite 白云石
taraxerane 蒲公英赛烷
tar barrel 焦油桶;柏油桶
tar base 柏油浇底层;柏油打底;焦油碱
tar based 焦油碱的
tar based composition 焦油碱成分
tar based compound 焦油碱化合物
tar base roof preservative 柏油打底的屋面防腐层
tar bearing basic refractory 含焦油的碱性耐火材料
tar binder 焦油沥青结合料;柏油结合料
tar-bitumen 焦油-石油沥青;焦油沥青混合料
tar-bitumen binder 焦油沥青结合料
tar-bitumen blend 焦油沥青—石油沥青混合料(一般指含沥青较多的)
tar-bitumen mixture 焦油沥青混合物
tar blast-furnace slag chip(ping)s 鼓风炉柏油层炉渣小块;鼓风炉柏油层炉渣碎片
tar-board 硬板板
tar boiler 焦油锅炉;柏油熔化炉
tar-bonded 柏油结合的
tar-bonded basic refractory 焦油结合碱性耐火材料
tar-bonded dolomite brick 沥青结合白云石砖
tar-bonded magnesite 焦油结合菱镁矿
tar-bound 柏油黏结的
tar-bound flexible pavement 柏油黏结柔性铺面
tar-bound surface 柏油结合面;焦油沥青碎石面层;焦油沥青结碎石路面
tar broken rock 柏油碎石
tar broken stone 柏油碎石
tar brush 焦油帚;柏油帚;柏油刷
tar bucket 柏油桶
tarbuttite 三斜磷锌矿
tar cancer 焦油致癌
tar carpet 柏油路面磨耗层
tar cement 焦油沥青胶粘剂;沥青水泥;柏油黏结料;柏油胶合料
tar cementing agent 柏油水泥黏结料
tar coat 柏油罩面;柏油盖层
tar-coated broken rock 柏油罩面的碎岩
tar-coated broken stone 柏油罩面的碎石
tar-coated chip(ping)s 柏油罩面小块;柏油罩面碎片
tar-coated chip(ping)s carpet 焦油石屑磨耗层
tar-coated crushed rock 柏油罩面的碎石
tar-coated iron blast-furnace slag chip(ping)s 炼铁鼓风炉炉渣碎片
tar-coated limestone 柏油罩面的石灰石
tar-coated pipe 涂柏油管
tar-coated road 柏油铺面(道)路;柏油路;焦油沥青罩面路
tar-coated road-metal 柏油罩面铺路石
tar-coated sand 柏油罩面砂
tar-coated slag 焦油矿渣
tar-coated tube 涂柏油管
tar-coated unit 柏油罩面部件;柏油罩面构件;柏油罩面设备;柏油罩面单元
tar coating 焦油沥青盖层;柏油盖层
tar composition 焦油组成;焦油成分
tar concrete 沥青混凝土;焦油沥青混凝土;柏油混凝土
tar concrete pavement 焦油混凝土路面;柏油混凝土铺面
tar-containing coal 含油煤
tar content 焦油含量
tar cooler box 焦油冷却箱
tar covering 焦油沥青罩面
tar crushed rock 拌柏油的碎石岩
tar crushed stone 拌柏油的碎石
tar dipping ladle 焦油勺;柏油舀勺
tar distillate 焦油馏出物
tar distillation 焦蒸馏
tar-distillation fraction 焦油馏出组分
tar-distillation plant 焦油蒸馏厂
tar-distillery waste(water) 焦油厂废水
tardive 迟发的
tardolomite brick 焦油(沥青)白云石砖
tar down 涂柏油
tar-dressing machine 浇柏油机;柏油浇灌机
tar dry penetration surfacing 柏油干渗透铺面材料
tar dry process penetration macadam 柏油干处理渗透碎石路面
tardy debtor 违约债务人
tar dyestuff 焦油颜料;焦油着色剂
tardyon 慢子
tare 容器空重;皮重;称瓶重量;车身重量;本身自重;包装重(量)(指皮重)
tare and draft 皮重及重量损耗折扣
tare and tret 扣除皮重计算法;皮重和添头
tare balance 药物天平;精密天平
tared 配衡的
tare effect 支架效应
tare gross 除皮重量

tare mass 自重(皮重)
tare mass of freight container 集装箱自重
tar emulsion 乳化沥青;焦油沥青乳液;乳化焦油沥青;柏油乳液
tare weight 容器重量;皮重;自重;车皮重量;包装重量(指皮重);包皮重量
tar extractor 除焦油器
tar felt 油毛毡;焦油沥青毡;防水毡
tar felt roof cladding 油毛毡屋顶覆盖层
tar felt roofing 油毛毡屋顶材料
tar filler 焦油沥青填(缝)料;柏油填(缝)料
tar finisher 柏油铺面机
tar flexible conduit 焦油软管
tar flux 焦油沥青稀释剂;柏油稀释剂
tar for road purposes 道路用柏油
tar gas 焦油气
target 信号圆板;指标;地物;靶牌;靶标;标板
target abeam 物标正横
target acquisition 目标探测;目标搜寻;目标捕获;物标搜索;物标录取;捕获目标
target acquisition gate 目标信号检测门
target acquisition radar 目标搜索雷达
target acquisition technique 目标探测技术
target actuation 靶激发
target adsorbate 目标吸附质
target aircraft 靶机
target (allowable) leak rate 指标泄漏率
target analysis 目标选择(分析);目标分析
target and training submarine 靶标潜艇
target angle 物标舷角;反弦角
target approach 进入目标
target approach chart 进入目标图
target area 作业区;空投区;目标区(域);靶板面积;靶面积
target area survey 目标区地形测量
target atom irradiation 靶原子辐射
target attribute 目标属性
target background 目标背景
target bearing 目标方位
target being tilt 靶标已倾斜
target braking 目的制动
target brief 目标简况
target burner 撞击式燃烧器
target butt 靶垛
target capacity 目标容量
target cathode 靶阴极
target characteristic identification 目标特征识别
target chart 目标图
target chemistry 靶化学
target code 目标码
target complex 目标群
target computer 目标(程序)计算机;特定程序计算机
target concentration 目标集中
target configuration 目标程序配置
target contract 目标(价)合同;合同指标
target cost 目标成本
target cost contract 目标成本合同
target cost control 目标成本控制
target cost decomposing 目标成本分解
target cost determination 目标成本测定
target cost implement 目标成本落实
target cost management 目标成本管理
target cost management of railway transport enterprise 铁路运输企业目标成本管理
target coverage 目标范围
target cross-section 目标有效截面;回波面积
target cut-off potential 靶截止电位
target cutoff voltage 靶截止电压
target damaged 靶标已损坏
target date 预定日期;目标日期
target decomposition 目标分解
target density 靶密度
target depth 目标水深;设计井深
target-designating system 目标标示系统
target designation 目标指示
target designation system 目标指示系统
target designator 目标指示器
target detect 目标检测
target detecting device 目标探测器
target detecting system 目标探测系统
target detection 目标探测
target detection technique 目标探测技术
target detection unit 目标探测装置

target detector 目标探测器
target diagram 电灯特性图
target direction 目标方向
target director post 目标指示站
target discrimination 目标识别;目标鉴别
target discriminator 目标鉴别器
target distribution 目标分布
target dividend per share 每股目标股息
target dossier 目标档案
target drone 靶机
target echo 目标回波;目标反射信号
target electrode 靶电极
target elevation 目标高度
target estimate 目标预算
target estimate type 轨道限额合同
target exposure 目标暴露量
target fade 目标信号消失;目标信号减弱
target fidelity 目标逼真度
target figure 预定数字
target flicker 靶信号闪变
target flow transmitter 靶式流量变送器
target following 目标跟踪
target for increased production 增产指标
target framework 靶标架
target glass 玻靶;靶玻片
target growth rate 目标增长率
target heater 靶面加热器
target homing 目标搜寻
target identifiability 目标的可识别性
target identification 目标识别;目标判别
target identification equipment 目标识别设备
target identification system 目标识别系统
target illuminating laser radar 目标照明激光雷达
target illuminating source 目标照明光源
target illumination 目标照度
target image 目标图像;靶像
target imaging 目标成像
target-indicating sight 目标指示瞄准具
target indicating system 目标指示系统
target indicator 照明弹;目标指示器
target information 目标信息
target in kind 实物指标
target instruction target word 目标指令
target intelligence 目标信息
target intensity 目标强度
target in transit 照准张标
target lamp 目标灯;灯塔;道岔信号灯
target language 目标语言
target length 目标长度
target level(l)ing rod 靶板水准(标)尺;标杆
target lifetime 目标寿命
target light 道岔信号灯
target line-method 照准线法
target location 目标定位
target lost 目标失踪
target machine 目标程序机
target mapping 标图
target marker 目标识别器
target marketing 目标行销
target material chart 目标资料图
target material graph 目标资料图
target mean strength 设计平均强度
target memory 目标存储器
target mesh 靶网
target mesh microphonics 靶网颤噪效应
target metal 靶极金属
target meter 靶式流量计
target model 空袭目标
target molecule 靶分子
target nucleus 靶核
target-object contrast 标板反差
target of air attack 空袭目标
target of budget 概算指标
target offset 对目标偏移的校正
target of transportation cost 运输成本指标
target output 产量指标
target pattern 攻击航线
target perfect 靶标完好
target phase 运行阶段;目的程序运行阶段;目标阶段
target photo 目标相片
target photograph 目标相片
target pinpoint 目标定点
target plotting 标图

target point 照准点
target (point) braking 目的制动【铁】
target population 人口指标
target position 预定位置
target position finding 目标测位
target positioning data 目标定位数据
target position map 目标位置图
target predicted position 目标预测位置
target present position 目标的现时位置
target price 预定价格;指标价格;标准价格
target prism 目标棱镜
target productivity 计划生产率
target profit 计划利润;目标利润
target profit pricing 目标利润定价
target profit rate 预定利润率
target profit transfer price 目标利润转移价格
target program(me) 目标程序
target programming 目标规划
target property 目标特性
target range 目标范围;打靶场
target range and bearing computer 目标距离和方位计算机
target rate 指标速率;目标收益率
target rate of growth 指标增长率;目标增长率
target rate of return 目标利润率
target-rate-of-return pricing 以目标利润率定价;照目标利润定价;目标收益率价格设定
target recognition 目标识别
target reflectivity 目标反射率
target reflector 目标反射器
target region 靶区
target responsibility system for environmental protection 环境保护目标责任制
target returns principle 目标利润原则;目标利润原理
target rod 靶尺;靶板水准尺;标杆
target routine 目标程序
target search 目标搜索
target seeker 目标跟踪头
target seeking 目标搜索
target seeking device 目标搜索装置
target-seeking guidance 自动跟踪
target seeking unit 目标搜索器
target selection 靶区优选
target selector 目标选择器
target setting 指标设定;目标整定;设定目标
target ship 靶船
target-sighting apparatus 目标跟踪仪
target signal 目标信号
target signature 目标特征图;目标特征
target simulation 目标模拟
target size 靶子大小
targets of environment(al) planning 环境规划目的
target speed 目的速度
target staff 靶板标尺;靶板水准尺
target stated in contract of fixed output 包产指标
target stratum 目的层位
target stratum depth 目的层深
target strength 预期强度;目标强度;期望强度;物标强度;(混凝土等的)试配强度
target swop 目标替换
target system 目标体系
target system of eco-agriculture 生态农业指标体系
target temperature 预定温度
target theory 靶子学论;靶子论;靶学说;靶理论
target tightening 目标集中
target timing 目标定时
target tissue 靶子组织
target track correction 目标跟踪校正
target tracking 目标跟踪
target-tracking filter 目标跟踪滤光器
target-tracking laser radar 目标跟踪激光雷达
target-tracking radar 目标跟踪雷达
target tree of water pollution control 水污染控制目标树
target true motion radar 目标真实运动显示雷达
target type flowmeter 靶式流量计
target type fluid energy mill 单喷式气流粉碎机;靶板式气流粉碎机
target value 指标值
target vane 目标照准器
target variable 目标变数;目标变量

target viscosity 目标黏滞度
target voltage 靶电压
target volume 目标体积
target water quality criterion 目标水质基准
target wire 照准(十字)丝
target word 目标字;目标词
target zone 目标区
tar-gravel roofing 焦油沥青砾石屋面;柏油砂石屋顶
tar-grouted macadam 灌焦油沥青碎石路
tar-grouted stone 薄焦油沥青碎石路;薄柏油薄浆的石料
tar-grouted surfacing 灌焦油沥青碎石面层;灌焦油沥青碎石路面;灌柏油碎石面层;灌柏油碎石路面;柏油浇灌碎石路面
tar heater 热柏油炉;柏油加热器
tar heavy oil 厚柏油;重质柏油
taria 细木镶嵌装饰
tariff 运价;租金表;资费表;关税;税则率;收费制;收费表
tariff act 关税法
tariff agreement 运价协议;运费协定;关税协定
tariff association 保险费率同盟
tariff barrier 关税壁垒
tariff bureau 关税局
tariff ceiling 关税最高限额
tariff concession 关税减让
tariff design 费率设计
tariff differentiation 差别税率
tariff diminution 关税减让
tariff discount 关税折扣
tariff duty 关税
tariff equalization tax 海关平衡税
tariff factory 保税工厂
tariff for tare carriage 空包装箱运价
tariff-free zone 自由关税区
tariff increase 运价增加
tariff index 运价指数
tariff in force 现行税率
tariff kilometerage 运价里程
tariff law 税法
tariffless 不收税的
tariff level 关税水平;收费水平
tariff list 资费表
tariff mileage 运价里程
tariff nomenclature 税则类别
tariff of abomination 被厌恶的关税
tariff of complete railway vehicle 铁路整车运价
tariff of goods classification 货物分等运价
tariff of luggages and parcels 行李包裹运价
tariff of railway container 铁路集装箱运价
tariff of railway part-load 铁路零担运价
tariff of regional railway 地方铁路运价
tariff on electricity 电费率
tariff parity 运价比价
tariff per axes kilometer 轴公里运价
tariff per passenger-kilometer 人公里运价
tariff per person-kilometre 人公里运价
tariff per ton-kilometre 吨/公里运价
tariff per vehicle kilometre 车公里运价
tariff policy 关税政策
tariff preference 关税优惠
tariff protection 关税保护
tariff quotas 关税配额
tariff rate 协定费率;运价率;运价降低;规定税率;规定费率;关税税率;税率
tariff rate of railway section 区段运价率
tariff-rate quota 关税率配额
tariff report 关税同盟
tariff revision 运价修订
tariff schedule 关税表;收费率表
tariff structure 价格结构
tariff subheading 分税目
tariff system 收费制
tariff wall 关税壁垒
tariff zoner 收费率区分器
Tarim marine trough 塔斯曼海槽
tar-impregnated 柏油浸渍的
tar-impregnated brick 沥青浸渍砖
tar-impregnated paper 柏油纸
tar-impregnated roofing paper 盖屋顶用柏油纸
tar impregnation 柏油浸透(物)
tarine 氨基乙磺酸
tariric acid 塔利酸
tar joint runner 柏油接缝填料

tar kettle 沥青锅;柏油烧锅
tar light oil 轻质柏油;轻焦油
tarmac 铺地沥青;柏油碎石(路)
tar-macadam 柏油沥青碎石路;焦油处理的碎石路;铺地用沥青;柏油碎石路;焦油路面材料
tar-macadam binding course 柏油石渣层
tar-macadam carpet 灌焦油沥青碎石磨耗层
tar-macadam mixer 焦油沥青碎石拌和机
tar-macadam pavement 焦油沥青(碎石)路面;柏油碎石路面
tar-macadam paver 焦油沥青碎石铺路机
tar-macadam paving 焦油沥青(碎石)铺面;焦油沥青碎石路面
tar-macadam plant 柏油碎石道路材料制造厂
tar-macadam surfacing 焦油沥青碎石路面
tar-mac runway 柏油跑道
tar mastic 焦油胶泥
tar medium oil 中性柏油;中焦油
tar membrane 焦油沥青层
tar mixer 柏油调合机;柏油调拌器;柏油拌和机
tar mixing machine 柏油拌和机
tar mixing plant 焦油沥青调和装置
tar mortar 焦油灰浆
tarn 山中小湖;山间水湖;冰斗湖
tarnish 光泽变暗;金属失光;晦光;雾斑;脱色;沾污;表面失去光泽;表面变色;锈斑;色斑
tarnish decolo(u)ration 无光泽
tarnish decolo(u)rization 表面变色
tarnished 有污点的;脱色的
tarnish film 氧化膜;锈膜
tarnishing 去光泽的;有污点的;失泽
tarnishing action 失泽作用
tarnishing oil 去光泽油
tarnishproof board 防晦暗板
tarnish resistance 抗蚀性;抗蚀力;抗溶蚀力
tar number 焦油值
tar oil 高黏重质原油;焦油油;煤焦油;松溜油;松焦油;粗焦油
tar-oil fraction 焦油馏分
tar-oil preservative 柏油防腐剂
tar-oil type preservative 焦油类防腐剂
tarp 舱盖布;防水帆布;柏油帆布;焦油帆布
tar pail 柏油桶
tar paint 焦油沥青涂料;柏油涂料;焦油涂料;焦油沥青漆
tar paper 焦油纸;沥青纸;柏油沥青纸;防潮(沥青)纸
tar paper cover 焦油纸覆盖(面);柏油纸覆面
tar paper subgrade treatment 柏油纸覆盖路基
tarpaulin 油布;焦油防雨布;焦油防水布;柏油防水布;篷布;船员;舱盖布;防雨布;水防水布;防水帆布
tarpaulin canvas 柏油帆布
tarpaulin paper 防水油纸;柏油防水纸;防潮纸
tar-paved 铺柏油的(路面)
tar pavement 焦油沥青碎石路;柏油铺面
tar paving 柏油铺面;焦油沥青铺面;焦油沥青面层;舱盖布
tar penetration coat 浇灌柏油层
tar penetration macadam 柏油渗透的碎石路面
tar petroleum asphalt mix(ture) 焦油石油沥青混合物
tar pitch 柏油;硬焦油脂;硬焦油;焦沥青(脂)
tar pitch clay pipe sealing compound 柏油沥青黏土管密封口合成物
tar pollution 焦油污染
tar pot 焦油沥青烧锅;焦油沥青罐;柏油锅;柏油罐
tar pouring rope 浇注柏油的绳索
tar pre-impregnating 柏油预浸
tar prepared roofing 柏油成品屋面材料
tar prepared sheet roofing paper 柏油处理过的薄型屋面材料纸
tar pre-saturated 柏油预饱和
tar-primed base 浇柏油沥青透层的基层
tar product 焦油沥青产物;焦油产品
tar production 焦油生产
tar protective coat(ing) 柏油保护层材料
tar pump 沥青泵;柏油泵
tar rag felt 柏油毡
Tarran 塔兰(一种装配房屋)
tarras cement 火山灰质水泥
tar ready sheet roofing paper 预制柏油型屋顶材料纸

tarred 涂柏油;涂柏油的
tarred board 焦油纸板
tarred chip(ping)s 涂柏油碎片
tarred coarse chip(ping)s 涂柏油的粗糙碎片
tarred composition roofing 涂柏油混合物屋顶材料
tarred cord 浸焦油麻绳
tarred cord dipped in asphalt 焦油麻绳;浸沥青焦油麻绳
tarred felt 油毡;油毛毡;油毡;油毛毡;涂柏油的屋顶油毛毡
tarred felted fabric 涂柏油的油毛毡织物
tarred filter 配衡滤器
tarred fittings 小油绳;小绳
tarred hemp rope 三股油麻绳
tarred joint runner 柏油填缝条
tarred limestone 涂柏油的石灰石
tarred macadam 焦油沥青碎石;柏油碎石
tarred macadam road 柏油碎石路
tarred macadam surfacing 焦油沥青碎石路面
tarred marline 细油麻绳;柏油麻绳
tarred oakum 焦油麻絮;焦油麻丝
tarred paper 焦油纸;柏油纸
tarred prepared roofing 涂柏油的预制屋顶材料
tarred prepared sheet roofing paper 涂柏油的预制薄板屋顶材料纸
tarred ready roofing 涂柏油的预制屋顶材料
tarred road 涂柏油路;柏油路
tarred rolled-strip roofing felt 涂柏油的屋顶卷材油毛毡
tarred roofing felt 屋面油毡;屋顶油(毛)毡
tarred rope 油麻绳;油浸绳;涂柏油麻绳;柏油麻绳
tarred rope sealing 涂柏油的密封材料;柏油麻绳止水
tarred slag 焦油沥青矿渣
tarred stone 焦油沥青碎石(路)
tarred tow 涂柏油的拖绳
tar refining 焦油沥青精炼
tar removal 脱焦油;除焦油
tar-resisting paint 抗柏油油漆
tarring 浸柏油;焦油化;涂柏油
tar road metal 柏油铺路碎石
tar-road-mix surface 路拌柏油混合料面层;路拌柏油混合料路面;路拌沥青混合料面层;路拌沥青混合料路面
tar roofing 柏油屋面
tar roofing felt 油毛毡
tar roofing product 柏油屋顶材料产品
tar rope sealing 柏油绳密封材料
Tarros type 塔罗斯型
tarry 焦油状的;涂柏油的
tarry cut 焦油馏分
tarry device 调时装置
tarry distillate 焦油馏分
tarry matter 煤胶物质
tarry smell 柏油臭味
tarry stool 煤焦油样粪
tarry valve 停留调时阀;定时阀;调时阀
tar sand 沥青砂;焦油砂;焦油沥青砂;柏油砂混合料
tar sandstone 沥青砂岩
tar saturant 柏油浸透的
tar-saturated 浸柏油的
tar saturation 柏油饱和状态
tar saturator 柏油浸透器;焦油浸出机
tar seal 沥青封彻;沥青封闭层
tar-seal pool 沥青封闭油藏
tar-seal trap 沥青封闭圈闭
tar seep 沥青苗
tar separation 除焦油
tar separator 焦油分离器
tar sheet 盖舱布;柏油帆布
tar sheet roof(ing) felt 柏油薄型屋顶材料纸
tarsia 拼花制品;镶木制品;嵌木制品;嵌木细工
tar slag chippings 柏油矿渣
tar slurry 焦油泥浆
tar soil stabilized base 焦油沥青稳定土基层
tar solution 焦油溶液
tar spirit 由木焦油蒸馏得到的强溶剂
tar spray can 浇柏油罐
tar spray car 浇柏油罐车
tar-sprayed 浇柏油的
tar-sprayed road 浇焦油沥青
tar sprayer 焦油喷射机;焦油沥青喷洒机;柏油喷洒机
tar spraying 焦油沥青表面处治;浇洒柏油;喷洒焦

tar spraying machine 油沥青；浇柏油；喷洒柏油
tar spraying machine 焦油沥青喷洒机；柏油喷洒机
tar spraying tank 焦油喷洒车；洒柏油柜车
tar spraying vehicle 柏油喷洒车
tar stabilization 焦油沥青稳定化；柏油加固处理
tar-stained wool 沥青毛
tar styrene paint 焦油苯乙烯涂料
tar surface 焦油沥青碎路面
tar surface dressing 焦油沥青表面处治
tar surface treatment 焦油沥青表面处理
tar surfacing 焦油沥青路面
tartan 合成橡胶材料（运动场铺地面用）
tartan twin 格子双晶
tartar emetic 酒石酸养锑钾；吐酒石
Tartarian polarity chron 鞑靼极性时
Tartarian polarity chronzone 鞑靼极性时间带
Tartarian polarity zone 鞑靼极性带
tartaric acid 酒石酸
tartaric tester 酒石测试器
tartaric tile 酒石瓦；酒石砖
tartaric tiling 酒石面砖；酒石盖瓦
tartaric value 酒石值
Tartary Strait 鞑靼海峡
tar trap 焦油收集阱；焦油阱
tartrate 酒石酸盐
tartrazine lake 酒石黄色淀
tartrazine yellow 酒石黄
tar-treated 柏油处治的；柏油处置的
tar-treated surface 焦油沥青处治面层；焦油沥青处治路面
tartufite 纤方解石
tarungar 碎石筐（又称石笼）
tar urethane 沥青氨基甲酸酯
tar urethane rubber 沥青氨基甲酸酯橡胶
tarus 屋脊筒瓦；筒形隆起屋顶坡面；复折屋顶筒形脊
tar use 焦油使用
tar value 焦油值
tarvan 载重汽车和铁路货车
tar vapo(u)r 焦油蒸气
tarvia 他维亚（一种用于筑路的沥青焦油）
Tarvialithic 柏油混合料（一种冷铺柏油混凝土混合料）
tarviated macadam 焦油沥青碎石路；柏油碎石
tarviated macadam road 柏油碎石路
tar-water 焦油冷浸剂
tar-water emulsion 焦油和水乳浊液
tar waterproof board 沥青防水纸板
tar well 焦油收焦器
tar yield 焦油收率；焦油产率
tar yield graduation 焦油产率分级
tas-de-charge 起拱层；基石；拱顶基石
taseometer 应力计
Tashtyp polarity superchron 塔什蒂普极性超时
Tashtyp polarity superchronzone 塔什蒂普极性超时间带
Tashtyp polarity superzone 塔什蒂普极性超带
tasimeter 温湿度电微压计；电微压计（测温度变化的）
task 任务；定额
task-ambient lighting 作业环境照明
task analysis 任务分析
task attribute 任务属性
task-based appraisal 以工作为依据的评价
task-based participation 以工作为依据的参与
task call 任务调用
task committee 工作委员会
task completion date 任务完成日期
task control block 任务控制块；任务控制程序段
task control card 任务控制卡片
task control program(me) 任务控制程序
task creation 任务建立
task data 任务数据
task definition 任务内容
task definition table 任务定义表
task description 任务说明书
task descriptor 任务描述符
task design 工作设计
task determiniation 任务确定
task dictionary table 任务解释表
task dispatch 任务调度程序
task dispatcher 任务调度程序
task equipment 专用设备
task execution memory 任务执行存储器
task family 任务族
task force 专题工作组；工作组；任务强制转移；特别工作组
task force on low and cold emission sources 低和冷排放源问题工作队
task group 专题工作组；工作组；特别工作组
task idea 任务观念
task illumination 专用照明；任务照明；作业照明
task input/output table 任务输入输出表
task interrupt control 任务中断控制
task letter of certificate 设计任务书
task macro 任务宏指令
task management 计件作业管理；任务管理；定量作业管理
task management program(me) 任务管理程序
task management system 任务管理系统
task master 工头；工长；监工
task memorandum 任务备忘录
task method of budgeting 按任务编制预算法
taskmix 作业混合
task name 任务名字
task of design 设计任务
task of making segment known 段挂号任务
task option 任务选择
task-oriented group 任务定向群体
task performance 工作特征；工作特性
task priority 任务优先级
task program(me) 任务程序
task queue 任务排队【计】；任务队列【计】
task quit 任务停止
task reassigment 任务再指定
task retry routine 任务再执行程序
task scheduler 任务调度程序
task scheduler queue 任务调度排队
task scheduling 任务调度
task selection 任务选择
task session 专题研究会；工作会议
task status 任务状态
task status iudex 任务状态索引
task supervisor 任务管理程序
task suspension 任务暂停；任务挂起
task switch 任务转换
task synchronization 任务同步
task termination 任务终止；任务结束
task time 工时定额；任务时间；定额时间
task timer 任务计时器
task variable 任务变量
task work 件工；包干工作；计件工作
Tasmania geosyncline 塔斯曼地槽；塔斯马尼亚地槽
Tasmanian bluegum 篮桉
Tasmanian oak 橡木（澳大利亚塔斯马尼亚产）
Tasmania seaway 塔斯曼海路
tasmanite 沸黄霞辉岩；油页岩；含硫树脂；塔斯曼油页岩
Tasman Pulp and Paper Co. Ltd. 塔斯曼纸浆和纸公司（新西兰）
Tasman Sea 塔斯曼海
taspinite 杂块花岗石
tassel 游檐木；缨子；流苏（饰）；穗子；垂花饰【建】；承梁木
tassel and slide 大装穗饰（窗帘拉绳上的）
tassel bush 流苏形灌木
tassel flower 流苏花
tassement polaire 极填料
taste and odo(u)r control 除味去臭
taste and odo(u)r removal 除味去臭
taste of water 水味
taste removal 除味
taste suppression 去味
taste test 舌感试验
tat 粗袋布
tatami 榻榻米（日本）；稻草垫席
tatan-barium white 钛钡白
tatanising 钛涂层增强
tatarian dogwood 红瑞木；红硬木
tatarskite 水硫碳钙镁石
tate-beche 并列单边带发送系统
tatenerian arch 悬链状拱
tatol evapotranspiration 蒸发蒸腾量
tat patti 粗袋布
tattering 拉紧
tattler 猛烈雷暴
tattle-tale 钻进速度记录仪
tattooing 刺花
tatty 湿帘（用以降低室内气温，保持潮湿的门帘或窗帘）
Tauber test 陶贝尔试验
Tau cross T形十字架
taula 毛石平台
taunt mast 高桅杆
tauochrone 等时差线
taupe 灰褐色
taupe brown 灰棕色
taupe gray 褐灰色
taurine 氨基乙磺酸
tauriscite 七水铁矾
tausonite 等轴锶钛石
taut 整齐的；严格的；拉紧的
taut band 紧带
taut band ammeter 紧带安倍计
tautening rod 紧索棒
taut helm 转舵吃力
taut-line cableway 紧索（缆）道
tautness 张紧度；紧固度
tautness meter 应变计；拉力计；紧度计；伸长计
tautochrone 等时曲线；等时降落轨迹
tautochronism 等时性
tautological implication 永真蕴涵
tautology 永真式；重言式
tautology rule 重言式规则
tautomer 互变异构体；互变体
tautomeric 互变异构的；互变的
tautomeric effect 互变效应
tautomeric form 互变异构形式
tautomeric formula 互变异构式
tautomeride 互变体
tautomerism 互变异构性；互变异构现象；互变（现象）
tautomerization 互变
tautomerizing 互变（现象）
tautonymy 重名关系
taut planning 突击计划
taut ship 装备良好的船；完备的船
taut-wire (测海上距离用的)张绳；张紧索
taut wire buoy array 紧缆浮标阵列
taut wire gear 海上基线测距仪
taut-wire printing 紧丝印刷（法）
taut-wire system 张紧索系统
taut-wire traverse 张绳测距法（海上测距用）
tave 侦察鱼群的网
tavelling crane 移动起重机
tavern 客寓；小酒店；酒馆
tavorite 羟磷锂铁石
taw 井下爆炸器
tawmawite 铬绿帘石
tawny 黄褐色的；茶色的
tawny or gray 黄褐色或灰色
tax 捐税；税(收)；税捐；税金；赋税
tax abatement 减税
taxability 纳税能力；应纳税；课税能力；可课税性；赋税能力
taxable 征税的；可征税的
taxable amount 应税金额
taxable base 应税基数
taxable capacity 可税能力；纳税能力；付税能力
taxable equivalent yeild 赋税等值收益率
taxable horsepower 收税马力
taxable income 应交税所得；课税所得；可征税收入；可税所得
taxable items 税目
taxable level 课税环节
taxable limit 课税限度；纳税起征线
taxable profit 应税利润；课税利润；可征税利润
taxable property 可征税财产
taxable unit 应纳税单位；课税单位
taxable value 计算税价格
taxable year 应税年度；课税年度
tax accounting 税务会计
tax accrued 应计税金
taxadjunct 未分类土壤；未划分土壤
tax adjustment 税收调整
tax administration 税务管理；税收管理
tax affidavit 免税声明
tax agency 税务机构
tax allocation with a period 捐税当期分配；赋税当期分配
tax allowance 税收优待
tax allowances for anti-pollution investments 税率优待反污染投资

tax and insurance 本金、利息、税金和保险费
tax anticipation bills 税收共付票据;赋税抵偿国库券
tax anticipation notes 赋税抵偿债券
tax apportionment 税收分配
tax appraisal 纳税鉴定
tax articles 应纳税商品
taxas 河船最高甲板室
taxas deck 河船最高甲板
tax assessment 征税估价
taxation 征税;纳税;税制;税收;税款
taxation bears hard 赋税沉重
tax(ation) bureau 税务局
taxation categories 税目
taxation of distribution link 分配环节课税
taxation system based on progressive rates 全额累进税制
taxation treatment 课税处理方法
tax at source 从源课税
tax audit 税务监察;税收审计
tax authorities 税务机关
tax avoidance 避税;避免纳税
tax base 征税基准;征税基数;征税标的;课税依据;课税基准;计税基数
tax basis 征税基数;课税标准
tax bearer 纳税人;赋税负担者;负税人
tax bearing 税负
tax bearing capacity 税负能力
tax bearing ratio 税负率
tax-benefit rule 课税受益原则;课税沾惠原则
tax bond 抵税公债
tax bracket 税收等级;税率等级;税级
tax burden 课税负担;税收负担;税负
tax burden ratio 税收率;税负率
tax categories 税种
tax certificate 税征
tax collecting 征税
tax collection 收税
tax collector 税务员;收税员;收税人
tax computation 税款计算
tax concession 减免税办法;税收优惠;税收减让(指对投资的鼓励)
tax corporate 课税法人
tax corporation 课税法人
tax criterion 课税扣除;课税津贴;纳税扣除;纳税额的减除;税收抵免;税款减除数;税捐信贷;税额减免;抵免;赋税宽减额
tax credit method 税收抵免法
tax credit scheme 免税方案
tax criteria 课税标准
tax cutback 税收削减
tax cut by law 依法减税
tax day 征税日期
tax declaration 纳税申报
tax deductible 可减免课税的
tax deduction 应税金额的扣除;课税免除;课税扣除;课税减免;纳税时可扣除的项目;纳税扣除;赋税扣抵
tax deduction method 已纳税款扣除法
tax deferral 延期纳税;延期缴税
tax delinquency 税收滞纳
tax delivery 上交税金
tax difference 课税差别;计税差别
tax differentiation 税收差别
tax differentiation on petroleum 对石油的差别计税
tax dirigation 干预经济的税收政策
tax dodger 偷税人;偷税漏税者;逃税人;逃避纳税者
tax dodging 漏税;偷税
tax due 到期应纳税款
taxed article 已税品
taxed commodity 已税商品
taxed product 已税产品
tax effect 课税影响;课税效果
tax effort 课征效率
tax equalization 纳税均等
tax equity 课税公平;纳税公平
taxes and duties 税费
taxes and levies 捐税
taxes and profits to be delivered to the sate 上交税利
taxes assessed by taxpayers' report 纳税人申报的税额
tax escape valves 逃税途径

taxes collected in advance 预收税款
taxes due 应缴税金
taxes during construction 建造期的税项
taxes estimating 税款预算
taxes in kind 用实物缴税
taxes on buildings 房产税
taxes on business 营业税
taxes on house 房产税
taxes on the possession of resources 资源占有税
taxes other than income 所得税以外其他捐税
taxes paid in kind 用实物缴税
taxes payable 应缴税金
taxes withheld 扣缴税捐
tax-evader 偷税漏税者
tax evasion 偷税漏税;偷税
tax exclusion 课税外项目;课税豁免;免税;不预课税;不予课税
tax exempt 免税债票;免税的
tax-exempt bond 免税票证;免税公债;免税债券
tax exempt exports 免税出口货物
tax-exempt export tax-free export 免税输出
tax-exempt fellowship 免税研究奖金
tax-exempt import 免税进口货物;免税进口
tax-exempt import tax-free import 免税输入
tax-exempt income 免税收益;免税收入
tax-exempt interest 免税利息
tax exemption 免征额;税收减免
tax exemption certificate 免税执照;免税证(明)书;免税凭证
tax exemption certification 免税证明
tax exemption system 免税制度
tax-exempt items 免税项目
tax-exempt items allowance 免税项目
tax-exempt organization 免税组织
tax-exempt property 免税房地产
tax-exempt scholarship 免税奖学金
tax farmer 包税人
tax farming 包税
tax ferret 逃税侦查员
tax for default 滞纳加征税款
tax forms 纳税申报表(格)
tax for regulating differential income 级差收入调节税
tax for the examination of deeds 验契税
tax fraud 偷漏税
tax-free 免税的;无税
tax-free allowance 免税宽减额;免税津贴
tax-free article 非课税品
tax-free bond 免税债券
tax free corporation 非课税法人
tax-free distribution 免税分配额
tax-free exchange 免税交换;免税交易;免税资产交换
tax-free exchanges and transfers 免税资产交换或转让
tax free export 免税出口货物
tax-free import 免税进口货(物);免税进口
tax-free income 免税所得
tax-free interest 免税利益
tax-free investment 免税投资
tax free liquidation 免税清算
tax-free or tax-refund 无税或退税
tax-free profit 免税利润
tax-free reserve funds 免税准备基金
tax-free trade zone 免税贸易区
tax-free transfer 免税资产转让
tax function 税务功能
tax fund 税源
tax haven 纳税隐藏所;免税港口;免税地区;逃税乐园;避税港;避税掩蔽所
tax haven activities 避税港活动
tax hike 赋税增加
tax holiday 免税期;赋税优惠期
taxi 滑走;出租小汽车;出租汽车
taxi cab 出租小汽车;出租汽车
taxi-channel 滑行水道;滑行航道
taxi-channel light 滑行水道灯
taxi-channel marker 机场标志板
taxi coach 大型出租汽车
taxi distance 滑行距离(飞机)
taxifolin 黄杉素
taxiing guidance system 滑行引导系统
taxiing speed 滑行速率
taxiing traffic 飞机滑行交通

taxilane 滑行道
taxi levy 税收
taximeter 计程器;里程计;计价表;车费计;车用计费器
tax in arrears 滞纳税款
tax incentive 税收减让(指对投资的鼓励);赋税刺激
tax incentives for investment 投资税收减让
tax incidence 课税归宿;赋税归宿
taxing 滑行;乘出租汽车
taxing authority 征税官员
taxing of agricultural land 对农业用地的课税
tax in kind 实物付税
tax inspection 纳税检查
tax investment 避税投资
taxi operating speed 出租汽车运行速度
taxi parking area 出租汽车停车场
taxi plane 出租飞机
taxirank 出租汽车站;出租汽车停车场;出租汽车停车处
taxi service 客运汽车出租
taxi sign 出租汽车标志
taxi stand 出租汽车站;出租汽车停车场
taxi strip 滑行道
taxite 斑杂岩
tax items 税目
taxitic-streaked pseudofluidal structure 斑杂条带状假流纹构造
taxitic structure 斑杂构造
taxi track 滑行道
taxi-track light 滑行道(照明)灯
taxiway 滑行跑道;飞机滑行道
taxiway light 飞机滑行灯;滑行道(照明)灯
taxiway lighting 滑行道照明
tax law 税法
tax lease 课税租赁
tax liability 应纳税金;纳税义务;赋税义务
tax lien 课税扣押权;课税留置权;税务扣押权
tax limit 课税限额
tax-loaded 负担税收者
tax loading 负担税收
tax loophole 税收漏洞;赋税漏洞
tax-loss benefit 亏损税款抵免;亏损利益
tax-loss carry-back 赋税亏损退算;纳税时亏损转回
tax-loss carry-forward 赋税亏损结转;纳税时亏损结转
tax map 税务地图
tax money 税款
tax obligation 纳税义务
Taxodiaceae 杉科
tax office 税务所
tax of vehicle and ship licence 车船使用牌照税
taxology 分类学
taxon 分类单元;分类群;分类单位;[复]taxal
tax on a contract 合同税
tax on a corporation 公司税
tax on aggregate income 综合所得税
tax on agriculture 农业税
tax on artificial persons 法人税
tax on automobiles 汽车使用税
tax on capital profit 资本利得税
tax on commodities 货物税
tax on communication 交通税
tax on consumption 消费税
tax on cost 原价税
taxon cycle 类群分化循环
tax on enterprise 企业税
tax on expenditures 消费税
tax on imports 进口税
tax on income 所得税
tax on income and profit 所得及利润税
tax on licenses 营业执照税
tax on mine 矿税
tax on net investment income 投资净收入税
taxonomic 分类(学)的
taxonomical group 分类群
taxonomic biochemistry 分类生物化学
taxonomic category 分类项目;分类阶元
taxonomic character 分类性状
taxonomic frequency rate 分类频率速度
taxonomic series 分类系列
taxonomic species 分类种
taxonomy 分类学
taxonomy grade 分类等级
taxonomy system 分类系统

tax on personal income 个人所得税
tax on product 产品税
tax on profit 利得税
tax on property 财产税
taxon range zone 分类单元延限带
tax on reappraisal of tax 资产重估税
tax on revaluation 重估价税
tax on separate income 分类所得税
tax on special consumption behavio(u)r 特种消费行为税
tax on stalls 摊贩业税
tax on the above-norm bonus 超限额奖金税
tax on the write-up 资产重估增值税
tax on transaction 交易税
tax on turnover 周转税
tax on undistributed profit 未分配利润税;未分配利润率
tax on value added 增值税
tax on writeup 重估增值税
tax paid in advance 预付税款
tax paper 纳税通知书
tax paradise 纳税隐藏所;避税港
tax participation 产业税参股
tax payable 应付税捐;应付税
tax payable account 应付税款账户
tax payable audit 应交税金审计
taxpayer 纳税人
taxpayer revolt 纳税人抗税
taxpayers consultant 纳税辅导
tax payment 支付税款;纳税;税款
tax payment date 纳税日期
tax payment in kind 实物纳税
tax payment place 纳税地点
tax payment receipt 纳税收据;纳税凭证;完税凭证
tax planning 税务计划
tax planning scheme 赋税计划
tax policy 税收政策
tax position 课税情况
tax price 课税价格
tax privilege 税收特惠
tax pyramiding 重叠计税
tax rate 税率
tax rate limit 税率限度
tax rate reduction 降低税率
tax rate schedule 税率表
tax rate structure 税率结构
tax rebate 减税退款;退税
tax receipts 税收收入
tax receivable 应收税款
tax receivable delinquent 应收滞纳税款
tax receiver 收税员;收税人
tax reduction 减税
tax reference price 征税参考价格;课税参考价格
tax reform(ation) 税制改革
tax refund 退税
tax refundable 可获发还之税款
tax refunded 已退税款
tax registration 纳税登记工作
tax regulation 税则;税收条例
tax reimbursement 退税
tax reimbursement for export 出口退税
tax-related provisions 有关税收的规定
tax relief 减免税;税款减免;税捐减免
tax remission 赋税减免;赋税豁免;赋税分配
tax repayment 预付税款
tax reserves 纳税准备金
tax reserve certificates tax home 纳税准备证书
tax retention ratio 纳税后利润留存比率
tax retirement 免税
tax returns 纳税申报单
tax revenue 税收收入;税收;赋税收入
tax revision 税制修改
tax revolt 反税
tax roll 课税清册;捐税清册;纳税清册
tax saving 节税
tax schedule 税率表
tax search 税务检查
tax selling 纳税抛售;免税售出
tax sensitivity 赋税敏感性
tax sharing 课税分成;赋税分担
tax shelter 将利润、收益用于投资等用途以推迟纳税;纳税隐藏所;逃税手段;赋税优惠;避税港
tax-sheltered annuities 避税年金
tax shield 税收挡板;赋税保护

tax shifting 赋税的转嫁
tax sparing 纳税饶让
tax sparing credit 免税信贷
taxs shield 节税手段
tax stamp 印花
tax stop clause 承ином人支付可能增加的财产税条款
tax structure 税结构;赋税结构
tax surcharge on imports 进口附加税
tax system 课税制度;税制
tax table 税率表
tax take 税收总额
tax technique 课税方法;课征方法
tax threshold 免缴税入息额;起征界限
tax timing 税款调节
tax title 税契
tax to be paid in advance in quarterly instalments 分季预缴税额
tax treatment 课征方式
tax unit 课税单位;纳税单位
tax value 课税价格
tax withheld 已扣缴预提税款;扣缴捐税
tax withheld certificates 扣缴税款凭证
tax write-off for all the expenditures 纳税时各项费用的冲销
tax year 课税年度;纳税年度
tax yield 赋税收入
Taylors approximation 泰勒近似
Taylor's bag filter 泰勒袋滤器
Taylor's circle 泰勒圆
Taylor's column 泰勒柱
Taylor's connection 泰勒接线法
Taylor-Scotson type automatic voltage regulator 泰勒斯科特自动调压器
Taylor's differential piece-rate system 泰勒差别计件工资制
Taylor's diffusion 泰勒扩散
Taylor's effect 泰勒效应
Taylor's equation 泰勒方程式
Taylor's expansion 泰勒展开式
Taylor's formula 泰勒公式
Taylor's four-roll mill 泰勒四滚筒磨
Taylor's functional organization 泰勒的职能组织
Taylor's graphic(al) construction 泰勒图解法
Taylor's instability 泰勒不稳定性
Taylor's linearization 泰勒线性化
Taylor's lino(leum) 泰勒亚麻油地毡
Taylor's method 泰勒拉丝法(封入玻璃管拉丝法)
Taylor's number 泰勒数
Taylor's Piecework Plan 泰勒论件计划
Taylor's principle 泰勒原理
Taylor's quotient 泰勒比值;速长比
Taylor's resilient 泰勒弹力摛纵机构
Taylor's series 泰勒系列;泰勒级数
Taylor's series expansion 泰勒级数展开
Taylor's stability chart 泰勒稳定图
Taylor's stability number 泰勒稳定数
Taylor's system of functional foremanship 泰勒的职能工长制
Taylor's theorem 泰勒定理
Taylor's vortex 泰勒旋涡
Taylor's vortex flow 泰勒涡流
Taylor's wave 泰勒波
Taylor's wire method 泰勒线法
Taylor-White process 泰勒怀特炼钢法
Taz glacial stage 塔兹冰期
tazheranite 等轴钙锆钛矿
Taz stage 塔兹冰阶【地】
tazza 喷水池台座
T bandage 丁字带
T-bar 丁字形钢;丁字铁;丁字钢
T-beam 丁字形梁;丁字梁;丁字架;T形梁
T-beam and slab construction 肋形板结构
T-beam bridge T形梁桥;单肋式梁桥
T-beam cross-section 丁字梁截面
T-beam floor T形梁楼板;丁字梁楼板
T-beam floor panel 丁字形梁楼板节间;丁字梁楼节间
T-beam footing T型梁柱脚;倒T型梁连接的底脚
T-beam girder 丁字梁桁架
T-beam rib T形梁肋
T-beam slab 肋形板
T-beam stirrup 丁字梁籀筋
T-beam with double reinforcement 复筋丁字梁
T-beam with single reinforcement 单筋丁字梁

T-bend 三通管;丁字弯头;T形支管
T-bevel T形斜角规
T-block T形大楼;T形块料
TBM 隧道掘进机
T-bolt 丁字形螺栓;丁字螺栓;T形栓;T形螺栓
T-branch T形支管
T-branch pipe T形管;T形分叉管;三通管
T-bulb-bar 球缘T形材
T-bulb steel 圆形丁字钢
T butt joint 丁字对接头
Tchahar taq 四方圆穹亭(莎桑尼亚建筑的)
Tchebycheffian rule 柴贝夫法则
Tchebycheff's rule 柴贝夫法则
T-circuit T形网络
T-clamp 丁字形夹头
T-coating 防反光膜
T-concrete beam 混凝土T形梁
T-connection 三通接头;丁形连接;T形接头
T-connection sheet pile T形接头连接板桩
T-crossing 丁字街;横穿通行
T cross-section T形断面
T-dike 丁坝;丁字突堤;丁字堤
tea 茶叶;茶剂
tea ash 茶灰
tea bag infusion paper 茶叶袋浸泡纸
tea bag paper 茶叶袋纸
tea bar 茶室;茶馆
teaberry 冬绿树;冬青树
tea-booth 茶亭
tea-box board 茶箱板
tea brick 茶砖
tea-caddy 茶叶筒;茶叶罐
tea canister 茶叶罐
tea-ceremony house 单栋茶室(日本)
teacherage 教员住宅;教师宿舍
teacher room 教室
teachers college 师范学院
teachers school 师范学校
teacher's training college 教师培训学院;师范学院
tea chest 茶叶筒
teach-in 专题讨论会
tea china 茶杯
teaching 教学的
teaching aid 教学用具
teaching amp 教学地图
teaching and playback robot 示教再现型机器人
teaching and research room 教研室
teaching assistant 大学助教
teaching block 教学楼;教学大楼
teaching building 教学大楼
teaching hospital 教学医院
teaching lab(oratory) 教学实验室
teaching microscope 教学显微镜
teaching model 教学模型
teaching plan 教学计划
teaching pool (室内的)教学游泳池
teaching principles 教学原则
teaching program(me) 教学计划;教学大纲;教学程序
teaching reinforcement 教学的加强
teaching replica 教学模型
teaching research group 教研组
teaching wing 教学侧楼
teach mode 示教状态;示教方式
tea-colo(u)red crystal 茶晶
tea cropper 茶叶收获机
teacup with rim shaped like Buddha's fingernails 佛指甲茶杯
tea cutter 茶叶切断器
tead cycle time 读周期时间
tea drier 焙茶机
tea dust 茶末
tea-dust glaze 茶叶末釉
tea estate 茶树种植园
tea factory 茶厂
tea field 茶园
tea finishing machine 茶叶精制机
tea finish machine 茶叶加工厂
tea firing machine 茶叶干燥机;焙茶机
tea-flake glaze 茶叶末釉
tea garden 茶叶种植园;茶室公园;设有茶点部的公园;茶园
teagle 卷扬机;绞盘机
teagle post 支柱

tea green 茶绿色
tea green glaze 茶绿釉
tea-green marl 茶绿泥灰岩
tea grove 茶树林
tea grower 茶农
tea house 茶馆
teak 柚木;麻栗树
tea kiosk 茶亭
tea kitchen 茶点厨房;茶房
teak plank 柚木板
teak tread 柚木踏板;柚木梯级;柚木踏梯
teak trimming 柚木装修
teak window 柚木窗
teakwood 麻栗木;柚木
teakwood decking 柚木甲板铺板
tea-leaf rolling machine 茶叶压实机
tea-leaf steaming machine 茶叶蒸制机
tea-leaf withering machine 茶叶干燥机
teallite 硫锡铅矿
team 工作班;群;班组
team blower 蒸汽喷雾机
team design 成套设计
team economic accounting 班组经济核算
team grade 队级
team hour 工组小时
teaming 联运(由开挖处移去挖土到岸上);兽力运输
teaming contractor 马车运输承包商
team management 小组管理
team-oriented management by objective model 以队组为中心的目标管理模型
teams and groups 班组
teamster 卡车驾驶员;货车驾驶员
team theory 集组理论;队论
team track 装卸线【铁】;货物装卸线【铁】;货车装卸线;车队装卸轨道
team truck 货车装卸队
team work 集体工作;合作;协作;协同动作;共同研究;联合动作;班组工作
team-work achievement 协作的成就
team-work project 协作设计;协作规划;协作方案
team yard 货车装卸场;货场
tea napkin 茶布
tea oil 茶油
tea packaging machine 茶叶包装机
tea packing machine 茶叶包装机
tea plant 茶树
tea plantation 茶园;茶坊;茶场
tea plant pruning machine 茶树剪修机
tea pocket forming and filling machine 茶叶制袋包装机
tea pot 茶壶
teapot-effect 茶壶效应
tea pot welcome spring 报春壶
teapoy 三脚几;三条腿装饰台;茶几
tea-processing equipment 茶叶加工设备
tear 破损;撕开;扯裂
tear and wear 磨损;损耗
tear-and-wear allowance 允许磨耗;磨损容差;磨损留量;容许磨耗
tearaway force 脱开力
tearaway load 脱开力;断开力
tear down 扯下;拆除;解体
teardown of engine 拆卸机器
teardown stand 拆卸台架
teardown time 拆卸时间
tear drop 表面圆斑;水滴
teardrop balloon 水滴状气球
teardrop body 流线型车身
teardrop set bit 胎体滴状突出式金刚石钻头
tearer 强烈风暴
tear factor 抗摩因数;抗裂系数
tear fault 捩断层;横推断层
tear fracture 撕断
tearing 釉面干燥裂纹;图像撕碎;撕破;撕裂的;扯裂
tearing away 塌落
tearing away of bank 河岸滑坡;河岸崩坍
tearing down 拆除;拆毁;拆卸
tearing energy 撕裂能
tearing failure 撕裂破坏
tearing foil 防爆膜
tearing force 撕力
tearing instability 撕裂不稳定性
tearing limit 抗扯裂极限
tearing method 分裂法

tearing mode 撕裂模
tearing mode instability 撕裂模不稳定性
tearing mode of crack 撕开型裂纹
tearing resistance 抗断强度;撕破强度
tearing rupture 扯裂
tearing strain 撕裂应变;断裂应力;扯裂应变
tearing strength 抗扯破强度;撕破强度;撕裂强度;扯裂强度
tearing-strength test 撕裂强度试验
tearing tendency 开裂倾向
tearing test 撕破试验;扯裂试验
tearing tester 撕裂检验器
tear in two 撕成两半
tear jaw 撕裂(试验)用夹头
tear line 撕裂线(包装袋)
tear-off lid 撕开盖
tea room 饮茶室;茶水间;茶室
tearout 撕取力;撕裂力;撕断力;扯下;撕下
tear plate 扁豆形花纹钢板
tear proof 抗扯裂的;抗撕裂的
tear resistance 抗扯性;抗扯裂性;耐磨力;防裂阻力;抗撕裂性
tear rupture 撕裂
tear-shaped 梨形的
tear speed 撕裂速率
tear stains 撕裂斑
tear strength 抗(撕)裂强度;抗扯强度
tear stress 撕裂应力
tear tape 撕条
tear test 撕破强度试验;撕裂试验
tear tester 撕裂试验机;撕裂度测定仪;扯裂试验机
teas 茶类
tease 去除表面瑕疵;梳理
tea seed cake 茶籽饼
teaseed oil 茶油
teasel raising machine 刺果起绒机
Teaser 受激辐射可调电子放大器
teaser (剧院前台拱后的)横幕;熔制工;台口檐幕;梳松机;扯麻机
teaser coil 梯塞线圈
teaser rate of interest 引诱利率
teaser transformer 梯塞式变压器;副变压器
tea services 茶具
teaser winding 梯塞绕组;可调绕组
tea set 梅形茶具;茶具
tease tenon 削宽榫;阶形榫
tea shook 茶箱板材
tea sifting and grading machine 茶叶筛分机
tea sineusis 茶园
teasing 油漆面缺陷的修理
tea sorting machine 茶叶分选机
teaspoon 茶匙
teaspoonful 茶匙
tea stalk 茶梗
tea stall 茶摊;茶亭;茶水站
tea station 茶站
tea steaming machine 茶叶蒸制机
tea stem 茶梗
tea strainer 茶滤器
teat 突出部;凸缘;凸出部分
tea table 茶桌;茶几
tea things 茶具
tea tray 茶盘
tea troll(e)y 茶食推车
tea wagon 茶几柜(有轮的);茶食推车;茶具柜
tea worker's lung 茶工尘肺
teazel 角柱
teaze tenon 阶形榫
teazle post 木屋角柱
tebam 读经台(犹太教堂)
tebi 埃及泥砖
tecassir 清真寺中的妇女楼座
technetium-ruthenium method 锝-钌法
technetron 场调管
technic 技巧;工艺;技术
technical 专业的;专门的
technical ability 技能
technical accident 技术性事故
technical administration 技术管理
technical adviser [advisor] 技术顾问
technical advisory 技术咨询
technical advisory work 技术咨询工作
technical agreement 技术协议
technical aircraft hangar 飞机库

technical analysis 工艺分析;工业分析;技术分析
technical analysis package 技术分析软件包
technical and economic appraisal 技术经济论证
technical and economic evaluation 技术经济评估
technical and economic target 技术经济指标
technical and organizational measures 技术和组织措施
technical and organization measures 技术组织措施
technical application 工程应用
technical appraisal 技术评价;技术评估;技术鉴定
technical appraisement 技术评估;技术鉴定
technical appraiser 技术鉴定者
technical arbitration 技术性仲裁
technical archives of aids-to-navigation 航标技术档案
technical assessment 技术性鉴定;技术评价;技术评定
technical assistance 技术援助
technical assistance and manufacturing agreement 技术援助和制造协定
technical assistance program(me) 技术援助方案
technical assistant 估量估计助理员(英国);技术助手;技术助理
technical atmosphere 工业(用)大气压力;工程大气压
technical azotification 工业固氮
technical barrier 工艺障;技术难关
technical barrier to trade 贸易的技术性壁垒
technical betterment 技术改造
technical breakthrough 技术突破
technical bulletin 技术公告
technical capability 技术性能
technical casing 技术套管
technical center 技术中心
technical ceramics 工业陶瓷;技术陶瓷
technical certificate 技术证明书;技术认证
technical characteristic 技术特征;技术特性;技术特色;技术特点;技术性能
technical circular 技术通报
technical class 技术等级
technical classification 技术分类
technical class of marine structure 港口水工建筑物等级
technical clearing house 技术情报交流中心
technical code 技术规范
technical code for work and acceptance 施工验收技术规范
technical coefficient 技术系数
technical coefficient of production 生产技术系数
technical cohesive strength 工程内聚强度
technical collaboration 技术协作
technical collaboration agreement 技术协作协定
technical command 技术领导
technical commission 技术委员会
Technical Commission for Climatology 气候学技术委员会
technical committee 技术委员会
technical committee for standardization 标准化技术委员会
technical competence 技术发展水平
technical condition 技术状况;技术条件
technical condition of logging 测井技术条件
technical constant 技术常数
technical consultant 技术咨询;技术顾问
technical consultation 技术咨询
technical consultation service 技术咨询机构;技术咨询服务
technical content 技术含量
technical control board 技术控制台
technical control center 技术控制台
technical controller 技术管理人员
technical control project 技术控制工程
technical cooperation 技术协作;技术合作
technical cooperation committee 技术合作委员会
technical co-ordination 技术协作
technical corrigendum 技术勘误表
technical cost function 技术价格函数
technical counsel service 技术顾问处;技术评议服务;技术咨询机构;技术咨询服务
technical criterion 技术标准;技术准则
technical crop 工艺作物
technical damage 机械损伤
technical data 详细的技术资料;工程技术数据;技

术资料;技术性能;技术数据
technical data acquisition 技术资料收集
technical data digest 技术资料文摘
technical data required for the pump 泵的性能参数
technical data sheet 技术数据表
technical dates 特殊日期
technical decline 技术性下跌
technical default 技术性违约
technical demand 技术人员需要量
technical demonstration 技术论证
technical description 技术说明
technical design 技术设计
technical design change 技术设计改变
technical design plan 技术设计方案
technical design review 技术设计审查
technical detail 技术细节
technical detection 技术检测
technical development 技术开发;技术进展;技术改造;技术发展
technical development plan 技术研制计划;技术发展规划
technical direction 技术指导;技术说明书
technical director 技术指导者
technical division 技术处
technical division of labo(u)r 技术分工
technical documentation 技术文件编制
technical drawing 计算绘图;技术设计图
technical economic 技术经济的
technical economic analysis 技术经济分析
technical economic comparison 技术经济比较
technical economic computation 技术经济计算
technical economic definition phase 技术经济条件确定阶段
technical economic demonstration 技术经济论证
technical economic evaluation 技术经济评估
technical economic index 技术经济指标
technical-economic index 技术经济指标
technical-economic index of blasting 爆破技术经济指标
technical-economic index of drainage 排水技术经济指标
technical-economic index of hoisting 提升技术经济指标
technical-economic index of loading 装岩技术经济指标
technical-economic index of transport 运输技术经济指标
technical economics 技术经济学
technical economy 技术经济
technical efficiency 技术效率
technical efficiency analysis 工艺效率分析
technical engineering of naval-base 海军基本技术工程
technical equipment 技术设备
technical error 技术误差
technical evaluation 技术鉴定
technical examination 技术检查
technical exchange 技术交流
technical expert 技术专家
technical expertise 技术鉴定;技术专长
technical explanation on construction details 技术交底
technical failure 技术事故;技术失效;技术故障
technical feasibility 技术可行性;技术可能性
technical feasibility analysis 技术可行性分析
technical feature 技术特性表;技术性能
technical fee 技术费用
technical final age 工艺伐期令
technical floor 设备层
technical force 技术力量
technical forecasting 技术预测
technical form of expression 工艺表达式
technical function 技术职能
technical gelatine 工业明胶
technical glass 工业玻璃;技术玻璃
technical grade 工业品等级;工业级
technical grading 工艺分级
technical group 技术组
technical guidance 技术指南;技术指导
technical guidance documents 指导性技术文件
technical handbook 技术手册
technical hangar 维修车间;维修库;维修棚
technical identification 技术验证
technical index 技术指标
technical information 技术资料;技术信息;技术情报
technical information document 技术情报文献
technical information exchange station 技术情报交换站
technical information file 技术情报资料
technical innovation 技术革新;技术发明;技术创新
technical innovator 技术革新者
technical input coefficient table 技术投入系数
technical inspection 技术检查
technical inspection check list 技术检查对照表
technical inspection location 技术检查站
technical inspection report 技术检查报告
technical inspection station at frontier 国境技术检查站
technical inspection work 技术检查工作
technical installation 技术设施
technical institute 工艺学院;工业专科学校;工程研究院;技术研究院
technical instruction 技术指南
technical intelligence 技术情报
technicality 学术性;专业性质;专业性;专门性;专门名词;技术性;技术细节
technicalization 技术化
technical know-how 专门知识;专门技术;专门技能;技术知识;技术诀窍;技术工艺知识;技术保密
technical know-how market 技术诀窍市场
technical knowledge 技术知识
technical labo(u)r 技能劳动
technical leader 技术领导
technical leadership 技术领导
technical lettering 工程字
technical level 技术水平;技术等级
technical level(l)ing 工程水准测量
technical liason 技术协作
technical life 技术寿命
technical limitation 技术限制
technical line 技术路线
technical literature 技术资料;技术文献
technical load 工艺负荷;技术性负载
technical(ly) training 技术培训
technical malfunction 技术故障
technical management 技术管理
technical management board 技术管理委员会
technical management of civil aviation 民用航空技术管理
technical management system of railway construction enterprise 铁路施工企业技术管理制度
technical manpower 技术力量
technical manual 技术指南;技术细则;技术手册;技术参考书
technical matter 技术问题
technical maturity 工艺成熟度
technical meaning 专门意义
technical means of exploration 勘探技术手段
technical measurement 技术测定
technical measures 技术措施
technical mechanics 工程力学
technical memorandum 技术备忘录
technical message 技术信息
technical monitoring in power plants 发电厂技术监督
technical monopoly 技术专利;技术垄断;技术独占
technical norm 技术指标;技术标准
technical note 技术说明;技术备忘表
technical obsolescence 技术陈旧
technical of injector method 注浆法工艺
technical operation 技术作业;技术操作
technical operation colo(u)r 技术操作颜色
technical operation process 技术操作过程
technical operations of passenger train 旅客列车技术作业
technical operation station 技术作业站【铁】
technical order 技术指示;技术条令;技术命令;技术规程
technical panel 技术委员会
technical paper 技术论文
technical parameter 技术参数
technical parameter in DC electric(al) method 电法勘探的技术参数
technical parameter of injector 注浆孔工艺参数
technical patrol 技术养护(队)
technical people 科技人员
technical performance 技术性能
technical performance index 技术性能指标
technical philosophy 技术原则
technical photography 技术摄影术
technical plan 技术计划
technical planning committee 计划委员会
technical plant 工业植物;技术植物;技术设备;机械化机组
technical plant offer 技术设备报价
technical play equipment 技术活动设备
technical policy for environmental protection 环境保护技术政策
technical possibility 技术可靠性
technical powder 工业粉末
technical power expense 工业用动力费用
technical press 技术刊物
technical principle 技术原则
technical problem 技术问题;技术难题
technical process 工艺过程
technical process of production 生产工艺过程
technical productivity of handling machinery 装卸机械技术生产率
technical progress 技术进步
technical property 工艺性能
technical proposal 技术建议(书)
technical provision 技术条件
technical publication 技术书刊;技术文件
technical pure 工业纯净;工业纯的
technical-pure grade 工业纯品
technical purity 工业纯度
technical qualification 技术资历
technical quality control 技术质量控制
technical rate of substitution 技术置换比
technical rationalization 技术合理化
technical reconnaissance 技术勘测
technical reconstruction 技术改造;技术重建
technical reference 技术参考资料
technical reform 技术改造
technical reformation 技术改造
technical reform of railway 铁路技术改造
technical regulation 细则;说明书;技术条件;技术条例;技术规范;技术规程
technical regulation on the design of sampling program(me) 采样方案设计技术规范
technical regulations for production 生产技术规程
technical renovation 技术革新
technical renovation of railway 铁路技术改造
technical report 技术报告
technical report instruction 技术报告说明书
technical representative 技术代表
technical requirement 技术要求
technical requirement explanation 技术交底
technical research division 技术研究部
technical research institute 技术研究所
technical research report 技术研究报告
technical reserves 技术储备(金)
technical resource 技术资源
technical responsibility 技术责任
technical review group 技术评审组
technical road patrol 养路队
technical rotation 工艺轮伐期
technicals 技术术语;特殊技术用语
technical safety measures 技术安全措施
technical sales excutive 技术销售主任;技术销售经理
technical schedule 工艺规程;技术规程
technical school 技术专科学校;技术学校
technical science 技术科学
technical scrutiny 技术审核
technical secondary school 中等专科学校
technical section 技术部门
technical service 技术整备;技术服务
technical service by medical and health workers 医疗卫生人员技术服务
technical service contract 技术服务合同
technical service department 技术业务部门
technical service group 技术勤务组
technical skill 专门技术;专门技能;技术专长
technical sodium citrate 工业柠檬酸钠
technical solution 技术方案
technical soundness 技术可靠性
technical specialist 技术专家
technical specification 技术条款;技术要求说明;技术条件;技术(规格)说明书;技术规格(书);技

术规范;操作规范
technical speed 技术速度【铁】;技术航速
technical staff 技术人员;技术干部;全体技术员
technical standard 技术规格(书);技术定额;技术标准
technical standard of road 道路技术标准
technical statement 技术方案
technical station 技术站
technical stoneware 工业用炻器
technical storage 技术储备
technical strength 实际强度
technical structure 技术结构
technical study 技术研究
technical study tour 技术考察
technical summary 技术总结
technical supervision 技术监督;技术监督
technical support team 技术支援小组
technical survey 工程测量
technical system 技术系统
technical target 技术指标
technical term 专业名词;专业词汇;专门术语;专门名词;专门技术名词;技术条件;技术名词;术语
technical test 技术考试;技术考查
technical through train 技术直达列车
technical tie-up 技术联合
technical tips 工业方法
technical title 技术职称
technical tolerance 技术容限;技术公差
technical total loss 推定全损
technical trader 期货商
technical train 技术列车
technical training cost for employees 职工技术培训费
technical transfer 技术转让
technical transfer commission 技术转让费
technical transformation 技术改造
technical transport 专门运输工具
technical trend documents 技术趋向文件
technical update 技术更新
technical vehicle 工程车
technical viewpoint 技术观点
technical visa 技术签证
technical visco(si)meter 工业黏度计;工程(用)黏度计
technical wear and tear 技术上消耗磨损
technical white oil 工业用白油;白油
technical word 专业用语;技术用语;术语
technical workers with specific grades 一定级别的技术工人
technical working group 技术工作组
technical work plan in railway transportation 铁路运输工作技术计划
technic assistance 技术援助
technic code 技术规范
technic condition 工艺条件
technic-economic analysis 技术经济分析
technic-economic comparison 技术经济比较
technic-economic comprehensive index 技术经济综合指标
technic economic strategy 技术经济对策
technician 技术专家;技术员;技术人员;技师;技工
technician control 技术控制
technician control desk 技术控制台
technician engineer 技师
technician on probation 见习技术员
technic index 工艺指标
technicist 技术员;技术人员;技师
technic management of railway construction enterprise 铁路施工企业的技术管理
technic metal 实验合金
technico-economic index 技术经济指标
technicolo(u)r 五彩的;天然色;彩色印片法;彩色
technicolo(u)red 色彩鲜艳的;彩色的
technic quality of railway construction enterprise 铁路施工企业技术素质
technic responsibility system 技术责任制
Technicrete 德克尼克利脱(一种防水防冻水泥硬化液)
technics 工艺学;工程学
technic transfer 技术转让
technique 工艺方法;技术设备;技术;技巧;技能;机助判读技术;机助判读技术;操作法
technique assessment 技术评价
technique development contract 技术开发合同

technique development expense 技术开发费
technique for project evaluation 项目评估技术
technique goods 技术商品
technique market 技术市场
technique of catch device on air 气体捕集器法
technique of city planning 城市规划技术
technique of display 显示技术
technique of manufacture 制造技术
technique of production 生产工艺
technique of simulation 仿真技术
technique of strain measurement 应变测量方法
technique of variance analysis 方差分析技术
techniques for extracting inclusion solution 包裹体溶液提取法
techniques for high temperature and high-pressure 高温高压技术
techniques of discriminant analysis 判别分析法
techniques of inclusion study 包裹体技术
techniques of pest control 病虫害防治技术
technique trade 技术贸易
technocausis 烙铁术
technocracy 专家管理
technocrat 科技主义者
techno-economic analysis 技术经济分析
techno-economic appraisal 技术经济论证
techno-economic comparison 技术经济比较
techno-economic evaluation 技术经济评估
techno-economic forecast 技术经济预测
techno-economist 技术经济学家
technogenous movement 人为激发运动
technologic(al) achievement 科学技术成就;技术成就
technologic(al) advance 科学技术发展;技术进步
technologic(al) and economic characteristics of railway 铁路技术经济特征
technologic(al) and economic research strengthening 加强技术经济研究
technologic(al) appraisal 技术鉴定
technologic(al) appraisement 技术鉴定
technologic(al) approach 技术表现手法
technologic(al) arrangement 工艺布置
technologic(al) article 技术论文
technologic(al) assessment 技术评价;技术评定
technologic(al) audit 技术审查;技术检查
technologic(al) borrowing 技术借取
technologic(al) breakthrough 技术性突破
technologic(al) capability 技术能力
technologic(al) chain 工艺流程;工艺过程;技术操作过程
technologic(al) challenge 技术挑战
technologic(al) change 科学技术进步;技术变化
technologic(al) change in agriculture 农业技术改革
technologic(al) characteristic 技术作业特性
technologic(al) coating 工艺性涂层;工业化涂层
technologic(al) coefficient 工艺系数;技术系数
technologic(al) condition 工艺条件
technologic(al) cooperation 技术合作
technologic(al) cost 工艺费用
technologic(al) data 技术资料
technologic(al) datum 技术基准
technologic(al) design 工艺设计
technologic(al) development 工艺发展;技术开发;技术发展
technologic(al) director 技术负责
technologic(al) documents 技术文件
technologic(al) economic indicator justice 技术经济指标合理
technologic(al) economic indicator unjustice 技术经济指标不合理
technologic(al) economics of energy 能源技术经济学
technologic(al) economic synthetic(al) evaluation 技术经济综合评议
technologic(al) environment 技术环境
technologic(al) factor 技术因素
technologic(al) fix 工艺安装
technologic(al) flow 工艺过程
technologic(al) flow sheet 工艺流程图
technologic(al) forecasting 技术预测
technologic(al) gap 技术(上的)差距
technologic(al) gas 工业气体
technologic(al) guidance 技术指导
technologic(al) hazard 工艺危害;技术危害

technologic(al) ideas 技术设想
technologic(al) improvement 技术改造
technologic(al) innovation 工艺革新;技术革新;技术创新
technologic(al) installation 技术设备
technologic(al) item 技术产品
technologic(al) know-how 技术知识
technologic(al) level 工艺水平
technologic(al) line 技术生产线
technologically feasible 工艺上可行的;技术上可行的
technologically induced growth 技术进步带来的增长
technologic(al) magnetization 技术磁化
technologic(al) management 技术管理
technologic(al) management of cargo handling 装卸工艺管理
technologic(al) manual 技术手册
technologic(al) material commodities 工艺材料矿产
technologic(al) mineralogy 工艺矿物学
technologic(al) model 技术模式
technologic(al) operation 工艺操作
technologic(al) package 一整套技术
technologic(al) parameter 工业参数;工艺参数
technologic(al) petrology 工艺岩石学
technologic(al) plastics 工艺塑料;高级技术塑料
technologic(al) procedure 工艺规程
technologic(al) process 工艺流程;工艺过程;流程;技术作业过程;技术操作过程
technologic(al) process automation 工艺程序自动化
technologic(al) processing assessment 加工技术评议
technologic(al) process of cargo handling 装卸工艺流程
technologic(al) process of trains at passenger station 客运站技术作业过程
technologic(al) production norm 生产技术参数
technologic(al) progress 工艺发展;科学技术发展
technologic(al) property 工艺性能;工艺特性
technologic(al) rating 技术作业定额
technologic(al) reformation project 计改项目
technologic(al) requirement 工艺要求
technologic(al) revolution 技术革命
technologic(al) school 工艺学校
technologic(al) science 技术科学
technologic(al) sophistication 技术上的高度精良化;采用尖端技术
technologic(al) specification 技术规范
technologic(al) sphere 技术圈
technologic(al) state-of-the-art 最先进技术水平
technologic(al) study 工艺研究
technologic(al) test of coal 煤的工艺性试验;煤工艺试验
technologic(al) trade-off 技术的权衡;技术权衡
technologic(al) transformation 技术改造;技术改革
technologic(al) unemployment 技术进步引起的失业
technologic(al) upgrade 技术升级
technologist 工艺学家;工艺人员;科技人员;技术人员;技师
technology 工艺(学);技术(学)
technology acquisition and development 技术引进与发展
technology and equipment 技术和装备
technology assessment 技术再评;技术评价;技术评估;技术评定
technology consulting contract 技术咨询合同
technology equipment for insulation glove 绝缘手套生产线
technology forecasting 技术预测
technology gap 技术差距
technology import 技术引进
technology input 技术引进
technology-intensive 技术密集型;技术密集的
technology-intensive industry 技术密集型产业
technology-intensive investment 技术密集的投资
technology life cycle 技术生命周期
technology negotiation 技术谈判
technology of automated cartography 自动制图工艺
technology of garden construction 造园技术

technology of infrared remote sensing 红外(线)遥感技术
technology of inorganic chemicals 无机物工艺学
technology of manufacturing anion generator 负离子发生器制造技术
technology of metals 金属工艺学
technology of microwave remote sensing 微波遥感技术
technology of nitrogen generator 氮气发生器技术
technology of production 生产技术
technology of roller coating for leather surface 辊印涂饰皮革表面技术
technology of waste disposal 废物处置技术
technology of wastewater treatment 废水处理工艺
technology of water quality stabilization 水质稳定技术
technology pilot project 技术控制工程
technology test 工艺试验
technology tester 工艺试验机
technology trade 技术贸易
technology trade license 技术贸易许可证
technology transfer 技术转让;技术推广
technology transfer agreement 技术转让协议
technology transfer expense 技术转让费
technology transfer program(me) 技术转让计划
technology transferring contract 技术转让合同
technology transformation 技术转移
technopolis 科技社会;科技城
technoscape 科学技术景观
technostructure 技术专家控制体制;技术管理阶层;技术(专家)阶层;技术专家体制
technotron 结型场效应管
Techrete system 德克尼利脱构造体系(采用预制墙和楼板)
Teclu burner 特克卢燃烧器;双层转筒燃烧器
tecnazene 四氯硝基苯
Teco 德科(一种专利的木材连接件)
tectate-imperforate 覆盖层无穿孔的
tectate-perforate 覆盖层具穿孔的
tectiform 盖形的;屋顶形的(建筑物)
tectite 熔融石
tectofacies 构造相
tectofacies map 构造相图
tectogene 构造带;深地槽;深凹槽
tectogenesis 构造运动;造山运动【地】
Tectona 柚木属
Tectona grandis 柚木
tectonic 构造的
tectonic accretion 大陆增生
tectonic accumulation 构造聚集作用
tectonic active area 构造活动区域
tectonic active belt 构造活动带
tectonic active region 构造活动区域
tectonic activity 构造活动性
tectonic adjustment 构造调整
tectonic analog(ue) experiment 构造模拟试验
tectonic analysis 构造分析
tectonic arkose 构造长石砂岩
tectonic axis 构造轴
tectonic background 构造背景
tectonic basin 构造盆地
tectonic bay 构造湾
tectonic belt 构造带
tectonic block 构造断块
tectonic breccia 构造角砾岩
tectonic characteristic 构造特征
tectonic chronology 构造年代
tectonic closed system 构造封闭系统
tectonic community 构造群落
tectonic creep 构造蠕动;构造蠕变
tectonic critical surface 构造临界面
tectonic crustal phenomenon 地壳构造现象
tectonic cycle 造山运动;造山循环;构造循环;构造旋回
tectonic cycles and movements 构造旋回及构造运动
tectonic deformation 构造形变;构造变形
tectonic deformation map 构造变形图
tectonic denudation 构造剥蚀
tectonic depression 构造洼地;构造凹陷
tectonic discontinuity 构造结构面
tectonic disturbance zone 构造扰动带
tectonic division map 构造分区图
tectonic domain 构造域

tectonic driving force 构造驱动力
tectonic earthquake 构造地震
tectonic element 构造要素;构造单元
tectonic emplacement 构造侵位
tectonic energy 构造能量
tectonic episode 构造序幕
tectonic erosion 构造侵蚀
tectonic eustatism 构造海面升降
tectonic event 构造事件
tectonic evolution 构造演化
tectonic evolution diagram 构造演化图
tectonic fabric 构造组构
tectonic facies 构造相
tectonic fan-delta complex 构造冲积扇—三角洲复合体
tectonic feature 构造特征
tectonic fissure 构造裂缝
tectonic fold 构造褶皱
tectonic foundation 构造基底
tectonic fracture 构造裂隙
tectonic framework 构造轮廓;构造格式;构造格架
tectonic framework map 构造格架图
tectonic generation 构造序次
tectonic geology 构造地质学
tectonic grade 构造级别;构造等级
tectonic hinge 构造支点带
tectonic joint 构造节理;构造缝
tectonic knot 构造结;构造节
tectonic lake 构造湖
tectonic land 构造陆地
tectonic landform 构造地形
tectonic land mass 构造地块
tectonic land subsidence 构造地面沉降
tectonic layer 构造层
tectonic lens 构造透镜体;构造扁豆体
tectonic lenticle 构造透镜体
tectonic level 构造面;构造层位;构造层次
tectonic line 构造线;大断裂线
tectonic-lithologic(al) columnar section 构造岩性柱状图
tectonic location 构造位置
tectonic location of seismic fault 地震断裂的构造位置
tectonic-magmatic map 构造岩浆图
tectonic magnitude 构造等级;构造等级
tectonic map 构造图;构造地质图;大地构造图
tectonic melange 构造混杂岩
tectonic-metamorphic map 构造变质图
tectonic metamorphism 构造变质(作用)
tectonic migration 大地构造迁移
tectonic migration forms 构造迁移形式
tectonic migration mechanism 构造迁移机制
tectonic migration regions 大地构造迁移区
tectonic mineral 构造矿物
tectonic mixture 构造混合体
tectonic model 构造型式;构造类型
tectonic model experiment 构造模型试验;构造模型实验;构造模拟实验
tectonic model map 构造模式图
tectonic moraine 构造冰碛物
tectonic motion 构造运动
tectonic movement 构造运动;地壳构造运动
tectonic movement direction 构造运动方向
tectonic movement intensity 构造运动强度
tectonic network 构造网
tectonic oil pool 构造油田
tectonic open system 构造开放系统
tectonic order 构造序次;构造级别;构造等级
tectonic order and tectonic generation 构造级序
tectonic origin of earthquake 地震的构造成因
tectonic outlier 构造外露层
tectonic outline map 构造纲要图
tectonic overpressure 构造附加压力
tectonic overprinting 构造叠加
tectonic-pal(a)eotopographic(al) map 构造古地理图
tectonic pattern 构造形式
tectonic phase 构造幕
tectonic plain 构造平原
tectonic plate 构造板块;地层
tectonic plate motion 地壳板块运动
tectonic plate movement 构造板块运动
tectonic pole 构造极地
tectonic process 构造作用;构造运动

tectonic profile 构造切面;构造剖面图;构造剖面
tectonic province 构造区
tectonic rebuilding 构造再造作用
tectonic relationship 构造关系
tectonic rock-controlling process 构造控岩作用
tectonic rotation 构造旋转
tectonics 营造学;地层构造;构造学;构造地质学;大地构造学
tectonic scale 构造级次
tectonic seaquake 构造海震
tectonic-sedimentary facies map 构造岩相图
tectonic semi-open system 构造半开放系统
tectonic setting 构造背景
tectonic setting of studied area 研究区构造位置
tectonic slide fault 构造滑动断层
tectonic spring 构造温泉
tectonic stillstand period 构造宁静期
tectonic strain 构造应变
tectonic strain effect 构造应变效应
tectonic strain energy 构造应变能
tectonic stress 地壳应力;构造应力;大地构造应力
tectonic stress analysis 构造应力分析
tectonic stress field 构造应力场
tectonic stress field at profile 剖面构造应力场
tectonic stress field control of ore 构造应力场控矿
tectonic stress field map 构造应力场图
tectonic stress history 构造应力作用过程
tectonic strike-line map 构造走向线图
tectonic style 构造样式
tectonic sublayer 亚构造层
tectonic superimposing 构造叠加
tectonic superimposition 构造叠加作用
tectonic syntaxis 构造系统的联合
tectonic synthesis 构造综合分析
tectonic synthesis method 构造综合法
tectonic system 构造系;构造体系
tectonic system control of ore deposition 构造体系控矿作用
tectonic systems and earthquake map 构造体系与地震图
tectonic systems controlling earthquakes 控震构造体系
tectonic terrace 构造阶地
tectonic transport 构造运移
tectonic type 构造形式
tectonic types of oreforming area 成矿区构造类型
tectonic unit 构造单元;大地构造单元
tectonic unmixing 构造不混合
tectonic valley 构造谷
tectonic weaving 构造编织
tectonic zone 构造带
tectonism 构造作用;构造变动
tectonism in crust 壳内构造运动
tectonism in matle 地幔地壳运动
tectonite 构造岩
tectonite fabric 构造岩组构
tectonization 构造作用;构造化
tectonization feature 构造变动特征
tectonization marker 构造变动标志
tectonization sequence 构造变动顺序
tectonized pinch out 构造变形尖灭
tectonoblastic 构造变晶的
tectonoclastic 构造碎裂的
tectonodynamics 构造动力学
tectono-eustatism 构造性海面升降运动;构造海面变动
tectonofacies map 构造相图
tectono-geochemical field 构造地球化学场
tectono-geochemistry 构造地球化学
tectono-geochemistry of igneous rock 火成岩的构造地球化学
tectono-geophysical field 构造地球物理场
tectono-karstic depression 构造岩溶洼地
tectono-magmatic belt 构造岩浆带
tectonomagnetism 构造地磁学
tectono-metallogenic condition 构造成矿条件
tectono-metallogenic process 构造成矿作用
tectono-metallogenic type 构造成矿类型
tectono-metamorphic belt 构造变质带
tectonometer 地壳构造测量仪
tectonophysics 地壳构造物理学
tectonosphere 构造圈;地壳构造层
tectonosphere theory 层圈构造说
tectono-stratigraphic(al) division 构造地层分区

tectono-stratigraphic(al) terrane 构造地层地体
tectono-stratigraphic(al) unit 构造地层单元
tectono-thermal event 构造热事件
tectorial 覆盖着的;构成覆盖物的;屋顶状的;房顶形的
tectorial membrane 覆膜
tectorium 疏松层
tectosequent 构造表象的
tectosilicate 架状硅酸盐;网硅酸盐
tectosome 构造相层;构造体
tectosphere 构造圈;构造层
tectostratigraphic 构造地层的
tectotope 构造境
Tecuta 德库塔(一种薄铜屋面板)
tedder 干草摊晒机;摊晒机;摊草机;翻草机
tedder shaft 翻草器轴
tedder side rake 侧向摊晒搂草机
tedding machine 干草干燥机
tedge 竖管;直立物
tee 丁字形;宝顶
tee bar 丁字钢;丁字形钢
tee beam 丁字形梁;丁字梁;T形梁;丁字架
tee beam and slab construction T形梁板结构;肋形板结构
tee-beam and slab structure 肋形板结构
tee-beam footing T形梁柱脚;倒T形梁连接的底脚
tee bend T字形接头
tee bevel 斜角规
tee bolt 丁字螺栓;T形螺栓
tee branch T形管;三通管
tee branch pipe 三通(管)
tee branch tube 三通(管)
tee building component T形组成部分;T形预制建筑构件
tee building member T形建筑构件
tee building unit T形建筑单元
tee cock 三通旋塞
tee component T形建筑构件
tee conduit 三通;丁字管
tee-connection T形接头
tee-connection sheet pile T形接头连接板桩
tee-drive unit 三路传动器
tee fitting T形配件
tee floor slab T形桥面板;T形楼板
tee fluid flow meter 三通液体流量计
tee girder T形大梁;丁字形大梁
tee grill T形格栅
tee handle T形手柄;T形把手
tee hanger 丁字形弹簧吊架
tee head T形柱头;T形支撑
tee-head bolt 丁字头螺栓
tee-head type T字头形
tee hinge 带铁扁担铰链;T形铰链
tee-hinge junction T形铰合连接
tee-hinge member T形铰合构件
tee-hinge panel T形铰合面板;T形铰合墙板
tee-hinge strap T形铰合铰带;T形铰合皮带
tee-iron T形钢;丁字形铁件;T形梁;丁字钢
tee-jet nozzle 平射喷嘴
tee joint 三通接头;丁字接头;丁字形接头;T形接头
tee junction 三通连接;T形接头
teem 倾注;倒出
teemer 浇筑工
teeming 铸件
teeming aisle 铸锭跨
teeming brick 浇钢砖
teeming crane 浇铸起重机;浇注起重机
teeming furnace 铸造电炉
teeming ladle 浇筑桶
teeming ladle crane 钢水包起重机
teeming lap 浇筑重皮;浇筑折叠
teeming line 浇筑线
teeming platform 铸锭台
teeming speed 铸速
teeming stage 浇筑平台
tee off 分线;分叉
tee-off current 分支电流
tee panel T形板
tee piece 三通;丁字管;T形片;T形接头
tee pipe 三通管;三通接头;丁字管;T形管;T管
tee pipe coupling 三通管接头
teepleite 氯硼钠石
Teepol 蒂普尔(一种引气剂)
tee profile T形型钢;丁字形钢

tee rail 工字轨
tee reducing on both runs 双头缩小三通
tee reducing on one run 一头缩小三通
tee reducing on one run and outlet 出口和一头小三通
tee reducing on outlet 支路异径丁字管;出口缩小三通
tee reducing on run 直路异径丁字管
tee rest 丁字形刀架
tee ruler 丁字尺
tee section T形截面;T形材;丁字钢
tee-slot 丁字形槽
tee-socket 承口三通管
tee spot welding 丁字点焊
tee square 丁字尺
tee steel 丁字钢;T形钢
teeter 上下摇摆
teeter chamber 搅拌室
teetered bed 搅拌床层;浅流化床层
teetering sand 悬浮沙
teeter-totter 跷跷板
teeth bar 齿条
teeth of a cogwheel 轮齿
teeth of a saw 锯齿
teeth of the bit 钻头的齿
teeth of the gale 逆风
teeth parts 齿轮元件
teeth per inch 每英寸齿数
teeth setting 锯齿偏侧度
teetotum 手转陀螺
tee tube 三岔管;丁字管
tee union 三通接头
tee unit T形单元
tee valve 三通阀
tee weld T形焊缝
tee welding 丁字焊;三通焊接
tee with side branch 侧口三通
Teflon 特氟隆(聚四氟乙烯)
Teflon-coating 特氟隆涂层(混凝土钢模接触面)
Tefloncompensator 聚四氟乙烯补偿器
Teflon film 特氟隆薄膜
Teflon guide strip 特氟隆导向条
Teflon heat exchanger 聚四氟乙烯换热器;氟塑料换热器
Teflon insert 特氟隆夹入物
Teflon neoprene bearing 特氟隆橡胶支座;氟板橡胶支座
Teflon pipe 特氟隆管
Teflon plate bearing 特氟隆板支座;氟板支座
Teflon plate-coated sliding bearing 特氟隆滑板支座
Teflon plunger 特氟隆柱塞密封
Teflon seal 聚四氟乙烯密封件
Tego 蒂戈铅基轴承合金
Tego film 蒂戈膜;酚醛树脂胶片;酚醛树脂胶膜
tegula 底瓦;屋顶盖瓦;瓦片状排列;瓦的;[复]tegulae
tegular 鳞片状的
tegurium 尖形墓顶
tehnic bronze 碲镍青铜
Te-Hua ware 德化窑瓷
teichoic acid 磷壁酸质
teichopsia 闪光暗点
Teijin Tetoron 帝特龙聚酯纤维
teil 菩提树
teineite 碲铜石
Teisen disintegrator 动力除尘器
tekaol 亚麻(油)厚油
Tekken type cracking test 铁研式抗裂试验
teklite 玻陨石
tekram 市场
Teks 泰科斯自钻孔螺钉
tektite 熔融石;铁毛矾石;玻璃陨体;玻璃陨石
tektonite 构造岩
telamon 男像柱;人像柱;[复] telamones
telargpalite 碲银钯矿
Telcoseal 泰尔科铁镍钴合金
Telcuman 泰尔铜锰镍合金
teleadjusting 远距调整
teleammeter 遥测电流计;遥控安培计;遥测安培表
teleautogram 传真电报机
teleautomatics 遥控自行装置学;远距离控制;遥控自动学;遥控自动技术;遥控机械学;自动遥控装置
telebarometer 远距离气压计;远距离气压表

telebinocular 遥测双目镜
telebit 二进制遥控系统
telecamera 摄像机;电视摄像机
telecar 遥控车;收发报汽车
telecast 电视广播
telecaster 电视广播机
telecentric iris 焦阑
telecentric light 远心照明
telecentric mirror 远心镜
telecentric optic(al) system 远心光学系统;远心光路系统
telecentric stop 远心光阑;焦阑
teleceptor 距离感受器
telechelic polymer 远螯聚合物
telechiric 远距离操纵手;远距离操纵的
telechirics 控制系统
telechron clock 遥控记时钟
telechron motor 遥控电动机
telechron timer 计时开关
telecinesia 感应运动
teleclinometer 井斜仪
telecom 电信
telecommand equipment 遥控设备
telecommand receiver 遥测指令接收机
telecommunication 远距离联系;远程通信;无线电通信;电信;长途通信;长距离通信
telecommunication access method 远程通信存取法;电信存取法
telecommunication and signal(l)ing building 通号综合楼
telecommunication and signal(l)ing workshop 通号车间
telecommunication authority 电信主管机关
telecommunication building 电信大楼
telecommunication bureau 电信局
telecommunication cable 通信电缆;电信电缆
telecommunication circuit 电信电路
telecommunication engineering 电信工程(学)
telecommunication environment 电信环境
telecommunication equipment 无线电通信设备;电讯设备;电信设备
telecommunication facility 通信设备;通讯工具
telecommunication interference 电信干扰
telecommunication laser 电信激光器
telecommunication line 电信电路;通信线路;电信线路
telecommunication management 通信管理
telecommunication management network 电信管理网
telecommunication optic(al) fiber 电信光纤
telecommunication pipe line 电信管线
telecommunication post bureau 邮电局
telecommunications 通信学
telecommunication satellite 通讯卫星;通信卫星;电信卫星
telecommunications satellite 实用通信卫星
telecommunication station 电信站;无线电通信站;无线电通信台
telecommunication system 远距离通信系统
telecommunication terms 电信术语
telecommunication time 远程通信时间
telecommunication tower 电信塔
telecommunication traffic 电信业务
telecommuting 电信办公
telecompass 远距离罗盘;遥示罗经;遥控罗盘;电视罗经
teleconference 远距离通讯会议;距离感受器;电信会议
teleconferencing 电信会议
teleconferencing network 电信会议网络
teleconnection 远距离联系
telecontrol 远距离控制
telecontrol board 遥控板
telecontrol center 运动控制中心
telecontrol frequency 远动用频率
telecontrolled power station 遥控发电站;遥控电站
telecontrolled substation 遥控变电所
telecontrol pumping house 遥控泵房
telecontrol relay 遥控继电器
telecontrol room 遥控室
telecontrol system 遥控系统
telecontrol system pulse generator 遥控系统脉冲发生器
telecopier 传真复印

telecounter 远距离计算机
telecoupler 共用天线耦合器
Telectal alloy 泰雷铝硅合金
electroscope 电传照相机
teledata processing 遥测数据处理
teledata processing system 遥测数据处理系统
teledepth 气压测深仪
teledip ga(u)ge 电子式浮尺
teleequipment 遥控装置
telefax 用户传真;光波传信法;公用电话网传真
teleferic 单塔式架空缆车
teleflex 转套;软套管
telefluoroscopy 远距荧光屏检查
telefork 叉式起重拖车
Telefunken beam antenna 德律风根定向天线
Telefunken process 德律风根法(陶瓷金属封接法的一种)
telega 俄罗斯式无簧四轮货车
telega(u)ge 遥控仪表;远距离测量仪(表);遥控仪;遥测计;方向计;遥测机
telegon 无接点交流自整角机
telegoniometer 遥远测角计;遥测角计;无线电测距仪;方向计
telegram 电报
telegram in plain language 明码电报
telegram multiple 分送电报
telegram retransmission centre 自动转报中心
telegram retransmission system 自动转报系统
telegraph 摇车钟;电码;电报机;电报
telegraph address 电报挂号
telegraph and communications center 电报及通信中心
Telegraph and Telephone International 国际邮政电报电话联合会
telegraph automatic system 自动转报系统
telegraph buoy 水下电缆浮标;水底电缆浮标;电缆浮标
telegraph cable 海底电缆;电报电缆
telegraph cable landing buoy 海底电缆接岸浮标
telegraph chain 车钟链
telegraph chute 挡板式溜眼
telegraph code book 电码书
telegraph cord (井场发动机的)远距离开关控制钢绳
telegraph department 电报局
telegraph exchange 用户电报;电报交换机
telegraph gong 机舱自动控制信号钟
telegraphic(al) address 电报挂号
telegraphic(al) code 电报电码
telegraphic(al) money order 电汇单
telegraphic(al) reply 回电
telegraphic(al) time signal 电报时号
telegraphic(al) transfer 电汇
telegraphic(al) transfer reimbursement 电报索汇
telegraphing 透印;透底
telegraph modulated wave 键调波
telegraph office 电报局
telegraphoscope 电传照相机
telegraph pole 电杆(电信用)
telegraph receiver 收报机
telegraph room 电信室
telegraph ship 电缆敷设船
telegraph signal 电报信号
telegraph switching equipment 电报交换机
telegraph transmitter 发报机
telegraph wheel 调节手轮(司钻台上)
telegraphy 电报
teleguide 遥导
telehoist 可伸缩起重杆;伸缩式起重机
telehygrometer 遥测湿度计
teleidoscope 液层扫描镜
teleindicating system 远距离指示系统
teleindicator 远距指示器;远距指示器
teleinformation 遥测信息
teleirradiation 远距离辐照
telekinesis 感应运动
telekinesy 远距离作用
telelight 远传信号灯
telemagmatic coalification 远程岩浆煤化作用
telemagmatic metamorphic 远程岩浆热变质作用
telemagnetic compass 电传磁罗经
telemanometer 遥测压力表
telemeasurement 远距离测量;遥测
telemeasuring data 遥测数据

telemechanical apparatus 遥动装置;遥控机械装置
telemechanical device 遥动装置;遥控机械装置
telemechanical installation 遥动装置;遥控机械装置
telemechanical receiver 远动接收机
telemechanical system 远动系统;遥控机械系统
telemechanical transmitter 远动发送机
telemechanical unit 远动装置;遥控机械装置
telemechanics 远动学;遥控力学
telemechanisation 远距离机械化
telemechanism 遥控机构
telemechanization 远动机化;远距离操纵;远动化
telemetacarpal 端掌骨
telemeteorograph 远距气象计;遥测气象计;气象遥测仪
telemeteorography 远距气象测定学;遥测气象仪器学
telemeteorometry 遥测气象学
telemeter 远距离电测装置;遥测装置;遥测仪;遥测器;遥测计;遥测发射器;遥测传送;光学测距仪;测远仪;测距仪
telemetered air monitoring 遥感空气监测
telemetered array of seismic stations 遥测地震台阵
telemetered control equipment 遥测控制设备
telemetered data 遥测数据
telemetered information 遥测到的信息
telemetered medium environment buoy 海洋环境遥测浮标
telemetered seismic net 遥测地震台网
telemetered signal 遥测信号
telemetered vibration data 遥测振动数据
telemeter equipment 遥测设备
telemetering 远距离测量;遥测
telemetering amplifier 遥测放大器
telemetering antenna 遥测天线
telemetering channel 遥测通道
telemetering commutator 遥测系换向器
telemetering counter 遥测计数器
telemetering current meter 遥测流速仪;遥测流量计;电报海流计
telemetering data 遥测数据
telemetering data receiving set 遥测数据接收机
telemetering depth meter 遥测水深仪;遥测深度仪
telemetering device 遥测装置;遥测设备
telemetering device of water-level indicator 水位遥测仪
telemetering device of water stage 水位遥测仪
telemetering environment(al) profiling system 遥测环境剖面装置
telemetering equipmentt 遥测装置
telemetering force balance type converter 遥测力平衡变换器
telemetering gear 遥测装置
telemetering hydrophone 遥测水下测音器;遥测水听器
telemetering installation 遥测装置
telemetering link 遥测线路
telemetering medium 远距离测量装置;遥测方法
telemetering monochromer 遥测单色计
telemetering oceanographic current meter 遥测海洋流速仪;遥测海流计
telemetering ocean profiling system 遥测海洋剖面装置
telemetering of hydrologic(al) data 水文数据遥测
telemetering oscillator 遥测振荡器
telemetering pickup 遥测传感器
telemetering rain ga(u)ge 遥测雨量计
telemetering receiver 遥测接收机
telemetering receiving apparatus 遥测接收装置
telemetering seismic station network 遥测地震台网
telemetering seismograph 遥测地震仪
telemetering seismometer 遥测地震计
telemetering sender 遥测发送器
telemetering sensor 遥测传感器
telemetering station 遥测站;遥测台
telemetering system 遥测系统;遥测计
telemetering technique 遥测(技)术
telemetering terminal equipment 遥测终端设备
telemetering totalizer 遥测加法器
telemetering transmitter 遥测发射机
telemetering water-level indicator 遥测水位指示器;遥测水位指示仪
telemetering wave buoy 遥测波浪浮标;波形遥测浮标;波浪遥测浮标

telemetering weather station 遥测气象站
telemeter level 测深水平仪
telemeter processor 遥测处理机
telemeter receiving set 遥测接收机
telemeter rod 测距尺
telemeter system 远距离控制方式
telemeter transmitting set 遥测发射机
telemetric apparatus 远距测定装置
telemetric circuit 遥测电路
telemetric control 遥测控制
telemetric data 遥测数据
telemetric data analyzer 遥测数据分析器
telemetric data converter 遥测数据转换器
telemetric data monitor 遥测数据监控器
telemetric data monitoring set 遥测数据监控器
telemetric data receiver 遥测数据接收机
telemetric data receiving-recording-scoring set 遥测数据接收与记录设备
telemetric data recorder 遥测数据记录器
telemetric data transmitter 遥测数据发射机
telemetric data transmitting set 遥测数据发射装置;遥测数据发射机
telemetric instrument 遥测仪器
telemetric integrator 遥测积分器
telemetric link 遥测线
telemetric oscillogram 遥测波形图
telemetry 远距离测量;远距测定法;遥测技术;遥测(技)术;遥测(方)法;测远术;测距术;测距法
telemetry-acquisition 遥测采集
telemetry and command antenna 遥测与指令天线
telemetry and command station 遥测与指令站
telemetry and command system 遥测指令系统
telemetry antenna 遥测天线;测距天线
telemetry automatic processing system 遥测自动处理系统
telemetry automatic reduction equipment 遥测自动还原设备
telemetry auto reduction system 遥测自动简化系统
telemetry band 遥测频带
telemetry blackout 遥测中断
telemetry building 遥测室
telemetry carrier 遥测载波
telemetry checkout system 遥测检查系统
telemetry code 遥测编码
telemetry code modulation 遥测编码调制
telemetry command 遥测指令
telemetry command system 遥测指令系统
telemetry computation 遥测计算
telemetry control 遥测控制
telemetry control and monitor 遥测控制与监听
telemetry data 遥测资料;遥测数据
telemetry data center [centre] 遥测数据中心
telemetry data compressor 遥测数据压缩器
telemetry data digitizer 遥测数据数字转换器
telemetry data evaluation system 遥测数据鉴定系统
telemetry data gate module 遥测数据门模件
telemetry data monitor 遥测数据监测仪
telemetry data network 遥测数据传递网
telemetry data receiver 遥测数据接收机
telemetry data reduction 遥测数据处理
telemetry data system 遥测数据系统
telemetry decommutation and display systems 遥测信息还原与显示系统
telemetry encoder 遥测编码器
telemetry equipment 遥测设备
telemetry equipment calibration 遥测设备校准
telemetry filter 遥测滤波器
telemetry frequency 遥测频带
telemetry fuze 遥测引信
telemetry ground station 遥测地面站
telemetry ground system 遥测地面系统
telemetry information 遥测信息
telemetry input system 遥测输入系
telemetry link 遥测线路;遥测线
telemetry mast 遥测桅
telemetry modulation system 遥测调制系统
telemetry net 遥测网
telemetry network 遥测台网
telemetry observation system 远程观测系统;遥测监测系统
telemetry parameter 遥测参数
telemetry processing 遥测处理

telemetry processing station 遥测处理站
telemetry processor 遥测数据处理机
telemetry readout 遥测读出
telemetry receiver 遥测接收机
telemetry reference 遥测基准脉冲
telemetry report 遥测报告
telemetry schedule 遥测程序
telemetry ship 遥测船
telemetry simulator 遥测模拟器
telemetry station 遥测站
telemetry system 遥测系统
telemetry terminal system 遥测终端系统
telemetry tracking, command and monitor station 遥测跟踪指令和监听站
telemetry tracking and command station 遥测跟踪和指令站
telemetry tracking system 遥测跟踪系统
telemetry transmitter 遥测发射机
telemetry unit 遥测装置
telemicroscope 遥测显微镜;望远显微镜
telemometer 遥测计
telemonitor 远程监控;遥远监视
telemonitoring 遥控监测
telemotion 无线电操纵
telemotor 液压操舵器;液压操舵机;遥控发动机;遥控电动机;遥测电动机;舵机操纵器
telemotor adjustment 油压操舵机调整
telemotor circuit 油压操舵机油管路
telemotor control steering gear 遥控液压操舵装置
telemotor gear 遥控机机构;遥控传动装置
telemotor oil 油压操舵机用油;液压传动操舵装置用油
telemotor receiver 液压传动操舵装置接收器
telemotor steering gear 油压推杆式操舵装置;遥控操舵装置
telemotor system 油压操舵系统;遥控液压操舵装置;遥控操舵系统
telenetwork 遥测台网
teleobjective 望远镜头
teleology 目的论
teleoperated 远距离操纵
teleoperation 电信经营业务
teleoperator 遥控操作器;遥控操纵装置;遥测操纵器
teleoroentgenography 远距离摄片
teleoseal 铁镍钴合金
telephone and computer information service 电话与计算机信息服务
telephone and telegraph 电话及电报
telephone area 电话用户区
telephone bid 电话投标单
telephone block system 电话闭塞系统
telephone book 电话号码簿
telephone booth 电话亭;电话间
telephone box 公用电话间;电话室;电话间
telephone buzzer 电话蜂鸣器
telephone cabinet 电话橱
telephone cable 电话电缆
telephone central office 电话总局;中央电话局
telephone circuit 电话线路
telephone communication 电话通信
telephone communication unit 电话通信装置
telephone company 电话公司
telephone concentration system 电话集中系统
telephone connection box 电话接头箱
telephone cross talk 电话串话
telephone directory 电话簿
telephone distribution box 电话分线盒
telephone duct 电话电缆管道
telephone engineering 电话工程学
telephone exchange 电话局;电话交换机
telephone exchanger 总机;电话交换台
telephone exchange room 电话总机室;电话交换室
telephone exchanging system 电话交换系统
telephone fixer 电话固定器
telephone for harbo(u)r district 港区电话
telephone installation 电话安装;电话装置
telephone installation work 电话安装工程
telephone insulator 通信线路绝缘子
telephone jack 电话插孔
telephone junction box 电话组线箱
telephone key 手键
telephone kiosk 电话间
telephone line 电话线路;电话线
telephone line conditioning 电话线调节

telephone link 电话线
telephone link conditioning 电话线调节
telephone network 市内电话网
telephone number 电话号码
telephone office 电话局
telephone operator 电话员;电话接线员
telephone outlet 电话出线口
telephone plant 电话设备
telephone plug 电话插座;电话插塞
telephone point 电话接线插口;电话测试点;电话连接点
telephone private branch exchange 用户小交换机
telephone rate 电话费
telephone receiver 电话耳机;听筒;收话器;受话器
telephone register 收报机
telephone repeater 电话中继器;电话增音机
telephone repeater station 电话中继站;电话增音站
telephone sender 电话记发器
telephone service 电话业务
telephone set 台式电话机;电话机
telephone shelf 电话隔板
telephone signal 电话信号
telephone socket 电话插座
telephone station 公用电话台
telephone subscriber 电话用户
telephone survey 电话调查
telephone switchboard 电话交换台
telephone switchboard room 电话总机室
telephone switching system 电话交换系统
telephone system 电话系统
telephone system for port dispatching 港口调度电话
telephone-telegraph switch 电话电缆转换开关
telephone traffic 话务量
telephone transmitter 话筒;送话器;船舶收发报话机
telephone trunk zone 长途电话电路段
telephone-type submarine cable 电话型海底电缆
telephonist 电话接线员
telephonograph 电话录音机
telephonometer 通话计时器
telephony 电话学
telephoto 传真照片;传真电报
telephoto effect 传真效果
telephotograph 遥拍照片;传真照片
telephotography 远距离照相术;照片传真术
telephoto lens 远摄镜头;远距离摄影镜头;摄远镜头
telephotolens 远摄物镜
telephotometer 远距光度计;遥测光度计;遥测光度表
telephotometry 遥测光度学;光度遥测术;光度遥测法
telephoto objective 摄远物镜
teleplotter 电话绘图器;电传绘迹器
telepneumatolitic way 远源气化方式
teleport 电信港
teleprinter 电传打印机;电打字机
teleprinter over multiplex 多工电传打字机
teleprinter room 电传打字电报机房
teleprocessing 远程处理;遥控加工;遥控处理;远程信息处理
teleprocessing system 遥控处理系统;远程信息处理系统
telepsychrometer 遥测干湿表
telepunch 遥控穿孔(机)
telepuppet 遥控机器人
telepyrometer 定时高温计
telequipment 遥控装置
teleradiography 远距离射线照相术;远距X线照相术
teleran 电视雷达导航系统;电视雷达导航仪;电视雷达导航
telereading 远距离读数
telerecord 电视录像
telerecorder 电视录像器;遥测记录器;遥测记录仪;自记式遥测仪
telerecording 电视录像;遥测记录
telerecording apparatus 录像机
telerecording bathyphotometer 遥测深水光度计
teleregister 显示牌价荧光屏
telerobot 遥控机器人
teleroentgenogram 远距X线照片
teleroentgenography 远距离摄片;远距X线照相术
telerun 遥控;远距离操纵
telesat 通讯卫星

telescope 望远镜瞄准镜;望远镜;套入
telescope and scale method 望远镜尺度法
telescope arm 望远镜;伸缩臂
telescope-axle mirror (工具经纬仪望远镜的)侧镜
telescope cal(l)ipers 内径测微器;光学测微仪
telescope cap 望远镜头帽;望远镜盖
telescope casing 望远镜套筒
telescope chimney 套筒式烟囱
telescope claw mounts 瞄准镜插座
telescope collar 望远镜调整圈
telescope constant 测距误差常数
telescoped bulkhead 伸缩式舱壁
telescoped ice 叠冰;层状冰
telescope direct 正镜【测】
telescope disk 望远镜镜片
telescope(d) joint 套管连接装置;伸缩管连接;套管接头;插接
telescoped ore deposit 远生矿床
telescoped pipeline 变径管道
telescope drivingsystem 望远镜驱动装置
telescoped shape 套叠形状
telescope-feed 套筒式进料;套筒式
telescope flint 望远镜用燧石
telescope flint glass 望远镜用火石玻璃
telescope fork 伸缩套筒叉
telescope holder 望远镜支架
telescope in normal position 正镜【测】
telescope in reversed position 倒镜
telescope in sequence 顺序伸缩
telescope jack 筒式千斤顶;套筒(式)千斤顶
telescope joint 套筒接头;套筒接合;套管连接;插接套管接头
telescope ladder 伸缩梯
telescope leg 伸缩腿架
telescope level 带望远镜的水准仪;瞄准镜水准器;水准仪
telescope level tube 望远镜水准管
telescope magnifier combination 望远镜放大镜组合
telescope mast 伸缩桅
telescope objective 望远物镜
telescope partition 伸缩隔墙
telescope piece 伸缩器
telescope pipe 套筒管
telescope port 望远镜观察窗口
telescope reverse 倒镜
telescope sealing of shiplift 升船机密封装置
telescope shade 望远镜色片
telescope structure 嵌入构造
telescope support 望远镜支架;伸缩支架
telescope table 伸缩桌
telescope tube 望远镜镜筒;套筒管
telescope with load 带载伸缩
telescope with reticle 带分划板的望远镜
telescopic 可伸缩的;套筒的;套管的;伸缩的
telescopic access tube 伸缩式接管
telescopic aerial 可伸缩天线
telescopic alidade 望远镜照准仪;带望远镜的方位镜
telescopic antenna 拉杆天线
telescopic aperture 望远镜孔径
telescopic arm 伸缩臂;套筒臂
telescopic astronomy 望远镜天文学
telescopic axle 套筒轴
telescopic baffle 伸缩挡板
telescopic band (伸缩式输送机的)伸缩带
telescopic bar 伸缩套杆
telescopic bellmouth 套筒式漏斗口;套筒式喇叭口
telescopic belt 伸缩皮带机
telescopic bogie 伸缩式装料车
telescopic boiler 伸缩式锅炉;长身锅炉
telescopic boom 伸缩式吊杆;伸缩臂
telescopic boom crane 能伸缩的吊杆起重机;伸缩式吊杆起重机
telescopic boom excavator 伸缩铲挖机
telescopic caisson 套筒式沉箱
telescopic cap 望远镜盖
telescopic centering 伸缩式膺架;伸缩式拱架;伸缩对中座
telescopic chimney 套筒式烟囱;伸缩式烟囱
telescopic chute 可回缩的脊椎式溜槽;套筒式卸(料)槽;套筒式倾卸槽;套筒式溜子;套筒式溜槽;套筒倾卸槽;套筒溜槽;套筒式卸槽;伸缩式倾斜槽;伸缩(式)溜)槽
telescopic conveyer 伸缩传送带

telescopic cover 伸缩罩
telescopic cradle 伸缩式支船架
telescopic crane 伸缩式起重机
telescopic cyclone 伸缩旋风器
telescopic cylinder 伸缩式油缸
telescopic cylinder of two-stages 两级伸缩缸
telescopic damper 伸缩式减震器
telescopic derrick 无缝吊杆；套叠钻塔；伸缩式钻架
telescopic dipper stick 烧勺伸缩柄
telescopic door 折门；伸缩式门
telescopic drawbar 伸缩拉杆
telescopic drive shaft 伸缩式驱动轴
telescopic duct 伸缩管
telescopic expansion joint 伸缩接头
telescopic extension lance 套管伸缩喷枪
telescopic feed 伸缩式套筒给进架
telescopic feeder 伸缩式进给装置
telescopic feed hammer drill 伸缩式推进凿岩机
telescopic feed screw 伸缩进刀螺旋
telescopic finder 检像望远镜；探测望远镜
telescopic flagpole 可拉ံ旗杆；可伸缩旗杆
telescopic floor shuttering center 伸缩地板模板拱架
telescopic fork 伸缩套筒叉
telescopic form(work) 滑升模板；套筒式模板；穿行式钢拱模板；伸缩式模板
telescopic funnel 伸缩烟筒
telescopic gangway 可伸缩跳板
telescopic gas-holder 套筒储气柜
telescopic gate （装在鱼道入口处的）叠合滑动门；（大坝的）叠合滑动鱼道入口门
telescopic girder 伸缩式大梁；伸缩梁
telescopic guard 机床导轨伸缩防护装置
telescopic guideway 可伸缩导轨
telescopic guiding carriage 伸缩导架
telescopic hand-held laser rangefinder 手持望远镜式激光测距仪
telescopic heliotrope 望远镜回照器
telescopic hoist 伸缩式起重机；伸缩吊
telescopic hoist boom 伸缩式吊臂
telescopic holder 多层储气罐
telescopic hood wagon 有伸缩式罩套的货车
telescopic hydraulic cylinder 伸缩套筒液压缸
telescopic hydraulic ram 套筒式液压顶杆；伸缩式液压顶杆
telescopic jack 套筒式千斤顶
telescopic jib 伸缩臂
telescopic jibbed excavator 伸缩支架挖土机
telescopic jib crane 伸缩式臂架起重机
telescopic joint 伸缩接缝；伸缩接头；套筒接合；套管接合；伸缩式连接；伸缩式接头；伸缩管接合
telescopic ladder 消防云梯；伸缩梯
telescopic lap splice 套接接头
telescopic leader 伸缩导杆；可伸缩的打桩机导向支柱
telescopic leg 伸缩支柱；伸缩柱
telescopic leg tripod 活腿三脚架
telescopic lifting mechanism 伸缩举重机构
telescopic lighthouse 可升降灯塔
telescopic light pole 套筒式灯杆
telescopic line 伸缩套管；伸缩式管线
telescopic link 伸缩式杆
telescopic loader 伸缩装卸机；伸缩臂式装载机
telescopic loading spout 可伸缩散装头
telescopic loading trough 伸缩式装载；伸缩式双层装载槽
telescopic manipulator 伸缩式机械手
telescopic mast 套管天线杆；伸缩柱；伸缩套管式天线；伸缩式钻塔；伸缩式桅杆
telescopic mast head 伸缩式桅顶
telescopic mesh refinement 筒式格栅清除
telescopic metal joist 可伸缩的金属支撑件；套筒式钢梁
telescopic meteor 望远镜流星
telescopic mount 套管式装配；伸缩式架座
telescopic multiple par shaft 万向接头伸缩轴；万向接头套管轴
telescopic optic(al) system 望远镜光学系统
telescopic performance 望远镜性能
telescopic photometer 天文光度计
telescopic pipe 伸缩管
telescopic plunger 伸缩柱塞
telescopic pole 伸缩式钻塔；伸缩式桅杆
telescopic position 望远镜位置

telescopic prop 伸缩支柱；可伸缩支柱
telescopic radiant 望远镜辐射点
telescopic ram 伸缩式油缸
telescopic range finder 光学测距仪；望远镜测距仪
telescopic rapier 伸缩剑杆
telescopic resolution 望远镜分辨率
telescopic revolving ladder 伸缩回转梯
telescopic riser 伸缩式竖管
telescopic rotary tower 伸缩回转塔架
telescopic scaffold board 伸缩脚手板
telescopic screw 套筒螺旋
telescopic settlement ga(u)ge 套筒式沉陷计；伸缩式沉降计
telescopic shaft 套管轴；伸缩轴；伸缩套筒；伸缩套管
telescopic shaft coupling 伸缩筒式联轴节
telescopic shock absorber 可伸缩减震器；筒式减震器；伸缩式减震器；套筒式减震器
telescopic shock strut 可伸缩减震柱；套管减震柱
telescopic sight 望远镜瞄准镜
telescopic sighting alidade 望远镜照准仪
telescopic speed 伸缩速度
telescopic spindle 套筒指轴
telescopic spout 可伸缩的下料口；套筒式下料口
telescopic spreader 伸缩式吊具；伸缩式吊架
telescopic staff 可伸缩式标尺；塔尺【测】
telescopic steel shuttering 套筒式钢模板
telescopic strut 套筒支柱；伸缩撑杆
telescopic style spring bar 伸缩表耳簧
telescopic support 伸缩支架；望远镜支架
telescopic (top)mast 伸缩桅
telescopic tower 伸缩塔架；套管天线杆；伸缩式塔身
telescopic tower crane 伸缩塔架式起重机
telescopic tripod 套管式三脚架；伸缩式三脚架
telescopic tube 伸缩套筒；伸缩(套)管
telescopic tube extensometer 望远镜管式伸缩仪
telescopic tube settlement ga(u)ge 伸缩式沉降测管
telescopic type absorber 伸缩筒型避震器
telescopic type damper 伸缩筒型避震器
telescopic valve 套筒排泥阀
telescopic vent 伸缩通风口
telescopic working platform 套筒工作台
telescopiform 可伸缩的；套叠式(的)
telescoping 纵转望远镜；可伸缩式；套管式；叠缩；迭生作用
telescoping arm 伸缩杆；伸缩臂
telescoping boom 可伸缩悬臂；套筒式吊杆
telescoping-boom crane 可伸缩吊杆起重机
telescoping casing 伸缩套管
telescoping chute 套筒式卸料槽；套筒式倾卸槽
telescoping connection 伸缩管式联结；伸缩(管)式连接
telescoping crawler excavator 伸缩式履带挖掘机
telescoping cylinder 伸缩缸
telescoping derrick 伸缩式桅杆起重机
telescoping drilling derrick 伸缩式桅杆起重架；伸缩式钻井架
telescoping excavator 伸缩式挖掘机
telescoping excavator boom 伸缩式挖掘机悬臂
telescoping form(work) 可伸缩模板；套筒式模板；滑升模板
telescoping ga(u)ge 可伸缩内径规；伸缩量规；伸缩规
telescoping jack 套筒千斤顶；双重螺旋千斤顶；双重螺旋起重器
telescoping jib 伸缩式吊杆
telescoping joint 伸缩式接头
telescoping loading ramp 伸缩式装货斜道；伸缩式装货滑道
telescoping mast 伸缩式立柱
telescoping of shuttle 货叉伸缩
telescoping outrigger 伸缩平衡重
telescoping rail 伸缩轨
telescoping range finder 测距仪
telescoping shore column 望远架支柱
telescoping shuttering 滑升模板；套筒式模板
telescoping sliding shutter 望远镜滑动开闭门
telescoping spreader 伸缩式吊具；伸缩式吊架
telescoping stabilizer 伸缩平衡锤；伸缩平衡重
telescoping tripod 伸缩式三脚架
telescoping tube 伸缩管；伸缩套筒；伸缩套管；分层沉降管
telescoping-type tremie 套筒式水下灌注混凝土导管

telescoping valve 套筒阀
telescopiny 嵌入过程
telescopy 望远镜学
telescreen 电视屏幕；荧光屏
teleseism 远震
teleseismic arrival 远震到时
teleseismic data 远震资料
teleseismic detection 远震探测
teleseismic distance 远震距
teleseismic epicenter location 远震震中定位
teleseismic event 远震事件
teleseismic magnitude 远震震级
teleseismic microseismic noise 远震微震噪声
teleseismic network 远震台网
teleseismic noise 远震噪声
teleseismic record 远震记录
teleseismic signal 远震信号
teleseismic station 远震台站
teleseismic wave 远震波
teleseismology 遥测地震学
teleseme 信号机
teleservice 用户终端业务
teleset 电视接收机
teleshopping 电信购物
telesignal(l)ing 遥信；遥测信号
telesignallization 远距离信号设备；遥测信号设备；遥测信号化
telesis 有目的的使用
telesmoke 烟气望远镜；黑烟读数仪
telesounder 遥测回声仪
telespectroradiometer 远距光谱辐射计
telespectroscope 远距分光镜；遥测分光镜
telespeedometer 遥测速度计
telestation 遥测台站
telestereoscope 立体望远镜；双眼望远镜；反光立体镜；体视望远镜
Telesterion 提列斯蒂利昂大会堂(公元前希腊)
telesthesia 超限感觉；超常感觉
telestudio 演播室；电视演播室
teleswitch 遥控开关；遥控键
telesyn 遥控同步机
telesynd 远程同步遥控装置；遥控同步装置
telesystem 望远镜系统
teletach(e)ometer 遥测转速计；遥测转速表；遥测流速仪；遥测流量计
teletext 文字电视广播；图文电视
telethermal deposit 远成热液矿床
telethermal ore deposit 深成低温热液矿床
telethermal process 远成热液作用
telethermograph 遥测温度计；遥测温度表；遥测温度自动记录器
telethermometer 电气寒暑表；遥感温度表；遥测温度计；遥测温度表；远距温度计
telethermoscope 遥测温度器；遥测温度计；遥测温度表
teletorium 演播室；电视演播室
teletorque 遥控力矩器；交流自整角机
teletraffic 长途业务
teletransmission 远程传送；遥测传送；电信传递
teletransmitter 遥测传送器
teletron 显像管；测试图显像管
teletube 显像管
teletube glass 显像管玻璃
teletype 电传
teletype control 远距离控制；远距离控制
teletype writer 电传打字机；打字终端
teletype writer terminal 终端电传打字机；电传终端机
teleutospore 真菌冬孢子
televimonitor 电视监视器
television 电视接收机
television aerial 电视天线
television amplifier 电视放大器
television and infrared observing satellite 电视和红外辐射观测卫星
television and radar navigation 雷达电视导航
television and radar navigation(al) system 电视雷达导航系统
television borehole camera 电视钻孔照相机
television broadcast band 电视广播频带
television broadcasting 电视广播
television broadcasting room 电视广播室
television broadcasting station 电视广播台
television broadcast satellite 电视广播卫星

television building 电视大楼
television cabinet 电视接收机箱
television cable 电视电缆
television camera 电视摄像机
television camera tube 摄像管
television car 电视车
television compass 电视罗经
television control 电视管理;电视控制
television control center 电视管理中心;电视控制中心
television counter 电视计算电路
television coverage 电视有效区
television-directed 用电视遥控的
television electronic microscope 电视电子显微镜
television engineering 电视工程
television equipment 电视装置;电视器材
television facsimile 电视传真
television field frame 电视场帧
television guider 电视导星器
television high voltage tester 电视高压测试器
television interface 电视接口
television interference 电视干扰
television map 电视地图
television method 电视法
television modulator 电视调制器
television monitor 电视监视器
television-monitored production 现场电视监测生产法
television monitoring system 电视监视系统
television non-linear distortion tester 电视非线性失真测试仪
television operating center 电视中心
television outlet 电视出线口
television picture tube 显像管
television plant 电视设备
television power amplifier 电视功率放大器
television projector 电视放映机
television radar air navigation 电视雷达航空导航仪;电视雷达导航系统
television radar navigation 电视雷达导航
television ranging system 电视测距系统
television receiver 电视接收机
television recording 电视录像
television reflect meter 电视反射功率计
television reflector lens 电视反射透镜
television room 电视(广播)室
television screen 电视屏
television set 电视机
television studio 电视演播室;电视广播室
television surveillance 电视监控
television system for simulation training 模拟训练电视系统
television telephone 可视电话
television telephone set 可视电话机;电视电话机
television testing oscilloscope 电视测试示波器
television torch 电视发射圆;电视发射塔
television tower 电视塔
television tracking 电视跟踪
television tracking system 电视跟踪系统
television transmitter 电视发射机
television transposer 电视传播机
television tuner 电视调谐器
television typewriter 电视电传打字机
television viewer 电视观众
televisor 电视接收机;电视机
televoltmeter 遥测伏特计;遥测电压表
televox 声控机器人
telewattmeter 遥测功率计
telewriter 打字电报机;电写机
telex 用户电报;电传
telex room 电传打字室
Telfener rack 特尔费纳式铁轨
telfer 电动高架单轨缆车;高架索道;电动缆车
telferage 高架缆索;电动缆车(系统)
telfer conveyer 缆索输送机
telfer line 高架索道
telfer railroad 高架索道
Telford 泰尔福装配式房屋;锥形块石
Telford base 大块石基层;泰尔福式基层;大石块基础;锥形块石基层;紧压石基层;大石基层
Telford foundation 泰尔福式基层
Telford macadam 大块石基层碎石路;泰尔福式碎石路
Telford pavement 大块石铺面;碎石路面;泰尔福式路面
Telford road 大石(块)基层道路
Telford stone 基层大块石
Telford subbase 大块石底基层;大块石基底
telfurane 替氟烷
telinite 结构镜质体
telite 微结构镜煤
Tellard 特拉德水泥着色剂
tellemarkite 钙铝榴石
teller (量尺的)长度标
teller console 用户服务台
Tellerette 特勒填料;惰性填(充)料
Tellerette packing 特勒花环型填料;特勒环填料
Teller Rosette 特勒花环填料
Teller Rosette packing 特勒花型填料
teller's desk 出纳台
teller system 用户服务系统
tellevel 液面高度计
telling technique system 技术交底制度
tellite 指示灯
telltable clock 指示时钟
telltale 液量指示器;指数器;指示器;警报器;示警装置;舵位指示器;应变杆;裂缝监测片;驾驶动作分析仪;登记机
telltale board 操作信号盘;控制信号盘
telltale cock 警号龙头;警报龙头
telltale compass 倒挂罗经
telltale cover flag 监视孔盖板
telltale device 信号装置
telltale float 水位浮标
telltale hole 信号孔;泄漏信号孔;警报孔;指示(器)孔
telltale pipe 显示管(溢水指示);溢水显示管
tell-tales 警告悬条标
telltale title 说明字幕
telltale tube 显示管(溢水指示)
telltale water level 警报水位
tellurantimony 碲锑矿
tellurate 碲酸盐
telluretted hydrogen 碲化氢
tellurian 地球居住者
telluric 地球的
telluric acid 碲酸
telluric acid anhydride 三氧化碲
telluric band 大气谱带
telluric chloride 四氯化碲
telluric current 地电;大地电流
telluric current method 大地电流法
telluric current prospecting 大地电流法勘察
telluric electric(al) field 大地电流场
telluric electric(al) field method 大地电流场比法
telluric electricity 土地电
telluric electricity field 大地电场
telluric electromagnetic sounding 大地电磁测探
telluric line 大气谱线;大气吸收线
telluric magnetic force 大地磁力
telluric oxide 三氧化碲
telluric water 地中水
tellurides 碲化物
tellurion 地球仪
tellurire 黄碲矿
tellurium 自然碲
tellurium-antimony method 碲—锑法
tellurium bromide 二溴化碲
tellurium chloride 二氯化碲
tellurium compound 碲化合物
tellurium deposit 碲矿床
tellurium dibromide 二溴化碲
tellurium dichloride 二氯化碲
tellurium dioxide 二氧化碲
tellurium donor complex 碲供体络合物
tellurium ga(u)ge 测碲计
tellurium hexafluoride 六氟化碲
tellurium lead 碲铅合金
tellurium monoxide 一氧化碲
tellurium ores 碲矿
tellurium oxychloride 二氯一氧化碲
tellurium tetrachloride 四氯化碲
tellurium-xenon method 碲—氙法
tellurobismuthite 碲铋矿
tellurohauchecornite 硫碲铋镍矿
telluroid 似地球面
tellurometer 精密测地仪;脉冲测距仪;脉冲测距器;微波测距仪;测距仪;无线电测距仪;雷达测距仪
tellurometer survey 微波测距
tellurometer traverse 微波测距导线
telluropalladinite 斜碲钯矿
tellurous bromide 二溴化碲
tellurous chloride 二氯化碲
telnet 远程登录
Telnic bronze 特尔尼克耐蚀青铜
teloalgnite 结构藻类体
teloclarite 微结构镜质亮煤
telocollinite 均质镜质体
telofusite 微结构丝煤
telogelinite 均匀凝胶体
telogen 调聚体;调聚剂
telogenetic fabric 表成组构
telognosis 远距诊断
telomere 调聚物;端粒
telomeric reaction 调聚反应
telomerization 调节聚合反应
telomerization reaction 调聚反应
telophase 分裂末期
telosemicollinite 均质半镜质体
telotaxis 趋触性
Telpak 宽频带通信通道
telpher 高架索道;电缆吊车;电动小吊车;电动缆车;电动单轨悬挂系统;电动单轨吊车
telpherage 高架索道运输;高架索道系统;架空电缆运输(法);索道运输系统;索道运输装置
telpherage line 高架索道;架空索道
telpher carrier 架空单轨起重葫芦
telpher conveyer 吊式输送机;电动缆索输送机;缆道输送机;索道输送机
telpher conveyer system 架空单轨输送系统;架空输送系统
telpher crane 电动小吊车
telpher hoist 架空单轨重车;架空单轨起重葫芦
telpher line 高架索道;电动缆车线路;架空单轨吊运线;电动缆车索道
telpher rail 架空单轨
telpher railroad 电气化铁路;电缆铁路
telpher railway 电气化铁路;电缆铁路
telpher track 架空单轨
telphoto 远距照相镜头
TEL process combined 离心喷吹 TEL 法
telquel 现状条款
telseis 遥测地震
Telsmith breaker 特尔史密斯破碎机
Telsmith crusher 特尔史密斯破碎机
Telsmith gyrasphere crusher 特尔史密斯型旋球式破碎机
Telsmith pulsator 特尔史密斯脉动筛
telstar 通信卫星
Telstar satellite 电星通信卫星
temagamite 碲汞钯矿
temblor 地震
temenos 寺院围墙;神庙区(古希腊)
temforming 马氏体形变处理
temo 卡车驾驶员
temoin 挖方土柱(标记挖方深度);土柱(标记挖方深度用);挖土深度标桩
temper 钢化;韧度回火;调合;回火
tempera 胶质壁画颜料;水粉画;水彩;蛋胶画
temperable 可回火的
tempera colo(u)r 水彩画颜料
Temperac quilt 坦姆扑拉克卷材
temperal distribution 瞬时分布
temperal homogeneous 时齐的
temperance 节制
tempera painting 水彩画
temperate 温和的(气候)
temperate and cold savanna(h) 温带和寒带稀树草原
temperate and cold scrub 温带和寒带密灌丛
temperate arid climate 温带干燥气候
temperate belt 温带和地带;温带
temperate climate 温带气候
temperate climate with summer rain 温带夏雨气候
temperate climate with winter rain 温带多雨气候
temperate continental climate 温带大陆性气候
temperate decideous forest 温带落叶林
temperate desert climate 温带沙漠气候
temperate ecology 温带生态学
temperate ecosystem 温带生态系统
temperate forest 温带森林

temperate forest and woodland 温带森林与林地
temperate forest biosphere 温带森林生物层
temperate forest climate 温带森林气候
temperate forest region 温带林带
temperate forest zone 温带林带
temperate glacier 暖冰川；温带冰川
temperate grassland 温带草原
temperate grassy climate 温带草原气候
temperate house 中温温室
temperate karst 温带岩溶
temperate lake 温带湖
temperate low belt 温带低压带
temperate marine climate 温带海洋性气候
temperate mixed forest 温带针阔混交林
temperate mixed forest region 温带混交林带
temperate monsoon climate 温带季风气候
temperate needle-leaf forest 温性针叶林
temperate periglacial 温带冰缘
temperate rainforest 温带雨林
temperate rainy climate 温带夏雨气候；温带湿润气候；温带多雨气候
temperate water carbonate 温带海水碳酸盐岩
temperate waters 温带水域
temperate westerlies 温带西风带；西风带；温带亚风林
temperate-westerlies index 温带西风带指数
temperate woodland 温带疏林
temperate zone 温带
temperate zone summer green forest cinnamon soil zone 温带夏绿林褐土带
temperatoplankton 温带浮游生物
temperator 恒温器；温控器
temperatuer detector 温度检测计
temperature 温度
temperature, salinity and current recorder 温度、盐度和流速记录仪
temperature above freezing 零上温度；凝固点以上温度；冰点以上温度
temperature above zero 零上温度
temperature-absolute zero 绝对温度零度
temperature acclimation 温度适应
temperature action 温度作用
temperature actuated pressure relief valve 温度控制卸压阀
temperature adaptation 温度适应
temperature adjusting device 温度调节装置
temperature adjustment 温度调节
temperature advection 温度平流
temperature alarm 过热报警；温升过热报警信号；温升报警信号；温感报警器；温度过高警报；温度报警钟
temperature aloft 高空气温
temperature altitude chart 气温高度图；温度高度图
temperature and humidity 温度和湿度
temperature and humidity datum 温湿度基数
temperature and humidity infrared radio meter 温度与湿度红外辐射计
temperature and time of storage 储[贮]藏温度和时间
temperature anomaly 温度异常；温度距平
temperature antiseptic effect 温度防腐作用
temperature approach 最低温差
temperature association 温带群落；温带群丛
temperature at debiteuse mouth 槽口温度
temperature at orifice 小眼温度
temperature at work area 作业地带温度
temperature at work place 作业地带温度
temperature average 平均温度；温度平均值
temperature balance 温度平衡
temperature band 温度带
temperature bar （混凝土的）温度筋；分布筋
temperature barrier 热障；热点；温度阻限
temperature bath 恒温池
temperature behaviour 温度表现
temperature below freezing 凝固点以下温度；冰点以下温度
temperature below zero 零下温度
temperature belt 温度带
temperature binding stress 温度约束应力
temperature boundary layer 温度边界层
temperature build-up 温度升高
temperature bulb 温度传感器；测温泡
temperature buzzer 温度警报器
temperature cabinet 定温箱

temperature calibration 温度标定
temperature calibration value 井温仪刻度值
temperature change 温度改变；温度变化；温度变动
temperature-change epitaxy 变温外延
temperature change of hydrothermal system 热液系统温度变化
temperature change resistance 耐温度变化性
temperature changing heat transfer 变温传热
temperature characteristic 温度特性
temperature characteristics 温度特性曲线
temperature chart 温度图
temperature checking of furniture lacquer 家具漆膜的冷裂试验
temperature classification 耐热等级；温度分类
temperature classify of spring 温泉的温度分类
temperature climate 温暖气候
temperature clock 测温钟
temperature coefficient 温度系数
temperature coefficient feedback 温度系数反馈
temperature coefficient of capacitance 电容温度系数
temperature coefficient of capacity 电容温度系数
temperature coefficient of delay 温度延迟系数
temperature-coefficient of electric (al) resistance 电阻温度系数
temperature coefficient of permeability 导磁性温度系数
temperature coefficient of permittivity 电容率温度系数
temperature coefficient of reactivity 反应性温度系数
temperature coefficient of refractive index 折射率温度系数
temperature coefficient of resistance 电阻温度系数
temperature coefficient of resonance frequency 谐振频率温度系数
temperature coefficient of resonant frequency 谐振频率温度系数
temperature coefficient of solubility 溶解度温度系数
temperature coefficient of voltage drop 管区降温度系数
temperature coefficient power regulation 温度系数功率调节
temperature colo(u)r 火色
temperature colo(u)ration 受热色变
temperature colo(u)r scale 色温标
temperature compensated 温度补偿的
temperature compensated capacitor 温度补偿电容器
temperature compensated crystal oscillator 温度补偿(石英)晶体振荡器
temperature compensated flasher 温度补偿闪光断续器
temperature compensated ga(u)ge 温度补偿片
temperature compensated pendulum 温度补偿摆
temperature compensated reference element 温度补偿基准元件
temperature compensated regulator 温度补偿调节器
temperature compensated thermist current meter 温度补偿热敏电阻海流计
temperature compensated Zener diode 温度补偿齐纳二极管
temperature compensating circuit 温度补偿电路
temperature compensating coil 温度补偿线圈
temperature compensating device 温度补偿装置
temperature compensating resistance 补偿电阻温度
temperature compensation 温度补偿
temperature compensation alloy 温度补偿用合金
temperature compensation depth 温度补偿深度
temperature compensation device 温度补偿器件
temperature compensation range 温度补偿范围
temperature compensation system 温度补偿系统
temperature compensator 温度补偿器
temperature composition diagram 温度组成图
temperature condition 温度条件；温度情况
temperature conditioning 温度调节
temperature conditioning chamber 温度调节室
temperature conductivity 温度传导率；导温率；导热性
temperature constancy 温度恒定

temperature contour 等温线
temperature contraction apparatus 温度收缩仪
temperature contraction of pipes 管道的温度收缩
temperature contraction stress 温缩应力；温度收缩应力
temperature contrast 温度差(异)；温度不均匀分布；不均匀温度分布
temperature control 温度控制；温differences控制
temperature control amplifier 温度控制放大器
temperature control chamber 定温室
temperature control circuit 温度调动电路
temperature control coating 温控涂层；调温层
temperature control equipment 温度控制装置；温度控制器；温度调节设备
temperature control for high temperature boiler 高温锅炉温度控制器
temperature controlled 温度控制的
temperature controlled cabinet 温度控制箱
temperature controlled chamber 温度控制室
temperature controlled crystal oscillator 温度控制晶体振荡器
temperature controlled package 温度调节组件
temperature controller 温度控制器；调温器
temperature control(ling) system 温度控制系统
temperature control loop 温度控制回路
temperature control material 温度控制材料
temperature control measure 温度控制措施
temperature control relay 温度继电器
temperaturecontrol sensor 温度控制传感器
temperaturecontrol system test set 温度控制系统测试设备
temperature control thermistor 控温热敏电阻器
temperature control valve 温度控制阀；温度调节阀
temperature conversion 温度换算
temperature conversion table 温度换算表
temperature correction 气温改正；温度校准；温度校正；温度改正；温度订正
temperature correction chart 温度修正图表
temperature correction value 温度改正值
temperature covariance 温度共分散
temperature crack(ing) 温度裂缝；温度开裂
temperature current curve 温度电流曲线
temperature curve 温度(变化)曲线；温度变动曲线；体温曲线；变温过程
temperature curve for melting 熔化温度制度
temperature curve of exhaust 排气温度曲线
temperature curve of melting 熔化温度曲线
temperature cycle 温度循环
temperature cycling test 冷热交变试验；温度循环试验
temperature damping 温度衰减
temperature decreament 温度减退
temperature-decreased pressure reducer 减温减压器
temperature decrease of air conditioning 空调降温
temperature defect 温度亏损
temperature deflection 温度翘曲；温度挠曲
temperature deformation 温度变形
temperature degeneration 温度致衰；温度减退
temperature density 温度密度
temperature departure 气温距平；温度距平
temperature dependence 温度关系；温度依赖性；温度相关性
temperature dependence of creep 蠕变的温度影响
temperature-dependent 随温度变化的；与温度有关的；随温度而变的
temperature-dependent behavio(u)r 温度依赖行为
temperature-dependent constant 温度相依常数
temperature-dependent resistor 热变电阻器
temperature depth curve 温深曲线
temperature depth profile curve map 温度—深度剖面曲线图
temperature depth salinity system 温盐深度测量装置
temperature detecting apparatus 温度检测装置
temperature detecting device 温度检测装置
temperature detecting element 高温计探头
temperature detecting unit 温度检测装置
temperature detector 感温器；检温器；热敏元件；温度传感器；探温计
temperature determination 温度确定；温度测定
temperature development 升温过程

temperature difference 气温差;温(度)差
temperature difference between supply and return water 供回水温差
temperature difference correction factor of envelope 围护结构温差修正系数
temperature difference heat engine 温差热机
temperature difference method 温差法
temperature difference module 温差组件
temperature difference quotient 温差商数
temperature difference resistance 耐温差性
temperature difference stress 温差应力
temperature differential 温差
temperature differential filter 温差过滤器
temperature differential weathering 温差风化
temperature diffuse scattering 温度漫散射
temperature diffusivity 温度扩散(系数);温度传导系数
temperature dispersion 温度分散
temperature display panel 温度显示板
temperature distillation chart 蒸馏温度曲线图
temperature distortion 受热变形
temperature distribution 温度分布
temperature distribution infrared radio meter 温度分布红外辐射计
temperature disturbance 温度扰动
temperature drift 温度漂移
temperature drift instability 温度漂移不稳定性
temperature drive 温度差
temperature drop 降温;温降;温度下降
temperature drop correction 温度降校正
temperature drop drop of temperature 温度降(低)
temperature drop fall 温降
temperature drop of groundwater 地下水温下降
temperature effect 温度作用;温度影响;温度效应;温差作用
temperature efficiency 温度效率
temperature efficiency index 温(度有)效指数;温度效率指数
temperature efficiency ratio 温度效率比
temperature element 热元件;温度元件
temperature entropy chart 温熵表;温熵图
temperature environment 温度环境
temperature equalization 温度平衡;温度均一化;温度均衡
temperature equalization and distribution device 温度均衡和分配装置
temperature equilibrator 温度补偿器
temperature equilibrium 温度平衡
temperature excursion 温度偏离额定值;超温
temperature expansion 热膨胀
temperature expansion coefficient 温度膨胀系数
temperature expansion of pipes 管道的温度膨胀
temperature exponent 温度指数
temperature extremal 温度极值
temperature extreme 温度极限
temperature factor 温度因素;温度因数;温度系数
temperature fall 温度下降;温度落差
temperature feedback 温度反馈
temperature field 温度场
temperature figure 温度分布图式
temperature float 温度浮沉子
temperature float technique 变温悬浮法
temperature fluctuation 温度脉动;温度升降;温度波动;温度变化;温带变幅
temperature fluctuation in water 水的温度波动
temperature fluctuation range 温度波动范围
temperature force 温度力
temperature forecast 温度预报
temperature frictional stress 温度摩擦应力
temperature front 温度界面
temperature funtion 温度函数
temperature ga(u)ge 温度计;温度表
temperature ga(u)ge pipe 温度计管
temperature ga(u)ge unit 温度感受元件
temperature grade 温度等级
temperature gradient 热阶度;温度梯度;温度陡度;梯温法
temperature gradient chromatography 温度梯度色谱法
temperature gradient furnace 梯温炉
temperature gradient instability 温度梯度不稳定性
temperature gradient method 温度梯度法;分段升温

temperature gradient paper chromatography 温度梯度纸色层法
temperature gradient technique 温度梯度技术
temperature gradient type incubator 温度梯度型保温箱
temperature gradient zone melting 温度梯度区熔
temperature gradient zone refining 温度梯度的区熔提纯
temperature gravity conversion table 温度比重换算表
temperature gravity graph 温度密度线示图
temperature hardened spectrum 温度硬化能谱
temperature head 热位差;温差
temperature height diagram 温度高度图
temperature history 温度随时间的变化
temperature hole 测温井
temperature homogeneity 温度均匀性
temperature humidity cycling 温湿循环
temperature humidity graph 温湿(曲线)图
temperature humidity index 温度—湿度指标;温湿指数
temperature humidity infrared radio meter 温(湿)度红外辐射仪;温湿度红外辐射计;温度—湿度红外辐射计
temperature hyperbola 温度双曲线
temperature hysteresis 温度滞后现象
temperature increase 温度增高;温升
temperature increment 温度增量;温度升高
temperature in degree 摄氏温度
temperature-independent 不随温度变化的;与温度无关的
temperature indicating coating 示温涂层
temperature-indicating compound 指示温度化合物
temperature indicating controller 温度指示控制仪
temperature indicating crayon 温度笔
temperature indicating device 温度指示器
temperature indicating instrument 温度测定仪
temperature-indicating paint 示温颜料;示温涂料;示温漆
temperature-indicating pigment 示温颜料
temperature indicating strips 温度指示条
temperature indication 温度指示
temperature indicator 温度指示器;温度计;示温器
temperature indicator adapter 温度指示器接头
temperature induce 温度诱导
temperature induced 温度诱致
temperature induced load 温度感生荷载
temperature influence 温度影响
temperature infrared radio meter 温度红外辐射计
temperature in situ 现场温度
temperature instability 温度不稳定性
temperature instrument 温度计
temperature insulation 温度绝缘
temperature insulation property 温度绝缘特性
temperature interval 温度区间;温度间隔
temperature inversion 逆温;气温逆增;温度转换;温度转化;温度逆增;温度倒布
temperature inversion layer 逆温层
temperature joint 温度缝;伸缩缝
temperature jump 温度跃升;温度跃迁;温度跃变
temperature jump distance 温度跃变距离
temperature jump method 温度突变法
temperature lag 温度滞后
temperature lamination 温度分层;温度层结
temperature lapse condition 温度减率条件
temperature lapse rate 温度直减率;温度递降率;温度递减率
temperature level 温(度)位;温度范围
temperature limit 应用温度范围;温度极限
temperature limitation 温度限制
temperature limit controller 温度限制控制器
temperature limited condition 温度限制条件
temperature limited diode 温度限制二极管
temperature limited region 温度限制范围
temperature limiting criterion 温度限制准则
temperature limiting relay 过热继电器
temperature load 温度荷重;温度荷载
temperature loading 温度力
temperature log 井温测井;温度录井;测温井
temperature loss 温度损失
temperature margin 温度限度
temperature maximum 温度极大值
temperature mean 平均温度;温度平均值

temperature measurement 温度测量;测温
temperature measurement method 温度测量方法
temperature measurement system 温度测量系统
temperature measuring 温度测定
temperature measuring device (大体积混凝土的)温度监测装置
temperature measuring element 温度测量元件;测温元件
temperature measuring equipment 测温装置
temperature-measuring hole 测温孔
temperature measuring meter 温度测量仪表
temperature measuring set 温度测量仪
temperature measuring thermocouple 温度量测热电偶
temperature meter 温度计;温度测量器;温度表
temperature microstructure 温度微结构
temperature minimum 温度极小值
temperature mixed corrosion 温度的混合溶蚀
temperature model of river water 河水温度模型
temperature modulated choke system 温度控制的阻风门系统
temperature-modulated forced-air heating 强迫式加热温度调节
temperature moisture diagram 温湿图解
temperature monitor 温度监视器;温度测量仪
temperature movement 温度伸缩(材料);温度变位(膨胀或收缩);温度胀缩;温度位移
temperature noise 热噪声;温度噪声;温度本底值
temperature normal 温度正常值
temperature number 温度准数
temperature observation 温度观测
temperature of a solid-state transition 固态转变温度
temperature of cement hydration 水泥水化温度
temperature of clarification 澄清点
temperature of combustion 燃烧温度
temperature of comparison 对照的温度
temperature of earth heat 地热温度
temperature of effluent 排放物温度
temperature of evapo(u)ration 蒸发温度;蒸发温度
temperature of exit gas 出口煤气温度
temperature of explosion 爆炸温度
temperature of freezing of petroleum 石油的凝固点
temperature of fusion 溶解温度
temperature of hydration 水化(作用)温度
temperature of ice 冰温
temperature of interpolation 内涵法推断的温度
temperature of inversion 转化温度
temperature of low-temperature treatment 低温处理温度
temperature of melting 熔融温度
temperature of mine 矿井温度
temperature of outgoing air 排风温度
temperature of reaction 反应温度
temperature of recrystallization 重结晶温度
temperature of reservoir 热储温度
temperature of return water 回水温度
temperature of seabed thermal water 海底热水温度
temperature of set(ting) 凝结温度;(水泥的)水化温度
temperature of solidification 固化温度;凝固温度
temperature of spring 泉水温度
temperature of storage 储[贮]藏温度
temperature of surface softening 表面软化点
temperature of the peak 峰温
temperature of use 操作温度;使用温度
temperature of wastewater 废水温度
temperature on hot face of lining 衬板受热面温度
temperature optimization 温度最适化
temperature oscillation method 温度振荡法
temperature overload relay 温度检测式过载继电器
temperature parameters of fan 风扇耐温参数
temperature pattern 温度影响范围;温度分布形式;温度场
temperature-phased anaerobic digestion 温相厌氧消化
temperature pick-up 温度传感器;测温传感器;温度上升
temperature plug 热电偶;测量温度用的热电偶
temperature poise 温度平衡
temperature press(ing) 温度压力

temperature pressure 温度压力
temperature pressure compensating computer 温度压力补偿运算器
temperature pressure curve 温压曲线;温度压力曲线
temperature probe 温度传感器;测温传感器
temperature profile 温度线图;温度剖面;温度分布线;温度(分布)曲线;温度分布(变化)图
temperature profile recorder 温度剖面记录仪;温度廓线自记仪
temperature program(me) 温度制度;温度程序
temperature programmed desorption 程序升温脱附
temperature programmed oxidation 程序升温氧化
temperature programmed work 程序升温操作
temperature program(me) for glass-melting 玻璃熔化温度制度
temperature programmer 程序升温器
temperature programming 温度程序设计;程序升温
temperature-proof 抗温度的
temperature property 温度特性
temperature province 温度区域
temperature radiation 热辐射;温度辐射
temperature radiator 温度辐射体
temperature rainy climate 温带季雨气候
temperature range 温度较差;温度极限;温度幅度;温度范围;温差范围
temperature range for crop plants 对作物的温度
temperature range limit 温度变化幅度;温度变化范围
temperature range of crystallization 析晶温度范围
temperature range property 温度变化上染性能
temperature rate characteristic 温度走时特性
temperature rating 温度额定值
temperature ratio 温(度)比
temperature reaction 温度反应
temperature rebar 温度钢筋
temperature receptor 温感器;温度感受器
temperature record 温度记录
temperature record and control 温度记录和控制
temperature recorder 自记温度计;温度记录器;温度记录计;温度记录仪
temperature recorder controller 温度记录控制器
temperature recorder panel 温度记录板
temperature recording controller 温度记录调节器
temperature recovery 温度恢复
temperature recovery curve 温度恢复曲线
temperature recovery factor 温度恢复系数
temperature reduction 温度下降;温度订正
temperature reductioner 降温器
temperature refraction 温度折射
temperature regime 温度状况
temperature regulating chamber 温度调节箱
temperature regulating equipment 温度调节设备
temperature regulation 温度调节
temperature regulation of melting 熔化温度制度
temperature regulator 调温器;温度控制器;温度调节器
temperature reinforcement 抗温钢筋;(混凝土的)温度筋;温度钢筋;温差钢筋
temperature relay 热(动)继电器;温度(控制)继电器
temperature resistance 耐温性;耐热性
temperature resistance coefficient 热阻(力)系数;电阻温度系数
temperature-resistant 抗温度的;耐热的
temperature-resistant coating 耐热敷层
temperature response 温度响应
temperature rise 温升;温度上升;升温
temperature rise coefficient 温升系数
temperature rise curve 温升曲线
temperature rise efficiency 升温效率;温升效率
temperature rise limit 温升极限
temperature rise of groundwater 地下水温上升
temperature rise of light 号灯温升
temperature rise period 升温期;升温阶段;温升周期
temperature rise ratio 温升比
temperature rise room 蒸汽养护室;升温室
temperature rise test 温升试验
temperature rod 温度测杆
temperature run 发热试验
temperature salinity and current report 温盐流

报告;温度含盐量及海流观测报告
temperature salinity curve 温盐曲线
temperature salinity diagram 温盐图解
temperature sample introduction system 升温进样系统
temperature saturation 温度饱和
temperature scale 温度计刻度;温度标尺;温度标;温标;位标
temperature schedule 温度制度;温度分派表
temperature section 温度断面(图)
temperature seiche 温度静振;温度假潮
temperature selector 温度选择器
temperature sending device 温度发送装置;传温器
temperature sensing 对温度变化灵敏性
temperature sensing bulb 温包
temperature sensing capsule 温度受感膜盒
temperature sensing device 热敏装置;温度敏感装置;温度敏感元件;温度传感器
temperature sensing element 感温元件;感受元件;热敏元件;温度敏感元件;温度受感元件
temperature sensing feed-back 温度敏感反馈
temperature sensing probe 温度敏感探针
temperature sensing system 温度探测系统
temperature sensitive 热敏的;温度敏感的;对温度灵敏的
temperature sensitive element 温度敏感元件
temperature sensitive paint 热敏油漆;示温涂料;示温漆;变色涂料;变色漆
temperature sensitive period 温度敏感期
temperature sensitivity 温度敏感性;温度灵敏度;变温误差
temperature sensitivity coefficient 温度敏感系数
temperature-sensitivity of pressure 压力的温度灵敏度
temperature sensitization apparatus 热敏感器
temperature sensor 热敏元件;温度感测元件;温度传感器;测温传感器
temperature sequence 温度序
temperature set(ting) (水泥的)水化温度
temperature setting dial 温度调节度盘
temperature shift 温度变化
temperature shock 热震;温度骤变
temperature shock tester 高低温冲击试验台
temperature shrinkage 冷(却收)缩;温度收缩
temperature slope 温度梯度;温度坡度
temperature span 温度差距
temperature spread 温度差距
temperature stability 热稳定性;温度稳定性
temperature stabilization 温度稳定性
temperature stabilizer system 温度稳定系统;温度稳定器系统
temperature state point 温度状态点
temperature steel 分布筋;温度伸缩钢筋;温度(钢)筋
temperature step 温度跃变
temperature strain 热应变;温度应变;温差应变
temperature stratification 热结层;温度分层
temperature stress 热应力;温度应力
temperature stress peak 温度应力峰
temperature stress rod 温度应力钢筋
temperature stress-strain diagram 温度应力应变图
temperature structure 温度结构
temperature sum 温度总和
temperature summation 积温
temperature sum rule 温度总和规律
temperature surge 温度波动
temperature survey 温度调查;温度观察;井温测量;温度测量;地温调查;地热测量
temperature survey hole 测温钻孔
temperature susceptibility 温度敏感性;温度敏感度
temperature susceptibility ratio 感温比
temperature susceptivity 温度敏感性;温度敏感度
temperature swing adsorption 变温吸附
temperature switch 温度开关
temperature telemeter 温度遥测仪
temperature test 温度试验
temperature tester 测温计
temperature thermostat 恒温器
temperature time combination 温度时间关系
temperature time curve 温度时间曲线
temperature time equivalence 温度时间等效
temperature time factor 温度时间因数
temperature tolerance 温度容限
temperature tracking 温度跟踪

temperature transducer 温度传感器;温度变换器;测温传感器
temperature transfer coefficient 导温系数
temperature transformation 相变温度
temperature transmitter 温度传感器
temperature transport 温度输送
temperature traverse 温度横向分布;温度变化
temperature treatment 定温处理
temperature tube 测温管
temperature type 温度类别
temperature type of thermal groundwater 地下热水温度分类
temperature uniformity 温度均匀性
temperature upper limit 温度上升
temperature valve 温度阀
temperature variation 温度变化
temperature variation curve 温度变化曲线;温度变动曲线;温度变动
temperature variation method 温变法
temperature vertical lag 温度垂直滞后
temperature vibration 热振动;热振波动;温度波动
temperature-viscosity chart 黏度温度曲线图
temperature-viscosity curve 温度黏度曲线;黏度温度曲线
temperature warm zone 温带林
temperature warping 温度翘曲;温度挠曲;温度变化引起的翘曲;温差扭曲
temperature warping stress 温度翘曲应力
temperature wave 温度波
temperature well 测温井
temperature (well) log(ging) 温度测井(曲线)
temperature wind 温度风
temperature zone 温度带
temper bending test 韧弯试验
temper brittleness 回火脆性
temper carbon 回火碳
temper colo(u)r 回火(颜)色
temper current 回火电流
temper draw 使回火
tempered 经过回火的
tempered air 调和空气
tempered drawing colo(u)r 回火色
tempered forged steel 回火锻钢
tempered glass 钢化玻璃;回火玻璃;韧化玻璃;热处理玻璃;强化玻璃
tempered glass door 钢化玻璃门;保暖玻璃门
tempered glass railing panel 钢化玻璃拦板
tempered glassware 强化玻璃器皿
tempered hardboard 高强耐水硬纤维板;热处理硬质纤维板;调质硬质纤维板
tempered hardening 回火硬度
tempered heat-adsorbing glass 吸热钢化玻璃
tempered martensite 回火马氏体
tempered martensite embrittlement 回火马氏体脆化
tempered oil 轴承表面防护油膜;调和油
tempered oil film 保护油膜
tempered paint 调和漆
tempered plate glass 强化玻璃;回火玻璃平板
tempered safety glass 回火安全玻璃;钢化安全玻璃
tempered service hardboard 热处理辅助硬质纤维板
tempered sorbite 回火索氏体
tempered steel 回火钢;还原钢;退火钢;淬硬钢;脆硬钢
temperer 热处理工;淬火工
temper graphite 回火碳
temper hardening 淬火硬化;二次硬化;回火硬化
temper hardening brass 回火硬化镍铝黄铜
tempering 型砂浸湿;混练;人工老化;调和;淬硬;淬火强化;回火
tempering air 冷却用空气;调温空气
tempering air damper 冷风门;调节风门
tempering apparatus 钢化设备
tempering backfire 回火
tempering bath 回火槽
tempering between plates 二板之间淬火
tempering bloom 钢化虹彩;钢化彩虹
tempering coil 冷却盘管;调温旋管;调温蛇管
tempering colo(u)r 回火色
tempering curve 回火曲线
tempering drawing 回火
tempering forge 回火锻炉
tempering furnace 回火炉;淬火硬化炉;钢化炉

tempering glass 钢化玻璃
tempering hardness 经过回火硬度;再生硬度;回火硬度
tempering house 进行回火的房子
tempering machine 调温机
tempering metal 中间合金
tempering mixer 混料机;搅料机
tempering mortar 搅拌砂浆(加水后)
tempering oil 回火油
tempering pan 盘式捏练机
tempering plant 回火车间
tempering point 回火温度
tempering quality 回火质量
tempering range 回火温度范围
tempering sand 回火砂;调质砂
tempering steel 回火钢
tempering stove 回火炉
tempering stressing 回火温度
tempering tank 混合桶;混合槽
tempering temperature 回火温度
tempering time 回火时间
tempering tower 回火塔;缓冲塔
tempering tub 捏泥盘
tempering water 调和水;工作水分
Temperite 氯化钙防冻剂(混凝土)
Temperite alloy 坦普莱特铅锡铋镉易熔合金
temper justice with mercy 宽严并举
temper medium 回火介质
temper number 钢化指数;回火后硬度;淬火指数
temper pass 平整道次
temper pass mill 表面平整机;表面轧光机
temper resistance 回火稳定性
temper rolling 表面光轧
temper sand 调匀砂
temper screw 调节螺杆;平衡螺杆(钢绳钻探用);调节螺丝;调节螺钉
temper-screw clamps 丝杠机构中的绳卡
temper the jar 平衡推进器给进(冲击钻进时)
temper tinting 加热着色
temper troostite 回火屈氏体
temper water 砂子调节用水;调质水
tempest 风暴天气;风暴;恶劣天气;暴风雨
tempestite 风暴雨沉积;风暴流沉积;风暴层
tempest-swept 风暴席卷
tempestuous storm 惊涛骇浪
tempforming 形变回火
temp grad 温度梯度
Tempietto (1502年意大利罗马圣比得教堂院内的)回廊式建筑物
tempil 测温剂
tempilac 坦皮赖克示温器
tempilaq 测温用示色液
tempil pellets 示温法
tempil stick 热色棒;测温(色)笔;热敏蜡笔;测温棒
tempinis 黄红色硬木(有深色斑的,菲律宾产)
Templar's Church at London 在伦敦的圣殿骑士教堂
Templaster 坦普拉斯特(几种建筑材料的商品名称)
template 型板;样型;样板;梁垫;框图模板;模片;土圭;透明绘图纸;垫石;承板;薄铁板;标准框;板型
template action 模板作用
template activity 模版活性
template beam 垫梁
template board 模版纸板
template cutter 模板切割器
template die 板模
template excavate 模板挖掘机
template eyepiece 样板目镜;轮廓目镜
template feeding machine 喂模机
template for anchor bolt 地脚螺栓模板
template for boring 镗孔样板
template forming 样板刀成型
template frame 放样架
template ga(u)ge 螺纹样板
template grinding 样板轮廓磨削
template grinding machine 样板磨床
template guide 模制导板
template hardware 模制小五金;金属样板
template impulse control 模板脉冲控制
template jig 模板式钻模
template layout 型板布置
template 属性单元
template matching 模板匹配;模板比较
template material 模板材料

template mechanism 模板机理
template method 样板法;量板法
template milling machine 样板铣床
template mo(u)lding 刮板造型
template of angle domain 角域量板
template of electric(al) sounding curve 电测深量板
template paper 模版纸
template platform 底盘式平台;导管架平台
template process 仿形铣齿法
templates and models 样板和模型
template shop 模板工场;放样(车)间
template strand 模股;模板股
template theory 模板学说
template with machined bushings 机器样板
temple 庙宇;庙;寺院;支撑
temple architecture 寺庙建筑;寺院建筑
temple building 庙宇建筑
temple city 庙宇城
temple complex 庙宇复合建筑
temple construction 庙宇建筑
temple court 庙宇庭院
temple enclosure 庙宇围墙
temple facade 庙宇正面
temple forecourt 庙宇前庭
temple front 庙宇前部
temple gate 山门
temple gateway 庙宇门道;庙宇入口
temple ground plan 庙宇地平面图
temple nucleus 庙宇中心
Temple of Abu Simbel 阿布辛贝勒神庙(古埃及)
Temple of Amon at Karnak 卡纳克的阿蒙神庙
Temple of Aphaia 阿发伊神庙(古希腊)
Temple of Aphaia at Aiyina 埃伊纳岛上的阿发伊神庙(古希腊)
Temple of Aphrodite Ourania 阿夫罗狄蒂—乌拉尼阿神庙
Temple of Apollo 阿波罗神庙(古希腊)
Temple of Apollo Epicurius 阿波罗神庙(古希腊)
Temple of Ares 阿维斯神庙(古希腊战神)
Temple of Athena at Syracuse 叙拉古城的雅典城神庙
Temple of Athena Polias at Priene 普里恩的波利阿斯雅典娜神庙
Temple of Bel at Palmyra 巴尔米拉古城的贝勒神庙
Temple of Demeter 德墨忒耳神庙
Temple of Demeter at Paestum 帕埃斯图姆的德墨忒耳神庙
Temple of G. T. at Selenius 塞勒涅斯的阿波罗神庙
Temple of Heaven 天坛
Temple of Hera 赫拉神庙
Temple of Hera Lacinia at Agrigentum 意大利阿格里真托的赫拉拉西尼亚神庙
Temple of Horus 太阳神庙(古埃及)
Temple of Horus/Edfu 埃德夫赫鲁斯太阳神神庙(埃及)
Temple of Isis 伊希斯神庙
Temple of Jupiter 朱庇特神庙
Temple of Jupiter Heliopolitanus at Baalbek 巴勒贝克的朱庇特神庙
Temple of Minerva 密涅瓦神庙
Temple of Nemesis 涅墨西斯神庙
Temple of Nemesis at Rhamnus 拉姆努斯的涅墨西斯神庙
Temple of Nike Apteros at Athens 雅典的奈基神庙
Temple of of Bacchus 巴斯克神庙
Temple of of Concorde at Rome 罗马城的康科迪娅神庙
Temple of of King Sethos I 塞斯奥王一世祠庙
Temple of Poseidon 波赛冬神庙
Temple of Poseidonat Paestum 帕埃斯图姆的波赛冬神庙
Temple of Romulus 罗慕路斯神庙
Temple of Saturn 萨宿恩神庙
Temple of Soul's Retreat 灵隐寺(杭州)
Temple of the Azure Clouds 碧云寺(北京香山)
Temple of the Large Leshan Buddha 乐山大佛寺
Temple of the Sleeping Buddha 卧佛寺(北京香山)
Temple of the Sun 太阳庙
Temple of Venus 维纳斯神庙(罗马)

Temple of Venus and Rome at Rome 罗马的维纳斯和罗马神庙
Temple of Vespasian 韦斯巴芬祠
Temple of Vesta at Rome 罗马的维斯太神庙
Temple of Vesta at Tivoli 蒂沃斯的维斯太神庙
Temple of Zeus at Olympia 奥林匹亚的宙斯神庙
Temple of Zeus at Athens 雅典的宙斯神庙
Temple on the Ilissus 伊利色斯神庙
temple plan 庙宇平面图
temple portico 庙宇柱廊;庙宇门廊
temple precinct 庙宇辖区
Temples of Egypt 埃及神庙
templet 样板;梁垫;模片;道尺;属性单元
templet assembly 模片组合
templet cutter 模片切缝器
temple terrace 庙宇林阴路堤;庙宇阳台
templet laydown 模片安装
templet matching 模板匹配
templet method 模片法
templet milling 靠模铣
temple tomb 庙宇坟场
temple tower 庙塔
templet shaping press 样板压机
templet shop 放样车间
temple with twin sanctuaries 有孪生殿的庙宇
templin chuck 楔形夹头(拉力试验用)
templum in antis 角柱间殿堂;角柱间空间
Templum vaticanum 梵蒂冈教堂
tempolabile 瞬时即变的
temporal 瞬时的;时间的
temporal absorption coefficient 季节性吸收系数
temporal adaptation 时间适应
temporal aeration 瞬时曝气
temporal and geographical limits of liability 赔偿责任的时效和地效
temporal and spatial correlations 时空相关
temporal and spatial factors 时空因数
temporal and spatial variation 时空变化
temporal average velocity 瞬时平均速度;瞬时平均流速
temporal change 时间变化
temporal coding 时间编码
temporal coherence 时间相干性
temporal coherent beam 时间相干束;时间相干光束
temporal correlation 时间相关
temporal dependence 时间依从
temporal dimension 时间因次
temporal distribution 时间分配;时间分布;时程分配;瞬时分布
temporal filter 时间滤波器
temporal fluctuation 瞬时波动
temporal frequency domain 时间频率域
temporal frequency spectrum 时间频率谱
temporal gain 时间增益
temporal gain control 时间增益控制
temporal hour 钟盘时间等分
temporal interference 时间干涉
temporal isolation 时间隔离
temporality 暂时性
temporal limit of liability 赔偿期限
temporally absorbed water 暂时吸收水
temporally coherent light 时间相干光
temporally spatial model 时空模型
temporal match 时间匹配
temporal mean velocity 瞬时平均速度;瞬时平均流速;时平均速度
temporal method 暂行办法
temporal parallax 时间视差
temporal pattern 时程分配型式
temporal radiometer 瞬时辐射计
temporal redundancy 时间冗余度
temporal registration 时间配准
temporal resolution 时间分辨率;时间分辨力
temporal response 时间响应
temporal sampling theorem 时间采样定理
temporal sequence 时间层序
temporal series 时间序列
temporal-spatial distribution of tidal current velocity 潮流流速时空分布
temporal spirit 时代性
temporal stage 时相阶段
temporal structure 瞬时结构
temporal summation 时间上的叠加

temporal trend 时间趋势
temporal value 时间价值
temporal variation 暂时的变化;瞬时变化
temporaries 暂时存储单元
temporarily erected 暂时安装的;临时装设的
temporarily extinguished 暂时熄火
temporarily lighted buoy 临时灯标
temporarily out of stock 暂时缺货
temporary 临时工;暂时(性)的;临时(的)
temporary acceptable daily intake 暂行每日容许摄入量
temporary access (road) 临时通道;临时便道
temporary accommodation 临时住所;临时住宅;临时通融
temporary adaptation 暂时适应
temporary adhesion 暂时黏合
temporary adjustment 暂时调整
temporary adminstrating line revenue 临管线收入
temporary admission 免税进口
temporary advance 暂付预支;暂记预支
temporary advance and sundry debtor 暂记欠款
temporary agreement 临时协议
temporary airfield 临时机场
temporary anchorage 临时锚地
temporary annuity 有限年金
temporary aquifer 临时含水层
temporary area 暂时区;数据暂时存储区
temporary arrangement 临时办法
temporary assembly 临时装配
temporary assignment 暂时分配;临时任务
temporary axis 瞬时轴
temporary backing 临时垫块;临时垫板
temporary ballast 暂时压载;临时压载
temporary bank protection 临时性护岸
temporary base level 暂时基准面;临时基(准)面
temporary bench mark 临时水准(基)点
temporary binding agent 临时结合剂
temporary block 工棚;临时房屋;临时建筑;工作单元块
temporary block in a recursive procedure 递归过程的工作单元块
temporary boarded fence 临时板围;临时板篱
temporary borrowing 临时借款
temporary boundary effect 瞬时边界影响
temporary bracing 临时支撑
temporary bridge 临时性桥(梁);临时桥;临时便桥;便桥
temporary bridge for construction 施工便桥
temporary budget 临时预算;短期预算
temporary building 暂设房屋;临时性建筑(物);临时性房屋;临时建筑(物)
temporary bulkhead 应急舱壁;临时(隔)舱壁
temporary bunker 临时煤舱
temporary buoy 临时浮筒;临时浮标
temporary cable 临时索
temporary canal 暂时渠;临时渠(槽)
temporary cessation of business 暂停营业
temporary check dam 临时谷场
temporary climax 暂时演替顶极;暂时顶极群落
temporary close 暂时关闭
temporary coating 工艺性涂层
temporary cofferdam 临时围堰
temporary column 临时支柱
temporary communication 临时交通
temporary concrete lining 临时衬砌混凝土
temporary condensation 暂时冷凝
temporary connection 暂时联系
temporary construction 临时性结构孔;暂设工程;临时(性)建筑(物);临时构筑物;临设工程
temporary convergence 暂时趋同
temporary credit 暂收款
temporary crown 暂用罩冠
temporary current 暂时流水;临时洋流;短暂流
temporary dam 临时围堰;临时(挡)水坝
temporary data set 暂时数据集;临时数据集
temporary deadlight 临时舷窗盖
temporary debt 临时欠款
temporary decking 临时路面
temporary deficit 暂时逆差
temporary deformation 瞬时变形
temporary deposit 暂时存款;暂存款
temporary derrick 临时性起重桅杆
temporary dike 临时堤
temporary disability 暂时丧失工作能力;暂进丧失工作能力
temporary discharge pipe 临时泄水管
temporary discharge piping 临时泄水管
temporary discharge tube 临时泄水管
temporary discharge tubing 临时泄水管
temporary disk 临时盘
temporary diversion tunnel 临时导流隧洞
temporary dry dock 临时干坞
temporary duty 临时(关)税
temporary duty certificate 临时关税证书
temporary dwelling 应急住宅;暂设住宅;临时简易住宅
temporary earth electrode 临时接地电极
temporary earths 临时接地器
temporary enclosure 临时性围墙
temporary encroachment 暂时侵占(堆放施工材料)
temporary equilibrium 暂时均衡
temporary erection 临时竖立;临时架设;临时安装
temporary erection plate 临时安装板
temporary erection support 临时安装支架
temporary error 暂时错误
temporary expansion 暂时膨胀
temporary export 暂时出口
temporary extinction 暂时干涸
temporary face 临时封板
temporary facility 临时设施
temporary facility expense 临时设施费
temporary facility fee 临时工程费
temporary failure 暂时失效
temporary fastening 临时固定
temporary fastening device 临时固定装置
temporary fence 临时性围栏
temporary file 暂时文件;临时文件
temporary filling 暂时性充填
temporary film 临时薄膜
temporary filter 临时过滤器
temporary firn-limit 暂时雪线
temporary fixture 暂时装置品
temporary flush board 临时挡水板
temporary flyover 临时天桥;临时立体交叉(多指临时人行桥)
temporary food service establishment 临时性食品服务设施
temporary food tolerance 暂行食品最大允许含量
temporary foreign capital 短期外资
temporary form(work) 临时模板
temporary fortification 临时性防御工事
temporary frame work 临时框架
temporary fuel 暂时燃料
temporary fund 临时资金
temporary gate 临时水闸
temporary ga(u)ge 临时文件;临时水尺
temporary geographic grid 临时地理坐标图
temporary grip 临时锚具
temporary ground 临时用地
temporary groundwater 季节性地下水
temporary (ground) water-table 暂时地下水面
temporary guest 临时宾客
temporary guide base 井口导向盘(海洋钻探用)
temporary habitat 临时生境;暂生境
temporary hardness 碳酸盐硬度;暂时硬度
temporary hard water 暂(时)硬水
temporary hearing defect 暂时性听力缺损【救】
temporary hotbed 临时温床
temporary housing 临时住房;临时性住宅
temporary hunting prohibited area 暂时禁猎区;临时禁猎区
temporary import 暂时进口
temporary income 临时收入
temporary induction 暂时性感应
temporary influence 暂时影响
temporary investment 暂时投资
temporary irrigating canal 临时灌溉渠
temporary irrigation 临时灌溉
temporary joint 临时结合;暂时结合;临时接头;临时接合;临时接缝
temporary joint fillet 临时性接缝带
temporary laying 暂时铺设
temporary leakage 暂时性渗漏
temporary library 暂时程序库
temporary light 临时灯标
temporary lighting 临时照明
temporary line 临时线路;便线
temporary lining 临时衬砌
temporary liquefaction 暂时液化
temporary load during construction 施工期荷载
temporary load(ing) 临时荷载;短时负载;短时荷载
temporary loan 暂时借款;临时贷款;短期贷款
temporary location 中间单元;暂时单元
temporary magnet 暂时磁体
temporary magnetism 暂时磁性;非永久磁性
temporary magnetization 暂时磁化;暂存剩余磁化强度
temporary maintenance expedient 临时维护措施
temporary marker 临时路标杯;临时标志;临时标记
temporary measure 临时措施
temporary memory 暂存器
temporary method 暂用方法;暂行方法
temporary mo(u)ld 一次铸型
temporary nail 临时钉
temporary navigation 临时通航
temporary navigation facility 临时通航设施
temporary navigation lock 临时船闸
temporary navigation structure 临时通航建筑物
temporary needs 临时需要
temporary notices to mariners 临时航海通告
temporary nursery 临时苗圃
temporary observation point 临时性观测点
temporary observation station 临时性观测站
temporary operator 临时值机员
temporary order 临时排序
temporary overproduction 虚假的生产过剩
temporary palace 行宫(古建筑)
temporary par of exchange 暂时平价
temporary partition 临时隔墙;临时隔断
temporary partnership 暂时合伙
temporary pass 临时性通路
temporary pasture 临时牧地
temporary pattern 简易模样;单件或小批生产的模型
temporary pavement 临时路面
temporary paving 临时线路;临时铺装
temporary payment 暂付款(项);临时款项
temporary perched water 临时(性)滞水;上层滞水
temporary pictorial display 暂时图像显示
temporary pier 临时墩
temporary plankton 暂时性浮游生物
temporary planting 假植
temporary platform 临时站台
temporary plug(ging) 暂时堵塞
temporary point 临时点
temporary pointer variable 暂时指示字变量
temporary poison 暂时毒物
temporary post 临时支柱
temporary power 临时电源
temporary power supply 临时供电
temporary prohibition 临时的禁止令
temporary project 临时性工程
temporary proprietorship account 暂记资本账户
temporary protective coating 临时防锈涂料
temporary provisions 暂行规定;暂行条款
temporary radiobeacon 临时无线电信标
temporary radiobeacon service 临时无线电信标业务
temporary realm 临时区域
temporary receipt 临时收据
temporary receipts account 临时收入账户
temporary recharge 暂时性补给
temporary register 暂存器
Temporary Regulations of Management for Famous Historic Sites and Scenic Spots 风景名胜区管理暂行条例
temporary regulation structure 临时性整治建筑物
temporary reinforcement 临时配筋
temporary relocation 暂时搬迁
temporary repair 临时修理;临时修复;临时性修理;临时性检修
temporary repair shed 临时修复库
temporary repair siding 临修线
temporary replaced by lighted buoy showing the same characteristics 临时替用同灯质的灯标
temporary replaced by unlighted buoy 临时代用无灯浮标
temporary replacement 临时更换零件
temporary residence 临时住宅;临时住处
temporary restaurant 临时性饭店;临时性餐厅
temporary rest position 待班台位【铁】

temporary rest track 待班线
temporary retaining board 临时挡板
temporary revenue 临时收入
temporary river 间歇河(流)
temporary river works 临时性河道工程
temporary rivet 临时铆钉
temporary road 便道;临时道路
temporary roofing 临时(的)屋面材料
temporary route 临时线路;临时路线
temporary rudder 临时应急舵;代用舵
temporary runway 暂用跑道;临时跑道
temporary rust protection 临时(性)防锈
temporary sample plot 临时标准地
temporary scaffold 简易脚手架
temporary scaffolding 临时脚手架
temporary seal (for joint) 临时封缝;接缝临时填封
temporary seismograph station 临时地震台站
temporary service 临时(性)服务;短期工作
temporary service fee 临时劳务费
temporary set 弹性变形;暂时定形;高弹性形变
temporary settlement 临时居民点
temporary shade house 临时遮阴棚
temporary shade plant 短期荫蔽植物
temporary shaft 临时竖井
temporary shed (临时性的)简易工棚
temporary sheet piling 临时板桩
temporary shelter 临时防空洞;临时隐避所;临时避难所
temporary shelter for workers 临时工房
temporary shiplock 临时船闸
temporary shoring 临时支撑
temporary shuttering 临时模板
temporary sign 临时标志
temporary signal 临时信号;临时讯标
temporary silo 移动沙丘
temporary site 临时用地
temporary site access ramp 临时通道
temporary site factory 临时工棚
temporary snow-line 暂时雪线
temporary speed restriction 暂时限速
temporary speed restriction section 临时限速区段
temporary spring 暂时泉
temporary stability 暂时稳定
temporary stage 临时舞台;简易脚手架
temporary staging 临时台架;临时脚手架;简易脚手架
temporary station 临时台站;临时测站
Temporary Statute of Plan Administrating Land-use for Construction 建设用地计划管理暂行办法
Temporary Statute of Supervision and Management of Mineral Resources 矿产资源监督管理暂行办法
temporary stop 临时停车
temporary stop signal 临时停车信号
temporary storage 中间存储(器);暂时储[贮]存;暂时储[贮]藏;暂时库容;暂时存储;暂存器;工作单元区;暂时蓄水量;临时存储;临时存贮;临时仓库
temporary storage area 暂储[贮]区
temporary storage array 暂存阵列
temporary storage control 暂时存储控制
temporary storage table 临时存储表
temporary storage yard 临时堆场
temporary storehouse 临时仓库
temporary strain 暂时应变
temporary stream 间歇河(流);季节性流水;季节性河流
temporary stress 暂时应力;临时应力;施工应力
temporary structure 临时工程;临时(性)建筑(物);临时结构
temporary stuff 临时工作人员
temporary subsidy 临时补贴
temporary substitute 替工
temporary support 临时支柱;临时支架;临时支护;临时支承;临时支撑
temporary support point 临时支点
temporary support works 临时支撑工程
temporary surface flow 临时性表流
temporary surfacing 临时木工
temporary table 工作单元表
temporary tabular display 暂时表形显示
temporary target 临时讯标
temporary task 临时任务

temporary terminal 临时码头
temporary text delay 正文暂时延迟
temporary threshold shift 暂时性阈移;暂时性阈移;暂时性阈移
temporary tidal station 临时测潮站
temporary tide station 临时潮位站
temporary timbering 临时支撑
temporary toe weight 坡脚临时压重(物)
temporary toilet 临时厕所
temporary tolerance 暂时最大允许含量
temporary torrent 暂时性洪流;季节性流水;季节性河流
temporary tour train 临时旅游列车
temporary track 临时线(路);临时轨道;临时便线
temporary trade 暂时贸易
temporary trailer park 拖挂车临时停车场
temporary traverse 临时导线
temporary troll(e)y 临时台车
temporary tunnel support 隧道临时支护
temporary underpinning 临时支撑工程
temporary unemployment 临时性无业
temporary unlighted buoy 临时无灯浮标
temporary valve 暂时阀
temporary variable 临时变量
temporary wall 封闭墙
temporary water 临时水源;季节性水
temporary water-bearing layer 暂时含水层
temporary water connection 临时接水
temporary water course 临时(过)
temporary water discharge pump room 临时排水泵房
temporary water flow 暂时性流水
temporary water sealing 暂时性止水
temporary water supply 临时供水
temporary water table 暂时水位
temporary way 临时道路;临时线(如建筑工地临时运料用)
temporary wayleaves 临时道路通行权
temporary weir 临时围堰
temporary weld 临时点固焊缝;临时点定焊缝
temporary welds 临时焊缝(接)
temporary wilting 暂时干枯;暂时调萎
temporary wood frame structures and tents 临时木结构及帐篷
temporary wood pile 临时木桩
temporary wood piling 临时木桩
temporary worker 临时工;替工
temporary work files 临时工作文件
temporary works 暂设工程;临时性工程;临时工程
temporary works expense 临时设施费
temporary yard 临时堆场
temp-stick 温度计
temse 筛子
tenability 支持能力
tenacious 有附着力的;黏着力强的;黏滞(的)
tenacious clay 黏土;强黏性土
tenacious consistency 韧性结持度
tenacious wood 坚韧木料
tenacity 极限张应力;韧性;韧度;热力性
tenacity damping factor 黏滞阻尼系数
tenacity of soil 土壤韧度
tenaculum forceps 单爪钳
tenail(le) 钳堡(筑城);凹角堡
tenancy 租赁;成员;(土地或房屋的)租用;租赁权
tenancy agreement 租约;租借合同
tenancy at sufferance 默许租赁
tenancy at will 不定期租赁
tenancy by the entirety 夫妻合租
tenancy condition 租赁条件;成员条件
tenancy for life 终身占有
tenancy from year to year 自动延期租赁
tenancy in partnership 合伙占有
tenant 租借人,租户;租地人;承租人;成员组;房客;租屋人;居住者
tenantable 可租赁的
tenant at will 不定期租户
tenant contribution 承租人租金外的支出
tenant dwelling 出租住房
tenant farmer 佃农
tenant handbook 承租条例;承租人手册
tenant house 出租房屋
tenant in severalty 独自拥有财产者
tenantless 无人租住的
tenantless room (未使用的)空闲房间

tenant maintenance 租户自行维修
tenant mix 租户组合
tenant right 租地权
tenantry 租赁;承租人
tenant services 为租户服务
tenant's fixtures 租户加装的设施
tenant's improvement 承租人添建筑
tenant's meter 住户电表;客房电表
tenant's right of first refusal 承租人首买权
tenaplate 涂胶铝箔
ten-bay 十个桥跨的;十开间的
ten billion 一百亿
ten-bit component 十位分量
ten-channel multispectral scanner 十通道多谱扫描仪
ten-chord spiral 十弦螺旋曲线
tend 趋势
ten-day mean 旬平均值
ten-day report 旬报
ten-days average humidity 旬平均湿度
ten-days average temperature 旬平均温度
ten days' plan 旬计划
ten-day star 时星;旬星
ten-day's wagon loading plan 旬间装车计划
ten days' wagon requisition plan 旬要车计划表
ten degree grade xylene 十度级二甲苯
tendency 趋向;趋势;倾向;动向
tendency chart 趋势图;变化图
tendency equation 趋势方程;倾向方程
tendency for flash rusting 闪锈倾向
tendency interval 趋势时距
tendency method 趋势法
tendency monitoring 趋向监测
tendency of channel change 河道变化趋势
tendency of exchange rate 汇率走势
tendency of market 市场价格趋势
tendency of persistence 持续性趋势;持续性倾向
tendency of river channel change 河道变化趋势
tendency of stage 水位变化趋势
tendency plane analysis 趋势面分析
tendency profile 趋势廓线
tendency to bubble 起泡倾向
tendency to corrosion cracking 腐蚀开裂倾向
tendency to cracking 开裂倾向;断裂倾向
tendency to flow 流动倾向
tendency to fracture 断裂倾向
tendency to reboil 再生气泡倾向
tendency towards devitrification 析晶倾向
tendency towards sliding 滑动趋势
tendentious face analysis method 趋势面分析法
tendentiousness 倾向性
tender 易倾斜的【船】;供应船;煤水车;投标书;提供者;提供;脆弱的;偿付品;偿付;附属船(舶);标件
tender acceptance 投标书的接受
tender advertisement 投标广告
tender allowance 投标津贴
tender back lug 煤水车后端定位铁
tender-based seaplane 舰基水上飞机
tender bogie 煤水车转向架
tender bolster 煤水车身承梁
tender bond 押标保证;投标担保;投标保证金;投标保函;投标保单
tender book 投标书
tender brake cylinder 煤水车闸缸
tender brake gear 煤水车制动器
tender chafing block 煤水车防擦板
tender clause 投标条款
tender clay 软质黏土
tender coal board 煤水车挡煤板
tender coal bunker 煤水车煤槽
tender committee 投标委员会
tender condition 投标条件
tender contractor 指名投标
tender coupler 煤水车车钩
tender date 投标日
tender design 招标设计;供投标用的设计;投标设计;投标计划
tender document 投标文件;招标文件;交单;标书(招标的)
tender draft gear 煤水车牵引装置
tender drain cup 煤水车放水杯
tender drawing 投标图样
tendered price 投标价格

tenderer 提出者;提供者;招标人;投标者;投标商;投标人
tenderers sheet 投标商名单;投标人名单
tender evaluation 评价;投标估价
tender for construction 投标承建
tender form 标书;投标格式;标书格式
tender for procurement 标购
tender for the construction of 投标承建
tender front tank lug 煤水车水箱前定位铁
tender green 嫩绿
tender guarantee 投标保证书;投标保函
tendering 准备投标的过程;招标;发包;报价
tendering by comparing offer 比价招标
tendering condition 投标条件
tendering date 投标日期
tendering documents 投标文件;招标文件
tendering for conveyance 托运
tendering form 承包形式
tendering on an international competitive basic 国际竞争性招标;国际竞争性投标
tendering period 投标期限
tendering procedure 招标程序
tendering procedure (adjudication) 投标手续
tendering ring 投标集团
tender intermediate safety chain 煤水车中间保险链
tender invitation 投标邀请
tender item 提供项目;投标项目
tenderizer 整平机
tender journal box 煤水车轴箱
tender locomotive 煤水机车
tenderness 易倾性
tender notice 招标公告
tender of abandonment 委付移交
tender offer 投标书承诺;投标出价;投标报价;投标开价;收购股权
tender of specified contractors 指定投标
tender opening 开标
tender opening date 开标日期
tender out 申请;提出
tender performance 投标履约情况;提出履行合同
tender period 投标期
tender plant 不耐寒植物
tender porcelain 软瓷
tender price 投标价(格);标价;中标价格;标底
tender price of project 工程标底
tender procedure 投标手续;投标程序;招投标程序
tender proposal 投标建议(书)
tender rejection 投标书的谢绝
tender result 投标结果
tender sheet 标单
tender ship 小稳性船;易倾斜的船;高重心船
tenders price 开标价格
tender step 煤水车脚蹬
tender subject 棘手问题;敏感问题
tender sum 投标总价;发包价格
tender system 投标制度;投标程序
tender tank 煤水车水箱
tender tank man hole shield 煤水车水柜人孔盖
tender to contract option 投标期间汇期权
tender tone 柔和色调
tender validity 招标有效期;招标有效期
tender vessel 小稳性船;易倾斜的船;高重心船
tender yarn 脆纱
ten-digit micrometer 十位数测微器
tend(ing) 抚育;维护保养
tending and cutting of forest 森林抚育采伐
tending and intermittent cutting of forest 抚育间伐
tending felling 抚育伐
tending of wood 森林抚育;抚育森林
tendonless release 松弛作用
tendon 预应力筋;钢丝束;钢筋束;力筋;锚索
tendon anchor 钢丝束锚具;力筋锚具
tendon and winding machine 卷管机
tendon crossing each other 相互交叉筋束
tendon duct 钢丝束通道
tendon layout 力筋布置图
tendonless prestressing 化学法预(加)应力;非张拉钢筋法预加应力
tendon method 预应力钢筋法
tendon profile 钢腱纵向布置;钢筋变形曲线图;预留孔洞;预留孔道
tendon tension meter 钢筋拉力测定仪

tendon wire 力筋用钢丝
tendril 卷须攀缘植物
tend to be balanced 趋向平衡
tenebrescence 磷光消失;变色荧光
Tenelon 高锰高氮不锈钢
tenement 公寓;经济公寓;地产;租用房屋;租用地;寓所;(公寓的)一套房间
tenemental 地产的;出租的
tenement block 经济公寓大楼;出租公寓;出租大楼
tenement building 公寓;经济公寓
tenement house 出租住宅;公寓;经济公寓;经济房屋;租借屋宇
tenement style 长排建造式
tener 大型聚光灯
tenesmic 下坠的
tenesmus 下坠
tenexinite 薄壁壳质体
tenfold 十倍
ten-fold mixer 十仓混合机
ten-footer 十英尺见方的小屋侧房
ten-frame-hive 十框蜂箱
tengchongite 腾冲矿
tengerite 水碳钙钇石
tenia (陶立克建筑的)挑檐处带形花边饰
ten-in-one odeometer 十联固结仪
Tenite 吞耐特(乙酸乙酯纤维素塑料)
ten(-membered) ring 十节环
ten million 一千万
ten-minute rating 十分钟最大功率额定值
tennantite 砷黝铜矿
tennatry 出租的财产
Tennessee marble 田纳西州大理石(美国)
Tennessee softwood 田纳西州软木材(美国)
tennis club 网球俱乐部
tennis court 网球场
tennis lawn 草地球场
tennis stadium (有观众座位的)网球场
tennoned into 榫进
tenon 雄榫;合型销;凸榫;榫(头)
tenon and mortise 阴阳榫;榫和榫眼;雌雄榫
tenon and mortise joint 雌雄榫接合;凸凹榫接合;榫接
tenon-and-mortise work 榫卯
tenon-and-slot mortise 榫槽接合
tenon-bar splice 棍板榫接;榫条拼接
tenon cutter 榫凿
tenon-cutting machine 制榫机;开榫机
tenon dowel joint 嵌销接合
tenoned 开榫的
tenoner 制榫机;开榫机;接榫者
tenon hole 榫孔
tenoning 开榫
tenoning machine 制榫机
tenon joint 榫接合
tenon jointing 用榫接合
tenon-making machine 制榫机;开榫机
tenon saw 开榫锯;夹背锯;榫锯;手锯
tenon through 穿斗
tenon tooth 榫齿
tenor 支票期限;矿石品位;欠债期限;品位;地质品位
tenor draft 限期汇票
tenorite 黑铜矿
tenor of ore 矿石品位
tenor trombone 高音长号
tenpenny 三英寸大钉
tenpenny nail 大钉;三英寸大钉
ten point road test 十项道路试验
tenpounder 十磅重的东西
ten-pounds gasoline 十磅汽油
ten-pounds ga(u)ge pressure 十磅表压力
tensammetry 表面张力电量法
ten sampling intervals per decade 十分度取样
ten's carry 十进位脉冲
tens complement 十进制补码
ten's dial 十分度盘
ten's digit 十进制数(字);十进位数
ten's driver 十进位驱动器
tensegrity system 无一定尺寸限制的结构
tenseness 紧紧度
tensibility 延性;可挠性;可展性;可拉伸性
tensible 能伸展
tenside 表面活性剂;表面活化剂
tensile 张力的;拉伸的;力的;抗张力的;受拉的;伸张
tensile-adhesion property 拉伸黏结性能

tensile and compression test 断裂和挤压试验
tensile area 拉力区;受拉区;受拉面积
tensile bar 拉力试棒;受拉钢筋
tensile bending stress 弯曲引起的拉应力;拉伸弯曲应力
tensile bond 抗拉砌合;抗拉黏结
tensile break 拉伸断裂
tensile capability 受拉张裂;张拉开裂;受拉开裂;抗拉性能
tensile capacity 拉伸容量
tensile carriage 拉力架
tensile circumferential stress 圆周拉应力;周向拉应力
tensile component 张拉组件;张拉构件;抗拉组件;抗拉构件
tensile-compressive stress partially reversed 不全反复拉压应力
tensile crack 张裂隙;张裂缝;张拉破裂;张断裂;拉张裂缝;拉伸裂缝;受拉裂缝
tensile cracking 拉伸断裂;张拉开裂;受拉开裂
tensile creep 拉伸徐变;拉伸蠕变
tensile creep test 拉伸蠕变试验
tensile deformation 拉伸变形;拉力变形
tensile elasticity 拉伸弹性;抗张弹性
tensile elongation 拉伸长度;伸长度
tensile failure 拉伸破坏;拉伸断裂;拉力破坏;拉断;受拉破坏
tensile fatigue test 张力疲劳试验;拉伸疲劳试验
tensile fault 张断层
tensile figure 伸展数;张力数;抗张值;抗拉强度指标
tensile fixation 张拉固定
tensile force 张力;拉力
tensile force-extension relation 张力拉伸关系式
tensile force ga(u)ge 拉力仪
tensile fracture 张破裂;张断面;拉伸破裂
Tensile fracture zone in South China Sea Basin 南海海盆张裂系
tensile ga(u)ge 张力计
tensile impact 张力冲击
tensile impact test 拉伸冲击试验
tensile intensity 拉力强度
tensile joint 受拉接头;受拉接合;受拉接点
tensile layer 张力层
tensile load(ing) 张力载重;拉伸荷载;拉伸负荷;抗拉荷载
tensile load test 拉力负载试验
tensile loop joint 环受张力连接;环受力连接;张力环连接
tensile machine 张力试验机
tensile member 受拉杆件
tensile membrane failure 受拉膜破损
tensile modulus 拉伸模数;拉力模量;抗张模量;抗拉模数;抗拉模量
tensile modulus of elasticity 拉伸弹性模量
tensile power 受张能力;受拉能力;抗张能量
tensile prestress 受拉预应力
tensile product 抗张积;抗拉积
tensile property 受(张)性能;拉伸性能;抗张性能;抗拉性能
tensile quality 抗张质量;抗拉质量;受张能力;受拉能力
tensile radial stress 径向张应力;径向拉应力
tensile recovery 拉伸回复力
tensile region 张力区;拉力区;受拉区域
tensile reinforcement 拉力钢筋;抗拉钢筋;受拉钢筋
tensile resilient modulus 抗压回弹量;弯拉回弹模量
tensile resistance 抗拉力
tensile rigidity 抗拉刚度
tensile ring 拉力环
tensile rupture 拉裂;拉断
tensile rupture test 拉伸破坏试验
tensile safe factor 抗拉安全系数
tensile sample 拉伸试样
tensile section 受拉断面
tensile shear crack 受拉剪力裂缝
tensile shear(ing) strength 拉伸剪切强度
tensile shear(ing) test 拉伸剪切试验
tensile shock test 张力冲击试验
tensile shrinkage stress 收缩引起的拉应力
tensile shrinkage stress crack 收缩拉应力裂缝
tensile-socket 张拉索接
tensile specimen 拉力试棒
tensile spilling strength test 劈裂抗拉强度试验

劈裂法试验;巴西法试验
tensile spilling test 拉裂试验;劈裂抗拉强度
tensile splitting force 受张分裂的力;受拉分裂的力
tensile splitting reinforcement 拉裂钢筋
tensile splitting strength 拉裂强度
tensile splitting strength test 拉裂强度试验;抗拉劈裂强度试验
tensile splitting test 拉裂试验
tensile spring 张丝
tensile steel 受拉钢筋
tensile strain 张胁变;张(拉)应变;拉(伸)应变;拉伸变形;抗拉应变;受拉应变
tensile strength 张拉强度;盛张强度;拉伸强度;拉力强度;抗张强度;抗拉强度;抗剪强度;伸张强度
tensile strength across the grain 横向抗拉强度(木材);横纹抗拉强度(木材);横断面抗拉强度(木材)
tensile strength and collapse resistance 拉伸和断裂强度
tensile strength in bending 弯曲抗拉强度
tensile strength limit 极限抗拉强度;拉力强度极限;抗拉强度极限
tensile strength of casing 套管抗拉强度
tensile strength of fibre [fiber] 纤维抗拉强度
tensile strength of malthoid 油毡拉力
tensile strength of rock 岩石抗拉强度;岩石的抗拉强度
tensile strength of wood perpendicular to grain 木横纹抗拉强度
tensile strength on bending test 弯曲试验的抗拉强度
tensile strength on direct test 轴向拉伸强度;直接拉伸强度
tensile strength parallel to the grain 木纹纵向抗拉强度;顺纹抗拉强度
tensile strength perpendicular to the grain 横纹抗拉强度(木材)
tensile strength test 抗张强度测定;抗拉强度试验
tensile strength tester 抗拉强度试验机
tensile strength test of rock 岩石(的)抗拉试验
tensile strength to weight ratio 抗拉强度—重量比
tensile stress 张应力;拉(伸)应力;抗拉应力
tensile stresses under restraint 限制作用下的拉应力
tensile stress-strain modulus 拉伸应力—应变模量
tensile structural plane 张性结构面
tensile structure 张拉结构
tensile system 受拉系统
tensile tectonite 张性构造岩
tensile test 张拉试验;拉伸试验;拉力试验;拉拉试验
tensile test bar 拉伸试棒
tensile test curve 抗拉试验曲线
tensile test diagram 拉力试验图
tensile tester for monofilament 单纤维强度试验仪;单丝抗拉强度试验仪
tensile tester for single strand 单丝抗拉强度试验仪;单纱强度试验仪
tensile testing machine 张力试验机;拉伸试验机;拉力试验机
tensile test piece 抗拉试件;受拉试件
tensile test specimen 拉伸试验试棒
tensile trajectory 受拉主应力轨迹线
tensile tripod 拉力架
tensile type load cell 张拉式压力盒
tensile type of deformation 伸长类变形
tensile wave 张力波;拉力波
tensile yield 拉伸屈服;扯断伸长率
tensile yield point 拉伸屈服点;张力屈服点;受拉屈服点
tensile zone 受张区;张拉区;拉力区;受拉区
tensility 延性;可展性;可拉伸性
tensil pull 拉力
tens(i)ometer 压差计;引伸仪;张力器;拉力计;张力表;拉伸试验仪;拉伸试验机;拉力计;拉力表;表面张力计
tensiometric measurement 张力测验
tension 拉伸,张力;张紧;拉力;牵力;膨胀力
tension adjuster 张力调节器
tension adjusting gear 张紧拉力调节装置;张力调节装置
tensional boundary 拉张边界
tensional deep fracture 张性深断裂
tensional deflection of drill tool 钻具扭曲度
tensional gasket 可受张的填料;可受拉的填料;膨胀的填料
tensional joint 张节理【地】
tensional joint fan of extrados 外弧张节理扇
tensional (steel) wire 张紧钢丝
tensional structural plane 张性结构面
tension analyer 张力分析器
tension anchorage 张拉锚碇
tension apparatus 拉紧装置;基线拉尺器
tension area 张拉面积;拉力区;抗拉面积;受拉区;受拉面积
tension arm 张力臂;力臂
tension at filter bag 滤袋织物过滤负荷
tension axis 张力轴
tension band 箍圈;张紧带
tension bar 拉(力)杆;拉尺杆;抗拉钢筋;受拉钢筋
tension bar loader 拉杆式装载机
tension bar type buoy 拉感式浮筒;拉杆式浮筒
tension bar type mooring buoy 拉条式系船浮标
tension beam 受拉(力)梁
tension block 张力轮
tension board 钢筋张拉台座
tension bolt 抗拉螺栓;拉力螺栓;拉紧螺栓;受力螺栓
tension boom 抗拉缘条
tension brace 支撑杆;拉杂;拉紧牵条;拉撑;牵条
tension bracket 张力架
tension brake-wheel 张力惰轮
tension bridle 张紧装置
tension brittleness 拉应力脆性;内应力脆性
tension butt strap 抗拉鱼尾板;抗拉平接板
tension by flexure 弯曲受拉
tension cable 受拉缆索
tension capacity 抗拉能力
tension carriage 拉力架;拉力器;紧索架
tension cavity 张力性空洞
tension chain 张紧链;受拉链;拉链
tension chord 拉弦杆;受拉弦杆
tension click 张力棘爪
tension click spring 张力棘爪簧
tension coefficient 张力系数
tension compensation regulation 张力补偿调节
tension compensator 张力补偿器
tension component 抗拉组件;抗拉构件
tension-compression bearing 拉压支座
tension-compression fatigue limit 拉压疲劳极限
tension-compression system 拉压系统
tension connection 受张连接;受拉连接;拉力连接
tension control 张力控制
tension controlled cylinder 给进调节液压缸(转盘钻进时)
tension correction 张力修正;拉力改正
tension crack 张裂,张(力)裂缝;张拉破裂;张裂缝;拉伸裂缝;拉裂缝;疲劳裂缝;受拉裂缝
tension cracking 受拉开裂;拉伸龟裂
tension cup 张力杯
tension curve 张力曲线
tension cut-off 张拉截面
tension deformation 受拉变形
tension detector 张力检测器
tension device 张拉装置
tension diagonal 斜拉杆;拉力斜撑;受拉斜杆
tension draft 张力牵伸
tension drilling 部分减压钻进
tension drum 张紧滚筒;(皮带机的)拉紧鼓筒
tension drum of conveyer 皮带机的张紧滚轮;输送机张紧滚筒
tension dryer 张力干燥器
tension dynamometer 张力功率计;张力动力计;张力测功计;张力仪;拉力测力计;拉力表
tensioned face 受拉面
tensioned highline 高架张索
tensioned in pairs 成对受张;成对受拉;成双张紧的;成对张紧的
tensioned rivet 受拉铆钉
tensioned side 受拉面
tensioned structure 牵拉结构
tensioned wire 张丝;受拉钢丝;拉制钢丝;受张钢丝
tension elasticity 拉伸弹性
tension electric(al) process 电流直接加热淬火法
tension element 抗拉构件;挠性件;受拉杆件
tension equipment 张紧装置
tensioner 张拉装置;张力器;拉紧器;张紧装置;张紧器;张紧轮;补偿器
tensioner of large type 大型拉紧装置
tensioner ring 胀环;扩张环
tensioner system 拉紧装置(深海钻井平台)
tension face 受拉面
tension factor 张力系数
tension failure 拉伸破坏;拉力破坏;受拉破坏
tension fault 正断层;张力断层;张断层
tension feeler 张力探测器
tension fiber 受拉纤维
tension field 张量场
tension field action 张力场效应
tension fillet 受缘条;受拉贴角条;受拉贴角架;受拉焊缝
tension finger 张力导纱器
tension fissure 张裂缝;张拉裂缝
tension flange 应力凸缘;受拉翼缘;抗拉法兰;拉翼缘;受拉梁翼;受拉法兰;受拉边
tension fold 张褶皱;拉张褶皱
tension foot 尺架
tension force 张力
tension fracture 张(性)破裂;张性断裂;张力裂缝;拉伸断裂;拉裂;拉裂;伸长破裂
tension-free 无拉力的;无张力的
tension-free pore water 无张力孔隙水
tension free rolling 无张力轧制
tension ga(u)ge 张力计
tension gear 张拉装置;牵引装置
tension gradient 张力梯度
tension grip 张拉夹具;拉力试验夹具
tension guide 张力导纱器
tension head 表面张力水头;张力水头
tension impact test 张力冲击试验
tension impulse 电压脉冲
tension increment 张力增量
tension indicator 张力指示器;张力计
tensioning 张力调整;张拉,拉紧
tensioning area 受张(力的)面积;受拉(力的)面积;抗拉面积
tensioning bar 拉杆
tensioning block 受张力块;受拉力块
tensioning bracket 张力托架
tensioning cable 受张力钢缆;受拉力钢缆
tensioning device 补偿器
tensioning element 预加应力元件
tensioning equipment 张力调整装置;补偿器
tensioning force 张力;拉力
tensioning insulator 耐力绝缘子
tensioning jack 用千斤顶张拉(钢丝);张力千斤顶;液压拉伸器;张拉千斤顶
tensioning lever 拉紧柄;拉紧杆
tensioning load 张拉荷载
tensioning loss 张力损失;拉力损失
tensioning machine 张拉机;拉伸机
tensioning method 施张力方法;施拉力方法
tensioning nut 张紧螺母
tensioning order 施张力次序;施拉力次序
tensioning plant 施张车间;施拉车间
tensioning process 受张过程;受拉过程
tensioning reinforcement 受张钢筋;受拉钢筋
tensioning rod 受张杆件;受拉杆件
tensioning rope 张力绳索
tensioning steel 受张力钢;受拉力钢
tensioning strand 受张绞合线;受拉绞合线
tensioning system 受张系统;受拉系统;绷紧系统
tensioning tendon 受张钢筋束;受拉钢筋束
tensioning value 张力值;拉力值
tensioning wedge 施张力楔形物;施拉力楔形物
tensioning wires in pairs 成对地施张力于钢丝;成对地施拉力于钢丝
tensioning yard 施张力工场;施拉力工场
tensioning zone 受张区域;受拉区域
tension insulator 耐张绝缘子
tension intensity 拉力强度
tension jack 张紧千斤顶;拉紧千斤顶
tension joint 拉力接合;受拉接头;受拉接合;受拉接点
tension leg platform 张力腿式生产平台;张力腿(式)平台
tension length 锚段长度;锚段
tension level(l)er 张力平整机
tension level(l)ing 张力平整
tension lever 拉紧柄;扳紧柄
tension line 张力线
tension link 张力调节器
tension load 张力荷载;拉伸负荷

tension magnet 节动磁铁
tension mat 风扇网
tension measuring device 张力测定仪
tension member 抗拉构件;受拉牵条;受拉构件;受拉杆(件);加强件
tension meter 张力器;张力计;拉力计
tension modulus 张力模量
tension-moisture curve 张力含水量曲线
tension nut 张拉螺母;张紧螺帽
tension of a spring 弹簧的拉伸
tension on belt 拉紧皮带
tension-open basin 拉开盆地
tension packer 张力封隔器
tension pair 牵力副
tension parallel to grain 顺纹拉力
tension parallel to grain of timber member 木料顺纹拉力
tension perpendicular to grain of timber member 木料垂直于木纹拉力
tension pick-up 张力传感器
tension piece 拉杆;缀条
tension pile 应力桩;张力桩;拉力桩;拉桩;抗拔桩;受拉桩
tension pile test 抗拔桩试验;拉力桩试验
tension pin 绷紧空心销
tension pole 张紧柱
tension pulley 张力盘;张力轮;张紧轮;受拉滑轮
tension pulley block 张紧滑轮
tension quantity 张量
tension rail 张力杆
tension raker 受拉斜桩
tension recovery 张力松弛
tension reducing mill 张力减径机
tension reel 张力卷筒;张力滚柱
tension region 拉力区;受拉区
tension-regulated contact wire 可调应力接触线
tension regulator arm 张力调节臂
tension regulator for bucket line 斗链张力调节装置
tension regulator stopper 张力调节器限位片
tension reinforcement 拉力钢筋;抗拉钢筋;受拉钢筋
tension resistance 抵抗张力;抵抗拉力
tension rib 受张肋;张力肋
tension rift 张力断裂
tension ring 张力圈;张拉环;拉力环
tension ring indicator 拉力环状指示计;拉环指示计;拉力环状指示针
tension rod 张杆;拉力杆;拉杆
tension roll(er) 张力辊;张力滚柱
tension roller system 张力辊系统
tension roof 悬挂式屋顶
tension roof structure 张拉屋盖结构
tension rope 张紧绳
tension servo 张力伺服
tension set 张力变形
tension shackle 弹簧张紧钩环
tension-shear basin 张扭性盆地
tension-shear fault 张扭性断层
tension-shearing 张扭性的
tension-shear structural plane 张扭性结构面【地】
tension sheave 张紧轮
tension side 受拉面;主动边;紧边;皮带主动边;受拉侧;受拉边
tension side of belt 皮带紧边
tension sleeve 箍圈;张拉套筒;受拉套管;带丝扣套环(钢丝绳用);螺纹套管
tension slide 张力滑板
tension solenoid drive 张力电磁铁驱动
tension specimen 拉力试杆;拉力试棒;受拉试件;测定抗拉强度的试件
tension splice 受拉拼接
tension spring 拉力弹簧;拉簧
tension stiffening 受拉劲化(混凝土)
tension strain 张应变;张拉应变;拉应变;受拉应变
tension strap 受拉带元
tension strength 抗张强度;抗拉强度
tension strength at break 拉断强度
tension strength test 张拉强度试验
tension strength tester 张拉强度试验机
tension stress 张应力;拉应力;拉伸应力;抗拉力;受拉应力
tension stringing 张力架线

tension structure 悬挂结构;充气结构;张拉结构;拉力结构
tension structure fabric 牵拉结构织物
tension suture 减张缝合法
tension switch 张力开关
tension tannage 张紧鞣法
tension test 张力试验;拉伸试验;拉力试验
tension test diagram 抗拉试验图
tension tester 张力试验机;张力测试仪;拉力试验机
tension test for loose gear 可卸零部件拉力试验
tension test sample 张拉(强度)试样
tension theory 张力学说
tension-thining 拉伸变薄
tension tie 受张系杆;受拉系杆
tension time index 张力时间指数
tension together 绑紧在一起;拉紧在一起
tension toggle operator 收紧器;绳索收紧器
tension tower 张力架
tension track structure 张性追踪构造
tension transformer 电压互感器
tension tripod 拉力架
tension-type boudin 张裂型石香肠
tension-type hanger 张力型油管挂;张力型油管挂
tension unit 张紧辊装置
tension vatiation method 张力变动方法
tension-velocity curve 张力速度曲线
tension washer 张力垫圈;张紧垫圈
tension wave 张力波;拉力波
tension wave front 张力波前沿;拉力波前沿
tension weight 拉紧重锤;张力重;平衡重;拉重;张力锤;张紧配重;平衡锤
tension weld 受拉焊缝
tension wheel assembly for concrete pole 水泥柱用棘轮
tension wheel assembly for lattice steel 桁架钢柱棘轮
tension wheel assembly for tubular steel pole 圆钢筋柱上用棘轮
tension wheel in tunnel 隧道内棘轮装置
tension winch 自动调节绞缆机
tension wire 张力线;张拉钢丝;引张线
tension wire alignment method 张线准直法
tension wood 应拉木;收缩性大易裂开和扭曲的木材;伸张木
tension wrapped fin tube 绕片式翅支管
tension wrench 转矩扳手;扭力扳手
tension yield point 拉力屈服点;受拉屈服点
tension zone 张拉区;受拉区;张力地带;拉力区;受张地带
tensiostats 张力保持器
tensity 张紧度
tens jumper 十位跳杆
tens jumper distance piece 十位跳杆隔片
tenslator 张力控制装置
tens of minutes indicator 十位分指示盘
tenso-friction 张力摩擦
tensometer 土壤湿度计;张力计;应变仪;伸长计;拉力计
tensometry 张力测量法
tensoplast 弹性胶布
tensor 伸张器;张力器;张量
tensor algebra 张量代数
tensor analysis 张量分析
tensor angle 张角
tensor bundle 张量丛
tensor calculus 张算;张量计算;张量分析
tensor conductivity 张量电导率
tensor contraction 张量约缩;张量短缩
tensor controller 张力控制器
tensor density 张量密度
tensor differentiation 张量微分
tensor-divergence 张量发散
tensor-divergence theorem 张量发散定理
tensor ellipsoid 张量椭球
tensor equality 张量等式
tensoresistance 张致电阻效应
tensor field 张量场
tensor force 张量力
tensor (force) interaction 张量力相互作用
tensor form 张量形式
tensor function 张量函数
tensor geometry 张量几何
tensorial set 张量集
tensor interaction 张量相互作用

tensor nature 张量性质
tensor neutralization 品位中和
tensor of polarizability 极化率张量
tensor of stress 应力张量;压应力张量
tensor operation 张量运算
tensor operator 张量算符
tensor permeability 张量导磁率
tensor polarization 张量极化
tensor potential 张量势
tensor product 张量(乘)积
tensor product algebra 张量积代数
tensor-product method 张量乘积法
tensor product representation 张量积表示
tensor property 张量特征
tensor quadric 张量二次曲面
tensor representation 张量表示
tensor space 张量空间
tensor subscript 张量足标
tensor sum 张量和
tensor susceptibility 张量磁化率
tensor-tensor effect 张量—张量效应
tensor transformation 张量变换
tensor-valued function 张量赋值函数
tensor viscosity 张量黏性
tensor well 张量势阱
tenso-shear 张量切变
tenso-shear boundary 张剪边界
tenso-shear joint 张剪性节理
tenso-shear structural plane 张扭性结构面【地】
tenso-sheer deep fracture 张剪性深断裂
tensotast 应变仪;张力仪
Tensovic 坦索维克(一种浸透树脂的加强木材)
ten's place 十位
tens star screw 十星螺钉
tent 帐篷;张开帐篷;架设帐篷;混凝土路养护篷;旅游帐篷
tentacle 触器;触毛;触感元件
tentaculitid extinction 竹节石绝灭
tentaculocyst 石针
tentage 宿营设备;帐篷;路面养护帐篷
tental housing 租赁房屋
tentation data 暂设数据;暂定数据;假设数据;假定数据;推测数据;暂用数据;预定数据;试用资料
tentative 暂行的;暂定(的);假定的;试行的;初步的
tentative annual report 试编年度报表
tentative assembling 工地装配
tentative assembly 试拼装
tentative booking 暂定船位;试定舱位
tentative budget 试编预算
tentative code 暂行规范
tentative control regime 试探控制方式
tentative control strategy 试探性的控制策略
tentative criterion 暂拟准则
tentative data 预定数据;暂设数据;假定数据;推测数据;试验性数据
tentative design criterion 草案性设计规范;草案性设计要求;试验性设计规范;试验性设计要求
tentative diagnosis 初步诊断
tentative function 试探函数
tentative idea 初步设想
tentative label 试探标号
tentative layout 试验性布置
tentative method 暂行方法
tentative modernistic 草案性现代化;试验性现代化
tentative program(me) 比较程序;暂行程序
tentative proposal 初步设想
tentative provisions 暂行条例;暂行规定
tentative route 临时路线
tentative routine 暂行程序
tentative schedule 试验性时间表
tentative site 比较坝址;暂定港址;暂定厂址;暂定坝址;比较港址;比较厂址
tentative sketch 草图
tentative specification 暂行规格;暂行规范;暂行规程;试验性(的)规格;试行规范;试行标准;暂行规定
tentative specification of harbo(u)r work 港口工程技术规范
tentative standard 试行标准;暂行标准;假设标准;试用标准
tentative steady compression process 假设稳定压缩试验法
tentative strategy 试探性策略

tentative technical specification 暂行技术规范;
暂行技术规程
tentative valuation 财产初步估计
tentative value 试用值
tent bed 帐篷式卧床;行军床
tent camp 露营
tent-dried 帐篷式干燥的
tented arch 帐形纹;帐篷式拱;四心外心桃尖拱
tented ice 拱形冰锥
tenter 展幅机;拉幅机;绷布机
tenter dryer 拉幅干燥框
tenter hook 保护钩;拉幅钩
tentering machine 展幅机;拉幅机;伸幅机
tentering machine with drying cylinders 滚筒烘干拉幅机
Tentest 天铁斯特绝缘纤维板
tent fly 帐篷盖
tent for cold area 寒区帐篷
tent-guy 帐篷索
tenth highest annual hourly volume 一年第十个最大小时交通量
tenth-normal 十分之一当量浓度的
tenth-normal solution 十分之一当量浓度
ten-thousandth micrometer 万分表尺
ten-thousand-Yuan norm 万元定额
tenth-power width 十分之一功率点宽度
tenth ring 十环
tenth-thickness value 十分之一厚度值
tenth-value layer 十分之一值层
tenth-value thickness 十分之一厚度值
tenting 斜张天幕;海冰受压拱起成帐幕状;海冰隆起
tenting an awning 斜张天幕
tent line 帐篷线
tent membrane 帐篷薄膜
tentonic contact 构造接触
tentorlum 幕;[复]tentoria
tent peg 帐篷桩(子)
tent poling 突然偏转(声波测井中,由于周波跳跃或仪器的停顿所产生的曲线突然偏转)
tent roof 锥形屋顶;帐篷屋顶
tent roofed shed 帐篷式堆场
tent-shaped clinker store 帐篷式熟料储[贮]库;帐篷式熟料堆场
tent-shaped pile 帐篷式料堆
tent-shaped roof 人字形屋顶
tent string hook 帐篷绳钩
tent structure 帐篷构造
tent system 帐篷系列
tent system for construction of high points 高处建筑帐篷体系
tent texture 帐篷结构
tent trailer 帐篷拖车
tent with alumin(i)um poles 铝管帐篷
tenuicutinite 薄角质体
tenuity 稀薄度
tenuity factor 稀薄因子
tenure 不动产占有(期);持有权;占有条件;所有权;(土地的)使用权
tenure of use 使用年限
ten-way valve 十通阀
ten wheeler 十轮卡车
ten-year flood 十年一遇洪水
ten years 十年
tepee butte 锥形残丘
tepee structure 帐篷构造
tepefy 微温;变温热
tepering gutter 宽度渐变的沟;楔形沟
tepetate 钙积层
tephigram 温熵图;熵温图
tephra 火山(喷发)碎屑
tephrite 碱玄岩
tephritoid 似碱玄岩
tephrochronology 火山灰年代学
tephroite 锰橄榄石
tephros 火山灰沉积物;火山灰
tephrosious 白灰色的
tepid 微热的
tepidarium 中温浴室(古罗马);温水浴室;温水浴池
tepid spring 微温泉
tepid water 微温水
T-equivalence T 等价
tera 太(1 太 $=10^{12}$)
teracainum 丁卡因
terahertz 太赫

teram 楼梯踏步端头旋涡形装饰;踏步涡饰
teraohmmeter 太欧姆表;超绝缘计
teratogenesis 致畸作用;畸形生成;畸形发生
teratogeny 畸形生成;畸形发生
teratron 亚微米波振荡器;亚毫米波振荡器
Terbeck gas burner 泰尔贝克煤气燃烧器
terbium boride 硼化铽
terbium bromide 溴化铽
terbium carbonate 碳酸铽
terbium nitride 氮化铽
terbium ores 铽矿
terbium oxide 氧化铽
terbium silicide 硅化铽
terbutryn 去草净
Tercod 特格德碳化硅耐火材料
terdiurnal tide 三分之一日潮
terebene 芸香烯;松节油与松油精之混合物(用作防腐剂,涂料等);松节油精
terebinth 松脂木
terebinthaceae 树脂科
terebinthinism 松节油中毒
Teredinidae 船蛆科
teredo 蛀船虫;海蛆
teredo navails 凿船虫;船蛆
teremkovite 捷辉锑银铅矿
terephthalate 对苯二甲酸酯;对苯二(甲)酸盐
terephthalic acid 对酞酸;对苯二(甲)酸
terephthalic aldehyde 对苯二醛
terephthalonitrile 对苯二腈
terephthaloyl 对苯二酰
terephthaloyl chloride 对酞酰氯
terephthaloyl dichloride 对苯二酰氯
terephthaloyldioxydibenzoyl chloride 对苯二酰二氧二苯甲酰氯
tergal 背的
tergeminate 三次双生的
tergite 背片
terhalide 三卤化合物
teriodide 三碘化物;三碘化合物
terlinguaite 黄氯汞矿
Terlite 特莱(一种粉煤灰陶沥青骨料)
term 学期;限期租用的地产;专门名词;期限;期间;条款;条件
Termacoust 德玛库斯特(一种木纤维板)
term analysis 光谱项分析
term bond 一次到期债券;定期债券
term-by-term 逐项的
term-by-term differentiation 逐项微分
term-by-term integration 逐项积分
term cash 现金支付
term day 支付日;结算日;付款日
term diagram 项图
term discount 费用折扣
termed saddle stone 山墙顶石
term end 期末
term hour 定时观测时
termierite 坛埃洛石
terminable contract 有限期合同
terminable interests 有限期的产权;可中止的产权权益
terminal 枢纽线端;远端;终期的;终端站;终点;接线柱的终端;接线柱[电];接线端;末端的;末端;特种码头;水陆交接点码头;水陆交接点车站;输入端;端子;头;端饰;端板
terminal access 终端访问
terminal access network 终端访问网络
terminal account 终结账户
terminal accuracy 末段引导精度
terminal acoustic(al) resistance 终端声阻
terminal address selector 终端地址选择口
terminal administration 终端主管部门
terminal aerodrome 终点飞机场
terminal aids 终端导航设备
terminal airport 终点航(空)站
terminal air traffic control system 终端区空中交通管制系统
terminal amino group 末端氨基
terminal amplifier 终端放大器
terminal analysis 末端分析
terminal anchor 端部锚固(压力管道的)
terminal anchor clamp for contact wire 接触线终端锚固线夹
terminal anchor clamp for messenger wire 承力索终端锚固线夹

terminal and trunkline 站线【港】
terminal angle 终点角【测】
terminal apron 终点停机坪(机场)
terminal area 终端区域;终点站(机场);连接盘;焊连接盘
terminal area control 终端区域管理
terminal area forecast 终点气象预报
terminal area impact point 终端地区弹着点
terminal area navigation system 终端区导航系统
terminal area radar control 终端区域雷达控制
terminal assembly 终端装置;接头装配;接头排;端接组件
terminal bar 结合堤;末端棒眼
terminal basin 终碛盆地
terminal bay 终端架
terminal bending moment 终端弯矩
terminal block 电缆管块;线夹;线弧;用户接线;接线盘;接线盒;接线板;接线排;端子排;端饰头;端饰块
terminal board 接线盘;接线盒;接线端子板;接线端(面)板;接线板;端子板架;端子板;分电器接线板
terminal bolt 端螺栓
terminal bond 终端接头;终端导接线;端键
terminal box 接线箱;接线板;终端盒;拉线盒;接线盒;端子箱;出线盒[电];分线箱
terminal box of lightning arrester 避雷器接线盒
terminal budding 劈顶芽抱
terminal building 码头房屋;街道端建筑;终点站建筑;终点站候车室;终点站房;港口建筑物;港口房屋;码头建筑(物);候机室;候机楼;航站大楼
terminal bushing 电极套;电极夹头
terminal button 终钮
terminal cable 终端电缆
terminal call 终端局通话
terminal called 被呼叫的终端
terminal capacity 终点容量;枢纽通过能力
terminal carboxyl group 末端羧基
terminal card 终止卡片
terminal center 终端中心
terminal charges 装卸费(用);港口费(用);落货费;结算
terminal check 最后校验;终端检查
terminal check valve 管端止回阀
terminal crossing 末端交叉
terminal cistern 终池
terminal clamp 终端线夹;漏板夹头;电极夹头;出线夹
terminal clearance capacity 枢纽限界容量
terminal cleft grafting 顶劈接
terminal cluster 终端线束;接线元件组;集线箱
terminal command 终点指挥部
terminal communication 终端通信
terminal complex 码头区
terminal component 终端设备成分
terminal computer 终端计算机
terminal computer communication 终端计算机通信
terminal condition 终端条件;结束状态;边界条件;界限条件
terminal connecting line 枢纽联络线
terminal connection 终端接线
terminal connector 终端线夹;终端连接器
terminal console 终端控制台
terminal constraint 终端约束
terminal contact 终端触点;端点接触头
terminal contract 定期合同
terminal control 终端控制;末端控制
terminal control area 航空终端站控制区
terminal control computer 终端控制计
terminal control device 终端控制设备
terminal control equipment 终端控制设备
terminal controller 终端控制器;终端控制单元
terminal control system 终端控制系统
terminal control unit 终端控制器;终端控制单元
terminal cost 站场费用
terminal cost for transport 运输的站场费用
terminal country 终端国
terminal cursor 终端游标
terminal cylinder 终柱
terminal date of an agreement 契约终止(日)期
terminal degree 终端绷紧(固定)度
terminal delay 终端滞后;终点迟滞;传送滞后
terminal deletion 末端缺失

terminal delivery 码头交货
terminal department store 终点站百货商店
terminal depot 码头型货运站；码头仓库
terminal destination 终端目的地
terminal devanning 区内拆箱
terminal device 终端装置；终端设备；末端装置
terminal difference 终端温差
terminal digit of chelicera 螯趾
terminal disinfection 终结消毒
terminal display mode 终端显示方式
terminal disposal of radwaste 放射性废物最终处置
terminal dissipation line 终端衰耗线
terminal dissipation zone 终端消散带（冰川）
terminal distributor 终端配电盘；终端分配器
terminal dock 港内系泊池；港口坞池
terminal drying 最终干燥；末期干燥；后期干燥
terminal dummy load 终端假负载
terminal ear 漏板耳朵
terminal edge 射入边
terminal effect 终端效应
terminal emulation 终端仿真
terminal emulator 终端仿真程序
terminal (end) 终端
terminal endpoint 端点
terminal Eocene event 始新世末期事件
terminal equipment 终端设备；终端机
terminal eruption 顶剖喷发
terminal exchange 终端局
terminal expense 晚期费用
terminal face 山麓断层面；端面；冰川抹端
terminal facet 端面
terminal facility 港坑口设备；油库设备；终点站设备；终点设施；港口设施；码头设施；码头设备
terminal fall velocity 下落终速（度）；终端沉速；临界沉降速度
terminal fanout 终端扇出端
terminal fault 终端故障
terminal feature 尾景；最后性能；尽头街景；尽头布置；街道尽端布置
terminal field 终点场
terminal figure 胸像柱；柱式胸塑像
terminal fitting 终端接头
terminal forging temperature 终锻温度
terminal front 终碛前缘
terminal function 终端机的功能
terminal function key 终端功能键
terminal graphic(al) plotter 终端绘图仪
terminal group 端基
terminal guidance 终端导引；终端导航；末段制导；航空终端站区导引
terminal guidance system 终端制导系统
terminal handling charges 装卸区操作管理费；装卸费（用）
terminal hardware 终端硬件
terminal head loss 期终水头损失
terminal hood 接线端盖帽
terminal hotel 机场、车站旅馆
terminal hour 终时
terminal hydrophilic group 末端亲水基；亲水端基
terminal hydroxyl 末端羟基
terminal identifier 终结标识符
terminal impact prediction 终端弹着预测
terminal impedance 终端阻抗；端阻抗
terminal impulse 终点油量；终点冲量
terminal incorporation 末站掺入；末端掺入
terminal indicator 终端指示器
terminal information control system 码头信息控制系统
terminal installation 终端装置；终端设备
terminal installation for data transmission 数据传输终端装置
terminal instrument procedure 终端仪表进场程序
terminal insulator 接线柱绝缘体
terminal interchange 道路终点枢纽
terminal interface 终端接口
terminal interface controller 终端接口控制器
terminal interface module 终端接口模件
terminal interface processor 终端接口处理机
terminal iron 终端交叉铁板
terminalization 终点化；码头枢纽化
terminal keyboard 终端键盘
terminal lamina 终板
terminal landing 终端停车平台
terminal landing system 终端着陆系统

terminal landmark 界标
terminal laser level 激光低能级
terminal launch facility 终端发射设施
terminal layer of drilling 终孔层位
terminal layout 枢纽布置（图）
terminal lead 终端引线
terminal ledge 终棱
terminal level 终端能级
terminal limit switch 终端限位开关；终端极限开关
terminal line 终线；码头线；顶线
terminal linkage 端键
terminal linkage board 端子连接板
terminal linking system 终端连接系统
terminal load 终端负载；末端荷载
terminal loop 迂回线；终点环线；枢纽环线
terminal loss 终端损失；终端损耗
terminal lug 终端接线片；接线片；端子衔套；耳接头
terminal machine 终端机
terminal market 农产品集散中心市场；码头市场；车站市场；终点市场；集散中心市场；期货市场；批发市场
terminal master 码头管理（人）员
terminal member 末端部件
terminal memory 终端存储器
terminal Miocene event 中新世末期事件
terminal module 终端模件
terminal moment 终端力矩
terminal monitor program(me) 终端监督程序
terminal moraine 终碛；终堆石；尾碛；边堆石
terminal motivation 终末激发
terminal movement 枢纽内运行
terminal node 终止节点
terminal nut 接线柱螺母
terminal of a cable linking box 电缆连接箱端子
terminal of an agreement 协议终止（日）期；协议期满日；满期日；合同终止（日）期；合同期满日；契约期满日
terminal office 终端局
terminal office for communication 通信架站
terminal of relay 继电架端
terminal of submerged pipeline 潜水管线末端
terminal of voyage 航程起讫点
terminal operation 装卸作业；集装箱装卸作业操作；码头作业；终点作业
terminal operator 终端操作人员；码头经营人；码头公司
terminal organs 首尾机构
terminal-oriented network 面向终端网络
terminal outfit 末站设备
terminal oxidation 末端氧化
terminal pad 焊盘；端子垫片
terminal pair 端子线对；端子对
terminal panel 终端面板
terminal parameter 终止参数；终值参数
terminal part 终部
terminal pedestal 胸座；胸像台；胸台；轴线尽端雕像台座
terminal performance objectives 终端操作目标
terminal phase 初生相
terminal phase guidance 终端导航
terminal pillar 终端柱
terminal pin 引线插头；尾销
terminal plate 终端片；终板；端子板
terminal platform 终端平台
terminal point 终点；接线点；端点
terminal point gun 端子连接器
terminal point of base 基线端点
terminal pole 终端杆
terminal polling 终端轮询
terminal pond 终端塘；端部压力池
terminal port 终端入口；终端出口；终端港；起讫港；枢纽港
terminal portion 终末部
terminal post 线头接栓；接线头端子；电极；终端立柱
terminal post cap 终端立柱帽
terminal potential 端电压；端电势
terminal pressure 终点压力；末端压力；端压力
terminal processor 终端处理机
terminal product inhibition 末端产物抑制
terminal protocol 终端协议
terminal pulleys 前后鼓轮
terminal pulse 终端脉冲
terminal pump station 终点泵站；末端泵站
terminal queue 终端队列

terminal radiation program(me) 终端辐射计划
terminal railway station 铁路终点站
terminal rate 装卸港口费率；装卸费率；装卸车站费率
terminal reaction 末道反应；终点反应
terminal receiving system 库前接受制度；外围仓库制度
terminal regeneration 终端再生
terminal reheat 终端再热
terminal reheat system 末端再热系统
terminal reliability 终端可靠性
terminal repeater 终端转播机；终端增音机
terminal repeater station 终端中继站
terminal representative 港方代表
terminal reservoir 终端压力池；输水管道末端压力水池
terminal resistance 终端电阻
terminal rod 终点杆
terminal room 终端室
terminal roundabout line 枢纽迂回线
terminal routing 指向运输终点的运输
terminal saddle 站口鞍座
terminal screw 固定螺丝；固定螺钉；接线螺柱；接线螺钉；端接螺钉
terminal section 终段；（防波堤等的）末端；端截面；堤头段
terminal segment of chelicera 螯钳
terminal seizing signal 终端占用信号
terminal server 终端服务器
terminal service 码头作业
terminal serviceability 最终耐用性
terminal serviceability index 最终耐用性指数
terminal set 终端机
terminal settling velocity 终末沉降速度；终端沉速；终沉降速度
terminals for power supply 电源端子
terminal shaft 终端轴
terminal share 终端摊额
terminal sheet （屋顶的）终端板
terminal shoe 终端垫板
terminal shunting 枢纽内调车
terminal signal 终点信号
terminal situation 终结情况
terminals of a unit block 组合端子
terminals of lager zero of relay racks 零层端子
terminals of zero level 零层端子
terminals on distributing board 分线盘端子
terminals on distribution board 分线盘端子
terminal source editor 终端源编辑程序
terminal spark plug end 火花塞端子
terminal speed 终速度；末速（度）
terminal square 站前广场
terminal stage 终了期
terminal state 终态
terminal-state manifold 终态流形
terminal station 终点（车）站；终端站；输油终点站
terminal statue 胸像柱；轴线尽端雕像（街道布置）
terminal sterilization facility 终点消毒站
terminal stopping device 行程开关；终端自动开关装置；到达自动开关装置
terminal storage 最终储[贮]存
terminal strap 接头条
terminal stria 终纹
terminal string 终端串
terminal strip 接线条；端子板
terminal stud 接线柱【电】；接线端头
terminal stud connector 柱螺栓线接器
terminal sub-control station 终端控制分站
terminal subscription 定期订费
terminal supervisory program(me) 终端管理程序
terminal support 终端支架
terminal surface 终端曲面
terminal switch 终端开关；终电开关；终点开关
terminal symbol 终止符号；终结符；终端符号
terminal synchrone 终端等时线
terminal system 终端系统
terminal tank （管道的）终端压力池；终点油库
terminal tankage 终端油罐储量
terminal taxiway 通侯机楼的滑行道
terminal telomer 终端调节剂
terminal temperature 终始温度；终点温度
terminal temperature difference 末端温差
terminal test bay 终端设备测试架
terminal throughout capacity 终点站通过能力

terminal time 装卸时间;终端时间
terminal-time indicator 结束时间指示器;终点时间指示器
terminal tower 终端塔
terminal traffic 终端业务;终点交通
terminal traffic control 终端交通管制
terminal transit 区内调运
terminal transmission equipment 终端传输设备
terminal transmission interface 终端传输接口
terminal transmitter 终端发射机
terminal transparency 终端透明度
terminal treatment plant 最终处理厂
terminal trunk 终端干线
terminal unit 终端设备;终端装置
terminal unsaturation 端基不饱和性
terminal user 终端用户
terminal value 终值
terminal value control 终值控制
terminal valve 进出口阀
terminal vector 终末向量
terminal velocity 临界速度;下沉速度;终端速度;终(点)速度;极限速度;末速(度);末端速度;稳定状态速度;收尾速度;端(点)速(度);落速
terminal-velocity dive 极限速度俯冲
terminal versatility 终端通用性
terminal vertex 悬挂点;端点
terminal very high-frequency omnirange 终端甚高频全向指向标;终端(站)甚高频全向(导航)信标
terminal very high-frequency omnirange area 终端甚高频全向信标区
terminal voltage 终端电压;高压电极电压;端(子)电压
terminal volume 码头货运量
terminal wave 终末波
terminal wheel 终端绳轮
terminal yard 站场;枢纽调车场;收尾(工)场
terminal zone 枢纽区
term in a symbolic expression 符号表达式的项
terminate 端接
terminate a contract 终止合同;解除合同
terminate cable 终接电缆
terminate contract before the data of expiration 提前终止合同
terminated level 终接电平
terminated line 终端线路;端接线
terminated stop 止门器(开启45度或90度)
terminate load 终端负载
terminate resistance 终端电阻
terminating 连接;接通;线头扎接;终接;终端
terminating block of main distributing frame 总配线架终端块
terminating circuit 终接电路
terminating continued fraction 有尽连分数
terminating decimal 有尽小数
terminating format 终端格式
terminating impedance 终端阻抗;负载阻抗
terminating load 终端负载
terminating machine 终端机
terminating Markov chain 可终止的马尔可夫链
terminating network 终端网络
terminating office 终端局
terminating ornament 端饰
terminating packet 终端信息包
terminating phase 终结相位
terminating plug 终端塞子
terminating queue 终结队列
terminating resistance 终端电阻
terminating resistor 终端电阻器
terminating set 端接装置
terminating signal(l)ing unit 终端信号装置
terminating symbol 终结符;结束符号
terminating traffic 到达交通
terminating traffic volume 到达交通量;集中交通量
terminating trigger 终端触发器
termination 终止;末端;终接装置;终端;解雇;结局(指网络模型);满期;期满
termination agent 终止剂;链终止剂
termination amount 终止款额
termination bar 收头棒
termination beam 钢柱单支下锚角钢
termination capacitor 端接电容
termination chargess 终止合同费用
termination codon 终止密码子
termination condition 结束条件

termination connection point 终端连接点
termination dead end 终点
termination environment 终端区
termination expenses 终止合同费用
termination factor 终止因子
termination impedance 终端阻抗
termination interrupt 结束中断
termination job processing 终端作业处理
termination light 终端灯
termination load type power meter 终端负载式功率计
termination message 终了信息
termination network 终端网络
termination notice 终止通知
termination of a block 分程序终止
termination of a contract 契约终止
termination of agency relationship 代理关系终止
termination of agreement 协定结束;协定期满;契约期满
termination of an agreement 契约终止
termination of barrier 壁垒的界限
termination of block 程序块终止
termination of business 停业
termination of contract 终止合同;解除合同;合同终止;合同期满;合同解除
termination of contract of carriage clause 运送合约终止条款
termination of employment 雇佣期满
termination office 终端局
termination of insurance 保险终止
termination of insurance status 保险失效
termination of procedure 过程终止;过程终结
termination of railway 路线终点【铁】
termination of report processing 报表处理终止
termination of risk 保险(责任)终止
termination of series errors 截断误差;级数误差
termination of task 任务终止
termination of tenancy 租借权终止
termination of term 满期
termination pay 解雇费
termination piece 终端件
termination point 端点
termination point of a vessel 容器的界限点
termination product 最终产品
termination queue overflow 结束排队溢出
termination rack 终端网络
termination rate method 终点比率法
termination reaction 终止反应
termination report 总结报告
termination routine 结束程序
termination schedule 解约附表
termination signal 终止信号
termination step 链终止阶段
termination symbol 终结符
termination value 解约价值
terminative time of fault growth 终止活动时间
terminator 终止子;终止剂;终止程序;终结符;端套管;终端适配器;终端连接器;终端负载连接器;终端负载;结束符;端子排;晨昏线
terminator codon 终止密码子
terminator/initiator 停启程序
terminator program(me) 终止程序
terminator record 终结符记录
terminolic acid 终油酸
terminology 专门名词;技术名词;命名法;术语学;术语
terminology of geographic(al) names 地名术语
terminology standard 术语标准
term insurance 定期保险
terminus 界石;界标;胸像台;界桩
terminus boundary marker 界桩
termipoint 端接
termitarium 白蚁穴;[复]termitaria
Termite 铅基轴承合金;特米特铅基轴承合金
termite 白蚁
termite clause 白蚁条款
termite control 白蚁防治
termite damage 白蚁危害
termite heap 白蚁堆
termite hill 白蚁窝;白蚁巢
termite inspection 白蚁检查
termite nest 白蚁巢
termite-proof 防白蚁(的)
termite proof cable 防蚁电缆

termite-proof concrete 防白蚁混凝土
termite protection 防白蚁
termite resistant cable 防蚁电缆
termite resisting 防白蚁的
termite runway 白蚁走道
termite shield 防蚁板(防潮层高度的金属悬挑板);防蚁罩;防白蚁垫片;白蚁挡板
termite treated 防白蚁的
Termitidae 白蚁科
term lease 定期租赁
term life insurance 定期人寿保险
term load 分期偿还贷款
term loan 项目定期贷款;中长期贷款;定期贷款
term mortgage 定期抵押
term of acceptance 承兑日期
term of charter 租船期限
term of contract 合同有效期;合同期限;契约履行期限
term of delivery 交付期限
term of discount 贴现期限;贴现期
term of expectancy 使用年限
term of hour 定期观测时间
term of insurance payments 保险费偿还期
term of life 工程寿命;使用期限
term of maintence 保修期限
term of office 任期
term of payment 支付条款;付款条件
term of service 任期;使用期限;使用期;服务期限;保修期限
term of tax 纳税期限
term of tender validity 招标有效期
term of validity 有效期(限)
term of validity of offer 要约的有效期限;出售的有效期限
term of voyage 航行期间
term of works 工期
termolecular mechanism 三分子机理
Termotile 德蒙太(一种方形白色绝热瓦)
term piece 尾端边木
terms and condition 条款;条件;费率表
terms of access 进入术语
terms of an agreement 协议条款
terms of contract 合同条款
terms of credit 信用证条件;信贷条款;债方条款;赊账条件;贷款条件
terms of delivery 交货条件;交割条件
terms of employment 雇佣条件;待遇(指对待、权利方面)
terms of estimate 预算项目;估算项目
terms of exchange 交换条件
terms of experiment analysis at lab 室内试验分析项目
terms of loan 贷款条件
terms of loan agreements 信贷协定的期限和条件
terms of payment 结算期;支付条件;支付期限;付款方式
terms of price 价格条件
term(s) of redemption 偿还期限
terms of reference 职权范围;职能区域;职能范围;任务大纲;权限
terms of sale 买卖条件
terms of service 服务条件
terms of settlement 解决条件;结算条件;清算条件
terms of shipment 装运条件;装船条件
terms of tender 招标条件
terms of trade 进出口交换比价;进出口比价;贸易条款;贸易条件
terms-of-trade effect 贸易条件效果
terms of trade indexes 贸易条件指数;进出口比价指数
term structure 期限结构
term structure of interest rate 利率的期限结构
term-term matrix 项相关矩阵
term time 法定支付期
termwise 逐项地
termwise differentiable 逐项可微的
termwise differentiation 逐项微分
termwise integrable series 逐项可积级数
termwise integration 逐项积分
ternary 三重性;三重的;三元的;三态的;三进制的;三个组成的;三变量
ternary acid 三元酸
ternary addition 三进制加法
ternary algebra 三元代数

ternary alloy 铅锡镉合金;三元合金
ternary alloy steel 三元合金钢
ternary arithmetic 三进制运算
ternary bilinear form 三元双一次形式
ternary blends 三组分混合料
ternary code 三元码;三进制码;三进制代码;三单元制码
ternary collision 三元碰撞
ternary colo(u)r 三元混合色
ternary complex 三元络合物
ternary complexes analysis method 三元络合物分析法
ternary composition 三元合金
ternary compound 三元化合物
ternary compound structure 三元化合物结构
ternary counter 三进制计数器
ternary cubic(al) form 三元三次型
ternary cycle 三循环;三次循环
ternary device 三进制设备
ternary diagram 三元图解;三元图
ternary distribution coefficient 三元分配系数;第三次分布系数
ternary distribution constant 三元分布常数
ternary dye 三元混合染料
ternary electrolyte 三元电解质
ternary ethylenepropylene rubber packing 三元乙丙胶垫
ternary eutectic 三元共晶
ternary expansion 三进展开
ternary fission 三裂
ternary form 三元型
ternary incremental representation 三进制增量表示法
ternary liquid systems 三元液体系统
ternary logic 三进制逻辑
ternary logic element 三进制逻辑元件
ternary mix 三种原料配料;三元配料
ternary mixture 三组分混合物
ternary multiplication 三进制乘法
ternary notation 三进制记数法;三进制
ternary number 三进制数
ternary number system 三进制数系;三进位数系
ternary phase 三元相
ternary phase diagram 三元相图
ternary plate 三合板
ternary polymerization 三元共聚
ternary quadratic form 三元二次形式
ternary quartic form 三元四次型
ternary relation 三元关系
ternary representation 三进制表示法
ternary scale 三进制标度
ternary sediment 三相沉积(物)【地】
ternary sequence 三进制程序
ternary set 三分点集
ternary solution 三元溶液
ternary steel 三元钢
ternary system 三元系;三元体系
ternary system ceramics 三元系陶瓷
ternary system piezoelectric(al) ceramics 三元系压电陶瓷
ternary tree 三叉树
ternary vapo(u)r-liquid equilibrium 三元蒸气液体平衡
ternary variable 三态变量
terne 包覆铅锡合金的低碳钢板;镀锡;镀铅铁板
terne alloy 铅锡合金
terne coating 镀铅锡合金
terned 镀锡的;镀铅的
terne plate 镀铅锡合金薄钢板;镀铅锡薄(铁)板;钢铁薄铅涂层;镀铅铁板
terne plating 镀铅锡合金层;铅锡镀层
tern schooner 三桅帆船
Teron 缔纶聚酯纤维(罗马尼亚)
terotechnology 设备综合工程学;设备使用保养技术
terpances 萜烷/甾烷
terpane 萜烷
terpanes content 萜烷含量
terpene 萜烯
terpene maleic anhydride condensation product 萜烯—顺(丁烯二酸)酐缩合物
terpene phenol resin 萜烯酚醛树脂
terpene resin 萜烯树脂
terpene resinite 萜烯树脂体
terpenic 萜烯(类)的

terpenic series 萜烯系
terpenoids 萜烯类
terphenyl 联三苯;三联苯
terpinene 萜品烯;萜二烯
terpinene maleic anhydride adduct 萜品烯—顺(丁烯二酸)酐加成物
terpineol 萜品醇
terpine resin 萜烯树脂
terpinolene 萜品油烯
terpinyl acetate 醋酸萜品酯
terpolymer 三元聚合物;三元共聚物
terpolymer finish 三元共聚物整理
terpolymerization 三聚作用
terra 陆地;土地
terra alba 白土;制管土;石膏粉;纯白生石膏;白色软黏土
terra calcis 石灰性土
terra cariosa 硅藻土;钙质细土;板状硅藻土;蚀余硅质岩
terrace 露天平台;露台;海底台地;平台;排屋;排房;水平埂;室外平台;地坪;地埂;阶地;台地
terrace access 阳台通道
terrace agriculture 梯田农业
terrace at entrance 入口平台
terrace awning 平台遮阳篷
terrace awning blind 阳台遮阳
terrace back edge 阶地后缘
terrace channel 阶地水沟;地埂槽沟
terrace clearance 离地距离;离地净空;离地净高;离地高度
terrace cover 阶地覆盖
terrace cropping 梯田种植
terrace cultivation 梯田耕作
terraced aisle (剧院等的)台阶式通道
terraced dwelling 排屋;阶梯状住房
terraced field 梯田
terraced funerary temple 阶状式葬礼庙宇
terraced garden 露台花园
terraced house 行列式房屋;梯阶式房屋;行列式住宅;台阶式住宅
terrace die 凸模
terraced land 梯田地;筑成梯田的地;地埂地
terraced mastaba(h) 平顶斜坡坟墓(古埃及)
terraced mortuary temple 露台式停尸殿堂
terrace door 阳台门
terrace door fitting 阳台门装配附件
terrace door furniture 阳台门装置
terrace door hardware 阳台门金属小五金;阳台门金属附件
terrace door hardware item 阳台门金属附件项目
terraced pavilion 榭
terraced pyramid 露台棱柱体
terrace drainage 平台排水
terraced riprap 台级形乱石堆;阶状抛石体
terraced roof 平台屋顶;平台屋顶;平屋顶
terraced street 阶梯街道
terraced temple 阶梯式庙宇
terrace dwelling unit 排屋居住单元
terrace edge 阶地外缘;阶(地)边缘
terrace element 阶地要素
terrace elevation 阶地高度;阶地高程
terrace epoch 阶地时期
terrace family dwelling 成排家庭住宅
terrace farmland 梯田
terrace fault 阶梯断层
terrace-field 梯田
terrace flexture 阶状挠曲
terrace formation 阶地形成;(河流的)阶地地层
terrace front edge 阶地前缘
terrace function 阶地函数
terrace furnace 梯台式炉
terrace garden 台地园;平台园;梯级园
terrace garden housing 台阶式花园住宅
terrace grade 阶地坡度;梯田坡度;地埂坡度
terrace gravel 砾石阶地
terrace heating 阶梯式采暖
terrace height 阶地高度;阶地高程
terrace house 毗连式住宅;街坊房屋;阶梯式房屋;联排式住宅;连栋房屋
terrace-house development 联式住宅区
terrace interval 阶地间距;梯田间距
terrace land 阶地;台地;梯地
terrace loading 阶梯荷载
terrace of river 河成阶地

terrace outlet 梯田排水口;梯田出水口
terrace outlet channel 阶地排水沟;梯田排水沟
terrace placer 阶地砂矿;阶地矿砂
terrace prevent 平台防护
terrace relationship 阶地关系
terrace restaurant 平台饭店
terrace ridge 阶地脊;台地脊线
terrace rim 阶地眉峰
terrace rood 平台式屋顶
terrace roof 平屋顶;平台屋顶
terracer 修梯田机
terraces 河流沉积残留砾石
terrace scarps 阶地陡坎
terrace sediment 阶地沉积物
terrace slope 阶地坡;阶地坡;上下坡踏步;台阶
terrace spacing 阶地间距;梯田间距
terrace spring 阶地泉
terrace step 平台(式)踏步;平台梯级
terrace structure 阶地构造
terrace surface 阶面
terrace system 阶地系统;梯田系统
terrace-temple 阶梯式庙宇
terrace tile 阶梯式砖瓦
terracette 小土滑阶坎
terrace type 阶地类型
terrace-type pit 梯级式露天矿
terrace walk 梯级步道
terrace wall 梯级式挡土墙;挡土墙
terrace wall type reforming furnace 梯台式墙型重整炉
terrace water channel 阶地(排灌)水渠;梯田(排灌)水渠
terrace waterproofing 平台防水
terrace width 阶地宽度;梯田宽度;梯坡田宽度
terrace winding 登坡盘道
terracing 梯层状轮廓;梯层状构造;修筑梯田;形成阶地;阶地状地形;阶地形成;梯田化;地埂耕作法;等高耕作;筑坛
terracing grader 筑埂平地机
terracing machine 梯田修筑机
terracing of the land on the slopes 坡地梯田化
terracing with sod 铺草皮作业
terra-cotta 琉璃砖;瓷砖;玻璃砖;混合陶器;红土无釉陶器;红陶;低温多孔赤陶;赤土(陶瓷);赤褐色;镶铺瓷砖;陶土制品;琉璃砖;赤陶(制陶器或建筑用,上釉或无釉的);焙烧黏土陶瓦;赤陶
terra-cotta architecture 陶瓷建筑
terra-cotta block 陶瓷砌块;玻璃砖块
terra-cotta block finish 陶砖饰面
terra-cotta block protection 陶砖护岸
terra-cotta clay 陶塑用黏土
terra-cotta conduit 陶土管
terra-cotta decoration 陶瓷装潢
terra-cotta decorative feature 陶瓷装潢特色;陶瓷装潢面貌
terra-cotta decorative finish 陶瓷装潢终饰
terra-cotta decorative fixture 陶瓷装潢装置
terra-cotta enrichment 陶瓷增添美化;陶瓷增添装饰
terra-cotta facing 陶瓷面层
terra-cotta lining 陶瓷衬料
terra-cotta lumber 密孔砖
terra-cotta ornament 陶瓷装饰品
terra-cotta ornamental feature 陶瓷装饰特色
terra-cotta ornamental finish 陶瓷装潢终饰
terra-cotta panel 陶瓷面板
terra-cotta partition (wall) 陶瓷分隔(墙)
terra-cotta pipe 陶管;陶瓦管;陶土管;陶制管
terra-cotta roof(ing) tile 陶瓷屋瓦
terra-cotta sewage conduit 陶瓦沟管
terra-cotta sewage pipe 陶制下水管;陶瓦下水管
terra-cotta slab 陶砖板
terra-cotta statue 陶瓷塑像
terra-cotta (sur)facing 陶瓷铺面(材料)
terra-cotta tile 陶瓦;陶土砖瓦;陶砖
terra-cotta tube 陶土管
terra-cotta vase 赤砖;赤陶花瓶
terra-cotta wall tile 陶瓷墙砖
terra-cotta warriors 兵马俑
terra-cotta work 陶塑工程
terra disiena 天然氧化铁
terra firma 硬土;稳固地基;陆地;旱地;硬土稳固地基;坚固地基;大地
terra fusca 淋溶棕色石灰土;脱钙棕色土

terragraph 地面立体测图仪
terrain 岩体出露区;领域;地域;地形;地体;地势;地层
terrain analysis 地物分析
terrain appreciation 地形判断;地物评价
terrain-avoidance radar 地形躲避雷达
terrain camera 地形摄影机
terrain clearance 相对航高;离地距离;离地净空;离地净高;离地高度
terrain clearance indicator 绝对测高计
terrain coefficient 地面利用系数
terrain condition 地形条件
terrain configuration 地表外形
terrain correction 地形修正;地形校正
terrain correction range 地改范围
terrain correction value 地形改正值
terrain correction value of farther distance 重区地改值
terrain correction value of medium distance 中区地改值
terrain correction value of nearby distance 近区地改值
terrain data translator 地形数据转换器
terrain description 地形描述
terrain details 地形碎部;地面细部
terrain echoes 地面杂乱回波
terrain effect 地形影响
terrain-effect-analysis program(me) 地形效应分析程序
terrain-effect-analysis routine 地形效应分析程序
terrain emboss 地形浮雕
terrain error 地形误差;地势误差
terrain estimation 地形评价;地面估测
terrain factor 地形因素
terrain feature 地形要素;地形特征;地面要素
terrain-following 地形跟踪(飞行)
terrain-following radar 地形显示雷达
terrain influence 地体影响
terrain intelligence 地形信息
terrain intelligence data 地形自动记录器
terrain interpretation 地形判读;地层鉴别
terrain line 天然界限;地性线;地表线
terrain map 地形图
terrain-mapping radar 地形测图雷达
terrain-matching system 地形匹配系统
terrain model 地形模型
terrain model panel 地形模型板
terrain morphology 地面起伏
terrain mo(u)ld alternation 地形模型修改
terrain movement 地层运动;地层移动
terrain nomenclature 地理通名
terrain of sea bottom 海底地貌
terrain orientation 地面定向
terrain pattern 地形模型
terrain point 地面点
terrain point spacing 地形点间距
terrain power spectrum 地面功率谱
terrain profile 地形断面图
terrain profile photography 地形剖面摄影
terrain profile recorder 机载地形记录器
terrain profilter 地形测绘仪
terrain profiling 地形断面记录
terrain radiation 大地辐射
terrain reduction 地形改正
terrain relief 地面起伏
terrain remote sensing 地形遥感
terrain rendering 地形绘制
terrain representation 地貌表示法
terrain sampling 地形采样
terrain scale 地形图比例尺;地形标尺
terrain scanning 地形扫描
terrain sensing 地形遥感
terrain sheet 地形图图幅
terrain slope 地形坡度
terrain slope measurement 地形坡度测量
terrain static correction 地形静校正
terrain study 地形调查
terrain surface 地面
terrain tract 地域
terrain-type map 典型地貌图
terrain unit map 地貌单元图
terrain utility coefficient 地物利用系数
terrain with high stem crops 高苗地
terral levante 陆上累范特风

terra miraculosa 红玄武土
terrane 岩层;地体
terra nera 黑土
terra nova wall plaster 墙面有色抹灰粉饰
terrapak 机械撞击
terrapin 龟鳖族;海龟类;淡水龟
terraplane 汽垫车;气垫车
terra ponderosa 重土
terra-probe 水下振砂器
terraqueous 水陆的
terrarium 小动物饲养园
terra rossa 红土(颜料);红色(石灰)土;含碳酸钙的铁质黏土;蚀余红土;赤铁矿质红黏土
terras 火山土;粗面凝灰岩
terras cement 火山灰水泥
terra sigillata 古代红或黑色陶器;烧后呈红色的黏土;封印土;雕塑用硅质黏土
terras soil 黏磐土
terrastatic pressure 地静压力
terratolite 密高岭土
terra verte 绿土
Terrawode 塔拉沃德(一种可受钉砌块)
terrazzo 威尼斯嵌镶;水磨石地面;磨石子;水磨石
terrazzo aggregate 水磨石集料;水磨石骨料
terrazzo asphalt tile 水磨沥青地面
terrazzo base 水磨石基础
terrazzo block 水磨石板
terrazzo capping 水磨石覆盖
terrazzo chip(ping)s 水磨石石屑
terrazzo cill 水磨石窗台
terrazzo coating 仿大理石面
terrazzo concrete 水磨石混凝土
terrazzo cope 水磨石墙帽;水磨石顶层
terrazzo coping 水磨石挡板;水磨石盖顶
terrazzo cove 水磨石凹圆挑檐;水磨石拱
terrazzo dado 水磨石墙裙;水磨石护壁板
terrazzo dividing panel 水磨石隔板
terrazzo dressing 水磨石饰面
terrazzo faced tread 面层为水磨石的梯级
terrazzo finish 水磨石的罩面处理;水磨石饰面;水磨石地坪;水磨石地面
terrazzo finished surface 水磨石饰面
terrazzo firma 坚硬的土壤;旱地
terrazzo floor cover(ing) 水磨石地面覆盖层
terrazzo floor(ing) 磨石(子)地面;水磨石地坪;水磨石地面(做法)
terrazzo floor(ing) finish 水磨石地板坪饰面;水磨石地板面饰面
terrazzo flooring tile 水磨石铺地砖
terrazzo grain 水磨石颗粒
terrazzo grinder 水磨石机
terrazzo grinding machine 水磨石打磨机
terrazzo jointless floor(ing) 无接缝水磨石地板
terrazzo layer 水磨石铺嵌工;水磨石工
terrazzo machine 水磨石机
terrazzo mix 水磨石混合料
terrazzo panel 水磨石块
terrazzo panelling 水磨石分块
terrazzo pavement 水磨石铺面
terrazzo plant 水磨石工厂;水磨石车间
terrazzo polishing 水磨石上光
terrazzo sill 水磨石窗台
terrazzo sink drop 水磨石冲洗台;水磨石水槽落水
terrazzo skirting 水磨石踢脚线
terrazzo slab 水磨石平板;水磨石面板
terrazzo stair tread 水磨石楼梯踏板
terrazzo surface 水磨石地面
terrazzo tile 水磨石平面;水磨石平板;水磨石砖;水磨石板;水磨石饰面砖
terrazzo tile machine 水磨石制造机
terrazzo tile press 水磨砖压机;水磨石(砖)厂
terrazzo topping 水磨石罩面
terrazzo trade 水磨石手工业;水磨石工场
terrazzo tread 水磨石踏板
terrazzo wall capping 水磨石墙顶盖
terrazzo wall cope 水磨石墙帽;水磨石墙顶盖
terrazzo wall tile 水磨石墙面砖;水磨石墙面板
terrazzo ware 水磨石器皿;水磨石制品
terrazzo work 水磨石作业;水磨石工作;水磨石工程
terreau 有机质
terre-capital 土地资本
terrene 岩体出露区;岩体;地域
terreous 土色的
terreous marble 含土大理石

terre pise 垒墙黏土;泥土建筑
terreplein 垒道;平顶土堤
terrestrial 陆地的;地球的
terrestrial adjustment 陆上调整
terrestrial airport 陆地机场
terrestrial alternation 地心引力
terrestrial ambient 大地环境
terrestrial ambient noise 大地环境噪声
terrestrial animal 陆生动物;陆栖动物
terrestrial animal community 陆生动物群落
terrestrial apparent resistivity 大地视电阻率
terrestrial atmosphere 地球大气
terrestrial attraction 地心吸力
terrestrial beds 大陆沉积层
terrestrial biologic(al) resources 陆地生物资源
terrestrial biosphere 陆生生物界;陆生生态系统;陆地生物圈;地球生物大气层
terrestrial biota 陆地生物
terrestrial bird 陆禽
terrestrial branch 地支
terrestrial camera 地面摄影机
terrestrial camera station 地面摄影站
terrestrial channel 地上信道
terrestrial coordinates 地面坐标;地球坐标;地理坐标
terrestrial coordinate system 地球坐标系;地面坐标系统
terrestrial cross bearings 陆标交叉方位
terrestrial current 杂散电流;地电流;大地电流
terrestrial deposit 陆相沉积(物);陆地沉积(物);陆成沉积(物);洞穴沉积
terrestrial deposition 地表沉积
terrestrial dimension 实地尺寸
terrestrial disturbing function 地球摄动函数
terrestrial dynamic(al) time 地球力学时
terrestrial ecology 陆上生态学;陆生生态学
terrestrial ecosystem 陆地生态系(统)
terrestrial ecosystem evaluation 陆地生态系统评价
terrestrial ecosystem hydrology model 陆地生态系统水文模型
terrestrial effect 地球弯曲影响
terrestrial electricity 地电;大地电
terrestrial ellipsoid 地球椭圆体;地球椭球体
terrestrial ellipsoid dimension 地球椭球尺寸
terrestrial environment 陆地环境
terrestrial environmental remote sensing 陆地环境遥感
terrestrial equator 地球赤道
terrestrial erosion 地表侵蚀
terrestrial eyepiece 正像目镜
terrestrial facies 陆相
terrestrial facies theory on origin of petroleum 陆相生油说
terrestrial flow 地面水流;地面水
terrestrial frozen water 地面冻结水
terrestrial geodesy 地面大地测量学
terrestrial geophysics 陆上地球物理学
terrestrial globe 地球仪
terrestrial gravitation 地球重力;地球引力
terrestrial gravitational constant 地球引力常数
terrestrial habitat 陆生环境
terrestrial heat 地热
terrestrial heat flow 大地热流量
terrestrial horizon 地平;岸线地平
terrestrial interface equipment 地面接口设备
terrestriality 陆地性
terrestrial latitude 地球纬度;地面纬度
terrestrial life 陆栖生物;土壤微生物;地下生物
terrestrial link 地面链路
terrestrial longitude 地球经度;地面经度;地理经度
terrestrial longitude line 地理经度线
terrestrial magnetic disturbance 地磁扰动
terrestrial magnetic field 地磁场
terrestrial magnetic pole 地磁极
terrestrial magnetism 地磁(学)
terrestrial mapping astronomical theodolite 地面测绘天文经纬仪
terrestrial meridian 天文子午线;地球经圈;地面子午线;地面子午圈
terrestrial microwave 地面微波
terrestrial movement 地动;大地运动
terrestrial navigation 陆标导航;地文导航;地面导航;地控导航
terrestrial non-thermal continuum radiation 地

球非热连续辐射
terrestrial object 陆标
terrestrial parallel 地球纬度圈；地面纬圈
terrestrial perturbation 地球摄动
terrestrial photogrammetric camera 地面摄影测量摄影机；地面测量摄影机
terrestrial photogrammetric mapping 地面摄影测量成图
terrestrial photogrammetry 地面摄影测量(学)
terrestrial photograph 地面像片；地面摄影照片；地面摄影像片
terrestrial photographic(al) mapping 地面摄影测图
terrestrial photographic(al) survey 地面摄影测量
terrestrial photo theodolite survey 地面摄影经纬仪测量
terrestrial plant 陆生植物
terrestrial plant theory on origin of petroleum 陆目植物生油说
terrestrial poles 地极
terrestrial polynometry 地面导线测量
terrestrial position determination 地面位置测定
terrestrial radiation 地球辐射(包括大气辐射)；地面辐射
terrestrial radiation balance 地球辐射平衡；大地辐射平衡
terrestrial radiation thermometer 地面辐射温度表
terrestrial reconnaissance 地面勘察
terrestrial reference 地面方向物；地标
terrestrial reference flight 地球基准飞行
terrestrial reference guidance 地球基准制导；地球基准导引
terrestrial refraction 地面折射；地面蒙气差；地面大气折射；大地折射
terrestrial root 地下根
terrestrial satellite 地球卫星
terrestrial science 地学；地球科学
terrestrial scintillation 地面闪烁；大型闪烁
terrestrial sediment 陆地沉积(物)；陆成沉积(物)
terrestrial soil 陆地土壤
terrestrial space 近地空间
terrestrial sphere 地球
terrestrial station 地面电台
terrestrial stereocamera 地面立体摄影机
terrestrial stereogram 地面摄影立体像对
terrestrial stereophotogrammetry 地面立体摄影测量；地面立体摄影测绘；地面立体测量(学)
terrestrial stereoplotter 地面摄影立体测图仪；地面立体测图仪
terrestrial stress 地应力
terrestrial surface radiation 地球表面辐射
terrestrial survey 地面测量
terrestrial survey camera 地面测量摄影机
terrestrial telemetric equipment 地面遥测装置；地面遥测设备
terrestrial telemetric station 地面遥测站
terrestrial telescope 正像望远镜；地面望远镜；大地望远镜
terrestrial topocentric coordinate system 站心地面坐标系
terrestrial triangle 地面三角形
terrestrial triangulation 地面三角测量
terrestrial water 大陆水
terrestrial water quality 地表水水质
terrestrial waters 地表水体
terrestrial waveguide 地球波导
terrestrial wind system 陆成风系
terre verte 绿土
terre verte pigment 地绿色颜料
terrier 地籍册；地产册
terrificientiherbosa 深沼草本群落
terrific sea 巨浪
terrigenous 陆源的；陆生的
terrigenous clastic rocks 陆源碎屑岩
terrigenous deposit 陆源物；陆源沉积(物)
terrigenous lake 陆源湖
terrigenous matter 陆源物质
terrigenous mineral 陆源矿物
terrigenous mud 陆源淤泥
terrigenous province 陆源区
terrigenous sand 陆源砂
terrigenous sediment 陆源沉积(物)
terriherbosa 陆地草丛；陆生草本群落

territorial air 领空
territorial classification 区域分组
territorial control 国土整治
territorial engineering 国土工程
territorial industry 地方工业
territoriality 地盘性；大陆性
territorial jurisdiction 领土管辖权；区域管辖；属地管辖
territorial limit 领土极限；领土范围
territorial management 国土管理
territorial maritime belt 领航带；领海带
territorial planning 国土计划；国土规划
territorial property 地产
territorial resources 国土资源
territorial rights 不同范围的利益
territorial sales budget 分区销售预算
territorial sampling 区域抽样(法)
territorial sea 领水；领海水域；领海
territorial waters 领水；领海水域；领海
territorial waters demarcation line 领海线
territorial yard 地区编组场【铁】
territory 国土；领土；领地；区域；地区；地盘；版图
territory economics 国土经济学
territory health quarantine 国境卫生检疫
territory lighting 地面照明
territory management program(me) 国土整治计划
territory of nationality 少数民族地区
territory planification 土地规划
territory planning 区域规划；领土规划
territory rights caluse 销售地区权利条款
territory under jurisdiction 领土
terro-metallic clinker 黑黏土缸砖
terron 草泥坯块
terrwicket 小梯田
terry 厚绒布；无线电测高计；雷达自动测高计
terry poplin 毛圈府绸
terry swatch 毛圈布布样
terskite 突硅钠锆石；水硅锆钠石
tert-butyl acetate 醋酸特丁酯
tert-butyl alcohol 叔丁醇
tert-butyl catechol 叔丁基邻苯二酚
tert-butylhydroperoxide 叔丁基化过氧氢
tert-butyl perbenzoate 过苯甲酸叔丁酯
tert-butyl peroxide 叔丁基过氧化物
tert-butylpyrocatechol 叔丁基邻苯二酚
Tertiarium 特蒂锡铅焊料
Tertiary 第三纪【地】；第三期；第三级
tertiary air 三次空气；三次风
tertiary air duct 三次风管
tertiary alcohol 叔醇
tertiary amine 三元胺；三级胺
tertiary amine value 叔胺值
tertiary arsenate 三代砷酸盐
tertiary arsine oxide 氧化砷
tertiary beam 三重梁；三级梁
tertiary branching 三次分枝
tertiary breaker 三级碎石机；三级破碎机
tertiary butanol 叔丁醇
tertiary butyl benzene 第三丁苯
tertiary butyl cyclohexane 第三丁基环己烷
tertiary butyl cyclopentane 第三丁基环戊烷
tertiary butylhypochlorite process 叔丁基次氯酸盐法
tertiary calcium phosphate 正磷酸钙
tertiary carbon atom 叔碳原子
tertiary cementite 三次渗碳体
tertiary circulation 局部环流；三级环流
Tertiary clay 第三纪黏土
tertiary cleaning 第三次精选
tertiary coil 三次线圈；第三线圈；第三级线圈
tertiary colo(u)r(s) 三次色
tertiary comminution 三级磨碎
tertiary creep 加速蠕动；加速蠕变；三重蠕变；三段蠕变；第三期蠕变；第三阶段蠕变
tertiary creep stage 加速蠕变阶段
tertiary crusher 细碎破碎机；三级碎石机；三级破碎机
tertiary crusher with rolls 三级辊式破碎机
tertiary crushing 三级破碎；第三级破碎
tertiary current 三次电流
Tertiary deposit 第三纪沉积层；第三纪沉积物
tertiary device 第三设备
tertiary effluent 三级出水

tertiary filter 三级滤池；三级过滤器
tertiary filtration 三级过滤
tertiary flo(a)tation 第三次浮选
tertiary float feed 三次浮选给料
tertiary Gothic 第三期哥特火焰式
tertiary grinding 三级粉碎；三级研磨；三次研磨；三次磨矿；第三级研磨
tertiary highway 三级公路
tertiary hydroperoxide 叔丁基过氧化氢
Tertiary igneous rocks 第三纪火成岩
tertiary industry 第三产业；三次产业
tertiary jig 三洗跳汰机
tertiary level of prevention 第三级预防
tertiary maturation ponds 三级塘
tertiary mixture 三元混合物
tertiary nitro compounds 叔硝基化合物
tertiary occupation 第三产业；服务性职业
tertiary oxidation pond 三级氧化塘
Tertiary period 第三纪【地】
tertiary point 三等点
tertiary pond 三级塘
tertiary production 三级产业(指服务业)
tertiary pyroelectricity 第三热电性
tertiary recrystallization 三次再结晶
tertiary reduction 三级压延；三级粉碎
tertiary reflector antenna 三级反射面天线
tertiary reflex 三级反射
tertiary rib (哥特式拱上的)中间肋
tertiary risks 投资低风险
tertiary road 三级路；三级公路；三级道路
tertiary sampler 三次取样器
tertiary screen 三道筛
Tertiary sediment 第三纪沉积物
tertiary sewage treatment 污水三级处理；三级污水处理；三级废水处理
tertiary sewage treatment method 污水三级处理法
tertiary sewage treatment of no-chlorination 非氯化污水水三级处理
tertiary solid 三元固体
tertiary spectrum 三重光谱
tertiary spiral 三生螺纹
tertiary split 三级风流分支
tertiary stabilization ponds 三级稳定塘
Tertiary structure 第三纪构造【地】
tertiary structure 三级结构
tertiary substation 三次变电所
Tertiary system 第三系
tertiary target 三类目标
tertiary to secondary centre circuit 三级至二级中心线路
tertiary treated wastewater 三级处理后的污水；三级处理的废水
tertiary treatment 三级处理；第三级处理
tertiary treatment for sewage 污水三级处理
tertiary treatment method 三级处理法
tertiary treatment pond 三级处理塘；三级处理池
tertiary triangulation 三级三角测量
Tertiary-type ore deposit 第三纪型矿床(浅成热液矿床)
tertiary voltage 三次电压；第三电压
tertiary washbox 三洗跳汰机
tertiary wastewater 三级废水
tertiary wastewater treating process 污水三级处理法；三级废水处理过程
tertiary waste(water) treatment 污水三级处理
tertiary winding 稳定绕组；三级绕组；三次绕组；第三线圈
tertibrach 三腕板
tertlary butyl acetate 乙酸叔丁酯
tert-pentyl acetate 醋酸特戊酯
tertschite 多水硼钙石
teruggite 砷硼镁钙石
Tervueren faience 锡釉陶器
terylen(e) 涤纶
terylen(e) ladder web 涤纶梯状织物(装百叶窗用)
terylen(e) rope 涤纶绳
terylen(e) tape 涤纶带
terylene wastewater 涤纶废水
Terzaghi's bearing capacity theory 太沙基承载力理论
Terzaghi's consolidation theory 太沙基固结理论
Terzaghi's rock load 太沙基岩石荷载
Terzaghi wash-point soil penetrometer 太沙基水冲式土壤贯入仪

Tesaar lens 特沙尔透镜
teschemacherite 碳铵石
teschenite 沸绿岩
tes facility 检测手段
teshirogitlite 钛金红石
tesitol 半熔酚醛树脂
Tesla 特斯拉
Tesla coil 高频火花检漏仪;特斯拉空心变压器
Tesla coupling 特斯拉耦合
Tesla current 特斯拉电流
Tesla-D'arsonval ammeter 特斯拉-达松伐耳安培计
Tesla-D'arsonval current 特斯拉-达松伐耳电流
Tesla transformer 特斯拉空心变压器;特斯拉变压器
tessarace 四面体的顶点
tesselate 花纹状嵌饰
tesselated mirror 嵌和反射镜
tessella 小块大理石
tessellate 花纹状嵌饰,镶缀成花纹的
tessellated 细工镶嵌的;棋盘格状的;分成方格的
tessellated macula 棋盘状斑
tessellated path 拼花路面
tessellated pavement 镶嵌成花纹的路面;罗马马赛克铺面;嵌石铺面;嵌成花纹路面
tessellated paving 嵌花铺面;罗马式镶嵌铺面;镶石铺面;嵌石铺面
tessellated work 镶嵌工艺;镶嵌细工
tessellation 嵌石装饰;棋盘形布置
tessera 镶嵌用彩色玻璃;嵌镶块;镶嵌物;嵌石铺面;[复]tesserae
tessera flooring 马赛克地面;马赛克楼面
tesseral 镶嵌物(似)的
tesseral harmonics 田谐函数;田调和函数
tesseral system 等轴晶系
tessha glaze 铁砂釉
test 检验;探试;试验;实验
testability 可检验性;试验性
testable hypothesis 可检验的假设
test accessories 试验辅助设备
testaceum 碎瓦片铺面
test adapter 试验附加装置
test address 测试地址
test adit 试验导洞;试挖横坑道
test again and again 反复试验
test air tunnel 试验风洞
test alignment 精度检查
test alphabetic 测试字母的
testament 遗嘱
testamentary trust 遗嘱信托财产
test amplifier 测试用放大器;测试用放大镜
test analysis report 检验分析报告;试验分析报告
test anchor 试验用地锚
test and adjustment of level 水准仪检验校正
test and adjustment of theodolite 经纬仪检验与校正
test and handling 试验与操作
test and maintenance truck 试验及维修车
test and operation group 试验与操作群
test and verify program(me) 测试验证程序
test antenna 试验天线
test anvil 测砧
test apparatus 试验仪器;试验工具
test application 试验的应用
test area 试验区
test arrangement 试验安排;试验布置;试验准备
test ascent 试验性上升
test assembly 试验装置;试验组件
test assignment 测试任务
test at constant load 定荷载试验;常荷载下试验
testator 立遗嘱人
testatrix 女立遗嘱人
test audit 抽查
test automation 测试自动化
test average 试验平均值
test balloon 试验气球
test bar 杆状试件;试验条;试验(杆);试验(棒)
test base 试验基地
test basin 试验池
test batch 试验批料;试验的一批
test bay 试验间;试车台
test beam 梁状试件;(混凝土)试件;试验梁
test beam mo(u)ld 试验梁模
test bed 试验台座;试验台架;试验台(面);试验床
test bed boat 水上试验站
test-bed firing 试车台点火

test bed instrument 试验台测试仪器
test-bed instrumentation 试验台测试仪器
test bed of cement 水泥试验台
test bench 试验台
test bench for tachometer generator 测速发电机试验台
test bench installation 试验台设备
test bench roller 试验台滚轮
test bias 试验偏倚
test bit 试验钻头
test block 试验台;试片
test board 检验板;试用配电盘;试验盘;试验板;测试台;测试板;测量台
test booster 测试放大器
test bore(hole) 试钻孔;试验(钻)孔
test bore pattern 试验孔布置形式
test boring 探坑;试钻;试(验)钻孔;探穴
test boring record 试井记录
test box 试验盒
test building 试验间
test burner 试验灯
test button 金属试样
test buzzer 测试蜂鸣器
test by digging 挖掘试验
test by dry mortar 硬练试验
test by elastic wave 弹性波探查
test by trial 探索性试验;尝试
test by wet mortar 软练试验
test cabinet 化验室;测试室
test cable 试验电缆
test cap 测试端头
test capacity 试验容量;试验能力
test car 试验汽车
test card 试验卡片;试验表格;测试纸;测试图表;测试卡;测试板
test carriage 测试车
test car technique 测试车观测法
test case 校验数据和条件
test case design 测试情况设计
test case generation 测试程序的数据和条件生成
test casting 取样铸件
test cavity 试验腔
test cell 试验电池;试验箱;试车间;测试用单元
test-cell instrumentation 试车间检测仪器设备
test cell run 试车台试马
test center 试验中心
test certificate 检验证书;检验证
test chamber 试验室;试验间;风洞试验段
test charge 试验电荷
test chart 草图;试验张;试验图;测试图(像);测试表;视力表
test check 试样检验;试验检查;抽样;抽查
test circuit 检查电路;试验环线;测试电路
test clamp 试验用夹子
test class 试验等级
test clerk 试验员
test clip 试线夹
test coach 检测车
test cock 验水旋塞;试验旋塞;试验栓
test code 试验指导书;试验码;试验规则;试验规范;试验法规
test coil 试验用线圈
test compaction 压实试验
test complex 综合试验设备
test computer 测试计算机
test concrete 试验混凝土
test condition 试验条件;实验条件;测试条件
test conductor 试验指导人
test cone 试验用测温锥;试验火锥
test connection 测试电路
test connector 测试用终接器
test console 试验控制台
test constant 试验常数
test contact 试验触点
test controller 测试员;测试器
test cord 测试塞绳
test core 圆柱试体;钻芯试件
test correction chart 弯液面校正表(氟氢酸测斜)
test coupon 附连试验板
test crew 试验机组
test criterion 检验判据;检验标准
test cube 立方(体)试块;立方试体;立方试件;混凝土(立方体)试块;混凝土立方块(体)强度试验;试块

test cube mo(u)ld 立方体试模(混凝土)
test cube strength 立方体试块(抗压)强度;立方体试件(抗压)强度
test cubicle 试验间
test current 试验电流;测试电流
test cut 试验开挖
test cycle 试验周期;试验循环
test cylinder 圆柱体试块;圆柱试体;柱状试样;圆柱形试块(混凝土)
test cylinder (jar) 试验缸
test cylinder of concrete 混凝土圆柱形试块
test data 校验数据;检查数据;试验数据;测试数据
test data generation 检查数据生成;测试数据生成
test data generator 试验数据发生;测试数据产生程序
test data memorandum 试验数据单
test data record 试验数据记录
test data recorder 试验数据记录器
test data reduction 试验数据换算
test data report 试验数据报告
test data sheet (数据)记录
test data system 试验数据系统
test date 试验日期
test deck 模拟试验组;测试卡片叠
test description of sample 试样说明
test desk 试验桌;测试台
test detail 试验细节(说明)
test detect function 测试检测操作
test detect routine 测试检测例行程序
test determination 试验测定
test development 用试验进行开发
test device for flexible pipes 软管试验槽
test device for rotary machine 回转机试验装置
test diagram 试验曲线
test direction team 试验指导小组
test director 试验领导人
test distance 试验距离
test distributor 试验分配器
test dive 试验性俯冲
test documentation 试验文件
test dosing 增量试验法
test-drill 试钻
test drill core 试样岩心
test drilling 钻屑试样
test drive 试验驾驶;试车
test driver 试车驾驶员
test driving 试验打桩
test dummy 试验模型
test duration 试验周期;测试持续时间;试验时间;试验持续期
test dust 试验用粉尘;试验尘
test duty table 试验责任表
test dye 试验用染料
test dyeing 试样染色
test dynamometer 试验用测力计
tested capacity 经检验定出的产量;试验能力
tested capacity of well 试验井出水率
tested data 测定数据
tested recipe 成熟配方
tested value 试验值
tested well capacity 井试验容量;井的试验容量
testee 测验对象
test efficiency 试验效率
test embankment 试验堤
test encoder 测试编码器
test engine 试验发动机
test engine data 试验发动机数据
test environment 试验环境
test equipment 测试工程师;测试仪器
test equipment engineering 试验设备工程学
test equipment error analysis report 试验设备误差分析报告
test equipment list 试验仪器清单
test equipment team 测试设备组
test equipment tester 测试设备检验装置
test equipment tool 测试设备工具
tester 检验仪;检验器;检查设备;试验者;试验员;试验仪器;测试器;试验计;试验机;测试人员;测试器;测定器;分析仪器;华盖(布道台、坟墓、大床等的)
tester for fiberboard strength 纤维板强度试验机
test error 试验误差
test evaluation 试验鉴定
test event 测试事件

test examiner 试验检查人
test excavation 试验开挖
test exposure 试验曝光
test face 试验工作面
test facility 试验装置
test facility and instrument 检测设备和仪器
test failure 试验时的损坏
test failure load 试验性破坏荷载
test fan 试验用功率吸收风扇
test-fence 曝晒架
test fence exposure 曝晒架曝晒
test figure 试验值;试验图表;试验数字;试验数据
test file 测试文件
test fill 试验填方
test filling 填充试验;试填
test film 试验片
test final selector 测试终接选择器
test fire 试验火
test firing 试验点火
test fittings 检测连接装置
test fixture 试验夹具
test fixture board 测试夹板
test flight 试飞
test flight worksheet 试飞工作单
test floor 试验台座;试验台面;测试台
test fluidized velocity 试验流化速度
test flume 试验(水)槽
test-fly(ing) 试飞
test for allocation 分配测试
test for bacteria 检验细菌
test for blank 空白区检查
test for clay content of aggregate 集料黏土含量试验
test for contamination 污染试验
test for correlation 相关试验
test for dangerous goods package performance 危险货物包装性能试验
test for equal variance 等方差试验
test for equilibrium 均衡性测试
test for expansive pressure 膨胀压力试验
test for fault 探伤试验
test for grading 评估试验;评级试验
test for homoscedasticity 同方差性检验
test for linearity 线性检验
test for miscibility with water 水可混(和)性试验
test formulation 试验配方
test for pollution 污染试验
test for reinforcing plate 补强板试验
test for resistance to penetration 插入阻力试验
test for serial correlation 序列相关检验
test for serial randomness 序列随机性检验
test for silt 泥沙含量分析
test for soundness 水泥安定性试验;安定性试验(水泥)
test for specific gravity of slip 泥浆比重测定
test for trend time series 动态趋势检验
test for upward or downward trend 升降趋势检验
test for weldability 可焊性试验
test frame 试验架
test frequency 试验次数
test fuel 试验燃料
test function 测试函数;测试功能
test gas 试验气体;测试燃气
test ga(u)ge 检测规
test ga(u)ge requirement 试验仪表要求
test generation 测试生成
test generator 测试(码)生成程序
test gimulator 测试模拟器
test glass 光学样板;试验(管);试验(杯);玻璃样板
test grid 检验格网
test groin 试验丁坝
test ground 测试场地
test groyne 试验丁坝
test hammer 检修用手锤;测试锤(非破损法测定混凝土强度)
test harness 试验用接线
test head 试验用闷头(钢管的);试验压头
test heat exchanger 试验换热器
test hole 勘探浅眼;勘察钻孔;试验钻孔;试验(性)孔;试孔;测试孔;钻探孔
test hole boring 勘探孔钻进
test hole drilling 勘探孔钻进;试孔钻探
test hole pattern 勘探孔布置
test-hole work 试孔钻探

test house 试验站
testibiopalladite 等轴碲钯矿
testification 举证
testify 作证
testimonial 证明书;鉴定书
testimonial by accountants 会计师证明
testimonial paper 证明文件
testimony 证据;证词
testimony of witness 人证
test impact pendulum 摆锤冲击试验
test index 试验指标
test indicator 试验指示器;测试指示器
test information 试验资料;试验情报
testing 测验;测试
testing aeroplane 试验用飞机
testing after completion 完井测试
testing age 试验龄期
testing agency 试验单位
testing aircraft 试验飞机
testing and recording car 测试记录车
testing and research engineering 试验研究工程
testing and scoring system 测验与记分系统
testing apparatus 测试仪器;检验仪器;试验装置;测试设备;试验仪器
testing appliance 试验设备
testing arrangement 试验布置
testing bed 试验台座
testing bedrock 测定基岩厚度及性质的浅孔钻进
testing bench 试验台架
testing block 试验台架;试块
testing body 试验车车身
testing box 试验箱
testing button 试验按钮
testing cab(in) 测试室
testing cable 测试电缆
testing car 试验车
testing certificate 试验证(明)书;试验检定(证)书;出厂证(书)
testing column 试验柱
testing committee 测试委员会
testing compression specimen 抗压试件
testing configuration 试验布置
testing control 检验控制;检查控制;测试控制
testing cost 测试费
testing crew 试验队
testing curve 试验曲线
testing data 试验资料
testing data of sample 样品测试日期
testing department 试验车间;试验部门
testing device 试验设备;测试装置
testing device for pantograph 受电弓试验台
testing digital module 测试数字模块
testing drain 沟道检验
testing drill hole 试验孔
testing engineer 试验工程师
testing equipment 试验设备;测试设备;检验设备
test in getting bumper crops 丰产试验
testing exemption 免检
testing expenses 试验费
testing facility 测试设备
testing farm 试验农场
testing field 试验工地;试验场(地)
testing fill 试验填方
testing findings 试验结果
testing gallery 试验平巷;试验廊道
testing goodness of fit 拟合优度检验;拟合良好性检验
testing grid 检验格网
testing ground 试验场
testing hammer 检验锤
testing hardware 测试硬件
testing head 试验水头
testing hypothesis of curvilinear relationship 曲线关系的假设检验
testing hypothesis of linear relationship 线性关系的假设检验
testing indicator 试验指示器
testing in hard rock 硬岩测试
testing institute 试验学会;试验研究院
testing instrument 检验仪器;试验仪器;测试仪器
testing jack 测试塞孔
testing jig 测试架
testing laboratory 化验室;试验室
testing level 测试电平

testing lever 试验杠杆
testing line 试验品系
testing linearity of regression 回归线性检验
testing liquid 试验液体
testing load 试验荷载;试验负载;试验负荷
testing machine 试验机
testing machine for material strength 力学强度试验机
testing medium 试验介质
testing method 试验方法;试验法;测定方法
testing method of sample 样品测试方法
testing method standard 测试方法标准
testing mortar 试验灰浆
testing mo(u)ld 试模
testing nethod for industrial wastewater 工业废水检验法
testing nominative 测试标准
testing of characteristic 特性试验
testing of dynamic(al) balance 动平衡试验
testing of electric(al) heating on backlites 后风挡玻璃电加热检测
testing of grids 铅板试验(蓄电池)
testing of materials 材料试验
testing of steel tape 钢尺检验
testing of steering gear 试舵
testing of well 试井
testing operator 测试人
testing panel 试验板
testing period 试验周期
testing pile 试验桩
testing piling 试验桩
testing pit 试验竖井
testing plant 试验设备;试验车间;测试设备
testing platform 试验平台
testing plug 试油水泥塞
testing point 试验点
testing position 试验位置;测试座席
testing press for props 支柱承载试验压机
testing pressure 试验压力
testing principle 试验原理
testing priority 试验优先次序
testing procedure 测试程序;试验方法;试验步骤;测试步骤
testing program(me) 试验程序
testing project for moisture transporting in aeration zone 饱气带水分运移测试工程
testing project of representative sample 代表性样品试验方案;代表性样本试验方案
testing range 试验范围;试验量程
testing record sheet 试验记录卡;试验记录表
testing regulation 试验法规;试验规程
testing reliability 试验可靠性;试测可靠性;测试可靠性
testing report 试验报告(单)
testing resctor 试验堆
testing result 试验结果
testing rig 试验设备;试验台;试验架
testing room 试验室
testing routine 试验常规;测试程序
testing sample 测试样品
testing schedule 试验大纲
testing screen 检验筛;试验(用)筛
testing sequence 试验程序;测试顺序
testing set 测试装置;试验设备;试验器;成套试验器
testing shaft 测试竖井
testing sheet 检验单;试验单
testing shop 试验车间
testing sieve 试验(用)筛;实验筛
testing sieve cloth 试验筛网
testing sieve cloth series 试验筛布系列;试验筛网系列
testing sieve cloth standard 试验筛网标准
testing sieve plate 试验筛板
testing sieve shaker 筛分试验振荡器;试验筛振动器
testing sieve standardization 试验筛的标准化
testing site 试验现场;试验地点;试验场地
testing size 试验粒度
testing skill 试验技术
testing socket 试验插座
testing software 测试软件
testing specification 试验细则;试验规程
testing specimen 测试标本
testing stand 试验台
testing standard 测试标准

testing station 试验站
testing strip 试验纸条
testing structure 试验性结构
testing surface 试验表面
testing switchboard 试验用配电板
testing system 试验系统
testing system flexibility 试验系统的柔度
testing table 检验表;试验表
testing tank 试验池
testing technique 试验技术
testing tension 试验电压
testing terminal 测试终端;测试端子
testing terminal for tunnel structural reinforcement 隧道结构钢筋监测端子
testing term of sample 样品测试项目
testing the system 系统测试
testing time 试验时间;测试时间
testing tool 测试工具
testing track 试验跑道
testing transformer 试验用变压器;试验变压器
testing tray 取样槽
testing tube 试管
testing tunnel 观测平硐
testing type of sample 样品测试类型
testing unit 试验装置;测试装置;测试设备
testing valve 试验阀
testing voltage 试验电压;测试电压
testing wedge 检验光楔
test in place 现场试验;工地试验
test input tape 测试输入带
test instruction 测试指令
test instruction debugging 测试指令调试
test instrument 测试仪器;试验工具
test instrumentation 测试仪表
test instrumentation system 测试仪表系统
test instrument board 试验仪表板
test instrument for coal dust explosion 煤尘爆炸试验仪
test interval 测试间隔时间
testis cerebri 四叠体下丘
test item 试验项目
test jack 测试插座
test jack panel 测试塞孔面板
test jar 试验缸
test key 押码;测试电键;密押
test kit 测试包
test lab(oratory) 测试实验室
test lamp 试验灯;测验灯
test landing 试验性着陆
test launcher 试验发射装置
test lead 试验引线;测试引线;测试端
test length 试验长度
test level 校核水准;试验水准;测试电平
test limitations 试验的限制
test limit 试验限度
test line 试验线路
test line generation procedure 测试行生成过程
test list 检验单
test load 试载;试验荷载
test-loaded 加式载的
test loading 试验装载;试验加荷
test loop 环行试验线;试验回路;测试回路;测试环线;测试环路
test loop antenna 试验环形天线
test loop road 试验环道
test lowering 水位下降试验(地下水)
test lug 连接试样块(与金属本体铸在一起);本体试块(铸造用语)
test machine 试验机
test machine for door open/close equipment 车门开闭装置试验器
test macro 测试宏指令
test mandrel 试验心轴
test manual 测试手册
test mark 检验标志
test-market 试推销上市
test marketing 市场试销
test material 试验材料
test material preparation 试验料制备;试料制备;试件制备
test matrix 检验矩阵;试验矩阵;测试矩阵
test-meter 试验仪表
test method 检测方法;探伤方法
test method for light(weight) aggregate 轻骨料混凝土试验方法
test methods for light(weight) aggregate concrete 轻集料混凝土试验法
test mile 试验时程(汽车);试验里程
test missile 试验性导弹
test mission 试飞任务
test mixer 试验混合器
test mix(ture) 试验拌和物;试验混合物
test mode 检查方式;试验模式;试验方式;测试状态;测试方式
test model 试验模型;试验模式;测试模型
test modulation 试验调制
test module 试验组件;测试模块
test module for rotational flexure fatigue 回转扭曲疲劳试验模型
test monitor 试验监测器
test natural angle of repose 天然休止角试验
test norm 试验标准
test number 试验号(码);试验次数;代号
test numeric 检验数字的;测试数字的
test-object 试标
test objective 试验目的
test observation 试验观察
test observatory 试验性观测台
test odd 奇次谐波测试
test of artificial soil 人工土壤测定
test of convergence 收敛性检验
test of customary characteristics 习惯特征的测验
test of drinking water 饮用水检验
test of fit 拟合检验
test of goodness of fit 拟合优度检验;适合性测定
test of grasping soil into a ball 握团试验
test of hypothesis 假设验证;假设检验
test of independence 独立性检验;独立性测验;独立性测定
test of long duration 长期试验
test of mechanical property 力学性能试验
test of normality 正态性检验
test of pozzolanicity 火山灰活性试验
test of precipitation infiltration 降水入渗试验
test of predictive power of a model 模型预测能力检验
test of qualification 合格试验
test of sand selection 选砂试验
test of sewage 污水试验;污水检验;污水检测
test of short duration 短期试验
test of significance 显著性试验;显著性检验;显著性测验;有效检验
test of significance for correlation 相关显著性实验
test of skewness and kurtosis 偏度峰度检验
test of smoothness 修平法检验
test of standard atmosphere pressure 标准大气压测定
test of subtense bar 基线横(测)尺试验
test of theodolite 经纬仪试验;经纬仪检测
test of the program(me) 程序检验
test of transit 经纬仪检验
test of trend 趋势检验
test of unusual use 非常用途测验
test of water self-purification 水质自净试验
test of weld point 焊接试验
test oil 试油
test OK 无故障
test on core 线芯试验
test on prototype bridge 原型桥试验
test on residue 残渣试验;残留物试验
test on residue accumulation 残留累积试验
test on sand 砂地试验
test on site 工地实验;现场试验;工地试验
test operating procedure 试验操作程序;试验步骤;测试操作过程
test operation 试运行
test organism 测定菌
test oscillator 测试振荡器;测试用振荡器
test outfit 试验设备
test over-flow condition 测试溢出条件;测试上溢条件
test pack 测试包
test packer 测试封隔器
test panel 检验样板;试验样板;试验台;试验盘试板;试板
test paper 试卷;试纸
test paper method 试纸(检测)法
test parachute 试验用降落伞
test parameter 试验参数
test paraphernalia 试验用附属装置
test particle 试验粒子
test pattern 试验图像;试验模型;测试图(像);测试码模式;标准图形(电测试用)
test pattern beam 测试图电子束;测视图电子束
test pattern evaluator 测试码模式评估程序
test pattern generation 测试码模式生成
test pattern generation program(me) 测试码模式生成程序
test pattern generator 试验图像发生器;测试图案信号发生器;测试码模式生成程序
test pattern sequence 测试码模式序列
test pencil 试验电笔;试验笔;试电笔
test performance 试验性能
test period 试验周期;试验期;试验阶段
test phase 试验阶段
test piece 试块;试(验)样板;试验片;试片;试件
test piece for welding 焊接试板
test piece model 试验样品
test piece mould 试件模具
test piece number 试样号
test pile 试桩;试验桩
test pile driving 打试桩;打桩试验
test piling 试验桩;试桩
test pilot 试飞驾驶员
test pipe 测水管
test pit 探坑;探井;试验竖井;试坑
test-pit digging 试挖
test pitting 挖探坑;坑探
test plan software 测试程序软件
test plate 检验片;试压板;试板;玻璃样板;试验盘(用于试验材料抗压性);试验板(地耐力试验用);光学样板
test platform 试验台
test plot 初步方案图;试验(小)区;试验田;试验计划;试验地
test plug 试验插塞
test point 试验点;测试点
test point scattering 试验点分布区
test pole 试线杆
test portion 试验部分;试料
test position 测量台
test prescription 试验配方
test press 打样机
test pressing 试验压力
test pressure 试验压力;测试压力
test pressure ga(u)ge 试验压力计
test prism 棱柱体试件
test probe 测试探头
test problem 检查问题;检查题目
test procedure 试验步骤;试验方法;试验法;试验程序
test process of corrosion test 溶蚀试验过程
test prod 测试用探棒;测试棒
test production 试验性生产
test profile 试验结果曲线的示意图;试验断面
test program(me) 试验计划;实验大纲;测试程序;检验程序;试验大纲
test program(me) package 测试程序包
test program(me) system 测试程序系统
test project 试验规划;试验设计
test protocol 试验规程
test pulse 试验脉冲
test pump 试验泵
test push button 试验用按钮
test rack 试验机架
test rain 测试雨量
test rain temperature 测试雨温度
test range 检验范围;测试范围
test-range tracking 试验靶场跟踪
test rate 试验速率
test reach 试验河段
test reactor 试验堆;试验用反应器
test reading 试验读数
test receiver 测试接收机
test recipe 试验配方
test record 试验声盘;试验记录;测试记录
test record sheet 试验记录单
test regulation 试验规则
test relay 试验继电器
test replacement 检查替代;测试替代
test report 试验报告
test report No. 试验报告编号

test report table 试验报告表
test request message 测试请求报文
test requirement 试验要求
test requirement manual 试验要求手册
test requirements document 试验要求;试验的技术要求
test requirement specification 试验技术规范
test result 试验成果
test result processor 测试结果处理器
test results of dredge pump 泥泵试验结果
test revision 复试
test rig instrument 试验台仪表
test rig panel board 试验台控制板
test-rig platform 试验台的平台
test ring 试验环线
test road 试验路;试车跑道
test rod 探棒;触针
test routine 检验程序;试验例程;试验程序
test rule 试验规则
test rules for inventory management 库存管理检查规则程序
test run(ning) 试运转;试运行;试样的制备;试验运转;试车;测试运行;小生产;环路试验
test run train 试运转列车
test runway 试飞用跑道;试飞跑道
test sailing 试航【航海】
test sample 试样;试验样品
test sample mo(u)ld 试验样品模具
test samples of bearing 持力层的土样试验
test sampling 试验取样
test schedule 试验进度表
test scheme 试验方案
test scope 测试示波器
test section 试验区段;试验截面;试验断面;试验段;试验部分;试线区段;测试区
test section of a wind tunnel 风洞试验段
test section of high-speed line 高速试验段
test selector 测试(用)选择器
test sequence 试验顺序
test series 试验系列
test session 试验阶段
test shaft 探井;试(验)井
test sheet 试验单
test shield 测试护套
test sieve 试验筛析;试验筛
test-sieve shaker 摇筛机
test sieve size 筛号
test sieve vibrator 摇筛机
test sieving 筛(分)析
test signal 测试信号
test signal generator 测试信号发生器
test signal oscillator 试验信号振荡器
test silo 试验发射井
test simulation 实验模拟
test simulator 试验模拟器
test site 试验基地;试验地点;试验场(所)
test skip 测试跳跃
test slab 板的试验;试验板
tests of financial position 财务状况测试
tests of profitability 利润率测试
tests of welded vessels 焊接容器试验
test soil 土壤试样
test solution 试液;试验溶液;测试溶液
tests on completion 竣工检验
tests on cores 岩芯测试
tests on site 现场试验
test space 测试空间
test specification 试验规格;试验规范;测试说明;测试规格;测试规范;测试范围
test specimen 拉力试棒;试样;试验样品;试件
test specimen mo(u)ld 试样模(子)
test specimen preparation 试验制备
test specimen tube 试样筒
test speed 试验速度
test speed record 试验速度的记录
test splice 试验连接;测验接头
test spring 试验弹簧
test stage 试验台
test stand 试验架;试验用架子
test standard 检验标准
test stand operation 试车台运转
test statement 测试语句
test station truck 试验车
test statistics 检验统计量;试验统计资料;测试统计量

test status output 测试状态输出
test step 测试步
test step by step 分步测试
test stone 试金石
test strength 试验强度
test structure 试验结构
test studies of rock and soil 岩土实验研究
test subroutine 测试子程序
test summary sheet 试验总表
test supervisor 试验负责人
test supervisor program(me) 测试用监控程序;测试管理程序
test support equipment 试验辅助设备
test support program(me) 测试支援程序
test switch 试验用开关;试验开关
test system 测试系统
test tank 观察柜;试验罐;试验池;试验槽
test tape 测试带
test tapping 试验
test-target 检验目标
test task 测试任务
test team 试验组
test tee pipe 试验T形管
test tee tube 试验T形管
test thrust chamber 推力试验室
test timer 测试计时器;测试定时器
test to destruction 破裂试验;破坏试验
test to failure 故障前试验
test tolerance 试验允差;试验公差
test tone 测试音调
test track 试验线路;试验车道;试验轨道;试验车道
test traffic 行车试验
test train 试验列车
test translator 检验翻译程序;测试转换器;测试翻译程序
test tree 标准木
test trench 探槽;试验坑;试坑;试槽
test trenching 挖探坑
test trial 复核试验
test trough 试验槽
test trunk 试验干线
test tube 风洞;试(验)管
test-tube brush 试管刷
test tube centrifuge 试管离心机
test-tube clamp 试管夹
test-tube holder 试管夹
test tube plant 试验工厂
test tube rack 试给架;试管架
test tube slant 试管斜面
test-tube stand 试管架
test-tube support 试管架
test tube without rim 不带边试管
test tube with rim 带边试管
test tube with side tube 带支管试管
test-type launcher 试验发射装置
test types of sight 视力检定表
testudinate 龟甲形拱顶的;拱状的
testudo 龟甲形大盾或屏风(古罗马);弓形屋顶(罗马建筑)
test under load 加载试验
test unit 试验单元
test-use 运用试验
test value 试用值
test valve 检查阀;试验阀
test vehicle 试验车辆;试验车
test vessel 试验容器;试验船
test wait 测试等待
test wall 试验墙
test water 检验(用)水;试验用水
test water meter 试验水表
test wave 试验波浪
test wedge 三角试片
test weld 试焊
test welding 试验焊接;试焊接
test well 勘探钻井;评价井;试(验)井;试孔
test well number 试验孔数
test wheel 试验轮
test with notched test piece 用凹槽试件进行冲击试验;凹槽试件冲击试验;缺口试样试验
test without packer 无封隔器测试
test word 押码
test work 试验工作
test working 试车(机器的);试(验)运转

test zone 试验区
tetartine 钠长石
tetartohedral 四分面的
tetartohedral form 四分面形
tetartohedrism 四分对称性
tetartohedron 四分面体
tetartohedry 四分面象;四分对称
tetartoid 五角三四面体
tetartopyramid 四分锥
tetartosymmetry 四分对称
tete-a-tete 双人椅子;面对面
tether 系链
tethered air diving system 管供式空气潜水系统
tethered balloon 系留气球;有系绳气球
tethered breakwater 用铁链拴系的浮体防波堤
tethered buoyant platform 带缆浮力平台
tethered float breakwater 浮球阵防波堤
tethered manned submersible 带缆载人潜水器
tethered mixed gas diving 管供式混合气潜水
tethered one-man atmospheric system 带缆单人常压潜水系统
tethered remotely operated vehicle 带缆遥控器
tethered submersible 带缆潜水器
tetherless submersible 无缆式潜水器
Tethyan metallogenic belt 特提斯成矿带
Tethyan molluscan realm 特提斯软体动物地理区系
Tethyan realm 特提斯区系
Tethys 古地中海;特提斯海(古地中海)
Tethys-Himalayan tectonic domain 特提斯—喜马拉雅构造域
Tethys ocean 特提斯海洋
Tethys seaway 特提斯海路
Tethys tectonic active zone 特提斯构造活动带
Tethys tectonic zone 古地中构造带
Tetmajer 蒂特迈杰硅青铜
tetometamorphism 异变态
Tetoron 蒂托纶聚酯纤维
tetraalkyl ammonium salt 四烷基季铵盐
tetra-atomic base 四价碱
tetra-atomic phenol 四元酚
tetra-atomic ring 四节环
tetra-auricupride 金铜矿
tetraazaphenalene 四氮杂萘并苯
tetrabarium aluminoferrite 铁铝酸四钡
tetrabasic acid 四元酸;四价酸
tetrabasic alcohol 四元醇
tetrabasic carboxylic acid 四羧酸
tetrabasic ester 四羧酸酯
tetrabenazine 丁苯那嗪
tetrabenzo-porphyrazine 四苯并紫菜嗪;四苯并四氮杂卟吩
tetraborane 四硼烷
tetraborate 四硼酸盐
tetraboric acid 四硼酸
tetraboron decahydride 十氢化四硼
tetrabromobisphenol 四溴双酚
tetrabromoethane 四溴乙烷
tetrabromofluore scein 四溴荧光素;曙光红
tetrabromophthalic anhydride 四溴酞酸酐;四溴(代)邻苯二甲酸酐
tetrabutylammonium iodide 碘化四丁铵
tetracaine 丁卡因;阿美素卡因
tetracalcium alumina cement 矾土水泥;四钙高铝水泥
tetracalcium alumina ferrite 铁铝酸四钙
tetracalcium aluminate hydrate 水化铝酸四钙
tetracalcium aluminoferrite 铁铝酸四钙
tetracalcium phosphate 磷酸四钙
tetracalcium trialuminate hydrate 水化三铝酸四钙
tetracene 丁省;并四苯
tetrachloride 四氯化(合)物
tetrachloro 四氯
tetrachloro aniline 四氯苯胺
tetrachloro benzene 四氯苯
tetrachlorobiphenyl 四氯联苯
tetrachloro compound 四氯化合物
tetrachloroethane 四氯乙烷
tetrachloroethylene 四氯乙烯
tetrachloroheptane 四氯庚烷
tetra chloroisoindolinone 四氯异吲哚啉酮
tetrachloromethane 四氯甲烷;四氯化碳
tetrachlorophenol 四氯苯酚;四氯酚
tetrachlorophthalic acid 四氯酞酸
tetrachloro-phthalic anhydride 四氯酞酸酐;四氯

(代)邻苯二甲酸酐
tetrachloro silicane 四氯硅烷
tetrachoric 四项的
tetracid 四酸
tetracid base 四酸价碱
tetracontane 四十烷
tetracosandienoic acid 二十四碳二烯酸
tetracosandioic acid 二十四烷二酸
tetracosane 二十四烷
tetracosane diacid 二十四烷二酸
tetracosanic acid 二十四酸
tetracosanoic acid 木焦油酸
tetracosendioic acid 二十四碳烯二酸
tetracosene diacid 二十四碳烯二酸
tetracosenic acid 二十四碳烯酸
tetracosenoic acid 二十四碳烯酸
tetracosyl 二十四基
tetracyclic coordinate 四圆坐标
tetra-cyclone 四级旋风分离器
tetrad 四元基;四位二进制;四个一组;四个一套;四分体;四次对称晶
tetrade canoic acid 肉豆蔻酸
tetrad effect 四素组效应
tetrad grouping 四素组分组
tetradic 四阶张量
tetrad segregation type 四分型
tetradymie 辉碲铋矿
tetraethylammonium 四乙胺
tetraethylammonium chloride 氯化四乙胺
tetraethylammonium iodide 碘化四乙胺
tetraethylene glycol 四甘醇
tetraethylene pentamine 四亚乙基五胺
tetraethyl lead 四乙铅;四甲铅;四工铅
tetraethyl lead blending of gasoline 汽油掺四乙基铅
tetraethyl lead poisoning 四乙铅中毒
tetraethyl lead pollution 四乙铅污染
tetraethyl orthosilicate 原硅酸四乙酯
tetraethyl propane-tetracarboxylate 丙四羧酸四乙酯
tetraethyl pyrophosphate 焦磷酸四乙酯
tetraethyl radio lead 四乙基放射铅
tetraferroplatinum 铁铂矿
tetrafluoethylene 四氟乙烯
tetrafluoethylene resin 聚四氟乙烯树脂
tetrafluoroethylene resin 四氟乙烯树脂
tetrafluoromethane 四氟化甲烷
tetrafurfuryl acrylate 四氢糠醇丙烯酸酯
tetragon 四角形;正方晶系;四重轴;四边形
tetragonal 正方形;正方晶(格);正方的;四方(的);四边形的
tetragonal cell 正方形单元
tetragonal crystal 四方晶体
tetragonal crystal system 四角晶系
tetragonal derrick 四脚钻塔
tetragonal dipyramid 四方双锥
tetragonal disphenoid 四方四面体
tetragonality 四方性
tetragonal lattice 四方晶格
tetragonal martensite 正方马氏体
tetragonal prism 正方柱;正方棱柱体;四角棱镜;四方柱
tetragonal pyramid 正方锥;四方(单)锥
tetragonal sphenoid 正方楔
tetragonal structure 四方晶体结构;四方晶结构
tetragonal system 正方晶系;四方晶系
tetragonal tetrad 四边形四分体
tetragonal trapezohedron 四方偏方(三八)面体;三方偏方面体
tetragonal trisoctahedron 四角三八面体
tetragonal zirconia polycrystal 多晶氧化锆
tetragonal zirconia polycrystal ceramics 四方氧化锆多晶陶瓷
tetrahalide 四卤化物
tetrahedral 四面体的
tetrahedral angle 四面(体)角
tetrahedral anvil 四面加压式砧模;四面顶锤
tetrahedral anvil ultra-high pressure and high temperature apparatus with tie rods 拉杆式四砧超高压高温设备
tetrahedral arrangement 四面体排列
tetrahedral block 四面体块
tetrahedral bonding structure 四面体键结构
tetrahedral cluster 四面体簇;四面体丛

tetrahedral configuration 四面体构型
tetrahedral congruence 四面线汇
tetrahedral coordinates 四面坐标
tetrahedral coordinate system 四面体坐标系
tetrahedral coordination 四配位数;四面体配位
tetrahedral defect 四面体缺陷
tetrahedral element 四面体单元
tetrahedral element characteristic 四面体单元体的特征
tetrahedral formula 四面体式
tetrahedral function 四面函数
tetrahedral garnet 四光榴石
tetrahedral group 四面体群
tetrahedral hybrid orbital function 四面体混合轨函数
tetrahedral interstice 四面体间隙
tetrahedral layer 四面体层
tetrahedral packing 四方体堆置
tetrahedral pore 四面体孔
tetrahedral research satellite 四面体研究卫星
tetrahedral sheet 四面体(晶)片
tetrahedral symmetric curve 四面对称曲线
tetrahedral symmetry 四面体对称性
tetrahedral theory 四面体假说
tetrahedral toolmaker's straight edge 四棱平尺
tetrahedral winding 四面体绕组
tetrahedrite 黝铜矿
tetrahedron 四脚锥体;四面体
tetrahedron block 四面体块体;(空心的)四面块体
tetrahedron block revetment 四面块(体)护岸
tetrahedron group 四面体群
tetrahedry 四分对称
tetrahexahedron 四六面体;二十四面体
tetrahydrate 四水合物
tetrahydric acid 四元酸
tetrahydric phenol 四元酚
tetrahydric salt 四酸式盐
tetrahydrofarfuryl acrylate 丙烯酸氢糠酯
tetrahydrofuran 氧杂环戊烷;四氢呋喃;丁撑氧
tetrahydrofurfuryl alcohol 四氢糠醇;四氢化呋喃甲醇
tetrahydronaphthalene 四氢化萘;四氢化萘
tetrahydrophthalic acid 四氢化邻苯二甲酸
tetrahydrophthalic anhydride 四氢化邻苯二甲酸酐
tetrahydrotrimethypicene 四氢化三甲基匹
tetraiodofluorescein 赤藓红
tetrakalsilite 四型钾霞石
tetrakis(ethoxymethyl)benzoguanamine 四乙氧甲替苯代三聚氰二胺
tetrakishexahedron 四六面体
tetrakishydroxymethyl phosphonium chloride 四羟甲基氯化锑
tetrakis(methoxymethyl)benzoguanamine 四甲氧甲替苯代三聚氰二胺
tetralite 特屈儿(一种烈性炸药)
tetramer 四聚物;四聚体
tetramerous 四重;四部分
tetramethrin 胺菊酯
tetramethyl ammonium 四甲铵
tetramethyl ammonium azide 叠氮化四甲基铵
tetramethylene 环丁烷
tetramethylene diamine 四亚甲基二胺
tetramethylene glycol 丁二醇
tetramethylol benzoguanamine 四羟甲替苯代三聚氰二胺
tetramethylol cyclohexanol 四羟甲基环乙醇
tetramethylolmethane 季戊四醇
tetramethyl orthosilicate 硅酸四酯
tetrammine 四氨络物
tetramorphism 四晶现象
tetramorphous 四晶形;四晶的
tetranatrolite 四方钠沸石
tetra-n-butoxy titanium 四正可一氧基钛
tetranitromethane 四硝基甲烷
tetra-pack 四角包
tetra paper 臭氧试纸
tetraphene 丁苯
tetraphenylboron sodium 四苯硼(酸)钠
tetraphenyl ethylene 四苯乙烯
tetraphenylprophin complex of mercury 汞的四苯基卟吩络合物
tetraphernl methane 四苯基甲烷
tetraphyllous 四叶的
tetraploid 四倍体;四倍的

tetrapod 钢筋混凝土四脚锥体;四脚锥体(防波堤用);四角支座;四角护岸块;四角防波石
tetrapod (hollow) block 四脚空心方块
tetrapod mo(u)ld 四脚锥体浇铸模
tetrapod wall 四脚锥体堆墙;四脚锥体堤
tetrapody 四标准英尺的单位
tetrapolar 四端
tetrapolarity 四极性
tetrapropylene-benzene sulfonate 四丙烯苯磺酸盐
tetrapylon 古建筑十字交叉通道的凯旋门
tetraquetrous 四棱角的
tetraselenodont 四月牙形的
tetrasilicane 十氢化四硅
tetrasilicon decahydride 十氢化四硅
tetrasome 四体
tetrasomy 四体性
tetrastoon 回廊庭院;四面有门廊或柱廊的宫院;[复]tetrastoa
tetrastrotium aluminoferrite 铁铝酸四锶
tetrastyle 建筑物正面有四根柱子的;四柱式门廊
tetrastyle colonnade 四柱式柱廊
tetrastyle temple 四圆柱式庙宇
tetrastylic portico 四柱式门廊
tetrastylos 建筑物正面有四根柱子的
tetrataenite 四方镍纹石
tetrathionate 连四硫酸盐
tetrathylammnonium hydroxide 氢氧化四乙铵
tetratogenesis 畸形发生
tetratohedral crystal 四分面晶体
tetratolite 密高岭土
tetrawickmanite 四方羟锡锰石
tetraxon 四轴针
tetrazo compound 双偶氮化合物
tetrdymite 碲铋矿
tetrode 四极真空管
tetrode transistor 四极晶体管
tetrodon poisoning 河豚中毒
tetrodotoxin 河豚毒素
tetrol 呋喃
tetroxide 四氧化物
tetryl 特屈儿(一种烈性炸药)
tetryl-azide detonator 特屈儿一重氮化合物引爆剂
tetrytol 特屈儿混合炸药
tewel 通风孔;烟筒
tex 软质纤维板
texalite 水镁石
Texas fever 德克萨斯热
Texas pink granite 德克萨斯州浅粉红色花岗岩(美国)
Texas red 德克萨斯红
Texas tower 雷达台;德克萨斯式塔
Texas triaxial design method 德克萨斯三轴设计法
Texas triaxial method 美国德克萨斯州三轴试验法
tex finishing 拉毛涂装法
texibond 聚乙酸乙烯酯类黏合剂
texitinite 结构木质体
texrope belt 三角皮带
texrope drive 三角皮带传动
text alignment 正文校准
text clipping 正文裁剪
text colo(u)r index 正文颜色索引
text creation 文本创建
text display 文本显示
text extent 正文范围
text figure 插图
text file 文本文件
text font 文本字体
text formatting program(me) 文本格式程序
textile 织物原料;纺织物
textile abrasion tester 织物耐磨试验仪
textile analysing glass 织物分析镜
textile belt 布皮带
textile block 花纹砌块
textile block style 织物纹样砌块风格
textile board 织板
textile carpet 织物地毯;机织地毯
textile ceramics 纺织用陶瓷
textile coating 织物涂布
textile college 纺织(工)学院
textile cone 织纹锥螺
textile department store 纺织品百货商店
textile design 织物设计
textile desizing wastewater 纺织脱浆废水
textile dressing 织物整理

textile dust 纺织物粉尘
textile dye 纺织染料
textile dyehouse wastewater 纺织印染厂废水
textile dy(e)ing 纺织品染色
textile dy(e)ing and finishing industry 纺织染整工业
textile dy(e)ing and finishing plant 纺织染整厂
textile dy(e)ing mill 纺织品染色厂
textile dyestuff 纺织品染料;纺织着色剂
textile dye wastewater 纺织染液废水
textile effluent 纺织流出水;纺织出水;纺织厂废水
textile engineering institute 纺织(工)学院
textile factory 纺织厂
textile fiber 纺织纤维(制品)
textile filtering medium 织物滤材
textile finisher 纺织加工厂
textile finishing 纤维板贴面;织物整饰
textile glass 纺织玻璃
textile glass fabric 玻璃纤维织物;玻璃丝布
textile glass fiber 纺织玻璃纤维
textile glass mat 玻璃纤维毡;玻璃丝毯
textile glass prepreg 纺织玻璃预浸料
textile glass staple fibre 玻璃短纤维
textile glass staple fibre product 定长玻璃纤维织品
textile glass tube 纺织玻璃管
textile-grade talc 纺织级滑石
textile hydrostatic pressure tester 织物耐水压试验仪
textile industry 纺织工业
textile industry wastewater 纺织工业废水
textile insulated cable 编织绝缘电缆
textile lacquer 织物用漆;纺织品用漆
textile machinery industry 纺织机械制造业
textile material 织物材料
textile mill 纺织厂
textile mill waste(water) 棉纺厂废水;纺织工业废水;纺织厂废水
textile-printing 织物印花
textile reinforcement 织物加固
textile screen cloth 网布
textile wastewater 纺织废水
textile wastewater effluent 纺织废水处理后出水
textile wastewater reuse 纺织废水回用
textile wastewater treatment 纺织废水处理
textile wrap(ping) 纺织品包裹材料
text index 正文索引
textinite 木质结构体
text library 目标程序库
text link 正文连接
text of a procedure 过程的正文
textolite 织物酚醛塑胶;胶布板;夹布胶木;层压胶布板;酚醛塑料布
textolite bearing 夹布胶木轴承
textolite gear 夹布胶木齿轮
text-oriented data base 面向原文的数据库
texto-ulminite 结构腐木质体;木质结构腐木质体
text paper 高级印刷纸
text path 正文路径
text processing 正文处理
text representation 正文表示
Textron Inc. 特克斯特隆公司(美国)
text segment 正文段
textual coating 织物涂层
textual error 原文错误;文字上的错误
textual statement 报表正本
textural adjustment of technologic(al) and e-conomics 调整技术经济结构
textural anisotropy 结构异向性;结构非均质性
textural anomaly 结构异常
textural behavio(u)r 织构行为
textural classification 组分分类;结构分类
textural classification of soil 土壤结构分类;土的质地分类
textural component of sedimentary rocks 沉积岩结构组分
textural criticism 考证
textural of geologic(al) working adjustment 调整工作结构
textural research 考证
textural soil classification 土组织分类;土壤结构分类
textural variation of orebody 矿体结构变化
texture 影纹;组织;质地;织物;织构,结构;纹理(木材岩石等的);网纹
texture analysis 纹理分析
texture and structure of ores 矿石结构(与)构造
texture anisotropy 织构
texture brick 粗纹砖;粗面砖
texture classes 质地分级
texture coating 凸纹涂料;凸纹厚浆涂料;浮雕漆
textured 有装饰性条纹的(粉刷);起纹理
textured board 有纹理的板;纹理板
textured brick 粗纹砖;织纹面砖;花纹砖;网纹砖
textured brick slab 花纹砖板
textured coarse concrete 表面粗纹理骨料;表面粗纹理集料;粗纹理骨料;粗纹理集料
texture depth 纹理深度
texture developed by hot working 织构化热加工
textured exposed concrete 花纹装饰混凝土
textured fabric 花式织物
textured fair-faced concrete 花纹饰面混凝土
textured fiber 变形纤维
textured filament 变形丝
textured finish 织纹饰面;织物状饰面;织纹状饰面;花纹饰面;花纹抹面
texture discrimination 纹理判别
textured material 织构材料
textured metal sheet 起纹金属板
textured paint 起纹理的涂料;凸纹漆;粗质地涂料
textured pattern 纹理图案
textured paving 起纹道面
textured pile 花式绒头(指地毯)
textured plywood 肌纹胶合板(具有不同机械加工表面);结构胶合板
textured surface 变形表面
textured tester 变形丝试验仪
textured tile 花纹砖
textured wall(ing) surface 花纹墙面
textured yarn 花色纱;结构纱;膨松纱;变形丝;变形纱(线)
textured yarn fabric 变形丝织物
textured yarn goniometer 织构测角计
texture effect 质地感
texture enhancement 纹理增强
texture-finished paint 凸纹漆;粗质地涂料;纹理饰面涂料
texture fundamental element of soil 土结构的基本单元
texture grade 质地等级;结构等级
texture grade of soil 土结构等级
texture gradient 纹理梯度
texture grading 质地分类
texture index 质地指标
texture index of asbestos 石棉织构指数
texture map indices 纹理图标志
texture maturity 结构成熟度
texture meter 结构深度仪
texture of argillaceous rocks 泥质岩结构
texture of asbestos 石棉织构
texture of cements 胶结物结构
texture of clastic grains 碎屑颗粒结构
texture of domainal cleavage fabric 域组构结构
texture of drainage network 河网疏密度;水文网疏密度
texture of fault rock 断层岩结构
texture of igneous rocks 火成岩岩石结构
texture of kerogen 干酪根结构
texture of metamorphic rocks 变质岩岩石结构
texture of microlithon 微劈石结构
texture of ores 矿石组成
texture of sedimentary rocks 沉积岩岩石结构
texture of soil 土壤组成;土壤质地;土壤结构
texture of wood 木材结构
texture paint 起纹涂料;起纹漆;纹理漆
texture (pebble) finish 凸纹涂料
texture profile 质地剖面
texture separate 质地分级
texture split 沿纹理劈裂的
texture split shakes panel 纹理裂缝屋面板
texture statistics 纹理统计
texture structure 纹理结构
texture synthesis 纹理合成
texture technique 织构技术;结构化技术
texture-tone analysis 结构-色调分析
texture type of soil 土结构类型
texturing 塑制纹理;抹面法;拉毛;花纹(混凝土表面的);织构化;花纹抹面

texturing effect 变形效应
texturing twist 变形捻度
texturing variables 变形变数
texturity 不缩标签
texturization 刻花;纹饰
tête de pont 桥头堡
T-filled weld 丁字形角焊
T-fillet weld 丁字接头角焊缝
T-fillet welding T 形接头贴角焊
T-fitting 三通;T 形支管
T-fixture 丁字形夹具
T-flip-flop 翻转触触发器
T-ga(u)ge T 型规划线
T-girder 丁字梁;丁字大梁;T 形大梁
T-grade separation T 形立体交叉;丁字形立交
T-groin 丁字形丁坝
thadeuite 磷镁钙石
Thai architecture 泰国建筑
Thai Imperial Palace 泰国皇宫(建筑)
Thailand-Vietnam-Burma old land 泰越缅古陆
Thai style 泰国式
thalamus 内室(古希腊建筑的)
thalasometer 验潮器
Thalassal 萨拉萨尔铝合金
thalassic 海域的;海洋的
thalassic deposit 深海沉积
thalassic rock 深海岩(石)
thalassium (of the oceans) 海水群落
thalassochemistry 海洋化学
thalassocracy 制海权
thalassocratic 海洋盛期的
thalassocratic period 最大海侵阶段
thalassocratic sea level 最大海侵面
thalassocraton 海洋克拉通【地】;大洋克拉通
thalassogenic 造海的;成海的
thalassogenic movement 造海运动;海底运动
thalassogenic sedimentation 海底沉积(作用)
thalassogensis 造海作用
thalassogeosyncline 深海地槽
thalassography 海洋学
thalassology 海洋学
thalassophile element 海水元素;海生元素;亲海元素
thalassophobia 海洋恐怖
thalassophyte 海生植物
thalassoplankton 海洋浮游生物
thalassostatic period 海面稳定期
thalenite 红钇石
thalfenisite 硫镍铁铊矿
thalidomide 反应亭
thallic 含正铊的
thallic bromide 三溴化铊
thallic fluoride 三氟化铊
thallic iodide 三碘化铊
thallic sulfate 硫酸铊
thallic trifulfide 三硫化二铊
thallium acetate 乙酸铊
thallium-beam clock 铊束流钟
thallium bromide 溴化铊
thallium chloride 氯化铊
thallium compound 铊化合物
thallium deposit 铊矿床
thallium detector 铊探测器
thallium glass 铊玻璃
thallium hydroxide 氢氧化铊
thallium iodide 碘化铊
thallium monochloride 一氯化铊
thallium monoxide 一氧化铊
thallium ore 铊矿
thallium oxide 一氧化铊;氧化铊
thallium ozone test paper 铊臭氧试纸
thallium poisoning 铊中毒
thallium pollution 铊污染
thallium tribromide 三溴化铊
thallium trifluoride 三氟化铊
thallium triiodide 三碘化铊
thallium trioxide 三氧化二铊
thallium yellow 铬酸铊;铊黄
thallofide cell 铊氧光电管
thallophyte 原植体植物;叶状植物
thallose 叶状植物
thallotoxicosis 铊中毒
thallous bromide 一溴化铊
thallous carbonate 碳酸亚铊

thallous chlorate 氯酸亚铊
thallous chloride 一氯化铊
thallous fluoride 一氟化铊
thallous hydroxide 氢氧化铊
thallous iodide 一碘化铊;碘化(亚)铊
thallous malonate 丙二酸铊
thallous oxide 一氧化铊;一氧化二铊
thallous phosphate 磷酸亚铊
thallous pyrovanadate 焦钒酸亚铊
thallous sulfate 硫酸亚铊
thallous sulfide 一硫化二铊
thallous sulphate 硫酸亚铊
thallus 菌体
thalposis 温感
thalweg 沿河谷渗流;最深谷底线【地】;谷(底)线;河流谷底线;河谷线;河道中泓线;海谷深泓线;海谷底线;深泓线;山谷或海底谷最底部联线
thanalocoenosis 生物尸积群
T-handle T字手柄形;丁字(形)把子;T形操纵杆;丁字形手把
T-handle pentagonal fixed socket wrench 丁字柄五角固定套筒扳手
T-handle square fixed socket wrench 丁字柄四方固定套筒扳手
T-handle swinging socket wrench 丁字柄活动套筒扳手
T-handle taper reamer 丁字柄锥形铰刀
Thanite 森乃特(硫氰乙酸异茨酯)
thank-you-madam 小脊或凹坑(俚语);(道路上的)浅沟
tharandite 镁铁白云石
thatch 用茅草盖屋顶;茅草;草屋顶
thatchboard 胶合茅草板;稻草板;草秸板
thatched cottage 茅屋顶村舍;茅屋
thatched pavilion 茅亭
thatched roof 茅草顶;草屋顶
thatched shack 茅草棚;草棚
thatched shed 茅草棚
thatcher 盖屋顶工(用稻草等);茅屋顶铺设工
thatcher's labo(u)rer 茅屋顶铺设辅助工
thatcher's server 茅屋顶铺设辅助工
thatch fiber 滑秸
thatching 盖茅屋顶;铺茅草顶
thatching reed 盖屋顶芦苇
thatching straw 盖屋顶稻草
thatch-making machine 编席机
thatch roof(ing) 茅草屋顶
thatchy 茅屋顶的
thaumasite 硅灰石膏
thaumatropy 结构转化
thaw 开冻,解冻
thaw bowl 融化盘
thaw bulb 融泡
thaw cave 融洞
thaw collapse 融陷;融沉
thaw collapsibility 融陷性;融化塌陷
thaw compression 融化压缩
thaw compression factor 融化压缩因子;融化压缩因数
thaw compression test 融化压缩试验
thaw consolidation 融化固结
thaw-consolidation test 融化固结试验
thaw depression 融冻洼地
thaw depth 融化深度
thawed condition 融化状态
thawed consolidation 融固结作用
thawed lake 解冻湖
thawed layer 融化层;融冻层
thawed reach 解冻河段
thawed soil 解冻土;融土
thawed zone 融区
thaw flowing 融流动
thaw frozen landslide 热熔滑坍;热熔滑坡;热熔滑动;热融滑坡
thaw house 融化间
thawing 融解;融化
thawing and freezing test 融冻试验
thawing apparatus 融冰装置;融冰器
thawing grooves 融沟
thawing hole 融解洞
thawing ice 融冰
thawing index 融化指数
thawing layer 融冻层
thawing pipe 融冻管

thawing pit 化冻坑
thawing season 解冻季节;融化季节
thawing shed 解冻库;化冻棚
thawing surface 融化面
thawing tank 融冻桶
thawing water 融水;融解水;融化水
thawing water irrigation 融水灌溉
thawing weather 解冻天气;冰雪融化天气
thaw lake 融陷湖;融沉湖;融沉湖
thaw line 融线
thaw of river 河流解冻
thaw of snow 融雪
thaw out 融化
thaw period 融雪期;融化期
thaw pipe 融化解冻管
thaw point 融点
thaw season 融季
thaw settlement 融沉降
thaw settlement basin 融冻沉陷盆地
thaw settlement depression 融冻沉陷洼地
thaw settlement funnel 融冻沉陷漏斗
thaw sink 融坑
thaw stable permafrost 融化稳定永冻土
thaw subsidence 融化下沉;融沉
thaw subsidence factor 融陷因子;融沉因数;融沉因数
thaw temperature 融解温度
thaw water 融化水;解冻水
thaw weather 融化天气
the ABC of architectures 建筑学入门
the absorption of technology 技术吸收
the acid content of soil 土壤酸含量
the actual site of action 实际作用部位
T-head 丁字头;丁字形柱头;丁字形支撑柱
the adaxial suface 近轴面上
T-head bolt T形槽用螺栓;大头螺栓;丁字头螺栓
T-headed T形头的
T-headed buttress dam 扶壁式坝;丁字形支墩坝
T-headed jetty 丁字形突堤码头;丁字形防波堤
T-headed spoke 丁字形辐
T-head pier 引桥式码头;丁字形码头;丁字码头;丁形首部码头;T形码头
T-head pier or L-shape pier T形码头或 L 形突码头
T-head post 丁字形柱
theaism 茶中毒
the allocating factor of service track 整备线配置系数
The American Institute of Planners 美国规划工作者协会
the American society for testing material cards ASTM 卡片
The American Society of Heating, Refrigerating and Air Conditioning Engineers 美国采暖、制冷与空调工程师学会
The American Society of Heating and Ventilation Engineers 美国采暖与通风工程师学会
the amount melted 熔化量
the anchor holds 锚吃得住
the ancient city of wiang Ta kan 万泰堪古城(泰国)
The Antique 古文物(泛指古希腊、古罗马遗物)
The Architects Collaboratives 建筑师协作组织(1964年格罗庇乌斯组织的)
the area of liquid phase 液相面积;气相面积
the arrangement of tubes 管子排列
The Asphalt Institute Pavement Design Method 地沥青协会路面设计法
theaspirone 茶香螺酮
the assured 被保险人
theater 剧场
theater air material area 战区空军物资区
theater depot 战区仓库
theater dimmer 剧场调光器
theater headquarters 战区司令部
theater-in-the-round (舞台设在公众座中央的)环座剧场
theater navy headquarters 战区海军司令部
theater of operations 战区
theater of war 战区
theater operational facility 战区作战设施
theater seating 剧场座位
theater war reserves 战区战备物资
theatre = theater

theatre architecture 电影院建筑;剧院建筑
theatre auditorium 电影院观众席;剧院观众席
theatre block 电影院大楼;剧院大楼
theatre building 电影院建筑;剧院建筑
theatre construction 电影院建筑;剧院建筑
theatre dimmer 电影院调光器;剧场调光器;具有调光器
theatre for three-dimensional film 立体电影院
theatre glass 观剧望远镜
theatre heating 电影院采暖;剧院采暖
theatre-in-the-round 露天圆剧场;(舞台设在公众座中央的)环座剧场;中心舞台剧场
Theatre of Marcellus at Rome 罗马的麦西剧院
theatre restaurant 有演出节目的餐馆
theatre seating 观众席
theatre seats 剧院座位
theatre sight lines 剧场视线
theatrical 舞台表演艺术;戏剧演出
theatrical building 观演建筑
theatrical designer 剧院设计师
theatrical gauze 剧院布景用纱网
theatrics 舞台效果;舞台表演艺术;戏剧演出
the authorities 当局
the authorities concerned 有关当局;主管机关
the average length of railway vehicles 车辆平均长度【铁】
the baker 百业之人
the basic policies of China's environmental protection 中国环保基本政策
the basket maker 打篮户
the battle of the brands 厂牌战
the beginning of a year 年头
the best demagnetizing field 最佳退磁场
the best demagnetizing temperature 最佳退磁温度
the best growing condition 最佳生长条件
the best varieties for 年优良品种
the bid and tender law 招标投标法
The Boeing Company 波音公司(美国)
the boss 主管
the breaching party 违约方
the butcher 百业之人
theca 珊瑚壳;珊瑚壁
the candlestick maker 百业之人
the capital obtained from insurance of government bonds 国债资金
the central and western regions 中西部地区
the central line of foundation trench 基槽中心线
The Central Office for International Rail Transport 国际铁路运输中央事务局;国际铁路运输中心事务所
the character of radiation 放射性质
the chemistry of submerged soils 水田土壤化学
the chief 主管
the choice of flow direction 流向选择
the classification of voltage 电压等级
the coldest month 最冷月份
the cold months of the year 十冬腊月
the committee on technical cooperation 技术合作委员会
the common-lead dating method 普通铅测年法
the competent authority 主管机关
the complete sets of equipment 成套装置
the construction of housing 居民住宅建设
the content of soil 土壤的成分
The Corporation of Lloyd 劳氏公司
the correct path 正路
the cultivation of trees 栽培树木
the Cultural Palace of the Nationalities 民族文化宫
the current pattern of spending 现行支出机构
the degree of processing and the conversion rate of agriculture products 农产品加工深度和转化率
the department concerned 有关部门
the determination of geologic(al) ages 地质年龄测量方法
The Dock, harbo(u)r Authority 船坞与码头管理(期刊)
the downtown district 繁华地区
the earth day 地球日
the economic growth rate 经济发展速度
the edge of a knife 刀刃
the effect of earth tides 固体潮影响

the effect of intermediate layer 中间层影响
the effect on gravity field 影响重力场因素
the English Channel 英吉利海峡
the enhanced integrated drive electronics 增强型集成传动电子线路
the entire profile 整个剖面
the equipment support drawing 设备支架图
the erosion caused by sloping surfaces 坡面侵蚀
the Estimates 财政收支概算(英国)
theetsee 马来黑漆
the expiration of the period for exemption 免税期满
the extent of actual control 实际控制的范围
the facade of a shop 门面
The Federation of Civil Engineering Contractors 土木工程承包商联合会
the fifth grade of minerogenetic prospect 五级成矿远景区
the fifth grade of oil-gas prospect 五级油气远景区
the fifth grade pf coal-forming prospect 五级成煤远景区
the fifth trophic level 第五营养级
the final solution to pest control 最终解决防治问题
the first and the fifteenth day of the lunar month 朔望
the first front 头场霜
the first generational pesticide 第一代农药
the first grade of minerogenetic prospect 一级成矿远景区
the first grade of oil-gas prospect 一级油气远景区
the first grade pf coal-forming prospect 一级成煤远景区
the first party 甲方
the first position of Gauss 高斯第一位置
the first trophic level 第一营养级
the Forbidden City 紫禁城
the forested in the humid temperate zones 湿润温带森林地区
the forestry and grass coverage 林草植被率
the forming of head 封头的成型
the fourth grade of coall-forming prospect 四级成煤远景区
the fourth grade of minerogenetic prospect 四级成矿远景区
the fourth grade of oill-gas prospect 四级油气远景区
the fourth trophic level 第四营养级
the framework of theory on site-dimensions 规模框架理论
theft and pilferage 偷窃
theft-prone 偷窃倾向的;扒窃倾向的
the functional basis for land utilization 土地利用的主要根据
the general price level 消费价格水平
the general table of soil classification in highway tunnels 公路规范中土分类总表
the geologic(al) retrieval and synopsis program(me) 地质检索和摘要程序
The Gorges Dam 三峡大坝
the Gothic style of architecture 哥特式建筑
the government procurement law 政府采购法
the grays 灰色派
the greatest step discharge rate 最大分级排水速率
the Great Lake 大西洋
the Great lakes 北美洲五大湖
The Great Mosque 大清真寺
the Great Wall 万里长城
the height of the transverse metacenter above the center of buoyancy 稳心在浮心上的高度
the highest demagnetizing temperature 最高退磁温度
the highest effective frequency 最高有效频率
the highest possible speed 可能最高车速
The Houses Parliament 国会大厦(美国)
the humus content of the soil 土壤腐殖质含量
Theim formula 齐姆公式
the inadequately employed 不完全就业人员
the income and expense report of transport revenue 运输收入进款收支报表
the income and profit ratio 收入利润率
the inflow way of water in gallery 坑道进水方式
the insured 被背书人
the Interim Regulations on the Fuel Oil Tax 燃油税暂行条例
The International Tanker Owners Pollution Federation Ltd. 国际油轮船东污染联合会
the International Year of the Quiet Sun 国际宁静太阳年
Theisen disintegrator 机械洗涤器
Theisen gas cleaner 动力除尘器;机械洗涤器
Theisen gas scrubber 机械洗涤器;动力除尘器
Theis formula 泰斯公式
theisite 西锑砷铜铅石
Theis well function 泰斯井函数
the K-Ar dating method 钾氩测年法
the K-Ca dating method 钾—钙测年法
the key technical indices 主要技术指标
the kinds of measuring geologic(al) ages 年龄测量方法种类
the labo(u)r expense of freight transport 货运人员工资
the labo(u)r expense of passenger transport 客运人员工资
the largest particles 最大颗粒
the largest source 最大来源
the law of constant heat summation 赫斯定律
the law of equivalent proportions 当量比例定律
the law permit 法律许可
the least square smoothing 最小二乘圆滑
Thelen evapo(u)rator 锡伦蒸发器
the length of distribution channel 销售渠道长度
Thelen pan 锡伦盘
the lever of finance and taxation 财政税收杠杆
the lightest element 最轻的元素
the limits to growth 增长的极限
the loose mass of the earth's crust 一层松散物质
the loose surface 疏松表面
the lowest available zinc 最少有效锌
the lowest price 最低价(格)
the Lu-Hf dating method 镥-铪测年法
thelweg 国界线水道的主航道中央线
the macerals classification of brown coals and lignites 国际褐煤显微组分分类
the macerals classification of hard coal international coal petrology committee 国际烟煤和无烟煤显微组分分类
the main design figures 主要设计指标
the mass spectrometers for isotope analysis 同位素分析质谱仪
thematic atlas 专题地图集
thematic cartography 专题地图制图学
thematic chart 专题海图
thematic data 专题地图资料
thematic information 专题地图资料
thematic interpretation of radar imagery 雷达图像主题判读
thematic land-use map 土地利用专题图
thematic map 专题(地)图;专门图;特种地图
thematic map of synoptic(al) representation 概括表示法专题地图
thematic map of synthetic(al) representation 合成表示法专题地图
thematic mapper 专题制图仪;专题成像仪
thematic mapping 专题制图;主题判图
thematic marine map 专题海图
the means of relay protection 继电保护方式
theme extraction 专题要素抽出
theme files 专题文件
theme processing 专题处理
the method for identification of minerals and rocks 岩矿鉴定方法
the method of Lugeon test 压水试验方法
the methods and instruments for isotope analysis 同位素分析方法及仪器设备
the microlithotypes of coal international coal committee petrology 国际煤岩委员会分类
the middle ages 中世纪
the middle of ship 舯
the middle of spring 仲春
the minimum-possible saturation concentration 最小可能饱和浓度
the mining plan did not ratify 开采设计采批准
The Ministry of Water Resources of CHina 中国水利部
themionic ionization ga(u)ge 电离真空规
themite welding 热焊(接)
the mode of action between grain and water 土粒与水的作用方式
the month of passenger flow 客流月
the month of regime data 动态资料的月份
the Monument to the People's Heroes 人民英雄纪念碑
themophore 蓄热器
the most conspicuous early sign of attack 受害初期最明显的迹象
the most disadvantageous combination of loads 最不利荷载组合
the most disadvantageous load 最不利荷载
the most favo(u)rable time interval of electromagnetic distance measurement 电磁波测距最佳观测时间段
the most important practical benefit 最重要的实际价值
the most important sources of energy 能量的最重要来源
the most probable velocity of molecular 分子最可几速率
the moving-range chart 移动差距图
thenardite 无水芒硝;天然无水芒硝
Thenard's blue pigment 钴蓝颜料;特纳德蓝
the national budget for this fiscal year 本年度国家预算
the natural disaster relief system 灾害救助制度
the nature's behavio(u)r 天行
the negative unit on negative unit 负地形单元上的负单元
the negative unit on positive unit 正地形单元上的负单元
the number of allocated passenger car 客车配属系数
the number of joint sets 节理组数
the number of pollution factors 污染要素数量
the number of years for a work 工作年限
the number of years set for allowable aquifer dewatering 含水层允许疏干年限
theodolite 经纬仪
theodolite adapter 经纬仪衔接器
theodolite adjustment 经纬仪校正
theodolite camera 摄影经纬仪
theodolite compass 经纬仪罗盘
theodolite goniometer 经纬测角仪
theodolite level 经纬水准仪
theodolite line 经纬仪导线
theodolite-magnetometer 磁偏角经纬仪
theodolite man 经纬仪测量员
theodolite note 经纬仪测量手簿
theodolite pier 经纬仪墩座
theodolite polygon 经纬仪导线
theodolite reading 经纬仪读数
theodolite stadia traverse 经纬仪视距导线
theodolite station 经纬仪测站
theodolite survey(ing) 经纬仪测量
theodolite telescope 经纬仪望远镜
theodolite traverse 经纬仪导线
theodolite traverse point 经纬仪导线点
theodolite traverse station 经纬仪导线站
theodolite triangulation 经纬仪三角测量
theodolite tribrach 经纬仪三角基座
theodolite trigonometric level(l)ing 经纬仪三角高程测量
theodolite with compass 罗盘经纬仪
theodolite with stadia wires 视距(丝式)经纬仪
the old-age insurance system 养老保险制度
theologeion (古剧场为圣职者设的)楼厢
theologic(al) control 流变调节
theologic(al) response 流变响应
theophrastite 施羟镍矿
theophylline ethylenediamine 茶碱乙二胺
theophylline meglumine 茶碱葡胺
theophylline sodium 茶叶碱钠
theophylline sodium glycinate 茶碱甘氨酸钠
the optimum isochron 最佳时线
the optimum of application 施用的最适宜时间
the optimum slope 最佳斜率
theorem of area moment 面积距定理
theorem of Bernoulli 伯努利定理
theorem of consolidation 固结定理
theorem of corresponding states 对应态定理
theorem of double ratio 复比定理
theorem of five moments 五弯矩定理;五力矩定理
theorem of four moments 四力矩定理;四弯矩定理

theorem of image sound 镜像原理
theorem of impulse 冲量定理
theorem of joint rotation 结点转动定理
theorem of Lagrange's mean 拉格朗日中值定理
theorem of least work 最小功定理;最小功能定理
theorem of mean 平均值定理;中值定理
theorem of minimum energy 最小能量定理
theorem of minimum potential energy 最小位能定理
theorem of polyhedra 多面体定理
theorem of reciprocal deflection 位移互等定理
theorem of reduction 简化定理;还原定理
theorem of superposition 叠加定理;叠加原理
theorem of the mean 中值定理
theorem of three-moments 三弯距原理;三弯矩定理;三力矩定理
theorem of virtual displacement 虚位移定理
theorem of virtual work 虚功定理
theorem on addition of probabilities 概率相加定律
theorem on multiplication of probabilities 概率相乘定律
the oremoval of soil material 土壤物质的移动
theorem prover 定理证明程序
theorem proving 定理证明
the ore reserve at percentage 拥有资源所占百分比
theoretic(al) abstraction 理论抽象
theoretic(al) acid consuming 理论耗酸量
theoretic(al) acoustics 理论声学
theoretic(al) air 理论空气(量)
theoretic(al) air for combustion 理论燃烧空气量
theoretic(al) air free density 理论最大容重;理论最大密实度
theoretic(al) amount of air 空气理论量
theoretic(al) amplitude 理论振幅
theoretic(al) analysis 理论分析
theoretic(al) arithmetic 理论计算
theoretic(al) assumption 理论假定
theoretic(al) astronomy 理论天文学
theoretic(al) astrophysics 理论天体物理学
theoretic(al) availability index 理论有效指数
theoretic(al) basis 理论基础
theoretic(al) basis of input-output 投入产出的理论基础
theoretic(al) bathymetrical datum 理论深度基准面
theoretic(al) boundary 理论极限值;理论边界
theoretic(al) calculation 理论计算
theoretic(al) cam curve 理论凸轮曲线
theoretic(al) capacity 理论生产能力;理论生产率;理论生产力;理论容量
theoretic(al) center of frog 辙叉理论中心
theoretic(al) chance 理论机遇
theoretic(al) chemistry 理论化学
theoretic(al) cleaned coal recovery 理论精煤回收率
theoretic(al) coefficient 理论系数
theoretic(al) combustion gas volume 理论燃烧气体体积;燃烧气体理论容量
theoretic(al) combustion temperature 理论燃烧温度
theoretic(al) composition 理论成分
theoretic(al) concept 理论概念
theoretic(al) cost 理论成本
theoretic(al) cost of living 理论生活费
theoretic(al) cost standard 理论成本标准
theoretic(al) crystallography 理论晶体学
theoretic(al) curve 理论曲线
theoretic(al) cutoff frequency 理论截止频率
theoretic(al) damping 理论阻尼
theoretic(al) cut-off point of bar 钢筋理论断点
theoretic(al) deflection coefficient 理论弯沉系数
theoretic(al) density 理论密度
theoretic(al) depreciation 理论折旧
theoretic(al) depth 理论深度
theoretic(al) derivation 理论推导
theoretic(al) design 理论设计
theoretic(al) dilution line 理论稀释线
theoretic(al) dimension 理论量纲;理论尺度
theoretic(al) discharge 理论流量
theoretic(al) distribution 理论分布
theoretic(al) distribution function 理论分布函数
theoretic(al) draft 理论抽力
theoretic(al) dry 绝对干燥(的)

theoretic(al) drying process 理论干燥过程
theoretic(al) duration of a significant interval 有效间隔理论持续时间
theoretic(al) econometrics 理论经济计量学
theoretic(al) elasticity 理论弹性力学
theoretic(al) elevation 理论高程
theoretic(al) environmental science 理论环境学
theoretic(al) equation 理论方程(式);理论公式
theoretic(al) error 理论误差
theoretic(al) error of optic(al) range-finder 光学测距机的理论误差
theoretic(al) flow 理论流量
theoretic(al) flow characteristic 理论水流特征
theoretic(al) flow rate 理论排量
theoretic(al) fluidized velocity 理论流化速度
theoretic(al) fluid mechanics 理论流体力学
theoretic(al) formula 理论公式
theoretic(al) formula in smoothing 修匀的理论公式
theoretic(al) formula of interfering wells 干扰井理论公式
theoretic(al) frequency 理论频数;理论频率
theoretic(al) frequency distribution 理论频率分布
theoretic(al) frog point 辙叉(心轨)理论尖端
theoretic(al) generalization 理论概括
theoretic(al) geodesy 理论大测量学
theoretic(al) glass composition 理论玻璃组成
theoretic(al) graphics 理论图学
theoretic(al) gravity formula 理论重力公式
theoretic(al) head 理论扬程;理论压头
theoretic(al) heat 理论热量
theoretic(al) heat consumption for glass melting 玻璃熔化理论热耗
theoretic(al) heat requirement 理论耗热量
theoretic(al) heel length 辙叉跟理论长度
theoretic(al) height 理论高度
theoretic(al) height of fall 瀑布的理论高度
theoretic(al) horsepower 理论功率
theoretic(al) ideal wave 理想波;理论(的)理想波
theoretic(al) indicator card 理论指示图;理论示功图
theoretic(al) lead 理论导程
theoretic(al) lead of turnout 道岔理论导程
theoretic(al) leasing price 理论出租价格
theoretic(al) length of lead rails 道岔理论导程
theoretic(al) length of turnout rails 道岔理论长度
theoretic(al) level 理论水平
theoretic(al) limit 理论极限
theoretic(al) limit load 理论极限荷载
theoretic(al) line 理论线;理论开挖线
theoretic(al) line capacity 线路理论通过能力
theoretic(al) location 理论位置
theoretically complete combustion 理论完全燃烧
theoretically highest tide level 理论最高潮面
theoretically lowest tide level 理论最低潮面
theoretically perfect plate 理想塔板
theoretically perfect plate tray 理想塔板
theoretically perfect tray 理论塔板
theoretically relieving capacity 理论释放能力
theoretic(al) margin 理论边缘;理论边限
theoretic(al) maximum concentration 理论最大浓度
theoretic(al) mean 理论平均值
theoretic(al) measure 理论测度
theoretic(al) mechanics 理论力学
theoretic(al) meteorology 理论气象学
theoretic(al) mix of concrete 混凝土理论配合比
theoretic(al) mode 理论众数;理论模式
theoretic(al) model 理想模型;理论模型
theoretic(al) models of variogram 变差函数理论模型
theoretic(al) module 理论模数
theoretic(al) morphology 理论形态学
theoretic(al) numerical aperture 理论数值孔径
theoretic(al) oceanography 理论海洋学
theoretic(al) optics 理论光学
theoretic(al) organic carbon 理论有机碳
theoretic(al) organic chemistry 理论有机化学
theoretic(al) output 理论产量;理论功率
theoretic(al) oxygen demand 理论需氧量
theoretic(al) parallax 理论视差
theoretic(al) parity 理论平价
theoretic(al) physics 理论物理学
theoretic(al) plate 理想塔板;理论(塔)板

theoretic(al) plate number 理论塔板数
theoretic(al) point frog 理论辙叉心
theoretic(al) point of frog (铁路的)辙叉交点
theoretic(al) point of switch 转辙器理论尖端
theoretic(al) potential flow 理论位流;理论势
theoretic(al) prediction 理论预测;理论推算;理论判断
theoretic(al) prediction method 理论推算法
theoretic(al) price 理论价格
theoretic(al) prices on a composite basis 复合理论价格
theoretic(al) prior information 理论的先验信息
theoretic(al) probability 先验概率;理论概率;事前概率
theoretic(al) probability distribution 理论的概率分配
theoretic(al) profile 标准断面;理论断面;理论剖面
theoretic(al) pump displacement 计算跨度;水泵理论排水量;泵的理论排水量
theoretic(al) quantitative approach 理论数量方法
theoretic(al) quantity of cable steel 缆索理论用钢量
theoretic(al) quarry-stone yield 理论成荒率
theoretic(al) raw meal consumption 理论料耗
theoretic(al) recovery per cent graduation of cleaned coal 精煤理论回收率%分级
theoretic(al) recurrence interval 理论重现期
theoretic(al) recurrence period 理论重现期
theoretic(al) relative frequency 理论相对频率
theoretic(al) resolving power 理论分辨率
theoretic(al) scale 理论比例尺
theoretic(al) screen cut 理论筛选分割
theoretic(al) sea level datum 理论深度基准面
theoretic(al) seismograph 理论地震图
theoretic(al) seismology 理论地震学
theoretic(al) separation coefficient 理论分离系数
theoretic(al) service life 理论使用寿命
theoretic(al) shape 理论图像
theoretic(al) size 计算尺寸;理论尺寸
theoretic(al) size of map 地图理论尺寸
theoretic(al) soil mechanics 理论土力学
theoretic(al) solubility equilibrium concentration 理论溶度平衡浓度
theoretic(al) solution 理论溶液
theoretic(al) source 理论根源
theoretic(al) source function 理论震源函数
theoretic(al) span 理论跨度
theoretic(al) specific impulse 理论比推力
theoretic(al) spectrum 理论谱
theoretic(al) speed 理论速度
theoretic(al) stage 理论级
theoretic(al) statistics 理论统计学
theoretic(al) stiffness 理论刚度
theoretic(al) stratigraphy 理论地层学
theoretic(al) strength 理想强度;理论强度
theoretic(al) stress distribution method 理论应力分布法
theoretic(al) thermal efficiency 理论热效率
theoretic(al) thickness of oil water transition zone 理论油水过渡带厚度
theoretic(al) throat 焊缝计算厚度;(焊缝的)理论厚度
theoretic(al) throat depth 焊缝理论厚度
theoretic(al) thrust coefficient 理论推力系数
theoretic(al) time curve 理论时间曲线
theoretic(al) time section of model 理论模型时间剖面
theoretic(al) toe length 辙叉趾理论长度
theoretic(al) torque 理论扭矩
theoretic(al) track capacity 线路理论通过能力
theoretic(al) traffic capacity 理论输送能力
theoretic(al) unloading capacity 理论卸船生产率
theoretic(al) value 理论值
theoretic(al) value of earth tides 固体潮理论值
theoretic(al) velocity 理论速度
theoretic(al) volumetric(al) efficiency 理论容积效率
theoretic(al) water power 理论水电出力;理论水电发电能力
theoretic(al) weight method 理论秤量法
theoretic(al) weight per meter pipe 管子每米理论重量
theoretic(al) yield 理论产量
The Organization of American States 美洲国家

组织
the organization of labo(u)r of railway enterprise 铁路企业劳动组织
theories of bureaucracy 行政组织理论
theories of light 光的二象性理论
theorization 理论化
theorize 建立学说;建立理论;理论化
theory 学说;理论
theory for large-scale systems 大系统理论
theory lower-bound 理论下界
theory model of quantification 数量化理论模型
theory of adaptation 适应理论
theory of alternate dominance 交互显性说
theory of anharmonic 交比理论
theory of anharmonic ratio 交比理论
theory of antecedent conflicts 先导冲突理论
theory of approximations 逼近论
theory of architecture 建筑学理论
theory of artificial (earth) satellite motion 人造地球卫星运动理论
theory of automatic control 自动控制理论
theory of balances 平衡理论
theory of barrels 圆筒理论
theory of bending 弯曲理论
theory of benefits 受益理论
theory of branching process 分支过程理论
theory of breaking 破坏理论
theory of buckling 弯折原理;弯曲原理
theory of categories 范畴论
theory of cement setting and hardening 水泥凝结硬化理论
theory of chances 概率论;变革理论
theory of colo(u)r vision 色觉学说
theory of comminution 粉碎理论;破碎理论
theory of comparative advantage 比较有利条件论;比较利益论;比较成本理论
theory of comparative cost 比较成本理论
theory of comparative productivity 比较生产力论
theory of concentration of metallogenesis 矿化集中(金属)区学说
theory of concrete creep 混凝土徐变理论;混凝土蠕变理论
theory of conjugate beam 共轭梁定理;共轭梁原理;共轭梁理论
theory of conjugate stress 共轭应力理论
theory of conjugation 共轭原理
theory of conscious activity 能动性
theory of consolidation 固结原理;固结理论
theory of constant energy of deformation 形变能守恒原理
theory of consumer's behavio(u)r 消费者行为论
theory of consumer's budget 消费者预算论
theory of consumer's choice 消费者选择论
theory of continental drift 大陆移动理论;大陆漂移(学)说
theory of continental growth 大陆成长理论
theory of continuous support 连续支承理论;连续支撑理论
theory of control 控制(理)论
theory of creep 蠕变原理;徐变理论;徐变力学
theory of creep buckling 徐变压曲理论
theory of cross ratio 交比理论
theory of crowd passage 人流理论;人群通行理论
theory of cultural landscape 文化景观论
theory of decision-making 决策理论
theory of decision management 决策管理理论
theory of deformation 变形原理
theory of demand 需求论
theory of determinats 行列式理论
theory of discrete supports 点支承理论
theory of discs 圆片原理;圆盘原理
theory of distance distortion 距离变形理论
theory of distortion 面积变形理论
theory of distortion of angles 角度变形理论
theory of distribution 分配理论
theory of double image measurement 双像量测理论
theory of dynamic(al) programming 动态规划理论
theory of ecologic(al) development 生态发展论
theory of ecologic(al) equilibrium 生态平衡论
theory of economic calculation 经济计算论
theory of economic development 经济发展论
theory of economic growth 经济成长论

theory of economic policy 经济政策理论
theory of elastically restrained beams 弹性的束梁原理
theory of elasticity 弹性理论
theory of elastic laminated system 弹性层状体系理论
theory of elasto-plastic deformation 弹塑性变形理论
theory of electrolytic dissociation 电离理论
theory of empathy 神人概念
theory of employment 就业理论
theory of engagement 啮合原理
theory of environment(al) adaptability 环境适应论
theory of epigenesis 发生论
theory of equation 方程论
theory of equilibrium 平衡理论
theory of error 误差理论
theory of evolution 进化论
theory of external economy 外部经济理论
theory of extremal statistics 极值统计理论
theory of extreme values 极值理论
theory of failure 破坏理论
theory of fields 域论
theory of fixed beams 固定梁理论
theory of flexure 挠曲理论
theory of floating bodies 浮体理论
theory of flow net 流网理论
theory of fluid flow 液流理论
theory of form and design 造型及设计原理
theory of function 函数论
theory of function of a complex 复变函数论
theory of functions of complex variable 复变函数论
theory of functions of real variable 实变函数论
theory of game 对策论;博弈论
theory of games of strategy 策略博弈理论
theory of general adaption syndrome 一般适合综合学说
theory of geographic(al) environment 地理环境说
theory of gradually varied flow 渐变流理论
theory of graph 图论
theory of groundwater dynamics 地下水动力学理论
theory of groundwater movement 地下水运动理论
theory of group 群论
theory of hardening 固化理论;硬化理论
theory of harmony 和谐论
theory of humus mutrition 腐殖质营养理论
theory of ideal plasticity 理想塑性理论
theory of image coding 图像编码理论
theory of induced cleavage 预裂隙理论;诱生劈理理论【地】;诱生裂隙理论
theory of inducement to invest 投资诱导论
theory of infiltration 下渗理论;入渗理论
theory of ionization 电离理论
theory of kinetic momentum 动能理论
theory of lattice analogy 格构类比理论;网络模拟理论
theory of limitation 极限论
theory of lubrication 润滑理论
theory of machines 机械原理;机械理论
theory of macrodistribution 宏观分配理论
theory of man-land relationship 人一地相关原理
theory of marginal productivity 边际生产力论
theory of material strength 材料强度理论
theory of maximum strain energy 最大应变能理论;最大变形能理论
theory of mechanics of materials 材料力学理论
theory of molecules 分子论
theory of Molodensky 莫洛金斯基理论
theory of mutation 突变学说
theory of network structure 网络结构理论
theory of non-linear vibration 非线性振动理论
theory of optimal control 最佳控制理论
theory of ordinal utility 序数效用论
theory of perturbation 摄动理论
theory of planning 规划理论
theory of plastic flow 塑性流动理论
theory of plasticity 塑性理论
theory of plastic-viscous flow 塑性黏滞流动理论
theory of plates 板体理论

theory of plate tectonics 板块构造理论
theory of population density 人口密度理论
theory of probability 概率论;或然率论
theory of profit upon alienation 利润让渡论
theory of proportional parts 比例部分论
theory of proportioning 混凝土配合比理论;配合比理论
theory of proportions 均衡理论;比例理论
theory of quantification 数量化理论
theory of quantification program(me) 数量化理论程序
theory of queues 排队(理)论
theory of queuing 排队(理)论
theory of radioactivity 放射论
theory of reaction rate 反应速率理论
theory of reciprocal deflection 变位互等理论
theory of reciprocity 互换理论;互换定理;位移互等定理
theory of redox 氧化还原理论
theory of reflection 反映论
theory of reflection and cognitive analysis 反映论与认识分析
theory of regionalized variables 区域化变量理论
theory of regulation by credit policy 信贷政策调节论
theory of relativity 相对论
theory of risk-bearing 风险负担理论
theory of risks 风险论
theory of rupture 破裂理论
theory of sampling 取样原理;取样理论;样品理论
theory of sampling distribution 抽样分布理论
theory of seismic method 地震法原理
theory of seismic rays 地震射线理论
theory of servomechanism 伺服控制论;伺服机构理论
theory of shallow shells 薄壳理论
theory of sheets 薄板理论
theory of shells 壳体理论
theory of similarity 相似性理论;相似论
theory of solids 固体论
theory of spline 样条理论
theory of stability 稳定性理论
theory of statistic(al) test 统计检验理论
theory of stochastic service system 随机服务系统理论
theory of storage 存储理论
theory of straight-line distribution 直线分布理论
theory of strain energy 变形能理论
theory of strength 强度理论
theory of strength of materials 材料力学理论
theory of structural inflation 结构性通货膨胀论
theory of structures 结构论
theory of subsidence 沉降理论
theory of super-sonic speed 超声波速理论
theory of surface area 表面积理论
theory of suspended separation 悬浮分离理论
theory of taxation 赋税学说;赋税论
theory of technique 技法论
theory of the balance of trade 贸易差额论
theory of the displacement of continent 大陆漂移(学)说
theory of the firm 厂商理论
theory of the internal origin 内因论
theory of the stages of economic growth 经济成长阶段论
theory of thin plate 薄板理论;薄板力学
theory of thin shell 薄壳理论;薄壳力学
theory of tidal channel 潮汐沟渠理论
theory of torsion 扭转理论;转矩理论
theory of transformation 变换论
theory of treatment after pollution 先污染后治理说
theory of two points 二点论
theory of ultimate equilibrium 极限平衡理论
theory of underdevelopment 经济欠发达理论
theory of unemployment 失业理论
theory of unit hydrograph 单位过程线理论
theory of urban design 城市设计理论
theory of vibration 振动理论
theory of volcano accumulation 火山堆积说
theory of washout by rain 预水冲刷理论;雨洗理论;雨伞冲刷理论
theory of welfare 福利论
theory of zero growth 零增长论

theory on autocontrol 自动控制学说
theory on economic development strategy 经济发展战略理论
theory on failure by bending 挠曲破坏理论;弯曲破坏理论
theory on macroregulation and control 宏观调节和控制理论
theory recovery 理论回收率
theory that those who has invested should be benefited 谁投资谁受益论
the other side of the river 对岸
the overall price level 物价总水平
The Overseas Economic Cooperation Funds 海外经济合作基金;海外经济协助基金
the parameters of radioactivity 放射性参数
the parameters used for calculation of ages 计算年龄的参数
the Paris Opera 巴黎歌剧院
the parties concerned 有关方(面)
the party concerted 当事人一方
the pattern of isotopic abundance 同位素丰度型式
the Pentagon 美国国防部五角大楼
the People's Insurance Company of China 中国人民保险公司
the percentage of crystalline phase 结晶质比率
the percentage of daughter minerals 子矿物比率
the percentage of gas phase 气相比率
the percentage of iron-rich matter 铁质比率
the percentage of liquid carbon dioxide 液态二氧化碳比率
the percentage of liquid phase 液相比率
The Piazza of S. Mark 圣罗马广场(位于威尼斯)
the plate pieces of spheric(al) storage tank 球形储[贮]罐瓣片
the pollution quota of pollutant 污染物污染指标
the porosity and structure of soil 土壤孔隙和结构
the positive unit on negative unit 负地形单元上的正单元
the positive unit on positive unit 正地形单元上的正单元
the Potala Palace 布达拉宫
the present stage 现阶段
the press 报刊
the price ask 喊价
the price asked 要价
the problem of employment 就业问题
the production of wood for various uses 生产各种需要的木材
the project of making some people unemployed obtain employment again 再就业工程
the proper authorities 有关当局
the public bidding system 招标投标制
the Pyramids 埃及金字塔
the quality of producing a greater yield 增加产量的一种品质
theralite 霞斜岩
theralite group 霞斜岩类
theralitic essexite 霞斜岩质厄塞岩
therapeutical gymnastic room 医疗体育室
therapeutic bathtub 治疗浴盆
therapeutic centre 治疗中心
the rated flow of fixture 卫生器具定额流量
the rate of fulfilment for daily passenger transport plan 日计划兑换率
the rate of gaseous products of denitrification 反硝化作用气态产物量
the rates against U.S. dollars 美元兑换率
the ration of cost can be used time to the whole time used 空费系数
the ratio of reserves class is not conform to the rules 储量比例不合要求
the ratio of sale to production 产销衔接状况
therblig 基本动作;(工艺操作中的)基本分解动作
therbligs 分解动作
the receipt of freight transport 货运票据
the receipt of passenger transport 客运票据
the refrom of the railway tariff system and the pricing of the transport industry 运价体制改革与运输企业定价
the Regional Oil Combating Centre 区域性防油污中心
the regulations on financial supervision 财政监督条例

the relative acid content of the soil 土壤相对含酸量
the Renaissance 文艺复兴
the right of investment, mortgage and guarantee 投资、抵押、担保权
the right way 正路
Therlo 色罗铜合金
therm 撒姆
thermae 温泉浴场;温泉
Thermae of Caracalla 卡拉卡勒温泉
Thermae of Dioletian 迪奥克兰汀温泉
Thermae of Nero 内罗温泉
Thermae of Titus 蒂图斯温泉
Thermae of Trajan 曲拉斯温泉
thermafiber blowing wool 矿渣吹制棉
thermal 热泡;热的;温热的
thermal aberration 热像差
thermal abrasion 热蚀
thermal absorber 吸热器
thermal absorption 热吸收
thermal acceptor 热受主;热受体;热接受器
thermal acclimation 热(的)适应
thermal accommodation coefficient 热适应系数
thermal action 温度作用
thermal activation 热激活;热激发;热活化
thermal activation energy 热激活能
thermal activity 热中子放射性
thermal addition 热力加成作用
thermal adhesion 热黏合;热黏附
thermal adhesive 热黏合剂;热胶粘剂
thermal adjustment time 热响应时间
thermal admittance 蓄热系数;热导纳
thermal aerobic filter 热气滤池
thermal after-effect 热后效应
thermal ag(e)ing 热(力)老化;热陈化
thermal agitation 热噪声;热骚动;热扰动;热搅动;热激发
thermal agitation noise 热激噪声
thermal agitation voltage 热骚动电压
thermal air current 热空气流
thermal alteration index 热变指数
thermal alteration of petroleum 石油的热演化
thermal amplitude decrement 热振幅衰减倍数
thermal analog computer 热模拟计算机
thermal analysis 热(学)分析
thermal analysis instrument 热分析仪器
thermal analyzer 热分析仪
thermal and moisture protection 隔热与防潮工程
thermal and photochemically enhanced Fenton reaction 热光化学强化芬顿反应
thermal anemometer 热风速仪
thermal anisotropy 热各向异性
thermal anomaly 热异常;热(的)反常
thermal arrest 热制动
thermal-arrest calorimeter 集热量热器
thermal asphalt 石油硬沥青
thermal assisted evaporative separation 热蒸发分离
thermal aureole 热晕轮;热接触变质晕
thermal background emission 背景热辐射
thermal baffle 遮热板;隔热板;隔热挡板;绝热隔板
thermal balance 高温天平;热天平;热平衡;热量平衡
thermal balance time 热平衡时间
thermal bardness 热硬度
thermal barrier 跃变层;热障;热垒;热激波;保温层
thermal barrier coating 热障涂层
thermal-bath 浴
thermal beam 热中子束
thermal behavio(u)r 热性能
thermal belt 高温区;高温带;热异常带;温度垂直带
thermal bending 温度弯曲(由温差引起的)
thermal black 热炭黑;热裂炭黑
thermal blooming 热晕
thermal blooming compensation 热晕补偿
thermal blooming spectroscopy 热聚焦光谱学
thermal bombardment 热轰击
thermal bond 热结合层;热结合(元件的)
thermal bonding 热黏合
thermal booster reaction 热引发反应
thermal boring 热(力)钻(孔);高温钻孔;热力钻探;热力钻进
thermal bounce 热反弹
thermal boundary 热障;热限

thermal boundary layer 热边界层;温度边界层;传热界面层
thermal break 热破坏;绝热物;热障
thermal breakage 受热破裂(玻璃的局部)
thermal breakdown 热逸散;热失控;热击穿;热分解
thermal break-off 热折断(挤压缺陷)
thermal break-through 热穿透
thermal breeder 热增殖堆
thermal breeder reactor 热中子增殖反应堆;热中子增殖堆
thermal breeding 热增殖
thermal bridge 热(电)桥(测无线电干扰用)
thermal broadening 热致宽
thermal broadening effect 热增宽效应
thermal buckling 热皱折;热失稳
thermal budget 热平衡
thermal buffer 热缓冲器
thermal bulb 测温筒;测温包
thermal bulb-type thermometer 热球式温度计
thermal burden rating 热负载额定值
thermal burn 热灼伤
thermal calculation 热(工)计算
thermal calibration 热校准
thermal camera 热感照相机
thermal capacitance 热容;热电容
thermal capacitivity 热电容量
thermal capacity 热容(量)
thermal capacity value 热容量值
thermal capture 热中子俘获
thermal cathode 热电阴极
thermal cell 热激电池
thermal center 热节;热中心
thermal change 热变
thermal character 热特性
thermal characteristic 热学特性;热工特性
thermal characteristic measurement 热特性测量
thermal characteristics of lake 湖水温度特征
thermal checking 过热生裂;热龟裂
thermal chemistry 热化学
thermal chlorination 热氯化
thermal chocking 热阻现象
thermal circuit-breaker 热胀断器;热继电器断路器
thermal circulation 热环流
thermal cleaning 热清洗;热清洁处理
thermal cleaning system for furnace 炉子热清扫系统
thermal climate 炎热气候;热型气候
thermal coagulation 热凝聚;热凝结
thermal coalescence 热聚结
thermal coefficient 热系数
thermal coefficient of conduction 导热系数
thermal coefficient of expansion 热(膨)胀系数
thermal coefficient of linear expansion 热线胀系数
thermal collapse 热破坏
thermal collector 热力除尘器
thermal column 加温塔(化工设备);热柱
thermal column beam 热柱中子束
thermal combustion 热力燃烧
thermal comfort 热舒适
thermal comfort index 热舒适指标
thermal comfort ventilation 热舒适通风
thermal comparator 导热比测器;热传导比测器
thermal compensation 热补偿
thermal compensation of pipe 管道热补偿
thermal compensator 热补偿器
thermal component 热分量
thermal compression evapo(u)ration 热压蒸发
thermal compressor 热压汽器;热力压缩机;喷射器压缩机
thermal computing element 热计算元件
thermal condition 热状态;热工制度
thermal conditioning of sludge 污泥热调节
thermal conductance 热传导;热导(率);导热系数;热电导;导热性
thermal conduction 热传导;导热性;导热
thermal conduction module 导热模件;导热模数
thermal conduction per unit area 单位面积热传导
thermal conduction rate 导热率
thermal conduction rating 导热率
thermal-conduction resistance 传导热阻
thermal conductivity 热导(系)数;热导;热传(导)性;热传导率;导热能力;导热率;导热度

thermal conductivity analyzer 热导分析器
thermal conductivity apparatus 导热率仪
thermal conductivity bridge 热导率桥
thermal conductivity cell 热导池;导热析气计
thermal conductivity cell detector 热导池检测器
thermal conductivity coefficient 热传导系数;导热系数
thermal conductivity coefficient test of frozen soil 冻土导热系数试验
thermal conductivity detector 热传导式探测器;热传导测定器;热导(池)检测器
thermal conductivity gas analyzer 热导式气体分析仪;热导式气体分析器
thermal conductivity ga(u)ge 热导真空计;热导真空规
thermal conductivity measuring apparatus by guarded plate (method) 平板法导热系数测定仪
thermal conductivity measuring apparatus by heat pulse 热脉冲法导热系数测定仪
thermal conductivity method 热导法
thermal conductivity of liquid 液体导热(性)
thermal conductivity of rock 岩石热导率
thermal conductivity tester by absolute plate method 直接平板法导热仪
thermal conductivity tester by comparative plate method 比较平板法导热仪
thermal conductivity vacuum ga(u)ge 热传导真空计
thermal conductor 导热材料;导热物质;热导体;导热体
thermal confinement 热约束
thermal conformer 温度随变生物
thermal consolidation 热固结
thermal constant 热物理常数;热常数;热学常数
thermal contact 热接点;热接触;热触点
thermal contact resistance 接触热阻;热接触电阻
thermal container 热保温集装箱;保温集装箱
thermal content 热含量
thermal contraction 热收缩
thermal contraction apparatus 温度收缩仪
thermal contraction crack 热缩裂隙
thermal control 热的控制
thermal control coating 热控涂层
thermal control coating material 温控涂料
thermal controller 热控制器
thermal control material 调温涂料
thermal control surface coating 调温表面涂层
thermal convection 热力对流;热对流
thermal convection cell 热流循环
thermal convection rainstorm 热对流暴(风)雨
thermal convection storm 热对流风暴
thermal convective current 热对流气流;热对流湖流;热对流海流
thermal conversion 热转换;热对流
thermal conversion of gases 气体的热转化
thermal conversion processes 热转化过程
thermal converter 热转换器;热交换器;热变换器
thermal converter system 热变换器制
thermal coulomb 热库仑
thermal couple 热电偶
thermal cracker 热裂化装置
thermal cracking 加热裂解;热(裂)解;热裂化;热分裂;温缩裂缝;热应力裂隙;热裂纹
thermal cracking gas chromatography 热裂解气相色谱(法);热解气相色谱(法)
thermal cracking of petroleum 石油热裂解
thermal cracking process 热裂化过程
thermal cracking unit 热裂化装置
thermal creep 热蠕变
thermal critical point 临界温度;热临界点
thermal critical range 热临界带
thermal cross 热电偶接点
thermal curable system 热固化系统
thermal current 热稳定电流;热(气)流
thermal current converter 热电交换器
thermal current rating 热额定电流
thermal curtain 热障
thermal curve 热力曲线
thermal cutoff 热断开;热熔断器;热中子截止
thermal cutout 热熔断路器;热断流器;温控开关;热熔保险器
thermal cutting 热切割
thermal cycle 热循环;热环流
thermal cycle efficiency 热循环效率

thermal cycling 热循环;温度循环变化
thermal cycling stability 热循环稳定性
thermal cycling test 热循环试验
thermal cyclization reaction 热环化反应
thermal dam 热障;热点
thermal damage 热损伤
thermal day 热日
thermal deaeration 加热除气;热力除气
thermal deaerator 热法脱气器;热法除气器
thermal death 热致死;热寂
thermal death equipment 热消毒器
thermal death inactivation point 热灭活点
thermal death point 致死温度;热致死点;热死温度;热毙点
thermal death-point 热死点
thermal death rate 热死率
thermal death time 热(致)死时间;热消毒时间;热寂时间
thermal decay time log 热中子衰变时间测井
thermal decomposer 热分解器
thermal decomposition 加热分解;热(分)解
thermal decomposition furnace 热解炉;热分解炉
thermal decomposition method 热解法;热分解法
thermal decomposition of wastes 废物热解
thermal decomposition product 热解产物
thermal defect 热缺陷
thermal deflection 热偏转
thermal defocussing 热散焦
thermal deformation 热形变;热变形
thermal degradation 热降解
thermal degradation zone 热降解作用带
thermal degrading 热降解
thermal deicer 热温除冰设备
thermal delay 热延时器
thermal demagnetization 热退磁;热去磁
thermal demagnetizer 热退磁仪
thermal denitration 热脱硝
thermal density current 温差异重流
thermal density flow 温度差异重流
thermal density fluctuation 热密度波动
thermal depolymerization 热解聚(作用)
thermal depopulation 热致粒子数减少
thermal depression 热力低压
thermal desalting 加热脱盐法
thermal design 热设计;散热设计
thermal design criterion 热力设计准则
thermal desorption 热脱附;热解吸
thermal desorption method 热解吸法
thermal desorption spectroscopy 热脱附谱(术)
thermal desorption spectrum 热脱附谱
thermal desorption substance 热解吸物
thermal destruction curve 热力破坏曲线
thermal desulfurization 加热脱硫
thermal detection element 热红外探测元件
thermal detection radiometer 热探测辐射计;热控测辐射计
thermal detector 热(中子)探测器;热检测器;热电探测器;电阻辐射热测定器;测温器
thermal deterioration 热损坏
thermal development 热显影
thermal difference 热差值
thermal diffusion 热扩散(作用)
thermal diffusion coefficient 热扩散系数
thermal diffusion column 热扩散柱;热扩散分离管
thermal diffusion constant 热扩散常数
thermal diffusion current 热扩散电流
thermal diffusion depth 热扩散深度
thermal diffusion factor 热扩散因子;热扩散因数
thermal diffusion flow 热扩散流
thermal diffusion fractionation 热扩散分离;热扩散分级
thermal diffusion length 热扩散距离
thermal diffusion method 热扩散法
thermal diffusion plant 热扩散工厂
thermal diffusion potential 热扩散势
thermal diffusion ratio 热扩散比
thermal diffusion system 热扩散系统
thermal diffusion tube 热扩散管
thermal diffusion zone 热扩散区
thermal diffusivity 热扩散性;热扩散系数;热扩散率;热扩散(作用);散热系数;散热率;导温系数
thermal dilatation 热膨胀
thermal dilatometer 热膨胀仪
thermal disadvantage factor 热不利因子

thermal discharge 热泻出物;热释放;热排放;温排水
thermal discoloration 热褪色
thermal disinfection 热力消毒
thermal dispersion 热散布
thermal display area 热显示区
thermal display zone 热显示区
thermal dissociation 热力离解;热离解(作用)
thermal distillation 热蒸馏
thermal distortion 热成畸变;热变形
thermal distortion temperature 热变形温度
thermal distress 热损坏
thermal distribution 热分布;温度分布
thermal disturbance 热扰动
thermal drain 热疏水
thermal drift 热泻出物;热漂移
thermal drill 火力钻机
thermal drilling 高温钻孔;热钻孔;热力钻探;热力钻进
thermal drilling of stone 岩石热力钻子
thermal drive 热驱动
thermal driven device 热驱动装置
thermal drop feed oiler 热差点滴加油器
thermal dryer 热干燥机;热力干燥机
thermal drying 热烘干
thermal dust collector 热式聚尘器
thermal dust precipitator 热式尘埃计
thermal effect 热效应;热骚动
thermal effectiveness 热处理效果
thermal effect of ultra-sound 超声波热效应
thermal efficiency 热效率;温度效率
thermal efficiency coefficient 热效率系数
thermal efficiency index 温度效率指数
thermal efficiency ratio 热效率比;温度效率比
thermal effluent 热排放物
thermal effusion 热隙透
thermal elasticity 热弹性
thermal electret 热驻极体
thermal electric(al) potential 热力电位
thermal electric(al) (power) plant 热电站
thermal electric(al) station 热电站
thermal electricity 热电
thermal electromotive force 热电动力
thermal electromotive potential 热电动势
thermal electron emission 热电子发射
thermal element 热元件;发热元件;温差电偶;热敏元件
thermal elimination 热消除
thermal ellipsoid 热椭圆体
thermal elongation 热伸长;受热延长
thermal emission 热发射
thermal emission ion source 热发射离子源
thermal emissive power 热发射功率
thermal emissivity 热发射率;热辐射系数
thermal endurance 耐热性;热稳定性;热寿命;热耐久性
thermal energy 热能
thermal energy analysis 热能分析
thermal energy balance 热能平衡
thermal energy consumption 热能消耗
thermal energy converter 热能交换器;热能变换器
thermal energy density 热能密度
thermal energy gap 热能隙
thermal energy input 热能输入
thermal (energy) neutron 热中子
thermal energy range 热能区
thermal energy yield 热能当量
thermal engine 热力发动机;热机
thermal engineering 热力工程;热工学
thermal enhancement 热增强作用
thermal enregy region 热能区
thermal entropy 热熵;温度熵
thermal environment 热环境
thermal environment of the earth's surface 地球表面的热环境
thermal equation of energy 热能方程
thermal equation of state 热状态方程
thermal equator 热赤道
thermal equilibration 热平衡
thermal equilibrium 热平衡;热力平衡
thermal equilibrium condition 热平衡条件
thermal equivalent 热当量
thermal equivalent of work 热功当量
thermal equivalent spectrum level 热当量光谱仪

thermal equivalent value 热当量值
thermal erosion 热蚀;热侵蚀
thermal etching 热蚀(法);热力学侵蚀;热浸蚀;热腐蚀
thermal evapo(u)rated thin film 热蒸发薄膜
thermal evapo(u)ration 热蒸发
thermal evolution 热演化
thermal examination 热工检查
thermal excavation 热力开挖
thermal exchange 热交换
thermal excitation 热骚动;热激励;热激发
thermal excitation energy 热激发能量
thermal exhaustion 热耗散
thermal expand of seawater 海水的热膨胀
thermal expansibility 热膨胀性;热膨胀率
thermal expansion 热伸长;热胀;热膨胀
thermal expansion analysis 热膨胀分析
thermal expansion and contraction 热性胀缩;热膨胀及热收缩
thermal expansion character 热膨胀特性
thermal expansion coefficient 热膨胀系数;热膨系数;膨胀系数
thermal expansion compatibility 热膨胀适应性;热膨胀的适应性
thermal expansion compensator 热膨胀补偿器
thermal expansion curve 热膨胀曲线
thermal expansion detection 热膨胀辐射探测
thermal expansion factor 热膨胀因数;热膨胀系数
thermal expansion instrument 热膨胀仪
thermal expansion measurement 热膨胀测定
thermal expansion mo(u)lding 热膨胀模塑成型
thermal expansion reactivity coefficient 热膨胀反应性系数
thermal expansion relay 热膨继电器;热膨胀继电器
thermal expansion test 热膨胀试验;热膨胀测定
thermal expansion valve 热膨胀阀;调温膨胀阀
thermal expansivity 热膨胀性;热膨胀系数;热膨胀率
thermal expansivity of gas volume 天然气体积热膨胀系数
thermal expansivity of oil volume 油的体积热膨胀系数
thermal expansivity of pore volume 孔隙体积的热膨胀系数
thermal expansivity of rock grain 岩石颗粒的热膨胀系数
thermal expansivity of water volume 水的体积热膨胀系数
thermal explosion 热爆炸;热爆
thermal exposure 热辐射量
thermal factor (室内环境的)冷热因素
thermal farad 热法拉
thermal fatigue 热疲劳
thermal fatigue fracture 热疲劳断裂
thermal fatigue test 热疲劳试验
thermal feedback 热反馈
thermal feedback effect 热隙透效应
thermal field 热场;温水场;温度场;热力场
thermal fisser 热中子裂变物质
thermal fission 热中子裂变
thermal fissionability 热中子作用下可裂变性
thermal fission factor 热中子裂变因子
thermal fission yield 热中子裂变产额
thermal fixation 热固着
thermal flame safeguard 火焰保护器
thermal flasher 热效闪光器
thermal flexure 热致弯曲;热弯(曲)
thermal flow 热流
thermal flowmeter 热流(量)计;量热式流量计;热式流量计;热量计
thermal flow study 热流值研究
thermal fluctuation 热涨落;热起伏
thermal fluid 热流体
thermal fluid heater 热载体加热器
thermal flux 热通量;热流量;热流
thermal foam 热泡沫
thermal focal spot 热焦斑
thermal foil 保温箔片
thermal force 热荷载;热力;温度力;温差力
thermal forming 热赋能
thermal fracture 热破裂
thermal fragment 热裂
thermal fragmentation 热裂
thermal front 暖锋【气】

thermal galvanometer 热效式电流计;温差电流计
thermal gas lens 热气体透镜
thermal gas lens measurement 热气测量;加热型气体透镜测量
thermal gasoline 热法汽油
thermal ga(u)ge 热敏电阻式压力计;热规
thermal generating station 热电厂;火力发电站
thermal generator set 火力发电机组
thermal geotechnics 热岩土工程学
thermal glass 耐急冷急热玻璃;低膨胀热稳定玻璃
thermal glow 热释光
thermal glow characteristic 热辉光特性
thermal glow curve 热辉光曲线
thermal glow peak 热释光峰
thermal gradient 热(塑)梯度;热阶度;热陡度;温度梯度;温度坡度
thermal gradient furnace 梯温炉
thermal gradient measuring probe 热梯度探头;热梯度探测仪
thermal gradiometer 热量梯度计
thermal graphite reactor 石墨慢化热中子反应堆
thermal gravimeter 热解重量仪
thermal gravimetric analysis 热解重量分析
thermal ground 热地;放热地面
thermal groundwater 地下热水
thermal halo diagram 热晕图
thermal halogenation 热卤化作用
thermal halo method 热晕法
thermal harm 热害
thermal head 热压头;热位差;热能头
thermal heat sealing 热封焊
thermal high 热高压
thermal history 热史;热历史
thermal hoop-stress 热箍应力
thermal horsepower 热马力
thermal hydrolysis 热水解
thermal hydrolytic dewatering technology 热水解脱水技术
thermal hygrometer 热湿度表
thermal hysteresis 热滞后现象;热滞(后);温度滞后
thermal ignition system 热点火系统
thermal image 热点像
thermal image heating method 成像加热法
thermal imager 热像仪
thermal imager common module 热像通用组件
thermal imagery 热影像;热成像
thermal imaging 热辐射成像;热成像
thermal imaging camera tube 热成像摄像管
thermal imaging device 热成像装置
thermal imaging response 热成像响应
thermal imaging system 热像仪;热成(像)系统
thermal imaging tube 热像管
thermal impact 热冲击
thermal impact resistance 抗热冲击(性)
thermal impedance 热阻抗
thermal impulse welding 脉冲热焊
thermal inactivation 热钝化
thermal inactivation point 热纯化点
thermal inactivation temperature 热纯化温度
thermal incineration 加热焚化;热焚烧
thermal incinerator 热焚烧炉;焚化垃圾炉
thermal index (人体感觉的)冷热指标
thermal induced optic(al) distortion 热感应光学畸变
thermal induced phase separation 热致相分离
thermal inductance 热感
thermal induction 热诱导;热感应
thermal inertia 热惯量;热贯性;热惰性
thermal inertia factor 热惰性指标
thermal inertia map 热惯量图
thermal inertia mapper 热惯量成象仪
thermal infrared 热红外
thermal infrared band 热红外波段
thermal infrared channel 热红外通道
thermal infrared detection array 热红外探测器阵列
thermal infrared distribution map 热红外分布图
thermal infrared image 热红外影像;热红外图像
thermal infrared imager 红外(线)热像仪
thermal infrared imagery 热红外影像;热红外图像
thermal infrared imaging camera 红外(线)热成像摄像机
thermal infrared imaging goggles 红外(线)热成像眼镜
thermal infrared imaging viewer 红外(线)热像观察仪

thermal infrared mapper 热红外成象仪
thermal infrared mapping system 热红外成像系统
thermal infrared multispectral scanner 热红外多谱段扫描仪
thermal infrared scanner 热能红外线扫描器;热红外扫描仪
thermal infrared scanning 热红外扫描
thermal infrared scanning data 热红外扫描数据
thermal infrared sensor 感温热红外传感器
thermal infrared spectrum 红外(线)热光谱
thermal infrared wave length 热红外波长
thermal infrared window 热红外窗口
thermal initiated polymerization 热引发聚合
thermal initiation 热引发(作用)
thermal injury 温度损伤
thermal instability 热不稳定性
thermal instrument 热效应仪器;热电效应仪表;电热仪表
thermal insulant 热绝缘体
thermal insulating cement 隔热水泥;绝热水泥
thermal insulating coating 隔热涂料
thermal insulating coefficient 绝热系数
thermal insulating course 隔温层;隔热层
thermal insulating cover 隔温覆盖层
thermal-insulating diving suit 保暖潜水服
thermal insulating door 保温门
thermal insulating efficiency 绝热效率
thermal insulating foam glass 绝热泡沫玻璃
thermal insulating glass 隔热玻璃
thermal insulating layer 隔温层;隔热层;保温层
thermal insulating material 隔热材料;绝热材料;保温材料
thermal insulating mortar 隔热砂浆;保温砂浆
thermal insulating painting 绝热涂料
thermal insulating pipe 保温管
thermal insulating plaster 绝热灰浆
thermal insulating property 绝热性质
thermal-insulating roof 隔热屋顶;保温屋面;保温屋顶
thermal-insulating suit 保暖服
thermal insulating tank car 保温槽车
thermal insulating tube 保温管
thermal-insulating value 绝热值
thermal insulation 隔热;绝热;热(的)绝缘;绝热材料;保温材料;保温
thermal insulation berm 保温护道【道】
thermal insulation blanket 绝热层
thermal insulation board 隔热板;保温板
thermal insulation course 隔温层
thermal insulation door 保温门
thermal insulation for building 房屋隔热
thermal insulation layer 绝热层
thermal insulation material 绝热材料;热绝缘材料;保温材料
thermal insulation mortar 保温砂浆
thermal insulation of ceiling 顶棚保温
thermal insulation of roof 屋顶隔热;屋顶保温
thermal insulation plate 隔热板
thermal insulation property 保温性
thermal insulation slab 隔热板;保温板
thermal insulation system 隔热系统
thermal insulation technique 绝热技术
thermal insulation thickness 保温层厚度
thermal insulation value 绝热值
thermal insulator 隔热物;绝热体;绝热器;热绝缘体
thermal intake length 热量引入区
thermal intensity 热强度;热力强度
thermal intensity of combustion chamber 燃烧室热力强度
thermal interaction 热互作用
thermal internal boundary layer 热内边界层
thermal inversion 逆温;热转化;温度转换;温度转化;温度逆增
thermal ionization 高温电离;热(致)电离
thermal ionization energy 热电离能
thermal ionization mass spectrometer 热电离质谱仪;热电离质谱计
thermal ionization sources 热电离源
thermal island 热岛
thermal island effect 热岛效应
thermal isolation 建筑隔热
thermal isomerization 热异构化
Thermalite 德玛莱特轻质绝缘材料(商品名)

thermalite 隔热材料;隔热板
thermalization 热化;热能化
thermalization coefficient 热化系数
thermalization column 热化柱
thermalization range 热化区
thermalization time 热化时间
thermalization time constant 热化时间常数
thermalize 热能化
thermalized neutron 热化中子
thermalized position 热化位置
thermalizing 热化
thermalizing collision 热碰撞
thermalizing column 热柱
thermalizing time 热化时间
thermal jet 热成急流
thermal jet engine 热力喷气发动机
thermal jet piercing drilling 热焰射穿钻孔
thermal jet piercing method 热焰射穿钻孔法
thermal junction 热结;热电偶接合
thermal karst 热岩溶;热力类岩溶;热喀斯特
thermal lag 热滞;热惯性
thermal lag of the oceans or the earth 海洋或地球的热滞后
thermal lake 热水湖
thermal land 热地
thermal layer 逆温层;热激波;热层;温跃层
thermal leads 热前置
thermal leakage coefficient 热漏泄系数
thermal leakage factor 漏热因数
thermal lens 热透镜
thermal lens effect 热透效应
thermal lensing compensation 热透补偿
thermal lensing spectroscopy 热聚焦光谱学
thermal lethal rates 热致死率
thermal level 耐热水平;热能级
thermal life 热寿命
thermal light 热光;室温指示灯;超温指示灯
thermal limit 耐温极限(材料的);发热极限(电机功率的);发热限制
thermal load 热荷载;热应力;热负载;热负荷
thermal load capacity 高温荷载能力;热负荷能力
thermal load scheme 热负荷计划
thermal locality zone 热异常区
thermal log 热测井
thermal logging 温度测井(曲线)
thermal loss 热消耗量;热损失;热(损)耗;热量损耗
thermal loss in weight 热失重
thermal low 热低压【气】
thermalloy 镍铜合金;耐热耐蚀合金;热合金;铁镍耐热耐蚀合金;铁镍热磁合金
thermal lubricator 热润滑器
thermally activated chemical vapo(u)r deposition 热活化化学气相沉积
thermally activated creep 热激活蠕变
thermally altered seawater 热变质废水
thermally chocked duct 不导热管道
thermally cracked gas 热裂解气
thermally depoled 热去极化
thermally disordered 热无序的
thermally driven 热驱动的
thermally engraved hologram 热雕全息照片;热雕全息图
thermally fissile Pu 热裂变钚
thermally foamed plastics 热发泡塑料
thermally induced flow 热导流
thermally induced refractive-index gradient 热感应折射指数梯度
thermally induced strain 热生应变
thermally induced stress 热生应力
thermally insulated 绝热的;隔热的
thermally insulated cladding panel 有隔热包层的墙板
thermally insulated door 保温门
thermally insulated shroud 隔热罩
thermally massive structure 大热容量结构
thermally metamorphosed metallization 热变质成矿作用
thermally reinforced weak soil 加热固结的弱土壤
thermally reversible expansion 可逆热膨胀
thermally sensitive 热敏的
thermally sensitive resistance 热敏电阻
thermally stable 热稳定的
thermally stimulated conductivity 热刺激电导率
thermally stimulated current 热刺激电流

thermally symmetrical design 热对称设计
thermally tuned valve 热调管
thermally unstable substance 热不稳定物质
thermal machine 热机
thermal magnetic breaker 热磁开关
thermal magnetizing material 热磁材料
thermal magnon 热磁振子
thermal man 热人(耐燃试验用人体模型)
thermal map 热图
thermal map of sea surface 海面热图
thermal mapping scanner 热成像扫描仪
thermal mapping technique 热测绘技术
thermal mass 热质量
thermal mass flowmeter 热质流量计
thermal measuring instrument 热学测试仪器
thermal mechanical working 热机械加工
thermal medium 热载体;热介质
thermal medium boiler 热载体加热炉;热载体锅炉
thermal medium heater 热载体加热炉
thermal medium return nozzle 热载体回流口
thermal medium storage tank 热载体储[贮]罐;热载体储罐
thermal medium vapo(u)rizer 热载体蒸发器
thermal metamorphic rock 热力变质岩
thermal metamorphism 热力变质(作用);热变质(作用)
thermal meter 热工仪表;热电式仪表;测热仪表
thermal method 热测法
thermal methods of oil recovery 油的加温回收法
thermal microphone 热线传声器;热送话器
thermal migration 热迁移
thermal mineral spring 热泉
thermal misadjustment 热力失调
thermal model 热模型
thermal moon 热运动
thermal mortar 保温砂浆
thermal movement 热移动(构件因温度变化而膨胀或收缩);热位移;热变位;温度变化
thermal naphtha 热裂化石脑油
thermal neutron absorption 热中子吸收
thermal neutron activation 热中子激活
thermal neutron beam 热中子束
thermal neutron calibration 热中子校准
thermal neutron capture cross section 热中子俘获截面
thermal neutron converter reactor 热中子转换反应堆
thermal neutron cycle 热中子循环
thermal neutron detector 热中子探测器
thermal neutron fission 热中子裂变
thermal neutron flux 热中子通量
thermal neutron flux standard 热中子通量标准
thermal neutron laminagraphy 热中子薄层照相法
thermal neutron leakage 热中子泄漏
thermal neutron leakage factor 热中子漏逸率
thermal neutron lifetime 热中子寿期
thermal neutron range 热中子区
thermal neutron reaction 热中子反应
thermal neutron reactor 热中子堆
thermal neutron spectrum 热中子谱
thermal nitrogen oxides 热氮氧化物
thermal noise 热噪声;热噪声;热杂波
thermal noise generator 热力探矿
thermal noise limited operation 热噪限制工作
thermal noise power 热噪声功率
thermal noise tester 热噪声测试器
thermal noise thermometer 热噪声温度计
thermal noise voltage 热噪声电压
thermal NO_x 热氮氧化物
thermal oiler 热加油器
thermal optic(al) distortion 热光畸变
thermal oscillation 热振荡
thermal oscillation frequency 热振荡频率
thermal output 热功率
thermal output of heat surface 加热面的热输出
thermal overload 热过载;热超负荷
thermal overload capacity 热过载容量
thermal overload protection 热过载保护
thermal overload relay 热过载继电器
thermal overrun 热超限
thermal oxidation 加热氧化;热氧化(作用)
thermal oxidative degradation 热氧化降解
thermal oxidization 热氧化
thermal oxidizer 废气燃烧炉

thermal paint 加热防腐漆;示温漆
thermal parameter 热力参数;热参数
thermal peak 温度峰值
thermal peaking 火电调峰
thermal penetrator 热力穿孔器
thermal performance 热特性
thermal performance design 热工设计
thermal pericyclic reaction 热周环反应
thermal perturbation 温度扰动
thermal photograph 热辐射摄影
thermal picture 热异常图像;热图像(红外线暗视器摄取的图像)
thermal piercing 热力穿孔
thermal piercing-drill 火焰穿孔钻
thermal pinch 热收缩
thermal pinch effect 热箍缩效应
thermal pipe 热导管(防冰装置的导管)
thermal pit 热坑
thermal pitch 石油硬沥青
thermal plastic elastomer 热塑性弹性体
thermal plasticity 热塑性塑性;热塑性
thermal plasticization 热塑炼
thermal plastic spacer 热塑性隔条
thermal plotter 热绘图器
thermal plug 热力偶塞
thermal plume 热柱;热烟流;热卷流
thermal pollution 热污染;热沾染;高温污染
thermal pollution control 热污染控制
thermal pollution from electric(al) power plant 电厂热污染
thermal pollution of water body 水体热污染
thermal pollution source 热污染源
thermal polymerization 热聚合(作用)
thermal polymerization of oil 油的热聚合(作用)
thermal potential 热位;热势;热力势
thermal potential difference 热势差
thermal power 发热量;火力;火电;热能;热力;热功率;热动力;热电
thermal power generation 火力发电
thermal power plant 火力发电厂;热电厂
thermal power station 火力发电站;火电站;火电厂;热力发电站;热电站
thermal precipitation 加热沉淀;热力沉降
thermal precipitation device 热沉降装置
thermal precipitation slide 热沉淀载片(尘末显微检查)
thermal precipitator 加热吸尘器;加热沉淀器;热沉淀器;电热取样计
thermal pressure 热压力;热压
thermal prestressing 加热法预加应力
thermal printer 热式打印机;热敏式打印机
thermal printing 热转印刷
thermal probe 热探针;热探头
thermal process 热处理;热(力)过程
thermal processing 热处理
thermal process parameter 热工参数
thermal profile 热剖面;温度曲线
thermal property 热(学)性质;热学性能
thermal property of basin 盆地热特征
thermal property of minerals 矿物热学性质
thermal property of rock and soil 岩土的热学性质
thermal property of water 水的热性质
thermal prospecting 热力探矿
thermal protection 防热;过热保护;热防护
thermal protection shield 热防护层
thermal protection suit 保暖服
thermal protection system 热防护系统;防热系统
thermal protection system material 防热系统材料
thermal protective coating 热防护层;热保护层;防热涂层;隔热层
thermal protective diving suit 保暖潜水服
thermal-protective shell 防热壳
thermal protector 过热保护装置
thermal pulse 热脉冲
thermal pulse method 热脉冲法
thermal pump 热力泵
thermal quality of snow 雪质
thermal quenching 热淬火
thermal radiating 热辐射的
thermal radiating material 热辐射材料
thermal radiation 热能辐射;热辐射
thermal radiation attenuating cloud 热能辐射衰减云
thermal radiation burn 热辐射灼伤

thermal radiation destruction distance 热辐射破坏距离
thermal radiation detector 热辐射探测器
thermal radiation injury 热辐射伤害
thermal radiation in the infrared 红外(线)热辐射
thermal radiation phenomenon 热辐射现象
thermal radiation shield 热辐射防护屏
thermal radiator 热辐射体;热辐射器
thermal radiometer 热辐射仪;热辐射计
thermal radio waves 热射电辐射
thermal radius 热半径
thermal randomness 热混乱度
thermal range 热区
thermal rating 耐热等级;热功率;热定额;热导测定
thermal ratio 回热度
thermal reaction 热化学反应;热反应
thermal reactivation 热法再生
thermal reactor 热(中心)反应堆;热反应器;热堆
thermal reactor shield 反应器热屏
thermal reclamation 热法再生
thermal reconnaissance 热辐射侦察
thermal recovery 热还原
thermal rectification 热精馏
thermal reduction of sludge 污泥热分解
thermal reduction reaction 热还原反应
thermal reduction time 热杀菌时间
thermal refined structure 调质处理结构
thermal refining 热精炼;调质处理
thermal reflowing 软熔发亮处理
thermal reflowing process 热力再流动法
thermal reforming 热重整
thermal reforming plant 热重整装置
thermal refractive index coefficient 折射率温度系数
thermal regeneration 加热再生;热回授
thermal regenerative cell 热再生电池
thermal regime 热状况;热体制;热力情况
thermal region 温泉区;高温区;热区
thermal regulation 热工制度
thermal relaxation 热弛豫
thermal relaxation time 热弛时间
thermal relay 热敏继电保护装置;热力继电器;热(动)继电器;温度继电器;温差(控制)继电器
thermal release analysis 热释分析
thermal releaser 热释放器
thermal relief 消除应力退火;限热;热控放气门
thermal relief gate 热控放泄门
thermal reservoir 热库
thermal resistance 有效热阻;抗热性;耐热性;热阻(率);热(敏)电阻;热变电阻
thermal resistance of fin 翅片热阻
thermal resistance to phase transition 相变热阻
thermal resistivity 热阻系数;热阻性;热阻率;导热抵抗力
thermal resistor 热敏电阻(器);热变电阻器
thermal resolution 热分辨率
thermal resource 热源
thermal response 热反应(系数);热(力)响应
thermal response simulation 热反应模拟
thermal result 热效应
thermal retardation 热迟延;保温
thermal retardation coefficient 相延迟温度系数
thermal reversibility 热能可逆性
thermal rig 火力钻机
thermal river 热水河
thermal rod effect 热棒效应
thermal run-away 热散发;热耗散;热逸溃;热逸散;热失控;热击穿
thermal scanner 热扫描仪
thermal scatter(ing) 热散射
thermal seal(ing) 热合
thermal sea power 海洋热能
thermal seepage 热渗眼
thermal self-action effect 热自作用效应
thermal self-focusing 热自聚焦
thermal sensation 热感觉
thermal sensibility 热灵敏性
thermal sensitive ceramic 热敏陶瓷
thermal sensitive effect 热敏效应
thermal (sensitive) element 热敏元件
thermal sensitive parameter 热敏参数
thermal sensitive stock 热敏原料
thermal sensitivity 热敏性;热敏度
thermal sensitizer 热敏剂

thermal sensor 热敏元件;热传感器
thermal separator 绝热分隔器
thermal setting adhesive 热固胶粘剂
thermal shadow 热阴影
thermal shaving foam dispenser 热刮泡沫分配器
thermal shield(ing) 热屏(蔽);热防护;屏蔽层
thermal shock 热震(荡);热际;热激波;热(冲)击;热波动;温度突跃;温度急变;温度冲击
thermal shock crack(ing) 热冲击裂缝
thermal shock effect 热冲击效应
thermal shock fatigue 热震断裂
thermal shock fracture 热冲击破断
thermal shock parameter 热震参数
thermal shock resistance 抗急冷急热性;耐热震性;热稳定性;抗热冲击阻力;抗热冲击(性)
thermal shock resistance measuration 抗热震性测定法
thermal shock resistance of coating 涂层抗震性能
thermal shock resistance test 抗热震试验
thermal shock resisting stoneware 耐热震炻器
thermal shock rig 热冲击试验设备
thermal shock shield 热冲击屏蔽
thermal shock test 热震试验;热冲(击)试验
thermal shock time current rating 热短时额定电流
thermal short circuit 热短路
thermal shrinkage 降温收缩;降热收缩;热皱缩;热(收)缩
thermal shrinkage differential 热收缩差异
thermal shrinkage stress 热收缩应力
thermal shrink packaging 热收缩包装
thermal shroud 隔热罩
thermal shunt 热分流器
thermal shut-down circuit 热击穿保护电路
thermal signal 热信号;热量信号
thermal simulation 热模拟
thermal simulation experiment 热模拟实验
thermal singularity 热奇异点
thermal sintering 热烧结
thermal siphon 热虹吸
thermal sleeve 隔热套管
thermal slope 热坡度
thermal slope wind 热梯度风
thermal small current meter 温差式微流速计
thermal softening 热软化
thermal source 热中子源;热源
thermal source of water pollution 水热污染源
thermal spalling 热震碎;热脱落;热剥落;热爆裂;受热剥落
thermal spectrum 热谱;热辐射谱
thermal spectrum analysis 热光谱分析
thermal spike 热峰;热钉;温度峰值
thermal splitting 热裂解
thermal spray 热喷雾;热喷镀
thermal spraying 热喷涂
thermal spring 热泉;温泉
thermal stability 耐热度;热稳定性
thermal stability curve 热稳定性曲线
thermal stability help out parameter 热稳定性辅助指标
thermal stability measuration 热稳定性测定法
thermal stability of raw meal nodules 生料球的热稳定性
thermal stability parameter 热稳定性指标
thermal stability test 热稳定性试验
thermal stabilization 热加固(法)
thermal stabilization method 气热加固法
thermal stabilizer 热稳定剂
thermal standard 灼热标准
thermal state 热态
thermal steady state 热稳定状态
thermal steering 热引导;热成引导
thermal still 热蒸馏釜
thermal stimulated luminescence 热激发光
thermal storage 蓄热
thermal storage capacity 蓄热量;蓄热能力
thermal storage characteristic 蓄热特性
thermal storage coefficient 蓄热系数
thermal storage effect 蓄热作用
thermal storage heating 蓄热式采暖
thermal storage heating system 蓄热式供热系统
thermal storage tank 蓄热水池
thermal storage valve 储[贮]热阀门
thermal storage vessel 储[贮]热容器
thermal storage wall 蓄热墙

thermal straightening 热矫直
thermal strain 热应变;热胁变;温差应变
thermal strain ga(u)ge 热应变计
thermal strain meter 热应变计
thermal stratification 加热分层;热分层;热成层;温度分层;温度层析作用
thermal stratification in lakes 湖泊热分层
thermal strength 热力强度
thermal stress 热应力;温度应力;温差应力
thermal stress analysis 热应力分析
thermal stress cracking 热应力开裂;热裂
thermal stress distribution 热应力分布
thermal stress duty 热应力功能
thermal stress fatigue 热应力疲劳
thermal stress ga(u)ge 热应力计
thermal stress in pipes 管子中的热应力
thermal stress meter 热应力计
thermal stress reduction 热应力缓和
thermal stress reduction factor 温度应力折减系数
thermal stretch(ing) 热拉伸;热伸长
thermal striae 热条痕
thermal structure 热(力)结构;热力构造
thermal substation 热力站
thermal substractive process 热抽除法
thermal sweep 热力蠕变
thermal switch 热控(制)开关;热动开关
thermal symmetry 热对称性
thermal system 热系统
thermal technique 热工艺
thermal telomerization 热调聚反应
thermal tempering 热强化
thermal tensioning 加热张拉
thermal test(ing) 耐热性试验;热(力)试验
thermal testing instrument 热学测试仪器
thermal test reactor 热中子试验反应堆
thermal texturing 表面烧落处理(混凝土);烧落(混凝土)
thermal thicket 热障
thermal thunderstorm 热雷暴
thermal tide 热力潮
thermal time constant 热时间常数
thermal time delay relay 热延时时继电器
thermal time lag 热延迟时间
thermal time relay 热时间继电器
thermal time scale 热时标
thermal titration 温度滴定
thermal touch 触敏温度计
thermal tower 热塔
thermal tracking 热跟踪
thermal transducer 热效换能器;热换能器
thermal (transfer) printing 热转移印花;热转印刷
thermal transition 热转化;热跃迁
thermal transmission 热传递;热(传)导;传热系数
thermal transmissivity 热传递系数
thermal transmittance 热辐射透过率;热传递;传热系数
thermal transmittance test 热传导试验
thermal transpiration 热流逸
thermal transport formula 热传递公式
thermal transport property 热传递性质
thermal trap 热收集器
thermal treatment 热加工;热处理
thermal treatment laboratory 热处理实验室
thermal trip 热效应断路器;热断路装置
thermal tube jacket 热炮管套
thermal tuning 热调谐
thermal tuning constant 热调谐常数
thermal tuning time constant 热调谐时间常数
thermal turbulence 热紊流;热湍流
thermal type 热动式
thermal type anemometer 热式风速表
thermal type galvanometer 热动式电流计
thermal type meter 热动式仪表;热电式仪表
thermal type microphone 热线式传声器
thermal type radiation detector 热辐探测器
thermal unbalance 热不平衡
thermal unit 热量单位;热单位
thermal updraft 热上升气流
thermal upset 热缩镦
thermal utilization factor 热中子利用因子;热利用率
thermal vacuum 热真空
thermal vacuum chamber 热真空容器
thermal value 热值;热当量

thermal valve 热阀;热动式调节阀
thermal vapo(u)riser 热气化器
thermal velocity 热速度
thermal velocity neutron 热能中子
thermal vibration 热振动
thermal viewer 热观察仪
thermal viscosity 热黏性;热黏度
thermal volatilization 热挥发作用
thermal volatilization analysis 热挥发分析
thermal voltage 热电压
thermal volume change 热体积变化
thermal vorticity 热涡度
thermal vorticity advection 热成涡度平流
thermal wake 热尾流
thermal walking 热位移
thermal wall 热点
thermal wallboard 阻热墙板;保温墙板
thermal wall epitaxy 热壁外延
thermal waste 热废水
thermal wastewater 含热废水;热废水
thermal water 热水
thermal water from very deep origin 超深源热水
thermal water pollution 热水污染
thermal wattmeter 温差电偶瓦特计
thermal wave 热波;温度波
thermal wave flowmeter 热波流量计(测污水流量用);热波测流计(测污水流量用)
thermal wave imaging technique 热波成像技术
thermal weathering 热力风化
thermal wedge 热楔
thermal weight change curve 热重(量)变化曲线
thermal weldability 热焊接性
thermal weld(ing) 热焊接
thermal wheel 热轮;转轮式换热器
thermal wind 热风;热成风
thermal wind equation 热成风方程
thermal window 热窗(戴克里先公共浴场的);浴场式窗
thermal zone 高温区;高温带;温度垂直带
thermantidote 冷风扇
thermarine recorder 深度温度计
therma room 温泉浴室
thermatomic black 热裂炭黑
thermatron 高频加热装置
thermautostat 自动恒温箱
thermcale 热泉
thermcutout 热保险装置
thermel 热电温度计(装有热电偶的)
thermel protection device 热保护装置
Thermex 德么克斯(一种绝热玻璃)
thermfiber blowing wool 矿渣吹制棉
thermic 热的
thermic anti-cyclone 热反气旋
thermic borehole 热法钻孔
thermic boring 热钻孔(在混凝土上钻孔的方法);热力钻孔;高温钻孔;热钻孔;热法钻探;热法钻孔
thermic copying 热敏复制术
thermic cumulus 热积云
thermic development 热显影
thermic drilling 高温钻孔
thermic element 热法还原铁
thermic equivalent 热当量
thermicon 红外(线)敏感摄像管;热敏摄像管
thermic process 热工过程
thermie 兆卡
thermilinear thermistor 热线性热敏电阻
thermindex 示温涂料;示温漆
therming 四面装饰
thermion 热离子
thermionic activity 热离子激活性;热离子活度;热放射效率
thermionic amplifier 电子管放大器
thermionic cathode 热离子阴极;热电阴极
thermionic cell 热离子电池
thermionic conduction 热离子传导;电子传导
thermionic controller 热离子调节器
thermionic conversion 热离子转换;热离子变换
thermionic converter 热离子交换器;热离子换能器;热离子变换器
thermionic current 热离子电流
thermionic detection 热离子检测
thermionic detector 碱火焰电离检测器;热离子(发射)检测器
thermionic device 热离子器件

thermionic discharge 热离子放电
thermionic effect 热离子效应
thermionic emission 热离子发射;热电放射
thermionic emission current 热离子发射电流
thermionic emission detector 热离子发射检测器
thermionic emission effect 热电放射效应
thermionic emitter 热离(子)发射体
thermionic fuel element 热离子燃料元件
thermionic generation 热离子发电
thermionic instrument 热离子仪器
thermionic multiplier tube 热阴极电子倍增管
thermionic oscillator 电子管振荡器
thermionic reactor 热离子反应堆
thermionic rectifier 热离子整流器
thermionic relay 热离子继电器;电子继电器
thermionics 热阴极电子学
thermionic source 热离子源
thermionic vacuum ga(u)ge 热离子真空计
thermionic vacuum tube 热离子真空管
thermionic voltmeter 热离子管电压表
thermionic work function 热离子功函数
thermischer 热效率
thermisopleth 变温等值线;等变温线
thermister 热控管;热元件;热敏电阻器;热敏电阻;热变阻器;热变电阻器
thermistor alloy 热敏电阻合金
thermistor anemometer 热敏式风速计
thermistor bead 球状热敏电阻
thermistor bolometer 热敏电阻式辐射热测量计;热敏电阻检测器;热敏(电阻)测辐射热计
thermistor bolometer detector 热敏电阻测辐射热仪
thermistor bridge 热敏电阻电桥
thermistor bridge circuit 热敏电阻桥形电路
thermistor compensation 热敏电阻补偿
thermistor composition 热敏材料
thermistor controller 热敏电阻检测器
thermistor detector 热敏电阻检测器;热敏电阻辐射热测量器
thermistor flowmeter 热变电阻流量计
thermistor ga(u)ge 热敏真空计;热敏电阻式压力计;热敏电阻式测量仪
thermistor heat detector cell 热敏电阻测热器
thermistor mount 热敏电阻座
thermistor nomogram 热敏电阻列线图
thermistor oscillator 热敏电阻振荡器
thermistor power meter 热敏电阻功率计
thermistor probe 热敏探示器;热敏电阻传感器
thermistor (temperature) sensor 热敏电阻传感器
thermistor thermometer 热敏电阻温度计;热敏电阻温度表
thermistor type motor protection 热敏电阻型电动机保护装置
thermistor-type thermostat 热敏型恒温器
thermistor vacuum ga(u)ge 热敏电阻真空计
thermit(e) 铝粉焊熔剂;热剂;热焊剂;铝热剂
thermit(e) bomb 高热剂燃烧弹
thermit(e) collar 高热轴环
thermit(e) crucible 热剂焊用坩埚
thermit(e) fusion welding 热剂铸焊
thermit(e) iron 铝热还原铁
thermit(e) joint 铝热焊接缝
thermit(e) metal 铝热焊金属粉
thermit(e) method 铝热法
thermit(e) mixture 铝热剂混合器;热剂
thermit(e) mo(u)ld 铝热铸型;热剂模
thermit(e) pressure welding 加压热焊剂
thermit(e) process 铝热剂法;铝热法
thermit(e) rail welding 钢轨热剂焊
thermit(e) reaction 铝热反应;热剂反应
thermit(e) splice 高温搭接焊;铝热剂钢筋搭接
thermit(e) steel 铝热焊钢粉
thermit(e) welded joint 热焊接头
thermit(e) welded rail joint 热焊钢轨接头
thermit(e) welding 热焊接术;铝热剂焊接法;铝热(接);铝热剂焊;铝镍焊;热剂焊
thermium 温泉群落
thermo 德蒙卢克斯(一种玻璃纤维夹层玻璃)
thermoacoustic(al) effect 光声效应
thermoacoustic(al) recording meter 热声自动记录仪
thermoacoustic array 热声阵
thermoacoustimetry 热声学法
thermoadsorption 热吸附作用

thermoaeroelasticity 热弹性;气动弹性力学
thermoammeter 热电偶安培计;热电流计;热电流表;温差电偶电流计;温差电偶安培计
thermoanalysis 热分析
thermoanelasticity 热滞弹性
thermoanemometer 热温式风速计;温热式风速计;温差式风速仪;温差(式)风速表
thermoartesian system 热自流系统
thermobalance 高温天平;热天平
thermobarograph 自记温度气压计
thermobarometer 虹吸气压计;虹吸气压表;温压表;温度气压计;温度气压表;温度测高计
thermobarometric(al) wave 温压波
thermobattery 热电池组;温差电池组
thermobilling 按燃气热值收费
thermobiology 高温生物学
thermobiosis 高温生活
thermobottle 热水瓶
thermocamera 热像照相机
thermocapillary movement 温差毛细管水运动
thermocatalysis 热催化
thermocatalystic detector 热催化检测器
thermocatalytic exhaust device 热催化排气装置
thermocatalytic polymerization 热催化聚合
thermocatalytic purification 热催化净化(作用)
thermocaustica 热灼剂
thermocauterectomy 烙除法;热烙除法
thermocauterization 热烙术
thermocautery 热烙术;热烙器
thermocell 热电池;温差电池
thermochemical calorie 热化学卡
thermochemical cycle 热化学循环
thermochemical destruction curve 热力化学致死曲线
thermochemical engine 热化学发动机
thermochemical equation 热化学方程;热化学反应式
thermochemical equilibrium 热化学平衡
thermochemical isotope effect 热化学同位素效应
thermochemical kinetics 热化学动力学
thermochemical method 热化学方法
thermochemical property 热化学性能
thermochemical reaction 热化学反应
thermochemical remanent magnetization 热化学剩余磁化
thermochemical theory 热化学理论
thermochemical treatment 热化学处理
thermochemistry 热化学
thermochor 分子体积与温度关系
thermochroic 反射热线性
thermochroism 反射热线作用
thermochromatic effect 热色效应
thermochromatic property 热色性能
thermochromatism 热色性
thermochromic display 热色显示
thermochromism 热致变色;热色现象;热变色现象
thermochromy 热色性
thermochrosis 反射热线作用
thermocirculator 热循环装置;热环流器
thermocline 斜温层;温跃层
thermocline layer 降温带;降温层;温度跃(变)层;水下温度突变层
thermocoagulation 热凝固(术)
thermocoax 超细管式热电偶;瑟莫科斯热电偶
thermocolorimeter 热比色计
thermocolo(u)r 变色温度指示;热变色;示温颜料;示温涂料;彩色温度标示;变色温度;示温色
thermocompensation capacitor 温度补偿电容器
thermocompression 热压
thermocompression bond 热压胶合
thermocompression bonder 热压接合器
thermocompression bonding 热压焊(接)
thermocompression distill 热压蒸馏法
thermocompression evapo(u)rator 热压缩蒸发器
thermocompression welding 热压焊(接)
thermocompressor 热压缩机;热压机
thermoconductivity 热传导率
thermocontact 热接触
thermocontroller 热敏控制剂
thermoconvection 热对流
thermocooling 温差环流冷却
thermocople junction 热电偶接点
thermocouple 热电偶;温差电偶
thermocouple ammeter 热电偶式安培计;温差电

偶安培计
thermocouple attachment 热电偶安装
thermocouple block 热电偶砖;插热电偶的砖
thermocouple cartridge heater 热电偶圆筒加热器
thermocouple comparator 热电偶比较仪
thermocouple converter 热变换器
thermocouple current meter 热电偶流速仪;热电偶海流计
thermocouple detector 热电偶检测器
thermocouple dipole 热电偶极子
thermocouple element 热电偶元件
thermocouple ga(u)ge 热电偶仪表;热电偶真空计;热电偶气压计
thermocouple generator 热电偶发电器
thermocouple instrument 热电偶仪表;热电偶式测试仪器
thermocouple junction 热电偶结
thermocouple material 热电偶材料
thermocouple measurement 热电偶测量
thermocouple metal 热电偶合金
thermocouple meter 热电偶仪表;热电偶式电表;热电偶电表;温差电偶仪表;温差电偶计
thermocouple microphone 热电偶传声器
thermocouple needle 热电偶针;热电偶计
thermocouple plug 热电偶塞
thermocouple probe 热电偶探头
thermocouple protection tube 热电偶保护管;热电偶护管
thermocouple pyrometer 热电偶高温计;热电高温计
thermocouple relay 热电偶继电器
thermocouple restorer 热电偶恢复器
thermocouple sensor 热电偶传感器
thermocouple sheath 热电偶套管
thermocouple simulator 热电偶模拟器
thermocouple thermometer 热电偶温度计
thermocouple thermometry 热电偶测温法
thermocouple tube 热电偶套管;热电偶管
thermocouple-type ga(u)ge 热电偶式真空计
thermocouple vacuum ga(u)ge 温差电偶真空计;热电偶(式)真空计
thermocouple wattmeter 热电偶瓦特计;热电偶功率计
thermocouple well 热电偶管;插热电偶的套管
thermocouple wire 热电偶线
thermocouple wire protection 热电偶导线保护
thermocrete 高炉熔渣
thermocuring 热固化
thermocurrent 热电流;热差电流;温差电流
thermocurrent meter 热电偶流速仪
thermocushion 热垫层
thermocut-off 热动关闭器;热保险装置
thermocutout 热保险装置;热断流器
thermocyclogenesis 热力气旋生成
thermode 热电极
thermodenaturation 热变性作用
thermodesorption 热解吸
thermodestruction 热破坏
thermodetector 温差探测器;热检测器;热检波器;热电检波器;测温仪;测温器;测温计
thermodialysis 热渗析
thermodifferential analysis 差示分析
thermodiffusion 热扩散(作用)
thermodiffusion coefficient 热扩散系数
thermodiffusion cycle 热扩散循环
thermodiffusion factor 热扩散系数
thermodiffusion fractionation 热扩散分级
thermodiffusivity 热扩散性
thermodilatometric analysis 热膨胀分析
thermodilatometry 热膨胀法
thermodilution 热稀释法
thermodilution curve determination 温度稀释曲线测定
thermodine 温变层
thermodissociation 热解离
thermodrill 热力钻孔;热力钻机
thermodrying 烘干
thermoduric 耐热的
thermoduric bacteria 耐热细菌
thermodynamic(al) 热动力的;热力学的
thermodynamic(al) acidity 热力学酸度
thermodynamic(al) activity 热力学活性;热力活性
thermodynamic(al) analogy 热力模拟
thermodynamic(al) analysis 热力学分析

thermodynamic(al) argument 热力学理论
thermodynamic(al) barrier 热障
thermodynamic(al) change 热力学转化
thermodynamic(al) chart 热力学图表;热力图
thermodynamic(al) circulation 热环流;热力循环;热力环流
thermodynamic(al) coefficient 热力系数
thermodynamic(al) concentration 热力有效浓度;热力浓度
thermodynamic(al) condition 热动力条件
thermodynamic(al) coordinate 热力学坐标
thermodynamic(al) criterion 热力学准则
thermodynamic(al) cycle 热力循环;热力学循环;热动力循环
thermodynamic(al) data 热力学数据
thermodynamic(al) definition 热力学定义
thermodynamic(al) degeneration 热力学上的退化
thermodynamic(al) dew-point temperature 热力学露点温度
thermodynamic(al) diagram 热力学图;热力图解
thermodynamic(al) disturbance 热力扰动
thermodynamic(al) driving force 热力驱动力
thermodynamic(al) duct 热力学管
thermodynamic(al) efficiency 热力学效率;热力效率
thermodynamic(al) energy 热力学能
thermodynamic(al) energy equation 热力学能量方程
thermodynamic(al) entropy 热力学的熵
thermodynamic(al) equation of state 热力学物态方程
thermodynamic(al) equilibrium 热力(学)平衡;热动(力)平衡
thermodynamic(al) equilibrium constant 热力学(平衡)常数
thermodynamic(al) equilibrium equation 热力学方程
thermodynamic(al) equilibrium state 热力学平衡状态
thermodynamic(al) flux 热力学通量
thermodynamic(al) function 热力(学)函数
thermodynamic(al) function of state 热力学态函数
thermodynamic(al) heating system 热力供热系统
thermodynamic(al) ice-bulb temperature 热力学冰球温度
thermodynamic(al) index 热工指数
thermodynamic(al) instrument 热工仪表
thermodynamic(al) investigation 热力试验;热力研究
thermodynamic(al) law 热力学定律
thermodynamic(al) limit 热力(学)极限
thermodynamically admissible process 热力学容许过程
thermodynamically stable phase 热力学稳定相
thermodynamic(al) medium 热工介质
thermodynamic(al) metamorphism 热动力变质
thermodynamic(al) method 热力学法
thermodynamic(al) model 热力模型;热动力学模型
thermodynamic(al) model of climate 气候的热力学模式
thermodynamic(al) parameter 热力学参数
thermodynamic(al) partition coefficient 热力学分配系数
thermodynamic(al) perfect degree 热力完善度
thermodynamic(al) potential 热力(学)势(能);热力势;热动力势
thermodynamic(al) pressure 热力学压力
thermodynamic(al) principle 热力学原理
thermodynamic(al) probability 热力学概率
thermodynamic(al) process 热力(学)过程
thermodynamic(al) property 热力(学)性质;热力学性能;热工性能
thermodynamic(al) propulsion system 热力推进系统
thermodynamic(al) quantity 热力量
thermodynamic(al) reasoning 热力学推论(计算吸附作用)
thermodynamic(al) relation 热力学关系
thermodynamic(al) restriction 热力学约束
thermodynamic(al) scale 热力学温标
thermodynamic(al) scale of temperature 热力学温标

thermodynamic(al) similarity 热力学相似
thermodynamic(al) solution theory 热力学溶液理论
thermodynamic(al) solution thereom 热力学溶液理论
thermodynamic(al) stability 热力学稳定性
thermodynamic(al) stability condition 热动力稳定性条件
thermodynamic(al) state 热力学状态
thermodynamic(al) stratification 热力层次
thermodynamic(al) study 热力学研究
thermodynamic(al) system 热力(学)系统;热力学体系
thermodynamic(al) table 热力学表
thermodynamic(al) temperature 热力学温度
thermodynamic(al) temperature scale 热力(学)温标
thermodynamic(al) test 热力学试验
thermodynamic(al) theory 热力学理论
thermodynamic(al) trap 热动力疏水器
thermodynamic(al) (type steam) trap 热动力(式)疏水器
thermodynamic(al) variable 热力学态函数;热力学变量
thermodynamic(al) wet-bult temperature 热力学湿球温度
thermodynamicist 热力学家
thermodynamics 热力学;热动力学
thermodynamics erosion 热力学侵蚀
thermodynamics isotope effect 热力学同位素效应
thermodynamics of nuclear materials 核材料热动力学
thermodynamics of reactions 反应热力学
thermodynamics of the nucleus 核热力学
thermoefficiency 热效率
thermoelastic after-effect 热弹性后效应
thermoelastic distortion 热弹性变形
thermoelastic effect 热弹(性)效应
thermoelastic inversion 热弹转换;热弹性逆转
thermoelastic inversion point 热弹性逆转点
thermoelasticity 热弹性(力学)
thermoelastic strain 热弹性应变
thermoelastic stress 热弹性应力
thermoelastic stress-strain law 热弹性应力—应变规律
thermoelastic theory 热弹性理论
thermoelectric(al) 热电的;温差电的
thermoelectric(al) actinography 热电日射表
thermoelectric(al) actinometer 热电日射表
thermoelectric(al) action 热电作用
thermoelectric(al) air conditioner 热电式空调装置;热电空调器
thermoelectric(al) air conditioning plant 热电式空气调节装置
thermoelectric(al) air conditioning system 热电式空调系统;温差电空调系统
thermoelectric(al) alarm 热电警报器
thermoelectric(al) baffle 热电挡板
thermoelectric(al) battery 热偶电池
thermoelectric(al) cell 热电式感温元件;温差电池
thermoelectric(al) ceramics 热电陶瓷
thermoelectric(al) circulating cooling water 热电循环冷却水
thermoelectric(al) colo(u)rimeter 热电色度计
thermoelectric(al) comminution 热电破碎
thermoelectric(al) comparator 热电比较器;热电比测器
thermoelectric(al) conversion 热电转换
thermoelectric(al) converter 热电转换器;热电交换器;热电变换器
thermoelectric(al) cooler 热电制冷器;热电冷却器
thermoelectric(al) cooling 热电制冷;热电冷却;温差电制冷
thermoelectric(al) cooling device 温差式冷却装置
thermoelectric(al) cooling module 热电制冷器
thermoelectric(al) coolor 温差制冷器
thermoelectric(al) couple 热电偶;温差电偶
thermoelectric(al) crystal 热电晶体
thermoelectric(al) current 热电流;温差电流
thermoelectric(al) cycle 热电循环
thermoelectric(al) device 热电装置;热电器件
thermoelectric(al) diagram 热电图
thermoelectric(al) diode 温差电二极管
thermoelectric(al) effect 热电效应;温差电效应;

温差电势效应
thermoelectric(al) element 热电元件
thermoelectric(al) equivalent 热电当量
thermoelectric(al) galvanometer 热电电流计
thermoelectric(al) generating station 热力发电站
thermoelectric(al) generation station 热力发电站
thermoelectric(al) generator 热电式发动机;热电发电机;热变换器
thermoelectric(al) heating 热电加热;热电供暖
thermoelectric(al) heating device 热电发热器
thermoelectric(al) heat pump 温差电热泵
thermoelectric(al) instrument 热电式仪表
thermoelectric(al) inversion 热电变换
thermoelectric(al) junction 热电偶接头
thermoelectric(al) law 温差电定律
thermoelectrically heterogeneous effect 热电非均匀效应
thermoelectric(al) material 热电材料
thermoelectric(al) measurement 热电测量;电阻测量
thermoelectric(al) method 热电势法
thermoelectric(al) nuclear battery 温差核电池
thermoelectric(al) photometer 热电光度计
thermoelectric(al) pick-up 热电拾取器
thermoelectric(al) pile 热电堆
thermoelectric(al) plant 热电厂
thermoelectric(al) potential 热电势
thermoelectric(al) power 热电能;热电功率;温差电势
thermoelectric(al) power generation 热力发电
thermoelectric(al) power generator 热力发电机;热变换器
thermoelectric(al) power plant 火力发电厂;热电厂
thermoelectric(al) power station 热电站;热电厂
thermoelectric(al) power system 热电动力系统
thermoelectric(al) property 热电性能;温差电性
thermoelectric(al) pyrometer 热电偶高温计;热电高温计;热电高温表;温差电偶高温计
thermoelectric(al) radiometer 热电偶辐射计
thermoelectric(al) refrigeration 热电制冷
thermoelectric(al) refrigeration system 热电式制冷系统;温差电制冷系统
thermoelectric(al) sensing element 热电传感元件
thermoelectric(al) sensor 热电式传感器;热电传感器
thermoelectric(al) series 热电序;热电特性表;热电次序
thermoelectric(al) solar cell 热电太阳能电池
thermoelectric(al) station 热电站
thermoelectric(al) thermometer 热电温度计;温差(电偶)温度计
thermoelectric(al) traction 热力电动牵引;热电牵引
thermoelectric(al) transducer 热电式传感器
thermoelectric(al) type 热电式
thermoelectric(al) winding system 电热缠丝机
thermoelectric(al) wire 热电导线
thermoelectricity 热电学;热电(性);热电现象;热电物质;温差电
thermoelectro analogy apparatus with conductive paper 导电纸热电模拟装置
thermoelectrochemistry 热电化学
thermoelectrode 热电电极
thermoelectrolumines 热激电致发光
thermoelectroluminescence 热激电致发光
thermoelectromagnetic 热电磁的
thermoelectromagnetic pump 热电磁泵
thermoelectrometer 热电计
thermoelectrometry 热电学法
thermoelectromotive 热电动的
thermoelectromotive force 热电动势;温差电动势
thermoelectromotive potential 热电动势
thermoelectron 热电子
thermoelectronic emission 热电发射
thermoelement 热敏元件;热电元件;热电偶;温差电偶
thermoemission reactor 热发射反应堆
thermoenergetic agent 高热能工作剂
thermoexpansion compensation angular ball bearing 热胀补偿向心球轴承
thermofiber blowing wool 矿渣吹制棉
thermofin 热隔层;传热翅片
thermofission 热分裂

thermofixation 热固化
thermoflash process 热闪蒸法
thermoflask 热水瓶
thermofluid enamel 热塑性瓷釉
thermofluid system 热流体系统
Thermofor 塞摩福流动床;载热固体
Thermofor clay burning process 塞摩福流动床白土烧再生过程
Thermofor continuous cracking plant 塞摩福流动床连续裂化车间
Thermofor continuous percolation 蓄热器连续渗滤
Thermofor continuous percolation process 塞摩福流动床连续渗滤过程
thermoforming 热塑成型;热压成型;热成型
thermoforming machine 热成型机
thermoforming machinery 热成型机械
thermoforming plastic sheet 热成型塑料板
thermoforming process 热成型过程;热成型工艺;热成型(方)法
thermofractograph 热分离色谱
thermofractography 热裂谱法
thermo-fracture 热裂解
thermofuse 热熔丝
thermofusion welding 热熔焊(接)
thermogalvanic corrosion 热电腐蚀;电化学腐蚀
thermogalvanometer 热电偶检流计;热电偶电流计;温差电偶检流计;温差电偶电流计
thermogas chromatography 热气相色谱法
thermogenerator 热偶池;热偶发生器
thermogenesis 热产生;生热作用;产热
thermogenetic 热产生的
thermogenic 热源的;生热的
thermogenic action 致热作用
thermogenic bacteria 产热细菌
thermogenic centre 生热中枢
thermogenic gas 热成因气
thermogenic 生热学
thermogenic soil 热成土
thermogenous 生热的
thermo-geotechnology 热岩土工程学
thermoglobe anemometer 热球式电风速计
thermogram 自记温度图表;热像图;热解曲线;热分析图;温谱图;温度自记曲线;温度图;温度记录图;温度过程线;温度分布图;差突分析曲线
thermograph 自记温度计;自记温度表;自记式温度计;自动温度器;记录温度计;红外(线)影像;热图像(红外线暗视器摄取的图像);热红外图;温度记录仪;温度记录器;温度记录仪
thermographacamera 热像照相机
thermograph correction card 温度计订正表
thermographic(al) inspection 热像检测;温度记录法检查
thermographic(al) process 热摄影法
thermography 热压凸印刷;热发影术;热熔印刷;热谱法;热敏成像法;热红外图像;温度记录法;温场照相术;差热分析;测温术;发黏温度记录;表面测温术
thermogravimeter 热重仪;热解重量仪
thermogravimetric analysis 热重(量)分析;热失重分析;热解重量分析
thermogravimetric curve 热重曲线;热失重曲线
thermogravimetry 热重法;热重(量)分析法;热重(量)分析;热失重法;热解重量分析术
thermogravitational column 热扩散柱
thermohaline 热盐的
thermohaline circulation 热盐环流;温盐环流
thermohaline circulation of the ocean 海洋的温盐环流
thermohaline convection 热盐对流;温盐对流;温度盐分合成对流
thermohaline current 温盐洋流;温盐海流
thermohaline environment 温盐环境
thermohaline field 温盐场
thermohaline structure 温盐结构
thermohardening 热硬化;热定形
thermohardening lacquer 热固性涂料
thermohardening resin 热固性树脂
thermohydraulic effect 热水力学效应
thermohydraulics 热工水力学
thermohydro conduction 热湿传导
thermohydrodynamic(al) 热流体动力的
thermohydrodynamic(al) equation 热流体动力方程

thermohydrodynamic(al) instability 热流体动力不稳定性
thermohydroforming 热液压成型
thermohydrograph 温湿度记录器
thermohydrometer 热比重计;温度比重计;温差比重计
thermohygrogram 温湿自记曲线;温湿(曲线)图
thermohygrograph 温湿计;温度湿度记录器;自记温湿度计;自记温湿度表
thermohygrometer 温湿表;温度湿度计
thermohygrostat 恒温恒湿器
thermohynamic condition 热力工况
thermoimage device 热成像器件
thermoinactivation 热灭活法
thermoindicating panel 温度指示屏
thermoindicator paint 热变涂料;示温涂料;示温漆;示温油漆;变色漆
thermointegrator 土壤积热仪;土壤积热器
thermoionic cell 热离子室
thermoionic emission 热电子发射
thermoionic emitting element 热离子元件
thermoionization 热电离
thermoisogradient 热等梯度
thermoisohyp 实际温度等值线
thermoisopleth 等温线;变温等值线
thermoistor 热控管
thermojet 热射流;热喷涂;热喷射;热力喷气发动机
thermojunction 热接点;热电偶接点;热电偶接头;热电偶接点;热电偶;温差电偶;温差电结
thermojunction ammeter 热电偶式安培计
thermojunction battery 热电偶电池组
thermojunction type 热电偶式
thermojunction type meter 热电偶式测量计
thermokarst 热岩溶
thermokarst pit 热喀斯特坑
thermokarst topography 热喀斯特地形
thermokinetic analysis 热动力学分析
thermokinetics 热动力学
thermolabile 不耐热的
thermolabile material 不耐热材料
thermolability 不耐热性
thermolability test 热不稳定试验
thermolamp 热灯
thermolator method 导热压铸法
thermolens effect 热透效应
thermolite 绝热平屋顶;高热炭丝红外线灯
thermolith 耐火水泥;耐火胶结材料
thermolization 热化
thermolize 表面热处理
thermolizing 表面热处理
thermolog(ging) 温度测井(曲线)
thermology 热学
thermoluminescence 加热发光;热致发光;热释光;热激发光;热发光
thermoluminescence ages 热发光年龄
thermoluminescence curve 热致曲线
thermoluminescence dating 热致发光判断年代法;热释光判断年代法
thermoluminescence dating method 热释光测年法
thermoluminescence detector 热释光探测器
thermoluminescence dosimeter 热致剂量计
thermoluminescence method 热发光方法
thermoluminescent 热释光的;热致发光的
thermoluminescent detector 热释光探测器
thermoluminescent dosimeter 热释光剂量计;热发光剂量计
thermoluminescent dosimetry 热释光剂量学
thermolysis 热解(作用);热分解;热放散
thermolysis curve 热解曲线
thermolysis of wood 木材热解
thermolysis water yield 热解水收率
thermolytic 热放散的
thermolytic dissociation 热解
thermolytic reactivity 热解反应性
thermomagnet 热磁
thermomagnetic 热磁(的)
thermomagnetic alloy 热磁合金
thermomagnetic converter 热磁换能器
thermomagnetic coversion 热磁转换
thermomagnetic effect 热磁效应
thermomagnetic generator 热磁发电机
thermomagnetic motor 热磁电动机
thermomagnetic oxygen analyzer 热磁氧分析仪
thermomagnetic phenomenon 热磁现象

thermomagnetic preparation 热磁选矿法
thermomagnetic property 热磁性
thermomagnetic transducer 热磁换能器
thermomagnetic treatment 热磁处理
thermomagnetism 热磁学;热磁性;热磁现象
thermomagnetization 热磁化
thermomagnetometry 热磁学法;热磁测定法
thermoman 热人(耐燃试验用人体模型)
thermomechanical analysis 热机械分析
thermomechanical analysis method 热机械分析法
thermomechanical analyzer 热机械分析仪
thermomechanical curve 温度—形变曲线
thermomechanical effect 热机械效应
thermomechanical generator 热机发电机
thermomechanical magnetic treatment 磁场形变热处理
thermomechanical method 热机械方法
thermomechanical property 热机械性质;热机械性能
thermomechanical puling 热力浆化法
thermomechanical treatment 形变热处理(法);热机械处理
thermomechanical vibration treatment 振动形变热处理
thermomechanical working for texturization 织构化热加工
thermomechanics 热变形学;热力学
thermomeion 高负温距平区
thermometal 热敏金属;温差双金属;双金属
thermometamorphism 热同素异形;热力变质(作用)
thermometer 寒暑表;温度计;温度表;体温计
thermometer adjustment 温度计调整
thermometer anemometer 温度计式风速表
thermometer bird 温度计鸟
thermometer boss 温度计套管;温度计接口
thermometer bulb 温度计小球;温度计水银球;温度计球;温度计泡;温包
thermometer calibration 温度计校正
thermometer collar 温度计套圈
thermometer conversion 温度计换算
thermometer correction 温度计校正
thermometer error 温度计误差
thermometer float 测温浮标
thermometer for water temperature 水温计
thermometer frame 温度计架
thermometer glass 温度计玻璃
thermometer guard unit 温度计保护装置
thermometer lag 温度计延迟
thermometer method 温度计法
thermometer microscope 测温显微镜
thermometer reader 温度计读数机
thermometer reading 温度计读数
thermometer scale 温标;温度计标度
thermometer screen 仪器百叶箱;温度表百叶箱
thermometer set 温度计组
thermometer shelter 仪器百叶箱;百叶箱温度计;百叶箱温度表
thermometer shield 温度计罩
thermometer shutter 百叶箱
thermometer sounding 温度测深
thermometer stem 温度计枢轴
thermometer support 温度计支架
thermometer tube 温度计管
thermometer type thermistor 温度计式热敏电阻
thermometer well 温度计套管;温度计插孔;温度计插池
thermometer with contacts 触点式温度计
thermometric analysis 量热分析;测温分析
thermometric conductivity 热导系数
thermometric constant 需热量
thermometric depth 温测深度;测温深度
thermometric diffusivity coefficient test of frozen soil 冻土导温系数试验
thermometric gas 测温气体
thermometric hydrometer 测温比重计
thermometric indicator 测温仪
thermometric level(i)ing 沸点气压高程测量
thermometric liquid 测温液体
thermometric property 测温性质
thermometric scale 温度标;温标
thermometric scale for agriculture 农业温标
thermometric sounding 温度测深
thermometric titration 热量滴定法;热滴定(法);温度滴定;测温滴定法
thermometrograph 自记式温度计;温度记录器;温度记录计
thermometry 计温学;计温(技)术;计温法;温度测量(法);测温学;测温(技)术;测温(法)
thermomicrograph 红外(线)显微照片;热显微照片
thermomicroscopy 热显微镜学
thermomineral water 热矿水
thermomodule 热电微型组件
thermomolecular ga(u)ge 热分子真空计
thermomolecular pressure 热分子压力
thermomolecular vacuum ga(u)ge 热分子真空计
thermomotive 热动力的
thermomotor 热发动机
thermomultiplicator 热倍加器
thermonatrite 水碱;一水碳酸钠
thermonegative 吸热的
thermonegative reaction 吸热反应
thermoneutrality 热中和性
thermonuclear 热核子;热核(的)
thermonuclear apparatus 热核设备
thermonuclear bomb 热核炸弹
thermonuclear charge 热核装料
thermonuclear chemical engineering 热核化学工程
thermonuclear condition 热核条件
thermonuclear device 热核装置
thermonuclear explosion 热核爆炸
thermonuclear fire 热核火力
thermonuclear fuel 热核燃料
thermonuclear fusion 热核聚变;热核反应
thermonuclear fusion reaction 热核聚变反应
thermonuclear machine 热核装置
thermonuclear neutron soure 热核中子源
thermonuclear power 热核动力
thermonuclear power plant 热核电站
thermonuclear reaction 热核反应
thermonuclear reaction rate 热核反应速度
thermonuclear reactor 热核反应堆;热核堆
thermonuclear reactor blanket 热核反应堆再生区
thermonuclear temperature 热核温度
thermonuclear transformation 热核转化;热核变化
thermonuclear weapon 热核武器
thermonucleat reactor 聚变反应堆
thermonucleonics 热核子学;热核技术
thermooptic(al) aberration 热光像差
thermooptic(al) coefficient 热光系数
thermooptic(al) constant 热光常数;热光常量
thermooptic(al) effect 热光效应
thermooptic(al) instability 热光不稳定性
thermooptic(al) property 热光性质;热光性能
thermooptic(al) stability 热光稳定性
thermooptics 热光学
thermoosmosis 热渗;热渗透(作用);热渗透(性)
thermooxidative stability 热氧化稳定性
thermooxidizing 热氧化
thermopaint 测温漆;热敏油漆;示温涂料;测温涂料;彩色温度标示漆
thermopair 热电偶
thermopalpation 温度差别按诊法
Thermopane 得蒙平(一种隔热玻璃)
Thermopane glass 隔热玻璃(一种专利产品)
Thermopane glazing 绝热框格玻璃窗
Thermopane unit 框格格玻璃构件(门窗上的)
thermoparticulate analysis method 热微粒分析法
thermopause 热大气层顶部;热成层顶
thermoperiodicity 温周期性
thermoperiodism 温周期现象;温变周期性
thermoperm alloy orthonik alloy 铁镍合金
thermophile 嗜温植物;高温(常指40℃以上);嗜热生物
thermophile bacteria 嗜热菌
thermophilic 嗜温植物;嗜温的;嗜热的;适温的
thermophilic activated sludge process 适温活性污泥法
thermophilic aerobic biological wastewater treatment 嗜热好氧生物废水处理
thermophilic aerobic digestion 高温好氧消化;亲热需氧消化(量)
thermophilic aerobic reactor 高温好氧反应器
thermophilic bacteria 喜温细菌;嗜热菌;适温细菌;耐热细菌
thermophilic cellulose decomposing 高温纤维素分解菌
thermophilic digestion 高温蒸煮;高温(厌氧)消化;嗜热消化
thermophilic digestion treatment 高温消化处理
thermophilic fermentation 高温发酵
thermophilic form 适热型
thermophilic hydrolysis-oxic contact oxidation process 高温水解好氧接触氧化法
thermophilic microorganism 嗜热微生物;适热微生物
thermophilic organism 喜温生物
thermophilic range 嗜热范围
thermophilic sludge digestion 污泥高温消化
thermophilic species 喜热树种
thermophilous 喜温植物;适温的
thermophilous organism 喜热生物;适热生物
thermophilous plant 好湿植物
thermophone 热(致)发声器;传声温度计
thermophoresis 热泳现象;热迁移
thermophosphor 热磷光体
thermophosphorescence 热发磷光
thermophotography 热照相术
thermophotovoltaic 热光伏电的
thermophotovoltaic cell 热光伏元件;热光电元件
thermophotovoltaic conversion 热光伏变换;热光电转换
thermophylactic 抗热的
thermophysical 热物理的
thermophysical property 热物理性能
thermophysics 热物理学
thermophyte 耐热植物;适温植物
Thermopil 德蒙皮尔(一种音响传声材料)
thermopile 热电偶;热电堆;温差电堆;温差电池
thermopile generator 热电堆发电机
thermopile stick 测温色笔
thermopile using Peltier effect 采用帕尔帖热电堆
thermoplast 热塑性材料;热塑性塑料;热范性
thermoplastic 加热软化;热塑性塑料;热塑性材料;热塑体;热塑(的)
thermoplastic acrylic 热塑性聚丙烯系纤维
thermoplastic acrylic resin 热塑性丙烯酸树脂
thermoplastic adhesive 热塑性黏合剂
thermoplastic backer 热塑性衬里
thermoplastic binding 热塑性装订;无线装订
thermoplastic bitumen product 热塑性沥青制品
thermoplastic box toe 热塑性内包头
thermoplastic cable 包耐热塑料电缆
thermoplastic cement 热塑性胶
thermoplastic coating 热塑性涂层
thermoplastic concentrate 热塑性塑料提浓物
thermoplastic-covered wire 热塑绝缘线
thermoplastic decoration 热塑装饰品
thermoplastic deformation 热塑性变形
thermoplastic display 热塑显示
thermoplastic elastomer 热塑性弹性体;热塑性弹性材料
thermoplastic enamel 热塑性瓷釉
thermoplastic ethylene-propylene diene monomer 热塑乙烯—丙烯二烯单体
thermoplastic family 热塑塑料族
thermoplastic fiber bonding 热塑性纤维结合力
thermoplastic fibre 热塑性纤维
thermoplastic film 热塑性胶片;热塑薄膜;热塑性塑料薄膜;热塑膜
thermoplastic floor(ing) tile 热塑地板砖;热塑性铺地板
thermoplastic flow 热塑流
thermoplastic glue 热塑性塑料胶
thermoplastic hologram 热塑全息照片
thermoplastic hot-melt adhesive 热塑热熔黏合剂
thermoplastic hubcap 热塑性塑料轮毂帽
thermoplastic insulated cable 热塑性绝缘电缆
thermoplastic insulating tape 热塑绝缘(胶)带【电】
thermoplastic insulation 热塑性塑料隔热
thermoplasticity 热塑性;热熔塑性
thermoplastic laminate 热塑性层压
thermoplastic layer 热塑性薄层
thermoplastic light valve 热塑料光阀;热塑光阀
thermoplastic material 热塑性材料;热塑料料;热塑材料
thermoplastic memory system 热塑存储系统
thermoplastic microsphere 热塑性塑料微球体
thermoplastic mo(u)lding compound 热塑性模塑料

thermoplastic nature 热塑性
thermoplastic nitrile-polyvinyl chloride 热塑腈聚氯乙烯
thermoplastic olefin 热塑性烯烃
thermoplastic phenolic resin 热塑性酚醛树脂
thermoplastic photoconductor 热塑性(光)导体
thermoplastic photoelectric conductor 热塑光电导体
thermoplastic plate 热塑性塑料板
thermoplastic polyamide 热塑性聚酰胺
thermoplastic polyester 热塑性聚酯
thermoplastic polymer mixture 热塑性聚合混合物
thermoplastic powder 热塑性粉末
thermoplastic powder coating 热塑性粉末敷层
thermoplastic protective tape 热塑防护带
thermoplastic puff 热塑性内包头
thermoplastic putty 热塑(性)油灰
thermoplastic resin 热塑(性)树脂
thermoplastic resin adhesive 热塑性树脂黏合剂;热塑性树脂胶
thermoplastic resin matrix composite 热塑性树脂复合材料
thermoplastic rubber 热塑性橡胶
thermoplastics 热塑塑料
thermoplastics bodied vehicle 热塑车身汽车
thermoplastic sheet 热塑薄型板;热塑(性塑料)片
thermoplastic spacer 热塑性隔条
thermoplastic synthetic(al) rubber material 热塑性合成橡胶材料
thermoplastic tape 热弹性带
thermoplastic tile 热塑性地面(饰面)砖;热塑性屋面料;热塑性铺地板
thermoplastic tiling 热塑(制)板
thermoplastic traffic marking material 热塑性路标材料
thermoplastic urethane fabric 热塑性聚氨酯织物
thermoplastic veil 热塑性纤维表面毡
thermoplastic vinyl tile 聚氯乙烯地面砖;热塑乙烯饰面砖
thermoplastic vulcanizer 热塑性硫化体
thermoplastic welding strip 热塑性焊条
thermoplegia 热射病
thermopneumatic control system 热气动控制系统
thermopolymerization 热聚合(作用)
thermopositive 放热的
thermopower 热能
thermoprecipitation 热沉淀
thermoprecipitinogen 热沉淀原
thermopress paint 热敏油漆
thermopress paint spraying 热压喷漆
thermoprobe 热探头;测温探针
thermoprotective material 热防护材料
thermoprotective sight glass 防温视镜
thermoptometry 热光学法
thermoquenching 热(浴)淬火
thermoradiography 热射线摄影术
thermoradiometry 放射量热法
thermoreceptor 热感受器;温度感受器
thermoreduction 铝热法
thermoregulation 温度调节
thermoregulator 调热器;热(量)调节器;温度控制器;温度调节器;调温器
thermorelay 热继电器;热电继电器
thermoremanence 热顽磁
thermoremanent magnetism 热剩余磁性;热剩磁
thermoremanent magnetization 热剩余磁化强度
thermoremanent remagnetization 热剩余再磁化
thermoresistance 热阻
thermoresistant 抗热的
thermoresistive effect 温阻效应
thermoroll forming 热辊成型
thermorunaway 热致破坏;热致击穿;热失控
thermorunning fit 热转配合
thermos 热水瓶;保温瓶
thermosalinograph 温盐测量仪
thermos bottle 热水瓶;保温瓶
thermos-bulk blowing machine 保温瓶吹泡机
thermoscope 验温器;测温锥;测温仪;测温器
thermoscopic bar 示温棒;测温棒
thermoscreen 仪器百叶箱;热屏(蔽)
thermosensitive 热敏的
thermosensitive ceramics 热敏陶瓷
thermosensitive effect 热敏效应
thermosensitive material 热敏材料

thermosensitive meter 热感测定仪
thermosensitive paint 热敏油漆;热敏涂料
thermosensitive resistor 热敏电阻器
thermosensitive transducer 热敏传感器
thermosensor 热敏元件
thermoset 热凝物;热固性材料;热固(性);热固树脂;热固化性
thermoset deflashing 热固性塑料去飞边
thermoset extrusion 热挤塑法
thermoseting plastics 热固塑料
thermoseting material 热凝材料
thermoset plate 热固性塑料板
thermosetting 热硬化;热凝的;热固性;热固化;热固(的)
thermosetting acrylic resin 热固性丙烯酸树脂
thermosetting adhesive 热固性黏结剂;热固性胶合剂;热固性胶粘剂;热固胶(合剂)
thermosetting cement 热固性黏合胶;热固性胶
thermosetting coating 热固性涂料
thermosetting composite 热固复合塑料;热固复合材料
thermosetting fibre 热固性纤维
thermosetting glue 热补胶火
thermosetting ink 热固性油墨
thermosetting insulation 热固绝缘
thermoset(ting) material 热固(型)塑料;热固性物质;热固(性)材料
thermosetting mo(u)lding compound 热固性模塑料;热固性模塑化合物
thermosetting mo(u)lding powder 热固成型粉
thermoset(ting) plastics 热固性塑料;热硬塑料
thermosetting plates 热固塑料印版
thermosetting polymer 热固性聚合物;热聚合物
thermosetting powder coating 热固粉末涂层
thermosetting powders 热固性树脂粉末
thermosetting property 热硬化性;热凝性;热固性
thermosetting resin 热固性树脂;热固树脂
thermosetting resin adhesive 热固性树脂黏结剂
thermosetting resin composite 热固性树脂复合材料
thermosetting resin matrix composite 热固性树脂基复合材料
thermosetting sand 加热硬化砂
thermosetting synthetic resin 热固性合成树脂
thermosetting varnish 热固性(油)漆
thermosetting yarn 热定形纱
thermoset varnish 热固性清漆
thermos flask 保温瓶
thermoshield 热屏(蔽)
thermosiphon 热虹吸蒸发器;热虹吸管
thermosiphon circulation 热对流循环法
thermosiphon cooling 热虹吸管冷却
thermosiphon cooling system 热虹吸冷却系统;温差环流冷却系统
thermosiphon reboiler 热虹吸管再煮锅
thermosiphon theorem 热对流原理
thermosistor 调temp器
thermosizing 热锤击尺寸整形
thermos method 蓄热法;保温法
thermosnap 热保护自动开关
thermosoil-hygrometer 热偶式土壤温度计
thermosol 热溶胶
thermosonde 热感探空仪;热感探测仪
thermosonimetry 热发声法
thermosounder 热感探测仪
thermosounder recorder 深度温度计
thermosphere 热电离层;热大气层;热(成)层
thermospray liquid chromatography-mass spectrometry 热喷雾液相色谱法—质谱法联用
thermostability 耐热性;热稳(定)性
thermostabilization 热稳定
thermostable 热稳定的
thermostage 春化阶段
thermostaking 热定标
thermostart 加热起动装置
thermostat 温度自动调节器;恒温箱;恒温器;恒温槽;热动开关;温度继电器;温度调节装置
thermostat air bath 恒温空气浴槽
thermostat blade 恒温器开关叶;恒温器断流计
thermostat drier 恒温烘箱
thermostated container 恒温箱
thermostat equipment 恒温设备
thermostat for refrigerator 冰箱用温度控制器
thermostat for vehicle 车辆恒温器

thermostat housing 恒温器壳体
thermostatic alarm 温度报警器;定温自动报警器
thermostatically controlled 恒温控制的
thermostatically-controlled bath 恒温浴
thermostatically controlled furnace 恒温调节炉
thermostatically controlled warmer 恒温控制加热器
thermostatically operated 恒温控制的
thermostatic apparatus 恒温装置
thermostatic bath 恒温浴;恒温槽
thermostatic bimetal 恒温双金属
thermostatic blending mixer 恒温混合搅拌器
thermostatic bottle 恒保温瓶
thermostatic carbureter 恒温化油器
thermostatic chamber 恒温室
thermostatic container 恒温容器
thermostatic control 恒温控制;恒温调节;温度自动控制
thermostatic control valve 恒温控制阀;恒温调整器
thermostatic cutout 恒温开关
thermostatic developing tank 恒温显影桶
thermostatic equipment 恒温设备
thermostatic evapo(u)rator 恒温蒸发器
thermostatic evapo(u)rator pressure regulator 恒温蒸发器压力调节器;恒温节流阀
thermostatic expansion valve 恒温膨胀阀;热力膨胀阀
thermostatic fire detection system 恒温火警系统
thermostatic heat transfer 恒温传热
thermostatic oven 恒温炉;恒温加热炉
thermostatic pyrometer 恒温高温计
thermostatic radiator valve 散热器的恒温阀
thermostatic reaction 恒温反应
thermostat(ic) regulator 恒温调整器;恒温调节器
thermostatics 静热力学;热静力学
thermostatic shower mixing valve 淋浴恒温混合阀
thermostatic steam trap 恒温式疏水器;恒温凝汽罐
thermostatic switch 恒温开关
thermostatic trap 恒温隔汽具
thermostatic valve head with remote sensor 内装遥控传感器的恒温阀门头
thermostatic water bath 电热恒温浴缸
thermostatic water blending mixer 恒温水混合器
thermostatic water-circulator bath 超级恒温水浴
thermostatic water valve 恒温水阀
thermostat irrigator 恒温冲洗器
thermostat oil cooler 恒温油冷却器
thermostat relay 恒温器继电器;恒温继电器
thermostat sensing element 恒温器传感元件
thermostat switch 恒温器开关
thermostat varnish 恒温稳定漆;耐热漆
thermostat water bath 恒温水浴锅
Thermosteel 德蒙斯蒂尔(一种绝热的钢屋面材料)
thermosteresis 热损失;热耗损
thermosteric anomaly 热比容偏差;比容异常
thermostriction 热致收缩
thermos vessel 恒温器
thermoswitch 热敏开关;热控(制)开关;热电偶继电器;双金属热膨胀开关
thermosynthesis 热合成
thermosyphon 热虹吸蒸发器;热虹吸器;温差环流系统
thermosyphon cooling system 温差环流冷却系统
thermosystaltism 温度性收缩
thermotactic 趋温的
thermotactic optimum 适温
thermotank 节温箱;节温柜;恒温箱;调温柜
thermotank system 自动调节温度装置
thermotape 热塑带
thermotaxic 趋温的
thermotaxis 向热性;趋温性;趋热性
thermotaxy 热排性;热排聚形
thermotechnical 热工的
thermotel 负载指示器
Thermotex 德蒙塔克斯(一种绝缘板)
thermotics 热学
thermotolerance 耐热性
thermotolerant 耐热的
thermotolerant bacteria 耐热细菌
thermotostat 自动恒温器
thermotracheotomy 热烙气管切开术
thermotropic 正温
thermotropic(al) liquid crystal polymer 热致液晶聚合物

thermotropic bacterium 适热细菌
thermotropic model 正压大气模式
thermotropism 向温性;向热性
thermotropy 热互变;热变性
thermotube 热管
thermovac evapo(u)rator 热真空蒸发器
thermovacuum 热真空
thermovacuum forming 热真空成型
thermovalve 热敏阀
thermovent 散热口
thermoviewer 热像图仪
thermoviscoplasticity 热黏塑性
thermoviscosimeter 热黏度计
thermoviscosity 热黏度
thermovision 热视
thermovoltaic 热伏打
thermovoltaic detection 热伏探测
thermovoltmeter 热电式伏特计;电电压计
thermovolumetric analysis 热容量分析
thermoweld 热焊(接);熔接
thermowell 热电偶套管;温度计套管;温度计插孔
thermsilid 硅铁耐酸合金
therm window 热窗(古罗马戴克利先公共浴场的)
the road of common property 共同富裕的道路
the role of foreign trade as a lever 外贸杠杆作用
the root-meansquare velocity of molecular 分子均方根速率
the root of the plant 植物的根
therophyte 一年生植物
the routing of water drainage 排水走向
Thersilion at Megalopolis 迈加洛波利斯的塞锡利翁(希腊)
therzolite 二辉橄榄岩
the sagger makers 匣户
the same stratigraphic level 同一地层标高
thesaurus 实库
the saved ship 获救船
the saving things 被救场
the second generational pesticide 第二代农药
the second party 乙方
the second position of Gauss 高斯第二位置
the second trophic level 第二营养级
Theseion 古希腊神庙
the selection of plant site 厂址选择
the selling party 卖方
the sequential occupancy release system 顺序占用解放系统
the shortest ultraviolet 波长最短的紫外线
the situation of mineral resources 矿产资源形势
the six draft readings 六面吃水
the smallest possible value 最小可能值
the Society of Naval Architects and Marine Engineers 造船及轮机工程试协会
the south-north seismotectonic zone 南北地震带
the special care and placement 优抚安置制度
the specifications of highway soil test 公路土工试验规程
the state of the art 现代工艺水平;目前工艺水平
the statute of reclaimation of land 土地复垦规定
the stone of a column 柱基础石;柱石
the strategy of developing the country by relying on science and education 科教兴国战略
The Swedish International Development Authority 瑞典国际开发署
the system of examination and approval of projects 项目审批制
the system of holding the legal person responsible for project 健全的项目法人责任制
theta angle 收敛角
theta pinch 角箍缩
the task of repairing and reinforcing the main dykes 干堤的修复加固任务
the tax to be paid each month 每月应纳的税款
the tax withheld from each payment 每次所扣的税款
the theory of railway tariff 运价理论
the thickness of the layer 土层的厚度
the third generational pesticide 第三代农药
the third grade of minerogenetic prospect 三级成矿远景区
the third grade of oil-gas prospect 三级油气远景区
the third grade pf coall-forming prospect 三级成煤远景区

the third party 丙方
the third trophic level 第三营养级
the three parties of the contract 合同三方当事人
the three-wastes 三废(指废水、废气、废渣)
the time of the beginning and the end for drought season 旱季起止时间
the time of the beginning and the end for rainy season 雨季起止时间
the tip of anything 物体的尖端
the tip of the axis 轴的末端
the tools and machines used in farm operations 耕作农机具
the total volume of a soil 土壤全部体积
the transfer of and changes in property rights of enterprises 产权划转和产权变动
the tributary area of water transport 水运腹地
the Tropic(al) of Capricorn 南回归线
the Tropic(al) of Cancer 夏至线
the twelve channels 正经
the twenty-four solar terms 二十四季节
the type of area cultivated 栽培面积的类型
the undersigned 署名人
the undertaker's 殡仪馆
The United Nations Whole System Mid-term Environmental Scheme 联合国全系统中期环境方案
the United States Comprehensive Soil Classification 美国土壤综合分类系统
the usable life of machine 机器的使用年限
the use of gradient centrifugation 利用分级离心法
the usual units of ages 年龄单位
the U Th-Pb dating method 铀-钍-铅测年法
the utilization ratio of arrival and departure track 到发线利用率
the utilization ratio of throat 咽喉利用率
The venin's generator 特夫南等效发生器
Thevenin's law 等效发生器法则
Thevenin's theorem 等效发电机定理;特夫南定理
the volume of gas phase 气相体积
the volume of liquid phase 液相体积
the washing away of soil by moving water 径流对土壤的冲刷
the weakest point 最弱点
the weakest side 最弱边
the west 西方
the west mountain forest soil area 西部山地森林土区
the whole of the works 整个工程
the wilds 荒野
the Winter Solstice 冬至
the withholding agent 扣缴义务人
the withholding income tax return 扣缴所得税报告表
the work of budgeting wages 劳动工资计划工作
the work of cultivating soil 土壤耕作工作
the world geomagnetic chart 世界地磁图
the world loss of cropland 世界范围的耕地损失
the world-wide navigational warning service 全世界航行警告业务
the year of regime data 动态资料的年份
the Yellow Sea 黄海
thiadiazin 代森环
thiakol rubber 丁氰合成橡胶
thiambutosine 二苯硫脲
thiamine 硫胺
thiamine chloride 氯化硫胺
thiamine disulfide 二硫胺
thiamine propyldisulfide 丙硫硫胺
thiamine tetrahydrofurfuryl 呋喃硫胺
thiamine tetrahydrofurfuryl disulphiele 呋喃硫胺
thiazine 硫氮杂苯
thiazole 间硫氮(杂)茂;间氮硫茂
thiazole dye 噻唑(结构)染料
thiazolidine 四氢噻唑;噻唑烷
thiazoline 硫氮戊环烯;噻唑啉
thick 浓的;密的;厚的;稠(密)的
thick activated sludge 浓缩活性污泥
thick and thin block 提琴式滑车
thick arch dam 厚拱坝
thick arch support lining with thin wall 厚拱薄边墙衬砌
thick beaded edge 串珠厚边
thick-bedded 厚层状;厚层的

thick-bedded method (涂层的)厚底方法
thick-bedded orebody 厚层矿体
thick-bedded structure 厚层状构造
thick-bedded texture 厚层状结构
thick board 厚木(板)
thick body cup-saucer 厚体杯碟
thick cement grout 稠水泥浆
thick-channel device 厚沟道器件
thick china 贴面(瓷)板
thick cloth producing equipment 粗厚织物生产设备
thick cloud 密云
thick coal seam 厚煤层
thick-coated arc welding rod 厚药皮焊条
thick-coated electrode 厚药皮焊条
thick contour 粗等高线
thick covering 厚药皮
thick cylinder 厚筒;厚壁筒
thick darkness 漆黑
thicken 增浓;增厚;增稠;变厚
thickened 增稠的
thickened concentrate 浓缩精矿
thickened deslimed pulp 浓密脱泥矿浆
thickened-edge (道路的)厚边
thickened-edge concrete pavement 厚边式混凝土铺面;厚边式混凝土路面
thickened-edge concrete slab 厚边式混凝土路面板
thickened-edge (cross) section 厚边式(横)断面;厚边路面横断面
thickened-edge design 厚边板设计;厚边路面设计
thickened-edge plate 加厚边缘的板
thickened-edge slab 厚缘板;厚边板
thickened-edge type concrete pavement 厚边式混凝土路面
thickened-edge type section 厚边型横截面;厚边型路面横截面
thickened fuel 稠化燃料
thick(ened) line 粗线
thickened oil 增稠油;厚熟油;调墨油;稠化油(料)
thickened portion of the slab 板的变厚部分
thickened product 浓密产品
thickened pulp 浓纸浆
thickened sludge 增稠污泥;浓污泥;浓缩污泥
thickened slurry 浓缩泥浆
thickened underflow 浓缩底流;浓泥;浓密底流
thickened waste-activated sludge 浓缩废活性污泥
thickener 沉降槽;增稠器;增稠剂;浓缩剂;浓缩机;稠化器
thickener clear zone 浓密机澄清区
thickener sludge zone 浓密机泥浆区
thickener tray 浓密机层
thickener tunnel 浓密机排砂坑道
thickener underflow 浓缩机底流;浓密机底流
thickener vat 浓缩机底槽(圆桶)
thickening 致密;增稠过程;浓缩;稠化(过程);变厚;变浓;变稠;增浓;增厚;增稠
thickening agent 增稠剂;浓密剂;稠化剂
thickening cone 圆锥浓缩机
thickening cyclone 浓密旋流器
thickening fibre 加厚纤维
thickening filter 浓缩滤器
thickening flushing 稠泥浆洗井
thickening material 增稠剂
thickening of activated sludge 活性污泥浓缩
thickening of mud 泥浆加稠;泥浆稠化
thickening of side wall 边墙加厚
thickening of sludge 污泥浓缩
thickening of slurry 泥浆浓缩
thickening process 增稠过程
thickening rate 增稠速度
thickening settling tank 浓缩沉淀池
thickening tank 浓缩槽;浓缩池
thickening technology 增密工艺
thickening time 增稠时间;浓化时间;稠化时间
thickening time of cement 水泥初凝时间
thickening time tester for oil well cement 油井水泥稠化时间测定仪
thickening time test of cement 水泥初凝时间测定
thickening treatment 浓缩处理
thickening-upward sequence 向上变厚层序
thickening washer 加厚垫圈
thicket 植丛;灌木丛;密(灌)丛;热带密灌丛;树丛;丛薄

thick filament 粗丝
thick-film 厚膜
thick-film amplifier 厚膜放大器
thick-film chip resistor 厚膜片电阻器
thick-film circuit 厚膜电路
thick-film column 厚膜柱
thick-film integrated circuit 厚膜集成电路
thick-film interconnection 薄膜连接电路
thick-film lubrication 厚油膜润滑;厚膜润滑
thick-film pressure transmitter 厚膜压力传感器
thick-film resistor 厚膜电阻器
thick-film rust preventive 厚膜防锈的
thick-film test 厚膜试验
thick first-year ice 当年厚冰(厚度一百二十厘米以上)
thick fog 一级能见度;浓雾
thick glue 稠胶
thick ice 厚冰
thick indifference class 甚无差异类
thick iron 多倍厚板
thick juice 稠汁
thick juice press 稠汁压滤机
thick juice sulfitation 稠汁亚硫酸化
thick lamellar structure 厚纹层构造
thick laminae 厚纹层状
thick layer 厚层
thick layer dredging 厚层开挖
thick layer excavation 厚层开挖
thick layer filling 厚层充填的
thick layer of fluid mud 厚浮泥层
thick layer sample 厚层样品
thick lift compaction 厚层压实
thick lift construction 厚层施工法
thick lift paving 厚层铺筑法【道】
thick line 粗线
thick liquid 浆
thickly inhabited 人烟稠密的
thick mo(u)lding compound 稠模塑料
thick mud 浓泥浆;稠泥浆
thick multiply plywood 多层厚胶合板
thickner 增稠剂
thickness 黏稠;密度;厚度;厚薄;稠密
thickness adjusting plate 厚度调节板
thickness after forming 成型之后厚度;成型后的厚度
thickness allowance against corrosion 防锈厚度留量
thickness along axial plane 沿轴面的岩层厚度
thickness and corrosion allowance 厚度和腐蚀裕度
thickness averaging method 厚度平均法
thickness by penetration 穿透法测厚
thickness by reflection 反射厚度;反射法测定的厚度
thickness change 稠度改变;厚度改变
thickness chart 厚度图
thickness-chord ratio 厚弦比
thickness coefficient 厚度系数
thickness control handle 铺设厚度控制杆
thickness design 厚度设计
thickness design method 厚度设计法
thickness deviation 厚度偏离率
thickness dial ga(u)ge 带千分表的厚度规
thickness drag 厚度阻力
thickness dummy 装帧样本
thicknesser 压刨(机);刨压机;刨板机;刨
thickness extension vibration mode 厚度伸缩振动模式
thickness feeler 量隙规;千分垫
thickness figure of mid-layer 中间层厚度图
thickness flexure vibration mode 厚度弯曲振动模式
thickness ga(u)ge 厚度量规;厚度计;厚度规;厚薄规;塞规;测厚仪;测厚计;测厚规
thickness index 结构数(路面强度指数);厚度指数
thicknessing machine 去厚机械刨(木工);刨板机;压刨(机);定厚机
thickness isogram of local interval 局部层间等厚图
thickness line 厚度线
thickness loss 厚度损耗
thickness measurement 厚度测量;厚度测定
thickness measurement with laser 激光测厚
thickness measurement with ray 射线测厚

thickness meter 测厚仪
thickness mo(u)lding 角饰(平面与下垂竖面相交处的);橡胶的底脚;檐口线脚(檐口下);厚线条饰;一组线脚的最后一道线脚;挑檐滴水板下的线脚;挑出物下的线脚
thickness of active layer 活动层厚度
thickness of apical plate of aquifer 含水层顶板岩层厚度
thickness of aquifuge 隔水层厚度
thickness of arch 拱的厚度
thickness of armo(u)r layer 护面层厚度
thickness of artificial recharge stratum 人工补给层厚度
thickness of ballast 道砟厚度
thickness of bearing stratum 持力层厚度
thickness of bed 岩层厚度;地层厚度
thickness of blasting layer 炸层厚度
thickness of caprock 盖层厚度
thickness of cascade calcareous-sinter precipitation 瀑水钙华厚度
thickness of cave sediments 洞穴沉积物厚度
thickness of coal formation 煤层的厚度
thickness of coal seam 煤层厚度
thickness of coating 油漆厚度;抹灰厚度;粉刷厚度
thickness of compacted layer 压实层厚度
thickness of compacted soil 加固土层厚度
thickness of Conrad discontinuity 康拉德面速度
thickness of construction 建筑厚度(路面)
thickness of cover 剥离厚度
thickness of covering 涂药厚度
thickness of covering strata 盖层的厚度
thickness of crust of weathering 风化壳厚度
thickness of cutting 切削厚度
thickness of dam 坝厚
thickness of deposit 沉积厚度
thickness of destructive zone of bottom bed 底板破坏带厚度
thickness of diffusion layer 扩散层厚度
thickness of dredged layer 挖层厚度【疏】
thickness of dredging layer 挖层厚度【疏】
thickness of drill tip 钻尖厚度
thickness of dry strata with oil 含油千层厚度
thickness of failure zone 破坏区的厚度
thickness of finish 粉刷厚度
thickness of formation 地层厚度
thickness of frozen soil layer 冻土层厚度
thickness of gas water transition zone 气水过渡带厚度
thickness of ga(u)ge 标准量规的厚度
thickness of glide mass 滑体厚度
thickness of gravel stuffing 填砾层厚度
thickness of grouting layer 注浆分层厚度
thickness of half value 半值厚度
thickness of high resistance layer 高阻层厚度
thickness of horse 断片厚度
thickness of ice 冰厚
thickness of individual bed 单层厚度
thickness of layer 层厚(度)
thickness of leaky aquifer 越流层厚度
thickness of lining 衬砌厚度
thickness of link plates 链板厚度
thickness of main desired layer 主要目的层厚度
thickness of marker horizon 标准层厚度
thickness of mud-stone flow 泥石流厚度
thickness of oil-gas transition zone 油气过渡带厚度
thickness of oil slick 油膜厚度
thickness of oil-water layer 油水同层厚度
thickness of oil-water transition zone 油水过渡带厚度
thickness of orebody 矿体厚度
thickness of overburden 积土厚度;剥离厚度
thickness of overlying strata 上覆地层厚度
thickness of parting 夹矸厚度
thickness of pipe 管厚;管壁厚度
thickness of pipe wall 管壁厚
thickness of plastering 抹灰厚度
thickness of plate 板的厚度
thickness of relative aquifuge 相对隔水层厚度
thickness of relaxation ring 松动圈厚度
thickness of retaining layer 保护层厚度
thickness of roadbed 道床厚度
thickness of rock stratum 岩层厚度
thickness of roof rock 盖层厚度

thickness of root face 钝边高度
thickness of sample 样品厚度
thickness of seam 缝道宽度;缝宽
thickness of sinter 泉华厚度
thickness of slab 板厚
thickness of sliding zone 滑动带深度
thickness of soft layer 软弱层厚度
thickness of soil layer 土层厚度
thickness of source layer 源层厚度
thickness of source rock 母岩厚度
thickness of spraying 喷射厚度
thickness of step 梯级厚度
thickness of stone shield 石盾厚度
thickness of strata with oil 油层厚度
thickness of stratum 地层厚度
thickness of stripping layer 剥离层厚度
thickness of tectonic layer 构造层厚度
thickness of the absorbed layer 吸附层厚度
thickness of the earth's crust 地壳厚度
thickness of the fifth layer 第五层厚度
thickness of the first layer 第一层厚度
thickness of the fourth layer 第四层厚度
thickness of the magnetic body 磁性体厚度
thickness of the nth layer 第 n 层厚度
thickness of the overburden 表土厚度
thickness of the ozone layer 臭氧层的厚度
thickness of the second layer 第二层厚度
thickness of the third layer 第三层厚度
thickness of thin mica 薄片云母厚度
thickness of thread 螺纹厚度
thickness of tooth 齿厚
thickness of underlayer 垫层厚度
thickness of water strata with oil 含油水层厚度
thickness parallel to axial plane at given point 某点平行轴面的厚度
thickness parameter parallel to axial plane 平行轴面厚度参数
thickness piece 厚薄规厚隙规
thickness range 板厚范围
thickness rank 厚度分级
thickness ratio of rudder 舵的厚度比
thickness recorder 厚度记录器
thickness recovery 产品厚度回缩
thickness reduction due to punching stretch 冲压拉伸减薄量
thickness setting scale 厚度标尺
thickness shear 厚度切变
thickness shear vibration electro-mechanical coupling factor 厚度切变振动机电耦合系数
thickness shear vibration mode 厚度切变振动模式
thickness taper 楔削厚度
thickness tester 测厚仪
thickness time 稠化时间
thickness tolerance 厚度公差;厚度负偏差
thickness tolerance of steel plate of pipe 钢板或钢管厚度负偏差
thickness vibration 沿厚度方向振动
thickness vibration electro-mechanical coupling factor 厚度振动机电耦合系数
thickness-width ratio 厚宽比
thick oil 稠油
thick oil wastewater 稠油废水
thick overlay 厚面层
thick paint 厚漆
thick panel 厚胶合板
thick paste 稠膏
thick plank 厚板
thick planting 密植
thick plate 厚板
thick plate glass 厚平板玻璃
thick polished plate glass 厚的抛光玻璃板
thick producing section 厚生产层
thick quartz ga(u)ge 厚石英测量仪
thick reinforced bar cutting machine 粗钢筋切断机
thick resistive bed 高阻厚层
thick rope 缆索;粗绳
thick rough cast plate glass 厚的粗浇平板玻璃;厚的未加工浇注玻璃板
thick rubber pad 厚胶垫
thick seam 厚地层
thick-section casting 厚壁铸件
thick set 密簇;稠密的
thick sheet glass 窗橱玻璃;厚板玻璃;橱窗玻璃;

三厚(窗)玻璃(5毫米厚窗玻璃)
thick shell 厚壳体;厚壳
thick-shell penstock 厚壁压力水管
thick-skinned structure 地壳厚构造
thick sludge 浓缩污泥
thick slurry 浓(泥)浆;变稠的料机
thick slurry process 浓浆法
thick smear 厚涂片
thick smear method 厚片法
thick source 厚源
thick space 厚衬铅
thick squall 强风暴
thick stick 杠子
thick string 粗绳
thick stuff 厚度为4~12(1英寸=0.0254米)英寸的原木材
thick suspended sediment 高浓度悬浮泥沙
thick target model 厚靶模型
thick-thin chart 等熵厚度图
thick undergrowth 茂密灌木丛
thick wall 厚墙;厚壁管;厚壁(的)
thick(-wall) article 厚壁制品
thick wall cavity 厚壁空洞
thick-wall chamber 厚壁电离室
thick(-wall) cylinder 厚壁圆筒
thick(-wall) cylinder theory 厚壁圆筒理论
thick(-wall) cylindric(al) shell 厚壁圆柱壳
thick walled 厚壁的
thick-walled castings 厚壁铸件
thick-walled cylinder by disperse tapered blocks 楔块分置式厚壁圆筒
thick-walled cylinder by seamless steel pipe 无疑钢管式厚壁圆筒
thick-walled fibre 厚壁纤维
thick-walled pipe 厚壁管
thick-walled sampler 厚壁采样器;厚壁取土器
thick-walled tube 厚壁筒;厚壁管(筒)
thick-walled vessel 厚壁容器
thick-wall steel cassete 厚壁钢制防爆盒
thick weather 阴天;能见度不好的天气;雾天
thick wire 粗丝
thief 取样;分流器
thief hatch 取样口;取样孔;取样盖
thief hole 取样口;取样孔
thief knot 平结
thief proof 防盗(法)
thief proof mesh 防盗网
thief proof net 防盗网
thief resistant lock 防偷窃锁
thief rod 泥浆取样杆;采样杆
thief sample 泥泵试样
thief sampler 粉末取样器;油舱底部水量测定器
thief sampling 泥泵取样
thief zone 漏水带;漏失(泥浆)带;渗水带
thief zone in a well 井中漏失区段
Thiele-Burrau regularization 蒂埃尔·布劳正规化
Thiele coordinates 蒂埃尔坐标
Thiele-Geddes method 蒂埃尔·盖德法
Thiele tube 均热管
Thiessen disintegrator 泰森洗涤机
Thiessen gas cleaner 泰森洗涤机
Thiessen gas scrubber 泰森洗涤机
Thiessen network 泰森格网
Thiessen polygon 泰森多边形
thievery 蚁贼
thieving tank 测定油舱底部水量
thigmotaxis 趋触性
thigmotropism 向触性
Thikbut 德邦板条(一种对接的屋面板条)
thill 辕;耐火黏土厚层;底黏土
thimble 锥心空心夹具;锥管;壳筒;护环;嵌环;铁制嵌轮;索眼心环;绳端心环;顶针;穿线环;Z形搅拌器
thimble assembly 套管组件
thimble bat 套筒底板;锥形垫板;耐火材料底板(窑具)
thimble chamber 套管型电离箱
thimble connector 套管连接器
thimble coupling 套管联轴节
thimble-eye 套筒眼
thimble filter 壳筒滤器
thimble-headed trysail 三角斜帆
thimble hook 套筒钩;套管钩
thimble ionization chamber 指形电离室

thimble joint(ing) 套接;套管接口;套管接合;套筒连接;套管结合;套管接头
thimble punch holder 套管冲头托架
thimbler 心环;电震极
thimbler eye splice 心环索眼插接
thimble seat 套管座
thimble shore 临时船尾撑住(下水时)
thimble slag car 渣罐车
thimble test 高温稳定性试验
thimble-tube boiler 套管锅炉
thimet 甲拌磷
thin 浅薄;薄的;变稀
thin air foil theory 薄翼理论
thin aluminum filter 薄铝滤光片
thin-arch dam 薄拱坝
thin as a piece of paper 薄如纸
thin away 变稀薄至消失;冲淡;尖灭
thin-base film 薄感光胶片
thin bed 薄层
thin-bedded 薄层状;下卧薄层的;有薄层基础的;基层薄的;薄层的
thin-bedded adhesive 薄底胶粘剂
thin-bedded fixing technique 薄底胶粘工艺
thin-bedded method 薄底方法
thin-bedded orebody 薄层矿体
thin bedded structure 薄层状构造
thin-bedded texture 薄层状结构
thin bladed knife 薄刃刀
thin blanket feed 薄层投料法
thin board 薄纸板
thin butt 粗端过磨的
thin cement grout 稀水泥浆
thin cement slurry 稀水泥浆
thin coal seam 薄煤层
thin coat 薄膜
thin coated electrode 薄包剂焊条
thin coating 薄质涂料
thin-coat plaster 罩面灰浆
thin concrete overlay 薄层混凝土罩面
thin consistency 稀稠度;稀释度;低黏度;低稠度;不稠;稀黏滞度
thin contour 细等高线
thin core bit 薄壁取芯钻头
thin core dam 薄心墙坝
thin corporation 虚弱公司;负债过多的公司
thin cutter 精细切碎机
thin-cut veneer 薄切割胶合板(饰面)
thin cylinder 薄壁圆筒
thin-cylinder formula 薄壁圆筒公式
thin dielectric(al) capacitor 薄介质电容器
thin down 逐渐稀薄;逐渐变弱;变细
thin drilling bit 薄壁钻头
thin drizzle 细毛毛雨
thin edge 薄边
thin-edged 薄边的
thin-edged knife 薄刃刀
thin emulsion carrier film 薄感光胶片
thin epoxy joint 薄层环氧接缝
thin equity 低资产净值
thin faced bit 薄壁(金刚石)钻头
thin feel 稀薄手感
thin-film 薄膜
thin-film band pass filter 薄膜带通滤光片
thin-film capacitor 薄膜电容器;薄膜电容
thin-film chromatography 薄层色谱(法)
thin-film circuit 薄膜电路
thin-film coating 薄膜涂层;薄膜镀层
thin-film coil-coated system 薄膜卷涂系统
thin-film column 薄膜柱
thin-film component 薄膜元件
thin-film composite structure 薄膜合成结构
thin-film cryotron 薄膜冷子管;薄膜低温管
thin-film crystal 薄膜晶体
thin-film deposition 薄膜沉积
thin-film detector 薄膜探测器
thin-film device 薄膜器件
thin-film diamond 薄膜金刚石
thin-film distillation 薄膜蒸馏
thin-film electrothermal component 薄膜电热元件
thin-film element 薄膜元件
thin-film epitaxy 薄层外延
thin-film evapo(u)ration technique 薄膜蒸镀法
thin-film evapo(u)rator 刮膜蒸发器;薄膜蒸发器
thin-film ferrite coil 薄膜铁氧体磁芯线圈

thin-film field-emission cathode 薄膜场致发射阴极
thin-film fixed bed reactor 薄膜固定床反应器
thin-film gas sensor 薄膜式气敏元件;薄膜气体传感器
thin-film heating test 薄膜加热试验
thin-film hologram 薄膜全息图
thin-film hybrid integrated circuit 薄膜混合集成电路
thin-film hydrogen sensor 薄膜氢传感器
thin-film image converter 薄膜变像管
thin-film infrared detector 薄膜红外探测器
thin-film insulated wire 薄膜绝缘电线
thin-film integrated circuit 薄膜集成电路
thin-film interferometer 薄膜干涉仪
thin-film ionics 薄膜离子学
thin-film laser 薄膜激光器
thin film layers 多层薄膜
thin-film light guide 薄膜光导管;薄膜光导
thin-film lubrication 薄油膜润滑;薄膜润滑
thin-film magnetic head 薄膜磁头
thin-film magnetic module 磁模组件;薄膜磁组件
thin-film magnetoresistive head 薄膜磁阻磁头
thin-film material 薄膜材料
thin-film memory 磁薄膜存储器;薄膜存储器
thin-film memory computer 薄膜存储式计算机
thin-film microelectronics 薄膜微电子学;薄膜微电子器件
thin-film microscopy 薄膜显微术
thin-film monolithic circuit 薄膜单片集成电路
thin-film notch filter 薄膜陷波滤波器
thin-film optic(al) coating 光学薄膜
thin-film optic(al) components 薄膜光学元件
thin-film optic(al) modulator 薄膜光调制器
thin-film optic(al) multiplexer 薄膜光复用器
thin-film optic(al) shutter 薄膜光学快门
thin-film optic(al) switch 薄膜光开关
thin-film optic(al) waveguide 薄膜光波导
thin-film optics 薄膜光学
thin-film osmosis desulfurization 薄膜渗透法脱硫
thin-film oven test 薄膜烘箱试验(地沥青);薄膜蒸发试验
thin-film passive network 薄膜无源网络
thin-film photocatalytic slurry reactor 薄膜悬浮液光催化反应器
thin-film photocell 薄膜光电管;薄膜光电池
thin-film photoconductor 薄膜光电导体
thin-film photodiode 薄膜光电二极管
thin-film photoresistor 薄膜光敏电阻
thin-film physics 薄膜物理学
thin-film prism 薄膜棱镜
thin-film processing 薄膜工艺
thin-film production machine 薄膜生产机
thin-film pyroelectric(al) detector 薄膜热电探测器
thin-film rectifier 薄膜整流器
thin-film resistance thermometer 薄膜电阻温度计
thin-film resistor 薄膜电阻器;薄膜电阻
thin-film scanner 薄膜扫描机
thin-film semi-conductor 薄膜半导体
thin-film semi-conductor detector 半导体薄膜探测器
thin-film sensor 薄膜传感器
thin-film single junction quantum interferometer 薄膜单结量子干涉仪
thin-film solar cell 薄膜太阳电池
thin-film solid-state image sensor 薄膜固体像传感器
thin-film spreading agent 薄膜分散剂
thin-film storage 薄膜存储器
thin-film strain ga(u)ge 薄膜式应变压力计;薄膜式应变片
thin-film switch 薄膜开关
thin-film technique 薄膜技术
thin-film technology 薄膜工艺学
thin-film thermistor 薄膜热敏电阻器
thin-film thermoelectrics 薄膜温差电学
thin-film thickness 薄膜厚度
thin-film thickness meter 薄膜厚度计
thin-film transducer 薄膜换能器
thin-film-transistor material 薄膜晶体管材料
thin-film water 固着水;薄膜水
thin-film waveguide bolometer 波导型薄膜测热器;波导式薄膜测辐射热计
thin-film waveguide spectrum analyzer 薄膜波导光谱分析仪

thin first-year ice 当年薄冰(厚度30~70厘米)
thin flame 焊焰
thin flat head rivet 扁平头铆钉
thin flat head socket shank rivet 扁平头半空心铆钉
thin flat nose pliers 薄扁嘴钳
thin flattened chips 扁石屑
thin flattish chips 扁石屑;薄片石屑
thin flowing consistency 稀浆稠度
thin fog 轻雾(霾);四级能见度;薄雾
thin foil honeycomb core 薄箔蜂窝芯子
thin foll 薄膜
thing 潮式呼吸
thin-gage goods 薄壁制品
thin ga(u)ge conduit 薄铁管
thin-ga(u)ge skin 蒙皮
thin ga(u)ge strip 薄标准钢条
T-hinge 丁字铰链;T形长尾铰链;丁字形铰链
T-hinge strap 丁字铰接
thin glass plate 薄玻璃板
thin glaze 薄釉
thin gold film resistor 薄金膜电阻器
thing personal 动产
thing pledged 担保物
thing possessed 占有物
thing real 不动产
things liable to duty 征税物品
things mortgaged 抵押物;抵押品
things personal 个人财产
things pledged 抵押物
things real 不动产
things to be declared 报关物品
thin hollow brick 薄空心砖
thin hologram grating 薄全息光栅
thin iron sheet 薄铁板
thinium 沙丘群落
think 认为
thin kerf 窄截槽(采矿)
thin-kerf bar 薄截盘
think factory 智囊团
thin knurled nut 滚花扁螺母
think tank 智囊团
thin lamellar structure 薄纹层构造
thin laminae 薄纹层状
thin-layer 薄膜层;薄层
thin-layer charging 薄层加料法
thin-layer chromatograph 薄层色层分析仪
thin-layer chromatographic densitometry 薄层色谱光密度定量法
thin-layer chromatography 薄色层分离法;薄层色谱(法);薄层分析法;薄层析(法)
thin-layer chromatography densitometer 薄层光密度计
thin-layer chromatography scanner 薄层色谱扫描仪
thin-layer chromatography-scanning spectrodensitometry 薄层色谱法—扫描光谱密度法联用
thin-layer deaerator 薄层脱气塔
thin-layer deposit 薄层沉积
thin-layer dredging 薄层浚挖
thin-layer electrochemistry 薄层电化学
thin-layer erosion 薄层冲刷
thin-layer evapo(u)rator 薄层蒸发器
thin-layer fill 薄层堆填;薄层填土
thin-layer filling 薄层充填的
thin-layer gel filtration 薄层凝胶过滤(法)
thin-layer microreactor 薄层微型反应器
thin-layer plate 薄层板
thin-layer plate densitometry 薄层板上光密度测定法;板上光密度测定法
thin-layer rod chromatography 薄层棒色谱法
thin-layer sample 薄层样品
thin-layer scanner 薄层扫描仪;薄层扫描器
thin lead 薄片条
thin lens 薄透镜
thin lift 薄层
thin liquid 稀液
thin liquid film membrane 液态薄膜状隔膜
thin list 薄表
thinly deposited copper sheet 电解沉积的纯铜薄片
thinly inhabited 人烟稀少的
thinly populated country 居民稀少地区
thinly stratified 分成薄层的;薄层叠成的

thin magnetic film 薄磁膜
thin margin 薄利
thin market 不景气的市场;交易稀少的市场;交易滞呆的市场
thin membrane 薄膜
thin membrane box structure 薄壁箱形结构
thin membrane dam 窄心墙坝;薄心墙坝
thin membrane mirror 薄膜镜
thin metal plate 薄金属板
thin mud 稀泥浆
thinnable 可变薄的
thinned array 稀疏天线阵
thinned array antenna 稀疏阵列天线;稀布阵列天线
thinned-debris-flow deposit 稀性泥石流沉积
thinned oil 稀释石油
thin(ned) section method 薄片鉴定法
thinned water 稀释用水
thin negative 弱反差
thinner 信那水;稀释剂;稀料;冲淡剂
thinner for acrylic baking coating 丙烯酸烘漆稀释剂
thinner for nitrocellulose finishes 硝基漆稀释剂
thinner for oleoresinous paint 油基漆稀料
thinner for traffic paint 道路标线漆稀释剂
thinner shaft 较薄的柱身
thinning 细化;稀释;疏枝法;变薄
thinning additive 稀漆剂
thinning agent 稀释剂;稀漆剂
thinning cycle 疏伐周期
thinning for tending 抚育性疏伐
thinning from above 上层疏伐
thinning limit 稀释限度
thinning machine 压薄机;轧薄机
thinning of the Arctic ice cap 北极冰盖逐渐变薄
thinning of the ozone layer 臭氧层逐渐变薄
thinning out 尖灭【地】
thinning out of strata 地层尖灭;地层变薄
thinning ratio 稀释比(率)
thinning tank 稀释槽;稀释罐
thinning-upward sequence 向上变薄层序
thin nose pliers 扁嘴钳
thin nut 薄螺母
thin oil 稀油;轻油
thinolite 岸钙华
thinophyte 沙丘植物
thin ore bed 矿层(煤层)薄
thin out 稀释;冲淡;减少;尖灭;变细
thin overburden 浅覆盖层;浅剥离层
thin overcast 较明朗的阴天
thin overlay 薄罩面层
thin pipe 薄壁圆筒
thin (pipe)line 小直径管线
thin placing lift 薄浇筑层(混凝土)
thin plank 薄板
thin planting 稀植
thin plaster 涂墙稀灰泥;稀墙泥
thin plate 薄板
thin-plate element 薄板元件;薄板构件;薄板单元
thin plate glass 薄板玻璃
thin plate lement 薄板构件;薄板元件
thin plate marking 薄板标志
thin plate mill 薄板扎机
thin plate orifice 薄孔板
thin plate structure 薄壁板结构
thin plate weir 薄板堰;薄壁堰
thin polymer film 聚合物薄膜
thin prism 薄棱镜
thin rain 稀疏薄雨
thin resin-base coating 树脂基薄涂层
thin ribbed plate 浅肋形板
thin rim type rotor 薄轮缘式转子
thin roof 薄壳屋顶
thin screen model 薄屏模型
thin screwdriver with wood(en) handle 木柄薄螺丝起子
thin seam 薄煤层
thin section 磨片;薄剖面;薄片
thin-section analysis 薄片分析;磨片分析
thin-section ball bearing 薄壁套圈的球轴承
thin-section casting 薄壁铸件
thin-section of cement paste 水泥净浆体透明磨片
thin segment 细段
thin-set 薄黏结层
thinset method 薄贴法

thinset terrazzo 薄层水磨石
thin shear web 抗剪薄腹板
thin sheet 钢皮
thin sheet flow 薄层水流
thin sheet glass 薄的平板玻璃;薄的窗玻璃;薄片玻璃
thin sheet metal 薄金属(片);薄板金属
thin-sheet mill 薄板轧机
thin sheet (plate) 薄钢板
thin-shell 薄壳(的)
thin-shell arch bridge 薄壳拱桥
thin-shell barrel 薄壳圆筒
thin-shell barrel roof 筒形薄壳屋顶
thin-shell concrete 薄壳混凝土
thin-shell construction 薄壳构造;薄壳结构
thin-shell cylinder 薄壳圆筒
thin-shelled 薄壳的
thin-shell element 薄壳单元
thin-shell penstock 薄壁压力水管
thin-shell precast 薄壁预件;预制混凝土薄壳;薄壳预制件
thin-shell precast concrete members 预制混凝土薄壳构件
thin-shell precast unit 薄壳预制(构件)单元
thin-shell reinforced concrete structure 锯齿形钢筋混凝土薄壳结构
thin-shell roof 薄壳屋顶
thin-shell structure 薄壳结构;薄壳构造
thin-shell surface 薄壳面
thin ship 瘦体船
thin-skin model 薄膜模型
thin skinned 薄皮的
thin-skinned ingot 薄皮钢锭
thin-skinned lining 薄壁炉衬
thin-skinned structure 薄皮构造
thin-slab 薄板;薄楼板
thin-slab construction 薄板构造
thin-slab structure 薄板结构
thin slice 薄片
thin slurry 稀料浆
thin soil 薄土;薄地
thin source 薄源
thin space 薄空铅
thin speaker 薄型扬声器
thin sphere 薄壳球体;薄壁球体
thin spot 薄弱点
thin spread 薄层撒布(碎石);薄层浇布;薄层分布
thin steel tube 薄壁钢管
thin stock 薄坯
thin-stock reel 小型型钢卷取机
thin stratum 薄层
thin strip process 薄带法
thin structure 薄形结构
thin surfacing 薄面层(沥青面层)
thin surfacing (coat) 薄层表面处治
thin tape 薄带
thin target 薄靶
thin tubesheet 薄管板
thin type speaker 薄型扬声器
thin vault 薄壳拱顶
thin veneer 薄饰单板;薄面板
thin viscosity 稀黏滞度
thin-wafer-channel multiplier 薄片通道式倍增器
thin-wall 薄壁
thin-walled 薄壁的
thin-walled bar 薄壁杆(件)
thin-walled(bed) beam 薄壁梁
thin-walled bearing 薄壁轴承
thin-walled bearing bush 薄壁轴承衬
thin-walled bit 薄壁钻头;薄壁(金刚石)钻头
thin-walled case 薄壁壳体
thin-walled casing tube 薄壁套管
thin-walled cavity 薄壁空洞
thin-walled cell 薄壁管胞
thin-walled chamber 薄壁电离室
thin-walled channel 薄壁槽钢;轻型槽钢
thin-walled conduit 薄壁涵管;薄壁管
thin-walled construction 薄壁结构;薄壁构造
thin-walled container 薄壁盛料器
thin-walled copper tube 薄壁铜管
thin-walled counter 薄壁计数管
thin-walled cylinder 薄壁圆柱体;薄壁圆筒
thin-walled driven sampler 打入式薄壁取土器
thin-walled fixing 薄墙壁铆件;自挤压螺旋铆件

thin-walled hollow brick 薄壁空心砖
thin-walled hollow product 薄壁中空制品
thin-walled magnesian crucible 薄壁氧化镁坩埚
thin-walled metal mo(u)ld 薄壁金属型
thin-walled mo(u)ld 薄壁结晶器
thin-walled penstock 薄壁压力水管
thin-walled pipe 薄壁管
thin-walled plaster 罩面灰浆
thin-walled pod 薄荚
thin-walled prefabrication 薄壁预制
thin-walled pressure vessel 薄壁容器
thin-walled profile 薄壁纵断面
thin-walled sampler 薄壁取样器;薄壁取土器
thin-walled sampling 薄壁取样法
thin-walled section 薄壁剖面;薄壁断面
thin-walled Shelby tube sampler 谢尔贝薄壁取土器
thin-walled shell 薄壳
thin-walled shoe and barrel sampler 束节式取土器
thin-walled soil sampler 薄壁取土(样)器
thin-walled sphere 薄壁球体
thin-walled stainless 薄壁不锈钢管
thin-walled steel drill rod 薄壁钢质钻杆
thin-walled steel section 薄壁型钢
thin-walled steel structure 薄壁钢结构
thin-walled steel tube 薄钢管;薄壁钢管
thin-walled structure 薄壁结构
thin-walled triangle weir 薄壁三角堰
thin-walled tube 薄壁取样器;薄壁管
thin-walled (tube) sampler 薄壁管取土器;薄壁取样管;薄壁管取样器
thin-wall fixing 自挤压螺旋铆件;薄墙壁铆件
thin-wall plaster 薄壁墙粉刷;罩面灰浆
thin-wall sampler 薄壁取样器
thin weak 淡
thin-webbed beam 薄腹梁
thin-webbed girder 薄腹大梁
thin-webbed girder bridge 薄腹梁桥
thin-webbed plate 薄腹板
thin-weir 薄壁堰
thinwin 斑驳朱古力棕色硬木(印度产)
thin-window 薄窗式的
thin-window counter 薄窗膜计数管
thin-window counting tube 薄窗计数管
thin-window image intensifier 薄窗图像增强器
thin-wire 细线
thin-woodfree printing paper 薄道林纸
thioacetamide 硫代乙酰胺
thioacetic acid 硫代乙酸;硫代醋酸
thioacetone 丙硫酮
thio-acid 硫代氧酸
thio-acid compound 硫代(酸)化合物
thio-alcohol 硫醇
thio-aldehyde 硫醛
thioallyl ether 烯丙基化硫
thioarsenite 硫亚砷酸盐
thiobacilli 产硫酸杆菌
thiobacillus 硫杆菌
thiobacillus denitrificans 脱氮硫杆菌
Thiobacillus ferro-oxidant 硫杆菌铁氧化剂;氧化铁硫杆菌
thiobacillus thiooxidant 氧化硫杆菌
thiobacillus thioparus 排硫杆菌
thiobenzaldehyde 苯甲硫醛
thiobenzoate 硫代苯甲酸
thiobenzophenone 二苯甲硫酮
thiocarbamoyl 氨基硫羰基
thiocarbamoyl 苯基芥子油
thiocarbanilide 均二苯硫脲;对称二苯硫脲
thiocarbomate 硫代氨基甲酸酯
thiocarbonyl chloride 二氯硫化碳
thiochrome 硫色素
thiocol based material 乙硫橡胶基材料
thiocyanate 硫氰酸盐;硫氰酸酯
thiocyanate value 硫氰酸值
thiocyanation 硫氰化作用
thiocyanato 氰硫基
thiocyanic acid 硫氰酸
thiocyanide 硫氰化物
thiocyanoacetates 硫氰乙酸酯类
thiocyanogen 硫化氰
thiocyanogen number 硫化氰值
thiocyanogen value 硫氰值;硫化氰值

thiocyclohexane 硫杂环己烷
thiodiphenylamine 硫代二苯胺
thioelaterite 硫弹性沥青
thiogenic 产硫的
thiogenic bacteria degradation 产硫菌降解
thioglocollic acid 硫代乙醇酸
thioindigo 硫靛蓝
thioindigo colo(u)r 硫靛颜料
thioindigo maroon 硫靛褐红
thioindigo pigment 硫靛颜料
thioindigo pink 硫靛桃红
thioindigo red 硫靛红
thiokerite 九硫沥青
thio-ketone 硫酮
thiokol 聚硫橡胶;乙硫橡胶
thiokol based material 乙硫橡胶基材料
thiokol tip 乙硫橡胶镶尖
thiol 硫醇
thiol acid 硫羟酸
thiolate 硫羟酸盐;硫醇盐;烃硫金属
thiolin 硫麻子油液
thiolinic acid 硫麻子油酸
thiolysis 硫解(作用)
thionaphthene 苯并噻吩
thionic acid 连硫酸
thionine 劳思氏紫
thionizer 脱硫塔
thionyl chlor-fluoride 氟氯氧化硫
thionyl chloride 亚硫酰(二)氯;氯化亚硫酰
thionyl hydrazine 亚硫酰阱
thiopexy 硫固定(作用)
thiophahate 土布散
thiophanate-methyl production wastewater 甲基硫菌灵生产废水
thiophane 噻吩烷
thiophene 硫茂;噻吩
thiophenol 苯硫酚
thiophenoxide ions 苯硫氧化物离子
thiophenyl acetone 苯硫基丙酮
thiophenylamine 吩噻嗪
thiophenylmercaptan 硫酚并菲
thiophil 亲硫的
thiophile element 亲硫元素
thiophilic bacteria 嗜硫细菌
thiophos 对硫磷
thiophosgene 硫光气
thiophosphate 硫代膦酸酯
thioplast 硫塑料;硫合橡胶;弹性硫塑料
Thioploca 辫硫菌属
thio-propenyl ether 丙烯硫醚
thio rubber adhesive 聚硫橡胶黏合剂
thiosauite 钙长石
thiosemicarbazide 硫代氨基脲;硫代氨基甲酰肼;灭鼠特;氨基硫脲
thiosemicarbazone 缩氨基硫脲
thiosemicardasone 安苯硫脲
thiosulfate 硫代硫酸盐
thiosulfate ion 硫代硫酸根离子
thiosulfite 硫代亚硫酸盐
thiosulfuric acid 硫代硫酸
thiosulfurous acid 硫代亚硫酸
thiosulphate 硫代硫酸盐
thiourea 硫脲
thiourea resin 硫脲(甲醛)树脂
thioxane 氧硫杂环己烷
third angle drawing 第三角画法
third angle projection 第三角投影(法);正投影;三次投
third angle projection drawing 第三角投影画法
third angle projection method 第三象限投影法
third angle system 第三象限法;第三角法
third anode 第三阳极
third brush 第三电刷
third-brush control 第三刷控制
third-brush regulation 第三刷控制
third category gassy mine 三级瓦斯煤矿
third category vessel 三类容器
third chord 第三弦杆
third circle goniometer 三圆测角仪
third class 三等(品);三等的
third-class conductor 第三类导线
third-class paper 三等证券
third-class pressure vessel 三类压力容器;三级压力容器

third-class road 三级公路
third-class room 三等舱
third-class triangulation 三等三角测量
third coat 第三次油漆层;第三次粉刷层;第三度漆
third cosmic velocity 第三宇宙速度
third-country currency 第三国货币
third deck 第三层甲板
third-degree burn 三度烧伤
third echelon 第三梯级
third engineer 三管轮【船】
third engineer's room 三管轮室
third-faucet configuration 第三龙头布置
third-faucet point-of-use device 第三龙头用水点
third-faucet unit 第三龙头装置
third firing 三次烧成;第三次煅烧
third fittings 第三道装修
third fixing 木工最后三道整修(房屋的)
third flag carrier 第三国船舶
third floor 四楼(英国);三楼(美国)
third gear 三挡传动;三挡齿轮;三挡;第三速度齿轮;第三排挡;第三挡
third generation 第三代
third generation computer 第三代计算机
third generation container 第三代集装箱
third generation container terminal 第三代集装箱码头
third-generation hardware 第三代硬设备
third generation software 第三代软设备;第三代软件
third-generation synthetic fibre 第三代合成纤维
third-grade gravity base station for exploration geophysics 物探重力三级基点
third grade structure 三级构造
third grade timber(3rd grade timber) 三级木材
third hand tap 三锥;第三锥
third harmonic 三次谐波;第三谐波;第三次谐波
third harmonic generation 三次谐波振荡;三次谐波生成
third harmonic tuning 三次调谐
third integral 第三积分
third land 第三个槽沟
third law of motion 第三运动定律
third law of thermodynamics 热力学第三定律
third leaky system 第三越流系统
third legs 第三对足
third-level address 三级地址;第三级地址
third-level addressing 三级地址
third level calibration 三级刻度
third-level storage 三级存储;第三级存储器
third liability 第三方责任
third market 三级市场(美国术语,指大宗股票的场外交易)
third mate 三副【船】
third mate's room 三副室
third Minamata disease 第三水俣病
third motion shaft 第三运动轴
third officer 三副【船】
third order 第三级的;第三等的
third-order aberration 三级像差
third-order aberration coefficients 三级像差系数
third-order aberration theory 三级像差理论
third-order benchmark 三等水准点
third-order correlation 三阶相关
third-order distortion 三级畸变
third-order equation 三次方程
third order goal 三阶目的
third order leveling field book 三等水准观测手簿
third-order level(l)ing 三等水准测量
third-order light mixing 三阶光混频
third-order reaction 三级反应
third-order reactor 三级反应器
third-order theory 三阶原理
third-order triangulation 三级三角测量;三等三角测量
third-order triangulation point 三等三角点
third part claim 第三者的索赔权
third party 第三者;第三人;第三方
third party audit 第三方审计
third party bill of lading 以第三者为装船人的提单
third party certification 第三方认证
third-party certification system 第三方鉴定制
third party charges on the consultant 第三方对咨询人的收费
third-party claim 索赔涉及第三方

third party claimant 第三方索赔人
third party inspection 第三方检验
third party insurance 第三者责任险;第三者保险;第三方保险
third party leasing 向第三者租用
third party liability 第三者责任
third party liability insurance 第三人责任险
third party motor vehicle liability insurance 机动车第三者责任险
third party shipper 以第三者为托运人
third party warranty 第三方担保书
third pinion 三轮齿轴;第三小齿轮
third pitch 二分之三百三十一度屋面坡度
third point flexural test 三分点挠曲试验
third point load(ing) 三分之一跨度点荷载;三分点荷载;三等分点荷载;第三点载重;三点加载
third pollution 第三污染
third power 立方;三次幂
third principal stress 第三主应力
third programme 第三期工程
third projector 第三投影器
third proportion(al) 比例第三项
third quarter 下弦【天文】
third quarter neap tide 下弦小潮
third quarter of the moon 下弦月
third quartile 第三四分位数
third radiation belt 第三辐射带
third rail 电动机车输电轨;供电滑接线;接触轨;第三轨(条);导电轨
third-rail below contact 三轨下部接触式
third rail collector shoe 受电靴
third rail ga(u)ge 第三轨轨距
third rail height 第三轨高度
third-rail middle contact 三轨中部接触式
third rail (supply) shoe 第三轨集电靴
third-rail system 三轨制
third rail top contact 三轨上部接触式
third-rail traction 第三轨条牵引
thirds off 不扣减三分之一
third speed 三挡速率;三挡速度;第三速度
third-speed gear 三挡;第三速度齿轮
third spudding date 三开日期
third-stage biological aerated filter 第三级曝气生物滤池
third-stage denitrification filter 第三级脱氮滤池;第三级反硝化滤池
third stage of creep 三段蠕变
third stor(e)y 四楼(英国)
third tap 精丝锥;三锥;三攻丝锥
third terrace 三级阶地
third to fourth-order leveling 三四等水准测量
third voltage range 第三级电压范围
third watch 第三班(船上值班表)
third way 第三种方法
third wheel 中间轮;接轮;三轮
third wire 第三线
third yield value 第三屈服值
thirl 连通道
thirling 两巷道掘通;联络小巷;掏槽落煤法;巷道掘通(隧道);石板瓦钻孔;连通道
thirst fever 缺水热
thirsty formation 漏失岩层
thirsty soil 干燥土
thirtieth highest annual hourly volume 一年第三十个最大小时交通量;年第三十位最大小时交通量
thirtieth traffic volume 第三十位交通量
thirty-degree piezoelectric(al) cut 压电三十度截法
thirty-five colo(u)r film 三十五毫米彩色胶片
thirty-seven cm brick wall 三七墙
thirty-twomo 三十二开
thirty-type 州字型
this account 本账户
this article 本文
this end up 此端向上
thisltes and throns 荆棘
this side up 此端向上;此边向上
Thistle board 地斯特尔石膏板(一种轻质纸面石膏板);轻质板;石膏板(排挡间的);纸面石膏板
thistle cutter 割蓟机
thistle funnel 长梗漏斗
thistle tube 长梗漏斗
this year 本年度
thitka 缅甸桃花心木
thitsiol 缅漆酚

thiuram 秋兰姆
Thiviers red 赛威尔红(含铁的石英质色料)
thixocasting 触融压铸;触变铸造
thixopectic liquid 触变胶液体
thixotrometer 触变黏度计;触变仪;触变计
thixotrope 触变胶
thixotropic 摇溶的;具有触变作用的;触变性的;触变的
thixotropic action 触变作用
thixotropic agent 摇溶剂;触变剂
thixotropic behavior 触变行为
thixotropic breakdown 触变破坏
thixotropic casting 触变浇注
thixotropic cement 触变水泥
thixotropic clay 触变黏土
thixotropic coating 触变型涂料
thixotropic coefficient 触变系数
thixotropic dispersion 触变分散体
thixotropic effect 触变效应
thixotropic fiberglass size 触变性玻璃纤维浸润剂
thixotropic flow 触变性流动;触变型流动
thixotropic fluid 触变黏土浆;触变性液体;触变(性)流体
thixotropic gel(ation) 触变凝胶(作用)
thixotropic hardening 触变硬化
thixotropic injection 触变灌浆
thixotropic level 触变等级
thixotropic liquid 触变性液体;触变液体
thixotropic liquid method 触变性液体法
thixotropic mud 触变泥
thixotropic paint 触变性涂料;触变型涂料;触变漆
thixotropic paste 触变性涂膏
thixotropic propellant 触变喷气燃料
thixotropic property 触变性(质)
thixotropic ratio 触变比
thixotropic resetting time 触变再现时间
thixotropic resin 触变型树脂
thixotropic sculptural property 触变可塑性
thixotropic sediment 触变沉积物
thixotropic sludge 摇溶污泥
thixotropic slurry 触变泥浆
thixotropic transformation 触变转化
thixotropic type paint remover 触变型脱漆剂
thixotropic vehicle 触变性漆料
thixotropic viscoelasticity 触变黏塑性;触变黏弹性
thixotropic viscosity 触变黏度
thixotropy 摇溶性;摇溶现象;振动液化;搅динамическ性;触变性;触变(体)
thixotropy index 触变指数
thixotropy period 触变周期
thizone 硫腙
thjalcusite 硫铊铁铜矿
thoft 横座板
thole 供祭品坛;还愿壁龛;橹桩;桨栓;圆屋(顶);壁龛;凹座
thole board 桨架板;桨耳板
tholeiite 拉斑玄武岩
tholeiite series 拉斑玄武岩系列
tholeiitic diabase 拉斑辉绿岩
tholeiitic dolerite 拉斑粒玄岩
tholeiitic glass 拉斑玄武玻璃
tholeiitic magma 拉斑玄武岩浆
tholeiitic olivine dolerite 拉斑橄榄粒玄岩
thole pin 橹桩;桨栓
Thollon prism monochromator 索仑棱镜单色仪
tholobate 圆顶房支承结构;穹顶支承结构;穹隆顶底座;圆屋顶座
tholos 房屋(圆形或部分圆形的);蜂窝形房屋;圆形房屋;圆屋顶(希腊古典风格);蜂窝形圆形陵墓;圆形建筑(物);[复]tholoi
tholos tomb 蜂窝形墓穴
tholus 蜂窝形墓穴
tholus type of tomb 蜂窝形的圆形陵墓
Thomas converter 碱性转炉
Thomas converter steel 碱性转炉钢
Thomas-Gilchrist process 碱性转炉炼钢法
Thomas iron 托马斯生铁;碱性转炉生铁
Thomas meal 托马斯碱性转炉渣粉
Thomas pig 碱性转炉生铁
Thomas pig-iron 碱性转炉铸铁;托马斯铸铁
Thomas slag 碱性转炉渣
Thomas steel 碱性转炉钢;托马斯钢
Thompson filter 汤普逊波器
Thompson machine 汤普逊摩擦试验机

Thompson process 汤普逊电弧焊接法
Thompson's deflector 张角偏转仪
thomsenolite 汤霜晶石;方霜晶石
Thomson circulation theorem 开尔文环量定理
thomsonite 杆沸石
Thomson measuring weir 汤姆逊量水堰
Thomson process 接触对焊法
Thomson's balance 汤姆逊平衡
Thomson's coefficient 汤姆逊系数
Thomson's cross section 汤姆逊散射截面
Thomson's effect 同质热电效应;汤姆逊效应;汤姆逊热电效应
Thomson's electrometer 汤姆逊静电计
Thomson's formula 汤姆逊公式
Thomson's galvanometer 汤姆逊电流计
Thomson's guard-ring 汤姆逊保护环
Thomson's heat 汤姆逊热
Thomson's inclined coil instrument 汤姆逊斜圈式仪表
Thomson's limestone 汤姆逊灰岩
Thomson's meter 汤姆逊式仪表
Thomson's method 汤姆逊法
Thomson's parabola apparatus 汤姆逊抛物线装置
Thomson's parabolas 汤姆逊抛物线
Thomson's scatter(ing) 汤姆逊散射
Thomson's scattering formula 汤姆逊散射公式
Thomson's type 汤姆逊式
thong 皮带
thorbastnaesite 水氟碳钙钍矿
thoreaulite 钽锡矿
thoria 氧化钍;钍土
thoria cathode 氧化钍阴极
thoria ceramics 氧化钍陶瓷
thoria gel 氧化钍凝胶
thoria gelatination 氧化钍凝胶化
thoria microsphere 氧化钍微粒
thoria-molybdenum cemet 钼钍陶瓷金属
thorian aeschynite 含钍易解石
thorianite 放射性氧化钍;方钍石
thoria refractory 氧化钍耐火材料
thoria refractory product 氧化钍耐火材料产品
thoriated 含钍
thoriated tungsten cathode 钍钨阴极
thoriated tungsten electrode 钍钨极
thoride 钍化物
thorite 钍石含量;钍石
thorium-bearing 含钍
thorium-bearing carbonatite deposit 含钍碳酸岩型矿床;含钍石碱性岩床
thorium bismuthide 铋化钍
thorium boride 硼化钍
thorium carbide 碳化钍
thorium channel window width 钍道窗宽
thorium chloride 氯化钍
thorium content anomaly 钍含量异常
thorium dioxide 氧化钍;二氧化钍
thorium emanation 钍射气
thorium emanation concentration value in soil 土壤钍射气浓度值
thorium emanation consistence 钍射气浓度
thorium-free optic(al) glass 无钍光学玻璃
thorium-lead age method 钍铅年代测定法
thorium nitride 氮化钍
thorium ore 钍矿
thorium oxalate 草酸钍
thorium oxide 氧化钍
thorium oxyfluoride 氟氧钍
thorium/potassium anomaly 钍钾异常
thorium series 钍蜕变系
thorium silicate 硅酸钍
thorium silicide 硅化钍
thorium sulfate 硫酸钍
thorium-uranium method 钍—铀法
Thorman tray 推液式条形泡罩塔板
thornbush 多刺高灌丛
thorn damage 刺伤
Thorne barker 桑恩袋式剥皮机
thorn forest 棘刺林;热带棘林;热带旱树林;热带旱生林
thorn hedge 刺篱;紫荆篱
thorn palm 刺棕榈
thorns 荆棘
thorn scrub 多刺高灌丛
Thornthwaite index 桑斯威特指数

Thornthwaite moisture index 桑斯威特湿润指数
Thornthwaite precipitation effectiveness index 桑斯威特降水效率指数
Thornthwaite's classification of climates 桑斯威特气候分类法
Thornthwaite's formula 桑斯威特公式
thorny plants hedge 刺篱
thorny problem 难题
thorny scrub 热带草生灌丛
thorny subject 难题
thorny under-growth 荆棘
thorogummite 羟钍石;钍脂铅铀矿
thoron 钍射气
thorosteenstrupine 胶硅钍钙石
thorotrase 胶质二氧化钍
thorotungstite 钍钨矿
thorough 通过墙身的顶砖
thorough bond 穿通砌合
thorough burning 完全燃烧;烧透
thorough cleaning 大扫除
thorough consolidation 完全凝固
thorough cut 通沟
thorough examination 彻底审查
thoro(ugh)fare 大道;公共航道;通行;大街;超高速公路;干道;内陆航道;通道;大路
thoroughfare insulator 穿越绝缘子;穿墙绝缘管
thoroughfare track 车站行车交通线
thoroughfare track of hump 驼峰迂回线
thorough index 全指数
thorough investigation 彻底调查
thoroughly air dried 完全烘干的
thorough repair 大修;彻底修理
thorough rivet 穿通铆钉
thorough roasting 充分焙烧
thorough way 直达道路
thoroughwort 穿心佩兰
thortveitite 钪钇石
thor-uraninite 钍铀矿
thorutite 钛钍矿
though-and-though sawing 毛板下锯法
though cut 路堑
though slurry 生料浆
though statistically significant 在统计上是显著的
thought 横座板
thought experiment 想像实验
thoughts of ecology 生态学思维
thousand board measurement feet 千量材尺
thousand feet board measure 千板英尺(木材计量单位)
thousand-flower glass 玻璃锦砖
thousand-foot board measure 千板英尺量材法;千板尺量度
thousand-grain weight 千粒重
thousand peaks in emerald colo(u)r 千峰翠色
thousand slates 千张石板瓦
thow(e)l 桨栓;橹栓
thow off fault 断层断距【地】
thow point 露点
thraquinone lake 茜素色淀
thrash 振摆;逆风浪航行;顶风浪前进
thrasher 打谷机;捶击机
thrashing 系统失效;颠簸;打散;捶击
thrashing floor 打谷场;打谷场院
thrashing ground 打谷场
thrashing machine 打谷机;捶击机
thrash out plan 经过研究讨论订出计划
thrash over 推敲;反复研究;重复研究
thrawn 挠曲的
thraw point 融化点
thread 线状物;线索;线道;线;纤维;细条纹;细丝;螺线;螺纹;螺丝;缕;连接丝扣;棉纱;麻纱;穿线
thread anchorage 螺杆端杆锚具
thread angle 螺纹牙型角;螺纹断面角
thread an unthread tree 对未穿线的树进行穿线
thread at both ends 螺旋柱
thread bar 冷轧螺纹钢筋;精轧螺纹锚;迪维达格锚固系统;螺杆
thread bolt 切削螺纹
thread bottom diameter 螺纹内径
thread brake 螺旋制动(器)
thread break detector 断线检测器
thread cal(l)ipers 螺纹卡尺
thread cap 螺纹帽
thread-changing joint 变丝接头

thread chaser 螺纹梳刀;螺丝花刀;螺丝钢板;螺丝扣(牙);扳牙
thread chasing lever 螺纹刻度手柄
thread chasing machine 螺纹车床;螺丝车床
thread contour 螺纹轮廓
thread control block 连接控制块
thread count 织物经纬密度;经纬密度
thread counter 线纹计数器
thread coupling 螺纹过渡连接件
thread crest clearance 螺纹顶部间隙
thread cutter 攻丝扳牙;螺纹刀具;丝板;车丝扳牙
thread cutting 切削螺纹;螺纹切削;车制螺纹
thread-cutting compound 螺纹切削油
thread cutting die 攻丝用扳牙
thread-cutting lathe 螺纹车床
thread cutting machines 螺纹加工机床
thread-cutting mechanism 螺纹切削结构
thread-cutting oil 螺纹切削油
thread-cutting screw 攻丝螺钉
thread-cutting stop 车螺纹停止器
thread-cutting tool 螺纹切削刀具;螺纹槽刀
thread dial 螺纹刻度盘
thread-dial indicator 螺纹千分表
thread die 切丝扳牙
thread dope 螺纹润滑脂;丝扣漆
thread driver 螺丝起子;螺丝刀
threaded 带螺纹的
threaded anchorage 螺纹锚固;螺纹锚具
threaded and coupled 螺纹连接
threaded attachment 螺纹连接
threaded bar 螺纹钢筋
threaded bar end anchorage device 螺丝端杆锚具
threaded bolt 螺纹栓;螺纹锚栓;螺栓;螺杆
threaded bush(ing) 螺纹衬套
threaded cast-iron fittings 螺纹铸铁管件
threaded chlorinated polyvinyl chloride plastic fitting 氯化聚氯乙烯螺纹塑料管件
threaded collar joint 螺纹接箍连接
threaded column 螺旋柱
threaded conduit 螺纹管
threaded connecting 螺纹连接
threaded connection 螺纹连接;丝扣连接;螺纹联结;螺纹接合
threaded connector 螺纹接头
threaded coupling 螺纹接头;螺纹束节
threaded drill pipe 有细扣钻杆
threaded drum 螺纹连接柱体
threaded end 螺纹管口;螺纹端;丝头
threaded fastener 螺纹紧固件
threaded file 链接文件;连接文件
threaded fit 螺纹配合
threaded flange 螺纹法兰;螺纹连接凸缘
threaded for union 螺纹活接头
threaded ga(u)ge 螺纹规
threaded glass globe 螺纹螺丝玻璃球
threaded hole 螺栓孔;丝孔
threaded hood 机匣罩
threaded hook 螺纹挂钩
threaded insert 螺纹嵌入件
threaded inspection opening 螺纹检查孔
threaded-in winding 拉入绕组;穿入式绕组
threaded joint casing 丝扣连接套管
threaded joint(ing) 螺纹接合;螺纹管接头;丝扣接头;螺纹接头
threaded kelly 有细扣方钻杆
threaded line pipe 螺纹管;来复管
threaded list 线索表;穿插表
threaded mandrel 螺纹心轴
threaded nail 螺纹钉
threaded nozzle 螺纹喷管
threaded nut 螺母
threaded one end 一端螺纹
threaded outlet (井点总管上的)螺纹接头
threaded part 螺纹部分
threaded pipe 螺纹管
threaded pipe connection 管子丝扣连接
threaded pipe end 螺纹口
threaded pivot 螺纹枢
threaded plate 套丝板
threaded plug 螺纹插头;螺纹插塞
threaded portion 螺纹部分
threaded projecting bar 螺纹伸杆
threaded protector 护丝;护箍
threaded rod 螺纹锚栓;螺纹杆

threaded shaft 螺纹轴
threaded sleeve 螺纹支管;螺纹套筒;螺纹套管;丝扣套筒
threaded sleeve joint 丝扣套筒接头
threaded-sleeve-type fitting 螺纹轴套型接头
threaded spindle 螺纹心轴
threaded spindle nose 螺纹主轴端
threaded stud 螺杆;双头螺栓
threaded tree 连接树形网络;穿线的树
threaded tube 螺纹管
threaded valve stem 螺栓阀杆
threaded wrought iron pipe 螺纹口熟铁管
thread end reamer 螺纹端铰刀
thread engagement 螺纹啮合
threader 螺纹加工机;螺纹铣床;螺丝磨床;螺丝车床
thread extractor 打皮辊花机
thread failure 丝扣折断
thread fit 螺纹配合
thread form 螺纹牙形
thread form character 扣型代号
thread forming machine 螺纹成型机
thread forming tool 螺纹成型刀具
thread fraise 螺纹铣刀
thread galvanometer 艾因托反文电流计
thread generator 螺纹加工机
thread glass 线纹饰玻璃
threadgoldite 板磷铝铀矿
thread grinder 螺纹磨床
thread grinding 螺纹磨削;磨具刻画线条
thread groove 螺纹槽
thread guide 导丝器
thread head 螺纹头
thread height 丝扣高度
thread indicator 螺纹指示器;螺距仪
threading 旋压螺纹;线纹装饰让;攻丝;刻螺纹;套丝;穿纱;穿入套管;穿入拱;穿线;车螺纹
threading chaser 螺纹刀具
threading chuck 攻丝卡盘
threading cutter 螺纹铣刀
threading dial 车螺纹指示盘
threading die 螺纹扳手;螺纹扳(牙);扳牙
threading hob 螺纹滚铣刀
threading lathe 螺纹车床
threading lubricant 攻丝润滑剂
threading machine 攻丝机;丝扣车床
threading nut 攻丝螺帽
threading-process die 螺纹扳牙
threading roll 螺纹滚花辊
threading speed 穿引速度
threading tool 螺纹刀具;螺纹车刀;挑扣刀
threading tool chaser 整修螺孔用丝锥;螺纹扳手
threading unit 螺纹车床
thread inside calliper 内螺纹卡尺
thread in signal 卷带信号
thread interval 螺纹距;螺距;十字丝间距
thread joint 螺纹接头
thread-lace scoria 纤维火山渣
thread length 螺纹长度
threadless 无螺纹
threadless binding 无线装订
threadless pipe 无螺纹管
threadlike 丝状
threadlike phase 螺线型相位
thread like surface flaw 泥釉缕
thread link 穿线链接
thread losing 脱扣
thread micrometer calliper 螺纹测微器
thread mill 螺纹铣刀
thread miller 螺纹铣床
thread milling 铣螺纹;螺纹铣削
thread milling attachment 螺纹铣床附件
thread milling cutter 螺纹铣刀
thread minimum tensile strength 丝扣最小抗拉强度
thread mounting 拧紧螺丝
thread of a screw 螺丝扣
thread of channel 中泓(线);河道主流线;河道中泓主流
thread of crystal face 晶面螺纹
thread of current 流线
thread off channel 河道中泓线
thread of oakum 捻缝麻絮
thread of stream 细流;主流线;主溜线【铁】;中泓线;河流主溜线;河流中心线;河道主流线;河道中

泓线；水流纹线
thread of the current 主流线
thread opening machine 开纺头机
thread out signal 倒带信号
thread peeling machine 旋风式螺纹铣床
thread pitch 螺纹距；螺距
thread plate 搓丝板
thread plug ga(u)ge 螺纹塞规
thread producing machine 螺纹加工机
thread profile 螺纹牙形
thread protector 螺纹护环
thread protector of drill tool 钻具护丝
thread rating 螺纹额定值
thread restoring tool 螺纹修整刀具；螺纹修复刀具
thread ring ga(u)ge 螺纹环规
thread roller 滚丝机
thread rolling 滚丝；搓丝；螺纹旋车
thread-rolling die 滚丝模；搓丝机
thread rolling feeder 螺纹滚轧机；螺纹滚压机
thread rolling machine 滚丝机；螺纹滚压机床；搓丝机
thread root 螺纹牙底；螺纹根
thread root casing wall cross section area 扣根管壁截面积
thread root diameter 螺纹内径
thread-rouing method 搓条法
thread sealant 螺纹填料
thread sealing mechine 塑料线烫订机
thread sewing machine 锁线机；串线机
thread slipping 滑扣
threads of per inch 每时扣数
threads of tectonic network 构造网线
threads per inch 每英寸螺线数；每英寸扣数
thread spike 螺纹道钉
thread spindle 螺纹锭
thread spinner 纺机
thread stab 刺扣
thread standard 螺纹标准
thread stay 螺旋撑条
thread steel nail 螺纹钢钉
thread sticking 黏扣
thread-stitching 串线订
thread stripping 螺纹滑扣
thread tap 螺丝攻；丝锥
thread taper 丝扣锥度
thread tapper 攻丝器
thread through 穿过
thread tooth length 丝扣齿合长度
thread top clearance 螺纹顶部孔隙
thread tree 穿线的树
thread type 螺纹形式
thread type of casing 套管扣型
thread up 拧上丝扣
thread-whirling 螺纹回转（法）
thread whirling machine 旋风式螺纹铣床
thread worm 蜗杆螺纹
thready pulse 丝状脉
threatened species 受威胁物种
threatening appearance of weather 险恶的天气现象
threatre-in-the-round 表演场设在观众坐席中央的剧院
three-abreast seats 三人并肩座椅
three-account system 三账户制
three-acting jack 三作用千斤顶
three-address architecture 三地址结构
three-address instruction format 三地址指令格式
three-address system 三地址制；三地址系统
three-aggregate bin 三集料库；三骨料库
three-air monitoring station 三气监测站
three-aisled 三条通道的（剧院观众座位间）
three-aisled basilica 三条通道教堂（古罗马）
three-ammeter method 三安培计方法
three-angle equation 三角变方程
three antis 三防（海、陆、空国防）
three-aperture interferometer 三孔干涉仪
three-apexed rotor （转子发动机的）三角转子
three arch bridge 三拱桥
three-armature direct current motor 三枢直流电动机
three-arm base 三脚架基座
three-armed protractor 三臂分度仪
three-arm mooring 三脚碇泊浮筒
three-arm pickling machine 三摇臂式酸洗机

three-arm protractor 三杆分度仪；三杆分度器；三杆定位仪；三臂分度器
three-arm rift system 三支裂谷系；三叉裂谷系
three-arm unloader 三臂卸卷机
three-aspect automatic block system 三显示自动闭塞系统【铁】
three-aspect signal 三显示信号；三位显示信号
three-aspect three-block system 三显示三闭塞区段系统
three-aspect two-block system 三显示二闭塞区段系统
three-attributes of colo(u)r 色彩的三项属性（指色相、明度、彩度）；颜色的三属性
three-axial ellipsoid 三轴椭球体
three-axial magnetometer 三轴磁强针
three-axis autopilot 三轴自动驾驶仪
three-axis control 三轴坐标控制
three-axis control system 三轴控制系统
three-axis-degree-of-motion trainer 三自由度陀螺练习器
three-axis gyro indicator 三自由度陀螺指示器
three-axis gyropilot 三轴稳定自动驾驶仪
three-axis gyroplatform 三轴陀螺平台
three-axis laser gyro package 三轴激光陀螺装置
three-axis mounting 三轴装置
three-axis reference 三轴基准
three-axis reference system 三轴参考系
three-axis simulator 三轴模拟器；三度空间模拟器；三度空间仿真器
three-axis stabilization 三轴稳定（作用）
three-axis stabilized platform 三轴稳定平台
three-axle bogie 三轴转向架
three-axle carrier 三轴运输机
three-axle gear 三轴传动装置
three-axle girder body 三轴桥梁绞车架
three-axle model 三轴型
three-axle rear-dump truck 三轴后卸卡车；三轴翻斗卡车
three-axle roller 三轴压路机；三轴串联（式）压路机
three-axle tandem roller 三轴串联滚压机；三轴串联（式）压路机；串联式三轮压路机；串联式三滚筒压路机
three-axle trailer 三轴挂车
three-axle truck 三轴货车；三轴卡车；三轴转向架
three-axle truck trailer 三轴载货挂车
three-axle vehicle 三轴汽车；六轮汽车
three-axle wagon 三轴货车
three-ball and trunnion universal joint 三球枢轴式万向节
three-bank cylinders 三组汽缸
three-bar bumper 三杆保险杠
three-bar curve 三杆曲线
three-base method 三基线法
three-batcher paver 三料斗摊铺机
three-bay 三开间的
three-beam 三射束
three-beam speckle pattern interferometer 三射束光斑图像干涉仪
three-beam tube 三柱管；三束管
three-bearing crankshaft 三轴承曲轴
three-bearing machine 三轴承电机
three-bearing pinion 三轴承小齿轮
three-bearing type 三轴承式
three-bed 三床（位）的
three-bed room 三床房间
three-bedroomed dwelling unit 三割卧室居住单元
three-bedroomed house 三个卧室的住房
three-bell tee 三承三通
three-bin system 三堆法
three-bladed 三叶的
three-bladed propeller 三叶螺旋桨
three-blade drag 三刃刮路机
three-blade hub 三叶毂
three-blade marine type propelle agitator 三叶推进桨式搅拌器
three-blade mixing paddle 三叶搅拌桨
three-blader 三叶螺旋桨
three-block four-indication automatic block signal 三闭塞四显示自动闭塞信号
three-blocks method 三大块法
three-body collision 三体碰撞
three-body problem 三体问题
three-body recombination 三体再化合；三体复合
three-boiling system 三级蒸发系统

three-bolt ventilation diving equipment 三螺栓通风式潜水装置
three-boom drift-jumbo 三臂平巷凿岩台车
three-bowl calendar 三轮砑光机
three-box section 三室箱形截面
three-box section girder bridge 三室箱形桥梁
three-bracket testing 三托架测试
three-brush bottle washer 三刷洗瓶机
three-brush generator 三刷发电机
three-burn 三次燃烧的
three-button (filter) plate 三钮板
three-by three-arrangement 三三制配置
three-by three-complex 三三制发射场综合设施
three-by-three-configuration 三三制配置形式
three-carburettor engine 三化油器发动机
three-car interaction 三车相互作用（交通流量理论）
three-cavity amplifier tube 三腔放大管
three-cavity klystron 三腔速调管
three-cavity mould 三开箱铸型
three-cavity mo(u)lding 三箱造型
three-cell 三腔式
three-cell furnace 三室炉
three-center curve 三圆心曲线
three-centered 三心的
three-centered arch 三心椭圆拱；三心平圆拱【建】；三心拱
three-centered arch dam 三心拱坝
three-centered circular arch section 三心拱形断面
three-centered circular arch support 三心拱支架
three-centered (compound) curve 三心复曲线
three-centred arch 三心圆拱
three-chamber air reservoir 三室储[贮]气筒
three-chamber mill 三仓磨
three-channel amplifier 三通道放大器
three-channel coal burner 三通道燃烧器；三风道喷煤管
three-channel coal fired burner 三通道喷煤管
three-channel colo(u)rimeter 三道比色计
three-channel impeller pump 三路叶轮泵
three-channel multipoint recorder 三道多点记录器
three-channel seismic discriminator 三通道地震鉴别器
three-channel stereo 三声道立体声
three-channel tracking receiver 三信道跟踪接收机
three-channel ultraviolet visible spectrophotometer 三道紫外—可见光分光光度计
three-chord bridge 三弦桥
three-circuit 三电路
three-circuit receiver 三调谐电路接收机
three-circuit transformer 三绕组变压器；三卷变压器
three-circuit tuner 三回路调谐器
three-cleft 三裂的
three-coat 三道抹灰
three-coat plaster 三层灰泥
three-coat plastering 三道抹灰
three-coat plaster work 三道抹灰工作
three-coat system 三层系列
three-coat three-bake 三涂三烘工艺
three-coat work 三道涂刷工作；三道抹灰作业；三道抹灰工作
three-coil dynamo 三绕组发电机
three-coil generator 三绕组发电机
three-coil regulation 三线圈调整
three-collar cross pipe 三盘四通；三盘十字管
three-collar tee fitting 三盘三通；三盘丁字管
three-colo(u)r 三原色；三彩（考古）
three-colo(u)r acousto-optic modulator 三色声光调制器
three-colo(u)r additive method 三色叠加法
three-colo(u)r camera 分色照相机
three-colo(u)r coefficient 三色系数
three-colo(u)r colo(u)rimetry 三色色度学
three-colo(u)r decoration 三色装饰
three-colo(u)r display 三色复原
three-colo(u)r red ware 三彩器皿
three-colo(u)r filter 三色滤光片
three-colo(u)r glaze 三彩釉
three-colo(u)r hologram 三色全息图
three-colo(u)r knob yarn 三色结节纱
three-colo(u)r photometry 三色光度学；三色光度测量；三色测光
three-colo(u)r recorder 三色录像机
three-colo(u)r reproducer 三色重现装置
three-colo(u)r separation 三色分离

three-colo(u)rs overglaze decoration 三彩釉上装饰
three-colo(u)r system 三色系统
three-colo(u)r tube 三色显像管
three-column core 三柱型铁芯
three-column transformer 三心柱变压器
three-compartment chute 三室溜槽
three-compartment compound mill 三仓磨机
three-compartment mill 三室磨碎机；三仓磨
three-compartment revolving door 三分隔(空间)转门
three-compartment ship 三舱制船
three-compartment steam drier 三格蒸汽干燥机
three-compartment subdivision 三舱不沉制
three-complementary colo(u)r 三分量；三部分组成的；三元的
three-component 三分量；三部分组成的；三元的
three-component accelerograph 三分量自记加速表
three-component accelerometer 三分量加速度仪；三分量加速度计
three-component anemometer 三分量风速仪
three-component concrete 三组分混凝土
three-component epoxy resin 三种成分的环氧树脂
three-component geophone 三分量检波器
three-component geophone array 三分向地震检波器台阵；三分量地震检波器台阵
three-component hydraulic binder 三组分的液压黏结剂
three-component magnetic survey 三分量磁测
three-component measuring system 三组分测量系统
three-component moving coill-type seismometer 三分量动圈式地震仪
three-component refiner 三元系区域精炼炉
three-component seismograph 三分向地震仪；三分量地震仪
three-component sequence 三元结构层序
three-component strong-motion accelerograph 三分量强烈加速度仪
three-component system 三组分系统
three-component tree 三合树
three-component unit 三组分单元
three-component zoom lens 三组式变焦距镜头
three-conductor cable 三芯电缆
three-cone bit 三锥牙轮钻头；三牙轮钻头
three-cone logging cable 三芯测井电缆
three-cone rock bit 三牙轮岩石钻头
three-cone tilting mixer 三维倾斜搅拌机
three-connections and the one level(l)ing (assuring that a construction site is connected to water/ power and roads/ and that the land is levelled before a building project is begun) 三通一平(水、电、道路要通、施工场地要平)
three-constant 三常数
three-constituent mixture 三组分混合料
three-contact plug 三心塞子
three-coordinates measuring machine 三坐标测量仪
three-cord-thread 三股棉线
three-core 三心
three-core block 三孔砌块
three-core brick 三孔砖
three-core cable 三芯电缆
three-core cell 三心单元
three-core conductor 三芯导线
three-core plastic wire 三芯塑料线
three-cornered 三棱的
three-cornered exchange 三角汇兑
three-cornered file 三角锉
three-cornered scraper 三角刮刀
three-cornered trade 三角贸易
three-corrections for horizontal direction 三差改正
three-cross 三线交叉
three-crystal electron interferometer 三晶体的电子干涉仪
three-cup anemograph 三杯风速仪
three-cup anemometer 三杯风速仪；三杯风速表
three-curve (calculation) method (室内照明计算的)三曲线划分法
three-cutter underreamer 三切削刃管鞋下扩孔器；三刮刀刃管鞋下扩孔器
three-cylinder compound locomotive 三缸复式机车

three-cylinder engine 三汽缸发动机；三缸发动机
three-cylinder locomotive 三缸机车
three-cylinder locomotive valve gear 三汽缸式机车阀动装置
three-cylinder marble machine 三辊制球机
three-cylinder reciprocating pump 三缸泵
three-cylinder two-stroke engine 三缸二行程发动机
three-cylinder washer 三筒洗机
three-cylindrical-die thread rolling 三圆柱模滚轧螺纹法
three-day pass 三天假期许可证
three-day sickness 三日热
three-day strength 三日强度(混凝土的)
three-decision problem 三判决问题
three-deck cage 三层罐笼
three-decker 三层甲板船；三层讲坛
three-decker rate system 三重甲板运费制
three-decker (tunnel) jumbo 三层舱客机
three-deck sinking stage 三层凿井
three-deck vessel 三层甲板船
three-deck vibrating screen 三层振动筛
three-deck washing classifier 三层式洗涤分级机
three-degree of freedom 三自由度
three-degree of freedom structure 三自由度建筑物
three-degree zone 三度带
three-D holographic memory 三维全息照相存储器
three-diamond 三菱形符号(货物标志)
three-diaphragm brake chamber 三膜片制动分泵
three-diaphragm element 三膜片元件
three-dimension 三维
three-dimensional 立体的；三向的；三维空间的；三维的；三度(的)
three-dimensional acquisition radar 三坐标探测雷达
three-dimensional analog computer 三维模拟计算机
three-dimensional animation 三维动画片
three-dimensional anomaly transforming to two-dimensional anomaly 三度异常化二度异常
three-dimensional architectural model 三维结构模型
three-dimensional area-covering structure 屋顶框架结构；三维顶盖结构
three-dimensional arrangement 立体结构；三维组合
three-dimensional array 三维数组
three-dimensional attack 三维空间攻击
three-dimensional autopilot 三通道自动驾驶仪
three-dimensional bar 三维杆件
three-dimensional bearing structure 三维承重建筑
three-dimensional bedding 立体花坛
three-dimensional bending 空间弯曲；三维弯曲
three-dimensional bin display system 三维面元显示系统
three-dimensional bin record 三维面元记录
three-dimensional biplane 三维双翼
three-dimensional body 三维物体
three-dimensional CAD project 三维计算机辅助设计项目
threedimensional camera 立体照相机
three-dimensional circuit 三维电路
three-dimensional classification 三维分类法
three-dimensional colo(u)r display 三维彩色显示
three-dimensional compaction 三向压制
three-dimensional composite multiple-purpose figure 三维万能组合图形
three-dimensional composition 三维组合
three-dimensional computer model 三维计算机模型
three-dimensional conductivity 三线电导率；三维电导率
three-dimensional configuration 三维形态
three-dimensional consolidation 三向渗透固结；三固结；三维固结
three-dimensional construction 三维建筑
three-dimensional (construction) polymer 立构聚合物
three-dimensional continuity 三维连续性
three-dimensional continuum 三维连续统；三维连续体
three-dimensional contouring 立体造型
three-dimensional control 三维控制

three-dimensional coordinates 三维坐标
three-dimensional crimp 三维卷曲
three-dimensional crosslinked molecule 体型交联分子
three-dimensional crosslinked network 三维交联网络
three-dimensional cross-slide system 空间交滑轴系；空间滑轨系统
three-dimensional curve 三维曲线
three-dimensional data 三维数据
three-dimensional data stereo-display 三维资料立体显示
three-dimensional deformation 三维变形
three-dimensional design 立体造型设计；空间设计；空间度量；三维设计
three-dimensional diffraction pattern 三维衍射图
three-dimensional digital image 三维数字图像
three-dimensional display 立体显示；三维显示
three-dimensional display system 三维显示系统
three-dimensional distribution 立体分布；三维分布
three-dimensional distribution of forces 三维力分布
three-dimensional domain 三维域
three-dimensional drape 三维悬垂
three-dimensional dynamic spectrum analyser 三维动态频谱分析仪
three-dimensional effect 立体作用；立体效果；空间效应；三维效应
three-dimensional elasticity analysis 三维弹性分析
three-dimensional electrode 三维电极
three-dimensional electrode method 三维电极法
three-dimensional electrode reactor 三维电极反应器
three-dimensional electronically scan-(ning) fix array radar 三坐标电子扫描固定阵雷达
three-dimensional element 三维部件；三维单元
three-dimensional equality 三向等长
three-dimensional extended orebody 三向延伸矿体
three-dimensional extent of fissure 三向裂隙比；三维裂隙比
three-dimensional fabric 立体织物
three-dimensional fault 三维断层
three-dimensional figure 三维图
three-dimensional film 立体电影；立体影片
three-dimensional finite element method 三维有限元法
three-dimensional fix 三维空间定位
three-dimensional flow 空间绕流；空间流；三元(水)流；三向流；三维流(动)
three-dimensional follow-up 三度空间追踪
three-dimensional frame 三向构架；三维构架
three-dimensional frame (d) load bearing structure 三维
three-dimensional frame(work) 三维空间框架；三维空间构架；空间构架
three-dimensional fringe 三维条纹
three-dimensional fundamental form 三维基本形
three-dimensional geodesy 三维大地测量学
three-dimensional geometry 立体几何学；三维几何学
three-dimensional graph 三维空间图
three-dimensional graphic operation 三维图形操作
three-dimensional grating 三维光栅
three-dimensional gravity graticule 三度重力量板
three-dimensional ground motion 三维地面运动
three-dimensional guidance 三维制导
three-dimensional holding plate feeder 三坐标夹板式送料装置
three-dimensional hologram 三维全息图
three-dimensional hologram grating 三维全息光栅
three-dimensional holograph 三维全息摄影
three-dimensional holographic(al) view 立体全息图
three-dimensional holography 三维全息术
three-dimensional homing 三维自导引
three-dimensional-hyperbolic tracking system 三维双曲线跟踪系统
three-dimensional image 三维影像；三维像
three-dimensional image coordinates 三维像点坐标
three-dimensional imager 三维成像器

three-dimensional imaging 三维成像
three-dimensional inclination stack velocity analysis 三维倾斜叠加分析
three-dimensional inelastic analysis 三维非弹性分析
three-dimensional interconnection 三维互联
three-dimensional inter penetration 三维穿透度
three-dimensional inversion 三维反演
three-dimensional isoparametric element 三维等参数单元
three-dimensional isotropy 三维各向同性
three-dimensional joint meter 三向测缝计
three-dimensional lattice 三维点阵
three-dimensional layer-built total-sun method 三向分层总合法
three-dimensional layout 立体设计;立体安排;立体展示;三维观测系统
three-dimensional linear drift 三维一次漂移
three-dimensional load-carrying structure 三维承重结构
three-dimensional log 三维测井
three-dimensional maneuver 三维机动
three-dimensional map 立体地图;三维地图;三维地图;三面图
three-dimensional marine survey 海上三维地震勘探
three-dimensional mathematic(al) model 三维数学模型
three-dimensional matrix 三维矩阵
three-dimensional measurement 空间度量
three-dimensional memory 三度存储器
three-dimensional method of geodesy 三维大地测量法
three-dimensional migration 三维偏移
three-dimensional migration in two steps 二步法三维偏移
three-dimensional migration profile of linking well 连井三维偏移剖面
three-dimensional migration series section 三维偏移系列剖面
three-dimensional migration velocity map 三维偏移速度图
three-dimensional milling 立体铣削
three-dimensional model 立体模型;三维模型;三维模式
three-dimensional model(l)ing 三维造型
three-dimensional model test 整体模型试验
three-dimensional module house building 三维模数式住房
three-dimensional motion 三维运动;三度运动
three-dimensional net structure 三维网络结构
three-dimensional network 三维网(络)
three-dimensional network structure 体型网络结构
three-dimensional non-linear analysis 三维非线性分析
three-dimensional normal distribution law 高斯度空间正态分布定律
three-dimensional nozzle 三元喷管
three-dimensional nucleus 三维晶核
three-dimensional optic(al) transform 三维光学变换
three-dimensional package 三维组装
three-dimensional panelling pattern 三维镶嵌风格;三维镶嵌式样
three-dimensional pantograph 立体刻模机
three-dimensional pantograph engraving machine 立体刻模铣床;立体缩放仪刻模铣床
three-dimensional parallel system 三维平行系统
three-dimensional parameter unit 三维参数单元
three-dimensional penetration 三维渗透
three-dimensional perception 立体感
three-dimensional perspective 三维透视图
three-dimensional perspective view map 三维透视图
three-dimensional photoelasticity 三维光弹性学;三维光致弹性;三向光弹性;三维光测弹性学
three-dimensional photo(graph) 立体照片
three-dimensional photography 立体摄影术;立体照相术;三维摄影
three-dimensional picture 立体图像;立体图片
three-dimensional pipeline 空间管线;三维管线
three-dimensional Pitot tube 三向皮托管;三维全静压管;三维空速管

three-dimensional plot 三维图像;三维空间立体图;三维绘图
three-dimensional polar curves 三维极坐标曲线
three-dimensional polycondensation 三向缩聚
three-dimensional presentation 三维表示(法)
three-dimensional prestressing 三维预加应力;三向预应力
three-dimensional printing 立体印刷
three-dimensional problem 三维问题;三维问题
three-dimensional profile 立体河床纵剖面
three-dimensional profiling 三维仿形加工;成型面加工
three-dimensional profiling machine 三向靠模铣床
three-dimensional program(me) 立体图程序
three-dimensional projection 立体放映
three-dimensional projection screen 立体投影屏
three-dimensional projector 立体放映机
three-dimensional purlin(e) 三维桁条;三维檩条
three-dimensional quadratic drift 三维二次漂移
three-dimensional radar 三维雷达
three-dimensional radial plastic flow 三维空间径向塑性流动
three-dimensional radial timebase 三维径向时基
three-dimensional random distribution 三维乱向分布
three-dimensional randomly orientated 三维乱向
three-dimensional ray tracing 三维地震射线跟踪
three-dimensional reconstruction 三维重现
three-dimensional rendering 三维绘制
three-dimensional rod 三维杆件
three-dimensional scanner 三维扫描仪
three-dimensional scattering 三维散射
three-dimensional scintigraphy 三维闪烁照相法
three-dimensional search and height finding radar 三坐标搜索与测高雷达
three-dimensional seeing 立体视觉;三维度视力
three-dimensional seepage (flow) 三维渗流
three-dimensional seismic model 三维地震模型
three-dimensional seismic ray system 三维地震射线跟踪
three-dimensional seismic raytracing 三维地震射线跟踪
three-dimensional seismic survey 三维地震勘探
three-dimensional shape of a cave 三度空间组合形态
three-dimensional simulation 三维模拟
three-dimensional sink 三维汇点
three-dimensional sketch 三维草图
three-dimensional slope angle model display 三维倾角模型显示
threedimensional solid 三维固体
three-dimensional solid chart 三维实体图
three-dimensional solid modeling 三维实体模型
three-dimensional sound 立体声
three-dimensional space 三维空间;三度空间
three-dimensional space frame system 三维空间刚架体系
three-dimensional space model 三维空间模型
three-dimensional space pattern 三度空间模式
three-dimensional spectral classification 三元光谱分类
three-dimensional-spherical tracking system 三维球面跟踪系统
three-dimensional state of deformation 空间变形状态;三维变形状态
three-dimensional state of stress 三维状态应力;三向应力状态
three-dimensional storage 三维存储器
three-dimensional strain 三维应变
three-dimensional streamer 三维电缆
three-dimensional stress 三维应力;三轴应力;三向应力;三度空间应力
three-dimensional stressing 三维应力
three-dimensional stress state 空间应力状态;三维应力状态;三向应力状态
three-dimensional stress system 三维应力体系
three-dimensional stretching 三维张开;三维拉伸
three-dimensional structural description 三维结构描述
three-dimensional structural system 三维构造系统
three-dimensional structure 立体结构;空间结构;体型结构;三向结构;三维结构;三度结构
three-dimensional supporting structure 三维支

承建筑;三维支承结构
three-dimensional surface 三元面;三维面
three-dimensional symbol 立体符号
three-dimensional system 空间体系;三向体系;三维体系;体系;三维空间体系
three-dimensional system of forces 三维作用力系统
three-dimensional tank 三维水池
three-dimensional tape controlled 三坐标带控制
three-dimensional target tracking acquisition radar 三坐标目标跟踪和指示雷达
three-dimensional tectonic map 三维构造图
three-dimensional template 立体样板
three-dimensional tensioning 三维张紧;三维拉紧
three-dimensional terrain coordinates 三维地形坐标
three-dimensional thin layer chromatography 三维薄层色谱法
three-dimensional tracking 立体跟踪;三维跟踪
three-dimensional transformation 三维变换
three-dimensional traversing 三维导线测量
three-dimensional treatment 立体处理;三维处理
three-dimensional trend analysis 三维趋势分析
three-dimensional triangulation 空中三角测量
three-dimension(al) turbulence 三元紊流
three-dimensional turn 三维转弯
three-dimensional unit 三维单元;立体道具(剧场的)
three-dimensional vector 三维向量
three-dimensional velocity 三维速度
three-dimensional velocity log 三速度测井;三维速度测井
three-dimensional vibration 三向振动
three-dimensional view 三维视图
three-dimensional viewing 立体观察
three-dimensional vision 立体观察法;立体图像
three-dimensional wallpaper 三维墙纸
three-dimensional water quality model 三维水质模型
three-dimensional wave equation migration 三维波动方程偏移
three-dimensional weave 三向织物
three-dimensional weight-carrying structure 三维载重结构
three-dimensional weighted migration 三维加权偏移
three-dimensional woven fabric 三维织物
three-dimensional X direction 三维X方向
three-dimensional Y direction 三维Y方向
three-dimensional Z direction 三维Z方向
three-dimensional zonation 三向地带性;三间地带性
three-dimension leadership model 三构面领导理论
three-directional optic(al) square 三向直角头
three-distinct layers 三个明显层
three-D log 三速度测井
three-drift hypothesis 三流假说
three-drum boiler 三汽包锅炉
three-drum boiler low-head 低头三锅筒锅炉
three-drum hoist 三滚筒提升机
three-drum rapid shifting hoist 三鼓式快速换卷扬机;三筒式快速换挡绞车;三筒式快速换卷扬机;三鼓式快速换挡绞车
three-drum reel 三鼓卷纸机
three-drum steam hoist 三鼓式蒸汽绞车
three-drum system 三鼓法
three-edge-bearing method of testing clay pipes 三支边试验陶管法
three-edge bearing test 三边支承试验;三边承重试验
three-edged rule(r) 三棱尺;三角尺
three-eighth rule 八分之三法则
three-elastic axes 三弹性轴
three-electrode 三电极的
three-electrode arc lamp 三电极弧光灯
three-electrode sounding 三极测深
three-electrode spark 三电极火花
three-electrode thyratron 三极闸流管
three-electrode vacuum tube 三极真空管
three-element 三元
three-element aerial 三单元天线
three-element antenna 三元天线
three-element arch 三心拱;三铰拱

three-element control 三元调节
three-element filter 三元件滤波器
three-element servosystem 三元件伺服系统
three-energy transient phenomenon 三能量瞬变现象
three-engined 三发动机的
three-engined plane 三发动机飞机
three-engine hook up 三引擎组合
three-entry connector 三引线连接器
three-equidistant supports 三等距离支点
three-exactly coincident rasters 三个精确一致的光栅
three-faced sawing 三面下锯法
three-factor analysis of variance 三因素方差分析
three-factors of earthquake prediction 地震预报三要素
three-family dwelling 三家合住房屋
three feet 喇叭皮带线盆
three-feet and six inches ga(u)ge 三英尺六英寸轨距
three-feet line 三英尺安全线(甲板舱口周围划的安全线)
three-feet line ga(u)ge 三英尺一英寸十二分之一的轨距
three-field generator 三磁场发电机
three-field system 三区制
three-figure compass card 三位度数罗经盘
three-figure method 圆周法
three-filter colo(u)rimetry 三滤光镜比色计
three-fingered calipers 三指内径规;三触针井径规
three-finger rule 三指定则
three-fixture bathroom 三件套浴室
three-flanged nail 三叶钉;三刃钉
three-flight stair(case) 三段楼梯;三跑楼梯
three-floored 三层楼面的
three flute bit 三槽铣刀
three-fluted drill 三槽钻头
three-foiled arcade 三叶形饰的连拱廊
three-foiled arch 三叶形饰拱
three-foiled ground plan 三叶形平面图
three-foiled tracery 三叶形窗花格
three-fold 三倍(的);三重
three-fold altarpiece 三重圣坛层
three-fold block 三轮滑车;三滑轮滑车组;三饼滑车
three-fold coincidence 三重符合
three-fold coordination 三配位
three-fold degeneracy 三重简并
three-folded yarn 三股线
three-fold representation 三重表达
three-fold statically indeterminate 三次超静定的
three-fold statically indeterminate structure 三次超静定结构
three-fold symmetry axis 三次对称轴线
three-foot rule 三英尺长折尺
three-forked controlled jump 三叉控制转移
three-forked jump 三叉转移
three-fork truck 三叉式装卸机
three-foundation weave 三原组织(平纹斜纹缎纹三个基本组织)
three-four-five rule 勾三股四弦五法则;三、四、五定直角法
three-fourth running down clause 四分之三碰撞(责任)条款
three-furnace process 三炉联炼法;三联法
three-Gods in white-and-blue 青花三星
three-grade system signal 三级制信号
three-grate combination cooler 三篦床复合冷却机
three-groove chucking reamer 三槽机用铰刀
three-grooved drill 三槽钻头
three-group method 三组法
three-group reactor 三群反应堆
three-groups of organisms 三类有机体
three-group theory 三群理论
three-grouser track shoe 三肋履带板
three-gun tube 三枪管
three-gun turret 三联炮塔
three-gyro platform 三陀螺稳定平台
three-gyro-stabilized platform 三陀螺稳定平台
three-halves power law 二分之三方律;二分之三次方定律
three-harness twill 三页斜纹
three-harrow unit 三耙组合
three-heading system 三平巷掘进法
three-head milling machine 三轴铣床

three-high 三辊的
three-high arrangement 三辊式配置
three-high berth 三层铺;三层床
three-high bloomer 三辊开坯机
three-high blooming mill 三辊式初轧机
three-high cogging mill 三辊式开坯机;三辊开坯机
three-high finishing train 三辊式精轧机组
three-high house 三辊式机座
three-high jobbing mill 三辊式型钢轧机
three-high Lauth plate mill 三辊劳特式钢板轧机
three-high mill 三辊轧机;三辊式轧机
three-high plate mill 三辊钢板轧机
three-high rolling stand 三辊式轧机机座
three-high rolling train 三辊式轧钢机组
three-high shape mill 三辊式型钢轧机
three-high straddle carrier 三层跨云车
three-high train 三辊式装置
three-high universal mill 三辊万能式轧机
three-hinged 三铰的
three-hinged arch 三铰拱
three-hinged arch bridge 三铰拱桥
three-hinged arch dam 三铰拱坝
three-hinged arched girder 三铰拱大梁
three-hinged arch-ribbed dome 三铰拱肋穹隆
three-hinged arch truss 三角式桁架;三铰拱桁架
three-hinged braced chain 三铰桁链
three-hinged frame 三铰钢架
three-hinged half-frame 三铰半框架
three-hinged rectangular frame 三铰矩形构架
three-hinged roof 三铰屋顶
three-hinged steel arch 三铰钢拱
three-hinged stiffening truss 三铰加劲桁架
three-hinged trussed arched girder 三铰桁架拱大梁
three-hinged trussed frame 三铰桁架构架
three-holder 有三个坑位的厕所
three-hole aperture 三孔光栏
three-hole wedge 三孔楔子
three-hop 三次反射
three-hour geomagnetic planetary index 三小时行星地磁指数
three-hour range index 三小时段指数
three-hundred and sixty degree nozzle rotation 三百六十度转动管嘴
three-industrial wastes 工业三废
three-in-hand 三驾马车
three-inland ship 三岛型(轮)船
three-in-one brake valve 三合一制动阀
three-in-one cell 三试件压力室;三试件压力盒
three-in-one drawbar 三杆牵引装置
three-in-one process 三合一措施
three-in-one unit 三用装置
three-input adder 全加器
three-input subtracter 全减法器
three-input switch 三输入开关
three-instantaneous processing 三瞬处理
three-intersection in space 空间后方交会
three-island berth 三岛式泊位
three-islander 三岛型(轮)船
three-jaw 三爪
three-jaw chuck 三爪卡盘;三卡瓦卡盘
three-jaw chuck with interchangeable false jaws 互换虎钳口式夹条
three-jaw collet chuck 三套爪夹盘
three-jaw concentric chuck 三爪同心夹盘
three-jaw concentric lathe chuck 三爪定心夹盘
three-jaw crusher 颚板破碎机;三爪破碎机
three-jaw draw-in type chuck 三爪内拉簧式夹盘
three-jaw drill chuck 三爪钻卡
three-jaw equalizing drive 三爪同心卸轮器
three-jaw hand operated chuck 三爪手动夹盘
three-jaw independent chuck 三爪分动卡盘
three-jaw independent lathe chuck 三爪定心夹盘
three-jaw lever-operated chuck 三爪杠杆夹盘
three-jaw self-centering chuck 自定心三爪卡盘
three-jaw universal chuck 三爪万能夹盘
three-jiont unit 三根钻杆组成的立根
three-kind rate meter 三种费率计算器
three-kpc arm 三千秒差距臂
three-lane 三车道
three-lane at a time paving 三车道一次铺设
three-lane dual carriageway road 双向三车道路
three-lane dual highway 三车道双行道
three-lane pavement 三车道路面

three-lane road 三车道路道;三车道公路;三车线道路
three-lane road surface 三车道路面
three-lane section 三车道断面
three-lane traffic handling 三车道交通运行
three-layer 三层
three-layer board 三层刨花板;三合板;三层结构板
three-layered chimney (由内烟道管、砖砌层及外套构成的)三层烟囱
three-layered panel 三层胶合板
three-layered panel facade 三层胶合板正面
three-layered roof(ing) slab 三层屋顶板
three-layered shell 三层壳
three-layered system 三层系统
three-layered wall panel 三层墙板
three-layer elastic system 三层弹性体系
three-layer mineral 三层矿物
three-layer particle board 三层碎料板
three-layer sandwich glass 三层安全玻璃
three-layers template method 三层量板法
three-layer structure 三层构造
three-layers type 三层型
three-layer system 三高度层系统
three-layer type structure 三层型结构
three-leaf 三叶
three-leaf filling twill 三页纬面斜纹
three-leaf folding shutter door 三扇折叠式百叶门
three-leaf pin hinge 三叶铰链
three-leaf twill 三页斜纹
three-leaf warp twill 三页经面斜纹
three-leaved bear trap dam 三叶式活动坝;三叶卧式闸门熊阱坝
three-leaved rose curve 三瓣玫瑰线
three-leg crane 三角式吊杆
three-legged 三腿的
three-legged backfill tamper 三脚回填夯锤
three-legged derrick 三腿摇臂起重机;三腿起重机;三腿吊杆;三足起重机;三腿钻塔;三脚钻塔;三脚钻架
three-legged dividers 三脚规
three-legged gravity escapement 三脚式重力擒纵机构
three-legged lewis 三脚起重爪
three-legged rammer 三脚夯锤
three-legged roller 三脚压路机
three-legged support 三脚支架
three-legged tamper 三脚夯锤
three-leg interchange 喇叭形立体交叉
three-leg jack-up rig 三脚顶升机具
three-leg mooring 三脚碇泊浮筒
three-legs intersection 三支交叉口;三路相交的交叉
three-leg-sling 三索套;三套索吊装装置
three-legs type core 三心柱铁芯(变压器)
three-length recording 三长度记录
three-lens 三透镜;三镜头
three-lens camera 三镜头摄影机
three-lens condenser 三透镜聚光器
three-lens corrector 三透镜校正系统
three-lens objective 三透镜物镜;三镜片物镜;三片物镜
three-lens system 三透镜系统
three-letter point 三字点(磁罗经)
three-letter signal 三字母信号(国际信号规则的)
three-level 三能级
three-level addressing 三级寻址;三级地址;三级编址
three-level basement 三层地下室
three-level building operational system 三平面建筑作业系统
three-level channel 三阶通路
three-level control network 三级防治网
three-level costate prediction 三级共态预测
three-level excitation 三能级激发
three-level fluorescent solid 三能级荧光固体
three-level generator 三能级振荡器;三级振荡器
three-level grade separation 三层(的)立体交叉
three-level jumbo 三层钻车;三层台车
three-level laser 三能级激光
three-level laser configuration 三能级激光结构
three-level laser material 三能级激光器材料;三能级激光材料
three-level light emitter 三能级光发射体
three-level operation 三平面作业

three-level record(ing) 三级记录
three-level return system 三级记录系统；三级返回系统
three-level section 三点折线断面；三点断面
three-level structure 三平面构筑物；三层式构筑物
three-level subroutine 三级子例行程序
three-level system 三能级系统
three-light-beam source 三束光源
three-light window 三扇窗；三截窗；三分格窗
three-limb core 三柱铁芯
three-limb tube 三臂管
three-line high draft system 三罗拉式大牵伸装置
three-line pavement 三车道路面
three-line target 三线靶标
threeling 三连晶
three-link hitch 三点悬挂装置
three-lip core drill 三缘空心钻头
three-lip keyway cutter 三面刃键槽铣刀
three-lipped drill 三缘钻头
three-lipped twist drill 三缘麻花钻
three-lobe bearing 三叶形轴承；三叶轴承
three-lobe chamber 三叶形汽缸
three-lobe epitrochoidal bore 三弧外旋轮线缸体
three-lobe pump 三叶片式水泵；三突瓣式水泵；三叶片式泵
three-lobe rotor 三弧旋转活塞（转子发动机的）
three-lobe tracery 三叶窗花格
three-loop system 三环系统
three-L's （驾驶员的）三注意
three-main routes of absorption 三种主要吸收途径
three-man module 三人舱
three-market 三方竞争市场
three-masted 三桅的
three-masted vessel 三桅船
three-master 三桅船
three-member cog 三角垛式支架
three-membered ring 三节环
three-member joint 三杆联结
three-minute glass 三分钟定时器
three-mirror system 三镜系统
three-miss blister fabric 三列凸纹针织物
three-missile-to-a-site complex 三发射架的发射场综合设施
three-mode control 三模控制
three-mode tracking 三模跟踪
three-moment equation 三弯矩方程；三力矩方程
three-moment method 三弯矩法；三力矩法
three-moment theorem 三弯矩定理
three-months inspecting shed 三月检库
three-motor all-electric shovel 三电动机电铲
three-motor drive 三发动机驱动
three-motor shovel 三动力组电铲；三电机电铲
three-motor tape-transport mechanism 三马达纸带传运机构
three-motor travel(l)ing crane 三发动机移动式起重机
three-motor type crane 三发动机式起重机
three-name paper 三方签署的票据
three-nave 三个中殿（教堂中的）；三个正厅（教堂中的）
three-necked bottle 三颈瓶
three-necked flask 三颈烧瓶
Three-North Shelterbelt 三北防护林带
three-numeral code group 三位数数码组
three-nutrient compound fertilizer 三元复合肥料
three o'clock-welding 横向自动焊
three-operating crane 三用起重机
three-over four curvilinear draft 三上四下曲线牵伸
three-over four roller drafting 三上四下罗拉牵伸
three-pack door 三肚板门；三镶板门
three-pack epoxy resin 三色环氧树脂
three-pack unit 三包堆（材料）单元
three-pair 三对；四楼房间
three-panel truss 三节间桁架
three-pan system 三锅系
three-paper test 三张纸试验
three-parameter method 三参数法
three-parameter model 三参数模型
three-part alloy 三元合金
three-part box 三节砂箱
three-part clock 三用钟
three-part flask 三开砂箱
three-part hydraulic binder 三种成分的水力黏合剂；三种成分的液压黏合剂
three-parties 三方当事人
three-part line 三股绳索；倍率为三的钢丝绳滑轮组
three-part line tackle 三股绳索滑轮组
three-part mold 三片的合模
three-part pattern 三开模
three-part rate schedule 三部收费制
three-part system 三元系
three-part tree 三合树
three-party draft 三名汇票
three-pass boiler 三回程锅炉
three-pass economic boiler 三回路经济锅炉
three-pass stove 三转热风炉；三通式热风炉
three-path amplifier 三通道放大器
three-pedestal base 三垛式座
three-pen recorder 三笔记录仪
three-person flat 三人公寓
three-phase 三相绕线式感应电动机；三相（的）
three-phase aerobic biological fluidized bed 三相好氧生物流化床
three-phase alternater 三相交流发电机
three-phase alternating current 三相交流电
three-phase alternating current induction motor 三相交流感应电动机
three-phase alternating current locomotive 三相交流电机车
three-phase alternating current motor 三相交流电(动)机
three-phase alternating current plasma generator 三相交流等离子发生器
three-phase alternating current supply 三相交流电源
three-phase alternating current traction 三相交流电牵引
three-phase alternator 三相交流发电机
three-phase arc furnace 三相电弧炉
three-phase arc welding 三相电弧焊
three-phase armature 三相电枢
three-phase asynchronous motor 三相异步电动机
three-phase autotransformer 三相自耦变压器
three-phase bar-winding 三相条绕组
three-phase biological fluidized bed 三相生物流化床
three-phase breaking time 三相断路时间
three-phase bridge 三相桥
three-phase cable 三相电缆
three-phase carbon arc lamp 三相碳精灯
three-phase change-over switch 三相转换开关
three-phase circuit 三相电路
three-phase commutator machine 三相整流子电机；三相换向电动机
three-phase commutator motor 三相换向电动机
three-phase connection 三相连接
three-phase current 三相电流
three-phase current limiting fuse 三相限流熔断器
three-phase current locomotive 三相电机车
three-phase current motor 三相交流电动机
three-phase current transformer 三相电流互感器
three-phase current vibrator 三相电路振荡器
three-phase design 三阶段设计
three-phase diagram 三相图
three-phase diagram of soil 土的三相草图
three-phase distribution 三相分布
three-phase driving 三相驱动
three-phase electric-arc furnace 三相电弧炉
three-phase electric-drive 三相电力传动装置
three-phase electromotive force 三相电动势
three-phase equilibrium 三相平衡
three-phase excitation 三相激励
three-phase exciter 三相励磁机；三相激磁器
three-phase fault 三相故障
three-phase Flotrol 三相恒流充电机
three-phase fluidization 三相流态化
three-phase fluidized bed 三相流化床
three-phase fluidized bed reactor 三相流化床反应器
three-phase four core cable 三相四芯电缆
three-phase four-wire installation 三相四线装置
three-phase four-wire system 三相四线制
three-phase froth 三相泡沫
three-phase full wave connection 三相全波接法
three-phase furnace 三相电炉
three-phase generator 三相发电机
three-phase heating 三相加热
three-phase hoist motor 三相起重电动机
three-phase induction 三相感应
three-phase induction motor 三相感应电动机
three-phase induction regulator 三相感应调节器
three-phase inverter 三相反向变流器
three-phase kilowatt-hour meter 三相电度表；三相电表
three-phase line 三相线路
three-phase load 三相负载
three-phase low frequency welder 三相低频焊机
three-phase machine 三相电机
three-phase magnetic amplifier 三相磁放大器
three-phase motor 三相电动机
three-phase network 三相网络；三相电力网
three-phase of soil 土壤三相
three-phase oil immersed transformer 三相油浸式变压器
three-phase one shot reclosing device 三相一次自动重合闸装置
three-phase peak load 三相尖峰负荷
three-phase potential transformer 三相电压互感器
three-phase power supply net 三相供电网
three-phase power transmission 三相电力输送
three-phase railway 三相电力牵引铁路
three-phase rectifier 三相整流器
three-phase rectifier welder 三相整流焊机
three-phase region 三相区
three-phase separation region sludge layer 三相分离区污泥层
three-phase separator 阳气水分离器
three-phase series commutator motor 三相串励换向电动机
three-phase series motor 三相串励电动机
three-phase seven-wire system 三相七线制
three-phase short-circuit 三相短路
three-phase short-circuit curve 三相短路曲线
three-phase shunt commutator motor 三相并励换向电动机
three-phase shunt motor 三相并激电动机
three-phase six-wire system 三相六线制
three-phase slip-ring induction motor 三相滑环式感应电动机
three-phase slip-ring motor 三相滑环式电动机
three-phase smelting furnace 三相电熔炉
three-phase soil 三相土
three-phase squirrel-cage asynthonous traction motor 鼠笼式三相异步牵引电动机
three-phase squirrel cage induction motor 三相鼠笼式感应电动机
three-phase squirrel cage motor 三相鼠笼式电动机
three-phase starter 三相起动器
three-phase stator 三相定子
three-phase switch 三相开关
three-phase synchronous generator 三相同步发电机
three-phase system 三相制；三相系
three-phase three-shot reclosing device 三相三次自动重合闸装置
three-phase three-wire system 三相三线制【电】
three-phase traction 三相牵引
three-phase transformer 三相变压器
three-phase two shot reclosing device 三相二次自动重合闸装置
three-phase unbalanced power 三相不平衡功率
three-phase voltage 三相电压
three-phase wattmeter 三相瓦特计；三相功率计
three-phase winding 三相绕组
three-phase with neutral wire 4 hole socket-outlet 三相四线制
three-phonon absorption 三声吸收
three-phonon process 三声过程
three-photon decay 三光子衰变
three-piece 三件一套的
three-piece bail 组合式耳环（三部分组成的）
three-piece core box 三开芯盒
three-piece cutting edge 三件式切削刃；三件式刀刃
three-piece flare fitting 三件式护口管接头
three-piece laminated safety sheet glass 三层胶合安全玻璃
three-piece pattern 三开模
three-piece toggle plate 三件式肘板
three-piece tree 三合树
three-piece windshield 三块式风挡

three-pin clevis link 三孔联板；三角调节板
three-pin connection strap 三孔联板
three-pin frame 三铰构架
three-pinned 三心的；三脚的；三节的
three-pinned arch 三铰拱
three-pinned arch bridge 三铰拱桥
three-pinned arched girder 三铰拱大梁
three-pinned frame 三铰构架；三铰框架
three-pinned half-frame 三铰半框架
three-pinned rectangular frame 三铰矩形框架
three-pinned roof 三铰屋顶
three-pinned steel arch 三铰钢拱
three-pinned truss(ed) frame 三铰桁架支撑的框架
three-pin plug 三心插塞；三脚插头
three-pin rigid frame 三铰钢架
three-pin steel lattice rigid frame 三铰钢格构刚架
three-pipe air conditioning system 三管空调系统
three-pipe line 三管线
three-pipe system 三管采暖系统
three-pipe water system 三管洽水系统
three-piping system 三管配置方式
three-pit liming system 三槽浸灰法
three-pit system 三槽法
three-pit system of liming 三槽浸灰法
three-place accuracy 三位准确度
three-plate clutch 三片离合器
three-plier 三行式铆钉
three-plug connector 三线插座
three-plug method 三弧法
three-plus-one address 三加一地址
three-ply 三重的；三层
three-ply board 三(合)夹板；三层板
three-ply built-up roof cladding 三层复合的屋顶覆盖层
three-ply butt 三折铰链；三折合页
three-ply riveting 三层铆接
three-plywood 三夹板；三合板
three-point arbitrage 三角套汇
three-point arbitrage inforeign exchange 三角套汇
three-point assay 三点检验
three-point attachment 三点联结
three-point attitude 三点着陆姿态
three-point bearing 三点支承
three-point bending 三点弯曲
three-point bending testing 三点抗折试验
three-point bit 三刃钻头；三刃钎头
three-point bond 三点联结
three-point circuit 三点电路
three-pointed arch 尖顶拱
three-point fix 后方交会；三点定位
three-point fix method 三点定位法；三标两角法
three-point free linkage 自由式三点悬挂装置
three-point guidance 三点法引导
three-point hitch 三点挂济
three-point intersection 后方交会；三点法交会
three-point intersection in space 空间三点交会法
three-point landing 三点着陆；三点落地(飞机)
three-point linkage 三点联结(机构)
three-point linkage mounting 装在三点悬挂装置上
three-point loading 抗弯负载
three-point lock 三点固定门锁
three-point method 三点法
three-point method of current ga(u)ging 测流三点法
three-point method of stream ga(u)ging 测流三点法
three-pointodontograph 三点画齿规
three-point perspective 三点透视
three-point plug 三脚插塞
three-point probe 三探针
three-point problem 三点法；三点问题；三垂线井下定向法；三标两角定位法
three-point projection 三点投影
three-point refueling tanker 三点加油机
three-point resection 三点后方交会法
three-point roller reamer 三牙轮扩孔器
three-point starter 三端起动器；三点起动器
three-point starting box 三端启动电阻箱；三点起动箱
three-point support 三点支承
three-point suspension 三点悬吊；三点悬挂
three-point switch 三点开关

three-point test 三点试验；三点测验
three-point tone controlling 三点控制法
three-point transmission 三点传动
three-pole 三极式
three-pole derrick 三脚井塔；三脚井架
three-pole double throw switch 三刀双投开关
three-pole knife switch 三极闸刀开关
three-pole of near source 近场源三极
three-pole single-throw switch 三刀单投开关
three-pole switch 三极开关；三刀开关
three-polyphosphoric sodium 三聚磷酸钠
three-portal facade 三门进入口正面
three-port slide 箱式活门；箱式滑阀
three-port slide valve 三通滑阀
three-port two stroke engine 三缸二行程发动机
three-position 三位
three-position control 三位控制
three-position four-way valve 三位四通阀
three-position modulation 三单元调制
three-position signal 三位式(显示)信号；三相信号机
three-position switch 三位置转换开关；三位开关
three-position valve 三位阀
three-post cruciform bollard 三柱十字柱
three posts 三柱桩(屋顶架)
three-pressure stage turbine 三级压力式汽轮机
three-primary colo(u)rs 三原色
three-prism spectrograph 三重棱镜摄谱仪；三棱镜摄谱仪
three-probe method 三探针法
three-product jig 三产品跳汰机
three-product separation 三物分离
three-pronged hoe 三叉锹
three-pronged program(me) 三点方案
three-prong plug 三脚插头；三相插头
three-prong rope grab 三尖端捞绳矛
three-purlin(e) beam 三架梁
three-quantities of reserves estimates 三量统计
three quarter 七分头；四分之三
three-quarter (attached) column 露出墙面四分之三柱径的柱壁；四分之三柱
three-quarter bat 大半头砖；七分头；四分之三砖
three-quarter block 四分之三块体
three-quarter bound 布皮脊装订
three-quarter brick 四分之三砖；大半头砖；七分头
three-quarter closer 收口用去角七分头；四分之三闭砖法
three-quarter design speed method 四分之三设计车速法(一种设计弯道超高的方法)
three-quarter floating axle 四分之三浮动轴
three-quarter hard 四分之三硬度
three-quarter hard annealing 增硬退火
three-quarter header 四分之三长丁砖；四分之三砖
three-quarter mo(u)lding 四分之三浇筑
three-quarter niche 四分之三壁龛
three-quarter plate watch 四分之三夹板表
three-quarter small column 四分之三小柱
three-quarter tile 四分之三瓦；四分之三砖
three-quarter track 四分之三履带车
three-quarter tracked 四分之三履带式的
three-quarter turn 回转270°的楼梯
three-quarter-turn stair(case) 三跑楼梯
three-quarter view 四分之三侧视；半侧视图
three-race mill 三环槽式磨碎机
three-rail railway 双轨距铁路
three-rail test track 三轨试验(导轨)装置；三轨实验(导轨)装置
three-rail top contact 三轨上部接触式
three-rail track 双轨铁路；双轨铁道；双轨距线路
three-ram pump 三柱塞泵
three-range 三量程的
three-range transmission 三挡齿轮
three-range winding 三列绕组
three-rate tariff system 三级税率系统
three-ratio gear 三级变速箱
three-ray interference 三射线干涉
three-recycle flow biological filter 三重环流生物流化床
three-reflector feed system 三反射面馈源系统
three-resource figure of spectrum anomaly 能谱异常三源图
three-roll bending machine 三辊弯扳机
three-roller bit 三滚轮钻头；三牙轮钻头
three-roller mill 三辊中速磨

three-roll grinder 三辊式磨料机
three-roll mill 三辊筒轧机；三辊式滚轧机
three-roll piercer 三辊穿孔机
three-roll random roller 三轴串联(式)压路机
three-roll roller 三辊压路机
three-roll tandem roller 串联式三轮压路机；串联式三滚筒压路机
three-roll-type coiler 三辊式卷取机
three-roll type machine 三辊辊式机械
three-roll unbender 三辊校直机；三辊矫直机
three-roomed 三室的
three-roomed apartment 三室公寓
three-roomed dwelling unit 三室居住单元
three-roomed flat 三室套房
three-roomed living unit 三室生活单元
three-rope suspension grab(bing) bucket 三索悬挂式抓斗
three-row 三行的
three-row bearing 三列轴承
three-row hatch 三排横列舱口；三列舱口
three-row ridger 三行起垄机
three-sample theory 三样本理论
three-sampling intervals per decade 三分度取样
three-screen sand 三筛砂
three-screen sizing unit 三筛式分级装置
three-screw base 三脚架基座
three-screw pump 三螺杆式(水)泵
three-sectional dock 三分段式浮船坞
three-section coil 三节线圈
three-section cut 三段掏槽
three-section floating dock 三段式浮船坞
three-section framing bent 三节排架；三段排架
three-series theorem 三数列定理
three/seven lime earth(3:7 lime earth] 三七灰土
three/seven lime earth rammed 三七灰土夯实
three-sheave block 三轮滑车
three-sheet mineral 三片层矿物；三晶片矿物
three-shift operation 三班作业
three-shift pouring 三班轮换浇注(混凝土的)
three-shift work 三班工作制
three-shift work day system 每日三班(工作)制
three-shift work(ing) 三班(工作)制；三班作业
three-ship element 三机小队
three-ship operating 三船作业
three-side(d) (bull)dozer 三向推土机；斜铲推土机
three-sided mold-board 三边床模型板；三边样板
three-side hung sections and top hung fanlight 三扇平开和一扇上悬腰窗
three-sider 三面形(体)
three-simultaneities [completion of a project simultaneously with the completion of the facilities for controlling waste water/ waste gas and industrial wastes] 三同时(指完成工程的同时,完成废水、废气、废渣的处理设施)
three-simultaneous rule 三同时原则
three-slit interference 三狭缝干涉
three-slot boiling head 三缝波淋头
three-slot burner head 三缝燃烧器
three-socket cross pipe 三承四通；三承十字管
three-socket cross tube 三承四通；三承十字管
three-soil-water regime 三种土壤水分状况
three-sources of water 三个水分来源
three-space 立体(的)
three-span 三跨度
three-span bridge 三跨桥梁
three-span cable supported bridge 三跨缆索吊桥
three-span continuous beam 三跨连续梁
three-span continuous bridge 三跨连续桥
three-span continuous girder 三跨连续大梁
three-span frame 三跨框架
three-speed door panel 三扇伸缩门板；三扇搭接门板
three-speed gear 三速齿轮；三连齿轮；三级变速装置
three-speed gear box 三速齿轮箱；三级变速箱
three-speed motor 三速电动机
three-speed transfer case 三速变速箱；三速分动箱
three-spout packing machine 三嘴包装机
three-spray gun 三喷嘴喷枪
three-square 截面成等边三角形；三等边灯心草
three-square file 三角锉(刀)
three-square oil stone 三角油石
three-square rule 三棱尺
three-square scale 三棱尺

three-square screw 三棱螺丝
three-square tool 三棱刀
three-stability 三稳定性
three-stable state device 三稳态设备;三稳定设备
three-stack(ed) 三烟囱的
three-stage 三级的
three-stage/6-yard marshalling station 三级六场编组站【铁】
three-stage air compressor 三级空气压缩机
three-stage biologic(al) treatment process for nitrogen removal 三段生物处理除氮工艺
three-stage blower 三级压缩机
three-stage compressor 三级压缩机
three-stage converter 三级变矩器
three-stage counter current extraction system 三级逆流抽提系统
three-stage coutercurrent reactor 三段逆流反应器
three-stage crushing 三级破碎;三级破碎
three-stage design 三阶段设计;三段设计
three-stage differentially pumped vacuum system 三级差动抽气真空系统
three-stage distillation 三级蒸馏
three-stage electrodialysis unit 三段电渗析装置
three-stage engineering approach 工程上三步骤研究法
three-stage gas sealing system 三级气密封系统
three-stage gas turbine 三级式燃气轮机
three-stage jumbo 三阶段隧道盾构
three-stage least squares estimates 三段最小平方估计
three-stage least squares method 三段最小平方法
three-stage mass spectrometer 三级质谱计
three-stage mixing 三段混炼
three-stage open-circuit crushing 三级开路破碎
three-stage parachute 三级降落伞
three-stage process 三阶段工艺;三段法
three-stage reduction 三段碎矿
three-stage refiner 三段区域精炼炉
three-stages fluidized bed dryer 三层硫化床干燥器
three-stages hot air conveying type dryer 三层带式热风干燥机
three-stage sintering 三段烧结
three-stage-six-yard bidirectional longitudinal marshalling station 双向三级六场纵列式编组站
three-stage/six yard marshalling station 双向纵列编组站【铁】
three-stages of ore reserves 矿山三级矿量(黑色冶金)
three-stage steam ejector 三级蒸汽喷射器
three-stage stereo-comparator 三像片盘立体坐标仪
three-stage supercharger 三级增压器
three-stage system 三级管网系统
three-stage/three-yard marshalling station 单向纵列式编组站【铁】;三级三场编组站【铁】
three-stage-three-yard unidirectional combined marshalling station 单向三级三场混合式编组站
three-stage torque converter 三级变扭器;三级转矩变换器;三级变矩器
three-stand tandem mill 三机座串列式轧机
three-star fix 三星定位
three-star problem 三星定位(法)
three-start screw 三头螺纹
three-stateable 可三态的
three-state buffer 三态缓冲器
three-state control signal 三态控制信号
three-state data output 三态数据输出
three-state driver 三态驱动器
three-state enable 三态使能;三态启动
three-state output 三态输出
three-states of matter 物质三态
three-station duplex transmission system 三局双工传输制
three-stator winding synchro 三定子绕组自动同步机
three-step action 三位作用
three-step control 三位控制
three-step decompression cycle 三级泄压循环
three-step design 三阶段设计
three-step distance relays 三段距离继电保护装置
three-step feed box 三级进给箱
three-step switch 三级开关
three-stimulate measurement method 三刺激实测法

three-stor(e)y 三楼
three-stor(e)yed 三层的
three-storied house 三层楼房(屋)
three-stranded rope 三股绳
three-strand rod mill 三线式型钢轧机;三线式棒材轧机
three-strand roller chain 三鼓式滚子链
three-strands double crossing-over 三线双交换
three-strand sennit 三股编绳
three-stub tuner 三短截线调谐器
three-tandem roller 三轴压路机
three-term control 三项控制
three-term controller 三项控制器
three-terminal 三端
three-terminal circulator 三通循环器;三通路循环器
three-terminal contact 三端子接点
three-terminal line 三端线路
three-terminal network 三端网络
three-term recurrence 三项递归
three-thread overlock 三线包缝机
three-thread wire method 三线测螺纹法;三钢丝法
three-through pump 三联泵
three-throw 三曲柄的
three-throw crank 三连曲柄;三拐曲柄
three-throw crank pump 三拐曲柄泵
three-throw crankshaft 三曲柄曲轴;三拐曲轴
three-throw high-speed pump 三缸高速泵
three-throw plunger pump 三柱塞泵;三联柱塞泵
three-throw pump 三缸泵;三冲程泵;三程泵
three-throw switch 三向转辙器;三开道岔
three-throw tap 三通旋塞;三通阀
three-throw turnout 三开道岔
three-throw turnout of contrary flexures 异向三开道岔
three-throw turnout of similar flexures 同向三开道岔
three-tier winding 三列绕组
three-time estimate 三种时间估计法;三时估计数;三点估计
three-times overload relay 三次过荷继电器
three-tine grapple 三叉抓斗
three-tine log grappler 三爪木材抓斗
three-tone plan position indicator 三音平面位置指示器
three-tone signal 三音信号
three-ton truck 三吨载货汽车
three-tooth ripper 三齿松土机;三齿耙路机
three-to-single-phase welding transformer 三相输入的单相焊接变压器
three-towered 三塔高的
three-transient-sinusoidal successive wave method 三瞬态正弦连续波法
three-tray thickener 三层浓密机
three-trees 三腿梯子
three-tripod traversing 三联脚架法(导线测量的)
three-truss bridge 三桁架桥
three-tube camera 三管式摄像机
three-types of triaxial test 三轴试验的三种类型
three-unbalanced power 三相不平衡功率
three-unders in coal mining 三下采煤
three-unique catalytic converter 三元催化净化器
three-unit code 三元码
three-unit regulator 三合一式调节器
three-valued logic 三值逻辑
three-value simulation 三值模拟
three-valve 三阀的
three-valve engine 三气门发动机
three-vaned 三叶片的
three-vaned centrifugal pump 三叶片离心泵
three-variable analysis 三变量分析
three-variable model 三变量模型
three-variable production 三变量生产
three-variance overhead analysis 三次差异制造费用分析;三差异制造费用分析;三差异间接费用分析
three-vector 三维向量
three-vidicon camera 三光导视像管摄影机
three-view drawing 三面图
three-voltmeter method 三伏特计法
three-W 三种轨道病害
three-wastes [waste water, waste gas and waste residue] 三废(指废水、废气、废渣)
three-wastes utilization and disposal 三废治理
three-watch system 三班轮值制

three-wattmeter method 三瓦特计法
three-wave mixing 三波混合
three-way 三通管接头;三向的;三路
three-way analysis of variance 三向方差分析
three-way analyzer 三用分析仪
three-way arch 三路门拱
three-way bulb 三光灯泡
three-way cable subway 三路电缆地下管道
three-way calling 三方呼叫
three-way catalytic converter 三元催化排气净化器
three-way catalytic reactor 三元催化反应器
three-way circulation 三方式循环
three-way cock 开关旋塞;三向龙头;三通旋塞;三通塞门;三通开关;三通活栓
three-way compensating valve 三向补充阀
three-way conduit tile 三路管道瓦筒
three-way connection 三通接头;三向连接
three-way connection (pipe) 三路管
three-way connector 三通接线夹
three-way control valve 三通控制阀;三路控制阀
three-way cord 三芯塞绳;三股软线
three-way coupling box 三通电缆分线盒
three-way cross 三交
three-way distributing chute 三股分配斜槽
three-way double observation 三程二次观测
three-way drag 三向刮路机
three-way dump body 三向倾卸车身;三向倾卸式车厢
three-way dump discharge 三向倾卸
three-way dumper 三向翻斗车
three-way dumping 三向倾卸
three-way dump trailer wagon 三向倾卸拖车
three-way dump truck 三向自卸汽车;三向卸载车;三向倾卸货车
three-way elbow 三向弯头;三通弯头
three-way electrochromatography 三路电色谱法
three-way emission control 三元催化净化
three-way fitting 三通管件
three-way flat slab 三向无梁楼板
three-way fuse plug 三线保险丝塞
three-way fuse socket 三线保险丝塞孔
three-way glass stopcock 三通玻璃栓
three-way hub 三面毂
three-way intersection 三路交叉口;三岔路口;三叉路口
three-way jack 三线插孔
three-way key 三路电键
three-way lamp 三向灯
three-way lattice(d) grid 三向花格网架
three-way layout 三元配置
three-way loop observation 三程循环观测
three-way magnetic valve 三通电磁阀
three-way piece 三路管
three-way pipe 三向弯头;三向弯管;三通管;三路管
three-way positioning method 三路定位方法
three-way radio 三用接收机
three-way receiver 三用接收机;三调谐电路接收机
three-way reinforcement 三向配筋;三向布筋
three-way reversing cock 三通换向旋塞
three-way set 三用机
three-way signal 三组交通信号;三面信号
three-way space grid 三向空间网架
three-way split former 三瓣式模筒
three-way stop cock 三路栓
three-way strap 丁字铁角;T形铁角;铁T角;三向系板
three-way support 三脚架
three-way switch 三向开关;三通开关;三路开关;三路道岔;三联开关;三开道岔;三点开关
three-way system 三元催化废气净化系统
three-way system of reinforcement 三向配筋制;三向配筋(体系);三向布筋制;三向布筋体系
three-way tap 三通开关;三通龙头;三向龙头;三路龙头
three-way tipper 三向倾卸装置;三向倾卸车
three-way tipping gear 三向倾卸机构
three-way tipping trailer 三向倾卸式挂车
three-way turnout 三开道岔
three-way union 三通管接头
three-way valve 三向阀;三通开关;三通阀;三路阀
three-way valve manifold 三通阀联箱
three-way van 三门厢式送货车
three-webbed box section 三腹板箱式断面
three-wedge bearing 三油楔轴承

three-weight batcher 三种成分重量配料计量器
three-well-defined horizons 三个具限分明的土层
three-wheeled 三轮的
three-wheeled all-purpose roller 三轮多用滚压机
three-wheeled automobile 三轮汽车
three-wheeled(-en) delivery van 三轮送货车
three-wheeled fork-lift truck 三轮铲车
three-wheeled motor vehicle 三轮汽车
three-wheeled static roller 三轮静力压路机
three-wheeled steam roller 三轮蒸汽压路机
three-wheeled taxi 三轮出租汽车
three-wheeled truck 三轮卡车
three-wheeled vehicle 三轮汽车
three-wheeled wagon drill 三轮车式钻机
three-wheeler 三轮小车;三轮(卡)车
three-wheel landing 三轮着陆
three-wheel landing gear 三轮起落架
three-wheel pipe cutter 三轮截管器
three-wheel plow 三轮犁
three-wheel pump 三轮泵
three-wheel roller 三滚筒滚压机;三轮压路机;三轮路碾;三滚筒压路机
three-wheel tractor 三轮拖拉机
three-wheel two axle roller 三轮双轴压路机;三轮双轴路碾
three-winding 三线圈;三绕组的;三绕组
three-winding autotransformer 三绕组自耦变压器
three-winding transformer 三线圈变压器;三绕组变压器
three-wing bit 三翼钻头;三刃钻头
three-wing drill bit 三翼钻头
three-winged 三翼的
three-winged drag bit 三翼刮刀钻头
three-wing revolving door 三扇式转门
three-wire 三线(的)
three-wire balancer 三线均压器
three-wire compensator 三相自耦变压器;三线补偿器
three-wire connection 三线接法
three-wire control 三线控制
three-wire direct current equipment 三线制直流装置
three-wire generator 三相发电机;三线发动机;三线发电机
three-wire installation 三线装置
three-wire leveling method 三线读数水平测量法
three-wire level(l)ing 三线水准测量;三丝水准测量
three-wire level(l)ing method 三线(性)读数水平测量法;三丝水准测量法
three-wire line 三线线路
three-wire measurement 三线测螺纹法
three-wire meter 三线制电量计;三线式仪表;三线式电度表
three-wire mooring 三线系留
three-wire rotary converter 三线变流机
three-wire strand 三股绞线
three-wire system 三线系统;三线制;三线测螺纹法
three-wire system generator 三线式发电机
three-wire transformer 三线制变压器
three-wire trunk 三线中继线
three-wire trunk line 三线式中继线
three-yarn nettle 三股油麻绳
three-yarn nettle stuff 三股油麻绳填塞料
three-year cycle 三年周期
three-zero-one resinous adhesive 三零一树脂胶
three-zones of card top 卡片顶部的三行区
thresh 捶击
thresher 脱粒机
thresh grain 脱粒
threshing 机械法除绳锈;脱粒
threshing barn 脱谷板棚
threshing bucket 打谷桶
threshing drum 打谷筒
threshing floor 禾场
threshing machine 脱粒机;脱谷机
threshing mechanism 脱粒装置
threshing mill 脱粒机
threshing outfit 脱谷装备
threshing performance 脱净率
threshing rig 脱粒机组
threshing set 脱粒机组
threshold 阈限;异常下限;最低限度;门限值;阈限;门口;挡堰;分界
threshold acceleration 阈加速率;阈加速度
threshold adjustment 限界调整;阈调整;临界调整;阈限调整
threshold agreement 界限协议
threshold amplifier 阈值放大器
threshold analysis 阈限分析;临界分析
threshold audiogram 听阈图;听力图
threshold behaviо(u)r of the reaction 反应阈性能
threshold character 阈值性状
threshold circuit 阈电路
threshold comb (电动走道的)终端梳形板
threshold concentration 入口浓度;下限浓度;阈限浓度;阈浓度;浓度极限;起始浓度
threshold condition 阈状态;阈值条件
threshold constraint 门限制约
threshold contraction 阈收缩
threshold contrast 阈值反衬;阈值对比度;阈限对比
threshold control 阈值控制
threshold current 阈电流;门限电流
threshold decoder 门限解码器
threshold decoding 门限解码
threshold degeneracy 阈值简并
threshold depth 阈限深度;海槛深度
threshold depth of oil generation 成油门眼深度
threshold detectability 听阈
threshold detection 阈值检测;阈监测
threshold detector 阈值探测器;阈值检测器;阈探测器;门限检波器
threshold dosage 阈量;阈剂量
threshold dose 阈量;阈剂量;监界剂量;极限用量
threshold dose inhibitor 低剂量抑制剂
threshold dose treatment 低剂量处理
threshold draft-proofer 门框防风罩;门框防风设施
threshold drift 阈漂移
threshold effect 低限效应;阈作用;阈(值)效应
threshold effect limit 阈值有效极限
threshold effect low value 阈值有效低值
threshold element 阈(值)元件;阈因素;门限元素
threshold energy 阈能;门限能量
threshold evaluation 阈评价
threshold extension demodulator 阈展解调器;门限扩展解调器
threshold factor 阈因子;临阈系数
threshold field 阈场
threshold fissioning material 阈裂变物质
threshold flow 冲动流量
threshold flow density 阈值通量密度;阈值流量密度
threshold for injury 造成伤害的最低限值
threshold for photoionization 光致电离阈
threshold frequency 阈频率;界限频率;门限频率
threshold friction 阈限摩擦;极限摩擦
threshold function 阈值函数;阈函数
threshold function table 阈值函数表
threshold gain 阈值增益
threshold gate 阈值元件
threshold gradient 阈限坡降;初始坡降
threshold howl(ing) 临振啸声
threshold illuminance 阈照度
threshold illumination 阈照度;临界照明;临界照度
threshold increment 阈限增量
threshold intensity 阈界强度
threshold inversion 阈值反转
threshold Jacobi's method 阈值雅可比法
threshold legibility 可读临界
threshold level 阈(值)电平;阈阈值;阈级;门限(信号)电平;门槛电位
threshold light (飞机场跑道的)入口界灯;着陆道标灯;(飞机场跑道的)入口标灯
threshold lighting 跑道入口灯
threshold limit 阈极限
threshold limit value 阈值;阈极限值;最高允许值;最低限值;临界阈值;安全限量
threshold limit value short-term exposure limit 短时接触阈限值
threshold limit valve ceiling 阈值限值上限
threshold-logic network 阈逻辑网络
threshold logiic(al) circuit 阈值逻辑电路
threshold luminance difference 阈亮度差
threshold magnitude 临界震级
threshold mode 阈值方式
threshold modulation 阈调节
threshold moisture content 阈含水率;临界含水率
threshold odor 阈恶臭
threshold odor concentration 阈气味浓度
threshold odor number 阈气味值
threshold odо(u)r 限定臭气浓度;阈臭;极限恶臭;气味浓度
threshold of adverse effect 有害作用阈
threshold of audibility 最小可听度;可闻度;听阈
threshold of detectability 检验阈;检测阈;检测能力阈;检核阈;听阈
threshold of discomfort 不舒适感觉阈;不适阈
threshold of feelings 感阈;感觉阀;可听上限;感觉阈值
threshold of grain motion 颗粒起动下限
threshold of hearing 听(觉)阈;听觉临界;听觉极限
threshold of illuminance 照度阈值
threshold of illumination 照度阈值
threshold of intelligibility 感觉阈
threshold of laser oscillator 激光振荡阈值
threshold of luminescence 发光界限
threshold of no adverse effect 无害作用阈(值)
threshold of nucleation 成核临界温度
threshold of pain 痛阈(声学术语)
threshold of reaction 反应阈值
threshold of sensitivity 灵敏度阈(值);灵敏度阈
threshold of specific effect 特异作用阈
threshold of temperature 感温阈;感温极限
threshold of tickle 痒阈(声学术语)
threshold of toxicity 毒性极限
threshold operator 阈值运算器
threshold pick-up 临界灵敏度
threshold plate 终端板(电动走道的);门槛
threshold potential 阈(值)电位;临界电位
threshold power 阈值功率;极限功率
threshold power density measurement 阈值功率密度测量
threshold pressure 阈限压力
threshold price 门槛价格;入门价格
threshold pumping 阈抽运
threshold quantity 阈值;临界量;门限值
threshold reaction 阈反应
threshold reduction 阈值减低
threshold region 阈域;阈区
threshold response point 临时特征曲线
threshold retention 始值保持量
threshold returns 阈界利润
threshold saturation 临界饱和(土壤在此饱和界限下产生水流);起始饱和下限
threshold seal(er) 门槛(密)封条
threshold sensitivity 阈灵敏度;临界灵敏度
threshold setting 阈值调整
threshold shift 阈值迁移;阈移位;阈移
threshold shift test 阈值旋转试验
threshold signal 阈值信号;临界信号;门限信号
threshold signal-to-interference ratio 阈信号干扰比
threshold signal-to-noise ratio 阈值信噪比
threshold size 定值粒度;临界尺寸
threshold sound intensity 阈值声强
threshold speed 阈流速
threshold state 临界状态
threshold strength 耐久极限
threshold substance 阈质
threshold switch 阈值开关
threshold temperature 阈温;临界温度;极限温度
threshold temperature of oil generation 成油门限温度
threshold time-series model 临界时间数列模型
threshold treatment 阈处理;入口处理;起始处理;低限处理
threshold tube 限制管;阈管
threshold value 限界值;阈(限)值;阈界值;门槛值
threshold value limit 阈值极限
threshold value of measurement 测量阈值
threshold value of organic carbon 有机碳门阈值
threshold value of track detector 径迹探测器阈值
threshold velocity 阈流速;泥沙起动流速;起动流速;初动速度
threshold vibration 阈振动;临界振动
threshold voltage 阈值电压;阈电压;临界电压;门限电压
threshold wavelength 限度波长;阈波长;临界波长
threshold wave number 阈波数
threshold weight 门限钻压
threshold worker 入门职工
threshold zonal filtering 阈值带滤波
threshold zone 入口(区)段;起始区域

threshold zone of tunnel 隧道入口段
thresold 限极
thribble 三联管
thribble board 三联管工作板;井架工作平台铺板;井架二层平台
thrice-cut 三次切割
thrid-degree erythema 三度红斑
thrift 节约;节省
thrift account 储蓄账户
thrift basin 节水池;省水池
thrift lock 储水闸;节水阀(门);节水船闸;省水船闸
thrift mate 万用连接器
thrift shop 旧货店(美国)
trihedral corner reflector 三面正交反射器
trihedral reflector 三面反射镜
thrill 震颤
thriving and powerful 富强
thriving and prosper 繁荣昌盛
thriving business 繁荣产业
throat 狭道;注入口;窄路;窄颈段;工作导槽;流液洞;孔颈;淡深;滑车上接近钩环眼部分;焊接厚度;弯曲处;缩口管;滴水管;滴水槽;滴水
throat angle 前端斜角(螺丝钢板)
throat area 咽喉区;临界截面面积(喷管的);临界截面积;颈截面;焊缝计算截面;喷口面积;喷管临界截面积
throat bolt 眼环螺栓
throat brace 掌状系杆
throat bullet 可调式尾喷管锥体;可调节的尾喷管锥体
throat cheek 流液洞侧壁砖
throat cover (block) 流液洞盖板砖
throat depth 焊口有效厚度;焊接厚度;焊喉深度;焊喉厚度;焊缝厚度
throat depth of fillet weld 角焊缝厚度
throat diameter 临界截面直径(喷管的);喉部直径
throat diameter of runner 转轮喉(口直)径
throated flume 窄颈测流槽
throat halyard 桁叉端吊索
throating 平刨刨口槽;壁炉喉部;烟道汇集口;滴水(槽凹口)
throating plate 披水板
throat inside burner 下煤减压装置;喷燃管内缩口
throat length 咽喉区长度
throatless chamber 圆筒形燃烧室
throat liner 喉部衬砌(泥水管的)
throat of (fillet) weld 凹角喉喉;角焊缝喉部(焊缝的最小厚度);填角焊缝喉长
throat of frog 辙叉咽喉;辙叉喉
throat of hinge 铰颈
throat of station 车站咽喉区;车站咽喉区【铁】
throat of threading die 螺丝扳牙导口
throat of tooth 齿喉
throat of weld 凹角喉
throat of yard 站场咽喉区;车站咽喉区;车场咽喉区【铁】
throat opening 下料口;空心搁板门框背面的缝口
throat opening area 喷管喉部面积
throat opening factor 通过量因数
throat operation 喉道区段营运
throat plane 喷管临界截面
throat points 咽喉道岔【铁】
throat pressure 临界截面压力
throat radius 喉道半径
throat ring 节流环(尾水管进口处);喉部环;入口环;填料衬环
throat (section) 临界截面
throat seizing 绳圈合扎
throat shape 喉道形态
throat sheek block 流液洞侧砖
throat-side block 流液洞侧砖
throats of crossing 广场入口;交叉口入口
throat switch 咽喉道岔【铁】
throat-tap flow nozzle 长颈式流量喷嘴
throat thickness 焊接厚度;焊喉厚度
throat thickness of fillet weld 角焊缝厚度
throat track 咽喉线
throat turnouts 咽喉道岔【铁】
throat velocity 临界截面速度;临界速度
throat width 入口宽度
thrombopenic purpura 陆地坏血病
thrombus 菱形
throne 宝座
throne hall 皇家大厅

throne room 皇家房间;觐见室
thron hedge 荆棘篱
Thronth Waite moisture index 桑思韦特土壤水分指数
thron woodland 多刺疏林
throttle 主气门;节流气门;节流活门;调速气门;操纵阀;风门;风量调节阀;扼流;节流;进汽量
throttle adjusting screw 风门调整螺钉
throttle apparatus 节流装置
throttle balance level 节气门平衡杆
throttle button 节流阀钮
throttle chamber 节流室
throttle channel 喷口调节通路
throttle circuit 牵引力控制电路
throttle coefficient 节流系数
throttle condition 进口参数
throttle control 节流控制;风门控制;扼流控制
throttle-control bell crank 风门控制直角杠杆
throttle-control bell crank lever 操纵节流直角杆
throttle control lever 节流控制杆
throttle control wire tube bracket 节汽操纵线管托架
throttle-control wire tube clip 风门控制线管夹
throttle curve 节流曲线
throttle damper 节流闸板
throttled cycle 节气循环
throttled directional valve 节流换向阀
throttled flow 节流流动
throttle disc 节流板
throttled nut 槽形螺母
throttle down 关小气门;降低流速
throttle down a well 降低井的流量
throttled steam 减压蒸汽
throttled surge chamber 阻尼式调压器;节流式调压室
throttle(d) valve 节流阀
throttle expansion 节流膨胀
throttle expansion valve 节流膨胀阀
throttle factor 节流系数
throttle flap 节流挡板
throttle flow 节流;进汽量
throttle flowmeter 节流式流量计
throttle-flow ratio 节流流量比
throttle full open 全节流
throttle gate 节流闸门
throttle governing 节流调速;节流调节
throttle governor 节流调速器;节流调节器
throttle grip 节流阀手柄
throttle hand control 节流手控制杆
throttle lever 节气阀杆;油门杆;节流门操纵杆;节流(阀)杆
throttle lever connection 节流杆接头
throttle lever grip 节流阀杠杆柄
throttle lever handle pin 节流杆手柄销
throttle lever latch 节流杆闩
throttle lever latch guide 节流杆弹键导承
throttle lever latch guide bolt 节流杆弹键导螺栓
throttle lever latch joint 节流杆弹键螺栓
throttle lever latch spring 节流杆闩弹簧
throttle lever quadrant 节流杆扇形板
throttle linkage 风门控制杆系;油门控制杆系;节气操纵杆系;节流联动装置
throttle lock 风门锁;油门锁
throttle loss 节流损失
throttle mechanism 节流机构
throttle nozzle 节流喷嘴
throttle opener 节气门开启器
throttle opening 阻力孔;开节流阀;节流孔
throttle orifice 节流圈;节流孔板
throttle pedal 节流阀踏板;风门踏板
throttle pipe 节流管
throttle plate 节流板
throttle plate screw 节流板螺钉
throttle pressure 节流压力;调速汽门前压力
throttle pressure control group 节流压力控制组;节流压力控制系统
throttle pressure reducing valve 节流减压阀
throttle process 节流过程
throttle regulating valve 节流调节阀
throttle regulation 节流调节
throttle regulator 节流调速器;节流调节器
throttle resistance 节流阻力
throttle rod cotter 风门杆开尾销
throttle runner 节流浇口

throttle setting 节流阀调整;风门装置
throttle shaft bushing 节流阀轴衬套
throttle slot 节流缝
throttle spindle 节流阀轴
throttle stand pipe 节流阀立管
throttle steam 节气阀调节蒸气;节流(蒸)气
throttle (steam) temperature 调速气门前气温;进气温度
throttle stem packing gland 节流杆填密压盖
throttle stop screw 节流阀制动螺钉
throttle switch 节流开关
throttle thermometer 节流测热计
throttle valve 减压阀;风门;节气阀;节流阀;阻流活门
throttle valve body 节流阀体
throttle valve casing 节流阀体
throttle valve gear 节流阀装置
throttle valve lever 节流阀杆
throttle valve shaft 节流阀轴
throttle valve stop 节流阀停止器
throttle washer 节流圈
throttling 油门调节;节气;节流;阻气
throttling action 节流作用
throttling adjustment 节流范围调整
throttling area 节流面积
throttling bar 节流条
throttling bend 节流弯管
throttling butterfly valve 蝶形节流阀
throttling calorimeter 节流量热器
throttling cock 节油门旋塞;节气门旋塞;节流旋塞;扼流栓
throttling coefficient 节流系数
throttling cooler 气体节流式制冷器
throttling discharge 排出口节流
throttling expansion 节流膨胀
throttling factor 节流系数
throttling flow meter 节流流量计;差压流量计
throttling governing 节流调节
throttling groove 节流槽
throttling loss 节流损失
throttling method 节流法
throttling of steam 总气门
throttling orifice 节流喷嘴;阻力孔;节流孔
throttling port 气门
throttling process 节流过程
throttling range 节制范围;节流范围;调节范围
throttling resistance 节流阻抗
throttling signal intersection 限流信号交叉口
throttling sleeve 节流短管
throttling thermometer 节流测热计
throttling turbine 节流调节式涡轮机;节流调节透平
throttling type flowmeter 节流式流量计
through air duct 通过式风管
through analysis 全面分析
through and through 原木平行下锯(法);直通的;完全
through and through sawing 直锯法
through arch 下承拱;通拱
through ball joint 通行球节
through ball rate 通球率
through ball test 通球试验
through band 绿波波宽;车辆通过两号志间区段的时间
through base 带槽底板
through beam 下承梁;连续梁;全通梁;通梁
through bias 直通偏流
through bill of lading 直达提单;联运提单;通过提单
through blade screwdriver 直刀身螺钉钻
through block connection (建筑物内的)公共通道
through bolt 细杆螺栓;贯穿螺栓;双头螺栓;穿螺栓
through bolt design 贯穿螺栓设计
through bond 贯通砌缝;贯墙丁砌;横向砌合;槽向砌合
through bonder 系石;横砌石
through bracket 贯穿机件的托座
through bridge 下承(式)桥;穿式桥梁
through bridge roadway above boom 下承桥面
through building freeway 穿越建筑物的高速干道
through button circuit 通过按钮电路
through call 直接通话
through cantilever bridge 下承式悬臂桥
through capacity 工作交换容量
through car 直达车;联运车

through car equivalent 直行车当量
through cargo 直达货(物);联运货物
through cargo arbitrary 直达货物增额;直达货物(运费)增额;联运货物附加费
through cargo clause 联运转船货条款
through cargo manifest 直达货物清单
through carriage 联运程运输;全程运输
through car unit 直行小车单位
through-carved work 镂花雕刻;镂空雕刻;镂花雕刻
through cave 穿洞
through chain conveyer 链板式输送机
through channel 直通通道
through characteristic 通过特性;穿透特性
through check 劈裂;穿裂;详龟;贯通裂纹;贯通裂缝;贯穿裂缝
through circuit 转接电路
through-circulation 连续环流
through-circulation dryer 通风干燥机
through city traffic 越城交通
through coach 直通客车
through cock 直旋塞
through condenser 穿心式电容器
through consignment note 直达货发货通知单
through consultation 通过协商
through container service port 集装箱枢纽港
through convoy 直达运输护送队
through coupling 直接耦合;直接联轴节
through court 对穿庭院
through crack 贯穿裂缝;裂穿;从上到下开裂;穿透裂缝;穿裂
through current 直通电流
through current bias 直通电流偏置
through curve 通过曲线
through cut 贯穿开挖;开切眼【采矿】;明挖;明路堑;明挖路堑
through cutting 贯穿切割
through cutting method 贯穿切割法
through deep fracture 贯通的深断裂
through dial(l)ing 直通拨号
through distribution 物资调运的全过程
through documents 直达单据;全程单据
through-double cap 全空心的盖
through dovetail 贯穿燕尾榫;露头燕尾榫
through drainage 贯穿水系
through drill device 穿心式夹具
through-drive 直接传动(的)
through-drive design 直接传动设计
through-dry 使干燥;彻底干燥
through-drying 实干
througher 联络小巷;小巷
through error 贯通误差(隧道的)
through express train 直达旅客特别快车
through expressway 直通街道
throughfall 直接降水(量)
through fault 穿越故障;穿性故障
through feed 贯穿进给
through-feed centerless grinding 贯穿进给无心磨削
through-feed grinding 贯穿进给磨法
through-feed method 贯穿磨法;贯穿进给法
through feed screw 空心螺旋立轴
through-fixing 贯穿式固定;穿式结合
through-flight 直达航班
through flow 贯通流动;径流式;壤中流;次表层流;穿透流;穿堂风;通流
through flow boiler 直管式锅炉
through-flow heater 通流加热器
through-flow turbine 贯流式水轮机
through flushing system 循环式冲洗系统
through freight 直达货运;联运运费;联运费;全程运费
through freight train 直达成组列车
through furnace 直通炉
through geosyncline 贯地槽
through girder 下承(桥)大梁;下承板梁
through girder bridge 下承桥;下承梁桥;穿式桁架桥
through glacier 分叉冰川
through-going shaft 全通轴;贯穿轴
through-going structure 穿透构造
through goods train 直达成组列车
through-hardening 全部硬化
through heating 穿透加热;穿透淬火
through highway 直达公路;过境公路;直通道路

through hole 通孔;透眼;金属化孔;透孔
through hole coating 通孔镀敷
through hole plating 通孔镀敷
throughing 完全的;彻底的
through insulator 穿越绝缘子
through jack 转接塞孔
through joint 直通接线装置
through-jointed frame 穿过式构架
through journey time 全程时间(公路工程)
through kilometrage 直达里程
through ladder 直通扶梯
through layer 次表层
through line 直通线;正线【铁】;贯通电线路;槽线
through lintel 贯通过梁;穿过梁;整墙过梁
through-living room 纯属临时居室
through lot 对穿地块(地块前后均临街道);直通地段
through lot yards 对穿地块的前院
through main line 直通干线
through manifest 联运舱单
through metal 金属支架
through midsole 直通垫外底
through migration 徒步游历
through milling 轴向滚铣
through-mortice 直通榫眼
through motor transportation 直达汽车运输
through movement 直达交通;全程运输
through multimode transport 全程多种方式联运
through neutralite filtration 中和过滤
through of belt 带槽
throughout 总处理能力;自始至终;始终
throughout like watering 进行生长全期浇水
throughout rate 直达运价率
throughout the growing season 整个生长期间
throughout time 解题时间
through parcels train 直达包裹列车
through part-load wagon 直达整零车
through pass 过道
through passenger flow 直通客流;通过客流
through passenger flow diagram 直通客流图
through passenger train 直通旅客列车;直达旅客列车
through path 直通通路
through pin 穿通销
through pipe piece 穿过舱壁的管道接头
through plane 贯通面【铁】
through plate 通连板
through plate center keelson 中央立龙骨
through plate girder 下承梁;下承式板梁
through position 转接台
through post 通柱
through product 筛下物;筛下产品
throughput 吞吐量;信息通过量;矿石处理量;解题能力;容许能力;吞吐能力;透光率;通过量;生产量;出料率;总收获量;物料通过量
throughput capacity 物料通过额;物料通过量;吞吐能力;通过能力【铁】;生产能力;吞吐量
throughput capacity of port 港口通过能力
throughput capacity of container berth 集装箱泊位通过能力
throughput capacity of rail siding 铁路装卸线通过能力
throughput capacity of storage 库场通过能力
throughput capacity of the transport network 运输网生产能力
throughput capacity of wharf 码头通过能力
throughput efficiency 通过效率
throughput homogenizing 通过式均化
throughput membrane resistance 过膜阻力
throughput number 排出流量数
throughput of column 塔中蒸气通过速度
throughput of screen 格栅通过量;滤网通过量
throughput of water 水流出;水流量;水的通过量
throughput power 传输功率
throughput rate 流过率;吞吐量;通过速率;生产率;物料通过速率
throughput volume 物料通过体积
throughput weight 物料通过重量
through quenching 淬透
through rail 直通轨
through rail-ocean rate 铁路、海路联运运价率
through railway diameter 贯通城市的铁路直径线
through rainfall 贯穿降水量
through rate 直通运价;直达运费;联运运费率;联运运价;联运费率;全程运费率;通过率;通过速率

through reamer 长铰刀
through repair 大修
through repeater 中间带电机
through retort 直通干馏炉
through road 直通道路;过境道路
through roadway 直通车道;过境车道
through rod 通杆;双头活塞杆
through-roof pipe 出屋顶管道;出层顶管道
through route 高速公路;直达路线;过境路线;通过进路
through-route sign posting 设置直达路线号志;设置直达路线标志
through(-running) train 直达列车
throughs 过筛物
through scavenging 直通清除
through section 流通截面
through service 过境检修
through shake 木材在年轮间开裂;透环流;(顺木纹方向的)贯透裂;贯通轮裂;穿透裂缝;木材年轮裂
through-shaped girder 双腹板梁
through shipment 联运;全程运输;直达运输
through signal 通过信号
through-solution hydration 通过溶液的水化
through span 下承跨;穿过式桥跨;穿过桥跨
through speed of passenger train 旅客列车直达速度
through station 通过站;中间站;直通式车站;贯通式车站
through stone 横砌砖石(穿透墙厚的);系石;贯石(贯穿墙壁而露其两端的长石);贯墙石;穿墙石
through street 直通街道;直达街道;直通大街
through structure 穿式结构
through supervision 直通监控
through survey 贯通测量
through switchboard 转接交换台
through tenon 透榫;贯通榫;穿透榫石;贯穿雄榫;穿榫
through terminal 跨线总站(指几条线路的总站);过境总站;直通枢纽;直达总站;通过式总站;通过式枢纽站
through the entire surface of the seed 通过整个种子表面
through-the-ice sounder 穿冰测深器
through-the-lens finder 分光镜式取景器;反光式取景器
through-the-wall air conditioner 穿墙式
through-the-wall handler 穿墙式机械手
through-the-wall-tile 穿墙式
through the year 终年
through threshing 脱粒干净
through ticket 联运票;通票
through track 直通线;直通线路;直通轨道;直轨道;贯通线;通行线(指铁路)
through track group facility 跨线设备
through traffic 直达运输;直达交通;过境交通;联运;全程运输;通过交通;过境交通
through-traffic highway 直达公路;过境公路
through traffic revenue 直通运输收入
through traffic road 过境交通道;过境道路;直通快速车道;直达交通高速公路;直达道路
through traffic roadway 过境交通(行)车道
through train 直通列车;直快列车;直达快车;直达车
through train orginated from several adjoining loading points 阶梯直达列车
through train originated from one loading train 始发直达列车
through train with empty cars 空车直达列车
through transit 全程运输
through transit club 联运保赔协会
through transmission 透射传输
through-transmission technique 穿透技术
through transport 直通运输;直达运输;联运
through transportation operation income 直通运行作业收入
through transport by land and water 水陆联运
through trip 过境出行
through truck route 过境货运路线
through truss 下承式桁架;下承桁架;穿式桁架;穿过式桁架
through truss bridge 穿式桁架桥;穿过式桁架桥;下承(式)桁架桥
through tubing perforating gun 过油管射孔枪
through type 穿过式;直通型

through type current transformer 贯通式变流器
through-type freight station 直通式货站
through-type goods station 通过式货运站
through-type goods yard 通过式货场
through-type passenger station 通过式客运站
through type pyrometer 贯通式高温计
through type shed 贯通式机车库
through type station 通过式车站
through-type transformer 单芯变压器；穿越式变压器
through-valley 贯通谷
through valve 贯通阀
through ventilation 穿堂风
through view 鸟瞰；眺望
through-voided 全空心的
through volume 通过容积
through voyage 直达航程
through-wall air conditioner 过墙式空调器
through-wall flashing 穿墙泛水
through-wall flashing piece 通过墙的泛水件
through wall flaw 穿透性裂纹
through-wall pipe 穿墙管
through-wall sleeve 穿墙套管
through wall type ultrasonic flowmeter 穿壁式超声波流量计
throughway 直通公路；直达道路；过境道路；直通街道；高速公路；高速干道
through (way) cock 直通旋塞
through way-town plan 城市方位物图
through (way) valve 直通阀
through weld 透焊缝；贯穿焊缝
through welding 焊透
through-wiring 全部布线
through yard 直通场【铁】；通过车场【铁】
throw 落差；拉出；曲柄行程；偏移度；偏心距离；喷程；抛掷；投掷；射程；断距；冲程
throw a double 连续工作两班
throw a lease 退租
throw at end of points 尖轨尾程
throwaway 抛弃；废品；临时利用件
throwaway bio-degradable plastic bag 可降解一次性塑料袋
throwaway bit 一次性钻头；不重磨车刀
throwaway chip 不重磨刀片
throwaway cutter 不重磨铣刀
throwaway drill 不重磨钻头
throwaway element 更换型元件
throwaway face cutter 不重磨硬质合金面铣刀
throwaway filter element 可更换过滤器组件
throwaway hob 不重磨滚刀
throwaway process of fume gas desulfurization 抛弃法烟气脱硫
throwaway share 不修复犁铧
throwaway tool 不重磨刀具
throwaway-type bag 一次性口袋
throwaway-type filter 一次性滤纸；一次性滤器
throwaway-type house 临时性房屋
throwback 切换
throw bearing 曲柄轴颈
throw blasting 扬弃爆破；抛掷爆破
throw clay 拉坯黏土；拉坯泥料
throw crank 推动曲柄
thrower 抛料机；喷洒器；拉坯工人；平舱机；抛掷物；抛射器；抛料机；投掷器；甩水圈
thrower belt conveyer 离心式皮带运输机
thrower belt unit 抛料带式输送机
throwering belt 抛掷式充填机皮带
thrower ring 润滑油圈
throw faulting 下落断层作用
throw-in 接通；接入；投入
throwing 制丝；拉坯成型法；抛粒；甩溅
throwing chamber 抛草机外壳
throwing circle 折叠圆面
throwing jet 喷浆机；水泥喷枪
throwing lever 闸杆
throwing line 抛缆
throwing marks 拉坯形成的条纹
throwing on 喷上
throwing out 气裂
throwing power 均镀能力；电镀本领；匀镀能力；泳透性；泳透力；涂覆力；布散能力
throwing stick 拉坯棒
throwing the chain 甩链子
throwing wheel 抛头

throw-in lever 起动杆
throw into action 接入
throw lathe 小型手摇车床
thrown fault 下落断层
thrown glaze 施釉法
thrown-on finish 涂上终饰
thrown-on plaster partition (wall) 喷涂灰浆的分隔墙
thrown silk 加捻丝线；捻丝
thrown slip 下落滑动
thrown solder 脱焊
throw of crankshaft 曲轴弯头；曲柄半径
throw of eccentric 偏心轮偏心度
throw of eccentricity 偏心轮行程
throw off 断开；丢弃；切断
throw off a belt 卸下皮带
throw of faults 断层断距
throw-off belt 卸载运输车；皮带式卸料机
throw-off carriage 卸料车
throw-off effect 断开效应
throw-off end 卸料端
throw off hook 开脱钩
throw-off shelf 卸料台；卸料架
throw off the rope 摘掉钢丝绳
throw of hook 摘钩
throw of lever 杆行程
throw of piston 活塞行程
throw of pointer 指针偏转
throw of pump 泵的冲程；泵冲程
throw of switch 尖轨动距；尖轨动程
throw of the crusher 压碎机的锤击；破碎机颚板冲程
throw of the point 轨尖摆幅
throw of traverse 排线器排幅
throw on curve 曲线几何偏移
throw on the rope 开动提升卷筒
throw-out 切断；抛出
throw-out bearing 顶推轴承；离合器分离轴承
throw-out clutch 常合式离合器
throw-out collar 推环
throw-out conveyer 抛掷式输送器
throw-out fork 换档叉
throw-out lever 止推杆；解脱子；停车杆
throw out of gear 摘开齿轮
throw-outs 废品
throw-out spiral 抛出纹；盘尾纹；输出螺旋线；输出带
throw-out switch 电磁离合器控制开关
throw-out tail 抛出尾纹
throw over 转接；换向；切换
throw-over blades and pick-ups 叶片(混凝土拌和机)
throw-over catch 回转挡
throw-over dress 斗篷
throw-over gear 跨轮架
throw-over motion 回转
throw-over plow 可换向犁
throw-over relay 固定双位继电器
throw-over switch 投掷开关；扳倒开关
throw rod 动作杆；扳动杆
throws 曲拐
throw screening 投掷过筛
throw shaft 扳动轴
throw shot cleaning 抛丸清理
throw transporter 震动运输机；推杆输送机
throw up a burr 打掉毛口
thrufeed method 贯穿进给法
thrum 线头；植绒帆布；绳屑；绳毛绒
thrum cap 绳头编帽(防碰撞用)
thrum mat 绳毛垫
thrumming 帆布浸底植绒
thruput 通过量
thru-roof pipe 出屋顶管道
thrus 尺寸不足的；过筛眼的；过网眼的
thrust 止推；推力；冲复作用；侧向压力
thrust action 猛插动作；猛推动作
thrust-anemometer 推力风速表
thrust arm 推力杆
thrust at springer 拱脚推力；拱端推力
thrust augmentation 推力增大
thrust-augmentation control 推力增加控制
thrust-augmented 推力增大的
thrust augmenter 推力增加器
thrust-augmenting nozzle 导流管(螺旋桨的)

thrust axis 推力轴
thrust balancing device 推力平衡装置
thrust ball bearing 止推球(滚珠)轴承；止推滚珠轴承；推力球轴承
thrust ball bearing with flat seat 平面推力球轴承
thrust bar 止推杆
thrust barrel 推力室
thrust bearing 轴向轴承；止推轴承；止推支座；推力座；推力轴承；推力止推轴承
thrust bearing disc 推力轴承板
thrust bearing load 推力轴承负荷
thrust bearing pad 推力轴承垫座
thrust bearing race 推力轴承的挡圈
thrust bearing resisting plate 推力轴承止推板
thrust bearing runner 转环；镜板；推力头
thrust bearing segment 推力轴瓦扇形块；扇形推力轴瓦
thrust bearing shoe 止推轴承瓦；推力轴瓦
thrust bed 推力试车台
thrust block 承推底块；斜撑块；止推支座；止推承座；逆冲推覆体；推力座；推力轴承；推力块；顶推垫块；冲掩底块；冲掩块；止推座；轨撑【铁】
thrust block concrete 止推混凝土管座
thrust block keep 推力轴承盖
thrust block masonry 止推圬工管座
thrust block of pipe 水管支墩
thrust block plate 止推座板
thrust block seat (ing) 止推座
thrust bolt 压紧螺栓
thrust borer 凿岩机；凿进机；冲击钻孔机；冲击钻机；冲击式钎子钻机；冲击式钎子
thrust boring 顶管法；冲击钻孔
thrust boring-auger method 冲击钻孔法
thrust boring machine 推管装置；顶管穿孔机
thrust box 止推轴承箱
thrust brake 轴向压力；止推装置
thrust brearing recess 推力轴承穴
thrust breast heart 斑纹心
thrust buildup 推力增大
thrust bush(ing) 推力套筒；推力衬套；衬套(涡轮钻具的)
thrust button 止推轴颈
thrust cap 推压盖；(导叶接力器的)端盖
thrust capacity 轴向负荷支承能力；轴向负荷承受能力
thrust center 推力中心
thrust chamber 火箭发动机；推力室
thrust chamber pressure 推力室压力
thrust chamber valve switch 推力室阀门开关
thrust coefficient 推力系数
thrust collar 轴扣；止推轴环；止推轴承定位环；止推环；推力环
thrust collar bearing 推力环轴承
thrust component 推力分量
thrust control 推力调节
thrust controller 推力调节器
thrust correction 推力修正
thrust curve 侧压力曲线
thrust-cutoff phase 推力截止阶段
thrust cylinder 推力室
thrust-decay period 推力衰减期
thrust deduction 推力减额
thrust deduction coefficient 推力减额系数
thrust deduction factor 推力减额系数
thrust deduction fraction 推力减额系数
thrust deflector 推力转向器；射流偏向器
thrust deflexion 推力转向；推力偏差
thrust deviation 推力偏差
thrust device 止推装置；推力装置；推进器
thrust direction 推力方向
thrust-direction control 推力方向调节
thrust diverter 推力转向器
thrust drag 推阻力
thrust drop 推力下降
thrust duration 推力持续时间
thruster 助推器；推进器；推冲器；船首推进器；侧推器
thruster assisted anchor(ing) 推力辅助抛锚
thruster assisted mooring system 推力器辅助下锚系统；侧推系泊系统
thrust error 推力误差
thrust face 压力面；推力面；受力面
thrust fault 逆掩断层；逆断层；逆冲断层；冲断层
thrust fault plane 冲断面
thrust fold oil-gas field 逆掩褶皱油气田

thrust (force) 轴向力;推力
thrust force of shield 盾构推力
thrust frame 发动机架
thrust gear 止推装置;推移机构;齿轮联动装置
thrust grading curve 推力分布曲线
thrust graph 推力图线
thrust horsepower 推力马力;推进马力
thrust increment 推力增量
thrust indication 推力指示
thrust indicator 推力指示器
thrust influence line 推力影响线
thrusting 逆掩作用
thrusting action 推压作用
thrusting force 侧向压力;推力
thrusting force of locomotive 机车推力
thrust jack 推力千斤顶;推力测定计;推进千斤顶
thrust journal 止推轴颈
thrust-kilogram 公斤推力
thrust level 发动机额定推力
thrust level control 推力调节
thrust-lift nozzle 推力升力喷管
thrust limiter 推力限制器
thrust line 轴向压力线;支撑线;推力作用线;推力线;标定基准线
thrust-line adjustment 推力方向调整
thrust line arch 推力线拱
thrust line vault 推力线拱顶
thrust load 纵向负荷;推力荷载
thrust loading 重量推力比;推力负荷
thrust-loss indicator 推力下降信号器
thrust margin 推力余量
thrust-measuring device 推力测量装置;测推力装置
thrust-measuring equipment 推力测量装置
thrust-measuring installation 推力测量装置
thrust metal 推力轴承座
thrust meter 推力计
thrust misalignment 推力偏心率
thrust moment 推力力矩
thrust motion (挖土机斗柄的)推压运动;推压动作
thrust nappe 逆冲推覆体;冲断推覆体
thrust nozzle 推力喷管
thrust nut 推力螺母
thrust of arch 拱推力
thrust of earth 土推力
thrust of pump 泵推力;泵的压头
thrust of soil 土推力
thrust of springer 拱底推力
thrustor 推力器
thrust orientation 推力方向性
thrust output 推力输出(量);输出推力
thrust pad 止推轴承调节垫;止推轴承衬;推力瓦块
thrust pickup 推力传感器
thrust pin 止推轴销
thrust plane 逆冲断层面;冲断面;冲断面
thrust plate 止推盘;止推挡板;止推板;逆冲推覆体;冲断层板块
thrust plunger 止推销;止动销
thrust point 推力点
thrust-pound 推力磅
thrust power 推进功率
thrust push 推力
thrust race 推力轴承的挡圈
thrust rail 轴向压力轨道;顶推导轨
thrust range 推力范围
thrust rating 推力定额
thrust recess 推力轴承穴;推承凹座
thrust recovery 推力恢复
thrust reduction 推力减额
thrust reserves 推力储备
thrust-reversal device 推力反向器
thrust reverser 制动推力器;反推装置;反推力器
thrust-reverser cascade 推力转向叶栅
thrust-reversing mechanism 推力逆转机构
thrust ring 圆屋顶支承环;止推环;推力环;推；土推环
thrust roller 挡轮
thrust runner 推力轴承滑道
thrust screw 压紧螺钉;压紧螺丝;止动螺钉;推力螺旋;推力螺钉
thrust section 推力室;冲断面
thrust segment 推力瓦块
thrust shaft 止推井;后మ井;推力轴
thrust sheet 止推挡板;迭掩盘;逆冲岩席;逆冲推覆体;冲断岩席

thrust shoe 止推瓦;推力瓦块
thrust side of piston 活塞推力侧
thrust slice 逆冲推覆体
thrust slip 冲断滑动
thrust slip fault 逆断层;冲断层
thrust spoiler 推力阻流片;推力扰流器
thrust stage 凸形舞台;三面观众的舞台;突出式舞台(半岛式舞台);半岛式舞台
thrust stand 推力试验台;发动机试验台架
thrust stiffness factor 抗推劲度系数;侧移劲度系数;侧移刚度系数
thrust strip 止推板
thrust structure 承受推力结构
thrust surface 推力面;推进器的受力面;受力面;冲断面
thrust taper roller bearing 锥形滚柱止推轴承
thrust terminator 止推器
thrust test 推力试验
thrust tool 刀具推力
thrust to weight ratio 推力重量比
thrust truck 直达货运路线
thrust-type earthquake 冲断层型地震
thrust unit 止推装置;推进装置;推动装置
thrust vector control 推力矢量控制
thrust ventilator 推力风机
thrust wall 顶推后座墙
thrust washer 止推垫圈;推力垫圈
thrust-washer key 止推环键
thrust web 推力受力元件;承推力壁
thrust-weight ratio 推力重量比
thrust wheel 支重轮
thrust yoke 推力轭
thru-wall flashing 过墙泛水;穿墙披水
thruway 直达道路;过境道路;快速公路
thucolite 古碳质岩
thud 重击声
thuja 侧柏;金钟柏
thuja koraiensis nakai 朝鲜崖柏
thujane 侧柏烷
thujanol 侧柏醇
thujene 侧柏烯
thujone 侧柏酮
thujorhodin 玫红黄质
thujyl alcohol 侧柏醇
Thule group 图勒群
thulia 氧化铥
thulium boride 硼化铥
thulium nitride 氮化铥
thulium ore 铥矿
thulium oxalate 草酸铥
thulium oxide 氧化铥
thulium silicide 硅化铥
thumb 外圆角砖
thumb bolt 指按(门窗)闩;元宝螺栓
thumb cleat 单挽钩
thumb control abjustable wrench 拇指控调扳手
thumb edge 抹圆角边
thumb head screw 扁头螺丝
thumb hole 拇指孔
thumb index 书边标目
thumb jump 按钮弹跳开关
thumb knob 门锁按钮把手
thumb knot 反手结
thumb latch 手铐式碰锁;指按锁栓;指按门闩;诺福克插销;门窗插销
thumb lever 操纵钮;拇指操纵杆
thumb-mark 拇指纹
thumb mo(u)lding 拇指模;扁圆的突线脚;拇指饰【建】;抹圆角线脚
thumbnail 简略的
thumbnail sketch 略图;极小略图;草图;目测图
thumb nut 元宝螺母;元宝螺帽;翼形螺母;翼形螺帽;指旋螺母;蝶形螺母;指旋螺帽
thumb-operated control stick 按动操纵杆
thumb pad 凹形垫
thumb piece 门锁按钮;翼形件
thumb pin 锹钉;图钉
thumb plane 窄小木工刨;指形刨;净面刨
thumb pressure oil can 抽压油壶
thumb print 拇指纹
thumb rings with various decorations 各色搬指
thumb rule 经验规则;经验法则;计算中近似法
thumb screw 屋顶钩住铅板的钩子;推拉窗锁位器;手拧螺丝;墙钩(使铅板贴到墙面上);翼螺钉;指

旋螺钉;拇拧螺丝;蝶形螺(丝)钉;元宝螺丝
thumb tack 揿钉;图钉
thumb-tack needle for subcutaneous embedding 揿针
thumb turn 转动门钮;指旋销
thumbwheel 拇指旋转控制器;拇指轮
thumbwheel controller 拇指旋转式控制器
thumbwheel decade switch 指轮十进制开关
thumbwheel switch 指轮开关
thump 砰然声;撞击声;键击;强冲击;低频噪声;捶
thumper 车载大锤
thump killer 键击噪声抑制器
thunder and lightning 雷电
thunder-arresting 避雷
thunderbolt 雷电;霹雷;落雷;雷电击毁
thunderclap 雷声
thunder cloud 雷雨云
thunder egg 玉髓的卵形结石
thunder-gust 雷电交加的暴风雨;伴有大风的雷暴雨
thunderhead 气象雷达站;雷暴云砧(雷暴前常见的)
thunder shake 木纹纵裂;电裂
thunder shower 雷阵雨
thunder squall 雷电交加的风暴;雷飑
thunder storm 雷雨;雷暴;锋面雷暴;雷暴雨
thunderstorm belt 雷雨区
thunderstorm cell 雷暴云泡
thunderstorm charge separation 雷暴电荷分离
thunderstorm damage 雷害;雷雨灾害
thunderstorm day 雷雨日;雷暴日
thunderstorm electricity 雷电
thunderstorm front 雷雨锋
thunderstorm locating device 雷暴定位装置
thunderstorm rain(fall) 雷雨;雷暴雨;雷暴降雨量
thunderstorm warning 雷暴警报
thunderstreak 雷电
thundery front 雷雨锋
thundery precipitation 雷雨云降水
thundery rain 雷暴雨
thundery sky 险恶天气
thurane 刚性聚氨酯泡沫
thuresite 微斜钠闪正长岩
thuringite 鳞绿泥石
thuringite rock 磷绿泥石岩
thurland 盐结壳地
thurm 用手工做圆木;小断层【地】;柜橱工艺;岩岬
Thurso flagstone group 瑟索板层岩群【地】
Thurstone temperament schedule 瑟斯顿个性测验表
Thurston's alloy 瑟斯顿铸造锌合金
Thurston's brass 瑟斯顿高锌黄铜
Thurston Scales 瑟斯顿尺度
Thurston's metal 瑟斯顿锡基轴承合金
Thury regulator 瑟雷调节器
Thury screw-thread 细牙螺纹
Thury system 高压直流制
thuya 侧柏;金钟柏
thuzic acid 侧柏酸
Thvera normal polarity subchron 斯维拉正向极性亚时
Thvera normal polarity subchronzone 斯维拉正向极性亚时间带
Thvera normal polarity subzone 斯维拉正向极性亚带
thwack 夯;捣实
thwacking 拍泥;拍打成型
thwacking frame 波形瓦坯模具
thwaite 开垦地
thwart 脚手架横撑木;横座板
thwart knee 横座板肘板
thwart stanchion 横座板支柱
thyme 百里香
thyme camphor 百里酚
thymele 剧院前舞台讲坛(古希腊)
thyme lemon oil 百里柠檬油
thyme linaloe oil 百里香油
thymene 百里烯
thyme oil 百里香油
thymic 百里香的
thymol 百里酚油;百里酚
thymol blue 百里酚蓝;百里酚磺酸
thymol-carboxylic acid 百里酚羧酸
thymol ethyl ether 百里酚乙醚
thymol iodide 百里碘酚
thymolphthalein 百里酚酞

thymolsulfonic acid 百里酚磺酸
thymolsulfonphthalein 百里酚磺酸;百里酚蓝
thymol urethane 氨基甲酸百里酚酯
thymol violet 百里酚紫
thymoquinone 百里醌
thymoquinone-oxime 百里醌肟
thymotic acid 百里酸
thymotic anhydride 百里酸酐
thymotinic acid 百里亭酸
thymotor 闸流管电动机
thymotrol 闸流管电动机控制
thyratron 闸流管
thyratron automatic voltage regulator 闸流管自动电压调整器
thyratron chatter detector 闸流管振动检测器
thyratron commutation 闸流管整流
thyratron control 闸流管控制器
thyratron control characteristic curve 闸流管控制特征曲线
thyratron controlled generator 闸流管控制发电机
thyratron counter 闸流管计算器
thyratron drive 闸流管驱动
thyratron gate 闸流管门
thyratron inverter 闸流管换流器;闸流管变流器
thyratron motor 闸流管电动机
thyratron-motor drive 闸流管电动机传动
thyratron rectifier 闸流管整流器
thyratron-rectifier drive system 闸流管驱动系统
thyratron relay 闸流管继电器
thyratron servo 闸流管伺服系统
thyratron stroboscope 闸流管频闪观测器
thyratron switch 闸流管开关
thyriothecium 盾状囊壳
thyristor 闸流管;硅控整流器;硅可控整流器;薄石英片整流器;半导体闸流管;半导体开关元件
thyristor installation 可控硅装置
thyristor inversion 可控硅逆变
thyristor inverter 可控硅换流器
thyristor supply 可控硅电源
thyristor valve 可控硅阀
thyrite 碳化硅陶瓷材料
thyrite arrester 泰利避雷器;塞利特压变电阻避雷器;非线性电阻避雷器
thyrite arrestor 塞利特压变电阻避雷器
thyrite exciter 非线性电阻励磁机
Thyrite resistor 碳化硅非线性电阻器;塞利特电阻器
thyrode 硅可控整流器;可挂硅整流器
thyroid 盾形
thyroma 二层后台向外开的门(古罗马);古代临街开的门
thyroneum 大门入口连柱廊的通道
thyrorion 从古希腊宫廷入口到柱廊的通道
Thysen-Emmel 蒂森·埃米尔高级铸铁
Tianqi porcelain 天启瓷
tibari 住宅柱厅(印度)
Ti-based oxides coating anode 钛基涂层电极
Tibetan pagoda 藏式塔
Tibetan style 西藏式
Tibetan-style lamaist temple 藏式喇嘛庙
Tibet inula root 藏木香
Tibisirie fiber 蒂比锡里纤维(圭亚那)
tiburitine 石灰华
Tic coated tip 碳化钛涂层刀片
tick 勾号;滴答声
tick day 决算日
ticker 振子;振动子;载波传声器;断续装置
ticker office 售票处
ticket 船票
ticket acceptance slot 车票出入机
ticket affairs office 票务室
ticket agency 售票代理处
ticket barrier 检票栏
ticket bezel 票口
ticket book 本票
ticket booth 售票房
ticket check 验票
ticket check machine 读票机
ticket-collector 收票员
ticket converter 卡转换器
ticket counter 售票台;售票柜台
ticket day 交易结算日;交割日的前一天
ticket devaluation machine 剪票机
ticketed contour 计曲线
ticket examination 检票

ticket examination gate 检票口
ticket hall 售票厅
ticket identification 识别票号
ticketing information area 售票问讯处
ticketing stall after normal time 补票亭
ticket jam 夹票
ticket lobby 售票厅
ticket office 客票房;票务室;票房
ticket office machine 票房机
ticket policy 标准保险单
ticket-reading gate 检票门
ticket readout unit 读票机
ticket room 售票室
ticket sampling 标签抽样
ticket selling 售票
ticket-selling system 售票系统
tickets storage 储票库
ticket-transfer office (换车、船、飞机的)签票房
ticket vending machine 自动售票机
ticket window 售票窗
ticking 枕套及褥罩织物
ticking sling 划关计数
tickler 再生线圈;记事本;初始器
tickler coil 反馈线圈
tickler-coil oscillator 电感反馈振荡器;反馈线圈式振荡器
tickler guide 弹性推针导管
tickler spring 推针弹簧
tick mark 小十字符;记号;抖动标记
tick note 交货单
tick-over 低速工作
Tico 梯科镍锰铜钢
Ticonal 蒂克纳尔镍铁铝磁合金
Ticonium 蒂克尼姆铸造齿合金
tic-tac escapement 滴答擒纵机构
tic-tac rhythm 钟摆律
tida 潮间(带)台状冰脚
tidal 潮汐间的;潮汐的
tidal acceleration 潮汐加速度
tidal accommodation 感潮码头
tidal action 潮汐作用
tidal air 潮流气
tidal alarm 潮警报
tidal amplification 潮型放大;潮汐放大
tidal amplitude 潮振幅(潮差的一半);潮汐振幅;潮汐变幅;潮幅
tidal analysis 潮汐分析;潮汐分析;潮位分析
tidal anoxia 换气性缺氧症
tidal area 潮浸地带
tidal arrow 潮流矢符(指潮流方向的箭头)
tidal atlas 潮汐图
tidal backflow 感潮倒灌
tidal backwater 回潮水;潮水顶托
tidal ball 潮汐信号球;报潮球
tidal bank 潮滩
tidal barrage 挡潮闸(门);挡潮堤(坝);防潮堰堤;防潮堰堤
tidal barrier 挡潮堤(坝)
tidal-bar trap 潮砂坝圈闭;潮坝圈闭
tidal basin 有防潮闸的港池;有潮港池;感潮港池;潮汐盆地;潮汐港池;潮坞;潮港池;潮船坞;封闭式港池
tidal bay 有潮海湾
tidal beach 潮滩
tidal bedding 潮汐层理;潮流层理;潮成层理
tidal belt 潮间带
tidal bench mark 潮位基准点;潮汐水准点;潮汐标记;验潮水准点;验潮标志
tidal berth 有潮泊位;感潮泊位
tidal boat 乘高潮进出港的船舶
tidal bore 涌潮;怒潮;虎潮;海啸;潮涌;潮水涌浪;潮水上涨;潮津波
tidal boundary layer 潮汐边界层
tidal bulge 潮涨
tidal calculation 潮汛计算
tidal capacity 潮量;潮汐容量
tidal chamber 高水舱
tidal change 潮位变化
tidal channel 有潮小海湾;感潮河道;感潮航道;潮沟;潮道
tidal-channel deposit 潮汐水道沉积
tidal channel facies 潮汐水道相
tidal chart 潮汐图
tidal circulation 潮汐环流

tidal clock 钟形示潮器;时钟式自记潮位计;潮汐信标;潮汐(钟表形)信标;潮位钟
tidal coefficient 潮汐系数
tidal compartment 有潮区;感潮区域;潮区
tidal compensation 潮汐补偿
tidal component 调和分量;潮波分量;分潮
tidal condition 潮汐情况
tidal constant 潮汐参数;潮汐常数
tidal constituent 分潮
tidal correction 潮汐校正;潮汐改正;潮汐订正;潮位订正
tidal crack 潮汐裂缝(冰的)
tidal creek 涨潮河滨;潮沟
tidal creek deposit 潮沟沉积
tidal current 潮汐水流;潮汐流;潮流
tidal current analysis 潮流分析
tidal current analyzer 潮流分析仪
tidal current atlas 潮流图集
tidal current bedding 潮流层理
tidal current chart 潮流图
tidal current curve 潮流曲线
tidal current cycle 潮流周期
tidal current denudation 潮流剥蚀作用
tidal current diagram 潮流图
tidal current ellipse 潮流椭圆
tidal current hour 潮流时
tidal current limit 潮流界(限)
tidal current observation 潮流观察;潮流观测
tidal current prediction 潮流预报
tidal current prediction table 潮流预报表
tidal current scour 潮刷;潮蚀;潮流冲刷;潮流冲蚀
tidal current signal 潮流信号
tidal current survey 潮汐测量;潮流测量
tidal current table 潮汛涨落表;潮流预报表;潮流表
tidal curve 潮汐曲线;潮汐过程线;潮位曲线;潮候过程线;潮候曲线
tidal cuspate foreland 潮成尖甲
tidal cycle 潮汐周期
tidal data 潮汐资料;潮汐数据;潮汐潮流资料
tidal datum 潮准线;潮汐基准面;潮位基准面
tidal datum plane 潮汐基准面;潮高基准面
tidal day 太阴日;潮汐日;潮日
tidal defence 防潮设施;防潮堤
tidal deformation 潮汐变形
tidal delta 感潮三角洲;潮汐三角洲;潮成三角洲
tidal delta facies 潮汐三角洲相
tidal-deltaic deposit 潮汐三角洲沉积
tidal-delta marsh 潮盐沼
tidal deposit 潮积物
tidal depositional cycle 潮汐沉积旋回
tidal diagram 潮汐图表;潮汐图
tidal difference 潮汐改正值;潮汐校正值;潮汐改正数;潮汐时差;潮高差;潮差
tidal discharge 潮流量
tidal divide 潮道分水岭;分潮岭
tidal dock 有潮码头;有潮港池;感潮码头;感潮港池;通潮闸坞;潮汐港池;潮汐港
tidal double ebb 双低潮
tidal double flood 双高潮
tidal drag 潮汐阻力
tidal drainage 潮式引流法
tidal drift 潮汐漂移;潮汐漂流物
tidal duration 潮历时
tidal eddy 潮汐漩涡
tidal effect 潮汐作用;潮汐影响;潮汐效应
tidal efficiency 潮汐效率
tidal electric(al) power generation 潮汐发电
tidal electric(al) power station 潮汐电站
tidal electric(al) power plant 潮汐发电站
tidal elevation 潮胀
tidal embankment 挡潮土堤
tidal energy 潮汐能(量);潮汐能;潮能
tidal energy resources 潮汐水能资源
tidal entry 乘潮进港
tidal envelope 潮汐包线
tidal epoch 迟潮时间;潮滞;潮相;潮汐迟角
tidal erosion 潮蚀;潮流冲刷作用;潮流冲刷
tidal establishment 候潮时差
tidal estuarial channel 潮汐河口航道
tidal estuary 有潮河口;潮汐河口
tidal evolution 潮汐演化
tidal excursion 潮程
tidal factor 潮汐因子
tidal fall 落潮;退潮

tidal feature 潮汐特征
tidal fever 潮热
tidal flap gate 防潮阀闸
tidal flap valve 潮汐瓣阀
tidal flat 水槽地;漫滩;潮汐滩(地);潮滩;潮坪;潮漫滩;潮成平地
tidal flat autochthonous organic matter 潮坪原生有机质
tidal-flat channel 潮坪水道
tidal flat deposit 潮滩沉积;潮浦沉积;潮坪沉积
tidal flat facies 潮浦相;潮坪相
tidal flat lagoon 潮坪泻湖
tidal flat lagoon deposit 潮坪泻湖沉积
tidal flood 涨潮
tidal flood current 涨潮流
tidal flood interval 涨潮间隔
tidal flood strength 涨潮强度
tidal flow 潮汐水流;潮汐流;潮流
tidal flow capacity 潮流能力
tidal fluctuation 潮位波动;潮位变化
tidal flush 潮流冲刷作用;潮流冲刷;潮流冲蚀
tidal force 引潮力;潮汐力;潮力
tidal force of the moon 月潮引潮力
tidal forecasting 潮汐预报
tidal forest 红树林;潮汐林
tidal formula 潮汐公式
tidal frequency 潮汐频率;分潮频率
tidal friction 潮汐摩擦
tidal funnel 喇叭形潮汐河口;潮汐河口
tidal funnel estuary 感潮喇叭形河口
tidal gate 挡潮闸(门);潮汐闸门;防潮闸(门)
tidal ga(u)ge 检潮标;潮位水尺;潮位计;潮水位计
tidal generation force 引潮力
tidal generator 潮汐发生器;潮汐发动机
tidal glacier 入海冰川;随潮流的冰山;濒海冰川
tidal gravimeter 潮汐重力仪
tidal gravity correction 潮汐重力改正
tidal harbo(u)r 有潮差港;感潮港;潮汐港湾;潮港;潮港
tidal harmonic analysis 潮汐调和分析
tidal height 潮位;潮高
tidal high water 高潮位
tidal hour 潮时
tidal hydraulics 潮汐水力学
tidal hypothesis 潮汐假说
tidal indicating buoy 潮位指示浮标
tidal indicator 示潮器;潮水指示器
tidal inequality 潮汐不等
tidal influx 进潮量;潮水倒灌
tidal information 潮汐情报;潮汐资料
tidal information service 潮汐情报工作
tidal inlet 有潮小海湾;感潮水道;感潮海湾;进潮口;潮水入口;潮流通道
tidal instability 潮汐不稳定性
tidal institute 潮汐研究所
tidal interval 潮汐间隙;潮汐间隔
tidal inundation 潮水泛滥
tidalite 潮汐沉积;潮流沉积;潮梭岩;潮汐岩;潮积岩
tidal lag 高低潮时差;潮滞;潮汐滞后
tidal lagging 潮汐迟滞
tidal lagoon 潮汐礁湖
tidal lamination 潮汐分层;潮汐层化;潮成纹理
tidal land 潮间地
tidal levee 挡潮堤(坝)
tidal level 潮位;潮水位;潮水水位
tidal lift 涨潮
tidal light 潮汐信号灯(在港入口处指示潮汐情况的灯);潮灯;报潮灯(标);潮汐指示灯
tidal limit 涨潮界;潮汐界限;潮区界(限)
tidal line 潮汐线
tidal load 潮压;潮汐压力;潮汐荷载
tidal lock 涨潮船闸;进港船门;入口船闸;挡潮船闸;潮闸;潮门;防潮闸(门)
tidal lockup 潮壅;潮水顶托;潮水顶托
tidal loop 等潮高线
tidal low water 低潮位
tidall-water coast 受潮汐影响的海岸
tidally induced residual current 潮汐余流
tidal mark 潮痕;潮标
tidal marsh 盐沼;潮沼(泽);潮汐沼泽地;潮水沼泽;潮泛之沼泽
tidal marshland 潮漫滩
tidal mechanism 潮汐发生仪;潮汐发生器
tidal meter 验潮仪;测潮仪;测潮计;潮汐表

tidal migration 潮汐推移
tidal mill 潮力站;潮力磨坊
tidal mixing 潮汐混合
tidal model 潮汐模型
tidal monitoring station 验潮站
tidal motion 潮流运动
tidal movement 潮汐运动;潮水流动;潮流运动
tidal mud deposit 落潮淤淤积层;潮汐淤积层;潮区淤积层;潮区淤泥沉积;涨落潮淤沉积;潮汐沉积物;潮区沉积物
tidal mud flat 潮区泥地;潮汐泥滩;潮区泥泞低地
tidal observation 验潮;潮汐观测;潮位观测;潮水观测
tidal observation rod 验潮水尺
tidal observation station 验潮站
tidal observatory 验潮站;潮汐站;潮汐观测站;潮汐观测所
tidal opening 引潮时间
tidal oscillation 潮位振荡
tidal outfall 潮汐溢水道
tidal outlet 出潮口;潮汐出口
tidal pattern 潮型
tidal peak 高潮
tidal period 潮汐周期
tidal perturbation 潮汐摄动
tidal phase 潮汐位相;潮位相
tidal phenomenon 潮汐现象
tidal plain 潮汐平原
tidal plane of reference 潮位基准面;潮位参考面
tidal platform ice foot 潮间台状冰脚;潮间冰阶;潮间(带)台状冰脚
tidal pole 验潮杆;验潮标;潮水水尺;潮水(标)尺;测潮水尺
tidal pool 潮坑;积水坑;潮水塘;潮水坑;潮间带水坑;潮池
tidal port 感潮港;候潮港;潮汐港
tidal portion 感潮段;感潮区;潮区段
tidal potential 潮汐位能;潮位(势);潮汐势(能)
tidal power 潮汐功率;潮汐能发电;潮汐能;潮力;潮汐发电功率;潮力
tidal power barrage 潮汐发电坝;潮汐坝
tidal power generating station 潮力发电站
tidal power generation 潮汐发电;潮力发电
tidal power plant 潮汐水电站;潮汐发电厂;潮汐电站
tidal power resources 潮汐水能资源
tidal power station 潮汐力发电厂;潮汐发电站;潮汐电站;潮水发电站
tidal power turbine 潮汐发电涡轮机
tidal prediction 潮汐预报;潮汐推算;潮位推算
tidal predictor 潮汐预报机
tidal pressure 潮压
tidal pressure ridge 潮压冰脊
tidal prism 进潮棱柱水体;进潮量;潮棱(柱体)
tidal progressive wave 潮汐推进波
tidal propagation 潮波推进
tidal quality 潮汐特性
tidal quay 有潮码头;高潮码头;感潮码头;海上码头;全潮码头(任何潮汐都可靠泊);潮汐岸壁(式)码头;潮区码头
tidal race 急潮(流);强潮流
tidal range 潮汐变化范围;潮汐变幅;潮位变幅;潮差
tidal range at neaps 小潮潮差;小潮差
tidal range at springs 大潮潮差
tidal ratio 潮升比(两地的)
tidal reach 有潮河段;感潮(河)段;潮汐河段;潮水区;潮区段
tidal record 潮汐记录
tidal reduction 潮汐订正;潮位订正;潮水修正
tidal regime 潮汐状况;潮汐特性;潮汐情况
tidal region 感潮区;潮汐区(域);潮汐地区
tidal-regulating structure 潮汐调节构筑物;潮汐节制结构物
tidal regulation 潮汐控制;潮汐调节;潮水调节
tidal report 潮汐报告
tidal resonance 潮汐共振
tidal resonance theory 潮汐共振说
tidal respiration 潮式呼吸
tidal rhythm 潮汐节律
tidal rip 激潮;潮头卷浪;潮隔
tidal rise 潮升
tidal river 有潮河;感潮河段;感潮;受潮汐影响的河道;潮汐河流;潮汐河道;潮水河
tidal river reach 感潮河段

tidal river station 潮水河(测)站
tidal rod 潮水(标)尺
tidal rotary current 旋转潮流
tidal salt marsh 潮盐沼
tidal sand ridge 潮沙脊;潮成沙脊
tidal sand ridge deposit 潮沙脊沉积
tidal sand wave 潮沙波
tidal scale 验潮水尺;验潮杆;验潮标;观潮水尺;测潮水尺
tidal scheme 感潮区建港计划
tidal scour 潮汐冲刷;潮水冲刷;潮刷;潮蚀;潮汐冲刷;潮流冲蚀
tidal scour delta 潮汐侵蚀三角洲
tidal sea 潮汐海
tidal section 潮汐段
tidal sediment 潮汐沉积物
tidal sequence 潮汐层序
tidal shallow 潮滩
tidal ship 乘潮进港船舶
tidal signal 潮汐信号;潮汐信号;潮水信号
tidal signal station 潮汐信号站;潮汐信号台;潮位信号站
tidal slough 潮汐沼泽
tidal spectrum 潮汐谱;潮汐频谱
tidal spring 多潮泉
tidal staff 验潮杆;验潮标;潮位水尺;潮水(标)尺;测潮水尺
tidal stage 潮汐相位;潮位;位相;潮位相
tidal stand 平潮;停潮
tidal station 验潮站;潮汐站;潮汐观测站;潮汐观测所;潮汐站
tidal stencil 潮型
tidal stirring 潮汐混合
tidal storage 蓄潮
tidal strain 潮汐应变
tidal stratification 潮汐分层;潮汐成层
tidal stream 感潮河;潮汐(河)流;潮流
tidal stream atlas 潮汐水流图集;潮流图集
tidal stream data position 潮流数据可查表的地方
tidal stream diamond 潮流菱形符号
tidal stream information 潮流资料
tidal stream table 潮流表
tidal stress 潮汐应力
tidal stretch 感潮河段;潮区
tidal surface 潮汐面
tidal surge 潮水涌浪;潮水猛涨
tidal suspension deposition 潮汐悬移质沉积
tidal swamp 盐沼;潮淹沼泽;潮汐沼泽
tidal swelling 潮胀
tidal synobservation 同步验潮
tidal system 潮系
tidal table 潮汐表
tidal theory 潮汐(学)说;潮汐理论;潮波理论
tidal tide interval 月潮间隙
tidal torque 潮汐扭矩
tidal traffic 潮汐状交通
tidal train 候潮机动班车
tidal undulation 潮振动;潮汐振动;潮位升降;潮位波动;潮水升降
tidal valve (干船坞坞闸的)潮阀
tidal velocity 潮流速度
tidal velocity difference 潮流速度差
tidal volume 进潮量;潮气量;潮流气量
tidal water 潮汐水面;潮水
tidal water level 潮汐水位;潮位;潮水水位;潮水高度
tidal waters 有潮水域;感潮水域;潮汐水域
tidal water volume 潮汐水量;潮水量
tidal waterway 感潮水道;感潮河道;潮汐航道;潮浸地带
tidal wave 异常高潮(位);浪潮;潮啸;潮汐波;潮波
tidal wave crest 异常高潮波顶
tidal wave crest line 潮波波峰线
tidal waves system 潮波系统
tidal wave theory 潮波理论
tidal wave velocity 潮波速度
tidal way 潮汐水道
tidal wedge 楔形潮沟口;潮楔谷;潮水楔
tidal weir 挡潮堰;潮堰
tidal wetland 潮汐湿地;潮湿地
tidal wharf 有潮码头
tidal wind 潮汐风;潮风
tidal work 抢低潮施工
tidal yielding 潮汐屈服

tidal zone 潮区;潮间带;潮差段
tidal zone biology 潮间带生物学
tidal zone corrosion (发生在潮位变化区的)干湿交替腐蚀
tidal zone organisms 潮间带生物
tiddle-shaped 提琴形的
tide 潮汐
tide age 潮龄
tide amplitude 潮振幅(潮差的一半);潮汐振幅;潮幅;半潮差
tide anchorage 候潮锚地
tide and half tide 半续潮
tide ball 潮汐信号球;报潮球
tide barrage 挡潮坝
tide barrier 防潮墙;防潮堤
tide batten 测潮杆
tide beach 海滩;潮滩;潮汐海滩
tide blow 激潮
tide bore 海啸;涌潮
tide-bound 受潮水阻碍;等潮水的;潮阻
tide box (水工模型用的)潮水箱
tide bulge 潮汐波;潮波
tide bulge chart 潮波图
tide clock 钟形示潮器;潮汐钟;潮汐信标
tide control 潮汐控制
tide-control apparatus 潮汐控制仪
tide correction 潮汐校正;潮汐改正
tide crack 潮致冰裂;潮汐冰裂
tide current 潮流
tide current meter 潮流计
tide curve 潮位曲线;潮位过程线;潮候曲线
tide cycle 潮汐周期
tide data 潮汐资料;潮汐数据
tide day 太阴日;潮日
tide delta 潮汐三角洲
tide-dominated delta 潮控三角洲
tide-dominated facies 潮汐控制相
tide duration 潮汐历时
tide ebb 落潮;退潮;落涨
tide effect 潮汐作用
tide ellipse 潮汐椭圆
tide embankment 防潮堤;海岸堤
tide epoch 涨潮时间;涨潮,大潮发生时间;潮期
tide factor 潮汐因子
tide flat 潮汐滩(地);潮坪;潮漫滩;潮成平地
tide flood 涨潮
tide fluctuation 潮汐涨落;潮水涨落
tide fluctuation curve 潮汐涨落曲线
tide-forming force 引潮力
tide-forming force of the moon 月潮引潮力;太阴引潮力
tide-free 无潮的;不受潮汐影响的
tide-free region 无潮区
tide-free river 无潮河流
tide-free sea 无潮海
tideful 被潮水淹没的
tide gate 急潮水段;急潮水道;挡潮闸(门);潮闸;潮汐闸门;潮门;防洪闸(门);防潮闸(门);防潮门
tide ga(u)ge 水标尺;验潮器;验潮仪;检潮标;水位计;潮位水尺;潮位器;潮位计;潮位杆;潮位标尺;测潮仪;测潮计;测潮表;潮水位计
tide-ga(u)ge house 潮水计房
tide-ga(u)ge station 潮汐站;潮位站;潮位测站;潮水站
tide-ga(u)ge telemetry 遥测验潮仪
tide-ga(u)ge well 验潮井;潮水计井
tide-ga(u)ging station 潮位站;潮位测站;验潮站
tide-generating force 起潮力;成潮力;引潮力
tide-generating force of the moon 太阴引潮力
tide-generating force of the sun 太阳引力潮
tide-generating machine 潮汐发生器;潮汐发动机
tide-generating mechanism 潮浪发生器;潮浪发生机构
tide-generating potential 引潮位;引潮势
tide-generating set 潮汐发电机组
tide generator 生潮装置;生潮机;潮汐发生器;潮汐发动机
tide glacier 入海冰川
tide harbor(u)r 有潮港
tide head 感潮界限;感潮河段上限
tide head limit 涨潮界
tide height 潮位
tide hole 观潮冰穴(冰中钻孔观测潮高);冰上验潮孔

tide house 验潮站
tide impulse 潮浪冲击
tide indicating buoy 潮流指示浮标
tide indicator 示潮器;潮汐信标;潮水指示器;潮高指示器
tide-induced electromagnetic field 潮感电磁场
tide information 潮汐资料
tide lag 潮汐滞后;潮时差
tide lagging 迟潮时间;潮时延后
tide land 潮漫滩;潮浸地带;潮间地;受潮地区;潮淹区;潮渡区;潮浸区
tideland spruce 锡特加云杉;锡特加云杉
tideless 无潮
tideless beach 无潮海滩
tideless coast 无潮海岸
tideless region 无潮区
tideless river 无潮河流;不受潮汐影响的河道
tideless sea 无潮海
tide level 潮位
tide level ga(u)ge 潮位尺
tide-level glacier 潮位冰川
tide limit 潮位界;潮区限;潮区界(限);潮界
tide line 涨潮线;潮汐线;潮水痕迹
tide lock 进港船闸;入口船闸;挡潮闸(门);挡潮船闸;潮闸
tide-lock 防潮闸(门)
tide machine 潮汐机
tide mark 潮标;涨潮点;潮汐尺;潮汐标尺;潮水痕迹;潮痕;最高高潮痕;潮标
tide marsh 潮沼(泽)
tide-meter 潮位计;验潮仪;潮汐仪;潮汐计
tide-motor 潮汐发动机
tideness river 不感潮河段
tide notes 海图上有关潮汐情况的注释;潮汐附注
tide observation 验潮;潮汐观测;潮水观测
tide of long period 长周期潮(汐)
tide of second order 第二级潮汐
tide-over run 紧急复印
tide phase 潮相位;潮汐位相
tide pole 验潮水尺;验潮杆;潮位水尺;潮位杆;潮位标尺;测潮杆;测潮标杆
tide pole reference point 验潮杆基准点
tide pool 满潮池;低潮池(落潮时露出的池);潮间带水坑;潮池
tide port 潮口
tide power 潮汐力
tide power barrage 潮汐发电坝
tide predicting machine 潮汐自动推算机;潮汐推算机;潮汐预报机
tide prediction 潮汐预报;潮位预报;潮候推算
tide predictor 潮汐推算机
tide pressure 潮压
tide prevention 防潮水
tide priming 潮时提前
tide-producing force 起潮力;成潮力;引潮力
tide-producing potential 引潮位;引潮势
tide race 急潮(流);强潮流
tide raising force 引潮力
tide range 潮幅;潮汐变化范围;潮位变幅;潮差
tide recorder 自记验潮仪;潮水位计
tide reducer 潮汐改正尺
tide register 验潮仪;潮位器;潮水位计
tide reproduction 潮汐重演
tide retard 迟潮时间
tide retardation 潮汐延迟;潮时后延
tide riding level 乘潮水位
tide rip 急潮(流);潮头卷浪;潮隔;潮激;激潮
tide rise 潮升
tide rock 半潮干出礁石;受潮岩;低潮露出岩
tide rode 顶流锚泊;顶流锚泊
tide roll 潮汐自记纸卷
tide sail 候潮开航
tide sail level 乘潮水位
tide sail stage 乘潮水位
tide sequence 潮汐序列
tide sets in 潮向内流
tide sets out 潮向外流
tide shaft 潮位标尺
tide signal 潮汐信号
tide signal station 潮汐信号站;潮汐信号台
tide species 潮汐种类
tide staff 验潮杆;水位水尺;潮位水尺;潮位尺;潮水位尺;测潮标杆
tide stand 平潮

tide station 验潮站;潮汐站;潮汐测站;潮位站;测潮站
tide stretch 感潮河段
tide synthesizer 潮浪合成仪
tide table 潮汐表
tide type 潮汐类型
tide variation 潮位变化
tide vibration 潮汐振动
tide-waiter 海关负责上船监视卸货的人员
tide walker 随潮漂流物
tide washed 潮汐冲刷
tide water 潮汐水面;潮水;侵陆潮水
tide water control forest 防潮林带
tide water glacier 随潮流的冰川;入海冰川;滨海冰川
tide(water) level 潮水位
tidewater mill 海运材制材厂
tide water railroad 潮水影像的码头铁路线
tidewater railroad station 海港车站
tidewater railway station 海港车站
tidewater red cypress 落羽杉
tide waters 感潮水域;潮汐水域
tide water volume 潮汐水量
tide wave 潮汐波;潮浪;潮波
tideway 潮汐水道;潮汐流槽;感潮段;潮路;受潮汐影响的部分(河流);潮流
tide weir 潮堰;防潮堰
tide wind 潮汐风
tidework 潮汐工程
tide zone 潮涨落地带
tide zone facies 潮区相;潮间相;潮间带相
tidology 潮汐学
tidy 罩布;小垫布;整齐的;杂物箱
tie 相持(统计学);条件;柱箍;轨枕;联系;联结;联结件;捆轧;孔间距;带
tie adzer 枕木削平机
tie adzing 枕木削整
tie adzing machine 比削枕木机
tie aligning 方正轨枕
tie angle bar 连接角钢
tie arched girder 系行拱大梁
tie back 锚碇装置;后拉条;牵索;拉条
tieback anchor 土锚杆;拉条地锚;拉杆式锚定装置;拉杆锚
tieback curtain 带窗帘钩的帘
tieback load 拉条荷载
tieback location 拉条位置
tieback method 锚拉挡土墙施工法
tieback string 回接管
tie back wall 锚杆挡墙;锚碇墙;锚碇挡墙
tie band 系带
tie bar 承拉螺栓连接钢筋;并连案;系杆;连接角钢;连岛沙洲;拉筋;拉杆;转向拉杆;拉杠
tie-bar joint with rope lashing 捆绑连接系杆;用捆绑连接的系杆
tie bar placer 拉杆压入机;拉杆铺设机
tie-bar print 组芯芯座
tie beam 连系梁;系梁;梁;联梁;拉梁;拉结梁;锚梁;舱盖压条;封舱门条;连接梁
tie-beam roof 系梁屋架;系梁屋顶
tie-bed (铺铁路的)碎石层
tie bed scarifier 枕基爬松器
tie block 接合砖
tie bolt 承拉螺栓地脚螺栓;连接螺栓;系杆;拉杆(混凝土路面接缝的);地脚螺栓;连接螺栓;拉紧螺栓;收集螺栓;贯穿螺栓
tie border 枕木钻孔机
tie-boring machine 枕木钻孔机
tie breaker 连线上的电流断路器
tie brick 拉结砖;束砖
tie buffer coat 过渡涂层
tie bus 联系母线
tie cable 联结电缆;连接电缆
tieceron 居间肋
tie chain 系紧链条
tie circuit 连接线;连接路线
tie clamping plate 连接夹板;固定夹板
tie coat 衔接(涂)层;结合层;黏结层
tie column 系杆加劲柱;构造柱
tie condition 连接条件
tie conductor 系统连接线;联络线;连接线;连接路线
tie cotter 连接楔;连接栓;连接销
tie crib 轨枕盒
tie cutter 切枕木机

tied aid 限制性援助;附带条件的援助
tied arch 弦系拱;系杆拱;拉杆拱
tied-arch bridge 系杆拱桥
tied arch frame 有拉杆的拱式构架;拉杆拱构架
tied-back bulkhead 锚碇板桩岸壁
tied-back sheet-piling wall 锚碇钢板桩岸壁
tied bulkhead 锚固的板桩墙;用拉杆加固的板桩岸壁;用拉杆固定的堵头;用拉杆固定的闷头
tied cantilever bridge 有拉杆的悬臂桥
tied cantilever construction 有拉杆的悬臂结构
tied column 加拉杆柱;加箍柱;有拉杆的柱;联系柱;构造柱
tied concrete column 混凝土系柱
tied cottage (公司专供退职员工的)特供住房
tied deposit 专用存款
tied dock floor 锚柱式坞底板
tied house (公司专供退职员工的)特供住房;附属商店
tied-in clause 附带条款
tied indicator 相联系的指示物
tie disc 系圆片
tied island 陆连岛
tie distance 系杆间距;轨枕间距;撑间距
tied joint 绑扎连接;绑扎联结
tied loan 限制性贷款;专项贷款;附条件的贷款;附带条件的贷款
tiedown 系泊(专指拖车住房在驻地的系固)
tie(-down) bolt 系紧螺栓;锚碇螺栓
tie-down diagram 栓系图
tie-down fitting 系紧装置;系固装置
tie-down insulator 悬垂形绝缘子;悬式绝缘子
tie-down point 栓系点
tie-down sheaves 系紧滑轮
tied pair 相间对
tied plan 枕木面;规枕面
tied rank 相持秩
tied retaining wall 拉杆挡土墙;锚式挡土墙;锚碇式挡土墙;锚碇挡墙
tied rib arch 有拉杆的肋拱
tie drilling machine 轨枕钻机
tied sheeting type 锚碇板桩式
tied sheeting wall 锚系板桩墙
tied sheet pile wall 锚碇板桩墙
tied sheet piling 锚固板桩
tied shop 特约商店
tied structure 系杆结构;有系杆的结构
tied-up 停泊(指系泊作业)
tied wall 锚碇坞墙;有拉杆的船坞墙;有拉杆的岸墙;锚碇岸壁
tie element 联系杆件
tief-cordierite 低温堇青石
tie for power shovel 电铲轨枕
tie gang 换轨工班
tie girder 系梁
tie hole 系孔
tie hoop 螺旋箍筋;联系钢箍
tie-in 使连结合成一体;联测;打结
tie-in constraint 关联约束
tie-in crew 接管班;接班
tie-in equation 关联方程
tie-in gang 接头班
tie-in line 接入线
tie-in party 接头班
tie-in point 邻接点;连接点
tie-in sale 搭配销售
tie inserter 枕木插入机
tie-in team 接头班
tie-intransformer 联络变压器
tie-in weld 接头班;死口焊口
tie iron 墙箍;系墙铁
tie laying machine 轨枕铺设机
tie length 轨枕长
tie line 系线;专用路线;直接连线;直达线路;直达通信线路;直达连接线;联络线;连线;连锁线;连结线;连接线(路);连接导线【测】;结线;接入线;加固连接线;数据自动传输装置;数据通信线路;对角线;转接线;直达耦合线路;扎线
tie-line-bias 特性曲线法(能源供应)
tie line closure 联络测线闭合
tie line method 系线测量法
tie line oscillation 联络线振荡
tie-link 数据自动传输装置
tiemanite 硒汞矿
tie measurement 联测

tie member 系紧构件;系件;系杆;拉杆
tie mill 条锯机
tiemolecule 连接分子
tien 殿(中国)
tienshanite 天山石
tie number 集装箱船的层号
tie nut 固定螺母
tie on 拧上
tie pad 轧道垫板;轨枕衬垫
tie piece 系紧梁;系梁;系筋;条状模型加固片
tie pile 拉桩
tie plate 固定板;垫板(钢轨);系板;缀板;木甲板牵板;锚板
tie plate (bar) 钢轨垫板
tie plate column 缀板柱
tie plate member 缀板构件
tie plate mill 钢轨垫板轧机
tie plug 轨枕塞孔栓;塞孔栓;道钉口木塞
tie point 测量终点;联结点;联测点;连接点;连测点;接合点
tiepoint coordinates 联结点坐标
tier 锚链舱内空的部位;绳卷的中空部位;多船并靠系浮;并靠系浮船
tier a cable 排链
tier antenna 分层天线
tier array 堆叠天线阵;单排天线阵
tier building 多层建筑(物);多层房屋(建筑)
tier bus 叠合汇流排
tierce point 尖拱;桃尖拱拱尖
tierceron 中间肋(穹的);居间拱肋
tierceron vault 居间肋拱顶
tiered and distributed system 分层分布系统
tiered array 分层天线阵
tiered burners 分层布置喷燃器;分层布置的喷燃器
tiered pile 成层料堆
tiered stone umbrella 多层石块伞形(建筑)
tiered warehouse facility 立体仓库
tie remover 起枕器
tie renewal 抽换枕木
tie replacing machine 轨枕抽换机
tierer 排锚链人
tie respacer 方枕器
tie respacing 方正轨枕
tie-ridge 沟横梗
tie ridger 起垄器
tiering 排列成层;排链;堆积成层;堆叠;分层布置
tiering height 码头堆货高度;堆货高度
tiering machine 堆货机;仓库用起重运输机
tier number 层号(集装箱)
tier number of container 集装箱装载层数
tie rod 拉杆;系拉杆;系杆;轨距联杆;联结杆;连接杆;转向拉杆;中心杆;拉杠;横拉杆;长螺栓
tie rod anchored retaining wall 锚杆式挡土墙
tie-rod arm 转向臂
tie-rod-ball nut 横联杆球头螺母
tie rod end 转向横拉杆球铰接头
tie rod spider 拉杆辐射架
tie-rod stator frame 拉杆式定子框架
tier of beams 横梁层列
tier of blocks 一排房屋
tier of corbel bracket 斗拱铺作(古建筑)
tier of seating 成排座位;(常指剧场中的)行列座位
tier of seats 座位排列
tier pile 同厚度宽度的木材垛
tier pole 排杆
tierra caliente 高山暖温带
tierra helada 高山永冻带
tierra templada 高山温带
tiers 脚手架框架层数;浇筑层段
tiers in market 市场层次
tiers-point arch 三分之一拱
tier structure 多层构架建筑
tier table 宝塔形台
ties 柱箍;钢筋箍
tie scoring machine 枕木开槽机;轨枕锯面机
tie sheet pile type 锚碇板桩式
tie spacer 轨距调整器;调整轨距器
tie spacing 轨枕间距
tie spreader 轨枕铺置机;轨枕铺装机;扩枕机
tie strap 支撑板
tie strip 横向重叠
tie-strut 系杆;支撑杆;连接杆
tie tamper 轨枕捣固机;枕木捣固器;轨枕捣固机
tie tamping machine 拉杆打夯机

tie-tod yoke 系杆轭
tie tongs 枕木钳;轨枕钳
tie trunk 联结干线;连接中断线
Tietze extension theorem 蒂策开拓定理
tie-up 捆绑;用来捆扎的东西;停顿;停泊待命;停泊处
tie-up basin 系留港池;停泊港;待命港池
tie-up capital 呆滞资本
tie-up facility 系船设施;系船设备
tie-up force 系留力;系船力
tie-up wharf 待命码头
tie wall 拉结墙;横墙
tie water 结合水
tie welding 带连接板焊接;连接板焊接
tie wire 扎丝;张线;扎线;钢筋扎丝;钢丝捆带;拉筋;捆扎钢丝;绑扎;绑扎钢丝;绑扎(用)钢丝;绑扎(用)铁丝
Tiffanian (stage) 帝方阶【地】
tiffany finish 蒂法尼涂饰法
Ti-foon 提风(一种防风雨窗)
tige 柱杆;柱身
Tiger board 泰格板(一种石膏板)
tiger down spot glaze 虎皮斑釉
tiger eye glaze 虎眼釉
tiger eye stone 虎眼宝石
tiger grain 虎斑纹理(木材)
tiger heart 斑纹心
tiger lily heart 斑纹心
tiger's eye 虎目石;虎眼石
tiger skin glaze 虎斑釉
tiger spot 虎斑
tiger wood 虎皮木材;美洲核桃木
tight 致密的;固着物;坚实和坚固;水密钻孔
tight adjustment 精调(整)
tight alignment 精确调准;精确调谐
tight and loose pulley 固定轮和滑动轮;固定轮和游滑轮
tight bedding 坚实基础;不透水基础
tight bend 急弯
tight binding approximation 紧束缚近似
tight black oxide 密黑氧化
tight (bore) hole 缩径钻孔
tight bricking 紧密砖砌
tight burning 致密烧结;完全烧结
tight butt joint 无间隙对接接头
tight cesspool 不渗漏污水池
tight circling approach 小航线进场
tight connection 紧密接合
tight constraint 紧约束
tight container 紧密容器;紧闭容器
tight coupling 紧耦合;紧密连接;紧接头;密耦合;密封连接;强耦合
tight credit 信用紧缩;紧缩信用
tight cure 充分硫化;彻底硫化
tight curve 小半径曲线;密集曲线
tight distribution system 不漏水配水系统
tight drill-hole 缩径钻孔
tight-edged 纸边收缩
tighten 绷紧;拧紧;使紧密;上紧
tighten auditing 强化审计
tightened inspection 严格检验
tightened sampling inspection 严格抽样检查
tightener 张紧装置;张紧器;张紧工具;拉紧工具;紧线器;紧轮;收紧器
tightener sheave 张紧滚轮;皮带张紧轮
tightener sprocket 张紧链轮
tightening 拉紧;紧密;上紧
tightening belt 张紧皮带
tightening bolt 固紧螺栓;拉紧螺栓;紧固螺栓
tightening chain 拉紧链
tightening coupling 拧紧接头
tightening device 拉紧装置;扣紧装置
tightening flap 密封用衬垫
tightening idler 张紧托滚
tightening indicator 张紧指示器
tightening key 紧固楔子;锁紧键;斜扁销调整楔;斜扁销
tightening nut 拉紧螺母
tightening of bearing 轴承拉紧
tightening of credit 紧缩信贷
tightening pulley 张力轮;张紧轮
tightening rail 张紧滑轨
tightening scale dial 张紧器刻度盘
tightening screw 固紧螺丝;固紧螺钉;扣紧螺丝;扣紧螺钉;紧固螺丝;紧固螺钉
tightening sequence 紧固顺序

tightening slide 张力滑板
tightening torque 拉紧转矩;拧紧扭矩
tightening up screw 固紧螺杆;拉紧螺杆
tighten the belt 张紧皮带
tighten the economy 紧缩经济
tighten the rope 绷紧绳索
tighten up 拉紧;扎紧;绑紧
tighten up the money supply 紧缩资金投放
tight financing policy 紧缩性财政政策
tight fit 牢配合;静配合;紧配合;紧密吻合;紧密贴合
tight fitting 密封连接
tight-fitting bolt 精制螺栓;紧配螺栓;紧固螺栓
tight fitting door 密闭门
tight flask 一般砂箱;固定砂箱
tight flux bath 浓溶剂槽
tight fold 紧褶皱;紧闭褶皱;紧闭型褶皱
tight formation 致密岩层;稳固岩层;低渗透率层;低渗率层
tight ga(u)ge 紧轨距
tight grid 不漏水配水系统;不漏水管网
tight head 钻杆盘根盒
tight hitched tool 紧悬钻具(钢绳成就钻)
tighting ear 拉紧吊耳
tight joint 致密接头;紧密结合;紧密接合;接头瞎缝;密缝
tight junction 紧密联结
tight knot 紧密木节;紧密节;(木材的)紧节;坚固木节;抽紧结
tight lattice 紧密栅格
tight-line pumping 泵到泵输送
tight-lining 紧索越障法
tight-lock coupler 密锁自动耦合器
tightly cemented formation 紧密胶结岩层
tightly coupled multiprocessor 紧密耦合多处理机
tightly-coupled processor 紧密结合处理机
tightly coupled system 紧密耦合系统
tightly jointed masonry 密缝圬工
tightly keyed structure 紧密锁结构
tightly packed 紧密包装的
tightly-pressed structure 挤压型经济结构
tightly wound arm 紧卷旋臂
tight manhole cover 检查井密盖;密封式入井盖板;密封式入孔盖板;密封式入井盖板
tight market 紧俏市场;供不应求的市场
tight monetary policy 紧缩信用政策
tight money 信用紧缩;银根紧缩;高息资金;高息借款;紧缩银根
tight money policy 紧缩银根政策;紧缩信用政策
tightness 严密度;致密度;紧缩感;紧密性;紧密(度);紧度;密封性;密封度;松紧度
tightness against dust 防尘密封度
tightness degree 气密度
tightness measuring instrument 密封性测定仪
tightness of concrete 混凝土的气密性
tightness of joints 接合密封性
tightness of the curtain 链幕密闭度
tightness test 紧密性试验;压力试验;气密(性)试验
tightness test of air box 空气箱密性试验
tightness test pressure 致密性试验压力
tightness to gas 气密性
tight nut 紧固螺母
tight pull 拉紧
tight pulley 固定轮
tight rate 紧定额
tight riveting 紧铆;紧密铆接;紧密铆合;防漏铆接
tight rock 致密岩石;致密岩
tight rope 绷紧绳索;绷索
tight sand 密实砂层;致密砂层;紧砂
tight seal 严密封闭;紧密封;紧封;紧密
tight seam 紧密焊缝
tight sheathing 镶拼底板(钉在筋筋或格栅上的);密封式板撑;密铺挡土板;密铺挡板;密铺衬板;封闭板撑;密铺望板
tight sheeting 连续式挖方支撑
tight shook 紧桶板料
tight shot 紧凑拍摄
tight side 紧面;紧边;切片光面;受拉段;单板整面;单板紧边
tight side of belt 皮带紧边
tight side of veneer 单板的紧面
tight silo 密闭青储[贮]塔
tight size 紧配合间隙;紧配尺寸
tight (sky) line 张紧式索道
tight soil 紧实土(壤);密实土

tight spigot 夹紧套筒
tight-spiral 紧螺旋
tight spot (隧道的)设计开挖线;计价线【铁】
tight stacking 密实堆积(木料等)
tight standard 严格标准
tight staunch and strong 紧密
tight stowage 紧密码垛
tight-strong seam 密固接缝;强固紧密焊缝
tight-strong weld 密实强固焊缝
tight structure 紧密结构
tight surface 致密路面;密封面
tight tolerance 紧密度容限
tight to petrol 对石油密封;对汽油密封
tight to surface water 对地面水的密封
tight tube fiber 紧套光纤
tight-tube optic(al) communication fiber 紧包光通信纤维
tight tube optic(al) fiber 紧套光纤
tight turn (道路的)急弯;急转弯
tight veneer 紧单板
tight weld 致密焊缝;密封焊接
tight wire 崩紧;紧系钢索;绷索;(钢制的)绑绳
Tiglian warm epoch 梯格林温暖期
tiglic acid 顺芷酸
tigna 木屋架拉梁;木屋顶系梁
tignum 建筑材料
tigroid substance 虎斑质
tigrolysis 虎斑溶解
TIG welding 惰性气体保护钨极焊
tikhonenkovite 水氟铝锶石
Tikhvin polarity superchron 蒂克赫维恩极性超时
Tikhvin polarity superchron zone 蒂克赫维恩极性超时间带
Tikhvin polarity superzone 蒂克赫维恩极性超带
tikker 自动收报机;振子;载波传声器
til 浪线;波形号
tilasite 氟砷钙镁石
tilde 否定号
tile 矩形耐火板;花砖;瓦(片);瓷砖;扁平砖(修补用)
tile-and-a-half (tile) 阔瓦
tile and slab press 砖瓦坯压制机
tile arch 砖瓦
tile bathroom 砖砌浴室
tile batten 挂瓦板条;挂瓦条
tile beam 砖梁
tile blue 瓦蓝(浅灰蓝色)
tileboard 压实纤维墙面板;仿釉面墙板;花砖式饰板;花砖式饰面拼板
tile body 砖体
tile ceiling 砖瓦顶棚
tile cement fixing method 水泥固定法
tile cementing agent 砖瓦粘合剂
tile chimney 砖砌烟囱
tile clay body 黏土砖体
tile cleaner 瓷砖清洗剂
tile cleaning agent 砖的清洗剂
tile clip 瓦夹;固瓦弹簧夹
tile colo(u)r 面砖色泽
tile compression strength 面砖的抗压强度;瓦的抗压强度
tile conduit 瓦管
tile construction method 砖建筑阀
tile contractor 空心砖承包人;瓷砖承包人;瓦铺砌承包人
tilecoping on gable 封檐瓦垄
tile course 瓦行
tile cradle 瓦管座;瓦基座
tile creasing 逐层砂层;逐层挑出屋面瓦;瓦出檐;砖墙外挑瓦顶;墙头挑出瓦
tile culvert 瓦管涵(洞);管涵;瓦管阴沟
tile cut across 横砌砖
tile cutter 砖瓦切割刀具;瓷砖刀
tile cutting machine 面砖切割机;瓦管切割机
tiled finish 面砖饰面
tile(d) floor 砖砌地面;花砖地面;瓷砖楼地面
tiled kitchen 砖砌厨房
tile drain 缸瓦排水管;排水瓦管;瓦管排水(沟);瓦沟
tile drainage 瓦管排水
tile draining machine 瓦管敷设机
tile(d) roof(ing) 瓦屋顶
tiled stair(case) 砖扶梯
tiled stove 砖砌烘箱
tile(d) valley 瓦天沟;瓦排水斜沟;瓦管屋面天沟;瓦排水(屋面)斜沟;瓦管天沟

tiled viewpoint 砖形视口
tile edging 瓦口;颌版
tile-end 瓦挡
tile envelope 陶土套管
tile-faced panel 贴砖镶板;贴砖护墙板
tile facing 面砖贴面
tile factory 制砖厂
tile feed 贴砖;进瓦
tile field 沥滤场;瓦管排水场;瓦管灌溉系统
tile fillet 瓦砌封角条;瓷砖嵌镶;瓷瓦泛水;瓦泛水
tile filter bottom 瓦管滤池底瓦管;瓦管滤池底;瓦管集水系统
tile finish 砖的精加工;面砖饰面
tile fitting 瓦工配件;砖瓦装配
tile fixer 瓦瓦固定材料
tile fixing 砖瓦铺设
tile fixing adhesive 固定砖胶合剂
tile fixing agent 砖定型剂
tile fixing medium 固定砖的介质
tile fixing work 砖的固定工作
tile floor 花砖地;瓷砖地;瓦铺地(面);花砖地面
tile floor cover(ing) 地砖覆盖
tile floor(ing) 地砖贴面地板
tile floor(ing) finish 铺装地面加工
tile format 砖瓦尺寸
tile ga(u)ge 瓷砖水尺
tile girder 砖梁
tile glazed coat(ing) 面砖加釉
tile glazing 砖瓦上釉
tile grinder 自动石板研磨机
tile grinding machine 面砖砂轮机
tile hammer 铺瓷砖用锤;打瓦锤;瓷砖锤
tile hanging 墙头瓦;竖瓦;挂瓦法;墙面挂瓦;屋瓦接合法;缠结瓦片
tile hook 挂瓦钩
tile hung wall 墙面吊瓦;贴面砖墙
tile jacket 陶土套管
tile joint 瓦管接合;瓦管接头
tile jointing composition 砖缝结合剂
tile jointing compound 砖缝接合混合物
tile jointing mass 砖连接块件
tile junction 瓦管接合
tile key 翻边
tile kiln 瓦窑
tile lath 挂瓦条
tile layer 瓦管铺设机
tile laying 埋瓦管;铺瓦;瓦瓦
tile laying machine 面砖铺设机
tile-like coating 仿瓷(砖)涂料
tile line 砖缝线
tile lining 瓷砖内衬;砖衬;瓦板内衬
tile lino(leum) 块状油毛毡;贴砖毡
tile lintel 砖过梁
tile lintel floor 空心砖加筋密肋楼板;陶土空心砖加筋的密肋楼板
tile listing 瓷砖泛水;瓦泛水
tile making 制瓦
tile making factory 面砖制造厂
tile masonry 砖瓦圬工;铺瓦圬工;铺瓦圬工
tile masonry wall 砖瓦工墙
tile masonry work 铺瓦圬工
tile mastic 铺贴面砖玛碲脂;铺贴面砖胶粘剂
tile nail 瓦钉
tile of wood 木瓦
tile ore 瓦铜矿;赤铜矿
tile-packed column 瓦片填充塔
tile packing 瓦填料
tile panel 砖面板
tile papered on the back 砖背贴纸瓷砖
tile papered on the front 砖面贴纸瓷砖
tile partition (wall) 砖隔墙;砖隔断
tile path 砖路
tile pattern 制砖模
tile paved surface 地砖面层
tile pavement 瓷砖地面
tile paving 砖铺面;砖路面;花砖面;铺瓦
tile peg 瓦栓钉
tile pick (瓦工用的)尖头锤
tile pin 瓦钉;木瓦钉;压瓦钉
tile pipe 瓦筒;瓦管
tile plant 砖瓦厂
tile press 面砖压制机;压砖机;瓦压机;瓷砖压制机
tiler 泥瓦工;泥水匠;瓦工;铺瓦工;瓦工;贴面砖工

tile red 瓦红(橙色)
tile repair screw 陶土管修理班
tile roof cladding 盖瓦顶
tile roof cover(ing) 瓦顶
tile-roofed house 瓦房
tile roof(ing) 瓦屋面;瓦屋顶;铺瓦;瓦层面;盖瓦屋面
tile roof sheathing 瓦屋顶覆盖
tiler work 贴面工程;贴面砖作业;铺瓦作业
tilery 瓦厂;装饰性砖瓦铺贴术;砖瓦装饰术;制瓦厂;瓦窑
tile sealant 砖瓦密封料
tile sealing composition 砖瓦密封组合物
tile sealing compound setting 砖瓦密封混合物
tile setting 砖瓦砌筑
tile setting adhesive 砖瓦砌筑粘结剂
tile setting medium 铺砖瓦介质
tile shell 空心砖外壳
tile size 花砖尺寸;瓦片尺寸;瓷砖尺寸
tile slabber 铺地面砖和墙面砖工人;瓦模工
tile stack 砖砌烟囱;砖瓦垛
tilestone 石板;石瓦
tile store 砖瓦货仓;砖瓦堆栈
tile strength 砖瓦强度
tile strip 挂瓦条
tile subdrain 瓦管暗沟;地下排水管;地下瓦管;地下瓦沟
tile tie 系瓦绳
tile tint 面砖色泽
tile trenching machine 瓦管管道挖掘机
tile tube 瓦管
tile tubing 瓦管道
tile-type mixer 可倾倒拌和机
tile underdrain 瓦管暗沟;地下排水瓦管
tile underdrainage 瓦管地下排水
tile valley 特制天沟瓦
tile walk 花砖人行道;花砖便道
tile wall 砖墙
tile wastewater-disposal lines 废水处置瓦管
tile with double interlock 双连销瓦
tile with non-slip pattern of short ribs 有防滑短的瓷砖
tile with round edge external corner 圆边外角砖;二面圆边外角(釉面)砖
tile with round edge opposite 两面圆方砖(釉面)砖
tilework 砖瓦厂;铺瓦工作;砖瓦;瓦作
tilia 椴树
Tilia vulgaris 欧洲椴木
tiling 炉衬修补;铺瓦工作;铺瓦;铺瓷砖;屋顶铺瓦;贴面砖;贴瓷砖
tiling batten 挂瓦板条;挂瓦条
tiling plaster 砌面砖用灰浆;面砖用灰浆;砌面用灰浆;石膏灰;胶凝石膏
tiling the plane 砌平面
tiling work 砖砌工作;铺瓦工作;贴砖工作;贴砖工程
tiliting axis 倾转轴
till 钱柜后;钱柜;漂积土;抽屉;冰碛物;冰碛(土)
tillable land 可耕地;适耕地;适耕地
tillage 耕作;耕种;耕地
tillage doesn't reach the B horizon 耕不到B层土
tillage equipment 耕作设备
tillage implement 整地机具,耕地工具;耕作农具
tillage land 可耕地;适耕地
tillage meter 耕耘机具测试装置
tillage methods in humid tropics 湿润热带地区的整地方法
tillage objective 耕作目的
tillage pan 耕作层;犁底层
tillage pattern 耕地图形
tillage requirements 耕作要求
till billow 冰碛波状丘陵
till countermanded 至撤销为止;撤销前有效
till cured concrete 硬化中的混凝土
tilled field gley soil 耕作地潜育土
tilled land 耕地
tilled soil 耕作土(壤)
tiller 耕作机;舵杆;舵柄;翻土机
tiller becket 舵轮系绳
tiller chain 舵链
tiller combine 联合耕作机
tiller grip 手柄端头
tiller head 舵柄头
tiller line 转舵索

tiller quadrant 舵扇
tiller rope 转舵索
tiller rope guide 转舵索导环
tiller rotro 耕耘旋转锄
tiller steering 舵柄操舵
tiller steering gear 舵柄转舵装置
tiller telltale 舵角指示器
tiller tie bar 舵柄连接拉杆
tiller type mixer 翻土式拌和机
tilleyite 粒硅钙石
tillite 冰碛岩
till-less agriculture 免耕农业
tilloid 类冰碛岩
till plain 冰碛平原
tilt 斜置;掀斜;倾斜;偏斜
tiltable 可倾斜的
tiltable antenna 仰角可调天线
tiltable plate telescope 带形射电望远镜
tiltable sunshade 花园太阳伞
tilt adjuster 倾斜度调节器
tilt adjustment 角调节;横向倾斜角调节;横向倾斜;倾斜改正;倾斜角校正;倾角改正;俯仰改正
tilt aerial photo 倾斜航片
tilt-and-bend shading 倾斜与弯曲阴影补偿
tilt and turn window 翻窗;斜窗;倾斜转动式窗
tilt angle 倾斜角;转动角;倾角;摆角
tilt angle control 倾角控制系统
tilt angle method 倾角法
tilt angle of photograph 像片倾(斜)角
tilt angle of photographic(al) axis 主光轴倾角
tilt arm 倾斜杆
tilt autosyn 倾斜自动同步机
tilt back angle 向后倾翻的位置;倾斜后角;后翻角
tilt-bed quenching car 倾翻式熄焦车
tilt block 翘起断块
tilt-block basin 掀断盆地【地】;掀断盆地;倾斜掀断盆地
tiltboard 倾斜台
tilt-boat 张设天幕的小艇
tilt boundary 倾斜角界;倾斜界;倾侧边界
tilt brace 斜撑顶杆
tilt cab(in) 倾翻式车厢;前倾式驾驶室
tilt calculator 倾斜计算装置;倾斜计算器
tilt cart 倾卸小车
tilt casting 倾斜浇筑(法)
tilt cast method 倾斜浇(注)法
tilt check mechanism 测斜机构
tilt check rotating nail 测斜转螺
tilt circle 倾角圆;倾角圈
tilt coil 倾斜线圈
tilt-compensation seismometer 倾斜补偿地震计
tilt control 倾斜振动控制系统;倾斜角度控制;倾斜度调整;舵角控制
tilt correction 倾斜校正
tilt corrector 倾斜校正器
tilt cylinder 转斗液压缸;倾倒用液压缸;翻斗唧筒;摆缸
tilt cylinder pin 倾斜汽缸销
tilt deck trailer 可倾斜拖车;倾卸式平板拖车;倾倒台拖车
tilt displacement 倾斜位移
tilt distortion 倾斜变形
tilt down 使向下倾斜;翻下
tilt dozer 斜推式推土机;斜铲(式)推土机;倾斜式推土机;拖挂式筑路机械;推挂式筑路机械
tilted 倾斜的
tilted anterior tomogram 正体倾斜体层摄片
tilted aquifer 倾斜含水层
tilted arch 上心拱
tilted blade 倾刀
tilted block 掀斜断块;掀斜地块;翘起地块
tilted block structure 倾斜块状式构造
tilted cylinder mixer 倾斜圆筒混合机;倾斜式筒形混合机
tilted disc check valve 倾斜盘防逆阀
tilted electron lens 倾斜电子透镜
tilted fault block 掀斜断块;翘起断块
tilted fault-block mountain 掀斜式断块山
tilted flume 倾斜水槽;倾斜测流槽
tilted garden parasol 花园太阳伞
tilted gold-leaf electrometer 倾斜金箔静电计
tilted iceberg 倾斜冰山
tilted interface 倾斜界面
tilted mixer 倾倒式混合器;倾斜搅拌器;倾斜式搅拌机

tilted model 倾斜模型;斜置模型
tilted mountain 倾斜山
tilted pad bearing 自位轴承
tilted parabola 倾斜抛物线
tilted parabola waveform 倾斜抛物线波形
tilted photo coordinate system 倾斜像片坐标系统
tilted photograph 倾斜像片;倾斜摄影航片
tilted-plate separator 倾板式分离器;斜板分离器
tilted position 倾斜位置
tilted shape 倾斜形状
tilted-slab cupola 斜板小圆屋顶
tilted-slab dome 斜板穹隆
tilted-slab roof 斜板屋顶
tilted-slab segment 斜板扇形体
tilted-slab structure 斜板结构
tilted stratum 倾斜地层;倾斜层
tilted tree 马刀树(即醉林)
tilted trough 斜槽
tilted turret 倾斜转动架
tilted-up 倾斜的
tilter 摇动装置;摇摆台;倾转架;倾卸架;倾卸车;倾架;倾翻机构;倾倒设备
tilter mixer 倾斜式混料器
tilt error 倾斜误差
tilt field 倾斜场
tilt finder 倾斜测定仪;倾斜测定器
tilt ga(u)ge 倾斜仪;倾度规
tilt graticule 角度分划板;倾斜分划板
tilth 耕作;耕过的土地;耕层
tilt hammer 跳动锤
tilt head (三脚架上的)倾斜头
tilth rotor 耕耘旋转锄
tilth top soil 熟化土壤
tilt hydraulic hammer 倾压油缸;倾斜水压扬汲机;倾斜水锤泵
tilt indicator 倾角指示器
tilting 掀动;翘起;翘动
tilting angle indicator 倾角指示器
tilting angle of upright 门架倾斜角【港】
tilting arrangement 倾动装置
tilting bearing 斜垫自位轴承;摇座;斜垫轴承
tilting bed 倾翻平台
tilting block 摆动箱体
tilting board 椽上封檐板条
tilting body 摆式车体
tilting box 斜板箱;翻卸箱;摆动箱体
tilting-box sampler 倾卸式取样机
tilting bucket conveyer 翻斗式输送机
tilting bucket rain ga(u)ge 倾斗式雨量计
tilting bulldozer 自调推土机;枢承推土机
tilting burner 倾斜式燃烧器
tilting cab(in) 倾斜式驾驶室
tilting cart 倾卸小车;倾卸货车;翻斗车
tilting chair 盘条翻转台;翻锭机
tilting concrete mixer 倾翻式混凝土搅拌机
tilting crucible furnace 倾动式坩埚炉
tilting-cup feeder 翻杯式给料器
tilting cylinder block 摆动缸体
tilting dam 顶部泄水坝;倾斜坝
tilting-deck cage 可翻卸罐笼
tilting device 倾动装置
tilting digester 回转式蒸煮器
tilting dipper bucket 倾翻铲斗
tilting disc filter 倾盘式过滤机
tilting door 卧倒门
tilting dozer 斜铲推土机;倾斜式推土机;倾翻式推土机
tilting drum 倾斜转筒;倾斜鼓筒
tilting drum batch mixer 倾筒式分批搅拌机;倾筒式分批拌和机
tilting drum concrete mixer 斜筒式混凝土搅拌机;斜鼓形混凝土搅拌机;倾斜式滚筒混凝土搅拌机;倾筒式混凝土搅拌机;倾角式混凝土搅拌机
tilting drum mechanism 倾转筒机构
tilting drum mixer 倾斜式搅拌机;斜鼓形搅拌机;倾筒式搅拌机
tilting easel 倾斜承影面
tilting equipment 倾动装置
tilting error 倾斜
tilting failure 倾倒破坏
tilting fillet 连檐(木);倾斜缘条;披水条;三角垫木;檐口垫瓦条;斜塔式闸门;镶边压条;倒角压头
tilting flagpole 倾斜旗杆

tilting float 倾斜式浮标
tilting floor 倾斜地板
tilting flume 可调坡度水槽;活动水槽
tilting force 倾斜力
tilting forehearth 倾斜式前炉
tilting frame 倾斜框架
tilting front end bucket 前倾铲斗
tilting front gate 可倾式前(斗)门
tilting furnace 罩式炉;可倾炉;倾转式炉子;倾注炉;倾斜式炉
tilting gate 翻转闸门;倾斜式闸门;倾倒式闸门;卧式闸门;卧倒门
tilting ga(u)ge 倾斜水尺;倾斜式水尺;倾斜度测量仪;倾斜测压管
tilting gear 倾翻机构;翻钢机;翻车装备
tilting gearing 倾斜装置
tilting glass cutting table 倾斜玻璃切台
tilting hangover 倾斜拖尾
tilting head press 摇头压榨机;上回转台压榨机;摆头印刷机
tilting hydraulic press 倾斜式液压机
tilting idler 倾翻惰轮
tilting jack 倾倒千斤顶
tilting ladle 可倾桶;倾动式前炉;倾动式浇包
tilting-lens rectifier 倾斜镜头纠正仪
tilting level 倾斜水平仪;微倾水准仪
tilting level screw 水准仪微倾螺丝;微倾水准仪螺丝
tilting lever 斜度调整杆;倾斜杆
tilting load(ing) 倾斜荷载;翘起荷载;倾覆荷载
tilting low-bed trailer 倾斜式平板拖车
tilting machinery 倾动装置
tilting manometer 回转式压力计;倾斜式压力计
tilting mechanism 倾筒式机械;倾翻机构;摆动机构
tilting method with transit 经纬仪倾角法
tilting micromanometer 倾斜式微压计
tilting milling spindle 斜铣轴
tilting mirror 转镜;倾斜镜
tilting mixer 斜鼓形搅拌机;可倾式搅拌机;可拌和机;可倾倒拌和机;倾斜式搅拌器;倾斜(式)搅拌机;倾斜式拌和机;倾筒式搅拌机;倾翻式搅拌机;倾斜式拌和机;锥形翻出料式搅拌机;倾斜式鼓形搅拌机
tilting moment 倾斜力矩;倾覆力矩;翘起力矩;翻倒力矩
tilting motion 倾倒
tilting mould 翻转模板
tilting nozzle 翘动管嘴
tilting of rail 轨身倾斜
tilting of strata 倾斜地层
tilting of wheels 车轮倾斜
tilting pad 倾斜垫
tilting-pad bearing 斜瓦轴承;倾斜轴瓦轴承;斜垫轴承
tilting pad thrust bearing 调心止推轴承
tilting pan 倾卸式料槽
tilting pass 翻转道次
tilting photo print 倾斜航片
tilting piece 铺瓦垫片;檐口垫瓦条;镶边压条;倒角压头
tilting pinch rolls 翻钢夹紧辊
tilting plate method 斜板法
tilting platform 倾卸平台
tilting platform trailer 倾卸式拖车
tilting pneumatic cell 翻动气压器
tilting position 倾斜位置
tilting pot buggy 翻料车
tilting propeller 倾斜式螺旋桨
tilting rail 斜坡轨道
tilting ram 倾翻顶杆
tilting range 倾斜范围
tilting reflector 上下可动反射镜
tilting reverberatory furnace 倾倒式反射炉
tilting ring tachometer 摆环式转速计
tilting roll(er) 摆动式辊;翻钢辊
tilting rotor 倾斜的水平旋翼
tilting saw table 斜置锯子桌;倾斜锯子桌
tilting screw 倾斜螺旋
tilting seat 可调整靠背斜度的座椅;转动式座椅;折叠式座(椅);可折座
tilting skip 倾斜料斗;倾斜料斗;翻倒箕斗
tilting sluice gate 倾卸式泄水闸门
tilting spillway gate 倾翘式溢水堰闸门;倾倒式溢流堰闸门
tilting spout 倾动流铁槽

tilting stability 倾覆稳定(性)
tilting steam trap 倾斜凝汽阀
tilting stiffness 抗翘刚度;抗翘起刚度
tilting structure 掀斜构造
tilting system 倾翻系统
tilting table 倾斜式升降台;倾斜工作台;摆动升降台
tilting table stop 倾斜台触止
tilting tank 倾翻桶
tilting test 倾斜试验
tilting track 升降机轨道;起重机轨道
tilting trailer 倾卸式拖车
tilting-type 可倾式
tilting-type drag scraper 拖式倾卸铲运机
tilting-type mixer 倾筒式搅拌机;可倾混铁炉;可倾混合炉
tilting up of the mast 桅杆的竖立
tilting vehicle 倾卸式运输车;倾卸货车
tilting weir 活动堰;可倾式堰
tilt ladle 倾转式浇包
tilt lever 转斗杠杆
tilt link 转斗连杆
tilt-loading trailer 倾斜装卸式拖车
Tiltman block 隔墙砌块(侧面有双槽榫接合)
tiltmeter 斜度仪;斜度计;倾斜计;倾斜仪;测斜仪;地面倾斜度测量仪
tilt mixer 行倾斜校正器;倾斜混波器;低频失真校正电路
tilt mo(u)ld 倾斜铸型
tilt-mo(u)ld billet 可倾模锭坯
tilt observation 倾斜观测
tilt of aerial camera 航空摄影机倾斜
tilt of wavefront 波前倾斜
tiltometer 斜度仪;斜度计;倾斜仪;倾斜器;倾角计
tilt pouring 倾转浇注
tilt-pour process 倾注法
tilt ram 斜置水锤泵;斜置撞锤
tilt relief valve 斜置安全阀
tilt rod fuze 倾斜杆状引信
tilt roof 斜屋面;斜屋顶;帐篷式屋顶;遮顶(船、车、帐篷、地摊等);篷顶
tilt sag 倾斜下陷
tilt sensor 测倾敏感元件
tilt slide rule 倾斜滑尺
tilt step 倾斜突变
tiltster 自动翻卸货机
tilt storm 倾斜暴
tilt stowing hatchcover 斜置式舱盖
tilt survey 倾斜测量
tilt switch 倾卸开关
tilt table test 倾斜台试验
tilt to coast 向岸倾斜
tilt to lake centre 向湖心倾斜
tilt tolerance 倾斜容限
tilt-top table 翻面桌
tilt to sea 向海倾斜
tilt-trailer 翻斗拖车;倾卸式拖车;倾卸式挂车
tilt trap 倾卸式阱
tilt tuner 斜调谐器
tilt-type mixer 可倾式搅拌机;可倾式拌和机
tilt-up 翘起装配;吊装构造
tilt-up compound 混凝土平浇板隔板
tilt-up concrete panel 倾斜混凝土镶板
tilt-up construction 板式建筑;就地平浇竖立施工法(混凝土墙板);比式建筑;立墙平浇建筑法;翘起结构;吊装结构;翻起法施工;翻立脱模施工
tilt-up construction method 预制构件装配施工法
tilt-up door 上翻门
tilt-up method 向上翻转竖立(施工)方法
tilt-up panel 翻起施工墙板
tilt-up structure 立墙平浇结构
tilt-up wall 倾斜的墙
tilt-up walling 立墙法
tilt vector 倾斜矢量
tilt voltage 倾斜电压
timber = lumber
timber anchor pile 木锚桩;木拉桩
timber and half hitch 圆材捆加半结;拖材结
timber and lumber standard 木材标准;木材规格
timber and lumber standardization 木材标准化
timber and space 肋距
timber-and-tile roof 木托瓦屋顶;木瓦屋顶
timber apron 木护坦;木底板;木挡板
timber arch 木拱
timber arch bridge 木拱桥

timber arch bridge of glued and laminated wooden strips 胶木板拱桥
timber arched girder 木拱大梁
timber architecture 木建筑
timber area calculator 材积专用计算机
timber assortment 木材种类;木材分类
timber baffle 木挡板
timber banister 木楼梯栏杆
timber bar 木格条;木棒;木杆
timber base 木材底板
timber base plate 木底板
timber basin 储[贮]木场;海湾储[贮]木池;木材港池;水上储[贮]木场
timber bath (tub) 木浴盆
timber batten 木挂瓦条;木板条
timber bead 木压条
timber beam 木梁;木槛
timber beam bridge 木梁桥
timber beam floor 木梁地板
timber bearing structure 木承重结构
timber beetle 蛀木筒虫
timber bent 木排架
timber bent sill 排架座木
timber bin 木箱;木仓
timber binder 木黏合剂
timber binding beam 木联结梁
timber binding joist 木联结格栅
timber block 木块
timber blocking 枕木梁;木垫块
timber board 木板
timber boarding 木板条;木隔板;木地板
timber board sheathing 木衬板;木望板
timber bob for hauling 运输原木小吊车
timber bollard 木标桩;木缆桩;木缆柱
timber bond 砌体用木条连接
timber borer 木料蛀虫
timber box caisson 木沉箱
timber box girder 木制箱式梁
timber brick 木块;木砖
timber bridge 木桥
timber bridge pier 木桥桩;木桥墩
timber bridge superstructure 木桥上部结构
timber broach 锥形尖头木材工具;木扩孔器
timber building 木结构建筑;木建筑;木房(屋)
timber building block module 木建筑砌块模数
timber bulkhead 木板桩岸壁
timber bundle 木排;成捆木材
timber cage cofferdam 木笼围堰
timber caisson 木沉箱
timber cap 帽木
timber cargo 木材货
timber carriage 木材运输车;木材拖车
timber carrier 运木船;木材船
timber cart 运木车
timber centering for stone arch bridge 石拱桥木拱架
timber centring 木拱架
timber chestnut soil black calcareous soil zone 草原草地栗钙土黑钙土带
timber chock 码头护木
timber church 木屋教堂
timber chute 木斜槽;木滑道;木材滑道;筏道
timber ciborium 木佛堂;木神龛;木华盖
timber cistern 木水塔;木制储[贮]水器
timber cladding 木板贴面
timber cleat 木楔;木屋架上承托檩条的三角木
timber cofferdam 木围堰
timber column 木柱
timber column bent 座架木墩
timber column for frame bent 座架排架木柱
timber component 木构件
timber composite pier 混合结构突堤码头;浮栈桥
timber compound beam 木叠(合)梁
timber concession 采伐木材证
timber connection 木接合;木材接合
timber connector 结合环;接木环;木结构连结件;木结构连结件;木结构结合件;木材接合件;木料结合器
timber construction 木结构;木建筑;木构造
timber construction type 木构造形式
timber counter ceiling 木材吊平顶
timber crayon 木蜡笔
timber crib 木屋;木笼;木框;木垛支架;木垛;填石木笼;牛栏;叠木笼

timber crib breakwater 木笼式防波堤
timber crib cofferdam 木笼围堰
timber crib dam 木笼坝;木框坝;木垛坝;木笼填石坝
timber-crib type cofferdam 木笼式围堰
timber cribwork 木笼构筑物(填石)
timber culvert 木质涵洞;木涵洞;木材涵洞
timber curing wastewater 木材加工废水
timber cutter 伐木工
timber cutting chain saw 伐木链锯
timber cutting curb (沉箱的)木刃脚
timber cutting saw 伐木锯
timber dam 木坝;木框坝
timber dapper 木材锯割机
timber dealer 木材商(人)
timber deck 木面板
timber deck cargo 木材甲板货
timber deck weir 木板堰;木面板堰
timber delivery tax 木材发运税
timber depot 储木场
timber derrick 木井架;木质人字起重机;木钻塔
timber dimension 木材尺寸
timber dog 骑马钉;木扒钉;蚂蟥钉;扒钉
timber dolly (打桩机的)木垫桩
timber dome 木穹隆
timber door 木门
timber door frame 木门框
timber door threshold 木门槛
timber dowel 合缝钉;木销钉;木榫
timber drawer 坑木回收者;坑木回收机
timber dryer 木材烘干设备;木材干燥设备;木材烘干炉
timber drilling derrick 木钻井架
timber drying kiln 木材干燥窑
timber drying plant 木材烘干设备;木材干燥厂
timber eaves gutter 木檐沟
timbered area 木材铺盖面积;木盖面积
timbered back 木支护顶板
timbered bottom 木支护井底车场
timbered drift 木支护平巷
timbered-horizontal cut and fill stoping 木支护的水平分层充填采矿法
timbered raise 木支护天井
timbered rill 木支护的对角工作面
timbered shaft 木支护竖井
timbered soil 林地土壤
timbered stope 有木支撑的工作面;方框支架工作面
timbered stopping method 支柱采矿法
timbered trench 加撑基坑
timbered tunnel 木撑隧道
timber elevator 木料升降机
timber engineering 木结构工程;木材工程
timberer 木材商人;木材工人;支架工
timber export trade 木材出口贸易
timber faced 木材护面的
timber facing 木装饰面;木镶面板
timber falsework 木工作架;木鹰架;木脚手架
timber fastener 木扣件;木接合件;木紧固件
timber felling 筏木
timber fence 木栅栏
timber fence post 木栅栏桩
timber fender 木栅;木围墙;木护舷;木护栏;护木;防撞护木
timber fillet 木嵌条;木砧
timber finish 木面层
timber flat roof 木平屋顶
timber float 木排;木筏
timber floating 木排浮运;木材浮运
timber floating dock 木质浮船坞
timber floating floor cover(ing) 木夹层地板覆盖层
timber floating floor(ing) finish 木夹层地板终饰
timber floating sorting yard 木材浮运分类场
timber floor base 木踢脚
timber floor cover(ing) 地板覆盖层
timber floor(ing) 木地面;木地板;木底板
timber floor joist 木楼板龙骨
timber floor of bridge 木桥面
timber flume 木槽;木渡槽
timber folded plate roof 木折版屋顶;木折板屋盖
timber folded slab roof 木折板屋盖;木折板屋顶
timber folding door 木折叠门
timber foot block 木垫条
timber footing (纵横搭接逐层排列的)木基础

timber forest 用材林;乔木林;原始森林
timber form factor 用材形数
timber form(work) 木模(板);木模工程
timber for window-frames 窗框用木
timber frame 木骨架;木构架
timber frame bent 座架木排架
timber frame construction 木构架构造;木构架建筑;木构架结构
timber framed 木结构建筑;木屋
timber framed brick construction 木构架包砖结构
timber-framed building 木构架建筑(物)
timber framed facade 木框架正面
timber framed floor 木架楼板;木构楼板;木构架地板
timber framed house 木构架房屋
timber framed infilling panel 木框填充墙板
timber framed mirror 木框镜子
timber framed wall 木框架墙
timber frame structure 木构架结构
timber framework 木结构框架;木骨架
timber framework building 木骨架房屋
timber framework wall 木框架墙
timber framing 木骨架;木骨架构造;木构架
timber framing shop 木支架加工车间
timber freeboard 木材干舷(运木材时的干舷高度)
timber fungus 木菌
timber gate 木制大门
timber gatehouse 木哨所;木警卫室;木门房
timber girder 木大梁
timber girder bridge 板梁木桥
timber grab 木材抓具;木材抓斗;木材吊具
timber grade 木材等级
timber grain 木纹
timber grapple 木材抓钩
timber grid 木格栅
timber grid footing 木格栅底脚
timber grill 木栅格
timber grillage 木格排;木格床;木垛
timber grillage footing 木格架底脚
timber grille 木格子窗;木栅栏
timber groin 木丁坝
timber ground 木材场地
timber ground floor 底层木地板
timber groyne 木丁坝
timber guide crib 导水木笼
timber guide wall 木导墙
timber handrail 木栏杆;木扶手
timber harbo(u)r 木材港
timber hauling vehicle 木材拖运车
timber head 肋骨顶材延长部;木缆桩;木导架;拖缆木桩;系缆柱
timber header 木丁头
timberheads 双柱系缆桩;双柱系船桩
timber height 用材高;枝下高(木材)
timber hipped-plate roof 四坡木板屋顶;折边无梁木板屋顶
timber hitch 系木绳套;圆扣结;系木索环;捆扎圆木绳结(木材水运)
timber hitch and half hitch 拖材结
timber hoist tower 吊机木塔
timber hollow floor filler 木空心地板填料
timber horizontal bracing 横向木支撑
timber house 木结构住宅;木房(屋)
timber hut 木营房;木棚屋
timber hydrolysis plant 木材水解厂
timber industry 木材工业
timbering 用木支撑;支撑工程;支撑;临时支撑;结构材;木支柱式支撑;木支撑;安装木板
timbering and shoring 木支撑及顶撑
timbering for excavation 开挖木撑
timbering gallery 木框支撑式巷道
timbering machine 支柱机;支架机;支架安装机
timbering material 支撑材料
timbering of a cut 挖方支撑
timbering of foundation trench 基槽支撑
timbering of trenches 沟槽木支撑
timbering residue 木屑
timbering set 木支撑体(隧道)
timbering sheeting with timber 木支撑
timbering sideboard 模板的侧木
timbering support 枝撑
timbering to deep trenches 深槽木撑
timbering work 木支撑工程
timber Ionic column 爱奥尼亚式木柱

timberjack 伐木工(人)
timber jack chain 木材拖索
timber jetty 木栈桥码头;木栈桥;木结构突堤
timber joint 木支架构件接头;木连接
timber joint connector 木构件接合件;木材接合件;木结构节点连接件
timber jointing method 木材接合法
timber joist 木托梁;木格栅
timber joist floor 木格栅地板
timber joisting 木格栅
timber joist roof floor 木格栅屋顶楼板
timber joist upper floor 木格栅上层楼板
timber knee 木肘材
timber knot 木节
timber lagging 木衬护;挡土板;挡水板;插板
timber lamella cupola 薄木片小屋顶
timber lamella dome 薄木板穹隆
timber land 林区;材用林地;森林;林地
timber lath and plaster 板条抹灰
timber lattice beam 木花格梁
timber latticed beam 格栅式木梁
timber latticed cupola 格栅式圆屋顶
timber latticed girder 格栅式大梁
timber lead 木导架
timber licence 木材采伐证
timber limit 林木界限;树木限界(纬度限界)
timber line 林线;林木线;树线;森林界线;森林分布线;树木线
timber-lined 用木材支护的
timber lined passage 木护板通道
timber lining 木护板;木衬板;木支架;木衬(砌)
timber lintel 木过梁
timber load bearing structure 承重木结构
timber load-carrying structure 载重木结构
timber loaded structure 有荷载的木结构
timber loader 木材装载机
timber load line 木材载重线(装运木材的满载吃水线)
timber load water-line 木材载重水线
timber longitudinal beam 木纵梁
timber man 支撑工;架子工;木工;木材商(人);支架工
timber market 木材市场
timber marking 林木打号;木材打印
timber mast 木支柱
timber mat 木假顶
timber matting 木排
timber mattress 柴排
timber measure 木材丈量
timber measurement 木材体积
timber meeting face 木质贴接面
timber meeting piece 止水木条
timber member 木构件
timber member connected by hard wood key 木构件的键结合
timber member connected by split ring and bolt 木构件用开环和螺栓结合
timber member connected with bolts and nails 木构件用螺栓和钉结合
timber mill 重型锯木机;制材车间;制材厂;锯木车间;锯木厂;木工厂;木材厂
timber mining 支柱采矿法;木材滥采;森林滥伐
timber mopboard 木踢脚板
timber mould 木模
timber moulding box 木砂箱
timber needle 木针
timber of bridge 木桥面
timber pack 木垛
timber packing 木平底船(沉箱下水船只)
timber panel(ling) 木配电板;木镶板
timber partition (wall) 木分隔墙;木隔墙;木隔板
timber pass 木滑道;木筏道;筏道
timber patent glazing bar 专利玻璃窗木嵌条
timber paving work 木材铺地工作
timber peg 木桩;木栓;木钉
timber permeable revetment 木制透水护坡
timber pickling 木材浸渍(一种防腐处理)
timber pier 木突堤;木栈桥码头;木码头;木墩
timber pile 木桩
timber pile anchorage 木桩锚碇
timber pile bent 桩式木墩;木桩排架
timber pile bent abutment 木桩排架桥台
timber pile dike 木桩丁坝;木桩堤
timber piled jetty 木桩突码头

timber piled wharf 木桩码头
timber pile groin 木桩丁坝
timber pile jetty 木桩式突堤；木桩式码头
timber pile-staging 木桩台
timber pile top ring 木桩顶铁箍；木桩顶钢箍
timber piling 木板桩
timber pillar 木柱桩；木柱
timber plain splice 木材平面拼接
timber plain web(bed) beam 木普通腹板梁
timber plain web(bed) girder 木普通腐败大梁
timber planking 木面板
timber plate girder 木板梁
timber pond 储木港池；木材港池；水上储[贮]木场
timber port 运木港（埠）
timber positive 全木的
timber post 木桩；木柱；木杆
timber prefabricated construction 木材预制建筑
timber preservation 木材防腐
timber preservation plant 木料防腐处理车间
timber preservative (agent) 木材防腐剂
timber principal post 木主杆
timber prismatic shell roof 木棱柱形薄壳屋顶
timber processing 木材加工
timber processing enterprises 林业加工企业
timber processing plant 木材加工厂
timber product 木材制品
timber product processing 木材产品加工
timber prop 木柱（采矿）；木支柱
timber protection coat 木材防腐保护层
timber pulpit 木讲坛
timber purlin(e) 木屋顶桁条
timber puttyless glazing bar 无油灰木制玻璃窗心条
timber raft 木排；筏基
timber rafter 木椽
timber rail 木栏杆；木围栏
timber rake 掘根耙
timber ramp 木坡道
Timber Research and Development Association 木材研究发展协会
timber resources 木材资源
timber revolving door 木旋转门
timber right 伐木权
timber rigid frame 木刚性结构；木刚性框架
timber road 木制道路；木质道路
timber rock-fill crib 填石木笼
timber roller shutter 木制卷升百叶窗
timber rolling shutter 木制卷升百叶窗
timber roof 木屋顶
timber roof bar 木顶梁（采矿）
timber roof cladding 木屋面覆盖物
timber roof cover(ing) 木屋顶覆盖层
timber roof floor 木屋顶地板
timber roofing 木屋顶材料；木屋面
timber roof sheathing 木屋顶盖板
timber roof spire 木屋顶尖端
timber roof truss 木屋顶桁架
timber rot 朽病；木材腐朽；木材腐烂
timber rubbing 木摩擦条
timber rubbing strip 护舷木
timber runner 木承辊；木滑道
timber runway 木滑道
timber saddle 木支座
timber sanitary cove 木制便桶盖；木制卫生凹圆形板
timber saw 木锯
timber sawmill 制材厂
timber scaffolder 木脚手架工
timber scaffold(ing) 木脚手架
timber scaffold pole 脚手架木杆
timber scram 木支护扒矿平巷
timber screed for plastering 用于粉刷的木刮板
timber scrub board 木擦洗板
timber seal 止水木条；木质止水条；木质水封；木止水条
timber seat 木座位；木底座
timber selling 木材贩卖
timber separator 木隔板
timber set 木支架；木棚；木井框
timber shaft 木桩支撑竖井
timber shave 双柄剥皮刀
timber sheathing 木盖板；木护套
timber shed 木棚；木材储[贮]存棚
timber sheeting 木板桩

timber sheet pile 木板桩
timber sheet-pile bulkhead 木板桩护岸
timber sheet-pile groin 木板桩丁坝
timber sheet-pile groyne 木板桩丁坝
timber sheet piling 木板桩
timber sheet piling cofferdam supported with guard timber piles at the outside perimeter 外围支撑导桩的木板桩围堰
timber sheet piling cofferdam with internal frame bracing 用内木框架支撑的板桩围堰
timber shell 木壳
timber shell cupola 木壳圆屋顶
timber shell dome 木壳穹隆
timber shell roof 木壳屋顶
timber ship 运木船
timber shirting 木踢脚
timber shutter(ing) 木百叶窗；木模壳；木模（板）；木板模
timber sill 木槛
timber sill of frame bent 座架木排架底木
timber single-post shore 单杆木支撑；单杆木支撑
timber size 木材尺寸；木支撑尺寸
timber skirting (board) 木踢脚板
timber sleeper 枕木；木枕
timber-sleepered track 木枕线路
timber slide 木滑道；木材滑道；滑木槽
timber sliding panel 木滑板
timber soffit 木拱腹
timber solid web(bed) beam 木实心腹板梁
timber sorting pond 水上木材分类场
timber sorting yard 木材分类场
timber source 木材资源
timber space frame 木空间框架
timber space load-bearing structure 木间隔承重结构
timber spacer 定位木块
timber spatial weight-carrying structure 木空间承重结构
timber spire 木螺旋形体；木尖塔；木尖端装饰
timber splicing 木材拼接；木材拼接（法）
timber spud 木草铲
timber staging 木脚手架；木台
timber stain 木材变色
timber stair(case) 木楼梯
timber staircase builder 木阶梯式建筑（物）
timber stanchion 木撑杆；木支柱
timber standard 木材计量标准单位（一百六十五立方英尺）
timber stand improvement 用材林改进法
timber-steel-rivet method 木材钢铆钉法
timber steeple 木尖塔（教堂上的）
timber step 木台阶；木踏步
timber stick 木杆
timber stiff leg derrick 木肢起重杆
timber stop 闸木；叠梁闸板
timber storage basin 水上储木场
timber storage shed 木材仓库
timber storage yard 木材堆（置）场
timber store 木堆栈
timber strength 木材强度
timber stress 木材应力
timber string 木楼梯斜梁
timber stripe 原条
timber structural system 木结构体系
timber structural technique 木结构技术
timber structure 木结构
timber structure drawing 木结构图
timber strut 木支柱；木支撑；木撑
timber strut framed bridge 木撑架桥
timber stud (房屋的)主要木框架；木立筋；木板墙筋
timber-stud partition (wall) 木立筋隔断；木柱隔墙
timber-stud wall 木柱墙壁
timber sub-floor 毛木地板；木桥面底层；木底层地板
timber support 木支架；木撑
timber support guniting 木支架喷浆
timber supporting structure 木下部结构；木支承结构
timber surfacing 木料表面修整；木料铺面
timber surround 木围绕物；木包裹层
timber suspended floor 木架空楼板；木料悬桥面
timber sway bracing 交叉木支撑
timber swing door 木制双开式弹簧门
timber switch sleeper 道岔木枕
timber system construction 预制木结构

timber terminal 木材码头
timber thickness 木材厚度
timber three-dimensional load-bearing structure 木空间承重结构
timber threshold 木门槛
timber tie 系木；枕木
timber tilted-slab roof 折叠式木质斜屋顶
timber-to-metal connection 木与金属连接
timber-to-timber connection 木与木连接
timber tower 木塔（楼）
timber trade 木材贸易
timber trailer 木拖车；木材搬运用拖车
timber transom 木横档（门窗的）；木摇头窗；木气窗
timber tread 木踏板（楼梯等的）
timber treating plant 木材处理车间；木材处理厂
timber tree 成材木
timber trespass 木材盗伐
timber trestle 木柱架；木排架
timber trestle bridge 木栈桥；木排架桥
timber troll(e)y 坑木载运车
timber truck 运木材车
timber truss 木桁架
timber truss bridge 木桁架桥
timber trussed girder 木桁架式大梁
timber truss frame 木桁架构架
timber truss girder 木桁架梁
timber type of construction 木结构类型
timber underframe 木托架；木底架
timber unitized unit 木砌块模数；木质通用构件；木质成套设备
timber valley gutter 木天沟雨水管；木斜沟槽
timber vault 木拱顶
timber volume 木材体积
timber wagon 运材车；木质货车
timber wall 木板壁
timber warehouse 木材仓库
timber washboard 木踢脚板；木洗衣板
timber water tower 木水塔
timber weight-carrying structure 木承重结构
timber weir 木堰（口）；木板堰
timber window 木窗
timber window shutter 木窗挡板；木百叶窗
timber without pith 无髓心木材
timber wood 建筑用材；用材木（直径3英寸以上）；建筑用木材；成材
timber work 木工作业；木材建筑物；木材加工厂；木材加工；木材工作；木材工厂；木结构；木材制品
timber working machine 木材加工机
timber working machine tool 木工机床
timber worm 蒸馏釜的蛇形管；木蛀虫；木蛀虫；木料虫
timber yard 储[贮]木场；木材场；木材码垛；木材堆（置）场
timber yield 木材产量
timbo 用作家具和内部细工的红木板材
timbre 音质；音品；音色
time accounts statistics 时间核算统计
time accuracy 时间准确度；时间读数精度
time action 定时作用
time address 时间地址
time-address signal(l)ing 时间寻址信号方式
time-adjacent fields 时间相邻场
time-adjusted investment 投资现值
time-adjusted rate of return 时间调整收益率；按时间调整报酬率
time-adjusted-return method 收益折现法
time-adjusted revenue 收入限值
time adjusting device 时限装置；时限调正装置；时间调整装置
time adjustment 时间调整
time advance 时间提前
time alarm 限时报警器；时限警报；时限警告信号
time-align technique 时间调整技术
time allowance 工时定额；宽限时间；放宽时间；时间的宽限
time allowance for acceleration and deceleration 加减速时分
time allowed for loading 装货时间
time amplitude 出没时幅
time analyser 时间分析器
time analysis chart 时间分析图
time analysis diagram 时间分析图
time and again 一再；反复
time-and-altitude azimuth 时间与高度方位

time and control code 时间控制码
time and date 时间及日期
time and events record 时间和事件记录
time and events recorder 时间与事件记录仪
time and frequency standard 时间和频率标准
time and location code 时间位置码
time and material pricing 按工时及材料定价法
time and materials 工料成本加和润费(建筑合同)
time and motion 工时标定
time and motion study 工时与动作分析;时间动作研究;查定;操作测定
time and movement study 时间动作研究;查定
time and piece rate 时件工资
time and space 时间与空间
time and space distribution of rainfall 降雨时空分布
time and temperature development 定时定温显影
time-angle 时间角
time arbitrage 时间套利
time area concentration diagram 集流图
time area curve 时间—面积曲线
time-area-depth analysis 暴雨面深分析
time-area-depth curve 时间—面积—降水深度曲线(暴雨的)
time-area diagram 时间—面积图;流时—面积图
time arrive guarantee 按时到达保证
time arrow 时间向量
time-assigned signal(l)ing 时间分配信号方式
time at inflection point 拐点处的时间
time at shot point 爆破点时间
time availability 时间利用率;时间可用度
time availability of rig-month 钻机台月利用率;台月利用率
time average 时间平均值;时间平均;时间均值
time-average beam 时间平均束流
time-averaged current 时间平均电流
time-averaged hologram 时间平均全息照片
time averaged temperature 时间平均温度
time averaged velocity 均时速度
time-averaged yield 时间平均产额
time-average hologram 时间平均全息图
time-average holographic interferometry 时间平均法全息干涉测量术
time-average holography 时间平均全息术
time-average intensity 时间平均强度
time-average interferometer 时间平均干涉仪
time-average interferometry 时间平均干涉度学
time-average measurement 平均时间测定;平均时间测量
time-average output 时间平均输出
time-average pattern 时间平均条纹图样
time average photography 平均时间照相
time-average product array 时间平均乘积阵
time-average sample 时间平均样
time-average shadow-Moire method 时间平均阴影条纹法
time-average velocity 时间平均速度
time averaging 求时间均;按时间平均
time-averaging operation 按时间取平均数
time axis 极轴;时轴;时间轴(潮汐曲线)
time-axis shift 时轴偏移
time azimuth 时角求方位法;时间方位
time azimuth table 时间方位表
time balance 时间摆
time ball 报时信号球;报时球
time band method 时区带法
time bandwidth product 时间带宽积
time bargain 期货交易;定期贸易
time barred 失时效
time base 时间坐标;时间基准;时基
time base accuracy 时基准确度
time base charging capacitor 时基充电电容器
time base circuit 时基电路
time base compression 时基压缩
time base control 时基控制
time base correction 时基校正
time base corrector 时基调整器
time base counter chain 时基计数电路
time base diagram 时基图;时基曲线图
time based optic(al) range finder 时基光学测距仪
time base error 时基误差
time base error correction 时基误差校正器
time base expansion 时基扩展
time base filtering 时基滤波

time base flutter 时基抖动
time base frequency 时基频率
time base generator 时基振荡器;时基发生器
time baseline 时基线
time baseline parallax method 时基线视差法
time-base oscillator 定时信号振荡器
time base power amplifier 时基功率放大器
time base pressing type spectrum analyzer 时基压缩式频谱分析仪
time base reference information 时基基准信息
time base repetition 时基重复频率
time base resolving 时基分解
time base stabilizing 时基稳定
time base stroke 扫描行程
time base sweep 时基扫描
time base sweep multivibrator 时基扫描多谐振荡器
time base unit 时基装置
time base voltage 时基电压
time base waveform 时基波形
time bearing display 时间方位显示器;时间方位显示
time behavio(u)r 时效效应;时间特性;时变性能
time bell 船钟;报时钟
time belt 时区
time between arrivals 到达时间间隔
time between overhaul 翻修时限;翻修间隔时间
time between overhual 大修期
time bias 时偏;时间补偿
time bias integrator 时偏积分器
time bias parameter 时间偏向参数
time bias setting 时偏整定
time bill 远期票据;定期汇票
time bill discount 远期票据贴现
time blanking 时间消隐
time-book 工作时间记录表簿
time bounded Turing machine 时间界限图灵机
time break 时间间隔;时断信号(地);定时开关;定时爆炸;爆炸时刻;爆破信号
time budget 时间预算;定期预算
time by equal altitude of a star 单星等高法测时
time by equal altitude of two stars 双星等高测时法
time by ex-meridian altitude of star 近子午圈恒星高度法测时
time by star transit 恒星中天测时
time calculation 时间计算
time calibration 时间校准
time card 工时记录卡(片);记时卡(片);出勤时间记录卡;工时卡;钟片;工作时间;考勤卡;记录卡(片)
time case 时间角色
time cell 时间间隔
time center 时间中心
time certificate 定期存款票据
time change 时间变更
time change component 定期检修的部件
time changeover 时间转换
time-changing flux 时变通量
time channel 时路
time character 时间特性
time-charactaristic 时值的
time character of secondary magnetic field 二次磁场的时间特性
time charges 计时费用
time chart 时间图表;时间表
time charter 租船按时计费;定期租船合同;定期船;定期船租
time charterer 期租人
time charterer's interest 期租人的股息
time charter hire 定期船租金;定期船雇佣
time charter on trip 航租
time charter on trip basis 航次期租
time charter party 定期租船合同
time check 时间检验;定期支票
time checker 记时员;记录时员
time check lamp 报时灯
time circuit 时限电路;时间电路
time classification 时代分类
time clerk 计时员
time clock 工时记录钟;时钟;定时时钟;定时器;考勤钟
time clock actuated controller 时钟驱动(信号)控制机
time clock feature 时钟功能

time closing 延时闭合
time code 时间密码;时间码;时间电码;时间代码;时间编码
time code control 时间码控制;程序控制
time code detector 时间码检测器
time code disable 时间码阻塞
time code gate pulse 时间码选通脉冲
time code generator 时间码发生器;时间编码发生器
time code head 时间码磁头
time code lamp driver 时间码指示灯驱动器
time code muting 时间码静噪
time code playback amplifier 时间码重放放大器
time code playback gain 时间码重放增益
time coder 时间编码器
time code reader 时码读出器;时间编码读出器
time code record amplifier 时间码录制放大器
time code record level 时间码录制电平
time code signal 时码信号;时间码信号
time code timing 时间码定时
time code track detecting level 时间码磁迹检测电平
time code translator 时间编码转换器
time code word 时间码字
time coding 时间编码
time coherence 时间码相干性
time-coherent light 时间相干光
time comparator 时间比较仪
time comparison 时间对比
time complexity 时间复杂性
time compressed single sideband system 时间压缩式单边带系统
time-compression 时间压缩
time compression coding 时间压缩编码
time-compression curve 时间压缩曲线
time compression multiplex(ing) 时间压缩(多路)复用
time-compression single sideband system 时间压缩式单边带系统
time-concentration curve 时间浓度曲线
time condensation method 时间缩短法
time connection 时间连接
time-consolidation curve 时间固结(关系)曲线
time-consolidation curve of soil 土的时间固结曲线
time-consolidation relation for clay 黏土时间—固结关系
time constant 时间常数;时间常量
time-constant method 时间常数法
time constant of controlled plant 调节对象时间常数
time constant of damping 缓冲时间常数
time constant of water passage 压力水管时间常数
time constant term 时间常数项
time-consuming 费时的;浪费时间;耗时的;耗费时间
time consuming software routine 费时的软件程序
time-continuing process 持续过程
time contour 等时线图
time contour map 交通等时线;等时线图;交通等时线图
time contract 定期交易合同
time-contrast-index curve 时间反差指数曲线
time control 曝光定时;时间控制;定时控制
time-controlled 定时控制的
time-controlled system 时钟控制系统
time-controlled traffic signal 定时交通信号
time controller 时间控制器
time control pulse 定时脉冲
time conversion 时间换算
time converter 时间变换器
time coordinate 时间坐标
time coordination 时间协同;时间上的协同
time corrected value 时间修正值
time correction 时间校正;时间改正
time correction circuit 时间校正电路
time-correction factor 时间改正因子
time-correlated observation 时间相关观测
time-correlated photograph 有时间记录的相片;标有时间记录图像
time correlation 时间相关;时间关联;时代对比
time correlation method 时间相关法
time correlation system 时间相关系统
time cost 工时成本;时间费用;时间成本(按时间计算的费用成本)

time cost estimates 时间成本估计
time cost trade-off 时间成本权衡法;时间费用权衡
time-count cycle 计时周期;计时循环
time counter 计时器
time counting distribution 时间计数分布
time course 时序;时间过程
time-creep curve 时间蠕动曲线
time critical process 时间临界进程
time cross-section 时间剖面图
time-current characteristic 时间电流特性
time-current test 时间电流试验
time-current threshold 时间电流临界值
time curve 时距曲线;时间曲线;时间关系曲线
time cut-off relay 限时断路继电器
time cut-out 时限自动开关;定时断路器
time-cycle 时间周期
time cycle controller 定时循环控制器
time-cycle signal 周期信号
timed 时控的;同步的
timed acceleration 定时加速度
timed breaker 定时断路器
time deformation 徐变变形;随时间而发生的变形;依时间而变的变形
time deformation curve 时间形变曲线;时间应变曲线
time delay 延时;落后;时延;时间延误;时间延迟
time-delay aeration 延迟曝气
time delay analyser 延时分析器;时延分析器
time delay analysis 延时分析
time delay characteristic 延时特性
time delay circuit 时延电路;延时电路
time delay correction means 时间延迟校正方法
time delay device 延时装置;时间继电器
time delay distortion 时间延迟失真
time delayed action 延时动作
time delay element 延时元件;延迟元件
time delay factor 时延因子
time delay fuse 延时熔断丝;延时保险丝
time-delay mechanism 延时装置
time delay network 时延网络
time delay of supplying current 供电延迟时间
time delay relay 缓动继电器;延时继电器
time-delay servo 延迟伺服机构
time-delay squib 延迟点火管
time delay subroutine 延时子程序
time delay switch 延时开关
time delay unit 延时器
time-delay valve 延时阀
time demodulation 时间解调
time de-multiplexer 时间分路器
time dependence 随时间变化;时间相关;时变;与时间有关
time-dependent 随时间变化的;时控的;时间依赖;时变的;与时间有关的
time-dependent behavio(u)r 时效特性;时随行为
time-dependent control 时变控制;程序控制
time-dependent deflection 随时间变化的挠度;时间相关挠度
time-dependent deformation 时间相关变形;徐变变形;与时间有关的变形
time-dependent effect 时间效应;含时效应
time-dependent factor 时间变量因素
time-dependent flow 非定常流;非常定流
time-dependent fluid 时间依赖性流体
time-dependent lensingn effect 时间相关透镜效应
time-dependent mechanism 时间相关机制
time-dependent motion 依时运动
time-dependent ocean current 依时海流
time-dependent perturbation 时间相关扰动
time-dependent process 随时间变化过程
time-dependent propagation 时间相关传播
time-dependent property 与时间依存的性质
time-dependent rebound 时间相关回跳
time-dependent signal program(me) 随时间变化的信号程序
time-dependent spectral density 与时间有关密度谱函数
time-dependent strength 强度随时间增加
time-dependent stress accumulation 时控应力累积作用
time-dependent stress analysis 时效应力分析;非定常应力分析
time-dependent stress relaxation 时控应力松弛作用

time-dependent system 时变系统;不稳定系统
time-dependent wave function 含时波动函数
time-dependent wind stress 依时风应力
time deposit 定期存款
time-depth chart 时间—深度图
time-depth conversion 时深转换;时间—深度转换
time-depth curve 时间—深度曲线;垂直时距曲线
time-depth graph 时间—深度图
time-depth plot analysis 时间—深度图分析
time derivation 时间导数
time derivative 按时间的导数
time derived channel 时间分支信道;时分信道;时分通道
time determination 测时
time development 定时显影
time-development current 非常定流
time deviation 时间偏差
time device 计时装置;计时器;时控装置
timed explosion 定时爆破
timed flow control 预定气流调节
timed hot water supply 定时供应热水
time diagram 天赤道平面图;时系图;时间图
time difference 时差
time difference in high-water 高潮时差
time difference in low-water 低潮时差
time difference of arrival 脉冲到达时间差
time difference of flood and ebb 涨落潮历时差
time difference of tide 潮汐时差;潮时差
time differencing method 时间差分法
time differential 对时间的微分
time digital converter 时间数字变换器
time dilatation 时间流逝变慢效应
time dilation 时间扩展;变慢效应
time dilution of precision 钟差精度因子;时间精度因子
time dimension 时间因次;时间限度;时间维度
time director 操作耗时控制器
time-discharge curve 时间流量曲线
time-discharge rate 定时放电率
time discount 贴现
time discounting 时间贴现
time discriminating circuit 时间鉴别电路
time discrimination 时间鉴别
time discriminator 鉴时器;时间鉴频器
time-displacement chart 位移量曲线图
time displacement curve 时间位移曲线
time displacement of flame 火焰时间位移;火焰传播速度
time displacement operator 时移算符
time display controller 时间显示控制仪
time display control switch 时间显示控制开关
time display unit 时间显示单元
time distance 时距;时间距离
time-distance basis 时距原理;时间距离原则
time-distance curve 时间距离(关系)曲线;时距(关系)曲线
time-distance diagram 时间距离图
time-distance graph 时距图;时间距离关系图;航程航时图;时距曲线
time-distance relay 时间距离继电器
time-distance trajectory 时间距离迹线
time distribution 工时分配;时间;时间分布
time distribution of runoff 径流时间分配
time diversity 时间分集
time dividing channel(l)ing 时分多路通信
time division 划分时间;时间区分;时间划分;时间分隔;时间分割;时分;分时通道
time division circuit 时间分割电路;时分电路
time division command synchronization 时分指令同步方式
time division common control 时间分割综合控制
time division control access 时间分隔控制存取
time division data link 时分数据链路
time division duplex 时分双工
time division jamming 时间选择性干扰
time division multiple access 时分多址连接方式;时分多址存取;时分多路访问;时分多路存取
time division multiplex 时分多路复用;时分多路传输;分时多路方式
time division multiplex bus 时分多路总线
time division multiplex communication 时间分割多路通信;时分多路通信
time division multiplexer 时分多路转换器;时分多路调制器;分时多路转换装置

time division multiplexing 时分复用;时分多路转接;时分多路转换;时分多路复用技术
time division multiplexing technique 时分复用技术
time division multiplex telegraph system 时间分隔多路电报制
time division multiplex transmission system 时分多路传输系统
time division multiplier 时间划分乘法器;时间分隔乘法器;时间分割乘法器;分时乘法器
time division radio relay system 时间分割式无线电中继系统
time division scrambler 时分保密机
time division switching 时间分割开关;时分开关;时分交换
time division switching system 时间分割制交换系统
time division system 时间分割制;时间分割系统;时分制通信;时分系统
time division telegraph 时分制电报
time division telemetry system 时分制遥测系统
time domain 时域;时间域;时间演变;时畴
time domain analysis 时域分析
time domain analysis method 时域分析法
time domain and frequency-domain 时域与频域
time domain coding 时域编码
time domain conversion 时域转换
time domain display 时畴显示
time domain distortion 时域失真
time domain electro-magnetic method 时间域电磁法
time domain equalizer 时域均衡器
time domain filtering 时间域滤波法
time domain induced polarization method 时间域激发极化法
time domain method 时域法
time domain multiple access system 时域多重访问系统
time domain processing 时间域处理
time domain reflectometer 时域反射仪;时域反射计
time domain response 时域响应
time domain study 时域研究
time domain waveform 时域波形
timed pulse 计时脉冲
timed pump system 定时泵制
time draft 定期票据;习惯期限的汇票;远期汇票
time drawdown curve 时间—降深曲线
timed separation 定时分隔
timed shipment 限期装船
timed signal service 报时信号服务
timed siphon 定时虹吸
timed spark system 定式火花式;定时火花放电装置
timed stop 定时车站
timed syphon 定时虹吸
timed unit 时控装置
time duration 历时
time duration of flood 洪水持续时间
time duration of test 试验时间
timed valve system 定时阀制
time edge effect 时间边缘效应
time effect 时效;时间作用;时间效应
time-effect curve 时间—效应曲线
time effect of rock pressure 围岩压力的时间效应
time-effect relationship 时间—效应关系
time effect treatment 时效处理
time efficiency 时间效率
time elapsed 经过的时间;所经时期
time element 延时元件;时间因素
time-element control 定时限控制
time element in grade reduction 降坡的时间因素
time-element relay 时限元件继电器
time element starter 定时机构起动器
time equalizer 时间均衡器
time error 时致误差;时间误差
time error correction 时间误差校正
time-event analysis 时间事件分析
time expand 延长时间;时间延长
time expansion 时间扩展
time-expired 期限终止
time exposure 定时曝光;长时间曝光
time exposure technique 定时曝光操纵法
time extension 工期延长
time factor 时间因子;时间因数;时间系数;时间利用系数

time factor curve 时间因子曲线
time-fall 时降
time field 时间域
time field calibration 时间场校准
time figure 时间注记
time filtering 时间滤波
time fire 定时引爆
time flow control 定时流量控制
time flow mechanism 时间流结构
time flutter 时间颤动(扫描不稳定性);扫描颤动
time for acceleration 加速时间
time for back-up 倒扣时间
time for bid submission 提交投标时间
time for completion 竣工期限;预计完成时间;预定完成时间;竣工时间
time for completion of contract 总工期
time for completion of whole works 总工期
time for escape 排泄时间;逸漏时间
time for fresh water circulation 清水循环时间
time for milling 套铣时间
time for mixing and circulating oil 混油循环时间
time for mounting well head 装井口时间
time for payment 支付时间;付款期(限)
time for possession 进场占用时间
time for tests 检验时间
time for the layout of work place 布置工作地时间
time frame 时间帧;时间量级;时间范围
time freight 快运货物;快递货件;快递货;定时货运;定期租船;定期货运
time frequency 时间频率
time-frequency address matrix 时频地址矩阵
time-frequency collision avoidance system 时间频率避碰系统
time-frequency domain 时间频率域
time-frequency filtering 时间频率滤波
time front 时间曲线
time function 时间函数
time function generator 时间函数发生器
time fuse 定时引信;定时引爆;定时导火线;定时保险丝;时间引信
time gain 节省的时间
time gain control 时间增益控制
time gate 时间选择器;定时启门电路;定时门;定时开启电路;定期开启闸电路;时间门
time gating 时闸;按时选通
time gear 定时齿轮
time gears 定时齿轮系
time gradient 时间梯度
time gradient coefficient 时间梯度系数
time graph 时距图;时距曲线;时间曲线图
timegrapher 校表仪
time handed in 交付时间
time harding law 时间硬化理论
time hardness 时间硬度
time harmonic (电压和电流波形的)时间谐波
time-haul 途中运输
time headway 前后两车之间的时距;车头时距
time-height section 时间高度部分;时高剖面
time hierarchy 时间谱系
time history 随时间变化;时间历程;时间经历;时间关系曲线图;时间关系曲线
time-history analysis 时间经历分析;时间关系分析
time-history analysis method 时程分析法
time-history file 时间历程文件
time-history form 时间关系式
time-history plot 随时间变化的曲线图
time-history recorder 时程记录仪
time-history response 历时反应;时间经历反应;时程反应
time homogeneous 时齐的
time-homogeneous process 时间均匀过程
time-hopping 时跳
time horizon 时间水平
time hour materials 工时和材料
time hull insurance 船舶定期保险
time image 时像
time-impulse 力—时积
time-impulse distribution 时钟脉冲分配
time in advance 提前期
time increment 时间增量
time increment index 时间增量指数
time-independent 与时间无关的
time independent motion 不依时间变化的运动
time-independent perturbation 不含时微扰

time-independent toxicity test 非时间制约毒性实验
time-indicating device 时间指示器
time indicating mechanism 报时机械装置
time indication 时间指示
time indices 记时
time-in port 停港时间;在港时间
time insurance 定期保险
time integral 时间积分
time integral of force 力的时间积分
time integral scale 时间积分比尺
time-integrated concentration 积时浓度
time-integrated spectra 时间积分光谱
time-integrating sampler 积时式取样器;积时式采样器;时间累积采样器;时间累积采样器
time-integrating sampling 积时式采样;积时法取样;积时法采样
time-integrating suspended sampler 时间累积悬移质采样器
time-integration method 时间累积法;时间积累法
time-integration sampling 积时式取样;积时式采样;积时法采样;积时法采样
time integration stage 时间积分级
time-intensity curve 时间—强度(关系)曲线
time-intensity graph 时间—强度(关系)曲线图
time interest earned ratio 已获得的计时利息比率
time-interleaving 时间重叠
time interval 时距;时距;时间间隙;时间间隔;间歇时间
time interval bell 时间间隔铃
time interval between following movement 列车追踪运转时隔
time interval between following movements in a station 站运行间隔时间
time interval between passenger trains' dispatching 旅客列车开行间隔
time interval between trains spaced by automatic block signals 追踪列车间隔时间
time interval between two adjacement trains at station 车站间隔时间
time interval between two opposing trains arriving at station not at the same time 不同时到
time interval between two stations 车站间隔时间
time interval between two trains at a station 车站内列车进站间隔时间
time interval comparator 时间比较仪
time-interval counter 时间间隔计数器
time interval error 时间间隔错误
time interval for two meeting trains at station 会车间隔时间
time interval for two trains dispatching in succession in the same direction 同方向列车连
time interval indicator 时间间隔测量器;时间间隔指示器
time interval maps 时间间隔地图
time-interval measurement 时间间隔测量
time interval measuring instrument 时间间隔测量仪
time interval method 隔时法;时间间隔平均法;时间间隔法
time interval of exposure 曝光间隔
time interval of observation 观测时间段
time interval of sounding 测深点时距
time interval principle 时间间隔原理
time-interval radiosonde 时距探空仪
time interval recorder 时间间隔记录器
time interval system 时间间隔制;时间间隔系统
time invariance 时间不变性;时不变
time-invariant 时不变;时间恒定的;不随时间变化的
time-invariant boundary 不随时间变化的边界
time-invariant channel 非时变信道
time-invariant filtering 时不变滤波
time-invariant model 时不变模型
time invariant operation 时间不变性运算
time-invariant regulator 时间恒定调整器
time-invariant seepage 稳定渗流
time-invariant system 时间不变系统;时不变系统;时间恒定系统;常系数控制系统
time jitter 时间跳动;时间起伏;时标抖动
timekeeper 工时记录员;记时员;记时仪;记录时员;计时器;计时机构;守时仪器;钟表;精确钟表机构;精确计时装置;计时员;记录员;时计
time keeping 工时记录;测时;保时;记时记录;工时测定;守时

timekeeping classification 时间记录分类
timekeeping instrument 计时仪器;测时仪器
timekeeping performance 守时性能
timekeeping system 时间记录系统
time lag 落后;时滞;时延;时间滞后;时间滞差;时间迟滞;时迟;时间延迟;延时;时间落后
time-lag action 时延作用;延时作用;时滞作用
time-lag device 时滞器件
time-lag disturbance 时滞引起的扰动
time-lag fuse 延时熔丝;惯性保险丝
time-lagged type 时间滞后式
time-lag network 时滞电路
time-lag of a catchment 集水时间滞后
time-lag over-current protection 时限过电流保护
time-lag overload relay 时限过载继电器
time-lag relay 时滞继电器;延时继电器;时限继电器
time-lag system 时滞系统;时迟系统
time-lag trip 延迟断开;延迟释放
time lapse 延时;慢速拍摄;时延;时间压缩;时间推移;时间流逝
time-lapse camera 时距照相机
time-lapse indicator 时间推移指示器
time-lapse interferometry 时间推移干涉量度术
time-lapse microcinematography 定时显微电影技术
time-lapse photography 延时摄影;慢速摄影;慢过程电影摄影法
time lead 时间超前
timeless functional architecture 长期有效的实用建筑
timelessness 永恒性;无时间性
timeless observation 与时间无关的观测
time-light curve 时间光通量曲线
time-like 类时
time-like direction 类时间方向
time-like interval 类时间隔
time-like path 类时路径
time-like vector 类时矢量
time limit 限定时间;完成日期;时间期限;极限时限;期限;时间限制;时间极限
time limit attachment 定时装置;时滞装置;时限附件
time limit for arbitration 仲裁的时间限制
time limit for blasting 爆破时限
time limit for filing claims 索赔期限;索赔期;索赔时数
time limit for payment of claim 索赔期限
time limit for submission of loan 提出索赔的时效
time limit method 时限法
time limit of offer 要约的期限;出售的期限
time limit of repair 修理期限
time limit of the lease 租借期限
time-limit parking 限制时间的停车
time (limit) relay 延时继电器;限时继电器
time-limit system 时限方式
time line 时线;时间界限;等时线
timeliness 及时性;合时
time-load curve 时间荷载曲线
time loading test 蠕变试验
time loan 定期贷款
time lock 自动定时锁;时间同步;时间锁定;定时锁
time log 工时记录
time loss 时间损失;时间损耗
time lost in waiting for 等待损失时间
timely and sufficient funding 及时,足额发放
timely completion 如期完工;按期竣工
timely rate of report filing 报告提交及时率
time magnification 时间放大倍数
time magnifying 时间标度的放大
time management 时间管理
time mapping 时间映射
time-mapping asynchronous simulation 时间映射异步模拟
time mark 时间标记;时标;记时标;时间记号
time marker 计时器;时间指示器
time mark generator 时标振荡器;时标发生器
time-marking device 时间标记装置
time mean speed 时间平均速度;定期贷款;时间平均车速;时间平均车速
time-mean velocity 时间平均速度;时均流速
time measurement 时间测量;测时
time measurement unit 时间衡量单位
time-measuring counter 测时计数器
time-measuring machine 计时机

time-measuring mechanism 测时机械装置
time measuring resolution 测时分辨力
time memory 时间记忆
time meridian 计时子午圈;时区子午线
time microscope 时间显微镜
time mode over-ride 超时过闸时限
time-modulated beam 时间调制束
time modulation 时间调制
time modulation technique 时间调制技术
time money 定期贷款
time movement 计时机构
time moving 时间移动
time multiplex 时间分割多路传输;时分制多路传输
time multiplexed system 时分多路复用系统
time multiplexing 时分复用
time multiplex techniques 时分多工技术
time-multiplex transmission 时分制多路传输
timenoguy 小吊绳;防纠索
time non-variant system 时间定常系统
time norm 时间定额
time normalization 时间归一化(法)
time number of magmatism 岩浆活动次数
time number of metamorphism 变质作用次数
time number of tectonization 构造变型次数
time observatory 时间观象台
time occupancy 时间占有率
time of action 动作时间
time of advent 到达时间;波至时间
time of aeration 曝气时间
time of application 施用时间
time of arrival 运到时间;抵达时间;到达时间;到车时刻;倒时;船舶抵港时间
time-of-arrival location 时差定位法
time of breakthrough 穿透时间
time of cave collapse 洞穴塌陷时间
time of closest point of approach 到达最接近点的时间
time of closure 关闭时间
time of coincidence 重合时间
time of coming out 刊出时间;出现时间
time of compilation 编制时间
time of completion 完工日期;工期;竣工时间;完成期限
time of concentration 集中时间;集流时间;汇流时间
time of concentration in conduits 管道中集流时间
time of concentration in pipe [tube] 管道中集流时间
time of connection delay 等待时间
time of consolidation 固结时间
time of curing 养护时间(混凝土)
time of day 最新流行的式样;由钟表示的时间
time-of-day clock 日时钟;日历钟
time of day pretime signal 多段式定时信号
time of delay 延迟时间;时间滞后;时间滞差
time of delivery 装船时间;交货时间;交货日期;交货期(限);交货季度;交付时间;送到时间;递交时间
time of departure 离开时间;开航时间;开船时刻;开车时刻;出港时间;出发时间
time of detention in transit 中转停留时间
time of developing 显像时间
time of discharge allowed 容许卸货时间
time of discharge used 实用卸货时间
time of drawdown 泄降时间;降落时间
time of duration 持续时间
time of duration of rainfall 降雨持续时间
time of earthquake at epicenter 震中震时
time of ebb 落潮时间
time of effect 有效期间
time of efficiency 有效期限
time of emptying 泄空时间;放空时间
time of entering water 入水时间
time of entry 进水时间
time of exposure 露光时间;曝光时刻;曝光时间
time of fall 落下时间;退水时间;水位降落时间;沉落时间
time of filing 归档时间
time of filling 充泥时间
time of final setting 终凝时间(水泥)
time of finishing works 完工时间
time of firing 燃烧时间
time of first induced earthquake 第一次发震时间
time-of-flight mass spectrometer 飞机时间质谱仪

time of flight of impulse 脉冲通过时间
time-of-flight spectrometer 飞机时间光谱计
time of floc formation 混凝体形成时间
time of flood peak 洪峰时刻
time of flow 管内流行时间;流动时间;洪水波传播时间;水流时间
time of flow observation 流量观测时间
time of fold formation 褶皱形成时间
time of (free) fall 自由降落时间
time offset method 单纯平移滞后演算法
time of ground concentration 地面集流时间
time of handing-in 交发时间
time of haul 搅(拌)运(输)时间(商品混凝土的);运送时间;运输时间;运行时限(商品混凝土搅拌车)
time of heating 加温持续时间
time of high water 高潮时
time of hoisting 提升工作时间
time of ignition 发火时间
time of incidence 倒时
time of infiltration by water pouring 灌水入渗时间
time of initial setting 初凝时间
time of integration 积分时间
time of intermittence 中断时间
time of investigation 研究时间
time of irradiation 照射时间
time of irrigation 灌水时间
time of lag 延滞时间;延迟时间;滞时
time of larger earthquake 较大地震时间
time of loading 装载时间
time of lockage 过闸时间
time of low entropy 低熵时代
time of low water 枯水期;低潮时
time of maximum water level 最高水位出现时间
time of metamorphism 变质时代
time of migration 运移时间
time of minimum water level 最低水位出现时间
time of mixing 搅拌时间;拌和时间
time of nautical twilight 航海晨昏蒙影时间
time of observation 观测时间
time of observation bias 观测偏差时间
time of occurrence 潜伏期;出现时刻;出现时间;发生时刻
time of occurrence at the epicenter 震中震时
time of one complete oscillation 全波周期
time of opening 开启时间;开发时间
time of operation 作用时间
time of origin 原发时间;起始时间;始点时间
time of oscillating 振荡时间
time of oscillation 振荡时间;摆动时间
time of passage 通过时间
time of passage through kiln 窑内停留时间;通过窑炉的时间
time of passage through kiln of material 物料在窑内停留时间;物料通过窑所需时间
time of payment 付款时间(限)
time of peak 峰现时刻;峰时
time of peak arrival 峰值到达的时间
time of penetration 渗透时间
time of perihelion passage 近日点时间
time of per loading 每次装载时间;每次荷载时间
time of permanency 持续时间
time of persistence 余辉时间
time of preparation and wind-up 准备与结束时间
time of primary migration 初次运移时间
time of primary migration peak 初次运移高峰时间
time of priming 冲水时间(水泵);起动时间
time of principal earthquake 主震时间
time of propagation 传播时间
time of pumping 抽水时间
time of receipt 收到时间
time of reception 接收时刻
time of recession 消落时间;退水时间;过时错误
time of recording 收录时间
time of reflection 反射时间
time of regression 海退时间
time of relaxation 松弛时间;弛张时间
time of releasing 分离瞬间
time of response 响应时间
time of rest 休息时间
time of rise 涨洪时刻;涨潮时间;增长时间;建立时间;起潮时刻;上升时间
time of run 运转时间
time of running in 下钻所需时间

time of sailing 开航时间
time of satellite perigee passing 卫星(通)过近地点时刻
time of seawater intrusion 海水入侵时间
time of secondary migration 二次运移时间
time of sedimentation 沉淀时间
time of set(ting) 凝结时间;凝固时间
time of setting in 涨潮时间
time of setting up a call 接通时间
time of shipment 交运时间;交货时间
time of ship's arrival 船舶抵港时间
time of shock at the origin 震源震时
time of shuttle movement 货叉伸出时间
time of side dressing 旁施时间
time of simultaneous observation 水位统测时间
time of slaking 消化时间;熟化时间;湿化时间
time of stripping 拆模时间
time of tectonic system formation 构造体系成生时期
time of test(ing) 试验时间;试验次数
time of the ground stress measuring 地应力测量时间
time of tide 潮时
time of tracer arrival 示踪剂到达时间
time of tracer drop 示踪剂投放时间
time of train arrival 到达时刻
time of train departure 开车时刻【铁】
time of transgression 海侵时间
time of transit 过渡时间;传播时间;运送时间
time of transmission 发播时刻
time of transport 运输时间
time of travel 行进时间;传播时间
time of travel curve 传播时间曲线(汇流时间对水位的关系曲线)
time of traveling 实际行驶时间
time of turnaround of freight car 货车周转时间
time of turnaround of locomotive 机车周转时间
time of turnaround of wagon 货车周转时间
time of unloading 卸荷时间;卸料时间
time of usage 使用时间
time of validity of a claim 索赔有效期
time of vibration 振动时间
time of warming 加温时间
time of water bursting 突水时间
time of water-level observation 水位观测时间
time of water sampling 水样采取时间
time of water source operation 水源地投产时间
time of wave arrival 波浪到达时间
time of well cleaning 洗井用的时间
time of working out 制定时间
time on daywork 小时工资率
time on target 追随目标时间
time on trips 起下钻时间
time opening 开始时间;定时断开
time optimal 时间最优的
time optimal control 时间最优控制
time optimal problem 时间最优问题
time optimal steering 最短时间操舵
time optimum control 时间最优控制
time optimum control system 快速最优控制系统;时间最优控制系统
time order 定期订单
time-ordered contraction 时序收缩
time-ordering product 时序乘积
time or distance selector switch 时间距离选择器开关
time-oriented sequential control 面向时间的顺序控制;时序控制
time origin 基准时间;时间基点
time-out 暂停;窝工时间;停工时间;时间已过;超时
time-out check 超时检验
time-out control 超时控制
time-out error 超时错;过时错误
time-output command 时间输出指令
time-output cycle 时间输出循环
time over 超时
time over-check 超时检验
time overcurrent 时限过电流
time-overcurrent protection 时限过电流保护
time overcurrent relay 时限过电流继电器
time over target 越过目标的时间
time parallax at a single station 单站时间视差法【测】
time parameter space 时间参数空间

time path 时间路径
time-path curve 时距曲线;时间距离(关系)曲线
time path with fluctuation 波动的时间路径
time pattern 时间模式;按时图案
time pattern control 时间程序控制
time pattern of cash flows 现金流量时间分布模式
time payment 定期支付;定期付款;分期付款
time payment debts 定期偿还的债务
time-payment method 计时工资方法
time-payment schemes selling 分期付款出售
time penalty 时间损失
time penalty clause 延期损失不负责条款
time period 时期;时间周期;时间极限;时限
time period for recovery of investment 投资回收期
time per loading 一次装载时间;装载一次的时间
time per piece 每件的时间
time phase 时相;时间相位;时间推移
time phase angle 时间相位角;时间相角
time-phase dispersion 时间相位散布
time phasing 分期实施
time phasor 时间相量
time piece 钟表;时计
timepiece lubricant 钟表润滑剂
time pin 定时针
time-place clustering 时间一地点集聚性
time plot 时距曲线向量图;时间曲线图
time polarity 时间极性
time polarity control 时间极性控制
time polarity control code 时间极性控制码
time policy 限期保险单;定期保险单
time port 定时孔(测斜仪)
time postpone 时间延迟
time prearranged assignment 时间预分配
time preassign 时间预分配
time-preassigned multiple address 时间预分配多址
time preference 时间优惠;时间偏好
time-preference rate 时差率
time-pressure curve 时间压力曲线
time-price differential 时间价格差
time process 时间过程
time profile 时间剖面
time-program(me) 时间程序
time program(me) command 时间程序指令
time program(me) control 时间程序控制
time-proof 经久的;耐用的
time proportional control 时间线性控制
time proportioning controller 时间比例控制器
time pulse 时间脉冲
time-pulse distributor 时钟脉冲分配器;时间脉冲分配器
time pulse generator 时钟脉冲产生器;时标脉冲发生器
time pulse relay 定时脉冲继电器
time purchase 分期付款购买
time quantization 时间量子化;时间量化
time quantum 时间量子;时间段;时间单元
time quantum method 时间量子法
time quenching 控制时间淬火;马氏体等温淬火;定时淬火
time queue 时间队列
time qunatized control 量化时间控制
time quota standard 时间定额标准
timer 限时器;计时员;计时器;跑表;时间传感器;定时仪;定时器
timer advance 定时器(提前发火)
time range 时间间隔;时间范围
time-range extending method 时距扩大法
time rate 计时工资率;计时工资标准;计时工资;时间率;时间变率
time rate control 时间比控制
time rate of change 时间变率
time rate of heat capacity 单位时间的热容量
time-rate system 计时收费制;计时工资制;计日工资制
time rating 连续运行功率;额定时间
time ratio 时间比
time-ratio-controlled chopper 时间比可控的斩波器
timer box 计时器箱
timer circuit 计时电路
timer clock 定时器时钟
timer command 定时命令
timer control 计时器控制;时钟控制

timer control bracket 定时控制托架
timer control routine 计时控制程序
timer control word 定时控制字
timer counter circuit 定时计算电路
timer cut-off 定时器控制停车
timer distributor 定时配电盘
time read 时间读数
time receiver 时号接收机
time receiving 收时
time reckoning 计时法
time recorder 时间记录器;定时记录器
time record(ing) 时间刻度;时间记录;时程记录
time record(ing) device 记录装置
time recovery error 时间恢复误差
time redundancy 时间冗余
time reference 时间基准;时间参数;时标
time reference coordinate 时间参考坐标
time reference line 时间基线
time reference pulse 时间基准脉冲
time reference scan(ning) beam 时基扫描波束
time reference system 时间基准系统
time reflection 时间反演
time reflex 时间反射
time register 计时器;时间寄存器
time regulator 时间记录器;时间调节器
time relationship 时间关系
time relay 限时继电器;时间继电器;定时继电器
time release 定时解锁器;限时释放器;延时释放;定时开关
time remaining 时间持续
timer equipment 定时装置
time required for train departing from station 列车出站时间
time require for train entering into station 列车进站时间
time resolution 时间分辨率;时间分辨
time resolution of uncertainty 不确定性时间分辨率;不确定性的时间分辨
time-resolved 时间分辨的
time-resolved emission spectrum 时间分辨发射光谱
time-resolved fluorescence 时间分辨荧光
time-resolved interference spectroscopy 时间分辨干涉光谱学
time-resolved microphotography 时间分辨显微照相术
time-resolved spectrograph 时间分辨摄谱仪
time-resolved spectroscopy 时间分辨光谱(学)
time-resolved spectrum 时间分辨谱;分时光谱;展时光谱图
time resolved X-ray diffraction 时间分辨 X 射线衍射
time-resolving image converter 时间分解变像管;时间分辨变像管
time resolving spectrometer 时间分辨分光光度计
time response 时间反响;响应时间;加速响应;时间响应;时间反应
time response spectrum 时间反应谱
time reversal 时间转换;时间反演
time-reversal invariance 时间反转不变性
time reversal test 时间互换测验
time reverse test 时间互换测验
time reversibility 时间可逆性;时间互换性
time reversible 装卸日数可调剂使用
timer expiration exit routine 计时终止出口程序
timer initiation facility 计时器启动设备;计时器初始化设施
timer interrupt 时钟中断
timer interrupt enable 允许定时中断
timer interruption 计时中断;定时中断
time risk 时间风险;时间的风险
timer lagging machine 插板机
timer lever stud 定时杆柱螺栓
timer multitransmission 定时多路发送
time rock correlation 时岩对比
time-rock unit 时岩单位;时间地层单位
timer paper 校仪表用纸
timer programmer 程序计时器
timer recorder 计时器
timer recording pulse 定时器记录脉冲
timer set 定时装置
timer shaft 定时器轴
timer-shaft cam 分电轴凸轮;分路轴凸轮
timer-shaft gear 分路轴齿轮;分电轴齿轮

timer supervision 计时器管理程序
timer timing pulse 计时器计数脉冲;定时器定时脉冲
time running of gyroscope 陀螺漂移
timer word 计时字;计时器用字
times 时代;次数;倍数
time sale 分期付款销售
time sample 时间抽样
time sampling 时间量化
times and becomes symbol 乘值符号
time saved 节省时间
time saving 节约时间;节省(的)时间;省时
time-saving analysis 省时分析
time-saving apparatus 省时装置
time-saving commodities 省时商品
time-saving device 省时装置
time-saving unit 省时装置
times between overhauls 大修间隔(时间)
time scale 时间量程;时间刻度;时间尺度;时间标度;时间比例;时标;调时制;时序表
time-scale calibration method 时间标法
time scale check 时标校验
time scale factor 时标换算系数;时标因子
time scale of scour and silting 冲淤时间比
time scale ratio 时间比尺(模型)
time-scale riverbed deformation 河床变形时间比尺
time scan(ning) 时间扫描
time schedule 工作时间进度表;工作进度(表);时间进度表;时间表
time schedule and liquidated damages 时间表和违约赔偿
time schedule (chart) 行车时间表;工作进度表;施工进度表;制造进度表
time-schedule control 日程控制;时间表控制
time schedule controller 时序控制器;时间表控制器;程序调节器
time schedule controlling 程序控制
time section 时间剖面;时间断面
time sections correlation 时间剖面对比
time selection 时间选择
time selection circuit 时间选择电路
time selective action 选择性延时动作;时间选择性作用
time selective fading 时间选择性衰落;时间选择衰落
time selector 时间选择器
times enlargement 放大倍数
time sense 时间分辨力
time-sensing element 时间传感元件
time sensor 时间传感器
time separation 时间间隔
time sequence 时间序列;时间顺序;时序
time-sequenced 排定时序的
time sequencing 时间顺序
time sequential 时序的
time sequential analysis 时序分析
time sequential routine 操作时序
time series 时序列;时系列;时间序列;时间系列;时间数列
time series analysis 时序分析;时间序列分析
time series data 时间序列数据
time series distribution 时间序列分布
time series forecasting 时间数列预测法
time series method 时间序列法
time series model 时间数列模型
time series of ocean measurements 海洋测量时间序列
time series of water quality determination 水质测定试剂序列
time series prediction 时间序列预测
time series trend 时间数列趋势
time series vector 时间序列向量
time served worker 满时工人
time service 授时;时间工作;时间服务;报时业务
time set 时间装定
time setting 时间装定;时间置位
time setting circuit 时间置位电路
time-settlement curve 时间一沉降曲线
time-settlement graph 时间一沉降曲线;建筑下沉曲线
time share 时间共享;时间区分;分时
time-share computer 时间划分计算机
time-shared 时间分配的;时分的;分时

time-shared amplifier 时间分配放大器;分时放大器
time-shared basis 分时制
time-shared bus 分时总线
time-shared computer 分时计算机
time-shared control 分时控制
time-shared file system 分时外存储器系统
time-shared monitor system 分时监督系统
time-shared multiplexing system 时间分割多路传输系统
time-shared operator 分时算符
time-shared programming 分时程序设计
time-share driver 分时驱动器
time-shared service 分时服务
time-shared system 分时系统
time-share estate 按时分享权(分享部分地产)
time-share monitor system 分时监督系统
time-share network 分时网络
time-share scanner 分时扫描器
time-share simulator 分时模拟器
time share tube 时间分割管
time-share use 按时分享用权
time sharing 分时制;时间分配;分用时间;分时;时间分享;时间分割;分时的
time-sharing accounting 分时统计;分时记账;分时会计
time-sharing allocation 分时分配
time-sharing allocator 分时分配程序
time-sharing application 分时应用
time-sharing arrangement 定时分享安排
time-sharing assembler 分时汇编程序
time-sharing base 分时基址
time-sharing circulation 分时environ
time-sharing computer 分时计算机
time-sharing console 分时操作控制台
time-sharing control system 分时控制系统;分时控制方式
time-sharing control task 分时控制任务
time-sharing design system 分时设计系统
time-sharing driver 分时驱动器;分时驱动程序
time-sharing driver CPU 分时中央处理机
time-sharing dynamic(al) allocator 分时动态分配程序
time-sharing environment 分时环境
time-sharing executive 分时执行(程序)
time-sharing executive system 分时执行程序系统
time-sharing firm 分时公司
time-sharing format 分时方式
time-sharing incremental compiler 分时可增编译程序
time-sharing industry 分时工业
time-sharing information system 分时信息系统
time-sharing interface terminal 分时接口终端
time-sharing interrupt 分时中断
time-sharing job control block 分时作业控制块
time-sharing laser 时间分割激光器;分时激光器
time-sharing meter reading 分时读数
time-sharing method 时间分配法;时间划分法
time-sharing mode 分时型
time-sharing monitor 分时监督程序
time-sharing monitor program(me) 分时监督程序
time-sharing monitor system 分时监督系统
time-sharing multiplexer channel 分时多路转接器通道
time-sharing network 分时网络
time-sharing operating system 分时操作系统
time-sharing operation 分时操作
time-sharing option 分时选择;分时方案
time-sharing polling 分时轮询
time-sharing priority 分时优先级
time-sharing process(ing) 分时处理;分时加工
time-sharing quantum 分时量
time-sharing radiometer 分时辐射计
time-sharing ready 分时就绪
time-sharing scanner 分时扫描器
time-sharing scheme 时间划分制;时间划分方案
time-sharing service 分时服务
time-sharing simulator 分时模拟器
time-sharing software 分时软件
time-sharing system 时间分配系统;分时制;分时系统
time-sharing system computer network 分时系统计算机网
time-sharing system network 分时系统计算机网络

time-sharing terminal 分时终端
time-sharing two frequency laser 分时双频激光器
time-sharing work 分时作业
time sheet 工时测定表;装卸时间表;工作时间记录表;工时卡;工时记录表;计工单
time shift 倒换时间
time shifting 时间位移
time-shift keying 时移键控
time-shrinkage curve 时间—收缩曲线
time shutter 曝光快门
time signal 按时信号;时间信号;时号;报时信号
time signal clock 报时钟
time signal generator 时间信号发生器
time signal in broadcasting 广播报时
time signal in international system 国际式时号
time-signal set 计时信号装置
time-signal station 报时站
time simulation 时间模拟
timesing 次数记录
times-interest earned 盈利对利息的倍数
time siphon 定时虹吸
time slice 时间片;时间分段;时间段
time slice end 时间片端
time slice interval 时间片间隔
time-slice map 等时切片剖面
time slice of three-dimensional migration 三维偏移的时间切片
time slice time-sharing 时间片分时
time slicing 时间限制;时间分片;时间分段
time slot 时隙;时间间隙;时间分隔法
time slot interchange 时隙互换
time slot number 时隙号码
time-slot pattern 时间槽模式
times of average amplitude 平均振幅倍数
times of back out 倒扣次数
times of bottom hole accident 井下事故次数
times of eruption 喷发次数
times of explosion 爆炸松扣次数
times of lacking in weight 欠重次数
times of low water 枯水期
times of mechanical accident 机械事故次数
times of milling 套铣次数
times of pipes moving 活动钻具次数
times of pumping slurry 灌浆次数
times of transportation 运输次数
times of turnover 周转次数
time solution 时域解
time sorter 时间选择器
time sorting 时间分选
time-space diagram 时空图;时距图;绿波运行图;时间距离图
time-space distribution of ozone 臭氧的时空分布
time space graph 时距图
Time-Space in Architecture 建筑时空论
time-space interval 时空间隔
time-space problem 时间空间问题
time-space relationship 时空关系
time-space trade-off 时空折中
time-space trajectory 时间距离迹线
time spacing of vehicles 时间车间距
time span 时间间隔;时间长度
time span of prediction 预测的时间间隔
time speed distribution 时间车速分布
time spent in preparatory and ancillary operations 准备和辅助工作时间
time split system 时间分隔制
time spread 时间差价买卖
times sign 乘号
time stability 时间稳定性
time stage curve 时间水位曲线
time stamp 记时打印机
time standard 工时定额;时间标准
time standard generator 时间标准发生器
time star 时星
time started 开始时间
time stationary process 时间平稳过程
time step 时间步长
time-stepping 时间步进的
time-stepping vehicle-specific simulation 变时定车仿真
time-step ratio 时间步长比
times-three register 乘三寄存器
time-strain diagram 时间—应变图
time-strain experiment 时间应变实验

time-stratigraphic(al) facies 时代地层相
time-stratigraphic(al) unit 年代地层单位;时代地层单位【地】
time-stratigraphic(al) facies 时间地层相
time-stratigraphic(al) unit 时间地层单位
time-stratigraphy 年代地层学
time stream transformation 时间流变换
time structure 时间结构
time structure of investment 投资的时间构成
time study 工时分析;工时(定额)研究;计时研究;时间研究;测时
time-subsidence curve 时间—下沉曲线;时间—沉降曲线;沉降过程曲线
time survivor curve 存活曲线
time sweep 时间扫描
time sweep flip-flop 时间扫描双稳态多谐振荡器
timesweep unit 时间扫掠器
time switch 控时断路器;计时开关;定时开关
time switch for automatic staircase lighting 自动楼梯照明计时开关
time synchronization 时间同步
time-synthesis schema 时间合成法
time system 报时系统;时制;时标系统
time table 行车时刻表;工作时间表;时间表;时刻表;列车时刻表
timetable comparing program(me) 时刻表比较程序
timetable editor 时刻表编辑器;时间表编辑
timetable file 时间表格文件
timetable handling 时间表管理
timetable holder 时刻表座架
time taken by program(me) 程序花费的时间
time taking 记时;时间读数
time-temperature control 时间—温度控制
time-temperature curve 时间—温度曲线;温时曲线;温度—时间曲线
time-temperature diagram 时间—温度图
time-temperature index 时间—温度指数
time-temperature indicator 时间—温度显示器
time-temperature relationship 温度—时间关系
time-temperature-transformation curve 时间—温度—形态转变曲线
time-temperature-transformation diagram 时间—温度转变图
time-term method 时项法
time-tested 作过运行时间试验的;经过时间考验的
time tick 时间分段信号;报时嘀答声
time ticket 计工单
time tie 时间互换
time to change concave 更换固定衬板时间
time to change mantle 更换可动衬板时间
time-to-climb 爬升时间
time to closest point of approach 最近会遇时间
time to coat ability 可涂面漆时间
time to count 起算时间
time-to-death 致命时间
time-to-digital conversion 时间数字转换
time-to-failure 故障总周期;破坏时间;损坏时间
time to go 到达所需时间
time to maneuver 试操作时间
time tone 报时音
time to peak 达冲时间;到达峰值时间
time to purchase 采购时间
time to sparkover 距离放电时间;接近闪络时间
time totaliser 总时计
time to turn 转向时间
time-toxicity curve 时间毒性曲线
time tracking indicator 真航迹指示器
time transformer 时间变换器
time transgressive 年序堆积层
time-transgressive unit 时间海侵单位
time transient 时间瞬态
time transmission 授时;播时
time-travel curve 时程曲线;时距曲线;时间行程曲线;时间距离(关系)曲线
time-travel diagram 时间行程图
time trend 时间趋势
time-trend analysis 时间趋势分析
time-undervoltage protection 时限低电压保护装置
time unit 计时单位;时间受控装置;时间单位;地质年代单位
time update 及时修改
time used 实际使用时间
time used for a day in port 船舶平均每次在港停

泊天数
time used per stay in port 船舶平均每次在港停泊天数
time utility 时间效用
time utility factor of dredge(r) 挖泥船时间利用率
time utilization rate in a project 疏浚时间利用率
time utilization rate of dredge(r) 挖泥船时间利用率
time vale of matching point 配合点的时间值
time value 时值;时间价值
time value factor 时间价值因素
time value of money 资金时间价值;货币的时间价值
time-variable 时间变数;时间变量
time-variable control 按时间控制;程序调节;变时控制
time-variable data 时间变化数据
time-variable impedance 时变变阻抗
time variance 工时差异;时间变异
time-variant 时变的
time-variant control system 时变控制系统
time-variant deconvolution in frequency domain 频域时变反褶积
time-variation 随时间变化;时间函数;时变
time-variation control 时间变化控制
time-varied gain 随时间变化的增益;时间变化增益
time-varied sensitivity 随时间变化的灵敏度
time-varying 随时间改变的;随时间变化(的);时变
time-varying amplitude 随时间变动的振幅
time-varying boundary condition 时间变化边界条件
time-varying capacitance 随时变化的电容
time-varying channel with delay spread 时变时延散布信道
time-varying coefficient 时变系数
time-varying coordinate 时变坐标
time-varying field 时变场
time-varying figure 时变图
time-varying filtering 时变滤波
time-varying flux 时变通量
time-varying force 随时间变化的力;时变力
time-varying gradient 随时间变化的梯度;时变梯度
time-varying level 时间变化标程
time-varying linear filter 时变线性滤波器
time-varying model 时变模型
time-varying network 时变网络
time-varying parameter 时间变换参数时变参数
time-varying parameter model 时变参数模型
time-varying power spectrum 时变功率谱
time-varying process 随时间变化过程
time-varying signal measurement 随时间变化的信号测量
time-varying sinusoidal gradient 随时间变化的正弦曲线梯度;时变正弦曲线梯度
time-varying system 时变系统;变系数控制系统
time vector 时间矢量
time-vector method 矢量分析法
time-velocity interpolation interval 时间—速度插值间隔
time vernier 时间游标尺
time wage 计时工资
time wage rate 计时工资率
time wage system 计时工资制
time waveform 时间波形
time-weighted 时间加权的
time-weighted absolute error 时间加权绝对误差
time-weighted average 时间加权平均值
time-weighted average concentration 时间加权平均浓度
time-weighted exposure level 时间加权曝露量
time-weighted squared error 时间加权平方误差
time width 时间宽度
time window 时间窗;时窗
time window length 时窗长度
time window method 时间窗口法
time work 计时工作;日工
time worker 计时工作者;计日工(人)
time-write interval 写入时间间隔
time yield 蠕变;徐变
time-yield of concrete 混凝土蠕变;混凝土的徐变;混凝土的蠕变
time zero 计时起点;时间零点
time zone 时区;等时区
time zone chart 时区图

time-zone disease 时区病
time-zone hand 时区针
timing 整步;校时;记时;计时;配时;调校;定时(的);调时;额定工时
timing adjusting shaft 定时调整轴
timing adjustment 定时调整
timing algorithm 定时算法
timing analysis 定时分析
timing analyzer 定时分析器;定时分析程序
timing and firing control 定时爆破控制
timing and sequencing control 定时程序控制
timing angle 计时角;定时发火变化角
timing back-pumping 定时回扬
timing belt 同步齿形带
timing belt pulley 同步皮带轮
timing bolt 定时螺丝;调整螺丝
timing box 计时箱
timing buoy 定时浮标
timing cam 定时凸轮
timing capacitor 时基电容器;时标电容器;定时电容
timing case 定时箱
timing center 时桄中心
timing chain 正时锤;定时链条;计时机构传动链条;定时链
timing-chain case 同步链传动箱
timing-chain connecting link 定时链连杆
timing-chain lock 定时链锁节
timing channel 定时通道;时间通道
timing chart 计时表;脉冲波形图
timing circuit 计时线路;计时电路;时标电路;定时电路
timing clock 时钟;定时钟;定时时钟
timing clutch 同步离合器
timing condenser 时基电容器;时标电容器
timing-consuming activity 耗时作业
timing contactor 延时开关;延时接触器;定时接触器
timing control 延时控制;时间控制;时间调节器;定时控制
timing controller 定时控制器;定时控制器;定时信号控制器
timing control reference 时间控制标记
timing correction 计时校正
timing current switch 定时电流开关
timing curve 定时曲线
timing cycle 定时循环
timing device 延时装置;计时装置;计时设备;计时器;时限装置;时间控制装置;定时装置;定时设备;定时器;报时设备
timing diagram 正时图;定时图
timing disc 定时盘
timing display 定时显示
timing drum 定时转筒;定时转鼓
timing-drum button setting 定时转鼓的阀钮装置
timing electronics 计时电子学设备
timing element 限时元件;时限元件;定时元件
timing equipment 计时装置
timing error 记时误差;计时误差;同步误差;同步错误;时间错误;定时误差;定时错误
timing extraction 定时抽取
timing fare system 计时票制
timing float 定时浮标
timing for applying pressure 加压计时
timing frequency 定时频率
timing ga(u)ge 定时计
timing gear 定时齿轮装置;正时齿轮;定时齿轮
timing-gear box 定时齿轮箱
timing-gear case 定时齿轮箱
timing-gear-case cover 定时齿轮箱盖
timing-gear chamber 定时齿轮室
timing-gear cover 定时齿轮盖
timing-gear gasket 正时齿轮衬
timing gears 定时齿轮系
timing-gear train 定时齿轮系
timing generator 定时信号振荡器;定时信号发生器;定时发生器
timing hole 定时孔
timing hot water supply 定时供应热水
timing ignition 正时点火
timing impulse 定时脉冲
timing index 定时标志;程序调节器的指示器
timing indicator 定时表
timing in econometric model 经济计量模型的时间选择
timing key 配时键

timing lamp 定时标灯;调时标灯
timingless back-pumping 不定时回扬
timing lever 定时杆
timing machine 校表仪
timing magnet 延时磁铁
timing mark 时间标志;定时记号
timing mark check 时标校验
timing mark sensor element 时标读出元件
timing master 定时主控
timing mechanism 定时机构
timing modification 更改正时
timing motor 计时电动机;时限电动机;定时电动机
timing network 定时网络
timing nut 摆轮调整螺母;定时螺母
timing of decision 决策(的)计时
timing of engine 发动机定时
timing of magneto 磁电机定时
timing of surcharge removal 卸载期
timing of tectonic movement 构造运动定时性
timing of variables 变量的时限
timing oscillator 定时振荡器;定时信号振荡器
timing out 定时已过;定时失效
timing period 延迟期间
timing piece 计时部件
timing plan 配时计划
timing pointer 定时指针
timing pulse 时标脉冲;定时脉冲
timing pulse distributor 定时脉冲分配器
timing pulse generator 时标脉冲发生器;定时脉冲发生器
timing pulse shaper 定时脉冲形成器
timing range 时间调节范围
timing ranger 时间telephone调节器
timing recovery 定时恢复
timing reference 时间基准标记
timing reference signal 时间参考信号
timing regeneration 定时再生
timing register 自动计时器;(自动的)计时器;定时寄存器
timing relay 时限继电器;时间继电器;定时继电器
timing resistor 时标电阻
timing routine 定时程序
timing sampling 脉冲调幅;时间量化;定时采样;定时采样
timing schedule 进度表
timing screw 调时螺丝
timing sector 正时扇形
timing separation 按时间区分
timing sequence 时标序列
timing sheet 计时卡片
timing signal 时标信号;时标同步;时标定时;定时信号
timing signal generator 定时信号发生器
timing switch 计时开关;定时开关
timing system 记时系统;时标系统;定时系统
timing tape 校时带;定时带
timing telegram 定时报文
timing track 同步轨道
timing uncertainty 计时不准
timing unit 定时装置
timing valve 延时阀
timing voltage 整步电压;同步电压;定时电压
timing washer 调时垫片;调时垫片;定时垫圈
timing watch 钟表机构(测斜仪)
timing wave 记时波
timing-wave amplifier 定时波放大器;计时信号放大器
timing-wave generator 定时信号发生器
Timken 铬镍钼耐热钢;蒂姆肯铬镍钼耐热钢
Timken film strength 蒂姆肯膜强度
Timken steel 蒂姆肯镍铬钼耐热钢
Timken X 蒂姆肯X耐热合金
timming filter 补偿滤色片
timoneer 舵手
Timonex 锑白(一种氧化锑)
Timor Sea 帝汶海
Timor trough 帝汶海槽
Timoshenko beam 铁摩辛柯梁
timothy 梯牧草
tin 洋铁罐;导脂铁片;白铁罐
tin adhesion 镀锡层的粘附
tin alloy 锡合金
tinalsiite 硅钛钙钾石
tin amalgam 锡汞合金

tin ash 锡粉;氧化锡;二氧化锡;锡灰(锡与铅的混合氧化物)
tin bar 锡条;锡块
tin-base babbit 锡基巴比合金;锡基轴承合金
tin-base babbit metal 锡基巴比合金
tin-base bearing 锡基轴承
tin-based alloy 锡基合金
tin-base white-metal linings 锡基白合金衬层
tin bath 锡浴;锡槽;成型槽
tin bath block 锡槽砖
tin bath bottom block 锡槽底砖
tin bath partition wall 锡槽空间分隔墙
tin bath siege 锡槽底砖
tinbender 钢板工;白铁工
tin bibromide 二溴化锡
tin bifluoride 二氟化锡
tin bisulfide 硫化锡;二硫化锡
tin brass 锡黄铜
tin bronze 锡青铜
tin bronze alloys 锡青铜合金
tin buddle 锡淘洗盘
tin burning 锡焊
tin-cadmium phosphide detector 磷锡镉探测器
tin-cadmium phosphide monocrystal 磷锡镉单晶
tincal 原硼砂;硼砂原矿;粗硼砂
tincalcite 钠硼钙石
tincalconite 天然硼砂;八面硼砂
tin can 锡罐
tin-canning 墙面板起拱;墙面板扭曲;屋面板起拱;屋面板扭曲
tin case 马口铁罐
tin chloride 氯化锡
tin-clad 镀锡的;包锡的
tin-clad fire door 包白铁皮防火门;金属包层防火门;包层防火门
tincle 粒迹
tin-coat 包锡
tin-coated drum 镀锡铁桶
tin-coated hurricane lantern 镀锡桅灯
tin-coated roofing bolt 镀锡瓦楞螺栓
tin-coated roofing screw 镀锡瓦楞螺丝
tin-coated steel 马口铁;镀锡钢
tin coated steel plate 镀锡钢板
tin-coated wire 镀锡线
tin coating 包锡;锡涂层;镀锡
tinct 色泽;色调;染料
tinction 染色
tinctorial 染色的;着色的
tinctorial power 着色能力
tinctorial property 着色性能
tinctorial substance 着色剂
tincture 酊剂
tincture odo(u)r test(ing) 界限臭味试验
tincture of feeling 感慨
tincture of iodine 碘酒
tindalo 从淡橙色到暗红色的硬木(菲律宾产)
tin-decorating finish 印铁漆
tinder 火绒;引火物;火种
tinderbox 易燃建筑物;火绒盒;导火线盒
tin dichloride 氯化锡;二氯化锡
tin diethyl 二乙基锡
tin difluoride 氟化亚锡;二氟化锡
tin dioxide 二氧化锡
tin discharge 锡拔染法
tin disk 洗槽(洗岩粉分样品用)
tin dross 锡渣
tine 松土齿;锄刀
tine adapter 耙齿插座
tine angle 齿倾角;齿尖倾角
tine bar 搂齿梁
tine contact 齿啮合
tined harrow 钉齿耙
tined loader 抓爪式装载机
tined wheel 齿销轮
tine harrow 齿耙
tine electroplating 锡的电镀
tine length 齿长度
tin enamel 锡釉
tine of formation trap 圈闭形成时间
tine plaque 锡斑;镀锡装饰板
tine plough 齿形犁
tine point 铲尖
tine potato digger 齿形马铃薯挖掘机
tine potato machine 齿状马铃薯挖掘机

tine rake 搂齿杆;齿耙
tine ram 抓齿转动油缸
tine shaft 搂齿梁
tine spindle 搂齿轴;搂齿梁
tin ethide 二乙基锡
tin exudation 锡汗
tin file 锡锉
tin finish 锡增量处理
tin fluoride 氟化亚锡;氟化锡
tin foil 锡纸;锡箔
tin-foil echo 锡箔条回波
tin foil electrode 锡箔电极
tin foiling 制锡箔
tin-foil paper 锡纸
tin-foil plate 锡箔片
tin-free bronze 无锡青铜
tin-free steel 无锡钢板
tin frit 锡晶料
t'ing 亭(中国)
ting beam 底脚梁
tinge 色泽;色头;色彩;带色的
tin glaze 锡釉
tin glaze tile 锡釉瓦
tingle 砌砖用线垫;金属软夹;钓子(砖工);线垫;固定夹片
tingle-tingle 澳洲桉树
tin group 锡组
tinguaite 细霓霞岩;丁古岩
tinguaite porphyry 丁古斑岩
tin horn 波状金属管
Tinicosil alloy 蒂尼科西尔合金
Tinidur 蒂尼杜尔耐热合金(钢)
tin-indium film 锡—铟薄膜
tin ingot 锡锭
tin iodide 碘化锡
Tinite 蒂尼特锡基轴承合金;蒂尼特锡基含铜轴承合金
tinkal 粗硼砂
tinker 小机器匠;修补工;小机工;补锅工人
tinkering 熔补
tinkertoy module 堆叠式微型组件
tin-kettle 白铁壶
tin-knocker 钣金工;白铁工
tin labeller 罐头贴标机
tin-lead bonding 锡焊
tin leaded bronze 锡铅青铜
tin-lead plating 镀锡铅合金
tin-lead solder 锡铅钎料;锡铅焊料
tin leaf 锡箔
tin leak(ing) 漏锡
tin-lined 衬锡(的)
tin-lined case 镀锡箱子;铅皮胎箱
tin-lined conduit 镀锡管
tin-lined lead pipe 衬锡铅管
tin-lined pipe 镀锡管;衬锡管
tin-lined tube 镀锡管
tin liner 锡衬垫
tin liquor 锡液(氯化亚锡溶液)
tin lode 锡矿脉
tin luster 锡光泽彩
tin machine 镀锡机
tinman 白铁工
tinman's shears 白铁皮剪钳;白铁工人剪钳
tinman's snips 美式白铁剪;白铁剪刀
tinman's snips chrome plated handle 电镀柄白铁剪
tinman's snips with contoured handle 环柄白铁剪刀
tinman's snips without spring 不带弹簧白铁剪
tinman's solder 铅锡焊料;软焊料
tinman's solder alloy 铅锡焊料合金
tinmen's bick iron 白铁工台砧
tin metal sheet 锡片
tin mill 镀锡薄钢板轧机
tin mine 锡矿山
tin monosulfide 硫化亚锡
tinned 罐装的;镀锡的
tinned conductor 镀锡导线
tinned copper 锡铜
tinned copper braid 镀锡铜编织层
tinned food 罐头食品
tinned iron 马口铁;白铁皮
tinned iron burr 镀锡铁圈
tinned iron sheet 马口铁

tinned plate 马口铁;镀锡钢板
tinned sheet iron 镀锡铁皮;白铁皮
tinned stainless-steel wire 镀锡不锈钢丝
tinned steel sheet 镀锡薄钢板
tinned wire 镀锌铁丝;镀锡铜线;铅丝;铅皮线;镀锡(导)线;铳线
tinner 锡矿矿工;白铁工
tinner's rivet 白铁工用铆钉
tin-nickel 锡—镍
tin-nickel coating 锡镍镀层
tinning 镀锡;包锡
tinning bath 镀锡槽
tinning line 镀锡工作线;镀锡机组
tinning plant 镀锡车间
tinning pot 镀锡槽
tinning rollers 镀锡辊带材
tinning stack 镀锡器具
tinning unit 镀锡机
tinny 锡的
tinny glaze 含锡釉
tin oil 锡油
tinol 锡焊膏;焊锡膏
tin opener 罐头刀
tin ore 锡矿石;锡矿
tin oxalate 草酸亚锡
tin oxide 氧化锡;锡白
tin oxide ceramics 氧化锡陶瓷
tin oxide electrode brick 氧化锡电极砖
tin oxide refractory 氧化锡耐火材料
tin oxychloride 二氯氧化锡
tin pants 防水衣裤
tin pewter 锡锑合金
tin phosphorus bronze band 锡磷青铜带
tin pickup 底部锡
tin pipe 白铁管
tin pipe with lead jacket 有铅管的白铁管
tin placer 砂锡矿床
tin plate 镀锌钢板;镀锡铁皮;镀锡钢板;镀漆铁皮;马口铁;白铁皮;镀锌板
tinplate bar 镀锡薄钢板坯
tinplate brad 镀锡平头顶
tinplate container 锡钢皮容器
tinplate ink 印铁油墨
tinplate manufacture 镀锡薄钢板生产
tinplate mill 马口铁轧机;镀锡薄钢板轧机
tinplate roof cover(ing) 屋面马口铁覆盖板
tinplate strip 马口铁条;镀锡薄钢带
tin-plating 镀锡
tin poisoning 锡中毒
tin powder 锡粉
tin-printing 印锡
tin printing paper 薄印刷纸
tin protosulfide 硫化亚锡
tin refiner 锡精炼厂
tin-refining plant 锡精炼设备
tin roofing 白铁皮屋面
tin saw 砖锯
tinsel 薄片碎玻璃(装饰用);锡铅合金;金属丝;金属片;金属箔;镀银的碎玻璃片(装饰材料);箔
Tinsel City 俗丽城(好莱坞的俚称)
tinsel cord 软线;箔线
tin shears 白铁剪刀
tin sheet 锡片;锡板
tin sheet iron 镀锡铁皮
tin shingle 马口铁屋面板;镀锡屋面板
tin side detection device 浮法玻璃着锡面检测装置
tin slab 锡锭
tin smelter 锡熔炼炉
tinsmith 锡匠;铁皮工;白铁匠;白铁工
tinsmith shears 铁皮剪子;铁皮剪刀
tinsmith snips 铁皮剪
tinsmith solder 锡铅软焊料
tinsmith's solder 锡焊料
tinsmith work 白铁工程
tinsmithy 白铁工工作间
tin snips 金属薄板剪;白铁皮剪;平头剪;铁皮剪
tin solder 锡钎料;锡料;锡焊剂;焊锡;软焊料
tin solder(ing) 锡焊
tin spar 锡石
tin spirit 锡醇
tin steamer 白铁蒸笼
tin stone 锡石
tin sulfate 硫酸亚锡
tin sulfide 硫化亚锡

tin sweat(ing) 锡珠;锡热析;锡汗
tint 线晕;浅色;色辉;色调
tin tack 锡钉
tintage 着色;染色;涂色;上色
tint base 配色浆
tint blue positive film 淡蓝阳片
tint colo(u)r 浅色
tint control 色度控制
tint correction 色调校正
tint' dye 染料
tinted concrete walk 着色混凝土人行道
tinted glass 有色玻璃;颜色玻璃;彩色玻璃
tinted glass block 颜色玻璃块
tinted glass brick 颜色玻璃砖
tinted glass wall 有色玻璃墙
tinted glass window 有色玻璃窗
tinted granolithic 有色人造石铺石(的)
tinted laminated glass 套色夹层玻璃
tinted lens 色辉透镜
tinted paint 有色油漆;有色涂料;彩色油漆;彩色涂料
tinted putty 有色油灰
tinted rough rolled glass 有色毛玻璃;有色粗轧玻璃
tinted shading 分层设色【测】
tinted spectacles 有色眼镜
tinter 染色器;色料;着色剂;染色剂;调色浆;素色幻灯片
T-intersection 丁字路口;丁字形交叉(口)
tin test 锡试验
tin tetrachloride 四氯化锡
tin tetraiodide 碘化锡
tinticite 白磷铁矿
tintinatie 硫铋锑铅矿
tinting 化色;糊版;起脏;调色;上色;浮脏;浮墨
tinting colo(u)r 调色剂;着色用染料
tinting efffect 着色效应;着色效力;着色效果
tinting material 着色剂
tinting method 分层设色法
tinting paste 着色浆;调色漆;调色浆
tinting pigment 着色颜料
tinting power 着色力
tinting shade 淡色;遮光器;有色玻璃
tinting strength 色度;着色强度;着色能力;着色力
tinting vehicle 着色料
tin titanium 钛化锡
Tintocrete 挺托(一种水泥喷涂材料)
tint of passage 通过色辉
tint(o)meter 色调计;色辉计;测色计;比色计
Tintopal 挺托巴尔(一种彩色半透明玻璃)
tint plate 色辉板;色板
tint-reducing power 消色力
tint retention 保色性
tin tricyclohexylhydroxide 三环己基氢氧化锡
tin triethyl 三乙基锡
tin trip 镀锡带钢
tint screen 网线片;色辉荧光屏;色调荧光屏
tint-tone 浅色调
tin-tube 锡管
tin tube filler 锡管充药器
tin tube shutter 锡管封口机
tint unevenness 色调不均
tinty 色彩不调和的
tin-vanadium yellow 钒锡黄
tin violet 锡紫
tinware 锡器;马口铁器皿
tin weighting 锡增量
tin welding 锡焊
tin whistle 瓦楞铁皮涵管
tin white 锡白
tin wire 焊锡丝
tinwork 锡工厂;锡制品;白铁制品
tiny clutch 超小型离合器
tiny crack 微裂纹;发裂
tiny horsepower hydraulic turbine 小马力水轮机
tinzalconite 三方硼砂
tinzenite 锰斧石;廷斧石
tin-zinc 锡—锌
tin-zirconium 锆化锡
tip 小费;翼尖;尖嘴;倾卸;倾倒;梭尖;端头;倒出;触点;插塞尖端;废物堆
tip angle 顶锥角
tip awn 顶芒
tip-awned variety 顶芒品种
tip-back 后倾;后翻
tip barrow 倾卸手(推)车
tip beam 倾倒车身转轴
tip bearing 调心轴承;枢支座
tip box 倾卸箱
tip brazing 刀片铜焊
tip bucket 倾斜戽斗
tipburn 顶烧
tip car 自动卸货车
tip cart 倾斜手推车
tip cavitation 车叶尖空泡现象
tip chute 溜槽;倾卸滑槽;倾斜消槽;倾斜滑槽
tip circle 顶圆
tip cleaner 喷嘴通针
tip cleaning 电极端清理
tip clearance 螺纹牙尖间隙;船体叶梢间隙;齿顶间隙
tip-clearance area 齿顶间隙面积
tip clearance vortex 车叶尖余隙涡流
tip cone 顶锥
tip cutting 顶插
tip diameter 小头直径;轮缘直径
tip die 顶芽
tip disaster 矸子堆灾难;矸子堆祸害;尾矿堆灾难;尾矿堆祸害;废石堆灾难;废石堆祸害
tip distance 轮冠距
tip downwards 向下翻倒;向下倾卸
tip easing 齿顶修边
tip eddy 桨叶尖漩涡;尖端涡流
tip edge 齿缘;齿顶棱边
tip end 小头(原木的);小端;原木销
tip engagement 极尖插入度
tip follower 尖顶从动杆
tip grade 桩端标高;桩尖标高;坡脚线
tip hardness 矸刀硬度
tip holder 焊条夹钳;喷嘴焊钳;导电嘴夹头
tiphium 池塘群落
tip-in 粘附插页
tip insertion 嵌漏嘴
tip jack 插孔;塞孔;尖头插座;单孔插座
tip-jet 叶端喷口
tip layering 顶压条
tip lead 塞尖引出线
tip leak(age) 顶部漏气
tip-leakage flow 顶部泄漏流量
tipler 翻斗水冲便桶;倾卸车
tipless bushing 无漏嘴漏板
tip line 断夹尖线
tip loading installation 散装货倾倒装舱装置
tip lorry 翻斗小卡车;自卸卡车;自动卸货车;翻斗卡车
tip loss 叶端损失
tip motor lorry 机动翻斗车
tip node 梢节点
tip of drain 排水板底
tip of drawing 绘图笔尖
tip of drill 钻尖
tip-off 解焊;脱焊;拆焊;分接头(电、水)
tip of nozzle 喷管出口截面
tip of pile 桩尖;桩端;桩(的)尖端
tip of plug 插塞尖
tip of propeller 推进器叶端
tip of stylus 绘图笔尖
tip of the needle 针尖
tip-on machine 粘插页机
tip orifice 喷头孔板
tip over 翻倒
tippable container 可倾翻集装箱
tipped 镶齿的
tipped bit 镶刃钻头;补强钻头
tipped drill 焊接钻头
tipped edge 镶硬质合金切削刃
tipped fill 卸土堆;倾土堆;倾卸堆;倾倒填土;抛填土;堆填料
tipped milling cutter 焊齿铣刀
tipped panoramic distortion 侧向全景畸变
tipped rockfill 倾填堆块;倾斜堆块;抛填堆块(体)
tipped steel 镶刃钎头
tipped stone rubble 抛填乱石;碎石堆积层;倾斜乱石
tipped tool 镶片刀具;焊接刀片车刀
tipper 镶尖装置;自卸汽车;自动倾卸车;倾卸装置;倾卸汽车;倾卸货车;倾卸车;倾卸机构;倾倒装置;翻斗车;翻倒装置;翻车机
tipper chassis 翻斗车车盘;翻斗车底盘
tipper-hopper 翻斗
tipper vehicle 倾卸式载重汽车;倾卸式运输车
tipping 装尖头;尾弯【船】;船尼下沉;翻倒;崩刃
tipping angle 倾斜角;倾角;倾翻角(度)
tipping apparatus 倾卸机
tipping arrangement 翻转设备
tipping at sea 抛海
tipping barge 倾卸甲板驳
tipping barrow 倾卸手推车;翻斗手推车
tipping bin 倾卸料仓;倾卸仓;倾卸斗;翻斗
tipping bin for non-tipping trailer 非倾翻拖车的倾翻箱
tipping bucket 转动式装土斗;倾卸斜斗;倾卸料斗;倾卸斗;倾斜斗
tipping-bucket flusher 翻斗式冲洗器
tipping bucket rain(fall) ga(u)ge 倾斗式雨量器;倾卸式雨量计;倾杯式雨量计
tipping-bucket rain-gauge 翻斗式雨量计
tipping cart 倾卸手推车;倾卸车(辆);倾斜手车;翻斗(手推)车
tipping chassis 倾翻式底盘
tipping condition 倾斜状态
tipping cradle 装料设备;倾卸车斗
tipping cup rain ga(u)ge 倾杯式雨量计;倒杯式雨量计
tipping device 翻卸机;倾卸装置;倾卸设备;倾翻装置
tipping door 卧式闸门;卧倒门
tipping gate 卧式闸门
tipping gear 倾卸装置;倾卸装置;翻转装置
tipping grate 卸渣炉算;可倾动式炉算;倾动炉算;摆动炉排
tipping height 仰填高度
tipping hod 倾卸桶
tipping hopper 卸料斗;倾卸漏斗;倾卸料斗;倾斜漏斗;倾料斗;斗式倾卸器
tipping jetty 卸货码头(散货);倾卸码头(突出式)
tipping ladle 倾倒式浇注桶
tipping launch 侧倾下水船
tipping line 倾翻线
tipping link 脱卸接头
tipping load 倾卸荷载;倾翻荷载
tipping loader 翻斗装载机
tipping lorry 翻斗卡车;自卸(汽)车;自倾卸卡车;自动倾卸车;倾卸车身运货车;翻斗卡车
tipping machine 包头机
tipping mechanism 倾卸机械装置;倾卸机构
tipping mixer 可倾式搅拌机;倾筒式搅拌机
tipping pitman 倾卸轴
tipping plant 倾卸设备
tipping point 白人趋避点;卸货点;卸货站;倾卸点;倾翻点
tipping-resistant coating composition 高触变性涂料;不倾出涂料
tipping service 倾卸装置
tipping site 倾置场
tipping skip 翻斗车;倾卸土斗车;翻转式箕斗
tipping space 垃圾倾倒场
tipping speed 倾翻速度
tipping stone rubble 抛填片石;抛填块石
tipping system 倾翻系统
tipping through mixer 倾卸式搅拌机
tipping top link 车厢倾翻用上拉杆
tipping trailer 倾卸拖车;翻斗拖车;倾斜铁拖车
tip(ping) tray 倾卸盘;倾翘洒水盘;倾卸盘
tipping tray conveyer 倾槽式运输机
tipping tray distributor 倾盘式配水器
tipping trough 倾卸槽;倾覆槽;翻斗槽
tipping trough conveyer 倾槽式运输机;倾槽式输送机
tipping trough mixer 斜筒搅拌机;倾槽式搅拌机;倾槽式拌和机
tipping truck 翻斗运货车;倾卸卡车;自卸(汽)车;自动倾卸车;倾卸车
tipping tube 溜管;倾卸溜管;倾卸管
tipping unit 倾卸设备;倾卸装置;翻卸装置
tipping vehicle 自卸(汽)车
tip(ping) wagon 自卸(汽)车;倾卸货车;翻斗料车;自卸货车
tip plate 嘴板
tip-plate process 梢板法(制漏板)
tipple 自动装卸装置;筛煤场;倒煤场;翻笼

tipple building 井口房
tipple dust control 卸货尘埃控制
tippler 翻车工;自卸卡车;翻车机;倾卸装置;倾卸机构
tippler lift 自动倾卸升降机
tip press 制漏嘴压机
tip protector 尖头保护器
tip pruning 打顶
tip radius 齿顶圆半径
tip relief 齿形修缘;齿顶修整
tip relief of tooth 尖角圆角
tip resistance 桩端阻力
tip retainer 焊条夹钳;端部固定件
tip shaft 倾倒车身转轴
tip sheet 内情通报(股票等)
tip side 接头端;尖ախ联线;触点端
tip speed 梢速;车叶尖速度
tip spreader 倾卸式撒布机(筑路用)
tip-stall 梢部失速
tip temperature 垃圾场温度
tip-tilt 倾角
tiptopite 廷磷钾铝石
tip truck 自动倾卸车;自卸垃圾车;自动倾卸卡车;翻斗卡车
tip turbine 叶端喷流透平
tip-up door 上翻门
tip up seat 翻椅;倾靠椅(剧场用);翻动椅
tip-up troll(e)y 卸料小车(输送的)
tip up wards 向上翻转
tip velocity 叶片顶部速度;周缘速度
tip vortex 翼梢漩涡;端头漩涡;车叶尖涡流
tip wrapper 端部包装物
tiradaet 黏结剂;黏合剂
tiragalloite 硅砷锰石
tirant 系梁;系杆
tirckling filter bed 滴滤床
tire 轮胎修补车间;轮胎;车胎;轮箍
tire air ga(u)ge 轮胎气压表
tire band 密封用衬垫
tire bender 弯胎机
tire-bending machine 轮胎弯机
tire brake 轮胎闸(作用在轮胎上的闸)
tire breaker strip 轮胎缓冲层
tire builder 轮胎成型床
tire capacity 轮胎载重量
tire casing 外胎
tire center (路面上的)轮胎印迹中心
tire chain 轮胎防滑(铁)链
tire contact area 轮胎与路面接触面(积)
tire cord 轮胎帘布
tire crane 汽车起重机
tire cut 轮胎割痕
tired gantry crane 轮胎龙门吊
tire dimension 轮胎尺寸
tire dozer 轮胎式推土机;轮式推土机
tired roller 轮胎式压路机
tire flexing 轮胎挠曲
tire ga(u)ge 轮胎气压计
tire grip 轮胎接触地面附着力
tire hop frequency 轮胎跳震频率
tire impression 轮胎花纹
tire inflation 轮胎充气
tire inflation kit 轮胎充气工具
tire inflation pressure 轮胎充气气压
tire iron 拆轮胎棒
tire-killing 轮胎磨耗
tire life 轮胎寿命
tire mill 轮箍轧机
tire noise 轮胎噪声;轮胎滚滑噪声
tire opener 轮胎撑开器
tire paint 轮胎漆
tire-pavement friction 轮胎路面间摩擦阻力
tire press 轮胎压机
tire pressure 轮胎压力
tire pressure ga(u)ge 轮胎压表;轮胎气压计;轮胎压力计
tire pressure watch 轮胎压力表
tire printing 轮胎印
tire pump 轮胎充气机;轮胎打气筒
tire rack 轮胎架
tire recapping mo(u)ld 轮胎修复硫化机
tire repair kit 补胎工具箱;补胎工具
tire rim 胎环
tire roller 轮胎(式)压路机
tire rolling 轮箍轧机
tire rolling mill 轮箍滚扎机
tire section 轮胎断面
tire shop 轮胎店
tire side wall 外胎边壁
tire slip 轮胎打滑
tire spoon 撬胎棒
tire sprinkling system 轮胎喷淋系统
tire steel 轮箍钢
tire thickness 轮箍厚度
tire toe-out 轮胎外倾
tire tread 轮胎着地面;轮胎凸纹;车轮踏面
tire tread design 轮胎花纹设计
tire vulcanizing machine 轮胎热补机
tire wearing surface 轮胎磨耗面
tiring 轮箍术
tiring house 化妆室(剧院内)
tiring room 化妆室(剧院内)
tirocinium 学徒身份;学徒期限
tirodite 锰镁闪石
T-iron T型钢;丁字铁;丁型铁
tiron 试剂
Tiros satellite 泰罗斯卫星
tirr 高沼面腐殖层
Tirrill burner 梯利尔燃烧器
Tirril regulator 梯瑞尔调整器
tirtank 拖拉坦克(集装箱)
Tischenko reaction 蒂森科反应
Tisco alloy 蒂斯科铬硅耐磨耐蚀合金
Tisco manganese steel 锰钢
Tisco Mn steel 蒂斯科耐磨锰钢
Tisco steel 高锰镍耐磨钢;蒂斯科耐磨镍锰钢
Tisco Timang steel 锰镍耐磨钢
Tiscrom 梯斯克拉姆铬锰钢
tisinaite 水硅钛锰钠石
tisodium phosphate 磷酸三钠
Tisserand criterion 蒂塞朗准则
Tissier's alloy 蒂塞西合金
Tissot's indicatrix 底索曲线
tissue 薄纸;薄毡(欧洲称法);薄绢
tissue culture 组织培养
tissue equivalent material 体模材料
tissue grasping forceps 持组织镊
tissue machine 薄纸造纸机
tissue paper 拷贝纸;绵纸;薄页纸;薄棉纸
tissue paper for walls 糊墙用薄页纸;糊墙用薄棉纸
tissue paper holder 卫生纸匣;卫生纸盒
tissue reconstruction 组织改建
tissue roentgen 物理当量伦琴
tissue space 组织间隙
tissue wrapping paper 薄包装纸
tit 小疙瘩;突击队
titabiotite 钛云母
titaisation 钛涂层增强
Titait 钛钨金属陶瓷
Titanal 蒂坦铝合金;蒂坦纳尔铝基活塞合金
Titanaloy 蒂坦钛铜锌耐蚀合金;蒂坦纳洛伊钛铜锌合金
titanate 钛酸酯;钛酸盐
titanate ceramics 钛质陶瓷;钛酸盐陶瓷
titanate porcelain 钛质瓷
titanaugite 钛辉石
titan-barium white 硫酸钡填充的二氧化钛
titanbiotite 钛黑云母
titanbruchevite 钇钛烧绿石
titan-calcium white 钛钙白
titanclinohumite 钛斜硅镁石;钛橄榄石
titan crane 轨行式重型锤头起重机;(自动的)巨型运输机;巨型起重机;巨大起重机;桁架式起重机;桁架桥式起重机
titandiopside 钛透辉石
titan-elpidite 钛斜钠锆石
titangarnet 钛榴石
titanhornblende 钛闪石
titanhydroclinohumite 钛斜硅镁石(钛橄榄石)
titania 氧化钛
titania brick 氧化钛砖
titania ceramics 氧化钛陶瓷;钛氧陶瓷
titania porcelain 钛质瓷
titania type electrode 钛型焊条
titania whiteware 钛质白色陶瓷
titanic 正钛;钛的
titanic acid 钛酸
titanic carbide 碳化钛
titanic chloride 四氯化钛
titanic hydroxide 原钛酸;氢氧化钛;钛酸
titanic iron ore 钛铁矿
titaniferous 含钛
titaniferous ferromanganese 含钛锰铁
titanit 蒂坦钛钨硬质合金
titanite 榍石
titanium acetylacetonate 乙酰丙酮钛
titanium alkoxide 烷氧基钛;钛烷氧基酯
titanium alloy 钛合金
titanium-aluminium alloy 钛铝合金
titanium-aluminium-molybdenum alloy 钛铝钼合金
titanium-aluminium-tin alloy 钛铝锡合金
titanium anode 钛阳极
titanium-barium pigment 钛钡颜料
titanium boride 硼化钛
titanium brown 钛棕
titanium butoxidel 丁氧基钛
titanium-calcium pigment 钛钙颜料
titanium carbide 碳化钛
titanium carbide coating 碳化钛涂层
titanium carbide-nickel cermet 碳镍化钛合金陶瓷
titanium carbide sintering alloy 碳化钛烧结合金
titanium carbonitride 碳氮化钛
titanium carbonitride coating 碳氮化钛涂层
titanium chloride 二氯化钛
titanium copper 钛铜
titanium copper alloy 钛铜合金
titanium diboride 二硼化钛
titanium dichloride 二氯化钛
titanium dioxide 钛白;二氧化钛
titanium dioxide activated carbon supported photocatalyst 二氧化钛活性炭负载型光催化剂
titanium dioxide catalyzed ozonation 二氧化钛催化臭氧氧化
titanium dioxide-coated activated carbon 二氧化钛涂覆活性炭
titanium dioxide coated mica 钛系珠光颜料
titanium dioxide film 二氧化钛膜
titanium dioxide heterogeneous photocatalyst 二氧化钛多相光催化剂
titanium dioxide photocatalyst 二氧化钛光催化剂
titanium dioxide photocatalytic oxidation process 二氧化钛光催化氧化工艺
titanium dioxide pigment 二氧化钛白粉颜料
titanium dioxide sensitized photo-oxidation 二氧化钛敏化光氧化工艺
titanium dioxide thin-film 二氧化钛薄膜
titanium dioxide thin-film electrode 二氧化钛薄膜电极
titanium dioxide waste 二氧化钛废料
titanium doped silica 掺钛二氧化硅
titanium electrode 钛电极
titanium electrolytic cell 钛电解槽
titanium equipment 钛制设备
titanium ethoxide 乙氧基钛;钛酸乙酯
titanium getter pump 钛泵
titanium green 钛绿
titanium gypsum 钛石膏
titanium heat exchanger 钛材换热器
titanium hydride 氢化钛
titanium hydride brazing 氢化钛烧焊
titanium hydroxide 氢氧化钛
titanium isopropoxide 异丙氧基钛;钛酸异丙酯
titanium-magnesium pigment 钛镁颜料
titanium master alloy 高钛合金;钛母合金
titanium mineral 钛矿物
titanium mother alloy 钛母合金
titanium niobate 铌酸钛
titanium nitride 一氮化钛;氮化钛
titanium nitride coating 氮化钛涂层
titanium ore 钛矿(石)
titanium oxalate 草酸钛
titanium oxide 氧化钛;三氧化钛
titanium oxide porcelain 氧化钛陶瓷
titanium paint 钛涂层
titanium peroxide 三氧化钛
titanium pigment 钛白粉;钛(白)颜料
titanium plate 钛材板片
titanium plate heat exchanger 钛材平板式换热器
titanium polymer 钛聚合物
titanium potassium oxalate 草酸钛钾

titanium pyrophosphotungstate 焦磷钨钛
titanium silicide 硅化钛
titanium silicon 钛硅合金
titanium stabilized 钛稳定的
titanium-sublimation pump 钛升华泵
titanium sulfate 硫酸钛
titanium tetrachloride 四氯化钛
titanium tetraethoxide 钛酸乙酯;四乙氧基钛
titanium-tin alloy 钛锡合金
titanium trioxide 三氧化钛
titanium white 二氧化钛;钛白(粉)
titanium white coat(ing) 钛白涂层;钛白刻图膜
titanium yellow 钛镍黄;钛黄
titanize 钛化
titanizing 钛化
titanmica 钛云母
titano-chromite 钛铬铁矿
titanomagnetite 钛磁铁矿
titanomagnetite content 钛磁铁矿含量
titanomagnetite ore 钛磁铁矿矿石
titanometric titration 钛盐滴定(法)
titanometry 钛盐滴定(法)
titanopyroxenite 钛辉石岩
Titanor metal 钛工具钢
titanosis 钛尘肺;钛病
titanous oxalate 草酸钛
titanspinel 钛尖晶石
titan yellow 钛镍黄;钛黄
titanyl sulfate 硫酸氧钛
titer 纤度;最小滴定量;脂酸冻点测定(法);脂酸冻点;定度;滴度;滴定率;滴定度;标准液
titer method 滴定率法
Tithonian 蒂托阶【地】
titian 金黄色的;橙红色的
Titius-Bode law 伯德定则
title 名称;物权;图标;所有权;称号;产权证;标题;标题
title abstract 产权证记录
title and risk 所有权及风险
title block 注记空白位置;工程图明细表(方框图表);图签;图廓外整饰
title company 担保所有权公司
title deed for house 房契
title deed (for land) 地契;地契;契据;物权契据;所有权契据;房契;土地证
title directory entry 标题目录输入
title division 标题部分
title domain 标题区
titled trough 斜低压槽【气】
title exception 保险单中出现的例外产权
title format 标题格式
title for technical personnel 技术职称
title index 标题索引
title insurance 所有权保险;产权保险
title match 锦标赛
title of airfield 机场名称
title of associated mineral 伴生矿物名称
title of chart 海图标题栏
title of drawing 图纸名称
title of geologic(al) report 地质报告名称
title of invention 发明名称
title of mineral commodities 矿产名称
title of ore deposits 矿床名称
title of ore districts 矿区名称
title of paragenesis mineral 共生矿物名称
title of part 部分名称
title of railroad 铁路名称
title of reserve-report 储量报告名称
title of ship 船舶物权
title of standard 标准的名称
title of survey area 测区名称
title of test report of mineral dressing ability and metallurgy 选冶试验报告名称
title of test report of ore physical ability 矿石物性试验报告名称
title of the chart 海图图名
title option 标题选择
title page 标题页;标题扉页
title panel 标题栏
title registration 所有权注册
title search 所有权审查;所有权调查
title sheet 图纸的标题页;具名图纸
title signature 标题书帖
titles to real estate 不动产所有权

title theory 所有权原理
title theory state 所有权理论州
title to land 土地所有权
title to property 物件所有权;财产所有权
title transfer 产权转移
title update 当前所有权记录
titling fount 标题字铅字
titling load 倾覆荷载
titrable acidity 可滴定酸度
titrand 标准滴定液;被滴定物
titrant 滴定用标准液;滴定剂;滴定标准液
titratable 可滴定的
titrate 滴定;被滴定液
titrating solution 滴定液
titration 滴定(法)
titration alkalinity 滴定碱度
titration analysis 滴定分析(法)
titration apparatus 滴定仪
titration cell 滴定池
titration coulometer 滴定电量计
titration curve 滴定曲线
titration curve of neutralization 中和滴定曲线
titration curves of waste 废水滴定曲线
titration detector 滴定检测器
titration end-point 滴定终点
titration error 滴定误差
titration exponent 滴定指数
titration in nonaqueous solvent 非水滴定(法)
titration kit 滴定器
titration method 容量法
titration of dissolved oxygen 溶解氧滴定
titration of halide 卤化物滴定
titration of total acidity 总酸度测定
titration test 滴定试验
titration thief 滴定阱
titrator 滴定仪;滴定器
titre 最小滴定量;滴度;滴定率;滴定度
titre method 滴定率法
titrimeter 滴定计
titrimetric analysis 容量分析(法);滴定分析(法)
titrimetric method 滴定法
titrimetry 滴定分析(法);滴定法
titroscope 滴定式酸碱计
titular 名义上的
titular possession 有权持有的财产
tivanite 羟钛钒矿
tivano 提旺诺风
tiza 钠硼钙石
Tizit 高速切削工具合金
tjaele 冻坡层;多年冻土;冻土
tjandi (印度8~14世纪的)墓塔
tjarle 冻土
T-jetty T形突堤码头
T-joint 丁字接头;T形接头
T-jointing 丁形连接
T-joist T型格栅
T-junction T形接缝;丁字路口;丁字形交叉(口);T字形三岔路口
T-junction box T形套筒
T-juncture 三通接头
tlalocite 水氯碲铜石
tlapallite 硫碲钙石
TLD 调谐液体阻尼器;调频晃动阻尼器
TMD 调频质量阻尼器
T-method 张力变动方法
TNT 黄炸药
TNT equivalent 梯恩梯当量
T-number T指数
T-nut 丁字形螺母
to act as a source of decaying material 作为腐烂物的来源
toadback handrail 蛤蟆臂形扶手(楼梯);蛤蟆背形(楼梯)扶手
to adopt measures suiting local condition 因地制宜
toad's-eye tin 蛙目锡石
toadstone 玄武斑岩
toadstool 毒菌
toal bit pressure 钻头总荷载
to allot a profit 分配利润
to-and-fro 往复的;来回的
to-and-fro (aerial) ropeway 往复式索道;往复架空索道
to-and-fro bending test 往复弯曲试验

to-and-fro method 来回方法
to-and-fro motion 往复运动
to-and-fro movement 来回运动
to-and-fro oscillations 来回震动;来回摆动
to and-fro ropeway 往复架空索道;往复索道
to and fro traffic 往复交通
Toarcian 多尔斯阶
to arrival at dredging site 抵达疏浚工地
to arrive 业经装船
toaster 烘炉;多士炉
toat 刨把;台刨把手;短刨把柄
toatal heat 变浓热
to attempt…rashly 轻率尝试
tobacco 烟草
tobacco-curing 烟的生产
tobacco goods 烟草货物
tobacco industry 烟草工业
tobacco jack 黑钨矿
tobacco juice 烟液
tobacco manufacture 制烟业
tobacco plant 烟叶
tobacco product 烟制品
tobacco-seed oil 烟草籽油
tobaccosis 烟草尘肺
tobacco smoke 烟草烟雾
tobacco tar 烟草焦油
tobacco tongs 烟斗点火钳
tobacco warehouse 烟页仓库;烟草仓库
tobacco waste(water) 制烟废水;烟草制品废水
to balance the book 结清账簿
to beat the world 打破世界纪录
to be built for full depth reclamation of thick lifts of asphalt 为厚层沥青整层翻修而造成的
to be changed depending on the site 随地更换
to be changed depending on the weather 随气候更换
to be deposited many centuries ago 在许多世纪前沉积下来的
to be designed as explosion proof 应按防爆设计
to be employed in agriculture 从事于农业工作
to be formed in the bodies of aminal and plant 存在于动植物体内
to be hard and shiny 硬而有光泽
to be in cash 有现款
to be kept upright 直立安放;竖立安放
to be leached from the soil 从土壤中淋溶出去
to be left off unfinished 半途而废
to be liquidated 被清算
tobelite 托铵云母
to be out of cash 缺款
to be out of shape 走样
to be protected from cold 防冷
to be put in file for reference 存档备查
tobermorite 雪硅钙石
tobermorite gel 纤维钙石凝胶;雪硅钙石胶
to be visible on plate 在板上可见
Tobias acid 托彼厄斯酸;吐氏酸
Tobin brass 托宾黄铜;铜锌锡青铜
Tobin bronze 托宾青铜;铜锌锡青铜
Tobin's tube 托宾通风管
toboggan 手撬;利用重力向下运东西的输送机
Tobol interglacial stage 托波尔间冰期
to break one's word 不守信用
to break the journey 中途下车
to break the world record 打破世界纪录
to break up of rock 岩石粉碎
to bring about a prosperous economy 繁荣经济
to bring potential into play 发挥潜力
to bring the initiative into play 调动积极性
to bubble up 使冒泡
to buckle outward 向外弯曲;向外凸出
to burn with a blue flame 在燃烧时发出蓝焰
to carry a big weight in 举足轻重
Tocco hardening 表面淬火
Tocco process 高频局部加热淬火法;高频加热淬火法
to centre to centre 上静点
to center to center 上静点
tochilinite 羟镁硫铁矿
tocicological assessment of pollution of freshwater ecosystem 淡水生态系统污染毒物评估
to come up to advanced world standards 达到世界先进水平
tocornakite 碘银汞矿

to cut down to grade 挖至设计标高
tod 繁茂处;灌木丛
toddler toy 木制拖行玩具
to declare the contract avoided 宣告合同无效
to depart from actual realities 缺乏实际根据
to distribute a profit 分配利润
to do a lot of damage 造成很大破坏
todorokite 钙锰矿;钡镁锰矿
to do very great damage to land 对土地造成极大破坏
to drain 通往下水道
toe 靴头;柱底;轴踵;趾;焊脚;焊缝外表面边界;焊边;墙趾;坡脚;材料或构件的凸缘
to each according go one's needs 按需分配
toe apron 坡脚护坦;坡底护堤
toe-ballasting rubble 镇脚石
toe basin 坝趾消能池;坝趾消力池;坝趾静水池;消力池
toe bead 底部膏条;周边挡板;下部维护侧板;踏板;密封填料;工作平台
toe bead wire 缘趾线环
toe beam L 形梁;带趾梁
toe bearing 止推轴承;立式止推轴承
toe blanket 底部铺盖;坝趾铺盖
toe block 辙叉趾间距铁;护脚块体;坡脚砌块
toe board (挡土墙的)趾板;围护侧板;周边挡板;下部维护侧板;踢脚板;踏板;搁脚板;矮挡板;下部围护侧板;脚手挡板
toe brake 脚踏闸
toe bund 堤脚棱体
toe circle 坡趾圆;坡脚圆
toe construction 堤脚施工
toe contact 小端齿接触;齿顶接触
toe conveyer 轴踵输送机
toe course 护脚砌片;坡脚砌层
toe crack 焊趾裂纹;焊缝边缘裂纹;边界裂缝
toe cut 底部掏捞;底部掏槽
toe cut-off wall 坡脚拦墙;坡脚截水墙
toed 钉铆钉的;钉斜钉的
toe damage 坡脚破坏
toed-in 内八字脚
toe ditch 投入河中;投入海中;投入沟中;挖沟
toed keystone 带榫的拱心石
toed material 钉斜钉的材料
toe(d) nail 斜钉
toe(d) nailing 斜钉法;斜叉钉法
toe dog 小撑杆
toed-out 外八字脚
toe drain 坝脚排水设施;坡脚排水;(坝渠岸堤的)背水面坡脚排水;坝趾排水(层)
toe drainage 滤水坝(趾)
toed voussoir 带榫拱石
toe end of frog 辙叉跟端
toe-end of tooth 轮齿趾端
toe excavation 坡脚挖方;坡脚开挖
toe extension 辙叉趾延长轨
toe fail 坝趾失效
toe failure 坡趾破坏;坡脚坍塌;坡脚破坏;坡脚滑坍;坡脚滑动
toe filter 坝址反滤层;护脚反滤层;坡脚(反)滤层;坝趾滤层;坝趾反滤层;坝趾倒滤层
toe guard 脚挡板
toehold 立足板条;坡脚稳定;抗滑趾
toe hold of dam 坝趾抗滑趾
toe hole 梯段底部炮眼;底部炮眼
toe-in 内八字脚;内向;前端内倾
toe index 转位
toeing 斜装;斜向;轮子斜向
toeing angle 侧舷安装角
toe-in ga(u)ge 前轮前束测定器;测量前轮前束装置
toe joint 斜连接;斜接头;齿接合;齿接
toe joint bolt 辙叉趾端连接螺栓
toe-jointing V 形接口;齿接
toe kerb 护坎
toe length of frog 辙叉趾长度
toe level of pile 桩尖标高
to eliminate black smoke 不冒黑烟
to eliminate waste 杜绝浪费
toe line 桩尖标高线;桩顶水平线;桩顶线;坡趾线
toe load of fastening 扣压力;扣件扣压力
toe mattress 护脚排
toe-nailed 斜钉钉法
toe-nailing wood-strip flooring 地板斜钉;木条地板斜钉

toe of bead 缘趾(轮胎的)
toe of beam 工字梁梁边
toe of bracket 肘板趾
toe of dam 坝脚;坝趾
toe of fill 路堤坡脚;填土坡脚
toe of fillet 角焊长度
toe of frog 辙叉趾
toe of hole 孔底;井底
toe of pile 桩尖
toe of retaining 挡土墙墙趾
toe of roadbed 道床坡脚
toe of side slope 坡脚
toe of slope 坡脚;斜坡底部;坝趾;坡面图
toe of structure 建筑物基脚
toe of the dam 坝脚
toe of the weld 焊缝边界
toe of wall 墙趾;岸壁前趾
toe of weld 焊趾
toe-out 外八字脚;前轮负荷束;外向
toe-out on turns 转弯外倾
toe piece 凸轮镶片
toe piston 脚踏按钮
toeplate 踢脚护(条)板
toe pressure 趾压力
toe pressure of foundation 基础趾压力
toe protection 护脚;坡脚防护;坝趾防护
toe protection of dam 坝趾防护
toe protection of slope 坡脚防护
toe protection stone 坝趾护石
toe protection unit 护脚块体
toe-punching 打脚号
toe resistance of pile 桩端阻力
toernebohmite 羟硅铈矿
toe scour 堤趾冲刷
toe slab 挡土墙趾板;趾板(挡土墙的);前趾板
toe spread 辙叉趾开口距
to establish business relation 建立业务关系
toe steel 桩脚附加钢筋;附加构造钢筋
toe stone 护脚石
toe stress 趾应力;坝趾应力
toe tip 脚尖
toe-to-toe drilling 竖直大炮眼的钻孔
toe trench 护脚坑;护底坑
toe trench method 沟渠填石护坡法
toe trench of levee 堤基防渗槽
toe wall 趾墙;坡脚墙;坝趾挡土墙;坝趾墙;坝趾齿墙
toe weight 坡脚压重;坝趾压重
toe weld 趾部焊缝
to excavate holes alongside the wreck for passing slings 沉船侧打千斤洞
to exist in both tropical and subtropical zones 存在于热带和亚热带
to exist in solution 存在于溶液中
toe zone 焊趾区
tofall 披屋
tofan 托范风暴
to fire an engine 启动发动机
tofranil 托法尼
to-from indicator 来去指示器
toft 小丘;花园地;宅基;屋基
toggery 服装商店;服装用品店;服装店
toggle 系素桩;肘环套接;肘节;拉钳;紧线套环;紧线鬼爪;曲肘;触发器;反复电路;拨动
toggle action 曲柄动作;肘杆曲柄动作
toggle actuator 拨转开关
toggle angle 肘节角
toggle arm 肘节臂
toggle bolt 套环螺栓;系墙螺栓;系墙;肘节螺栓;中环螺栓;固定铁箍的螺栓;肋节栓
toggle brake 肘节制动器;套环制动器;带伸缩元件制动器
toggle chain 肘节链
toggle chuck 肘环套接卡盘
toggle circuit 触发器
toggle clamp 肘节式夹钳;铰接夹
toggle clip 夹扣
toggle clip closure 夹扣封闭器
toggle clutch 肘节式离合器
toggle connection 弯头连接
toggle control key 切换控制键
toggle coupling 弯头接合
toggle crusher 肘节压碎机
toggle draw die 双动柱模
toggle drawing press 肘杆式拉深压力机

toggle flip-flop 计数触发器;反转触发器;触发器
toggle frequency 反复频率
toggle grip 肘节柄
toggle hook 肘节钩;(系在绳索末端的)索眼钩;带栓的搭钩
toggle jaw crusher 肘节颚式破碎机
toggle joint 肘接;曲柄杠杆装置;弯头连接;弯头接合;肘节接合;肘节
toggle-joint brake 肘节闸;肘接制动器
toggle jointing 弯头接合
toggle-joint lever 肘节杆
toggle-joint press 曲柄压机
toggle-joint riveter 肘节铆接机
toggle-joint riveting machine 肘节铆机
toggle joint unit 曲柄杠杆装置
toggle lever 肘节杆;曲柄连杆
toggle-lever arm 曲杆臂
toggle-lever grip gear 带有套环的保护装置
toggle lever press 曲柄压机;曲柄杠杆压机
toggle link 肘节杆;肘杆;曲柄连杆
toggle linkage 肘节链系
toggle locking 肘杆锁定的
toggle mechanism 系墙螺栓;肘节机构;曲柄杠杆装置;曲柄杠杆机构;弯头接合
toggle nut 肘节螺母
toggle pin 肘接销钉
toggle plate 肘环套接板;肘板;肋板;推力板;套环
toggle plate press 肘板压力机
toggle press 肘节式压榨机;肘杆式压印机;肘节压砖机;曲臂压砖机;曲柄压机
toggle puncher 手动冲床
toggler 栓紧装置
toggle rate 计时频率
toggle rod 肘杆
toggle seat 肘座位置;肘节支座;肘板承座
toggle speed 计时速度;启动速度;反复速度
toggle switch 肘式开关;肘节开关;可逆肘头拨动式开关;跷动式小开关;触发器;拨动开关;拨动(式)小开关;扳手开关;扳钮开关
toggle type jaw crusher 肘节式颚式破碎机;复杂摆动式颚式破碎机
toggle-type tablet machine 肘杆压片机
toggle valve 肘形阀门
to give change 找零钱
to give up halfway 半途而废
to give way to navigation 退避让航
to go into full construction 开足马力施工
to go into full production 开足马力生产;满负荷生产
to go out of form 走样
to have common interests 共同的利益关系
to have electricity, gas, running water and steam heating system 有水、电、暖、煤气设备
toich 年泛滥湿地
toil 苦力;苦活
toilet 洗涤;盥洗间;卫生间;厕所
toilet and shower 厕所及淋浴室
toilet-and-shower room 盥洗淋浴室
toilet articles 梳妆用具
toilet bowl 大便器;抽水马桶;厕座;便纸
toilet cabinet 化妆柜
toilet compartment 盥洗室
toilet cubicle 厕所间;厕所隔断
toilet cubicle door 厕所隔间门
toilet dam 大便挡水装置
toilet enclosure 厕所隔断;卫生间中的小隔断
toilet equipment 卫生间设备
toilet facility 盥洗室设备;盥洗设备
toilet fan 盥洗室风扇;盥洗室通风机
toilet faucet 盥洗室龙头
toilet flushing 盥洗室冲洗;便池冲洗;冲水便桶
toilet flushing cistern 盥洗室冲洗水箱
toilet flushing pipe 厕所冲洗管
toilet flushing water 厕所间冲洗水
toilet pan 厕所间水槽;厕所间水盆
toilet paper 卫生纸
toilet paper dispenser 卫生纸分配器
toilet paper holder 卫生纸架;卫生纸匣;卫生纸盒;便纸架
toilet partition 卫生间隔板;厕所隔断
toilet partition wall 卫生间隔墙
toilet-pot 盥洗室用器皿
toilet roll holder 手纸架
toilet room 盥洗室;化妆室

toilet seat 马桶座圈;便桶座圈
toilet seat lid 马桶盖
toilet set 梳妆用具;洗手间成套器皿
toilet soap 香皂
toilet stall 盥洗室小分隔间
toilet table 化妆台;梳妆台
toilet trailer 流动厕所;厕所拖车
toilet unit 盥洗室设备;卫生间单元
toilet ware 洗手间器皿;化妆物品
toilet waste water 盥洗室污水
toilet waste water pipe 盥洗室污水管
toilet water 化妆用香水
to impregnate with creosote 杂酚油灌注
to improve the environment 改善环境
to influence the water supply of the soil 影响土壤水分的供应
toise 突亚斯(法国长度单位)
Tokai Accident Control Center 东海交通灾害控制中心(日本)
token 行车凭证;象征性的;代用币;标志
token bus network 令牌传送网络
token card 记号卡片;特征卡(片);标记卡片
token deliverer 凭证递交机
token export 象征性输出
token import 象征性输入;象征性进口
tokenism 象征性雇佣
token money 代用货币;符号货币
token passing 令牌传送
token payment 象征性偿付
token ring 令牌环
token-ring bus 令牌总线
token-ring network 令牌环网;令牌环网
toki 炻器
tokonoma 壁龛(日本房屋的)
tokuthon 低毒硫磷
Tokyo datum 东京大地基准面
Tokyo type smog 东京型烟雾
tolane 二苯乙炔
tolane dibromide 二溴化二苯乙炔
tolbachite 托氯铜石
tole 涂漆锡器;涂漆马口铁皮
Toledo 托莱多(一种可折叠钢架)
tolerable 可容忍的
tolerable backlash 容许齿隙游移
tolerable concentration 可容许浓度
tolerable defect 容许疵病
tolerable deviation 容许偏差
tolerable distortion 容许畸变
tolerable fluctuation 容许漂移
tolerable frequency variation 容许频率变化
tolerable level 可容许量
tolerable limit 容许限(度);容许范围;容限
tolerable range 容许限(度)
tolerable risk 容许危险性
tolerable settlement 容许沉降(量)
tolerable viscosity 容许粘度
tolerable water supply quality 可容许供水水质
toleragen 耐受原
tolerance 限差(范围);允许公差;裕度;公差;宽容度;耐性;耐受性;耐受;耐量;容许误差;容许量;容许间隙;容许的误差;容许;容限;容忍度;容差;忍耐力;容限
tolerance allowance 容许
tolerance analysis 容差分析
tolerance and fit 公差及配合;公差和配合
tolerance chart 公差图表
tolerance chimera 耐受性嵌合体
tolerance class 公差等级
tolerance clearance 间隙公差;配合间隙
tolerance control 公差控制;偏差控制
tolerance degree 容许程度
tolerance deviation 容许偏差;容差离差
tolerance distribution 容许分布
tolerance distribution model 允许量分配模型
tolerance domain 容许域;容许量范围
tolerance dose 耐受剂量;容许剂量
tolerance ecology 耐污生态
tolerance factor 容许系数
tolerance flux 容许通量
tolerance for aberration 像差容限
tolerance for back-silting 回淤富裕深度;回淤富余深度
tolerance for indistinctness 不清晰度容限
tolerance for sand up 回淤富裕深度;回淤富余深度
tolerance frequency 允许频率;容限频率;容计频率
tolerance interval 容许区间
tolerance level 允许的辐射级;宽容度
tolerance level in production 生产宽容度
tolerance limit 公差限度;公差极限;耐受限;耐受极限;容许限(度);容许界限;容许极限;容差极限;容限
tolerance limit median 平均容许极限量
tolerance measurement 公差量测
tolerance number of defects 容许缺陷数目
tolerance number sequence 公差数号顺序
tolerance of dimension 尺寸公差
tolerance of fit 配合公差
tolerance of parallelism 平行度公差
tolerance of salinity 耐盐度
tolerance of set pressure 调定压力公差
tolerance of shade 耐阴性
tolerance on fit 配合公差
tolerance on hardness 容许硬度
tolerance pollution capacity 耐污能力
tolerance range 公差范围;公差带;容许限(度);容许量范围;容许界限;容许范围;忍耐范围;存活界限
tolerance region 容许区域
tolerance response curve 忍受响应曲线
tolerance species 耐污物种;耐污染物种;耐受物种
tolerance specification 公差规格说明
tolerance standard 公差标准;容许量标准
tolerance stipulated 规定公差
tolerance system 公差制(度)
tolerance test 公差配合试验;耐药量试验;耐量试验;容耐试验
tolerance to climate 气候耐受性
tolerance to heat 耐暑
tolerance to high altitude 高原耐性
tolerance unit 公差单位
tolerance value 公称值;容许残留值
tolerance zone 公差带;公差带;尺寸公差带
tolerant deviation 容许偏差
tolerant dose 容忍剂量
tolerant grasses 难以防除杂草
tolerant plant 耐阴植物
tolerant species 耐污物种
tolerant to drought 耐旱的
tolerant to heat 耐热(的)
tolerant tree 耐阴树
tolerated dose 耐药量
tolerated structure 暂准建筑物
tolerate P-levels 耐磷
toleration 耐受性
toleration ecology 耐性生态学
tolerator 杠杆式比长仪
tolerific 产生耐受性的
to level out 趋向平缓
to level the ridges before planting 播前把垄整平
tolilic acid 二甲基乙醇酸
toll 通行税;通行费;长途电话
toll area 收通行税地区【道】
toll bar 拦门;收费拦门;通行税征关卡;收通行税卡【道】;收费门
toll barrier 收费处路障
toll board 长途交换台
toll booth 收税亭;收费亭(道路);收费所
toll-booth area 收通行费地点【道】
toll bridge 征收桥;征收卡;收费桥
toll cable 长途线;长途通信电缆;长途电缆
toll call 收费长途电话;长途呼叫
toll call telephone 长途电话机
toll center 中央长途电话局;长途电话总局;长途电话中心
toll charges 通行税;道路税(养路费)
toll circuit 城际通道
toll collection 征税;征收费用;收费设备
toll collector 收费者;收费人
toll communication 长途通信
toll communication center 总枢纽
toll communication network 长途通信网
toll communication optic(al) cable 长途通信光缆
toll control 长途控制
toll dues 通行费
toll enquiry position 长途查询台
toll enrichment 收费浓缩
toller 收费者
toll facility 收通行费的设施【道】;收费设施
toll final selector 长途终接器
toll-free 免税;免费;免税服务费
toll-free bridge 免费桥
toll-free call 免费电话
toll-free highway 免税公路;免费公路
toll-free hotline 免费热线
toll-free number 免税数
toll-free road 免费道路
toll gate 收税卡;通行税征税关卡;收费卡口;收费门
toll highway 收税公路;收费公路
toll house 收费亭;通行税征税所;通行税征税所;通行收费处;收费所;收费处
toll keeper 通行税征税员
toll line 长途线;长途电话线
toll man 过境税收人;通行税征税员
toll network 长途通信网
toll office 长途局;长途电话局
toll of road maintenance 养路费
toll on transit 通行税
toll plaza 通行税广场
toll road 收税道路;收费(公)路;收费道路
toll road bonds 收费公路公债
toll service 长途业务
toll station 长途电话局
toll supervisors console 收费监督控制台
toll switchboard 长途交换台;长途交换机
toll tax 通行税
toll telephone 长途电话
toll telephone circuit 长途电话线路
toll terminal 长途终端局;长途电话局直通用户线
toll terminal loss 长途终接损耗
toll through 道路税(过桥税);过桥费
toll traffic volume 收费交通
toll train 长途专用设备
toll trunk line 长途中继线
toll tunnel 收费隧道
toll turn 牲畜市场税
toll-type interchange 收费立体交叉
tollway 收费公路;收税道路
tolserol 甲苯丙醇
Toltec architecture 托尔克建筑
toluene 甲苯
toluene bromide 溴甲苯
toluene chloride 氯甲苯
toluene concentrating agent 甲苯浓缩剂
toluene dichloride 二氯甲苯
toluene di-isocyanate 二异氰酸甲苯;甲苯二异氰酸酯
toluene dilution ratio 甲苯稀释比(率)
toluene formaldehyde resin 甲苯甲醛树脂
toluene halide 卤甲苯
toluene method 甲苯法
toluene resin 甲苯树脂
toluene sulfochloride 甲苯磺酰氯
toluenesulfonamide 对甲苯磺酰甲基亚硝酰胺
toluenesulfonate 甲苯磺酸盐
toluene sulfonic acid 甲苯磺酸
toluidine 甲苯胺
toluidine blue 甲苯胺蓝
toluidine maroon 甲苯胺褐红
toluidine red (由煤焦油制得的)有机红颜料;甲苯胺红
toluidine toner 甲苯胺有机调色剂
toluidine yellow 甲苯胺黄
toluol sulfonamide 甲苯磺酰胺
toluol sulfonanilide 甲苯磺酰(替)苯胺
tolu tree 吐鲁胶树
toluylene 二苯乙烯
toluylene blue 甲苯蓝
toluylene diamine 甲代苯二胺
toluylene orange 甲苯橙
toluylene red 中性红;甲苯红
tolyl aldehyde 甲苯醛
tolylene diamine 亚苄基二胺;甲苯二胺
tom 支撑;支柱;倾斜粗洗淘金槽
tomahawk 修整铆钉梢的铁锤;石斧
tomahawking 修整铆钉梢
to make into fertilizer 做成肥料
to make loans 发放贷款
to make public 公布
tomb 墓;坟墓
tombac 顿巴克黄铜;德国黄铜
tombak 含锌黄铜;顿巴克黄铜
tomb architecture 坟墓建筑

tombarthite 稀土;汤硅钇石
Tombasil 顿巴希尔耐磨硅黄铜
tomb chamber 墓室;墓穴
tomb chamber pyramid 金字塔形墓室;棱锥形墓室
tomb chapel 坟地小教堂
tomb-chest 柜式石棺;墓柜
tomb ciborium 坟墓祭坛华盖;坟场龛室
tomb-mosque 坟地伊斯兰教堂
tomb niche 墓壁龛
Tomb of Agamemnon 亚加米农陵墓
Tomb of Emperor Zhu Yuanzhang 明孝陵(南京)
tombolo 连岛沙洲;连岛坝;沙颈岬;陆连岛
tombolo cape 沙颈岬
tombolo cluster 连岛沙洲群;复式陆连岛;复合连岛沙洲
tombolo series 连岛沙洲群;复式陆连岛;复合连岛沙洲
tomb slab 坟墓石板
tombstone 墓石;墓碑
tombstone in block 墓碑石块料
tombstone in slab 墓碑石块料
tomb tholus 圆形陵墓
tome 册
tomichite 砷钛钒石
tomite 藻煤
tommy 实物工资制;定位销钉
tommy bar (套筒扳手的)旋转手柄;螺丝旋杆;螺丝旋棒;铁杆;插杆
Tommy bar nut 带孔螺母;带孔螺帽
Tommy hole 扳手孔
tommy screw 贯头螺丝;贯头螺钉;虎钳丝杠
tommy-shop 面包店;厂内商店
tommy wrench 螺旋头部用活动杆;套筒扳手
to mobilize one's enthusiasm 调动积极性
tomogram 层析 X 射线照片
tomographic scanner 层析 X 射线扫描器
tomography 断面 X 射线照相术;层析 X 射线照相法;层析 X 射线照相法;层面 X 射线照相术
Tomonaga-Schwinger's equation 朝永振一郎施温格尔方程
tomophotography 层析 X 射线影屏照相术
tomosynthesis 层析 X 射线照相组合
to move away from navigation channel 退槽让航
tompoming 搓涂
Tomson's type 汤姆森方式
tom-tom scouring machine 锤式精练机
tom-tom washer 锤式洗染机
ton 冷吨;短吨
tonal anomaly 色调异常
tonal boundary (航测地形的)色调界限
tonal brightness 色调亮度
tonal content 色调含量
tonal control 色调控制
tonal controller 色调控制器
tonal density 色调密度
tonal difference 色调差
tonal distortion 灰度畸变;色调失真
tonal fidelity 色调逼真度
tonal gradation 色调梯级;色调等级
tonalite 英云闪长岩;石英闪长岩
tonalite porphyry 英云闪长斑岩
tonal range 灰度范围;色调梯度;色调范围;色调度
tonal relationship 色调对比
tonal scale 色标
tonal signature 色调特征
tonal value 影像明晰度;亮度级;色调
tonal value shift 色调值偏差
tonal wedge 色楔;半色调楔
Tonawanda Pine Punus strobes 北美乔松
ton burden 负载吨
ton by ton delivered 按吨交货
ton comparable 吨位丈量
tondal 吨达
ton-day of cargo in storage 货物堆存吨天数
tondino 半圆饰;圆盘形图;圆盘形浮雕
tondo 圆画;圆浮雕;圆形浮雕
tone 影调;行情;市场的景况;色头;色光;色调;风格
tone and gray scale 色调和灰阶
to nearer 交与持票人
tone-burst generator 单音脉冲发生器
tone chamber 音室
tone channel 音道
tone code ranging 音码测距法

tone colo(u)r 音色
tone control 音质控制
tone-control aperture 色调控制孔径
tone copy 色彩复制
tone correction 音调校正
tone dialing 按钮拨号
tone distortion 色调失真
tone-down 柔和;使柔和
tone filter 音频滤波器
tone generator 音频振荡器;音频发生器
tone homogeneous degree 色调均匀程度
tone intensity 彩色强度
tone keyer 单音调制器
tone-kilometers of goods per kilometer of railway line 货物运输密度
tone level 音平
tone-line 色调线
tone-line technique 色调线法
tone localizer 音频振幅比较式定位器;等信号着陆信标
tone modulation 音调制
tone modulator 音调制器
tone of gray 灰色色调;黑白亮度等级;半色调等级
tone of the market 市场的景况
tone plate 色调板
tone quality 音质
toner 显像粉;有机调色剂;油漆调色工;增色剂;静电复印粉;调色浆;调色剂;色原;色剂;色粉
toner brown 色原棕
toner carrier 色剂载体
toner concentrate 色剂浓缩
toner container 色剂容器
toner dispense gear 墨粉分配传动机构
tone receiver 信号音接收器;音频接收机
toner empty sensor 墨粉不足传感器
tone rendering 色调再现
tone reproduction 影调再现;色调再现
tone reversal 色调改变
toner layer potential 墨粉层电位
toner mark 色剂标记
toner metering roller 墨粉调节辊
toner particles 增色剂微粒
toner-receptive coating 可着色的涂层
toner yellow 色原黄
tone sense 音感
tone separation 影调析离
tone signature 色调特征
tones of goods originated 货物发送吨数
tones of grey 半色调等级;黑白亮度等级;灰色色调
tone soft gray 黑白亮度级
tone telemetry 音调遥测术
tone to detect 受检音源
tonet reservoir 色剂容器
tone tuning 调音
tone value 色调
tone wedge 色调楔
tone wedge set 色调光楔组
ton force 吨力(合 9807 牛顿)
Tonga Development Bank 汤加开发银行
Tonga plate 汤加板块
tong arm 钳臂
Tonga trench 汤加海沟
tongbaiite 桐柏矿
tong bucket 吊钳平衡锤
tong crane 钳式吊(车)
tong dies 管钳板牙
Tongguan ware 铜官窑器
tong keys 管钳板牙
tong mark (玻璃片的)夹钳痕;挂钩痕迹;钢化玻璃的夹痕
tong operated mechanism 夹钳开闭机构
tong pipe up 用管钳拧上管子
tongs 抓下凹槽;夹子;夹钳;火钳
tongs furnace charger 夹钳加料机
tongs hanger line 吊钳绳
tongs jaw 钳子牙口;钳子夹板;钳颚
tong space 大钳安置高度
tong-test ammeter 夹线式电流计;夹表;钳式安培计
tong torque controller 拧管扭矩指示器
tong torque ga(u)ge 扭矩仪(机械大钳用);大钳扭矩量测仪
tong torque indicator 拧管扭矩指示器
tong-type ammeter 钳式电流表
tongue 岩枝;岩舌;高度尺;熔岩舌;曲流朵体;牵引架;凸缝;凸出部;榫舌;舌状物;舌簧;赤龙;斧柄榫舌;岸岬
tongue amperemeter 舌簧式电流表
tongue and groove 企口拼接;榫槽式;舌榫;舌槽(接缝);榫槽
tongue and groove block 企口砌块
tongue and groove board 企口板
tongue and groove connection 企口连接;舌槽接合
tongue-and-grooved 企口的;有榫槽的
tongue and groove edge 榫槽接合边;企口边
tongue and groove face 榫槽面
tongue and groove floor(ing) 企口楼板;企口地板
tongue and groove joint 企口接缝;舌榫式接头;舌槽式接合;榫槽接合;企口接合;企口缝
tongue and groove joint of sewer 管道企口接口
tongue and groove labyrinth 迷宫密封;舌槽迷宫式;舌槽迷路
tongue and groove matched 拼花的;企口的
tongue and groove material 榫槽接合板材
tongue and groove method 闭口式孔型轧制法
tongue and groove pass 直孔闭孔型;闭口孔型
tongue and groove plank 嵌口接合板
tongue and groove rolling 闭口孔型轧制
tongue and groove seal contact face 榫槽密封面
tongue and groove sheet pile 企口板桩
tongue and groove siding 企口(侧)墙板;榫槽接合墙板;企口接合墙板
tongue-and-lip joint 楔形榫槽接合
tongue bar 尖铜条;尖铜棍(墙壁加固用)
tongued and grooved 舌与槽的
tongued and grooved block 企口块体
tongued and grooved block flooring 企口块地板
tongued and grooved brick 企口砖
tongued and grooved ceiling 企口顶棚;企口天花板
tongued and grooved deal 企口接合木铺板
tongued and grooved edge 企口侧边
tongue(d) and groove(d) flange 榫槽面法兰
tongued and grooved flange 密封法兰;舌槽法兰
tongued and grooved floor 企口地板
tongued and grooved form 企口块模板
tongue(d) and groove(d) joint 企口缝;企口接口;舌槽式接合;槽榫接合
tongued and grooved plane 企口刨
tongued and grooved plywood 企口胶合板
tongued and grooved sheet pile 企口板桩;锁口板桩
tongued and grooved shingle 企口墙面板;企口屋顶板
tongued and grooved siding 企口墙板
tongued and grooved surface 榫槽面;舌槽面
tongued and grooved timber 舌槽榫
tongued and grooved wood 企口木料
tongued and grooved wood sheet pile 企口块木板桩
tongue depressor 压舌板
tongued flange 带凸台的法兰;阳凸缘
tongued floor cover(ing) 铺舌榫地板
tongued floor(ing) 铺舌榫地板
tongue(d) miter 舌榫斜拼合;舌榫斜接;斜棒拼接
tongued washer 凸边垫圈
tongue file 小圆锉
tongue floor 企口铺板;企口地板
tongue forceps 舌钳
tongue grafting 舌接
tongue-groove block 企口块体
tongue-groove flooring 企口块地板
tongue-hill 舌状丘
tongue joint 企口接口;榫接;榫齿接合;舌榫接合
tongue joint with lug 带舌接合杆
tongue joint with movable lug 带活舌接合杆
tongue-like basin 舌状盆地;舌状流域
tongue-like distribution 舌状分布
tongue-like fold 舌状褶皱
tongue method 舌形试样法
tongue miter 斜拼合;企口斜角缝
tongue of flame 火苗
tongue of landslide 滑坡舌部
tongue of warm air 暖空气舌
tongue piece 舌片
tongue (rail) 尖轨;道岔尖轨;铁道岔尖轨
tongue rail made of special section rail 特种断面尖轨

tongue roll 凸舌轧辊
tongue scraper 整平机;刮舌板;舌板式铲运机
tongue-shaped 舌状的
tongue-shaped glacier 舌状冰川
tongue-shaped sand flood 舌状积沙
tongue shell 舌贝
tongue socket 绳头打捞筒
tongues of albic horizon 漂白层
tongue switch 转舌道岔;尖轨转辙器;尖轨转换器
tongue tear strength 舌形撕破强力
tongue tear test 舌形试样撕破强力试验;舌形扯破试验
tongue tip 舌尖
tongue(-type) switch 尖轨转辙器;舌形转辙器
tongue valve 活瓣阀
tongue wedge 舌片式槽楔
tongue weld 嵌合锻接
tonguing 企口接合;振动拍摄;舌榫接合
tonguing and grooving machine 制榫机
tonguing chisel 企口凿
tonguing cutter 榫槽铣刀
tonguing plane 槽刨;榫舌刨
tonic receptor 张力感受器
tonic train signal(l)ing 声列信号
tonify without causing stagnation 补而不滞
toning 调色
toning blue 调色蓝
tonite 吐纳特(一种烈性炸药)
ton-kilometer 吨/公里
ton-kilometers charged 计费吨公里
ton-kilometers hauled 运行吨公里
ton-kilometers of goods 货物吨公里
ton-kilometers of transportation 运行吨公里
ton-kilometers operated 运营吨公里
ton-kilometre 吨/公里
Tonk's fitting 可调节承板的金属支架
Tonk's strip 通克条(书柜内承受搁板的可调镀锌铁条)
ton-mile 吨/英里
ton-mileage 吨/英里;吨/英里
tonnage 量吨;军舰排水量;加工压力;吨(位);吨数
tonnage admeasurement 吨位丈量
tonnage board 量吨开口盖板
tonnage breadth 量吨宽度
tonnage capacity 量吨容积;容积吨
tonnage certificate 吨位证书
tonnage CO$_2$ equivalent 以吨计的二氧化碳当量
tonnage code 吨位丈量规范
tonnage coefficient 吨位系数
tonnage-day in port 停泊吨位天
tonnage deck 量吨甲板;量度甲板
tonnage deduction 量吨减额
tonnage depth 量吨深度
tonnage displacement 排水吨位
tonnage dues 吨位税;吨位费;吨税
tonnage dues certificate 吨税证
tonnage dues memo 吨税缴款单
tonnage duty 吨税
tonnage equivalent unit 标准箱(集装箱)
tonnage exemption 免除吨位
tonnage freight train 重载货物列车
tonnage-grade curve 吨一品位曲线
tonnage hatch 量吨开口;量吨舱口
tonnage law 吨位丈量法;吨位法
tonnage length 量吨长度
tonnage mark 吨位标志
tonnage measurement 吨位丈量
ton(nage) mile 吨/海里
tonnage noting 载重计算法
tonnage of cargo handled 操作量(指装卸)
tonnage of cargo transferred 货物装卸量
tonnage of exempted and open spaces 免除和开敞部位吨位
tonnage of freight arrived 货物到达吨数
tonnage of freight dispatched 货物发送吨数
tonnage of freight transported 货物运送吨数
tonnage of lacking in weight 欠重吨数
tonnage of load volume 载货容积吨数
tonnage of ship 船舶吨位;船舶吨数
tonnage of traction 牵引吨数
tonnage of vessel 船舶容积
tonnage opening 量吨开口
tonnage oxygen 工业用氧
tonnage passed by 通过吨位;通过吨数

tonnage per hour 吨位/小时
tonnage plan 吨位图
tonnage pool 货运联营(以吨计)
tonnage rate for hauling 运价
tonnage rating (of traction) 牵引定数
tonnage scale 吨位运费标准
tonnage schedule 吨位运费标准
tonnage section 丈量截面
tonnage standard 吨位丈量规范
tonnage tax 吨税
tonnage train 加开货物列车
tonnage utilization ratio 载重量利用率
tonnage value 船价(按吨位计算)
tonnage well 量吨井孔
tonne 公吨
tonne coal equivalent 吨煤当量
tonne per day 吨每日
tonne per year 吨每年
tonne petroleum equivalent 吨石油当量
tonner 吨级
ton of coal equivalent 煤当量吨
tonofilament 张力原丝;张力细丝
ton of refrigeration 冷吨;冷冻吨;冷藏吨重
tonogram 张力图
tonograph 张力描记器
tonometer 音调计;张力计;气压计
tonoplast 液泡形成体;液泡膜
tonoplast vacuole 液包膜【给】
tonoscope 音波振动描记器
tonotaxis 趋张力性
to note from the label 从标签得知
tonotransducer 张力换能器
tonotron 雷达显示管
ton-profit 每吨利润
tons carried 货物运送吨数
tons conveyed 货物运送吨数;货物发送吨数
tons deadweight 载重吨
ton sea mile 吨/海里
ton slates 称重石板瓦;称重石板瓦
tons loaded of goods 货物装载吨数
tons loaded per wagon 货车静载重
tons of cargo delivered 出库吨数
tons of dry material 干土吨
tons of dry sludge solid 干污泥固体吨数
tons offered of goods 货物待运吨数
tons of goods originated 发送吨数
tons of goods received 货物到达吨数;到达吨数
tons of goods transported 货物运送吨数
tons originated of goods 货物起运吨数
tons per centimeter 吃水吨数/厘米
tons per centimeter immersion 每厘米浸水吨数;每厘米吃水吨数
tons per day 每日吨数;日吨数;吨/日
tons per hatch-hour 舱时量
tons per inch 英寸吨;每英寸排水吨数
tons per ship-hour 船时量
tons per square inch 吨平方英寸
tons received of goods 货物到达吨数
tons registered 登记吨位【船】
tonstein 高岭石泥岩夹矸;白土石
tons transported 货物运送吨数
tons unloaded of goods 货物卸车吨数
ton weight 重量吨
tool 用具;工具;机具;机床
toolability 修补性
tool adjustment 刀具调整
tool and cutter grinding machine 工具磨床
tool angle 刀具角;刀尖角
tool apron 刀座
tool arm 刀臂
tool back clearance 刀具纵向后角
tool backlash movement 退刀动作
tool back rake 刀具纵向前角
tool bag 工具袋;工具包
tool bar 横木;通用机具架;通用机架
toolbar assembly 通用机架总成
tool bar bulldozer 通用机械式推土机
tool bar carrier 通用机架
tool bar clamp 刀杆夹具
toolbar cultivator 通用机架式中耕机
toolbar lister 开沟犁
toolbar machine 装在通用机架上的机具
tool base clearance 刀具最小后角
tool belt frog 工具套

tool bit 割刀;刀头;刀具
tool-bit holder 刀夹
tool block 刀架
tool box 工具箱
tool-box-door hinge 工具箱门铰链
tool-box-door latch 工具箱门闩
tool-box-door latch hasp 工具箱门闩扣
tool-box on cross slide 横向滑板刀箱
tool-box slide 刀箱滑板
tool cabinet 工具室;工具箱
tool car 工具车;检修车
tool carbide 刀具硬质合金
tool carriage 拖板;大刀架
tool carrier 送钻头工;刀架
tool-carrier chassis 自动底盘
tool-carrying device 刀架;车床刀架
tool case 工具箱
tool ceramics 工具陶瓷
tool charpening 刀具刃磨
tool chest 工具箱
tool chuck 刀夹
tool clamp 工具夹;刀夹
tool clearance 刀具后角
tool compartment 工具箱
tool control 工具管理
tool crib 工具仓库
tool cupboard 工具柜
tool design 工具设计
tool drawer 工具抽斗
tool drawing 工具图
tool dresser 钻具修理工;磨轮修整器
tool dressing 钻具修整;刀具磨锐
tool dynamometer 刀具功率计
too lean 过稀
tooled ashlar 凿纹方石;(平行的)柱槽琢面石
tooled by Crandall 用凿石锥加工
tooled finish 用工具修琢面层(石料);琢成面;凿石面;琢有平行槽的石面;琢纹饰面;假錾石饰面
tooled finish of stone 平行琢石面;(平行的)琢石面
tool edge 切削刃
tooled joint 压缝;用专门工具勾灰浆缝
tooled margin of stone 斧琢石边
tooled surface 凿石面;琢石面
tooled work 用凿刀修琢的石工
tool ejector 工具拆卸器
tool engineering 工具学
tool equipment 工具设备
tooler 阔凿;石工用宽凿
tool escape 退刀槽
tool extractor 钻具打捞器;工具拔出器;捞钩;拔除工具
tool fabrication change sheet 工具制造更改单
tool face 刀面
tool-faced masonry 琢石砌体
tool-facing masonry 琢面圬工
tool failures 工具失效
tool file 刀具文件
tool flank 刀具侧面
toolframe 通用机架;刀具架
tool ga(u)ge 工具量规;刀具样板;刀具量规;刀具检查器
tool geometry 刀具几何形状
tool grab 捞钩;打捞工具
tool grinder 工具磨床;车刀磨床
tool grinder sharpener 工具磨机
tool grindery 磨刀间
tool grinding 工具磨削
tool-grinding department 磨刀车间;磨工车间
tool-grinding installation 工具磨削装置
tool-grinding machine 工具磨床
tool guide 刀架导轨
tool head 刀具主轴箱
tool-head feed screw 刀架进给螺杆
tool-head slide 刀架滑台
tool-head traversing screw 刀架进给螺杆
tool holder 刀把;刀具柄;刀夹;刀杆;绘图头
tool holder of lathe 车床刀杆
toolholder slide 刀架滑台
toolholder support 刀夹支架
toolholding 刀具夹紧
toolholding device 刀夹具
toolholding slider 刀夹滑板
tool hold key 刀夹钥匙
toolhouse 工具室;工具棚;工具间;工具房

tool house foreman 码头工具管理员
toolie 钻具修理工
tooling 封面压花;凿面;用刀具切割加工;压缝;上嵌缝料;砂浆接缝压光;平行线凿石工艺;勾灰缝
tooling actuator 设备制动器
tooling chart 机床调整卡片
tooling cost 刀具加工成本
tooling for refinery maintenance 炼油厂检修设备用的工具
tooling layout 刀具设计
tooling resin 工具树脂
tooling system 刀具系统
tooling time 修整时间
tooling zone 刀具调整区域
tool-in-use system 刀具工作系统
tool joint 钻具接头
tool joint box (锁链接的)母接头
tool joint pin 公接头
tool joint screw grab 接头丝扣打捞爪
tool kit 工具箱;工具匣;成套工具
tool-lapping machine 刀具研磨机
tool life 工具寿命;工具使用寿命;刀具耐用度
tool lifter 自动抬刀装置
tool list 工具清单
tool lorry 工具车
tool magazine 多刀刀座;刀库;刀具库;刀具仓
toolmaker 工具制造者;刀具制造厂
toolmaker lathe 工具车床
toolmaker's button 钮形孔距规
toolmaker's ga(u)ge 工具样板
tool maker's lathe 工具车床
toolmakers measuring microscope 工具测量显微镜
toolmaker's microscope 工具显微镜;工具显微测定器
toolmaker's shop 工具车间
toolmaker's straight edge 刃口平尺
toolmaker's vise 万向虎钳
toolmaking 工具制造
toolmaking lathe 工具车床
toolman 刀具工
tool management 工具管理
toolmanufacturer 刀具制造厂
tool mark 工具商标;刻痕;工具痕;切削具切痕;刀痕
tool master 高精度工具
tool material 工具材料
tool mechanism 工具机构
tool microscope 工具显微镜;工具显微测定器
tool mounting 刀架
tool nipper 送钎工
tool nose 刀尖
tool of know ledge 丁字尺(俚语)
tool operation 工具操作
tool orthogonal rake 刀具前角
tool outfit 工具装备(件);成套工具箱;成套工具
tool pad 工具架;组合工具
tool pan 刀盘
tool pick-up 自动抬刀架
tool pits 刀头
tool position 刀具位置
tool post 刀架;工具架
tool-post collar 刀架套圈
tool-post for front and rear slide 前后两用刀架
tool-post for rear cross slide 后横刀架用刀座
tool-post grinder 刀架磨头
tool-post ring 刀架圈
tool-post rocker 刀架球形垫圈
tool-post screw 刀架螺旋
tool-post wrench 刀架扳手
tool preseting measuring instrument 刀具预调测量仪
tool presetter 刀具预调仪
tool presetting equipment 刀具调整设备
tool profile grinding machine 工具曲线磨床
tool pusher 钻工领班
tool rack 工具架
tool recess 退刀槽
tool-relief mechanism 自动抬刀机构
tool resistance 刀具阻力
tool rest 刀架
toolrest slide 刀架滑板
toolroom 工具间;工具室;工具车间;工具房
toolroom lathe 工具车床
toolroom machine 工具机

toolroom microscope 工具显微镜;工具显微测定器
tool saddle 刀架鞍板
tools cost 工具费
tool section 刀条
tool set 全套工具;成套工具
tool setter 刀具调整工
tool-setting 工具调整;刀具调整;刀具安装;调刀
tool-setting biagram 刀具安装图
tool setting ga(u)ge 调刀千分表
tool set-up 机床的调整
tools factory 工具厂
tools for assembly and disassembly of bogie 转向架拆卸工具
tools for cleaning rust 除锈工具
tools for downhole operation 井下作业工具
tools for unitized goods (货物的)成组工具
tool shank 工具柄部
tool shank for endmill 端铣刀刀柄
tool shank for rough boring tool 粗镗刀柄
tool sharpener 工具磨具;磨工具的设备;刀具磨;砂轮;刀具磨床
tool sharpening machine 刀具刃磨机
toolshed 工具房;工具棚
tool shelf 工具盘
tool shiner 钻具修理工
tool shop 工具车间
tool side 贴模边
tool side clearance 刀具横向后角
tool side rake 刀具横向前角
tool slide 切削进给刀架;刀架滑台;刀架滑鞍
tool slider 刀具滑座
toolsmith 工具锻工
tools of rock mechanics 岩石力学测定工具
tool specification 刀具规格
tool spindle mechanism 刀具旋转轴机构
tools register 工具底账;工具登记簿
tool stand 工具台
tool standard 工具标准
tool standardization 工具标准化
tool station 刀具厂
tool steel 工具钢
tool stones 优质金刚石
tool storage room 工具库
tool support 刀架
tools workshop 工器具车间
tool table 工具台
tool tear 刀具折裂
tool tempering 工具淬火
tool test 工具试验
tool tester 工具试验机
tool thrust 切削力
tool tip grinding machine 刀头磨床
tool tip material 刀头材料
tool tips 陶瓷刀具
tool trailer 工具车
tool up 装备加工机械
tool van 工具车
tool vehicle 工具车
tool wagon 工具车
tool window 工具传递窗
tool withdrawal mark 退刀伤痕
tool work 工具学
tool world 工具世界
tool wrench 钻挺扳手
tool write-up card 工艺卡(片)
toon 红椿木(东印度群岛和澳洲产的)
to order 空白抬头
tooth 细齿纹;咬合
tooth adapter 齿座
tooth addendum 齿顶高
tooth angle 研磨角;齿角
tooth-angling mechanism 搂楼倾斜机构;齿倾角调整机构
tooth annulus 环状齿圈
tooth axe 齿斧
tooth bar (松土机的)耙齿座杆;搂齿梁
tooth base 齿座;斗齿体
tooth bearing 轮齿接触面
tooth block 齿形utta
tooth breadth 齿宽
tooth brush holder 牙刷架
toothbrush holder tile 挂牙刷瓷砖
tooth bucket 带齿铲斗
tooth caliper 测齿长尺

tooth chamfer 齿轮齿倒棱
tooth chamfering machine 齿轮倒角机
tooth chisel 齿凿
tooth chisel finish 齿凿面
tooth chisel finish of stone 齿凿石面
tooth clamp 齿夹
tooth clearance 齿顶与齿根间隙
tooth contact 齿面接触
tooth core 齿缘;齿型心
tooth correction 轮齿修正
tooth coupling 齿式联轴器;齿式联轴节
tooth crest 齿顶
tooth crown 齿轮轮周;齿冠
tooth crushing pressure 轮齿压碎压力
tooth curve 齿向曲线
tooth cutting 切齿
tooth cutting machine 切齿机
tooth dedendum 齿根高
tooth depth 齿高
tooth diametral pitch 径节距
toothed bar 齿条;齿杆
toothed bar type self-loading forage box 齿杆式自装饲料拖车
toothed beater 钉齿式逐稿轮
toothed belt 锯齿形皮带
toothed bit 带齿钻头
toothed chain 齿链
toothed chain wheel 齿链轮
toothed clutch 齿轮离合器;齿形离合器
toothed concave 钉齿式凹板
toothed connector 齿板连接件;齿状式结合环;齿结合环
toothed coupling 齿形联轴器;齿轮联轴节
toothed crown 齿顶缘
toothed cutter 齿形切割刀
toothed disc mill 齿盘锯
toothed disk drawframe 锯齿盘式并条机
toothed drive belt 齿状传动带
toothed flywheel 齿圈飞轮
toothed forceps 有齿镊
toothed gear 齿轮
toothed gear drive 齿轮传动
toothed gearing 齿装置;齿轮传动装置
toothed gear type coupling 齿轮形联合器
tooth edge miller 齿边倒圆铣刀
toothed harrow 钉齿耙;齿耙
toothed harrow maintenance 钉齿耙保养
toothed jaw 齿式夹块
toothed link 齿板
toothed lock 啮合锁
toothed lock washer 齿式锁紧垫圈
toothed nail 带齿钉;刺钉
toothed ornament 齿形装饰
toothed plate 带齿连接板;齿板
toothed quadrant 扇形齿轮
toothed rack 齿条;齿轨
toothed rack push bench 齿条式推床
toothed rail 齿轨
toothed railroad 齿轨铁路;齿轨铁道
toothed railway 齿轨铁路;齿轨铁道;防滑齿轨铁道
toothed retractor 有齿拉钩
toothed ring 齿圈;齿环;齿状接合环
toothed ring connector 齿状结合环;齿结合环;齿环连接器
toothed ring dowel 有齿接环;齿环接钉
toothed roll crusher 带齿辊碎机;齿辊(式)破碎机
toothed roller 爪形镇压器;钉齿辊;齿形镇压器
toothed roller bit 牙轮钻头
toothed roller drilling 牙轮钻进
toothed scoop 带齿铲斗
toothed scoop shovel 齿斗式机铲
toothed scoop shovel excavator 带齿铲斗挖土机;齿斗式挖土机;齿斗铲土机
toothed scraper 齿形矿耙
toothed segment 扇形齿板;齿轮扇形
toothed shears 齿形剪断机
toothed sill 齿形消力槛;齿槛
toothed sleeve 有齿套筒
toothed tyre 花键接合的轮胎
toothed wheel 齿缘轮;齿轮
toothed-wheel gear 齿轮联动装置
toothed-wheel shaper 刨齿机
toothed-wheel shaping machine 刨齿机
tooth effect 啮合效果;啮合作用

tooth end 齿顶点
toother 齿形砖墙;齿接砖;齿接砖墙
tooth error 齿形误差
tooth face 齿面
tooth factor 齿轮系数
tooth fillet 齿根过渡曲面
tooth fillet radius 齿根圆角半径
tooth flank 齿面;齿根面;齿根侧面
tooth form 齿形
tooth-form factor 齿形因数;齿廓因数
tooth ga(u)ge 齿规
tooth gear 齿轮传动装置;齿轮
tooth generating 齿轮滚铣
tooth geometry 轮齿几何学
tooth grinding 锉牙
tooth head 齿顶高
tooth holder 斗齿插座;齿座齿夹;齿座;齿夹
toothing 使表面粗糙;留马牙槎(砖墙);刻齿;锉齿;装齿;直槎;留槎;磨齿;马牙槎【建】;齿轮连接;齿接;齿形接渣口
toothing bond 齿接
toothing course 齿形缝道(砌筑砖墙预留的)
toothing of brick wall 砖砌体齿形待接插口;砌砖体留齿插口;砖墙留槎;砌砖体齿形接口;留接砖墙
toothing of stone 石砌体留齿插口;石砌体预留齿形接口;石砌体齿形接口;待接石
toothing plane 开槽刨;带齿刨刀;锯齿刨;齿状刨;齿形槽刨
toothing stone 齿形接砌石块
tooth joint 槽齿联结;槽齿接合
tooth key 齿销
toothless drag harrow 无齿拖板;无齿耙
tooth load 轮齿负载
tooth lock washer 齿锁紧垫圈
tooth marks 走刀痕迹
tooth mesh 轮齿啮合
tooth of cutter 刀齿
tooth ornament 齿状饰;犬齿饰;中凸的四瓣花饰的线脚;齿形装饰
tooth outline 齿形;齿廓
tooth overlap 轮齿重叠
tooth parts 齿形各部
tooth pin 齿销
tooth pitch 齿距;齿节
tooth plane iron 梳形刨刀
tooth point 齿顶
tooth point shovel 单尖松土铲
tooth pressure 齿压
tooth profile 齿形;齿廓
tooth profile angle 齿形角
tooth punching machine 冲齿机
tooth rack 齿条
tooth rake 齿耙
tooth ratio 电机槽宽;槽齿比
tooth rest 支齿点;刀齿支片;刀齿支承板
tooth ripple 齿状波纹;齿形波纹;齿纹波
tooth-roller cracker 齿辊破碎机
tooth root fillet 齿根圆角
tooth-rounding hob 倒角滚刀
tooth scarifier 齿式松土机;齿式翻土机;齿式翻路机
tooth section 齿割面
tooth sector 转向齿轮;齿弧
tooth-shoe cutter 齿形管鞋
tooth sides fit of spline 在键齿侧配合
tooth space 齿隙;齿距;齿槽
tooth space indicator 齿距指示器
tooth-space micrometer 齿距千分尺
tooth spacing 齿槽空间距
tooth-spacing angle 齿间角
tooth spiral 齿旋
tooth surface 齿面
tooth swage set 锯齿侧向错齿量
tooth thickness 齿厚
tooth thickness micrometer 齿厚千分尺
tooth tip 斗齿尖;齿顶
tooth-tip leakage 齿尖漏磁;齿尖泄漏
tooth-tip reactance 齿间漏抗
tooth tolerance 轮齿公差
tooth top 齿顶
tooth to tooth composite error 齿间组合误差
tooth trace 齿间曲线;齿轨迹
tooth type coupling 齿式联轴节
tooth wearing height 牙齿磨损高度

tooth wearing mode 牙齿磨损模式
tooth wearing rate 牙齿磨损速度
tooth wheel rim 齿轮缘;齿轮辋;齿形轮缘
tooth width 齿宽
top 最高的;极点;桅盘;上升到顶;顶上;顶篷;顶(部);此端向上;上部
top address 顶地址
top aggregate size 最大集料尺寸;最大骨料尺寸
top aileron 上部副翼
top airfoil 上翼面
top air inlet louver 上部进气百叶窗
top all previous records 超过历史最高水平
top and bottom film box 放映机上下的胶片盒
top and bottom pilot tunneling method 水下导洞法
top and bottom process 顶底炼镍法;分层法
top-and-bottom saw 上下锯机
top-and-butt joint 大小头拼合
top and end grinding machine 磨头磨尼机
top angle 上部角钢(钢梁上部设置的);顶尖角;顶端角钢(钢梁上部设置的)
top annular plate 顶部环板
top antenna 车顶天线
top application manure 追肥
top area 顶面面积
top armo(u)r 桅盘围
top arrangement 炉顶装置
topartron 托帕特朗射频质谱仪
top assembly drawing 顶部装配图
top a tank out the easy way 兜底倒出
top available frequency 可用频率上限
to pay the penalty 缴罚款
topaz 金黄宝石;黄玉;黄晶;黄宝石;托帕石
topaz granite 黄玉花岗岩
topaz greisen 黄玉云英岩
topazization 黄玉化作用
topaz muscovite greisen 黄玉云母云英岩
topazolite 黄榴石
topaz quartz 黄晶
topaz quartz greisen 黄玉石英云英岩
topaz rock 黄玉岩
top backing up roll 上支承辊
top balance jewel 上摆钻
top ballast 面砟[铁];上层道砟
top-band pattern 上口护条样
top bar 上梁;上层(粗)钢筋;上鼠笼条;上层线棒;顶面钢筋;顶拉钢筋
top base 上基座;上基面
top beam 系杆;系椽;系梁;上梁;顶梁
top beam roof 顶梁屋顶
top bearing 上盖轴承
top bearing adjusting nut 上轴承调节螺母
top bearing cover 上轴承端盖
top bearing housing 上轴承箱
top bearing lock nut 上轴承锁紧螺母
top bearing shaft nut 上轴承轴螺母
top bearing shell 上轴瓦
top beater 上击肥轮
top bed 上层;顶层
top bell 顶料钟
top belt 上段运输带
top bench 上部阶段
top-benching 上梯段式开采法
top bench tunnel(l)ing 正台阶开挖
top bevel 上磨面;顶斜面
top bevel angle 顶斜角
top blade 上刀片
top blast 面吹
top-blast refining 顶吹氧精炼
top block 拱顶石;顶滑轮;(吊杆的)顶滑车
top-blower oxygen convertor 顶吹氧气转炉
top blowing oxygen 顶吹氧
top-blown converter 顶吹转炉
top-blown rotary converter 顶吹转炉
top board 上压板
top boiler connection 锅炉上接头
top boom 上弦(杆);顶弦;顶桁
top boom junction plate 上弦连接板
top boom member 上弦杆构件
top boom plate 上弦杆板
top boot 顶罩
top border 上图廓
top bow 顶弯梁
top-bow rest 顶弯梁架

top-bow separator 顶弯梁柱
top-bow socket 顶弯梁座
top box 上型箱;上箱
top bracing 上平联;上拉条;顶拉条;顶撑
top bracker 车顶托架
top branch 顶部分支
top brass 上部轴瓦
top brick 顶部用砖
top burner 灶面燃烧器
top burning kiln 顶燃式隧道窑
top-burning tunnel kiln 顶燃式隧道窑
top burton 顶滑车组
top camber 上曲面
top canch 顶
top cap 用再生橡胶补胎胎;石层之上的土层(采石场);盖顶;试样帽
top car clearance 电梯厢上部净空
top carriage 上架
top carriage trunnion bearing 上架耳轴轴承
top case 上型箱
top casting 上(浇)铸;顶铸
top cavity 上模巢
top ceiling 舱顶天花板
top center 上止点;上死点
top center heading 提前上死点
top-center indicator 上死点指示器
top centre heading 提前上死点
top chalk 上层白垩
top chamber 上闸室
top charging 炉顶装料
top check of caisson 沉箱顶板
top chisel 钳工錾
top chock 上轴承座
top chord 上弦杆;上弦;顶弦
top chord force 上弦杆力;顶弦力
top chord of truss 桁架上弦杆
top chord panel 上弦节
top chord plate 上弦杆板;顶弦板
top clamp 上夹钳;车顶夹
top clamping bracket 车顶托架
top clamping plate 顶层压紧板
top class 最高级
top-class optic(al) component 一级光学部件
top cleaner 甜菜顶清除器
top clearance 端部间隙;顶部间隙;电梯厢上部净空;齿端部间隙
top clearance of tooth 齿顶间隙
top clipping 剪梢
top closing 拱顶封合;隧道顶部合拢;顶部合拢(隧道)
top closure 炉顶封闭
top coat 面涂;面漆;面层涂料;面层;外涂层;涂面漆;涂层;顶面层;顶层(指油漆涂料等);表面涂层
topcoating 面漆;外层涂料;顶面涂饰;涂面漆
topcoat operation 顶层操作;面层操作
top colo(u)r 表面色;面色
top column 悬臂顶柱
top compartment 顶层;顶室
top concrete 面层混凝土;顶层混凝土
top concrete layer 上层混凝土
top condition 最佳条件;良好状态
top cone 高炉炉头;上部锥体;顶锥
top connection 上接头
top-contact rail 顶接触导轨;顶部接触轨
top contraction 顶层收缩
top contruction 炉顶结构
top couch (roll) 上伏辊
top course 抗磨表层;顶层铺砌;顶层;面层
top course concrete 上层混凝土
top course of tank block 上层池壁砖
top course tile 上瓦层(紧挨屋脊);顶层瓦
top cover 顶盖板;顶盖;顶板岩层
top cover labyrinth 上密封盖
top cross 顶交
top cross beam 装卸桥顶部十字架
top-cross test 顶交测验
top crown wheel 上小钢轮
top crown wheel screw 上小钢轮螺钉
top curve 上口线
top cut 顶端竖截(橡木);顶部截断;中槽;顶挖槽;顶面掏槽;顶部切削;顶部截槽
top-cut excavation method 反台阶形挖法
top cut method 顶部开挖法

top cutting 上部截槽
top cylinder cup 上端汽缸碗
top daylighting 顶部采光
top dead center 上止点;上死点
top dead-center indicator 上止点标记
top deck 上层车厢;顶甲板
top depth of cemented annulus 水泥环顶深度
top depth of cementing plug 水泥塞顶深度
top depth of gravel pack 砂砾充填段顶深
top depth of producing horizon 生产层顶深
top devit 塔顶吊杆
top diameter 圆木梢径;木材梢径;梢径;端端直径
top diaphragm plate 上隔膜板
top die 上膜具
top digit 高位数;高数位
top discharge 顶部出料
top disposal 表面处置
top distance 齿冠距
top distributing gear 炉顶布料装置
top documents 顶层文件
top dog 顶钩
top door rail (门的)上部导轨
top doup 上绞综
top-down 自顶向下的
top-down analyzer 自顶向下分析算法
top-down analysis 自上而下分析
top-down approach 自上而下近似;自顶向下(分析)法
top-down construction method 逆筑法
top-down design 自上而下设计;自上而下的设计;自顶向下设计
top-down effect 下行效应
top-down fast-back parser 自顶向下快速分析程序
top-down goal-oriented analysis procedure 自上而下面向目标分析步骤
top-down method 自顶向下(分析)法;逆作法
top-down parser 自顶向下分析程序
top-down parsing 自上而下剖析;自顶向下(分析)法
top-down parsing algorithm 自顶向下分析算法
top-down planning 自顶向下规划
top-down process 自顶向下分析过程
top-down recognizer 自顶向下识别算法
top-down system 自顶向下方式
top-down technique 自顶向下分析技术
top drainage 地表排水
top draining 地表排水
top drawing 顶视图
top dressing 过筛土(种植草皮用);浇注;土面施肥;敷面;表面处治
top dressing compound 表施肥料;顶肥;表面处理材料;敷面材料
top drift 上导坑
top-drive hydraulic power head 液压马达动力头
top driven 顶传动的
top-driven centrifuge 顶部驱动离心机
top-drive pneumatic power head 风动马达动力头
top-drive power head 动力头
top-drive press 上传动压力机
top drive tachometer 高速转速计
top drop anchor 下锚
top-drying 表干
top dust hood 顶罩
tope 卒塔婆(印度佛塔);佛塔(印度)
top efficiency 最高效率
tope guide 游览指南;圆顶塔导游人
topek 爱斯基摩人夏季住的小屋
Topeka and Santa Railway Company 艾奇逊—托皮卡—圣菲铁路公司
Topeka grading 托皮卡式级配
Topeka type asphaltic concrete 托皮卡型沥青混凝土
top elevation 顶面高程
tope-mound 印度塔;圆顶塔
tope-mound base 圆顶塔基座;圆顶塔基础
top end 顶端;端梢
top end cap 管顶盖
top-end computer 高档计算机
top-end facility 高档次设备
top-end-lop 枝梢头木
top end rail 上横梁;上端梁(集装箱)
top end transverse member 上端梁(集装箱)
top entering propeller agitator 顶伸式螺旋桨搅拌器

top entering (type) agitator 顶伸式搅拌器
top entry 顶伸入;顶端插入
top erosion 顶部冲刷
tope shrine 圆顶塔神龛
top extension 顶伸柱
top face 铺浆面
top failure 顶板崩落
top fastener 车顶夹
top fastening bolt 顶螺栓
top feed 顶部加料
top-feed baler 顶喂捡拾压捆机
top feeder head 顶冒口
top-feed furnace 顶部加料炉
top-feed hammer crusher 上部给料锤式粉碎机
top-feed oxygen bessemer furnace 顶吹氧气转炉
top felt 上毛布
top felt press 上毛毯压榨
top fermentation 顶发酵;表层发酵
top fiber 顶橼纤维;上层纤维
top filler 炉顶加料器
top filling 炉顶加料;填充上限
top fill up 装满
top-fired kiln 顶烧窑
top firing 顶部加热
top flange 上翼缘;上翼;上弦;上凸缘
top flange plate 上缘板;上翼(缘)板
top flap 上窗盖
top flask 上型箱
top flight 优秀的;高级的;最高一层房屋
top floor 顶层(指楼房);采板层楼面
top flow line 面层流线
top fluid 上部液体
top-fluid condenser 两气循环高温级冷凝器
top-fluid turbine 前级工质透平
top-for-bottom 底部朝上
top form 顶部模板;外模;上模
top former 上方喷浆成型器
top frame 顶面导架;顶框;顶骨架;顶层构架
top frame member 上部框架构件
top frame of gate 闸门的顶部构架
top fuller 圆锤;上用圆凹楔形锤
top-gaining 最大增重的
topgallant mast 上桅;顶桅
top gas 炉顶煤气
top gas exhaust port 炉顶排气孔
top gate 进水闸(门);顶浇浇口
top gate leaf 进水闸门门扇
top gatherer 甜菜顶收集器
top gating 顶注式浇铸系统
top gear 高速挡;高速齿轮;末挡(速度)
top gear acceleration 最高齿加速度
top gear ratio 最高齿轮速比
top-gear speed on gradient 上坡时的高挡行驶速度
top girth 梢端干围
top glazed 顶部上釉
top grade 高级
top-grade concentrate 上等精矿
top grade variety 高级品种
top-grade wood carving tool 高级木雕刻刀
top grain 顶层粒面
top grasping belt 抓顶皮带
top grass 高草
top grassland 高草草原
top grating 顶部格子板
top groove wear 顶槽磨损
top growth 上长
top gudgeon 上舵钮;上舱钮
top guide 前导承;上导头
top guide bar 上导杆
top guide bearing 上导轴承
tophaceous 砂质的;沙质的
top hamper 上部累赘船具
tophan box 离心罐
top handling attachment 顶吊架
top hang window 吊头窗
top hat 钢壳;帽形钢锭缺陷;缩顶;上帽
top hat annealing furnace 罩式退火炉
top hat beam 高顶帽形梁
top-hat furnace 罩式炉
top hat kiln 钟罩窑;钟罩式窑;高帽窑
top hat section 扎帽形截面
top hat-section stringer 管形截面桁条
top-hat structure 顶部加劲结构;顶部加筋结构

top head 上盖
top head drive bit 动力头钻机
top heading 上导坑;上半断面;顶导洞;顶部导坑;顶部导洞(隧道的);上部超前掘进法
top heading and bench method 坑台法;顶导坑及平台法;顶撑及平台法
top-heading method 上半断面开挖法;上半断面超前施工法;顶撑法;上导坑法;上导洞法(开挖隧道);上部超前掘进法
top heading-under head bench 支承顶推法
top heater 末级加热器
top heaviness 树冠长势过重
top-heavy 上重下轻;投资过度
top-heavy load 上部重的荷载
top height 林分高
top-height diameter 林分高大木直径
tophet 多飞特合金
tophet alloy 托非特镍铬电阻合金
top hinge 上铰链;顶铰
top-hinged casement 上旋窗;上悬窗
top-hinged inswinging window 内开旋窗;顶悬内开撑窗;上悬内开窗
top-hinged outswinging (窗扇的)上悬外撑
top-hinged outswinging window 上悬外开窗;顶悬外开撑窗
top-hinged sash (window) 上旋窗
top-hinged swinging door 翻门
top-hinged swinging window 上悬窗
top hinged window 上旋窗;上悬窗;铰链装在顶上的窗
top hinge of arch 拱顶铰
top hog 顶部挠曲;顶部拱曲
top hold-down 上支架;上部压板
top holder 车顶架
top hole 上部孔段;顶孔
top-hole choke 上孔阻风门;上孔节气门
top hole pressure 井口压力;出口压力(压气升液器的)
top hole temperature 井口温度
top hooker 块石装载工
top hopper 茎叶箱
top hung 上铰链推窗;上悬式(窗);顶悬
top-hung casement 上悬窗
top-hung sash 上旋窗;上悬窗
top-hung window 上旋窗;吊窗;上悬窗
top-hung window opening inwards 内开上悬窗
top-hung window opening outwards 外开上悬窗
topiary 修整树形的;修剪艺术;植物整形
topiary art 修剪艺术;整形艺术
topiary garden 整形树木园
topiary tree 整型树
topiary work 修树工作;修枝工作;修剪整形(树木的)
topic 论题;课题;题目
topical action 局部作用(量)
topical application 局部施用
topical diagnosis 部位诊断
topical type 局限型
topical words 主题词
topic map 专题地图
topic of development 研究课题
topic sentence 标题句子
top idler 上托辊
top illumination 顶部照明
top inlet 顶部进风道;顶部进气道
top inwall line 炉身水平线
top iron 顶弯梁
top iron and screw 盖铁
top jack 上提综杆
top jaw 上部凸轮;分离凸轮
top joint 顶连接;顶部接头
top kerf 顶面掏槽;顶(采矿)
top knife 上刀片
top knife block 上刀片滑块
top labo(u)r 高级技工
top ladder 桅杆;吊臂;悬臂
top land 端环梢脊
top landing 上部出车台
top lantern 枪灯
top lap 搭接部分(屋面板);上研磨盘;搭接长度;顶超
top lateral bracing 纵向上平联系;上弦横向水平支撑;上弦侧向支撑;上平纵联;上部水平支撑;顶横支撑

top laterals 纵向上平联
to play a decisive role 举足轻重
to play a special role 特殊作用
top layer 顶层
top layer diversion 表层引水;表层取水
top leaf 上部窗扇;上部门扇;护顶层;上分层;顶上叶
top leaf of spring 顶簧片
top level 顶面标高;顶面高程
top level die-filling 上面装模
top-level environment 顶层环境
top lever 顶装合合杠杆;外操纵杆;上操纵杆;上提综杆
toplift 千斤索;顶索
top lifter 提顶臂
top lift frame 顶吊框架
top lift(ing) spreader (起吊集装箱的)挂顶吊架;吊具顶吊
top lift sling 吊索底吊
top light 桅(顶)灯;顶光;桅杆顶灯;天窗;电视塔顶灯
top lighting 顶部采光;顶部照明
top limit 最高限额;上限
top line 背线
top lined eaves 顶部铺盖的屋檐
top line of teeth 齿顶线
top line recker (悬挂装置的)上抬杆铰接环
top link adjustment (悬挂机构的)上拉杆调节
top link joint 上拉杆铰链
top-link sensing 上拉杆传感
top liquid level 顶部液面
top load 尖峰负荷
top-loaded vertical antenna 顶部加载竖直天线
top loading 装在上面
top-loading balance 上皿式天平
top-loading belt 上部装载带型运输机
top-loading belt conveyer 上载式带式运输机
top loading filter 顶部加料式过滤机
top loading kiln 顶装窑(由窑顶部装坯)
top loading of explosive 顶部装药
top-loading washer 顶装式洗涤机
top locking pintle 顶部锁紧销钉
top-lock lift 顶销提臂
top-lock-lift hole cap 顶销提臂孔盖
top-lock nut 上端防松螺母
top log 顶段原木;梢端原木
top longitudinal bracing 上弦纵向水平支撑
top main plane 上主翼
topman 采板工人;工地杂工;地面上工作的建筑工人
top management 主管;最高管理阶层;企业首脑人物上层管理
top manager 高级管理员;首席经理
top margin 上图廓;顶边
top mark 顶部标志;顶标(志)
topmark buoy 顶标浮标
topmast 中桅(三节桅);上桅;顶桅
topmast hounding 中桅桅肩;上桅桅肩
top mat 上层钢筋网
top material 顶杆
top mesh 上层钢筋网
top moor layer 沼泽地表层
top mop 顶涂层;平屋面顶层涂沥青
topmost 最高处
topmost stor(e)y 最高层;楼顶层
top mo(u)ld half 凸模
top mounted rod 顶插棒
top mulch 顶层覆盖物
top nailing 顶部钉固
top needle 上盖针
top nipper 上钳板
top nipper arm 上钳板臂
top node 顶端节点
top noise 局部噪声
top-notch 最高质量
top-notch product 第一流产品
top nozzle 上喷嘴
top nut 上螺母
topoangulator 像片测倾仪
topo beacon 地形测量标志
topocenter 站心
topocentric aberration 站心光行差
topocentric Cartesian coordinate 站心直角坐标
topocentric celestial coordinate system 站心天球坐标系
topocentric coordinates 站心坐标;站心球面坐标

topocentric coordinate system 站心坐标系;站心坐标系
topocentric equatorial coordinates 站心赤道坐标
topocentric horizon 视地平
topocentric libration 站心天平动
topocentric origin 站心原点
topocentric right ascension 站心赤经
topocentric terrestrial coordinate system 站心地面坐标系
topocentric velocity 站心速度
topochemical polymerization 局部化学聚合
topochemical reaction 局部化学反应
topochemistry 局部化学
topocline 地形渐变群;地形递变;地理倾差
top of boring 孔顶
top of climb 升限
top of crest 波峰顶
top of cutting slope 堑顶
top of dam 坝顶
top of descent 降限
top of drain 排水板顶
top of embankment 路堤顶
top of engine room 机舱顶
top of floor 肋板上缘;地板表面
top-of-form set 表头设置;表头格式设置
top of foundation 基础顶部;基础顶面
top of furrow 沟头
top (of grade) 坡顶
top of hump 脊峰;山峰
top of layer 层顶
top of main space 机舱顶
top of mast clamp 杆顶抱箍
top of oil horizon 油层顶部
top of piston 活塞顶
top of process(ing) interval 处理井段最小深度
top of rail 轨面;轨顶
top of roadbed 路基顶面
top of rock bed 岩层顶层
top of roof rock 盖层顶界
top of saddle 鞍座顶
top of shoulder 肩顶
top of side slope 边坡顶
top of slab 板顶面
top of sleeper 轨枕面
top of slope 坡顶;边坡脚;边坡角
top of soil sample 土样顶面
top of source oil window 生油窗顶界
top of stack 栈板的顶
top-of-stove ware 玻璃烹饪器皿
top of stroke 冲程上限;冲程顶
top of superstructure 上层建筑顶
top of tank wall 池顶
top-of-the-atmosphere fluxes 大气顶部通量
top of the flood 涨潮顶点
top of the slag 渣面
top of tooth 齿尖
top of weir 堰顶
top of well 井口
Topogon lens 弯月形镜头;小孔径宽视场镜头
topogram 内存储信息位置图示
topograph 分域图
topographer 地形测量员
topographer-assistant 见习地形测量员
topographic(al) 地形学的;地形的;地形测量的
topographic(al) adjustment 地形图接边
topographic(al) adolescence 地形青年期
topographic(al) and coastal survey 地形岸地测量
topographic(al) and cultural indication 地形地物标志
topographic(al) and geomorphic(al) condition 地形地貌条件
topographic(al) annotation 地形调绘
topographic(al) appreciation 地形判断
topographic(al) base (map) 地形底图
topographic(al) basis 地形底图
topographic(al) bathymetric(al) map 海陆地形图
topographic(al) beacon 地形测量标志
topographic(al) cartography 地形图制图学
topographic(al) change 地形变化
topographic(al) characteristic 地形特征
topographic(al) classification 地形类别;地形分类
topographic(al) climax 地形顶极;地形演替顶极
topographic(al) condition 地形情况;地形状况;地形条件

topographic(al) condition of spring emerging 泉出露的地形条件
topographic(al) contour 地形等高线
topographic(al) control 图根控制
topographic(al) control point 地形控制点
topographic(al) control survey(ing) 地形控制测量
topographic(al) correction 地形校正;地形修正;地形改正;地形订正
topographic(al) crest 棱线
topographic(al) curl effect 地形涡度效应;地形旋度效应
topographic(al) data 地形资料;地形测量资料
topographic(al) data bank 地形数据库
topographic(al) data center 地形数据中心
topographic(al) data digitizer 地形资料数字化器
topographic(al) deflection 地形垂线偏差
topographic(al) details 地形细图;地形细节;地形细部
topographic(al) diagnosis 部位诊断
topographic(al) distribution 地形分布
topographic(al) divide 地形分水线;地形分水岭;地形分界界;地形分界线;地表分水线
topographic(al) drafting 地形图;地形绘图
topographic(al) drawing 地形图;地形绘图
topographic(al) effect 地形影响;地形效应
topographic(al) engineer 地形测量工程师
topographic(al) entity 地形要素
topographic(al) estimation 地形评价
topographic(al) expression 地形表示(法)
topographic(al) factor 地形因素;地形系数
topographic(al) feature 地下情况;地形;地貌;地形要素;地形特征;地形特性;方位物;显著地形
topographic(al) feature of riverbed 河床地形特征
topographic(al) field book 地形测量手簿
topographic(al) film 地形测图航摄胶片
topographic(al) form 地形形状;地形特征;地形
topographic(al) high 高地
topographic(al) identification 地形判读
topographic(al) index 地形指数
topographic(al) infancy 地形幼年期;幼年期【地】
topographic(al) influence 地形影响
topographic(al) information 地形资料;地形信息
topographic(al) information processing 地形资料整理
topographic(al) irregularity 地形不规则
topographic(al) isostatic correction 地形图内容地壳均衡校正;地形地壳均衡校正
topographic(al) isostatic deflection 地形均衡垂线偏差
topographic(al) isostatic reduction 地形地壳均衡改正
topographic(al) latitude 地理纬度;大地纬度
topographic(al) legend 地形图符号
topographic(al) level(l)ing 水准地形测量;地形水准测量;地形高程测量
topographic(al) level(l)ing line 地形测量水准线
topographic(al) lineament 地形线性体
topographic(al) loading effect 地形负载效应
topographic(al) longitude 大地经度
topographic(al) manuscript 地形图底稿
topographic(al) map 地形图;地形测图
topographic(al) map content element 地形图要素
topographic(al) map of construction site 工点地形图
topographic(al) map of urban area 城市地形图
topographic(al) mapping 绘制地形图;测制地形图;地形绘图;地形测绘
topographic(al) mapping with electronic tach(e)ometer 电子速测仪测图
topographic(al) mapping with plane table 平板仪测图
topographic(al) mapping with transit 经纬仪测图
topographic(al) map revision 地形图修测
topographic(al) map subdivision 地形图分幅
topographic(al) map symbol 地形图图标;地形测量站;地形图图式
topographic(al) masses 地形质量
topographic(al) matching guidance film 地形匹配制导胶片
topographic(al) materials 地形资料
topographic(al) maturity 地形壮年期;壮年地形
topographic(al) meridian 地理子午线;地理子午圈

topographic(al) meter 地形测量仪
topographic(al) model 地形模型
topographic(al) model panel 地形模型板
topographic(al) mo(u)ld alternation 地形模型修改
topographic(al) note 地形测量手簿
topographic(al) obstacle 地形障碍
topographic(al) old age 老年地形;地形老年期
topographic(al) original map 地形原图
topographic(al) parallel 地理纬度平行圈
topographic(al) parameter 地形参数
topographic(al) pattern 地形图图形;地形类型
topographic(al) plan 地形平面图
topographic(al) plane-table survey 平板仪地形测量
topographic(al) planimetry 地物测图
topographic(al) plot 地形测图
topographic(al) plotting 编绘地形图;地形描绘
topographic(al) point 地形点
topographic(al) point spacing 地形点间距
topographic(al) portrayal 地形形态描述
topographic(al) position finding 地形测量定位
topographic(al) preparedness level 成图状况
topographic(al) profile 地形纵断面;地形剖面
topographic(al) profile map 地形剖面图
topographic(al) profile photography 地形剖面摄影
topographic(al) quadrangle 地形图梯形图幅
topographic(al) reconnaissance 地形踏勘;地形勘察;地形勘测
topographic(al) reduction 地形改正
topographic(al) reform design 地形改造设计
topographic(al) relief 地形起伏
topographic(al) relief grade 地形起伏程度
topographic(al) remo(u)ld of garden 园林地形改造
topographic(al) scale 地形图比例尺;地形标尺
topographic(al) sheet 地形图图幅
topographic(al) sketch 地形略图
topographic(al) slope 地形坡度
topographic(al) staff 地形测量标尺;地形尺;地形测量人员
topographic(al) street system 顺应地形布置的道路网;依地形布置的道路网;地形道路网
topographic(al) survey(ing) 地形测量;地形调查;地貌调查
topographic(al) survey of bridge site 桥址地形测量
topographic(al) surveyor 地形测量员
topographic(al) survey recorder 地形测量记录页
topographic(al) symbol 地形图符号;地形符号
topographic(al) tachometric telemeter 地形快速测距仪
topographic(al) terms 地形测量术语
topographic(al) triangulation 图根三角测量
topographic(al) unconformity 地形不整合
topographic(al) ware 风景陶瓷器皿;风景画陶瓷器皿
topographic(al) youth 地形幼年期;幼年期
topographic(al) zoning 地势分带
topographic-geologic(al) map of coalfield 煤田地形地质图
topographic-geologic(al) map of exploration area 勘探区地形地质图
topographic-geologic(al) map of mine district 矿区地形地质图
topographic-geologic(al) map of ore deposit 矿床地形地质图
topographic-geologic(al) map of ore district 矿区地形地质图
topographist 地形学家;地形测量员
topography 局部记载;金属表面形貌学;地形学;地形描述;地形描绘;地形测量学;表面形态测量学
topography note 地形测量手簿
topography of basin bottom 盆地底部地形
topography of bridge site 桥位地形
topoinhibition 局部抑制
topologic(al) Abelian group 拓扑阿贝耳群
topologic(al) algebra 拓扑代数
topologic(al) algebraic systems 拓扑代数系
topologic(al) classification 拓扑分类
topologic(al) complex 拓扑复合形
topologic(al) consistency 拓扑相容性
topologic(al) convergence 拓扑收敛

topologic(al) critical point 拓扑临界点
topologic(al) cross section 拓扑截面
topologic(al) data 拓扑数据
topologic(al) degree 拓扑度
topologic(al) description 拓扑描述符
topologic(al) descriptor 拓扑描绘子
topologic(al) design 布局设计
topologic(al) dynamics 拓扑动力学
topologic(al) editing 拓扑编辑
topologic(al) embedding 拓扑嵌入
topologic(al) extension 拓扑扩张
topologic(al) feature 拓扑要素
topologic(al) field 拓扑域
topologic(al) groupoid 拓扑广群
topologic(al) groups 拓扑群
topologic(al) immersion 拓扑浸入
topologic(al) insertion 拓扑嵌入
topologic(al) invariance 拓扑不变性
topologic(al) irreducibility 拓扑不可约性
topologic(al) lattice 拓扑格
topologic(al) limit 拓扑极限
topologically conjugate 拓扑共轭的
topologically connected components 拓扑连通分支
topologically corresponding 拓扑对应的
topologically equivalent spaces 拓扑等价空间
topologically identical model 拓扑恒等模型
topologically invariant 拓扑不变的
topologically irreducible 拓扑不可约的
topologically nilpotent element 拓扑幂零元
topologically solvable group 拓扑可解群
topologic(al) manifold 拓扑流形
topologic(al) matrix 拓扑矩阵
topologic(al) method 拓扑方法
topologic(al) nilpotent operator 拓扑幂零算子
topologic(al) pair 拓扑偶
topologic(al) phenomenon 拓扑现象
topologic(al) polyhedron 拓扑多面体
topologic(al) product 拓扑乘积
topologic(al) relation 拓扑关系
topologic(al) relationship 拓扑关系
topologic(al) ring 拓扑环
topologic(al) semigroup 拓扑半群
topologic(al) simplex 拓扑单形
topologic(al) solid sphere 拓扑球
topologic(al) space 拓扑空间
topologic(al) sphere 拓扑球面
topologic(al) structure 拓扑结构
topologic(al) substitute 拓扑代换
topologic(al) sum 拓扑和
topologic(al) transformation 拓扑变换
topologic(al) transformation group 拓扑变换群
topologic(al) triangle 拓扑三角形
topologic(al) unit 拓扑单位
topologic(al) upper limit 拓扑上限
topologic(al) viewpoint 拓扑观点
topology 外形学;拓扑;地志学
topology of network 网络拓扑
topology of the system 系统布局
topology tolerance 布局容差
topomap 地形图
topometer 地形测斜仪
topometric survey(ing) 地形测量
topometry 地形测量学
topomineralogy 标型矿物学
top one-wayfired soaking pit 单侧上烧嘴均热炉
toponomy 地名学
toponym 地名;部位名称
toponymic index 地名索引
toponymy 地名学;部位命名法
top opening 顶孔
top-opening baler 顶倾式捡拾压捆机
top opening freezer 开顶式冰柜
top operator 栈算符
topophysis 部位效应
top oriented management 高级适向管理
toposcopy 局部检查
toposequence 地形系列
topostratigraphic(al) unit 地形地层单位
topostratigraphy 地形地层学;地方地层学
topotactic polymerization 局部定向聚合
topotaxy 同构变晶
topotype 地区型;地模标本
top-out 需求顶点;上部建筑;上层建筑;封顶

top overhaul 大修
top paint 面漆
top panel 上面板
top parallel entry 上部平行平巷
top pass 上排孔型(轧制用语);上轧槽;顶部焊道
topped crude 脱烃石油;拔顶原油
top pendant 吊重短索(挂绞辘用)
topper 千斤调整索;拔顶装置
top performance 最优性能
top petrol tank 上置式油箱
top phase 上相;顶相
top picking 挑顶(隧道工程)
top piling 上层码垛
top pinch roll 上夹送辊
topping 注满;蒸去轻油;截冠;混凝土面层;前置;堤顶填高;齿顶变尖;拔顶;上部
topping boom 吊杆
topping coat 塌涂;三道粉刷中的第二道;面层下的毛抹灰
topping compound 面层胶混合料;面层胶混合料
topping finish 面层修整
topping governor 主控调速器;极限调速器
topping heater 拔顶加热器
topping joint 面层叠缝
topping lift 吊扬索;千斤索;顶索
topping lift bail 连接在千斤索拉端的三角眼铁
topping lift block 千斤索滑车
topping lift bull rope 小千斤;千斤索升降绳
topping lift chain 千斤链
topping lift eye bracket 千斤索眼板座
topping lift fall 千斤索拉端
topping lift purchase 千斤索绞辘
topping lift winch 千斤索定位绞车
topping line 千斤索
topping mechanism 切削装置
topping mix(ture) 面层配合料
topping of dyke 堤坝的顶面覆盖
topping off 装满;装舱操作;舱顶装载
topping of road 路面;铺筑路面;道路面层
topping out 主体完工;封顶
topping-out ceremony 上梁庆宴;封顶仪式(建房)
topping paint 水线漆
topping pipe still 拔顶管馏器
topping plant 前置机组;初馏装置;拔顶厂
topping power 前置功率
topping roll 上轧辊
topping rope 千斤调整索
topping slab 面层板;平板面层
topping still 粗馏塔;初馏塔;拔顶塔
topping unit 蒸馏装置;拔顶设备
topping up 舱内惰气加压;补充惰性气体;液体补充;添油;充气
topping-up pump 注液泵;充液泵;充气泵;补给泵
topping wall 顶墙
topping winch 千斤索绞车
top pintle 上舵栓;顶部销钉
top pipe branch 顶部支管
top pipe inlet 顶管入口;顶管进口
top pipe junction 顶管接头;顶管连接处
top piston ring 顶活塞压环
top pivot 上枢轴
top-pivoted window 上翻窗;上悬窗
top plate 钢板弹簧舌片;顶板
top-plate angle 顶片角
top-plate cutting angle 顶片锯切角
top-plate screw 顶板螺钉
top platform 顶部平台
top plating 顶边列板
topple 主轴俯仰现象;倾覆
topple axis 主轴俯仰轴;倾覆轴;倒转轴
toppling (collapse) 崩塌
toppling failure 倾覆破坏
top plug 上塞;上胶塞
top point 最高点
top point of vertical curve 竖曲线顶点
top-pour 上浇;顶浇
top pouring 上铸(法);上浇;顶浇
top pouring through pencil gates 顶浇雨淋式内浇道
top-pour ladle 倾注桶;倾动式浇包
top pressure 最高压力;上压力;顶部压力
top price 最高价(格);高价
top print 型芯记号
top priority 最优先

top priority lending rate 最优贷款利率
top prism 上部三棱镜
top product 塔顶产品
top projection 尾柱顶端
top pruning 打顶
top pulley 上滑车
top pulverizer 甜菜顶切碎器
top quality 最优质的
top quality equipment 最优质设备
top-quality rubber 优质橡胶
top rafter 顶椽
top rail 柱顶线盘;(枢架顶上的)水平构件;上横档;栏杆扶手;上冒头;上栏;上档(门窗的);(门的)上框;顶横杆;桅楼扶手
top rail of door 门717711;上冒头
top rail to fabric 网顶横杆
top rake 削角;纵坡角;前倾角
top reflector 上部反射层
top reflux 塔顶循环回流;塔顶回流
top register 栈顶寄存器
top reinforcement 架立钢筋;上部钢筋
top removal 打尖
to press through 绞漆;挤净
top rest 上刀架
top retaining wall 上挡墙
top ring 炉头钢圈;炉顶钢圈;顶圈
top-ring groove 顶活塞环槽
top riser (楼梯的)最高一级踏步面板;顶冒口
top road bridge 桁架顶车路桥;上承式桥;上承式公路桥;上承桥
top rod 上部钻杆
to produce fruits and vegetables 生产水果和蔬菜
to produce more food 生产更多的粮食
to produce various food 生产各种食物
top roll 急流顶漩;急流顶溢;上轴;上辊;顶部旋滚
top-roll assembly 上轧辊组合部件
top-roll balance 上轧辊的平衡
top-roll-balance arrangement 上轧辊平衡装置
top-roll-balancing mechanism 上轧辊机构
top-roll counterweight balance arrangement 上轧辊重锤式平衡装置
top roller 拉边机;拉边辊;顶端滚柱
top-roll hydraulic balance arrangement 上轧辊液压式平衡装置
top-roll spring balance arrangement 上轧辊弹簧式平衡装置
to promote economic property 繁荣经济
top root ratio 冠根比率
top rot 顶腐
top-rounding 上层环行(路)
to provide a complete set of services 提供成套服务
to provide a solid foundation 提供牢固的基础知识
to provide staff 配备人员
top-running bridge 沿轨道上翼缘行走的桥架(起重机)
top-running crane beam 上承式吊车梁
top-running stacking crane 支承桥式堆垛起重机
tops 最初馏分;顶馏分;拔头油
top sag 顶部下垂
topsail 中桅帆;上桅帆
topsail halyard bend 上桅扬帆结
top sample 顶部取样
top sand 粗砂;粗粒砂
top sash 顶部窗格;顶部窗扇
top scalping deck 上层筛面
top science 尖端科学
top seal 顶部密封
top sealer 灌缝剂;顶部防水材料
top sealing 灌缝
top secret 绝密
top secret file 绝密文件
top section 拱顶截面;拱顶段;顶截面
top segment 抄针盖板
top-serrated knife (切割器的)上面刻齿动刀片
top set 上分支
topset bed (三角洲的)顶积层
topset deposit 顶积层沉积
topset slope 顶积坡
tops from crude distillation 原油拔顶气;拔顶气
top shadow line 屋面板上黑色砂砾条
top shear blade 上剪片
top shed 上梭口
top shield 进水闸(门);上部防护屏;隧洞顶部防护板;隧道顶盾
top shielding 上部屏蔽
top shooting 挑顶爆破
topside 舷侧;外舷部;上面;水线以上部分
topside arm 车顶侧柱
topside black 水线上部舷侧用黑漆
topside frame 顶边肋骨
topside lighting 顶部侧光
topside line 船舷顶线;船边顶线
topside paint 干舷漆;船舷漆
topside planking 舷侧外板
topside plate 顶边列板
topside plating 舷侧外板
topside rake 顶侧倾度
topside scantlings 顶边船材尺度
topside sounder 上层探测器;顶层探测器;电离层探测器
topside sounding 顶外探测
topside strake 舷侧顶板外板;副舷侧厚板
topside tank 顶边舱;舷侧水柜;上部油舱
topside view 俯视图;顶视图
top sieve 上筛
top size 最大尺寸
top-size aggregate 集料最大尺寸;骨料最大尺寸;最大骨料尺寸;最大尺寸的集料
top skimmer 最表层鱼
top slab 顶板
top slat 车顶铰接机构
top-slat iron 车顶铰接机构杆
top-slice method 下向水平分层崩落法
top slicing 下行分层崩落采矿法
top slicing and caving 下向水平分层崩落采矿法
top slicing by rooms 下向水平分层崩落采房法
top slicing combined with ore caving 下向水平分层崩落采矿法
top slicing cover caving 下向水平分层顶板崩落采矿法
top-slicing mining 下向分层崩落采矿法
top-slicing with inclined slices 下向倾斜分层崩落采矿法
top slide 顶部滑板
topsman 桅盘瞭望水手;桅楼瞭望水手
top-socket 车顶弯梁座
top sod 面层草皮;表层草皮
topsoil 土壤耕作层;土的上层;耕土;土壤上层;上层土(壤);浮面土;表土;表层土(壤);沃土
topsoil cover 表土保护层
top soil erosion 表土侵蚀
topsoil excavation 表土开挖
topsoiling 用填砂筑路
topsoil planting 土皮植草法
topsoil replacement 表土更换;换表层土
topsoil road 天然砂土路
topsoil stripping 表土剥离;挖除表土
topsoil surfacing 表土面层;天然土面层;天然土铺面
top speed 最高速率;最高速度;最大转速;最大速度;全速
top spit 粗废料
top spit from quarries 采石场粗废料
top sprays 上喷嘴组
top squeeze molding machine 上压式造型机
top steam 最大蒸气压(力);顶部吹气;釜顶蒸汽
top steel feed roll 上传送钢辊
top step 顶踏级;顶踏步;顶踏脚;顶踏阶
top-stitched seam 明缝
topstone 压顶石;盖顶石;顶层石
top story (英国住房的)顶层房间;楼顶层;最高层;顶层
top strap 车顶条
top stratum 覆盖层;顶层【地】;地表岩层;[复] strata
top structure 顶部结构
top sun louver 顶遮光板
top support 车顶托架
top surface 顶面;上表面;顶面面积
top surface bubble 玻璃带表面气泡(玻璃等的)表面气泡
top-surface camber 上曲面
top-surface dummy joint 顶面假接头
top surface error 顶面误差
top surface of liquid 液体的上部表面
top surge 涌浪最高水位;最高涌浪
top surge water level 涌浪最高水位;最高升水位
top suspension 上悬式
top suspension (basket) centrifuge 上悬式离心机
top swage 上型模;上陷型模
top switch 顶开关
top sword 上压刀
top tackle 吊挂卫板
top tank 日用油箱;日用油柜;日用水柜;上箱
top tank air 罐顶空气;上层气体
top tapping standard riddling machine 顶霹式标准振筛机
top teeming process 上注法
top telescope 顶端望远镜;顶部望远镜;顶副镜
top temperature 炉膛温度
top tension 上线拉力
top tension wire 顶张拉钢丝
top terminal 高压电极
top terminal landing 顶层电梯平台
top the lifts 升起吊杆
top the oil 拔油顶
top the oil sand 钻开油层
top thrust 最大推力
top tie 顶部拉杆
top tier 顶层
top tight 顶部系统
top-timber 顶木梁;顶肋材
top tone 墨色;表光
top tower control 塔顶调节
top trace 上部记录线;(地震图的)上部记录曲线
top track (推拉隔断上的)顶轨
top transom 栏杆上档
top transversal strut 上横撑
top transverse strut 上横撑
top tread roller 托链轮
top trench 横浇口
top tumbler 上鼓轮;上导轮
top turbine 前级透平
top union 冠接点
top unloader 上部卸载机(青储[贮]塔)
top up 装满
top usable frequency 可用频率上限
to put an end to waste 杜绝浪费
to put into effect 付诸实施
to put into practice 付诸实施
to putty 打腻子
top velocity 最高速度
top veneer 胶合板面层
top view 鸟瞰图;正面观察;上视图;顶视(图);俯视图
top wall 上盘;顶盘
top wall of intrusion 岩体顶盖
top washing 顶部清洗
top water 顶部水;顶层水;冲水;表层水
top water level 最高蓄水位;最高水位;水库最高蓄水位
top water sprayer 上喷水器(在被轧制带材的上面)
top wave 顶波
top wear 顶面磨耗
top wearing surface 磨耗层(道路)
top weight 垛顶部货重量
top width 顶宽;坝顶宽度
top wind bracing for truss bridge 桁架桥上风撑
top wing 上翼(堆石坝截水墙的)
top wing tank 顶边舱
top wire 上网
top with gum 涂胶;上胶
topwood 梢头材
top wool 毛条
topwork 整顶工作
top working roll 上工作辊
top yeast 表层酵母
tor 突岩
Toran 托兰(一种近程无线电定位系统)
toran 临时拱门(印度);大门甬道(通向庙宇的);高耸庙门(印度)
Toran method 道兰法
torbanite 页岩;藻烛煤;块煤;苞芽油页岩
Torbar 托把(一种冷加工钢筋)
torbernite 铜铀云母
torch 火炬(信号);火把;焊接灯;氢氧焊枪;气炬;手电筒;电筒;吹焰器
torch-applied asphalt felt 热熔油毡;喷灯施工油毡
torch-applied method 热熔法;喷灯铺毡
torch assembly 火炬装置
torch atomizer 火炬喷射器
torch blower 火焰吹管

torch braze welding 焊炬钎焊
torch brazing 焊炬钎焊；钎焊；气炬硬焊；气炬钎焊；吹管硬焊
torch burner 喷灯
torch cooky 火炬饼
torch corona 火焰状电晕
torch cutting 气炬烧割；气炬切割；气割
torch cutting edge 火焰割边
torch cutting outfit 火焰切割器具；割炬
torcher 为石板屋顶塞泥灰的工人
torchere 落地台灯；明火灯架；设于地板上的上射灯光；装饰的壁灯座
torch firing 喷烧
torch hardening 火焰淬火
torch head 焊枪管；气焊焊头；气焊嘴；喷灯头；炬头
torch igniter 火舌式点火器；火炬式点火器
torching 瓦间嵌灰泥；喷灯除漆；屋顶嵌灰泥；烧去旧漆；尾板嵌灰泥；喷灯除旧漆
torching roofing felt 热熔油毡
torch lamp 喷灯
torch light bulb 电珠
torch lighter 火炬点火器
torch machine 喷灯机
torch oil 喷灯油
torchon paper 粗面水彩画纸
torch peat 蜡脂泥炭；高树脂泥炭
torch pipe 喷射管；喷灯
torch-resistant steel 耐喷烧钢
torch soldering 气焊；气焊（接）
torch weld 气焊焊缝
torch welding 气炬焊；气焊（接）；气焊；吹管焊接
torch with non-variable head 固定头焊枪
Tordon 毒莠定除草剂
torductor 扭矩测量仪
tore 圆环形凸线脚；环面；座盘饰；管环
to reach all-time high 达到历史最高水平
to reduce the change of errors 减少出错
torendrikite 镁钠闪石
tore of reflection 反射环
toreutic 金属浮雕的
toreutics 金属浮雕术；金属浮雕工艺；浮雕术
Torfoleum 托福轮板（为一种绝缘板）
tori 凸圆线脚
toric lens 复曲面透镜
toric lenticular 复曲面凸凸透镜
toriconic(al) head 折边的锥形封头；带折边的锥形封头
toriconic(al) reducer 带折边的锥形变径段
toric smoothing machine 复曲面细磨机
toric surface 复曲面
torii 神社牌坊（日本）；[复]torii
toripherical 准球形的
torise load zasts 念珠状负荷模
torispheric(al) head 碟形封头；带折边的球形封头
Torkret pump 单缸液压（混凝土）泵
tormenter 台口侧幕；回声防止幕；消声隔片；边幕
tormentor 边幕；消声隔片；回声防止幕；台口侧幕
tormentor tower （舞台的）第一道侧幕背面塔架
tornadic thunderstorm 龙卷雷暴
tornado 旋风；陆龙卷（风）
tornado belt 陆龙卷区；龙卷区
tornado cellar 避风窖
tornado cloud 管状云
tornado dust collector 旋风式集尘器；龙卷风（式）集尘器
tornado echo 龙卷回波
tornado lantern 风灯
tornado prominence 龙卷日珥
tornadotron 旋风管
tornaria 柱头幼虫
tornebohmite 硅稀土石
torn fiber 粗糙面；（粗锯解成的）毛面
torn grain 撕裂纹
torn surface 有擦痕的表面；裂痕面；磨损面
torn to pieces 撕碎
torn-up 磨耗的
torn-up surface 开裂的路面；磨损的路面
toroid 环形磁路；超环面
toroidal 螺旋管形；环形的；曲面
toroidal bend 螺旋线形弯曲
toroidal coil 环形磁路
toroidal coordinates 圆环坐标
toroidal core 圆环柱心；环形磁芯
toroidal-core current transformer 带环状铁芯的电流互感器
toroidal electromagnetic type X-ray tube 环形电磁型 X 射线管
toroidalintake guide vanes 超环面进口导向叶片
toroidal lens 曲面透镜；超环面透镜
toroidal magnetic circuit 环形磁路
toroidal magnetic field 环形磁场；环向磁场
toroidal membrane 喇叭形薄膜；喇叭形底面；环形薄膜；环形面层
toroidal memory 环形存储器
toroidal nozzle 环形喷管
toroidal reactor 环状堆
toroidal recording spot 环形记录斑点
toroidal shell 圆形壳；圆或椭圆形壳；椭圆形壳
toroidal solenoid 环式螺线管
toroidal spot 环形斑点
toroidal surface 超环面
toroidal swirl 漩流
toroidal swirl chamber 漩流式燃烧室；涡流式燃烧室
toroidal transformer 环形变压器
toroidal type parametron 环形参数器
toroid shell 环壳
toroid structure 超环面结构；复曲面结构
toroid worm pipe 螺旋管
toromatic transmission 变扭传动装置
Toros-Anatolia epsilon structural system 托罗斯—安纳托里亚山字型构造体系
torose 多瘤的
torose load cast 下突重荷模
torpedo 信号雷管；响墩信号；响墩；鱼雷形装置；油井爆破药筒；电鳗
torpedo bit 不取心金刚石钻头
torpedo boat 鱼雷（快）艇
torpedo-boat depotship 鱼雷快艇母舰
torpedo boat on board 舰载鱼雷艇
torpedo-boat stern 鱼雷艇式船尾
torpedo craft 鱼雷艇；鱼雷舰艇
torpedo crown 不取心金刚石钻头
torpedo danger area 鱼雷危险区域
torpedo danger zone 鱼雷危险区
torpedo depot 鱼雷库
torpedo depot boat 鱼雷储藏船
torpedo depot ship 鱼雷储藏船
torpedo detection sonar 鱼雷探测声呐
torpedo exercise head 操雷头
torpedo experimental establishment 鱼雷试验站
torpedo gravel 尖砾石
torpedo gunboat 鱼雷炮舰
torpedo head magazine 鱼雷弹头储[贮]藏库
torpedoing 爆炸
torpedo man 爆破工
torpedo mine 鱼雷或水雷；主动式水雷
torpedo net boom 防雷网杆
torpedo range 鱼雷演习区（域）
torpedo recovering search boat 鱼雷回收搜索艇
torpedo recovery boat 鱼雷回收船；捞雷船
torpedo recovery vessel 鱼雷回收船
torpedo retriever 鱼雷回收船
torpedo sand 粗砂；粗粒砂
torpedo shop 鱼雷工厂
torpedo sinker 铅鱼（流速仪用）；鱼雷型测锤
torpedo speed boat 鱼雷快艇
torpedo splice 鱼雷式铰接器
torpedo stern 鱼雷艇式船尾
torpedo storage ship 鱼雷仓库船
torpedo storehouse 鱼雷仓库
torpedo switch 间接开关
torpedo testing barge 鱼雷试验驳船
torpedo testing range 鱼雷试验靶场
torpedo-type weight 鱼雷型测锤；铅鱼
torpedo warhead locker 鱼雷弹头舱
torpedo workshop 鱼雷工场
torpex 铝末混合炸药
torque 转矩；扭距；回转力矩；偏振光面旋转效应
torque actuator 扭矩液动机
torque allowance 转矩值容许误差；转矩允许误差；转矩容许误差
torque amplifier 转矩放大器；增矩器
torque amplitude 转矩振幅
torque angle 转矩角
torque-angle curve 功角特性曲线
torque arm 转矩臂；扭矩臂
torque at peak horsepower 最大马力转矩
torque at rated load 额定负载转矩
torque axis 扭矩轴
torque balance 扭力天平；扭力计；扭力秤
torque balance converter 扭矩平衡变换器
torque-balance device 扭矩平衡装置
torque-balance system 扭矩平衡法
torque bar 方钻杆
torque-biasing full time four-wheel drive 扭矩差动型常时四轮驱动
torque bimorph 转矩双压电晶片
torque chain 扭矩传递链
torque characteristic 转矩特性；扭转性能；扭转特性
torque coefficient 扭矩系数
torque coefficient of pump impeller 泵叶轮扭矩系数
torque compensation 扭矩补偿
torque compensator 扭矩补偿器
torque control 扭矩调节
torque controller 扭矩调节器
torque control method 转矩控制法；扭矩控制法
torque control wrench 转矩调节扳手
torque conversion 变扭；转矩转换；转矩变换；转矩转变；扭矩变换
torque converter 液力变矩器；液力变矩器；液力变矩器；转矩变换器；力变换器；扭矩转换器；扭矩变换器；变矩器；变矩器
torque converter charging fluid 液力变矩器用液
torque converter charging pressure 液力变矩器控制压力
torque converter charging valve 液力变矩器控制阀
torque-converter control lever 液力变矩器控制杆
torque-converter drive 变矩器驱动（装置）
torque converter drive clutch 变矩器驱动离合器
torque-converter fluid 转矩变换器液体；变矩器液体；液力变矩器用液
torque converter hold valve 变矩器控制阀
torque-converter housing 液力变矩器箱；液力变扭器壳
torque converter inlet relief valve 变矩器输入安全阀
torque converter leakage 液力变矩器渗漏
torque-converter level tell-tale 液力变扭器
torque converter lever 液力变矩器操纵杆
torque converter liquid 液力变矩器用液
torque-converter locomotive 变扭器机车
torque converter oil pressure ga(u)ge 液力变矩器油压表
torque converter oil temperature telltale 液力变矩器油温警告灯
torque converter outlet relief valve 液力变矩器输出安全阀
torque converter planetary drive 变矩器行星传动装置
torque-converter power unit 液力变扭器传动
torque-converter reactor 转矩变换器反应器；变矩器反应器；液力变矩器反应器
torque converter recirculation pump 液力变矩器循环泵
torque converter size 液力变矩器尺寸
torque converter speed drive 变矩器加速阀门
torque converter stall point 变矩器失速点
torque converter stall ratio 变矩器失速比例
torque converter tank 液力变矩器油箱
torque-converter temperature telltale 液力变扭器液温信号灯
torque-converter transmission 液力变扭器传动；变扭器传动
torque creep 转矩滞缓；转矩蠕动；扭转打滑
torque current 转矩电流
torque current constant 转矩电流常数
torque curve 扭矩曲线
torque curve of hydraulic coupling 液压耦合器扭矩曲线
torque detector 扭矩测定器
torque diagram 扭矩图；扭矩特性曲线图
torque divider 转矩分配器；分动器
torque divider planetary set 转矩分配器星形装置
torque divider transmission 转矩分配器传动（装置）
torque divider transmission control valve 转矩分配器传动控制阀
torque division 转矩分配
torque drop 扭矩降

torque dynamometer 转矩计;扭矩测力计
torque efficiency 转矩效率;机械效率
torque failure 扭转破坏
torque fluctuation diagram 扭矩波动图
torque force 扭转力;扭力
torque free motion 无扭转运动
torque ga(u)ge 扭力计;扭力表;扭矩仪;扭矩计;扭矩表
torque gradient 转矩梯度;比力矩
torque grading curve 转矩分布曲线
torque horse power 转矩马力
torque indicating (hand) wrench 转矩指示扳手
torque-indicating wrench 测力扳手
torque indicator 转矩指示器
torque inertia ratio 转矩惯性比
torque insulator 转矩隔绝器
torque level 扭力计;扭矩仪
torque limit 转矩极限
torque limitation 转矩限制;扭矩限度
torque limiter 转矩限制器;扭矩限制器
torque-limiter clutch 扭矩限制离合器
torque link 扭接杆件
torque-load characteristic 扭矩—负载特性曲线
torque load(ing) 转矩负载;扭矩负载
torque loop 转矩环路
torque loss 转矩损耗
torque machine 大扭矩机械
torque magnetometer 转矩式磁力计
torque master 转矩检测装置;转矩传感器
torque measurement 转矩测量;转矩测量
torque measuring device 转矩测量装置
torque-measuring gear 测扭矩设备
torque measuring wrench 转矩指示扳手
torquemeter 转矩计;扭力仪;扭矩计;扭矩表;扭矩仪;扭矩计;扭矩表;旋转力矩测量器;扭力测定仪;扭矩器;扭矩测量仪
torquemeter chart 扭矩测量曲线图
torque metering device of rotary kiln 回转窑转矩测量装置
torque moment 扭力矩;扭矩;转矩
torque motion 转矩操作机构
torque motor 旋转电机;转矩电动机;力矩马达;扭矩马达;慢速电动机;合闸用电动机;陀螺修正电动机
torque multiplication 扭矩增大比
torque noise 转矩噪声
torque of cutter head 刀头扭矩
torque output 转矩输出;扭力输出(量);输出扭矩
torque peak 转矩峰值
torque performance 扭矩特性
torque pickup 力矩传感器;转矩传感器
torque pillar 转矩柱
torque pipe 扭力管
torque property 扭矩性能;扭矩特性
torque pulsation 转矩脉动
torquer 转矩发送器;转矩产生器;力矩器;加扭器;扭力测定仪;扭矩计;扭矩表
torque rating 计算扭矩扭率;额定扭矩
torque ratio 扭矩比
torque reaction 转矩反作用;扭矩反作用;反转力矩
torque-reaction joint 反作用转矩节
torque-reaction rod 反作用转矩杆
torque-reaction stand 转矩测量台
torque recorder 扭矩记录器
torque reducer 减扭器;扭矩降低器
torque reduction 扭矩减少
torque regulator 转矩调整器
torque rheometer 转矩流变仪
torque ripple 转矩脉动
torque rise 扭矩增大
torque rod 扭转杆;扭矩杆;回转杆;方钻杆
torque-rod bracker 扭矩杆托架
torque-rod end pin 扭矩杆端销
torque rod joint 扭矩连接
torque rod lower pin 扭矩杆下销钉
torque rod mounting parts 扭矩杆的固定零件
torque rod rubber bushing 扭矩杆橡胶衬套
torque rod suspension 扭矩杆悬挂
torque rod upper pin 扭矩杆上销钉
torque scatter 扭矩的偏差范围(紧固件等的拧紧)
torque selector 扭矩选数器
torque selsyn 力矩式自整角机
torque sensing coupling 扭矩传感联轴节
torque sensor 扭矩传感器

torque setting 调扭矩
torque sewer pump 转矩污水泵
torque spanner 转矩扳手;扭力扳手;扭矩扳手
torque-speed characteristic 转矩速度特性曲线
torque-speed curve 转矩速度特性曲线
torque spider 扭力辐
torque split 扭矩分解
torque spring 扭矩弹簧
torque stand 扭矩测台
torque stay 转矩管
torque summing member 转矩相加器
torque synchro 力矩式同步机;扭矩传递同步器
torque tension test 扭矩张力试验
torque test 扭力试验;扭矩测定
torque tester 扭力试验仪
torque-to inertia ratio 扭矩—惯矩比
torque transducer 转矩传感器;扭矩传感器
torque transfer 扭力传递装置
torque transmitter 扭矩传递装置
torque tube 转矩管;扭力管;扭矩管;万向轴管
torque-tube ball 转矩管球接
torque tube drive 扭力管传动;扭矩管万向节传动
torque-tube flange 转向管凸缘
torque-tube flowmeter 扭矩管式流量计
torque-tube propeller shaft 转矩管螺旋桨轴
torque-tube type suspension 转矩管式悬置
torque-twist curve 扭矩—加捻曲线
torque-twist diagram 扭矩—扭转角曲线图(零件的)
torque-type magnetometer 转矩式磁力仪
torque type power meter 扭力式功率计;扭力测功器;扭力(测)功率计;转矩式功率计
torque-type visco(si)meter 扭矩式黏度计;扭转黏度计;回转式黏度计;旋转黏度计
torque-vibration damper 扭振减震器;扭转振动阻尼器
torque viscosity 扭矩黏度
torque-winding diagram 力矩卷绕图
torque wrench 限力扳手;转矩扳手;扭力扳手;扭矩扳手
torque yield 扭矩屈服限
torquing of nuts 转矩拉紧螺母
torquing rate 加矩速率
torquing unit 攻螺丝的机械化工具
torr 托(真空压强单位)
torrefy 烘烤;焙烧
Torrens title 州产权登记法
torrent 激流;洪流;山溪;山洪;奔流
torrent control works 急流控制工程;山洪控制工程
torrent deposit 急流沉积
torrent flume 急流槽
torrential 急流的;洪流的
torrential burst 山洪暴发
torrential cross-bedding 急流交错层
torrential cross-bedding structure 急流交错层理构造
torrential downpour 暴雨
torrential flood 湍流;山洪;暴发洪水
torrential flow 奔流;急流;激流;奔奔流
torrential rain 骤雨;倾盆大雨;暴雨
torrential rain strength 暴雨强度
torrential regulation 急流整治
torrential river 急流河;湍急河流;山溪
torrential stream 急流河;湍急河流;山溪
torrential wash 急流冲刷
torrential wash way 急流冲刷道
torrent of rain 倾盆大雨;骤雨
torrent rapids 急湍;急滩
torrent works 山溪整治设施
Torres Strait 托雷斯海峡
torreyite 羟锌镁矾
torric (土壤的)干燥状况
Torricellian barometer 水银气压计
Torricellian law 托里拆利定律
Torricellian theorem 托里拆利定理
Torricellian tube 托里拆利管
Torricellian vacuum 托里拆利真空
Torricelli's theorem 托里拆利定理
torrid area 热带地区
torrid environment 热带环境
torrid zone 热带
Torrington needle bearing 冲压外座圈滚针轴承
torrmenter 回声防止幕
Torrox 干燥氧化土

torsade 编织装饰;螺旋形线脚;绳索式线脚
torsal line 挠点线
torsal plane 挠切面
torsal point 挠切点
torse 可展(曲)面;扭花环装饰
torsel 旋涡花形装饰;游檐木;墙中梁端垫块;承梁木
torshear bolt 扭剪螺栓
torsibility 抗扭性
torsimeter 扭矩计
torsiogram 扭振图;扭力图
torsiograph 扭力计;扭振自记器;自记扭力计;(转)振(动)记录仪;扭振自(动)记(录)仪;扭力记录仪
torsiometer 转矩计;扭力计;扭矩计
torsion 扭转;挠率
torsional 扭转的
torsional amplitude 扭转幅度
torsional analogy 扭转比拟
torsional anemometer 扭转式风速计
torsional angle 扭转角;扭角
torsional balance 扭力天平;扭力秤;扭矩秤;扭秤
torsional balance beam 扭秤臂
torsional balancer 扭转振动阻尼器;扭振减震器;扭力平衡器
torsional bar 扭力杆;扭力棒
torsional bar anchor 扭力杆支座
torsional bar belt-force-limiter 扭杆式带负荷限制器
torsional bar end plug 扭力杆端部塞子
torsional bar front retainer 扭力杆前挡块
torsional bar housing (inner) 扭力杆内套筒
torsional bar housing (outer) 扭力杆外套筒
torsional bar independent suspension 扭杆弹簧式独立悬挂
torsional bar rear retainer 扭力杆后挡块
torsional bar replacer 扭力杆拆卸工具
torsional bar retainer 扭力杆限制块
torsional bar retaining cap screw 扭力杆固定螺钉
torsional bar seat 扭杆座
torsional bar spacer 扭力杆调整套筒
torsional bar spring 扭杆弹簧
torsional bar spring anchor plug 扭杆弹簧固定塞
torsional bar spring anchor plug cover 扭杆弹簧固定塞盖
torsional bar spring tension 扭杆弹簧扭转
torsional bar stabilizer 扭力杆稳定器
torsional bar support 扭力杆支座
torsional bar suspension 扭杆悬架
torsional bar torsional lever 扭杆
torsional bearing 扭转轴承
torsional bending 扭转弯曲
torsional box girder 抗扭箱形纵桁
torsional box structure 抗扭盒式结构
torsional bracing 扭力拉条
torsional braid 扭辫
torsional braid analysis 扭带分析;扭辫分析
torsional braid tester 扭力带试验机
torsional break 扭力破坏;扭坏
torsional buckling 扭转屈曲;扭屈;扭曲
torsional buckling load 由扭矩引起的临界荷载
torsional capacity 抗扭能力;扭张力;容许扭转负荷
torsional circle 扭环
torsional coefficient 扭转系数
torsional coil spring 扭力盘形弹簧
torsional compliance 扭力柔度
torsional constant 扭转常数;扭力常数
torsional couple 扭转力偶;扭力偶
torsional crack 扭转开裂
torsional cracking 扭转裂缝
torsional creep 扭转蠕变;扭蠕变
torsional creeping crack 扭转蠕变裂纹
torsional crimping process 扭转卷曲法
torsional critical speed 扭转临界速度
torsional curve 扭矩—转角曲线
torsional cyclical load triaxial apparatus 扭转周期荷载三轴仪
torsional damper 扭振阻尼器;扭振减振阻尼器
torsional damping arrangement 扭力减震器
torsional deflection 扭转变形
torsional deformation 扭转形变;扭转变形;扭曲变形
torsional displacement 扭转位移
torsional distribution graph (车身的)扭曲变形分布图

torsional divergence 扭转发散
torsional drum coefficient 扭鼓常数
torsional drum correction 扭鼓改正
torsional drum correction value 扭鼓改正值
torsional dynamometer 扭转式测力计;扭力计;扭力功率计;扭力测力器;扭力测功率计
torsional eccentricity 扭转偏心
torsional effect 扭转效应
torsional elasticity 抗扭弹性;扭转弹性
torsional electrometer 扭转静电计
torsional endurance 扭转疲劳限度
torsional endurance limit 扭转疲劳极限;扭转持久极限
torsional exciter system 扭振激发装置
torsional failure 由扭矩引起的破坏;扭转失效;扭转破坏;扭损;扭坏
torsional fatigue 扭转疲劳
torsional fatigue failure 扭转疲劳破坏
torsional fatigue strength 扭转疲劳强度
torsional fatigue test 扭转疲劳试验;扭疲劳试验
torsional fault 扭转断层
torsional flexibility 扭转伸缩性;扭转挠性
torsional-flexural buckling 弯扭屈曲;扭转挠曲
torsional flow 扭转流动
torsional flutter 扭转颤振
torsional force 抗扭力;扭力;转动力
torsional formula 扭转公式
torsional fracture 扭转断裂;扭力破裂
torsional frequency 扭转振动频率;扭转频率;扭振频率
torsional galvanometer 扭转式检流计;扭转电流计
torsional ga(u)ge 扭力表;扭矩计;扭矩表
torsional gravimeter 扭转重力仪;扭秤
torsional ground motion 扭转地面运动
torsional gum 扭振橡胶弹簧
torsional head 扭秤头
torsional hygrometer 扭转湿度计
torsional impact test 扭力冲击试验
torsional indicator 扭转指示器;扭力指示器;扭力计;扭矩指示计
torsional instability 由扭矩产生的不稳定性
torsional joint 扭节理;扭合接头
torsional laminar displacement 扭转层状位移
torsional level(l)ing suspension 扭力杆调平式悬架
torsional load(ing) 扭转荷载;扭转负荷;扭力荷载
torsionally braced 扭转支撑的
torsionally coupled building 扭转耦联建筑物
torsionally flexible coupling 扭转弹性联轴节
torsionally rigid 扭转抗扭的
torsionally stiff 不易扭转的
torsional magnetostriction pickup 扭转磁致伸缩传感器
torsional member 抗扭构件
torsional membrance (推力轴承的)弹簧油箱
torsional meter 扭力仪;扭矩计;扭力功率计;扭力表;扭矩计;扭矩表
torsional method 扭转法
torsional mode 扭转振型;扭转振动;扭转模态;扭振形式
torsional mode delay line 扭转式延迟线
torsional modulus 扭转模量;扭转变形模数;扭振模数;扭曲模件;扭力模数
torsional moment 转矩;扭力矩;扭力矩;扭矩
torsional oscillation 扭转振荡;扭振;扭曲振荡
torsional-oscillation damper 扭摆缓冲器;扭转摆动衰减器
torsional oscillation method 扭转振动法
torsional pairing 扭转配对
torsional pendulum 扭转摆;扭摆
torsional pendulum method 扭摆试验
torsional piston ring 扭转活塞环
torsional problem 扭转问题
torsional prop 扭力支柱
torsional recilience 扭转弹性变形
torsional recovery test 扭(转)复原试验
torsional reinforcement 抗扭配筋;扭抗钢筋;扭力钢筋
torsional relaxation 扭力松弛
torsional resilience 扭转回能
torsional resistance 抗扭(阻)力;抗扭(能力);扭(转)阻力;扭曲度;阻距
torsional resonant column apparatus 扭转共振柱仪

torsional response 扭转响应
torsional restraint 扭转约束
torsional ribbon 扭丝
torsional rig 扭转试验台
torsional rigidity 抗扭劲度;抗扭刚度;扭转刚度
torsional ring 扭环
torsional rubber mount 抗扭橡胶支座
torsional scale 扭秤
torsional seismograph 扭转地震仪
torsional seismometer 扭转地震仪;扭转地震计;扭摆地震计
torsional shaft 扭转轴;扭力轴
torsional shake 扭转振摆
torsional shear 扭转剪切;扭转剪力;扭剪(切)
torsional shear apparatus 扭剪(切)试验装置
torsional shearing stress 扭剪(切)应力
torsional shear strength 扭剪(切)强度
torsional shear test 扭转剪切试验;扭转剪力试验;扭剪(切)试验
torsional shock absorber 扭震防冲器
torsional spectrum 扭转谱;扭转反应谱
torsional spring 抗扭弹簧;扭转弹性;扭转弹簧;扭力弹簧;扭簧
torsional stability 扭转稳定性
torsional steel 抗扭钢;扭钢
torsional stiffness 抗扭劲度;抗扭刚度;扭转刚度
torsional stiffness graph (车身的)扭转
torsional strain 扭(转)应变
torsional strength 抗扭强度;扭转强度;扭断强度
torsional stress 抗扭应力;扭转应力;扭应力;扭曲应力
torsional suspension 抗扭吊架;扭悬;扭式杆式悬架
torsional tensor 扭曲张量
torsional test 扭力测试;抗扭试验;扭转试验;扭曲试验;扭力(强度)试验
torsional test diagram 扭转试验图
torsional tester 抗扭试验器;扭转试验机
torsional testing machine 扭转试验机;扭力试验机
torsional twist crimping 扭转加捻卷曲
torsional vibration 扭转振动;扭曲振动;摇旋振动
torsional vibration balancer 扭振平衡器
torsional vibration damper 扭振减振器;扭转振动减震器;扭振平衡器;扭力消震器
torsional vibration pick-up 扭转振动传感器
torsional viscometer 扭转黏度计
torsional visco(si)meter 扭力粘度计
torsional wave 扭转波;扭力波;扭波
torsional wire 扭线;扭力丝
torsional wire apparatus (测试橡胶的)扭丝仪
torsional (wire) visco(si)meter 扭丝式黏度计
torsion angle 扭转角
torsion bar for vehicles 车辆转力杆
torsion-bar spring 扭转轴弹簧
torsion-bar suspension 扭转悬置
torsion-bending bimoment 扭弯双力矩
torsion-bending moment 扭弯力矩
torsion-box structure 扭抗盒式结构
torsion bracing 扭力拉杆
torsion-braid analyzer 扭力带分析仪
torsion buckling 扭曲
torsion coefficient 挠曲系数
torsion-cone consistometer 扭力圆锥稠度计
torsion degree 扭转度
torsion failure 扭转破坏
torsion-fibre 扭转纹
torsion flex suspension 扭曲悬挂
torsion force balance 阻力天平
torsion fracture 岩芯自卡被扭成螺旋形裂隙
torsion-free 无扭转的
torsion impact test 冲击扭转试验
torsion indicator 示扭器
torsion interferometer 测扭干涉仪
torsion-less 不扭转的
torsion machine 绕簧机
torsion member 抗扭构件;受扭构件
torsionmeter 扭转仪;扭力计;扭矩计;扭矩测量仪
torsion micrometer 微扭力计
torsion module 挠模
torsion modulus 扭转模量
torsion of a curve 曲线的挠率
torsion oscillation 扭摆;扭转振荡
torsion problem 扭转问题
torsion-proof 防扭绳;防扭的
torsion reinforcement 扭力钢筋;抗扭配筋

torsion ring shear apparatus 环形试件扭剪仪;环式扭剪仪
torsion rod 受扭杆;扭杆
torsion rod spring 受扭杆簧
torsion stiffness 抗扭刚度;抗劲度
torsion strength 抗扭强度;扭转强度
torsion stress 扭应力
torsion-string galvanometer 扭线电流计
torsion test 抗扭试验
torsion tester 抗扭试验器;扭力试验仪
torsion testing machine 扭曲试验机
torsion type 旋转型的
torsion-type magnetometer 悬丝式磁力仪
torsion visco(si)meter 扭转式黏度计
torsiversion malposition 扭转错位
torslograph 扭力自动记录仪
torso 裸体躯干雕像;螺旋形柱;扭曲形柱
torsoclusion 扭转压法
torso mountain 残丘;剥蚀残山
Torsteel 托尔钢(一种螺纹圆钢筋)
tort 民事侵权行为;侵权行为
tortile 扭卷;扭曲的;盘绕
tort liability 侵权行为的赔偿责任
tortoise plastron 龟板
tortoise shell 龟甲
tortoise shell figure 铁水面龟纹
tortoiseshell pattern 龟背纹
tortoise-shell ware 玳瑁斑纹器皿
tortoise shell wood 龟甲木
tortuosity 扭曲度;曲折
tortuosity coefficient 沟路曲折系数
tortuosity factor 迂回因素;弯曲系数
tortuosity of porous medium 多孔介质的弯曲率
tortuosity ratio 弯曲比
tortuous channel 曲折河道;弯曲水道;弯曲河槽;弯曲航道
tortuous flow 缭绕流;曲折流;紊流;弯曲水流
tortuous passage 迂回通道;曲折通路
tortuous path 曲径;弯路
tortuous river section 弯曲河段
tortuous stream 弯曲水流
tortuous voids 迂回空隙
torture 使翘曲
torture track 破坏性试验道
toruloid 念珠状的
torulose 念珠状的
torulus 小圆凸
to run the risk 冒险
torus 座盘饰;圆枕;轮环;纹孔塞;椭圆环;凸圆线脚;[复]tori
toruselement 安全环套
torus ring 紧圈;锚环
torus roll 包铅圆筒形挡风雨板;圆凸卷边接缝
torus shell 环壳
torvane 小型十字板剪切仪;小型十字板仪
toscanite 斑苏粗安岩
tosecan 平行划线器
toseki 日本陶石
to select a good time 择期
to sell at a discount 减价出售
to sell to disadvantage 亏本销售
tosher 小渔船;专运木材小船
TOSHIBA 东芝
Toshiba non-linear resistor 东芝非线性电阻器
toshnailing 埋头倾斜钉入
to specifications 按照说明书
to sponsor 主办
tossing board 抖动板
tossing tub 摇洗槽
to stain 上色
tosudite 迪间蒙石
to suit measures to local condition 因地制宜
tosyl 甲苯磺酰
total 总共;总的;合计;全部的;完全的
total ablation 总消融量(冰川)
total ablation rate 总消融
total aborbing surface 总吸收表面
total absorbed dose 总吸收剂量
total absorption 总吸音能力;总吸收(量);全吸收
total absorption capacity 总吸收能力
total absorption coefficient 总吸收系数
total absorption spectrometer 总吸收分光计;全吸收分光计
total absorptivity 总吸收率

total acceleration 总加速度
total acceleration time 总加速时间
total acceptance angle 全接受角
total access time 总存取时间
total accumulation 总累积量
total accumulation rate 总累积率
total accuracy of gravity survey 重力测量总精度
total accuracy of sounding 测深总精度
total accuracy of terrain correction 地形改正总精度
total acid 总酸
total acidic group 总酸性基
total acidity 总酸度
total acidity of soil 土壤总酸度
total acid number 总酸值
total acid value 总酸值
total acoustic(al) power 总声功率
total active soil force 总主应力
total activity 总活度
total actual spending 实际支出总额
total admittance 总导纳
total adsorption 总吸附
total adsorption capacity 总吸附能力
total adsorption density 总吸附密度
total adsorption of oxalate 总草酸盐吸附
total adsorptive capacity 总吸附(容)量
total aggregate 施工骨料总量;总集料;总骨料
total agricultural output 农业总产值
total air 总空气量;总含气量
total air-borne fraction 污染物在大气中的总份额
total air content 总含气量
total air defense 总体防空
total air-gas mixture 空气—燃气化合物
total air pollution 总空气污染
total air pressure 全风压
total air quantity 总风量
total algae density 总藻类密度
total alkali 总碱
total alkali consumption 总碱耗
total alkalinity 总碱度
total alkalinity of soil 土壤总碱度
total allowable catch 许可总鱼获量
total allowance 总余量;总留量
total all the expenditures 计算全部支出
total alpha-radiation survey 总阿尔发(放射性)测量
total alpha-radioactivity survey 总阿尔发(放射性)测量
total ammonia nitrogen 总氨氮
total amount 总数;总量;总额;总计
total amount control of pollutant 污染物总量控制
total amount index 总量指标
total amount indicatrix 总量指标
total amount of air 总空气量
total amount of anion 阴离子总和
total amount of bacteria 细菌总数
total amount of cation 阳离子总和
total amount of cloud 总云量
total amount of erosion 总侵蚀量;侵蚀总量
total amount of estimated investment 概算投资总额
total amount of heat 总热量
total amount of investment 投资总额
total amount of pollutant discharge standard 污染物排放标准总量
total amount of pollution emission control planning 排污控制总量规划
total amount of resources 资源总量
total amount of soluble metal 溶性金属总量
total amount of wages 工资总额
total amount of weld fumes 焊接发尘量
total amount on board 载货总量
total amounts of reactants 反应物总量
total amplification 全振幅
total amplitude 正负峰值间总幅度;全振幅
total analysis 全分析;全量分析
total and adjustable price contract 调值总价合同
total and permanent disability 完全和永久丧失工作能力
total angular momentum 总角动量
total angular momentum quantum number 总角动量矩数
total annual charges 全年总费用
total annual cost 全年总费用

total annual expense 全年总费用
total annual goods transport 年货运总量
total annual person travel 年客运总量
total annual yield 年总产量
total-annular eclipse 全环食
total apparent adsorption density 总表观吸附密度
total area 总面积
total area cut 总切削面积
total area of cross-section 毛截(面)面积
total area of emittance 总发射度面积
total area under cultivation 耕作总面积
total argon 总氩量
total arsenic 总砷
total artificial recharge 总人工回灌量
total ascent time 上升总时间
total ash 总灰分
total ashed water solubles 总水
total asphalt pavement 全厚式沥青铺面
total assembly 总装配
total assets 总资产;资产总额
total assets turnover 资产总额周转率
total association 全相联
total attenuated reflection 总衰减反射
total attenuation 总衰减
total available alkali 总有效碱容量
total available chlorine 总有效氯
total available energy 总有效能量
total available resources 现有资源总量
total average 总平均值
total average loss 全损
total backlash 总间隙
total bag sampling system 总量取样法
total bandwidth 总带宽
total bank height 堤岸总高
total basal area 总断面积
total base number 总基数;总底数;总碱值
total beam 总束流
total bed load 底沙总量
total bed sediment 底沙总量
total bed sediment discharge 底沙输移总量
total benefits 总受益收入
total beta-radiation survey 总贝塔(放射性)测量
total beta-radioactivity survey 总贝塔(放射性)测量
total bicycle parking ratio 自行车存放率
total binding energy 总结合能(核的)
total biological oxygen demand 总生物需氧量
total biomass of marine organism 海洋有机体的总生物量
total biomass of terrestrial plant 陆生植物的总生物量
total bit 总位数
total bit load 钻头总荷载
total bitumen 纯沥青(含)量
total blast nozzle area 风嘴总面积;风口总有效截面积
total body burden 总体负荷(量)
total body radiation 全身辐射
total brightness 总亮度
total budget rate 总收支率(指冰川)
total buffer intensity 总缓冲强度
total building density 总建筑密度
total building system 综合建筑系统;建筑体系
total burn-off 总耗油量
total by-products 副产品总产量
total cadmium 总镉
total calendar hours 日历总时数
total calorific value 总发热量
total capability for load 总供电能力;总负载能力
total capacitance 总电容
total capacity 总容量;总能力;总库容
total capacity for water 总含分量
total capacity of a well or spring 井或泉的总容量
total capacity of groundwater reservoir 地下水库总库容
total capacity of well 井的总容量
total capacity reservoir 总库容
total capcity of a well 一口井的总产量
total capital 总资本;资本总额
total capital profit ratio 总资本利润率;资本总额与利润比率
total capital turnover 资本总周转率
total capital turnover rate 资本总额周转率
total carbamate residue 总氨基甲酸酯残基

total carbohydrate 总碳水化合物
total carbon 总碳(量)
total carbonate concentration 总碳酸盐浓度
total carbonates 碳酸盐总量
total carbon dioxide capacity 二氧化碳总量
total carbonic acid 总碳酸;全碳酸
total car ratio 汽车存放率
total case annealing 完全相变退火
total casing length 套管累计长
total casing weight 套管总重;套管累计重量
total cation 阳离子总和
total centre-of-mass energy 总质心能量
total chain yield 总链产额
total channel anomaly 总道异常
total channel length 河网总长度
total channel storage 总槽蓄量;河槽总蓄水量;河槽调蓄总量
total channel window width 总道窗宽
total characteristic (curve) 总特性(曲线)
total charge 总电荷;总装药量
total check 累加校验
total chemical analysis of soil 土化学全分析试验
total chemical oxygen demand 总化学需氧量
total chemical potential energy approaches 总化学势能计算法
total chlorine 全氯
total chlorine or nitrogen 氯或氮的总量
total cholrine 总氯
total chromatographable organics 总可色谱有机物
total chromium 总铬
total clearance 总间隙
total cloud cover 总云量
total cloudiness 总云量
total coefficient of elongation 总延伸系数
total coliform bacteria population 总大肠杆菌群
total coliform concentration 总大肠杆菌浓度
total coliform detection 总大肠杆菌检测
total coliform population number 总大肠杆菌群数
total coliforms 总大肠杆菌
total collapse 总塌陷量;总湿陷量;全毁
total colo(u)r difference 总色差;全色差
total column ozone depletion 臭氧柱消耗总量
total combustion method 总燃烧法;完全弧烧法
total commission 全部佣金
total community oxygen consumption 总群落耗氧量
total compaction 总压实量
total complementary energy 全余能
total complement measurement 补体总量测定法
total components 总成分
total composite error 总合成误差
total composition 全量组成
total compressibility 总压缩系数
total compression load on joint contact surface 连接接触面总压紧荷载
total concentrates 精矿总产量
total concentration 总浓度
total concentration distribution graph 总浓度分布图
total concentration of Magnesium 镁的总浓度
total concrete stress 混凝土总应力
total condenser 全凝器
total conductivity 总电导率
total configurational entropy 总组合熵
total confining pressure 总周围压力
total connected electric(al) load 总连接电荷
total consumption 总消耗量
total consumption burner 全消耗性喷灯;全消耗型燃烧器
total consumption expenditures 全部消费支出
total consumption of coal 总煤耗
total contain alkali volume 总合碱量
total contaminant 总污染物
total contaminant loading 总污染物负荷
total contaminant quantity control 总污染物量控制
total contamination 总污染
total content 总含量
total content of organic carbon 总有机碳含量
total continuous period of test 试验总延续时间
total contract sum 工程合同总造价
total control counter 总控制计数器
total control over revenue and expenditures 统收统支

total conversion error 总转换误差
total conversion time 总转换时间
total cooling effect 总冷却功效;总冷却效应
total cooling power 总冷却功率
total copper 总铜
total core length 岩芯总长度
total core recovery 总岩芯采取率
total correction 总改正量
total correction of error 误差总校正;误差总改正
total correction of star's altitude 星体高度总改正量
total correction of sun's altitude 太阳高度总改正量
total correlation 总相关;全相关
total correlation coefficient 总相关系数
total correlation factor 全相关因子
total cost 总值;总造价;总费用;总成本;全部费用
total costing 总成本核算
total cost of building 总造价
total cost of drill 钻探总成本
total cost of engineering 工程总投资
total cost of heavy repair 大修总费用
total cost of investigation terms 调查项目总费用
total cost of loading/discharging 装卸总成本
total cost of material 材料总成本
total cost of non-factor inputs 非要素投入总成本
total cost of upkeep 维修总费用
total cost used 实际总成本
total counter 总计数器;总计数器
total count ground gamma-ray survey 地面伽马总量测量
total-count measurement 总计数测量
total-count method 总计数法
total count rate standard deviation map 总计数率标准偏差图
total-count surveys 总计数测量
total coverage 总建筑(占地)面积系数(总建筑占地面积和建筑用地总面积之比)
total cover-degree 总盖度
total creep 累积变形;累积蠕变;总蠕动;总蠕变
total critical load 总临界荷载
total critical pressure 总临界荷载
total cross-section 总截面;全截面
total crown 总路拱
total cultivated area 耕地总面积
total cumulation 全累积
total cumulative output value of industry 累计工业总产值
total cumulative resultant displacement 总合成位移
total current 总电流
total current regulator 总电流调节器
total curvature 总曲率;全曲率;弯道曲率总和
total curve 累积曲线
total cyanide 总氰化物;氰化物总量
total daily production 总日产量
total daylight factor 全采光系数
total dead time 总迟延时间
total deadweight 总载重量
total death rate 总死亡率
total debts 借记总数
total debt to total assets 负债总额对资产总额的比率
total decompression time 减压总时间
total decrease 保留地让出总量
total deflection 总挠度;总变形
total deformation 总变形量;总应变;总变形
total deformation length 总变形长度
total delay 总延时;总延迟
total delivery head (水泵的)总扬程;总送水扬程;总输水水头;供水总水头
total demand 总需求(量)
total demonstrated reserves 累计探明储量
total departure 总偏差;总横距;横距和
total deposition 沉积总量
total depreciation 折旧总额
total depth 总水深;总深(度);全齿高
total depth of bell 钟口总深
total depth of shallow wells 浅井总深度
total depth of vertical shaft 竖井总深度
total derivative 全导数
total design concept 总体设计概念
total design data planning 总体数据规划
total design engineer 总体设计工程师
total development cost 开发总成本

total deviation 总偏差;累积偏差
total deviation intensity 全弯曲强度
total diamond loss 金刚石总消耗量
total difference equation 全差分方程
total differential 全微分
total differential equation 全微分方程
total differential head 全压头
total diffuse reflectance 全漫反射比
total diffuse reflection 全漫反射
total digestible nutrients 总消化营养;总消化养分
total dimension 总尺寸
total disability 完全丧失劳动力
total discharge 总流量;排放总量
total discharge control of pollutant 排污控制总量
total discharge head 出口总压力
total discharge of pumping wells 抽水孔总流量
total discharge of sediment 总输沙量
total discharge volume 总排水体积
total dispersion 全色散
total displacement 总移位;总位移;总排水量;总排(放)量
total disposable income 可支配收入总额
total disposable means 可支配资力总量
total dissolvable manganese 总可溶解锰
total dissolved carbonate 总进入碳酸盐
total dissolved carbonate concentration 总溶解碳酸盐浓度
total dissolved carbonate species 总溶解碳酸盐物种
total dissolved chromium 总溶解铬
total dissolved copper 总溶解铜
total dissolved gas 总溶解气体
total dissolved gas concentration 总溶解气体浓度
total dissolved inorganic carbon 总溶解无机碳
total dissolved ions 总溶解离子
total dissolved materials 总溶解物质
total dissolved metal 总溶解金属
total dissolved metal concentration 总溶解金属浓度
total dissolved nitrogen 总溶解氮
total dissolved organic carbon 总溶解有机碳
total dissolved organic matter 总溶解有机物
total dissolved phosphorus 总溶解磷
total dissolved salt 总溶解盐
total dissolved sediment 总溶解沉积物
total dissolved solid 总溶解固体;溶解性总固体;可溶性固体总量;溶解质总量
total dissolved solid value ppm 溶解物总量
total dissolved substance 总溶解物质
total distance indicator 总距离指示器
total distortion 总失真
total diversion efficiency 总输水效率
total dollar return 投资的总现金收益
total domestic water consumption 总生活耗水量
total dosage attack 总剂量攻击
total dose-stay time 总剂量停留时间
total draft 总阻力
total drag 总阻力
total drawdown 总降深
total drawing deformation 拉拔总变形量
total drift 总漂移;总漂沙量;总漂流
total drilling control 钻井工程全面控制
total drilling cost 钻井总费用;总成本【岩】
total drilling footage 总钻进尺
total drilling footage of big diameter borehole 大孔径钻总进尺
total drilling footage of drilling 钻探总进尺
total drilling footage of land drilling 土钻总进尺
total drilling footage of machine drilling 机钻总进尺
total drilling hours 全井总台时
total drilling months 总月数【岩】
total drilling time 总钻探时间
total drop (断层的)总落差
total dry biomass 总干生物质
total drying time 总烘干时间;总干燥时间
total dry matter 总干物质
total dry residue 总干燥余渣
total dump 总转储;全部转储
total dust 总尘量
total dynamic(al) discharge head 总动力出流水头
total dynamic(al) error 总动态误差
total dynamic(al) head 总动力(力)压头;总动力水头
total eclipse 全食

total eclipse of the Sun 日全食
total ecosystem 全生态系统
total effective temperature 有效温度总和
total efficiency 总效率
total effluent nitrogen concentration 总污水含氮浓度
total effluent oxygen demand 总污水需氧量
total effluent phosphate concentration 总污水磷酸盐浓度
total elastic-plus-creep recovery strain 弹性徐变总回复应变
total electric(al) charge 总电荷
total electrochemical potential energy 总电化学势能
total electrolyte 总电解质
total elongation 总延伸(量)
total emission 全发射
total emission control 总排污量控制;总排放控制;排污控制总量
total emission standard 总体排污标准;总排放标准
total emission strength 总发射强度
total emissive power 总发射能力
total emissivity 总发射率
total employment 总就业人数;就业总数
total energy 总能量
total energy balance 总能量衡算
total energy concept 全能量系统
total energy consumed 能源消费总量
total energy consumption 全能耗
total energy equation 总能量方程(式)
total energy flow 总能流
total energy flux 总能通量
total energy method 总能量法;全能量法
total energy of waves 波浪总能量
total energy plant 全能电站
total energy residual curve for wide frequency band 宽频带总能量剩余曲线
total energy system 总供热系统;全能量系统
total enthalpy 总热焓
total enthalpy gain 总增焓
total entropy 总熵
total environment 总环境;全环境
total environmental control 总环境控制
total environment(al) emission 环境总排放物
total environmental emission 总环境排放量
total environmental quality index 总环境质量指数
total environment(al) resistance 环境阻力
total environmental value 总环境评价
total equity 全部股本
total equivalent amount 总当量
total eradication 总体杀灭
total erosion 总侵蚀量
total error 总误差
total error of division 节距总误差
total error of tooth spacing 齿距总误差
total esterified fatty acid 酯化脂肪酸总量
total estimate 总估计
total estimated cost 总投资;总概算造价;总概算
total estimated settlement 总预估沉降量
total evapo-transpiration 总散发量
total evapo(u)ration 总蒸发量
total evapo(u)ration of watershed 流域总蒸发量
total excess profit 总超额赢利
total exchangeable bases 盐基交换总量
total exchange capacity 总交换能力
total excursion 全行程;位移峰峰值
total excursion of an oscillating quantity 振动量的全行程
total expansion of tank volume 罐body容积全变形值
total expected value of decision maker 决策者的期望值
total expenditures 全部开支
total expenses 总费用
total expense of prospecting and exploration 地质勘查总费用
total export-import volume 进出口总额
total extension 总伸展;总伸长
total extinction 全消光
total fabric 全组构
total factor productivity 全要素效率
total failure 工作中断;完全失效
total failure strength 完全破坏时强度
total feed 总进给
total feedback 总反馈

total field 总场
total filterable residue 总滤过性残渣
total financing 融通资金总额
total financing-to-value ratio 融资总额与资产价值之比
total fixed cost 总固定成本
total fixed solid 总固定固体
total flash-off 烧化余量
total flexibility matrix 总柔度矩阵
total float 总富裕(时间)(网络进度计划中的);全多余;实际工作时差值
total flood duration 洪水总历时
total floor area 总建筑面积;展开面积
total floor depth 楼板总厚度
total floor space 楼面总面积;地板总面积
total flow 总消耗量;总流量
total-fluoride adjustment buffer 总氟化物调节缓冲器
total flux 总涌水量;总通量;总流量
total footage 总进尺;全井总进尺
total force 合力
total fracture density 总裂缝强度
total free calcium oxide 总游离氧化钙量
total free dissolved carbohydrate fraction 总溶解游离碳水化合物分数
total free lime 游离氧化钙总量;总游离石灰量
total free surface 总自由面(浮选)
total frequency 累频;累积频率;频率
total frequency curve 频率累计曲线
total frequency curve of tidal level 潮位累积频率曲线
total (frequency) response curve 总频率响应曲线
total friction head 总摩擦损失;总摩擦(水头)损失
total fuel consumption 总燃料消耗量
total fusion ages 全熔融年龄
total gain 总增益
total gas-oil factor 总油气比
total gas-oil ratio 总油气比
total gear reduction 总传动比;齿轮总减速
total general bacteria 一般细菌总数
total generation 总发电量;总产量
total generation rate 总的产生速度
total geomagnetic intensity 总地磁场强度
total Gibbs energy 总吉布斯能量
total goal 整体目标
total gradient current 总梯度流
total grain surface 总颗粒表面
total graph 全图
total gross floor area 总建筑毛面积
total gross traffic 总运输量
total gross weight 总毛重
total growth 总生产量
total gyroscopic drift 陀螺总漂移值
total half-life 总半衰期
total half-width 全半宽
total haloacetic acid 总卤代乙酸
total halogen concentration 总卤素浓度
total hardness (水的)总硬度
total hardness of water 水的总硬度
total harmonic distortion 总谐波失真
total harmonic distortion level 总谐波失真电平
total harmonic ratio 总谐波比
total haul 总运距
total haulage 总运距
total head 总压头;总水头;总高差;全值
total head differential 总扬程差;总水头差
total head efficiency 总压头效率
total head (grade) line 总能头线
total head gradient 总水头梯度;总水头比降;总水力梯度;总水力比降
total head line 总水头线
total head loss 总水头损失
total head of dredge pump 泥泵总扬程
total head pressure 总压头
total head tube 全压头管;总压头管
total head value 全压头值
total heat 总热量;总热;热函;全热
total heat balance 总热平衡
total heat consumption 总耗热量
total-heat content 总热函
total-heat-entropy diagram 总热熵图
total heat flux 总热负荷;总热流量;总热通量
total heat history 总热历程
total-heating surface 总传热面;总受热面

total heat(ing) value 总热值
total heat load 总热负荷
total heat of evapo(u)ration 总蒸发热量
total heat of solution 总溶解热;溶解热
total heat radiation pyrometer 全辐射高温计
total heat requirement 总需求热量
total heat transfer 总热传递
total-heat-transfer coefficient 总传热系数
total heat transfer rate 总传热率
total heave 总平错;总隆起量
total heavy metal 总重金属;重金属总量
total height 总高(度)
total height of cage 罐笼全高
total heterotrophic bacteria 总异养菌
total heterotrophic bacteria count 异养菌总数
total hoisting capacity 额定起重量
total holding time 全部占用时间
total holdup 总滞留量
total horizontal water pressure 总水平水压力
total humic acid 总腐殖酸
total humic acids content percent graduation 总腐殖酸含量分级
total humidity 总湿度
total hydraulic gradient 总水力坡降线
total hydraulic head 总水头
total hydraulic head on damaged part 破口总水压力
total hydrocarbon 总烃;总碳氢化合物
total hydrocarbon content 总含烃量
total hydrostatic(al) end force 流体静压轴向力
total identifiable chlorinated hydrocarbon 总可鉴别的氯化烃
total illumination 总照度
total image run-out 全像偏离
total immersion corrosion test 全部浸没腐蚀试验
total immersion test 全浸试验
total immersion thermometer 全浸温度计
total impedance 总阻抗
total impellent force 总推力
total impurities 总杂质
total incineration 完全焚烧
total income 总收益
total increment 总增量;总生长量
total indebtedness 负债总额
total indicator 总和指示器
total indicator reading 指针读数
total industrial and agricultural output 工农业总产值
total industrial demand for each commodity 每种商品的产业部门总需求
total industrial output value 工业总产值
total inflow 总入流(量);总来流量;总进水量
total inflow of river network 河网总入流量
total influence value of water table 总水位削减值
total information 总信息
total information system 综合信息系统
total inhibition point 总抑制点
total inorganic carbon 总无机碳
total inorganic dissolved solid 总溶解无机固体
total inorganic nitrogen 总无机氮
total input 总投入;总输入信号;统一输入信号
total inspection 全面检查
total installed capacity 总装机容量
total installed horsepower 总设备马力
total insured value 总保险值
total intake 总摄入量;总进水量
total integral 整体
total interception loss (植物对降水的)总截留损失(量)
total internal reflection 全内反射
total internal reflection cavity 全内反射腔
total internal reflection light fibre 全内反射导光纤维
total internal reflection resonator 全内反射共振器
total interstage flow 总级间流量
total inversion 全反转
total invested capital 投入资本总额
total investment 总投资(额);全部投资
total investment of grouting (灌浆的)总投资
total investment of recharge works 回灌量
total investment of water source construction 水源地建设总投资
total ion concentration 总离子浓度
total ion current 总离子流

total ion current profile 总离子流剖面图
total ion exchange capacity 总离子交换容量
total ionic concentration 离子总浓度;离子总量;矿化度
total ionic strength adjustment buffer 总离子强度调节缓冲器
total ionization cross section 总电离截面
total ion monitor 总离子记录器
total ions chromatogram 总离子色谱图
total ions detector 总离子流检测器
total iron 总铁;全铁
total iron binding capacity 总铁结合力
total iron combining power 总铁结合力
totaliser 累加器
total isotopic composition 总同位素成分
totality 总体;总数
totalization 综合化
totalizator 累积计数器;累积计算器
totalize 总计
totalizer 总雨量计;累加器;累器;累积计算器;累积计数器;加法器;累计计数
totalizing 总计
totalizing instrument 积分仪;求和仪
totalizing key puncher 总计键盘穿孔机
totalizing meter 累计计数
totalizing punch 总计穿孔机
totalizing puncher 总码穿孔机
totalizing wattmeter 总功率表;累计功率计
total Kieldah nitrogen 总基耶达氮
total labo(u)r force 劳动力总数
total latent heat 总潜热
total latitude 纵距和;总纵距
total lead 总铅(量)
total lead angle 总提前角
total lead time 总提前时间
total leakage 总泄漏耗损;总渗漏量
total leakage flux 总漏磁通
total leak test 总漏泄试验
total length 总长(度);全长
total length closing error of traverse 导线全长闭合差
total length of adit 平峒总长度
total length of bridge 桥梁全长
total length of connecting traverse 附合导线总长
total length of exploratory trench 槽探总长度
total length of fault zone 断层带总长度
total length of line network 线网总长(度)
total length of long casing thread 长扣套管丝扣总长度
total length of short casing thread 短扣套管丝扣总长度
total length of sound wave logging well 声波测井总长度
total length of static cone penetration test 静触探总进尺
total length of survey line 测线总长
total length of synthetic(al) logging well 综合测井长度
total length of track 股道全长
total length of turnout 道岔全长
total length of water pressure test segment 压水试验总长度
total length of well pipe 井管总长度
total level width 总能级宽度
total liabilities and networth 负债总额及净值;负债及净值总数
total liability 总负债
total lift 总提升高度;总升力;总起升高度;全扬程(装卸桥);全升力
total lifting area 总升力面积
total light amount of heat releasing light 热释光总光量
total light flux 总光通量
total light output 总发光量
total light yield 总发光量
total linear swelling and shrinkage 胀缩总率
total linear swelling shrinkage 总线性胀缩(率)
totalling 相加;总和
totalling meter 求积计数器
total liquid volume 液相总体积
total load demand 总负荷需要量;总负荷需求
total load discharge 总输沙量
total load(ing) 总负载;总荷重;总荷载;总负荷;全沙

total loading area 总荷载区
total loading of pollutant 污染总负荷
total loading time 总装填时间
total load rejection 全部弃荷
total lockage time 过闸总时间
total longitudinal strength 纵总强度
total lorries 运输车数
total loss 总损失;总损耗;亏损总额;全损;全漏失;全部损失
total loss control 全面损失控制
total loss cover 全损险
total loss coverhulls 船舶全损险
total loss of prestress 预应力全消失
total loss of soft magnetic material 软磁性材料总损耗;软磁材料总损耗
total loss of vessel only 只保船舶全损
total loss on ignition 总灼烧损失;灼失总量;燃烧总热损失
total loss only 仅负完全损失之责(指保险);仅保全损;全损险
total loss only and general average 全损险及共同海损
total loss only by total loss of vessel 根据本船全损的全损险
total lot amount 总批量
total lumen efficiency 总流明效率
total luminous flux 总光通量
total lunar eclipse 月全食
totally additive 完全可加性的
totally additive set function 完全加性集函数
totally bounded 完全有界的
totally bounded distribution 完全有界分布
totally bounded metric space 完全有界度量空间
totally bounded set 预紧集
totally bounded uniform space 完全有界一致空间
totally buried 完全淹没的
totally continuous 完全连续的
totally differentiable function 完全可微函数
totally differetiable 完全可微的
totally disconnected 不连通的;完全不连通的
totally disconnected closed set 完全不连通闭集
totally disconnected compact groups 完全不连通紧群
totally disconnected graph 全不连通图;完全不连通图
totally disconnected group 完全不连通群
totally disconnected metric space 完全不连通度量空间
totally disconnected set 完全不连通集
totally disconnected spaces 完全不连通空间
totally discontinuous function 完全不连续函数
totally distributed traffic control system 全分散式交通控制系统
totally enclosed 全封密的;全封闭(的);全风化冰脊;完全封闭的
totally enclosed cabin 全闭式舱
totally-enclosed cockpit 全密闭式座舱
totally enclosed fan cooled 全封闭风扇冷却
totally enclosed fan cooled squirrel cage motor 全封闭风扇冷却鼠笼式电动机
totally enclosed fan-cooled type machine 全封闭风扇式电机
totally enclosed fan cooling motor 全封闭扇冷式防爆电动机
totally enclosed frame 全封闭外壳
totally enclosed lifeboat 全密封式救生艇
totally enclosed machine 全封闭式电(动)机
totally enclosed motor 全封闭电动机;防爆电(动)机;全封闭式电(动)机
totally enclosed non-ventilated 全封闭不通风的
totally enclosed pipe-ventilated 全封闭管子通风式
totally enclosed type 全封闭式
totally enclosed type induction motor 全封闭式感应电动机
totally enclosed vertical elevator 全封闭垂直提升机
totally enclosed water cooled 全封闭水冷却式
totally enclosured motor 全封闭式电(动)机
totally enclosured motor lifeboat 全封闭式救生艇
totally enclosured motor propelled survival craft 全封闭式机动救助艇
totally enclosured type 全封闭式
totally enclosure treatment facility 全封闭处理设施
totally geodesic 全测地的

totally geodesic hypersurface 全测地超曲面
totally hyperbolic 全双曲型
totally hyperbolic differential equation 全双曲型微分方程
totally imaginary field 全虚域
totally-inclosed cockpit 全闭式座舱
totally isotropic 全迷向的
totally isotropic subspace 全迷向子空间
totally matric algebra 全部矩阵代数
totally ordered 全序的
totally ordered set 全序集
totally porous packing 全多孔型填充剂
totally reflected 全反射层的
totally sealed environment 完全密封的环境
totally self-checking circuit 全自检查电路
totally unimodular matrix 全单位横矩阵;完全单位模矩阵
total machine control system 整个机器控制系统
total magnetic field 总磁场
total magnetic force 总磁力
total magnetic intensity 总磁力强度;总磁场强度
total magnetic intensity chart 总磁强图
total magnetic susceptibility 总磁化率
total magnification 总放大倍数(显微镜)
total make 总出量
total management 综合管理;全面管理
total management system 综合管理系统;全盘管理系统;全面管理系统
total manganese 总锰
total mask 全面罩
total mass 总质量
total mass flux 总质量流量
total mass of S/R machine 有轨巷道堆垛重量
total mass stopping power 总质量阻止本领
total maximum daily load 日最大污染物负荷总量
total mean increasing 总平均生长量
total mean increment 总平均生长量
total mean power 总平均功率
total mean square error of elevation difference 高差全中误差
total measurable benefit 可测总收益
total measure 全测度
total measurement (volume) 总容积
total measuring error 测量总误差
total mechanical impedance 总机械阻抗
total mechanical ventilation 全面机械通风
total melting loss 熔炼总损耗
total mercury 总汞(量)
total metabolism 总代
total metal concentration 总金属浓度
total metal content 金属全量
total metal contents in soil 总金属含量
total metal input 总金属输入量
total metals 金属总产量
total microbial biomass 总微生物生物质
total mileage 总行驶里程
total mineralization 总矿化度
total mineralization of groundwater 地下水总矿化度
total mineral resources 矿产资源总量
total mineral resources prediction 矿产资源总量预测
total minutes-miles of congestion 拥挤总分——英里
total mixed liquor volatile suspended solids 总混合液挥发性悬浮固体
total-mixing cycle 总混炼周期
total mix(ture) 总混合物
total mobilization 总动员
total modulation 完全调制
total moisture 总水分;总湿度;全水分
total moisture content 总含水率
total moisture sample 全水分煤样
total molar concentration 总质量摩尔浓度
total molybdate0reactive phosphorus 总钼酸活性磷
total monomer 总单体
total months 总月数
total mortality rate 总死亡率
total motion resistance 总行驶阻力(车辆)
total movement 终位移
total net ablation rate 总净消融量
total net accumulation rate 总净累积率
total net additional cost 费用净增加数总额
total net budget 总净收支

total net charge 总装药量
total net earning 盈余总额
total net inflow volume 总(的)净流入量;净入流总量;净来水总量;净进水总量
total net lifting capacity 总的净提升能力
total net profit 盈余总额
total net weight 总净重
total network travel passenger/time 线网乘行人次总数
total net worth 净值总额
total nitrogen 总氮
total nitrogen oxide 总氧化氮;总氮氧化物
total noise spectrum 总噪声谱
total non-carbonate carbon 总非碳酸盐碳
total non-filterable residue 总非过滤性残渣
total non-residue 总非残渣
total normal force 总法向力
total normal pressure 法向总压力
total normal stress 法向总应力
total number 总号
total number density 总数密度
total number of bacterial colony 菌落总数
total number of calls 船舶停泊总艘次数
total number of grid point 格点总数
total number of heterotrophic bacteria 异养菌总数
total number of junction 节点总数
total number of samples 样本总数
total number of stations 总站数目
total number of wagons handled 货车在港作业车次数;火车在港作业车次数
total numbers labo(u)r efficiency 全员劳动效率
total obscuring power 总蒙蔽本领
total observation accuracy of gravimeter 重力仪观测总精度
total of longest consecutive jobs 需时最长的各项连续工作
total olefins 总烯含量
total on board 载货总量
total online program (me) and information control system 全部机内联用程序和信息控制系统
total opening 全开度
total operating cost 总运行成本
total operating profit 总生产赢利
total operating time 总运行时间;总计操作时间
total operation expense 总操作费用
total operation time 总工作时间
total order 全序
total ordering relation 全序关系
total ores 原矿总产量
total organically bound chlorine 总有机结合氯
total organic bromine compound 总有机溴化合物
total organic cabon analyzer 有机碳总量分析仪
total organic carbon 有机碳总量;总有机碳(量)
total organic carbon analyzer 总有机碳量分析仪
total organic carbon analyzer for seawater 海水总有机碳测定仪
total organic carbon content 总有机碳含量
total organic chlorine 总有机氯
total organic chlorine compound 总有机氯化合物
total organic component 总有机组分
total organic dissolved solid 总溶解有机固体
total organic halide 总有机卤化物
total organic halide formation potential 总有机卤化物生成势
total organic halogen 总有机卤素;总有机卤素量
total organic halogen concentration 总有机卤素浓度
total organic halogen formation potential 总有机卤素生成势
total organic loading 总有机负荷
total organic matter 总有机物
total organic nitrogen 总有机氮
total organics concentration 总有机物浓度
total organo-chlorine compound 总有机氯化合物
total organohalogen 总有机卤素
total outflow 出流总量
total outlay 支出总额
total output 总生产率;总动力;总出量;总产量;总产出
total output of energy 出口总能量
total output panel 总输出功率记测板;总输出电屏;总功率记测板
total output value 总产值;生产总值

total output value of agriculture 农业总产值
total overall reaction 全部反应
total overburden pressure 总上覆压力
total overshoot 总超调量
total over shoot offset 总过冲偏移量
total oxidant 总氧化剂
total oxidant concentration 总氧化剂浓度
total oxidation 完全氧化
total oxidation process 完全氧化法
total oxides of nitrogen 总氮氧化物量
total oxidizable nitrogen 总可氧化氮
total oxygen 总氧
total oxygen-consuming capacity 总耗氧量
total oxygen demand 总需氧量
total oxygen demand automatic recorder 总需氧量自动记录仪
total oxygen utilization rate 总耗氧量速率
total ozone column 臭氧气柱总量
total ozone content 臭氧总含量
total ozone deviation 臭氧总量离差
total ozone mapping psectrometer 臭氧总量绘图系统
total ozone oscillation 臭氧总量起伏
total ozone unit 总臭氧单位
total paid-in capital 实缴股本总额
total parallax 总视差;全视差
total particle concentration 总颗粒浓度
total particulate matter 总颗粒物
total particulate phosphorus 总颗粒磷
total passing 总体搬运
total-passing gradation 全部通过筛号的集料等级;全部通过筛号的骨料等级
total path loss 总路径损耗
total peak corrected value 总和峰校正值
total peak distortion 总峰值失真
total penetration 总贯入量
total penetration of pile 桩的总贯入量;桩的入土深度
total percent of ash 总灰分百分含量;总灰分
total permeability 总磁导率
total persulfate nitrogen 总过硫酸盐氮
total phosphate 总磷酸盐
total phosphorus 总磷;全磷
total phosphorus concentration 总磷浓度
total phosphorus level 总磷水平
total phosphorus load 总磷负荷
total phosphorus model 全磷模型
total pipe-line mileage 管线总英里数
total pitch error 节距总误差
total plan 总体规划
total plant capacity 总电站装机容量
total plate count 总板式计数
total polarization error 总极化误差
total pollutant 总污染物
total pollutant loading 总污染物负荷
total pollutant quantity control 总污染物量控制
total pollution loading 总污染负荷
total pore space 总空隙;总孔隙
total pore volume 总孔隙体积
total porosity 总透气量;总气孔率;总孔隙率;总孔隙度;全孔隙度
total port area 港域总面积
total potential 总位能
total potential energy 总势能
total power 总功率
total power consumption 总功耗
total power density 总动力密度
total-power loss 总功率损耗
total power of station 电站总功率
total-power receiver 总功率接收机
total-power requirement 总需要功率;功率总需要量
total-power telescope 总功率望远镜
total precedence grammar 全优先文法
total precipitation 总降水量;总降水;全年总雨量
total precipitation of storm 暴雨总量
total-prefabricated construction 全装配式建筑
total-prefabricated structure 全装配式结构
total pressure 总压强;总压(力);全压
total pressure difference 总压力差
total pressure drop 总压(力)降
total pressure drop of pool 油藏总压降
total pressure efficiency 全压效率
total pressure gradient 全压力增encrypt率;全压力梯度
total pressure loss 总压力损失

total pressure method 总压法
total pressure probe 总压力探头;总压力传感器;总压力测管;全压管
total-pressure probe 总压传感器
total-pressure recovery coefficient 总压力恢复系数
total pressure tensor 总压强张量
total pre-stressing force 总预应力
total-presure at exit 出口总截面压力
total price 总价
total price index of living cost 生活费用价格总指数
total probability 总概率;全概率
total probability formula 全概率公式
total probability theorem 全概率定理
total processing time 总处理时间
total production 总产量
total production cost 生产总成本
total production of society 社会总产值
total productive capacity 总生产能力
total productive maintenance 全面生产性维修
total product of society 社会总产值
total products 成品总产值
total-products forecast 总生产量预测
total profile length 剖面总长度
total profit 总利润;利润总额
total profit and loss for the previous year 去年全年总损益
total project cost 全部工程费用
total project duration 总工期
total propeller-disk area 螺旋桨旋转盘总面积
total propeller-width ratio 螺旋桨总宽度比
total proper motion 总自行
total pseudo-ordering 全伪序
total pump(ing) head (水泵的)总扬程;总抽水扬头;总扬程;水泵总扬程
total push 总推动
total qualitative analysis 全定性分析
total quality assurance program(me) 全面质量保证
total quality charges 质量收费
total quality control 全面质量管理;总质量控制;全面质量控制
total quality control circle 全面质量管理圈
total quality control system 全面质量管理制
total quality management 全面质量管理
total quantity consumed 消费总量
total quantity control of pollutant discharge 总量控制;排污总量控制
total quantum number 总量子数
total radiant power 总辐射功率;全辐射功率
total radiation 总辐射量;全辐射
total radiation dosage 总辐射剂量
total radiation dose 总辐射剂量
total radiation power 总辐射功率
total radiation pyrometer 全辐射高温计
total radiation radiometer 全辐射辐射计
total radiation temperature 全辐射温度
total radiator 全辐射器;全波辐射器
total rainfall 总雨量
total rare earth 总稀土
total rare earth content 稀土总量
total rate of mining and dressing recovery 采选综合回收率
total rate of mining dressing and melting recovery 坏选冶综合回收率
total reaction 全反应;总反应
total reactive force 总冲力;总反作用力
total reactive phosphorus 总活性磷
total receipts 收入总额
total receiver point number 检波点总数
total receiving water pollutant budget 总纳污水体污染物平衡
total recoverable phenolic compound 总可回收苯酚化合物
total reduced sulfur 总还原硫
total reduction 总减速;总传动比
total reduction capacity 总还原能力
total reduction gear ratio 总传动减速比
total reduction ratio 总减速比
total reflectance 全反射比
total reflecting resonator 全反射共振腔
total reflection 全反射
total reflection condition 全反射条件
total reflection factor 全反射系数

total reflection layer 全反射层
total-reflection long capillary cell in absorption spectrophotometry 全反射长光程毛细管分光光度法
total-reflection method 全反射法
total reflection of wave 波的全反射
total reflection prism 全反射棱镜
total reflection refractometer 全反射折射计
total reflectivity 全反射率
total reflectometer 全反射仪;全反射计
total reflector current 反射极总电流
total reflux 全回流
total-reflux operation 全回流操作
total refraction 全折射
total regression 总回归;全响应
total regression curve 总回归曲线
total reinforcement area 钢筋总(截面)面积
total relaxation 总松弛
total release 总排(放)量;排放总量
total relief 体视率
total requiremant 总需(要)量
total reserves 总储(存)量;总储藏量
total reserves of the production area 产地总储量
total reserve system 全额准备制
total reservoir capacity 总库容
total reservoir storage 水库总容量
total residence time 总停留时间
total residual bromine 总残溴
total residual chlorine 总余氯(量);总残余氯
total residual oxidants value 总残氧化剂量
total residue 总筛余;总残渣;总残物;总残留量
total residue of evapo(u)ration 蒸发残渣总量
total residue on ignition 灼烧残渣总量
total resistance 总阻力;总电阻;全电阻
total resistance of locomotive 机车总阻力
total resistance of wagon 车辆总阻力
total resolution efficiency 总分辨效率
total resource recovery 总资源回收
total resources 总资源(量)
total response 总响应
total resultant pressure 总的合压力
total retention time 总停留时间;总保留时间
total retirement benefits 退休总津贴
total revenue 营业收入总额;总收益
total reversal solidification 倒向凝固
total rig power 钻机总功率
total rig time 钻探总台时;钻机总台时
total rise of a roof 屋面总高度
total risk 全损险
total risk factor 总危险度
total river discharge 总流量
total rubber 总胶量
total run 总行程;总长度(椽梁跨度水平距离);椽木水平跨度
total runoff 总流量;总径流(量)
total sag 总垂度
total salable expenditures 总推销费
total sales method 销货总额法
total salinity 总盐度;总含盐量;全盐量
total salt 全盐量
total salt content 总含盐量
total salt determination 全盐量测定
total sample 全部试样;全部样品
total sample volume 土样总体积
total sampling method 全采样法
total scanning time 总扫描时间
total scattering 总散射;全散射
total scattering coefficient 总散射系数;全散射系数
total scattering probability 总散射概率
total scope of architecture 全面建筑观
total sediment 总泥沙量;泥沙总量
total sediment discharge 总输沙率;总输沙量
total sediment inflow 输入泥沙总量
total sediment load 总含沙量;泥沙总量
total sediment nitrate demand 总底泥需硝酸盐量
total sediment outflow 输出泥沙总量
total sediment transport capability 总输沙能力
total sediment weight 泥沙总重
total sensitivity 总灵敏度
total sensitivity of colo(u)r film 彩色片总感光度
total separation 除尘效率
total settlement 总沉降量
total shaft horse-power 总轴马力
total shaft resistance 总侧面阻力

total shear 总剪力;总剪切
total shearing force 总剪切力
total shearing resistance 总抗剪力
total shearing stress 总剪切应力
total shift 总移距
total ship-days in port 船舶停泊总艘天数
total shrinkage 总收缩
total silver concentration 总银浓度
total site area 总建筑基地面积
total site construction area 总建筑基地面积
total size 总范围
total sky brightness 累积天空亮度
total slip 总滑距
total sludge solid 总固体污泥
total snow load 总雪荷载
total social product 社会总产品
total soil-moisture stress 土壤水分总应力
total soil water potential 土壤水总潜力
total solar eclipse 日全食
total solar radiation 太阳总辐射
total solid 总固体量;固体总量
total solid capacity 总固体含量
total solid concentration 总固体浓度
total solid content 总固体含量
total solid matters 总固体物
total solid residue 总固态残渣
total solids 总固体物;总固体;总干物质
total solids by drying 总干固物(干燥法测得)
total solids content 总固形物含量
total solids in wastewater 废水中总固体量
total solids of water 水的总固体
total solids test 总固体测定
total solid yield 总干物质量
total soluble cadmium concentration 总溶性镉浓度
total soluble cations 总溶性阳离子
total soluble chemical oxygen demand 总溶性化学需氧量
total soluble matter 总溶性物质;总溶解物;总可溶物;总溶度
total soluble phosphorus 总溶性磷
total solubles 总可溶物
total soluble salt 总溶性盐
total soluble solid 总溶性固体;总可溶物
total solution 全溶液;全部解法
total sorbed amount 总吸附量
total sound absorption 总吸声率
total space (城市的)综合空间;全空间
total spacing 总间隔
total span 总跨度;总臂距
total specific ionization 比电离
total spectral emissivity 总频谱发射率
total stacking fold 总叠加次数
total state 总态
total static head 总静压头;总静力高差
total station optic(al) electronic tach(e)ometer 全站型电子速测仪
total station optic(al) electronic tachometric theodolite 全站型光电速测经纬仪
total station optic(al) electronic tachymeter 全站型电子速测仪
total station optic(al) electronic tachymetric theodolite 全站型光电速测经纬仪
total steel area 钢筋总面积
total step iteration 整步迭代法
total stiffness matrix 总刚度矩阵
total stocks 总储[贮]存量
total stock value 库存价值总额
total storage (capacity) 总库容;总储藏量;总蓄水量;水库总容量
total storage capacity when earthquake is induced 发震时总库容
total storage space 库场总面积
total stor(e)y space 楼层总面积;楼层空间
total storm precipitation 一次暴雨降水量
total strain 总应变
total strain field 总应变场
total strain theory 总应变理论;全量理论
total stream load 河流总输沙量;河流泥沙总量
total strength 总强度
total stress 总应力
total stress analysis 总应力分析
total stress approach 总应力法
total stress cell 总应力盒

total stress cell method 总应力盒法
total stressing force 总应力
total stress method 总应力法
total stress path 总应力路径
total stress strength index 总应力强度指标
total stress strength parameter 总应力强度参数
total stretching force 总拉伸力
total structural intensity 结构总强度
total suction 总吸水量
total suction head 总吸升水头
total sulfur 总硫;全硫
total sulfur content 总硫量;总含硫量;含硫总量
total sulfur content graduation 全硫含量分级
total sulphur content 总硫量;总含硫量;含硫总量
total sum 总价;总厚度;总额
total-sum compressibility 总压密量
total sum of squares 总平方和;平方总和
total sum of trip 出行总量
total supply 总供水量
total surface 总面积;总表面
total surface area 总表面积
total surface dose 总表面剂量
total suspended load 总起重量
total suspended material 悬移质总量
total suspended matter 总悬浮物
total suspended particle matter 总悬浮颗粒物
total suspended particles 总悬浮微粒;总悬浮颗粒
total suspended particulate index 悬浮颗粒总量指数;总悬浮微粒量指数
total suspended particulate matter 总悬浮颗粒物
total suspended particulates 总悬浮微粒;总悬浮颗粒;悬浮固体总量
total suspended sediment 悬移质总量;悬浮泥沙总量
total suspended solid concentration 总悬浮固体浓度
total suspended solids 悬移质总量;总悬浮固体(量)
total swelling and shrinkage rate 胀缩总率
total symmetry 全对称
total synthesis 全部合成
total system 总(体)系统;大板体系;全系统
total systematization 总体配套系统化
total system capacity 总系统容量
total system cost 全系统费用
total system optimization 整体最优化;整体最佳化
total tankage 总容量
total taxes 总税额
total telegraph distortion 总失真
total temperature 总温;滞止温度
total tender value of project 项目的总投标价值
total terminal centres (运输机组的)总长度
total testing time 总计试验时间
total theater 全台剧场;全面剧院(建筑大师格罗派斯1927年设计的)
total thermal analysis 总热分析;热全分析
total thermal power 总热功率
total thermal remanent magnetization 总热剩余磁化强度
total thickness 总厚度
total thickness of coal seam 煤层累计厚度
total thickness of coat film 涂膜总厚度
total thickness of oil-gaseous section 含油气段的总厚度
total thickness of pavement 路面总厚度
total thickness of workable coal seam 可采煤层累计厚度
total throw (断层的)总落差
total thrust 总推力
total tidal discharge 潮流总量
total time 总时间;总计时间
total time of passenger's travel(l)ing 旅客旅行总时间
total time of treatment 总处理时间
total times of coring 取芯总次数
total titratable acidity 总酸度
total tolerance 总容许限度
total tonnage 总吨位
total torque 总转矩
total traffic 总报务量
total traffic ton-miles 总运输吨哩
total train braking force 列车总制动力
total train resistance 列车总阻力

total transfer 总计传递;全转移;全额转让
total transform 全变换
total transmission factor 全透射因数
total transmittance 总透射率;总透过系数;总透光度
total transparent length 透光总长度
total transportation expense 总运费
total transportation volume 总运输量
total transport cost 总运费
total transport expense 总运费
total transport of mass 总质量运输
total travel 总行程
total travel passenger/time for mass transit 公交乘行人次总数
total travel time 总行程时间
total treatment 完全处理
total (treatment) system 综合处理系统
total trihalomethane formation potential 总三卤代甲烷生成势
total trim 总调整
total tuyere area 风口总有效截面积;风口总面积
total ultimate reserves 总工业储量
total unfilterable residue 总非过滤性残渣
total unsaturates 总不饱和物
total upstream erosion 上游总冲刷量;上游泥沙流失总量
total value 总值
total value increment percent 总价格生长率
total value industrial and agricultural output 工农业总产值
total value industrial and agricultural production 工农业总产值
total value of fixed assets to be depreciated 应计折旧的固定资产总值
total value of import 进口总值
total value of production 生产总值
total value of production of railway transportation 铁路运输总产值
total value of social commodities 社会商品价格总额
total value of trade 贸易总值
total values of output 总产值
total variable cost 总变动成本
total variation 总变差;全变差
total ventilating power 总通风功率
total ventilation pressure 总通风压力
total viable count 总活菌计数
total viable numbers 总活菌数
total view 整体展望;全景
total void ratio 总空隙率;总开空隙率
total volatile acids 总挥发酸类
total volatile basic nitrogen 挥发性盐基总氮
total volatile solids 挥发性固体物总量;总挥发性固体
total volatile suspended solid 总挥发性悬浮固体
total voltage 总电压
total volt-ampere 视在功率;总功率
total volume 总容积;总体积;全容积
total volume of construction 施工总体积
total volume of cycle system 循环系统总体积
total volume of ebb tide 落潮主槽
total volume of excavated soil 挖土机总量
total volume of exploring mining and exploratory trench 坑槽探方量
total volume of expulsive liquid 排出液体的总体积
total volume of flood 涨潮总量
total volume of flood tide 涨潮总量
total volume of import and export trade 进出口贸易总额
total volume of imports and exports 进出口贸易总额
total volume of orebodies 矿体总体积
total volume of railway freight 铁路运输量
total volume of rain 总雨量
total volume of rainfall 降雨总量
total volume of silt 泥沙总量
total volume of soil sample 土样总体积
total volume of storm 暴雨总量
total volume of system 系统总体积
total volume of the soil aggregate 土壤团粒总体积
total volume of trade 贸易总量;贸易总额
total volume of voids 总孔隙体积
total volume to be dredged 拟挖总方量
total vorticity 总涡量

total wage bill 工资总额
total wage plan 工资总额计划
total wages 总工资;工资总额
total wagon-hours in port 货车在港总停留车时;
  火车在港总停留车时
total wall thickness 墙的总厚度
total water/cement ratio 总水灰比
total water consumption 总耗水量
total water content 总水量
total water content instrument 总含水量测试仪
total water ga(u)ge 总水柱高度;水头测量总单位压力
total water immersion test aggregate 浸水试验骨料
total water requirement 总需水量
total watershed contaminant load 总流域污染物负荷
total water-soluble carbohydrate 总水溶性碳水化合物
total water solubles 总水溶物
total water supply 总供水量
total water yield of mine 矿坑总涌水量
total wave force 总波浪力
total wave load 总波力
total wear of rail head 轨头总磨损
total weight 总重(量);总容重;全重(量)
total weight of drill tool 钻具总重
total weight of gravel 砾石总重量
total weight of hard filth 硬垢总量
total weight of pot filth 锅垢总量
total well capacity 井的总容量
total wet density 总湿密度
total wheel base 总车轴距;全轴距
total wheel base of bogie assembly 转向架全轴距
total width 总宽(度)
total-width of roadway 路基全宽度
total-width ratio 总宽(度)比
total wind force 总风力
total work 总功
total work cost 总生产成本;总工作费(用);总工程造价
total working weight 总工作重量
total yard performance 编组场作业总性能
total yearly output 年总产量
total yield 总收率;总收获量;总产量;总产出
total yield of water sources 水源地总出水量
total zinc 总锌
totameter 流量计
totara 新西兰罗汉松
totating 转动的
tote 总数;台刨把手
tote bin 有腿货箱
tote box 搬运箱;运输斗;有腿货箱
totem 图腾
totem-pole 图腾柱;推拉输出电路;物像柱(北美印第安人屋前刻的)图腾柱
totem-pole amplifier 图腾柱放大器
totem post (刻有图腾像的)图腾柱
tote pan 装运箱;托盘
toter 导弹装载起重机
tothed bar type crop loader 齿杆式作物装载机
to the layout line 按照划线
to the order of 某一订货人
to the satisfaction of the engineer 使工程师满意
to the weather 外露部分(屋面板)
tot lot 幼儿园游戏场
totter 动摇
Toucas 塔卡斯铜镍合金
touch 联系;接触;触式;触力;触感
touchable 可触的
touch ahead 缓慢前进
touch and stay 弯靠与停靠
touch and stay clause 停留条款
touch ar a port 弯靠某港
touch arc starting 接触式引弧
touch at 弯靠(中途在一港口暂时靠停)
touch at a port 在某港湾靠
touch bottom 接触到实质;触底;擦浅【船】
touch button 灵敏按钮
touch call 按键(式)呼叫
touch catch 触试门扣;弹子扣
touch coating 局部修饰用漆;局部修补用漆
touch core 无芯头芯
touch down 转入排水状态;着地

touch-down dispersion 着陆点散布
touch down point 接地点
touch-down zone 接地区;着陆区
touch-down zone elevation 着陆区最高点
touch dry 表干;指触干
touched horn of pantograph 刮弓
touch end travel 接触端位移
touch feedback 接触反馈
touch gun 补漆用喷枪
touch hole 点火孔
touching angles 对顶三角形
touching button 触钮
touching each other 相互重叠的枝条
touching point of bit 钻孔定位
touching up 修整;表面润饰;润色;找补
touching warmth 步履感热;触觉感热
touch key 触键
touch magnetization 接触磁化
touch-me-not 凤仙花
touch paper 火药纸;导火纸;火硝纸(导火用)
touchplate-controlled automatic door 接触板控制自动门
touch-retract 接触再后退引弧
touch sanding 用砂轻磨
touch screen 触摸屏;触敏显示器
touch sensor 接触式传感器
touchstone 黑燧石;试金石;检验标准
touch switch 接触式开关;接触开关
touch tabulator key 按钮制表键;按键式制表机键
touch tack 接触粘着性
touch the bottom 擦底
touch the runway 接触跑道
touch-tone data service 按键式数据服务
touch-tone dialing 按钮式音频拨号
touch-tone telephone 按钮式电话
touch-tone type key 触发声式键
touch trigger probe 触发式测头
touch up 修整
touch-up paint 修补漆
touch-up painting 补漆
touch welding 触焊;接触式焊接
touchwood 朽木;引火木;火绒用朽木;火绒
tough 韧性的
tough alloy steel 韧性合金钢;韧合金钢
tough alumina 块状氧化铝;不纯氧化铝
tough bronze 韧青铜
tough cake 精铜(含铜约99%)
tough cathode 电解纯铜
tough clay 硬质黏土
tough consistency 韧性结持度
tough (copper) cake 韧铜
tough digging 坚韧岩层掘进
toughen 韧化(作用)
toughened ceramics 增韧陶瓷
toughened glass 钢化玻璃;韧化玻璃;淬火玻璃;强化玻璃;铅丝玻璃;夹层玻璃;防弹玻璃;安全玻璃
toughened glass railing panel 钢化玻璃拦板
toughened glass (suspension) insulator 钢化玻璃绝缘子
toughened glass ware 钢化玻璃器皿;强化玻璃器皿
toughened plate 钢化玻璃板
toughened polished plate glass 钢化打磨过的玻璃板
toughened polystyrene 韧化聚苯乙烯
toughened resin matrix 韧性树脂基体
toughened safety glass 钢化安全玻璃;钢化安全玻璃
toughened safety glass lens 钢化安全玻璃镜片
toughened water ga(u)ge glass 钢化水位计玻璃
toughened zirconia ceramics 韧化氧化锆陶瓷
toughening 韧化处理;韧化(作用);增韧
toughening degree 钢化程度
toughening rubber 韧性橡胶
toughening treatment 钢化处理
tough fracture 韧性断裂;韧性断口
tough hardness 韧硬度
tough iron 韧性铁(可锻铸铁);韧铁
toughlay 耐磨的(绳索)
tough lead 韧铅
tough metal 韧性金属
toughness 坚韧性;坚韧度;粘稠性;韧性;韧度;弯曲性
toughness-brittleness transition 韧性-脆性转变
toughness coefficient 韧性系数;韧度系数

toughness factor 抗冲击系数;韧性因数;韧性系数;韧性率;韧度因子;韧度因素;韧度系数;韧度率
toughness hardness 韧性硬度
toughness index 黏稠度指数;韧性指数;韧度指数
toughness number 韧性指数
toughness of impact 冲击韧性
toughness of material 材料韧性
toughness of road metal 道砟韧度
toughness of road metal test 道砟韧度试验
toughness plastics 韧性塑料
toughness requirements 韧性要求
toughness test 黏稠度测定;韧性试验;韧度试验;冲击载荷强度试验
toughness testing 黏稠度测定
toughness value 韧性值
tough (pitch) copper 韧铜
tough plastic clay 韧塑黏土
tough rubber sheath 韧性橡皮护皮
tough rubber sheathed cable 坚韧橡皮绝缘电缆;韧性橡皮绝缘电缆
tough rubber sheathed rubber insulated cable 韧性橡胶护套绝缘电缆
tough steel 韧性钢
tough value 韧度
tough water 黏稠水
tough wood 硬木;韧木
tougor pressure 膨压
toukul 圆形茅草独室房屋(非洲)
tou kung 斗拱
Toul faience 托尔锡釉精陶
touple 转床刀具
tourbillon watch 旋转机构表
tourelle 出挑塔楼;滚动装置;突墙(墙)角塔
tourer 旅行机
tour gaging 巡测
tour garden 游赏庭园;环游式庭园
tour harbo(u)r 旅游港
tourie 吸收罐
tourill 吸收罐
touring bus 游览公共汽车
touring car 游览车;旅客车
touring map 游览图
touring plane 游览机
touring route 游览路线
tourism 旅游业;旅游事业
tourism and resort zone 旅游度假区
tourism building 旅游建筑
tourism environment 旅游环境
tourist 游客;旅游者;旅游路;旅客
tourist agency 旅游局
tourist attraction 对观光者吸引力;对游客的吸引力;旅游胜地
tourist base 旅游基地
tourist bureau 旅游局
tourist bus 旅客车
tourist cabin 旅游客舱;旅游轮
tourist camp 旅游营地;拖挂式活动住房营地
tourist capacity 游客容量
tourist cave 旅游洞穴
tourist circuit 游览路线
tourist city 旅游城市
tourist class 经济舱;普通客轮
tourist court 汽车游客旅馆
tourist facility 旅游设施
tourist function 旅游功能
tourist geography 旅游地理学
tourist home 旅游者住处;旅客之家;旅游住宅
tourist hotel 旅游旅馆
tourist house 旅游借宿处;旅游借宿屋
touristic zone 游览区;旅游地带
tourist industries 旅游事业
tourist industry 旅游业
tourist map 游览图;旅游图;旅游地图
tourist plane 游览机
tourist pollution 旅游污染
tourist regional environment 旅游区域环境
tourist resort 旅游胜地
tourist road 游览路;游览道路;风景区道路
touristry 旅游
tourist ship 旅游客仓;旅游船
tourist site 游览地
tourists novelty 旅游用品
tourist spot 旅游区
tourist terrace 旅游者观光台

tourist tower 观光塔
tourist trade 旅游业
tourist train 旅游列车
tourmaline 电气石；碧硒
tourmaline deposit 电气石矿床
tourmaline ga(u)ge 电气石计
tourmaline granite 电气石花岗岩；电气花岗岩
tourmaline greisen 电气云英岩；电气石云英岩
tourmaline muscovite greisen 电气石云母云英岩
tourmaline quartz greisen 电气石英云英岩
tourmaline schist 电气片岩；电气石片岩
tourmalinization 电气石化作用；电气石化
tourmalite 电英岩
Tour-Marshall test 图尔—马歇尔弯曲试验
tournadozer 轮式推土机
Tournai marble 图尔奈大理石（比利时）
Tournaisian (stage) 杜内阶【地】
tournament 比赛图
tournament loading 分散加载方式
tournament selection sort 比较选择分类
tournament sort 重复选择分类
tournament sorting 比赛排序；分类分类
tournapull 轮式牵引车
tournatractor 轮式牵引车
Tournay brass 陶奈黄铜
tourniquet 压脉器
tour of observation 考察
tour report 班报表
tourstry attraction 旅游胜地
tour survey of economy 经济考察
tour terminal 旅游码头
Tourun Leonard's metal 轴承用韧性锡青铜
Toussaint's formula 杜赛公式
to vacuum pump 接真空泵
to vary from subhumid to arid 在半湿润至干旱范围内的变化
tow 麻屑；牵引缆绳；丝束；拖曳
tow abreast 旁拖；绑拖
tow-abreast metering 并排限流
towage 牵引作业；牵引费；牵引；拖轮费；拖航；拖带费用；拖带费；拖曳
towage clause 拖航条款；拖带条款
towage contract 拖航合同；拖带契约
towage dues 拖带费
towage firm 驳运公司
towage lien 拖船费留置权
towage service 拖带行业
towage transportation 拖带运输
tow aircraft 拖航机
tow alongside 旁拖；绑拖
towan 沿岸沙丘
toward breeze 顺风
toward low potential from high potential 由高势区向低势区
toward west 向西（的）
tow at side 绑拖
tow-away zone 违章泊车；(不准停车的)拖车区
tow bar 拖杆；拖挂装置
tow beam 拖网拱梁
towboat 拖船；拖驳；拖轮
towboat and barge status system 拖轮及驳船动态显示系统
towboat transport 拖驳运输
tow breaking operation 解体作业【船】；船队解体作业
tow car 救援车
tow carding engine 短麻梳麻机
tow chain 拖车链
tow cleaner 麻屑清理机
tow collecting machine 丝条集束机
tow-component glass 二元组分玻璃
tow conveyer 牵引式输送机；牵引输送机
tow-course construction 双层筑路（法）
tow dimensional code 二组码
towed array surveillance system 拖曳式阵列监视系统；拖曳式基阵监视系统
towed asphalt paver 拖式沥青混合料摊铺机
towed blade grader 拖式平地机
towed boat 拖驳
towed cable length 拖缆长度
towed flexible barge 柔性液囊
towed grader 拖式平地机
towed hopper type gritter 拖挂式料斗式铺砂机
towed load 牵引荷载；拖挂荷载；拖带荷载

towed machine 拖曳的机器
towed manned submersible 拖曳载人潜水器
towed material 纤维束材料
towed moving machine 拖式割草机
towed offset 拖曳偏位
towed paver type aggregate spreader 拖曳式铺路骨料撒布机
towed pneumatic-tired roller 拖挂轮胎压路机
towed ripper 拖式松土机
towed roller 拖挂压路机；拖式压路机；拖式碾压机
towed sampler 拖曳式采样器
towed sand spreader 拖挂铺砂机
towed scarifier 路犁；耙土机；拖式翻土机
towed scraper 牵引式铲运机；拖式铲运机
towed sheep-foot roller 拖式羊足碾
towed ship 被拖船
towed smooth drum roller 拖式光碾压路机
towed speed 拖曳速度
towed sprinkler 牵引式喷灌机
towed target 拖靶；拖
towed-target aircraft 拖靶机
towed-type chassis 拖挂式底盘；拖挂式车身架
towed-type grit spreader 拖挂式平地机
towed-type gritter 拖挂式铺砂机
towed-type mixer 拖挂式拌和机
towed-type road brush 拖挂式扫路机
towed-type sand spreader 拖挂式撒砂机
towed-type sheep's-foot roller 牵引式羊足碾
towed-type vibrating sheep's foot roller 牵引式振动羊足碾
towed unit 被拖曳单元
towed vehicle 挂车；拖曳车；拖挂车
towed-vehicle skid test 拖挂车滑溜试验
towed vibratory roller 拖挂振动压路机
towel airer 毛巾干燥器
towel bar 毛巾挂；毛巾杆
towel bar fitting 毛巾挂零件
towel cabinet 毛巾橱；毛巾柜
towel clamp towel forceps 布巾钳
towel clip 巾夹
towel dispenser 毛巾分放架
towel forceps 巾钳
towel hanger 毛巾架
towel holder 毛巾架
towel hook 毛巾钩
towel ladder 毛巾架
towel motion 毛巾起圈装置
towel-pull-out rod 抽拉式毛巾杆
towel rack 毛巾架
towel rail 毛巾架；毛巾杆
towel ring 毛巾环；毛巾挂环
towel supply shelve 毛巾搁架
towel tree 树形毛巾架
tow equipment 顶推设备
tower 杆塔；望楼；塔设备；塔器；塔架(输电线路)
tower abrasion mill 塔式磨粉机
tower absorption system 塔式吸收系统
tower acid 塔酸；塔式法硫酸
tower acid system 塔酸回收系统
tower ager 塔式汽蒸机
tower antenna 铁塔(式)天线
tower arch 塔形拱
tower base 塔基；塔底；塔底座
tower basin 塔池
tower beacon 锥形标；塔标
tower biologic(al) filter 塔式生物滤池
tower bioreactor 塔式生物反应器
tower bivane 塔形双翼风向标；塔式双翼风向标
tower block 塔式建筑
tower block building 塔式建筑
tower block facade 塔式房屋正面
tower block rising line 塔式建筑压升水管
tower body 塔体；塔身
tower bolt 重型门闩；短套筒插销；管形插销；管销；管插销；短筒插销；大型汽缸盖长螺栓
tower boom 塔臂
tower bottom 塔底
tower bottom pressure 塔底压力
tower bottoms 塔釜液；塔底产物(残油)；塔底残油
tower brace 高架支撑
tower bracing 高架撑架；高架支撑
tower bridge 塔形桥楼
tower bucket 塔架内提升斗
tower-builder 建塔者；建塔工人

tower building 塔楼；摩天楼；造标；塔式建筑
tower burst 塔上爆炸
tower cable excavator 管式缆索挖土机；塔式缆索挖土机；塔架缆索挖土机
tower cable way 高架索道
tower cellar 塔楼地下室
tower clock 塔式时钟；塔钟；屋顶钟
tower column 桥塔立柱；桥梁立柱
tower communication 塔台通信
tower concentrator 塔式浓缩器
tower concrete spouting plant 高架混凝土喷射机
tower control center 塔台控制中心
tower cooler 冷却塔；塔式冷却器；塔冷却
tower core oven 立式烘芯炉
tower counter-flow 塔逆流
tower crane 塔式起重机；塔吊
tower crane erection equipment 塔式起重装配设备
tower crane with inner mast 转柱式塔式起重机
tower crane with slewing cat head 塔顶回转式塔式起重机
tower crane with slewing upper platform 上回转平台式塔式起重机
tower crib 塔式木笼
tower crown 钻井起重机顶部
tower derrick 塔式起重机；塔式起重架；塔式起重桅杆
tower diameter 塔径
tower door 机膛盖板
tower dryer 塔式干燥机；干燥塔；塔形干燥器
tower dump 混凝土料斗
tower dwelling 塔楼；塔式住宅
towered 高耸的；有塔的
towered fire-temple 有塔的火神庙
towered gateway 塔式门道
tower evapo(u)ration 塔式蒸发
tower evapo(u)rator 塔式蒸发器
tower excavator 拖铲挖掘机；塔式挖土机；塔式索斗挖掘机
tower filling 添塔料
tower floor 高楼楼层
tower footing 塔基
tower-footing impedance 杆塔接地阻抗
tower for cable-stayed bridge 拉索桥塔
tower foundation 塔架基础
tower foundation post 塔架基础桩
tower frequency 塔台频率
tower furnace 塔式炉；竖式炉
tower gantry 吊塔；塔吊
tower gantry crane 高架塔式起重机；塔式门式起重机；塔式龙门起重机；塔架门式起重机
tower gas cooler 塔式气体冷却器
tower girder 塔架大梁
tower girt 塔架围梁
tower grillage 塔架格床
tower gutter tile 塔沟瓦
tower head 塔架顶
tower height 塔高；视标高
tower hoist 塔式提升机；塔式升降机；塔式起重机；塔式卷扬机；塔架提升机；塔吊
tower hoist bucket 塔式提升机料斗
tower house 中世纪城堡
towering 高耸厦景
towering cumulus 塔状积云
tower-installed 安装塔架；安装井架
tower intake 进水塔；塔式取水口
tower internals 塔内件
tower jib crane 塔式挺杆起重机
tower karst 岩溶孤峰；塔状喀斯特
tower landform 塔状地貌
tower launcher 塔式发射架
tower leaching 塔淋浸出；塔浸
tower leg 塔架支柱；塔腿
tower lighting 体育场高塔照明
tower loader 前端式装载机；塔式装载机
tower loading 杆塔计算负载；铁塔负载
tower machine 拖铲挖掘机；塔架式挖掘机
tower man 塔台指挥人员；区号标志管理员；望火员
tower mast 格子桅；塔式灯桅
tower mill 塔式粉碎机
tower-mounted 塔上架设的；安装塔架；安装井架
Tower of Babel 巴别塔（古巴比伦建筑）
tower officer 塔台飞行指挥员
Tower-of-hanoi function 梵塔函数

Tower-of-hanoi problem 梵塔问题
Tower of London 伦敦塔
tower of silence 静堡
Tower of Winds 钟楼(古希腊);风塔(古希腊)
tower packing 塔填充物;塔填料
tower pickle 塔式酸洗机
tower pier 塔架;塔形桥墩;塔架(式)桥墩
tower pincers 胡桃钳
tower pit 塔架底坑(混凝土)
tower pivot 塔架支枢
tower plant 高塔式拌和厂
tower platform 塔架平台;钻井平台;钻架平台;塔式平台
tower portal of bridge 桥头堡
tower process 塔法
tower purification 塔式脱硫
tower purifier 塔式净化器
tower radiator 塔形辐射器
tower reactor 塔式反应器
tower reclaiming system 塔酸回收系统;塔式回收系统
tower restaurant 高楼菜馆;高楼饭店;高楼餐馆
tower roof 塔顶
tower scaffold(ing) 塔楼脚手架
tower scrubber 塔式洗涤器
tower section 机膛
tower shaft 塔身
tower shaft kiln 塔式立窑
tower-shaped church 塔式教堂
tower shield facility 塔台防护设施
tower silo 塔式仓;塔式筒仓;圆筒粮仓;圆筒仓
tower skirt 塔裙
tower slewing crane 塔式回转起重机;塔式悬臂起重机
tower's liability clause 拖带责任条款
tower sludge 塔内淤渣;塔内污泥
tower span 吊桥的桥垮;塔式架间距;塔架间跨度;塔架间距
tower spotting 杆塔定位
tower stair(case) 高楼梯;上塔楼梯
tower still 塔式蒸馏器
tower stor(e)y 高楼楼层
tower stove 立式烘干机
tower structure 框架式结构;塔式结构;塔式建筑物
tower substructure 塔架下部结构
tower superstructure 塔式上部建筑;塔式建筑的上部建筑
tower support 高压电线支架;塔式支座;塔架支座
tower system 塔组;塔式挖掘机开采法
tower telescope 太阳塔;塔台望远镜;塔式望远镜
tower top 塔顶
tower top pressure 塔顶压力
tower top temperature 塔顶温度
tower tray 塔盘;塔板(化)
tower trickling filter 塔式滴滤池
tower truss 铁塔桁架
tower-type antenna 塔式天线
tower-type batch drier [dryer] 塔式分批干燥机
tower type batch plant 塔式称量装置
tower-type fermentor 塔式发酵罐
tower-type grain drier 塔式谷物干燥机
tower-type headframe 塔式井架
tower-type Koepe winder 塔式戈培提升机
tower-type launch pad 塔式发射台
tower-type structure 塔型结构
tower unloader 塔式卸料机
tower wagon 高架检修车;梯车;塔车
tower wall(ing) 塔墙
tower wall material 塔壁料
tower washer 洗涤塔;塔式洗矿机;塔式洗涤器;湿式净气塔
tower washing 塔式洗涤
tower weight 塔重
tower winding 塔式提升
tower with constant batter 均匀倾斜的塔
tow finisher card 短麻末道梳麻机
tow formation 顶推驳船队队形
towhead 嫩柳覆盖的沙洲
tow-hook 拖钩
towing 修整坯体边缘;修边;拉纤;拖航;拖带
towing accumulator 拖缆缓冲器
towing aft 吊拖
towing aircraft 拖航机
towing airplane 拖航机

towing anchor at short stay 拖锚
towing and general service launch 拖带和公务交通艇
towing and salving clause 拖救条款
towing angle 拖曳角;拖角
towing apparatus 拖带设备
towing appliance 拖运设备;拖带设备
towing arch 拖缆承梁;拖网拱梁
towing arrangement 拖带设备
towing a sling 拖关(装卸货物)
towing bar 拖缆承梁
towing basin 船模拖曳试验(水)池;船模(型)试验池
towing beam 拖缆承梁
towing bitt 拖缆桩
towing block 拖缆掣架
towing bollards 拖缆桩
towing bridge (带拖钩的)拖带桥楼
towing bridle 龙须缆(船);拖索;拖缆勒头
towing bridle cable 龙须缆(船)
towing buoy 雾中拖标
towing cable 牵索
towing cable winch 拖缆卷扬机;拖缆绞盘
towing capstan 牵引绞盘;拖缆绞盘
towing carriage 拖车
towing chain 拖带链
towing channel 拖曳试验槽(航模);船模拖曳试验槽
towing charges 拖欠费用;拖带费用;拖船费用
towing device 拖曳装置;拖带装置
towing dynamometer 牵引力测力计
towing dynamometer test 拖挂牵引试验
towing efficiency 拖带效率
towing engine 拉纤机车;拖缆绞车
towing equipment 牵引设备;拖拉设备
towing eye 挂钩;连接环;牵引环
towing fish-plate 拖航三角板
towing force 牵引力;牵力;拖力(船舶);拖力
towing fork 拖引叉
towing gallows 拖缆承梁
towing gear 拖曳装置;拖曳设备;拖挂连接装置
towing handle 拖杆
towing hawser 拖缆
towing head 拖缆桩
towing helicopter 拖航直升机
towing hook 拖钩;牵引钩
towing hook platform 拖钩台(架)
towing knee 顶推架
towing lantern 拖带灯
towing launch 港内拖轮;拖轮
towing light 拖带灯
towing line 拖缆
towing loop 拖圈
towing machine(ry) 拖缆绞车;牵引装置;拖拉机械
towing method 牵引法(船)
towing model 拖挂模型
towing off line 拖离用拖缆(油轮)
towing of raising wreck 沉船拖航
towing operation 拖运作业;拖挂作业;随航作业
towing pad eyes 拖带眼板
towing painter 拖带小艇的绳或小链条
towing pass 曳船道;牵引道
towing path 拉纤道;纤道
towing pendant 拖缆端短索
towing performance 拖曳性能
towing plane 牵引的飞机
towing point 牵引点
towing pole 拖挂杆
towing port 拖缆孔
towing post 拖缆桩;拖缆柱
towing power 曳引力
towing rail 拖缆承梁
towing resistance 拖带阻力
towing resistance measurement 拖动阻力测定
towing rig 拖带设备
towing ring 牵引环
towing rope 拖缆;拖绳
towing rope winch 牵绳绞盘
towing shackle 牵引钩环;拖缆卸扣;拖航卡环
towing signal 拖带信号
towing spar 雾中拖标
towing speed 拖船航速;拖行速度
towing strap 曳索
towing tank 模型试验池;拖曳水池;拖曳水槽;船

模拖曳试验水池;船模拖曳试验槽
towing tension 拖力
towing test 牵引试验
towing the bow 拖头
towing the stern 拖尾
towing the waist 拖腰
towing timber 拖缆桩
towing to and from the site 往复拖运
towing track 拉纤铁路
towing traction 牵引(力)
towing trial 拖带试验
towing tripper 拖缆钩掣
towing trol(l)ey 牵引车;拖车;曳引车
towing trough 船模拖曳试验槽
towing tug 拖轮
towing unit 拖曳机具;拖曳单元;牵引装置;拖带船队
towing vehicle 曳引汽车;曳引车;牵引车
towing vessel 拖引船;拖船
towing voyage 拖带航程
towing warp 拖船票
towing waterline mark 拖航水标志
towing winch 曳引汽车;曳引绞车;牵索绞车;拖缆绞车
towing wire 拖索
towing wire cable 拖曳钢缆
towing yoke 连接叉
tow-island ship 两岛式船
tow line 拖缆;曳引索;纤绳;拖绳;船缆
towline 拖索;拖缆;拖船索
towline toggle 拖缆销
tow-man concept 二人管理制度
town 城镇建筑师;市镇;城镇
town and country construction 城乡建设
town and country maintaining and construction taxes 城乡维护建设税
town and country planning 城乡规划;城市及郊区规划;城郊规划
town architect 城镇建筑
town architecture 城镇设计
town area 城镇用地领域;城镇用地范围
tow-navigation lock 双室闸门
town block 城镇街区;城镇街坊
town border station 城市燃气门站;煤气城边点;煤气降压站;煤气城边站;城市门站
town boundary 城市界线
town built in blocks 棋盘式城市建设
town car 市内公共汽车
town cartography 城市图制图学
town center 城市中心
town chapel 城镇礼拜堂;城镇小教堂
town church 城镇教堂
town configuration 城镇轮廓;城镇外形
town development 城市建设
town development committee 城市发展
town district 城市地区;城市区域
town drain 城镇排水沟
town drainage 城镇地面排水;城市排水;市镇排水;城镇排水
town drainage system 城镇排水系统
town drain pipe 城镇排水管
town driving beam 市内行车束光
town dweller 区镇居民;城镇居民
town enclosure 城市界限;城市围栏
town engineering 城镇建设工程
town expressway 城镇快速道路
town extension 城市扩展
town fireproof plan 城市防火计划;城市防火规划
town fog 城市雾
town forest 城镇林带
town fortification 城市设防;城镇防御工事
town garden 城市花园;城市园林;城市公园
town gas 家用煤气;城市燃气;城市煤气;城市供气;民用煤气
town gateway 城门通道
town hall 公共集会的大礼堂;市政厅;(公共集会的)大会堂
town hall complex 市政厅建筑群
town heating 城市供热
town house 城镇住房;城镇住宅;城市住宅;城市住房;排房;花园洋房
townhouse dwelling 联立城镇住房
town improvement 城市改善

town in a plain 平原城市
town landscape 城市景观;城市风景
Town (lattice) truss 方格(式)桁架
townless highway 不经过集镇的道路
townlet 小镇;小城镇
town level 城市水平
town load 城镇道路
town-lot drilling 城区钻探;城区钻进
town main 城市总管;城市煤气总管
town map 城镇图
town of first administrative importance 一级行政中心
town on a table mountain 台地城镇(指美籍意大利建筑师索莱里所建议的方案)
town plan 城镇规划;城市规划图
town planer 城市规划者;城市规划人员
town-plan inset 城市插图
town planning 城镇规划;城市规划
town planning act 城镇规划条例;城市规划法(规);城市规划条例;城市规划法(规)
town planning and administration 城市规划管理
Town Planning and Administration Bureau 城市规划管理局
town planning authority 城市规划权威;城市规划当局
town planning institute 城市规划机构;城市规划学会
town planning restriction 城市规划限制
town planning study 城市规划研究
town planning survey 城市计划调查;城市规划测量
town planning system 城市规划系统
town planning theory 城市规划理论
town planning works 城市规划工作
town power supply 城镇供电
town refuse 城镇垃圾;城市垃圾
town refusing 城镇垃圾
town renewal 城市改造;城市改建
town road 城镇道路;市区道路;市街
town rubbish 城市废物;城市垃圾
town sauna bath 城市桑拿浴室
town scape 城镇景观
townscape plan(ning) 城市景观规划;市容规划
Townsend characteristic 汤森德特性曲线
Townsend coefficient 汤森德系数
Townsend current 汤森德电流
town sewage 城镇污水
town sewage treatment 城市污水处理
town sewage treatment plant 城市污水处理厂
townsfolk 城市居民
township 乡;镇;区;市区
township and village enterprise development zones 乡镇企业开发区
township and village enterprises 乡镇企业
township corner 市镇范围
township enterprises 乡镇企业
township highway 市镇公路
township line 市镇界线
township road 乡镇道路;乡公路;市镇道路
township-run enterprises 乡镇企业
township system 市镇道路系统
town site 城市建设基地;城市用地;城址;城镇位置;城市计划用地
town-site survey 城区测量
townsman 市民
town square 城市广场
town station 市内车站;城市车站
town street 市内街道;市街
town sunniness 市镇白天亮度
town surface drainage 城镇地面排水
town survey(ing) 城镇测量
town symbol 城镇符号
town/township enterprises 乡镇企业
town traffic 城镇交通
town wall 城墙
town waste 城市垃圾;市废物
town water 城市用水
town water demand 城镇用水量
town water requirement 城镇用水量
town water supply 城镇供水;城镇给水;城市供水;城市给水
towny 城市生活的;城镇里的
tow opener 短麻开松机
tow out 拖开
tow packing 麻屑填料

tow-part stator 分半定子
tow-part system 双元系统
tow path 纤道;牵道;牵船路
tow pendant 拖缆索
tow-piece mo(u)ld 对开模
tow-plane-wave aluminium-coated blazed hologram 二平面波镀铝闪烁全息图
tow-plus-one address instruction 二加一地址指令
tow-point distribution 二点分布
tow pole 拖挂杆
tow rail 拖网拱梁
tow ring layout 驱动环节设计
tow rope 拖绳
tow rope crane 缆索起重机
tow rope horsepower 拖缆马力
tow rope power 拖缆力
tow rope resistance 拖带阻力;拖曳阻力
tows 短麻屑
tow sack 麻袋
tow scutcher 麻屑加工机;打麻器
tow shackle 牵引钩环
tow shaker 麻屑抖落器;筛麻机;短亚麻除尘机
tow speed log 拖速仪
tow spinning frame 短麻细纱机;短麻精纺机
tow-stage casting 两段铸造
tow steering light 操舵用的灯
tow-step die 重合模
tow tank 船模(型)试验池
tow target 拖
tow-target aircraft 拖靶机
tow target system 拖靶装置
tow the head 拖头
tow the midship 拖腰
tow-tier exchange system 双重汇率制度
tow-to-top 丝束直接成条
tow tractor 曳引式拖拉机
tow trailer 牵引的拖车
tow transformer 丝束成条机
tow truck (将车辆拖走的)救援车;牵引卡车
towveyor 输送器
tow way 牵引道
tow yarn 短麻纱线
Toxement 托斯科水泥(一种防潮层水泥)
toxer dryer 塔形干燥器
toxic 有毒(性)的;毒剂
toxic action 毒性作用
toxic alage 有毒藻类
toxicant 有毒物;毒药;毒物;毒素
toxicant exposure 染毒
toxicant exposure cabinet 染毒柜
toxicant exposure chamber 染毒室
toxicant filtering tank 滤毒罐
toxicant monitoring 毒物监测
toxicant protective system 防毒系统
toxicants concentration 毒剂浓度
toxic cation 有毒阳离子
toxic chemicals 有毒化学品
toxic chemical waste 有毒化学废物
toxic chemical wastewater 有毒化学废水
toxic chian 木船船底用沥青防污漆
toxic concentration 毒物浓度
toxic concentration low 低毒性浓度
toxic constituent 毒性成分
toxic contamination 毒物污染
Toxicc Substance Control Act 有毒物质控制法令(美国)
toxic detonation product 毒性爆炸物
toxic dilution zone 毒物稀释带
toxic discharge 有毒排放物;有毒废水;毒物排放
toxic dose 中毒剂量;毒性剂量
toxic dose low 低毒性剂量
toxic effect 毒性效应;毒效
toxic effect zone 毒作用带
toxic element 毒性元素
toxic environment 有毒环境
toxic extraction procedure 毒物萃取法
toxic fish poisoning 毒鱼中毒
toxic fume 有害气体;毒烟;毒雾
toxicfurs of caterpillar 毛虫毒毛
toxic gas 有害气体;有毒气体;有毒材料;毒气
toxic ground level concentration 地面毒物浓度
toxic hazard 中毒危险;毒性危害;中毒灾害
toxic honey poisoning 有毒蜂蜜中毒
toxic index evaluation of water environment 水环境毒物指示评价
toxic industry 有毒工业
toxic industry waste 有毒工业废物
toxicity 毒性;毒度
toxicity analysis 毒性分析
toxicity analyzer 毒性分析仪
toxicity assessment 毒性评价;毒性评估
toxicity bioassay 毒性生物测试
toxicity characteristic 毒性特征;毒性特性
toxicity characteristic leaching procedure 毒性特性沥滤法
toxicity data 毒性数据
toxicity elimination 消除毒性;清除毒性
toxicity elimination and management strategy 排毒力与管理对策
toxicity emission factor 毒性排放系数
toxicity equation 毒性平衡
toxicity equivalent 毒性当量
toxicity equivalent conversion factor 毒性当量换算系数
toxicity evaluation 毒性评估
toxicity experiment of pesticide 农药室内毒力测定
toxicity hazard zone 毒性危害区;毒性风险区
toxicity identification 毒性鉴定
toxicity identification evaluation 毒性鉴定评估
toxicity index 毒性指数;毒力指数
toxicity indices 毒理指标
toxicity investigation evaluation 毒性调查评估
toxicity level 毒性水平;毒性级;毒性标准
toxicity mode 毒性型;毒性类
toxicity of effluent 废水的毒性
toxicity of metal dust 金属粉屑的毒性
toxicity of metal ion 金属离子毒性
toxicity of pesticide 农药毒性
toxicity of pesticide residue 农药残毒
toxicity of pollutant 污染物的毒性
toxicity on inhalation 吸入毒性
toxicity reduction evaluation 毒性衰减计算
toxicity spectrum 毒性谱
toxicity test 毒性检验;毒力测定;毒性试验
toxicity test of fishes 鱼类毒性试验
toxicity threshold 毒性阈;毒性界限;毒性极限;毒力阈
toxicity versus fish 对鱼的毒性
toxic level 毒性水平
toxic limit 耐毒限度;耐毒极限;毒性极限
toxic liquid 有毒液体
toxic load 毒物负荷
toxic material 有毒物质;有毒材料
toxic metal 有毒金属;毒性金属
toxicodendron 干毒漆树叶
toxicoderma/toxicodermia 中毒性皮病
toxicodermitis/toxicodermatitis 中毒性皮炎
toxicological analysis 毒物分析
toxicological assessment of pollution 污染毒物评估
toxicological index 毒物指数
toxicological index of river water 河水毒物指数
toxicological index of water 水的毒物指数
toxicological pollution index 毒物污染指数
toxicological test 毒物测试
toxicological testing 毒性检验
toxicologic index 毒理学指标
toxicology 毒物学;毒理学
toxicopexis 毒物中和
toxic organic compound 有毒有机化合物
toxic organic contaminant 有毒有机污染物
toxic organic material 有机毒物;有毒有机物
toxic organic pollutant 有毒有机污染物
toxic organics 有毒有机物
toxic organics removal 除有毒有机物
toxic organic substance 有毒有机物
toxicosis 中毒
toxic paint 有毒涂料;有毒油漆
toxic persistant material 持久性毒物
toxic pollutant 有毒污染物;含毒性污染物质;毒性污染物
toxic pollutant concentration 有毒污染物浓度
toxic pollution 毒物污染;毒物污染
toxic potency 毒效力
toxic residue 残毒
toxic salt reduction 毒性盐类浓度降低
toxics control 毒物控制
toxic smog 有毒烟雾;毒雾

toxic smoke 毒气;毒烟
toxic soil 毒土
toxic solvent 有毒溶剂
toxics point source discharge 有毒物质点污染源排放
toxic substance 有毒物质;毒性物;毒物
toxic substance pollution 有毒物质污染
toxic tolerance 耐毒性
toxic-treated board 毒化处理板材(防虫蛀)
toxic unit 毒性单位;毒度单位
toxic value 毒度值
toxic vapo(u)r indicator 有毒蒸气指示剂
toxic waste 有毒废水;有毒废物
toxic waste disposal 有毒废物处置;毒性废水处理
toxigenic 产生毒素的
toxin 毒素
toxin amount 毒素量
Toxkan Dao tectonic knot 托什干构造结【地】
toxogonin 双复磷
toy-lot 儿童游戏场地
toy shop 玩具店
toy theater 木偶剧院;木偶剧场
T-panel 丁字镶板;丁字墙板;T 形墙段
T-pattern array 横测线(T 排列)
T-Pee lateral 三角柱面体支管
T-piece 三通管;T 形管;三向接头
T-pipe 丁形管;T 形管;丁字形管;三通管;丁字管节
T-pipe fitting T 形管配件
T-plate T 形(钢)板;铁 T 角;丁字形板
T-post 丁字形(支)柱
trab 墙板
trabeated architecture 檐梁式建筑(学)
trabeated construction 楣式构造;柱顶横檐梁式结构;横檐梁式结构
trabeated style 柱顶横檐梁式;横梁式
trabeated system 柱顶横檐梁式结构系统
trabeation 柱顶上部;柱顶盘;柱顶横檐梁;横梁式结构
trabecula 小梁;分隔带;径列条;横隔片
trabeculae 萨列阿横梁
trabes 天花板格栅梁(古罗马)
trabs 墙板
trabuk (alloy) 特拉布克锡镍合金
Tracaba method 特拉科巴式体系建筑法
trace 追踪;轨迹;记录道;描图;描绘;痕迹;曲线图;微量雨;微量;扫迹
traceability 跟踪能力;可追溯性;可追查性;可寻迹性;示踪能力
traceability control 检寻控制
traceable 可示踪的
traceable cost 可追踪成本;可追溯成本
traceable paper 描图纸
trace amount 痕量;微量
trace analysis 痕量分析;微量分析
trace arsenic 微量砷
trace bit 跟踪位
trace cadmium 微量镉
trace chemistry 痕量化学
trace command 跟踪指令
trace component 痕量组分
trace concentration 痕量浓度;微量浓度
trace constituent 痕量成分
trace contaminant 痕量污染物;微量污染物
trace contamination 痕量污染;微量污染
trace contour 描绘等高线
traced drawing 底图
trace diagram 描迹图
trace drawing 描好的图
trace element 痕量元素;微量元素;分散元素
trace element analysis 痕量元素分析
trace element anomaly 分散元素异常
trace element concentration 微量元素浓度
trace-element fertilizer 微量元素肥料
trace element of oil 石油的痕量元素
trace element pollution 痕量元素污染;微量元素污染
trace elements analysis 微量元素分析
trace elements in seawater 海水微量元素
trace enrichment 痕量富集
trace equalization 道间均衡
trace facility 跟踪功能
trace fine sand 少量细砂
trace flow 跟踪流程;信号流跟踪器
trace flue 火焰回路

trace fossil 遗迹化石;足迹化石;踪迹化石;痕迹化石
trace gas 痕量气体;微量气体
trace gas technique 微量气体技术
trace heavy metal concentration 微量重金属浓度
trace horizon 地平线
trace impurity 痕量杂质;微量杂质
trace infrared system 红外(线)跟踪系统
trace ingredient 微量成分
trace inhibitor 痕量抑制剂
trace inorganic element 微量无机元素
trace interpretive program(me) 追踪解释程序
trace interval 扫描线时间间距
trace lead 微量铅
trace level 痕量级;痕迹量级;微量级;示踪量
trace-level activity 微量放射性
trace level organics 微量级有机物
trace linearity 扫描直线性
trace material 痕量物质;微量物质
trace measurement method 阿尔法径迹测量法
trace metal 痕量金属;微量金属
trace metal analysis 痕量金属分析
trace metal concentration 微量金属浓度
trace metal contaminant 微量金属污染物
trace metal input 微量金属排入
trace metal pollution 痕量金属污染;微量金属污染
trace mineral 微量无机物
trace mineral supplement 微量矿物质添加剂
trace normalized 道归一化
trace nutrient 微量营养
trace of a matrix 矩阵的迹
trace of precipitation 微量降雨;微量降水
trace of rain 微雨量
trace of rainfall 微雨量
trace of rake 扫寻痕纹(饰面抹灰)
trace of the plane 平面痕迹
trace organic 少量有机质
trace organic element 微量有机元素
trace organic matter 痕量有机物
trace organic pollutant 微量有机污染物
trace organics 微量有机物
trace packet 跟踪包
trace paper 描图纸
trace phosphorus 微量磷
trace point 迹点
trace pollutant 痕量污染物
trace pollution 痕量污染;微量污染
trace program(me) 跟踪程序;追踪程序;示踪程序
trace quantity 迹量;痕量;示踪量
tracer 曳光器;追踪装置;追踪指标;追踪者;追踪程序;跟踪机构;跟踪程序;迹线;描图员;描图器;描图工具;描记器;描绘器;描迹笔;货物短缺调查;航迹自画器;曲线记录装置;示踪装置;示踪物(质);示踪器;示踪剂;示踪程序
tracer activity 示踪物的放射性
tracer analysis 示踪分析
tracer range 跟踪范围
tracer atmospheric diffusion experiment 示踪扩散实验
tracer atom 示踪原子
tracer atom analysis 示踪原子分析
tracer chemistry 示踪化学
tracer composition 曳光剂
tracer compound 示踪化合物
tracer control 仿形控制
tracer-controlled lathe 靠模车床;仿形车床
tracer-controlled machine 仿形控制机床
tracer-controlled milling machine 仿形控制铣床
tracer curve 示踪剂曲线
tracer detector 示踪探测器
tracer determination 示踪测定
tracer diffusion 示踪原子扩散
tracer diffusion coefficient 示踪扩散系数
tracer dilution technique 示踪剂稀释技术
tracer-displacement technique 示踪物置换法
tracer dose 示踪剂量
tracer dye 示踪染料;地下水流动示踪染色料
trace recorder 微雨计
trace record 记录道
tracer edge 切削导刃
trace reflex 痕迹反射
tracer element 示踪元素
tracer experiment 显迹实验;示踪实验;示踪扩散实验;显迹试验;示踪试验
tracer finger 仿形器指销;仿形器针头

tracer flow method 示踪剂测流法;示踪测流法
tracer free 无示踪物的
tracer gas 空气流动示踪气体;检测气体;示踪气体
tracer gas instrument 示踪气体测定仪
tracer gas technique 示踪气体法
trace(r) glass 示踪玻璃
tracer grain 示踪颗粒
tracer head 跟踪头;随动磁头
tracer heating 示踪装置加热
tracer housing 仿形口罩
traceried opening 花格窗口
traceried rose window 玫瑰花格窗
tracer injecting hole 试剂投放孔
tracer isotope 示踪同位素
tracerlab 示踪物实验室
tracer label 示踪标志
tracer-label(l)ing 示踪物标
tracer laboratory 示踪实验室
tracer level 痕迹量;示踪量级;示踪量
tracer-level activity 示踪量级放射性
tracer liquid 检测液体
tracer material 示踪物质;示踪材料
tracer method 示踪元素法;示踪法
tracer milling machine 靠模铣床
tracer mixture 曳光剂
tracer molecule 示踪分子
tracer needle 描形针
trace rotating mechanism 扫描线转动机构
trace rotation system 径向扫描旋转系统
trace routine 跟踪程序;检测程序
tracer pin 仿形器指销;仿形器针头
tracer point 靠模指
tracer pulse technique 示踪剂脉动法
tracer receive probe 示踪剂接收探头
tracer receiving hole 试剂接收孔
tracer releaser 示踪剂投放器
tracer roller 靠模滚子
tracer routine 跟踪程序
tracer scale 示踪物量
tracer-scale extraction 示踪量萃取
tracer solution 示踪溶液
tracer spool valve 仿形阀
tracer study 示踪元素研究
tracer stylus 仿形器指销;仿形器针头
tracer technique 示踪技术;示踪法
tracer test(ing) 示踪元素试验;示踪法试验;示踪试验
tracer unit 仿形装置
tracer wheel 描迹轮
tracery 曲线窗花格;窗花格
tracer yarn 标志纱
tracery decorating art 花格窗装饰艺术
tracery filling 窗格镶填;窗格镶嵌
tracery frieze 镂花格檐壁
tracery motif 花格窗花纹图案
tracery ornamental art 窗花格装饰艺术
tracery parapet 镂空女儿墙
tracery pattern 窗花格风格;窗花格格局;窗花格图案;窗花格花样
tracery wall 花墙
tracery window 雕花格窗;石雕花格窗
tracery with four-leaf-shaped curves 四叶窗花格(哥特式的)
traces 遗迹
trace sediment 痕迹沉淀
trace sensitivity 轨迹灵敏度
trace sequential 道编排
trace slip 形迹滑距;交迹滑距;平行滑距
trace-slip fault 形迹滑距断层;交迹滑动断层
traces of a shot 炮记录道数
traces of wide line stack 宽线叠加道数
trace standard certified of agricultural chemicals 合格农药残留标准
trace standard of agricultural chemical 农药残留标准
trace standard of agricultural chemistry 农药残留标准
trace standard of pesticides 农药残留标准
trace substance 痕量物质
trace survey 线索调查
trace table 跟踪表
trace time 扫描时间;扫掠时间
trace-truck 带拖车的卡车
trace vector 跟踪向量

trachea 木导管
tracheal sac 气囊
tracheary element 管状分子
trachelion 柱颈(希腊陶立克式柱的)
trachelium 柱头饰(陶立克柱);柱颈饰
Trachelomonas granulata 颗粒囊裸藻
Trachelomonas lacustris 湖生囊裸球藻
tracheolar tapetum 反光气管层
tracheophyte 导管植物
Trachidermus fasciatus 松江鲈鱼
trachite 粗面岩
trachitic tuff 粗面质凝灰岩
trachitic welded agglomerate 粗面质熔结集块岩
trachitic welded breccia 粗面质熔结角砾岩
trachitic welded tuff 粗面质熔结凝灰岩
trachographic(al) relief 山形法绘图
trachoma 沙眼
trachomatous pannus 赤膜下垂
trachyandesitic glass 粗安玻璃
trachybasalt 粗玄岩;粗面玄武岩
trachy basaltic glass 粗玄武玻璃
trachy discontinuity 不规则面不连续;凹凸不整合面【地】;凹凸不整合
trachy dolerite 粗面粒玄岩
trachyic glass 粗面玻璃
trachy mugearite 粗粒面橄榄粗安岩
trachyte 火山岩;粗面岩
trachyte andesite 粗面安山岩
trachyte for cement 水泥用粗面岩
trachyte group 粗面岩类
trachyte prophyry 粗面斑岩
trachytic 粗面状;粗面岩的;粗面的
trachytic andesite 粗面安山岩
trachytic lava 粗面熔岩
trachytic porphyry 粗面斑岩
trachytic texture 粗面结构
trachytic tuff 粗面凝灰岩
trachytiod 似粗面状
trachytoid texture 似粗面结构
trachy unconformity 不规则面不整合
tracing 线路图追索;线路图寻迹;显踪;追踪;故障寻找;跟踪术;跟踪法;描图;描记法;描绘;蒙绘;透写;示踪
tracing analysis 曲线分析
tracing arm 绘图臂
tracing attachment 描图附件
tracing backing 跟踪垫
tracing bar 描图杆
tracing boundary method 界限追索法
tracing carriage 绘图笔架
tracing cloth 透明纸;描图纸;描图布;绘图透明蜡布
tracing curve 描迹曲线
tracing device 描绘装置
tracing distorsion 描纹失真
tracing distortion 描纹畸变;描绘畸变;示踪畸变;包线失真
tracing error 跟踪误差;描纹误差
tracing experiment 示踪试验
tracing feed 仿形进给
tracing filtering 跟踪滤波
tracing for checking lines of sounding 检查透写图
tracing groundwater 找地下水
tracing head 跟踪头
tracing house 描图室
tracing instrument 描图仪(器)
tracing joints 追踪节理
tracing line 吊绳
tracing linen 描图布;描图麻布
tracing loss 描纹损耗
tracing machine 描图机
tracing method 追踪法
tracing migration 追踪迁移
tracing milling 仿形铣
tracing needle 描图针;描迹针
tracing of contour lines 描绘等高线
tracing of flow pattern 流态示踪
tracing of sounding lines 测深线透写图
tracing paper 描图纸
tracing paper method 透明纸法
tracing pattern 曲线形态
tracing pen 自记笔;描迹笔尖;描迹笔
tracing point 轨迹点;描迹针;描迹点;描绘点
tracing probe 故障探寻器
tracing room 描图室

tracing route 追索路线
tracing routine 追踪程序;跟踪例行程序
tracing rut 车辙
tracing sediment discontinuity 追踪沉积间断
tracing seismic reflected interface 追踪地震反射面
tracing sheet 描图纸;绘图片
tracing stand 透写台
tracing stereometer 立体测图仪
tracing table 描图桌;透写桌;测绘台
tracing table alongside the apparatus 仪器绘图桌
tracing technique 示踪技术
tracing valve 仿形阀
tracing wheel 描线轮;描迹轮
Tracitex 蔡司塔斯科纸(一种描图纸)
track 线路;轨道;小道;印制线;音轨;轨检车;股道【铁】;跟踪航行;路线;路径;路轨;跨距;径迹;记录带导道;记录槽;火车轨道;航路【空】;航迹【航海】;纹道;搜寻线;磁轨;磁道;船路
trackability 可跟踪性
track accessories 轨道附件
track address 道地址;磁道地址
track address register 跟踪地址寄存器
track adhesion 履带与地面的附着力;磁迹附着
track adjuster 轨道调整器;链轨进度调整器
track adjuster ball check valve 履带调节球形止回阀
track adjuster relief valve 履带调节安全阀
track adjusting 轨道调整
track adjusting cylinder 履带调节汽缸
trackage 铁路使用费;线路里程;里程;铁路线;铁道路线
trackage classified by speed 按速度分类的线路
trackage classified by traffic volume 按运量分类的线路
track ag(e)ing 径迹老化
trackage rights 线路使用权;过轨使用权
track air compressor 拖带式空气压缩机
track alignment 线路中心线;轨道方向
track along parallel 等维航线
track analysis 径迹分析
track-and-guidance laser radar 跟踪制导激光雷达
track angle 轨迹角;航向角;航迹角
track annealing 径迹退火
track area 股道区
track around the crest of hump 驼峰迁回线
track assembly 履带装置;轨道装置;轨道结构
track autoradiography 径迹放射自显影
track axis 铁路轴线
track balance 磁道平衡
track ball 控制球
track ballast 道作
track ballast cleaning 清洗道作
track bar 拨道棍
track barrow 有轨手推车
track beam 轨梁
track bed 道床【铁】
trackbed drain 道床排水沟
track belt 链板固定带
track block 自动闭锁;履带蹄块;履带板;区间闭塞
track blockade 线路遮断
track blocking 线路闭锁
track block section 闭塞区间
track boat 纤拉的船
track bolster 转向架承梁
track bolt 线路螺栓;轨条螺栓;轨道螺栓;路轨接缝板螺栓;接头螺栓;(钢轨接头的)夹板螺栓
track bond 钢轨接合;钢轨对接
track booster 轨道降压器
track brace 轨距拉条
track brake 轨制动器;轨道制动(器)
track-brake jaws 轨道制动夹钳
track broom 扫路机;电车轨道扫寻
track buckling 轨道鼓出
track bushing 履带销套
track cable 载重索;轨道索;缆道;承重索;承重钢丝绳;承载索道
track cable joint 承载索接头
track cable scraper 索拉耙铲机
track calibrate 磁道校准
track capacity 线路通过能力【铁】;股道容量;铁路线路通过能力;磁道容量
track capacity of line 线路容量【铁】
track car 轨道车
track carrier 履带提架

track-carrier roller 链轨托轮
track center 轨道中心
track center distance 轨道中线距;线间距;轨道中心距离;履带中心距(离)
track centerline 轨道中(心)线
track centers 轨道中(心)线
track center to center 轨道中(心)线
track center-to-center spacing 轨道中心距
track centre distance 线间距;轨道中心距离
track centre line 轨道中(心)线
track chain conveyer 轨道链式输送机;链轨式输送机
track chamber 径迹室
track change control 轨道变化控制
track chart 航路图;航迹图
track circuit 轨道电路【铁】
track-circuit signalling 轨道电路信号
track classification 线路等级
track clear 轨道空闲;轨道空闲【铁】
track clearance 车道间隙;轨道净空(保持轨道上的一定空间);规划建筑线
track clearer 轨道清理机;排障器;拔草板
track-closed post 线路封闭标志
track closure 线路封锁
track components 轨道部件
track condition table 轨道条件表
track connection box 轨道连接箱
track construction 铺轨
track construction base 铺轨基地
track construction train 铺轨列车
track container crane 轨道式集装箱起重机
track control device 运行轨迹控制装置
track crane 轨道式起重机;轨道起重机
track crawling 航线徐变
track creeping 轨道爬行
track crossing 平交道口【铁】;铁路与铁路交叉
track cross level 轨道水平;左右水平
track crossover 路轨交叉;渡线【铁】
track current maintenance 线路经常维修
track curvature radius 线路曲率半径;轨道曲率半径
track curve 弯道
track curve radius 弯道半径
track damage 线路病害
track defect indicator car 查道车
track defect 线路病害
track deformation 轨道变形
track density 轨道密度;磁道密度
track depression 路线降低;降低轨道
track descriptor record 磁道描述符记录
track deterioration 线路病害
track deviation 航迹偏差
track deviation limit 航迹偏离极限
track diagram 轨道图
track diamonds 主切削金刚石(在冲洗槽主动边的)
track disorder 轨道变形
track distortion 轨道变形;磁道畸变
track distribution 航迹分布
track district 工务段
track ditch 线路边沟
track dividing 轨道划分
track division 工务段
track doubling 修建第二线
trackdown 跟踪目标
track drainage 线路排水
track-drill 钢轨钻
track drive 履带式传动;履带传动
track duty 磁迹占空比
track dynamic stabilizer 轨道动力稳定机
tracked air-cushion vehicle 有轨气垫车;有轨气垫式快车
tracked bulldozer 履带式推土机
tracked crane 履带式起重机
tracked loader 链轨式装载机
tracked vehicle 履带式车辆;有轨车辆;履带运行车辆
track element 磁道单元;磁道单元
track elevation 提高轨道
tracker 跟踪装置;跟踪器;跟踪机构;纤工
tracker ball 轨迹球;跟踪球;球形跟踪器
track eradicator 轨迹消除器
track error 磁道错误
tracker wire 跟踪线
track etching 径迹蚀刻
track etching analysis 径迹蚀刻分析(法)

track etching dosimeter 径迹蚀刻剂量计
track expansion 线路扩建
track experiment 轨道试验
track extension 线路延长
track facility 线路设施;线路设备
track failure 线路故障
track fault 线路病害
track filter 轨道滤波器
track fixed stake 线路固定标桩
track flag 信号旗
track follower 磁道跟随器
track following servosystem 磁道跟踪伺服系统
track for container trains 集装箱列车作业线
track for damaged wagon 损坏车停留线
track for either direction working 双向行车线路
track-foreman 线路领工员
track for locomotive running 机车走行线
track format 信息道格式;磁道格式
track for storing locomotive 机车存放线;机车停留线
track frame 履带托轮架;履带台车架;履带架
track galvanic effect 轨道生电现象
track gang 养路道班
track gap 磁道间隙
track ga(u)ge 轨距规;轨距尺;轨距;轨道尺;履带中心距(离);道尺
track ga(u)ge meter 轨距测量仪
track geometry 轨道几何形位
track geometry measurement 轨道几何测量
track geometry tolerance 轨道几何尺寸容许偏差
track grade 轨道坡度
track gradient 线路坡度
track group 履带组合;线束【铁】;磁道组;编组场线束
track guide 履带导板;航路引导
track guiding guard 履带定向护板
track guiding wheel 履带导向轮
track hand 养路工
track haulage 有轨运输
track history 全程轨迹
track hold 跟踪保持;磁道挂起;磁道保持
track homing 跟踪寻的
track hopper 轨行式漏斗;轨道卸料斗
track identifier 磁道标识符
track idler 履带诱导轮;履带惰轮
track in 镜头推进
track index 信息道索引;道变址;磁道检索
track-induced dynamics 轨道相关动力学
tracking 追踪;跟踪术;跟踪法;跟踪;漏电痕迹;拉纤;航迹跟踪;铺轨;统调;随纹
tracking accuracy 跟踪准确度;跟踪精度
tracking after step 步法追踪
tracking and control centre tracking and control center 跟踪控制中心
tracking and data acquisition 跟踪和数据获取;跟踪和数据获得
tracking and data acquisition system 跟踪和数据获取系统;跟踪和数据获得系统
tracking and data relay satellite 跟踪与数据中继卫星
tracking and data relay satellite system 跟踪与数据中继卫星系统
tracking-and-guidance 跟踪制导
tracking-and-guiding radar 跟踪制导雷达
tracking and illuminating laser radar 跟踪照明激光雷达;激光跟踪和照明雷达
tracking antenna 跟踪天线
tracking antenna system 跟踪天线系统
tracking area 跟踪区
tracking axis 跟踪轴
tracking backlog work 追踪积压工作
tracking ball 跟踪球
tracking beam 跟踪波束
tracking camera 跟踪式照相机;跟踪摄影机
tracking capacitor 统调电容器
tracking circuit 跟踪电路
tracking computer 跟踪计算机
tracking control 统调控制
tracking counter 跟踪计数器
tracking coverage 跟踪范围
tracking cross 跟踪十字光标;十字光标
track(ing) data 跟踪数据
tracking data processor 跟踪数据处理机
tracking device 跟踪装置

tracking display system 跟踪显示系统
tracking distorbion 随纹失真
tracking drift buoy 跟踪漂流浮标
tracking driven device 跟踪传动装置
tracking encoded theodolite 跟踪编码经纬仪
tracking equipment 跟踪设备
tracking error 循迹误差;随纹误差
tracking feed 跟踪馈电
tracking filter 跟踪滤波器
tracking flare 跟踪闪光
tracking frequency 跟踪频率
tracking generator 跟踪发生器
tracking information 跟踪信息;跟踪情报
tracking instrument 跟踪仪
tracking jack 铺轨千斤顶
tracking jitter 跟踪回波起伏
tracking joint 履带接头;轨道接头
tracking lag 跟踪滞后
tracking laser radar 跟踪激光雷达
tracking lidar 跟踪激光雷达
tracking mark 追踪标志;跟踪标志
tracking meter 循迹表
tracking mirror 循迹反光镜;跟踪反射镜
tracking mode 跟踪式;跟踪方式
tracking mode switch 跟踪模开关
tracking module 跟踪组件
tracking network 跟踪网
tracking noise 跟踪噪声
tracking oscillator-analyzer 跟踪振荡器分析器
tracking point 跟踪瞄准点
tracking power 跟踪本领
tracking preprocessor 跟踪预处理机
tracking problem 跟踪问题
tracking program(me) 跟踪程序;追踪程序
tracking radar 跟踪雷达;监视雷达
tracking range 跟踪距离;跟踪范围
tracking rate 跟踪速率
tracking receiver 跟踪接收机
tracking resistance 抗电弧径迹性
tracking rope 纤绳
tracking routine 跟踪程序
tracking ruts 轨辙车印;铁轨辙印
tracking sensitivity 跟踪灵敏度
tracking ship (导弹靶场的)遥测船
tracking signal 跟踪信号
tracking stand 测标台
tracking station 追踪站;跟踪站;跟踪台
tracking study 跟踪研究
tracking symbol 跟踪符号
tracking system 跟踪装置;跟踪系统
tracking telemetry and command station 跟踪遥测和指令船
tracking telescope 跟踪望远镜
tracking television 跟踪电视
tracking velocity 跟踪速度
tracking window 跟踪窗
track inspecting and measuring car 轨道检测车
track inspection 轨道检测;轨道检查
track inspection car 轨道检查车
track inspection railcar 轨道检查车
track inspector 巡道工
track insulation 轨道绝缘
track intermediate repair 线路中修
track intersection 线路交叉
track-in-track 径迹中的径迹
track irregularity 轨道不平顺
track irregulating 轨道偏移
track jack 轨道(式)千斤顶;起轨器;起轨机;起道机
track joist 托轨梁
track jumper cable 轨道跨接电缆
track junction 线路连接
track junction point of port 港区铁路接轨点
track kilometrage 线路公里里程
track label 磁道标号
track layer 铺轨工;履带式车(辆);养路工;履带式拖拉机;铺轨工人
track-layer machine 铺轨机
track layer's hammer 铺轨锤
track laying 轨道铺设;铺轨;磁迹时间对准
track-laying and girder erecting machine 铺轨架桥机
track-laying car 铺轨车
track-laying crane 铺轨机;铺轨起重机;铺轨吊车
track-laying depot 铺轨基地

track-laying gang 铺轨队
track-laying gantry 铺轨龙门架;铺轨机;龙门式铺轨机
track-laying machine 铺轨机
track-laying machinery 铺轨机械
track-laying machine with single-cantilever 单悬臂(式)铺轨机
track-laying tractor 铺轨牵引车
track-laying train 铺轨列车
track-laying type asphalt finisher 履带式沥青路面整修机
track-laying type combined paver 履带式联合铺路机
track-laying type long-boom crane 履带式长臂吊车
track-laying type overhead loader 履带式高位装载机;履带式过顶装载机
track-laying vehicle 履带式牵引车;履带式车辆;铺轨车
track layout 轨道配线;轨道连接
track lead 钢轨引接线
track leading from receiving yard to hump 驼峰推峰线【铁】
track length 股道长度;径迹长度
track length of railway station 铁路车站线路长度
trackless 非履带的;无轨的
trackless earthmoving 非履带挖运土;无轨挖运土
trackless haulage 轮胎式运输;无轨运输
trackless mine 无轨矿井
trackless mining 无轨采矿
trackless mining area 无轨采区
trackless traffic 无轨交通
trackless transportation 无轨运输;无轨交通
trackless trolley 无轨电车
trackless troll(e)y bus 无轨电车
trackless tunnel(l)ing 隧道无轨掘进
trackless vehicle 无轨车辆
track level 线路水准
track lever 起轨杆
track lifter 起轨器
track lifting 起道
track lifting and lining machine 起轨拨道机;起拨道机
track lifting device 起道装置
track lifting jack 轨道顶升升千斤顶
track lifting truck 轨道起重车
track lifting winch 起轨绞车
track lifting work 起道作业
track line 线路长度;运转钢索;架空线(路);架空轨道;航迹线;船首方位测探
track liner 撬棍;拨道机;拨道杆
track lining 拨道
track lining device 拨道器
track link 链节;履带轨节
track link assembly 履带链节组合
track (link) pin 履带销;链轨销
track link rail 履带链节导轨
track loader 装车机;履带式起重机
track load spectrum 轨道荷载谱
track lowering 落道
track maintenace 线路养护
track maintenance division 工务段
track maintenance equipment 线路维修设备;养路维护设备
track maintenance equipment siding 养路设备专用线
track maintenance force 养路工作队
track maintenance section 养路工区
track maintenance statistic chart 线路维修统计表
track maintenance tools 养路工具
trackman 养路工;纤路;拖船道;田径运动员;护路员工;线路工
track map 线路图【铁】
track mark 磁道标记
track marking 线路标桩
track mat 挠性地垫
track maul 道钉锤
track measuring car 轨道检查车
track mechanical analysis 轨道力学分析
track mileage 铁路英里里程
track miles 股道总延长
trackmotor car 轨道自动车
track-mounted 履带的;链轨式
track-mounted drill 履带式钻机

track-mounted jumbo 有轨钻车
track-mounted troll(e)y 轨行跑车
track-mounted wagon drill 履带式自动车
track navigation computer 航迹导航计算机
track normally open 正常关闭的线路
track number 轨道数;股道号码
track obstruction 线路障碍;线路故障
track occupancy 线路占用
track occupancy indication 线路占用表示
track occupation 线路占用
track occupation time 股道占用时间
track occupied 轨道占用
track of a cyclone 气旋路径
track of perturbation 摄动轨道
track of typhoon 台风路径;台风轨迹;飚风轨迹
track on turnout side 岔道侧;铁道岔道
track opening 铁道入口孔
track orbit 移动轨道
track oriented decrossing 线路别疏解
track out 镜头后移
track packing 轨道垫实
track pad 履带垫;履带板
track pan 轨道给水槽
track panel 轨排
track panel with concrete tie 混凝土轨枕轨排
track path 履带轨迹
track patrolling 轨道巡查
track permanent deformation 轨道永久变形
track pin 链轨销;履带销
track pit 车轮辙
track pitch 铁轨分岔;履带节距;道(间)距
track plan 线路平面(图);线路布置图
track planter 履带式机械
track plotter 航迹记录器【船】;航迹绘图仪
track pointer 轨道指示器
track polishing 导轨抛光
track post 线路标志
track predictor 航迹预绘
track profile 线路纵断面(图);轨道前后高低
track protected equipment 线路定期防护设备
track protection 轨道保护
track quality index 轨道质量指数
track rail 铁轨
track rail bond 轨端连接线;钢轨导(电)接线;铁轨连接器;道机导电接头
track raiser 扛轨器;起道机
track raising 起坡;上坡路;升坡
track raising device 起道机
track raising operation 起道作业
track reactor 轨道电抗器
track record (房地产开发商的)经营记录
track recorder 航迹记录器【船】
track recovery 道恢复;轨道恢复
track relay 轨道继电器;拨轨继电器
track relay transformer 轨道受电变压器
track renewal 换轨;线路更新;线路大修;轨道更换
track-renewal train 换轨列车
track replacement 线路更换;轨道更换
track residual deformation 轨道残余变形
track resistor 轨道电阻器
track resolution 跟踪分辨率
track return circuit 路轨回流电路
track-return power system 路轨回流供电制
track return system 轨道回流系统
track revolver 磁轨旋转器
track rigidity 线路刚度
track ring 凸轮环;定子环
track road 纤道;拖船道
track rod 轨距杆
track-rod ball plug 横联杆球塞
track-rod bolt 横联杆螺栓
track roller 履带压路机;履带滚轮;支重轮;履带支重轮;链轨支重轮
track roller frame 履带支重轮架
track roller shaft 履带支重轴
track rolling profile 变坡线路纵断
track rope 运输索;装载索;跑索;承载索道;承载(钢)索
track sand 轨道防滑砂;刹车用砂;制动用砂
tracks by parallel shifting apart 线路平行错移;股道平行错移
track scale 车辆磅秤;轨道衡;称量车;车辆秤
track scale track 轨道衡线
track scanning meter model 径迹扫描仪型号

track seal 履带密缝;履带密封
track section 线路区段;养路工区;轨道区段
track-sectioning cabin 轨道分段室
track section insulator 轨道区段绝缘节
track segment 轨道区段
track selection 道选择
track selector 道选择器
track-sensitive target 径迹敏感靶
track set in paving 暗轨铁路;暗轨铁道;暗轨
track setting 轨距调整
track sheet 材料流程单
track shifter 移轨器;移轨机;移道机;搬道机;扳道机
track shoe 履带板;链板;履带瓦;链轨板
track shoe width 履带板宽度
trackside 线旁;沿线
trackside box 轨旁盒
trackside circuit 轨旁电路
trackside equipment 线路设备
trackside facility 沿线设施
trackside isolator 轨旁隔离开关
track sign 线路标志
track signal 声迹信号
track sign post 铁路路标
tracks in surface 铁轨状况良好
track skeleton 轨排;轨框;轨节;架线车
track skids 轨道垫木
track skim for frost heaving roadbed 冻害垫板
track skims 轨道垫块
track slack 履带松弛
track slewing 移动轨道
track slip 履带滑转;滑转;链轨滑转;打滑
track spacing 轨距;股道间距;搜寻线间距
track span 铁道跨距;股道长度
track spike 道钉;铁轨道钉;线路道钉;铁路道钉
track split 轨道分开
track stake 轨道桩
track standard 轨道类型;轨道标准
track steam crane 轨道蒸汽吊
track stiffness 轨道框架刚度;轨道刚度
track storage effect 轨道电路蓄电效应
track strengthening 线路加固
track stringer 铁路纵轨枕;轨道纵梁;轨道梁;桥面引车道下的纵梁
track structure 追踪式构造;轨道结构
track supervisor 养路领工员;养路分段长
track surface 导轨面
track survey in borehole 钻孔径迹测量
track survey in water 水中径迹测量
track survey on bog 沼泽径迹测量
track survey on desert 沙漠区径迹测量
track survey on lake bottom 湖底径迹测量
track survey on outcrop area 露头区径迹测量
track survey on river bottom deposit 河底沉积物径迹测量
track survey on snow layer 雪层上径迹测量
track switch 轨道开关
track switching 轨道转换
track system 履带系统;轨道系统;薪金线制度;铁道路线
track tamper 轨道夯实器
track tamping 轨道垫实
track tamping machine 轨道夯实机
track telling 轨道辨别
track tension 履带拉紧;履带张力
track tension adjusting gear 履带张力调节装置;履带张力调节器具
track tensioning 履带张拉
track theostat 轨道变阻器
track tie 轨枕
track-time curve 线路时间曲线
track-to-track moving time 道间移动时间
track towing conveyer 轨道牵引式输送机
track tractor 履带拖拉机;轨道牵引车;汽车列车
track transformer 轨道送电变压器;轨道变压器;拨轨变压器
track transformer box 轨道变压器箱
track transmitter 跟踪发射机
track treadle 轨道接触器
track twist 三角坑【铁】
track-type excavator 履带式挖土机;履带式挖掘机
track-type jumbo 有轨钻车
track-type loader 履带式装载机;履带式装运机;履带式装料机
track-type loading shovel 履带式装载铲

track-type tractor 履带式拖拉机
track-type undercarriage 履带式运载装置;履带式底盘
track under the hump 驼峰下线路
track up 前进跟踪
track vacancy detection 轨道空闲状态检测
track vacancy monitoring 轨道占用状态监视
track vehicle 履带式车
track velocipede 轨道三轮车
track-velocity curve 线路速度曲线
track walker 巡路人员;巡道工;铁道护路员
trackway 行迹;轨道【铁】;纤道
track way of gravity davit 重力式吊艇杆上滑轨
track well (两站间的)轨道凹槽
track wheel 承重轮;履带轮
track wheel flat dump car 轨道平板倾卸车
track wheel wood dump wagon 轨道木倾卸车
track-while-scan 跟踪搜索;扫描跟踪
track-while-scan program(me) 跟踪扫描程序
track-while-scan radar 扫描跟踪雷达;搜索跟踪雷达
track width 轨道宽度;履带宽度;道宽;磁道宽
track winch 起轨绞车
track with cant 有超高的轨道;设超高的线路
track with check rail 带护轨的轨道
track without cant 无超高线路
track works 轨道工程
track wrench 轨道扳手;螺栓扳手
Tracoba 特拉科巴式体系建筑
Tracoba system 特拉科巴预制构造系统
tract 广阔地面;土地;束;地带;发展用地
tractability 精细性
tractable 易处理的(材料等)
tract builder 购地建屋经营者
tracer driven carrier 拖行小车
tract house 地区性住宅;(设计相同的)屋屯住宅;大量生产类同房子;一片相同的房屋
tract index 土地记录索引
tracting material 摹图材料
traction 曳引力;拉法;拉法;拖曳搬运;推移作用;推移;涂料收缩;牵引
traction(al) force 推移力
traction(al) load 推移质;推移荷载
traction(al) velocity 推移速度
traction bandage 牵引绷带
traction battery 车辆用电池;车辆电池;牵引动力电池
traction booster 驱动轮加载装置
traction-booster system 驱动轮加载系统
traction bow 牵引弓
traction bracing 牵力顺联
traction by pulling and extension 拔伸牵引
traction cable 牵引索;牵引绳
traction calculation 牵引计算
traction capacity 牵引生产力;牵引能力;拖运能力;推移能力
traction clutch 传动离合器;行走传动离合器
traction coefficient measurement 附着系数测定
traction control 传动操纵
traction control line 牵引控制线
traction control unit 牵引力调节装置
traction crusher 牵引式碎石机
traction current 牵引电流
traction current system 牵引电流系统
traction device 驱动轮加载装置
traction district 牵引区段
traction ditcher 牵引式挖沟机
traction drill 牵引式钻机
traction drill(ing) machine 拖拉式钻机;牵引式钻机
traction drive 行走轮驱动;槽摩擦轮传动;牵引
traction drive belt 行走部分的传动皮带
traction drive lift 曳引驱动电梯
traction driven 车拖式的
traction driven mower 行走轮驱动式割草机
traction-driven rotary sweeper 牵引驱动旋转扫路机
traction drive shaft 驱动轮传动轴
traction duster 行走轮驱动的喷粉机
traction dynamometer 牵引力测力计
traction earthing system 牵引接地系统
traction element 牵引构件
traction engine 牵运机;牵引机;牵引发动机;牵引车
traction equipment 牵引装置;牵引设备
traction feed 牵引供电

traction fleet 动力车队
traction frame with pulley attachment 滑轮牵引架
traction gear 牵引装置
traction gear box 行走部分变速箱
traction generator 牵引发电机;电力牵引系统
traction grader 牵引式平土机;牵引式平地机
traction horsepower 牵引马力
traction indicator 牵引力计
traction lift 卷扬提升机
traction line 牵引索
traction load 牵引荷载;牵引负荷;底移质
traction locomotive 牵引机车
traction machine 牵引式机械
traction machinery 牵引机械
traction mass 牵引质量
traction meter 牵引计
traction method 拖曳方法
traction modulus 牵引模数
traction motor 牵引马达;牵引电动机
traction (motor) current 牵引电流
traction motor fan 牵引电机风扇
traction motor suspension 牵引电机悬挂
traction gear motor test stand 牵引电机试验台
traction motor voltage 牵引电压
traction mower 牵引式割草机
traction network 牵引网
traction noise 拖曳噪声
traction of tracks 轨道摩擦力
traction permeameter 拉力磁导计;牵引磁导计
traction phenomenon 牵引现象;拖曳现象
traction pole 牵引柱;牵引杆
traction power 曳引力;拉力;牵引功率;牵引动力
traction power lighting substation 牵引动力照明变电所
traction power of coupler 车钩牵引力
traction reduction 牵引复位
traction resistance 牵引阻力
traction return cable 牵引回流电缆
traction return current 牵引回流
traction return current circuits 牵引回流电路
traction return rail 牵引回流轨
traction roller 牵引辊
traction rope 拖曳索;牵引索;牵引绳;拖曳吊索（空中索道用,使吊车移动）
traction sheave elevator 卷扬提升机
traction shovel 牵引式挖土机;牵引式铲土机;牵引铲
traction speed lever 前进速度调节杆
traction sprayer 行走轮驱动的喷雾机
traction spring 牵引缓冲弹簧
traction steel cable 牵引钢索;牵引钢丝绳
traction stress 牵引应力
traction substation 牵引变压所;牵引变电所
traction suspension rope 牵引吊索
traction system 牵引系统
traction system dynamics 牵引系统动力学
traction table 牵引床
traction test 牵引试验
traction thickener 周边传动式浓缩机
traction thrust 推挽力
traction tongs 牵引钳
traction transformer 牵引变压器
traction transformer for three phase three winding connection 三相三绕组接线牵引变压器
traction-transmitting belt 牵引传送皮带
traction transport 牵引搬运;拖曳搬运;推移质
traction tread 牵引轮胎
traction tube 牵引管
traction-type cell 牵引型电池
traction-type dynamometer 拉力表
traction(-type) elevator 牵引式升降机
traction type hydraulic coupling 牵引型液力耦合器
traction-type tyre 牵引式轮胎
traction unit 牵引单元
traction variable-speed control 行走部分无级变速器操纵杆
traction wheel 主动轮;机车主动轮;驱动轮;牵引式轮
traction winch 牵引绞车
tractive 牵引的
tractive ability 牵引能力
tractive armature type relay 衔铁上升式继电器
tractive capacity 牵引能力;推移能力

tractive characteristics 牵引特性
tractive coefficient 牵引系数
tractive connection 牵引连接
tractive current 牵引流;拖曳水流;拖流;贴底泥沙流
tractive efficiency 牵引效率
tractive effort 牵引作用;牵引效力;牵引力
tractive effort at adhesion limit 受黏着限制的牵引力
tractive effort at the wheel rim 轮周牵引力【铁】;轮辋牵引力
tractive effort developed by diesel engine 柴油机牵引力
tractive effort of a locomotive 机车牵引力
tractive effort-speed curve 牵引力—速度曲线
tractive element 牵引构件
tractive force 曳引力;推移力
tractive force balance 牵引力平衡
tractive force chart 牵引力图
tractive force-current characteristic curve 牵引力电流特性曲线
tractive force diagram 牵引力平衡图
tractive force limit 牵引力极限
tractive force method 牵引力图解法
tractive force of train 列车牵引力
tractive force sensor 牵引力传感器
tractive force variation 牵引力变动
tractive load 牵引荷载
tractive machine 拖曳车
tractive magnet 牵引电磁铁
tractive performance diagram 牵引特性曲线
tractive power 曳引力;牵引能力;牵引功率
tractive power supply system 牵引供电系统
tractive resistance 牵引阻力
tractive stress 界面切向应力
tractive substation 牵引变电所
tractive tonnage calculation 牵引计算
tractive unit 牵引机组
tractive viscosity 牵伸黏度
tract loan 分小地块建小户住宅的贷款
tractolift 叉车兼牵引车
tractometer 工作测定表
tractor 牵引器;牵引机;拖拉机
tractor airscrew 拉进式航空螺旋
tractor-allied equipment 牵引设备
tractor articulated roller 拖式路滚
tractor attachment 拖拉机附属设备
tractor automatic arc welding machine 拖车式自动电焊机
tractor backhoe 牵引式反向铲
tractor-based equipment 牵引设备
tractor-bucket machine 牵引斗式装载机
tractor cable winch 卷扬机
tractor car 摆渡车
tractor-carryall unit 拖拉铲运机
tractor clutch 拖拉机离合器
tractor-compressor 牵引压力机
tractor cram-shell bucket 曳车蟹壳桶
tractor crane 牵引车起重机;拖拉机起重机
tractor cultivator 机引中耕器;机引圆盘耙
tractor ditcher 机引挖沟机
tractor dog 牵引夹头
tractor dozer 拖拉式推土机;拖拉机式推土机;推土拖拉机;推土机;履带式推土机
tractor-dragged 拖拉机牵引
tractor-drawing 拖拉机牵引的
tractor-drawn 拖拉机牵引
tractor-drawn equipment 拖拉机牵引设备
tractor-drawn multi-rubber-tyre roller 拖拉机曳引多轮胎压路机
tractor-drawn plough 机引犁
tractor-drawn polydisk plow 机引垂直圆盘犁
tractor-drawn roller 拖拉机曳引压路机
tractor-drawn rotavator 机引滚动铲
tractor-drawn scraper 拖拉牵引(式)铲运机
tractor-drawn train 拖挂列车
tractor-drawn two unit articulated roller 拖式双联路碾
tractor-draw-plate 拖拉机牵引板
tractor drill 拖拉机钻机
tractor-driven 用拖拉机驱动的
tractor duster 机引喷粉器
tractor elevator 牵引式提升机
tractor engine 拖拉机发动机
tractor forklift 牵引式叉车装载机

tractor fuel 拖拉机用燃料;低锌烷燃料;发动机用轻油
tractor gate 履带式闸门;环轮高压闸门;拖拉式闸门
tractor grader 拖拉机平地机;牵引式平地机
tractor grip lug 拖拉机履带蹄齿
tractor haulage 拖拉机拖运
tractor-hauled 拖拉(式)拖运的;拖拉机牵引
tractor-hitched sprayer 机引喷雾机
tractor hoist 拖拉机起重绞车
tractor hydraulics 拖拉机液压系统
tractor hydraulic system 拖拉机液压系统
tractorization 拖拉机化
tractor jumbo 履带式钻车;牵引式钻车
tractor kerosene 拖拉机煤油
tractor ligroin 拖拉机用粗汽油
tractor link 拖拉机拉杆(悬挂装置的)
tractor loader 履带式铲斗车;牵引装载机;牵引式挖掘装载机;拖拉机装载机
tractor-loader-digger unit 牵引式装载挖土联合机
tractor luboil 拖拉机润滑油
tractor-mounted 履带式
tractor-mounted blade 拖拉机装具铲刀
tractor-mounted elevator 装在拖拉机上的升运器
tractor-mounted excavator 拖拉机载挖土机
tractor-mounted pump 拖拉机载泵;装在拖拉机上的泵
tractor-mounted sweeper 拖拉机载扫路机
tractor-mounted trench excavator 拖拉机载挖沟机
tractor-mounted vibrating blade 装在拖拉机上的振动式刮刀;装在拖拉机上的振动式铲板
tractor oil 拖拉机润滑油;拖拉机用油;拖拉机油
tractor-operated trench hoe 拖拉机操纵挖沟机
tractor operator 拖拉机手;拖拉机驾驶员
tractor plant 拖拉机厂
tractor plough 机引犁;拖拉机犁
tractor power take-off 牵引力启动
tractor prader 拖式平路机
tractor-propelled 拖拉机驱动的
tractor-propelled harrow 机引耙
tractor propeller 牵引式螺旋桨
tractor-pulled carrying scraper 拖拉机牵引的二轮铲运机
tractor-pulled sheep's-foot 牵引式羊足碾(拖拉机)
tractor pulley opening 拖拉机皮带盘孔
tractor revolving crane 拖拉机旋转起重机
tractor roller 牵引式压路机;机引压土器
tractor rope winch 拖拉机绞盘
tractor scraper 自行式铲运机;拖拉机铲土机;拖拉机拖曳铲运机;拖拉机拖曳铲土斗;拖拉机铲运机
tractor-semitrailer 牵引车及半拖车;牵制式半拖车;拖拉机式半拖车;拖拉机半挂车
tractor shoe 履带板;拖拉机履带片
tractor shovel 履带式铲斗车;拖拉机式铲土机;拖拉机铲土车;拖拉机铲运机;拖拉机铲土机;推土拖拉机
tractor shovel backacter 牵引式反铲挖土机
tractor snow plough 拖式(双犁)除雪机
tractor sprayer 拖拉机喷雾器
tractor station 拖拉机站
tractor sweep 平顶铲
tractor throttle setting 拖拉机油门调节
tractor traction 拖拉机牵引
tractor trailer 牵引车挂车;拖运车;拖拉机挂车;拖拉机拖车;牵挂式运粮车
tractor-trailer engagement 牵引车与挂车的连接
tractor-trailer trucking 牵引式挂车运输
tractor train 挂车列车;拖引列车
tractor tread 拖拉机履带
tractor-truck 半挂汽车;拖拉机手扶拖拉机;拖车;拖车头
tractor-truck for semitrailer 拖半挂车的牵引车
tractor tug 牵引船
tractor type 牵引机型
tractor wagon 拖车;拖运料车;拖拉机挂车
tractor with trailer 带拖车拖拉机
tractrix 曳物线;等切面曲线
tractrix horn 曳物线喇叭
tractus pyramidalis lateralis 锥体侧束
tractus triangularis 三角束
tradable amount 可交易的金额;可交换的金额
tradable emission rights 可以转让的排放权
tradable goods 可外贸货物;贸易财物
trade 信风;工种;行业;通商
tradeable permit 可交易许可证
trade acceptance 贸易承兑汇票

trade acceptance draft 商业承兑汇票
trade access 贸易机会
trade account 贸易项目;贸易科目;贸易账户
trade accounts payable 应付贸易账款;应付货款
trade accounts receivable 应付贸易账款;应收货款
trade act 贸易法
trade adjustment 贸易调整
trade adjustment assistance 贸易调整救济
trade agency 贸易代办处
trade agreement 贸易协定;同业协议;通商协定
trade agreement rate of duty 贸易协定税率
trade air 信风空气
Trade and Development Board 贸易和发展理事会
Trade and Development Committee 贸易及发展会议
trade and industry 贸易与工业
trade-and-payment agreement 贸易支付协定;贸易和支付协定
trade and payment liberalization 贸易和支付自由化
Trade and Tariff Act of 贸易及关税法
trade area 商业区;贸易区
trade assets 营业资产
trade association 专业组织;同业公会;同业工会
trade association guild 同业公会
trade association specification 工业标准说明书;行业协会说明书
trade away 卖掉;贴旧换新
trade balance 贸易平衡;贸易差额;外贸收支情况
trade barrier 贸易障碍;贸易壁垒
trade bid 贸易平衡;贸易差额
trade bill 贸易汇票;商业汇票
trade bloc 贸易同盟
trade block 贸易集团
trade board 劳资协商会(英国);贸易促进会
trade books 商业账簿
trade (brand) name 商标名;商品名
trade burdens 贸易吨位
trade burning appliance 商品燃烧用具
trade capital 营业资本
trade capital movement 贸易资本流动
trade caravan 商队
trade cartel 贸易卡特尔
trade center 贸易中心
trade channel 贸易渠道
trade character 商品个性标志
trade charges 贸易费用
trade circular 商业报单
trade city 贸易城市;商业城市
trade coating 市售涂料
trade commissioner 商务专员
trade concession 贸易减让
trade contact 贸易往来
trade contract 贸易契约;贸易合同
trade control 进出口控制;贸易管制
trade council 同业公会;同业工会
trade creation 贸易开拓;贸易创造
trade creation effect 增加贸易作用
trade credit 交易贷款;贸易信贷;商业信贷;贸易信用
trade credit and advances 贸易信贷及预付款
trade credit on exports 出口贸易信贷
trade creditor 销货客户
trade crew 贸易商行职员;贸易商行雇员
trade cumulus 信风积云
trade currency 贸易通货
trade current 贸易风暖流
trade custom 贸易惯例
trade cycle 景气变动;经济危机周期;贸易周期;贸易循环;商业循环
trade deficit 贸易逆差;贸易赤字
trade development 业务发展
traded goods 贸易商品
trade discount 批发折扣;同业折扣;同行折扣;商业折扣;商业减价
trade discrimination 贸易歧视
trade dispute 贸易争议
trade distorting deuice 贸易干扰手段
trade distortion 贸易异常现象;贸易畸形发展
trade diversion 贸易转移;贸易转向
trade diverting aid 贸易转向援助
trade diverting effect 贸易转向效果
trade dollar 贸易美元
trade drift 信风漂流;商业意义;贸易风暖流

trade economy 贸易经济
trade edition 普及版;输出版
trade effluent 工业污水;工业排水;工业废水
trade effluent officer 工业排污监察员
trade effluent pump 工业排污泵;工业排水泵
trade exhibition 商品交易展览会
trade expansion 贸易扩展
Trade Expansion Act 扩张贸易法;扩大贸易法(美国);贸易扩展法
trade expenses 营业费(用)
trade fair 交易会;贸易交易会;商品展览会;商品交易会
trade finance 贸易金融
trade financing 贸易融资
trade fixture 经营设施(属于租户的)
trade flow 贸易流量
trade fluctuation 贸易波动
trade foreign exchange earnings 贸易外汇
trade foreman 专业工长;工种工长
trade for interior work 室内工程行业
trade framework 贸易体制
trade friction 贸易摩擦
trade furnace 商品炉子
trade gang 贸易商行职员
trade gap 贸易差额
trade goods 外贸货物
trade granite 片麻岩
trade group 贸易集团
trade grouping of rocks 石材商品分类
trade guild 同业公会
trade hall 工业会所;工会会所
trade hazard 贸易危险;贸易风险
trade identity 贸易恒等式
trade impact on environment 贸易对环境的影响
trade-in 以旧物折价换取新物;贴旧换新;折价物
trade-in allowance 贴旧换新折价
trade indifference curve 贸易无差别曲线
trade information 贸易情况
trade in pollution permit 排污许可证交易
trade-ins 折价物
trade-ins of plant and equipment 厂房和设备的折价物
trade-in transaction 贴旧换新交易
trade-in value 作价价值;折价价值;折算价值;贴旧换新价值
trade inventories 贸易库存
trade inversion 信风逆温
trade investment 业务性投资;商业投资
trade invisible trade 贸易外收支
trade journal 专业刊物;贸易杂志
trade law 贸易法
trade leads 经济情报
trade liabilities 商业负债
trade liberalization 贸易自由化;贸易放宽
trade license 贸易许可证
trade liquid 工业废液
trade loss 交易损失;商业性损失
trade magazine 行业杂志;贸易杂志;专业杂志
trademan 技工;手工工人
trade management 行业管理
trade mark 牌子;牌号;商标
trademark act 商标法
trademark infringement 侵犯商标权;侵犯商标专用权
trademark law 商标法
trademark leasing 商标转让
trademark method 商标方法
trademark privileges 商标专有权
trademark protection 商标保护
trademark registration 商标注册
trademark registration administration 商标注册管理
trademark registration fee 商标注册费
trademark registration law 商标登记法
trademark regulations 商标管理法规
trademark system 商标系统
trade matrix 贸易比例矩阵
trade method 贸易方式
trade mission 贸易访问团
trade multiplier 贸易乘数
trade name 商品名称;商号;店名;厂商名称;品名;商业名称
trade negotiation 贸易谈判
trade-off 物货交换;折中选择;折中(方案);折中;交替换位;卖掉;比较评定;比较;权衡利弊;权衡法;权衡
trade-off between risk and rate of return 风险与报酬率比益
trade-off curve 选择曲线
trade-off process 权衡过程
trade-off of the benefit 收入和成本间的比较
trade-off study 比较研究;折中研究
trade-off value 权衡值
trade organization 同业组织
trade partner 贸易伙伴;贸易对象
trade party 贸易一方;商业参与者
trade payable 应付贸易账款
trade payable turnover period 应付贸易货款周转期;应付货款周转期
trade payment 贸易收支
trade policy for environmental protection 环境保护贸易政策
trade post 商栈
trade practice 贸易习惯;贸易惯例
trade preference 贸易优先权;贸易优惠
trade premises 工商业房产
trade price 批发价(格);同行价格
trade promotion 贸易促进
trade promotion centre 贸易促进中心
trade prospect 贸易前景
trade protectionism 贸易保护主义
trade protection society 交易保护社
trade protocol 贸易议定书;贸易协定书;贸易草约
trade-quota agreement 贸易配额协定
trader 小商人
trade reference 贸易查询资料;商业信用查询;商业备咨
trade refuse 工业垃圾;工业废物;工厂垃圾
trade refusing 工业废料
trade register 商业登记
trade regulation 贸易条例
Trade Relations Association 贸易关系协会
Trade Remedy Assistance Office 贸易救济服务处
trade representative 贸易代表;商务代表
trade restriction 贸易限制;贸易管制;交易管制
trade rights 贸易权利
trade risks 贸易风险
trade route 贸易线路;贸易路线;贸易航路;通商路线;商路
trader's price 居间价格
trader's schedule 贸易者一览表
trade safeguarding 通商保护法
trade safe-guarding act 贸易保护法;通商保护法
trade sale 批发
trade sales paint 市售涂料
trade sanction 贸易制裁
trade secret 贸易秘密;商业秘密;商业机密;厂商的制造秘密
trade sewage 工业污水
trade sewerage 工业污水
trade share 贸易份额
trade size 商业尺寸
tradesman 技工;工匠
trade strategy of mineral commodities 矿产贸易对策
trade study 比较研究
Trades Union Congress 工会联合会(英国);贸易协会联合会
trade surplus 贸易盈余;贸易顺差
trade tax 交易税;贸易税
trade team 从事贸易人员
trade term 贸易用语;贸易术语
trade test 技术考试;技术考查
trade tyre 折价轮胎
trade treaty 通商条约
trade turnover 贸易额
trade union 同业工会;工会;行业公会
trade up 标高价格
trade usage 贸易习惯;贸易惯例
trade use 零售用
trade value 贸易值
trade volume 贸易量;贸易额
trade wall 贸易壁垒
trade war 贸易竞争
trade warehouse 贸易货栈
trade waste 工业废液
trade waste sewage 工业废水
trade waste surcharge 工业废液附件罚款

trade waste water 工业废水;工业污水
trade water 工业用水
trade-weighted average 贸易加权平均数
trade-weighted depreciation 贸易加权贬值
trade-weighted exchange rate 贸易加权汇率
trade wind 贸易风;信风
trade-wind belt 信风带
trade-wind circulation 信风环流
trade-wind cumulus 信风积云
trade-wind current 信风海流
trade-wind desert 信风沙漠
trade-wind equatorial trough 信风赤道槽
trade-wind front 信风锋
trade-wind inversion 信风逆温
trade-wind region 信风区
trade-wind system 信风系统
trade-wind zone 信风带
trade year 贸易年度
trade zone 贸易区;贸易带
trading 即刻获利
trading account 营业账户
trading advantage 贸易有利条件
trading association 同业公会
trading center 交易中心
trading certificate 商业执照
trading company 贸易公司;商行
trading concern 贸易商号
trading corporation 贸易公司
trading crowd 交易团体
trading currency 贸易通货
trading desk 交易专柜
trading down 经营低挡商品增加销售量
trading estate 商业用地;计划工业区
trading firm 贸易公司
trading floor 交易场所
trading house 贸易商行
trading limit 交易限额;贸易区域;贸易范围;航行区域
trading market 贸易市场
trading mechanics 贸易机构
trading-off resources 权衡资源
trading on equity 举债经营
trading on the equity 借款扩大资本
trading operation 交易
trading partner 商业合伙人
trading place 交易场所
trading port 贸易港;通商口岸;商港;商埠
trading post 交易所;贸易站
trading post system 交易柜台制
trading profit 销货毛利;营业利润;交易利润
trading record sheet 交易记录表
trading schedule 贸易表
trading service product 服务性项目
trading stamp paper 贸易用商标纸
trading statement 销售表
trading suspended 暂停交易
trading town 商业城市
trading under license 许可制贸易
trading unit 交易单位
trading up 经营高档商品提高利润率;升格销售
trading volume 交易量;交易额
trading warehouse 贸易货栈
trading warranty 营运区保证
tradition 传统
tradition aeration 传统曝气(法)
traditional activated sludge aerobic biochemical treatment 传统活性污泥好氧生化处理
traditional activated sludge process 传统活性污泥法
traditional aeration 传统曝气(法)
traditional agroecosystems 传统农业生态系统
traditional architecture 传统建筑
traditional bioassay 传统生物测定
traditional biologic(al) prevention and treatment 传统生物防治
traditional cartographic(al) manner 传统制图方法
traditional ceramics 传统陶瓷
traditional Chinese curved roof 大屋顶
traditional Chinese garden 中国传统园林
traditional Chinese medicine department 中医科
traditional Chinese medicine wastewater 中药废水
traditional Chinese painting 国画
traditional Chinese painting of mountains and water 中国传统山水画
traditional Chinese style 中国传统风格
traditional coagulant 传统混凝剂
traditional concept 传统观念
traditional construction 传统式施工;传统式结构
traditional costing 传统成本计算
traditional culture 传统文化
traditional culture street 古文化街
traditional debugging 传统调试
traditional decision theory 传统决策理论
traditional decoration 古彩;传统装饰
traditional delineation 白描
traditional design 传统设计
traditional development strategy 传统发展战略
traditional exterior circulation fluidized bed 传统外循环流化床
traditional farming 传统农业
traditional floculant 传统絮凝剂
traditional form 传统形式
traditional form of building 建筑的传统形式
traditional garden building 传统园林建筑
traditional general cargo ship 传统杂货船
traditional health care 传统医学
traditional ideas 传统观念
traditionalism 传统主义
traditional line drawing 白描
traditionally established fishery 传统性渔业
traditionally used by shopkeepers to mark prices 码子
traditional management 传统管理
traditional material 传统材料
traditional measurement 传统方法测定
traditional medicine 传统医学
traditional microbiological analysis 传统性微生物分析
traditional name 惯用名
traditional oceanography 传统海洋学
traditional point sources 传统点污染源
traditional pollutant 传统污染物
traditional porcelain 传统瓷
traditional roller 传统压路机;传统滚筒;传统辊筒
traditional sludge drying bed 传统污泥干化床
traditional style 传统风格
traditional technique 传统方法
traditional water treatment 传统水处理
tradition of building 建筑惯例;建筑传统
traethyl silicate 硅酸四乙酯
traffic 运输;业务量;交通;通信;报务;通信量
trafficability 可通行性;通行能力;通过能力;车辆通过性;行车能量;交通能量
trafficability characteristic 通过性
trafficability condition 能通行情况;交通能量条件;车辆通行情况
trafficability map 通行程度图
trafficability of bridge 桥梁通过能力
trafficability of highway 公路通行能力
trafficability of navigation channel 航道通过能力
trafficability of waterway 航道通过能力
trafficable 可通行的
traffic accessibility in city 市区交通通达度
traffic accident 车祸;行车事故;交通事故
traffic accident appraisal 交通事故评价
traffic accident rate 交通事故率
traffic accident record 交通事故记录
traffic activated signal 车辆触发信号
traffic-actuated control 交通传动控制
traffic-actuated controller 交通传感控制机;交通传动控制机;车动控制器;遥感信号控制机
traffic-actuated detector 交通传动器
traffic-actuated signal 交通传动信号;车动信号
traffic actuated signal control 车辆感应信号控制
traffic-actuated speed controller 车动速率控制器
traffic adaptive network signal timing program (me) 交通网自适应信号配时程序
traffic-adjusted master controller 交通自调式主控制机
traffic-adjusted signal 交通自动调节信号
traffic administered by computer and teletype writer 计算机和电传通信业务管理
traffic aesthetics 交通美学
traffic aisle 交通通道
traffic analysis 交通分析
traffic and accident loss 交通及事故损失
traffic architecture 交通建筑(学);交通建筑艺术(处理)
traffic area 行车范围;运量吸引面积;通车地带
traffic area index 交通面积指数
traffic arrow 交通箭头标志
traffic artery 运输要道;运输干线;运输干道;交通要道;交通干线
traffic assignment 运量分配;交通分配
traffic at intersection 交叉口交通
trafficator 转向指示器;汽车方向指示器;方向转向指示器;(汽车的)方向指示器
traffic attractor 交通吸引点
traffic ban 交通封锁
traffic behavio(u)r 交通现象;交通动态;交通行为
traffic block 运输堵塞;交通阻塞;交通中断;交通拥塞;交通堵塞
traffic block working 行车闭塞法
traffic blow 行车冲击力;交通冲击;车轮的冲击
traffic boat 工作艇;交通艇
traffic bottleneck 道路卡口;运输咽喉;交通线上的薄弱环节;交通狭窄路段;交通瓶颈(地段)
traffic bound 行车碾压
traffic-bound macadam 车结碎石路
traffic-bound method(of construction) 车结筑路法
traffic-bound road 交通拥挤的道路;交通紧张的道路;车结路;交通频繁道路;车压(实)路(面)
traffic bureau 交通局
traffic button 交通路钮;路钉(路面划线用的圆头钉)
traffic capacity 运输能力;业务容量;交通运输能力;交通通过能力;交通容量;交通能量;通行能力;通过能力;输送能力
traffic capacity curve 通过能力曲线
traffic capacity estimate 通过能力估计
traffic capacity study 通过能力研究
traffic-carrying 负载交通(的)
traffic-carrying deck 行车桥面
traffic carrying street 交通频繁的大街;城市通衢
traffic catchment area 吸引范围【铁】;运输吸引范围
traffic category 交通类别
traffic cell 交通流量单元
traffic census 交通调查;交通统计
traffic-census chart 交通调查表
traffic-census point 交通调查点
traffic centre[center] 交通中心
traffic channel 业务信道
traffic channelling 交通渠化;交通分流
traffic circle 大转盘;交通转盘;交通环形枢纽;环形交通枢纽;环形交叉路;环形交叉(道)口;环形交叉
traffic circulation 交通运行;交通循环
traffic circulation map 交通运输图
traffic classification 交通分类
traffic clearance 行车净距;车轮净距
traffic clearance indicator 电路通畅指示器
traffic coating 路标涂料
traffic coefficient 交通量系数
traffic-compacted road 车压(实)路(面);车结路;交通密集路
traffic compaction 交通密集
traffic components 交通组合
traffic composition 交通组成;车辆组成
traffic concentration 交通密度;交通流密集;交通集中
traffic condition 运输条件;运输情况;交通状况;行车条件;交通条件
traffic condition bad 交通不利
traffic cone 交通锥标;锥体交通标志;锥形交通路标
traffic conflict count board 交通冲突计数板
traffic-congested area 交通拥挤地区
traffic congestion 信号拥挤;交通拥塞;交通拥挤;交通堵塞;通信量拥挤
traffic congestion area 交通拥挤地区
traffic constable 交通警(察)
traffic constitution 交通结构
traffic container for door-to-door transport 挨户送货货箱
traffic control 行车调度;运输调度;交通控制;交通管制;交通管理
traffic control and surveillance 交通控制与监视
traffic control area 交通管制区
traffic control barrier 路栏(交通管制)
traffic control centre 交通管制中心
traffic control computer 交通管理计算机
traffic control department 调度处

traffic control device 交通管制设施;交通管制装置;交通管制设备;交通管理设施;交通管理设备
traffic control in dam area 坝区交通管制
traffic controller 行调;调度员
traffic controller's office 调度室调度所
traffic controller's office on railway 铁路调度室
traffic control light 交通控制信号灯
traffic control measures 交通控制措施
traffic control of combined system 联合式交通管理
traffic control procedure 交通管理程序
traffic control regulations 交通管理规则
traffic control signal 交通管制信号
traffic control signal mark 通行信号标
traffic control signal station 通信信号台
traffic control system 行车调度系统;交通控制系统;交通管制系统;交通管理系统
traffic control technology 交通控制技术
traffic convergence 车流汇合
traffic conversion equation 车辆换算公式
traffic conveyed free 免费运输
traffic coordinate centre 交通调度中心
traffic cop 交通警(察)
traffic cordon count 区界交通出入量调查
traffic corridor control system 交通走廊控制系统
traffic count 交通流量观测;交通量计数;交通计数;交通调查
traffic counter 交通计数器
traffic counter loop 环状车辆计数器
traffic count station 运量观测站;交通量观测站
traffic court 交通法庭
traffic created wave 航行波
traffic creep 交通蠕动
traffic cut 交通流线交叉
traffic data 通信量数据
traffic data processing 交通数据处理
traffic deadly accident 交通死亡事故
traffic decking 桥梁车行道
traffic deck surfacing 混凝土路面;水磨石路面
traffic delay factor 交通延迟系数
traffic demand 运输需要;交通需求
traffic demand administration 交通需求管理
traffic density 运输密度;交通密度;通信业务量密度;车辆密度
traffic department 运输局;交通科;交通局;运输科
traffic destination 交通终点
traffic detector 交通检测器;交通传感器;交通探测器
traffic diagnosis 交通鉴定分析
traffic diagram 通信图
traffic diffusion 交通扩散;车队散布
traffic digital control 数符交通控制
traffic directing 交通指挥
traffic direction 行车方向;运行方向
traffic direction block 交通指向牌;交通指路牌
traffic direction sign 运动方向标
traffic direction stud 交通指向钮
traffic discharge 交通输出量;交通量;通车量
traffic dispatch 运输调度
traffic dispatching order 调度命令
traffic dispersion 交通离散;车队散布
traffic distribution 运量分配;交通分配;交通分布
traffic disturbance 交通干扰;交通矢调;交通混乱
traffic divergence 交通分流
traffic divergency 交通分流
traffic diversion 交通疏散;交通改道;交通分散;货物运输变更;变更运输
traffic diversion curve 交通分散曲线
traffic-diversion line 分流线路
traffic divided blocks 交通隔离墩
traffic-dividing line 分流线路
traffic division 交通分区;调度科
traffic dynamic(al) theory 交通动力学理论
traffic dynamics 交通动力学
traffic element 交通组成部分;交通要素;交通单元
traffic engineer 交通工程师
traffic engineering 交通运输工程;交通工程(学)
traffic environment 交通环境
traffic equivalence factor 交通当量系数
traffic estimate 运量预测;交通估计
traffic estimating 交通量估计;交通估计
traffic expense 车务费用
traffic facility 交通设施
traffic factor 交通因素
traffic feeder line 集散运输线

traffic figures 运量观测数据
traffic film 交通渣膜
traffic filter 信号滤波器
traffic flange 路标法兰
traffic flow 客货流通量;信号流;交通运输流量;交通流(量);交通量;车流方向
traffic flow control 交通流控制
traffic flow density 交通流量密度;交通强度;货流密度
traffic flow diagram 交通流量图
traffic flow feature 交通流特性
traffic flow map 交通流量图
traffic flow measure 交通流测量
traffic flow model 交通流模型
traffic flow security 通信码流保密
traffic flow sheet 交通流量表
traffic flow theory 交通流理论
traffic fluctuation 交通量增减
traffic forecast 运输预测;交通预测;交通量预测
traffic-free area 车辆禁行区;禁止通车区
traffic-free encave 禁止通车区;车辆禁行区
traffic-free precinct 无车辆来往地区;禁止通车区;车辆禁行区
traffic function 交通功能
traffic generation 交通始发;交通产生;交通发生
traffic generator 交通始发点;交通始发处
traffic grooming 业务量疏导
traffic growth 运量增长;交通增长
traffic growth-rate 交通增长速率;交通增长速度
traffic guardrail 交通护栏
traffic guidance 交通指挥;交通管理;航路指示;通信指挥
traffic guidance system 交通指挥系统
traffic guide light 交通指挥灯;红绿灯;交通指示灯
traffic handling 交通处理
traffic hazard 险段;交通险阻;交通风险;滩险
traffic hazard warning signal 安全航行信号
traffic hub 交通中心(城市);交通枢扭
traffic impact 交通压力
traffic impedance 交通阻抗
traffic improvement 交通改造
traffic income 交通进款
traffic index 交通指数(表示荷载重复作用的)
traffic-induced ventilation 车行(引起的)通风
traffic information 交通信息
traffic injury 交通工具损伤
traffic inquiries 通信询问
traffic installation 交通设施
traffic instruction 运输指示
traffic integrator 交通积分器
traffic intensity 交通强度
traffic interchange 互通式立(体)交(叉)
traffic interference 交通干扰
traffic interruption 行车中断;交通阻断;交通中断
traffic intersecting 行交叉【铁】
traffic intersection 行车交叉
traffic in transit 过境运输;过境交通
traffic investigation 交通调查
traffic island 交通(安全)岛
traffic isle 交通岛;交通安全岛
traffic jam 交通阻塞;交通拥塞;交通拥挤;交通堵塞;车辆拥挤
traffic jam area 交通拥挤地区
traffic jams and congestion 交通阻塞
traffick 通行车辆
trafficker 买卖者;策划者
traffickist 唯交通论者
traffic lane 行车道;交通分道;通车车道;车道
traffic lane line 车道线
traffic law 交通法律
trafficless 无交通往来的
traffic level 交通程度;交通水平
traffic lever 定向挺杆
traffic light 交通红绿灯;交通控制灯;交通管理色灯;交通灯光信号;红绿灯
traffic lighting 交通照明
traffic limitation 行车限制
traffic line 运输线路;运输线;交通线路;交通线;道路网;车道线
traffic line marker 划线机(交通)
traffic line marking 交通线标划
traffic line marking machine 交通线标划机
traffic line paint 交通线漆;交通线涂料
traffic load 行车荷载;交通荷载;汽车荷载;使用荷载;使用负载;传输负载;行车量;交通负荷;通信量
traffic loading 交通负荷
traffic management 运输管理;交通管理
traffic management device 交通管理设施
traffic management procedure 交通管理程序
traffic manager 车务经理;车务处长
traffic map 交通图
traffic map of research area 工作区交通位置图
traffic mark 交通车辆标线;交通线桥;交通线标
traffic marker 交通路标
traffic marking 交通标线
traffic master plan 交通总规划
traffic maze 交通混乱
traffic measurement 交通量计数;话务测量
traffic measurement device 交通量计量设施;交通量计算装置
traffic measurements 交通量计算
traffic mirror 道路反光层
traffic mishap 交通事故
traffic mode 交通方式
traffic mode classification 交通方式划分
traffic model 交通模型;通信模型
traffic model split 交通方式划分
traffic monitoring 交通监视;交通监督;交通监测
traffic monitoring system 行车监督系统;交通监测系统;通信监督系统
traffic movement control 运输调度
traffic movement phase 通行信号显示;通行信号时间
traffic movement survey 交通动向调查
traffic need 交通需求
traffic network 交通系统;交通网(络)
traffic network planning 交通系统规划;交通网规划
traffic network study tool 交通网络法
traffic noise 交通噪音;交通噪声;船舶噪声
traffic noise control 交通噪声控制
traffic noise index 交通噪声指数;交通噪声指标
traffic noise management 交通噪声
traffic nuisance 交通公害
traffic offence 违反交通规章事件;交通违法
traffic officer 交通警官
traffic of port 港口吞吐量
traffic opening 通航口(桥);通车口(桥隧)
traffic operation 行车作业;交通营运
traffic operation office 行车室;运转室【铁】
traffic order 调度命令
traffic organization 交通组织服务;交通组织
traffic origin 交通起点
traffic pacer 绿波交通
traffic paint 路面标志油漆;路标漆;交通漆;马路划线涂料;马路标线漆;道路划线漆;马路划线漆
traffic paint abrasion tester 路标漆耐磨试验器
traffic pan 车压硬土层;车轨磐
traffic parking-transit problem 交通停车过境问题
traffic passing through dam 过坝运量
traffic patrol(ling) 交通巡逻
traffic patrolman 交通巡逻警察
traffic pattern 交通型式;交通路线图;交通量变化图;交通类型;起落航线
traffic peak 运输高峰;交通高峰
traffic permit 通行证(车辆)
traffic phase 行车信号显示;交通信号相
traffic pilot 多路转换器
traffic place 交通广场
traffic place map of survey area 测区交通位置图
traffic plan 运输计划
traffic planning 交通规划;道路交通规划
traffic playground 交通公园
traffic point city 交通枢纽城市
traffic police 交通警(察)
traffic pollutant 交通性污染物
traffic pollution 交通污染
traffic pollution sources 交通污染源
traffic ponding 交通拥塞
traffic pool 运输联营组织
traffic position map 交通位置图
traffic post 交通指挥岗
traffic potential 运输潜力
traffic pressure 交通压力
traffic priority 行车优先权
traffic profile 交通量变动图
traffic prognosis 交通预测
traffic program(me) 运输方案
traffic prohibited 禁止通行;禁止交通

traffic prohibited sign 禁止通行标志
traffic promotion 运输业务促进
traffic psycology 交通心理学
traffic railing 交通栏杆;行车栏杆
traffic rate 通信量速率
traffic records 运量观测记录;交通记录
traffic regulation 运输调整;交通管理;运输规程;交通规则;交通法规
traffic rehabilitation 交通改善
traffic relation 运输关系
traffic requirement 交通需求;运输需要
traffic research and development network 交通科研网
traffic research image processing 交通研究图像处理
traffic resistance 交通通畅度
traffic responsive control 交通感应控制
traffic responsive metering 交通感应式限流(控制)
traffic returns 运输统计报告;运输统计报表;运输利润;交通统计表
traffic road 通车道路
traffic rotary 环形交叉
traffic roundabout 环形交叉
traffic route 交通路线;运输路线;业务量路由
traffic rows 交通行列
traffic rules 交通规则
traffics 运量
traffic safety 行车安全;交通安全
traffic safety device 交通安全设施;交通安全措施
traffic safety education 交通安全教育
traffic safety facility 交通安全设施
traffic schedule 运行计划
traffic sealing 交通压密(路面)
traffic segregation 交通分流;交通分隔
traffic separation line 通航分隔线
traffic separation scheme 交通危险区略图;分道通航制;分道航行制
traffic separation zone 通航分隔带
traffic separator 交通分隔带
traffic shed 候车棚;停车棚
traffic sign 交通路牌;交通记号;交通标志
traffic signal 交通信号;交通控制信号灯;交通灯光信号;红绿灯交通信号;红绿灯
traffic signal control 交通信号管理;交通信号控制
traffic signal controller 交通信号控制机
traffic signal installations 交通信号装置
traffic signal lamp 交通信号灯
traffic signal(l)ing device 交通信号装置
traffic signal optimization program(me) 交通信号最优化程序
traffic signal speed sign 交通信号系统速率标志
traffic signal system 交通信号系统
traffic signs and signals 交通标志和信号
traffic simulation 交通模拟
traffic simulator 行车模拟器
traffic situation 交通场所
traffic snarl 交通混乱
traffic span 通航跨度(桥的);通行跨度(桥隧的)
traffic speed 交通速率
traffic square 交通广场
traffic stability 交通稳定性
traffic stagnation 交通停滞
traffic station 交通站
traffic statistics 交通统计
traffic stop 交通中断;交通停车处
traffic stream 交通流量;交通流;车流
traffic stream line 交通流线;车流线
traffic stream model 交通流模型
traffic strip 快车道;分车带
traffic stripe 中央车道;路面交通标线
traffic stripping 划车道线(路面上);路面上划车道线
traffic structure 交通构筑物;运输结构;交通结构
traffic stud 路钉;交通钉
traffic study 运输研究;交通研究
traffic supervision 车辆监视
traffic surface 通行路面
traffic surveillance 交通监视;交通监理;交通监控
traffic surveillance and control system 交通监控系统;车辆监视控制系统;车辆监测控制系统
traffic surveillance system 交通监视系统
traffic survey 交通观测;交通调查
traffic-switching intersecting 行调交叉【铁】
traffic system 交通系统
traffic technology market network 交通技术市场信息网络系统
traffic terminal point 交通枢纽点
traffic test 行车试验
traffic testing period 交通试验期;行车试验期
traffic through (port) 港口吞吐量
traffic topping 道路路面
traffic tower 交通指挥塔
traffic transport industry 交通运输业
traffic trend 运输发展趋势
traffic tube 车行隧道
traffic tunnel 交通隧洞;交通隧道
traffic turn-over 运输周转量
traffic unit 运输单位;运量单位
traffic variation 交通变动
traffic victim 交通事故受害者
traffic volume 交通密度;运量;交通容量;交通流(量);交通量;通车量
traffic volume assignment 交通量分配;交通量分布
traffic volume computer 交通量计算机
traffic volume count board 交通量计数板
traffic volume counter 交通量记录仪;交通量计数器
traffic volume forecast 运量预测;交通(流)量预测
traffic volume observation 交通量观测
traffic volume observation station 交通量观测站
traffic volume of intersection 交叉口交通量
traffic volume per second 秒交通量
traffic volume prediction 运量预测
traffic volume prognosis 交通量预测
traffic volume survey 交通量观测;交通量调查
traffic volume trends 交通量发展趋势
traffic volume variation 交通量变化图
traffic warden 交通管理人员
traffic wave theory 交通波动理论
traffic way 公路;交通路线;车道;交通要道;道路
traffic weighing station 运输量过磅站
traffic without shipping documents 无票运输
traffic zone 交通管制段;运输区;交通小区;交通频繁地区;交通分区;交通地带
traffic zoning 交通区划;交通分区
traffic-zoning paint 交通区划漆;交通区划涂料
Traffolyte 特拉福利特纸(一种墙纸)
Trafford tiles 特拉福德牌石棉水泥瓦(一种波纹石棉水泥瓦)
traf-o-line marker 交通划线机
tragacanth 黄胶蓍树胶
tragacanth gum 黄蓍胶
tragantine 可溶性淀粉
tragedy of commons 公用全悲剧
tragedy of the commons 公用资源悲剧
traied plough 牵引式犁
T-rail T形钢轨;平底钢轨
trail 星象迹线;小路;小径;临时道路;后续;痕迹;牵引杆;尾迹;拖着走;拖曳;串集;余波;散步道;漫步路
trailable 挤脱的
trailable force 挤脱力
trail batch 混凝土试块;小量试拌
trail-behind 牵引式的
trail bike 轻便摩托车
trailblazer 路径导向标;创新者
trail board 蝴蝶板;船首饰板
trailbreaker 创新者
trail builder 斜板推土机;开路推土机;拖挂式筑路机械
trail cable (电动机械的)拖尾电缆
trail car 挂车
trail crane 电动有轨吊车
trailed air-blast sprayer with engine drive 机动牵引式鼓风喷雾机
trailed decoration 泥浆饰物;泥浆画
trailed disk plow 牵引式圆盘犁
trailed implement 牵引式农机具
trailed model 牵引型
trailed plow 牵引式犁
trailed ripper 牵引式松土机
trailed vibrating roller 拖挂振动式压路机
trailer 自航耙吸式挖泥船;挂车;拖车
trailer air compressor 拖车式空(气)压(缩)机
trailer aircraft 拖靶机
trailer axle 拖车轴
trailer block 随附信息组
trailer body 拖车车厢;拖车车体
trailer bogie 拖车转向架
trailer brake 拖车制动器
trailer brush 拖挂扫路车
trailer bus 牵引式公共汽车;双节公共汽车
trailer camp 拖车营(地);拖车式活动住房营地;拖车住户集中地;拖车居住营地
trailer car 拖车
trailer card 后续卡片;尾部卡片
trailer carrier 运载拖车
trailer chassis 拖车车架;拖车车盘;拖车底盘;底盘车
trailer coach 活动住房拖车
trailer coach home 拖车式活动住房
trailer coach (home) park 拖车式活动住房营地
trailer coach space 拖车式活动住房营地
trailer compactor 拖挂夯实机
trailer compressor 拖挂空气压缩机
trailer concrete pump 拖挂混凝土泵
trailer connector 拖车接头;拖车挂钩
trailer construction 拖车构造
trailer converter dolly 拖带挂车;台车
trailer coupling 挂车拖接器;拖挂装置;拖车拖接器;拖车联结器
trailer court 拖车式活动住房营地;拖挂组成的营区;拖车营(地);拖车停放场
trailer crane 拖挂式起重机;拖车式起重机;随车(式)起重机;随车(式)吊
trailer deck 挂车甲板
trailer distributor 拖挂式分布机;拖挂式洒布机
trailer-drawn train 挂车列车
trailer dredge(r) 拖斗挖泥船
trailer dump 翻斗挂车
trailer dwelling 汽车拖带的住房;拖车住房
trailer edge 后刃
trailer-erector 竖直拖车;安装拖车
trailer excavator 拖挂挖土机
trailer fifth wheel 拖车接轮
trailer grit spreader 拖挂铺砂机
trailer gritter 拖挂铺砂机
trailer gritting machine 拖挂铺砂机
trailer-hauling tractor 牵引车
trailer hitches 拖车系紧装置
trailer hitch for vehicle 车辆牵引索
trailer house 汽车拖带的住房;用汽车拖带的活动住房;拖车房屋
trailer interchange 拖车交接
trailerist 拖挂挂车的汽车驾驶员
trailerized 拖挂化的
trailerized tank 拖挂式油罐车
trailer label 尾部标志
trailer landing gear 拖车起落架
trailer launcher 拖车式发射装置
trailer launching platform 拖车式发射台
trailer light coupling 挂车灯接头
trailer load 集装箱拖运架整车货物
trailer-loading elevator 拖车装载升运器
trailer mixer 拖拌和机
trailer mobile home park 拖车式活动住房营地
trailer-mounted 装在拖车上的
trailer-mounted capstan 装在拖车上的绞盘
trailer-mounted container 拖载运的集装箱;拖车载运的集装箱
trailer-mounted drill 拖车装钻机
trailer mower 牵引式割草机
trailer nozzle 马蹄形喷嘴
trailer oil distributor 喷油拖车
trailer on flat car 平车运装;拖上平板车;铁路挂车平车
trailer on flat car/container flat car service 拖车上平板车和集装箱上平板车运输
trailer park 拖挂车停车处;挂车停放场;旅行车停车场(美国);拖车营(地);拖车停放场;拖车式活动住房营地
trailer park sewer 拖车式活动住房驻地污水管道
trailer pintle eye 挂钩孔;拖车挂钩孔
trailer platform 牵引式平板车;拖车平台
trailer pump 拖拖泵车;移动式消防泵;装在拖车上的消防泵
trailer record 后续记录;尾部记录
trailer-rig 拖车式钻机
trailer road-sweeping machine 拖式扫路机;拖挂式扫路机
trailers and mobile homes 篷车拖车和活动房屋
trailer sanitation station 拖挂式活动住房卫生站
trailer saw 拖挂锯车
trailer scraper 拖挂式刮土机

trailer ship 被拖船;挂车船
trailer site 拖挂的活动住房营地
trailers-on-flat-car 挂车驮运
trailer sprayer 牵引式喷雾机
trailer-spreader 拖挂式撒铺车;拖挂式撒布车
trailer standing area 挂车停放场
trailer suction 耙吸
trailer test equipment 拖车试验装置
trailer tipping device 拖车倾翻装置
trailer tire [tyre] 拖车轮胎
trailer towability 牵引挂车的能力
trailer transfer point 拖车转运点
trailer truck 汽车列车;拖车转向架;带有平板拖车的卡车;双节卡车
trailer type 拖车式
trailer-type concrete mixer 拖车式混凝土搅拌机
trailer-type extinguisher 拖车式灭火机
trailer-type machine 拖挂式机械;牵引式机械
trailer-type method 拖挂方法
trailer-type overrun brake 拖车超速闸
trailer wagon 牵引小车;拖车;翻斗挂车
trailer with adjustable drawbar height 牵引杆高度可调式挂车
trailer with canvas cover 帆布顶篷拖车
trailer yard 挂车编组场
trail formation 纵队队形
trailgate 拖在船模后的尾流栅
trail herding 定线放牧
trail hitch 拖引叉
trailing 泥浆绘制图案;牵引;顺向行驶;从动;成串拖集;拖尾
trailing action 车辆的转向销作用
trailing aerial 拖曳天线
trailing antenna 下垂天线;拖曳天线
trailing axle 后轮轴
trailing bar 拖板
trailing black 后沿变黑
trailing bogie (车辆的)牵引转向架
trailing box 拖车
trailing broadcaster 牵引式撒播机
trailing buoy antenna system 拖曳式浮标天线装置
trailing cable 牵索;尾曳电缆;拖电缆;拖曳电缆
trailing carriage 后方台车
trailing conveyer 接运输送机
trailing coordinates 轨迹坐标
trailing edge 尾边;下降沿;下降边;叶片后缘;后缘;后沿;随边
trailing-edge pulse time 脉冲后沿时间
trailing-edge wedge 楔形后缘
trailing end 终端;后端
trailing form(work) 滑模
trailing frog 并线器
trailing gas shield 尾气保护
trailing habit 蔓延习性
trailing head 耙头(耙吸式挖泥船的)
trailing hopper suction dredge(r) 耙吸(式)挖泥船
trailing idler 从动空转轮
trailing implement 牵引式农具
trailing leaf 节流挡板
trailing line 拖管索
trailing load 牵引总重;牵引重量;牵引质量;牵引载荷;拖挂重量
trailing of a point 挤岔
trailing of a switch 挤岔
trailing plant 蔓性植物
trailing plate expansion joint 后缘板伸缩缝
trailing point 反向转辙器;顺向道岔【铁】;背向道岔
trailing pointer 背向转撤器
trailing pole tip 后极尖;磁极后端
trailing portion 舵后部
trailing scoop grader 拖挂式铲斗平地机
trailing shock wave 尾激波
trailing smear 拖尾;拖影
trailing spit 岛后沙坝
trailing spud 后桩
trailing suction double hopper dredge(r) 双舱耙吸式挖泥船
trail(ing) suction dredge(r) 尾部吸入挖泥船;尾部吸入式挖掘船;拖曳式吸泥船
trailing suction dredge(r) with side suction tube 边耙挖泥船
trailing suction dredge(r) with stern well 艉耙挖泥船

trailing suction hopper dredge(r) 吸扬式开底挖泥船;自航耙吸式挖泥船;耙吸(式)挖泥船
trailing suction sidecasting dredge(r) 边抛耙吸(式)挖泥船
trailing switch 背向道岔;反向转辙器
trailing tilting spud 拖曳倾卸桩【疏】
trailing train 后行列车
trailing transient 后沿瞬变特性
trailing truck 拖曳式卡车;拖曳运货车
trailing vortex 后缘涡流;尾涡
trailing vortex lines 后缘涡流线
trailing wake 尾流
trailing wall 导流构筑物
trailing web 后腹板
trailing weight 牵引质量
trailing wheel 后轮;从轮;从动轮
trailing white 后沿变白;拖白
trailing wire 下垂天线
trailing-wire aerial 拖垂天线
trailing-wire antenna 拖曳天线
trailing zero 后补零
trail line 退曳线
trail-marking pheromone 标迹信息素
trail mix 混凝土的试拌
trail net 拖网
trail of fault 断层迹
trail of precipitation 降水迹
trail of the fault 断层痕迹
trail road 小路;临时通道;试用道路
trail rope 拖绳
trails of precipitation 雨幡
trails of rain 雨幡
trail spark plug 后置火花塞
trail transfer car 拖车
trail-type 牵引式;拖车式
train 序列;载重车队;列车;火车;导火药
trainability 可训练性
train acceleration 列车加速度
train accident 列车事故
train accommodation 接车【铁】
train air signal valve 列车号态气阀
train and vehicle ferry 火车和汽车轮渡
train apparatus 列车设备
train appreciation 列车增值
train approach warning device 列车渐近报警器
train arrival and departure operation 列车到发技术作业
train arrived ahead of scheduled time 列车早到
train arrived behind scheduled time 晚点到达【铁】
train assembly station 调车场;车辆编组站;铁路列车编组站;编车站【铁】
train attendant 列车员
train automatic follow 列车自动跟踪
train axis 转动轴
train bar 轮系夹板
train block system 行车闭塞制;行车闭塞系统;行车闭塞法
train boat 列车渡轮
train-borne 列车载的
train-borne ATP/ATO equipment and software 车载ATP/ATO设备及软件
train-borne equipment 车载设备
train brake 真空制动(器)
train braking distance 列车制动距离
train calculation length for long term 远期列车计算长度
train calling 列车呼叫
train capsized 列车倾覆
train carrier 火车轮渡;火车渡船
train class 列车等级
train classificationl train formation 列车编组
train clearing siding 列车洗刷线
train coasting 列车滑行
train collision 列车冲突
train composition list 列车编组顺序表
train conductor 列车长
train configuration 列车编组
train configuration type 列车编组型式
train configuration unit 列车编组单元
train consist list 列车编组顺序表
train consist report 列车编组顺序表
train construction speed 列车结构速度
train control 列车控制;列车调度

train control automation 铁路行车自动化
train controller 列车调度员
train control relay 列车控制继电器
train control section 调度区段
train control unit 列车控制单元
train conveyer 列车输送带
train crew district 列车段
train crossing 列车交错;列车会车;列车回车
train crossing plan on intermediate station 中间站列车会让计划
train deceleration 列车减速度
train delay 列车延误
train de luxe 高级列车;特别列车
train density 列车密度
train departure and arriving time 列车到发时刻
train derailment 列车出轨
train describer 列车记录器
train despatcher 列调【铁】
train destination 列车去向
train destination indicator 列车到达站指示器
train diagram for automatic block signals 追踪运行图
train diagram for double-track 比线运行图【铁】
train diagram in pairs 成对运行图
train diagram not in pairs 不成对运行图
train diagram recorder 运行图描绘仪【铁】;运行图记录仪【铁】
train disgram for single-track 单线运行图
train disinfecting siding 列车消毒线
train disinfecting track 列车消毒线
train dispatched ahead of scheduled time 列车早开
train dispatched in succession 连发列车
train dispatcher 列调【铁】;列车调度员;火车调度员;铁路调度员
train dispatching order 调度命令
train dispatching section 调度区段
train dispatch schedule 列车调度计划
train distribution plan 列车分配计划
train disturbance 列车故障
train drifting 列车滑行
train dynamic(al) characteristic 列车动力性能
train-earth information receiving system 车—地信息接收系统
trained fruit tree 整形果树
trained personnel can make professional progress 人才的成长
trained worker 专家
trainee 受训者;学员;培训学员;受训人
trainee cartographer 见习制图员
train entering wrong track 列车进入异线
trainer 教练员;教练设备;培训教员;数字逻辑演算装置;电子培训设备
train ferry 列车轮渡;列车渡轮;列车渡船;火车轮渡;火车渡轮;火车渡口
train ferry terminal 火车轮渡码头
train-ferry track 铁路轮渡线路
train-ferry traffic 列车轮渡运输
train fire 列车火灾
train flow 列车流
train formation 车辆编组
train formation plan 列车编组计划;列车编成计划
train frequency 群频(率)
train general testing device 列车静调设备
train grapher 自动记录器
train gross load (tons) 列车总重
train gross ton kilometers 列车叫重吨公里
train-ground information coupler 车—地信息耦合
train-ground information receiving system 车—地信息接收系统
train-ground radio system 车—地无线电系统
train-group 车组
train-group number 车组号
train guard 运转车长
train hauling out-of-ga(u)ge goods wagon 超限列车
train hauling part-load traffic 零担货物列车
train headway 列车间距
train hospital 列车医院
train hour 列车小时
training 训练;整治;教育训练;培训;导治;导流
training adviser 培训顾问
training aid 训练辅助器材
training all workers and staff 全员培训

training analyst 训练分析员
training area 训练区;训练场
training area files 训练区文件
training attachment 训练配属
training balloon 训练气球
training bank 导水堤;导流河岸;导流堤岸
training base 训练基地
training bay 培训区
training bill 训练大纲
training bulletin 训练公报
training captain 训练机长
training center 培训中心;训练中心
training chart 训练图
training class 训练班
training classifier 训练分类器
training class of environmental protection 环境保护训练班
training college 师范学院
training command 训练部
training computer 训练计算机
training cost 培训费用
training course 培训科目;整治水道;训练班;实习课;导流水道
training cross-section 导流断面
training cruise 航行实习
training curve 导流曲线
training dam 顺坝;导流坝
training department 训练部
training design 培训设计
training device 训练设备;训练器材
training device man 训练器材保管员
training dike 导流丁坝;导水堤;导流坝;围堤;顺坝;导堤
training dike at head of central bar 洲头顺坝
training dike at mountain brook 溪口导坊
training dike at river mouth 河口导流坝;河口导流坝
training dike at tail of central bar 洲尾顺坝
training embankment 顺坝;导流堤
training evaluation 训练评定
training expense 培训费用
training experimental group 训练实验组
training facility 训练设施
training field 训练场
training film 教学电影
training for depth 通行深度疏导;疏浚
training for discharge 疏导;导流
training for promotion 晋升训练
training for sediment 治沙;泥沙疏导;导沙
training function 培训职能
training idler 导向轮;调心托辊
training in road sense 交通安全教育
training in rotation 轮训
training institute for environment(al) pollution control 环境污染控制进修学校
training iterative method 训练迭代法
training jetty 导水突堤;导流突堤;导流丁坝;导流坝
training kitchen 培训厨房
training launch control center 训练发射控制中心
training levee 导水堤;导流堤
training light signal 灯光导标
training load 训练负荷
training loop 培训回路
training management 训练管理
training manual 训练手册;培训手册
training map 教学地图
training memorandum 训练备忘录
training mission 训练任务
training mole 导流突堤;导流堤岸
training objective 训练目标;培养目标
training of channel 河流疏浚;河道疏浚;渠道疏浚
training of entire staff 全员培训
training officer 培训工作人员
training off job 脱产培训
training of local personnel 培养本国人才
training of mole 导流突堤;导流堤岸
training of sediment 泥沙疏浚
training-of-trainers workshop 教员学习班
training oneself 炼己
training on the job 岗位培训;在职训练;在职培训
training order 训练命令
training performance indicator 培训业绩指标
training period 训练期
training plan(ning) 培训计划;训练计划

training pool 游泳训练池
training pool hall 游泳训练馆;室内游泳训练池
training program(me) 训练计划;训练程序;培训程序
training range 训练靶场
training reactor 教学用反应堆
training regulation 训练条令
training report 训练报告
training schedule 训练进度表
training scheme 培训方案
training school 专科学校
training school for nurses 护士培训学校
training set 训练集
training ship 教练船;培训船;实习船
training site 训练现场;训练基地
training specification 培训规范
training-spur dike 导流丁坝
training standard 训练标准
training stereoscope 训练立体镜
training structure 整治建筑物;导流建筑物;导治建筑物
training system 导流系统
training time 训练时间;培训时间
training wall 束流坝;导水墙;导(流)墙;导(流)堤;导流壁;导流坝;顺坝
training works 治导工程;整治建筑物;导流工程;导治工程
training work shop 培训工场
training year 训练年度
train integrated management system 列车集成控制系统
train interval 列车间隔
train interval running 时间间隔行车【铁】
train-kilometer 列车公里;列车走行公里
train lead locomotive kilometers 本务机车走行公里
train left behind booked time 晚点发车
train left behind scheduled time 晚点发车
train lighting dynamo 火车照明用发电机
train line 列车风管;列车线
trainline cut-out cock 列车管路截断塞门
trainline release 全部制动管路缓解
train list 列车编组顺序表
train list information after departure 列车确报;确报
train list information in advance 列车预报
train list information in advance and before departure 列车预确报
train list information reporting station 确报站
train loader 装车机;不摘挂列车装载机
train load(ing) 列车载重;火车荷载;列车重量;铁路荷载;整车货(物);列车装载能力;列车载重量
train locomotive 列车机车
train log 火车运材
train make-up 火车编组
train make-up station 火车车辆编组站;调车(编组)场;编组站【铁】
train making-up 火车编组
train man 列车乘务员
train manifest 列车编组顺序表
train meter 偏差测定计(测量目标实际位置的偏差)
train momentum 列车动能
train monitoring and tracing function 列车跟踪功能
train movement against the current of traffic 双线反向行车
train movement monitoring program(me) 列车运行监督及最终程序
train movement recording equipment 行车记录设备
train net load (tons) 列车载重
train net ton-kilometre 列车载重吨公里
train noise 火车噪声
train number 列车车次;车次号【铁】
train number identification system 车号识别系统【铁】
train number indication 车次表示【铁】
train number indicator 车号指示器【铁】
train number record 列车号记录
train of barges 驳船队列;驳船队;船队
train of cars 汽车队列
train of empty stock 空车列车
train of equipment for road construction 筑路机械队列

train of gearings 齿轮系
train of impulses 脉冲系列;脉冲串
train of irregular waves 不规则波列
train of lacking in weight 欠重列车
train of loading 一系列车轮荷载(测试桥梁用)
train of loads 荷载分布图
train of locks 梯级(式)船闸
train of mechanism 机构系统
train of one class 同级列车
train of oscillations 振荡列;振荡串
train of paving plants 铺筑机列(道路)
train of powder 导火线
train of pulses 脉冲链;脉冲群;脉冲序列
train of pusher barges 顶推驳船队
train of rollers 轧辊机列;压路机组
train of skywave 序列天波
train of surfacing plants 路面铺筑机组
train of toothed wheels 齿轮系
train of waves 波列;波群
train of wheels 齿轮系
train oil 鱼油
train operating room 运转室【铁】
train operating routing 列车运行交路
train operating yard 运转场
train operation 列车运行
train operation accident 行车事故
train operation adjustment 行车调整;列车运行调整
train operation and control automation 铁路行车指挥自动化
train operation autoplotter 列车运行自动绘图仪
train operation organization 行车组织
train order 行车命令
train oriented decrossing 列车种类别疏解
train overcoming the gradient by momentum 列车动能闯坡
train overturned 列车倾覆
train parking for ferry 列车待渡场
train passing in opposite direction 对向列车通过
train path 列车运行线;列车进路
trainphone 列车电话
train pipe 列车间管;列车管
train police 乘警
train position indication 列车位置表
train printer 列式打印机
train propelled by locomotive 列车推送运行
train-protecting signal 护车号志
train protection 列车防护
train pulled forward by locomotive 列车牵引运行
train pulse 序列脉冲
train pushed forward by locomotive 列车推送运行
train-pushing engine 顶推机车
train radio set 列车无线电台
train reassembly yard 列车改编场
train receiving and delivery at junction station plan 分界站交接列车计划
train reception and departure 接发列车
train recovering time lost 列车恢复正点
train register book at station 车站行车日志
train regulation 列车调整
train reserves 列车备用
train resistance 列车阻力;船队阻力
train resuming its normal running 列车恢复正常运行
train reversal 列车折返
train reverse operation 列车折返
train route 列车进路
train running at gaining speed for making up time lost 列车赶点运行
train running at reduced speed 列车减速运行
train running before scheduled time 列车早点
train running behind scheduled time 列车晚点
train running circulation 列车运行交路
train running delay 列车运缓
train running in fleet 列车成队运行
train running in reverse direction 反向行车【铁】
train running notice 行车通告
train running on scheduled time 列车正点(运行)
train running resistance 列车运行阻力
train running time in block section 列车区间运转时间
train schedule 列车时刻表
train schedule chart 铁路列车运行图
train service number 列车服务号

train-set 列车组;车组;车底【铁】
train set leaded 车列牵出
train-settling cost per hour 列车停留小时成本
train set transferred 车列转线
trains hauling above normal tonnage 超轴牵引（指铁道）
train shed 车棚(火车);列车棚
train signal 行车信号(机)
train simulation rationale 列车模拟原理
train skip 列车越站
train spaced by automatic block signals 追踪列车
train spacing 列车追踪间隔;列车间隔
train spacing point 分界点【铁】
train spacing point with distribution tracks 有配线分界点
train spacing point without distribution tracks 无配线分界点
train specific braking force 列车单位制动力
train specific tractive force 列车单位牵引力
trains per hour 每小时列车开行对数
train staff 火车路签;铁路乘务员;路签
train-staff instrument 路签机
train staff lamp 路签灯
train's technical speed 列车技术速度
train-stock 车列【铁】
train stock reverved 列车保留
train stop 转角限制器;止车器
train stop mark 停车标志
train succession time 列车间隔时间
train sway 列车摇晃
train test line 试车线
train tonnage 列车重量
train tonnage rating 列车牵引定数
train travel(l)ing speed 列车旅行速度
train trip 列车行程
train tunnel 铁路隧道
train type printer 列式打印机
train unit 列车组;列车
train up-turned 列车倾覆
train value 列值
train waiting for a path 列车等线
train waiting for a receiving track 列车等级
train washing equipment 列车清洗设备
train washing siding 列车洗刷线
train wave 列波
trainway 铁路线
train way-side coupler 车—地信息接收器【铁】
train weight 列车重量
train wheel 火车车轮
train wheel bridge 上夹板
train wheel bridge foot 上夹板位钉
train wheel bridge support 上夹板托
train withdrawn from schedule 列车停运
train working diagram 列车运行图
train working graph 列车运行图
train working plan 列车工作计划
train working plan on classification 编组站列车工作计划
train working program(me) 列车工作方案
train working schedule 列车时刻表
train working timetable 列车时刻表
trait 品质;特点
Trajan's column 图拉真纪功柱(古罗马)
trajectile 抛射物
trajectories of principal stresses 主应力轨迹线
trajectory 轨迹线;流线;抛射线;弹道
trajectory analysis 轨迹分析;轨道分析
trajectory bucket 挑流反口弧;挑流鼻坎
trajectory bucket type energy dissipater 挑流式消能戽;鼻槛挑流式消能结构
trajectory cable 正交钢索
trajectory case 轨道角色
trajectory derivative 轨道导数
trajectory determination laser 测轨道激光器
trajectory measurement 弹道测量
trajectory method 轨迹法;抛射线法
trajectory of electron motion 电子运动轨迹
trajectory of grit particles 杂粒运动轨迹
trajectory of jet 射流航迹;射流轨迹
trajectory of prestressing force 预应力轨迹线
trajectory of principal stress 主应力轨迹
trajectory of strain 应变轨迹
trajectory of stresses 应力轨迹;应力迹线
trajectory of stretching force 拉伸力轨迹

tram 正确位置;正确调整
tram car 市内电车;有轨电车;矿车;电车;煤车
tramcar motor 电车电动机
tram depot 有轨电车停车站;有轨电车停车场;有轨电车车库
tramegger 高阻表;迈格表
tramline 有轨电车线(路)
trammel 长径规;旋转筛;渔网;游标卡尺;梁规;横木规;划割圆形玻璃设备;长臂(圆)规;椭圆规
trammel drain 网状排水(盲)沟
trammel(l)ing 滚筒洗矿;滚筒筛选;转筒筛选
trammel net 双重刺网
trammel roll 管壁减薄扎机
trammel screen 旋转式圆筒筛
trammer 机车;推车工;推车
tramming 用量规量;乘有轨车;有轨运输
tramp 不定期货船
tramp cargo 不定期货运物资
tramp discharge 沉料排出口
tramp element 夹入元素;残存元素
tramper 夯实器;不定期船;不定航线货船
tramper cargo 不定期货运物资
tramp glass 玻璃屑
tramping 不定线的海上运输;不定期运输
tramp iron 废铁;杂铁;孔内金属落物;混杂铁块
tramp iron detector 夹杂铁的探测器
tramp-iron magnetic separator 铁质夹杂物磁力分离器
tramp-iron rejector 金属夹杂物分离器;铁质夹杂物排除器
tramp-iron removal 去掉夹杂铁
tramp-iron remover 铁质夹杂物排除器
tramp-iron separator 夹杂铁分离器;夹杂铁捕集器
tramp-iron separator 磁吸法分离废铁机
tramp-iron trap 夹杂铁收集器
trampling method 踩法
tramp material 混入物
tramp metal detector 夹杂金属检查器
tramp navigation 不定线航行
tramp oil 液中油
tram pole 电车线杆
tramp-pick 掘土撬杆
tramp ship 不定期货船;不定航线货船
tramp shipping 不定期运输
tramp species 舶来物种
tramp steamer 流动货船;不定期货船;不定航线货船
tram rail 吊车索道;运料车轨(道);电车轨(道);有轨电车轨道
tram road 架空索道;电车路;电车道;有轨电车路
tram rope 运输钢丝绳
tram service 有轨电车交通
tram stop 电车站
tram trammels 椭圆量规
tramway 缆道;架空索道;轻轨道;电车(轨)道
tramway bridge 电车路线桥
tramway dead-end terminal (电车道的)尽头式终点站
tramway for cutting 采伐轨道
tramway for felling 采运木轨道
tramway mast 索道立柱
tramway portage 缆车盘运
tramway rail 电车轨道
tramway subway 地下电车道
tramway switch 有轨电车转换器;有轨电车转辙器
tramway system 电车道系统
tramway terminal 电车终点站
tramway track 电车轨道;有轨电车线(路)
tramway type equipment 简单悬挂【电】
tranche 份额;分批发行
tranche CD 分档存单
tranche certificate of deposit 分档存单
tranche drawing 份额提款(国际基金组织贷款)
tranche policy 份额政策
Trancor 特兰科尔合金
tranfer time 中转时间
tranformation reaction 变态反应
T-range jack T型张拉千斤顶(张拉单跟钢丝束)
transient program(me) table 临时程序表
tranish film 锈膜
tranisonic aircraft 跨音速机
tranport subsidy 运输补贴
tranquil 平稳的
tranquil arc 平稳电弧

tranquil flow 缓流;平流;稳流;妥流
tranquil flow reach 缓流河段
tranquility 泊稳
tranquility condition 泊稳条件
tranquility conditioning 船舶泊稳调节
tranquility of harbo(u)r basin 泊稳
tranquil(l)ing tank 稳流箱
tranquillityite 静海石;宁静石
tranquil(l)izer 镇静剂;增稳装置
tranquil regime 平稳状态
tranquil rest area 安静休息区
tranquil state 稳定态
trans 反式
transacetylation 乙酰转移作用
transacter 事务处理系统
transaction 横断面;学报;交易;会报;事项处理;事项;事务处理;事务;动态实体;成交价
transactional analysis 交流分析
transaction backout 事项复原
transaction-based routing 基于事务处理的路由选择
transaction basis 交易行为基础
transaction card 报单卡片
transaction center 流通中心
transaction code 事项码;事项处理码;事务代码
transaction command security 事项命令安全性;事项处理命令安全性;事务处理命令保密性
transaction control program(me) 事务控制程序
transaction control system 事务控制系统
transaction cost 交易成本
transaction data 更新数据;交易数据;事务数据;变动性数据
transaction data set 事项数据集;事务数据集
transaction data set record block 事项数据集记录块;事务处理数据组记录块
transaction display 事项处理显示;事务显示
transaction equation 费歇尔氏交换方程式
transaction exposure 交易风险;交易暴露
transaction file 细目文件;远行外存储器;更新文件;具体事务文件;事务文件;动态实体文件
transaction flow 交易流量
transaction for account 记账交易
transaction identifier 事务标识器
transaction journal 事项处理日志;事务日志;事务记录
transaction key 更新关键字
transaction listing 事务清单
transaction load balancing 事项处理均衡装入;事务荷载平衡
transaction log 事项登记;事务记录
transaction logging 事务处理记录
transaction management system 事务管理系统
transaction margin 交易界区
transaction motive 交易动机
transaction of near delivery 近期交易
Transaction of the American Society Journal of Civil Engineers 美国土木工程师协会会刊
transaction-oriented application 面向事项处理应用
transaction processing program(me) 事务处理程序
transaction processing system 事理系统
transaction program(me) 事项程序;事务处理程序
transaction record 事项(处理)记录;事务记录;动态实体记录
transaction record header 事项(处理)记录头标;事务处理记录标题
transaction record number 事务记录号
transaction restart 事项处理重新启动
transaction selection menu 事务处理选择表
transaction services 事项处理服务程序;事项处理服务;事务处理业务
transactions in foreign assets 外国资产的往来
Transactions of American Society of Civil Engineer 美国土木工程师协会会刊
transactions velocity 交易速度
transactions velocity of circulation 交易流通速度
transaction tape 修改带;更新数据;事务带;动态实体带
transaction tax 交易税
transaction value 交易价格
transaction with credit 信用交易
transaction work sheet 交易计算工作底稿
trans-addition 反式加成(作用)

trans-adhesive map type 透明注记剪贴字
transamination 转氨作用;氨基转移(作用)
transamine 反苯环丙胺
transargonomic structure 超氩结构
transatlantic 横渡大西洋的
trans atlantic geo-traverse 横大西洋地质断面调查
transatlantic lane 横渡大西洋航线
transatlantic port 大西洋彼岸港口
transatlantic telegraph cable 横渡大西洋海底电缆
trans atlantic trade 北大西洋航线
trans-atmospheric pollution 越界空气污染
transaudient 传声
trans-backland 中转腹地
transbasin diversion 跨流域引水;跨流域调水
transbeam 横梁
transboarder bridge 交界桥
transbooster 可调扼流圈
transborder 交界处(位于国境)
transborder bridge 渡桥;边界桥;跨界桥
transborder disposal of hazardous waste 危险废物的越界处理
transboundary air pollution 越界空气污染
transboundary movement 越境转移
transboundary river basin management 越境流域管理
transboundary transport of pollutant 越境污染物输移
transboundary waters 越界水域;跨越国界水域
trans-butene diacid 反丁烯二酸
trans-butene dioic acid 反式丁烷二酸
trans-butenediol 反丁烯二酸
transcation 更新数据
transceiver 无线收发两用机;无线电收发两用机;收发信机;收发设备;收发器;收发两用机;收发报机
transceiver circuit 收发两用电路
transceiver data link 收发器数据传送装置
transcend 超越
transcendence 超越(性)
transcendental 超越的
transcendental basis 超越基
transcendental curve 超越曲线;超越曲面
transcendental element 超越元素
transcendental equation 超越方程
transcendental estimation 超越估计
transcendental factor 超越因数;超越因素
transcendental field extension 超越扩张域
transcendental function 超越函数
transcendentality 超越性
transcendental number 超越数
transcendental surface 超越曲面
transcend the actual level of productive forces 超越实际生产力水平
transchannel effect 传递通道效应
transcocrystallized Al-Si alloy 过共晶铝硅合金
transcoder 代码转换机;变码器
transcolumn effect 传递柱效应
trans-compound 反式化合物
transconductance 跨电导;跨导
transconductance bridge 测量电子管互导的电桥
transconfiguration 反式构型
transcontainer 跨国(跨海)集装箱;远洋集装箱;国际大型标准集装箱
transcontinental 横贯大陆的
transcontinental air-mass 横跨大陆气团
transcontinental geographysical survey 横贯大陆的地球物理测量
transcontinental line 横贯大陆干线
transcontinental railroad 横贯大陆铁路
transcope 转绘仪器
transcriber 信息转换器;转点仪;再现装置;抄录器
transcribing machine 播放机
transcript 抄本;正式文本;副本
transcription 改编;抄录
transcription device 转录设备
transcription map 转录图
transcription section 地名室
transcription system 地名转写法
transcrystalline 横晶的;穿晶(的)
transcrystalline corrosion 穿晶腐蚀
transcrystalline crack(ing) 穿晶裂缝;穿晶开裂;穿晶裂纹
transcrystalline failure 跨晶断裂;穿晶断裂
transcrystalline fracture 穿晶断裂;穿晶断口

transcrystallization 横结晶
transcrystallization structure 晶变结构
transcurium 超锔
transcurrent 横冲的
transcurrent fault 横推断层;横冲断层;平移断层;侧移断层
transcurrent motion 横推运动
transcurrent thrust 横冲断层
transcycle geosyncline 串旋回地槽
trans-derivative 反式衍生物
transdifferentiation 分化转移
trans-divide diversion 流域间分流
transduce 传感
transduced element 转导因子
transduce measuring unit 换能器测量装置
transducer 换能器;换流器;通信系统;传输系统;测量变换器;转录程序;转换器;传感器;振子;四端系统;传送器;传感器;变换器
transducer array 换能器阵
transducer baseline 换能器基线
transducer beam angle 换能器波束角
transducer directivity 换能器指向性
transducer dissipation loss 换能器耗散损失
transducer element 传感元件
transducer equivalent spectrum level 换能器等效谱强级
transducer gain 换能器增益
transducer loss 换能器损失
transducer phase shift 传感器相移
transducer sensitivity 传感器灵敏度
transducer submergence 换能器吃水
transducer supply 传感器电源;变送器电源
transducer translating device 发送器变换装置
transducing piezoid 压电换能石英片;压电换能晶体;换能用压电晶体
transductant 转导子
transduction 转导
transductor 磁放大器;饱和电抗器
transductor regulator 磁放大器调节器
transect 横切;样条;横断面
transecting section method 不平行断面法
transection 横切切开;横切(面);横断
transection glacier 分叉冰川
trans-effect 反式效应
trans-elimination 反式消去作用;反式消除
trans-eluvial landscape 过渡性残积景观
transenna 教堂围屏;庙堂前栅栏围墙
transept 建筑翼部;耳堂
transept aisle 侧廊(教堂十字交叉部分的)
transeptal aspsis 教堂端半圆形
transeptal basilica 带翼厅的长方形基督教堂
transeptal chapel 教堂翼部礼拜堂
transept chapel 小教堂(从教堂十字交叉处进入的)
transeptless 无建筑翼部
transept square 十字形建筑平面交叉的方形部分
trans-esterification 酯转移
transet 动圈式电子控制仪
trans-etherification 醚交换
Trans-Eurasian seismic belt 横贯欧亚地震带
trans Europe container express 贯欧集装箱快车
Trans-Europe-Express 横贯欧洲特别快车
trans Europe express merchandises 贯欧快运业务
transafaunation 转变宿主
transfer 移交;转运;转移;转向装置;转接;转换;连续自动(化);换乘;换距;让与;切换;输送;传输;传递;变换
transferability 可转让性(财产);移动性;可移性;可动性;可传递性
transferable 可转移的;可传递的
transferable account 可转拨账户
transferable and divisible letter of credit 可转让分割信用证
transferable and separable letter of credit 可转让与可分割信用证
transferable bond 可转让债券
transferable credit 可转让信用证
transferable deposit 可转让存款
transferable discharge permit 可转让的排污许可证;容许输水流量
transferable goods 可移动财产
transferable instrument 可转让证券
transferable letter of credit 可转让信用证;可转让信用证
transferable loan facility 可转让贷款融资

transferable shares 可转让的股票
transferable water amount 可调水量
transfer address 转移地址;传送地址
transfer administrative functions to lower levels 体制下放
transfer admittance 转移导纳;传递导纳
transfer agent 转移剂;转让代理人
transfer algorithm 传送算法
transfer and stamp taxes 产权转移与印花税
transfer and transport 迁移运输
transfer apparatus 传递设备
transfer appropriation account 转账拨款账户
transfer arm 机械手
transfer arrangement (横移轧件用的)拖运装置
transfer at death 遗产转让
transfer-at-sea gear 海上传送装置
transfer bank 移送台架
transfer bar 传动杆
transfer bar mechanism 推杆机构
transfer bed 机动台架
transfer behaviour 传递特性
transfer belt 传送带
transfer between orbits 轨道过渡
transfer bin 转载仓
transfer board 转接交换台
transfer bogie 传送小车
transfer bolt cam 转栓螺栓闸刀
transfer bond 传递黏结力(混凝土和先张钢筋束间的黏结应力);传力黏结
transfer book 过户册
transfer box 转运箱;转接分线箱
transfer bridge 交界桥;渡桥;跳板
transfer buggy 转运车;传送小车
transfer bus 转接母线;转接汇流条;换接汇流排;切换母线
transfer busbar 转换母线
transfer by belt conveyer 皮带转载
transfer by shuttle tram 梭车转载
transfer cal(l)ipers 移测卡钳;移测卡规
transfer canal 输水渠;调水渠;传送管道
transfer car 运输车;转运小车;转运车;中转车;输送车;传送车;搬运车
transfer card 转移卡片;转账卡片;控制转移卡;传送卡
transfer cart 运送车
transfer car with rack pushing mechanism 齿条传送顶车机
transfer case 分动箱;变速箱
transfer-case breather 分动箱通气管
transfer-case cover 分动箱盖
transfer-case cover gasket 分动箱盖垫密片
transfer-case drain plug 分动箱放油塞
transfer case flange 变速箱凸缘
transfer-case idle shaft 分动箱中间轴
transfer-case mounting cushion 分动箱座的软垫
transfer-case mounting spring 分动箱座弹簧
transfer-case to front axle 分动箱至前车轴的传动轴
transfer-case to intermediate axle shaft 分动箱至中间车轴的传动轴
transfer-case to pillow block shaft 分动箱至主轴枕的传动轴
transfer cell 转送热室
transfer certificate 让与证书
transfer certificate of title 所有权转让证书
transfer chain 传送链条;传递链
transfer chamber 转送室;转接室;加料室;传递室
transfer characteristic 转移特性;传输特性;传动特性;传递特征;传递特性
transfer characteristic correction 转移特性校正;传输特性校正
transfer characteristic curve 转移特性曲线
transfer charges 装卸区堆场操作费;过户手续费;换装费
transfer check 转账支票;转移校验;传送校验;传送检验;传输检查
transfer chute 转运溜井;转运溜槽;传送溜槽
transfer circuit 传送电路
transfer coating 移膜涂饰;转移涂布
transfer coefficient 转移系数;转换系数;转化系数;传递系数
transfer coefficient of element 元素过渡系数
transfer column 传力柱;抽柱
transfer command 转移指令;转移命令

transfer commission 转让佣金;转让手续费
transfer company 转运公司
transfer condition 移交条件;转让条件
transfer constant 转换常数;迁移常数;替续常数;传送常数;传输常数;传递常数
transfer contact 转换触点;转换触点;切换触点
transfer container 转运容器
transfer control 转移控制;传送控制器;传送控制
transfer control instruction 传送控制指令;转移控制指令
transfer conveyer 转载运输机;转运传送机
transfer cost 转运费;转移成本
transfer crane 运送吊车;装卸桥;龙门起重机;搬运起重机;搬运吊车
transfer crane system 集装箱搬运吊车方式
transfer current 转移电流
transfer current ratio 转移电流比
transfer curve 转移曲线;转换曲线;传递曲线;传递函数曲线
transfer cylinder 递纸辊筒
transfer day 转名日;过户日
transfer decoration 移图印花装饰;转印装饰法;贴花
transfer deed 转让证书;过户契据;财产转让证书
transfer delay 转换延时;传输延迟
transfer depot 转运车站;折返站
transfer device 转换装置;转点仪
transfer diaphragm 转料隔板
transfer die 自动模
transfer distance of adsorption component 吸附组分迁移距离
transfer distance of bacteria 细菌迁移距离
transfer documents 转账凭证
transfer door 转运门
transfer drift 转运平巷
transfer earning 转移收益
transfered heat 传递热
transferee 受让方;受让者;受让人
transferee of stock 受股人
transfer efficiency 转移效率;转换效率;合金过渡系数;传递效率
transfer electrode 转移电极
transference 转移作用;转让;转卖;让渡
transference cell 交换细胞;迁移电池
transference number 迁移数
transference number of ions 离子迁移数
transference of heat 热(气)传递;热转送
transference of load 荷载传递
transfer energy 转换能
transfer equation 转移方程;转换方程
transfer equilibrium 相平衡
transfer equipment 转移设备;转换装置
transferer 转让方
transfer expenditures 转移性支出
transfer extraction 转换抽提
transfer facility 传递设施;转运设施
transfer factor 转移系数;换乘系数;传输常数
transfer fee 过户费;转运费;转让费(用)
transfer feed 连续自动送料;连续自动送进
transfer film 贴花薄膜
transfer finishing 移膜涂饰(法)
transfer floor 中转(换乘)层
transfer form 转让书
transfer forming 连续自动送进成型;传递模塑(法)
transfer frequency matrix 转移频率矩阵
transfer function 转移函数;转换函数;迁移函数;传递函数
transfer-function amplifier 传输函数放大器
transfer-function coefficient 传递函数系数;传导函数系数
transfer function measuring instrument 转移函数测量仪
transfer function model 变换函数模式
transfer function operator 传递函数算子
transfer fund 划拨基金
transfer funds to 腾出资金
transfer gallery of conveyer 输送机廊道
transfer gantry 龙门吊车
transfer gear 传动齿轮;移送机;转运装置;传导齿轮
transfer gear box 齿轮传动箱;变速箱
transfer girder 传递大梁;托梁;传力梁
transfer glass 光学玻璃块
transfer gradient 转移特性梯度
transfer grid 篦条运输机

transfer grille 送气格栅
transfer hatch 转移舱口
transfer heat 热迁移;迁移热
transfer homomorphism 转移同态
transfer hopper 转运仓;接转漏斗;输送斗;传递料斗
transfer house 接转机房
transfer hub 中转枢纽
transfer impedance 转移阻抗;跨点阻抗;传输阻抗;传递力阻抗
transfer-in box 转入箱
transfer income 转让收入
transfer-in cost 转入成本
transfer ink 转印油墨;转写油墨
transfer instruction 转移指令
transfer instrument 转换式指示仪表
transfer interpreter 传送译印器;传输翻译器
transfer investment audir 转出投资的审计
transfer jack 转接塞孔
transfer joint 转移时的临时连接
transfer key 转接电键;传送键
transfer ladle 转运浇包;铁水罐车;输送浇包
transfer lag 转移延迟;换算延迟;传递迟滞
transfer laminator 转移涂层机;转移胶合机
transfer law 传递定律
transfer ledger 过户账目
transfer length 传递长度
transfer length of prestress 预应力传递长度
transfer lettering system 图文转印材料
transfer line 连续自动式生产线;连续生产线;接送线;输送管路;传输管线
transfer linearity 转换直线性;转换线性;传输直线性
transfer line exchanger 转油线换热器
transfer locus 传递函数轨迹图
transfer lorry 转运货车
transfer loss 转移损耗;传递损失
transfer machine 自动线;连续自动工作机床;传送机;传递机
transfer-matic 自动线
transfer matrix 转移矩阵
transfer matrix method 传递矩阵法
transfer means 输送工具
transfer mechanism 移送机构;转账机构;拖运机构
transfer medium 传送介质;转换介质;代用介质;传送媒体;传递介质
transfer memory 转移存储
transfer method 移植法;转载方法;转移法;进位法;迁移取样法;迁移法
transfer mobility 传递迁移率
transfer mode 换乘方式
transfer model 迁移模式
transfer mode of plotter 绘图机传动方式
transfer moratorium 延期偿付;停止偿付
transfer motion 传动
transfer mo(u)ld 传递式模具
transfer mo(u)lding 压铸;转移造型;连续自动送进成型;传递模塑(法)
transfer mo(u)lding pressure 传递模塑压力
transfer network 转移网络
transfer node 换乘节点
transfer note 转科记录
transfer number 过户编号
transfer of accounts 转账
transfer of advanced technology 先进技术转让
transfer of a position line 船位线转移
transfer of appropriation 转拨款项
transfer of axes 坐标轴变换;坐标系转换
transfer of building line 轴线投测
transfer of building property right 房产权转移
transfer of cargo to another storage 转库
transfer of colo(u)r 沾色
transfer of conditioned reflex 条件反射传递
transfer of control 控制转移
transfer of control card 控制转移卡;程序执行启动卡
transfer of coordinates 坐标传递
transfer of development right 开发转让权
transfer of domicile registration 转户
transfer of drilling weight 钻压转移;钻压传递
transfer of elevation 标高传递
transfer of equipment 设备转移
transfer of financial resources 资金转拨;转移资金
transfer of fire 火力转移
transfer of forces 力的传递
transfer of foreign exchange risk 外汇风险转移

transfer of funds 资金转拨
transfer of inputs 投入的转移
transfer of land-use rights 土地使用权转让
transfer of lighting indication 灯光转移
transfer of line and grade 轴线和倾斜度校正
transfer of load 荷载传递;荷载转移
transfer of luggages and parcels 行李和包裹的中转
transfer of manpower 抽调劳动力
transfer of moment 力矩的传递
transfer of momentum 动量转移
transfer of money 资金转移
transfer of names 转名
transfer of pollutant 污染物的转移
transfer of prestress 预应力传递
transfer of profits 让利
transfer of property 所有权转移
transfer of quota 转让配额
transfer of refuse 垃圾(的)转运
transfer of resources 资源转让
transfer of rolling stock 车辆互用
transfer of shares 让股
transfer of skill 技术转让
transfer of technique with compensation 有偿转让技术
transfer of technology 技术转让;技术传授
transfer of title 产权让渡(土地);转让所有权;所有权的移交
transfer of wagon directly from one country railway to another 原车过轨
transfer of water in soil system 土壤中水传递
transfer of wind load 风载传递
transferometer 传递函数计
transfer operation 转移操作;传送操作
transfer optics 传递光学系统
transferor 转让者;转让人;让与人;让股人;让渡人(权利或财产);出让人;出让方
transfer orbit 转移轨道
transfer orbit sun sensor 转移轨道太阳传感器
transfer order 汇兑;导向指令
transfer (ownership) 过户
transfer package 转让一揽子
transfer padding 转移轧液
transfer pair 变换对
transfer paper 复写纸
transfer partition 隔仓板
transfer passage 移注液;放出槽
transfer passenger 换乘乘客
transfer passengers' waiting room 中转旅客候车室
transfer path 转移轨迹
transfer payment 转让支付;转拨款项;转账性支付;转账付款;转移支付;转让性付款;转拨付款
transfer payment slip 转账支票
transfer peak 传递峰值
transfer pipe 传动管
transfer pipe method 移管法
transfer pipet(te) 移液细管;移液吸(移)管
transfer piping 输送管(道)
transfer pit 转道沟
transfer plate 传送板带
transfer platform 换乘站台;移动台架;转向平台;中转站台
transfer point 中转地点;过渡点;转点【测】;转载点;转运站;转运点;转移点;转刺(点)【测】
transfer point of belt conveyer 胶带输送机转运点
transfer port 转口港;中转港(口);输送孔;输气孔;传送孔
transfer posting machine 转换定位机
transfer price 转让价格;转账价格;内部调拨价格
transfer prices among commercial enterprises 商业内部调拨价格
transfer pricing 内部调拨定价
transfer printing 移图印花;贴花印刷
transfer probability matrix 转移概率矩阵
transfer process 移图印花法;传递过程;转移过程
transfer-prohibited signal 转接禁止信号
transfer pump 转运泵;输油泵;输送泵;调驳泵;传送泵;驳运泵
transfer pumping unit 输送泵车间
transfer railway 通过铁路
transfer raise 转运溜井
transferral 转移;传递
transfer rate 转移(速)率;转换率;输移率;输送

transfer rate of information 率;传送速率;传送率;传输速率;传输率
transfer rate of information 信息位传送速率
transfer ratio 转移系数;传送系数;传送比
transfer reaction 转移反应
transferred account 转账
transferred account card 转账卡
transferred acquisition 转让取得
transferred air duct 转移风管
transferred air grill(e) 转移风口
transferred arc 转移弧;过渡电弧
transferred credit 转移信用
transferred electron generator 电子传递发生器
transferred group 转移群
transferred information 转移信息
transferred labor 转移人工
transferred position line 移线;转移(的)位置线
transferred title 转移产权
transferred value of the means of production 生产资料转移价值
transferred volume 换乘量
transferred wind 转移风
transfer register 送风调节器;传送寄存器
transfer relay 切换继电器
transfer request 传送请求
transfer request handling 转移请求处理
transfer resistance 转移电阻;传热阻
transferring course 转移横列
transferring data in storage 主存储内的数据传输
transferring mass model 转质模型
transferring (of rail) (钢轨的)移换
transferring oil 输送油
transferring passenger 转机旅客
transferring powers to units at lower levels 放权
transferring roll(er) 转移辊
transferring the tow 接拖
transferring water 调用水量;分流水量
transfer roll 中间辊
transfer routing 小运转交路【铁】
transfer saving 划拨储蓄款
transfers between associated banks 联行往来
transfer scheme 转移方案;读出和记录电路
transfer section 传送部分
transfer sequence 传输序列
transfer service 转接业务
transfer shed 周转仓库
transfer sheet 传送记录纸;复制纸
transfer signal 转移信号
transfer skirt 对接裙
transfer slide 转运溜槽
transfer slip 转账传票
transfer speed 传输速度
transfer statement 变换语句
transfer station 转运站;转输站;转换处;中转站;换乘站;传送站;传输站
transfer steamer 中转列车轮渡
transfer stitch 转移线圈
transfer strength 传力强度;传递强度
transfer strip 切换片
transfer sump 转运水仓
transfers under military grants 按照军事援助的转让
transfer surface 传递面
transfer switch 转换开关;转接开关
transfer switchboard 转接交换台
transfers with retained life estate 保留生前权益财产转让
transfer system 转运系统;传送系统;传输系统;传递系统
transfer table 移车台;转运台;转移盘;转移表;转换台;横移车;转换表;转车台
transfer table vector 转移表向量
transfer tape technique 带迁移技术
transfer tax 转让税;交易税
transfer technique 迁移取样技术
transfer tensor 辐射转移张量
transfer terminal 中转码头;换运枢纽;换装码头
transfer test 传送检验
transfer time 转移时间;切换时间;传送时间;传输时间
transfer time of byte flow 位流传输时间
transfer to another track 转线
transfer track 装卸线【铁】;转运轨道;转运道;中转线;交换车停留线
transfer track for depot 出入段线

transfer track for locomotive 机车换挂线
transfer track service 转线调车
transfer trade 转手贸易
transfer traffic volume 转移交通量
transfer train 小运转列车
transfer transaction 转让交易
transfer trip 转接释放
transfer trip cable 连跳电缆
transfer trip cable splice 连跳电缆接头
transfer tub 斗式转载列车
transfer turn table ribbon feeder 转台自动进给装置
transfer-type heat exchanger 传递式热交换器
transfer-type semi-trailer 转运式半拖挂车
transfer unit 传质单元
transfer value 转移价值
transfer valve 移注阀;输送阀;输入阀;放出阀
transfer vector 转移向量
transfer vector address 传送向量地址
transfer vector storage 传送向量存储器
transfer velocity 转移速度;传输速度
transfer voltage 转移电压
transfer voltage ratio 转换电压比
transfer voucher 转账凭证;转账凭单
transfiguration 变形
transfilter 变压器滤波器
transfinite 超限
transfinite incompleteness 超限不完全性
transfinite induction 超限归纳法;超穷归纳法
transfinite numbers 超限数
transfixation 贯穿;刺通
transfixion 贯穿术
transfixion pin 贯穿针
transflow valve 横流阀
transfluence 越流
transfluxor 多孔磁芯存储转换器;多孔磁芯存储器;多孔磁芯;磁通转换器
transfluxor memory 多孔磁芯存储器
transform 转变;折算;反式;变换式;变换;变革;反式异构体
transformability 可变换性
transform analysis 变换分析
transformant 转化体
transformat 传送格式;传输格式
transformation 转形变异;转换;转化;转变;改造;改变;换变;代换;重排作用;变换
transformational analysis of factorial data 析因数据的变换分析
transformational criterion 变换分析准则
transformational semantics 变换语义
transformation apparatus 转换仪器;改造机
transformation base 转换基面
transformation by reciprocal direction 倒方向变换
transformation calculus 变换微积分
transformation chain 中转变链
transformation coefficient 归化系数;变换系数
transformation computation 换算
transformation constant 转化常数;变换常数
transformation cross-section 换算截面
transformation curve 转换曲线
transformation definition language 变换定义语言
transformation dislocation 相变位错
transformation energy 变换能量
transformation equation 转换方程;变换方程
transformation factor 转化系数
transformation family 变换族
transformation formula 变换公式
transformation framework 变换结构框架
transformation function 转换函数;变换函数
transformation gain 变换增益
transformation granite 改造花岗岩
transformation group 变换群
transformation hysteresis 变换滞后
transformation induced plasticity 相变诱起塑性;高强度及高延性
transformation induced super plasticity 相变诱发的超塑性
transformation instrument 转换仪器
transformation load 换算荷载;变换荷载;临界荷载
transformation loop 变换回路
transformation loss 变换损耗
transformation matrix 转置矩阵;变换矩阵
transformation member 变换构件
transformation model 变换模型

transformation of an air mass 变性气团
transformation of atmospheric nitrogen 大气氮的转变
transformation of axis 轴的变换
transformation of city 城市改造
transformation of coefficient 系数的变换
transformation of colo(u)r 色彩调节
transformation of coordinates 坐标变换
transformation of data 资料变换;数据变换
transformation of energy 能的转化
transformation of flood wave 洪水波变形
transformation of function 函数变换
transformation of geologic(al) variable 地质变量的变换
transformation of inverse hyperbolic tangent 反双曲正切变换
transformation of logarithmic 对数变换
transformation of parabola 抛物线的变换
transformation of pollutant 污染物输移;污染物的转化
transformation of river pattern 河型转化
transformation of scale 比例变换;比尺变换
transformation of series 级数变换
transformation of similitude 相似变换
transformation of square root 平方根变换
transformation of tensor 张量变换
transformation of tide wave 潮波变形
transformation of variables 变数(的)变换;变数代换
transformation of wave 波浪变形
transformation on variables 变量的变换
transformation operator 变换算子;变换算符
transformation parameter 变换参数
transformation period 转化周期;转变期;变换周期
transformation plasticity 相变塑性
transformation point 相变点;转变点;变变点;变换点
transformation principle 转化因素
transformation program(me) 变换程序
transformation range 转变范围
transformation rate 转换率;转化率;变换速率;变换速度;变换率
transformation ratio 油气转化率;变比
transformation ratio of chloroform bitumen A 氯仿沥青A的转化率
transformation ratio of hydrocarbon 烃转化率
transformation ratio of kergen 干酪根转化率
transformation reaction 转变反应
transformation relation 变换关系
transformation resistance 变换电阻
transformation rule 变换规则
transformation series 放射性系;放射系列;变换系
transformation space 变换空间
transformation stage 转化阶段
transformation stress 相变应力
transformation substance 转化物质
transformation table of commodity information 矿产信息转换表
transformation table of geologic(al) information 地质信息转换表
transformation temperature 相变温度;相变点;转变温度;变换温度
transformation temperature range 转变温度范围;转变温度区
transformation tensor 变换张量
transformation theory 变换论;变换理论
transformation-toughened zirconia 相变增韧氧化锆
transformation toughening 相变增韧
transformation twin 转变双晶;转变孪晶
transformation unknown 变换未知数
transformation voltage ratio test 变压比试验
transformative transducer 变换传感器
transform boundary 转换边缘;转换边界
transform diagram 转移图
transformed air-mass 变性气团
transformed area 换算面积
transformed catenary 变形悬链线
transformed coordinate system 变换坐标系
transformed cracked section 按开裂换算的截面
transformed (cross-)section 换算截面;变换横截面
transformed current 变性海流
transformed field equations 变换场方程
transformed flow net 转换流网;变换流网

transformed lattice structure 变换点阵结构
transformed network 变换网(络)
transformed nucleus 变换原子核
transformed phase composition 变换相成分
transformed print 转换像片
transformed rock 变质岩
transformed schist 变质片岩
transformed section 换算截面;转换截面
transformed section method 变换截面法
transformed static load 换算静载(重)
transformed theory 变换论
transformed value 变换了的数值
transformed wave 变相波
transform equipment 变电设备
transformer 互感器;大货车;变压器;变量器;变换器
transformer action 变压作用;变压器作用;变压器效应
transformer active power 变压器有效功率;变压器有功功率;变压器消功功率
transformer and switch room 变电室
transformer array 变压器阵列
transformer bank 变压器组
transformer barrier tube 变压器绝缘套筒
transformer bay 变压器间
transformer board 变压器用纸板
transformer box 变压器箱
transformer bracket 变压器支架
transformer bridge 变压器电桥;变换器电桥
transformer build 变量器构型
transformer bus 变压器母线
transformer-busbar scheme 变压器母线组接线
transformer bushing 变压器进线套管
transformer capacity 变压器容量
transformer case 变压器外壳
transformer C-core reeling machine 变压器C型铁芯卷绕机
transformer center tap 变压器中点抽头
transformer coil 变压器线圈
transformer coil winding paper 变压器线圈缠绕用纸
transformer connection 变压器接线组
transformer container 变压器容器
transformer control room 变电控制室
transformer core 变压器铁芯
transformer-coupled amplifeir 变压器耦合放大器
transformer-coupled direction finder 变压器耦合测向仪
transformer-coupled load 变压器耦合负载
transformer-coupled oscillator 变压器耦合振荡器
transformer-coupled power amplifier 变压器耦合功率放大器
transformer-coupled pulse amplifier 变压器耦合脉冲放大器
transformer-coupled stage 变压器耦合级
transformer coupling 电感耦合;变压器耦合
transformer-distribution station 变配电站
transformer divider 变压器分压器
transformer drying methods 变压器干燥法
transformer efficiency 变压器效率
transformer electromotive force 变压器电动势
transformer-fed deflector coil 变压器馈电偏转线圈
transformer feedback 变压器反馈
transformer filter 变压器滤波器
transformer for electric(al) furnace 电炉用变压器;电炉变压器
transformer for withstand voltage test 耐压试验变压器
transformer gallery 变压器廊
transformer hall 变压器室
transformer house 变电室
transformer inductance 变压器电感
transformer kiosk 变压器亭
transformer lead 变压器引线
transformer line unit connection 变压器线路组结线
transformer load loss 变压器有载损失;变压器负载损耗
transformer locomotive 变流机车
transformer loss 变压器损耗
transformer matching 变压器匹配
transformer modulation 变压器调制
transformer noload losses 变压器空载损耗
transformer normal tapping ratio 变压器额定变比
transformer oil 变压器油
transformer oilproof board 变压器用防油纸板
transformer oil resistance 耐变压器油性
transformer overcurrent trip 变压器过流跳闸装置
transformer overload 变压器过载
transformer pillar 变压器柱
transformer pit 变压器洞室
transformer pole 变压器杆
transformer protective screen 变压器保护屏
transformer pulse response waveform 变压器脉冲响应波形
transformer ratio 变压系数;变压比;变流比
transformer ratio-arm bridge 变量比率臂电桥
transformer-ratio bridge 变压比电桥
transformer ratio test 变压比试验
transformer reactive power 变压器无功功率
transformer read-only store 变压器只读存储器
transformer/rectifier unit 整流变压器单元
transformer regulation 变压器调节
transformer relay 变压器继电器
transformer room 变压器室;变压器房;变电间
transformers 变形金刚
transformer-secondary voltage 变压器次级电压
transformer sheet 变压器钢板
transformer shell 变压器外壳
transformer silicon steel sheet 变压器矽钢片;变压器硅钢片
transformer size 变电站容量
transformer stamping 变压器冲片
transformer station 变电站
transformer steel 变压器钢
transformer (sub)station 变电站;变电所
transformer substation of harbo(u)r 港区变电所
transformer substation of port area 港区变电所
transformer switch 变压器开关
transformer tank 变压器箱;变压器柜
transformer tapping 变压器抽头
transformer testing instrument 变压器测试仪
transformer-thermal converter voltmeter 变压器热转换器伏特计
transformer tuning capacitor 变压器调谐电容器
transformer turbine 涡轮变压器
transformer turn-to-turn insulation test 变压器匝间绝缘试验
transformer-type ammeter 互感器式安培计
transformer type coupling filter 变压器型耦合滤波器
transformer undercurrent trip 变压器欠流跳闸装置
transformer underload trip 变压器欠载跳闸装置
transformer utilization factor 变压器利用系数
transformer vault 变压器室;变压器窖
transformer voltage ratio 变压器电压比;变压器变比
transformer winding 变压器绕组
transformer with natural cooling 自然冷却式变压器
transformer with split windings 分裂变压器
transformer with tap changer 带有抽头变换器的变压器
transformer yard 变压器场
transform fault 转换断层
transform formula of Kossakin 库沙金变形公式
transforming agent 转化因素;转化介质
transforming amplitudes 变换振幅
transforming Dupuit well formula 变换的裴布衣水井公式
transforming factor 转化因子
transforming function transformation function 变换函数
transforming meshwise 网状变换
transforming principle 转变因素
transforming printer 航空照片纠正仪
transforming section 变换段
transforming station 变电站;变电所
transforming valve 减压阀
transform integrals 变换积分
transformism 变成论
transform layer 变换层
transform operator 变换算子
transform operator pair 变换算子对
transform optics 变换光学
transform pair 变换对
transform phase multiimpule method 改变相位多脉冲法
transform plate boundary 转换板块边界【地】
transform processing 变换域处理
trans-frontier pollution 超国境污染;越境污染
transfuse 移注
transfusion 过熔作用;倾注
transfusion way 过熔作用方式
transgranular 晶内;穿晶(的)
transgranular crack 粒间贯通裂纹;穿晶裂纹
transgranular fracture 粒内破裂;晶穿断裂
transgressing continental sea 进浸大陆海
transgressing sea 海侵地区
transgression 海进;亲和性转移;超覆;超度
transgression of sea 海侵;侵陆海
transgression phase 海进相
transgression regression 海侵海退
transgressive 海浸的;海进的
transgressive cycle 海侵旋回
transgressive deposit 海侵沉积物
transgressive facies 海侵相;海进相
transgressive intrusion 贯层侵入;不整合侵入
transgressive overlap 海侵超覆;海进交错;海进超覆;入侵超覆
transgressive phase 海侵相;海进相
transgressive reef 海侵礁;超覆礁
transgressive rhythm 水进韵律
transgressive sediment 海侵沉积物
transgressive sequence 海侵层序;海进层序;水进式层序
transgressive series 海侵岩系
transgressive sill 海侵底脊
transgressive succession 海进序列
transhipment entry 转运报单
transhipment free entry 免税品进口报关单
transhipment manifest 转运舱单
transhipment permit 转运许可证
transhipment shipping bill 货物转船清单
transhorizon propagation 超地平传播
transience 暂时性
transient 暂态的;暂时性的;过渡的;瞬态声;瞬态;瞬时的;瞬态值;瞬变(的)
transient ablation area 暂时消融区
transient acceleration 瞬时加速度
transient accumulation area 暂时堆积区
transient action 瞬时作用
transient action force 瞬时作用力
transient aerodynamics 瞬变空气动力学
transient agent 暂时性毒剂
transient analysis 暂态分析;瞬时分析
transient analyzer 过渡过程分析器;瞬态分析器;暂态分析仪;瞬变(特性)分析器
transient area 暂存区;暂批存储区;非常驻程序区
transient area control table 过渡区控制表
transient area descriptor 过渡区描述符
transient armature current 定子瞬变电流
transient behavio(u)r 暂态特性;瞬时状态;瞬时特性;瞬变行为;瞬时性态
transient biont 暂时生物
transient boiling 瞬态沸腾
transient boiling range 过渡沸腾区
transient buildup 瞬态结构
transient bursting water 瞬时突水
transient calculation 瞬态计算
transient calibration 瞬时校准
transient calibration pulse 瞬时校准脉冲
transient-cause forced outage 暂态性故障
transient cavity 瞬时气穴;瞬时空穴
transient circuit 瞬态电路
transient coherence 瞬时相干性
transient condition 过渡工况;瞬态条件;瞬变条件;瞬变工况;瞬变状态
transient contact 过渡性接触
transient control executive area 过渡控制执行区
transient creep 过渡蠕变;瞬态蠕动;瞬时徐变;瞬时蠕变
transient creep curve 趋稳蠕变曲线
transient current 暂态电流;过渡电流;瞬态电流;瞬时电流;瞬变电流
transient current efficiency 瞬时电流效率
transient current offset 瞬态电流补偿
transient current velocity 瞬时流速
transient curve 过渡曲线;瞬态曲线
transient data 过渡过程数据;瞬态数据;瞬变数据
transient-decay current 瞬态衰减电流
transient decay rate 瞬时衰变率
transient deflection 瞬时弯沉

transient deformation 瞬时形变;瞬时变形
transient delay 过icze延时;瞬时延迟
transient deviation 瞬态偏差
transient digitizer 瞬态数字转换器
transient dipole moment 瞬时偶极矩
transient directory 暂驻目录
transient discharge 瞬时排放
transient displacement 瞬时位移
transient distortion 瞬态失真;瞬态畸变;瞬时畸变
transient distribution 瞬态分布
transient disturbance 瞬变扰动
transient-divided reset 瞬时分段复位
transient driving force 瞬态推力
transient earth resources phenomenon 瞬态地球资源现象
transient effect 暂态效应;瞬变作用;顺变作用
transient electromagnetic instrument 过渡场电法仪
transient equilibrium 暂时平衡;过渡平衡;瞬间平衡
transient error 暂态误差;过渡误差;瞬态误差;瞬时误差
transient event 暂现事件
transient excitation 瞬态激振
transient experiment 瞬态实验
transient failure 瞬时失效
transient fault 瞬態故障;瞬时故障
transient feedback 暂态反馈;软反馈
transient field current 瞬态励磁电流
transient field method 瞬变场法
transient-flight data 瞬时飞行数据
transient flight path 瞬变航迹
transient floating point gain control 瞬时浮点增益控制
transient flow 瞬变流(动);非定常流;不稳流动;不稳定流动
transient flow permeability 瞬时流动透气性;瞬变流透气率
transient flux-linkage 瞬间磁链
transient force 瞬态力;瞬时力
transient-free 无瞬变的
transient function 瞬时函数
transient Gaussian process 瞬态高斯过程
transient generation 瞬变产生
transient generator 暂态发电机
transient geomagnetic variation 瞬地磁变
transient growth 瞬时生长
transient hardness 暂时硬度
transient harmonics 瞬间谐波
transient heat conduction 瞬态热传导;不稳定传热
transient heat conduction equation 瞬时热传导方程
transient heat flow 瞬时热传递
transient heating condition 过渡热状态
transient hot channel factor 瞬态热管因子
transient hotel 中转旅馆;暂住性旅馆;暂时性旅馆
transient hot spots 瞬变热点
transient housing 瞬时住房
transient image 瞬息图像
transient impedance 瞬变阻抗
transient indicator 瞬時指示器;瞬变过电压指示器
transient intermediate 瞬时中间产物
transient internal voltage 瞬变电动势
transient irradiation 瞬时辐照
transient laser behavio(u)r 瞬态激光特性
transient laser response 瞬变激光响应
transient library 暂驻程序库
transient liquid sintering 过渡液压烧结
transient load 瞬载;瞬时荷载;瞬时负荷;瞬变荷载
transient loading 临时荷载
transient lodging facility 临时居住设施
transient lunar phenomenon 月球暂现现象
transient magnetic variation 瞬磁变化;瞬磁变
transient measurement 瞬变测量
transient melting 瞬时熔化
transient method 暂法;瞬态法;瞬变场法
transient microbe 过渡微生物
transient micropulsation signal 瞬微脉动信号
transient moment 瞬时力矩;过渡期内力矩
transient motion 暂态运动;瞬态运动;瞬时运动;瞬变运动;不稳定运动
transient nature 瞬时性质
transient noise 瞬态噪声
transient occupancy 临时使用;临时居住
transient off time 瞬时断开时间

transient operation 瞬间操作
transient oscillation 瞬态振荡;瞬时振荡
transient output 不稳定功率
transient overload current 瞬态过载电流
transient-overload forward-current limit 瞬态过载反向电流限制
transient-overload reverse-voltage limit 瞬态过载反向电压限制
transient overshoot 瞬时过调量;瞬时过冲
transient overshooting 瞬时过冲
transient overspeed 瞬态超速
transient overvoltage 瞬态超压
transient overvoltage counter 瞬时过电压计数器
transient part 瞬变部分
transient peak 合闸尖峰电流;瞬间峰值
transient peak-inverse voltage 瞬态反相峰值电压
transient peak loading 瞬时峰荷
transient peak value 瞬时峰值
transient performance 瞬态特性
transient period 过渡周期;瞬变周期
transient phase distortion 瞬态相位失真
transient phenomenon 暂时现象;过渡现象;瞬变现象
transient plasma 瞬间等离子体;不稳定等离子体
transient point 变态点
transient population 暂留人口;暂住人口
transient pore pressure 瞬态孔隙压力
transient power 瞬态功率
transient pressure 过渡压力;瞬态压力;瞬时压力;瞬变压力
transient pressure pulse 瞬变压力脉冲
transient problem 瞬态问题
transient process 过渡过程;过渡步骤;瞬态过程;瞬变过程
transient program(me) 暂用程序;过渡程序;临时程序
transient program(me) area 过渡程序区
transient program(me) library 过渡程序库;临时程序库
transient program(me) table 过渡程序表
transient propagation delay 瞬态传播延迟时间
transient property 瞬变特性
transient pull 瞬时牵引力
transient pulse 瞬态脉冲
transient pumping test 瞬时抽水试验
transient radiation damage 瞬时辐照损伤
transient radiation effects 瞬时辐照效应
transient reactance 瞬态电抗;瞬变电抗
transient reaction 瞬变反应
transient reactor test facility 瞬态反应堆试验装置
transient recorder 瞬态记录器
transient recording 瞬态记录
transient recovery voltage 瞬时恢复电压;瞬变恢复电压
transient regime 瞬变状态
transient region 过渡区
transient release 瞬时释放
transient residential building 临时居住建筑
transient resistance 瞬态电阻
transient resonance 瞬态共振;瞬时谐振
transient response 过渡反应;瞬态响应;瞬态反应;瞬时反应;瞬变响应
transient response analysis 瞬态响应分析
transient-response characteristic 瞬变过程特性;非定常过程的特征
transient response data 瞬态响应数据
transient-response efficiency 瞬变负荷效率
transient response model 瞬态响应模型
transient rod 瞬变棒
transient routine 过渡例行程序
transient saliency 瞬态凸极性
transient saturation 瞬态饱和
transient seepage flow 不稳定渗流
transient (self-) recorder 暂时自动记录器
transient service 临时检修;小修
transient shock wave 瞬变激波
transient short-circuit 瞬态短路
transient situational disturbance 短暂情境障碍
transient solution 瞬时解
transient sounding method 瞬变测深法
transient-speed characteristic 瞬变过程中的转速特性
transient stability 瞬态稳定;动态稳定性
transient-stability limit 动态稳定极限

transient stability margin 暂态稳定储备余量;暂态稳定储备系数;瞬时稳定储备系数
transient stability power limit 瞬态稳定功率极限
transient stability region 瞬时稳定区
transient stage 瞬态级
transient state 暂态;过渡(状)态;瞬态;瞬时状态;非稳定状态
transient state analysis 瞬态分析
transient state stability 瞬态稳定度
transient state temperature 暂态温度
transient state vibration 瞬态振动;非定常振动
transient statistics 瞬变统计学
transient stop exploration 暂缓勘探
transient storage area 暂驻存储区
transient strain 瞬态应变;瞬时应变
transient stress 瞬态应力;瞬时应力;瞬变胁强
transient stress-energy modulus 瞬态应力能模量
transient study 瞬态研究
transient suppressor 瞬变抑制器;电涌抑制器
transient surface deflection 瞬时路面挠度;瞬时表面变形
transient swing 暂态波动;瞬时摇摆
transient system deviation 瞬态系统偏差
transient technique 瞬变技术
transient temperature 瞬时温度;瞬变温度
transient temperature distribution 瞬态温度分布
transient temperature gradient 过渡温度梯度
transient temperature rise 瞬态温升
transient term 暂时项
transient thermal blooming 瞬变热晕
transient thermal defocusing 瞬变热散焦
transient thermal impedance 瞬态热阻抗
transient thermal response 瞬时热效应
transient thermal selffocusing 瞬变热聚焦
transient thickening 暂时稠化;暂时变稠
transient time 过渡时间;瞬态时间;瞬变时间
transient time-constant 瞬变时间常数
transient torque 瞬时转矩
transient tracer concentration 顺变示踪浓度
transient tracers in the ocean 海洋中的瞬变示踪剂
transient trajectory 瞬变轨道
transient trip 瞬变过程解扣
transient unusable reserves 暂不能利用储量
transient variation 瞬时变化;顺变
transient vibration 暂态振动;瞬态振动;瞬时振动;瞬变振动
transient voltage 过渡电压;瞬态电压;瞬时电压;瞬变电压
transient voltage regulation 瞬态电压调整
transient wave 瞬时波;瞬变波
transient waveform 瞬变波形
transient well test 不稳定试井
transient X-ray source 暂现X射线源;暂态X射线源
transillumination 迆光试验;透照法;透射(法);透穿照射
transim 船用卫星导航装置
transimpedance 跨阻抗;跨导倒数
transinformation 转移信息;跨信息;传送信息;传输信息;传递信息
transinformation content 传送信息量
transinformation rate 传送信息速率
trans-interchange 反位转移作用;反式交换
transire 沿岸运输许可证;货物放行证
trans-isomer 反式异构体
trans-isomeride 反式异构体
trans-isomerism 反式异构现象
transistance 跨阻抗作用;晶体管作用;晶体管效应
transister = transistor
trans-isthmian highway 横穿海峡公路
transistor 晶体管;半导体管
transistor ag(e)ing 晶体管老化
transistor-alpha meter 晶体管放大系数测定器
transistor amplifier 晶体管放大器
transistor array 晶体管阵列
transistor biasing 晶体管偏置
transistor bottom view 晶体管引线连接
transistor characteristics 晶体管特性
transistor chip 晶体管薄片
transistor chopper 晶体管斩波器
transistor clipping circuit 晶体管限幅电路
transistor curve tracer 晶体管特性曲线描绘仪
transistor encapsulation welder 晶体管封焊机
transistor frequency multiplier 晶体管倍频器

transistor gain 晶体管增益
transistor inverse power supply 晶体管逆变电源
transistorized measuring amplifier 晶体管式测量放大器
transistorized memory 晶体管存储器
transistorized millivoltmeter 晶体管毫伏表
transistorized multimeter 晶体管万用电表
transistorized ultrasonic thickness indicator 晶体管超声波测厚仪
transistorized version 晶体管化线路
transistorized voltage regulator 晶体管稳压器
transistor life 晶体管寿命
transistor limiter 晶体管限幅器
transistor locator 晶体管测位器
transistor meter 晶体管测试器
transistor noise coefficient 晶体管噪声系数
transistor notation 晶体管符号
transistor oscillator 晶体管振荡器
transistor package 晶体管外壳
transistor radio 晶体管收音机
transistor rectification arc welder 晶体管整流弧焊机
transistor relay 晶体管继电器
transistor scaler 配置晶体管的脉冲计数器
transistor switch 晶体管换接器
transistor switching circuit 晶体管开关器
transistor tester 晶体管测试仪
transistor transmitter 晶体管发射机
transistor trigger 晶体管触发器
transistor video amplifier 晶体管视频放大器
transistor voltmeter 晶体管伏特计
transit 纵转【测】;转口;转接;中转(客货运输);中星仪;中天;过境;过渡;经纬仪;通过;串视
transit and bearing 串视线和方位
transit-and-stadia survey 视距经纬仪测量
transit area 转运区(域);中间集结区
transit authority 运输管理机构
transit axis 水平轴;传动轴
transit beacons 叠标(导航)
transit bearing 叠标方位;串视方位
transit bond 过境保函
transit call 转接呼叫;渡越呼叫
transit camera 摄影经纬仪
transit cargo 转运货;中转货物;过境货(物)
transit car without resorting 无调中转车
transit car with resorting 有调中转车
transit center 转接中心
transit charges 中转费用;中转费;过境费
transit circle 子午仪;子午环
transit clause 运输条款;转运条款
transit cold store 中转性冷库
transit company 转运公司;中转公司
transit compass 转镜仪;转镜经纬仪;经纬仪罗盘
transit concrete mixer 混凝土运拌车;混凝土运送拌和车
transit crane 全转式起重机
transit curve 汇流曲线
transit declaration 过境申报单
transit declinometer 经纬仪式磁偏计
transit depot 转运仓库;临时仓库
transit distribution facility 中转运调设备
transit Doppler positioning 子午仪卫星多普勒定位
transit dues 过境税;通行税;通过税
transit duty 过境税;通行税;通过税
transite 石棉水泥防火板;石棉水泥板;波纹石棉水泥瓦
transit entry 过境报关手续
transit equipment 转接设备
transiter 中天记录器
transit exchange 转接站;转接局;变换站
transit expressway 过境快速公路;过境快速道路
transit facility 转接设备
transit fee 过境费
transit field method 瞬变场法
transit flow 转速通量
transit form 转报底
transit formality 过境手续
transit godown 临时仓库;码头仓库
transit goods 转口货物
transit harbo(u)r 转口港
transit imaging 瞬变成像
transit improvement program(me) satellite 子午仪改进卫星
transit instrument 经纬仪;中星仪

transit interest 中转利息
transit intersection field book 经纬仪交会法测量观测手簿
transition 转变段;过渡;换接过程;换接;渡越;变迁
transition account 跨期分摊账户
transition aeration sludge 过渡曝气污泥
Transitional 中石器时代
transitional 过渡的;瞬变
transition(al) altitude 过渡高度
transition(al) architecture 过渡式建筑
transition(al) area 过渡使用地区;过渡区
transitional area from syncline to anticline 向斜到背斜的过渡区
transition(al) association 过渡性植物群丛
transition(al) basin 过渡港池
transition(al) bed 过渡阶段河床;过渡阶段河槽;过渡层
transition(al) belt 过渡(地)带;过渡传送带
transition(al) bog 过渡性沼泽
transition(al) boiling 过渡沸腾
transition(al) boundary 过渡边界
transition(al) capacitance 过渡电容
transition(al) card 过渡卡
transition(al) circular curve 过渡曲线
transition(al) climate 过渡性气候;过渡型气候
transition(al) coding 过渡编码
transition(al) coil 过渡线圈
transition(al) condition 过渡状态
transition(al) cone 过渡锥
transition(al) contact 过渡接触
transitional credit 过渡性信贷
transition(al) crust 过渡壳
transition(al) curve 过渡曲线
transition(al) delta-plain 过渡性三角洲平原
transition(al) delta-plain deposit 过渡性三角洲平原沉积
transition(al) depth 过渡水深
transition(al) effect 过渡效应
transition(al) element 过渡元素
transition(al) element analysis 过渡元素分析
transition(al) environment 过渡环境
transition(al) facies 过渡相【地】
transition(al) facies zone 过渡相带
transition(al) fiber 过渡光纤;过渡光纤
transition(al) fit 过渡配合
transition(al) fittings 过渡配件
transition(al) flight 过渡飞行
transition(al) flow 过渡流
transition(al) flow regime 过渡流态
transition(al) form 过渡形式;过渡型;过渡类型;过渡形状
transition(al) frequency 过渡频率
transitional grade (zone) 缓和坡段
transitional gradient 缓和坡段
transitional grading 缓坡处理;(融合地形景色的)坡面修整
transition(al) hopper 过渡斗仓
transition(al) housing 过渡室
transition(al) joint 过渡连接(汽液)
transition(al) lattice 过渡晶格;过渡点阵
transitional layer 转变层
transition(al) leading marks 过渡导标
transition(al) length 过渡长度
transitional lift 瞬变升力
transition(al) line 过渡线
transition(al) map 过渡图
transition(al) measure 过渡措施;缓和措施
transition(al) metal 过渡金属元素;过渡金属
transition(al) metal carbide 过渡金属碳化物
transition(al) metal ions 过渡金属离子
transition(al) metal nitride 过渡金属氮
transition(al) moor 过渡性泥炭沼泽
transitional neighbourhood 演变中的邻里
transition(al) part 过渡段
transition(al) payment 过渡性支出
transition(al) period 过渡周期;过渡时期;过渡期
transition(al) phase 过渡相;过渡阶段
transition(al) phenomenon 过渡现象
transition(al) point 过渡点
transition(al) population 过渡群体
transition(al) primer 过渡涂层
transition(al) profile 过渡段水面曲线;渐变剖面
transitional provision 过渡性条款
transition(al) range 过渡区;过渡距离

transition(al) range marks 过渡导标
transitional reach 河道过渡段
transitional reach outside entrance 口门外衔接段
transition(al) redox environment 过渡氧化还原环境
transition(al) regime 过渡水流动态
transition(al) region 过渡区域;过渡区
transition(al) regulation 过渡调节
transition(al) resistance 过渡电阻
transition(al) resistor 过渡电阻
transition(al) response 过渡特性;暂态响应
transition(al) rock series 过渡岩系
transition(al) rough zone 过渡粗糙区
transition(al) season 过渡季节
transition(al) section 过渡截面;过渡断面;过渡段;过渡部分
transition(al) series of series 过渡岩系
transition(al) settlement 过渡性住宅区
transition(al) slab 过渡路面板;渐变板
transition(al) sleeve 过渡套管
transition(al) soil 过渡性土壤
transition(al) span 过渡孔
transition(al) stage 过渡阶段;渐变阶段
transition(al) state 过渡状态;过渡态
transition(al) state of flow 过渡流态
transition(al) stops 过渡停止位置
transition(al) structure 过渡组织;过渡结构
transition(al) style 过渡风格;过渡式
transition(al) surface 过渡表面;缓和表面
transition(al) swamp 过渡性沼泽
transition(al) time 过渡时间
transition altitude 转换高度
transitional track 转换线
transitional tropic(al) Atlantic air-mass 热带大西洋过渡气团
transitional tropic(al) Gulf air-mass 热带墨西哥湾过渡气团
transitional tropic(al) maritime air-mass 热带海洋过渡气团
transitional tropic(al) Pacific air-mass 热带太平洋过渡气团
transitional tropic(al) superior air-mass 热带过渡高空气团
transitional type 过渡类型
transitional type delta 过渡型三角洲
transitional type formation 过渡型沉积建造
transitional water 过渡水
transitional waters 过渡水域;过渡水区;中间水区
transitional yard 过渡庭院
transitional zone 过渡区域;过渡区;过渡段;过渡地区;过渡地带;过渡带
transitional zone deposit 过渡带沉积
transitional zone extension 过渡区延伸段
transition arch 过渡发券;过渡拱
transition assertion 变迁断言
transition belt 给料皮带机
transition between energy level 能级间的跃迁
transition between gradients 不同坡度间的过渡段;竖向缓和曲线
transition between process states 各进程状态之间的转换
transition card 转移卡;转换卡片;控制转移卡;控制转换卡
transition cell 转变点电池
transition chamber 进料室;进料端窑头罩
transition circular curve 缓和曲线
transition coding 变换编码
transition coefficient 跃迁系数
transition coil 分流线圈
transition condition 暂态;暂时状态
transition count test method 变换计数测试法
transition coupling 过渡配合
transition course 过渡层
transition curvature 缓和曲率
transition curve 过渡曲线;转变曲线;缓和曲线变更率;缓和曲线
transition curve location 缓和曲线测设
transition curves between gradients 竖向缓和曲线
transition curve with varied length 不等长缓和曲线
transition energy 转变能;临界能量
transition experiment 转换实验
transition fishplate 异型鱼尾板
transition fittings 特种接头

transition flare 喇叭形过渡段
transition flow 渐变流;瞬变流(动);不稳(定)流动
transition frequency 跃迁频率;过渡频率;交界频率;交叉频率
transition front 相变前沿
transition heat 转换热;转化热;转变热
transition housing 进料室;进料端窑头罩
transition impedance 转移阻抗
transition in cylindric(al) shells 柱状壳体的过渡
transition interval 转变间隔
transition jet 缓和射流
transition joint 异型接头;中间接头
transition knuckle 折边
transition lake 转变中的湖泊
transition law 转移律
transition layer 过渡层;转变层
transition layer of salinity 盐度跃(变)层
transition layer of temperature 温度跃(变)层
transition length 过渡长度;转变长度;缓和曲线长度
transition level 转换高度
transition lighting 过渡照明
transition lime 过渡石灰
transition line 转移线
transition line analogy 传输线模拟
transition link 转换连接
transition loss 转变损失;渐变段损失;渡越损失
transition mast 转换柱
transition material 半透水材料;转换材料
transition matrix 转换矩阵
transition moment 跃迁矩
transition network grammar 转移网络文法
transition nozzle 变径喷管
transition of double socket 双承异径管;双承大小头管节;双承大小头
transition of electron 电子迁移
transition of shots 镜头转换
transition of single socket 单承大小头
transition order 转变次序
transition patching 临时(性)修补
transition phase 转变阶段
transition pipe 过渡管段;大小头管段;异径管;转换导管;渐变管
transition point 转运点;转换点;转变点;临界点
transition polymerization 逐步聚合
transition position 转变位置
transition potential 转变点电位
transition probability 跃迁几率;跃迁概率;转移概率
transition probability law 转移概率律
transition probability matrix 转移概率矩阵
transition probability system 转移概率系统
transition process 转变过程
transition radii on gradients 竖向缓和曲线半径
transition range 转变范围
transition rate 跃迁率
transition ratio 转换比(率)
transition region 过渡区;渐变区;渐变段;渡越区
transition rock 过渡岩层
transition rule 转移规则
transition section 转换段;过渡区;渐变段
transition slab 桥头搭板;过渡性路面板
transition slab at bridge head 桥头搭板
transition soil 过渡土(层)
transition spiral 螺旋形过渡曲线;螺旋形缓和曲线;缓和曲线
transition stage 转变状态
transition state 中间状态;变换状态
transition state theory 转变态理论
transition symbols 转移符号
transition system 转换系统
transition table 转换表;传递表
transition tangent 缓和切线
transition temperature 转脆温度;转变温度;脆性转变温度;变迁温度
transition through forbidden band 禁带跃迁
transition tip 换接电极头
transition tube 渐变管
transition-type folds 过渡型褶皱
transition wood moor (soil) 过渡森林沼泽土
transition zone 过渡区段;渐变段;转变区;转变带;渐变区;缓和段;渡越区
transition zone of cross section 断面渐变段
transition zone of curves 缓和曲线段
transition zone of curve widening 加宽缓和段
transition zone of illumination 光过渡

transitive 过渡的;传递的
transitive closure 传送闭包;传递闭包
transitive closure of relations 关系的传递闭包
transitive dependency 传递依赖;传递相关性;传递从属性
transitive digraph 可迁有向图
transitive graph 传递图
transitive relation 传递关系
transitive relationship 传递关系
transitive system of index numbers 可迁指数体系
transitivism 互易感觉
transitivity 可迁性;可传递性;传递性
transitivity of strategic equivalence 策略等价的传递性
transit lane 过境交通专用车道
transit letter of credit 转口信用证
transit level 经纬水准仪
transit line 视线;照准线;经纬仪导线
transit line productivity 交通线路生产率
transit load profile 公共交通载客量图
transit loss 转接损耗
transit magnetism 非永久磁性
transit mall 有公交的步行街;公交步行街
transit man 导线测量员;经纬仪测量员;测量员
transit marks 叠标(导航);导标
transit meridian 子午圈
transit method 中天法
transit micrometer 子午仪测微器
transit-mixed concrete 车拌混凝土;路拌混凝土;运送(车)拌和混凝土;运输搅拌混凝土;运拌混凝土;车运搅拌混凝土
transit-mixed concrete batching plant 车拌混凝土批量配料设备
transit mixer 移动式搅拌机;运送搅拌车(混凝土);运送拌和机(混凝土);自动搅拌机;混凝土运输搅拌车
transit-mixer truck 混凝土搅拌车
transit-mixer vehicle 搅拌式运料车(混凝土);混凝土运输车
transit-mixing concrete plant 车拌混凝土设备
transit-mixing railcar 有轨运送拌和车
transit-mixing truck 搅拌运料车;搅拌车;商品混凝土搅拌车
transit mixture 在途搅拌混合料
transit navigation system 子午仪导航系统
transit network 转接网;传输网(络)
transit note 经纬仪测量手簿;测角计算用表
transit office 转接局
transitory 瞬息;瞬变
transitory decomposition 暂时分解
transitory frozen earth 季节性冻土
transitory frozen ground 季节性冻土
transitory frozen soil 季节性冻土
transitory income 一时性收入;暂时收入;临时收入
transitory loading 瞬时荷载
transitory over voltage 瞬间超电压
transitory pruning 初步修剪
transitory starch 过渡淀粉
transitory target 瞬间目标
transitory variation 暂时变化
transit passenger 过境旅客
transit pass for imports 进口过境单
transit period 运到期限
transit period of goods 货物运到期限
transit period of parcels 包裹的运到期限
transit pipeline 总管线;转输管线
transit plug 塞套
transit point 转接点;经纬仪测站
transit polygon 经纬仪导线
transit port 转口港;中转港(口);中转港(口);过境港
transit press 暂时压力
transit pressing 暂时压力
transit pressure 暂时压力
transit proceed-to-send signal 转接着手发码信号
transit raceway 传动滚道
transit radio telescope 中天射电望远镜
transit railway 运境铁路;通过铁路
transit rate 中转费率
transit register 转接寄存器
transit ride 乘公共车辆出行
transit riders 公共交通客流
transit road 过境道路
transitrol 自动频率微调管

transitron 负跨导管
transitron oscillator 负跨导管振荡器
transit route 运输路线;转接路由;中转线路;过境运输路线;过境路线
transit routing 转接路由
transit rub 装运擦痕
transit rule 坐标增量配赋法
transit satellite 子午仪卫星;子午卫星
transit seizing signal 转接占用信号
transit shed 转运货棚;周转仓库;中转货棚;港口仓库;临时堆栈;码头货棚;码头仓库;前方货场;前方仓库;通栈
transit shed and yard 一线库场
transit square 型架经纬仪;坐标经纬;工具经纬仪
transit stadia method 经纬仪视距法
transit stadia traverse 经纬仪视距导线
transit stadia traverse field book 视距经纬仪导线测量手簿
transit state 过境国
transit station 转接局;经纬仪测站
transit stop 公共交通停车站
transit storage 转运仓库;过境保管
transit stream length 河流过境段长度
transit survey 经纬仪测量
transit tax 通行税
transit telegram 转报
transit telescope 中天望远镜
transit temperature 转化温度
transit terms 中转条款;过境运输条款
transit theodolite 转镜经纬仪;经纬仪
transit time 运输时间;中天时;中天时间;切换时间;通过时间;渡越时间;传播时间
transit time effect 切换时间效应;渡越时间效应
transit time heating 渡越时间加热
transit-time measurement 传播时间量测(声波等)
transit time microwave diode 渡越时间微波二极管
transit time mode 渡越时间模式
transit time oscillation 渡越时间振荡
transit-time tube 速调管
transit-time ultrasonic flowmeter 过渡时超声波流量计
transit track 直通线路
transit trade 转口贸易;过境贸易;通过贸易
transit traffic 过境运输;过境交通;过轨运输;通过交通
transit traffic line 过境运输线路
transit train 中转列车;直通列车
transit traverse 经纬仪导线
transit traverse point 经纬仪导线点
transit traverse station 经纬仪导线站
transit triangulation 经纬仪三角测量
transit tribrach 经纬仪三角基座
transit trigonometric level(l) ing 经纬仪三角高程测量
transit trip 中途列车出行;公共交通出行
transit truck 运送车;汽车式搅拌车
transit trunk 转接中继线
transit tube 地下铁道
transit unit line capacity 交通单元线路容量
transit usage map 公共交通客流图
transit vehicle 过境车辆
transit velocity 通过速度
transit vernier 经纬仪游标
transit visa 过境签证
transit warehouse 周转仓库
transit without visa 无签证过境
transit with stadia wires 视距(丝式)经纬仪
transit working 转接工作
transit working area 暂时工作区
transit yard 直通(车)场【铁】;通过(车)场【铁】
translate 货币折换;翻译
translate and edit 编译
translated channel 转换通路
translated distribution 平移分布
translated negative exponential distribution 平移负指数分布
translating circuit 译码电路
translating equipment 转换设备
translating gear 中间齿轮;变换齿轮
translating names in original 原文式译名
translating relay 变换继电器
translating unit 转换器
translation adjustment 折算调整
translation adjustment for fiscal year 财政年度

货币折算调整
translational 平移的
translation(al) algorithm 平移算法
translation(al) axis 平移轴
translation(al) cam 平移凸轮
translation(al) coefficient 平移系数
translation(al) component 移动分量;平移分量;线位移分量
translation(al) control 平移控制;移动控制
translation(al) coordinates 平移坐标
translation(al) elastic stiffness 平移弹性刚度
translation(al) element 平移要素
translation(al) energy 平移位能;平移能;直线运动能量;进动能量;平动能(量)
translational fault 直移断层
translation(al) flow 平移流
translation(al) frequency 平移频率
translation(al) gliding 直移滑动
translation algorithm 翻译算法
translation(al) group 平移群
translation(al) invariant 平移不变式
translation(al) invariant metric 平移不变度量
translation(al) landslide 平移式滑坡;直移滑坡
translation(al) lattice 平移晶格;平移点阵
translation(al) loss 平移损失
translation(al) mode 平移振型
translational moment 侧移弯矩
translation(al) motion 平移;平动;平移运动;直移运动
translation(al) movement 平移运动;直移运动
translation(al) operation 平移运行
translation(al) operator 平移算子
translation(al) parameter 平移参数
translational partition function 平动配分函数
translation(al) plane 平移面
translation(al) principle 平移原理
translation(al) response 平移反应
translational rigidity 侧移刚度
translation(al) seismograph 平移地震仪
translation(al) shell 可移动的壳体;平移壳
translational shell of double curvature 可移动的双曲线率壳体
translation(al) slide 平移滑动
translational slip 平移滑移
translational speed 移动速度
translation(al) state 移动状态;平动态
translational stiffness 侧移劲度;侧移刚度
translational stiffness coefficient 侧移劲度系数;侧移刚度系数
translational stiffness factor 侧移劲度系数;侧移刚度系数
translation(al) surface 可移动的表面;平移曲面
translation(al) surface generation 过渡曲线生成
translation(al) surface shell 平移面壳
translation(al) symmetry 平移对称
translation(al) symmetry element 平移对称元
translation(al) system 平移系统
translation(al) time lag 平移时滞
translation(al) vector 平移矢量
translation(al) velocity 平移速度;转移速度
translation(al) wave 移进波;平移波;推进波
translation cable 横过大西洋电缆
translation cam 直动凸轮
translation circuit 转接电路;转换电路
translation continental margin 移ളa大陆边缘
translation description language 翻译描述语言
translation exception 转换失效;转换故障
translation exposure 折算风险暴露
translation field 选择器线弧
translation form 翻译形式
translation gain 转换增益
translation gain or loss 折算损益;外币折算损益
translation-invariant function algebras 平移不变函数代数
translation language 翻译语言
translation loss 平动损失;放声损失
translation memory 译码存储系统;译码存储器;翻译存储器
translation noise 转播噪声
translation of axes 坐标轴(的)平移
translation of coordinate axis 坐标轴转换;坐标轴平移
translation of data 数据平移
translation of rigid body 刚体平移

translation program(me) 译码程序
translation rate 转换率
translation risk 外币折算风险
translation specification exception 翻译说明异常
translation table 转换表;翻译表
translation to decimal fraction 化为十进位
translation value 移动数值
translation vector 移动向量;平动矢量
translation wave 推进波;推进波
translator 转移器;转换器;转发器;传送器
translator station 转播站;转播台
translator unit 传送部分;翻译部分
translatory 平移的
translatory derivative 移动导数
translatory fault 直移;平移断层
translatory flow 平移流
translatory motion 平移
translatory resistance 平移阻力;平动阻力;平顶阻力
translatory speed 转换速度
translatory velocity 平移速度
translatory wave 移进波;平动波;推进波
translauncher 转移发射装置
trans-level combination 超层次结合
transliterate 直译
transliteration 直译
transliteration key 译青表
translithospheric fracture 超岩石圈断裂
transloader 装运机
translocated injury 易位损伤
translocation 运输(作用);易位;移位(作用);转移;联测定位;输导作用
translocation mode 联测定法
transloeated herbicide 内吸转移的除草剂
translot 横槽
translucence 半透明性;半透明度;半透彻度;半透彻
translucence liquid 半透明液体
translucence water 半透明水
translucency 半透明性;半透明度;半透明;半透彻度;半透彻
translucent alumina 半透明氧化铝
translucent alumina ceramics 半透明氧化铝陶瓷
translucent attritus 半透明暗煤
translucent beads 半透明球
translucent body 半透明体
translucent ceiling 半透明顶棚
translucent china 半透明瓷
translucent coating 半透明涂层;半透明罩面;半透明涂料
translucent colo(u)r 半透明颜料;半透明色(料)
translucent concrete 半透明(的)混凝土;预制的混凝土和玻璃板组合
translucent concrete construction 半透明混凝土结构;半透明混凝土组合结构
translucent concrete floor 半透明混凝土楼板
translucent concrete window 玻璃砖混凝土窗
translucent copy 半透明拷贝
translucent diffuser 半透明散光罩
translucent door 半透明门
translucent enamel 透光珐;半透明珐琅
translucent film 半透明漆膜
translucent fused silica glass 半透明石英玻璃
translucent glass 半透明玻璃
translucent glass paper 半透明玻璃纸
translucent glaze 半透明釉
translucent humic matter 透明分解腐植物质
translucent lighting 半透明照明
translucent mask 半透明罩
translucent material 半透明材料
translucent medium 半透明介质
translucent mosaic 半透明镶嵌
translucent origin 半透明原稿
translucent paint 半透明漆
translucent panel 透明板;半透明板
translucent paper 半透明纸
translucent plastic sheet 半透明塑料板
translucent projection screen 半透明投影屏幕
translucent PVC panel 半透明聚氯乙烯板
translucent rubber 半透明橡胶
translucent screen 半透明屏
translucent sheet 半透明纸
translucent soap 半透明皂
translucent surface 半透明表面
translucent vitreous silica 半透明石英玻璃

translucid 半透彻的
translucidus 透光云
transmarginal inhibition 超限抑制
transmarine port 海外港口
transmat 反射软片
transmembrane action potential 跨膜动作电位
transmembrane gradient of hydrogen ion 氢离子跨膜梯度
transmembrane potential difference 跨膜电位差
transmembrane pressure 过膜压力
transmigration 反式迁移作用;迁徙
transmiss-gear ratio 转速比
transmissibility 蓄水层输出水能力;可透性;可传性;可传动性;导水性;传输率;传递率;传递比
transmissibility coefficient 可透性系数;导水系数
transmissibility of vibration 振动可传性
transmissibillity coefficient of aquifer 含水层导水系数
transmissible 可传送的
transmissible credit 可转让信用证
transmissiometer 透射计
transmission 透射比(率);透射;透过;透光;通信;输电;传送;传射;传递;传播;发射;播送
transmission adaptor 传输适配器;传送转接器;传输衔接器
transmission agent 传动系统
transmission and distribution 输配电
transmission and distribution line 输配电线
transmission and information exchange system 传输和信息交换系统
transmission and receive control equipment 发射和接收控制设备
transmission anomaly 异常传输
transmission apparatus 传输设备
transmission arm 传动臂
transmission array 传输阵
transmission assembly 变速箱总成
transmission attaching screw 变速器接合螺钉
transmission attenuation 传输衰减;传播衰减
transmission band 传送频带;传输频带;传动刹车带
transmission bar 传动杆
transmission-beam method 透射法
transmission bearing retainer 传动轴承保护圈
transmission belt 传动皮带;传动带
transmission belting 传动带
transmission block 传送数据块;传送块;传输字组;传输块
transmission block character 传输块字符
transmission box 传动箱;变速箱
transmission brake 传动系制动器
transmission bridge 馈电电桥;传输电桥
transmission by formite 杂物传播
transmission by hand 经手传播
transmission capacity 输电容量;输电能力;输电量;渗透能力;渗漏能力;传送能力;传输容量;传播能力
transmission cartridge 传爆药筒
transmission case 传动箱;齿轮箱;变速箱
transmission case cover 传动箱盖
transmission case drain plug 传动箱放油塞
transmission category 传输范畴
transmission cavity 发射机谐振腔
transmission chain 传动链
transmission channel 传输信道;传输通路;传输通道;发送信道
transmission characteristic 传输特性;传递特征;传递特性
transmission characteristics tester 传输特性测试仪
transmission circuit method 传输线路法
transmission clutch 传动离合器
transmission code 传输码;传输代码
transmission code violation 传输代码破坏
transmission coefficient 传导系数;转移系数;透射系数;渗透系数;传输系数;传播系数;穿透几率
transmission coefficient of detection 检波传输系数
transmission colo(u)r 透射色
transmission condition 传输条件
transmission constant 透水常数;传递常数;传导常数
transmission control 传动控制;传送控制;传输控制
transmission control character 传送控制符;传输控制字符;传输控制符号

transmission control code 传送控制码
transmission control layer 传输控制层
transmission control procedure 传送控制程序;传输控制规程
transmission control protocol 传输控制协议
transmission control protocol/internet protocol 传输控制和互联网协议
transmission control station 传输控制站
transmission control unit 传输控制器
transmission control valve 传送控制阀
transmission cooling 传动冷却
transmission copying 传输复制;传感拷贝
transmission cost 输电费用;输电成本;传输费用
transmission countershaft 逆转轴;传动箱中间轴;传动副轴;变速器中间轴
transmission countershaft gear 传动副轴齿轮
transmission countershaft low speed gear 传动副轴低速齿轮
transmission countershaft reverse gear 传动副轴回动齿轮
transmission countershaft second speed gear 传动副轴中速齿轮
transmission crank 传动曲轴
transmission cross section 透射截面
transmission curve 透射曲线;输电曲线;传输曲线;传动曲线;传递曲线
transmission cycle 传输周期
transmission data set 传输数据组
transmission delay 传输延迟;传动延迟
transmission densitometer 透射光密度计
transmission density probe 透射式密度探测器
transmission depth 透射深度
transmission device 传递装置
transmission device of load 荷载传送装置;荷载传递装置
transmission diagram 传动图
transmission diffraction 透射衍射法
transmission distance 传输距离
transmission distortion measuring set 传输失真测量仪
transmission dynamometer 传送测力计;传动式测力计;传动功率计;传动测功计
transmission dynode 透射倍增极
transmission echelon 透射梯
transmission efficiency 输电效率;传送效率;传输效率;传动效率;传递效率
transmission efficiency measuring set 传输效率测量器
transmission electron diffraction 透射电子衍射
transmission electron image 透射电子图像
transmission electron microscope 透射电子显微镜;穿透式电子显微镜
transmission electron microscope method 透射电镜法
transmission electron microscopy 透射电子显微术;透射电子显微镜检查法;透射电子技术
transmission energy converter 发射换能器
transmission engineering 传动工程
transmission equipment 输电设备;传输设备;传递装置
transmission equivalent 传输衰耗等效值;传输当量
transmission error 传输误差
transmission error control 传输差错控制
transmission experiment 透射实验
transmission explosive 传爆炸药
transmission extension 传输扩展
transmission extension drive 延伸传输驱动
transmission fabric 传动带用布
transmission facility 输电设施;传输设施;传输设备
transmission factor 传导系数;透射因子;透射因素;透射系数;传输因子;传输因素;传输因数;传递系数;传播系数
transmission fatigue 传递性疲劳
transmission film 透射膜
transmission filter 透射滤波器
transmission first and reverse speed gear 变速器头挡及倒车齿轮
transmission fluoroscope 透射荧光镜
transmission fork 拨叉;变速叉
transmission for lifting gear 提升设施传动
transmission forward-reverse lever 前进倒车换挡杆
transmission frequency 输电频率;传输频率
transmission frequency bandwidth 传输频带宽度

transmission frequency characteristic 传输频率特性
transmission gain 增益;传递增益
transmission gate 传输门
transmission gate circuit 传输门电路
transmission ga(u)ge 透射式测厚计
transmission gear 变速齿轮;传动装置;传动齿轮;传导齿轮
transmission gear box 传动齿轮箱;齿轮变速箱
transmission gear cover 传动齿轮盖
transmission gearing 变速装置;传动装置
transmission gear ratio 齿轮减速比
transmission grating 透视光栅;透射光栅;透镜光栅
transmission grease 减速箱润滑脂;变速器润滑油脂
transmission grid 输电网
transmission group 传输组
transmission group identifier 传输组标识符
transmission guard 传动装置护板;传动装置护罩
transmission header 传送报文头;传输头标;传输报头
transmission high energy electron diffraction 透射高能电子衍射
transmission housing 变速箱壳
transmission hydraulic control 传动液压控制
transmission interface 传输接口
transmission interface converter 传输接口转换器
transmission interruption 传输中断
transmission lag 传输延迟
transmission length 传力长度;传递长度;预应力筋锚固长度(混凝土)
transmission level 透射级;传输电平;发射电平
transmission level measuring set 传输电平测试器
transmission level meter 传输电平表
transmission light 透射光
transmission light intensity 透过光强度
transmission light interference microscope 透射光干涉显微镜
transmission limit 传输限度;传输极限
transmission line 输油管线;输电线;输电线路;输电线(路);输水管线;馈(电)线
transmission line adapter 传输线衔接器
transmission line admittance 传输线导纳
transmission line amplifier 传输线放大器
transmission line attenuation 传输线衰减
transmission line cable 传输线电缆
transmission line calculator 传输线计算机
transmission line characteristic impedance 传输线特性阻抗
transmission line constant 传输线参数
transmission line control 传输线控制
transmission line control block 传输线控制块
transmission line controlled blocking oscillator 传输线控制间歇振荡器
transmission line corridor 出线走廊
transmission line current 传输线电流
transmission line distortion 传输线畸变
transmission line efficiency 传输线效率
transmission line error 传输线误差
transmission line hit 传输线瞬断
transmission line impedance 传输线阻抗
transmission line level 传输线信号电平
transmission line loss 传输线损耗
transmission line model 传输线模型
transmission line noise 传输线上的干扰;传输线(路)噪声
transmission line of information 信息传输线
transmission line oscillator 传输线振荡器
transmission line parameter 传输线参数
transmission line power 传输线功率
transmission line pylon 输电塔架;输电塔桩
transmission line reflection 传输线路反射;传输线反射
transmission line reflection coefficient 传输线反射系数
transmission line regulator 传输线调整器
transmission line speed 传输线路传输速率
transmission line stabilized oscillator 传输线稳频振荡器
transmission line switching 传输线路开关
transmission line theory 传输线理论
transmission line tower 拉线塔;输电线塔;输电塔桩;输电塔架
transmission line trademark 输电线牌号

transmission line traffic capacity 传输线通话能力
transmission line transducer loss 传输线换能器损失
transmission line voltage 传输线电压
transmission link 传输线路;传输链路
transmission lock 变速齿轮箱锁
transmission loss 配水损失;声透射损失;传输耗损;隔音量;馈电损耗;透射损失;输气损耗;输电损失;输电损耗;传送损耗;传输损失;传输损耗;传声损失;传递损失;传播损耗;传输衰减
transmission loss distance 传输损耗距离
transmission lube system 传输润滑系统
transmission lubricant 传动装置润滑剂
transmission lubricating valve 传动润滑油阀
transmission main 输水干管
transmission main shaft 传动主轴;变速器主轴
transmission mast 输电杆;输电线杆
transmission mast for power 输电电杆
transmission matrix method 透射矩阵法
transmission measurement 透射测定法
transmission measuring equipment 传输测量设备
transmission measuring set 传输测试仪;传输测量器
transmission mechanism 透射机理
transmission media 传输手段
transmission medium 传送介质;传输媒质;传输媒体;传输介质
transmission mode 传输模式;传输模;传输方式;传输波型;波模
transmission mode bit 传输方式位
transmission mode of short-wave 短波传播模式
transmission modulation 传输调制
transmission monitor 主监察器;传输监视器
transmission-mounted drive line brake 传动装置驱力制动器
transmission network 输电网;传输网(络)
transmission neutralizer valve 传输中和器阀
transmission of economic growth 经济增长的传递
transmission of electricity 输电
transmission of element 元素递变
transmission of energy 能量输送;能量传输;能量传递
transmission of heat 传热;热的传递;热传递
transmission of humidity 渗潮
transmission of induced resonance 诱导共振传递
transmission of light 光透射;光的透射
transmission of load 荷载传递
transmission of motion 运动传递;传动
transmission of pathogen 病原传播
transmission of pressure 压力传送;压力传递
transmission of reading 读数传送
transmission of sound 声音传播;声的透射;声的传递;声传输;传声
transmission of wastewater 废水输送
transmission of water 输水
transmission of water regime information 水情传递
transmission oil 汽车传动油;传动油;传动润滑油
transmission oil cooler 变速器冷油器
transmission oil sump 变速器油箱
transmission output shaft 驱动轴;传动轴
transmission packet 发送消息包
transmission panel 输电仪表板
transmission parameter 传输参数
transmission path 传输通道;传输路径;传递通路;传播途径;传播路径
transmission pattern 传播方式
transmission performance 传输性能
transmission performance rating 额定传输性能
transmission photocathode 透射光电阴极
transmission pipeline 输送管(道)
transmission plant 输电设备
transmission pole 输电杆;输电线杆
transmission polynomial 传输多项式
transmission power 传动功率
transmission power train 传动装置[铁]
transmission preprocessor 传输预处理机
transmission pressure 输电电压;传输电压
transmission primaries 发送原色
transmission print 晒图
transmission priority 传输优先级
transmission pulse 发射脉冲
transmission pump relief valve 输送泵减压阀;输送泵安全阀

transmission quality 传输品质
transmission ram 传动唧筒
transmission range 夜视距离;传输距离;传输范围;传播距离
transmission range selector lever 输送范围换向杆;输送范围换位杆
transmission rate 垂直渗透率;传送率;传输速率;传导率
transmission ratio 转速比;速比;传送比;传动比;传递比
transmission regulator 传输调节器
transmission report file 发送报告文件
transmission reverse gear 变速器倒挡齿轮
transmission reverse idler gear 传动用回动空转轮;传动回动空转轮
transmission reverse speed shift fork 倒挡换挡义
transmission rod 传动杆
transmission rope 传动绳;传动钢丝绳
transmission route 传输路由
transmission rubber belting 传动胶带
transmission rule 传递规则
transmission screw 传动螺杆
transmission secondary emission 穿透式二次发射
transmission security 传输保密
transmission service 传输服务
transmission service profile 传输服务剖面图
transmission shaft 传动轴
transmission sliding gear 传动滑动齿轮
transmission sound 声音传递
transmission spectrometry 透射光谱法
transmission spectrum 透射谱
transmission speed 传送速度;传输速度;传动速度;传递速度;发信速度
transmission standard 传输标准
transmission star coupler 变速器星形耦合器
transmission statement 传输语句
transmission strap 传动带
transmission subsystem 传输子系统
transmission subsystem interface 传输子系统接口
transmission subsystem profile 传输子系统概述文件
transmission support bracket 变速器支架
transmission synchromesh 传动同步配合;传动同步接合
transmission system 馈电系统;输电系统;传输系统;传动系统;传递系统;发送系统;输气管线系统
transmission table 传输表
transmission target 透射靶
transmission terminal card 传输终端卡片
transmission test 传动试验
transmission test bay 传输测试架
transmission time 传送时间;传输时间;发送时间
transmission topography 透射形貌学
transmission tower 输电(线路)塔;拉线塔输电(杆)塔
transmission trunking 传输集群
transmission type 传送类型
transmission type cavity frequency meter 传输式空腔频率计
transmission-type electron microscope 透射式电子显微镜
transmission type frequency meter 传输式频率计
transmission type ga(u)ge 传递式仪器
transmission-type radio isotope ga(u)ge 透射式放射同位素测量仪
transmission unit 透射单位;传输单位;传递单位
transmission velocity 传送速度
transmission wave 传射波
transmission window 传输窗
transmission yoke 传动轭
transmissive 可发射;可传动的
transmissive exponent 透射指数
transmissive optic(al) disc 透射光盘
transmissive viewing screen 透过式屏幕
transmissive wave 滤过波;透过波
transmissivity 输水能力;透射系数;透射(率)比(率);透光度;通过能力;输能能力;导水系数
transmissivity tester 透射率检验仪
transmissometer 浑浊度仪;透射表;大气透射仪
transmissometry 透射测量仪
transmit 转达;透光;传送;传输
transmit amplifier 发送放大器
transmit-buffer-empty 发送缓冲器空闲
transmit control block 传输控制块;发送控制块

transmit data register 发送数据寄存器
transmit data set 传送数据集;传输数据组
transmit decrepitation thermometer 传导爆裂测温仪
transmite data register 传输数据寄存器
transmit efficiency of shutter 快门透光率
transmit flow control 传输流量控制;发送数据流控制;发送流控制
transmiting receiving device of high frequency track circuit 高频轨道电路收发器
transmit leg 传输引线;发送方
transmit modulator 发送调制器
transmit operation 发送操作
transmit-receive 收发两用
transmit-receive cavity 收发开关空腔谐振器
transmit-receive direction 发射接收方向
transmit-receive path 发射接收路径
transmit receiver 收发两用机
transmit-receive switch 收发开关
transmit receive unit 收发装置
transmit side 传输端
transmit status 发送状态
transmit symbol 发送符
transmittability 传输能力
transmittal-carrier operation 载波发送制
transmittal letter 附函
transmittal method by mean sea-level 平均海面传递法
transmittal mode 传送方式;传输方式
transmittance 透射比(率);透光率;透光度;透射性;透射系数;透明度;热透射;传透系数
transmittance meter 能见度测量仪;透射表;大气透射计
transmittance method 透射法
transmittance of lens 镜头透射率
transmittance of light(ing) 透光度;烟雾透光率
transmittancy 相对透射比;透射率;透光度
transmitted beam 透射束;透过光速;发射波束
transmitted-data circuit 发送数据线路
transmitted diagnostic ultra-sonoscope 穿透式超声诊断仪
transmitted efficiency 传输效率
transmitted energy 透射能;发射能量
transmitted flux 传输通量
transmitted information 传输信息
transmitted information transmitter data 发送信息
transmitted intensity 透射强度;透射
transmitted light 透射光
transmitted lighting 透射照明
transmitted light intensity 透射光强度
transmitted load 间接荷载;传递荷载
transmitted medium 传导介质
transmitted mode 传送方式
transmitted power 传输功率;传递功率;发射(声)功率
transmitted power level 发射功率电平
transmitted radiation 漏过辐射
transmitted ratio 传送比
transmitted side band 发射边带
transmitted sound 透射声;发射声
transmitted spectrum 透射光谱
transmitted wave 折射波;透过波;透射波;发射波
transmitted work 透射光观察工作
transmitter 传送机;传递质;传送者;传感器;传送者;变送器;发送器;发送机;发射机;发报机;发话器
transmitter amplifier 发射机放大器
transmitter aperture 发射天线孔径
transmitter card 发送电路板
transmitter clock 发送时钟
transmitter collimating optics 发射器准直光学装置
transmitter data 发送数据;发送储息
transmitter distributor 发射机分配器;发信器
transmitter empty 发射机空载
transmitter holding register 发送器保持寄存器
transmitter installation 发送装置
transmitter modulator 发射机调制器
transmitter monitor 发射机监控器
transmitter off 关闭发送机;发送器断电
transmitter on 发送器接通
transmitter optic(al) connector 发射机光连接器
transmitter ready 发送准备就绪;发送准备
transmitter-receiver 收发报机;收发器;收发两用机;发送接收机
transmitter receiver probe 发送接收探头

transmitter-receiver serial parallel module 串行并行收发器
transmitter-receiver set 收发装置
transmitter relay 发射机继电器
transmitter responder 应答发射机
transmitter signal element timing 发送信号单元定时
transmitter start code 发信起动码;发送器启动码;发送起始码;发送开始码
transmitter substance 传递质
transmitter synchro 自同步发射机
transmitter telescope 激光发射管;发射望远镜
transmitting 发射
transmitting amplifier 发信放大器
transmitting and receiving dipole 发送与接受振子
transmitting and receiving radio set 无线电收发站
transmitting antenna 发射天线
transmitting aperture 发射孔径
transmitting band directivity factor 发送频带方向系数
transmitting band response 发送频带响应
transmitting bank 通知银行
transmitting boundary 透射边界;传输边界
transmitting branch 发信支路
transmitting capacity 输电容量;输电能力;输电量;传输能力;发送能力
transmitting chain tester 发射链路测试器
transmitting circuit 发信电路
transmitting core 传输铁芯
transmitting coverage 发射有效区
transmitting crystal 发射晶体
transmitting data register 发送数据寄存器
transmitting density 透射密度
transmitting directivity 发射方向性
transmitting directivity factor 发送方向系数
transmitting efficiency 发送效率;发射效率
transmitting element 传递元件;发送单元;发射元件
transmitting equipment 发送设备
transmitting film 透光膜
transmitting filter 发信滤波器
transmitting fork filter 发送音叉滤波器
transmitting frequency 发射频率
transmitting frequency converter 发射频率变换器
transmitting gear 传动齿轮
transmitting instrument 变送器
transmitting interference 传输干扰
transmitting jigger 发射振荡变压器
transmitting line of sounder 测深仪零线;测深仪发射线
transmitting loop loss 传输回路损耗
transmitting low pass filter 发信低通滤波器
transmitting magnetic compass 传输磁罗经
transmitting magnetic moment 发射磁矩
transmitting mechanism 传动系
transmitting medium 传导介质;传播媒质
transmitting motion 传动
transmitting objective lens 发射物镜
transmitting of water regime information 水情传递
transmitting optics 发射光学装置;发射光学
transmitting oscillator 送波器;发射振荡器
transmitting paraboloid 抛物面发射天线
transmitting power 发射功率
transmitting power level 发射功率电平
transmitting power monitor 发射功率监测器
transmitting power response 发送功率响应
transmitting probe 发射探头
transmitting pulley 中间滑轮
transmitting ratio 传动比
transmitting ray 发射线
transmitting reduction 透光缩小
transmitting relay 传输继电器;发送继电器
transmitting response 发送响应
transmitting rheostat 传动变阻器;传导变阻器
transmitting selsyn 自动同步传感器
transmitting set 传输装置;发送机;发射机组;发射机;发报机(组);发射机;发报机组
transmitting signal conversion equipment 发信信号转换设备;发信信号转换装置
transmitting signal element timing 发信信号单元定时
transmitting spectrum factor 发送频谱因数
transmitting star-coupler 传递星形耦合器
transmitting station 发射台;发报台;发报局;播音站

transmitting system 发送系统;发射装置;变送系统
transmitting time 发送时间
transmitting tissue 引导组织
transmitting transducer 传送变换器;发射换能器
transmitting tube 发射管
transmitting voltage 发送电压
transmitting zone 传导带
transmityper 光电导航信号发射机,导航信号发送机
transmodalist 国际联合运输人
transmodulator 转化调制器
transmogrify 使完全变形
trans-mountain 穿山的
trans-mountain diversion 跨流域引水;跨流域调水;穿山引水
trans-mountain diversion project 穿山引水工程;跨流域引水工程
trans-mountain diversion works 穿山引水工程
trans-mountain line 越山管线
trans-mountain pipeline 越山管线
trans-mountain tunnel 穿山隧洞;穿山隧道
trans-mountain water diversion 跨山引水
transmultiplexer 传输多路复用器;复用转换器
transmutation 所有权的转移;所有权的让与;原子核嬗变;质变;变质
transmutation glaze 变色釉
transmutation product 转变产物;人造同位素
transmutative force 引起变形的力
transmuted wood 变性木材
transnational 跨国的
transnational agreement 跨国契约
transnational business strategy 跨国经营战略
transnational company 跨国公司
transnational corporation 跨国(家)公司
transnational enterprise 跨国企业
transnational operations 跨国经营
transnational tourist agency 跨国旅游机构
transnatural 超自然
trans-Neptunian planet 海外行星
transnormal 超常规;超常的
transnormal inverse matrix 异常逆矩阵
transnoundary pollutant 越境污染物
transobuoy 浮标式自动气象站
transoceamic communication 海外通信
transoceanic aeroplane 横渡大洋飞机
transoceanic cable 越洋电缆
transoceanic cable system 越洋电缆系统
transoceanic communication 远洋通信
transoceanic distance 越洋距离
transoceanic flight 越洋飞行
transoceanic link 越洋线路
transoceanic navigation 越洋航行;大洋航行
transoceanic sonde 越洋探空仪
transoceanic steamer 远洋轮(船)
transoceanic (submarine) cable 越洋海底电缆
transoceanic traffic 远洋交通
transoceanic vessel 远洋船
trans-ocean tanker 远洋油船
transocean (transport) 跨海运输
transoid conformation 反式构像
transolver 控制同步机
transom 摇头窗;腰头窗;连接环节;门顶窗;横楣;舷板;上亮子;船尾横材;中横框;气窗;门窗中槛;楣窗;横梁(车辆的)
transom bar 中槛;中贯挡;门窗横挡
transom beam 船尾梁
transom board 船尾板
transom bolt 横梁螺栓
transom bow 平头前倾船首;方形船头;方形船首
transom brace 横梁拉条
transom bracket 气窗托架
transom catch 气窗插销
transom chain 气窗启闭链条;气窗链
transom cleat 系缆角铁;拉绳挂钩
transom drive 船尾机驱动
transom(e) 气窗;门顶窗;横梁;中贯挡;横挡
transom flap 尾板;船尾均衡翼板
transom floor 尾肋板;船尾横骨架
transom frame 船尾构架;艉板肋骨;气肋骨;带气窗的门框
transom lifter 气窗垂直启闭器;气窗梃销
transom light 气窗
transom pier 门式墩;门式桥墩
transom plate 船尾板
transom pylon 门式桥塔

transom stern 方形船尼;方尼型
transom window 楣窗;气窗;有横楣分隔的窗;带横档的窗
transonde 越洋探空仪
transonic 跨声速的
transonic acceleration 跨音速加速度
transonic aerodynamics 跨音速空气动力学
transonic barrier 跨声速障
transonic boundary 跨声速边界层
transonic cascade 跨音速叶栅
transonic characteristic 跨音速特征
transonic compressor 跨音速压气机
transonic effect 跨声速效应
transonic flow 跨音速流动
transonic flow field 跨音速流场
transonic Mach number 跨声速马赫数
transonic nozzle 跨音速喷管
transonic range 跨音速范围
transonics 跨音速空气动力学
transonic speed 跨音速;跨声速
transonic stream tube 跨音速流管
transonic test section 跨音速试验段
transonic thermal blooming 跨声速热晕
transonic velocity 跨音速
transonic wind tunnel 跨音速风洞
transopaque mineral 透明—不透明矿物;半透明矿物
transosonde 越空探测;平移探空仪
transownership 跨所有制
transpacific 横渡太平洋的
trans Pacific trade 太平洋航线
transparant illumination 透射照明法
transparence 透明性;透射度;透光度
transparency 明晰度;透射度;透明性;透明(物)体;透明软片;透明(度);透光度;投影薄膜
transparency area 透明区
transparency cabin 透明舱
transparency canopy 透明舱盖
transparency carrier 透明基板
transparency celluloid 透明赛璐珞
transparency cellulose 透明纤维素
transparency ceramics 透明陶瓷
transparency clarity 透明度
transparency colo(u)r 透明色料
transparency conducting electrode 透明传导电极
transparency conducting layer 透明导电层
transparency conductive electrode 透明导电电极
transparency conductive film 透明导电膜
transparency cover 透明盖
transparency cup grease 透明杯脂
transparency cutting oil 透明切削油
transparency degree 透明度
transparency dielectric(al) layer 透明电介质层
transparency display window 透明显示窗
transparency dye 透明染料
transparency electrode 透明电极
transparency electro-luminescent cell 透明场致发光电池
transparency emulsion 透明乳液
transparency enamel 透明色釉
transparency ferro-electric(al) ceramics 透明铁电陶瓷
transparency film 透明薄膜
transparency glass 透明玻璃
transparency glass plate 透明玻璃板
transparency glaze 透明釉
transparency-grey-scale 透明灰度梯度
transparency horticultural glass 温室玻璃
transparency ice 透明冰
transparency insulating coating 透明绝缘敷层
transparency lacquer 透明亮漆
transparency layer 透明层
transparency leather 透明革
transparency liquid 透明液体
transparency manifold 透明复写纸
transparency material 透明物质
transparency medium 透明媒质
transparency meter 透明度仪;透明度计
transparency mirror 透光反射镜
transparency mode of operation 透明运行方式
transparency negative 透明底片
transparency nose 头部整流罩
transparency-nucleus model 透明核模型
transparency of atmosphere 大气透明度

transparency of seawater 海水透明度
transparency of water 水的透明度
transparency paramagnetic substance 透明顺磁材料
transparency photocathode 透明光电阴极
transparency plastic model 透明塑料模型
transparency ratio 透明比率
transparency reactor 透明反应器
transparency reflector 透明反光镜
transparency scale 透明分度盘
transparency scheme 透明方案
transparency screen 透明屏
transparency silica glass 透明石英玻璃
transparency sky cover 透光云量
transparency swelling 透明膨胀
transparency varnish 透明漆
transparency velvet 透明丝绒
transparency vitreous silica 透明石英玻璃
transparency window 透明窗(采用各种透明材料的窗)
transparency zone 透明带
transparent 可穿透的;透明的;透彻的
transparent adhesive varnish 透明胶性漆
transparent alumina ceramics 透明氧化铝陶瓷
transparent attritus 透明碎屑
transparent beryllia ceramics 透明氧化铍陶瓷
transparent body 透明体
transparent box package 亮格箱
transparent carrier 透明基板
transparent celluloid 透明赛璐珞
transparent cellulose 透明纤维素
transparent ceramics 透明陶瓷
transparent/clear 透明/清
transparent coating 透涂层;透明涂层;透明膜
transparent colo(u)r 透明色料;透明颜料
transparent conductive coating 透明导电涂料
transparent crystal 透明晶体
transparent cup grease 透明杯脂
transparent dielectric(al) layer 透明介质层
transparent dye 透明颜料
transparent enamel 透明磁釉;透明瓷釉
transparent ferro-electric(al) ceramics 透明铁电陶瓷
transparent film 透明模;透明胶片;透明薄模
transparent filter 透明滤光片
transparent finish 透明涂面
transparent flexible swing door 透明柔性弹簧门
transparent frit 透明玻璃料;纯玻璃料
transparent fused silica 透明石英玻璃
transparent glass 透明玻璃
transparent glass ceramics 透明微晶玻璃;透明玻璃陶瓷
transparent glass door 透明玻璃门
transparent glass fibre reinforced plastics 透明玻璃纤维增强塑料
transparent glaze 透明釉
transparent hologram 透明全息图
transparent horticultural glass 透明温室玻璃
transparent hose 透明蛇皮管;透明软管
transparent ice 透明冰
transparent ink 透明油墨
transparentizing 透明化
transparent lacquer 透明漆;透明喷漆;透明亮漆
transparent lamp 透明灯(泡)
transparent layer 透明层
transparent length 透光长度
transparent liquid level ga(u)ge 透光式液面计
transparent magnesium alumin(i)um spinel ceramics 透明镁铝尖晶石陶瓷
transparent magnetic glass 透明磁性玻璃
transparent material 透明物质;透明材料
transparent medium 透明介质
transparent mirror 单向镜(从亮室看是镜,从暗室看是玻璃);单面镜;透明镜
transparent negative 透明负片;透明底片
transparent nose 头部整流罩
transparent organism 透明生物
transparent original 透明原稿
transparent overlay 透明薄模
transparent oxide ceramics 透明氧化物陶瓷
transparent painting 涂透明漆
transparent paper 玻璃纸;透明纸
transparent paramagnetic substance 透明顺磁材料
transparent parchment 透明羊皮纸

transparent passage(way) 开敞通道;透明通道
transparent photocathode 透明光电阴极
transparent pigment 透明颜料
transparent plastic building unit 透明塑料建筑构件
transparent plastic model 透明塑料模型
transparent plastic pipe 透明塑料管
transparent plastic roof cover(ing) 透明塑料屋面覆盖层
transparent plastics 透明塑片料;透明塑料
transparent plastic sheet 透明塑料片;透明塑料板
transparent plastic tile 透明塑料瓦
transparent plate glass 透明厚玻璃
transparent polycrystalline alumina ceramics 透明多晶氧化铝陶瓷
transparent polycrystalline ceramics 透明多晶陶瓷
transparent positive 透明正片
transparent positive original 幻灯母片;底图
transparent reactor 透明反应器
transparent scale 透明刻度盘;透明度盘
transparent shade 透明遮阳棚
transparent sheet 透明纸
transparent side wall 透明边墙
transparent sky cover 透光云量
transparent tape 透明带
transparent thoria ceramics 透明氧化钍陶瓷
transparent tracing paper 透明描图纸
transparent transmission 透明传输
transparent triforium 透光拱廊(教堂侧廊上的)
transparent varnish 透明清漆
transparent vitreous silica 透明石英玻璃
transparent window 透明观察孔线;透明窗(采用各种透明材料的窗)
transparent yttria ceramics 透明氧化钇陶瓷
transparticle effect 传递颗粒效应
transpassive region 超钝化区
transpassivity 超钝度
transpercolation hydric 过渡型湿润状况
transpex 透明聚甲基丙烯酸甲酯
transpierce 戳穿
transpiration 蒸腾(作用);流逸;散发率
transpiration(al) current 蒸腾流
transpiration(al) steam 蒸腾流
transpiration coefficient 蒸腾系数;散发系数
transpiration-cooled 蒸发冷却的
transpiration-cooled blade 蒸发冷却式叶片
transpiration cooling 蒸发冷却法;蒸发冷却
transpiration depth 年散发量
transpiration effection leaching fractions 蒸腾对淋溶等级的作用
transpiration efficiency 蒸腾效率
transpiration loss 蒸腾损失
transpiration observation 蒸腾量观测
transpiration rate 蒸腾速度;蒸腾比率
transpiration ratio 蒸腾率;蒸腾比;散发率
transpiration test 蒸腾作用的测定
transpiring moisture 蒸发水分
transpirometer 蒸腾计
transplant a cutting 插条
transplant(ation) 移植
transplanter 栽植机
transplanting 移植;栽植
transplanting semi-mature trees 移植半熟树
transplanting width 栽植幅度
transplant test 转厂试验
transplutonics 钚后元素
transplutonium 超钚(元素)
transplutonium element 超钚元素
transplutonium isotope 超钚同位素
transpolar 通过极地
transpolarizer 铁电介质阻抗
transpond 转发
transponder 应答器;应答发射机;转发器;雷达反射器;脉冲转发器;脉冲收发两用机;发射机应答器;发射机
transponder antenna 应答器天线
transponder beacon 应答式无线电指向标;应答机信标;应答(式无线电)信标;应答(器)信标
transponder buoy 应答浮标
transponder control unit 应答机控制部分
transponder dead time 应答器寂静时间
transponder radar beacon 应答器式雷达信标
transponder reply efficiency 应答器效率;应答器的应答效率

transponder set 应答器;发射机应答器设备
transponding radar beacon 应答式雷达信标
transport 运送;走带机构;转运;水陆两用车;输送;传送走带机构;传递率
transportability 易运输性;可运输性;可便携性
transportable 可运输的;可迁移;可搬运的
transportable auger 移动式输送螺旋
transportable basket drying stove 可移篮式烘干炉
transportable black-top plant 轻便沥青拌和设备
transportable boiler 列车电站锅炉;便运锅炉
transportable breakwater 可移动防波堤
transportable communication system 移动式通信系统
transportable console 流动控制盘
transportable control 流动控制
transportable copy 可传送复制品
transportable darkroom 移动式暗室
transportable defense radar 移动式防御雷达
transportable detection radar 移动式探测雷达
transportable earth station 移动式地面站
transportable electronic shop 移动式电子修理所
transportable equipment 移动式设备
transportable fallout detector 轻便式沉落微尘探测器
transportable ground beacon 移动式地面信标
transportable ground communication station 移动式通信地面站
transportable harbo(u)r 临时拼装式流动港口
transportable height finder 移动式测高仪
transportable multiplex communication center 移动式多路通信中心
transportable MW radio communication 移动式微波无线电通信设备
transportable reactor 可运输反应堆
transportable receive-only station 移动式只收站
transportable rocket fuel purification system 移动式火箭燃料净化装置
transportable satellite communications link terminal 移动式卫星通信线路终端机
transportable satellite communication terminal 移动式卫星通信终端设备
transportable satellite earth station 移动式卫星地球站
transportable set 移动式设备
transportable station 移动式电台
transportable tent 移动式帐篷
transportable tower crane 移动式塔式起重机;移动式塔吊
transportable transformer 移动式变压器
transportable transmit and receive station 移动式收发站
transportable tropospheric scatter communication equipment 移动式对流层散射通信设备
transport added value 运输增加值
transport address 传输地址
Transport Advisory Committee 运输咨询委员会
transport agency 运输代理
transportainer 运输货箱
transport aircraft 运输机
transport air cushion vehicle 运输气垫车
transport airplane 运输机
Transport and Communication Commission 运输及通信委员会
transport and communication facility 交通运输设备
transport and communication trade 交通运输贸易
transport and deposition of acidifying pollutant model 酸化污染物输移与沉淀模型
transport-and-dumping bridge 运输排土桥
transport and loading vehicle 运输装载车
Transport and Road Research Laboratory 运输与道路研究实验室(英国)
transport and storage by ion technique 电离技术储运
transport and storage in controlled atmosphere 气调储运
transport and storage in reduced pressure 减压储运
transport and supply column 运输队
transport apron 运输机停机坪
transport area 运输卸载区域
transport artery 运输动脉
transportation 运输;迁移;输运;搬运(作用)
transportation account code 运输账户代码;运输会计代码
transportation administration 运输管理
transportation agent 运输船上运输部门负责军官代理人
transportational 运输的
transportation allowance 交通津贴
transportation amount 搬运量
transportation and communication 交通运输车辆
transportation and communication expense 交通运输费用
transportation and erection equipment 运输与安装设备
transportation and erection of lining 衬砌运输和安装
transportation and handling procedure 运输和装卸过程;搬运装卸过程
transportation and telecommunication intelligence 运输与电信情报
transportation average 平均运输量
transportation burden 运输量
transportation by leaps and bounds 跃移
transportation by load haul-dump equipment 铲运机运输
transportation by railway 铁路运输
transportation by traction 拖曳搬运
transportation capacity 运输能力;通行能力
transportation center 运输中心
transportation chain 运输系统
transportation charges 运输费(用)
transportation clearance 运输尺寸
transportation column 运输队
transportation company 运输公司
transportation competency of stream 河流输送能力
transportation congestion 运输充塞
transportation control 运输队;运输调度
transportation control card 运输控制卡片
transportation-control depot 运输控制仓库
transportation control number 运输控制编号
transportation control point 运输调度站
transportation corridor 运输走廊
transportation cost 运输成本;运输费(用)
transportation cost and fee plan in terms of cost items 分要素编制的运输成本费用计划
transportation cost and fee plan in terms of main items 按主要项目编制的运输成本费用计划
transportation cost plan 运输成本计划
transportation data system 运输数据系统
transportation demand 运输需求;交通需求
transportation demand analysis 运输需求分析
transportation demand and supply 运输供求关系
transportation demand theory 运输需求理论
transportation department 运输部门
transportation deposit 搬运沉积
transportation depot 运输站
transportation device 运输设备
transportation direction 搬运方向
transportation distance 搬运距离
transportation distance of compressed air 压气输送距离
transportation economics 运输经济学
transportation emergency 运输非常状态
transportation engineer 运输工程师
transportation engineering 运输工程(学)
transportation equipment 运输工具;交通设备
transportation facility 运输设施;运输工具
transportation facility on one side of building 在建筑物一旁的运输设施
transportation finish 运输工具涂料
transportation firm 运输公司
transportation forecasting model 交通预测模式
transportation group 运输大队
transportation hub 交通中心(城市);交通枢纽
transportation-in 运入运费
transportation industry 运输业
transportation institute 运输学院
transportation insurance 运输保险
transportation intelligence 运输情报
transportation investigation 运输量调查
transportation investment 运输投资
transportation junction 交通结点;运输枢纽
transportation leakage 许可运输漏损率;运输中的漏损
transportation lifeline 运输线

transportation limitation 运输极限
transportation management 运输管理
transportation management information system 运输管理信息系统
transportation manifest 远洋运输舱单
transportation map 交通运输图
transportation marketing 运输市场
transportation means 运输工具
transportation medium 运输工具;运输方式;运输手段;搬运介质
transportation mode 运输方式;搬运方式
transportation model 输送模型
transportation motor pool 运输汽车基地
transportation movement 运输管理
transportation network 交通运输网;运输网
transportation networks theory 运输网路理论
transportation of bed load 河床推移质的输移;河床推移质的输送;推移质输移
transportation of dangerous goods code 危险货物运输规则
transportation of debris 泥沙输移;沉积物输移
transportation of excavated material 挖出土方的运输
transportation officer 运输船上运输部门的负责军官
transportation of flood current 洪流搬运作用
transportation of goods 货物运输
transportation of groundwater 地下水搬运作用
transportation of lake 湖泊搬运作用
transportation of materials 材料运输
transportation of merchandises 货物运输
transportation of ocean current 洋流搬运作用
transportation of passengers 客运
transportation of radioactive material 放射性物质运输
transportation of refuse by rail road 铁路输送垃圾
transportation of refuse on ship board 船只运垃圾
transportation of reguse by motor-lorry 运货汽车输送垃圾
transportation of rivers 河流搬运(作用)
transportation of running water 流水搬运
transportation of sedimentary 沉积物搬运
transportation of sediments 河沙搬运;泥沙搬运;沉淀物的移运;沉淀物运移
transportation of small goods 小搬小运
transportation of tidal current 潮汐搬运作用
transportation of wave 波浪搬运作用
transportation of wind 风的搬运作用
transportation of wood 运材
transportation on the site 现场运输;施工现场运输
transportation operating agencies 运输管理机构
transportation optimization technique 运输最佳化技术
transportation order 运输命令
transportation out 销售运费;销货运费
transportation outlays 运费支出
transportation parameter 交通参数
transportation passing through the dam 过坝运输
transportation pipeline 输送管线
transportation plan 运输计划
transportation planning 运输规划;交通运输规划;交通规划
transportation policy 运输方针;运输政策
transportation pollution 交通公害
transportation pollution sources 交通运输污染源;交通污染源
transportation pool 运输车场
transportation poor 交通弱者
transportation power 搬运力
transportation priorities 运输优先权
transportation problem 运输问题
transportation process 运输过程;泥沙输移过程;输沙过程
transportation-produced noise 运输噪声
transportation production 运输生产
transportation program(me) 运输规划
transportation program(ming) 运输规划
transportation quality 运输质量
transportation railroad 铁路运输
transportation range 运输距离
transportation rate 运价
transportation rate of suspended load 悬沙输沙率
transportation report and control system 运输报告与控制系统

transportation request 运输申请
transportation research 运输研究
transportation resource 运输方法
transportation revenue 运输收入
transportation revenue accounting audit 运输收入会计核算审计
transportation revenue audit 运输收入审计
transportation revenue management audit 运输收入管理的审计
transportation revenue plan accomplishment audit 运输收入计划完成情况审计
transportation road under shaft 井下通路
transportation route 运输路线;运输路径
transportation routine 运输线
transportation safety 运输安全
transportation service accessory 运输附属服务设施
transportation services 运输服务
transportation source of water pollution 水污染交通源;上交通运输污染源
transportation station 运输站
transportation subsidies 交通费补贴
transportation supply 运输供给
transportation supply mode 运输供给模式
transportation survey 运输调查;运输量调查;运量观测
transportation system 运输系统;运输方式;港口集疏运系统;交通系统
transportation system management 交通系统管理
transportation tank 油罐车
transportation tax 运输税
transportation technology 运输技术
transportation terminal 运输终点站
transportation terminal questionnaire (method) 交通站点旅客咨询(调查)法
transportation tongs 运夹具
transportation tons 运输吨数
transportation tool 套管送入工具
transportation turbidity current 浊流搬运作用
transportation velocity 输送流量
transportation ventilation of tunnel 隧道运营通风
transportation volume 运输量
transportation within the project site 工地运输;场内运输
transportation work 运输功
transportation workshop 运输专题讨论
transportation zone 运输区
transport attendant 运输管理工
transport automobile 运输汽车
transport axis 交通中轴线
transport band 运输带
transport barge 运输驳
transport basket 运输容器
transport box 集装箱;输送容器
transport by colloidal solution 呈胶体熔液搬运
transport by complex 呈络合物搬运
transport by haloids 呈卤化物搬运
transport by lighter 驳船运输
transport by steam power 蒸汽动力运输
transport by true solution 呈真熔液搬运
transport by water 水运
transport by water of pollutant to water courses 污水流入水道
transport capacity 输送量;输沙能力
transport capacity of bed load 推移质输移量;推移质输沙量;底沙输送量
transport capacity of stream 河流运输能力
transport capacity of suspended load 悬移质输送能力
transport-capacity vs traffic-volume curve 运量适应图
transport car(t) 运输车;转运车
transport case 运输箱;包装箱
transport cask 运输桶
transport caterpillar tractor 运输用履带拖拉机
transport circuit 传送电路
transport combine 运输联合企业
transport combined of merchandises 货物联运条约
transport command 运输指挥部
Transport Committee 运输委员会
transport community 运输共同体
transport company 搬运公司
transport competency 运送能力;载运能力;运送

能力
transport competent 输沙能力
transport competition 运输竞争
transport concentration 输沙浓度
transport condition 运输条件
transport connection 传输连接
transport connection endpoint identifier 传输连接终端点识别器
transport container 运输箱;运输容器;包装箱
transport control center 运输控制中心;运输调度中心
transport controller 运输调度员
transport control number 运输控制编号
transport control system 运输控制系统
transport conveyance 交通工具运输
transport cooperation between railway departments and enterprises 路企运输协作
transport coordination 运输协作
transport corps 运输队
transport cost of goods article 货物品名别运输要求
transport cost of luggages and parcels 行李包裹运输成本
transport cost of railway line section 铁路区段运输成本
transport cost of rolling stock 机车车辆运输成本
transport crane 转运吊机;运输用的起重机;运输起重机;转运起重机;大跨度装卸桥;装卸桥
transport delay 传输延迟
transport delay unit 传送延迟器
transport density of freight traffic 货物运输密度
transport development theory 运输发展理论
transport distance for calculating tariff 货物运价里程
transport documents 运输单据
transport drift 运输平巷
transport duct loss 供气管道损失
transport earning linked revenue 运输进款挂钩收入
transport economics 运输经济学
transport economy 交通运输经济
transported bridge 运送式桥
transported capacity 搬运能力
transported clay 运积黏土
transported deposit 运积物;运积矿床;输移沉积物
transported energy of wave 传递波能;波的传递能量
transported form of ore-forming elements 成矿元素搬运形式
transported grab crane 桥式抓斗起重机
transported overburden anomaly 运积物异常
transported quantity 运输量
transported rate 搬运速率
transported regolith 搬运表土
transported soil 运积土;冲积土(壤);搬运土
transportee 运输者
transport efficiency 迁移效率
transport enterprise assets added value rate 运输企业资产增值率
transport enterprise capital profit-tax rate 运输企业资金利税率
transport enterprise cost profit rate 运输企业成本利润率
transport equation 迁移方程
transporter 运输装置;运输者;运输机;运输船;输送器;散装货装卸机;运送机;运输车
transporter bridge 桥式起重机;运送式桥;单塔式架空缆车桥;运输机桥
transporter crane 桥式起重机;行车;装卸桥;桁架桥式起重机
transporter-erector 运输及竖直安装车
transporter-erector-launcher 运输
transport ergonomics 运输环境工程学
transport grab crane 抓斗式装卸桥;抓斗式装卸重机
transporter tower 运输塔
transporter truck-tractor 运输牵引车
transporter type apron 运输工具转运式客机坪
transport expenditures 运输支出
transport expense 运送费;运输支出
transport expense audit 运输支出审计
transport factor 迁移率;输送因数
transport film 输片
transport fixed installations 运输固定设备

transport flowchart 运输流程图
transport form 转运形式
transport fuel 运输用燃料
transport function 传送功能
transport gas 输运气体
transport grade of radioactive package 放射性包装件运输等级
transport gross output 运输总产值
transport in bulk 散运
transport income 运输进款
transport income all the railway 全路运输进款
transport income dued to senior administrative agency 欠缴上级运输进款
transport income payable 应缴运输进款
transport index 运输指数
transport infrastructure 运输基本设备结构
transporting 运输
transporting action of running water 流水搬运作用
transporting capacity 运送能力;输送能力;搬运能力
transporting condition assessment 交通运输评议
transporting condition 运输条件
transporting distance 输送距离
transporting equipment 运输设备
transporting erosive velocity 输移冲刷流速;搬运侵蚀速度
transporting means for building construction 施工运输方式;施工运输工具
transporting mechanism 搬运机构
transporting outer size 运输时外形尺寸
transporting power 运送能力;输送功率;搬运能力
transporting species 运载体
transporting speed 运送速度
transporting stream 运输流
transporting system 运输系统
transporting velocity 运送速度;输送速度;输送流速;搬运速度(泥沙的)
transport input and output survey 运输投入产出调查
transport interface data unit 传输接口数据装置
transport kernel 迁移影响函数
transport law 运输法
transport layer 传送层;传输层
transport layer protocol 传输层协议
transport limitation 运输限制
transport line 运输线
transport linkage 运用铰链机构
transport liquidated revenue 运输清缴收入
transport machine 运输机
transport machinery 运输机械
transport market 运输市场
transport market and economic theories related 运输市场及相关经济理论
transport mean free path 输运平均自由程
transport mechanism 运载机构;迁移机构;传送机构
transport medium 传播介质
transport mileage 运输里程
transport mixer 搅拌车(混凝土)
transport model 运运模型;传输模式
transport module 传输模块
transport monitor system 运输监控系统
transport movement 运输
transport movement control 运输调度
transport movement control center 运输调度中心
transport network 运输网;传送网;传输网(络)
transport network control 传输网络控制
transport network data unit 传输网络数据装置
transport network endpoint 传输网络点
transport No. 运输号码
transport node 交通枢纽站;交通枢纽点
transport number 迁移数
transport of a slick 浮油迁移
transport of both travellers and goods 客货运输
transport of debris 河道砂石运移;砂石搬运
transport of dissolved organic compound 溶解有机化合物输移
transport officer 运输管理员
transport of groundwater contaminants 地下水污染物输移
transport of groundwater pollutants 地下水污染物输移
transport of hazardous materials 物品的运输
transport of moisture 水汽输移

transport of momentum 动量输移;动量输送
transport of part-load goods 零担货物运输
transport of pollutant 污染物的迁移
transport of pollutant by erosion 污染物侵蚀转移;污染物的侵蚀转移
transport of pollutant in soil 污染物在土壤中的迁移
transport of pollutants 污染物迁移
transport of silt 泥沙运移;泥沙移移;泥沙输送;泥沙搬运;河道泥沙移运
transport of wagon-load goods 整车货物运输
transport of water 输水
transporton 运送子
transport on own account 自用运输
transport operating earning of railway bureau 铁路局的运输营业收入
transport operation 运输业务
transport operator 运输经营者;运输部门
transport organization of ships 船舶运输组织
transport over inland waterways 内河运输
transport package 运输包装;包装
transport packaging 运输包装
transport pan 运输槽
transport park 运输车场
transport performance 运输性能
transport phenomenon 迁移现象
transport piece 迁移部分
transport plane 运输机;运输飞机
transport point 运输所
transport process 输运过程
transport process of pebbles 卵石输移过程
transport product 运输产品
transport production quota 运输生产定额
transport program(me) 运输方案
transport property of molecules 分子运动性
transport protocol 点到点传输协议
transport protocol data unit 传送协议数据单元;传输协议数据装置
transport quota 运输定额
transport rate 转移率;输移率
transport rate of bed load 底沙输移率
transport rate of bed sediment 底沙输移率
transport rate of suspended load 悬移质输沙率
transport ratio 透移率
transport regulation 运输管制
transport relaxation distance 输移滞后距离
transport revenue payment bill 运输收入交款单
transport revenue receivable from subordinate unit 下级欠缴运输进款
transport rocket 运载火箭
transport rotorcraft 运输直升机
transport route 运输线路
transport runway 运输机用跑道;运输机用的跑道
transport scheduling 运输计划图表的编制
transport sector expense 运输部门支出
transport semitrailer 运输用半挂车
transport service 运输业;传输业务
transport service access point 传送服务访问点
transport service revenue 客货运服务收入
transport ship 运输船;输送船
transport speed 行驶速度
transport station 传输站
transport station address 传输站地址
transport station address space 传输站地址空间
transport station header 传输站标题
transport statistics 运输统计
transport stream 运输车流
transport stress 传送应力
transport structure 运输结构
transport submarine 运输潜水艇
transport surgeon 运输船军医
transport symbol 运输符号
transport system 运输体系;热能液流系统
transport target from loss to profit achieved 扭亏增盈运输利润指标
transportation charges 运费
transportation cost 运费
transportation expense 运费
transport tax and addition 运税税金及附加
transport tensor 转移张量
transport terminal activity 运输终点站
transport terminal function 传送终端功能
transport to stock storage 搬至仓库储存
transport tractor 运输用拖拉机

transport trade 运输业
transport traffic regulating group 运输交通调度组
transport transmission 运载传递
transport troll(e)y 运输车
transport truck 运货车
transport tunnel 运输隧洞;运输隧道
transport under controlled temperature 控温运输
transport undertaking 运输事业
transport unit 传送装置;传送延迟器;传输装置
transport user 运输用户;运输车用户
transport vehicle 运输器;运输机;传送装置;运输设备;运输车辆
transport vehicle park 运输车辆停放场
transport vesicle 运转泡;运输小泡
transport volume survey 运量调查
transport water 跳汰机冲水
transport with ice between goods 夹冰运输
transport within station 站内运输
transport work ticket 运输工作通行证
trans-porvincial or trans-regional branches 跨省区分行
transposable elements 转位因子
transpose 线路交叉;转置;调换
transpose capacitor 交叉电容器
transposed conductor 换位导线
transposed differential operator 转置微分算子
transposed dyadic 转置并向量
transposed equation 转置方程
transposed integral equation 转置积分方程
transposed linear mapping 转置线性映射
transposed mapping 转置映射
transposed matrix 转置矩阵;转置伴随矩阵
transposed operator 转置算子
transposed pair 交叉线对
transposed representation 转置表示
transposed storm 移置风暴;移置暴雨
transposed strip 换位导线
transposed transmission line 换位输电线路
transpose matrix 矩阵传置
transpose of a matrix 转置矩阵;矩阵转置;矩阵的转置
transpose of a relation 关系的转置
transposer 移项器;换位器
transposing gear 中间齿轮
transposition 易位;移置;移项;移位;转置;转位;换位;数字倒置;对换;调换;导线交叉(法);反接;变换;反位
transposition adjustment 移置调整
transposition arm 交叉臂
transposition arrangement 交叉配置
transposition block 换位线接板;换位接线板
transposition board 交叉板
transposition changeover box 交叉转换盒
transposition circuit 交叉电路
transposition error 交叉误差;换位误差
transposition foliation 置换面理;置换剥理;置换剥离
transposition index 交叉指数
transposition insulator 交叉绝缘子;换位绝缘子
transposition limit 移置限制
transposition line 换位线(连续梁图解法)
transposition method 相似流域比拟法;移置法
transposition number 交叉序号
transposition of storm 风暴移置;暴雨移置
transposition of unit hydrograph 单位过程线移置
transposition pattern 交叉图
transposition pin 交叉用绝缘子直螺脚
transposition section 交叉区
transposition system 移位制;交叉制式;导线交叉制式
transposition tower 换位塔;导线交叉塔;导线交叉杆
transposition weighing 换位置称法
transposition wire 换位用导线
transposon 转位子
transpression 转换压缩;转换挤压
transput 传输
transput declaration 传输说明
trans-quantitative method 转化定量法
transradar 变换雷达
transrectification 阳极检波;交换整流
transrectification characteristic 阳极检波特性;板极检波特性曲线
transrectification factor 阳极检波系数;检波系数;

换流因数
transrectifier 电子管检波器;板极检波器
transregional cooperation 跨区协作
transregional purchase of commodities 跨区收购
transregional sales of commodities 跨区交易
transregional supply of commodities 跨区供应
transring 横环
transsection 横切面
transseptal fibers 横隔纤维
transship 转运;驳运;驳船
transship cargo 转船货;中转货物
transship entry 中转货进口报单
transshipemt centre 中转中心
transshipment 转运;转船运输;转船;中转(客货运输);货物换装;驳载;驳运
transshipment acitivity 转运机构
transshipment and recording 使用车计划
transshipment area 转运区(域)
transshipment bill of lading 转运提单;转船提单
transshipment by lying a part of goods unloaded in a part-load wagon 坐车中转
transshipment by unloading goods from one part-load wagons to goods section then loading into another 落地中转
transshipment cargo 转运货物;转船货
transshipment cargo rate 联运运价
transshipment charges 转船费;中转作业费
transshipment crane 转运起重机;输送起重机
transshipment delivery order 转船栈单;转船交货单
transshipment entry 转船货报关单
transshipment facility 装卸设备
transshipment from one part-load wagon to another directly 过车中转
transshipment gallery 栈桥
transshipment harbo(u)r 换向港
transshipment note 转船通知
transshipment of cargo 货物装卸
transshipment of intermodal cargo 联运货物转载
transshipment of part-load goods 零担中转作业
transshipment of through cargo 联运货物转载
transshipment part-load wagon 中转整零车
transshipment permit 转船许可证
transshipment point 转载点;转运站;货物转运站
transshipment port 中转港(口);换向港
transshipment shed 转运货棚
transshipment shipping bill 转船货装船单
transshipment station 中转站;换装站
transshipment surcharge 转船附加费
transshipment terminal 中转枢纽;中转码头;换装枢纽;换装码头
transshipment track 换装线【铁】
transshipping of cargo at waterfront 岸边装卸货物
transshipping of goods at waterfront 岸边装卸货物
transship train 小运转列车
transspecific 特异性转移
transsuper aerodynamics 跨音速和超音速空气动力学
trans-tactic 有规反式构形
transtage 中间级
transtainer 移动式集装箱吊运车;特兰斯泰纳
transtainer system 高架换装机法
transtat 可调变压器;自耦变压器
transtension 转换拉伸
transtensional basin 转换拉张盆地
transthermy 透热法
transtidal wave 超潮波
transtmitted-carrier operation 载波发送制
transtrum 屋架横木(古罗马);横梁(古罗马建筑);[复]transtra
transtympanic electrode 穿鼓膜电极
transudate 漏出液;漏出物
transudation 渗出作用
transudation seeping 漏出
transuranic element 越铀元素;超铀元素
transuranic waste 超铀废物
transuranium 超铀
transuranium element 铀后元素;超铀元素
transuranium metals 铀后金属
transuranium waste 铀后废物
transus temperature 转变温度
Transvaal jade 绿钙铝榴石

Transvaal old land 德兰士瓦古陆
transvar 可调定向耦合器
transvar coupler 可调定向耦合器;定向可变耦合器
transversal 贯线;截线;横截(线)
transversal beam 横梁
transversal brace 横向支撑
transversal bracing 横向支撑
transversal branch 横支
transversal bubble 照准部水准器;横向水准仪;横向水准器
transversal coast 横向海岸;横海岸
transversal coastline 横向海岸线
transversal cofferdam 横向围堰
transversal crack 横向裂纹;横向裂缝
transversal current 横向水流
transversal discrepancy 横向差值;横向不符值
transversal displacement 横向位移
transversal dune 横向沙丘;横向沙垄
transversal earthquake 横震
transversal electromagnetic force 横向电磁力
transversal erosion 横向侵蚀
transversal extension 横向引长;横向扩张
transversal fault 横推断层
transversal filter 横向滤波器
transversal force 横向力
transversal inclinometer 横向倾斜仪;横向倾斜计
transversality 相截性;横截性
transversal level 横轴水准器;横向水准仪;横向水准器
transversal line 横轴线
transversal load(ing) 横向荷载
transversally swing migration 横向摆动迁移
transversal magnification 横向放大率
transversal mass 横质量
transversal position 横向位置
transversal pre-stress steel reinforced concrete mast 横腹杆预应力钢筋混凝土支柱
transversal rip 横向裂痕;横向裂缝
transversal river 横向河
transversal scale 复比例尺
transversal section 横断面
transversal shiplift incline 横向斜面升船机
transversal shrinkage 横向收缩
transversal stabilization 横向稳定
transversal strain 横向变形
transversal stream 横向河
transversal stress 横向应力
transversal surface 横截面
transversal switch 水平开关
transversal table 横移车
transversal thrust 横向推进;横向推动;横冲断层
transversal turbulent diffusion coefficient 横向紊动扩散系数
transversal type district station 横列式区段站
transversal type passing station 横列式会让站
transversal ventilation 横向式通风
transversal wave 横波
transversal wave velocity 横波速度
transversal zoning 横向分带
transverse 横的;横隔;横断;横的
transverse abutment 横向拱座系统;横向支座系统;横向桥台
transverse acceleration 侧向加速度
transverse acoustic(al) mode 横向声学模式
transverse airflow 横向气流;横向空气流动
transverse aisle 横过道
transverse anisotropy 横向各向异性
transverse anterior line 横前线
transverse apparent resistivity curve 横向视电阻率曲线
transverse arch 横向拱;横拱
transverse architrave 横线脚
transverse-arch kiln 横拱窑
transverse arch of foot 横弓
transverse arm 横臂
transverse arrest 横阻
transverse artery 横贯干线
transverse axis 横坐标轴;横轴线;横轴
transverse baffle 折流板;横向挡板
transverse bar 与海岸约成直角方向延伸的沙洲;横向沙洲;横向沙坝;横向潜洲;横沙洲;横沙坝;横贯沙洲
transverse bar deposit 横向沙坝沉积
transverse barrel vault 横筒式拱顶;横向拱顶

transverse basin 横切盆地
transverse bead notched bend test 横向焊道缺口弯曲试验
transverse beam 横臂;横向梁;横梁
transverse beam for timber truss bridge 木桁架桥横梁
transverse beam travelling-wave tube 横向柱行波管
transverse bearing 向心轴承;径向轴承
transverse belt conveyer 横带式皮带输送机;横带输送机;横向胶带输送机;横向带式输送机
transverse bending 横向弯曲
transverse bending action 横向弯曲作用
transverse bending moment 横向弯矩
transverse bending stiffness 横向弯曲劲度
transverse bending strength 横向抗弯强度;横向抗挠强度
transverse (bending) test 横向弯曲试验
transverse bending testing machine 横向弯曲试验机
transverse bent 横向排架;横向构架;横排架
transverse berthing velocity 横向靠船速度
transverse block-faulted region 横向块断区
transverse bow 横向支撑;横向凸凹边缘
transverse bow propeller 船首横向推进器
transverse brace 横向支撑;支柱;横向竖联;横木;横拉筋;横撑
transverse brace system 横向连接系统;横撑系统
transverse bracing 横向支撑;横拉条;横撑
transverse bracket 横向肘板;横向托座;横向牛腿
transverse buckling 横向屈曲
transverse bulkhead 横隔墙;横隔舱;横隔壁;横舱壁
transverse bulkhead stiffener 横舱壁加强筋
transverse cable 横向钢丝绳
transverse cable force 横向系缆力
transverse center line 横向中心线
transverse center of buoyancy 浮心横向位置
transverse center of gravity 重心横向位置;横向重心
transverse centre line 横向中心线
transverse channel 横向渠道
transverse chart 横轴海图;横切墨卡托投影图;横切渐长投影图
transverse check 横向检验
transverse chordal tooth thickness 横向弦齿厚
transverse circulating current 横向环流
transverse circulation 横向环流
transverse cleavage belt 横向劈理带
transverse cleavasse 横向冰隙
transverse coastline 横式海岸线
transverse cofferdam 横向围堰
transverse component 横向分量;横向分力
transverse compression wave 横向压缩波
transverse conductivity 横向传导率
transverse conduit 横向管
transverse construction joint 横向工作缝;横向施工缝
transverse contraction 横向收缩
transverse contraction joint 横向收缩缝
transverse control 横向控制
transverse conveyer 横向输送机;横向运送机;横向运输机
transverse copying 横向仿形
transverse correction 横向修正
transverse correlation coefficient 横相关系数
transverse corrugated 横向瓦楞状的;横向波状的
transverse coupling 横向耦合
transverse crack 横向裂纹;横向裂缝;横向开裂;横裂缝;横裂
transverse cracking 横裂
transverse crawler travel(l)ing gear 横向履带爬行机构
transverse crevasse 横向裂纹;横向裂缝
transverse cross-section 横截面
transverse crown bar 横顶杆
transverse crown stay 横顶杆
transverse current 横向水流;横向电流;横向潮流;横流
transverse current force 横向水流力;横向潮流力
transverse current force coefficient 横向水流力系数
transverse current microphone 串联传声器
transverse current velocity 横向流速

transverse curvature 横向曲率
transverse cutting 横切
transverse cutting bridge 横向切割机桥架
transverse cutting geological profile 横切地质剖面图
transverse cutting machine 横向裁剪机
transverse cutting profile 图切剖面图
transverse cylindric(al) conformal projection with two standard meridians 双标准经线等角圆柱投影
transverse cylindric(al) orthomorphic chart 横切墨卡托投影图;横切渐长投影图;横轴墨卡托投影(地)图
transverse cylindric(al) orthomorphic projection 横轴圆柱正形投影
transverse cylindric(al) orthomorphic projection chart 横轴圆柱正形投影图
transverse deformation 横向变形
transverse deviation 横向偏差
transverse diameter 横径
transverse diameter of pelvic inlet 入口横径
transverse diaphragm 横隔板
transverse differential protection 横联差动保护;横差保护
transverse diffusion 横向扩散
transverse diffusion tensor 横向扩散张量
transverse dike 横堤;跳水坝;挑流坝;丁坝
transverse dip component versus depth diagram 横向倾角分量—深度图
transverse direction 横向
transverse disk 横盘;暗板
transverse dislocation 横向位错;横变位
transverse dispersion 横向弥散
transverse dispersion coefficient 横向弥散系数
transverse displacement 横向位移
transverse distribution 横向分布
transverse distribution of load 荷载横向分布
transverse Doppler effect 横向多普勒效应
transverse drainage 横向排水
transverse drainage for bridge surface 桥面横向排水
transverse drainage method 横向排水法
transverse drain ditch 横向排水沟
transverse dredging 横挖【疏】
transverse dredging method 横挖法
transverse dummy joint 横向假接缝
transverse dune 横向沙丘;横向沙垄
transversed valley 横谷
transverse earthquake 横向地震
transverse edge effect 横向端部效应
transverse effect 横向效应
transverse elasticity 横向弹性
transverse elastic test 横向弹性试验
transverse electromagnetic wave 横电磁波
transverse electromechanical coupling factor 横向机电耦合系数
transverse engine 横置发动机
transverse equator 横轴投影赤道
transverse expansion 横向膨胀
transverse expansion joint 横向伸缩缝;横向膨胀缝
transverse extension 横向引长;横向伸长;横向扩张
transverse extensometer 横向应变计;横向伸缩仪
transverse facial cleft 面横裂
transverse failure 横向断裂
transverse fault 横断层
transverse feed 横向进给
transverse fiber 横向纤维
transverse field 横向场
transverse field travelling-wave tube 横向场行波管
transverse fillet weld 正面角焊;横向贴角焊缝;横向角焊缝
transverse-film bolometer 横膜辐射热测量仪
transverse fissure 横向裂纹;横向裂缝;横向接缝;横裂缝;横裂
transverse fissure detector 横裂缝检查器
transverse flexural strength 横向抗弯强度
transverse flow 横向水流;横向流(动);横流
transverse flow oil separator 横向流除油器
transverse flow velocity 横向流速
transverse flux magnet 横向磁力线磁体
transverse fold 横褶皱;横跨褶皱
transverse force 剪力;横力
transverse force determination 确定横向(剪)力

transverse fracture 横向破裂
transverse frame 横架;横向构架;横肋骨;横骨架
transverse framed ship 横骨架船
transverse frame system 横骨架结构
transverse framing 横骨架结构
transverse furrow 横沟
transverse gallery 横向平巷;横洞
transverse gear 横动装置
transverse girder 横梁;横向桁材;横向大梁;横大梁
transverse gradient 横向梯度;横向坡度;横向比降;横比降
transverse graduation 斜分尺
transverse graticule 横轴投影经纬网;横轴投影格网
transverse grooving 横向抗滑槽纹(路面)
transverse hall 横厅
transverse heat conduction 横向热传导
transverse heating 横向射频加热;横向加热
transverse holding ground 横截河埂储[贮]木场
transverse illumination 横向照明
transverse impact modulator 对冲型元件
transverse impedance 横向阻抗
transverse incision 横切口
transverse inclination 横倾
transverse incline shiplift 横向斜面升船机
transverse inclinometer 横向倾斜计
transverse interchange of sediment 泥沙横向交换
transverse interference 横向干扰
transverse interval 横向间距;横间隔宽度
transverse isotropy 横向同性;横向各向同性;横各向异性
transverse jetty 横栈桥
transverse joint 横向接头;横向接缝;横节理;横缝
transverse joint cutter 横向切缝机;切缝机
transverse joint edge 横向接缝边
transverse joint spacing 横缝间距
transverse latitude 横轴投影纬度
transverse launching 横向下水
transverse length extension vibration mode 横向长度伸缩振动模式
transverse level link 横杆连接杆
transverse lever support 横杠托
transverse lines 横线
transverse load 横向荷载;横向负载;横向负荷
transverse load distribution 荷载横向分布
transverse loading 横向荷载
transverse longitude 横向投影经线
transverse-longitudinal cutting 横向纵切
transverse lumber 横材
transversely arranged frame 横向布置的构架;横向布置的框架
transversely distributed line load 横向分布线荷载
transversely hinge-connected slab method 铰接板法
transversely isotropic material 横向同性材料
transversely loaded 横向承载的;受弯曲荷载的;受横向荷载的
transversely traversable 可横越的;横向可通过的
transverse magnetic mode 横磁模式
transverse magnetic wave 横磁波
transverse mass 横向质量
transverse member 横向构件;横向撑;横梁;横臂;挺杆;吊杆(起重机的)
transverse Mercator chart 横轴墨卡托投影(地)图;横切墨卡托投影图;横切渐长投影图
transverse Mercator projection 横轴墨卡托投影;横切墨卡托投影;横切渐长投影
transverse Mercator projection chart 横轴墨卡托投影(地)图
transverse Mercator projection grid 横轴墨卡托投影格网
transverse meridian 横轴投影经线
transverse metacenter 横稳心;横稳定(倾)中心
transverse metacentric height 横向定倾中心高度;横稳心高度
transverse metacentric radius 横向稳心半径
transverse microscopic dispersion 横向微观弥散
transverse migration 横向迁移
transverse mode 横模
transverse mode control 横向模腔
transverse mode-controller 可控横模
transverse model datum 横向模型基准面
transverse module 端面模数
transverse modulus of elasticity 剪切弹性模数;剪切弹性模量;横向弹性模量

transverse modulus of rupture 弯曲强度极限
transverse moment 横向力矩
transverse momentum 横向动量
transverse motion 横向运动
transverse movement 横向运动
transverse movement of material 泥沙横向运动
transverse moving blade shaper 横向移动刮刀式整形机
transverse neutral axis 横向中和轴
transverse number 横构件特征数
transverse optic(al) modulator 横向光调制器
transverse oscillation 横向震荡;横向振荡;横向摆动
transverse outlet 出口横径
transverse oval cupola 横向椭圆屋顶
transverse oval dome 横向椭圆穹隆
transverse parallax screw 横向视差螺旋
transverse parking 横向停车
transverse part 横部
transverse pavement line 横道线;道路横向标线
transverse permeability 横向透水性;横向渗透性
transverse phase curve 横向相位曲线
transverse piezoelectric(al) effect 横压电效应
transverse pile-cap girder 横向桩头帽梁
transverse pipe 横向管
transverse pitch 横向中心距;横向节距;齿面
transverse pitch of rivets 铆钉横距
transverse plane 横向偏转板;横剖面
transverse plank 横板
transverse pole 横轴投影极点
transverse position 横向位置
transverse pressure 横向压力
transverse pressure angle 横压力角
transverse-pressure gradient 横向压力梯度
transverse prestress 横向预应力
transverse process 横突
transverse profile 横剖面图;横剖面
transverse-profile angle 端面齿形角
transverse profile of lithologic(al) phase 岩性岩相横剖面图
transverse profilometer 横向验平仪
transverse profilometer truss 横向验平析架
transverse projected area 正面投影面积
transverse projection 横轴投影;横投影;横投影
transverse projection parallel 横向投影纬线
transverse pull 横向拉力;横向拉拔
transverse radius of curvature 平行圈曲率半径;纬度圈曲率半径
transverse reciprocating blade 横向往复式刀片
transverse recording 横向记录
transverse reinforcement 横向钢筋
transverse relaxation 横向松弛
transverse rhumb line 横轴投影恒向线;横轴投影等变形线
transverse rib 横肋
transverse ribs clast stripes 横向肋状碎屑带
transverse rib slab 横向排肋(厚)板
transverse ridge 横脊(线);横向山脊
transverse ridge-rib 横向脊肋
transverse rigidity 横向刚性;横向刚度
transverse ripple mark 横向波痕
transverse rocking force 横向摇摆力
transverse rod 横杆
transverse rolling 横轧
transverse roof 横向屋顶
transverse roof bar 横顶杆
transverse roof stay 横顶杆
transverse round bar 横圆拉杆;横圆棒
transverse rupture strength 横断面视图
transverse scale 斜线读微尺;横线读数
transverse scan 横向扫描
transverse scoring machine 横切机
transverse scour mark 横向冲刷痕
transverse screed 横向整平板;横向整平机
transverse screed tension 横向整平拉力
transverse screw 横向螺旋;横向螺丝
transverse sea 横向浪
transverse seam 金属薄板的横接缝;横向缝;圆周接缝;横接缝;横缝
transverse seam welding 横向缝焊(接)
transverse section 横截面;横剖面;横断面图;横断面
transverse section figure 横断面形状
transverse section of grotto 溶洞横剖面图
transverse section shape 横断面形状

transverse seiche 横向静振
transverse sensitivity 横向灵敏度
transverse sensitivity ratio 横向效应系数;横向灵敏系数;横向灵敏度比
transverse shaft 横轴
transverse shaper 横列机
transverse shear 横向剪切;横向剪力;横切力
transverse shearing stress 横向剪应力
transverse shear testing 横切剪力试验
transverse shear wave 横切边波
transverse shift 横移
transverse shrinkage 横向收缩
transverse shrinkage rate 横向线缩率;横向收缩系数
transverse sidewalk grade 人行道横坡
transverse sleeper 枕木;轨枕;横轨枕
transverse slide 横向滑杆;横向滑板
transverse slipway 横向滑道
transverse slope 横向坡降;横向坡度;横向降;横坡度;横坡;横比降;起拱路面横坡
transverse slope of bridge deck 桥面横坡
transverse slope of water surface 水面横比降
transverse slot 横向缝隙
transverse snapping 横向瓣断
transverse sound velocity 横声速
transverse space 横距
transverse spacing 横间距
transverse span wire 横撑力素
transverse spheric(al) aberration 横向球面像差
transverse spreading blade 横向铺料板
transverse spring 横置弹簧;横弹簧
transverse stability 横向稳定性;横稳性【船】
transverse stabilization cable 横向稳定钢缆;横向稳定钢索
transverse stack 横向叠加
transverse stage 横式舞台
transverse stay 横撑条;横向拉条;横向拉杆
transverse steel 横向钢筋
transverse stereographical projection 横球面投影
transverse stiffener 横向加劲构件;横向加强构件;横向加劲杆;横刚材
transverse stiffening 横向加强;横向加固
transverse stiffness 横向劲度;横向刚性;横向刚度
transverse strain 横向应变
transverse strain moment 横向应变矩
transverse strength 弯曲强度;抗弯强度;挠曲强度;横向强度
transverse stress 弯曲应力;横向应力
transverse stressing 横向应力
transverse stretching 横向张拉;横向拉伸
transverse strut 横撑;横杆
transverse stylolitic seams 横向缝合线
transverse submerged arc welding 横列双丝并联埋弧焊
transverse surface 横截面;横断面
transverse sweep 横向扫肃
transverse system 横骨架结构
transverse system of framing 横骨架式
transverse table 横移台;横移车
transverse tamping 横向夯实
transverse tangent cylindric(al) conformal projection 等角横切圆柱投影
transverse tangent cylindric(al) equidistant projection 横轴等距离切圆柱投影
transverse tangent cylindric(al) equivalent projection 横轴等面积切圆柱投影
transverse tectonic fissure 横向构造缝
transverse tendon 横向筋
transverse tension crack 横向受拉裂缝
transverse tensioning 横向张力
transverse test 抗弯试验;抗变试验
transverse thickness profile 横向厚度剖面
transverse thrust 横推断层;横冲断层
transverse thruster 横向推进器
transverse thrust unit 侧推装置
transverse tie bar 横向系杆
transverse tissue 横列组织
transverse tooth thickness 横向弦齿厚;横向齿厚
transverse transport of bedload 底沙横向输移
transverse transport of sediment 横向输沙;横向泥沙输移
transverse tube 横向管
transverse tubule 横小管;横管
transverse tubule system 横管系统

transverse tunnel vault 横向圆筒拱顶
transverse type of shop 横列式车间
transverse valley 横谷
transverse vein 横脉
transverse velocity 横向速度
transverse velocity gradient 横向速度梯度
transverse ventilation 横向通风
transverse vertical plane 横垂直平面
transverse vibration 横向振动
transverse vibrator 横向振捣器
transverse wagon vault 横向筒形拱顶
transverse wall 横隔墙;横向外墙;横切壁;横墙
transverse warping 横向弯翘;横向翘曲
transverse water flow 横向水流
transverse wave 横波;地震横波
transverse web 加强肋骨;横向桁材
transverse web frame 横强肋骨
transverse wedge 横楔
transverse weld 横向焊缝
transverse white bands of nails 指甲白色横纹
transverse wind force coefficient 横向风力系数
transverse wing wall 横向翼墙
transverse wire of the cloth 金属丝网纬线
transversing gear 横动机构
transversing mechanism 横动机构
transversion mutation 颠换型突变
transversum 横骨
transverter 转换器;换能器;换流器;变压整流机;变换器
transveyer 运送机;输送机
trans-vinylation 乙烯基转移作用
transvoal jade 钙铝榴石
trans-watershed diversion 跨流域引水
transwitch 硅开关
transyte 窄门道(中世纪建筑);窄走廊(中世纪建筑)
tranuranic element 铀后元素
tranylcypromine 反苯环丙胺;苯环丙胺
trap 陷阱;陷波器;油陷;油圈闭;止回器;阱;截流湾;活板;圈闭;曲颈管;汽水阻;疏水器;收集器;存水弯管;存水弯;存水器;存水井;沉沙井;沉沙槽;分离器;返水弯管;捕捉;捕获;捕获器;捕俘;暗色岩
trap above unconformity 不整合上的圈闭
trap address 陷阱地址;捕获地址
trap amplifier 陷波放大器
trap arm 疏水器臂;存水弯高度
trap-ash 暗色岩灰
trap below and above unconformity 不整合上下圈闭
trap bit 陷阱位
trap bottom wagon 活底货车;活底车;底开门车
trap cellar 舞台下活动门小室
trap circuit 吸收滤波器;吸收电路
trap coil 陷波线圈
trap dike 暗色岩墙
trap dip 疏水器的最低部位
trap door 通气门;上翻门;陷阱门;落板装置;落板口;活板门;调节风门;天窗;水平翻门;底板孔口盖;仓门
trapdoor fault 翻转断层
trap door on roof 屋顶天窗;屋顶活门
trap-down 下落断片
traped particle 俘获粒子
trap effect 陷阱效应
trap efficiency 拦沙效率;截淤效率
trap efficiency of reservoir 水库淤沙效率;水库截留(泥沙)效率
trap elevator 舞台(活动)地板门电梯
trap energy 陷阱能
trape notch 梯形凹口;过水口;梯形堰;梯形缺口
trapeze 经纬度格;梯形
trapeze hanger 悬挂托架
trapezial 斜方形的
trapeziform 斜方形的;不等四边形
trapezium 斜方形;梯形;不规则四边形;不等(边)四边形
trapezium-type 梯字型
trapezium yard 异腰梯形车场;梯形车场
trapezohedral tetartohedron 偏方四分面体
trapezohedron 偏方三八面体;四角三八面体
trapezoid 斜方形的;梯形;不规则四边形;不等边四边形
trapezoidal 梯形的
trapezoidal abutment 梯形桥台

trapezoidal arch 梯形拱
trapezoidal bay 梯形开间;梯形排架间距
trapezoidal belt 梯形皮带
trapezoidal boudin 梯形石香肠
trapezoidal box girder 梯形箱式大梁;梯形箱梁
trapezoidal brick 梯形砖
trapezoidal bucket 梯形料斗;梯形桶
trapezoidal buttress dam 梯形坝
trapezoidal canal 梯形截面河
trapezoidal channel 梯形截面槽;梯形渠道;梯形河槽
trapezoidal chute 梯形溜槽
trapezoidal cleavages 梯形劈理
trapezoidal concrete box girder bridge 梯形混凝土箱梁桥
trapezoidal cross-section 梯形横截面;梯形(横)断面
trapezoidal cut 梯形掏槽
trapezoidal deflection 梯形偏转
trapezoidal distortion 梯形失真;梯形畸变
trapezoidal ditch 梯形边沟
trapezoidal divisor 梯形分割绘制法
trapezoidal eaves gutter 梯形檐沟
trapezoidal excavator 梯形挖掘机
trapezoidal flume 梯形水槽;梯形渡槽;梯形测流槽
trapezoidal formula 梯形公式
trapezoidal frame 梯形构架;梯形结构;梯形刚构
trapezoidal frame bridge 梯形刚架桥
trapezoidal generator 梯形脉冲发生器;梯形波发生器
trapezoidal girder 梯形大梁
trapezoidal groin 梯形丁坝
trapezoidal groyne 梯形丁坝
trapezoidal gutter 梯形檐槽;梯形排水沟
trapezoidal hydrograph 梯形过程线
trapezoidal integration 梯形积分法;梯形积分
trapezoidal jig 梯形跳汰机;梯形淘汰机
trapezoidal law 梯形定律(应力分布的)
trapezoidal leading mark 梯形导标
trapezoidal load 梯形荷载
trapezoidal masonry 砌多角石圬工;多角石砌圬工
trapezoidal metal plate 梯形金属板
trapezoidal metal sheet 梯形金属薄板
trapezoidal method 梯形法
trapezoidal modulation 梯形调制
trapezoidal notch 梯形槽口
trapezoidal notch weir 梯形槽口(量水)堰
trapezoidal pass 梯形孔型
trapezoidal pattern 梯形图
trapezoidal piece 梯形段;梯形构件;梯形零件;梯形件;梯形块;梯形片
trapezoidal profile 梯形剖面;梯形断面;梯形侧断面
trapezoidal projection 梯形图幅投影;梯形投影
trapezoidal pulse 梯形脉冲
trapezoidal purlin(e) 梯形檩条;梯形桁条
trapezoidal reinforced concrete channel 梯形钢筋混凝土槽
trapezoidal rib 梯形肋
trapezoidal rigid frame 梯形刚架
trapezoidal roof truss 梯形屋架
trapezoidal rule 折梯形总计法;梯形规则;梯形法则
trapezoidal sandwich panel 梯形复合板;梯形夹芯板
trapezoidal section 梯形断面
trapezoidal-shaped belt 梯形皮带
trapezoidal share 梯形犁铧
trapezoidal shock pulse 梯形冲击脉冲
trapezoidal slab 梯形厚板
trapezoidal speed-time curve 梯形速度—时间曲线
trapezoidal steel box girder 梯形的箱形钢梁
trapezoidal steel box girder bridge 梯形钢箱梁桥
trapezoidal strap 三角皮带
trapezoidal tear test 梯形撕裂试验
trapezoidal terrace 梯形阶地
trapezoidal thread 梯形丝扣;梯形螺纹
trapezoidal thread ga(u)ge 梯形螺纹规
trapezoidal thread tap 梯形丝锥
trapezoidal three-layer(ed) panel 梯形三层胶合板
trapezoidal trim 梯形门窗贴脸;梯形饰条
trapezoidal truss 梯形桁架
trapezoidal truss frame 梯形桁架式构架
trapezoidal-type section 梯形(横)断面
trapezoidal unit 梯形构件
trapezoidal vault bay 梯形拱顶跨度;梯形拱顶开间

trapezoidal voltage 梯形电压
trapezoidal waveform 梯形波(形)
trapezoidal weir 梯形堰;梯形槽口(量水)堰
trapezoidal wood lath(ing) 梯形灰板条(抹灰用)
trapezoidal worm 梯形蜗杆
trapezoid body 斜方体
trapezoid bucket 梯形铲斗
trapezoid cross-section 梯形截面
trapezoid formula 梯形公式
trapezoid line 斜方线
trapezoid method 梯形试样法
trapezoid overflow pipe 梯形溢流管
trapezoid pattern data buffer 梯形图案数据缓冲存储器
trapezoid pattern data register 梯形图案数据寄存器
trapezoid pedestal 裙形垫底
trapezoid rule 梯形规则
trapezoid slot 梯形槽
trapezoid tear strength 梯形试样撕破强力
trapezoid truss 梯形桁架
trap field 飞靶射击场
trap filter 陷波器;陷波滤波器
trap formed by a single curved fault 由单一弯曲断层形成的圈闭
trap formed by intersection of a salt plug barrier with fault 盐栓遮挡—断层圈闭
trap formed by intersection of fault with fold 由断层与褶皱交切断层形成的圈闭
trap formed by intersection of structural "nose" with a fault 由断层与构造鼻交切形成的圈闭
trap formed by several intersecting faults 由若干交切断层形成的圈闭
trap formed by two intersecting faults 由二交切断层形成的圈闭
trap for oil 油捕集器
trap frequency 陷波频率
trap gate 拦沙闸(门);截淤闸门
trap-hole 陷阱
trap in 陷入
trap instruction 陷阱指令
trap level 自陷级
trap loader 活底装载机
trap mask 陷阱屏蔽
trap mine 触发地雷
trap model 陷阱模型
trap on the migration line 在运移路线上的圈闭
trap operation 陷阱操作
trap out 陷出
trap outlet 疏水器出口;存水弯出口
trap pattern 陷阱型
trapped 截留的
trapped air 存气;内部气泡;残存空气;滞流空气;裹入的空气
trapped air cushion vehicle 空气润滑气垫船
trapped air in old ice 古冰气泡中的空气
trapped-air process 吹塑薄膜挤出法
trapped air void 存气空隙
trapped argon 捕获氩
trapped bathroom gull(e)y 浴室排水沟;浴室集水沟
trapped dust 捕集粉尘
trapped electromagnetic wave 封闭电磁波
trapped electrons 捕获电子
trapped fluid 截面液;被困油液
trapped fuel 存油
trapped gull(e)y 带水封雨水口
trapped humidity 残存湿度(指气体中含有的水分)
trapped magnetic field 捕集磁场
trapped mode 集中振荡模
trapped moisture 残存潮湿;残留潮湿(指多孔固体材料中含有的水分)
trapped oil 被困油液
trapped oxide 夹杂氧化物
trapped particle dynamics 捕获粒子动力学
trapped particle instability 俘获不稳定性;捕获粒子不稳定性
trapped-program(me) interrupt 捕捉程序中断
trapped radiation 俘获辐射
trapped rare gases 捕获稀有气体
trapped species 俘获组分
trapped state 受限制姿态;捕获状态
trapped water 埋藏水
trapped wave 陷波;拦获波

trapper 捕捉器;通风口值班工人(采矿)
trappide 暗色岩
trapping 陷阱型;引导传播;入陷转移;湿套印性;叠印效果;捕集;捕捉;捕俘
trapping apparatus 捕集仪
trapping centre 陷获中心
trapping constant 陷获常数
trapping corner 致死角;截留角
trapping cross-section 捕集截面
trapping effect 俘获效应
trapping index 陷扰指数
trapping instability 俘获不稳定性
trapping layer 陷获层;吸收层;阻挡层
trapping level 陷获级
trapping mode 陷阱方式;入陷状态;捕捉状态
trapping muck 卸碴
trappings 装饰;服饰
trapping spot 陷点
trapping structure 拦沙建筑物
trapping temperature 捕集温度
trapping theory 陷波理论
trap pit 水封阴井
trappoid 暗色岩状
trappoid breccia 暗色岩状角砾岩
trap points 防护道岔;阻止道岔;保护道岔
trap pressure 气液分离器内压力
trap primer 存水弯充水;存水弯原有存水
trap rejection 陷波频率抑制
trap rock 两层沥青间的垫石;暗色岩;深色火成岩
trap sandstone 暗色砂岩
trap screw 存水弯螺旋栓;存水弯的螺丝塞
trap seal 存水弯水封;水封深度
trap sealed high below unconformity 封闭于高部位的不整合面下圈闭
trap sealed lower below unconformity 封闭低于不整合面的圈闭
trap setting 设陷阱;捕捉装置;捕捉设置
trap siding 避难线【铁】
trap standard 水封标准
trap state 捕捉状态
trap switcher 陷波开关
trap test 收集性能试验
trap theory 陷阱理论
trap track 避难线【铁】
trap-tufa 暗色岩灰
trap tuff 玄武凝灰岩
trap type of oreforming structures 成矿构造圈闭类型
trap under unconformity 不整合面下圈闭
trapunto 提花结纬凸纹布
trap valve 滤阀;除污阀
trap vector 捕获向量
trap weir 疏水器的最高部位;翻板堰;翻板坝
trap window 活板窗
trap with removable lower part 底部可拆的存水弯
trascoro 唱诗班席位(西班牙教堂的)
trash 垃圾;夹杂物;废物
trash baffle 胸墙;拦污墙
trash bar 拦污栅;刮土板机;拦污条
trash boom 拦污栅;拦污浮排
trash bund 防蚀埝
trash burning 废物燃烧
trash can 垃圾箱;垃圾筒;储[贮]尘箱;金属制垃圾箱
trash chain 杂草清除链
trash chamber 排杂器壳;除杂室
trash chute 垃圾滑槽;泄水滑槽;垃圾道;倒垃圾管;倒垃圾槽;废物滑槽;废料滑槽
trash clearance 耕耘机机架地隙
trash closet 垃圾间
trash cultivator 中耕除草机
trash cutter 杂草切碎器;废茎残叶切除机
trash disposal 污物处置
trash drum 草屑清除滚筒
trash dump 垃圾堆
trash eliminator 杂质分离器;除草残茬机
trashery 废物;垃圾
trash extractor 残屑分离机
trash farming 残留耕作;残茬耕作
trash gate 垃圾闸门
trash-intercepting device 拦污设施;拦污设备
trash link 除杂输送带条杆
trash mover 清除杂质输送器
trash plow 除草犁

trash pump 排污泵
trash rack 粗筛;细密拦污栅;拦污栅;拦废物栅;垃圾格栅
trash rack cleaner 拦污栅清洗器
trash rack loss 拦污栅(水头)损失
trash rack rake 拦污栅;拦污栅清理耙;清污耙
trash rack well 沉污井;拦污井
trash rake 杂草残茬楼耙
trash rake hoist 清污起重机
trash receptacle 废料桶
trash-removal device 除污装置;除污设备
trash remover 杂草与残茬抛杂器
trash roll 排杂辊
trash screen 拦污栅;拦废物筛;垃圾筛网;垃圾网
trash separator 除杂器
trash sluice 排污闸
trash sluice of fishway 鱼道清污
trash truck 垃圾车
trashway 泄污道;排污道
traskite 硅钛铁钡石;硅铁钡钛矿
trasour 插图室
trass 火山灰;粗面凝灰岩;浮石凝灰岩;浮石火山灰
trass cement 火山灰水泥;火山灰(质)水泥
trass cement mortar 火山灰水泥砂浆;火山灰质水泥砂砂;火山灰质水泥砂浆
trass clinker ratio 火山灰—熟料比
trass concrete 火山灰混凝土
trass lime cement mortar 火山灰石灰水泥灰泥;火山灰石灰水泥灰浆
trass lime mortar 火山灰石灰灰泥;火山灰石灰灰浆
trass lime powder mix(ture) 火山灰石灰粉状混合物
trass mortar 火山灰砂浆;浮石火山灰砂浆
trass Portland cement 火山灰硅酸盐水泥;火山灰波特兰水泥
trass powder 火山灰石灰粉末
Tratado de Cooperacion Amazonica 亚马孙合作条约组织
tratol 氨基甲酸百里酚酯
Traube's murmur 奔马律
trauma 外伤;创伤
trauma by other people 他伤
trauma due to a fall 坠损
traumaticin 马来乳胶溶液
trautwinite 钙铬榴石
Trauwood process 特劳伍德线材电流加热法
Trauzl method 特劳泽(铅柱试验)法
Trauzl test 特劳泽试验
trave 小横梁;藻井
travel 行程;移动;水平行程;冲程
travelable 可移动的
travel accommodation 旅行舱位
travel advance 预支差旅费
travel agency 旅游局;旅行社
travel allowance 差旅津贴
travel and entertainment expense 旅行及交际费
travel arm 方向测量尺
travel behavio(u)r 出行行为
travel blank 运送空白
travel brake 输送制动;运行制动
travel bureau 旅行社
travel cable 拉索;牵引索
travel chart 工序卡
travel collar 滑动套筒
travel cost method 旅行费用法
travel demand forecasting 出行(交通)需求预测
travel-dependent control 移动相关控制
travel distance 移距
travel down the kiln 物料顺窑移动
travel end limiter 运行终端限速器
traveller = traveler
travel(er) gantry 移动式龙门架
travel expense 旅差费
travel forecasting systems 运输预测系统
travel forward 向前运动
travel frequency 行程频率;移动频率
travel gear 传输装置;传输齿轮
travel ghost 移动叠影
travel(l)ed distance 行走距离
travel(l)ed lane 行车车道
travel(l)ed portion 行车部分
travel(l)ed soil 运积土;转积土
travel(l)ed way 行车道
travel(l)er 左右移动的幕(剧院舞台);桥式起重

机;移动式起重架;移动式起重机;(起重机的)行车大梁;移动式脚手架;活动运物架;钢丝圈;旅客;双开幕;移动式门架;滑马
travel(l)er burnout 钢丝圈烧毁
travel(l)er chatter 钢丝圈颤动
travel(l)er curtain (可拉到舞台两旁的)横拉幕;双开幕
travel(l)er gantry 移动式龙门架
travel(l)er horse 帆索架
travel(l)er sprinkler 行喷式喷灌机
travel(l)er's scissors 旅行剪刀
travel lift 移动式吊车;龙门起重机
travel line 运行线;路线
travel(l)ing agitator 移动式搅拌器
travel(l)ing allowance 出差津贴
travel(l)ing and miscellaneous expense 旅差费
travel(l)ing angle 移动角
travel(l)ing anticyclone 移动性高压;移动性高(气)压
travel(l)ing apron 输送机胶带
travel(l)ing (arch) centering 移动式拱架
travel(l)ing arm 滑动臂
travel(l)ing auditor 巡回查账员
travel(l)ing autodoffer 移动式自动落纱机
travel(l)ing bags 旅行包
travel(l)ing bar 推料机的推杆;游荡性沙洲;移动性沙洲;移动沙洲
travel(l)ing (-bar) grizzly 活动棒条筛
travel(l)ing barrel 活动(泵)缸
travel(l)ing batcher and mixer 混凝土汽车;混凝土流动配料搅拌车
travel(l)ing beach 移动海滩
travel(l)ing bed (排स装置等的)活动底
travel(l)ing bed adsorber 移动床吸附器
travel(l)ing belt 运输皮带;运输带;移动式输送机;胶带输送机;输送机胶带;传送带
travel(l)ing belt conveyer 皮带输送机
travel(l)ing belt feeder 运输带给料机;运输带给矿机
travel(l)ing belt screen 运输带筛
travel(l)ing bituminous mixing plant 移动式地沥青(混合料)拌和设备;移动式地沥青(混合料)拌和厂
travel(l)ing block 游动滑车(组);游车;移动滑车;动滑轮;动滑车
travel(l)ing block guard 游动滑车护罩
travel(l)ing block machine 砖块输送机;铺地砖机
travel(l)ing bogie 运输小车
travel(l)ing box 运输箱
travel(l)ing boxtype solar cooker 旅游箱式太阳灶
travel(l)ing breaker plate 循环移动的破碎板
travel(l)ing bridge 桥式起重机;高架吊架;移动桥;移动吊架
travel(l)ing bridge crane 移动式龙门起重机;移动式龙门吊;移动式吊桥;移动桥式吊车;桥式起重机;桥式吊
travel(l)ing-bridge scraper 移动桥式刮泥机
travel(l)ing bridge sludge collector 移动桥式集泥器
travel(l)ing bucket crane 移动式抓斗起重机
travel(l)ing cable 运行索;移动电缆;随行电缆
travel(l)ing cableway 移动式缆道
travel(l)ing cableway crane 移动式缆索起重机
travel(l)ing car 移动小车
travel(l)ing carriage 旅行车辆;车架;移动台;施工挂篮
travel(ling) cavitation 行移汽蚀;行进空化
travel(l)ing centering 移动式拱架;移动式鹰架;移动式脚手架
travel(l)ing charge 随行装药
travel(l)ing cleaner 移动式清垢器
travel(l)ing clock 旅行钟;旅行闹钟
travel(l)ing cold-mix plant 移动式冷拌和厂
travel(l)ing column vertical boring machine 柱立式镗床
travel(l)ing comfort 运行的平稳性
travel(l)ing concrete mixer (plant) 混凝土搅拌车;移动式混凝土搅拌机
travel(l)ing contact 滑动接点;可动接点;活动接触;动接触
travel(l)ing counting glass 游动式织物密度镜
travel(l)ing crab 移动式料斗;移动式绞车;移动起重机;移动绞车
travel(l)ing cradle 移动式摇架;移动式吊篮

travel(l)ing crane 移动桥式吊车;移动起重机;移动式起重机;桥式起重机;起重机行车
travel(l)ing crane and car shed of harbo(u)r 港区流动机械库
travel(l)ing cyclone 移动性气旋
travel(l)ing deflectometer 移动式弯沉仪;移动弯沉仪
travel(l)ing depression (low) 移动性低压;移动性低(气)压
travel(l)ing derrick 移动式人字起重机
travel(l)ing derrick crane 移动式转臂起重机;移动转臂起重机
travel(l)ing detector 行波探测器
travel(l)ing dispensary 巡回医疗队
travel(l)ing distributor 移动式撒布机;移动(式)喷油车;移动(式)喷洒器;移动洒油车
travel(l)ing dryer 移动式干燥机
travel(l)ing dune 移动沙丘;流动沙丘
travel(l)ing earthquake wave 地震行波
travel(l)ing electrode 移动电极;移动电极
travel(l)ing engine 汽车发动机
travel(l)ing environment 旅行环境
travel(l)ing expense 旅费;差旅费
travel(l)ing expense for business 业务差旅费
travel(l)ing feeder 移动式给料机;带式移动给矿机
travel(l)ing fixture 随行夹具
travel(l)ing flights sludge collector 移动刮板集泥器
travel(l)ing foam plug 移动式泡沫挡帘
travel(l)ing forge 轻便锻炉
travel(l)ing forge crane 锻造用移动式起重机
travel(l)ing form construction method 活动模板施工法
travel(l)ing form paver 移动式摊铺机;移动式铺路机
travel(l)ing forms 滑模;移动式模板;挂篮;活动式模板
travel(l)ing formwork 移动模板;活动模板;滑动式模板
travel(l)ing frame 移动式框架
travel(l)ing framework 移动式脚手(构)架;移动式构架
travel(l)ing furnace 输送带式炉
travel(l)ing ganger 巡视工长
travel(l)ing gantry 移动式行车;移动式龙门架;移动式门架
travel(l)ing gantry crane 移动式门式起重机;移动式龙门起重机;移动式龙门吊;移动龙门吊架
travel(l)ing gantry tower 自走式门架起重机
travel(l)ing gear 移动齿轮
travel(l)ing gear brake 行走机构制动器
travel(l)ing grate 移动式炉排;移动式炉箅;炉排;移动炉箅;转动炉箅;活动炉箅
travel(l)ing grate cooler 移动箅式冷却器
travel(l)ing grate heat exchanger 移动箅式热交换器;炉箅子加热器
travel(l)ing grate preheater 移动箅式预热器
travel(l)ing-grate retort 移动炭化炉
travel(l)ing grate rotary kiln 带移动箅子的回转窑
travel(l)ing-grate sintering 移动炉栅烧结
travel(l)ing grate (stoker) 链条炉排
travel(l)ing grate stoker boiler 链条炉
travel(l)ing grid 活动格网
travel(l)ing grizzly feeder 移动式格筛输送机
travel(l)ing guide carriage 移动小车
travel(l)ing guiding runways 运行导轨
travel(l)ing guiding wheels device 运行导向轮装置【机】
travel(l)ing hair 移动丝
travel(l)ing hatch-way beam 舱口活动梁
travel(l)ing head grinder 磨头纵向移动的磨床
travel(l)ing head shaper 移动刀架式牛头刨
travel(l)ing head veneer jointer 刀头转动式单板铣边机
travel(l)ing heater 移动加热器
travel(l)ing heater method 移动加热法
travel(l)ing high 移动性高(气)压
travel(l)ing hoist 移动式起重机(械);移动(式)卷扬机;移动式滑车;移动起重机
travel(l)ing hopper 料斗式卸料车
travel(l)ing hydraulic motor 液压行走马达
travel(l)ing induction coil 移动感应圈;移动感应圈
travel(l)ing information system 旅游信息通报系统

travel(l)ing irrigation 自走式喷灌
travel(l)ing jack 吊运小车
travel(l)ing jib crane 移动挺杆起重机;移动(式)悬臂起重机;移动式臂架起重机
travel(l)ing library 流动图书馆
travel(l)ing lighting gallery 移动式照明平台;移动式照明通道;灯光渡桥
travel(l)ing light spot 移动光点
travel(l)ing limit angle 移动极限角
travel(l)ing lizards 防控拉索
travel(l)ing load 移动荷载;移动负载;活动荷载;动荷载
travel(l)ing low 移动性低(气)压
travel(l)ing low depression 移动性低压
travel(l)ing luffing crane 移动式俯仰起重机
travel(l)ing man 移动车驾驶员
travel(l)ing mast 移动式桅杆
travel(l)ing mechanism 行驶机构;迁移机构
travel(l)ing merchants 行商
travel(l)ing microscope 移测显微镜;移测显微镜
travel(l)ing mixer 移动式搅拌机;移动式拌和机;移动搅拌机;活动式拌和机
travel(l)ing mixer method of proportioning 混凝土搅拌车配料法
travel(l)ing mixing machine 移动式搅拌机
travel(l)ing mixing plant 移动式搅拌站;移动式搅拌厂;移动式拌和设备
travel(l)ing motion 移动行程
travel(l)ing motor 行走电动机;移动式电动机
travel(l)ing motor crab 电动移动起重机
travel(l)ing motor hoist 移动式电动起重机
travel(l)ing mould 活动式模子
travel(l)ing nut 移动螺母
travel(l)ing of S/R machine 有轨巷道堆垛水平运行
travel(l)ing paddle mixer 移动式桨叶搅拌机
travel(l)ing pan filter 转盘过滤机;动盘滤机
travel(l)ing pedestal 行驶车座
travel(l)ing period of grinding ball 磨球的回转周期
travel(l)ing phase 移动相位;前进相位
travel(l)ing pile driver 移动式打桩机
travel(l)ing pipe shield 移动式管垫【给】
travel(l)ing plant 移动式机器;移动式装置;移动式设备
travel(l)ing plant mixing 移动式厂搅
travel(l)ing platform 移动台座;自动步道(输送式)
travel(l)ing plough 移动耙料机
travel(l)ing portable jib crane 移动式悬臂吊车
travel(l)ing portable rig 移动式轻便钻机
travel(l)ing portal jib crane 门式移动悬臂起重机;移动门架式悬臂起重机
travel(l)ing port system 活动舷门系统
travel(l)ing pressure wave 运行性压力波
travel(l)ing probe 移动探针
travel(l)ing public 乘客
travel(l)ing range 行程范围;运行范围;运转范围
travel(l)ing rate 行进速率
travel(l)ing reclaimer 移动式取料机
travel(l)ing resistance 行走阻力
travel(l)ing roller 移动(式)滚筒;移动滚轮;移动滚子
travel(l)ing roller table (横行的)输送辊道
travel(l)ing rope 拖曳绳;运行索
travel(l)ing rotary plow feeder 移动式叶轮给料器
travel(l)ing runways 运行轨道【机】
travel(l)ing runways level 运行轨道高程
travel(l)ing salesman 行商;巡回推销员;旅行推销商
travel(l)ing saw 移动式锯
travel(l)ing scaffold(ing) 移动式脚手架;移动脚手架
travel(l)ing scale 活动秤
travel(l)ing schedule 旅行日程
travel(l)ing scissors 旅行刀具
travel(l)ing scraper 移动式刮板机;桥式刮泥机
travel(l)ing screen 直接测量水速的幕帘;旋转式垃圾筛;移动式格栅;移动筛;转筛;活动测流隔板;活动箅栅;带筛;测速幕
travel(l)ing screen method 移动屏(测流速)法;屏移法
travel(l)ing seismic wave 地震行波
travel(l)ing service 旅行服务
travel(l)ing sheave (在轨道上的)滑动滑车

（组）；游动滑轮；游动滑车
travel(l)ing shiploader 移动装船机；移动式装船机
travel(l)ing shot 移动摄影
travel(l)ing shuttering 移动(式)模板
travel(l)ing shuttering work 移动式模板
travel(l)ing sidewalk 自动步道(输送式)；活动人行道
travel(l)ing sieve 抖动筛
travel(l)ing skip 移动式卸料(箕)斗
travel(l)ing slewing stacker 移动式旋臂堆料机
travel(l)ing slit 移动狭缝
travel(l)ing solvent method 移动溶剂法
travel(l)ing speed 运行速度；行走速度；行进速率；行进速度；行车速度；旅行速度【铁】
travel(l)ing speed end limiter 运行终端限制器
travel(l)ing spider 游动吊卡；活动卡瓦；活动夹具；活动夹持器
travel(l)ing spray booth 移动式喷雾室
travel(l)ing sprinkler 移动式喷灌机
travel(l)ing stacker 移动式堆料机
travel(l)ing stage 移动式舞台；移动式工作平台；移动式工作台
travel(l)ing stair(case) 移动楼梯；移动式楼梯；自动楼梯；自动扶梯
travel(l)ing stairway 移动梯；自动扶梯
travel(l)ing standard 巡检用标准仪器
travel(l)ing stay 随行刀架
travel(l)ing steady 随行刀架
travel(l)ing surge 行进涌浪
travel(l)ing table 移动工作台
travel(l)ing table mould 移动式平台模板
travel(l)ing thermocouple 滑环热电耦
travel(l)ing thread 行程
travel(l)ing thresher 移动式脱粒机
travel(l)ing tidal wave 涨潮行波
travel(l)ing time 实际行驶时间；行移时间
travel(l)ing time table 列车运行时分表
travel(l)ing tower 移动塔架；移动(式)塔架(缆道)
travel(l)ing tower crane 移动塔式起重机；移动式塔式起重机；移动式塔吊
travel(l)ing trader 客商
travel(l)ing tripper 移动式卸料装置；移动式倾卸装置；移动卸料器；S 形卸料小车
travel(l)ing troll(e)y 滑车；手推小车
travel(l)ing type suspended weigh-batcher 移动式悬挂定重称料机
travel(l)ing unloader 移动式卸货机
travel(l)ing unloading tower 缆道移动塔；塔式装船桥；移动(式)卸货塔
travel(l)ing valve 移动阀门；游动凡尔；游动阀
travel(l)ing velocity 行进速度；旅行速度【铁】
travel(l)ing velocity of photo 影像移动速度
travel(l)ing water distributor 移动式配水器
travel(l)ing wave 行进波；行波；移动波；进行波；前进波；推进波
travel(l)ing-wave accelerator 行波加速器
travel(l)ing-wave aerial 行波天线
travel(l)ing-wave amplifier 行波管放大器；行波放大器
travel(l)ing-wave antenna 行波天线
travel(l)ing-wave cavity 行波共振腔
travel(l)ing-wave coefficient 行波系数
travel(l)ing-wave confinement 行波约束
travel(l)ing-wave ferromagnetic amplifier 行波型铁磁放大器
travel(l)ing-wave field 行波场
travel(l)ing-wave waveguide 行波波导
travel(l)ing-wave helix 行波螺旋波导
travel(l)ing-wave klystron 波速调管；行波速调管
travel(l)ing-wave linac 行波直线加速器
travel(l)ing-wave line 行波线
travel(l)ing-wave linear accelerator 行波线性加速器；行波直线加速器
travel(l)ing-wave maser 行波微波激射器
travel(l)ing-wave multiplebeam klystron 行波多束速调管
travel(l)ing-wave optic(al) parameter oscillation 移动波光参量振荡
travel(l)ing-wave oscillation 行波振荡
travel(l)ing-wave oscillograph 行波示波管；行波示波器
travel(l)ing-wave oscilloscope 行波示波器
travel(l)ing-wave parametric amplifier 行波参量放大器

travel(l)ing-wave photomultiplier 行波光电倍增管
travel(l)ing-wave phototube 行波光电管
travel(l)ing-wave power amplifier 行波功率放大器
travel(l)ing-wave protection 行波保护
travel(l)ing-wave ratio 行波比
travel(l)ing-wave resonator 行波谐振器
travel(l)ing-wave separator 行波分离器
travel(l)ing-wave single mode laser 单模行波激光器
travel(l)ing-wave structure 行波结构
travel(l)ing-wave system 行波系统
travel(l)ing-wave theory 行波学说
travel(l)ing-wave tube 行波管
travel(l)ing-wave tube amplifier 行波管放大器
travel(l)ing-wave tube oscillator 行波管振荡器
travel(l)ing-wave type oscillation 行波型振荡
travel(l)ing-wave type parametric amplifier 行波型参量放大器
travel(l)ing weigh hopper 移动式称料斗；移动称料斗
travel(l)ing wheel 行走轮；行走滑轮；行轮；走轮
travel(l)ing wheel-heed roll grinder 轴磨床
travel(l)ing winch 移动式绞车；移动绞车
travel(l)ing winged stacker 移动翼式堆料机
travel(l)ing wire 移动丝；动丝
travel loader 侧向叉车
travel map 游览图
travel mechanism 移动机构
travel mixer 移动式搅拌机；移动式拌和机；自动推进式黏土拌和机
travel mode 交通方式
travel motor 移动电动机
travel of elevator 电梯行程
travel of escalator 自动扶梯行程
travel of grout 水泥浆扩散
travel of lift 电梯行程
travel of plunger 柱塞行程长度
travel of sand 泥沙推移
travel of sound 声音行程
travel of wave 波向
travel on official business 因公出差旅行
travel parameter 行程参数
travel-pass shaving method 对角线剃齿法
travel path 传播途径；传播路径
travel pattern 运行模式
travel performance test 行驶性能试验
travel plant 移动式装置；移动式设备
travel-plant method 移动式设备拌制法(混凝土)
travel-plant mixing 移动式设备拌制法(混凝土)
travel pod 远程转场外挂油箱
travel position 运输状态
travel range 行程范围
travel rate 移动速率
travel rate of flood wave 洪水波传播速度
travel-reversing switch 终点转换开关
travel rope 拉索；牵引索
travel service 旅行社
travel shot 移动摄影
travel speed 行程车速；运行速度；运行速
travel speed lever 行驶速度控制手柄
travel switch 行程开关
travel time 时距工资；路程时间工资；行程时间；走时；旅行时间；路途时间；途用时间；地震走时；传播时间；出行时间
travel-time anomaly 走时异常
travel-time chart 行程时间表
travel-time curve 行程时间曲线；走时曲线；汇流时间曲线；时距曲线；地震走时曲线；传播时间曲线(汇流时间对水位的关系曲线)；时程曲线
travel time delay 行程时间延误
travel-time difference 走时差
travel time inversion 走时反演
travel time map 行程时间图
travel-time model 移动时间模型
travel time of flood wave 洪水波传播时间
travel time ratio 行程时间比
travel time residual 行程时间残数；地震走时差
travel timetable 走时表
travel trailer 旅行拖挂车；游览拖车
travel trailer park 拖车式活动住房营地
travel unit 运载工具；运载单元
travel utilization ratio 行程利用率；行程利用率

travel way 行车路线
travel wire 动丝
traversability 可遍历性
traversable gallery 横穿廊道
traversable ramp 可横越的坡道
traversable roof 横通屋顶
traversal 穿程；遍历
traversal vibration 横向振动
traverse 横切线；横截线；横放；横断；横渡；横穿风；曲线航行；切前；切割；往复移动(横向)；导程【测】；穿程；测量导线；遍历
traverse adjustment 平衡测量；导线平差；导线调整；导线测量平差
traverse angle 导线转折角；导线折角；导线角
traverse angular closure 导线角度闭合差
traverse axis 水平轴
traverse bed 摇臂横动床面
traverse bed of radial drilling machine 摇臂钻横动床面
traverse carriage 横动滑架
traverse closed between previously fixed control points 附合导线【测】
traverse closure 导线闭合
traverse control 横转控制；导线测量控制
traverse course 图上路线；导线路线
traverse dihedral angle 横上翘角
traverse error of closure 导线闭合差
traverse feed 横进刀
traverse gear 横向传动齿轮；横动装置
traverse geologic(al) map 路线地质图
traverse guide rail 横向导轨
traverse gyro(scope) 方位陀螺
traverse layout 导线布设(方案)
traverse leg 导线边
traverse leveling field book 高程导线手薄
traverse line 导线边；导线【测】；方向线
traverse linear closure 导线全长闭合差
traverse loop 导线环
traverse measurement 移动测量；导线量测；导线测量
traverse measurement network 导线测量网
traverse measurements 导线测量成果
traverse method 移测法；导线法；穿越法
traverse multi-bucket excavator 横向多斗挖掘机
traverse net 导线网
traverse net plan 导线网扩展图
traverse network 导线网；导线控制网
traverse network adjustment 导线网平差
traverse network plan 导线网扩展图
traverse node point 导线节点
traverse of geologic(al) observation 地质观察路线
traverse parallax 横视差
traverse parallax screw 横视差螺旋
traverse path 导线路线
traverse pit 转车台坑
traverse planer 滑枕水平给进式牛头刨床
traverse (plane) table 小平板仪
traverse plus point 导线加桩点
traverse point 横断点；导线点
traverse positionable crane 横移式起重机(沿甲板横向)
traverser 转盘(铁路)；活动平台；横过物；横撑；横臂；横梁；横件；移车台；转动发射装置；转车台；活动平台；导线测量员
traverse resistance 横向阻力
traverse rod 窗帘轨；(挂窗帘等的)水平滑杆；窗帘杆
traverser pit 转盘地坑；转车台坑
traverse sailing 之字形航法；曲折航行计算法；曲线航行法
traverse sampling 测线采样
traverse scan 扫描数字化器
traverse segment 导线节
traverse servo power amplifier 方位伺服功率放大器
traverse side 导线边
traverse sketch 导线测量略图；导线草图
traverse speed 小车运行速度
traverse station 导线站；导线段；导线点
traverse station mark 导线站标志；导线标志；导线点标志
traverse stiffener of steel web 钢腹板横向加劲肋
traverse survey(ing) 地形测量；导线测量
traverse survey network 导线测量网

traverse table 车辆转盘【铁】；测量用小平板；移动工作台；移车台；转盘；折航表；往复式磨平机平台；导线测量表；转车台；铁路转盘
traverse tables 导线测量用表
traverse target 导线觇标【测】；导线觇板【测】
traverse trench 横截沟
traverse unloading elevator 横向卸载升运器
traverse wind 逆吹风；挡路风
traversing 横切；描绘拱块；横向进刀；定导线网；导线测量；遍历
traversing base 横行底座
traversing bridge 横穿桥；横移式活动桥；横移滑动桥；伸缩桥；可伸缩的活动桥
traversing caisson 横动闸门；横动沉箱式船坞闸门；(船闸的)推拉式闸门；推拉式坞门
traversing chute 摆动落煤管
traversing conveyer 横向输送器
traversing crane 桥式吊(车)
traversing device 排线装置
traversing dock-gate 横动坞门
traversing feed 纵向进给
traversing gate 横拉闸门；横动闸门
traversing gear 横动装置；横动齿轮
traversing guide 方向导轨
traversing handwheel 方向手轮
traversing head 瞄准镜头
traversing jack 旋转千斤顶；横移式起重器；横动千斤顶
traversing lever 横杆
traversing mandrel 横移心轴
traversing mechanism 旋转机构；横动机构
traversing method 导线法；导线测量法
traversing network 导线测量网
traversing pinion 方向机齿轮
traversing probe 横向移动探针
traversing screw jack 滑座螺旋起重器；横式螺旋起重器
traversing slipway 移动式船台；横向滑道；横向滑船道
traversing speed 横移速度
traversing table type 工作台移动型
travertine 钙华；凝灰石；石灰华；石灰质松石
travertine column 钙华柱；石灰华柱
travertine facing 钙华贴面
travertine lining 钙华镶衬
travertine marble 钙华大理石
travertine slab 钙华石板；钙华厚片
travertine texture 仿石灰华饰纹
traviated 分成开间的；横向分隔的(天花板)
traviated ceiling 横分隔天花板
travis 梁格；天花板格
Travis tank 特拉维斯池；水解池
travolator 输送机自动步道
travopolye crop rotation system 草田轮作制
trawl 曳网；水文施测器
trawl anchor 沙地锚
trawl beam 拖网承梁
trawl board 拖网板；水下观测板；水文拖测(器)板
trawl boat 拖网鱼船
trawl buoy 拖钩浮标
trawler 拖网渔船；拖网船
trawler boat 拖网渔船
trawler-drifter 拖网兼漂网渔船；拖网—漂网渔船
trawler factory ship 鱼获物加工船
trawler fleet 拖网渔船队
trawler-seiner 拖网—围网渔船
trawler ship 拖网渔船
trawl gallow 网板架【船】
trawl head 桁架网；拖网架；水文拖测网
trawling 拖网作业
trawling efficiency 拖曳效率
trawling gear 拖网装置
trawling light 拖网灯
trawl net 拖网
trawl sampling 拖网采样
trawl winch 拖网绞车
trawolater 自动步道(输送式)
traxcavator 履带式挖掘机；重型履带式拖拉机；大型履带拖拉机；装载机；汽车式挖掘机
tray 淋浴；料盘；浅盘；塔板【化】；吊货盘；大型货盘
tray accumulator 盘式蓄电池
tray aerator 盘式曝气器
tray and compartment drier 箱式干燥器
tray boot 料盘中心槽

tray burner 盘架炉
tray cap 塔盘泡罩；分馏塔盘的泡罩
tray ceiling 盘形顶棚；槽形顶棚
tray column 盘式塔；多层塔；板式塔
tray conveyer 盘式输送机；槽式运输机；吊盘输送机；槽式输送机
tray deck 塔盘板
tray design 塔盘设计
tray down-spout 塔盘下流管
tray dryer 盘架干燥器；盘架干燥炉；盘式烘燥机；盘式干燥器
tray drying 盘式干燥
tray dumper quench conveyer 料盘式淬火传送带
tray efficiency 塔板效率；板效率
tray elevator 托盘升送机；托架提升机；吊盘提升机
tray flatness 塔盘水平度
tray floor 塔盘
tray frame 盘架
tray frame construction 盘式车架结构
tray freezing method 托盘式冻结法
tray gradient 塔盘压力降；塔盘梯度
tray heater 蒸馏塔盘框式加热器
tray hydraulics 塔盘水力学
tray internal materials 塔盘内件材料
trayle 蔓叶花饰；(英国古代用的)葡萄串形装饰
tray lift 吊盘提升机
tray lowerer 托盘降送机
tray mo(u)ld 盘式模型
tray of column 塔盘
tray oscillation 塔盘振荡
tray outside diameter 塔板外径
tray rail 盘子搁架
tray ring 塔盘环
tray riser 塔盘蒸汽上升口；塔盘升气管
tray roll(er) 支承辊；过渡辊台
tray salver 托盘
tray scrubber 盘式洗涤器
tray separator 盘式磁选机
tray shape deaerator 盘式除氧器
tray spacing 塔盘间距
tray stiffer 塔盘支承
tray support 塔盘支承
tray support ring 塔盘支撑环
tray thickener 料盘式增稠器；多层增稠器；多层浓缩机；多层浓缩槽
tray-top table 盘桌
tray tower 盘式塔
tray-type bed-load sampler 盘式推移质采样器；盘式河底推移质采样器
tray-type deaerator 浅盘式除气器
tray type frame 盘式车架
tray-type gas scrubber 盘式气体洗涤器
tray-type sampler 盘式采样器
tray-type separator 盘式分选器
tray weir 塔盘堰；塔母堰
treacherous passage of sharp-bend pattern 急弯型险滩
treacherous quicksand 难处理流沙
treacherous weather 变化莫测的天气；变化不稳定的险恶天气
treacle mo(u)lding 凸缘线脚；突圆边饰
treacliness 黏滞度；黏滞性
tread 支撑面；轮距；楼梯踏步板；楼梯踏步；楼梯踏板；龙骨长度；级宽；刃宽；胎面；胎伤；踏面；踏步；踏板；车辙
tread arc width 胎面弧宽
tread band 车轮触轨面
tread bar 轮胎抓地齿
tread bar contact area 胎面花纹接地面积
tread base 胎面底层
tread board 踏腰线板板；踏板
tread brake unit 踏面制动单元
tread brake unit without spring-loaded brake 未配弹簧制动器的踏面制动单元
tread brake unit with spring loaded brake 配有弹簧制动器的踏面制动单元
tread braking 踏面制动
tread cap 胎面冠部
tread caterpillar 履带
tread chord width 胎面弦宽
tread circle 切轨圆
tread concavity 胎面压扁
tread configuration 胎面外层
tread contour 胎面花纹

tread contour plate 胎面样模
tread cover(ing) 踏步覆盖物
tread cushioning layer 胎面底层
tread depth 胎面花纹深度
tread design 胎面花纹设计
tread detachment 胎面剥离
tread drive dental engine 脚踏牙钻车
treaded spade 脚踏助力铲
treader 表土细碎机
tread ga(u)ge 胎面厚度规
tread grinder 楼梯踏步板磨光机
tread groove 胎面花纹沟
treading warmth 脚接触热
treadle 轨道接触器；驱动踏板；踏板
treadle bowl 踏杆转子
treadle disc 踏蹬
treadle drilling machine 脚踏钻床
treadle drive 踏板驱动
treadle grate 踏综杆栅子
treadle hammer 踏锤；踏板锤
treadle lathe 脚踏车床
treadle loom 踏杆式织机
treadlemill 踏板传动式试验台；脚踏传动式磨
tread length 踏步长度；踏板长度
treadle operated valve 踏板操作阀
treadle pedal 踏腰线板板
treadle valve 踏板操纵的阀门
tread lug 胎面突起花纹
tread material 踏板材料
tread mill 脚踏轧机；踏旋器
treadmill test 平板试验
tread nosing 踏步凸边
tread of rear 后轮距
tread of travel(l)ing wheel 行走轮轮距
tread of wheel 车轮踏面
tread pattern 轮胎胎面花纹；轮胎花纹
tread pattern groove 胎面花凹纹
tread plank road 踏板路
tread plate 履带链条；(桥梁伸缩缝的)盖板；花纹钢板；防滑条；防滑板；网纹板；防滑金属踏步板
tread ply 胎面花纹层
tread pressure 轮胎着地面压强
tread projection 踏面突沿
tread radius 胎面半径
tread reinforcement fillet 胎面花纹加强筋条
tread return 探头踏步板；踏板延侧面；踏板端翼
tread rig 胎面横向花纹
tread rise 踏步步高；踏步高
tread road 铁路
tread roll 车轮凸缘和踏面的辊轧辊
tread roller 履带托轮；履带导轮
tread rubber 轮胎面橡胶
tread run 踏步级宽；踏步宽度
tread separation 胎面脱壳
tread surface 踏步面(楼梯)
tread taper 踏面斜度
tread thickness 胎面厚度
tread tile 踏步地砖
tread tractor 履带式拖拉机；大型拖拉机
tread truck 轮距
tread void 胎面花纹沟
treadway (电动走道的)载人踏板；临时铺板路；跳板道；车行板
treadway bridge 军用桥；应急桥；轮距桁梁式桥；双车道浮桥
tread wear 胎面花纹磨损
tread-wear life 胎面磨损寿命
tread wheel 脚踏陶车
tread width 楼梯级宽；胎面宽度；踏步宽；踏板宽度；楼梯踏板宽度
treasure-house 宝库；金库；国库
treasurer 出纳员；总会计师；会计(指人员)；司库；财务长
treasure room 贵重物品保管室；贵重品储[贮]藏室；贵重货舱
treasuruy function 库务功能
treasury 金库；宝库
treasury bill 国库证券；国库券
treasury bond 国库券；库存债券；库存公司债；财政长期债券
treasury currency outstanding 财政部发行的货币
Treasury Department 财政部
treasury note 国库券
Treasury of Atreus at Mycenae 阿特柔斯王陵墓

Treasury of Sikyon 锡科人宝库
Treasury of the Athenians 古雅典人宝库
Treasury of the Siphaians 锡弗诺岛人宝库(位于希腊德尔菲)
treasury overdraft 国库透支
treasury security 国库券
treasury stock 库存股份
treasury stock account 库存股票账户
treasury view 财政部观点
treatability 可处理性
treatability factor 可处理性系数
treated 经处理的
treated annulus 经防渗处理的隧道环;经处理的环带
treated bituminous surface 用沥青处理过的表面
treated chemical metal cleaning wastewater 处理过的化学金属清洗废水
treated clay 活化白土
treated domestic wastewater 处理过的生活污水
treated drying oil 加工干性油;处理干性油
treated earth 活化白土
treated effluent 处理过的污水;处理过的废水
treated effluent contaminant 处理过的污水污染物
treated effluent reuse 处理过的废水再用
treated felt 已处理过的毛毡
treated gem 加工过的宝石
treated grinding wheel 处理砂轮
treated industrial waste 处理过的工业废物
treated industrial wastewater 处理过的工业废水
treated joint 处理过的接缝
treated lens 镀膜镜头
treated oil 精制油;经过处理的油;加工油
treated pigment 处理过的颜料
treated pile 浸渍木桩;防腐处理过的木桩
treated plot 经处理过的小区
treated pole 防腐处理的电杆;浸渍电杆
treated rail-tie 防腐枕木
treated ring 经处理的环带
treated roofing 浸渍过的屋顶材料
treated rubber 精制橡胶
treated secondary municipal wastewater 处理过的城市二级废水
treated seeds 已处理的种子
treated sewage 已处理污水;处理过的污水
treated sewage discharge 处理过的污水排放
treated sewage effluent 处理水后出水
treated sleeper 经过防护处理的轨枕;防腐枕木;防腐处理过的枕木
treated steel 加特种元素钢
treated surface 表面处置层
treated tie 经过防护处理的轨枕;防腐处理的枕木;防腐处理的枕木
treated timber 防腐处理的木料;已加工木料;经过防护处理的木材;防腐处理木材;防腐处理的木材
treated timber sleeper 油枕
treated wastewater 处理过的污水;处理过的废水
treated wastewater discharge 处理过的污水排放
treated water 已处理水;净化水;经处理水;处理过的水
treated water conduit 清水导管
treated water outlet 净化水出口
treated water pump 清水泵
treated water storage 处理过的水储[贮]存池;处理过的水库
treated water tank 清水池
treated wood 经过防护处理的木材;经处理木材;阻燃木材;防腐木材
treated wooden sleeper 油枕
treated yarn 处理纱
treater 净水器;精制器;浸渍器;纯净器;处理器
treating coal liquefaction wastewater 净化煤液化废水
treating column 精制塔;处理塔
treating cylinder 木材处理加压筒
treating effect 处理效果
treating efficiency 处理效率
treating equipment 净化设备
treating of erratic high-grade sample 特高品位样品处理
treating plant 净化处理设备;处理车间;木材处理厂
treating pond 净化池
treating process 精制过程;处理过程
treating solution 精制用溶液
treating surface water and ground water 治理地表水和地下水

treating tower 精制塔
treating water 处理水
treatise 专题论文;论文
treatise on dam 坝论
treatment 加工;急救;待遇(指对待、权利方面);处置;处理
treatment agent cost 处理剂费
treatment and disposal of radioactive waste 放射性废物处理和处置
treatment and disposal of radioactive wastewater 放射性废水处理和处置
treatment at different stages 分期处理
treatment before hardening of concrete 混凝土硬化前凿毛;混凝土硬化前处理
treatment building 手术房屋;治疗房屋(医院)
treatment by chlorinated copperas 绿矾氯化处理(水的)
treatment by dilution 稀释处理
treatment charges 选矿费;处理费(用)
treatment class 经营措施等级
treatment condition 处理条件
treatment cost 处理成本
treatment cycle 循环处理(过程)
treatment effect 处理效果
treatment efficiency 处理效率
treatment element 处理单元
treatment equipment for the wastewater 排水处理装置
treatment equipment of garbage 垃圾处理装置
treatment equipment of sludge 污泥处理设备
treatment facility 处理设备
treatment flowsheet 处理流程
treatment for earthquake-proof of foundation soil at seismic region 地震区地基土的抗震处治
treatment in accordance with local condition 因地制宜
treatment in accordance with seasonal condition 因时制宜
treatment in squeezing ground 挤压地层处理
treatment kinetics 处理动力学
treatment loss 精制损失;处理损失
treatment maladjustment 处理失调
treatment measure of seepage deformation 渗透变形防治措施
treatment measures 防治措施
treatment method 处理方法
treatment methods of rock slope 岩坡破坏的防治措施
treatment mode 处理模式
treatment of atomic waste 原子工业废料处理
treatment of a waste with another waste 以废治废
treatment of coagulated sludge 混凝沉淀污泥处理
treatment of cultivated soil 耕作土壤处理
treatment of dam foundation 坝基处理
treatment of data 资料整理;资料整编;数据加工;数据处理
treatment of domestic refuse 生活垃圾处理(法)
treatment of electroplating effluent 电镀废水处理
treatment of elevation 立面处理
treatment of emulsion 乳化液处理
treatment of facade 立面处理
treatment of hazardous waste 有害废料处理
treatment of ice pavements 结冰路面的处理
treatment of industrial residue 工业废渣处理
treatment of industrial solid waste 工业固体废物处理
treatment of industrial wastes 工业废物处理
treatment of insular species 隔离区处理
treatment of karst foundation 岩溶地基处理
treatment of karst subgrade 岩溶地基处理
treatment of make-up water 补充水处理
treatment of mud 泥浆处理
treatment of municipal refuse 城市垃圾处理
treatment of municipal sewage 城市污水处理
treatment of nitrogen oxides 氮氧化物治理
treatment of ore-dressing wastewater 选矿废水处理
treatment of radioactive wastes 放射性废物处理
treatment of radioactive wastewater 放射性废水处理
treatment of reclaimed water 回收水处理(循环再利用的废水处理)
treatment of refuse 垃圾处理

treatment of sewage 污水处理
treatment of soil steam 土壤蒸气处理
treatment of solid wastes 固体废物处理
treatment of space 空间处理
treatment of tannery wastes 制革工业废水处理
treatment of three-types of wastes 三废治理
treatment of wastes 废水处理
treatment of wastewater 废水处理
treatment of wastewater containing heavy metal 重金属废水处理
treatment of water 水处理
treatment plant 处理车间;净水厂;处理厂
treatment plant effluent 污水处理厂出水;处理厂出水
treatment process 处理过程处理工艺;处理方法
treatment process of leachate 渗沥水处理工艺
treatment rate 处理速率
treatment rate of domestic sewage 城市污水处理率
treatment rate of wastewater 废水处理率
treatment room 处理室
treatment sewage effluent 处理水后出水
treatment system 处理系统
treatment tank 精制罐;污水处理池;处理罐;处理池
treatment technology 处理工艺;处理技术
treatment time 处理时间
treatment train 处理序列
treatment yard for productive wastewater 生产废水处理厂
treat with salt water 盐水处理
treaty 协定;合约;合同;条约
treaty articles 合同条款;协定条款
treaty condition 合同条件;协定条件
treaty contents 协定内容;合同内容
treaty obligation 条约义务
Treaty of Brussels 布鲁塞尔条约
treaty of commerce and navigation 通商航海条约
treaty particulars 协定摘要
treaty port 通商口岸
treaty protection 条约保障
treaty ratification 条约批准;条约的批准
treaty valid 条约生效
treaty wording 协定文本;合同文本
treble 三重的;三倍(的)
treble back gear 三重后齿轮装置;三重齿轮传动装置
treble block 三饼滑车
treble cloths 三层织物
treble control 高音控制
treble cut 三纹
treble damage 三倍于原物的损害赔偿;三倍赔偿
treble frequency 高音频率
treble kiln floor 三层干燥床
treble loudspeaker 高音传声器
treble planking 三层船壳板铺板法
treble ported slide valve 三通闸阀
treble pump 三联泵
trebler 三倍倍频器
treble riveted butt joint 三行铆钉对接
treble riveted joint 三行铆钉接合
trebles 三粒级煤
treble-slot 三隙缝
treble tariff 三重税
trebling 木船首部抗冰强度板
trebling circuit 三倍频电路
trebly 三重地
trecento 意大利艺术时期
trechmannite 三方硫砷银矿
trechometer 轮转计
tree 柑橘树;林木;乔木
tree age 树龄
tree algorithm 树算法
tree allocated processor 树形网络分派信息处理机;树形网络分派的信息处理机
tree and branch architecture 树形一分支结构
tree approach 求树法
tree archive 树木档案库
tree automaton 树状自动机
tree balling machine 树苗挖掘机
tree bark 树皮堵漏处理
tree belt 林带;树带;绿化带
tree belt area 保护林带区
tree binary form 树状二元型
tree-bordered paseo 绿茵大道

tree-branch pipeline 树枝状管网
tree breaker 伐树机
tree breeding 树苗育种
tree burst 树木碎裂
tree butchery 滥伐森林
tree caliper 树径尺
tree canopy 树冠覆盖面
tree census 树木调查
tree class 级别树
tree classification 树分类【计】;树木分类
tree clearer 灌木切除机
tree climate 林区气候
tree clubmoss 树石松
tree code 树码
tree code decoding 树码译码
tree compass 直径铗
tree computation of height 树形结构计算
tree copy 树复制
tree cover 林木植被
tree-crop cycle 三种作物的轮作周期
tree crown 树冠
tree crown scale 树冠直径量尺
tree crusher 树木轧碎机
tree crystal 树枝状晶体
tree crystalization 树状晶体
tree cutter 推树机;除根机;伐木机
tree database management system 树型数据库管理系统
tree deletion 树删除程序
tree derivation 树形图
tree diagram 树状图解;树形图
tree diagram of conditional probability 条件概率的树形图
tree diagraph 树图
tree disease 树木病害
tree dozer 推树机;除根机;伐木机
tree entanglement 树木阻塞
tree entry 树形登记项
tree enumeration of labelled 带标号树的枚举
tree farm 林场;私营林场
tree feller 伐木锯
tree felling 采伐
tree felling saw 伐木锯
tree form 树形
tree form-factor 树木形数
treeframed view 树状框视图
tree frog 雨蛙
treegarden irrigation 园林灌溉
tree graph 树图
tree grate 树池保护格栅;树木围栅
tree groin 植树拦砂丁坝
tree guard 树干保护套栏;树木护栏
tree guide frame 采油树导向架
tree-gule 树木黏料
tree-hand section 徒手切片
tree heath 树木石南
tree height 树高
tree-height formula 树高公式
tree height measurement 树高测量
tree height of three 三层树形结构
tree height reduction algorithm 树形结构简化算法
tree hierarchy 树形层次
tree house 巢屋;建在地面以上或树上的房子
treeing 树枝状组织;树枝状晶体;丛生;不规则金属淀积
treeing breakdown 树枝状击穿
tree initiation voltage 树形发生电压
tree injection 树木注射
tree injector 树木注射器
tree lace 胶线
tree lawn 绿化(地)带;绿化草地
tree layer 树层
treelcaf design 木叶纹
tree-like 枝状;枝晶的;树枝状(的)
tree-like crystal 枝晶
tree-like drainage 树状河网;树状水系
tree-like structure 树形结构
tree-like transition table 树状转移表
tree limit 树木限界(纬度限界);树木界限
tree line 林木线;树线;树线;森林线;森林界线
tree-lined footpath 林阴小道
tree litter 枯枝落叶层
tree marking 林木打号
tree matrix 树型矩阵;树矩阵

tree moss 树上的苔藓
tree mo(u)ld 树模
tree mover 挖树机
treenail 滴珠饰;木橛;定位销钉;圆锥饰;木堵头;木钉;木栓;木键
treenail wedge 插在木钉两端的楔子
Treene interstade 特伦间冰阶
tree network 树形网络
tree node 树节点
tree nursery 苗圃;树苗圃;树苗床
tree of economic value 经济林木
tree of heaven 臭椿
Tree of Jesse 耶西树(表示基督的家系)
tree of procedure activation 过程激励树
tree of the Gods 臭椿
tree ordering 树形排序
tree peony 牡丹
tree peony figure 牡丹纹
tree pink 树石竹
tree pipe system 树枝状管网
tree plan 植树方案
tree plantation 林场
tree planter 植树机
tree planting 植树
tree planting auger 植树螺旋钻
tree-planting day 植树节
tree planting machine 植树机
tree planting protection 植树保护
tree pointer 树指示字
tree polynomial 树多项式
tree poppy 树罂粟
tree program(me) 植树方案
tree prop 树木支柱
tree-pusher 推树器
tree representation 树状表达式
tree resistance 树木被车辆撞断阻力
tree-retard method 树枝挂淤法;树枝挂淤法
tree retards 树枝挂淤;植株滞流;树丛滞流(河道);植树滞流挂柳
tree ring 年轮(树木);树年轮;树木年轮
tree ring climatology 年轮气候学
tree-ring dating method 树轮测年法
tree rose 树冠蔷薇
trees 树木
tree savanna 稀树干草原
tree saver 采油树内护套
tree saw 树锯
treescape 树景;树木景观
tree search 树搜索
tree search for an immobile hider 对不动隐藏者的树形搜索
tree search method 树形寻优法
tree-seed drill 造林播种机
tree seedlings 树苗
tree seed orchard 林木种子园
tree shaped 单干的
trees have been planted to fix the sand 封沙育林
tree shears 树木剪切;修枝剪
tree sign 森林符号
tree site 有林地
tree sorting 树形排序
tree species 树种
tree species selection 树种选择
tree stage 成材树阶段
tree stake 护树桩
tree stinger 推树机
tree stock 砧木
tree structure 树状结构;树形结构
tree stump 树桩;树墩
tree survey 树木生态观察;测树
tree system 树枝形配电方式;树枝配电方式;树配电方式;树形系统;分枝配电方式;分支配电方式
Treetex 特拉特克斯板(一种绝缘板)
tree top 树顶;树冠
tree topology 树形布局
tree trunk 树干
tree-type distribution system 树枝状配水系统
tree volume table 立木材积表;树木材积表
tree-walk 树径
tree wall 树墙
tree wart 树瘤
tree well 树坑
tree wire 跨树保护线
trefoil 品字形;三瓣形花饰;三叶形;三叶饰

trefoilapsis 三叶形拱点
trefoil arch 三叶形拱;三叶拱;三心花瓣拱
trefoiled arcade 三叶连拱廊
trefoiled ground plan 三叶状平面图
trefoiled tracery 三叶窗饰;三叶式花格窗
trefoil flower 三叶花饰
trefoil ornament 三叶饰
trefoil-shaped tower (block) 三叶状塔
trefoil stage 三叶期
trefolate ceramic heat exchanger 三叶形陶瓷热交换器
treillage 格子结构;花格架;格子墙;格子篱;格子架;格沟
treis 天花板格
treithyltin 三乙基锡
trellis 格子结构;格子;格构;花架;藤架
trellis arch 攀藤拱架
trellis bond 格子结构连接
trellis bridge 格构桥;桁梁桥
trellis casing 格子罩;格构罩
trellis code 格式码
trellis column 格形柱
trellis cupola 格构圆屋顶
trellis design 方格纹
trellis diagram 格图;树形图
trellis dome 格构穹隆
trellis drainage 格状水系;格状排水系统
trellis drainage pattern 格子水系;格状水系
trellis dune 格状沙丘
trellised arch 攀藤拱架
trellised drainage 格状排水系统
trellised drainage pattern 交织水系
trellised drainage system 交织水系
trellised fence 方格篱笆
trellised verandah 花格走廊
trellis gate 格状闸门
trellis girder 格式大梁;格梁;格构大梁
trellis masonry wall 间孔砖墙;格形砌筑墙
trellis mast 格式桅
trellis member 格构构件
trellis mode 格状水系模式
trellis mo(u)lding 格子饰的线脚
trellis pattern 格子型
trellis post 格形柱
trellis tower 格状起重机;格形塔;格构塔
trellis web 格状梁腹;格构式梁腹
trellis window 格构窗;花格窗
trellis work 网格结构;花格制品;格构细工;格构构架;格构工作;格子细工;格构工程
trellis work bridge 桁梁桥;格构桥
Tremadoc Slates 特雷马多克板岩
tremble 振颤
trembler 电震板;自动震动器;自动振动器;振动子;振动器;振颤片;振颤片;断续器;电铃
trembler bell 电铃
trembler blade 振颤片
trembler coil 振颤线圈
trembler firing device 振动式发火装置
trembler spring 蜂鸣器弹簧
trembling barrel 磨光滚筒
trembling bell 振颤式电铃;电铃
trembling moor 振沼
trembling motion 狄拉克振动
trembling poplar 欧洲山杨
tremendous potentiality 巨大潜力
tremendous sea 巨浪
tremie 混凝土导管(用于水下灌注混凝土);水下浇灌;水下混凝土导管;导管(指灌注水下混凝土用的)
tremie concrete 水下导管灌筑混凝土;水下混凝土
tremie concrete connection 水下混凝土连接
tremie concrete dowel 水下灌筑混凝土榫槽;水下灌注混凝土榫槽
tremie concrete seal 水下混凝土止水层;水下混凝土封底
tremie concreting 导管浇注混凝土法
tremie consolidating 混凝土稠度
tremie construction 导管法施工
tremie(d) concrete 导管浇注混凝土;导管注的(水下)混凝土
tremied concrete base 水下灌筑混凝土基础;水下灌注混凝土基础
tremie(d) concrete wall 导管灌注混凝土墙
tremie dike 导流堤;导流坝
tremie method 水下混凝土法;导管施工法;导管

法(水下用导管灌筑混凝土法)
tremie pipe 灌筑水下混凝土用漏斗管;混凝土漏斗管;混凝土导管(用于水下灌注混凝土);水下混凝土导管;水下灌注混凝土导管;水下灌注混凝土导管;(指灌注水下混凝土用的)导管
tremie placing 导管浇筑(法)
tremie seal 水下封底;水下混凝土止水层
tremie slab 导管法浇制(混凝土)板
tremie tube 水下混凝土导管
tremie tube concrete 水下导管浇注混凝土;水下管浇混凝土
tremie tube formwork 导管浇注水下模板
tremolabile 不耐震的
tremolite 透闪石;天然硅酸钙镁
tremolite asbestos 透闪石石棉
tremolite hornfels 透闪石角岩
tremolite marble 透闪石大理岩
tremolite quartzite 透闪石石英岩
tremolite schist 透闪石片岩
tremolite serpentine schist 透闪蛇纹片岩
tremolo 震音装置
tremometer 微震仪;微地震仪
tremor 小震;震颤;颤动(地震)
tremorgram 震颤描记图
tremorine 震颤素
tremor tardus 慢震颤
tremostable 耐震的
tremulor 震颤仪
tremulous tongue 舌颤动
trenail 定缝钉销钉;木栓;木堵头;木钉
trench 沟壑;沟槽;沟;路槽;两侧坡度很陡的海沟;海槽;堑壕;地沟;舱内货物的通风道
trench abyssal plain 海沟底平地
trench and feather edge method 半槽式断面法(路面工程的)
trench axis 海沟轴
trench back 堑壕后崖
trench backfill compactor 沟壕回土夯实器;回填夯实器
trench backfill(ing) 沟槽回填;挖沟回填
trench bottom 沟槽底;堑壕底
trench box 沟槽掘进防护箱;移动式沟槽支撑;基槽箱;地沟箱
trench brace 沟槽支撑;槽撑
trench bracing 沟壁横撑
trench bucket 挖沟机斗
trench car 壕车
trench compactor 墙基夯实机;蛙式打夯机
trench conduit 槽式管线
trench conveyer 地沟运输机
trench cover 管沟盖板;沟盖(板)
trench cut method 挖沟支撑法
trench cutting 挖沟;开管沟
trench cutting boom 挖沟杆
trench-cutting machine 挖沟机
trench digger 开沟机;掘沟机;挖沟机
trench digging 挖沟
trench-digging attachment 挖沟装置
trench-digging machine 挖壕机
trench drain(age) 地下排水沟;沟槽排水;沟槽式排水;排水沟;排水暗沟;明沟排水
trench duct 电缆通道;电缆沟;电缆槽
trenched 挖成沟的;堑形的,沟状的;挖有沟槽的
trenched cone 沟形锥体
trench edge 堑壕边缘
trencher 挖壕人;沟渠开挖机;开渠机;开沟机;掘沟器;渠道开挖机;挖沟机;挖沟工人;反铲
trencher bucket 挖沟机铲斗
trencher ladder 联式挖沟机
trencher tooth 挖沟铲齿
trench excavating machine 挖沟机
trench excavation 开挖沟槽;挖掘管沟;挖沟工程;挖沟;挖方支撑
trench excavator 挖坑机;挖壕机;挖沟机;挖沟机
trench extrapolation 趋势外延法
trench fault 沟状断层;沟伏断层;堑形断层
trench fill (concrete) 沟槽混凝土回填
trench filler 平沟机;填沟机
trench fill revetment 伐道喷砌面
trench flame thrower 堑壕喷火器
trench-forearc geology 海沟弧前地质学
trench-forming shovel 挖沟铲;反铲沟成形铲
trench grader 沟槽整修平土机;沟槽整修平地机
trench-grave 地穴墓;穴式墓

trench hoe 反向铲;锹式挖沟机;挖沟锹;反向铲土机;反铲挖沟机
trench hoe attachment 挖沟机反铲设备
trench incinerator 槽式焚化炉
trenching 开沟;开槽;挖沟;挖方支撑;挖槽;槽探
trenching area 挖沟区
trenching bucket 挖沟铲斗
trenching chain 挖沟切土链
trenching exploration 槽探
trenching hoe 反铲挖掘机;挖沟铲;反铲
trenching initial geologic(al) logging 探槽原始地质记录
trenching machine 开沟机;掘挖沟机;挖沟机;成槽机
trenching machine of the bucket elevator type 提升斗式挖沟机;斗式挖沟机
trenching machinery 挖沟机械
trenching plane 槽刨;沟槽刨;开槽刨
trenching plant 挖沟机
trenching plough 犁式挖沟机
trenching spade 掘沟用铲
trenching works 挖沟工程
trench jack 沟槽横向支撑用的千斤顶;沟槽液压斜撑
trenchless construction 不开挖基槽而建造的建筑物;不开槽施工
trench-like depression 槽状凹地
trench line 堑壕线
trenchliner 铲式挖沟机;斗式挖沟机
trench method 沟填法;开沟施工法;堑壕法;挖沟法;沉埋法;槽式断面法
trench method of blasting 挖沟爆破法
trench of subsidence 下沉凹槽;沉降海沟;沉降海槽
trench opening 沟壕上口
trench periscope 堑壕潜望镜
trench pipe 沟内管道
trench pit 保水沟
trench plain 海沟平原
trench planting 沟状栽植
trench-plough 深耕犁
trench plowing 超深耕
trench refilling 沟槽还土
trench roller 沟槽压路机;沟槽路碾;沟槽滚压机
trench rooted 扎根深的
trench sampler 刻槽取样器
trench sampling 刻槽取样;探槽采样;槽沟取样
trench section 沟槽断面;槽式断面
trench sheeting 沟槽护板;槽沟挡板
trench sheeting and bracing 沟槽钢板桩与撑杆
trench sheetpile 沟槽板桩
trench shield 壕沟挡壁构件;沟槽掘进防护板;盾构;移动式沟槽支撑;炸开孔异坑筑路堤法
trench-shooting 炸开孔异坑筑路堤法
trench shoring 沟壁支撑
trench shoring system 沟槽支撑系统
trench shuttering 沟槽支撑;堑壕护板
trench side 沟壁
trench silo 沟式地窖;沟式青储[贮]窖
trench slope 路堑边坡
trench-slope basin 海沟大陆坡盆地
trench sowing 畦播
trench spoil 挖沟弃土
trench surplus 挖沟余土
trench system 堑壕体系
trench test 槽探
trench timbering 沟槽支撑;挡土支护
trench timber sheeting 护沟木板桩
trench-tomb 地穴墓;穴式墓
trench transplanting 沟栽;畦沟移植
trench-trench transform fault 沟—沟式转换断层
trench-type construction 槽式建筑法(路面)
trench-type tunnel construction 槽式隧道施工(法)
trench wall 沟壁;槽壁
trench width 沟槽宽度
trench with lagging 衬砌板桩沟槽
trench with lateral lagging 水平支撑板沟槽
trench works 挖沟作业;挖沟工作;挖沟工程
trend 走向[地];锚链与船首尾向所成的角;趋势;倾向;动向
trend adjusted indexes 趋势调整指数
trend adjusted time series 调整长期趋势后的时间数列
trend adjustment for average method 平均数趋势调整法

trend analysis 走向分析;趋势分析
trend analysis method 趋势分析法
trend analysis model 趋势分析模型
trend-based forecasting 趋势预测法
trend bucker 始终赢利的企业
trend component 趋势成分
trend computation 趋势计算
trend correction method 趋势更正法
trend curve 趋势曲线
trend cycle 趋势循环
trend-cycle ratio method 趋势循环比率法
trend effect 趋势效果
trend extrapotation 趋势外推法
trend forecast(ing) 趋势预测
trend in development 发展趋势
trend line 走向线;趋势直线
trend map 趋势图
trend model 趋势模型
trend mothod 趋势法
trend of a fault 断层走向
trend of channel 航道走向
trend of coast 海岸走向
trend of deep fracture 深断裂走向
trend of development 发展趋势
trend of fold axis 褶皱轴向
trend of mountain chain 山脉走向
trend of railway line 线路走向
trend of river system 河系走向
trend of stock price 股价走势
trend of the market 行情
trend of the tilt 倾斜方向
trend of the times 时间价值
trend of time series 动态趋势
trend percentage 趋势百分率
trend ratio 趋势比率
trend ratio method 趋势比率法
trend record 趋势记录
trend surface 趋势面
trend surface analysis 趋势面分析
trend test 趋向试验;倾向检验
trend towards urban development 城市发展趋向
trend value 趋势值
trend variable 趋势变量
trength excavation 打直井
trenh extrapolation 趋势外推法
Trenner's automatic pipet 特化内尔氏自动吸管
Trente biotic index 特伦特生物指数
Trent economic water quality model 实用特伦特水质模型
Trentonian (stage) 特伦顿阶[地]
trepan 环钻;线锯;圆锯;钻井机(架);凿井机;钻矿机;开孔;打眼
trepanation and drainage 钻孔引流法
trepan borer 套孔机
trepan boring 套孔
trepan boring machine 套孔机
trepan chisel 冲击钻头
trepang 海参
trepang agate 海参玛瑙
trepang blue 海参青
trepang glaze 海参釉
trepanner 穿孔机;(钻矿井井筒的)大直径钻井设备;打眼机
trepanning 打眼;穿孔;套孔;开孔
trepanning drill 套料钻;套孔锥
trepanning machine 打眼机
trepanning method 圆槽释放法
trepanning tool 套孔刀
trepan-sunk shaft 大直径钻头钻凿井眼
trephination and irrigation 钻孔冲洗术
trephine 环钻;线锯;圆锯
treponema 密螺旋体
treppe 阶梯现象
treppeniteration 楼梯迭代;梯子迭代
treptomorphism 等化学变质作用
trermostatically controlled X-ray film dryer 电热恒温 X 线胶片干燥箱
tresaunce 通道(中世纪建筑);窄门道(中世纪建筑)
tresaunte 僧院回廊
Tresca criterion 特雷斯卡准则
Tresca yield criterion 特雷斯卡屈服准则
Tres Grande Vitesse 法国高速列车
trespass 非法侵入
trestle 柱架;栈桥;栈道;高凳;台架;叉架;凹底车

trestle and pier 栈桥式码头
trestle approach 栈桥引道
trestle bay 栈桥桥跨;栈桥跨距
trestle bent 栈桥排架;高架桥排架
trestle bent span 栈桥桥跨
trestle board 制图板;绘图板;大图板;图板;大绘图板
trestle bridge 栈桥;高架桥;架柱桥;排架桥
trestle cable excavator 高架索道挖土机;高架缆索挖土机
trestle cap 栈桥帽木
trestle concreting 混凝土浇灌架
trestle crane 龙门起重机;门式起重机;门架起重机
trestle dam 支架坝;栈桥式坝
trestle excavator 高架缆索挖土机
trestle flume 栈桥式渡槽;高架渡槽
trestle for pipe 杆桥;管道栈桥;管桥
trestle frame 栈桥构架
trestle horizontal closure 栈桥平堵截流
trestle hydraulicking 装架冲挖
trestle ladder 自立支架梯
trestle pile 支架桩
trestle road along cliff 栈道
trestle scaffold(ing) 排架式脚手架;高架脚手架
trestles core 转芯架
trestle stand 栈桥架;栈桥;架柱桥
trestle stringer 栈桥纵梁
trestle structure 栈架结构;排架结构
trestle table 支架桌;搁板桌;搁架绘图板
trestle tower 栈桥支墩;栈桥塔座
trestle track excavator 高架索道挖土机;高架轨道挖土机
trestle tree 排架柱木;桅顶纵桁
trestle type column 栈架式柱;架柱
trestle unloading siding 栈桥卸车线
trestle work 栈架工程;栈桥;鹰架;栈台;搭排架工程
tresuante 走廊
tret 损耗补贴
Tretol 特拉托尔(一种玛蹄脂,水泥快凝剂等建筑材料)
trevis 天花板格
Treviso faience 特勒维索陶器(意大利)
Treviso porcelain 特勒维索软瓷(意大利)
trevor 暗褐花岗岩
trevorite 镍磁铁矿
trexan 氨克生
triable 可试验的
triac 三端交流开关
triacetate 三乙酸盐;三醋酯纤维;三醋酯纤维
triacetate-base film 三醋酸酯基胶片
triacetate metrical membrane 三醋酸材料膜过滤器
triacetate rayon 三醋酯人造丝
triacetate staple fibre 三醋酯短纤维
triacetin 甘油三乙酸酯;醋酸甘油
triacetyl glycerine 三醋酸甘油酯
triacid salt 三酸式盐
triacontahedron 三十面体
triacontane 三十烷
triacontanedioic acid 三十烷二酸
triacontanoic acid 三十烷酸
triacontanol 三十烷醇
triacontyl 三十基
triacontylene 三十碳烯
triactic 三同立构
tri-actional screw jack 三作用千斤顶
triad 三重组;三元基;三位组;三同步发射台组;三数组;三色组;三价元素;三个一组;三点荧光组
tria-data 三角测量资料
triad axis 三重轴;三次对称轴
triad group 三元组群
triad homotopy set 三元组同伦集
triadic 三人组合的
triadic mean 三点平均
triadic product test 三产品测试
triadic type 三支式
triadimefon 粉锈宁
triad of the gyro 三陀螺组
triaene 三叉体
triage 筛余
triakisoctahedron 三角八面体;三八面体
triakistetrahedron 三角三四面体
trial 试验;实验;审理;三个一组
trial-and-error 累试法
trial-and-error learning 尝试错误学习

trial-and-error method 拼凑法;试误法;试算法;试错法;试拌校正法;尝试误差法;反复试验法;试配法;尝试法;尝试错法
trial-and-error method for proportioning 尝试错法配料
trial-and-error method of blending aggregates 掺和集料试配法;掺和骨料试配法
trial-and-error observation 试探性观测
trial-and-error phase 试验阶段
trial-and-error procedure 选配法;试错法;逐步试凑法;试算法;试凑步骤;试凑法;尝试错法
trial-and-error process 试算法
trial-and-error solution 择试解法;逐次逼近解法;试凑法解
trial-and-error test 试探试验
trial array 试验点阵
trial asphalt 试用地沥青
trial assembly 试装(配)
trial at anchor 系泊试车
trial balance 试算表
trial balance after closing 结账后试算表
trial balance before adjustment 调整前试算表
trial balance before closing 结账前试算表
trial balance of subsidiary ledger 明细账科目余额试算表
trial balance of totals 全额试算表
trial balance of totals and balances 全额及差额试算表
trial balloon 测验风速
trial bar 检验杆
trial batch 试验性拌和;试拌(和);小量试拌
trial-batch method 批试法;试配(料)法;试抖(配料)法;试拌配料法;试拌法
trial-batch mixing proportion 试拌配料(比)
trial batch of concrete 混凝土试件;混凝土小量试样;混凝土小量试件
trial board 试航委员会
trial bore 试验孔
trial borehole 试钻孔;试钻洞;初探钻孔
trial boring 探穴;探坑;探查孔;试钻;初探钻孔
trial boring tool 试钻工具
trial captain 试航船长
trial charge 试验装药
trial charging 试充电
trial coat 试涂层;试制涂层;试验涂层
trial combination of aggregates 集料试配比;骨料试配比
trial concrete mix 混凝土试拌;试验性配料
trial condition 试航状态;试车状态
trial course 试操纵航向;船速校验线上航向
trial curve 试验曲线;试错曲线
trial cut 试切
trial design 探讨设计方案;试设计
trial-developing pumping test 开采试验抽水
trial dimension 试算尺寸
trial discharge standards of industrial three wastes 工业三废排放试行标准
trial displacement 试航排水量
trial division 试除
trial divisor 试用除数
trial dredging 试挖【疏】
trial drilling 初探钻孔
trial drive 试车
trial driving 试桩;试验打桩
trial dyeing 试染
trial erection 试装配;试安装
trial exchange 试用交换局
trial exposure 试拍
trial face 试验工作面
trial fill(ing) 试填土;试填充
trial flight 试验飞行
trial frame 试镜架
trial free 无偿试用
trial function 试探函数
trial furnace 试验炉
trial grade line 试用坡度线
trial ground 试验场
trial heading 勘探导洞;探洞;试验项目
trial heading excavation 探洞开挖
trial hole 样洞;探孔;探坑;试钻孔;试验性钻孔;验孔;试孔
trial horse-power 试车马力
trial implementation 试行
trial injecting water 试验性压水

trial inspection 初步检查
trial installation 试验性装置
trialism 三元论
trial jar 试瓶
trialkoxyparaffin 三烷氧基烷烃
trialkylaluminium 三烷基铝
trialkylamine 三烷基胺
trialkylated 三烷基化的
trial lens case 测试镜片箱
trial license 试车执照
trial load 试验荷载;试置荷重;试载
trial load analysis 试荷载分析
trial-load method 荷载试配法;试载法;试算荷载法
trial lot 试验路段;试验段
triallyl phosphate 磷酸三烯丙酯
trial making-up 试装配
trial maneuver 试操纵
trial-manufacture 试制
trialmanufacture on a small batch basis 小批试制
trial map specimen 地图试用图样
trial match 试匹配
trial material 试用材料
trial measurement 试验性测量
trial method 试算法;试算;试配法
trial mix 试用混合料;试验配合;试拌配合比;试拌和;试拌
trial mix design 试拌法设计
trial mixing method 试验配合法
trial mix series 试拌组次
trial mixture 试拌混合料
trial number 试验数字
trial(-on) 试用
trial operation 试运行
trial order 试验订单;试购
trial output 试车功率
trial pavement 试验路面
trial pavement section 试路段
trial period 试用期;试用阶段;试验周期;试车时期
trial pile 试桩;试验桩
trial piling 试验打桩
trial pit 样洞;浅探井;探坑;探井;试坑;初探浅井
trial plant 试验厂;实验厂
trial point 试测点
trial predication diagram 试航预测图
trial production 试生产
trial produce 试生产
trial-produce on a small batch basis 小批试制
trial production 试产;产品试制
trial production of new products 新产品试制
trial proportioning 试验配合比
trial pumping 试验抽水;抽水试验
trial race 试车行程
trial range 船速校验线距离;标柱间距离
trial record 试记录
trial regulation 试行规定
trial regulation works 试验性整治工程
trial road study 试用路线研究
trial rod 探尺
trial run 临时生产;试运转;试运行;试生产;试航【航海】;试车;试验
trial run cost 试运转费用
trial running 试运行
trial running expense 试运转费用
trial sale of new product 新产品试销
trial section 试验段
trial-section method 试验路段法
trial seeding 试播
trial shape of axis 轴线试用形状;拱轴线试用形状
trial shot 试验爆破
trial site 试用地点
trial slip 试验纸条
trial solution 试探解;尝试解
trial speed 试验速度;试航速度;试车速度;试操纵速度
trial square 验方(角)尺
trial standards of industrial three wastes discharge 工业三废排放试行标准
trial station 试验站
trial steel work fixing 钢梁结构试装配
trial stretch 试验范围
trial strip 试航【航空】
trial substitution 试换法
trial system 试验系统
trial table 试验表;试算表
trial test 预试验;预试;探索性试验;试探性试验;初

步试验;初步实验
trial-test investigation 探索性试验分析
trial-test well 试探井
trial transmission 试输电
trial trench 试槽
trial trench method 试槽法
trial trim line 破舱试算水线
trial trip 试车行程
trial trip rating 试运转功率
trial-trip train 试运转列车
trial tunnel 探洞
trial tunnel excavation 探洞开挖
trial value 试用值;试验值
trial vector 试验矢量;试探向量;试探矢量
trial vehicle 试验车
trial voyage 试航航程;试航【航海】
trial wedge method 滑楔试算法
trial weld 试验焊接
tri-amine 三元胺
triamorph 三晶形物
triamterene 氨苯喋啶
triamyl borate 硼酸三戊酯
trianco 振动台(混凝土)
triangle 三角形;三角板;三角
triangle antenna 三角形天线
triangle axiom 三角形公理
triangle bag method for odo(u)r sensing measurement 三点比较式气味袋法
triangle bar 三角形支承棒
triangle belt 三角皮带
triangle caisson 三角形沉箱
triangle classification of soils 土的三角坐标分类法;三角坐标土壤分类法
triangle closing error 三角形闭合(误)差;三角闭合差
triangle closure 三角形闭合(差);三角闭合(差)
triangle computations 三角形解法
triangle condition equation 三角形条件方程
triangle connection 三角接线法
triangle coordinate classification system 三角坐标分类系统
triangle core wire 三角芯股芯丝
triangle crossing 三角形交叉(口)
triangle defect 三角形缺陷
triangle equation 三角方程
triangle error of closure 三角形闭合(误)差
triangle excavating process 三角开挖法
triangle file 三角锉(刀)
triangle generation 三角波发生
triangle generator 三角波脉冲发生器
triangle gin 三脚起重机
triangle grid 三角形网格
triangle hoisting gallows 三角起重架
triangle impact concavities 三角形撞击坑
triangle impulse generator 三角形脉冲发生器
triangle inequality 三角不等式
triangle junction 三角形联结
triangle law 三角定律
triangle load distribution 三角形荷载分布
triangle mesh 三角网;三角形网眼
triangle mesh fabric 三角形网络钢丝网
triangle-mesh wire fabric 三角孔钢丝网;三角形钢丝网;三角钢丝网
triangle method 三角法
triangle moment 三角矩
triangle noise 三角形分布噪声
triangle of doubt 示误三角形
triangle of error 误差三角形;示误三角形;闭合误差三角形
triangle-of-error method 示误三角形定点法
triangle of forces 力三角形
triangle of maximum pressure 压力三角形
triangle of moments 力矩三角形
triangle of position 定位三角形;船位三角形
triangle-planting 三角形植树
triangle pole 三角杆
triangle profile 三角异形
triangle property 三角性质
triangle scale 三棱尺
triangle-section crawler shoe 三角履带板
triangles for chart work 海图作业三角板
triangle shooting 三角形爆破(地震勘探)
triangle solution concavity 三角形溶蚀坑
triangle square 三角板

triangle strand construction rope 三角形股钢索
triangle strength 三角形图形强度
triangle table 三角形表
triangle test 三点试验法
triangle tie 三角撑
triangle tile 三角面砖
triangle track 折返三角线;三角线【铁】
triangle trade 三角贸易
triangle type junction terminal 三角形枢纽
triangle wave 三角波
triangle waveform 三角波形
triangulable 可三角剖分的
triangular 三角形圆屋顶;三角形的;三角的
triangular acceleration 三合促进剂
triangular alluvial plain 三角形冲积平原
triangular arch 三角拱
triangular arched corbel-table 三角形拱支承的挑出层
triangular arched window 三角形拱窗
triangular array 三角形阵列;三角形台阵
triangular array station 三角台站
triangular asperities technique 三角形凸点密封技术
triangular association scheme 三角结合方案
triangular base 三角基座
triangular basis 三角形基;三角基
triangular bastion 三角形突堡
triangular beacon 三角岸标
triangular beam 三角形梁
triangular belt 楔形带;三角带
triangular box section beam 三角匣形梁;三角箱形梁
triangular bracing 三角形拉撑
triangular buoy 三角浮标
triangular caisson 三角形沉箱
triangular cam 三角凸轮
triangular cast-iron end piece 三角形铸铁端片
triangular chain 三角链
triangular chart 三角坐标分类图;三角图
triangular chart for soil classification 土分类三角图
triangular chisel 剁刀
triangular classification chart 土壤分类三角坐标图;三角坐标分类图
triangular classification of soil 土壤三角形坐标分类(法)
triangular coarse file 粗三角锉
triangular column 三角形柱
triangular compasses 三脚规;三角规
triangular compensation trade 三角补偿交易
triangular configuration 三角联结
triangular-cored optic(al) fiber 三角芯光纤
triangular corner seal 三角形角密封片
triangular crib groin(e) 三角形木笼丁坝;三角填石木笼丁坝
triangular crib groyn(e) 三角形木笼丁坝;三角填石木笼丁坝
triangular cupola 三角形装饰
triangular cut 三角形掏槽
triangular dam profile 三角形坝体断面
triangular dam to provide a horizontal crest 坝顶三角洲人字头
triangular decomposition 三角形分解;三角分解(法)
triangular decoration 三角形装潢
triangular dee 三角形D盒
triangular design 三角设计
triangular diagram 三角形图表;三角形图样;三角形图解;三角图解;三角图
triangular distortion 三角畸变
triangular distribution load(ing) 三角形分布荷载
triangular distribution model 三角形分布模型
triangular division method 三角区划法;三角划分法
triangular divisor 三角形分割绘制法
triangular dome 三角形穹隆;三角底穹隆
triangular duct 三角形风道;三角风管
triangular elastic wedge 三角形弹性楔
triangular element 三角形部件;三角元;三角形单元
triangular (element) distribution 三角形分布
triangular equation 三角方程
triangular equivalent projection 三角等积投影
triangular error 三角形闭合(误)差
triangular exploration grid 三角形勘探网
triangular (eye) plate 三角眼铁;三角眼板

triangular facet 三角面
triangular facet of fault 断层三角面
triangular factorization 三角分解(法)
triangular fibre-length distribution 三角形纤维长度分布
triangular file 三角锉(刀)
triangular fillet 三角形嵌条;三角垫木
triangular finishing pass 三角精轧孔型
triangular flag 三角旗
triangular flask 三角瓶
triangular flight 三角航线飞行
triangular form 三角形;三角形式
triangular frame 三角格构;三角形构架;三脚架
triangular frame dike 三角框架堤
triangular framing 三角形框架;三角形构架;三角形格构
triangular fret mo(u)lding 三角格子饰线脚
triangular full loading 三角形(分布)全荷载
triangular function generator 三角形函数发生器
triangular gate 三角形内浇道
triangular girder 三角形大梁
triangular grafting 三角接
triangular graph 三角形统计图表
triangular graph paper 三角坐标纸
triangular grating 三角形光栅
triangular grid 三角网
triangular ground plan 三角形底层平面
triangular guide 三角形导轨
triangular hollow-web beam 三角形空腹梁
triangular hydrograph 三角形过程线
triangular inequality 三角不等式
triangular island 三角岛
triangularity 成三角形
triangularization 三角形化
triangularized form 三角化形式
triangular junction 三角岔道
triangular kneading block 三角形捏合块
triangular laser 三角形激光器
triangular lattice 三角形栅格
triangular latticed construction 三角形格构建筑
triangular lifting eye 三角形吊孔
triangular load 三角形荷载
triangular load distribution 三角形荷载分布
triangular loop 三角形斤斗
triangular matrix 三角形矩阵;三角矩阵
triangular measuring weir 三角形测算坝
triangular mesh 三角网篦;三角孔网;三角形网格;三角网
triangular method 三角形法;三角法
triangular misclosure 三角形闭合差
triangular motion 三角形运动
triangular net 三角形网
triangular network 三角形网络;三角形网格;三角网
triangular noise 三角波噪声
triangular notch 三角形槽口;三角凹口
triangular notch weir 三角形缺口堰;三角堰;三角形槽口量水堰
triangular number 三角形数
triangular oil groove 三角油槽
triangular oil stone 三角油石
triangular operator algebra 三角形算子代数
triangular oreblock method 三角形块段法
triangular organization 三三制
triangular oscillating blade 摆动三角刀
triangular panel 左三角翼;三角形板
triangular parachute 三角形降落伞
triangular pediment 三角形山墙;三角形人字墙
triangular phase diagram 三角形相图
triangular phase plot 三角形相位图
triangular pigsty 三角形垛式支护
triangular-pitch 三角形节距
triangular planting 三角形栽培;三角(形)植树
triangular plate 三角片
triangular plot showing relative amount of anions 阴离子含量三角图
triangular plot showing relative amount of cations 阳离子含量三角图
triangular plotting scale 绘图用三棱尺
triangular point 三角点
triangular porch 三角形入口处;三角形门廊
triangular Pratt truss 三角形普拉式屋架;三角形普拉特式屋架;三角形普拉特桁架
triangular prism 三面体棱镜;三棱柱;三棱镜;三角棱镜

triangular profile 三角形轮廓;三角形外形
triangular-profile weir 三角形剖面堰
triangular pulse 三角形脉冲;三角波脉冲
triangular pyramid 三棱锥(体)
triangular pyramid space grid 三棱锥网架
triangular pyramid-type packing 三角金字塔形填料
triangular rabbet joint 三角凸凹榫接头;三角凸凹榫接合;三角榫槽接合;三角舌槽接合;三角槽舌接合;三角凹凸槽接头
triangular reflector 三角形反射器
triangular road camber 三角形路拱
triangular roof truss 三角形屋架
triangular routes 三角航线
triangular rule 三角定规
triangular ruler 三角尺
triangular rust scraper 三角除锈刮刀
triangular scale 三棱尺;三角比例尺
triangular scraper 三角刮刀
triangular section 三角形截面
triangular section bar 三角形杆
triangular-section step 三角形踏步
triangular section to top of 坝顶三角洲人字头
triangular segmental conductor 三角截面导体
triangular shape 三角形模型;三角形外形;三角形状
triangular shell 三角面壳;三角薄壳
triangular shell structure 三角面壳体结构
triangular side ditch 三角形边沟
triangular signal 三角波信号
triangular slab 三角岩板;三角石板;三角平板
triangular slip road 三角形岔道
triangular soil classification chart 三角土壤分类图;土壤分类三角图
triangular soil classification system 土壤三角坐标分类法
triangular solid scraper 实边三角刮刀
triangular space 三角区;象限
triangular spacing 三角形栽植
triangular spade 三角铲
triangular spring 三角簧
triangular stacking fault 三角形堆垛层错
triangular steel post 三角形钢柱
triangular steering control arm 三角形方向操纵杆
triangular stone 三棱石
triangular strand wire rope 三角股钢丝绳
triangular stress distribution 三角形应力分布
triangular structure 三角形结构
triangular stump 三角形矿柱
triangular support system 三角形支架法;三角形支撑体系
triangular symmetric curve 三角对称线
triangular system 三角形体系;三角形系统;三角形方程组;三角系统
triangular thread 三角螺纹;V 形螺纹
triangular three-mirror ring cavity 三镜式三角形环形共振腔
triangular tie 三角形支撑
triangular tile 三角形面砖;三角形瓷砖
triangular tongue and groove joint 三角企口接合;三角榫接;三角榫槽接合
triangular tooth 三角齿
triangular track 折返三角线;三角(回)车线【铁】
triangular trade 三角贸易;三国间的三角贸易
triangular trim 三角形装饰;三角形边饰
triangular truss 三角形屋架;三角形桁架;三角桁架
triangular-truss bridge 三角桁架桥;三弦桥
triangular-type gate 三角形闸门
triangular unit 三角形构件
triangular velocity distribution 三角形速度分布
triangular voltage sweep 三角波扫描
triangular washer 三角垫圈
triangular wave 三角形波;三角波
triangular waveform 三角波形
triangular-wave oscillator 三角波发生器
triangular-wave polarography 三角波极谱法
triangular web 三角形腹杆;三角形腹板
triangular weir 三角堰;三角形量水堰;三角形槽口量水堰;三角量水堰
triangular window 三角窗
triangular wire 三角金属线;三角金属丝
triangular wood lath(ing) 三角形灰板条
triangulated area 已布设三角网地区
triangulated bracing 成三角形的拉撑
triangulated girder 桁架大梁
triangulated graph 三角形剖分图
triangulated grid framework 有三角格栅的结构框架
triangulated height 三角高程
triangulated lattice 三角形格构
triangulated level(l)ing 三角高程测量(法)
triangulated roof truss 三角桁架;三角孔屋架
triangulated strip 三角锁系
triangulated system 成三角状系统
triangulated truss 三角形桁架
triangulated welded arch 三角焊接拱
triangulateration 三角三边测量;边角测量
triangulateration network 边角网
triangulation 三角格构桁架;三角剖分;三角测量;分成三角形
triangulation adjustment 三角网平差;三角测量校正;三角测量平差
triangulation arc 三角测量弧
triangulation balloon 三角测量气球
triangulation base(line) 三角测量基线
triangulation benchmark 三角高程测量点
triangulation camera 三角测量摄影机
triangulation chain 三角网系;三角锁
triangulation control 三角控制
triangulation (control) network 三角控制网
triangulation control point 三角控制点
triangulation data 三角测量数据
triangulation datum 三角测量基准(点)
triangulation diagram 三角网布设略图
triangulation field squad 三角测量作业组;三角测量外业组
triangulation height 三角测量高程
triangulation-listening-ranging sonar 三角听音测距声呐
triangulation location 三角测量定位法
triangulation loop 三角测量锁环
triangulation mark 三角点标志;三角测量标记
triangulation method 三角形法;三角测量法
triangulation net(work) 三角测量网
triangulation operation 三角测量作业;三角测量
triangulation pillar 三角点标石
triangulation point 三角点
triangulation problem 三角剖分问题
triangulation program(me) 三角网解算程序
triangulation publication 三角点一览表;三角点成果表
triangulation radar system 三角测量雷达系统
triangulation reconnaissance 三角测量选点
triangulation routine 三角形例行程序
triangulation signal 三角测量觇标
triangulation station 三角点;三角测站;三角测量站;三角测量标点
triangulation station mark 三角测站标志;三角测站标
triangulation strip 三角锁系
triangulation supported method 三角形支持法
triangulation survey 三角网测量
triangulation survey for bridge construction 桥梁施工三角(网)测量
triangulation (survey) station 三角网测站
triangulation system 三角系
triangulation target 三角测量觇标
triangulation tower 三角测量觇标
triangulation-traverse network 三角导线混合网
triangulation work 三角测量
tri-apsidal 有三个半圆形室的;三个半圆形室
tri-apsidal chevet (教堂的)三个半圆形内室
tri-apsidal church 带三个半圆室的教堂
triarch 三原型
triarimol 嘧菌醇
triaromatics 三环芳香烃
triarticular 三节的
triarticulate 三节的
triarylmethane 三芳基甲烷
triarylmethane colo(u)ring matters 三芳基甲烷染料
triaryl methyl 三芳甲基
triaryl phosphate 磷酸三芳酯
Triassic 三叠系
Triassic period 三叠纪
Triassic sandstone 三叠纪砂岩
Triassic system 三叠系
triaster 三星体
triatic 由三部分形成的
triatic stay 临时支索;横牵索;桅间索;水平支索
triatomic acid 三价酸
triatomic phenol 三元酚
triatomic ring 三元环
triaxial 三轴的;三维的
triaxial accelerometer 三轴加速计
triaxial apparatus 三轴压力试验仪;三轴仪;三轴压缩仪
triaxial borehole deformation ga(u)ge 三轴钻孔形变仪
triaxial cable 三轴电缆;三线电缆
triaxial cell 三轴压力传感器;三轴仪压力室;三轴仪压力盒;三轴仪;三轴仪压力室;三轴压力盒;三轴测压仪;三向压力传感器
triaxial ceramics 三组分陶瓷
triaxial chamber 三轴压力室
triaxial chart 三轴图;三角坐标图
triaxial compaction apparatus 三轴压实仪;三轴压缩试验仪
triaxial compaction test machine 三轴压实仪
triaxial compression 三轴压缩
triaxial compression apparatus 三轴压缩试验仪;三轴压缩仪;三轴压力试验仪
triaxial compression strength 三轴抗压强度
triaxial compression test 三轴压缩试验;三轴压力试验;三轴向压力试验;三轴抗压试验
triaxial compression test apparatus 三轴压缩试验仪
triaxial compressive strength 三轴向抗压强度
triaxial compressive stress 三轴压应力
triaxial confining pressure 三轴限制压力
triaxial deformation 三轴形变
triaxial diagram 三轴图;三角坐标图
triaxial distribution 三轴分布
triaxial ellipsoid 三轴椭圆体;三轴椭球体
triaxial equipment 三轴仪
triaxial extension test 三轴引张试验;三轴伸长试验;三轴拉伸试验
triaxial extrude pipe machine 三轴式挤压制管机
triaxial fabric 三轴向织物
triaxial flux-gate magnetometer 三轴饱和磁力仪
triaxial forced balance accelerator 三轴力平衡式加速度计
triaxial forced balance accelerometer 三轴平衡式加速度计
triaxiality 三轴性;三轴
triaxiality factor 三轴向因数
triaxial load cell 三轴向负载传感器
triaxial loading 三轴加载;三轴加荷;三向加载
triaxial mount 三轴装置
triaxial orientation 三轴定向
triaxial photoelastic measurement 三轴光弹量测法
triaxial porcelain 三组分瓷器
triaxial pottery 三组分陶瓷
triaxial pressure 三向压力
triaxial prestress 三轴预应力;三向预应力
triaxial reference system 三轴参照系
triaxial roving fabric 三轴向无捻粗纱布
triaxial seismograph 三轴地震仪
triaxial seismometer 三轴地震仪
triaxial shear 三轴剪切;三轴剪力
triaxial shear apparatus 三轴向剪切仪;三轴剪切试验仪;三轴剪力仪;三轴剪力试验仪
triaxial shear equipment 三轴向剪变仪;三轴剪切仪;三轴剪切试验仪;三轴剪力仪;三轴剪力试验仪
triaxial shear test 三轴剪切试验;三轴剪力试验
triaxial state of stress 三轴应力状态;三向应力状态
triaxial strain 三轴应变
triaxial strain cell 三轴应变计
triaxial strain ga(u)ge 同轴应变计
triaxial strength 三轴强度
triaxial stress 三轴应力;三轴向应力;三向应力
triaxial stress field 三轴应力场
triaxial stress state 三轴应力状态
triaxial stress-train plot 三轴应力-应变图
triaxial system 三轴压力试验法;三轴试验(方)法
triaxial tandem roller 三轴串联(式)压路机
triaxial tensile test 三轴拉伸试验
triaxial tension 三轴向张力
triaxial test 三轴受压试验;三轴试验
triaxial testing apparatus 三轴试验仪
triaxial testing cell 三轴试验容器
triaxial (test) method 三轴试验(方)法
triaxial test of reiteration compacting 反复加荷

三轴试验
triaxial utraviolet dye laser 三轴紫外染料激光器
triaxial weaving 三轴向织造
tri-axis locator 三轴定位器
triaxon 三轴骨针
triazanaphthalene 三氮杂萘；三氮杂苯
triazene 三氮烯
triazene paper 三氮烯纸
triazenyl 三氮烯基
triazine-containing azo-dye 含三嗪偶氮染料
triazine triol 三聚氰酸
tribal economy 部落经济
tribalism 部落文化
tribal name 部落名称
tribal society 部落社会
triband 三重频带
tribar 三角柱体(混凝土块用于防波堤)；山字块体；三柱块体(混凝土)
tribar method 三杆法(分辨率检查)
tribasic acid 三元酸；三碱酸
tribasic carboxylic acid 三元羧酸
tribasic ester 三元酸酯
tribasic lead sulfate 三碱式硫酸铅
tribasic magnesium phosphate 正磷酸镁
tribasic zinc phosphate 磷酸锌
tribenzylamine 三苄胺
tri-bit encoding 三位编码
tri-blade-drag bit 三刮刀钻头
triblet 心棒；心轴
tribo-abhesion separation 摩擦—黏附分选
tribocouple 摩擦电偶
triboelectric(al) action 摩擦带电作用
triboelectric(al) attraction 摩擦带电吸引力
triboelectric(al) behavior 摩擦带电性能
triboelectric(al) bond 摩擦电键
triboelectric(al) characteristic 摩擦带电特性
triboelectric(al) charge 摩擦电荷
triboelectric(al) charging 摩擦带电；摩擦充电
triboelectric(al) contact 摩擦带电接触
triboelectric(al) double layer 摩擦电双层
triboelectric(al) effect 摩擦带电效应
triboelectric(al) imaging 摩擦带电成像
triboelectric(al) means 摩擦带电方法
triboelectric(al) property 摩擦带电性能
triboelectric(al) relationship 摩擦带电关系
triboelectric(al) series 摩擦电序列
triboelectricity 摩擦电；摩擦发电
triboelectrification 摩擦起电
triboelectrostatic gun 摩擦静电喷枪
triboelectrostatic powder spray 摩擦静电粉末喷涂
triboeletric affinity 摩擦电亲和力
triboemission 摩擦发光发射
tribology 摩擦学
tribology in iron and steel works 钢铁加工摩擦学
triboluminescence 摩擦发光
triboluminescent 摩擦发光的
tribometer 摩擦力测量仪；摩擦力测定仪；摩擦计
tribometry 摩擦力测量术
tribophysics 摩擦物理学
triboplasma 摩擦等离子体
tribrach 三脚架；三脚台
tribranch 三角基branch
tribromamine 三溴胺
tribromaniline 三溴苯胺
tribromide 三溴片；三溴化合物
tribromoacetic acid 三溴乙酸
tribromo-benzene 三溴苯
tribromo-compound 三溴化合物
tribromo-dichloroethane 三溴二氯乙烷
tribromoethane 三溴乙烷
tribromophenol 三溴苯酚
tribromophenol bismuth 三溴酚铋
tribromophenyl 三溴苯基
tribunal 显要席(古罗马教堂)；法庭；法官席
tribunal of commerce 商事法院
tribune 论坛；后殿；演讲台；圣楼；廊台；看台(古罗马赛马场)；讲坛
tribune column 观礼坛柱；讲坛柱
tribune niche 观礼壁龛；讲坛壁龛
tributary 支路；支流；河流支流；从属的；附属的
tributary area 支流流域面积；汇水面积；支流区域；附属地区
tributary area of harbo(u)r 港口腹地(使用港口的内陆)

tributary area of port 港口属地；港口腹地(使用港口的内陆)
tributary arm 支汊
tributary basin 集水区；支流流域；流域
tributary cause 旁因
tributary channel 支渠
tributary data terminal equipment 从属数据终端设备
tributary ditch 支沟；进水沟
tributary drain 排水支管；支流
tributary drainage area 支流流域
tributary flow 支流
tributary glacier 支冰川
tributary inflow 支渠入流；支流流入(水)量；支流来水
tributary inlet 支流口
tributary junction 支流汇合处
tributary lending 为专门用途的借贷
tributary loss 支线损耗
tributary modem 从属调制解调器
tributary mouth 支流河口
tributary path 从属道路；附属道路
tributary receiver 支路接收装置
tributary river 支流
tributary station 从属站；附设局；辅助站；分站
tributary stream 支流
tributary system 支流水系
tributary unit 支路单元
tributary unit group 支路单元组
tributary unit pointer 支路单元指针
tributary valley 支流河谷
tributary waterway 支水道；支流水道
tributoxyethyl phosphate 磷酸三丁氧基乙酯
tributyl 三丁基
tributyl aconitate 丙烯三羧酸三丁酯
tributyl borate 硼酸三丁酯
tributyl citrate 柠檬酸三丁酯；柠檬酸丁酯
tributyl phosphate 亚磷酸三丁酯
tributyl phosphite 亚磷酸三丁酯
tributyl tin acetate 乙酸三丁基锡
tributyltin fluoride 氟化三丁基锡
tributyl tin oxide 氧化三丁基锡
tricalcium 三钙盐
tricalcium aluminate 铝酸三钙
tricalcium aluminate hydrate 水化铝酸三钙
tricalcium disilicate 二硅酸三钙
tricalcium phosphate 磷酸三钙
tricalcium silicate 硅酸三钙
tricalcium silicate hydrate 水化硅酸三钙
tricam connecter 三凸轮接头
tri-camera method 三镜航摄仪摄影测图法
tri-camera photography 三机照相
tricapability 三重能力
tricar 三轮车；三轮汽车；三轮机器脚踏车
tricarballylic acid 丙三羧酸
tricarbimide 异氰尿酸
tricarbonate 三碳酸盐
tricarboxylic acid cycle 三羧酸循环
trice 拉起；瞬息；吊索
tricen fiber point 三个着丝点的
tricerores 三角形蜡孔
trichalcite 丝砷铜矿
tricharged 三电荷
trichasium 三岐式
trichatrophia 毛折
trichilemma 毛膜
trichion 发际中点
trichite 晶星；针状晶体；发晶；发雏晶
trichlorgnate 胺硫磷
trichloride 三氯化物
trichloroacetic acid wastewater 三氯乙酸废水
trichlorobenzene 三氯苯
trichloro-butyl malonate 丙二氯苯三丁基酯
trichloroethylene 三炔化三氯；三氯乙烯
trichloroethyl phosphate 磷酸三氯乙酯
trichloromethane 氯仿；三氯甲烷
trichloromonofluoromethane 一氟三氯甲烷
trichobezoar 毛粪石
trichoblast 毛丝体
trichobotheria 毛点
trichochromogenic 毛色产生的
trichoid sensillum 毛状感受器
trichome 藻丝
trichome hydathode 毛状排水器

trichopore 毛孔
trichorrhea 骤脱毛发
trichosiderin 发铁色素
trichothallic 毛状菌体
trichotillomania 毛边套
trichotomy 三分法
trichroic 三色的
trichroic coating 三向色镀层
trichroism 三原色性；三向色性；三色性；三色现象
trichromat 三色器
trichromate 三铬酸盐
trichromatic 天然色；三元色白；三原色的；三色的
trichromatic analysis 三色分析
trichromatic coefficient 三原色系数
trichromatic colo(u)rimeter 三原色色度计；三色比色计
trichromatic coordinates 三原色坐标
trichromatic diagram 三色图
trichromatic edition 三色版
trichromatic image 三原色图像
trichromatic optic(al) separator 三色光学分离器
trichromatic photography 三色照相术
trichromatic printing 三原色印花
trichromatic response 三色响应
trichromatic response theory 三色响应理论
trichromatic specification 三色标志特性
trichromatic system 三原色系统；三色制；三色系统
trichromatic theory 三色说；三色理论
trichromatic unit 三原色单位
trichromatism 三色原理；三色性；三色现象
trichromatopsia 正常色觉
trichrom-emulsion 三色乳胶
trichromic acid 三铬酸
trichromscope 三色镜
trichter cathode 漏斗形阴极
tricing line 拉绳；控制绳
tricity 三联城市；三核都市
trick 镜面刻度线；特技；舵工的值班时间
trick drum 提花滚筒
trick effect 特技效果
trickle 细流；滑流
trickle bed 涓流床；滴流床
trickle bed reactor 滴流床反应器
trickle charge 连续补充充电；涓流充电；点滴式充电法
trickle charger 小电流充电器；涓流充电器；弱电持续充电器
trickle collector 涓流集热器
trickle cooler 淋式冷却器；阶式冷却器；水淋式冷却器
trickle cooling 淋式冷却
trickle deaerator 滴流式脱气器
trickle-down theory 流下理论
trickle drain 溢流管；滴流排水(管)
trickle feed 补充细流量给料
trickle filter bed reactor 滴滤床反应器
trickle flow hydrodesulfurization 滴流式加氢脱硫(法)
trickle hydrodesulfurization 滴淋氢化脱硫；滴流加氢脱硫(法)
trickle irrigation 细流灌溉；滴灌
trickle irrigation system 细流灌溉
trickle irrigator 滴灌器
trickle reactor 滴流式反应器
trickle tube 滴水管
trickling condenser 滴漏冷凝器
trickling cooler 水淋冷却器；水淋冷凝器
trickling cooling plant 冷水塔；喷洒冷却装置
trickling-down effect 渗透效应
trickling filter 洒滴滤池；生物滤器；生物滤池；渗滤器；滴滤池；滴滤池；滴漏滤池
trickling filter bed 散水滤床
trickling filter bed reactor 滴滤床反应器
trickling filter distilate 滴滤池渗出物
trickling filter plant 滴滤装置
trickling filter process 散水滤床法；滴滤过程；滴滤法；滴滤池法
trickling filter recycle 滴滤循环
trickling filter solid contact process 滴滤池-固体接触工艺
trickling filtration (process) 滴滤法
trickling towers 塔式滴滤池
trickling towers effluent 塔式滴滤池废水
trickling water 渗流水；渗漏滴水；滴灌水

trick lock 暗锁;数字转锁;对字锁
trick paint 纹理油漆;图案颜料
trick photography 特技摄影
trick recognition 特征识别
trick scene 旋转舞台;特技场景;特技布景
trick shot 特技镜头
trick valve 蒸汽机滑阀;阿伦滑阀
trick wheel 控制轮;提花轮
trick wheel bit 提花轮片
trick work 特技品
tricky business 艰巨工程;棘手任务
triclazate 三环酯
triclinic 三斜的;三斜
triclinic base 三斜底面
triclinic cell 三斜晶胞
triclinic crystal 三斜晶体
triclinic fabric 三斜组织;三斜组构
triclinic feldspar 三斜长石
triclinic hemihedral class 三斜半面体类
triclinic holohedral prism 三斜多面柱镜
triclinic holoihedral class 三斜全面体类
triclinicity 三斜度
triclinic prism 三斜柱体;三斜柱
triclinic symmetry 三斜对称
triclinic system 三斜晶系
triclinium 餐厅(古罗马三面用餐躺椅的);躺椅(古罗马圆餐桌椅的)
triclinochloritoid 三斜硬绿泥石
triclinogibbsite 三斜三水铝石
tricolor system 三原色法
tri-colo(u)r 三色的;三色旗
tricolo(u)r amaranth 雁来红
tricolo(u)r camera 三管彩色摄像机
tricolo(u)r chromatron 三色色标管;单枪栅控式彩色显像管
tricolo(u)r darkroom lamp 三色暗房灯
tricolo(u)r direct view tube 直观式三色管
tricolo(u)r red lantern 三色灯
tricolo(u)r red light 三色灯
tricolo(u)r emulsion 三色乳胶;三色感光乳剂
tricolo(u)r filter 三色滤色器;三色滤光片
tricolo(u)r glaze 三彩釉
tricolo(u)r kinescope 三色显像管
tricolo(u)r picture tube 彩色显像管
tricolo(u)r recption 三色接收
tricolo(u)r with china-ink ground 墨地三彩
tricolo(u)r with tender yellow 浅黄三彩
tri-column pier 三柱式桥墩
tri-component 三分量
tri-component magnetic survey in borehole 井中三分量磁测
tri-composition foundation 三合土基础
triconch 三边为半圆的四方平面布置形式
triconch choir 中央方形三面半圆形建筑中的歌唱班席位
triconch church 中央方形三面半圆形教堂
tricone bit 三牙轮钻头;三锥齿轮钻头
tricone bit with tungsten carbide inserts 三牙轮硬合金球齿钻头
tricone compartment mill 三锥式分室;三圆锥室磨碎机
tricone mill 三锥式球磨机;三锥式磨机
tricone rock bit 三牙轮凿岩钻头;三锥轮凿岩钻头;三牙轮岩石钻头;三角岩芯钻头
tricone roller bit 三锥滚轮钻头
triconodont 三锥牙;三锥齿
tricorn bit 三角钻头
Tricosal 特立库塞尔(一种水泥防水硬化液)
tricosane 二十三烷
tricosane diacid 二十三烷二酸
tricosanic acid 二十三酸
tricosanol 二十三醇
tricosendioic acid 二十三碳烯二酸
tricosene diacid 二十三碳烯二酸
tricosenoic acid 二十三烯酸
tricosoic acid 二十三酸
tricosyl 二十三基
tricosylacetic acid 二十三基乙酸
tricot weft insertion machine 衬纬经编机
tricresol 三煤酚
tricresyl phosphate 亚磷酸三甲苯酯;磷酸三甲苯酯
tricritical point 三重临界点
tricrotism 三波
tricrotous 三波的

tricrystal 三晶
tricusidal quartic 三尖点四次线
tricuspid 三尖拱;三尖头的;三尖瓣的
tricuspid cap 三角冠
tricyanic acid 三氰酸;三聚氰酸
tricycle 三轮机车;三轮摩托车;三轮机;三轮车
tricycle gear 三轮起落架
tricycle landing gear 三轮式起落架
tricyclene 三环烯
tricyclic aromatics 三环芳香化合物
tricyclic compound 三环化合物
tricyclic hydrocarbon 三环烃
tricyclic naphthenes 三环环烷
tricyclic ring 三环核
tricycloalkane 三环烷烃
tricyclohexylmethane 三环己基甲烷
tricyclopentadienide 三环戊二烯化物
tricyclopentadienyl-butoxy uranium 三环戊二烯基丁氧基铀
tricyclopentadienyl-cyclohexyliloxy-uranium 三环己氧基铀
tricyclopentadienyl-ethoxyuranium 三乙氧基铀
tricyclovetivene 三环岩兰烯
tricyle-type row-crop tractor 三轮型中耕拖拉机
tridane 十三烷基苯
triddling 机动车滴高液
tridecandioic acid 十三烷二酸
tridecane 十三烷
tridecane diacid 十三烷二酸
tridecane dicarboxylic acid 十三烷二羧酸
tridecane phosphonic acid 十三烷基膦酸
tridecanoic acid 十三酸
tridecanoyl 十三酰
tridecene 十三烯
tridecene dicarboxylic acid 十三烯二酸
tridecyl amine 十三胺
tridecyl benzene 十三烷基苯
tridecyl cyanide 十四腈
tridecylendioic acid 十三烯二酸
tridecylene 十三烯
tridecylenediacid 十三烯二酸
tridecylenic acid 十三烯酸
tridecylic aldehyde 十三醛
tridecyl sulfate 十三硫酸盐
tridecyne 十三块
trident 三叉鱼叉;三叉曲线;三叉戟式测距仪;三戟
trident maple 三角枫
trident of Newton 牛顿三叉线
trident-shaped dock 叉式港池;分岔式港池
trident spear with barbs 带倒刺三剌鱼叉
trident-type dock 三叉式港池;三叉式船坞
T-ridge roof 丁字脊屋顶
tridiagonal matrix 三对角(线)矩阵
tridimension 三维
tridimensional 立体的;三维的;三度的;三向的
tridimensional data 三度空间数据
tridimensional map 三维地图
tridimentional measurement 立体测量
tridiphenylmethyl 三联苯甲基
tri-drag-blade bit 三刮刀钻头
tridrum boiler 三筒式锅炉
triductor 三次倍频器
tridymite 鳞石英
tridymite heroynitite 鳞英铁尖品岩
tridymus 三联胎
tridyne 三合气
Trief process 特利夫湿磨矿渣水泥法
trielcon 三爪钳
trielectrode arrangement 三极测量法
triennial 三周年纪念
trienol 瑞士产合成桐油
trier 检验者;试验者;试验员;试验仪表;试验物;试验机;试料
triethane chloride 不均三氯乙烷
triethanolamine 三乙醇胺
triethanolamine oleate 油酸三乙醇胺酯
triethanolamine salt 三乙醇胺盐
triethide 三乙基金属
triethylaluminium 三乙基铝
triethylantimony 三乙锑
triethyl-bismuth 三乙基铋
triethyl-bismuthine 三乙铋
triethylchlorotin 三乙基氯化锡
triethyl citrate 柠檬酸三乙酯

triethyl cyanurate 三聚氰酸三乙酯
triethylene 三乙烯
triethylene glycolbismethyl-acrylate 二缩三乙二醇双甲基丙烯酸酯
triethyl-gallium 三乙镓
triethylic borate 硼酸乙酯
triethyl phosphate 磷酸三乙酯
triethyl-silicane 三乙基甲硅烷
triethyl-silicon 三乙基甲硅烷
triethyltin chloride 氯化三乙基锡
trifactor classification 三因分类
triferrous carbide 碳化三铁
Triffin Plan 特里芬计划
trifilar suspension 三点悬置
trifilar wire 三股线
trifluoride 三氟化物
trifluorochloromethane 三氟氯甲烷
trifluoro-compound 三氟化合物
trifluoromethane 三氟甲烷
trifluralin 三氟硝接
triflux 三工质热交换器
trifoil cross-section 三叶形截面
trifoliate 三叶的
trifoliate orange 枳
trifolium 三叶线
trifora 侧廊上的楼层(教堂)
trifora arch 侧廊上楼层的拱(教堂)
triforium 教堂拱门上拱廊;三拱式
triforium arcade 侧廊上楼层的拱廊(教堂)
triforium arch 侧廊内侧纵向三拱(教堂)
triforium gallery 侧廊内侧上方的三个拱洞拱(教堂)
trifoveolate 三穴的
trifrequency airborne electromagnetic system 三频航电仪
trifuel engine 三燃料发动机
trifuel pipe-line engine 三燃料管线站用发动机
trifurcating box 三芯线终端套管;三路接线盒;三支电路接线盒
trifurcating joint 三心电缆与三根单心电缆的接线盒
trig 楔子;制动三角木;刹车
triga 三驾双轮马车;三马战车雕饰;[复]triage
trigamma function 三伽马函数
trigatron 引燃管;充气触发管
trigatum 三角测量基准(点)
trig card 三角点成果表
trigdatum 三角测量基准(点);三角测量资料;三角测量数据;[复]trigdata
trigeminy 三联现象;牵联
trigger 引发物;引发剂;制轮器;止滑器;起动设备;启动器;触发信号;触发器;触发;掣子;掣板;扳机;闸柄
trigger action 制动效应;触发作用
trigger action relay 触发继电器
trigger amplifier 触发脉冲放大器;触发放大器
trigger bit 带钩钻头
trigger blocking oscillator 触发式间歇振荡器;触发间歇振荡器
trigger bolt 锁舌锁住钮;制动锁栓
trigger button 触发按钮
trigger cam 制轮凸轮
trigger carrier 触发器座架
trigger chain reaction 引发物连锁反应
trigger circuit 起动线路;起动网路;同步起动电路;触发电路
trigger clear terminal 触发器清零端
trigger control 触发控制
trigger decoder 触发译码器;触发解码器
trigger delay 触发延迟
trigger delay circuit 触发延迟电路
trigger delay multivibrator 触发脉冲延迟多谐振荡器
trigger discriminator 起动脉冲鉴频器;触发脉冲鉴别器
triggered blocking generator 可控间歇振荡器;触发间歇振荡器;触发间歇发生器
triggered display system 触发显示系统
triggered gaps 电花隙避雷器
triggered multivibrator 触发式多谐振荡器
triggered numerical recording set 触发数字记录装置
triggered spark gap 触发火花隙
triggered teleseismic recording equipment 触发远震记录设备
triggered-time base 等待式扫描电路;触发式时基

triggered-time clock 触发信号钟
trigger electrode 触点电极
trigger equipment 触发设备
trigger extension 扳机延伸杆
trigger finger 扳机状指
trigger flip-flop 计数触发器
trigger gas 触发气
trigger gate 触发闸门；触发选通脉冲；触发门
trigger gate voltage 触发选通脉冲电压；触发控制极电压
trigger gear 扳机机构
trigger generator 触发脉冲发生器；触发脉冲发生器
trigger housing 扳机座
trigger indicator 触发指示器
triggering 触发；触发
triggering earthquake 触发地震
triggering effect 触发效应
triggering energy 触发能
triggering gap 触发间断
triggering level 启动电平；触发电平
triggering mechanism 触发机制；触发机理；触发机构
triggering mechanism 引发机理
triggering of earthquake 地震的触发；地震触发
triggering pulse 触发脉冲
triggering sensitivity 触发灵敏度
triggering signal 触发信号
triggering structure survey 发震构造调查
triggering system 触发系统
triggering threshold 触发阈值
triggering value 触发值
trigger input 触发脉冲输入
trigger input circuit 触发输入电路
trigger inverter 触发脉冲变换器
trigger level 触发电平
trigger mechanism 击发机构；扳机机构
trigger mode 触发模式
trigger piece 闸柄；扳柄
trigger pin 扣机销
trigger point 扳机点
trigger point latch 触发点闩锁
trigger price mechanism 控制价格结构
trigger pull 扳机力
trigger pulse 触发脉冲
trigger pulser 触发脉冲器
trigger pulse signal 触发脉冲信号
trigger recognition 触发器识别
trigger reflector 脉冲反射器；触发脉冲反射器
trigger register 触发器式寄存器；触发寄存器
trigger relay 触发继电器
trigger release 触发释放装置
trigger release spray pump 扳机喷雾器
trigger scope 触发显示器
trigger select 触发选择
trigger selector 触发选择器
trigger selector switch 触发脉冲选择开关
trigger shaft 击发杆
trigger shaper 触发脉冲形成器
trigger sharpener 触发脉冲锐化器；触发峰化器
trigger source 触发源
trigger spark 点火器火花；触发点火花
trigger squeeze 压扳机
trigger starter 触发起动器
trigger sweep 触发式扫描；触发扫描
trigger switch 触发开关
trigger-timing pulse 定时触发脉冲
trigger tube 触发管
trigger unit 触发器
trigger valve 触发管
trigger voltage 起动电压；启动电压；触发电压
trigger winding 触发绕组
trigger word 触发字
trig horizontal direction and list 三角点水平方向及成果表
trigistor 三端开关器件
triglist 三角点成果表；三角测量成果表
trig loop 制动圈
triglyceride 甘油三酸酯；三酸甘油酯
triglyceride fatty acid 脂肪酸三甘油酯
triglycerol 缩三油
triglycidyl isocyanurate 异氰脲酸三缩水甘油酯
triglyeride 甘油三酯
triglyph 三槽板陶立克柱式；三槽板浅槽饰；三陇板（陶立克柱式的特征之一）；三联浅槽饰；三槽板

triglyph frieze 三陇板檐壁
trigodgenite 三方蓝辉铜矿
trigon 三角形；测时三角规
trigona 三角
trigonal 三角的；三方
trigonal aspect 三分方位
trigonal bipyramid 三方双锥
trigonal bipyramid face 三方双面锥
trigonal biyramid 三角双锥
trigonal crystal 三角系晶体；三方结晶
trigonal (crystal) system 三角晶系
trigonal field 三角场
trigonal frame 三脚架
trigonal lattice 三角系点阵
trigonal network 三角形网(络)
trigonal point 三分方位点
trigonal prism 三方柱
trigonal pyramid 三方单锥
trigonal system 三方晶系
trigondodecahedron 三角十二面体
trigone 三棱；三角
trigonia grit 毛粒状粗砂岩
trigonid 下三角座
trigonite 砷锰铅矿
trigonometer 平面直角三角形计算工具；三角学家
trigonometric(al) approximation 三角近似法
trigonometric(al) beacon 三角测量觇标
trigonometric(al) calculation 三角测量计算
trigonometric(al) chain 三角锁
trigonometric(al) cofunctions 三角余函数
trigonometric(al) control 三角控制；三角测量控制
trigonometric(al) control point 三控制点；三角控制点
trigonometric(al) coordinates 三角坐标
trigonometric(al) curve 三角曲线；三角函数曲线
trigonometric(al) data 三角测量数据
trigonometric(al) definite integral 三角定积分
trigonometric(al) device 三角函数装置
trigonometric(al) equation 三角方程
trigonometric(al) expression 三角式
trigonometric(al) fixed point 已知三角点；三控制点
trigonometric(al) frame(work) 三角控制网
trigonometric(al) function 三角化形式；三角函数
trigonometric(al) function forecast method 三角函数预测法
trigonometric(al) height traversing 三角高程导线测量
trigonometric(al) identify 三角恒等式
trigonometric(al) integral function 积分三角函数
trigonometric(al) interpolation 三角内插法
trigonometric(al) interpolation polynomial 三角插值多项式
trigonometric(al) level(l)ing 三角水准测量；三角高度测量；三角高程测量(法)；三角测量高法
trigonometric(al) level(l)ing network 三角高程网
trigonometric(al) parallax 三角视差
trigonometric(al) point 三角点
trigonometric(al) point description 三角点说明；三角点之记
trigonometric(al) polynomial 三角多项式
trigonometric(al) ratio 三角比
trigonometric(al) ray-trace method 三角光线追迹法
trigonometric(al) ray tracing 三角光线追迹
trigonometric(al) relation 三角关系
trigonometric(al) series 三角级数
trigonometric(al) station 三角点；三角(测量)标点；三角网测站
trigonometric(al) substitution 三角代换
trigonometric(al) sum 三角和
trigonometric(al) survey 三角测量
trigonometric(al) system 三角系统
trigonometric(al) table 三角函数表
trigonometry 三角学
trigonulum 三角室
trigonum 三角形镶嵌物；三角
trigonum collaterale 侧副三角
trigonum dorsale 背侧三角
trigonum durum 背侧三角
trig point 三角点
trigram 三字母组
tri-grid 三向网格的
trig station 三角测站

triguaiacyl 三基
trihedral 三面形(体)
trihedral angle 三面角
trihedral prism 三面体棱镜；三角棱镜
trihedral (tetrahedral) tool-maker's straight edge 三棱直尺
trihedral vertex 三面顶点
trihedron 三面形(体)；三面体
trihemellitic acid 苯偏三酸
tri-hinges 三联铰
trihydrate 三水合物
trihydric alcohols 三元醇
trihydric salt 三酸式盐
trihydrocalcite 三水碳酸钙；三水碳钙石
trihydrol 三水分子
triiodide 三碘化物
triiodo-compound 三碘化合物
tri-iron tetroxide 铁黑
triisoamylamine 三异戊胺
triisobutene 三聚异丁烯
triisobutylene 三聚异丁烯
tri-isopropanolamine 三异丙醇胺
trijunction 三线交点
trikalsilite 三型钾霞石
triketopentane 戊烷三酮
trilamellar membrane 三层膜
trilaminar structure 三片层结构
trilateral 三角形；三边形；三边的
trilateral agreement 三边协定
trilateral crosssection 三角形截面
trilateral figure 三边形
trilateral trade 三边贸易
trilateration 长距离三角(边长)测量；三边测量
trilateration method 三边测量法
trilateration net 三角测量网
trilateration network 三边测量网；测边网
trilateration survey 三边测量
trilead cable 三线电缆
trilete 三裂缝的
trilete aperture 三裂口
trilete marking 三射痕
trilete scar 三裂痕
tri-level 三层的
tri-level grade separation 三层式立体交叉；三层立体交叉
trilinear 三线性的；三线的
trilinear chart 三角折线图；三线图；三角坐标图；三角坐标分类图
trilinear coordinates 三线坐标；三角坐标
trilinear degrading stiffness 三线性退化刚度
trilinear form 三线性形式
trilinear idealized hysteresis loop 三线性理想化滞流图；三线性理想化滞流环线
trilinear line coordinates 三点线素坐标
trilinear method 三线法
trilinear point coordinates 三线点素坐标
trilinear primary curve 三直线基线
trilinear survey 三点后方交会法
trilinolenin 亚麻精
trilith(on) 巨石纪念碑；三石塔(两石柱上架石梁的纪念碑)；石柱上架石梁的纪念门
trill 三连晶
trilling 三连晶
trillion 百亿亿
Trillor vibrator 特里洛牌振动捣实器
trilobal 三叶形
trilobated 三裂片的
trilobated leaf 三裂叶
trilobation 三裂
tri-lobe cam 三突齿凸轮；三凸角形凸轮；三工作边凸轮
trilobed wheel 三叶轮
trilobite limestone 三叶虫灰岩
trilock type 三齿防转式
trilocular 三室的
trilogarithm 三重对数
trilogy 三联
trim 修坯；修剪；纵倾【船】；装饰物；门头线；门贴脸；门窗贴脸；去除飞边；区标；潜艇浮力；平衡度；平舱；贴脸；首尾吃水差；吃水差(船首尾的)；镶边框；装饰；清整
trim a budget 削减预算
trimaceral microlithotype 三组分显微煤岩类型
trim and stability booklet 纵倾和稳定计算书

trim angle 纵倾角(船的)
trimaran 三体船
trimask 多层彩色蒙片;三次掩蔽(的)
trimask structure 三次掩蔽结构
trim ballast 整修铺用碎石;平舱压载
trim band 带箍
trim bronze 装饰青铜
trim budget 削减预算
trim by bow 艏倾;首吃水大于尾吃水
trim by head 艏倾;首吃水大于尾吃水
trim by stern 艄倾;尾倾;首吃水大于尾吃水
trim by the bow 首倾【船】;船首前倾
trim by the head 首倾【船】;船首前倾
trim by the stem 首倾【船】;船首前倾
trim by the stern 后坐;后倾【船】;艉倾;尾吃水大于首吃水;船尾纵倾度
trim control 平稳控制
trim cooler 调温冷却器
trim correction 纵倾调整;吃水差调整
tri-m-cresylphosphate 磷酸三间甲苯基酯
trim curve 纵倾曲线
trim cutter 切边机
trim diagram 纵倾计算图表
trim die 切边冲模
trim-edge 饰边;型条边
tri-media filter 三层滤料滤池
tri-media polishing filter 三层滤料终端处理滤池
tri-medium bed 三滤料滤床
trimellitate 偏苯三酸盐
trimellitic acid 偏苯三酸
trimellitic anhydride 偏苯三酸酐
trim enamel paint 装饰性瓷漆
trimenon 三个月
trimer 三体;三聚物
trimeric benzene 三聚苯
trimeric sodium phosphate 三聚磷酸钠
trimerite 硅铍钙锰石;三斜石
trimerization 三聚
trimerous 三对的
trimesic acid 均苯三酸;苯均三酸
tri-met 三镜头航空摄影
trimetal 三金属;三层金属轴承合金
trimetal detector 三元合金探测器
trimetallic catalyst 三金属催化剂
tri-metallic strip 三金属带
trimetaphosphate 三偏磷酸盐
trimethylamine 三甲胺
trimethyl ammonium chloride 氯化三甲铵
trimethyl borate 硼酸三甲酯
trimethylene 环丙烷
trimethylen-edinitrilo-tetraacetic acid 丙二胺四乙酸
trimethylene glycol 亚丙基二醇
trimethyl phosphate 磷酸三甲酯
trimethyltetrahydronaphthalene 三甲基四氢化萘
trimethylthiophene 三甲基噻吩
trimetric 斜方的
trimetric drawing 正三轴测图
trimetric projection 正三测投影;三度投影;正三轴投影
trimetric system 斜方晶系
trimetrogon 三镜头摄影仪;三镜头航摄机
trimetrogon camera 三镜头航空摄影机;三镜头航空照相机
trimetrogon charting 三镜照相制图
trimetrogon mapping 三镜头航空摄影测图
trimetrogon method 三镜头航摄机摄影法
trimetrogon photogrammetry 三镜头航空摄影测量学
trimetrogon photograph 三镜头空中照相;三镜头摄像片;三镜头航摄照片
trimetrogon photography 三镜照相;三镜头航摄机摄影术
trimetrogon set 三镜头摄影像片
trim fabricator 饰条制造工
trim for interior work 内部装饰条材
trim hardware 小五金装饰件;装饰小五金
trim-heel regulating system 纵横倾调节系统【船】
trim hole 前装炮眼;爆破泄放孔;修整孔洞
trim in 镶进;嵌入;镶入
trim indicator 纵倾指示器
trimix diving 氦—氮—氧潜水
trimix saturation diving 氦—氮—氧饱和潜水

trim joist 托梁;承接梁
trim knob 调整片钮
trimline 毁林线
trimmability 可微调性
trim maker 饰条制造工
trim mark 光边角线
trimmed block 装饰混凝土砌块
trimmed displacement 纵倾排水量
trimmed edge 铣削过的外底边
trimmed estimator 修整统计量
trimmed hedge 修剪的丛篱;装饰篱笆
trimmed in bunkers 煤舱平舱;燃料舱平舱
trimmed joist 承接梁;托梁;楼梯洞口短格栅;次梁;楼梯洞口短栅栏
trimmed opening 饰边门洞;有贴脸的门洞;镶边框的孔口
trimmed rafter 装饰的椽木;装饰的椽子;装饰椽子;整饰过的椽
trimmed sheet 精整薄板
trimmed size 纸张光边尺寸;切齐寸
trimmed spur 修切山嘴
trimmed surface 修琢面
trimmed tunnel 修饰隧道;修饰过的隧道;衬砌隧洞
trimmed value 平衡值;调整值
trimmed waterline 纵倾水线
trimmer 周边炮眼;修整器;修整机;修平机;异形墙砖;集料器;活挡头;横截圆锯机;切刀;切边工;千斤格栅(美国);平舱器;平舱机;平舱工人;微调电容器;托梁;调整器;调整片;承梁;承接梁;安全矿灯剔丝
trimmer arch 炉前拱;承梁拱;壁炉前拱;托梁拱;矢高小的(石)拱
trimmer beam 托梁;承接梁
trimmer capacitor 微调电容器
trimmer condenser 微调电容器
trimmer conveyer 移动式输送机
trimmer joint 多牙榫接;托梁接头(与小梁相接处);镶榫接头;镶榫接合
trimmer joist 楼梯格栅;活挡头;托梁
trimmer plank unit 托梁垫板构件
trimmer resistor 微调电阻器
trimmers 边炮眼
trimmer saw 修整锯
trimming 修整;修坯;消除危石;镶边;整平;找顶;规格剪切;截齐;剪切;去毛刺;去飞边;切屑;微调;挖底;打枝;打侧杈;飞边;镶边框;装饰;清整
trimming algorithm 裁剪算法
trimming allowance 修边余量
trimming and curling 剪卷
trimming and mounting diagram 相片镶嵌略图
trimming bit 整修铣刀
trimming blast 修整爆破
trimming charges 平舱费
trimming cloth 修饰用布料
trimming condenser 补偿电容器
trimming condition 纵倾状态
trimming cut 切边
trimming data 修整数据
trimming deflashing 修边
trimming device 修整装置
trim(ming) die 精整冲模;修边模;冲模
trimming dust 铣削革屑
trimming filter 校正滤波器;补偿滤波器
trimming function 平衡作用
trimming gear 翼尼修整装置
trimming glass capacitor 玻璃微调电容器
trimming glassglaze potentiator 玻璃釉微调电位器
trimming guide 导片裁切器
trimming hatch(way) 匀货舱口;灌补舱口
trimming hole 修正孔;匀货舱口;灌补舱口
trimming joist 整饰横梁;楼梯格栅;加劲格栅;千斤格栅;托梁;支托托梁的格栅;过梁
trimming knife 修边刀;切边刀
trimming knife machine 修边机
trimming lathe 修整用车床
trimming machine 修坯机;滚边机;截锯机;剪边机;切边剪机;挖底机;缝边机;刨版边机;修整机;木料裁角机;剪切机
trimming method 裁剪方法
trimming moment 纵倾力矩;平衡力矩
trimming of loose rock 清除浮石
trimming of shoulder 路肩修整
trimming of slope 斜坡修整;边坡修整
trimming of working face 工作面整修

trimming operation 平舱作业
trimming out 打枝;打侧杈
trimming piece 微拱模架
trimming plate 切边钢板;切削去边缘的钢板;切去边缘的板
trimming press 修整压力机;修边压力机;整形压力机;切边机;冲压压力机
trimming process 校正过程
trimming pump 纵倾平衡系统泵
trim(ming) punch 精整冲头;切边冲头
trimming rafter 整饰的椽;千斤椽
trimming saw 修枝锯;修剪锯
trimming shears 修整剪切机
trimming shop 整修车间;整修工场
trimming slivers 切边整割机
trimming speed 修整速度
trimming tab 配平调整片
trimming table 纵倾计算图表
trimming tank 压载水舱;纵倾平稳水舱;尖舱
trimming tool 修整工具
trimming tree 整形树
trimming wheel 修整砂轮
trimming yard 整修工场
TRIMO [modern, mobilia, modello] 三 MO 设计方案
trim on even keel 平吃水
trimorphic flower 三形花
trimorphism 三形;三态性;三晶
trimorphous 三形的
Trimorphous metal 三晶型金属
trimotor 三发动机
tri-motored airplane 三发动机飞机
trim packing 车厢内衬
trim paint 装饰漆;门窗漆
trim plate 调整片
trim saw 修整锯;小木锯;修枝锯
trimscript 切标
trim sheet 纵倾计算表
trim size 地图或海图纸修整后的尺寸;实际尺寸
trim stone 镶边石
trim stop 切边定位法
trim system 平衡装置
trim tab 调整片
trim-tab control 平衡配平片操纵
trim-tab servo 调整片随动机构
trim tank 平衡水舱
trim the top of column 塔顶回流
trim tight surface 密封面
trim type bucket 清舱抓斗;耙集式抓斗
trim valve 微调阀;调整阀
trinacriform 三尖形
trinalgon 硝酸甘油
trinal layer filter 三层滤池
trinary 三元的
Trinasco 特里奈斯库沥青液(一种修补地面和屋面的沥青液)
Trinazzo 掺大理石屑的彩色沥青地面
trincomalee 亭可马里暗红色硬木
tringle 杆子;帐杆;挂帘横杆;狭直条饰;挂帘子的横杆;模杆
Trinidad and Tobago Oil Co. 特立尼达和多巴哥石油公司
Trinidad asphalt 特立尼达(地)沥青;特立尼达天然沥青
Trinidad lake asphalt cement 特立尼达沥青胶泥
Trinidad pitch 特立尼达天然沥青
Trinidad Pitch Lake 特立尼达沥青湖
Trinidad refined asphalt 特立尼达精制沥青
tri-nip press 三压区压榨
Trinitian (stage) 特林尼特阶【地】
trinitramine 三硝胺
trinitrate 三硝酸酯;三硝酸盐
trinitrides 叠氮化物
trinitrin 三硝酸甘油酯
trinitrocellulose 火棉
trinitro-compound 三硝基化合物
trinitrocresol 三硝基煤酚
trinitroglycerin 三硝酸甘油酯
trinitrol 三硝油
trinitro-methylamine 三硝甲胺
trinitron 三镜筒观测镜
trinitrotoluene 黄色炸药
trinitrotoluene [TNT] 梯恩梯炸药;三硝基甲苯
trinitrotoluene poinsoning 三硝基甲苯

trinity 三位一体
Trinity House 引航工会(英国)
trinity ophiolite 三位一体蛇绿岩
Trinity Series 三一统【地】
trinket 最上帆
trinodal quartic 三结点四次线
trinodal seiche 三节假潮
trinol 垂诺耳
trinol [TNT] 黄色炸药
trinomen 三名
trinomial 三项式
trinomial distribution 三项分布
trinomial equation 三项方程
tri-n-propyl phosphate 磷酸三正丙酯
trinuclear 三环的
trinucleated 三环的
trio 三组;三辊(拉伸机)
trioctahedral 三八面体的(结晶结构);三八面层
trioctyl phosphate 磷酸三辛酯
triode 三极管
triode amplifier 三极管放大器
triode argon detector 三极氩检测器
triode clipper 三极管削波器
triode detection 三极管检波
triode detector 三极管检波器
triode driver 三极管激励器
triode electron gun 三极管电子枪
triode generator 三极管振荡器
triode gun 三极式电子枪
triode ion ga(u)ge 三极管电离真空计
triode laser 三极管气体激光器
triode mixer 三极管混频器
triode oscillator 三极管振荡器
triode time selector 三极管时间选择器
triode transmitter 三极管发射机
triolein 三油精
Triolith 特里奥利特防腐剂(一种木材防腐剂)
triolocation 三位置
trio mill 三辊式轧机
Trioptic 三元万能测长机
triose 丙糖
triovariant system 三变体系
trioxide 三氧化物
trioxysulfotungstate 三氧硫钨酸盐
trip 出行;周程;矿车列车;解扣机构;航次(指船);倾斜器;倾翻器;摘纵机构转动;跳闸装置;断开装置;短途旅行;单航次;扳机卡榫;安全开脱器;接合机构
trip account 航次账单
tripack 三层彩色片;三层版
tripack colo(u)r film 三层彩色胶片
tripack film 三层膜
tripack material 三层感光材料
trip aim 出行目的
trip alarm 跳闸报警
trip amplifier 停堆放大器
triparatite seismic network 三合地震台网
trip arm 松放杆(犁安全器的);定捆杆
triparted 三深裂的
triparted array 三联台阵
triparted force 三组分力
triparted observation 三点观测
triparted seismic network 三联地震台网
triparted vault 三个成一组的穹顶
triparties currency agreement 三通货协定
tripartite 三个一组的;三组分的;三深裂的;分成三部分的
tripartite agreement 三方协议
tripartite arch 三个一组的拱;三连拱
tripartite arrangement 三方经营
tripartite array 三合台阵
tripartite concave grating 三重凹面光栅
tripartite debts 三角债
tripartite idler 三节托辊
tripartite indenture 三联合同
tripartite inspection system 三检制
tripartite logarithmic response spectrum 三对数坐标反应谱
tripartite net observation 三合台网观测
tripartite presbytery 三合一长老席位
tripartite response spectrum 三坐标反应谱
tripartite stair(case) 三跑楼梯;三梯段楼梯
tripartite station 三向测站
tripartite transaction 三方交易

tripartite vault 三个组成的穹顶;三部分组成的穹顶;三轴穹顶;三个一组穹顶
tripartite window 三连窗
tripartition 分裂三部分
trip assignment 出行分配
trip attraction 出行吸引
trip beam 带安全器的犁辕
trip block 自动停止进刀挡块
trip bolt 紧固螺钉;紧固螺栓
trip bridge 吊桥
trip bus 脱扣母线;跳闸母丝
trip button 解扣按钮
trip cam 释放凸轮
trip casing spear 脱开式套管打捞器;套管打捞筒
trip change 列车交换
trip charter 航租;航次租船;定程租船;承租契约
trip choice 出行选择
trip circuit 解扣电路;脱扣电路
trip classification 出行分类
trip coil 脱扣线圈;跳闸线圈;自动开关释放器;解扣线圈
trip consumption of time 出行耗时
trip contour map 交通等密线图
trip cost 出行费用
trip destination 出行终点
tripdial 里程计;里程数
trip discharger 翻卸装置
trip distance 出行距离
trip distribution 行程分布;出行分布
trip distribution prediction 出行分布预测
trip dog 脱档;绊担;自动爪;自动停车器;止制钩;止动钩;解扣;脱扣钩;跳闸装置;跳挡装置
Tri-pedal 特利板(一种不滑的铁路面)
trip end 交通汇集点;行程终点
trip end model 出行终点模型
trip endpoint 出行端点
trip engine 扳动发动机
triperchromic acid 过氧铬酸
tripericlinal chimera 三层周缘嵌合体
tripestone 弯硬石膏
trip feeder 推车机
trip feeding 推车
trip-free 自动跳闸;自动解扣
trip-free circuit breaker 不跳闸断路器
trip free relay 自由释放继电器
trip-free release 自动释放
trip-free type 自由脱扣型
trip gear 脱开机构;跳闸装置;跳动装置;断路装置;扳动装置;扳动机构;接合机构
trip generation 出行生成;出行发生
trip generation method 交通发生法
trip halyard 补助帆索
trip-hammer 杵锤;夹板落锤
triphane 锂辉石
triphase 三相(的)
tri-phase current 三相电流
tri-phase soil 三相土
triphasic potential 三相电势
tri-phenarsazine chloride 氯化三吩砒嗪
triphenyl 三苯基
triphenylamine 三苯胺
triphenyl antimony 三苯锑
triphenylmethane 三苯甲烷
triphenylmethane dye 三苯甲烷染料
triphenylmethane dyestuff 三苯甲烷染料
triphenyl phosphate 磷酸三苯酯
triphenyltin fluoride 氟化三苯锡
trip hitch 脱钩式联结器
trip holder 夹紧模座
trip hook 自动脱开钩
triphosgene 三光气
triphylite 磷酸锂矿
tripinnate 三回羽状的
trip intensity 出行强度
trip interchange model 行程交换模型;出行交换模型
trip key 开路电键;断路电键
triplane 三翼机
triplasy 三分式
triple 三重的;三元组;三层;三倍数;三倍量;三倍(的)
triple access 三重访问
triple acting jack 三作用千斤顶
triple acting pump 三动泵

triple action paddle mixer 三作用桨叶式搅拌机
triple action presses 三动式压力机
triple action skid control brake system 三级降压防滑制动系统
triple agency 三代店(即代购代销代营店)
triple air lock 三道卸料闸门;三道锁风阀
triple annealing 三重退火
triple aplanat 三层消球差镜
triple arch 三连拱
triple arch dam 三拱坝
triple-arched 三连拱的
triple articulation arch 三铰拱;三跨拱;三跨相连的拱;三节拱
triple-band filter 三频带滤波器
triple band-saw 三联带锯
triple-barrel carburetor 三筒汽化器
triple-barrel motor 三燃烧室发动机
triple beam balance 三梁天平
triple bed room 三床位房间
triple belt 三层皮带;三层胶合皮带
triple bevel scarve 三头斜榫
triple block 三饼滑车
triple board platform 二层台(三根钻杆组成立根高度)
triple body 三通阀体
triple bottom 三层底
triple box section 三腔式断面;三箱式断面
triple braided 三层编包线
triple buff 三折布抛光轮
triple bunk 三层铺
triple busbar 三母线
triple busbar system 三母线系统
triple cable 三股绞花;三芯电缆
triple cambered aerofoil 三曲翼面
triple camera 三镜头摄影机
triple cancellation 三系斜杆
triple carriageway road 三块板道路;三幅(式道)路
triple case 三重壳
triple-casement window 三连窗;三扇窗
triple-cavity mo(u)ld 三穴模具;三腔模;三孔模型
triple cavity process 三腔模成型法
triple chain conveyer 三链式输送机
triple chain wire 三线网
triple-change gear 三级变速箱
triple channel 三重沟道
triple chrome piston ring 三面镀铬型活塞环
triple circuit jack 三线塞孔
triple circuit system 三管路系
triple clamp 三角轧头
triple-clamp bolt 三角轧头螺钉
triple cleaning 三次精选
triple cloud points 三浊点
triple coaxial transformer 三路同轴线变压器
triple coil 三重线圈
triple coincidence 三重符合
triple-coincidence switch 三重联合开关
triple collision 三重碰撞;三元碰撞;三体碰撞
triple-concentric cable 三芯同轴电缆
triple condenser 三透镜聚光器
triple conductor 三芯导线;三股绞线
triple control 三重控制
triple-conversion receiver 三重变频接收机
triple copy 第三份
triple cordon 三单干形
triple-core cable 三芯电缆
triple correlation 三重相关
triple correlation coefficient 三重相关系数
triple correlator 三重相关器
triple corrugated sheet iron 三瓦楞铁皮
triple cotton-covered 三层纱包的
triple counterpoint 三重对位
triple course 三层瓦(檐口处)
triple (course) surface treatment 三层式表面治;三层表面处治;三层表面处理
triple covered wire 三层包线
triple-cropping 三作
triple cross 罗马宗教十字架;三连十字架
triple cross-compound turbine 三轴多缸汽轮机
triple-crystal diffractometer 三晶衍射仪
triple curve 三相曲线
triple-cut quartersawn 三重叠切割锯木
triple cylinder 三通筒
triple-cylinder gasket 三通筒垫密片
triple-deck 三层的

triple deck chip screen 三层木片筛
triple-deck concentrating table 三层摇床
triple deck dryer 三层烘缸
triple deck dry screen 三层干筛
triple-decker 三层式道路；三层式立体交叉；三层道路交叉
triple-deck grade separation structure 三层立交桥；三层立体交叉结构
triple-deck screen 三层筛
triple-deck vibrating screen 三层(式)振动筛
triple detection 双超外差接收
triple diagonal matrix 三对角(线)矩阵
triple dial timer 三表轨计时器
triple diffuser 三重喉管(化油器的)
triple diffusion process 三次扩散工艺
triple-diffusion technique 三次扩散技术
triple discharging gate 三道卸料闸门
triple distilled water 三级蒸馏水
triple diverty 三重分集
triple drive 三重驱动
triple-drum scraper hoist 三滚筒耙矿绞车
triple drum tamper 三滚筒压实机
triple dye 三重染色
triple effect 三效
triple-effect evapo(u)ration 三效蒸发
triple-effect evapo(u)ration concentration 三效蒸发浓缩
triple-effect evapo(u)rator 三用蒸发器；三效蒸发器
triple-engined airplane 三发动机飞机
triple-engined type 三发动机式
triple-engine layout 三发动机布局
triple entry 三平巷
triple-entry room-and-pillar mining 三进路房柱式采矿法
triple-entry system 三平巷掘进系统
triple-error correcting code 三重错误校正码；三差校正码
triple excited state 三重激态
triple-expansion 三次膨胀
triple-expansion engine 三胀式蒸汽机；三涨式蒸汽机；三级膨胀式蒸汽机；三级膨胀发动机
triple-expansion steam pump 三级膨胀蒸汽泵；三膨胀式蒸汽泵
triple extension ladder 三节式可延伸梯
triple filament method 三灯丝方法
triple-fired furnace 三段式炉
triple flap grate 三道翻板阀
triple flashing light 三连闪光灯
triple-flight auger 三头螺旋螺旋
triple-flight stair(case) 三跑楼梯；三梯段楼梯
triple-float-type seaplane 三浮筒式水上飞机
triple flow turbine 三排汽口涡轮机
triple flue boiler 三燃烧管锅炉
triple-fluked anchor 三爪锚
triple-fluted bit 三刃铣刀
triple focusing 三重聚焦
triple focusing mass spectrometer 三重聚焦质谱仪
triple-fold disappearing stair(case) 三折隐梯
triple-folded hornreflector 三折喇叭形反射器天线
triple form 三元组形式
triple frequency 三倍频
triple frequency harmonic 三次谐波
triplefunction sonar 多功能声呐
triple gate 三态门
triple gate feeder 三道闸门喂料机
triple gear 三联齿轮；三合齿轮
triple geared drive 三联齿轮传动；三级齿轮驱动
triple-glazed units 三层玻璃窗
triple glazing 三重玻璃
triple grating polychromator 三层光栅多色仪
triple-grid amplifier 三极管放大器
triple grip anchor 三爪锚
triple grips 三向连接铁板
triple grouser 三组锚碇桩
triple grouser shoe 三组锚碇桩座；三齿履带板
triple gyrocompass 三陀螺罗经
triple half-wave filter 三半波滤光片
triple hand tamper 三人手持夯
triple-headed capstan 三卷筒绞盘
triple heading 三平巷
triple-heading room-and-pillar mining 三平巷房柱采矿法

triple head-lock and double side-lock tile 三向顶嵌合与双向嵌合的瓦片
triple helix 三股螺旋
triple-hinged 三铰链的；三铰接的
triple-hinged arch 三铰拱
triple-hinged arched girder 三铰(接)拱梁
triple-hinged arch-ribbed dome 三铰拱肋穹隆
triple-hinged half frame 三铰半框架
triple-hinged portal frame 三铰门架
triple-hinged rectangular frame 三铰矩形框架；三铰矩形构架
triple-hinged roof 三铰屋顶
triple-hinged steel arch 三铰钢拱
triple-hinged trussed arched girder 三铰桁架拱梁
triple-hulled ship 三体船
triple-hulled vessel 三体船
triple humped curve 三峰曲线
triple-hung sash window 三扇式上下启闭窗
triple indexing center 三重分度头
triple integral 三重积分
triple integration 三重积分法
triple interaction 三因子交互影响
triple interferometer 三次干涉仪
triple interleaved parity 三重插入奇偶检误
triple interlocking tile 三向嵌合的瓦片
triple interpolation 三向内插
triple ion 三离子物
triple jet 三孔喷嘴
triple-jet experiment 三射流实验
triple joint 三头连接
triple junction 三重接头；三重点；三联点
triple junction of plates 板块三结合点
triple junction structure 三向联结构造
triple labyrinth seal 三层迷宫式挡风圈
triple laminate 三层胶合板；三层夹板；三夹板
triple lancet window 三尖顶拱窗
triple-lap pile 三叠接桩
triple launcher 三联发射装置
triple-layer coating 三层镀膜
triple-layer grid 三层网架
triple-layer winding 三层绕组
triple-leaf standard 三层板铲柄
triple-length 三倍长度
triple-length working 三倍长工作
triple mast 三脚电杆；三脚天线脚
triple mat 三元复合毡
triple mirror 三垂面反射镜
triple mode catalytic converter 三作用催化转换器
triple mode emission control system 三方式排气净化系统
triple modular redundancy 三重模块冗余度
triple modulation beacon 三调制信标
triple mould 三重模板
triple-mounted 三联装的
triple mounting 三联装架
triple mower 三刀割草机
triplen 三次谐波序列
triplener 三合滤波器
triple net lease 三净租赁(财产税、保险费和维修费均由租户承担)
trip length 行程长度
triple offset 三级起步时差(交通)
triple paper cup dispenser 三层式纸杯配出器
triple pass 三程
triple pen 三机掩体
triple petticoat bell shaped insulator 三裙碗式绝缘子
triple-pinned 三销接的
triple-pinned truss(ed) frame 三销架式构架
triple pipe 三重管
triple-pipe chemical churning process 三重管旋喷法
triple piston type pump 三汽缸泵
triple plane swivel joint 三面回转接头
triple plate clutch 三片式离合器
triple ply belt 三层皮带；三层胶合皮带
triple point 三重点；三相点；三态点
triple point depression curve 三相点降低曲线
triple point of water 水的三相点温度
triple point path 三相点轨迹
triple-point refueling 三点加油
triple-pole 三极(的)
triple-pole double throw 三刀双掷
triple-pole fuse 三极熔丝

triple-pole knife switch 三极闸刀开关
triple-pole on-off switch 三刀开关；三极开关
triple-pole single throw 三刀单掷
triple-pole switch 三极开关
triple-post lift 三柱式车库用举升器
triple precision 三倍精度
triple precision number 三倍精度数
triple prism 三棱镜
triple-prism reflector 三个棱镜反射器
triple probe 三探针
triple product 三重积
triple-product convolution 三重积对合；三重积卷积
triple pulley block 三级滑车组
triple pulse coding 三脉冲编码
triple pump 三联泵；三缸式(水)泵
triple-purpose 三用的
triple-purpose breed 三用品种
tripler 三重器；三倍器；三倍频器；乘三装置
triple ram preventer 三闸板防喷器
tripler circuit 三倍电路
triple reaction 三重反应
triple reduction gear(box) 三级减速器
triple reduction gear(ing) 三级减速装置
triple redundancy 三重冗余
triple-redundant 三重余度的
triple reflection 三次反射
triple-reset system 三重置系统
triple response 三重反应
triple rib tire 三环花纹轮胎
triple-rivet(ed) joint 三行铆钉；三行铆接
triple rivet(ing) 三行(式)铆钉
triple rocker 三重摇杆
triple rod extensometer 三杆应变计
triple roll crusher 三滚筒压缩机；三滚轴破碎机；三辊破碎机
triple roller 三组镇压器
triple-roll(er) mill 三辊磨
triple-roll roller 三滚筒压路机
triple root 三重根
triple-round edge 三圆边(釉面砖)
triple round tile 三向呈圆边的面砖
triple row 三行的
triple rows line seeding 带状三行条播
triples 三联式拖车方式
triple-safe air brake system 三合安全气制动系
triple-safe boom hoist 三保险横臂吊车
triple salt 三聚盐；三合盐
triple scalar product 三重内积；三重标积
triple scale 三重刻度；三重标度
triple scan(ning) 三重扫描
triple scattering 三重散射
triple-scoop feeder 三斗给料机
triple screw 三纹螺旋；三螺旋桨
triple screw propeller 三推进器
triple screw propeller ship 三推进器船
triple-screw ship 三螺旋桨船
triple seal 三重密封
triple sealing ring 三层式密封环
triple seal piston ring 三重密封活塞环
triple seal-tab 三叶密封片
triple seat 三人并肩座椅
triple segmental baffle 三弓形折流板
triple-shear 三剪力；三系剪刀
triple shift 三班(工作)制；三班
triple-shock intake 三激波进气扩散段；三激波进气口
triple-silo complex 三发射井的发射场综合设施
triple single pole motorized isolator 成组三台单极电动隔离开关
triple sleeve antenna 三层重叠高增益天线
triple smoothing 三次修匀；三次平滑化
triple solid laminated partition 三层实心层叠隔墙
triple-speed gramophone 三转速电唱机
triple spiegel 三角柱镜
triple-spiral ring packing 三头螺旋环填料
triple-spool 三转子
triple-squirrel-cage rotor 三鼠笼转子
triple-stacked 三层的
triple-stage pump 三级泵
triple staining 三重染色法
triple stand 三脚架
triple star 三合星
triple start worm 三头蜗杆
triple state 三相(的)

triple stereoset 三重立体组
triple strand roller chain 三列套筒滚子链
triple stripe 三线条纹
triple-stroke deep well pump 三冲程深水泵
triple-stub transformer 三短线变量器
triple-substituted 三代的
triple substitution 三元取代
triple superheterodyne 三次变频超外差接收机
triple superphosphate 三过磷酸钙
triple superphosphate fertilizer 三元过磷酸钙肥料
triple surface treatment 三层表面处理
triple suspension safety hook 三爪安全提引钩
triplet 三重线;三重的;三元级;三同步发射台组;三联体;三角形链;三件一套;三合透镜;三个一组;三电台机组;弹丸
triple table 三元组表
triple tandem strander 三级串联捻胶机
triple tangent 三重切线
triple tariff meter 三部收费电度计
triplet bending magnet 三重偏转磁铁
triplet charm 三重色
triplet code 三联体密码
triple telescopic boom 三节伸缩吊杆
triplet glass 三合透镜
triple thread 三线螺纹;三头螺纹;三片螺纹
triple (threaded) screw 三头螺纹
triple-throat carburetor 三腔化油器
triple throughing roll(er)（输送带的）三合一槽式滚轴
triple throw plunger pump 三行程柱塞泵
triplet hypothesis 三联体假说
triplet interval 三重线间隔
triplet lens 三片三组镜头;三合透镜;三合镜
triplet method 三片法
triplet model 三片模型
triplet objective 三合物镜
triplet of three-spatial variable 三空间变量体
triple tombolo 三连岛沙洲
triplet piston pump 三联式活塞泵
triplet quadrupole 三元四极透镜
triple-track 三声道
triple trailer 三联拖车
triple transit echo 三次行程回波
triple travel signal 三次行程信号
triple treat sonar 综合声呐站
triplet-singlet 三重线—单谱线
triplet spectrum 三重线谱
triplet splitting 三劈裂
triplet state 三重线态;三重态
triplet theory 三联体理论
triplet thin lens 三合薄透镜
triple tube 三重管
triple-tube core barrel 三重岩芯管;三管岩芯筒
triple-tuned coupled circuit 三重调谐耦合电路
triple-tuned transformer 三绕组调谐变压器
triplet unit impulse 三元脉冲
triple turbine 三级涡轮机
triple turret 三联炮塔
triple twill 三层斜纹
triple type cable 三芯型电缆
triple-unit 三合单元
triple valve 三通开关;三通阀
triple-valve body 三通阀体
triple-valve cap 三通阀盖
triple-valve piston 三通阀活塞
triple vector product 三重向量积
trip level 断路电平
triple Venturi（化油器的）三重喉管
triple venturi carburetor 三重喉管式化油器
triple warp weave 三线捻办
triple weave 三层织法
triple wedge 三楔边
triple weighing batcher 三种成分重量配料计量器;三斗称重分批配料器
triple-window 三层窗
triple wire 三丝铜网
triple withdrawal refiner 三重回收区域精炼炉;三重回收提纯器
triple-working switches 三动道岔
triplex 三重;三元三件物体;三部;三倍(的)
triplex apartment 三层楼公寓
triplex autolanding 三重操纵系统自动进场着陆
triplex backfill tamper 三腿回填夯
triplex board 三层板

triplex building 三住户楼房;有三套住房的房屋;三居室的楼房
triplex cable 三芯电缆
triplex camera 三镜头照相机;三镜头摄影机
triplex carburetor 三联式汽化器
triplex chain block 三重链滑车
triplex control 三重控制
triplex controller 三重控制器
triplex cutter 三面刀盘
triplex derrick 三杆吊车
triplex design tubular derrick 三重伸缩式管子钻塔
triplex drilling hook 三角钻钩
triplexer 三通天线转发开关;三工器
triplex glass 夹层玻璃;三层玻璃
triplex hook 三爪提引钩
triplex house 三联式住宅
triplexing 三联炼钢法;三部法(冶金的)
triplex net winch 三联式网机
triplex plunger pump 三柱塞泵
triplex-process 三联炼钢法;三联法
triplex pump 三联式泵;三级泵;三缸式(水)泵;三缸泵
triplex rake classifier 三耙式分级机
triplex receptacle outlet 三孔插座出线口;三孔插座出口线
triplex reciprocating pump 三缸往复泵
triplex reflector 三个棱镜反射器;三次反射镜
triplex row 三行
triplex safety glass 三层式防碎玻璃
triplex sheepfoot roller 三滚筒羊脚压路机
triplex silk convered wire 三层丝包线
triplex single action pump 三缸单作用泵
triplex spindle 三联(销)轴
triplex system 三工制
triplex-system(ed) aircraft 三重操纵系统飞机
triplex winding 三重绕组;三分绕组
triplicate 一式三份;三次重复;重复三次
triplicated plots 三次重复区
triplicate ratio circle 三重比圆
triplication 增至三倍;三倍量
triplication design 三次重复试验设计
triplication formula 三倍量公式
triplicity 三重;三倍
trip line 脱扣线;浮锚拉索
triplite 氟磷铁锰矿;氟磷锰石
triplography 三重造影术
triploid 三倍体
triploidite 羟磷锰石
triploidy 三倍性
triplopia 三重复视
triply conjugate system of surfaces 三重共轭曲面系
triply connected curve 三连通曲线
triply-discharging 三位的
triply isothermal-conjugate system of surfaces 三重等温共轭曲面系
triply linked 三重链接
triply linked tree 三重链接的树
triply orthogonal family of surfaces 三重正交曲面族
triply orthogonal system of surfaces 三重正交曲面系
triply periodic function 三周期函数
tri-ply wood 三夹板;三合夹板
trip magnet 断路电磁铁
trip maker character(istic)出行者特征
trip map 路程图
trip mechanism 跳动装置;分离装置
trip meter 短距离里程表
trip mileage of inspection on aids-to-navigation 航标巡检里程
trip mode 出行方式
trip number 航次数;车次号【铁】
tripod 三脚台;三脚架;三脚插头;三角支撑物;鼎
tripod adaptor 三脚架适配器
tripod angle 三面角
tripod base 三脚架基座
tripod bolt 三脚螺栓;三脚架穿钉
tripod-borne 装在三脚架上的
tripod clevis 三脚架 U 形环
tripod crane 三支腿起重机;三脚起重机
tripod derrick 三脚起重机;三脚架桅杆起重机;三脚起重机;三脚支架吊车
tripod dolly 三脚矮橡皮轮小车

tripod drill 三脚钻机;三脚架式钻机
tripod fate 三杆闸机
tripod for base measurement 轴杆架
tripod head 三脚架头
tripod-head level(l)ing in base measurement 轴杆头水准
tripod jack 三脚千斤顶
tripod landing gear 三轮起落架
tripod leg construction 三脚机架结构
tripod legs 三脚架的腿;三脚架腿
tripod loader 三脚架式垛草机
tripod magnifier 三脚放大镜
tripod mast 三脚桅杆
tripod-mounted 三脚架安装的;装在三脚架上的
tripod mount(ing)三脚架
tripod pendulum type batch centrifugal 三足离心机
tripod planting 三脚式植树挂浆法;三脚架式植树挂浆;等边三角形岸边植柳
tripod puller 三脚架拔桩机
tripod receptacle 三脚架插孔
tripod regulator 三脚架调节器
tripod rest 三脚架
tripod rotary drill 三脚架回转式钻机
tripod sheave wheel 三脚吊滑轮
tripod sheer 三脚起重机
tripod shoe 三脚架铁尖
tripod socket 三脚架底座
tripod stack 三脚架码垛机
tripod table 三脚桌
tripod undercarriage 三脚起落架
trip off 跳开
tri-point rock drill 三脚架式凿岩机;三脚钻机;三脚架式钻岩机
tripolar 三极的
tripolar coordinates 三极坐标
tripolar electro magnet 三极电磁铁
tripolar lead-cobalt battery 三极式铅钴电池
tripoli 硅藻土;硅藻岩;松软硅质岩
tripoli earth 风化硅石
tripoli-powder 硅藻土粉
tripoli-powder brick 硅藻土粉砖
tripoli-powder concrete 硅藻土粉混凝土
tripoli-powder slab 硅藻土平板
tripolite 硅藻石;硅藻土;风化硅石
tripolite earth 板状硅藻土;风化硅石
tripolycyanamide 三聚氰胺
tripolymer 三聚体
tripolyphosphate 三聚磷酸盐
tri-porosity method 三孔隙度组合法
tripotassium phosphate 磷酸三钾
trip-out 停机;跳闸断路;切断;跳出
trip-out rate 跳闸率
trip-out torque 跳闸力矩
trip-over lever（翻转犁的）翻转控制杆
trip-over stop 跳档
trippable 可剥离的(墙面覆盖物)
trip paddle 脱扣控制杆;松开控制杆;脱扣踏板;脱扣浆片
trip pawl 解扣装置爪
trip peak 三重峰
trip pedal 控制踏板
tripped drop hammer 打桩锤;落锤
tripper 卸料装置;卸料器;开底装置;脱扣装置;跳开装置;安全器;自动倾卸输送机
tripper body 翻斗车身
tripper bus 高峰小时使用的公共汽车
tripper car 翻斗车;倾卸车
tripper coal yard 倒煤场
tripper conveyer 自动倾卸(式)输送机
tripper-conveyer gallery 自动倾卸输送机廊道
tripper per day 每天往返运行次数(货车)
tripper man 卸车工
tripper pump 扫舱泵
tripper trailer（带式运输机的）卸料小车
tripper type conveyer 有卸料器的皮带运输机;倾卸式运输机;带卸料车的皮带输送机
trip pile driver 松扣打桩机
trip pin 脱扣销;松放销
tripping 脱扣;跳开
tripping bar 跳动杆;钩杆;脱扣杆;跳闸杆
tripping battery 断路器用蓄电池
tripping bracket 防歪斜衬板;防倾肘板;防颤肘板
tripping cam 脱开凸轮;跳动凸轮

tripping car 翻斗车;自卸车;自动卸料车;自动倾卸车
tripping centre 漂心
tripping circuit 跳闸回路;跳开电路
tripping coil 脱扣线圈;跳闸线圈
tripping current 脱扣电流;跳闸电流
tripping device 摘钩装置;摘钩设备;倾卸装置;脱扣装置;水流紊动激励器;释放装置;翻车器;解扣装置
tripping frequency 跳闸频率
tripping gear 脱钩装置;倾卸装置;松开机构
tripping impulse 触发脉冲
tripping lever 离合杆;起动杆;分离杆
tripping line 拉放阀;锚抓拉索;收帆索
tripping link 滑钩
tripping magnet 解扣电磁铁;切断电磁铁
tripping mechanism 倾卸机械装置;倾卸机构;脱扣机构;脱钩机构;跳闸机构;倾翻机构;抛掷机构
tripping off 跳开
tripping operation 跳闸
tripping out 跳开
tripping palm 助钩角
tripping pulse 脱扣脉冲
tripping relay 脱扣继电器;跳闸继电器
tripping speed 解扣速度
tripping spring 制动弹簧
tripping voltage 跳闸电压
trippkeite 软砷铜矿
trip post jack 解脱式顶柱
trip pressure 启闭压力
trip production 出行(交通)产生
trip production method 交通生成法
trip purpose 出行目的
trip radius 出行半径
trip relay 切断继电器
trip rider 跟车工;列车跟车工;矿车跟车工
trip roller 离合杆滚轮;排出滚轮
trip rope 放开制动器的活索;桩锤活索;操纵杆拉绳
trip-rope control 接合杆绳索控制
tripropellant 三元推进剂
tripropylborate 硼酸三丙酯
tripropylene 三聚丙烯
triprotic acid 三元酸
trip scale 脱扣秤
trip service 普通检修
trip setting 脱扣整定值
trip shaft 解扣轴;绊轴;扳动装置轴
trip shank 带弹回安全器的铲柄
trip shore 临时船尼撑住(下水时)
trip signal 解扣信号;脱扣信号
tripsis 研碎;研磨
trip speed 出行速度;脱扣速度
trip spindle 解扣轴
trip-spotting hoist 列车调度绞车
trip spring 解扣弹簧;释放簧
trip standard 带松脱装置的铲柄
trip survey 出行调查
trip switch 遥控开关;控制开关
trip table 出行表
tripteral 具有三个侧翼或侧柱的
trip termini map 旅程起讫点图
trip test 解扣试验;跳闸试验;跳脱试验
trip time 往返时间;起下钻时间;跳闸时间;断开时间;出行时间
trip time space 出行时距
tripton 水中的浮屑;非生物悬浮细粒;非生物性悬浮物
trip trigger 解扣触发器
triptych 三个相连的雕刻;三张相连的图画;三折书牒;三间隔形式;三联画;三幅一联画或雕刻
tripuhyite 锑铁矿
trip value 跳闸值
trip valve 自动停止阀
trip wire 在船模前面装置的促进乱流的金属卷线
trip (wire) switch 跳闸开关
trip worm 脱扣蜗杆
triquetra 三角形饰
triquetrous 三角形的;有三角形断面的
triradiata 三辐体;三射的
triradiate calcareous spicule 三射石灰骨针
triradiate ridge 三射脊
triradiate slit 三射缝
triradiation 三向辐射
trirectangular quadrilateral 三直角四边形

trirectangular spheric(al) triangle 三直角球面三角形
trirectangular trihedral 三直角三面形
triregional 三部的
trirhombohedral 三菱面体的
triroll ga(u)ge 三滚柱式螺纹量规
tri rotor pump 三转子泵
trisalt 三酸盐
trisantia 窄门道(中世纪);窄走廊(中世纪)
trisaturated 三相饱和的
trisazo compound 三偶氮化合物
trisazo dye 三偶氮染料
trisecant 三重割线;三度割线
trisecant curve 三重割曲线
trisect 三等分
trisected 三剖面的
trisection 三等分
trisector 三等分器
trisectrix 角的三等分线;三分角;三等分角线;[复]trisectrice
triserial 三列的
trisha 脚踏三轮车
trishaw 脚踏三轮车;三轮车
trishore 三脚支撑
trisilane 丙硅烷
trisilanyl 丙硅烷基
trisilicate 三硅酸盐
trisimint 水合硅酸铝镁
trisimulus filter 三原色滤色片
triskelion 三支形饰;三腿形饰;[复]triskelia
triskelion cross section 三弯叶形截面
tri-slab 三T形板;三肋板
trislope screen 三层坡度减小筛
trislot 三槽
tri-slot nut 三槽螺母
trisnitrate 三硝酸酯;三硝酸盐
trisoctahedron 三角八面体;三八面体
trisoctahedronal layer 三八面体层
trisodium 三钠
trisodium cellulose 纤维素三钠
trisome 三体
trisomic ratio 三体比率
trisomy 三体性
trisonic 三音速
trisonic aerodynamics 三种声速范围的空气动力学
trisonic range 三种音速范围
trisonics 三音速气动力学;三声速空气动力学
trisonic test 三音速试验
trisonic wind tunnel 三速风洞
trispast 三轮滑车;三饼滑车
trisphaeridine 三球定
trispool 三转子
trisquare 曲尺
tristable 三稳态
tristate 三州间地区;三态
tristate bus driver 三态总线驱动器
tristate bus line 三态总线
tristate control 三态控制
tristate controller 三态位式控制器
tristate logic 三态逻辑
tristearate 三硬脂酸盐
tristearin 硬脂;甘油三硬脂酸酯;三硬脂酸甘油酯
tristetrahedron 三四面体;三角三四面体
tristichous 三列的
tristigmatic 三柱头的
tristimulus 三色激励
tristimulus coefficient 三色激励系数
tristimulus colo(u)rimeter 三色源色度计;三色激励比色计;三激源色度计;三刺激色度计
tristimulus colo(u)rimetry 三色激励测色法
tristimulus designation 三色指示;三色标示
tristimulus diagram 三色图
tristimulus filter 三色激励滤色片
tristimulus light filter 三激光过滤器
tristimulus system of colo(u)r specification 三部分定色系统
tri-stimulus value 三刺激数值;三色激励值
tristramite 水磷钙铀矿
trisul 三叉饰
trisulfate 三磺酸盐
trisulfonic acid 三磺酸
trisulphate 三硫酸盐
trit 三进制数位
tritan 三苯甲烷

tritane carboxylic acid 二苯甲基苯甲酸
tritangent plane 三重切面
tritanomalous vision 色弱
tritanopia 第三色盲
tri-tee beam 三T形梁
tri-tee slab 三T形板
triternate 三回出叶的
triterpene 三萜(烯)
triterpenoid 三萜系化合物
tritet oscillator 多谐晶体振荡器
trithilite 垂塞拉特
trithioacetaldehyde 三聚乙硫醛
trithioacetone 三聚硫酮
trithiocyanuric acid 三聚硫氰酸
trithion 三硫磷
trithionic acid 连三硫酸
trithioozone 臭硫
tritiate 氚化
tritiated 氚化了的
tritiated adenosine 氚标记腺苷
tritiated arginine 氚化精氨酸
tritiated compound 含氚化合物;氚化物;氚标记化合物
tritiated hydrocarbon 氚化烃
tritiated thymidine 氚标记胸腺嘧啶核苷
tritiated titanium target 氚化的钛靶
tritiated water 氚化水
tritiation 氚化作用
tritide 氚化物;氚化合物
tritioboration 氚硼化
tritium 氚;超重氢
tritium bearing waste 含氚废物
tritium breeding materials 氚增殖材料
tritium dating 氚年代测定法;氚测年法
tritium-helium method 氚—氦法
tritium-labeled 氚标记的
tritium labeled amino-acid 氚标记氨基酸
tritium-labelled compound 氚标记化合物
tritium labilization 氚活化
tritium method 氚法
tritium oxide 氧化氚
tritium raito 氚比
tritium target 氚靶
tritium unit 氚单位
tritol 垂陶耳
tritolyl phosphate 磷酸三甲苯酯
tritomite 锥稀土矿;硼硅铈矿
tritomite-(Y) 硼硅钇矿
Triton 美人鱼雕像;半人半鱼海神;人头人身鱼尼的海神美人鱼雕像(希腊神话)
triton 氚核
Triton block 梯恩梯药块;梯恩梯炸药块
triton magnetic resonance 氚核磁共振
tritopine 三陶品
tritoxide 三氧化物
tritriacontane 三十三烷
tritriacontyl 三十三基
tritriated waste 含氚废物
tri-truck 三轮卡车;三轮汽车;三轮货车
tritubercular 三结节的
triturate 研制剂;磨碎物;磨碎;捣碎
triturated clay 研成粉状黏土;粉状黏土
triturating 研制过程
triturating machine 研制机
trituration 研制剂;研磨;研粉作用;磨碎
triturator 研钵;捣碎器;捣碎机
tri-twist flower glaze 三捻花釉
trityl bromide 三苯甲基溴
trityl cellulose 三苯甲基纤维素
triumphal 凯旋(式)的
triumphal arch 教堂大拱门;凯旋门
triumphal column 凯旋柱
triumphal gate(way) 凯旋门
triumph arch 凯旋门;教堂大拱门
triundulate 三波曲的
triunsaturated glyceride 三不饱和酸甘油酯
triuranium 三铀
trivacancy 三空位
trivalent alcohol 三元醇
trivalvular 三瓣的
trivariant 三变的
trivariate normal distribution 三变量正态分布
trivat 割绒刀
trivector 三向量

tri-vent press 三压区沟纹压榨
trivet 割绒刀;三脚架;矮金属脚架;三脚座
trivet table 三腿桌;三脚桌
trivette 割绒刀
trivial bundle 平凡丛
trivial digraph 平凡双图
trivial family 平凡族
trivial graph 平凡图
triviality 平凡性
trivialization 平凡化
trivialize 平凡化
trivial knot 平凡纽结
trivial leak 小漏
trivial line bundle 平凡线丛
trivial microbundle 平凡微丛
trivial name 土名;俗名
trivial nomenclature 习惯命名法
trivial place 平凡位
trivial proximity 平凡邻近
trivial sheaf 平凡层
trivial solution 明显解;平凡解
trivial space 平凡空间
trivial subgroup 平凡子群
trivial uniformity 平凡一致性
trivial valuation 平凡赋值
trivial vector bundle 平凡向量丛
trivium 三学科(中世纪学校的);三道体区
trivoltine 三化的
trivoltinism 三化性
tri-wall corrugated board 三层双面波纹纸板
triyl magnesium chloride 氯化三苯甲基镁
trizonal 三带的
troble shooting data 追查故障用的数据
trocar 套针
trochanter 转节
trochanterian 转子的
trochanteric 转子的
trochanterion 转节点
trochanter major 大转子
trochanterplasty 转子成型术
trochanter spur 转节距
trocheameter 车轮转速计
trochilus 凹圆线;凹环形线脚
trochlear fovea 滑车凹
trochoid 旋轮线;余摆线;滑车关节;陀螺形的;枢轴状;枢轴关节;次摆线【数】;车轴关节;摆线
trochoidal analyzer 次摆线分析器
trochoidal curve 余摆线
trochoidal focusing mass spectrometer 余摆线聚焦质谱仪
trochoidal form 余摆线波形
trochoidal mass analyzer 次摆线质谱(分析)仪;摆线质量分析仪;余摆式质谱分析仪
trochoidal mass spectrometer 余摆线质谱仪;余摆线型质谱计
trochoidal motion 摆线运动
trochoidal orbit 余摆线轨道
trochoidal profile 摆线波剖面
trochoidal ripple 摆动小波
trochoidal rotor 旋轮线转子
trochoidal rotor housing 旋轮线型缸体
trochoidal theory 摆动理论
trochoidal track 旋轮线轨迹
trochoidal type machine 旋轮线转子机
trochoidal wave 余摆线波;余摆线;深海波;次摆线波;摆线波;摆动波
trochoidal waveform 摆动波形
trochoid pump 次摆线泵
trochometer 里程计;里程表;车程计;路程计
trochotron 余摆线管;电子注开关管
troctolite 橄长岩
Trodaloy 铜铍合金
troegerite 砷铀矿
troegrite 济砷铀云母
troffer 暗灯槽;天花板凹槽;灯槽;槽形支架
Trofil 特罗菲尔(一种聚乙烯纤维)
trogtalite 硬硒钴矿;方硒钴矿
troilite 陨硫铁
Troisier's node 信号结
trolit 聚苯乙烯绝缘材料
trolitul 聚苯乙烯塑料材料
troll 拖钓
trolled as soon as placed 随打随抹光
trolleite 羟磷铝石

troller 拖钓船
troll(e)y 缆车;空中吊运车;滑触线;倾倒式货车皮;台车;手推小车;斗车;触轮;成层盆地;载重滑车
troll(e)y base 手摇车底座【铁】;受电轨座
troll(e)y batcher 触轮式分批称料斗
troll(e)y batcher plant 吊式分批投料机
troll(e)y beam 吊车梁
troll(e)y block 游动滑车(组);起重机小车滑轮组
troll(e)y boom 无轨电车吊杆
troll(e)y bucket 小车戽斗
troll(e)y bus 无轨电车;电车(指无轨电车)
troll(e)y bush 汇流环的滑轮套管;电车滑轮轴衬
troll(e)y-bus motor 无轨电车电动机
troll(e)y car 无轨电车吊运车;无轨电车;电车;触轮式电车
troll(e)y car shed 电车检修场
troll(e)y coach 无轨电车
troll(e)y collector 滑线式集电器
troll(e)y conveyer 吊运式输送器;吊运式输送机;吊式输送机;吊空输送机;链轮式架空输送机;空中悬吊输送机
troll(e)y cord 吊车绳索;空中吊运车绳索
troll(e)y crane 悬挂式起重机;架空起重机
troll(e)y fall 快递吊货索
troll(e)y for caisson 沉箱台车
troll(e)y for turning use 翻身小车
troll(e)y-frog 接触电线线岔;电车线交叉
troll(e)y girder (桥式的)小车行进大梁;小车大梁;桥吊的小车行驶大梁
troll(e)y guide 推车轨
troll(e)y haulage 架空线电机车运输;架空无轨运输
troll(e)y head 集电头;受电头;电触轮滑触靴
troll(e)y hoist 悬挂式起重小车;电动小吊车
troll(e)y jib 无轨电车导电杆
troll(e)y-jib tower crane 小车塔式起重机
troll(e)y ladder 推移梯
troll(e)y ladle 悬挂式浇包;单轨吊包
troll(e)y line 有轨电车架空线;架空线(路);架空电线
troll(e)y load 小车荷载
troll(e)y locomotive 架线式电(力)机车;滑触式电(力)机车;触轮式电气机车;触轮式电力机车
troll(e)y motor 电车电动机
troll(e)y-pivot 集电器杆支枢;受电器杆支枢;受电杆轴销
troll(e)y pole 集电器杆;集电杆;电车触轮杆;触轮杆;受电杆
troll(e)y pole catcher 跳停器
troll(e)y-pole current collector 滚轮杆接电器
troll(e)y post 触轮臂
troll(e)y rail 手推车小铁轨
troll(e)y rope 空中吊运车缆索;电车缆索;装卸桥的小车牵引索;架空吊车索
troll(e)y-shield globe 触轮托罩
troll(e)y shoe 集电靴;滑式式集电器
troll(e)y span 小车轨距
troll(e)y system 缆车系统;空中吊运系统
troll(e)y track 门轮滑轨;小车轨道;滚轮滑轨;高架导轨
troll(e)y train 电车
troll(e)y travel (桥式起重机的)小车行程;吊运车行程;轮式集电器;桥吊的小车行驶
troll(e)y travel(l)ing speed 小车运行速度
troll(e)y wheel 滚轮;电车触轮;触轮
troll(e)y-wheel harp 触轮托罩
troll(e)y wire 有轨电车架空线;架空接触线;滑接线;电车架空线;电车电线;触线
troll(e)y wire guard 电机车架空线保护
troll(e)y wire locomotive 电气机车
troll(e)y wire network 触线网
troll(e)y wire suspension 电车线悬挂装置
troll(e)y wire termination 简单悬挂下锚
Troll flower 金莲花
Trollhatten furnace 特罗哈顿电炉
trolling 拖钓作业;拖捕
trolling line 拖钓绳
trolly 小车
tromantadine 醋胺金刚烷
Trombe wall 特隆布墙
trombone 可调节 U 形波导节;可调 U 形同轴线
tromechanical signal controller 机电式信号控制机
trometamol 氨基丁三醇
tromethamine 氨基丁三醇
trommel 转筒筛;回转筛;滚筒筛

trommel screen 旋转式转筒筛;转运筛;筒筛;选矿筒;滚筒筛
trommel screening 滚筒筛;转筒筛
trommel washer 转筒洗筛
Tromolite 特罗莫赖特烧结磁铁
tromometer 微地震测量仪
Tromp area diagram 特伦普面积曲线图
Tromp blast 特伦普井筒注水通风法
Tromp curve 选粉机效率曲线
Tromp cut-point 特伦普分选点;实际分选比重
tromp cut-point 分离密度;分配密度
trompe 穹隆的一角的形状
Trompeter zone 免压圈
Tromp process 特伦普介质分选法
trona 天然苏打;天然碱
trona deposit 天然碱矿床
Trona process 天然碱流程
trondhjemite 奥长花岗岩
trone 特朗秤
trone weight 特朗衡器
troop barge 军队运输驳船
trooper 军队运输船
troopship 运输舰;运兵船;军队运输船
troops topographic(al) service 部队测绘勤务
troop test 野外试验;部队试验
troop train 军用列车
troop truck company 运输汽车连
troostite 锰硅锌矿
tropadyne 超外差(式)电路
tropal 衬里织物
trope 奇异切面
Tropenas converter 侧吹酸性转炉
tropept 热带始成土
trophic analysis 营养分析
trophic state index 营养状况指数
trophic state prediction 营养状况预测
trophic symbiosis 滋养共生
tropholytic zone 营养分解层
trophophase 营养期
trophy 战利品雕饰;奖品
trophy case 珍藏品橱;珍藏品匣;奖品橱;奖品匣
trophy room 锦标存放室
tropic(al) 热带的
tropic(al) air 热带空气
tropic(al) air-mass 热带气团
tropic(al) air (mass) fog 热带气团雾
tropic(al) alpine srub 热带高山灌丛
tropic(al) alpine tundra 热带高山冻原
tropic(al) animal 热带动物
tropic(al) anti-cyclone 热带高压;热带反气旋
tropic(al) aquarium 热带水族馆
tropic(al) area 热带区带
tropic(al) Atlantic air-mass 热带大西洋气团
tropic(al) aubalpine forest 热带亚高山森林
tropic(al) automatic weather station 热带自动气象站
tropic(al) black clay 热带黑黏土
tropic(al) black earth 热带黑土
tropic(al) bleach 热带漂白粉
tropic(al) bleached soil 热带漂白土
tropic(al) brown soil 热带棕色土
tropic(al) calm zone 热带无风带
tropic(al) cell 热带环流
tropic(al) century 回归世纪
tropic(al) climate 热带性气候;热带气候
tropic(al) climatology 热带气候学
tropic(al) closed circulation 热带闭合环流
tropic(al) cloud cluster 热带云团
tropic(al) cloud forest 热带云雾林
tropic(al) continental air-mass 热带大陆气团
tropic(al) convergence 热带辐合带
tropic(al) crop 热带作物
tropic(al) curing 高温养护法;高温养护混凝土
tropic(al) curvature 向性弯曲
tropic(al) cyclone 热带气旋;热带气旋
tropic(al) cyclone landfall 热带气旋登陆地
tropic(al) cyclone program(me) 热带气旋计划
tropic(al) cyclone-prone area 易于受热带气旋侵袭的地区
tropic(al) day 酷热日;太阳日;常用日
tropic(al) depression 热带低(气)压
tropic(al) desert 热带荒漠
tropic(al) desert climate 热带沙漠气候
tropic(al) designed 为热带设计的

tropic(al) diarrhea 热带腹泻
tropic(al) disease 热带病
tropic(al) disturbance 热带扰动
tropic(al) earth 热带土
tropic(al) easterlies 热带东风带
tropic(al) ecology 热带生态学
tropic(al) ecosystem 热带生态系统
tropic(al) environment 热带环境
tropic(al) exposure testing 热带曝晒试验；热带暴露试验
tropic(al) ferriallitic soil 热带铁铝土
tropic(al) finish 热带漆
tropic(al) fish 热带鱼
tropic(al) fishes 热带鱼类
tropic(al) floral realm 热带植物地理区系
tropic(al) floristic subregion 热带植物亚区
tropic(al) forest 热带森林
tropic(al) forest ecosystem 热带森林生态系(统)
tropic(al) forestry action plan 热带森林行动计划
tropic(al) forestry action program(me) 热带森林行动计划
tropic(al) forest soil 热带森林土
tropic(al) forest zone 热带林带
tropic(al) freeboard 热带干舷
tropic(al) freshwater load line 热带淡水载重线；热带淡水满载吃水线
tropic(al) freshwater load line mark 热带淡水载重线标志；热带淡水满载吃水线标志
tropic(al) front 热带锋
tropic(al) fruit culture 热带果树栽培
tropic(al) fruit (tree) 热带水果
tropic(al) grass 热带高密草原
tropic(al) grassland 热带草原
tropic(al) grasslands climate 热带草原气候
tropic(al) Gulf air-mass 热带墨西哥湾气团
tropic(al) higher high water 回归较高高潮；回归高高潮
tropic(al) higher high water interval 回归高高潮间隙
tropic(al) higher low water 回归较高低潮；回归高低潮
tropic(al) high water 回归高潮
tropic(al) high water inequality 回归高潮不等
tropic(al) hospital 热带医院
tropic(al) hurricane 热带气旋；热带飓风
tropic(al) insulation 热带绝缘
tropicalization 热带适应性
tropicalization test 高温湿度试验；热带试验；热带气候适应性试验
tropicalize 热带气候处理
tropicalized 热带化的
tropicalized packing 耐热包装
tropic(al) karst 热带岩溶
tropic(al) lake 热带湖泊
tropic(al) level 营养级
tropic(al) life zone 热带生物带
tropic(al) line 回归线
tropic(al) load line 热带载重线；热带满载吃水线
tropic(al) load line mark 热带载重线标志
tropic(al) lower high water 回归低高潮
tropic(al) lower low water 回归低低潮
tropic(al) lower low water interval 回归低低潮间隙
tropic(al) low water 回归低潮
tropic(al) low water inequality 回归低潮不等
tropically packed 热带包装的
tropic(al) marine air-mass 热带海洋气团
tropic(al) marine climate 热带海洋气候
tropic(al) maritime air(-mass) 热带海洋气团
tropic(al) meteorology 热带气象学
tropic(al) monsoon 热带季风
tropic(al) monsoon climate 热带季风气候
tropic(al) monsoon forest 热带季风林
tropic(al) month 回归月；分至月
tropic(al) ocean 热带海洋
tropic(al) ocean climate 热带海洋气候
Tropic(al) of Cancer 北回归线
Tropic(al) of Cancer tide 回潮
Tropic(al) of Capricorn 南回归线；冬至线
tropic(al) outfit 热带装备
tropic(al) Pacific air-mass 热带太平洋气团
tropic(al) pack 热带包装
tropic(al) packing 防热防潮包装

tropic(al) pedocalic soil 热带钙层土
tropic(al) plant 热带植物
tropic(al) radiator 热带冷却器
tropic(al) rain 热带雨
tropic(al) rain climate 热带雨气候；湿热多雨气候
tropic(al) rain forest 热带雨林
tropic(al) rainforest climate 热带雨林气候
tropic(al) rain(y) climate 热带多雨气候
tropic(al) range 大潮高高潮和大潮低低潮的潮高差；回归大潮(潮)差；回归潮差
tropic(al) red earth 热带红土
tropic(al) red loam 热带红壤土
tropic(al) region 热带地区
tropic(al) revolution 分至周
tropic(al) revolving storm 热带旋转风暴；热带风暴
tropic(al) river 热带河流
tropic(al) savanna(h) 热带疏林
tropic(al) savanna(h) climate 热带草原气候；热带稀树草原气候；热带大草原气候
tropic(al) scrub 热带密灌丛
tropic(al) seas 热带海域；热带海区
tropic(al) shallow water 热带浅水
tropic(al) shelterwood system 热带伞伐作业
tropic(al) soil 热带土
tropic(al) storm 热带风暴
tropic(al) superior air-mass 热带高空气团
tropic(al) switch 热带用开关
tropic(al) temperature lamination 热带温度层结
tropic(al) tidal current 回归潮流
tropic(al) tide 回归潮；分至潮；热带潮
tropic(al) timber 热带材
tropic(al) to subtropic(al) subzone 热带一亚热带地区
tropic(al) tree 热带树木
tropic(al) trial 热带试验
tropic(al) tropopause 热带对流层顶
tropic(al) trough 赤道槽
tropic(al)-type driver's cab(in) 热带型驾驶舱
tropic(al) upland rice soils 热带旱地水稻土
tropic(al) use 热带用途；热带使用
tropic(al) velocity 回归大潮流速；回归潮流速
tropic(al) vortex 热带涡漩
tropic(al) water 热带水
tropic(al) water line 热带载重线
tropic(al) waters 热带水域
tropic(al) wear 热带服装
tropic(al) weather 热带气候
tropic(al) wet and dry climate 热带草原气候
tropic(al) wet climate 热带雨林气候
tropic(al) wind observation system 热带风观测系统
tropic(al) wood 热带木材；热带材
tropic(al) woodland 热带疏林；热带林地
tropic(al) year 回归年；分至年；民用年
tropic(al) zone 热带区带；热带海区；热带地区
tropic(al) zone tropic(al) belt 热带的
tropicopolitan 热带地方菌；全热带的
tropics 热带地区
tropolone 草酚酮
tropolones 草酚酮类
tropometer 旋转计
troponparasite 转主寄生
tropopause 对流层顶
tropopause break 对流层顶中断
tropopause chart 对流层顶图
tropopause funnel 漏斗状对流层顶
tropopause inversion 对流层顶逆温
tropophilous 湿旱生的
tropophyte 湿旱生植物
tropophytia 湿旱生植物群落
troposcatter(ing) 对流层散射
troposcatter receiver 对流层散射接收机
troposphere 对流圈；对流层
tropospheric attenuation 对流层衰减
tropospheric chemistry 对流层化学
tropospheric dust suspension 对流层悬浮(土)
tropospheric effect 对流层效应
tropospheric error 对流层误差
tropospheric fading 对流层衰落
tropospheric fallout 对流沉降物
tropospheric life-time 在对流层停留时间
tropospheric mode 对流层振荡模
tropospheric ozone 对流层臭氧

tropospheric ozone pollution 对流层臭氧污染
tropospheric ozone station 对流层臭氧观测站
tropospheric propagation 对流层传播
tropospheric radio duct 对流层无线电波道；对流层无线电波导
tropospheric radio propagation 对流层电波传播
tropospheric refraction 对流层折射
tropospheric refraction correction 对流层折射校正；对流层折射改正
tropospheric refraction effect 对流层折射效应
tropospheric refraction reduction 对流层折射改正
tropospheric residence time 在对流层停留时间
tropospheric scatter communication 对流层散射通信
tropospheric scatter(ing) 对流层散射
tropospheric scatter radio communication system 对流层散射无线电通信设备
tropospheric scatter transmitter 对流层散射发射机
tropospheric superrefraction 对流层超折射
tropospheric wave 对流层波
troptometer 扭力计；扭力仪；扭转仪；测扭计
trosted paint 无光泽油漆
trot 系泊小船的浮筒绳索
trot line 捕蟹系绳
trot mooring 三臂系船设施
trottoir 步道；人行道
trouble analysis 故障分析
trouble and failure report 故障事故报告
trouble and health insurance 事故及医疗保险
trouble and waiting time 事故及停待时间
trouble area 不稳定岩层区；不稳定岩层段
trouble back jack 故障信号塞孔
trouble blinking 故障闪烁信号；事故指示信号
trouble block 故障块
trouble car 检修车
trouble chart 故障图；事故图表；事故记录卡片
trouble clearing 故障排除；排除故障
trouble detection and monitoring 故障检测和监控
trouble detector 故障探测器
trouble finder 故障寻迹器
trouble-free 无困难；可靠的；无故障的
trouble-free coating 可靠涂层；安全涂层
trouble-free landing 无故障着陆
trouble-free life 无故障使用期
trouble-free oil clutch 无故障的浸油离合器
trouble-free operation 无干扰运行；安全运转；连续工作；无故障运转；无故障运行；无故障操作；顺利运行
trouble-free performance 无故障性能
trouble-free running 正常运转；正常运行；无故障运转
trouble-free service 无故障使用；安全工作；安全服务
trouble-free wheel motion 车轮顺利运行
trouble history card 故障记录卡片
trouble-hunting 寻找故障
trouble indicator 故障指示器
trouble lamp 检修灯；故障灯；故障指示灯；故障警报灯；探查灯
trouble light 故障灯；临时灯
trouble light for inspection 检查事故灯
trouble-locating 故障检寻；故障检索；故障查找
trouble-location problem 故障定位问题；故障定位题目
trouble man 故障检修员；检修员；检修人员
trouble-proof 无故障的；不发生故障的；安全的
trouble proof-saving 预防故障的
troubleproof spot 无故障处
trouble record 故障记录
trouble recorder 故障记录器
trouble removal 故障排除；事故处理
troubles and tool 事故与工具
trouble-saving 预防故障
troubles free time 事故解案时间
troubles happening time 事故发生时间
troubleshoot 寻找故障；故障寻找；故障检修；检查故障；检修故障；排除故障；调试；查找故障；消除故障
trouble-shooter 故障检修员；检修员；检修人员
trouble-shooter program(me) 查错程序
trouble shooting 故障探寻；故障探测；故障解除；故障检查；故障分析；故障查找；查找故障；故障检修；检查故障；解决麻烦；检修；排除故障；消除故障
trouble-shooting data 故障检查数据；检修指南

trouble-shooting manual 故障查找手册;故障检修手册
trouble-shooting time 故障寻找时间
trouble signal 故障信号
troublesome freight 不易照管的货物
troublesome impurity 难除杂质
troublesome soil 难处理土
troublesome taste 讨厌的味道
troublesome weed 难根除的杂草;难除掉杂草
troublesome zone 易出事故地带;破碎带;复杂地层
trouble spot 故障点;出故障处;易出故障处
troubles predicting 故障提示
troubles waste time 事故损失时间
trouble tracer 故障检寻器
trouble unit 故障单元;故障部件
trouble waters for sailing 航行困难的水域
trough 中间包;谷沟;沟;曲线上极小值;商业周期低潮;电缆槽;槽线;槽沟;槽地;槽;凹点
troughability 成槽(可能)性
trough aloft 高空槽
trough and paddle mixer 桨叶式搅拌机;桨叶式拌和机
trough-and-paddle (type) mixer 槽桨式(混凝土)搅拌机
trough approach 槽形引道
trough axis 向斜轴
trough band conveyer 槽带输送机;槽形胶带运送机;槽形胶带输送机
trough beam 双山形梁;槽形梁
trough belt 槽带;槽形带;槽式皮带
trough-belt conveyer 槽形胶带运送机;槽形胶带输送机;槽式皮带输送机;槽带式运输机;槽式运输机
trough bend 溜槽转弯;深海槽;拗槽
trough block 槽形砌块
trough bridge 下承桥
trough casting 中间罐浇铸;中间包浇铸
trough-chain conveyer 刮链式输送机;槽形链板输送机
trough classifier 耙式分料器
trough column 槽形柱
trough compass 长盒罗针
trough connection 槽接头
trough conveyer 槽式输送机;槽式运输机;输料槽;槽形输送机
trough core 向斜中心;岩芯槽;槽核
trough crab 装料槽起重绞车
trough cross-bedding 槽形交错层
trough cross bedding structure 槽状交错层理构造
trough cross set 槽形交错层组
trough depth of wave 波谷深度
trough distributor 布水槽
trough downwarped extracontinental basin 海槽型陆外下陷盆地
trough drop 波幅(波谷低于静水面之下的高度)
trough earthquake 海沟地震
troughed 槽形的
troughed alumin(i)um sheet 槽形铝板
troughed belt 槽形带;槽形皮带
troughed belt conveyer [conveyor] 槽板带式输送器;槽形胶带输送机;槽式胶带输送机
troughed block 槽形块;沉陷断块
troughed bulkhead 槽形舱壁
trough-edge 槽谷缘
troughed plate 带熟料兜槽的箅板
troughed profile 槽形外形;槽形轮廓
troughed roller conveyer 槽形辊子输送机
troughed roof cladding 槽形屋面板;槽形屋顶覆盖层
troughed rubber belt shuttle conveyer 槽形皮带穿梭式输送机
troughed sheeting 槽形薄板
trough fault 槽状断层;槽形断层;凹槽断层
trough feeder 供料槽
trough floor 槽形铺面
trough floor slab 槽形楼板
trough flow 槽状流
trough for air conditioning unit 空调器安装槽
trough for steering chain 舵链槽
trough-freight 过境货运
trough-freight traffic 过境货物运输
trough-freight transportation 过境货物运输
trough girder 槽形大梁;(支承铁轨的)槽形梁
trough grate 盆状架棚;槽形炉箅

trough grate furnace 槽形箅炉
trough grate tile 槽形炉箅砖
trough grinding-polishing machine 槽式磨光机
trough gutter 箱形水槽;排水沟;槽沟;檐沟;槽形天沟;槽形沟
troughing 做槽材料;沟槽;槽型钢材;槽形截面;槽钢
troughing belt 运输槽
troughing conveyer 槽式输送器;槽形皮带输送机
troughing distribution system 槽式配水系统
troughing idler 槽形托辊;槽形支承辊(输送带用);槽形布置托滚
troughing plate 槽形钢板;槽纹镶板;槽形钢板
troughing roller 凹槽滚轴(槽带输送机的);凹槽滚轮
troughing rolls 槽形输送带滚子
trough in westerlies 西风槽
trough iron 槽铁;宽口钢梁;槽钢
trough keel 槽形龙骨
trough lavatory basin 盥洗室洗涤盆
trough lift 槽式提升机;带有承船厢的升船机;承船槽式升船机
trough lighting fitting 灯槽装置
trough-like basin 槽形盆地
trough limb 向斜翼
trough line 向斜底线;商业周期低潮线
trough mixer 搅拌槽;槽形混合器;槽式搅拌机;槽式混合器;槽式拌和机
trough of long period wave 长周期波谷
trough of low-pressure 低压槽
trough of syncline 向斜槽
trough of tidal wave 潮汐波谷
trough of wave 波谷
trough-paddle mixer 槽桨式(混凝土)搅拌机
trough plate 槽形板;槽形钢板
trough plate girder bridge 槽形板梁桥
trough pressure 波谷压力
trough pressure diagram 波谷压力图
trough reflector 槽式反射器
trough-ridge system 槽脊系统
trough roller 槽型滚筒;槽形滚筒;槽滚
trough roof 双山头屋顶;双山墙屋顶;槽形屋顶;M形屋顶
trough section 沟槽(节)段;槽式截面;槽形断面
trough separation 槽式分离
trough-shaped 槽式的;凹形的;槽形的;槽式断面
trough-shaped earth pushing process 槽形推土法
trough-shaped girder 槽形梁
trough-shaped paving 槽式断面铺砌
trough-shaped submarine valley 槽形海底谷
trough shaped valley 槽形谷
trough-shoulder 槽谷肩
trough slab 槽板
trough spillway 槽式溢洪道
trough steel 槽钢
trough-style pulp beater 槽式打浆机
trough tile 槽瓦;槽形瓦
trough tip wagon 倾斗车;槽式翻斗车
trough truck 油槽车
trough truss bridge 桁架式下承桥
trough-type bin 槽形料仓
trough-type blending bed 槽形混合料床
trough-type concrete distributor 槽形混凝土浇灌机
trough-type distributor 槽型分配器
trough-type mixing machine 槽形混合机
trough-type sheeting 槽形薄板
trough urinal 小便池;小便槽
trough valley 槽形(河)谷;槽谷
trough vault 长方形反水槽式穹顶;凹槽拱顶;倒槽式拱
trough wash basin 洗涤木盆
trough washer 洗矿槽;槽洗机
trough water closet 槽式便池
trough zone 波谷带
trousered undercarriage 减阻罩起落架
Trouton's coefficient of viscous traction 特鲁顿黏性引力系数
Trouton's constant 特鲁顿常数
Trouton's elongational viscosity 特鲁顿伸长黏度
Trouton's law 特鲁顿定律
Trouton's ratio 特鲁顿比值
Trouton's rule 特鲁顿法则
Trouton's viscosity 特鲁顿黏度
trow 木槽;扁首方尾平底驳

trowel 泥刀;抹子;抹刀;镘刀;镘;花铲子;瓦刀;铁抹子
trowelability 可抹性
trowelable 可镘抹修平的;可镘抹整平的
trowelable coating 镘涂涂料
trowel adhesive 高黏度黏结剂;高黏度黏合剂
trowel application 镘刀用法;镘刀涂抹
trowel-applied coat 镘涂层;涂抹层
trowel-applied finishing 抹平;抹光;镘光
trowel-applied seamless floor(ing) material 镘抹无缝地面材料
trowel coating 抹涂层;镘涂层
trowel elevator 刮板升运器
trowel finish 压光面;赶光面;抹子压光;抹光面;镘平表面;镘抹面
trowel finishing 镘抹平;镘平;抹光;镘光
trowel finish in plastering 抹灰压光面
trowel hand 抹灰工
troweling 铁抹赶光
troweling course 找平层
troweling machine 镘光机
trowel joint 抹子缝
trowel(l)ability 可抹性;抹光性能
trowel(l)ed face 抹光面;镘平面
trowel(l)ed finish 抹面;镘平;镘光
trowel(l)ed joint 用镘刀刮出一定深度的灰缝;抹子缝
trowel(l)ed stucco 抹灰;抹子拉毛
trowel(l)ed surface 镘平面;抹光面;镘光表面
trowel(l)ing 镘(刀涂)光;抹涂;镘平;抹灰;抹光;镘涂
trowel(l)ing consistency 镘涂稠度;涂抹稠度
trowel(l)ing machine 抹光机;抹平机;镘光机
trowel man 抹灰工;瓦工
trowel mark 镘抹痕迹
trowel-mastic 高黏度玛琋脂;抹光用玛琋脂
trowel off 用镘刀涂抹
trowel plaster 镘涂粉刷
trowel test 抹平试验
trowel tool 洒布器;抹胶刮刀;撒布器
trowel trades hydrated lime 泥工行业用熟石灰
Trower cornice 特罗韦牌檐口玻璃条(每节3英尺长,内装条形灯)
Troy ounce 金衡盎司
troy system 金衡制(衡量宝石、金、银之制)
troy weight 金衡制(衡量宝石、金、银之制);金衡
TR probe 发送接受探头
trub 冷却残渣
Trube's correlation 特鲁贝对比
trubulation 湍流搅动
truck (升降机门槛与井口门框间的)过桥;小商品;运货车;载重汽车;载重车;卡车;旗杆帽;无盖货车;桅帽;运货汽车(美国)
truck access pad 进车坪
truckage 载重汽车运输费;卡车运费
truck agitator 汽车搅拌机;搅拌器汽车;拌和车;汽车式拌和机
truck-air truck 陆空联运
truck and rail van 载重汽车和铁路货车
truck axle 转向架轴
truck back hoe 汽车式反(向)铲
truck bearing spring 车架轴承簧
truck berth 卡车车位
truck body 运货汽车车身;载重车车身;卡车身;车体
truck body panel 卡车车体外板
truck bolster 转向架承梁;车架承梁
truck box 载重车厢
truck brake cylinder 转向架闸缸
truck cab 载重汽车驾驶室
truck capacity 载重汽车载重量;卡车载重量
truck car 转向架车辆
truck chamber kiln 钟罩式窑;活底窑;台车窑;车式窑
truck chassis 载重汽车底盘;载货汽车底盘;卡车底盘
truck column 拱架柱
truck combination 货车组合
truck concrete mixer 汽车式混凝土搅拌机
truck-concrete mixer 混凝土搅拌车
truck contact 手车触头
truck convoy 载重汽车队
truck crane 轮胎起重机;汽车式起重机;汽车起重机;汽车吊;起重汽车;随车(式)起重机;随车

(式)吊
truck creel 转向筒子架
truckcrete 混凝土搅拌运料车
truck crop 蔬菜
truckc size 载重汽车规格
truck delivery 卡车运送;卡车送货;卡车递送;货车递送
truck diesel 载重车用柴油机
truck diesel engine 卡车用柴油发动机
truck drier 车载干燥器;车式干燥器
truck driver 卡车驾驶员
truck-dumped riprap 卡车倾卸的乱石堆
truck dumper 卡车倾卸机
truck dump piler 卡车倾卸堆垛机
truck elevator 装车升运器
truck engine 卡车发动机;筒形发动机
truck entrance 运货汽车入口
trucker 载重汽车驾驶员;易货易货者;货物交换者;搬运工;送货工;卡车驾驶员
truck evapo(u)rator 载货汽车蒸发器
truck excavator 汽车挖土机
truck farm 蔬菜农场
truck farming 庭院经济
truck fee 卡车运费
truck-ferry 汽车轮渡
truck fill stand 装车台
truck firm 汽车运输公司
truck fleet 卡车队
truck frame 载重汽车车架;履带架
truck garden 蔬菜农场
truck grader 卡车式平地机;汽车式平地机
truck ground scale 地磅
truck haulage 卡车运输
truck hauling 卡车载运;卡车拖运;汽车载运
truckhead 汽车货运终点站
truck high 汽车装卸台高度
truck-high platform 与卡车高度相同的装车站台
trucking 卡车载运;汽车运输
trucking activities 汽车载运活动范围
trucking area 汽车货运服务区
trucking center 汽车运输中心
trucking company 汽车货运公司
trucking company bill of lading 汽车公司提单
trucking expense 卡车运费;搬运费(用)
trucking facility 汽车运输用器材
trucking industry 卡车运输业
trucking operation 汽车货运业务
trucking terminal 汽车运输总站;汽车运输终点站
truck jack 车辆起重器
truck journal box 转向架轴箱
truck jumbo 台车盾构
truck kiln 台车式窑
truck lane 运货汽车车道;载重汽车车道;卡车道;货车专用车道
truck laying 轨道敷设
truck leasing firm 载重汽车出租公司
truck lever 转向架制动杠杆
truck licence plate number 卡车登记牌
truck lift 自动装卸机
truck lifter 车式提升机
truck light 装在电杆转向架上的灯;桅冠灯
truck line 主航道线;汽车运输公司
truck liner 卡车周转箱;班车箱
truck-load 货车载装(物)
truck-load bin 装车料仓
truckload cargo 整车货(物)
truck loader 自动装载机;自动装卸机;装车机;轮胎式装载机
truck loading 车辆荷载;载重汽车荷载;车载重;货车荷载
truck-loading facility 装车设备
truck-loading hopper 装车料斗
truck-loading platform 装车台
truck-loading rack 装车架
truckload rate 整车运价
truckman 汽车运输业者;搬运工
truckmaster 汽车技术员
truck-mixed concrete 汽车搅拌混凝土;路拌混凝土;商品混凝土;车上搅拌混凝土;车拌混凝土;运输搅拌混凝土;拌和车拌制的混凝土
truck mixer 卡车式搅拌机;混凝土搅拌(运输)车;混凝土搅拌输送车;混凝土拌和车;汽车式拌和机;车载搅拌机;车拌和机
truck mixer method 混凝土车拌法;车拌法

truck mixer plant 混凝土拌和车;汽车式搅拌机
truck mixing 卡车拌和(混凝土);卡车拌制
truck mixing method 混凝土车拌法;车拌法
truck-mounted 车载式
truck-mounted attachment 车载装置
truck-mounted auger 车装麻花钻
truck-mounted bucket loader 车载桶式装载机
truck-mounted concrete pump 车载混凝土泵;混凝土泵车
truck-mounted crane 汽车起重机;装在汽车上的起重机;汽车式起重机;汽车吊;车载吊车
truck-mounted digging attachment 车载挖掘装置
truck-mounted distributor 车载洒布机;车载洒布机
truck-mounted ditcher 挖沟车;汽车式沥青喷布机
truck-mounted dragline 车载索铲
truck-mounted drill 凿岩台车
truck-mounted drilling rig 钻车;车装钻机;车载钻机;汽车式钻机;钻车
truck-mounted excavator 挖土车;车载挖土机
truck-mounted hammer 打桩车;车载夯锤
truck-mounted hoist 车装绞车
truck-mounted jumbo 汽车台钻;汽车式钻堡
truck-mounted loader 车载装载机
truck-mounted mixer 混凝土拌和车;汽车式搅拌机
truck-mounted mortar mixer 车载砂浆拌和机
truck-mounted power shovel 车载动力铲
truck-mounted pressure distributor 车装自动沥青喷洒机;(装在汽车上的)自动沥青喷洒机
truck-mounted rig 车装钻探设备
truck-mounted shovel 轮胎式挖掘机;汽车式挖土机;汽车式铲土机
truck-mounted slick line reel unit 试井车
truck-mounted steam generator 蒸汽车
truck-mounted tipping unit 卡车倾卸设备
truck-mounted tower crane 汽车起重机;汽车塔吊;汽车式塔起重机
truck-mounted water well drilling rig 车装水井钻机
truck move 汽车输送
truck organization 汽车运输队
truck paint 卡车漆
truck payload 货车有效荷载
truck percentage 货车百分比
truck plant 载货汽车制造厂
truck platform 汽车装卸台
truck-platform height 与卡车底板相同的高度
truck platoon 汽车运输排
truck population 载重汽车保有量
truck push/pull for bogies 转向架推拉小车
truck ramp 汽车运输斜坡道
truck rent expense 卡车租费
truck roller 轮胎式压路机
truck route 货车路线
truck scale 载货汽车磅秤;货车地秤;货车磅;汽车地磅;地秤(汽车)
truck schedule 载重汽车一览表
truck scraper 自走式铲运机
truck shortage 载重汽车不足
truck shovel 汽车铲土机;装在汽车上的挖土铲;轮胎式挖掘机
truck side 卡车侧
truck side bearing 闸瓦托
truck snow plough 卡车雪犁
truck spacing 转向架间距
truck spreader 装在卡车上的撒布机
truck spring 车架弹簧
truck spring hanger 车架弹簧吊架
truck stop 载货汽车停车场;卡车服务站
truck stripe 转向条纹
truck system 实物工资制
truck system supported by air cushion 气垫有轨系统
truck tank 油槽汽车
truck terminal 运货汽车终点站;卡车站;货车终端
truck tire 载货汽车轮胎
truck tire casing 载重汽车轮胎外胎
truck-tire remover 载货汽车轮胎拆卸工具
truck ton 载重汽车吨位
truck to shovel ratio 车铲比
truck towing rate 拖运率
truck track 运货汽车轮链
truck tractor 运输货车牵引车;载重牵引车;轮胎式拖拉机;卡车拖拉机;汽车拖拉机
truck tractor for highway 公路两用牵引车

truck-tractor for semi-trailers 半拖挂车用牵引车
truck tractor for yard 场地两用牵引车
truck-tractor semitrailer 载重牵引车半挂车
truck traffic 货车交通
truck trailer 卡车拖挂车;载重拖车;载货挂车;卡车拖车;货运拖挂车;牵引车后挂车
truck-trailer combination 卡车拖车组合;牵引车及挂车
truck train 汽车运输列车
truck-transit mixer 汽车式搅拌机;混凝土搅拌运料车;混凝土运送搅拌车
truck transportation 卡车运输
truck transport company 汽车运输连
truck turn-around 汽车运输往返
truck turntable 运货车转台
truck-type bulk transporter 散货载重汽车;散装运货车
truck-type highway 卡车公路
truck-type switchboard 车架式配电盘
truck-type switchgear 车架式开关装置
truck tyre 载重轮胎;货车轮胎
truck tyre noise 卡车轮胎噪声
truck unloading 汽车卸载
truck unloading hopper 汽车卸料斗
truck wash(down) 载重汽车冲洗
truck weighing scale 汽车地磅
truck wheel 载重汽车车轮;运货车轮
truck wheelbase 转向架轴距
truck window pane broken 车窗玻璃破损
truck wind-screen broken 车风窗玻璃破损
truck with bottom dump body 底卸式货车
truck with crane 载重汽车附起重机
truck with end dump body 后卸式卡车;后卸式货车;尾卸式自卸(汽)车
truck with lift 汽车式起重机
truck with trailer 带拖车的运货汽车
trudellite 易潮石
true absorption 真吸收
true absorption coefficient 真吸收系数
true abundance 真丰度
true acidity 真实酸度
true activation energy 真活化能
true add 原码加
true adsorption 实际吸收;实际吸附
true adsorption density 实际吸附密度
true age 真实年龄
true air speed 真风速;真空速
true airspeed calculator 真空速计算器
true airspeed computer 真空速计算器
true airspeed indicator 真空速表
true airspeed meter 实际空速计
true air temperature 真气温
true alarm base 智能探测器底座
true alarm heat sensor 智能热探测器
true alarm photoelectric(al) smoke sensor 智能光电烟雾报警器
true alarm photo sensor 智能光电探测器
true altitude 真高度
true altitude above the rational horizon 地心高度
true altitude above the sensible horizon 地面真高度
true amplitude 真幅;真方位(角)
true amplitude recovery 真振幅恢复
true amplitude section 真振幅剖面
true angle of internal friction 真内摩擦角
true angular distance 真角距
true annealed sheet 全退火薄钢片
true annual interest 实际年息
true annual ring 真年轮
true anomaly 真近点角
true aperture 实际孔径
true arch 纯拱
true area 真实面积
true aryl nitro-compound 真芳基硝基化合物
true average 真实平均数
true axial angle 真光轴角
true azimuth 真方位(角);天体真方位
true bacteria 真细菌
true-banked turn 正常盘旋
true bearing 真象限角;真航向;真方位(角)
true bearing adaptor 真方位指示装置
true bearing of landmark 陆标真方位
true bearing rate 真实方位变化率
true bearing unit 真方位指示装置;真方位显示装置

true binary notation 二进制原码表示法
true boring cutter 精加工镗刀
true bottom 真底板
true breeding 不分离的
true brightness 真实亮度
true brinelling 击蚀
true bruising 击蚀
true carry 真实进位
true cedar 真雪杉
true cedar cedrus liabnotica 真杉
true centrifugal casting 精确离心铸造
true cleavage 真劈理
true coal 纯煤
true coal combustion heat 纯屯煤燃烧热
true code 原码;原料原码;实代码
true coefficient 真系数
true coefficient of friction 真摩擦因数
true coefficient of thermal expansion 真热膨胀系数
true cohesion 真黏着力;真黏聚力;真黏结力;真凝聚力
true colo(u)r 真色度;纯水色
true colo(u)r fidelity 彩色逼真(度)
true colo(u)r image 真彩色图像
true colo(u)rity unit 真实色度单位
true colo(u)r unit 真色度单位
true complement 真补数;实补码;补数;补码
true cone tip resistance 真锥头阻力;真锥尖阻力
true conic(al) projection 真圆锥射影
true constant 真实常数;真常数
true contact 真实接触
true continuous absorption 真连续吸收
true coordinates 真坐标
true copal 真柯巴树脂
true correlation 真实相关
true cost 真正成本;实际成本
true counter-current flow 真逆流流动
true course 真航向
true creep 实际蠕变
true critical pressure 真临界压力
true critical property 真临界性质
true critical shear stress 真临界剪应力
true critical temperature 真临界温度
true date-time group 真日时组
true degree of dissociation 真离解度
true density 实际密度
true depression angle 真俯角
true depth 真深度
true deviation 真实偏差
trued for alignment 线向调整
trued for level 调整标高;调整高程
trued for position 调整位置
true difference 真实差数
true diffusion international 技术扩散
true diorite 真闪长岩
true dip 真倾斜;真倾角;真倾度
true dip angle 真倾角
true direction 真方向;真方位(角)
true discount 实际贴现率;实在折扣;实在贴现;实际折扣
true distance 真距离
true dormancy 真正休眠
true dynamite 真硝甘炸药
true earning 真正收益;实际收益
true earth pressure 实际土压
true elastic strain 纯弹性应变
true elevation of stone monuments 柱石面真高
true empty-running speed 实际空转速度
true environment 真实环境
true equator 真赤道
true equilibrium 真实平衡;真平衡
true equilibrium price 真正平衡价格
true equinox 真春分点
true erosion 实际侵蚀量;实际冲刷量
true error 真误差;真差;测量误差
true evergreen forest 真常绿林
true exit 真出口
true expression 真表达式
true extract 真正浸出物含量
true factor analysis 真因子分析
true-false test 真伪试验
true fault 真误差;真实故障
true fibre 粗细均匀的纤维
true firs 真枞

true fluidity 真流动性
true folding 真褶皱
true form 原码形式;原料原码;真实形态;实际形状
true formation resistivity 真岩层电阻率;地层真电阻率
true fraction 真分数
true fresco 壁画技术(壁画技术中一种)
true frictional resistance 真摩阻力;真摩擦力
true frost injury 真正的霜害
true fungi 真正真菌素
true ga(u)ge length 正确隔距长度
true ground speed meter 真地速测速仪;地磁计程仪
true groundwater velocity 真地下水流速;地下水真速度;地下水真流速;地下水实际流速
true growth rate 真生长率
true hardness 真硬度
true heading 真首向;真方向
true heading computer set 真方向电子计算机
true heat capacity 真热容量
true height 真实高度;实际高度;实际高程
true height of the barometer 气压表真高度
true hemp 粗细均匀的麻类
true homing 真返航向;真自动引导
true horizon 真水平;真实水平;真地平线;真地平(圈)
true horizon plane 真地平面
true horizontal 真水平的
true horsepower 指示马力;实际马力
true humus calcareous soil 典型腐殖质碳酸盐土
true image 直接像
true income 真正所得
true inflation 真正通货膨胀
trueing 校正;调整;矫直;整平
trueing unit 精密修整装置
trueing-up 平整木面
true internal friction 真内摩擦
true jade 翡翠
true joint 真缝
true laterite 典型砖红壤
true length 实长
true libration 真天平动
true lifetime 真实寿命
true line of bearing 真方位线
true liqduid 真液体
true lode 真矿脉
true loess 真黄土
true lubrication 真润滑
true luminance 真实亮度;真实发光率
true luminance information 真实发光率信息
true magnetic charge 真磁荷
true marine fishes 纯海洋性鱼类
true maximum 真实极大
true meadow 真草甸
true mean 真实平均数
true mean dispersion 真正平均散布
true mean place 真平位置
true mean temperature 真平均温度
true mean value 实际平均值
true median 真实中位数
true meridian 真子午线
true middle latitude 真中分纬度
true midnight 真午夜
true model 等大模型
true monthly mean precipitation 真月平均降水量
true motion 真运动;绝对运动
true motion display 真运动显示器;真运动显示;绝对运动显示器
true motion indication 真运动指示
true motion indicator 真运动显示器;目标真运动指示器
true motion radar 真运动雷达;绝对运动雷达
true motion radar presentation 雷达真运动显示;雷达绝对运动显示
truemotor load 热马力
true nature 本性
trueness 真实性
trueness of shape 外形的真实性;形状的真实性
trueness to type 典型性
true net energy for maintenance 维持真净能
true noon 真正午;真午
true normal at the observer 测站真法线
true north 真北
true north heading 真北

true object 实物
true obsidian 真黑耀岩
true ohm 实际欧姆;电磁制欧姆
true origin 坐标原点;绝对原点
true output 实际出力
true perpendicular 真垂线
true picture 真实图像
true place 真位置
true plane 校准平面
true plot 真实图表;真实读数;绝对运动图;对地运动图
true plot diagram of situation 绝对运动图
true plotting 绝对运动作图;对地运动作图
true point mutation 真点突变
true pole 真极
true porcelain 瓷器
true porosity 真气孔率;真孔隙率;实际空隙率
true porphyry 真斑岩
true position 真位置;真船位
true power 有效功率
true precession 真实进动
true price 真正价格;实际价格
true prime vertical 真东西圈
true profit 真正利润;实际利润
true purity 真纯度
true pyramid 真棱锥体
true quotient field 真实商域
truer 整形器;校准器;砂轮修整工具
true radio bearing 真无线电方位
true rake 实际前角
true rake angle 真前角
true range 真射程
true range indication 真实航程指示
true rate of interest 实际利率
true reactivity 真实反应性
true refuse density 垃圾真密度
true regression plane 真回归面
true relay 实际延迟
true representation 真实的表现;真表示
true resin 真树脂
true resistance 真电阻;欧姆电阻
true resolution 实际分辨率
true rib 真肋
true rising and setting 真出没(太阳、月亮)
true rolling 真正滚动;纯滚动
true-rolling bit 纯滚动牙轮钻头;牙轮钻头
true rotation 真自转周
true-running 正确运行
true sample 原试样;代表性岩石标本
true scale 真比例尺
true screw 定(螺)距螺旋桨
true section 真实断面图
true semidiameter 真半径
true sensitivity 真实灵敏度
true September equinox 真秋分点
true shale thickness in coal formation 夹矸的真厚度
true sidereal time 真恒星时
true side-wall friction 真侧壁摩擦力
true side-wall friction resistance 真侧壁摩阻力
true size 真实尺寸
true skin 真皮
true slime 真矿泥
true slip 真实滑移;真断距【地】;实滑距
true slip ratio 真实滑脱比
true slump 正确坍落度;正常坍落度;实际坍落度;正坍落(混凝土)
true slump type 正常坍落型
true soil 土层
true solar day 真太阳日;视太阳日
true solar time 真太阳时;视太阳时;视时
true solution 真溶液
true solvent 真溶剂
true south 真南
true specific gravity 真比重
true specific gravity of pure coal 纯电煤真比重
true specific weight 真比重
true speed vector 真速度向量
true steady state 真稳态
true steel 共柝钢
true steppe 真草原
true stiffness 真实刚度
true strain 真实应变;真实变形;实际应变
true stress 真应力;真实应力;实际应力

true stress-strain curve 真应力一应变曲线
true subgrade 符合设计要求的路基;完全符合设计要求的路基
true sun 真太阳;视太阳
true supply price 实际供给价格
true table 真值表
true tailings 最终尾矿
true temperature 真实温度;实际温度
true temperature drop 真温度降
true tensile strength 最大拉伸强力
true thickness 真厚度
true thickness of coal formation 煤层的真厚度
true time 真时
true-time operation 实时运算;实时操作
true-to-scale 符合比例尺的
true-to-shape 形状正确的;真实形状的
true to size 符合真实尺寸;真实尺寸;尺寸准确的
true total temperature 真总温度
true track 真航迹向;真方向
true tracking 真运动;绝对运动
true tracking indicator 真运动显示器;绝对运动显示器
true tracking radar 真运动雷达
true triaxial apparatus 真三轴仪;三向主应力三轴仪
true triaxial shear apparatus 真三轴剪力仪
true type 理想类型
true undersize recovery 实际筛下回收率
true-up 安装正确;校准;调准;整形
true value 真实值;真值;真正价值;实际价值
true vapo(u)r pressure 真蒸气压(力)
true vapo(u)r process 真气相过程
true variability 实际变异性
true vault 真实拱顶
true vector 真矢量
true velocity 真速度;真实速度;实际流速
true vernal equinox 真春分点
true vertex 真顶点;真奔赴点
true vertical 真地平经圈;真垂直线;真垂线
true vertical photograph 理想垂直摄影像片
true viscosity 真黏度;真实黏度
true vocal cord 真声带
true volume 真体积;真实体积;实际体积
true volume measure 实际材积度
true weight 实重
true width 真宽度;真厚度
true wind 真风
true wind direction 真风向
true wood 心材;真木
true yellow 纯黄(色)
true yield 实际产率
true yield point 真屈服点
true zenith 真天顶
true zenith distance 真天顶距;顶距;实际天顶距
true zero 真零点
trug 灰浆槽;水槽
truing 修整;整平
truing course 整平层
truing device 修整装置
truing diamond 修整用金刚石刀
truing diamond holder 修整用金刚刀架
truing face 修正面
truing plane 整形刨;长刨;细刨;平刨
truing tool 整修工具;校准工具
truing up 校准
truite 百级碎
tru-lay rope 预搓丝股的钢索
tru-lay wire rope 不松散钢丝绳
T-rule 丁字尺
trullo 圆顶房屋(干垒毛石的)
truly asynchronous 真异步式
truly eluvial landscape 正残积景观
trumbone 长号
trumeau 窗间墙;(两窗或两门之间的)间柱;(两窗或两门之间的)间壁;门中间石柱
trumming materials 修整材料
trumpet 漏斗形浇口;漏斗形的;喇叭形的;喇叭管
trumpet arch 跨屋隅的小拱;喇叭形拱
trumpet assembly 中心注管;中心浇口
trumpet bell 浇注漏斗
trumpet brass 管乐器黄铜
trumpet buoy 音响浮标;雾号浮标
trumpet cooler 管式冷凝器
trumpet guide 喇叭形导纱器
trumpet inlet 喇叭形进口

trumpet interchange 喇叭形立交
trumpet intersection 喇叭口交叉口
trumpet joint 融叭形管接头
trumpet junction 喇叭形接头
trumpet log 号角测井
trumpet metal 管乐器黄铜
trumpet-shaped ruby laser 喇叭状红宝石激光器
trumpet spillway 喇叭口溢洪道
trumpet-type interchange 喇叭形立体交叉;喇叭形互通式立交
trumpet vault 喇叭形穹顶
trump stone 主景石
trump tree 主景树
truncate 截尾;截头;截断;截短;平头;缩补;舍弃
truncated 剖面剥蚀的;平切
truncated address 短缩地址
truncated angular arch 截顶锥角拱;锥角拱;截顶角形拱
truncated anomaly 切余异常
truncated anticline 削蚀背斜
truncated bowstring truss 平截弓形桁架;平端弓形桁架
truncated buoy 截锥形浮标
truncated cone 平截锥;圆台;截锥(体);截圆锥;截头圆锥(体);截顶锥
truncated cone banking 锥坡;截锥坡
truncated cone shape grinding media 截锥形研磨体
truncated conic(al) roller 截锥形磨辊
truncated corner 修圆转角;削圆转角
truncated cylinder 平截筒柱;斜截筒柱;截柱
truncated distribution 截止分布;截尾分布
truncated Doppler profile 截多普勒轮廓
truncated earnings function 截尾收入函数
truncated edge 钝边
truncated error 截断误差
truncated exponential frequency 截断指数频度
truncated float 截锥形浮子
truncated Fourier transformation 截断傅里叶变换
truncated frequency distribution 截尾频率分布
truncated function 截尾函数
truncated gate 胸墙下闸门;半截闸门
truncated Gaussian 高斯截短曲线
truncated glume 截形颖片
truncated headland 削蚀的岬角
truncated landform 屑蚀地形
truncated length 截尾长度
truncated linear model 截尾线性模型
truncated miter gate 胸墙下的人字闸门
truncated negative exponential distribution 截尾负指数分布
truncated normal decision distribution 截尾正态决策分布
truncated normal distribution 截尾正态分布;截断正态分布
truncated octahedron 截顶八面体
truncated paraboloid 截尾抛物面
truncated picture 截像
truncated podzol 剥蚀灰壤
truncated Poisson distribution 裁截泊松分布
truncated posterior distribution 截尾后验分布
truncated prism 斜截棱柱;截棱柱
truncated profile 剥蚀剖面
truncated pyramid 斜截头角锥;斜截棱锥;菱台;截头角锥体;截棱锥;截角锥(体);截顶角锥体;平顶金字塔
truncated roof 平塔顶
truncated sample 截断样本
truncated soil 剥蚀土壤
truncated spectrum 截断频谱
truncated spur 削断山嘴
truncated tetrahedron 平截四面体
truncated tip 锥头电极
truncated tip electrode 锥头电极
truncated triangular decomposition 截顶三角分解(法);舍项三角分解(法)
truncated two-stage least square 截尾二阶段最小平方法
truncate leaf 截形叶
truncating and rounding 截尾和舍入
truncation 削平;截尾;截舍;截断;截除;舍项;舍位;舍去
truncation error 截断误差;舍项误差;舍位误差;不含入误差

truncation frequency 截断频率
truncation function 截断函数
truncation method 截尾法;平切法;舍项法;舍位法
truncation on assignment 赋值时的切断
truncation specification 截断规定
truncation test 截断试验
truncation for trap 削截圈闭
truncation width 截断宽度
truncheon 插尺
truncus inferior 下干
trundle 矮轮手推车;手车;小轮;转轴颈;无盖货车
trundle bed (可推入大床下的)矮床;套床
trunk 中继线;象鼻管;箱子;总线;总管(气、水等);主干;旅客衣箱;集群;母线;膨胀阱;围壁;围蔽室;筒形;树干;大旅行箱;粗干材
trunk amplifier 干线放大器
trunk answer from any station 分机代答中继来话
trunk boiler 筒形火管锅炉;筒形火管锅炉
trunk bulkhead 围舱壁
trunk buoy 链筒浮筒;筒形浮标
trunk busy 中继占线
trunk cabin 半显露舱室
trunk cable 干线电缆;长途通信电缆;长途电缆;主干电极电缆
trunk call 长途呼叫;长途电话
trunk call office 长途电话局
trunk chain 主链
trunk circuit 局内线
trunk code 长途电码
trunk conference telephone 干线会议电话
trunk connection 长途接线
trunk conveyer 主运输机;主要巷道运输机;主输送机
trunk coupling unit 干线耦合单元
trunk deck 箱形凸起甲板;围壁室甲板;凸起甲板
trunk deck beam 凸起甲板横梁
trunk deck vessel 箱形甲板船;有围壁室的甲板船
trunk depth 支承表面至喷嘴之间的距离
trunk dial 箱形字盘;箱式表盘
trunk directory inquiry 长途电话局查询台
trunk dispatching telephone 干线调度电话【铁】
trunk distribution frame 长途配线架
trunk duct 主干管;通风总管
trunked hatchway 围壁室舱口
trunked mobile communication system 集群移动通信系统
trunked telephone connect 集群电话互联
trunked telephone connector 集群电话互联器
trunk elevator 装货电梯
trunk engine 无十字头发动机;筒状活塞发动机;单动式发动机
trunk equipment 中继设备
trunk exchange 长途电话局
trunk feeder 喂料干管;主馈电线;主馈电路;主电路;供料干管;供电干线
trunk for light 采光围壁
trunk glacier 主干冰川
trunk greening 树干绿化
trunk grid 干线系统;干线网
trunk group 中继线群;干线组
trunk group area 多局制市内电话网
trunk hatch 主升降口
trunk highway 干路;公路干线;干线公路;干线道路
trunk hunting 自动选中继线;干线寻我;长途自动寻线
trunking 管槽;通风管;通风道;线槽;中继方式;中继
trunking capacity 集群能力
trunking communication 集群通信
trunking diagram 中继线方式图
trunking loss 管道损失
trunk junction circuit 长途中继电路
trunk length 体长
trunk lid (汽车的)后行旅箱盖
trunk lid key 行李间钥匙
trunk lift 管式升降机
trunk line 线主管道;油管干线;主干路;管道干线;干线;长途线
trunk line circuit 长途线路
trunk line duct 干管;干渠
trunk line heating furnace 干线加热炉
trunkliner 干线客机
trunk line route 主航道
trunk line waterway 水道干线
trunk main 主干线;中断干线;中继线;管道干线

给水主干管;大直径管;主下水道;排水干管;干管
trunk mat 行李舱地板垫毯
trunk mooring buoy 链筒系泊浮筒
trunk network 干线网
trunk-offering final selector 长途插入终接器
trunk-offering selector 长途接入选择器
trunk of funnel 外烟囱
trunk or main part of a breakwater 堤干(防波堤)
trunk pipe line 管道干线
trunk piston 筒状活塞;筒形活塞;筒裙式活塞
trunk piston brake cylinder 柱状活塞闸缸
trunk plank 包装棉花压板
trunk platform 无栏杆拖车
trunk position 长途台
trunk railway 干线【铁】;干线铁路;铁路干线;铁道干线
trunk-record circuit 长途记录电路
trunk-record position 长途记录台
trunk river 干河
trunk road 港口主干道;干线公路;干线道路;干线道路;干路;县干道
trunk road of port 港口主干道
trunk roadway 主要运输道
trunk room 行李间;行李房;箱子储藏室
trunk rot 干腐
trunk route 主航线;干线
trunk route decca 台卡跟踪测距导航系统
trunk sanitary sewer 污水干管
trunk sewer 主下水道;主排水管道;干沟;污水总管;污水主管;污水干管
trunk shoot 树干枝条
trunk siding 主气专用线
trunk snap 断干
trunk stem 主干
trunk stream 主流;干流;干河
trunk subsystem 中继子系统
trunk switchboard 长途交换台;长途交换机
trunk system 干线系统;干线排水系统;干管系统
trunk to a space 通道围阱
trunk transmission line 主输电线
trunk tree 树干腐朽病
trunk ventilator 围壁通风筒
trunk vessel 有围壁室的船
trunk way 烟道
trunk wrapping 缠裹树干(园林)
trunnel 木销钉;大木钉;木钉
trunnion 枢轴;枢销;耳轴
trunnion axis 水平轴;枢轴
trunnion axis of theodolite 经纬仪水平轴
trunnion bascule bridge 枢轴仰开桥
trunnion bearing 空心轴轴承;枢轴轴承;耳轴承;枢轴支座
trunnion bed 炮耳座
trunnion bend 耳轴弯头
trunnion brass 耳轴铜衬
trunnion carrier 管套座
trunnion connected drive 空心轴传动连接
trunnion cross 活节十字头
trunnion discharge mill 中心卸料器;空心轴卸料磨机
trunnion feed mill 空心轴喂料磨机;耳轴加料磨
trunnion-hoisting gantry 耳轴吊架
trunnion joint 耳轴式万向接头
trunnion liner 空心轴衬板
trunnion lining 空心轴衬里
trunnion mill 耳轴磨
trunnion-mounted 转轴安装的
trunnion (mounted) mill 轴颈支承的磨机;耳轴磨
trunnion mounting 耳轴支架
trunnion of drag arm 耙臂耳轴
trunnion pier 仰开桥桥墩
trunnion pipe 耳轴管
trunnion ring 耳轴环
trunnion screen 筒筛
trunnion seat 炮耳座
trunnion support 叉轴支座;耳轴支座
trunnion tip wagon 耳轴式自卸车
trunnion-type bascule bridge 耳轴式竖旋桥
trun-off thyristor 可关断晶体闸流管
Truscon floor 德鲁斯康式楼板
Truscon paste 德鲁斯康牌防水糊
Truscon precast unit 德鲁斯康牌预制构件
truscottite 特水硅钙石;白钙镁沸石
truss 桁架;桁材

truss action 桁架作用
truss analogy method 桁架比拟法
truss-and-beam bridge 桁架梁桥
truss arch 拱形桁架;桁拱
truss bar 构架杆;桁架杆;弯曲钢筋;桁架钢筋
truss beam 桁梁;平行弦桁架;桁架梁
truss beam bridge 桁架梁桥
truss bolt 锚栓;地脚螺栓;桁架螺栓
truss boom 桁架伸臂
truss bow 桁架用接环
truss brace 桁架支撑
truss bridge 桁架桥
truss bridge with subdivided panels 再分式桁架桥
truss-buttress dam 桁架式支墩坝;桁架支墩坝
truss camber 桁架拱度;桁架反拱;桁架反挠度
truss chord 桁弦
truss cupola 桁架圆屋顶
truss depth 桁架高度
trussed arch 拱形桁架;桁构拱
trussed arch aqueduct 桁架拱式渡槽
trussed arch bridge 桁架拱桥
trussed arched girder bridge 拱形桁架梁桥
trussed arch girder 桁架式大梁;构成拱梁;桁架拱梁
trussed arch portal frame 桁拱门架
trussed arch with three-hinge 三脚桁架拱
trussed bar 桁架钢筋
trussed beam 桁杆;桁架式梁;桁架梁;平行弦桁架;桁架梁
trussed beam bridge 桁架梁桥
trussed beam plough 吊挂式犁
trussed beam with diagonals and suspenders 带有斜杆和吊杆的桁架式梁
trussed bent 桁构架桥
trussed-box-girder bridge 箱形桁架桥
trussed bridge 桁架桥;桁构桥
trussed camber 桁架起拱
trussed cantilever arm 桁架式悬臂
trussed center 构架式拱架(托)
trussed frame 桁架式构架;桁架构架;桁构构架
trussed girder 平行弦桁架;构成桁架梁;桁架式大梁;桁架大梁;桁构大梁
trussed joist 桁架式格栅;桁架托梁
trussed member 桁架杆件;构架杆件;桁构杆件
trussed opening 桁架门孔;桁架式洞口
trussed partition (wall) 桁架式构造的隔墙;格构式间壁;木隔墙;框架式隔墙;桁架式隔墙;桁架间壁
trussed pole 桁构式杆;桁构式柱;桁架式柱;桁构式桅杆;桁构电干
trussed portal 桁架式门座
trussed post 桁架式柱;桁构支柱
trussed principal member 桁架主杆件
trussed purlin(e) 桁架式檩条;桁架梁;格构式檩条;加劲檩条(拉杆);桁架檩
trussed rafter 桁架式椽条;桁架椽;桁架式人字木
trussed rafter principal 构成主椽;构成主梁;桁架主椽
trussed rafter roof 桁架椽屋顶;人字形桁架屋顶
trussed ridge roof 桁架脊屋顶;桁架式双坡屋顶;格构式双坡屋顶
trussed rod 圆钢斜撑;圆钢拉杆
trussed roof 桁架屋顶
trussed spacing 桁架间距
trussed span 桁架跨度
trussed steel arch bridge 钢桁架拱桥
trussed-steel beam 钢桁梁;桁梁
trussed steel joist 构成钢格栅;桁架式钢格栅;桁架钢格栅
trussed stringer 构成纵梁;桁架纵梁;桁架式纵梁
trussed structure 桁架式结构
trussed system 桁架体系;桁架系统
trussed-wall opening 桁架式洞口;桁架式斜屋顶;桁架式开洞
truss frame 桁架
trussframed 构架的;桁架的
truss-frame plow 架式犁
truss girder 梁构桁架
truss-head rivet 大圆头铆钉
truss-head screw 大圆头螺钉
truss height 桁架高度
truss hoop 桁箍
trussing 捆扎;构架系统;杆系;桁架系统;桁构系统;桁架构件
trussing piece 捆扎件

truss in space 空间桁架
Trussit 多孔(拉制的)钢丝网
truss jack-up 桁架自升式(平台)
truss joint 桁架节点
truss leg 桁架柱
truss member 桁架杆件
truss panel point 桁架接点
truss post 桁架中柱;桁架式柱;桁架支柱
truss principal 桁架主杆件;人字木;承重桁架
truss rod 桁架拉杆;桁架系杆;桁架对角系杆
truss rod system 桁架杆系统
truss spacing 桁架间距
truss span 桁架跨度
truss-stiffened suspension bridge 用桁架加劲的吊桥;桁架加劲吊式桥
truss structure 桁架结构
truss theory 桁架理论
truss timber 屋架木料
truss type model 桁架式模型
truss up 捆绑
truss vertical 桁架竖杆
truss web 桁架腹杆
truss wing 桁架臂杆
truss without sloping members 无斜杆的桁架
truss with parallel chords 平行弦桁架;梁式桁架
truss with sub-divided panels 再分格桁架;再分节间桁架
truss with sub-ties 有分(系)杆的桁架;K 形桁架
truss with triangular web opening 三角形网格桁架
trusswork 桁架结构
truss yoke 桁箍
trust 信托;信任;托拉斯
trust account 信托会计;信托账户
trust accounting 信托业务会计
trust ageement 信托协议
trust agent 信托代理商
trust-and-agency fund 信托及代理基金
trust and banking company 信托银行公司
trust and investment business 信托投资业务
trust and savings bank 信托和储蓄银行;信托储蓄银行
trust assets 信托资产
trust banking 信托银行业务
trust bonds 信用证券;信托债券
trust business 信托业务
trust busting 取缔垄断
trust certificate 信托证书
trust charges 信托费
trust company 信托公司
trust consultancy corporation 信托咨询公司
trust corpus 信托资产本值
trust deed 信托书;信托契据;委托书;财产信托证书
trust deed for surveying mark 测量标志委托保管书
trust deposit 信托存款;信托保证金
trusted program(me) 受托程序
trustedtite 方硒镍矿
trustee 移交财产或管理权给受托人;受托人;受托(管理)人
trust(ee) bank 信托银行
trustee council 理事会
trustee in bankruptcy 破产信托人;破产受托人;破产管理人;破产财产管理人
trustee investment 受托人投资
trustee savings bank 信托储蓄银行
trusteeship 信托制度;托管
trustee's sale 受信托人销售
truster 信托人
trust estate 信托财产;受托管之房地产
trust fee 信托费
trust function 信托职能
trust fund 信托资金;信托基金
trust fund account 信托基金账户
trust fund appropriation 信托基金拨款
trust fund bureau 信托金库
trust fund ledger 信托基金分类账
trust fund program(me) 信托基金方案
trust indenture 信托书;信托合约
trust institution 信托机构
trust instrument 信托证书
trust insurance 信托保险
trust inter-fund transaction 信托基金户间往来
trust investment 投资信托业;信托投资

trust letter 信托书
trust obligation 信托合约
trust of housing loan bonds 住宅贷款信托债券
trustor 信托人
trust port 托拉斯港
trust process 信托手续
trust property 信托财产
trust purchasing 信赖性购买
trust receipt 信托收据
trust region 信任区
trust remainder 信托财产的指定继承人
trust revolving fund 信托周转基金
trust service corporation 信托服务公司
trust service system 信托服务系统
trust shop 信托商店
trust stock company 信托股份公司
trust territory 托管(领)地
trust unit 信托单位
trustworthy product 信得过产品
truth 真实性
truth function 真值函数
truth model 真实模型
truth set 真值集合
truth table 真值表
truth table computer 真值表计算机
truth table generation testing 真值表产生试验
truth table method 真值表法
truth value 真值;真伪值
truth-value function 真值函数
truth value matrix 真值矩阵
truth value of proposition 真值命题
try 尝试
try amplifier 试验放大器
try-and-error method 逐次逼近法;试凑法
try cock 试验旋塞;超位检验栓
try engine 试机器
try ga(u)ge 校验量规
try hole 高炉探尺孔;探测孔
trying a wheel 箍轮
trying contract 结卖合同
trying iron 料钎;玻璃液取样铁棒
trying out tool 工具试验
trying plane 细刨;平刨;大刨;长刨
try(ing) square 验方角尺;验方角规;矩尺;直角尺;曲尺
try-net 试验网
tryoff 铸型试合箱
tryout in typical cases 典型试验
try plane 平刨;大刨;长刨
trysail 斜桁纵帆
try square 曲尺;验方(角)尺;直角切削工具;直角尺
tryst 集合所;约会处;市场
try state 试验位置
Tsar's palace 沙皇宫廷
tschermakite 钙镁闪石;镁钙闪石
tschermigite 铵明矾
tscherwinskite 贫硫沥青
tschirwinskite 含硫沥青
T-section 丁字形截面;T形截面
T-section girder 丁形梁;T形大梁
T-section steel T形钢;丁字形钢
T-shaped T字形;T形截面的;T形的
T-shaped entrance and exit 穿廊式出入口
T-shaped forehearth T形料道
T-shaped girder T形大梁
T-shaped pier T形码头
T-shaped retaining wall 丁字形挡土墙
T-shaped rigid frame bridge T形刚构桥;T形刚架桥
T-shaped road junction 丁字街
T-shaped truss rigid frame bridge 桁式T形刚构桥
T-shore 丁字形支撑;T形支撑
tsilaisite 钠锰电气石;锰电气石
tsingtauite 青岛岩
T-slab 单肋板;T形板
T-slot bolt 丁字形螺栓
T-slot cutter 丁字槽铣刀
T-slot (milling) cutting 丁形槽铣削
T-slot piston 带T形槽活塞;丁形槽活塞
T-slot screw T字槽螺丝
T-socket 丁形套管
T-square frame T方形架
T-square (ruler) 丁字尺

TSS Network 分时系统计算机网
T-steel T形钢;丁字形钢
T-steel bar T形钢杆;T形型钢
T-steel dowel T形钢暗榫
t-stop T光圈
T-strap T形接板
tsumcorite 砷铁锌铅石
tsumebite 绿磷铅铜矿
tsumoite 楚南铋矿
tsunami 津浪;海啸;地震津波;地震海啸
tsunami barrier 海啸护堤;地震海啸防波堤;防潮大堤
tsunami earthquake 海啸地震
tsunami engineering 海啸工程
tsunami flooding risk analysis 海啸危险分析
tsunami ga(u)ge 海啸仪
tsunamigenic earthquake 海啸地震
tsunami height distribution function 海啸高度分布函数
tsunami instrument 海啸仪
tsunami numerical modeling 海啸数值模拟
tsunami potential 海啸势
tsunami prediction 海啸预报
tsunami protection breakwater 海啸防波堤;地震海啸防波堤
tsunami run-up 海啸行近
tsunami source 海啸源
tsunamite 海啸岩
tsunami warning 地震警报
tsunami warning system 地震海啸警报系统
tsunami warning system in the Pacific 太平洋海啸警报系统
tsunami wave 海啸波(浪)
T-support T形支撑(架)
Tsushima current 对马海流
tsuzuic acid 粗租酸
Tswett's method 茨维特法
T system 横管系统
T-test T检验
T tile T形面砖;T形瓦
T-trail T形轨
T-trap T形带
T-tube 丁字管
T tubule 横小管;横管
T-type T形的
T-type equal pressure valve 工字形无压阀门
T-type highway 货运公路
T-type intersection 丁字交汇;丁字形交叉(口)
T-type tooth collect bar 钉齿校正框
tub 圆形容器;浴缸;坐浴(浴)盆;木桶;木盆;桶;大金属桶
tuba 管状云;大旱
tubage 插管法
tubain 土芭树脂
tubal 管的
tubal static mixer 管状静态混合器
tubbable 可放入桶中的
tub basket 桶筐
tub bath 盆浴
tubber 双尖镐
tubbing 孔壁支护用丘宾筒;井壁支护用丘宾筒
tubby 瓮声
tub chair 桶式软垫椅
tub coating 槽式涂布机
tub cock 浴盆龙头;澡盆旋塞;澡盆龙头
tub colo(u)ring 槽法着色
tube 筒体;管子式管道;管(子);花冠筒;电子管;地下道;灯管
tube adapter 管子适配器;管接头
tube agglutination test 试管凝集试验
tube alley 管沟
tube amplifier 电子管放大器
tube-and-coupler scaffold(ing) 管子和扣件联结的脚手架;钢管脚手架;扣件式钢管脚手架;单管脚手架
tube-and-coupler shoring 钢管支撑;钢管支承组件;带联结器的钢管支撑
tube and fin radiator 肋片管式散热器
tube and pipe of alumin(i)um alloy 铝合金圆管
tube and tank process 管罐裂化过程
tube-anemometer 管状风速表
tube arrangement 管线布置;管系布置;管路布置;管道布置
tube (array) 列管

tube-axial fan 轴流风扇;管装式轴流风机;轴流式风扇;轴流式风机
tube axial-flow fan 圆筒式轴流通风机
tube bank 管束
tube base 管底
tube base coil 插入式线圈;插拔线圈
tube beader 管子卷边工具;管子卷边器
tube bedding 管道垫层
tube (belt) conveyer 管状胶带输送机
tube bender 弯管机
tube bending machine 弯管机
tube blower 吹灰器;冲灰器
tube booster 传爆管
tube brush 管刷
tube bunch 管束
tube bundle 管群;管簇
tube bundle arrangement 管束布置
tube bundle heat exchanger 管束式换热器
tube bundle of tube-type heat exchanger 列管式换热器的管束
tube buoy 管状浮标
tube caisson foundation 管桩基础;管柱基础
tube canal 管沟
tube cap 管帽
tube carrier 供料机旋转管托架
tube casing 管围壁
tube ca(u)lker 管子卷边工具
tube cavitation 管的气蚀
tube cell 粉管
tube center distance 管中心距
tube characteristic 电子管特性
tube checker 管测试仪
tube clamp 管夹板
tube clarifier 斜管澄清池
tube cleaner 管内除垢器;管子清洁器;洗管器;洗管机;清管器
tube cleaning 洗管
tube clearance 管眼余隙
tube clip 管卡;管夹
tube closing 管的封密
tube coefficient 管子系数
tube coil 盘香管;盘管;管圈
tube conductor 空心导线
tube connector 管子接头
tube connect(tion) 管接头
tube construction 管形结构
tube construction with different diameter 异径井身
tube construction with same diameter 同径井身
tube conveyer 输送管;管式气动传送器;管输送机
tube core 管心
tube corrosion 管子腐蚀
tube count 计数管计数
tube couples 塑料管接头
tube coupling 管节;管接头
tube cradle 管子吊架
tube cross-section area 隧洞横截面面积
tube current division 电子管内的电流分配
tube cutter 截管器;截管机;切管器;切管机
tube cutting machine tool 管子切削机床
tube delivery gate 管道放水(斗)门
tube derrick 管子钻塔
tube diameter 管直径
tube diffuser 扩散管
tube ditch 管沟
tube door 管子门
tube(d) railway 地下铁道
tube drain 管道排水
tube drainage 管道排水
tube drainage system 管道排水系统
tube drawing 拔管;管材拉拔
tube drawing bench 拉管台;拔管机
tube-drawing machine 制管机;拉管机
tube-drawing process 拉管法
tube drift 管塞(子);打桩器
tube drive twister 积式式退解捻捻线机
tube drive unwinding 积式式退解
tube driving 打管桩
tube driving rig 打管桩设备
tube drop 涵管跌水
tubed tyre 充气轮胎
tube duct 管沟
tube efficiency 电子管效率
tube electric(al) furnace 管形电炉

tube electrometer 电子管静电计
tube end belling press 管端扩口压力机
tube end cutting machine 管端切割机
tube end expanding machine 管端扩粗机
tube end sizing press 管端定径压力机
tube end swaging press 管端锤头压力机
tube expander 扩管器;胀管器;辘管器
tube extractor 管道萃取器;捞管器
tube extruder 制管机
tube face 管泡表面;灯泡座;灯管座
tube fitter 管道装配工
tube fittings 管接头;管子配件;管道配件
tube flaring kit 管子扩口器
tube flaring tool 管子扩口工具
tube-float 管状浮标
tube floor (slab) 圆孔空心楼板
tube flow 管流
tube foot 管足
tube-formed bottle 管制瓶
tube-forming machine 管材成型机
tube-forming mill 管材成型机
tube for precipitation 沉淀管
tube for upstream migration 回游管道(水坝上溯鱼类)
tube frame 管制框架;管框架
tube freight traffic 管道输送
tube freight traffic system 管道输送系统
tube-friction 管道摩阻(力)
tube furnace 管形炉
tube furnace pyrolyzer 管炉热解器
tube fuse 熔丝管;管状保险丝
tube generator 电子管振荡器
tube germination 管萌发
tube hanger 管子吊架
tube head 管头
tube holder 灯座;管座;管座
tube holes on tube sheet 管板管口
tube ice 管状冰;空心筒形冰
tube impedance 电子管内阻
tube inlet orifice 管口节流圈
tube-in-sheet evapo(u)rator 板管式蒸发器
tube-in-sleeve alidade 转镜照准仪
tube-in-tube 筒中筒
tube-in-tube condenser 套管式冷凝器;套管冷凝器
tube-in-tube heat exchanger 套管式热交换器
tube-in-tube structural system 筒中筒结构体系
tube-in-tube structure 筒中筒结构;套管式结构
tube-in-tube system 筒中筒结构体系
tube joint 管接头
tube junction 管接头
tube lamp 管形灯泡
tube lane 管道通路
tube leak 管子泄漏
tube leakage 管子泄漏量
tubeless amplifier 无电子管放大器
tubeless tyre 无内胎轮胎;无管轮胎
tube-life charts 管使用时间图表
tubelining 彩色泥浆勾轮廓(面砖装饰方法);彩色泥浆勾画轮廓
tube machine 制管机
tube mast 管桅(杆)
tube mechanism 料筒机构
tube method 恢复水位法
tube mill 转筒研磨机;制管机;制管厂;轧管机;管磨机;管磨;管材轧机;钢球磨煤机;筒磨机
tube mill head 管磨机磨头
tube milling 管磨粉碎机
tube-mounted ceiling 管架顶棚;管架天花板
tube nest 管束;列管;集成管
tube newel 管子旋梯中柱;管子楼梯端柱
tube nipple 管接头
tube noise 管内噪声;电子管噪声
tube nucleus 管核
tube of flow 细流;流水管;流管
tube of force 力线管;力束;力管;电力线管
tube of glass fibre reinforced plastic 玻璃钢管
tube of magnetic induction 磁感应管
tube orifice 管孔
tube oscillator 电子管振荡器
tube parameter 电子管参数
tube parts 配管部件
tube pass 管程
tube pass partition 管程分程隔板
tube photometer 管光度计

tube pitch 管心距;管间距
tube placing 埋管;管段安放
tube planting 筒状容器栽植
tube plate 管板
tube (plate) settler 异向斜管沉淀池【给】;异向斜板沉淀池【给】
tube plug 管塞(子)
tube pointer 管夹头锻制机
tube press 管压机;内胎硫化热压器
tube press dies 管式冲压压力机
tube pressure-regulating valve 管形调压阀
tube pulling machine 拨管器
tube purlin(e) 管子檩条;管子桁条
tube push bench 顶管机
tube radiator 管式散热器
tube railing 管子栏杆
tube railing fittings 管子栏杆零件;管子栏杆附件
tube railroad 地下铁道
tube railway 地下铁道;地下管道铁路;地下铁路
tuber calcanei 跟结节
tubercle 锈瘤;小块茎;突起
tubercle formation 形成结节
tubercle of sellae 鞍结节
tubercular corrosion 点腐蚀;斑点腐蚀;斑点玻璃
tuberculated pipe 结瘤管;翅片散热管
tuberculation 管瘤;结管瘤;结节;结核
tubercule 管子密封环;管瘤;铁锈结节
tuberculin precipitation 沉淀结核菌素
tuberculoidin 沉淀结核菌素
tuberculopneumpconiosis 尘肺结核
tuberculosis hospital 结核病院
tuberculosis sanatorium 结核病疗养院
tube reamer 扩管器
tube resistance furnace 管形电阻炉
tube ring 电子管振动噪声
tube-rolling mill 制管厂;管材轧机
tuberose 块茎状的
tube roughness 隧洞粗糙度
tuberous 块茎状的
tuber zedoariae 莪苓
tube sample 管装土样
tube sample boring 取样钻孔(土的);取土样钻孔;取土样孔;薄管取样钻探
tube sampler 管状取样器
tubes and pipes 管材
tube scale 管垢
tube scaler 刷管器
tube scope 管内检查镜
tube scraper 管子刮刀;刮管器;刮管刀
tube screen 管状筛
tube screw machine 管子螺丝车床
tube sealing 封管
tube seam 管接缝;管缝
tube searcher 管道检查器
tube seat 管座
tube section 管段;管子断面
tube settler 斜管沉淀池
tube sewer 管沟
tube shaft 尾轴
tube shears 管料切断机
tube sheet 管形铁板;管板
tubesheet holes 管板孔
tubesheet lining 管板衬里
tubesheet manufacture 管板制造
tube sheet support 管板支承板
tube sinking 管段沉放;缩口
tube-sinking cast-in-situ pile 沉管灌注桩
tube sinking method 沉管法
tube skeleton 管子支架;管子骨架
tube skin temperature 炉管表面温度
tube socket 管子钟口;管子承口;管承窝;管座
tube soil sampler 管状取土器
tube solenoid 圆筒螺旋管
tube sorting machine 分管机
tube space 管距;管际空间
tube spacing 管距;管间距
tube spanner 管子扳手;管钳
tube spinning 管表变薄旋压
tube-spring manometer 管状弹簧压力计
tube static mixer 管式静态混合器
tube station 地下铁道车站;地铁车站
tube stay 管牵条
tube still 管状加热炉

tube-still heater 管式蒸馏釜加热器
tube stop(per) 管塞(子)
tube straightener 管材矫直机
tube straightening machine 管材矫直机
tube straightening press 管材矫直机
tube stretching machine 管材拉伸矫直机
tube stretch reducing mill 管材张力减轻机
tube structure 管筒(式)结构;筒形结构;筒体结构
tube support (钻塔内的)钻杆垫架;管支柱;定位管支撑
tube support plate 管支承板
tube test 试管试验
tube tester 电子管试验器;电子管测试仪
tube-to-tube sheet 管子对管板
tube-to-tube sheet joint 管子对管板的连接
tube-to-tubesheet joint load 拉脱力
tube-train 地下铁道列车
tube trench 管形地沟;管沟
tube tunnel 管状隧道;管沟
tube turbining 涡轮清管
tube turn U 形管
tube-type 电子管型号;管式
tube-type and wire electrofilter 管式电除尘器
tube-type axial fan 管式轴流风机
tube-type bag 管式袋
tube-type beam splitter 管式分束镜
tube-type boiler 管式锅炉
tube-type brake 管式制动器
tube-type buoy 管式浮标
tube-type camera 管式照相机
tube-type centrifuge 管式离心机
tube-type clarification process 管式澄清法
tube-type classifier 管式分级机
tube-type cooler 管式冷却器
tube-type cracking furnace 管式裂解炉
tube-type delivery gate 管式输水斗门
tube-type diffuser 管式扩散器
tube-type drier 管式干燥器;管式干燥机
tube-type drive twister 管式加捻机
tube-type dryer 管式干燥器;管式干燥机
tube-type electric(al) furnace 管式电炉
tube-type electrometer amplifier 管式静电计放大器
tube-type explosive press 管式爆炸压力机
tube-type extractor 管式提取器
tube-type feeder 管式加料机
tube-type filter 管式过滤器
tube-type float 管式浮标
tube-type format 管式结构
tube-type ga(u)ge 管式气压计
tube-type heater 管式加热器
tube-type heat exchanger 列管式换热器
tube-type heating furnace 管式加热炉
tube-type jet dy(e)ing machine 管式喷射染色机
tube-type line strander 管式绞线机
tube-type micrometer 管式测微仪
tube-type mill head 管式磨头
tube-type mill(ing) 管式磨
tube-type mill liner 管式磨衬里
tube-type mill shell 管式磨外壳
tube-type oscillator 管式振荡器
tube-type pneumatic dryer 直管气流式干燥器
tube-type precipitator 管式除尘器
tube-type radiator 管式散热器
tube-type resistance furnace 管式电阻炉
tube-type sample 管式取样品
tube-type sampler 管式取样器;管式采样器
tube-type scaffold(ing) 管式脚手架
tube-type settler 管式沉降器;管式沉淀器
tube-type socket 管式承窝
tube-type still 管式蒸馏釜
tube-type telescope 管式望远镜
tube-type turbine 管式水轮机
tube-type (type) furnace 管式炉
tube-type type generator 管式发生器
tube-type valve 管式阀
tube-type visco(si)meter 管式黏度计
tube union 联管节;管接头
tube upsetting machine 管端镦厚机
tube upsetting press 管端镦厚机
tube valve 管阀;管形阀;管式阀;管式调压阀
tube vice 管子虎钳;管虎钳
tube voltage drop 管压降;电子管电压降
tube voltmeter 电子管伏特计;电子管电压表

tube wall 管壁
tube wall thickness 管壁厚度
tube wall thickness ga(u)ge 管壁厚度计
tube wastage 管子漏失量
tube wear 管子磨损
tube-welding machine 焊管机
tube-welding mill 焊管机
tube well 管井;机井
tube well irrigation 管井灌溉
tube well pump 管井泵;深井泵
tube winding 缠绕管
tube within a tube 双壁管
tub file 敞开文件
tub-gardening 桶栽园艺
tubing 装管;管道;管材;软绝缘管;铺设管道;配管;纺织玻璃管
tubing anchor 油管锚
tubing and casing rollers 管材和套管滚压器
tubing and sucker rod socket 油管和抽油杆打捞筒
tubing arrangement 管线布置
tubing back-off 倒开油管
tubing bender 弯管机
tubing bleeder 油管泄放阀
tubing block 油管滑车
tubing board 塔上起下钻工作平台
tubing brick 管砖;空心砖;筒砖
tubing calipers 管子卡尺
tubing catch 打捞筒【岩】
tubing catcher 油管夹持器;捞管器
tubing clamps 油管卡;钻管托夹
tubing clip 管夹
tubing connecting links 油管接环
tubing convey perforating gun 接油管射孔枪
tubing coupling 油管接箍;管件连接接头
tubing coupling tongs 管件连接钳
tubing depth 油管下入深度;管通深度;套管深度
tubing dog 油管打捞矛;套管打捞矛
tubing drill 管钻;捞砂筒;泥泵
tubing elevator 管式提升机;油管吊卡;套管式提升机
tubing feeder 管式给料机;管道给料机
tubing for casing completion 油管完成
tubing frame 管框架;管构架
tubing hanger 油管夹持器
tubing head 管头;油管头
tubing-head adapter flange 油管头法兰
tubing head pressure 管头压力
tubing heat exchanger 管式热交换器
tubing hook 油管大钩
tubing joint 油管接箍
tubing lamp 充气灯管;管灯
tubingless completion 小井眼完成;无油管完井
tubing line 油管吊绳;提管绳
tubing liquid level detecting 测油管液面
tubing loop 管圈
tubing machine 装管机械
tubing method of cementing 导管注水泥法;导管灌浆法
tubing mill 管磨机
tubing packer 油管封隔器;套管盘根盒
tubing perforator 管式凿岩机;管式射孔器;套管冲孔器
tubing plan 管道平面图
tubing plug 油管堵塞器
tubing program(me) 埋管程序
tubing pulling unit 管子起拔器
tubing pump 管式抽油泵;管式泵
tubing record 下油管记录
tubing ring with wedges 油管卡套
tubing rotator 油管转动器
tubing socket 油管打捞筒;套管打捞筒;打捞筒
tubing spear 油管打捞矛;套管打捞矛
tubing spider 油管卡盘
tubing swivel 油管旋吊头
tubing thread 油管丝扣
tubing thread form 油管扣型
tubing tongs 油管吊钳
tubing with external upset ends 外加厚油管
tubing with internal upset ends 内加厚油管
tubing with plain ends 内平油管
tub lead trap 浴盆下水管存水弯
tubler key 套管扳手
tub mixer 槽式搅拌机
tub orchard 桶栽果园

tuboscope 钻杆探伤器
tub plant 桶栽植物
tub scouring machine 槽式洗涤机
tub-sizing 槽法施胶
tub sizing machine 槽式施胶机
tub tap 澡盆龙头
tubular 管材;管状(的)
tubular air cooler 管式空气冷却器
tubular air heater 管式空气加热器;管式热风器
tubular air preheater 管式空气预热器
tubular annulus 环状间隙(双层管子之间或钻杆与孔壁之间的)
tubular arch bridge 筒形拱桥
tubular arched girder 管制拱形大梁
tubular arrester 管型避雷器
tubular atomizer 管式喷雾器;同心缝隙喷雾器
tubular axis 管轴;管式轴
tubular axle 空心轴
tubular beam 管状梁
tubular bearing structure 钢管承重结构
tubular biofilm reactor 管状生物膜反应器
tubular biological filter 管式生物滤池
tubular bit 管状钻头
tubular boiler 管式锅炉;火管锅炉
tubular boom 管状吊杆
tubular borer 管钻头
tubular-bowl centrifuge 管式转筒离心机;管式离心机
tubular-bowl clarifier 管式澄清器
tubular-bowl ultra-centrifuge 管式超速离心机
tubular box spanner 管套筒扳手
tubular brace 管支撑
tubular brick 管砖;空心砖;筒砖
tubular bridge 管桁桥;管造桁架桥;管子桥;管状桥;管格桥
tubular budding 管状芽接
tubular building 管状建筑物
tubular butt seal 管状的平焊接
tubular cage extension 车挡板的管状加高部分
tubular (cage) recuperator 鸟笼式换热器
tubular calandria 管式加热器
tubular capacitor 管状电容器;管形电容器
tubular cascade evapo(u)rator 管式级联蒸发器
tubular case 圆柱状盒
tubular casing 管形套
tubular ceramic capacitor 管式陶瓷电容器
tubular chassis 管制汽车底盘
tubular clarifier 管式净化器
tubular coil 管式线圈
tubular collection electrode 管式除尘极
tubular collector 管式收集器
tubular collimator 视准管
tubular colonnade foundation 管桩基础
tubular column 管柱
tubular column drilling method 管柱钻孔法
tubular column foundation 管柱基础
tubular compass 管式罗针;管式罗盘;管式罗经
tubular concrete 管内浇灌混凝土
tubular condenser 管式冷凝器
tubular conduits on the surface 明设电线管
tubular construction 箱形结构;管状建筑;管式结构;钢管结构;筒形结构
tubular container 管状容器
tubular continuous digester 管式连续浸煮器
tubular contraction of visual field 管状视野缩小
tubular cooler 管状散热器;管式冷却器;单筒冷却机
tubular corolla 管状花冠
tubular cross bar 十字形铁圆管
tubular cross member 管制横向构件
tubular cyclone collector 管式旋风除尘器
tubular cyclones 管式旋风分离器;筒状旋风分离器
tubular cyclone separator 管状旋风分离器
tubular data 表列资料
tubular deduster 管式收尘器
tubular derrick 管状桅杆;管式塔;钢管钻塔
tubular diagonal (rod) 管子斜杆件
tubular diesel pile hammer 筒式柴油机打桩锤
tubular discharge lamp 管状放电灯;条带形灯;管状白炽灯
tubular dowel 圆锚筋
tubular drawing pen 管式绘图笔
tubular drawing point pen 管形储[贮]水绘图笔
tubular dryer 管式干燥器;管式干燥机
tubular drill 管钻

tubular dust collector 管式吸尘器;管式收尘器;管式集尘器;袋式集尘器
tubular electric(al) furnace 管状电炉
tubular electric(al) heater 管式电热器
tubular electrode 管状焊条
tubular electrode precipitator 管式电除尘器
tubular electrostatic precipitator 管式静电除尘器
tubular elevator 管式升运器
tubular equipment 管状设备
tubular evaparator 管式蒸发器
tubular exchanger 管状换热器;管式热交换器
tubular extraction unit 管状分离机组;管状抽吸机组
tubular extractor 管状分离器;管状抽提器
tubular extract ventilation unit 管状抽气式通风设备;管状抽气式通风机组
tubular fan 管道形通风机
tubular-film 管形薄膜
tubular filter 管式过滤器
tubular flexible conveyer 管式挠性输送机
tubular float level indicator 管状浮标液面指示器
tubular flow 管式流
tubular flower 筒状花
tubular flow reactor 管式流反应器;管式流反应堆;管流式反应器
tubular-flow reactor 管道式流动反应器
tubular fluorescent lamp 荧光管灯
tubular foraminous wall reactor 管状多孔壁氧化反应器
tubular foundation 管状基础
tubular frame 管道支架;圆管框架;圆管结构;管式框架;管架;钢管框架
tubular frame pile driving plant 管架打桩机
tubular frame saw 管架锯
tubular framework 管构桁架
tubular furnace 管式炉
tubular furniture 钢管家具
tubular fuse 管式保险丝
tubular gasket 管型垫带
tubular gas lens 管状气体透镜
tubular ga(u)ge 管材量规
tubular girder 管腹工字梁;管式大梁;管式大梁;管腹工字梁;钢管横梁;空心板梁;筒形梁
tubular goods 管类材料;管材
tubular graphitic heat-exchanger 浮头列管式石墨换热器
tubular grip handle 管状捏手把
tubular guide 管形导轨
tubular hacksaw frame 管式钢锯架
tubular hacksaw frame chrome plated 管式镀铬钢锯架
tubular handle 管子手柄
tubular handrail 管状扶手
tubular heater 管式加热器
tubular heat exchanger 管式换热器;列管式换热器
tubular heating coil 加热盘香管;蛇管式加热器
tubular heat intercharger 管式内换热器
tubular hexagon box wrench 六角形套筒扳手
tubular housing 管状外壳
tubularia 筒螅;水螅
Tubularia crocea 水母类
tubular incandescent lamp 管状白炽灯
tubular incandescent light 管形白炽灯
tubular inside micrometer 管式内径千分尺
tubular intermediate column 管状中(间支)柱
tubular key 管形螺丝扳手;套管扳手
tubular lamp 管式灯
tubular lamp bulb 管形灯泡
tubular lamp post 管子灯柱
tubular lattice boom 管式格构吊杆
tubular lattice mast 管制格构桅
tubular leg 管子支柱;管子腿
tubular level 管状水准器;管形水准器;水准管
tubular linear induction motor 管形直线感应电动机
tubular line lamp 管式白炽灯
tubular lock 管形锁
tubular mast 套管天线杆
tubular membrane 管状膜;管式膜
tubular metal construction 管状金属结构
tubular metal space frame 金属管空间框架
tubular nose 管形鼻
tubular orebody 管状矿体
tubular oven 管式加热炉
tubular ozone generator 管式臭氧发生器

tubular part 短管
tubular particle (chip) board 管孔木屑板; 管孔刨花板; 管孔碎料板
tubular pile 管桩
tubular pile quay 管桩码头
tubular pile wharf 管桩码头
tubular piling foundation 管柱桩基础; 管柱基础
tubular plate staybolt 炉管炉板设备螺撑
tubular pole 钢管电杆; 空心电车杆; 管式电线杆
tubular post 管状柱
tubular preheater 管式预热器
tubular product 管状制品; 管材; 管状产品
tubular prop 管柱
tubular pump 管式泵
tubular radiator 管状散热器; 管式散热器
tubular rail 管式栏杆
tubular reactor 管状反应器; 管式反应器; 管式反应堆
tubular reverse osmosis 管状反渗透
tubular rheostat 管状电阻
tubular rivet 管形铆钉; 空心铆钉; 空心管形铆钉
tubular rod 管式抽油杆
tubular rotary cooler 筒式回转冷却器
tubular saw 开孔锯; 内径锯; 孔壁修整工具; 管钻
tubular scaffolder 管子脚手架安装工
tubular scaffold(ing) 钢管脚手架; 管子脚手架; 金属管脚手架; 管子脚手架; 管式脚手架
tubular scaffolding in alumin(i)um alloy 铝合金管子脚手架
tubular screw conveyer 管式螺旋运输机
tubular screw pump 管式螺旋泵
tubular section 管状钢材; 管状截面; 管状钢材; 管状构件
tubular shape 管式
tubular-shaped house 管状房屋
tubular skeleton 管件构架; 管子骨架
tubular sole 底部管梁
tubular space structure 管式空间结构
tubular specimens 管子标本
tubular splice 管子连接件; 套管连接
tubular spring 管状泉
tubular stage 管制脚手架
tubular steel 钢管
tubular steel bar 钢管棒
tubular steel barrow 钢管手推车
tubular steel bridge 钢管桥
tubular steel column 钢管柱
tubular (steel) derrick 钢管井架; 钢管钻塔
tubular steel folding chair 钢管折椅
tubular steel frame 钢管框架
tubular steel grid dome 钢管网穹窿
tubular steel handle 钢管手柄
tubular steel ladder 钢管梯子
tubular steel lighting column 钢管灯柱
tubular steel lighting column with bracket 带托架的钢管灯柱
tubular steel mast 钢管桅(杆)
tubular steel pier 管状钢闸墩; 管状钢桥墩; 管状钢墙墩; 管状钢墩柱; 钢管墩
tubular steel prop 钢管支柱
tubular steel scaffolding 钢管脚手架
tubular steel string 钢管楼梯斜梁
tubular steel support 钢管支承
tubular steel system 钢管体系; 钢管系统
tubular steel table 钢管桌
tubular steel tower 钢管塔; 钢管塔架
tubular steel truss 钢管桁架
tubular steel window 钢管窗
tubular strand 空心股线
tubular stranding machine 管式捻股机
tubular structure 管状结构; 管状建筑物; 管状构造; 管筒(式)结构; 钢管结构
tubular structure system 圆筒形结构系统; 管制的结构系统
tubular strut 管形支柱
tubular support 管支承; 管架
tubular surface 管状曲面
tubular surface condenser 管式冷凝器
tubular switch 管形开关
tubular tank 管式罐
tubular texture 管状结构; 片状结构
tubular tie 管拉杆
tubular towing pole 管子拉杆
tubular traverse 管形排线器
tubular turbine 贯流式水轮机
tubular type closer 管式合绳机
tubular-type column 管形柱
tubular type electrolytic condenser 管状电解电容器
tubular type precipitator 管式除尘器
tubular type radiator 管式散热器
tubular type strander 管式捻股机
tubular tyre 管式轮胎
tubular vapo(u)r cooler 管状蒸气冷却器
tubular ventilator 管状通风器
tubular visual field 管状视野
tubular washer 管状洗涤器
tubular water turbine with extended axis 轴伸贯流式水轮机
tubular welded frame scaffold 焊接管构脚手架; 管状焊接框架脚手架
tubular well 管井; 机井
tubular wire 管状焊丝
tubular wire-stranding machine 管式绞线机
tubular wire winder 管形绕线机
tubular xenon lamp 管形氙灯
tubular yoke 管式牛轭
tubulate 有管的; 管状的; 焊管脚; 装管
tubule 小管
tuck 褶缝; 船舷至船尾过渡区; 船尾突出部
tuck and pat pointing 嵌凸缝; 凸嵌勾缝; 凸扁嵌缝
tucker 装填; 折页刀; 填塞
tuck-in 嵌入卷边; 塞入部分; 卷入部分
tucking blade 折页刀
tucking board 电缆沟壁板; 撑板; 塞入式挡土板
tucking frame 插板撑架
tuck-in selvage 折入边
tuck in the hook 针钩集圈
tuck joint 凸嵌(灰)缝
tuck joint pointing 凸嵌勾缝
tuck net boat 网兜渔船
tuck plate 尾端外板; 呆木
tuck-pointed joint 嵌凸缝
tuck-pointer trowel 勾灰刀
tuck pointing 砖砌体嵌凸缝; 勾凸缝; 凸嵌勾缝; 凸嵌
tuck-pointing joint 凸嵌(灰)缝
tuck point joint filler tool 嵌凸缝工具
tuck stone 挂钩砖; 玻璃熔窑护砖
tuck timber 尾立材
tuck-under 自动俯冲趋势(在临界马赫数附近时的)
tuck-under parking 楼下停车
tuck-up 自动上升趋势
tuck wall 间隙砖层
tuck weave 皱纹组织
Tudor accumulator 都德式蓄电池
Tudor apple 都德式球形饰
Tudor arch 尖顶拱门; 都德式拱
Tudor architecture 都德式建筑
Tudor ceiling 都德式顶棚
Tudor flower 都德式三叶饰; 都德式花饰; 三叶花饰
Tudor leaf 都德式叶饰
Tudor ornament 都德式饰件
Tudor rose 都德式玫瑰饰
Tudor style 都德式风格【建】
Tudor window 都德式窗
tuekite 硫锑镍矿
tufa 硅华; 凝灰岩
tufa cement 凝灰岩水泥
Tufboard 都夫板(一种木纤维墙板)
tuff 凝灰岩; 凝灰石
tuffaceous 凝灰质的
tuffaceous agglomerate 凝灰质集块岩
tuffaceous bioclastic limestone 凝灰质生物碎屑灰岩
tuffaceous boulder 凝灰质巨砾岩
tuffaceous breccia 凝灰质角砾岩
tuffaceous cement 凝灰岩水泥
tuffaceous chemical sedimentary rock 凝灰质化学岩
tuffaceous coarse sandstone 凝灰质粗砂岩
tuffaceous cobblestone 凝灰质粗砾岩
tuffaceous conglomerate 凝灰质砾岩
tuffaceous dolomite 凝灰质白云岩
tuffaceous facies 凝灰岩相
tuffaceous fine sandstone 凝灰质细砂岩
tuffaceous granulestone 凝灰质细砾岩
tuffaceous gyprock 凝灰质石膏岩
tuffaceous halite rock 凝灰质盐岩
tuffaceous limestone 凝灰岩; 凝灰质石灰岩
tuffaceous marl 凝灰质泥灰岩
tuffaceous mudstone 凝灰质泥岩
tuffaceous shale 凝灰质页岩
tuffaceous siliceous rock 凝灰质硅质岩
tuffaceous siltstone 凝灰质粉砂岩
tuffaceous slate 凝灰质板岩
tuffaceous texture 灰结构
tuff agglomerate-lava 凝灰集块熔岩
tuff ball 泥球
tuff breccia 凝灰角砾岩; 火山灰角砾岩
tuff cement 凝灰岩水泥
tuff cone 凝灰锥; 凝灰岩锥
tuffcrete 凝灰岩(水泥)混凝土
tuffed rug 绒头地毯; 簇绒地毯
tuff for cement 水泥用凝灰岩
tuff for glass 玻璃用凝灰岩
tuffite 沉凝灰岩; 层凝灰岩
tuff lava 凝灰熔岩; 熔结凝灰岩
tuff loam 凝灰岩壤土
tuff of grass 草丛
tuff rock 凝灰岩
tuff sandstone 凝灰质砂岩
tuffstone 砂屑凝灰岩
tufftride 扩散渗氮
tufftride method 碳氮共渗法
tuff-turbidite 凝灰浊流岩
tuff volcanic rock 凝灰火山岩
tuff volcano 凝灰火山
tuft 毛撮; 毛簇
tufted 绒头; 簇绒; 簇生的
tufted carpet 立毛地毯; 簇绒地毯
tufted setae 毛簇
tufted shoot 丛生嫩枝条
tufting 簇绒(法); 装饰
tufting machine 簇绒机
tuft mockado 充丝绒
tuft observation 丝丛法观察
tuft pull strength 绒头抗拔强力
tug 拉拔; 牵引; 拖轮; 悬吊式挂车(运材)
tug and barge fleet 拖驳船队
tug and barge-train 拖驳船队
tug and lighter fleet 拖驳船队
tug arch 拖缆拱架; 拖缆承梁
tugarinovite 氧钼矿
tug assistance 拖轮协助
tug-barge combination 拖轮驳船队; 拖带船队; 拖驳船队; 顶推驳船队
tug-barge pushing fleet 顶推驳(船)
tug-barge towing fleet 拖带船队
tug boat 拖船; 拖轮; 工作拖轮
tugboat stern 拖轮式船尾
tugboat transport 拖驳运输
tugee 被拖船
tuggage 拖带费; 拖带
tugger 矿用轻便小绞车; 拖拉器
tugger hoist 轻便绞车; 拖拉式卷扬机; 绳索卷扬机
tugging 牵引感; 拖
tug hire 拖船租金
tug light 拖船灯
tugman 拖轮船员
tugmaster 拖轮船长; 拖船船长; 大拖车; 大牵引车
tug pulley (钢绳冲击钻的)拉绳轮; (钢绳冲击钻的)导向轮
tug rim 钢绳传动凹轮
tug-tender 拖曳交通; 拖轮兼驳渡船
tugtupite 硅铍铝钠石
tug wheel 拖尾轮
tuhualite 硅铁钠石
Tukey's gap test 图基间隙检验
Tukey's pocket test 图基快速检验
Tukey's quick test 图基快速检验
Tukey statistic 图基统计量
Tukon hardness 图康显微硬度
Tukon tester 图康显微硬度计; 图康试验器
tulameenite 铜铁铂矿
tulip poplar 北美鹅掌楸
tulip tree 北美鹅掌楸
tulip valve 漏斗形阀
tulipwood 郁金香木; 鹅掌楸木; 南美红木
tulle 薄纱(丝质或尼龙)
tumblast 转筒喷砂; 抛丸滚筒
tumble 鼓转; 倾覆; 倒扳
tumble axis 倾覆轴

tumble bay 净水池;回旋池;消能池;静水池
tumble blender 转鼓式掺和机;桶混机
tumble-bug 滚动刮土机
tumble-bug scraper 滚动刮土机
tumble card 分部倒用卡片;部分倒用卡
tumble curing 转动硫化
tumbled diamond 磨圆金刚石
tumbled-in course 对角砌层;斜嵌砖层;嵌砌砖层
tumble-down 就要倒坍的(建筑物);破烂的(如房子等);倒塌的;待修品
tumble dryer 转筒式烘干机,转筒式干燥器;转筒式干燥机,转笼烘燥机
tumble enamelling 转鼓涂漆
tumble gate 翻板闸门;倾倒闸门;卧倒闸门;卧倒门;倒拱式闸门;倒翻闸门
tumble home 立面大部分向内倾斜的房屋;舷缘内倾;内倾
tumble-home bow 撞角型船首
tumble-home bulwark 内倾舷墙
tumble house 内倾
tumble in 内倾
tumble polish 滚转抛光
tumble polishing 转鼓抛光;滚筒抛光(法)
tumbler 转滚;转臂;驱动轮;倾翻机构;平底无脚玻璃杯;倒扳开关;大玻璃杯;齿轮换向器;翻转器;扳倒开关
tumbler barge 自卸驳(船)
tumbler bearing 摇座;铰式支座;球支座
tumbler blowing machine 压饼—吹制成型机
tumbler bolt 伞头螺栓
tumbler cup blowing machine 压饼吹杯机
tumbler gear 三星齿轮;摆动换向齿轮
tumbler holder 转动轮子手柄;玻璃杯格架
tumble(r) lock 杠杆锁
tumbler seat 折叠式座(椅)
tumbler switch 双联开关;双连开关;凸件起动开关;顺逆换向开关;倒扳开关;翻转开关;拨动式开关
tumbler test 转鼓试验
tumbler-type washing machine 转鼓式洗涤机
tumbler washer 转笼式洗衣机
tumbleweed 风滚草
tumbling 滚筒清理法;滚动抛光;滚动;翻转作用
tumbling action 翻滚作用
tumbling barrel 滚转筒;滚磨筒;清理滚筒;串筒
tumbling barrel finishing 转鼓涂装
tumbling barrel method 转鼓除锈法;滚筒清洗法
tumbling barrel process 转鼓涂装;滚筒涂漆(法)
tumbling bay 回旋池;滚水坝消能池;跌落检查井;消能池;跌落窨井
tumbling bay junction 跌水窨井
tumbling body 滚动体
tumbling box 滚磨筒
tumbling-bug course 内倾砖层;斜嵌砖层;嵌砌砖层
tumbling crusher 转筒式粉碎机;转鼓式粉碎机
tumbling door 分泥门;分泥板【疏】;翻泥门
tumbling flow 倾泻水流
tumbling hammer type rapper 挠臂锤式打捶机
tumbling-in 内倾;向内倾斜;嵌进(木材)
tumbling mark 滚转痕
tumbling mass 转动体
tumbling mill 转鼓式研磨机;滚筒式磨机;滚磨机
tumbling mixer 滚筒拌和机;转筒混合器;转筒拌和机;转鼓(式)混合器;翻转式拌和机
tumbling regime 跌水情况
tumbling rod 摆动杆
tumbling shaft 凸轮轴
tumbling sheave 翻身滑车
tumbling-type 滚动式
tumbril 二轮弹药车
tumour hospital 肿瘤医院
tumulus 钟状火山;古坟;坟丘;熔岩肿瘤;[复]tumuli
tun 烟囱身;大桶;顿
tuna 金枪鱼
tuna and bonito clipper 金枪鱼及鲣船;鲔鱼及渔船
tunable 可调谐;可调
tunable antenna 可调谐天线
tunable birefringent filter 可调谐双折射滤光器
tunable cavity 可调谐腔;可调空腔谐振器
tunable coherent light source 可调谐相干光源
tunable-dye laser 调色激光器
tunable dye laser with an ultra-short cavity 超短腔可调谐染料激光器
tunable echo box 可调回波共振器
tunable filter 可调谐滤光器
tunable laser 可调谐激光器
tunable laser oscillator 可调谐激光振荡器
tunable metal film filter 可调金属膜滤光器
tunable optic(al) bandpass filter 可调光学带通滤波器
tunable preamplifier 可调谐前置放大器
tunable varactor diode 电调谐变容二极管
tuna boat 捕金枪鱼船
tuna clipper 金枪鱼船
tuna long liner 金枪鱼延绳钓船
tundish 漏斗;集灰盘;中间流槽;中间罐;中间包;浇口盘
tundish bottom 中间罐底
tundish brick 有孔巨型浇钢用砖;中间盛钢桶耐火砖
tundish casting 中间罐浇注
tundish lining brick 中间盛钢桶衬砖
tundra 苔原;冻原;冻土地带;冻土带
tundra climate 苔原气候;冻原气候
tundra community 冻原群落
tundra ecosystem 冻原生态系统
tundra landscape 苔原景观
tundra soil 冻原土;冰沼土
tundra soils in the arctic area 北极冰沼土
tundra vegetation 冻原植物;冰沼植物;冰原植被
tundrite 碳硅钛钠石
tundrite-(Nd) 碳硅钛钕钠石
tuned amplifier 调谐放大器
tuned-anode 板极调谐
tuned-anode coupling 板极调谐耦合
tuned-base oscillator 基极调谐振荡器
tuned cavity 空腔共振器
tuned circuit 调谐电路
tuned-collector oscillator 集电极调谐振荡器
tuned coupled circuit 调谐耦合电路
tuned damper 动力吸振器
tuned-emitter oscillator 发射极调谐振荡器
tuned filter 可调滤波器
tuned grid circuit 调栅电路
tuned liquid damper 调频液体阻尼器;调频晃动阻尼器
tuner mass damper 调频质量阻尼器
tuned plate amplifier 板极调谐放大器
tuned plate oscillator 板极调谐振荡器
tuned-radio-frequency receiver 高放式接收机
tuned-radio-frequency transformer 射频调谐变压器
tuned sloshing damper 调频液体阻尼器;调频晃动阻尼器
tuned transducer system 调谐转换系统
tune in 调入;调准
Tunelite 特纳水泥(一种快速凝固水泥)
tune-out 解谐
tuner 高频头;调谐器
tuner drum 频道选择器转鼓
tuner screw 调节螺钉
tune-up 调准;检修(发动机)
tune-up inspection 调整检查
tune-up procedure 调整程序
tune-up test 调整试验
tune up the mill 磨机调整
tungalloy 钨系硬质合金
tungar 钨氩管;充电整流管
tungar bulb 钨氩整流管
tungar charger 钨氩充电器
tungar rectifier 钨氩整流器
tungar tube 吞加管
tungate 桐油制成的催干剂;桐油衍生的金属皂
Tungcrete 桐漆(一种仿石油漆)
Tungelinvar 腾格林瓦合金
tung oil 桐油
tung oil stain 桐油污渍
tung oil tree 油桐;桐树;桐油树
tung oil varnish 桐油清漆
tung seed 桐油籽
tung stand oil 熟桐油
tungstate 钨酸盐
tungstate based water treatment agent 钨酸盐基水处理
tungstate ceramics 钨酸盐陶瓷
tungstate glass 钨酸盐玻璃
tungstate radicle 钨酸根
tungstates 钨酸盐类
tungsten 钨
tungsten alloy 钨合金
tungsten and molybdenum group elements 钨钼族元素
tungsten-antimony-gold-quartz vein deposit 钨锑金石英脉矿床
tungsten arc 钨电极弧
tungsten arc cut 钨极电弧切割
tungsten arc inert-gas cutting 惰性气体保护钨极电弧切割
tungsten arc inert-gas welding 惰性气体保护钨极电弧焊
tungsten arc lamp 钨丝弧光灯
tungsten arc melting 钨弧熔化
tungsten arc welding 钨极电弧焊;钨焊条电弧焊;钨弧焊
tungsten blue 钨青铜
tungsten boride 硼化钨
tungsten bronze 钨青铜
tungsten bronze structure 钨青铜结构
tungsten bronze type structure 钨青铜型结构
tungsten bulb 钨丝灯泡
tungsten bur 钨钢牙钻
tungsten carbide 碳化钨
tungsten-carbide alloy 碳化钨合金
tungsten-carbide ball 碳化钨球
tungsten-carbide bit 硬合金钻头;碳化钨(合金)钻头
tungsten-carbide blade 碳化钨刀片
tungsten-carbide boring bit 碳化钨硬质合金钻头
tungsten-carbide bur 碳化钨牙钻
tungsten-carbide button bit 碳化钨硬质合金珠齿钻头
tungsten-carbide ceramics 碳化钨陶瓷
tungsten-carbide chisel 碳化钨钻头;碳化钨硬质合金一字形钻头
tungsten-carbide coating 碳化钨敷层
tungsten-carbide composite 碳化钨复合材料
tungsten-carbide composition 碳化钨制品
tungsten-carbide coring drilling 硬质合金钻进
tungsten-carbide drilling 硬合金钻进
tungsten-carbide drill steel 碳化钨硬质合金钢钎
tungsten-carbide insert 镶齿;碳化钨硬质合金刀片
tungsten-carbide insert bit 镶齿钻头
tungsten-carbide powder 碳化钨粉末
tungsten-carbide roll 碳化钨硬质合金轧辊
tungsten-carbide-tipped bit 镶硬质合金钻头
tungsten-carbide tipped cutting teeth 碳化钨硬质合金截齿
tungsten-carbide-tipped drill rod 镶硬质合金钻杆
tungsten-carbide tipped inserts 碳化钨合金镶嵌块
tungsten-carbide-titanium carbide-cobalt alloy 钨钛硬质合金
tungsten-carbide tool 炭化钨
tungsten chrome steel 钨铬钢
tungsten-chromium steel 钨铬钢
tungsten-chromium tool steel 钨铬工具钢
tungsten clip 钨夹
tungsten coil furnace 钨丝炉
tungsten contacts 钨质电接触器材
tungsten copper 钨铜
tungsten-copper composite 钨铜复合物
tungsten-copper composition 钨铜制品
tungsten-copper contacts 钨铜电接触器材
tungsten-copper-nickel composition 钨铜镍制品
tungsten-copper-nickel contacts 钨铜镍电接触器材
tungsten-coring drilling 硬质合金钻进
tungsten diboride 二硼化钨
tungsten dioxide 二氧化钨
tungsten disc 钨盎
tungsten disulfide 二硫化钨
tungsten electrode 钨极;钨电极
tungsten emitter 钨发射体
tungsten filament 钨丝;钨灯丝
tungsten filament lamp 钨丝灯
tungsten-free cemented carbide 无钨硬质合金
tungsten fuse 钨熔丝
tungsten halogen lamp 卤钨灯;钨丝卤化灯;钨卤灯泡
tungsten heater 钨丝加热器
tungsten helix 钨螺旋线
tungsten hexachloride 六氯化钨
tungstenic 含钨
tungsten inclusion 夹钨
tungsten inert gas welding 钨极惰性气体保护电

弧焊
tungsten inert gas welding machine 钨极惰性气体保护弧焊机
tungsten insert bit 镶碳化钨钻头
tungsten-iodide lantern 碘钨灯
tungsten iodine lamp bulb 碘钨灯泡
tungsten iron 钨铁
tungstenite 硫钨矿;辉钨矿
tungsten lake 磷钨颜料
tungsten lamp 钨丝灯(泡);钨灯
tungsten light 钨丝灯
tungsten lighting 钨丝照明
tungsten minerals 钨矿物
tungsten-molybdenum steel 钨钼钢
tungsten-molybdenum thermocouple 钨-钼热电偶
tungsten nickel 钨镍合金
tungsten nitride 氮化钨
tungsten ore 钨矿(石)
tungsten oxide 氧化钨
tungsten particle 钨粒
tungsten pentexide 五氧化二钨
tungsten point 钨接点
tungsten powder 钨粉
tungsten quartz halogen lamp 石英卤素灯
tungsten rod 钨条
tungsten salt 钨酸
tungsten sensitivity 钨丝灵敏度
tungsten silicide 硅化钨
tungsten steel 钨钢
tungsten sulfide 硫化钨
tungsten target 钨靶
tungsten tetrachloride 四氯化钨
tungsten-titanium carbide 钨钛碳化物
tungsten trifluoride 三氟化钨
tungsten trioxide 三氧化钨
tungsten whisker 钨质须簧
tungsten wire furnace 钨丝炉
tungsten zirconium 锆化钨
tungstic acid 钨酸
tungstic ocher 钨赭石;钨华
tungstite 钨华
tungstomolybdic pigment 钨钼酸系颜料
tungstosilicic acid 钨硅酸
tung tree 油桐;桐油树
tungum 铜康(一种铜基合金)
tungusite 硅钙铁石
Tungus sea basin 通古斯海盆
tuning 调谐;电流谐振
tuning block 调谐组件
tuning category 类别调整
tuning circuit 调谐电路
tuning dial 调谐刻度盘
tuning fit 推入配合
tuning fork 音叉
tuning-fork frequency control 音叉频率控制
tuning fork grizzly 音叉式棒条筛
tuning fork set 连续音叉
tuning-fork transducer 音叉换能器
tuning-fork type 音叉式
tuning fork type ammeter 音叉式安培计
tuning for radio-beam 定向接收调谐
tuning hammer 定弦器
tuning lamp 调谐指示灯
tuning probe 可调探针
tuning rod 音杆
tuning screw 调整螺钉;调节螺钉
tuning-stub antenna 带调谐短截线天线
tuning unit 调谐单元
T-union 三通接头
Tunis crochet 突尼斯棱纹钩编花边
Tunis grass 突尼斯禾草
tunisite 碳钠钙铝石
tunking fit 推入配合
tunnel 隧道(又称隧洞);岁收;地道;虫道
tunnel abutment 隧道墙座
tunnel access 隧道引道
tunnel advance speed 隧道推进率
tunnel a hill 凿山隧道
tunnel alignment 隧道线形
tunnel alley 轴隧道
tunnel annealing furnace 隧道式退火炉
tunnel annunciating device 隧道通知设备
tunnel approach 隧道引道
tunnel aqueduct 隧道渠

tunnel arch 隧道拱(圈)
tunnel arm support and registration 隧道支撑臂与定位
tunnel atmosphere 隧道空气
tunnel axis 隧道纵轴线
tunnel battery locomotive 坑道用电机车
tunnel beach 槽状海滩
tunnel-bearing grease 轮机脂
tunnel (belt) conveyer 隧道式输送机
tunnel beneath ocean 海底隧道
tunnel blasting 隧洞爆破;隧道爆破
tunnel (bore) hole 隧道钻孔
tunnel borer 全断面掘进机;隧道钻进机;隧道挖凿机;隧道开凿机;隧道掘进机
tunnel boring jumbo 巨型隧道盾构;巨型隧道掘进机
tunnel boring machine 掘岩机;隧道钻(进)机;隧道挖凿机;隧道联合掘进机;隧道开凿机;隧道掘进机
tunnel boring machine for full section 全断面隧道掘进机
tunnel burner 火道式燃烧器
tunnel canal 隧道运河
tunnel cap 槽形泡罩
tunnel car 坑道运输车
tunnel carriage 隧道开掘车
tunnel center line 隧道中心线;隧道主轴线
tunnel chamber 洞室
tunnel chamber for drying 隧道式干燥室
tunnel clearance 隧道限界;隧道净空;隧道净断面;隧道建筑界限;隧道建筑界线
tunnel cllinker 隧道衬砌用缸砖
tunnel communication 隧道通信
tunnel concrete work 隧道混凝土工程
tunnel concreting train 隧道混凝土浇灌车
tunnel conduction 隧道导电
tunnel construction 隧道施工;隧道建筑;隧道工程
tunnel construction clearance 隧道建筑界
tunnel construction control 隧道施工控制
tunnel construction ga(u)ge 隧道建筑限界
tunnel construction with shield driven method 盾构法隧道施工
tunnel contour 隧道断面轮廓
tunnel control area 隧道控制区
tunnel control room 隧道控制室
tunnel conveyer 隧道输送机;地下运输机
tunnel cooler 隧道式冷却器
tunnel cost 隧道造价
tunnel cross profiling 隧道横断面测量
tunnel cross river 过江隧道
tunnel cross-section 隧道截面;隧道(横)断面
tunnel cross-section convergence 隧道断面收敛
tunnel cross-section survey 隧道横断面测量
tunnel cryotron 隧道冷子管
tunnel curing chamber 隧道式养护窑
tunnel curing kiln 隧道式混凝土蒸汽养护窑
tunnel current 隧道电流
tunnel cut-off 用隧道缩短线路长度;隧道取直;隧道裁弯取直;隧道裁弯
tunnel deterioration 隧道病害
tunnel development 隧峒开拓
tunnel diagonal deformation 隧道对角变形
tunnel diagram of prism 棱镜展开图
tunnel diameter 洞径
tunnel diesel locomotive 隧道内用柴油机车
tunnel diode 隧道二极管
tunnel disposal 坑道处置
tunnel diversion 隧洞导流;隧道导流
tunnel diverter 隧道式泥沙分流设施;分流隧道
tunnel drainage 隧道排水
tunnel drainage facility 隧道排水设备
tunnel dryer 隧道窑(式)干燥器;隧道(式)干燥机;隧道干燥窑
tunnel drier for bobbins 隧道式干燥机
tunnel drill 隧道用凿岩机;架式风钻;隧道钻(进)机;隧道凿岩机
tunnel drill hole 隧道钻孔
tunnel drilling car 隧道钻孔车;隧道打眼车
tunnel drivage 隧洞掘进
tunnel driving 隧洞掘进;隧道掘进
tunnel driving method 隧道开挖法;隧道掘进法
tunnel drying 隧道干燥
tunnel drying oven 隧道式干燥窑
tunnel effect 隧道效应

tunnel effect element 隧道效应元件
tunnel element 沉埋管段
tunnel emergency light 区间紧急照明灯
tunnel emission 隧道发射
tunnel emitter amplifier 隧道射极放大器
tunnel enamelling furnace 隧道式搪烧炉
tunnel entrance pretension 隧道洞口伪装
tunnel equipment 隧道坑内设备
tunnel erase 隧道擦除
tunnel erosion 漏斗状侵蚀;地下洞蚀
tunnel escape 轴隧应急出口;轴隧太平洞;地轴弄应急出口
tunnel excavation 隧道开挖
tunnel excavation method 隧道开挖方式
tunnel excavation speed 隧道推进率
tunnel excavator 隧道挖掘机
tunnel exploration engineering 坑探工程
tunnel face 隧道口;隧道开挖面;隧道工作面;掌子面;隧道面
tunnel fan 隧道通风机
tunnel-fiber 隧道纤维
tunnel finish 隧道饰层;隧道内装修
tunnel fire hazard 隧道火灾
tunnel flat 轴隧平台
tunnel floor 隧道底板
tunnel fork 隧道分岔段
tunnel for luggage and postbag 行包邮政地道
tunnel form escape 轴隧式船尾(浅水船)
tunnel form stern 轴隧式船尾(浅水船)
tunnel forms with accelerated curing 加热养护隧道模板
tunnel form(work) 隧道模(板)
tunnel for passenger 旅客地道
tunnel for pedestrians 地下步行道
tunnel for transporting luggage and postbag 行包邮政地道
tunnel for utility mains 总干管隧道;水电管线廊道
tunnel frame 轴隧肋骨
tunnel front 隧道口;隧道工作面
tunnel furnace 隧道窑;隧道式烘炉;隧道式退火炉
tunnel geometry 隧道几何形状
tunnel grating measurement and control system 风洞光栅测控系统
tunnel grease 尾轴轴承承滑脂
tunnel guard 隧道警卫人员
tunnel haulage 隧道碎渣搬运
tunnel haulage system 洞内运输卡车
tunnel heading 隧道平巷;隧道工作面;隧道导坑;掌子面
tunnel headroom 隧道净高
tunnel horizontal deformation 隧道水平变形
tunnel hygiene 坑道卫生
tunnel in clayey stratum 黏土层隧道
tunnel in deep length 深埋隧道
tunnel indicators and alarms 隧道标志与报警
tunnel in earth 土质隧道
tunnel in medium depth 中埋深隧道
tunnel intake 隧道式进水口
tunnel intersection 隧道交叉(口)
tunnel invert 隧洞底;隧道底拱;隧道仰拱;隧道底拱;隧道底板;隧道倒拱
tunnel invert elevation 隧道内底高程
Tunnelite 特纳莱牌水泥(一种快硬水泥)
tunnel jointer 支撑工
tunnel jumbo 隧道掘进钻车;隧道掘进用钻机
tunnel junction 隧道结
tunnel kiln 隧道窑
tunnel kiln car 转窑运料车;隧道窑运料车
tunnel kiln plant 隧道窑工厂
tunnel kiln sponge iron process 隧道窑海绵铁生产工艺
tunnel kiln with dryer 一次码烧隧道窑
tunnel lehr 连续式退火窑
tunnel laser 隧道激光导向
tunnel(l)ed aqueduct 隧道导水管;地下输水道
tunnel(l)ed oceanic outfall 隧道式污水排海口
tunnel(l)ed outfall 隧洞式排污口
tunnel(l)ed winding 埋入式绕组
tunnel lehr 连续式退火窑;隧道式退火炉
tunnel length 隧道长度;洞长
tunnel(l)er 挖隧道的人;隧道工人;掘进机;隧道掘进机
tunnelless underground laying 无沟地下铺设
tunnel lighting 隧道照明
tunnel lighting design 隧道照明设计

tunnel lighting metal pipe 区间金属线槽安装
tunnel liner 隧道衬砌
tunnel(l)ing 坑道作业;掘进;隧道现象;隧道施工;隧道开挖;隧道掘进;隧道贯穿;隧道工程;暗挖法
tunnel(l)ing boring machine 联合掘进机
tunnel(l)ing by pipe jacking 隧道顶管施工法
tunnel(l)ing construction method 隧道施工法
tunnel(l)ing crew 隧道掘进工作班
tunnel(l)ing current 隧道效应电流
tunnel(l)ing equipment 隧道掘进设备
tunnel(l)ing footage 开拓进尺(煤矿)
tunnel(l)ing form(work) 隧道模板
tunnel(l)ing gang 隧道掘进队;隧道工程队
tunnel(l)ing law 隧道开挖法则
tunnel(l)ing leaky mode 隧道效应漏模
tunnel(l)ing machine 隧道开挖机;隧道掘进机
tunnel(l)ing operation 隧道作业;隧道掘进作业;隧道掘进开挖;隧道掘进
tunnel(l)ing plant 隧道掘进装置;隧洞掘进装置
tunnel(l)ing quality index 隧道围岩质量指标
tunnel(l)ing rate 隧道掘进速率;隧洞掘进速率
tunnel(l)ing record 隧道掘进记录
tunnel(l)ing shield 隧盾;隧道防护支架;隧道盾构;盾构
tunnel(l)ing speed 隧道掘进速度
tunnel(l)ing through 隧道掘通
tunnel(l)ing transport 平巷掘进运输
tunnel(l)ing variation with pressure 隧道随压力的变化
tunnel lining 隧道被覆;隧道衬砌
tunnel lining geometry 隧道衬砌几何形状
tunnel lining machine 隧道衬砌机
tunnellite 四水硼锶石
tunnel loader 隧道装载机
tunnel location 隧道位置;隧道定位
tunnell-type sediment diverter 隧道式泥沙分流设施
tunnel luminaire 隧道灯具
tunnel machine 隧道掘进机
tunnel machinery 隧道掘进机械
tunnel maintenance and protection 隧道养护
tunnel maintenance sub-distribution box 区间隧道维修配电箱
tunnel mapping 隧道踏勘
tunnel media 隧道介质
tunnel monitoring measurement 隧道监控量测
tunnel (monoindication) obstruction signal 隧道遮断信号
tunnel muck 隧道挖出土石料;隧道挖出土方;隧道弃渣;洞渣装运机
tunnel muck hauling 出渣
tunnel name plate 隧道铭牌
tunnel net 长袋网
tunnelnet paint 渔网漆
tunnel of love 公园供游人穿行的隧道;公园供游人穿行的暗道
tunnel of tunnel 隧道修复
tunnel of utility mains 公用干管隧洞
tunnel opening 轴隧出入口;隧道口
tunnel operation 隧道建造;隧道掘进
tunnel operation and maintenance 隧道营运及维修
tunnel operator 隧道操作人员
tunnel oven 隧道窑
tunnel pack(ing) 隧道衬垫
tunnel patrol car 隧道巡逻车
tunnel personnel 隧道员工
tunnel piercing 隧道掘进
tunnel plant system 风洞装置系统
tunnel platform 轴隧平台
tunnel plug 隧道堵头;隧道堵头;隧道堵塞段
tunnel portal 隧道口部;隧道口;隧道洞门;(隧道的)洞门;峒门
tunnel post 隧道标
tunnel profile 隧道纵断面
tunnel recess 轴隧尾室;轴隧端室;鹅颈槽(集装箱)
tunnel reconnaissance 隧道勘测
tunnel reconstruction work 隧道改建工程
tunnel remodelling 隧道扩大改造
tunnel resistance 隧道阻力
tunnel resistor 隧道电阻器
tunnel ring 隧道管节;隧道圆环;隧道圈;隧道拱圈;隧道衬砌环

tunnel roof 巷道顶板;隧道顶板
tunnel roof settlement 隧道拱顶下沉
tunnels at shallow depth 浅埋隧道
tunnel (screw) propeller 轴隧推进器
tunnels driven from portal 隧道口部推出的盾构
tunnel sealing 隧道防水层
tunnel section 隧道横断面
tunnel service classification 隧道使用分类
tunnel set 隧道棚子
tunnel shaft 中间轴;隧道竖井;隧道井
tunnel shaft bearing 中间轴承
tunnel shaft bearing pillow block 支撑轴承
tunnel shape 隧道形状
tunnel shed 棚洞
tunnel shield 隧盾;隧道铠框;隧道防护板支架;隧道盾构
tunnel shield work 盾构工程
tunnel shovel 隧洞用机铲
tunnel side wall 隧道边墙
tunnel sign 隧道标
tunnel signal 隧道信号
tunnel site selection 隧道选址;隧道位置选定
tunnel skip 隧道料车
tunnel slot 隧道槽;闭合槽
tunnels on steep slopes 陡坡隧洞
tunnel spillway 溢洪隧洞;隧道排水槽
tunnel spiral 隧道螺旋线
tunnel spoil 隧道挖出的泥石;隧道弃土;隧道开挖的弃土
tunnel steel form(work) 隧道钢模板
tunnel stool 中间轴承台
tunnel stoving machine 隧道式熏白机
tunnel structure 隧道结构
tunnel structure ga(u)ge 隧道限界;隧道建筑界限
tunnels under city street 城市地下隧道
tunnels under flat ground 平坦地层隧道
tunnel supervisory system 隧道监控系统
tunnel support 坑道支护;隧道支护;隧道支撑;隧洞支撑
tunnel support with principal rafter 背顶式支护
tunnel surrounding rock 隧道围岩
tunnel survey(ing) 隧道测量;坑道测量
tunnel terminal 坑道头部
tunnel test 隧道试验;风洞试验
tunnel through mountains and cut across ridge 凿山劈岭
tunnel tramway 地下电车道
tunnel transition 隧道渐变段
tunnel-tray cap 槽形泡帽
tunnel trench 地下壕堑
tunnel triangular network 隧道三角网
tunnel triangulation network 隧道三角网
tunneltron 隧道管
tunnel trunk 隧道洞身;轴隧通道;地轴弄应急出口
tunnel (tube) 隧洞
tunnel tube section 隧道管段;管式隧道断面
tunnel-type discharge carrier 隧道式排出物运载工具;地下排水道
tunnel-type penstock 隧道式高压引水管
tunnel-type sediment diverter 泥沙分流隧道;泥沙分流隧道;隧道泥沙分流设施;隧洞式泥沙分流设施;分流隧道
tunnel-type tail race 隧道式泄水道
tunnel-type tray 槽形泡帽塔盘
tunnel-type tray cap 长方形泡罩
tunnel under no pressure 无压隧洞;无压隧道
tunnel under pressure 有压隧洞;有压隧道
tunnel under stockpiles 料堆隧道坑式输送
tunnel vault 隧道穹顶;隧道拱顶;筒形屋顶;圆筒拱顶;筒形穹顶;筒形拱顶
tunnel-vaulted 圆筒形拱顶的
tunnel-vault(ed) roof 筒形拱顶
tunnel ventilation 隧道通风
tunnel ventilation fan 隧道通风机
tunnel ventilation system 隧道通风系统
tunnel ventilator 隧道通风机
tunnel vision 管状视
tunnel voltage 隧道电压
tunnel wall displacement 隧道墙体位移
tunnel washing operation 隧道清洗操作
tunnel water 隧道涌水
tunnel well 轴隧泄水阱;轴隧坑;地轴弄污水井
tunnel width 洞宽
tunnel wind 隧道风

tunnel wind friction 隧洞风摩擦
tunnel windings 隧道绕组
tunnel without cover 明洞
tunnel works 隧道工程;洞身工程;隧道掘进工作
tunny boat 捕金枪鱼船
tunoscope 电眼
tun wagon 酒桶车
tup 撞锤;锤体;冲面
tupelo 美国紫树
tupeloe 山荣黄木(北美洲产)
tupelo gum 紫树
tuperssuatsiaite 钠铁坡缕石
tupid 铁锤
tup impact 落锤冲击试验
tuple 重数
tup pallet 锤下钢垫
tupper 砌砖辅助工;小工
tuque 绿至橄榄色硬木(委内瑞拉产)
turacin 羽紫素
Turam method 土拉姆法(定源式双线框交流电法)
turanite 羟钒铜矿
Turbadium 船用锰黄铜
turbid 浑浊的;混乱的
turbid circulating water 浊循环水
turbid current 浑浊水流
turbid density flow 浑水异重流
Turbide 特比德烧结耐热合金
turbid flow 乱流;浑浊水流
turbidimeter 浊度计;浊度表;浑浊度仪;浑浊度测定仪;比浊计
turbidimeter fineness 浑浊度仪测定的细度;浊度计细度
turbidimeter method 比浊计法
turbidimetric 浊度计的
turbidimetric analysis 浊度分析;浊度测定;比浊分析
turbidimetric apparatus 浊度测量仪
turbidimetric assay 浊度测定法
turbidimetric method 浊度测定法;比浊法
turbidimetric sulfate determination 硫酸盐比浊测定(法)
turbidimetric titration 比浊滴定(法)
turbidimetry 浊度测定法;浊度测定;测浊法;比浊法
turbid inflow 浑水入流;浑水流入口
turbidite 浊流岩;浊流层;浊积岩
turbidite basin 浊积盆地
turbidite facies 浊流相
turbidite fan 浊积扇;深海扇
turbidite mud 浊积泥
turbidite rhythm 浊积岩韵律
turbidite sedimentation 异重流沉积作用;浊积作用
turbidite sequence 浊流层序
turbidity 浊度;浑浊性;浊浑;浑浊度
turbidity coefficient 浊度系数;混浊系数;浑浊系数
turbidity current 悬浮质流;异重流;浊流;浊流;混浊流
turbidity current denudation 浊流剥蚀作用
turbidity current deposit 浊流沉积;浊流沉淀
turbidity current deposition 浊流沉积作用
turbidity current deposit of lake 湖泊浊流沉积
turbidity current deposit of marine 海洋浊流沉积
turbidity-detection with laser light 浑浊度—激光检测
turbidity dynamics 浊流动力学
turbidity effect 浊度影响
turbidity factor 浊度系数;混浊因子;混浊因素;浑浊度因素
turbidity fan 浊流扇;浊流扇
turbidity flow 浊流;浑浊流
turbidity fluctuation 浊度变化
turbidity index of drinking water quality 饮用水水质浊度指标
turbidity indicator 浑浊度指示剂;浑浊度指示计
turbidity inflow 污水流入量
turbidity level 浊度水平
turbidity limestone 浊积灰岩
turbidity maximum zone 最高混浊区
turbidity measurement 浊度测定;浑浊度测定法
turbidity mechanism 浊度机理
turbidity meter 浊度计;浊度表
turbidity method 浊度法;比浊浊法
turbidity monitor 浊度监测器;浊度监测计
turbidity of atmosphere 大气的浑浊度

turbidity of seawater 海水浊度
turbidity of water 水蚀度;水的浊度;水的浑浊度
turbidity plume 羽状混浊水团
turbidity plumes 浑浊水团
turbidity point 浊度点
turbidity reduction 降浊;浑浊度降低
turbidity removal 浊度去除;浑浊度去除;除浊
turbidity removing efficiency 除浊效率
turbidity scale 浊度标
turbidity screen 浑浊层
turbidity sediment 浊流沉积;浊流沉淀
turbidity size analysis 浊流粒径分析;浊流粒度分析
turbidity standard 浊度标准液
turbidity test 浊度试验
turbidity titration 浊度滴定(法)
turbidity transmitter 浊度变送器
turbidity unit 浊度单位
turbidity value 浊值;浊度值
turbidity water 浑浊水
turbid layer 浑浊层
turbid layer flow 浊层流
turbid layer transport 浊层搬运
turbid matter 浑浊物质
turbid medium 浑浊溶剂
turbidness 浊度;浑浊
turbidometric titration 浊度滴定(法)
turbidometry 浊度分析法
turbidostat 恒浊器
turbid river 浊水河流
turbid water 浊水;浑水
turbid water fixed-bed model 浑水定床模型
turbid-water spring 浊水泉
turbination 倒圆锥形
turbine 叶轮机;轮机;汽轮机;涡轮机;透平机;透平
turbine aerator 平板型曝气器;涡轮充气器
turbine agitator 涡轮式搅拌器
turbine air vent pipe 水轮机通气管;水轮机进气管
turbine axial thrust 水轮机轴向推力
turbine ball bearing cover 涡轮滚珠轴承盖
turbine beam 涡轮机支承梁
turbine bearing 涡轮机轴承
turbine blade 轮机叶;汽轮机叶片;涡轮(机)叶片;透平叶片;透平机桨叶;水轮机叶片
turbine blade cooling 涡轮机叶片冷却
turbine blower 涡轮(式)鼓风机
turbine brake 涡轮制动器
turbine bucket vibration 涡轮机叶片振动
turbine bypass 水轮机旁通管
turbine case 涡轮机壳;水轮机罩
turbine casing 涡轮外壳;涡轮(式)涡壳;涡轮机(机)壳
turbine cavitation 涡轮机气蚀
turbine-centrifugal foam breaker 涡轮式消沫离心机
turbine centrifugal pump 涡轮式离心泵
turbine chamber 水轮机室;水轮机(机)坑
turbine characteristic 水轮机特性
turbine compressor 涡轮式压缩机;涡轮式压气机
turbine constant 水轮机常数
turbine control panel 水轮机控制盘
turbine cover 水轮机盖
turbine cover cap 涡轮盖帽
turbine cover packing 涡轮罩填密
turbine cover screw cap 涡轮盖螺旋帽
turbine cylinder 涡轮壳;涡轮机壳体;透平壳
turbine diaphragm 涡轮机导流隔板;涡轮导流片
turbine direct drive 涡轮机直接传动
turbine disc 涡轮机转轮;涡轮(圆)盘
turbine discharge 水轮机过流量
turbine discharge temperature control 涡轮出口温度控制
turbine draft tube 水轮机尾水管
turbine drill 涡轮钻具
turbine drilling 涡轮钻进
turbine drill rod 涡轮钻机钻杆
turbine-drive 涡轮传动
turbine-driven 涡轮机驱动
turbine-driven fan 涡轮通风机
turbine-driven pump 叶轮机驱动泵;涡轮传动泵
turbine drum 涡轮机转鼓
turbine dynamometer 涡轮功率计
turbine efficiency 涡轮机效率;水轮机效率
turbine electric(al) drive 涡轮机电力传动
turbine electric(al) generator 涡轮发电机
turbine-electric(al) locomotive 涡轮电动机车

turbine engine 涡轮发动机;透平机
turbine entry duct 涡轮入口导管
turbine exhaust pipe 涡轮排气管
turbine-exit temperature 涡轮出口温度
turbine fan 涡轮(式)风扇机;透平式风机
turbine flo(a)tation 叶轮浮选;涡轮浮选
turbine floor (水电站的)水轮机层
turbine flowmeter 涡轮式流量计;涡轮流量计;涡流流量计;水轮式流量计
turbine flow per second 涡轮秒流量
turbine foundation 涡轮机底座
turbine gallery 水轮机廊道
turbine gas absorber 叶轮气体吸收器;涡轮气体吸收器
turbine gas purifier 涡轮气体净化器
turbine gate 水轮机闸门;水轮机阀门
turbine-geared propulsion unit 涡轮机减速推进装置
turbine generator 涡轮发电机;水轮发电机;汽轮发电机;透平发电机
turbine generator room 涡轮发电机舱
turbine generator set 涡轮发电机组
turbine go-devil 透平式清管器
turbine guard valve 水轮机保护阀
turbine guide-wheel 水轮机导轮
turbine hall 透平机房;汽轮机房;水轮机房
turbine house 涡轮机室
turbine housing 涡轮机壳体
turbine idling 水轮机空转
turbine inlet bend 水轮机送水弯管;水轮机进水弯管
turbine inlet temperature 涡轮进口温度
turbine inlet temperature control 涡轮进口温度控制
turbine inlet valve 水轮机进水阀
turbine input 水轮机输入功率
turbine inspection platform 水轮机检查平台
turbine installation 涡轮机装置
turbine jet 喷射式涡轮
turbine jet engine 涡轮喷气机
turbine like structure 涡轮状构造
turbine lip 水轮机盖
turbine locomotive 涡轮机车
turbine losses 涡轮机损耗
turbine machinery 涡轮机械
turbine main shaft 水轮机主轴;水轮机大轴
turbine meter 旋转式煤气表;涡轮式水表;涡轮式流量计;水轮式流量计
turbine mixer 叶轮式搅拌机;叶轮式混合器;叶轮搅拌器;立轴搅拌机;桨叶式搅拌机;涡轮式搅拌机;涡轮(式)混合器;涡轮式拌和机;涡轮搅拌机;涡轮混合机
turbine mixer with stationary diffusion ring 带固定扩散环的涡轮混合器
turbine motor 涡轮马达
turbine name-plate capacity 涡轮机标志能力
turbine noise 涡轮噪声
turbine nozzle 涡轮导向装置
turbine nozzle exit velocity 涡轮导向器气体流出速度
turbine nozzle vane 涡轮导向器叶片
turbine oil 汽轮机油;涡轮机油;透平油
turbine operating board 水轮机操作盘
turbine outlet 涡轮气体出口
turbine outlet pressure 涡轮出口气体压力
turbine output 涡轮机功率;水轮机输出功率;水轮机出力
turbine performance 涡轮性能
turbine performance chart 水轮机性能图表
turbine performance curve 涡轮机性能曲线;涡轮机运行曲线
turbine pig 透平式清管器
turbine piping 水轮机管路系统
turbine pit 透平室;水轮机(机)坑
turbine pit liner 水轮机坑衬砌
turbine power 透平动力
turbine-powered automobile 透平汽车;燃气轮机汽车
turbine-powered glider 涡轮动力滑翔机
turbine power plant 涡轮动力装置
turbine pressure ratio 涡轮压力比
turbine propeller 涡轮螺旋桨
turbine-propeller combination 涡轮螺旋桨组合
turbine-propeller engine 涡轮螺旋桨发动机
turbine proper 水轮机本体

turbine propulsion 涡轮推进
turbine protective device 涡轮机保安设备
turbine pump 再生泵;汽轮泵;涡轮(式)泵;透平泵;水轮泵
turbine radial(-flow) 径流式涡轮
turbine reaction 涡轮反作用度
turbine regulation 涡轮机调整;水轮机调节
turbine relief valve 涡轮机空放阀
turbine room 涡轮机室;涡轮机舱;水轮机室
turbine rotor 汽轮机转子;涡轮机转轮
turbine runner 水轮机转轮
turbine scraper 透平式清管器
turbine self-regulation factor 水轮机自调因数
turbine servomotor 水轮伺服电动机;涡轮机继动器
turbine set 涡轮机组
turbine setting 水轮机安装(高程)
turbine shaft 涡轮轴;水轮机轴
turbine shaft gland 水轮机轴密封套
turbine shape meter 涡轮式水表
turbine shape mixer 涡轮式搅拌器
turbine ship 汽轮机船;涡轮机船
turbine shroud 涡轮壳体
turbine shroud ring 涡轮壳环
turbine sifter 叶轮筛
turbine speed control 涡轮转速调节
turbine stage 涡轮机级
turbine station 涡轮机站
turbine stator 涡轮机固定部分
turbine stay ring 水轮机座环
turbine steamer 汽轮机轮船;汽轮机船
turbine stirrer 叶轮搅拌机;涡轮搅拌机
turbine surgical instruments 风动手术机械
turbine tanker 涡轮油船
turbine test code 水轮机试验规则
turbine transport 涡轮发动机运输机
turbine tubing 水轮机管路系统
turbine (tunnel) propeller 环围整流推进器
turbine-type 涡轮式
turbine-type agitator 涡轮式搅拌机;叶轮式搅拌器;汽轮式搅拌器;透平式搅拌器
turbine-type deep well pump 涡轮式深水泵
turbine-type impeller agitator 涡轮型叶轮搅拌器
turbine(type) intersection 涡轮式立体交叉
turbine-type meter 涡轮式水表
turbine-type mixer 涡轮式搅拌器;强制式搅拌机;强制式拌和机
turbine-type rotary intersection 涡轮式环形交叉
turbine-type watermeter 涡轮式水表
turbine valve 水轮机阀
turbine venting 水轮机通气
turbine vessel 涡轮船
turbine wheel 涡轮叶轮;涡轮机叶轮;涡轮
turbine wicket gate 水轮机导叶
turbining 涡轮清垢;水轮机工况运行
Turbiston 特比斯通高强度黄铜
turbo 透平
turbo-accelerator 涡轮加速器
turbo-air separator 涡轮式风力选粉机
turbo-alternator 涡轮交流发电机
turbo-alternator plant 汽轮交流发电厂
turbo-axial fan 涡轮轴流式扇风机
turbo-blower 离心鼓风机;涡轮式空压机;透平式风机;涡轮式通风机;涡轮式鼓风机;叶轮鼓风机;汽轮鼓风机;涡轮鼓风机
turbo-blower characteristics 涡轮增压器特性曲线
turboblown diesel engine 涡轮增压柴油机
turbo burner 涡轮燃烧炉;涡轮燃烧器
turbo-car 燃气轮机汽车
turbo-charge 涡轮增压;透平增压
turbo-charged diesel 涡轮增压柴油机
turbo-charged diesel engine 透平增压柴油机
turbo-charged piston engine 涡轮气活塞发动机
turbocharger 汽轮增压器;涡轮增压器;透平增压器
turbocharger compression ratio 涡轮增压器压缩比
turbocharger compressor 涡轮增压器压缩机
turbo-charger running defects 涡轮增压器运行故障
turbo-charger speed 涡轮增压器转速
turbocharging 涡轮机增压;涡轮增压
turbo-charging installation 涡轮增压装置
turbo-circulator 涡轮循环泵
turbo-classifier 循环空气分级机;透平式选粉机
turbo-compound aero-engine 涡轮活塞式混合发动机

turbo-compounded diesel 涡轮增压柴油机;复合式柴油机
turbocompound engine 涡轮复式发动机
turbo-compressor 叶轮压缩机;离心涡轮压缩机;汽轮压缩机;涡轮式压缩机;涡轮式压气机;涡轮离心压缩机
turbo-convertor 涡轮变流器
turbocopter 涡轮发动机直升机
turbo-coupling 涡轮联轴节
turbocyclone 涡轮旋风器
turbodiesel 涡轮柴油机
turbo-dissolver 涡轮型溶解机
turbo-distributor 涡轮分配器
turbo-dozer 涡轮推土机
turbodrier 涡轮式干燥器
turbo-drill 涡轮钻具;涡轮钻机
turbodrill housing 涡轮钻具外壳
turbo-drilling 涡轮钻掘;涡轮钻进
turbo-drilling rig 涡轮钻机
turbo-drive 涡轮机传动
turbo-driven 涡轮驱动的
turbo-driven compressor 涡轮压缩机
turbo-driven supercharger 涡轮驱动增压器
turbo-dynamo 涡轮发电机;直流水轮发电机;涡流发电机
turbo-electric(al) 涡轮电力的;涡轮电动的
turbo-electric(al) drive 涡轮机电力推进装置;涡轮电力驱动
turbo-electric(al) installation 涡轮机电力装置
turbo-electric(al) locomotive 涡轮机电气机车
turbo-electric(al) propellant 透平电动推进
turbo-electric(al) propulsion 涡轮发电动力装置
turbo-electric(al) ship 涡轮机电力推进船
turbo-engine 涡轮发动机
turboexciter 涡轮励磁机
turbo-exhauster 涡轮排气机
turbo-expander 涡轮膨胀机;透平式膨胀机
turbo-extractor pump 涡轮排水泵
turbofan 内外函喷气发动机;涡轮式风扇;涡轮式风扇;涡流鼓风机;涡轮式通风机
turbofan power plant 涡轮风扇动力装置
turbofan transport 涡轮风扇式运输机
turbo-fed 涡轮泵供油的
turbofed gas generator 涡轮气体发生器
turbo-feed pump 涡轮给水泵
Turbo-float tray 波纹浮动塔盘
turbofurnace 旋风炉膛
turbo-generator 汽轮发电机;涡轮发电机;涡流发电机
turbo-generator room 涡轮发电机舱
turbo-generator set 汽轮发电机组
turbo-generator unit 涡轮发电机组
turbogrid 叶轮式格子
turbogrid plate 叶轮式栅格板;叶轮式格子塔盘
turbogrid tower 穿流板塔
turbogrid tray 叶轮式栅格分馏塔盘;穿流栅板
turbohearth 涡轮敞炉
turbo-impeller 涡轮机叶轮
turbo-inverter 涡轮反用换流器
turbojet 喷气式发动机
turbojet engine 涡轮喷气发动机;涡轮喷气机
turbojet version 涡轮喷气型
turbolator 扰流子
turbo-liner 涡轮螺浆式客机
turbomachine 涡轮机
turbomachine design performance 涡轮机设计性能
turbomachinery 涡轮机
turbomachinery compressor 涡轮机压气机
turbomachine test stand 透平机械试验台
turbomat 汽轮机自动控制
turbo-mill 涡轮粉碎机;涡轮研磨机
turbo-mixer 立轴拌和机;汽轮式混合器;涡轮式混合器;涡轮式搅拌机;叶轮式搅拌机;叶轮式混合器;涡轮式搅拌机;涡轮混合器
turbomolecular pump 涡轮分子泵
turbonada 狂风;大雷雨
turbonator 汽轮发电机
turbonite 胶纸板
turbo-packer 涡轮式包装机
turbopause 紊流层顶;湍流层顶
turbo-power 透平动力;涡轮动力
turbo-power unit 涡轮机动力装置
turboprop 涡轮螺桨发动机

turboprop boat 涡轮螺旋桨船身式水上飞机
turboprop engine 涡轮螺旋桨发动机;涡轮推进机
turboprop performance 涡轮螺旋桨发动机性能
turboprop transport 涡轮螺旋桨运输机
turbopump 汽轮泵;涡轮泵;透平泵
turbopump assembly 涡轮泵组
turbopump injection 涡轮泵压式供油
turboramjet 涡轮冲压组合喷气发动机
turbo-ram jet engine 涡轮冲压式喷气发动机
turboram rocket 涡轮冲压火箭
turbo-reciprocating engines 涡轮往复蒸汽机联合装置
turbo-refrigeration machine 封闭式透平制冷机
turbo-refrigerator 涡轮式制冷机;透平制冷机
turbo-regulator 涡轮调节器
turbo-rotor 涡轮转子
turbo-scavenging blower 涡轮扫气泵
turbo-separator 旋风分离器;循环空气(式)选粉机;透平式选粉机;叶轮除渣器
turboset 汽轮机组;汽轮发电机组
turboshaft 涡轮轴;涡轮动力轴
turboshaft-compressor 涡轮轴压缩组
turboshaft engine 涡轮轴发动机
turbosphere 紊流层;湍流层
turbo-starter 涡轮起动机
turbo-steam separator 蒸汽旋风分离器
turbostratic structure 混层结构
turbo-supercharged engine 涡轮增压机;涡轮增压发动机
turbo-supercharger 涡轮增压器
turbotrain 燃气轮列车
turbo-tray dryer 涡轮盘式干燥机
turbotrol 汽轮机控制
turbo-type 涡轮式
turbo-type supercharger 涡轮式增压器
turbo-unit 汽轮发电机组;涡轮发电机组
turbo-visory equipment 涡轮机监控设备
turbo-washer 涡轮洗矿机
turbo-wheel 透平叶轮
turbo-winder 涡轮卷纬装置
turbstile 旋转栅门
turbulence 扰动;紊流;紊动性;湍动;骚乱;动荡
turbulence accident 涡流事故
turbulence broadening 湍流致宽
turbulence cell 湍流元
turbulence chamber 压气输送水泥混合室;湍流式燃烧室
turbulence (current) 湍流
turbulence decay 紊流衰减
turbulence diffusion 紊流扩散
turbulence diffusion equation 紊流扩散方程
turbulence element 湍流元
turbulence energy 涡流动能;湍流能;湍动能
turbulence factor 紊流因子
turbulence flow 无旋流
turbulence-free flow 无紊流;无湍流
turbulence heating 湍动加热
turbulence indicator 紊流指示器;紊流计
turbulence induced noise 紊流感应噪声
turbulence-induced vibration 紊流引起的振动
turbulence intensity 紊流强度;紊流度;紊动强度
turbulence inversion 湍流逆温;湍流逆流
turbulence level 湍流强度;湍流级;湍流度
turbulence loss 扰动损失
turbulence method 湍流法;湍动法
turbulence motion 紊流运动
turbulence muffle burner 涡流隔焰炉烧嘴
turbulence noise 湍流噪声
turbulence number 湍流度
turbulence promoter 湍流促进器
turbulence pump 涡流泵
turbulence rate 紊流率
turbulence resistance 紊流阻抗
turbulence scale 紊流度;湍流尺度;湍流标度
turbulence separation 紊流分离(现象);湍流分离(现象)
turbulence spectrum 紊流谱;湍谱;湍动谱
turbulence theory 紊流理论;湍流理论
turbulence transition 湍流转变
turbulent 紊流的;湍流的
turbulent air 湍动空气
turbulent airflow 湍动气流
turbulent amplifier 紊流放大器
turbulent annulus 湍流圈

turbulent arc 飘移电弧;漂移电弧
turbulent atmosphere 紊流大气
turbulent attack 紊流腐蚀
turbulent ball tower 湍球塔
turbulent ball tower dedusting 湍球除尘
turbulent ball tower scrubber 湍珠洗涤
turbulent boundary layer 紊流边界层;湍流边界层
turbulent boundary layer noise 湍流附面层噪声
turbulent buffeting 紊流抖振
turbulent buoyant jet 紊流浮射流
turbulent burner 湍流燃烧器;紊流煤粉燃烧器
turbulent burning velocity 紊流燃烧速度
turbulent coagulation 湍流聚结法
turbulent coefficient 紊流迁移系数;湍流系数
turbulent component 湍流分量
turbulent condition 紊流状态;紊流态;湍流状态
turbulent contact tower 湍球塔
turbulent convection 紊乱对流;湍流对流
turbulent current 乱流;紊流;湍流
turbulent damping 紊动性衰减
turbulent decay 湍流衰减
turbulent density current 紊动异重流
turbulent density fluctuation 紊流密度波动;紊流密度变动
turbulent diffusion 涡流扩散;紊流扩散系数;紊流扩散;湍流扩散模式;湍流扩散
turbulent diffusion coefficient 紊动扩散系数;湍流扩散系数
turbulent diffusion equation 紊流扩散方程
turbulent diffusion process 紊流扩散过程
turbulent diffusivity 紊流扩散系数;湍流扩散系数
turbulent drag 紊流阻力
turbulent dynamo 湍流发电机
turbulent eddy 湍流漩涡;湍流涡
turbulent energy 湍流能量;湍流能;湍动能
turbulent energy spectrum 湍流能谱
turbulent entrainment 紊流夹卷;湍流夹卷
turbulent exchange 紊流扩散;紊流交换;紊动转换;湍流交换
turbulent exchange coefficient 湍流交换系数
turbulent field 湍流场
turbulent-film evapo(u)rator 机械搅拌薄膜蒸发器;动膜式蒸发器
turbulent fin tube 紊流式翅片管
turbulent flame 紊流火焰;湍流火焰
turbulent flow 混流;紊流;紊动水流;湍流动;湍流
turbulent flow burner 紊流燃烧器;湍流燃烧器
turbulent-flow condition 紊流条件
turbulent flow contact absorber 湍流接触吸收器
turbulent-flow core 湍流核
turbulent flow preheater 湍流预热器
turbulent flow regime 紊流流态
turbulent flow resistance 紊流阻力
turbulent fluctuation 紊流脉动;紊动;湍流脉动;不规则脉动
turbulent fluid 湍动流体
turbulent fluidized bed 湍动流化床
turbulent flux 涡流通量;紊流扩散率;湍流通量
turbulent heat exchange 紊流热交换
turbulent heating 湍流加热
turbulent heat transfer 紊动传热
turbulent integral scale 湍流积分尺度
turbulent intensity 紊流强度;紊动强度;湍流强度
turbulent inversion 湍流逆温;湍流逆流
turbulent jet 紊动射流
turbulent kinetic energy 紊流动能;湍流动能
turbulent kinetic energy dissipation rate 湍流动能耗散率
turbulent kinetic energy equation 湍流动能平衡方程
turbulent layer 沸动层;流沙层;湍流层;紊流层
turbulent level 紊流度;湍流强度;湍流级;湍流度
turbulent Lewis number 紊流刘易斯数;湍动刘易斯数
turbulent line width 湍流谱线宽度
turbulent magnetic field 湍动磁场
turbulent mantle convection 湍动地幔对流
turbulent medium 紊动媒质
turbulent method 湍流法;湍动法
turbulent mixer/reactor 涡流混合反应器
turbulent mixing 紊流搅拌;湍流混合;湍流混合
turbulent mixing chamber 湍动混合室
turbulent mixing zone 湍流混合区
turbulent motion 扰动;涡旋运动;湍动;紊流运

动;紊动;湍流运动;湍动
turbulent movement 湍流运动
turbulent noise 湍流噪声
turbulent number 湍流度
turbulent open channel flow 明渠纹流;紊动明渠流
turbulent pattern 漩涡式
turbulent perturbation 湍流扰动
turbulent plasma 湍动等离子体
turbulent Prandtl number 紊流普朗特数
turbulent process 湍流过程
turbulent promoter 湍流促进器
turbulent propagation 紊流传播
turbulent range 紊流区
turbulent rate 紊流速率
turbulent reach 紊流段
turbulent region 紊流区;湍流区
turbulent resistance 紊流阻力;湍流阻力
turbulent Reynolds stress 湍动雷诺应力
turbulent roughness 紊动粗糙度
turbulent scale 湍流尺度;湍流标度
turbulent scattering 湍流散射
turbulent Schmidt number 湍动施密数
turbulent sea 激浪
turbulent separation 紊流分离(现象);湍流分离(现象);湍流边界层分离
turbulent shear 湍流剪力
turbulent shear flow 紊流剪流;紊动剪切流
turbulent shear force 湍流剪切力
turbulent shear stress 湍流切应力
turbulent (skin) friction 紊流摩擦;紊流表面摩擦
turbulent skin-friction coefficient 湍流表面摩擦因数
turbulent smooth 紊动光滑的
turbulent spectrum 湍谱
turbulent speed 紊流速度
turbulent stream 紊动水流
turbulent strength 紊动强度
turbulent stress 雷诺应力;紊流应力;湍流应力
turbulent structure 湍流结构
turbulent suspended composite velocity 湍流悬浮分速
turbulent texture 流状结构
turbulent throttle 涡节流
turbulent tidal current 湍动潮流
turbulent trail 紊流尾迹;湍流尾迹
turbulent transfer 湍流输送
turbulent transitional 紊流过渡的
turbulent type mud flow 紊流型泥石流;湍流型泥石流
turbulent velocity 湍流速度;紊流速度;湍速
turbulent velocity distribution 紊流流速分布;湍流速度分布
turbulent velocity field 紊流速度场
turbulent viscosity 湍流黏滞性
turbulent vorticity equation 紊流涡旋方程
turbulent wake 紊动尾流;湍流尾流;湍动尾流
turbulent wave 湍动波
turbulent wind 旋风;湍流风
turbulent wind structure 湍动风结构
turbulent zone 紊流区;湍流区
turbulivity 涡流度;紊流度;湍流系数;湍流度
Turc formula method 陶凯公式计算法
tureen 盖艂
tur(e)ing unit 整形修整装置
turf 泥煤;赛马场;粗泥炭;草皮;草根土
turf area 草地(面积)
turfary 泥炭田
turf bed 泥岩层;泥炭层
turf bottom 万能犁体
turf cofferdam 草土围堰
turf court 草地网球场
turf cutting 切草簇
turf dam 草土坝;草土小坝
turf density 草丛密度
turf drain 铺草排水沟;松节瓦管;长满草皮的水沟;草坪排水沟
turfed area 铺草(皮)面积
turf edging knife 草皮切边刀
turfed hair grass 发草
turfed slope 草坡;草皮边坡
turfed spade 草皮铲
turf grand stand 赛马场大看台
turf grass 草皮草
turf growth retardant 草皮抑长剂

turfing 铺草皮
turfing area 铺草(皮)面积
turfing by seeding 播籽植草
turfing by sodding 铺草皮的草地
turfing plough 草皮犁
turf moor 沼泽;泥沼地;泥炭沼(泽)
turf mount 泥炭丘
turf muck block 营养土块
turf peat 泥煤
turf piercer 草地疏松透气器
turf planting 翻草皮栽植
turf process 草炭过程
turf reconditioning 草坪更新;草场更新
turf revetment 草皮护坡
turf slope 草皮坡
turf sod 草皮
turf sod cutter 草皮铲;铲草皮器
turf sod stockpile 草皮堆垛
turf sprinkler 草地喷水器;草地喷灌器
turf sprinkler system 草地喷洒系统;草地喷水系统
turf strip 条形草地
turf stripping 草皮铲剥
turf surface 草皮地面
turf swamp 泥炭沼泽
turf wall 草皮土墙
turfy 含泥炭的;多草的;草皮的
turfy soil 生草土;草炭土
turgesency 膨胀性
turgidity 膨胀性
turgite 水赤铁矿
turgor pressure 涨压
Turgo turbine 土尔戈式涡轮机;土尔戈式透平;土尔戈式水轮机
turguoise 绿松石
turjaite 云霞黄长岩
Turkey blue 土耳其蓝;钴铬绿
Turkey red 土耳其红;铁红
Turkey red oil 土耳其红油
turkey stone 细粒磨石
turkey tails 火鸡尾山墙顶饰
Turkey umber 土耳其棕土;土耳其赭土
Turkish-Aegean plate 土耳其—爱琴海板块
Turkish architecture 土耳其建筑
Turkish bath 土耳其浴室
Turkish carpet 土耳其地毯
Turkish-Central Iran-Gangdise plate 土耳其—中伊朗—冈底斯中间板块
Turkish closet 土耳其厕所
Turkish gallotannin 土耳其种没食子丹宁
Turkish minaret 土耳其式尖塔(伊斯兰教寺院的)
Turkish mortar 土耳其砂浆
Turkish plate 土耳其板块
Turkish (type) mosque 土耳其式清真寺
Turk Loydu 土耳其船级社
Turkoman 土库曼装饰织物
Turk's head 牡蛎属头巾式饰结;穆斯林巾式刷子
Turk's head rolls 特克头轧辊
turmeric 姜黄
turmeric paper 姜黄纸;姜黄试纸
turmeric test paper 姜黄试纸
turmoil 骚乱;动乱
turn 一轮;一回;一次载重量;转动;转变期;匝;轮次;车削;转弯
turnable 转盘;转车台
turnable bay 回旋池
turnable bridge 平旋桥;平开桥
turnable gate 转动式门;转动式闸门
turnable infrared laser 可调红外激光器
turnable polish 滚转抛光
turnable press 转盘压(砖)机
turnable roller 回转滚子
turnable support 旋转支座
turnable table 旋转工作台
turn about 调向;调头;船舶卸货所需时间;转向
turn-about plow 翻转犁
turnadozer 转轮式推土机
Turnall 特纳尔牌(为多种石棉水泥制品)
turn and bank indicator 转弯倾斜指示器
turn-and-slip indicator 转弯倾斜指示器
turn angle 测量转角度;转角;转测角
turnapull scraper 拖式铲运机
turnarocker 拖拉后卸式搬运机
turnaround 调头;船在港内周转时间;正常停工检修;回车线;回车场;车道;周转期;来回程;街道

交叉处的转盘;检修期;船只进港
turnaround area 转盘地带
turnaround condition 反转条件
turnaround depot 折返站
turnaround documents 周转资料;周转文件
turnaround engine house 机车折返库
turnaround factor of container 集装箱周转率
turnaround line 折返线
turnaround loop 回车道;回车场;转向环道
turnaround machine 钢锭回转机;回转机
turnaround number of storage capacity 库场容量周转次数
turnaround of a ship 船舶在港时间
turnaround of rolling stock 车辆周转
turnaround of unit 装置检修;装置工作周期
turnaround pad 回车坪
turnaround plan 检修计划
turnaround point 机车折返点
turnaround point of vehicle 车辆转向点
turnaround specialist 周转专家(能使未完工项目或无利可图的项目变为有利可图的人)
turnaround speed 周转速度;周转速率
turnaround station 折返站
turnaround system 机车折返制;巡回系统;周转系统
turnaround table 调头转台
turnaround taxiway 回旋滑行道(飞机)
turnaround time 整个运算时间;轮转周期;来回时间;解题周期;往返航次时间;调头时间;反向传送时间;周转时间;换向时间
turnaround time of goods section 货位周转时间
turnaround velocity 周转速度
turnaround wye 折返三角线;转向三角线;三角线【铁】
turn a tap 拧龙头
turn away 转向一边
turnback 折回;手工制管鞋
turn-back capability 折返能力
turn back flow 折流
turn-back of train 列车折返
turn-back station 折返站
turn-back station of passenger train stock 车底折返站
turn-back track 折返线;机车折返线
turn-ban 不准转弯
turn barrel 转筒
turn-batch mixing drum 倾倒式搅拌机
turn bench 钟表车床
turn berth clause 停泊条款
turn branch switch 铁路支线转辙器
turn bridge 转桥;平旋桥;平开桥
turn-broaching machine 车拉床
turnbuckle 拉线螺丝;伸缩螺杆;螺旋扣;螺丝扣;螺丝接头;紧索螺套;花兰螺扣;花兰螺栓;松紧螺丝扣;反正扣
turnbuckle closure 螺套封闭器
turnbuckle rig 混合结构钻塔的钻机
turnbuckle screw 旋扣螺丝;紧线螺丝;接合螺丝;松紧螺丝扣;花兰螺丝
turn button 旋转式按钮;窗插销
turn by banking 只压坡度的转弯
turn-cap 风帽;旋转帽;烟囱风帽
turn circle radius 转弯半径
turn control 转弯操纵
turn controller 转弯驾驶杆
turn coordination 转弯协调
turn creel 转动式筒子架
turn crossing 可调头交叉口;允许掉头的交叉口
turn down 可翻折的;折叠式的;旋小;摺;不受理;量程比
turn-down lamp 变光灯泡
turn-down range 调节范围
turn-down ratio 操作弹性;极限负荷比;调节比
turn-down region 调节范围
turn-down rig 翻料装置
turn down rims 翻边的
turn-down wiring 下行线路
turn-down zone 调节范围
turned and bored joint 车镗接头
turned bolt 紧配螺栓;光制螺栓;精制螺栓
turned-down coupling 斜角接箍;倒角接头
turned finish 车削面光洁度
turned groove 车制槽
turned hardening 变温锻炼

turned joint 车削接头
turned knee 折肘
turned-over capital 周转资金
turned position 转动位置
turned sorts 倒放铅字
turned wooden articles 车床加工的木配件(如栏杆柱等)
turned work 旋木工件;车制工件
turn effect 匝效应
turner 转架;锅铲;搅动器;车削工件;车工
turnerite 独居石
Turner method 特纳法
turner scalerometer 金属划痕硬度试验计;回跳硬度计
Turner's yellow 车工铅黄;意大利船黄;氯化铅黄;特纳黄
turnery 旋制品;车削工艺;车削车间;车削产品;车工工艺;车床工厂
turn foot roller 转足羊毛碾
turn formation 转换队形
turn from deficits to profits 扭亏增盈
turn green 返青
turn grey 变成灰色;调成灰色
turn hillside into terraced fields 坡改梯
turn home 拧紧
turn in 向里弯曲;下班;系上(绳索)
turn inbound 转向导航台
turn indicator 转弯指示器;转数表;匝数计
turning 弯角;车旋;车削外圆;车削(加工);车床修坯;转弯
turning ability 转向能力;车削性能
turning agency 复代理
turning an arch 造拱;起拱
turning angle of flow 水流转角
turning area 转弯处;调头处;回车场;航途转弯区;调头区(船舶)
turning axle 旋转轴
turning bar 转向杆;壁炉条;壁炉拱的支承铁杆
turning basin 回旋水域;回水池;回船(港)池;掉头水;调港池;船舶基地;回船(港)池;船舶调头港池;转船池;转头水域
turning basin circle 转头水域
turning berth 掉头区;调头泊位
turning block 转动块;转动滑块
turning-block linkage 回转导杆机构
turning bolt 扣子;搭扣;长插销;锁扣
turning bridge 旋开桥
turning buoy 转向浮标;转弯浮标
turning by ahead astern engine 进倒车掉头
turning by lathe 车床加工
turning by pulling the bow 拖头掉头
turning by pulling the bow and pushing the stern 拖头顶尾掉头
turning by pulling the stern 拖尾掉头
turning by pushing the bow 顶头掉头
turning capacitor 可调电容器
turning capacity 最大转弯能力
turning center 转向中心;车削加工中心
turning circle 旋回圈;转车盘;回转圈;回旋圈;回船场;回车盘;转弯圆;转船池;过渡曲线
turning (circle) trial 回转试验
turning control 调谐旋钮
turning corner 转角
turning couple 旋转力偶;扭力偶;倾覆力偶
turning crane 旋转式起重机;旋臂(式)起重机;回转式起重机
turning curvature 转弯曲度
turning diameter 调头区直径;转弯直径;回转直径;调头直径
turning direction 回转方向
turning dolphin 调头系缆墩
turning dynamometer 车削测力仪
turning effect 转动效应
turning effort 转动作用;转动力;回转力
turning engine 转车机;盘车机
turning equipment 转向设备【铁】
turning error 转向误差;调谐误差
turning force 回转力
turning-fork oscillator 音叉振荡器
turning gear 转向装置;转轴装置;转车机;盘车装置
turning gouge 旋木用车刀;圆凿(子);弧口旋凿
turning handle 转动柄;转动手柄
turning head 多刀转塔;车削刀架
turning hole 铰削孔

turning-in 向里弯曲
turning in heavy sea 大风浪中掉头转向
turning interval 旋回周期;回转周期;回转间隔
turning invisible subsidy to open subsidy 暗补变明补
turning jack 转盘
turning joint 旋转接头;旋转接合;活动关节
turning labyrinth 旋转迷宫
turning lane 回车道
turning lathe 旋床;车床
turning leverage 回转效应
turning-lifting gate 转升门
turning locus 回转轨迹
turning locus for combination vehicle 连接车回转轨迹
turning machine 车床
turning machinery of swing bridge 旋开桥的转动机械
turning mat 变成没有光泽
turning mill 车床
turning moment 扭转力矩;旋转力矩;转向力矩;转矩;回转力矩
turning moment diagram 转矩图
turning motion 回转运动
turning movement 旋转运动;转弯运行;倾覆力矩
turning of aggregate 集料转运
turning off steam 关闭暖气
turning of rail 钢轨调头
turning of tide 转潮;潮汐转流
turning-on 整经
turning on steam 开放暖气
turning out device 转出工具
turning-over board 造型平板;翻箱板
turning pair 旋转力偶;回转对偶
turning path 转弯小径;转弯轨迹
turning performance 转弯特性;回转性能
turning period 回转时间
turning piece 砌拱弧形模板;简便拱架;门闩控制钮把手;筑小拱的弧形板;轮转盘;单拱架
turning pin 铅管开口塞;细木管楦;锥形木塞;硬木旋塞
turning place 转向处;回车场;调头处
turning plate 水准尺垫;标尺座;转盘;尺垫【测】
turning plow 翻土犁
turning pocket 袋形回车场
turning point 旋转点;转折点;转机;转点;商业周期转折点;导线转点;反转点;转向点
turning point convergence zone 反转点会聚区
turning point on a curve 曲线的转向点
turning-point pin 尺桩【测】
turning point test 转向点检验
turning power 转动能力;回转力
turning quality 回转性能
turning radius 转弯半径;旋回半径;旋回半径;转向半径;回转半径;回车半径
turning radius of ship 船舶回转半径
turning radius trial 回转半径试验
turning range marks 转向叠标
turning rate control 转动速率控制
turning roadway 转弯车道
turning roadway terminal 转向车道岔口
turning roll 转向辊;滚轮架
turn(ing)-round time 周转时间;换向时间
turning rule 转弯规则
turning saw 绕锯;弧形锯;弧锯;曲线锯
turning scale 车前碎屑
turning shop 车工车间
turning short round 就地旋回
turning short round by anchor 抛锚掉头
turning shovel 火铲
turning site 转弯部位
turning slide 回转滑阀
turning slider crank mechanism 旋转滑块式曲柄导杆机构;旋转滑块曲柄机构
turning slide valve 转动滑阀
turning speed 转弯速率
turning station 转折点;转坡点;转点
turning strickle 车板
turning test 转向试验
turning the gear over 改变吊杆位置
turning time 转弯时间
turning tool 转动工具;车刀
turning to pasture 放牧

turning track 转向轨距
turning track width 转向轨迹宽度
turning traffic 转弯交通
turning tube type liquid level ga(u)ge 旋转管式液位计
turning unit 旋转部件
turning value 转向值
turning valve 旋转挡板;转阀;回转阀;换向阀
turning vane 转动叶片;导向装置;导向叶片;导流叶(片);导流片;导流板
turning-vane steering gear 转翼式液压操舵装置
turning wheel 转轮;转轮;车刀
turning width 转弯宽度
turning with the aid of current 利用流力掉头
turning zone 商业周期转折期
turn insulation 线匝绝缘
turn in synchronism 同步转动
turn into 变成
turn into dive 转入俯冲
turn into the line 转入管道
turn inward 转向机场
turnkey 成套项目;总承包;全工程统包;全部包建的工程承包方式;转动钥匙;交钥匙方式;总控键;监狱的看守
turnkey agreement 交钥匙工程协议书
turnkey bid 全包建投标;交钥匙投标
turnkey building construction 整套承包建筑(包设计施工和包工包料)
turnkey construction 交钥匙工程
turnkey contract 整套合同;整套承包合同;全承包合同;全部承包合同;总承包;交钥匙合同;承包工程
turnkey contracting 建设项目总承包
turnkey cost estimate 整套承包估价
turnkey delivery 承包安装及启用
turnkey factory 全部包建的工厂
turnkey franchising 交钥匙特许
turnkey front panel 转键前面板
turnkey handover 建筑竣工移交;整套承包移交
turnkey housing construction 整套承包住宅建筑
turnkey industry 统包工业
turnkey investment 交钥匙投资
turnkey job 承包工程之至投产使用;交钥匙工程;全承包工程
turnkey leasing 交钥匙租赁(专指租给低收入住户的公共住房)
turnkey microcomputer system 交钥匙微机系统
turnkey-mutual help-new construction 总承包—互助—新住房建设
turnkey-new construction 总包整套新建公共住房;交钥匙新建公共住房
turnkey operation 交钥匙工程项目;成套包建项目;交钥匙作业
turnkey project 成套包建项目;交钥匙项目;交钥匙工程项目;交钥匙工程;包建工程;全承包设计;全承包工程
turnkey public housing 总承包公共住房
turnkey rehabilitation 总承包公共住房的修复
turnkey system 整套系统;承包系统;包用系统;转键系统
turnkey terms 交钥匙条件
turnkey type building 交钥匙型建筑;总承包型建筑
turnkey type of contract 整套承包合同
turn knob 球形把手;(门锁的)旋转球柄;门锁控制钮把手;旋钮
turn left 左转弯
turn light 转向灯
turn lock 回船闸
turn marking 转向标志
turn mechanism 旋转机构
turnmeter 转率计;回转速度指示器;回转计
turn-miller 立式车铣床
turn number 匝数
turn of bilge 船底边弯曲部分;舭部弯曲处
turn-off 切断;停用;断开;转开
turn off current 断路电流
turn-off delay time 断开延迟时间
turn-off point from main sequence 主序反向点
turn-off pulse 熄火脉冲
turn off the runway 转弯离开车道
turn-off time 断开时间;断时间
turn-of-nut method 旋转螺母方法
turn of speed 速力
turn of the market 市场变化

turn of the tide 潮汐转流;转潮;涨落潮更替;再次涨潮
turn of tidal current 转流;转潮流;潮流转向
turn of tidal stream 转流;潮流转向
turn of tide 潮汐转流
turn-on 接通;扭开;开启;使通导;通路
turn-on delay 通导延迟
turn on investment 投资回收
turn on time 接通时间;接入时间
turnout 向外弯曲;向外;切断;分道(叉);线叉;会船区;渠道分叉口;道岔【铁】;产额;岔道;分水处;分岔
turnout back-passing the point 顺向道岔【铁】
turnout branch 道岔侧线
turnout center distance 道岔中心距离
turnout closure 道岔密贴
turnout design 道岔设计
turnout face-passing the point 对向道岔
turnout for work 出勤
turnout from curved track 曲线道岔
turnout guard rail 道岔护轨
turnout in a block section 区间道岔
turnout lane 避车道;港湾式停车道
turnout locking 道岔锁闭
turnout main 道岔主线
turnout mast 道岔柱
turnout number 道岔号数
turnout of traffic 转弯避让其他机车
turnout passing-passing the point 顺向道岔【铁】
turnout rate 转换率
turnout sleeper 岔枕
turnout survey 道岔测量
turnout through speed 过岔速度
turnout tie 岔枕
turnout track 岔道线【铁】;牵出线【铁】;分叉道;转辙
turnout wall 翼墙;侧墙
turnout zone 道岔区
turnover 让渡;倾覆;翻身;翻面;翻板;营业额;移交;周转量;周转;工程维修费;临时投资额;交易额;年龄级周转;回转循环;倾覆;翻层现象
turnover board 翻转底板
turnover capacity 转运能力;周转能力
turnover capacity of harbo(u)r 港口吞吐量
turnover capacity of storage space 库场通过能力库存物资周转量;库场利用率;库场货物周转量
turnover cartridge 翻转式拾声头
turnover days of circulating capital 流动资金周转天数
turnover door 上拉式叠库门;翻转门
turnover elevator 倾翻式升运器
turnover forms 翻转模板;翻模板
turnover frequency 过渡频率;交界频率;交叉频率
turnover job 大修
turnover jolt squeeze mo(u)lding machine 转板式震压造型机
turnover labo(u)r 替工
turnover mechanism 翻转机构
turnover mo(u)lding 翻转造型
turnover number 转化数;转换数
turnover of animal herds 畜群的周转
turnover of cargo stored 货物平均堆存期
turnover of current assets 流动资产周转率
turnover of finished goods 制成品周转率
turnover of form 模板周转
turnover of freight traffic 货物周转量
turnover of funds 资金周转
turnover of goods 货物周转量
turnover of investment 投资周转额
turnover of locomotive 列车机车全周期时间
turnover of material in stock 储存材料周转率
turnover of net working capital 净运用资本周转率
turnover of parcel transportation 行包周转量
turnover of passenger traffic 旅客周转量
turnover of payment 应付账款周转率
turnover of ratio of accounts payable 应付账款周转率
turnover of special marshalling locomotive 专用调车机车全周转时间
turnover of storage 货物平均堆存期
turnover of total assets 总资产周转率;资产总额周转率
turnover of total liabilities and net worth 负债及资本净值总额周转率
turnover of total operation assets 经营资产总额周转率;经营资本周转率
turnover of water 水的翻转
turnover of working funds 流动资金周转率
turnover per capita 人均周转额
turnover plow 翻土犁
turnover radius 转向点半径
turnover rate 换水率(池水循环一次所需时间,以小时计);转换率;周转率;周转率常数;净值周转率
turnover rate constant 周转率常数
turnover rate of current capital 流动资金周转率
turnover rate of funds 资金周转率
turnover rate of personnel 人员变动率
turnover rate of standard current funds 定额流动资金周转率
turnover rate of whole circulating funds 全部流动资金周转率
turnover ratio 周转比率
turnover (ratio) fixed assets 固定资本周转率
turnover ratio of gross and net working capital 运营资本总额及净额周转率
turnover ratio of net worth 净值周转率
turnover ratio of receivable account 应收账款周转率
turnover (ratio) of total liabilities 负债总额周转率
turnover ratio of total liabilities and net worth 负债总额及净值周转率
turnover ratio of total operating assets 运营资产总额周转率
turnover ratio of working capital 流动资本周转率
turnover rig 翻面装置
turnovers 流动者
turnover shuttering 翻转模板
turnovers of commercial and service trades 服务行业的流转额
turnover study 周转率调查
turnover table mould 翻台模架
turnover tax 周转税;交易税;商品周转;按营业额征收之税项
turn over taxes to the state 上交税费
turnover temperature 转换温度
turnover time 周转时间;更新时间;换向时间
turnover time of freight train per week 货车周转时间
turnover type entry guide 翻转式导口
turnover volume of passenger transported 客运周转量
turnover zone 移动区
turn performance 转弯性能
turn-picture control 图像转换控制
turn piece 锁闩控制钮把手
turnpike 收税栅;征收通行费的高速公路;铁路跨线桥;收税(高速公)路;收费高速公路
turnpike road 主干路;收税路;收税道路;收费高速汽车公路
turnpike stair(case) 螺旋式楼梯
turnpin 黄杨木捣塞;硬木旋钉
turn place 车辆掉头处;车辆调头处
turnplate 旋转台;转台;转盘;转车台;回转盘
turn plate of rail way 轨道转盘
turnplate high incline 转盘式斜面升船机
turn plow 翻土犁;铧式犁
turn point 转向点
turn power on 接通电源
turn prohibition sign 禁止转向标志;禁止转弯标志
turn-pull button 转拉电钮
turn radius 转弯半径
turn rail 钢轨调头
turn rate 旋转率
turn ratio 线匝比
turn regulation 限制转弯
turn restriction 转弯限制
turn round 运转;周转;掉头
turnround diagram of passenger train stock 旅客列车车底周转图
turnround of a ship 船的周转
turnround period 船在港内周转时间
turnround port 中转港(口)
turn round quickly 快速转动
turnround time 往返时间
turnround time of passenger train set 旅客列车车底周转时间
turns 手动车床
turn scrap material to good account 利用废料
turnscrew 螺丝刀;旋扣螺丝;改锥;螺丝起子;旋凿;螺丝起子;传动丝杠
turn separator 匝间间隔片;匝间垫条
turn sheet 转台板
turn sieve 翻转筛
turn sign 急弯标志
turn-signal control lamp 转向信号指示灯
turn signal flasher 转向信号闪光灯
turn signal lamp 转弯示向灯
turn signal switch 扭转信号开关
turn(s)-insulating 匝间绝缘
turn slots 转弯车带
turn space 车辆掉头处
turns per inch 捻密度;每英寸转数
turns ratio 匝数比;变压器匝比
turn stair(case) 回转式楼梯
turnstile 旋转式栅栏;转栅;转式栅门;回转栏(门);绕杆;十字回转门
turnstile antenna 挠杆式天线;绕杆式天线
turnstile crane 转臂(式)起重机
turnstile feed 绕杆式馈源
turnstile gate 转千式闸机
turntable 转车盘;转台;转动台;转车盘;回转压砖机;回转盘;回旋盘
turntable antenna mount 转台式天线座架
turntable doffing mechanism 转台式落条筒机构
turntable engine 转车台发动机
turntable mounting 转盘架
turntable mumble 转盘噪声
turntable of shiplift 升船机转盘
turntable-pit 转盘坑
turntable slewing ring 转盘式回转支承
turntable support 转台支架
turntable switch 转盘型道岔
turntable tip 转台翻车机
turntable track 转盘线
turntable type slewing supporting device 转盘式旋转支承装置
turntable undercarriage 转车台台底盘
turn the country green 绿化祖国
turn the soil 翻土
turn tidal stream 转潮流
turn to port 向左转舵;左转弯
turn to starboard 向右转舵
turn-to-turn 匝间
turn-to-turn fault 匝间故障;匝间短路
turn-to-turn insulation 匝间绝缘
turn-to-turn short (circuit) 匝间短路
turn tread (楼梯的)回转踏步;扇形踏步;斜踏步;转弯踏板
turn trough 转槽
turn turtle 翻船
turn up 旋入;折起;在甲板集合;开大;掘起;翻卷边
turn up edging machine 翻边机
turn-up wiring 上行路线
turn volume 吞吐量
turn wastes into wealth 变废为宝
turnwrest plough 转壁双向犁
turnwrest swing plow 摆杆式转壁双向犁
turn yellow 泛黄
Turonian 土仑阶【地】
Turonian-Senonian transgression 土仑—赛诺期海浸
turpentine 松脂;松节油
turpentine fiber 松节油纤维
turpentine lacquer 松节油喷漆
turpentine medium 松节油溶剂
turpentine oil 松节油
turpentine oil-resisting test 耐松节油试验
turpentine-orchard 松脂林
turpentine separator 松节油澄清槽
turpentine soot 松烟
turpentine stain 松节油脂着色;松节油瑕疵
turpentine substitute 松节油代用品
turpentine varnish 松节油清漆
turpentine vehicle 松节油溶剂
turpidometer 浮沉测粒计;浮沉测定颗粒仪
turquoise 土耳其玉;绿松石
turquoise blue 绿蓝色;青绿色;绿松色
turquoise blue crackle glaze 宝石蓝纹片釉
turquoise deposit 绿松石矿床
turquoise glaze 孔雀绿釉
turquoise green pale 淡碧绿色

turret 转塔;转动架;六角转头;角架;楼塔;塔楼;六角刀架
turret block 六角刀架
turret clock 塔钟
turret clock striking cams 塔钟打点凸轮
turret clock trains 塔钟打点轮系
turret coal cutter 塔式截煤机
turret deck 坛甲板
turret deck vessel 坛甲板船
turret dome 角楼圆顶
turret door 转塔门
turret drill chuck 六角车床占夹
turreted roof 塔楼屋顶;角楼屋顶;有塔楼屋顶;有角楼屋顶;塔状屋顶
turreted steamer 塔式汽船
turreted tower 角楼
turret graver 多头刻图仪
turret hatch 转塔顶门
turret head 六角头
turret ice 竖立冰;侧立冰
turret index 六角刀架转位
turret lathe 转塔车床;六角车床
turret-like 塔楼状的
turret miller 带六角头回转铣床
turret multiple punch 转塔式压力机;转塔式冲床
turret press 转塔式压力机
turret punch press 六角形零件压冲机;转塔式六角冲床;六角零件压力机
turret ring gear 转塔环形齿轮
turret step 塔楼梯级;旋转楼梯踏步;塔楼踏步;三角形踏步
turret terminal 按钮端
turret tuner 旋转式频道选择器;回转式调谐器
turret type feeder 转台型喂料器
turret type pump 塔式泵
turricula 塔状物
turriculate 有小塔的
turriculated 塔楼的;塔楼状的;角塔式的
turriform 塔形
turtle 龟标;海龟
turtle back 折流板;疤眼;龟背;船尼防浪损的拱形架
turtle-back beam 凸起甲板横梁
turtle back poop 鲸背式船尼楼
turtle geometry 龟标几何
turtle graphics 龟标制图
turtle's calipash 鳖裙(色釉品种之一)
turtle stone 龟背石;泥土内的大圆石
turtoning boom rig 连杆索具
Turugart tectonic segment 图噜喀尔特构造段
Tuscan 托斯卡式建筑(古罗马)
Tuscan arch 托斯卡式拱(古罗马)
Tuscan base 托斯卡式柱基(古罗马)
Tuscan capital 托斯卡式柱帽(古罗马)
Tuscan column 托斯卡(式)柱(古罗马);托斯卡柱
Tuscan column base 托斯卡式柱基(古罗马)
Tuscan floor 托斯卡(式)楼板(古罗马)
tuscanite 硫硅钙钾石
Tuscan order 托斯卡(式)柱式(古罗马);托斯卡式柱型(古罗马);托斯卡柱建筑
Tuscan red 托斯卡红;铁红色料
Tuscan style 托斯卡式【建】;托斯卡柱型
tusche 制版墨;黑墨(汁)
tusionite 硼锡锰石
tusk and tenon joint 镶尖榫接头
tusking 装齿形物;墙牙岔;马牙接搓
tusk nail 斜钉
tusk nailing 斜钉(法);斜向钉钉子;多牙钉;齿形钉
tusks 墙面牙石
tusk tenon 加劲凸榫;多牙尖榫;长牙镶榫
Tussah (wild) silk 柞蚕丝(织物)
tusses 犁尖;墙面接搓突石;石块牙砌;墙面牙石
tussock 生草丛;草图;草丛
tussock grass 生草丛;丛生禾
tussock plant 生草丛植物
tussock tundra 生草丛冻原
tussocky 多丛草的;丛草状的
tutamen 保护器
tutanaga 锌铁皮;白铁皮
Tutane 氖丁烷
Tutania alloy 锡锑铜合金
Tutania metal 图塔尼阿锡锑铜合金
tutenag(ue) 白铜
tutin 土亭
tutorial light 指导灯

tutorial mode 个别教学方式
Tuttle lamellae 塔特尔纹
tut work 计件工件
tuyere 鼓风管嘴;(熔炉的)喷气管嘴;吹风管嘴
tuyere action 风嘴工作
tuyere arch 风口拱墙
tuyere area 风口区
tuyere belt 风口带钢箍
tuyere block 风口冷却器
tuyere block bottom 带风口大砖的转炉底
tuyere bottom 风嘴底
tuyere brick 鼓风管嘴耐火砖;风嘴耐火砖;风口砖
tuyere cap 风口盖
tuyere connection 风口接头
tuyere cooler 风口冷却器
tuyere cooling plate 风口冷却板
tuyere head 风嘴帽
tuyere notch 风嘴孔
tuyere pipe 风管
tuyere plug 风口塞
tuyere type burner 风口型燃烧器
tuyere valve 风口阀
tuyere velocity 风口速度
tuyers 多孔风管
tvalchrelidzeite 硫砷锑汞矿
TV broadcast station 电视广播台
TV building 电视大楼
TV camera 电视摄像机
TV cheque monitoring system 电视检查监控系统
tveitite 氟钇钙矿
TV field strength meter 电视场强仪
TV method under well 井下电视法
TV modulator 电视调制器
TV plant 电视设备
TV radar display 电视雷达显示
TV recording 电视录像
TV room 电视室
TV sensor 电视传感器
TV set 电视接收机;电视机
TV studio 电视播放室;电视演播室
TV telephone 电视电话
TV theater 电视剧院
TV torch 电视塔
TV tower 电视塔
TV tracker system 电视跟踪系统
TV tracking 电视跟踪
TV traffic control system 电视交通管制系统
TV transmission network 电视传输网
TV transmitter 电视发射机
TV wall controller 电视墙控制器
Twaddell's degree 特瓦德尔度(液体比重表示法之一)
Twaddell's hydrometer 特瓦德尔比重计
Twaddle gravity scale 特沃德尔比重标度
Twaddle scale 特沃德尔比重标
Twaddle's hydrometer 特沃德尔比重计
T-wall T形墙;T形挡土墙
tweed 粗花呢
tweel 限流闸板;炉闸门
tweel blocks 门砖
tweendeck 甲板间
tweendeck cargo space 甲板间货舱;二层舱
tweendecker 多层甲板船
tweendeck frame 甲板间肋骨
tweendeck height 甲板间高度
tweendeck hold 甲板间舱
tween-deck ship 双层甲板船
tweendeck tonnage 甲板间舱吨位
tweendeck tonnage section 甲板间吨位截面
tweeter 高音喇叭;高音喇叭
tweeter (speaker) 高频扬声器
tweeter-woofer 高低音两用喇叭
tweezers 镊子;夹钳;小钳子;挟子
T-weld 丁字接头焊缝;T形焊
T-welding T形焊接
twelve-bay 十二开间
twelve bolt ventilation type diving equipment 十二螺栓通风潜水装置
twelve-cylinder 十二缸
twelve-direction mixing unit 十二路混合设备
twelve edge 十二边
twelve(-membered) ring 十二节环
twelve-mile limit 十二海里界限
twelvemo 十二开本

twelve note music 十二分音符法
twelve o'clock welding 仰焊
twelve-ordinate scheme 十二纵标格式
twelve-point scanner 十二点扫描器
twelve point socket 十二方套筒
twelve point socket wrench 十二方套筒扳手
twelve point sphere 十二点球
twelve principles of efficiency 效率十二原则
twelve pulse rectifying 十二脉冲整流
twelve punch 十二穿孔
twelve to seven calcium aluminate 七铝酸十二钙
twelve-wire 十二根钢丝的
twelve-wire cable 十二股(金属丝)缆绳
twelve-wire jack 张拉十二根钢丝的预应力千斤顶
twentieth highest annual hourly volume 一年第二十个最大小时交通量
twenty day biochemical oxygen demand 二十天生化需氧量
twenty day BOD 二十天生化需氧量
twenty-eight day strength 二十八天强度(混凝土)
twenty-eight lunar mansions 二十八宿
twenty equivalent unit 英尺标准集装箱
twenty-five points moving average 二十五点移动平均
twenty-foot equivalent unit 二十英尺当量单位;集装箱换算箱;标准箱(集装箱);集装箱标准单位(20英尺长集装箱的载重量);换算箱(集装箱)
twenty-four cm brick wall 单砖墙
twenty-four fortnightly periods 二十四节气
twenty-fourmo 二十四开
twenty-four solar terms 二十四节气
twenty-high roll mill 二十辊冷轧机
twenty-inch map 二十五英寸地图
twenty-one points moving average 二十一点移动平均
twenty-sided polygon 十二边形
twere 风口
twice 两次;两倍;二次
twice a week 每周两次
twice continuously differentiable 二次连续可微
twice-daily ga(u)ge reading 每日两次水尺读数;每日两次观读水位
twice firing 二次烧成
twice-horizontal frequency oscillator 水平双频振荡器
twice-laid rope 再生绳
twice-laid stuff 再生材料
twice-ploughed fallow 复耕休闲地
twice-pressed brick 二次压制砖
twice storied cordon 两侧多层单干形
twice the straight-line depreciation rate 双倍直线法折旧率
twicher 粉刷边缘用泥刀
twiddling line 小船横舵柄绳
twifold 两倍的;双重的
twig 小枝;树枝
twig cutting 插枝
twig fence 篱笆围栏;篱笆
twigs of the chaste tree 荆条
twilight 曙暮光
twilight arch 蒙影光弧
twilight area 过渡区域;过渡地区;半衰落区
twilight brightness 曙暮光亮度
twilight correction 曙暮光改正
twilight emission 曙暮光发射
twilight industry 弱小企业
twilight spectrum 曙暮光光谱
twilight zone 城镇中衰退地带;微明带;极夜地区;蒙影地带;曙暮光区;半阴影区
twill angle 斜纹角度
twill backed 斜纹衬里
twill backing 斜纹背织物
twill canvas 加料帆布
twilled double warp 形成斜纹的双向翘曲
twilled herringbone weave 人字形编织;人字形斜纹组织
twilled mat 斜纹方平织物
twilled weave 斜纹编织
twill fabric 斜纹织物
twills 斜纹织物
twill set 斜列式植针法
twill shape screen 斜纹滤筛网
twill (type) screen 斜纹筛网;斜纹滤筛网
twill weave 斜纹组织;斜纹编织

twin-action brake 双作动式制动器
twin angle probe 双角探头
twin arc welding 双弧焊接
twin arithmetic(al) units 双运算器
twin arrangement 并排安装
twinax 屏蔽双绞馈线
twin axial cable 双同轴电缆
twin axis 共轭轴;双晶轴
twin axles 共轭;双轴;共轭轴(线)
twin-ball float 双球式浮标;双球浮子
twin band 孪生带;双晶带
twin band mill 双带锯机(制材厂);复式带锯制材厂
twin-bar bumper 双杆保险杠
twin barrel carburetor 双腔化油器
twin barrel tube 双腔管;双筒管
twin-batch mixing drum 双拌式搅拌筒
twin-batch pav(i)er 双拌(式)混凝土摊铺机
twin bead 双沿口
twin beam bridge crane 双梁桥式起重机
twin beam bridge type grabbing crane 双梁桥式抓斗起重机
twin-beam model 双射束模型
twin beam oscilloscope 双束示波器
twin beams 并置梁;双梁
twin-bearing screen 双轴承筛分机
twin-bedded room 双人房间
twin beds 一对双人床
twin beds guest room 双床间客房
twin-belt conveyer 双带式输送机
twin-belt take-off machine 双皮带牵引出机
twin-belt vanner 双带式淘洗机
twin berth 双人住舱
twin blades 双叶片;双翼
twin-block concrete sleeper 双块混凝土轨枕
twin-boat-type airplane 双船身式飞机
twin-boom aircraft 双尾撑飞机
twin-boom carriage 双臂凿岩台车
twin-boom drilling rig 双架钻机
twin-boom stacker 双悬臂堆料机
twin boom tracklayer 双臂铺轨机
twin booster 对偶助推器
twin-bore tunnel 双孔隧道
twin boundary 孪晶面边界;孪晶界面;双晶面面;双晶间界
twin bowls 双便池
twin-box girder 双箱形截面大梁
twin-box-shaped rigid frame 双联箱形刚架
twin bracket 双环板;定位双环
twin branch type antenna 双枝形天线
twin-break contact 双断接点
twin brick 双砖;方砖
twin bridge 并列桥
twin bucket 双斗
twin-bulkhead tanker 双纵向舱壁油轮
twin bundled lines 双分裂导线线路
twin cable 平行叠置双电缆;双芯电线;双芯电缆;双股钢丝绳
twin-cable aerial tramway 双线架空索道
twin-cable ropeway 有轨电车轨道;平行架空索道;双线架空索道
twin-cable system 双电缆制;双电缆系统
twin-cable tramway 有轨电车轨道;平行架空索道
twin cage hoist 双吊篮吊车
twin calculating machine 双联计算机
twin camera 双镜头摄影机;双镜摄影机
twin camshaft chain drive 双凸轮轴的链传动装置
twin cam shaft type 双凸轮轴式
twin cam tracks 双层三角针道
twin cantenary wire dropper clip 双承力索吊弦线夹
twin cantilevers 双悬臂
twin cantilever shell 蝶形双悬臂壳体
twin carbon arc 双碳极电弧
twin carbon arc brazing 间接碳弧钎焊
twin-carbon arc lamp 双碳极弧光灯
twin carburetor system 双化油器系统
twin casting 双流浇铸
twin catenary wire anchor clamp 双承力索锚固线夹
twin cathode 双阴极
twin-cathode ray beam 双电子束
twin(-cell) culverts 双孔涵管;双孔涵洞
twin center 双晶中心
twin-chain elevator 双链斗式升运器

twin chamber 双线闸室;双联闸室;并联闸室
twin chamber electrostatic precipitator 双室静电除尘器
twin-channel 双通道的;双通道;双渠;双路
twin channel system 双信道制
twin check 双重校验;双重校对;比较校验
twin chime unit 双套信号钟装置【铁】
twin-choke carburetor 双风门化油器
twin chute 双叉溜槽
twin-circular mill 双平行圆锯制材厂
twin cities 孪生城市
twin coil 双线圈
twin coiler 双条圈条器
twin columns 双柱联体;双柱;并置柱;并合柱;同基双柱;双柱联体
twin compartment tank 两室储罐
twin compass system 双罗盘系统
twin-component methacrylate compound 双组分甲基丙烯酸酯化合物
twin compressor 复式压缩机
twin-concentric cable 双芯同轴电缆
twin conductor 平行双芯线;双芯电缆;双芯导线;双股导线
twin conduit 成对管道
twin-cone bit 双牙轮钻头
twin cone crusher 双圆锥破碎机
twin-cone drum 双锥式卷筒
twin-cone mixer-granulator 双锥式混合制粒机
twin contact 双头接点;双接点;双触点
twin contact wire 双接触线
twin contact wire clamp 双接触线接头线夹
twin contact wire clip 双接触线定位线夹
twin contrast 孪晶对比
twin controller 双联控制器
twin controls 双重操纵系统
twin-core 双芯
twin(-core) cable 双芯电线;双芯电缆
twin core conductor 双芯导线
twin-core current transformer 双芯电流互感器
twin-core shot firing cable 双芯起爆电缆
twin crane 双联起重机
twin crankshaft engine 双曲轴发动机
twin cross-section 双断面
twin crystal 双晶体;双晶
twin culvert 双孔涵洞
twin-cylinder 双筒联生
twin cylinder air-cooled engine 双缸风冷式发动机
twin cylinder drilling engine 钻探用双缸发动机
twin-cylinder engine 双缸发动机
twin cylinder mixer 双筒混合机;V形混合机
twin cylinder piston pump 双缸活塞泵
twin cylinder pump 双缸泵
twin cylinders 双汽缸;双筒联生;双筒联体
twin derrick posts 龙门式起重柱
twin-derrick system 双桅杆起重设备
twin-detector scanner 双探头扫描机
twin-diaphragm jig 双膈膜夹具跳汰机;双隔膜跳汰机
twin-diaphragm pump 双膈膜泵
twin die 双出口挤泥机
twin diesel 双缸柴油机
twin digitato 重对羽状的
twin-disc-bearing 双圆盘轴承
twin-disc resilient drive 双盘弹性传动器
twin dishwashing sink bowls 双洗碟盆水盘
twin-disk colter 双圆盘开沟器
twin disk cutters 双蝶滚刀
twin docks 双船坞
twin dosing tanks 双投药池
twin-double 两张双人床客房
twin drift 双平巷
twin-drill 并列钻;双钻
twin drive 双驱动(器);双电动机驱动;双传动
twin-drive belt 双传动皮带
twin-drum dryer 双滚筒干燥机;双辊干燥机
twin drum drive 双筒驱动
twin drum mixer 双筒拌和机
twin drum sorting 双鼓分类法
twin-drum type concrete paver 双筒混凝土摊铺机
twin drum winch 双(卷)筒扬机;双筒绞车
twin dry plate clutch 双片式摩擦离合器
twin-dual roll crusher 四辊轧碎机
twin dust collector 二级捕尘器
twine 麻线;盘绕;双股线;搓

twine can 绳筒
twine carrier 绳盒
twin edge grinder 双边研磨机
twin edger 双圆锯裁边机;双裁边锯
twined wire 缠丝
twine guide plate 导绳板
twine holder cleaner 绳夹清洁器
twine holder mechanism 绳夹
twine holder plate 绳夹板
twine keeper 绳夹
twine knife 割绳刀
twin elbow 双肘管;双弯头;双弯管
twin electric(al) horn 双警报器
twin electrode 双芯焊条
twine machine 制线机
twin engine 双联发动机;双发动机
twin-engined 双发动机的
twin-engined bus 双发动机公共汽车
twin-engined drive 双发动机驱动
twin-engined freighter 双发动机货机
twin-engined monoplane 双引擎单翼机
twin-engined plane 双发动机飞机
twin-engined single-shaft system 双机单轴式
twin-engined tractor 双引擎拖拉机
twin engine vehicle 双发动机汽车
twin entry 双平行平巷布置
twiner 搓绳机;缠绕植物
twine-tying mechanism 麻绳捆扎机构;麻绳打捆机构
twine unit 麻绳机
twin exhaust pipes 双排气管
twin fabric disk 双橡胶纤片
twin failure 双重故障
twin fan 双联风扇
twin feed 双面进料
twin feeder 双线馈线
twin feeder clamp 双线馈线线夹
twin feeder wire 双线等距线
twin-feed spiral 双头螺旋
twin fiber cable 双芯光缆
twin-filament lamp 双丝灯
twin fillet wolt 双面角焊缝
twin filter 双滤油器
twin-finned 双垂直尾翼的
twin-finned airplane 双垂尾飞机
twin-finned layout 双垂尾布局
twin flapper-and-nozzle valve 双喷嘴挡板阀
twin flashing yellow lights 双闪黄光灯
twin flexible cord 双芯软绳;双股软线
twin flight locks 双线多级船闸
twin float 双子浮标;双球浮子;双联浮标;双浮筒
twin-float seaplane 双浮筒式水上飞机
twin-flood lighting 两向泛光照明;双向泛光照明
twin floods 双生洪水;双峰洪水
twin flue 双烟道;双联火道
twin formation 双晶形成
twin former 双网成形装置
twin former type thermoforming machine 双端型热成型机;双成型头型热成型机
twin frame lift 双柱式车身底架支撑型举升器
twin-frame (saw) mill 双框锯厂;双框锯制材厂
twin fuel engine 复式燃料泵;双联燃料泵
twin fuel pump 双联燃料泵
twin furnace 双炉膛
twin-furnace boiler 双炉膛锅炉
twin-fuselage configuration 双机身布局
twin gates 双扇门
twin-geared press 双边齿轮驱动压力机
twin geminate 双干树
twing habit 蔓草类型
twin girders 双大梁;并置大梁
twin gliding 双晶滑移
twin gliding plane 孪晶滑移面;双晶滑移面
twin grating spectrograph 双光栅摄谱仪
twin-grid rectifier 双栅整流器
twin grinder 双面研磨机
twin grinding 双面研磨
twingrip 双支承;双夹钳;双柄
twingrip bath(tub) 双手把浴盆
twin grip type centerless grinder 双支承砂轮无心磨床
twin ground plate 双面研磨的镜玻璃;双面研磨玻璃板

twin ground plate glass 双重研磨的平板玻璃
twin-gun attachment 双联机枪装置
twin-head auger 双头螺钻机；双头螺旋钻机
twin-head grinder 双头研磨机
twin-head holder 双刀头刀夹
twin-head horizontal rotary drill(ing) machine 双头水平螺旋钻机
twin headspan wire 双横承力索
twin headspan wire clamp 双横承力索线夹
twin-head submerged arc welding machine 双头浸入式电弧焊机
twin-head tandem arc-welding machine 双头串联电弧自焊机
twin-head type sloping lobe 双头鹿角形
twin-heater system 双加热器体系
twin heavy-duty paste mixer 双罐厚浆混合机
twin hive 双群箱
twin hood crane 双钩吊车
twin horn cleat 双羊角
twin house 成对房屋；拼连的两所房屋
twin hull boat 双体船
twin hull craft 双体船
twin-hulled dredge(r) 双体挖泥船
twin-hull flying boat 双船身式水上飞机
twin hulls 双船体
twin-hull semi-submersible derrick barge 双体半潜式钻探船；双体半潜式钻井船；双体半潜式起重船；双身半潜式钻探船；双身半潜式钻井船；双身半潜式起重船
twin hull ship 双体船
twin hull trawler 双体拖网渔船；双体拖网船
twin-hull unit 双体平台
twin hull vessel 双体船
twin hydraulic jack 双液压千斤顶
twin ignition 双火花塞点火
twin images 孪生像
twin image separation 双像分离
twining axis 孪晶轴
twining dislocation 双晶位错
twining movement 缠绕运动
twining plant 缠绕植物
twining stem 缠绕茎
twining stem mo(u)lding 绕茎卷须线脚饰；双茎式装饰
twin-input reduction gear 双主动齿轮减速齿轮
twin-input single-output gear 双主动齿轮单出轴齿轮
twin installation 双联装置
twin insulator strings 双联绝缘子串；双连绝缘子串
twin ionization chamber 双层电离室
twin islet 双岛
twin jack 双塞孔
twin-jaw crusher 双颚式轧碎机；双颚式破碎机
twinjet 双喷气发动机
twin-jet nozzle 双喷嘴
twin jetties 成双导流堤；双导流堤；双边突堤；双边靠船码头
twin jib tracklayer 双臂铺轨机
twin king pile 双主柱桩
twinkler 闪光体
twinkling 瞬息；闪烁；闪光星
twin lamella 双晶片
twin-lamp socket 双灯座
twin landing gear 双轮起落架
twin laser 双激光器
twin lateral 孪支流
twin lateral stream 双生侧面河
twin launcher 双联发射装置
twin launching device 双联发射装置
twin law 孪晶律
twin layout 双岸式布置
twin-layshaft configuration 双中间轴式结构
twinlead 双芯引线
twin-lead cable 平行双电缆
twin-lead type feeder 平行双馈线
twin-leaf bascule bridge 双翼衡重式仰开桥；双叶竖旋桥；双翼竖旋桥；双翼衡重仰开桥
twin leaves 双生叶
twin lens 双镜头
twin-lens camera 双镜头照相机
twin lever 双节臂
twin lift container crane 双箱式集装箱起重机
twin lifts 双线升船机；双线电梯；双线降船机
twin lift spreader 双吊式吊具；双吊吊架

twin-lift transporter container crane 双吊式集装箱装卸桥
twin line 平行传输线；双线电路
twin line preheater 双板列预热器
twin liquid single system injection 双液单系统注浆
twin liquid twin system injection 双液双系统注浆
twin load circuit 双负载电路
twin-lobe cam 双凸角凸轮
twin lock 双线船闸；双室闸门；双室船闸；复式船闸；双船闸
twin-log floating breakwater 双木浮式消波堤
twin loudspeaker 双频带扬声器
twin low-oblique transformation 双浅倾转换
twin masts 龙门桅(杆)
twin method 双晶法
twin missile carrier 双管火箭发射架
twin mo(u)ld 双模
twin-mo(u)lding machine 两箱造型机
twin multi-stage locks 双线多级船闸
twinned bars 双钢筋
twinned binary code 孪生二进制码
twin(ned) (crystal) 孪晶；双晶
twinned depression 孪生低气压
twin(ned) flight locks 双线梯级船闸；复线梯级船闸
twinned glacier 双股冰川
twinned grooves 双纹；重纹
twinned single rigid rapiers 双层刚性单剑杆
twin needle 双头钩针
twinning 孪生作用；孪生；孪晶作用；孪晶现象；双晶现象；双股绞合；成对
twinning axis 孪生轴；孪晶轴
twinning deformation 孪生变形
twin(ning) law 双晶律
twin(ning) plane 孪晶面；孪生面；双晶面
twinnite 特硫锑铅矿
twin northlight frame 双北向采光框架
twin opening 双联孔
twin opening subway 双联孔人行隧道
twin pack 两种成分(的)；双单元；双组件；双包装
twin packer 双密封器；双封隔器；双层堵塞器
twin-pack system 两个组合系统
twin-pan intensive mixer 双盘强烈混合机
twin paradox 时钟佯谬
twin-part methacrylate compound 双异丁烯酸化合物
twin pav(i)er 双鼓式混凝土搅拌摊铺机；双鼓式(混凝土)铺路机
twin pen recorder 双笔记录装置
twin photocell 双光电管
twin photogrammetric camera 双镜头测量摄影机；双镜测量摄影机
twin photograph 双镜头摄影照片
twin pier 双线码头；双桥墩；成对窗间墙
twin pin clevis link 双孔联板
twin-pinion single-output reduction gear 双主动齿轮单出轴减速齿轮
twin pipe culvert 双联管式涵洞
twin-pipe direct return system 双管直接回水(供暖)系统
twin pipe elbow 管子的双弯头
twin-pipe hot water downfeed system 下给式双管热水系统
twin-pipe hot water upfeed system 上给式双管热水系统
twin pipeline 双管线
twin-pipe system 双管系统
twin plane 双晶面
twin-plane boundary 孪晶面边界
twin plane reflection 孪晶面反射；双晶面反射
twin planting 双干树
twin plate clutch 双片式离合器
twin-plate process 双面研磨法
twinplex 四信路制；双路制；双路移频制
twinplex plotter 双投影器绘图仪
twinplex projector 双投影器
twin pointer 孪生指针
twin polisher 双面抛光设备
twin polishing 双面抛光
twin post auto lift 双柱式汽车举升器
twin post lift 双柱升降机
twin pot system 双槽系统
twin-powered drive 双动力驱动

twin-powered excavator 双动力挖土机
twin power plant 双岸式水电站
twin preheater 双列预热器
twin press 双压区复式压榨
twin (pressure) vessel conveyer 双仓输送机；双仓泵
twin-prism separator 双棱镜分离器
twin prism square 双棱镜直角器
twin-probe snow-density ga(u)ge 双探针雪密度计
twin projection plotter 双投影绘图仪
twin propeller 双推进器；双螺旋桨的
twin pug 双轴拌和机；双向拌泥机
twin pug mill 双捏拌和机
twin pulley block 双联滑轮组；双滑轮组
twin pulse code 双脉冲编码
twin pumps 双缸泵；双水泵；双联泵
twin rail runway 双轨滑道
twin-ram-jet target 双冲压喷气发动机靶机
twin ramp 双跳板
twin ring ga(u)ge 双联环规
twin ring structures 孪生环形构造
twin river 孪生河川；双支流；孪生河流
twin rocker shovel 双翻铲机
twin-roll 双滚筒的；双滚筒
twin roll breaker 双辊破碎机
twin roll crusher 双辊破碎机
twin-roll drum dryer 双辊干燥机；双转鼓式干燥机；双辊滚动烘干机
twin roller 双轧机；双滚柱
twin-roller chain 双列套筒滚子链
twin roller stretching machine 双辊拉伸机
twin roller type 双滚轮式
twin roll feeder 双辊喂料机
twin rolling mill 二辊式轧机
twin-roll press 双辊加压；双辊压滤机
twin-roll toilet tissue dispenser 双卷卫生纸配出器
twin-roll vari-nip press 双辊变距压滤机
twin room (有两张单人床的)双人房间
twin rope drum 双绳卷筒
twin rope grab 双绳抓斗
twin rotor condenser 双动片电容器
twin-rotor continuous fluxing mixer 双转子连续塑化混合机；双辊连续塑化混合机
twin rotor crusher 双转子破碎机
twin rotor engine 双缸转子发动机
twin rotor hammer crusher 双转子锤式破碎机
twin rotor hammer mill 双转子锤式磨
twin-rotor housing 双缸体
twin-rotor layout 双旋翼式布局
twin rotor roll crusher 双转子辊式破碎机
twin-rotor side-intake-port engine 双缸端面进气转子发动机
twin rotor system 双旋翼系统
twin-row 双行(的)；双列的
twin row ball-bearing slewing ring 双行滚珠轴承圈
twin row radial engine 双排星型发动机
twin-rudder ship 双舵船
twin runner type turbine 双转轮式涡轮机；双转轮式水轮机；双工作轮涡轮机；双工作轮水轮机
twin saw-tooth frame 双锯齿形框架
twin scavenge pump 双扫气泵
twin-screed finisher 双括板平整器
twin screw 双推进器；双螺旋桨(的)
twin-screw and nut 双螺杆与螺母
twin-screw and single-rudder ship 双车舵船
twin-screw and triple-rudder ship 双车三舵船
twin-screw and twin-rudder ship 双车双舵船
twin-screw bulk cement trailer 双螺旋送料散装水泥拖车
twin-screw conveyer 双管螺旋输送机；双管绞刀输送机
twin-screw extruder 双螺杆挤出机
twin-screw extrusion compounding 双螺杆挤出复合
twin-screw feeder 双管绞刀喂料机
twin-screw mixer 双螺杆混合机
twin-screw polymerization reactor 双螺杆聚合器
twin-screw propeller 双螺旋桨
twin-screw pugmill 双螺杆叶片式拌和机
twin-screw pump 双螺旋泵
twin-screw sand classifier 双螺旋分砂机
twin-screw ship 双螺旋桨船

twin-screw steamer 双推进器船舶
twin-screw streamer 双螺旋桨轮船
twin-screw vessel 双螺旋桨船
twin seal 双密封
twin-seal tapping sleeve 双密封穿孔套管
twin segment 孪生段
twin serial camera 双镜头连续摄影机；双镜连续摄影机
twin-shaft 双轴的
twin-shaft continuous-mix pugmill 双连续拌和叶片式搅拌机
twin-shaft mixer 双轴搅拌机
twin-shaft paddle mixer 双轴桨板搅拌机；双轴式桨叶拌和机；双轴搅拌机
twin-shaft pier 双柱式桥墩；双柱式墩；双柱墩
twin-shaft preheater kiln 双立筒预热窑
twin-shaft screen 双轴筛
twin sheet piles 双板桩
twin-shell blender 双壳拌和器；双壁式掺和机；V形混合机
twin shell mixer 双筒混合机
twin ship lock 双联船闸
twin sideband 双边带
twin-side heading method 双侧壁导洞法
twin simple catenary 双接触的简单悬挂
twin simplex pump 双联单缸泵
twin-single pump 单作用双联泵；单作用双缸泵；双联单作用泵
twin six 水平对置式十二缸；双六汽缸
twin-six engine 双六缸发动机；水平对置式十二缸发动机
twin-skeg stern 双导流尾鳍
twin-skip system 双箕斗提升系统
twin slab culvert 双孔板式涵洞
twin slab drainage culvert 双联板排水涵洞
twin sleepers 双轨枕
twin sliding window 双重滑移窗
twin sloid tyres 双实心轮胎
twin socket 双插座
twin spans 双跨(度)
twin-spar construction 双梁结构
twin spark ignition 双火星塞点火
twin spark magneto 双发动机火磁电机
twin-speed power take-off 双速功率输出轴
twin spiral water turbine 双蜗壳水轮机；双螺形水轮机
twin spot 双生斑
twin-stage 双级
twin-stage grinding 两级粉磨
twin-stand mill 双机座轧机
twin-station fixture 双位夹具
twin steam engine 双缸蒸汽机
twin steamer 双螺旋桨轮船
twin-steer 双转向装置
twin stopped ladle 双塞棒钢包
twin storms 双生暴雨；双峰暴雨
twin strainer 双联滤器
twin-strand mill 双线式轧机
twin stream 双系列
twin stream preheater 双列预热器；双股流型预热器
twin striation 孪晶条纹；双晶纹；双晶条纹
twin subcarrier 双副载波制
twin surface grinder 双端面磨床
twin-tandem 成双串联(焊接)
twin-tandem wheel 双串式轮
twin tank furnace 双池窑炉
twin tanks 双槽；两舷水柜
twin tape transporter 双带运输装置
twin target 双机目标
twin-tee filter 双T形滤波器
twin tenons 夹口榫；双凸榫；双榫；双雄榫；双榫舌
twin-T filter 双T滤波器
twin theater 双演出场的剧院
twin throttle 双节流阀
twin tie plates 辙后双垫板；辙叉跟垫板
twin ties 双轨枕
twin timber mast 双木桅杆；双木柱
twin tires 双轮胎
Twintite 汀太特(一种专利的蒸汽活塞阀)
twin T network 双T形网络
twin torch 双头喷灯
twin-towered 双塔的
twin-towered facade 双塔式建筑物正面
twin-towered gatehouse 双塔门楼

twin towers 双塔；双列预热器塔；双塔式建筑
twin-track airway 双航路
twin-track procedure 双航路航线飞行规则
twin track recording 双轨记录
twin-track tunnel 双线隧道
twin tractor 双联拖拉机
twin-trailer (road) train 双挂车汽车列车
twin transistor 双晶体管
twin transposition cable 双绞电缆
twin travel(l)ing crane 双轮同时移动起重机；双轮移动起重机
twin tube clamp 双管线夹
twin tube clamp with double eye 双耳双管线夹
twin tube clamp with eye 单耳双管线夹
twin-tube conveyer 双管输送机
twin-tube heater 双管加热器
twin-tube tunnel 双管形隧道
twin tubing 双气管
twin tunnels 双座隧道；双孔隧道；复线隧道；双隧道
twin turbine 双转轮式涡轮机；双转轮式水轮机；双流式水轮机；双流型涡轮机；双流型汽轮机
twin-turbine torque converter 双涡轮液力变矩器
twin turbo-charger 双涡轮增压器
twin turret rolling-up machine with slitters 带纵切刀双位回转卷绕机
twin-twin landing gear 双复式着陆架
twin-twisted 双股扭绞的
twin-twisted bar reinforcement 双股扭合钢筋；双股螺纹钢筋
twin-twisted bars 双股扭合钢筋；双股螺纹钢筋
twin-twisted reinforcement 双绞钢筋；扭结钢筋
twin-twisted round bars 双螺纹面钢筋；双绞圆钢筋
twin-twisted strand 双绞股线
twin-twisted wire strand 双股扭绞钢丝束；双股绞钢丝束
twin tyres 双轮胎；双料轮胎
twin valve 双出口阀
twin vessel conveyer 双仓泵
twin vibrator 双振动器
twin vibratory pile hammer 双筒振动式打桩锤
twin vibratory roller 双筒振动压路机；双筒振动(路)碾
twin volute pump 双蜗壳式泵；复式螺旋泵
twin-volute water pump 双蜗壳水泵
twin-walled spacecraft 双壁航天器
twin water turbine 双流式涡轮(机)；双流式透平；双流式汽轮机；双工作轮涡轮机；双工作轮水轮机
twin wave 孪生波；双生波
twin-webbed 双腹板的
twin-webbed plate girder 双腹板梁
twin-webbed T-beam 双腹板T形梁
twin well 双井
twin wheel loading 双轮负荷
twin-wheel nose gear 双轮前起落架
twin wheel twin nozzle type turbine 双轮双喷嘴式水轮机
twin winding head 双缠绕头
twin wire 双线缆；双股线
twin-wire arch 双丝弓
twin-wire control 双线控制；双线操纵
twin wire parallel power submerged arc welding 双丝并列埋弧焊
twin wire series power submerged arc welding 双丝串联埋弧焊
twin wood(en) mast 双木桅杆；双木柱
twin worm 双蜗杆
twin zone furnace 两段式加热炉
T-wire 塞尖引线
twirl 扭转，快速转动
twist 三角坑；扭曲；绞合；扭转(度)；扭弯；捻度；捻；挠曲；歪曲；水龙卷；搓合；搓；编织
twist-against-twist 反向加捻
twist amplifier 扭转放大器
twist-and-contraction meter 捻度试验仪
twist and steer control 联合操纵系统
twist angle 旋转角
twist a tail 施加压力
twist auger 螺旋钻；麻花钻
twist balance 捻度平衡
twist balance index 捻度平衡指数
twist bar 来复线；扭杆
twist barrel 金属缠制枪管
twist bit 螺纹钻；麻花(活)钻头；蛇形钻；螺钻
twist bit twist drill 麻花钻头

twist blades 扭转叶片
twist boat 扭船式
twist boundary 扭转晶界；扭曲型间界
twist center 铰钮中心；扭轨记录
twist centre of member section 杆件断面扭转中心
twist chair 扭椅式
twist change over region 捻度转换区
twist coil chain 扭节链
twist constant 捻度常数
twist contraction 捻缩
twist contraction rate 捻缩率
twist counter 测扭仪；测扭器
twist course 扭曲试验跑道
twist direction 加捻方向
twist distribution 扭转分布
twist drill 螺旋钻头；麻花钻；扳钻
twist-drill cutter 麻花钻槽铣刀
twist drill ga(u)ge 螺旋钻径规；麻花钻径规
twist drill grinder 麻花钻研磨器
twist drill grinding machine 麻花钻头刃磨机
twist drill spear 打捞工具
twisted acacia 扭曲相思树
twisted asbestos yarn 合股石棉线
twisted auger 取土样的麻花钻；螺旋钻；取土麻花钻
twisted bar 螺旋钢筋；螺纹钢筋；扭转钢筋；扭杆
twisted bevel 扭曲边
twisted bit 螺旋钻
twisted blade 扭转车叶；扭叶片；扭曲叶片
twisted board 扭曲板
twisted cable 绞合电缆
twisted cable ornament 绞合缆索装饰
twisted colo(u)red body 绞色胎
twisted column 螺旋形柱；绞绳柱；绞绳形柱；麻花形柱；涡卷圆柱
twisted conductor 分层绞线
twisted connecting rod 扭转连杆
twisted cord 绞合软线；双绞软线；双绞软线
twisted curve 挠曲线
twisted deformed bar 扭转变形钢筋
twisted effect 扭曲效应
twisted fabric 冷扭钢筋网
twisted fiber 扭曲纤维；合股纤维
twisted-fibred 扭转纹的
twisted fillet 螺纹形嵌条
twisted fringe 花纹边；麻花条纹
twisted grain 交错木纹；(木材的)扭转纹理；扭曲木纹
twisted-grained 扭转纹的
twisted growth 木材节疤；扭转纹；木材扭曲纹
twisted joint 扭接
twisted knotless net making machine 捻结织网机
twisted-lead transposition 扭线换位
twisted line 绞线；扭绞线
twisted magnetic field 扭绞磁场
twisted pair 扭绞双股电缆；扭绞双线电缆；扭绞对；扭绞二线电缆；双芯绞合电缆
twisted-pair cable 双纽绞电缆；双芯绞合电缆；双绞线；对绞电缆
twisted-pair feeder 双绞式馈线
twisted-pair line 双组线；双绞线；缠绕双线
twisted-pair wire 双绞线
twisted pair wiring 双芯绞合线；扭绞二股线
twisted plate 弯曲板
twisted rod 螺纹杆；螺纹钢筋
twisted rope 螺旋钢索；搓捻绳索
twisted rope frieze 绳纹壁缘；麻花壁缘
twisted shell bit 螺壳木钻头
twisted shielded pair cable 双绞屏蔽电缆
twisted silver fir 扭枝银枞
twisted sleeve joint 金属套管扭接；扭接套管接头
twisted square bar fabric 扭转钢筋网
twisted steel 螺旋形钢筋；螺纹钢(筋)；扭转钢筋
twisted steel bar 螺纹钢筋
twisted steel mat 螺纹钢筋网
twisted steel sheet 螺纹钢筋网
twisted stratum 曲折地层；扭曲地层
twisted straw 搓合的禾杆
twisted straw deafening 搓合的禾杆隔声材料
twisted thread 捻线
twisted thread canvas 双线帆布
twisted waveguide 扭曲波导管
twisted weave 捻织
twisted winding 绞合绕组
twisted wire 绞合线；扭绞钢丝；麻花线

twisted wire anchor 弯钩锚固铁件
twisted wire cloth 金属绞丝布
twisted wire tie 金属缆拉索
Twisteel 扭纹钢(一种钢筋牌号)
twist equalizer 扭型均衡器
twister 拧结器;陆龙卷(风);绞扭器;加捻装置;捻线机;受扭晶体;缠绕物
twister bimorph 扭转双层元件
twister feeder 绞合馈电线
twister pinion 扭结小齿轮
twister ring 捻线钢领
twister winder 帘子线倍拈机
twist factor 捻系数;捻度系数
twist-fibrillated tape 捻裂法原纤化薄膜条
twist flat drill 麻花平钻
twist-free 不扭转的
twist fringing device 捻边装置
twist gear 捻度齿轮
twist gimlet 螺纹锥;麻花(手)钻;手摇钻;螺旋锥;螺旋手锥
twist grip 可转动把手
twist grip throttle control 把手油门掣(摩托车)
twistiness 弯曲性
twisting 加捻;扭曲;扭转
twisting action 扭转作用
twisting apparatus 测挠仪
twisting bar 扭绞导条
twisting box 扭转导板
twisting buckling 屈曲;扭曲
twisting buckling load 扭曲荷载
twisting channel 游荡性水道;蜿蜒水道;蜿蜒河道;摆动河道;摆动河床
twisting coefficient 扭曲系数
twisting count 捻数
twisting couple 旋转力矩;扭转力矩;扭转力偶
twisting degree 扭转程度;扭曲程度
twisting die 扭模
twisting effort 扭力
twisting error of tower 塔标扭转误差
twisting error of tripod 脚架扭转误差
twisting failure 扭转破坏
twisting force 扭力
twisting frame 捻丝机;捻线机;并线机
twisting-in 捻接法
twisting instability 扭曲不稳定性
twisting joint 扭接
twisting load(ing) 扭转荷载;扭转负荷;扭荷载
twisting machine 绞线机
twisting meander system 曲流系统
twisting mechanism 拧结机构
twisting modulus 扭转模量
twisting moment 扭力力矩;捻矩;旋转力矩;转矩;扭矩;扭力矩
twisting moment diagram 扭矩图
twisting moment distribution 扭矩分布
twisting motion 扭绞运动
twisting number 捻度
twisting of wire 钢丝绳的扭转
twisting pass 扭转孔型
twisting point 扭点
twisting resistance 扭阻力;扭转力
twisting rigidity 扭曲刚度;抗扭刚度
twisting rod 扭杆
twisting section 蜿蜒河段
twisting stick 花兰螺栓
twisting stiffness 扭曲劲度
twisting strain 扭转应变;扭应变
twisting strength 扭曲强度;抗扭强度;扭转强度
twisting stress 扭转应力;扭应力
twisting test 扭转试验;扭曲试验;扭力试验
twisting tester 扭力试验器
twisting torque 捻矩
twisting unit stress 单位扭应力
twisting vibration 扭转振动
twisting wire 扭转的钢丝绳
twist irregularity 捻度不匀
twist joint 绞接(头);扭接(头);扭绞连接
twistleaf nordmann fir 扭叶高加索冷杉
twist-less 不扭转
twist level 捻级
twist link type chain 扭节式胎链
twist-link type tyre chain 扭节式胎链
twist-lint chain 麻花链环
twist liveliness 捻度不稳定性

twist-lively 捻皱缩
twist lock 旋锁;转锁;扭锁
twist lock pin 扭销
twist lock span 扭锁销
twist machine 拧转式放炮器
twist mapping 扭转映射
twist migration 捻度转移
twist moment 扭矩
twist motion 扭转运动
twist multiplier 粘度系数;捻度系数
twist nut 螺纹螺母
twist-off 扭断;扭转断裂;扭开
twist-off closure 拧开瓶盖
twist-off on the bottom 钻头切削部分损坏(因压力过大或转速过高的)
twist of pillar 观测台扭曲
twist of rifling 膛线缠度
twist of track 三角坑【铁】
twist-on explosives 炸药捻线
twist opening 旋流风口
twistor 磁扭线
twistor feeder 绞合馈电线
twistor storage 磁扭线存储器装置
twist outlet 旋流风口
twist pair cable 对绞电缆
twist pair type telephone cable 扭绞四芯电话电缆
twist pin 扭销
twist propagation 捻回传递
twist retraction 捻缩
twist rotation 扭转
twist rug 加捻经纱地毯
twists and turns 曲折
twist setter 定捻器
twist shrinkage 捻缩
twist stress relaxation 扭力松弛
twist system 换位制
twist takeup 捻缩
twist test 抗扭试验
twist tester 检捻器;捻度试验机;捻度测试仪
twist testing machine 捻度试验机
twist tube threading 捻管生头
twist type piston ring 扭曲式活塞环
twist warp 扭转翘曲
twist waveguide 扭旋波导管;扭波导
twist web 麻花钻心
twist wire weave 捻织网
twisty road 迂回道路;盘旋道路
Twitchell reagent 特威切耳试剂
twitcher 拉抹;角边抹子
two-acting hinge 双向铰链
two-action line 双向传输线
two-action trunk 双向中继线
two-address 两地址;二地址;双地址
two-address architecture 二地址结构
two-aerial synchronized satellite 双天线同步卫星
two-aircraft element 双机组
two-aisle 两楹间房屋
two-aisle building 两披肩房;双走廊房屋;带两披肩的房屋
two-aisle cabin 双过道客舱
two-anchor mooring 双锚系泊;双锚锚泊
two-and-a-half brick wall 六二墙;两砖半墙
two-and-a-half-inch map 二英寸半地图
two-and-two check 双经双纬格子花纹
two-and-two twill 二上二下斜纹
two-aperture electronic lens 双孔电子透镜
two-aperture lens 双孔透镜
two-aqueous phase extraction 双水相萃取
two-armature 双电枢
two-armature direct current motor 双枢直流电动机
two-armature generator 双电枢发电机
two-armed spider 双臂支架
two-armed spiral pattern 双臂漩涡图样
two-arm kneader 双臂搅拌机
two-arm mooring 用双锚固定的系锚
two-arm yoke 双臂轭
two-array 二极排列
two-articulated arched frame 双铰拱形框架
two-articulated flat arched girder 双铰平拱大梁
two-articulated flat parabolic arched girder 双铰平抛物线拱大梁
two-articulated gable(d) frame 双铰山墙式框架;双铰人字形框架

two-articulated parabolic arched girder 双铰抛物线拱大梁
two-articulated segmental arched girder 双铰扇形拱大梁
two-aspect automatic block 二显示自动闭塞
two-axis autopilot 双通道自动驾驶仪
two-axis car 两轴车
two-axis equivalent circuit 双轴等效电路
two-axis hitch 双轴悬挂装置
two-axis laser gyroscope 双轴激光陀螺仪
two-axis linkage 双轴悬挂装置
two-axis spot wobble 两轴光点颤动
two-axis synchronous machine 双轴同步电机
two-axle bogie 两轴转向架;双轴转向架;二轴转向架
two-axle car 二轴车
two-axle carrier 双轴车架
two-axled coach 两轴客车
two-axled wagon 两轴货车
two-axle engine 双轴式发动机
two-axle full trailer 双轴全挂车
two-axle gear 双轴传动机构
two-axle motor unit 双轴牵引车
two-axle scraper 双轴铲运车;双轴耙运机
two-axle semitrailer 双轴半挂车
two-axle tandem roller 双轴串联式压路机;双轮压路机
two-axle trailer 双轴挂车
two-axle truck 两轴转向架;双轴货车
two-axle truck trailer 双轴货挂车
two-band 双频段
two-band filter 双带滤波器
two-band model 双带模型
two-band picture 双带图像
two-band receiver 双波段接收机
two-bank 双排的
two-bank engine 双列(汽缸)发动机
two-bank radial engine 双排星型发动机
two-banks opposing spray pattern 对喷
two-bank stage pump 双排阶段泵
two-barrel carburetor 双室气化器
two-barrel transit tube 双孔交通隧道
two-base method 双基线
two-base range finder 双基测距计
two-batch truck 装两批料卡车;双货厢卡车;双拌载重车
two-bath developer 双浴显影剂;双液显影
two-bath development 双浴显影
two-bath fixation 双槽定影
two-bath method 两浴法(化工过程);两次涂沥青法
two-bath process 双浴法
two-bay 两开间;双跨(度)
two-bay chapel 两开间小教堂
two-bay frame 两跨框架
two-bay gable(d) frame 双跨山墙式框架;双跨人字形框架
two-bay portal frame 双跨桥门架
two-bay spillway 两跨溢洪道;双闸孔溢洪道;双堰孔溢洪道
two-beaked anvil 两角铁砧
two-beam 双束
two-beam accelerator 双束加速器
two-beam diffraction 双束衍射
two-beam finisher 双括板整平器
two-beam holography 双束全息术
two-beam interference 双光束干涉;二束干涉
two-beam interference microscope 双光束干涉显微镜;双束干涉显微镜
two-beam interference microscopy 双光束干涉显微术
two-beam interference pattern 双光束干涉图
two-beam interference spectrometer 双光束干涉分光计
two-beam interferometry 双光束干涉量度学;双光束干涉测量学
two-beam pulsed accelerator 双束脉冲加速器
two-beam spectrograph 双光束摄谱仪
two-beam storage ring 双束存储环
two-beam tube 双束管
two-bearing computer 双方位计算机
two-bearing machine 双轴承型电机
two-bearings and run between 双角测向法;双并方位标
two-bed demineralizer 双床脱矿质器;双床水软

化器
two-bed guest room 双人客房；双床间客房
two-bed hydrodesulfurization 双床加氢脱硫
two-bed room 双人卧室
two-bedroom unit 两卧室单元
two-bed sleeping room 双人卧室
two-bed system 双级系统；双床装置
two-bench stripping 双梯段剥离
two-bench-type seat 双人座椅
two-berth room 双人舱；双铺舱
two-bin inventory system 两箱存储系统；两堆存货系统
two-bin system 两级存货管理制；双箱制；双接收制；分存控制法
two-bit-time-adder 双拍加法器
two-bladed 双叶的
two-bladed bit 双刃钻头
two-bladed fan 双叶风扇
two-bladed propeller 双叶推进器
two-blade drag 双刃刮路机
two-blade propeller 双叶螺旋桨
two-blade shutter 双百叶片快门
two-boat trawler 双拖网渔轮；双拖网渔船
two-body force 二体力
two-body interaction 二体相互作用
two-body orbit 二体轨道
two-body problem 二体问题
two-body satellite 双体卫星
two-boiling system 双沸系统设备
two-bolt lock 双栓锁；双舌锁；双销锁；双重锁
two-boom drill jumbo 二臂凿岩台车
two-boom rock drill 二臂凿岩台车
two-boom stope-jumbo 双臂采矿台车
two-boreholes electromagnet wave method 双孔电磁波法
two-boson 双玻色子
two-bowl calender 双辊压延机
two-bowline 双套结
two-bowl scraper 双斗刮土机；双斗铲运机
two-box spitzkasten 双箱角锥分级机
two-brick wall 两砖墙
two-brush generator 二刷发电机
two-bulb type resistance thermometer 双球式电阻温度计
two-busbar regulation 双汇流条调整；双母线调整
two-by-four 二乘四英寸木枋
two-by four method 二比四板材建造法
two-by-two taxiing 双机跟进滑行
two-cage hoisting 双罐笼提升
two-cage rotor 双鼠笼转子
two-canal recorder 双记录仪
two-can coating 两罐装涂料；双组分涂料
two-car bank 双梯厢组
two-card module 双卡模块
two-car garage 双车车库
two-carrier loss 双载波损耗
two-carrier theory 二载流子理论
two-car tandem dumper 双翻式翻车机；双车串联翻车机
two-car test 双车试验
two-casement window 双扇窗
two-cats in kiln colo(u)r 窑彩双猫
two-cavity amplifier tube 双腔放大管
two-cavity block 双孔砌块
two-cavity die 双型腔压型
two-cavity klystron 双腔速调管
two-cavity mould 双型腔铸型
two-cavity oscillator 双腔振荡器
two-cell partition 两区分割
two-cells one board horizontal form 两鼓一板水平式线(护舷)
two-cells one board vertical form 两鼓一板垂直式线(护舷)
two-cells one panel in horizontal form 二鼓一板水平式
two-cells one panel in vertical form 二鼓一板垂直式
two-centered arch 双心拱；二心拱
two-centred compound curve 二圆心复合曲线
two-centre shell model 双中心壳模型
two-chamber air reservoir 双室储[贮]气筒
two-chamber brake 双室式制动器
two-chamber electrostatic precipitator 二室静电除尘器

two-chamber filter 双腔滤波器
two-chamber lock 双室船闸；双室闸门；复式船闸
two-chamber mill 两仓磨；双仓磨
two-chamber separator mill 带选粉机的双仓磨
two-chamber vacuum furnace 双室真空炉
two-channel 双通道
two-channel ARQ equipment 二路自动请求重发设备
two-channel duplexer 双道天线收发转换开关
two-channel lump sorting machine 双道矿块分选机
two-channel monopulse radar 双通道单脉冲雷达
two-channel noise figure 双通道噪声系统
two-channel radiometer 双通道辐射计
two-channel receiver 双通道接收机
two-channel recorder 双道记录器
two-channel resolver 双通道旋变变压器
two-channel selsyn 双通道自整角机
two-channel switch 双信道开关
two-channel tracking receiver 双信道跟踪接收机
two-chord method 双弦法
two-circle goniometer 双圈测角器；双圆测角仪
two-circle reflecting goniometer 双圈反射测角仪
two-circuit 双路；双回路；双槽路
two-circuited cubic 双环道三次线
two-circuit hydraulic braking system 双线路液力制动系
two-circuit method 双电路法
two-circuit microprocessor 双电路微处理机
two-circuit prepayment meter 两路预付电度计；双路预付电度计
two-circuit receiver 双调谐电路接收机
two-circuit reception 双调谐电路接收
two-circuit resonant absorber 双路共振吸声器
two-circuit switch 双路开关
two-circuit transformer 双绕组变压器
two-circuit tuner 双回路调谐器
two-circuit winding 双路绕组
two-coat 双面涂布；涂二道漆
two-coat lime plaster 石灰砂浆抹面二遍法
two-coat metallic finish 两层金属饰面
two-coat paint 双层油漆
two-coat plaster 双层抹灰；两度抹灰
two-coat plastering 二道抹灰作业；两道抹灰
two-coat plaster work 两层抹灰工作；二道抹灰工作
two-coat system 两层作业法(涂涂料或油漆)
two-coat work 二层工作；二道抹灰工作；两层抹灰；双层铺筑法；双层铺砌法(路面)；二层细工；双层涂漆；双层抹灰
two-coil instrument 双线圈仪表
two-coil magnetic detector 双线圈磁感应检测器
two-coil method 双线圈法
two-coil selective cracking 双炉选择性裂化
two-collar tee fitting 双盘丁字管
two-colo(u)r automatic sheet-fed offset press 对开自动双色胶印机
two-colo(u)r detector array 双色探测器阵列
two-colo(u)r diagram 两色图
two-colo(u)r electronic distance measuring instrument 双色激光测距仪
two-colo(u)r hologram 二色全息图
two-colo(u)r image 双色像
two-colo(u)r indicator 双色指示剂
two-colo(u)r knop yarn 双色结子花线
two-colo(u)r laser distancer 双色激光测距仪
two-colo(u)r laser range finder 双色激光测距仪
two-colo(u)r photography 双色摄影术
two-colo(u)r photometry 两色光度测量；双色测光
two-colo(u)r presentation 两色显示
two-colo(u)r process 双色制版法；双色过程；双色复制法
two-colo(u)r pyrometer 双色温度计；双色高温计
two-colo(u)r radiolocation 双色无线电定位
two-colo(u)r reproduction 双色重现；双色再现
two-colo(u)r reticle 双色调制盘
two-colo(u)r sheet fed offset press 对开双色胶印机
two-colo(u)r tracker 双色跟踪装置；双色跟踪仪
two-colo(u)r tube 双色显像管；双色管
two-column core 双柱型铁芯
two-columned pier 双柱式(桥)墩
two-column planer 双柱龙门刨(床)
two-column radiator 双柱散热器；双柱暖气片
two-column strategy 二行战略

two-column tariff system 二栏税则
two-compartment ball mill 两仓球磨机；双仓球磨机
two-compartment bin 双室料箱
two-compartment cell 双室电解槽
two-compartment coating 两罐装涂料
two-compartment compound mill 双仓复式磨机
two-compartment floodability 两舱进水不沉性
two-compartment jig 两段式跳汰机
two-compartment mill 双仓磨；双室式磨机
two-compartment mixing drum 双室拌和筒
two-compartment oil tank 双层油罐
two-compartment septic tank 双格化粪池；双舱化粪池
two-compartment ship 两舱制船
two-compartment silo 双室立仓
two-compartment slope 双格斜井
two-compartment subdivision 二舱不沉制
two-compartment submersible 双舱式潜水器
two-compartment tank 双格储槽；双舱储槽；双室池窖
two-compartment tube mill 双转筒研磨机
two-compass crossing method 罗盘交汇法
two-component 两部分(的)；两种成分(的)；双构件(的)；双组分
two-component accelerometer 双向加速度仪；双向加速度计；双向加速度表
two-component adhesive 双组分黏合剂
two-component alloy 二元合金
two-component coating 双组分涂料
two-component concrete 双组分混凝土
two-component decay 双组分衰变
two-component electromagnetic log 双分量电磁计程仪
two-component eutectic 二元共晶
two-component flow 双组分流
two-component glue 双组分胶粘剂；双组分黏结剂
two-component gun 双口喷枪
two-component lens 双合透镜
two-component lognormal model 二组分对数模型
two-component model 二组元模型
two-component pallograph 两向振动仪；双向测振器
two-component particle 双成分粒子
two-component sealant 双组分密封膏
two-component sequence 二元结构层序
two-component spray apparatus 双组分喷涂机
two-component spray gun 双组分喷枪；双口喷枪
two-component spraying 双组分喷涂
two-component system 双构件系统；二成分系统；双组分系；双组分体系；二元物系
two-computer back-to-back system 双计算机背对背系统
two-concrete floor 双层混凝土地面
two-condition cable 双流电缆
two-condition code 双态码
two-conductor 双导体
two-conductor line 双导线线路
two-conductor shielded cable 双线屏蔽电缆
two-conductor wiring 双线布线
two-cone bit 二牙轮钻头；双锥牙轮钻头；双牙轮钻头
two-connector manifold 有双歧接头的歧管
two-container varnish 两罐装清漆；分装清漆
two-contingency system 双偶然性体系
two-control 双路操纵的
two-coordinate neutron diffractometer 二坐标中子衍射仪
two-core 双活性区
two-core block 双孔砌块
two-core cable 双芯电缆；双芯线
two-core fixture wire 双芯电器引线
two-core per bit 每位两个磁芯；双磁芯存储方式
two-core per bit system 每位两磁芯系统
two-core switch 双磁芯开关
two-core transformer 双芯变压器
two-core voltage regulating transformer 双铁芯调压变压器
two-counter machine 双计数器计算机；二计数器机器
two-couple 双力偶
two-couple method 双力偶法
two-course 双层的
two-course beacon 双向信标

two-course concrete floor 二次施工的混凝土地面;双层混凝土地面
two-course concrete pavement 双层混凝土铺面;双层混凝土路面
two-course method 双层铺筑法;双层铺砌法(路面)
two-course net 双列网眼
two-course pavement 双层铺面;双层路面
two-course radio range 双航道无线电指向标
two-course range 双航向信标
two-crew cockpit 双座座舱
two-crop-paddy 双季稻
two-crystal spectrometer 双晶体分光计
two-cup edging machine 双压盖盐磨边机
two-cusped arch 小双尖拱;内拱弧梅花雕饰的小双尖拱;双尖拱
two-cut tapping system 双线割胶制度
two-cycle 二冲程;二冲程循环
two-cycle coast 双旋回海岸
two-cycle diesel engine 二冲程柴油机
two-cycle engine 二冲程发动机
two-cycle generator 双周波发电机
two-cycle goniometer 双圈式晶体界面角测量计
two-cycle method 二次弯矩分配法;二次循环法
two-cycle principle 二程循环原理
two-cycle scheme 双循环电路
two-cycle valley 双循环谷
two-cylinder compound jacquard 双花筒复合提花机
two-cylinder compound locomotive 双缸复合机车
two-cylinder dresser 双烘筒上浆机
two-cylinder electron lens 双筒电子透镜
two-cylinder locomotive 双缸机车
two-cylinder pump 双缸泵
two-cylinder steam engine 双汽缸蒸汽机
two-cylinder turbine 双缸式涡轮机
two-cylindrical lens 双筒透镜
two-daylight press 双板压榨机
two-decision problem 双判决问题
two-deck classifier 双台分粒器;双板分粒机
two-deck drawplate oven 双层屉烤炉
two-deck dryer 双层干燥机;双层烘缸
two-decked ship 双层甲板船
two-decker 双层甲板船;双层车
two-deck(er) cage 双层罐笼
two-deck jumbo 双层钻车;二层钻车
two-deck transport 双层运输机
two-dee cyclotron 双D盒回旋加速器
two-dee system 双D盒系统
two-degree discontinuity 二维不连续面
two-degree-freedom gyro 二自由度陀螺(仪)
two-degree of freedom laser gyroscope 二自由度激光陀螺(仪)
two-degree(s) of freedom gyroscope 双自由度陀螺仪
two-demineralization system 双级除盐装置
two-derrick boom cargo handling 联杆吊货法
two-derrick cargo handling 双吊杆装卸
two-design point aircraft 双设计点飞机
two-diameter piston 双级活塞
two-digital group 二位组
two-digit code services 两位数代码业务
two-dimension 二维
two-dimensional 两维图;两维轮廓图;两维;双向的;二向的;二维的
two-dimensional δ function 二维得尔塔函数
two-dimensional action 两维作用
two-dimensional aligned 二维定向
two-dimensional angle domain 二度角域
two-dimensional array 二维阵列;二维数组
two-dimensional array grammar 二维阵列文法
two-dimensional autocorrelation 二维自相关
two-dimensional average 二维平均值
two-dimensional biharmonic equation 二维双调和方程
two-dimensional body 二维体
two-dimensional boundary layer 二维边界层
two-dimensional chromatography 两向纸色谱法;两向色谱(层)法;二维色谱法
two-dimensional code 二维码
two-dimensional comb function 二维梳状函数
two-dimensional complex autocorrelation 二维复自相关
two-dimensional composite figure 二维组合图形
two-dimensional compression method 双向压缩法
two-dimensional concatenation 二维连接关系
two-dimensional consolidation 两向固结;双向渗透固结;双向固结;二向固结
two-dimensional consolidation test 双向固结试验
two-dimensional constriction 双向收缩
two-dimensional contraction 两向收紧
two-dimensional convolution 二维卷积
two-dimensional coordinate system 平面坐标系(统)
two-dimensional copying 二坐标仿形
two-dimensional correlation 二维相关
two-dimensional cross correlation 二维互相关
two-dimensional current 二元流
two-dimensional curve 二维曲线;平面曲线
two-dimensional development method 双向展开法
two-dimensional display 平面显示器;二维显示
two-dimensional distribution 二维分布
two-dimensional drag 二维阻力
two-dimensional drawing 两维绘图
two-dimensional elasticity 两维弹性;二维弹性
two-dimensional element 二维单元
two-dimensional ensemble 二维集
two-dimensional equation 二维方程
two-dimensional extended orebody 二向延伸矿体
two-dimensional extension 二向扩展
two-dimensional extent of fissure 二向裂隙比;二维裂隙比
two-dimensional finite element method 二维有限元法
two-dimensional flow 双向流;二元水流;二元流;二向流(动);二维流(动)
two-dimensional flow net 二维流网
two-dimensional fluid 二元流体
two-dimensional Fourier transform 二维傅里叶变换
two-dimensional frame 二向框架
two-dimensional fundamental form 二维基本形
two-dimensional gas 两维气体;二维气体
two-dimensional Gaussian function 二维高斯函数
two-dimensional grating 二维光栅
two-dimensional gravity graticule 二度重力量板
two-dimensional ground motion 二维地面运动
two-dimensional hologram 二维全息图
two-dimensional hologram grating 二维全息光栅
two-dimensional holographic image 二维全息图像
two-dimensional image 二维像
two-dimensional irrotational flow 二维无旋流
two-dimensional isotropy 二维各向同性
two-dimensional jet 二维射流
two-dimensional laser 二维激光
two-dimensional laser scanning system 二维激光扫描系统
two-dimensional lattice 二维点阵
two-dimensional layout 二度空间装配
two-dimensional linear drift 二维一次漂移
two-dimensional linear shift in-variant system 二维线性平移不变系统
two-dimensionally scanned laser sensor 二维扫描激光传感器
two-dimensional map 二维地图
two-dimensional mathematical model 二维数学模型
two-dimensional matrix 两维矩阵
two-dimensional maze search 二维迷宫查找
two-dimensional memory 线送存储器;二维存储器
two-dimensional method 两元法;双向法
two-dimensional mode 二维模式
two-dimensional model 二维模型
two-dimensional Mohr circle 二维莫尔圆
two-dimensional motion 二维运动
two-dimensional normal distribution 二维正态分布
two-dimensional nozzle 二维喷管
two-dimensional nucleation theory 二维成核理论
two-dimensional optic(al) radar 二维光雷达
two-dimensional paper chromatography 双向纸色谱;双向纸层析;二维纸上色谱法
two-dimensional parallel flow 平面平行流
two-dimensional phase 二维相
two-dimensional photoconductor array 二维光电导体阵列
two-dimensional photometry 二元测光
two-dimensional picture 二维图形
two-dimensional plane model 二维平面模型
two-dimensional plane stress 二维平面应力
two-dimensional plot 双向图
two-dimensional positioning system 双向定位系统
two-dimensional pressure 二维压力
two-dimensional problem 两维问题;平面问题;二维问题
two-dimensional projection 二维平面投影
two-dimensional pseudo isostatic anomaly 二维准均衡异常
two-dimensional quadratic drift 二维二次漂移
two-dimensional random process 二维随机过程
two-dimensional random walk 二维随机游动;二维随机行走
two-dimensional rectangle function 二维矩形函数
two-dimensional reinforcement 二维增强材料
two-dimensional run-length encoding 二维行程编码
two-dimensional scan 二维扫描
two-dimensional scan(ning) technique 双向扫描术
two-dimensional section model 二维剖面模型
two-dimensional seismic model 二维地震模型
two-dimensional signal 二维信息
two-dimensional space 二维空间
two-dimensional spatial filter 二维空间滤波器
two-dimensional spectral classification 二元光谱分类
two-dimensional square lattice 二维方格
two-dimensional state of stress 二向应力状态
two-dimensional statistic(al) method 二维统计法
two-dimensional steady uniform flow 二向稳定等速流
two-dimensional steady water quality model of non-boundary river 无界河二维稳定水质模型
two-dimensional strain 二维应变
two-dimensional stress 二向应力等速流;二向应力
two-dimensional stress state 二维应力状态
two-dimensional stress system 二维应力系统;二维应力体系
two-dimensional strike slip fault 二维走向滑动断层
two-dimensional surface wave 二维面波
two-dimensional system 平面体系;二维体系;平面系
two-dimensional systematic sampling 二维等距抽样
two-dimensional tidal current 二维潮流
two-dimensional time 二维时间
two-dimensional topographic(al) data 二维地形数据
two-dimensional trend analysis 二维趋势面分析
two-dimensional turbulence 二维湍流;二维湍流
two-dimensional Turing machine 二维图灵机
two-dimensional ultrasonic tomogram 两维超声声像图
two-dimensional vector 二维矢量;二度矢量
two-dimensional water quality simulation system 二维水质模拟系统
two-dimensional wave 二维波
two-dimensional wave spectrum 双向波谱
two-dimensional weighted moving average 二维移动加权平均
two-dimension prestress 双向预应力
two-dining kitchen 两居室并带厨房、餐室的住宅
two-direction 双向
two-directional 双方向的
two-direction(al) compression 二向压制
two-directional excitation 双向游振;双向激振
two-directional focusing 双向聚焦
two-directional load(ing) 双向荷载
two-direction linear stepping motor 双向直线步进电动机
two-direction pressing 双向压制
two-direction self-aligning ball thrust bearing 双向自动调整止推滚珠轴承
two-direction thrust bearing 双向推力轴承
two-disc bit 双圆盘钻头
two-disc tensioner 双盘张力器
two-disk bit 双圆盘钻头
two-disk thermistor bridge 双片式热变电阻桥

two-door 双门;双摺门
two-door car 双门小客车
two-door (type) body 双门车身
two-dot chain line 双点画线;双点划线
two-double biological aerated filter 折叠曝气生物滤池
two-draft scale 双拉盘式轨道衡
two-drogue refueling 双锥套式加油
two-drum crab 双筒提升机构;双筒卷扬机
two-drum drag scraper hoist 双卷筒铲运提升机
two-drum hoist 双滚筒卷扬机
two-drum reel 双鼓卷纸机
two-drum slusher 双滚筒耙矿机
two-drum steam hoist 双卷筒蒸汽提升机
two-drum winch 双筒卷扬机;双筒绞车
two-drum winder 双鼓卷纸机
two-dwell motion 双停歇期运动
two-edged 两刃;双锋
two-edged knife 双面刀
two-effect evapo(u)rator system 双级蒸发装置
two-eighth method 八分之二法
two/eight lime earth 二八灰土
two-electrode chamber 双电极室
two-electrode gun 双极枪
two-electrode system 双电极系统
two-electrode tube 二极管
two-electron donor 双电子供体
two-electron shift 双电子转移
two-electron spectrum 双电子谱线
two-electron system 双电子系
two-element air ejector 双组空气抽逐器
two-element arch 单铰拱
two-element electronic system 双元件电子系统
two-element feed control 双元件馈给控制
two-element field 二元域
two-element interferometer 双元干涉仪
two-element lens 两片型透镜
two-element microphone 双元传声器
two-element regulator 双元件调节器
two-element relay 二元继电器
two-end cheese winding 双股并线络筒
two-end feeding 两端喂料
two-engine approach 双发动机进场着陆
two-engined vehicle 双发动机车辆
two-engine landing 双发动机着陆
two-engine nacelle 双发动机舱
two-engine rubber-mounted crane 双发动机汽车吊
two-engine rubber-mounted execavator 双发动机汽车挖土机
two-engine rubber-mounted slewing crane 双发动机减振悬臂起重机;双发动机减振回转起重机
two-engine rubber-mounted telescopic crane 双发动机减振伸缩臂起重机
two-engine system 双发动机系统
two-entry system of mining 双平巷掘进系统
two-ethyl butyl acetate 乙酸异己酯
two-ethyl hexoate 异辛酸酯;异辛酸盐;乙代己酸盐
two-ethyl hexyl acetate 乙酸异辛酯;醋酸异辛酯
two-ethyl hexyl acrylate 丙烯酸-2-乙基己酯
two-excimer laser 双激元激光器
two-faced plush cloth 双面绒布
two-factor analysis of variance 二因素方差分析
two-factor analysis of variance model 二因素方差分析模型
two-factor interaction 二因素相互作用
two-factor theory 双因素理论
two-factor variance analysis 双因素方差分析
two-family detached dwelling 独立双户住宅
two-family duplex 独立双户住房;双联住宅
two-family dwelling 独立双户住房;双联住宅
two-family house 两户住宅;双联式住宅;双户住宅;二联式住宅
two-fan cleaning 双风扇复选
two-feed knitter 双路进线针织机
two-field induction machine 双磁场异步电机
two-field picture 两场合成图像
two-field system 二圃制农业
two-filament bulb 双丝灯泡
two-film theory 双膜理论;双模理论
two-fixture bathroom 两件套浴室
two-fixture lavatory 两件套卫生间
two-flame burner 叉形火焰焊矩
two-flanged hollow shaft 双法兰空心轴

two-flanged shaft 双法兰轴
two-flank gear rolling tester 双面啮合检查仪
two-flight screw 双螺纹螺杆
two-flight stair(case) 两梯段式楼梯;两跑楼梯
two-floor 两层楼
two-floored 双层楼的
two-floor garage 双层汽车库
two-floor (height) panel 两层楼高的墙板
two-floor kiln 双层干燥床
two-floor stage 双层舞台
two-flow condenser 双流程冷凝器
two-flow core 双流程堆芯
two-fluid 双流体
two-fluid cell 两液电池
two-fluid chemical consolidation process 双液化学加固土壤法
two-fluid manometer 双流体压力计;双液面流量计
two-fluid mode 双流体模式
two-fluid model 双流体模型;二流体模型
two-fluid nozzle 两液喷嘴;气动雾化喷嘴;双相喷嘴
two-fluid process 双液灌浆法;双液法
two-fluid reactor 双流体反应堆
two-fluid spray nozzle 双流体喷嘴
two-fluid system 双流体系统;双工质系统
two-fluid theory 两流体理论;二流体理论
two-flute bit 双槽钻头
two-flying whales in kiln colo(u)r 窑彩双鲸鱼
two-fold 两倍的;双重(的)
two-fold axis 二重轴
two-fold axis of symmetry 二重对称轴
two-fold-classification 双重分类
two-fold coincidence 双重符合
two-fold degeneracy 二重简并(度)
two-fold diffraction 双重衍射
two-fold door 双摺门;双折门
two-fold frequency system 双频制
two-fold-gang 双联
two-fold node 二重节
two-fold purchase 双倍四滑轮滑车;双倍滑轮组;双倍复滑车
two-fold symmetry 双重对称
two-fold tackle 双倍滑轮组
two-fold window 双折窗(扇);双摺窗(扇)
two-fold yarn 双股线
two-foot 两英尺长的尺
two-force member 二力构件;两力构件
two-for-one twister 倍捻捻线机;倍捻机
two-for-one twisting machine 倍捻机
two-frame gyroscope 双框架陀螺仪
two-frame laser gyroscope 双座激光陀螺仪
two-frequency beacon 双频信标
two-frequency channel 双频率通路
two-frequency converter 双频直流变换器
two-frequency dialling 双频拨号
two-frequency duplex 双频双工制
two-frequency echo-sounder 双频回声测深仪
two-frequency fish finder 双频鱼探仪
two-frequency gas laser 双频气体激光器
two-frequency method 双频率法
two-frequency selection 双频拨号
two-frequency signal 双频信号
two-frequency signal generator 双频信号发生器
two-frequency signal(l)ing 双频制信号
two-frequency signal receiver 双频制信号接收机
two-frequency sounder 双频测深仪
two-full tidal cycles 两个全潮汐周期
two-gang 双联的
two-gang condenser 双联电容器
two-gang plough 双组犁
two-gang plow 双组犁
two-gang saw 双排锯
two-gang variable capacitor 双联可变电容器
two-gap klystron 双隙速调管
two-gate feeder 双闸门给料机;双口供料器;双口给料器
two-gate loading chute 双口装载溜槽
two-glass test 两杯试验
Two-Goals 一控双达标政策
two-grades accountability system 双级责任制度
two-grate combination cooler 两篦床复合式冷却机
two-grid 双栅的
two-group constant 双群常数
two-group critical mass 双群临界质量
two-group diffusion theory 双群扩散理论

two-group model 双群模型
two-group theory 双群理论
two-group treatment 双群处理
two-gun oscillograph 双电子枪示波器
two-gyro pendulous gyrocompass 双转子摆式罗经
two-half hitch 两半结
two-hand control 双手操纵
two-handed control 双手控制器
two-handed hammer 大锤
two-handed saw 拉锯;大锯;双人操作锯
two-handed work 双人工作
two-handled knife 双柄刮皮刀
two-head automatic arc welding machine 自动双弧焊机
two-headed wrap 双磁头卷带方式
two-head grinder 双头研磨机
two-heading entry 双平巷
two-height creel 双层纱架
two-high bloomer 二辊式初轧机
two-high (cogging) mill 二辊式轧机
two-high cold reduction mill 两辊冷轧机
two-high mill 二辊式机座
two-high plate mill 二辊式中厚板轧机
two-high pull-over mill 二辊周期式薄板轧机
two-high reversing beam mill 二辊可逆式钢梁轧机
two-high reversing blooming mill 二辊可逆式初轧机
two-high reversing-finishing mill 二辊可逆式精轧机
two-high reversing mill 二辊可逆式轧机
two-high reversing-plate mill 二辊可逆式中厚板轧机
two-high roller level(l)er 两排辊子的矫直机;二重式多辊矫直机
two-high rolling mill 二辊式轧机
two-high straddle carrier 双层跨运车
two-high universal mill 二辊式万能轧机
two-hinged 双铰的
two-hinged arch 两铰拱;双铰拱
two-hinged arch bridge 双铰拱桥
two-hinged arch truss 双铰拱桁架
two-hinged braced arch 双铰弓弦拱
two-hinged frame of one bay 双铰单跨框架
two-hinged frame of one span 双铰单跨框架
two-hinged portal frame 双铰门式刚架
two-hinged portal frame bridge 双铰门式刚架桥
two-hinged reinforced concrete arch 双铰钢筋混凝土拱
two-hinged rigid frame 双铰刚架
two-hinged structure 双铰结构
two-hinge frame 双铰框架;双铰构架;双铰刚架
two-Hi universal type slabbing mill 二辊式万能板坯初轧机
two-hole coupler 双孔耦合器
two-hole directional coupler 双孔定向耦合器
two-holer 双出口厕所
two-hole wedge 双孔楔子
two-hole whirl plate 双孔旋流片
two-hologram method 双全息图法
two-hop-E 第二E层波
two-hop transmission 二次反射传输
two-horned anvil 两角铁砧
two-horse chariot 双马牵引的四轮车;纪念碑上的双马战车塑像
two-humped camel 双峰驼
two-hundred kg samples coke oven test 二百公斤焦炉炼焦试验
two-image photogrammetry 双像摄影测量学
two-impulse trajectory 双冲轨道
two-independent ignition 双源独立点火
two-in-hand winding 双路叠绕法
two-in-one motor 二合一电动机
two-in-one oedometer 双联固结仪
two-input adder 双输入加法器;二输入端加法器;半加(法)器
two-input gate 双输入门
two-input module 二输入端组件
two-input servo 双输入伺服系统
two-input subtracter 半减(法)器
two-input switch 双输入开关
two-integrator system 双积分器系统
two-jaw(ed) chuck 两爪卡盘;两爪夹盘;双卡瓦卡盘

wo-jaw grab 双瓣(式)抓斗
two-jet carburetor 双喷管化油器
two-jet flapper valve 双喷嘴挡板阀
two-joint drive 双万向节传动
two-knob lock 双头开关器
two-lane 双车道
two-lane bridge 双车道桥(梁)
two-lane canal 双航道运河;双线运河
two-lane channel 双线航道
two-lane highway 双车道公路
two-lane road(way) 双车道道路
two-lane canal 双船道运河
two-lane single carriage 双车道单幅车行道;单幅双车道
two-lane traffic 双车道交通;双车交通
two-lane tube 双车道隧管
two-lane tunnel 双车道隧道
two-lap lapping machine 双盘研磨机
two-large divisions 两大部分
two-layer 双层(的)
two-layer ballast 双层道砟
two-layer cast stone 双层铸石
two-layer coating 双层镀膜
two-layer columnar joints 双层柱状节理
two-layer construction 双层施工;双层构造
two-layered 双层的
two-layered cylinder 双层筒
two-layered problem 双层次问题
two-layered space frame shell 双层空间壳架
two-layered texture 双层结构
two-layer electrode 双层电极
two-layer equation 双层式(路面设计)公式;双层式
two-layer fabric 双层织物
two-layer flow 双层流
two-layer formula 双层式(路面设计)公式
two-layer laminates 双层层压制品
two-layer lattice 双层晶格
two-layer level bridge 双层桥
two-layer metallic effect paint 双层金属闪光漆
two-layer mineral 双层矿物
two-layer mode 二层模式
two-layer model 两层模型
two-layer ocean 两层模式大洋
two-layers of wood(en) joists 双层木格栅
two-layer spiral coil 双层螺旋式线圈
two-layer structure 两层构造
two-layers type 双层型
two-layer system 双层系统
two-layer theory 双层体理论
two-layer type structure 二层型结构
two-layer weave 双层织造
two-layer winding 双层绕组
two-lay sling 双吊索
two-leading shoe brake 双助势蹄制动器
two-leaf 双折的;双叶饰
two-leaf door 两扇门
two-leafed door 双折门;双摺门
two-leafed window 双折窗;双摺窗
two-leaf flexible swing door 双折弹簧门
two-leaf hinged shutter door 双折铰接百叶门
two-leaf sliding shutter door 双折推拉百叶门
two-legged lewis 双腿起重爪
two-legged singlephase transformer 双柱式单相变压器
two-legged transformer 双柱式变压器
two-leg lavatory 双柱洗脸盆
two-leg mooring 用双锚固定的系锚
two-leg pipe derrick 双腿管制钻塔
two-leg propeller strut 人字尼龙扶架
two-leg sling 双索链钩;两索套
two-legs of breech pipes 两叉连管
two-leg strut 双支脚尼轴架
two-lens 双透镜
two-lens objective 双透镜物镜
two-lens ocular 双透镜物镜;双透镜目镜
two-lens system 双透镜系统
two-level 双层的
two-level address 二级地址【计】
two-level addressing 二级寻址;二级定址
two-level allocation 二级分配
two-level building operational system 双层施工体系
two-level energy system 二能级系统
two-level flocculation and sedimentation tank 双层絮凝沉淀池
two-level fork junction 双层式Y形立体交叉
two-level formula 两层公式
two-level garage 双层车库
two-level grammar 二级文法
two-level junction 双层式立体交叉;双层立交
two-level logic 两级逻辑;双电平逻辑
two-level medium 二能级介质
two-level microprogram(me) control 二级微程序控制
two-level mold 双层塑模
two-level nonreturn system 双位无回程系统
two-level operation 双层次施工;双层次操作
two-level priority interrupt 二级优先中断
two-level recording 二级记录;二级记录
two-level return system 双位有回程系统;双级归零记录系统;二级归零记录系统
two-level rotor 双层转子
two-level roundabout junction 双层式环形立体枢纽;双层式环形立体交叉
two-level scheme 双级方案
two-level signal 双亮度信号灯
two-level stage 双层舞台
two-level subroutine 二级子例行程序;二级子程序
two-light candlestick 双蜡烛烛台
two-light frame 两扇窗框
two-light headlamp 双光前大灯
two-light lamp 双光灯泡
two-light window 李窗;双窗
two-linear array 二数组法
two-line clamshell 双绳抓岩机;双绳多瓣式抓岩机
two-line grab (bucket) 双绳吊桶;双绳抓斗
two-line ground 双线接地
two-line mixtures 两个品系的混合
two-line string up 双绳提升
two-linked 双铰接的
two-lip end mill 双刃端面铣刀
two-lip end milling cutter 双面刃端铣刀
two-lip hand ladle 双嘴手杓
two-lip ladle 双嘴勺
two-liquid centrifuge 两液离心分离机;双液离心机
two-lobe blower 双叶转子鼓风机
two-lobe cam 双凸块凸轮
two-lobe chamber 双弧工作室
two-lobed epitrochoidal housing 双弧外旋轮线缸体
two-lobed epitrochoidal working chamber 双弧外旋轮线工作室
two-lobed epitrochoid rotary mechanism 双弧外旋轮线旋转机械
two-lobe epitrochoidal bore 双弧长短幅外旋轮线汽缸
two-lobe epitrochoidal shape 双叶长短辐圆外旋轮线形状
two-lobe pump 双叶片水泵;双突瓣式泵;双凸瓣式水泵
two-lobe rotary pump 双凸轮旋转泵
two-lock recompression chamber 双闸门减压舱
two-log washer 双转轴细矿机
two-loop sampling valve 双环管进样阀
two-loop servomechanism 双回路伺服机构
two-loop temperature controller 双环温度控制器
two-luminous points of equal intensity 两个等光强的发光点
two-machine jumbo 双机台车
two-machine operation 双机操作
two-magnet extraction system 双磁铁引出系统
two-magnification rangefinder 双放大率测距仪
two-main system 双干管制;双干管系统;双干管式
two-man crew 双人乘员组
two-man crew station 双人乘员组位置
two-man cross-cut saw 双人锯
two-man foxhole 双人散兵坑
two-man ladle 抬勺
two-man policy 二人管理制度
two-man riprap 两人搬得动的抛石
two-man rule 双人制;二人管理制度
two-man saw 双人锯
two-man steel shelter 双人钢架掩蔽部
two-mass oscillator 双料振荡器;双重振荡器
two-mass vibrator 双料振荡器
two-master 双桅船
two-media photogrammetry 双介质摄影测量(学)
two-men optic(al) system 双人光学瞄准系统
two-mesh filter 双节滤波器
two-meter folding rule 两米折尺
two-meters contour interval 两米等高距
two-meter steel tape line 两米钢带尺
two-meter tape line 两米带尺
two-methods of control and five aspects of improvement 两管五改
two-metre folding rule 两米折尺
two-metre steel tape measure 两米钢带尺
two-metre tape measure 两米带尺
two-mica gneiss 二云母片麻岩
two-mica granite 二云花岗岩
two-mica quartzite 二云母石英岩
two-mica quartz schist 二云母石英片岩
two-mica schist (leptynite) metabasite formation 二云母片岩—变基性岩建造
two-mill type of grinding 两级粉磨;双机粉磨流程
two-mirror anastigmat 双镜消像散镜
two-mirror objective 双镜物镜
two-mirror system 双反射镜系统
two-mode turbulence 二模式湍动
two-mold machine for forming brick 双模制砖机
two-months vote 二月法投票表决
two-motion selector 两级动作选择器
two-motion switch 两级动作开关;双动作选择器
two-motor car 双电动机车
two-motor drive 双动力驱动
two-motored airplane 双发动机飞机
two-movable jaw crusher 可卸双颚板破碎机
two-neck bottle 双颈瓶
two-needle bar 双针床
two-noded vertical vibration 双节点垂直振动
two-noded vibration 双节点振动
two-nozzle gun 双喷口喷枪
two-oblique photograph 双镜头倾斜航摄照片
two-oblique photography 双镜头倾斜摄影术
two-out-of-three system 三取二系统
two-oxide ferroelectrics 双氧化物铁电体
two-pack 双杆件(的);双副;分包装
two-package 双组分
two-package catalyst urethane coating 双组分催化聚氨酯涂料
two-package formulation 双组分配方
two-package polyol urethane coating 双组分多元醇聚氨酯涂料
two-package polyurethane coating 两罐装聚氨酯涂料
two-pack coating 双包装涂料
two-pack component system 双组分系
two-pack system 双组分式;双包装
two-pack varnish 两罐装
Two-Pagoda 双塔寺(中国泉州开元寺)
two-pair front 三楼前房
two-panel door 双腹板门
two-panelled door 双镶板式门;双腹板门;双格板门;两格式门;两镶板式门
two-pane sash 双格窗框
two-parameter 双参数
two-parameter adaptive control system 双参数自适应控制系统
two-parameter control 双参数控制
two-parameter flow 双参数流动
two-part 分包装;双杆件(的);两部分(的);两种成分(的)
two-part adhesive 分售式黏合剂;双组分黏结剂;双组分胶粘剂;二元黏合剂
two-part bearing 分轴承
two-part buoy system 双浮标系统
two-part cracking 两部分裂化
two-part experiment 二部实验
two-part formulation 两料配制;两料配方;二元配制;两元配方
two-part glue 两合胶
two-part line 回绕绳;回绕线
two-part line tackle 双轮复滑车
two-part mo(u)ld 两箱造型
two-part mo(u)lding 两箱造型
two-part paint 双组分涂料
two-part prepayment meter 两部预付两度计
two-part putty 双料配制的灰泥
two-part sealant 双组分密封膏;双组分密封材料;双组分密封胶
two-part step-rate prepayment meter 两部阶跃

预付电度计
two-part step tariff 两部制费率
two-part tariff meter 两部制收费煤气表
two-part up and over door 双折上翻门
two-party draft 双名汇票
two-pass 双行程
two-pass boiler 双回程锅炉
two-pass condenser 双流式凝汽器；双路冷凝器；双道冷凝器；双程冷凝器
two-pass cooling 双路冷却
two-pass grate 二段式链箅机
two-pass horizontal regenerator 双通路卧式蓄热室
two-pass superheater 双流程蒸汽过热器
two-pass surface treatment 双层表面处理
two-pass type 双程式
two-path 双通道
two-path circuit 双路电路
two-path cylinder dryer 分路式烘筒
two-path distance measuring system 双路测距装置
two-path feedback 双路反馈
two-pedal control 双踏板控制
two-pedal operation 双踏板操纵
two-peg method 两点校正法【测】
two-pen chart recorder 双笔图表记录器
two-pendulum gravimeter 双摆重力仪
two-pendulum method 双摆法
twopenny nail 四英寸大钉；一英寸号钉
two-pen potentiometer continuous recorder 双笔电位计式连续记录器
two-pen recorder 双笔记录器
two-pen recording thermometer 双针记录式温度计
two-period cumulative multiplier 两期积累乘数
two-person car 双座汽车
two-person flat 双人公寓
two-person game 双人博弈
two-person zero sum game 两人零和对策；二人零和博弈
two-phase 两相；双相（的）
two-phase additive 两相添加剂
two-phase aeration 两级曝气
two-phase aggregate 二相集合体
two-phase alloys 双相合金
two-phase alternating current circuit 两相交流电路
two-phase alternating current servo 二相交流伺服系统
two-phase anaerobic digestion 两相厌氧消化
two-phase anaerobic digestion system 两相厌氧消化系统
two-phase anaerobic fluidized bed reactor 两相厌氧流化床反应器
two-phase anaerobic-oxic process 两相厌氧—好氧工艺
two-phase biological contact oxidation 两相生物接触氧化
two-phase biological fluidized bed 两相生物流化床
two-phase coherent diffractogram 双相共格衍射图
two-phase composite 两相复合材料
two-phase connection 两相接法
two-phase contactor 两相接触器
two-phase current 两相电流；双相电流
two-phase design 两次设计；二阶段设计；二次设计
two-phase digestion process 两相消耗法
two-phase equilibrium 两相平衡；双相平衡
two-phase five-wire system 两相五线制
two-phase flow 两相流；两相水流；两相流（动）；两相混合流；双相（水）流；二相流
two-phase four-wire system 两相四线制
two-phase generator 两相（交流）发电机；二相发电机
two-phase ideal sand 二相理想砂土
two-phase induction motor 二相感应电动机
two-phase ion exchange column 两相离子交换柱；二相环
two-phase loop 二相环
two-phase method 二段法
two-phase morality 双相死亡现象
two-phase operation 两相运行
two-phase permalloy 两相透磁钢
two-phase plug-flow reactor 两相推流式反应器

two-phase region 双相区
two-phase relay 双相继电器
two-phase reservoir 两相热储
two-phase sampling 两相取样
two-phase selsyn 两相自整角机；两相自动同步机
two-phase separation 两相分离法
two-phase shift keying 二相移键控
two-phase soil 二相土
two-phase system 两相制；二相制；二相系统
two-phase three-wire system 两相三线制；二相三线制
two-phase titration method 两相滴定法
two-phase transformer 双相变压器
two-phase wet unit weight 二相湿容重
two-phonon excitation 双声子激发
two-photon absorption 双光子吸收
two-photon fluorescence method 双光子荧光法
two-photon fluorescence monitor 双光子荧光监测器
two-photon holography 双光子全息术
two-piece 两半的；两部分的；二片
two-piece apex seal 双片式径向密封片
two-piece bearing 可调轴承；拼合轴承；对开轴承
two-piece calotte 双护盖
two-piece can 两片罐
two-piece chassis 双节底盘
two-piece drop system 双管道向下供暖系统（上部分配集中供暖）
two-piece glue-jointed panel 双片胶粘板
two-piece housing 剖分式外壳；拼合式外壳
two-piece laminated insulating glass 双层叠合绝缘玻璃板
two-piece laminated safety sheet glass 双层黏合安全玻璃
two-piece link 拼合式链节；部分式链节
two-piece mo(u)ld 双件模
two-piece panel 双层板
two-piece set 两构件支架；两件套
two-pier elbow draft tube 双墩肘形尼水管
two-pin escapement 双销擒纵机构
two-pin frame 双枢框架；双铰框架
two-pinned 双铰接的
two-pinned arch 两铰拱；双铰拱
two-pinned arch bridge 双铰拱桥梁
two-pin plug 两脚插头；双柱插头；双线插头
two-pin rigid frame 双铰刚架
two-pin socket 双铰承窝；双插销座
two-pin strap 双孔联板
two-pipe culvert 双管涵洞
two-pipe direct return system 双管直接回水（供暖）系统
two-pipe direct water downfeed system 下给式双管热水供暖系统
two-pipe heating system 双管供暖系统；双管采暖系统
two-pipe hot water downfeed system 下给式双管供暖系统；双管下给式热水供暖系统
two-pipe hot water heating 双管热水供暖
two-pipe hot water system 上给式双管热水系统
two-pipe hot water upfeed system 上给式双管热水（供暖）系统；双管上给式热水供暖系统
two-pipe meter 双管计
two-pipe plumbing 双管卫生工程管道
two-pipe rising system 双管道向上送热供暖系统（下部分配集中供暖）
two-pipe steam heat-supply network 双管制蒸汽热网
two-pipe system 双管道排水系统；双管供热系统；双管系统；双管供水系统；双管供暖系统；双管采暖系统
two-pipe water system 两管制水系统
two-pitch airscrew 双距螺旋桨
two-pitch propeller 双位变距螺旋桨
two-pit tannage 二槽鞣法
two-place interceptor 双座截击机
two-plane balancing 双面平衡
two-plane theory 双面说
two-plane theory of chiasma 双面交叉说
two-plane type 双面式
two-plane-wave hologram 双平面波全息图
two-plate clutch 双片式离合器
two-plate press 双板压榨机
two-plug cementation 双木塞注水泥（法）
two-plug-in reducing pipe 双插渐缩管

two-plug-in reducing piping 双插渐缩管
two-plug-in reducing tube 双插渐缩管
two-plug-in reducing tubing 双插渐缩管
two-plus-one address 二加一地址
two-ply 两层板；两合板；双重；双股；双层(胶合)板
two-ply belt 双股皮带
two-ply bridge floors 双层木板桥面
two-ply laminating machine 双贴层压机
two-ply riveting 双层铆接
two-ply stock 双重纸
two-ply tyre 双层轮胎
two-ply vacuum exhaust station 双真空排气台
two-plywood 薄层镶面胶合板
two-ply yarn 双股线
two-pocket bore 双套钻
two-pocket magazine grinder 双袋库式磨木机
two-point 双接点；双点；二接点
two-point arbitrage 两点套汇；两地套汇
two-point arbitrage in foreign exchange 两角套汇
two-point bit 一字形钻头；一字形钎头；双刃钻头
two-point boundary value problem 两点边值问题
two-point breaker 双触点断电器
two-point cam 双夹瓣三角
two-point detection 两点检查法
two-point discrimination 两点辨别
two-point distribution 两点分布
two-point focus setting 双点调焦
two-point form 两点式
two-point lifting 两点悬吊法
two-point linkage 两点悬挂装置
two-point loading 两点加载
two-point method 两点法
two-point method of stream ga(u)ging 两点测流法
two-point mooring 两点系泊；双浮筒系泊；双浮标系泊
two-point perspective 二点透视；两点透视
two-point press 双点压力机
two-point problem 两点问题
two-point recorder 双点记录器
two-point resolution 两点分辨率
two-point rod extensometer 二点杆式伸缩仪
two-point spark plug 双电极火花塞
two-point suspension scaffold 两点摇摆式脚手架；两点悬吊式脚手架
two-point test 两点试验法
two-polar 双极的
two-pole 两极
two-pole field circuit breaker 双断口灭磁开关
two-pole knife switch 双极闸刀开关
two-pole of near source 近场源二极
two-pole plug 二眼插头；二极插头
two-pole switch 双极电门；双极开关
two-pool plan 双池布置（潮汐发电）；双池式布置
two-port 四端；二端对；双口
two-port junction 两端口接头
two-port network 二端对网络
two-port parameter 四端网络参量；四端参量
two-port system 二端对系统
two-port transducer 四端网络传感器
two-port valve 二通阀
two-port waveguide-junction 双引出线波导管连接
two-position 二位式；双位置
two-position action 双位控制作用；双位置作用
two-position control 开关控制；两位调节；双位（置）控制；双位调节
two-position controller 两位控制器；双位控制器；双位调节器
two-position differential gap 双位差隙
two-position differential gap action 双位差隙作用
two-position differential gap control 双位差隙控制
two-position diffrential control 二位置微差控制
two-position discontinuous temperature control 双位不连续温度控制
two-position four-way valve 二位四通阀
two-position mode 双位制
two-position propeller 双位变距螺旋桨
two-position relay 二位置继电器；双位继电器
two-position signal 二位信号机；双位信号；双位式信号机；双重位置式信号；二位式标志
two-position system 双位制
two-position temperature controller 双位温度控制器

two-position valve 双位阀;二位阀
two-position viewfinder 双焦取景器
two-position winding 双层绕组
two-post (car) lift 双柱汽车举升器
two-pot coating 两罐装涂料
two-pot system 双壶系统;双锅系统
two-price 二重价格
two-price house 两价商店
two-price system 两价制
two-prism device 双棱镜仪
two-prism square 双棱镜直角器
two-probe arrangement 双探针装置
two-product test 双产品试验
two-projector method of relative orientation 单独像对相对定向
two-prong claw hook 双叉爪钩
two-prong rope grab 双叉式钢绳捞矛
two-prong sharp retractor 锐双齿牵开器
two-proton decay 双质子衰变
two-proton radioactivity 双质子放射性
two-pulse 双脉冲
two-pulse canceler 双脉冲对消器
two-pulse counting circuit 双脉冲计数电路
two-purpose plough 两用犁
two-push-down tape machine 两下推带机器
two-quadrant multiplier 二象限乘法器
two-quarter brick 半砖
two-quarters 二分之一砖;半砖;半头砖
two-race mill 双球程磨机
two-rail railway 双轨铁路;双轨铁道
two-rail scraper 双轨刮料机
two-rail single car railway 双轨条单车缓行器
two-rail surface track 双轨地面铁路
two-rail system 双轨系统
two-raker 双耙侧向搂草机
two-ram hydraulic steering gear 双缸式液压操舵装置
two-range 双距离
two-range Decca 双程台卡导航系统;双距离式台卡导航系统
two-range winding 两平面绕组;双排绕组
two-ranks of concrete pile curtains 双排混凝土桩防渗墙
two-rate meter 两种计数的测量表;双额电表
two-rate register 双价电度累计装置
two-rate two-part prepayment meter 双价率两部电度预付计
two-ray interference 双光束干涉
two-reaction machine theory 双反应电机理论
two-reaction method 双反应法
two-reaction theory 双反应论
two-redundant code 二冗余码
two-region 双区
two-region reactor 双区堆
two-resistance theory 双阻力理论
two-resonance maser 双共振微波激射
two-resonator klystron 双腔速调管
two-resonator klystron amplifier 双腔速调管放大器
two-revolution printer 二回转印刷机
two-revolving field method 双旋转磁场法
two-ribbon flare 双带耀斑
two-ring piston 双环式活塞
two-ring storage system 双环存储装置
two-rod piston 双活塞杆的活塞
two-roll calender 双辊压延机
two-roll chain driven crusher 链传动双辊破碎机
two-roll coater 双辊涂漆机
two-roll crusher 双滚轴破碎机
two-roll embossing machine 双辊压花机
two-roller 双辊磨
two-roller bearing 双滚柱轴承
two-roller bit 双牙轮钻头
two-roller skew-rolling mill 二辊式斜轧机
two-roll grizzly 双辊轴筛
two-roll machine 双辊磁选机
two-roll mixing warm-up mill 双辊混合加温磨机
two-roll piercer 二辊穿孔机
two-roll press 双辊挤压机
two-roll screen 双辊筛
two-room camera 明室照相机;明室暗室合制版照相机
two-room house 两室住宅
two-rope grab hoist 双绳抓斗绞车

tworotor 双缸转子内燃机
two-row ball bearing 双列球轴承
two-row cylinder 双排汽缸
two-row elevator-digger 双行挖掘升运器
two-row hatch 两横列舱口;两列舱口
two-row machine 双行作业机具
two-row method 双排炮孔爆破法
two-runner turbine 双转轮式涡轮机;双转轮式水轮机;双流式涡轮(机);双流式透平;双流式汽轮机;双工作轮涡轮机;双工作轮水轮机
two-runway field 双跑道机场
two-sample test 双样本检验
two-sashed window 双扇窗
two-scale 双刻度;双标度
two-scale frequency meter 双刻度周波计
two-scale method 双刻度法;双标度法
two-scale notation 二进制记数法
two's complement 二进制补码;二的补码
two's complement arithmetic 二的补码运算
two's complement number 二进制补码数
two-screed finisher 双括板整平机
two-screed finishing machine 双样板修整机
two-screen expanding wall 双屏伸缩隔墙
two-screw chip feeder 双螺旋喂料器
two-screw comparator 双螺旋坐标量测仪
two-screw pump 双螺杆泵
two-seat 双座的
two-seat cabin 双座舱
two-seated 双座的
two-seated version 双座型
two-seater 双人座车;双座汽车
two-seater airplane 双座飞机
two-seater monoplane 双座单翼机
two-seat version 双座型
two-section 双节
two-section choke 双节抗流圈
two-section choke coil 双节扼流圈
two-section filter 双节滤波器
two-section group retarder 两段式线束缓行器
two-section income statement 两部式收益表
two-section method 双段法
two-sector growth model 两个部门增长模型
two-sect upflow biological aerated filter 两段升流曝气生物滤池
two-sets of turnouts facing each other and laid on the same side the original track 同侧对向道岔
two-sets of turnouts facing each other and liad on opposites sides of the original track 异侧顺向道岔;异侧对向道岔
two-sets of turnouts trailing each other and laid on the same side the original track 同侧顺向道岔
two-shaft free power turbine engine 双轴自由燃气轮机
two-shaft hammer mill 双轴锤式粉碎机
two-shaft method 双标尺法
two-shaft orientation 两井定向
two-shaft orientation method 双井定向法
two-shaft pier 双柱式(桥)墩
two-shaft pug-mill concrete mixer 双轴混凝土捏拌机
two-sheave block 双轮滑车
two-sheeted covering 双叶覆盖
two-sheet map 双页地图
two-sheet mineral 二片层矿物;双层结构矿物
two-shields ball bearing 双保护垫的滚珠轴承
two-shift 两班制
two-shift operation 两班制运行
two-shift work 两班制;两班作业
two-ship element 双机组
two-shipper 双机
two-shoe brake 双蹄式制动;双瓦制动器
two-shoe external contracting brake 双蹄外缩式制动器
two-shore method 双管化学灌浆法
two-shot 中近景;双人特写镜头;双镜头拍摄;双镜头
two-shot chemical grouting 双液化学灌注
two-shot grout(ing) 双液灌浆
two-shot solution grout(ing) 双液灌浆
two-sided alternative 双边替换
two-sided bottom switchyard 双面井底车场
two-sided coating 双面涂布
two-sided cutting 双面掏槽

two-sided exponential 双侧指数
two-sided exponential distribution 双侧指数分布
two-sided finish 双面抛光
two-sided goods 双边货物
two-sided ideal 双边理想
two-sided identity 双边单位元
two-sided invariant linear functionals 双不变线性泛函
two-sided invariant mean 双不变平均
two-sided invariant pseudometric 双不变伪度量
two-sided inverse 双边逆元
two-sided knit goods 双面针织物
two-sided Laplace transform 双边拉普拉斯变换
two-sided manifold 双边流形
two-sided module 双边模
two-sided regular representation 双侧正则表示
two-side dressing 双面修整
two-sided sampling plans 双限抽样检验方案
two-sided search 两侧搜索
two-sided spectra 双边谱
two-sided spectral density 双侧谱密度
two-sided surface 双侧曲面
two-sided swing parting 双面甩车调车场
two-sided target 双面靶
two-sided test 双侧检验
two-sided town gateway 双侧城镇通道
two-sided unit 双边器件
two-sided vector space 双侧向量空间
two-sided waxing 双面涂蜡
two-sided wedge-shaped brick 两侧楔形砖
two-sided zero 双边零
two-sided Z-transform 双边Z变换
two-side hung sections 二扇平开
two-side mosaic 双面嵌镶幕
two-side receiving coil 双滑动臂调谐线圈;双边接收线圈
two-sides 两面
two-side structural sealant glazing 两边结构密封镶嵌玻璃
two-side welding 双面焊(接)
two-signal method 双信号法
two-signal selectivity 双信号选择性
two-signal test method 双信号测试法
two-site assay 双位测定
two-slag practice 双渣熔炼
two-sled 双橇
two-slit diffraction 双缝衍射
two-slit interference 双缝干涉
two-slit interferometer 双缝干涉仪
two-sloe 双层外底
two-sludge sequencing batch reactor process 双污泥序批间歇式反应器工艺
two-socket reducing pipe 双承异径管;双承渐缩管
two-solution developer 双液显影
two-solution fixation 双槽定影
two-solvent gradient 双溶剂梯度
two-solvent process 双溶剂过程
two-source dead time determination 双源静寂时间测定法
two-source frequency keying 双源频率键控
two-source supply 双水源供水
two-source system 双水源系统
two-space dimensions filter 二维空间滤波器
two-span 双跨
two-span beam 双跨梁
two-span bridge 双跨桥
two-span cable supported bridge 双跨缆索吊桥
two-span continuous beam 两跨连续梁
two-span girder 双跨大梁
two-span heavy load 双孔重载
two-spark distributor 双火花塞分电器
two-spectra binary 双谱分光双星
two-speed 双速的
two-speed adapter lever 双速接头杆
two-speed blower 双速鼓风机
two-speed clutch 双速离合器
two-speed controller 双速控制器
two-speed drive 双速传动
two-speed feed 双速供料
two-speed final gear 双速主减速齿轮
two-speed gear 双速齿轮变速装置
two-speed gearbox 双速齿轮变速机;双速减速机
two-speed motor 双速电动机
two-speed overhead countershaft 双速天轴

two-speed pump 两级泵
two-speed pump-turbine 双速水泵水轮机
two-speed resolver 双通道旋转变压器
two-speed reversing gear 双速齿轮逆转装置
two-speed slip-ring motor 双速滑环式电动机
two-speed synchro 双通道自整角机
two-sphere 二维球
two-spindle borer 双轴镗床
two-split head 双隙头
two-spool 双转子的
two-spool compressor 双转子压缩机；双转子压气机；双轴压气机
two-spool engine 双转子发动机；双路式发动机
two-spool gas generator 双转子燃气发生机
two-spool gas turbine 双路式燃气轮机
two-spout oil can 双口油桶
two-spout unit packing machine 双嘴包装机
two-stable position 双稳位置；双稳定状态
two-stacked array 双层天线阵
two-stack plumbing 双管卫生设备
two-stage 二级(梯级、平台等)；两级；二阶段
two-stage acceleration 二级加速
two-stage activated sludge process of night soil 粪便二段活性污泥处理
two-stage addition 两级附加；两级掺加
two-stage aeration 两级曝气法；两级曝气；两段曝气；二级曝气
two-stage aeration tank 两级曝气池；两段曝气槽
two-stage ages 二阶段年龄
two-stage agitation 两段搅拌
two-stage air cleaner 两级空气洁净器
two-stage air compressor 两级空气压缩机；双级空气压缩机
two-stage air sampler 两段空气采样器
two-stage air washer 双级空气喷淋室
two-stage Aitken estimation method 两阶段艾特肯估计法
two-stage Aitken estimator 两阶段艾特肯估计量
two-stage anaerobic biofilter 两级厌氧生物滤池
two-stage anaerobic treatment 两级厌氧处理
two-stage bidding 两阶段招标；两段投标(国际承包工程方式之一)；两段式招标
two-stage bids 两步投标
two-stage biofilter 两级生物滤池；两段生物滤池；二级生物滤池
two-stage biological aerated filter 两级曝气生物滤池
two-stage biological contact oxidation 两级生物接触氧化
two-stage biological treatment process 两级生物处理工艺
two-stage breaking 两级破碎
two-stage carbonation 两段碳酸化作用
two-stage cascade generator 双级级联起电机
two-stage catalytic converter 两级催化转化器
two-stage cementing 双级注水泥
two-stage centrifugal pump 两级离心泵
two-stage clarification 两段澄清
two-stage coal combustion process 两段煤燃烧过程
two-stage combustion 两级燃烧；两段燃烧法；二段燃烧
two-stage combustion engine 双阶段燃烧发动机
two-stage comminution 两级粉碎；两级破碎
two-stage compression 双级压缩
two-stage compressor 二级压缩机；两级压缩机；二级压气机
two-stage control 双级控制
two-stage controller 两级调节器
two-stage converter 两级转换器；双级液力变矩器
two-stage cooler 两级冷却；二级式冷却器
two-stage crusher 两段破碎机；双级破碎机；两段破碎机
two-stage crushing 两级破碎；二次破碎
two-stage curing 两阶段养护；二阶段养护；两级养护
two-stage cyclone 二级式旋流器
two-stage dam 两级跌水坝；二级坝
two-stage de-airing extruder 双级真空挤泥机
two-stage demineralizer 两级去矿物质器
two-stage design 两阶段设计
two-stage desulfurization 两段脱硫
two-stage digester 两级消化器；两段消化器
two-stage digestion 二级腐化；二级消化

two-stage digestion tank 二段消化池
two-stage disgestion 两段消化作用
two-stage ejector 两级喷射器
two-stage electron image converter 二级电子变像管
two-stage electrostatic image intensifier 二级静电像增强器
two-stage estimation 两段估计
two-stage experiment 两阶段实验
two-stage filter 两级过滤器
two-stage firing 两级燃烧；二阶段烧成法
two-stage firing process 二步煅烧
two-stage-four-yard marshalling station 单向混合式编组站(铁)；二级四场编组站(铁)
two-stage-four-yard unidirectional combined marshalling station 单向二级四场混合式编组站
two-stage gas turbine 双级式燃气轮机
two-stage grinding 两级粉磨
two-stage grinding system 二级粉磨系统
two-stage growth curve 二阶段增长曲线
two-stage hoisting 二级提升
two-stage holdback device 两级锁定机构
two-stage hybrid reactor 两级混合床反应器
two-stage image intensifier 二级像增强器
two-stage injector 复合喷嘴
two-stage intermittent aeration membrane bioreactor 两级间歇曝气膜生物反应器
two-stage jaw crusher 二级颚式轧碎机；二级颚式破碎机
two-stage joint 两级接缝；双重防水节点
two-stage least squares 二段最小平方法；两段最小平方方法
two-stage least squares estimator 两段最小平方估计量
two-stage least squares residuals 两段最小平方剩余残差
two-stage lift 两级提升器
two-stage liquifaction 两段液化
two-stage mass spectrometer 双级质谱计
two-stage melting process 两段熔化法
two-stage method 两阶段(方)法
two-stage microscope 双级显微镜
two-stage mixing 二次搅拌；二阶段搅拌
two-stage oxidizing process 两级氧化工艺
two-stage phase 二次相
two-stage plasticator 双级塑炼机
two-stage pneumatic-conveyor dryer 两段气流输送干燥器
two-stage policy 两阶段策略
two-stage precipitator 双级电吸尘器；双区电除尘器
two-stage prediction method 两阶段预测法
two-stage pressure-gas burner 两级压力气体燃烧器
two-stage pretest estimator 两段预先检验估计量
two-stage principal component method 两段主分量法
two-stage process 两阶段程序；两步法
two-stage proportioning 两级配料
two-stage pump 两级水泵；两级泵；双级泵；二级泵
two-stage rectifier 二步纠正仪
two-stage reduction 两级粉碎；两级破碎
two-stage reduction gearing 两级减速装置；两级减速齿轮
two-stage regeneration 两段再生
two-stage regression method 两段回归法
two-stage regulator 双级调节器
two-stage road construction 两阶段筑路(法)
two-stage rotary air compressor 二级旋转式空气压缩机
two-stage sampling 两阶段抽样法；两级抽样；二重抽样；二级抽样；二次抽样；二层抽样
two-stage seal 双重嵌封
two-stage separator 两级分离器
two-stage shredder/granulator 双阶撕碎/成粒机
two-stage single-action compressor 两级单作用压缩机
two-stage sodium ion exchange 二级钠离子交换
two-stage stationary air compressor 二级固定式空气压缩机
two-stage supercharger 二级增压器
two-stage superheater 二级过热器
two-stage suspension 二级悬挂
two-stage system 两级管网系统
two-stage tendering 两阶段招标；两段招标

two-stage-three-yard unidirectional combined marshalling station 单向二级三场混合式编组站
two-stage torque converter 二级扭矩变换器
two-stage trickling filter 两级滴滤池；两级滴滤池
two-stage tube 双级管
two-stage turbocharger 二级涡轮增压器
two-stage turbocharging 双级涡轮增压
two-stage type filter 双层式滤尘器
two-stage valve 两级阀
two-stage vented extruder 两段式排气挤出机
two-stage vortex cyclone 双级涡流旋风分离器；双级涡流旋风除尘器
two-stage well point system 二级井点系统
two-star inertial guidance 双星座天文惯性导航
two-stars navigation 双星导航
two-star stellar inertial guidance system 双星天文惯性制导系统
two-state 双态
two-state circuit 双态电路
two-state control 双态控制
two-state device 双态装置
two-state Markov crossings 两态马尔可夫跨越
two-state process 双态过程
two-state system 双态系统
two-state variable 二态变量；二状态变量
two-state high-head storage scheme 两层高位储存系统
two-station sole press 双位外底压合机
two-stator induction machine 双定子感应电机
two-stator-rotor motor 双定转子电动机
two-step action 双位控制作用；双位控制动作；双位作用
two-step closure 二段关闭
two-step cross beam 双步梁
two-step crystallization 二步结晶法
two-step current protection unit 二段式电流保护装置
two-step decompression cycle 两级泄压循环
two-step design 二阶段设计
two-step die 两级冲裁模
two-step diffraction migration 二步法绕射偏移
two-step distance protection characteristic 两段距离保护特性
two-step exercise test 阶梯试验；二阶运动试验
two-step footstool 两级踏凳(台)
two-step heat treatment 两阶段热处理
two-step injection system 双级喷射系统
two-step loan 两步贷款
two-step method 二步法
two-step microscopy 双级显微术
two-step nozzle 双级喷管
two-stepped tenon 双重榫
two-step photoionization 二级光电离
two-step plus inverse time-lag 两段加反时限
two-step process 两步处理法
two-step Raman scatter 两级拉曼散射
two-step read 两步读出
two-step reduction 两步缩小法
two-step relay 两级动作继电器
two-step side-pilot tunnel(l)ing method 两阶段侧壁导坑法
two-step three-dimensional envelope migration 二步法三维包络偏移
two-stop batch(ing) plant 两级分批配料设备
two-stor(e)y 两层楼
two-stor(e)y aquifer 双层滞水池
two-stor(e)y bent 两层排架；双层排架
two-stor(e)y cliff 双旋回海崖；双层海崖
two-stor(e)yed 双层的
two-stor(e)yed aquifer 双层含水层
two-stor(e)yed church 双层教堂
two-stor(e)yed corridor 双层廊
two-stor(e)yed forest 两层林
two-stor(e)yed surrounding aisle 周围双层侧廊；周围双层耳房
two-stor(e)y faces 双层割面
two-stor(e)y garage 双层车库
two-stor(e)y pavilion 阁
two-stor(e)y sedimentation tank 双层沉降槽；双层沉淀池
two-stor(e)y septic tank 双层化粪池；双层化粪池
two-stor(e)y settling tank 双层沉淀池；二层沉淀池
two-stor(e)y silo 双层库
two-stor(e)y stage 双层舞台

two-stor(e)y tank 双层储箱；双层池
two-stor(e)y valley 谷中谷；双循环谷
two-storied cliff 双层海崖
two-storied gate 双层门
two-storied house 两层楼房子
two-storied truss bridge 双层桁架桥
two-strand anneal and pickle line 双线退火和酸洗作业线
two-strand coil conveyer 带卷双路运输机
two-strand line 双线加工作业线
two-strand roller chain 双股滚柱链
two-stream approximation 二流近似
two-stream hypothesis 二星流假说
two-stream instability 二束不稳定性；双流不稳性
two-stream preparation plant 双流程选矿厂
two-striped scolopsis 双带眶棘鲈
two-strip structure 两条结构
two-stroke 二冲程
two-stroke cycle 二冲程循环
two-stroke diesel engine 二冲程柴油发动机
two-stroke engine 二种程式发动机；二冲程发动机
two-stroke marine diesel engine 二冲程船用柴油发动机
two-stroke mix(ture) 两冲程拌和(料)
two-stroke petrol engine 两冲程汽油发动机
two-stroke vibrating conveyer 两冲程振动式输送机
two-stub transformer 双杆变压器
two-stub tuner 双短截线调谐器
two-subcarrier system 双副载波制；双副载波系统
two-supply circuit 双电源电路
two-surface lens 二反射面透镜
two-symbol 二进位的
two-system power supply 双系统供电
two-table machine 双台成型机
two-tailed test 双尾检验；双侧检验
two-tape Turing machine 两带图灵机
two-tenth depth method 零点二水深法
two-terminal device 二端器件
two-terminal line 两端线路
two-terminal network 二端网络
two-terminal pair network 二端对网络
two-terminal switch 双端开关
two-terminal system 两站制；两端点制
two-theodolite observation 双经纬仪观测
two-third point 上三分点
two-thirds formula 三分之二公式
two-thirds rule 三分之二规则
two-thousand metres above sea level 海拔高达二千米
two-thread linking machine 双线套口机
two-thread overlock machine 双线包缝机
two-thread worm 双头螺纹螺杆
two-thread yarn twist tester 双股纱线捻度试验仪
two-throw crank 双拐曲柄
two-throw crankshaft 双弯曲轴
two-throw pump 双吸泵
two-throw switch 双通开关；两掷开关
two-tier 双排；双层
two-tier bridge 双层桥楼【船】
two-tier cambox 双层三角座
two-tiered pilastered facade 双层壁柱正面
two-tier exchange rate 双重汇率
two-tier foreign exchange market 双重外汇市场
two-tier market 双层市场
two-tier price 双重价格
two-tier price system 两价制
two-tier rhombic 双层菱形天线
two-tier winding 两平面绕组
two-tip torch 双嘴割炬
two-tone 双色
two-tone colo(u)r 阴阳色；双色
two-tone detector 双音检波器
two-tone diaphone 双声低音雾号；双音发生器
two-tone dyeing 双色调染色
two-tone finish 双色调涂饰剂；双色(喷)漆
two-tone finished leather 双色调涂饰革
two-tone leather 双色革
two-tone method 双音信号法
two-tone paint 两色油漆；双色油漆
two-tone stripe 双色条纹
two-tone system 双音系统
two-tone vehicle 双色彩汽车
two-tone working 双音工作

two-toning 双色调涂饰
two-tool machine 双刀创齿机
two-to-one 二比一乱石墙(两份石头一份灰浆)
two-to-one principle 二对一原理
two-to-one ratio 二对一比率
two-to-one slope 二比一边坡
two-tower facade 双塔(建筑物)正面
two-trace method 双线扫描法；双迹法
two-track 双轨
two-track road 双线道路
two-track suspension pulley 双线悬吊滑轮
two-track tape 两道带
two-transducer flaw detector 双探头探伤仪
two-transit fix method 双经纬仪定位法
two-transit observation 双经纬仪观测
two-tray (type) thickener 双层浓密机
two-trunnion mill 双耳轴磨机
two-tube electronic lens 两管电子透镜
two-tube tunnel 双管隧道
two-tuples 二元组
two-turn stellarator 双叶仿星器
two-turn winding 双匝绕组
two-type pile 二型桩
two-types of chemical balance 两种化学天平
two-unit 双机组
two-unit grinding 两级粉磨；双机粉磨流程
two-unit system 双组系统
two-unit tester 双样试验机
two-unit triaxial test apparatus 二联三轴试验仪
two-valley model 双能谷模型；双谷模型
two-value 双值；二值
two-value capacitor motor 双值电容式电动机
two-valued 二值的
two-valued condition 双值状态
two-valued function 双值函数
two-valued logic 二值逻辑
two-valued output signal 双值输出信号
two-valued variable 双值变量；二值变数；二值变量
two-valve concrete pump 双阀混凝土泵
two-variable computer 双变数计算机；双变量计算机
two-variable function generator 双变量函数发生器
two-variable matrix 双变量矩阵
two-variable single equation estimator 两段最小平方估计量
two-variance analysis 两差异分析
two-variance overhead analysis 两差异制造费用分析
two-vee 双 V 粒子
two-vertical cable planes 双竖索面
two-voltmeter method 双伏特计法
two-walled 双墙的
two-walled sheet-pile cofferdam 双排板桩箱形围堰
two-wall sheet-pile structure 双排板桩墙结构
two-wall sheet-piling cofferdam 双排板桩围堰
two-water action 双向水流洗涤方式
two-wattmeter method 双瓦特计法
two-waveband 双波段
two-waveband radio set 双波段接收机
two-wave coupling 二波耦合
two-wavelength dye laser 双波长染料激光器
two-wavelength microscope 双波长显微镜
two-wavelength microscopy 双波长显微术
two-way 双向的；双向；双频道；双路；二通
two-way accelerator 双向加速器
two-way account 两用账户
two-way alternate 双向交换
two-way alternate communication 双向交替通信
two-way alternative plow 翻转式双向犁
two-way amplifier 双向放大器
two-way analysis fo variance 两因素方差分析
two-way anchor 双向锁定
two-way automatic private line 双向自动专用线路
two-way balanced traffic 双向平衡运行(船闸或航道)
two-way beam 双向梁
two-way bit 鱼尾形钻头；双翼钻头；双刮刀钻头
two-way block footing 两向块状底脚；两向块形底脚；双向块形基脚
two-way block foundation 两向块状基础
two-way break-before-make contact 双向先断后合触点
two-way cable conduit 双向电缆道；双向线路；双向电路

two-way canal 双向运河；双线运河
two-way channel 双行航道；双向信道；双向通道；双向航道；双线航道
two-way chute 裤叉溜子；双向卸料槽；双向卸料；双向槽；双叉溜槽
two-way circuit 双向的线路；双向电路
two-way clamp 双向箝位
two-way classification yard 双向编组场【铁】
two-way cock 双通开关；两路旋塞；两通旋塞；双向旋塞；双向龙头
two-way column 双向角柱；双向柱
two-way communication 双向交往；双向通信；双向沟通
two-way communication channel 双向沟通渠道
two-way concrete slab 双向钢筋混凝土板
two-way conduit tile 双向电缆道
two-way contact 切换接点；切换触点；双向触点
two-way contactor 双向接触器
two-way control 双向控制
two-way control chart 双向管理图；复式管理图
two-way controlled transmission 双向受控式传送
two-way control valve 两通控制阀
two-way conveyer 双向运输机
two-way cord 双芯塞绳
two-way cross 双交
two-way curved arch bridge 双曲拱桥
two-way curved arch tile 拱波
two-way cylinder 双作用油缸；双向油缸
two-way detector array 二维探测器阵列
two-way device 双向装置
two-way discharge chute 裤衩溜子
two-way distribution 双向分布
two-way distributor 双路分配器
two-way dozer 双向推土机
two-way drag 双向刮路机
two-way drainage 双向排水
two-way drive 双向传动
two-way dump body 双向倾卸车身
two-way dumper 双向翻斗车；双向倾卸车
two-way dumping 双向翻倾
two-way dump trailer 双向倾卸拖车；双向倾卸挂车
two-way dump truck 双向倾卸货车
two-way duplex circuit 双向双工电路；双向单工电路
two-way dynamic microphone 双路动圈传声器
two-way electrochromatography 两向电色谱法
two-way elution technique 双向淋洗技术
two-way entrance-exit 双行出入口
two-way exit-entry visa 双程出入境签证
two-way feed 双向馈电
two-way feeding chute 双通供料溜槽
two-way flat slab 双向无梁楼板
two-way flat slab floor 双向板楼板；双向平板楼面
two-way flow 双向流
two-way flow of message 双向交流信息
two-way flow pumps 双联泵
two-way footing 加筋底脚；双向配筋基脚；双向受力基脚
two-way fuse plug 双线保险丝插塞
two-way fuse socket 双线保险丝插座
two-way glass 单向透明玻璃镜
two-way grid 双向网格
two-way grillage 井形床格
two-way half-duplex circuit 双向半双工电路
two-way hydraulic control 双向式液压操纵
two-way hydraulic shock absorber 双向液压减振器
two-way hypothesis 二向假设
two-way inlet 双向出入口
two-way intercommunication system 双向内部通话装置
two-way intercom system 双向内部通信系统
two-way joist 双向格栅；双向梁；双向龙骨；双向格(式搁)栅
two-way joist construction 双向格栅构造；双向龙骨构造
two-way laser array 二维激光阵列
two-way lattice(d) grid 双向格构网架
two-way launching 双滑道下水
two-way layout 二元配置
two-way level(l)ing 纵横向调平
two-way limit switch 双向限位开关
two-way link 双向链路

two-way lock 双向船闸;双线船闸
two-way loudspeaker 双路扬声器
two-way merge sort 二路归并分类
two-way mirror 双向镜
two-way observation 双程往返观测
two-way operation 双向运行;双向操作
two-way pallet 双向镘灰板
two-way paper chromatography 双向纸上色层分析法
two-way phase switcher 双向移相器
two-way power feeding 双边供电
two-way prestressed slab 双向预应力板
two-way price quotation 双向报价
two-way pulse transmission 双向脉冲传输
two-way pump 可逆泵;双向泵
two-way quotation 包括买价和卖价的双向报价单
two-way radial sector accelerator 双向径向扇加速器
two-way radio 交直流两用无线电设备;无线电对话机;双向无线电设备;收发两用无线电设备
two-way radio communication 双向无线电通讯
two-way radio system 双向无线电通信系统
two-way ram 双作用活塞
two-way ramp 双向匝道【道】;双向坡道
two-way receiver 交直流两用收音机;双频道接收机
two-way reinforced 双向配筋的
two-way reinforced concrete 双向配筋混凝土;双向配筋(的钢筋)混凝土;双向钢筋混凝土
two-way reinforced concrete slab 双向钢筋混凝土板
two-way reinforced footing 双向配筋基础;双向配筋基脚
two-way reinforced foundation 双向配件基础
two-way reinforced slab 双向配筋的板
two-way reinforcement 双向配筋;双向加筋;双向钢筋
two-way repeater 双向增音器;双向增音机
two-way reversing switch 双向转换开关
two-way ribbed slab 双向肋平板
two-way ribs 双向密肋
two-way road 双向道路;双线公路
two-way road system 双线道路系统
two-way route 双程航线
two-way shot 双向爆破
two-way side dump car 两侧倾卸车
two-way signal 双组信号;双面信号
two-way simplex connection 双向单工连接
two-way simplex operation 双向单工操作
two-way simultaneous 双向并行
two-way simultaneous communication 双向同时通信
two-way slab 双向配筋混凝土板;四边支承板;双向板
two-way slab system 双向板体系
two-way slope 双向坡度
two-way snow plow 双向雪犁
two-way sort 双向分类
two-way space-framed steel structure 双向空间钢网架
two-way space grid 双向空间网架
two-way speaker 双频道扬声器
two-way split former 对开式模筒
two-way spout 双向卸料槽
two-way spread 双向排列
two-way spring butt 双向弹簧铰
two-way stitch transfer 双线移圈花纹
two-way stop sign 双向停车标志
two-way strap 叉形连接板
two-way stretch 双向拉伸
two-way stretch resonator 双向可调谐振腔
two-way sulky plow 双向乘式犁
two-way switch 双路开关;双向开关;双投开关;双头开关;双路开关;双联开关;双控开关
two-way symmetric accelerator 双向对称加速器
two-way synchrotron 双向同步加速器
two-way system 双系统;双向配筋系统;双向体系
two-way system of reinforcement 钢筋双向制;双向配筋体系;双向配筋法
two-way tap 两路龙头;双路开关
two-way television 双向电视
two-way television relay 双向电视中继系统
two-way three-step ship lock 双线三级船闸
two-way time aligned 双程时间对齐
two-way tipper 双向倾卸车;双向翻斗车

two-way trade 双轨贸易;双边贸易
two-way traffic 双向行车;双向通信业务;双向交通(量);双线交通
two-way transit 双向过境
two-way transmission 双向传输
two-way travel time 两端传播时间
two-way trunk line 双向中继线
two-way truss system 双向桁架系统
two-way tumbler switch 双连开关
two-way valve 两通阀;双向阀;双通活门;双通阀;二通阀;二路阀
two-way variable displacement pump 双向变量泵
two-way vertical compressional wave 双向垂直压缩波
two-way video 双向视频
two-webbed 双腹板的
two-webbed beam 双腹梁
two-webbed plate girder 双腹板大梁
two-webbed T beam 双腹板 T 形梁
two-wedge bearing 双油楔轴承
two-week and three months inspection shed 双周—三月检车库
two-week inspecting shed 双周检车库
two-weight shaft plumbing 双荷重竖井对中
two-wheel axle 两轮轮轴
two-wheel brake 双轮制动器
two-wheel drive planter 双轮驱动式播种机
two-wheeled chassis 双轮底盘
two-wheeled ploughing tractor 两轮犁耕拖拉机
two-wheeled tractor 双轮拖拉机
two-wheeler 两轮半拖车;挂车;两轮挂车
two-wheel frame plow 双轮架式犁
two-wheel garage jack 二轮修车起重器
two-wheel handcar 双轮手推车
two-wheel landing gear 双轮起落架(飞机)
two-wheel plough 双轮犁
two-wheel road roller 二轮压路机
two-wheel rubber-tired trolley 双轮气胎电车
two-wheel scraper 双轮刮路机;双轮铲运机
two-wheel tractor 两轮拖拉机
two-wheel trailer 两轮拖车;两轮挂车
two-wheel trailer booster 双轮后推助力器
two-wheel unit 单轴双轮走行
two-wheel wagon drill 双轮钻探车
two-winding transformer 双线圈变压器;双绕组变压器
two-window glue-jointed two-piece panel 双窗铰接预制墙板
two-window office 双窗办公室
two-window one-piece panel 整块双窗墙板
two-window panel 双窗墙板
two-wing building berth 双翼船台
two-winged building 有两翼的建筑物
two-winged stope 双翼回采工作面
two-wing slotting cutter 双叶刨槽铣刀
two-wire 双线的;双线
two-wire antenna 双线天线
two-wire carrier system 双线载波制
two-wire channel 双向通道;二线制通道;二线信道
two-wire circuit 两线电路;双线回路;双线电路
two-wire circuit system 双线电路制
two-wire connection 二线连接
two-wire former 双网成型装置
two-wire jack 张拉两根钢丝的千斤顶
two-wire line 二线制电路
two-wire loop terminal 双线回路端子
two-wire (paper) machine 双网造纸机
two-wire repeater 双线增音器;二线转发器;二增音机;二线式转发器
two-wire switch 双联开关
two-wire system 两线制;双线方式;双线电路;二线制
two-wire-wide strands 双排股线
two-wire winding 双线绕组
two-withe 两层隔墙;双重隔墙;两重烟道隔板
two-word bead 二字珠
two-word list element 二字链表元
two-year ice 两年冰
two-zone air-conditioning system 双区空调系统
two-zoned-core 双区堆芯
two-zone drawing system 双区拉伸体系
two-zone furnace 双温区炉
two-zone high drafting 双区大牵伸
two-zone tubular furnace 双区管式炉

tw-pulse timer 双脉冲定时器
T-wrench 活络扳手;T形手扳钳;丁字柄套筒扳手;丁字把手
twyer(e) 风口
Twyman interferometer 特环曼干涉仪;台曼干涉仪
Twystron 行波速调管
tybat(e) 孔口边顺砌石块
tychite 杂芒硝;硫碳镁钠石
tychocoen 随遇生物
Tychonoff space 完全正则空间
Tychonoff theorem 吉洪诺夫定理
tychoplankton 偶然浮游生物
tycoon 实业界巨头
Tyeh limestone 大冶灰岩【地】
tyf(h)on 压缩空气或蒸汽发音器;大喇叭
tygon tube 聚乙烯管
tygoweld 环氧树脂复合黏合剂
tying bar 陆连沙坝;连岛沙洲
tying clause 相连条款;附带条款
tying contract 连锁契约;搭卖合同;附有条件的合同;附条件的合同
tying device 打捆装置
tying-down of capital 资金积压
tying frame 结经机
tying fund 连锁基金
tying in 联测法;结经
tying-in machine 结经机
tying machine 扎捆机;打捆机
tying mechanism 捆扎装置;打捆装置
tying point 联结点;连接点;联系点
tying products 附加商品
tying stich 重结线迹
tying up capital 备用资金
tying-up place 锚地;系留地;系留码头;系泊地
tyler 泰勒架空索系
Tyler mesh 泰勒筛;泰勒标准筛号
Tyler scale 泰勒筛号尺寸
Tyler screen 泰勒筛;泰勒标准筛
Tyler series 泰勒制筛序
Tyler sieve 泰勒筛
Tyler standard grade scale 泰勒粒级标准
Tyler standard mesh 泰勒标准筛网
Tyler standard screen 泰勒标准筛
Tyler standard screen scale 泰勒标准筛制;泰勒标准筛分标度
Tyler standard series 泰勒制标准筛序;泰勒标准筛
tylosis 导管内的侵填体
tylosoid 拟侵填体
tylote 两头圆针
tymapanitic note 鼓音
tymp 水冷铁铸件
tympan 门楣中心;山墙饰内三角面部分;拱圈与拉梁间弧形部分
tympanic canal 鼓阶
tympanicity 鼓响性
tympanic lip 鼓唇缘
tympanic resonance 鼓音
tympanitic 鼓响的
tympanosquamous fissure 鼓鳞裂
tympanum 振动膜;山墙饰内三角面
tympanum electrode 鼓岬电极
tympanum enclosed by arch 拱圈与门头间的弧形部分
tympanum enclosed by pediment 人字山墙与门头间的三角部分
tympany 鼓音
Tyndall cone 廷德尔亮锥
Tyndall effect 廷德尔效应;丁达尔效应
Tyndall flowers 廷德尔冰花
tyndallimeter 悬体测定计;廷德尔计
tyndallimetry 悬体测定法;廷德尔法
tyndallization 廷德尔灭菌法
Tyndall light 廷德尔光
Tyndall meter 廷德尔计;廷德尔测尘计
Tyndallometer 廷德尔悬浮体浓度计;悬浮体浓度(测定)仪
Tyndallometry 廷德尔悬浮体测定法
Tyndalloscope 廷德尔悬浮体浓度计
Tyndall phenomenon 廷德尔现象;丁达尔现象
Tyndall scattering light 廷德尔散射光
Typar 蒂帕尔聚丙烯纺丝黏合织物
type 形式;型号;型别;样式;类型;类别;模式;铅字;典型
type 1 batch 第一类批处理

type 1 important construction 一类建筑物
type 1 programming 第一类程序设计
type 2 batch 第二类批处理
type 2 counter model 二型计数模型
type 2 distribution 二型分布
type 2 error 二型误差
type 2 general construction 二类建筑物
type 2 programming 第二类程序
type 2 sampling 二型抽样
type 2 superconductor 第二类超导体
type 3 programming 第三类程序设计;第三类程序
type 4 programming 第四类程序设计
type A appliance A 型燃具;无烟道燃具
type A finish 木纹饰面(拼缝镶板做模板);A 型饰面;木纹饰面;A 类饰面
type and feature of microstructure and fabric 显微构造类型和性质
type and grade of lime 石灰的种类和等级
type and property of fault 断层类型和性质
type and property of folds 褶皱类型和性质
type and property of primary structure 原生构造类型和性质
type and property of synsedimentary structure 同沉积构造类型和性质
type and source of organism 有机体类型和来源
type approval 形式批准;产品定型
type-approval inspection 定型检查
type approval test 工业性试验;型式试验;定形试验
type area 模式地区;典型地区;标准地区
type array 类型数组
type association 类型结合
type attribute 类型属性
type A wave 连续波
type ball 球形字锤
type B appliance B 型燃具;开启式有烟道燃具;开启式有烟道燃具
type bar 打印杆
type bar printer 杆式打印机
type batch 成批打印
type B error 取伪
type B finish 光滑或浅木纹饰面;B 类饰面
type Boolean 布尔型
type C appliance C 型燃具;室内封闭式燃具
typecast 浇铸铅字
type certificate drawing 打字校准图
type C finish C 类饰面;特光滑饰面
type charter party 标准租船合同
type classification 体型鉴定
type compatibility 类型相容性;类型兼容
type constancy 类型的稳定性
type conversion 类型转换;类型变换
type crane 轮胎式起重机
type crossing 型交叉
type culture 模式培养
type curve 标准曲线
type curves of well function 井函数标准曲线
type-curve solution 标准曲线解
type declaration 类型说明
type definition 类型定义
type design 定型设计
type D finish D 类饰面
type distribtution 类型分布;典型分布
type E finish E 类饰面
type evaluation 形式评价
type exchangeable ones 可换的离子类型
type face 打印页
typeface 版面
type figure of sounding curve 电测深曲线类型图
type forty-five degree bend 一型四十五度弯头
type fossil 标准化石
type genus 典型属
type grade 体型等级
type grading 典型级配
type grading curve 典型级配曲线;类型分级曲线
type ground plan 典型底层平面图
type house 定型房屋;定型设计房屋
type I and II probabilities 一型和二型概率
type identifier 类型标识符
type I error 第一类错误
type II error 第二类错误
type impression control 印痕控制器
type in 打入
type in quadruplicate 打一式四份
type intensity 类型强度

type IV cement 低热水泥
type K expansive cement K 型膨胀水泥(硫铝酸盐型)
type list 类型表;简单变量表
type locality 模式地点;模式产地;典型地区;标准产地
type map 类型图
typematic key 重复作用键
typematic key stroke 键的重复启动
type matter 文字说明
type metal 印刷合金
type M expansive cement M 型膨胀水泥(硅酸盐型)
type number 型号
type of a constant 常数类型
type of activity 活动类型
type of adsorbates 吸附质类型
type of age distribution function 年龄分布函数类型
type of alarm 报警方式
type of analogy tests 模拟试验的种类
type of anchorage 锚固形式
type of antenna of borehole electromagnetic wave method 井中电磁波法天线类型
type of attachment 连接形式;附件类型
type of aquifer 含水层类型
type of bar 沙坝类型
type of bar of continental shelf outer edge 大陆架外缘坝类型
type of barrier 遮挡类型
type of basin 盆地类型
type of beach 海滩类型
type of beach ridge 滨岸堤类型
type of bearing pile 支承桩的类型
type of benchmark 水准标石类型
type of blasting material 爆破材料类型
type of bollard heads 系船柱柱头形式
type of bond (砖的)砌合形式
type of borehole electromagnetic wave method 井中电磁波法类型
type of bottom switchyard 井底车场形式
type of breaker 破波类型
type of building 建筑物类型;房屋类型
type of business 企业类型
type of calculating formula 计算公式类别
type of calderas 破火山口类型
type of car 车辆类型
type of car ferry 汽车渡船型
type of cave formation 洞穴成因类型
type of cave rocks 洞穴岩石类型
type of cement 水泥种类;水泥类型
type of cementation 胶结类型
type of chemical well-cleaning liquid 化学洗井液种类
type of close packing 紧密堆积方式
type of cloud 云型
type of coal fissure 煤裂隙型;煤的裂隙类型
type of coal gasification 煤气化型
type of coal metamorphic 煤变质作用类型
type of coal seam roof 煤层顶板类型
type of coast 海岸类型
type of coastal platform 海岸台地类型
type of coastal terrace 海岸阶地类型
type of colo(u)r 颜色种类
type of colo(u)r center 色心种类
type of combined mining 联合开采方式
type of commercial enterprises 商业企业种类
type of complex frequency spectrum in rock and minerals 岩矿石的复频谱类型
type of compliance tests 符合性测试的类型
type of compounds 化合物类型
type of compression curve 压缩曲线类型
type of concrete 混凝土类型
type of configuration in induced polarization method 激电法电极装置类型
type of configuration of electric(al) sounding 电测深法装置类型
type of configuration on electric(al) profiling 电剖面法装置类型
type of connection surface 连接面形式
type of construction 结构类型;建筑形式;建筑模式;建筑类型
type of continental shelf 大陆架类型
type of contract 合同类型;包工方式

type of coordination polyhedron 配位多面体类型
type of coral reef 珊瑚礁类型
type of corrosion inhibitor 缓蚀剂种类
type of creep curve 蠕变曲线类型
type of crustal architecture and structures 构壳结构构造类型
type of crust of weathering 风化壳类型
type of crystal lattice 晶格类型
type of crystallization 结晶作用类型
type of curve equation 曲线方程的类型
type of dam 坝型;坝的类型
type of deep fracture 深断裂类型
type of delta 三角洲类型
type of density current 密度源类型
type of desert 荒漠类型
type of diamond bit 金刚石钻头类型
type of diapiric structure 底辟构造类型
type of dilution 矿石贫化分类
type of discharge-drawdown curve 涌水量水位降深曲线类型
type of distributary of estuary 河口支流类型
type of drawings 制图类别;图纸类别
type of dredging plant 疏浚设备类型
type of drilling equipment 钻探设备类型
type of drilling machine 钻机类型
type of dry and damp soil base 土基干湿类型
type of dynamic(al) sounding 动力触探种类
type of earthquake 地震类型
type of economic organization 经济组织的类型
type of economic results 经济效果的种类
type of electric(al) prospecting curve 电测曲线类型
type of electric(al) sounding curve 电测深曲线类型
type of electrode 焊条牌号
type of elution 洗脱方式
type of engineering 工程类型
type of erosion(al) groove 侵蚀沟类型
type of estuary 河口类型
type of exogenic process 外营力类型;外力地质作用类型
type of exogenic process anomaly 外成作用异常类型
type of exploration line 勘探线类型
type of exploration tunnel 探矿坑道类型
type of exploration tunnel(l)ing equipment 坑探设备类型
type of exploration tunnel(l)ing instrument 坑探仪表类型
type of exploratory grid 勘探网类型
type of explosion crater 爆破漏斗形式
type of expression 表达式的类型
type of fabrication 制作类别;建造形式
type of facies 相类
type of fault rock 断层岩类型
type of feed 装填方式
type of fender 护舷类型
type of field shear test 野外剪切试验种类
type of field source in DC electrical prospecting 直流电法场源类型
type of field source in electric(al) prospecting 电法勘探场源类型
type of filter 过滤器种类
type of filter framework 过滤器骨架种类
type of filter net 滤网种类
type of finishing 饰面(工程)类别
type of flood plain elevation 河漫滩高度类型
type of flow 流动形式;流动方式
type of focus 聚焦形式
type of foliation 面理类型
type of force 力的种类
type of forest 林型
type of foundation 基础形式;基础类型
type of foundation soil 地基种类
type of foundation subsidence 地基沉降类型
type of frequency distribution 频率分布类型
type of frost-damage 冻害类型
type of fumarole 喷汽孔类型
type of gas lift valve 气举阀形式
type of gently dipping bed and massive rocks 平缓块状岩层型
type of geochemical anomaly 异常种类
type of geochemistry survey 地球化学调查分类
type of geologic(al) data 地质数据的类型

type of geologic(al) variable 地质变量的类型
type of geomagnetic pole 地磁极类型
type of geophysical prospecting method 物探方法类型
type of geosyncline 地槽的类型
type of geosyncline-platform 地槽地台的类型
type of geothermal system 地热系统类型
type of gravity anomaly 重力异常类型
type of gravity tectonics 重力构造类型
type of ground spectral data 地物波谱数据类别
type of groundwater 地下水的类型
type of groundwater balance 地下水均衡类型
type of groundwater regime 地下水动态模型
type of groundwater regime curve 地下水动态曲线类型
type of groundwater reservoir 地下水库类型
type of groundwater sources 地下水源地的类型
type of harbor work 海港工程种类
type of heading 巷道形式
type of heat hazard source 热害源类型
type of heavy compensation equipment 升沉补偿设备类型
type of hot-water 热水型
type of hydrocarbon source rock 烃源岩类型
type of hydrogeologic(al) borehole 水文地质钻孔类别
type of hydrogeologic(al) survey 水文地质调查种类
type of hydrothermal eruption 水热喷发形式
type of image point 映射点类型
type of impactite 撞击岩类型
type of impact structures 撞击构造类型
type of inclined bed and fractured rocks 倾斜破碎岩层型
type of inclusion 包裹体类型
type of index numbers 指数型
type of induced earthquake 诱发地震类型
type of infiltration field 渗流场的类型
type of intercepting ditch 截水沟类型
type of interference in electric(al) shield 电屏蔽干扰类型
type of investigation program(me) 勘察项目种类
type of investigation work 已有调查工作性质
type of island 岛屿类型
type of isomorphous substitution 类质同像代替种类
type of isotope ray 同位素射线种类
type of joints 连接形式；结合类型；接缝类型
type of karst shape 岩溶形态类型
type of kerogen and its characteristic 干酪根类型及其特征
type of lagoon 泻湖类型
type of landing nipple 油管工作筒类型
type of lava flow 熔岩流类型
type of layout chart 观测系统类型
type of leakage 渗漏类型
type of line 线形
type of lineation 线理类型
type of lithojacies anomaly 岩相异常类型
type of lithologic(al) anomaly 岩性异常类型
type of load 荷载形式；荷载类型
type of loading test 荷载试验种类
type of locomotive 机车类型
type of luminescence of minerals 矿物发光性种类
type of luster 光泽种类
type of magmatic formation 岩浆建造类型
type of magnetic field in electromagnetic method 电磁法的激励场类型；电磁法的磁场类型
type of magnetic pole 磁极类型
type of magnetism of specimens 标本磁性类型
type of manganese nodule 锰结核类型
type of manhole 人孔类型
type of map 图类
type of map folding 地形折叠形式
type of marine faue 海洋动物区系型
type of mathematic(al) simulation 数学模拟的类型
type of mating electric(al) machine 配套电机型号
type of message 消息类型
type of metal ion 金属离子类型
type of metalliferous sediments 含金属沉积物类型
type of metamorphism 变质作用类型
type of migmatization 混合作用类型
type of mineral deposit 矿床类型

type of mineral deposit-hydrogeology 矿床水文地质类型
type of mineral water 矿物中水的类型
type of mining 开采方式
type of models 模型类型
type of morphologic(al) survey 地貌调查分类
type of motion of pal(a)eomagnetic field 古地磁场活动分类
type of mud-stone flow 泥石流种类
type of name 名字类型
type of natural building materials 天然建筑材料种类
type of natural vegetation 天然植被类型
type of neotectonic movement 新构造运动分类
type of nodule 结核类型
type of observation net and points 观测网点类型
type of observation point 观测点类型
type of offset 错动方式
type of offshore rig 海上钻探装置类型；海上钻井装置类型
type of oil field water 油田水类型
type of open mining 露天开采方式
type of operand field 操作数域的类型
type of operation 经营方式
type of ore 矿石类型
type of ore anomaly 矿异常类型
type of orebody structures 矿体构造类型
type of ore-containing structures 容矿构造类型
type of orefield structures 矿田构造类型
type of ore losses 矿石损失分类
type of ore losses ratio 矿石损失率分类
type of oscillation 振荡类型
type of outcrop 露头类型
type of outer harbor engineering 港外工程类型
type of pal(a)eomagnetic pole 古地磁极类型
type of pal(a)eoseismic trace 古地震遗迹类型
type of parent rock 母岩类型
type of payment 付款方式
type of penetration sounding machine 触探机类型
type of piles 桩的类型
type of pit engineering 坑探工程种类
type of planar boundary of mineral deposit 矿床平面边界类型
type of plate 板块类型【地】
type of point survey signal 观测标类型
type of polarization of rocks and minerals 岩矿石的极化类型
type of poles 极的类型
type of polymorphic transition 同质多像转变类型
type of population 人口类型
type of pore geometry 孔隙结构类型
type of pore in rock 岩石的空隙类型
type of pore space 孔隙空间类型
type of preflush 前置液名称
type of pressure charging 增压形式
type of probe 探头种类
type of probe of pressuremeter 旁压仪探头种类
type of production place 产地类型
type of prospecting area 普查地区类型
type of prospective ore deposits 远景矿产类型
type of pseudo-chromatism 假色种类
type of pump 泵型号
type of pumping test 抽水试验类型
type of rail joint 钢轨接头形式
type of reef 生物礁类型
type of regime observation point 动态观测点的类型
type of regional geophysical survey 区域地球物理调查类型
type of remanent magnetization 剩余磁化强度类型
type of reservoir 储集层类型
type of reservoir of petroleum and gas 油气储集层类型
type of retaining wall 挡土墙类型
type of retarder 缓行器类型
type of river mouth bar 拦门沙形式
type of river runoff 河流的径流类型
type of rock anomaly 岩石异常类型
type of rock heat 岩温型
type of rock sample 岩石样品类型
type of rock void 岩石的空隙类型
type of roof 屋顶类型

type of route 路线类型
type of salinization 盐渍化类型
type of sample 样品类型
type of sample examination 样品检验种类
type of samples 样品种类
type of Sea 海类型
type of seabed thermal water 海底热水类型
type of seal 密封形式
type of sedimentary facies 沉积相类型
type of sedimentary formation 沉积建造类型
type of sedimentary rock 沉积岩类型
type of seepage deformation 渗流变形类型
type of semi-solid rocks 半坚硬岩石类
type of seismic calamity 震害类型
type of seismic curve 地震曲线类型
type of seismic remains 遗迹类型
type of seismic section 地震剖面类型
type of service 业务类型
type of shaft 井型
type of shaft sealing 轴封形式
type of shear strength index 抗剪强度指标类型
type of shear test of rock mass 岩体剪切试验种类
type of sheet pile structure 板桩结构类型
type of ship 船舶类型
type of ship's employment 船舶营运方式
type of slip tectonite 滑构造岩类型
type of slope 边坡类型
type of slope failure 土坡稳定破坏类型
type of soft rocks 松软岩石类
type of soil 土质类型；土壤种类；土壤类型；土的种类
type of soil anomaly 土壤异常类型
type of soil sample 土壤样品类
type of soil sampler 取土器种类
type of soil water 土中水的类型
type of solid bitumen 固体沥青类型
type of solid rocks 坚硬岩石类
type of sonde 电极系类型
type of space lattice 空间格子类型
type of spring 泉水类型
type of stabilization pond 稳定塘类型
type of stabilizer 稳定剂种类
type of stack gas plume 烟缕形式
type of steels 钢种；钢材牌号
type of stereographic(al) projection 立体投影方式
type of stream sediment samples 水系沉积物样品类型
type of stress and strain curve of rock mass 岩体应力应变曲线类型
type of structural anomaly 构造异常类型
type of stratigraphic(al) cross-section 地层剖面类型
type of structural trace 构造形迹类型
type of structure 形式；结构类型
type of subsurface water anomaly of bed rock 基岩地下水异常类型
type of support 支护方式
type of supporting 支撑类型
type of surface 路面类型
type of swab 抽子形式
type of tectonic layer 构造层类型
type of tectonic movement 构造运动类型
type of tectonic stress field 构造应力场类型
type of tectonic system 构造体系类型
type of tectonic system controlling earthquakes 构造体系控震类型
type of tectonic system map 构造体系图类型
type of tectonite 构造岩类型
type of terrace texture 阶地结构类型
type of terrestrial heat field anomaly 地热田异常类型
type of texture bond 结构连结类型
type of tide 潮汐类型
type of towers 舰标类别
type of tracer 示踪剂种类
type of track 轨道类型
type of traction 牵引种类【铁】
type of trading 贸易方式
type of traffic 运输类型；业务种类
type of trouble 事故种类
type of turbine 水轮机形式
type of turnout rails 道岔钢轨类型
type of twinning 双晶类型
type of underground mining 地下开采方式

type of underground structure 地下结构物类型
type of variables 变量(的)类型
type of vegetation 植被(类)型
type of vehicles 机车车辆类型
type of vertical boundary of mineral deposit 矿床垂向边界类型
type of vibration 振动类型
type of virtual geomagnetic pole 虚地磁极类型
type of volcanic craters 火山口类型
type of volcanic vent 火山口形式
type of wastewater 污水类型;废水类型
type of water-bearing medium 含水介质类型
type of water body 水体类型
type of water quality 水质类型
type of water quality standard 水质标准的类型
type of water sample 水样品类型
type of weathered zone 风化带类型
type of weathering 风化作用类型
type of well engineering hazard 水井工程病害类型
type of well logging equipment 测井仪器型号
type of well tube 井管类型
type of wind-sand damage 风沙危害类型
type of wind-sand landform 风沙地貌类型
type of wire twine 缠丝种类
type of work in production 工种
type one counter model 一型计数模型
type out 打出
typeover 改写
type permitted 允许形式
type-plan of floor 典型的楼层平面图
type pressure 轮胎压力
type procedure 类型过程
type profile 模式剖面
typer 印刷装置;印刷机
type reaction 典型反应
type real 实数类型
type record 记录类型
type region 典型地区;标准地区
typer ribbon 色带
type sample 标样;典型试样;标准样品
type sash 局部转轴窗
type-script 打印原稿;打印件
type section 模式剖面;定型剖面;典型剖面;标准剖面;标准断面
type selection 选型
type selection load 选型负荷
type series 模式组
type-series container 批准箱型集装箱
type setter 排字机
typesetting 排版
type S expansive cement S 形膨胀水泥
type ship 型船;母型船
type species 模式种;典型种;标准种
type specification 类型说明
type-specificity 型特殊性
type specimen 模式标本;典型试件;标准样品
type standard 体型标准
type standardization 定型标准化
type statement 类型语句
type statian 模式迹
type-stick 透明注记
type symbol 标型符号
type technology 形式技术
type test 形式试验;定型试验;典型试验
type tree 标准木
type two 45 degree bend 二型 45 度弯头
type-undivided ore 类型未分矿石
type wash 去油墨剂
typewriter table 打字机桌
type X gypsum wallboard X 型滞火石膏墙板;特种防火石膏板
type X lath 特种防火石膏板条
typhoid fever 伤寒
typhon 古埃及神庙
typhoon 台风;大喇叭
typhoon anchorage 防台锚地
typhoon course 台风行程
typhoon damage 台风损失
typhoon detective line 台风警戒线
typhoon evacuation team 台风撤运小组
typhoon eye 台风眼
typhoon forecasting 台风预报
typhoon imminent warning 台风紧急警报
typhoon information 台风消息
typhoon movement 台风动向
typhoon path 台风路径
typhoon rain 台风雨
typhoon rainstorm 台风暴雨
typhoon recurvature 台风转向
typhoon region 台风地区
typhoon signal 台风信号
typhoon source 台风源地
typhoon squall 台风飑
typhoon storm surge 台风暴潮
typhoon surge 台风涌浪
typhoon track 台风路径
typhoon urgent warning 台风紧急警报
typhoon warning 台风警报
typhoon zone 台风区
typical 典型的
typical adjustable Ro-Ro shore ramp 典型可调式滚装岸上坡道
typical analysis 典型分析
typical apartment building 典型的公寓建筑
typical basic invariant 标准基本不变量
typical basis 典型值
typical bond 典型砌合
typical building 标准建筑
typical case 典型事例;典型
typical characteristc 典型特征
typical characteristics of untreated domestic sewage 典型未经处理的家庭生活污水特性
typical compacting method 标准击实法
typical complete equipment 定型的整套装置
typical composition 典型成分
typical construction 典型建筑;典型构造;典型工程
typical creep curve 标准蠕变曲线
typical cross-section 典型横剖面;标准横断面
typical cube 标准立方试块;典型的立方块强度试验;标准混凝土立方试块
typical curve 标准曲线
typical design 定型设计;典型设计;标准设计
typical design data 典型设计数据
typical detail 定型详图;典型详图
typical detail drawing 定型详图;标准详图
typical drawing 定型图;标准图
typical drawings adopted 选用通用图
typical element 定型构件
typical environmental condition 标准环境条件
typical environmental solution 标准环境溶液
typical example 典型
typical feature 类型特征
typical feature of simulated processes 仿真过程的典型特征
typical fixed Ro-Ro shore ramp 典型固定式滚装岸上坡道
typical fleet 标准船队
typical floor plan 高层建筑中的标准层平面;典型楼层平面图;标准层平面(图)
typical floors 标准楼层;典型楼层
typical flowsheet 典型流程
typical form 典型形式;定型模板;标准模板
typical gradation curve 典型粒度曲线;典型级配曲线
typical grading 典型级配
typical groundwater 标准地下水
typical highway vehicle 公路标准车
typical house 定型房屋
typical hydrograph 典型过程线
typical instance 典型事例
typical investigation 典型调查
typical joint 典型接头;典型接缝
typical joint arrangement 典型接缝布置
typical ladder dimensions 典型爬梯尺寸
typical large compartment cage 典型大罐笼
typical layout 示范布置;典型布置
typical length 标准长度
typical line 特征线
typicallized form 定型模板
typical load(ing) 典型荷载
typical mapping film 标准航摄胶片
typical meadow 点型草甸;典型草甸
typical model 定型
typical module 典型模式
typical mooring pattern 典型系统模式
typicalness 典型性
typical odo(u)r 典型气味
typical organic pollutant 特性有机污染物
typical panel arrangement 典型墙段布置
typical parts 典型件
typical pattern 典型样式
typical per capita wastewater flow 每人典型废水量
typical pollutant 特性污染物
typical problem 典型问题;标准问题
typical products in mass production 大批量产品
typical profile 典型纵剖面;典型剖面
typical project 样板工程
typical rates of water use 典型用水量
typical river reach 典型河段
typical sample 典型样式;典型样品;典型样本;典型土样;典型试样
typical section 典型截面;典型断面
typical section of road 典型路段
typical shear value 标准抗剪(力)值
typical shell 典型的薄壳
typical ship dimensions 典型船舶尺寸
typical ship type 标准船型
typical side slope 典型边坡
typical solution 标准溶液
typical space 典型空间
typical speed 标准速度
typical spiral nebula 典型旋涡星云
typical stacking height 标准堆高
typical stacking method 典型堆料法
typical standard anchor 典型紧固件;标准型紧固件
typical storm 典型暴雨
typical test 典型试验
typical thickness 标准厚度
typical throughput 典型吞吐量
typical tie rod 典型拉杆
typical turbine generator unit bay 典型的水轮发电机机组段
typical uncharged solute 标准无电荷溶质
typical upper floor plan 典型的上层楼平面图
typical value 均值;平均数;代表值
typical velocity 标准速度
typical vessel ramp arrangement 典型的船舶跳板布置图
typical water treatment condition 标准水处理条件
typical weekday 典型工作日
typical width 标准宽度
typical year 典型年
typification 定型;典型化
typified drawing 定型图
typified form 定型模板
typing 分型(法)
typing room 打字室
typist 打字员
typist's error 打字错误
typist's mistake 打字错误
typist's room 打字间;打字室
typochemical element 典型化学元素;标志元素
typodont 矫正咬合器
typographic(al) mark 印刷标记
typology 类型学
typomorphic(al) 标型的
typomorphic(al) facies 标型相
typomorphic(al) mineral 标型矿物
typomorphic(al) peculiarities 标型特征
typomorphic characteristic of mineral 矿物标型特征
typomorphism of minerals 矿物的标型
typotron 高速(字)标管
tyraline 一品红;苯胺红
tyrannopolis 消费城市(专制统治的)
tyre = tire
tyre bar 轮箍钢条
tyre bead 轮胎缘;胎圈
tyre bead toe 轮胎缘趾
tyre bender 弯胎机
tyre boot 胎内衬片
tyre brake 轮胎制动器
tyre branding iron 轮胎烙印铁
tyre breaker 轮胎隔层
tyre builder 轮胎成型床
tyre building machine 轮胎装配床;轮胎成型机
tyre burst 轮胎爆裂;爆胎
tyre canvas 轮胎帆布
tyre capacity 轮胎承载能力;车胎承载能力
tyre carcass 轮胎胎壳;轮胎帘布层
tyre casing 外胎

tyre cement 衬胎胶
tyre center （路面上的）轮胎印迹中心
tyre chain 轮胎防滑铁链；轮胎防滑（铁）链；胎链；胎垫
tyre chain adjuster 胎链调整器
tyre chain for cross-country operation 野外用轮胎防滑链
tyre-chain pliers 胎链钳
tyre-chain repairing pliers 胎链修理钳
tyre-chain repair link 胎链备修链节
tyre chains 防滑轮胎链
tyre clawbar 轮胎撬棍
tyre cold patching cement 冷补胎胶
tyre contact area 轮胎接触面积
tyre cord 轮胎箍绳
tyre cover 外胎
tyre crane 汽车起重机；轮胎起重机；轮胎吊
tyre creep 大窑滚圈滑动
tyre cut 轮胎割痕
tyre demounter 轮胎拆卸器
tyred light tramway 轻型胶轮电车线路
tyre dozer 轮胎式推土机
tyred tractor 轮胎式拖拉机
tyred tractor shovel 轮胎式铲运机；轮胎式拖拉挖土机；轮胎式拖拉铲土机；轮胎铲斗车
tyre engraving 胎面花纹
tyre fabrics 帘子布
tyre fastening 轮胎固着法；紧轮箍圈
tyre fender 车轮挡泥板；轮胎碰垫；轮胎护舷
tyre finish 轮胎漆
tyre flap 胎内衬带
tyre force 轮胎作用力
tyre fork 装轮胎叉
tyre ga(u)ge 轮胎气压计；轮箍规
tyre girdle 轮胎防滑套爪
tyre grip 轮胎附着力
tyre gripping pattern 轮胎防滑花纹
tyre hand air pump 轮胎打气筒
tyre heater 轮胎热压器
tyre hot patch 轮胎热补；热补胎胶
tyre impression 轮胎印痕
tyre inflater 轮胎充气机
tyre inspection 轮胎检查
tyre iron 拆轮胎棒
tyre last 轮胎耐久性
tyre lathe 轮箍加工机床
tyre lever 轮胎撬板
tyre life 轮胎寿命
tyre load 轮胎荷载
tyre lock 胎锁
tyre mounted asphalt finisher 轮胎沥青路面整修机
tyre-mounted crane 轮胎式起重机
tyre-mounted tower crane 轮胎式塔式起重机
tyre mounting 轮胎架
tyre mounting and dismounting machine 轮胎装卸机
tyre noise 轮胎噪声；轮胎滚滑噪声
tyre of conventional construction 普通轮胎
tyre of spray gun 喷枪轮箍
tyre opener 轮胎撑开器
tyre patch 补胎片
tyre-pavement friction 轮胎和路面的摩阻力
tyre-penetration factor 轮胎深入系数
tyre press 轮胎装胎压床；轮胎压机；轮箍压机
tyre printing 轮胎印
tyre protective chain 轮胎防护链
tyre pump 轮胎打气筒；轮胎打气泵
tyre regroover 轮胎（式）压槽器
tyre repair kit 补胎工具
tyre rim 轮胎钢圈
tyre roller 轮胎压路机；轮胎式压路机；轮胎压路机
tyre scuffmark 轮胎擦印
tyre sealer 补轮胎胶
tyre section 轮胎断面；轮胎花纹
tyre slip 轮胎打滑
tyre soles process 胎面翻新
tyre steel 轮箍钢
tyre storage rack 轮胎存放架
tyre strap 轮胎衬带
tyre stripper 轮胎拆卸器
tyre thickness 轮箍厚度
tyre-to-pavement contact area 轮胎与路面接触面（积）
tyre tourniquet 轮胎密封衬带
tyre tread 轮胎凸纹；轮胎花纹
tyre tread life 轮胎行驶里程
tyre tread pattern 轮胎花纹
tyre-tru(e)ing machine 轮胎正圆机
tyretskite 蒂羟硼钙石
tyre tube 内胎
tyre valve 轮胎气门嘴；轮胎阀
tyre-valve cap 轮胎阀帽
tyre vulcanizer 轮胎硫化机
tyre wheel 罩胎轮
Tyrian purple 泰尔红紫
tyrian purple 皇紫
tyro 新手
Tyrol 提罗喷漆枪（一种手持的喷涂彩色水泥砂浆的喷涂器）
Tyrolea 提罗尔抹灰（一种防水抹灰）
Tyrolean finish 拉毛罩面；粗面罩面；提罗尔防水砂浆罩面
tyrolean plaster 喷灰浆
tyrolite 铜泡石（天蓝石）；丝砷铜矿
tyrosine 酪氨酸
tyrrellite 硒铜钴矿
Tyrrell's hook 提勒耳氏钩
Tyrrhenian Sea 第勒尼安海
Tyseley alloy 泰泽利饰用铸锌合金
tyuyamunite 钙钒铀矿；钒钙铀矿
Tyvek 蒂维克纺黏型聚丙烯纤维
T zone T 带

U

U-abutment U 形岸墩；U 形桥台
U-antenna U 形天线
ubac 山阴【气】；背阳坡
U-bar 槽钢
Ubbelohde drop point 乌伯娄德滴点
Ubbelohde liquefying-point test 乌伯娄德液化点试验
Ubbelohde melting point 乌伯娄德熔点
Ubbelohde visco(si)meter 乌伯娄德黏度计
U-beam U 形梁
U-bearing carbon-siliceous rock 含铀碳硅质岩
U-bend U 形弯头；马蹄(形)弯头；回填弯段
ubethetical sale 不道德售货
ubiquist 随遇植物
ubiquitous organism 随遇生物
ubiquitous surface wave 随遇面波
ubitron 波荡射束注入器
U-block U 形混凝土过梁；U 形砌块
U-boat 潜水艇
U-bolt 钩键；钩环；马蹄(形)螺栓；骑马螺栓；槽形螺栓；U 形螺栓
uca 招潮属
uchucchacuaite 硫锑锰银铅矿
U-clamp U 形压板；U 形(管)夹
U-content anomaly map of survey area 测区铀含量异常图
U curve 淬火截面硬度分布曲线
Udden grade scale 尤登粒级
Udden-Wentworth grade scale 伍登—温德华斯粒级标度
udell 冷凝水接受器
udent 湿新成土
udert 湿变性土
udic 土壤长湿状
udoll 湿软土
udometer 雨量计
udomograph 自记雨量计
udox 湿氧化土
U-dozer U 形铲推土机
U-duct U 形烟道；U 形通风管道
udult 湿老成土
U-end frame U 形支腿
U-expansion bend U 形膨胀弯管
U-factor 热损失系数
uferflucht 离岸浮游生物
U-flame furnace 马蹄形火焰炉
U-form U 形的
Uganda mahogany 乌干达桃花心木
ugandite 暗橄白榴岩
U-ga(u)ge U 形(压力)计；U 形检漏器
Ugine Sejournet process 玻璃润滑剂高速挤压法
ug-in unit 插件
uglies 非理想建设
ugly and threatening weather 险恶气候
ugly sea 险恶海面
ugly sky 阴沉天空
ugly weather 阴沉天气
ugrandite 铬钙铁榴石
U-hanger U 形吊钩
Uhde sulfur recovery process 乌德硫磺回收过程
uhligite 锆钙钛矿；钙锆钛矿
U-hook U 形钩
Uhuru catalogue 乌呼鲁 X 射线源表
uinal 乌纳尔
Uinta formation 温塔岩组
uinta(h)ite 硬沥青；天然硬沥青
U-iron U 形铁；水落管卡；槽铁；U 形水管卡
U-joint U 形接头
ukay 哥伦比亚木材分类
UK barrel 英制石油产品量桶
Ukena colo(u)rimeter 乌肯纳比色计
UK gallon 英制加仑(1 英制加仑=4.546)
uklonskovite 水钠镁矾
Ulakhan polarity superchron 阿拉克罕极性超时
Ulakhan polarity superchronzone 阿拉克罕极性超时间带
Ulakhan polarity superzone 阿拉克罕极性超带
Ulatisian 乌拉梯斯期

Ulbricht globe 乌布利希球形光度计
Ulbricht sphere 积算球；乌布利希球
ulcer 远极孔
U-leathering ring U 形皮碗
uleine 乌勒因
Ule's colo(u)r standard 乌利型染色标准
Ule's scale of colo(u)r 乌利水色计
ulexite 钠方硼解石；硼钠钙石；硼钠方解石
Ulfsunda faience 乌尔夫宋达陶器(瑞典)
U-link U 形插塞；U 形连接环；U 形链
ullage 油罐油面上部的空间；液舱耗损量；罐顶空气；漏损(量)；空距；空挡；减量；缺量；途耗
ullage board 膨胀余位尺；膨胀余位测板
ullaged condition 漏损情况
ullage foot 膨胀余位测尺
ullage ga(u)ge 膨胀余位测尺
ullage hole 膨胀余位测孔
ullage of expansion 膨胀余位
ullage plate 膨胀余位测孔盖板
ullage plug 膨胀余位测孔栓塞
ullage port 膨胀余位测孔
ullage report 空距表(油轮)
ullage rocket 气垫增压火箭
ullage rule 测量罐空的量尺
ullage scale 液面标尺；膨胀余位测尺
ullage space (储罐油面上面的)气相空间
ullage stick 液面标尺；膨胀余位尺；膨胀余位测板
ullage table 空距表(油轮)；膨胀余位表；测油尺
ullage tape 液面标尺
Ullmal alloy 厄马尔合金
ullmannite 辉锑镍矿；锑硫镍矿
Ulloa's ring 布格晕
ulmain 棕腐殖无结构镜煤
Ulmal 尤马尔铝合金
ulmarium 榆树苗圃
ulmic 棕腐质的
ulmic acid 棕腐酸；赤榆酸
ulmification 棕腐化作用
ulmin 棕腐酸；赤榆树脂
ulmin brown 铁棕(色)
ulminite 腐木质体
ulminium 尤耳明铝合金
ulmin vegetable jelly 棕腐质
ulmohumic acid 赤榆腐殖酸
Ulmus campestris 英国榆
Ulmus tonkinensis 越南榆
ulnar area 中域
ulnar loop 正箕
ulotrichous 波纹发的
Ulpian basilica 古罗马法院
ulrabasic appinite 超基性暗拼岩
Ulsterian 乌耳斯特阶【地】
ulterior chartered freight 以后的租船运费
ultex 整块双焦点镜
ultimate 最终的；最后的；终结的；根本的；临界；极限(的)
ultimate amount of wear 极限磨耗量
ultimate analysis 元素分析；最后分析；极限分析；基本元素分析
ultimate analysis of coal 煤元素分析
ultimate angle of soil slope 土坡的极限坡角
ultimate aseismic capacity 极限抗震能力
ultimate base level 终极基准面；海面基准面
ultimate bearing capacity 极限承载量；极限承载(能)力
ultimate bearing capacity of foundation 地基极限承载力
ultimate bearing capacity of pile 桩极限承载力
ultimate bearing capacity of single pile 单桩极限承载力
ultimate bearing capacity of soil 土壤极限承载力
ultimate bearing moment 极限弯距
ultimate bearing power 极限承载(能)力
ultimate bearing pressure 极限支承压力；极限(承载)压力
ultimate bearing property 极限承载能力
ultimate bearing quality 极限承载能力
ultimate bearing resistance 极限承载(能)力

ultimate bearing stress 极限支承应力
ultimate bearing value 极限承载力值
ultimate bearing value of foundation soil 地基土极限承载力值
ultimate bending capacity 极限弯曲能力
ultimate bending moment 极限弯曲力矩；极限弯距
ultimate bending strength 极限弯曲强度；极限抗弯强度
ultimate bend stress 极限抗弯应力
ultimate biochemical oxygen demand 最终生化需氧量；总生化需氧量；极限生化需氧量
ultimate biodegradation 终极降解
ultimate biological oxygen demand 最大生化需氧量
ultimate bond stress 极限黏合应力
ultimate boundedness 毕竟有界性
ultimate breaking load 极限破坏荷载；极限断裂荷载
ultimate breaking strength 极限断裂强度
ultimate buffer 终极侵冲性
ultimate capacity 最大容量；最大能量；最大能力；总功率；极限容量；极限能力；极限量
ultimate carbonate hardness 极限碳酸盐硬度
ultimate carbonate hardness method 极限碳酸盐硬度法
ultimate carcinogen 终致癌物
ultimate carrying capacity 极限承载(能)力
ultimate carrying capacity of single pile 单桩极限承载力
ultimate carrying capacity of station 车站最终通过能力
ultimate carrying resistance 极限承载(能)力
ultimate column resistance 柱的极限强度
ultimate column stability 柱的极限稳定性
ultimate column strength 柱的极限强度
ultimate composition 元素组成
ultimate compressive load 极限抗压荷载
ultimate compressive strength 极限抗压强度；抗压极限强度；极限压缩强度
ultimate compressive stress 极限压缩应力
ultimate condition 极限条件
ultimate consistency 极限稠度
ultimate consumer 最终用户；最后消费者
ultimate consumer market 最后消费者市场
ultimate control 最后控制
ultimate cost 最终成本；最高费用；总成本
ultimate creep coefficient 极限徐变系数
ultimate current 最终电流
ultimate curvature 极限曲率
ultimate damping 极限阻尼；极限衰减
ultimate death 最终死亡
ultimate deflection 极限形变
ultimate deformation 极限变形
ultimate demand 最终需求
ultimate deposition 最终淤积(量)
ultimate depth 最终深度
ultimate design 最终设计；极限设计
ultimate design method 极限状态设计法；极限强度设计(方)法
ultimate design resisting moment 极限设计抗力矩；极限设计抗弯矩
ultimate destabilization 极限不稳定性
ultimate destination 最终目的地；最终到达(港)
ultimate detection limit 最大检测限度
ultimate discharge 极限排放
ultimate disposal 最终处置；最终处理；最后处理
ultimate disposal of industrial waste 工业废物最终处理
ultimate disposal of radwaste 放射性废物最终处置
ultimate disposal of sludge 污泥最终处置
ultimate distribution 最终分配；最终分布
ultimate earthquake 极限地震
ultimate elongation 最大伸长；极限延伸值；极限延伸率；极限伸长(率)；断裂伸长(率)
ultimate elongation in percent 极限伸长率
ultimate elongation percentage 极限伸长率
ultimate equilibrium 极限平衡
ultimate erosion base level 终极侵蚀基准面

ultimate error 极限误差
ultimate error of sampling 抽样最大误差
ultimate expansion 极限膨胀
ultimate expenses 最高费用
ultimate factor of safety 极限安全系数;最大安全系数
ultimate failure 极限破坏
ultimate filter 最后过滤器;终端过滤器;终点过滤器
ultimate flexural capacity 极限弯曲应力;极限弯曲能力
ultimate flow 最大流量
ultimate gain 最大增益
ultimate height of soil slope 土坡的极限高度
ultimate holding company 最终控股公司
ultimate infiltration capacity 稳定下渗容量;稳定下渗能力;稳定入渗容量;稳定入渗能力
ultimate installation 最终装机(容量)
ultimate interception range 最大截击距离
ultimate interest 最终利益
ultimate interlock strength 锁口极限强度
ultimate landform 最终地形;终地形
ultimate lateral capacity 极限侧向承载力
ultimate liability 主要责任
ultimate life 最终寿命;极限寿命
ultimate limit(ing) state 极限状态;承载能力极限状态
ultimate limit(ing) switch 极限限制开关
ultimate lines 驻留谱线
ultimate load 最大荷载;最大负载;最大负荷;极限荷载;极限负荷
ultimate load analysis 极限荷载分析
ultimate load bearing capacity 极限荷载承载能力
ultimate load design 极限荷载设计
ultimate load design method 极限荷载设计法
ultimate loading 极限荷载;最大荷载
ultimate loading capacity 最终承载力
ultimate load method 极限荷载法
ultimate load of pile 桩的极限荷载
ultimately controlled variable 最终受控量;最终被控变量
ultimate maximum 极限最大值
ultimate median tolerance limit 起始半数致死浓度
ultimate mixture 最终混合料
ultimate moment 极限力矩;极限力矩;破损力矩
ultimate moment capacity 极限力矩承载能力
ultimate motor fuel 最优汽车汽油;最优发动机燃料
ultimate mutual solubility 无限互溶性
ultimate net loss 最终实际损失;最终纯损;最后实际损失;最后实际赔偿费
ultimate net retention 最终实际自留
ultimate objective 最极目标
ultimate output 最终产量;最大功率;极限产量
ultimate oxidation 极限氧化
ultimate oxygen demand 最终需氧量;极限需氧量
ultimate oxygen demand index 极限需氧量指数
ultimate particle 基本粒子
ultimate particle size 极限粒度
ultimate passive earth pressure 极限被动土压力
ultimate penetration 极限针入度;极限贯入度
ultimate pH 极限酸碱度;极限 pH 值
ultimate pile load 桩的极限荷载
ultimate pinna 末级羽片
ultimate plain 终平原
ultimate plate load capacity 板荷载极限承载力
ultimate point 极限点
ultimate position 终了位置
ultimate potential recovery 最终潜在可采储量
ultimate pressure 极限压强;极限压力
ultimate pressure of a pump 泵的极限压力
ultimate principle 极限原理
ultimate production 最终产量;总产量;极限产量
ultimate production aircraft 最新生产型飞机
ultimate pullout capacity 极限抗拔力
ultimate quantity of value 最后价值量
ultimate range 最大距离;极限范围
ultimate recovery 最大复原;最终开采量;最终回采率;最终采收率
ultimate reserves 最终储量
ultimate resilience 极限回弹力;极限弹力;极限冲击韧性
ultimate resistance 极限应力;极限抗力;极限电阻;试桩强度
ultimate resistance strength 极限抗力强度
ultimate rock strength 岩石的极限强度

ultimate run 最大航程
ultimate sampling unit 最后取样单位
ultimate seismic-resistant capacity 极限抗震能力
ultimate sensibility 最高灵敏度
ultimate set 相对伸长;最终固化度
ultimate settlement 最终沉陷(量);最终沉降(量)
ultimate shaft resistance 极限侧阻力标准值
ultimate shear 极限剪力
ultimate shear(ing) resistance 极限抗剪力
ultimate shear(ing) strength 极限抗剪强度;极限剪切强度
ultimate shear(ing) stress 最大剪应力;极限(抗)剪应力
ultimate shear(ing) moment 最大剪力矩
ultimate shrinkage 极端收缩;完全收缩
ultimate sink 终端散热器
ultimate size 极限尺度
ultimate slope angle 极限坡角
ultimate slope design 最终边坡设计
ultimate sludge disposal 污泥最终处置
ultimate solvency 最大偿债能力;最大偿付能力
ultimate spacing pattern 最大排泄面积
ultimate stage 极限阶段
ultimate standard water demand 最大标准需水量;总标准需水量
ultimate state 极限状态
ultimate state design 极限状态设计
ultimate static resistance 极限静抗力
ultimate storage 最终储存量;最大储存量
ultimate storage area 最后储藏区
ultimate strain 极限应变;极限形变
ultimate stream 最大水流
ultimate strength 终值强度;极限强度;强度极限
ultimate strength design 极限强度设计
ultimate strength design method 极限状态设计(法);极限强度设计(方)法
ultimate strength method 极限强度计算法
ultimate strength method of design 极限强度设计(方)法
ultimate strength of rupture 最终破裂强度;极限破坏强度
ultimate strength theory 极限强度理论
ultimate strength value 强度终值
ultimate stress 最大应力;极限应力
ultimate stress and strain 极限应力和应变
ultimate stressed state 极限应力状态
ultimate subsidence 最终沉陷(量);最终沉降(量)
ultimate sulfur 元素分析硫
ultimate supporting capacity 极限支承压力
ultimate swelling 顶端膨大
ultimate temperature 极限温度
ultimate temperature rise 极限温升
ultimate tensile strain 极限延伸值;极限拉伸变形
ultimate tensile strength 抗拉极限强度;极限拉伸强度;极限抗张强度;极限抗拉强度
ultimate tensile strength of fiber 极限纤维拉伸强度
ultimate tensile stress 最大抗拉应力;极限张应力;极限拉应力
ultimate tension 极限张力;极限拉伸;极限拉力;极限抗拉强度
ultimate tip resistance 极限端阻力标准值
ultimate torque 极限扭矩
ultimate total dosage 中毒剂量
ultimate use of water 水的最终用途
ultimate-use temperature 最终使用温度;极限使用温度
ultimate vacuum 极限真空
ultimate value 最后值;最大值;极值;极限值
ultimate value of deflection 极限形变值
ultimate vertical bearing capacity of a pile 单桩竖向极限承载力标准值
ultimate void ratio 极限孔隙比
ultimate wealth owner 最终财富所有者
ultimate weight-carrying capacity 极限荷载能力;极限荷重能力
ultimate wilting 永久凋萎
ultimate wilting point 永久凋零点;极限凋萎点
ultimate working capacity 极限工作量
ultimate yield 最终产率;最终产量
ultimatum 最后通牒;最后结论
ultimo 前月;上月

ultimogenitary 最后生的
ultisol 老成土
ultor 最高压级;高压阳极;第二阳极
ultra-abyssal bottom 超深海底
ultra-accelerator 超加速器;超促进剂
ultra-achromatic lens subminiaturization 超消色差透镜
ultra-acid(ic) rock 超酸性岩(石)
ultra-acoustics 超声波学
ultra-albanite 超沥青
ultra-audible 超声(波的)
ultra-audible frequency 超声波频率
ultra-audion circuit 超再生电路
ultra-audion oscillator 超再生振荡器;超三极管振荡器
ultra-audio wave 超声波
ultra-basic 超基性的
ultra-basic complex 超基性杂岩体
ultra-basic glass 超基性玻璃
ultra-basic igneous rock 超基性火成岩
ultra-basic lava 超基性熔岩
ultra-basic magma 超基性岩浆
ultra-basic metamorphic rocks 超基性变质岩类
ultra-basic rock 超盐基性岩;超基性岩
ultra-basic rocks 超碱岩类;超基性岩类
ultrabasite 异辉锑铅银矿
ultra-bornological 超有界型
ultra-bright 超亮
ultra-calan 超卡兰
ultra-cataclasite 超碎裂岩
ultra-centrifugal 超速离心的;高速离心的
ultra-centrifugal cell 超离心元件
ultra-centrifugal method 超离心法
ultra-centrifugal mill 超离心研磨机
ultra-centrifugal sedimentation 超离心沉积
ultra-centrifugation 超(速)离心法
ultra-centrifuge 超(速)离心机
ultra-centrifuge method 超离心机法
ultra-centrifuger 超级离心机
ultra-centrifuging 超离心技术
ultra-chromatography 超色谱法;荧光色谱法
ultra-clay 细粒黏土;超黏(土)粒;超微黏粒
ultra-clean 特净;超净(的)
ultra-clean air system 超净空系统
ultra-cleaner 高效除杂机
ultra-clena coal 特净煤
ultra-close-range photogrammetry 超近摄影测量
ultra-coarse 超粗
ultra-coating 超硬包覆层
ultra-cold neutron 超冷中子
ultra-cold storage freezer 超低温冰箱
ultra-conductor high-intensity magnetic separator 超导强磁场磁选机
ultra-connected 超连通的
ultraconservative 极端保守
ultra-crackate 超加氢裂化产物
ultra-cracking 超加氢裂化
ultracryotomy 冰冻超薄切片术;超薄冷冻切片
ultra-crystalline 超微晶
ultra-deep 超深的
ultra-deep dive 超深潜水
ultra-deep exploration 超深勘探
ultra-deep fracture 超深断裂
ultra-deep mining 超深开采
ultradense zircon brick 高致密锆英石砖
ultra-directional microphone 超(指)向传声器
ultra-divisor 超因子
ultra-dry bulb hygrometer 超干球湿度计
ultra-duralumin 超硬铝
ultra-dyne receiver 超外差(式)接收机
ultra-emulsifier 超声波乳化器
ultra-fast 超快
ultra-fast amplifier 超速放大器
ultra-fast coherent phenomenon 超速相干现象
ultra-fast computer 超高速计算机
ultra-fast curing 超速固化
ultra-fast light pulse 超速光脉冲
ultra-fast neutron 超快中子
ultra-fast phenomenon 超快现象
ultra-fast pinch 超快箍缩
ultra-fast pulse 超快脉冲
ultra-fast recovery photodiode 超速复原光电二极管;超快复元光电二极管
ultra-fast relaxation 超速弛豫

ultrafax 电视高速传真;电视传真电报
ultra-fiche 超小平片;超缩微平片;超缩微胶片
ultra-filter 超细过滤器;超滤子;超滤器;超滤集;超级滤网;高速滤池
ultra-filtrate 超滤液;超滤滤液
ultra-filtration 超滤作用;超过滤法;超滤(法);超过滤(作用)
ultra-filtration apparatus 超滤设备
ultra-filtration crucible 超滤坩埚
ultra-filtration dynamic membrane 超滤动态膜
ultra-filtration electrodeionization technology 超滤—电去离子工艺
ultra-filtration inorganic membrane 超滤无机膜
ultra-filtration membrane 超滤膜
ultra-filtration membrane bioreactor 超滤膜生物反应器
ultra-filtration membrane filtration 超滤膜过滤
ultra-filtration membrane surface 超滤膜面
ultra-filtration monitor 超滤监测器
ultra-filtration process 超过滤法
ultra-filtration reactor 超滤反应器
ultra-filtration system 超滤系统
ultra-filtration treatment 超过滤处理
ultra-fine 过细的;超细的
ultra-fine ceramic powder 超细陶瓷粉末
ultra-fine classifier 超细粒分级机
ultra-fine colloidal antimony oxide 超微细胶体氧化锑
ultra-fine dust 特细粉末;超细微粒;超细粉末;超细粉尘
ultra-fine electrolytic powder 超细电解粉末
ultra-fine enameled wire 特细漆包线
ultra-fine equiaxil grain 超细等轴晶粒
ultra-fine fiber 超细纤维;超滤器;超滤器
ultra-fine fiber glass 超细玻璃纤维
ultra-fine filtration 超滤
ultra-fine focus 超细焦点
ultra-fine fraction 超细颗粒
ultra-fine glass 超细玻璃
ultra-fine glass fabric 超细玻璃丝布
ultra-fine glass fibre 超细玻璃纤维
ultra-fine glass wool 超细玻璃棉
ultra-fine grain 超细颗粒;超细晶粒;超微粒
ultra-fine grain ceramics 超细粒度陶瓷
ultra-fine grain developer 超精细显影液;超细粒显影剂
ultra-fine-grain emulsion 超微粒乳剂
ultra-fine grain size 超细晶粒度
ultra-fine grinder 特细粉碎机
ultra-fine grinding 超细研磨;超细粉碎;超细粉磨
ultra-fine grinding mill 超细粉磨机
ultra-fine grinding technology 超细粉磨工艺
ultra-fine ion exchange resin 超细粉末离子交换树脂
ultra-fine material 超细材料
ultra-fine particle 超细粒子;超细颗粒
ultra-fine pore 超细孔
ultra-fine powder 超细粉(末);超细粉料
ultra-fine pulverizer 特细研粉机
ultra-fines 超细粉;超细颗粒;超细粉末;特细粉末
ultra-fine sand 微细砂
ultra-fine silver emulsion plate 超微粒银盐干版
ultra-fine soot 特细煤烟;超微烟灰
ultra-fine titanium dioxide 超微细二氧化钛
ultra-fine titanium dioxide pigment 超微细二氧化钛颜料
ultrafine zinc 超细粉末锌
ultra-fining 超精炼
ultra-fining heat treatment 超细化热处理
ultra-fining process 超精炼过程
ultra-finish 极超精加工
ultra-fitration theory 超滤学论
ultra-forming 超重整(过程)
ultra-fort steel 超强强钢;超强钢
ultra-gamma ray 超伽马射线
ultra-gravity wave 超重力波;短周期重力波
ultra-hard 超硬;特硬的
ultra-hard coating 超硬涂层
ultra-hard high speed steel end mill 超硬高速钢立铣刀
ultra-hard material 超硬材料
ultra-hard ray 超硬射线
ultra-harmonic 超调和(的)
ultra-harmonic oscillation 超谐振荡

ultra-harmonic resonance 超谐共振
ultra-harmonics 超(高次)谐波;高次谐波
ultra-harmonic vibration 超谐振动
ultra-high access memory 超高速存取存储器
ultra-high-altitude 超高空
ultra-high-altitude photograph 超高空摄影片
ultra-high build coating 超厚(膜)涂层
ultra-high-early-strength cement 超早强水泥
ultra-high-early-strength Portland cement 超高早强硅酸盐水泥;超高早强波特兰水泥
ultra-high-energy accelerator 超高能加速器
ultra-high-energy beam separator 超高能束流分离器
ultra-high energy particle 超高能粒子
ultra-high-field low-flutter cyclotron 超高场低颤回旋加速器
ultra-high frequency 特高频;超高频
ultra-high frequency aerial 特高频天线
ultra-high frequency amplification 超高频放大
ultra-high frequency amplifier 超高频放大器
ultra-high frequency antenna 特高频天线
ultra-high frequency band 特高频带;超高频带
ultra-high frequency bandpass filter 特高频带通滤波器
ultra-high frequency bridge 超高频电桥
ultra-high frequency carbon film resistor 超高频碳膜电阻
ultra-high frequency carrier 特高频载波
ultra-high frequency cathode ray tube 超高频阴极射线管
ultra-high frequency channel 特高频频道
ultra-high frequency converter 特高频变频器;超高频变频器
ultra-high frequency correlator 超高频相关器
ultra-high frequency electricity 超高频电
ultra-high-frequency electromagnetic field 超高频电磁场
ultra-high frequency generator 超高频发生器
ultra-high frequency handheld transceiver 超高频手持式收发机
ultra-high frequency heterodyne wave meter 特高频外差波长计;超高频外差频率计
ultra-high frequency jammer 超高频干扰器
ultra-high frequency land mobile radio communication equipment 超高频陆地移动无线通信设备
ultra-high frequency mini-trunking radio system 超高频小型中继无线通信系统
ultra-high frequency mobile transceiver 超高频移动式收发机
ultra-high frequency multiplex telephone facility 超高频多路复用电话设备
ultra-high frequency oscillator 超高频振荡器
ultra-high frequency poker 超高频探测棒;超高频探测器;超高频振荡器
ultra-high frequency power booster 超高频功率升高器
ultra-high frequency power signal generator 超高频功率信号发生器
ultra-high frequency preamplifier stage 特高频前置放大级
ultra-high frequency protection 超高频保护
ultra-high frequency radio set 超高频无线电设备
ultra-high frequency radio telephone equipment 特高频无线电话设备
ultra-high frequency receiver 特高频接收机;超高频接收机
ultra-high frequency resonance frequency meter 超高频谐振式频率计
ultra-high frequency signal generator 超高频信号发生器
ultra-high frequency sound device 超声波测云器
ultra-high frequency super-conducting magnetometer 超高频超导磁强计
ultra-high frequency sweep signal analyzer 超高频扫描信号分析仪
ultra-high frequency test receiver 超高试验接收机;超高频测试接收器
ultra-high frequency tetrode 超高频四极管
ultra-high frequency transformer 超高频变压器
ultra-high frequency translator 特高频转播机;特高频差转机
ultra-high frequency transmission line 特高压输电线路

ultra-high frequency transmitter 特高频发射机;超高频发射机
ultra-high frequency triode 超高频三极管
ultra-high frequency tuner 特高频调谐器;超高频调谐器
ultra-high frequency tuning device 特高频调谐装置
ultra-high frequency tuning mechanism 特高频调谐机构
ultra-high frequency ultrasonic wave 超高频率超声波
ultra-high frequency ultra-sound 超高频超声波
ultra-high frequency vision transmitter 特高频图像发射机
ultra-high frequency wave 超高频电波
ultra-high frequency wavelength 超高频波段
ultra-high frequency wave meter 超高频波长表
ultra-high molecular weight polyethylene 超高分子量聚乙烯
ultra-high power 超高功率
ultra-high power arc furnace 超高功率电弧炉
ultra-high power electrode 超高功率电极
ultra-high pressure and high temperature device with two recessed dies 凹形对顶砧超高压高温装置(简称凹砧装置)
ultra-high pressure and high temperature technique 超高压高温技术
ultra-high pressure apparatus 超高压装置
ultra-high pressure device with two recessed dies 凹形对顶砧超高压装置
ultra-high pressure measurement 超高压测量
ultra-high pressure mercury lamp 超高压汞灯
ultra-high pressure mercury vapo(u)r lamp 超高压汞灯
ultra-high pressure sintering 超高压烧结
ultra-high pressure technique 超高压技术
ultra-high pressure vessel 超高压容器
ultra-high purity 超高纯度
ultra-high purity metal 超高纯金属
ultra-high purity reagent 超高纯试剂
ultra-high purity water 高纯水;超高纯水
ultra-high quality filter effluent 超高质滤池出流物
ultra-high quality filtration 超高质过滤
ultra-high rate filtration 超高速过滤
ultra-high resistance 超高电阻
ultra-high resistance and resistivity tester 超高电阻及电阻率测试仪
ultra-high resolution 超高分辨率
ultra-high resolution bathymetry 超高分辨测深术
ultra-high sand concentration 超高含砂浓度
ultra-high-speed 极高速;超高速;特高速
ultra-high speed camera 超高速摄影机
ultra-high-speed computer 超高速巨型计算机
ultra-high-speed copier 超高速复印机
ultra-high speed extrusion 超快速挤压
ultra-high speed facsimile 超高速传真机
ultra-high-speed particle 超高速粒子
ultra-high speed photographic instrument 超高速相器械;超高速照相仪器
ultra-high speed photography 超高速摄影术
ultra-high speed pulse repeater 超高速脉冲再生器
ultra-high speed radiography 超高速放射线相术
ultra-high speed railway 超高速铁路
ultra-high speed sewing machine 超高速缝纫机
ultra-high speed streak camera 超高速扫描摄影机
ultra-high speed transient recorder 超高速瞬态记录仪
ultra-high speed x-autoradiography 超高速X放射自显影术
ultra-high strength 超高强度
ultra-high strength concrete 超高强混凝土
ultra-high strength filament 超高强度长丝
ultra-high strength high-modulus fiber 超高强度高模量纤维
ultra-high strength steel 超高强度钢
ultra-high-sulfur 超高硫
ultra-high temperature press 超高温压机
ultra-high vacuum 超高真空
ultra-high vacuum chemical vapo(u)r deposition 超高真空化学气相沉积
ultra-high vacuum coater 超高真空镀膜机
ultra-high vacuum meter 超高真空计
ultra-high-vacuum monochromator 超高真空单色仪

ultra-high vacuum pump 超高真空泵
ultra-high vacuum seal 超高真空密封
ultra-high-vacuum system 超高真空系统
ultra-high vacuum technology 超高真空工艺学
ultra-high vacuum valve 超真空阀
ultra-high velocity impact 超高速碰撞
ultra-high viscosity mixer 超高黏度混合器
ultra-high voltage 超高压【电】
ultra-high voltage electron microscope 超高压电子发射显微镜
ultra-hypabyssal facies 超浅成相
ultra-hyper-bolic 超双曲线
ultra-hyper-bolic differential equation 超双曲线微分方程
ultra-hyper-pyrexia 超高热
ultra-ideal 超理想
ultra-infinte point 超无穷远点
ultra-intense beam 超强束
ultra-ionization potential 超电离电位
ultra-large 特大的
ultra-large crude carrier 超型油轮；超巨型油轮；超大型油轮
ultra-large crude oil carrier 超大型油轮
ultra-large oil carrier 超型油轮
ultra-large product carrier 超大型成品油船
ultra-large reefer carrier 超大型冷藏船
ultra-large type 超大型
ultra-lattice 超格
ultra-ligation 超扎法
ultra-light aircraft 超轻型飞机
ultra-light alloy 超(级)轻合金
ultra-limit 超(级)限
ultra-linear 超线性
ultra-linear amplifier 高线性放大器；超线性放大器
ultra-linear amplifier circuit 超线性放大电路
ultra-linearity 超直线性
ultra-long period 超长周期
ultra-long period seismograph 超长周期地震仪
ultra-long range 超远程声的
ultra-long range sonar 超远程声呐
ultra-long spaced electric(al) log 超长距电测井；超长电极距测井
ultra-long spaced sonde 超长电极系
ultra-long spaced sonde log 超长电极系测井
ultra-long spacing electrical log 超长极距测井
ultra-long wave 超长波
ultra-low birefringence fiber 超低双折射光纤
ultra-low carbon stainless steel 超低碳不锈钢
ultra-low-carbon steel 超低碳钢
ultra-low cement castable 超低水泥浇注料
ultra-low density polyethylene 超低密度聚乙烯
ultra-low drift amplifier 超低漂移放大器
ultra-low expansion glass ceramics 超低膨胀微晶玻璃
ultra-low expansion mirror telescope 超低膨胀率镜望远镜
ultra-low expansion modified fused silica glass 超低膨胀石英玻璃
ultra-low frequency 特低频；超低频
ultra-low frequency amplifier 超低频放大器
ultra-low frequency dualtrace oscilloscope 超低频二踪示波器
ultra-low frequency ocean wave recorder 超低频自记测波仪
ultra-low frequency oscillator 超低频振荡器
ultra-low frequency oscilloscope 超低频示波器
ultra-low frequency phase shifter 超低频移相器
ultra-low-friction cylinder 超低摩擦汽缸
ultra-low-loss fiber 超低损耗光纤
ultra-low-melting alloy 超低熔合金
ultra-low noise travelling wave tube 超低噪声行波管
ultra-low pressure reverse osmosis membrane 超低压反渗透膜
ultra-low refractive index optic(al) glass 超低折射率光学玻璃
ultra-low temperature 超低温
ultra-low temperature ball valve 超低温球阀
ultra-low temperature freezer 超低温冰箱
ultra-low temperature glue 超低温胶
ultra-low temperature pump 超低温泵
ultra-low temperature resistant adhesive 耐超低温胶粘剂
ultra-low temperature storage 超低温保存

ultra-low temperature thermistor 超低温热敏电阻器
ultra-low temperature throttle valve 超低温闸阀
ultra-low viscosity oil 超低黏度油
ultra-low volume 超低容量；超低剂量
ultra-low volume aerial application 超低容量飞机喷药
ultra-low volume spray(ing) 超低容量喷药；超低剂量喷药
ultra-low weight portable power-driven sprayer 超轻量手提式电动喷雾器
ultra-lumin 超硬铝
ultra-macroion 超巨型离子
ultra-mafic 超铁镁质的；超镁铁质的
ultra-mafic extrusive rock group 超镁铁喷出岩类
ultra-mafic glass 超碱玻璃
ultra-mafic nodule 超铁镁质结核
ultra-mafic structure 超镁铁岩构造
ultra-mafite 超铁镁岩；超镁铁岩
ultra-mafites 超镁铁岩类
ultra-magnifier 超磁放大器
ultramarine 群青(色)
ultramarine ash 群青色的灰
ultramarine blue 绀青；深蓝；群青蓝；群青(色)
ultramarine brown 群青棕
ultramarine green 群青绿
ultramarine trade 海外贸易
ultramarine violet 群青紫
ultramarine yellow 群青黄
ultra-matic drive 超自动传动装置
ultra-matic stitch selector 超自动线迹
ultra-membrane filtration 超膜滤
ultra-metamorphic ore-forming process 超变质成矿作用
ultra-metamorphic rocks 超变质岩
ultra-metamorphism 超变质作用
ultra-metamorphosed 超变质的
ultra-mfic rock 超基性岩
ultra-micro 超微(的)；超微量
ultra-microacervatio 超微量凝聚
ultra-microanalysis 超微(量)分析
ultra-microassay 超微量测定；超微分析
ultra-microbalance 超微量天平
ultra-microbe 超微生物
ultra-microcamera 超微型照相机
ultra-microchemical manipulation 超微量化学操作
ultra-microchemistry 超微(量)化学
ultra-microcrystal 超微细晶；超微结晶
ultra-microcut 超薄切片
ultra-microearthquake 超微(地)震；极微小地震
ultra-microelectrode 超微电极
ultra-microelectrophoresis 超微量电泳
ultra-microelement 超微量元素
ultra-microfiche 超微缩照片；缩微照片；特超缩微胶片
ultra-microfigure 超缩微像片
ultra-microfluorophotometer 超显微荧光光度计；超微量荧光光度计
ultra-microfossil 超微化石
ultra-micromethod 超微量法
ultra-micrometer 超微计；超级测微仪；超(级)测微计
ultra-micromotor 超微型电机
ultra-micron 超微细粒；超微粒(子)
ultra-microphotography 超微型照相术
ultra-micropipet 超微量吸管
ultra-micropolarimeter 超微量偏振计
ultra-micropore 超微孔(隙)
ultra-micropowder 超微粉末
ultra-microsampling 超微量取样；超微量采样
ultra-microscope 超显微镜；超高倍显微镜；超倍显微镜
ultra-microscopic 超显微；超微型；超光度计
ultra-microscopic amicroscopic 超显微镜的
ultra-microscopic cracking 超微裂纹
ultra-microscopic dust 超微粉尘；超微尘(粒)
ultra-microscopic granule 超显微镜粒
ultra-microscopic inclusion 超微观包裹体
ultra-microscopic microorganism 超微型微生物
ultra-microscopic organism 超显微(镜)生物
ultra-microscopic scale 超显微构造尺度
ultra-microscopic view 超微观
ultra-microscopy 超显微术；超显微镜检查；超倍显微镜检验法

ultra-microsome 超微粒体
ultra-microspectrophotometer 超微量分光光度计；超显微分光光度计
ultra-microstructure 超微结构
ultra-microtechnique 超微量技术；超微技术；超微工艺
ultra-microtitrator 超微滴定器
ultra-microtome 超薄切片机
ultra-microtomy 超薄切片术；超薄切片法
ultra-microwave 超微波
ultra-microwave interferometer 超微波干涉仪
ultra-miniature 超缩微
ultra-miniature camera 超小型摄影机
ultra-mizer 超微细粉碎机
ultra-modern 超现代化(的)；最新(式)的
ultra-modern ship 最新式差
ultra-mylonite 超糜棱岩化；超糜棱岩
ultra-mylonitic texture 超糜棱结构
ultra-Neptunian planet 海外行星
ultra-nsonic transducer 超声波传感器
ultra-optimeter 超精度光学比较仪；超级光学计；超光测仪
ultra-orthoflow fluid catalytic cracking process 超正流流化催化裂化过程
ultra-oscilloscope 超声波示波器；超短波示波器
ultra-paque microscope 超暗显微镜
ultra-particle colloid grinder 超微粒胶体磨
ultra-perm 超坡莫高透磁合金；超导磁铁镍钼铜合金
ultra-phonic 超声的
ultra-phonics 超声波学
ultraphosphate 过磷酸
ultraphotic 超视；超光
ultraphotic ray 视外线；不可见射线；不可见光线
ultra-photometer 超光度计
ultra-plane microscope projection lens 超平场显微镜投影镜头
ultra-plankton 超微型浮游生物
ultra-poor products 超差品
ultra-porcelain 超高频瓷
ultra-portable 超轻形式；超便形式；极轻便的
ultra-portable camera 超小型电视摄像机
ultra-power 超幂
ultra-powerful Nd-Fe-B permanent magnetic alloy 超强钕铁硼永磁合金
ultra-precise 超精密(的)
ultra-precise displacement sensor 超精密位移传感器
ultra-precise laser distancemeasuring instrument 超精密激光测距仪
ultra-precise laser rangefinder 超精密激光测距仪
ultra-precise measurement 超精密测量
ultra-precision 超精度
ultra-precision display 超精度显示器
ultra-precision gear 超精密齿轮
ultra-precision manufacture 超精度生产
ultra-pressure mercury lamp 超高压汞灯
ultra-product 超积
ultra-project meter 超级投影仪；超精度投影(光学)比较仪
ultra-protective trade policy 超保护贸易政策
ultra-pure 超高纯；超纯的
ultra-pure hydrogen generator 超纯氢发生器
ultra-pure material 超纯物质
ultra-pure metal 超纯(度)金属
ultra-pure silicon tetrachloride 超纯四氯化硅
ultra-pure water 超纯水
ultra-purification 超提纯
ultra-purified water 超纯水
ultra-purity 超纯度
ultra-quick continuous flushing light 超快连闪光
ultra-quick flushing light 超快闪光
ultra-quick light 超快闪光
ultra-quinine 超奎宁
ultra-radio frequency 超无线电频；超射频率
ultra-rapid cinematograph 高速电影摄影机
ultra-rapid cinematography 高速电影摄影术
ultra-rapid developer 超速显色剂
ultra-rapid flasher 超高速闪光灯
ultra-rapid flash photography 超速闪光摄影；超高速闪光摄影
ultra-rapid flow 超速水流；高速水流；高速水道
ultra-rapid hardening cement 超快硬水泥；特快硬水泥

ultra-rapid high pressure ga(u)ge 超高速高压规
ultra-rapid lens 超高速透镜;超大孔径物镜
ultra-rapid vulcanization 超速硫化
ultra-rays 宇宙(射)线
ultrared heating 红外(线)加热
ultrared radiation 红外线辐射
ultrared ray 红外(线)
ultrared ray video microscope 红外(线)视频显微镜
ultrared sweeping temperature measuring instrument 红外(线)扫描测温仪
ultra-relativistic dispersion relation 超相对论性色散关系
ultra-relativistic electron 超相对论性电子
ultra-relativistic particle 超相对论性粒子
ultra-relativistic theory 超相对论理论
ultra-rkose 超长石砂岩
ultra-safe 极其安全的;特别安全的
ultra-safe lending 绝对安全贷款
ultra-scan laser densitometer 超精度扫描激光密度计
ultra-scnic scaling 超声波洁治术
ultrasecure laboratory 极安全的实验室
ultra-selective cracking process 超选择转化法;超选择性裂化法
ultra-sensitive 高灵敏度的
ultra-sensitive amino-acid analyzer 超灵敏氨基酸分析仪
ultra-sensitive clay 超灵敏黏土
ultra-sensitive exposure meter 高灵敏度曝光表
ultra-sensitive hydrogen detector 超灵敏度氢检测器
ultra-sensitive impact measurement device 超灵敏冲击测量仪
ultra-sensitive magnetometer 超灵敏磁强计
ultra-sensitiveness 超灵敏性
ultrasensitive pressure ga(u)ge 高灵敏度压力计
ultra-sensitive seismometer 极灵敏地震计;超灵敏度地震计;超灵敏地震计
ultra-sensitive super-conduction cryogenic magnetometer 高灵敏度低温超导磁力计
ultra-sensitivity 特高灵敏度;超灵敏度
ultra-short cavity laser 超短腔激光器
ultra-short feedback 超短反馈
ultra-short infrared pulse 超短红外脉冲
ultra-short laser pulse 超短激光脉冲
ultra-short light pulse 超短光脉冲
ultra-short light pulse laser 超短光脉冲激光器
ultra-short-lived nuclide 超短半衰期核素
ultra-short-period binary 超短周期双星
ultra-short pulse 超短脉冲
ultra-short pulse laser 超短脉冲激光器
ultra-short-range radar 超短距测距雷达
ultra-short sound wave 超短声波
ultra-short synchronized laser pulse 超短同步激光脉冲
ultra-short time film evapo(u)rator 离心式瞬时薄膜(蒸发)浓缩器
ultra-short wave 超短波
ultra-short wave antenna 超短波天线
ultra-short wave broadcasting emitter transmitter 超短波广播发射机
ultra-short wave communication 超短波通信
ultra-short wave diathermy 超短波透热
ultra-short wave diathermy apparatus 超短波透热机
ultra-short wave oscillator 超短波振荡器
ultra-short wave pulse laser 超短波脉冲激光器
ultra-short wave radio set 超短波无线电设备
ultra-short wave radio station 超短波无线电台
ultra-short wave receiver 超短波接收机
ultra-short wave transmitter 超短波发射机
ultra-short wave vacuum tube voltmeter 超短波真空管电压表
ultra-short wave wireless telephone 超短波无线电话机
ultra-short whisker 超短针状单晶
ultra-sima 超硅镁层
ultra-simatic layer 超硅镁层
ultra-smooth finish 极光滑的修整
ultra-smooth powder coating 超高平整性粉末涂料
ultra-soft 超软
ultra-soft clay 超软弱黏土
ultra-some 超微体

ultra-sonator 超声波振荡器
ultrasonic 超音速的;超声波动的;超声波的;超声(波)
ultrasonic absorption coefficient 超声波吸收系数
ultrasonic activation 超声波活性化
ultrasonic agglomeration 超声波凝聚
ultrasonic agglomerator 超声波凝结器
ultrasonic agitation 超声波搅拌
ultrasonic airlift loop reactor 超声膜提式环流反应器
ultrasonic algae removal 超声波除藻
ultrasonically activated extruder 超音速驱动的挤压机
ultrasonically induced oxidation 超声波诱发的氧化
ultrasonically promoted reaction 超声波加速反应
ultrasonically tested area 已经探伤部分
ultrasonic amplitude transformer bar 超声变幅杆
ultrasonic anaerobic biochemical process 超声波—厌氧生化法
ultrasonic anemometr 超声波风速计
ultrasonic angle beam testing 超声波斜角探伤试验
ultrasonic apparatus for material testing 超声波材料试验仪
ultrasonic apparel machinery 超声波成衣机械
ultrasonication 超声提取法;超声波作用;超声波旋律;超声波破碎
ultrasonic atomizer 超声波雾化器
ultrasonic atomizer apparatus 超声波雾化器
ultrasonic attenuation measuring device 超声波衰减测量仪
ultrasonic attenuation technique 波衰减技术
ultrasonic bath 超声波浴
ultrasonic bath basin 超声波浴盆
ultrasonic beam 超声波束;超声波射束
ultrasonic bedload detector 超声波推移质测定器
ultrasonic bedload sampler 超声波推移质取样器
ultrasonic biological contact oxidation 超声波—生物接触氧化
ultrasonic bond 超声波键合;超声波焊
ultrasonic bonder 超声波键合器
ultrasonic bonding 超声波黏结
ultrasonic bonding machine 超声波焊接机
ultrasonic bond meter 超声波结合强度测定计
ultrasonic Bragg cell 超声波布拉格盒
ultrasonic Bragg diffraction 超声波布拉格衍射
ultrasonic brazing 超声波钎焊
ultrasonic brazing unit 超声波钎焊装置;超声波焊接装置
ultrasonic bubble camber 超声波凝聚
ultrasonic camera 超声波摄像机
ultrasonic carrier 超声波载波
ultrasonic carver 超声波加工机
ultrasonic casting 超声波注浆
ultrasonic cavitation 超声波空化现象;超声波空化
ultrasonic cement analyzer 超声波水泥分析器
ultrasonic checking 超声波检验
ultrasonic chemical process 超声波化学工艺
ultrasonic chemistry 超声波化学
ultrasonic cleaner 超声波洗净机;超声波清洁器;超声波清洗机;超声波清洁器;超声波除垢器
ultrasonic cleaning 超声波洗涤;超声波清洗;超声波净化
ultrasonic cleaning agent 超声消化剂;超音清洁剂
ultrasonic cleaning bath 超声波洗净槽
ultrasonic cleaning equipment 超声波清洗装置
ultrasonic cleaning machine 超声波清洗器
ultrasonic cleaning process 超声波清洗工艺
ultrasonic cleaning set 超声波清洗装置
ultrasonic cleaning tank 超声波清洗箱
ultrasonic cleaning unit 超声波零件清洗装置
ultrasonic clearing machine 超声波清理机
ultrasonic coagulation 超声波凝聚;超声波凝结;超声波聚结
ultrasonic coat thickness ga(u)ge 超声波涂层厚度计
ultrasonic communication 超声波通信
ultrasonic component 超声波元件
ultrasonic concentrator 超声波浓度仪
ultrasonic concrete tester 超声波混凝土测试仪
ultrasonic continuous wave spectrometer 超声波连续波光谱仪
ultrasonic control 超声波控制
ultrasonic converter 超声波变换器
ultrasonic counter 超声波计数器

ultrasonic crack detection 超声波裂纹检测
ultrasonic crack detector 超声波探伤仪
ultrasonic cross grating 超声波交叉光栅
ultrasonic cross-sectional imaging unit 超声波截面显像仪
ultrasonic crystal 超声波晶体
ultrasonic current 超声流
ultrasonic current meter 超声波流速仪;超声波海流计
ultrasonic cutting 超声波切割
ultrasonic dedusting 超声波除尘
ultrasonic defect detector 超声波探伤仪
ultrasonic defectoscope 超声波检验器
ultrasonic deferring line glass 超声波延迟线玻璃
ultrasonic degradation 超声波降解(作用);超声波断层造影法
ultrasonic degreasing 超声波脱脂
ultrasonic dehydration 超声波脱水
ultrasonic delay-line 超声波延迟线
ultrasonic delay-line glass 超声波延迟线玻璃
ultrasonic delay-line memory 超声波延迟线存储器
ultrasonic delay time-temperature coefficient 超声波延迟时间温度系数
ultrasonic densimeter 超声波比重计
ultrasonic depolymerization 超声波解聚合作用
ultrasonic depth finder 超声波深度测试仪;超声波回声探伤仪;超声波回声探伤仪;超声波测深器;超声波测深计
ultrasonic destruction 超声波破坏
ultrasonic detecting 超声波探伤检查
ultrasonic detection 超声波探伤;超声波探测(法)
ultrasonic detector 超声波探测器;超声波检测仪
ultrasonic detector for density of mixed fuel 超声波混合燃料浓度检测仪
ultrasonic detectoscope 超声波探伤器
ultrasonic device 超声波设备;超声波器件
ultrasonic dewatering technology 超声波脱水技术
ultrasonic diagnostic apparatus 超声波诊断仪;超声波透热设备
ultrasonic diagnostic method 超声波诊断法
ultrasonic diagnostic scanner 超声波诊断扫描仪
ultrasonic dicing 超声波切割
ultrasonic diffraction microscope 超声波衍射显微镜
ultrasonic digital water level ga(u)ge 超声波数字水位计
ultrasonic disinfection 超声波消毒
ultrasonic disintegration 超声波破解
ultrasonic disintegrator 超声波粉碎器
ultrasonic dispersion 超声波色散;超声波分散
ultrasonic dispersion machine 超声波分散机
ultrasonic Doppler effect 超声波多普勒效应
ultrasonic Doppler method testing system 超声波多普勒法检测系统
ultrasonic Doppler technique 超声波多普勒技术
ultrasonic Doppler testing system 超声波多普勒检测系统
ultrasonic drill 超声波钻
ultrasonic driller 超声波钻机
ultrasonic drilling 超声波钻孔;超声波钻法;超声波打孔
ultrasonic drilling machine 超声波钻床;超声波打孔机
ultrasonic dust removal 超声波除尘
ultrasonic echo 超声波回声
ultrasonic echogram 超声波回波图;超声波回波描记图
ultrasonic echo-impulse digital thickness meter 超声波回波脉冲数字式测厚仪
ultrasonic echo sounder 超声波回声测深仪;超声波回声测深器;超声波回声测深仪
ultrasonic echo sounding 超声波测深(法)
ultrasonic echosounding device 超声波测深装置
ultrasonic effect 超声波效应
ultrasonic electrostatic sprayer 超声波静电喷涂机
ultrasonic emitter 超声波发生器
ultrasonic emulsification 超声波乳化(作用)
ultrasonic emulsion breaking 超声波破乳(化)
ultrasonic energy 超声波能
ultrasonic engineering 超声波工程学
ultrasonic erosion 超声波腐蚀
ultrasonic examination 超声波探伤;超声波检查
ultrasonic examination of welds 焊接的超声检验

ultrasonic excitation 超声波激发
ultrasonic extraction 超声波萃取;超声波抽提
ultrasonic extractor 超声波萃取器;超声波抽提器;超声波拔桩机
ultrasonic/Fenton oxidation-coagulation process 超声波/芬顿氧化—混凝工艺
ultrasonic field 超声波场
ultrasonic film thickness ga(u)ge 超声波膜厚计
ultrasonic filter 超声波滤波器
ultrasonic flaw-detecting machine 超声波探伤机
ultrasonic flaw detection 超声波探伤
ultrasonic flaw detector 超声波探伤仪;超声波探伤器
ultrasonic flow 超声波速流;超声波流
ultrasonic flow detector 超声波测流器
ultrasonic flow inspection 超声波流量检查
ultrasonic flow measurement 超声波流测(法)
ultrasonic flow measuring instrument 超声波量测量仪表
ultrasonic flowmeter 超声波流量计
ultrasonic fluctuation velocimeter 超声波动测流计
ultrasonic flue dust elimination 超声波烟道除尘
ultrasonic fluorometer 超声波荧光计
ultrasonic fluviograph 超声波水位计
ultrasonic forming 超声波成形
ultrasonic frequency 超音频;超声波频率;超声波频
ultrasonic frequency generator 超声波发生器
ultrasonic ga(u)ge 超声波探伤仪;超声波计时器
ultrasonic ga(u)ging method 超声波测流法
ultrasonic generator 超声波发生器
ultrasonic grating 超声波光栅
ultrasonic grating constant 超声波常数
ultrasonic handpiece 超声波机头
ultrasonic hand-washing apparatus 超声波洗手装置
ultrasonic hologram 超声波全息照片;超声波全息图
ultrasonic holography 超音全息照相术;超声波全息照相术;超声波全息照相法;超声波全息术
ultrasonic homogenization 超声波乳化(作用)
ultrasonic humidifier 超声波加湿器
ultrasonic hydrophone 超声波水听器
ultrasonic image camera 超声波成像照相机
ultrasonic image visualization technique 超声波像目视技术
ultrasonic imaging 声成像;超声波成像
ultrasonic imaging device 超声波成像仪
ultrasonic imaging method 超声波显像法
ultrasonic imaging technique 超声波成像技术
ultrasonic immersing 超声波浸渍
ultrasonic immersion technique 超声波浸没探伤技术
ultrasonic impulse transmitter 超声波脉冲发射器
ultrasonic inspection 超声波探伤;超声波探测(法);超声波检验;超声波检查
ultrasonic inspection system 超声波测试装置
ultrasonic interferometer 超声波干涉仪
ultrasonic interferometry 超声波干涉度学
ultrasonic internal threading unit 超声波加工内螺纹装置
ultrasonic irradiation 超声波照射;超声波辐照
ultrasonic irradiation enhanced degradation 超声波辐照强化降解
ultrasonic lapping machine 超声波精研机
ultrasonic leak detection 超声波检漏
ultrasonic leak detector 超声波检漏器
ultrasonic leak locator 超声波泄漏定位仪
ultrasonic lens 超声波透镜
ultrasonic level controller 超声波液位控制器
ultrasonic level ga(u)ge 超声波液位计;超声波水准仪;超声波料面计
ultrasonic level indicator 超声波料位指示器
ultrasonic level measuring instrument 超声波物位测量仪表
ultrasonic level meter 超声波液位计;超声波水准仪;超声波料位计
ultrasonic level switch 超声波液位开关
ultrasonic light diffraction 超声波致光衍射
ultrasonic light modulator 超声波光调制器;超声波调光器
ultrasonic light valve 超声波光阀
ultrasonic luminescence 超声波发光
ultrasonic machine tool 超声波加工机床
ultrasonic machining 超声波加工

ultrasonic machining unit 超声波加工设备
ultrasonic marigraph 超声波自记潮位计
ultrasonic material level meter 超声波物位计
ultrasonic measurement 超声波测量
ultrasonic-mechanic multiplicate polisher 超声波机械复合抛光机
ultrasonic memory 超声波存储器
ultrasonic metal inspection 超声波金属探伤;超声波金属检查
ultrasonic metal welding 超声波金属焊接
ultrasonic meter 超声波流量计;超声波测速仪;超声波测量仪
ultrasonic method 超声波法
ultrasonic method of material testing 超声波材料试验法
ultrasonic method of measuring 超声波测量法
ultrasonic micrometer 超声波测微仪
ultrasonic microphone 超声波传声器
ultrasonic microscope 超声波显微镜
ultrasonic microtome 超声波切片机
ultrasonic mixing 超声波混合
ultrasonic mode changer 超声波振动模式变换器
ultrasonic modulation cell 超声波调制元件
ultrasonic modulator 超声波调制器;超声波调节器
ultrasonic monitor 超声波监测器
ultrasonic motor 超声波马达
ultrasonic multiple grating 超声波多重光栅
ultrasonic nebulization 超声波喷雾吸入法
ultrasonic nebulizer 超声波雾化器
ultrasonic negative-ion generator 超声波负离子发生器
ultrasonic obstacle detector 超声波障碍探测器
ultrasonic oscillation 超声波振荡
ultrasonic oscillator 超声波振荡器
ultrasonic output meter 超声波输出计
ultrasonic oxidation method 超声波氧化法
ultrasonic particle monitor 超声波粒子监测器
ultrasonic pen pointer 超声波指示笔
ultrasonic pen table 超声笔感应板
ultrasonic phenomenon 超声波现象
ultrasonic photocatalytic degradation 超声波—光催化降解
ultrasonic photodiffusion 超声波光电扩散
ultrasonic physics 超声波物理学
ultrasonic plastic welder 超声波塑料焊接机
ultrasonic power 超声波功率
ultrasonic pressure ga(u)ge 超声波压力规
ultrasonic pretreatment 超声波预处理
ultrasonic probe 超声波探针;超声波探头
ultrasonic process 超声波压铸法;超声波法
ultrasonic processing 超声波处理
ultrasonic pulse 超声波脉冲
ultrasonic pulse generator 超声波脉冲发生器
ultrasonic pulse method (测混凝土强度的)超声波脉冲(测强)法
ultrasonic pulse method for strength 超声波脉冲测强法
ultrasonic Q-switch 超声 Q 开关
ultrasonic quartz 超声波石英
ultrasonic radar 超声波雷达
ultrasonic radiation 超声波辐射
ultrasonic radiator 超声波辐射器
ultrasonic rail-defect detector car 超声波钢轨探伤车
ultrasonic rail-flaw detector 超声波钢轨探伤仪
ultrasonic rail inspection 超声波钢轨探伤
ultrasonic rail inspection car 轨道超声波检测车;超声波钢轨探伤车
ultrasonic rail tester 超声波钢轨探伤器
ultrasonic rail-testing train 超声波测验钢轨列车
ultrasonic ranging 超声波测距
ultrasonic rapid sector scanner 超声波快速扇形扫描仪
ultrasonic rebound combined detector 超声—回弹综合检测仪
ultrasonic receiver 超声波接收机
ultrasonic reflection technique 超声波反射法;超声波反射技术
ultrasonic reflection test 超声波反射试验
ultrasonic reflectoscope 超声波探伤器
ultrasonic refractoscope 超声波折射检验器
ultrasonic remote control 超声(式)遥控
ultrasonic remote welding 超声波遥焊
ultrasonic rock drill 超声波凿岩机

ultrasonics 超声波学;超声波空气动力学
ultrasonic scaler 超声波除垢器
ultrasonic scaling apparatus 超声波除垢器
ultrasonic scanner 超声波扫描器
ultrasonic scanning 超声波扫描
ultrasonic scanning image system 超声波扫描成像系统
ultrasonic scanography 超声波扫描照相术
ultrasonic sealing 超声波热合密封;超声波热合;超声波焊接(法)
ultrasonic seam welding 超声波缝焊
ultrasonic seismology 超声波地震学
ultrasonic sensing device 超声波感应设备
ultrasonic sensing equipment 超声波感应设备
ultrasonic sensing unit 超声波感应设备
ultrasonic sensor 超声波传感器
ultrasonic separation 超声波分离
ultrasonics for viscosity measurement 超声波法黏度测定
ultrasonic shipboard wave height sensor 船用超声波高计;船用超声波测波仪
ultrasonic soldering 超声波软钎焊;超声波软焊;超声波钎焊;超声波焊接(法)
ultrasonic soldering iron 超声波钎焊烙铁
ultrasonic soldering unit 超声波钎焊装置
ultrasonic sonochemistry 超声波声化学
ultrasonic sound 超声
ultrasonic sounder 超声波探测器
ultrasonic sounding 超声波探测(法)
ultrasonic sounding apparatus 超声波检测仪
ultrasonic sound pulse 超声波脉冲
ultrasonic source 超声波源
ultrasonic space grating 超声波空间光栅
ultrasonic spectrometry 超声波光谱学;超声波测谱学
ultrasonic spectroscope 超声波分光镜
ultrasonic spectroscopy 超声波频谱学
ultrasonic spectrum 超声谱
ultrasonic spot welder 超声波点焊机
ultrasonic spot welding 超声波点焊
ultrasonic spot welding machine 超声波点焊机
ultrasonic standing wave 超声波驻波
ultrasonic static painting 超声波静电涂漆
ultrasonic sterilization 超声波灭菌法
ultrasonic stimulation 超声波增强
ultrasonic stress analyzer 超声波应力分析仪
ultrasonic strobolux 超声波频闪观测器
ultrasonic stroboscope 超声波频闪观测器
ultrasonic switch 超声波开关
ultrasonics technique 超声波技术
ultrasonic television camera tube 超声波电视摄像管
ultrasonic temperature measurement 超声波测温法
ultrasonic temperature measuring apparatus 超声波测温仪
ultrasonic test equipment 超声波探伤设备
ultrasonic tester 超声波检验器;超声波测试器
ultrasonic test(ing) 超声波试验;超声波检测;超音频探伤;超声波探伤(法);超声波检验;超声波测试;超声波探伤试验
ultrasonic testing apparatus 超声波试验仪器
ultrasonic testing equipment 超声波测试设备
ultrasonic testing method 超声波试验法
ultrasonic testing of concrete 超声波混凝土检验
ultrasonic therapy 超声波透热法
ultrasonics thermometer 超声波温度计
ultrasonic thickness ga(u)ge 超声波厚度计;超声波厚度仪;超声波测厚计
ultrasonic thickness indicator 超声波厚度测定器;超声波测厚器
ultrasonic thickness measurement 超声波厚度测量;超声波测厚
ultrasonic thickness measuring unit 超声波测厚仪
ultrasonic thickness meter 超声波测厚仪;超声波测厚计
ultrasonic thickness tester 超声波测厚计
ultrasonic tomographic apparatus 超声波断层摄影诊断仪
ultrasonic tomography 超声波切面诊断仪;超声波层析 X 射线摄影机;超声波 X 线断层照相术
ultrasonic train 钢轨探伤车
ultrasonic transducer 超声波传感器;超声波探头;超声波换能器

ultrasonic transducer system 超声波换能器系统
ultrasonic transmitter 超声波发射器
ultrasonic travelling wave 超声波行波
ultrasonic treatment 超声波处理
ultrasonic treatment unit 超声波处理装置
ultrasonic type detector 超声波检测仪;超声波探测器
ultrasonic underwater communication 超声波水下通信
ultrasonic vehicle detector 超声波车辆检测器
ultrasonic vibration 超声波振动
ultrasonic vibration dyeing 超声波振荡染色
ultrasonic vibration generator 超声波振荡器
ultrasonic vibrator 超声波振动器
ultrasonic videocorder 超声波录像仪
ultrasonic visco(si)meter 超声波黏度计
ultrasonic visualization 超声波显像法
ultrasonic walking aid 超声波助行仪
ultrasonic washbasin 超声波洗面器
ultrasonic washer 超声波洗涤器;超声波洗涤机
ultrasonic washing 超声波清洗
ultrasonic washing machine 超声波清洗机
ultrasonic wastewater treatment 超声波废水处理
ultrasonic water level measuring equipment 超声波水位仪
ultrasonic wave 超声波
ultrasonic wave cleaner 超声波清洗机
ultrasonic wave height meter 超声波测量高度计
ultrasonic wavemeter 超声波频计
ultrasonic wave nondestructive testing 超声波无损探伤法
ultrasonic wave quadrant section diagnostic meter 超声波扇形切面诊断仪
ultrasonic web-break detector 超声波卷筒纸断裂指示器
ultrasonic welding 超声波焊接(法);超声波焊
ultrasonic welding inspection 超声波焊缝检查
ultrasonic welding machine 超声波焊接机
ultrasonic wire drawing 超声波拔丝
ultrasonic working 超声波加工
ultra-sonogram 超声波图;超声波记录图;声像图
ultra-sonograph 超声波影像;超声波图记录仪;超声波频录像
ultra-sonography 超声波照相法;超声波显像探伤法;超声波图像术;超声波图像检查;超声波扫描术;超声波频记录器
ultra-sonoholography 超声波全息术
ultra-sonomicroscope 超声波显微镜
ultra-sonoscope 超声波仪;超声波图示仪;超声波探伤仪;超声波探测仪;超声波计
ultra-sonoscopic method 超声波示波诊断法
ultra-sonoscopy 超声波显像;超声波显示技术;超声波图像
ultra-sonotomography 超声波降解
ultra-sonovision 超声波显示装置;超声波电视
ultra-sophisticated 超精密;超尖端的
ultra-sound 超声波;超声
ultra-sound camera 超声波照相机
ultra-sound computerized tomography 超声波计算机断层摄影术
ultra-sound diathermy 超声波透热法
ultra-sound disinfection 超声波消毒
ultra-sound Doppler's method 超声波多普勒法
ultra-sound emitting crystal 超声波发射晶体
ultra-sound excitation 超声波激发
ultra-sound injury 超声波损伤
ultra-sound pattern 超声波图像
ultra-sound tomogram 超声波切面显像
ultra-sound transmitting transducer 超声波发射换能器
ultra-spatial element 超空元素
ultra-spectrophotometry 超度分光度测定法
ultra-speed 超高速
ultra-speed centrifuge 超速离心机
ultra-speed highway 超高速公路
ultra-speed streak camera 超高速图像拖尼摄像机
ultra-speed train 超高速列车
ultra-speed welding 超高速焊接
ultra-spheric(al) function 超球函数
ultra-spheric(al) polynomial 超球多项式
ultra-stability 超稳定;超稳定性;超高稳定性
ultra-stable system 超稳定系统
ultra-sterile 超无菌的
ultra-strength 超强度;超高强度

ultra-strength material 超强度材料;超高强度材料
ultra-strength steel 超高强度钢
ultra-strong magnetic field 超强磁场
ultra-structural change 超微结构变化
ultra-structural organization 超微结构;超微结构
ultra-structural technology 超微结构技术
ultra-structure 超显微镜看出的结构;超显(微)结构
ultra-subharmonic oscillation 超亚谐振荡
ultra-sweetening 超级脱硫
ultrated 红外(线)的
ultratelescopic meteor 微流量
ultra-temperature 超高温
ultra-thermometer 超级温度计;超高温温度计;限外温度计
ultra-thermostat 超级恒温器
ultra-thin 超薄型(的)
ultra-thin cellulose acetate membrane 超薄醋酸纤维膜
ultra-thin fibre 超细纤维
ultra-thin film 超薄(薄)膜
ultra-thin-film capacitor 超薄(薄)膜电容器
ultra-thin laser 超薄激光器
ultra-thin material 超薄材料
ultra-thin membrane 超薄膜
ultra-thin polished section 超薄光薄片
ultra-thin section 超薄切片
ultra-thin sectioning 超薄切片法
ultra-thin solid carbide slotting saw 超薄型固体硬质合金开槽锯
ultra-thin tinplate 超薄镀锡薄钢板
ultra-thin volume excitation 超薄容积激发
ultratine particulate organic matter 超细颗粒有机物
ultra-tone generator 超音频发生器
ultra-trace 超痕量
ultra-trace amount 超示踪量
ultra-trace analysis 超痕量分析
ultra-trace element 超痕量元素
ultra-turbid water 高浊度水
ultraudion 三极管回授式检波器
ultra-vacuum 超真空
ultra-vacuum vessel 超真空容器
ultraviolet 紫外线的
ultraviolet absorbent 紫外线吸收剂
ultraviolet absorbent glass 吸收紫外线玻璃
ultraviolet absorber 紫外线吸收剂
ultraviolet absorbing filter 紫外吸收滤光镜
ultraviolet absorbing glass 吸收紫外玻璃;吸收紫外线玻璃
ultraviolet absorption 紫外(线)吸收(作用)
ultraviolet absorption curve 紫外吸收曲线
ultraviolet absorption detector 紫外吸收检测器
ultraviolet absorption device 紫外吸收装置
ultraviolet absorption ozone monitor 紫外吸收臭氧监测仪
ultraviolet absorption spectrogram 紫外吸收光谱图
ultraviolet absorption spectrometry 紫外线吸收光谱法
ultraviolet absorption spectrophotometry 紫外吸收光谱法;紫外吸收分光光度学
ultraviolet absorption spectroscopy 紫外吸收光谱分析;紫外吸收分光法
ultraviolet absorption spectrum 紫外(线)吸收光谱
ultraviolet adsorbance 紫外吸光度
ultraviolet alpine sun lamp 紫外线太阳灯
ultraviolet analysis 紫外分析
ultraviolet analyzer 紫外线检偏镜
ultraviolet and infrared ray lamp 紫外线红外线灯
ultraviolet and visible detector 紫外光及可见光检测器
ultraviolet A radiation 紫外线 A 波动
ultraviolet astronomy 紫外天文学
ultraviolet background 紫外线背景
ultraviolet band 紫外线谱带;紫外线波段;紫外区;紫外波段
ultraviolet barrier 光屏蔽剂
ultraviolet camera 紫外线照相机
ultraviolet catastrophe 紫外线灾难
ultraviolet cell 紫外线试泡;紫外光电元件
ultraviolet communications 紫外线通信
ultraviolet curing 紫外线固化
ultraviolet degradation 紫外线降解
ultraviolet densitometry 紫外显像测密术

ultraviolet detection 紫外线探测
ultraviolet detector 紫外线检测器;紫外检测器;紫外监测器
ultraviolet disinfection 紫外线消毒
ultraviolet dosimeter 紫外线剂量仪;紫外线剂量计
ultraviolet electrodeless discharge lamp 紫外无极放电灯
ultraviolet energy 紫外线能量
ultraviolet enhanced ozonation process 紫外强化臭氧氧化工艺
ultraviolet exposure 紫外线照射(法)
ultraviolet filter 紫外线滤光镜
ultraviolet fluorescence 紫外荧光
ultraviolet fluorescence meter 紫外线荧光仪
ultraviolet gas analyzer 紫外线气体分析器
ultraviolet generater 紫外线发生器
ultraviolet germicidal lamp 紫外线杀菌灯
ultraviolet glass 紫外线玻璃
ultraviolet illumination 紫外线照射(法)
ultraviolet image 紫外线成像;紫外(图)像
ultraviolet imagery 紫外成像
ultraviolet induction 紫外线诱导
ultraviolet infrared film curing apparatus 紫外—红外线漆膜固化装置
ultraviolet inhibitor 紫外线抑制剂
ultraviolet injury 紫外线损伤
ultraviolet instrumentation 紫外仪器
ultraviolet instrumentation method 紫外仪器法
ultraviolet intensity 紫外强度
ultraviolet irradiation 紫外线照射(法)
ultraviolet lamp 紫外线灯;紫外(光)灯
ultraviolet laser 紫外线激
ultraviolet lens 紫外光镜头
ultraviolet light 紫外线(灯);紫外光;紫外辐射
ultraviolet light activator for resin polymerization 紫外线光敏树脂固化器
ultraviolet lighting 紫外线光明
ultraviolet light recorder 紫外光记录器
ultraviolet light screening agent 光屏蔽剂
ultraviolet light source 紫外(线)光源
ultraviolet light treatment 紫外线处理
ultraviolet light vacuum ga(u)ge 紫外线真空计
ultraviolet luminescence 紫外线发光
ultraviolet luminescent method 紫外光发光法
ultraviolet meter 紫外光计
ultraviolet microscope 紫外线显微镜;紫外光显微镜
ultraviolet microscopy 紫外显微法
ultraviolet monitor 紫外监测器
ultraviolet mutagism 紫外诱变
ultraviolet nitric oxide spectrometer 紫外一氧化氮分光计
ultraviolet organic pollution monitor 紫外线法水质有机污染监测仪
ultraviolet oscillograph 紫外线示波图;紫外线示波器
ultraviolet-oxidation system 紫外一氧化法
ultraviolet photoelectron spectrometry 紫外光电子谱法
ultraviolet photoelectron spectroscopy 紫外线光电子光谱法;紫外光电子能谱学
ultraviolet photograph 紫外线照片
ultraviolet photography 紫外线照相(术);紫外线摄影(术);
ultraviolet photometer 紫外光光度计;紫外光度计
ultraviolet photometric titration 紫外光度滴定
ultraviolet photometry 紫外光度法
ultraviolet photomicrography 紫外显微照相术
ultraviolet photostat 紫外直接影印照片
ultraviolet power meter 紫外照度计
ultraviolet protective property 抗紫外线性;防紫外线性
ultraviolet radiation 紫外线照射(法);紫外(线)辐射
ultraviolet radiation damage 紫外辐射损伤
ultraviolet radiation detector 紫外辐射探测器
ultraviolet radiation intensity 紫外辐射强度
ultraviolet radiation sources 紫外线辐射源
ultraviolet radiator 紫外线灯
ultraviolet range 紫外线区(域)
ultraviolet ray 紫外(线)
ultraviolet ray adsorbance 紫外射线吸光度
ultraviolet ray ageing 紫外线老化
ultraviolet ray flaw detector 紫外线探伤器
ultraviolet (ray) intercepting glass 防紫外线玻璃

ultraviolet ray range 紫外线范围
ultraviolet ray resisting paper 紫外线防护纸
ultraviolet rays transmission 紫外线透射
ultraviolet (ray) transmitting glass 紫外线玻璃
ultraviolet ray treatment 紫外线照射(法)
ultraviolet receptor 紫外光感受器
ultraviolet recorder 紫外记录仪
ultraviolet recording paper 紫外线感光记录纸
ultraviolet reflectance 紫外反射
ultraviolet reflectance spectrometry 紫外反射光谱法
ultraviolet region 紫外线区(域);紫外光区
ultraviolet scanner 紫外扫描器
ultraviolet screener 紫外线屏蔽(剂);防紫外线剂
ultraviolet screening agent 紫外线屏蔽剂;防紫外线剂
ultraviolet self-mattering process 紫外自遮摄影法
ultraviolet-sensitive image tube 紫外敏感成像管
ultraviolet sensitizer 紫外敏化剂
ultraviolet signature 紫外线特征
ultraviolet source 紫外源
ultraviolet spectrogram 紫外线光谱图
ultraviolet spectrograph 紫外光谱仪;紫外光谱
ultraviolet spectrography 紫外光谱法
ultraviolet spectrometer 紫外线光谱仪;紫外分光计
ultraviolet spectrometry 紫外分光术;紫外分光法
ultraviolet spectrophotometer 紫外(线)分光光度计
ultraviolet spectrophotometric method 紫外分光光度法
ultraviolet spectrophotometry 紫外线光谱光度计;紫外分光光度学;紫外分光光度测量;紫外分光光度测定法
ultraviolet spectroscope 紫外光分光镜
ultraviolet spectroscopy 紫外(线)光谱学;紫外线光谱法
ultraviolet spectrum 紫外(光)谱
ultraviolet stability 抗紫外线(光)稳定性
ultraviolet stabilizer 紫外(线)稳定剂;防紫外线剂
ultraviolet sterilization 紫外线杀菌;紫外线灭菌法
ultraviolet sterilizing 紫外线消毒
ultraviolet sun radiation 紫外太阳辐射
ultraviolet surveillance 紫外线监视
ultraviolet target radiometer 紫外目标辐射计
ultraviolet technique 紫外线技术
ultraviolet telescope 紫外线望远镜;紫外望远镜
ultraviolet television microscope 紫外线电视显微镜
ultraviolet transmission 紫外线穿透性
ultraviolet transmittance 紫外线透过率
ultraviolet treatment 紫外线处理法
ultraviolet vidicon 紫外视像管
ultraviolet-visible absorption spectrometry 紫外—可见光吸收光谱法
ultraviolet-visible colo(u)rimetric monitor 紫外—可见光比色监测器
ultraviolet-visible detector 紫外光可见光检测器
ultraviolet-visible effluent monitor 紫外—可见光流出物监测器
ultraviolet-visible spectrophotometer 紫外—可见光分光光度计
ultraviolet-visible spectrophotometry 紫外—可见光分光光度法
ultraviolet wavelength 紫外线波长
ultraviolet zone 紫外区
ultra-vires 超越权限;越权
ultra-vires act 超过法定权限的行为
ultra-viscometer 超声波黏度计
ultra-viscoson 超声波(振动式)黏度计
ultra-visible 超显微镜的
ultra-visuscope 超视镜
ultraviolet photoelectron spectrometer 紫外光电子分光计
ultraviolet photometry 紫外光度测量
ultra-vulcanian 超武尔卡诺式;强火山作用
ultra-wear-resistant material 超耐磨材料
ultra-white 超白
ultra-white region 超白区;白外区
ultra-wide-angle camera 特宽角摄影机
ultra-wide-angle lens 超广角镜头
ultra-wide band signal generator 超宽频带信号发生器
ultra-wide dynamic(al) range 超宽度动态范围
ultra-X ray 超 X 射线

ultromotivity 自动力
Ultron 厄尔管;波导耦合正交场放大管
ulvospinel 钛铁晶石;钛尖晶石
umangite 红硒铜矿
umbellate 伞形的
umbellate form 伞形
umbellifer 伞形植物
umbelliform 伞形的
umbellule 小伞形花序
umber 赭色(的);棕土;赭土色;铁锰质土
umber glue 赭色
umbilic(al) connector 临时管道及电缆连接器
umbilic(al) type diving equipment 管供式潜水装具
umbilicate 凹陷的
umbilicate type 凹陷型
umbilic cord 操纵缆
umbo 壳顶
umbonate 凸形的
umbozerite 硅钍钠锶石
umbra 阴影;全影;本影;暗影
umbra area 阴影区
umbraculiferous 伞盖状的
umbrage 树阴
umbral dot 本影点
umbral eclinse 本影食
umbrascope 烟尘浊度计
umbrella acacia 伞形相思树
umbrella aerial 伞形天线
umbrella agreement 一揽子协议
umbrella antenna 伞状天线;伞形天线
umbrella arch 隧道护拱;隧道护板
umbrella article 总括条款
umbrella collector 伞状收集器
umbrella convention 总公约
umbrella core 伞状型芯;伞状泥芯
umbrella creel 伞形粗纱架
umbrella effect 雨伞效应;阳伞效应;伞形效应
umbrella form 伞形(结构)
umbrella hood 伞形罩
umbrella insulator 伞形绝缘子
umbrella patent 要求全面保护的专利
umbrella project 一揽子项目;大型项目
umbrella prop 伞形支柱
umbrella pulley 伞形滑轮
umbrella roof 挑篷;伞形(屋)顶;中柱挑出长方形屋顶;单柱伞形屋顶;车站棚顶
umbrella shape 伞形
umbrella shaped 伞状的
umbrella-shaped roof 伞形屋顶
umbrella shell 伞形薄壳;伞形屋面壳
umbrella-shoe enamelled metal reflector lamp 搪瓷伞形罩灯
umbrella stand 有伞形顶的看台;箭筒
umbrella-stand type intersection 米字交汇
umbrella stull 伞形支架
umbrella system 伞形修枝
umbrella tent 伞形帐篷
umbrella tree 伞状树;伞形树
umbrella type 华伞型;伞型;伞形
umbrella-type alternator 伞形水轮发电机;伞形交流发电机
umbrella type (hydraulic) generator 伞式水轮发电机
umbrella type hydrogenerator 伞式水力发电机
umbrella vault 伞形拱顶
umbrella vent 伞形通风筒
umbrept 暗始成土
umbric epipedon 暗色表层
umformer 换流机
umia(c)k 皮筏
umite 硅镁石
Umkehr effect 逆转效应
Umklapp process 倒逆过程;反转过程
Umklapp scattering 反转散射
U-mode record 不定长方式记录
umohoite 钼铀矿
umpire 仲裁长;评判人;裁判员
umpire analysis 仲裁分析
umpire examination 仲裁检验
umpoverished rubber 失弹性橡胶
umptekite 碱闪正长岩
unabated 未减的
unaberrated image 无像差图像

unable-deliverable cargo 无法交付货物
unable-deliverable goods 无法交付货物
unable to pay 无支付能力的
unable to supply 不能供应
unabsorbed 未吸收的
unabsorbed cost 未吸收成本;未纳入成本的支出;未分配成本;未吸收的成本;待摊成本
unabsorbed expenses 未吸收费用
unabsorbed gas 未被吸收的气体
unaccelerated 未加速的
unaccelerated condition 无加速条件
unaccelerated sulfur vulcanizate 无促进剂的硫化胶
unacceptable 不能接受的
unacceptable condition 不可接受的条件
unacceptable date 不能接受的期限
unacceptable product 不合格品
unacceptable risk 不可接受的危险性
unaccepted product 不合格产品
unacclimated activated sludge 未驯化的活性污泥
unaccommodated 无设备的
unaccompanied 无伴奏的
unaccompanied luggage 非随身携带的行李;不随身行李
unaccompanied shipment 非随载运行李
unaccomplished 无技术的
unaccountability 无法解释
unaccountable 无法解释的
unaccounted 未予说明的
unaccounted for 不可预见费;(工程概算、预算中的)备用费
unaccounted-for loss 未计及的损失
unaccounted for water 未明水量;不收水费
unacked time 延迟确认时间
unacknowledged information transfer service 无确认式信息传递服务
unackowledged 未公开承认的
unactivated state 未活化态;非激活状态
unactuated 未开动的
unadaptable through passenger train 无改编通过旅客列车
unadapted 未被采用的
unadapted river 不相称河(流)
unaddressable storage 非可编址存储器;不可寻址存储器
unaddressed 无地址的
unadept 外行
unadjustable 不可调节
unadjusted 未平差的;未调节的
unadjusted credits 未调整贷项
unadjusted debits 未调整借项
unadjusted eye 肉眼
unadjusted return on investment method 未调整投资回收方法
unadjusted trial balance 结账前试算表;结算前试算表;调整前试算表
unadjusted trial statement 调整前试算表
unadmitted assets 非公认资产;不计列资产
unadorned 没有装饰的;无装饰的;未加装饰的
unadulterated water 纯水
unaerated 未充气的
unaerated lagoon 未曝气池
unaffected 未受影响的
unaffected area 原质区(未受热影响区)
unaffected base metal 未影响的母材
unaffected by temperature fluctuations 不受温度波动影响
unaffected zone 未受影响区;未受热影响区
unaflow 单流
unaged property value 老化前性能数值
unaided eye 目视;肉眼
unaided viewing 肉眼观察
unaired room 不通风房子
unaka 残丘(块)
unakite 绿帘花岗岩
unaligned bundle 不定位光纤束
unallied 无关系的
unallocate 重新分配地址
unallocated household 无房户
unallocated logical storage 不可分配逻辑存储器
unallocated physical storage 不可分配物理存储器
unallotted apportionment 未分发的经费
unallotted household 无房户
unallowable 不许可的
unallowable character 非法字符

unallowable code 非法代码
unallowable code check 非法代码检查
unallowable cost 不许可费用
unallowable digit 非法数位;不允许数位
unallowable instruction 非法指令【计】
unallowable instruction digit 非法指令数字
unalloyed 纯金属的;非合金的
unalloy steel 非合金钢;纯钢
unalloy steel plate 非合金钢板
unaltered 未改变的
unaltered belt 不变带
unaltered rock 未变质岩(石);新鲜岩石;未蚀变岩石;完整岩石
unaltered sample 未蚀变样品
unaltered zone 不变带
unambiguous 明显的;非二义性的;不含糊
unambiguous right-linear stochastic grammar 非含混右线性随机文法
unamortized expenses 未摊销费用;待摊费用
unamortized interest during construction 未摊还建造期利息
unanchored 非锚碇的
unanchored sheet pile quaywall 悬臂板桩岸壁
unanchored sheet piling 无锚板桩;悬臂板桩;非锚碇板桩
unanimity rule 一致性法则
unanimous agreement 一致同意
unanimous verdit 一致裁定
unannealed 未退火的;不退火
unannealed track 未退火径迹
unannounced 未经宣布的
unanswerable 无法回答的
unanswered 无回答的
unanticipated load 非预期荷载
unanttuated forward wave 无衰减前向波
unappealing 无吸引力的
unappeasable 无法平息的
unapplicable 不适用的
unapplied cash 未指定用途的现金
unapplied expenses 未分配费用
unapplied income 未分配收入
unappreciated 未受赏识的
unapprehended 未被领会的
unapproachable 无可匹敌的;不可接近的
unappropriated 未占用的
unappropriated budget surplus 未分配预算结余
unappropriated estimated surplus 未支配预计盈余额
unappropriated surplus 未指拨盈余
unapproved 未经同意的
unapt 未必的
unarguable 无可争辩的
unarmed 保险
unarmed explosive train 未解除保险的传爆系统
unarming 未解除保险
unarmored 无装甲的
unarmored body 无装甲车体
unarmored wheeled vehicle 无装甲的轮式车辆
unartificial 非人工的
unartistic 非艺术的
unary 一元关系
unary array 一元数组
unary Boolean operator 一目布尔算符
unary form 一元形式
unary linear regression model 一元线性回归模型
unary minus quadruple 一目减四元组
unary operation 一元运算;一元操作;一目运算;单目运算
unary operator 一元(运)算符;一目运算符;单项算符
unary system 一元系(统);单组分系统
unassembled 未装配的;未接合的
unassembled chassis 未装合底盘
un-asserted claim 未确定的索赔
un-assessed authorization 未分摊的核定款项
unassignable 不得转让的
unassisted flight 无动力推进飞行
unassisted gravity return 重力回水(锅炉)
unassociated 无联系的;未缔合的
unassociated gas 不含油的天然气
unassorted 未分级的;不混合
unattached 无关联的
unattackable 不易锈蚀的
unattacked 未锈蚀的;未腐蚀的

unattended 未加以注意的;无人值班;无人看守
unattended boiler 自动化锅炉
unattended engine-room 无人操纵机舱
unattended equipment 自动设备
unattended equipment area 无人管理设备区域
unattended level crossing 无人看守道口
unattended lighthouse 无人看守的灯塔
unattended lightship 无人看守的灯船
unattended machinery space 无人机舱
unattended operation 无人运行;无人看守的作业
unattended pumper 自动抽油井
unattended pumping station 无人管理自动泵站
unattended relay station 无人值守中继站
unattended remotely controlled power station 无人值班的遥控电站
unattended repeater 无人增音机
unattended repeater station 无人增音站
unattended station 无人值守站;无人值守台
unattended substation 无人值班的变电所
unattended time 待修时间
unattened station 无人管理的车站
unattenuated 未变稀的;不衰减的
unattenuated forward wave 无衰减前向波
unattenuating wave 等幅波
un-audited financial statement 未审核的财务报表;未审核之财务报表
un-audited voucher 未审核凭单
unaugmentable matching 不可扩张匹配
unauthorized access 越权存取
unauthorized access to information 对信息的未授权访问
unauthorized construction 违章建筑
unauthorized edition 翻印版
unauthorized item 未核定的补给品项目
unauthorized project 未经批准的项目
unauthorized use 违令使用
unauthorized use of water 非授权用水
unauthorized user 非法用户
unauthorized vehicle 未经审验许可使用的车辆
unavaibility 没有利用率
unavailability 无效(性);无供应;不能作业时间;不能利用;不可利用性
unavailable 不能利用的;不可得的
unavailable days 不能作业天数
unavailable energy 无用能;无效能;不可能
unavailable layer 无用层
unavailable moisture 无效水分
unavailable part 不可用部分
unavailable reserves 尚难利用的储量
unavailable second 不可用秒
unavailable soil moisture 无效土壤水分
unavailable soil water 无效水(土内)
unavailable time 不能作业时间;不可用时间
unavailable water 不可给态水
unavoidable 不可避免的
unavoidable accident 不可避免的事故
unavoidable cause 不可避免的原因
unavoidable cost 固定成本;不可控(制)成本
unavoidable delay 不可避免的延迟;不可避免的迟延
unavoidable hazards 不可避免的危险
unavoidable lost-time 不可避免的时间损失
unavoidable risk 不可避免的风险
unavoidable stoppage 不可避免的间断;不可避免的耽搁
unawakened 未被激发的
unazotized 不含氮的
unbacked 无靠背的
unbacked shell 高强型壳;不填砂壳型
unbaffled 无挡板的
unbaffled cylinder 无导流片汽缸
unbaffled rod bundle 无扰流棒束
unbaked 未烘烧的;未熟的
unbaked brick 未烘干砖
unbalance 平衡差度;失去平衡;失衡;不平衡;不均衡
unbalance coefficient of cargo handled at the port 吞吐量不平衡系数
unbalance coefficient of locomotive to shed 进车不平衡系数【铁】
unbalanced 未结束的;不平衡的
unbalanced acceleration 未被平衡的加速度
unbalanced addition 不均衡增加
unbalanced assignment problem 不平衡分配问题

unbalanced attenuation 不平衡衰减
unbalanced attenuator 不平衡衰减器
unbalanced bid (quotation) 未列分项成本的标价;不扎实标价;毛估标价;不平衡报价
unbalanced book 未结算账簿
unbalanced brake 不平衡制动器
unbalanced bridge 不平衡电桥
unbalanced budget 不平衡预算;不均衡预算
unbalanced capacity 不平衡容量
unbalanced case 不平衡状况
unbalanced centrifugal acceleration 未被平衡的离心加速度
unbalanced centrifugal force 未被平衡的离心力
unbalanced circuit 不平衡电路;不对称回路
unbalanced class of procedures 规则等级的不平衡
unbalanced coefficient of cargo throughput 港口生产不平衡系数
unbalanced coefficient of reservoir inflow 入库不平衡系数
unbalanced construction 不平衡结构;不均衡结构
unbalanced current 不平衡电流
unbalanced cutter chain 不平衡截煤机链
unbalanced data link 不平衡数据链路
unbalanced development 不平衡发展
unbalanced economic growth 不平衡经济增长
unbalanced elevation 不平衡超高
unbalanced error 非平衡误差;不平衡误差
unbalanced factor 不平衡因素;不平衡系数
unbalanced factor of passenger flow section 客流断面不均匀系数
unbalanced feeder line 不平衡馈电线
unbalanced finance 不平衡(的)财政
unbalanced flow 紊流
unbalanced food chain 不平衡食物链
unbalanced force 不平衡力
unbalanced fuel pump 不均衡供油燃料泵
unbalanced growth 非平衡增长;不平衡生长;不均衡增长
unbalanced head 不平衡水头
unbalanced heating system 不均衡采暖系统;不均衡供暖系统
unbalanced hoisting 不平衡提升
unbalanced increase 不平衡增长
unbalanced indicator 不平衡指示力
unbalance direction coefficient 方向不均衡系数
unbalanced lane movement 不平衡车道行使;不均衡车道运行
unbalanced line 不平衡线路;不平衡传输线
unbalanced load 不平衡负载;不平衡负载;不均衡荷载;不均衡负荷
unbalanced-load meter 不平衡负载指示仪表;不平衡电路用电度表
unbalanced-load protection system 不平衡荷载保护方式
unbalanced mass 不均衡质量;不均衡体
unbalanced merge sort 不平衡合并分类;不平衡分配分类
unbalanced modulator 不平衡调制器
unbalanced moment 不平衡弯矩;不平衡力矩
unbalancedness 不平衡性
unbalanced network 非平衡网络;不平衡网络
unbalanced output 非对称输出;不平衡输出
unbalanced peak hours 不平衡的高峰时间
unbalanced phase 不平衡相位
unbalanced polyphase system 不平衡多相制
unbalanced pressure 不稳定压力;不平衡压力
unbalanced pump 不平衡(的)泵
unbalanced reaction 不平衡性反应
unbalanced resistance 不平衡电阻
unbalanced response 不平衡反应
unbalanced ring 不平衡环
unbalanced rudder 不平衡舵
unbalanced section roll bar 不对称断面轧材
unbalanced sediment transport 不平衡输沙
unbalanced shed 不清晰梭口
unbalanced shothole 不平衡爆破的炮眼
unbalanced site 不平衡位置
unbalanced speed 不平衡速度
unbalanced superelevation 不平衡超高;欠超高
unbalanced system 不平衡制;不平衡系统
unbalanced three-phase circuit 不平衡三相电路
unbalanced threep-hase load 不平衡三相负荷
unbalanced threep-hase voltage 不平衡三相电压
unbalanced-throw screen 惯性筛

unbalanced to ground 对地不平衡
unbalanced traffic 不平衡运输;不平衡交通(量)
unbalanced transformation 不平衡变换
unbalanced transportation problem 不平衡运输问题
unbalanced type 非平衡型;不平衡型;不均衡型
unbalanced valve 不平衡阀;不平衡的阀
unbalanced vane pump 不平衡型叶片泵
unbalanced voltage 不平衡电压
unbalanced weight conveyer 偏重输送机;不平衡输送机
unbalanced weight vibrator 偏心振动器
unbalanced wire circuit 不平衡有线电路
unbalance force 力的不平衡
unbalance model 非平衡模型
unbalance static 静不平衡
unbalance system economics 非平衡系统经济学
unbalance uranide method 不平衡铀系法
unballast 卸除压载;无压载
unballasted 未铺砂石的;未铺石渣的
unballasted deck 未铺石渣的桥面
unballasted floor 无渣桥面
unballasted track 无渣轨道;无渣板式轨道
unbanked 未堆起来的
unbanked attitude 无坡度姿态
unbarred road crossing 无栏木道口
unbated 未减少的
unbeaconed 无标志的
unbeaten pulp 生纸浆
unbeaten track 未开辟地区
unbeknown 未知的
unbender 矫直机
unbender unit 端头扳直机
unbending column 坚硬直柱;笔直的柱
unbending test 反复弯曲试验
unbeneficiated ore 原矿;未经精选的矿石
unbenzoled petrol 无苯汽油
unberthed passenger 统舱旅客
unberthing 离泊
unbeveled end 无坡口
unbeveled face 未开坡口面
unbias(s)ed 不偏的;无偏(差)的;不加偏压的
unbias(s)ed check 无偏(差)检验
unbias(s)ed confidence interval 无偏置信区间
unbias(s)ed critical region 无偏临界区域
unbias(s)ed error 不偏误差;无偏误差
unbias(s)ed estimate 无偏估算,无偏估计;无偏差估计量
unbias(s)ed estimation 无偏估计;无偏差估量
unbias(s)ed estimator 无偏估计量
unbias(s)ed ferrite 非磁化铁
unbias(s)ed form 非偏置形式
unbias(s)ed importance sampling 无偏重性抽样
unbias(s)ed important sampling 无偏重性抽样
unbias(s)ed mean 无偏平均数
unbias(s)edness 不偏性;无偏性
unbias(s)ed ratio estimator 无偏比估计量
unbias(s)ed regression estimator 无偏回归估计量
unbias(s)ed sample 不偏试样;准确的样本;无偏样本
unbias(s)ed statistics 无偏统计(值);无偏统计(量)
unbias(s)ed test 无偏试验;无偏(差)检验
unbias(s)ed variance 无偏方差;正态均方差
unbilled gas 未计价燃气量
unbilled sales 未结账销量
un-billed works 未开账单工程
unbitt 解缆
unblank 非空格
unblanketed 无阴影的
unblanking 增辉
unblanking circuit 正程增辉电路
unblanking mixer 开锁混频器;开启混频器
unbleached 未漂白的;未经漂白的
unbleached rags 未漂破布浆
unblemished 无疵的
unblended 未混合的;未掺和的
unblended asphalt 未掺配(过的)地沥青
unblended octane number 掺和前辛烷值
unblended oil 未调和油料
unblind 截断符号;无效符号
unblock 开启;开放;解块;非阻塞
unblocked level 开启电平

unblocked record 非成组记录;非成块记录
unblocked road 未封锁的道路
unblocked set 未楔固支架
unblocking 消除堵塞;解除闭塞
unblown 未增压的;尚未盛开的;未受风吹的
unboarded derrick 无围护钻塔;无围护井架
unbodied 无定形的;无实体的;脱离实际的
unbodied oil 非聚合油;非干性油
unboiled silk 未脱胶丝
unbolt 松开;打开;旋开螺栓
unbolt detergent 未复配洗涤剂
unbolted 粗糙的;未上螺栓的;卸掉螺栓的;未筛过的
unbonded 未砌合的;未黏合的;未黏着的;未胶结的
unbonded beam 无黏结梁;无约束梁
unbonded concrete overlay 隔离式混凝土加厚层
unbonded design 无黏结(梁)设计;按无握裹效应设计
unbonded felt 原棉毡;未合成的棉毡
unbonded fuel element 未黏合的燃料元件
unbonded glass wool 未经数值黏结的玻璃棉
unbonded mat 无粘毡
unbonded member 无黏结构件;无黏结预应力构件
unbonded overlay 不黏连加铺层
unbonded post-tensional prestressing concrete slab 未胶结的后张预应力混凝土板
unbonded post-tensioned slab 未灌浆固结的后张预应力混凝土板
unbonded post-tensioning 无黏结后张法;不灌浆的后张预应力
unbonded prestressed bar 无黏结预应力钢筋;未黏结的预应力钢筋(后张法中)
unbonded prestressed concrete 无黏结预应力混凝土
unbonded prestressed reinforcement 不黏着预应力筋
unbonded prestressing 后张法预应力
unbonded pretensioning 无黏结后张法
unbonded reinforcement 未绑扎钢筋
unbonded sand 不含黏结剂型砂
unbonded strain ga(u)ge 非黏合应变计
unbonded tendon 无黏结预应力钢丝束;无黏结钢筋丝(束);无黏着力筋
unbonded wool 未黏合的玻璃棉
unboosted 无助推的
unbordered 无边的
unbound aggregate 无黏合料骨料;无结合料集料;未黏结集料;无黏结骨料
unbound base 无结合料路基
unbound crushed aggregate base 非结合破碎集料基层
unbounded 没有根据的
unbounded construction 非黏结构造
unbounded covering surface 无界覆盖面
unbounded function 无界函数
unbounded gate delay 门的无界延迟
unbounded ocean 一望无际的海洋
unbounded set of real numbers 无界实数集
unbounded solution 无界解
unbounded space 无界空间
unbounded stream 无约束水流;无界水流
unbounded variable 无界变量
unbounded wave 无限制波
unbound (road) base 无结合料基层
unbound state 未结合状态
unbound water 非结合水
unbrace 放松
unbraced 未加支撑的;不加撑的
unbraced column 无支撑柱
unbraced excavation 无支撑开挖
unbraced frame 无支撑框架
unbraced length 自由长度;无支撑长度
unbraced length of column 柱的自由长度;无支撑的柱长
unbraced wall 无支撑墙;完整墙
unbranched-chain hydrocarbon 无支链烃
unbranched river 无分叉河流
unbranched stream 无分叉河流
unbranded gasoline 无商标汽油
unbreakable bond 不可裂开的胶结;强力胶合
unbreakable glass 不碎玻璃
unbreakable slate 不碎石板
unbridgeable 无法架桥的
unbridled competition 盲目竞争
unbroadened resonance 无展宽共振

unbroken 未破坏的;完整的
unbroken bulkhead 无开口舱壁
unbroken curve 连续曲线;完整曲线
unbroken formation 完整岩层;不破碎岩层
unbroken furrow 未碎垡片
unbroken ice 未破碎冰
unbroken line 实线
unbroken material 完整的材料
unbroken package 未开封包装
unbroken rail 完整的钢轨
unbroken wall 完整墙
unbroken wave 连续波;未碎波;未破碎波
unbuffered 无缓冲(装置)的
unbuild 毁坏
unbuilt 未兴建的;未建筑的
unbuilt area 待建地区;无建筑物地区;未建区域;未建面积
unbuilt detergent 无助洗剂的洗涤剂
unbunched 未聚束
unbundle 分类交易
unbundling 非附随
unbundling technology 分门别类的技术
unburdened 无荷载的
unburnable 烧不掉的
unburned 未烧过的;未燃烧的
unburned block 未经煅烧的块材
unburned brick 未烧的砖;不烧砖
unburned clay 未焙烧的黏土
unburned clay brick 非烧结黏土砖
unburned combustible 未烧尽的可燃物,未烧尽的可燃物
unburned fuel 未燃烧的燃料;未燃燃料
unburned gas 未燃烧气体;未燃气体
unburned high-alumina brick 不烧高铝砖
unburned hydrocarbon 未燃烃;未燃碳氢化合物
unburned mixture 未燃烧混合物
unburned part 未燃部分
unburned refractory 不烧耐火材料
unburned refuse 未燃垃圾;不可燃垃圾
unburned stopper 不烧塞头
unburned tile 未经煅烧的砖
unburnt 欠火的;未烧过的;未燃的
unburnt brick 欠火砖;未烧透砖
unburnt gas 不完全燃烧气体
unburnt lime 未焙烧透石灰
unburnt plain roof(ing) tile 未烧透的屋面平瓦
unburnt product 不烧制品
unburnt solid 未燃固体
unburnt tile 未烧透砖;欠火砖
unbuttoning 拆卸;拆开;解扣
unbypassed 无旁路的
uncage 释放;松开;松锁;放出笼来
uncaging section 松销机构(测斜仪)
uncalcined gypsum 生石膏
uncalendered 未经压光处理的
uncalibrated analysis 未校准分析
uncalibrated gain control 未校准增益控制
uncallable bond 不能提前收回的公司债
uncallable condition 不能提前偿还的条件
uncalled capital 未收资本;未收股款;未缴股本
uncalled condition 不适宜的条件
uncalled shares 未收股款
uncalled subscription 未催缴股款
uncallow 表土
uncambered bottom 平底
uncancelled residual 未被抵偿的残差
uncanned 无外壳的;无护罩的
uncap 打开盖子
uncapped 开盖的
uncapped fuse 未装雷管导火线
uncased 未包混凝土的;无套壳的;无套管的;无外壳的
uncased-bored pile 无套管钻孔灌注桩
uncased (bore) hole 无套管钻孔;未下套管钻孔;无井管孔
uncased concrete pile 无套管式混凝土桩;钻孔灌注桩
uncased crossing 不加套管穿越
uncased drill hole 无套管钻孔
uncased pile 无套管钻孔灌注桩;无套管桩;无壳套桩
uncased piling 打无壳套桩
uncased well 无套管井;未加固水井
uncashed check 未兑现支票

uncast 未定角的
uncatalog 未编目;解除目录;解除编目;目录上没有
uncatalog file 非目录式文件
uncatalogued file 非编目文件
uncemented 未胶结的
uncemented fault 未胶结断层
uncemented rock 未胶结岩石;松散岩石
uncemented tile 外孔瓦管;多孔瓦管
uncentering 未聚于一点;拆御拱架;拆除拱架
uncentralized capital 非集中性资金
uncentralized expenditures 非集中性支出
uncertain 不确定的
uncertain degree of measurement 测量不确定度
uncertain delivery lag 不确定延迟交货
uncertained risk analysis 不确定风险分析
uncertain environment 不确定环境
uncertain factor 不(确)定因素
uncertain genetic circular features 成因不明环形体
uncertain hydrodynamic environment 不确定水动力环境
uncertain hydrodynamic factor 不确定水动力因素
uncertain hydrodynamic parameter 不确定动水参数
uncertain interpretation 不确定性解释
uncertain knowledge 不确定知识
uncertain planning horizon 不确定计划水平
uncertain pollutant sources 不确定污染源
uncertain quantity 不确定量
uncertain rate 不确定汇率
uncertain region 不确定区(域);不确定范围;不肯定区域;不可辨区
uncertain relation 不定关系
uncertain sweep width 不确定扫描宽度;变扫描宽度
uncertain system 不确定系统
uncertainty 不确定性;不确定度;不肯定性;不可信度;不可靠;不可测性;不精确(性);不定因素;不定(性)
uncertainty analysis 不确定性分析;不确定分析
uncertainty condition 测不准条件;不确定情况
uncertainty decision 不定型决策
uncertainty element 不定元
uncertainty ellipse 不定性椭圆
uncertainty event 不确定事件
uncertainty function 不定性函数
uncertainty of exchange rate 外汇波动
uncertainty of measurement 测量不确定度
uncertainty of objective 目标的不确定性
uncertainty of plotting 绘图误差
uncertainty of relative orientation 相对定向不定性
uncertainty principle 测不准原理;不确定性原理;不定性原理
uncertainty relation 不定关系;测不准关系;不确定性关系
uncertainty relationship 不确定关系
uncertain wind 不定向风;变向风
unchain 解开锁链
unchamfered butt 平头双接;不削边铰链
unchamfered joint 无坡口接头
unchangeable 不能改变的
unchangeable joint 冻结接头【铁】
unchanged 未改变的,未变的
unchanging benchmark 固定水准基点
unchanging grade 不变等级
unchannelized intersection 非渠化式交叉;非渠化式交通交叉口
uncharge 未充电的;不付费用的;卸载;解除负荷;无负荷
uncharged 未充电的;解除负荷的;不付费用的;卸货;无荷载的;无电荷的;不带电(荷)的
uncharged battery 未充电蓄电池
uncharged demolition target 未装药的爆破目标
uncharged molecule 无电荷分子
uncharged particle 不带电粒子
uncharged state 未充电状态
uncharged surface 不带电表面
uncharted 未经测绘的;图上没有(标明)的
uncharted area 无图地区
uncharted port area 未测图港区
unchartered 不合规则的
unchecked 未(经)校核的;未经核实的
unchecked development 盲目发展
unchecked mosaic 无控制点镶嵌图
unchecked spot elevation 原始高程

unchipped surfacing 不撒石屑面层
unchlorinated drinking water 未加氯处理的饮用水
unchock 除去楔子
unchoked running system 开放式浇注系统
unchoking effect 消扼流作用;去扼流作用
unchopped radiometer 非调制辐射计
unchuted landing 无减速伞着陆
uncivilized 未开发的
unclaimable land 无法开垦的土地;不可利用的土地
unclaimed balance 不动户余额
unclaimed cargo 无人认领货物
unclaimed dividend 未付红利;未领股利
unclaimed field 未改良土地
unclaimed freight 没人认领的物品
unclaimed goods 无主货物
unclaimed wage funds 未付工资
unclaimed wages 未领工资
unclamp 松开(夹子等);松开
unclamped elevator 开式提引器
unclasp 松开
unclassed 未归类的
unclassed ship 无级船
unclassified 类别不明;无密级的;未分类的
unclassified excavation 未经分类的开挖;不分类开挖
unclassified feed 混合入选给料;不分级入选给料
unclassified miscellaneous vessel 不入级杂用船
unclassified reserves 未分级储量
unclassified river 未分类河流
unclassified road 不列级公路
unclassified store 其他物料;不分类存料
unclassified stream 未分类河流
unclassified technical order 一般性技术规程
unclassified track 未识别的目标回波
unclean bill of lading 不清洁提(货)单
unclean surface 不清洁表面;不洁表面
unclean water 污染水
unclear number 质量数
unclog 清除阻碍;清除障碍物;清除油污
unclose 打开
unclosed 未关的;非闭合的;未结束的
unclosed contour 未封闭等高线;不闭合等高线
unclosed dock 开敞式港池
unclosed force polygon 非闭合力多边形
unclosed ground 开阔地
unclosed land 公共土地
unclosed level(l)ing line 支水准路线
unclosed traverse 支导线;开导线;不闭合导线【测】
unclothed 无覆盖的
uncoated 无覆盖的;未涂面层的
uncoated asbestos-cement pipe 未涂层石棉水泥管
uncoated asphalt rag felt 面层未涂沥青的粗油毡
uncoated cast-iron pipe (未涂保护层的)光(面)铸铁管
uncoated chip(ping)s 不拌沥青的石屑
uncoated electrode 无药焊条
uncoated flat product 无镀层钢板制品
uncoated laser 不涂膜激光器
uncoated lens 不涂膜透镜
uncoated lightweight aggregate 无釉面轻集料
uncoated paper 无涂层纸
uncoated steel pipe 未涂层钢管
uncoated tar(red) rag felt 面层未涂柏油的粗油毡
uncoated tinplate base 镀锡原板
uncoaxiality 不同轴性
uncoded 未编码的
uncoded word 未编码字
uncode input 编码输入
uncoil 开捆;开卷;解开
uncoiler 开卷机
uncoiler leveller 开卷—校直机
uncoiling machine 开卷机
uncoiling reel 开卷机
uncoiling the wire rope 放开钢丝绳
uncoiling unit 开卷机
uncoilling reel 折卷机
uncoil-stand 退卷装置
uncollected premium 未收保费
uncollected voucher 未收凭证;未收凭单
uncollectible 无法收集的
uncollectible account 坏账;呆账
uncollectible check 无法收款的支票;无法兑付的支票

uncollectible debt 坏账;呆账
uncollectible note receivable 无法收现的应收票据
uncollectible receivable account 无法收现的应收账款
uncollectible tax 无法征收的税款
uncollided 未经碰撞的
uncollided radiation 未经碰撞辐射
uncollimated 未经准直的
uncolo(u)red 无色的;未着色的;未加颜色的
uncombined 未化合的;非结合的;非化合的;不结合的
uncombined carbon 游离碳
uncombined lime 游离石灰;游离氢氧化钙
uncombined oxide 非结合态氧化物
uncombined water 游离水;未结合水
uncombined water in concrete 混凝土中的未结合水
uncombusted rubbish 不可燃垃圾
uncombustible rubbish 不可燃垃圾
uncomfortable 不相符的;不舒服的;不适的
uncomfortable loudness level 不适响度级
uncommercial 非商业性的
UN Commission Transnational Corporation 联合国跨国公司委员会
uncommitted 不承担责任的;不承担义务的
uncommitted amount 未支配数额;未支配款项
uncommitted contract 不受约束合同
uncommitted surplus 未拨用盈余
uncommon condition 异常情况
uncompacted 未压实的;不密实的
uncompacted concrete 未压实混凝土;未捣实混凝土
uncompacted fill dumped 非密实填土堆积
uncompacted filling 不密实填土;不密实填料
uncompacted fresh concrete 未压实新拌混凝土;未捣实新拌混凝土
uncompacted gravel 不密实充填砾石
uncompacted rock 不密实岩(石)
uncompacted rockfill 未压实堆石(体)
uncompacted soil 未压实土;未压密土
uncompacted tip 不压实的垃圾场
uncompahgrite 辉石黄长岩
uncompensated 无补偿的
uncompensated acidosis 非代偿性酸中毒
uncompensated alkalosis 非代偿性碱中毒
uncompensated amplifier 无补偿放大器
uncompensated attenuator 未补偿衰减器
uncompensated basin 非补偿盆地
uncompensated compass 非补偿罗经
uncompensated flowmeter 无补偿流量计
uncompensated force 离心力
uncompensated ionization chamber 无补偿电离室
uncompensated load 不平衡荷载
uncompensated loss 未补偿损耗
uncompetitive 无竞争的
uncomplemented 未结合补体的
uncompleted anticline 不完整背斜
uncompleted building works 未完建筑施工工程
uncompleted construction 未完施工;未完建筑工程;未完成的工程;未竣工工程
uncompleted construction project 未完工程项目
uncompleted contract 未完成履行的契约;未完成(的)合同
uncompleted orbit 未满轨道
uncompleted project construction 未完成工程
uncompleted transaction 未完成交易;未完成会计事项
uncomplexed ions 未络合离子
uncomplexed metal ions concentration 未络合金属离子浓度
uncomplicated 简单的
uncomposed 未分解的
uncompounded cement 纯胶浆
uncompressed vegetable fiber 疏松的植物纤维
uncompromising 不妥协的
unconcerned 无关的
uncondensable 不冷凝的
uncondensed 不凝缩的;不凝结的
uncondensed polycyclic aromatic 未稠合多环芳烃
uncondensible 不可冷凝的
unconditional 无条件的
unconditional acceptance 无条件承付;无条件承兑
unconditional bank guarantee 无条件的银行保函

unconditional branch 无条件转移
unconditional convergence 无条件收敛;非条件收敛
unconditional delivery 无条件交货
unconditional endorsement 无条件背书
unconditional grants 不指定用途的财政补贴
unconditional inequality 无条件不等式;非条件不等式
unconditionality 无条件性
unconditional jump 无条件转移;无条件跳转;非条件转移
unconditional jump instruction 无条件转移指令
unconditional letter of credit 无条件的信用证
unconditional liquidity 无条件流动能力
unconditionally stable 无条件稳定
unconditional obligation 无条件的义务
unconditional operation 无条件(的)交易
unconditional optimum 非条件最优
unconditional probability 无条件概率;非条件概率
unconditional stability 无条件稳定
unconditional stability criterion 无条件稳定准则
unconditional statement 无条件语句【计】
unconditional transfer 无条件跳转;无条件传送
unconditional transfer of control 控制无条件转移;无条件控制转移
unconditioned 无条件的
unconditioned inhibition 非条件抑制
unconditioned reflex 无条件反射
unconditioned zone 非空气调节区
unconfident 不保密
unconfined 无约束的;无限制的;无限的;无侧限的
unconfined aquifer 自由含水量;自由含水层;潜水含水层;无压含水层;无承压含水层;非承压地下水
unconfined aquifer turned from confined aquifer 承压转无压含水层
unconfined avalanche 无约束雪崩
unconfined blast 敞口吹风
unconfined charge 无约束药包
unconfined compression 无侧限压缩
unconfined compression apparatus 无侧限压缩装置;无侧限压缩仪
unconfined compression strength 无侧限压缩强度;无侧限抗压强度
unconfined compression test 无侧限压缩试验;无侧限压缩强度试验;无侧限抗压强度试验
unconfined compressive strength 无侧限压缩强度;无侧限抗压强度
unconfined concrete 无约束混凝土;无侧限混凝土
unconfined condition 无约束条件
unconfined-confined water 潜水承压水
unconfined cylindric(al) sample 无侧限圆柱体试样
unconfined deformation modulus 无侧限变形模量
unconfined explosion 无限制爆炸
unconfined flow 无压流;非限制流;无压流
unconfined groundwater 自由地下水位;无压地下水;非限制地下水;非承压(地下)水
unconfined mixture 无限制混合物
unconfined pad bearing 无侧限板式支座
unconfined seepage flow 无压渗流
unconfined shear strength 无侧限抗剪强度;无侧限剪切强度
unconfined shot 无约束爆破
unconfined specimen 无侧限试件
unconfined steady flow in well 井内自由稳定流
unconfined stream 无约束(水)流
unconfined swelling 无侧限膨胀
unconfined tensile strength 无侧限抗拉强度
unconfined water 未压水;非承压水
unconfined water basin 潜水盆地
unconfined water table 自由潜水面;无压地下水位
unconfirmed 未证实的;未经确认的
unconfirmed credit 未确认的信用证;非保兑信用证
unconfirmed letter of credit 非保兑信用证;未确认的信用证;不保兑信用证
unconfirmed offer 虚盘
unconformability (岩石的)不整合(性);不一致的状态或性质
unconformability of overlap 超覆不整合
unconformable 不整合的
unconformable contact 不整合接触(面)
unconformable fault 不整合断层
unconformable interface 不整合界面
unconformable oil-gas field 不整合油气田
unconformable stratification 不整合层理
unconformable stratum 不整合地层
unconformable to national conditions 不合国情的
unconformity 不整合性(岩石的);不整合
unconformity barrier 不整合遮挡
unconformity iceberg 混成冰山
unconformity icebery 分层冰山
unconformity interface 不整合界面
unconformity of metamorphic terrain 变质地层不整合
unconformity pool 不整合油藏;不整合油
unconformity spring 不整合泉
unconformity stratum 不整合地层,[复]strata
unconformity trap 不整合圈闭
uncongealable 不可冻结的
uncongealable cell 不冻电池
uncongealable dynamite 不冻结的炸药
unconjugated acid 非共轭酸
unconnected 不连续的;不连接的
unconnected course 分离层
unconnected depositor 仅在存款
unconnected graph 非连通图
unconnected shaft 不连续的轴
unconscionable contract 无理合同;极不公平的合同
unconsciousness 不省人事
unconscious selection 无意识选择;非意识选择
unconserved air conditioning plant 冷却水不回收的空调设备
unconsistency 不一致性
unconsolidated 未固结的;松散的;非固结的
unconsolidated backfill 未固结回填;非固结回填料
unconsolidated clay 未固结黏土
unconsolidated coverage 松散覆盖层
unconsolidated deposit 未固结沉积物;未固结沉积层;松散沉积(物);松软沉积(物)
unconsolidated-drained shear test 不固结排水剪切试验
unconsolidated-drained test 不固结排水试验
unconsolidated formation 松散层
unconsolidated glacial deposits 松散冰碛物
unconsolidated ice cover 松散冰盖
unconsolidated investment 不全并计算的投资;不合并计算的投资
unconsolidated layer 疏松层
unconsolidated material 未固结物质;未固结土料;松散材料;疏松物质;疏松沉淀物
unconsolidated mud 未回结淤泥
unconsolidated overburden 松散覆盖层
unconsolidated post-tensioned slab 未固结的后张混凝土板
unconsolidated rock 松散岩体
unconsolidated sediment 未固结沉积物;松散泥沙;松散沉积(物);松软沉积(物)
unconsolidated silt and fine sand type 未胶结的粉细砂岩
unconsolidated soil 未固结土
unconsolidated soil material 不固结土质
unconsolidated strata 松散地层
unconsolidated surface layer 未固结表层
unconsolidated track 未压实轨道
unconsolidated-undrained test 不固结不排水试验
unconsolidated-undrained triaxial compression strength 不固结不排水三轴抗压强度;不排水三轴剪切强度
unconsolidated-undrained triaxial shear strength 不固结不排水三轴剪切强度
unconsolidated-undrained triaxial test 不固结不排水三轴试验;未固结不排水三轴试验
unconstrained 无约束的
unconstrained beam 无约束梁
unconstrained chain 不限定链条
unconstrained damping layer 无约束的阻尼层
unconstrained minimization 无约束极小化
unconstrained optimization 无约束最优化
unconstrained shrinkage 无约束收缩
unconstrained system 无约束系统
unconsumable products 不可消费的产品
uncontainerized cargo 非集装箱化货
uncontaminated 洁净的;无杂质的;未受污染的;未被污染的
uncontaminated air 洁净空气;未污染的空气
uncontaminated field 未污染土地
uncontaminated oil 洁净油
uncontaminated river 未受污染河流
uncontaminated soil 结净土壤;洁净土壤;未污染土(壤)
uncontaminated stream 未受污染河流;未被污染的河流
uncontaminated water 洁净水;无杂质水;未污染的水;未污染的水
uncontaminated zone 未污染带;未被污染区
uncontamination 非污染
uncontinuity 不连续性
uncontinuous 不连续的
uncontinuous change 阶段变速;间断变化
uncontinuous construction 不连续施工(法)
uncontracted weir 非收缩堰;不收缩堰
uncontrollability 不可控性
uncontrollable 无法控制的;不可控(制)的;不可调节的
uncontrollable condition 失去操纵的状态
uncontrollable cost 不能控制的成本;不可控制成本
uncontrollable diesel generator set 不可控柴油发电机组
uncontrollable expenses 不易控制费用;不能统制费用;不能控制的费用;不可控制费用
uncontrollable locomotive 非支配机车【铁】
uncontrollable maneuver 失去控制的动作
uncontrollable marketing variables 不可控销售变数
uncontrollable spin 不可控制的螺旋
uncontrollable system 不可控系统
uncontrolled 无(法)控制的;非受控;非控制的;不受控制的;不加管制的
uncontrolled area 无控制地区
uncontrolled barrage 无控制的拦阻射击
uncontrolled clearance 无节制地清除;无节制地采伐森林
uncontrolled combustion 无控制燃烧
uncontrolled combustion phase 无控燃烧期
uncontrolled company 非受控公司
uncontrolled crack 无控制的裂纹
uncontrolled crest spillway 自由洪峰溢洪道
uncontrolled development 无控制的发展
uncontrolled disintegration 无约束冰崩
uncontrolled diving 失控潜水
uncontrolled dumping 无控制倾倒
uncontrolled emission 无节制地排放
uncontrolled engine 无净化装置发动机
uncontrolled enterprise 非受控企业
uncontrolled flight condition emergency 失去控制的紧急情况
uncontrolled flooding 自由淹灌;无控制泛滥
uncontrolled flow 不可调节流;自由流动;无控制流(动);失控流动
uncontrolled fragments 非预制破片
uncontrolled intersection 无管制交叉口
uncontrolled mosaic 无控制点镶嵌图
uncontrolled nuclear transformation 不受控核变化
uncontrolled oscillation 无控振荡
uncontrolled pedestrian crossing 无控制人行横道
uncontrolled pedestrian cross walk 无控制人行横道
uncontrolled prices 非控制价格
uncontrolled spillway 无控制溢洪道
uncontrolled spillway weir 自流式溢流堰
uncontrolled storage 无控制蓄水
uncontrolled tipping 无节制地倾弃
uncontrolled traffic system 非控制交通系统
uncontrolled unit 非受控单位
uncontrolled urbanization 城市化失控
uncontrolled variable 不可控制变数;不可控变量;不可调变量
uncontrolled vehicle 未装废气净化装置的车辆
uncontrolled ventilation 自由通风;无控制通风
uncontrolled weir 自流(式)溢流堰;无控制堰
unconvectional energy resource 非常规能源
unconvenanted 不受公约约束的
unconventional 非规范的;非常规的
unconventional fuel 非常规燃料
unconventional railway 非常规技术铁路
unconventional vessel 非常规容器
unconventional water resources 非传统水资源
unconvertdr lime 未转化石灰
unconverted 无变化的;未转化的
unconverted monomer 未聚合的单体
unconvertible loan stock 不能兑换的债券

uncooked 未煮过的
uncooled motor 非冷却式发动机
uncooled nozzle 非冷却喷管
uncooled piston 非冷却活塞
uncooled reactor 无冷却剂堆
uncooperative 不合作的
uncoordinated 未测坐标的；无组织的；未经协调的
uncoordinated borehole 未测坐标的钻孔
uncored brick 无孔砖
uncork 未加塞的
uncorrectable error 无法校正的错误；不可校正的错误
uncorrected 未修正的
uncorrected acceleration 未校正加速度记录
uncorrected delay 未校正延迟
uncorrected error 漏校正错误
uncorrected field 未修正场
uncorrected gear 非修正齿轮
uncorrected reading 未校正的读数
uncorrected retention volume 未校正保留体积
uncorrected spectrum 未修正频谱；未改正频谱
uncorrecting 航向倒算；反修正
uncorrelated 不相关的
uncorrelated function 不相关函数
uncorrelated increment 不相关增量
uncorrelated independent random variable 不相关独立随机变量
uncorrelated noise reduction 不相关噪声削减
uncorrelated products 不相关产物
uncorrelated random erros 不相关随机误差
uncorrelated random force 不相关随机力
uncorrelated random variables 不相关随机变量；不相关的随机变量
uncorrelated reaction products 不相关反应产物
uncorrelated sample 不相关样本
uncorrelated variable 不相关变量
uncorroded 未受腐蚀的
uncorrupted 未腐败的
uncorruptible 不易腐蚀的
uncountable 不可数的
uncountable for loss 未计入损失
uncountable set 不可数集
uncounted 无数的
uncounted for loss 未计入损失
uncounted for water 未计水量
uncouple 脱开
uncoupled 解耦合
uncoupled axle 不连轴
uncoupled bond 非耦联键
uncoupled charging 非耦合装药
uncoupled differential equation 不耦联微分方程
uncoupled electron 非耦联电子
uncoupled equation of motion 不耦联运动方程
uncoupled factor 非耦合系数
uncoupled lever chain 起钩杆链
uncoupled mode 非耦合模态；非耦合振荡模
uncoupled oscillations 解耦合振荡
uncoupled particle 解耦合粒子
uncoupled wheel 游滑轮
uncoupler 解耦联剂；锻钩联剂
uncoupling 列车解钩；解钩；无关联的；脱开联轴节；拆开
uncoupling chain 联结器分离链；车钩提杆链
uncoupling device 摘解器；架空索道摘解器
uncoupling gear 放松机构
uncoupling hammer 下跌的桩锤
uncoupling lever 脱钩操纵杆
uncoupling lever bracket 车钩提杆座
uncoupling lever chain 分离杆链
uncoupling lever hanger 车钩提杆吊
uncoupling lever shaft 分离杆轴
uncoupling phenomenon 解耦合现象
uncoupling rigging 脱钩机构
uncoupling rod 车钩提杆
uncoursed 乱砌的；不分层的
uncoursed ashlar masonry 错列琢石圬工
uncoursed masonry 堆砌圬工；不成层圬工；不成层乱砌毛石
uncoursed random rubble 不分层蛮石乱砌圬工；不分层乱石圬工
uncoursed rubble 不分层毛石；不成层乱砌圬工；乱砌毛石
uncoursed rubble masonry 堆砌粗石圬工；不分层乱石圬工；不成层毛石圬工

uncoursed rubble stone masonry 粗石圬工
uncoursed square rubble 不分层方块(蛮)石圬工
uncoursed work 不成层圬工
uncover 干出(落潮时露出)；揭露
uncover and make opening 剥露和开口
uncovered 露天的；未冲销；无遮盖的；无覆盖的
uncovered area 无图地区
uncovered burner 无罩燃烧器
uncovered cable 无包皮电缆
uncovered canal 明渠
uncovered digester 露天消化池
uncovered drain 明渠
uncovered floor 无面层的毛地板；无面层的毛楼板
uncovered goods 未保险货物
uncovered platform 露天站台
uncovered reservoir 露天水库；露天水池
uncovered rock 明礁
uncovering of foundation 基础露面
uncovering plough 翻地犁
uncover work 剥离工作
uncowled 无盖的
uncracked 未裂开的；未裂化的
uncracked asphalt 未裂化沥青
uncracked condition 无裂缝状态
uncracked elastic condition 不开裂弹性状态
uncracked hydrocarbon 未裂化烃
uncracked residue 未裂化残油
uncracked zone 未开裂区
uncreeped 不皱的
uncreosoted 未用杂酚油处理的
uncropped areas 非耕作区
uncropped location 荒地
uncrossed cheque 非划线的支票
uncrushable 压不碎的
uncrushed 未经破碎的
uncrushed gravel 非破碎砾石；未破碎砾石；天然风化砾石
uncrushed material 完整的材料；未轧碎的材料
uncrystallizable 不能结晶的
uncrystallized 非晶的；未结晶的
unction 油腻性；涂油
unctuarium 浴疗室
unctuous 油性的
unctuous clay 油性黏土；重黏土
unctuousness 油腻的
unctuous paint 油质涂料
unctuous soil 松软肥沃的土壤
uncultivated 未开垦的
uncultivated field 未耕地
uncultivated land 荒地；未开垦地；未耕地
uncultured 未耕作的
uncupping of gasholder lift 储气罐塔节脱挂
uncured 未处治的
uncured glass wool 树脂固化玻璃棉
uncurtained 幕被拉起的；没有幕的；帘被拉起的；没有帘的；无遮蔽的；无窗帘的
uncustomed 未经海关通过的
uncustomed goods 未报关货物
uncut 未经琢磨的；未雕刻的
uncut bimetallic balance 未切口双金属摆轮
uncut diamond 未琢磨的金刚石；粗金刚石
uncut edges 不切边装订
uncut loop carpet 不剪绒的圈绒地毯；毛圈绒地毯
uncut modillion 未琢磨的托饰
uncut sheet glass 未选裁的板玻璃
uncut stone 未琢磨金刚石
undaform 浪蚀(底)地形
undamaged 无损坏的；未破坏的；完整无损的；完好的
undamaged recovery 无损回收
undamaged ship 未受损船
undamaged value 完整无损价值
undamped 无阻尼的；不衰减的
undamped analytical balance 无阻尼的分析天平
undamped control 无阻尼控制；不稳定调节
undamped forced vibration 无阻尼的强迫振动
undamped free vibration 无阻尼自由振动
undamped free vibration frequency 无阻尼自振动频率
undamped frequency 无阻尼频率
undamped mode 无阻尼模式
undamped motion 非阻尼运动
undamped natural circular frequency 无阻尼自振圆频率

undamped natural frequency 无阻尼固有频率；无阻尼自然频率
undamped natural period 无阻尼自振周期
undamped oscillation 连续振荡；无阻尼振荡；无衰减振荡；等同摆动；持续振荡；不衰减振动
undamped oscillator 无阻尼振荡器
undamped pendulum 无阻尼摆
undamped period 无阻尼周期；无阻尼振荡周期
undamped period of oscillation 等幅振荡周期；等幅摆动周期
undamped vibration 无阻尼振动；无衰减振动
undamped wave 无阻尼波；无衰减波；无衰耗波；等幅波；等波幅；不衰减波
undark 夜明涂料
undated 未注日期的
undated check 无日期支票
undated cheque 未注日期支票
undated endorsement 无日期背书
undated reference to standard 不注明日期的引用标准
undated securities 不定期证券
undathem 浪蚀岩层；浪蚀底岩层；浪成地层
undathem facies 浪成岩相；浅海岩相
undaturbidite 浪积浊流沉积；浪成浊积岩
undecagon 十一边形
undecalcified 未脱钙的
undecanal 十一醛
undecandienoic acid 十一碳二烯酸
undecane 十一烷
undecanoic 十一烷的
undecanoic acid 十一酸
undecanoic amide 十一酰胺
undecanonitrile 十一腈
undecanoyl 十一酰
undecap 十一烯酸
undecendioic acid 十一碳烯二酸
undecene 十一碳烯
undecene diacid 十一碳烯二酸
undecene dicarboxylic acid 十一碳烯二羧酸
undecenyl 十一碳烯基
undecidability 不可判定性
undecidable 不可判定的
undecimal 十一进制的
undecked 无装饰的
undecked boat 无甲板船
undecked ship 无甲板船
undecked vessel 无甲板船
undeclared cargo 未向海关申报的货物
undeclared goods 未报关货物
undeclared identifier 未说明标识符
undecomposable 不可分解的
undecomposed 未析出的；未分解的
undecomposed explosive 未变质炸药
undecorated 未经装饰的
undecyl 十一基
undecyl alcohol 十一醇
undecylamine 十一胺
undecylenate 十一碳烯酸盐
undecylene 十一碳烯
undecylenic acid 十一烯酸
undecylenic alcohol 十一碳烯醇
undecylenyl acetate 醋酸十一碳烯酯
undecylic acid 十一酸
undecylic alcohol 十一醇
undecylic aldehyde 十一醛
undecyl mercaptan 十一硫醇
undecyl phenol 十一基苯酚
undecyl-polyoxyethylene-ether-alcohol 十一基聚氧乙烯醚醇
undecyl thiophane 十一基噻吩烷
undecyl thiosulfate 十一基硫代硫酸盐
undecyndioic acid 十一碳炔二酸
undecyne 十一炔
undecyne diacid 十一碳炔二酸
undecyne dicarboxylic acid 十一碳炔二羧酸
undecynic acid 十一炔酸
undefended building 未设防建筑物
undefended city 未设防的城市
undefiled 未污染的
undefine 未定义
undefined 下定义的；没有定义的
undefined assets 不确定性资产
undefined course of river 迁徙河流
undefined format 未定义形式

undefined instruction 未定义指令
undefined length block 不定长信息组
undefined-length record 不定长记录
undefined record 未定义记录;未定义界记录;不定长记录
undefined scale 自由比例尺
undefined structure 无定型结构;无定形结构;非稳定结构
undefined symbol 无定义符号;未定义符号
undefinite boundary 未定国界
undefinite coastline 不定岸线
undeflected 未偏转的
undeflected beam 未偏转束
undeformable 不变形的
undeformed 无变形;无应变;未剥皮的;未变形的
undeformed representation 无变形表示法
undeformed state 无变形状态
undegraded 未经碰撞的
undegraded beam 未慢化束
undegraded material 未降解物质
undelayed 未延迟的
undelayed channel 无迟延信道
undeliverable articles 无法交付物品
undeliverable cargo 无法交付货物
undeliverable check 无法交付的支票
undeliverable goods 无法交付货物
undeliverable import goods 无法交付的进口货
undelivered article 未交付的物品
undelivered cargo 未提取货物
undelivered cargo list 催提货单
undemarcated 未分界的
undemonstrable 无法论证的
undeniable 不可否认的
undensified 不密实的
undensified rock 不密实岩(石)
undependent variable 独立变数;独立变量
undeposit cash 未存款
undepreciated balance 未提折旧余额
undepreciated balance of the cost 未折旧原价余额
undepreciated cost 未(提)折旧成本
undepreciated currency 未贬值的货币
undepreciated value 未贬值的价值;未折旧价值
under a 9-hours photoperiod 在九小时的光周期下
underabsorbed burden 少分配的成本及费用;少分配制造费用
underabsorbed overhead 少分配制造费用
under-ag(e)ing 凝固不足;硬化不足;时效不足
under a head of 在……水压下
under a legal incapacity 缺乏法律上的行为能力
under all conditions 在所有条件下
under allowance 尺寸下偏差
under annealing 不完全退火
underapplied expenses 少分配的费用
underapplied manufacturing expenses 少分配制造费用
underapplied manufacturing overhead 少分配制造费用
under average 受损
under axial loading 在轴向荷载下
underbaked 欠烘的;未烘透的
underbalance 欠平衡;平衡不足
under ballast 底渣
underbar burner 炉算下点火燃烧器
underbead covering 内部裂纹
underbead crack 焊道下裂纹;焊注下裂纹;熔珠下裂纹;焊珠不足的裂缝;焊道下裂纹;焊根裂纹
underbead crack susceptibility 焊道下裂纹敏感性
underbead crack test 焊道下裂纹试验
underbeam 下梁
under beam girder 梁下纵行架;梁下桁(梁)
under beam tonnage 梁下吨位
underbed 下端;底面;垫层;底座;底灰;底层;基础板;底垫
under-belly equipment 机腹下悬挂农具
underbid 递价过低
under-bidding 投低价标;投低标价
under-billing 少算运费
under blanket 底衬
underblower 算下送风机
under blowing 底风;底吹
underboarding 垫板;衬(底)板
underbody 物体下部;水下船体;船体水下部分;车身底座
under bond 保税

under-bound finish 基料不足的涂膜
under-braced 自下支撑的;支撑不足
under-bracing (电杆的)杆根横木;帮桩;下联;下(部)支撑;下撑
under branch 下部枝条
underbreak 欠挖
underbreak in tunnel excavation 隧道欠挖
underbridge (立体交叉的)下穿(式)桥;跨线桥;桥下
underbridge clearance 跨线桥桥下净空;桥下净空;下穿(式)桥桥下净空
underbright 水平线附近云隙中射出
underbrush 小树丛;林下灌丛;矮丛林
underbuilding freeway 穿过建筑物下层的高速街道
underbunching 电子聚束不足
under-burned brick 未烧透砖;欠火砖
under-burned clinker 欠烧熟料
under-burning 欠火;欠烧;生烧
under-burnt 欠烧的;欠火的;未烧透的
under-burnt brick 欠火砖
under-burnt lime 欠烧石灰
under burnt material discharged from shaft kiln 立窑跑生料
undercalcined 欠烧的
under canvas 驶帆
undercapacity 强度不足;生产率不足;非饱和容量
undercapacity factor 欠载系数
undercapacity operation 开工不足
under-capitalization 资本过少
undercapitalize 投资紧缩;投资不足
undercarriage 履带行走装置;行走机构;机脚;起落架(飞机);底座;底架
undercarriage arrester 起落架拦阻装置
undercarriage bay 起落架舱
undercarriage bracing 起落架拉条
undercarriage brake 起落架制动器;起落架制动装置
undercarriage cushioning 起落架减震
undercarriage door 起落架舱门
undercarriage indicator 起落架状态指示器
undercarriage jack 底架千斤顶
undercarriage springing 起落架弹性装置
undercarriage track 起落架轨距
undercarrier 不足量运送人
undercart 起落架(飞机)
under-car temperature 客车下温度
under-car ventilation 客车下通风
undercasing 底箱(铸造用)
undercast 高空阴天
undercharge 充电不足
undercharged airmail correspondence 免收航空附加费的航空函件
undercharging 非饱和充电
under-chassis 底盘;机脚;起落架(飞机);底部框架
underclass 下层居民;贫民;经济最底层
underclay 煤层下黏土;底黏土(层)
underclay limestone 底黏土灰岩
underclearance 下部间隙;下部空隙;桥下净空
undercliff 滑坡阶地;滑坡坍塌形成的阶地;岩脚坡;崖下阶地;阶地;滑动崖脚坡;副崖
undercloak 檐口卷边层;黏土砖层;檐口处垫瓦;下部覆盖层;基层;垫层挡水板;垫层
under clothing burn 衣下烧伤
undercoat 着色底漆;中涂层;底釉;底衬;打底层
undercoater 面漆下涂层;罩面涂层基层
undercoater for external use 外用底层抹灰
undercoat glaze 衬底釉
undercoating 底漆;内涂层;上底漆;底涂层
undercoating for waterproofing 防水层底子
undercoating lacquer 头道漆
undercoating material 底层抹灰材料
undercoating paint 底层涂料;打底漆;头道漆;底涂涂料
undercoating varnish 打底清漆
undercoation 底涂层
undercoat mixed plaster 底层抹灰混合砂浆
undercoat mix(ture) 底层灰浆
undercoat of paint 底层涂料
undercoat stuff 底层涂料
under colo(u)r removal 底色去除
under-commutation 欠换向
under-compacted 欠压密
under-compensated meter 补偿不足电度计
under-compensated 欠补偿的
undercompensated integral control 欠补偿的积分控制

undercompensation 欠补偿;不完全补偿;补偿不足
undercompound 欠复绕;欠复励(的);欠复激
undercompound excitation 欠复励
undercompound generator 欠复励发电机
undercompound winding 欠复激绕组
under compression 在压缩条件下;受压
underconpaction facies 欠压实相
underconpaction zone 欠压实带
underconsolidated clay 欠固结黏土
underconsolidated soil 欠固结土
underconsolidated soil deposit 欠固结土沉积
underconsolidation 未充分固结
under construction 修建中;正在施工;在施工中;在建筑中;在建造中;在建工程;施工中的;地下建筑
underconsumption 低消耗;消耗不足;消费压制;消费不足
undercontact rail 底面接触的导轨;底接触导轨
under continuous headway 前后保持安全距离陆续通过(车、船等)
under control 情况正常
under controlled conditions 在控制条件下
undercool 过度冷却(指使液体冷到凝固点以下而不凝结)
undercooled 过冷的
undercooled graphite 过冷石墨
undercooled liquid 过冷液体
undercooled steam 过冷蒸汽
undercooling 过冷
undercooling point 过冷点
undercorrected lens 校正不完全透镜
undercorrection 改正不足;校正不足
undercounter dumbwaiter 柜台下小型运货升送机
undercounter refrigerator 柜台下冰箱
undercoupling 耦合不足;欠耦合
undercourse 衬层;檐口垫底层
under cover 有遮盖的
under-cover payment 贿赂
under-cover storage 带顶储库;非露天储存
undercritical 次临界的
under-croft 地窖;地下室;穹窿状地下室
undercrossing 下穿式立交;地道;下穿交叉(路);桥下道路;地下通道
undercrown surface 活塞顶内表面
under crown wheel 下小钢轮
undercure 固化不足;欠塑化;欠硫化;欠固化;欠处理
undercured 未充分养护的;未硬结的
undercured concrete 欠养护混凝土
undercuring 养护不够;欠养护;欠熟;欠处理;硬化不足
undercurrent 底流;低电流;下层流;海底潜流;潜在势力;潜流;电流不足;低电流;暗流;暗潮
undercurrent circuit breaker 低电流断路器;欠电流断路器
undercurrent protection 低电流保护装置;欠电流保护
undercurrent relay 低电流继电器;欠(电)流继电器
undercurrent release 欠(电)流释放;低电流释放
undercut 底部截槽;齿轮根切;沟割;向下凹陷;下植;下冲;下边部分被削掉的;咬边;落底;截槽;潜挖;平板玻璃小缺口;淘蚀;掏槽;倒凹的;不足伐
undercut bank 凹蚀岸
undercut door 截底门
undercut price 削价
undercut side wall 凹圆线脚边墙;带凹圆线脚边墙
undercut slope 淘蚀岸坡;底切岸坡;凹岸;暗夏坡;底切坡
undercut-slope bank 底切坡岸
undercut swing saw 落地式摆锯
undercut tenon 燕尾榫;合角榫;欠削榫舌
undercutter 下刻机;截煤机;凹形挖掘铲
undercutting 刨削 T 形槽;凹割;基切;切底(作用);淘刷;掏蚀(作用);地下截根;底切(作用);底部冲刷(作用);低切馏分;道砟破底;暗掘;切入或挑出
undercutting cold front 下切冷锋
undercutting jib 下截盘
undercutting machine 破底清挖道砟机械
undercutting method 下切法(隧道)
undercutting of riverward toe of levee 河堤迎水坡脚掏刷
undercutting of track 落道
undercut(ting) slope 底切坡

undercutting slope bank 底切坡岸
undercutting tool 平面切刀
undercutting turning tool 沉割车刀
undercut trimmer saw 脚踏截锯机
undercut weld 底切焊接
under-cylinder pan 滚筒下谷粒盘
underdamp 不完全衰减;弱衰减
underdamped circuit 低衰减振荡电路
underdamped response 欠阻尼响应
underdamped system 欠阻尼体系;低阻尼系统
underdamping 弱衰减;不足阻尼;阻尼不足;弱阻尼;欠阻尼;不完全衰减
underdeck 甲板下
underdeck cargo 甲板下货物;舱内货(物);舱货
underdeck clause 舱内条款
underdeck equipment 甲板下设备
underdeck girder 梁下桁(梁);甲板桁材
underdeck machinery 舱内机械
underdeck operation 舱内作业
underdeck shipment 舱内装货
underdeck space 甲板下舱容
underdeck stowage 舱内装载
underdeck tonnage 甲板下吨位;下甲板吨位
underdeck tonnage section 甲板下吨位剖面
underdense 欠密的
under-deposit corrosion 垢下腐蚀
under-depreciation 折旧不足
underdesign 欠安全的设计;安全系数不足的设计
under designed 未有足够的安全系数的
underdetermined 欠定的
underdetermined system 欠定组
underdetermined system of equations 欠定方程组
underdetermined value 欠定值
underdeveloped area 不发达地区
underdeveloped country 不发达国家
underdeveloped region 未充分开发地区;不发达区域;未充分开发的地区
underdevelopment 显影不足;显像不足;欠发达状态
under-dispersion 低扩散
underdraft 轧件下弯
underdraft furnace 下抽炉
underdrag 下侧牵引
underdrain 地下排水管;阴沟;盲沟;地下渠(道);地下沟渠;地排水管;暗渠;暗沟
underdrainage 地下排水;暗沟排水
under-drained dock floor 底板下排水式坞底板
underdrain manifold 底排水支管
underdrain network 地下排水管网
underdrain pipe 地下排水管
underdrain system 地下排水管网
underdrain system design 地下排水系统设计
underdrain tile 底部排水管道
underdraught 下压
underdrawing 在屋面瓦下嵌灰泥
under-drawn 向上铺对
underdrive 下传动;减速传动;低档;传动迟缓
underdriven 下面传动的;底部驱动的
underdriven logframe sawing machine 从下部开动的架锯机
underdriven roll 空转辊
underdriven vertical edger 下传动的立辊轧机
underdrive(-type) double action drawing press 底传动式双动拉伸压力机
undereamer 扩孔器;套管靴下扩孔器
undereamer cutter 套管靴下扩孔器的推出式切刀
undereamer lug 套管靴下扩孔器的推出式切刃
undereaming bit 套管靴下扩孔钻头
underearth 地面下土层;底黏土
underearth-eaves course 檐下瓦层
under easy steam 缓速航行
under-eaves course 檐口垫瓦
under-eaves tile 檐口垫瓦
underemployed 半失业
under-employed land 未完全被利用的土地
underemployment 就业不足
underemployment equilibrium 就业不足的均衡状态
under-enumeration 漏查
underestimate 估计过低;估计不足;低估
underestimation 低估价(的);估计不足
underevaluation 低估
underexcitation 欠激励(状态)
underexcitation generator 欠激励发电机
underexcitation protection 欠激励保护(装置)

underexcited compound generator 欠复激发电机
underexhaustion 未发好
underexpanded nozzle 欠膨胀喷管
underexpansion 不完全膨胀
under-expenditures 节余
under exploitation 开发不足
underexpose 未充分照射
under-exposed 欠曝光
under-exposure 感光不足;曝光不足;欠曝光;显像不足;照射不足
under-exposure portion 曝光不足部分
under-exposure region 曝光不足区
under external pressure 在外压作用下
underfeature 小起伏地形;小地物
underfeed burning 不给料式燃烧
underfeed combustion 下部给料燃烧方式;下给料式燃烧
underfeed furnace 下饲式炉膛;下加热式炉;下供燃料炉膛;下部加(燃)料炉;火下加煤煤气发生器
underfeed(ing) 喂料不足;供料不足;地下补给;底部进料;补给不足;供料不充分
underfeed stoker 火下加煤机;下饲式加煤机;下方给煤机
underfeed system 下行式热水系统;下给式热水系统
underfelt 底层油毡;衬底油毡;垫油毡
under field conditions 在自然条件下
under field factors 在自然条件下
underfill 焊接处的凹陷;未焊满;未充满;不满
underfilled pass 未充满孔型
underfilling 底层填料;未充满
underfill method 炸开孔穴填筑路堤法
under-film corrosion 膜下腐蚀
underfire air 炉箅下的进风;一次空气
underfired 欠烧的;欠火的;未烧透的;从下部燃烧的
underfired brick 未烧透砖
underfired furnace 下部燃烧炉
under-firing 欠热;欠火;欠烧;生烧
under-firing gas 加热煤气
under-fishing 捕捞不足
underfit river 不相称河(流)
underfit stream 不相称河(流)
underflood 暗潮
underflooding 地下水升高;部分洪淹
underfloor 地面下的;地板下的
underfloor cable duct 地板下电缆渠
underfloor cable duct installation 车底电缆槽安装
underfloor culvert 闸门涵洞;闸底涵洞
underfloor discharge chute 地下卸料溜子
underfloor drainage system 地板下排水系统
underfloor duct 地板下通风管道;地下(电缆)管道;地面暗管道;地板下线渠;地下电缆管沟
underfloor engine 安装车厢地板下发动机;底架下的发动机;车地板下安装的发动机
underfloor exhaust duct 楼板下排风管道;地板下排风管道
underfloor extraction 地下卸料;地下出料
underfloor filler 地板下面的填料
underfloor fresh air duct 地板下送风管道;楼板下送风管道
underfloor heating 地板下供热;地板下供暖;地板采暖
underfloor heating by electricity 楼板电热采暖
underfloor heating cable 地板下面的电热线
underfloor heating installation 地板下面的供热装置
underfloor heating system 地板下面的供热系统
underfloor horizontal furnace 地板下的卧式炉
underfloor installation 车底安装
underfloor lifting line 架车线
underfloor pump 底板下排水泵
underfloor raceway 地板下线渠;楼板线槽;楼板下电路;地下电缆管道;地板下线槽
underfloor safety vision screen 安装车厢地板下安全反光镜
underfloor socket 地板下面的电线插座
underfloor stopping 地板下的填塞物
underfloor venthole 坑内换气口
underfloor warming 地板下面采暖
underfloor warming installation 地板下面的采暖装置
underfloor warming system 地板下面的采暖系统
underfloor wheel lathe 不落轮镟床;地坑式车轮镟床

underfloor wiring 地板下面布(电)线
underfloor wiring system 地下电缆系统
underfloor wiring work 地板下布线工程
underflow 沉沙;浓浆;浓泥;冰下水流;下溢;海底潜流;潜流;地下水流;底流;槽下水流
underflow baffle 地下水流挡板;底流挡板
underflow conduit 潜流(水)道
underflow-density 底层流密度
underflow-density current 底部异重流
underflow flow 潜流水流
underflow gate 底流式闸门
underflowing irrigation 地下水流灌溉
underflow mark 底流痕
underflow of classifier 筛出物;分选器的分出物
underflow pump 浓浆泵;低流泵
underflow turbidity current 底部浊浊流
underflow water 潜流水;伏流水
under fluorescent light 在日光灯下
under focus 弱焦点;欠(聚)焦
underfoot 托换基础;加强基础;少计总额
underfoot coating 地面涂料
underframe 底座;车身底板;底(构)架;底盘架;底框;
underframe clearance 犁架高度
underframe cross tie 底架横撑;底架横梁
underframe pivot 底架上的心盘
underfrequency 低于额定频率
underfrequency protection 低频率保护装置
under full steam 全速航行
under-gate conveyance 门下输水
underga(u)ge 短尺寸;尺寸不足
under girder 纵大梁
undergirdth 底部拉网束带
under given conditions 在特定条件下
underglaze 用于涂油前的;釉底的;上釉前着色的
underglaze blue 青花
underglaze blue and underglaze red covered jar 青花釉里红盖罐
underglaze blue bowl 青花碗
underglaze blue jar with sanskritdesgn 青花梵文罐
underglaze blue vase with fruit and floral sprays 青花折枝花果纹瓶
underglaze blue vase with pagoda shaped cover 青花塔式盖瓶
underglaze brown painting on white ground 白地釉下赭彩
underglaze colo(u)r 釉下彩;釉底颜料;釉底色彩
underglaze decoration 釉下陶瓷花饰;釉下彩饰;釉下装饰;釉底装饰
underglaze red 釉里红
undergoing deformation 在变形中
under grade 质量不高
undergrade crossing 铁路在下交叉
undergrade lumber 劣等木料
undergrade wood 等外材
undergraduate 大学肄业生;本科学生
undergraduate course 本科
undergrate blast 箅下送风;炉底进风
undergrate chamber 箅下风室
undergrate compartment 箅下隔仓
under grazing 轻度放牧
undergrinding 研磨不足
underground 地下
underground aerodrome 地下机场
underground airfield 地下(飞)机场
underground air passage 地下通风道
underground air-raid shelter 地下防空洞
underground alignment 坑道定线;地下放线;地下定线(测量)
underground anchor 地下锚碇装置
underground asbestos mining 地下石棉矿
underground barn 地下粮库
underground basilica 长方形地下会议厅(古罗马)
underground basin 地下水流域
underground bin 地下储仓
underground blast-hole drilling 地下坑道爆破钻进
underground blasting 坑内爆破;地下爆破
underground block 地下建筑物
underground bookstore 地下书店
underground borehole 坑道钻孔
underground boring 坑道钻探;坑道钻进;井下钻进
underground bud 地下芽
underground building 地下房屋;地铁建筑;地下

建筑
underground bunkering 井下装仓
underground burial chamber 地下墓室
underground cable 管道电缆;地下光缆;地下电缆
underground cable box 地下电缆接线盒
underground cable channel 地下电缆槽
underground cable entry 地下电缆入口
underground cable terminal box 地中电缆盒
underground canal 暗渠;地下渠(道)
underground car park 地下停车场;地下车库
underground cavern 地下岩洞;地下洞穴
underground cavity 地下洞室
underground cellar 地窖;地下室
underground cemetery 地下公墓
underground chamber 地下室;地下洞室
underground channel 地下渠(道)
underground circuit 地下线路
underground cistern 地下水窖;地下液槽;地下蓄水池
underground clay mining 地下采土
underground coal gasification 煤炭地下气化;煤地下气化;煤的地下气化
underground coal mining district 井下采煤区
underground collapse 地下坍陷
underground collector 地下接触轨集电器
underground commercial centre 地下商业中心
underground communication conduit 地下电信电缆管道
underground complex 地下综合建筑
underground condition 地下情况;坑道状态;地下条件
underground conductor 地下线;地下导体
underground conduit 电缆通道;电缆沟;电缆槽;地下公用管道
underground conduit net(work) 地下管道网;地下水道网
underground conduit system 地下管渠系统
underground connecting triangle for shaft plumbing 井下定向联系三角形
underground construction 地下建筑
underground construction in rock 地下岩层施工;岩石下的地下工程
underground construction machinery 地下工程施工机械
underground construction survey 洞内施工测量
underground construction work 地下工程
underground contact 地下接触
underground contour map 地下等深线图
underground conveyer 地下运输机
underground corrosion 地下腐蚀
underground course 地下砌层
underground crossing 地下跨越
underground crusher 井下破碎机
underground dam 潜坝;地下截水墙;地下防水墙;地下挡水墙;地下坝
underground department store 地下商场
underground development 地下开拓
underground dewatering 地下疏干
underground diaphragm (墙面的)地下防水层;地下防水墙
underground diaphragm wall 地下防渗墙
underground digging 地下挖掘
underground disposal 排入地下;地下处置;地下处理
underground disposal by pressure 地下压入处理
underground disposal of industrial wastewater 工业废水地下处置
underground distribution 地下配电;地下布线
underground drain 地下排水沟;暗沟;地下排水管
underground drainage 地下排水管道;地下水系;地下疏干法;地下排水
underground drainage area 地下流域面积
underground drainage pipe system design 地下排水系统设计
underground drain pipe 地下排水管
underground drain system 地下排水系统
underground drill 坑道钻机
underground drill-hole 地下钻孔;坑道钻孔
underground drilling 地下钻探;坑道钻进;井下钻进
underground drilling machine 坑道钻机
underground drinking water source 地下饮用水源
underground duct 地下管道
underground dumping car 井下自卸卡车

underground dwelling 地下窑居
underground economy 地下经济
underground electric generating station 地下发电站;地下发电厂
underground engineering 地下工程
underground engineering geology 地下工程地质学
underground engineering of waste disposal 地下排污工程
underground environment 地下环境
underground erection chamber 地下安装室
underground erosion 潜蚀;地下潜蚀
underground excavation 地下巷道;地下开挖
underground excavator 井下挖掘机
underground exploration 地下勘探;地下勘察
underground explosion 井内爆破;地下爆炸;地下爆破
underground explosion breccia 隐爆角砾岩
underground expressway 地下快速道路
underground external plaster work 地下建筑外层抹灰
underground fan 井下通风机
underground filling device 地下装油装置
underground filtration 地下过滤
underground filtration field 地下过滤场
underground flooding 地下注水
underground flow 渗流;潜(水)流;地下(水)流
underground flow distance 地下水距离
underground formation 地下形成(油、气的)
underground free way 地下通道
underground gallery 地下廊道
underground garage 地下车库
underground gasification 煤炭地下气化;地下气化(作用)
underground gas storage 地下燃气库
underground geologic(al) documentation 地下地质编录
underground getting 地下开采
underground glory-hole method 地下放矿漏斗开采法
underground haulage 地下运输
underground headquarter 地下指挥所
underground heated pipeline 埋地热油管道
underground height measurement 井下高程测量
underground holding tank 地下储油箱;地下储水箱
underground hopper 地下料箱;地下料斗;地下储[贮]料仓
underground horizontal control survey 井下水平控制测量;井下平面控制测量
underground hospital 地下医院
underground hotbed 地下温室
underground hydrant 地下消火栓;地下消防栓;地下式消防栓
underground hydraulic power plant 地下水电厂
underground hydraulic power station 地下水电站
underground ice 地下冰
underground industry 地下工业
underground injection 地下注水
underground installation 地下敷设;地下装油装置;地下(装油)设施
underground interchange 地下互通式立体交叉;地下交通枢纽
underground irrigation 地下灌溉
underground junction 地下结合
underground karst form 地下岩溶形态
underground lake 地下湖;暗湖
underground lake deposit 地下湖积物
underground laying 地下敷设
underground leakage 地下渗漏;地下渗流
underground level(l)ing 坑道水准测量;地下水准测量
underground line 地下管线;埋地线;地下线路;地下铁路线;地铁线路
underground line in shallow 浅层地下铁道
underground loader 井下装载机;地下装载机
underground loading facility 地下装油设施
underground locomotive 地下铁道火车头;地下铁道牵引机车
underground lowering 地下水下降
underground magazine 地下炸药库
underground main 地下总管;地下干管
underground mall 地下商业街
underground mark 地下标志;地下标记;地下标点
underground market 地下商业街;地下商场

underground metal mining 金属矿开采
underground milling 地下放矿漏斗开采法
underground mining 地下开采
underground mining production exploration engineering 地采生产勘探工程
underground movement 地下移动
underground nuclear explosion 地下核爆炸
underground nuclear power station 地下核电站
underground nuclear test 地下核试验
underground nuclear waste storage 地下核废料封存库
under ground observation laboratory 地下观测室
underground oil depot survey 地下油库测量
underground oil reservoir 地下储油仓
underground oil storage 地下油库;地下石油储存
underground opening 地下巷道;地下孔洞;地下洞室
underground opening stability evaluation 峒室稳定性评价
underground operation 地下作业
underground palace 地下宫殿
underground park(ing) (garage) 地下停车(场)
underground parking lot 地下停车场;地下停车库
underground passage(way) 地(下)通道
underground path 地道
underground penstock 地下进水管;地下输气管;地下输水管
underground percolation 地下渗漏;地下渗流
underground perforated drainpipe system 地下花管排水系统
underground pipe 地下管(道)
under-ground pipe comprehensive design 管线综合设计
underground pipe-driving survey 顶管测量
underground pipe entry 地下管道入口
underground pipeline 地下管线
underground pipeline engineering 地下管道工程
underground pipeline finder 地下管线探测仪
underground pipelines comprehensive design 管线综合设计
underground pipeline survey 地下管线测量
underground (pipe) net(work) 地下管(道)网
underground pipe system 地下管系
underground piping 地下管道
underground pit 地下坑道
underground plant 地下工厂;地下电站;地下电厂;地下厂房
underground pollution 地下污染
under ground portion 地下部分
underground power house 地下(发)电站;地下(发)电厂
underground power plant 地下(发)电厂;地下(发)电站
underground power station 地下(式)电站;地下(式)电厂
underground pressure 井下压力;顶板压力;地压
underground project 地下(工程)项目
underground prospecting engineering survey 井探工程测量
underground public utility 地下公用设施
underground quarry 地下采石场
underground railroad 地下铁道
underground railroad station 地下铁道车站
underground railway 地下铁道;地铁
underground railway profile 地下铁道纵断面
underground railway survey 地下铁道测量
underground railway train 地下铁道列车
underground recharge 地下回罐
underground recording 地下记录
underground reservoir 地下水库;地下含水层
underground residential distribution 地下住宅配电系统
underground residue 地下水储量
underground restoration 地下还原处理
underground retention time 地下水滞留时间
underground return flow 地下水回流
underground rhizome 地下走茎
underground river 地下暗河;伏流;暗河;地下河(流)
underground rock cavern design 地下岩石洞室设计
underground room 地下室
under ground root 地下根
underground runoff 地下径流

underground runoff modulus map 地下径流模数图
underground seepage loss 地下散失
underground service 地下供电系统;井下工作;井下作业
underground service conductor 地下引出线
underground service entry 矿井下作业进口;井下作业进口;地下作业进口
underground service road 地下辅助道路
underground shaft 盲井;地井
underground shelter 地下掩(护)体;地下人员掩蔽所;地下防空洞;防空洞
underground shopping arcade 地下商场
underground shopping center 地下街道;地下购物中心
underground shovel 地下掘土机;地下挖土铲;坑内动力铲;井下机铲;井下铲斗装载机
underground silo 地下青储窖
underground solute evaluation method 地下溶质评估法
underground source of drinking water 地下饮用水源;饮用水地下水源
underground space 地下空间
underground space usage 地下空间利用
underground spring 地下泉
underground sprinkler system 埋管式喷灌系统
underground stage plumbing 井下分段摄影法
underground station 地下车站
underground stem 地下茎
underground storage 地下蓄水;地下水库;地下(水)储盐;地下储罐;地下储藏
underground storage of energy 地下储能
underground storage of liquefied natural gas 地下存储液化天然气
underground storage of LNG 地下存储液化天然气
underground storage of nuclear wastes 地下存储核废料
underground storage of oil 地下储油
underground storage of water 地下储藏水
underground storage tank 地下储罐
underground storage water 地下蓄水
underground storehouse 地下仓库
underground storeroom 地下仓库
underground stratification 地下成层现象
underground stream 地下河(流);伏流(河);暗河
underground street 地下商业街
underground structure 地下结构(物);地下建筑物;地下构筑物
underground structure contour map 基准层构造图;地下构造等高线图
underground stucco 地下外层抹灰
underground substation 地下变电站;地下变电所
underground surge tank 地下平衡箱;地下均压箱
underground survey(ing) 坑道测量;地下(结构)测量
underground survey map 矿井测量图
underground surveyor 地下测量员
underground tank 地下油箱;地下水箱;地下水池
underground terminal 地下枢纽
underground town 地下商业街;地下城;地下市镇
underground trackless mining 地下无轨采矿
underground traffic 地下交通
underground traffic artery 地下交通干道
underground transportation 地下运送
underground traverse 矿坑导线;井下导线;地下导线
underground traversing 井下导线测量;地下导线测量
underground tube 地铁
underground tube ties 地下管线
underground tube ties system 地下管架系统
underground tunnel 地下隧洞;地下隧道
underground tunnel gate 地下廊道卸料门
underground unconformity 地下不整合
underground undercrossing 地下铁道穿越
underground undermanager 负责井下工作的副矿长
underground utilities 地下管网
underground utility 地下管线;地下公用事业;地下公用管道设施(水、电、气等的);地下公共设施
underground utility line 地下管网布线
underground valve chamber 地下气门室;地下阀门箱
underground vault 地窖

underground ventilation pipe 地下通风管
underground ventilation plant 地下通风机
underground waste deposit 废物的地下堆存
underground waste disposal 地下废物处置;地下废料处置
underground water 潜水【水文】;地下水
underground water basin 地下水域
underground water behaviou(u)rs observation 地下水动态观测
underground water conservation district 地下水保护区
underground water course 地下水道;潜流水道;地下河道
underground water divide 地下水分水线;地下水分岭;地下水分界
underground water flow 地下水流
underground water head 地下水压头
underground water influence 地下水影响
underground watering 浸润灌溉;底土灌溉
underground water level 地下水(水)位;潜水面(指地下水)
underground water loss 地下水损失
underground water parting 地下(水)分水线;地下分水岭;地下水分界;地下分水结
underground waterproofing 地下防水
underground water resources 地下水资源
underground watershed 地下流域;地下分水岭;地下水分界;地下分水界;地下分水结
underground water storage 地下储水库
underground water supply 地下水供应;潜水供应
underground water system 地下水系统
underground water table 地下水位以下;地下水位
underground working 地下开采
underground working-out section area extent 地下采空区范围
underground works 地下工厂;地下工程;坑道作业;井下作业
underground workshop 地下厂房
underground yield 地下水流量;地下水产(水)量
undergrove 下层树林
under growth 林下植物;林下灌木;林下草层;树下植被;树下草地;下层林丛
undergrowth planting 下木栽植
underguard 下部护板
under hair 底绒
under half steam 半速航行
underhammer 下置式击锤
underhand 下向梯段式的;正台阶式的;俯采式的
underhand bench 下向梯段;正台阶
underhand blasting 下向梯段爆破;正台阶式爆破;俯采式爆破
underhand double stope 下向双翼回采工作面
underhanded 人手不足的
underhand longwall 下向梯段式长壁回采
underhand single stope 下向单翼回采工作面
underhand stope 下向回采;俯采工作面
underhand stope bench 下向回采梯段
underhand stoping 下行梯段回采法
underhand work 下向梯段凿岩;下向梯段开挖;下向梯段掘进;下向梯段回采;下向采掘;正台阶式凿岩;正台阶式开挖;正台阶式掘进;正台阶式回采;俯采式凿岩;俯采式开挖;俯采式掘进;俯采式回采
under hardening 淬火不足
underhead crack 焊面下裂缝
underheated 加热不足的
underheating 加热不足
underhole 下部掏槽
underhorn 下角
underhung 自下承接的;靠轮拉动的(拉门);自下支承的
underhung crane 高架移动下悬起重机;悬挂式吊车;悬挂式单梁吊车;下悬起重机
underhung door 扯门;靠轮子在轨道上滚动的拉门;悬挂式推拉门
underhung rudder 吊舵;铲型舵
underhung spring 悬簧
under-ice diving 冰下潜水
under-ice reef blasting 冰下炸礁
under-ice saturation dive 冰下饱和潜水
under-ice tunnel 冰下隧洞
under-ice work 冰下作业
underimproved land 未充分利用的土地

underimprovement 改善不足
under individual tree canopies 不同树冠层下
under-inflation 打气不足
under inlet 输入不足
underinsurance 不足额保险;不足保险;保险不足
underinsure 保险价值过低
under interlayer 胶合层不足
under investment 投资不足
underkeel 龙骨下的
underkeel clearance 船底以下的水深;龙骨下富裕水深;龙骨下富余水深
under laboratory conditions 在实验室的条件下
under-laid 垫层下面的
underlap 遮盖不足;重叠不足;欠连接;提前还船;不重叠
under lapped valve 负遮盖的阀
underlay 铺在下面;衬底层;底基层;衬(垫)底;安置在下面;下衬;倾斜余角;铺底;地毯垫层;地毯衬垫;从下面支撑;衬底
underlayer 下覆岩层;下卧层;下垫层;垫(底)层;底基层;底垫;衬底材料
underlay foil 垫底薄铁皮
underlaying geologic(al) formation 岩床地质层系
underlaying material 衬底材料
underlayment 垫平层;衬料;衬垫之物;找平层;木垫板;防水漏层;防水层;衬底(层)
underlay mineral (上浇沥青屋面的)砂砾
underlay paper (混凝土路基下的)衬垫纸
underlay shaft 下盘斜井
underlay sheet(ing) 衬垫薄板
underlease 转借;转租;分租
underlessor 重出租人
underlet(ting) 分租
under-lever 下置杠杆
underlie 下覆;倾斜余角
underlier 下盘斜井
underlie shaft 下盘斜井
underline 腹线;强调(线)
under-line bridge 铁路在下交叉
underlined character 下加线字符
underlined platform 加下险地台
underlining felt 屋面油毛毡;屋面衬垫油毡;衬底油毡;层面衬底
underlining felt of roof 屋面衬(垫油)毡
under load 承荷载的;负载过轻;欠载;不满荷载;不满负荷
underload breaker 欠载断路器
underload circuit breaker 轻载断路器
underloaded 欠载的;轻载;轻载的;负载不足(的);负荷不足;非满载的;轻荷的
underloading 装料不足;负载不足
underload refractoriness 荷重耐火度
underload refractory 荷重软化温度
underload relay 轻载继电器;欠载继电器;负载不足继电器
under low moisture conditions 在低湿条件下
under lustred 光泽不够的
underlying 下覆的;垫底焊
underlying alluvial soil 下覆冲积土
underlying alluvium 下覆冲积层
underlying bed 下覆(岩)层;下垫层
underlying bedrock 下覆基岩;底部基岩
underlying bonds 高级债券
underlying cause of death 主要死因;基本死因
underlying characteristics grammar 基础表征文法
underlying clay layers 下卧黏土层
underlying clay strata 下卧黏土层
underlying company 子公司;附属公司
underlying cost 基本成本
underlying country rock 下覆围岩
underlying course 底基层;垫层
underlying deposit 下覆沉积
underlying detail 分层设色底图碎部
underlying documents 原始凭证
underlying fill 下覆填土层
underlying financing 原有贷款
underlying foundation soil 下卧地基土;下覆地基土
underlying geologic(al) formation 岩床地质层系;下覆地质层组
underlying graph 底图;基础图;基本图
underlying ground 底(层)土
underlying layer 下卧层;下覆层
underlying layer of soil 下卧层土
underlying mass 下覆体

underlying material 下覆材料
underlying metal 底层金属
underlying of ore body 矿体下界面
underlying phreatic flow 下覆潜流
underlying price 基础价格
underlying rock 母岩;下覆岩石;下垫岩石;底岩;底层岩石
underlying sand layer 下卧砂层
underlying sandstone 下卧砂岩
underlying sand stratum 下卧砂层
underlying saturated clay strata 下卧饱和黏土层;下覆饱和黏土层
underlying securities 附属公司证券
underlying sediment 下覆沉积
underlying shaft 下盘斜井
underlying soil 下覆泥沙;底层土;下覆土(壤)
underlying stratum 下卧层;下覆岩层;下覆地层;下覆层
underlying surface 下覆面;下垫面;基面;伏面
underlying system 下覆系统
underlying tax 直接税
underlying topography 下垫面地形
underlying wall rock 下部围岩
under-maintained 维护不足
under-maintenance 维修不足
undermanned 人员不足的;人手不够的
undermargined 保证金不足
under mechanical stress 在机械应力下
undermelting 底融
undermentioned 下述的
undermine 潜挖;掏槽;洞穴;底部挖掘;暗掘
undermining 基蚀;下部淘空;潜挖;挖坑道;淘刷;淘底冲刷;底切;底部冲刷(作用)
undermining area 采空区
undermining blast(ing) 坑内爆炸;坑道爆破;潜挖爆破
undermining flow 潜水流
undermining method 暗挖法
undermining of levee 堤防淘底
undermining sublevel 分段下面拉底
undermining-torrent 侵蚀野溪
undermixing 混合料的不均匀性;搅拌不足;搅拌不匀;混合不足;混合不均;拌和不足
undermoderated lattice 慢化不足栅格
undermoderated mixture 慢化不足混合物
undermoderated system 慢化不足体系
undermoderation 慢化不足
undermodified 变质处理不足的
under modulated 欠调制的
undermodulation 欠调制;调制不足
undermounted 车架下悬挂的
underneath drive 下部传动
underneath the roof 屋顶下面
underneath type 下置式
under nitrogen blanket 在氮气层之下
under-nodularizing 球化不良
under normal operation 在正常条件下
undernourished profile 供砂不充分剖面
under offer 在出售中
under one's hand and seal 签名盖章
under operation condition 在工作条件下
underpaid 工资微薄的
underpainting 叠印的底色;底色
underpan 底盘
underpart 下部结构;非重要构件
underpass 地槽;下穿线;高架桥下通道;铁路在下交叉;下穿交叉(路);人行地道;桥下通道;地下通道;地下铁路;地下过道;地道
underpass approach 地道引道
underpass bridge 地道桥
underpass deflector 下水道改流装置
underpass deflector 地下水道导流装置
underpass grade separation 下穿铁路立体交叉
underpass-type tunnel(l)ing method 地道式隧道施工法
underpavement 下层铺面;下层路面;路面下层
underpay 不足额的工资
underpayment 付款不足
underpeopled 人口稀少的
underpick 下投梭
underpickling 欠酸洗;酸洗不够
under-pier 码头下的
underpin 加固基础;从下面支撑;筑基础;由下向上做支撑;由下向上支护井筒

underpinned pier 托换座墩
underpinned pile 托换桩
underpinner 支撑木
underpinning 基础托换(法);基础加固托换法;托(换柱)基;托换基础(法);托换(技术);上向连接井壁补
underpinning by vibroflo(a)tation 振冲法托换
underpinning engineering 托换工程
underpinning foundation 托换基础
underpinning in economic theory 经济理论基础
underpinning post 托换支柱(隧道坑道);隧道托柱;托换柱(隧道)
underpinning technology 托换技术
underpinning to column foundation 托换柱基
underpinning wall 加强基墙
underpinning with cylinders 筒井从下加固基础;用井筒托换基础
underpinning work 托换基础工程;支撑工作
underpitch groin 威尔士拱;交叉穹梭;交插穹梭
underpitch vault 高低交叉穹顶;交插穹顶;隧道拱顶
underpit operation 井下作业
underplant 下木;树下矮灌木
underplanted 设置不全的;栽植于下的
underplaster 暗装混合龙头
underplate 底盘;底座;基础;底板;垫板
under platform 下层站台
underpopulation 人口稀少;人口稀少区
underpopulation area 人口稀少地区;人口过稀区
underpopulation front 人口稀少区边缘
underpopulation zone 人口稀少地区;人口过稀区
underport 底孔
under port burner 小炉下口燃烧器
underpour type gate 下泄式闸门;下射式闸门;下冲式闸门;底泄式闸门
under power 低功率
under-powered 功率不够的
under power protection 低功率保护装置
underpower protection device 低功率保护装置
underpower relay 功率不足继电器
underprediction 预测偏低
underpressing 模锻不足
underpressure 受压下;低于大气压力;底部压力;稀薄,压力不足;真空计压力;降压;欠压;低压(力);抽空
under-pressure connection 有压接户;带气连接
under-prestressed 处于预应力状态的;预加应力的
under-prestressed beam 低预应力
under price 定低价格
underpriming 注油不足;浇注不足;起动注油不足
underproduce 生产供不应求;产量不足
underproduction 减产;生产不足;产量不足
underproof 不合标准的;低于标准的;不合格的;标准强度以下的;被试验的
underprop 用立柱加固;顶撑;撑柱;撑住
under proper moisture conditions 在适当的潮湿情况下
under protest 不报
underpunch 下部穿孔;低位穿孔
under-purlin(e) insulation 檩下保温层
under-purlin(e) lining 檩下衬;檩下内衬
underquenching 淬火不足
underquote 低报价格
underrange 次范围;不欠值;不定值
underrate 混等率;估计过低;低估
under reach 缩短动作
underreaching protection 继电器失灵保护装置
under-reamed 扩孔的
under-reamed bored pile 扩底钻孔桩
under-reamed foundation 底部扩大式基础;扩底基础
under-reamed pier 扩底墩
under-reamed pile 扩底桩
under-reamer 孔底扩大器;井下扩眼钻头
under-reamer bit 扩底钻头
underream(ing) 套管靴下扩孔;扩孔,井下开眼;孔内扩孔;扩大不足的钻孔
under-reaming bit 下向扩孔钻头;扩孔钻头
underrefining 精炼不足;欠精炼
under registration 记录不足;漏登
under-reinforced 加筋不足的;补强不足的;配筋不足的;低配筋的
under-reinforced beam 适筋梁低配筋梁
under-reinforced bonded beam 弱配筋梁
under-reinforced column 低配筋柱

under-reinforced concrete 配筋不足的混凝土
under-reinforced concrete structure 少筋混凝土结构
under-reinforced design 低筋设计
under-reinforced ratio 低配筋比(率)
under-reinforced section 配筋不足断面
under repair 修理中;在修理中
under-reporting 漏报
under reserves 保留追索权;保留追偿权
under review 审阅中
under-ridge gable tile 平脊瓦
under-ridge tile 脊(底)瓦;平脊瓦;脊下瓦
under-river shield 江底盾构
under-river tunnel 江底隧道
underroasting 焙烧不足
under-roll 欠压
underroof 下木
underrun 缺额材积;欠载运行;潜流;底面通过;底流;低于估计的产量;负载运行
underrun error 欠载运行出错
under runner 下磨盘转动的磨粉机
under-running stacker crane 仓库码垛机;下通式塔式起重机
underrunning third rail 底接触第三轨
underrusting 正在生锈;底锈;膜下锈蚀
under sail 驶帆
under sail and steam 机帆并用
undersample 采样不足
undersampling 采样过疏
undersanded 含砂过少的;含砂量不足的;掺砂太少的
undersanded concrete 含砂过少的混凝土;贫砂混凝土
undersanded mix(ture) 贫砂拌和料;少砂混合料;少砂拌和物
undersanding 含砂不足(混凝土)
undersand lake 沙下湖
undersaturated 欠饱和的;不饱和的
undersaturated basaltic glass 不饱和的玄武质玻璃
undersaturated condition 不饱和状态
undersaturated developer 不饱和显影剂
undersaturated exciter 欠饱和励磁机
undersaturated fluid 欠饱和流体
undersaturated liquid 欠饱和液体
undersaturated magma 不饱和岩浆
undersaturated oil pool 不饱和油藏
undersaturated rock 不饱和岩
undersaturated syenite aplite 不饱和正长细晶岩
undersaturated syenite pegmatite 不饱和正长伟晶岩
undersaturated with gas 气体欠饱和的;气体不饱和的
undersaturation 欠饱和(现象);不饱和;未饱和
underscouring 下冲
under screen push bar 纱窗下推杆
underscreen water 筛下水
underscribe 雕合(接头);合缝接头
underscrub 下层密灌丛
undersea 在海中(的);在海下面进行的
undersea boat 潜水艇
undersea cable 海底电缆
undersea coal field 海底煤田
undersea communication and detection system 水下通信和探测系统
undersea delta 潜三角洲;水下三角洲
undersea detection 海下探测
undersea device 海底设备;潜水装置;水下设备
undersea exploration 海底勘探
undersea exploration ship 水下探测船
undersea fan 海底扇
undersea feature 海底地形
undersea fiber-optic cable 海底光缆
undersea fiber-optic connector 海底光纤连接器
undersea habitat 水中生境
undersea jetty 潜堤
underseal 封闭腻子;基层处治;底封;防蚀涂层
undersea laser radar 水下激光雷达
undersea light wave cable 海底光缆
undersealing 底封;密封不足;密封不严
undersealing asphalt 填充混凝土地面板洞的沥青;封底沥青;密封沥青
underseal work 底封;基层处治
underseam 深部煤层;底部煤层
undersea mine 海底矿山

undersea mineral resources 海底矿产资源
undersea mining 海底采矿;水下开采
undersea mining technology 海底采矿技术
undersea oil production plant 海底集油装置
undersea pipeline 海底管路;水下管路;海底管道
undersea prospecting 海底勘探
undersea ranging 海下测距;水下测距
undersea rescue vehicle 潜水艇救艇
undersea ship 水下舰艇
undersea simulation system 海下模拟系统;水下模拟系统
undersea sonar mapping 海底声呐测图
undersea surveillance 水下监视
undersea table-top mount 海底桌状山;海底平顶山
undersea technology 海底技术
undersea television 水下电视
under seat heater 座下温暖器
undersea tunnel 海底隧道
undersea tunnel survey 海底隧道测量
under seeded method 深播法
under seeding 深播
underseepage 下方渗流;地下渗漏;地下渗流;底部渗流
undersell 抛售物资;抛售货物
underselling 廉价出售
under-sensitive 灵敏不足
underserrated 底面刻齿的
underserrated section 底刃面刻齿的动刀片
underserviced 公共设施不足
underserviced city 公共设施不足的城市
underset 下层逆流
undershaft benchmark 井下水准点
undershaft level(l)ing 井下水准测量
undershaft trigonometric level(l)ing 井下三角高程测量
under shed 下开口
undershelve 底柜
under shield 下部挡板;挡泥板
undershoot 下冲;欠调量;欠程;低插;负脉冲(信号);负尖锋
undershoot distortion 下冲失真
undershooting 落地过低(飞机)
undershoot landing 落地过低(飞机)
undershot 下冲的;下射的
undershot elevator 底部输送手运器
undershot gate 下泄式闸门;下射式闸门;下冲式闸门
undershot head gate 下泄式渠首闸门;下射式引水闸门
undershot turbine 下冲式水轮机
undershot-type wasteway 下冲式废水渠
undershot-type waterway 下冲式废水道
undershot water wheel 下射式水轮(机);下击式水轮;下冲水轮;下射式水轮(机);底射式水轮
undershot wheel 下击式水轮;下冲式水轮;底射式水轮;低水位水轮
undershrub 小灌木;矮灌木
underside 隐蔽面;下侧;底面;车身底面;车身底板;反面;下面
underside end (给进液压缸的)下端
underside groove 背面坡口
underside of arch 拱的内弧面;拱的内侧
underside of piston crown 活塞顶内侧
underside of the piston head 活塞头下部
underside of vault 拱顶内侧(面)
underside view 底部图
undersieves 筛下料
undersignated area 未计划用地
undersigned 下列签字人
undersigned area 未计划用地;待定用地
under simulated commercial conditions 在模拟的商品生产条件下
undersink 洗气池下面
undersink cold tap unit 洗气池下冷水龙头
undersink system 厨房水盆下系统
undersize 过筛物料;过筛量;逊径;过小;尺寸过小;尺寸不足;负公差尺寸
undersize aggregate 逊径集料
undersize core 小于标准直径的岩芯
undersized 欠浆的;欠胶的;筛下的;尺寸过小的;尺寸不足的
undersized bearing 小尺寸
undersized sample 容量不足的样本
undersize in the oversize 筛上物中的筛底料

undersize material 筛底料;尺寸不足的物料
undersize of aggregate 粒径不足集料;粒径不足骨料
undersize (of hole) (孔的)缩径
undersize part 小尺寸的零件
undersize particle 过筛小颗粒料;过细颗粒
undersize percentage 筛下料百分率
undersize percentage of lump coal 块煤下限率
undersize recovery 筛下产品回收率
undersize section 尺寸过小断面
undersize sieve 细筛;草籽筛
underslating felt 屋顶瓦下油毡;油毡衬层
under sluice 底部泄水孔;底部闸门;冲砂闸;下泄式水闸;底部泄水闸;底部泄水道
under-sluice canal 底部泄水渠
under-sluice channel 底部泄水渠
undersluice section 底部泄水渠
under-sluice tunnel 底部泄水隧洞
underslung 吊起的;吊着的;下悬式;悬挂的
underslung alternate plow 架下悬挂键式犁
underslung bridge crane 悬挂式桥式起重机
underslung car frame 下悬式车架;悬挂式电梯缆绳架
underslung charger 吊车式加料机
underslung charging crane 悬挂式加料起重机;悬臂式加料吊车
underslung chassis 下置车架底盘
underslung conduit 悬吊管
underslung conveyer 悬挂式输送机;悬挂式运输机
underslung crane 悬挂式起重机
underslung hoist 下悬式起重机
underslung jib 悬吊截盘
underslung monorail system 悬挂单轨系统
underslung pipe 悬吊管
underslung radiator 下悬式散热器
underslung rocket unit 机身下的火箭发动机
underslung spring (位于车轴下的)悬挂式钢板弹簧
underslung suspension 悬挂式安装法
underslung track 悬挂轨道
underslung trolley 悬挂小车
underslung tube 悬吊管
underslung tubing 悬吊管
underslung type 下吊式
underslung water tank 车下水箱
underslung worm 下置蜗杆
underslung worm drive 下置式蜗杆传动
undersoil 亚土层;下层土;底(层)土
undersoil water 底土水
undersonic 次声的
undersonic frequency 次声频
under spall 下剥
under specified conditions 在特定条件下
underspeed 降低速度;速度不足
undersquare engine 长行程发动机
understable 欠稳定的
understaffed 人员不足的;人手不够的
understain 浅染
understanding 掺砂太少
understatement 少报
under steam 用蒸汽机航行
understeer 转向不足;转弯不足(汽车);对驾驶盘反应迟钝
understeering 不足转向(车辆)
understock 存货不足
under-stocking 不足立木度
under-stoke 自下给(燃)料
under-stoker 自下上煤机
understo(re)y 下层(矮生)植物;下木;下层(林)木
under strata formation 下层地层构造
understratum 下层地层;下(部地)层;地层的下层
under-stream period 开工期
under-strength 强度不够的
understress 在受力时;在应力状态下;承受应力;应力不足
understressed 应力不足的;加压不足
understressed concrete 应力不足的混凝土
understressing 低应力锻炼法;应力不足
understretched 处于张拉中的
understructure 下层结构;下层构造;下部结构;下部建筑;底座结构
under study 研究中
under-study plan 接替计划
under suction state 按吸入状态
undersulfated cement 低硫水泥

under-sulfonate 磺化不足
under superelevation 欠超高
under-supply 供应不足
undersurface 下面;底面
undersurface collection 底层表面集电
under-surface filling 液面下灌浆
under-surface indicator 液面下指示器
underswing 负脉冲(信号)
under symbiotic conditions 共生条件下
undersynchronous 低于同步的
undersynchronous braking 低于同步制动
undertake 承揽;承诺;承担;承包
undertake financial responsibility 承担经济责任
undertaker 营业者;企业主;承包者;承包商;承包人;承办人
undertake responsibility 承担责任
undertake transfer of personnel 人员流动
undertake treatment within a prescribed limit of time 限期治理
undertaking 计划;承包;事业;承揽
undertaking invenue 事业收入
undertaking syndicate 企业联合组织
undertamping 没有夯实;夯实不足
under taping 反贴边
undertenant 转租承租人
undertension 降低电压;欠电压;电压不足
undertensioned 处于拉伸中
under test 在试验中;试验中
under the auspices of 由……主办
under-the-cover modem 集成调制解调器;分立式调制解调器
under the erosion(al) basis 侵蚀基准面以下
under the influence natural forces 在自力的影响下
under the lee 在下风中
under the proper condition 在适当条件下
under these circumstances 在这种情况下
under the wind 下沉风;在下风的
under-thickness tolerance 厚度负公差
underthroating 檐板下作成滴水槽;檐板滴水凹圆线
underthrow distortion 欠幅失真
underthrust 下冲断层
underthrust earthquake 俯冲断层地震
underthrust fault 俯冲断层
underthrust plate 下冲板块;俯冲板块
undertighten 紧密不足;拉得不紧;拧紧不够;绑得不紧
under-tile 底瓦;垫瓦
undertint 褪色;淡色;柔和色调;浅色
under-tipped face 坑道斜入口面
undertone 淡色;浅色;潜在倾向;底彩;低音;低调
undertone tint 底彩
undertonguing 丝扣欠紧
undertow 底流;下层逆流;离岸流;回卷;退浪;沿坡降下;回卷流;底溜
undertow current 回卷(底)流;底层逆流;底层反向流
undertrack hopper 轨下料箱
undertrack screw conveyer 轨下螺旋输送机
undertrading 利用资产过低
under treatment 不足的处理;处理不足
under-tree sprinkling 树下喷水
under-trowelling 抹光不足
under upper deck tonnage 上甲板下吨位
underutilization 利用不足;未充分利用
underutilization manpower 未充分利用的人力
underutilized land 未充分利用的(土)地
underutilized lane 低利用车道
undervaluation 低估价(的);估计过低;估计不足;低估价值;评价过低
undervaluation duty 低估税
undervalue 低估价格;评价过低
undervalued 定值低的
under-veneer 带贴面板的毛地板
undervibration 振捣不足(混凝土);振捣不够;振动不足;振荡不足
under-voltage 欠(电)压;电压不足
under-voltage circuit breaker 欠压断路器;欠压电路器;低电压断路器
under-voltage condition 欠压状态
under-voltage detector 欠压检测器
under-voltage no-close release 低电压不闭合释放装置;欠压无闭合断路器
under-voltage open phase relay 低电压开相继电器
under-voltage protection 欠压保护;低电压保护

under-voltage relay 欠压继电器；失压继电器；低电压继电器
under-voltage release 欠压脱扣器；欠压释放；欠压断路器；低电压跳闸；低电压释放
under-voltage release unit 低电压释放装置
under-voltage reverse-phase relay 低电压反相继电器
under voltage supervision 低压监测
under-voltage tripping 低电压跳闸；低电压释放
under-voltage warning device 电压不足信号器
underwall 底帮
underwash(ing) 河床冲刷；河底涮刷；淘刷
underwate formwork 水下立模
underwater 在水下的；水线以下；水下（的）
underwater acoustic 水声学；水下声
underwater acoustic(al) communication set 水声通讯设备
underwater acoustic(al) communication system 水下音响通信系统
underwater acoustic(al) measurement 水下测量；水声学测量
underwater acoustic(al) system 水声系统
underwater acoustic(al) telemeter 水声遥测器
underwater acoustic(al) test facility 水声测试设施
underwater acoustic(al) transducer 水声换能器
underwater acoustic(al) wave 水下声波
underwater acoustic(al) waveguide 水下声道
underwater acoustic(al) communication apparatus 水声通信机
underwater acoustic(al) communication set 水声通信机
underwater adhesive 水下黏结剂
underwater aerator 水下曝气器
underwater air shield arc welding 水下气体保护电焊
underwater ambient noise 水下环境噪声
underwater antenna 水下天线
underwater antifouling coat 水下防污漆
underwater appendages 水下附体
underwater apron 水下护坦；水下护脚
underwater arc cutting 水下电割
underwater arc cutting welding torch 水下焊割两用焊枪
underwater arc welding 水下电(弧)焊(接)
underwater bank 水下洲滩；水下浅滩
underwater bank protection 河岸水下保护
underwater bar 水下沙坝
underwater barometer 水下压力表
underwater battery plotting room 水下武器指挥室
underwater bed level(l)ing 水下机床整平
underwater bed tamping 水下机床夯实
underwater bell 水中雾钟
underwater blasting 水下爆炸；水下爆破；水下爆扩；水底爆炸
underwater blasting cutting 水下爆破切割
underwater blasting gelatin(e) 水下爆破胶质炸药
underwater blasting hole 水下爆破孔
underwater blasting research 水下爆炸研究
underwater blast injury 水下爆炸伤
underwater blast wave 水中冲击波
underwater block erection 水下块体安装
underwater block setting 水下方块安装；安装水下方块
under water body 船体水下部分；水下船体
underwater bore-hole blasting 水下钻孔爆破
underwater boring vessel 水下钻探船
underwater breakwater 水下防波堤
underwater breathing apparatus 潜水呼吸器；水下呼吸装置；水下呼吸器
underwater breathing organism 水下呼吸生物
underwater bucket 水下抓捞斗
underwater (bull)dozer 水下挖泥机；水下推土机
underwater buoyant bridge 潜水浮桥；水(面)下浮桥
underwater burst 水下爆炸
underwater cable 海底送电线；水下电缆；水底电缆
underwater camera 水下照相机；水下摄影机
under water camera observation 海底摄影观测
underwater cartridge chopper 水下元件切断机
underwater cleaning 水下除污
underwater coat(ing) 水下油漆
under water coating compound 水下施工涂料；湿表面施工涂料
underwater communication 水下通信
underwater communications system 水下通信系统
underwater concrete 水下混凝土
underwater concrete mix 水下混凝土(配合)混合料；水下混凝土配合比
underwater concreting 水下浇筑混凝土；水下浇注混凝土；水下浇灌混凝土；水下混凝土浇筑；水下灌筑混凝土；水下灌注混凝土
underwater concreting box 水下混凝土浇注箱
underwater connection 水下连接
under water construction 水下混凝土施工；混凝土水下施工
underwater consumable-electrode water-jet cutting 水下熔化极水喷射切割
underwater craft 潜水船；水下舰艇
underwater crane 水下升降机；水下吊机
underwater cross-section survey 水下横断面测量
underwater current 潜流；水中暗流
underwater cutter 水下割草机
underwater cut(ting) 水中切割；水下切割(氢氧焰)
underwater cutting blowpipe 水下(切割)割炬
underwater cutting equipment 水下切割设备
underwater damage 水下部分破损；船体水下部分损伤
underwater danger 水下障碍物
underwater debris clearing 水下清渣
underwater defect detecting 水下探伤
underwater defence 水下防御
underwater demolition 水下拆除
underwater demolition team 水下爆破队
underwater deposit 水下沉积(物)
underwater desilting 水下清淤
underwater desilting explosion 水下排淤爆破
underwater detection 水下探测
underwater detection and classification system 水下探测与分析系统
underwater detection equipment 水下探测设备
underwater detector 水下雷管；水下引爆器
underwater diffuser 水下扩散器
underwater digging 水下挖掘
underwater dike 潜堤；潜坝
underwater discharge pipe 水下排泥管
underwater discharge pipeline 水下排泥管线
underwater dismantlement 水下拆除
underwater display 水下显示
underwater dobie blasting 水下裸露爆破
underwater Doppler navigation system 水下多普勒导航系统
underwater drill(ing) 水下钻探；水下钻井；水下钻进
underwater drilling and blasting ship 炸礁船；钻爆(炸礁)船；水下钻孔炸礁船
underwater drilling vessel 水下钻探船
underwater dune 水下沙丘
underwater eagre 海底涌潮
underwater earthquake 水下地震
underwater electric(al) pump 浸没式电泵
underwater electrode 水下焊条
underwater emission 水下发射
underwater engineering 水下工程
underwater environment 水下环境
underwater excavation 水下挖土；水下挖掘；水下挖方；水下开挖(工程)
underwater excavator 水下挖掘机
underwater execution 水下施工
underwater exhaust 水下排气
underwater explosion 水下爆炸；水下爆破
underwater explosion cutting 水下爆破切割
underwater explosion injury 水下爆炸伤
underwater explosion-making hole 水下爆破孔
underwater explosion research 水下爆炸研究
underwater fairleader 水下导缆器
underwater fast-hardening concrete 水下快干混凝土
underwater fault detection 水下探伤
underwater fill 水下填筑；水下填土；水下填方
underwater fire-cracker welding 水下鞭炮焊
underwater firing 水下发射
underwater fittings 船体水线下装置
underwater flexible pipe 水下挠性管
under water floating launching way 水下浮动滑道(船)
underwater foundation 水下基础
underwater foundation bed level(l)ing 水下基床整平
underwater foundation-bed tamping 水下基床夯实
underwater foundation trench excavation 水下基槽开挖
underwater garage 水下车库
underwater gradient 海底坡度；水下坡度
underwater habitat 潜水员水下居住舱；水下住所；水下栖所；水下居住舱
underwater highway 水底公路
underwater holography 水下全息摄影
underwater hull 水下部分船体；船体的水下部分
underwater hull TV camera 水下船身电视摄像机
underwater hydraulic excavator 水下液压挖掘机
underwater hydraulic-operated cutter drive gear 水下液压铰刀驱动装置
underwater ice 水下冰；水内冰
underwater illumination 水中照明；水下照明；水下照(明)度
underwater illumination intensity 水中照明度
underwater individual coat 水下分段油漆
underwater inspection 水下检查
underwater instrument 水下仪器
underwater kit 潜水装具；水下装备
underwater lamp 水下灯
underwater landslide 水下滑坡
underwater laser 水下激光器
underwater laser device 水下激光装置
underwater laser radar 水下激光雷达
underwater laser surveying system 水下激光测量系统
underwater laser television 水下激光电视
underwater light 水下照明灯
underwater lighting 水下照明
underwater line 水下线
underwater listening device 水声设备
underwater listening post 水下监听站
underwater longitudinall-section survey 水下纵断面测量
underwater loudspeaker 水下扬声器
underwater maintenance 水下保养
underwater manipulator 水下机械手
underwater mechanic 水下机械军士
underwater microphone 水下传声器
underwater mining ship 水下采矿船
underwater missile firing ship 水下导弹发射舰
underwater missile test evaluation center 水下导弹试验与鉴定中心
underwater navigation 水下航行；水下导航
underwater nephelometer 水下浊度计
underwater noise 水下噪声
underwater nuclear burst 水下核爆炸
underwater object detecting set 水下目标探测器
underwater observation 水下观测；水下观察
underwater obstacle 水下障碍物
underwater oil search ship 水下探油船
underwater oil tank 水下储[贮]油库；水下储[贮]油罐
underwater operation 水下作业
underwater operation group 水下作业大队
underwater optic(al) fiber cable 水下光缆
underwater ordnance locator 水下兵器定位器
underwater ordnance research boat 水下兵器研究试验艇
underwater outfall 水下出口
underwater oxygen decompression 水下吸氧减压法
underwater paint 防水漆；水下施工涂料；水工漆
underwater parking building 水下停车库
underwater patching 水下修补
underwater pedestrian tunnel 水底人行地道
underwater pelletizing 水下造粒
underwater performance 水下性能
underwater petroleum seep sensor 水下油渗推测器
underwater photogrammetry 水下摄影测量术；水下摄影测量(学)
underwater photograph 海底照片
underwater photographic(al) inspection 水下摄影检查
underwater photography 海底照相；水下摄影(术)
underwater physiology 水下生理学

underwater picture 水下图像
underwater pile driving 水下打桩
underwater pipe ditch 水下管沟
underwater pipe ditch excavation 水下管沟开挖
underwater pipeline 水下管线;水下管路;水下管道
underwater plasma welding process 水下等离子体焊接法
underwater platform 水下平台
underwater plug 海底岩颈;水下岩颈
underwater position fixing 水下电氧切割
underwater positioning 水下定位;水流定位
underwater positioning apparatus 水下定位装置
underwater positioning device 水下定位装置
underwater positioning tool 水下定位装置
underwater positioning unit 水下定位装置
underwater projectile 水下投射体
underwater propulsion noise 水下推进器噪声
underwater protection 水下防护
underwater pump 水老鼠;潜水泵
underwater radiac set 水下放射性检测仪
underwater range finding 水下测距
underwater ranging 水下音测
underwater ranging battery 水下音测队
underwater reef blasting 水下炸礁
underwater reserves 水下储量
underwater revetment 水下护坡;水下挡墙
underwater ridge 潜埂
underwater rigging 水下索具
underwater robot 水下自动装置;水下遥控继续装置;水下机器人
underwater rock drill 水下凿岩机
underwater rockfill 水下抛填
underwater rockfill dam 水下堆石坝;水下填石坝
underwater rock pinnacle 水下尖礁
underwater salvage 水下打捞
underwater sample 水下土样;水下试样
underwater sand ridge 水下沙脊
underwater sawing 水下锯割
underwater scattering meter 水下散射计
underwater schlieren 水下纹影
underwater scraper 水下刮土机
underwater screen 水下筛;水中滤网
underwater seal 水下封税;水下封层
underwater search equipment 水下搜索设备
underwater security advance warning 水下安全预警
underwater sediment 水下沉积(物)
underwater seismics 水下地震勘探
underwater self-homing device 水下导引头
underwater sensors 水下传感器
underwater servicing device 水下辅助装置
underwater servicing unit 水下辅助装置
underwater setting coating 水下固化涂料
underwater ship 潜艇;潜水船
underwater shooting 水下爆炸
underwater sight-seeing boat 观光潜水器
underwater signal(l)ing 水下(声)信号
underwater sill 潜槛
underwater slipways-end 水下滑道
underwater slope 水底坡度;水下边坡
underwater soil 水底土壤
underwater solifluction 水下泥流
underwater sonic communication gear 水下电话机
underwater sonic gear 水下水声装置
underwater sound 水中声;水声
underwater sound channel 水下声道
underwater sound communication set 水下电话机
underwater sound communication system 水声通信系统
underwater sound device 水下音响装置
underwater sound equipment 水声设备
underwater sound gear 回声测深仪;水下回声测深仪;水声设备
underwater sound integrated measuring set 水声综合测量仪
underwater sound laboratory 水声研究室
underwater sound measurement 水声测量
underwater sound-pressure microphone 水下声压式传感器
underwater sound projector 下声发射器;水下声发射器;水声发射器
underwater sound system 水声系统
underwater sound telephone 水下音响电话

underwater sound transducer 水下水声换能器;水声换能器
underwater sound wave 水下声波
underwater sport 潜水运动
underwater spring 海底泉
underwater squeezing explosion 水下挤淤爆破
underwater stage decompression 水下阶地减压法
underwater stereoscopic television 水下立体电视
underwater stop valve 水下断流阀
underwater storage 水下储[贮]存
underwater storage tank 水下储罐
underwater structure 海底构造;水下结构;水下构筑物
underwater surveillance 水下监视
underwater survey 水下勘测;水下测量
underwater survey work 水下测量作业
underwater tank 水下储[贮]罐
underwater target identification and delay system 水下目标识别与延时系统
underwater target simulator 水下目标模拟器
underwater task light 水下作业灯
underwater telemetry 水下遥测(术)
underwater telephone 水下电话
underwater television 水下电视
underwater television camera 水下电视摄像机
under water television observation 水下电视观测
underwater test 水下试验
underwater test range 水下试验靶场
underwater topographic(al) map 水下地形图
underwater topographic(al) survey 水下地形测量
underwater topography 海底地形;水下地形
underwater torch 水下焊炬
underwater-to-ship 潜对舰
underwater towed craft 水下拖曳器
underwater tow vehicle 潜水拖艇;水下拖曳器
underwater tracking system 水下跟踪系统
underwater transducer 水下换能器;水声换能器
underwater transducer array 水声换能器基阵
underwater transparency meter 水下透明(度)计
underwater trench 水下管沟
underwater trench excavation 水下管沟开挖
underwater tube 水下沉管
underwater tunnel 水底隧道;海底隧道;海底坑道;水下隧洞;水下隧道
underwater TV camera 水下电视摄像机
underwater untethered submersible 无系缆式潜水器
underwater vegetation 水生植物;水生植被
underwater vehicle 海底作业车
underwater viewer 水下窥视器
underwater visibility 水下能见度
underwater vision 水下视觉
underwater wall 水下墙
underwater wave 水下波
underwater weapons station 水下武器站
underwater weight of a wreck 沉船水中重量
underwater weight of residual cargo 余货水中重量
underwater weight of residual mud 余泥水中重量
underwater welding 水中焊接;水下焊(接)
underwater well 水下(油)井
underwater window (游泳训练或电视装置用的)水下窗
underwater works 水下构筑物;水下工程
underwater zero 水下爆炸中心
underway 下穿道;在航;水下穿道;水底通道;非系泊船舶;正在进行中的
underway bottom sampler 走航底质取样器
underway lamp 航行灯
underway replenishment ship 航行补给船
underway transfer rigging 海上补给传送索具
under-wear 潜水内衣
under weigher hopper 秤下漏斗
underweight 重量不足(的);低于额定重量;分量不足
underweight bit 金刚石镶嵌不足的钻头
underwind 下卷式
underwinding 下卷式卷绕
under-window spandrel (上下层窗槛之间的)窗台墙;(上下层窗槛之间的)矩形墙
under-window spandrel panel (上下层窗槛之间的)窗台板;(上下层窗槛之间的)矩形板
underwing 翼下
underwood 矮树丛;下层林丛;下(层)木

Underwood chart 安特乌得图表
Underwood distillation method 安特乌得蒸馏计算法
underwood growth 下层林丛
underwood planting 林下种植
underwork 偷工;省工;根基;下部结构;下部建筑;支持结构;附属性工作;支承结构
underworker 辅助工人
underworkings 地下巷道;井下巷道
underwrite 包销
underwrite insurance for 办理保险
underwriter 证券承销商;保险业者;保险商;保险人;保险公司
underwrite risks 立约保险
underwriter's guarantee 承包者保证书;承包商保证书;承包人保证书;保险人保证书
underwriter year 保险业务年度
underwrite transportation insurance 经营运输保险业务
underwriting 保险业;包销
underwriting account 保险账目;保险账户
underwriting agreement 包销协议书;包销协议
underwriting commission 包销佣金
underwriting fee 包销费用
underwriting limit 承保限额
underwriting pool 包销联营
underwriting profit 承保收益
under-yielding 产量低的
undesignated area 未计划用地;待定用地
undesigned 没有计划的;设计的安全系数不足的
undesirable constituents 不良成分
undesirable deformation 不合要求的变形
undesirable geologic(al) phenomenon 不良地质现象
undesirable material 不合格材料
undesirable tendency 不良倾向
undesired carry 非理想进位
undesired frequency 寄生频率
undesired plant 讨厌的植物
undesired signal 干扰信号
undesired sound 噪声;噪音
undesulfured 未脱硫的
undetectable 不能探测的
undetected error 漏检错误;未查出的误差;未被检出的误差
undetected error rate 漏检故障率;漏检错误率;未发现误差率
undetected failure 未发现的故障
undetection 未检出
undeterminable 未定的;不可测定的
undeterminable loss 无法测定的损失;不可测定的损失
undeterminate 未测定的
undeterminate boundary 未定界限
undeterminate coefficient 未定系数
undeterminate coefficient of equation 方程待定系数
undetermination 待定
undetermined 未(确)定的;待定的
undetermined boundary 未定国界
undetermined coefficient 待定系数
undetermined function 不定函数;待定函数
undetermined multiplier 未定乘数;待定因子
undetermined parameter value 待求参数值
undetermined value 未定值;非已知数
undeveloped 未开发的;不发育的
undeveloped area 未开发区(域);处女地
undeveloped country 未开发国家
undeveloped head 未开发水头
undeveloped land 没有建筑物的荒地;没有建筑物的空地;未开发地;未开发的地区
undeveloped resources 未开发资源
undeveloped shower 未展开簇射
undeveloped water power 未利用水能;未开发(的)水能
undeveloped waters 未开发水域
UN Development From Business Edition 联合国发展论坛商业版
undevelopment surface 不可展面
undewriter 包销人
undible fish 非食用鱼
undifferentiated 未分化的;分化不良型的
undifferentiated igneous rock 未分异火成岩
undifferentiated marketing 无差异行销

undifferentiated race 未分化型族
undifferentiated type 未分化型
undifferentiation 未分化
undiffracted light 非衍射光
undigested 未被市场吸收的
undigested residue 未消化残余物
undigested sludge 未消化污泥；生污泥
undigested stool 不消化粪
undiked reach 无坝河段
undiluted 未稀释的
undiluted effluence 未稀释废水
undiluted engine oil 未稀释的发动机油
undiluted epoxy resin 未稀释的环氧树脂
undiluted wastes 未稀释废水
undiluted wastewater 未稀释废水
undiminished 等幅的
undirected graph 无向图
undirected tree 双向树；不定向树
undirectional approach 双向逼近法；不定向逼近法
undirectional flow 不定向流
undirectional motion 不定向运动
undischargeable claim 非可解除债务
undischarged 未卸下的
undischarged bankrupt 未解除债务之破产人
undischarged cuttings 未排出的岩粉
undisclosed 未透露的
undisclosed agency 不公开的代理
undisclosed factoring 不公开的代理融通；不分开的代理融通
undisclosed partner 秘密合伙人
undisclosed principal 不公开的委托人
undiscovered 未(被)发现的
undiscovered loss 未发现的损失；未发觉的损失
undiscovered reserves 未发现储量
undiscovered reservoir 未探明的热储
undiscovered resources 未经发现资源
undiscovery resources 未发现资源量
undisplaced joint 未变位的接头
undissociated molecule 未解离分子
undissolved 不溶(解)的
undissolved matter 非溶解物质
undissolved residue mineral 不溶残余矿物
undissolved solids 不溶解固体
undissolved substance 不溶物(质)
undistillableheavy residue 不能蒸馏的重残油
undistilled 未蒸馏的
undistinguishable 无特征的；不能区分的
undistored power 不失真功率
undistorted 无失真的；不失真的
undistorted ground 未经挖动的断面；未经挖动的土地
undistorted hologram 无畸变全息图
undistorted image 无失真图像；不失真图像
undistorted image formation 无畸变成像
undistorted linear scale 正态线性比(例)尺
undistorted model 正态模型；无扭曲立体模型；无畸变模型
undistorted picture 无失真图像
undistorted power output 不失真功率输出
undistorted ship model 正态船模
undistorted signal 不失真信号
undistorted stereomodel 无畸变立体模型
undistorted stereopair 无畸变立体像对
undistorted transmission 无失真传输
undistorted view 无畸变视图
undistorted viewing 无畸变观察
undistorted wave 无失真波；无畸变波
undistributable reserves 不可分配准备金
undistributed profit 未分配盈利；未分配利润
undistributed profits audit 未分配利润审计
undisturbance 平静
undisturbed 未经扰的；原状的；未受扰的；未扰动的
undisturbed ages 未扰动年龄
undisturbed bedrock 原状基岩
undisturbed clay 原状黏土；未扰动黏土
undisturbed core 原状岩芯
undisturbed disk-shaped clay 原状蝶形黏土
undisturbed disk-shaped clay sample 未扰动蝶形黏土样
undisturbed explosive 未分解炸药；未变质炸药
undisturbed field 未受扰动的地热田
undisturbed flow 稳流；未乱流；未扰动流
undisturbed forest 未受扰林

undisturbed foundation 未扰动地基
undisturbed ground 原生土(壤)
undisturbed homogeneous sample 未扰动均质土样
undisturbed jet 无扰动射流；未扰动射流
undisturbed line-of-sight system 无干扰瞄准系统
undisturbed model 正态模型
undisturbed motion 无扰运动
undisturbed Newtonian viscosity 未扰动(的牛顿)黏度
undisturbed-one output 未扰动1输出
undisturbed orbit 无摄轨道
undisturbed plasma 未受扰动等离子体
undisturbed readout 不干扰读出
undisturbed rock 原状基岩
undisturbed sample 原状样品；原状土样；原状土；原状标本；无扰动试样；无扰动试样；不扰动土样
undisturbed sample boring 原状土样钻取
undisturbed sample of soil 原状土壤标本；未扰动土样
undisturbed sand 原状砂
undisturbed sand sample 未扰动砂样；原状砂样
undisturbed sand sampling 取未扰动砂样
undisturbed settling 未扰动沉降；原状沉淀；未扰动沉淀
undisturbed soil 原状土(样)；未扰动土(壤)；未搅动土；不扰动土
undisturbed soil curve 原状土曲线
undisturbed soil sample 原状土样；原状土样；无扰动试样；未搅动土样；不扰动土样
undisturbed soil sampler 原状取土器
undisturbed state 原状态；未扰动状态
undisturbed strata 未变动岩层
undisturbed stream 未扰动水流；未扰动流
undisturbed strength 原状强度
undisturbed structure 未扰动结构
undisturbed sun 非扰动太阳
undisturbed water 未扰动水
undisturbed water level 平静水面；无波浪水面；未扰动水(平)面
undisturbed-zero output 未扰动0输出
undisturbed-zero output signal 未干扰0输出信号
undivided carriageway 无分隔带的车行道
undivided highway 无分隔带公路；不分线公路
undivided ore 未分品级矿石
undivided profit 未分配的利润
undivided road 无左右行车线的道路；无分隔带道路；不分隔道路
undivided shaft 整杆；整柱
undivided share 不可分割的份儿
undivided surplus 未分配盈余
undivided two-way road 无分隔带的双向车道
undo 反向交易
undocking 出坞；离码头
undograph 自记测波仪
undoing cover or hedge 解除抛补或避险
undoped 无掺杂的
undoped cutback 不掺杂的稀释(沥青)
undoped flow system 无掺杂气流系统
undoped material 非掺杂材料
undoped single crystal 非掺杂单晶体
undor 广义旋量
undosed 未经照射的
undrained 无排水管路的；无排水设施的；不排水的
undrained analysis 不排水分析
undrained compression test 不排水压缩试验
undrained condition 不排水条件
undrained creep 不排水蠕变
undrained cycle load 不排水循环荷载
undrained lake 不排水湖；不流通湖
undrained load condition 不排水荷载条件
undrained load cycle 不排水荷载循环
undrained loading 不排水加载
undrained movement 不排水沉降
undrained quick shear test 不排水快剪试验
undrained settlement 瞬时沉降
undrained shear 不排水剪切；不排水剪力
undrained shear strength 不排水抗剪强度
undrained shear strength of soil 土的不排水剪切强度
undrained shear test 不排水抗剪试验；不排水剪切试验
undrained slope 未疏干边坡
undrained state 不排水状态

undrained test 不排水试验
undrained trench 无排水设施的路槽
undrained triaxial compression test 不排水三轴压缩试验
undrained triaxial test 不排水三轴试验
undrain settlement 瞬时沉降
undrawn tow 未拉伸丝束
undrawn yarn 未拉伸丝
undressed 未修整的；未经处理的；未加工的
undressed casting 未清理铸件
undressed kid 未加工的小山羊皮
undressed lumber 未抛光的木材；原木；毛材
undressed timber 未刨木料；原木
undressed weld 未加工焊缝
undressing 卸装
undressing cab(in) 化妆室
undressing cubicle 化妆室
undressing room 更衣室
undried 未干燥的
undried pellets 未干燥的小片
undrillable 不能钻的；不可钻的
undrilled 未开钻的
undrilled porved 未钻即可证实的储量
undrinkable water 非饮用水
un-drinking water quality standard 非饮用水质量标准
unduation grade 起伏坡度
undue 过分的；未到期的；不正当的
undue bill 未到期票据
undue debt 未到期债务
undue deflection 挠度过大
undue delay 不应有的延误；不应当的延迟
undue note 未到期票据
undue wear 过度磨损；过度磨耗
undulance 波动
undulant 波动的
undulant fever 波浪热
undulant temperature 波浪式温度
undulant wavelike 波浪形的
undular 波状的；波纹的；波浪形的
undular hydraulic jump 波形水跃
undular jump 不完全水跃；波状水跃；波动水跃
undular mo(u)lding 波纹线脚
undulate(d) 波状的；波浪形的；波动的
undulated beam injector 波动射束注入器；波荡射束注入器
undulated fiber 波浪形纤维
undulated ipecac 波纹吐根
undulated nappe 脉动水舌；起伏水舌；波动水舌
undulated parabolic cylinder 波状抛物型筒壳
undulated series winding 波绕组
undulated sheet iron 瓦垅薄钢板；瓦楞铁皮；波形铁皮；波纹铁皮；波纹铁
undulated type reducing valve 波纹管式减压阀
undulated vault 波纹状拱顶
undulate fold 波状褶皱
undulate folding 波状褶皱
undulating 波浪形；微起伏
undulating anticline 波状背斜
undulating area 丘陵区
undulating cable 波形钢索
undulating coast 高低不等的海岸；起伏海岸
undulating current 波动电流；波荡电流
undulating facade 有起伏的(建筑物)正面
undulating flow 有波纹的水流
undulating grade 起伏坡度；起伏(不定)坡度；不定坡度
undulating ground 丘陵地；起伏地；波状土地；波状平原
undulating land 起伏的土地；波状土地
undulating light 脉动光灯；强弱变光；定闪光
undulating line 波动线；波状线
undulating membrane 波纹膜；波形膜；波动膜
undulating mo(u)lding 波浪形饰的线脚；波形线条；波形线脚
undulating oceanographic(al) recorder 波动海洋记录仪
undulating orbit 波荡轨道
undulating pulse 波状脉
undulating quantity 波动值
undulating-ribbon structure 波浪形带状结构
undulating road 起伏道路
undulating surface 波纹状表面；波状表面
undulating terrain 丘陵地区；丘陵地段；丘陵地带

undulating topography 起伏地形；波状地形
undulating tracery 波浪形花格窗；曲线窗花格
undulating track 起伏线路
undulating wire 波纹截面钢
undulation 起伏运动；起伏不平度；地形起伏；波状起伏；波浪形磨耗；波浪形；波动；表面不平整度
undulation period 波动周期
undulator 波纹印报机；波纹器；波纹机；波动器；波荡器
undulatory 波形的；波动的
undulatory extinction 波状消光；波动消光
undulatory jump 波状水跃
undulatory motion 波状运动；波动；波荡运动
undulatory movement 波状运动；波形运动
undulatory theory 波动学说；波动理论
undulatus 浪云；波状云
unduly long-haul traffic 过远运输
unduly short-haul traffic 过近运输
undundled hardware 零星硬件；散装硬件
undundled service 零星服务
undurable goods 软商品；非耐用货物
undusted clinker 无粉尘熟料
undy mo(u)lding 波浪形饰的线脚；波形线条；波形线脚；波纹饰
undynamic(al) force rotary scraper 无动力回转刮刀
uneared premium 剩余保险费
unearned 非勤劳所得的
unearned advantage 非劳力利益
unearned burden 非营业费用
unearned dividend 分外股利；分外股息
unearned income 未得到的收入；分外收益；非营业收益；非营业收入；非劳动收入；非工资性收入
unearned increment 土地自然增价；非劳力增值
unearned profit 分外利润；非营业利润
unearned revenue 未得到的收入；分外收入
unearned surplus 分外盈余；非营业盈余
unearth 揭露；掘出；发掘
unearth cultural relics 发掘文物
unearthed 未接地的；采掘出的
unearthed relics 出土文物
unearthed voltage transformer 不接地电压互感器
uneconomic 浪费；不经济的
uneconomicalness 不经济
UN Economic Commission for Africa 联合国非洲经济委员会
UN Economic Commission for Latin American 联合国拉丁美洲经济委员会
uneconomic line 亏损线路
uneconomic site 不经济坝址
unedged 毛边的
unedged timber 毛边木料
unedited format 非编辑格式
unelastic 刚性的；非弹性的
unelastic behavio(u)r 非弹性状态
unelastic bending 非弹性弯曲
unelastic consolidation 非弹性固结
unelastic effect 非弹性效应
unelasticity 非弹性
unelastic lateral buckling 非弹性侧向压曲
unelastic resistance 非弹性抗力
unelastic scattering 非弹性散射
unelastic torsion 非弹性扭曲
unembanked 无护堤的
un-embanked alluvial land 未筑堤的冲积土地
unemployed 未加使用的；失业的；未就业的
unemployed capital 闲置资本；游资
unemployed labor force 失业劳动力
unemployed ship 闲置船舶；停航船泊
unemployed shipping 无营运业务的
unemployed workers in motion 不在职流动职工
unemployment 失业
unemployment benefit 失业救济(金)；失业补助金
unemployment compensation 失业救济(金)；失业补助；失业补贴；失业补偿
unemployment compensation board 失业救济委员会
unemployment compensation funds 失业补偿金
unemployment dole 失业津贴
unemployment equilibrium 失业均衡
unemployment fund 失业基金
unemployment insurance 失业保险
unemployment insurance benefits 失业保险救济

unemployment insurance expenses 失业保险费
unemployment insurance statistics 失业保险统计
unemployment insurance tax 失业保险税
unemployment pay 失业救济金；失业补贴
unemployment rate 失业率
unemployment relief 失业救济
unemployment reserves 失业储备金
unencapsulated 未密封的
unencapsulated source 未封装源
unenclosed arcade 不能关闭的拱廊
unenclosed exterior stair(case) 室外明楼梯；不封闭室外楼梯
unenclosed opening 不封闭孔洞
unenclosed plaza 不能关闭的广场；敞开的广场
unenclosed porch 不能关闭的廊
unenclosed relay 敞开式继电器
unenclosed stair(case) 不封闭楼梯
unencumbered 没有抵押的；可自由支配的
unencumbered balance 未支配余额
unencumbered property 没有负担的财产
unending circulation 连续环流；闭合环流
unendorse check 未背书支票
unenforceable 不能实施的
unenforceable contracts 不能执行的合同
unenforced 未实施的
unenriched 未装饰的；未浓集的；未富化的
unenriched uranium 未浓缩的铀
unenumerated articles 零星货物(少于贸易中常用单位额度)
unequable 未调匀的；不调匀的
unequal 不等的
unequalable 不能相比的
unequal absorption coefficient 不等吸收系数
unequal activation center[centre] 不等活性中心
unequal addendum gear 不等齿顶高齿轮
unequal addendum system 不等齿顶制；不等齿顶高齿轮系
unequal angle 不等角
unequal angle iron 不等边型材；不等边角钢；不等边角铁
unequal-arm scale 不等臂秤
unequal cavity acceleration 不等腔加速
unequal class intervals 不等组距
unequal clusters 容量不等的群体
unequal coefficient fo reflection 不等反射系数
unequal coincidence counting 不等符合计数
unequal compressibility 不等压缩性
unequal crossing over 不等交换
unequal-degree-of saturation method 不等饱和度法
unequal depression 不均匀沉陷
unequal disintegration constant 不等衰变常数
unequal division 不等分裂
unequal draught 不均匀压下
unequal emission rate 不等发射率
unequal energy dissipation 不等能量耗失
unequal fillet weld 焊脚不等的角焊缝
unequal fission 不等分裂
unequal fission process 不等裂变过程
unequal gable roof 不等长双坡屋顶
unequal harmonic vibration 不等谐振动
unequal helix structure 不等螺旋线结构
unequal impact parameter 不等碰撞参数
unequal impulse 不等脉冲
unequal intervals 不等区间
unequal irradiation effect 不等辐射效应
unequalized profit rate of capital 资本未平均化的利润率
unequal leg 长短腿；不等长翼；不等长边；不等长腿
unequal (leg) angle 不等肢角钢；不等边角钢
unequal-leg angle iron 不等肢角钢
unequal-leg bulb-angle 不等肢圆头角钢
unequal-leg fillet welds 不等边角焊缝
unequal lines 分布不均的抓痕
unequal load(ing) 不均匀荷载；不均匀负载
unequally distributed 不均匀分布的
unequally spaced 非等间距的；不等步长的
unequally spaced data 非等间距数据
unequally spaced data point 非等间距数据点
unequally spaced intervals 非等间距区间
unequally spaced teeth 不等距轮齿
unequally-tempered scale 不均匀温标；不等程音阶
unequal melting point 不等熔点
unequal orbital motion 不等轨道运动

unequal oscillation frequency 不等振荡频率
unequal perturbation 不等微扰
unequal pointed arch 不对称尖拱
unequal precision observation 不等精度观测
unequal pressure 不等压力
unequal probability 不等概率
unequal productive capacities 不同等的工作能力
unequal pulse 不均脉
unequal quantization effect 不等量子化效应
unequal relative velocity 不等量相对速度
unequal replication 不等重复
unequal sample sizes 不等样本含量
unequal sections 不等边型钢
unequal segmentaion 不等分裂
unequal sensibility array 不等灵敏度组合
unequal settlement 不均匀下沉；不均匀沉箱；不均匀沉陷；不均匀沉降；不等(量)沉陷
unequal-sided 不等边的
unequal sides 不等边型钢
unequal sized clusters 容量不等的群体
unequalsized retinal image 不相等的网膜像
unequal span 不等跨
unequal stop 不对称触止
unequal stress 不等应力
unequal subclass 不等子类
unequal subclass numbers 不等子类含量
unequal subsidence 不均匀沉陷；不等量沉陷
unequal tangential velocity 不等切向速度
unequal treaty 不平等条约
unequal triangle 不等边三角形
unequal twill 单面斜纹
unequal valent 异价
unequal variance 不等方差
unequigranular 粒径不等的；颗粒不匀的
unequilateral tube 不等边管
unequilateral turnout 单式不对称道岔【铁】
unequilibrated type 不平衡型
unequilibrium crystallization 非平衡结晶作用
unequipped 未装备的
unequivalent exchange 不等价交换
unerected 未安装的
unerring 无误的
unerwater-surveillance system 水下监视系统
unescorted convoy 无护航的海上运输队
unescorted flight 无护航飞行
unessential 不重要的
unessential element 非必要元素
unestablished 未建成的
unestablished coenosium 未建成群落
unestablished flow 未定型流
unesterified 未酯化的
unesterified fatty acid 未酯化脂肪酸
unetched fibre 未侵蚀纤维
unetherified 未醚化的
unevacuable 无法后送的
unevaporated 未蒸发的
uneven 参差状；参差不齐的；不匀整；不匀；凹凸不平的
uneven brightness 光泽不匀
uneven caving 不均匀崩落
uneven coating 镀层不均
uneven coefficient of aquifer 含水层不均匀系数
uneven colo(u)r 不匀染色；不匀匀色
uneven distribution 厚薄不匀；分布不均；不均匀分布
uneven dye 不匀染色
uneven face wear 唇面不均匀磨损(钻头)
uneven fracture 参差(状)断口；不平整断口
uneven front 不整齐前沿
uneven frost heaving 不均匀冻胀
uneven ga(u)ge 不均匀厚度
uneven glaze 疙瘩釉
uneven grain 不均匀木纹；木材的不均匀纹理；不匀纹理；不均匀纹理；不(均)匀颗粒；不均匀晶粒
uneven ground 不平地；起伏地；不平整地面
uneven heating 不均匀加热
uneven injection 不均匀注入
uneven irradiation 不均匀辐照
uneven keel 首尾吃水不等
uneven length code 不等长电码
uneven lines 分布不均的抓痕
uneven load 不均匀荷载
uneven lustre 光泽不匀
unevenly-compressible foundation 不均匀压缩

地基
unevenness 不齐；不平整性；不平整度；不平坦(性)；不平(度)；不均匀性；不(均)匀度
unevenness in height at joint 钢轨接头上下错牙
unevenness in texture 质地不匀
unevenness of application 撒药不均匀
unevenness of rail at joints 接头轨端不平
uneven number 奇数
uneven number of teeth 单数齿数
uneven paralleled bars 高低杠
uneven ripening 成熟不均匀
uneven river bed 不平整河底
uneven rolling 碾压不匀
uneven running 不匀转动；不平稳运转
uneven seabed 不平整海底
uneven selvedge 不匀布边
uneven settlement 不均匀沉陷；下沉不匀；不均匀沉降
uneven shed 不清晰梭口
uneven sizing 定径不匀
uneven snow coverage 不均匀积雪
unevenspan greenhouse 不等屋面温室
uneven surface 粗糙(表)面
uneven temper 不均匀钢化；不均匀淬火
uneven temperature distribution 不均匀(的)温度分布
uneven terrain 起伏地
uneven texture 不匀质地；不匀结构；不均匀纹理
uneventful 无重大事故的
uneven topography 凹凸地形
uneven track 不平整轨道
uneven wear 非均匀磨损；不匀磨损；不匀磨耗
unexcellent law of environment 环境非优定律
unexcited 欠激
unexpanded composition cork 不膨胀成分构成的软木
unexpanded pure agglomerated cork 不膨胀的纯聚结软木
unexpected 不可预报的
unexpected accident 意外事故
unexpected expenses 意外支出；意外开支
unexpected failure 意外的损坏
unexpected halt 意外停机；突然停机
unexpected pay 意外支出；不可预见费(用)
unexpected property loss 财产意外损失
unexpected risks 意外风险
unexpended 未用完的
unexpired 未尽的
unexpired cost 未消耗成本；未耗成本
unexpired expenses 未满期费用；未耗用费用
unexpired insurance 未到期保险
unexpired insurance premium 未过期保险费
unexpired patent 生效的专利
unexpired premium 未到期保险费
unexpired term 有效期未满；未满期限
unexplained deviation 未解释的变量
unexploded 未爆炸的
unexploded bomb 未爆炸弹
unexploded cartridge 拒爆药包；未爆药卷
unexploded charge 瞎炮；盲炮
unexploded explosive 残余炸药
unexploited 未(经)开发的
unexploited area 未开发区(域)
unexploited waters 未开发水域
unexplored 未经勘察的
unexplored area 未经探察的地区；未勘查地区
unexplored region 未勘查地区
unexplosive 防爆的；非爆炸性的；不爆炸的
unexposed 未经照射的
unexposed film 未曝光胶片
unexposed film reel 未曝光胶卷
unexpressed 未表达的
unexpressive 未能表达原意的
unextended compound 无油胶料
unextinguished discount on funded debt 未偿清长期债款的折扣；摊余债券折价
unfaced 表面未经加工的；不整面的
unfaced asphalt rag felt 未整面的沥青粗油毡
unfaced board 表面未加工(玻璃)板
unfaced surface 未光面
unfacetted 无磨面的
unfactored takeoff distance 未经修正的起飞滑跑距离
unfadable 不褪色的

unfading 不褪色的
unfailing performance 可靠性能
unfailing service 可靠服务；无故障的工作
unfair 超出弹性界限的；不公正的
unfair competition 非公平竞争；不正当竞争；不公平竞争
unfair dismissal 不公正的解雇
unfair distribution of income 分配不公
unfair goods 不公正的产品
unfair hole 不通孔
unfair labor practice strike 不公正做法的罢工
unfair list 不公平名单
unfair price 不合理价格
unfair trade 非公平交易
unfair trade practices 不公正的商业做法
unfashioned 未成型(加工)的
unfasten 解缆；解开；放松
unfaulted 无断层的
unfaulty condition 无故障状态
unfavo(u)rable 不利的；不适宜的
unfavo(u)rable balance of deficit 逆差
unfavo(u)rable balance of trade 贸易入超；逆差；入超；外资逆差
unfavo(u)rable condition 不良情况
unfavo(u)rable condition for car rolling 溜车不利条件【铁】
unfavo(u)rable defect 不利的缺陷
unfavo(u)rable exchange 不利汇率
unfavo(u)rable factor 不利因素
unfavo(u)rable geologic(al) condition 不良地质条件；不良地质情况
unfavo(u)rable geologic(al) features 不良地质现象
unfavo(u)rable geology 不良地质
unfavo(u)rable influence 不利影响
unfavo(u)rable leverage 负效应
unfavo(u)rable physical condition 不利的自然条件
unfavo(u)rable variance 负差异；逆差异；不利差异
unfavoured state 不利状态
unfavoured transition 不利跃迁
unfeasibility 非可行性
unfeasible state 非可行状态
unfed 未加燃料的
unfeed 无报酬
unfelt earthquake 无感地震
unfenced 没有篱笆的
unfenced machinery 无防护机械
unfermentable 不能发酵的
unfettered 不受约束的
unfibered 非纤维质的
unfibrated 非纤维组织的
unfibred 非纤维质的
unfill 曝光不足
unfilled 无充填的；未填充的
unfilled aperture 分立孔径
unfilled cog 未充填垛式支架
unfilled corner 缺角
unfilled crater 未填满的弧坑
unfilled fissure 未充填裂隙
unfilled level 未满能级
unfilled order 未执行的订单；未交货订单；未定货单；未定订货单
unfilled orders 未发货订单
unfilled porosity 透气度；非饱和孔隙度
unfilled section 欠缺断面
unfilled state 未满态
unfilmed 未敷膜的
unfiltered 未滤过的；未过滤的
unfiltered water 未过滤水
unfinished 毛面的；未修整的；未完工的；未完成的；未(精)加工的；粗加工的
unfinished bolt 毛面螺栓；粗制螺栓
unfinished book 未完成的著作
unfinished bottom 未修饰的外底
unfinished building 未完工建筑物
unfinished building space 未完工的建筑场所
unfinished business 未了事宜
unfinished construction 未完工程
unfinished edition 未完版
unfinished goods 半制品
unfinished hex nut 粗制六角螺母
unfinished leather 未整饰的皮革
unfinished nut 粗制螺母

unfinished product 未成品；粗制品
unfinished round pin 粗加工圆销
unfinished screw 粗制螺丝
unfinished section 未轧完品
unfinished sheet 粗加工薄钢板
unfinished voyage 未完航次
unfinished washer 粗加工垫圈
unfinished work 未完工作；未成品
unfinned fuel element 无肋片燃料元件
unfired 未燃烧的；未经焙烧的；非直接火焰加热的；不直接接触火的；不烧的
unfired brick 砖坯；未烧砖(坯)；未经煅烧的砖；不烧砖；不经过煅烧的砖
unfired ceramic body 陶器坯
unfired clay 不经煅烧的黏土
unfired density 生坯密度
unfired explosive 拒爆残药；剩药
unfired fusion welded pressure vessel 非直接火焊接压力容器
unfired hole 拒爆炮孔
unfired material 欠烧物料
unfired pressure vessel 非直接火压力容器；非受火压力容器；不用火烧的压力容器
unfired semi-stable dolomite refractory 不烧半稳定性白云石耐火材料
unfired steam generating vessel 非直接火蒸汽锅炉
unfired steam generator 非燃式蒸汽发生器
unfired strength 生坯强度
unfired tube 非点火管
unfished ground 未开发的渔场；未捕海区
unfished stock 未捕群
unfissured rock 无裂缝岩石
unfit currency 不合格货币
unfit dwelling 不宜居住的住宅
unfitness of butt joint 错边量
unfitness of butt weld 焊缝对口错边量
unfit parts box 损坏零件箱
unfix 松开
unfixed 非固定的；不固定的
unfixed bar 不固定沙洲
unfixed chrome 未结合的铬
unfixed material 机动材料
unfixed point 未定点；不定点
unfixed sand 流沙(也称流砂)；未固定沙
unfixed soil 非固定土壤
unflanged 无凸缘的
unflared 无喇叭口的
unflared landing 无拉平的着陆
unflatness 不平度；不平坦
unflattened roll 理想刚性辊
unfluted 不开槽的；无凹槽的
unfluted shaft of column 无槽柱身；无槽立柱
unfluxed asphalt 未加稀释油的沥青
unfluxible 不熔化的；不可溶的
unfoamed crosslinked gel 不发泡交联凝胶
unfocused laser 未聚焦激光器
unfocused processing 未聚焦处理
unfolded 展开的
unfolded chain section 无折叠链段
unfolding 伸展
unfolding drawing 展开图
unforced rotary kiln 正常工作的回转窑
unforcement 不强化
unforeseeable event 不能预料的事件
unforeseen 不可预报的
unforeseen circumstance 意外情况
unforeseen demand 未预见用水量
unforeseen event 料不到的事件；不可预见的事件
unforeseen expenses 意外开支；临时开支；不可预见费(用)
unforeseen incident 意外事故
unforeseen items 不可预见项目
unforeseen loss 不可预见的损失
unforeseen site condition 未能预见现场条件
unforeseen works 不可预见工作；不可预见工程
unformatted 无格式的；非格式化
unformatted display 非格式化显示
unformatted file 无格式文件；非格式文件
unformatted image 非格式化图像
unformatted input-output 无格式输入输出
unformatted mode 非格式化方式
unformatted read 无格式读
unformatted record 无格式记录
unformatted request 非格式化请求

unformatted system services 非格式化系统服务程序
unformatted write 无格式写
unformed 未成型(加工)的;不用模板的
unforseen demand 未预见需求
unforseen event 非常事态
unfortified 我增强的;未设防的;不稳定的
unfortunate variance 不利差异
unframed door 无框门;无框门;板条门
unframed glass 无框玻璃
unfree meander 强制型曲流
unfree variation 不自由变异;不自由变分
unfree water 吸着水;固着水;死水;非自由水
unfreezable drain valve 防冻排液阀
unfreezable tap 防冻龙头
unfreeze 解冻
unfreezing 解卡(钻具或钻杆);解冻;冻融作用
unfreezing port 不冻港
unfrequent inquiry 非经常性调查
unfrozen 不冻的
unfrozen estuary 不冻河口
unfrozen lake 不冻湖
unfrozen port 不冻港
unfrozen river 不冻河流
unfrozen soil zone 不冻土区
unfrozen stream 不冻河流
unfrozen water content 未冻(结)含水率
unfrozen water surface 未封冻水面
UNF thread 统一标准细牙螺纹
unfueled part 非燃料元件
unfulfilled 未完全的
unfulfilled obligations 未完成的义务;未履行的义务;未尽义务
unfunded debt 流动债务;非抵押债务;非长期债务
unfused 未熔化的
ungaite 奥长英安岩
ungalvanized 未镀锌的
ungalvanized iron and steel 未镀锌钢铁
ungalvanized plate 未镀锌薄板
ungalvanized steel plate 未镀锌钢板
ungalvanized wire 未镀锌线
ungarbled 未经筛分的
ungated 无大门的;无闸门的
ungated flow 敞开泄流
ungated level crossing 无道口栏木的公路与铁路交叉;无栏木道口
ungated spillway 无闸门溢洪道
unga(u)ged basin 无资料流域;未施测流域;未设站流域
unga(u)ged lime plaster 非罩面石灰砂浆;纯石灰砂浆
unga(u)ged local inflow 未施测区间入流;未控制区间入流
unga(u)ged river 未施测河流
unga(u)ged stream 未施测河流
unga(u)ged watershed 未施测流域
ungear 脱离啮合
ungeared 无齿轮啮合的
ungeared engine 无减速器发动机
ungemachite 菱碱铁矾;碱铁矾
ungerade 奇态
unglazed 无光的;没有装玻璃的;素烧的;未上釉的;未装玻璃的
unglazed crucible 无釉坩埚;素烧坩埚
unglazed earthenware 素烧陶器
unglazed extruded tile 无釉挤压砖
unglazed lightweight aggregate 无釉轻集料
unglazed pipe 未上釉(瓦)管
unglazed porcelain 无釉瓷;素烧瓷;素瓷
unglazed pressed tile 未上釉的压制方砖
unglazed printing paper 无光道林纸
unglazed roofing tile 无釉瓦;素烧瓦
unglazed roof tile 无釉屋面瓦
unglazed structural facing unit 无上釉建筑用饰面块;未上釉的结构用砌面块
unglazed tile 滞光瓷砖;无釉(面)砖;素烧瓷砖
unglazed ware 无釉陶瓷器皿
ungovernable 难控制的;难调节;无法控制的
ungoverned 无调节的
ungraded 不符合标准的;不合格的;未分级的;劣质的;未整平的;次级(的);不分级的
ungraded aggregate 无级配骨料;非级配集料;非级配骨料
ungraded pole line 同高架空线路

ungraded position 未定级职位
ungraded products 次品
ungraded river 未均衡河流
ungraded stream 未均衡河流
ungrained plate 未研究板材
ungrained surface 未研磨表面
ungraphitised carbon 非石墨化碳
ungrazed area 未放牧地区
ungrease 未加润滑脂润滑的
ungreenable black 不泛绿精元
ungridded 无栅的
unground 未磨过的;未打磨的
ungrounded 没有基础的;无事实根据的;未接地的;不接地的;不接地
ungrounded bridge 不接地电桥
ungrounded supply system 不接地系统;不接地供电系统
ungrounded system 不接地制;不接地系统
unground material 未经研磨的材料
unground resin 未磨碎树脂
ungrouped data 未分组资料;未归类资料
ungrouped regression model 不分组回归模型
ungstenic 钨的
unguarded 无防护的
unguarded flange way 辙叉有害空间
unguent 滑油;润滑材料
unguentum 软膏
ungula 蹄状物
ungulic acid 蹄酸
unhairing and scudding knife 去毛和净面刀
unhairing beam 去毛用刮皮板;刨皮砧板
unhairing cylinder 去毛机刀轴
unhampered flow of traffic 无阻碍车流;畅行交通
unhandy 难操纵
unhardenable 不能硬化的
unhardened 未硬化的;未淬火的
unhardened concrete 未硬结混凝土;未硬化的混凝土
unhardened facility 无原子防护的设施
unhardening 未硬化的
unharmed 未受伤害的
unharmful 无害的
unharming 无害的
unharmonic 非谐
unharmonic oscillator 非谐荡振器
unhealth 不良的
unhealthy 不卫生的;不健康的
unhealthy area 不卫生地区
unhealthy condition 不健康状态
unhealthy effect 不良影响
unhealthy tendency 不良倾向
unheated 未加热的
unheated air drier 常温气流干燥机;未加热空气式干燥器
unheated air drying 未加热空气干燥;非热风干燥
unheated building space 无取暖建筑空间
unheated greehouse 不加热的暖房;不加热的温室
unheated lime 未加热的浸灰池
unheated line 不加热管线
unheeded 未受注意的
unherded 无人放牧的
unhewn 未切削的;未琢凿的;未砍削的
unhindered (gravity) discharge 无阻碍排料
unhindered terminal settling velocity 自由终沉速度
unhitch 摘钩;解结;脱钩
unhitched 摘钩
unhook 摘开;摘钩;脱钩
unhulled 未脱壳的
unhumified 未腐殖化的
unhumified organic matter 未腐殖化的有机物;未腐解有机物质
un-hump yard 非驼峰编组站
unhydrated 未水化的
unhydrated cement 未水化的水泥
unhydrated ion 未水合离子
unhydrated plaster 干灰膏
unhydrolyzable nitrogen 非水解性氮
unhydrous plaster 干灰膏;未水化灰浆
unhygienic 不卫生的
unialignment 单一调整
unialignment tool 单一调整工具
uniaxial 一轴的;单轴的
uniaxial anisotropic material 单轴各向异性材料

uniaxial anisotropic matter 单轴各向异性物质
uniaxial anisotropy 单轴各向异性
uniaxial compression 单轴(向)压缩
uniaxial compression test 单轴压缩试验;单轴抗压试验
uniaxial compressive strength 单轴抗压强度
uniaxial compressive strength test 单轴压缩强度试验;单轴抗压强度试验
uniaxial compressive test 单轴压缩试验;单轴抗压试验
uniaxial crystal 一轴晶;单轴晶(体)
uniaxial cyclic loading 单轴循环加载
uniaxial deformation 单轴向变形
uniaxial dynamic(al) analysis 单轴向动力分析
uniaxial eccentricity 单(向)偏心
uniaxial hot pressing 单向热压
uniaxial interference figure 一轴晶干涉图
uniaxiality 单轴性
uniaxial load 单轴加载;单轴荷载
uniaxial loading 单轴向荷载
uniaxial longitudinal strain 单轴伸缩应变
uniaxially 单向地
uniaxial mechanical pressure 单向机械压力
uniaxial negative character 一轴负光性
uniaxial negative crystal 负单轴晶体
uniaxial orientation 单轴取向
uni-axial pantograph 极坐标缩放仪
uniaxial positive character 一轴正光性
uniaxial pressing 单向加载;单向加压
uniaxial pressure 单向压力
uniaxial rotational oblate ellipsoid 单轴旋转扁球体
uniaxial rotational prolate ellipsoid 单轴旋转长球体
uniaxial saturated compressive strength of rock 岩石单轴饱和抗压强度
uniaxial shear strength 单轴抗剪强度
uniaxial state of stress 单轴(向)应力状态;单向应力状态
uniaxial strain 单轴向应变
uniaxial strain test 单轴(向)应变试验
uniaxial stress 单轴(向)应力;单向应力
uniaxial stress and temperature cycling test 单轴应力一温度循环试验
uniaxial stress field 单轴应力场
uniaxial stress-strain curve 单轴应力一应变曲线
uniaxial tensile strain 单轴拉伸应变
uniaxial tensile strength 中心受拉强度;单轴抗拉强度
uniaxial tensile strength test 单轴抗拉强度试验
uniaxial tension 单轴向张力;单轴向拉伸
uniaxial tension test 单轴拉伸试验
uniaxial test 单轴向试验
uniaxial torpedo 单轴测斜仪
uniboom seismic sound source 单响地震声源
unibus 单母线
unibus system 单总线式
unicast 单播
unicast process 陶瓷型法
unicell block 单砌块;单孔砌块;单槽砌块
unicellular 闭孔
unicellular plastics 单孔塑料;闭孔泡沫塑料
unicellular rubber 闭孔泡沫橡胶
unichannel 单一河槽
unichassis 单底盘;单层底板
unichoke 互感扼流圈
unichrome process 光泽镀锌处理
unicity 单一性
uniclinal 单斜的;单倾
uniclinal fold 单斜褶皱
uniclinal structure 单斜构造
unicline 单斜
unicode state assignment 单码状态分配
unicoil 单线圈
unicoil winding 单线圈绕组
unicolo(u)r 单色的
unicolo(u)red marble 单色大理石
unicomponent system 一元系;一元素
uniconductor 单导体
unicontrol 统调;同轴调谐;单向控制;单向调整;单钮调谐
unicounter machine 单一计数器械
unicouple 单耦合
unicoupler 单耦合器
unicube 立方体单元

unicursal curve 有理曲线；单行曲线
unicuspid 单尖
unicycle 独轮车；单轮脚踏车
unicycle graph 单圈图
unicyclic graph 单圈图
Unidal 铝镁锌系形变铝合金
unidan mill 多仓式球磨机
unidentate ligand 单齿配位
unidentifiable 无法鉴别的
unidentifiable cargo 无标志货物
unidentifiable parameter 不可识别的参数
unidentifiable point 不明显点
unidentified 未鉴定的；未鉴别的
unidentified material 未鉴定的材料
unidentified pip 未识别出的目标尖头信号
unidentified product 未经鉴定的产品；未鉴定的产物；组成未明的产物
unidentified sight 未识别的物体
unidentified source 未证认源
unidentified target 未识别的目标
unidentified track 未识别的目标回波
unidiameter joint 等直径连接套管
unidied time 统一时间
unidigit 一位数
unidimensional 一次元的；线性的；一维的；一度空间的
unidimensional consolidation 一维固结；单向压密；单向固结
unidimensional flow 一维流
unidimensional hologram 一维全息图
unidimensional magnification 一维放大
unidimensional search 一维寻查
unidimensional space 一维空间
unidirection 单向；单方面
unidirectional 单向性的；单（方）向的
unidirectional air flow 单向气流
unidirectional antenna 单向天线
unidirectional bending loading 单向弯曲加载
unidirectional bus 一维总线
unidirectional chained list 单向链接表
unidirectional channel 单向信道
unidirectional combined type marshalling station 单向混合式编组站【铁】
unidirectional conductivity 单向电导率
unidirectional conductor 单向导线；单向导体
unidirectional coupler 单向耦合器
unidirectional current 直流电；单向电流
unidirectional direction finder 单向测向器
unidirectional engine 单转向发动机
unidirectional fabric 单向纤维织物
unidirectional fibre 单向纤维
unidirectional flow 直流；单向水流；单向流（动）
unidirectional flow of information 单向信息流
unidirectional flow ventilation 单向流通风
unidirectional glass cloth tape 单向玻璃布带
unidirectional glass tape 单向玻璃带；玻璃纤维无纬带
unidirectional laminate 单向层压板
unidirectional line 单向运行线路
unidirectional loading 单向加载
unidirectional loading four-pair sliding-face cubic(al) ultrahigh pressure and high temperature device 单向加载四对斜面式立方体超高压高温装置
unidirectional loading four-pair sliding-face cubic(al) ultrahigh pressure device 单向加载四对斜面式立方体超高压装置
unidirectional loading link type cubic(al) ultrahigh pressure device 单向加载铰链式立方体超高压装置
unidirectional loading multiple anvil ultrahigh pressure device 单向加载多压砧式超高压装置
unidirectional loading three-pair sliding-face cubic(al) ultrahigh pressure device 单向加载三对斜面式立方体超高压装置
unidirectional log-periodic(al) antenna 单向对数周期天线
unidirectional longitudinal type marshalling station 单向纵列式编组站【铁】
unidirectional magnetic field 单向磁场
unidirectional martensite 定向马氏体
unidirectional membrane rupture valve 单向破膜阀
unidirectional microphone 单向麦克风；单向扩音器；单向传声器
unidirectional mill 单向轧机
unidirectional motion 单向运动
unidirectional moving-coil trigger 单向动圈触发器
unidirectional prepreg machine 单向预浸机
unidirectional pressure 单向压力
unidirectional pulse-amplitude modulation 单向脉幅调制
unidirectional pulses 单向脉冲
unidirectional radio direction finder 无线电单向定向器；单向无线电测向仪
unidirectional relay 单向继电器
unidirectional replication 单向复制
unidirectional ring laser 单向环形激光器
unidirectional rotation 单向转动
unidirectional roving fabric 单向无捻粗纱布
unidirectional runway 单向起降的跑道
unidirectional rupture diaphragm valve 单向破膜阀
unidirectional search 单向搜索
unidirectional settlement 单向沉陷
unidirectional solidification 定向凝固；单向凝固
unidirectional switch 单向开关
unidirectional tape prepreg 单向带预浸料
unidirectional tension 单向拉伸
unidirectional tow prepreg 单向丝束预浸料
unidirectional traffic 单向交通；单向运行
unidirectional traffic flow 单向车流
unidirectional transducer 单向转换器；单向传感器
unidirectional transmission 单向传输
unidirectional transversed type marshalling station 单向横列式编组站【铁】
unidirectional turbine 单转向的涡轮机
unidirectional voltage 整流电压；单向整流电压；单向电压
unidirectional water head 单向水头
unidirectional wind 单向风
unidirectional working 单翼开挖面；单翼工作面
unidrive gear box 单独传动齿轮箱
uni-engraver 刻图工具
unification 一致；统一化；通用化
unification in parallel 并行一致化
unification of the currency 统一币制
unification set 一致集合
unified 统一的
unified allocation of resources 统一调配
unified approach 统一方法
unified approach in our external dealings 统一对外的原则
unified architectural details 建筑统一详图
unified aseismic performance index 统一抗震性能指数
unified average social rate of profit 统一的社会平均盈利率
unified base 联合机座；单一机座
unified base wages 统一的工资底数
unified boiler control 锅炉联合控制
unified bonds 统一债券
unified brand 统一牌号
unified budget 综合预算；统一预算
unified bus 单一总线
unified classification system of soil and rock 岩土统一分类系统
unified coarse thread 统一标准粗牙螺纹
unified construction 统一施工；统一建设；统建
unified credit policy 统一的信贷政策
unified curve number of log 测井曲线统一编号
unified deficit 统一赤字
unified development 统建
unified distributed material 统配物资
unified diversity 多样统一
unified earthquake magnitude 统一震级
unified electric(al) machine theory 统一化电机理论
unified equipment 统一标准装置
unified extra-fine thread 统一标准超细牙螺纹
unified field theory 统一场论
unified fine thread 统一标准细牙螺纹
unified foreign exchange rate 统一外汇汇率
unified gyratory crusher 单面进料的圆锥式破碎机
unified housing development 统建住房
unified income tax 统一所得税
unified industrial and commercial tax 工商业统一税
unified interpretation 统一解释
unified joint economic legal relation 统一连带经济法律关系
unified large market 统一大市场
unified law 统一法
unified leadership 统一领导
unified magnitude 统一震级
unified management 统一经营；统一管理
unified member 统一构件
unified model 统一模型
unified net of Doppler stations and zero-order traverses 多普勒导线联合网
unified network 统一网络
unified nomenclature 统一命名
unified number of log interpretation 测井解释统一编号
unified plan 综合规划
unified plan and management at different levels 统一计划分级管理
unified planning 统一计划；统一规划；统筹
unified planning and management at different levels 统一规划领导分级管理
unified planning and overall arrangement 统筹安排
unified planning project 整体规划方案
unified power system 统一电力系统
unified precast element 统一的预制构件；通用预制构件
unified precast reinforced concrete structural element 钢筋混凝土统一预制构件
unified program(me) planning 联合规划法
unified ratio of reserves 统一的准备金率
unified regulations 统一规范
unified scale 统一标度
unified screw thread 统一标准螺纹
unified soil classification system 统一土壤分类制；土的统一分类（法）
unified standard 统一标准
unified state control over income and expenditures 统收统支
unified structure element 统一构件
unified thread 统一螺纹；标准螺纹
unified transfer 统一调度
unified transfer tax credit 财产转移税统一抵免额
unified wage scale 统一工资标准
Unified World Geodetic Datum 统一世界大地基准
unifier 一致化置换；一致化算子
unifies soil classification system 统一土壤分类系统；统一土壤分类体制；统一土壤分类体系
unifilar electrometer 单丝静电计
unifilar helix 单股螺旋线
unifilar magnetometer 单丝磁力仪
unifilar suspension 单线悬挂
unifilar variometer 单丝磁变计
unifilar winding 单线绕组
unifile volume 单文件卷
unifloe tank furnace 纵火焰池窑
uniflote 单体浮箱（拼装式趸船的单元）
uniflow 单向水流；单向流动的；单向流（动）；单流的；直流；顺流
uniflow boiler 强压给水式锅炉；单流（式）锅炉
uniflow condenser 单流式冷凝器
uniflow cooling 单向流动冷却
uniflow engine 单流（蒸汽）机；单流（式）发动机
uniflow rotary dryer 顺流转筒式烘干机
uniflow scavenging system 轴流单向扫气系
uniflow steam engine 单流式蒸汽机
uniflow tank furnace 单向流动池窑
uniflow type compressor 顺流式压缩机；单流式压缩机
uniflow type condenser 单流式冷凝器
unifluenced 不受干扰的
uniflux condenser 单流式冷凝器
uniflux fractionating tray 长条泡罩分馏塔盘
unifluxor 匀磁线
uniflux tray 单流向式塔板；单流式塔盘
unifocal lens 单焦透镜
uniform 均匀的；均一的；统一的
uniform acceleration 匀加速度；均匀加速度；等加速度
uniform accounting 统一会计（制度）
uniform accounting system 统一会计制度
uniform-admission turbine 全周进汽汽轮机
uniform aerated concrete 均匀充气混凝土

uniform air-stream 均匀气流
uniform amplitude 恒定振幅;等幅
uniform apparent flow 均匀表观流动
uniform approximation 一致近似法;一致逼近;均匀近似法
Uniform Arbitration Act 统一仲裁法(美国)
uniform array 等距(离)天线阵
uniform atomizing 均匀雾化
uniform attack 均匀侵蚀
uniform automatic data processing system 标准自动资料处理系统
uniform beam 均匀梁;等截面梁
uniform beam bridge 等截面梁桥
uniform bed 均质层
uniform bill of lading 统一提单
Uniform Bills of Lading Act 统一提单法(美国)
uniform bound 一致有界
uniform boundedness principle 一致有界原理
uniform brand 统一品牌
uniform bridge 均一电桥;等截面电桥
uniform brightness 均匀亮度;等亮度
uniform broadcasting 均匀撒播
uniform budget 统一预算
uniform building code 统一建筑法规;统一建筑规范
uniform buoyage (system) 统一浮标系统
uniform cantilever beam 等截面悬臂梁
uniform cap 制服帽
uniform channel 均匀河槽;等断面渠道
uniform chart 等刻度记录纸
uniform chromaticity chart 均匀色度图
uniform chromaticity coordinates 均匀色度坐标;等色差色度坐标;等色差三色系数
uniform chromaticity scale 均匀色度标
uniform chromaticity scale chart 均匀色品图
uniform chromaticity scale diagram 均匀色品图;均匀标度色品图
uniform chromaticity scale system 均匀色品系
uniform chromaticity spacing 均匀色度间距
uniform chromaticity triangle 均匀色三角
uniform chromaticness subspace 均匀色子空间
uniform circular load 均布圆形荷载;环形均布荷载
uniform circular motion 匀速圆周运动;等速圆周运动
uniform code 均匀代码;统一规范
uniform coefficient 匀度系数
uniform colo(u)r scale 均匀色标
uniform colo(u)r space 均匀(彩)色空间
uniform column 等截面柱
uniform combustion 均匀燃烧
Uniform Commercial Code 全国统一商业准则(美国);统一商法
uniform commercial industrial appraisal report 工商统一标准评估表
uniform compaction 均匀压实;均匀压密
uniform components of delay 均匀延误(部分)
uniform components of queue length 均匀排队长度部分
uniform components of stop rate 均匀停车率
uniform compression 均匀压缩
uniform concrete 匀质混凝土
uniform condition 一致条件
uniform construction 均匀结构
uniform construction index 统一建筑指标
uniform contact 均匀接触
uniform continuity 一致连续;均匀连续性
uniform convergence 一致收敛;均匀收敛
uniform cooling 均匀散热
uniform corrosion 整体腐蚀;均匀腐蚀
uniform cost-accounting 统一成本会计
uniform cost accounting standards 统一成本会计标准
uniform cost control 统一成本控制
uniform costing 统一成本核算法
uniform cost system 统一成本制度
uniform coverage 有同类地图资料地区
uniform crop 同龄林
uniform cross flow 均匀横流
uniform cross-section 不变截面;恒定断面;等交叉截面;等截面
uniform cross-section area 等截面面积
uniform cross-section pile 等截面桩
uniform current 均匀电流;等速流
uniform current density 均匀电流密度

uniform curvilinear motion 等速曲线运动
uniform customs 统一惯例
uniform customs and practice 统一习惯和做法
uniform customs for credits 信用证统一惯例
uniform decompression 等速减压
uniform deflection test 等挠曲试验
uniform delivered price 单一到货价
uniform delivered pricing 统一运送订价法;统一发货定价法
uniform demand 均匀需求
uniform density 均匀充实;均匀密度;均衡分布密度
uniform density lens 均匀密度透镜
uniform design 均匀设计
uniform development 等速发展;等速发育(地形)
uniform dielectric 均匀电介质
uniform diffuser 均匀漫射体;均匀漫射面;均等扩散面
uniform diffuse reflection 均匀漫反射
uniform diffuser face 均匀扩散面
uniform diffuse transmission 均匀漫透射;均匀扩散传输
uniform diffusion 均匀漫射
uniform diffusion theory 均匀扩散理论
uniform dilatancy 均匀膨胀
uniform discharging technique 匀速卸料技术
uniform (distributed) load 均布荷载
uniform distribution 一致分布;均匀配水;均匀分布
uniform effluent standard 统一排放标准
uniform elastic material 均匀弹性材料
uniform elongation 均匀伸长
uniform emission standard 统一排污标准
uniform encoding 线性编码;均匀编码
uniform equivalent annual cost criterion 等量年成本标准
uniformer 变流机
uniform ergodic theorem 一致遍历定理;均匀遍历定理
uniform expansion 均匀膨胀
uniform feeding device 均匀给料装置
uniform feeding mechanism 均匀给料装置
uniform field 均匀场
uniform field Kerr cell 均匀场克尔盒
uniform flow 均匀流;等速流;等流速;等流
uniform flow depth 均匀水流深
uniform flow equation 等速流方程
uniform flow formula 等速流公式
uniform flow of traffic 均匀货流
uniform flow rate 固定流量
uniform fluid 均匀流体
uniform fluid flow 均匀流体流动
uniform flux 均匀通量
uniform force 均匀力
uniform force square pole 方形恒力杆
uniform fraction 均匀馏分
uniform function 单值函数
uniform ga(u)ge 均匀厚度
uniform general charter 标准件杂货租船公司
uniform geometry technique 规则形状技术
uniform grade 均匀坡度;不变坡
uniform gradient 均匀坡度
uniform gradient series 等差系列
uniform grading 均匀级配
uniform grain 匀晶
uniform granite 均质花岗岩
uniform gravel 粒径均匀的砾石;均匀砾石
uniform-grid method 均匀格网法
uniform hardening 均匀淬火
uniform hashing 均匀散列
uniform hazard 均匀危险因素;均匀风险
uniform hemisphere 均匀半球
uniform illumination 均匀照明
uniform impedance 均匀阻抗
uniform infiltration from surface 有地表均匀渗入
uniform in size 统一尺寸
uniform interface 一致接口
uniform internal diameter 均匀一内径
uniform investment 均衡投资
uniform invoice 统一分配;统一发票
uniformitarianism 均变说
uniformity 匀细度;一致性;均匀性;均一性;同一性
uniformity coefficient 匀质系数;均质系数;均匀(性)系数
uniformity coefficient of sand 砂的均匀系数
uniformity distributed load 均布荷载

uniformity factor 均匀系数;匀质因数;匀质系数;匀度系数
uniformity modulus 均质模数;均变模数
uniformity of colo(u)r 色泽均匀(性)
uniformity of fuel spray 燃料喷雾的均匀性
uniformity of illuminance 照度均匀度
uniformity of illumination 照明的均匀性;照明均匀度;高级分布均匀度;亮度分布均匀
uniformity of light 光的均匀性
uniformity of structure 构造的均质性
uniformity of texture 结构的均质性
uniformity of type 体型一致
uniformity of washing aggregate 洗砂石均匀程度
uniformity varying amounts method of depreciation 规则变数折旧法
uniformity varying load 等变荷载;等变负荷
uniformization 单值化;统一化;均匀化
uniformization transformation 均匀化变换
uniform Lambertion distribution 均匀朗伯分布
uniform landing 均匀着落
uniform law 全国统一法律(美国)
uniform layer 均质层
uniform level of liability 统一责任标准
uniform liability principle 同一责任制
uniform liability system 统一责任制
uniform light 均匀光(线);均匀照明
uniform lightness-chromaticness scale 均匀色度标
uniform lightness scale 均匀明度标(尺)
uniform limit partnership act 统一有限合伙条例
uniform line 均匀线
uniform live load 均布活荷载;均布动荷载
uniform load(ing) 均匀荷载;均匀负载;均匀负荷;均匀加感;均匀荷载;均布负载
uniform load(ing) bit 均匀负载钻头;均匀负荷钻头
uniform luminance 均匀亮度
uniform luminance area 均匀亮度面积
uniformly accelerated motion 匀加速运动;等加速运动
uniformly almost periodic function 一致殆周期函数
uniformly approximated 一致逼空间
uniformly best constant risk estimator 一致最佳不变风险估计量
uniformly best distance power test 一致最佳距离功效检验
uniformly better decision function 一致较好决策函数
uniformly bounded variation 一致有界变差;均匀有界变差
uniformly boundness 一致有界
uniformly convex space 一致凸空间
uniformly decreasing motion 等减速运动
uniformly definite subspace 一致定子空间
uniformly diffusing surface 均匀散射面;均匀漫射面
uniformly distributed 均匀分布的
uniformly distributed function 均匀分布的函数
uniformly distributed linear load 均布线荷载
uniformly distributed load 均匀分布荷载;均(匀分)布荷载
uniformly distributed loading 均匀载荷
uniformly distributed partial load 均布局部荷载
uniformly distributed static load 均布静载
uniformly distributed stress 均布应力
uniformly equicontinuous set 一致等连续集
uniformly equivalent distance 一致等价距离
uniformly equivalent function 一致等价函数
uniformly feasible direction 一致可行方向
uniformly graded 均匀级配的
uniformly graded aggregate 均匀级配集料;均匀级配骨料
uniformly graded sand 均匀级配砂
uniformly graded soil 同一级配的土壤;均匀级配土
uniformly increasing load 均匀变化荷载
uniformly loaded 均布荷载的
uniformly locally compact space 一致局部紧空间
uniformly locally connected space 一致局部连通空间
uniformly minimum risk 一致最小风险
uniformly minimum variance 一致最小方差
uniformly mixed system 均匀混合系统
uniformly modulated random process 均匀调制的随机过程
uniformly monotone mapping 一致单调映射

uniformly more powerful test 一致较大功效检验
uniformly most powerful confidence region 一致最大功效置信区域
uniformly most powerful invariant test 一致最大功效不变检验
uniformly most powerful test 一致最佳检验;一致最大功效检验
uniformly most powerful unbias(s)ed tests 一致最大功效无偏检定
uniformly open mapping 一致开映射
uniformly optimal plan 一致最优方案;均匀最优方案
uniformly optimal plan for moving target 动目标的一致最优方案
uniformly primitive recursive function 一致原始递归函数
uniformly progressive flow 均匀行进流
uniformly retarded motion 匀减速运动;等减速运动
uniformly rotating fluid 匀速旋转流体
uniformly sedimentation 均匀沉淀;均匀沉降
uniformly settling 均匀沉淀;均匀沉降
uniformly shaped blade 等截面叶片
uniformly sized grains 尺寸匀称的颗粒
uniformly spaced 均匀间隔的;等间距的;等间隔的
uniformly variable motion 匀变速运动
uniformly varying distribution load 线性变化分布荷载
uniformly varying load 匀变分布负荷;均匀变化荷载;均变荷载;等变荷载;等变负荷
uniformly varying stress 等变应力
uniformly yielding foundation 均匀沉陷的基础
uniform magnetization 均匀磁化
uniform market 统一市场
uniform market prices 统一市场价格
uniform mass 均匀质量
uniform mat 均匀底板;等厚度基床;等厚度筏基
uniform material 均匀物料;均匀泥沙;均匀材料
uniform mixing 均匀搅拌;均匀混合
uniform mix(ture) 匀质混合物
uniform modulus 均匀模量
uniform moisture content 均匀的含水率
uniform moment 均匀弯矩
uniform monetary limitation of liability rule 统一赔偿自然限额规则
uniform mortar 匀质砂浆
uniform motion 匀速运动;均匀运动;等速运行;等速运动
uniform motion cam 等速运动凸轮
uniform norm 一致范数
uniform open channel flow 均匀明渠流
uniform particle material 均匀颗粒材料
uniform particle size 均匀粒径
uniform particle source 均匀颗粒源
uniform particulate size 均匀粒度
uniform pattern 均匀样本
uniform peripheral pressure 周边均布压力
uniform pirce 统一报价
uniform pitch 等齐缠度膛线;等螺距
uniform pitch screw 等距(离)螺纹螺杆
uniform plane-polarized electromagnetic wave 均匀平面偏振电磁波
uniform plane wave 均匀平面波
uniform point source 均匀点源;均匀点辐射源
uniform policy conidtions 统一保单条件
uniform porosity 均匀密集气孔
uniform potential flow 均匀势流
uniform power spectrum 匀匀能谱
uniform practices 统一工作制度
uniform pressure 均匀压力;等压力
uniform-pressure-drop valve 定差减压阀
uniform pressure expander 等压撑涨器
uniform price 统一价(格)
uniform price auction 统一价格拍卖
uniform pricing code 统一价格代码
uniform principles 统一原则
uniform probability design 均匀概率设计
uniform product 粒度匀齐的产品
uniform profile 均质剖面
uniform quality 均质
uniform quantization 均匀量化
uniform radiation 均匀辐射
uniform rainstorm pattern 均匀暴雨型
uniform random number 均匀(分布)随机数
uniform random sampling 均匀随机取样

uniform random variable 一致随机变量
uniform rate 统一运费率
uniform rate of profit 统一利润率
uniform receipt method 统一收款法
uniform recharge over a plane 面状均匀补给
uniform rectilinear translation 匀速直线平移
uniform reference 一致基准
uniform regulated discharge 均匀调节放水
uniform regulation 统一条例
uniform residential landlord and tenant act 房主与房客关系统一法案
uniform resistance 均匀阻力
uniform resonance 一致共振
uniform resource identifier 统一资源定仪符
uniform resource indicator 通用资源指示程序
uniform resource locator 统一资源定位器
uniform risk spectrum 统一危险性谱
uniform road slab 等厚路面板
uniform roughness of surface 表面一样糙度
uniform rules 统一规则
uniform rules for a combined transport documents 复合运送单据统一规则
Uniform Rules for Collection 托收统一规则
Uniform Rules for Foreign Exchange Contracts 外汇契约统一规则
Uniform Sales Act 统一货物销售条例(美国)
uniform sales laws 一致销售法
uniform sample 均匀样品;均匀样本
uniform sampling 均匀取样;均稀抽样
uniform sampling fraction 一致抽样比
uniform sand 均质砂;均匀砂
uniform sandstone 均质砂岩
uniform scale 一致标度;等分度盘;等分标尺
uniform scale interpolation 等分标尺内插
uniform schedule 等分标度
uniform scores test 一致分数检验
uniform section 均匀断面;等截面
uniform section area 等截面面积
uniform-section blade 等截面叶片
uniform-section conduit 等截面管
uniform section pipe 等截面管
uniform section tube 等截面管
uniform sediment 均匀泥沙
uniform series 均匀系列;等额系列
uniform series compound amount factor 定额序列复利因子
uniform series present worth factor 定额序列现值因子
uniform settlement 均匀沉降
uniform settling 均匀沉淀
uniform shallow water 均匀浅海
uniform shock wave 均匀激波
uniform sidereal time 平恒星时
uniform size aggregate 均匀粒径骨料;等粒径集料;等粒径骨料
uniform-sized package 统一尺寸包装
uniform size of tree seedlings 均匀树苗
uniform slag 均匀渣
uniform slope 均一坡度;均匀斜度;均匀坡;坡度;不变坡
uniform soil 均质土(壤)
uniform sound distribution 声均匀分布
uniform source of light 均匀光源
uniform space 一致空间;等距(离)空间
uniform speed 匀速
uniform stability 一致稳定性
uniform standard 统一标准
uniform standards of professional appraisal practice 职业评估操纵统一标准
uniform steady flow 均匀稳定流
uniform stiff cohesive soil 均匀的硬黏性土
uniform strain 均匀应变
uniform strain shearing strength tester 等应变剪切强度测试仪
uniform strength 等强度;均匀强度;均一强度
uniform strength beam 等强度梁
uniform stress 均匀应力;均布应力
uniform string 均匀弦
uniform subsidy 职工制服补贴
uniform supercharge 均匀超载
uniform surcharge 均匀超载
uniform suspension 均匀悬浮(物)
uniform sweep rate 线性扫描率
uniform system 统一制度;统一系统;均一体系;

全林渐伐作业法
uniform system of maritime buoyage 国际助航浮标统一系统;国际统一海上浮标系统;国际通用浮标制;国际海上浮标统一系统
uniform taper(ing) 均匀的锥形收缩
uniform temperature 恒温;一定温度;恒定温度;等温
uniform temperature distribution 均匀的温度分布
uniform test procedure 统一试验法;统一试验程序;标准试验法
uniform tests 统一试验
uniform texture 结构均匀
uniform thickness 均匀厚度;等厚度
uniform-thickness concrete slab 等厚式混凝土板;等厚混凝土(路面)板
uniform-thickness slab 匀厚平板;等厚平板
uniform thread 统一标准螺纹
uniform tightness 均匀紧密度
uniform time charter 统一定期租船合同;标准论时租船契约
uniform tracing method 等速跟踪观测法
uniform tracing technique 等速跟踪观测法
uniform traffic regulations 统一交通法规
uniform turbulence 均匀涡流;均匀紊流;均匀湍流
uniform turbulent flow 均匀紊流;均匀湍流
uniform twist 等齐缠度膛线
uniform twist rifling 等齐缠度膛线
uniform type of rocks 均质岩石;均一岩层
uniform underflow 均匀底流
uniform varying motion 等变运动
uniform vehicle code 统一车辆规则;统一车辆法规
uniform velocity 匀速;均匀速度;等速(度)
uniform velocity motion 等速运动
uniform vendor and purchaser risk act 统一房地产交易风险法案
uniform vorticity 均匀旋度
uniform vorticity fluid 均匀涡旋流(体)
uniform wage points 统一工资分
uniform (water) depth 均匀水深
uniform wear 均匀磨损;均匀磨耗
uniform wind field 均匀风场
uniform yielding foundation 均匀沉降的基础
uniframe 与车身制成一体的车架
unifrequency 单频率
unifrequent 单频率的;单频率
unifunctional circuit 单功能电路
unifunctional pipeline 单功能流水线
unifying composition 一致置换合成
Unigel 尤尼杰尔炸药
unignited basis 未灼烧基
unigraph design storm procedure 单位线设计暴雨法;设计暴雨单位线法
uni-grinder 单辊磨
uniguide 单向波导管
unihabitable 不适于居住的
unihibited 不受禁止的
unijunction 单结
unijunction transistor 单结晶体管
Unilac 统一蜡克板(一种表面涂清漆的硬质板)
unilacunar node 单叶隙节
unilaterably connected 单侧连通的
unilateral 单向性;单向的;单方面(的);单侧向;单侧(的);单边的
unilateral amplifier stage 单向放大级
unilateral antenna 单向天线
unilateral arrangement 单边协议
unilateral array 单向阵列
unilateral bearing 单向方位;单侧方位
unilateral circuit 单向电路;不可逆电路
unilateral component 单侧支
unilateral conduction 单向导电(性)
unilateral conductivity 单向电导率;单向导电性;单向导电率;单向传导性
unilateral conductor 单向导体
unilateral contact line powering 单边供电
unilateral continuity 单向连续;单侧向连续性
unilateral contract 单项合同;单方面承担义务的合同
unilateral control 单向控制
unilateral deformation of riverbed 河床单向变形
unilateral denunciation 单方面废止;单方面废除
unilateral diffusion 单向扩散
unilateral digraph 单侧有向图
unilateral direction conduction rectifier 单向导

电整流器
unilateral discharge 单侧卸料
unilateral gear 单侧齿轮
unilateral heating 单面加热
unilateral illumination 单侧照明
unilateral impedance 单向阻抗
unilateral import quota 自由配额
unilateralization 单向化
unilateral lighting 单侧照明
unilateral limit 单侧极限
unilateral load 单侧荷载
unilaterally aligned standard 单边看齐标准
unilaterally compound pitting 单侧复纹孔式
unilaterally connected graph 单向连通图
unilateral matching 单向匹配
unilateral method 一侧观察法
unilateral mistake 单方错误
unilateral moving dislocation 单向移动错位;单向位移错位
unilateral network 单向网络
unilateral observation 一侧观察
unilateral parking 城市道路单边停车;道路单侧停车;道路单边停车;单边停车
unilateral prohibition of waiting 一边禁止停车制
unilateral repudiation 单方拒付
unilateral repudiation of a debt 单方不履行债务
unilateral risks 风险单一
unilateral splitting 单向分岔
unilateral spotting 一侧弹着观察法
unilateral stream 一侧支流水系
unilateral surface 单侧曲面
unilateral switch 单向开关
unilateral system 单向制;不对称制
unilateral system of parking 单边停车制
unilateral thickness ga(u)ge 单侧面厚度计
unilateral tolerance 单向公差
unilateral tolerance method 单向公差法
unilateral transducer 单向转换器;单向换能器;单向传感器
unilateral transfer 片面转移;片面移转
unilateral transmission 单侧传动
unilateral undertaking 单方面承诺
unilateral waiting 城市道路单边停车;一边禁止停车制;道路单侧停车;道路单边停车;单边停车
unilateral winding 单向绕法
uniline 单一路线;单行;单相线路;单线;单列
unilinear development 直线发展
uniline message 单线信息
unilluminated hemi-sphere 暗半球
unilobate delta 单叶三角洲
unilocular 单房的
uniloy alloy 铬镍合金
Unimate 通用机械手
unimeter 多刻度电表;伏安
unimixing 均匀混合
unimodal 单峰的
unimodal current rose 单向水流玫瑰图
unimodal distribution 单峰分布
unimodal function 单峰函数
unimodal hydrograph 单峰过程线
unimodality 单峰性
unimodel frequency curve 单峰频率曲线
unimode magnetron 单模磁控管
unimodular 单位模的
unimodular matrix 么模矩阵;单模矩阵
unimodular property 么模特性
unimodule 单组件
unimodulus matrix 么模阵
unimolecular elimination 单分子消除
unimolecular film 单分子膜
unimolecular reaction 单分子反应
unimolecule 单分子
unimpaired 未受损害的
unimpaired capital 未动用资本
unimpaired visibility 未削弱视界
unimplemented 未实施的
unimpregnated 未浸染的
unimproved airstrip 未铺筑道面的简易机场
unimproved area 未经平整的储存区
unimproved footing 未处理地基
unimproved land 未处理土地;未开发地;未改善土地
unimproved runway 未铺筑道面的跑道;土跑道
unimproved type 未改良型

unimproved value 没有增加的价值
unincorporated area 非市政区
unincorporated association 非公司团体
unincorporated business income 非公司工商业组织收入
unincorporated enterprise 非公司企业
unincorporated firm 非公司厂商
unincorporated government enterprise 非公司组织的政府企业
unincorporated private enterprise 非公司私营企业
unindorsed check 未经背书支票
uninflammable 不易燃烧的;不易燃的
uninflated 未升高的;未加压的;未充气的
uninhabitable 不适合居住的
uninhabited island 荒岛;无居民的岛
uninhabited oasis 无人烟绿洲
uninhibited fuel 无抑止剂汽油
uninhibited oil 未加抑制剂的油品
uninjured 未受损伤的
uninked ribbon 无油墨色带
uninoculated 未经变质处理的
uninodal 单节的
uninspected 未经检验的;未经检查的
uninstrumented strip 无设备的跑道
uninsulated 未绝缘的;不保温的
uninsulated conductor 裸电线;裸导线;无绝缘导线
uninsulated line hanger 非绝缘吊弦
uninsulated overlap 非绝缘锚段开关
uninsulated overlapped section 非绝缘搭接区
uninsulated pipe 未保温的管道
uninsulated surface 非绝缘面
uninsulated transition mast 非绝缘转换柱
uninsulated wire 无绝缘电线
uninsulation 非绝热
uninsurable 不能保险的
uninsurable risk 不可担保的风险;不可保风险;不可保的险
uninsurable title 不能保险的房地产权
uninsured 未保险的
uninsured deposit 未保险的存款
uninsured loss 意外损失
unintegrable 不能积分
unintelligent terminal 非智能终端
unintelligibility 不清晰性;不清晰度
unintended inventory 非意愿的存货
unintentional non-linearity 固有的非线性
unintentional poisoning 事故性中毒
unintentional pollution 无意污染;无意识污染
unintentional radar interference 无意的雷达干扰
uninterchangeability 不可互换性
uninterpreted name 不解释名
uninterruptable power system 不间断电源
uninterrupted 不中断的;不停的;不(间)断的
uninterrupted concreting 混凝土连续浇筑(法);混凝土连续浇注(法)
uninterrupted cycle of erosion 完整侵蚀循环
uninterrupted duty 不间断工作方式
uninterrupted flow capacity 不间断车流量
uninterrupted power supply [UPS] 不间断(供电)电源;不中断电源
uninterrupted power system 不间断电力系统
uninterrupted-run 连续生产
uninterrupted service 无事故运行
uninterrupted space 连续空间
uninterrupted traffic flow 不间断车流
uninterrupted travel time 连续行程时间;不间断行程时间
uninterrupted working time 连续作业时间
uninterruptible power supply 连续供电
uninterruptible power system 不间断电力系统
uninterruption power sources [UPS] 不停电电源
union 中间接头;联合群落;联合目录;联合;活接头;并集
unionarc welding 磁性焊剂二氧化碳保护焊
Union Bank of Switzerland 瑞士联合银行
union bend 接头弯管;活接弯头;弯管接头;螺纹弯头
union body 联结体;连接体
union bound 一致限
union cap 螺纹压帽
union carbide 联合碳化物;合成碳化物
union carpet 双面绒头地毯
union catalog(ue) 联合目录
union chuck 双爪卡盘

union clip 连接夹(铸铁槽的);连接套箍
union cloth 交织织物
union cock 连接龙头;有管接头的水龙头;联结旋塞;接头旋塞
union colo(u)rimeter 联合比色计
union cone 联管锥
union coupling 联管节;联管器;管接头;管接合
union coupling sleeve 联接套管;连接衬套;连接套管
union cross 中间十字接头
union database 联合式数据库
Union de Banques Suisse 瑞士联合银行
union depot 联合车站
union elbow 弯头(接头);中间弯头;联管肘管;活接弯头
union ell L 形管接头
union fitting 管接头
union flange 管接法兰;管节法兰;法兰盘连接
union function 并集函数
union fund 工会基金
union gear 联杆吊货装置
union hook 组合吊货钩
union-hose connector 软管中间接头
union intersection principle 交并原理
unionized 未游离的
unionized ammonia 未离子化氨
union joint 管(子)接头;管节连接;管接合;联节管;联管节;连接器接头;喇叭管连接;活接头;连接管
union kraft 夹柏油防潮纸;双层夹沥青包装纸
union lever 连接杆
union liner 活接头垫圈
union link 结合杆
union management cooperation 劳资合作
union-melt powder 合熔焊粉
union melt process 联合熔接法
union-melt weld 埋弧焊
union-melt welding 埋弧自动焊;合熔式自动电焊;焊剂层下自动焊
union metal 铅基础土金属轴承合金
union mortgage clause 标准抵押权条款
union nipple 联管锥;接头;接管接头
union nut 连接螺套;联管螺帽;接管螺帽;管接螺母;联螺母;联管螺母;接管螺母;活接头螺母;接合螺帽
union of graph 图合并
Union of International Architects [UIA] 国际建筑师协会
Union of International Engineering Organizations 国际工程机构联合
union of sets 集合的联合;集合的并
union of symbol 联合符号
union paper 夹柏油防潮纸
union piece 连接零件
union pipe joint 管接头
union purchase rig 联杆吊货索具
union purchase system 联杆吊货法;双索单钩吊货系统;双杆联吊装卸方式
union reducer 渐缩管接头;缩接;缩管接头
union ring 联管节
union rope socket 联合型绳卡
union screw 管接头对动螺纹
union(-screw) coupling 联管节;联管接
union-screwed bonnet 螺纹压盖
union socket 联管套节
union special double lockstitch machine 双线锁缝机
union spigot 联管锥
union station 总(车)站;联运站;联合车站
union stud 联管节柱螺栓
union suit 工作服
union swivel 旋转联管节;活接头
union swivel end 转环连接端;联管节旋转端
union tail 联结尾管;连接尾管
union tee 三通活接头;T形管接头;中间三通接头;连接丁字管;活接三通;三通管接头
union thimble 活接头套圈;互套心环;嵌圈
union three way cock 联合三通旋塞;连接三通旋塞
union type 联合类型
union unt 接头螺套
union vent 合用排气管
union wage 规定工资
union wrench 管接头扳手;联管节扳手
union yarn 混纺纱

unioresistant glass 不透紫外线玻璃
uniparous branching 单出聚伞状分枝式
uniparous cyme 单歧聚伞花序
uniparted hyperboloid 单叶双曲面
unipath 单通路；非公用控制部件
uniphanar bending 平面弯曲
uniphanar filament 单平面灯丝
uniphase 单相(的)
uniphase amplifier 单相放大器
uniphase output 定相输出；单相输出
unipivot 单支轴；单枢轴(的)
unipivot instrument 单枢轴型仪表
unipivot support 单点支撑式
uniplanar 同平面的；单切面；单平面的
uniplanar bending 平面弯曲；同面弯曲
uniplanar flow 单面流
uniplane 单平面
unipod 独脚架
unipolar 含同性离子；单向的；单极(性)的；单极(向)
unipolar armature 单极电枢
unipolar conduction 单极传导
unipolar device 单极器件
unipolar dynamo 单极(发)电机
unipolar field effect transit 单极场效应晶体管
unipolar generator 单极发电机
unipolar group 单极群
unipolar induction 单极性感应
unipolarity 单极性
unipolar lead 单极导程
unipolar machine 单极电机
unipolar magnetic region 单极磁区
unipolar non-equilibrium conductivity 单极非平衡电导
unipolar pulse 单极性脉冲
unipolar sunspot 单极黑子
unipolar transistor 单极性晶体管；单极晶体管
unipolar transmission 单极传输
unipole 无方向性天线
unipole antenna 无方向性天线；单极天线
unipotential 等势(的)；单势；单电位
unipotential cathode 等电势阴极
unipotential electrostatic lens 单电位静电透镜
unipotential focus system 单电位聚焦系统
unipotential gun 单电位电子枪
unipotential lens 单电位透镜
uniprocesser system 单处理系统
uniprocessing 单(机)处理
uniprocessing system 单处理系统
uniprocessor 单(片)处理机
uniprocessor operating system 单处理机操作系统
uniprocessor system 单处理机系
uniprogrammed 单道程序的
uniprogrammed control system 单道程序控制系统
uniprogrammed system 单道程序系统
unipump 整体电动泵；组合泵
unipunch 点穿孔(器)；单元穿孔；单穿孔器；单穿孔机
unipunch pit sinking 单孔沉井
unique 独特的
unique code 专用化码；专用代码
unique continuation 唯一开拓
unique continuation theorem 唯一开拓定理
unique data system 唯一数据系统
unique decomposition theorem 唯一分解定理
unique equilibrium composition 孤平衡成分
unique equilibrium distribution 孤平衡分布
unique existence theorem 唯一存在定理
unique factor 单一因子
unique factorization 唯一析因
unique factorization domain 唯一析因整环
unique factorization theorem 唯一析因定理
unique feature 特色；独特特点
unique file 独立文件
unique flow 特定水流
uniquely colo(u)rable graph 唯一可着色图
uniquely definable point 明显地物点；明视地物点
uniquely defining class 唯一定义类目
uniquely ergodic 唯一遍历的
uniqueness 唯一性；独特性；独特的；单一性
uniqueness condition 唯一性条件
uniqueness factor variance 唯一因子方差
uniqueness of decision situation 决策形势的唯一性

uniqueness of judgements of probability 概率判断的唯一性
uniqueness of name 名字的唯一性
uniqueness of solution 解的唯一性
uniqueness principle 唯一性原理
unique proximity 唯一邻近
unique slip surface 单一滑动面
unique solution 唯一解(法)
unique uniformity 唯一的一致性
unique value 唯一值
unique wear-resistant white cast iron 高效抗磨白口铸铁
unique wedge design 独特楔设计
uniradiate 单放射形的
unirecord 单记录
unirecord block 单记录块
uniroller 单辊磨；单辊混炼机
Uniroyal 二氯萘酯(一种杀虫剂)
unirradiated 未经照射的
Unisan 统一马桶(一种用化学液体的马桶)
Uniseco 统一塞科(一种装配式住房牌号)
uniselector 旋转式寻线器；旋转式选择器；单动作选择器
uniselector distribution frame 旋转式选择器配线架
uniserial 同系列的；单系列；单排的
uniserial circle 单行环
uniserial crochet 单行趾钩
uniseriate 单列的
uniseriate ray 单列的射线
uniset 联合装置；单体机
uniset console 专用控制台；单体机控制台
unishank cementing machine 组合内底涂胶机
unishank construction 组合内底的结构
unishank process 组合内底制作工艺
unishear 手提电剪刀；单剪机
uniside layout main tracks 正线一侧式布置
unisilcates 单硅酸盐类
unislip propeller 等滑脱推进器
unisolated 未隔开的；未绝缘的
unisparker 单一发火器；单火花发生器
unissued bond 未发行债券
unissued capital stock 未发行股本
unissued mortgage bonds 未发行抵押债券
uni-stable state circuit 单稳态电路
unistor 不对称电阻
unistrand 单列(的)；单股
unistrand elevator 单列提升机
unit 基金单位；机组；单元(的)；单位
unit absorber 吸声附件；吸声构件；组合吸声器
unit acceleration constant 机组加速度常数
unit acceleration force 单位加速力
unit acceleration force curve 单位加速力曲线
unit activity 单位活度
unit address 装置地址；设备地址；单元地址；部件地址
unit air-conditioner 单位空(气)调(节)器；整装式空气调节器；窗式空气调节器
unit air heater 暖风机(组)
unit amount 单位(当)量
unit amplification 单级放大
unit amplifier 组合放大器
unit amplitude 单位振幅；单位幅角
unit and group bases of depreciation accounting 个别及分类折旧
unit angle 单位角
unit antenna 单位天线
unit apparent energy consumption 单位视在能耗【铁】
unit area 单位面积
unit area acoustic impedance 声比阻抗；单位面积声阻抗
unit-area capacitance 单位面积电容
unit area fee 单位面积收费
unit-area impedance 单位面积阻抗
unit area loading 单位面积荷载
unit area mechanical admittance 单位面积的力学导纳
unitarian hypothesis 一元论；一元假说
unitarian theory 一元论学说
unitarity 幺正性；统一性
unitarization of economic indicators 经济指标单一化
unitary 单元的；单式
unitary air conditioner 整体式空气调节器；单元式

空调机；整装式空气调节器；窗式空调机
unitary air conditioning system 单元式空调系统；单元式空调设备；单元式空调方式
unitary body 整壳式车身；全承载车身
unitary budget 统一预算
unitary code 一位代码；单代码
unitary concentrate 单位精矿回收
unitary control 独立控制
unitary controller 独立控制器
unitary controller interface 独立控制器接口
unitary declaration 单(一)说明
unitary decuplet 幺正十重态
unitary elasticity 一致弹性
unitary elasticity of demand 需求的单一弹性；需求弹性为一
unitary extension 一元(的)拉伸
unitary group 酉群
unitary-hydrograph method 单位水文过程线法
unitary impedance converter type 单阻抗变换器形式
unitary income tax 综合所得税
unitary matrix 酉矩阵；幺正矩阵；单式矩阵
unitary metal 单位金属回收
unitary multiplet 幺正多重态
unitary octet 幺正八重态
unitary operation 单值运算
unitary operator 幺正算子；幺正算符；单式算子
unitary ore 单位原矿回收
unitary packing 单件包装
unitary product 单位成品回收
unitary rate 单一汇率
unitary ratio 单位换算比
unitary representation 酉表示；幺正表示
unitary restriction 酉限制
unitary scattering factor 幺正散射因数
unitary space 酉空间；单元空间
unitary spin 幺旋
unitary stream profile 单线生产方式
unitary subgroup 酉子群
unitary symmetry 幺正对称
unitary symmetry theory 幺正对称理论
unitary symplectic group 酉辛群
unitary system 一元化制度；单元系(统)；单一体系
unitary tangent bundle 酉切丛
Unitary thermal polymerization 一元式气体热叠合过程
unitary transformation 酉变换
unitary transformation group 酉变换群
unitary transform of image matrix 影像矩阵酉变换
unitary vector space 酉空间
unit assemblage 总成
unit assemble 分段装配法
unit assembly 总成
unit assembly drawing 分总图；部件组合图；部件装配图；部分组合图
unit assembly shop 单元组装车间
unit at the grass-roots level 基层单位
unit automatic exchange 小型自动电话交换台；自动电话交换设备；内部自动电话交换机
unit ball 单位球
unit barge 分节驳船
unit batch can 单元料罐
unit bay 机组段
unit bearing assembly 带座球面轴承
unit binormal 单位副法线
unit bit 装置位
unit bit hydraulic horse power 钻头比水马力
unit bit weight 单位钻压
unit block 组匣；组合
unit block assembly rack 组合架
unit block interlocking 组匣式联锁
unit-block system 组匣系统；单元块制
unit block type all-relay interlocking 组合式电气集中联锁
unit body 个体建筑(建筑群中的)；单元体
unit bolt nut 螺栓螺母组合
unit bond 直缝砌砖法
unit breaking force 单位制动力
unit brightness 单位亮度
unit-built block 工厂预制房屋
unit-built building 工厂预制房屋
unit-built house 工厂预制房屋
unit cable 组合电缆
unit capacity 装置处理量；机组容量；单位容量；单

机容量
unit card 目录卡
unit cast 整体铸造;整体浇筑的
unit casting 单一(品种)铸件
unit cell 晶胞;基本晶格;单元干电池;单位晶格;单位晶胞;单室浮选机
unit cell dimension 晶胞大小
unit-cell vector 单位晶格矢量
unit character 单位性状
unit charge 单位电荷
unit chassis-body construction 底盘车体成套结构
unit check 设备检验;部件校验
unit circle 单位圆
unit clamp 专用夹具
unit clean 组件清洁
unit coefficient 单位系数
unit complex forcing function 单位复抗力函数;单位复抗剪函数
unit compressive stress 单位抗拉应力
unit computation 单元计算
unit conditioner 单位空(气)调(节)器
unit configuration 装置构型
unit connection 单元连接;成组连接
unit connector 组合连接器
unit constructed conveyer 单元组合式输送机
unit construction 元件组合建筑;组件结构;整体结构;独立装置;单元结构;成套组合体;分段组装
unit construction bridge 元件组合桥;装配式桥(梁);标准构件桥
unit construction bridge system 元件组合桥系统
unit construction computer 组件式计算机;部件式计算机
unit construction model 整体结构模型
unit construction mo(u)ld 整体结构模
unit construction of machine tools 机组编制系统
unit construction principle 组合结构原理
unit construction system 组合结构系统
unit consumption 单位消耗
unit consumption of material 材料单位消耗量
unit-contactor type 单元接触器式
unit contact pressure 单位接触压力
unit content 机组容量
unit content of cement 单位水泥用量
unit contribution 单位贡献
unit control 单元控制;单位控制;部件控制
unit control block 部件控制board
unit control error 单机控制误差
unit control information 部件控制信息
unit control module 部件控制模块
unit control room 机组集中控制室
unit control word 部件控制字
unit conversion 单位换算
unit conversion value 单位换算价值
unit conveyer 联合输送机
unit cooler 独立制冷机;直膨式冷却器;冷却机组;冷风机;设备冷却器;单元制冷器;单位冷却器
unit cost 单位投资;单位造价;单位成本;单位产品成本;单位
unit cost depreciation method 单位成本折旧法
unit cost in place method 构件成本估价法
unit cost of loading/discharging 装卸单位成本
unit cost of material 单位材料成本
unit cost of production 生产单位成本
unit cost of products 产品单位成本
unit cost per ton-kilometre 吨/公里单位成本
unit cost technique 单位成本技术
unit couple 单位力偶
unit creep 单位徐变
unit crest width 单位波峰宽度
unit cross-sectional area 单位(横)截面积
unit crystal 单晶体;单晶
unit cube 单位立方体
unit current repair 总成小修
unit deformation 单位形变;单位变形
unit-delay operator 单位延迟算子
unit deletion 设备删除
unit demand of electricity 机组自用电
unit depreciation 个别折旧(法);单件折旧
unit design 组合设计;总体设计;装置设计;单元设计;单体设计
unit diagnostics 部件诊断(程序)
unit die 组合压铸型;组合压铸模;成套模
unit direct digital control system 装置直接数字控制系统

unit discharge 单位流量;单宽流量
unit displacement 单位位移
unit displacement method 单位位移法
unit distance 天文单位距离;单位距离;单位间距
unit distribution 单位分布
unit dotter 刻点仪
unit double impulse 双元脉冲
unit doublet 双元
unit doublet function 二导阶跃函数
unit draft 单机牵引
unit drill 分组式条播机
unit drive 单位传动;单独驱动;单独传动
unit drive clutch 单独驱动离合器
unit dryer 单元干燥器
unit dry weight 干单位重;干容重
unit dry weight of sediment 泥沙干容重
unit dry weight of soil 土的干容重
unite 合一
unite as one 团结一致
United Economic Commission for Europe 欧洲联合经济委员会
united equipment 设备机组
united expansive agent 复合膨胀剂
united inches (玻璃、嵌板等的)半周长;综合尺寸
United Kingdom Atomic Energy Authority 英国原子能管理局
United Kingdom Chemical Information Service 英国化学情报处
United Mortgage Bankers of America 美国联合抵押银行协会
United Nation Department of Economic and Social Affair 联合国经济社会事务部
United Nation Economic and Social Council 联合国经济社会理事会
United Nation Organization Atomic Energy Commission 联合国原子能委员会
United Nations 联合国
United Nations Association of USA 美国联合国协会
United Nations Building 联合国总部大厦
United Nations Bureau of Technical Assistance Operations 联合国技术援助业务局
United Nations Capital Development Fund 联合国资本开发基金
United Nations Center for Housing, Building and Planning 联合国住房、建设与规划中心
United Nations Centre for Economic and Social Information 联合国经济社会资料中心
United Nations Centre for Human Settlements 联合国人类居住中心
United Nations Centre for Industrial Development 联合国工业中心
United Nations Centre for Natural Resources, Energy and Transport 联合国自然资源、能源和运输中心
United Nations Commission on Human Settlements 联合国人类住区委员会
United Nations Commission on International Trade Law 联合国国际贸易法律委员会
United Nations Conference on Environment and Development 联合国环境与发展会议
United Nations Conference on Human Settlements 联合国生境人类住区会议;联合国人类住区会议
United Nations Conference on New and Renewable Sources of Energy 联合国新能源及可再生能源会议
United Nations Conference on Prevention of Desertification 联合国防止沙漠化会议
United Nations Conference on Science and Technology for Development 联合国科学技术开发会议
United Nations Conference on the Human Environment 联合国人类环境会议
United Nations Conference on Trade and Development 联合国贸易和发展会议
United Nations Convention on the Law of the Sea 联合国海洋公约
United Nations development plan 联合国十年发展计划
United Nations Development Program(me) 联合国开发计划署
United Nations Disaster Relief Office 联合国救灾协调专员办事处;联合国救灾处

United Nations Economic and Social Commission for Asia and Pacific 联合国亚太经济社会
United Nations Economic Commission for Asia and Far East 联合国亚洲及远东经济委员会
United Nations Economic Commission for Europe 联合国欧洲经济委员会
United Nations Economic Development Administration 联合国发展局
United Nations Educational and Scientific and Cultural organization 联合国教(育)科(学)文(化)组织
United Nations Educational and Scientific and Cultural organization Division of Water Science 联合国教(育)科(学)文(化)组织科学部
United Nations Educational Organization 联合国教育发展组织
United Nations Educational Scientific and Culture Organization 联合国科教文组织
United Nations Environmental Conference 联合国环境会议
United Nations Environmental Education Conference in Tbilisi 第比利斯国际环境教育会议
United Nations Environmental Fund 联合国环境基金
United Nations Environmental Program(me) 联合国环境规划委员会
United Nations Environment Program(me) 联合国环境规划署
United Nations Food and Agriculture Organization 联合国粮农组织
United Nations Framework Convention on Climate Change 联合国气候变化框架公约
United Nations Fund for Population Activities 联合国人口活动基金
United Nations Habitat and Human Settlement Foundation 联合国生境和人类住区基金会
United Nations Headquarters Building 联合国大厦
United Nations High Commissioner for Refugees 联合国难民事务高级专员办事处
United Nations Human Environment Conference 联合国人类环境会议
United Nations Industrial Development Fund 联合国工业及发展基金
United Nations Industrial Development Organization 联合国工业及发展组织
United Nations Information Organization 联合国情报组织
United Nations Information Service 联合国情报服务站
United Nations Institute for Training and Research 联合国训练研究所
United Nations Laws of the Sea Conference 联合国海洋会议
United Nations Layout Key for Trade Documents 联合国贸易单据固定格式
United Nations Office of Public Information 联合国救灾协调专员办事处
United Nations Organizations 联合国组织
United Nations plan 联合国计划
United Nations program(me) of technical assistance 联合国技术援助方案
United Nations Relief and Rehabilitation Administration 联合国善后救济总署
United Nations Research Institute for Social Development 联合国社会发展研究所
United Nations Revolving Fund for Natural Resources Exploration 联合国自然资源勘探循环基金
United Nations Scientific Committee on the Effects of Atomic Radiation 联合国原子辐射影响科学委员会
United Nations Secretarial Building 联合国大厦
United Nations serial number 联合国产品编号
United Nations Social Defence Research Institute 联合国社会防护研究所
United Nations Special Commission on Environment 联合国环境特别委员会
United Nations Special Fund 联合国特别基金
United Nations Standards Coordinating Committee 联合国标准调整委员会
United Nations Statisitical Office 联合国统计局
United Nations Technical Assistance Administration 联合国技术援助管理处
United Nations Technical Assistance Board 联

合国技术援助理事会
United Nations Water Conference 联合国水事会议
United Nations World Intellectual Property Organization 联合国世界知识产权组织
united paleocontinent 联合古陆
united regenerator 连通式蓄热室
United States Advisory Committee on Weather Control 美国天气控制咨询委员会
United States Agency for International Development 美国国际开发署
United States Agricultural Research Center 美国农业研究中心
United States Agricultural Research Service 美国农业研究服务局
United States Army Corps of Engineers 美国陆军工程兵
United States bed material sampler 美国式底质采样器
United States Bureau of Reclamation 美国垦务局
United States Bureau of Standards 美国标准局
United States Bureau of the Census 美国调查局
United States Clean Water Act 美国清洁法
United States Coast and Geodetic Survey 美国海岸和大地测量勘查局
United States Coast Guard 美国海岸警卫队
United States Department of Agriculture 美国农业部
United States Department of commerce 美国商务部
United States Department of Energy 美国能源部
United States Environmental Protection Agency Water Quality Criteria 美国环境保护局水质基准
United States Environment and Resources Council 美国环境和资源委员会
United States Environment Protection Agency 美国环境保护局
United States Environment Protection Council 美国环境和资源委员会
United States Federal Power Commission 美国联邦电力委员会
United States gallon 美国加仑
United States Geographic Board 美国地理委员会
United States Geologic(al) Survey 美国地质勘测局
United States Hydraulics Laboratory 美国国家水力试验所
United States Hydrographic(al) Office 美国水文办公室
United States Inter-Agency Committee on Water Resources 美国资源局委员会
United States Joint Army-Navy Specifications 美国陆海军联合规范
United States League of Savings Associations 美国储蓄联合会
United States Maritime Administration 美国航运总署
United States Maritime Commission 美国航务委员会
United States National Bank 美国国家银行
United States National Bureau of Standards 美国国家标准局
United States Navy Hydrographic(al) Office 美国海军水道测量部
United States of American 美利坚合众国
United States Office of Saline Water 美国咸水办公室
United States of Soil 美国土壤局
United States of the Budget 美国预算局
United States Patent 美国专利
United States Pharmacopoeia 美国药典
United States President's Water Resource Policy Commission 美国总统水资源政策委员会
United States President's Water Resource Policy Consulting Commission 美国总统水资源政策咨询委员会
United States President's Water Resource Protection Commission 美国总统水资源保护委员会
United States Public Health Service 美国卫生署
United States saving bond 美国储蓄债券
United States Soil Conservation Service 美国土壤保护局
United States Standard Ga(u)ge 美国标准线规；美国标准规
United States Standards 美国标准

United States Steel Sheet Gage 美国薄钢板规格
United States Weather Bureau 美国气象局
united structure 联合构造
united-type fender 分片组合式碰垫
united water conservation district 联合水源保护
unit efficiency 装置效率；机组效率；分部效率
unit electric(al) flux 单位电通量
unit electrostatic charge 单位静电荷
unit element 单位元素；单位元
unit elongation 线应变；延伸率；单位伸长
uni-temper mill 双二辊式平整机；复二辊式平整机
unit engineering 装置工程
unit equipment 组合装置；组合设备
unit erection at berth 船台单元安装
uniterm 单元名；单一项目
uniterm design 单项工程设计
uniterm system 统一术语系统；单项系统
unit error 单位误差
unit essential equipment 单元主要设备；单位主要装置
unit tangent 单位切线
Unitex 统一特斯克板（一种硬质板牌号）
unit exception 设备异常；部件异常
unit extension 线应变；单位伸长
unit fabricator 单元制作者；单元制作厂
unit factor 装置因素；装置因数；单元系数
unit factor clause 单元因子子句
unit Faraday tube 单位法拉第管
unit feed 单位进刀量
unit feeder 单馈线
unit file 单位文件
unit flow 单位流量
unit flow rate 单宽流量
unit force 单位力
unit for chemical measurement 化学检测单位
unit-foreman 班组工长；工段工长
unit for interior work 室内工程用的型材
unit frame 整体式机架；同架
unit function 单元函数；单位函数
unit furniture 成套家具
unit furniture kitchen 配有家具的厨房单元
unit gain 单位增益
unit gap 单位间隔
unit generalized load 单机的综合负载
unit generator connection 发电机单元连接
unit graph 单位（水文）图；么线；单位（过程）线
unit green area 单位绿地
unit ground pressure 单位土壤压力
unit head 动力头
unit head boring machine 动力头式镗床
unit head broaching machine 动力头式拉床
unit head drilling machine 动力头式钻床
unit head machine 组合头钻床
unit head milling machine 动力头式铣床
unit head tapping machine 动力头式攻丝机
unit head turning machine 动力头式车床
unit heater 供暖机组；个别式供暖机组；暖风机；单位散热器；单位供暖机；单个式加热器；组合式供暖器；单位加热器；单位供暖器；单位采暖装置
unit heater-high-static type 高静压式暖风机
unit heater-low-static type 低静压式暖风机
unit heater with centrifugal fan 离心式暖风机
uni-thick 等厚材料；均匀厚度
unit-holder 投资单位持有者
unit horsepower 单位马力
unit humid weight of soil 土壤湿容重
unit hydrograph 么线；单位（水文）过程线；单位流量（水文）曲线（图）
unit hydrograph analysis 单位过程线分析
unit hydrograph application 单位过程线应用
unit hydrograph duration 单位过程线历时
unit hydrograph method 单位线法；单位（水文）过程线法
unit hydrograph synthesis 单位过程线合成
unit improvement factor 装置改进因数
unit impulse 单位脉冲
unit impulse function 单位脉冲函数
unit impulse response 单位脉冲响应；单位脉冲反应
unit in charge of construction 施工单位
unit income statement 单位损益计算书
unitin convoy 运输队
unit-independent fire extinguishing system 单元独立灭火系统
unit information 单元信息；部件状态信息

uniting of states 态的合并
unit injector 组合式喷射器；整体式喷油器
unit in kind 实物单位
unit-in-place method 按分项费用估算总造价法
unit installation schedule 机组安装进度
unit installed capacity 单位装机容量
unit insulator 单元绝缘子
unit interface 部件接口
unit interlock test 综合联动试验
unit interval 单位信号时间；单位时间（间隔）；单位时段；单位区间；单位间隔
unit investment cost 单位投资
unit investment quota 单位投资（定）额；单位产品投资额
unitised train 单元列车【铁】
unitization 联合经营；单元化；成组化
unitization device 成组工具（货物）
unitization of cargo 货物装运单元化；货物装运成组化
unitize 积木化
unitized 组合的；组成的；统一的；通用化的；成套（的）
unitized application 成组施肥
unitized bathroom unit 统一化浴室（连厕所）单元；标准化浴室单元；规格化浴室单元
unitized body 组合车身
unitized cargo 成组货物
unitized cargo-handling 成组装卸
unitized cargo tons 成组吨
unitized channel storage 通用的通道存储器
unitized construction 组织结构
unitized design 组合设计；通用化设计
unitized exchanger 组合式换热器
unitized goods transportation 成组运输
unitized joint 整体接头；联铸接头
unitized kitchen unit 统一化厨房单元
unitized machine 统一化电机；成套机器
unitized mill 单台传动开炼机
unitized press 自给式平板机
unitized shipment 集装箱运货
unitized stacking of cargo 成组堆垛
unitized substructure 组件底座
unitized tooling 组合加工
unitized train 单元列车【铁】
unitized transport 成组运输
unitized unit 组合单元；统一化单元
unitized unit method 统一化房屋单元建造法
unitizing 成组
unit kitchen 单元式厨房
unit labor cost 单位劳动成本
unit labor requirement 单位必要劳动量
unit labo(u)r cost 单位人工成本
unit lateral contraction 单位横向收缩
unitlaterally truncated distribution 单侧截尾分布
unit lateral pressure 单位侧压力
unit layouts of treatment plant 处理厂单元布局
unit length 单位长度
unit life depreciation method 按各项资产不同使用年限分别折旧法
unit lift thickness 单层铺土厚度
unit limitation of liability 按单位的责任限制
unit line 单位线
unit line connection 单元结线
unit load capacity 单位承载力
unit load cargo 成组货物
unit load device 成组装运设备
unit load handling 单元化运输
unit load(ing) 单位荷载；机组负荷；单元装运法；单位负荷；单位负载；成组货件；成组运输；货物装载单位
unit loading factor 单位负荷因数；单元负荷系数；单位负荷因素
unit load method 单位荷载法；成组运输法
unit load S/R machine 单元型有轨巷道堆垛起重机
unit load system 单位装载运输体制；单元装载运输方式；单元储[贮]存系统；成组运输系统
unit lock 整装门锁；单锁
unit locomotive moving kilometres 单机走行公里
unit locomotive moving ratio 单机走行率
unit longitudinal deformation 单位纵向变形
unit machine construction 组合机结构
unit magnetic mass in the electromagnetic system 电磁制中的单位磁质量
unit magnetic pole 单位磁极

unit magnification 等大
unit maker 装配单元预制厂
unit masonry 单位砌块;砌体单位;砌块
unit masonry work 砌筑单元
unit mass 单位质量
unit matrix 单位矩阵
unit measurement 单位量度
unit melter 单元窑;单元式池窑
unit membrane 单位膜
unit method 计件法
unit method of depreciation 计件折旧法
unit mix 单位配合比
unit module 单位模数
unit mo(u)ld 整体模(具);单塑模
unit mount 组件安装
unit nail value 单位握钉值
unit normal 单位法线
unit number 单位数
unit objectives 基层单位目标
unit of acceleration 加速度单位
unit of account 计账单位;计算单位
unit of acoustic(al) absorption 吸音单位
unit of activity 活度单位
unit of adjustment 调整单位
unit of angle domain topography 角域地形单元
unit of area 面积单位
unit of bond 砌层单元;砌合单元;砌层段
unit of capacity 容量单位
unit of coal 单位煤量
unit of compiling 编图单位
unit of concentration 浓度单位
unit of construction 构件;建筑单元
unit of cost 成本单位
unit of currency 货币单位
unit of enumeration 计数单位
unit of equipment 整套装置;整套设备
unit of error 误差单位
unit of fit 配合单位
unit of force 力(的)单位
unit offset 单位跃变
unit of geographic(al) division 地理区划单位
unit of geologic(al) mapping 填图地质单位
unit of heat 热(量)单位
unit of higher order 高(要)价单位
unit of illumination 照明单位;照度单位
unit of investigation work 已有调查工作单位
unit of issuing contract 发包单位
unit of length 长度单位
unit of light 光的单位
unit of luminous intensity 发光强度单位
unit of mass 质量单位
unit of material volume 材料体积单位
unit of measure 测量单位
unit of measurement 计量单位
unit of measurement of product 产品计量单位
unit of observation 观察单位
unit-of-output method 生产单位法
unit of ozone 臭氧单位
unit of power 功率单位;人力
unit of produced kind 生产的实物单位
unit of production 产量单位
unit of production depreciation method 生产单位折旧法;产量单位折旧法
unit of quantity 数量单位
unit of resistance 阻力单位
unit of sampling 抽样单位
unit of securities trading 证券成交单位
unit of structure 构造单元;构件;结构单元
unit of sunken tube 沉埋管段
unit of survey 调查单位
unit of taxation 纳税单位
unit of time 时间单位
unit of traffic 交通(折算)单位;客货运输单位(吨公里或旅客公里)
unit of transfer 传送单位;传输单位
unit of value 价值单位
unit of value added 增值单位
unit of volume 体积单位
unit of water measurement 量水单位
unit of weight 重量单位;权数单位;单位权
unit one 第一个工件
unit oneth kilometric indicating drum 个位公里里程记录鼓

unitool 统一刀盘
unit operating cycle 单位操作循环开关
unit operating equipment 单元操作设备
unit operating expense 单位营业支出
unit operation 单元处理;单元操作;操作一致;操作过程
unit operation(al) of cargo handling 操作工序(指装卸)
unit operation cost 单位经营成本
unit operations of chemical engineering 化工单元操作
unit operation ton 工序吨
unit operator 单位算子
unit optimization 装置最优化
unitor 连接器
unit output 机组出力;单位出力
unit outside system 制外单位
unit package 最小单位包装
unit packing list 装箱单;集装箱装箱单;成组货物装箱单
unit packing machine 单元式包装机
unit peak discharge 单位峰值流量
unit peak flow 单位洪峰
unit performance 设备性能
unit period 单位时段
unit permeance 单位磁导率
unit plan 单位平面(建筑物平面构成的基本单位)
unit plane 单位面
unit plant 单元机组;单元布置电站
unit pole 单位磁极;单极
unit pollutant loading rate 单位污染物负荷率
unit potential 单位电势
unit powder consumption 单位耗药量
unit power 单位功率
unit power clutch 单独驱动离合器
unit power mounting 发电机组组件安装
unit power output 单机出力
unit power plant 动力机组;成套发电设备;成套动力装置
unit-power sighting telescope 一倍瞄准望远镜
unit pressing 单位压力
unit pressure 单位压力
unit price 单位造价;单价
unit price analysis 单价分析
unit price contract 单价合同;单价发包合同;单价包干;按单价承包
unit price for work 工程单位造价
unit price marine fill 填海单价
unit price method in working drawing budget 施工图预算的单价法
unit price of standard coal 标准煤单价
unit pricing 报单价;计件定价;单价与总价分标法
unit process 基本过程;单元作业;单元过程;单元反应;单元处理;反应过程
unit processes in chemical synthesis 互感单元过程
unit processes of wastewater 废水处理单元过程
unit process of cargo handling 操作过程(指装卸)
unit process of wastewater 废水单元处理
unit processor 单处理机
unit process selection 单元过程选择
unit-producing yard 单元体生产场地
unit production cost 单位生产成本
unit production method 单位生产法
unit profit 单位利润
unit profit sharing 单位利润分配
unit program(me) 单位程序
unit project 单位工程
unit-pulse signal 单位脉冲信号
unit pump 组合泵;整体式电动泵;单位泵;成套泵设备
unit quantity 单位量
unit quantity of electricity 单位电量
unit rainfall 单位雨量;单位时间雨量;单位降雨
unit rainfall duration 单位降雨历时
unit rate 单位率
unit rate contract 单价合同
unit rate of flow 单位流量
unit rate of marine fill 填海单价
unit rates and prices 单价和合价
unit reaction coefficient of rock 岩石单位抗力系数
unit record 单位记录;单位记录;穿孔卡片

unit record device 单元记录装置;单元记录设备;单位记录设备
unit record equipment 单位记录装置;穿孔卡片设备
unit record file 单位记录文件
unit record pool 单位记录组;单位记录池
unit record principle 单位记录原则;单位记录法
unit record routine 单位记录程序
unit record system 单位记录系统;单位记录系统
unit record volume 单位记录容量;单位记录卷宗
unit rectangular function 单位矩形函数
unit refrigeration system 单元式制冷系统
unit regional value 单位区域价值
unit region value estimation method 单位区域价值估计法
unit repairing method 总成修理法
unit replacement 总成更换
unit replacement system 设备修理方法
unit replacement type 单元更换性
unit representation 单位表示
unit residual elongation 单位残余伸长
unit resistance 电阻系数;单位阻力;单位抗力;单位电阻
unit response 单位响应;部件响应
unit road system 集装路运系统
unit rotation 单位转角
unit runaway speed 机组飞逸转速;单位飞逸转速
unit runoff 径流模数;单位径流(量)
units 元件
unit sample 单位样本;单位取样;单位抽样
unit sampling 单位抽样
unit sand 统一砂;标准砂
unit segment 单位线段
unit-selector switch 单位选择器开关
unit sensitivity 单位灵敏度
unit separate regulation 单机调节
unit separation 单元分隔;单位分隔
unit separator 单元分离;单元分隔符;单位分隔符
unit separator character 单元分隔字符
unit service factor 装置运转率
unit settlement 单位沉降量
units for calculating products 产品计量单位
units for fast drying 快干设备
units for reserve calculations 储量计量单位
unit shaft resistance 单位侧向阻力;单位侧面阻力
unit shaft system 基础轴制
unit shear 单位剪力;单剪(力)
unit shear bolt 单剪螺栓
unit shearing stress 单位剪切应力
unit shear joint 单剪接合
unit shear resistance 单位抗剪强度
units in operation 经营单位
unit size 单元尺寸
unit skin friction 单位表面摩擦力
units of concentration of pollutant in environment 环境污染物浓度单位
units of satisfaction 满足程度单位
unit solid angle 单位立体角
unit space 单位空间
unit spacing 单位间距;机组间距
unit speed 单位转速;单位速率;(水轮机的)比速
unit sphere 单位球面
unit square matrix 单位方阵
unit status information 部件状态信息
unit step 单位阶跃
unit step function 单位阶跃函数;单位阶梯函数
unit step response 单位阶跃响应;单位阶梯响应
unit store 零件库
unit storm 单位暴雨
unit strain 单位应变
unit-stranded cable 单元式绞合电缆
unit stranding machine 单元式绞缆机
unit stratotype 单位层型
unit strength 单位强度
unit stress 单位应力;单位面积应力
unit string 单位串
unit strut 预埋构件
unit subgrade support 单位地基承载力;地基单位承载力
unit substation 单位变电所;成套变电站;成套变电所
unit surface conductance 单位面积传导
unit swelling strain 单位湿胀应变
unit switch 组合开关;单元开关

unit switch control 单元开关控制
unit switch controller 单元开关控制器
unit symbol 单位符号
unit system 成套灶具;单元系(统);单位制
unit tank furnace 单元式池窑
unit tax 单位税
unit tensile stress 单位抗拉应力
unit tension 单位拉力
unit tensor 单位张量
unit terminal building (航空站的)分枝站屋
unit test 单体试验;单体测试
unit testing 单元测试;单位测试
unit thermal conductance 单位热传导
unit time 单位时间
unit time step 单位时间步
unit time study 单位时间研究
unit tow 固定编组顶推驳船队
unit train 专列货车;运煤列车;专列火车;(直达的)固定编组列车;单元列车【铁】
unit transformer 单元厂用变压器
unit transistor 单位晶体管
unit transport 单位运输
unit treatment 单元处理
unit trial operation 综合试运行
unit triangle 单位三角形
unit triplet 三元
unit tube 单位管
unit tube of flux 通量的单位管
unit-twin cable 单式对绞电缆
unit type 设备型号;单元式
unit type air filter 单元式空气过滤器
unit-type cable 单位式电缆
unit type collector 单元型集水器;单元型集尘器
unit type crank shaft 组合曲轴
unit-type die-casting die 整体模铸模具
unit-type drill head 组合钻床钻头
unit-type drydock 整体浮船坞
unit type optic(al) fiber cable 单位式光缆
unit type signal mechanism 组合式信号机构
unit type vent 单式屋顶通气孔
unit-tyre vulcanizer 单元外胎硫化器
unit undertaking project 建设单位
unit under test 被测部件
unitune 单钮调谐
uni-tuning 同轴调谐;单钮调谐
unit value 单位(价)值
unit value bias 单价偏度
unit value index 单价指数
unit value index of exports 出口单位价值指数
unit value index of imports 进口单位价值指数
unit vector 单位向量;单位矢量
unit vector field analysis 单位向量场分析
unit vent 合用排气管
unit ventilation 分区通风
unit ventilator 通风机组;单位通风器;组件式通风器
unit voltage 单位电压
unit volume 单位体积;单位容积
unit volume adsorber 单位容积吸附剂
unit volume change 单位体积变化;单位容积变化
unit volume expansion 单位体积膨胀
unit volume soil 单位体积土
unit volume variation 单位体积变化
unit vulcanizer 固定模平板机;单元硫化机
unit water acceptance 单位吸水量
unit water content 单位体积含水量;单位含水量
unit water quantity 水耗率
unit water use 单位用水量
unit weight 么重;重度;容重;单位重;单位权;单位(长度)重量
unit weight haulage force 单位重牵引力
unit weight in dry state 干容重
unit weight in saturated state 饱和容重
unit weight of debris flow 泥石流容重
unit weight of freshwater 淡水容重
unit weight of frozen soil 冻土容重
unit weight of mud-stone flow body 泥石流容重
unit weight of rock 块石重度
unit weight of saturated gravel 饱和砾石容重
unit weight of saturated soil 饱和土容重
unit weight of seawater 海水容重
unit weight of soil 土的重度;土(的)容重
unit weight of soil constituents 土壤成分单位重量;土粒实容重
unit weight of soil grains 土粒容重

unit weight of solid constituents 土粒实容重
unit weight of solid particles 土粒容重
unit weight of water 水的容重
unit weight power 单位重功率
unit weight test 容重试验
unit width 单位宽度
unit width leakage 单宽渗漏量
unit wind load 单位风载
unit wire fabric 单块钢丝网;单层钢丝网
unity 均一;统一性;同质;单一;单位;单数;一致;单楼
unity coupling 整体耦合;联合耦合;完整耦合
unit years 机组运行年数
unity feedback 全反馈
unity feedback system 单位反馈系统
unity formula 整数釉式
unity gain amplifier 单一增益放大器;单位增益放大器
unity gain bandwidth 单位增益频宽;均一增益带宽
unit yield 单位产水量
unity in variety 多样化的统一
unit(y) matrix 单位矩阵
unity of composition 构成的统一性;组成的统一性
unity of design 设计的统一性
unity of economic movement 经济运动的统一性
unity of form (建筑物各部分的)设计的协调;形式统一
unity of interest 同等利益
unity of invention 发明的单一性
unity of person 单一法人
unity of plan 统一图案
unity of tectonic movement 构造运动统一性
unity power factor 整功率因数
unity power-factor test 功率因数等于一的电机试验
unity ratio equation 联合比值方程
unity slope 单位斜率
unity step method 单位步法
unity theorem 单性定理
univalence 一价
univalent game 一价的对策
univariable experiment 单变量试验
univariant 单变(度)的
univariant equilibrium 单变平衡
univariant system 单变系
univariate 单变量的
univariate analysis 单变量分析
univariate distribution 一元分布;单变量分布
univariate relaxation 单变化松弛
univariate spline 一元样条函数
univariate statistics 一元统计学
universal 广用的;万向的;万能的;通用的
universal adhesive 万能胶
universal agency 总代理商
universal agent 总代理商;总代理人;全权代理人;全权代表;全球代理人;全能代理人
universal air-operated clamping fixture 一般气动夹具
universal algebra 泛代数
universal algorithm 通用算法
universal amplification 通用放大
universal amplification curve 通用放大曲线
universal amplifier 通用放大器
universal and socket joint 球窝节
universal angle block 万能角规
universal angle plate 万向转台
universal apparatus 全能仪器;万能仪;通用仪(器)
universal arc suppressing reactor 通用消弧电抗器;通用熄弧电抗器
universal asychronous receiver transmitter 通用异步收发两用机;通用异步收发器
universal attachment 万用附件;万能附件
universal autocollimator 通用自动平行光管
universal autograph 全能自动测图仪
universal automatic computer 通用自动计算机
universal automatic map compilation equipment 全能自动测图仪
universal automation 普遍自动化
universal background 普遍背景
universal back rest 万能后支架
universal ball 万向节球
universal ball joint 万向球关节;球形万向节;球形万向接头;万向球铰接;万向球度
universal bar 通用条

universal beam 通用梁
universal beam bed 通用预应力张拉台
universal beam mill 万能式钢梁轧机铣床
universal bed 通用基础;通用底盘
universal bench mill 万能台式铣床
universal bender 万向弯管机;万能弯管机
universal bevel 万能曲尺;通用斜角规
universal bevel protractor 万能量角器;万能活动量角规
universal blower 通用鼓风机
universal blowpipe 万用焊接吹管
universal board 通用插件板
universal borer 万能镗床
universal boring and drilling machine 万能镗钻两用机(床)
universal boring head 万能镗刀盘
universal boring machine 万能镗床
universal bound 泛界
universal box 通用盒子构件
universal bridge 万用电桥;通用电桥
universal buffer 广域缓冲剂
universal building crane 通用建筑施工吊车
universal bulk carrier 通用散(装)货船
universal bulk ship 通用散(装)货船
universal bulldozer 通用推土机;通用型推土铲
universal bundle 通用丛
universal buoyage 国际通用浮标;统一浮标;通用浮标
universal burner 通用燃烧器;通用喷燃器
universal button box 通用按钮箱
universal calculator 通用计算器
universal calender 通用压延机
universal calorimeter 通用量热器
universal camera 通用摄影机
universal camera control system 万能摄影操纵仪
universal camera microscope 万能照像显微镜
universal card 通用插件
universal carrier 通用运输工具
universal case opener 通用开表壳器
universal cementing agent 万能胶粘剂;通用胶粘剂
universal center block 活节中心块
universal center pin 万向节中心销
universal centertype grinding machine 万能中心磨床
universal centre block 活节中心块
universal chain 通用链
universal chain control 一般连锁控制
universal characteristic curve of hydraulic torque converter 液力变矩器通用特性曲线
universal checking and testing stand for alternating current/direct current meter [AC/DC] 交直流电表万能检验台
universal checking and testing stand for relay 继电器万能检验台
universal chrome-plated vernier angle protractor 镀铬游标万能角度尺
universal chuck 联动夹盘;万用夹盘;万能卡盘;万能夹盘;万能夹具(台);通用卡盘;三爪夹盘
universal circular saw 万能圆锯
universal clamp 广用夹;万能夹持器;通用管箍
universal class 全类
universal claw 万能拆装器
universal closure 全称闭包
universal coal cutter 万能截煤机
universal cock 万向龙头;万能龙头
universal coefficient theorem 万有系数定理
universal coil 通用式线圈
universal column 通用柱
universal combined punching and shearing machine 万能联合冲剪机
universal combustion burner 自动通风煤气炉;通用燃烧器
universal command 通用指令
universal communications switching device 通用通信转换装置
universal compact group 万有紧群
universal comparator 全能比长仪;通用比长仪
universal compass 万能两脚规
universal compensator 万能补偿器
universal compressor 万能调整器
universal compressor tractor 通用压缩机拖拉机
universal compulsator 万能强制器
universal computer 通用计算机
universal connecting rod aligner 通用连杆对准器

universal connection 万向连接
universal connector 万向接头
universal constant 通用常数
universal contact 万能接头
universal container 通用集装箱
universal control equipment 通用控制阀
universal controller 万用控制器
Universal Convention of Post 万国邮政公约
universal coordinated time 通用协调时间
universal cord circuit 通用塞绳电路
universal counter 通用计数器
universal coupler 通用联轴节
universal coupling 万向联轴器;万向(联轴)节;万向接头;万能耦合器;万能联轴器;通用联轴节
universal covering 通用覆盖
universal covering group 通用覆盖群
universal covering manifold 通用覆盖流形
universal covering space 通用覆盖空间
universal covering surface 万有覆盖面;通用覆盖面
universal crane 全向起重机;万能装卸机;万能起重机;通用起重机
universal curve 万有曲线;通用曲线
universal cutter 普通刀具;通用刀具
universal cutter and tool grinder 万能工具磨床
universal cutter grinder 万能工具磨床
universal cutting pliers 剪线钳;万能剪钳;断线钳
universal data channel 通用数据信道
universal data voice multiplexing 通用数据话音多路复用
Universal Decimal Classification 国际十进制分类法
universal decimal classification 通用十进制分类法
universal decision element 通用判断元件
universal design theory 通用的组织设计理论
universal developer 万能显影液
universal die set 通用模架
universal digital autopilot 通用数字自动驾驶仪
universal digital cartographic(al) data 通用数字地图数据库
universal digital computer 万用数字计算机;通用数字计算机
universal digital recording machine 万能数字记录机
universal dividers 万能分线规
universal dividing head 万能分度头
universal domain 万有域
universal drafting machine 全能绘图仪;全能绘图机;万能制图机;万能绘图机
universal drawbar 万能挂钩
universal drawing 通用图
universal drier 万能式干燥机
universal drill jig 万能式钻模
universal drive shaft 万向传动轴
universal drive shaft guard 万向传动轴护罩
universal driving shaft 联动轴;万向传动轴
universal ear insert 通用耳塞
universal effective set 全有效集
universal electric(al) meter 万能电表;万能表
universal electric(al) motor 通用电动机
universal electric(al) tool 万能电动工具
universal electron microscope 万能型电子显微镜
universal engineer tractor 通用工兵牵引车
universal engine stand 发动机万能试验台
universal enveloping algebra 通用包络代数
universal equalizer 多频(率)音调补偿器
universal equipment 通用设备
universal equivalent 一般等价物
universal equivalent form 一般等价形态
universal excavator 万能挖土机;万能挖掘机;通用挖土机;通用挖掘机
universal extensible joint 可伸缩万向节;可伸缩万向接头
universal extension pulley 万能牵引滑轮
universal facepiece 通用面罩
universal face plate 万能花盘
universal fairleader 万向导缆器
universal fatigue machine 通用疲劳试验机
universal fatigue testing machine 万能疲劳试验机
universal film camera 全能胶片摄影机
universal filter 万用滤波器;通用滤波器
universal finisher 通用表面整机;通用表面处理机
universal finishing stand 万能式精轧机座
universal fixture 万能夹具(台);万能工具台
universal floating tool holder 万能浮动工具夹头
universal focus lens 固定焦距透镜

universal forceps 通用钳
universal fork-lift 万能叉式提升机;万能叉车
universal forklift truck 万能装卸机
universal frequency 通用频率
universal fuel tank kit 通用燃料箱组
universal function 通用处理函数
universal function generating unit 通用函数发生器部件
universal function generator 通用函数发生器
universal galvanometer 通用电流计
universal gas chromatograph 通用气象色谱仪
universal gas constant 普适气体常量;通用气体常数
universal ga(u)ge 万能(量)规
universal gear lubricant 通用齿轮油;通用齿轮润滑脂
universal gear tester 万能测齿仪
universal gear testing machine 万能齿轮试验机
Universal Geodetic Datum 统一大地基准(面)
universal glue 万能胶
universal gravitation 万有引力
universal gravitation constant 万有引力常数
universal grease 通用润滑脂;通用润滑剂
universal grinder 万能磨机;万能磨床
universal grinding 万能磨削
universal grinding attachment 万能磨床附件
universal grinding machine 万能研磨机;万能磨床;通用研磨机;通用磨床
universal grinding machine external 万能外圆磨床
universal grip pliers 多用卡钳
universal handle 万能手柄
universal handling dolly 万能维护车
universal hanger 万向挂架
universal head and column miller 万能升降台式铣床
universal high speed refrigerated centrifuge 万能高速冷冻离心机
universal hoist 通用绞车
universal hydrometer 通用比重计
universal I-beam 通用工字钢梁
universal implication graph 全隐含图
universal inclinometer 万向倾斜计
universal index center 万能分度中心
universal indexing head 万能分度头
universal indicator 万能指示剂;通用指示剂
universal indicator paper 通用试纸
universal indicator plant 通用指示植物
universal input or output 通用输入或输出
universal instability 漂移不稳定性
universal instruction set 通用指令组
universal instrument 全能(经纬)仪;普用仪;通用量测仪表;通用测量仪表;地平经纬仪
universal interconnecting 通用连接装置
universal interconnecting device 通用连接装置
universal internal grinder 万能内圆磨床
universal internationa ga(u)ge 通用国际量规
universal intervalometer 全能曝光控制器
universal joint 球窝铰链;万向节球;万向节;万向接头;万能(管)接头
universal-joint assembly 万向接头装置
universal-joint bearing 万向节轴承
universal-joint bearing housing 万向节轴承衬套
universal-joint bearing gasket 万向节轴承衬
universal-joint block 活节销
universal-joint boot 万向节护罩
universal-joint bushing 万向节衬套
universal-joint bushing and ring spider 万向节衬套及球十字架
universal-joint bushing plate 万向节衬套盖
universal-joint cap 万向节盖
universal-joint casing 万向节壳
universal-joint casing spring 万向节壳弹簧
universal-joint center ball pin 万向节中心球销
universal-joint centering sphere 万向节中心球
universal-joint center ring 万向节中心环
universal-joint cross 万向(联轴)节十字头
universal-joint (cross) bearing 万向接头轴承
universal-joint cross grease fitting 万向节十字轴油嘴
universal-joint cross relief valve 万向节十字轴溢油阀
universal-joint cross trunnion 万向节十字轴
universal-joint drive 万向节传动
universal-joint driving 万向节传动

universal-jointed shaft 连接轴;万向节轴
universal-joint flange 万向节凸缘
universal-joint flange yoke 万向节凸缘轭
universal-joint fork 万向节叉
universal-joint housing 万向节壳体
universal-joint housing cap 万向节壳盖
universal-joint jaw 万向节爪
universal-joint knuckle 万向接头关节
universal-joint knuckle retainer 万向节衬圈
universal-joint lubricant 万向接头润滑脂
universal-joint lubricating nipple 万向节加油嘴
universal-joint mechanism 万向节机构
universal-joint needle bearing 万向节针式轴承
universal-joint oil seal cap ring 万向节油封帽环
universal-joint pin and cotter 活节销及制销
universal-joint pressure lubricator 万向节压力润滑器
universal-joint roller 万向节滚柱
universal-joint shaft 万向节轴
universal-joint slip yoke 万向节滑叉
universal-joint snap ring pliers 万向节弹簧圈装卸钳
universal-joint spider 万向联轴节十字头
universal-joint spider bushing ring 万向节十字叉衬环
universal-joint spindle 万向(连)接轴
universal-joint splined shaft 万向节花键轴
universal-joint transmission flange 万向节传力凸缘
universal-joint trunnion 万向节耳轴
universal-joint trunnion block 万向节耳轴座
universal-joint yoke 万向节轭
universal key 万能键;通用键
universal keyboard 通用键盘
universal keyset 通用键组
universal kneading machine 通用(型)捏和机;通用捏练机
universal knee-and-column miller 万能升降台式铣床
universal knife 万能刀
universal Kriging estimator 泛克立格估计量
universal Kriging matrix 泛克立格矩阵
universal Kriging method 泛克立格法
universal Kriging variance 泛克立格方差
universal Kriging weighting coefficient 泛克立格权系数
universal laboratory extruder 实验室万能挤出机
universal ladle brick 万能弧形耐砖
universal lathe 万能车床
universal law 一般规律;普遍规律
universal leg 万能支腿
universal legatee 全财产继承人
universal level 圆(形)水准器;通用水准器
universal lever 万向操纵杆;万能杠杆
universal lifting gear 万能提升机
universal link 万向连接杆;通用线路
universal loader 万能装卸机
universal logic module 通用逻辑组件
universal lubricating grease 通用润滑剂
universally acknowledged 公认的
universally-jointed axle 万向节连接轴
universally valid formula 永真式;全称永真公式
universally valid premise 全称永真前提
universal machine 万能机床;万能工作机械;通用机械;通用机(器)
universal machine tool 通用机床
universal machine works 通用机械厂
universal magnetometer 全能磁力仪
universal mains unit 通用整流电源
universal manipulator 万能机械手
universal mapping property 通用映射性
universal material testing machine 万能材料试验机;通用材料试验机
universal matrix retainer 万用成型片座固定器
universal measuring instrument 通用测量仪表
universal measuring machine 万能测量显微镜
universal measuring microscope 万能测定显微镜
universal member 通用杆件
universal meter 万用电表;万用表;万能电表;通用电表;通用表
universal method 通用方法
universal method for point layout in double model 全能法双模型布点
universal method for point layout in single

model 全能法单模型布点
universal method of photogrammetric mapping 全能法(摄影)测图
universal method of photogrammetry 全能法(摄影)测图
universal metre 多用电表
universal microprobe analyzer 通用微探针分析器
universal microscope 万能显微镜
universal mill 万能轧机;万能铣床;万能磨粉机;万能滚节机
universal mill beam 万能轧机轧出的工字钢梁
universal miller 万能铣床;通用铣床
universal milling and boring machines 万能铣镗机床
universal milling attachment 万能铣削装置
universal milling machine 广用万能铣床;万能铣床;通用铣床
universal mill plate 轧制的非标准钢板;齐边钢板
universal mill stand 万能式轧机机座
universal mixer 通用型混合机;通用搅拌机;通用混合机
universal mobile excavator 通用挖土车;通用挖土机
universal mobile telecommunication system 通用移动通信系统
universal model 一般模型
universal money 世界货币
universal motor 交直流两用电动机;通用马达;通用电动机
universal mo(u)lding machine 通用制模机
universal mount 通用座架;通用支架
universal multiple bomb rack 通用复式炸弹架
universal multipole 通用多极
universal naming convention 统一命名约定
universal navigation beacon 通用导航信标
universal navigation computer 通用导航计算机
universal network 通用网络
universal numbering system 通用支数制
universal oil pressure bale press 通用油压打包机
universal oil product 通用油产品
universal operation table 万能手术台;万能操作台
universal output 常用出力
universal output transformer 通用输出变压器
universal overhead traction frame assembly 万能组合式牵引床
universal parallel groove clamp 分支线夹
universal peripheral controller 通用外围控制器
universal personal telecommunication 通用个人通信
universal pH indicator 大范围酸碱指示剂;大范围 pH 指示剂
universal photo 相片镶嵌图;全能法(摄影)测图
universal photometer 通用光度计
universal pH test paper 万用试纸
universal pile driver 通用打桩机
universal pile driving plant 通用打桩设备;通用打桩机;万能打桩机
universal piling plant 通用打桩机
universal pipe joint 万向联管节;万向节管;万能管接头
universal pipe wrench 自由管钳
universal piston vice and press 万能活塞压钳
universal Planck law 广义普朗克定律
universal plane 万能刨
universal planer 万能刨床
universal plate 钢条;扁钢;万能板材;齐边钢板;齐边柱厚板
universal pliers 通用钳;万能钳
universal plotting instrument 万能绘图仪;全能测量仪;万能绘图机
universal plotting machine 全能测图仪
universal plotting program(me) 通用制图程序
universal plotting sheet 通用空白海图;通用定位纸
universal polar stereographic(al) grid 通用极球面投影坐标网;通用极球面投影格网
Universal Postal Union 万国邮政联盟
universal postprocessor 通用后处理程序
universal potentiomerer 通用电位计
universal precision microtome 万能精密切片机
universal pressure 通用压力
universal pressure boiler 通用压力锅炉
universal primer 万能底漆;通用引物;通用底漆
universal processor 通用处理程序
universal product code 通用货单代码;通用产品代码
universal products 通用产品
universal program(me) 通用程序
universal property 通用性质
universal protractor 万能角度尺
universal provider 百货店
universal puller 万能式拆卸器;万能拉出器;万能拆卸机
universal pulley 万能滑轮
universal pulling machine 万能拔桩机
universal pump 通用泵
universal quick change chuck 多功能快换夹头
universal radial drilling machine 万能摇臂钻床
universal radical drill 万能摇臂钻床
universal receiver 通用接收机
universal relation 全域关系
universal relation assumption 泛关系假设
universal relaxation time 通用的弛豫时间
universal relieving lathe 万能铲齿车床
universal research microcope 万能研究显微镜;综合显微镜
universal resonance curve 通用谐振曲线
universal reversing roughing mill 万能可逆式粗轧机座
universal ring-type crusher 环式通用破碎机;通用环式破碎机
universal rocker arm drilling machine 万向摇臂钻床
universal rod 万向杆
universal rod body 万能杆件
universal rod holder 通用钻杆夹持器
universal roller 通用滚压机;通用辊压机
universal roller tension unit 通用型辊式张力装置
universal rolling 万能轧制
universal rotary microtome 万能手摇切片机
universal roughing mill 万能式粗轧机座
universal saddle 万能鞍架
universal saw 万能锯
universal saw bench 万能锯台;通用圆锯台
universal sawing machine 万能锯断机
universal saw table 万能锯台
universal Saybolt visco(si)meter 通用塞氏黏度计
universal scraper 万能刮刀
universal screw-key 万能螺丝把
universal screw wrench 活扳手;万能螺旋扳手
universal seamer 万能封罐机
universal seam welder 纵横两用缝焊机;万能滚焊机;万能缝焊机
universal seismograph 通用地震仪
universal serial bus 通用串行总线
universal set 全集;万有集;通用接收机
universal set-square 万能三角板
universal setting ga(u)ge 万能调整仪
universal shaft 万向转轴
universal shape 宽缘型钢
universal shaper 万能牛头刨床
universal shaping machine 万能牛头刨床
universal shaving machine 通用削匀机
universal shearing and punching machine 万能剪冲床
universal shears 万能剪切机
universal sheet-erosion equation 通用片蚀方程
universal shovel 全回转通用单斗挖掘机
universal shunt 通用分流器;艾尔顿分流器
universal signal 通用符号
universal sketchmaster 全能相片转绘仪
universal slabbing mill 万能式板坯初轧机
universal sliding microtome 万能滑动切片机
universal slip joint 可伸缩万向节
universal slotting machine 万能铲床
universal socket 万向套节;万向节座
universal socket wrench 万能套筒扳手
universal software 通用软件
universal soil loss equation 通用土壤流失公式
universal space 通用空间
universal spare parts 通用配件
universal spindle milling attachment 万能立铣装置
universal stage 旋转台;万能旋转台;万能旋转台;通用旋转台;费氏旋转台
universal stains 通用着色剂
universal staking machine 通用拉软机
universal stand 广用台;万能支架;万能台;通用台(架);通用架
universal standard 通用标准
universal standard data 通用定额资料
universal starching and drying machine 通用上浆干燥机
universal steel centering 通用钢拱架
universal steel plate 齐边钢板;万能钢板;通用钢板
universal stereo-plotter 全能立体测图仪
universal stimulus display instrument 多用刺激显示仪
universal stirring machine 多用搅拌机
universal stratigraphy 统一地层学
universal strength tester 万能强度试验机
universal structural mill 万能式轨梁轧机
universal submarine simulator 通用潜艇模拟器
universal successor 全财产继承人
universal supplementary search light 万能辅助探照灯
universal support 通用载体
universal support rod 万用支撑杆
universal surface ga(u)ge 万能平面规
universal surface with linear vectogram 具有线性向量图的万有曲面
universal switch 万能开关;通用开关
universal switchboard 通用交换台
universal asynchronous receiver transmitter 通用同步接收发送器
universal synchroscope 通用同步示波器
universal system for information in science and technology 全球科学技术情报系统
universal table 万能工作台
universal tapered ring 通用楔形管片环
universal tensile testing machine 万能拉力试验机
universal term on hydrogeology 水文地质通用名词术语
universal terrestrial camera 全能地面摄影机
universal test ammeter 通用试验安培计
universal test bench 通用试验台
universal tester 万能试验机;万能试验仪;万能测试仪
universal test indicator 万能试验指示器
universal test(ing) machine 万能试验机;全能试验机;通用试验机
universal theater 通用剧场
universal theodolite 全能经纬仪;通用经纬仪
universal thread counter 通用纱线支数计
universal thread grinder 万能螺纹磨床
universal thread grinding machine 万能螺纹磨床
universal tilting table 万能倾斜台
universal time 格林威治时(间);格林威治平时;极动世界时;世界通用时间;世界时
universal time coordinated 世界协调时
Universal Time Coordinated System 协调世界时系统
universal time two 季差世界时
universal tinter 通用着色器;通用色浆
universal tinting colo(u)r 一般着色
universal tongs 万能钳(子)
universal tool maker's microscope 万能工具显微镜
universal tool miller 万能工具铣床
universal tool milling machine 万能工具铣床
universal tool post 通用刀架
universal torn tape system 通用撕断纸带系统
universal torsion tester 万能扭力试验机
universal track ga(u)ge 万能轨距尺;万能道尺
universal tractor 通用拖拉机;通用式拖拉机
universal transit 全能经纬仪;通用经纬仪
universal transmission function 通用透射函数
universal transported loader 万能传送装卸机
universal transport vehicles 通用运输车辆
universal transverse Mercator coordinate 通用横轴墨卡托投影坐标
universal transverse Mercator projection 通用横轴墨卡托投影
universal trocar 多用套针
universal tube drill 万能管形钻
universal turbulence constant 通用湍流常数
universal Turing machine 通用图灵机
universal turning tool 万能车刀
universal twisting frame 通用式捻线机
universal two-axis slant bed machine 双轴斜床身万能机床
universal type 通用式
universal type ship 通用船
universal underwater mobile robot 通用水下移

动式机器人
universal upsetting welder 万能顶锻焊机
universal validity 普遍有效性
universal valve 万向阀
universal valve lifter 通用起阀器
universal V-die 万能 V 形弯曲模
universal vehicular radio 通用车载无线电设备
universal velocity profile 通用速度分布图
universal viewfinder 万能取景器;通用取景器
universal vise 万向虎钳;万能虎钳
universal wear tester 通用式耐磨试验仪
universal wheel wright machine 万能制轮机
universal winder 通用式络筒机
universal wood turning lathe 万能木工车床
universal woodworker 通用木工机床;多能木作机
universal woodworking machine 万能木工机
universal worktable 万能工作台
universal wrench 活络扳手
universal yarn numbering balance 通用式纱支测定天平
universal yoke 万向接头叉头
universal zoning 通用分带序列
universe 总体;全域
universe of discourse 个体域
universe of fuzzy sets 模糊集论域
universe point 通用点
universe population 全域总体
universe set 全域
universe value 全域值
university 综合大学;大学
university campus 大学城
University church at Salzburg 塞尔茨堡的大学教堂
university city 大学城
university function 学院功能
university ground 大学校园
university library 大学图书馆
University of Capetown process 开普敦大学工艺
university of science and technology 科技大学
university teaching hospital 大学教学医院
university zone 大学区
univibrator 单稳(态)多谐振荡器
univis oil 乌尼伏斯油
univocal resolution of force 力的单一(分)解
univoltage 单电压;单电位
univoltage lens 单电压透镜;单电位透镜
univolume file 单卷文件
uniwafer 单圆片
uniwave signal(l)ing method 单频信号法
unjammable lidar 不受干扰激光雷达
unjammed 未卡死的
unjointed concrete pipe 松接(式)混凝土管道
unjoint plywood 未拼接胶合板
unjust action 不正当行为
unjust disbursement 不正当支出
unjustifiable deviation 不正当绕航
unkempt 未雕琢的
unk face 凹面(石)
unkilled steel 不完全脱氧钢;沸腾钢
unknown 未知量;性质不明露头;未知因素;未知的
unknown accumulation 未知的积累
unknown aerial target 未识别的空中目标
unknown condition 未知条件
unknown constant 未知常数
unknown contact 接触关系不明
unknown crystal 未知晶体
unknown error code 未知错误指示码
unknown factor 未知因素
unknown function 未知函数
unknown geothermal area 未知地热区
unknown loss 未知损失
unknown number 未知数
unknown parameter 未知参数
unknown periodic component 未知周期分量
unknown point 未知点
unknown prime number 未知素数
unknown pulse 未知脉冲
unknown quantity 未知数;未知量
unknown sample 未知样品
unknown solution 未知液
unknown state 未知状态
unknown term 未知项
unknown the geologic(al) condition of ore deposition is bed 成矿地质条件不清
unknown track 未识别的目标回波
unknown trees 无名杂树
unknown value 未知值
unknown variance 未知方差
unlabel(l)ed 未(作)标记的
unlace 解开(带子)
unladen 空车;未载货的
unladen state 空车状态
unladen static deflection 空载静变位
unladen vehicle weight 空车重量
unladen weight 空车重量;空重;无载重(量)
unlagged endogenous variable 非滞后内生变量
unlagged piping 未保温的管道
unlapped 未覆盖的
unlatching of pile warp 毛圈高度控制
unlawful 非法的
unlawful act 非法行为
unlawful activity 非法活动
unlawful continuance 非法继续占用
unlawful conversion 非法变更
unlawful deviation 不合法绕航
unlawful dumping 非法轻弃
unlawfulness 非法
unlawful trading 非法交易
unlayered 未着色的
unlayered sheet 非分层设色图
unleachable 不可浸出的
unleached wood ashes 未滤溶木灰
unleaded 无铅的;未加铅的
unleaded antiknock additive 无铅抗爆添加剂
unleaded blend 无铅掺和料
unleaded fuel 无铅燃料
unleaded gasoline 无铅汽油;无铅汽油;不加铅汽油
unleaded octane rating 无铅辛烷值
unless otherwise mentioned 除另有说明外
unless otherwise noted 除另有说明外
unless otherwise stated 除另有说明外
unleveed river 无堤防河道
unleve(l)ed 未置平的
unleve(l)ling colo(u)r 不匀染料
unleve(l)ling dye 不匀(性)染料
unlevel luffing 非水平变幅
unlevel shade 不匀色泽
unlicensed 没有执照的;无执照的
unlicensed pilot 无执照的引航员
unlicensed transmitter 无执照发射机
unlifted wall 上升壁
unlifting pressure 上升压力
unlighted 不发光的
unlighted beacon 不发光导标;昼标(无灯光)
unlighted black paint 无光黑漆
unlighted buoy 无光浮标
unlighted mark 昼标
unlighted navigation mark 不发光航标
unlikelihood ratio 非似然比
unlike-signed correlation 异号相关
unlimed 除石灰的
unlimited 无限的
unlimited ceiling 无限升限
unlimited check 不限额支票
unlimited company 无限公司
unlimited compatibility 无限相容性
unlimited competitive bidding 无限竞争性投标
unlimited decimal 无限小数;无尽小数
unlimited duty-free storage 无限制的免税仓库
unlimited flow 无限流动
unlimited function 无界函数
unlimited general average guarantee 共同海损无限保证书
unlimited guarantee 无限担保
unlimited image number 无限次映射
unlimited liability 无限责任
unlimited liability company 无限责任公司
unlimited mortgage 无限制抵押
unlimited partnership 无限责任合伙企业;无限合伙
unlimited pattern area 无限制花型范围
unlimited thickness aquifer 无限厚度含水层
unlimited traffic source 无限活源
unline 上行线路
unlined 未镶面的;未衬砌的;无衬的;不镶衬的
unlined canal 无衬砌渠道
unlined landfill 非密封垃圾填埋
unlined metallic pipe 未衬里金属管
unlined metallic tube 未衬里金属管
unlined tank 无衬里水池
unlined tent 单帐篷
unlined tunnel 毛洞无衬砌隧洞;无衬砌隧道
unlined wall 无衬砌墙
unlining 分离
unlink 卸开;摘钩;解链;解环(作用);松开;使脱开;上行线路
unliquidated 未付的
unliquidated account 未清账目;未清账款;未清账
unliquidated claims 未偿偿债务;未偿债权
unliquidated damages 未确定损害赔偿额;未清算的损失
unliquidated debt 未清(偿)债务;未偿债务
unliquidated obligations 未偿债务
unlisted 未登名上册的
unlisted assets 账外资产
unlisted securities 未上市证券
unlisted stock 未上市股票
unlit triforium (gallery) 无照明的教堂拱廊
unload 卸除;卸(载);卖掉;倾销;起货;无载(荷)
unload a fixture 从夹具取下加工件
unload antenna 无载天线
unload cargo 卸货
unload characteristic 无载特性(曲线)
unload chord 不荷载弦杆
unload chute 卸料槽
unload curve 无载曲线
unload deadweight 空载时间
unloaded 无荷载的;无载的
unloaded antenna 无负载天线;不加载天线
unloaded characteristic 无负载特征
unloaded chord 无载弦;无荷载弦
unloaded circuit 无载电路
unloaded draft 空载吃水
unloaded energy 无载能量
unloaded hole 空炮眼;未装炸药的爆破孔;无药爆孔
unloaded/load ratio 空载/满载比
unloaded mechanism 无载机构
unloaded Q 空载 Q 值;无负载品质因数
unloaded quanlity factor 无负载品质因数
unloaded radial tire run-out 无荷载轮胎径向偏心
unloaded radial wheel run-out 无荷载车轮径向偏心
unloaded sag 无荷弧垂;无荷驰度;无荷垂度
unloaded spring 无载弹簧
unloaded track 不受载轨道
unloaded vulcanizate 纯硫化胶
unloaded waveguide 无载波导
unloaded weight 空车重量;无载重(量)
unloader 卸载机;卸料机;卸货器;卸货机;卸载器;卸板机;减荷器
unloader airslide system 风动斜槽卸料系统
unloader arm 卸粮伸出管
unloader blower chopper 卸载切碎吹送机
unloader boom 横梁式出料机
unloader box 卸载箱
unloader clutch 卸载装置离合器
unloader knife 卸料刮刀
unloader lifter 卸载起重机
unloader pump 卸货机泵
unloader's cutting auger 卸载机切碎螺旋推运器
unloader tower 卸货(机)塔
unloading 卸(去负)荷;卸料;卸货;卸载;排污;生物膜脱落;拆箱;放空
unloading apron 卸货停机坪
unloading area 卸货区;卸载场;释荷面积
unloading arm 卸料臂
unloading arrangements at main station 主要站卸车安排
unloading ashore 卸岸
unloading assembly 卸料设备
unloading auger 卸载螺旋运输机;卸货用螺旋输送机;卸荷钻;螺旋卸载机;螺旋卸料机;螺旋式输煤机
unloading auger clutch lever 卸粮螺旋离合器接合杆
unloading bare rib 裸拱卸架
unloading bay 卸货河湾;安装间
unloading belt conveyer 卸载皮带输送机
unloading bridge 卸货栈桥
unloading bulkhead 卸货推板
unloading by ice melt 冰盖卸荷运动
unloading capacity 卸载容量;卸货能力;卸货能力
unloading charges 卸货费(用)
unloading conveyer 卸载传送机

unloading crane 卸料吊车
unloading crew 卸货班
unloading curve 卸货曲线;卸荷曲线;卸载曲线
unloading cycle 卸载循环
unloading device 卸货装置;卸载装置;卸料设备;卸料设备;卸料机;取出机构
unloading dock 卸货码头
unloading efficiency 卸货效率
unloading elevator 卸载升运器
unloading equipment 卸载设备;卸货设备
unloading fish shelter 卸鱼棚
unloading form a wreck 从沉船上卸货
unloading gang 卸货班
unloading gang plank 卸货跳板
unloading gantry 卸矿栈桥
unloading gear 卸载设备;卸载传动装置
unloading hose 卸液软管
unloading installation 卸车装备
unloading joint 除荷节理
unloading line 卸油管路;卸油导管
unloading machine 卸砖机;卸载机
unloading machine with chain buckets 链斗卸车机
unloading mechanism 卸载装置
unloading modulus 卸荷模量
unloading of biotic film 生物膜脱落
unloading party 卸货班
unloading period 卸货时间;卸货期
unloading place 卸货处;卸货地点
unloading plan by stations 站别卸车计划
unloading plant 卸货设备
unloading platform 卸货平台;卸货作业台;卸货站台;卸货平台;卸车站台
unloading point 卸载点;卸货点
unloading port 卸运港;卸货港(口)
unloading position 卸载位置
unloading pressure relay 卸去负载压力继电器
unloading protecting relay 卸载保护继电器
unloading pulley 无载轮
unloading ramp 卸货斜台;卸货跳板
unloading rig 卸矿塔;塔式卸料机
unloading schedule 卸料时间表
unloading screw 螺旋出料机
unloading siding 卸车线
unloading skid 卸料台架
unloading speed 卸载速度
unloading station 卸货站
unloading suction dredge(r) 吹泥船
unloading team 卸货班
unloading terminal 卸货码头
unloading test 卸荷试验
unloading the film 卸下底片
unloading time 卸货时间
unloading tower 装卸桥
unloading track 卸货轨
unloading trough 卸载槽
unload(ing) valve 卸载阀;卸荷阀;放泄阀;减压阀;减荷阀
unloading well 自喷(油)井
unloading wharf 卸货码头
unloading yard 卸货处;卸货场;卸车场
unload lateral resistance 无荷载的横向阻力
unload loss 无载运转的损失
unload pointer 从地址计数器中减去;出空指针;出空指示字
unload side 非负荷端
unload weld 非承载焊缝
unlocated household 不定居户
unlock 开锁;松开;拔去销钉
unlocked area 非堵塞截面
unlock gimbal 随动万向架
unlock indicator 失锁指示器
unlocking cam spring 释放凸轮簧
unlocking click 释放爪
unlocking click spring 释放棘爪簧
unlocking click yoke 释放棘爪杆
unlocking click yoke spring 释放棘爪杆簧
unlocking dual control 非联锁双重控制
unlocking finger 释放爪
unlocking finger spring 释放爪簧
unlocking key 钥匙
unlocking lever nut 释放杆螺母
unlocking lever spring 释放杆簧
unlocking lever spring clip 释放杆夹

unlocking lever tube 释放杆套
unlocking link 开锁环
unlocking link stem 开锁环心杆
unlocking pallet 释放叉瓦
unlocking resistance 释放阻力
unlocking rod 开锁杆
unlocking shoe 开锁杆套
unlocking wheel 释放轮
unlocking wheel click 释放轮棘爪
unlocking wheel friction spring 释放轮磨擦簧
unlocking wheel holder screw 释放轮夹持螺钉
unlocking wheel operating lever 释放轮操作杆
unlocking wheel operating lever spring 释放轮操作杆簧
unlocking wheel plate 释放轮片
unlocking wheel plate distance piece 释放轮压板隔片
unlocking yoke cam 瞬跳杆凸轮
unlocking yoke driver 瞬跳杆驱动轮
unlocking yoke driving wheel stud 瞬跳杆传动轮柱
unlocking yoke ring 瞬跳杆衬圈
unlocking yoke spring 瞬跳杆簧
unlocking yoke spring click 瞬跳杆棘爪簧
unlocking yoke spring maintaining plate 瞬跳杆簧压片
unlocking yoke stud 瞬跳杆柱
unlocking zone 开锁区域
unlock signal 松开信号;释放信号
unlock statement 开锁语句【计】
unlubricated 未润滑的
unlubricated friction 干摩阻力;干摩擦(力);无润滑摩擦
unlubricated surface 无润滑表面
unlubrication friction 干摩阻力;干摩擦(力)
unmachinable 不能机械加工的
unmachined 未进行机械加工的
unmade ground 崎岖不平地面;未经挖动的地面;原始地面
unmade path 小径
unmade road 崎岖道路;失修道路
unmanageable 难(以)控制的;难(以)加工的;难管理的
unmanaged environment 难驾驭的自然环境
unmaned factory 无人工厂
unmanned 无人居住的;无人管理的;无人看守的;无人(驾驶)的;无人操作的
unmanned aerial surveillance system 无人操纵空中监视系统
unmanned barge 无人驳
unmanned capsule 无人舱
unmanned compartment 无人舱
unmanned control system 无人控制系统
unmanned crossing 无人看守道口
unmanned engine 无人驾驶机车
unmanned engine room 无人机舱
unmanned excavator 无人驾驶的挖掘机
unmanned free-swimming submersible 无人自航式潜水器
unmanned helper engine 无人驾驶辅机
unmanned lighthouse 无人看守的灯塔
unmanned lightship 无人看守的灯船
unmanned machinery space 无人机舱
unmanned module 无人驾驶舱
unmanned observation balloon 无人观测气球
unmanned operation 无人操作
unmanned orbital multifunction satellite 无人多用途轨道卫星
unmanned power station 自动化电站;无人管理电站
unmanned recovery vehicle 无人回收运载工具
unmanned remotely-operated vehicle 无人遥控潜水器
unmanned roving vehicle 无人驾驶自动车
unmanned seismic observatory 无人地震监测站
unmanned silo 无人发射井
unmanned station 无人台站;无人测站;无人操作台
unmanned submersible 无人潜水器
unmanned substation 无人变电所
unmanned surveillance equipment 自动监测仪;无人操纵监视仪器
unmanned underwater vehicle 无人潜水器
unmanned vehicle 无人潜水器
unmanufactured wood 原木;未加工木料

unmapped 地图未标出的;未制图的
unmapped area 无图地区
unmapped physical storage 非映像物理存取
unmapping area 未填图区
unmarked 未标记的
unmarked area 未标志区
unmarked end 未标注端
unmarked non-terminal 未标记非终结符
unmarked pole 不记名磁极
unmarked terminal 未标记终结符
unmarketable goods 不对路产品
unmarketed product 非销售产品
unmarred 未损伤的
unmarried print 双片
unmask 去屏蔽;无屏蔽
unmasticated rubber 未塑炼生胶
unmatch 未匹配
unmatchable 不可匹配
unmatched load 不匹配负载
unmatched seal 不匹配封接
unmatching 不匹配
unmatch record check 不匹配记录检验
unmatured 未熟的
unmatured debt 未到期债务
unmeasurable 无边际的;不能测量的;不可估量的;不可测量的
unmeasured 未测量的
unmeasured loss 其他损失;未计入损失;未测定损失
unmeasured sediment discharge 未测输沙量
unmeasured value 未测定值
unmechanized hump 非机械化驼峰【铁】
unmechanized hump yard 非机械化驼峰站【铁】
unmeet 不合适的
unmelted 未熔化的
unmelted charge 未熔炉料
unmendable 不可修理的
unmerited 无功而得的
unmetalled 未铺路面的;无路面的
unmetalled road 无路面道路;土路(无硬质路面的)
unmetamorphosed 未变质的
unmetered apartment 无水表的公寓
unmetered consumption 未计量封闭水
unmethodical 不讲方法的
unmicellized 未胶束化的
unminable 不可开采的
unminable protective safety territory 压矿
unmined 未开采的
unmined bed 未采层
unmined cause 未采原因
unmined of natural area or military area 自然保护区或军事禁区
unmineralized froth 未矿化泡沫
unmixed 未混合的;不混合的
unmixed batch capacity 未拌和的拌重量;未拌料重量;未拌料容量的拌容量;未拌料重量;未拌料容量
unmixedness 不混合度
unmixed zone 未混合区
unmixing 离析;未搅拌的;未混合的;不混溶
unmoderated 未慢化的
unmoderated reactor 无慢化剂堆
unmodifiable 不可改变的
unmodified 未改性的;未变性的
unmodified aircraft 未改型飞机
unmodified cellulose 未改性的纤维素
unmodified dam 未改建的坝
unmodified instruction 基本指令;无修改指令;未修正的指令;未修改指令;非修改指令;不可更改指令
unmodified isotherm equation 不变吸附等温方程
unmodified polystyrene 净聚苯乙烯
unmodified resin 净树脂;未改性树脂
unmodified titration curve 不变滴定曲线
unmodulated 未调节的
unmodulated beam 未调制射束
unmodulated carrier 未调制载波
unmodulated groove 无声槽
unmodulated noise 未调制噪声
unmodulated signal 未调信号
unmodulated track 无声声道
unmodulated waves 未调波
unmonitored 无监控的
unmonitored control system 开环控制系统;无监

视的控制系统
unmonitored polluted sources 未监测到的污染源
unmoor 离浮筒;双锚泊改单锚泊;拔锚
unmoradanted 未经媒染的
unmortgaged assets 无抵押资产;未抵押资产
unmounted 未上炮架的;未安装的
unmounted drill 无架凿岩机;手持式凿岩机;手持式风钻
unmovable data set 非移动数据集
unmuffled type furnace 无马弗炉
unmulched 未覆盖的
unnamed control section 无名控制段
unnamed drawing 无名图纸
unnamed segment 无名图段
unnavigable 不(能)通航的
unnavigable dam 不通航坝
unnavigable reach 不通航河段
unnavigable river 不通航河流
unnavigable stream 不通航河流
unnavigated waters 不通航水域
unneat beading 花边
unnecessary fixed assets 不需用的固定资产
unnecessary function 不需要的机能
unnecessary means of livelihood 非生活必需品
unneutralized 未中和的
unnormalized floating-point number 非规格化浮点数
unnormalized number 非规格化数
unnormalized relation 非规范关系
unnotched 无缺口的;无凹槽的
unnotched bar 光面钢筋;无槽杆件
unnotched specimen 无切口试样
unnotched type bearing 无滚珠槽的轴承
unnumbered 无数的;未编号的
unnumbered acknowledge 非编号确认;非编号回答
unnumbered command 无编号指令;非记数指令;非编号命令
unnumbered frame 非编号帧
unnumbered information 非编号信息
unnumbered response 非编号回答
unnumerable 无数的
unobligated appropriation 未承付的拨款
unobligated balance 不转结的年终预算结余;不结转的年终预算结余
unobservable 未观察到的;不可观测的
unobservable chance factor 不可观察的随机因子
unobservable quantity 不可观测量
unobservable random disturbance 不可观测的随机扰动
unobservable random variable 不可观测的随机变量
unobservable value 不可观测的值
unobserved 未观察到的
unobserved year 缺测年份
unobstructed 无阻挡的
unobstructed access 无障碍的入口;无障碍的通路
unobstructed capacity (集装箱的)净内容积;无阻碍容量
unobstructed crest of spillway 自由溢流堰顶;无闸堰顶
unobstructed influx 自由进流
unobstructed sight 无阻视线
unobstructed space 无阻碍空间
unobstructed spillway 无闸门溢洪道
unobstructed view 无障碍的景观;无障碍的视野
unobstructed vision 无障碍视界
unobstructive 小巧精致的
unobtainable tone 未接通信号
unobviousness 非显而易见性
unoccupied 未被占用的
unoccupied area 未占领地区
unoccupied land 空地
unoccupied level 空能级;未占能级
unoccupied line 空闲线路
unoccupied point 未设站点
unoccupied time 无酬时间;未占线时
unoccupied track 空闲线路
unode of a surface 单切面重点
unofficial 非正式的
unofficial agreement 非正式协商;非正式协定;非正式契约
unofficial market 非官方市场
unofficial method 非法定方法

unoil 脱脂;除油
unoiled 未涂油的
unopened 未开(小)口的;未开封的;未开放的
unopposed 无反对的
unordered 无序的
unordered basis 无序基
unordered pair 无序对
unordered symbol 无序符号
unordered table 无序表
unorganic 无机的
unorganized 无结构的
unorganized air supply 无组织进风
unorganized drainage 无组织排水
unorganized exhaust 无组织排风
unorganized natural ventilation 无组织自然通风
unoriented 无定向的;未定向的;非取向的
unoriented cobordism 无定向配边
unoriented current 无定向流
unoriented fabric 颗粒无定向组构
unoriented fibre 未取向纤维
unoriented rigion 未定向区域
unoriented specimen 非定向标本
unoriented welded tuff 无定向熔结凝灰岩
unoriginal 无独创性的
unornamented 未加装饰的
unorthodox form 非正式方式(合同、契约等的)
U-notch U 形槽口
unoverlap fetching 无重叠取数
unoxidizable 不锈的;不可氧化的
unoxidizable alloy 不锈合金;不可氧化的合金
unox process 污水纯氧曝气工艺;纯氧曝气法
unpack 信息分开;取货作业;拆包;分开
unpackage cargo 非包装货
unpackaged 未包装的;散装的
unpackaged sand 散装砂
unpackaged transfer 分件转让
unpacked 未包装的
unpacked cargo 无包装的货物;非包装货
unpacked decimal 分离十进制表示;非压缩十进制
unpacked format 分出形式
unpacked goods 无包装的货物
unpacked technology 不成套技术
unpack format 分离形式;分离格式
unpacking 拆箱
unpacking equipment 开箱设备
unpaged 未标页码的
unpaid 拒绝兑付;未付的;不受报酬的
unpaid account 未付讫的账款
unpaid appropriation 无偿调拨;无偿拨款
unpaid area 非付费区
unpaid balance 未付差额
unpaid bill 未付票据
unpaid check 未付支票
unpaid claims 未付赔款
unpaid declared dividend 未付已宣布股利
unpaid dividend 未付股息
unpaid expenses 未付费用
unpaid note 未偿票据
unpaid returns 未付资金报酬
unpaid seller 未收货款的卖主
unpaid stock 未付股本
unpaid tax 未付税款
unpaid transfer of resources 平调
unpaid transmission 无偿转拨
unpaid transmission of fixed assets 固定资产无偿转拨
unpaid up capital 未付资本;未付股本
unpaid up capital account 未付资本账(目)
unpaid wage 未付工资
unpainted 未油漆的
unpainted steel 无覆层钢
unpaired 无对的
unpaired terrace 不对称阶地
unpalletized cargo 非货盘化货物;非货板化货物;非成组货物
unparallel 不平行的
unparallel banks method 不平行边坡法
unparalleled surface 不平行表面
unpasteurized 未灭菌的
unpatented 未获专利权的
unpaved 未铺面的;未铺装的;未铺砌的;未铺面的
unpaved airfield 无铺筑面机场
unpaved airstrip 未铺筑道面的简易机场
unpaved strip 无铺筑道面的简易机场

unpedimented 无山墙的
unpeeled 未剥落的;未剥皮的
unpeeled log dump 未剥皮的原木堆
unpeeled wood 未剥皮木材
unpenetrability 不穿透性
unpeopled 人口稀少的;无人居住的
unperforated 未穿孔的
unpermeability 不透水性
unperturbed 未扰动的
unperturbed end-to-end distance 无干扰的端间距
unperturbed orbit 无摄动轨道;未扰动轨道
unperturbed random coil 无干扰无规线圈
unperturbed resonator 无激励谐振器
unpicked 未拣过的
unpickled spot 酸洗斑点
unpieced bulkhead 整体舱壁
unpiercement diaper 非刺穿底辟
unpigmented 不加颜料的;未着色的;无颜料的
unpigmented base coat 无颜料的底涂层
unpigmented cold water paint 不加颜料的冷水溶性涂料
unpigmented water-carried paint 不加颜料的水溶性涂料
unpiler 卸垛机
unpiloted 无人驾驶的
unpiloted ignition 未引火着火
unpitched sound 噪声
unplaced 未被安置的
unplaned timber 未刨的木材
unplanned 计划外的;无计划的
unplanned explosion 意外爆炸
unplanned freight traffic 计划外运输
unplanned investment 非计划投资
unplanned market 非计划市场
unplanted sludge drying bed 无设施的污泥干化床
unplastered 未抹灰的;未粉刷的
unplastered ceiling 未抹灰的顶棚
unplastered ceiling plate 未抹灰的顶棚板
unplastered gypsum ceiling plasterboard 未粉刷的顶棚石膏板
unplastered gypsum wallboard 未粉刷的石膏墙板
unplastered plate 未粉刷的石膏板
unplastered wall 未抹灰的墙
unplasticised polyvinyl chloride 硬聚氯乙烯
unplasticized 未增塑的
unplasticized polyvinyl chloride 未增塑聚氯乙烯
unplasticized PVC 未增塑聚氯乙烯
unplated 无涂层的
unpleasant smell 臭味;不愉快气味
unpleasant taste 讨厌的味觉;臭味
unpledged assets 未抵押资产
unploughed land 未耕土地;未耕地
unplowed land 未耕土地
unplugged formation 非密封性岩层;不致密岩层
unplumbed 未用垂球校准的;未经探测的
unpoisoned fuel 未中毒燃料
unpolarizable electrode 不极化电极
unpolarized 未极化的
unpolarized burst 非偏振暴
unpolarized component 非偏振成分
unpolarized continuum 非偏振连续谱
unpolarized electrode 非极化电极
unpolarized light 非偏振光
unpolarized particle beam 无偏振粒子束
unpolarizing 去偏振
unpolished 未抛光的;未磨光的
unpolished rice 糙米
unpollute 未被污染
unpolluted 未(受)污染的;未被污染的
unpolluted cooling water 未被污染的冷却水
unpolluted core 未污染岩芯
unpolluted freshwater lake 未被污染的淡水湖泊
unpolluted freshwaters 未被污染的淡水水体
unpolluted river 未受污染河流;未(被)污染的河流
unpolluted stream 未污染的河流;未受污染河流
unpolluted water 未(被)污染的水
unpolluted zone 未污染带;未被污染区
unpollution 未(受)污染;非污染
unpolymerized 未聚合的
unponderable 不可衡重的
unpopulated area 无人口区
unpowered 无发动机的
unpowered approach 无动力进场着陆
unpowered ascent 无动力上升

unpowered bogie 无动力转向架
unpowered control 无助力操纵
unpowered conveyer 无动力输送机
unpowered glide 无动力推进的下滑
unpowered homing weapon 无动力自导引武器
unpowered rudder 无助力操纵方向舵
unpractical 不切实际的;不切合实际的
unpracticalness 不切合实际性
unpractised 未实行过的
unprecedented 无先例的
unprecedented rise 空前涨价
unpredictability 不可预知性;不可预测性
unpredictable 无法预见的;不可预报的
unpredictable behavior 不可预测行为
unpredictable event 不可预测事件
unpredictable factor 不可预见因素;不可预测的因素
unpredictable items 不可预测项目
unpredictable works 不可预见工作;不可预见工程
unprepared field 无道面的机场
unprepared strip 无铺筑道面的简易机场
unpressured 非受压的
unpressurized 不加压的;非承压的
unprestressed 非预应力的
unpriced 未定价的;未标价的
unpriced bill of quantities 不带报价的工程量清单
unpriced proposal 不带报价的建议书
unprimed 未涂底漆的;未涂底色的
unprimed metal substrate 未涂底漆的金属底材
un-priming 不启动;未浇透层;未灌水;停止虹吸
unprocessed 未加工的
unprocessed book 未加工图书
unprocessed data 未处理数据
unprocessed information 未处理信息
unprocessed material 未加工材料
unprocessed mix 未加工的混合料
unprocessed timber 未加工木材
unproduction well 干井
unproductive 非生产性的;无结果的
unproductive area 非生产面积
unproductive boring 无效钻探
unproductive building 非生产性建筑
unproductive capital 非生产性资本
unproductive construction 非生产性建设
unproductive construction investment 非生产性建设投资
unproductive consumption 非生产性消费
unproductive cost 非生产费用
unproductive credit 消费信贷;非生产信贷
unproductive development 非生产的开拓工作
unproductive downtime 非生产性停歇
unproductive expenditures 非生产性支出;非生产性开支
unproductive expenses 非生产性费用
unproductive growth 不结果实枝
unproductive labo(u)r 非生产性劳动;非生产劳动
unproductive labo(u)rer 非生产劳动者
unproductive land 不毛(之)地
unproductive phase 非生产期
unproductive project 非生产性计划
unproductive projects in capital construction 非生产性基本建设
unproductive time 辅助工序时间;辅助工时;非有效工时;非生产时间
unproductive wage 非生产工资
unproductive working time 非生产性工作时间
unprofessional 非专业(性)的
unprofessional conduct 非专业行为
unprofitable firm 亏损企业
unprofitable line 亏损线路
unprogrammed decision 非程序化的决策
unprospected area 未勘探区
unprotect 无保护
unprotected 没有防护的;无保护的;未(加)保护的
unprotected beach 无防护海滩
unprotected cargo 无包装的货物
unprotected circuit 无保护电路
unprotected concrete pad 无防护的混凝土衬垫
unprotected corner 无传力设备的开ành工作
unprotected engine 无防寒设备的发动机
unprotected goods 无包装的货物
unprotected level crossing 无防护设备的平面交叉;无防护道口
unprotected metal construction 无防护金属结构
unprotected opening 不防火洞口

unprotected reversing thermometer 无保护颠倒温度表
unprotected river bed 无护岸河床;无防护的河床
unprotected shoulder 无保护层的路肩
unprotected site 无防护发射场
unprotected system file 非保护系统文件
unprotected terminal 无掩护码头
unprotected tool joint 未补强接头;无箍接头
unproved 未被证明的
unproved reserves 基本探明储量
unproven area 未探明区;未勘探地区
unprovided 未做准备的
unpruned 未剪枝的;未修剪的
unpublished 未发表过的;未发表的
unpublished data 未发表的数据资料;未出版的数据资料
unpublished data sources 未出版的数据资料线索
unpublished documents 未出版文献
unpublished guides to manuscript collection 未出版的馆藏稿本指南
unpublished manuscript 未发表过的原稿或手稿
unpublished materials 未发表的资料
unpublished report 未出版的报告
unpublished research materials 未出版的科研资料
unpulverized 未磨成粉状的
unpumped 未抽出的
unpunched 无孔的;未穿孔的
unpunched gasket 无孔密封垫
unpunished 未受处罚的
unpurified 未精制的;未纯化的
unpurified air 未净化空气
unputrefied 未腐烂的
unqualified 无资格的;无条件的;不合乎需要的;不合格的
unqualified certificate 不合格证明
unqualified endorsement 不受限背书
unqualified man 普通工(人)
unqualified material 经检查不合格材料
unquantifiable benefits 不能计量的收入
unquantized picture 未量化的图像
unquarried 未开采的
unquote 引用结束
unrammed 未夯实的;未捣实的
unrammed concrete 未夯实混凝土;未捣实混凝土
unrated 未率定的;未经定的
unrated ga(u)ging station 临时测站;未率定测站;未建立水位—流量关系的测站
unravel 放(工作钢绳)
unraveling (路面的)松散剥落;剥落;(混凝土)崩解
unreacted 不反应的;未反应的
unreacted radical 未反应的基团
unreacted state 未反应状态
unreactive aggregate 无反应骨料;惰性集料;惰性骨料
unreactive hydrocarbon 惰性碳氢化合物
unread 未经阅读的
unrealistic 不切合实际的
unrealistic assumption 不现实的假定
unrealizable assets 不可变(现)资产
unrealized appreciation 未实现增值
unrealized appreciation from valuation of assets 未实现资产重估增值
unrealized gains or losses 未实现损益
unrealized gross margin 未实现毛利
unrealized gross profit 未实现毛利
unrealized income 未实现收益;未实现收入
unrealized increment by appraisal of fixed assets 未实现固定资产估价增值
unrealized profit 未实现利润
unrealized revenue 未实现收入
unreasonable 不合理的
unreasonable arrangement of exploratory engineering 不合理的勘探工程布置
unreasonable cutting slope 不合理削坡
unreasonable delay 无理的拖延
unreasonable demand 不合理需求
unreasonable economic action 不合理的经济行为
unreasonable error 不合理误差
unreasonable interest rate 不合理(的)利率
unreasonable loading on wall face 坡面不合理加载
unreasonable price 不合理价格
unreasonable requirement 不合理要求
unreasonable reserve ratio or distribution 储量比例或布局不合理

unreasonable risks to health 对健康的过分风险
unreasonable rules and regulation 不合理的规章制度
unreasonable taxation 不合理税收
unreasonable use of water 水的不合理使用
unreceipted 未签收的
unreceipted note 未签收的票据
unreceived 未收到的
unreciprocal transfer 非互惠性转让
unreclaimable waste 不可修废品
unreclaimed 未收回的
unreclaimed field 未开垦地
unreclaimed region 未开垦地区
unrecognizable character 不可识别(的字)符
unrecognized 未被认识的
unrecorded 未记录的;未登记的
unrecorded book 未登记的图书
unrecorded deposit 未入账存款
unrecorded expenses 未入账费用
unrecorded income 未入账收入
unrecorded investment balance 未回收投资差额
unrecorded liabilities 未入账负债
unrecorded revenue 未入账收入
unrecoverable error 不可校正(的)错误
unrecoverable reserves 不可采储量
unrecoverable transaction 不可恢复事项(处理);不可恢复事务处理
unrecovered 未恢复的
unrecovered cost 未收回成本;未能收回的成本;未回收成本
unrecovered oil 残留石油
unrecovered strain 非弹性的应变;不可恢复(的)应变
unrectified 未纠正的
unrectified mosaic 未纠正镶嵌图
unrectified negative 未纠正底片
unrectified photograph 未纠正相片
unred gas 未计价燃气量
unreduced 未约化的;未还原的
unreduced design 非简约设计
unreduced matrix 未约矩阵;不可约矩阵
unreduced salt cake 未还原芒硝
unreducible 不可简化的;不可还原的
unreefed 未收口的
unreel 开卷;解轴;退绕;拆卷
unreeve 拉回;将绳从滑车抽出;卸下钢绳
unrefinable crude oil 不适于炼制的原油
unrefined 未精制的;未精炼的
unrefined information 未经提炼的情报
unrefined material 未精制材料
unrefined oil 未精制油
unrefined oil matter 未精制油料
unrefined sugar 粗制糖
unreflected 无反射的
unreflected reactor 无反射层堆
unrefueled 未加燃料的
unregister 未挂号
unregistered 未注册的
unregistered assets 未登记资产
unregistered bond 无记名债券
unregistered company 非注册公司
unregistered securities 非注册证券
unregularity 不规则性
unregulated 未调整的;未调节的
unregulated discharge 未调节的流量
unregulated felling 非规定采伐
unregulated flood flow 未调节洪流
unregulated polymer 无规聚合物
unregulated reaches of river 未整治的河段
unregulated rectifier 不稳压整流器;非稳压整流器
unregulated river 未整治的河流
unregulated (river)bed 未整治的河床
unregulated stream 未整治的溪流;未整治的河流
unregulated variable 不可调变量
unregulated voltage 未调节电压
unreheated 无加力燃烧室的
unreheated bloom 不再热的初轧方坯
unreinforced 无(配)筋的;无钢筋的;未加固的;不加固的
unreinforced bitumen (polymer) sheet 无胎沥青聚合物片材
unreinforced brick building 无筋砖建筑物
unreinforced brick masonry 无筋砖砌体

unreinforced concrete 无(钢)筋混凝土；素混凝土
unreinforced concrete pavement 素混凝土路面；无筋混凝土路面
unreinforced concrete slab 无筋混凝土路面板
unreinforced masonry 无筋砌体
unreinforced masonry panel walls 无配筋的砖石幕墙
unreinforced matrix 无增强基材
unreinforced nozzles openings 未补强的接管开孔
unreinforced pavement 无筋混凝土路面
unreinforced seal 无骨架密封
unreinforced section 不配筋断面
unreinforced surface 无筋路面
unreinforced weld 未增强的焊接
unreinforcement 未加固
unrelated business income 营业外收益；非本行业的收益
unrelated colo(u)r 孤立色
unrelated frequencies 无关频率
unrelated perceived colo(u)r 孤色；无关知觉色
unrelaxed modulus 未ының弛模量
unrelaxed modulus of elasticity 非松弛的弹性模数
unreliability 不可靠性；不安全性
unreliability of data 数据资料的不可靠性
unreliability of standard deviation 标准差的不可靠性
unreliable 不可靠的
unreliable estimator 不可靠估计量
unreliable observation 不可靠观测值
unreliable water 不可靠的水
unrelieved stress 残余应力
unremittance 不间断性
unremitted profit and loss 未达损益
unremitting 无间断的
unremovable 不能移去的
unremunerative 获利不多的
unremunerative service 无报酬的业务
unremunrative terms 无利可图的条件
unrendered 未抹灰的；不着色的
unrenewable 不可再生的
unrenewable resources 不可再生资源
unrepairable defective product 不可修复的残次品
unrepairable waste 不可修废品
unrepaired damage 未修的损伤
unrepeatable read 不可重读
unrepresentative 无代表性的
unrepresentative item 非代表性项目
unrequired current transfer 无偿现金转移
unrequited exports 无偿出口
unreservedly acceptance 无保留验收
unresisted rolling 无阻力横摇
unresolved 悬而未决的
unresolved complex mixture 未分解的络合物混合物
unresolved complex mixture of hydrocarbon 未分解的烃类络合物混合物
unresolved echo 不清晰回声(波)
unresolved peaks 未分辨出的峰
unresolved resonance 未分辨的共振
unresolved source 不可分辨源
unresonant 非谐振的
unresonant system 非谐振系统；非共振系统
unresounding 无回响的
unresponsive 无反应的
unrest 动荡
unrestrained beam 无约束梁
unrestrained joint 非约束节点
unrestrained land-use 无限制额土地利用
unrestrained member 非约束构件
unrestraint 无限制
unrestricted 无约束的；无限制的；不限制的；不加管制的
unrestricted air-bleb 无限制空气喷口
unrestricted flow 自由流动；无约束水流；无约束流；无限制流动
unrestricted grouting 无限制灌浆
unrestricted intersection 无管制交叉口
unrestricted invariant 无条件的不变式
unrestricted irrigation 无限制灌溉
unrestricted mostfavoured nation clause 无条件最惠国条款
unrestricted negotiation 公开议价
unrestricted operating licence 无限制操作许可证
unrestricted plastic flow 无制约塑性流动；无限制塑性流动
unrestricted random sample 无限制随机样品；无限制随机样本；非约束随机样品；非约束随机样本
unrestricted random sampling 无限制随机采样；无条件限制随机抽样
unrestricted speed 未加限制的速度
unrestricted spread 自由伸展
unrestricted surplus 未拨用盈余
unrestricted variable 无约束变量
unrestricted visibility 无限制能见度；无限远能见度
unresulted metallurgic(al) technique 冶炼技术未解决
unresulted productive condition 开采条件未解决
unretarded 未延迟的
unretarded hemihydrate gypsum plaster 快凝熟石膏灰浆
unretentive 无保持力的
unretouched photograph 未修版相片
unretouched photographic(al) reduction 未修版相片缩小
unrewind 未重卷；不重绕
unribbed 无肋条的
unribbed slab 无梗肋板
unripened 未熟的
unripened viscose 未熟黏胶
unrippled 无波纹的
unrisen 未升起的
unriven 未撕开的
unrivet 拆除铆钉
unriveting 拆除铆钉
unroasted 未经焙烧的
unroll 开卷
unrolled 退卷的；铺开的
unroller station 开卷站
unrolling of container 容器的展开
unroofed 露天的；无屋顶的
unrooted tree 无根树
unrotten 未腐烂的
un-round 不圆(的)
unrounded 不舍入
unrounded data 未四舍五入的数据
unruled card 无线卡
unruled quadric 非直纹二次曲面
un-run 不熔化；不运行
unrusted 未生锈的
unsafe 不安稳的；不安全的
unsafe act 不安全动作
unsafe building 不安全房屋；不安全建筑物
unsafe condition 不安全条件；危险条件；不安全情况
unsafe depth foundation protection 桥梁浅基防护
unsafe drinking water 不安全饮用水
unsafe error 偏危险误差；偏不安全误差；不安全误差
unsafe file 不安全文件
unsafe foundation depth 不安全的基础埋置深度
unsafe fuel 不安全燃料
unsafe range 不安全范围
unsafe reset 不安全复位
unsafe sector 不安全区
unsafe structure 不安全建筑物；不安全结构
unsafetied 未加保险丝的
unsafety 不安全性；不安全
unsalable 没有销路的；卖不掉的
unsalable goods 滞销商品；滞销货
unsalable product 没有销路的商品；没有销路的产品
unsalinization 非盐渍化
unsalutary 无益健康的
unsalvageable 不能回收的
unsampled 未取样的
unsampling 不取样
unsanded gypsum 无砂石膏灰料；无砂石膏灰浆
unsanded gypsum plaster 无砂石膏灰涂抹；无砂石膏灰粉刷
unsanitary 不卫生的
unsanitary water 不卫生的水
unsaponifiable matter 不能皂化物
unsaponifiable oil 不能皂化的油
unsaponified 未皂化的
unsatisfactory 不能令人满意的
unsatisfactory report 损失赔偿要求报告
unsatisfied 未满足的；不满意的
unsatisfied chemical bond 不饱和化学键
unsatisfied report 未最后定稿的报告
unsaturable 不能饱和的
unsaturated 未被饱和的；非饱和的；不饱和的
unsaturated acetal 不饱和缩醛
unsaturated acid 不饱和酸
unsaturated acrylic hydrocarbon 不饱和无环烃；不饱和链烃
unsaturated affinity 不饱和亲和力
unsaturated air 未饱和空气；不饱和空气
unsaturated alcohol 不饱和醇
unsaturated aliphatic aldehyde 不饱和脂肪醛
unsaturated aliphatic hydrocarbon 不饱和脂肪烃
unsaturated alkyd resin 不饱和醇酸树脂
unsaturated bond 不饱和键
unsaturated capacity 非饱和容量
unsaturated charging 非饱和充电
unsaturated chlorinated aliphatic hydrocarbon 氯化不饱和脂肪烃
unsaturated clay 非饱和黏土；不饱和黏土
unsaturated coefficient 不饱和系数
unsaturated compound 不饱和化合物
unsaturated conductivity 未饱和传导性；未饱和传导率；非饱和传导性
unsaturated current transformer 不饱和电流互感器；不饱和变流器
unsaturated derivative 不饱和衍生物
unsaturated dibasic acid 不饱和二元酸
unsaturated dive 不饱和潜水
unsaturated diving 非饱和潜水
unsaturated fatty acid 不饱和脂肪酸
unsaturated fatty hydrocarbons 不饱和脂肪族烃
unsaturated flow 非饱和土壤水分流动；非饱和(水)流；不饱和流
unsaturated green time 非饱和绿灯时间
unsaturated hydrocarbon 不饱和烃；不饱和碳氢化合物
unsaturated hydrocarbon of petroleum 石油的不饱和烃
unsaturated iron binding force 未饱和铁结合力
unsaturated lactone 不饱和内酯
unsaturated link 不饱和键
unsaturated linkage 不饱和键
unsaturated logic 非饱和逻辑；不饱和逻辑
unsaturated logic circuit 非饱和逻辑电路
unsaturated material 不饱和物质
unsaturated mineral 不饱和矿物
unsaturated monomer 不饱和单体
unsaturated neopentylglycol polyester 不饱和新戊乙二醇聚酯
unsaturated oil 不饱和原油
unsaturated permeability 未饱和透水性；不饱和渗透力；不饱和渗透度
unsaturated polybasic acid 不饱和多元酸
unsaturated polyester 不饱和聚酯(树脂)
unsaturated polyester coating 非饱和聚酯涂料；不饱和聚酯涂料
unsaturated polyester concrete equipment 不饱和聚酯混凝土设备
unsaturated polyester paint 不饱和聚酯(树脂)漆；不饱和聚酯涂料
unsaturated polyester putty 不饱和聚酯腻子
unsaturated polyester resin 不饱和聚酯树脂
unsaturated polyester varnish 不饱和聚酯漆
unsaturated rock 未饱和岩；不饱和岩
unsaturated rubber 不饱和橡胶
unsaturated side chain 不饱和支链
unsaturated silicon hydride 不饱和硅氢化合物
unsaturated soil 非饱和土(壤)；不饱和土(壤)；不饱和水分土壤
unsaturated solution 不饱和溶液
unsaturated standard cell 未饱和标准电池；不饱和标准电池
unsaturated state 不饱和状态
unsaturated steam 未饱和蒸汽；不饱和蒸汽
unsaturated synthetic(al) rubber 不饱和合成橡胶
unsaturated value 不饱和值
unsaturated vapo(u)r 未饱和汽
unsaturated void count 未饱和空隙数量
unsaturated zone 含气层；通气层；充气带；非饱水带；非饱和区；非饱和带；不饱和带；包气带
unsaturates 不饱和物
unsaturation 未饱和；不饱和(的性质或状态)
unsaturation phenomenon 不饱和现象
unsaturation test 不饱和度试验
unsawn timber 原木；粗材
unscathed 未受损失的

unscheduled 计划外的
unscheduled consumption 计划外消耗
unscheduled interruption 事故停电
unscheduled maintenance 计划外维修；出错维修；非预定维修
unscheduled manintenance time 出错维修时间
unscheduled shutdown 计划外停工
unscheduled use 计划外用途
unscientific 不科学的
unscoured silk 未脱胶丝
unscrambler 矫正器；辨声器；包装品自动推送台
unscrambling bed 炉子受料架
unscrambling grid 装料台的格栅
unscreened 无屏蔽的；没有用帘幕遮住的；未分级的；未筛（过）的；未屏蔽的；未经粒度分级的；未过筛的
unscreened gravel 混砂砾石；未过筛砾（石）；未过筛石料
unscreened lumps 未筛分的块
unscreened ore 原矿
unscreened sand 未过筛砂
unscrew 拧下螺丝；扭松螺丝；拧松；松螺钉
unsealed 未密封的；未加时的；非密封的
unsealed control gap 非密封调整间隙
unsealed joint 无封缝料的接缝；未封缝；不封缝
unsealed sample 未加封土样
unsealed source 非密封源
unsealing 开启；开封；未涂封闭底漆；未密封
unseamed 无缝的
unsearchable 无从探索的
unseasoned 未风干的；未干燥的
unseasoned lumber 未处理过的木材；新伐木材；未干燥（木）材
unseasoned timber 未处理过的木材；新伐木材；未干燥（木）材
unseasoned wood 未处理过的木材；新伐木材；未干燥（木）材
unseat 离开阀座
unseaworthiness 不适航
unseaworthy 不适航的
unseaworthy ship 不适航的船
unsecured 无担保的；未包装好的
unsecured account 无担保的账户；无担保项目
unsecured advance 无担保贷款
unsecured bond 无担保债券
unsecured creditor 无抵押担保的债权人
unsecured liability 无担保负债
unsecured loan 信用贷款；无押品货物；无担保贷款
unseeded 未加籽晶的【地】
unsegregated 未分离的
unselected 未经选择的
unselected clay 未选黏土
unselected core 未选磁芯
unselected marker 非选择性标记，不选择的标记
unselective absorption 无选择吸收
unseparated 无隔壁的
unserambler 自动堆垛台
unserviceability 不合用程度；使用不可靠性；不实用性
unserviceable 无法利用的；不合用的；不适用的；不能使用的；无用的
unserviceable brake 失灵制动器
unserviced 无人看管的
unserviced hung ceiling 不敷设管道的吊平顶
unserviced suspended ceiling 不敷设管道的吊平顶
unset 未镶嵌的；清除，未装配的；未凝固的；不固定的
unset concrete 新浇混凝土；新拌混凝土；未凝结混凝土
unset rim 凸边
unset saw 未装配的锯
unsetting 防沉扶壁
unsettled weather 多变天气；不稳定天气
unsevered filament 未切断长丝
unsewerage beam 无沟梁
unsewered 无下水道的；无沟渠的
unsewn binding 无线装订
unshaded 无遮蔽的（窗等）
unshaded area 无阴影区；非阴影区
unshaped 未成形的；不成型的
unshaped refractory 不定型耐火材料；不定形耐火材料
unshaped refractory product 不成型的耐火制品
unshapely 外形不良的
unshared 未平分的

unshared control unit 非共用控制器
unshared electron pairs 非共用电子对；未共用电子对
unshared lane 分用车道；非合用车道
unsharp 不清晰的
unsharp image 不清晰图像，不清晰影像
unsharp line 模糊线
unsharpness 不清晰度
unshatterable 不碎的
unsheared 未修剪的
unsheathed deck 无覆盖甲板
unsheathed wall 没有抹灰的墙；清水墙
unsheltered 无保护的
unshielded 无屏蔽的；无防护的；无保护的
unshielded arc welding 无保护层电弧焊
unshielded telescope 无屏蔽望远镜
unshielded twisted pair 无屏蔽双绞线对；非屏蔽双绞线
unshielted 无遮挡物的
unshift 不位移
unship 卸除舾装；从船上卸货
unshipped goods 未运出货物
unshock 无激波
unshocked 无激波的
unshored 无支撑的；不加撑的
unshored composite beam 无支撑叠合梁
unshorting 消除短路
unshrinkable 防收缩的；不收缩的
unshrinkable material 不收缩材料
unshrinkable steel 不收缩钢
unshrinking 不收缩的
unshrouded impeller （半）开式叶轮
unshrouded wheel 无围带的叶轮
unshuffle 反移
unshunted 未并联的
unsifted 未经详细检查的；未过筛的
unsifted plaster 未筛灰泥
unsighted 未经检验的
unsignalized intersection 无信号交叉口；非信号灯交叉口
unsigned 未签字的
unsigned copy 未签名本
unsigned integer 无符号（整）数
unsigned integer format 无符号整数格式
unsigned word 不带符号的字
unsigned word degree operand 不带符号的次数操作数字【计】
unsilvered plate 非镀银板
unsinkable lifeboat 不沉救生艇
unsintered 未烧结的
unsized 无填料的；无浆的；未筛分的；未过筛的；未分大小的；不分大小的
unsized coal 未筛分的煤
unsized ore 未过筛矿石；未分级矿石
unsizing 未分大小
unskilled 不熟练的
unskilled labor 粗工作；生手工人；小工；壮工；简单劳动；普通工（人）；不熟练工（人）；不熟练工人
unskilled operative 不熟练操作者
unskilled steel 沸腾钢
unskilled work 粗活工作；非技术工作
unskilled worker 不熟练工人
unskirted 无边缘的
unslacked lime 未消化石灰
unslaked 生的（石灰）；未消解的；未消毒或沸化的
unslaked and ground quicklime 未消解的磨细生石灰
unslaked lime 未消化石灰；生石灰
unslaked lime pile 生石灰柱
unslaved 无磁修正的
unsling 解开吊货绳圈
unslotted 无槽的
unslotted rotor 无槽转子
unslotted rotor turbogenerator 无槽转子汽轮发电机
unslugged 无磁滞的
unsluggish 无磁滞的
unsmeltable 不能熔炼的
unsmoothed 不平滑的；不光滑的
unsoaked 未浸透的
unsodded 未铺草皮的
unsodded land 未铺草皮地
unsodded slope 未铺草皮的边坡
unsoftening rock 非软化岩石

unsoil 除去表土；剥离表土
unsoiled 未沾土的
unsolder 焊开；拆焊
unsoldering 脱焊
unsolicited contract proposal 未邀请投标
unsolicited keyin 任意型键入
unsolicited message 任意型信息；非请求信息
unsolicited offer 主动报价；卖方自报价
unsolicited response 非请求响应
unsolicited technical alternative 不要求的技术替代方案
unsoluability 不可解性
unsoluble 不溶解的
unsoluble material 不溶物（质）
unsolvable 不能溶解；不可解的
unsolvable labeling procedure 不可解标记过程
unsolvable node 不可解节点
unsolvable remanent analysis 不溶残余分析
unsonsolidated soil 松散土（壤）
unsophisticated 无特殊装备的
unsophisticated buyer 没有经验的顾客
unsorted 未经分类的；未分选的；未分级的
unsorted aggregate 未筛分料料；未筛（分）骨料
unsorted ash 统灰；未筛分煤灰
unsorted table 未整理表
unsound 欠烧的；未消化透的；稳定性差的；不安定的；不致密的；不坚实的
unsound aggregate 不安定集料；不安定骨料，安定性不良的集料；安定性不良的骨料
unsound casting 不致密铸件
unsound cement 变质水泥；不健全水泥；不安定水泥；安定性不良的水泥
unsound construction 不牢固的结构
unsound deformation 不安全变形
unsounded cement 变质水泥
unsound grain 不完善粒
unsound joint 不坚固接头
unsound kernel 不完善粒
unsound knot 朽节；木料死节；活节；松软节；腐朽木节；不硬木节；不健全木节
unsoundness 无固定体积；不健全性；不安定性
unsound plaster 不安定灰泥；不良灰泥
unsound steel 不合格钢
unsound wood 腐朽木（材）
unspaced 无间隙的
unspaced butted weld 无间隙对接焊缝
unspecialized 无特定功能；非专业化的
unspecific 非特定的
unspecificness 无特殊
unspecified 非特定的
unspecified bit rate 未规定比特率
unspecified item 未定项目；不可预见项目
unspecified tree 无名杂树
unspent 未抽提完的
unspent allocation 未用完的拨款
unspent balance 未用完款项
unspent margin 未支用余额
unspinnable 不旋转的
unsplit 不裂口的
unsplit bedpiece 整体底板
unsplit bedplate 整体底板；整体底座
unsplit casing 整体外壳；整体机外壳
unsplit crankcase 整式曲柄箱；整体式曲轴箱
unsplit frame 整体机座
unsplit pattern 整模
unspoiled mode 不失真模式
unspoilt 未受破坏的
unspool 快钻
unspooling （卷扬机等）放缆退绕
unsprung 未装弹簧的
unsprung axle 簧下轴
unsprung parts 非悬挂部分；非破裂部分
unsprung weight 簧下重量；簧下重力；未安弹簧承重；未安弹簧的承重量
unsquared 非直角的；非方形的
unstability 不稳定性；不稳定（度）；不安定性
unstability constant of complex ion 络合离子的不稳定性常数
unstability of base line 基线不稳
unstabilized 未加固的；不稳定的
unstabilized crude 不稳定原油
unstabilized crude oil 不稳定原油
unstabilized presentation 非稳定显像
unstable 不稳定的

unstable air mass 不稳定气团
unstable arc 不稳定电弧
unstable area 不稳定区;不稳定地区
unstable association 不稳定组合
unstable atmosphere 不稳定大气
unstable atmospheric condition 不稳定大气条件
unstable balance 不稳定平衡
unstable bed load 不稳定底质
unstable channel 动床;不稳定河槽
unstable characteristic 不稳定特性
unstable chemical equilibrium 不稳定化学平衡
unstable clay 不稳定黏土
unstable community 不稳定群落
unstable complex 不稳定络合物
unstable component 非稳定环节
unstable compound 不稳定化合物;易分解化合物
unstable condition 不稳定条件;不稳定状态
unstable constant 不稳定常数
unstable constitution 不稳定组分;不稳定状态;不稳定的构造
unstable construction 不稳定结构;不稳定构造
unstable coordinates 不稳定坐标
unstable crack growth 失稳裂纹扩展
unstable crack propagation 裂缝不稳定开展;失稳裂纹扩展;不稳定裂纹扩展
unstable creep 不稳定蠕动
unstable distribution 不稳定分布
unstable dollar 不稳定币值
unstable dredge-cut 不稳定挖槽
unstable effect 不稳定效应
unstable element 不稳定元素
unstable employment 不稳定就业
unstable equilibrium 不稳(定)平衡;不稳定均衡
unstable exchange market 不稳定外汇市场
unstable filter 不稳定滤波器
unstable flow 不稳定流
unstable form 不稳定态
unstable formation 不稳定岩层;不稳定地层
unstable frame 不稳定框架
unstable gasoline 不稳定汽油
unstable ground 不稳(定)地面
unstable humus 不稳定腐殖质
unstable hydrocarbon 不稳定烃
unstable isotope 放射性同位素;不稳定同位素
unstable layer 不稳定层
unstable load 不稳定荷载
unstable lubrication 非永久润滑
unstable margin 不稳定陆缘区
unstable market 不稳定市场;不稳定的行情
unstable mineral 不稳定矿物
unstable motion 不稳定运动
unstable movement 不稳定运动
unstable nuclear 不稳定核
unstable operation 不稳定运转;不稳定工作
unstable optic(al) resonator 非稳定光学共振腔
unstable oscillation 不稳定振荡
unstable oxidation product 不稳定氧化产物
unstable particle 不稳定粒子
unstable phase 不稳定相
unstable protobitumen 不稳定原沥青
unstable radical 不稳游离基
unstable region 不稳定区
unstable regulation 不稳定调节
unstable relict matter 不稳定残积物
unstable relict mineral 不稳定残余矿物;不稳定残积矿物
unstable reservoir delta 水库不稳定三角洲
unstable residue 不稳定残积物
unstable resonator 非稳定共振腔;不稳定谐振腔
unstable ring resonator 不稳定环形谐振腔
unstable roof 不稳定顶板
unstable running 不稳定运转
unstable salt pair 不稳定盐对
unstable servo 不稳定伺服系统
unstable shock 不稳定激波
unstable slope 不稳定边坡
unstable soil 不稳定土(层);不稳定土壤
unstable solution 不稳定解
unstable stability 负稳定性;负稳定法
unstable state 不稳固状态;不稳定状态;非稳态
unstable stratification 不稳定层理【地】
unstable stratum 不稳定地层
unstable structure 非稳定性结构;不稳定结构;不稳定的构造

unstable system 不稳定系统;不稳定体系
unstable type gravimeter 不稳定式重力仪
unstable velocity 不稳定流速
unstable wave 不稳定波
unstable zone 不稳定区;不稳定地带
unstacker 拆垛机
unstacking 拆垛
unstaffed level crossing 无人看守道口
unstained 无污点的;无瑕疵的
unstained smear 不染色涂片
unstandardized technique 非标准法
unstarred non-terminal 未加星号非终结符;不带星号的非终结符号
unstart 未起动
unstated 未声明的
unstationary star 不稳定星
unstationary vibration 不固定振动
unstayed 无支撑的;未加固的;未加支撑的
unstayed flat heads and covers 无支撑平封头和盖板
unsteadiness 易变;不稳固性;不稳定(性);不定常
unsteady 不稳固的
unsteady air flow 非稳定气流
unsteady boundary layer 非稳定边界层;不稳定边界层
unsteady combustion 不稳定燃烧
unsteady conduction 不稳定的热传导;不稳定传导
unsteady current 不稳定电流;变量流
unsteady diffusion 不稳定扩散
unsteady dispersion 不稳定分散体系
unsteady distillation 不稳定蒸馏
unsteady drag 非定常流阻力;不稳定阻力
unsteady feed 不稳给料;不稳定的喂料
unsteady film boiling 不稳定膜态沸腾
unsteady flow 非稳(态)流(动);非稳恒流动;非稳流;非定常流;非常流;不稳定(动);变流
unsteady flow formula of boundary well 边界井的非稳定流公式
unsteady flow pattern 非定常流型;不稳定流型
unsteady force 非稳定力
unsteady heat conduction 不稳定导热
unsteady heat regenerator 不稳定回热器
unsteady heat transfer 不稳定传热
unsteady humus 不稳定腐殖土
unsteady load 不稳定荷载
unsteady mass transfer 不稳定传质
unsteady motion 不稳定运动
unsteady movement 不稳定运动
unsteady non-uniform flow 非恒定不均匀流;非定常变速流;非稳定均匀流;不稳定变速流
unsteady open channel sewer flow 不稳定明渠污水流
unsteady pumping test 非稳定流抽水试验
unsteady pumping test at fixed drawdown 非定流定降深抽水
unsteady pumping test at fixed quantity 非稳定流定流量抽水
unsteady rate 不稳定产量
unsteady resistance 非定常流阻力;不稳定阻力
unsteady riverbed 不稳定河床
unsteady running 不稳定运转
unsteady sea water purging 非稳常海水净化
unsteady seepage 不稳定渗流
unsteady-stage condition 非稳定工况
unsteady state 非稳态;不定态;非定常状态;不稳固状态;不稳(定)状态
unsteady state flow 不稳定状态流动;非稳态流动
unsteady state heat conduction 非稳定传热
unsteady state heat transfer 非稳态传热
unsteady state thermal anomaly 非稳态热异常
unsteady state transfer of heat 非稳态传热
unsteady uniform flow 等速不稳定流
unsteady wave 非定常波;不稳定波
unsteady wind 不稳定气流;不稳定风
unsteamed concrete 非蒸(汽)养(护)混凝土
unsteered 未操纵的
unstemmed shot 无炮泥爆破
unsterilized 未杀菌的
unstiffened 未变硬的;未加强的;未加劲的
unstiffened edge slab 未加劲边跨板
unstiffened member 未加劲构件
unstiffened plate 无加劲板
unstiffened suspension bridge 无加劲式吊桥;未加劲悬索桥

unstiffened type 未加劲式
unstitched 未缝合的
unstocked blank 无立木地
unstocked forest land 无林地
unstoried 未载入史册的
unstow 卸空
un-straight 不直
unstrained 无应变的;未应变的;未拉紧的;未过滤的;未变形的
unstrained medium 无应变介质
unstrained member (构架的)不受力杆;不受荷元件;无应变的构件;不受力构件
unstrained pile 自由桩
unstrained pile head 自由桩头
unstratified 无层理的;非层状的;不成层的
unstratified bed 不分层床
unstratified drift 非层状堆积物;非层状冰碛;不成层冰碛
unstratified filter bed 不分层的滤床
unstratified lake 无层结湖
unstratified mineral deposit 不成层矿床
unstratified rock 非层状岩;不成层岩
unstratified soil 非成层土;非层状土
unstratified structure 不成层构造
unstrengthened 未加强的
unstressed 无应力的;未受应力的;不受应力的
unstressed member (构架的)无应力杆件
unstressed pore 无受力孔隙
unstressing 无应力
unstriated 无纹的;无横纹的
unstring the block 解下游动滑车;松开游动滑车
unstriped 无纹线的
unstripped gas 原料气体
unstructured data item 非结构(化)数据项
unstructured dining 不固定式用餐
unstructured fiber 无结构纤维
unstructured program(me) 非结构式程序
unstuck 未系住的
unstudded 无挡的
unstudded cable 无挡锚链
unstudded shortlink chain 无挡短环链
unstuffing 取货作业;拆箱;掏箱
unstuffing and unsealing 未进行回填封闭
unsubdued 未减低的
unsubmerged 不淹没的
unsubmerged spur 不淹丁坝
unsubmerged weir 不淹没堰
unsubscribed capital stock 未认购股本
unsubscripted 无下标的
unsubscripted variable 无下标变量
unsubscripted variable name 无下标变量名
unsubstantial 不坚固的
unsubstituted 未取代的;未被取代的
unsubstituted aromatics 未取代的芳香烃
unsuccessful bidder 未得标人;招标落选者;未得标商;不中标者
unsuccessful call 不成功的呼叫;失败呼叫
unsuccessful execution 失败的执行
unsuccessful party 败诉的一方
unsuccessful plan 不成功的计划;失败计划
unsuccessful tender 未中标者
unsuccessive grain 不连续颗粒
unsued time 关机时间
unsuitable 不适宜的;不合适的
unsuitable containerizable cargo 不适合装箱货
unsuitable fill 不适用的回填材料
unsuitable packing 不适当的包装
unsuitable runway 不适用的跑道
unsuitable value 不适合值
unsuitable water 不适用的水
unsuited to needs 不合乎需要的
unsulfated cement 未加石膏的水泥
unsulfonated 未磺化的
unsullied 未受玷污的
unsunned 不见阳光的
unsupercharged 未超载的;无增压装置的;不增压的
unsupervised 无人监督的
unsupervised classification 无监督分类法;非监督分类
unsupervised crossing 不设交通岗的交叉口
unsupervised image classification 未被监视的图像分类
unsupported 无支柱的;无支撑的;未经证实的;自由的;无支承的

unsupported adhesive 无载体胶黏剂
unsupported area 未支护区
unsupported back 未支护顶板
unsupported back span 无支护顶板跨度
unsupported barrel 自由悬挂卷筒
unsupported catalyst 无载体催化剂
unsupported depth 无支撑深度
unsupported distance 悬空距离；自由长度；无支撑间距
unsupported ferric oxide ceramic membrane 无支撑氧化铁陶瓷膜
unsupported flank 无依托翼侧
unsupported height 悬空高度；自由高度；无支撑高度
unsupported length 自由长度；无支承长度；无支撑长度
unsupported space 悬空距离
unsupported span 无支护跨度
unsupported width 悬空宽度；自由宽度；无支护跨度；无支撑宽度
unsuppressible-literal 不可删文字
unsuppressible-replicator 不可删重复符
unsuppressible-suppression 不可删删除
unsurfaced 未铺(路)面的
unsurfaced asphalt rag felt 未涂沥青的粗油毡
unsurfaced exposed concrete 表面不暴露的混凝土
unsurfaced road 未铺路面的道路；(无路面的)土路
unsurfaced soil base 未铺(路)面的土基
unsurfaced strip 无铺筑道面的简易机场
unsurfaced subgrade 无路面路基；土路基
unsurfaced tar(red) rag felt 未涂柏油的粗油毡
unsurpassed 未被超过的
unsurveyed area 无实测资料地区；未测量区
unsurveyed coastline 未勘测岸线；草绘岸线
unsurveyed river 未勘测河流
unsurveyed shoreline 未勘测海岸线
unsurveyed stream 未勘测河流
unsusceptible 不敏感的；不易感受的
unsuspected 未知的
unswept pumping 固定频率泵浦
unsymmetric(al) 不对称的
unsymmetric(al) alkene 不对称烯属烃
unsymmetric(al) angular milling cutter 不对称角度铣刀
unsymmetric(al) attenuator 不对称衰减器
unsymmetric(al) balance 不对称平衡网络；不对称平衡
unsymmetric(al) beam 不对称梁
unsymmetric(al) bending 不对称弯曲
unsymmetric(al) circuit 不对称回路；不对称电路
unsymmetric(al) cross-section (道路的)不对称横断面
unsymmetric(al) crown 不对称路拱
unsymmetric(al) cruciform pier 不对称的十字形墩台
unsymmetric(al) curve 不对称曲线
unsymmetric(al) diaphragm 不对称膜片
unsymmetric(al) dispersion 不对称色散
unsymmetric(al) distribution 不对称分布
unsymmetric(al) distribution curve 不对称分布曲线
unsymmetric(al) double angle milling cutter 不对称双角铣刀
unsymmetric(al) double curve turnout 单式不对称道岔【铁】；异向双开岔道【铁】
unsymmetric(al) double curve turnout in the same direction 单式同侧道岔
unsymmetric(al) effect 不对称效应
unsymmetric(al) equilibrium 不对称平衡
unsymmetric(al) facade 不对称立面
unsymmetric(al) fold 不对称褶皱
unsymmetric(al) footing 不对称基础；不对称柱脚
unsymmetric(al) form 非对称造型
unsymmetric(al) grading 不对称分品连接
unsymmetric(al) ion 不对称离子
unsymmetric(al) joint 不对称接合
unsymmetric(al) kernel 不对称核
unsymmetric(al) laminate 不对称层板
unsymmetric(al) load(ing) 不对称荷载；不对称负载；不对称负荷；偏压
unsymmetric(al) matrix 非对称矩阵
unsymmetric(al) monomer 不对称单体
unsymmetric(al) multivibrator 不对称多谐振荡器
unsymmetric(al) operation 不对称运行(状态)

unsymmetric(al) oscillation 非对称振荡；不对称振荡
unsymmetric(al) peak 不对称峰
unsymmetric(al) plan 不对称平面
unsymmetric(al) plate girder 非对称板梁
unsymmetric(al) potential 不对称电位；不对称电势
unsymmetric(al) profile contour 不对称齿廓
unsymmetric(al) reagent 不对称试剂
unsymmetric(al) response 非对称响应；不对称反应
unsymmetric(al) rhythm 不对称韵律
unsymmetric(al) roof 不对称屋顶
unsymmetric(al) scalar 不对称标量
unsymmetric(al) short-circuit 不对称短路
unsymmetric(al) spreading 不对称扩张
unsymmetric(al) spring 不对称的弹簧钢板
unsymmetric(al) street 不对称街道
unsymmetric(al) stretch 不对称伸缩
unsymmetric(al) structure 不对称结构
unsymmetric(al) three-throw turnout 不对称三开道岔
unsymmetric(al) three-way turnout 不对称三开道岔
unsymmetric(al) transducer 不对称换能器
unsymmetry 不对称(性)；不对称现象
unsystematic(al) error 非系统性误差
unsystematic(al) measurement errors 非系统测量误差
unsystematic(al) risk 非系统性风险；可避免的风险
untack 分开
untagged 未加标签的
untamed 未受抑制的
untamped 未填实的；未夯实的
untamped backfill 未夯实的回填土
untanned 未经过鞣制加工
untapered 无尖削的
untapered pile 无锥度桩；无斜度桩；等(截)面桩
untapped 未开发的
untapped aquifer 未排放蓄水层
untapped reserves 未开发(的)资源
untapped resources 未开发资源
untarred 未浸沥青的
untarred jute spun yarn 未浸沥青的黄麻细纱
untarred road 非柏油路
untarred tow 没有涂焦油的麻屑；没有涂焦油的拖绳
untaxed 免税的；未完税的；不负担不重的；不征税的
untempered 未调和的(石灰等)；未(淬)回火的
untenantable 不适宜居住的；非租赁的
untenanted 无人租住的
untensioned 不张拉的；松弛的；不拉紧的
untensioned bar reinforcement 不张拉钢筋；普通钢筋；未加拉力的钢筋
untensioned rail fastening 无弹性扣件
untensioned reinforcement 不受拉钢筋；非张拉钢筋
untensioned reinforcement bar 非预应力钢筋
untensioned steel (reinforcement) 不张拉钢筋；普通钢筋；未加拉力的钢筋
untensioned wire 未加拉力钢筋
unterminated 无端接的
unterminated line 无端接线
untested 未经试验的
untethered atmospheric submersible 无缆常压潜水器
untethered diving 无缆式潜水
untethered remotely-operated vehicle 无缆遥控潜水器
untethered self-propelled vehicle 无缆自航式潜水器
untethered unmanned vehicle 无缆遥控潜水器
untethered vehicle 无缆式潜水器
unthinned 不薄的；不稀释的
unthread 使松脱；不通
unthreaded hole 光孔；无螺纹孔
unthreaded pipe 无车扣管子
unthreaded portion 无螺纹部分
unthreaded tube 未车扣管子
unthreshed 未脱粒的
untie 松开
untied loan 不附条件的贷款
untied pair 不同对
untight 漏的；未密封的；不紧密的
until further notice 在另行通知以前

untilled land 未耕地
untilted negative 无倾斜的底片
untimbered 无支撑的；无坑木支的
untimbered rill 无支护对角工作面
untimbered rill method 无支护倾斜分层充填开采法
untimbered rill stoping 无支护倾斜分层充填法回采
untimbered tunnel 无坑木支撑的隧道
untime 不定时
untimed prompt shipment 不限期即期交运
untimely 不适时的
untimely frost 不合时令的霜冷
untipped 不镶刀的
untipped steel bit 不镶刃的钢钻头
unto buyer 以购货人为受货人
unto collecting bank 以代收银行为受货人
untontaminated core 未污染岩芯
untouched 原状的；未触动过的
untouched edge 未切边
untouched measurement 非接触式测量
untoward effect 不良反应
untraceable cost 不可溯成本
untrained 未受训练的；未经培训的
untrained driver 无经验的驾驶员
untranslated storage 非转换存储；不转换存储器
untransposition 未换位
untrapped 不密封的；不吸收的；不截留的
untreated 未经处理的；未浸渍的；未处理的
untreated aggregate 未处治集料；未处治骨料；未处理骨料
untreated calcium sulphate 天然石膏
untreated cloth 未处理的织物
untreated coil 未处理的线圈
untreated effluent 未处理排放物；未处理过的污水；未处理废水
untreated fabric 未处理的织物；素布
untreated foundation 未处理地基；天然地基
untreated gasoline 未处理汽油
untreated glass fabric 未处理的玻璃织物
untreated gravel road 未经(黏合料)处理的砾石路；未处理砾石路
untreated joint 不处理接缝；外露接缝
untreated loading index 未处理过的负荷指数
untreated oil 未处理原油；未处理的油
untreated pole 未浸渍电杆；未经(防腐)处理的电杆
untreated quality 处理前质量
untreated river water 未处理过的河水
untreated sewage 未处理过的污水
untreated sleeper 未处理过的枕木
untreated sludge 未处理过的污泥
untreated surface 未经处理的面层；未经处理的路面
untreated timber 未经防腐处理的木材
untreated timber pile 未经(防腐)处理的木桩
untreated timber sleeper 素枕
untreated wastes 未处理废水
untreated wastewater 未经处理的废水；未处理过的污水
untreated water 原水；未处理水；未处理(过)的水
untreated wood pole 未经处理的木杆
untrimmed floor 无镶边装修楼板；无托梁楼板；无装饰格栅的楼板
untrimmed quarry stone 毛石料
untrimmed size 未切边的规格
untrimmed timber 未去皮木材
untrivial solution 非零解
untrue 不真实的
untubed well 未下套管的井
untunable magnetron 不可调磁控管
untuned 未调谐的；不调谐
untuned aerial 未调谐天线
untuned amplifier 不调谐放大器
untuned antenna 不调谐天线
untuned blade 不调频叶片
untuned circuit 未调谐电路；非调谐电路
untuned rope 未调谐的长反射器
untwine 解开(缠绕物)
untwine a station bottleneck 疏解车站咽喉
untwining 疏解
untwining at separate grade crossing 立体交叉疏解【铁】
untwining for approach line 进出站路线疏解
untwining for leading lines 线路疏解
untwining for train types 列车类别疏解
untwinned martensite 非李晶马氏体
untwist angle 反扭角

untwisted blade 非扭曲叶片
untwisted wire rope 无捻钢丝绳
untwisting 解开(缆绳)
untwisting machine 解拧机;松拈机
uniformity coefficient 不均匀系数
ununiform loops 毛环不齐
unusable 不适用的;不能使用的
unusable diamond 不能使用的金刚石
unusable glass 无用玻璃
unused balance 结存
unused balance of letter of credit 信用证未用余额
unused bit 未用位
unused code 禁用码;不适用码
unused combination 非法组合
unused command 非法命令
unused fixed assets 未使用固定资产
unused fund 未动用资金
unused heat 未利用热量
unused land 未用地;未利用的(土)地
unused land tax 未利用土地税;土地闲置税
unused name 未用名
unused portion of letter of credit 信用证未用部分
unused resources 未利用资源
unused storage location 未用存储单元
unused time 待修时间
unused water 没用过的水;未用过的水;未利用水
unused water energy 未利用水能
unused zone 未用地区;未利用的(土)地;不利用的地区
unusual 稀有的;例外的;不普通的
unusual and non-recurring item 特殊和偶然发生的项目
unusual attitude maneuver 异常姿态机动
unusual casualty reserves 非常灾害准备
unusual deformation 不正常变形
unusual drought 异常干旱
unusual exposure 异常照射
unusual flood 特大洪水;十年一遇洪水;非常洪水
unusual flood peak 非常洪峰
unusual increase 异常增加
unusual loads combination 非常荷载组合
unusual loss 非常损失
unusualness 异常
unusual service condition 异常工况;不正常工作条件
unusual sound 异响
unusual stage 非常水位
unusual structure 非寻常建筑物;罕见的建筑物
unusual weather 异常天气;反常天气;反常的气候
unvalued 无价值的
unvalued policy 未定值保险单;不定值保险单;保险金额未确定保单;保险金额未定保单
unvariable expenses 不变费用
unvariable point 不变点
unvarnished 未修饰的;未油漆的
unvaulted 非拱顶的
unvegetated 无植被的
unveil of a statue 新雕像揭幕仪式;雕像揭幕仪式
unvented 未放气的
unvented appliance 无烟道的燃具
unvented circulator 直接燃烧循环采暖器
unvented gas appliance 无烟道的煤气灶具
unvented space heater 无烟道采暖炉
unrented type 开放式燃(气用)具
unventilated 未加通风的;不通风的;没有通风设备的
unverifiable 无法考证的;无法检验的;无法核实的;不能证实的
unverified 未证实
unvernalized 未经春化处理的
unvibrated hairspring 未定长游丝
unvital 无活力的
unvouched 未被证明的;未加证实的
unvulcanized 未硫化的
unvulcanized rubber 非硫化橡胶
unwanted 不需要的
unwanted absorption 有害吸收
unwanted material 不需要的材料;不常用的材料
unwarranted claim 没有理由的索赔
unwarned 未受警告的
unwarp 解开包装
unwarranted 无根据的;未经授权的
unwarranted acceptance 不当接受
unwarranted claim 无根据的索赔
unwarranted losses 不应有的损失

unwashed 未清洗的;未冲刷的
unwatched 无人看守的;无人管理的
unwatched light 无人看守的灯标
unwatched lighthouse 无人看守的灯塔
unwatched lightship 无人看守的灯船
unwatered 缺水的;无水的
unwater(ing) 泄水;排干;疏干;抽水;放空水
unwatering borehole 降水孔
unwatering coefficient 疏干系数
unwatering conduit 泄水孔底;放水管
unwatering gallery 泄水廊道;排水廊道
unwatering method 排水方法
unwatering of lock chamber 水闸室泄水
unwatering pipe 排水管(道)
unwatering pit 排水坑
unwatering pump 排水泵
unwatering sump 排水坑
unwatering system 排水系统;脱水系统;疏干系统;放空水系统
unwatering yield 疏干开采量
unweathered 新鲜的(指岩石的风化程度);未(经)风化的
unweathered rock 新鲜岩石;未风化岩石
unweathered soil 未风化土
unweighable 不可称量的
unweighted 未加权的
unweighted analysis of variance 不加权方差分析
unweighted mean 未加权平均数
unweighted noise 未加权噪声
unweighted pair-group method analysis 非加权配对组法分析
unweighted regression analysis 不加权回归分析
unweighted signal-to-noise ratio 未加权信噪比;不加权噪比
unweldable 不可焊接的
unwelded 未焊好的
unwelded joint 接缝
unwheeling 架车
unwheeling repair 架修【铁】
unwholesome demand 不健康需求
unwieldy 难控制的;难操纵
unwieldy working piece 笨重工件
unwindase 解链酶
unwinder 开卷机;退卷机;拆线机;拆卷机
unwind(ing) 退卷;解退;消除程序修改;展开程序;解绕;程序展开;拆卷
unwinding coiler 开卷机
unwinding equipment 导开装置
unwinding installation 退绕装置;退卷装置
unwinding length 退绕长度
unwinding reel 开卷机;折卷机
Unwind's formulae 恩云公式(结构设计用)
unwind the call stack 解退调用栈
unwindy 无风的
unwired 无金属丝的
un-wooded 无树木的
unworkability 难加工性
unworkable boundary of ore body 矿体难采边界线
unworkable coal seam 不可采煤层
unworkable concrete 和易性差的混凝土
unworkable harness 无法治理
unworkable machine-hours 装卸机械非完好台时
unworkableness 难加工性
unworked 未用过的;未加工的
unworked country 未采区
unworked crystal 未加工晶体
unworked grease 未使用过的润滑脂
unworked penetration 静止针入度;工作前针入度
unworldliness 超脱性
unworn 未受磨损的
unworthiness 不适用性
unworthy 无价值的
unwound core 未嵌绕组铁芯
unwoven cloth 无纺布
unwoven glass fiber tissue 非织造玻璃纤维(薄)毡
unwrapping tool wire wrap stand 绕线机解线工具
unwritten dictates 不成文规定;不成文法
unwritten law 习惯法;不成文法
unwritten law of sea 海上不成文法
unwritten rules 不成文规定
unwrought 毛坯(木材);未制造的;未(经)加工的
unwrought log 未经加工的木材
unwrought timber 毛材;未加工的木材;未加工

木料;未经加工的木材
unyawed 无偏航的
unyielding 硬的;坚固的;稳定的;不可压缩的;不屈服(的);不沉陷的
unyielding abutment 无沉陷桥台
unyielding foundation 无沉陷的地基;不沉陷地基
unyielding support 不可压缩支架;不沉陷支座
unyielding surface 不沉陷的面
unzoned 未划分区域的
unzoned lens 简单透镜
U ore of dolomite type 白云岩型铀矿石
U ore of granite type 花岗岩型铀矿石
U ore of lamprophyre type 煌斑岩型铀矿石
U ore of limestone type 灰岩型铀矿石
U ore of sandstone type 砂岩型铀矿石
U ore of volcanic-rock type 火山岩型铀矿石
U-packing U 形断面盘根
upalite 针磷铝铀矿
up-and-down 起伏不平的(道路);起落
up-and-down alternating excavating process 上下轮换开挖法
up-and-down anchor 立锚
up-and-down blanking process 上下落料法
up-and-down cut shears 下切式剪切机
up-and-down draught kiln 倒焰窑
up-and-down fall 小关吊货索;舱口吊货索
up-and-down indicator 走时指示器;紧松指示器;松紧指示器
up-and-down locks 安全锁
up-and-down maneuver 升降机动
up-and-down method 上下法
up-and-down motion 上下运动;垂直方向往复运动
up-and-down movement 上下运动;垂直(往复)运动
up-and-down pull 上下作用的垂直力
up-and-down pull on the drawbar 上下推动牵引钩
up-and-down rod 水下联杆;直立杆;上下联杆
up-and-down screw 升降螺丝
up-and-down soke 上舵轮柄
up-and-down traffic 上下行交通;上下对开行车
up-and-down train routes 双进路
up-and-down trains 上下行列车
up-and-down welling 升降流
up-and-down locks 上下位锁
up-and-out option 见涨即止期权
up-and-over door 上卷门;翻门;上翻门
up-and-over door of the roll-up type 卷升式上翻门
up-and-over door of the swing-up type 摆式上翻门
up-and-over garage door 上盖式车库门
up behind 停拉
up-boom dredge 上挖式挖掘船
upbound 上行;上行的;上水的
upbound boat 上行船(舶);上水船(舶)
upbound cargo flow 上行货流
upbound commerce traffic 上水运量
upbound dredging 逆流施工【疏】
upbound entry time 上行进闸时间;上水进闸时间
upbound exit time 上行出闸时间;上水出闸时间
upbound fleet 上行船队
upbound journey 上水航程
upbound sailing 上水行驶
upbound steamer 上水船(舶)
upbound tows 上水顶推驳船队;上行船队;上水驳船队
upbound traffic 上水运量
up(-bound) train 上行列车
upbound vessel 上行船(舶);上水船(舶)
upbow 上弓
up brow conveyer 上山运输机
upbuilding 泥沙淤积;泥沙积聚
up-building of delta 三角洲增长;三角洲淤长
upcast 回风井;小炉上升道;排气坑道;上抛的;上风井
upcast air 上行风流
upcast fan 上行式通风机;上风井口扇风机
upcast fault 逆断层;上投断层
upcast header 上升管联箱
upcast shaft 竖向排气管;通风井;通风管
upcast side 上投侧
upcast slip 上投滑距
upcast ventilator 朝上通气筒;向上通风筒

up-chain 上行链路
upcharge 附加费用
upcheck 试验合格的分数
upcoast 上行海岸
up-coiler 卷绕机；上卷机；地上卷取机
up-conversion 增频转换；增频变频；升频转换；上转换；上变频
up-conversion phosphor 向上转换型磷光体
up-converter 升频器；上转换器；向上变换器；上变频器
up-converting laser 升频转换激光器
up-counter 升计数器；上计数器
up-country 内地；在内地
upcurrent 向上游(的)；上升气流
up curve 上升曲线
upcurved 上挑的；反弧的
upcurved spillway bucket 溢流堰消能鼻坎
up cut 上切式
up-cut grinding 同轴向磨削
up-cut milling 仰铣
up-cut shears 上切式剪切机
updatable data 可更新数据
updatable microfilm 可更新缩微胶片
update 修正；更新
update and forecast 更换及预测
update cycle 更新周期
updated 更新的；校正的
update data 最新资料
updated constant 更新常数
updated-record mark 更新记录标记
updated version 修改方案
update file 更新文件
update inconsistency 更新矛盾；更新冲突
update information 修正信息
update install 更新安装
update intent 更新意向；更新意图
update mark 更新标记
update mode 更新状态；更新方式
update number 更新号
update-only recovery 不可恢复的更新
update-record mark 更新过记录标志
update routine 修改程序
update transaction 更新事项(处理)；更新事务(处理)
update usage mode 更新使用方式
updating 更新；校正；使近代化
updating and alteration funds 更新改造基金
updating equipment 设备更新
updating formula 校正公式
updating maintenance 更新维护
updating of file 修改文件
updating plan 现代化计划
updating program(me) 修改程序；现代化计划
updating rate of real assets 固定资产更新率
updating routine 修改程序
updating technique 最新技术；最新工艺；适时校正法；适时修正法
updating technology 最新工艺
updating training 现代化训练
up digging 上挖
up digging excavator 仰掘机
updip 向上倾斜；(地层的)逆倾斜
updip attachment 上倾附着物
updip wedge out trap 上倾尖灭圈闭
updip wedge out type pool trend 上倾尖灭型油气藏趋向带
updip well 背斜井
up direction 上行方向；上坡方向
up-Doppler 上行多普勒
up-down asymmetry 上下不对称
up-down counter 加减计数器；升降计数器
up-down error 上下误差
up-down flip-flop 控制增减触发器
up-down operation on incline 上下坡作业
up-down reversal 上下颠倒
up-down traverse of the arm 臂部升降
updraft 上流式的；上风式的；向上排气；上抽的；向上通风；向上抽风；上曳气流
updraft boiler 塔式锅炉
updraft carburet(t)or 上吸式化油器；上风汽化器
updraft combustion 上抽式燃烧
updraft furnace 向上通风炉；直焰炉
updraft grate 上风炉箅
updraft hood 上吸式拔气罩

updraft kiln 向上抽风窑；直焰窑
updraft sintering machine 鼓风烧结机
updraft stream 向上气流
updraft type heater 直焰管式加热炉；竖形圆筒式加热炉
updraft velocity 上曳气流速度
updraft ventilator 趄上通风筒；上风道式通风机
up-draught 上抽的；上曳气流；上吸；上升气流
up-draught drying fan 鼓风干燥风机
up-draught flue 上抽烟道
up-draught gas producer 上抽煤气发生炉
up-draught intermittent brick kiln 上抽风式间歇砖窑
up-draught kiln 上抽风式窑；直焰窑；升焰窑
up-draught producer 上抽发生炉
up-drawing sheet glass process 平板玻璃垂直引上法
up-drawing tube machine 垂直拉管机
up-drawing tube process 垂直上拉法
up-draw machine 垂直引上拉管机
up-draw process 垂直引上法；垂直(连续)引上制法
up-draw sheet process 上拉法
updrift 逆向推移
updrift side 上游方向
up-elevator 上偏的升降舵
up-end 倒立；竖立；作大修改；吊桶朝上
up-ended 横置的；顶锻
upender 翻料机
upend forging 冲挤锻造
upend(ing) test 局部镦粗试验
up-face 上向工作面
upfaulted 上投断层的
upfaulted block 上投断块；上盘断块；断层上块
upfeed distribution 向上供送系统；泵压供水系统；下行上给式
upfeed distribution system 扬水式配水系统
upfeed down type 下行上给式
upfeed head tank system 上给下分式压力水箱系统
upfeed heating system 下分式采暖系统；上行式采暖系统
upfeed method 上给下分式
up feed or downfeed two pipe system 上供或下供式双管系统
upfeed system 下分式系统；上行下给式
upfeed system of heating 上行下给式供热；上分式供暖系统；上给下分式供暖系统
upfield shift 高磁场位移
upfloated 浮起的
upflow 上(升气)流；上流；竖流
upflow anaerobic biofilter 升流厌氧生物滤池
upflow anaerobic blanket reactor 升流厌氧污泥床反应器
upflow anaerobic filter 升流厌氧滤池
upflow anaerobic sludge bed 升流(式)厌氧污泥床
upflow anaerobic sludge bed process 升流厌氧污泥床法
upflow anaerobic sludge bed reactor 升流厌氧污泥床反应器
upflow anaerobic sludge blanket 上流式厌氧污泥床；升流(式)厌氧污泥床
upflow anaerobic sludge-fixed film bioreactor 升流厌氧污泥—固着膜生物反应器
upflow baffle 上流滤器
upflow basin 升流式池；上流式池
upflow biofilm filter 升流式生物膜池
upflow biofiltration reactor 升流生物过滤反应器
upflow biological aerated filter 升流曝气生物滤池
upflow biotower 升流生物塔
upflow clarifier 升流澄清池；上流澄清池
upflow coagulation 竖流凝聚(作用)；升流式凝聚(作用)
upflow composite biofilter 升流复合生物滤池
upflow contact clarifier 竖流(式)接触澄清池；升流(式)接触澄清池
upflow contactor 上向流接触器
upflow continuous digester 升流式连续蒸煮锅
upflow continuous dissolver 上流式连续溶解器
upflow expansion neutralizing filter 升流膨胀中和滤池
upflow filter 压力式滤清器；上流式滤清器；上流过滤器

upflow filtration 升流(式)过滤
upflow filtration process 升流式过滤法；升流过滤工艺
upflow fixed bed reactor 升流固定床反应器
upflow fixed film anaerobic bioreactor 升流固着膜厌氧生物反应器
upflow fluid-catalyst unit 上流流体
upflow fluidized bed 升流流化床
upflow furnace 顺流式烟道采暖炉
upflow limestone bed 升流石灰床；上流石灰床
upflow packed bed reactor 升流填料床反应器
upflow pebble bed clarifier 升流式卵石澄清池
upflow principle 上流原理
upflow reactor 升流反应器
upflow regeneration 逆流再生
upflow roughing filter 升流粗滤池
upflow roughing filtration 升流粗滤
upflow sand filter 升流式砂滤池；上向流砂滤器
upflow sludge bed 升流污泥床
upflow sludge bed-filter 升流污水床滤池
upflow sludge bed reactor 升流污泥床反应器
upflow sludge blanket 升流污泥床
upflow sludge blanket filter 升流污泥床滤池
upflow sludge blanket reactor 升流式污泥反应器
upflow system 上流式系统
upflow tank 上流(式)沉淀池
upflow tower 升流塔
upflow-type central furnace 升流式集中(采暖)炉
upflow velocity 升流速度
upfold 向上褶皱；隆皱；背斜褶皱
up former 上网挂浆
upfront 上前舱
upfront investment 预先投资
upglide 上滑
upglide cloud 上滑云
upglide motion 上滑运动
upglide surface 上滑面
upgrade 上坡道；加浓；提级；提高品位；提高矿物的品位；提高等级；升坡；升级；上坡；上限；高级
upgrade basin 顺流盆地
upgrade belt conveyer 升坡皮带输送机
upgraded design formula 先进的设计公式
upgraded line 改造线路
upgrade existing line 既有线改建与改造
upgrade existing track 既有线改建与改造
upgrade industries 产业升级
upgrade pumping station 上坡升压泵站
upgrader 提高质量装置
upgrade resistance 上坡阻力
upgrade run 上坡运行
upgrade section 升坡段；上坡区段；上坡坡段；上坡地段
upgrade set 高层集；上升集合
up gradient 上坡道
upgrading 提高质量；培训提升；提高品位；提高等级；叠积作用
upgrading and updating of product 产品升级换代
upgrading existing facility 改善现有设施
upgrading of boiler weld 锅炉焊缝加强
upgrading of human resources 提高人力资源的质量
upgrading training 升级训练
upgrading wastewater treatment plant 浓集污水处理厂
upgust 上升阵风
uphand sledge (hammer) 手用大锤
uphead welding 仰接焊
upheaval 隆起；动乱；动荡；土壤隆起；地层激变
upheaval coast 隆起海岸
upheaval island 上升岛
upheaval of land 土地隆起
upheave 上升
upheaved region 上升区
uphill 上坡的
uphill and downhill travel 轴向窜动；上下窜动
uphill belt conveyer 升坡皮带输送机
uphill casting 下铸法；底铸(法)
uphill conveyer 上山运输机
uphill diffusion 上坡扩散；富集扩散
uphill flow 上坡流
uphill furnace 倾斜炉
uphill grade 向上坡度
uphill gradient 升坡；上向坡度；上坡(坡度)

uphill grouting 仰灌;底灌
uphill line 上行管道;上坡管道
uphill quenching 上坡淬火
uphill road 上坡路
uphill running 反浇
uphill running gate 底注浇口(内浇口向上倾斜)
uphill screening 斜筛
uphill section 上坡坡段
uphill shaker 上行式振动运输机
uphill side 向上的坡面;山坡面;上坡侧
uphill slope 升坡;上山坡;上坡;向上的斜坡
uphill sloping profile 陡坡剖面
uphill teeming 下铸法
uphill welding 向上焊接;上坡焊
uphold principle 坚持原则
uphole 上向炮孔;向上倾斜的炮眼
uphole boring 仰孔钻进;上向钻进
uphole detector 井口检波器;上向孔测定器
uphole drilling 仰孔钻进;上向钻进;上向孔凿岩
uphole geophone 井口拾振器;井口检波器
uphole method 上孔法
uphole record 井口记录
uphole shooting 井口地震测井;炮井地震测井;上向孔爆破
uphole synthesized seismic record 井口合成地震记录
uphole time 升口时间;(初至波)到达井口时间
uphole velocity 升口速度
uphole work 仰孔作业;上向钻孔工作;上向炮眼作业
upholster 摆设;软座软垫制作;布置;装潢;装饰
upholstered 装潢过的
upholstered armchair 沙发椅
upholstered chair 软垫座椅
upholstered seat 软席
upholsterer 家具商;家具工;室内装修商
upholstering 室内装潢;装饰
upholstery 蒙皮材料;室外装饰;室内装饰业;室内装饰品(台布、窗帘、地毯等);室内装潢;车内装潢;室内装饰专业;室内装饰材料
upholstery fabric 家具装饰织物;家具布
upholstery for vehicle 车辆内装饰品
upholstery material 装潢材料
uphroe 天幕吊板
U-pipe U 形管
U piston pump U 形活塞泵
upkeep 养护费(用);拉修;维护;维持费(用);保养
upkeep and mending 房屋修缮;房屋维修;保养与维修
upkeep cost 维修费(用);维护费(用);保养费(用)
upkeep efficiency 维修费效率
upkeep expenses 维修费(用)
upkeep obligation 保养维修义务;维修契约
upkeep of equipment 设备维修
upkeep undertaking 维修任务
up-ladder 上升算子
upland 高原;高地;山地
upland area 陆域
upland bog 高地沼泽;高地泥泽
upland catchment 高地流域;上流流域
upland catchment water 高地汇水
upland channel 陆上冲沟;山沟
upland connection 陆上联系
upland cultivator 山坡地中耕机
upland down movement 升降运动
upland erosion 高地侵蚀
upland field 旱地
upland forest soil 高地森林
upland grassland 高地草原
upland lake 高地湖泊
upland meadow 山地草原
upland moor 高沼地;高沼;高地沼泽
upland nursery 旱秧田
upland peat 高地泥炭
upland plain 高位侵蚀面;高平原;山地平原;台地;高原
upland plow 坡地犁;山地犁
upland property 陆上地产
upland rice 陆稻;旱稻
upland rice nursery 旱秧田
upland rice soils 旱地水稻土
upland river 高地河流
upland river section 高地河段
upland soil 高原土(一种天然砂土混合物);高地土壤
upland stream 高地河流
upland surface water 高地地表水
upland swamp 高地沼泽
upland tropic(al) rainforest 高地热带雨林
upland vegetation 高地植被
upland water 地表水;高地来水;上游(来)水
upleap 逆断层
up-leg 上升段
up level 高电平
uplift 向上水压力;隆起;上托力;上升;上举;上浮(力);上顶压力;上顶力;上拔力;浮力
uplift cell 扬压力计;上托力计;浮托力计
uplift coast 上升海岸
uplifted block 上升断块;上升地块
uplifted (continental) shelf 上升陆架
uplifted horst 上升地垒
uplifted peneplain 上升准平原
uplifted plain 上升平原
uplifted reef 上升礁
uplifted shelf 上升陆架
uplifted side 上升侧
uplift force 浮力;上托力;上升力;上举力;上顶力
uplifting force 向上力
uplifting period 上升期
uplift(ing) pressure 向上压力;上顶压力;浮托力
uplifting stage 上升阶段
uplift intensity factor 上托力强度系数
uplift of pile 桩的上拔力
uplift pile 抗拔桩;上升桩;回升桩;浮托桩
uplift pole 隆起极
uplift pressure 浮托力;向上压力;扬压力;浮力
uplift pressure on dam 坝的上覆力(上举力)
uplift reduction coefficient 扬压力折减系数;扬水压力折减系数
uplift resistance 耐风揭性
uplift stress 反应力
uplift test 抗拔试验
uplift water pressure 上浮水压力
up-line 上行线;上行路线;入站线
up-link 上行链路;上行线路
up-link and down-link contributions 上行线路下行线路干扰量
up-link protection ratio 上行线路保护率
up-link signal 上行线路信号
upload 向上作用的负载;上载;向上作用的荷载
uploading air cleaner 顶部进气的空气滤清器
upload station 上行注入站
uplock 上位锁
up lock cylinder 上锁作动筒
uplooper 立式活套成形器
up main line 上行正线
up main track 上行正线
up maneuver 上升动作
up milling 迎铣;仰铣;逆铣
up movement 向上行动;向上运动;向上动作
upon a wind 迎风航驶
upon the spot 用现货
up or down an alteration of price clause 加价或减价条款
up-over 上向凿井;自下向上的凿井
up-over borehole 上向钻孔
up-packing press 上压式压力机
up peak 电梯乘客上行高潮
upper 上面的;上面
upper acceptance value 合格判断值上限;上限接受值
upper accumulator 高位累加器;上限累加器
upper aerofoil 上翼面
upper aileron 上部副翼
upper-air 高空;高层空气;高层大气
upper-air analysis 高空分析
upper-air chart 高空图
upper-air circulation 高空环流
upper-air climate 高空气候
upper-air climatology 高空气候学
upper-air contour 高空等高线
upper-air current 高空气流
upper-air cylinder gasket 上气筒垫密片
upper-air data 高层大气资料
upper-air disturbance 高空扰动
upper-air front 高空锋
upper-air hood 上部空气分布罩
upper-air map 高空天气图
upper-air mass 高空气团
upper-air monitoring network 高层大气监测网
upper-air observation 高空观测
upper-air report 高空气象报告
upper-air route 高空航路
upper-air sounding 高空探测;大气上层探测
upper-air sounding station 高空探测站
upper-air station 高空站
upper-air synoptic(al) station 高空气象(观测)站
upper-air temperature 高空气温
upper-air throttle 高空节流阀
upper airway 高空航路
upper alternate depth 上共轭水深
upper anchor bracket 上承锚底座
upper and lower groups design 高低组设计
upper and lower intervention limit 官价上下限
upper and lower limit alarm mechanism 上下限报警机构
upper and lower investment limit project 限上、限下项目
upper and lower limit 上下限
upper and lower limits of official rate of exchange 官方汇率上下限
upper and lower rail 上下冒头
upper annealing temperature 退火温度上限
upper anti-cyclonel 高空反气旋
upper apex 顶部
upper apex of fold 褶皱顶
upper approach 上游引航道
upper aquifer 上游含水层
upper-arm circumference 上臂周;臂围
upper asymptote 上渐进线
upper atmosphere 高空大气(层);高层大气;上层大气
upper atmosphere data 高层大气资料
upper atmosphere research satellite 高层大气研究卫星
upper atmospheric aerodynamics 高层大气空气动力学
upper atmospheric chemistry 高层大气化学
upper atmospheric dynamics 高层大气动力学
upper atmospheric photochemistry 高层大气光化学
upper atmospheric physics 高层大气物理学
upper atmospheric pollution 高层大气污染
upper atmospheric research rocket 高层大气探测火箭
upper atmospherics 大气上层
upper atmospheric thermodynamics 高层大气热力学
upper back 上背
upper baffle plate 上部隔板
upper ball cover 上轴承盖
upper ball race 上座圈
upper band 上带
upper bank 上游堤岸
upper bar 上鼠笼条;上层线棒
upper basement 上层地下室
upper basin 高库;上游水库
upper beacon light 上信标灯
upper beam 顶梁;上横梁
upper bearing 上轴瓦;上轴承
upper bearing bracket 上导轴承架
upper beater 上击肥轮
upper bed 上层床;上端;上层
upper bench 上台阶
upper bend 向上弓;向上拱;斜拱;鞍部
upper berth 上铺
upper binder 上层接结
upper blade 上刀片
upper block 上导轮
upper blown converter 顶吹转炉
upper boiler 上汽锅
upper bolster 上模座
upper boom 上弦(杆);上杆
upper boom junction plate 上弦杆连接板
upper boom plate 上弦杆板
upper border 上图廓
upper bosh line 炉腹顶线;上炉腹线
upper bottom 上底
upper bottom layer (梁的)下部的上排钢筋
upper bound 上限;上界
upper boundary 上限界;上界;上边线

upper bound depth of coal formation 煤层顶界面的深度
upper bound depth of shale in coal formation 夹矸顶界面的深度
upper bound depth of the water bearing formation 含水层顶界面深度
upper bound expression 上界表达式
upper bounding method 上界法
upper bound method 上限法
upper bound of symbol 上界符号
upper bound solution 上限解；上界解(法)
upper bound theorem 上限定理
upper brace 上部平台
upper bracket 上支架；到顶
upper bracket-arm 上机架支臂
upper bracket for cantilever 腕臂上底座
upper bracket for double track cantilever 双腕臂上底座
upper brake shaft bearing 手制动轴上导架
upper branch 上子午圈
upper branch of meridian 上子午线
upper branch of the observer's meridian 观测者子午线的上子午圈
upper break 上部缺失；上部间断
upper brick paving 立式砖块铺砌；立砌砖块铺砌；立砌砖块路面
upper bridge 上驾驶台；上层驾驶台
upper bubble 垂直度盘水准器
upper bunk 上床铺；上铺；上板铺
upper buoyancy chamber 上部浮圈
upper cage 上鼠笼
upper calendar 鞋面压延机
upper camber 上弧；上曲面
Upper Cambrian series 上寒武统【地】
upper canal reach 运河上游河段
upper canopy 上冠
upper canvases 上帆布输送带
upper cap 上部盖
upper cap nut 上螺母
Upper Carboniferous (series) 上石炭统【地】
upper case 大写(字)体
uppercase alphabetic character 大写字母
uppercase type 大写字体
upper cat bar 上推插销
upper central series 上中心列
upper chain 上链
upper chamber 上闸室
upper change point 上限点
upper chapel 大建筑物内部楼上小教堂
upper cheek plate (压碎机的)上颚板
upper choir 唱诗班的上席位
upper chord 上弦
upper chord junction plate 上弦杆连接板
upper chord member 上弦杆
upper chromosphere 色球高层
upper church 教堂的楼层
upper circle 经纬仪上盘；楼厅后座；楼厅；恒显圈；上盘；上规；三楼楼厅
upper citadel 主城堡
upper city 城市繁华区
upper clamp (经纬仪的)上盘制动螺旋
upper class 上类；上部
upper clearance 上方净空；上部净空
upper cloud 高空云
upper cloud layer 高空云层
upper cloud point 上浊点
upper clutch 上部离合器
upper coat 上涂层
upper coating 涂末道漆
upper cofferdam 上游围堰
upper coffer-wall 上游围堰
upper coil 上层线圈
upper cold front 高空冷锋【气】
upper colonnade 上柱列节理
upper column 上立柱
upper communication antenna 上部通信天线
upper Concordia intercept 一致曲线上交点
upper conductor 上导线
upper confining bed 上覆隔水层
upper connecting rod 上拉杆
upper contact 上接头
upper control arm 上置定位臂
upper control limit 控制上限；上界限；上控制(界)限

upper convex type 上凸型
upper cord bracing 上弦支撑
upper cord member 上弦杆件
upper core plate 上堆芯板
upper core support barrel 上部堆芯围筒
upper couch (roll) 上伏辊
upper coupling structure 上部连接结构
upper course 上(砌)层；上游河段；上游(段)
upper course-type 上游式(河流)
upper cover (电机的)上盖；上层覆盖；(水轮机)顶盖；上机壳
upper cover plate 上盖板
upper crankcase 上曲轴箱
Upper Cretaceous 晚白垩世【地】
Upper Cretaceous sandstone 上白垩砂岩【地】
Upper Cretaceous series 上白垩统【地】
upper critical 上临界的
upper critical cooling rate 上部临界冷却速度
upper critical cooling velocity 上临界冷却速度
upper critical depth 上部临界深度
upper critical speed 上限界速度
upper critical velocity 临界流速上限；上临界流速
upper croft 教堂拱门上的走廊；教堂拱门走廊
upper crown-length 上冠长
upper crown wheel 上小钢轮
upper crust 上地壳(层)；上部地壳
upper culmination 上中天
upper current 上层气流；高空气流；上层流
upper current conducting cap 陀螺顶电极
upper curtate 高部；上区段；上部穿孔区
upper curve 上曲线
upper cut 细切削；细加工
upper cut-off frequency 通带上限；上限截止频率
upper cylinder bore wear 上止点附近的汽缸壁磨损
upper cylinder half 上缸；上半汽缸
upper cylinder lubricant 汽缸上部润滑油
upper cylinder lubrication 上部汽缸润滑
upper cyma 上袅
upper dead center 上死点；上止点
upper deck 上(层)甲板；上层车厢；上层(舱)
upper-deck beam 上甲板梁
upper-deck cabin 上舱
upper-deck charge 上部分段装药
upper-deck hold 上层舱
upper-decking 铺上层桥面
upper-decking level 上部装罐水平
upper-deck of bridge 上层桥面
upper deep 上深槽
upper delta plain 三角洲上部平原
upper delta-plain deposit 上三角洲平原沉积
upper depression 上凹型
upper derivate 上导出数
upper derivative 上导数
upper derived function 上导函数
upper development road 上部回采平巷
upper deviation 偏差上限；上偏差
Upper Devonian series 上泥盆统【地】
upper die plate 上模板
upper differential coefficient 上导数
upper dimension bound 上维界
upper discharge 上面排料
upper door rail 门的上横挡；门的上冒头
upper draft tube 尾水管上段
upper drive mechanism 上部驱动装置
upper drum 上卷盘
upper edge 上切面
upper edge board 上边板
upper elastic limit 弹性上限
upper elevator screw bushing 升降机螺旋上端套
upper emulsion 上层乳剂
upper end 上端
upper end buoy 中沙内端浮标
upper end cover 上端盖
upper end of dredge-cut 挖槽上端
upper end of evapo(u)rator 蒸发器的上端
upper end of reach 河段上断面；河段上端
upper end of the schedular system 分级税制上层税率
upper (end-)shield 上半端罩
upper entry 上部平巷
upper envelope 上包络
upper epidermis 上表层
upper erengy band 上能带
upper estimator 上估计值

upper exhaust duct 上排风管
upper explosive limit 爆炸上限
upper extreme 上端
upper extreme point 上端点
upper face 上翼面；上表面
upper face height 上面高
upper failure plane 上层破裂面
upper falsework 上层脚手架；上部脚手架
upper fan 上扇
upper fascia (柱顶横梁的)顶条
upper feed rake 上喂入齿耙
upper felt 上毛毯
upper fiber 上缘纤维
upper fill 路堤上层(填土)
upper fillet and fascia 上枋
upper fixed jaw face 上固定颚板面
upper flammable limit 可燃上限
upper flange 上凸缘；上翼缘
upper flange junction plate 上翼缘连接板
upper flange of girder 梁上翼；大梁上翼缘；板梁上翼缘；上翼缘
upper flange plate 上翼板
upper flat plate 上平板
upper floor 楼上；上层楼面
upper floor ground plan 上层楼平面图
upper floor hearth 上层楼的壁炉地面
upper floor silo 上层小库
upper floor wall 楼层墙壁；上层楼的墙壁
upper flow regime 上部水流动态；高流态；射流状态
upper flue 上烟道
upper frame 顶框；顶架
upper frame rail 车架上梁
upper freezing-point curve 上凝固点曲线
upper frequency 上频率
upper-frequency limit 频率上限
upper-frequency limit of audibility 可闻声频上限
upper frigid zone 上寒带
upper front 高空锋
upper gallery 上层楼走廊
upper gangway 上部平巷
upper garment 上装
upper gasket 上部衬
upper gate 上闸(间)门；上游闸门
upper gate bay 上闸首
upper gate recess 上游闸门门槽
upper gauze 上层网
upper generator bracket 上机架
upper glass plate 上玻璃片
upper greensand 上海绿石砂层
upper grid plate 上格板
upper grip 上夹头
upper-growth 上木
upper guard cam 上护针三角
upper guard sill 上游护槛；上护槛
upper gudgeon 闸门顶枢；上耳轴；顶枢轴
upper guide 上导向器
upper guide bearing 上导轴承
upper guide-metal 上导轴瓦乌金
upper guide vane ring 上导叶轮
upper half 上半部
upper half-bearing 上半轴瓦
upper half casing 上缸
upper half crank case 上半部曲柄箱
upper half mean length 上半部平均长度
upper half method 上半断面超前施工法
upper half plane 上半面
upper half-power frequency 半功率频率上限
upper harmonic 高谐音；高(次)谐波
upper hatch(way) 上舱口
upper head 上闸首
upper header 上联箱；上集管(箱)
upper-header pipe 高架管道
upper head plate 上端板
upper hemispheric(al) flux 上半球光通量
upper high 高空高压【气】；高空反气旋
upper hind neck 上颈
upper hitch point 牵引装置下挂结点；上拉杆联结点
upper hopper door 小泥门
upper horizon soil 上层土(壤)
upper horizontal flue 上横气道
Upper Huronian 安尼米基系
upper-hybrid-resonance absorption 上混合共振吸收
upper hybrid wave 高混杂波

upper idler 上惰轮
upper idler roller 上惰轮辊筒
upper idler roller shaft 上惰轮辊筒轴
upper idler roller tire 上惰轮辊筒胎
upper impeller 上搅拌翼
upper incipient lethal temperature 上限起始致死温度
upper ineffective frequency 上界失效频率
upper inside weave 四层织物的第二层组织
upper insulator gasket 上部绝缘密封垫
upper intake passage 上部进路
upper integral 上限积分；上积分
upper inversion 高层逆温；对流层顶逆温
upper ionized layer 上电离层
upper joist floor 上层格栅板；上层格栅楼面
upper junction plate 上连接板
Upper Jurassic period 上侏罗纪【地】
Upper Jurassic series 上侏罗统【地】
upper laser level 激光高能级
upper lateral bracing 上弦横撑；上弦横向水平支撑；上部水平支撑；上部横向支撑
upper latitude 上纬度
upper latticed plate 上格状板
upper layer 上层
upper layer cloud 上层云
upper layer of ballast 面渣
upper layer of soil 上层土(壤)
upper layer of the ocean 海洋上层
upper leaf 水闸上游门扉；水闸上部门扉；上部矿层
upper leakage layer 上越流层
upper ledge 上框
upper level 高能级；上部水平
upper-level anti-cyclone 高空反气旋
upper-level chart 高空图
upper-level cyclone 高空气旋
upper-level disturbance 高空扰动
upper-level high 高空反气旋
upper-level loading 加载上限
upper-level of water 上层水位
upper-level ridge 高空脊
upper-level trough 高空(低压)槽
upper-level wind 高空风
upper lift drum 上部提升筒
upper limb 天体上边缘；上缘；上翼；上边缘
upper limit 最高限额；最高极限；上限(尺寸)；上极限尺寸
upper limitation of years frozen soil layer 多年冻层上限
upper limiting filter 上限滤波器；低通滤波器
upper limiting frequency 上限频率
upper limit of normal 正常值上限
upper limit of plasticity 塑性上限
upper limit of pressure 压力的上限
upper limit of production 生产上限
upper limit of size boxed dimension 最大尺寸
upper limit of ultimate strength 极限强度顶点
upper limit value 上限值
upper link 上拉杆
upper loading arch 上承拱
upper lobe 上部凸角
upper lock 上(游)船闸
upper lock approach 上(游)引航道；船闸上游引航道
upper lock arm 上锁臂
upper lock gate 上游闸门；上闸首；上闸门
upper lock head (船闸的)上游闸首
upper longitudinal bracing 上弦纵撑；上弦纵向水平支撑
upper longitudinal strut 上纵撑
upper low 高空低压
upper main sequence 上主星序
upper mantle 上地幔；地幔上部
upper mantle structure 上地幔结构
upper margin 上图廓；上边
upper marginal area 上部边缘区
upper mast 上部桅
upper mast table 上桅台
upper memory area 上端内存
upper memory blocks 上位内存块
upper meridian transit 上中天
upper millstone 上磨盘
upper mirage 上现蜃景【气】
Upper Mississippian 晚密西西比世
upper miter gate 上闸首人字闸门

upper mixing layer 高空混合层
uppermost 最上面的；最高的
uppermost complete deck 最上全通甲板
uppermost continuous deck 最上连续甲板
uppermost crust 最上部地壳
uppermost full length deck 最上统长甲板
uppermost layer 最上面的地层；表层；表面层；最上层
uppermost levels of the atmosphere 大气的最高层
uppermost phase 最大相位
uppermost stor(e)y 最高层
upper motion 上盘转动
upper mo(u)ld 上部模板
upper mounting plate 上平台
upper nappe profile 水舌上缘线
upper nicol 上偏光镜
upper nut 上螺母
upper ocean 海洋上层
upper oil-header 上集油箱
upper oil-pan 上油盘
upper ordinate set 上纵标集
Upper Ordovician period 上奥陶纪【地】
Upper Ordovician series 上奥陶统【地】
upper panel 上翼片
upper part 上部；上半部
upper partial 泛音
upper particle size limit 粒径上限
upper part of auditorium 观众厅上空
upper part of lobby 门厅上空
upper part of stage 舞台上空
Upper Permian period 上二叠纪【地】
Upper Permian series 上二叠统【地】
upper photic zone 上透光层
upper pickup 滑车系统提升至最高点
upper pintle 上枢轴；闸门顶枢
upper pivot 闸门顶枢
upper plane-bed facies lamellar structure 上部平底相纹理构造
upper plastic limit 高可塑限；塑性上限；上塑限
upper plastic moisture content limit 塑性上限含水量
upper plate (断层的)上盘；上(夹)板；顶壁；顶板
upper plate slow motion screw 上盘微动螺旋
Upper Pleistocene 晚更新世【地】
upper plenum 上空腔
upper plug 上塞
upper plunger 上冲杆
upper pole 上极
upper pole face 上极面
upper pole piece 上极面
upper pond 上游水池；(船闸)上游水库
upper pond level (船闸的)上游水位；上游水池水位
upper pool 高库；上游(进)水池；上蓄水池；上深槽；地面水塘；(船闸)上游水库；船闸前池；上游水塘
upper pool level (船闸的)上游水位
upper pool of lock 船闸上游水面
upper pool pipe 上游进水管
upper portion 上部
upper port sill 上首门槛
upper pour point 最高倾点；最高流动点
upper precision 精确度上限
upper pressure limit 压力上限
upper prestressing level 上层预应力标高
upper principal purlin(e) 上金檩
upper prop (坑道内伸缩顶杆的)内杆；上坑柱
Upper Proterozoic Subera 上元古亚代【地】
Upper Proterozoic Suberathem 上元古亚界【地】
upper punch 上冲头；上冲杆
upper purlin(e) tiebeam 上金枋
upper quadrant arm 上向臂板
upper quadrant signal 上向臂板信号机
upper quadrant single-arm semaphore signal 上向单臂板信号机
upper quartile length 上四分位长度
upper quire 唱诗班的上席位
upper race ring 上座圈
upper rafter 脑椽
upper rail 上导槽；上部轨；上冒头
upper ram 上锤头
upper range light 后灯标
upper range value 上限值
upper reach 上游河段；上游(段)

upper reinforcement 上部钢筋
upper reservoir 上游水库；上蓄水池；上池
upper residential floor 上面居住楼层
upper residential stor(e)y 上面居住楼层
upper ridge 正脊；高空脊
upper right side 右上方
upper rim ray 上边光
upper ring 上环
upper river 上游(河段)
upper river course 上游河道
upper roll 上辊
upper roller 托链轮
upper round head (大头坝的)上游圆顶
upper row 上列
upper sample 上层(土)样；上部试样；上层(取的)试样
upper sample of water 上层水样
upper sash 气窗窗框
upper seam 上部矿层
upper section 上半断面
upper seeding 浅播
upper segment 上节；上部砌块
upper semicontinuity 上半连续点
upper semicontinuous 上半连续的
upper semicontinuous decomposition 上半连续分解
upper semicontinuous function 上半连续函数
upper semi-lattice 上半格
upper semi-modular lattice 上半模格
upper shaft 上端轴
upper shaker screen 上部摇动筛
upper sheathing 上覆he
upper shed 上开口
upper shell 上筒体
upper shield 上罩；上盖
upper shift 上移
upper shoe 上模座
upper shore 高海岸带
upper side (断层的)上盘；上面；上端；上侧
upper side band 高频带；上边带
upper side band spectrum 上边带谱
upper side band wave 上边带波
upper side flat 上边滩
upper side frequency 上边频
upper side of deck beam 甲板梁上缘
upper side rail (集装箱)上侧梁；顶部桁材
upper side stone 上托钻
upper side track 上边声迹
upper sieve 上筛
upper sieve extension 上筛的延长部分
Upper Silurian period 上志留纪【地】
Upper Silurian series 上志留统【地】
Upper Sinian series 上震旦统【地】
upper skirt 上部裙筒
upper slide rest 上刀架；上部(滑动)刀架
upper slitter 上刀
upper slope 上翼坡
upper soil layer 土壤表层；上层土层
upper spacer 上隔板
upper span 上翼展
upper span wire 横跨上定位索
upper speed cone 高速锥
upper speed course 上速程
upper speed ring 上部调速环
upper spider 上机架
upper spindle 上轴
upper stage 前期；上一阶段
upper state 高能状态；上态；上能态
upper state relaxation 上能态弛豫
upper state symbol 上态符号
upper stay column (贯流式水轮机)上固定支柱
upper stay plate 上部缀板
upper steam cylinder gasket 上汽缸垫密片
upper steam cylinder jacket band 上部汽缸套带
upper steam passage 上部汽路
upper stem seal 上阀杆密封
upper stock 上舵杆；上等木材；舵杆头
upper stone monument damaged lower stone monument perfect 柱石已损坏盘石完好
upper stor(e)y 上面楼层；楼上；上一层
upper stor(e)y ground plan 上层楼平面图
upper strake 上层列板；上层建筑
upper stratosphere 高平流层
upper stratum 上层

upper stream 上游河段
upper stream course 河流上游段
upper stream water 上游水
upper string 上弦
upper structure 上部(回转)结构
upper strut 上部支撑
upper subbase 底基上层;上层路基;底基(层的)上层
upper sum 上和
upper support 上支架
upper support for frames 框架上部支撑
upper supporting roller 上托辊
upper support of bucket ladder 斗架上支架
upper surface 上翼面;上端
upper surface aileron 上翼面副翼
upper surface brake 上翼面减速板
upper surface of rail 铁轨顶面
upper surge tank 上调压井
upper suspension arm 上悬挂臂
upper swing jaw face 上摆动颚板面
upper tank 上集水槽
upper tectonic level 上部构造层次
upper terrigenous formation 上部陆源建造
upper three-blade rotor 上三叶轮增压器
upper threshold 上阈限
upper thrust wall 上游承推墙
uppertidal 潮上的
upper tier 上层
upper tracer 上扫描线
upper track 上道
upper track wheel 履带支托轮;托带轮
upper transit 上中天
upper transmission 上传动
upper treadle motion 上提综杆
upper tree layer 高树层
upper trench 上深槽
upper triangular matrix 上三角矩阵
upper-triangulation matrix 上三角形矩阵
Upper Triassic period 上三叠纪【地】
Upper Triassic series 上三叠统【地】
upper troposphere 对流层上部
upper trough 高空槽
upper tumbler 上(部)导轮;上部轮筒;上鼓轮
upper turntable 上回转台
upper turret station 上部炮塔位置
upper tweendeck 上甲板间;上二层舱
upper value 上方值
upper valve 上置式气门
upper valve cap 上部阀盖
upper valve gear housing 顶罩汽门室罩
upper valve seat 上端阀座
upper variation of tolerance 上偏差
upper wall (断层)上盘;顶壁
upper wall halo 上盘晕
upper warning limit 上警戒限
upper water 上游水;上层水;地面水
upper water box 上水箱
upper water course 河流上游
upper wheel 上锯轮
upper whorl 上轮
upper wig-wag 上摇摆轮
upper wind 高空风;上层风
upper wind chart 高空风图
upper wind girder 上风联杆
upper winding-bar 上层绕组铜条
upper wind observation 高空风观测
upper wing 上翼
upper wing wall 上游翼墙
upper wire 上网
upper wishbone link 上部叉形联杆
upper workings 上部采场
upper works 干舷;上层结构;水线上部建筑;上层建筑
upper yield point 塑性上限;上屈服点
upper yield point of plasticity 塑性上限
upper yield stress 上屈服应力
upper yoke 上铁轭
up price clause 加价契约条款
up projection 仰视图;垂直投影
up-pull 上拔
up-quenching 上淬(法);分级等温淬火
up-quenching crack 升温裂纹
upraise drift 上坡平巷
upraise room 上向矿房

upraise shaft 上向掘进的井筒
upraising 上向掘进
uprate 提高功率
uprated 提高功率的;大功率的
uprated engine 强化后的发动机;大功率发动机
uprating 提高定额;提高出力
uprating capacity 提高出力的潜力
uprating project 提高出力的设计
uprating treatment plant 提高定额处理厂
up rent potential 提租的可能额度
upright 直立(的);立式;立杆;竖立的;竖的;顶柱;垂直构件
upright barrel hook 立桶钩
upright batching plant 立式配料机
upright board 竖立式板
upright breakwater 直立式防波堤
upright brick 立砖
upright bucket type steam trap 浮筒式疏水器
upright casting 立浇
upright chamber 立式蓄热室
upright coal seam 直立煤层
upright column 立柱
upright condenser 立式冷凝器
upright conduit 竖管
upright converter 立吹炉
upright cordon 直立单干形
upright core 直(立)铁芯
upright course 立砌层;竖(立)砌层
upright cylinder 直立式汽缸
upright dial ga(u)ge 立式指示表
upright diamond 直立菱形
upright diapositive 正阳片
upright dive 正俯冲
upright drill(er) 立式钻床;直立式钻
upright-drilling machine 立式钻床
upright engine 立式发动机
upright fold 直立褶皱;垂直褶皱
upright freezer 立式冻结柜
upright galvanometer 垂直式电流计
upright granulator 立式成粒机
upright growth 直立枝
upright horizontal fold 直立水平褶皱
upright hull-building 正模造船法
upright image 竖立影像
uprighting 直立法;孔修正
uprighting tool 镗孔工具
upright joint 直立节理
upright letter 罗马字体;正体字
upright lever 垂直杠杆
upright lock 竖锁
upright method 直立法
upright of a boring machine 镗床柱
upright of frame 构架立柱;刚架立柱;框腿;框架立柱
upright of stiff frame 刚架立柱
upright panel 竖钉对接护墙板
upright picture 垂直图片
upright pipe 竖管
upright piping 竖管
upright plunging fold 直立倾伏褶皱
upright position 直立位置;垂直位置
upright post 立柱;脚手架立柱;竖杆
upright projection 侧视图;垂直剖面图;垂直投影
upright pump 立式水泵
upright radiator 立式冷却器
upright regenerator 立式蓄热室
upright resistance (桩的)抗拔力
uprights 工作架直柱
upright shaft 立轴
upright sitting position headrest 垂直座位头靠
upright spout 冲天流
upright stanchion 立柱;立杆;支柱
upright stem 直(立)船首柱
upright still 立式蒸馏釜
upright support 垂直支柱
upright tube 竖管
upright tubing 竖管
upright type 正体字
upright vacuum cleaner 立式真空吸尘器
upright ventilating slit door 竖式百叶门
upright wall 直立外堤;直立墙
upright (wall) breakwater 直墙式防波堤
uprise 上坡;直立管;升起
uprise conduit 上升竖管

uprise of salt 盐分上升;返盐
uprise pipe 上升竖管;上升立管
uprise piping 上升竖管
uprise tube 上升竖管
uprise tubing 上升竖管
uprising 向上;上向掘进
uprising gas 上升气体
upriver 上游方向;上水
upriver stepped face 堰坝上游阶形面
U-profile belt conveyer U 形胶带输送机
U-profile glass 槽形玻璃
uproot 根除;连根拔起;铲除
uprooting 根绝;挖出
uproot weeds 铲除杂草
up-run 上行(制气)
uprush 波浪上溅;波浪上冲;强上升流;上涌(船在波浪作用下的一种运动);上爬;上冲(流);垂直急流;冲岸浪;冲岸波(浪);向上奔流
uprushing wave 上滚波;上爬波
UPS charges 包裹递送费
up separator pressure of orifice meter 测气孔板上分离器压力
upset 上端局设备;弯曲伸缩率;联络小巷;镦锻;顶锻;颠倒;倒转;翻覆;压折
upset a balance 打破平衡
upset allowance 顶锻留量
upset bolt 冷镦螺栓;膨径螺栓;头部镦粗螺栓;镦锻螺栓
upset butt welder 电阻对焊机
upset butt welding 电阻对(接)焊
upset butt welding machine 电阻对焊机
upset current 顶锻(时)电流
upset current off time 无电顶锻时间
upset cutter 顶锻部有深槽的插刀
upset end 镦粗的一端;螺旋轴端;(管子的)加厚端;镦端头
upset-end casing 端部加厚套管
upset-end joint 管子的粗接头;加厚接头
upset force 顶锻力
upset for threading 为车制螺纹的镦粗
upset frame (砂的)箱框;砂箱框
upset kiln condition 混乱的窑况
upset metal 镦锻金属件
upset of steel bar 钢筋镦头
upset operation 不正常操作
upset pass 立轧道次
upset piping 端部加厚油管
upset pressure 顶锻压力
upset price 最低拍卖限价;拍卖底价
upset rivet 膨径铆钉;头部镦粗铆钉
upset rod 螺旋轴端
upset shackle 反装卸扣
upset speed 顶锻速度
upsetter (冷)顶锻机;平锻机
upset the apple cart 打乱计划
upset time 顶锻时间
upset(ting) 倾覆;缩锻;顶锻;冷镦;镦粗;顶锻
upsetting angle 倾覆角
upsetting arm 倾覆力臂
upsetting couple 倾覆力偶
upsetting device 镦锻设备;顶锻机构
upsetting die 镦锻模;顶锻模
upsetting force 顶(锻)压力;顶锻力
upsetting lever 倾覆力臂
upsetting machine 镦断机;平锻机;锻粗机;顶锻装置;顶锻机
upsetting moment 倾覆力矩;倾侧力矩
upsetting point 倾覆点
upsetting pressure 顶锻压力
upsetting stress 顶锻应力
upsetting temperature 顶锻温度
upsetting time 顶锻时间
upset tubing 端部加厚油管
upset weld(ing) 电阻对焊
upset welding joint 对接焊缝
upset wrinkles (管子的)镦厚损伤
up-shaft 气流向上排出的通风井;往上通风的竖井;向上通风竖井
upshear 逆剪切
up-shift 加速;换高速挡
up-shift range 加速幅度;高速换挡范围
upshot-type burner 上喷式燃烧器
up side 上行方向;上方方面;上面
upside-down 倒转的;倒置

upside-down and laterally reversal 上下左右全反转的
upside-down channel 倒转水道;倒转沟渠
upside-down charging 倒装
upside-down construction 先拱后墙施工法;逆作施工
upside-down firing 覆烧
upside-down karren 倒石芽
upside-down motion 倒转运动;倒动
upside-down mounting 倒装(法)
upside-down pile 粗头朝下的桩
upside-down prestressing jack 反装预应力千斤顶
upside-down roof 谷式屋顶;倒置屋面;倒置屋顶;反向屋面
upside of tonnage deck 量吨甲板上面
upside roll 上旁滚
up-sizing 放宽规格;放大尺寸
upslide 上投
upslide surface 上滑面
upslip 上滑距
upslope 上坡
upslope flow 上坡气流
upslope fog 上坡雾;山坡雾
upslope motion 上滑运动
upslope runway 上坡跑道
upslope time 上升时间;电流渐增时间
upslope wind 上坡风
up-splitting 上分岔
upspout 升流管
upspring 生长
upstage 末级;舞台后方;舞台后部;后台
upstair(case) 楼上(的)
upstand 竖柱;泛水
upstand at parapet 女儿墙勒脚
upstand beam 下承梁;倒梁;直立梁;凸形梁;上翻梁;倒 T 形梁
upstand diaphragm beam 上隔板梁
upstanding beam 直立梁;上凸梁;倒梁;反梁
upstand string 上翻边梁
upstand T-beam 倒 T 形梁
upstand T-beams slab 倒 T 形梁板;凸形梁板
upstart 窗台
up-station 上端局;高点
up-stop 上限动块;上止机构
upstream 逆流的;向上游(的);上游;上流;朝向河的上游
upstream anchor 上游锚
upstream apron 上游铺砌;上游铺盖;上游护坦
upstream area 上游地区
upstream batter 迎水坡;迎水面坡度;上游坡;上游面坡度
upstream bay 上游河湾
upstream blanket 上游铺盖
upstream-bound fish 上溯鱼
upstream chamber 进水室
upstream channel 上游河槽
upstream cofferdam 上游围堰
upstream control 上游控制
upstream dam survey 库区调查;坝上游区调查
upstream deck (坝的)上游面板
upstream deposit 河流上游沉积
upstream economizer 逆流式省煤器
upstream elevation 上游高程
upstream end 进水端
upstream end of culvert 涵洞入口
upstream extremity 上游端
upstream face 上游面;迎水面
upstream facing 上游护面
upstream fill (坝的)上游楔形体;上游填筑体;上游填土体
upstream floor 防渗底板;上游铺砌;上游护坦;上游底板;防渗铺盖;坝前护坦
upstream flow 河口涨潮流;上游流量
upstream garage (船闸的)上游停泊区
upstream ga(u)ge 上游水位标
upstream heel 迎水面坡脚
upstream impoundment 上游蓄水
upstream injection 逆流喷射
upstream leaf 上游向闸门扇
upstream level 上游水位高程
upstream line 吸引管线;吸入管线;接纳管线;上游管线
upstream lock 上游船闸
upstream maximum water-level 上游最高水位

upstream measuring section 上游测量断面
upstream method 上游法
upstream migrant 溯河洄游
upstream migrating 向上游洄游
upstream migration 溯河洄游
upstream minimum water-level 上游最低水位
upstream nose 桥墩分水尖;桥墩分水头;闸墩分水尖;闸墩分水头
upstream nosing 闸墩分水头;上游分水尖;上游鼻端;闸墩上游鼻端(护墩防冲、破冰等用)
upstream pier nose 桥墩上游桩
upstream pier nosing 桥墩上游分水尖;桥墩上游端
upstream ponding 上游蓄水量
upstream pressure 上游压力
upstream pressure of orifice meter 测气孔板前压力
upstream profile at crown 拱圈上游剖面线
upstream pumping station 上游泵站
upstream pumping unit 逆流泵站;上游泵站;上流泵站
upstream radius of crest 迎水面堰顶曲线;坝顶上游面曲率半径
upstream range 上游断面线
upstream reach 上游河段
upstream restriction 上流节流
upstream retarding reservoir 上游滞洪水库;上游蓄洪水库
upstream sailing 上水行驶
upstream separator 升流式选粉机
upstream shell 上游坝壳
upstream side 吸入侧;入口方向;上游侧;上游边;上流侧
upstream side of the injector 喷嘴入口端
upstream slanting filter 上游斜反滤层
upstream slope 迎水坡;迎水面坡度;上游坡(面);堤的上游面
upstream slope of dam 坝的迎水(面)坡(度);坝的上游面坡度
upstream slope protection 上游护坡
upstream sources of pollution 上游污染源
upstream spray pattern 逆喷
upstream station arrangement 首部式布置(水电站)
upstream surge basin 上游调压井;上游调压池
upstream table 上游水位
upstream toe 上游坝址;迎水面坡脚;上游坝脚;上游坝脚
upstream toe (of dam) 上游坝趾;坝的上游坡脚
upstream total head 上游总水头
upstream traffic 上水运输;上水(货)运量
upstream transitional section 上游衔接段
upstream vessel 逆流船;上行船(舶);上水船(舶)
upstream voyage 上水航行
upstream wall 上游坝墙
upstream water 上游(水);上层滞水
upstream water-level 上游水位高程;上游水位
upstream water line 上游水位线
upstream water source 上游水源
upstream well 逆地下水布置的井
upstream wing wall 上游翼墙
upstream zone 上游区
upstriker 上行式
upstroke 上行(冲)程;上升冲程
upstroke heater 上行式打手
upstroke hydraulic press 上行液压压力机
upstroke pin type beater 上行式钉式打手
upstroke press 上行式压力机;上行式平板机
upstructure 沿构造方向
upsurge 上涌浪
upsurging 上涌
upsurging swell 上涌隆起
up survey crystal volume 上测晶体体积
upswell 膨胀;隆起
upswept frame 弓形框架;上弯构架
upswing of sales 销售额增加
up symbol 升符号
uptake 咽喉管;(通向室外的)烟道;摄取;上风井;垂直上升管道
uptake and partitioning 吸收和分配
uptake breeching 上风井;上升烟道
uptake conduit 上升管渠
uptake dose 吸收剂量
uptake flue 上升烟道

uptake header 垂直联箱
uptake of oxygen 氧的吸取;氧的吸收
uptake of water by plant 植物的水分吸收
uptake pipe 上升管
uptake piping 上升管
uptake rate 摄入率
uptake rate coefficient 摄取速率常数
uptaker damper 上升道闸板
uptake shaft 上升排风道
uptake tube 上升管;引流竖管
uptake tubing 上升管
uptake ventilator 烟道通风器
uptake zone 吸收区(域)
up the hole sampling 自上而下取样
up the hole testing 自上而下试验
up throat 上行咽喉
upthrow 隆起;(逆断层)上升垂直位移;上投距;上投;上冲
upthrow fault 逆断层;上投断层;上升断层
upthrown block 上升盘
upthrow side 上投侧;上侧面
upthrust 仰冲断层;向外突出;上推;上冲
upthrust bearing 上推力轴承
upthrust block 上升盘
upthrust fault 上冲断层
upthrust fault scarp 上冲断层崖
upthrusting 垂直顶升
upthrust type propeller 上压式螺旋桨
uptilt 上击;翻成侧立状态
up-time 作业时间;正常运行时间;可工作时间;工作时间;周期上限;正常工作时间
up to an aggregate amount of 总金额
up-to-date 现代化的;最新(式)的;直到现在的;入账日
up-to-date directory 最新的电话号码本
up-to-date information 最新情报;最新信息
up-to-date maintenance of maps 地图(集)更新
up-to-date map 最新地图
up-to-dateness 现代化程度
up-to-date price 现时价格
up-to-date product 最新的产品
up-to-date style 现代(样)式;最新的
up-to-date type 最新式
up to grade 合格
up-to-standard 合乎标准(的)
up-to-standard rate 合格率
up-to-standard water 可用的水
up-to symbol 直到符号
up to the mark 达到标准
up-to-the-minute 最新式的;最近的
up-to-the-minute styling 最新式样
uptown 近郊的;住宅区;在近郊;向近郊;市中心较高处
uptrace 上升记录
up track 上行线
up traffic 上行
uptrain 上行车;上行列车
uptrain direction 上行方向;上坡方向
up travel stop 上行停止装置
up-trend 上涨趋势
up-trip 电梯上升
upturn 价格回升;防水板上翻部分
upturned 向外翻的
upturned beam 向上翻的梁;朝上梁
upturned bed 倒转层
upturned eaves 上翘屋檐;飞檐
upturned strata 倒转层
upturning 上翻
uptwister 上行捻线机
U pump U 形泵
upvaluation 升值
upwall 上向工作面
upward 向上(的)
upward-acting door 向上开的门;上闸门;向外开的门;上开门
upward-acting plate 上压板
upward adjustment 向上调整
upward adjustments of prices 调高物价水平
upward advance 上向掘进
upward air outlet 向上送风口
upward-angled spur dike 上挑丁坝
upward annular velocity of drilling mud 钻井泥浆在环状空间的回流速度
upward atmospheric radiation 向上大气辐射

upward attack 向上攻击
upward bias 向上偏斜；上偏
upward borehole 向上钻孔
upward break 上升脱离目标
upward broom 上升帚形
upward camber 朝上弯曲
upward capability 上升能力
upward-closing slide gate 上向关闭式滑板闸门
upward compatibility 向上相容性；向上兼容性
upward compatible 向上兼容的
upward component 向上分量
upward continuation 向上延拓
upward continuation integral 向上延续法；向上连续积分
upward continuation method of data 数据向上延拓法
upward continuation of gravity anomalies 重力异常向上延拓
upward conveyer 向上运的运输机
upward course 上升航向
upward cumulation 向上累积
upward current 上向流；上升（水）流
upward current classifier 上升（水）流分级机
upward current jig 上升水流跳汰机
upward current of air 上升气流
upward deflexion 向上偏转
upward diffusion cloud chamber 向上扩散云室
upward displacement 向上取代
upward draft 向上通风；上向通风；上升气流
upward drill-hole 向上钻孔
upward drilling 向上钻进；向上钻孔侵蚀；上向凿岩；池炉的穿透性侵蚀
upward facing 面向上
upward flight 上升飞行
upward-flow 向上流动；上向流；上升流
upward-flow chromatography 上向流色谱法
upward-flow evapo(u)rator 升膜蒸发器
upward-flow floc-banket clarifier 上流式泥渣层澄清池
upward-flow sand filter 伊梅迪娅砂滤池
upward-flow sedimentation tank 竖流式沉淀池
upward fluid velocity 冲洗液流返速度
upward force 向上力
upward gradient 上向坡度
upward gradient test 上流梯度试验
upward hole 上向孔
upward hydraulic pressure 向上水压力；扬压力；浮托力
upward hydrostatic(al) pressure 向上静水压力；扬压力；浮托力
upward identification light 上部标志灯
upward irrigation 浸润灌溉；底水灌溉
upward landscape 仰视景观
upward leakage recharge 顶托补给
upward leg 升弧段
upward-looking bowl sundial 仰釜日晷
upward-looking echo sounder 上视回声探测仪
upwardly-inclined conveyer 上斜式输送机
upwardly-inclined weld 倾斜仰焊
upward migration from source rock 自母岩向上运移
upward mining 向上开采；上向采矿
upward mobile 向上变动性
upward movement 上向运动；上升运动
upward movement of water 水（分）上升运动
upward ozone transfer 臭氧的向上输送；臭氧的上升输送
upward path 上升航迹
upward percolation 向上渗透；上向渗滤
upward-pointing groin 上挑丁坝
upward-pointing hole 上向钻孔
upward-pointing spur dike 上挑丁坝
upward pressure 向上压力；扬压力；上压力；浮托力；反向压力
upward price trend 涨价趋势；价格趋涨
upward pull 上向抽力
upward reference 向上引用
upward revaluation (币值)升值
upward revaluation of gold 黄金升值
upward revaluation surplus 重估价升值(盈余)
upward roll 上升侧滚
upward run 上行充电带
upward side 上方
upward slicing 上向分层采矿法

upward spin 上升螺旋
upward stability 上稳定性
upward stroke 上冲程
upward substitution effect 向上替代影响
upward system 上升系统
upward system of ventilation 向上通风系统
upward tendency 上涨的趋势
upward terrestrial radiation 向上地球辐射
upward the river 朝向河的上游
upward total radiation 向上全辐射
upward trajectory 上升段轨迹
upward transport 向上运输
upward transverse ventilating 向上横流式通风
upward trend 看涨(趋涨)
upward trend analysis 向上趋势分析
upward trend of price 涨风
upward velocity (冲洗液的)回流速度；上向速度
upward ventilation 向上通风；上向通风；上升(式)通风
upward ventilation system 向上通风系统
upward view 仰视图
upward water 上升水
upward water flow 上升水量；上浮水流
upward welding 向上焊接
upward welding in the inclined position 上坡焊
upward welding in the vertical position 向上立焊
upward zero crossing 上跨零点
upwarp 向上挠曲；上拱；翘起
upwarp deflection 向上翘曲
upwarp-downwarp structure 垅洼构造
upwarped structure 翘起构造
upwarping 向上弯曲；隆起；向上翘曲；翘起
upwarping folded belt 隆褶带
upwarping folded region 隆褶区
upwarping region 隆起区
upwarping zone of East China sea shelf margin 东海陆架边缘隆起带
upwash 升流
upwell 向上涌
upwelling 涌升流；上涌（船在波浪作用下的一种运动）
upwelling current 上升流
upwelling-current metallogenic model 上升洋流成矿模式
upwelling flow 上升(水)流
upwelling hot flow zone 上升热流带
upwelling oceanic current 上涌洋流；上翻洋流
upwelling phenomenon 上涌现象
upwelling region 涌流区；隆起区；上涌地区；上升流区
upwelling sites 上涌地点
upwelling water 上涌水；上升流
upwind 逆风；顶风；上风(方)向
upwind course 逆风航程
upwind difference formula 上风差分公式
upwind effect 上吹效应
upwind gear 迎风一侧的起落架
upwind landing 逆风着陆
upwind leg 逆风边
upwind position 迎风位置
upwind range 逆风航程
upwind takeoff 逆风起飞
upwind taxiing 逆风滑行
upwind threshold 逆风跑道入口
upwind turn 逆风转弯
upwind velocity 逆风流速
upwind wind source of pollution 逆风污染物
urac 脲醛树脂黏合剂
uralborite 乌硼钙石；乌拉尔硼钙石
Uralian emerald 绿钙铁榴石；乌拉尔祖母绿
Uralian (stage) 乌拉尔(阶)【地】
uralite 纤维闪石；纤闪石；水泥石棉板
uralite-diabase 纤闪辉绿岩
uralitization 纤闪石化
uralitophyre 纤闪粉岩
uralkyd 氨基甲酸乙酯改性醇酸(树脂)
Ural Mountains 乌拉尔山脉
uralolite 水磷钙铍石
Ural-Volga basin 伏尔加—乌拉尔盆地
uramphite 铵铀云母
uranami bead 熔透焊道
uranami welding electrode 底层焊条；低层焊条；打底焊条
uran-apatite 铀磷灰石
uranate (重)铀酸盐

urancalcarite 水碳钙铀矿
urania 氧化铀；二氧化铀
urania crucible 氧化铀坩埚
urania green 巴黎绿
urania sol 二氧化铀溶胶
uranic acid 铀酸
uranic chloride 四氯化铀
uranic oxide 二氧化铀
uranide(s) 铀系元素
uraniferous 含铀的
uraniferous coal 含铀煤
uraniferous lignite 含铀褐煤
uraninite 沥青铀矿；晶质铀矿；天然氧化铀
uranin yellow 荧光素钠；荧光黄
uranite 云母铀矿；铀云母(类)
uranium 铀云母类矿物
uranium-234 age method 铀234年代测定法
uranium-234 to U238 age method 铀234~U238年代测定法
uranium-235 铀235
uranium acetate 乙酸双氧铀；醋酸铀酰；醋酸双氧铀
uranium acetyl acetonate 乙酰丙酮铀
uranium-actinium 锕铀
uranium advantage marked limit 铀有利性指标限
uranium age 铀年龄
uranium amalgam 铀汞齐
uranium and gold-bearing conglomerate deposit 含金铀砾岩矿床
uranium-and-heavy-water reactor 铀重水反应堆
uranium antimonide 锑化铀
uranium antimony catalyst 铀锑催化剂
uranium bar 铀棒
uranium barium oxide 氧化钡铀
uranium-base fuel 铀基燃料
uranium-bearing 含铀
uranium-bearing fluorite vein deposit 含铀萤石脉矿床
uranium-bearing material 含铀物质
uranium-bearing mineral 含铀矿物
uranium-bearing ore 含铀矿
uranium-bearing quartz vein deposit 含铀石英脉矿床
uranium boride 硼化铀
uranium button 铀块
uranium carbide 碳化铀
uranium channel window width 铀道窗宽
uranium complex 铀络合物
uranium compound 铀化合物
uranium concentrate 铀浓缩物；铀精矿；浓缩铀
uranium consumption 铀消耗
uranium content 铀含量
uranium content anomaly 铀含量异常
uranium content meter 铀含量计
uranium cycle 铀循环
uranium decay series 铀蜕变系
uranium deposit 铀矿床
uranium deposit of penta element type 五元素型铀矿床
uranium-deuterium lattice 铀重水栅格
uranium dicarbide 二碳化铀
uranium dioxide 二氧化铀
uranium enriching plant 铀浓缩工厂
uranium enrichment 铀浓缩；富铀化
uranium equivalence of potassium 钾的铀当量
uranium equivalence of radium 镭的铀当量
uranium equivalence of thorium 钍的铀当量
uranium-equivalent of being measured element 待测元素铀当量
uranium family 铀系
uranium fission 铀裂变
uranium fuel cycle 铀燃料循环
uranium fuel element 铀燃料元件
uranium-fuelled reactor 铀堆
uranium fuel rod 铀燃料棒
uranium glass 铀玻璃
uranium-graphite lattice 铀石墨栅格
uranium-graphite reactor 铀石墨堆
uranium halide 卤化铀
uranium hexafluoride 六氟化铀
uranium intensification 铀盐加厚法
uranium isopropoxide 异丙氧基铀
uranium-isotope age 铀同位素年代
uranium jacket 铀外壳
uranium lattice 铀栅格

uranium-lead age method 铀铅年代测定法
uranium-lead dating 铀铅测年
uranium-lead two-stage model 铀—铅二阶段模式
uranium makeup 铀补给
uranium mine 铀矿
uranium mineral 铀矿物;铀矿石
uranium mine waste residue 铀矿渣
uranium mining 铀矿开采
uranium monobromotrifluoride 三氟一溴化铀
uranium monocarbide 一碳化铀;碳化铀
uranium monoiodotrifluoride 三氟碘化铀
uranium mononitride 一氮化铀
uranium monosulfide 一硫化铀
uranium monoxide 一氧化铀
uranium nitride 氮化(物)铀
uranium ocher 铀华
uranium ore 铀矿(石)
uranium ore processing plant 铀矿加工厂
uranium ores 铀矿
uranium oxalate 草酸双氧铀
uranium oxide 氧化铀;三氧化铀
uranium oxychloride 二氯二氧化铀
uranium-oxygen mineral 铀氧矿物
uranium peroxide 过氧化铀
uranium perspective area marked limit 铀远景区指标
uranium-phosphorite deposit 含铀磷块岩矿床
uranium pile 铀堆
uranium poisoning 铀中毒
uranium-potassium anomaly 铀—钾异常
uranium purification wastes 铀提纯废水
uranium-radium decay series 铀镭衰变系
uranium-radium series 铀镭系
uranium-radon-radium water 铀氡镭水
uranium-radon water 铀氡水
uranium reactor 铀(反应)堆
uranium reconnaissance 铀矿普查
uranium recovery 铀的回收
uranium red 铀红
uranium reducing method 铀还原法
uranium refinery 铀精炼厂
uranium reserves 铀储量;铀储备
uranium resonance 铀共振
uranium rod 铀棒
uranium-rod lattice 铀棒栅格
uranium salt 铀盐
uranium-sandstone-conglomerate deposit 含铀砂岩砾岩矿床
uranium series 铀系
uranium series element 铀系元素
uranium series method 铀系法
uranium silicide 硅化铀
uranium slug 铀棒
uranium tetrachloride 四氯化铀
uranium tetrafluoride 四氟化铀
uranium-thorium anomaly 铀—钍异常
uranium-thorium-lead age method 铀—钍—铅年代测定法
uranium-thorium-lead method 铀—钍—铅法
uranium toning 铀盐调色法
uranium trioxide 三氧化铀
uranium vein 铀矿脉
uranium water 铀水
uranium water lattice 铀水栅格
uranium yellow 铀黄;重铬酸钠
uran-mica 铀云母
uranmicrolite 铀细晶石
uranoaeschynite 铀易解石
uranoanatase 铀锐钛矿
uranochalcite 铀钙铜矿
uranocher 铀硫酸盐
uranocircite 钡铀云母
uranogenic lead 铀铅
uranography 星图学
uranohydrotorite 铀水钍石
uranoid 铀系元素
uranol 铀试剂
uranolepidite 铀铜矿
uranology 天体学(论文)
uranometrical survey 铀量测定
uranometric survey 铀量测量
uranometry 天体测量学
urano-organic ore 有机铀矿
uranophane 硅钙铀矿

uranophanite 硅钙铀矿
uranopilite 铀矾
uran ore 铀矿
uranosite 针硅铀矿
uranospathite 水铝铀云母
uranosphaerite 纤铀铋矿;纤铋铀矿
uranospinite 钙砷铀云母;砷钙铀矿
uranothallite 铀碳钙石
uranothorianite 铀方钍矿
uranotile 硅钙铀矿
uranotungstite 钨铀华
uranous 亚铀的
uranous-uranic oxide 八氧化三铀
uranphyllite 铜铀云母
uranpyrochlore 铀烧绿石
uranthorianite 铀方钍矿
Uranus 尤拉纽斯镍络合金钢
uranvitriol 铀铜矾
uranygl acetate 醋酸双氧铀
uranyl 双氧铀
uranyl acetate 乙酸双氧铀;醋酸铀酰
uranyl chloride 铀酰氯;氯化双氧铀
uranyl nitrate 硝酸双氧铀
uranyl propionate 丙酸铀酰
uranyl salts 铀酰盐类
uranyl silicate 硅酸铀
Urastone 乌拉石(多种石棉水泥制品)
urb 城市区域(美)
urban 城市街区;城市街坊
urban aerophotogrammetry 城市航空摄影测量
urban aesthetics 城市美学;市容
urban affair program(me) 城市事务计划
urban afforestation 城市绿化
urban agglomerates 城市集团
urban agglomeration 城市聚集体;城市集结;城市集聚;城市群
urban aggregate 城市集合体
urban air 城市空气
urban air blanket 城市空气覆盖层
urban air pollution concentration 城市空气污染浓度
urban air pollution model 城市空气污染模式
urban air pollution source 城市空气污染源
urban ambient air quality 城市环境空气质量
urban amenity 城市美观;城市环境舒适
urban and community redevelopment 城镇重建
urban and regional planning 都市与区域规划
urban and rural construction 城乡建设
urban and rural planning 城乡规划
urban appearance 市容
urban arboriculture 城市树木栽培
urban architecture 城市建筑(物)
urban area 市区(道路);城(市市)区;城市地区
urban area concentration 市区浓度
urban area congestion 市区拥挤
urban area district environment 市区环境
urban (area) extension 市区扩展
urban (area) land 市区用地
urban area planning 市区规划
urban area pollution 市区污染
urban area pollution sources 市区污染源
urban area study 城市地区研究
urban area to city area 市区面积比
urban art 城市(景观)艺术
urban arterial highway 市区干路;市区干道
urban atmosphere 城市大气
urban atmosphere pollution 城市大气污染
urban attraction 城市吸引力
urban authorities 城市管理机构
urban basic frame 城市基本结构
urban beauty 城市美观
urban branch road 城市支路
urban bridge 城市桥梁
urban building code 城市建筑法规
urban bus 市内公共汽车
urban business 城市商业
urban business spots 城市商业网点
urban capacity 城市容量
urban cell 城市单元
urban center 市中心(区);城市中心(区)
urban centered 以城市为中心的
urban chapel 城市小教堂
urban character 城市性质
urban chemical pollution 城市化学污染

urban church 城市教堂
urban clearance 城市清理;城市清洁
urban clearway 市区超速道路;城市高峰时禁止停车的道路;市郊高速公路;高峰禁止交通道路
urban climate 城市气候
urban climatology 城市气候学
urban commitment program(me) 城市住房修建计划
urban community 城市社区
urban complex 复合城市
urban complexity 城市综合体
urban concentration of population 人口向城市集中
urban configuration 城市轮廓;城市外貌
urban congestion 城市过密
urban connector 城市连接地区
urban conservation 城市(历史)保护
urban construction 城市建筑;城市建设
urban construction waste residue 城市建渣土
urban control survey 城市控制测量
urban core 城市核心;市中心区;城市中心
urban cosmetology 城市整容术
urban crisis 城市危机
urban critic 城市问题评论家
urban culture 城市文化
urban damage 城市灾害
urban decay 城市退化;城市衰败
urban decoration 城市装饰
urban desert 城市沙漠
urban design 城市设计;城区设计
urban detailed planning 城市详细规划
urban development 城市开发;城区发展;都市发展;城市改造;城市发展
urban development action grant 城市发展实施方案
urban development area 城市扩展区;城市发展区
urban development committee 城市开发委员会
urban development pattern 城市发展模式
urban development planning 城市发展规划
urban development strategy 城市发展战略
urban diagnosis 城市调查分析
urban disaster 城市灾害
urban disaster prevention 城市防灾
urban discharge 城市排污
urban disease 城市疾病
urban dispersion 城市疏散
urban district 市区;市内;城(市地)区
urban district planning 市区规划;城区规划
urban district sprawl 市区扩展;城市扩展
urban district waste 城市垃圾;城市废物
urban domestic pollution 城市生活污染
urban domestic sewage 城市生活污水
urban domestic water 城市生活用水
urban drainage design 城市排水系统设计
urban drainage district 城市排水区
urban drainage engineering 城市排水工程
urban drainage planning 城市排水规划
urban drainage stormwater management model 城市排暴雨水管理模型
urban drainage system 城市下水道;城市排水系统
urban drainage system management model 城市排水系统管理模型
urban drift 城市人口流入;城市人口集中
urban dwellers 城市人口;城市居民
urban dwelling 市区住宅
urban dynamics 城市动态
urban earthquake hazard protection 城市防震
urban ecological assessment 城市生态评价
urban ecological benefit 城市生态效益
urban ecological economics 城市生态经济
urban ecological forecasting 城市生态预测
urban ecological planning 城市生态计划;城市生态规划
urban ecological regulation 城市生态调控
urban ecological suitability 城市生态适宜度
urban ecology 城市生态学
urban economic benefit 城市经济效益
urban economics 城市经济学
urban economic system 城市经济体系
urban economy 城市经济
urban ecosystem 城市生态系统
urban effective area 城市功能影响区
urban element 城市要素
urban energy 城市能源

urban engineering 市政工程(学);城市工程(学)
urban enterprise 城镇企业;城市企业
urban environment 城市环境
urban environmental assessment 城市环境质量评价
urban environmental capacity 城市环境容量
urban environmental decay 城市环境恶化
urban environmental forecasting 城市环境预测
urban environmental hydrogeology 城市环境水文地质学
urban environmental management 城市环境管理
urban environmental planning 城市环境规划
urban environmental pollution 城市环境污染
urban environmental protection 城市环境保护
urban environment hygiene 城市环境卫生
urban environment quality 城市环境质量
urban erosion rate 市区土壤流失率
urban escape hatch 逃避城市枯燥生活的处所
urban expansion 市区膨胀;城市膨胀;城区扩展
urban explosion 市区膨胀;城市扩张;城市爆炸
urban express traffic 城市快速交通
urban expressway 城市快速道路;城市高速公路
urban expressway net(work) 城市快速道路网
urban extension 市区扩展;城市延伸;城市扩展
urban external communication 城市对外交通
urban fabric 城市建筑;城市构成
urban facility 市政设施;城市设施
urban ferry 城市轮渡
urban finance 城市财政
urban flood defence 城镇防洪;城市防洪
urban forest zone 城市树林地带
urban form 城市形式;城市形态
urban framework 城市结构;城市骨架;城市格局
urban freeway 城市高速公路
urban freight movement 城市内货流;城市内物流
urban fringe 市边缘区;市郊;城市边缘;边缘地区
urban function 城市职能;城市功能
urban garden 城市花园
urban gas utilities 城市煤气公司
urban geography 城市地理(学)
urban geology 城市地质学
urban green 城市绿化;城市绿地
urban group 城市群
urban growth 城市增长;城市生长;城市成长
urban hall 市政大厅;市政大厦
urban health 城市卫生
urban heating 城市供热
urban heat island 城市热岛
urban heat island effect 城市热岛效应
urban hierarchy 城市分级;城市等级
urban hierarchy planning 城市体系规划
urban highway 市区公路;城市道路
urban history 城市史
urban homesteading 城市定居
urban homogeneity 城市均质性
urban hotel 城市旅馆
urban hydrology 都市水文学;城市水文学
urban improvement area 城市备用地区
urban industrial pollution 城市工业污染
urban industrial society 城市工业群集
urban industrial structure 城市工业结构;城市产业结构
urban industrial system 城市工业体系
urban industry district 城市工业区
urban information system 城市信息系统
urban infrastructure 城市基础设施
urban inhabitant 城市居民
urban inhibitor 城市居民
urban investigation 城市调查
urbanism 城市规划;都市化;城市性
urbanist 城市规划专家;城市规划者
urbanite 城市居民
urbanity 城市的状况或文化特性
urbanization 市区化(主要指农用地逐渐变成住宅用地);都市化;城市化
urbanization control area 市区化控制区;市区化调整区域(规划中对建设加以抑制的地区,即非市区化地区)
urbanization factor 城市化因素
urbanization of population 人口都市化;城市人口集中
urbanization promotion area 市区化区域(规划成为市区的地区)
urbanization rate 都市化速率;都市化程度

urbanization trend 城市化趋势
urbanize 城市化
urbanized agglomeration 城市化集聚区
urbanized area 都市化区域;都市化面积
urbanized standard 城市化水平
urban labo(u)r force resource 城市劳动力资源
urban labo(u)r protection 城市劳动保护
urban land 城市土地;市区(土)地
Urban Land Institute 城市土地学会
urban land roll 城市地籍
urban landscape 城市风景;城市景观;城市景色
urban land-use evaluation 城市用地评价
urban layout 城市布局
urban library 城市图书馆
urban life style 城市生活方式
urban lighting 城市照明
urban living-space 城市居住空间
urban living standard 城市生活标准
urban local planning 城市小区规划
urban man 城市居民
urban map 城市地图
urban mapping 城市制图
urban mass 城市人口;城市群众
urban mass transit system 城市大运量交通系统
urban mass transportation 城市公共交通
urban mechanics 城市设备;城市装置
urban microclimate effect 城市微气候影响
urban-mileage 市内行驶里程
urban mobile system 市区交通运输系统
urban modernization 城市现代化
urban morphology 城市形态学;城市(结构)形态;城市组织
urban mortality 城市死亡率
urban motor vehicle permissible noise standard 城市机动车辆噪声标准
urban motor vehicle traffic 城市汽车交通;城市机动车辆交通
urban motorway 城市高速公路
urban multi-purpose system 城市多功能系统
urban museum 城市博物馆
urban nature 城市性质
urban network 城市网络;城市街道网;城市网架
urban noise 城市噪声
urban noise management 城市噪声管理
urban non-point source pollution control 城市非点源污染控制
urban nonstaple food production base 城市副食品生产基地
urbanoid 具有大城市特性的;大城市的;具有城市特点的
urbanologist 都市学专家;城市学专家;城市规划专家;城市规划者
urbanology 都市学;城市学
urban open space 城市园林绿地;城市园林保护地
urban orphanage 城市孤儿院
urban outbound traffic 城市对外交通
urban overall layout 城市总体布局
urban overall planning 城市总体规划
urban palace 城市宫(指文化宫、少年宫等)
urban park 城市公园
urban passenger flow 城市客流
urban pathology 城市病理学
urban pattern 城市形态;城市形式
urban personality 城市气质;城市个性
urban physical and ecological determinism 城市自然生态决定论
urban physical pollution 城市物理污染
urban pipeline engineering 城市管线工程
urban place 城市空间
urban planner 城市规划人员
urban planning 市政规划;都市规划;城市规划
urban planning authority 城市规划管理部门
urban planning commission 城市规划委员会
urban planning institute 城市规划设计院
urban planning norm and criteria 城市规划规范
urban planning of disaster management 城市防灾规划
urban planning of population 城市人口规划
urban planning study 城市规划研究
urban planning theory 城市规划理论
urban planning vocabulary 城市规划语言
urban plantation quota 城市绿地定额
urban policy 城市政策
urban pollutant 城市污染物

urban polluted river 城市污染河流
urban polluted river water 城市污染河水
urban pollution control planning 城市污染控制规划
urban pollution model 城市污染模式
urban pollution source 城市污染源
urban polygon 城市导线【测】
urban pool 城市游泳池;城市水池
urban population 市区人口;城市人口
urban population age composition 城市人口年龄构成
urban population census 城市人口普查
urban population composition 城市人口结构
urban population density 城市人口密度
urban population differentiation 城市人口异质性
urban population flow 城市流动人口
urban population structure 城市人口结构
urban power network 城市电力网
urban problem 城市问题
urban property 市区房地产
urban public open space 城市公共绿地
urban public services 城市公共设施
urban public traffic 城市公共交通
urban public transport 城市公共交通运行调度
urban public transportation 城市公共交通
urban public transport enterprise 城市公共交通企业
urban public transport fare structure 城市公共交通票价制
urban public transport passenger flow forecast 城市公共交通客流预测
urban public transport passenger flow survey 城市公共交通客流调查
urban public transport planning 城市公共交通规划
urban public transport sign 城市公共交通标志
urban public transport system 城市公共交通系统
urban public utility 城市公用事业
urban public utility surtax 城市公用事业附加税
urban qualitative theory 城市定性理论
urban radiation force 城市辐射力
urban rail transit 城市高速铁路运输;城轨交通
urban railway 市区铁路;城市铁路
urban rapid transit 城市高速交通
urban rapid transit system 城市快速转运系统
urban real estate tax 房地产税
urban reconstruction 城市改建
urban redevelopment 城市改造;都市重建;城市再发展;城市复兴;旧城改造;城市重建;城市改建
urban redevelopment work 城市再开发工作;城市复兴工作
urban refuse 城市垃圾
urban refuse collection 城市垃圾收集
urban refuse incinerator 城市垃圾焚化炉
urban region 城市地区
urban regional stability 城市区域稳定性
urban rehabilitation 城市整建;城市修复
urban relocation 城市搬迁
urban renewal 旧城改建;市区改造;城市重建;城市更新;城市改造
urban renewal area 城市更新区域;城市更新面积
urban renewal housing 城市更新住房
urban renewal plan(ning) 城市更新规划
urban renewal project 城市更新建设项目
urban renewal zone 城市重建区
urban requirement 城市要求;城市需求
urban residential area 城市住宅区
urban residential construction density 城市居民建设密度
urban resident 城市居民
urban revitalization 城市复苏
urban road 城市道路
urban road bridge 城市道路桥
urban road construction 城市道路建设
urban road hierarchy 城市道路体系
urban road network 市区线网;城市道路系统;城市道路网
urban road network density 城市干道网密度
urban road system 城市道路系统;城市道路网
urban runoff 城市(地面)径流
urban runoff model 城市径流模式
urban runoff pollutant 城市径流污染物
urban runoff pollution 城市径流污染
urban runoff quality 城市径流水质
urban runoff simulation model 城市径流模拟模式

urban-rural continuum 城乡连续区;城乡连续体理论
urban ruralization 城市乡村化
urban-rural transport 城乡运输
urban safety minimum 城市安全基准
urban sampling area 城市采样区
urban sanitation 城市下水道设备;城市环境卫生
urban sauna bath 城市蒸汽浴室
urban scale 城市规模
urban scenic spot 城市风景区
urban science 城市科学
urban sculpture 城市雕塑
urban section 市区
urban sediment control 城市泥沙防治
urban service boundary 城市服务设施范围
urban servitude 城市框架使用权
urban settlement 都市集居;城市居住区;城市居民点
urban sewage 城市污水
urban sewage discharge 城市污水排放
urban sewage effluent 城市污水处理后出水
urban sewage treatment 城市污水处理
urban sewage treatment plant 城市污水处理厂
urban sewer 城市下水道
urban sewerage 城市排水
urban sewerage system 城市下水道系统
urban shelter belt 城市防护带
urban shrinkage 城市萎缩
urban site 城市遗址
urban skeleton 城市网架;城市轮廓
urban sludge 城市污泥
urban slums 陋巷;城市贫民区;城市贫民窟
urban social and cultural determination 城市社会文化决定论
urban social benefit 城市社会效益
urban social planning 城市社会规划
urban society 城市社会
urban sociology 城市社会学
urban soil pollution 城市土壤污染
urban solid wastes 城市固体废弃物
urban space 城市空间
urban spatial determinism 城市空间决定论
urban spatial distribution 城市空间分布
urban sprawl 城市自发蔓延;城市无计划发展;市区扩展;城市扩展;市区伸延;市区建设;都市延伸;城市扩张;城市不规划发展
urban spread 城市延伸
urban square 城市广场
urban station 城市车站
urban stormwater 城市暴雨水
urban stormwater runoff 城市暴雨径流
urban strategy plan 城市战略规划
urban street 市区街道;城市街道;城市道路
urban street map 城市街道图
urban stress 城市压力;城区压力
urban structure 城市结构
urban study 城市研究
urban suburb 城市郊区
urban supply demand relationship 城市供需关系
urban supply network 城市供应网
urban surface subsidence 城市地面沉降
urban survey 城市调查;城市测量
urban swimming bath 城市浴场;城市游泳池
urban system 城市体系
urban system analysis 城市系统分析
urban telephone 市区电话
urban theater 城市剧院
urban throughfare 城市干道
urban through road 城市过境道路
urban topographic(al) survey 城市地形测量
urban tourism 城市旅游
urban traffic 市区运输;市区交通;城市交通
urban traffic control/bus priority system 城市交通控制/公共汽车优先系统
urban traffic control system 城市交通控制系统
urban traffic noise 城市交通噪声
urban traffic planning 城市交通规划
urban traffic pollution 城市交通污染
urban traffic survey 城市交通观测
urban tramway network 城市有轨电车网
urban transit 城市交通
urban transit traffic 城市过境交通
urban transport 市内运输;市内交通
urban transportation 城市运输;市区运输

urban transportation plannin 城市(交通)运输规划
urban transportation plan system 城市交通运输规划系统
urban transport modes 城市运输模式
urban traverse surveying 城市导线测量
urban trend 城市趋势
urban triangulation surveying 城市三角测量
urban tunnel 市区隧道;城市隧道
urban type industry 城市型工业
urban underground railway 城市地下铁道
urban unity 城市联合体
urban utility 城市公用事业
urban utility tunnel 城市市政隧道
urban villa 城市别墅
urban wall 城墙
urban waste 城市废物
urban waste treatment 城市废物处理
urban wastewater 城市污水;城市废水
urban wastewater management 城市污水管理
urban wastewater reclamation 城市废水的回收
urban wastewater reclamation and reuse 城市废水的回收和再利用
urban wastewater reuse 城市废水的再利用
urban water conservancy 城市水利
urban water consumption 城市水耗;城市用水量
urban water environment 城市水环境
urban water facility 城市给水设备
urban water management 城市水管理
urban water pollution 城市水污染
urban water resources 城市水(资)源
urban water supply 城市供水;城市给水
urban water supply engineering 城市给水工程
urban water supply network 城市供水管网
urban water supply quality 城市供水水质
urban water supply source 城市供水水源
urban water supply system 城市供水系统;城市给水系统
urban wind 城市风
urban zone 城市(地)带
urbicidal 对城市起毁灭作用的
urbiculture 城市特有的习俗;都市福利;都市生活;城市生活中特有的习俗和社会问题
Urbino ware 乌尔比诺陶器(意大利)
urbiphobia 城市恐惧症
urbmobile system 市区交通运输系统
Urca process 乌卡过程
urceolate 瓮形的
urchin 海胆
urdite 独居石
urea 尿素石
urea adduction 尿素络合
urea adhesive 尿素树脂黏合剂
urea anhydride 氨基氰
urea formaldehyde 脲醛
urea formaldehyde adhesive 脲醛胶黏剂;脲醛胶;脲醛黏合剂
urea formaldehyde foam 脲醛泡沫
urea formaldehyde glue 脲醛胶
urea formaldehyde plastics 脲醛塑料
urea formaldehyde resin 脲醛树脂
urea formaldehyde resin adhesive 脲醛树脂胶
urea melamine formaldehyde resin 尿素三聚氰胺甲醛树脂
urea nitrate 硝酸尿素
urea oxalate 草酸脲
urea peroxide 过氧化脲
urea plant 尿素装置
urea plastics 脲醛塑料
urea resin 尿素树脂;脲醛树脂
urea resin adhesive 尿素树脂黏合剂;尿素树脂黏合剂
urea resin foam 尿素树脂泡沫(材料)
urea resin glue 尿素树脂胶
urea resin laminated plastics 尿素树脂层压塑料
urea resin mo(u)lding compound 尿素树脂造型合成材料
urea resin varnish 脲醛树脂清漆;尿素树脂清漆
urea resin wood adhesive 脲醛树脂木材黏合剂
urea synthetic resin adhesive 尿素合成树脂胶
urea unit 尿素装置
urea varnish resin 尿素清漆树脂
ureilite 橄榄(易变辉石)无球粒陨石
urethanated oil 氨基甲酸乙酯(化)油
urethane 氨酯基;氨基甲酸(乙)酯(类)

urethane adhesive 氨基甲酸(乙)酯黏合剂
urethane alkyd 聚氨酯改性醇酸(树脂);氨基甲酸乙酯改性醇酸(树脂)
urethane bond 氨基甲酸(乙)酯键
urethane coated fabric 聚氨酯涂层织物
urethane coating 尿烷涂料;氨基甲酸(乙)酯涂料
urethane elastic fiber 氨基甲酸酯弹性纤维;氨纶
urethane elastic fiber production wastewater 脲烷弹性纤维生产废水
urethane elastomer 聚氨酯弹性体;氨基甲酸乙酯人造橡胶
urethane flexible foam internal pipe-cleaning device 聚氨酯软性泡沫管内清管器
urethane foam 氨基甲酸乙酯泡沫
urethane foam coating 氨基甲酸(乙)酯泡沫涂层
urethane foam filler 脲烷泡沫填缝料
urethane foam panel 氨基甲酸乙酯泡沫塑料板
urethane foam roofing board 聚氨酯泡沫屋面板
urethane-impregnated pad 氨基甲酸乙酯浸透垫板
urethane insulation 聚氨酯绝热材料;聚氨酯泡沫填充料
urethane lacquer 尿烷清漆
urethane laminated fabric 聚氨酯层合织物
urethane linseed oil 尿烷亚麻籽油
urethane oil 氨基甲酸(乙)酯(化)油
urethane paint 尿烷油漆
urethane plastics 聚氨酯塑料
urethane polymer 氨基甲酸乙酯聚合物
urethane putty 氨基甲酸(乙)酯腻子
urethane resin 尿烷树脂;聚氨酯树脂;氨基甲酸(乙)酯树脂
ureyite 隙铬石
urgeiss 古片麻岩
urgency principle in dispatch 紧迫性原则
urgency priority 紧急优先权
urgency rate 迫切率
urgency rating 紧急比率
urgency signal 紧急信号
urgent evacuation time 紧急疏散时间
urgent interrupt 紧急中断
urgently needed expenditures for key construction projects 急需安排的重点支出
urgent marine information broadcast 紧急海上情报广播
urgent message 紧急文件;急电
urgent meteorological danger report 紧急气象报告
urgent navigational danger report 紧急航行危险报告
urgent order 紧急指令
urgent remedial work 紧急补救工作
urgent repair 紧急修理;紧急抢修
urgent signal 紧急信号
urgent telegram 紧急电报;急电
urgranite 古花岗岩
uricite 尿环石
Uriconian rocks 尤里康岩层
urinae praeparatum 淡秋石
urinal 小便器;小便所;小便斗;小便池
urinal bowl 小便器
urinal channel 小便槽;排尿槽
urinal flushing 小便池冲洗
urinal flush valve 小便槽冲水阀
urinal gutter 小便池;小便槽
urinal range 一排小便槽
urinal spreader 小便器
urinal stall 小便器;小便隔间
urinal strainer 小便器滤网
urinal trap 小便器存水弯
urinal trough 小便槽
urinal water closet 大小便器
urinal with internal trap 虹吸式小便池
urinal with trap 带存水弯的小便器;带存水弯的小便池
urinary 尿池;小便所;小便池
urine-repellent 小便槽的挡板
urine stain 小便污垢
urinestone 黑色方解石
Urinettes 乌里奈特便器(一种公厕中女用便器)
urn 瓮
urn pit 骨灰瓮坑
Uroglena americana 美洲辐尼藻
urokon sodium 醋碘苯酸钠

urological department 泌尿科
urology laboratory 泌尿化验室
uropore 尿孔
urotroloine 乌洛托品
urotropine wastewater 乌洛托品废水
ursaenate 乌萨烯酸盐
Ursa minor 小熊星座
ursol 对苯二胺
ursolic acid 熊果酸
urstromtal 冰蚀宽谷（冰川拔削作用）
urstromthal 冰缘宽谷
urtite 磷霞岩
urunday 乌兰迪木（热带美洲产的红棕色硬木）
urusene 漆烯
urushi 漆
urushic acid 漆酸
urushi silk 漆丝
urushi tallow 漆蜡
urushi wax 漆蜡
urushi work 涂大漆
urushi-ye 用挥发性漆加深颜色的印刷品
urushoil 漆油；漆酚
urushoil resin coating 漆酚树脂涂料
urusite 纤钠铁矾
urvantsevite 软铋铅钯矿
usability 适用性
usability factor of runway 跑道可用系数
usability of electrode 焊条的工艺性
usability test 工作性能试验
usable aperture 有效孔径；可用孔径
usable area 有效面积
usable area of dwelling 住房使用面积
usable capacity 有效库容；工作库容；可使用库容
usable cargo space 有效载货容积
usable common open space 可使用的共用敞地
usable deadweight 有效载重量；有效载吨位
usable diamond 能用金刚石
usable dimensions of rear body 车厢有效尺寸
usable floor area 可用地板面积；使用面积
usable flow 有效流量
usable frequency 使用频率
usable frequency band 使用频带
usable height 有效高度
usable horsepower 有用马力；有用功率
usable-in future reserves 暂不能利用储量
usable iron ore 适用铁矿石
usable length 有效长（度）；可用长度
usable length of lock 船闸（的）有效长度
usable length of track 股道有限长度
usable life 适用期；有效期
usable logs 可用图表；合格图表
usable name of evolution stage 演化阶段名称
usable open space 可使用的敞地
usable power 有用功
usable rate of fire 实际射速
usable rate of geologic(al) report 地质成果利用率
usable rate of reserves 储量利用率
usable reflection 有效反映；有效反射
usable resources cumulative 可用资金积累额
usable room area 房屋使用面积
usable runway 可用的跑道
usable size 有用尺寸
usable stone 有用金刚石
usable storage 有效库容；可用储量；可使用库容
usable value 可用值；使用价值
usable velocity sensitivity 有效速度
usable width 有效宽度
usable yard 可使用的庭院
usage analytic(al) data of oil field water 油田水常规分析数据
usage bit 使用位
usage classify of spring 温泉的用途分类
usage coefficient 利用系数
usage count 使用计数
usage factor 利用系数；使用因子
usage factor of capital 资本金利用率
usage mode 使用方式
usage of groundwater 地下水利用
usage rate 利用率
usage report 耗用报告
usage temperature 使用温度
usage time 使用时间
usance 支付期票期限；期票支付时间
usance bill (draft) 远期汇票

usance bill rate 远期汇票贴现率
usance letter of credit 远期信用证
US and British system 美英制
US Army Corps of Engineers 美国陆军工程师团
US screw thread 赛勒螺纹
US system of environmental management 美国环境管理体制
US Three Mile Island nuclear power plant incident 美国三里岛核电站事故
US Department of Housing and Urban Development 美国住房及城市开发部
US dollar[US $] 美元
US Dollar devaluation 美元贬值
useable floor area 使用建筑面积
useable motor vehicle 机动车使用量
useable recording error 可用记录误差
useable source of drinking water 可饮用水水源
use age 使用时期
use and occupancy insurance 固定开支保险；停业保险；使用与占据保险
use attribute 使用属性
use before taking over 移交前使用
use bit 使用位
use capability 使用可能性
use capacity 使用能力
use capital 信用资本；使用资本
use characteristic 使用性能
use charges for postponed wagon sheet 货车篷布延期使用费
use charges for wagon sheet 货车篷布使用费
use coefficient 利用系数
use-cost 使用成本
use-cost difference 使用成本差
use count 使用计数
U-section steel pile U形钢板桩
use curve 利用曲线
used air 余风；废空气
used area 占用区
used automobile tire 旧汽车轮胎
used auto-tire 废旧汽车轮胎
used beat 用过的热
used car 半新车
used catalyst 用过的催化剂
used diamond 用过的金刚石
used fixed assets 已使用过的固定资产
used grease 用过的润滑脂
used heat 废热；余热
use-dilution test 实用稀释试验
use district 可用区域
used lime 用过的石灰
used lining 废炉衬
used liquid 用过的液体；废液
used material 使用材料
used oil 用过的油；废（机）油
used oil disposal 废油处置
used oil reclaimer 废油再生器
used oil regenerator 废油再生器
used packaging supplies in storage 库存已用包装物
used property 已使用财产
used rail 旧轨
used stone 用过的金刚石
used thickness 使用厚度
used train 使用车【铁】
used up bill of lading 已用提单
used (waste) heat 余热；废热
use factor 利用因数；利用系数
use fee 使用费
use-forbidden product 禁用产品
use foreign funds 利用外资
useful aperture 有效孔径
useful aquatic plant 有用的水生植物
useful area 有效面积；有效断面；可用面积；使用面积
useful area of screen 筛的有效面积
useful arts 工艺；手艺
useful beam 有用束流
useful capacitor 有效电容
useful capacity 有效容重；有效容量；载货吨位
useful casualty levels 有效杀伤程度
useful chamber length （船闸）闸室的有效长度
useful component 有用组分
useful compression ratio 有效压缩比
useful cross-section 有效截面；有效（横）断面

useful cross-sectional area 有效截面（面）积
useful deadweight 有效载重量；有效载吨位；载货吨位
useful diameter 有效直径
useful discharge 有效流量
useful earthing 有效接地
useful economic life 有效经济使用期
useful effect 有用效应；有用效果；有效作用
useful efficiency 有效功率；利用效率
useful energy 有用能
useful enlargement 有效放大倍率
useful exposure range 有效曝光范围
useful floor area 可用地板面积；使用面积
useful flow 有效流量；可用流量
useful flux 有效磁通
useful forest insect 森林益虫
useful function 有效机能
useful grate area 有效炉算面积
useful head 有效水压；有效水头；资用水头
useful heat 有效热；可利用的热
useful height 有效高度
useful height of a cupola 冲天炉有效高度
useful horsepower 有效马力；有效功率
useful index state 有利标志状态
useful information 有用信息
useful information coefficient 有用信息系数
useful investment 有用投资
useful length 有效长（度）
useful length of the chamber （船闸）闸室的有效长度
useful length of track 股道有效长（度）
useful life 有效期（限）；使用寿命；使用期（限）
useful life cycle 可使用寿命周期
useful life estimation 使用年限估定
useful life of device 设备使用年限
useful life of equipment 设备合理使用年限
useful life period 使用寿命期
useful life (span) 有效寿命
useful lift 有效升力
useful limit stress 工作极限应力
useful load 飞机装载量；有用荷载；有效载重量；有效载吨位；有效荷载；有效负载；有效负荷；作用荷载；资用荷载；工作载荷；实用负载
useful loading 有效负载；有效负荷
useful luminous flux 有效光通
useful magnification 有效倍率
useful materials 有用物资
useful mineral 有用矿物
useful modulation 有效调制
usefulness 效用；有用性
usefulness of accounting data 会计数据的用途
usefulness of a project 工程效益
useful output 有效输出；有用功率；有效产量
useful output at terminal 发电机出线端净出力
useful output of coupling 联轴器端净出力
useful physical life 使用寿命
useful picture area 有效图像面积
useful power 有用功耗；有效功率
useful power input 有效输入功率
useful pressure 有用压力
useful quantity 有效量
useful range 有效（测量）范围
useful range of leading marks 导标有效作用距离
useful refrigeration effect 有效冷量
useful screen area 有效屏幕面积
useful shoulder 有效路肩
useful signal 有用信号
useful space 有效空间
useful steam consumption 有效汽耗
useful storage 有效库容；可用蓄量；有效蓄水（量）；可用蓄量有效库容
useful surface 有效面
useful test 标准检定
useful thermal power 有用热功率
useful time 有用时间；使用寿命；使用年限
useful torque 有效转矩
useful transmission 有用透射
useful vacuum degree 使用真空度
useful value 使用价值
useful voltage 有效电压
useful volume 有效容积
useful volume of holder 储气罐有效容积
useful water 有效水分
useful width 有效宽度

useful width of the chamber (船闸)闸室的有效宽度
useful work 有用功;有效功;有效工作
useful working area 有效工作面积
use in fertilizer 作为肥料使用
useless 没用的;无用的
useless heat loss 无效热耗
useless land 无用土地
use level 用量;配合量
use no hooks 不可用钩;不得用钩;请勿用钩
use occurrence 引用性出现
use of aerial photograph 利用航测照片
use of agricultural labor 使用农业劳动力
use of agrochemical 农业化肥的使用
use of animal waste 牲畜粪便利用
use of blanks 空格的用法
use of emulsifying agents 乳化剂使用
use of funds 资金运用;资金使用
use of funds in a scattered way 分散使用资金
use of geologic(al) map and data 利用地质图和地质资料
use of geologic(al) map and information 利用地质图和地质资料
use of insecticide 使用杀虫剂
use of labor 使用劳动力
use of mineral commodities 矿产品用途
use of momentum 动能利用
use of names 名字的用法
use of patent 行使专利权
use of private carriage or special passengers train 包车
use of reclaimed water 回收水用途
use of revenue to finance expenses 以收抵支
use of scan(ning) electron microscope 扫描电子显微镜的使用
use of soil investigation 利用土壤调查
use of soil survey 利用土壤调查
use of the new technology 应用新技术
use of topographic(al) map 利用地形图
use of waste heat 废热利用
use-oriented language 面向用户的语言
use-pattern 使用方式
USEPA water quality criterion 美国环境保护局水质基准
use percentage 使用率
use property 使用性能
use-proved product 使用合格的制品
use quality 使用质量
user 用户;使用者
user account 用户记事
user agent 用户代理
user area 用户区(域)
user-assistance software 支持用户软件
use ratio 消耗率;使用率
user attribute data set 用户属性数据集
user auditor 客户审计师
user authorization file 用户特许文件;用户授权文件
user benefit 用户效益
user benefit economic analysis 使用者得益经济分析
user chain 用户链
user charges 使用费;用户费用
user class 用户级
user-coded virtual memory 用户编码的虚存
user collection charges 用户垃圾费
user command 用户命令
user communication interface 用户通信接口
user coordinate 使用者坐标
user cost 用户费用;用户成本
user datagram protocol 用户数据协议
user data set 用户数据组
user-defined 用户指定的
user-defined library 用户确定的程序库
user-defined subsystem 用户定义子系统
user-defined tariff 用户确定的关税
user density 用户密度
user-design command 用户设计命令
user-determined form 用户决定形式
use recycle paper 使用再生纸
user effort 用户花费
use reliability 使用可靠性
user equipment 用户设备
use requirement 使用要求
user equivalent range error 用户等效距离误差

user error procedure 用户错误处理过程
use-restriction plan 限量供水计划
user exit routine 用户出口程序
user facility 用户设备
user fee 使用费;用户费
user file 用户文件
user file directory 用户文件目录
user flag 用户标记
user friendly 方便用户的
user friendship 用户友谊
user graphic file 用户图形文件
user group 用户组;用户群;用户集团
user handbook 用户便览
user header label 用户首标
user identification 用户识别符
user identifier 用户标识符
U-series disequilibrium dating method 不平衡铀系列测年法
use rights 使用权
user interface 用户界面;用户接口
user interface management system 用户界面管理系统
user label 用户标号
user label exit routine 用户标号出口程序
user label handling 用户标号处理
user label information 用户标号信息
user lamp 用户指示灯
user level protocol 用户级协议
user library 用户(程序)库
user loop 用户回路
user management 使用单位主管
user man-machine interface 用户友好人机界面
user map 用户图
user microprogrammability 用户微程序控制性
user microprogrammer 用户微程序编制器
user microprogrammer processor 用户微程序编制加工器
user micro-programming 用户微程序设计
user mode 用户状态
user needs 用户需求
user network interface 用户网络接口
user node interface 用户节点接口
user number 用户代号
user of finance 资金使用者
user of funds 资金使用者
user of reclaimed water 回收水用户
user of service 使用人员
use rollers 在滚子上移动
user operation 用户操作
user option 用户任选项
user-oriented end product 提供用户的最后成果
user-oriented product 用户定向产品
user-oriented system 面向用户系统
user own coding 用户扩充工作编码
user partition 用户分区
user port function 用户端口功能
user privileges 用户特权
user processor 用户处理机
user profile 用户分布图
user programmable data acquisition system 用户可编程度数据采集系统
user program(me) 用户(可编)程序
user program(me) area 用户程序区
user program(me) error 用户程序错误
user program(me) list 用户程序表
user programming 用户的程序设计
user range error 用户距离误差
user requirement specification 用户需求规格书
user's demand 用户需求
user security verification routine 用户安全检验程序
user's guide 用户指南
user side 用户端
user's inspection to be final 以用户检验为最后依据
user's manual 用户手册
user's needs 用户需要;用户需求
user's specification 用户标准
user stack 用户栈
user stack pointer 用户栈指示器
user status table 用户状态表
user survey 用户调查
user task scheduling 用户任务安排
user terminal 用户终端

user terminal system 用户终端系统
user test 用户试验
user time 用户时间
user time-sharing 用户分时
user totaling exit routine 用户总出口程序
user-to-user circuit 用户到用户线路
user-to-user communication 用户对用户通信
user-to-user connection 用户间的连接
user-to-user service 用户对用户业务
user trailer label 用户尾部标记
user transducer 用户换能器
user view data 用户意图数据
user volume 用户卷宗
user water supply 用户供水
user writer 用户写入程序;用户记录器
uses type of coals 煤的应用类型
use tax 使用税
use technology 使用技术
use up 消耗掉;用尽
use up soil nutrient 消耗土壤养分
use value 使用价值;实用价值
use without compensation 无偿使用
US expenditures weight 以美国支出为权数
use-yellowing 使用泛黄
use zone plan 使用分区图
use zoning 功能(分)区;用途分区;使用分区制
US Forest Service 美国森林局
US gallon 美制加仑(1加仑=3.785升)
U-shape beam U形梁;槽开梁
U-shaped 马蹄形的;马鞍形;U形的
U-shaped abutment U形桥台
U-shaped bed 槽形台座
U-shaped bend (钢筋的)U形弯钩
U-shaped block 马蹄形块
U-shaped bolt U形螺栓
U-shaped brake 复合闸
U-shaped enamelling furnace 马蹄形搪烧窑
U-shaped glaciated valley U形冰蚀谷
U-shaped glass 槽形玻璃
U-shaped magnet 马蹄形磁铁;马蹄铁
U-shaped pipe U形管
U-shaped piping U形管
U-shaped rotary bit U形螺旋式钻头
U-shaped section U形断面
U-shaped stirrup U形箍筋
U-shaped stud U形龙骨
U-shaped tube U形管
U-shaped tubing U形管
U-shaped valley 槽形(河)谷;U形谷
U-shape stirrup U形箍筋
US hardness 美制硬度
ushers room 传达室;服务员室;引座员室
ushkovite 水磷镁铁石
US Housing Act 美国住房法
using advance technology 使用先进技术
using a phenolsulfuric acid colo(u)rimetric method 用苯酚磺酸光电比色法
using capacity of chemical liquid for well cleaning 化学洗井液用量
using capacity of preservation 防腐蚀剂用量
using computers to plan 使用计算机规划
using free-cultivation properly 适当使用免耕法
using non-metallic elements 非金属元素使用
using ordinary equipment 使用一般设备
using the money from the deposits for loans 以存支贷
using workers as cadres 以工代干
usn(in)ic acid 地衣酸
Usonian Houses 优素尼雅住宅(建筑大师赖特设计的一种纯美国风格住宅)
usovite 氟铝镁钡石
u-spin 幺旋
US sieve series 美国筛系
US standard screw thread 美制螺纹
U stage 万用旋转台
U-stair 两跑平行楼梯;双跑平行楼梯(指无梯井的双跑平行楼梯)
ustalf 干淋溶土
U-staple U形钉
U-steel U字钢;槽钢;U形钢
ustent 干新成土
ustert 干变性土
ustic 土壤水分偏干(状态)
U-stirrup U形箍筋;开口箍

ustoll 干软土
ustox 干氧化土
ustult 干老成土
usual condition 通常条件
usual despatch 正常装卸速度
usual discount 普通折扣
usual dispatch 正常装卸速度
usual element 常用元素
usual interest 普通利息
usual interest rate 普通利息率
usual life 活化寿命；适用期
usual load 一般荷载；普通荷载
usual loading condition 正常负荷状况
usually humid in tropic(al) area 热带地区多潮湿
usual market requirement 正常市场需求量
usual method of measuring discharge 测流常用法
usual name 俗名；常用名
usual packing 常规包装
usual practical 常用实用单位
usual practice 惯例
usual profit 通常的利润
usual residence 习惯居所
usual risks 一般(风)险；通常险别
usual service condition 正常工作条件
usual terms 普通条件
usucaption 凭时效取得财产权
usufruct 用益权
usufruct of immovable property 不动产用益权
usufructuary 享有用益权的人；有用益者
usufructuary right 用益权
usurious interest 高息
usurious loan 高利贷
usuriousness 放高利贷
usurpation 非法使用
usurpation of a franchise 非法使用特许权
usury 高利贷
usury ceiling 高利率限值；最高允许利率
US Water Convervation Laboratory 美国水保护实验室
US weighted purchasing power parity 以美国为权数的购买力比价
utahite 黄钾铁矾
Utah-type milking pool 犹他式挤奶站
utensil 用具；器皿；器具；厨房用具
utensil rack 炊具架
utensil room 工具间
utensil sink 厨房水槽
utensil storage 器具库
U-Th mix anomaly 铀—钍混合异常
Utica shale 尤蒂卡页岩
U-tie U形连接件；U形系墙铁
utilance 光通利用率
utilidor 保温管道
utiliscope 工业电视装置
utilitarian building 办公房屋；商业房屋
utilitarian classicism 古典实利主义
utilitarian element 可代用构件
utilitarianism 功利主义
utilities engineering 管网综合
utilities trench plan 公用设施地沟平面图
utility 公用工程；效用；有用；应用；公用事业(设备)；实用(性)
utility air 公用气源
utility aircraft 通用飞机
utility analysis 效用分析
utility architecture 通用建筑；实用建筑
utility area 生活区
utility armored car 通用装甲车
utility availability 公用设施条件
utility billing 公用事业
utility boat 工作艇；通用小艇
utility boiler 电站锅炉
utility bucket 通用铲斗；散装物铲斗
utility certificate 实用新型
utility circle (公用设施)利用范围圈
utility circuit 生活用电电路
utility closet 公用事业壁橱
utility cloths 实用织物
utility command 使用命令
utility company 公用事业机构
utility conduit 公用事业管线
utility consumption 动力消耗量(指水、电、汽)；动力电消耗量
utility control console 使用控制台；操作控制台
utility control facility 实用控制设备；实用程序控制设施
utility core 公用设施中心
utility cost 公用设施费用；公用事业费；通用费用
utility curve 效用曲线
utility debug 应用调试(程序)；实用调试(程序)
utility debugger 实用调试程序
utility design 实用设计
utility device list 使用设备表；实用设备表
utility diversion 设施迁改
utility dolly 通用砧
utility element 公用设施项目(如水、电、煤气管线等)
utility engineering 工程管线综合；综合管网
utility equipment 公用(事业)设备
utility facility 公用(事业)设施
utility facility program(me) 实用设备程序；实用服务设施程序
utility factor 利用率
utility factor of berth 泊位利用率
utility factor of dredge(r) 挖泥船利用率
utility factor (of equipment) 设备利用率；设备利用系数
utility factor of the position 台位利用系数【铁】
utility function 效用函数；公用功能；实用功能
utility gallery 公用事业管线廊道；公用隧管
utility gas 城市煤气；煤气公司煤气；公用煤气
utility generation 公用发电
utility glassware 日用玻璃器皿
utility grade wood 可用材
utility graph 应用图
utility green 利用绿地；资用绿地
utility index 效用指标
utility key 应用键
utility lamp 通用灯
utility landing craft 通用登陆艇
utility landing ship 通用登陆舰
utility level 效用水平
utility light 通用工作灯
utility line 公用事业管线；水电管线
utility logger system 实用记录系统
utility map 公用事业图
utility mapping 公用事业图制图
utility maximization 效用极大化
utility maximizing rule 效用最大化法则
utility-measurement experiments 效用测度实验
utility model 效用模型；(实用)新型；实用模型
utility model patent 实用模型专利
utility operator 辅助工
utility or other enterprise fund 公用事业或其他企业基金
utility package 应用程序包；实用程序包
utility payment 效用报偿
utility (pipe) line 公用工程管线；公用事业管线
utility piping 公用设施管线；公用设施敷管；公用事业管线
utility pole 电线杆
utility pouch 工具袋
utility power 公用电源
utility power-boat trailer 运载机动船的挂车
utility program(me) 应用程序；公用事业程序；实用程序(设计员)；辅助程序
utility programmer 应用程序员
utility protection 埋设物防护
utility raceway 设备管线；设备管道
utility radio communication 实用无线通信
utility radio transmitter 实用无线电发射机
utility rate 公用事业费率
utility ratio 应用率；实用率
utility refuse 服务行业废品；公用事业垃圾；公用事业废物；服务行业废物
utility room 杂品存储室；杂用室；杂务间；家庭用具存放室；附属生活设施间
utility routine 实用例程；实用(例行)程序
utility run 设备管路；设备管线
utility scale 效用尺度
utility service 服务行业；公用事业；公用服务事业；提供劳务
utility session 实用对话
utility shaft 设备竖井
utility software 实用软件
utility software package 实用软件包
utility structures 公用事业建筑
utility survey 公用设施调查；公用事业工程测量
utility system 应用系统；公用事业系统；实用系统
utility tax 公用事业税
utility theory 效用理论
utility theory of value 效用价值论
utility time 实用时间
utility tower 公用事业设备塔
utility tractor 通用拖拉机
utility tree 效用树；应用树；应力树
utility truck 公用载重汽车
utility tunnel 管线廊道
utility type 实用型
utility type unit 大型机组；电站型机组
utility undertaker 公用事业承揽者
utility undertaking 公共设施；公共服务事业
utility unit 效用单位
utility value 利用价值
utility vehicle 公用车辆
utility versus money loss 效用亏损与货币亏损比较
utility ware 日用器皿
utility waste 公用事业垃圾；公用事业废物；服务行业废物
utility wastewater 公用事业废水
utility window (公用建筑)采光窗
utility yard 杂务院
utilizability 可利用性
utilizable capacity 兴利库容
utilizable discharge 资用流量；可(利)用流量；可利用的排出物
utilizable flow 可用流量
utilizable regulation 兴利调度
utilization 运用；规格化；利用
utilization area 使用面积
utilization coefficient 使用系数
utilization coefficient of a blast furnace 高炉利用系数
utilization coefficient of light energy 光能利用率
utilization coefficient of open-hearth furnace 平炉利用系数
utilization coefficient of strokes (压力机)行程次数利用系数
utilization cut 放牧折减率
utilization efficiency 利用效率
utilization efficiency of fertilizer 肥料利用率
utilization engineer 公用工程工程师；使用维护工程师
utilization equipment 公用事业设备；设备利用；用电设备；日用电器；电气设备
utilization factor (焊机的)暂载率；利用因素；利用因数；利用系数；利用率
utilization factor of cargo handling machinery 装卸机械利用率
utilization factor of container volume 箱容利用率
utilization factor of man-hours 工时利用率
utilization factor of man-hours in loading-discharging 装卸工时利用率
utilization factor of runoff 径流利用系数
utilization factor of shothole 炮眼利用率
utilization factor of waste reclamation 废物资源利用率；废物回收利用率
utilization heat rate 利用热耗
utilization hours in a year 年利用小时
utilization of a coast 海岸利用
utilization of a loan 使用贷款
utilization of archives 档案利用
utilization of budget 执行预算
utilization of capacity 开工率；能力利用
utilization of debris 利用废料；废物(残渣)利用；废品利用；废料利用
utilization of digester gas 消化池气(体)利用
utilization of energy 能量利用
utilization of equipment 设备利用
utilization of forest resources 森林资源利用
utilization of fund 资金的运用
utilization of gangue 煤矸石利用
utilization of heat 热(的)利用
utilization of hour 工时利用
utilization of labo(u)r 劳动力使用
utilization of liquid sludge 液体污泥利用
utilization of local resource 本地资源的利用
utilization of manpower 人力的利用
utilization of natural depth 利用自然水深
utilization of nature 利用自然
utilization of petroleum 石油综合利用
utilization of refuse water 废水利用

utilization of resources 资源利用
utilization of site 场所的利用
utilization of sludge 污泥利用
utilization of solar energy 太阳能利用
utilization of solid wastes 固体废物利用
utilization of space 空间利用;场所的利用
utilization of the three-wastes 三废(综合)利用
utilization of tidal power 潮汐力利用
utilization of waste 废物利用;废品利用
utilization of waste heat 废热利用率
utilization of waste material 废料利用
utilization of wastewater treatment 污水处理利用
utilization of water 水的利用
utilization of water power 水力利用
utilization of water resources 水资源利用
utilization plan of storage space 场地利用计划
utilization policy of water resources 水资源利用方针
utilization rate 利用率
utilization rate for the semiprocessed materials 下料利用率
utilization rate of equipment 设备利用率
utilization rate of equipment and installation of dredge(r) 挖泥船设备利用率
utilization rate of installed equipment 安装设备利用率
utilization rate of static D/W tonnage 船舶的实载率
utilization rate of storage yard 库场利用率
utilization rate of warehouse 仓库利用率
utilization rate of working hours 工时利用率
utilization ratio 利用率;利用比;使用率
utilization ratio of civil aviation aircraft 民航飞机利用率
utilization ratio of the seating capacity of an aircraft 客座利用率
utilization results of energy 能源利用效果
utilization value 利用价
utilization variance 利用差异
utilize 运用;利用
utilized bandwidth (实际)使用带宽
utilized bandwidth ratio 使用带宽比
utilized capacity 利用的容量
utilized cost 利用成本
utilized efficiency of equipment 设备利用率
utilized efficiency of transport equipment 运输设备利用率
utilized flow 利用流量
utilized head 利用水头
utilized inflow 利用流入量
utilized river reach 已利用河段;开发河段
utilized river section 已利用河段
utilized river stretch 已利用河段;开发河段
utilize economically 经济使用
utilize foreign capital 利用外资
utilize to the capacity 利用设备达到最大能力
utilizing advanced technology 吸取先进技术
utilizing meteorologic(al) date 利用气象资料
utilizing water capacity 水的利用量
utilizing water content 水的利用含量
utiloy 镍铬耐酸钢
ut infra 如下所述
utisol 老成土
utmost 最高的;最远的;最大的;极度的
utmost dispatch 最快装卸速度
utmost good faith 绝对诚信
utopia 理想国
Utopian schemes of city planning 空想城市规划

U-trap 虹吸管;U形弯管;存水弯(头);U形存水弯
ut supra 如上所述
uttering false coin 行使伪币
uttering forged documents 行使假文件
U-tube U形管
U-tube aeration U形管曝气
U-tube aeration system U形管曝气装置
U-tube densitometer U形密度计
U-tube ga(u)ge U形管压力计
U-tube manometer U形管压力计
U-tube type manometer U形管压力计
U-turn U形转折;U形弯头;调头
U-type U形的
U-type barometer U形压差计
U-type crane U形起重机
U-type manometer U形压力计
U-type retaining wall U形挡土墙
U-type superposed folds U形叠加褶皱
UU-test 不固结不排水三轴试验
uvala 干宽谷;灰岩盆;溶蚀洼地
U-vale 干宽谷
U-valley U形谷
U value 总传热系数;U值
uvanite 钒铀矿
uvarovite 绿榴石;钙铬榴石
uvaser 远紫外激光器
uviofast 抗紫(外)线的
uviol glass 透紫外线玻璃
uvioresistant 抗紫(外)线的;不透紫外线
uvite 钙镁电气石
U-vortex 马蹄形涡流
U-washer 开口垫圈;U形垫圈
uytenbogaardtite 硫金银矿

V

vacamatic 真空自动式
vacamatic pump 真空自动泵
vacamatic transmission 真空自动控制变速箱
vacamatic valve 真空自动阀
vacancy 空缺额,缺额;空闲;空位;空格点;空房;缺位
vacancy 空闲;空位;空格点;空房;缺位;缺额
vacancy allowance rate 空房率
vacancy and rent loss 空房与收租损失
vacancy area 空白区域;空地区
vacancy chromatography 空穴色谱法
vacancy clause 空屋条款
vacancy concentration 缺位浓度
vacancy concentration of oxygen 氧离子缺位浓度
vacancy condensation 空位凝聚
vacancy defect 空位缺陷
vacancy diffusion 空位扩散
vacancy disc 空位盘
vacancy-dislocation interaction 空位错位的相互作用
vacancy factor 空房损失因素
vacancy loop 空位环
vacancy losses 房屋空闲损失
vacancy mechanism 空位机理
vacancy migration 空位移动
vacancy mobility 空位迁移率
vacancy pair 空位偶;空位对
vacancy rate 闲置率;空闲率;空缺率;空房率;空房比(一定地区中未使用房屋数占总住宅数的比率)
vacancy rate homeowner 非出租房屋闲置率
vacancy rate in rental housing 出租住房空闲率
vacancy ratio 住房闲置率
vacancy solid solution 缺位固溶体
vacancy source 空位源
vacancy test(ing) 空白试验
vacancy tracer exchange 空位示踪原子互换
vacancy wind 空位风
vacant 空位的
vacant building land 闲置的建设用地
vacant building site 空置建筑地盘
vacant electron site 电子空位
vacant field 闲田
vacant ground 空地
vacant house 空房
vacant job 空缺的职位
vacant land 空(闲)地;闲田;空置(的土)地
vacant land tax 土地闲置税
vacant lattice position 晶格中空位
vacant lattice site 晶格内空穴;点阵空位
vacant level 空级
vacant line 虚线
vacant lot 空地(出卖)
vacant position 空缺职位
vacant possession 空房
vacant premises exemption 空置物业豁免
vacant room 未使用房间;(未使用的)空闲房间;空房;未住人房间
vacant run(ning) 空转;无载运转;无载运行
vacant site 空置地盘(建筑);空位
vacant space 空舱位
vacant state 空态;未满状态
vacant tenement 空置住宅
vacant terminal 空间端
vacated cell 腾空单元
vacation 休假;漏涂点;假期;迁出
vacation allowance 休假津贴
vacation and holiday pay 假期和节日工资
vacationer 度假者
vacation home 度假屋
vacation hotel 假日旅馆;假日旅店
vacation house 假日宿舍
vacationist 度假者
vacation land 旅游胜地;度假胜地
vacation paradise 度假胜地
vacation pay 假期工资
vacation village 度假村
vacation with pay 带薪休假
vaccenic acid 异油酸

vaccigenous 产生菌苗的
vaccination 接种疫苗
vaccination of water 水的稳定处理
vaccinization 连续接种
vaccinogenous 产生菌苗的
vacillate 动摇
vacion pump 钛泵
vacor 灭鼠优
vac pump 真空泵
Vacquir tangential method 外奎尔切线法
vacreation 真空杀菌
vacreator 真空杀菌器
Vacseal pump 水压封闭泵
vacsorb 真空吸附
vacsorb pump 真空吸附泵
vacuate 抽稀;抽成真空
vacuation 抽成真空
vacuator 真空浮渣池;抽空装置;抽空机
vacu-forming 真空造型;真空模铸
vacuity 真空度;空容积;膨胀空隙
vacujet 真空捕尘凿岩机
vacuo 真空
vacuo-forming 真空造型
vacuo heating 真空加热
vacuo-junction 真空热电偶
vacuolar degeneration 空泡变性
vacuolar membrane 液泡膜
vacuolate 形成空泡
vacuolation 形成空泡;析稀(作用);空泡形成
vacuole 析稀胶粒;液泡;真空泡;空泡
vacuole formation 形成空泡
vacuole membrane 液泡膜
vacuolization 形成空泡;空泡形成
vacuolus 液泡
vacuometer 负压计;真空计;低压计
vacu-pad 真空轧染
vacuscope 真空计;真空仪
vacuseal 真空密封
vacu-therm instant sterilizer 真空加热直接杀菌器
vacu-therm pasteurizer 真空加热巴氏杀菌器
vacuum 真空
vacuum absorbing tape buffer 真空吸带缓冲器
vacuum accessories 真空设备用附件
vacuum advance control valve 真空点火提前控制阀
vacuum advancer 真空提前器;真空点火提前装置
vacuum air pump 真空(气)泵;真空抽气机;真空抽气泵
vacuum and blow process 真空吸料吹制法
vacuum and heating drying 真空加热干燥
vacuum and pressure air pump 抽空压气两用泵
vacuum and pressure relief assembly 真空和压力安全装置
vacuum and sub-atmospheric system 真空系统
vacuum anemometer 真空风速计;真空风速表
vacuum annealing 真空退火
vacuum apparatus 真空装置;真空设备;真空器
vacuum arc degassing process 电弧真空脱气法
vacuum arc furnace 真空电弧炉
vacuum arc furnace melting 真空电弧熔化炉
vacuum arc melting 真空电弧熔炼
vacuum arc melting process 真空电弧熔化法
vacuum area 真空区域
vacuum arrester 真空避雷器
vacuum arresting method 真空除尘法
vacuum asphalt 真空制沥青
vacuum assisted 真空加力的
vacuum assisted front disc 真空助力式前轮盘式制动器
vacuum assisted hydraulic brake system 真空助力液压制动系统
vacuum assisted mo(u)lding 真空辅助成型
vacuum assisted resin injection process 真空辅助树脂注射工艺
vacuum assisted sludge dewatering bed 真空辅助污泥脱水床
vacuum-assist pouring 负压浇筑
vacuumatic pressure ga(u)ge for ammonia 氨用真空压力表
vacuum augmenter 真空增强器
vacuum autoloader 真空自动加料机
vacuum automatic spark advance device 真空式自动提前点火装置
vacuum automatic tissue processor 自动真空组织处理机
vacuum back 真空压平板【测】;真空靠背
vacuum back-pumping 真空回扬
vacuum bag 真空袋
vacuum bag mo(u)lding 真空袋模制法;真空袋模塑成型
vacuum bag process 真空袋模法
vacuum bag technique 真空袋模塑
vacuum baking 真空烘焙(法)
vacuum balance 真空天平
vacuum baling 真空压捆
vacuum balloon 真空气球
vacuum bearing 真空轴承
vacuum behavio(u)r 真空行为
vacuum bell jar 真空钟罩
vacuum belt filter 真空带式过滤器
vacuum blast 真空吸入喷砂清理;真空喷砂
vacuum blasting 真空喷砂法
vacuum blower 吸入式输运器;真空吸送器;真空抽风机
vacuum blowing 真空吸料吹制
vacuum board machine 真空纸板机
vacuum bolometer 真空测辐射热计
vacuum bond 真空结合
vacuum bonding 真空黏合
vacuum boost brake 真空加力式制动器
vacuum booster 真空助力器
vacuum booster pump 预抽真空泵;真空助力泵
vacuum boring 真空钻进
vacuum bottle 热水瓶;保温瓶
vacuum box 吸水箱;真空室;真空框
vacuum brake 真空助力制动器;真空制动(器);真空增加制动阀
vacuum brakehose 真空制动器软管
vacuum braking system 真空制动系统
vacuum brazing 真空钎焊;真空焊接
vacuum break 破坏真空
vacuum breakdown 真空击穿
vacuum breaker 止回阀;单向阀;真空阻断器;真空断路器;消解水管真空设施;真空破碎机;真空破坏器;真空断流器;真空(调节)阀
vacuum breaker float 破坏真空浮子
vacuum breaking 真空破坏
vacuum break(ing) valve 真空破坏阀
vacuum bubble 真空泡;空泡
vacuum building 真空厂房
vacuum bulb 真空灯泡
vacuum bypass 真空侧管
vacuum calibration bath 真空校准浴
vacuum capacity 真空量
vacuum capping 真空封瓶
vacuum car 真空清洁车;卫生车
vacuum carburet(t)or 真空汽化器
vacuum carburization 真空碳化;真空渗碳
vacuum carburizing 真空渗碳(法)
vacuum case 真空室
vacuum casting 真空铸造;真空注浆成型;真空吸铸;真空外壳;真空模铸;真空浇铸;真空电流铸造机;减压注浆
vacuum cement 气密胶;真空黏结剂;真空气密水泥;真空封泥
vacuum centrifugal pump 真空离心泵;离心式真空泵
vacuum chamber 真空箱;真空室;真空罐
vacuum chamber aperture 真空室孔径
vacuum chamber dimensions 真空室尺寸
vacuum chamber wall 真空室壁
vacuum chilling 真空冷却
vacuum chlorinator 真空加氯机
vacuum chuck 真空吸盘;真空夹头;真空夹盘
vacuum circuit breaker 真空断路器
vacuum circulation system 真空循环式

vacuum clarifier 真空澄清器
vacuum cleaner 真空除尘器;真空清洁器;真空净化器;真空除尘器
vacuum cleaner alloy 真空净化器用合金
vacuum cleaner fan 真空吸尘抽风机
vacuum cleaning 真空吸尘;真空除尘
vacuum cleaning apparatus 真空净化设备
vacuum cleaning installation 真空吸尘装置;真空净化设备
vacuum cleaning line 真空除尘管(线)
vacuum cleaning plant 除尘设备;真空吸尘设备;真空清洁设备;真空清洁机具
vacuum cleaning system 真空吸尘系统
vacuum clear 真空净化
vacuum coating 真空镀膜(法)
vacuum cock 真空活塞
vacuum coefficient 真空系数
vacuum cold trap 真空冷阱
vacuum collector 真空罩;真空集尘器
vacuum column 真空(蒸馏)塔;真空箱
vacuum compression method 真空加压法
vacuum compressor 真空压缩机
vacuum concentration 真空浓缩
vacuum concrete 真空吸水处理混凝土;真空(法)混凝土;真空处理混凝土;真空作业混凝土
vacuum concrete method 混凝土真空吸水(处理)法
vacuum concrete pipe 真空处理混凝土管
vacuum concrete process 混凝土真空吸水(处理)法;真空混凝土法
vacuum condensation 真空缩合;真空冷凝
vacuum condenser 真空冷凝器
vacuum condensing point 真空冷凝点
vacuum condition heat transfer 真空热传输
vacuum cone dryer 真空锥形干燥器
vacuum connection 真空系统的管道连接
vacuum contact 真空电接触器材
vacuum container 真空集装箱
vacuum control 真空控制
vacuum control check-valve 真空止回阀
vacuum control clutch 真空控制离合器
vacuum controlled 真空控制的
vacuum controlled advance 真空提前控制
vacuum controlled clutch 真空控制式离合器
vacuum controlled economizer 真空省油器
vacuum controlled gearshift 真空控制换挡
vacuum controlled gear shifting 真空控制式齿轮变速
vacuum controlled ignition advance 真空控制式点火提前
vacuum controlled sparking timing 真空点火正时
vacuum controlled system 真空控制系统
vacuum control modulator valve 真空控制电磁阀
vacuum control piston 真空控制活塞
vacuum control unit 真空控制装置
vacuum control valve 真空控制阀
vacuum conveyer 真空吸入式输送器;真空输送机
vacuum conveyer tube 真空输送管
vacuum cooler 真空式冷却器
vacuum cooling 真空冷却
vacuum copy-holder 抽气压平稿图案;真空晒相架
vacuum corer 真空岩芯提取器;真空取样管;真空采样器;真空采芯管
vacuum core sampler 真空采样器
vacuum correction 真空校正
vacuum corrector 真空调整器
vacuum crane 真空吊车
vacuum crimper 真空卷边器
vacuum crystallization 真空结晶
vacuum crystallizer 真空结晶器
vacuum cup 真空吸盘;真空吸杯;保温杯
vacuum cup crane 真空吸盘式升降机
vacuum cup lifter 真空吸盘升降台
vacuum curve 真空曲线
vacuum cylinder 真空筒;真空缸
vacuum cylinder and plunger assembly 真空缸及柱塞总成
vacuum cylinder gasket 真空汽缸垫密片
vacuum cylinder oil 真空汽缸油
vacuum cylinder piston spring 真空汽缸活塞弹簧
vacuum cylinder push rod guide 真空汽缸推杆导管
vacuum cylinder reinforcing plate 真空汽缸加强板
vacuum dash pot 真空缓冲筒

vacuum deaeration 真空排气(法);真空除气
vacuum deaerator 真空除气器;真空除氧器
vacuum deairing 真空脱气
vacuum deairing concrete 真空去气混凝土
vacuum deairing machine 真空练泥机
vacuum decker 真空浓缩机
vacuum decrepitation extraction method 真空热爆浸取法
vacuum decrepitation gas measurement method 真空爆裂气体测量法
vacuum deep drawing 真空深层抽拔
vacuum defoamation 真空脱泡
vacuum degasifier 真空脱气器
vacuum degasing 真空除气;真空除气
vacuum degree 真空度
vacuum dehydration 真空脱水
vacuum deposited 真空沉积的
vacuum deposited circuit 真空淀积电路
vacuum deposited coating 真空镀敷法
vacuum deposition 真空喷涂法;真空镀膜;真空淀积;真空沉积
vacuum deposition of magnesium 真空镀镁
vacuum desiccator 真空干燥器;真空保干器
vacuum desorption 真空解吸
vacuum device 真空设备
vacuum dewatering 真空脱水
vacuum dewatering method 真空排水法
vacuum die casting 真空压力铸造
vacuum diffusion 真空扩散;真空浸提
vacuum diffusion pump 真空扩散泵
vacuum diffusion welding 真空扩散焊接
vacuum dilatometer 真空膨胀计
vacuum discharge 真空排放
vacuum dissipator 真空喷雾器
vacuum distance 真空距离
vacuum distillate 减压馏分
vacuum distillation 真空蒸馏;减压蒸馏
vacuum distillation plant 真空蒸馏装置;真空蒸馏设备
vacuum distillation process 真空蒸馏过程
vacuum distillation range 减压馏程
vacuum distillator 真空蒸馏器
vacuum distilling 真空蒸馏
vacuum distilling apparatus 真空蒸馏器
vacuum distilling column 真空蒸馏塔
vacuum distilling flask 真空蒸馏瓶
vacuum down 真空度降低
vacuum drainage method 真空排水法;真空排水(降低地下水位)施工法
vacuum drainage pipe system 真空排水管道系统
vacuum drain method 真空预压排水法
vacuum draw 真空抽吸
vacuum drilling 真空钻进
vacuum drive gyroscope 真空陀螺仪
vacuum dropper 轧花机的气吸式集棉筒
vacuum drum dryer 真空鼓式干燥机;真空转鼓烘燥机
vacuum drum model 真空成型立体模型
vacuum dryer 真空式干燥机;真空干燥器
vacuum drying 真空干燥的;真空干燥(法)
vacuum drying apparatus 真空干燥设备
vacuum drying chamber 真空干燥箱
vacuum drying of timber 木材真空干燥法
vacuum drying oven 真空干燥(烘)箱;真空(干燥)炉
vacuum drying pan 真空干燥罐
vacuum dust catcher 真空除尘器
vacuum dust-collector 真空收尘装置
vacuum dyeing 真空染色
vacuumed clay 已脱气的泥条
vacuum ejector 真空喷射器
vacuum electric(al) furnace 真空电炉
vacuum electric(al) insulation 真空电绝缘
vacuum electron beam welding 真空电子束焊
vacuum embossing 真空压花
vacuum encapsulation 真空封装
vacuum end cap 真空封帽
vacuum engine 真空发动机
vacuum engineering 真空工程
vacuum envelope 真空密封外壳
vacuum environment 真空环境
vacuum epitaxy 真空外延
vacuum equipment 真空设备
vacuum evapo(u)rated film 真空镀膜

vacuum evapo(u)rating 真空蒸镀;真空喷镀
vacuum evapo(u)ration 真空蒸发(作用);真空喷镀;减压蒸发
vacuum evapo(u)ration coating 真空蒸发涂敷金属;真空蒸镀金属;真空蒸镀
vacuum evapo(u)ration technique 真空蒸发技术
vacuum evapo(u)ration technology 真空蒸发工艺
vacuum evapo(u)rator 真空蒸发器
vacuum exhaust 真空排气
vacuum exhauster 真空脱气机
vacuum expander 真空膨胀器
vacuum exposure frame 真空曝光架;抽气晒版机
vacuum extens(i)ometer 真空伸长计
vacuum extract(ion) still 减压蒸馏提取器;真空吸出器;减压抽出器
vacuum extractor 真空提取器;抽真空装置
vacuum extruder 真空挤压机
vacuum extrusion press 真空挤压机
vacuum factor 真空因数
vacuum fan 抽风风机;真空抽气机;真空排气扇;真空风扇;抽气扇
vacuum feed 真空供给
vacuum feed tank 真空给油箱
vacuum feed(-water) pump 真空给水泵
vacuum field 真空场
vacuum filament lamp 真空白炽灯
vacuum fill 真空充填法
vacuum filler 真空填充器
vacuum filling 真空填充
vacuum filling machine 真空填充机
vacuum filter 吸滤器;真空(吸)滤器;真空滤池;真空过滤器;真空过滤机;抽空过滤斗
vacuum filtration 真空过滤(作用)
vacuum filtration process 真空过滤过程;真空过滤法
vacuum filtration yield 真空过滤产水量
vacuum finishing 真空后缩聚
vacuum firing 真空退火;真空烧成
vacuum firing furnace 真空炉
vacuum fittings 真空配件
vacuum flash 真空快速
vacuum flash distillation 真空闪蒸
vacuum flasher 真空闪蒸塔
vacuum flashing 真空闪蒸;减压闪蒸
vacuum flash unit 真空自蒸发装置
vacuum flash vaporization 真空闪蒸
vacuum flash vaporizer 真空闪蒸器;真空快速蒸发器
vacuum flask 真空瓶;热水瓶;保温瓶
vacuum flo(a)tation 真空漂浮法;真空浮选(法)
vacuum flo(a)tation process 真空上浮法;真空气浮法;真空浮选法;减压浮选法
vacuum fluctuation 真空漂移
vacuum fluorescent lamp 真空荧光灯
vacuum flush toilet 真空冲洗大便器
vacuum flush toilet system 真空冲洗式大便器;真空冲洗便池系统
vacuum follower instrument 真空随动装置
vacuum follower pump 真空随动泵
vacuum form 真空模板
vacuum form concrete 真空模板混凝土
vacuum formed 真空成型的
vacuum forming 真空成型(法)
vacuum forming equipment 真空成型设备
vacuum forming machine 真空成型机
vacuum fractionating distilling column 真空精蒸馏塔
vacuum fractionator 真空精蒸馏塔
vacuum frame 真空晒版架
vacuum freeze dryer 真空冷冻干燥器
vacuum freeze drying 真空冷冻干燥;冷冻真空干燥
vacuum-freezing vapo(u)r compression 真空冷冻气压压缩
vacuum-freezing ejector adsorption method 真空冷冻喷射吸附法
vacuum-freezing process 真空冷冻法
vacuum fuel feed 真空供油
vacuum fuel tester 真空及燃油泵输出压力测试仪
vacuum fumigation 真空熏蒸
vacuum furnace 真空炉
vacuum furnace container 真空炉容器
vacuum furnacing 真空炉冶炼;真空炉烧结;真空炉加热
vacuum fusion 真空熔融;真空熔化

vacuum fusion analysis 真空熔融分析
vacuum fusion chromatography 真空熔融色谱法
vacuum fusion gas analysis 真空熔化气体分析
vacuum fusion gas chromatography 真空融熔气相色谱;真空熔融色谱法
vacuum fusion method 真空熔化法
vacuum fusion technique 真空熔融技术
vacuum gas oil 真空瓦斯油
vacuum gas sampler 真空式气体取样器
vacuum ga(u)ge 流体压强计;真空(压力)计;真空规管;真空测定计;真空表;气体压力计
vacuum ga(u)ge control circuit 真空计控制电路
vacuum ga(u)ge for dredge pump 泥泵真空表
vacuum ga(u)ge pipe 真空计管
vacuum ga(u)ge pressure 真空计压力
vacuum gear shift(ing) 真空变速
vacuum generator 真空发生器
vacuum governor 真空调速器;真空调节器
vacuum grease 真空脂;真空润滑
vacuum grinding method 真空球磨法
vacuum grip device 真空吸盘
vacuum group 真空抽气机组
vacuum grown crystal 真空成长晶体
vacuum guide 真空导杆
vacuum guide servo 真空导向器伺服
vacuum gypsum treatment machine 石膏真空处理机
vacuum handling 真空处理
vacuum handling of cement 水泥真空输送
vacuum hardening 真空硬化
vacuum head 吸入侧真空;真空头
vacuum heater 真空加热器
vacuum heating 真空(式)供暖;真空加热;真空采暖
vacuum heating system 真空供暖系统;真空采暖系统
vacuum heat treatment 真空热处理
vacuum high temperature furnace 高温真空炉
vacuum hood 真空罩
vacuum horn 真空报警器
vacuum hose 真空软管
vacuum hot extraction 真空加热提炼
vacuum hot pressing 真空热压成型
vacuum hot stage microscope 真空温台显微镜
vacuum hydroextractor 真空脱水机
vacuum hydrosolver 真空式水溶性胶体搅拌机
vacuum ignition advancer 真空点火提前装置
vacuum impregnated 真空浸渍的
vacuum impregnating 真空油浸;真空浸渍
vacuum impregnation 真空浸透
vacuum impregnation equipment 真空浸渍设备
vacuum incandescent bulb 真空白炽灯泡
vacuum incandescent lamp 真空白炽灯
vacuum indicator 真空(度)指示器
vacuum inductance furnace 真空感应炉
vacuum induction 真空感应
vacuum induction casting 真空感应铸造
vacuum induction furnace 真空感应(电)炉
vacuum induction melting 真空感应熔化
vacuum induction melting furnace 真空感应电熔炉
vacuum induction remelt(ing) 真空感应炉重熔法
vacuum infiltration 真空侵入
vacuuming 真空吸尘;真空处理
vacuuming metalling 真空镀膜
vacuum injection 真空注射
vacuum installation 真空装置
vacuum insulated tank 真空绝热箱
vacuum insulation 真空绝缘;真空隔热
vacuum intake 真空吸料
vacuum intensifier 真空增强器
vacuum interferometer 真空干涉仪
vacuum interrupter 真空隔离开关;真空断路器
vacuum intervalve coupling 真空管间耦合
vacuum investing machine 真空包埋机
vacuum investment casting 真空失蜡浇铸
vacuum ionization chamber 真空电离室
vacuum ionization detector 真空电离检测器;真空电离测定器
vacuum ionization ga(u)ge 真空电离计;电离(式)真空计
vacuum ionization manometer 电离真空压力计
vacuum ionization sensor 真空电离传感器
vacuumization 真空处理
vacuum jacketed column 真空套塔

vacuum jacketed container 真空瓶
vacuum jacketed flask 真空套瓶
vacuum jar 真空瓶
vacuum jar with tap 带塞真空瓶
vacuum jet 真空喷射泵
vacuum jet device 真空喷射装置
vacuum jet package 真空喷射装置
vacuum junction 真空热电偶
vacuum kettle 真空锅;真空釜
vacuum kneader 真空捏合机
vacuum ladle 真空铸罐
vacuum lamp 真空灯(泡);真空白炽灯
vacuum latrine 真空(式)大便器
vacuum lead 抽真空管道
vacuum-leaf filter 叶片式真空过滤器;真空叶滤机
vacuum leak 真空漏泄
vacuum leak detector 真空泄漏检测仪;真空漏气指示器;真空检漏仪
vacuum leak hunting 真空检漏
vacuum leak rate 真空漏气速率
vacuum level 真空能级;真空度
vacuum lift 真空吸盘
vacuum lifter pad 提升用真空吸盘;真空升降机缓冲器
vacuum lifting 真空起重;真空吸附提升法;真空提升
vacuum lifting beam 真空提升梁
vacuum lifting equipment 真空吸盘提升设备
vacuum lifting gear 真空升降装置;真空吊货装置
vacuum lift system of lubrication 真空提升润滑系统
vacuum lightning arrester 真空避雷器
vacuum lightning protector 真空避雷器
vacuum light speed 真空光速
vacuum light velocity 真空光速
vacuum limiter 真空限制器
vacuum line 真空管(线)
vacuum line technique 真空管道操作
vacuum lining 真空衬炉
vacuum loader 真空装货机
vacuum loading 真空装料
vacuum lock 真空闸;真空锁;真空栓
vacuum machine 真空机械
vacuum magnetron sputtering coating glass 真空磁控溅射涂膜玻璃
vacuum man (升运未经养护的)预制板的工人
vacuum manifold 真空支管;真空歧管
vacuum manifold for dust removal 真空除尘器;真空(除尘)歧管;真空除尘集合管
vacuum manometer 真空压力计;真空压力表;真空测压计
vacuum mat 真空(操作)模板;真空吸水式覆盖;真空吸水板;真空吸垫
vacuum measurement 真气测量;真空测量
vacuum measuring instrument 真空测量仪表
vacuum melt 真空熔制
vacuum melting 真空熔融;真空熔炼
vacuum melting method 真空熔炼法
vacuum melting technique 真空冶炼技术
vacuum membrane distillation 减压膜蒸馏
vacuum metallizing 真空金属涂层;真空镀膜(涂装法);真空镀金;真空镀敷金属;真空沉积金属膜
vacuum metallizing coating 真空镀(金属)膜涂料
vacuum metallizing equipment 真空喷涂金属设备
vacuum metallurgy 真空冶金(学)
vacuum meter 真空计;真空表;低压计
vacuum method 真空预压法
vacuum method of drainage 真空排水法
vacuum method of preloading 真空预压法
vacuum method of testing sand 真空砂试验(方)法;砂的真空试验(方)法
vacuum method wellpoint 真空法抽水井点
vacuum mixer 真空搅拌机;真空混合机
vacuum mixing 真空搅拌
vacuum modulator 真空式调节器
vacuum modulator pressure 真空控制液压
vacuum motor 真空马达
vacuum mo(u)lded relief map 真空成型立体图
vacuum mo(u)ld(ing) 真空成(模)(板);真空压塑;真空模塑(法);真空(压塑)成型
vacuum multistage evapo(u)rator 真空多级蒸发器;多级真空蒸发器
vacuum negative pressure 真空负压
vacuum normal pressure method 真空常压法
vacuum nozzle 吸气嘴

vacuum of dredge pump 泥泵真空
vacuum oil 真空油
vacuum oil penetrator for hemp core 麻心真空浸油器
vacuum oil strainer 真空滤油机
vacu(um)ometer 低压计;真空(测定)计
vacuum operated 真空操纵的
vacuum operated clipper 真空驱动式剪毛器
vacuum operated clutch 真空作用离合器
vacuum operated load reducer 真空下降负荷限制器
vacuum operated sheet piler 真空吸盘的垛板机
vacuum operated spark control 真空控制点火定时装置
vacuum optic(al) bench 真空光具座
vacuum oven 真空炉
vacuum oven method 真空炉测湿法
vacuum overhead 真空塔顶馏出物;真空塔顶产物;减压蒸馏塔顶馏分
vacuum package 真空包装
vacuum packaging 真空包装
vacuum packed 真空包装的
vacuum packer for bag 袋真空包装机
vacuum packing 真空填充物;真空填充法
vacuum packing method 真空填充法
vacuum pad 真空夹砖器;真空垫
vacuum pan 真空盘
vacuum panel 真空模(板)
vacuum pan salt 真空锅制盐
vacuum pan water 真空锅水
vacuum paraffin embedding method 真空浸蜡包埋法
vacuum partial condenser 真空(蒸馏塔)部分冷凝器
vacuum passage 真空通路
vacuum pencil 真空笔
vacuum performance 真空性能
vacuum photocell 真空光电管;真空光电池
vacuum photoemissive cell 真空光电辐射元件
vacuum phototube 真空光电元件;真空光电管
vacuum pick-up 真空排粉;真空式检拾器
vacuum pipe(line) 真空管道;真空导管
vacuum pipeline conveyer 真空管道输送机
vacuum pipe-still 减压管式蒸馏装置
vacuum piping 真空管道;真空导管
vacuum piston 真空活塞
vacuum piston plate 真空活塞板
vacuum plant 真空装置
vacuum plate 真空板
vacuum plating 真空电镀
vacuum plenum 真空压力通风系统
vacuum plodder 真空压条机
vacuum plugging kit 真空充填器械包
vacuum pneumatic conveyer 抽吸式气动输送机
vacuum polarization 真空极化
vacuum polymerization 真空聚合
vacuum pot 真空锅
vacuum potting 真空浸渍
vacuum power brake 真空机动闸
vacuum power brake unit 真空增力制动器
vacuum power cylinder 真空动力汽缸
vacuum power gear change 真空助力换挡
vacuum power gear shift 真空助力换挡
vacuum power shift 真空换挡
vacuum preconcentrator 真空预浓缩器
vacuum preloading 真空预压;真空固结法
vacuum preloading method 真空预压法
vacuum prepared sample 真空制备的试样
vacuum preservation 真空保藏法
vacuum press 真空压制成形
vacuum press-in coating 真空浸涂
vacuum pressing 真空压制
vacuum pressing and casting 真空铸造
vacuum pressure 真空压力;真空度;负压
vacuum pressure die casting 真空压铸
vacuum pressure ga(u)ge 真空压力(两用)表;真空压力计;真空压力表
vacuum pressure impregnation 真空压力浸渍;真空压力浸涂(法);真空压力浸漆
vacuum pressure indicator lamp 真空压力指示灯
vacuum pressure process 真空加压法
vacuum pressure pump 真空压力泵
vacuum pressure valve 真空压力阀
vacuum printing frame 真空晒版机;抽气晒版机

vacuum process 真空作业;真空(密实)法;真空处理(法);减压铸造法
vacuum processed 真空加工的
vacuum processed concrete 真空作业混凝土;真空法混凝土;真空处理混凝土
vacuum production 抽成真空
vacuum profile 真空式剖面
vacuum pug 真空挤泥机
vacuum pug mill 真空搅拌机;真空练泥机;真空干碾机
vacuum pug mixer 真空泥料搅拌混合器
vacuum pump 真空抽气机;真空泵
vacuum pump adapter 真空泵接头
vacuum pump compressor 真空泵抽空压气机
vacuum pump for vent 排气真空泵
vacuum pumping 真空排气;抽真空(操作)
vacuum pumping equipment 真空抽气设备
vacuum pumping system 真空抽气系统;抽真空系统
vacuum pump lift 真空泵扬程
vacuum pump line 真空泵管道
vacuum pump oil 真空泵油
vacuum pump power supply 真空泵电源
vacuum pump rocker arm 真空泵摇臂
vacuum pump transporter 真空泵运输机
vacuum purge 真空驱气
vacuum purification 真空净化
vacuum quantometer 真空冲击电流计
vacuum receiver 真空脱气罐;真空接受器
vacuum recharge 真空回灌
vacuum recrystallization 真空精馏
vacuum rectifier 真空整流器
vacuum rectifier tube 真空整流管
vacuum rectifying apparatus 真空精馏装置
vacuum redistillation 真空再蒸馏
vacuum-reduced 降低真空
vacuum reduction 真空还原
vacuum refining 真空精馏;真空精炼
vacuum refining furnace 真空精炼炉
vacuum refrigerating machine 真空致冷机;真空制冷机;真空冷冻机
vacuum refrigerating process 真空冷冻法
vacuum refrigeration 真空制冷;真空冷冻
vacuum regulator 真空调节器
vacuum regulator valve 负压防止阀
vacuum relay 真空继电器
vacuum relay valve 真空式自动转换阀门;真空继动阀
vacuum release valve 真空泄漏阀;真空解除阀
vacuum relief valve 真空卸荷阀;真空(泄)漏阀;真空限制阀;真空释放阀;真空解除阀;真空保压阀;真空安全阀
vacuum removal method 真空排气法
vacuum rerun 真空再蒸馏
vacuum reserve tank 备用真空箱
vacuum residue 减压渣油
vacuum residuum 减压渣油
vacuum resistance 耐真空性
vacuum resistance furnace 真空电阻炉
vacuum retarding mechanism 真空减速机构
vacuum retort 真空蒸馏甑
vacuum return 减压重蒸馏
vacuum returnline device 真空回水装置
vacuum returnline heating system 真空采暖系统
vacuum return line system 真空回水系统
vacuum return pipe 真空回水管
vacuum rig 真空试验台
vacuum rolling 真空轧制
vacuum rotary filter 真空回转过滤器
vacuum safety valve 真空安全阀
vacuum sampler 真空取样管
vacuum sampling method 真空取样法;真空采气法
vacuum sampling tube 真空取样管
vacuum saturated 真空饱和的
vacuum saturating 真空饱和法
vacuum screen printing machine 真空丝网印花机
vacuum seal 真空封接
vacuum-sealed process 薄膜负压造型法
vacuum seal(tight) 真空密封
vacuum seal wax 真空密封蜡
vacuum seamer 真空封罐机
vacuum seasoning 真空干燥(法)
vacuum sensing line 真空信号输送管
vacuum separation 真空分离

vacuum servo 真空自动控制;真空助力
vacuum servo brake 真空助力制动器;真空补充制动器
vacuum servo hydraulic brake system 真空伺服液压制动系统
vacuum settle 真空调整
vacuum sewerage system 真空排水系统
vacuum sewer cleaner 真空吸污车;真空吸粪车
vacuum shelf dryer 真空干燥柜
vacuum shelf tray dryer 真空盘架干燥器
vacuum shutter(ing) 真空模板
vacuum silage 真空青储[贮]
vacuum sintering 真空烧结;减压烧结
vacuum sintering furnace 真空烧结炉
vacuum sizer 真空填料器
vacuum slot 真空吸嘴
vacuum snap-back forming 真空反吸成型
vacuum solar telescope 真空太阳望远镜
vacuum solenoid 真空电磁阀
vacuum solidification 真空固化
vacuum space 真空容积;真空空间
vacuum spark advance mechanism 真空点火提前机构
vacuum spark advancer 真空点火提前装置
vacuum spark control 真空控制点火提前机构;低压控制点火
vacuum spark discharge 真空火花放电
vacuum spark ion source 真空火花离子源
vacuum spark technique 真空火花法
vacuum spectrograph 真空摄谱仪
vacuum spectrography 真空光谱法
vacuum spectrometer 真空分光计
vacuum sphere 真空球
vacuum spray evapo(u)rator 真空喷雾蒸发器
vacuum spraying 真空喷涂
vacuum spray painting method 真空喷涂法
vacuum stability 真空稳定性
vacuum stable 在真空条件下保持性能稳定的
vacuum stage 真空级
vacuum steam distillation 真空蒸汽蒸馏
vacuum steam heating 真空式蒸汽供暖
vacuum steel 真空钢
vacuum still 真空蒸馏器;真空蒸馏釜
vacuum storage 真空储[贮]藏
vacuum stripping system 真空清舱系统
vacuum stuffer 真空充填机
vacuum sublimation 真空升华
vacuum suction 负压吸引
vacuum suction board 真空吸板
vacuum suction drainer vehicle 真空吸水排水车
vacuum suction filter 真空吸滤
vacuum suction plate 真空吸气板
vacuum suction process 真空抽吸成型
vacuum suction roll 真空吸辊
vacuum supply 抽真空
vacuum support 真空支承
vacuum suspended type 真空浮动式
vacuum sweeper 真空吸尘器;真空清扫机;真空除尘器
vacuum switch 真空(控制)开关
vacuum switch gear 真空开关装置
vacuum switch tube 真空开关管
vacuum system 真空设备;真空系统
vacuum system testing 真空系统试验
vacuum take-down 真空牵拉装置
vacuum tank 整流器外壳;真空箱;真空室;真空容器;真空柜;真空罐;真空池;真空槽;减压箱;抽气排出筒
vacuum tank spreader 真空罐式液肥喷洒机;真空罐式洒布机
vacuum tape guide 真空导向器
vacuum technique 真空技术
vacuum technology 真空工艺
vacuum telescope 真空望远镜
vacuum tempering apparatus 真空加湿器;真空回火装置
vacuum test 真空检验
vacuum testing 真空试验
vacuum test machine 真空试验仪
vacuum thermal method 真空热还原法
vacuum thermal reduction 真空热还原
vacuum thermal refining 真空热精炼
vacuum thermobalance 真空热天平
vacuum thermocouple 真空热电偶

vacuum thermometer 真空温度计
vacuum thermopile 真空温差电堆;真空热电堆
vacuum tight 真空密封;真空紧密的
vacuum tight chamber 真空密闭室
vacuum tight container 真空密封容器
vacuum tight furnace cover 真空密封炉盖
vacuum tight housing 真空罩
vacuum tight mo(u)ld 真空模子
vacuum tightness 真空气密性;真空密封;真空密闭;真空度
vacuum tight retort 真空密闭罐
vacuum tight sheath 真空密封外壳
vacuum tight window 真空密封窗
vacuum topping 真空分馏;减压拔顶蒸馏
vacuum tower 真空蒸馏塔
vacuum tower telescope 真空塔式望远镜
vacuum track 真空料道
vacuum transfer 真空输送
vacuum transfer vessel 真空转移容器
vacuum trap 真空凝气瓣;真空冷阱;真空阱;真空隔气具
vacuum treated 真空处理(的)
vacuum treated concrete 真空处理(的)混凝土;真空吸水处理混凝土
vacuum treated steel 真空处理钢
vacuum treating 真空处理
vacuum treatment 真空处理
vacuum triaxial test 真空三轴试验
vacuum triaxial test apparatus 真空三轴试验装置;真空三轴试验仪
vacuum triaxial test device 真空三轴试验装置
vacuum triaxial test unit 真空三轴试验装置
vacuum trip device 低真空保护装置
vacuum truck 真空油槽车;真空吸尘车
vacuum-tube 真空管;电子管
vacuum-tube accelerometer 真空管加速计;电子管加速计
vacuum-tube adapter 电子管适配器
vacuum-tube ag(e)ing 真空管时效
vacuum-tube ammeter 电子管安培计
vacuum-tube amplifier 真空管放大器;电子管放大器
vacuum-tube arrester 真空管避雷器
vacuum-tube bridge 电子管参数电桥
vacuum-tube characteristic 真空管特性
vacuum-tube circuit 电子管电路
vacuum-tube clipping circuit 电子管削波电路
vacuum-tube coating 真空管敷层
vacuum-tube converter 电子管变频器
vacuum-tube curve tracer 电子管特性图示仪
vacuum-tube detector 真空管检波器;电子管检波器
vacuum-tube direct current millivoltmeter 电子管直流毫伏表
vacuum-tube electrometer 真空管静电计;电子管静电计
vacuum-tube frequency converter 电子管变频器
vacuum-tube frequency multiplier 电子管倍频器
vacuum-tube generator 真空管振荡器
vacuum-tube glass 真空管玻璃
vacuum-tube Gm-meter 电子管跨导计;电子管互导测量仪
vacuum-tube hearing aid 真空管助听器
vacuum-tube keyer 真空管键控器;真空管电键器
vacuum-tube keying 真空管键法;电子管键控发报
vacuum-tube mica 电子管云母片
vacuum-tube microammeter 电子管微安计
vacuum-tube millivoltmeter 电子管毫伏表
vacuum-tube modulator 电子管调制器
vacuum-tube noise 真空管噪声
vacuum-tube oscillator 真空管振荡器;电子振荡器
vacuum-tube powermeter 电子管功率计
vacuum-tube pulser 电子管脉冲发生器
vacuum-tube receiving set 电子管接收机
vacuum-tube rectifier 真空管整流器;电子管整流器
vacuum-tube rejuvenator 电子管复活器
vacuum-tube relay 真空管继电器;电子管继电器
vacuum-tube socket connection 电子管管座接线
vacuum-tube switch 真空管开关
vacuum-tube tester 电子管测试器
vacuum-tube test set 真空管测试器
vacuum-tube transducer 真空管换能器
vacuum-tube transmitter 真空管发送器;电子管

发送器
vacuum-tube-type high-frequency converter 电子管式高频变换器
vacuum-tube video amplifier 电子管视频放大器
vacuum-tube voltage regulator 真空管电压调整器;电子管式稳压器
vacuum-tube voltammeter 电子管伏安计
vacuum-tube voltmeter 真空管电压表;电子管电压计;电子管伏特计;电子管电压表
vacuum-tube wavemeter 电子管波长计
vacuum tubing 真空管
vacuum tumbler dryer 真空转筒式烘干器
vacuum tweezer 真空镊子
vacuum type automatic bottling machine 真空式自动装瓶机
vacuum type drum filter 真空转鼓过滤机
vacuum type insulation 真空式热绝缘
vacuum type pneumatic micrometer 真空式气动测微计
vacuum type tensiometer 真空式张力计
vacuum type transfer tongs 真空转移钳
vacuum ultraviolet 真空紫外线
vacuum ultraviolet photon 真空紫外光子
vacuum ultraviolet radiation 真空紫外辐射
vacuum ultraviolet region 真空紫外线区域
vacuum ultraviolet spectrometer 真空紫外光谱仪
vacuum ultraviolet spectroscopy 真空紫外光谱(学)
vacuum under nappe 水舌下的真空
vacuum unit 真空装置;真空元件
vacuum unloader 真空式卸料机;气力卸载机
vacuum unloading 真空卸料
vacuum unloading system 真空卸料系统
vacuum unload machine 真空卸载机
vacuum up 真空度升高
vacuum valve 真空管;真空阀;负压活门
vacuum valve circuit 真空管电路
vacuum valve receiver 电子管收音机
vacuum vaporization 真空汽化
vacuum vapo(u)r 蒸气供暖
vacuum vapo(u)r deposition method 真空镀积法
vacuum vapo(u)r plating 真空蒸发镀膜
vacuum variable capacitor 真空可变电容器
vacuum ventilation 真空通风
vacuum ventilation system 真空(式)通风系统
vacuum vessel 真空室;真空容器;真空瓶
vacuum volatilization 真空挥发
vacuum watch 真空表
vacuum water sucker 真空吸水机械
vacuum wavelength 真空波长
vacuum welding 真空焊接
vacuum welding by electron beam 电子束真空焊接
vacuum well 真空井
vacuum well point 真空井点
vacuum wetting 真空湿润
vacuum window glazing 真空玻璃
vacuum windshield cleaner 真空驱动式风窗刮水器
vacuum windshield wiper 真空刮水器
vacuum windshield wiper control 真空控制风挡刮水器
vacuum windshield wiper pivot kicker 真空风挡刮水旋轴
vacuum windshield wiper swing arm 真空风挡刮水摆臂
vacuum wiper (汽车)真空式(窗玻璃)刮水器
vacuum xanthate mixer 真空黄化混合机
vacuum X-ray spectrometer 真空X射线分光计
vadose 渗流
vadose area 渗流区
vadose belt 上层滞水带
vadose epoch 渗流期
vadose feature 渗流地形
vadose karst 包气带岩溶
vadose pisolite 渗流豆粒
vadose region 渗流区
vadose solution 渗溶
vadose spring 渗流泉;上层滞水泉;包气带泉
vadose through dike 坝体渗流
vadose water 循环水下水;悬着水;重力水;含气带水;土壤水;渗流水;上层滞水;地下水位线之上的地下水;包气带水
vadose-water discharge 渗流水流量;渗流水出流;充气带滞水溢出;非饱和区土壤水流量

vadose water zone 中间带;渗水带【地】
vadose zone 非饱和层;含气层;通气层;渗流区;渗流带;渗流层;上层滞水带;上层渗水带;包气带
vaesite 方硫镍矿
va excape clause 反强卖实例
vaeyrrynenite 红磷锰铍石
vagabond current 地电流
vagaries of climate 多变的气候;变化无常的气候
vagile aquatic animals 漫游水生动物
vagile-benthon 河底漫游动物
vagile diatom 漫游硅藻
vagile organism 底栖游牧生物
vagina 胸像台座的上部
vaginula 基鞘
vagrant benthos 游移底栖生物;底栖游移生物
vagrant colo(u)rs 游彩
vagrant current 无定向流
vague copy 模糊不清的复制本
vague image 模糊像
vaguio 碧瑶风
vail 遮掩物;遮掩现象;低垂
val 瓦尔
valais wind 瓦拉伊斯风
valance 窗帘(盒);织物装饰条;帐幔;帘帐顶饰;花布(用作窗帘等);窗帘板;窗户顶部短帘
valanced 装有窗帘框架的;装有布帘的;装有帷幔的
valance lighting 窗帘盒顶部泛光照明;(带屏蔽框和屏蔽膜的)照明灯管;槽灯
Valanginian stage 凡兰吟阶【地】
vale 断裂谷
valence 原子价;化合价;评价强度
valence band 价带
valence charge 价电荷
valence crystal 共价晶体;价键晶体
valence effect 价效应
valence electron 价电子
valence energy coefficient 价能量系数
valence isomer 价异构体
valence model 化合价模型
valence mumber 价数
valence rule 价数规划
valence shell 价壳层
valence state 原子价态
valence vibrations 价振动
valencianite 冰长石
valency = valence
Valensi system 瓦伦西系统
valent 化合价的
Valentian series 瓦伦特统【地】
Valentian stage 瓦伦特阶【地】
valentinite 锑华
valent weight 当量
valerate 戊酸盐
valerene 戊烯
valerianate 戊酸盐
valerianic acid 戊酸
valeric chloride 戊酰氯
valeric 戊酸
valerolactam 戊内酰胺
valerolactone 戊内酯
valeronitrile 戊腈;丁基氰
valeryl chloride 戊酰氯
valethamate bromide 苯戊溴铵
valet room 随从室
valet towel holder 毛巾衣物架
valgoid 外翻状的
Valhalla (北欧神话中的)沃丁神接待战死者英灵的殿堂
valid address 有效地址
valid argument 有效论证
valid as form 自……起生效
validate 确认;使有效;使生效
validate the claim 使索赔有效
validating product 确认样品
validation 验证;确认
validation error 合法性错误
validation of a user's identity 确认用户身份的合法性
validation of the model 模型合理性论证;模型合理化论证
validation of title 法权验证
validation of training 培训效果检验
validation phase 审批阶段
validation procedure 验证程序

validation specification mode 核实规范说明方式
valid ballot 有效票
valid certificate 有效证书
valid combination 有效组合
valid conclusion 有效结论
valid contract 有效合同;有效约
valid documents for evidence 有效照明文件
valid drawing area 有效作图区
valid evidence 确凿的证据
valid formula 有效公式;永真公式
valid genus 有效属
validity 有效(性);有效度;永真性;真实性;合法性;确实性
validity check 效果检验;有效性检验;有效性检查;真实性检验;确实性检查
validity examination 有效性检验
validity limit 有效极限
validity of accelerated weathering 加速耐候试验的真实性
validity of a claim 索赔的有效性
validity of an argument 变元的有效性
validity of an award 裁决的有效性
validity of assumptions 假定的真实性;臆测确实性
validity of bids 投标有效期
validity of contract 合同的有效性;契约效力
validity of credit 信用证的有效性
validity of formula in fuzzy logic 模糊逻辑公式的有效性
validity of prediction 预测的正确性
validity of quasi-experimental analysis 拟实验分析的有效性
validity of tenders 投保期限;投标有效性;投标(书)有效期
validity of ticket 客票有效期限
validity of treaty 合同效力
validity period 有效期
validity problem 永真问题
validity rate 正确率
validity stamp 有效印章
validity test 有效测试
validity to bids 投标有效期
valid memory access 有效存取
valid memory address 有效存储地址
valid name 有效名
valid obligation 有效债务
valid operation 有效操作
valid parity check 有效奇偶校验
valid period 有效期限
valid period of standard 标准龄期
valid period of the offer 报价有效期
valid time 有效预见期
valinche 吸量管
valise 旅行袋
vallate papilla 筑垒术;(古建筑)壁垒
vallation (古建筑)壁垒;筑垒术
valleriite 墨铜矿
Vallette method of gap grading 瓦利特(混凝土)配合比设计法;瓦利特间断级配法(混凝土骨料级配法)
valley 谷值;沟谷;屋面天沟;屋顶排水沟;山谷;凹陷;凹部
valley area 流域面积
valley axis 河谷轴线
valley basin 谷盆
valley beam 檐槽梁托架
valley board 屋面斜沟支承板;斜(天)沟底板;屋顶斜沟(排水槽)底板;天沟底板
valley bog 低沼带(地);低位沼泽
valley bottom 谷底
valley brake 峰下缓行器
valley breeze 谷(微)风;
valley bridge 河谷桥梁
valley bulge 河谷凸起
valley catchment area 流域
valley channel 凹槽;天沟檐槽
valley clay roof(ing) tile 天沟瓦;斜沟陶土瓦
valley consumption (交通)最小负荷;河谷耗水量
valley consumptive use 河谷耗水量
valley contour 河谷形态
valley contraction 河谷缩窄
valley cross-section 山谷(横)断面;河谷(横)断面
valley current 最小电流
valley cut 天沟端瓦
valley dam 峡谷坝;河谷坝

valley deposit 河谷沉积物
valley depth 谷深
valley edge 谷缘
valley element 河谷要素
valley environment 流域环境
valley fen 谷沼泽地；河谷沼泽
valley fill 谷底堆积物；河谷填积；河谷堆积
valley-fill trap 河谷填充圈闭
valley flak trap 谷翼圈闭
valley flank 谷边；山谷边坡
valley flashing 斜天沟泛水；天沟披水；天沟防雨板；天沟泛水
valley flashing piece 斜沟泛水；天沟泛水
valley flat 谷地小平原；谷底小平原；谷底平地；漫滩地；河漫滩；河谷边滩；海漫滩
valley flat facies 河漫滩相
valley flat sediment 谷底沉积物；河谷沉积
valley floor 谷底
valley floor plain 谷底平原；河谷洪泛平原；河谷边滩
valley flow 低峰值流量
valley fog 谷雾
valley form 河谷形状；河谷地形；河谷地貌
valley formation 河谷形成
valley girder 天沟梁
valley glacier 谷冰川；高山冰川；山谷冰川
valley gradient 河谷坡降；河谷坡度；河谷比降
valley gravel 河谷砾石
valley-gravely placer 河谷砂矿
valley gutter (上大下小的)斜天沟；斜沟槽；阴戗天沟；天沟槽
valley harnessing 流域治理
valley head 谷源；谷头
valley iceberg 船坞形冰山
valley-in-valley 谷中谷；叠谷；双循环谷
valley-in-valley structure 谷中谷构造；叠谷构造
valley jack 斜承重椽；天沟次椽木；斜沟小椽条
valley jack rafter (四坡屋顶的)面坡椽；斜承重椽；天沟次椽条；斜沟小椽条；斜沟(处抬高的)椽(子)；斜天沟椽
valley length 谷长度
valley line 沿溪线【道】；最深谷底线【地】；谷线；河(流)谷(底)线；河谷线；合水线；深泓线；山谷线；谷(底)线
valley loess 山谷黄土；谷地黄土
valley margin 河谷边缘
valley meander 河谷曲流
valley migration 河谷迁移
valley of corrugation 瓦楞槽
valley of elevation 隆起谷；背斜谷
valley of loessland 黄土区河谷
valley of subsidence 向斜谷；沉降谷
valley pattern 谷型；河谷形(式)
valley plain 河谷平原
valley point 谷值点；谷点
valley point current 谷值电流
valley point voltage 谷值电压
valley post 天沟支柱
valley precipitation 山谷凝雾；山谷降水
valley project 流域开发项目；流域开发工程；流域规划
valley rafter 天沟椽；斜角椽；斜沟椽；斜沟(处的)椽(子)；(四坡屋顶的)沟椽
valley relief effect 山谷地形影响
valley roof 带斜沟的屋面；带天沟的屋面；带天沟的屋顶；有天沟的屋顶；带斜沟屋顶；带排水沟屋顶
valley roof tile 屋面天沟瓦
valley route 山谷线
valley sediment 河谷沉积物
valley sediment deposit 河谷沉积物
valley sediments 河谷泥沙
valley shape 河谷形状
valley-shaped warm anomalies 谷状暖异常
valley shifting 河谷迁移
valley shingle 天沟屋面板；斜沟木瓦
valley shoulder trap 谷肩圈闭
valley side 谷边
valley side batten 天沟两侧挂瓦条
valley side slope 谷边坡度
valley side trap 谷侧圈闭
valley sink 谷中穴；谷形洼地；长条形坑穴
valley slab 天沟板
valley slope 天沟坡度；谷坡；河谷坡降；河谷坡度；河谷比降
valley slope rock 谷壁岩石

valley soaker 天沟平口泛水
valley soffit 天沟底面
valley soil 谷地土壤；谷底土壤
valley-span antenna 山谷天线
valley spring 谷(边)泉；侵蚀泉
valley storage 谷槽蓄水；流域蓄水量；河谷蓄水(量)；河槽蓄水
valley system 谷系
valley temple 河谷庙
valley terrace 河谷台地；河谷阶地
valley tile 槽瓦；斜沟瓦；天沟瓦；槽形瓦
valley tolerance 凹度容差
valley topography 河谷地形
valley tract 河谷地区
valley train (狭)谷边碛；谷碛
valley trenching 河谷下切
valley type debris flow 河谷型泥石流
valley wall 谷壁
valley way 山谷道路
valley-wide flood 全流域性洪水
valley width 谷宽；河谷宽度
valley wind 谷风
Vallez filter 瓦利兹过滤机
Vallis Alpine 阿尔卑斯月谷
Vallis Inghirami 英希拉米月谷
Val mineral separator 维尔型选矿机
valonea 橡碗
valonia 橡碗
Valoniaceae 法囊藻科
valorem 按价
valorize 限价；政府限价
valproic acid 丙戊酸
valpromide 丙戊酰胺
Valray 瓦尔莱合金
valsoid 华尔兹形的
Valton floor 瓦尔登地板
Valton girder 瓦尔登大梁
valtropine 瓦托品
valuable advice 宝贵意见
valuable cargo 有用物资；贵重货(物)；高价货物
valuable clause 价值条款
valuable constituent 有价成分
valuable content 有价成分含量
valuable deposit 有用矿物床
valuable discovery 有价值的发现
valuable goods 贵重货(物)
valuable information 有价值的情报
valuable mineral 有用矿物
valuable papers 有价证券
valuable rock 有用矿石
valuables 贵重物品
valuable suggestion 有价值的建议；宝贵意见
valuation 估价；计价；评价
valuation account 计价账户；评价账户
valuation after release system 先放后核制
valuation allowance 计价备抵；备抵估价
valuation at cost 成本估价法
valuation at cost market 按成本的市价估价
valuation at market price 按市场价格估价
valuation charges 从价收费；超值费
valuation clause 船舶估价条款
valuation form 货物估价单
valuation judgment 价值判断
valuation method of depreciation 估价折旧法
valuation of fund application 资金运用的估计
valuation of geologic(al) variable 地质变量的取值
valuation of goods produced 货物估价
valuation of import 进口估价
valuation of insurable interest 保险权益定值
valuation of inventory 存货估价
valuation of leisure time 空闲时间的估价
valuation of property 财产估价
valuation of skilled resource flow 人才流动的评价
valuation of special drawing right 特别提款权估值
valuation of variation 变更的估价
valuation paper 货物估定价格单
valuation principle 估价法则；估价原理
valuation process 评价程序；赋值过程
valuation reserves 估价准备
valuation ring 赋值环
valuation vector 赋值向量
valuation work in process 在制品估价
valuator 估价人；估价者；赋值设备；标量设备
valuator device 求值设备；赋值设备；标量设备

valuator mode 定值设备操作方式
valubreeder 钍铀增殖堆
value 价值；数值；渗透性值
value-added 增值；附加值
value-added by manufacture 加工增值
value-added carrier 增值载波通信；加值载波器
value-added index 增值指数
value-added network 附加价值网络业务；增值网络
value-added of railway transport 铁路运输增值
value-added rates 附加价值率
value-added service 增值业务
value-added statement 增值表
value-added system 增值系统
value-added tax 增值税
value-added tax function 加值功能
value-added tax system 增值税制
value after the taking 土地征用后价值
value analysis 工程经济分析；价值分析
value as in original policy 按原保险单价格
value assessment 价值评估
value assurance 价值保证
value at cost 按成本定价
value at the frontier 边境价值
value barter trade 金额易货贸易
value basis 价值基础
value before the taking 土地征用前价值
value-begetting process 价值形成过程
value bill 估价单；价值单
value call 值调用；调值
value class 数值等级
value coefficient 价值系数
value composition 价值构成
value-composition of capital 资本的价值构成
value contrast 价值对比
value control 工程经济分析；价值控制
value-cost contract 建筑承包补偿合同；建筑承包补偿合约
value-creating substance 形成价值的实体
value criteria for environment(al) decision 环境决策价值准则
valued 贵重的
value date 结算日；起息日
value declared 申报价值
value determination 价值确定
value distribution 值分布
value distribution theory 值分布理论
valued policy 确定保险单；定值保险单
value engineering 工程经济学；工程经济分析；价值工程(学)
value engineering approach 价值工程方法
value engineering discipline 价值工程规划
value engineering philosophy 价值工程原理
value engineering program(me) 价值工程规划
value expression 价值表现
value for coefficient of variation 变异系数值
value forecasting 价值预测
value for interpolation 插入数值
value-free 不涉及价值
value function 价值函数
value function over time stream 在时间流上的价值函数
value group 值群
value harmony 明度谐调
value improvement 价值改善
value in account 账款
value in collection 托收价值
value index number 值指数；价值指数
value indicator 价值指标
value in pledge 抵押品的价值
value in security 抵押品的价值
value insured fee 保价费
value insured transport 保价运输
value integration 数据积分法
value in use 使用价值
value judgment 价值判断
valueless 没有价值的；无价值的
value number 实价率；计算值接收数
value of a day's labo(u)r 日劳动价值
value of a fuzzy game 模糊对策值
value of a game 对策的值
value of a hiding strategy 隐藏策略的值
value of all stocks 全部存货的值
value of assessment 课税价值
value of assimilative capacity of sewage efflu-

ent 污水处理后出水同化容量值
value of assimilative capacity of wastewater effluent 污水处理后出水同化容量值
value of atmospheric pressure 大气压力值
value of business sale 商业销售价值
value of capital stock 股本值
value of cargo 船货价值
value of chromatism 色差值
value of coalition 联盟的值
value of colo(u)r 彩色浓淡程度
value of commodity 商品价值
value of conditional simulation 条件模拟值
value of construction output 建筑业产值
value of construction work done as a subcontractor 分包者建筑完工价值
value of consumer services 服务消费品价值
value of credit 信贷值
value of currency 币值
value of demand 需要价值
value of diminishing of fertility 土壤肥力递减值
value of division 分划值;分格值
value of drawdown 水位降深值
value of drawdown designed 设计的水位降深值
value of drift 漂移值
value of drift of conditional simulation 条件模拟漂移值
value of drift of non-conditional simulation 非条件模拟漂移值
value of each contribution 投资价值
value of every commodity 每一商品价值
value of exchange 交换价值;交换价格
value of expression 表达式之值
value of extreme high contiguous deviation 特高邻差值【测】
value of fuel stocks 燃料库存价值
value of game 对策值;竞赛值
value of geologic(al) fruit 地质成果价值
value of geostress 地应力值
value of goods on the way 在途货物价值
value of gravity 重力值
value of human capital 人力资本的价值
value of hydraulic jump 水跃值
value of import and export 进出口贸易额
value of imports 进口额
value of imports by country of origin 进口货物国别价值
value of inferior boundary 下限值
value of insurable interest 保险价值
value of insurance 保险价值
value of labo(u)r power 劳工价值
value of land 土地价值
value of latitude 纬度值
value of machinery 机械设备价值
value of magnification 放大值
value of materials in stock 原材料库存价值
value of merchandise production 商品产值
value of money 现金价值;货币价值
value of new construction put in place 新建筑完工量价值
value of non-conditional simulation 非条件模拟值
value of output per unit of labor 单位劳动产值
value of purchase 收购额
value of residual volumetric(al) deformation 容积残余变形值
value of risk 风险代价;保险值
value of safe water pressure in the early days 早期安全水压值
value of salvaged materials 废物利用的价值
value of series 级数值;级数和
value of service 服务价值
value-of-service consideration 服务价值研究
value of set(ting) 调整值;安装值
value of sewage irrigation 污水农灌价值
value of simulation 模拟值
value of snow load 雪荷载值
value of stale hydrogen isotope in water 水的氢稳定同位素值
value of stale oxygen isotope in water 水的氧稳定同位素值
value of standard sample 标准样数值
value of statistical indicant 标志值
value of statistical item 标志值
value of stock 存货值
value of swelling shrinkage of foundation soil 膨胀土地基的胀缩量
value of symbol 符号值
value of the product 产品价值
value of the reply paid voucher 预付回报凭单的款项
value of the stock of assets at constant price 按不变价格计算的资产现有价值
value of the turn 周值
value of threshold 阈值
value of time saved 节省的时间价值
value of total cost 全部费用价值
value of trade 贸易额
value of uncompleted construction 未完工程价值
value of unit 数值单位
value of water level rising again 地下水位回升值
value of water pressure decrease 水压降低值
value of water temperature drop per unit distance 单位距离的水温下降值
value of wholesale trade 批发贸易额
value of widening the horizontal distance of the inner rail and outer rail on curvey within structural clearance 建筑接近限界曲线内、外侧水平加宽值
value of wind load 风荷载值
value on 签发
value on demand 需要价值
value parameter 值参数
value part 值部分
value per acre 每英亩收获量价值
value policy 实价保险
value produce 价值产品
value-put-in-place estimate 由定价估计价值
valuer 估价员;评价者
value received (支票)如数收讫
value relation 价值关系
value research 价值研究
value revolution 价值革命
values at sheet corners 图廓点坐标值
value securities 有价证券
values of amplitude factor 振幅因子值
values of definite magnitude 一定量的价值
values of dip angle of interface 界面段倾角值
values of error limit 误差限制值
values of error vector 误差向量值
values of factor K K系数值
values of geographic(al) coordinates 地理坐标值
values of the structural parameters 结构参数值
values of velocity factor 速度因子值
values of water quality 水质值
values of waveform of theoretic(al) record 理论记录波形值
value substitute 价值补偿
value substitution 值替换
value system 价值系统;价值观
value system design 价格系统设计
value theory 价值学说
value today 今日起息;即日结算;即日收;当日交割;当日交付
value-to-service 服务价值
value trade-off 价值权衡
value view of ecologic(al) economics 生态经济价值观
value-weight ratio 价值重量比率
valuewise independence of performance variable 操作变量的值态无关
value zero 零值
valuta 可使用的外汇总值
valut vertex 拱顶端
valvate 瓣裂的
valve 裂片;活门;活瓣;壳瓣;气门;电子管;阀(门);瓣膜
valve acceleration 阀加速度
valve action 整流作用;活门作用;阀作用
valve actuating gear 活门执行机构;活门活动装置;(发动机)气门传动机构;气阀机构
valve actuation 阀动
valve actuator 阀执行器;阀动器
valve adapter 电子管转接器
valve adjuster 阀门调节器
valve adjusting 气门调节
valve adjusting ball stud 气门调节圆头螺钉
valve adjusting nut 阀调准螺母
valve adjusting screw 气门间隙调整螺钉
valve adjusting washer 可调式阀座垫圈;阀座调节垫圈
valve adjustment screw 阀门调节旋钮
valve air hammer 有阀冲击器
valve air relief 自动放气阀
valve amplifier 电子管放大器
valve and valve gear 阀及阀动装置
valve arrangement 活门装置;气门布置
valve arrester 电子管避雷器
valve assembly 闸板组件
valve at inlet to penstock 压力钢管进口阀
valve at turbine inlet 水轮机进口阀
valve authority 阀权度
valve bag 包装纸袋;自封(口)袋;阀密封袋
valve bag filling machine 自封袋装填机;闭式袋包装机;灌口自封袋包装机
valve bag opening 自封袋插口
valve bag packer 自封袋包装机
valve bailer 带阀抽砂筒
valve ball 阀球
valve barrel 阀套;阀缸
valve base 阀座
valve base ga(u)ge 管底规
valve base pin 电子管管脚
valve beam 阀门梁
valve block 给油阀组;给水阀组;控制部件;阀组箱;阀体;阀锁
valve body 阀体;阀身
valve body attachment 阀体连接头
valve body cap screw 阀体盖螺钉
valve bonnet 阀罩;阀帽
valve bore 阀门孔
valve bounce 阀跳
valve bowl 阀箱
valve box 活门室;气门室;阀(门)箱;阀(门)室;阀盒
valve box cleaner 阀盒清扫器
valve bronze 阀青铜
valve bucket 气瓣活塞
valve buckle 阀框;阀的紧固螺栓
valve buffer 滑阀缓冲器
valve bulb wall 电子管壁
valve bush 阀衬
valve bushing 阀衬;阀套;阀门导管
valve cage 滑阀匣;气阀室;阀箱;阀盒
valve cage washer 阀柜垫圈
valve cam 阀凸轮
valve cap 气门盘;阀门顶;阀盖
valve case 阀芯座
valve casing 阀箱;阀壳
valve chamber 活门室;气门室;阀室;阀门井
valve characteristic 电子管的特性
valve checker 电子管检验器;电子管测试器
valve chest 滑阀胸;滑阀腔;阀箱;阀室;阀壳体
valve chest gasket 阀柜垫垫片
valve chest gland 阀柜压盖
valve circle 阀圆
valve circuit 阀门控制系统
valve clack 阀舌;阀瓣
valve clearance 气门间隙;阀余隙;阀门间隙
valve clearance adjustment 气门间隙调整;阀隙调整
valve closet 抽水马桶冲洗阀;(杠杆)阀冲水厕所
valve-closure member 关阀器
valve cock 龙头阀;阀栓
valve collar 阀(门)环
valve complement 电子管组
valve cone 阀锥(体)
valve conic(al) face 气阀圆锥面
valve control 阀(门)控制
valve control amplifier 阀门控制放大器
valve controlled 阀门控制的
valve-controlled servo 阀门控制伺服机构
valve controller 阀门控制器
valve control mechanism 阀门控制机构
valve control pillar 阀门操纵杆
valve converting tool 换阀工具
valve cord 阀绳
valve core 气门芯;阀芯
valve core housing 气门芯套
valve core removing tool 阀芯拆卸工具
valve cotter 气阀制销
valve coupling 电子管耦合;带止逆阀接头;阀联结
valve cover 阀套;阀盖
valve cover base 阀盖底(部)

valve cover gasket 阀盖垫密片
valve cover plate 阀室盖
valve cup 阀座
valve cup grab 阀座打捞器
valve cup ring 阀座环
valve current 电子管电流
valve current divider 换流阀分流器
valve cutter 气门刀具
valve cylinder 阀筒
valved cracker cell 阀控裂解炉
valve deck 阀盖
valve designation system 电子管型号命名制度
valve detector 电子管检波器
valve diagram 阀动图
valve diameter 阀径
valve dimension 电子管外形尺寸
valve discharger 阀门式卸料器
valve disk 阀盘;阀叶;阀碟;气门头
valve driving arm 阀驱动臂
valved window 有气门的窗;落地窗;玻璃窗
valve effect 单向导电性;阀效应
valve elbow 带阀弯头
valve electrocardiograph 电子管式心动电流描记器
valve electrode 整流电极;电子管电极;阀电极
valve envelope 电子管外壳
valve extractor 气门拆卸器;起阀器
valve face 阀面
valve face angle 气门工作面角度
valve facing 磨阀面
valve filament circuit 电子管灯丝电路
valve fishing tool 气门吊装工具
valve flap 阀舌;阀盘;阀瓣
valve flow performance 活门流量性能
valve follower 气门推杆;阀推杆
valve for air pump 气泵阀
valve for general use 通用阀
valve for vehicle tire 车辆轮胎气门
valve gallery 阀门廊道;阀廊
valve galvanometer 电子管检流计
valve gapper 气门间隙调整器
valve gate 闸板
valve-gate nozzle 阀门喷嘴
valve gear (发动机)气门传动机构;阀(动)装置;阀动齿轮
valve-gear link 阀装置连接
valve generator 电子管振荡器
valve grinding 磨阀
valve grinding compound 磨阀物;阀研磨剂
valve grinding machine 阀门磨床
valve grinding sand 磨阀砂
valve guard 阀柱护套;阀片升程限制器;阀门限程器;阀挡
valve guide 气阀导管;气门导管;阀门导管;阀导(承轨);阀导承
valve guide bushing 阀导衬套
valve guide cleaner 气门导管清扫器
valve guide clearance 阀片升程限制器升限;阀挡间隙
valve guide expansion reamer 阀导承可胀铰刀
valve guide ga(u)ge 气门导管测量表
valve guide puller 气门导管拆卸工具
valve guide reamer 阀导承铰刀
valve guide remover 气门导管拆卸器
valve guide retainer 阀导承护圈
valve guide seal 气门导管密封装置
valve guide shoulder 气门导管凸肩
valve gumming 阀面填缝
valve handle 阀门手轮;阀杆;阀柄
valve head 开关头;气门头;阀头;阀盖
valve heat dissipation 电子管热扩散
valve heater 电子管加热器
valve holder 管座;阀座
valve hood 阀帽;阀盖
valve house 开阀室;阀室
valve housing 阀箱
valve housing cover 阀室盖
valve housing cover packing 阀室盖填密
valve hummer 电子管蜂音器
valve-in-block engine 汽缸上装气门的发动机
valve-in-head 顶置气门
valve-in-head engine 顶置气门发动机
valve-in-head scavenging 头阀扫气;顶置阀扫气
valve injection 阀注射
valve in landers 装舱阀

valve inner chamber 阀内室
valve inner spring 阀内簧
valve input capacitance 电子管输入电容
valve input conductance 电子管输入电导
valve input impedance 电子管输入阻抗
valve installing tool 气门安装工具
valve integrator 电子管积分机
valve jacket 阀套;阀盒
valve key 气门扳手;气阀制销;阀钥匙;阀(门)键;阀门扳手
valve knock 阀敲击
valve lag 活门迟关;气门迟关
valve lagging 阀迟关
valve land 阀面
valve lap 阀余面
valve lash 气门间隙
valve-lash control 气门间隙调整
valve lead 导气程
valve leakage 阀漏失
valve leather 阀皮
valveless 无活瓣的;无气门的;无阀门的
valveless air hammer 无气门冲击器
valveless amplifier 无电子管放大器
valveless engine 无气门发动机;无阀式发动机
valveless filter 无阀滤池
valveless filtering pool 无阀滤池
valveless gravity filter 无阀重力滤池
valveless metering pump 无阀计量泵
valveless motor 无气门发动机
valveless pump 无阀泵
valveless tire 无气门轮胎
valve lever 气门杆;阀(门)杆;阀(门)操纵杆
valve lever bushing 阀杆衬套
valve lever pin 阀杆销
valve lift 阀门升程;阀门开启高度;阀开度
valve lift area 阀片升程面积
valve lift curve 阀升程曲线
valve lift diagram 阀升距图
valve lifter 气门拆卸工具;阀挺器;阀挺杆
valve lifter guide 起阀器导管
valve lifter roller 起阀器滚子
valve lifter spring 起阀弹簧
valve lifting cam 分配轴凸轮
valve lift pin 阀顶针
valve liner 阀衬
valve link 气门杆;阀杆
valve location 气门的布置;阀位
valve loss 阀面损耗;阀门泄漏
valve management program(me) 阀门操纵程序
valve manipulation 阀门操纵
valve mask 阀门罩
valve metal 阀用铅锡黄铜
valve microphonic effect 电子管颤噪效应
valve milling machine 阀门铣床
valve motion 阀动
valve movement 活瓣运动
valve needle 阀针
valve noise 电子管噪声;阀门噪声
valve nomenclature 电子管命名法
valve nominal capacity 电子管额定容量
valve nut 活门螺母
valve of cargo pipeline 货油管阀门
valve of communicating branch 交通支瓣膜
valve oil 阀油
valve oil can 带阀油壶
valve oil shield 阀防油罩
valve opening 活门开度;阀通口;阀门开(启)度;阀孔
valve operating mechanism 阀操作机构
valve operating piston 阀动活塞
valve operating stud 阀的操作钮
valve operation mechanism 阀动机构
valve operation test 阀动作试验
valve outer chamber 阀外室
valve outer chamber cover 阀外室盖
valve outer spring 阀外簧
valve overlap 气阀重叠
valve packing nut 阀填密螺母
valve pad 阀垫
valve panel 管座
valve parameter 电子管参数
valve parasitic capacitance 电子管寄生电容
valve passage 瓣口;阀门通道
valve passage resistance 阀道流动阻力

valve performance 阀性能
valve phase indicator 阀(门)相位指示器
valve pin 真空管脚;气阀制销;阀销
valve pipece 阀座
valve piston 阀门柱塞;阀门活塞
valve pit 阀井
valve plate 配流盘;浮阀塔板;阀片;阀板
valve play 气门间隙;阀隙
valve plug 气门顶杆;阀塞
valve plunger 阀柱塞
valve pocket 阀套;阀腔
valve-point 阀点
valve-point curve 阀点曲线
valve-point internal efficiency 阀点内效率
valve-point performance curve 阀点性能曲线
valve port 气门口;阀口
valve port diameter 阀芯口径
valve position 阀位置
valve positioner 阀(门)定位器
valve position-flow characteristic 阀位—流量特性
valve position indicator 阀位指示器
valve positioning servo loop 阀门定位伺服回路
valve position switch 阀定(位)开关
valve position transducer 阀位发送器
valve potentiometer 真空管电势计
valve pot type piston pump 阀箱式活塞泵
valve probe 电子管探示器
valve products 阀类产品
valve pump 阀面泵;阀式泵
valve rangeability 阀可调节范围;阀调节能力
valve reactor 换流阀电抗器
valve receiver 电子管收音机;电子管接收机
valve rectifier 电子整流管;电子管整流机
valve redundancy 阀的冗余
valve refacer 阀面磨光机;阀门磨光机;磨气门机
valve-refacer machine 磨阀面机
valve refacing machine 磨阀面机
valve regulation 阀调节
valve relay 电子管继电器;阀继电器
valve removal 卸阀器
valve remover 气门拆卸工具;起阀器;除气门机
valve reseater 阀门复位器
valve reseating 阀座重修整
valve reseating tool 装气门机
valve resistance 阀流阻力
valve resistance arrester 阀电阻避雷器
valve retainer cotter pin 气门弹簧托盘锁销
valve retainer lock 气门弹簧托盘锁扣
valve ring 活门环;阀圈
valve ring grab 阀圈打捞器
valve rock 阀摇杆
valve rocker (气)阀摇杆
valve rocker arm (气)阀摇臂
valve rocker arm bushing 阀摇杆衬套
valve rocker arm joint 阀摇臂接头
valve rocker arm shaft bracket 阀摇臂轴托架
valve rocker arm shaft spring 阀摇臂轴弹簧
valve rocker arm spring 阀摇臂弹簧
valve rocker arm support 阀摇臂支架
valve rocker arm washer 阀摇臂垫圈
valve rocker cover 阀摇杆盖
valve rocker gear 气面摇杆装置;气阀摇杆装置
valve rod 阀柱;阀杆
valve rod guide 阀杆导夹;阀杆导套
valve rod timing chain 阀柱定时链
valve rubber 橡皮垫
valve sack 自封口袋
valve sack-filling machine 灌阀自封袋包装机
valves and fittings 阀及其配件
valve seal 阀门装置水封;气门密封件;阀止水;阀门作用;阀门止水;阀(门密)封
valve seat 阀(门)座
valve seat angle 阀座斜角
valve seater 阀座刀具
valve seat grab 起阀座器
valve seat grinder 阀座磨床;磨气门座机
valve seat holder 阀座托架
valve seating 阀座
valve seat insert 气门座圈;阀座(密封)圈
valve seat insert replacing tool 阀座装卸工具
valve seat puller 拔阀座器
valve seat reamer 阀座铰刀
valve seat refacer 阀座磨光机
valve seat ring 阀座环

valve set 换流阀组;电子管组
valve setting 气门调节;阀位给定;阀调整;阀的装配
valve shaft 气门杆;阀轴
valve shaft guide 气门导管
valve shaft lever 阀轴杆
valve shield 气门挡板;电子管屏蔽
valve silencer 滑阀机构消音装置
valve size 阀门尺寸
valve skid 阀门制输器
valve socket 电子管座;阀座
valves of liquid level ga(u)ge 液位计阀组
valve spanner 气门扳手
valve speed 阀门启闭速度;阀开闭速度
valve spindle 截阀阀梗;阀杆
valve spindle guide 阀杆导程
valve spool 阀槽
valve spring 气门弹簧;阀(弹)簧
valve spring cap 气门弹簧托盘;阀簧座圈
valve spring cap lock 阀簧底座销
valve spring chamber 气门弹簧腔
valve spring collar 阀簧座环
valve spring compressor 气门弹簧压缩工具
valve spring cotter 气门弹簧销
valve spring cover 气门弹簧盖
valve spring cover gasket 阀簧盖衬
valve spring cup 气门弹簧座
valve spring damper 气门弹簧;阀簧减振器
valve spring force 气门弹簧力
valve spring holder 气门弹簧座圈
valve spring housing 气门弹簧罩
valve spring key 气门弹簧键
valve spring lifter 气门弹簧起弹器;气门弹簧拆卸器
valve spring relay 阀簧继电器
valve spring remover 气门弹簧拆卸工具;阀弹簧拆卸器
valve spring retainer 气门弹簧座
valve spring retainer clamp 气门弹簧座锁片
valve spring retainer key 阀簧底座键
valve spring retainer lock 气门弹簧锁销
valve spring retaining collar 气门弹簧座圈
valve spring seat 气门弹簧座;阀簧座
valve spring seat key 气门弹簧座锁键
valve spring seat pin 气门弹簧座锁销
valve spring surge 气门弹簧颤动
valve spring tester 气门弹簧试验器
valve spring tool 气门弹簧拆装工具
valve spring washer 气门弹簧垫圈
valve spring wire 阀簧钢绳;阀弹簧钢丝
valve stabilizer 电子管稳定器
valve staff 阀杆
valve station 阀室【石化】
valve steel 阀钢
valve stem 气门杆;气门蒂片;阀门塞;阀杆
valve stem bush 阀杆衬套
valve stem ga(u)ge 阀杆规
valve stem guide 阀导杆
valve stem lock nut 阀杆防松螺母
valve stem nut 阀杆螺母
valve stem packing 阀杆填料
valve stem packing leakage 阀杆密封漏泄
valve stem pin 阀杆销
valve sticking 阀胶着
valve stroke 阀门行程;气门行程;阀行程;阀门冲程;阀冲程
valve surge 气门颤动
valve surge damper 气门弹簧减振器;阀簧减振器
valve surging point 气门颤动点
valve-switching mechanism 阀转换机制;阀转换机构
valve system 活门系统;阀系统
valve tap 阀打捞锥
valve tee 带阀三通
valve tension spring 气门弹簧
valve test 电子管试验
valve tester 电子管试验器
valve the gas 放气
valve timing 阀定时
valve timing adjustment 活门定时调准
valve timing gear 阀定时装置
valve tool 气门保修用机具
valve tower 进水塔;取水阀塔;浮塔;分水塔
valve train 阀门控制系统
valve transmitter 电子管发射机
valve travel 活门行程;阀行程

valve tray 浮阀塔盘
valve type 电子管式的;阀门式
valve-type arrester 阀式避雷器
valve-type hydro-percussive tool with positive acting 阀式正作用冲击器
valve-type instrument 电子管式测试仪器
valve-type piston 阀式活塞
valve-type(surge) arrester 阀式避雷器
valve-type track circuit 阀式轨道电路
valve unit 阀门元件
valve voltage divider 换流阀分压器
valve voltage drop 管压降
valve voltage regulator 电子管稳压器
valve voltmeter 电子管伏特计;电子管电压计;电子管电压表
valve washer 滑阀垫板;阀座垫圈;阀门垫圈
valve wattmeter 电子管瓦特计
valve with ball seat 球座阀;球阀座
valve with handwheel 有操纵轮的阀门
valve with inclined stem 斜轴阀;斜杆阀
valve with spring core 弹簧芯阀
valve with straight stem 直轴阀;直杆阀
valve wrench 气门间隙扳手;阀门扳手
valve yoke 阀座
valving 阀系;阀门的配置;装设阀门
valvula 小瓣
valvular 瓣状的;活门的;有阀的
valvula spiralis 螺旋瓣
VAM buttress thread casing VAM 偏梯形扣套管
VAM buttress thread coupling VAM 偏梯形扣接箍
vamp 补片
vamping stitch 补钉线迹
vamure 墙顶路;屋顶女儿墙通道
van 有篷运货汽车;轻型货车;淘选铲;大篷货车;铲头矿物试验;簸分机
vanadate 钒酸盐
vanadia 含钒矿物
vanadic 钒的
vanadic acid 钒酸
vanadic ocher 钒土
vanadic oxide 五氧化二钒
vanadic salts 钒盐
vanadic sulfate 硫酸氧钒
vanadic sulfide 硫化钒
vanadine 钒土
vanadinite 钒铅矿
vanadinizing 渗钒
vanadiomagnetite 钒磁铁矿
vanadiomagnetite ore 钒磁铁矿矿石
vanadium additive 钒添加剂
vanadium alloy 钒合金
vanadium-aluminium alloy 钒铝合金
vanadium analyzer 钒含量分析仪
vanadium attack 钒腐蚀
vanadium-bearing bituminous rock 含钒沥青质岩
vanadium-bearing phosphorus rock 含钒磷质岩
vanadium boride 硼化钒
vanadium bronze 钒青铜;钒黄铜
vanadium carbide 碳化钒
vanadium cast iron 加钒铸铁
vanadium catalyst 钒催化剂
vanadium chloride 氯化钒
vanadium contact process 钒接触法
vanadium dichloride 二氯化钒
vanadium dioxide 二氧化钒
vanadium family element 钒族元素
vanadium ion 钒离子
vanadium iron 钒铁(合金);钒钢
vanadiumism 钒中毒
vanadium metal 金属钒
vanadium nitride 一氮化钒;氮化钒
vanadium ore 钒矿(石)
vanadium oxide 氧化钒
vanadium oxytribromide 三溴氧化钒
vanadium oxytrifluoride 三氟氧化钒
vanadium pentasulfide 硫化钒
vanadium pentoxide 五氧化二钒
vanadium pentoxide dust 五氧化二钒尘粉
vanadium pentoxide fume 五氧化二钒烟
vanadium pigment 钒系颜料
vanadium pollution 钒污染
vanadium porphyrin 钒卟啉
vanadium powder 钒粉
vanadium rutile 钒金红石

vanadium sesquioxide 三氧化二钒
vanadium silicide 硅化钒
vanadium slag 钒渣
vanadium stainless steel 钒不锈钢
vanadium steel 钒铁;钒钢
vanadium sulfate 硫酸氧钒
vanadium sulfide 硫化钒
vanadium tetrachloride 四氯化钒
vanadium tetraoxide 四氧化二钒
vanadium-tin yellow 钒锡黄
vanadium-titanium magnetite ore 钒钛磁铁矿矿石
vanadium trifluoride 三氟化钒
vanadium trioxide 三氧化二钒
vanadium-zirconium blue 钒锆蓝
vanadium-zirconium yellow 钒锆黄
vanadous 亚钒的
vanadous bromide 三溴化钒
vanadous chloride 氯化亚钒;二氯化钒
vanadous fluoride 三氟化钒
vanadyl chloride 氯化氧钒
vanadyl sulfate 硫酸氧钒
vanadyl trifluoride 三氟化氧钒
vanalite 蛋黄钒铝石
vanalium 钒铝合金
van carrier 跨运车;集装箱跨(运)车
vancometer 纸张透油性测量仪;光泽度仪;(测量纸的)透油计
van container 货箱集装箱;大型(货运)集装箱
vandalism 故意破坏的行为;人为故意破坏
vandalism of aids-to-navigation 故意破坏航标
vandal proof 防装饰被损坏的;防破坏
vandal proof equipment 防盗设备
Van Den Berg reaction 范等壁氏反应
vandenbrandite 绿铀矿
vandendrisscheite 橙黄铀矿
Van der pol's oscillator 文特波振子
Van der Waals forces 范德华力
Vanderwerp recuperator 范德韦帕式熟料冷却机
vandex 混凝土防水剂(其中的一种)
Van Dorn water sampler 范多恩采水器
vandyke 锯齿边;锯齿形装饰
Vandyke brown 科隆棕颜料;铁棕(色);深褐色;范戴克棕(颜料)
Vandyke pieces 铅笔刨花;锯齿边料
Vandyke red 天然红土;范戴克红
vane 叶片;叶轮;轮叶;瞄准板;视准器;传感片;风信旗;风环;(转杯风速表的)风杯
vane anemometer 翼式风速计;翼式风速表;翼轮风速仪;叶片式风速计;叶片风速表
vane angle 叶轮角
vane apparatus 十字板剪切仪;十字板剪力仪;导叶装置
vane armature 叶片式电枢
vane attenuator 片型衰减器
vane-axial cyclone 轴向叶片旋流分离器
vane-axial fan 叶轮式轴流风机;翼式轴流风扇;叶片式轴流风扇;导叶式轴流风机;叶片式轴流扇风机
vane borer 涡轮式钻土机;涡轮(式)钻(孔)机;十字板仪
vane borer constant 十字板常数
vane cascade 叶栅
vane channel 叶片间流道
vane compressor 叶片式压缩机;叶片式压气机
vane-control fan 带导向器风机
vane-control motor 叶片控制电机
vane curvature 叶轮曲率
vane dam 叶片坝
vaned diffuser 叶片扩压器
vaned diffuser system 导流片扩散系统
vaned disk agitator 盘式叶轮搅拌机
vaned drum 带叶片的转筒
vaned drum crystallizer 带叶片的转筒结晶器
vane diameter 十字板头直径
vane diffuser 叶片扩压器
vane dike 翼坝;(叶)片坝
vaned rotating disk 导流转盘
vane efficiency 叶片效率
vane exit angle (螺旋桨的)叶片出口角
vane feeder 叶片式给料机;螺旋加料器;螺旋加料机;分格轮(式)喂料机
vane flow meter 叶片式流量计;螺旋流量计
vane grill(e) 叶片式格栅
vane(hydraulic) motor 叶片液压马达;叶片式油

马达;叶片(式)液压电动机;叶轮液压马达
vaneless 无叶片的
vaneless diffuser 无叶片式扩散器;无叶扩压器
vaneless space 无叶片空间
vaneless-vaned diffuser 无叶—有叶混合式扩压器
vane motor pump 叶轮电动泵
vane motor rotary actuator 叶片式液压马达旋转促动器
vane number 叶片数
vane oil motor 叶片式油压马达
vane oil pump 叶片油泵
vane penetrometer 十字板贯入仪
vane pitch 叶栅栅距
vane pump 叶片泵;叶轮泵;括板泵;滑片泵
vane relay 翼片继电器;扇形继电器
vane rotor 叶片转子
vane-screw propeller 叶螺旋推进器;翼轮推进器
vane setting angle 叶片装置角;叶片安装角;叶片安放角
vane shear apparatus 十字板剪切仪;十字板剪力仪;十字仪;十字板剪切试验机;十字板剪力试验机
vane shear borer 十字板剪切仪;十字板剪力仪
vane shear strength 十字板抗剪强度;十字板剪力强度
vane shear test 十字板剪切试验;十字板剪力试验
vane shear tester 十字板剪切仪;十字板剪力仪
vane shear test of soil 土壤十字板剪力试验
vane solidity 叶片充实度
vane spreader 叶片式撒布机
vane starter 轮叶起动机
vane strength 十字板抗剪强度
vane strength value 十字板抗剪强度值
vane test (土壤的)十字板(剪切)试验
vane thickness 叶片厚度
vane tip 叶片尖端
vane twist 叶片的扭曲度
vane-type air vibrator 叶片型气力振动器
vane-type axial-flow compressor 叶轮轴流式压气机
vane-type baffle 叶片式折流板
vane-type blower 叶片式鼓风机;叶片式通风机
vane-type centrifugal compressor 叶轮离心式压气机
vane-type electro-hydraulic gear 叶片式电动液压传动装置
vane-type fuel transfer pump 叶片式输油泵
vane-type gasmeter 螺旋翼轮表;螺旋流量计
vane-type hydraulic motor 叶片式液压电动机
vane-type instrument 叶片式仪表
vane-type motor 叶片式液压马达
vane-type nozzle block 叶片式喷嘴组
vane-type oil pump 叶片式油泵
vane-type pump 离心泵;活叶泵;叶轮泵;叶片式泵
vane-type relay 叶片形继电器
vane-type resonator 扇形谐振腔
vane-type rotary air motor 叶片式风动马达
vane-type supercharger 叶片式增压器;转叶式增压器
vane-type vacuum pump 叶片式真空泵
vane valve 叶片阀
vane water ga(u)ge 叶轮水表
vane water meter 叶轮(式)水表
vane wattmeter 扇形功率计
vane wheel 翼轮推进器;叶轮
vane-wheel method 叶轮法
vane-wheel type flowmeter 叶轮式流量计
vane-wheel type water flow observer 叶片式水流观测器
vane-wheel type water meter 叶轮式水表
van freight container 大型集装箱
vang (斜桁)支索;张索
vang purchase 侧牵索滑车组
vanguard 先驱;先锋
vanguard system 前导系统
vanilla 香草
vanillin 香草醛
vanishing axis 没影轴
vanishing cream base 消失性乳膏基质
vanishing door 隐入墙中的拉门;隐门
vanishing line 消视线;消失线;没影线;合线
vanishing man concept 代替人力概念
vanishing of thread 退刀纹

vanishing plane 消失面
vanishing point (透视图的)投影点;没影点;消视点;消失点;灭点;合点
vanishing point check 合点检查
vanishing point condition 没影点条件;合点条件
vanishing point control 灭点控制;合点控制
vanishing point controller 主合点控制器
vanishing point method 没影点方法
vanishing point of stability 稳定性消失点
vanishing target 隐靶
vanishing tide 失潮
vanishing trace 合线
vanity basin 女用盥洗盆;梳妆台式洗面器
van line 汽车运输公司;长途运输公司;长途搬运公司
vanmeersscheite 万磷铀矿
Van Mises criterion 冯米塞斯准则
Van Mises distribution 冯米塞斯分布
vannal 扇域的
vannal region 扇域
vanner 整理矿砂机;皮带溜槽;淘选带;淘矿机;带式淘洗机
vanner concentration 带式摇床精选
vanning 装箱;集装箱装箱作业;货物装箱;淘选;淘矿(法)
vanning and devanning capacity 拆装能力
vanning and devanning depot 集装箱拆装场
vanning jig 振筛(淘汰机);机械振筛;跳汰机
vanning machine 淘矿机
vanning order 装箱指示
vanning report 装箱报告单
van of the advance 高速列车的先驱
vanoxite 复钒矿
van pool (集装箱)空箱堆场;合乘篷车
vanpooling 大篷车共乘
Van radiation belt 范艾伦辐射带
van semi-trailer 货柜半拖车
van ship 货车轮渡;火车渡船
Van Slyke aminonitrogen method 范斯莱克氨基测定法
vanthoffite 无水钠镁矾
Van Tongeren type classifier 涡流旋风式选粉机
van trailer 篷式挂车;厢式挂车
van truck 厢式货车
vanuralite 钒铝铀矿
vanuranylite 黄钒铀矿
Van Veen grab 范文咬合采泥器
van vehicle 运货车;有盖货车
vanway 通向房屋的车道
vapoil 气化油
vapometallurgical process 气化冶金方法
vapometallurgy 挥发冶金;气化冶金
vaporarium 蒸发室
vaporific 蒸气状的
vaporise = vaporize
vapotron 蒸发冷却器;高频发射管
vapo(u)r 蒸气;气化液体;气化物;水蒸气;雾气
vapo(u)rability 蒸发性;气化性
vapo(u)r absorption 蒸气吸收
vapo(u)r adsorption process 蒸气吸附法
vapo(u)r-air mixture 蒸气空气混合物
vapo(u)r-air ratio 蒸气空气混合比
vapo(u)r area ventilation 蒸气区域通风
vapo(u)rarium condition 蒸气参数
vapo(u)rarium filter 蒸气过滤器
vapo(u)r arlum 蒸气浴(疗室)
vapo(u)r ating dish 蒸发皿
vapo(u)ration 气化作用
vapo(u)ration parameter 蒸气参数
vapo(u)rator 蒸发器;(内燃机)汽化器
vapo(u)r balancer 蒸气平衡器
vapo(u)r balancing mechanism 蒸气平衡器
vapo(u)r balancing unit 蒸气平衡器
vapo(u)r barrier 隔气具;阻凝层;蒸气屏层;蒸气隔板;蒸气防护栅;隔气层;防蒸气渗透涂层;防湿层
vapo(u)r barrier film 蒸气隔膜
vapo(u)r barrier foil 蒸气阻层薄片
vapo(u)r barrier membrane 防潮气阻层
vapo(u)r barrier sheet(ing) 蒸气阻层薄板
vapo(u)r barring 蒸气阻凝
vapo(u)r bath 蒸气浴
vapo(u)r blanket 蒸气层
vapo(u)r blast 蒸气喷砂
vapo(u)r blasting 蒸气喷射;气水砂清理;喷气清理

vapo(u)r blast operation 蒸气喷砂处理
vapo(u)r blast process 蒸气喷砂清理法
vapo(u)r capacity 蒸气容量
vapo(u)r cavity 蒸气空泡
vapo(u)r chart 蒸气图
vapo(u)r check 隔气带
vapo(u)r chimney 排气道;蒸气囱
vapo(u)rchoc 蒸气枪
vapo(u)r cleaning 蒸气净洗
vapo(u)r-coated mirror 气化涂膜反光镜
vapo(u)r cock 油舱疏气阀
vapo(u)r composition 蒸气组成
vapo(u)r compression 蒸气压缩(法)
vapo(u)r compression cycle 蒸气压缩循环
vapo(u)r compression evapo(u)rator 蒸气压缩蒸发器
vapo(u)r compression heat pump 蒸气压缩热泵
vapo(u)r compression machine 蒸气压缩机
vapo(u)r compression method 蒸气压缩法
vapo(u)r compression process 蒸气压缩过程
vapo(u)r compression refrigerating system 蒸气压缩制冷装置;蒸气压缩制冷系统;蒸气压缩制冷方式
vapo(u)r compression refrigeration cycle 蒸气压缩制冷循环
vapo(u)r compression refrigeration system 蒸气压缩式制冷系统
vapo(u)r compression refrigerator 蒸气压缩式制冷机
vapo(u)r compression-vacuum distillation process 蒸气压—真空蒸馏法
vapo(u)r compressor 蒸气压缩机
vapo(u)r concentration 蒸气浓度;水蒸气密度;水气浓度
vapo(u)r condensation 蒸气凝结
vapo(u)r condenser 蒸气冷凝器;凝汽器
vapo(u)r condition 蒸气参数
vapo(u)r container 蒸气收集器
vapo(u)r content 蒸气含量
vapo(u)r-cooled generator 蒸发冷却发电机
vapo(u)r-cooled transformer 蒸发冷却变压器
vapo(u)r-cooling 蒸发冷却
vapo(u)r corrosion inhibitor 蒸气腐蚀抑制剂
vapo(u)r corrosiveness 气相腐蚀
vapo(u)r curable coating 气体固化涂料
vapo(u)r cure 蒸气养护;蒸气处冶
vapo(u)r curve 蒸气曲线
vapo(u)r cushion 蒸气垫
vapo(u)r cycle 蒸气循环
vapo(u)r cycle cooling 气相循环致冷;气相循环冷却;蒸发循环冷却
vapo(u)r degreaser 蒸气去油机
vapo(u)r degreasing 蒸气脱脂;蒸气去垢;蒸气除油
vapo(u)r delivery tube 蒸气配送管
vapo(u)r density 蒸气密度;气密度
vapo(u)r density apparatus 蒸气密度器
vapo(u)r density bulb 蒸气密度球
vapo(u)r density method 蒸气密度法
vapo(u)r deposited coating 蒸气沉积涂层法;气沉积涂层法
vapo(u)r deposited screen 蒸发荧光屏
vapo(u)r deposition 真空镀膜;气相沉淀;气相淀积法
vapo(u)r deposition method 蒸气淀积法
vapo(u)r deposition process 气相沉积法
vapo(u)r diffusion 蒸气扩散
vapo(u)r diffusivity 蒸气扩散率
vapo(u)r discharge lamp 蒸气放电灯
vapo(u)r discharge tube 蒸气排出管
vapo(u)r disengagement 蒸气分离
vapo(u)r-dividing head 蒸气分配蒸馏头
vapo(u)r dome 蒸气圆顶室
vapo(u)r effect 蒸气效力
vapo(u)r eliminator 蒸气消除器;蒸气分离器
vapo(u)r engine 气化液体燃料发动机
vapo(u)r enlarge lamp 气体放大灯
vapo(u)r enrichment 蒸气增浓
vapo(u)r escape 蒸气漏泄
vapo(u)r exchanger 蒸气换热器
vapo(u)r exhaust fan 抽油烟风扇
vapo(u)r expansion chamber 蒸气膨胀室
vapo(u)r explosion 蒸气爆炸
vapo(u)r extraction device 抽气装置

vapo(u)r extractor 蒸气提取器;抽气器
vapo(u)r-filled 蒸气充满式
vapo(u)r-filled thermoelectron type rectifier 充蒸气热电子式整流器
vapo(u)r-filled thermometer 充气温度计
vapo(u)r filter 蒸气过滤器
vapo(u)r flow 蒸气流
vapo(u)r flux 气体焊剂;水气通量
vapo(u)r fractometer 气相分离计
vapo(u)r-free 蒸气无凝结的
vapo(u)r galvanizing 气化镀锌
vapo(u)r-gas mixture 蒸气—气体混合物
vapo(u)r generation 蒸发生
vapo(u)r glaze 蒸气(沉积)釉
vapo(u)r-heated 蒸气换热的
vapo(u)r-heated double-rotary dryer 蒸气加热双转筒式干燥机
vapo(u)r heater 蒸气加热器
vapo(u)r heating 低压蒸气供暖;(低压)蒸气采暖
vapo(u)r heating equipment 蒸气供暖装置
vapo(u)r heating system 低压蒸气供暖系统
vapo(u)r holdup 蒸气截留量
vapo(u)r honing 喷气清理
vapo(u)rific inhibitor 挥发性防腐蚀剂
vapo(u)rimeter 蒸气计;蒸气压力计;挥发度计;气压计;气压表
vapo(u)rimetric method 蒸气测定法
vapo(u)r impermeability 蒸气不渗透性
vapo(u)r imperviousness 蒸气不渗透
vapo(u)ring extract process 气提法
vapo(u)ring liquid extinguisher 液雾灭火器
vapo(u)r inhalation 蒸气吸入
vapo(u)r injection 蒸气喷射
vapo(u)r insulation 蒸气绝缘;蒸气隔绝
vapo(u)rised kerosene burner 打气煤油炉
vapo(u)r riser 蒸发器
vapo(u)rish 蒸气状的
vapo(u)rising fuel 气化燃料
vapo(u)rium 含气量
vapo(u)rizability 气化性
vapo(u)rization 蒸气作用;挥发作用;气化(作用)
vapo(u)rization adjustment 气化调节
vapo(u)rization coefficient 蒸发系数
vapo(u)rization cooling 蒸发冷却
vapo(u)rization curve 液—气平衡线
vapo(u)rization efficiency 蒸发效率;气化效率
vapo(u)rization energy 气化能
vapo(u)rization heat 蒸发热
vapo(u)rization latent heat 蒸发潜热
vapo(u)rization loss 气化损失
vapo(u)rization of filament 灯丝蒸发
vapo(u)rization rate 气化速率
vapo(u)rization room 蒸发室
vapo(u)rization system 蒸发系统
vapo(u)rization zone 气化区域
vapo(u)rized 蒸发了的
vapo(u)rizer 喷雾器;蒸发器;汽化器
vapo(u)rizer tube 燃油蒸发管
vapo(u)rizing burner 气化燃烧器
vapo(u)rizing carburet(t)or 蒸发式化油器
vapo(u)rizing chamber 蒸发室;蒸发室
vapo(u)rizing coil 蒸发盘管
vapo(u)rizing combustion chamber 气化式燃烧室
vapo(u)rizing combustor 气化式燃烧器
vapo(u)rizing fuel 气化燃料
vapo(u)rizing liquid extinguisher 液雾灭火器;气化液体灭火器
vapo(u)rizing oil 气化油
vapo(u)rizing point 沸点
vapo(u)rizing property 气化特性
vapo(u)rizing rate 气化速率
vapo(u)rizing surface 蒸发面
vapo(u)rizing temperature 蒸馏温度;汽化温度;喷雾温度
vapo(u)rizing tube nipple 气化管乳头
vapo(u)rizing unit 气化装置
vapo(u)r jacket 蒸气夹套
vapo(u)r lamp 蒸气(放电)灯;气灯;放电灯
vapo(u)r lift pump 蒸气提升泵
vapo(u)r-liguid-solid technique 气液固工艺
vapo(u)r line 蒸气管线;蒸气管路
vapo(u)r-liquid equilibrium 气液平衡
vapo(u)r-liquid extraction 气液抽提
vapo(u)r-liquid interface 气液界面

vapo(u)r-liquid nucleation 气液成核
vapo(u)r-liquid process 气液交换过程
vapo(u)r-liquid ratio 气液比
vapo(u)r-liquid separator 气水分离器;气液分离器
vapo(u)r-liquid-solid growth 气液固生长
vapo(u)r-liquid-solid process 气体—液体—固体法
vapo(u)r liquor 蒸气液
vapo(u)r lock device 防止管内蒸气积聚装置
vapo(u)r lock(ing) 气塞;气阻;气锁;蒸(气)气塞;(发动机)气障;气封
vapo(u)r locking tendency 气阻倾向
vapo(u)r losses 蒸气损失
vapo(u)r meter 蒸气仪表
vapo(u)r method 蒸气降温法;蒸气(压)法;气浴法
vapo(u)r migration 蒸气迁移;蒸气渗透
vapo(u)r mist 蒸气薄雾
vapo(u)r mixing ratio 水气混合比
vapo(u)r nucleus 气核
vapo(u)rometer 蒸气压力计
vapo(u)rous 蒸气状的;蒸气饱和的;气态的
vapo(u)rous water 气态水
vapo(u)r penetration 蒸气渗透;蒸气渗透
vapo(u)r permeability 透气性;蒸气渗透性;透湿性
vapo(u)r-permeable 透气的
vapo(u)r permeance (木材的)蒸气渗透系数;透湿系数
vapo(u)r pervious material 透蒸气材料
vapo(u)r phase 气态;蒸气相;气相
vapo(u)r phase association 气相缔合
vapo(u)r phase chromatography 气相色谱分离法
vapo(u)r phase cooling 蒸发冷却
vapo(u)r phase cracked gasoline 气相裂化汽油
vapo(u)r phase deposition 气相沉积
vapo(u)r phase diffusion 气相扩散
vapo(u)r phase dy(e)ing 气相染色
vapo(u)r phase dye laser 气相染料激光器
vapo(u)r phase fugacity 气相易逸性
vapo(u)r phase granular activated carbon 气相颗粒活性炭
vapo(u)r phase gum 气相胶
vapo(u)r phase inhibitor 气相阻化剂;气相抑制剂;气相(膜)防锈剂;气相缓蚀剂
vapo(u)r phase inhibitor paper 气相抑制纸
vapo(u)r phase interference 气相干扰
vapo(u)r phase lubrication 气相润滑
vapo(u)r phase mixture 气相混合物
vapo(u)r phase nitration 气相硝化
vapo(u)r phase operation 气相操作
vapo(u)r phase oxidation 气相氧化
vapo(u)r phase oxidation process 气相氧化法
vapo(u)r phase polymerization 气相聚合
vapo(u)r phase process 气相过程
vapo(u)r phase purification 气相纯化
vapo(u)r phase reaction 气相反应
vapo(u)r phase reactor 气相反应器
vapo(u)r phase rearrangement 气相重排
vapo(u)r phase refining 气相精炼
vapo(u)r phase sintering 气相烧结
vapo(u)r phase spectrum 气相频谱
vapo(u)r phase synthetic(al) diamond 气相合成金刚石
vapo(u)r phase system 气相系统
vapo(u)r phase treatment 气相处理
vapo(u)r pipe 油舱通气管;蒸气管;透气管
vapo(u)r plating 真空镀膜;金属蒸发镀膜;气化渗镀
vapo(u)r-plating process 蒸气镀金法
vapo(u)r plume 蒸气卷流
vapo(u)r pocket 蒸气包;气囊
vapo(u)r point 蒸发点
vapo(u)r polishing 蒸气抛光
vapo(u)r pressure 蒸气压强;蒸气压(力);气压(强);水气压(力)
vapo(u)r pressure at pumping temperature 泵送温度下的液体蒸压
vapo(u)r pressure chart 蒸气压图表
vapo(u)r pressure constant 蒸气压常数;气压常数
vapo(u)r pressure curve 蒸气压力曲线
vapo(u)r pressure deficit 蒸气压差;水气压力差;饱和差
vapo(u)r pressure gradient 水气压梯度;蒸气压(力)梯度
vapo(u)r pressure index 蒸气压指数

vapo(u)r pressure inhibitor 蒸气压抑制剂
vapo(u)r pressure isotope effect 蒸气压同位素效应
vapo(u)r pressure lowering 蒸气压下降
vapo(u)r pressure method 水气压法
vapo(u)r pressure of water 水的蒸气压
vapo(u)r pressure osmometer 蒸气渗透压力计
vapo(u)r pressure osmometry 蒸气压渗透法
vapo(u)r pressure potential 蒸气压力势
vapo(u)r pressure ratio 蒸气压力比
vapo(u)r pressure test 气压试验;蒸气压测定
vapo(u)r pressure type 气压式
vapo(u)r-proof 气密(的);不漏蒸气的
vapo(u)r-proof barrier 防蒸气渗透隔层;防水气层
vapo(u)r-proof connection 气密连接
vapo(u)r-proof course 隔气层
vapo(u)r-proof curtain 气密幕
vapo(u)r-proof engine 防气式电机
vapo(u)r-proofing material 防湿材料
vapo(u)r-proof layer 防潮层
vapo(u)r-proof machine 气密式电机
vapo(u)r-proof membrane 防水气薄膜;气密薄膜
vapo(u)r pump 蒸气泵;扩散泵
vapo(u)r quantity 蒸气量
vapo(u)r reaction 气态反应
vapo(u)r recirculating still 蒸气循环式气液平衡试验器
vapo(u)r recirculation 蒸气循环
vapo(u)r recompression 蒸气再压缩
vapo(u)r recovery 蒸气回收
vapo(u)r recovery system 蒸气回收系统
vapo(u)r recovery unit 蒸气回收装置
vapo(u)r rectification process 蒸气分馏过程
vapo(u)r rectifier 蒸气精馏器
vapo(u)r-reflux contacting 蒸气与回流液接触
vapo(u)r release 蒸气排放
vapo(u)r release rate 蒸气释放速率
vapo(u)r removal 蒸气分离
vapo(u)r resistance 蒸气阻力;透湿阻力
vapour-resistant 耐湿的;耐蒸气的
vapour-resistant board 防蒸气石膏板
vapo(u)r resistivity 蒸气扩散阻力;蒸气阻率
vapo(u)r return 蒸气再发;气相回流管
vapo(u)r riser 导气管
vapo(u)r sampling 水气采样
vapo(u)r sampling rod 蒸气进样杆
vapo(u)r-saturated 蒸气饱和的
vapo(u)r scrubber 蒸气清洗装置
vapo(u)r seal 蒸气封闭;气封
vapo(u)r seal foil 蒸气密封薄片
vapo(u)r seal membrane 气密薄膜;蒸气密封膜
vapo(u)r seasoning 蒸气干燥(法)
vapo(u)r sensing 蒸气感测
vapo(u)r-set 蒸气凝结的
vapo(u)r solid 气化固体
vapo(u)r source 蒸气源
vapo(u)r space 蒸气空间
vapo(u)r-space volume 蒸气空间的容积
vapo(u)r state 气(化状)态
vapo(u)r steam blaster 蒸气喷射器
vapo(u)r stream 蒸气流
vapo(u)r substrate interface 蒸气衬底间界面
vapo(u)r system 蒸气系统;蒸气式
vapo(u)r table 蒸发表
vapo(u)r tension 蒸气张力;蒸气压(力);气压
vapo(u)r tension meter 蒸气压计
vapo(u)r testing apparatus 蒸气测试器
vapo(u)r-tight 气密(的);不漏气(的)
vapo(u)r-tight material 不透气材料;气密材料
vapo(u)r-tight tank 不渗器箱
vapo(u)r trail 雾化尾迹;水气尾迹
vapo(u)r transfer 蒸气迁移
vapo(u)r transfer coefficient 蒸气转移系数
vapo(u)r transfer rate 透湿量
vapo(u)r transmission 蒸气渗透;透湿性
vapo(u)r transmission property 蒸气透过性
vapo(u)r transport 水气输送
vapo(u)r trap 蒸气捕集器
vapo(u)r treatment 蒸气处理
vapo(u)r tube 蒸气管
vapo(u)r tube of flask 蒸馏瓶侧管
vapo(u)r uptake 蒸气上升管
vapo(u)r vacuum pump 蒸气真空泵
vapo(u)r vane 汽轮叶片

vapo(u)r velocity 蒸气速度
vapo(u)r vessel 蒸气容器
vapo(u)r viscosity 气体黏度
vapo(u)r volume 蒸气体积;蒸气容积
vapo(u)r volume equivalent 蒸气体积当量
Vaqueros formation 瓦克罗斯层【地】
var 乏(无功率单位);无功伏安
varactor 可变电抗器;变(容)抗器;变容二极管
varactor diode 变容二极管
varactor diode for parametric(al) amplifier 参放变容二极管
varactor diode parametric amplifier 变容二极管参量放大器
varactor frequency multiplier 变容管倍频器
varactor multiplier 变容管倍频器
varactor multiplier chain 变容管倍频链
varactor tuner 电调谐高频头
varactor tuning 变容管调谐
Varamat lens 法拉马特镜头
V-arch V 形拱
vardarac 伐尔达尔风
varec 海藻灰
varech 海草
varek 海藻
vargueno 装饰性书写柜
var hour 乏尔小时
var-hour meter 乏尔小时计
variability 可变性;差异度;变异性;变异度;变率;变化性;变度
variability index 可变性指数
variability law 变异定律
variability of annual precipitation 年降雨变化率;年降水变化率
variability of annual rainfall 年降雨变化率;年降水变化率
variability of runoff 径流变异性;径流变率
variability of stream flow 河道流量的变化性
variability of waves 波的可变性;波的变化性
variable 可变量;计量值;变数;变量;变化的
variable accelerated motion 变加速运动
variable acceleration 可变加速度;变加速度
variable action 可变作用
variable address 可变地址
variable adjacency matrix 可变邻接矩阵
variable air volume air conditioning system 变风量空气调节系统
variable air volume apparatus 变风量装置
variable air volume device 变风量装置
variable air volume system 变风量系统;变风量方式
variable air volume unit 变风量装置
variable-amplitude fatigue test 变幅疲劳试验
variable-amplitude recording 变幅记录
variable analysis 变量分析
variable angle 可变角度
variable-angle nozzle 可变角度喷嘴;可转喷嘴;可调导叶
variable-angle reflection 变角反射
variable annuity 可变年金;不定额年金;变动年金
variable aperture flowmeter 变孔口流量计
variable aperture seismic array 变孔径地震台阵
variable aperture shutter 可变开度光阀
variable architecture processor 可变结构处理器
variable area 可变截面;可变面积
variable area channel 可变面积槽
variable area exhaust nozzle 可变截面排气喷管
variable area flowmeter 转子流量计;可变面积流量计;可变截面流量计;变截面流量计
variable area flowrator 变截面流量计
variable area nozzle 截面可调喷管
variable area propelling nozzle 可变截面推力喷嘴
variable area record 变面积记录
variable area record section 变面积记录剖面
variable area track 面积调制声道
variable assembly line 变交流水线
variable attenuator 可变衰减器
variable-audio-frequency radiosonde 变声频无线电探空仪
variable ballast 可变压载
variable ballast tank 可调压载水舱;可变压载水舱
variable-band-width filter 可变频宽滤波器
variable bank 变动岸
variable bank line 不定岸线
variable baseline interferometer 变基线干涉仪

variable beam attenuator 可变光束衰减器
variable being explained 被解释变量
variable bias 可变偏压
variable binary scaler 变系数二进制分频器
variable binding 可变联结;变量结合
variable bit 可变比位
variable block 可变块
variable block format 可变(程序)块格式
variable bound 可变约束;可变上下界
variable-boundary layer 变界层
variable budget 临时预算;可变预算;弹性预算;变动预算
variable buoyancy device 可变浮力装置
variable camber 变曲面
variable camber gear 变弧装置
variable-capacitance amplifier 变容放大器
variable-capacitance diode 变容二极管
variable-capacitance parametron 变电容参变管
variable-capacitance transducer 变容转换器;变容换能器
variable capacitor 可变电容(器)
variable capacity 可变容量
variable capacity pump 可调流量泵
variable capital 可变资本
variable carrier 变幅载波
variable cavity 可调空腔谐振器;可变空腔
variable ceiling 多变云幂高
variable-cell method 变胞法
variable center arch dam 变中心拱坝
variable channel 不稳定水道;不稳定河槽;变化河槽;变动河床;变动河槽
variable charges 可变费用;非固定费用;变动费用
variable clause 可变条款
variable coefficient (可)变系数;变数系数
variable coefficient method 变量系数法
variable coil 可变电感线圈
variable color polarizing 可变色偏振
variable colo(u)r filter 变色滤光器
variable column 变量列
variable compression engine 可变压缩发动机
variable compression ratio 可变压缩比
variable compression ratio engine 变压缩比发动机
variable concentrated load 可变集中荷载
variable concentration 不定浓度
variable condenser 可变电容(器)
variable condenser block 可变电容器组
variable-conductance network 变导网络
variable connector 可变连接指令;可变连接器;可变连接符;可变连接点
variable contour interval 可变等高距
variable contrast 可变反差
variable contrast paper 可变反差照相纸;可变反差相纸
variable control block area 可(变)控制块区
variable cost 可变价格;可变价值;可变费用;可变成本;变动费用;变动成本
variable costing 可变成本法;变成本计算
variable cost method 变动成本法
variable cost pool 变动成本积累
variable cost ratio 可变成本比率;变动成本比率
variable costs 非固定费用
variable coupler 可调耦合器
variable coupling 可变电磁耦合
variable cross-section 变截面
variable cross-section method 变截面解法
variable cross-section wind tunnel 可变截面风洞
variable current 无定向洋流;不规则洋流
variable cycle 可变周期
variable cycle operation 可变周期操作
variable cycle pumping 可变循环泵送;变循抽油
variable daily limit 可调整的每日价格限度
variable data 可变数据
variable declaration 变量说明
variable declaration part 变量说明部分
variable deductions 变动减除数
variable-definition 变量定义
variable delay line 可变延迟线
variable delay pulse generator 可变延迟脉冲发生器
variable deletion 变量删除
variable-delivery mud pump 变量泥浆泵
variable-delivery pump (可)变流量泵;变量(输送)泵
variable density 变密度

variable-density channel 密度调制声道
variable-density disk 可变密度盘
variable-density filter 可变密度滤光片;变密度滤光器
variable-density log 变密度测井
variable-density logger 声波全波列测井仪
variable-density record 变密度记录
variable-density record film 变密度记录胶卷
variable-density record section 变密度记录剖面
variable-density seismic screen display 可变密度屏幕显示
variable-density system 疏密制;变密度制
variable-density track 密度调制声道
variable-density wind tunnel 变密度风洞
variable depth attachment 不同深播种附加装置
variable depth opener 变深度开沟器
variable depth section frame 可变高度截面的框架
variable depth sonar 可变深度声呐;变深声呐;深度可调声呐
variable design point method 变设计点法
variable device of vibratory element 振动可变装置
variable diagonal pulley 可变交叉皮带轮
variable diameter depth of hole 变径深度
variable diameter lifted bed 变径阴双层双容浮床
variable diameter number of hole 变径次数
variable diaphragm 可变光阑
variable-difference method 变量(值)差分法
variable-diffusion lens 变漫射透镜
variable dilution sampling 可变稀释采样法
variable dimension 可变尺寸
variable discharge 变化排水量
variable discharge permit 变量排污许可证
variable discharge turbine 可变量涡轮机;可变量透平;变流量式涡轮机;变量涡轮机
variable discriminator seismograph 可变鉴别器地震仪;带可变鉴别器的地震仪
variable dispersion equation 不定扩散方程
variable displacement device 可变位移式仪器
variable displacement hydraulic motor 变量液压电动机
variable displacement hydraulic pump 变量液压泵
variable displacement motor 可变排量油马达;变量马达;变量电动机
variable displacement pump 可变容积泵;可调流量泵;变容式泵;变量泵;变排量泵
variable domain 可变域;变域
variable draw motion 变速牵拉装置
variable drive 无级变速器
variable-drive power takeoff 变速传动式动力输出轴
variable drop device 变播量排种装置;变量排种器
variable-drop mechanism 播量调节机构
variable duration impulse system 脉宽调制系统
variable earnings program(me) 可变收入计划
variable eccentric wheel 可调偏心轮
variable electronic shutter 可变速电子快门
variable electrostatic contribution 不定静电分布
variable element 可变参数;变量参数
variable elements of semi-variable cost 半变动成本的变动部分
variable elevation 可变标高;可变高程
variable elevation beam 可变仰角波束
variable elevation beam-antenna 可变仰角射束天线
variable engine-transmission unit coupling 发动机传动机组的可变连接
variable error 可变误差;可变错误;不定误差;变量误差
variable evaluation 可变估值;可变估价;变动估价
variable expansion 可变膨胀;变膨胀
variable expansion valve 调节膨胀阀
variable expenditures related to traffic volume 与运输量有关的变动支出
variable expenses 可支出;可费用;变动费用
variable expenses ratio 变动成本比率
variable factor 可变因子;可变因素;可变系数
variable factor input 变动要素投入
variable factory over head 变动间接费用
variable factory overhead cost 变动的工厂制造费用
variable field 可变域;变(量)场
variable field length 可变域长度
variable floor load 变楼面荷载
variable flow 不均匀流;变(速)流

variable flow control 量调节;变流量调节
variable flow gate 变流闸门
variable-flow pump 可变流量泵;变流量泵;可调流量泵
variable-flux reactor 可变中子通量反应堆
variable flux voltage variation 变磁通调压
variable focal length 可变焦距
variable focal length lens 变焦距透镜;可变焦距镜头
variable focus 可变焦距
variable focus condenser 变焦距聚光镜
variable focus lens 可变焦(距)透镜
variable focus lens system 可变焦透镜系统
variable focus objective 可变焦物镜
variable focus pyrometer 变焦点高温计
variable force 可变力;不定力;变量的力;变力
variable format 可变格式;变化格式
variable format input 可变动式输入
variable free-stream velocity 变自由流速度
variable frequency 变频
variable frequency electric(al) drive 变频电气传动
variable frequency EPR spectrometer 变频电子顺磁共振波谱仪
variable frequency generator 可变频率振荡器
variable frequency induced polarization 变频激发极化法
variable frequency induced potentiometer 变频感应电位仪
variable frequency induction tool 变频感应测井仪
variable frequency oscillator 频率可变振荡器;变频振荡器
variable frequency power sources 变频电源
variable frequency pulse 变频脉冲斜向探测
variable frequency sinusoidal excitation 变频率正弦波激振
variable frequency speed regulation 变频调速
variable frequency synchro generator 变频同步发电机
variable frequency technique 变频技术
variable frequency trigger generator 可变频率触发振荡器
variable frequency type pile driver 变频式振动沉桩机
variable frequency vibration test 变频振动试验
variable frequency vibrator 变频振动器
variable fuel orifice carburetor 可变燃料口式汽化器
variable function 可变函数
variable function generator 可变函数生成程序;可变函数发生器;可变功能发生器
variable gain 可变增益;可变放大;可变收益
variable gain amplifier 可变增益放大器
variable gain daming 可变增益阻尼
variable gain device 可变增益装置
variable gap 可变狭缝;可变间隙
variable-ga(u)ge wheelset 轮对的可调节式量规
variable geometry aircraft 可变形飞行器
variable geometry nozzle 可变叶角喷嘴
variable geometry wing 变几何形机翼
variable grade channel 变坡渠
variable-grade spring 变量泉
variable gradient 可变梯度;变比降
variable grid leak 可变栅漏
variable head 可变压头;变水头
variable head permeability test 变水头渗透试验
variable-head permeameter 变水头渗透仪
variable identification 变量标识符
variable identifier 变量标识符;变量辨识器
variable immersion hydrometer 变容(式)比重计
variable impedance power meter 可变阻抗功率计
variable import levies 进口差额税
variable import levy 可变动的进口税征收税;非固定进口税;变动进口税征收额
variable incidence 可变倾角
variable inclination 可变坡度;不定坡度;变坡
variable in decision situation 决策形势中的变量
variable indicant 标志变异指标;变异标志
variable inductance 可变电感器
variable-inductance accelerometer 可变电感式加速度计
variable-inductance pick-up 可变电感拾音器;变感拾音器
variable-inductance transducer 变感换能器
variable inductor 可调电感器;可变电感线圈;可变电感器

variable information 可变信息;可变情报
variable information file 可变信息文件;可变信息存储器
variable inlet 可调进气口
variable inlet duct 可变进气管
variable inlet guide vanes 可调进气导片
variable inlet vane 可调进气叶片
variable instantaneous unit hydrograph 变动瞬时单位过程线
variable integer 可变整数
variable intensity 可变强度
variable intensity beam 可变强度光束
variable intensity device 可变强度式仪器
variable-intensity log 变强度测井
variable intensity pyrometer 可变强度高温计
variable interest rate 可变利率
variable interia 可变惯量
variable intermittent duty 变载断续工作方式
variable interval digitizer 可变间隔数字化器
variable interval reinforcement 变时型强化
variable in the model 模型内的变数
variable label 可变标号
variable labo(u)r cost 可变人工成本
variable-length 可变长度
variable-length block 可变长度块
variable-length code 可变长度码;可变长度编码;变长度码
variable-length command 可变长度命令
variable length computer 可变字长计算机
variable length data 可变长数据
variable-length data item 可变长(度)数据项
variable-length data package 可变长(度)数据包
variable-length encoding 变长编码
variable-length feed 可变长度卡片馈送
variable-length format 可变长格式
variable-length instruction 可变长度指令
variable-length instruction format 可变长度指令格式
variable-length instruction set 变长指令系统
variable-length pulse 可调宽度脉冲
variable-length record 可变长(度)记录;变长记录
variable-length record file 可变长记录文件;变长记录文卷
variable-length record format 可变长记录格式
variable-length record sorting 可变长记录分类
variable-length record system 可变长记录系统
variable-length table 可变长数据表
variable level filter 变水位过滤器
variable levies 可变税额;差额税;变动的征税
variable levy 差价税
variable levy system 差别征税率
variable lift device 变升力装置
variable light-filtering device 渐变光学滤光装置
variable linear differential transformer 可变线性差分变压器
variable linear system 可变线性系统
variable load 可变荷载;可变负载;可变负荷;变载(荷);变(动)荷载;变动负载;变动负荷
variable load plant 调峰电厂
variable logic 可变逻辑
variable loss and profit 变动损益
variable Mach number nozzle 可变马赫数喷管
variable magnetic flux geophone 变磁通式地震检波器
variable magnetic impedance geophone 变磁阻式地震检波器
variable magnetic-resistance type transducer 变磁阻式传感器
variable magnification 可变放大率
variable magnification lens 可变放大率透镜;变放大率透镜
variable marker counter 可变距离标志
variable mass 变质量
variable mass dynamics 变质量动力学
variable-mass system 变质量系
variable master clock 可变主钟
variable maximum controller 可变最长绿灯控制器;可变最长绿灯控制机
variable mesh size 可变网距;可变步长
variable metric algorithm 可变尺度算法
variable metric method 可变计量法;变尺度法
variable microcycle timing 可变微周期定时
variable micrologic 可变微逻辑;可变微程序控制

逻辑
variable moderator reactor 可变慢化剂反应堆
variable module model 变模量模型
variable modulus of elasticity 可变弹性系数;可变弹性模数;可变弹性模量
variable moment of inertia 可变惯性矩
variable money-capital 可变货币资本
variable motion 可变运动
variable mu 可变放大系数
variable multiplier 变量乘法器
variable multiplier unit 可变乘法器
variable mu pressure transducer 变导磁系数压力转换器
variable-mutual conductance 变互导
variable-mutual conductance tube 变跨导管
variable mutual conductance valve 变互导管
variable mu tube 可变放大系数管
variable mu valve 可变放大系数管
variable name 变量名(字)
variable name conflict 变量名冲突
variable nebula 变光星云
variable(-note) buzzer 可调蜂鸣器
variable nozzle 可调喷嘴;可变喷嘴;变截面喷嘴
variable numbering 可变编号
variable occurrence data item 可变重复次数数据项
variable of integration 积分变数;积分变量
variable operating cost 可变营运费(用);可变动营运费用
variable optic(al) attenuator 可变光衰减器
variable optic(al) density 可变光学密度
variable order 可变指令;可变顺序
variable order-quantity system 可变订货量系统
variable or flexible plan 可变偿还法
variable oriented berth 多方位泊位
variable oriented quay 多方位码头
variable oriented wharf 多方位码头
variable-orifice damper valve 变孔调节阀
variable-orifice flow controller 可变孔板流量控制器
variable oscillator 可变频率振荡器
variable output coupler 可变输出耦合器
variable overhead 变动管理费
variable overhead variance 可变间接费差异
variable pacer clock 可变定速时钟
variable-paper-speed recorder 可变纸带速率记录器
variable parameter 可变参数;可变参量;变量参数;变化参数
variable parameter channel 变参信道
variable part 可变部分
variable path-length cell 可变光程样品池;变光程样品池
variable path transmitter 可变光程器
variable per cent 变动百分率
variable per hour 小时变量
variable perspective camera 可调倾角摄影机
variable phase plate 可变相位板
variable phase shifter 可移相器
variable phase signal 可变相位信号
variable phasing sequence 可变相序
variable pitch 可变螺距;可变节距;变深度;变螺距
variable-pitch auger 变螺旋推运器;变螺距螺旋
variable-pitch blade 变螺距叶片
variable-pitch grid 变节距栅极
variable-pitch propeller 变螺距推进器;变距推进器;可调螺距螺旋
variable-pitch reversing propeller 变距可逆螺旋桨
variable-pitch screw 变(螺)距推进器
variable-pitch turbine 转桨式水轮机;可调桨叶角度的水轮机;卡普兰(式)水轮机;变(节)距水轮机
variable point 可变小数点数据
variable-point-mo(u)ld 变点模型
variable-point numeral 变点数
variable-point representation 可变小数点表示;可动点基数表示法;变点表示
variable point representation system 可变小数点记数制;变点表示系统
variable polarization 可变偏振
variable-pole-speed motor 变极变速电机
variable power 可变放大率;可变倍率
variable power eyepiece 可变放大率目镜
variable power lens 可变放大率透镜
variable power system 变焦度系统;可变焦度系统
variable power telescope 变倍望远镜
variable pressure 不定压;变压;变动压力

variable pressure accumulator 变压式蓄热器
variable pressure capillary viscometer 变压毛细管黏度计
variable pressure drop 可变压降
variable pressure operation 变压运行
variable pressure transducer 变压式传感器
variable prices 变化的价格
variable profile antenna 变轮廓天线
variable profit 可变利润
variable profit ratio 变动利润比率
variable propeller 可变螺距推进器;可变螺距螺旋桨
variable property element 变性质元素
variable proportion 变动比例
variable pump 变量泵
variable quadri-correlator 可变自动相位控制器
variable quantity 可变(参)量;变量
variable radio-frequency radiosonde 变射频无线电探空仪
variable radio object 变射电体
variable radio source 变射电源
variable radius arch 变半径拱
variable radius arch dam 变(半)径拱坝
variable radius theory 变径理论
variable radix 可变基数;可变底数
variable range 可调范围;可变范围;不定范围;变量域;变量范围;变程
variable range marker 可变距标;活动距标
variable range ring 可变距离圈
variable rate 变动比例
variable rate CDs 变动利率定存单
variable rate certificates of deposits 变动利率存单;可变利率存单
variable rate contract 可变利率合同
variable rate gilt-edged bonds 可变利率的金边债券
variable rate gilts 变动利率优良证券
variable rate mortgage 可变利率抵押(贷款)
variable rate of cost 成本变动率
variable rate of production 可变产量
variable rate pricing 可变利率作价
variable rate stock 可变利率证券
variable rate suspension 变刚度悬架
variable ratio 可变比
variable ratio pantograph 可变比例缩放仪
variable ratio reinforcement 变比型强化
variable ratio relay 变比继电器
variable-ratio transformer 可变比变压器
variable reach 易变河段
variable reactor 可变电抗器
variable recoil 可变后座
variable recoil control rod 节制杆
variable record length 可记录长度
variable reduction 可变缩拍
variable reduction method 可变简约法
variable reference 变量引用
variable regime 变动状态;变动情势
variable region 可变区
variable relation 变量关系
variable reluctance 可变磁阻(抗)
variable-reluctance detector 变磁阻检波器
variable-reluctance electromagnetic pendulum seismograph 变磁阻电磁摆地震仪
variable-reluctance microphone 变磁阻微音器;电磁传声器
variable reluctance pick-up 可变磁阻拾音器;变磁阻拾震器
variable-reluctance seismograph 变磁阻地震仪
variable reluctance stepper motor 可变磁阻步进电动机
variable reluctance transducer 可变磁阻传感器;变磁阻传感器
variable repeated load 可变反复荷载;变复荷载
variable reserves 可变准备;变动准备
variable resistance 可变阻力;可变电阻
variable-resistance accelerometer 可变电阻式加速度计
variable-resistance arm 可变电阻臂
variable-resistance calipers 变阻式井径仪;变电阻式测径仪
variable-resistance helium thermal bridge 变热阻氦热桥
variable-resistance level transducer 变阻式液位传感器
variable-resistance pick-up 变阻拾音器;变阻拾震器;变电阻拾声器
variable-resistance transducer 变阻传感器
variable resistance type flowmeter 变阻式流量计
variable resistor 电阻箱;变阻器;可变电阻器
variable resonator 可调谐振腔
variable response model 变动响应模型
variable response unit 电子响应装置
variable risk 可变风险
variable sampling 可变抽样法;变动抽样法
variable sampling bias 可变采样偏移
variable sampling precision 可变采样精度
variable's attribute 变量属性
variable scope 变量作用域
variable section 可变截面
variable section bar 变截面杆件;变断面杆件
variable section schottky barrier gate 变截面肖特基势垒栅
variable section surge chamber 变断面调压室
variable selection 变量选择
variable selectivity 可变选择度
variable sensitivity 可变灵敏度
variable sequence robot 可变程序机器人
variable series 变量数列
variable shutter 可变快门;可变光闸
variable sign 可变指示标志
variable size item 可变大小项
variable size node 变大小的节点;可变尺寸节点
variable slope 变坡;变比降
variable-slope pulse modulation 变斜率脉冲调制
variable slope tilting flume 变坡陡槽
variable sound absorber 可变吸音体
variable source area 不定污染源区
variable source area simulator 不定污染源区模拟器
variable source model 动态流域模型;不定污染源模型;变动水源模型
variable space 变量空间
variable-spacing interferometer 变距干涉仪
variable specific heat 不定比热;变比热
variable speed 可变速度;变转速;变速(度)
variable-speed axle generator 变速轴发电机
variable-speed belt feeder 变速胶带给料机
variable-speed blower 变速鼓风机
variable-speed chopper 变速斩光器;变速调制器
variable-speed chopper blade 变速调制器叶片
variable-speed chopper disk 变速斩波盘
variable-speed cone 变速锥
variable-speed control 可变速度控制器;变速控制
variable-speed control system 可变车速控制系统
variable-speed damper 变容消振器;变速消振器
variable-speed device 变速器
variable-speed drive 变速驱动装置;变速驱动
variable-speed electric(al) submersible pump 变速潜油泵
variable-speed engine 变速发动机
variable-speed fluid drive 无级液力变矩器;变速液压传动
variable-speed gear 变速齿轮;变速装置
variable-speed generator 变速发电机
variable-speed hydraulic governor 液压无级调速器
variable-speed inclined apron feeder 变速倾斜裙板喂料机
variable-speed indicator 变速标志
variable-speed induction motor 变速感应电动机
variable-speed lowering 变速下降
variable speed metering pump 变速计量泵
variable-speed model 变速模型
variable-speed modem 变速调制解调器
variable-speed motion 不等速运动;变速运动
variable-speed motor 调速电动机;变速马达;变速电动机;变速机
variable-speed planer 变速刨床
variable-speed plate feeder 变速板给料机
variable-speed pump 变速泵
variable-speed reducer 变速减速机
variable-speed scanning 变速扫描法;变速扫描
variable-speed shaft 变速轴
variable-speed sign 可变速度显示牌
variable-speed transmission bearing 变速传动轴承
variable-speed turbine 变转速汽轮机;变速汽轮机
variable-speed unit 无级变速装置;无级变速器
variable-speed wheel gear 变速轮
variable spring 变流量泵
variable spring hanger 变量弹簧吊架
variables sampling 变量抽样法
variable stage 灵活舞台
variable standard 可变动定额;可变标准器
variable standardization 变量标准化
variable star 变星
variable state 不稳定状态
variable stator blade 可调导叶
variable stator vane 变距静叶;变距定子叶片
variable step size 可变网距;可变步长
variables test 变量测试
variable stiffness method 变刚度法
variable stop 可调光栏
variable storage 可变储[贮]存量
variable stroke control 行程的无级调节
variable stroke engine 可变冲程发动机;变冲程(式)发动机;变程发动机
variable stroke feeder 变冲程给料机
variable stroke plunger pump 变冲程柱塞泵
variable stroke pump 变(冲)程泵;变排量泵
variable structure computer 可变结构计算机
variable substitution 变量误差;变量替换
variable supply 变动供给
variable symbol 可变符号;变量符(号)
variable system 变量系统
variable temperature cryostat 可变温度低温恒温器;变温低温恒温器
variable temporary duty 变载短时工作方式
variable terms 变项
variable-thickness arch 变厚度拱
variable-thickness arched dam 不等厚度拱坝
variable-threshold decoding 变门限解码
variable thrust 可调推力;可变推力
variable-thrust engine 可变推力发动机
variable-thrust motor 可调推力发动机
variable-thrust unit 可调推力发动机
variable time 可变时间
variable time increment 可变时间增量
variable time scale 可变时间比率;可变时标
variable time-setting relay 可调时间继电器
variable timing unit 可变定时装置
variable-tolerance-band compaction 可变裕度带压缩;可变容差数据精简
variable torque driving system 变转矩驱动系统
variable torque motor 可变转矩电动机
variable transconductance tube 变跨导管
variable transformer 可调变压器;可变比变压器;调压变压器
variable translating transponder 变频转发器
variable transmission 变速传动
variable traverse motion 变幅往复运动
variable unit hydrograph 变动单位(过程)线
variable utilization 可变利用度
variable-value control 变值控制
variable valve 可调节阀
variable vane pump 变量叶片泵
variable variate 可变异的
variable varied flow 不稳定变速流
variable vector 变(量)向量
variable velocity 可变速度;变速(度)
variable-velocity scanning 变速扫描
variable-velocity volute casing 变速蜗壳
variable viscosity 可变黏度
variable visibility 多变能见度
variable voltage 可调电压;可变电压
variable voltage control 变压控制
variable voltage generator 可调电压发电机;可变电压发电机
variable voltage variable frequency 变压变频
variable voltage variable frequency control equipment 变压变频控制装置
variable voltage variable frequency inverter 变压变频逆变器
variable voltage variable frequency inverter fan 变压变频逆变器风扇
variable volume pump 变流量泵
variable water level method 变水头法
variable waterway 变迁型河段
variable wavelength phase microscope 可变波长相位显微镜;变波长相位显微镜
variable weather 易变天气
variable weight 可变权
variable weighting index number 可变组成指数
variable wind 不定(向)风
variable window length 可变窗长

variable window length correlation method 可变窗长对比法
variable word length 可变字长;代码可变长度
variac 自耦(调压)变压器
variamine blue 变胺蓝
variamine blue B base 变胺蓝 B 色基
variamine blue salt 变胺蓝盐
variamine blue salt B 原色重氮盐蓝
variance 自由度;离散;特许施工证;差异;方差;变异数;变量
variance account 差异账户
variance analysis 偏差分析;差异分析;方差分析;变异分析
variance between primary sample units 初级抽样单位间的方差
variance components method 方差分量法
variance contribution 方差贡献
variance correction coefficient 方差校正系数
variance-covariance matrix 积差阵;方差—协方差矩阵;方差—积差矩阵
variance-covariance propagation law 方差—协方差传播律
variance distribution 方差分布
variance-exchange-rate system 自由浮动汇率制度
variance for stratified sampling 分层抽样方差
variance from planned cost of materials 材料计划成本差异
variance from standard method 标准差异法
variance function 方差函数
variance heterogeneity 方差的不一致性
variance in calculation 计算差异
variance inflation factor 方差膨胀因子
variance in price 价差
variance in quantity 量差
variance integration 偏差积分
variance in wages 工资差额
variance law 方差律;变分律
variance-length curve 变异长度曲线
variance-mean ratio 方差平均值比率
variance of a probability distribution 一个概率分布的方差
variance of classified data 分类数据方差
variance of conditional simulation 条件模拟方差
variance of deviation ratio 偏差比的方差
variance of estimation error 估计误差的方差
variance of non-conditional simulation 非条件模拟方差
variance of population 总体方差
variance of probability distribution 概率分布方差
variance of proportion 比例的方差
variance of ratio estimate 比率估计值的方差
variance of the Poisson's process 泊松过程变量
variance percentage 方差百分比
variance proportion 方差比例
variance rate 变化(速)率
variance ratio 方差比(率);变化率
variance-ratio distribution 方差比分布;F 分布
variance ratio test 方差比率检验;方差比检验
variance-ratio transformation 方差比变换
variance-reducing technique 方差缩减方法;方差缩减技术
variance-regression analysis 方差—回归分析
variance report 差异报告
variance stability 方差稳定性
variance test 方差检验
variance within primary sample units 初级抽样单位内的方差
variance with the contract 与合同有出入
variant 统计变式;差异的;不同的;变(异)体;变异的;变化的
variant character 异体;变数字符
variantional design 异体设计
variantly colo(u)red 杂色的
variant name 别名
variant of two borehole electromagnetic wave method 井中电磁波法工作方式
variant part 变体部分
variant pump 变速泵
variant scalar 变量标量
variant structure 变体结构
variant train diagram 分号运行图
variant train working diagram 方案运行图
variate 差动测量;变值
variate difference analysis 变差分析

variate-difference method 变量差分法;变差法
variate-value 变量值
variation 二均差;变异;变数;变迁;变更;变分;变度;变动;变差
variational 变化的
variational approach 变分(的方)法
variational calculus 变分学
variational curve 变分曲线
variational derivative 变分微商;变分导数
variational difference scheme 变分差分格式
variational equation 变分方程
variational formula 变分公式
variational formulation 变分公式
variational index 变化性指数
variational inequality 月运动不等性;变分不等式
variational integral 变分积分
variational method 变差法;变分法
variation along dip direction 沿倾向变化
variation along strike direction 沿走向变化
variational orbit 变分轨道
variational parameter 变分参数
variational principle 变分原理;变分法则
variational problem 变分问题
variational semi-diurnal component 变异半日分潮
variational semi-diurnal constituent 变异半日分潮
variational sensitivity 变分灵敏度
variational series 变异数列
variational wave function 变分波函数
variation amplitude of orebody 矿体变化幅度
variation analysis 变动分析
variation and control 变异和控制
variation between years 年际变化
variation calculation 变分计算
variation calculus 变分法计算
variation chart 地磁磁变图
variation chart in grade 品位变化曲线图
variation clause 灵活条款;可变条款
variation coefficient 变化因数;变化系数;变分系数;变动系数;变差系数;不均匀系数
variation coefficient of grade 品位变化系数
variation coefficient of unit weight 体重变化系数
variation curve of water table 水位波动曲线
variation diagram 变异图解
variation equation 变分方程
variation factor 不均匀系数;变化因数;变化系数
variation flow analysis 变动工艺流程分析
variation from norm cost 定额成本差异
variation from quality offered 品质与报价不符
variation from quality ordered 品质与订单不符
variation from standard cost 标准成本差异
variation from standard method 成本差异法
variation in abilities 工作能力差别
variation in average 平均数变动率
variation in bidding conditions 投标内容的变化
variation in brightness 亮度变化
variation in channel depth 航道水深变化
variation in discharge 流量变化
variation in earth's temperature 地温变化
variation in level of supports 支座(面)变位
variation in load 荷载变化;负载变化
variation in precipitation 雨量变化;降雨(量)变化;降水量变化
variation in sawing 锯偏;脱离锯线的部位
variation in scale 比例尺差异
variation in sensitivity 灵敏度变化
variation in speed 速度变化
variation in storage level 储存量变化
variation in temperature 温度变化
variation in temperature stresses 温度应力变化
variation intensity of orebody 矿体变化程度
variation in traffic flow 交通量变化
variation in velocity 速度变化
variation in voltage 电压变动
variation in wastewater flow rates 废水流量变化
variation in water level 水位变化
variation in water table 水位变化
variation in yearly traffic 年交通量变化
variation iteration 变分迭代法
variation magnetic field 变化磁场
variation map of syngenetic mineral 自生矿物变化图
variation margin 价格变动保证金;变动保证金
variation method 变分法
variation movement 回归性运动

variation of aids 航标变动;航标异动
variation of annual ore reserves 年度矿量变动
variation of blanking level 消隐电平变化
variation of brightness 亮度变化
variation of coefficient 变差系数
variation of compass 磁针变化
variation of contact wire 接触线畸变
variation of contract 合同变更
variation of coordinate method 坐标变换法
variation of development reserves 开拓矿量变动
variation of elements 根数变值法;参数变值法
variation of extraction reserves 回采矿量变动
variation of extremeness value method 极值变分法
variation of fault attitudes 断层产状变化特征
variation of field intensity 场强变化
variation of flow velocity along its course 沿程流速变化
variation of geologic(al) reserves 地质储量变动
variation of gloss 表面光泽不均匀
variation of horizontal pressure 水平压力变化
variation of instrument reading 仪表读数偏差
variation of latitude 纬度变化
variation of linear type 线性变化
variation of mining stope ore reserves 采场矿量变动
variation of nugget-effect type 块金效应型变化
variation of orebody 矿体变化性
variation of orebody occurrence 矿体产状变化
variation of orebody thickness 矿体厚度变化
variation of ore grade 矿石品位变化
variation of ore order 矿石品级变化
variation of ore type 矿石类型变化
variation of ore unit-weight 矿石体重变化
variation of parabolic(al) type 抛物线形变化
variation of parameter 参数的变更;参数变值法;参数变化
variation of poles 极移
variation of precipitation 雨量变化;降雨(量)变化
variation of prepared reserves 采准矿量变动
variation of price 价格变化;价格变动
variation of production reserves 生产矿量变动
variation of quantity 工程量变更
variation of radiant energy 辐射能变化
variation of rainfall 雨量变化;降雨(量)变化
variation of random type 随机型变化
variation of reservoir storage 库容变化;水库蓄水变化
variation of risk 风险变更
variation of season ore reserves 季度矿量变动
variation of sign 正负号替换;号的变更;符号变更
variation of stress 应力变化
variation of tape sensitivity 胶带不均匀度
variation of the declination 磁偏角变化
variation of tolerance 公差外
variation of twinkling type 跃迁型变化
variation of vertical 垂线变化
variation of works 工程变更
variation on contract 变更合同
variation orbit 变分轨道
variation order 变更设计文件;修改通知书;修改通知单;工程变更通知(单);变更命令
variation order procedure 变更命令程序
variation parameter 变化参数
variation parameter system 变参数系统
variation per day 每天变化量;日变量
variation per hour 每小时变化量
variation per minute 每分钟变化量
variation property of orebody 矿体变化性质
variation range 波动范围;变化区间;变化范围;变动范围
variation range of orebody 矿体变化范围
variation rate of orebody 矿体变化速度
variation rate of spring discharge 泉的流量变化率
variation rule 变化规律
variation sequence method 变异序列法
variation sequence of mineral size 矿化规模变异序列
variation series 变异数列
variations from plumb 垂直偏差量;偏离铅锤量
variations of gravity 重力变化
variations of water table 地下水位变化
variation tone 变调音
variation vector field 变分矢量场

variation within a year 年内变化
variator 温度补偿器;伸胀缝;伸缩接缝;变速器
varicolo(u)red 杂色的
varicolo(u)red clay 杂色黏土
varicolo(u)red subclay 杂色亚黏土
vari-distant rigidity rib 变间隔加固肋
varidrive 变速马达
varied 各种各样的;多种多样的;不同的
varied-angle glossmeter 变角光泽计
varied clay 层状黏土
varied curve 邻曲线;变曲线
varied discipline 各专业
varied flow 不均匀流;变速流
varied flow function 变流函数
varied irradiance level 变化的辐照度
varied length jib 可变长度起重臂
varied mineral nutrient conditions 不同矿物质营养条件
varied usage 变化使用
variegate 游离色;颜色缤纷;颜色斑驳;弄成杂色;发花
variegated colo(u)r 杂色;斑纹着色
variegated copper ore 斑铜矿
variegated glass 大理石纹玻璃
variegated glaze 斑纹釉
variegated leaf 斑叶
variegated marble 有彩斑大理石;杂色大理石
variegated of yellowish green 黄绿色的杂色图案
variegated position effect 花斑型位置效应
variegated sandstone 杂色砂岩
variegated variety 斑叶品种
variegated ware 斑纹(装饰)器皿;斑纹玻璃
variegate-leaved plant 斑叶植物
variegation 彩斑(现象)
varies in shape from linear to oval 直线形至椭圆形
varietal characteristic 品种特性
varieties and designs 花色品种
variety 品种;品质;多样性;多样化;变种
variety certification 品种检定
variety collection 品种收集
variety competition 品种竞争
variety control 品种控制
variety hall 杂技场;曲艺场
variety of soil 土壤变种
variety of the white fleshed group 白色果肉品种
variety plot 品种区
variety reduction 品种简化;品种减少
variety sander 棱角砂光机
variety saw 多用锯
variety shop 杂货店
variety store 杂货店
variety test 品种试验
variety test plot 品种试验小区
variety theatre 杂耍剧场
variety trial 品种试验
varifocal 变焦距;可变焦距(的)
varifocal erecting system 变焦正像系统
varifocal lens 变焦距透镜;可变焦距镜头
varifocal lens binocular 变焦距透镜双筒望远镜
varifocal Moire zone plate 可变焦距莫尔波带片
varifocal objective 变焦物镜
varifocal optical system 变焦光学系统
varifocal scanning system 变焦扫描系统
varifocal viewfinder 万能取景器
varifocus lens 变焦透镜;变焦物镜
Varignon's theorem 瓦利农(计算力矩和)定理
varigradation 差异均夷作用
varigrained 多种粒径的;多种粒度的
varimax method 最大方差法
varimax rotation factor solution 方差最大旋转因子解
varimin method 最小方差法
varindor 可变电感器;变感器
varing head 变水头
varing magnetic field 变化磁场
vari-nip press 可变辊隙压滤机
variocoupler 可变耦合器
variodencer = variodenser
variodenser 可变电容(器);变容器
Variogon lens 法里奥贡镜头
variogram 变量图
variogram of conditional simulation 条件模拟变差函数

variogram of non-conditional simulation 非条件模拟变差函数
variograph 变压计;变量计
variohm 可变电阻器;变阻器
Vario Klischograph 电子刻版机
variola 天花
variole 球颗
variolite 球颗玄武岩
variolitic 球粒玄武岩的
variolitic texture 玄武球颗结构;球颗结构
variolosser 可控损耗设备;可变损耗器
variometer 可变电感器;爬升率指示器;磁偏差计;变压表;变线仪;变感器
variometer bar 扭秤横杆
variometer coil 可变电感器线圈
variometer of mutual inductance 互感式可变电感器
variometer rotor 变感转动线圈
variometer stator 变感器固定线圈
varioplex 可变多路传输器;变路转换器;变工制
varioscale projection 变比例投影
various 各种各样的;多样的
various colo(u)r chlorinated rubber boottopping paint 各色氯化橡胶水线漆
various colo(u)r chlorinated rubber enamel 各色氯化橡胶瓷漆
various colo(u)r epoxy ester primer 各色环氧酯底漆
various colo(u)r epoxy ester putty 各色环氧酯腻子
various colo(u)r polyurethane enamel 各色聚氨酯瓷漆
various colo(u)r acrylic baking enamel 各色丙烯酸烘干瓷漆
various colo(u)r acrylic gloss coating 各色丙烯酸有光涂料
various colo(u)r acrylic matt coating 各色丙烯酸无光涂料
various combination 不同组合
various daily necessaries 日常必需品
various drilling condition 各种钻探情况
various extractants 各种提取剂
various hoisting devices 各类提升钻具
various household supplies 日用杂货
various interests 各种权益
various jobs 杂工
various matrixes 各种胎体(金刚石钻头)
various moisture contents 不同水分含量
various nitrogen carriers 不同氮载体
various permanent 各种永久性
various revisions 各种修改
various system alternatives 各种系统备选方案
various uses 各种用途
various walks of life 各行各业
various way solenoid valve 多通电磁阀
variplotter 自动作图仪;自动曲线绘制器;可变绘图仪
Variscan 华力西阶【地】
Variscan cycle 华力西旋回
Variscan-Indochina period 华力西—印支期
Variscan orogeny 华力西造山运动
variscite 磷铝石
variscope 彩色摄影器
variset 可变装置;可变框架结构
varisized 各种大小的
vari-slide 滑杆成形机
varislope screen 多层坡度加大筛
vari-speed drive 无级变速转动装置;无级变速传动(装置)
vari-speed feeder 变速给料机
varister 压敏电阻;可调电阻器;变阻器;可变电阻;调节电阻器;非线性电阻;变阻二极管
varister-compensated circuit 变阻器补偿电路
varister rectifier 变阻整流器
varistor = varister
varistructure 可变结构
varistructured system 可变结构系统
vari-sweep technique 变频带扫描技术
varisymbol 变符板
Varitone Mahogany 瓦利value花岗岩(一种带有黑灰色斑点的暗红色的)
varitran 自耦变压器;接触调压器
varitrol 自动调节系统
varitron 变光管
varley slide 杠杆式电阻箱

Varley's loop method 华莱回路法
Varley's loop test 华莱回路测验
varmeter 乏尔计
Varnene 瓦内内漆(商品名,一种油质着色剂)
varnish 假漆;清漆;粉饰
varnish base 清漆漆基;清漆底子;漆底
varnish black 清漆用炭黑
varnish brush 漆刷;清漆毛刷
varnish coater 涂清漆机
varnish coating 清漆涂层;涂清漆
varnish colo(u)r 清漆色;清漆涂料;涂清漆色
varnish cover 清漆涂层
varnish cure 清漆熟化;清漆固化
varnish dryer 清漆干燥剂;清漆催干剂
varnish-drying 清漆干燥
varnished 浸渍过的;上(清)漆的
varnished bias tape 涂漆料纸带;黄蜡带
varnished cambric 漆皮细麻布;漆布;涂清漆纱布
varnished-cambric covered cable 漆布绝缘电缆
varnished-cambric insulated cable 漆布绝缘电缆
varnished car (火车的)卧车;特别快车;(火车的)客车
varnished cardboard 漆纸板
varnished cloth 漆布
varnished cloth insulation 漆布绝缘
varnished cloth tape 绝缘布带;浸漆布带;漆布带
varnished cotton tube 涂漆套管
varnished fabric 浸漆织物;漆布
varnished glass cloth 玻璃漆布
varnished glass fabric 涂漆玻璃织物;玻璃(纤维)漆布
varnished glass tube 玻璃纤维漆管
varnished insulation 浸漆绝缘
varnished leather 漆革
varnished paper 浸渍绝缘纸;漆纸
varnished pressboard 浸渍漆板;漆纸板
varnished silk 浸漆丝绸;浸漆丝绸
varnished silk tape 绝缘丝带;浸漆丝带
varnished skeleton cannoned hour hand 凸起空心时针
varnished skeleton cannoned minute hand 凸起空心分针
varnished skeleton hour hand 凸起时针
varnished skeleton minute hand 凸起分针
varnished sleeve 漆(套)管
varnished tape 漆(布)带
varnished tube 浸漆绝缘管;黄蜡套管
varnished wire 漆包线
varnisher 清漆工人;油漆工人;涂清漆工
varnish extender 油漆调和料
varnish film 漆膜
varnish finish 清漆涂层
varnish for building construction purposes 房屋建筑用的清漆
varnish for deadening sound 隔音清漆;隔声清漆
varnish for foil 金属薄片用清漆
varnish formation 漆膜形成
varnish formulation 清漆组成
varnish for typographic(al) ink 凸版油墨用调墨油
varnish gum 清漆树脂
varnishing 涂漆;涂(刷)清漆;刷清漆;上清漆
varnishing brush 清漆刷
varnishing door 隐门
varnishing line 没影线
varnishing machine 浸(渍)漆机
varnishing point 没影点;遁点
varnish kettle 清漆罐;清漆锅
varnish lacquer 清漆;上清漆
varnish linseed oil 清漆亚麻仁油
varnish maker 清漆制造厂
varnish maker's and painter's naphtha 油漆制造和施工用石脑油;清漆与油漆用石脑油;漆用溶剂油
varnish make's and painte's naphtha 漆用石脑油
varnish maker's naphtha 清漆用石脑油
varnish manufacture 清漆制造
varnish medium 清漆介质
varnish oil 清漆(用)油
varnish paint 清漆(涂料);树脂性瓷漆
varnish paper 浸渍绝缘纸;浸渍绝缘;涂漆绝缘纸
varnish pot 清漆罐;清漆锅
varnish raw material 清漆原材料
varnish remover 脱漆剂;除清漆剂;除漆剂
varnish resin 清漆(用)树脂

varnish rubbing 清漆磨退
varnish silk 浸漆绸
varnish silk tape 绝缘绸带；浸漆绸带
varnish soybean oil 制清漆豆油
varnish spray gun 喷漆枪
varnish stain 清漆(型)着色剂；有色清漆
varnish system 清漆类别
varnish test 清漆试验
varnish-treated tape 漆带
varnish-tree 漆树；清漆树
varnish tube 浸漆纤维管
varnish-type paint 清漆
varnish vehicle 清漆媒介液；清漆媒介剂
varnish work 涂刷清漆
varnsingite 钠长楣伟晶岩
Varotal lens 法罗泰尔镜头
Varsailles garden 凡尔赛庭园
varulite 绿磷锰钠石
varve 纹泥；季候泥；成层黏土
varve(d) clay 明显分层的黏土；纹泥；带状黏土；季候泥(黏土)；成层黏土
varved dating method 纹泥测年法
varved glacial clay 冰河纹泥
varved structure 缟状构造
varve slate 纹板岩
vary directly as 和…成正比例
varying accelerated motion 变加速运动
varying acceleration 变加速度
varying-amplitude test 变振幅试验
varying-area channel 变截面管道
varying capacitive tape reader 变容式纸带输入机
varying capacity 可变容量
varying climatic and soil conditions 不同的气候和土壤条件
varying cross-section 变截面
varying density atmosphere 变密度大气
varying duty 可变负载；可变负荷；变化负载；变负荷；变动负载；变动操作条件；在变动负载下运行
varying function 变函数
varying length string 改变长度字符串
varying load 可变负载；不定荷载；不定负载；不定负载(荷)；变化负载；变动负载；变(动)负载；变动负荷；变量荷载
varying loading 变量荷载
varying optic(al) density 可变光学密度
varying parameter 可变参数
varying permeability trap 渗透性变化造成的圈闭
varying plant population 不同植物的群体
varying rigidity 变刚度
varying rigidity method 变刚度法
varying salt concentration 不定盐浓度
varying scale 可变比例尺
varying speed motor 变速马达；变速电动机
varying stress 变(化)应力
varying-stress amplitude 变应力振幅
varying time step 变动时间步长
varying-voltage control 变压控制
varying-voltage generator 变压发电机
varying-voltage rectifier 变压整流器
varying water head 变水头
vary inversely as 和……成反比例
vary off-line 使脱机
vary on-line 联机修改；使联机
vary-power telescope 可变倍率望远镜
vary-volume blower 变容式鼓风机
vary within-wide limits 变化幅度大
vasa-murrhina glass 大理石纹玻璃
vas collaterale 侧副管
vascular tracheid 导管状管胞
vasculum 采集器
vase 花瓶；瓶饰；瓶(形物)
vase carpet 瓶饰地毯
vased-wall (日本式)被服壁
vase-form 杯状的
vaseline 凡士林；矿脂；软石脂
vaseline gauze 凡士林纱布
vaseline oil 凡士林油
vase of Emperor Tangtai Zhong 唐太宗青花瓶(瓷器名)
vase of national amorous feeling 民族风情瓶
vase-painting 瓶饰涂饰
Vase rug 伐斯地毯
vase-shaped crown 杯状型树冠
vase with ancient chinese characters 古文字瓶 (瓷器名)
vase with dance design 舞瓶(瓷器名)
vase with enamel floral design 珐琅彩花卉纹瓶
vase with panda design 熊猫瓶(瓷器名)
vase with small mouth and pointed base 小口尖底瓶(瓷器名)
vase with washer shaped month 洗口瓶
vase with white prunus blossom design 白梅瓶 (瓷器名)
vashegyite 纤磷铝石；水羟磷铝石
vasiform 脉管状的
vasogen 矿脂氧化物
vast ice-floe 巨浮冰块
vast scale 大量的
vast scheme 庞大计划
vast stretches of paddy fields 成片的稻田
vat 木桶；大桶；大槽
vat black 还原黑
vat blue 还原蓝；瓮蓝
vat brilliant violet 瓮亮紫
vat colo(u)r 瓮染料
vat colo(u)ring 瓮染
vat dark blue 瓮暗蓝
vat deep printing black 瓮深黑
vat dye (stuff) 还原染料；瓮(染)染料
vaterite 六方碳钙石；球霰石
vatflation 增值税引起的通货膨胀
vat golden orange 还原金橙
vat golden yellow 还原金黄
vat grass green 还原草绿
vat grass yellow 还原草黄
vat green 还原绿
vat grey 还原灰
Vatican 梵蒂冈教廷
vat leaching 槽浸
vat lined 抄合
vat orange 还原橙色
vat powder 保险粉
vat printing assistant 瓮印染辅剂
vat red 还原红
vat red brown 还原红棕
vat scarlet 还原大红
vat sulfur dye 还原系列硫化染料
vatting 还原染色；瓮染
vat waste 沉积残渣
vat water disinfection 缸水消毒
vat yellow 还原黄；瓮黄
vaucanson chain 钩环节链
vauclusian spring 龙潭
vaudaire 沃丹尔风
vauderon 沃丹尔风
vaudeville theater 杂耍剧场
vaugnerite 暗花岗闪长岩
vaul-covered 拱顶覆盖的
vault 拱(形圆)顶；拱穹；金库；穹顶；钱箱；保险库；保管库房
vault abatement 拱顶压低
vault abutment 拱顶支座
vault action 拱顶作用
vault anchor 拱拉杆
vault apex 拱顶尖
vault apex block 拱顶的顶尖块
vault arch 拱门
vault axis 拱顶轴线
vault back 拱顶的拱背；拱顶外拱面
vault basilica 有拱顶的长方形建筑物
vault basilica church 有拱顶的长方形教堂
vault bay 穹隆开间
vault block 拱顶的楔形块
vault bond 拱顶的砌合
vault capping 拱顶的压顶
vault cash 库存现金
vault ceiling 拱顶顶棚；穹隆顶棚
vault center line 拱顶轴线；拱顶中心线；拱顶轴线
vault Christian basilica 有拱顶的长方形基督教堂
vault compartment 拱顶的隔舱
vault construction 拱顶建筑；拱结构
vault construction material 拱顶建筑材料
vault coping 拱顶的压顶
vault corridor 有拱顶的走廊
vault cover 拱顶盖
vault crown 拱顶尖
vault crown block 拱顶的顶尖块
vault depth 拱顶高度
vault door 保险库门；地窖门
vault ecclesiastical basilica 教堂中有拱顶长方形会堂
vaulted 穹隆状的；拱状的；盖有拱顶的；有圆顶的；穹顶的
vaulted area 拱顶下面积
vaulted basement 穹顶地窖
vaulted building 有拱顶的房屋
vaulted ceiling 拱顶顶棚；拱形顶棚；穹隆顶棚
vaulted chamber 有拱顶的小室
vaulted church 有拱顶的教堂
vaulted concrete 拱顶混凝土
vaulted dam 拱坝；双曲拱坝
vaulted edifice 有拱顶的大厦
vaulted hall 有拱顶的大厅
vaulted interior 有拱顶的内厅
vaulted plate 拱形板
vaulted roof 拱顶屋面；穹隆屋顶；圆屋顶；拱状屋顶；拱形屋顶
vaulted shell 穹壳；双曲壳体
vaulted slab 拱形板
vaulted structure 拱顶结构
vaulted walk 有拱顶的走道
vaulted work 拱顶砌筑工作；穹顶工程
vault extrados 拱顶外拱面；拱顶沟背
vault facing 拱顶饰面
vault form(work) 拱模板
vault grid 拱形格栅
vault haunch 拱顶的拱簸
vault head 圆拱顶盖
vault impost 拱顶的墩座
vaulting 穹顶工程
vaulting capital 支持穹隆的柱顶；拱顶扁倚柱头；穹隆柱顶
vaulting cell 穹隆开间
vaulting cone 拱形锥
vaulting course 穹隆的拱脚石层；穹隆的拱脚横带
vaulting engineering 拱顶工程
vaulting horse 鞍马；拱柱；拱墩
vaulting intrados 拱顶内拱面
vaulting masonry 穹顶圬工；筒拱圬工
vaulting masonry work 拱顶砌筑
vaulting pillar 筒拱柱
vaulting shaft 支穹柱身；拱顶支柱；筒拱柱；承肋柱；装饰性支柱
vaulting surface 拱顶表面
vaulting tile 穹隆用空心砖
vault key(block) 拱顶石
vault light 地下室采光；穹隆照明；地下室照明；地下室顶光；地下室灯
vault line 拱顶轮廓
vault lining 拱顶衬砌
vault-of-Heaven vase 天球瓶
vault painting 拱顶油漆
vault pattern 拱顶形式
vault pressure 拱顶压力
vault rib 穹隆拱肋；穹顶肋
vault rise 拱顶高
vault roof 拱顶；穹隆屋顶
vault severy 拱顶分隔块
vault shuttering 拱顶模架
vault soffit 拱腹；穹顶底面
vault span 拱顶跨度
vault spandrel 拱顶肩部
vault springer 拱顶脚石
vault style 拱顶风格
vault theory 穹隆理论
vault thickness 拱顶厚度
vault thrust 拱顶推力
vault top 拱顶顶；拱顶端
vault top block 拱顶的端块；拱顶的顶块
vault type 拱式
vault-type transformer 地下室型变压器
vault vertex 拱顶顶
vault vertex block 拱顶的端块；拱顶的顶块
vault with dentate springing lines 有齿形起拱线的拱顶
vauquelinite 磷铬铜铅矿
vauxhall bevel 古式斜边(小于七点五度)
vauxite 蓝磷铝铁矿
V-B consistometer 维勃稠度仪；VB仪
V-beam sheeting 波纹板；V形折板
V-belt 三角皮带；V形皮带
V-belt conveyer 三角皮带输送器

V-belt drive 三角胶带传动;V 形胶带传动
V-belt pulley 三角皮带轮
V-belt transmission V(ee)-belt drive 三角皮带传动
V-blade scraper with multiple drawoffs 多级抽泥的 V 形板刮泥器
V-blender V 形混合器
V-block 三角槽板;V 形块;元宝铁
V-block brake 楔形制动器
V-bottom 砂底船
V-bottom boat 尖底船
V-bottom hopper V 形底斗仓
V-box V 形分级箱;V 形箱
V-brick 多孔(空心)砖;竖孔空心砖;V 形空心砖
V-bucket carrier V 形斗链式提升机
V-bucket conveyer V 形斗式输送机
V center V 心
V-clamp V 形夹
V-column V 形柱
V-connection 开式三角形接法;V 形接法
V-crack V 形裂缝
V-crimp asbestos cement sheet 半波石棉瓦
V-cut 楔形掏槽;楔形开挖;V 形切割;V 形割法;V 形开挖
V-depression V 形凹槽
V-door V 形门
V-drag 拖式 V 形刮铲
V-drain V 形排水沟;V 形排水槽;V 形边沟
V-drainage V 形排水工程
V-drill 带尖钻
vear car 手推两轮货车;两轮拖车
veatchite 水硼锶石
veatchite-A 三斜水硼锶石
Vebe apparatus 维比稠度仪;维比仪(一种混凝土稠度测定仪)
Vebe consistency meter 维比稠度计
Vebe consistometer 维比稠度计;贝氏振动稠度计
Vebe consistometer test 维比稠度计试验
Vebe degree 维比稠度
Vebe meter 维比稠度计
Vebe specimen 维比试件
Vebe test 维比振动稠度试验
Vebe test method 维比试验法
Vebe time 维比时间
Vebe value 维比值
Vebe vibratory table 维比振动台(用于测定碾压混凝土稠度)
Vectian 阿普第阶
vectogram 向量图;矢量图
vectograph 偏振立体照片;偏振立体图;偏振光体视镜
vectograph method 偏振光立体观察法
vectography 偏振像片;偏振立体摄影术
vecton 矢量粒子
vectopluviometer 雨量计;风向雨量计
vector 矢量;向量;向径;引向目标;制导;媒介动物;传病昆虫
vector action 矢量运算
vector-active structure system 矢量有效结构体系
vector addition 向量加法;矢量加法;矢量和
vector admittance 矢量导纳;复导纳
vector algebra 向量代数;矢量代数
vector analysis 向量分析;矢量解析;矢量分析
vector angle 向量角;矢量角
vector area 矢面积
vector arithmetic 向量运算
vector-averaged current meter 矢量平均海流计
vector balancing 矢量平衡
vector-borne disease 媒介传染病
vector boson 矢量玻色子
vector calculator 向量计算器
vector calculus 矢算;矢量计算
vector canceller 矢量对消器
vector change 向量变换
vector component 分矢量
vector computer 向量计算机;矢量计算机
vector continue display mode 连续矢量显示法
vector control 向量控制;矢量控制
vector controller 矢量控制器
vector correlation 矢量相关
vector correlation coefficient 向量相关系数
vector coupling coefficient 矢量耦合系数
vector current 矢量电流;复数正弦电流
vector damping 矢量阻尼

vector diagram 向量图;矢量图
vector diagram of light vibration 光振动矢量图解
vector displacement 矢量位移
vector display 矢量显示
vector distribution 向量分布
vector drawing 向量图
vector drawing method 向量绘图法
vectored attacks 导向攻击
vector(ed) interrupt 向量中断;定向中断
vector element 矢元
vector equation 向量方程;矢量方程
vector expression 矢量式
vector field 向量场;流形上的向量场;矢量场
vector flux 矢(量)通量
vector force 矢量力
vector function 向量函数;矢量函数;矢函(数)
vector function generator 矢量函数发生器
vector ga(u)ge 定向测雨器
vector general 通用矢量
vector generation 向量产生
vector generator 向量产生器;矢量发生器;矢量产生器
vector graphics 向量图形学
vector group 向量群;矢量组
vector gunsight 向量瞄准器
vector height 矢高
vectorial 向量的;矢(量)的
vectorial addition 向量相加;矢量加法
vectorial adjustment 向量平差
vectorial angle 向量角;矢量角
vectorial astrometry 矢量天体测量
vectorial diagram 矢量图
vectorial field 向量场;矢量场
vectorial intensity method 矢量强度法
vectorial interpretation method 矢量解释法
vectorial line 有向线
vectorial mean 矢量平均值
vectorial metabolism 向量代谢
vectorial orbital constant 矢量轨道常数
vectorial photoelectric(al) effect 矢量光电效应
vectorial property 矢量特性
vectorial relation 矢量关系
vectorial resultant 矢量合成;合矢量
vectorial structure 方向构造
vectorial track method 矢量轨迹法
vectorial transport 向量运输
vectorial triangle 矢量三角形
vector impedance 矢量阻抗
vector inclination method 矢量倾角法
vector inequality 向量不等式
vectoring 引导
vectoring address 定向地址
vectoring aircraft 引导机
vectoring computer 引导数据计算机
vectoring doctrine 引导原则
vectoring error 引导误差
vectoring information 引导数据
vectoring instructions 引导指示
vectoring phase 引导阶段
vectoring post 引导哨
vectoring procedure 引导程序【计】
vectoring station 引导站
vector instruction 矢量指令
vector intensity method 矢量强度解释法
vector interaction 矢量相互作用
vectorization 向量化
vector length 向量长(度);矢量长度
vector locus 矢量轨迹
vector loop 矢量环
vectorlyser = vectorlyzer
vectorlyzer 矢量分析器
vector magnetocardiograph 向量心磁仪
vector Markov processes 向量马尔可夫过程
vector matrix 向量矩阵;矢量矩阵
vector measure 矢量测度
vector mechanism 矢量机构
vector meson 矢性介子
vector meter 矢量计
vector method 矢量法
vector mode 向量运算方式;向量(方)式
vector model 矢量模型
vector moment 向量矩
vector multiplication 矢(量)乘法
vector notation 矢量记号

vector of acceleration 加速度矢量
vector of generalized coordinates 广义坐标矢量
vector of observational correction 观测校正向量
vector of residuals 剩差向量
vector of strain 应变向量
vector of unit length 单位向量
vector of unknown parameter 未知参数向量
vector operation 向量运算;矢量运算
vector operator 向量算子;向量算符;矢算子;矢量运算子
vector-oriented 向量性
vector plotted cartographic(al) data 矢量制图数据
vector plotting 矢量绘图
vector-point diagram 矢点图
vector point plotting scanning mode 点扫描矢量绘图方式
vector polygon 向量多边形;矢(量)多边形
vector potential 向量位;向量势;矢(量)位;矢(量)势
vector power 矢量功率
vector problem 矢量问题
vector process 矢量过程
vector product 向量积;矢量(乘)积;矢积;叉积
vector projection 向量投影图;矢量投影图
vector property 矢量性质
vector quantity 矢量;向量
vector rain ga(u)ge 定向雨量计
vector ratio 向量比;矢量比
vector registor 向量寄存器
vector representation 向量表示;矢量表示(法)
vector rotation 矢量旋转
vector scan 矢量扫描
vectorscope 矢量显示器
vector separation 矢量分隔;矢量分解
vector servomechanism 矢量跟踪系统
vector set 矢量集
vector sight 矢量瞄准器
vector solution 矢量解法
vector space 向量空间;矢量空间;矢量空间
vector steering 推力矢量控制
vector stress 矢量应力
vector stress diagram 矢量应力图
vector subtraction 向量减法
vector sum 向量和;矢(量)和
vector sum excited linear prediction 矢量和激励线性预测
vector the aircraft 从地面引导飞机
vector theorem 矢量定理
vector to raster 矢量到光栅
vector tube 矢量管
vector-valued criteria 矢性判据;多目标判据
vector-valued objective function 矢性目标函数;多目标函数
vector variable 向量变量
vector variant 向量不变式
vector voltmeter 矢量伏特计;矢量电压表
vector wave 矢量波
vector wave equation 矢量波动方程
vector wave function 矢量波动函数
vectron 超高频频分析仪
vedette(boat) 舰载鱼雷艇;哨艇
vee V 形的
vee-angle 坡口角度
vee belt 三角带;V 形皮带
vee-blade scraper V 形刮片定泥机
vee-blender V 形混合器
vee block 三角槽铁
vee compressor V 形压缩机
vee crack 扩展裂缝;V 形混合器
vee crossing 人字辙叉
vee-cut V 形切割
veeder counter 测程仪【航海】;测程器;速力计
vee drain V 形排水沟;V 形排水槽
vee drainage V 形排水(沟)工程
vee-grooved pulley 三角皮带轮;V 形皮带轮
vee gutter V 形(排水)槽;三角形槽
vee joint V 形水平勾缝;V 形接缝
vee joint pointing V 形勾缝
veenered wall 砌有彩色面砖的内墙
veenite 维硫锑铅矿
vee notch 三角槽口;V 形缺口;V 形槽口
vee-notch weir 三角槽口堰
vee plate V 形极板
vee-port valve V 形孔口阀

vee pulley 三角皮带轮；V 形槽轮
veer 转方向；风向顺转；转向
veer aft 风向后移
veer and haul 放松和拉索
veer cable 锚链松出
veer from the line 船离队
veering 转向；顺转
veering of wind 风向顺转
veering wind 顺转风
vee roof V 形屋面；V 形屋顶
vee-rope V 形绳索
veer out cable 放缆
veer rope 松出绳索
vee-shaped batten V 形板条
vee slide V 形滑座；V 形导槽
vee support V 形支座
vee thread V 形螺纹；三角螺纹
vee tool V 形工具
vee tooled joint V 形勾缝
vee-trough V 形槽
Veevic 维易剂（一种油粉涂层和油漆清洗剂）
vee way V 形导轨
vega 低平原
vegetable adhesive 植物性黏合剂；植物胶黏剂；植物基料黏合剂
vegetable and fruit processing wastes 蔬菜水果加工废物
vegetable base 植物碱
vegetable basket construction 菜篮子工程
vegetable basket project 菜篮子工程
vegetable black 烟黑颜料；油烟；植物黑；松烟（黑）；灯黑
vegetable black pigment 植物黑颜料
vegetable butter 人造黄油
vegetable cellar 蔬菜地窖；菜窖
vegetable charcoal 木炭；炭
vegetable chemistry 植物化学
vegetable cork 隔热软木板；软木衬料
vegetable cover 植物覆盖（层）
vegetable crop 蔬菜（作物）
vegetable down 木棉
vegetable drying oil 干性植物油
vegetable dust 植物性填料；植物性粉尘
vegetable dye 植物染料
vegetable farm 菜园
Vegetable Farming Institute 蔬菜栽培研究所
vegetable fat 植物油（脂）
vegetable fatty acid 植物脂肪酸
vegetable fault 草籽毛
vegetable fiber 植物纤维
vegetable fiber board 木纤维板；植物纤维板
vegetable fiber material 植物纤维材料
vegetable filler 植物性填料
vegetable filling material 植物填充材料
vegetable garbage 植物性垃圾
vegetable garden 菜园
vegetable gelatin 植物胶
vegetable glue 树胶；植物胶
vegetable gum 植物胶
vegetable hair 植物纤维
vegetable industry 蔬菜加工业
vegetable insecticide 植物杀虫剂
vegetable insulation 植物绝热；植物隔声
vegetable ivory 象牙棕榈；植物象牙
vegetable jelly 植物冻
vegetable kingdom 植物界
vegetable layer 植被层
vegetable lifter 蔬菜挖掘器
vegetable lubricant 植物润滑油
vegetable matter 植物质；植物性物质
vegetable mo(u)ld 腐殖土（壤）；植物性松软沃土
vegetable oil 植物油
vegetable oil for industry purpose 工业植物油
vegetable oil tank car 食油罐车
vegetable origin 植物成因
vegetable parchment 植物羊皮纸；硫酸纸；假羊皮纸
vegetable pigment 植物颜料
vegetable plot 菜地
vegetable poison 植物毒
vegetable preparation room 蔬菜配制室；配菜室；蔬菜加工间
vegetable preparation tank 蔬菜洗涤池；洗菜池
vegetable processing 蔬菜加工
vegetable rubber 植物橡胶

vegetable silk 木棉；丝状植物纤维
vegetable sink 洗菜盆；洗菜池
vegetable sink unit 洗菜盆
vegetable soil 耕作土；植物土；腐殖土；种植土；熟化土壤；菜园土
vegetable sorter 蔬菜分级机
vegetable storage 储［贮］菜库；储［贮］菜仓
vegetable store 储［贮］菜库；储［贮］菜仓
vegetable-syntan tannage 植物合成鞣剂鞣法
vegetable tan liquor 植鞣液
vegetable tanning material 植物鞣料
vegetable tanning waste 植鞣沉淀
vegetable tar 黑油；植物焦油
vegetable theory 植物成因论
vegetable turpentine 植物质松节油
vegetable waste 蔬菜废物；蔬菜废料
vegetable wax 植物蜡；树蜡
vegetal cover 植物覆层；植物覆盖；植被
vegetal debris 植物残屑
vegetal discharge 植物散发耗水量；植被排水量
vegetal invasion of water 水草在水体中大量繁殖
vegetalization 植物化
vegetal period 植物生长期
vegetal retardance 植被阻滞
vegetated(stabilized) shoulder 植草加固路肩
vegetated waterway 植草水道
vegetation 植物；增殖体
vegetational cover 植被
vegetational growth 植被
vegetational plant geography 植被地理学
vegetational type 植被群落
vegetational zonation 植被地带性
vegetation(al) zone 植物带
vegetation and ecosystem mapping 植被和生态系统制图
vegetation anomaly 植物异常
vegetation bed 植物丛
vegetation belt 植被带
vegetation bending strength 植被弯曲强度
vegetation block 种植地段
vegetation chart 植被类型图
vegetation classification 植被类别
vegetation climate 植物气候
vegetation coast 种植海岸；植被海岸
vegetation community 植物群落
vegetation continuum 植被连续体；植被的连续
vegetation continuum concept 植被连续群概念
vegetation continuum index 植被连续群指数
vegetation cover 植被覆盖；植被被覆
vegetation coverage 植被覆盖度
vegetation cover strength 植被强度
vegetation damage 植物损害
vegetation damaging condition 植物受灾情况
vegetation disintegrator 植被切碎机
vegetation feature 植被
vegetation fibre 植物纤维
vegetation form 植被（类）型
vegetation growth 草木增长
vegetation loss 植物损失；植被损失
vegetation management 植被管理
vegetation map 植生图；植被（类型）图
vegetation monitoring 植被监测
vegetation nature 植物性质
vegetation of pal(a)eoclimate zone 古气候带植被
vegetation pattern 植被符号
vegetation period 营养期；植物生长期；植被生长期
vegetation plant geography 植被地理学
vegetation profile 植被剖面图
vegetation protection 植被保护
vegetation region 植被区域
vegetation sample 植物样品
vegetation sample type 植物样品类型
vegetation sampling locality 植物采样位置
vegetation sampling organ 植物采样器官
vegetation sampling period 植物候期
vegetation sampling position 植物采样部位
vegetation sampling record 植物采样记录
vegetation season 植物生长季节；植被生长季节
vegetation shielding influence 植被屏蔽影响
vegetation shore 植被海岸
vegetation stratum 植被层
vegetation study 植被研究
vegetation subsystem 植被亚系统
vegetation survey 植物测量；植被调查；植被测量

vegetation system 植被系统
vegetation tank 植被箱；生长箱
vegetation type 植被种类
vegetation unit 植被单元
vegetation vitality 植物生活力
vegetation zone 植被地带；海拔高的植被区
vegetative barrier 植物性屏障
vegetative breakdown 植被分类
vegetative classification 植被分类
vegetative cover（利于植物生长的）表土层；植物铺盖；植被
vegetative cycle 营养周期
vegetative euryhaline 营养广盐性
vegetative eurytherm 营养广温性
vegetative lining 植草衬护
vegetative practice 植被保护措施
vegetative protection 植被保护
vegetative sand control 植物固沙
vegetative screens 植被保护带
vegetative stenohaline 营养狭盐性；生长狭盐性
vegetative stenothermy 营养狭温性；生长狭温性
vegetative treatment 植被保护
vegeto-alkali 植物碱
vegetolene mineral matter 植物成因矿物质
vehicle 运载体；运载工具；展色剂；载体；连接料；客车；交通（运输）工具；漆料；调漆料；车辆；赋形剂；飞船
vehicle access 车辆引桥；车辆进出道；车辆出入口
vehicle accident rate 车辆事故率
vehicle actuated signal 车辆感应信号
vehicle actuated signal control 车辆感应信号控制
vehicle actuated traffic control 车动交通控制；车动交通管理
vehicle actuated traffic signal 车动交通信号
vehicle actuated traffic signal installation 车动交通信号装置
vehicle and civil architecture 船舶车辆及民用建筑型材
vehicle and implement washer 车辆与机具清洗设备
vehicle and passenger ferry 车客渡轮
vehicle and vessel license tax 车船使用牌照税
vehicle appearance 车容
vehicle area 车轴投影面积
vehicle axis system 车辆坐标系
vehicle axle 车轴
vehicle base 媒体介体；中间介体
vehicle battery 车辆用蓄电池
vehicle bridge 车辆渡桥
vehicle capacity 车辆载重（量）；车辆容量
vehicle cargo ship 车辆运输船；车辆货船
vehicle carrier 车辆运载船
vehicle chassis 车辆底盘
vehicle circulating availability 车辆周转率
vehicle circulation 车辆流通
vehicle classification count 车辆分类计数
vehicle clearance 车辆限界
vehicle compass 车载罗盘
vehicle composition 车辆组成
vehicle condition 车况
vehicle congestion rate 车辆拥挤度
vehicle construction 车辆制造；车辆构造
vehicle contour 车辆轮廓
vehicle control 车辆控制
vehicle control dynamics 车辆操纵动态特性
vehicle control scanning system 车控扫描系统
vehicle control system 飞行器控制系统
vehicle cost 车辆购置费
vehicle count 车辆计数
vehicle counter 车辆计数器；交通车辆计数器
vehicle crane 汽车起重机
vehicle cross passage 车辆横道
vehicle curb weight 车辆自重
vehicle currency 周转货币
vehicle-day 辆/日
vehicle deck 车辆甲板
vehicle deck of ship 船的车辆甲板
vehicle delay rate 车流迟滞率
vehicle design criteria 车辆设计标准
vehicle detecting equipment 侦车设备
vehicle detector 侦车器；车辆探测器；车辆感应器
vehicle detector pad（交通调查用）车辆感知板
vehicle diesel engine 车用柴油机
vehicle dimension 车辆尺寸

vehicle direct current kilowatt-hour meter 车用直流电度表
vehicle disinfection track 车辆消毒线
vehicle display 车内显示器
vehicle disturbance dynamics 车辆干扰动态特性
vehicle during the peak period 高峰车
vehicle dynamic(al) behavio(u)r 车辆动态特性
vehicle emission 车辆废气排出;机车排气;车辆排出物;汽车排放物
vehicle emission rate 车辆放射速度
vehicle engine 车辆发动机
vehicle engine photoelectric(al) igniter 车辆发动机光电点火器
vehicle entrance 车辆进出道;车辆出入口
vehicle excise tax 车辆税
vehicle exhaust 机车排气;汽车废气
vehicle exhaust analyzer 汽车排放分析仪
vehicle exhaust emission 车辆排气
vehicle exhaust monitoring equipment 机动车排气监测设备
vehicle exhaust nitrogen oxides monitor 机动车排出氮氧化物监测仪
vehicle exhaust system 车辆排气系统
vehicle extension 车辆绿灯延长时间
vehicle extension period 车辆放行延长时间
vehicle ferry 车辆渡船
vehicle finish 车辆(用)面漆
vehicle fleet 车群;车队
vehicle fleet control system 车队控制系统
vehicle flow 车流
vehicle flow paths 车辆流线
vehicle-for-hire 待运交通工具;待雇车辆
vehicle for repair label 车辆修理标签
vehicle frame section area 车辆截面面积
vehicle ga(u)ge 机车车辆界限;车辆界限;车辆界限
vehicle ga(u)ge device 限界门
vehicle gross weight 车辆毛重
vehicle guidance system 车辆制导系统
vehicle headlamp 车辆前灯
vehicle headlighting 车辆前灯灯光
vehicle hold 车辆舱【船】
vehicle-hour 辆/时
vehicle identification and location 车辆鉴别与定位
vehicle impact 行车冲击力
vehicle injury 交通工具损伤
vehicle inspection 车辆检查
vehicle-kilometer 车公里
vehicle-kilometer cost 车公里成本
vehicle kinematic envelope 车辆动态包络线
vehicle length 车辆长度
vehicle level 车辆等级
vehicle licensing 车辆许可驾驶标准
vehicle license fee 车辆执照费
vehicle lift 车辆电梯
vehicle lift gate 提升式闸门
vehicle lighting 车辆灯光
vehicle line capacity 车辆线路容量
vehicle load(ing) 车辆荷载
vehicle maintenance 车辆保养
vehicle manufacturer 车辆制造厂家
vehicle mass ratio 火箭质量比
vehicle message signs 车辆信息符号
vehicle microphone 车用扩音器
vehicle mile 车英里(1英里=1.609344千米)
vehicle mileage 车辆里程
vehicle-minute 车分
vehicle mix 车型混合率;车辆组成
vehicle mixing test 漆料混合试验
vehicle module 行车信号调制
vehicle monitoring radio communication system 车辆监控无线电通信系统
vehicle-mounted 装在车上的
vehicle-mounted generator unit 车载发电机组
vehicle noise 车辆噪声
vehicle of mixed construction 混合结构的车辆
vehicle of stain 展色料
vehicle operating cost 车辆运行费;车辆营运费
vehicle operation 车辆营运
vehicle outline 车辆轮廓线
vehicle overhang 车辆伸出部分
vehicle owner mail questionnaire method 车主函询调查法
vehicle parking 车辆停车(处)
vehicle path 车辆占用宽度

vehicle performance diagram 车辆行驶性能曲线图
vehicle pitching 行车(前后)颠簸
vehicle platoon 车队
vehicle pontoon 车辆运输驳船;车辆浮码头
vehicle-position-determining system 车辆定位系统
vehicle possession 车辆拥有量
vehicle power-supply unit 车辆电源装置
vehicle protection period 护考时间
vehicle rack container 汽车集装箱
vehicle radiator cap 车辆散热箱盖
vehicle radio 车辆收音机
vehicle radio station 车载电台
vehicle ramp 车行斜坡道
vehicle recall 车辆请求绿灯
vehicle recorder 车用播音机
vehicle reflector 车辆反光镜
vehicle registration 车辆登记
vehicle repair depot 汽车修理基地
vehicle repair track 修车线【铁】;车辆修理线
vehicle repair track at station 站修线
vehicle resistance 车辆阻力
vehicle response time 汽车反应时间
vehicle roll susceptibility 汽车侧倾敏感度
vehicle row load 车辆行列荷载
vehicle running board 车辆踏脚板
vehicle saturation level 车辆饱和状态
vehicle saturation point 车辆饱和点
vehicle scale 车辆磅秤
vehicle seat 车座
vehicle serial 汽车列车
vehicle shift 车班
vehicle-shift kilometers 车班行程
vehicle shunting device 车辆移动小车
vehicle size 车辆大小
vehicle spacing 车辆间隔
vehicles per lane per hour 每车道每小时通行车辆数
vehicle spring 车辆弹簧
vehicle station 车载电台
vehicle stopping distance 停车距离;车辆停止距离
vehicle stream 车流
vehicle subway 车行隧道
vehicle suspension spring 车辆悬挂弹簧
vehicle suspension system 车辆悬挂系统
vehicle tax 汽车运输税;车辆税
vehicle thrust 车辆推力
vehicle tilt 车辆倾斜(度)
vehicle-to-vehicle communication 车辆间通信
vehicle-track dynamics 车辆轨道动力学
vehicle-track model 车辆轨道模型
vehicle tractive capacity 车辆牵引能力
vehicle traffic 车流;车辆交通
vehicle traffic stream 车辆交通流
vehicle travelway 车行道
vehicle trip 车辆行程;汽车行程;车辆出行
vehicle tripod 车用三脚架
vehicle tube 车轮内胎
vehicle tunnel 车辆隧道
vehicle turbine 运输式透平
vehicle-turning area 车辆转弯面积;车辆转弯区
vehicle-turning radius 车辆转弯半径
vehicle type 车型
vehicle washdown(yard) 洗车场
vehicle washing 车辆冲洗
vehicle washing area 洗车场
vehicle weight 车重;汽车重量;车辆重量
vehicle wheel 车轮
vehicle wheel load 车辆轮载
vehicle wheel spoke 车轮辐条
vehicle wheel track 车辆轮迹带
vehicle with dual-ga(u)ge wheelsets 双轨距车辆
vehicular access 车辆出入口
vehicular access shaft 车辆入口竖井
vehicular accident 车辆事故
vehicular breakdown 车辆抛锚;车辆损坏
vehicular bridge 车行桥;公路桥(梁)
vehicular bump-integrator 车载式颠簸累计仪
vehicular cargo 车辆运载货物
vehicular disposal 用车运出
vehicular ecological system 飞船生态系统
vehicular elevator 载汽车电梯
vehicular equipment 车用设备

vehicular gap 车间(净)距;车辆间净距
vehicular gas turbine 运输式燃气轮机
vehicular horizontal clearance 行车横向净空
vehicular lane 车行道
vehicular language 沟通语言;交际语言
vehicular lift 载汽车电梯
vehicular live load 行车荷载;车辆可变荷载;车辆活荷载
vehicular load 车辆荷载
vehicular parking area 停车场
vehicular phase 车辆信号相
vehicular pollution 交通公害;车辆污染
vehicular radio 流动无线电设备;车载(式)无线电设备
vehicular stall 停车位
vehicular statistics 车辆统计
vehicular subway 地下道路;车行隧道
vehicular traffic 车辆通行;车辆交通
vehicular traffic tunnel 道路隧道
vehicular travelway driveway 车行道
vehicular tunnel 行车隧道;公路隧洞;公路隧道;车行隧道
vehicular undercrossing 下穿式立交
vehicular vertical clearance 行车竖向净空
vehiculary 供车辆通过的;车辆的
veil 遮掩现象;超薄毡;表面毡
veiled decoration 暗花
veiling 翳影;流挂;喷丝涂装法
veiling brightness 干扰性亮度;遮掩性亮度
veiling glare 杂光;光帷眩光;光幕眩光
veiling luminance 伞形照明
veiling reflection 干扰性反射;遮掩性反射;光幕反射;光膜反射
veil of cloud 云盖;云层
veil of glass fiber 玻璃纤维面层料
vein 岩脉;裂痕;脉纹;岩石纹理;纹理
vein bitumen 矿脉沥青;脉沥青
vein breccia 脉状角砾岩
vein deposit 脉矿床
vein dyke 岩墙岩脉;脉状岩墙
veined 有纹理的
veined ebony 纹理乌木;纹理黑檀木
veined gneiss 脉状片麻岩;脉状片麻岩
veined marble 脉纹大理石;有纹理的大理石
veined mercury deposit 脉状汞矿床
veined paper 花纹纸
veined pipe 有纹铸铁管
veined wood 纹理木(材)
vein end 脉梢
veiner 小V形槽;木纹凿
vein filling 脉质
vein filling of solid bitumen 固体沥青岩脉
vein fissure 脉裂隙
vein fissure water 脉状裂隙水
vein graphite 脉状石墨
vein gypsum 脉石膏
veining 结疤;形成纹理;铸型破裂;包砂;纹理
veining structure 纹理组织
vein intersection 两矿脉交叉点
veinite 脉混合岩【岩】
vein lead-zinc deposit 脉状铅锌矿床
vein material 脉质
vein matter 脉质
vein meshes 脉网
vein orebody 脉状矿体
vein pyrite deposit 脉状硫铁矿床
vein quartz 脉石英
vein quartz for flux 溶剂用脉石英
vein quartz for glass 玻璃用脉石英
vein rib 脉脊
vein rock 脉岩
veinstone 脉石
vein structure 脉状组织;脉状构造
vein system 脉系
vein texture 脉状结构
vein-tin 脉锡
vein-type copper deposit 脉状铜矿床
vein-type fissure water 脉状裂隙水
vein-type gneiss 脉状片麻岩
vein-type lead-zinc deposit 脉状铅锌矿床
vein-type orebody 脉状矿体
vein-type pitchblende-sulfide deposit 脉状沥青铀矿硫化物矿床
vein-type pyrite deposit 脉状硫铁矿床

vein-type structure 脉状构造
vein wall 脉壁
vein water 脉状水
veiny 脉状;多纹理的
Vela 船帆星座
Vela pulsar 船帆座脉冲星
velardenite 钙黄长石
velarium 露天剧场的帐篷;天篷;天幕
Veld(t) 韦尔德草地;草原
velet-finish glass 细毛面玻璃
velikhovite 氮硫沥青
velinvar 镍铁钴钒合金
vellum 仿牛皮纸
vellum bookcloth 充皮书面布
vellum glaze 缎光釉;半无光釉
vellum paper 厚牛皮纸
Vellux 维拉克斯复合织物(商品名)
velocimeter 速度计;水内声速计;声速计;测速仪;测速器;测速计
velocimetry 速度测量学
velocipede 轻便轨道三轮车
velocitron 质谱仪;电子灯
velocity 速度
velocity aberration 速度偏差
velocity adjustment 速度调整;初速修正量
velocity along rotor diameter 速度沿旋翼直径的分布
velocity along route 沿航线速度
velocity amplitude 速度(振)幅
velocity analysis 速度分析
velocity analyzer 速度分析器
velocity anisotropy 速度各向异性
velocity anomaly 速度异常
velocity anti-resonance 速度反共振
velocity-area measurement of discharge 流速断面测流法
velocity-area method 流速一面积法;速度一面积法
velocity at a point 测点流速
velocity at impact 冲击速度
velocity at measuring point 测点流速
velocity-azimuth-depth assembly 流速一流向一水深装置
velocity-azimuth display 速度一方位显示器
velocity band 速度范围
velocity-breaking step 跌水
velocity change 流速变化;速度变化
velocity circulation 速度环量
velocity coefficient 流速系数;速度系数
velocity comparison indicator 速度比较指示器
velocity compensation 速度补偿
velocity compensator 速度补偿器
velocity complex 速度合成
velocity component 速度分量;分速度
velocity-compounded impulse stage 复速冲动级
velocity-compounded stage 复速级
velocity-compounded turbine 复速级式汽轮机
velocity-compression cycle 冲压循环
velocity constant 速度常数
velocity contour 速度分布图;等流速线
velocity contour section 速度等值线剖面
velocity control 航速控制;速度控制
velocity control knob 速度控制旋钮
velocity-controlled system 控速调节系统;速度调节系统
velocity controller 速度控制器
velocity control servo 速度控制伺服机构
velocity control voltage 速度控制电压
velocity correction 速度校正;速度校正
velocity correlation 流速相关
velocity curve 速度曲线
velocity defect law 流速亏损律
velocity deficiency 流速差;速度差值
velocity-depth distribution 速度一深度分布
velocity-depth relation(ship) 流速一水深关系
velocity descontinuity 地震间断面
velocity detector 速度地裂检波器;测速计
velocity determination 速度计算;速度测定
velocity diagram 速度(分布)图
velocity discontinuity 速度不连续(性)
velocity dispersion 速度色散;速度频散;速度弥散(度);速度扩散;速度分散
velocity-distance curve 速距曲线
velocity-distance relation 速距关系
velocity distortion 速度畸变

velocity distribution 流速分布;速度分布
velocity distribution calculation 速度场计算
velocity distribution diagram 流速分布图;速度分布图
velocity distribution function 速度分布函数
velocity distribution in pipe 管道流速分布
velocity distribution law 速度分布定律
velocity distribution pattern 速度分布型
velocity disturbance 速度干扰
velocity effect 速度效应
velocity ellipse 速度椭圆
velocity ellipsoid 速度椭球
velocity equivalent 等效速度
velocity equivalent diameter 等流速当量直径
velocity error 速度误差
velocity error coefficient 速度误差系数
velocity error compensator 速度误差补偿器
velocity error constant 速度误差常数
velocity error correction 速度误差校正
velocity exploration 流速查勘;速度分布测定
velocity factor 流速系数;速度系数
velocity feedback 速度反馈
velocity feedback loop 速度反馈回路
velocity field 流速场;速度场
velocity field of water head drawdown 水头降速场
velocity field vector 速度场向量;速度场矢量
velocity filter 速度滤波器;扇形滤波
velocity filtering 速度滤波
velocity flowmeter 流速流量计
velocity fluctuation 流速脉动;流速波动;流速变动;速度起伏;速度脉动
velocity focusing 速度聚焦;按速度聚焦
velocity-focusing mass spectrograph 速度聚焦质谱仪
velocity formed pressure 速度形成的压力
velocity formula 速度公式;流速公式
velocity function 速度函数
velocity function of fund 资金周转函数
velocity gain 速度增量
velocity ga(u)ge 速度计;测速计
velocity ga(u)ging 流速测量;流速测定;测流速
velocity gauging 流速观测
velocity geophone 速度检波器
velocity gradient 速度梯度;流速坡(度);流速比降;速度梯度;速度变率;波速梯度
velocity gradient tensor 速度梯度张量
velocity graph 速度曲线
velocity gravitational air flow 空气重力流速
velocity head 流速水头;水头;速位差;速度水柱;速度(水)头;速度能高;动压水头
velocity head coefficient 流速水头系数;速头系数
velocity head correction coefficient 流速水头校正系数
velocity head correction factor 流速水头校正系数;速头改正因数
velocity head current meter 速头流量计;速头海流计;速头测速计;速头测流计
velocity head flowmeter 速头流量计;速头海流计;速头测速计;速头测流计
velocity head graph 速度能高图
velocity head of a retarder location 制动能高【铁】
velocity head of cars 车辆速度能高
velocity head of retarder location 减速器的制动能高
velocity head rating 车辆速度能高的测定
velocity head rod 速头测杆
velocity head-tacheometer 速头转速计
velocity/height ratio 速度高度比(值)
velocity hodograph method 速度矢端线法
velocity hump crest of retarder 制动能高【铁】
velocity hydrograph 流速过程线
velocity impact hardening 速度冲击硬化
velocity increase 速度提高
velocity increment 速度增量
velocity index 速度指数
velocity index of rock mass 岩体速度指数
velocity indicator 速度指示器
velocity in diffuser pipe 扩散管流速
velocity inertial navigation system 速度惯性导航系统
velocity inertial system 速度惯性系统
velocity information 速度信息

velocity in high velocity-layer 高速层速度
velocity inlet region 速度助流区
velocity in low-velocity-layer 低速层速度
velocity in pipes 管内流速;管流速度
velocity intake region 速度助流区
velocity integrator 速度积分器
velocity interface 速度界面
velocity interval 速度范围
velocity inversion 速度反演;速度反常
velocity inversion procedure 速度反演方法
velocity jamming 速度干扰
velocity jump 速度跃变;速度跳跃;速度不连续
velocity lag 速度滞后
velocity lag error 速度滞后误差
velocity layer 流速层;速度层
velocity level 速度级
velocity limit 速度极限
velocity limiting servo 速度限制伺服机构
velocity logging 速度测井
velocity made good 航迹速度
velocity match(ing) 速度匹配
velocity measure in bore hold 钻孔流速测量
velocity measurement 流速观测;流速测量;速度测量;测(流)速
velocity measurement at surface 表面流速测量
velocity measurement of satellite system 卫星测速系统
velocity measuring chronograph 测速计时仪
velocity measuring device 速度测量装置
velocity measuring duration 测速历时
velocity measuring dynamo 测速用发电机;测速电机
velocity measuring gyroscope 测速陀螺仪
velocity measuring mark 测速标
velocity measuring point 流速测点
velocity measuring vertical 测速垂线
velocity memory 速度存储器
velocity meter 速度(水)表;速度流量计;速度测量器;测速表
velocity method 速度法
velocity method duct sizing 速度递减求径法
velocity metre 测速表
velocity microphone 压差传声器;振速传声器;速率式传声器
velocity misalignment 速度偏差
velocity misalignment coefficient 速度偏差系数
velocity model 速度模型
velocity-modulated beam 速度调制束
velocity-modulated oscillator 速度调制振荡器;速调振荡器
velocity-modulated tube 速度调制管
velocity-modulated valve 调速管
velocity-modulating effect 速度调制效应
velocity-modulating generator 速调式振荡器
velocity modulation 调速;速度调制;速调系统
velocity modulation bunching 调速聚束
velocity modulation coefficient 速度调制系数
velocity modulation tube 调速管
velocity modulation type buncher 速度调制型聚束器
velocity monitor(ing) 速度监控
velocity negative feedback 速度负反馈
velocity node 速度节点
velocity noise 速度分布噪声
velocity of approach 行近流速;引െ速度;临近速度;临近流速;进场速度;接近速度;渐近流速;趋近速度;驶进速度
velocity of ascension 上升速度
velocity of breaking movement 止动流速
velocity of capillary rise 毛细上升速度
velocity of concentration dilution of tracer 示踪剂浓度稀释速度
velocity of consolidation 固结速度
velocity of creep 爬升速度;慢行速度;蠕变速度
velocity of crystallization 结晶速度
velocity of currency 现金流通速度
velocity of currency in circulation 纸币流通速度
velocity of deposit 淤积速度
velocity of detonation 爆炸速度;起爆速度
velocity of diffusion 扩散速度
velocity of discharge 泄水速度;流出速度;排水速度;排放速度;排出速度;出流流速;出口速度
velocity of discharge of groundwater 地下水流速
velocity of dislocation 错位速度

velocity of dislocation climb 错位攀移速度
velocity of dislocation glide 错位滑移速度
velocity of electromagnetic wave 电磁波速度
velocity of electrons 电子速度
velocity of energy 能速
velocity of entry 进入速度
velocity of escape 出流速度;逸出速度;脱离速度;拖离速度;逃逸速度
velocity of exhaust 排汽速度;排出速度;出口速度
velocity of falling bodies 落体速度
velocity of filtering 滤速
velocity of filtration 滤速
velocity of flow 流速;流动速度;水流速度
velocity of flow grit chamber 沉沙池流速
velocity of flue gas 烟道气流速
velocity of following 牵连速度
velocity of groundwater flow 地下水流速
velocity of hardening 硬化速度
velocity of increase 增长速度
velocity of infiltration by water pouring 灌水入渗速度
velocity of lifting movement 扬动流速
velocity of light 光速
velocity-of-light cylinder 光速柱面
velocity-of-light radius 光速半径
velocity of limit 极限流速
velocity of longitudinal wave 纵波波速
velocity of money 资金周转率
velocity of movement 运动速度
velocity of mud circulation 泥浆流动速度
velocity of mud-stone flow 泥石流流速
velocity of neotectonic movement 新构造运动速度
velocity of neutrons 中子速度
velocity of offing wave 海面波速
velocity of overland flow 坡面漫流速度;地表漫流速度
velocity of partial sedimentation 部分沉降速度
velocity of penetration 贯入速度
velocity of permeability 渗透速度
velocity of pressure wave 纵波波速
velocity of progress 前进速度;传播速度
velocity of propagation 流speed;扩散速度;传播速度
velocity-of-propagation meter 传播速度计
velocity of propagation of sound 声音传播速度
velocity of recession 退行速度
velocity of retreat 出流速度;排流速度;排流流速;(水工建筑物下游的)退水流速
velocity of rise of water 水位上涨速度
velocity of rotation 旋转速度;自转速度
velocity of sea floor sediments 海底沉积层的速度
velocity of sea-surface wave 海面波速
velocity of seawater intrusion 海水入侵速度
velocity of seismic wave 地震波速度
velocity of seperation 分离速度
velocity of servosystem 速度伺服系统
velocity of shear wave 横波波速
velocity of slope 斜坡移动速度
velocity of sound 声速(度)
velocity of subsidence 下沉速度
velocity of the Nth layer 第 N 层速度
velocity of tracer injection 示踪剂注入速度
velocity of translation 传播速度
velocity of turnover 周转速度
velocity of variation 变化速度
velocity of vertical line 垂线流速
velocity of water flow in pipe 管内水流速度
velocity of water level rising again 地下水位回升速度
velocity of wave 波速
velocity of wave advance 波浪推进速度;波的传播速度
velocity of wave decay 波浪衰减速度
velocity of wave group 波群速度
velocity of wave motion 波动速度
velocity of wave propagation 传波速度;波浪传速度;波(的)传播速度
velocity of wave travelling 波浪推进速度
velocity of whirl 涡流速度
velocity on a vertical 垂线流速
velocity package 加速器;加速舱
velocity pattern 速度场
velocity perturbation 速度波动
velocity photogram 速度记录

velocity photograph 速度图
velocity pick-up 速度拾震器;速度传感(器)
velocity plan 速度图
velocity potential 流速势;速度势
velocity potential function 流速位函数;流速势函数;速度势函数
velocity power spectral density 速度功率谱密度
velocity pressure 速度头;动压(力);流速压力;速度压力;风压
velocity profile 流速分布图;速度剖面(图);速度轮廓;速度廓线;速度分布图
velocity pulsation 流速震动;流速脉动
velocity rate 速度比率
velocity rating flume 流速仪率定(水)槽;流速仪检定(水)槽
velocity ratio 速度比(率);速(长)比;潮流振幅比
velocity ratio of the tides 潮速比
velocity recovery phase 速度恢复期
velocity redistribution 流速再分布
velocity-reducing dust collector 降速集尘器
velocity reducing steps 消能梯级;消能台阶;梯级跌水
velocity reduction method (风道)速度递减法
velocity refrigerator 速度式制冷机
velocity regulation 速度调整
velocity relation curve 流速关系曲线
velocity resolution 速度分解
velocity resonance 速度谐振;速度回声;速度共振
velocity response 速度响应;速度反应
velocity response envelope spectrum 速度反应谱曲线;速度反应谱包线
velocity response spectrum 速度响应谱;速度反应谱
velocity reversal 速度反向
velocity rod 浮杆;浮标;速度标杆;测(流)速杆
velocity rod correction 流速杆校正数;测速杆校正
velocity scale 流速比例;速度比例
velocity scan(ning) 速度扫描
velocity seismograph 速度地震仪
velocity seismometer 速度地震计
velocity selector 速度选择器
velocity sensitive pendulum 速度摆
velocity sensitive transducer 速度传感器
velocity sensitive trigger 速度式触发器
velocity sensitivity 速度灵敏度
velocity sensor 速度感传器
velocity separator 速度分离器
velocity servo 速度伺服(机构)
velocity servomechanism 速度限制的伺服机构
velocity servosystem 速度限制的伺服系统
velocity shear 速度切变
velocity shear boundary 速度切变界限
velocity shock 速度冲击
velocity shooting 速度测井
velocity slip 速度滑脱
velocity slope 速度曲线斜度
velocity sorting 速度分选;速度分类
velocity-sorting principle 速度分选原理
velocity space 速度空间
velocity space instability 速度空间不稳定性
velocity spectrograph 速度谱仪
velocity spectrum 速度(振)谱
velocity spectrum curve 速度谱曲线
velocity spread 速度分布
velocity stage 速度(分)级
velocity-stage impulse turbine 速度级冲动式汽轮机
velocity staging 速度级次变化
velocity strain 速度应变
velocity structure 速度结构;速度构造
velocity surface 速度面
velocity survey 速度测量
velocity survey data 速度测量数据
velocity through ports 过港速度
velocity-time diagram 时间—速度图
velocity to prevent erosion 防冲流速;不冲刷流速
velocity to prevent plant growth 防止长草流速;不长草流速
velocity to prevent sedimentation 防止淤积流速
velocity towards destination 向终点速度
velocity tracking 速度跟踪
velocity transducer 速度换能器;速度变换器
velocity traverse 速度横向分布
velocity triangle 速度三角形

velocity turbine 冲击式涡轮机
velocity type column air gage 流速式水柱气动量仪
velocity-type flowmeter 转子式流速仪;流速型流量计;速度式流量计
velocity-type governor 速度型调节器
velocity-type seismometer 速度式地震计
velocity variation 流速变化;速度变化
velocity-variation amplifier 速度放大器
velocity-variation oscillator 速变振荡器
velocity-variation tube 速度调制管;速调管;变速管
velocity vector 流速矢量;速度向量;速度矢量
velocity vector diagram 速度矢量图解
velocity versus altitude graph 速度高度曲线图
velocity vertical 测速垂线
velocity voltage 速度电压
velodrome 自行车比赛场;(室内的)摩托车比赛场;赛车场
velodyne 调速发电机
velograph 流速计;速度记录仪;速度计
velometer 轮船调速器;调速器;速度计;测速仪;测速器;测速表
velour 棉绒;丝绒;天鹅绒;拉绒织物
velour finishing 绒面处理
velours gros-grain 粗横梭绒
velours renaissance 复兴丝绒
velum 缟状云;(罗马剧场观众席的)帐篷
velum interpositum 中间帆
velum transversum 横帆
vel-velam fiber 维尔维兰纤维
velvet 绒状的;天鹅绒;丝绒
velvet carpet 割绒地毯;剪绒地毯;天鹅绒地毯
velveteen 棉绒;绒布
velvet finish 仿麂皮整理
velvet-finish glass 珍珠光玻璃;细毛玻璃;丝光玻璃
velvet knife 割绒刀
velvet-like finish(ing) 仿天鹅绒饰面;仿天鹅绒涂装法
velvet pile 割绒
velvet tapestry carpet 天鹅绒挂毯
velvety 柔软的;天鹅绒似的
velvety flock paper 天鹅绒短绒纸
Vema deep 维玛海渊
Vema fracture zone 维马破裂带
Vema trench 维玛海沟
V. E. michev discrimination table V. E. 米赫也夫鉴定表
venac 聚乙酸乙烯酯树脂
vena contract 收缩断面;缩脉
vena contracta 收缩断面;射流紧缩
vena hemi-azygos 半奇静脉
venamul 共聚体黏合剂;聚乙酸乙烯酯乳液
venasquite 镁硬绿泥石
vendancy 翠绿
vendaval 文达瓦尔风
vendee 买主;买方
vendee's lien 买主的留置权
vender 售货机
vending cafeteria 自助餐馆
vending machine 投币式售货机;自动售货机;售货机
vendor 供货商;卖主;卖方;售货机;包退包换
vendor charges backs 退货
vendor inspection 卖主检查
vendor lease 卖主租约
vendor parts index 售主部件索引
vendor parts list 售主部件一览表
vendor parts modification 售主部件更改
vendor parts number 售主部件号
vendor performance index 卖主债效指数
vendor provisioning parts breakdown 售主备件分类
vendor rating 卖主评级;卖主规格
vendor rating system 厂商评核制度
vendor reliability 卖主可靠性
vendor's account 卖主账(户)
vendor's credit 卖主信贷
vendor shipping configuration 卖主装运配置
vendor shipping instruction 卖主装运说明书;售主装运说明书
vendor's lien 卖主扣押权;卖主(对货物的)留置权;卖方留置权
vendor's market 卖方市场
vendor's shipping documents 卖方航运文件;售主装货单据

vendor standard settlement program 卖主标准结算程序
veneer 镶面;镶板;胶合;木纹面板;木饰面板;贴面板;饰面基层;饰面板;单板;风化盖层;薄木板;薄单板;(海蚀平台上的)薄层沉积
veneer adhesive 胶合板胶黏剂
veneer backing 贴面板衬背
veneer base 石膏饰面基层
veneer board 镶板;夹板;胶合板;贴面板
veneer bolt (可在车床上去皮的)短原木
veneer bonding adhesive 饰面板胶黏剂
veneer brick 饰面砖;镶面砖
veneer case 胶合板箱;薄板箱
veneer cementing agent 饰面板黏结剂
veneer clippers 单板切断机;单板剪裁机;单板铡刀
veneer computer 单板计算机
veneer construction 镶板结构
veneer core 单板芯板
veneer core plywood 单板芯胶合板;薄芯板的胶合板
veneer crown 罩冠
veneer crown grafting 切接
veneer cutter 单板切削机
veneer cutting 镶板切割
veneer cutting machine 截夹板机;镶面板切割机
veneer cutting method 单板切割法;单板锯割法
veneer cutting saw 单板锯
veneer door 胶合板门;夹门;贴面板门
veneer drier 单板干燥机
veneered 镶面的;贴面的;镶板的
veneered brick 护面砖;釉面砖;陶面砖
veneered construction 胶合构造;饰面构造;镶面构造;镶边构造;贴面结构
veneered door 镶面门;镶胶合板门
veneered facade 用饰面板处理的房屋正面
veneered flake board 贴面碎木板
veneered panel 镶板;胶合镶板
veneered particle board 贴面碎木板
veneered plastics board 镶塑料板
veneered plywood 贴面胶合板;饰面胶合板
veneered stock 胶合板(早期叫法)
veneered strip-glued board 细木工板
veneered structure 镶脂红色淀
veneered wall 镶板墙;空心墙;镶面墙
veneered wall tie 镶面墙系铁
veneered with brick 用砖镶面;贴面砖;砖镶面
veneered wood 胶合木材
veneer faced particle board 薄木片贴面的刨花板
veneer glue 胶合板胶
veneer grade 单板等级
veneer hammer 贴面锤
veneer holddown 层板压块
veneering 胶合板;面板贴敷;装饰板;薄料;胶合片料;木板胶合;贴面;贴胶合板面层;薄片胶合
veneering hammer 胶合锤
veneering masonry(work) 饰面圬工
veneering press 胶合用压力机
veneering work 饰面工作;黏贴胶合板工作
veneer jointer 单板刨光机;单板接缝机
veneer knife 镶面板刨刀;胶合板刨刀
veneer knife grinder 单板磨刀机
veneer lathe 旋板机
veneer lathe rotary peeling machine 单板旋床
veneer log 胶合板木方;单板原木
veneer machine 截(切)夹板机
veneer of mortar 灰浆胶层
veneer of soil 表土层;表层土(壤)
veneer of the crust 地壳表层
veneer of wall 镶墙面;贴面墙
veneer on 罩面
veneer panel 胶合镶板;镶板
veneer panel matching 贴面板拼花
veneer peeling machine 截夹板机;剥镶板机
veneer planer 胶合板刨床
veneer plaster 表层饰面灰;石膏饰板;护面灰;高强度抹灰材料
veneer plaster lath 饰面抹灰条板
veneer reeling machine 卷板机
veneer retainer 罩冠固位体
veneer rotary lathe 单板旋切机
veneer saw 单板锯;镶板锯;胶木板锯;胶合板锯
veneer sewing machine 单板缝合机
veneer shaver 镶面板刨刀;胶合板刨刀
veneer sheet 层板
veneer slicer 单板切片机;单板刨切机
veneer slicing 镶板切制
veneer stone 镶面石板
veneer stone anchoring 石板锚碇法
veneer stone facing system 石板贴面做法
veneer tie 镶面墙系铁;镶面板系铁;腰线;镶面系墙件;镶板墙拉杆;墙饰面层锚固件;墙带
veneer tile 外墙砖;外墙面砖
veneer trimming bit 木板修饰钻头
veneer wall 镶面墙
veneer wall tie 镶面系墙件;镶板墙拉杆;墙饰面层锚固件
veneer with paper backing 纸背贴面板
veneer wood 镶木;胶合板;贴面板
venenation 中毒
venerable building 古建筑物
Venesta 威尼斯塔牌胶合板(一种著名高级胶合板)
Venetian arch 威尼斯拱
Venetian awning 威尼斯式遮篷
Venetian ball 威尼斯玻璃球
Venetian blind 威尼斯百叶窗;活动百叶窗;软百叶帘;软百叶窗;百叶帘
Venetian blind direct energy converter 威尼斯百叶窗式直接换能器
Venetian blind effect 威尼斯百叶窗效应
Venetian blind interference 同波道干扰;条状波形干扰
Venetian blind multiplier 威尼斯百叶窗形倍增器
Venetian blind shutter 威尼斯百叶窗式快门
Venetian cloth 威尼斯缎纹织物
Venetian cover 威尼斯通风盖
Venetian dentil mo(u)lding 威尼斯排齿饰线脚
Venetian dentils 威尼斯排齿饰
Venetian door 威尼斯(百叶)门;联窗门;两侧供采光的门;采光门
Venetianed 有软百叶窗的
Venetian finish 威尼斯整理
Venetian glass 威尼斯玻璃
Venetian gondola 威尼斯小划船
Venetian lace 威尼斯花边
Venetian lake 胭脂红色淀
Venetian light 威尼斯式灯具;威尼斯式窗
Venetian mast (装饰街市的)彩色柱
Venetian mirror 无框镜;威尼斯式镜
Venetian mosaic 水磨石板;威尼斯镶面板;威尼斯镶嵌;威尼斯嵌摄;威尼斯马赛克;仿大理石镶嵌物
Venetian motif 威尼斯风格
Venetian pink 威尼斯淡粉红
Venetian pointed arch 威尼斯尖拱
Venetian red 褐红色;铁红;合成铁红;威尼斯红(主要成分为氧化铁)
Venetian Renaissance style 威尼斯文艺复兴
Venetian rose 威尼斯玫瑰红
Venetian school 威尼斯学派;威尼斯画派(出现于文艺复兴时代)
Venetian ship-bottom paint 基料溶解型船底漆
Venetian shutters 软百叶窗;活动百叶窗;威尼斯百叶窗;遮阳窗;固定百叶窗
Venetian sumac 威尼斯漆树
Venetian terrazzo 威尼斯水磨石
Venetian topping 威尼斯式面层
Venetian velvet 威尼斯丝绒
Venetian white 威尼斯白(色颜料)
Venetian window 附两侧窗的窗子;威尼斯(式)窗;三尊窗
Venetian yellow 威尼斯黄
Veneto-Byzantine style 威尼斯一拜占庭风格建筑
Venezuela asphalt 委内瑞拉(地)沥青
Venezuela mahogany 委内瑞拉桃花心木
venezuelan 白珍珠
Venezuelan crude(oil) 委内瑞拉原油
Venice blue 威尼斯蓝
Venice glass 威尼斯玻璃皿
Venice green 威尼斯绿
Venice turpentine 威尼斯松节油
Venice white 威尼斯白
Vening-Meinesz isostatic system 维宁—曼乃兹均衡说
Vening-Meinesz sea three pendulum apparatus 维宁—曼乃兹海洋三摆仪
venite 脉状混合岩
Venn diagram 文丘里图;维恩图
venom 毒液
venom filter 毒液过滤器
venose 多筋络的
vent 油舱通气管;孔;气孔;点炮小眼;出口(指水、气、汽、油等);放气口;放空(口);通气口;通风孔
ventage 通风孔;出口;小孔;漏口;孔隙;通风系;通气口
vent agglomerate 火山口集块岩
ventainer 文泰纳式通风集装箱
vent air 通风空气
vent air filter system 通风空气过滤系统
vent and air-conditioning equipment 通风空调设备
vent area 通风面积;送风口面积
vent bar 通风栅
vent block 有通风孔砌块
vent branch 通气支管
vent brick 通风孔砖
vent bushing 点火螺塞
vent cap 排气管罩;通气孔盖;通气(管)帽
vent capping 通气孔盖
vent cavity 通风空腔
vent channel 通气槽
vent cloth 排气布
vent cock 放气旋塞
vent condenser 排气冷凝器;通风凝结器
vent connection 排气接管;放气接管
vent connector 烟囱连接管;通气孔连接装置
vent court 通风天井
vent cover 通气孔塞
vent cowl 通风斗
vent decorative appliance 通风式装饰气灯
vent depth 裂纹深度
vent duct 通风道;通风管(通道)
vented appliance 带通风孔的用具;带烟囱的灶具
vented battery 有通气孔的蓄电池
vented-bore laser 出气孔型激光器
vented chuck 带通气孔的钎座
vented circulator 排烟循环器;有烟囱管的循环器
vented exhaust 通风排气;导出排气
vented form 透水模板;透气模板;通气模板
vented fuel element 排气式燃料元件;透气式燃料元件
vented gas appliance 带烟囱的煤气灶具
vented gas heater 带排气筒的煤气采暖器
vented lithium cell 通气锂电池
vented manifold 通风歧管
vented recessed heater 隐蔽通风式采暖炉
vented reservoir 开放式储油器
vented screw 排气螺杆
vented tank 通风的储桶
vented type 半密闭式燃(气用)具
vented-type drill 排气式凿岩机
vented wall furnace 有烟囱壁炉
venter posterior 后腹
vent fabric 通风织物
vent fan 吸气扇;排气器;通风扇
vent finger 通风道隔离片
vent flap 通风折翼;通风盖
vent flashing 通风防水板
vent flue 排气管;通气管
vent gas 排出气(体)
vent gas cooler 排气冷却器
vent grating 通风格子
vent groove 通气槽
vent gutter 通风道
vent gutter in hold 舱内货物的通风道
vent header 通风横管;通风管联箱
vent hole 排气孔;通气孔;放气孔;通风孔;指孔
venthole 出烟孔
vent hood 通气罩
vent hose 通风软管
ventiduct 地下通风道;通风管;通风道
ventifact 风磨石;风棱石
ventil 消音键
ventilate 通风;换气;装通风设备;使通风;使换气
ventilated 通风的;风冷的
ventilated area 通风面积
ventilated bin grain drier 通风箱式谷物干燥机
ventilated box 通风箱
ventilated box car 通风式箱式货车;通风棚车
ventilated by forced draught 强制通风
ventilated cabin 通风舱

ventilated cargo 通风货物
ventilated casing 通风罩
ventilated ceiling 通风天花板;通风顶棚
ventilated chamber 通风柜
ventilated commutator 通风式整流器;风冷整流子;风冷式换向器
ventilated compartment 通风舱
ventilated container 通风(式)集装箱
ventilated drying oven 通风烘箱
ventilated enclosure 通风机壳
ventilated facade 通风的建筑物立面
ventilated flat roof 通风的平屋顶
ventilated flight clothing 通风飞行服
ventilated gas 通风气体
ventilated insulation 通风绝热
ventilated light fitting 通风照明设备
ventilated lobby 通风门廊;通风休息
ventilated loft 通风阁楼
ventilated manhole cover 通气人孔盖
ventilated motor 通风式电动机;风冷电动机
ventilated pit closet 通气厕所
ventilated rib 通风筋
ventilated roof 架空通风屋面;通风屋顶
ventilated shelter 通风式防毒掩蔽部
ventilated shock 通气冲击
ventilated silo grain drier 通风谷物干储[贮]塔
ventilated sliding panel 通风滑板
ventilated suit 通风服
ventilated thermometer 通风温度计
ventilated totally-enclosed motor 全封闭通风式电动机
ventilated transport 通风运输
ventilated wind tunnel 通气式风洞
ventilating 通风作业;通气;通风(的)
ventilating acoustic(al) ceiling 通风吸声天花板;通风吸声顶棚
ventilating air 流通空气;通风风流
ventilating appliance 通风设备
ventilating area 通风面积
ventilating arrangement 通风布置
ventilating bead 通气珠状花缘;窗户挡风条;下扯窗盖条
ventilating block 玻璃通风砖;通风砌块
ventilating blower 通风机扇
ventilating brick 空心通风砖;通风砖;通风空心砖;多孔通风砖
ventilating cabinet 通风橱
ventilating capping 气帽;通风罩;通风帽
ventilating cavity 通风(气)洞
ventilating ceiling 通风顶棚;通风天花板
ventilating ceiling board 通风顶棚板
ventilating ceiling sheet 通风顶棚薄板
ventilating ceiling system 通风天花板系统;通风顶棚系统
ventilating chamber 通风室
ventilating chimney 通风烟囱
ventilating cock 通风旋塞
ventilating column 通风压差
ventilating cover 透气盖
ventilating cowl 通风帽
ventilating damper 通风风闸;风门
ventilating damper in wall 墙内风门
ventilating device 换气装置;通气设备;通风装置;通风设备
ventilating disk 通风片
ventilating dryer 通风(式)干燥机
ventilating dry kiln 通风干燥窑
ventilating duct 通风(管)道
ventilating ducting 通风管道
ventilating duct spacer 通风沟隔片
ventilating entry 通风平巷
ventilating equipment 通风设备
ventilating eyebrow 屋顶通气窗
ventilating facility 通风设备
ventilating fan 排气风扇;排风机;通风机;通风(风)扇
ventilating fixture 通风装置
ventilating flap 通风活盖;通风瓣
ventilating flow sedimentation tank 竖流沉淀池
ventilating flue 通风道
ventilating fret 通风饰;通风回纹饰
ventilating furnace 通风炉
ventilating glass block 通风玻璃块;通风玻璃砖
ventilating glass brick 通风玻璃砖
ventilating grill(e) 通风格子窗;通风格窗;通风格栅
ventilating hole 换气口
ventilating installation 通风装置;通风设备;送风设备
ventilating jack 通风管罩;通风管帽
ventilating lay-in ceiling 通风的镶装顶棚
ventilating light fittings 通风照明配件
ventilating light(ing) fixture 通风照明装置
ventilating line 通气管道;通风管道
ventilating machinery 通风机械
ventilating machine 通风机
ventilating method 通风方法
ventilating motor 通风用电动机
ventilating network 通风网路
ventilating opening 通气口;通气孔;通风口;通风孔
ventilating panel 通风孔盖板;通风仪表盘
ventilating piece 通风管道配件;通风部件
ventilating pipe 通风管(道)
ventilating pipe tray 通气管座;通风管座
ventilating pit 通风孔
ventilating plant 通风设备
ventilating plant room 通风设备室
ventilating position 通风地点;通风位置
ventilating pressure 换气压力;通风压力
ventilating relief cock 通风安全旋塞
ventilating ridge 通风屋脊
ventilating ridge tile 通风脊瓦
ventilating scoop 通风罩
ventilating screw 通风螺钉
ventilating section 通风段
ventilating shaft 通风井;通风道;通气井;(隧道的)通风竖井;风井
ventilating shutter 通风百叶窗
ventilating skylight 通风天窗
ventilating slit 通气孔隙
ventilating stack 排放烟囱;通风竖管
ventilating station 通风机房;通风测量站
ventilating structure 通风构筑物
ventilating suit 通风服
ventilating system 通风系统
ventilating tile 通风瓦
ventilating tissue 通气组织
ventilating tower 通风塔
ventilating trunk 通风(主)道;通风筒;通风干路;通风干道
ventilating tube 通风管(道)
ventilating unit 通风装置
ventilating vane 通风机叶片
ventilating window 气窗;(气)洞
ventilation 换气;排气;排风;透气;通风(法)
ventilation air 通风风流
ventilation and air-conditioner room 通风空调机房
ventilation and air-conditioning equipment 通风空调设备
ventilation and pressurization system 通风增压系统
ventilation and temperature control system 通风及调温系统
ventilation aperture 通风口
ventilation appliance 通风设备
ventilation area 通风范围;通风面积
ventilation breather 通气器
ventilation brick 通风(空心)砖
ventilation building 通风构筑物;通风机房;通风房屋
ventilation by extraction 抽气通风
ventilation by forced draft 压气通风;气压通风;对流通风
ventilation cabinet 通风柜
ventilation cap 通风帽
ventilation car 通风车
ventilation casement 气窗(扉);通风窗扉;通风窗
ventilation cell 通风间
ventilation chimney 通风管;通风道
ventilation circuit 换气线路;通风线路;通风回路
ventilation coefficient of mixed layer 混合层通风系数
ventilation column 通风管柱
ventilation conditioner 通风调节装置
ventilation conduit 通风管
ventilation control 通风控制
ventilation cowl 通风帽
ventilation cross-cut 通风石门
ventilation current 通风流
ventilation cycle 通风周期
ventilation design 通风设计
ventilation device 通风装置;通风设备
ventilation device of kiln 窑内通风设备
ventilation diver 通风式潜水器
ventilation drying unit 通风干燥机
ventilation duct 通风管道;空气通道
ventilation ducting 通风装置
ventilation during construction 施工(期)通风
ventilation engineering 通风工程
ventilation equipment 空调设备;通风设备
ventilation equipment depreciation apportion and overhaul charges 通风设备折旧摊销及大修费
ventilation equivalent 通气当量
ventilation exhaust line 通风排气管路
ventilation exhaust system 排风系统
ventilation facility 通风设备
ventilation fan 通风机;换气扇
ventilation fence 通风栅
ventilation flap 通风挡板
ventilation foreman 通风技术员
ventilation function 通风功能
ventilation gallery 通风平巷
ventilation grill(e) 通风格子窗
ventilation heat loss 通风耗热量
ventilation hole 通风孔
ventilation hood 通风罩;通风帽
ventilation in biologic(al) filter 生物滤池通风
ventilation index 换气指数;通风指数
ventilation in high rate (biological) filters 高负荷生物滤池通风
ventilation installation 通风装置;通风设备
ventilation lateral 通风平巷
ventilation layout 通风平面图
ventilation level 通风水平;通风平巷
ventilation light fitting 通风照明装置
ventilation light(ing) fixture 通风照明装置
ventilation line 通风管线;通风管道
ventilation load 换气负荷;通风负荷
ventilation loss 通风损失(量);通风损耗(量)
ventilation louver 放气窗;放气孔;百叶风口;通风百叶窗
ventilation luminaire (fixture) 通风照明装置
ventilation machine room 通风机房
ventilation map 通风图
ventilation method 通风方法
ventilation modulus 通气模量
ventilation network 通风网
ventilation network simulator 通风网路模拟装置
ventilation noise 通风噪声
ventilation number 通风次数
ventilation of coal pits (煤矿的)矿井通风
ventilation of exploratory tunnel 勘探坑道通风
ventilation of mill 磨内通风
ventilation of mine 矿井通风
ventilation of one-way and opposite angles 单翼对角式通风
ventilation of refrigerated wagon 冷藏车通风
ventilation of residences 住宅通风
ventilation of two-way and opposite angles 双翼对角式通风
ventilation of underground zone 地下区间通风
ventilation opening 通风孔;通风口
ventilation orifice 通风口
ventilation pavilion 风亭
ventilation-perfusion ratio 换气灌注比率
ventilation pipe 通风管(道);通风立管;风筒
ventilation pipe network 通风管网
ventilation pipe on roof 屋顶通风管
ventilation piping 通风管
ventilation plan 通风系统图;通风管布置图
ventilation planning 通风设计
ventilation plant room 通风机房
ventilation plug 通风塞
ventilation port level 通风管拉杆
ventilation pressure 通风压力
ventilation psychrometer 通风式湿度计
ventilation quantity 通风量
ventilation rate 换气率;换气次数;通风率;通风量;通风换气次数
ventilation requirement 换气量
ventilation resistance 通风阻力
ventilation ridge 通风屋脊

ventilation road 通风坑道
ventilation sample 通风空气试样
ventilation scheme 通风方案
ventilation scoop 通风罩
ventilation section 通风区段
ventilation shaft (隧道的)通风竖井;通风井道;通风井
ventilation sheet 通风板
ventilation shield 通风罩
ventilation shutter 通风百叶窗;百叶风口
ventilation slot 通风孔;通风槽
ventilation stack 气流管道;通风管(道);通风道
ventilation standard 通风标准
ventilation station 通风站;通风测量站
ventilation stopping 挡风墙
ventilation supply line 通风吸入管路
ventilation survey 通风测量
ventilation system 通风装置;通风系统;通风方式
ventilation system diagram 通风系统示意图
ventilation system for public area of station 车站公共区通风系统
ventilation system for sectional tunnel 区间隧道通风系统
ventilation system map 通风系统图
ventilation system of mines 矿井通风系统
ventilation system station tunnel 车站隧道通风系统
ventilation test 换气试验
ventilation tile 通风砖
ventilation tower 通风塔
ventilation trunk 通风道
ventilation tube 通风管
ventilation tubing 通风筒;通风管
ventilation tunnel 通风洞
ventilation unit 通风装置
ventilation valve 通风阀;风阀
ventilation vent 通气孔
ventilation vent cover 通风孔盖
ventilation volume 换气量;通风容积
ventilation window 通风窗
ventilation within mill 磨内通风
ventilation work 通风工程
ventilative diving 通风式潜水
ventilative diving equipment 通风式潜水工具
ventilator 通风口;通风窗;通风管;通气筒;通风机;通风装置;通风筒;通风器;通风机;扇风机;风扇;风机
ventilator air inlet 通风机空气入口
ventilator blade 通风机叶片
ventilator blower switch 通风机开关
ventilator blue 风斗蓝色漆
ventilator box 通风箱
ventilator cap 通风器帽
ventilator clamp bracket 通风器夹托架
ventilator clamp pin 通风器夹销
ventilator clamp slide 通风器夹滑板
ventilator clamp thumb screw 通风器夹紧翼形螺钉
ventilator coaming 通风筒围板;风斗座
ventilator coaming cover 风斗座罩
ventilator control 通风机控制机构
ventilator cover 通风机罩
ventilator cowl 通气管罩
ventilator cowling 通风罩
ventilator dash drain 通风机排风口
ventilator derrick 通风吊杆柱
ventilator door 通风器门
ventilator duct 通风筒;通风器导管
ventilator fan 通风风扇
ventilator frame 通气窗框
ventilator handle (车厢)通风窗手柄
ventilator hood 通风机罩;通风筒顶罩;通风器罩
ventilator link 通风器联杆
ventilator link bracket pin 通风器联节托销
ventilator loop 污水透气管回路
ventilator of roof area 车顶通风窗
ventilator of ro/ro ship 滚装船通风装置
ventilator outlet pipe 通风器出气管
ventilator outlet pipe air cleaner 通风器出气管滤气器
ventilator pipe 通风器管
ventilator pipe hood 通风管罩
ventilator room 通风机间
ventilator scoop 通风口;通风斗
ventilator valve 通风量控制阀

ventilator valve spring 通风阀簧
ventilator valve spring seat 通风阀簧座
ventilator with water spray 喷水通风器
ventilatory function 通气功能
ventilatory reserve precentage 通气储量百分比
Ventilock 文蒂罗克移窗(一种专利商品)
ventilometry 换气测量法
venting 泄走;铸型出气;扎气眼;排走
venting area 排气面积
vent(ing) channel 浇铸通气孔;通气槽
venting duct 通气管;通风管
venting equipment 通风设备
venting grill(e) 通风格子窗
venting installation 通风装置;通风设备
venting layer 排气层
venting loop 污水通气管路;污水透气管回路
venting panel 通风仪表盘;通风控制盘;防爆阀
venting pin 通气针;通气棒
venting pipe 通气管;通风管
venting plant room 通风设备室
venting quality 透气率;透气性
venting ridge tile 通风脊瓦
venting shaft 通风竖管;通风竖井
venting sliding window 通风滑动窗
venting stack 排气竖管;排气烟囱;通风烟囱;通风竖管
venting stave 通气板条;(储[贮]存混凝土、矿砂石等用的)通风板条
venting system 通风系统;排气系统;透气系统
venting tile 通风屋脊
venting tray 通风管座
venting unit 放散管
venting valve 通风阀
venting window 气窗;通风窗
venting wire 通气针
vent knife 通风槽刀片
vent level 通风平巷
ventlight 气窗;通风窗;透气亮子
vent line 通风系统;出油路
vent lining 通风管内衬
vent louver 气窗;通风窗
vent main 通气干管
vent manifold 通风歧管
vent network 通气网
vento di sotto 索托风
vent of eruption 喷溢口;喷发口;喷发道
vent opening 通气孔;通风口;出气孔
vent-opening eruption 开口喷发
vent opening in eaves soffit 檐底通风口
vent panel 通风孔盖板
vent peg 阀塞;通气孔塞;通风孔塞
vent pin 阀栓
vent pipe 出气管;排气管;透气管;通风管
vent pipe flashing 通风管的防水板
vent(pipe)line 排出管线;排出管路;放泄管线
vent pipe tile 通风管瓷砖
vent plug 阀塞;通气孔塞
vent port 送风口
vent pressure vacuum valve 通风系统的压力调节阀
vent purity control 通气口纯度控制
ventral branches 腹侧支
ventral brush 尾刷
ventral centerline 机腹中心线
ventral commissure 腹侧联合
ventral door 机腹舱门
ventral fins 后机身下翼
ventral tank 机腹油箱
vent relief valve 旁通溢流阀
vent rod 通气针
vent sash 腰头气窗;透气亮子;通气窗
vent screen 通风筛
vent scrubber 通气洗涤器
vent segment 隔片;通气槽片
vent shaft 通气井;通气管;通风井
vent sleeve 通风筒
vent spacer 通风槽片
vent stack 泄烟道;通气竖管;通气主管;通气竖立管;通气竖管;通气立管;放散管;放空烟道;放空烟囱
vent stake 烟道
vent stem 孔杆
vent stopper 通风管塞子;火门塞
vent suit 通风服

vent system 排气系统;通气系统;通气管系统;通风(管)系统
vent to the atmosphere 放空
vent trap 透气阀
vent tube 排气管
vent tunnel 通风洞
venture 投机(物);风险
venture account 投机买卖账户
venture accounting 冒险事业会计
venture business 投机企业;冒险企业;风险企业
venture capital 冒险(性)投资;投资(资本);投入的资本;投机资本
venture capital investment 风险投资
venture capital of new technology 新技术风险投资
venture enterprise 风险事业
venture evaluation and review technique 风险评审法
venture expansion fund 企业发展基金
venture investment 风险投资
venture making under risks 风险型决策
venture of competition 竞争风险
venture profit 风险盈利
venturer 冒险者
venture worth 风险价值
Venturi 喉管;流速管;缩喉管;细腰管;文丘里(又称文氏)
Venturi absorber 文丘里吸收器
Venturi air break 文丘里空气断开
Venturi assembly 文丘里管总成
Venturi batter 文丘里管斜度
Venturi blower 文丘里吹风管
Venturi chimney 文丘里式烟囱
Venturi conduit 文丘里管
Venturi diffuser 文丘里扩散器
Venturi draghead 文丘里耙头
Venturi dust scrubber 文丘里除尘器
Venturi dust trap 文丘里管式集尘器
Venturi effect 狭管效应;文丘里效应
Venturi ejector 文丘里喷射器
Venturi engine 文丘里式发动机冷却
Venturi feed tube 文丘里喂入管
Venturi flow control valve 文丘里流量控制阀
Venturi flowmeter 文丘里(管式)流量计
Venturi flume 文丘里(量)水槽;文丘里式量水槽;文丘里(测流水)槽
Venturi gas scrubber 文丘里雾化洗涤器;文丘里气体吸收器;文丘里涤气器
Venturi jet 文丘里型喷射器
Venturi loader 文丘里细腰式装药管
Venturi loop conduit 文丘里绕闸充水管道
Venturi-meter 文丘里水表;文丘里流速计;文丘里流量计;文丘里计量管;文丘里测量管
Venturi-meter coefficient 文丘里流量系数
Venturi-meter variable 文丘里流量计变数
Venturi mouthpiece 文丘里管嘴
Venturi nozzle 文丘里(管)喷嘴
Venturi nozzle scrubber 文丘里净气器
Venturi pipe 文丘里管
Venturi scrubber 文丘里洗涤器;文丘里涤气器;文丘里洗气器
Venturi scrubber cyclone separator unit 文丘里洗气—旋风式分离装置
Venturi scrubbing 文丘里洗气法
Venturi spreader 文丘里管形洒布机
Venturi stack 文丘里式烟囱
Venturi throat 文丘里喉管
Venturi tube 流速管;文丘里(测流量)管;文丘里流量计
Venturi tube mixer 文丘里管混合器
Venturi tube thermometer 文丘里管温度计
Venturi type expansion nozzle 文丘里扩散喷管
Venturi type fermentor 文丘里式发酵罐
Venturi type loader 文丘里式装药器
Venturi vertical upward gas flow scrubber 文丘里竖立式向上气流洗涤器
Venturi washer 文丘里管
ventury type expansion nozzle 超音速喷嘴
vent valve 放气阀;通流阀;排油阀;排气阀;通气阀;通风阀
vent valve guide 通风阀导承
vent valve piston 通风阀活塞
vent valve seat 通风阀座
vent valve spring 通风阀簧
vent wax 通气蜡

vent window 排烟窗
vent wire 气眼针;通气针
Ventzke scale 芬茨克标度
venue 犯罪地点;集合地点;会场
Venus 启明星;维纳斯女神;维纳斯雕像
Venus' hair stone 发金红石
Venwood 文伍德牌百叶窗(一种活动百叶窗)
veraier panel 精调整盘
veramon 凡拉蒙
veranda(h) 游廊;阳台;外廊;敞廊;走廊;凉台;廊
veranda(h) awning blind 阳台外廊遮篷
verbal 口头的;按字面的
verbal agreement 口头协议
verbal approval 口头批准
verbal command 号令
verbal contract 口头约定;口头合同
verbal notification 口头通知
verbal order 口头命令;口头订货
verbal scale 文字比例尺
verbal system 通话设备
verbal translation 直译
verbal treatment 口头处理
verbatim record 全文记录
verbilki-Tver porcelain 维比尔基—特维尔硬瓷器
Verchojanskij-Korakoje geosyncline 维尔霍扬斯克—科里亚克地槽
verdan 通用计数式计算机
verdant zone 无霜带
verd(e) antique 古铜绿;古绿石;杂蛇纹石
verdelite 电气石
verde salt 无水芒硝
Verdet constant 维尔德常数
verdigris 乙酸铜;碱性碳酸铜;碱性醋酸铜;铜锈;铜绿
verditer 铜盐颜料;铜绿
verditer blue 碱式碳酸铜蓝;铜蓝
verdure 青绿
verfication of additive independence 可加独立性的检验
verge 路肩;接近;屋顶端边;山墙檐口;山墙檐边(突瓦);山墙突瓦;边缘;边沿;摆轮心轴
verge board 挑檐封檐板;山头封檐板;挡风板;封檐板;博风板
verge course 山墙砖压顶;山墙瓦;用挑砖砌的墙压顶;檐口滴水瓦;悬山屋檐板;山墙顶部侧砌的砖
verge cutter 路肩修理器;切边器
verge dub-footed escapement 冕状爪式擒纵机构
verge escapement 冕状轮擒纵机构
verge fillet 山墙连檐木;山墙挂瓦条;山墙檐口嵌条;山墙搏风板
verge glass 罗盘面玻璃
verge mo(u)ld(ing) 山墙檐边突瓦下的线脚;山墙檐口线脚
vergence 构造转向;聚散度
vergency 聚散度
verge of road 公路路边;路面预留地;路边预留地
verge-perforated 边缘穿孔的
verge-perforated card 边缘穿孔卡
verge processing 边缘加工
verge-punched 边缘穿孔的
verge-punched card 沿边穿孔卡片
verge purlin(e) 山墙口桁条;檐口桁条;边桁
verge rafter 檐口椽;(古建筑)檐椽
verge ring 压坏
verge tile 脊瓦;檐(口)瓦
verge tilting 山墙挑瓦
verge trimmer 路肩整平器;边缘整平器;边缘修整器
verge turf 边缘草地
verging ray 分歧射线
verglas 雨凇冰
vericon 直像管
veridian 绿色颜料
verifiability 可检验性;可核实性;能证明
verifiable 可检验的
verifiable secure operating system 检验可靠的操作系统
verification 验证;校验;校核;鉴定;检验;检定;核实;核对;确认
verification and validation 检验和确认
verification by model test 模型验证试验;模型试验验证
verification by sampling 抽样检验
verification by test and scanning 通过抽查核对

verification by test and scrutiny 用抽查法证实;抽样验证
verification certificate 验(货)单
verification data 验证用数据
verification expression 验证式
verification mark 验证标记;检定标记
verification mode 验证方式;校验方式;核对方式
verification note 验单
verification of accounts 核对账目;账目核对
verification of alternatives and assumptions 方案及设想的验证
verification of assumptions 假设检验
verification of balance 核对余额
verification of cash 现金核对
verification of conformity 合格证明
verification of current meter 流速仪测流的检验
verification of footing and posting 总计及过账的检验
verification of forecast 预报验证;核对预测
verification of mathematic(al) model 数值模拟验证;数学模型试验
verification of model(ing) 模型验证;模拟验证
verification of risk aversion assumption 厌恶风险假设的检验
verification of thermometer 温度计校正
verification of water quality model 水质模型检验
verification relay 校核继电器;监控继电器
verification report 论证报告
verification scope 正确范围
verification setting 验证设置
verification sheet 复核表
verification technique 验证技术
verification terminal 验证用终端
verification terminal display 验证终端显示
verification terminal keyboard 验证终端键盘
verification test 证实试验
verification theorem 检验定理
verified by 复核(人)
verified copy 经核证的抄本;经核实的副本
verified correct 核对无误
verified output 落实产量
verified unit 核对单位
verifier 验证员;证明者;校验机;校对员;校对机;检验机;校对机;核对员;校对机;穿孔校对机
verify 验孔;证实;检孔;核对
verify a statement 证明检验报表;核对清单
verify error 检验误差
verify hypothesis 验证假设
verifying attachment 检验装置
verifying count of material received 核实运到原料数
verifying device 检验装置
verifying unit 校验装置;校对装置;检验装置
verify quantity of goods 查对货物数量
verify the turnover of materials stored 核库
verify unit 校验装置
verkhovodka 季节水
vermeil 镀金的银;朱红色;镀金的铜
vermex concrete 隔音混凝土
vermi biofilter 蚯蚓生物滤池
vermicelli 粉丝
vermicelli making equipment 粉丝制造设备
vermicular cast iron 蠕墨铸铁
vermicular clay 蠕虫状黏土
vermicular quartz 熔蚀石英
vermicular texture 蠕虫状影纹;蠕虫状结构
vermicular work 虫蚀状装饰;虫蚀状饰面
vermiculate 使成虫蚀状的装饰
vermiculated 虫迹状凿面的
vermiculated dressing 虫蚀状琢面;虫蚀状錾琢
vermiculated mosaic 细纹精刻的马赛克
vermiculated mottle 网纹斑
vermiculated pattern 虫蚀状图案
vermiculated plaster finish 虫蚀粉饰面
vermiculated rustic work 虫蚀状粗面(石)工;虫蚀粗琢石面工作;虫蚀状朴木房子
vermiculated work 虫蚀状装饰;虫蚀状饰面;虫蚀纹镶工;虫蚀纹雕刻
vermiculation 虫蚀状饰面;虫蚀状饰面;虫迹饰【建】
vermiculite 蛭石
vermiculite absorbent plaster 蛭石吸声粉刷
vermiculite acoustic(al) plaster 吸声蛭石灰泥;吸声蛭石粉刷

vermiculite aggregate 蛭石骨料
vermiculite brick 蛭石砖
vermiculite concrete 蛭石混凝土
vermiculite concrete block 蛭石混凝土块
vermiculite concrete brick 蛭石混凝土砖
vermiculite concrete screed 蛭石混凝土刮板
vermiculite deposit 蛭石矿床
vermiculite diopside 透辉石
vermiculite gypsum plaster 蛭石石膏抹灰(层);石膏蛭石灰浆
vermiculite insulation in bags 袋装蛭石声绝缘
vermiculite loose fill insulation 蛭石松散填充绝热材料
vermiculite mortar 蛭石砂浆;膨胀蛭石灰浆
vermiculite of exfoliation 蛭石的片状剥落
vermiculite plaster 蛭石抹面灰泥;蛭石灰浆
vermiculite plaster finish 蛭石灰泥抹面;蛭石砂浆抹面
vermiculite plastering 蛭石灰浆抹面
vermiculite slab 蛭石板
vermiculite sound absorbent plaster 吸声的蛭石灰膏
vermiculite sound-control plaster 声控蛭石粉刷;声控蛭石灰泥
vermiculus 硫化汞
vermiform filling defect 虫蚀状充盈缺损
vermil(l)ion 朱红色(的);银朱;朱砂
vermil(l)ion ate 朱砂代用品
vermilion cinnabar 朱砂
vermil(l)ion column 朱漆柱
vermil(l)ionette 人造银朱;赛银朱;朱砂代用品
vermil(l)ion paint 银朱涂料;朱红涂料
vermil(l)ion red 朱红(色);朱红淀
vermil(l)ion stamping ink 红印色;红印泥
vermin 害鸟;虫
verminal 虫的
vermin-proof 防虫(的);防虫蛀鼠咬的
vermin-proof mesh 防虫网
vermin-resistant 抗虫害的
Vermont marble 弗蒙特大理石
Vermont slate 有色石英岩;弗蒙特板岩
vernacular 地方的;本地语;本地的
vernacular architecture 地方建筑;乡土建筑;土风建筑
vernacular arts 乡土艺术;地方工艺
vernacular construction 乡土建筑;地方建筑
vernacular construction method 本乡本土的建筑方法
vernacular dwelling 民居
vernacular form 乡土味造型
vernacular housing 民居
vernadite 水羟锰矿
vernadskite 块铜矾
vernal aspect 春季相
vernal breeze 春风
vernal circulation 春季环流
vernal equinox 春分(点)
vernalization 春化作用;春化处理
vernal plant 春季性植物
vernal point 春分点
Verneuil furnace 韦尔讷伊式炉
Verneuil method 韦尔讷伊(单晶培育)法
vernier 游标尺;游标
vernier accuracy 游标精度
vernier acuity 游标对准视觉锐度
vernier adjustment 游标调整
vernier arm 游标臂
vernier bevel protractor 游标斜分度规;游标活动量角器;活动游标量角器
vernier cal(l)ipers 游标测径器;游标卡尺;游标测径规
vernier cal(l)ipers gage 游标卡尺的量具测径规
vernier cal(l)ipers 游标测径器;游标卡尺;游标测径规
vernier cal(l)ipers model roller type 滚轮式游标卡尺
vernier cal(l)ipers with depth gage 带深度计游标卡尺
vernier cal(l)ipers with dial indicator 带百分表游标卡尺
vernier cal(l)ipers with scriber 带有划线器的游标卡尺
vernier card of repeater 分罗经游标卡
vernier circle 游标盘;上盘

vernier circle of theodolite 经纬仪上盘
vernier clock 游标钟
vernier closure 游标读数闭合差
vernier closure meter 游标闭合计
vernier compass 游标罗盘(仪)
vernier condenser 微调电容
vernier control 游标调节;微调控制;微调
vernier coupling 游标式联轴器;挠性联轴节
vernier cursor 游标
vernier cutoff 游标发动机关车
vernier depth gage 游标深度尺;游标测高规
vernier device 游标装置
vernier dial 游动度盘;游标(刻)度盘
vernier division 游标刻度;游标分划;游标标度
vernier element 游标元件
vernier engine 游标发动机;微调发动机
vernier error 游标误差
vernier eyepiece 游标目镜
vernier gage 游标规;游标尺
vernier gage control 厚度的精调器自动控制
vernier gear tooth gage 游标齿厚规
vernier graduation 游标刻度
vernier height gage 游标高度规;游标高度尺
vernier labyrinth gland 游标式迷宫汽封
vernier level 游标水准器;游标水准管
vernier line 游标线
vernier micrometer 游标千分尺;游标测微器
vernier microscope 游标显微镜
vernier of sextant 六分仪游标
vernier plate 游标盘
vernier potentiometer 游标电位计
vernier protractor 显微半圆角度规;游标量角器;游标分度器
vernier quadrant 游标扇形(刻)度盘
vernier-rate control 速率微调控制
vernier reading 游标读数
vernier reading manometer 游标读数压力计
vernier reading microscope 游标显微镜
vernier-read instrument 游标读数器
vernier resolver 微动分解仪
vernier rocket 微调发动机
vernier scale 游标刻度盘;游标(尺)
vernier sextant 游标(式)六分仪
vernier slide callipers 游标滑动卡尺
vernier slide zero 游标零点
vernier tacheometer 游标视距仪
vernier theodolite 游标经纬仪
vernier transit 游标经纬仪
vernier tuning 游标微调;游标调谐
vernier weight indicator 精密重量指示器;精密指重表
vernier zero 游标零分划;游标零点
vernitel 精确数据传送装置
vernix 涂剂;清漆;护漆
vernolic acid 环氧油酸
vernonia oil 斑鸠菊油
Vernon Shale 维尔农页岩
Verona brown 维罗纳褐色
Verona green(earth) pigment 维罗纳绿色颜料
veronal 巴比土钠
Verona marble 维罗纳大理石
Verona yellow 车工铅黄;意大利铅黄;氯化铅黄;佛罗那黄
veronese green 水化三氧二铬绿颜料
verplanckite 水硅钡锰石
Verre grave 维涅牌玻璃(一种磨砂玻璃)
Versailles garden 凡尔赛庭园
Versailles Palace Park 凡尔赛宫苑
Versailles Treaty 凡尔赛宫(法国)
versant 坡度;山坡;山侧;边坡
versatile 万能的;通用的;多用途的;丁字着生的
versatile acoustic(al) material 多效吸音材料
versatile additive 多用途添加剂
versatile automatic data exchange 多用(途)自动数据交换机
versatile automatic test equipment 万能自动测试设备
versatile borehole surveying instrument 万能钻孔测斜仪;多用钻孔测斜仪
versatile digital computer 通用数字计算机
versatile dredge(r) 多功能挖泥船
versatile furnace 通用炉
versatile pile frame 多用途桩架;多功能桩架
versatile precision-rolling mill 多用途精密轧机;多品种精密轧机
versatile pulse shaper 通用脉冲形成器
versatile shaping machine 多功能牛头刨床
versatile spindle 万向轴
versatile system of signaling 可变式信号系统
versatile tamper 多功能捣固机
versatile underwater submersible 通用潜水器
versatile underwater vehicle 通用潜水器
versatile ventilator 多能通风机
versatility 通用性;多用性;多面性;多功能性;多方面适应性
versatility index 多方面适应性指标
versatility of farm tools 农业机械通用性
versatron 通用晶体管指示仪表
versed sine 正矢
versene 维尔烯
versicolo(u)r(ed) 红色的;斑色的;多色的;杂色的
versiliaite 氧锑铁矿
versine 正矢
versine equalizing instrument 正矢校平仪
versine inspection 正矢检查
versine shock pulse 正矢冲击脉冲
version 倒转术;方案;变型;版本
version number 版本号
version up 改订
versurae 舞台建筑的左右侧翼
vertaplane 垂直起落飞机
vertebrated catheter 分节导管
vertebrate fossil 脊椎动物化石
vertebratus 脊状云
vertex 底尖;角顶;尖轨理论尖端;天顶;台风路线转折点,顶;奔赴点,[复]vertices
vertex angle 顶角
vertex block 顶端砌块
vertex diopter 镜顶屈光度
vertex dioptric power 镜顶屈光度
vertex-disjoint path 点不重的路径
vertex equation 顶点方程
vertex focal length 镜顶焦距
vertex formula 镜顶公式
vertex hinge 顶端铰链
vertex hog 顶端弯拱
vertex in a graph 一图形中的顶点
vertex joint 顶端接头
vertex of a cone 锥顶
vertex of a conic 二次曲线的顶点
vertex of angle 角的顶点
vertex of a quadric 二次曲面的顶点
vertex of a triangle 三角形的顶点
vertex of curve 曲线顶点
vertex of great circle 大圆顶点
vertex of mirror 镜顶
vertex of negative lens 负透镜顶点
vertex of surface 面顶点
vertex of triangle 三角形顶点
vertex power 镜顶屈光度
vertex sag 顶端下垂
vertex stone 拱顶石头
vertex velocity 顶点速度;峰点速度
vertial-plane coupler 曲面车钩
vertible gulf 不可逾越的鸿沟
vertical 纵向的;直立的;帧;立式;立杆;铅直的;铅垂(方向);竖的;地平经圈;垂直线;垂直(的);垂直
vertical abutment joint 竖向接缝
vertical acceleration 竖向加速度;垂直加速
vertical accelerator 直立式加速器
vertical accelerometer 垂直加速计
vertical accent 突出竖线条的建筑(哥特式);竖线条建筑
vertical accretion (沉积物)向上增长;竖向淤积;竖向堆积;垂直淤积;垂直加积作用
vertical-accretion deposit 垂向加积沉积
vertical accuracy 高程精度
vertical activator of cement 漩涡式水泥活化器
vertical adjustment 高程平差;垂直校正;垂直调整
vertical aerial 竖天线
vertical aerial effect 竖天效应
vertical aerial photograph 水平航空相片;垂直航摄照片
vertical aerotriangulation 高程空中三角测量
vertical air duct 竖直风道;垂直风道
vertical air elevator 垂直气力提升机
vertical air-lock tube 竖向气闸管
vertical alignment 纵坡设计线;竖面线形;纵断面设计线【道】;竖向线位;竖向定线
vertical alignment of road 定â道路竖向线
vertical amplification ratio 竖向放大比
vertical amplifier 帧信号放大器;垂直扫描讯号放大器
vertical ampoule forming machine 立式安瓿机
vertical analysis 纵向分析;静态分析
vertical anchor 锚柱;竖锚;垂直锚碇
vertical anchor for pile 桩的垂直锚碇
vertical and horizontal arrangement 纵横排列
vertical anemometer 铅直风速表
vertical anemoscope 铅直风速仪
vertical angle 仰角;直角;高度角;竖(直)角;垂直角
vertical angle benchmark 三角高程测量点;垂直角测定基准标记
vertical angle field book 垂直角观测手簿
vertical angle of mark 标身垂直角
vertical angle of view 垂直视角
vertical angles 对顶角;对等角
vertical antenna 竖天线;垂直天线
vertical aperture correction 垂直孔栏校正
vertical armature 垂直衔铁
vertical arrangement 垂直排列
vertical ascent 竖向上浮
vertical assembly 竖式装配
vertical assembly building 垂直装配间
vertical astigmatism compensator 垂直像散补偿器
vertical axial current meter 竖轴流速仪
vertical axis 立轴;纵坐标;纵轴;竖(向)轴;垂直轴
vertical axis of revolution 旋转的竖直轴;竖直转轴
vertical axis sector gate 竖轴扇形闸门
vertical axle load 垂直轴荷载
vertical baffled mixing basin 竖式挡板混合池
vertical band saw 立式带锯
vertical band sawing machine 立式带锯机
vertical bank 垂直倾斜
vertical bar 竖井;竖向钢筋;垂直钢筋;竖直筋;竖尺;垂直筋
vertical barge lift 垂直式升船机
vertical bar generator 垂直条信号发生器
vertical bar magnetism 垂直软铁磁性
vertical bar of truss 桁架竖杆
vertical bar window 直棂窗
vertical batter stud 竖向支撑;竖撑条
vertical beam 竖梁
vertical beam ditcher 竖架挖沟机
vertical beam method 垂直探伤法
vertical bed 直立岩层
vertical bed type milling machine 直立床身式铣床
vertical bend 尖轨垂直弯曲
vertical bending machine 立式弯板机
vertical bending stress 垂直弯曲应力
vertical bending vibration 竖直挠曲振动
vertical bicycle stand 立式自行车(车)架
vertical binder 蹬筋
vertical biotic zonation 垂直生物分带
vertical blade 竖直叶片;垂直叶片
vertical blanking 帧回扫消隐;垂直消隐;垂直熄灭
vertical blanking generator 场消隐脉冲振荡器
vertical blast 垂直爆破
vertical blast mat forming aggregate 立吹薄毡机组
vertical blind 竖向百叶帘;垂直百叶窗
vertical blind ditch 竖向盲沟
vertical blind slat 百叶窗的垂直百叶板;百叶窗的垂直板条
vertical blowing process 立吹法;垂直喷吹法
vertical body break 辊身沿轴线垂直面折断
vertical boiler 立式锅炉
vertical bolt 竖插销;竖栓
vertical bond 竖直对缝砌
vertical boom ditcher 竖臂(式)挖沟机
vertical boom excavator 立臂式挖土机
vertical boom trencher 立臂(式)挖沟机
vertical boom trenching machine 立臂(式)挖沟机;竖臂(式)挖沟机
vertical boom type trencher 直立臂式挖沟机
vertical boom type trenching machine 竖杆式挖沟机
vertical boring and turning machine 立式镗车床
vertical boring machine 立式钻探机;竖式钻孔机
vertical boring mill 立式镗床
vertical bottle depth-integrating sampler 竖瓶积深式采样器

vertical boundary of aquifer 含水层垂向边界
vertical-bowl mixer 竖斗式拌和机
vertical brace plate 垂直撑板
vertical bracing 竖直支撑;竖向支撑;垂直支撑;垂直联条;垂直连接系;垂直剪力撑
vertical branch 直立枝;垂直分支
vertical branching 垂直支管
vertical breakwater 直立式防波堤
vertical bridging 高程加密
vertical broaching machine 立式拉床
vertical bucket elevator 竖式料斗提升机;垂直斗式提升机
vertical bucket mixer 竖斗式搅拌机
vertical bunker 垂直料仓
vertical burner 立式燃烧器
vertical calciner 立式分解炉
vertical calibration goniometer 垂直校准测向器;垂线校准测向器
vertical camera 立式照相机
vertical cantilever element 竖直升降机部件;垂直悬臂梁;垂直悬臂部分
vertical carrying current air classifier 垂直载流空气分级机
vertical cassette form mo(u)lding 成组立模成型
vertical cast brick panel 竖直浇注砖镶板
vertical casting 立浇;垂直浇注
vertical casting machine 立式连铸机
vertical cat bar 竖插销
vertical cavity 垂直孔洞
vertical cavity surface emitting laser 垂直空腔表面发射激光器
vertical cell 垂直孔
vertical cell container ship 立式格栅集装箱船
vertical cement kiln 水泥立窑;立式水泥窑
vertical cement kiln plant 立窑水泥厂
vertical centering 垂直对中
vertical centering alignment 中心垂线定线;垂直中心调准
vertical centering control 竖直中心调节;垂直定中调节
vertical center keelson 中内龙骨;直立中央内龙骨
vertical center line 垂直中心线
vertical center of gravity 垂直重心
vertical center hung pivot window 竖式中悬窗
vertical center hung window casement 竖直中悬窗
vertical center of buoyancy 垂直浮心
vertical centrifugal casting 立式离心铸造
vertical centrifugal casting machine 立式离心铸造机
vertical centrifugal machine 立式离心机
vertical centrifugal mill 立式离心磨
vertical centrifugal pump 立式离心泵
vertical chain 竖排水管;垂挂链
vertical chamber oven 立式炉
vertical channel 垂直通道
vertical character tabulation 垂直印刷控制字符
vertical check 垂直校验;垂直检验
vertical checkout 垂直测试
vertical chucking multi-spindle 立式多轴夹盘车床
vertical chute 竖直溜管;垂直溜槽
vertical circle 竖(直)圈;竖直(刻)度盘;地平经圈;垂直圆;垂直度盘;方位圈
vertical circle level 垂直度盘(指示)水准器
vertical circle level tube 垂直度盘水准管
vertical circular buoyant jet 垂直圆形浮射流
vertical circular grit settling tank 立式圆形沉砂池
vertical circulating 垂直循环
vertical circulation 垂直流通;铅直环流;垂直交通;垂直环流
vertical circulator 垂向环流
vertical city 空间发展城市;摩天楼城市
vertical cladding 墙面覆盖;垂直覆盖;垂直外墙挂板
vertical cladding panel 垂直覆盖板
vertical clamp (经纬仪)竖盘制动
vertical clamp screw 竖直制动螺旋
vertical classifier 纵向分级器;纵向分类器;立式选粉机;立式分级机
vertical cleaner 立式清净装置
vertical clearance 净高;竖向净空;垂直净空;垂直间隙;垂向净空
vertical cleat 垂直割理
vertical cliff 垂直崖壁;陡直岩壁
vertical climatic belt 垂直气候带

vertical closer 立式捻绳机
vertical closing machine 立式捻绳机
vertical closure 立堵截流
vertical closure method 立堵截流法
vertical coefficient of eddy diffusion 垂直涡流扩散系数;垂直涡动扩散系数
vertical coiler 立式卷取机
vertical collimation 垂直视准误差
vertical collimator 垂准器;垂准仪;竖准器
vertical column 垂直柱;大气柱
vertical combination 垂直统一管理;纵向联合;纵向合并
vertical combination merger 纵式联合
vertical common-axis array 直立共轴装置
vertical common-plane array 直立共面装置
vertical comparator 竖直比长器;垂直比测器
vertical compliance 垂直顺性
vertical component 竖直分量;竖直分力;竖向分量;垂直分量;垂直部分
vertical component of force 垂直分力
vertical component of magnetic field 磁场垂直分量
vertical component of pressure 垂直分压力
vertical component of secondary field 二次场垂直分量
vertical component seismogram 垂直分量记录
vertical component seismograph 竖向地震仪;垂直(分量)地震仪
vertical composite coal mill 立式组合煤磨机
vertical compound steam turbine 竖轴复式汽轮机
vertical compound turbine 叠置式多缸汽轮机
vertical compression 直压
vertical compressive force 垂直方向压力
vertical compressive stress 垂直压应力
vertical compressor 立式压缩机
vertical concrete spiral 竖轴混凝土蜗壳
vertical configuration 垂直地形
vertical coning 水锥;垂直角锥
vertical consistency 纵的一致性
vertical consolidating 纵向合并
vertical construction joint 垂直建筑缝;竖向施工缝;垂直施工缝
vertical contact comb 纵向触簧梳;垂直触簧梳
vertical contact pin 顶插棒
vertical contact stress 竖向接触应力
vertical contour of mining area 矿区高程控制
vertical control 高程控制;垂直控制
vertical control datum 高程控制基(准)面
vertical control network 高程控制网
vertical control point 高程控制点;控制点
vertical control station 高程控制点
vertical control survey 高程控制测量方法;高程控制测量
vertical control system 高程控制系统
vertical convection 铅直对流;垂直对流
vertical convection mixing 垂直对流混合(作用)
vertical convective mixing 垂直对流混合(作用)
vertical converter 竖式吹炉
vertical conveyer 立式输送机;立式运输机;垂直升降机
vertical conveying 垂直输送
vertical cooler 立筒式冷却器;立式冷却器;立式冷却机
vertical coordinate 纵坐标(轴);竖坐标;垂直坐标
vertical coordinate assessment 纵坐标法评价;纵坐标法分析
vertical copy miller 立式靠模铣床;仿型立式铣床
vertical cordon 单杆形果树;直立单干形
vertical core 垂直孔;垂直型芯
vertical core oven 立式烘芯炉
vertical core rockfill dam 芯墙式堆积坝;垂直芯墙堆石坝
vertical coring 垂直型芯;垂直取岩芯;垂直钻孔
vertical coring block 垂直空心砌块;垂直穿孔块
vertical coring brick 空心砖;深孔砖;垂直(孔)空心砖
vertical coring clay brick 垂直空心黏土砖
vertical coring clay brick panel 垂直空心黏土砖板
vertical coring engineering brick 垂直空心的工程用砖
vertical coring light(weight) brick 垂直空心轻质砖
vertical corrosion 垂直磨蚀
vertical counting accelerometer 垂直计数加速计
vertical counting circuit 垂直计数电路

vertical coupling 垂直联轴器
vertical coverage 垂直视线;垂直视界
vertical-coverage pattern 垂直作用范围
vertical crack 直裂纹
vertical crank 直立拐肘
vertical creep 垂向徐变
vertical cross-hair 竖丝
vertical crossing 垂直交叉
vertical cross-section 垂直截面;垂直断面
vertical cross-section of coal seam 煤层垂向横断面图
vertical cross tube boiler 立式横管锅炉;立式旋管锅炉
vertical crystal puller 垂直拉晶机
vertical curb 直立式路缘石;立缘石(侧石)
vertical curing chamber 立式养护窑
vertical current 垂直流
vertical current washer 垂直水流洗选机
vertical curtain 竖幕
vertical curvature 竖向曲度
vertical curve 纵断面曲线;竖曲线
vertical curve at crest of hump 峰顶竖曲线
vertical curve at sag 凹形竖曲线
vertical curved blade turbine impeller 垂直弯叶片涡轮桨
vertical curve location 竖曲线测设
vertical curve radius 竖曲线半径
vertical curve radius of lifting bridge 吊桥竖曲线半径
vertical cut 纵切;竖直切削;竖向挖方;竖向缺口;垂直侵蚀;垂直开挖
vertical cutter 立式切割器
vertical cutter head 垂直刀架
vertical cutting 立插
vertical cylinder 立式汽缸
vertical cylinder dryer 立式滚筒干燥机
vertical cylindric(al) furnace 圆形竖炉
vertical cylindric(al) tank 立式圆柱形筒
vertical damping 短轴阻尼;垂直阻尼
vertical damp-proof course 垂直防潮层
vertical datum 高程基准(面)
vertical decision matrix 垂直决策矩阵法
vertical declinometer 垂直测斜仪;垂直测斜器
vertical deep well pump 立式深井泵
vertical definition 竖直析像力;垂直清晰度
vertical deflecting electrode 垂直致偏电极
vertical deflection 帧偏转;高低提前量;竖向弯沉;竖向挠度;垂直偏转
vertical deflection amplifier 垂直致偏放大器
vertical deflection cell 垂直偏转元件
vertical deflection component 垂线偏差(分量)
vertical deflection mirror 垂直偏转镜
vertical deflection module 垂直偏转组件
vertical deflection oscillator 垂直偏转振荡器
vertical deflector 垂直偏转板
vertical deflector coil 垂直偏转线圈
vertical deformation 竖向变形;垂直形变;垂直扭曲;垂直变形
vertical denticulation 垂直小牙饰
vertical depression 垂直下降
vertical depth of inclined well 斜井垂直深度
vertical derrick 直塔
vertical design 竖向设计
vertical development 竖井开拓
vertical deviation 高低偏差;垂直偏差;垂线偏差
vertical dial 垂直式拨号盘
vertical diameter 垂直直径
vertical diaphragm cell 竖隔膜电池
vertical differential chart 垂直差值图
vertical diffusion 垂直扩散
vertical diffusion coefficient 竖向扩散系数
vertical digester 立式蒸煮锅;立式发酵罐
vertical dilution of precision 高程精度因子;垂直精度因子
vertical dimension 垂直尺度;垂直尺寸
vertical dimensioning 垂直标注
vertical diplopia 垂直复视
vertical dip slip 竖直滑距
vertical direction 竖直方向;垂直方向
vertical discharge line 垂直排料管
vertical disc dryer 立式蝶形干燥机
vertical disk type root cutter 立式盘刀块根切碎机
vertical dislocation 竖向位错;竖向错位;垂直位错;垂直错位;垂直错动;垂直变位

vertical dispersion 垂向弥散
vertical displacement 竖向位移;垂直位移;垂向位移
vertical displacement measurement 垂直位移测量【岩】
vertical displacement of double layered system surface 双层体系表面垂直位移
vertical disposition 纵排列
vertical distance 竖直距离;垂直距离
vertical distance between ship bottom and channel bed 船底至水底的垂直深度
vertical distance of traverse point 导线点垂距
vertical distribution 竖直分布;竖向分布;垂直分布;垂向分布
vertical distribution angle 刚性角
vertical distribution curve 垂直分布曲线
vertical distribution curve of flow velocity 流速垂直分布曲线
vertical distribution curve of light 垂直配光曲线
vertical distribution of light 光的垂直分布
vertical distribution of ozone 臭氧的铅直分布
vertical distribution of sediment 泥沙的垂向分布
vertical disturbing acceleration 垂直干扰加速度
vertical disturbing force 垂直方向扰动力
vertical divergence angle 垂直扩散角
vertical diversification 纵向多样化
vertical division 竖向分割;垂直划分
vertical division block 竖向分区
vertical division of climatic 气候的垂直分带性
vertical double action presses 立式双动冲床
vertical double plane transverse cable arrangement of cable-stayed bridge 斜拉桥垂直双面索纵向布置
vertical down-film condenser 立式降膜冷凝器
vertical down-film cooler 立式降膜冷却器
vertical down welding 向下立焊
vertical drain (沼泽填土中的,供排水用)砂柱;竖直排水;竖沟
vertical drainage 纵沟;竖向排水;竖沟;垂直排水
vertical drainage system 垂直式排水系统
vertical drain(pipe) 垂直排水管
vertical drain system 垂直排水系统
vertical drawing 垂直引上
vertical drawing machine 垂直引上机
vertical drier 立式干燥机
vertical drift 倾斜角
vertical drill 立式钻床;立式凿岩机;垂直钻
vertical drilling machine 立式钻床;方柱立式钻床
vertical drilling system 垂直钻探系统
vertical drop spillway 直落式溢洪道
vertical dryer 立式干燥机;立式烘干机;竖式干燥器
vertical dynamic(al) convergence 垂直动态会聚
vertical dynamic(al) convergence correction 垂直动态收敛校正
vertical earth pressure 竖向土压力
vertical earth rate 竖向地球速率
vertical easement curve 垂向缓和曲线;竖向介曲线
vertical eddy 竖直漩涡;垂向涡流;垂直涡流
vertical eddy viscosity coefficient 垂直涡流黏滞系数
vertical edge grinding machine 立式磨边机
vertical edger 立辊轧边机
vertical edging roll 立辊
vertical elasticity 铅垂弹性
vertical electric(al)-driven reciprocating pump 立式电动往复泵
vertical electrode 立焊焊条;垂直电极
vertical electrolytic surface grinding machine 立式电解平面磨床
vertical electrostatic precipitator 立式静电除尘器
vertical elevation 垂直高程
vertical elevator 垂直升降机;垂直提升机;垂直电梯
vertical engine 直立式发动机;立式发动机
vertical entrance and exit 垂直出入口
vertical epitaxial reactor 立式外延反应器
vertical erosion 向下侵蚀;向下冲刷;下蚀作用;下切作用;垂直侵蚀
vertical error 高程误差;垂直误差
vertical escape exit 竖向疏散口
vertical escapement 垂直擒纵机构
vertical exaggerated model 垂直放大模型
vertical exaggeration 垂直扩大;垂直放大
vertical exaggeration ratio 竖向放大比
vertical exit 垂直安全出口;竖向出口;竖向通道;竖向疏散口

vertical expansion joint 竖向伸缩缝
vertical exploration well 加深探井
vertical exploratory opening 勘探竖井
vertical extension 竖向延伸
vertical extent 竖向延伸量;垂直延伸
vertical exterior wall cladding 垂直外墙挂板
vertical extractor with fan 立式风扇排气器
vertical extruder 垂直挤出机
vertical extrusion process 挤压法
vertical fabric splitting machine 立式布机
vertical face 立面;竖直面;垂直面
vertical face breakwater 直立式防波堤;直墙式防波堤;铅垂面防波堤
vertical faced wharf 直立式码头
vertical face plate 立式花盘
vertical facies sequence 垂向层序
vertical fading 竖直衰落
vertical fan marker beacon 扇形指点标
vertical fault 竖向断层;垂直断层
vertical fault scarp 竖向断层崖
vertical feed 升降进给;垂直进刀;垂向输卡片;垂向馈送;垂向传送
vertical feed(ing) 垂直馈送;垂直送料
vertical feed opening 垂直进料口
vertical-fiber brick 凸纹砖
vertical-fiber lug brick 竖凸沿缸砖;竖凸沿砖;竖筋砖
vertical field balance 竖直场强磁力仪;垂直场强磁力仪
vertical field magnetometer 竖直场强磁力仪;垂直场强磁力仪
vertical field-strength diagram 垂直场强图
vertical file 立式档案柜
vertical fillet weld 立焊角焊缝
vertical film head 垂直薄膜磁头
vertical filter 竖流滤池;垂直滤波器
vertical filtration 竖流滤法
vertical fine boring machine 立式精密镗床
vertical fine motion drive 垂直微动螺旋
vertical fire tube boiler 立式火管锅炉
vertical firing shaft 发射竖井
vertical fissure 垂直裂隙;垂直裂缝
vertical flaring draft pipe 直锥形尾水管
vertical flaring draft tube 直锥形尾水管
vertical flat-blade turbine impeller 垂直平叶片涡轮桨
vertical flight gyroscope 倾斜稳定陀螺仪;垂直飞行陀螺仪
vertical flocculator 竖式絮凝池
vertical flow 竖向流
vertical flow basin 竖流池
vertical flow constructed wetland 垂直流人工湿地
vertical flow cupboard 垂直层流通风橱
vertical flow drier 垂直气流式烘干机
vertical flow dry electrostatic precipitator 气流垂直运动干式静电吸尘器
vertical flow electrostatic precipitator 垂直气流静电集尘器
vertical flow grit chamber 竖轴式沉砂池
vertical flow precipitator 垂直气流集尘器
vertical flow press 垂流式压机
vertical flow sedimentation tank 竖流沉淀池;竖轴式沉淀池
vertical flow settling basin 直流式沉淀池
vertical flow settling tank 竖流式沉淀池
vertical flow tank 垂直流向池;竖流式沉淀池
vertical flyback period 帧回扫时间
vertical flyback voltage 垂直回扫电压
vertical focusing 垂直聚焦
vertical fold 倾竖褶皱;竖直褶皱;垂直褶皱
vertical force 垂直(分)力;竖直(分)力
vertical force instrument 倾斜自差调整器;(磁罗经)倾差仪
vertical force magnetometer 竖直场强磁力仪;垂直场强磁力仪
vertical force of earth magnetism 地磁垂直分力
vertical force variometer 垂直磁力仪
vertical format 竖向形式;垂直格式
vertical format unit 垂直走纸格式控制器;垂直输纸格式控制器
vertical forming shoe 垂直摆锻锤头
vertical frame(d) gate 竖框式闸门
vertical frame saw 立式排锯;立式框锯;竖锯

vertical free face 垂直临空面
vertical frequency 帧频;半帧频
vertical frontal line 正垂线
vertical frontal plane 正垂面
vertical front haulm stripper 前置立式除茎叶器
vertical furnace 直立炉;立窑
vertical fusion 竖炉熔炼;竖炉熔化
vertical galvanometer 垂直式电流计
vertical gas flow dry precipitator 垂直气流干沉淀器
vertical gate 垂直闸门
vertical gating 垂直浇口;缝隙式浇口
vertical ga(u)ge 直立式水尺;竖直水尺;垂直水尺
vertical geothermal gradient 垂直地温梯度
vertical glazing 垂直玻璃窗
vertical glued laminated timber beam 垂直胶合木梁
vertical glue joint 垂直黏结
vertical gradient 铅直梯度;垂直梯度;垂直方向梯度
vertical gradient freeze method 垂直温度梯度凝固法
vertical gradient of gravity 重力垂直梯度
vertical gradient of magnetic anomaly 磁异常垂向梯度
vertical gradient of pollutant 污染物的铅直梯度
vertical gradient of ZZ 垂直梯度值
vertical grain 径切面;竖向纹理;四分开板材;径切纹理;径面纹理;垂直木纹
vertical grained 四开锯木法;垂直木纹的
vertical-grained interval 等高线间距
vertical grained lumber 径切纹理(木)材
vertical grain timber 直纹理木材
vertical granulator 立式成粒机
vertical gravity take-up 垂直重锤拉紧装置
vertical greening 垂直绿化
vertical grid 垂直格栅
vertical grid ordinate assessment 纵坐标法评价;纵坐标法分析
vertical grille 竖芯
vertical grinder 立式磨床
vertical grinding dispersion machine 立式研磨分散机
vertical grinding machine chucking internal 夹盘立式内圆磨床
vertical ground motion 竖向地面运动
vertical ground particle velocity 垂直地面质点速度
vertical guide 垂直导轨
vertical guide idlers 垂直导轮
vertical gust 垂直突风
vertical-gust recorder 垂直突风记录仪;垂直气流加速度计
vertical gyro 垂直陀螺仪
vertical gyroscope 垂直陀螺仪
vertical hair 十字纵丝;竖丝;十字丝线;垂直丝
vertical hammer 立式锤
vertical handling 垂直搬运
vertical handling system 垂直装卸方式
vertical hanger 竖吊杆
vertical head 静水压力
vertical head sander 立转带式砂光机
vertical height 垂直高度
vertical helical spring 垂直(螺旋形)弹簧
vertical helix type inferential meter 垂直螺翼式流量表
vertical hinged door 平开门
vertical hinge revolving (门轴的)竖轴旋转
vertical hodograph of convert reflection wave 转换反射波垂直时距曲线
vertical hodograph of direct longitudinal wave 直达纵波垂直时距曲线
vertical hodograph of multiple reflected longitudinal wave 多次反射纵波垂直时距曲线
vertical hodograph of reflected longitudinal wave 反射纵波垂直时距曲线
vertical hodograph of refracted p-wave 折射纵波垂直时距曲线
vertical hodograph of transmitted convert wave 透过转换波垂直时距曲线
vertical hodograph of transmitted longitudinal wave 透射纵波垂直时距曲线
vertical hold 垂直同步
vertical-hold control 竖直同步调整;帧同步调整;垂直同步调整
vertical hole 竖孔;垂直(钻)孔;垂直浆缝

vertical hole blasting 垂直眼爆破
vertical hollow shafting two speed motor 立式空心轴双速电机
vertical (or horizontal) tandem extruder 垂直/水平串挤出机
vertical hunting 图像上下摆动
vertical hydraulic lift 立式液压起落机构
vertical illumination 垂直照明;垂直(面)照度;垂直光照
vertical illuminator 垂直照明装置;垂直式照明器
vertical impact crusher 立式冲击破碎机;离心破碎机
vertical impact pulverizer 竖冲舂粉机
vertical incidence 垂直入射
vertical-incidence sound absorption coefficient 垂直入射吸声系数
vertical-incidence transmission 垂直入射传输
vertical-incidence transmission loss 垂直入射隔声量
vertical incident solar rays 太阳直射
vertical incision 垂直切口
vertical index 垂直指标;垂直度盘指标
vertical index bubble 垂度盘指示水准器
vertical index error 竖盘指示差;垂直度盘指示差
vertical index level 高程水准器;天顶水准器;垂直度盘指示水准器
vertical index screw 垂直度盘(指标)微动螺旋
vertical industrial pump 竖式工业泵
vertical infiltration 垂直渗入
vertical ingot 垂直锭料
vertical injection mo(u)lding machine 垂直铸模机
vertical in-line pump 竖式连续泵
vertical inserted curtain wall 立装式幕墙
vertical insert mo(u)lding machine 垂直加料模塑机
vertical instability 垂直不稳定(度)
vertical (intake) recharge 垂直补给量
vertical integration 纵向一体化(指跨国公司的垂直型经营);纵向联合;纵向结合
vertical intensity 垂直强度;垂直磁强
vertical intensity variometer 竖直强度磁力仪;垂直磁强仪
vertical intercept 垂直支距;垂直截距
vertical internal test signal 垂直扫描插入测试信号
vertical interval 垂直距离;垂直间距;垂直间隔;等高(线间)距
vertical interval keyer 垂直扫描间隔定时器
vertical introduced current air classifier 垂直引入气流分级机
vertical inverter 垂直偏转换向器
vertical iron 垂直软铁
vertical iron ladder 垂直铁梯
verticalism 垂直度;垂直性
verticality 竖直(度);垂直状态;垂直性;垂直(度);垂直状态
verticality of mast 立柱垂直度
verticality survey 垂直度测量
vertical jalousie 竖式固定百叶窗
vertical joint 直缝;竖直缝;竖(向)缝;垂直节理;垂直(接)缝
vertical joint face 竖向接缝面
vertical journal 枢轴颈
vertical judder 垂直位移
vertical keel 竖龙骨
vertical kerf 竖截槽
vertical kiln (水泥)立窑;立式烘炉;竖井炉
vertical kiln cement 立窑水泥
vertical kiln clinker 立窑熔渣
vertical kiln lime 立窑石灰
vertical kiln stone 立窑砌石
vertical knee and column milling machine 立式升降台铣床
vertical knee-and-column type miller 立式升降台铣床
vertical knife divider 立刀式分禾器
vertical knife lever shears 立刀杠杆式剪切机
vertical ladder 直爬梯;直立梯;竖梯;垂直梯
vertical ladder ditcher 竖向联斗挖沟机
vertical ladder furnace 垂直管炉
vertical lagging 垂直护板
vertical lake current 垂直湖流
vertical laminar flow 垂直平行流;垂直层流
vertical laminated wood 垂直层积木
vertical lamination 木材垂直叠合法;垂直叠合(木材)
vertical lapse rate of air temperature 气温垂直递减率
vertical lathe 立式车床;立车
vertical launching 垂直发射
vertical layer radiographic X-ray apparatus 垂直断层X线机
vertical laying 铅垂敷设
vertical layout 竖向定位;垂直式布置
vertical lead 垂直提前量
vertical lehr 垂直退火炉
vertical letter(ing) 立体字;长体字;直体字【计】
vertical leveling mirror 垂直调平镜
vertical lift 垂直提升(器);垂直升降机;垂直升船机
vertical-lift air-craft 垂直起飞飞机
vertical-lift bridge 竖向升降桥;升降桥;垂直升降桥;直升桥;提式升降桥
vertical-lift door 垂直升降门;直升门
vertical-lift fixed wheel gate 直升式定轮闸门
vertical-lift gate 垂直提升(式闸)门;直升式闸门;船坞插板门
vertical-lifting weir 垂直闸门堰
vertical-lift lock gate 直升式船闸闸门
vertical-lift span 立升跨度;直升跨度
vertical-lift weir 垂直升降堰
vertical light component 光(波)垂直分量
vertical light distribution 垂直配光
vertical lighting 纵向照明(设备)
vertical limb 垂直度盘
vertical lime burner 竖式石灰窑
vertical lime kiln 石灰立窑
vertical line 纵向线;铅垂线;竖线;垂直线
vertical linearity 帧幅直线性;垂直线性
vertical linearity control 竖直线性调节
vertical line of measuring points 测点垂线
vertical line of sight 垂直瞄准线
vertical link 垂直拉杆
vertical-load-carrying space frame 承受竖向荷载的空间框架;竖向荷载空间框架
vertical load increment 垂直荷载增量
vertical load(ing) 竖向荷载;垂直荷载;垂直负载;铅直荷载
vertical load system 竖向荷载系统
vertical location of center buoyancy 浮心垂直位置
vertical locking spring 垂直锁定弹簧
vertical longitudinal of orebody for reserve calculation 矿体储量计算垂直投影图
vertical longitudinal projection calculating reserves 储量计算垂直投影图
vertical longitudinal projection of orebody 矿体垂直纵投影图
vertical longitudinal section of coal seam 煤层垂直纵断面图
vertical long period seismograph 垂直向长周期地震仪
vertical loop 窗扇纵向滑轮;纵向回路
vertical louver 竖式百叶窗(帘);竖向百叶(窗)
vertically 竖直地;垂直地
vertically adjustable end wall 垂直可调节挡墙
vertically anchored platform 张力腿(式)平台
vertically corrugated bulkhead 垂直波形舱壁
vertically divided mold 垂直分型面铸型
vertically exaggerated model 竖向变态模型
vertically fired pit 中心换热式均热炉;垂直供热的均热炉
vertically mixed estuary 垂直混合河口
vertically moving biofilm process 垂向移动生物膜法
vertically offset rhomboid system 垂直偏差补偿斜方棱镜系统
vertically opening shutter door 垂直开启式百叶门
vertically perforated brick 竖孔空心砖
vertically pivoted shutter 竖向枢轴式百叶窗
vertically pivoted window 竖轴式旋窗;立转窗
vertically polarized electromagnetic wave 垂直偏振电磁波
vertically polarized wave 垂直偏振波
vertically sliding 垂直拉拉;垂直滑落
vertically sliding balanced sash 双垂直平衡滑动窗
vertically sliding door 直升拉门;立式拉门
vertically sliding sash 上下推拉窗扇;提拉窗扇
vertically sliding sash window 垂直活动的窗扇
vertically sliding window 垂直推拉窗
vertically split 垂直裂缝
vertically stacked loops 多层环形天线
vertically steady permeability coefficient 垂向稳定渗透系数
vertically steady quantity of infiltration 垂向稳定渗入流量
vertically stiff bridge 竖向刚劲式桥
vertically suspension door 纵吊门;竖吊门
vertically travel(l)ing derrick crane 竖向移动式转臂吊机
vertical machine 立式机床;立式电机
vertical machining center 立式加工中心
vertical magnetic balance 垂直磁力仪
vertical magnetic intensity 垂直磁力强度
vertical magnetic recording 垂直磁记录
vertical magnetic recording head 垂直磁记录头
vertical magnetic strength 垂直磁力强度
vertical magnetic transporter 竖向磁力运输机
vertical magnetization 垂直磁化
vertical magnetometer 垂直磁力仪
vertical management 垂直管理
vertical mattress suture 垂直褥式缝合
vertical meeting rail 竖直的碰冒头;门窗相遇的边梃
vertical member 立杆;竖向构件;竖杆;垂直构件;垂直杆件
vertical merger 纵向合并
vertical metal spiral setting 竖向钢壳装置;竖轴式金属蜗壳装置
vertical microprogramming 垂直微程序设计
vertical migration 垂直迁移;垂直洄游
vertical migration in the reservoir 在储层内向上运移
vertical mill 立磨
vertical miller 立辊轧机
vertical milling and boring machine 立式升降台铣镗床
vertical missing 垂直偏差
vertical mixer 竖向搅拌机;立式搅拌机
vertical mixing 垂直混合(作用)
vertical mode 垂直方式
vertical monitoring control network 高程监测控制网
vertical motion 垂直运动;竖向运动;升降设备;上下运动;垂直移动
vertical motion seismograph 垂直运动地震仪
vertical motion simulator 垂直运动模拟器
vertical mo(u)ld 竖模
vertical mo(u)lding 竖线条
vertical movement 竖向运动;垂直运动;垂直移动
vertical movement associated with horizontal displacement 伴有水平位移的垂直运动
vertical moving bed 竖式移动床
vertical moving pump 竖式移动泵
vertical mulching 竖向覆盖层;竖向防护层;垂直覆盖
vertical multispeed motor 立式多速电动机
vertical multistage furnace 立式多段燃烧炉
vertical multi-tubular boiler 立式多管锅炉
vertical multi-tubular steam boiler 立式多管蒸汽锅炉
vertical navigational clearance 竖向航行净空
vertical normal stress 垂直法向应力
vertical nozzle 垂直接管
vertical obstacle 垂直障碍物
vertical obstacle sonar 竖直障碍声呐
vertical obstruction 垂直挡板
vertical occurrence 直立
vertical off-normal contacts 上升离位触点
vertical offset 竖向支距;竖向支距;垂直截距;垂直断错
vertical offset electrode holder 垂直偏位电极座(点焊用语)
vertical one-tube system 竖向单管系统
vertical opening 井筒;垂直坑道;竖向通道
vertical open rod single action presses 立式单臂单动冲床
vertical ordinate 纵坐标
vertical organization structure 垂直组织机构
vertical orientation 垂直定向
vertical oscillating transformer 垂直扫描变压器用变压器
vertical oscillator 垂直扫描振荡器;场扫描振荡器
vertical outfall 垂直排放
vertical output amplifier 垂直输出放大器
vertical output transformer 垂直输出变压器

vertical overhead contact system 多边形架空接触网
vertical paddle mixer 立式搅拌机;垂直浆搅拌机
vertical palmette 直立单干形
vertical panel 竖直镶板
vertical parabola 抛物线形帧信号
vertical parallactic slide 上下视差滑尺
vertical parallax 纵视差;高低位差;上下视差;垂直视差
vertical parallax screw 竖直视差螺旋;上下视差螺旋
vertical parallax slide 上下视差滑尺
vertical parallel plate 垂直并联板
vertical parity check 垂直奇偶检验;垂向奇偶校验
vertical parting flaskless mo(u)lding 垂直分型无箱造型
vertical partition wall 隔断墙
vertical passageway 竖向通道
vertical pass point 高程过渡点
vertical pawl 上升爪
vertical pendulum 竖直摆;竖向摆
vertical perforation 竖孔
vertical permeability 垂向渗透率
vertical permeability coefficient 垂直渗透系数
vertical permeability of leaky aquifer 越流层垂向渗透系数
vertical phasing 帧定相
vertical photo control point 相片高程控制点
vertical photograph 竖直航摄照片;竖向摄影;垂直航摄照片
vertical photograph coverage 有垂直摄影资料地区
vertical photographic(al) image 垂直摄影相片
vertical photography 竖直(航空)摄影;垂直(航空)摄影
vertical picture 垂直图片
vertical pile 直桩;竖桩;垂直(板)桩
vertical piling reaction 竖向打桩反力
vertical pincushion distortion 垂直枕形畸变
vertical pipe 立管;竖管
vertical(pipe)line 立管;垂直管道
vertical pipe reaction column 立管式反应柱
vertical pipe suspension load sampler 竖管悬移质采样器
vertical pipe type suspension load sampler 立管式悬移质采样器
vertical piping 竖管;垂直管涌
vertical piston 竖活塞
vertical pivoted 沿垂直轴旋转的
vertical pivoted door 垂直旋转门;垂直枢轴式门
vertical pivoted sash 竖轴窗扇
vertical pivoted window 竖轴窗扇;立旋窗;竖轴式旋窗
vertical pivot hung window 竖轴窗扇
vertical plane 竖向平面;竖面;垂直(平)面
vertical plane machine 立式刨(床)
vertical plane of projection 垂直投影面
vertical plane projection of coal bed 煤层立面投影图
vertical planer 立式刨(床)
vertical-plane radiation pattern 垂直面辐射图形
vertical planing machine 立式刨(床)
vertical planning 竖向布置(图);垂直刨法
vertical planting 垂直绿化
vertical plate-bending machine 立式弯板机
vertical plate planter 垂直排种盘式播种机
vertical play 上下游隙
vertical plumb 垂线
vertical plunger cold chamber die-casting machine 立式冷室压铸机
vertical plunger pump 立式柱塞泵;方式柱塞泵
vertical pneumatic conveyer 垂直气力输送机
vertical point 天底点
vertical pointer 垂向指针
vertical polarization 垂直极化;垂直极化
vertical polarized shear wave 垂直极化横波
vertical position 悬垂位;立焊位置;垂直位置;垂向位置
vertical position of welding 立焊位置
vertical position welding 立焊
vertical post 竖杆
vertical pouring 立式浇注
vertical preferences 垂直优惠
vertical prepressure compacting 垂直加压法
vertical press 立式压力机;垂直压力机
vertical pressure 竖向压力;垂直压力;纵向压力

vertical pressure generator 立式充压型静电加速器
vertical pressure leaf filter 垂直压力叶滤机
vertical prestress(ing) 竖向预(加)应力
vertical prevention leakage 垂直防渗
vertical principal stress 垂直主应力
vertical prismatic coefficient 竖菱形系数
vertical probe 垂直探测器
vertical process camera 立式制版照相机
vertical profile 垂直剖面;纵剖面
vertical profile line 侧垂线
vertical profile map of road surface 路线纵断面图
vertical profile of ozone 臭氧铅直分布图
vertical profile plane 侧垂面
vertical projecting sign 竖直挑出的招牌
vertical projection 正面投影;竖直投影;垂直投影
vertical projection of orebody 矿体垂直投影图
vertical projection plane 垂直投影面
vertical prop 立柱
vertical propagation 垂直传播
vertical propel shaft 垂直驱动杆;垂直驱动轴
vertical pug mill 立式练泥机
vertical pulse 垂直脉冲
vertical pump 立式泵
vertical radial plane 垂直径向平面
vertical ramp 发射台
vertical range 垂直幅度
vertical ratio analysis 纵向比率分析
vertical reaction 竖直反力;垂直反力
vertical reaction wheel 垂直反滚轮
vertical recharge coefficient 垂直补给系数
vertical recording 垂直录声;垂直(磁)记录
vertical rectangular prismy forward calculation method 直立矩形棱柱法
vertical redundancy check 垂直冗余校验;垂向冗余码校验
vertical redundancy test 垂直冗余试验
vertical reference line 垂直基准线;垂直角的基准线
vertical reflection 垂直反射
vertical reflectivity 垂直反射率
vertical refraction coefficient 垂直折光系数
vertical refraction error 垂直折光差
vertical regenerator 立式蓄热室
vertical register 垂直寄存器
vertical reinforcement 竖向(配)筋;竖向钢筋;垂直钢筋
vertical reinforcing 纵向加劲;纵向增强
vertical reinforcing steel 竖向钢筋
vertical reset 垂直复位
vertical resilience 铅垂弹性
vertical resistance 垂直抗性
vertical resolution 竖直析像力;垂直析像能力;垂直分解度;垂直分辨率;垂直分辨力
vertical resolution wedge 垂直分辨率光楔;竖直分辨力检验楔
vertical resultant 垂直合力
vertical retort (直)立式炭化炉;竖式蒸馏器
vertical retort method 竖罐蒸馏法
vertical retort process 竖罐提炼法
vertical retort smelting 竖罐熔炼
vertical retort tar 竖式蒸瓶煤沥青
vertical retort zinc smelter 竖罐炼锌厂
vertical retrace 帧回描;竖直回描;垂直回扫;垂直回描
vertical retrace ratio 垂直回描率
vertical return 竖直回(水)管;垂直回管
vertical reverse temperature gradient technique 垂直反温差法
vertical revolving arm mixer 立式搅拌机
vertical revolving circle 立运仪
vertical rib 纵肋
vertical ridge for gable roof 垂脊
vertical ring drilling 垂直平面径向孔钻进
vertical ring-roller mill 立式环辊磨
vertical rise cable 竖立电缆;垂直电缆
vertical rise of concrete per hour 混凝土浇注每小时竖升量
vertical rise of escalator 自动扶梯垂直高度
vertical riser cable 立面上升电缆
vertical rising conveyer 垂直提升输送机
vertical-rising type of lock gate 直升式船闸闸门
vertical rod 立杆
vertical roller 直轴旋辊
vertical roller of fairleader 直立式导缆滚轮

vertical roller sluice gate 垂直辊轴泄水闸门
vertical roll(er) 立辊
vertical roof bracing 屋顶纵向支撑
vertical root 垂直根
vertical rotary feeder 立式旋转叶轮喂料机
vertical rotary planer 立式转刨床
vertical rotary planing machine 立式转刨床
vertical rotating dryer 立式转盘烘砂炉
vertical run 垂直敷设(电缆);垂直管道
vertical sag curve 凹形竖曲线
vertical sand drain 排水砂井;竖向排水砂井;垂直排水砂井;垂直排水矿井
vertical sand drying oven 立式烘砂炉
vertical sand pile 垂直砂桩
vertical sash 框格窗;竖式滑窗;上下扯窗;吊窗;竖式窗框;竖拉窗扇
vertical saw 垂直式往复锯;竖直式圆锯
vertical saw-frame 竖直锯框
vertical scale 垂直刻度;纵坐标;纵向比例尺;纵标度;竖直比例尺;竖向比例;竖向比尺;垂直比(例)尺
vertical scanning 垂直扫描
vertical scanning frequency 垂直扫描频率
vertical scanning generator 竖直扫描发生器;垂直扫描振荡器
vertical scanning interval 垂直扫描时间
vertical scarf 竖式嵌接
vertical scour 竖向冲刷
vertical screen 垂直屏蔽
vertical screw conveyer 立式螺旋输送机
vertical screw elevator 垂直螺旋提升机
vertical screw mixer 立式螺旋混合机
vertical screw pump 立式螺旋泵
vertical screw type conveyer 垂直螺旋(式)输送机
vertical scrolling 垂直卷动
vertical scrubber 立式洗涤器
vertical seam 竖缝
vertical seam welding 竖向缝焊
vertical sea wall 直立式海堤;直立式防波堤;直立式海塘
vertical section 纵剖面;铅直截面;竖直断面;竖向断面图;竖剖面;竖截面;垂直剖面;垂直断面
vertical section method 垂直断面法
vertical section of coal seam 煤层垂向断面图
vertical section of engineering geology 工程地质纵断面图
vertical sector 垂直光弧
vertical sedimentation basin 竖流沉淀池;立式沉淀池
vertical sedimentation tank 立式沉淀池;竖流沉淀池
vertical sediment concentration distribution 泥沙竖向浓度分布;垂向泥沙浓度分布
vertical seepage test method 垂直渗透试验法
vertical segregation 垂直分开式
vertical seismic coefficient 竖向地震系数;垂直地震系数
vertical seismic profile method 垂直地震剖面法
vertical seismic profile record 垂直地震剖面记录
vertical seismograph 垂向地震仪
vertical seismometer 垂直地震计
vertical selector 上升选线器
vertical separation 纵断距;垂直线间距离;垂直间距;垂直间隔
vertical separator 立轴涡流分离器;立式分离器;竖直同步脉冲分离器;垂直同步脉冲分离器
vertical service pipe 用户立管
vertical service run 用户立管
vertical service space 垂直服务通道
vertical settling basin 竖流沉淀池
vertical settling bunker 立式沉淀仓
vertical settling tank 竖流(式)沉淀池
vertical shading 垂直阴影;垂直黑斑
vertical shading signal 竖直补偿信号
vertical shadow angle 直立阴影角
vertical shaft 立轴;立井;竖轴;竖井;竖管;竖坑;导井
vertical shaft alternator 立轴式交流发电机
vertical shaft arrangement 竖轴式布置
vertical shaft axial flow full-admission turbine 竖轴轴流整圆周进水式水轮机
vertical shaft centrifugal pump 立式离心泵
vertical shaft current meter 竖轴流速仪
vertical shaft furnace 竖式燃烧炉

vertical shaft globe valve 立轴球阀
vertical shaft grinder 竖轴磨床
vertical shaft hammer crusher 立轴锤式破碎机
vertical shaft kiln 立窑
vertical shaft liner 立轴衬套
vertical shaft mixer 立轴式搅拌机；立轴(混凝土)拌和机
vertical shaft motor 垂直型电动机
vertical shaft pump 立式水泵
vertical shaft seal 立轴密封装置
vertical shaft water turbine 立轴式水轮机；竖轴式涡轮机；竖轴式透平；竖轴式水轮机；竖轴式汽轮机
vertical shaft winder 竖井提升机
vertical shaper 立式牛头刨床；插床
vertical shaping machine 立式牛头刨床
vertical shear 竖向剪切；竖直剪切；垂直切力；垂直剪切；垂直剪力
vertical shear beam model 竖向剪切梁模型
vertical shearing 垂直剪切作用
vertical shearing stress 垂直剪(切)应力
vertical shear method 竖向剪切法
vertical shear model 竖向剪切模型
vertical shears 立式剪床
vertical shear strength 垂直抗剪强度
vertical shear stress 垂直剪应力
vertical shear test 垂直剪切试验
vertical sheathing 竖桩板；竖复板
vertical sheeting 竖板桩
vertical sheeting for excavation 垂直挖方支撑
vertical sheet process 垂直引上法
vertical shift 帧偏移；垂直移位；垂直位移；垂直偏移
vertical shingling 木板瓦墙；挂墙面板；竖向护墙板；竖挂木板瓦
vertical ship lift 垂直升船机
vertical ship lift with counterweight 平衡重式垂直升船机
vertical ship lift with float 浮筒式垂直升船机
vertical ship lift with suspension gear 平衡重式升船机
vertical shock 竖向震动
vertical shoring 竖直支撑
vertical shrinkage 竖向收缩率
vertical shutter 垂直式拉门
vertical shuttering 垂直模板
vertical side ship 直舷船
vertical sides of opening 洞口侧壁
vertical siding 外墙覆面层；竖向墙板；竖钉墙板；垂直披叠板
vertical sieve tray 垂直筛板
vertical sight distance 纵向视距；竖向视距
vertical silo 立仓
vertical single pipe heating system 垂直单管采暖系统
vertical single-spindle mo(u)lder 立式单轴木工铣床
vertical sketching 垂直相片转绘
vertical sketch master 垂直相片转绘仪；反光转绘仪
vertical skip hoist 立式翻斗加料器；垂直提升加料机
vertical slat 纵向百叶板条
vertical slatted blind 竖向板条百叶窗
vertical slide 高程滑尺；垂直滑板
vertical slide door 吊门；竖门
vertical sliding door 竖(向推)拉门
vertical sliding sash 立式移窗；上下移窗
vertical sliding sash-window 垂直推拉窗；竖拉窗扇
vertical sliding shutter door 竖拉百叶门
vertical sliding steel shutter door 竖拉卷帘式铁门
vertical sliding units 垂直滑动组合板
vertical sliding wheel 垂直滑轮
vertical sliding window 竖向滑动窗；竖上拉窗；提拉窗
vertical sliding window sash 竖拉窗扇
vertical slip 竖直滑尾
vertical slip form 竖向滑(动)模(板)；垂直滑(动)模(板)
vertical slotting machine 插床
vertical slow motion screw 垂直微动螺旋
vertical sludge blanket clarifier 立式污泥澄清池
vertical soft iron 垂直软铁
vertical soil friction force 土的垂直摩擦力；垂直土摩擦力
vertical sounding 立面探测；垂直电测法
vertical space 垂直空间

vertical spacer 垂直定距块
vertical spacing 垂直线间距离
vertical sparring 竖立货舱护板
vertical spectrograph 竖直摄谱仪
vertical speed 垂直方向速度
vertical speed indicator 爬升率指示器；上升速度指示器
vertical spin 垂直绕转
vertical spindle 立轴；立式摘锭
vertical spindle centrifugal pump 立轴式离心泵
vertical spindle cutter 立式切茎器
vertical spindle disk crusher 立轴圆盘碎矿机
vertical spindle disk grinder 立式磨芯机
vertical spindle dry dock drainage pump 竖轴干船坞排水泵
vertical spindle grinder 立柱磨床
vertical spindle mill 立轴磨煤机
vertical spindle milling 立磨粉磨
vertical spindle milling attachment 立铣附件
vertical spindle motor 立式电动机
vertical spindle paddle mixer 竖轴叶片搅拌机
vertical spindle pump 立轴式泵
vertical spindle reciprocating table grinding machine 立式往返工作磨床
vertical spindle rotary grinder 立轴回转磨床
vertical spindle rotary planing machine 立式转刨床
vertical spindle rotary table grinding machine 立式回转工作台磨床
vertical spindle surface grinder 立式平面磨床
vertical spindle-type shredder 垂直轴刀式茎秆切碎器
vertical spinning 立式离心铸造
vertical split 垂直分画面
vertical split casing 垂直剖分式机壳
vertical split cross culvert (船闸)垂直向分流廊道
vertical spring-pivot hinge 门底弹簧；竖向弹簧支枢铰链；竖轴弹簧铰链
vertical stability 竖直稳定(性)；竖向稳定(性)；竖向稳定(度)；垂直稳定(性)；垂直稳定(度)
vertical stabilizer 垂直稳定翼
vertical stack 垂直堆积；垂直叠加
vertical stacked extractor 垂直堆积式萃取器
vertical stack section 垂直叠加剖面
vertical staff 竖尺
vertical starter 竖向起动器；垂直起动器
vertical static convergence magnet 垂直静会聚磁铁
vertical stations 站线；船体横剖线
vertical stave fan 立式风筒
vertical steam boiler 立式蒸汽锅炉
vertical steam engine 立式蒸汽机
vertical steel facing 竖向钢板护面
vertical steel ladder 垂直钢梯
vertical steel lining 竖向钢板衬砌
vertical steel surfacing 竖向钢板镶面
vertical stem 直船首柱
vertical stem globe valve 立轴球阀
vertical step wave 垂直阶跃波
vertical stiffener 竖向加劲肋；竖(向)加劲肋；竖向加劲材；垂直加劲条
vertical stiffening 竖向加劲
vertical still 竖式蒸馏釜
vertical stirrup 箍筋；垂直蹬筋；竖直箍筋；竖向箍筋；垂直钢箍
vertical storage hopper 垂直储[贮]料斗
vertical stove 立式烘干机
vertical straightening wheel 垂调直轮
vertical straight side single action presses 立式单动直边冲压机
vertical stratification 垂直结构；垂直分层；垂直层理
vertical stratigraphic(al) separation 铅直地层距离距
vertical strength 垂直强度
vertical stress 垂直应力
vertical stress component 垂直分力
vertical stretching 垂直伸展
vertical strip camera 连续垂直摄影机；竖直式连续摄影机；垂直摄影机
vertical strip door 垂直板条门
vertical stripe 直纹
vertical strip storage 机组作业线上垂直移动带材的备用量
vertical structural member 垂直构件
vertical structure 垂直结构；竖直结构

vertical structure of atmosphere 大气的铅直结构
vertical strut 垂直支柱
vertical strut member 竖杆件
vertical strut truss 垂直支柱构架
vertical stuffing box 垂直填料压盖
vertical subheading 纵标目
vertical submerged anti-corrosion pump 立式耐腐蚀液下泵
vertical submerged pump 立式液下泵
vertical submerged type 直升浸没式；直立浸没式；直浸没式
vertical subsurface flow constructed wetland 垂直潜流人工湿地
vertical suction type 直升吸水式；直立吸水式
vertical sunshading board 垂直遮阳板
vertical support 立式支座
vertical supporting member 垂直支承构件
vertical surface 竖直面
vertical surface aerator 竖流表面曝气器
vertical survey 高程测量；高差测量；竖向测量；垂直测量
vertical survey by intersection 交会高程测量
vertical suspension 垂直悬挂
vertical suspension installation support 垂直悬吊安装支架
vertical sweep 竖直扫描；垂直扫描
vertical synchronization 竖直同步；垂直同步
vertical synchronizing pulse 竖直同步脉冲
vertical synchronizing signal 垂直同步信号
vertical sync inverter 垂直同步倒相器
vertical sync separator 垂直同步分离器
vertical system 垂直系统
vertical system of combined exploratory opening drilling 垂直坑钻结合系统
vertical system of exploring opening 垂直坑道系统
vertical system of leadership 垂直领导
vertical system of shading 直照晕渲法
vertical table 垂直表
vertical tableflap flocculation tank 竖流折流式絮凝池
vertical tabulate 垂直列表
vertical tabulation 直列制表
vertical tabulation character 垂直列表字符；垂向制表字符
vertical tabulator key 垂向制表键
vertical tail 垂直尾翼
vertical takeoff and landing 垂直起落
vertical takeoff and landing aircraft 垂直起落飞机
vertical takeoff and landing plane 垂直起落飞机
vertical tangent screw 垂直微动螺旋
vertical tank 立式箱；立式罐；立式容器
vertical tapered draft pipe 正锥形尾水管
vertical taper joint grinder 立式磨口机
vertical technology transfer 纵向技术转移
vertical teeth 上升齿
vertical temperature gradient 温度垂直梯度；垂直温度梯度
vertical temperature profile radiometer 温度垂直剖面辐射计；垂直温度剖面辐射计；垂直温度廓线辐射仪
vertical test stand 垂直试验台
vertical text 垂直文本
vertical thermosiphon reboiler 立式热虹吸式重沸器
vertical thickness 铅直厚度；垂直厚度
vertical thread 纵丝；竖丝
vertical three-dimensional migration section 垂直三维偏移剖面
vertical throat canal 垂直流液槽
vertical throw 垂直落差
vertical tilework 垂直铺面砖工程
vertical tiling 竖向挂瓦；墙体贴面砖；挂瓦；竖向贴砖；竖瓦；垂直挂瓦
vertical tipped bit 直镶式钻头
vertical toe 垂直拖拉
vertical tolerance 容许超深；垂直公差
vertical tooth rack 立齿条
vertical total stress 垂直总应力
vertical tower 垂直提升塔架；垂直吊篮；架空装置(轴向提升平台用)
vertical tower launcher 塔式发射架
vertical track buckling 竖向轨道鼓曲
vertical traffic 垂直交通

vertical transfer 纵向转移;垂直投影
vertical transfer register 垂直移位寄存器
vertical transition 垂直跃迁
vertical transit system 垂直运送系统
vertical translation 垂直移动
vertical transmission 垂直传播
vertical transportable retort 竖式移动甑
vertical transportation 垂直运输
vertical transportation system 垂直运输设备;垂直运输系统
vertical transport coefficient 铅直输送系数
vertical transporter 提升机;垂直运输机
vertical transverse migration 垂直穿层运移
vertical trapezoidal sluice 竖式梯形泄水槽
vertical travel 垂直行程;垂直提升高度
vertical travel(l)ing formwork 垂直移动模板
vertical travel(l)ing shuttering 垂直移动模板
vertical tray conveyors 竖向槽式运输机
vertical triangulation 竖直三角测量
vertical triple action press 立式三动冲床
vertical trunking 下料管(混凝土浇注用)
vertical truss 立式桁架;立式构架
vertical tube 竖管
vertical tube bundle units 立式管束单元
vertical tube condenser 立管冷凝器
vertical tube evapo(u)rator 立管(式)蒸发器;竖管(式)蒸发器
vertical tube manometer 直管式压力表
vertical tube zone-freezing unit 立管区域凝固装置
vertical tubing 竖管
vertical tubular boiler 立式管式锅炉
vertical tubular furnace 立式管式炉
vertical turbine 立式涡轮机;立式透平;立式水轮机;立式汽轮机;竖轴式涡轮机;竖轴式透平;竖轴式水轮机;竖轴式汽轮机
vertical turbine aerator 竖流涡轮曝气器
vertical turbine pump 竖式叶轮泵
vertical turbulence 垂直紊动
vertical turbulent 垂直湍流
vertical turbulent buoyant jet 垂直紊动射流
vertical turbulent diffusion 垂直湍流扩散
vertical turbulent diffusion coefficient 垂直紊流扩散系数
vertical turbulent viscosity 垂直紊流黏滞度
vertical turn 竖直转弯
vertical turning technology 垂直切削工艺
vertical turret lathe 立体转塔车床;立式转塔车床
vertical twin-body trawl net 垂直双体拖网
vertical twisty biochemistry reactor 垂向折流生化反应器
vertical two-pipe system 竖向双管系统
vertical two-piping system 竖向双管系统
vertical type breakwater 直墙式防波堤
vertical type bucket 立式料罐
vertical type comparator 立式比较仪
vertical type condenser 立式冷凝器
vertical type evapo(u)rator 立式蒸发器;直立式蒸发器
vertical type feed water heater 立式给水加热器
vertical type high speed pump 立式高速泵
vertical type pipe making machine 立式制管机
vertical type pump 立轴式泵
vertical type shading device 垂直遮阳装置
vertical type(shaft) motor 立式电动机
vertical type speed change gear 立式减速机
vertical type speed reducer 立式减速机
vertical type stranding machine 立式捻股机
vertical untying 立体疏解
vertical up-or-down movement 垂直上下运动
vertical upstream face 垂直上游面
vertical up welding 向上立焊
vertical utility line 公用设施垂直管线
vertical utility run 公用设施垂直管线
vertical variation 铅垂变化;垂直磁变;垂直变化
vertical variation of sedimentation 沉积垂向变化
vertical vegetational zonation 植被垂直地带性
vertical velocity 竖向速度;垂直速度
vertical velocity coefficient 垂线流速系数
vertical velocity curve 竖(向)流速曲线;深度一速度曲线
vertical velocity curve method 竖直速度曲线法
vertical velocity distribution 垂线流速分布
vertical velocity distribution diagram 垂直速度分布图;垂直流速分布图
vertical velocity gradient 铅直速度梯度;竖直速度梯度;垂直速度梯度
vertical velocity of water surface 水面垂直速度
vertical velocity sensor 垂直速度传感器
vertical veneer slicer 立式单板刨切机
vertical ventilator with fan 立式风扇通风机
vertical vessel 竖立式容器;直立设备;立式容器
vertical vessel supported by skirt 裙座支承的立式容器
vertical vibrating screening centrifuge 立式振动离心脱水机
vertical vibration 竖向振动;垂直振动
vertical vibration mo(u)lding process 立式振动成型法;垂直振动成型法
vertical view 垂直取景;立面(俯视)图;俯视图
vertical visibility 垂直能见度;垂直可见度
vertical-vision-light door 带观察窗门;镶嵌竖窄条玻璃的门
vertical volute spring suspension 竖锥形弹簧悬置
vertical vorticity 铅垂涡量;垂直旋度;垂直涡度
vertical wainscot 竖向护墙板
vertical wall 直(立)墙
vertical wall breakwater 直墙(式)防波堤;直立墙防波堤
vertical wall-mounted flagpole 直立墙头旗杆
vertical wall set flagpole 直立墙头旗杆
vertical wall slab 竖直铺墙面板
vertical wall structure with perforated face 带开孔面板的直立墙结构
vertical washing machine 立式清洗机
vertical water-collection layout 竖向集水布置
vertical water tube boiler 立式水管锅炉
vertical water vapo(u)r gradient 水汽垂直梯度
vertical water velocity 竖向水流速度
vertical wave 立波;竖波
vertical wear of rail head 轨头垂直磨损
vertical wear on wheel tread 轮缘垂直磨损
vertical wear surface 垂直磨损面
vertical web 舱壁垂直桁
vertical web member 垂直翼板构件;垂直腹杆
vertical web stiffener 竖向腹板加劲肋
vertical wedge cut 竖向V形开挖
vertical weir 直立堰;垂直堰
vertical weld(ing) 立焊;竖焊;垂直焊(接)
vertical well 竖井
vertical well biologic(al) reactor 竖井式生物反应器
vertical well of water distribution 冷却塔配水竖井
vertical wheel 导线立轮
vertical whirl 立轴漩涡;竖向漩涡;垂直旋转
vertical whirler 立式旋转烘版机;立式烤版机
vertical winding machine 立式缠绕机
vertical windlass 起锚绞盘
vertical window 竖窗
vertical wind tunnel 直立式风洞
vertical wire 竖直丝;竖丝
vertical wraparound 光标的虚拟环线移动
vertical yoke current 竖直偏转电流;垂直偏转电流
vertical zonality 垂直分带性;垂直地带性
vertical zoning 垂直分带
vertical zoning of water system 水系统竖向分区
vertices of diagraph 有向图顶点
verticity 向磁极性
verticraft 直升(飞)机
verti-former 竖式造纸机
vertiginous current 旋流
vertigo 眩晕
vertijet 垂直起落喷气式飞机
vertimeter 上升速度计;垂直计
vertiplane 直升(飞)机
vertisol 转化土;变性土
vertometer 焦距测量仪;焦距计;焦度计;屈度计
vertoro 变压整流器
vertrep 垂向补给
vertumnite 水羟硅铝钙石
verved clay 季候泥黏土
very abrasive rock 高研磨性岩石
very active area of new tectogenesis 新构造运动极活跃地区
very angular 棱角明显的
very bed visibility 很坏能见度
very close 很密闭的
very close drift ice 非常密集流冰群
very close pack ice 密集流冰;非常密集浮冰群
very cloudy sky 密云天空;多云天空
very coarse-columned aggregate 极粗柱状团聚体
very coarse-crystalline texture 极粗晶结构
very coarse-granular aggregate 极粗团粒状团聚体
very coarse-granular psamitic texture 极粗粒砂状结构
very coarsely crystalline 极粗晶质
very coarse pebble 极粗砾
very coarse sand 极粗砂
very compact soil 极坚实土
very complete 极完整
very dark of colo(u)r 颜色极深;颜色极暗
very deep 特深埋
very dense 极密实的
very dense sand 极密实砂
very destructive to trees 对树非常有害
very difficult soluble 极难溶的
very disastrous 激震
very early universe 极早期宇宙
very end of the late Gothic style 哥特式风格的最末期
very fine clay 极细黏土
very fine columnur aggregate 极细柱状团聚体
very fine crumb aggregate 极细团粒状团聚体
very fine-crystalline texture 极细晶结构
very fine-granular aggregate 极细团粒状团聚体
very fine-granular psamitic texture 极细粒砂状结构
very fine grinding 超细粉磨
very finely crystalline 极细晶质
very fine particle 超细颗粒
very fine particulate organic matter 极细颗粒有机物
very fine pebble 极细砾
very fine prismatic aggregate 极细棱柱状团聚体
very fine sand 极细砂
very fine sediment 极细泥沙
very fine silt 极细粉沙
very fine wood particles 很小的木质颗粒
very fine wood particles removed by a saw 锯木时掉下木质颗粒
very firm 极坚固
very fissured 严重裂隙化
very friable 极脆弱
very good quality 质量很好
very good source rock 很好烃源岩
very good visibility 优良能见度
very hard 极硬
very hard washing 很难选
very heady seaway 汹涛海面
very heavy rain 特大雨
very high altitude 极高空
very high altitude method 特大高度定位法
very high bit rate digital subscriber line 甚高比特率数字用户线系统
very high density plasma 超高密度的等离子体
very high frequency 极高频(率);甚高频;超短波
very high frequency AM transceiver 甚高频调幅收发两用机
very high frequency and ultra-high-frequency direction finder 甚高频与超高频测向器
very high frequency antenna array 甚高频天线阵
very high frequency band 极高频段;甚高频段
very high frequency bandpass filter 甚高频带通滤波器
very high frequency channel 甚高频频道
very high frequency circuit 甚高频电路
very high frequency data link jammer 甚高频数据线路干扰器
very high frequency direction finder 甚高频测向仪
very high frequency direction finder system 甚高频测向系统
very high frequency direction finding 甚高频定向
very high frequency homing adapter 甚高频归航仪
very high frequency jammer 甚高频人为干扰发射机
very high frequency linear amplifier 甚高频线性放大器
very high frequency link 甚高频传输线路
very high frequency navigation system 甚高频导航系统
very high frequency omnidirectional radio range 甚高频全向无线电信标

very high frequency omnidirectional range 甚高频全向无线电指向标
very high frequency omnirange 甚高频全向指向标
very high frequency omnirange distance measuring equipment 甚高频全向信标测距设备
very high frequency oscillator 甚高频振荡器
very high frequency preamplifier stage 甚高频前置放大级
very high frequency radio beacon 甚高频无线电信标
very high frequency radio lighthouse 甚高频无线电灯塔
very high frequency radio range 甚高频无线电指向标
very high frequency radio station 甚高频电台
very high frequency radio telephone device 甚高频无线电电话设备
very high frequency radio telephone equipment 甚高频无线电电话设备
very high frequency receiver 甚高频接收机
very high frequency single channel receiver 甚高频单波道接收机
very high frequency single channel two-way radio equipment 甚高频单波道双向无线电设备
very high frequency solid-state TV 甚高频固态电视
very high frequency solid-state TV tuner 甚高频固态电视频道选择器
very high frequency station 甚高频电台
very high frequency station location marker 甚高频通信站位置标记
very high frequency television IF strip 甚高频电视中频部分
very high frequency television receiver 甚高频电视接收机
very high frequency television receiving equipment 甚高频电视接收设备
very high frequency transmission 甚高频传输
very high frequency transmitter 甚高频发射机
very high frequency tuner 甚高频频道选择器；甚高频调谐器
very high frequency tuning mechanism 甚高频电视调谐机构
very high frequency turret tuner 甚高频旋转式频道选择器
very high frequency TV tuner 甚高频电视频道选择器
very-high-level language 甚高级语言；超高级语言
very high load 特重货件；特大荷载
very high magnetic field 超强磁场
very high mountain 极高山
very-high-output fluorescent lamp 甚高输出荧光灯
very high rates of pressing 高能束压制
very high resolution 甚高分辨率
very high resolution radiometer 甚高分辨率辐射仪；甚高分辨率辐射计
very high resolution scan(ning) radiometer 甚高分辨率扫描辐射计
very high resolution spectroscopy 高分辨力光谱学
very high salinity water 极高盐水
very high sea 狂涛；八级浪
very high-speed 特高速
very high speed computer 甚高速计算机；超高速计算机
very high strength concrete 特高强混凝土
very high vacuum 甚高真空
very high workability 极高可加工性；极高和易性
very important person system 特种车辆优先信号系统
very insoluble cation 极不可溶性阳离子
very intense neutron source 超强中子源
very large boulder 超巨砾
very large bulk carrier 巨型散货船
very large cross-section 特大断面
very large crude carrier 巨型油轮（通指十五万～三十万吨级）；超级油轮
very large data base 超大型数据库
very large landslide 特大型滑坡
very large oil carrier 巨型油轮（通指十五万～三十万吨级）
very large oil-ore carrier 巨型石油矿石两用船
very large product carrier 巨型成品油轮
very large-scale integration 甚大规模集成电路；超大型计算题目；超大规模集成电路
very large-scale problem 超大型（计算）题目
very large ship 巨轮
very large wheel load 超大轮载
very light physical labor 极轻体力劳动
very long baseline interferometer 甚长基线干涉仪
very long baseline interferometry 甚长线干涉测量；射电干涉测量（法）；超长线干涉测量；超远基线干涉测量法
very long instruction word 超长指令字
very long oil 醇酸油
very-long range 超远程声的
very-long range radar 超远距测距雷达
very long term 极长期
very loose 极松散的
very loose sand 极松散砂
very low density polyethylene 超低密度聚乙烯
very low density sampling 甚低密度采样
very low frequency 极低频；甚低频；超低频
very low frequency antenna 超长波天线
very low frequency communication 甚低频通信
very low frequency electro-magnetic instrument 甚低频电磁仪
very low frequency highpotential test 甚低频高压试验
very low frequency jammer 甚低频人为干扰发射机
very low frequency receiver 甚低频接收机
very low frequency video signal 甚低频视频信号
very low income housing repair loans 极低收入家庭住房修理贷款
very low-strength field 特低场
very low-strength rock 强度极低的岩石
very-low-traffic roadway 极轻量交通道路
very low workability 极低和易性
very-narrow-beam receiver 甚窄束接收器；高方向辐射接收器；极窄束接收器
very old hummock 多年堆积冰；多年冰丘
very old ice 多年堆积冰
very open drift ice 非常疏流冰群
very open-grained pig iron 极粗晶生铁；伟晶生铁
very open ice 全分离流冰
very open pack ice 甚疏浮冰群；非常疏浮冰群
very organic soil 富含有机质
very pale of colo(u)r 颜色很浅；颜色很淡
very poorly drained 排水极劣
very poor visibility 很差能见度
very quick-setting cement 速凝水泥
very rough discontinuity 非常粗糙的结构面
very rough sea 六级浪；巨浪
very rounded 极圆状
very seedy glass 小气泡多的玻璃
very severely swelling-shrinkage foundation 严重胀缩地基
very shallow water wave 极浅水波
very short distance 很短的距离
very shortly attenuate 极短渐狭的
very short persistence phosphor 超短余辉荧光粉
very short wave 甚短波；超短波
very slightly fissured 极轻微裂隙化
very slightly soluble 极微溶解的
very slightly swelling-shrinkage foundation 轻微胀缩地基
very slow 很慢的
very slow permeability 极慢渗透
very slow sweep voltage wave 极慢扫描电压波
very small fracture 微冰隙
very soluble 极可溶的
very stable resistor 高稳定度电阻器
very sticky 极黏滞
very stiff 极坚硬的
very stiff mixture 特干硬混凝土拌和料；特干硬混凝土混合料
very strong acid 极强酸
very strong earthquake 强烈地震
very strong odo(u)r 极强恶臭
very thick bedded 巨厚层状
very thick bedded structure 巨厚层状构造
very thick platy aggregate 极厚片状团聚体
very thin bedded 极薄层状
very thin bedded structure 极薄层状构造
very thin platy 极薄片状团聚体
very unstable area 极不稳定地区
vesatile laser instrument for experiments 多用途激光实验仪
vesica 两端尖的椭圆形；梭饰；垂鱼饰；(耶稣像周围的) 光轮
vesicant 糜烂剂
vesicant war gas 糜烂性毒气
vesica-piscis 哥特式双尖卵状图形（古建筑）；椭圆光轮；(古建筑哥特式的)双尖圆形；双圆光轮
vesicle 气孔；小穴；小气泡；小泡；泡囊
vesicle fluid 水泡液
vesicle liquid 水泡液
vesicle liquor 水泡液
vesicular 有气泡的；微泡的；多泡状；多孔状；多孔的
vesicular aggregate 多孔集块；多孔骨料
vesicular basalt 多孔状玄武岩
vesicular film 微泡（法）胶片
vesicular lava 多孔（状）熔岩
vesicular nature 起泡性；多孔性
vesicular porosity reservoir 多孔型热储
vesicular process 微泡法
vesicular structure 气孔构造；多泡状构造；多孔构造
vesicular texture 多孔状结构
vesicular tissue 泡沫状组织
vesiculation 形成气泡
vesignieite 钒钡铜矿
vesorcinol 二羟甲苯
vessel 舰；容器；船(只)；船舶
vessel age 船龄
vessel agent 船舶代理商；船舶代理人
vessel approach to berth 船舶靠泊
vessel arrival and departure report 船舶到达及离岸报告
vessel barometer 球管气压计
vessel body 船体
vessel body design 船体设计
vessel body stress 船体应力
vessel bond 船舶抵押证书；船舶保证书
vessel buckling under external pressure 外压容器的失稳
vessel ceiling 船舶铺仓
vessel class 容器类别
vessel code 船舶代号
vessel constrained by her draught 限于吃水的船
vessel control center 船舶控制中心
vessel course 筒节
vessel dismantling 船舶装卸
vessel element 导管分子
vessel flange 容器法兰；设备法兰
vessel form 船式
vessel generator 船舶发电机
vessel husband 船舶代理人
vessel in distress 遇险船；遇难船
vessel inertial navigational system 船舶惯性导航系统
vessel lock 船闸
vessel material 容器材料
vessel member 导管分子
vessel mix 船舶编组情况
vessel model basin 船模（型）试验池
vessel model tank 船模（型）试验池
vessel mooring conditioning 船舶泊稳调节
vessel mooring condition 船舶停泊条件
vessel movement 船舶动态业务
vessel movement report 船舶动态报告
vessel movement reporting system 船舶动态报告制；船舶动态报告系统
vessel name 船名
vessel navigation 船舶导航
vessel noise 船舶噪声
vessel not under command 失去控制的船；失控船
vessel operating cost 船舶营运成本
vessel operating expenditures 船舶营运费用
vessel operator 船舶营运者
vessel pattern 船型
vessel perforation 导管穿孔
vessel pneumatic pump 仓式气压泵
vessel port 容器开口
vessel position report 船位报告
vessel reality 容器本体
vessel restricted in her ability to maneuver 操纵能力受限制的船
vessel sailing resistance 船舶航行阻力
vessel segment 导管分子
vessel sewage 船舶污水
vessel side 容器侧
vessels in service 在册船舶

vessel size 船型
vessel's number 船舶呼号
vessels per hour 艘/时
vessel squat calibration test 船舶航行吃水校准试验
vessels subjected to external pressure 承受外压的容器
vessels subjected to corrosion 经受腐蚀的容器
vessels subjected to ambient pressure 常压容器
vessel surveillance system 船舶监视系统
vessel test plate 容器试板
vessel ton 容器吨
vessel towed 被拖船
vessel traffic center 船舶交通(管理)中心
vessel traffic management 船舶交通管理
vessel traffic management center 船舶交通管理中心
vessel traffic management service 船舶交通管理系统
vessel traffic management system 船舶交通管制系统;船舶交通管理系统
vessel traffic reporting system 船舶交通报告系统
vessel traffic service 船舶交通服务
vessel traffic system 船舶交通系统
vessel turnaround 船舶周转
vessel underway 在航船
vessel unit 导管分子
vessel wall 容器壁
vessel waste 船舶废弃物
vessel wastewater 船舶废水
vessel-water relative velocity 船水相对速度
vest 赋予
Vesta 维士达牌玻璃
vested 法定的;既得的
vested benefit 归属利益
vested capital 已投资本;投入(的)资本
vested in possession 现有的权利
vested interests 既得利益(集团)
vested proprietorship 投入资本;确定资本
vested remainder 既得的地产指定继承权
vested right 固有权利;既得权利
vestee 不动产业主
vestiary 藏衣室
vestibule 挡风玻璃;(客车两端的)出入小室;窑前室;廊室;门厅
vestibule basin 过渡港池;前港池(有闸港池口门外的水域)
vestibule coach 带通道的客车
vestibule dock 过渡港池;前港池(有闸港池口门外的水域)
vestibule door 避风门;门厅门
vestibule school 新工人培训学校;技工学校
vestibule train (各车厢相通的)连廊列车
vestibule training 现场以外的训练;职前训练
vesticular 前庭的;前厅的
vestige 遗迹;痕迹
vestigial 残留的
vestigial analysis 痕量分析
vestigial element 痕量元素
vestigially period 意大利早期文艺复兴时期(15世纪)
vestigial sideband 残余边带;残留边带
vestigial sideband amplitude modulation 残余边带调幅
vestigial sideband demodulator 残留边带解调器
vestigial sideband emission 残留边带发射
vestigial sideband filter 残留边带滤波器
vestigial sideband modulation 残留边带调制
vestigial side band transmission 残余边带传输
vestigial sideband transmitter 残留边带发射机
vestigial sludge 剩余污泥
vestigiofossil 足迹化石
vestiluled 有门厅的
vesting 归属;赋予权利;保留退休金的权利
vesting of right 授予权利
vesting order 财产委托命令
vestopal 聚酯树脂
vestopal binder 聚酯树胶黏合料
vest-pocket calculator 袖珍计算器;袖珍计数器
vest-pocket camera 袖珍照相机
vest-pocket edition 袖珍版
vest-pocket garden 街心公园
vest-pocket housing 填空性住房
vest-pocket park 小块绿地;小(型)公园;街心公园;小游园
vest-pocket receiver 小型接收机
vestry 祭具;教堂备用室;祭服室;法衣室;(教堂)祭具室
vestured pit 附物纹孔
Vesuvian 维苏威式;白榴石
Vesuvian garnet 白榴石
vesuvianite 符山石
vesuvianite sharn 符山石矽卡岩
Vesuvian-type eruption 维苏威型火山喷发
Vesuvius Volcano 维苏威火山
veszelyite 磷锌铜矿
veteran 老手;富有经验的人
Veterans Administration 退伍军人管理局
veterans administration loans 退伍军人管理局住房贷款
veterans administration mortgage 退伍军人管理局住房抵押贷款
veterans's tax exemption 退伍军人财产免税权
veteran worker 老工人;熟练工(人)
veterinary clinic 兽医诊所
veterinary general hospital 兽医总医院
veterinary station 兽医站
veto in detail 部分否决权
veto power 否决权
Vezin cement sampler 维琴水泥取样装置
v-feel 速度感觉
V-groove V形小槽
V-grooved 带V形槽的
V-grooved and tonguing V形槽胶合板
V-grooved core cable 骨架型光缆
V-grooved plywood V形槽胶合板
V-grooved wall panel V形墙板
V-groove pulley 三角皮带轮带;V形皮带轮带
V-grooving and tonguing V形企口;V形雌雄榫
V-guide way V形导轨
V-gully V形沟
V-gutter 屋面斜排水沟;V形(水)槽;V形截面雨水槽;V形沟
VHF-DF 甚高频测向仪
VHF direction finder 甚高频测向仪
via 经由
Via appia antica 阿庇亚街道
viability 生活力
viability index 出生存活指数
viability of sample 样品生活力
viable count method 活菌计数法
viable proposal 可行性建议
viable site 有利坝址;经济坝址
via circuit 转接电路
viaduct 谷架桥;高架铁桥;高架铁路;高架道;高架桥;架空行车道;旱桥;栈道;跨线桥
viaduct and beam crane 高架长臂(堆煤)起重机
viaduct bent 旱桥排架
viaduct expressway 高架快速公路
viaduct pier 高架桥墩;旱桥(桥)墩
via face 沿工作面
via hole 借用孔;通路孔
vial 小瓶;药水瓶;管制瓶;管形瓶;水准器玻璃管;玻璃瓶
vialog 路面平整度测量仪;路程仪;路程计;测程仪
viameter 路面平整度测定仪;路面平整度测验仪;路面平整度测验计;路程计;计距器;计程仪;计程器;车程计;测距器
via pin 通路引线
via Suez Canal shipping line 苏伊士运河航线
viatic 道路的
vibrability (混凝土混合料)振动密实性
vibracast 浇捣混凝土
Vibracork 维布拉减振材料(一种减振材料)
viballoy 镍钼铁弹簧合金
vibramat 弹性玻璃丝垫
vibrameter 振动计;振动式计量器;测震仪
vibrapack 振动子换流器
vibrashock-isolating unit mounting 减震器隔震器
vibra shoot 振动槽
vibratable mix 可振动混凝土混合料
vibrated and tamped concrete 振动夯实混凝土
vibrated bed 振动床;搅动床(层)
vibrated brick panel 振动砖壁板;振动(过的)砖墙板
vibrated bulk concrete 振捣大体积混凝土
vibrated coarse concrete 振实的粗骨料混凝土
vibrated column 振实(的混凝土)柱
vibrated concrete 振实混凝土;振捣(过的)混凝土
vibrated concrete pile 振实混凝土桩
vibrated concrete pipe 振捣混凝土管
vibrated fluidized bed 振动流化床
vibrated frequency of controllable source 可控震源的振动频率
vibrated fresh concrete 振捣过的新浇混凝土
vibrated into place 振动装料
vibrated joint 振实的接缝
vibrated lining 振动打炉衬
vibrated mass concrete 振捣大体积混凝土
vibrated mortar cube test 振动成型砂浆立方试块试验;振动成型砂浆立方试件试验
vibrated optimum value 振动最宜值;振动最佳值;振捣最佳值
vibrated platinum 振动铂电极
vibrated rockfill 振捣堆石
vibrated rock fill dam 振实的堆石坝
vibratile 振动的
vibratile compacter 振动压实机
vibratility 振动性
vibrating 振动的;振捣
vibrating air separator 振动空气分选机
vibrating and finishing machine 振动抹面机
vibrating arm 振动杆
vibrating armature 振动电枢
vibrating backfill compactor 填土振动压实器
vibrating ball mill 振动球磨机
vibrating bar grizzly 振动式棒条筛;振动格筛
vibrating base-plate 振动底片;振动底板
vibrating base-plate compactor 振动平板压实机
vibrating batch charger 振动投料机
vibrating batching unit 振动配料单元
vibrating beam 震捣梁;振动梁;振捣梁;梁式振动器
vibrating bell 闹铃
vibrating bit 振动钻钻头
vibrating blade 振动刀
vibrating board 振动板;平板振动器;平板振动机;平板振捣器
vibrating board compactor 振动板压实器;振动板夯实机
vibrating body 振动体
vibrating boom 抖动杆
vibrating booster 振动子升压器
vibrating break 振动断续器
vibrating breaker 振动式断续器
vibrating capacitor 动态电容器
vibrating case 振动箱
vibrating casting 振动浇注
vibrating center 振动中心
vibrating centrifuge 振动(式)离心机
vibrating chute 振动输送料斜槽;振动流槽;振动溜槽;抖动式滑槽
vibrating chute transporter 振动斜槽输送器
vibrating circuit 振动电路
vibrating circuit pipeline 振动输料管道
vibrating circular pipeline 振动式循环管线
vibrating coating 振动抹浆
vibrating coil 振动线圈
vibrating coil method 振动线圈法
vibrating comb 振动梳
vibrating compaction 振实;振浮压实;振浮压密;振动夯实
vibrating compaction beam 振动(捣实)梁
vibrating compaction settlement ratio 振动压实沉降率
vibrating compactor 振动压路机;振(动压)实机
vibrating compactor block making machine 振实式砌块制造机
vibrating compactor mixer 振实拌和机
vibrating concrete float 振动抹灰板;混凝土振平器;混凝土振捣整平器;混凝土平板振捣器;混凝土表面振捣
vibrating contact 振动接点;振动触点
vibrating contact voltage regulator 振动接触式电压调整器;振动触点式电压调整器
vibrating controller 振动控制器
vibrating conveyer 振动式运输机;振动(式)输送机
vibrating conveying machine 振动式输送机
vibrating core 振动芯
vibrating core knockout machine 振动脱芯机
vibrating core process 振动芯模法
vibrating coring tube 振动取样管

vibrating crystal 振荡晶体
vibrating cylinder 振动缸
vibrating dampener 振动减低器
vibrating decorer 振动脱芯机
vibrating dehydrator 振动式脱水机
vibrating detector 振动检波器
vibrating device 振动装置
vibrating-dewatering screen 振动脱水筛
vibrating diaphragm 振动膜片;振荡光阑
vibrating disk 振动盘
vibrating dozer 振动(式)推土机
vibrating drill 振动钻
vibrating drilling 振动凿岩
vibrating drum-type roller 振动滚筒式压路机;振动鼓形碾;筒式振动路碾
vibrating earth borer 振动式黏土机
vibrating explosion 振动爆炸
vibrating extractor 振动拔桩机
vibrating feed bin 振动箱
vibrating feeder 振动喂料器;振动喂料机;振动式进料器;振动(式)给料机;振动加料器
vibrating filter 振动式过滤器;振动式过滤机
vibrating fine screen 振动式筛滤
vibrating finisher 振动抹面机;振动拌面器
vibrating finishing screed 振动式整平板
vibrating float 振动抹灰板
vibrating force 振动力;起振力
vibrating frame 振动筛;振动盘;振动框;振动式淘汰盘
vibrating galvanometer 振动电流计
vibrating grate 振动选矿筛
vibrating grate cooler 振动篦式冷却器
vibrating grid 振动栅
vibrating grinding mill 振动式研磨机
vibrating grizzly 摇筛机;振动(铁)栅筛;振动(条)筛;振动格筛
vibrating grizzly feeder 振动式给料器;振动格栅进料器;振动格筛进料器
vibrating hammer 振动(桩)锤
vibrating hammer method 振锤测定土壤含水率法
vibrating hammer test 振锤测定;振锤试验
vibrating hand-rolle 手扶振动辊;振动式手扶压路机
vibrating head 振动头
vibrating-impacting boring machine 振动冲击钻孔机;振动冲击成孔机
vibrating-impacting roller 振动冲击压路机
vibrating impacter 振冲器
vibrating isolation 隔振
vibrating jet method 振荡射流法(测动表面张力)
vibrating jigger 振动敲平锤
vibrating joint cutter 振动式混凝土切缝机
vibrating load 振动荷载
vibrating loader 振动式装载机
vibrating machinery 振动器;振动机械
vibrating mark 抖动标记
vibrating materials handling equipment 振动式材料输送设备
vibrating measuring 振动测量;振动测定
vibrating mechanical rectifier 振动机械整流器
vibrating mechanism 振动机构
vibrating membrane 振动膜
vibrating mesh 振动筛
vibrating metal 金属振动板
vibrating meter 振动计;示振仪;示振器;示振计
vibrating method 振动法
vibrating mill 振动磨碎机;振动球磨机;振动磨(机);振动粉碎机
vibrating mirror deflector 振动镜偏转器
vibrating mixer 振动拌和机
vibrating motor 振动马达;振动电动机
vibrating mo(u)ld 振动模具;振动模板
vibrating mo(u)lding machine 振动造型机
vibrating needle 振动磁针;振动棒;插入式振捣器;摆动磁针仪;振动杆;振捣棒
vibrating pan compactor 振动盘式压实机
vibrating part feeder 气流振动供料机
vibrating pebble mill 振动式乱石粉碎机;振动砾磨机
vibrating-pendulum gravimeter 振摆式重力仪
vibrating petrol engine 振动汽油发动机
vibrating pile driver 振动(式)打桩机
vibrating pile driver-extractor 振动沉拔桩机
vibrating pile hammer 振动打桩锤

vibrating piling hammer 振动打桩锤
vibrating pipe 振动溜管
vibrating plate 振动板;(切缝机的)振动切入板;振动片
vibrating plate compactor 振动板捣实器;平板振动压实器;振动式板压机;振动板压实器;振动(板)夯实机;平台式振动压实器;平板振动压实机;板式振捣器
vibrating plate furnace 振底炉
vibrating plate rammer 板振动夯压器;振动式夯板器
vibrating plate visco(si)meter 振动片黏度计
vibrating platform 振动装载台
vibrating platinum microelectrode 振动铂微电极
vibrating pneumatic loader 振动式风力装载机
vibrating poker 振动探棒
vibrating poker burn 振捣斑伤
vibrating press 振动压机
vibrating pulverizer 振动粉碎机
vibrating quartz crystal 振荡石英晶体
vibrating rammer 振动夯
vibrating reagent feeder 振动给药机
vibrating reed 振簧;振动片
vibrating-reed amplifier 振簧(式)放大器
vibrating-reed analyzer 振簧分析器
vibrating-reed electrometer 振簧(式)静电计
vibrating reed escapement device 弹性片调速器
vibrating-reed frequency sensor 振簧式频率传感器
vibrating-reed gyro 振簧式陀螺
vibrating-reed indicator 振簧指示器
vibrating-reed instrument 振簧式仪表
vibrating-reed magnetometer 振簧式磁强计
vibrating-reed method 振簧法
vibrating-reed oscillator 振簧振荡器
vibrating-reed rectifier 振簧式整流器
vibrating-reed relay 振簧式继电器
vibrating-reed tachometer 振簧式转速计
vibrating-reed type 振动片式
vibrating-reed viscometer 振簧式黏度计
vibrating regulator 振动调节器;起振力调节装置
vibrating replacement process 振冲法
vibrating riveter 振动铆钉机
vibrating rod oiler 振动杆加油器
vibrating roller 振动(式)压路机;振动式路碾;振动滚路机;振动镇压器;振动碾(式)振动辊压机
vibrating roller trailer 拖带式振动压路机
vibrating sample magnetometer 振动探针式磁强计
vibrating scalper 振动筛机
vibrating screed 振捣器镘面板;振动式整平板;振动刮板;长板振动器;振动找平机
vibrating screen 摇筛机;振动筛(分机);摆动筛
vibrating screen centrifuge 振动筛离心机
vibrating screen classifier 振动筛分机
vibrating screen dissolver 振动筛溶解器
vibrating screen with circular movement 圆行程的振动筛
vibrating screen with eccentric drive 偏心振动筛
vibrating scrubber 振动擦洗机
vibrating sensor 振动传感器
vibrating shake-out grid 振动落砂架
vibrating shake-out machine 振动落砂机
vibrating shakeout with eccentric drive 偏心振动落砂机
vibrating sheepsfoot roller 振动式羊脚压路机;振动式羊脚滚筒
vibrating shoe 振动靴
vibrating sieve 振动筛
vibrating slab compactor (混凝土)板振实器
vibrating sliding table press 振动式滑盘压机
vibrating slope method 振动坡法(测混凝土和易性)
vibrating smoother 振动式刮路机;振动式整平机;振动抛光机
vibrating spear 振动棒;振捣棒;振荡锤
vibrating spiral elevator 振动式螺旋提升机
vibrating spring 振动弹簧
vibrating squeeze mo(u)lding machine 振压(式)造型机
vibrating stressing bed 振动式加力台;振动式加力床
vibrating string 振动弦(线)
vibrating string accelerometer 振弦加速度计
vibrating string extensometer 振动弦式应变仪;

振弦式应变计;振动钢弦式应变计;振动钢弦式伸缩计
vibrating string gravimeter 振弦式重力仪
vibrating string strain meter 钢弦式应变仪;钢弦式应变计
vibrating string strainometer 振动弦式应变仪
vibrating stroke 振动冲程
vibrating system 振动系统
vibrating table 振动台;平台振捣器
vibrating tamper 振动捣固机;振动夯;振动板
vibrating tandem roller 振动式双轮压路机
vibrating target 振动靶
vibrating tele-hygrometer 弦式遥控温度计
vibrating test bench 振动台
vibrating test(ing) 振动试验
vibrating tine 振动叉
vibrating tool 振动工具
vibrating tray feeder 振动盘式给料机
vibrating tray reactor 振动盘反应器
vibrating trench compactor 振动式沟堑压实机;沟堑振实机
vibrating trestle 振动支架
vibrating trough 振动槽
vibrating trough batcher 振动槽式分批箱
vibrating trough conveyer 振动槽式输送机;振动输送槽
vibrating tube 振动溜管
vibrating tube mill 振动(式)管磨机
vibrating type automatic voltage regulator 振动式自动电压调整器
vibrating type circuit breaker 振动式断路器
vibrating type rectifier 振动式整流器
vibrating type regulator 振动式调节器
vibrating type track tamper 振动式轨道夯实机
vibrating vacuum dewatering process 振动真空脱水工艺
vibrating vacuum process 振动真空法
vibrating void-forming mandrel 振动穿孔芯棒
vibrating wedge 振动楔
vibrating wire 振动弦(线)
vibrating wire cell 振弦式压力盒
vibrating wire extensometer 振丝伸长计
vibrating wire indicator 振丝应变计;钢弦式应变仪;钢弦式应变计
vibrating wire piezometer 振弦式渗压计
vibrating wire porewater pressure meter 钢弦式孔隙水压力仪
vibrating wire(pressure) cell 钢弦式(土)压力盒
vibrating wire scanner 振动线扫描器
vibrating wire strain ga(u)ge 线振应变仪;振动钢丝应变仪;振弦式应变仪;振动金属丝应变计;钢弦式应变仪;钢弦式应变计
vibrating wire stressmeter 振弦应力计
vibrating wire tele-hygrometer 振弦式遥控湿度计
vibrating wire transducer 线振式换能器
vibration 振动;抖动;颤动
vibration abatement 振动抑制
vibration ability 振动能力
vibration absorber 减振器;消振器;吸振器;振动阻尼器;振动吸收器;振动衰减器;炉管消振装置;缓冲器
vibration-absorbing base 吸振基础;减振基础
vibration-absorbing foundation 防振基础
vibration-absorbing layer 减振层
vibration-absorbing measure 消振措施
vibration-absorbing mitten 防振手套
vibration-absorbing mount 减振架
vibration absorption 减振
vibration-absorptive material 减振材料
vibration acceleration level 振动加速级
vibration-actuated feed 振动式推进装置;振动给进
vibrational 振动的
vibrational acceleration 振动加速度
vibrational band 振动谱带
vibrational behavio(u)r 振动状态;振动性能
vibrational casting 振动铸造
vibrational compaction 振动致实
vibrational coupling 振动耦合
vibrational deactivation 振动钝化
vibrational degrees of freedom 振动自由度
vibrational energy 振动能(量)
vibrational energy level 振动能级
vibrational entropy 振动熵
vibrational excitation 振动激励

vibrational fine structure 振动微组结构;振动微细结构
vibrational instability 振动不稳定性
vibrational lattice 振动点阵
vibrational level 振动(能)级
vibrational line 振动线;振动谱线
vibrational load 振动荷载
vibrational method 振动方法
vibrational mode 振动模式
vibrational model 振动模型
vibrational property 振动特性
vibrational quantum number 振动量子数
vibrational relaxation 振动松弛
vibrational resilience 振动态回弹性
vibrational resonant disk 振动片
vibrational rotation 振动转动
vibrational-rotational band 振动转动(谱)带
vibrational-rotational energy 振动转动能量
vibrational-rotational spectrum 振动转动光谱
vibrational-rotational transition 振动转动跃迁
vibrational source 振动源
vibrational spectrum 振动(光)谱
vibrational state 振动(状)态
vibrational stress 振动应力
vibrational structure 振动结构
vibrational sum rule 振动求和定则
vibrational transition 振动跃迁
vibration amplitude 振动强度
vibration amplitude recording system 振幅记录系统
vibration analysis 振动分析
vibration analysis and deviation concept 振动分析和偏差概念
vibration analyzer 振动分析仪;振动分析器
vibration-and-shock safety switch 振动防护开关
vibration and shock testing 振动和冲击试验
vibration attenuation 振动衰减(量)
vibration attenuation curve 振动衰减曲线
vibration axis 振动轴
vibration balancing 振动平衡
vibration ball mill 振动球磨机
vibration band 振动带
vibration behavio(u)r 振动特性
vibration bit 振动钻头
vibration calculation 振动计算
vibration characteristic 振动特性
vibration characteristic factor 振动特征因数;振动特征系数;振动特性因数;振动特性系数
vibration characteristic temperature 振动特性温度
vibration check calculation 振动检验计算
vibration chute 振动槽;抖动式滑槽
vibration compaction 振实;振动压密;振动密实;振动法压实
vibration constant 振动常数
vibration control 振动控制
vibration control design 防振设计
vibration control system 振控系统;减振装置
vibration conveyer 振动式传送机
vibration criteria 振动评价标准
vibration curve 振动曲线
vibration damage 振动损害;振动破坏
vibration dampener 减振器
vibration dampening characteristic 振动衰减特性
vibration damper 减振装置;减振器;缓振器
vibration damper fly wheel 减振飞轮
vibration damper fly wheel cushion 减振器轮垫
vibration damper pilot 减振器导杆
vibration damper plate 减振板
vibration damping 振动阻尼;振动衰减
vibration damping material 振动材料
vibration deflection mode 偏移振型
vibration detecting laser apparatus 激光振动探测仪;激光振动检测仪
vibration detection system 振探测系统
vibration detector 振动指示器;振动检测器;振动传感器;拾振器
vibration diaphragm 振动膜
vibration direction 振动方向
vibration disease 振动病
vibration displacement 振动位移
vibration displacement level 振动位移级
vibration drill 振动钻机;振动设备
vibration drilling 振动钻眼;振动钻探;振动钻进;振动凿岩

vibration due to traffic 车辆引起的振动
vibration effect 振动效应
vibration effect of explosion 爆破振动效应
vibration energy 振动能
vibration energy level 振动能级
vibration environment 振动环境
vibration error 振动误差
vibration exciter 振子;振动式励磁机;振动激振器;激振器
vibration exposure time 振动经受时间;振动暴露时间
vibration failure mode of rock and soil 岩土振动破坏方式
vibration fatigue 振动疲劳
vibration fatigue machine 振动疲劳试验机
vibration fatigue test 振动疲劳试验
vibration feeder 振动喂料器;振动喂料机;振动式供料器;振动式给料器
vibration flanking transmission 振动侧向传输
vibration flow 振动流动
vibration-flow method 振动流度法;V-F 稠度测定法
vibration force 振动荷载;振动力
vibration foundation 振动基础
vibration fracturing 振动破裂
vibration-free 无振
vibration-free seat 防振座
vibration frequency 振动频率;波动频率
vibration frequency analyzer 振动频率分析仪;振动频率分析器
vibration frequency meter 振动频率计
vibration from blasting 爆炸振动
vibration galvanometer 振动式检流计
vibration ga(u)ge 振动仪;测振计
vibration generator 振子;振动发生器;激振器;起振器
vibration generator system 振动发生装置;激振器系统
vibration gravimeter 振动(式)重力仪
vibration grinding 振动研磨
vibration grizzly 振动格筛
vibration hammer 振动锤
vibration harmonic 谐和振动
vibration hazard 振动故障
vibration hazard for citizen 振动公害
vibration head 振动器插入棒
vibration impedance 振动阻抗
vibration indicator 振动绝缘体;隔振体;振动指示器;振动仪
vibration instrumentation 振动仪器
vibration insulating foundation 隔振基础
vibration insulation 振动隔绝;振动绝缘;隔振垫
vibration insulation material 防振材料;隔振材料
vibration insulator 隔振体;振动绝缘体
vibration intensity 振动强度
vibration investigation 振动研究;振动调查
vibration isolating mounting unit 减震组合
vibration isolating suspension 隔震悬柱
vibration isolation 防振;振动绝缘;隔振垫;隔振
vibration isolation counter-measures 隔振措施
vibration isolation effectiveness 隔振效率;隔振效果
vibration isolation equipment 防振装置
vibration isolation material 隔振材料
vibration isolation mounting 隔振装置
vibration isolator 隔振装置;隔振体;隔振(器);隔振垫
vibration laboratory 振动试验室
vibrationless 不振动的;无振动的
vibration level 振动等级
vibration level meter 振动级计
vibration limit 振动限度;振动界限
vibration liquefaction of sand 砂土振动液化
vibration load 振动荷载
vibration machine 振动机
vibration machines calibration 振动器校准
vibration magnet meter 摆动磁针仪
vibration measurement 振动测量;振动测定
vibration measurement instrument 振动测试仪
vibration measurement with laser 激光振动测量
vibration measurer 测振计;振动计;振动仪;测振仪
vibration measuring apparatus 振动测定仪
vibration-measuring device 振动测量器
vibration measuring equipment 振动测量设备

vibration measuring set 振动测量仪
vibration mechanics 振动力学
vibration meter 振动仪;振动(测量)计;测震仪;测震计;测振仪;测振计
vibration method 振动法
vibration mill 振动式磨矿机;振荡磨
vibration milling 振动磨矿(法)
vibration mode 振动方式
vibration model density 振动方式密度
vibration monitor 振动监测仪;振动监测器;振动探测计;振动检测器
vibration monitoring 振动监测
vibration mo(u)lding 振动成型
vibration mo(u)lding machine 振动制模机
vibration mount 隔振支座
vibration mounting 防振支座;振动支座
vibration multiplier 振动乘法器
vibration neutralizer 振动平衡器
vibration nodal point 振动节点
vibration node 振动结;振动节点;振动波节
vibration noise 振动噪声
vibration of amplitude 振幅
vibration of foundation 基础隔振;基础振动
vibration of molecule 分子振动
vibration overlap 振动重叠
vibration performance 振动性能;振动特性
vibration period 振动周期
vibration physiological effect 振动生理效应
vibration pickup 拾振仪;拾振器;振动传感器;测振器
vibration pile-driver 振动打桩机
vibration pile-driver extractor 振动沉拔打桩机
vibration piling 振动沉桩
vibration plane 振动面
vibration plus pressure 振动加压
vibration pollution 振动污染
vibration pollution forecasting 振动污染预测
vibration potential 振荡电位
vibration power meter 振动式功率计
vibration probe 振动探头
vibration problem 振动问题
vibration programmer 振动机构
vibration-proof 防振的;耐振的;抗振的
vibration-proof equipment 防振设备
vibration-proof foundation 耐振地基;防振基础
vibration-proof material 防振材料
vibration-proof rubber 橡胶防振制品;防振橡胶
vibration-proof structure 防振结构
vibration-proof trench 防振沟
vibration protection 防振
vibration puddling 振动捣实;振捣
vibration pump 振动泵
vibration ramming 振动紧实
vibration recorder 振动记录仪
vibration recording equipment 振动记录仪
vibration regime 振动制度
vibration regulator 激振调节器
vibration resistance 抗振性
vibration-resistant 抗振(的)
vibration ride environments 振动乘坐环境
vibration rig 振动试验机
vibration roller 振动滚路机;振动压路机;振动式滚压机
vibration rotation spectrum 振转光谱;振(动)转(动)光谱
vibration screen 振动筛
vibration-sensitive receptor 振动感受器
vibration separation 振动分级
vibration service lamp 耐振灯泡;防振灯
vibration severity 振动(强)烈度;振动激烈性
vibration shaping 振动成型
vibration sieve 振动筛
vibration simulator 振动模拟器
vibration source 振动源
vibration spectrum analyzer 振动频谱分析器
vibration staking 振荡式拉软
vibration staking machine 振荡式刮软机
vibration stone crusher 振动碎石机
vibration stopper 炉管消振装置
vibration-stopper gum 防振橡皮
vibration strain pickup 振动应变测定器
vibration strength 振动强度
vibration survey 振动测量
vibration survey system 振动测量系统

vibration syndrome 振动综合征
vibration system 隔振系统
vibration table 振动台
vibration tachometer 振动式转速计
vibration temperature 振动温度
vibration tempering 振动回火
vibration test(ing) 振动试验；耐振性试验
vibration testing device 振动试验装置
vibration testing machine 振动试验机
vibration testing of components 组件耐振性试验
vibration test stand 振动试验台
vibration theory 振动理论
vibration time 振动时间
vibration tolerance 振动容限
vibration transducer 振动传感器
vibration transmissibility 振动传递率
vibration transmission 振动传递；振动传播
vibration triaxial apparatus 振动三轴仪
vibration-type freguency meter 振动式频率计
vibration velocity 振速
vibration velocity level 振动速度级
vibration visco(si)meter 振动黏度计
vibration washer 振动洗涤器
vibration wave 振波；振动波
vibration welding 振动熔接
vibrative 振动的
vibratode 振动器极
vibratom 振动球磨机；振动磨(机)
vibrator 振子；振弦；振动式铆钉枪；振动器；振捣器；振打器
vibrator bar 振动条；振动筛条
vibrator beam 振动棒
vibrator coil 振动线圈；火花断续线圈
vibrator compaction 振动器振实
vibrator compactor 振动捣实器
vibrator contact 振动器接触头；振动器触点；振荡器触点
vibrator conversion 振子转换
vibrator converter 振动子换流器
vibrator cultivator 振动式中耕机
vibrator cylinder 振动头；振动器套筒
vibrator cylinder with flexible shaft 带软轴的振荡器
vibrator for concrete pavement 混凝土路面振捣器
vibrator frame 模板框
vibrator generator system 振动发生装置
vibrator head 振动头
vibrator inverter 振动式逆变器；振动变流器
vibrator-jetting deep compaction 深层振动压实
vibrator knee valve 振动器膝盖阀
vibratormeter 振动计；振动式计量器
vibrator mill 振荡磨
vibrator motor 振动器电动机
vibrator phase advancer 振动相位超前补偿器
vibrator pile hammer 振动式桩锤
vibrator plate 振动器板
vibrator power pack 振簧式变压装置
vibrator power supply 振动器电源；振动换流器供电
vibrator recording oscilloscope 振子记录示波器
vibrator sunk pile 振沉桩
vibrator supply 振动供料
vibrator trench compactor 振动式沟堑压实机；沟堑振实机
vibrator type converter 振动子式变换器
vibrator type inverter 振动器型反向变换器
vibrator type rectifier 振动式整流器
vibrator unit 振动器组
vibrator vehicle (物探和地震勘探用)振动车
vibratory 振动的
vibratory absorption process 振动吸收法
vibratory arc surfacing 振动电弧堆焊
vibratory base pan compactor 底盘振实器
vibratory base plate compactor 底板振实器
vibratory base slab compactor 底板振实器
vibratory beam (混凝土摊铺机的)振捣梁
vibratory belt cleaner 振动式皮带清扫器
vibratory block making machine 振动式砌块制造机
vibratory bowl feeder 振动杯进料器
vibratory bowl hopper feeder 振动圆盘送料器
vibratory bullfloat 振动长抹子；振动长镘刀；振动大抹子；振动大镘刀
vibratory centrifuge 振动离心机；振荡离心机
vibratory chip screen 木片振动筛

vibratory chute 振动输送槽
vibratory cleaning 振动清理
vibratory compacting 振动压实(法)
vibratory compacting roller 振动夯实机
vibratory compaction 振实；振动压实(法)；振动压密；振动成型
vibratory compaction settlement ratio 振动压实沉降率
vibratory compactor 振实器；振动式压实机
vibratory concrete mixer 振动式混凝土搅拌机
vibratory concreting 混凝土振动浇筑
vibratory conversion 振动转换
vibratory converter 振动变流器；振动变换器
vibratory conveyer 振动(式)输送机；振动(式)输送器
vibratory conveying machine 振动式输送机
vibratory cooler 振动冷却器
vibratory core 振动芯
vibratory cylinder 振动汽缸
vibratory device 振动装置
vibratory discharge system 振动卸料系统
vibratory disk mill 振动圆盘筛
vibratory drill 振动钻
vibratory drilling 振动钻探；振动钻进
vibratory drilling method 振动法【岩】
vibratory driver 振动打桩机
vibratory driving machine 振动打桩机
vibratory earth borer 振动钻土机；振动式土钻
vibratory equipment 振动设备
vibratory explosion 振动爆炸
vibratory feeder 振动(式)送料器；振动(式)给料机；振动(式)给料器
vibratory feed unit 振动给料斗
vibratory finishing beam 振动式整平尺；振动式整平器
vibratory finishing machine 振动式整平机；振动抛光机
vibratory float 振动式整平器
vibratory force 振动力
vibratory fracturing 振动断裂
vibratory grinding 振动粉磨
vibratory grinding mill 振动式研磨机
vibratory grizzly 振动铁栅筛；振动条筛
vibratory gyroscope 振动陀螺仪
vibratory hammer 振动锤
vibratory hammer test 振锤测定；振锤试验
vibratory hand float 手提式振动抹平刀
vibratory hand trowel 手提式振动抹平刀
vibratory head 振动头
vibratory hopper feeder 振动料斗送料器
vibratory impact 振动冲击
vibratory impulse 振动脉冲
vibratory in-line hopper feeder 直进式振动送料器
vibratory joint cutter 振动锯缝机
vibratory leacher 振动浸析器
vibratory linear shear apparatus 振动式直线剪切机
vibratory liquefaction 振动液化
vibratory liquidation test 振动液化试验
vibratory load 振动荷载
vibratory method 振动法
vibratory method of compaction 振实法(松软地基处理方法之一)
vibratory mill 振动球磨机
vibratory milling 振动球磨
vibratory mixer 振动式拌和机；振动搅拌机
vibratory mortar mixer 振动式砂浆拌和机
vibratory motion 振动运动；振动
vibratory motor 振动电机；产振马达
vibratory movement 振动
vibratory pile driver 振动沉桩机；振动打桩机
vibratory pile driving 振动打桩
vibratory pile driving and extracting frame 振动沉拔桩架
vibratory pile extractor 振动拔桩机
vibratory pile hammer 振动(打)桩锤
vibratory piston corer 振动活塞取样管
vibratory piston sampler 振动活塞取样管
vibratory plate compactor 振动平板夯；振动式平板压实机；振动压实板；振动捣实器
vibratory plate visco(si)meter 振板式黏度计
vibratory power 振动力
vibratory press 振动式压力机
vibratory pressing 振动压制；振动加压
vibratory pressure 振动压力

vibratory rammer 振动冲击夯
vibratory rate gyroscope 振动角速度陀螺仪；振动角速度传感器
vibratory response 振动响应
vibratory roller 振动碾；振动压路机；振动式碾压机；振动(式)路碾；振动辊压机
vibratory roller screed 振动辊压刮平器
vibratory rolling 振动滚压
vibratory rotary table machine 振动转盘机
vibratory screed (混凝土)振动整平机
vibratory screen 振动筛
vibratory screen with elliptic(al) movement 椭圆形行程的振动筛
vibratory-sensitive receptor 振动感受器
vibratory-shakeout machine 振动落砂机
vibratory sheepsfoot roller 振动式羊脚压路机；振动式羊脚滚筒
vibratory shock load 振动冲击荷载
vibratory sintering 振动烧结
vibratory sliding table press 振动式滑盘压力机
vibratory sluice 振动溜槽
vibratory smoother 振动式刮路机；振动式整平机；振动抛光机
vibratory smoothing screed 振动式整平板
vibratory smoothing wheeled roller 振动式滚压机；振动式光面辊
vibratory soil compacting plate 振动式土壤压实板
vibratory soil compaction 土壤振动夯实法
vibratory soil compactor 振动式土壤夯具；振动式土壤压实器
vibratory spiral pipe elevator 振动式螺旋管提升机
vibratory steel roller 振动钢路碾；振动钢路滚
vibratory stimulation 振荡激励
vibratory strain ga(u)ge 振弦应变计
vibratory strength 振动强度
vibratory stress 振动应力
vibratory stressing bed 振动式加力台；振动式加力床
vibratory stress relief 振动消除应力法
vibratory string 振动钢弦
vibratory table 振动台
vibratory tamper 振动(冲击)夯；振捣板
vibratory tandem roller 振动式双轮压路机
vibratory test(ing) 振动试验
vibratory testing machine 振动试验机
vibratory tie tamper 振动式砸道机；振动式拉杆夯实机
vibratory track 振动料道
vibratory trough conveyer 槽形振动输送机
vibratory tube mill 振动式轧管机；振动式管磨机
vibratory unit 振捣器
vibratory weigh scale 振动计重量秤
vibratory weight 激振体重量；激振器重量
vibratron 振敏管；电磁共振器
vibrex pile 振动灌注桩
vibrio 弧菌
Vibrio fluvialis 河流弧菌
vibro 振捣
vibro-assisted jig 动筛式跳汰机
vibrobatten 振动板
vibro-bench 振动台
vibro-boring 振动钻探；振动钻进
vibrocap 电雷管
vibrocast (用超声波或高频声波的)振动压铸
vibrocast concrete 振实混凝土；振动浇筑混凝土；振捣混凝土
vibro-casting 振动密实成型
vibro-casting concrete 振动灌浆混凝土
vibro-casting process 振动压铸法
vibro-classifier 振动分级机
vibro-compacted 振动压实的
vibro-compact fuel element 振动增实燃烧元件
vibro-compaction 振动压实；振冲挤密
vibro-compaction method 振实法
vibro-compaction slab press (混凝土)板振动压实机
vibro-compaction zone 振动增实带；振动加密区
vibro-composer 振动填实砂桩
vibro-compressor method 振实法
vibro-concrete pile 振动混凝土桩；振实混凝土桩
vibro-core cutter 振动岩芯提取器；振动岩芯切割器
vibro-corer 振动取土器；振动取样器；振动取(岩)芯器
vibro-coring 振动采取岩芯；振动取样

vibrocribble 振动粗筛
vibrocs 铁氧体磁致伸缩振动子
vibro-dampers 振动吸收装置(在机器下面)
vibro-damping mount 隔振基座
vibro-densification 振动压实(法);振动压密法
vibrodrill 振动钻机;振动设备
vibro-drilling 振动钻井;振动钻孔;振动钻探;振动钻进
vibro-drive extractor 振动打/拔桩机
vibro-driven anchor 振动挤入锚
vibro-driver 振动打桩机
vibro-driver extractor 振动打/拔桩机
vibroenergy mill 双向振动磨
vibroenergy separator 振动能分级器
vibro-extrusion method 振动挤出法
vibrofeeder 振动进给器;振动给料器;振动给料机
vibrofinisher 振动整平机;振动整面机;振动轧平机;振动平整机
vibroflo(a)tation 振动水冲法;振浮压实(法);振浮压密;振冲;冲水振实法
vibroflo(a)tation machine 振动浮选机
vibroflo(a)tation method 振动水冲振实法
vibroflo(a)tation process 灌水振动(压实)法
vibroflo(a)tation-type point-bearing concrete pile 振浮重压式混凝土端承桩
vibroflot 振动压实载重;振浮压实器;振动器
vibroflotating process 振浮压实法
vibroflot machine 振浮压实机
vibrogram 振动图
vibrograph 自记示波计;自记示振仪;振动显示器;振动(记录)仪;振动计;示振仪;示振器;示振计;测振仪
vibrographing 振动测量
vibrogrinding 振动磨矿(法)
vibro-grouted aggregate concrete 振动灌浆混凝土
vibrohammer 振动桩锤
vibrohammer pile driving 振动沉桩
vibro-hoe 振动锄
vibrohydropressed concrete pipe 振动挤压混凝土管
vibroimpact 振动冲击
vibroimpacting 振冲
vibroll 振动压路机
vibrolode 振动器极
vibrometer 振动仪;振动(测量)计;测振计
vibromill 振动球磨机;振动磨碎机;振动磨(机)
vibromixer 振动混合器
vibromotive force 起振力
vibro-mo(u)ld 振动铸造
vibro-mo(u)lding 振动成型
vibronic 电子振动的
vibronic coupling 振动耦合
vibronic meter 电子振动计
vibronic spectrum 振动光谱
vibropack 振动子整流器;振动子换流器;振动变流器
vibro-pickup 振动传感器
vibro pile 振动式灌注桩;振动(灌注)桩
vibro-pile driver 振动打桩机
vibro-pile hammer 振动打桩锤
vibropiling 振动打桩
Vibroplat 外布洛牌振捣器(一种附着式振捣器)
vibroplate 振动板;平板振动机;平板(式)振动器;平板(式)振捣器
vibroplatform 振动台;平台振捣器
vibroplex 振动电键
vibro-pulverization 振动粉碎(作用)
Vibropyl 外布洛牌振捣器(一种附着式振捣器)
vibrorammer 振动夯;振动捣实器;振动夯;振动器;振动板
vibro-record 振动记录;振动自记计
vibro-record drawing 振动记录图
vibro-replacement 振捣置换法;振冲置换
vibro-replacement stone column 振捣置换碎石桩;振冲碎石桩
vibroroller 振动式路碾;振动碾(机);振动压路机
vibroroller compacting 振动碾压(法)
vibrorolling 振动滚压;振动碾(法);振动滚轧
vibro-rotary drilling 振动回转钻进
vibros 铁氧体磁致伸缩振动子
vibroscope 振动仪;振动波探漏仪;示振仪;示振器;示振计;测震仪
vibroseis 振动地震
vibroseismic survey 振动地震量测
vibroseis source 连续振动源

vibroshears 高速振动剪床
vibroshock 振冲
vibroshock absorber 减振器
vibrosieve 振动筛
vibrosinking 振动沉桩法;振动沉柱法
vibrosinking pile 振动沉桩
vibrosmoothing trowel 振动抹平镘板;振动粉光抹子
vibro-soil compactor 土壤振动压实器
vibro-sounding rod 振动测深杆
vibrospade 振动铲
vibrostand 振动(试验)台;平台振捣器
vibrostand of structure 结构振动试验台
vibroswitch 振动(电)开关
vibrotamper 振动夯
vibrotamping 振捣
vibrotaxis 振动趋性
vibrotechnique 振动技术
Vibrotex machine 振荡洗涤机
vibrotron 压敏换能器;振敏管;电磁共振器
vibrotrunk 振动洗选槽
vibrovacuum dewatering 振动真空脱水
vibro washer 振荡洗涤机
vibtrating plate extractor 振动板式萃取器
Vicalloy 维卡合金
vicarage 牧师住宅
vicarious 代理的;错位的
vicarious authority 代理职权
vicarious responsibility 转成责任
vicarious trial and error 代替性尝试错误
Vicat apparatus 维卡仪(用于水泥稠度试验)
Vicat consistency apparatus 维卡稠度仪
Vicat consistency needle 维卡稠度针
Vicat hydraulic index 维卡指数
Vicat index 水硬系数
Vicat needle 维卡针;维卡检验计(测定水泥硬化程度)
Vicat needle apparatus 维卡仪;维卡针式测试器
Vicat needle test 维卡针入度试验
Vicat penetrometer 维卡透度仪;维卡透度计;维卡贯入仪(测定炸药塑性用)
Vicat softening point 维卡软化点
Vicat test 维卡计试验;维卡仪试验(测定水泥稠度用)
vice 老虎钳;夹紧;虎钳; = vise
vice bench 虎钳台;钳桌;钳工(工作)台;锉工台
vice captain room 副船长室
vice chairman 副主席;副董事长
vice clamp 螺丝扳手;虎钳夹;虎钳
vice clamp vibrator 附着式模外振捣器;钳式外部振捣器
vice grip 虎钳夹口
vice grip wrench 虎钳夹口扳手
vice jaw 虎钳口
vice minister 副部长
vicenary 二十进制的
vice pipe 通话管
vice-presidency 副董事长职位
vice president 副总裁;副会长;副董事长
vice proper 固有缺陷
vice screw 虎钳螺杆
vice stair(case) 螺旋楼梯
vice table cage 钳工箱
vice table caisson 钳工箱
vice versa 打颠倒
vicidity 黏稠性
vicinage 邻居;附近地区
vicinal 邻位;邻晶
vicinal face 邻晶面;近真面
vicinaloid 似邻接面
vicinal plane 邻接面
vicinal surface 邻位面
vicinity 近处;邻近
vicious circle 恶性循环
vicious cycle 恶性循环
vicious cycle of agroecosystem 农业生态恶性循环
vicious inflation 恶性通货膨胀
vicious spiral 恶性循环
vicissitude 变迁
Vicker 维氏硬度计
Vicker method 维氏法
Vicker's diamond hardness 维氏(金刚石)硬度
Vickers diamond hardness number 维氏金刚石硬度值

Vicker's diamond hardness test 维氏金刚石硬度试验
Vicker's diamond hardness tester 维氏金刚石硬度计;维氏钻石硬度计
Vickers diamond pyramid hardness test 维氏金刚石棱锥硬度试验
Vicker's diamond pyramid hardness tester 维氏钻石压头硬度计
Vicker's hardness 维氏硬度;维克硬度
Vicker's hardness number 维氏硬度值
Vicker's hardness scale 维氏硬度表
Vicker's hardness tester 维氏硬度计;维克硬度计
Vicker's harness test 维氏硬度试验
Vicker's indentation crack length 维氏压痕裂纹长度
Vicker's indenter 维氏压痕器
Vicker's method 维氏法
Vicker's pyramid hardness testing machine 维氏硬度试验机
Vicker's unit 维氏硬度单位
victaulic coupling 卡箍联结器
victaulic joint 管道接头(安装后能变动几度仍保持不漏水)
Victaulic pipe 维克斯利克型接头管子;卡箍管子
victim 受难者;受害者;受害人
Victoria blue 维多利亚蓝;碱性艳蓝
Victoria fast violet 维多利亚坚牢紫(一种紫色酸性染料)
Victoria green 维多利亚绿
Victorian architecture 维多利亚式建筑
Victorianism 维多利亚女王时代的风格
Victorian style 英女皇维多利亚时代建筑形式(1837~1901年);维多利亚式
Victoria-Simplex 维多利亚简型板(一种防火隔声楼板)
Victoria violet 维多利亚紫
Victoria yellow 维多利亚黄
Victoric joint 维克托利克型接头
Victory 调色油(俗称维力油)
victory gateway 凯旋门
victory memorial 胜利纪念碑
victory pillar 凯旋柱
victulic coupling 环带管接头
victulic joint 管道接头;管道接口
victual(l)er 供应船;运粮船
victual(l)ing 装载粮食
victual(l)ing bill 船用保税食物报关单
victual(l)ing expenses 餐费
victual(l)ing house 餐馆;饭店
victual(l)ing port 给养供应港
victual(l)ing quay 供应码头
victual(l)ing terminal 供应码头
victual(l)ing wharf 供应码头
victuals 船舶装载食物
video 影像;录像;图像
video acquisition system 视频获取系统
video amplifier 视频放大器
video band 视频波段
video bandwidth 视频带宽
video beam 像束
video beam television 像束电视
video cable 工业电视电缆
video camera 电视摄像(机)
video cassette recorder 盒式录像机
videocast 电视广播
video channel 视频通道
video character generator 视频时逝录像机
video characteristic 电视设备在视频带内的特性
video circuit 图像电路
video coding technique 视频编码技术
video colo(u)r bar selector 视频彩条选择器
video colo(u)r synthesizer 彩色视频合成器
video computer 录像计算机
video control signal 视频控制信号
video converter 视频变换器
videocorder 录像机
video correlator 视频相关器
video croobar 视频矩阵
video data display system 视频数据显示系统
video data interrogator 显示数据询问器
video data terminal 显示数据终端;数据显示终端
video disc system 光盘系统
video display 视频显示器;视频显示
video display board 图像显示板

video display terminal 视频显示终端;电视屏幕
video dissolve 图像叠化
video distribution switching unit 视频分配开关单元
video domain 视频频域
video engineer 电视工程师
video envelope 视频包络
video equipment 录像设备
video file 可视文件
video frequency 视频(率)
video(frequency) exchanger 视频
video-gain control 雷达回波强度控制
video gap 视频间隙
videograph 高速阴极射线印刷机
videograph echo sounder 示像回声测深仪
video graphic adapter 图像适配器
video graphics array 视频图形适配器
video guard band 视频保护带
video head 录像头
video head azimuth angle 视频磁头方位角
video image 视频图像
video image acquisition 可视图像采集
video input 图像输入
video insertion 视频插入
video jackfield 视频插孔板
video link 视频波道
video magnetograph 视频磁像仪
video map 视频地图
video mapper 视频成像仪
video-mapping transmitter 光电地图发送机
video measurement 视频测试
video memory 图像存储器
videometer 视频表;测试用电视接收机
videometer calibration 视频表校准
videometer drive 视频表驱动
video mixer 图像混合器
video/mono comparator 视频单片坐标量测仪
video on demand 点播电视
video output 图形输出
video package 图像信息程序包
video parameter 视频参数
video phase stabilizer 图像相位稳定器
video-phone 可视电话;电视电话
video picture 显像;视频图像
video plate 视频片
videoplayer 电视录放机;放像机
video point locator system 视频定位器系统
video printer 图像印刷机
video processing 视频处理
video processor 视频信息处理机;视频处理器
video recorder 录像器;录像机;视频信号记录
video recording 录像;显像记录
video recording device 显像记录装置
video routing switcher 声频视频程序转换器
video screen 电视屏幕
video sign 荧光灯广告牌;视频符号
video signal 目标信号;视频信号
video socket 视频插座
video system 录像系统
video telephone 可视电话;电视电话;显像电话
video telephone system 电视电话系统
video terminal 可视终端;视频终端
video test signal generator 视频测试信号发生器
videotex 信息传输机;视频数据检索系统
videotext 视像文本系统;可视图文;视频电文;视频电报
videotext picture 可视正文图形
videotext technology 可视图文技术
videotheque 录像带资料室
video track 图像跟踪
video transfer characteristic 视频传递特征
video transformer 视频变压器
video transmission 视频发送
video transmitter 电视发射机
video tube 电视接收管
video unit 视频装置
video wall 视频壁
video waveform 视频波形
vidicon 光导摄像管;视像管
vidicon camera 光导摄像管摄像机;摄像管摄像机
vidicon camera tube 光导摄像管
vidicon detector 光导摄像管检测器
vidicon telescope 视像管望远镜
vidicon tube 视像管

vidion camera system 光导照相机系统
vidpic 电视图像
Vienna blue 维也纳蓝
Vienna Convention for the Protection of the Ozone Layer 保护臭氧层维也纳公约
Vienna Definition Language 维也纳定义语言
Vienna lime 维也纳石灰
Vienna metallic-Cement amalgam alloy 维也纳金属接合剂汞合金
Vienna Postal Savings Bank 维也纳邮政储蓄银行
Vienna regulator 维也纳式校准器
Vienna Secession movement 维也纳分离派艺术运动
vierbein (防波堤用)四对称圆锥体钢筋混凝土管
Vierendeel box-section bracing 空腹箱形截面联结系
Vierendeel girder 空腹梁;框架梁;空腹杆;维林特大梁;带上下桁条的梁;空腹(式)大梁
Vierendeel girder bridge 空腹梁桥
Vierendeel member 空腹构件
Vierendeel pole 空腹杆
Vierendeel truss 空腹桁架;维林特桁架;连框桁架
Vietnam mountains 越南山地
Vietnam old land 越南古陆
view 景色;视形;视图;对景图
view angle 视角
view borrowing 借景
view camera 观察照相机;观测摄影机;叠影式摄影机
viewdata 影像数据系统;数据电视检阅
viewdata graphics 图像数据图形
viewdata index page 图像数据索引页面
viewdata modem 图像数据调制解调器
viewdata network 图像数据网
viewdata operation 图像数据工作
viewdata page 图像数据页面
viewdata phone 图像数据传真电话机
viewdata receiver 图像数据接收器
viewdata service 图像数据业务
viewdata signal 图像数据信号
viewdata system 图像数据系统
viewdata transmit-receive system 图像数据发送一接收系统
viewdisk 录像盘
view distance 视距
viewed from the drive 从传动方向看
view-endowed site 观赏地点;取景地点
viewer 观察者;观察器;观测者;取景器
view-factor 视角因数
view field 视场
view file plotting 视图文件绘图
view finder 观测器;反光镜;观察器;取景器;探视器;寻景器
viewfinder eye 瞄准器小孔
viewfinder eyepiece 取景器目镜
viewfinder hood 寻景器遮光罩
viewfinder indication of stop in use 取景器光圈
viewfinder mask 取景器像框
viewfinder monitor 寻像监视器
viewfinder parallax 检影视差
viewfinder picture 寻像器图像
viewfinder release button 取景释放钮
viewfinder window 取景窗口
view flying 视察飞行
view form above 俯视图
view from back 后视图
view front 前视图
viewing 观察
viewing angle 视野角;观察角;视线张角;视场角
viewing aperture 取景孔
viewing booth 观察室
viewing condition 观察条件;观测条件
viewing data 视图数据
viewing deck 观景台;瞭望台;观察台
viewing desk 观察桌
viewing distance 观察距离;视距
viewing equipment 观察
viewing field 视野
viewing floor 观光楼(层);瞭望楼(层)
viewing frustum 视见平截头体
viewing gallery 观光廊;观光走廊
viewing gondola 瞭望台;观光台
viewing head 观察头
viewing hole 观察孔

viewing indicator 目测指示器
viewing lens 取景物镜;取景镜头
viewing line 视线;观测线
viewing magnifier 观察放大镜
viewing microscope 观察显微镜
viewing mirror 观测镜;视镜
viewing objective 观察物镜
viewing optic(al) system 光学观测系统;观察光学系统
viewing optics 观察光学系统
viewing perspective 观测透视
viewing platform 瞭望台;观光台
viewing port 观察口;观察孔
viewing prism 观察棱镜;观测棱镜
viewing ratio 观察帧高比;视距比
viewing ray 观测光线;表面上的点与观察口中心点的连线
viewing room 审片室;预光室
viewing scale 观察比例尺
viewing screen 荧光屏;观察屏;电视屏
viewing slot 观察缝;观察口;观察孔
viewing spectacles 观测镜
viewing stor(e)y 观光楼(层);瞭望楼(层)
viewing system 观察系统;观测系统
viewing time 显示时间
viewing tower 瞭望塔
viewing transformation 观察变换;视见变换
viewing unit 观测装置
viewing window 观察窗;观察孔
view in opposite place 对景
view integration 意图综合
view integration process 意图综合过程
vie with one another to raise prices 竞相提价
view modeling 视图模型化;视图建模;意图模型化【计】
view modeling approach 意图模型法
view modeling technique 意图模型技术
Viewnafa computerized marine navigation system 维尤纳夫大型计算机组合导航系统
view of efficiency 效率观点
view of environment 环境观
viewphone 可视电话
view plane distance 视平面距离
viewpoint 观(察)点;视(图)点;对景位置;对景点
viewpoint movement 视点移动
viewport 视口;视区;(图形的)视见区;视窗
view reference point 视见参考点
view restructuring 视图重构
views and sizes 外表和尺寸
view screen 离心式扫雨器
view spot 景点
view surface 视图面
view-up 视图正视方向
view volume 视见约束体
view window 观景窗
viferral 吡啶与三氯甲醛聚合物
vigezzite 维铌钙矿
vigia 险礁;可疑礁滩;海图示警险区
vigilance man 不漏情况报告员
vigil basin 观察性流域
vignette 装饰图案;都德式葡萄饰;虚光;晕映图像;蔓叶花饰;葡萄饰【建】
vignetted band (岸边水域用)渐晕带
vignetted dot contact film screen 渐晕接触网目片
vignetted screen 渐晕网目片
vignetting 渐晕;烧结彩饰
vignetting at the center of the field 视场中心的渐晕
vignetting effect 渐晕效应
vignetting factor 遮光系数;渐晕因数
vignetting filter 渐晕校正滤光镜
vignetting mask 渐晕光栏
vignetting pattern 渐晕图案
vignetting stop 格晕光阑;渐晕光栏
vignite 磁性铁矿
vignole's rail 丁字(形钢)轨;阔脚轨
vigorous 猛烈的
vigorous agitation 剧烈振动;强烈搅拌
vigorous disinfection 强消毒(剂)
vigorous flush(ing) 强冲洗
vigorous growth 旺盛生长
vigorous wash 强冲洗
Vigoter effect pseudocolor encoding 维格特效应假彩色编码

vigour 生活力
Vigreaux column 维格勒柱
vihara 佛教寺院；印度教庙宇
viitaniemiite 氟磷铝钙钠石
vikingite 维硫铋铅银矿
viking log 舷边计程仪
Viking spacecraft 海盗航天飞船
villa 月票居民区；郊区别墅；别墅
villadom 别墅和居住别墅的人们
Villafranchian 维拉弗朗阶【地】
Villafranchina 阿斯蒂阶
villa garden 别墅庭院；庄园
village 村（庄）
village and town finance 乡镇财政
village and town run industry 乡镇工业
village building lot 乡村建筑地段
village church 乡村教堂
village environment 村落环境
village exchange 农村（自动）电话局
village forest 乡村林
village forest field 乡村林地
village geography 村落地理学
village green 村庄广场；村庄活动绿地
village hospital 疗养院；乡村疗养院；农村医院
village house lot 乡村庙宇地段
village landscape 村庄景观
village road 村庄道路；乡村路
village road bridge 乡村道路桥梁
village shop 乡间的店铺
village square 乡村广场
village-type arrangement 农村式布置
village-type grouping 农村式配合；农村式组合；农村式布置
village urbanization 乡村城市化
villagization 农村化
villa lot 别墅用地
villamaninite 黑硫铜镍矿
villa marina 海滨房屋；海滨房舍；海滨别墅
villanette 小别墅
Villard effect 维拉德效应
Villari effect 维拉里效应
Villari reversal 维拉里倒逆
villa rustica 乡村别墅
villatic 乡村的；别墅的
villiaumite 氟盐
villiersite 镍滑石
villus 绒毛
vilnite 硅灰石
vimana 印度方尖庙；婆罗门寺院上的塔
vimsite 维羟硼钙石
vinaceous 红色的
vinaceous figure 葡萄状花纹
Vinal Haven granite 维纳尔海文石；浅灰细纹花岗岩
vinayil 聚乙酸乙烯酯雷乳液黏合剂
vincentite 文砷钯矿
Vincent press 模锻摩擦压力机
Vinculum 文库兰牌预制楼板（商品名）
vinculum 大括号；线括弧；系带；纽带
vinculum longum 长纽
vindictive damage 惩罚性赔偿
Vindobonian 文多奔阶
vine 藤子；藤本植物
vine black 软质炭黑；葡萄藤黑；葡萄色
vine black pigment 葡萄黑颜料
vine crop 蔓生植物
vinegar 醋
vinegar bottle 醋瓶
vinegared lotus root 醋藕
vinegar eel 醋线虫
vinegar essence 醋精
vinegar meter 醋量计
vinegar milk 醋乳
vinegar worm 醋线虫
vine leaf 葡萄树叶形饰
vine plant 蔓草作物
vine rack 藤茎分离筛
vine rubber 藤本橡胶
vinery 葡萄园；葡萄温室
vine separator rod 除藤茎秆
vine spring 拨藤蔓弹性杆
vinette 叶形与卷须形饰；蔓叶花饰
vineyard 葡萄园
vineyard pole 葡萄园撑杆
Vinhatico （热带美洲产）有杂色线纹的橙黄色硬木

Vinikin formula 维里金公式
vinilex mixed paint 乙烯树脂调和漆
vinogradovite 白钛硅钠石
vinoleum 乙烯布；乙烯地毯
vinometer 酒精比重计
vinovo porcelain 维诺沃瓷器（意大利）
vinrez 乙酸乙酯共聚体乳液黏合剂
vinsol 纯木质素
Vinsol agent 文沙剂
vinsol resin 松香皂树脂；松香聚合物
viny butyral resin 乙烯丁醛树脂
vinyl 乙烯基
vinyl acetal 乙烯醇缩乙醛
vinyl acetal resin 乙烯醇缩乙醛树脂
vinyl acetate 乙烯基乙酸酯；乙酸乙烯（酯）；醋酸乙烯酯
vinylacetate 醋酸乙烯
vinyl acetate acrylic latex 乙酸乙烯—丙烯酸乳胶
vinyl acetate-acrylk ester 乙酸乙烯—丙烯酸酯共聚
vinyl acetate emulsion paint 乙酸乙烯酯乳化漆
vinyl acetate-ethylene copolymer 乙酸乙烯—乙烯共聚物
vinyl acetate finish 醋酸乙烯酯涂面料
vinyl acetate-maleic anhydride copolymer 乙酸乙烯—顺丁烯二酸酐共聚物
vinyl acetate plastics 醋酸乙烯酯塑料
vinyl acetate resin 乙酸乙烯（酯）树脂；醋酸乙烯树脂
vinyl acetate resin coating 乙酸乙烯酯树脂涂料
vinyl acetate-styrene copolymer 乙酸乙烯—苯乙烯共聚物
vinylacetic acid 乙烯基醋酸
vinylacetylene 乙烯基乙炔
vinyl adhesive 乙烯树脂粘合剂；乙烯基黏合剂
vinyl and acrylic acid adhesive 乙烯基和丙烯酸胶黏剂
vinyl-asbestos composition 乙烯基石棉组成
vinyl-asbestos compound 乙烯基石棉化合物
vinyl-asbestos floor cover(ing) 乙烯基石棉地板覆盖层；乙烯基石棉楼板覆盖层
vinyl-asbestos floor(ing) finish 乙烯基石棉地板饰面；乙烯基石棉楼板饰面
vinyl-asbestos mass 乙烯基石棉团块
vinyl-asbestos material 乙烯基石棉材料
vinyl-asbestos sheet 乙烯基石棉薄板
vinyl-asbestos tile 乙烯基（树脂）石棉瓦乙烯基石棉砖；乙烯基石棉板；石棉聚氯乙烯板
vinylated alkyd resin 乙烯基改性醇酸树脂
vinylated oil 乙烯基改性油
vinylation 乙烯化作用
vinyl base 乙烯基踢脚板
vinylbenzene 乙烯基苯；苯乙烯
vinyl blend 乙烯掺和剂
vinyl bonding adhesive 乙烯基黏合剂
vinyl butyrate 乙烯基丁酸酯；丁酸乙烯酯
vinylcabtiret 聚乙烯绝缘软质电缆
vinyl cabtyre cable 乙烯绝缘软性电缆
vinyl carbazole 乙烯基咔唑
vinyl cellulose 乙烯纤维素
vinyl cementing agent 乙烯基黏合剂
vinyl chloride 乙烯氯；氯乙烯
vinyl chloride-acetate copolymer 氯乙烯—乙酸乙烯酯共聚物；氯醋共聚物
vinyl chloride floor sheet 氯乙烯地面卷材
vinyl chloride incinerator 氯乙烯焚化炉
vinyl chloride latex 氯乙烯乳胶
vinyl chloride monomer 氯乙烯单体
vinyl chloride organosol 氯乙烯有机溶胶
vinyl chloride pipe 氯乙烯管
vinyl chloride plastics 氯乙烯塑料
vinyl chloride plastisol 氯乙烯塑性溶胶
vinyl chloride plywood 氯乙烯胶合板；氯乙烯夹板
vinyl chloride poisoning 氯乙烯中毒
vinyl chloride resin 氯乙烯树脂
vinyl chloride resin coating 氯乙烯树脂涂料
vinyl chloride resin enamel 氯乙烯树脂亮漆；氯乙烯树脂瓷漆
vinyl chloride resin paint 氯乙烯树脂漆
vinyl chloride resin varnish 氯乙烯树脂清漆
vinyl chloride rubber 氯乙烯橡胶；聚氯乙烯橡胶
vinyl chloride synthetic rubber 氯乙烯合成橡胶
vinyl chloride tar 氯乙烯焦油
vinyl chloride-vinylidene chloride resin 氯乙烯—偏(二)氯乙烯树脂

vinyl chloride wire 氯乙烯绝缘线
vinyl-coated 涂刷乙烯基的
vinyl-coated screening 涂塑窗纱
vinyl-coated sheet 乙烯基塑料覆面薄（钢）板
vinyl-coated steel 聚氯乙烯复合钢板
vinyl-coated steel plate 涂氯乙烯钢板
vinyl coating 乙烯基涂料
vinyl composition tile 乙烯基树脂复合面板
vinyl compound 乙烯(系)化合物；乙烯基化物
vinyl copolymer 乙烯系共聚物；乙烯基共聚物
vinyl cover 乙烯塑料罩
vinyl covered cord 乙烯基包裹线（用盐化乙烯基树脂包皮电线）
vinyl-covered gypsum wallboard 乙烯树脂覆面石膏墙板
vinyl covered steel plate 涂氯乙烯钢板
vinyl cube 乙烯基立方体
vinyl cyanide 乙烯基氰；丙烯腈
vinyl cyclohexane 乙烯环己烷
vinyl disk 乙烯塑料盘；塑料盘
vinyl duct 乙烯基（塑料通风）管道
vinyl elastomer 乙烯系弹料
vinyl emulsion 乙烯乳液
vinyl emulsion paint 乙烯(系)乳化漆
vinylene 乙烯撑
vinyl ester 乙烯基酯
vinyl ester of versatic acid 叔碳酸乙烯酯
vinyl esters 乙烯酯类
vinyl ether resin 乙烯基醚树脂
vinyl ethers 乙烯醚类
vinyl ethyl alcohol 乙烯基乙醇
vinyl ethylene 乙烯次乙基
vinyl ethyl ether 乙烯基乙醚
vinyl extrusion 乙烯压出品
vinyl fiber 乙烯纤维
vinyl film 乙烯基胶片；乙烯薄膜
vinyl floor base 乙烯基踢脚板
vinyl flooring 乙烯基树脂楼面板；乙烯基（塑料）铺地料
vinyl flooring cover(ing) 乙烯基地板覆盖层；乙烯基楼板覆盖层
vinyl flooring finish 乙烯基地板终饰；乙烯楼板终饰
vinyl flooring tile 乙烯基铺地砖
vinyl fluoride 乙烯基氟；氟乙烯
vinyl fluoride resin 氟乙烯树脂
vinyl foam 泡沫聚氯乙烯
vinyl foam cushioning 乙烯基泡沫衬垫
vinyl foils 乙烯箔
vinyl formic acid 乙烯基甲酸
vinyl glue 乙烯缩醛树脂胶
vinylglycollic acid 乙烯基乙醇酸
vinyl group 乙烯烃
vinyl group of plastics 乙烯类塑料
vinyl halide 卤乙烯
vinyl halide type 卤乙烯型
vinyl hose 乙烯基软管
vinylidene chloride 乙烯叉二氯
vinylidene chloride plastic 偏二氯乙烯塑料
vinylidene chlorite resin 亚乙烯氯化树脂
vinylidene fluoride 偏二氟乙烯
vinylidene resin 亚乙烯树脂；偏二氯乙烯树脂
vinyl insert(ion) 乙烯镶嵌物
vinyl insulated wire 塑料电线
vinyl iodide 乙烯基碘
vinylit 乙烯基树脂
vinylite 聚乙酸乙烯酯树脂
vinylit resin 乙烯基树脂
vinylit sheet 乙烯基树脂软片
vinyl ketones 乙烯基甲酮类
vinyl lacquer 乙烯基底漆
vinyl latex base 乙烯胶乳基
vinyl leather 乙烯基皮革
vinyl liner 乙烯密封衬垫；乙烯基衬里
vinyl medium 乙烯树脂介质
vinyl membrane 乙烯薄膜
vinyl methylether 乙烯甲醚
vinylmethylketone 丁烯酮
vinyl monomer 乙烯系单体
vinyl nonyl ether 乙烯基壬基醚
vinylog(ue) 插烯物
vinylogy 插烯（作用）
vinyloid sheet 乙烯薄板
vinylon 聚乙烯醇缩醛纤维；维复纶

vinylon canvas 维尼纶帆布
vinylon rope 维尼纶绳
vinyl overlay 乙烯基涂面层
vinyl paint 乙烯基树脂油漆;乙烯系涂料;乙烯基树脂涂料
vinyl panel 乙烯镶板
vinyl paste 乙烯基树脂糊
vinyl perchloride adhesive solution 过氯乙烯胶液
vinyl perchloride anti-corrosion paint 过氯乙烯防腐漆
vinyl perchloride enamel 过氯乙烯瓷漆
vinyl perchloride fire retardant paint 过氯乙烯防火漆
vinyl perchloride marking paint 过氯乙烯标志漆
vinyl perchloride matt enamel 过氯乙烯无光瓷漆
vinyl perchloride putty 过氯乙烯腻子
vinyl perchloride sea water resistance primer 过氯乙烯耐海水腐蚀底漆
vinyl perchloride semigloss enamel 过氯乙烯半光瓷漆
vinyl perchloride soft varnish 过氯乙烯软性清漆
vinyl perchloride varnish 过氯乙烯清漆
vinylphenol 乙烯基苯酚
vinylphenyl acetate 乙酸乙烯苯酯
vinyl pipe 乙烯(基)塑料管
vinyl plastic floor cover(ing) 乙烯基塑料地板覆盖层;乙烯基塑料楼板覆盖层
vinyl plastic floor(ing) finish 乙烯基塑料地板终饰;乙烯基塑料楼板终饰
vinyl plastics 乙烯(基)塑料;乙烯烃塑料
vinyl plastic sheet 乙烯基塑料片
vinyl plastisol 乙烯塑料溶胶
vinyl polymer 乙烯基聚合物;聚氯乙烯基
vinyl polymerization 乙烯聚合作用
vinyl pyrrolidone 乙烯基吡咯烷酮
vinyl resin 乙烯基树脂
vinyl resin medium 乙烯树脂介质
vinyl resin paint 乙烯基树脂油漆
vinyl resin vehicle 乙烯树脂车辆;乙烯树脂载体
vinyl sheet flooring 乙烯基卷材地面
vinyl sheet(ing) 乙烯薄板
vinylsiloxane rubber 乙烯基硅橡胶
vinyl sol paint 乙烯基溶胶涂料
vinyl sol pattern paint 乙烯基溶胶美术漆
vinyl solution 乙烯基溶液
vinyl solution coating 乙烯基溶液型涂料
vinyl steel plate 乙烯基饰面钢板
vinylstyrene 二乙烯基苯
vinylsulfone dyes 乙烯砜染料
vinyl sulfone dyestuff 乙烯砜染料
vinyl sulfonic acid 乙烯磺酸
vinyl-surfaced gypsum 乙烯基树脂(覆面)石膏板
vinyl tape 聚氯乙烯绝缘带
vinyl thioether 二乙烯基硫醚
vinyl tile 乙烯基树脂瓦(管);乙烯基树脂铺地;乙烯基瓦板;乙烯基铺地砖;乙烯基板
vinyltoluene 乙烯基甲苯
vinyl trichlorosilane 乙烯基三氯硅烷
vinyl-type vapo(u)r barrier 乙烯基树脂型防潮层;乙烯基树脂型汽隔
vinyl vehicle 乙烯载体
vinyl versatate 叔碳酸乙烯酯
vinyl wall cladding 乙烯塑料壁纸
vinyl wall covering 乙烯基树脂墙壁覆盖物;乙烯系墙布
vinyl wall facing 乙烯护墙面
vinyl wall lining 乙烯墙衬
vinyl wall paper 乙烯基墙纸
vinyon 聚乙烯塑料;聚氯乙烯塑料
vioform 蔚欧仿
Viogt effect 佛克脱效应
viol 锚索拉绳
viola calamlneria 芦叶堇菜
violaceous 紫罗兰色的
violaite 铁镁辉石
violan(e) 青辉石
violanthrene 蒽烯紫
violanthrone 紫蒽酮
violarite 紫硫镍矿
violate a contract 违反合同;违约
violate a law 违犯法律
violate a law of environmental protection 侵犯环保规定
violation of regulations 违章

violator 违反者
viol block 开口滑车
violence 猛烈性;强烈;暴力
violent 猛烈的
violent earthquake 强震;强烈震动;大地震
violent flood 暴洪
violent in toxicity 剧毒物
violent oscillatory motion 大幅度振动
violent shock 烈震;强震
violent stirring 剧烈搅拌;强烈搅拌
violent storm 十一级风;猛烈风暴;大风暴;暴风(雪)
violescent 带紫罗兰的
violet 紫色(的);紫罗兰色;紫
violet alizarin(e) lake 紫茜素色靛
violet-black 紫黑色
violet-blue 紫蓝色
violet-brown 紫褐色
violet-grey 紫灰色
violetish blue 紫色蓝;带紫头的蓝色
violet lake 紫色淀
violet paste 蓝油
violet photocell 紫光电池
violet pigment 紫色颜料
violet quartz 紫石英
violet ray 紫外线;紫射线
violet red 紫红(色)
violet ultramarine 紫群青
violet wood 紫(色)硬木
violle 外俄耳
VIP lounge 贵宾休息室
viral contamination 病毒污染
viraru (中美洲产)有红条纹的红灰色硬木
virazon 维拉丛风
Virchow's node 信号结
virescence 淡绿
virgae 杆状连锁沟
virgal precipitation trails 雨幡
virgation 分支
virgilite 硅锂石
virginal groove 哑纹
virginal humidity 原始湿度
virgin alloy 原始合金
virginal overburden pressure 原始上覆压力
virginal stress 天然应力
virgin ammonia liquor 粗氨水
virgin area 原始(林)区
virgin coil 空白纸带卷
virgin compaction 原始压密
virgin compression curve 原始压缩曲线;初压曲线;初始压缩曲线
virgin consolidation 原始压密
virgin copper 天然铜
virgin curve 原(始)曲线;初始曲线
virgin diamond 新金刚石
virgin field 未采矿区
virgin flow 天然径流
virgin forest 原始森林;原生林;未开发的森林
virgin formation pressure 原始地层压力
virgin glass fiber 原玻璃纤维
virgin ground 处女地;生荒地;未开垦地
virgin growth 原生材;原始林
Virginia initiative plant process 弗吉尼亚创始的工厂工艺
Virginian creeper 五叶爬山虎【植】
Virginian fence 犬牙形篱笆
Virginian pine 弗吉尼亚松
Virginia silver 弗吉尼亚银合金
virgin isotropic consolidation curve 原始各向等压固结曲线
virgin kerosene 直馏煤油
virgin land 生荒地;未开垦地;处女地
virgin material 纯净(原)材料;纯净物料;新料;原材料;新生材料;未使用材料;半成品
virgin medium 空白媒体;未用媒体;未用介质
virgin metal 原(生)金属
virgin naphtha 直馏粗汽油
virgin oil 直馏油
virgin paper tape 无孔纸带
virgin paper-tape coil 无孔纸带卷
virgin parchment 羊皮纸
virgin plastics 原生塑料
virgin rock 原生岩石
virgin sand 原始砂;未采油层

virgin soil 未垦土壤;原生土(壤);未开垦地;生荒地土壤;处女地
virgin state 初始状态
virgin state of stress 原始应力状态
virgin stock 直馏油料
virgin strength 原始强度
virgin stress 原始应力
virgin voyage 首航;首次航行;第一航次
virgo 弗戈镍铬钨系合金钢
virgule 斜线号
virial coefficient 维里系数
virial mass 位力质量
virial theorem 克劳修斯维里定理;位力定理
virial-theorem mass 位力定理质量
viride nitens 亮绿
viridescent 淡绿色的;带绿色的
viridian 翠绿色(颜料)
viridian green 古telligencerer
Viridian pigment 维利迪安颜料
viridity 翠绿;碧绿
Virmel engine 弗梅尔发动机
virocapsomer 壳粒
virola (用于镶板)美洲深红色木材
virology laboratory 病毒化验室
virtu 古董;艺术品
virtual acceptance 事实上的承兑
virtual access method 虚拟存取法;虚拟存储法
virtual address 虚(拟)地址;立即地址
virtual addressing 零级定址
virtual angle of friction 有效摩擦角
virtual argument 虚变元
virtual asymptotic(al) line 假渐近线
virtual axis 虚轴
virtual base(line) 有效基线
virtual beam method 虚梁法
virtual call 虚拟呼叫;虚调用
virtual call capability 虚呼叫功能
virtual call control 虚调用控制
virtual call facility 虚呼叫设备;虚调用手段
virtual cathode 虚阴极;假想阴极
virtual center of gravity 虚重心
virtual channel 虚通路
virtual channel logic 虚拟通道逻辑
virtual channel network 虚拟通道网络
virtual circle 虚圆;假圆
virtual circuit 虚拟线路;虚拟电路
virtual circuit control program(me) 虚拟线路控制程序
virtual common memory 虚拟公用存储器
virtual computer 虚拟计算机
virtual condition 实效状态;实际状态
virtual connection 虚连接
virtual console 虚拟控制台
virtual container 虚容器
virtual coupling 虚假偶合
virtual current 有效电流
virtual cycle 虚循环
virtual cylindrical gear 假想圆柱齿轮;当量圆柱齿轮
virtual decimal point 假定的小数点
virtual decision value 虚判决值;虚判定值
virtual declarator 虚说明符
virtual declarer 虚说明词
virtual deficiency 虚亏格
virtual deformation 虚拟变形;虚变形;虚变位;假变形;潜变形
virtual degree 假次数
virtual discharge 虚流量
virtual displacement 虚位移;虚拟排水量;虚变位;假位移
virtual displacement field 虚位移场
virtual displacement law 虚位移定律
virtual displacement principle 虚位移原理
virtual displacement theorem 虚位移定理
virtual duration 假想持续时间
virtual earth 虚(接)地;假接地
virtual-earth buffer 虚接地隔离器
virtual energy 虚能
virtual energy level 虚能级
virtual environment 虚拟环境
virtual external force 假想外力
virtual field 虚拟场
virtual field boundary 虚场边界
virtual file 虚拟文件
virtual file store 虚拟文件库

virtual flow 有效流量
virtual focus 虚焦点
virtual force 虚力
virtual form 虚拟形式
virtual friction-angle 视在摩擦角
virtual geomagnetic north pole 虚地磁北极
virtual geomagnetic north pole position 虚地磁北极位置
virtual geomagnetic pole 虚地磁极
virtual geomagnetic pole position 虚地磁极位置
virtual geomagnetic south pole 虚地磁南极
virtual geomagnetic south pole position 虚地磁南极位置
virtual grade 虚坡(度);机力坡度
virtual gradient 虚坡
virtual gravity 虚重力
virtual ground 虚假接地
virtual head 有效水头
virtual height 虚高;有效高度;视在高度;实际高度
virtual horizontal line 假想水平线
virtual image 虚像;可见(影)像;平面虚像
virtual image holographic(al) stereomodel 虚像全息立体模型
virtual image stereomodel 虚像立体模型
virtual impedance 虚阻抗
virtual indexed sequential access method 虚拟索引顺序存取法
virtual inertia 虚拟惯性
virtual injector 虚注入器
virtual interface 虚拟接口
virtualization 虚拟化
virtual leak 虚漏;假漏
virtual length 虚长(度);实际长度
virtual level 虚拟级;虚能级;假想电平
virtual load 虚载;虚荷载;有效荷载
virtual loading 虚荷载
virtual long-range coupling 虚假远程耦合
virtual loss of metacentric height 稳心高度减少值
virtual lower bound 虚下界
virtually safe dose 实际安全剂量
virtual machine 虚拟计算机;虚拟机
virtual machine kernal 虚拟机核心
virtual machine security 虚拟机安全性
virtual machine technique 虚拟机技术
virtual management 虚存管理
virtual map 虚拟图
virtual map file 虚地图文件
virtual mapping 虚图;虚拟图像
virtual mark 虚测标
virtual mass 虚质量;有效质量;视质量
virtual mass effect 附加质量效应;虚质量效应
virtual mass of liquid 液体虚质量
virtual mass of soil 土的虚质量
virtual memory 虚(拟)存储器
virtual memory address 虚拟存储器地址
virtual memory addressing 虚拟存储寻址
virtual memory allocation 虚拟存储器分配
virtual memory computer 虚存计算机
virtual memory concept 虚存概念
virtual memory executive system 虚存执行系统
virtual memory hardware 虚存硬件
virtual memory machine 虚存计算机
virtual memory management 虚存管理
virtual memory mechanism 虚存机构
virtual memory pointer 虚存指示字
virtual memory strategy 虚存策略
virtual memory structure 虚存结构
virtual memory system 虚存系统
virtual memory technique 虚存技术
virtual meridian 虚子午线
virtual mode 虚拟方式
virtual model 虚拟模型
virtual moment 虚力矩
virtual moment of inertia 虚转动惯量
virtual network 虚拟网络
virtual number of teeth 虚齿数
virtual operating system 虚拟操作系统
virtual orbital 虚轨道
virtual pair 虚对
virtual parameter 虚参数
virtual partitioned access method 虚拟分区存取法
virtual path 虚(拟)通路
virtual peripheral 虚外围设备
virtual photon field 虚光子场
virtual pitch 虚螺距
virtual plane 虚平面图
virtual plan-position reflectoscope 消视差平面位置显示器
virtual point 虚拟点;虚点
virtual point source 理论点源
virtual point source model 虚点源模式
virtual point source of pollution 虚拟点污染源
virtual potential 虚电势
virtual power 虚功(功)率;潜在功率
virtual preconsolidation pressure 虚拟前期固结压力;假定前期固结压力;视预固结压力
virtual pressure 虚压;有效压强;有效压力
virtual printer 虚拟打印机
virtual process 虚过程
virtual processor 虚拟处理机
virtual profile 虚纵断面
virtual quantum 虚量子
virtual rating 虚额定值;有效率;实际能力
virtual reactor 虚反应堆
virtual reality modeling language 虚拟现实建模语言
virtual record 虚记录
virtual resistance 虚电阻
virtual result 虚结果
virtual route 虚路由
virtual row 虚行
virtual screen 虚拟屏幕
virtual segment 虚段
virtual segment structure 虚段结构
virtual sequential access method 虚拟顺序存取法
virtual size 实效尺寸
virtual slope 虚坡;虚比降;假定比降;实际有效坡度
virtual sound source 虚声源
virtual source 虚拟源
virtual space 虚拟空间
virtual stage 虚拟级
virtual state 虚态
virtual steepness 虚陡度
virtual storage 虚拟存储器
virtual storage system 虚拟存储系统
virtual strain 虚应变
virtual stress 虚应力
virtual structure 虚结构
virtual target 视在目标
virtual technique 虚拟技术
virtual telecommunications access method 虚拟远程通信存取方法
virtual temperature 虚温
virtual terminal 虚拟终端
virtual terminal network 虚拟终端网络
virtual terminal protocol 虚拟终端协议
virtual ton kilometres hauled 实际输送吨公里
virtual ton-kilometres worked 实际工作吨公里
virtual translation 虚平移运动
virtual unbundling 虚拟释放
virtual unit address 设备虚址
virtual upper bound 虚上界
virtual value 有效值
virtual velocity 虚速度;实际速度
virtual viscosity 有效黏(滞)性;有效黏度
virtual waiting time 虚等待时间
virtual work 虚功
virtual work equation 虚功方程
virtual work law 虚功定律
virtual work principle 虚功原理
virtual work theory 虚功理论
virtual zero point 虚零点
virtual zero time 假想零时
virtue 优点
virtues and defects 优缺点
virtuosic 艺术家的;专家的
virtuoso 古玩收藏家;艺术品鉴赏家
virulence 毒性
virulent 有毒力的;剧毒的
virus 病毒
vis 螺旋式楼梯
visa 签证;背签
visa card 签证卡;通行卡
visa-granting office 签证机关
visalgen 测试图发生器
visamminol 齿阿米醇
visa passport 签护照
visavis S形长椅
vis-à-vis 面对面
visbreaker 减黏裂化炉
visbreaking 减黏(轻度)裂化;减黏
viscid 黏性的;黏滞的;半流体
viscid consistency 黏性稠度
viscidity 黏滞性;黏质
viscidity of soil 土的黏性
viscid sublayer 黏滞底层
viscin thread 黏丝
visco calculator 黏温性质计算尺
visco corder 黏度流变仪
viscoelastic 黏弹性的
viscoelastic analysis 黏弹性分析
viscoelastic behavio(u)r 黏弹性(性)状;黏弹性(性)质;黏弹性能
viscoelastic body 黏弹(性)体
viscoelastic correspondence principle 黏弹性对应原理
viscoelastic creep 黏弹性蠕变
viscoelastic cross effect 黏弹性横向效应
viscoelastic damper 黏弹性阻尼器
viscoelastic damping 黏弹性阻尼
viscoelastic deformation 黏弹性变形
viscoelastic design 黏弹性设计
viscoelastic dispersion 黏弹性分散体
viscoelastic flow 黏弹性流动
viscoelastic fluid 黏弹(性)流体;麦克斯韦流体
viscoelastic fluid system 黏弹性流体系统
viscoelastic ground 黏弹性地面
viscoelastic half-space 黏弹性半空间
viscoelasticity 黏(滞)弹性
viscoelastic layer 滞弹性层;黏弹性层
viscoelastic lithosphere 黏弹性岩石圈
viscoelastic material 黏弹性物质;黏弹性材料
viscoelastic medium 黏弹性介质
viscoelastic model 黏弹性模型
viscoelastic modulus 黏弹性模量
viscoelastic plate 黏弹性板
viscoelastic property 黏弹性(性)质;黏弹性能;黏弹特性
viscoelastic region 黏弹区
viscoelastic response 黏弹性反应;黏弹响应
viscoelastic response analysis 黏弹性反应分析
viscoelastic soil 黏弹性土
viscoelastic solid 黏弹固体
viscoelastic strain 黏弹性应变
viscoelastic stress 黏弹性应力
viscoelastic structure 黏弹性结构
viscoelastic substance 黏弹性物质
viscoelastic theory 黏弹性理论
viscoelastometer 黏弹计
viscoelastoplastic 黏弹塑性的
viscoelastoplastic soil 黏弹塑性土
viscogel 黏(性)凝胶
viscograph 黏度曲线(通常指黏温曲线)
viscoid 黏性体;黏丝体
viscoinelastic fluid 黏性非弹性流体
viscolloid 黏性胶体
viscometer 流变计;黏度计;黏滞计
viscometer gage 黏度计式真空计
viscometric 测定黏度的
viscometric analysis 黏度测定分析
viscometric function 测黏函数
viscometric titration 黏度滴定
viscometry 黏度测定术;黏度测定学
viscomill 高黏度磨机
viscoplastic 黏塑性的
viscoplastic deformation 黏塑性变形
viscoplastic flow 黏塑流;黏塑性流动
viscoplastic fluid 黏塑性流体
viscoplasticity 黏塑性
viscoplasticity method 直观塑性法
viscoplastic material 黏塑性材料
viscoplastic principal strain difference 黏塑主应变差
viscoplastic property 黏塑性能
viscoplastic soil 黏塑性土
viscoplastic solid 黏塑性固体
viscoplastoelastic 黏塑弹性
viscorator 连续记录黏度计
viscoscope 黏度指示器
viscoscope soil 黏弹性土
viscoscope stress 黏弹性应力
viscose 纤维黏胶;黏胶液;黏结胶

viscose artificial thread 黏胶人造丝
viscose binder 黏胶黏合剂
viscose cellar 黏胶窟
viscose cellulose 黏胶纤维素
viscose cord fabric 黏胶帘布
viscose fiber 纤维胶；黏胶纤维
viscose filament yarn 黏胶长丝
viscose film 黏胶薄膜
viscose flow 黏性流动
viscose glue 胶水；黏胶
viscose grinder 黏胶研磨机
viscose high tenacity yarn 高强力黏胶丝
viscose paper 黏胶纸；糊胶纸
viscose process 黏胶法；黏性过程
viscose pump 黏胶泵
viscose rayon 黏胶（丝）
viscose rayon fiber 黏胶人造纤维
viscose rayon tow 黏胶丝束
viscose rayon yarn 黏胶人造丝
viscose ripening 黏胶熟成
viscose ripening tank 黏胶熟成桶
viscose sheet for wrapping 包装用黏胶纸
viscose silk 黏胶人造丝
viscose solution 黏胶溶液
viscose sponge 黏胶海绵
viscose staple fiber 黏胶短纤维
viscose tank 黏胶桶
viscose trap 黏胶杂质分离器
viscose wastewater 黏胶废水
viscose yarn 黏胶纱
viscosifier 增黏剂
viscosimat 黏度自动调节器
visco(si)meter 黏度（测定）计；黏度计；黏滞计
visco(si)meter for cylinder torsion(al) method 扭筒法黏度计
visco(si)meter of bitumen 沥青黏(滞)度仪；沥青黏(滞)度计
visco(si)meter of mud 泥浆黏度计
visco(si)meter proportions 黏度计比率
visco(si)meter tip 黏度计尖头
visco(si)metric 测黏度
visco(si)metric analysis 黏度测定（分析）
visco(si)metric titration 黏度滴定
visco(si)metry 黏度法；黏度测量法；黏度测定术；黏度测定（法）；测黏法
visco(si)metry coefficient 黏滞系数
visco(si)metry factor 黏滞系数
visco(si)metry test(ing) 黏滞性试验
viscosine 暗色残油
viscosity 黏黏性；黏(滞)度；黏稠度；黏性（系数）
viscosity abnormality 黏度反常性
viscosity adjuster 黏度调整器
viscosity alarm recorder 黏度警报记录器
viscosity analyzer 黏度分析器
viscosity anomaly 黏度反常
viscosity average 黏度平均值
viscosity average molecular weight 黏度（平）均分子量
viscosity blending chart 黏度掺和线图
viscosity boundary 黏滞边界
viscosity breaker 减黏裂化炉
viscosity breaking 黏度降低；减黏裂化；降低黏度；降黏
viscosity coefficient 黏度系数；黏滞系数；黏性系数
viscosity constant 黏度常数；黏滞常数
viscosity control 黏度控制；黏度调节
viscosity control agent 黏度控制剂
viscosity controller 黏度控制器
viscosity-controlling agent 黏度调节剂
viscosity conversion 黏度换算；黏度转换
viscosity conversion table 黏度换算表
viscosity correction 黏度校正
viscosity cup 黏度杯
viscosity curve 黏度曲线（通常指黏温曲线）
viscosity deformation 黏度变形
viscosity-density constant 黏重常数
viscosity-density ratio 比密黏度；黏度密度比
viscosity diagram 黏度曲线图
viscosity effect 黏度效应；黏度性效应
viscosity equation 黏度方程
viscosity factor 黏度因素；黏度系数；黏滞因素；黏度因数；黏度系数；黏性系数
viscosity flow 层流；黏(滞)流
viscosity fluctuation 黏性波动
viscosity fluid 黏滞（性）流体
viscosity force 黏性力；黏滞力
viscosity friction 黏滞摩擦
viscosity function 黏度函数
viscosity gage 黏度真空计黏度仪；带字盘的黏度计
viscosity gradient 黏度梯度
viscosity-gravity chart 黏度比重图
viscosity-gravity constant 黏度—重度常数；黏度—比重常数
viscosity-gravity number 黏度重力数
viscosity increaser 增黏剂
viscosity index 黏度指数
viscosity index blending value 黏度指数调和值
viscosity index chart 黏度指数线圈
viscosity index constituents 黏度指数组分
viscosity index extension 黏度指数延伸；黏度指数扩展
viscosity index extention 黏度指数延伸值
viscosity index figure 黏度指数
viscosity index improver 黏度指数改善剂
viscosity indicating controller 黏度指示控制器
viscosity indicator 黏度指示器
viscosity law 黏度定律
viscosity loss 黏度降（低）
viscosity manometer 黏度压力计
viscosity measurement 黏度测量
viscosity meter 黏度计
viscosity modifier 黏度调节剂；黏度改进剂
viscosity modulus 黏滞模量
viscosity monitor 黏度监视器
viscosity number 黏滞度数；黏度（值）；比浓黏度
viscosity of asphalt 沥青黏(滞)性；沥青黏(滞)度
viscosity of bitumen 沥青黏(滞)性；沥青黏(滞)度
viscosity of coal-oil mixture 煤油混合燃料黏度
viscosity of liquid 流体黏滞度
viscosity of mud 泥浆黏度
viscosity of oil field water 油田水黏度
viscosity of petroleum 石油的黏度
viscosity of slip 料浆黏度
viscosity of slurry 料浆流动度
viscosity of suspending medium 悬浮溶剂的黏滞度
viscosity of wash liquid 冲洗液稠度
viscosity of water 水的黏度
viscosity pole 黏度极
viscosity pole hight 黏度极高度
viscosity pole line 黏度极线
viscosity pour point 黏度倾点
viscosity pyrometer 黏性高温计
viscosity ratio 黏性比；黏度比
viscosity recorder 黏度记录器；黏度自记仪
viscosity recording controller 黏度记录控制器
viscosity reducing 黏度降低
viscosity reducing agent 减黏剂
viscosity reductant 减黏剂
viscosity regulator 黏度稠节剂
viscosity remover 去黏剂
viscosity resistance 黏滞阻力；黏度阻力
viscosity retention (滑油的)黏度保持性
viscosity sensor 黏度探测器
viscosity slope 黏度变化斜率
viscosity stabiliser 黏度稳定剂
viscosity stability 黏度稳定性
viscosity standard oil 黏度标准油
viscosity sublayer 黏滞性亚层；黏性性副层
viscosity tachometer 黏性转速计
viscosity-temperature calculator 黏—温性质计算尺
viscosity-temperature characteristic 黏度—温度关系特性
viscosity-temperature chart 黏—温线图；黏度—温度曲线图
viscosity-temperature coefficient 黏—温系数；黏度—温度系数
viscosity-temperature curve 黏度—温度曲线
viscosity-temperature dependency 黏—温关系
viscosity-temperature graph 黏—温关系图
viscosity-temperature number 黏—温度值
viscosity-temperature relation 黏性温度关系
viscosity term 黏性项
viscosity test 黏滞性试验；黏度测定；黏(滞)度试验
viscosity theory 黏度学说；黏性性理论
viscosity tube 黏度管
viscosity unit 黏度单位
viscosity value 黏度值
viscosity wave 黏滞波
viscosity-yield curve 黏度收率曲线
Viscountess slate 韦康蒂斯石板瓦
viscous 黏稠；黏滞（的）
viscous air filter 黏滞滤气器；黏滞空气过滤器
viscous behavio(u)r 黏滞性质
viscous bitumen 黏沥青
viscous body 黏性体；黏滞体
viscous boundary layer 黏滞边界层
viscous cement grout 黏性水泥浆液
viscous clay suspension 黏性黏土悬浮液
viscous conductance 黏滞传导
viscous consistency 黏滞结持度；黏滞稠度
viscous coupled seismometer 黏滞耦合地震计
viscous creep 黏滞蠕变
viscous crude oil wastewater 黏性原油废水
viscous damper 黏性阻尼器
viscous damping 黏滞阻尼；黏滞衰减；黏性阻尼；黏性衰减
viscous damping coefficient 黏滞阻尼系数
viscous damping constant 黏滞阻尼常数
viscous damping factor 黏滞阻尼系数
viscous debris flow 黏性泥石流
viscous decay 黏滞衰减
viscous deformation 黏滞变形；黏性变形
viscous demagnetization 黏滞性退磁
viscous displacement 黏滞位移
viscous dissipation 黏滞耗散；黏性分散
viscous dissipation function 黏性耗散函数
viscous drag 黏滞曳力；黏性阻力
viscous-drag gas-density meter 黏滞阻力气体密度测量器
viscous effect 黏滞作用
viscous-elastic behavio(u)r 黏弹特性
viscous-elastic material 黏弹性材料
viscous emulsion 黏性乳化
viscous equation of motion 黏性流体运动方程；黏滞运动方程
viscous factor 黏滞因素；黏滞因数；黏滞系数
viscous fermentation 黏滞发酵
viscous fillers 黏品装填机
viscous filter 黏滞过滤器
viscous flow 滞流；黏滞流(动)；黏性流(动)
viscous flow air meter 滞流气体流量计
viscous flow equation 黏滞流(动)方程
viscous flow leak 黏性流渗漏孔
viscous flow leak hole 黏性流漏孔
viscous flow material transfer mechanism 黏滞流(动)传质机理
viscous flow permeability 黏滞性液流渗透率
viscous flow region 黏性流层
viscous fluid 黏滞(性)流体；黏性流体；黏稠液
viscous fluid flow 黏性流动；黏性液流动
viscous fluid system 黏性流体系统
viscous force 黏(滞)力；黏性力
viscous fracture 黏性破坏
viscous friction 液体层流阻力；黏滞摩擦；黏性摩擦
viscous friction coefficient 黏滞摩擦系数
viscous friction stress 黏滞摩擦应力
viscous gruel state 黏性粥状
viscous heat 黏性热
viscous heating 黏性发热
viscous heavy oil 高黏重质原油
viscous hysteresis 黏性磁滞
viscous impingement filter 黏滞撞击式滤尘器；黏性冲击式收尘器（取尘样用）
viscous impingement type filter 黏状式过滤器
viscous keying 黏性键固
viscous layer 黏滞层
viscous liquid 黏性液体；黏滞液体
viscous liquid flow 黏性液流动
viscous liquid sample 黏性液体试样
viscous loss 黏滞损失
viscous lubrication 黏性润滑；完整油膜润滑；稠油润滑
viscous magnetization 黏滞磁化
viscous melt-phase 黏熔相
viscous motion 黏滞运动
viscous mud 黏泥
viscous mud-flow 黏性泥石流
viscous mud-stone flow 黏性泥石流
viscousness 黏滞性
viscous neutral oil 黏中性油
viscous oil 高黏性原油；黏性油

viscous oil tank car 黏油罐车
viscous oily liquid 黏性油状液体
viscous packing 高黏充填
viscous paint 稠涂料
viscous pill 高黏液
viscous pour point 黏滞倾点
viscous precipitate 黏性沉淀
viscous pressure 黏滞压力
viscous processing 黏浆处理法
viscous property 黏性
viscous recoil 黏性回弹
viscous refractory 黏性耐火材料
viscous regime of flow 黏性流状态
viscous remanent magnetization 黏滞剩余磁化强度
viscous resistance 黏滞阻力
viscous shear 黏性剪力
viscous similarity 黏滞相似性
viscous slag 黏性渣
viscous slurry 黏性料浆
viscous state 黏性状态;黏流态
viscous strain 黏滞应变
viscous stress 黏滞应力
viscous stress tensor 黏滞应力张量
viscous sublayer 黏性底层
viscous thermoelastic material 黏性热弹材料
viscous trailing vortex 黏滞后缘涡流
viscous type air cleaner 黏附式空气净化设备
viscous type damper 黏性阻尼器
viscous vortex 黏性涡流
viscous water 黏着水
viscous yielding 黏性屈服;黏性变形
vise 老虎钳;签证;台钳
Visean transgression 韦宪期海浸
vise bench 虎钳台
vise clamp 台钳夹;虎钳夹
vise-grip 手虎钳;夹口虎钳
vise grip pliers 大力钳
viseite 磷方沸石;水硅磷钙石
viser 遮光板
vise tool (机械手的)钳夹抓手
vishnevite 硫酸钙霞石
visibility 可视性;可见性;可见度;能见度;视界;视见度
visibility angle 能见角
visibility chart 能见度图表;视界图
visibility code 能见度电码
visibility condition 能见情况;能见度条件;通视条件
visibility curve 可见度曲线;能见曲线;明视度曲线;视见曲线
visibility detector 能见度测试仪
visibility distance 可见度距离;能见距离;能见范围;明视距离;通视距离
visibility exceptional 能见度异常好
visibility factor 可见度因数;能见度因数;视见因数;视度系数
visibility function 可见度函数;能见度函数;视见函数;发光度函数
visibility function modification glass 视见函数修正玻璃
visibility glossmeter 鲜明度光泽计
visibility good 能见度好;七级能见度
visibility hypothesis 可见性假设
visibility indicator 能见度指示器
visibility limit 能见极限;能见度
visibility measuring set 能见度测量仪
visibility meter 可见度测定计;能见度仪(表);能见度(测定)计;视度仪
visibility moderate 六级能见度;能见度一般
visibility of precipitates 沉淀可见度
visibility of satellite 卫星能见度;卫星可见期
visibility on curves 弯路上的视距【道】;曲线上的视距
visibility poor 能见度不好;能见度不定;五级能见度
visibility range 能见距离;能见范围
visibility ratio 发光度
visibility reduction 可见度降低
visibility region 可见区域
visibility restriction 通视障碍
visibility scale 可见度标度;能见(等)级;视级
visibility test 能见度试验
visibility test after fracture 破碎后能见度试验
visibility unlimited 能见度极好
visibility variable 能见度不定

visibility very good 能见度很好;八级能见度
visible 显而易见的;可(看)见的;能见的;明显的
visible absorption spectrometry 可见光吸收光谱测定法
visible aids to navigation 视觉助航设施
visible air whistle 蒸汽汽笛
visible and infrared spectroscopy 可见光与红外光线光谱法
visible and infrared spin scan radiometer 可见光和红外自旋扫描辐射计
visible and near-infrared response 可见和近红外光谱响应
visible and near infrared waveband 可见光近红外波段
visible anti-reflection coating 可见增透膜
visible arc 可见扇形区;明弧
visible-area map 能见区地图
visible beam laser 可见光束激光器
visible bearing 可见方位
visible binarier 可见双星
visible cable 电视电缆
visible complementary filter 可见光补色滤色镜
visible concrete surface 混凝土外露面
visible condition 可见条件
visible crack 可见裂纹;可见裂缝
visible diameter 目估冠幅
visible display 可见显示
visible distance 可见(度)距离;能见距离
visible dye 染色渗透液
visible edge 可见边
visible emissivity 可见发射率
visible face 显示面
visible filter 可见光滤光片
visible flame operation 明火操作
visible horizon 可见水平;可见地平(线);视地平(线)
visible image 可见(影)像;视像
visible impurity 可见杂质
visible infrared dichroic mirror 可见红外分光镜
visible-infrared spectrometer 可见红外分光计
visible infrared spin-scan radiometer 可见光与红外自旋扫描辐射计;可见红外自旋扫描辐射仪
visible laser 可见(光)激光器
visible light 可见光
visible light radiation 可见光辐射
visible light sensitive photocatalyst 可见感光催化剂
visible light sensor 可见光传感器
visible light transducer 可见光传感器
visible light transmission 可见光穿透率
visible line 可见谱线;外形线;轮廓线
visible mark 可见标志
visible marker 可见标记
visible means 有形财产
visible meter 可见度测定计
visible mineral 露头矿物
visible noise 可见噪声
visible oil flow gage 目测滑油流量计;示滑油流量计
visible optic(al) isolator 可见隔离器
visible ore 露头矿脉;露头矿石
visible output 可见输出信号;通气检验器
visible part of spectrum 可见光谱段
visible penetrant 易见渗入剂
visible pits 可见麻点
visible plume 可见度烟缕
visible point 亮点
visible radiation 可视辐射;可见(光)辐射
visible range 可见光谱区;可见范围;能见距离
visible range monitor 可见光监测器
visible ray 可见射线;可见光线
visible reading 视觉读数
visible record computer 可见记录计算机
visible record equipment 显露式记录设备
visible recording 可见记录
visible reduction 可见度降低
visible region 可见光(谱)区
visible remote sensing 可见光遥感
visible reserves 视储量
visible search 能见检索
visible sharpness 可见清晰度
visible side 显示面
visible signal 可见信号;视觉信号
visible sky 可见天空(指从采光口可看到的那部分天空)

visible sound method 超声波显像探伤法
visible spectral remote sensing 可见光谱遥感
visible spectrometer 可见分光计
visible spectrometry 可见光谱法;可见分光法
visible spectrophotometric method 可见分光光度法
visible spectro-photometry 可见分光光度学
visible spectroscopy 可见光谱学
visible spectrum 可见光谱
visible structure 外露结构;可视建筑物
visible subsidies 有形补贴
visible sunrise 视日出
visible sunset 视日没
visible supply 可见仓储;可估量的供应
visible surface 可视表面;直观表面;显见面
visible trade 有形贸易
visible tree height 目估树高
visible ultraviolet laser 可见光和紫外激光器
visible ultraviolet spectrophotometric analysis 可见紫外分光光度分析
visible ultraviolet spectrum 可见紫外光谱
visible under-face 可视的底面
visible underside 可视的底面
visible-vertical filing system 显露式垂直归档制度
visible volume rendering 可见体绘制
visible warning system 可见预告系统;可见警告系统
visible wave 可见波
visible wavelength 可见波长
visible wavelength oceanic-colo(u)r sensor 可见光波长水色传感器
visible wavelength scanner 可见波长扫描器
visible wavelength sensor 可见光波长传感器
visible writing recorder 可见笔记录
visicode 视符号
visigraph recorder 示波记录器
visilog 仿视眼;仿视机
vis inertia 惰性
vision 目力;视力
visional motion of glacier 冰川的视运动
vision amplifier 图像放大器
vision area 视区
visionary city 未来城市
vision automatic gain control 图像放大器自动增益控制
vision bandwidth 图像信号带宽
vision carrier 图像载波
vision carrier frequency 图像载频
vision carrier spacing 图像载波间隔
vision channel 图像信号通路
vision circuit 图像电路
vision clearance 视距
vision cloth 含半透明纱的舞台幕布
vision control 图像控制
vision crosstalk 图像通道串扰
vision detector 图像检波器
vision distance 视能见距离
vision driver 图像激励器
vision driver stage 图像激励级
vision electronic recording apparatus 电子视觉记录仪;电子录像机
vision facility 图像设备
vision fading 图像消失;图像衰减
vision frequency 图像频率
vision frequency range 图像频率范围
vision intensity 可见强度
vision interference 图像干扰
vision interference limiter 图像干扰限制器
vision intermediate frequency 图像中频
vision intermediate frequency-modulator 图像中频调制器
vision klystron 图像速调管
vision light 观察窗;可外望的玻璃窗
vision-light door 带观察门
vision line 视线
vision mixer 图像混合器;视频
vision mixer control panel 图像混合控制台
vision mixing system 图像混合系统
vision mixture operator 图像混合操作员
vision modulation 图像调制
vision modulator 图像调制器
vision monitoring equipment 图像监视设备
vision on sound 图像信号对伴音的干扰
vision output power 图像输出功率

vision panel 透视窗板
visionproof barrier 防视装置;挡视线装置
visionproof device 防视装置;挡视线装置
visionproof glass 毛玻璃;防视玻璃
visionproof louver 遮挡百叶;防视百叶
vision radio frequency signal 图像射频信号
vision receiver 电视接收机
vision resolution 视力分辨率
vision screen(ing) 视线遮蔽
vision signal characteristic 图像信号特性
vision signal detector 图像检波器
vision strip 外露密封条
vision supervisor 图像控制技术人员
vision switcher 图像切换器
vision switching matrix 图像切换矩阵
vision transmitter 图像发射机
vision transmitter monitoring equipment 图像发射机测试柜
vision transmitter output 图像发射机输出
vision transmitter output power 图像发射机输出功率
vision transmitter power 图像发射机功率
visioplasticity 可视塑性
visit a node 访问一个节点
visiting card 名片
visiting gang 巡查班
visiting gauging section 巡回测流断面
visiting party 巡查班
visiting professor 客座教授;访问教授
visiting room 会客室
visiting scholar 访问学者
visiting team 巡查班
visiting traffic 过访性交通
visitor 访问者
visitor entrance 来宾入口
visitor room 来宾室;游客休息室
visitor's car 来宾用车;游客用车
visitors stand 观礼台
visitor traffic 旅客交通(量)
visitor waiting room 旅客候机室;旅客候车室
visitor walk way 旅客通道
vis major 不可抗力(法)
visors 遮阳板;护目镜;遮光棚
visor tin 双晶锡石
visotest 数字式校表仪
vissilk 纤维胶;黏胶丝
vista 远景;前景;视景线
vista clearing 树景整治;扩大视野
vista line 夹景;风景线
visual 直观;目视
visual access 目视可见
visual accommodation 视度调节
visual accumulation tube 直观淤积管
visual accumulation tube size analyzer 视测积存管颗粒分析器
visual achromatism 可见光消色差性;视消色差
visual acquisition 目视识别
visual acuity 目视的敏锐性;目视分辨率;目视敏锐度;视敏读;视亮度;视力敏锐度;视觉敏锐角;视觉敏锐度
visual adaptation 视觉舒适
visual aids 直观教具;直观教材;目视助航设备;视觉航标
visual aid to navigation 目视导航辅助设备
visual alarm 光报警信号;可见报警信号;视觉警报;视觉报警
visual alarm signal 视觉警报信号
visual alarm unit 灯光报警器;目视报警器
visual alignment 目视调整
visual alignment generator 测试图发生器
visual amplification 视放大率
visual analysis 目视分析;(街景设计的)视觉分析
visual and dimensional check 外观尺寸检查
visual angle 视野张角;视界角(度);视角
visual angle method 视角法
visual appearance 肉眼鉴定;肉眼鉴别
visual appraisal 目视判断
visual approach 目视渐进
visual approach slope indicator system 目视渐进坡度指示系统
visual arts 视觉艺术
visual astrophotometry 目测天体亮度
visual attribute 可见属性
visual-aural radio range 视听无线电指向标

visual-aural range 可见可听范围;声影显示范围
visual axis 视准轴;视轴(线)
visual balance 视觉平衡
visual band imagery 可见光谱段图像
visual beacon 能见信标;视觉航标
visual bearing 目测方位
visual binaries 目视双星
visual brightness 视觉亮度
visual broadcast 电视广播
visual call sign 视觉呼号
visual carrier 图像载波
visual cell 视觉细胞
visual center 视中心;视觉中枢;目视中心
visual characteristic 视觉特征;视觉特性
visual chart 目视地标飞行图
visual check 直观检查;目视检验;目视检查;肉眼检查;外观检查
visual classification of kerogen 干酪根光学分类
visual colo(u)rimeter 目测比色计;目视色度计;目视比色计;视觉色度计
visual colo(u)rimetry 目视色度测量;目视比色法;视感测色
visual comfort 视觉舒适感
visual comfort probability 视觉舒适概率
visual communication 图像通信;视觉通信
visual communication system 可见通信系统
visual comparison method 目测比较法
visual concrete 外露(骨料)混凝土
visual confirmation 目视判定
visual contrast 视觉对比
visual control 直观控制;目视操纵;视力控制
visual corridor 视线走廊;视觉走廊
visual counting technique 视测计数技术
visual course indicator 目视航向指示器
visual cut-off 可见截止点;图像截止
visual cut-out 图像截止
visual data 目测资料
visual data acquisition 可见数据集合
visual defect 外观疵点;视觉缺陷
visual defect in mo(u)ldings 模塑品外观疵点
visual degree 视度
visual demodulator 图像解调器
visual design 可视设计;视觉对象设计(指图表、阵列、标志、包装等的设计)
visual detection 图像信号检测
visual detector 可视指示器
visual diameter 视直径
visual digital display 目视数字显示
visual direction finder 目视探向仪;目视测向仪
visual display 直觉显示器;目视显示器;可视显示;可见显示;能见显示;目视显示器
visual display board 显示屏
visual display terminal 显示器;直觉显示终端;直观显示终端
visual display unit 显示装置;视觉显示装置
visual distance 目视距离;目测距离;视距
visual distance reception 视距信号接收
visual docking guidance system 飞机停靠目视引导系统
visual doubles 目视双星
visual edge match comparator 目视边缘匹配坐标仪
visual effect 视觉效果
visual effect of plants 植物的视觉效果
visual efficiency 目视敏度
visual end point 目视极点
visual environment 视(觉)环境
visual error 视觉误差
visual estimated grade 目测品位
visual estimation 目测
visual examination 目视检查;肉眼鉴定;肉眼鉴别;外观检查;外表检查;外表观察;表观检验;表观检查;外部检查
visual exploration technique 肉眼勘察技术
visual exposure meter 视像曝光计
visual factor 视觉因素
visual fatigue 视觉疲劳
visual field 可视区域;可见区;视野;视界;视场
visual field diagram 视野图
visual fix 目视定位
visual flight condition 目视飞行条件
visual flight rule weather 目视飞行规则天气
visual focusing 目视对光;视厚聚焦
visual freehand projection 目测随手方法

visual fusion 视熔
visual-fusion frequency 视熔频率
visual ga(u)ge 视规
visual grading 外观分级
visual ground sign 地面定向标
visual guidance of traffic 交通的视觉引导
visual guide 目视引导
visual Herschel effect 赫歇尔效应
visual horizon 可见地平线;视地平(线)
visual illusion 视错觉
visual image 可见图像;视觉像
visual impact 感观影响;视觉反应
visual impedance meter 阻抗显示装置
visual impression 视觉印象
visual-indicating ga(u)ge 刻度量规
visual indication 直观指示;视觉标示;直观显示;可见指示;可见信号;目视指示
visual indicator 指示剂;目测指示器;玻璃液面计
visual information 可视信息
visual-infrared-plane-parallel dichroic mirror 可见红外平行平面分光镜
visual inquiry 直观询问;可见询问
visual inquiry station 直观询问台
visual inspection 检视;直观检查;目视检查;目力检查;目检;目测(检查);肉眼鉴定;肉眼鉴别;肉眼检查;外观检验;外观检查;外部检查;外表检查
visual inspection visually inspect 目测外观检查
visual intensity 可见强度;可见光强度
visual interpretability 目视判读性
visual interpretation 目视判释;目视判读;目视解译
visual interpretation method 目视解译方法
visual intrusion 视线干扰
visuality 可见度
visualization 形象化;显像;目视观察
visualization method 显像法
visualization system 可视化系统
visualize 直观化;目测检验
visualized 直视的;直观的
visualizer 显像仪;观察仪;观测仪
visualizing hierarchical structure 可视化分层结构
visual landing direction indicator 目视降落方向指示装置
visual language 可视语言
visual lien of position 目视位置线
visual light 可见光
visual light cloud picture 可见光云图
visual light curve 目视光变曲线
visual line 视线
visual line of position 目测位置线;位置视线;定位视线
visual luminosity 目视(发)光度
visually inspect 目测检查
visual magnitude 目视星等;目视量;目测星等
visual measurement 目测
visual meteorological condition 目视气象条件;目测气象条件
visual method 可视测量法;目视法;目测法;视觉法
visual microphotometer 视觉显拋光度计
visual monitor 直接监护器;目测监测器
visual navigation 目视导航
visual navigation mark 视觉航标
visual object line 可见轮廓线
visual observation 直观研究法;目视观察;目视观测;目测(法);肉眼观察;外部观察
visual observation system 目力监测系统
visual obstruction 视线障碍物
visual operation console 目视操作控制台
visual optic(al) instrument 目视光学仪器
visual optic(al) pyrometer 目视光学高温计
visual orientation 目视定向
visual path 视径
visual perception 视觉
visual perception area 视觉区
visual performance 视觉功能
visual photometer 目视光度计;目测光度计
visual photometry 主观计光术;直视光度学;可见光光度测定;目视光度学;目视光度测量(法);目镜测定;视觉测光
visual photomtry 目视测光
visual picture 直观图形;直观图像;直观曲线;可见图像;目视图像
visual plane 视觉平面
visual polarimeter 目视偏振计
visual pollution 视觉污染

visual position 目测位置
visual problem 视觉问题
visual process 视觉过程
visual programming 可视编程
visual purple 光紫质;视紫红
visual pyrometer 目测高温计
visual radio direction finder 目测式无线电测向仪;目测波浪无线电测向仪
visual range 可视距离;可视范围;可见射程;目测范围;通视范围;视距;视程
visual ray 可见射线;可见光;维生射线
visual ray pyramid 视线棱锥
visual reading 直观读数;视读数
visual readout 可见读出;目视读出
visual reception 记录接收
visual record 直观记录
visual recorder 可见记录器
visual reflection factor 视觉反射系数;视觉反射率
visual relief map 视立体图
visual representation 直观表示
visual resolution 目视分辨率;目力分辨率;目镜分辨率;视力分辨率;视觉分辨率
visual resolving power 目视分辨率;视觉分辨率
visual response 视觉反应
visual-righting reflex 视觉翻正反射
visual rights 视觉权利
visual scale 光学标度;光学标尺
visual scan 可见扫描
visual scanner 光扫描器;可见扫描器
visual scene 图像景色
visual score 目测评分
visual search time 目视搜索时间
visual sector 可见扇形区;通视范围
visual selection 目测选种
visual sensation 视觉感
visual sensitivity 视觉灵敏度
visual sensitivity characteristic 视觉灵敏度特性
visual sensitometer 目测感光计
visual-sensory area 视觉区
visual sharpness 可见清晰度;视晰度;视觉清晰度
visual ship handling training simulator 直观操船训练模拟装置
visual signal 形示信号;指示信号;直观信号;光信号;可视信号;视觉信号;目视信号
visual signal transmitter 图像信号发射机
visual soil classification 土的肉眼分类
visual space 视空间
visual spectrophotometer 目视分光光度计
visual standard irradiance 可见标准辐照度
visual star 光学星
visual stimuli 视觉效应
visual storm signal 风暴信号
visual structure 可视结构
visual supervisory control 目视监督控制器
visual system 目视系统
visual target 可见目标
visual task 目视工作
visual telephone 可视电话;视频电话;电视电话
visual telescope 目视望远镜;望远镜
visual terminal 直观终端
visual test 目视检查
visual thermometry by estimating gas-liquid ratio 目测气液比测温法
visual threshold 视域;视觉极限
visual titration 目测滴定法
visual trace 目视跟踪
visual tracer 目测示踪剂
visual tracking 目视跟踪
visual training 视力训练
visual training aids 视力训练辅助器材
visual tube parallax 瞄准镜视差
visual tuning 目视调谐;目测调谐
visual tuning indicator 目视调谐指示器
visual type 目视式
visual ultravoilet chromatogram analyzer 可见紫外色谱分析器
visual variable 视觉变量
visual violet 视紫
visual wave data 目测波浪资料
visual wave height 目测波高
visual zenith telescope 目视天顶仪;目测天顶筒
vis-ultraviolet chromatogram analyzer 可见光—紫外色谱分析仪
visus 视力

vis viva equation 活力方程
vis viva formula 活力公式
vitaglass 维他玻璃(一种能透过紫外线的玻璃);透紫外线玻璃
vital action 生活作用
vital circuit 安全电路
vital communication line 交通要道
vital factors index 活力因素指数
vital force 生机;生活力
vital function 生活功能;生活机能
vital hardiness 生活抵抗力
vitalight 紫外线
vitalight lamp 紫外线灯
vital interest 切身利益
vitalism 活力论
vitality 生命力
vitality index 活力指数
vitality test 活力检验
vital optimum 活力最适度
vital part 要害部位;重要部件
vital phenomenon 生活现象
vital red 活性红
vitals 要害部件
vitals of a motor 马达的主要部件
vital statistics 人口统计
vitavite 莫尔道熔融石
vitiate contract 使合同无效
vitiated air 污浊空气
vitiated air chimney 废气通风烟囱
vitiated air duct 废气通风道
vitiated air floor duct 地面排泄废气风道;楼面排泄废气风道
vitiated air floor flue 排放废气风道
vitiated air grate 排放废气炉箅
vitiated air grid 排放废气格栅
vitiated air hole 废气洞
vitiated air line 废气管路
vitiated air opening 废气孔
vitiated air pipe 废气管
vitiated air shaft 废气排出井筒
vitiated expired air 排泄浊气;污浊排泄气
vitiate of water 水污染
vitiliginous 白斑的
vitiligo 白色斑;白斑病
viton gasket 氟化橡胶衬垫
Viton sheet 真空黑橡胶板
vitrail 门窗彩花玻璃
Vitrea glass 维脱利牌玻璃
vitrein 玻璃蛋白
vitreosil 熔凝石英
vitreous 琉态;玻璃状(的);玻璃质(的);玻璃态的
vitreous aggregate 玻璃质集料;玻璃质骨料
vitreous body 玻璃器;玻璃液;玻璃体
vitreous bonded crystalline ceramics 玻态黏合结晶陶瓷
vitreous brick 釉面砖;琉璃砖
vitreous carbon 玻璃碳;玻璃态石墨
vitreous china 玻璃陶瓷;透化瓷;陶器;玻璃搪瓷;玻化瓷
vitreous-china sanitary ware 瓷质卫生器;瓷质卫生瓷
vitreous clinker 玻璃状熔结块;玻璃态熟料
vitreous coating 涂釉层;搪瓷涂层
vitreous copper 辉铜矿
vitreous copper ore 辉铜矿石
vitreous cristobalite 琉态方晶石
vitreous cutter 玻璃体切割器
vitreous degeneration 玻璃样变性
vitreous electricity 正电;玻璃电荷;玻璃电
vitreous electrolyte 玻璃电解质
vitreous enamel label 搪瓷釉商标
vitreous enamel 玻璃釉;透明搪瓷;搪瓷釉;搪瓷珐琅;搪瓷;玻化釉
vitreous enamel coating 搪瓷法
vitreous enamel finish 珐琅外部修饰;釉瓷外部修饰
vitreous enameling sheet 玻璃搪瓷用薄板
vitreous enamelled building panel 上釉瓷的建筑镶板
vitreous enamelled product 搪瓷制品
vitreous enamelled steel 上釉瓷的钢材
vitreous enamelled steel wall panel 搪瓷墙板
vitreous evacuated container 透明真空箱
vitreous fertilizer 玻璃肥料

vitreous fiber 玻璃质纤维
vitreous floater 玻璃体浮游物
vitreous fluid 玻璃状(体);玻璃状废液
vitreous fracture 似玻璃状断口;玻璃状断面;玻璃状断口
vitreous fragment 玻屑
vitreous fused silica 透明石英玻璃
vitreous glaze 玻璃釉
vitreous humor 玻璃状液(体);玻璃液
vitreous infusion suction cutter 玻璃体灌注吸出切割器
vitreous insulating material 玻璃绝缘材料
vitreous lamella 玻璃层
vitreous lava 玻璃状熔岩
vitreous luster 玻态光泽;玻璃光泽;玻化光泽彩
vitreous material 透明材料;玻璃状材料;玻璃质材料
vitreousness 玻璃状态
vitreousness photometer 透明光度计
vitreous opacity 玻璃(状)体混浊
vitreous phase 玻璃相
vitreous phosphorus 玻璃状磷
vitreous rock 玻质岩
vitreous sand 玻璃质砂;色砂
vitreous sanitary ware 玻化卫生瓷
vitreous silica 熔融石英;透明石英;玻璃状石英
vitreous silica fiber 高硅氧玻璃纤维;玻璃质硅石纤维
vitreous slag 玻璃状炉渣;玻璃状矿渣;玻璃质炉渣
vitreous solder glass 透明焊料玻璃
vitreous solid 玻璃质固体
vitreous state 琉态;透明(状)态;玻璃(状)态
vitreous surfacing 表面玻璃化处理
vitreous texture 玻璃状结构;玻璃质结构
vitreous tile 釉瓷瓦;琉璃瓦
vitreous tile flooring 釉面砖楼地面
vitreous ware 透明瓷器;瓷器;玻璃器具
vitreous white arsenic 玻态砒霜
vitreous whiteware body 玻璃质白色瓷器
vitrescence 玻璃质化;玻(璃)态
vitrescent 成玻璃的;玻态的
vitreum 玻璃体
vitric 富玻璃质的;玻璃(质)的
vitric agglomerate 玻璃集块岩
vitric breccia 玻璃角砾岩
vitrics 玻璃制造法;玻璃器(类);玻璃状物质;玻璃制品
vitric tuff 玻璃凝灰岩;玻化凝灰岩
vitric tuff lava 玻屑凝灰熔岩
vitrifaction way 玻化方式
vitrifiable body 易玻化坯体;玻璃体
vitrifiable clay 易玻璃化黏土
vitrifiable colo(u)r 玻化颜料
vitrification 缸化;透明化;陶化;玻璃(固)化;玻化
vitrification brick 缸化砖
vitrification of organism 活质玻璃化
vitrification point 玻(璃)化温度
vitrification range 玻璃化温度范围;玻化(温度)范围
vitrified 成玻璃的;玻璃化的
vitrified abrasive 陶瓷砂轮;陶瓷结合砂轮
vitrified ashes 玻璃化的灰
vitrified block 陶瓷块
vitrified bond 黏土烧结;陶瓷结合剂;玻璃质黏合剂;玻璃质结合物
vitrified brick 釉面陶砖;釉面缸砖;陶砖;瓷砖;玻璃砖
vitrified ceramic drain 釉面陶质排水管
vitrified china 陶器;玻化瓷(器)
vitrified cinder 玻璃化的煤渣
vitrified clay 上釉陶土;上釉黏土;易熔化黏土;缸化黏土;玻璃砖
vitrified-clay conduit 缸瓦管
vitrified-clay culvert 上釉涵管
vitrified-clay drain tile 上釉陶土泄水管
vitrified-clay pipe 陶(土)管;上釉陶土管;釉面陶(土)管;缸瓦管;玻化黏土管
vitrified-clay pipe culvert 上釉(陶)管涵洞
vitrified-clay piping 缸瓦管
vitrified-clay tile 缸管;陶瓷瓦(砖);瓷砖;玻化瓦
vitrified-clay tube 缸瓦管
vitrified-clay tubing 缸瓦管
vitrified clayware 釉面陶土管
vitrified clinker 半熔融熟料
vitrified deck floor surfaced with road pavement 铺路缸砖

vitrified enamel 搪瓷
vitrified fort 玻璃化壁垒
vitrified grinding stone 瓷胶结砂轮
vitrified paving brick 铺面瓷砖;铺路缸砖
vitrified pipe 陶(制)管;陶土管;釉面陶土管;釉面瓦管;上釉瓦管;玻化管
vitrified resistor 涂釉电阻器;玻化电阻器
vitrified rock 玻璃化岩石
vitrified sewer pipe 缸瓦排水管;缸瓦下水管
vitrified sewer piping 缸瓦下水管
vitrified sewer tube 缸瓦下水管
vitrified sewer tubing 缸瓦下水管
vitrified silica 玻璃石英
vitrified slag 玻璃化矿渣
vitrified texture 玻璃质结构
vitrified tile 玻璃瓦;瓷砖;陶土管;缸瓦管;釉面瓦管;陶瓦管;陶土瓦;上釉瓦管;玻化瓷砖
vitrified tube resistor 玻化管电阻器
vitrified wall 玻璃化墙壁
vitrified wheel 陶瓷结合砂轮
vitrified whiteware body 玻璃化白色瓷器
vitriform 玻璃状的
vitri-fusin 闪炭;镜煤
vitrify 使玻璃化;玻璃化
vitrifying clay 易熔瓷土
vitrifying of wastes 废物玻璃化
vitrifying point 易熔点;玻化点
vitrina 玻璃状物质;半透明物质
vitrinertite 微镜惰煤
vitrinertoliptite 微镜惰壳质煤
vitrinite 镜质组;镜质体
vitrinite kerogen 镜质组干酪根
vitrinite reflectance 镜质体反射率
vitrinization 镜煤化
vitrinoid group 镜质组
vitriol 硫酸盐;矾(类)
vitriolic 硫酸的
vitriolization 硫酸盐化;硫酸处理
vitriol peat 硫酸盐泥炭
vitriol plant 硫酸厂
vitriphyric texture 显微玻基斑状结构
vitrite 微镜煤
vitriying clay 易熔黏土
vitrobasalt 玻璃玄武岩
vitro ceram 玻璃陶瓷
vitro-ceramic 玻璃陶瓷制品
vitroclarite 微镜亮煤
vitroclastic 玻璃状
vitroclastic structure 玻璃状构造;玻璃碎屑构造
vitroclastic tuff 玻璃凝灰岩
vitro-crystalloclastic psammitic texture 玻岩屑砂状结构
vitro-crystalloclastic tuff 玻晶屑凝灰岩
vitrodetrinite 镜屑体;碎屑镜质体
vitrodurite 微镜暗煤
vitrofusinite 镜质丝质体
vitrofusite 微镜丝煤
vitrolite 彩色玻璃面砖;瓷板;瓷砖
vitro-lithic psammitic texture 玻岩屑砂状结构
vitromicrolitic 玻璃显微晶质(的)
vitron 石英玻璃结构单元
vitropatic 玻基状
vitrophyre 玻(基)斑岩
vitrophyric 玻基斑状
vitrophyric glass 玻基斑状玻璃;斑状玻璃
vitrosemifusinite 镜煤半丝炭体
vitruvian opening 维特鲁威士门窗
vitruvian scroll 波形卷涡饰;波状涡纹
vitruvian scroll ornament 波状涡纹装饰
vitruvian wave scroll 波状涡纹;波形涡卷饰
vittae 油道
vitusite 磷铈钠石
viuga 浮加风
vivacious 多年生的
vivarium 鱼塘;人工环境动植物园
viver and forestry conservation 水利及森林资源保护
vivianite 蓝铁矿
vivianite earth 蓝铁矿
vivid green 鲜绿色
Viviparus quadratus 湖螺
vivo 柱身;自然条件下
vivosphere 生物圈;生存空间
V-joint V 形接头;V 形缝

V-jointed panel V 形镶板
V-jointer V 形连接
V-jointing V 形接头
V-joint pointing V 形勾缝
V-junction V 形交叉;V 形道路枢纽;V 形接头
V-ladders 人字形梯线
vladimirite 针水砷钙石
Vlaslov equation 符拉索夫方程
vlasovite 硅锆钠石
Vlasov-Maxwell equation 符拉索夫—麦克斯韦方程
V-line ceiling panel V 形吊顶板
V-line wall panel V 形吊墙板
vlotaism 伏打电学
vlucanization activator 硫化活化剂
vlume-and-flow controller 给进和冲洗液调节器;容积流量调节器
V-mo(u)lding V 形线饰
V-notch V 形切口;V 形槽口
V-notch Charpy specimen V 形缺口冲击试件
V-notched weir V 形槽间;三角堰;三角形槽口量水堰
V-notch weir V 形槽间;三角堰;三角形槽口量水堰;V 形缺口堰;V 形槽堰
vocabulary 词汇量;词汇表;词汇;词典;词表
vocabulary equivalence declaration 词汇表等价说明
vocabulary of terms 名词术语
vocabulary test 词汇测验
vocal cavity 声腔
vocal command 声指令
vocal function 发声机能
vocal input 声输入
vocal level 声电平
vocal music 声乐
vocal print 声纹
vocal room 发声室
vocation 业务
vocational aptiuted tester 职业才能测试仪
vocational assessment 人事考核
vocational education 职业教育
vocational rules 业务规章
vocational school 职业学校
vocational study 业务学习
vocational training 专业培训;职业培训
vocational training college 职业训练学院
vocational training for the employees 职工脱产培训
vocational work 业务
vocoder 声音编码器
vocodoer 声码器
vodas [voice-operated antisinging device] 音控防鸣器
voe 小海湾
voelckerite 氧磷灰石
V-of a derrick 钻塔正面
vogad 响度级调整装置
Vogel-Fulcher relation 沃格尔—福歇尔关系式
Vogel-Ossag viscosimeter 沃格尔—奥萨格(毛细管)黏度计
Vogel's red 沃格尔红
vogesite 闪辉正煌岩
voglite 碳铜钙铀矿
voice acceptance terminal 声音接收终端
voice-actuated terminal 音触终端
voice answer back 声应答装置
voiceband 声音频带
voice carrier 声信号载波
voice channel multiplexer 话路多路复用器
voice coder 声码器
voice coil 声圈
voice coil motor 音圈电动机
voice compression 声音压缩
voice control 声控
voice control technique 声控技术
voice data communication 音频数据通信
voice distress switch box 音响呼救开关盒
voice-excited vocoder 声激励声码器
voice-frequency cable 音频电缆
voice-frequency carrier 音频载波
voice-frequency circuit 音频电路
voice-frequency keying 音频键控
voice-frequency pulsing 音频脉冲发生
voice-frequency repeater 音频增音机
voice-frequency ringing 音频振铃
voice-frequency selective calling 音频选叫

voice-frequency terminal 音频终端
voice-frequency terminal equipment 音频终端设备
voice-frequency terminating equipment bay 音频终端架
voice input 声音输入
voice modulation 声频调制
voice-operated control 声音操作控制
voice-operated device 语控装置;音频控制设备;声音工作器
voice-operated device antisinging 语控防鸣器
voice-operated gain adjusting device 声控增益调节器;响度级调整装置;声音工作增益调整装置
voice-operated regulator 声音操作调节器
voice-operated switch 声音操作开关
voice output 声音输出(装置)
voice output scanner 声音输出扫描器
voice pipe 传话筒
voice print 声印
voice scrambler 声音扰频器
voice security telephone system 保密电话系统
voice synthesizer 声音合成器
voice terminal 声音终端
voice transmitter-receiver 收发话机
voice tube 传声管;传话筒
voice unit 声音单元
voicing 调声
void 孔隙;空隙;空泡;空洞;空的;空白(点);巨洞;缺乏
voidable contract 可取消合同;单方约束合同;可撤销的合同;可变为无效的合同
voidage 空穴现象;空隙量;空隙度;空位(现象);空洞率
voidage rate 枯竭率
void aggregate 填隙骨料
voidal lithophysa structure 空心石泡构造
voidal volcanic mud ball structure 空心火山泥球构造
void behind segments 盾尾净空
void between balusters 栏杆空档
void-cement ratio 隙灰比;孔隙与水泥容积比;孔隙水泥比;水泥隙灰比;水泥孔隙比;水泥空隙比
void-cement ratio method 隙灰比法
void-cement ratio method of proportioning material 隙灰比配料法
void channel 多孔管道
void coefficient 空穴系数
void coefficient of reactivity 反应性空位系数
void column 空柱
void compartment 空舱
void content 孔隙容积;孔隙率;孔隙量;孔隙含量;空隙含量;气孔率
void contract 无效合同;失效契约
void determination 孔隙测定;空隙测定
void documents 过期票券
voided check 已作废支票;作废支票
voided concretion 空心结核
voided contract 已作废合同
voided slab 空心板
voided slab bridge 空心板桥
void effect 空穴效应
void factor 孔隙率;空隙系数;空隙率;空隙比;孔隙系数
void filler 孔隙填充物
void-filling aggregate 填充性骨料;填隙集料;填隙骨料
void-filling capacity 填隙能力;填隙量
void-filling content 填隙量
void-filling gel 填隙凝胶量(水泥净浆颗粒间的空隙)
void former 孔隙形成器;留孔模
void fraction 空隙率;空隙度;截面含气量
void-free 密实的;无气孔的;无孔隙的;无空隙的
void-free cast method 无空隙浇注法
void-free fiber 密实纤维
void generator 空段发生器
void index 孔隙指数
void in shotcrete 隐蔽空隙;喷混凝土空隙
void item 空(白)项
voidless concrete 无孔混凝土
voidless mass 无孔物质
voidless matter 无孔物质
voidless substance 无孔物质
void measurement apparatus 孔隙测定仪;空隙测定仪

voidmeter 空隙测定仪
void method of proportioning 孔隙配料法;孔隙配合法;空隙配料法;混凝土孔隙配料法
voidness ratio 空心度
void of paste 跑浆
void of soil 土壤孔隙
void of weld 焊槽
void plugging 孔隙堵塞
void plywood 中空胶合板
void pocket 气囊
void ratio 孔隙比;孔洞率;空隙比;粉体空隙比
void ratio change 孔隙比变化
void ratio in densest state 最小孔隙比
void ratio in loosest state 最大孔隙比
void ratio of soil 土的孔隙比
void ratio-pressure curve 孔隙比—压力(关系)曲线
void ratio-strength curve 孔隙比—强度曲线
void ratio test 孔隙比试验
void ratio-time curve 孔隙比—时间曲线
void-reducing capacity 减隙能力;减少孔隙能力
void relation 空(白)关系
voids and hungry spots (指混凝土)蜂窝麻面
voids and pits (指混凝土)蜂窝麻面
voids content 空隙;气泡
voids filling theory 填充理论
void-solid ratio 门窗—墙面积比
void space 孔隙;空隙体积;空隙(空间)
void space in packing system 堆积中空位
void spacing factor 气泡间隔系数
void-strength curve 孔隙—强度曲线;空隙量—强度曲线;空隙比—强度曲线
void system 孔隙系统
void test 孔隙测定;空隙试验;空隙测定;气孔测定
void transaction 空的事务处理;空白事项
void-transaction log 空事务处理记录;空白事项登记记录
void transposition 虚换位;空换位
void value 没有(价)值
void volume 孔隙体积;空隙体积;空隙容积;空隙量;空体积;气孔容积
void water 孔隙水;空隙水
void water pressure 孔隙水压(力)
Voight model 伏格特模型
Voigt effect 伏格特效应
Voigt model 伏格特模型
Voigt notation 伏格特符号
Voigt rheological model 伏格特流变模型
Voigt's boundary lines 伏格特界线
Voigt solid 伏格特固体
voile 巴里纱
voilent earthquake 强烈地震
voilent storm 暴風雨
Voith-Schneider propeller 直翼推进器;平旋轮推进器
Voith-Schneider ship 平旋推进船
voiture 轻便马车;轻便敞篷汽车
voiturette 双座小型汽车
volage current transformer 变压变流器
volatibility test 易失性测试
volatile 挥发(性)的
volatile acid 挥发(性)酸
volatile acid alkalinity 挥发酸碱度
volatile acid concentration 挥发性酸浓度
volatile activity 挥发性物质的放射性
volatile alkali 挥发性碱
volatile ammonia 挥发性氨
volatile aromatics organics 挥发性芳香族有机物
volatile capital flows 不规则的资本流动
volatile cargo 易挥发性货物
volatile chlorinated aromatics 挥发性氯化芳香族
volatile chlorinated hydrocarbon 挥发性氯化烃
volatile combustible 挥发物(质)
volatile combustible matter 挥发可燃物
volatile component 挥发(性)组分;挥发(性)成分
volatile compound 挥发性化合物
volatile constituent 造气剂;挥发性组分;挥发性成分
volatile contaminant 挥发性污染物
voidtile contaminant analysis 挥发性污染物分析
volatile content 挥发分含量;挥发物含量
volatile corrosion inhibitor 挥发性防蚀剂
volatile covering 挥发性覆盖层
volatile diluent 挥发性稀释剂
volatile dissolved solid 挥发性溶解固体
volatile distillate 挥发性馏分

volatile dryer 挥发性催干剂
volatile element 挥发元素
volatile evapo(u)ration 挥发性蒸发
volatile evapo(u)ration residue 挥发性蒸发残渣
volatile fatty acid 挥发性脂肪酸
volatile fatty acid degradation 挥发性脂肪酸降解
volatile file 易变文件
volatile flammable liquid 挥发性易燃液体
volatile fluid 挥发液;挥发(性)油;挥发性流体
volatile flux 挥发(性)组分;挥发(性助)熔剂;挥发性成分
volatile fraction 挥发部分
volatile fuel 挥发性燃料
volatile grade 挥发(程)度
volatile halogenated organics 挥发性卤代有机物
volatile halogenated organics compound 挥发性卤代化合物
volatile halohydrocarbon 挥发性卤代烃
volatile hydrocarbon of existing kerogen 干酪根中挥发性烃
volatile hydroxybenzene 挥发性羟基苯
volatile inflammable liquid 挥发性可燃液体
volatile inhibition 气相防锈性
volatile inorganic compound 挥发性无机化合物
volatile insecticide 挥发性杀虫剂
volatile lacquer 挥发性面漆
volatile liniment 挥发性搽剂
volatile liquid 挥发性液体
volatile liquid hydrocarbon 挥发性液态烃
volatile loss 挥发损失;挥发减量
volatile loss from batch 配合料挥发量
volatile market 不稳定市场;不稳定的行情
volatile material 挥发性材料;挥发物(体);挥发性液体
volatile matter 易挥发物;挥发性物质;挥发物(质);挥发分
volatile matter analysis 挥发分分析
volatile matter yield 挥发分产率
volatile matter yield of floated coal 浮煤挥发分产率
volatile matter yield of raw coal 原煤挥发分产率
volatile memory 易失性存储器
volatile metal 挥发性金属
volatile mineral spirit 松香水;挥发性石油溶剂油;200号溶剂汽油
volatile mixed liquor suspended solid 挥发性混合液悬浮固体
volatileness 挥发性
volatile oil 挥发(性)油;芳香油
volatile organic acid 挥发性有机酸
volatile organic analysis 挥发性有机物分析法
volatile organic biogenic substance 挥发性有机生物剂
volatile organic carbon 挥发性有机碳
volatile organic chemicals 挥发性有机化学物
volatile organic chemicals wastewater treatment 挥发性有机化学物废水处理
volatile organic compound 挥发性有机化合物
volatile organic compound control technology 挥发性有机化合物控制技术
volatile organic compound emission inventory 挥发性有机化合物排放物清单
volatile organic compound level 挥发性有机化合物含量
volatile organic contaminant 挥发性有机污染物
volatile organic halogen 挥发性有机卤素;挥发性有机卤化物
volatile organic pollutant 挥发性有机污染物
volatile organics 挥发性有机物
volatile organohalogen compound 挥发性有机卤素化合物
volatile pattern 气化模
volatile phenol 挥发(性)酚
volatile phenol compound 挥发性酚化合物
volatile phosphorus compound 挥发性磷化合物
volatile poison 挥发性毒物
volatile pollutant 挥发性污染物
volatile price 飞腾性价格;变动剧烈价格
volatile product 挥发性产品
volatile residue 挥发性残渣
volatile rust-proofing agent 挥发性防锈剂
volatile short-term fund 不稳定的短暂基金
volatile solid 挥发性固体
volatile solid content 挥发性固体含量

volatile solvent 挥发性溶剂
volatile spacing agent 挥发增孔剂
volatile storage 易失性存储器
volatile substance 挥发性物质;挥发物体;挥发物(质)
volatile sulfur 挥发硫
volatile suspended matter 挥发性悬浮(固体)物
volatile suspended solid 挥发(性)悬浮固体
volatile suspended solids accumulation 挥发性悬浮固体累积
volatile suspended solids concentration 挥发性悬浮固体含量
volatile suspended substance 挥发性悬浮物
volatile thinner 挥发性稀释剂;挥发性冲淡剂
volatile total solids 总挥发性固体
volatile treatment 挥发处理;全挥发水处理
volatile varnish 挥发性清漆
volatile vehicle 挥发性展色剂;挥发性调漆料
volatility 易变性;易失性;挥发性;挥发度;变更率
volatility coefficient 挥发系数
volatility index 挥发度指数
volatility of spot rate movement 即期汇率反复变动
volatility of the file 文件的变动性
volatility product 挥发性乘积;挥发度积
volatility range 挥发范围
volatility resistance 耐挥发性
volatility test 挥发性试验;挥发度测定
volatilizable 可挥发的
volatilization 挥发作用
volatilization analysis 挥发分析
volatilization loss 挥发损失
volatilization method 挥发法
volatilization of alkalies 碱挥发
volatilization of pesticide 农药挥发
volatilization reaction 挥发反应
volatilization roasting 挥发焙烧
volatilization test 挥发性试验;挥发度测定
volatilization test(ing) 挥发试验
volatilize 挥发;使挥发
volatilized silica 硅粉(配制高强度混凝土外加剂用)
volatilizer 挥发剂;挥发器
volatilizer premix burner system 挥发预混式燃烧器装置
volatilizing loss 挥发损耗
volborthite 水钒铜矿
volcanello 内火山锥;寄生熔岩锥
volcanic 火山(作用)的
volcanic accident 火山突变
volcanic action 火山作用
volcanic activity 火山活动(性);火山地热活动
volcanic agglomerate 火山圆砾;火山块集岩;火山集块岩
volcanic agglomerate breccia 火山集块角砾岩
volcanic aggregate 火山岩集料;火山岩骨材
volcanic-allothigenou fragment 火山异源碎屑
volcanic and subvolcanic orefield structure 火山次火山型矿田构造
volcanic arenite 火山砂屑岩
volcanic ash 火山灰
volcanic ash bed 火山灰层
volcanic ash cloud 火山灰云
volcanic ash flow 火山灰流
volcanic ash reservoir 火山灰热储
volcanic ash soil 火山灰土壤
volcanic basin 火山盆地
volcanic belt 火山带
volcanic belts type 火山带型
volcanic block 火山岩块;火山块;火成岩块体
volcanic bomb 火山弹
volcanic bomb agglomerate 火山弹集块岩
volcanic bomb breccia 火山弹火山角砾岩
volcanic breccia 火山砾;火山角砾(岩)
volcanic breccia tuff 火山角砾凝灰岩
volcanic building material 易爆发性建筑材料
volcanic burner 在孔底加温液体的装置
volcanic chain 火山山脉
volcanic chimney 火山管
volcanic cinder 火山岩屑;火山渣
volcanic-cinder agglomerate 火山渣集块岩
volcanic-cinder breccia 火山渣火山角砾岩
volcanic clastic rock 火山碎屑岩
volcanic clay 火山黏土

volcanic cloud 火山云
volcanic coast 火山(海)岸
volcanic collapse 火山塌陷
volcanic conduit 火山喉管
volcanic cone 火山锥
volcanic conglomerate 火山砾岩
volcanic construction(al) material 易爆发性建筑材料
volcanic cycle 火山旋回
volcanic debris 火山碎屑
volcanic debris sandstone 火山岩屑砂岩
volcanic deposit 火山堆积(物)
volcanic detritus 火山岩屑
volcanic dome 火山丘;火山穹
volcanic dust 火山尘
volcanic-dust tuff 火山尘凝灰岩
volcanic earth 火山灰
volcanic earth from Pozzuoli 波苏埃里火山土
volcanic earthquake 火山地震
volcanic ejecta 火山喷发物;火山喷出物;火山抛出物
volcanic embryo 雏火山
volcanic eruption 火山喷发;火山爆发
volcanic eruption chipping 火山喷发碎屑
volcanic eruption period 火山喷发期
volcanic exogenic fragment 火山质外生碎屑
volcanic explosion 火山爆发
volcanic foam 浮岩
volcanic focus 火山源
volcanic-fracture structure 火山断裂构造
volcanic fragment 火山凝灰岩
volcanic fragmental facies 火山碎屑岩相
volcanic gas 火山气(体)
volcanic geology 火山地质学
volcanic geothermal area 火山地热区
volcanic glass 黑耀岩;火山玻璃
volcanic graben 火山地堑
volcanic gravel 火山砾(石)
volcanic hair 火山毛
volcanic hazard 火山灾害
volcanic horst 火山地垒
volcanic intensity 火山烈度
volcanic island 火山岛
volcanic island arc 火山岛弧
volcanicity 火山活动
volcanic lake 火山湖
volcanic landform 火山地貌
volcaniclastic 火山碎屑的
volcanic layering 火山陷沟
volcanic mechanism 火山机制
volcanic mechanism and its superposed fracture zone 火山机构及叠加其上的断裂带
volcanic mechanism and structure 火山机制与构造
volcanic mountain 火山
volcanic mud 火山泥
volcanic mud ball 火山泥球
volcanic mud ball structure 火山泥球构造
volcanic mud ball tuff 火山泥球凝灰岩
volcanic mudflow 火山泥流
volcanic muds and sands 火山泥和火山砂
volcanic neck 岩颈;火山颈
volcanic obelisk 碑状火山颈
volcanic orifice 火山口;火山孔
volcanic outer edge bar 火山外缘坝
volcanic phenomenon 火山现象
volcanic pipe 岩筒
volcanic pisolite 火山豆石岩
volcanic products 火山喷发物;火山喷出物
volcanic projectile 火山喷发物;火山喷出物
volcanic rain 火山雨
volcanic rent 火山裂罅
volcanic resource 火山资源
volcanic rift zone 火山裂谷带;火山断裂带
volcanic rock 火山岩(质)
volcanic rock phases 火山岩相
volcanic rock series 火山岩系列
volcanic rubble 火山碎屑
volcanic rumbling 火山鸣
volcanics 近地表火成岩
volcanic sand 火山砂
volcanic scoria 火山熔渣;火山灰渣;火山岩渣
volcanic seamount 海底火山
volcanic seaquake 火山海震
volcanic-sedimentary agglomerate 火山沉积集块岩
volcanic-sedimentary breccia 火山沉积角砾岩
volcanic-sedimentary chemical rock 火山沉积化学岩
volcanic-sedimentary clastic rocks 火山沉积碎屑岩类
volcanic-sedimentary clastic texture 火山沉积碎屑结构
volcanic sedimentary deposit 火山沉积矿床
volcanic-sedimentary mudstone 火山沉积泥岩
volcanic-sedimentary sandstone 火山沉积砂岩
volcanic-sedimentary siltstone 火山沉积粉砂岩
volcanic shield cluster 盾形火山
volcanic shoreline 火山岸线
volcanic sink 火山沉陷口;锅状火山口;火山塌陷;火山沉陷(区)
volcanic slope 火山成因坡
volcanic spine 火山栓
volcanic spring 火山(温)泉
volcanic structure 火山构造
volcanic sublimate 火山升华物
volcanic subsidence 火山沉陷(区)
volcanic theory 火山说;火山理论
volcanic thunder 火山雷鸣
volcanic thunderstorm 火山雷暴
volcanic treatment 孔内加热油井增产法
volcanic tremor 火山脉动
volcanic tuff 火山凝灰岩
volcanic type mud-stone flow 火山型泥石流
volcanic vent 火山(喷)口;火山道
volcanic water 火山水
volcanic zone 火山带
volcanism 火山作用;火山现象;火山活动
volcanized rock 火山石
volcannello 子火山
volcano 火山
volcano arc 火山弧
volcano chain 火山链
volcano cluster 火山群
volcanogene hydrothermal ore deposit 火山热液矿床
volcanogene magmatic ore deposit 火山岩浆矿床
volcanogenic 火山成因的
volcanogenic deposit 火山沉积
volcanogenic mineral deposit 火山成因矿床
volcanogenic mineralization 火山成矿作用
volcanogenic sedimentary deposit 火山沉积成因矿床
volcano-geologic(al) cross-section 火山地质剖面图
volcano-geologic(al) map 火山地质图
volcano-geologic(al) sketch 火山地质素描图
volcano hypothesis 火山假说
volcano island 火山岛屿
volcanology 火山学
volcano of unperiodical eruption 不定期喷发火山
volcano plug 火山塞
volcano seamount 火山海底山
volcano-sedimentary phase 火山沉积相
volcano-sedimentary-type boron deposit 火山沉积型硼矿床
volcano-subvolcanic hydrothermal ore deposit 火山次火山热液矿床
volcano-tectonic depression 火山构造洼地
volcano-tectonic line 火山构造线
volcano-tectonic map 火山构造图
volcano-tectonic rift zone 火山构造裂隙带
volcano-tectonic subsidence 火山构造沉陷
volchonskoite 富铬脱石;铬蒙脱石
volclay 高钠膨润土防水料
vole drill 竖式旋转冲击钻机
volet 三折书牒的一块板
Volgian stage 伏尔加阶【地】
Volhard's solution 沃尔哈德硫氰酸钾溶液
Volhard titration 沃尔哈德滴定
Volkmann's canal 福尔克曼氏管
Volkmann's retractor 福尔克曼氏拉钩
volkovskite 沃硼钙石
volley 齐发爆破
volleyball court 排球场
volley theory of nerve impulse 冲动排放论
Vollmer test 伏尔默试验
volometer 万能电表;电压电流表;伏安表
volscan 立体扫描(器)
Volspray 沃尔喷漆枪(一种油漆喷涂设备)
volt 伏(特)
voltage 电压;伏特数
voltage adjuster 调压器;电压调整器;电压调节器
voltage-ampere curve 伏安曲线
voltage amplifer 电压放大器
voltage amplification 电压放大
voltage amplification coefficient 电压放大系数
voltage amplifier 电压放大器
voltage-amplitude-controlled clamp 电压振幅箝位控制电路
voltage and current transformer 互感器
voltage and level meter 电压电平表
voltage and number of outgoing lines 出线电压和回路数
voltage balancer 均压器【电】
voltage balance relay 电压平衡继电器
voltage below-level 电压过低
voltage between layers 层间电压
voltage between lines 线(间)电压
voltage between phases 相(间)电压
voltage box 分压箱
voltage breakdown method 电压击穿法
voltage build-up rate 电压增长率
voltage changer 电压变换机;电压变换器;变压器
voltage circuit 电压回路
voltage circuit break protection 电压回路断线保护(装置)
voltage clamp 电压箝(位)
voltage clamping circuit 钳压电路
voltage clamp method 电压箝位法
voltage class 电压等级
voltage coder 电压编码器
voltage coefficient 电压系数
voltage coil 电压线圈
voltage comparator 电压比较器
voltage-comparison encoder 电压比较编码器
voltage compensation 电压补偿
voltage compensation method 电压补偿法
voltage compensator 电压补偿器
voltage conditioner 电压调节器
voltage constant 电压常数
voltage constant of logger 测井仪电压常数
voltage contrast 电压对比
voltage control 电压调整
voltage-controlled 电压控制的
voltage-controlled amplifier 压控放大器
voltage-controlled attenuator 压控衰减器;电压控制衰减器
voltage-controlled avalanche oscillator 电压控制雪崩管振荡器
voltage-controlled crystal oscillator 压控晶体振荡器
voltage-controlled current source 电压控制电流源
voltage-controlled oscillator 压控振荡器;电压控制振荡器
voltage-controlled phase shifter 压控移相器;电压控制相移器
voltage-controlled quartz crystal oscillator 压控石英晶体振荡器
voltage-controlled resistance 电压可控电阻
voltage-controlled variable delay line 压控可变延迟线
voltage-controlled voltage source 电压控制电压源
voltage controller 电压控制器
voltage control multichrome screen 电压控制多色穿透屏
voltage corrector 电压校正器
voltage crest 压峰
voltage-current characteristic 电压电流特性;伏安特性(曲线)
voltage-current converter 电压电流转换器
voltage-current dual 电压电流对偶
voltage decoder 电压解码器
voltage-dependent 压敏(电阻)
voltage-dependent resister 变阻器
voltage detector 电压检测器
voltage difference 电压差
voltage differential relay 电压差动继电器
voltage divider 分压器
voltage-divider current 分压器电流
voltage-dividing capacitor 分压电容器
voltage division 分压【电】
voltage division by time division 时间分割式分压

voltage double circuit 倍压电路
voltage doubler 电压二倍器;倍压器
voltage-doubler rectifier 二倍压整流;倍压整流器
voltage doubler tube 倍压整流管
voltage-doubling circuit 倍压电路
voltage doubling rectifying circuit 倍压整流电路
voltage drift 电压漂移
voltage drive 电压激励
voltage drop 电压降;压降【电】
voltage dropping resistor 降压电阻器
voltage effect 电压效应
voltage efficiency 电压效率
voltage encoder 电压编码器
voltage excursion 电压偏移
voltage factor 电压因数
voltage failure of main circuit 主回路失压
voltage feed 电压馈电
voltage feed antenna 电压馈接天线;电压馈电天线
voltage feedback 电压反馈
voltage feedback amplifier 电压反馈放大器;电压回授放大器
voltage feedback ratio 电压反馈比
voltage field 电压场
voltage follower 电压输出器;电压跟随器
voltage follower circuit 电压输出电路
voltage gain 电压增益
voltage generator 电压发生器;发动机电动势
voltage grade 电压等级
voltage gradient 电压梯度;电压陡度
voltage grading electrode 分压电极
voltage/hertz limiter 电压/频率限制器
voltage impulse 电压脉冲
voltage indicator 电压指示器
voltage influence 电压影响
voltage-invariant base-width 不随电压变化的基区宽度
voltage jump 电压突变;电压陡变
voltage lead 电压引线
voltage leap 电压突变
voltage level 电压电平
voltage level gain 电压电平增益
voltage leveller 电压电平器
voltage limit 电压极限
voltage limitator 电压限制器
voltage limiter 限压器;电压限制器
voltage loop 电压波腹
voltage loss 电压损失
voltage measurement 电压测量
voltage measurement determination 电压测定分析
voltage meter 电压表
voltage-mode logic circuit 电压模逻辑电路
voltage multiplier 仪表扩程器;电压倍增器;倍压器
voltage-multiplier circuit 倍压器电路
voltage-multiplying circuit 电压倍增电路;倍压电路
voltage mutual inductor 电压互感器
voltage node 电压波节
voltage of microphonic effect 颤噪电压
voltage of transformation 变换电压
voltage-operated 电压运行的
voltage-output element 电压输出元件
voltage parameter 电压参量
voltage pattern 电位起伏
voltage pen pointer 电压指示笔
voltage per unit band 单位频带电压
voltage phasor 电压相量
voltage probe 电压探针
voltage protection 电压保护装置
voltage pulsating factor 电压脉动因数
voltage pulsation 电压波动
voltage pulse 电压脉冲
voltage quadrupler 电压四倍器
voltage range 电压范围
voltage range multiplier 电压量程倍增器;仪表扩程器
voltage rating 额定电压;标定电压
voltage ratio 电压比
voltage-ratio method 电压比值法
voltage-ratio regulation 变压比调节
voltage-ratio type relay 电压比式继电器
voltage recovery 电压恢复
voltage recovery rate 电压恢复速度
voltage recovery time 电压恢复时间
voltage reducing device 电击防止装置
voltage reference 电压基准

voltage reference amplifier 电压基准放大器
voltage reference diode 标准稳压管
voltage reference element 电压参考元件
voltage reference tube 基准电压管;电压基准管
voltage reflection coefficient 电压反射系数
voltage-regulating equipment 稳压器;电压调整器
voltage-regulating potentiometer 调压电位计
voltage-regulating relay 电压调整继电器;电压调节继电器
voltage-regulating transformer 电压调整变压器
voltage regulation 调压;电压调节;电压变动
voltage regulation coefficient 电压稳定系数
voltage regulation coil 电压调节线圈
voltage regulation factor 电压调整率;电压变动率
voltage regulation of indication lamp 表示灯调压
voltage regulation of signal lamps 信号灯调压
voltage regulator 稳压器;调压器;电压稳压器;电压调整器;电压调节器
voltage regulator armature 电压调节器附件
voltage regulator circuit 电压调节电路
voltage regulator contact bracket 电压调整器接触点托架
voltage regulator contact screw 电压调节器接触螺钉
voltage regulator diode 稳压二极管
voltage regulator for vehicle 车辆电压调节器
voltage regulator panel 电压调整器面板
voltage regulator tube 调压管
voltage relay 电压继电器
voltage resonance 串联谐振
voltage response 电压响应
voltage-responsive circuit 电压响应电路
voltage responsive relay 电压继电器
voltage restraint 电压限制器
voltage ripple 电压纹波;电压脉动
voltage rise 压升;电压升(高)
voltage saturation 电压饱和
voltage saturation effect 电压饱和效应
voltage scanning 电压扫描
voltage-selected switch 电压选择开关
voltage selecting switch 电压选择开关
voltage-sensitive 压敏
voltage-sensitive effect 压敏效应
voltage-sensitive gate 电压敏感门
voltage sensitivity 电压灵敏度;变压误差
voltage series 电压系列
voltage sets 电压组
voltage-sharing 均压
voltage-sharing capacitor 电压均分电容器
voltage-sharing resistor 电压均分电阻器
voltage slider 滑线式调压器
voltage source 电压(电)源
voltage source inverter 电压源变换器
voltage specification 电压规格
voltage spread 最高和最低电压差
voltage stability 电压稳定性;电压稳定度
voltage stabilization 电压稳定
voltage-stabilized source 稳压电源
voltage stabilizer 常压调整器;稳压器;电压调整器;电压调整器
voltage-stabilizing 稳压
voltage-stabilizing circuit 稳压电路
voltage stable element 电压稳定元件
voltage stable negative resistance 电压稳定的负阻
voltage standard 电压标准
voltage standing wave ratio 电压驻波比
voltage standing wave ratio indicator 电压驻波比指示器
voltage summer 电压求和器
voltage supply equipment 供压设备
voltage surge 电压浪涌;电压冲击;冲击性过电压
voltage swing 电压摆幅;电压摆动
voltage switch 电压开关
voltage tapping device 电压分接头装置;电压抽头装置
voltage-time characteristic curve 电压—时间特性曲线
voltage-time curve 电压—时间曲线;伏秒特性曲线
voltage-time-to-breakdown curve 击穿电压与击穿时间(关系)曲线
voltage-to-digital converter 电压数字信息变换仪;电压数字变换器
voltage to earth 对地电压
voltage to frequency converter 电压频率变换器;电压频率变换器

voltage to neutral 对中性线电压;对中点的电压
voltage-to-pulse rate converter 电压脉冲重复频率变换器
voltage-transfer characteristic 电压传输特性
voltage transformer 压互;电压互感器;变压器
voltage transformer comparator 变压器的比较仪
voltage-triggered type 电压触发型
voltage tripler 三倍压电路;电压三倍器
voltage tripler rectifier 三倍压整流
voltage-tunable magnetron 电压调谐磁控管
voltage-tunable tube 电压调谐管
voltage-tuning oscillator 调压式振荡器
voltage-turns ratio 伏圈比
voltage type telemeter 电压式遥测计
voltage-variable capacitor 可变电压电容器;电压可变电容器
voltage-variable resistor 随电压变化的电阻器
voltage wave 电压波
voltage winding 电压绕组
voltage withstand 耐电压
voltaic 电流的
voltaic arc 电弧
voltaic cell 伏打电池
voltaic couple 伏打电偶
voltaic current 伏打电流
voltaic effect 伏打效应
voltaic electricity 伏打电
voltaic photo-detector 伏打光检测器
voltaic pile 伏打电堆
voltaite 绿镁铁矾;绿钾铁矾
voltammeter 库仑计;电量计;电压电流表;电压电流两用表;伏安(计);伏安表
voltammetric membrane electrode system 电量膜电极系统
volt-ammetry 伏安法
voltammogram 伏安谱图
voltamoscope 伏安器
volt-ampere 伏特安培
volt-ampere hour 伏安(小)时
volt-ampere-hour reactive 伏安时电抗性的
volt-ampere meter 伏安计;伏安表
voltascope 千分伏特计;伏安计示波器组件
Volta series 置换序列
volta's law 伏打定律
volt box 分压箱;分压器
volt cell 伏打电池
volt circuit 电压电路
volt-current characteristic 伏安特性(曲线)
Voltera integral equation 伏泰勒积分方程
Volterra dislocation 伏尔特拉位错
Volterra equation 伏尔特拉方程
Volterra integral equation 伏尔特拉积分方程
Volterra series 伏尔特拉级数
volt ga(u)ge 电压表
voltimeter 万能电表
voltite 电线被覆绝缘物
volt-line 伏特线
voltmeter 电压计;电压表;伏特计;伏安计;伏安表
voltmeter-ammeter method 电压电流表法
voltmeter commutator 伏特计换挡器
voltmeter indicator 伏特计指示器
voltmeter multiplier 伏特计倍增器
voltmeter sensitivity 伏特计灵敏度
volt-milliampere meter 伏特毫安计
volt-ohmmeter 电压—电阻表;伏特欧姆计
volt-ohm-milliam meter 伏欧毫安表
voltohmyst 伏特欧姆计
voltolization 电聚
voltolization process 电聚过程
voltol oil 高压放电聚合油;电聚合油
voltol process 电聚过程
volt-second 伏秒
voluble herb 绕缠的草本;缠绕草本
voluble shrub 藤本;绕缠的灌木;缠绕灌木
volume 航空旅客量,一组;容积;体积;声量;产量;册
volume absorption 体(积)吸收
volume absorption coefficient 体积吸收系数
volume absorption term 体积吸收项
volume allocation 卷宗分配
volume analysis 体积分析;产量分析
volume analysis method 体积分析法
volume and flow controller 容量与流速控制器
volume and flow direction figure 流量流向图

volume attenuation coefficient 体积衰减系数
volume-averaged flux 体积平均通量
volume backscattering return 体积后向散射
volume backscattering strength 体积反向散射强度
volume balanced slip 容积平衡式伸缩节
volume balance model 容量平衡模型
volume basis 容积基位;体积法;按体积
volume batcher 按体积比配器;体积比配料器
volume batching 按体积分批配合;体积配料法;体积配合法;体积比配料;按体积比配料;按体积比配合拌料;按容积比例配料
volumebolometer 脉容量测量器
volume bottles 体积缓冲瓶
volume bound 体积界(限)
volume box 套筒扳手
volume capacity 容量
volume-capacity ratio 拥塞系数
volume cargo 体积(大的)货物
volume center 体积中心
volume change 变形;体积变化;体积变更
volume change modulus 体变模量
volume changes in the soils 土壤的体积变化
volume charge 体(积)电荷;从量收费
volume charge density 体电荷密度
volume clearance 容积余隙
volume coefficient 容积系数;体积系数
volume coefficient of container 箱容系数
volume coefficient of petroleum 石油的体积系数
volume composition 体积组成;体积比;按容积计的成分
volume compressibility 体积压缩系数
volume compression 体积压缩
volume compressor 强度压缩器;体积压实机;声量压缩器
volume concentration 体积浓度;体积密度
volume concentration of impregnant 浸渍(材)体积率
volume conductance 体积电导率
volume conduction 容积导电;体积导电
volume conductivity 体电导率
volume conductor 容积导体
volume constancy 体积固定性;体积不变性
volume-constant diving suit 固定容积式潜水服
volume-constant dry diving suit 固定容积干式潜水服
volume content 体积含量
volume content of system 系统容量
volume contour 等容线
volume contraction 体积收缩
volume control 音量控制;卷控制;容量调节;体积控制;体积调节;声量控制
volume controller 音量控制器;增益调节器
volume conversion factor 体积换算系数
volume conversion factor of crude oil 原油体积换算系数
volume conversion factor of natural gas 天然气体积换算系数
volume coordinates 空间坐标
volume cost 体积成本;数量成本;单位体积成本
volume-cost-profit analysis 产量—成本—利润分析
volume-cost-profit relationship 量—本—利关系式
volume curve 容量曲线;容积曲线
volume-curve method 曲线求积法
volume damper 风量调节阀;风量调节板;风量调节器
volume decrease 体积缩减;体积缩小
volume-decrease potential 体积减少量
volume defect 体积缺陷
volume deformation 体积形变
volume density 重度;容积密度;体(积)密度;堆积密度
volume-density controller 流量密度(信号)控制机
volume-depth distribution curve 容积—深度分布曲线
volume design hour 交通量设计小时
volume diffusion 体(积)扩散
volume dilatation coefficient 体膨胀系数
volume dilational vibration 体积胀缩振动
volume dilatometer 体积膨胀计
volume discharge 容积流量
volume discount 数量折扣
volume-displacement meter 容积流量计
volume distortion 声量失真
volume distribution 体积分布

volume district 按建筑容积分区指定的地区
volume divergence 体散度
volume dose 积分剂量;体积剂量
volume dosing unit 容积计量供料装置
volume drying 整体干燥;成堆干燥
volume effect 体积效应
volume efficiency 容量系数;容积效率;体积效率;声量效率
volume elasticity 容积弹性;体积弹性
volume elastic modulus 容积弹性模数;容积弹性模量
volume electrostrictive effect 体积电致伸缩效应
volume element 卷宗元素;积分元素;体积元;体积单位
volume emission 容积发射;体积发射
volume emission rate 体积发射率
volume emissivity 体发射率
volume energy 体积能
volume entry 卷登记项
volume excitation 容积激发
volume exemption 容积免检
volume expander 音容控制器;声量扩展器
volume expansion 音量扩大;体胀率;体积膨胀;体积扩胀;声量扩展
volume expansion coefficient 体(积膨)胀系数
volume expansion coefficient of resilience 体积膨胀回弹系数
volume expansion meter 体积膨胀计
volume factor 体积因子;体积因素;体积系数
volume factor of crude at the present pressure 目前压力下原油体积系数
volume factor of crude at the primary pressure 原始压力下原油体积系数
volume factor of formation water 地层水体积系数
volume filling 容量装填法
volume fissure ratio 体积裂隙率
volume flask contraction of shotcrete (用长颈瓶测定的)水泥浆体积收缩
volume flocculation 容积絮凝
volume flow 容积流量;体积流量
volume flowmeter 容积流量计
volume flow rate 容积流率;体积流率
volume flux 体积通量
volume foam 浮石
volume for information interchange 信息交换存储媒体
volume formal 体积克式浓度
volume formality 体积克式浓度
volume fraction 容积分数;体积分数;体积分率;体积百分率
volume fraction of aggregate 集料体积分数
volume fraction of solid 固体体积比
volume free energy 体积自由能
volume frequency curve 容积频率曲线
volume gain 容积增加;体积增加
volume ga(u)ge 容积计
volume governor 容积调节器
volume graph 体积图
volume header label 卷宗头标
volume hologram 体积全息图
volume hologram memory 体(积)全息(图)存储器
volume holographic(al) element 体积全息元件
volume implant (放射源)体积植入
volume increase 体积膨胀
volume increment 材积生长
volume index 卷(末)索引;容积指数;物量指数;体积指数;体积指标
volume index of capital investment 投资物量指数
volume indicator 音量表;容积指示器;声量指示器
volume information 卷宗信息
volume initialization 卷的初始化
volume in-situ 下方【疏】
volume inspection 体积检查
volume insulation leakage 体绝缘漏泄
volume integral 容积积分;体积积分
volume integration method 体积积分法
volume in volume 体积百分数
volume ion density 体离子密度
volume ionization 体积电离度;体电离
volume ionization density 体电离密度
volume karst factor 体积岩溶率
volume label 卷宗标记;卷标签;卷标(号)
volume layout 卷布局
volume leakage 绝缘泄漏;绝缘漏泄

volume length mean diameter 体积长度平均径
volume level 强度级;体积级;响度级
volume lifetime 体积寿命
volume limit 容积极限
volume limiter 响度限制器
volume loading 容积负荷
volume loading of aeration tank 曝气池污泥负荷
volume loading rate 容积负荷率
volume loading rate in aeration tank 曝气池污泥负荷率
volume location database 卷定位数据库
volume loss 容积损失
volume magnetizing coefficient 体积磁化系数
volume magnetostriction 体积磁致伸缩
volume magnetostrictive effect 体积磁致伸缩效应
volume mark 卷标记
volume-mass relationship 容积质量关系
volume-mass relationship for sludge 污泥体积与含量关系
volume mean diameter 体积平均径
volume mean diameter of particle 颗粒体积平均(直)径
volume measurement 容积测量;体积计量
volume meter 量器;容积计;体积计;风量计
volume method 体积(估价)法;体积估算法;体积比配料法;体积(比)配合法
volume method of concrete mix 混凝土配合比体积设计法
volume method of concrete mix design 混凝土体积比设计法;混凝土按体积配合法
volume method of estimating cost 体积估算法(一种用体积毛估建筑造价的方法)
volume migration 迁移人口总数;迁移量
volume mix 容积混合;体积比配料法;体积比配合法
volume mix method 容积配料法
volume modulus 体积模量
volume modulus of elasticity 体积弹性模数;体积弹性模量;弹性容积模数
volume mo(u)lded per shot 一次压注容量;一次压注模制的体积量
volume name 卷名
volumenometer 容积计;排水容积计;体积计;视密度计
volumenometer method 容重测定法
volumenometry 折量法
voluntary geometry 体积几何
volume number 卷号
volume of activation 活化量
volume of air 空气体积
volume of air passengers 货物体积
volume of arrived goods 货物到达量
volume of bailer 提捞倘容积
volume of blast 鼓风量
volume of boiler 锅炉容积
volume of building 建筑体积
volume of business 营业额;业务(容)量;交易量;成交额
volume of cargo 货物量;船货体积
volume of cargo handled at port 港口货物吞吐量
volume of cargo in storage 货物堆存量
volume of cargo transferred 货物装卸量
volume of cargo transferred at berth 泊位作业量
volume of compartment 舱室容积
volume of construction 施工量
volume of continue carrying freight 货物接运量
volume of continue carrying passengers 旅客接运量
volume of corrosion solution 溶蚀液体体积
volume of current 电流容积
volume of cut(ting) 挖方量
volume of cylinder 缸容积
volume of dam 坝体积
volume of depression cone 降落漏斗体积
volume of digestion chamber 消化室容积
volume of discharge 流量
volume of displacement 排水(容)量;排水容积;排水体积
volume of dissocial chamber 游离室容积
volume of domestic wastewater 生活污水量
volume of dredging 疏浚量
volume of earthwork 土方(工程)量
volume of earthwork for ground level(l)ing 场地整平土方量
volume of ebb 退潮水量;落潮主槽;落潮量

volume of ebb tide 落潮总量
volume of equalization basin 调节池容积
volume of equipment 设备容积
volume of excavation 挖方体积;挖方量
volume of excavation work 挖方工程量
volume of expansion 声量扩大
volume of exports 出口量
volume of filling 填方工程量
volume of flood 涨潮(水)量;进潮总量;洪水总量
volume of flood tide 涨潮量
volume of flow 流量;径流总量;水流总量;水流流量
volume of freight handled 吞吐量
volume of freight traffic 货运量;货物运输量
volume of garbage 垃圾量;垃圾体积
volume of goods flow 货流量
volume of goods moved 货物运输量
volume of goods traffic 货运量
volume of goods transport 货物运输量
volume of gravel stuffing 填砾体积
volume of groundwater in storage 地下水储量
volume of hole radius 孔半径体积
volume of ice 冰量
volume of ice cone 冰锥体积
volume of immature source rock 未成熟烃源岩的体积
volume of inclusion 包裹体体积
volume of incoming and outcoming freights 吞吐量
volume of intergranular space 粒间孔隙量
volume of loading and discharging of ports 港口装卸量
volume of marketable goods 可供销售的商品数量
volume of mature source rock 成熟烃源岩的体积
volume of movement 运送量
volume of oil and gas bed 油气层容积
volume of ore block 矿块体积
volume of orebody 矿体体积
volume of organic matter 有机物容量
volume of output 产量
volume of overmature source rock 过成熟烃源岩的体积
volume of parcel dispatch and arrival and transshipment 行包发送、到达、中转量
volume of passenger flow 客流量
volume of passenger traffic 旅客周转量
volume of passenger transported 客运量
volume of pollutant 大量污染物
volume of pondage 调节容量
volume of pores 孔隙体积;孔隙量
volume of potassium absorption 吸钾量
volume of premium 保费数额;保费量
volume of production 生产量;产量
volume of purchase 收购量
volume of rail freight 货运量
volume of receipts 收入量
volume of recharge water 补给水量体积
volume of reservoir 库容;水库容量;水库容积
volume of river flow 河川径流总量
volume of river runoff 河川径流总量
volume of runoff 径流(总)量
volume of sales 销售量
volume of scintillation chamber 闪烁室体积
volume of sediment after compaction 沉积物压实后体积
volume of sedimentary rock 沉积岩体积
volume of sediment before compaction 沉积物压实前体积
volume of sending and carrying passengers 旅客运送量
volume of sending goods 货物发送量
volume of sending passengers 旅客发送量
volume of sending passengers by railway bureau 铁路局旅客发送量
volume of sending passengers by the railway 全路旅客发送量
volume of soil cave 土洞体积
volume of soil stripped 剥土体积
volume of solids 固体容积;土粒体积
volume of solution per unit time 单位时间溶液体积
volume of sound 声量
volume of source rock 烃源岩的体积
volume of spoil 弃土量
volume of stocks 储备量

volume of storm 暴雨(总)量
volume of streamflow 河水流量;河川径流总量
volume of stroke 缸的有效容积
volume of surface runoff 地面径流总量
volume of tank 箱容量;罐容量
volume of tapering 楔角量
volume of the circular flow 周转量
volume of the mineral substance (岩石中)矿物质体积
volume of the pore space 孔隙总体积
volume of the solid substance 土壤颗粒实有体积
volume of the water injected 压入水量
volume of timber 材积
volume of total 总体积
volume of total investment 投资总规模;投资总额
volume of trade 交易量;商业量
volume of trade bills 贸易汇票额
volume of traffic 运转量;交通量
volume of transaction 成交额
volume of transfer 换乘量
volume of transport 运输量;交通运输量;货物运输量
volume of trap 圈闭容积
volume of unitized cargo 成组货物量
volume of unit to be evacuated 装置抽气量
volume of ventilation air 通风量
volume of vessel 容器容量
volume of void in soil 土中孔隙体积
volume of voids 孔隙体积;空隙体积
volume of volcanic products 喷出物体积
volume of water 涨潮流量;水量
volume of water discharging on the ebb tide 退潮水量
volume of water entering on the flood tide 涨潮水量;进潮量
volume of water in soil 土中水的体积
volume of wood 材积
volume of work 工作量
volume of world exports 世界输出总量
volume of world trade 世界贸易量
volume on board 装舱方量;上方(指船放泥量);船方【疏】
volume percent(age) 容量百分比;容积百分数;容积百分率;容积百分比;体积百分比;体积百分数;体积百分率
volume percentage content 体积百分含量
volume percentage of fiber[fibre] 纤维体积率
volume percent of solids 固体体积百分率
volume per unit of depth 岩屑比;单位进尺的(岩粉)体积
volume photoconductivity 体积光电导性
volume photoeffect 体积光电效应
volume photoelectric(al) effect 体积光电效应
volume pit (测定流量用的)容积量斗
volume polarization 体极化
volume preparation 成卷准备
volume-preserving mapping 保体积映射
volume produce 批量生产;大量生产;成批生产;材积收获
volume-produced 大量生产的
volume production 批量生产;大量生产;成批生产
volume-production concrete mixer 大容量混凝土搅拌机
volume profile 流量断面图
volume pump 计量泵;定量泵
volume quota 定量配额
volume range 音程;声量范围
volume rate 运费率
volume rate of discharge 体积排放流量
volume rate of flow 流量;容积流率;体积流量
volume rating of flow 流量;容积流率
volume ratio 容积比(率);体积比率
volume ratio of truck to dipper 车铲容积比
volume receptor 容量感受器
volume recombination 体积复合
volume recombination rate 体积复合速率
volume recorder 声音自记器
volume recovery 容积回收率;体积回收率
volume reduction 容积减小;体积减小
volume reduction of solid wastes 固体废物减容
volume regulation 材积调节法
volume regulator 增益调节器;体积随变生物;体积调节器
volume resistance 体积阻力;体(积)电阻

volume resistivity 体(积)电阻系数;体(积)电阻率
volume reverberation 体积混响
volume scattering 体积散射
volume scattering coefficient 体积散射系数
volume scattering function 体积散射函数
volume scattering strength 体积散射强度
volumescope 容积测定仪;气体体积计;体积计
volume selector 容量选择器
volume sequence number 卷序号
volume serial number 卷(宗)序列号;卷顺序号;卷连续号
volume set 卷宗集合
volume shape factor 体积形状因子;体积形状系数
volume shrinkage 容积收缩;体积收缩
volume shrinkage mass 体积收缩量
volume solid 固体含量;体积固体分;含固量
volume sound range 声量范围
volume specific heat 容积比热(值)
volume specific thrust 容积比推力
volume-speed-density surface 流量—车速—密度曲面
volume stability 体积稳定(性)
volume stability under high temperature 高温体积稳定
volume stabilization 体积稳定(性)
volume state 卷宗状态;(矿物的)晶胞体积
volume sterilization 体积消毒
volume strain 体积应变
volume strain ga(u)ge 体积应变计
volume structure 体结构
volume surface mean diameter 体积面积平均径
volume-surface ratio 体积与表面比
volume surrounding 体外层
volume susceptibility 体积敏感度;体积磁化率
volume swelling 体积溶胀
volume switch 音量开关
volume switch procedure 卷开关过程
volume symbol 立体符号
volume table 材积表
volume table of contents 卷目录表
volume tank 小容积罐
volume target 体积目标
volume temperature coefficient 体积温度系数
volumeter 容量表;容量计;容积计;体积计
volume test 卷(宗)测试;容量测试;电路制度;大量(数据)试验;大量(数据)检验
volume theory 体积原理;体积理论
volume ton 容积吨
volume-to-surface-area ratio 体积表面积比
volume transport 体积运输
volumetric(al) 容量的;容积的;体积的
volumetric(al) absorption 体积比吸湿率;体积比吸收量
volumetric(al) analysis 容量分析(法);容积分析;体积分析;滴定分析(法)
volumetric(al) analysis method 体积分析法
volumetric(al) analysis test 容量分析试验
volumetric(al) apparatus 容量装置;容量器皿;容量(分析)仪器;测容量仪器
volumetric(al) base 按体积
volumetric(al) batch box 体积比配料箱;体积投配箱;体积配料比;体积分批箱;按体积配料斗
volumetric(al) batcher 容积配料斗;体积比配料箱;体积配料斗;按体积(比)配料器;按体积比配料斗
volumetric(al) batching 按体积比投配
volumetric(al) batching equipment 体积投配设备;按体积投配设备
volumetric(al) behavio(u)r 容积性质;体积状态
volumetric(al) blender 定容混合器
volumetric(al) box 按体积比配料斗
volumetric(al) capacity 容(积)量
volumetric(al) change 容积变化;容积变更;体积变化
volumetric(al) chromatography 体积色谱法
volumetric(al) coefficient 容积系数;填充系数;体积系数
volumetric(al) coefficient of water absorption 体积吸水率
volumetric(al) composition 容积组成
volumetric(al) compressibility 体积压缩性
volumetric(al) concentration 体积浓度
volumetric(al) concentration percentage 容积浓度百分率

volumetric(al) content 容积;体积含量
volumetric(al) content of system 系统容积
volumetric(al) contraction 体积收缩
volumetric(al) control speed 容积调速
volumetric(al) correction factor 容量校准因素
volumetric(al) cylinder 量筒;容筒
volumetric(al) deformation 体积变形
volumetric(al) determination 容量分析测定法;容量测定;体积测定
volumetric(al) dewater method 体积排水法
volumetric(al) diameter 容积直径
volumetric(al) displacement 体积排量
volumetric(al) displacement meter 容量置换计;体积流量计
volumetric(al) district 容积区域
volumetric(al) effect 体积效应
volumetric(al) efficiency 强度比率;容量效率;容量系数;容积效率;体积效率;充气效率
volumetric(al) efficiency of mine air 矿井通风容积效率
volumetric(al) efficiency of pump 泵容量效率
volumetric(al) efficiency of vessel 船舶容积效率
volumetric(al) elasticity 容积弹性
volumetric(al) error 滴定误差
volumetric(al) estimation method 体积估计法
volumetric(al) expansion 体积膨胀
volumetric(al) expansion coefficient 体(积膨)胀系数
volumetric(al) expansion of soil 土的体胀
volumetric(al) factor 容量因数;容积系数;填充系数;体积系数
volumetric(al) feed 容积计量喂料
volumetric(al) feeder 容量送料器;容积计量喂料机;定体积给料器
volumetric(al) filler 容量装填机;容量注入机;定量填装机
volumetric(al) filling device 容量装填装置
volumetric(al) filling machine 容量装填机
volumetric(al) flame spread 体积火焰扩散
volumetric(al) flask 量瓶;容量瓶;比重瓶
volumetric(al) flow 容积流量;体积流动
volumetric(al) flow meter 容积式流量计
volumetric(al) flow rate 容积流率;体积流量
volumetric(al) fragmentation of rock 岩石体积破碎
volumetric(al) glass(ware) 量器;量杯;玻璃量器;容量玻璃器皿;(有容量刻度的玻璃器皿)玻璃量具
volumetric(al) governor 容积控制器;流量调节器
volumetric(al) grains 等体积颗粒
volumetric(al) heat capacity 容积热容量;体积热容(量)
volumetric(al) heat exchanger 容积室换热器
volumetric(al) heat release rate 容积放热率
volumetric(al) liquid filling machine 液体计量灌注器
volumetric(al) loading 容量负荷;容积负荷
volumetric(al) loading rate 容积负荷率
volumetric(al) loss 容积损失
volumetrically stable 容积稳定的
volumetric(al) measurement 容积测量;容积测定;按容积测量
volumetric(al) measurement of discharge 容积法测流;容积法测流
volumetric(al) measuring device 容积计量装置
volumetric(al) meter 容积式煤气表;容积式流量表;容积式水表;容积式流量计
volumetric(al) method 体积比配料法;容量(测流)法;(测量混凝土空气量的)容积法;容积测流法;体积(配合)法;体积比配合法
volumetric(al) method of batching 体积配料法
volumetric(al) method of drainage flow per unit 单位排流容积法
volumetric(al) method of measuring discharge 容积法测流
volumetric(al) method of reservoir rock 储集层体积法
volumetric(al) method of sedimentary rock 沉积岩体积法
volumetric(al) metre 测容量仪器
volumetric(al) mixing 体积配合;体积混合
volumetric(al) modulus 容积模数;容积模量
volumetric(al) modulus of elasticity 体积弹性模数;体积弹性模量

volumetric(al) moisture content 体积含水率;体积含水量
volumetric(al) molar concentration 容量摩尔浓度;容量克分子浓度;体积摩尔浓度
volumetric(al) packaging 立体包装
volumetric(al) part 体积百分数
volumetric(al) performance 体积动态
volumetric(al) photoelectric(al) absorption index 体积光电吸收指数
volumetric(al) pigment concentration 体积颜料浓度
volumetric(al) pipet(te) 吸移管;移液吸(移)管;移液管;容量吸移管
volumetric(al) precipitation method 容量沉淀法;沉淀滴定法
volumetric(al) procedure 容量法
volumetric(al) production 体积产量
volumetric(al) proportion 容积比(率);体积比(率)
volumetric(al) proving 容积校准
volumetric(al) radar 立体雷达;空间显示多目标雷达;三度空间雷达
volumetric(al) rate 容积率
volumetric(al) rate of reaction coefficient 容量反应速度系数
volumetric(al) ratio 容积比(率);体积比
volumetric(al) reduction 体积减小
volumetric(al) refrigerating capacity 单位容积制冷量
volumetric(al) regulator 容积控制器;容积调节器
volumetric(al) sediment concentration scale 体积含沙量比例
volumetric(al) shrinkage 体缩率;体积收缩
volumetric(al) shrinkage limit 体积收缩极限
volumetric(al) shrinkage of oil 石油体积收缩率
volumetric(al) shrinkage rate 体缩率
volumetric(al) shrinkage test 体积收缩试验
volumetric(al) solids 体积固体
volumetric(al) solution 滴定(用)液
volumetric(al) specific gravity at air dried state 风干重度
volumetric(al) specific heat 体积比热
volumetric(al) standard 标准容器;容量标准
volumetric(al) state of stress 体积应力状态
volumetric(al) storage 容积储[贮]存量
volumetric(al) strain 容积应变;体积应变
volumetric(al) strain energy 体积应变能
volumetric(al) stress 容积应力;体积应力
volumetric(al) swell 体积膨胀
volumetric(al) theory 体积学说
volumetric(al) thermal expansibility 体积热膨胀率
volumetric(al) titrimetry 容量滴定法
volumetric(al) ton 容积吨;体积吨
volumetric(al) tonnage 容积吨位
volumetric(al) type diaphragm pump 容量型隔膜泵
volumetric(al) water content 体积含水量
volumetric(al) water ga(u)ge 容积式水表
volumetric(al) water meter 容积式水表;体积式水表
volumetric(al) water pump 容积式水泵
volumetric(al) weight 重度
volumetry 容量分析(法);容量法;容量测定
volumette 重复定量滴管
volume unit 容积单位;体积单元;体积单位;声量单位;单位体积
volume unit indicator 响度单位计
volume unit meter 声量计
volume vacaney diffusion 体积空位扩散
volume variance 产量差异
volume velocity 体积速度
volume viscoelasticity 体积黏弹性
volume viscosity 体积黏度
volume voltameter 容积电量计;体积电量计
volume-volume concentration factor 体积—体积浓缩系数
volume water absorption 体积吸水
volume weight 重度;容积重量;容积比重
volume weight by large-size sample 大体重
volume weight by small size sample 小体重
volume weight of ore 矿石体重
volume weight of soil 土壤重度;土的重度
volume weight of wood 木材重度
volume weight ratio 体积重量比

volume weight tube 体重管
volume withdrawal 体积采出量
volume yield 体积得率;体积产率
voluminal compressibility 体积压缩性
voluminal space filling factor 体积充满度
voluminosity 宏大;容量度;容积度;盘绕
voluminous 盘绕
voluminous powder 多孔粉末
volumometer 排水容积计;容量计;体积计
voluntary alienation 自愿转让
voluntary appropriation 任意指拨;任意拨款
voluntary appropriation of fixed assets 固定资产无偿转拨
voluntary arbitration 自愿仲裁;任意仲裁
voluntary assistance program(me) 自愿援助计划;自愿援助方案
voluntary association 民间社团
voluntary bankruptcy 自愿申请破产;自行申请破产;申请破产
voluntary ceiling 自动限额
voluntary chain 自愿联合公司;自愿连锁系统
voluntary compliance agreement 自愿(性)协议
voluntary contribution 自愿贡献
voluntary conveyance 无偿转让;自愿转让;无偿技术转与
voluntary cooperation 自愿合作
voluntary depreciation method 任意折旧法
voluntary export restraint 出口自动限制
voluntary export restriction 出口自动限制
voluntary grantee 无偿受让人
voluntary indemnity 自动赔偿
voluntary insurance 自愿保险;自动保险;任意保险
voluntary interrupt 随意中断
voluntary lien 自愿留置权
voluntary liquidation 自愿清理
voluntary loss 故意损失
voluntary movement 随意运动
voluntary observing ship 自愿观测船
voluntary payment 自动支付;提前偿付
voluntary prepayment 提前偿还
voluntary reserves 任意准备
voluntary restraint 自愿约束
voluntary restriction of export 自动限制出口
voluntary salvage 自愿救助
voluntary salvaging ship 自愿救助船
voluntary savings 自愿储蓄
voluntary security service 居民联防
voluntary societies 自愿组织
voluntary stranding 自愿搁浅;自动搁浅;抢滩
voluntary system of certification 自愿认证制
voluntary tax 自愿税收
voluntary transfer 自动转让
voluntary trust 自由信托
voluntary waste (对他人不动产的)故意损害
voluntary withdrawal 自动撤销
volute 螺旋形;螺旋饰品;集气环;盘涡形;蜗壳;涡饰;涡形(花样);涡螺;涡卷的;扶手涡旋饰
volute angle 蜗壳角
volute capital 旋涡形柱顶;旋涡形柱头
volute casing 涡壳;线形外壳;蜗形机壳;蜗线形外壳
volute casing cover 联结蜗壳底盘
volute chamber 螺旋室;环流室;蜗壳(腔);涡室
volute collector 蜗壳式收集器
voluted 旋涡形的
volute diffuser 蜗壳扩压器
volute gable 旋涡形山墙饰
volute helix 旋涡形线饰
volute ornament 旋涡形装饰
volute pump 螺旋(形水)泵;蜗壳泵;涡轮泵
volute siphon 涡流虹吸
volute spring 锥形弹簧;螺旋(形)弹簧;涡旋(形)弹簧
volute spring compressor stud 锥形簧压缩柱螺栓
volute syphon 涡流虹吸
volute throat 蜗壳喉部
volute tongue 蜗壳舌部
volute type 螺旋式
volute type air classifier 螺旋式风力分级机
volute with casement 楼梯扶手螺旋形弯头;扶手涡旋饰
volution 螺旋形
volvele 日月升落潮汐仪
Volvit 佛尔维特合金
volyer 网兜渔船

volynskite 碲铋银矿
vomax 电子管电压表
vomitive 大门
vomitorium 运动场看台入口；剧场看台入口；(剧场或运动场的)出入口
vomitory 看台入口；运动场入口；剧场入口
vomitus marinus 航海呕吐
vommon variable 公用变量
Von Arx current meter 冯阿克斯流速计
Von der Wehl steamer 冯丹士林蒸箱
Vondrek smoothing method 冯特克平滑法
Von Hofe eyepiece 冯霍夫目镜
Von Hofe modified eyepiece 冯霍夫改进型目镜
Von Karman constant 冯卡门常数
Von Mises yield criterion 米塞斯屈服准则
Von Neumann architecture 冯诺伊曼系统结构
Von Neumann ratio 冯诺伊曼比率
Von Zeipel method 冯蔡佩尔方法
voog 晶洞
Vorarlberg school 富拉伯格建筑格式
Vorce diaphragm cell 伏尔斯隔膜电解槽
vor refining 真空氧化精炼
vortair surface aeration 表面旋气式曝气
vortair surface aerator 表面旋气式曝气器
vortex 漩涡；旋涡；涡动；湍流漩涡
vortex action 旋涡作用；涡流作用
vortex adjustable centrifugal fan 涡流可调离心风机
vortex agitator 旋涡搅拌器；涡动搅动器；涡动搅拌器
vortex air cleaner 涡流空气滤净器
vortex air diffuser 旋涡式空气散流器
vortex amplifier 旋涡放大器
vortex axis 旋涡轴线
vortex band 涡流带；涡流层
vortex box 旋涡净气器
vortex breakdown 涡旋破碎
vortex breaker 旋涡破坏器
vortex burner 旋风燃烧器
vortex cage meter 涡流笼式流量计
vortex cast 旋涡铸型；旋涡模
vortex cavitation 涡流空蚀
vortex cavity 涡流区
vortex cell 旋流窝
vortex chamber 旋涡室
vortex-chamber grinding of cement clinker 水泥熟料的转桶研磨
vortex circulation and recirculation 涡流反复循环
vortex classifier 涡流分粒器
vortex cleaner 涡流除渣器
vortex cloud system 涡旋云系
vortex combustion chamber 旋涡式燃烧室
vortex cone 旋锥体
vortex conveyer dryer 旋风气流干燥器
vortex core 旋涡中心；涡流束；涡核
vortex crushing 回旋破碎
vortex density 涡旋密度；涡流密度
vortex distribution method 涡旋分布法
vortex electrocoagulation-flo(a)tation-contact filtration process 涡流电凝聚—气浮—接触过滤组合工艺
vortex equation 涡流方程
vortex excitation 涡流激励
vortex-excited oscillation 涡流激动振荡
vortex field 涡旋场；涡流场
vortex filament 涡(旋)线；涡(旋)丝；涡流流线
vortex finder 涡形定向器
vortex firing system 涡流式燃烧系统
vortex flow 涡流；涡旋流动
vortex flow grit 旋流沉砂池
vortex flow jet mixer 旋流混浆器
vortex flowmeter 漩涡流量计
vortex fluid amplifier 涡流型放大元件
vortex flux 涡通量
vortex-free discharge 无旋涡水流
vortex-free flow 无旋(涡)流
vortex-free motion 无旋涡流动；无旋(涡)运动
vortex gate 整流栅
vortex generator 旋涡发生器；扰流器
vortex grit separator 涡流砂粒分离器
vortex grit washer 涡流洗砂器
vortex hypothesis 涡流假说
vortex-induced vibration 涡致振动；涡列引起的振动
vortex intensity 旋涡强度；涡流强度

vortex interference 涡流干扰
vortex-interference effect 涡流干扰效应
vortex lattice 涡流栅
vortex-like flow mark 涡状流痕
vortex line 旋涡线；涡线
vortex loss 涡流损失
vortex melting furnace 涡流熔化炉
vortex meter 旋涡流量计
vortex motion 涡流运动；涡动
vortex of slurry 料浆搅拌
vortex pair 旋涡偶；涡对；涡动副
vortex path 涡流轨迹
vortex pattern 涡流图
vortex periodic 周期旋涡
vortex pipe 涡形管
vortex piping 涡形管
vortex plane rolling-up 涡面卷起
vortex point 旋涡点
vortex pot 涡斗
vortex precession flow meter 旋进流量计
vortex pump 旋涡泵；漩流泵；涡动泵
vortex rain 涡旋雨
vortex reactor 涡动反应器
vortex ring 环形涡流；涡轮；涡(流)环
vortex-ring cascade 涡流环叶栅
vortex row 涡列
vortex scrubber 冲击式除尘器
vortex separation 涡流分离
vortex-shaped discharge pattern 涡流型卸料方式
vortex shedding 旋涡区；旋涡尾迹；涡流效应；旋涡分离；旋涡脱落；涡流脱落；涡流发散
vortex shedding flowmeter 涡流式流量计
vortex shedding frequency 旋涡分离频率
vortex shedding mechanism 旋风流动机构
vortex sheet 涡面；涡流片；涡层
vortex-siphon spillway 漩涡虹吸溢洪道
vortex sound 涡流声
vortex-stabilized arc 涡流稳定弧
vortex streak 旋涡条纹
vortex street 涡漩迹；涡列；涡街；涡道
vortex street stability 涡道稳定性
vortex strength 涡强；涡流强度
vortex strip 涡流带
vortex structure 旋涡构造；旋卷构造
vortex structure chain 连环式旋扭构造
vortex suppressing basin 消涡池
vortex surface 涡面
vortex theory 涡流理论
vortex thermometer 涡流温度计；涡管温度表
vortex time 涡旋时间
vortex trail 漩涡尾
vortex train 涡列
vortex trunk 漩涡螺栓
vortex tube 涡(形)管；涡流管
vortex tube ejector 涡流管喷射器
vortex tube heat-protective clothing 旋风制冷防热服
vortex tubing 涡形管
vortex type flow 涡流
vortex value 涡动值
vortex vector 旋涡矢量
vortex viscosity 涡黏度
vortex zone 涡流区；旋流区
vortical activator of cement 旋涡式水泥活化器
vortical flow 涡流
vortical motion 旋涡运动；涡旋运动
vorticity 旋度；涡旋状态；旋涡强度；涡流强度；涡量；涡度；涡动性
vorticity center 涡度中心
vorticity equation 涡流方程；涡量方程；涡度方程
vorticity field 涡场
vorticity kinematics 涡流运动学
vorticity motion 涡旋运动
vorticity source 涡源
vorticity transfer 涡旋转移；涡度输送
vorticity transport 涡旋转移；涡旋传递
vorticity-transport hypothesis 涡量输运假设
vorticity transport theory 涡度输送理论
vortrap 旋流分级器
Vorwohler cement 福尔沃水泥
Voss connector 伏斯接缝销钉
Voss's polariscope 伏斯偏振镜
Vostochy Port 东方港(俄罗斯)

vostro account 来户；你方账户(国际汇兑)
votage hunting 电压摆动
votage pencil 电压笔
votator 螺旋式热交换器；螺旋式换热器
vote 投票
vote in the negative 反对票
voter assessment 投标评定
voting 选举；投票
voting circuit 表决电路
voting element 表决元件
voting logic 阈值逻辑
voting power 投票权
voting right 投票权
voting selector 表决选择器
voting table 投票信托
voting trust 投票信托
voting trust certificate 股权信托证
votive stupa-mound 奉献浮屠
votive tope-mound 奉献浮屠
voucher 转账单；证书；证人；会计凭证；凭证；凭单；收条；收据；担保人
voucher account 凭单账户
voucher audit 凭单审核
voucher check 凭单支票
voucher check system 凭单支票制
voucher deduction 凭单扣款
vouchered invoice 已核准发票
voucher for disbursement 支出凭单
voucher for receipt 收入凭单
voucher index 凭单索引
voucher jacket 凭单夹
voucher journal 凭单日记账
voucher of clearing 结算凭证
voucher payable 应付凭单
voucher payable account 应付凭单账户
voucher payable register 应付凭单登记簿
voucher record 凭单记录
voucher register 应付凭单登记簿；凭单登记簿
vouchers attached 凭证附后
voucher schemes 凭证制度
vouchers for business receipts 营业收款凭证
voucher system 应付凭单制；付款凭单制；凭单制度
vouching 以付款凭单核对证明
vouching account 以付款凭单(核对)证明的账户
vouching procedure 核付程序
voussoir 楔形拱石；楔块；拱楔块；拱石楔块；拱石；砌拱楔块
voussoir arch 楔形拱；楔块拱；砖石砌拱；砌块拱；分块拱
Voussoir beam 伏萨尔梁(岩石压力试验用)
voussoir brick 楔(形)砖；楔形拱砖；拱用砖；拱楔砖
voussoir joint 拱形砌缝；拱块(砌)缝
voussoir key 楔形键；拱斜石；拱楔块键
voussoir stone 楔形石
voyage 航行；航海家；航次(指船)；航程
voyage aboard 出国旅行
voyage account 航次账单
voyage charter 租船航次；租船按航次计费；航租；航次租船；航次期租；定程租船；程租；承租契约
voyage charter party 航次租船合同
voyage clause 航次条款；航程条款
voyage in stages 航程分阶段
voyage insurance 航次保险；航程保险
voyage letter 航次报告
voyage number 航次数；航次(指船)
voyage period 航行周期
voyage policy 航次保险单；航程保险单
voyage repair 航次修理
voyage report 航次报告
voyage round the world 环球航行
voyage schedule 航期表
voyal 锚索拉绳
voycharter 定程租船
voyol 锚索拉绳
vozhminite 砷硫砷镍矿
V-pad 三角垫
V-panel V形镶板
V-port valve 带(有)V形柱塞的阀
V-process 薄膜负压铸造
vrbaite 硫砷锑铊矿；硫砷汞铊矿
VR ditch VR式沟
V-rest V形支架
V-roof 内坡屋顶；蝶形屋顶；V形屋面；V形屋顶
vroom 加速时发出的声音

V-rope V形钢索
V-section 三角槽形断面
V shape 叉式
V-shape corrugated plate V形波纹板片
V-shaped V形的
V-shaped arch 尖形弓
V-shaped column V形柱
V-shaped dyke 鱼嘴(治鱼工程中的)
V-shaped ditch V形边沟
V-shaped guide 棱柱形导轨;三面导槽
V-shaped joint V形勾缝;V形接头;V形缝
V-shaped pier V形墩
V-shaped reinforcement V形钢筋
V-shaped roof V形屋顶
V-shaped sweep 箭形锄铲
V-shaped valley V形谷
vsicosity-temperature modulus 黏(一)温模数
V-slot V形槽
vsotskite 硫钯矿
V-stiffened wire lath V形加劲肋钢丝网条板
V-strap V形带钢
V-stripper V形卸料器
V-suffix chain 非多脉冲链
V-sweep V形铲
V-thread 管螺纹;三角螺纹;V形螺纹
V-thread screw 三角形螺旋;V形螺旋
VTMS 船舶交通管理系统
V-tool 三角凿刀;V形工具
V-tooled joint V形凿缝;三角凿嵌缝
V-trap 虹吸管
V-type V形的
V-type diaphragm pump V形隔膜泵
V-type fan belt V形风扇皮带
V-type mixer V形混合机
V-type motor V形发动机
V-type notch V形缺口
V-type refractometer V形折射仪
V-type ripper V形松土机;V形耙土机
V-type snow plough V形雪铧犁
V-type step pulley V形锥轮;V形塔轮
V-type thread 六十度V形螺纹;V形螺纹
vuagnatite 羟硅铝钙石
vug(g)(hole) (岩石中的)空穴
vuggy rock 多孔岩(石);多孔石
vulanerable spot 薄弱环节
vulcalose 同硬橡皮一样的绝缘材料
Vulcan 桩锤
Vulcan coupling 弗尔康联轴节

vulcanian eruption 火山喷发
vulcanic 火山作用的
vulcanicity 硬化作用;硫化作用;火山作用;火山现象
vulcanisate 硫化橡胶;硫化产品
vulcanised synthetic(al) rubber 硫化合成橡胶
vulcanism 火山现象
vulcanite 胶木;硬橡皮;硬橡胶(皮);硬化橡胶;硫化橡胶;碲铜矿
vulcanite roofing 用硬质弹性沥青片铺盖屋面;硬橡胶屋面
vulcanizable material 硫化材料
vulcanization 热补轮胎;橡胶硫化作用;硬化作用;硫化(作用)
vulcanization accelerator 硫化促进剂
vulcanization activator 硫化促进剂
vulcanization coefficient 硫化系数
vulcanization machine with side plate 侧板式硫化机
vulcanization of rubber 橡胶硫化
vulcanization reaction 硫化反应
vulcanization retarder 硫化延迟剂
vulcanizator 硫化剂
vulcanize 使硬化;使硫化
vulcanized 硬化的;硫化的
vulcanized agent 硫化剂
vulcanized butyl rubber 硫化异丁烯橡胶
vulcanized cable 硫化橡胶电缆
vulcanized fiber 钢纸;硫化纤维;硬化纸板;硬化纤维(板);硬化绝缘板
vulcanized latex 硫化胶乳
vulcanized oil 硫化油
vulcanized paper 硬化纸板;钢纸
vulcanized rubber 硬橡胶;硫化橡胶
vulcanized rubber cable 硫化橡胶包皮电缆
vulcanized rubber insulated cable 硬橡胶绝缘电缆
vulcanized rubber sheet 硫化橡胶片材
vulcanized synthetic(al) rubber 硫化合成橡胶
vulcanizer 橡胶硫化器;硫化器;硫化剂;硫化机;硫化罐;热补机
vulcanizing 硫化(的)
vulcanizing agent 硬化剂;硫化剂
vulcanizing apparatus 硫化器
vulcanizing boiler 硫化罐
vulcanizing chamber 硫化室
vulcanizing heater 硫化锅;加热硫化机
vulcanizing machine 硫化机;补胎机
vulcanizing period 硫化时间
vulcanizing press 加压硫化机

vulcan jockey 带偏心夹的联接装置
vulcanorium (海洋中脊与岛弧之间的)过渡构造带
vulcan powder 伏尔甘炸药
vulgar establishment 平均朔望潮高潮间隙;朔望高潮间隙
vulgar fraction 普通分数
vulgarian 暴发户
vulgaris 寻常的
vulnerability 易损(坏)性;脆弱性
vulnerability analysis 易损性分析;弱点分析
vulnerability function 易损性函数
vulnerability of structure 结构易损性
vulnerability to accidents 易损伤性;防事故性
vulnerability to jamming 抗扰性
vulnerable 脆弱的
vulnerable accessories 易损附件
vulnerable area 薄弱部分
vulnerable cargo 易损货物
vulnerable component 脆弱部件
vulnerable parts 易损零件;易损件
vulnerable period 易损期;易颤期
vulnerable phase 易损期
vulnerable species 渐危种
vulnerable spot (堤防)险工;弱点;脆弱部位
vulnerable system 易损系统
vulnerable to pollution 易受污染损害的
vulnerable to sunlight 不耐阳光的
vulnus 创伤
vulpinite 鳞硬石膏
vulsellum forceps 双爪钳
vultex 硫化橡胶浆
vultite 混合防滑罩面材料(由沥青乳液、水泥、砂和水组成)
vuonnemite 磷硅铌钠石
vuorelainenite 沃钒锰矿
V-valley V形谷
V-value 光学玻璃色散值;V值
V-weld V形焊缝
vycol glass 硼硅酸耐热玻璃
vycor 高硅耐热玻璃;纯石英玻璃
vycor glass 硼硅酸耐热玻璃;维克玻璃;高硼硅酸盐耐热耐蚀玻璃;高硅氧玻璃;石英玻璃
vycor tube 高硅氧玻璃管
Vynitop 聚氯乙烯涂层钢板;涂聚氯乙烯钢板
vys(e) 螺旋(式)楼梯
vys(e) stair(case) 有中柱的螺旋式楼梯
vyuntspakchkite 羟硅铝钇石

W

W 306 alloy 铜锡锰系标准电阻丝合金
Waalian interglacial stage 瓦勒间冰期
Waalian warm epoch 瓦尔温暖期
wabble 行程不匀；陀螺式摇摆；变度
wabble pump 摆动式泵
wabbler 摇动头；摆动器
wabbler mechanism 摇动机构
wabble saw 偏心圆锯；摇摆圆锯
wabble shaft 滚转轴
wabbly wheel 摆动车轮
wacke 玄武岩风化物；玄武土；玄土；玄砂石；瓦克岩
wackestone 粒泥状灰岩；粒泥石灰岩
wad 泥条；（封匣钵口用）耐火泥条；锰土；填塞；填料；潮坪；潮成平地
wadboard packer 板式装填器
wad box 填装器；团块压制盒；手动挤泥条机
wad clay 锰土；氧化的锰土
wad cutter 靶孔；冲孔型弹丸
wadded 制花样线脚模
wadden island 沙洲岛
wadding 衬料；填塞；填料；填塞物
wadding card 絮棉梳理机
wadding filter 填塞过滤器；填料过滤器
wadding picks 填充纬
wadding strip 填塞条
wadding warp 填充经
wadding weft 填充纬
wade across a river 涉水过河
waded wagons in transit 接运通过重车
waded wagons received at junction station for unloading 接入自卸重车
wadeite 钙钾板锆石；硅锆钙钾石
Wadell's sphericity factor 沃德尔球形度系数
wader 岸禽类
Wadhurst clay 瓦德赫斯特黏土
wadi 干涸河床；干谷；旱谷；沙漠中的绿洲
wadi deposit 旱谷沉积
wadi facies 旱谷相
wading 涉水
wading bird 涉禽
wading depth 涉水深度
wading equipment 涉水量测设备
wading measurement 涉水测深
wading measurement of discharge 涉水测流
wading pool （供儿童玩乐的）浅水池；儿童浴池；涉水池
wading rod 浅水测深杆；涉水测深杆；测深竿；测深杆；测流杆
wadi system 干谷型渠系
wadite 锰土
wadi-type depression 干谷（型）洼地；干谷低地；旱谷型洼地
wadsleyite 瓦兹利石
Wadsworth mounting 沃兹沃思装置
Wadworth prism 瓦德乌斯棱镜
wady 干涸河床；旱谷；沙漠中的绿洲
wady system 干谷型渠系
wady type depression 干谷型洼地
Waechtler's gold purple 威奇特勒紫金色料
Waelz method 华尔兹回转窑烟化法
wafer 干胶片；极板；大晶片；薄片；半导体薄片
wafer board 华夫板；木屑板
wafer bonder 薄膜片接合器
wafer breaker 分片器
wafer butterfly valve 对夹式蝶阀
wafer channeltron image intensifier 薄片渠道像增强器
wafer core 易割冒口芯片
waferer 压块机
wafering hay 制干草块
wafer machine 压块机
wafer map 晶片图
wafer matrix 晶片矩阵
wafer press 草饼压制机
wafer prober 晶片探测器；晶片检测器
wafer process 晶片加工
wafers 华夫刨花
wafer scriber 圆片划线器；划片机

wafer separator 分片器；分片机
wafer socket 冲压管座；饼形管座
wafer switch 晶质开关；晶片开关；夹层开关
wafer test 晶片试验
wafer type silent check valve 薄片型消音防逆阀
waffle arm 张力调整臂
waffle ceiling 方格天花板；格子平顶
waffle design 格纹设计图案
waffle die 格状校平模
waffle effect 格纹防滑效应
waffle floor 华夫型楼板；格纹桥面；格纹楼面；双向密肋楼板；轻型混凝土楼板；（华夫饼干式的）格纹桥面；网格式楼盖；密肋式楼板
waffle floor panel 地板格纹镶板；楼板格纹镶板
waffle floor plate 地板格纹平板；楼板格纹平板
waffle floor slab 地板格纹平板；楼板格纹平板
waffle footing 格栅基础；密肋式基础；网格式基础
waffle grid 网格；格栅
waffle iron 对开式铁芯
waffle-like pattern 网格状图案
waffle panel floor 格纹镶板楼板
waffle pattern 格纹图案
waffle-shaped grid 华夫型格式板
waffle slab 密肋楼板；格纹楼板；双向密肋楼板
waffle-slab floor 格纹平板地板；井式楼盖；格纹平板楼板；华夫型楼板；沉箱式楼板
waffle slab form 双向密肋模板
waffle soffit 格纹挑檐底面
waffle type 华夫饼式（表面有小格子）；格纹式
waffle weave 蜂巢组织
W. African kino 西非吉纳树胶
waft 浮动
wafter 风扇；转盘风扇
wag 上下移动
wage accrual 本期应付工资
wage administration 工资管理
wage advance plan 预付工资
wage allocation sheet 工资分配表
wage and price control 工资和物价管制
wage and related expenses 工资附加费
wage and salary administration 工资管理
wage and salary disbursement 工资与薪金支付额
wage and salary in kind 实物工资
wage and salary payment 工资与薪金支付
wage arbitration 调停工资纠纷
wage assignment 工资指定
wage bill 工资单
wage bracket 工资等级
wage-bracket table （按工资计算的）简易税额表
wage-by-age system 年资工资制
wage by piece 计件工资
wage capital 可变资本
wage category 工资级别；工资类别；工资种类
wage ceiling 工资最高额
wage contract 工资合同
wage control 工资管理
wage cost 工资成本
wage curve 薪资曲线
wage cut 削减工资
wage cutting 工资削减
wage day 工资发放日；发薪日；发放工资日
wage deduction 工资扣除；扣发工资
wage determination 工资决定
wage differential 工资差别
wage differential method 工资差额法
wage disparity 工资差别
wage earner 工资劳动者；依靠工资为生者
wage earnings 工资收入
wage exponent 工资指数
wage fraction 工资率
wage freeze 工资冻结
wage funds expense statement 工资基金支出表
wage funds income statement 工资基金收入表
wage garnishment 扣发工资
wage growth 工资增长
wage incentive 鼓励工资；工资刺激；奖励工资
wage incentive plan 奖励工资计划
wage income 工资收入

wage increase 工资增长
wage index 工资指数
wage inflation 工资膨胀
wage in railway transport enterprise based on efficiency 铁路运输企业工效挂钩
wage kikes 提高工资
wage labo(u)r 雇佣劳动者；工资劳动者
wage ledger 工资分类账
wage level 薪金标准；工资水平
wage lien 工资留置权
wage packet 工资袋
wage pattern 标准工资等级
wage payable audit 应付工资审计
wage payroll 工资单
wage per hour 小时工资
wage plan 工资计划
wage ranking system 工资等级制度
wage rate 工资标准；工资率；工资单价
wage rate paid 实付工资率
wage rate variance 工资率差异
wager boat 竞赛艇
wager policy 信用保单
wages 薪水；薪金
wages and benefits 待遇（指物质报酬）
wages and salaries 工资及薪金
wages and salaries book 工资薪金账簿
wages book 工资账簿
wage scale 工资级差；工资等级表；工资等级；工资标准
wage scale system 工资级差别
wages clerk 工资员
wages department 工资部门
wages-earner 以工资为生者
wages for piece work 计件工资
wages fund 工资基金
wages fund management 工资基金管理
wages fund plan 工资基金计划
wage sheet 工资表
wages in cash 现金工资
wages office 工资部门
wages or salaries 基本工资
wages per day 日工资
wages summary 工资汇总表
wage standard 工资标准
wage structure 工资结构
wage surcharge 工资附加费
wage system 工资制（度）
wage system based on ability 能力工资制
wage system based on post technique in railway enterprise 铁路企业岗位技能工资制
wage variation clause 工资浮动条款
wage worker 工资劳动者
wagging vibration 垂直摇摆振动
wag(g)on 旅行车；篷车
wag(g)on drill 移动式钻机
Wagner earth device 华格纳接地装置
Wagner ground 华格纳接地
wagnerite 氟磷镁石
Wagner pot 瓦格纳盆
Wagner's alloy 瓦格纳锡基合金
Wagner's casting machine 瓦格纳金属型铸造机
Wagner'scode 瓦格纳码
Wagner's earth connection 瓦格纳接地线路
Wagner's earth device 瓦格纳接地装置（一对阻抗的连接地点）
Wagner's fineness 瓦格纳细度
Wagner specific surface 瓦格纳比表面积
Wagner's reagent 瓦格纳试剂
Wagner's solution 瓦格纳试剂；瓦格纳溶液
Wagner's theory 瓦格纳氏学说
Wagner's turbidimeter 瓦格纳（浑）浊度计（测水泥细度）
Wagner surface area 瓦格纳比表面积
wagon 运货车；货车
wagon accelerator 车辆加速器
wagonage 运货车；货车运输
wagon arrester 阻车器；停车器
wagon average kilometre per day 货车平均日车

公里
wagon axle 车轴;车辆轴
wagon balance 地磅;车辆秤
wagon boiler 车式锅炉
wagon booster-retarder 矿车速度控制器
wagon box 车厢
wagon-box dryer 车厢式干燥机
wagon-box rivet head 扁头形大直径铆钉头
wagon brake 货车制动器
wagon bridge 公路桥(梁)
wagon building channel 车架槽钢;造车槽钢
wagon carrier truck 汽车运载车
wagon carrying trailer 装货车的拖车
wagon ceiling 筒形天花板;筒形顶棚
wagon control computer 车辆控制计算机
wagon coverduck 车篷帆布
wagon crane 车载吊车;起重车;车辆起重机
wagon day 货车日
wagon density 货车密度
wagon depreciation expenses 货车折旧费
wagon drill 移动式钻机;钻探车;轮胎式钻机;汽车钻机;汽车式钻机;车钻;车装钻机
wagon drill frame 车载钻机;钻探车架
wagon dumper(hoist) 翻卸货车机
wagon elevator 装车升运器
wagonette drill 钻车
wagon flow route diverse 变更运行路径
wagon flow scheduling controller 计划调度
wagon flow table 车流表
wagon flow to be resorted 改编车流
wagon for bearing goods weight 负重车
wagon for spraying salt 撒盐车
wagon for the carriage of traffic in powder form 粉末运输货车
wagon for the transport of milk 牛奶运输车
wagon for the transport of spent nuclear fuel 核废料运载车
wagon gear 拖车底盘部分
wagon-group used to straddle goods 跨装车组
wagon-headed 筒形穹的;圆筒形的;拱形的
wagon-headed ceiling 筒形天花板;斜(顶形)天花板
wagon-headed dormer 斜顶形老虎窗;筒形老虎窗
wagon-headed vault 筒形拱顶;筒形屋顶;筒形穹顶;筒形长拱;斜顶拱顶
wagon hoist 吊货车机;车氟起重机
wagon hopper 底卸式载货车
wagon jack 车辆起重机
wagon kilometer 货车(走行)公里
wagon kilometers per day 货车日公里
wagon-lit 卧车(铁路)
wagon load 货车荷载
wagon loaded with out-of-ga(u)ge goods 超限车
wagon-load goods loaded or unloaded not at goods station 途中装卸
wagon-load goods unloaded at two or more stations 整车分卸
wagon loading capacity utilization rate 货车载重力利用率
wagon loading elevator 货车装车升运器
wagon loading equipment 货车装载设备;装车设施
wagon loading list 货车装载清单
wagon loadings 装车数
wagon loadings not completed in monthly transport plan 落空车数
wagon loading tower 货车装载塔架
wagon-mounted air drill 车载风钻
wagon-mounted drill 钻车;凿岩台车;车钻
wagon number 车号
wagon operation times 货车作业次数
wagon plate 车厢钢板;筒形钢板
wagon postponed use charges 货车延期使用费
wagon requisition plan table 要车计划表
wagon retarder 车辆减速器;车辆缓冲器[铁]
wagon road 货车通路;行车道;车通道;货运道路
wagon roof 桁架椽子屋顶;斜顶形屋顶;拱形屋面;筒形屋顶
wagon scale 货车轨道衡
wagon-scraper 机动铲运机;汽车铲运机
wagon scrubbing and decontaminating point 货车洗刷除污所
wagon seal(ing) 货车施封
wagon-set 车组
wagon shaker 卸车机械;货车振动机

wagon sheet 货车篷布;篷布
wagon sheet controller 篷布调度员
wagon sheet dispatching plan 篷布回送计划
wagon shifter 货车车皮调编机
wagons loaded and unloaded locally 自装自卸车
wagons loaded locally and delivered at junction station 自装交出车
wagon spout 装车口
wagon stage 有轮布景架;车台;有轮舞台
wagon tipper 翻车机
wagon tipping plant 翻车机设备
wagon tippler 翻车设备
wagon tippler hoist 货车翻卸吊车
wagon top 斜顶;筒形顶;车顶
wagon train 运货列车
wagon traverser 货车转车台
wagon truck 厢式载重汽车;有篷载重汽车;有棚运货车
wagon turnround time 货车周转时间
wagon-type dryer 车厢式干燥机
wagon-type spreader 拖车式撒肥机
wagon unloader 货车卸货机;卸车机
wagon unloader pump 货车卸货机泵
wagon unloading machine 铁路车辆卸货机
wagon unloadings 卸车数
wagon vault 筒形拱顶;筒形屋顶;筒形穹顶;斜穹隆屋顶
wagon-vaulted roof 筒形拱屋顶
wagon vibrator 货车振动器
wagon washing line 货车洗刷线
wagon washing plant 货车洗刷所
wagon way 货车道路
wagon weigh bridge 货车轨道衡
wagon weigh(ing) machine 称车机
wagon weight detector 车重检测器
wagon with opening roof consisting of sliding pan 滑顶式货车
wagon with roller roof 辊道顶棚式货车
wagon with sliding roof and sides 滑动车顶车门式货车
wagon works 车辆厂
wagon yard 停车场
wagtail 分隔片;分隔条;吊窗隔条;隔条
Wahl correlation 沃尔对比
waif 无主材
wain 运货马车
wain house 货车仓库;货车车库
wainscot 装饰墙壁用瓷砖;装饰墙壁用材料;装饰板;木护墙;护墙(板);护壁(板);墙裙;壁板
wainscot(ing) cap 护壁板压顶
wainscot(ing) oak 护壁板栎木;护墙板橡木
wainscot(ing) rail 护墙板镶条
wainscot oak 橡木壁板;护壁橡木
wainscot rail 下墙板镶条
wainscot(t)ing 装壁板;镶护墙板;墙裙板;镶护墙板;护墙板(材料);护墙板安装法
wairakite 斜钙沸石
wairauite 铁钴矿
waist 光束腰(收敛部分);缩颈;收敛;池窑卡脖
waist anchor 备用主锚;备用船首大锚
waist board (船上)临时铺板
waist breast 中横缆;船中部横缆
waist cloth 吊铺围布
waist cutter spindle 腰窝铣刀轴
waist-deep 齐腰深的
waisted bolt 束腰螺栓
waisted cauldron 腰沿釜
waisted specimen 腰形试件;细腰形试块
waist hawser(line) 腰横缆;船中部横缆
waist-high 齐腰高的
waisting crack 拦腰裂开
waist-level finder 俯视取景器
waist level viewfinder 近腰部取景器
waist-level viewing 俯视观测
waist of Gaussian beam 高斯光束腰
waist plate 腰部舷板
waist radius 光束腰半径
waist rail 舷墙栏杆
waist setting attachment 腰窝铣刀
waist sheet 锅炉托板
waist sheet angle 锅炉托板角
waist sheet angle liner 锅炉托板角铁垫
waist shot 半身镜头
waist slab (楼梯)梯段板

waist stroke 斜舷板
waist trimmer 腰窝铣削铣刀
waist-trimming cutter 腰窝铣削铣刀
waist-trimming machine 腰窝铣削机
wait acknowledge 等待应答
wait before transmission 传输延迟;传输前等待
wait before transmit positive acknowledgement 确认传输延迟;确认传输前的等待
wait bit 等待位
wait condition 等候条件;等待条件
wait for call 等待呼唤
wait for dialtone before dialing 拨号前等候拨号音
waiting 等待;短时停车;待时
waiting area 罐车排队区;等候处
waiting bay 短时停车弯
waiting berth 等命泊位;待泊泊位
waiting buoy 待命浮筒;待泊浮筒
waiting charges 待时费
waiting cost 等待费用
waiting for a job within an enterprise 待岗
waiting for berth 等候泊位
waiting hall 候诊室;等候厅;候车厅;候车室
waiting in cement time 候凝时间
waiting insurance 停车保险
waiting lane 避车道;短时停车道
waiting line 等待线;排队
waiting line approach 排队法
waiting line model 等待线模型
waiting line problem 等待线问题
waiting line theory 排队(理)论
waiting list 排队名单;申请人名单;等待名单;队列清单;等待表
waiting loop 等待循环
waiting message indicator 待发信息指示器;待发信息数指示器
waiting meter (短时)停车收费表
waiting on cement 水泥候凝时间
waiting on cement time 水泥候凝时间
waiting on cement to set 候凝期
waiting passenger indicator (电梯)候乘人指示器
waiting period 备耕期
waiting position 待料位置
waiting post 待命系船墩
waiting room 休息室;旅客等候室;候诊室;候机室;候船室;候车室;等候室
waiting room for mothers with children 母子候车室
waiting space 等候处;等待空间
waiting state 等待状态
waiting system 等待(制)系统
waiting time 等候时间;等待时间;等待期(间);待时时间;待命时间
waiting time distribution 等待时间分布
waiting time for lockage 待闸时间
waiting vehicle 待行车辆
waiting zone 罐车排队区
wait line problem 排队问题
wait list 等待表
wait place 短时停车处
wait routine 等待程序
wait state bit 等待状态位
wait state of process 进程的等待状态
wait subroutine 延时子程序;等待子程序
wait subroutine call 延时子程序调用
wait time limit 等待时间限制
waive a claim 放弃要求
waived protest 放弃抗议
waive duty 免收税款
waiver 弃权证;放弃书
waiver clause 弃权条款;放弃条款
waive right of claim 放弃索赔权利
waiver of conditions 放弃要件
waiver of inspection of goods 放弃验货权
waiver of lien 留置权弃权声明书
waiver of premium 免交保险费;放弃奖金;放弃保险费
waiver of regulations 放宽条例
waiver of the right of indemnity 放弃索赔权(利)
waiver system 放弃承运权制度
waive the right 放弃权利
waiving interest 免付利息
wakabayashilite 锑雌黄
wake 航迹【航海】;尾流;尾迹;船迹

wake analysis and control 尾流分析与控制
wake boundary 尾流边界
wake coefficient 尾流系数；伴流系数
wake current 航迹流；尾流；船行流；船迹流
wake depression 尾随低压
wake drag 尾流阻力
wake dune 尾随沙(丘)；尾沙丘
wake efficiency 尾流效率
wake energy 尾流能量
wake factor 尾流系数；伴流系数
wakefieldite 钒钇矿
Wakefield sheet(ing) pile 组合企口板桩
Wakefield sheet piling 韦克菲德板桩
wake flame 尾焰
wake flow 尾流；伴流
wake flow theory 尾流理论
wake-following typhoon 尾随台风
wake front 伴流前沿
wake gain 伴流增量；伴流推力
wake-induced vibration 尾流引起的振动
Wakeman buff 威克曼黄石
wake pattern 伴流图
wake percentage 伴流率；伴流系数
wake rate 伴流速率；伴流速度
wake resistance 尾随阻力；尾流阻力；伴流阻力
wake shape 尾流形状
wake speed 伴流速率；伴流速度
wake stream 尾流；伴流
wake stream theory 尾流理论
wake strength 尾流强度
wake structure 尾流结构
wake-survey method 尾迹测量法
wake temperature 尾流温度
wake traverse method 尾迹移测法
wake value 伴流系数
wake velocity 尾迹速度
wake wall 防浪(胸)墙
wake zone 尾流区
walchowite 褐壤树脂
Walden inversion 旋光转化作用；瓦尔登转化
waldheimite 铁钠逸闪石
Waldo hardness tester 沃尔多硬度试验机
wale 腰梁；围图；条状隆起部；船舷上缘；撑梁；长垣；板桩导梁；横枋木(用于开挖支撑)
wale count 纵行密度
wale deflection 纵行歪斜
wale girder 横挡(大)梁
wale knot 绳头结；绳端结
walentaite 瓦伦特石
wale piece 横档；横夹木(码头等的)横撑；横撑板
waler 模板横撑；横枋木(用于开挖支撑)
waler-rod-type tie 横撑杆型拉杆；横撑杆型系杆
Wales basin 威尔士盆地
wale shore 船边撑木
wale slippage 纵行转移
wale thread plaiting 纵行添纱
walf checks 华夫格
waling 支腰挡；支撑；挖槽横撑；支腰梁；支横档；开挖横撑木；围图；水平木；导梁【港】
waling for excavation 开挖横撑木
waling guide beam 板桩导梁
waling stripe 横夹木；横撑水平联杆；支撑；横板条；围图；水平连杆；围檩
waling timber 横撑木
walk a boundary 步测边界线
walkaround 车间内弄堂；车间内走廊；人行栈桥
walk back 向后游动；倒转绞盘
walk-behind roller 手扶振动碾
walk-behind single drum 手扶(式)单轮压路机
walk board 跳板；走道板
walk dowm 信息丧失
walked-on finish 耐踩踏的面饰
walked-on roof covering 可行人的屋盖
walker 索拉铲土机
Walker casting wheel 沃尔克型圆盘铸锭机
walker crusher 迈步式破碎机
walker dragline 步行式拉铲挖掘机
walker excavator 跨步式挖掘机；迈步式挖掘机；步行式挖土机；步行式挖掘机
Walker interlocking joint 沃克(混凝土板)啮口接头
walkerite 镁针钠钙石
walker knot 桶梁结
walker river 游荡性河流；不定床河流
Walker-Steel's (swinging beam) hardness tester 沃克—斯蒂尔(摆杆)硬度计
walker stream 游荡性河流
walkie fork lift truck 手扶式升降搬运车
walkie-hearie 步听机
walkie-lookie 便携式(光导摄像管)摄像机；手提式电视摄影机
walkie-talkie 携带式无线电话机；对讲机；小谈机；步话机；报话机；背包电台
walk-in 简易门诊所
walk-in box 小冷库
walk-in closet (人能走进去的)大壁橱；进入壁柜
walk-in freezer 冷藏库；小型冻结间；冷冻间
walking area 步行范围；步行区
walking arrangement 行走装置；行走设备
walking bar 摆动梁；摆动轴
walking beam 摇梁；振动梁；工作梁；活动梁；平衡梁；步进式炉底；步进梁；摆动梁
walking beam continuous furnace 步进式连续炉
walking-beam conveyer 步进式输送机；步进式输送机
walking beam furnace 步进式(加热)炉
walking-beam kiln 步进梁式窑；步进式隧道窑
walking beam saddle pin 摆梁骑马钉
walking beam suspension 平衡梁悬挂装置
walking beam-type furnace 步进式连续加热炉
walking block 砌墙块石
walking boss 主管；监工
walking crane 轨形起重机；步走式起重机；移动式吊车；执行起重机；活动吊车；步行式吊车
walking crusher 行走式破碎机；迈步式破碎机
walking crushing plant 移动式破碎机；行动式碎石设备
walking cultivator 手扶中耕机
walking digger 跨步式挖掘机
walking distance 步行距离(圈)；走行距离；很短的距离
walking dragline 行动式拉铲挖掘机；跨步式索铲；迈步式索斗铲；步走式索斗铲；步行式拉铲挖掘机；行动式索斗挖掘机
walking dragline excavator 迈步式挖掘机；行走式索斗挖掘机；行动式索斗挖掘机
walking dragline stripper 移动式拉铲表土剥离机；行走式拉索脱模机
walking dredge(r) 行走式索斗式挖泥船；行走式挖泥机；移动式挖泥船
walking driver 移动式打桩机
walking excavator 行走式挖土机；行动式挖土机；移动式挖掘机；步行式挖土机；步行式挖掘机；移动式挖土机
walking foot sinkage 行走机械脚的沉陷量
walking ganger 巡视工长
walking gear 行走装置；步行装置
walking habit 步行习惯
walking jaw 移动式颚板
walking line 楼梯斜路步准线；步走线；楼梯行走线
walking machine 迈步式车辆；步行机
walking mechanism 行动式装置；行动式机构；步行机构；移动式机体
walking movement 步行运动
walking patrol 地上巡逻
walking pit prop 移动式坑柱；移动式矿井支柱
walking platform 步行平台
walking plough 步犁
walking potato digger 手扶式马铃薯挖掘机
walking prop 行走式支架；自移式支架
walking ram 移动式活塞夯实机
walking reciprocating mower 往复式手扶割草机
walking rotary cultivator 手扶旋耕机
walking scoop dredge(r) 移动式斗式挖泥机；移动式挖土机；移动式铲斗挖泥机；移动式铲斗挖泥船；步行式挖土机；步行式铲斗挖泥机
walking single-bottom plough 单体步犁
walking speed 步行速度
walking sphere 步行距离圈
walking spring-tooth cultivator 手扶式弹齿中耕机
walking sprinkler 移动式喷灌器；移动式喷灌机
walking spud 进位桩；艇定位桩
walking-spud system dredge(r) 支柱移动式挖泥船
walking strip 步行路带
walking support 移动式支架；移动支撑；迈步式支架
walking surface 步行地面；步行地区
walking survey 步行测漏
walking thread 运行路线
walking time 步行时间
walking tractor 手扶(式)拖拉机
walking trip 步行出行
walking unit 步行装置
walking vehicle 步行机
walking way 人行道
walking way rainway gutter 人行道雨水沟
walk-in refrigerator 小型冷藏间；小冷库；进入冷藏间
walk length of sound 声波波长
walkoff angle 离散角
walk-off-loss 逸散损失
walk-off-mode 离散模
walk on left sign 行人靠街左走标志
walk out 蠕变；罢工
walkover 人行桥
walk-over survey 踏勘
walk path 人行道；散步器；步道
walk plank 人行道铺板
walk-range 步行范围
walk-rate 步行率
walk route 运行路线；步行路线
walk-through 人行地道；地下步行道
walk-through abreast stall 并列通过式挤奶台
walk-through closet (能双向进入的)大壁橱
walk-through delivery van 穿越步行道搬运车；穿越步行道送货车
walk-through inspection 实地检测
walk-through stall 通过式挤奶台
walk tracks 查修路轨
walk type mower 手扶割草机
walk-up 无电梯的
walk-up apartment 无电梯公寓；多层公寓
walk-up apartment house 无电梯公寓楼
walk-up building 无电梯楼房；无电梯的大楼
walk-up domestic block 无电梯的住房；无电梯的住宅楼
walk-up domestic building 无电梯的居住楼房
walk-up residence block 无电梯的住家楼房
walk-up residence building 无电梯的居住楼房
walk-up stor(e)y 无电梯楼层
walkway 走道；过道；人行(通)道；天桥；步桥
walkway bracket 人行道托架
walkway clearance 过道宽度
walkway pad 通道垫
walkway test 踏板试验
wall 墙壁溶蚀；沟壁；间隔层；墙壁；索端之结节；长(壁)工作面
wallaba (产自热带美洲的)红棕色硬木
wall abutment tile 墙接合砖
wall accretion 炉瘤；炉结；炉壁结瘤
Wallace ga(u)ge 华莱氏硬度计
Wallachian orogeny 瓦拉赤造山运动
Wallach transformation 瓦拉赫转变
wall action 壁面作用；洞壁影响；边壁作用
wall adhesion 墙黏着力；墙附着力
wall admittance 墙壁声导纳
wall air conditioner 壁式空气调节器；壁式空调器
wall air diffuser 墙内空气扩散器；墙内空气扩散管
wall anchor 格栅锚件；墙内锚定装置；墙锚栓；墙锚；墙钩
wall anchorage 墙内锚固
wall-anchored 锚碇的
wall and window section 墙窗断面
wall aperture 墙洞
wall arcade (不开洞的)装饰拱廊；实心(连环)拱廊；实心连拱(廊)
wall arch 加固拱；壁拱；墙拱
wall area 墙(壁)面积
wall area index 墙面积指数
wallastonite marble 硅灰石大理岩
wall-attached chimney 附墙烟囱
wall-attached jet 附壁射流
wall-attachment amplifier 附壁型放大元件；附壁放大器
wall-attachment effect 附壁效应
wall attachment jet 贴附射流
wall base 踢脚板；墙基
wall basel 墙座
wall beam 墙上支托；墙(托)梁
wall bearer 墙身支座
wall-bearing 用墙承重的；墙承重；承重墙
wall-bearing construction 承重墙结构；墙承重结构
wall-bearing partition 承重隔墙

wall-bearing structure 墙承重结构
wall bed 壁内折叠床;壁床;墙基;壁龛床
wall belfry 钟楼墙垣
wall between windows 窗间墙
wall blank 无窗墙;光墙
wall block 墙砌块
wall board 建筑纸板;壁板;墙(面)板
wallboard corner bead 板墙角压条;板墙护角;墙板角护条
wallboard factory 壁板制造厂
wallboard hammer 墙板锤
wallboard installation 墙板安装
wallboard machine 壁板纸机
wallboard masonry 砖砌墙板
wallboard nail 墙板钉
wallboard switch 墙上开关
wall body 壁体
wall bolting 洞壁锚栓支护
wall bonded to piers 砌入柱墩的墙
wall bore machine 壁装钻床
wall box 墙上安装梁的凹穴;暗线箱;暗线盒;墙托架;墙上梁座;墙内电器盒
wall brace 墙斜拉索;墙斜撑
wall bracing 墙支撑
wall bracket 壁灯;墙装托架;墙(上)托架;壁支架
wall bracket crane 井场悬臂起重机;墙装吊架
wall brander 沿墙栏栅;沿墙格栅
wall-breaker 墙式防波堤
wall breaking 刷帮
wall breakthrough 墙壁穿通
wall-breakwater 直墙式防波堤;墙式防波堤
wall brick 墙砖
wall building block 墙砌块
wall building component 墙壁建筑构件
wall building member 墙壁建筑构件
wall building of mud 泥浆造壁
wall building property of mud 泥浆造壁作用
wall building tester of mud 泥浆造壁性能测验仪
wall building tile 砌墙砖
wall building unit 墙壁建筑构件
wall built-up 炉壁结块
wall bushing 穿墙套筒;穿墙套管
wall cabinet 墙面柜
wall cabinet with mirror door 带镜门壁橱
wall cake 孔壁泥皮;井壁泥皮
wall calendar 挂历
wall carpet 墙挂毯
wall casing 孔壁管;井壁管
wall cast in situ 现场浇灌墙壁
wall cavitation 孔壁扩大;孔壁洞穴;井身扩大;井壁洞穴
wall cavity 墙洞;墙身空腔
wall channel 墙槽沟
wall channeler 墙壁凿缝机;砖石墙勾缝刀
wall chart 挂图;海图挂图
wall chase 管线槽;墙槽
wall cladding 墙面覆盖层
wall cladding adhesive 墙面覆盖层黏结剂
wall cladding cementing agent 墙面覆盖层黏结剂
wall cladding panel 墙壁装饰板;墙壁护板
wall clamp 墙卡子;墙拉杆;墙撑杆
wall cleaner 孔壁刮洗器;孔壁刮刀;孔壁钢丝刷;井壁清洁器;井壁刮洗器;井壁刮刀;井壁钢丝刷
wall-cleaning guide 孔壁清理钢丝刷;井壁清理钢丝刷
wall clearance 孔壁间隙;孔壁(与钻具之间的)间隙;井壁(与钻具之间的)间隙
wall clock 壁钟;墙钟;挂钟
wall clock movement 挂钟机芯
wall closet 装在墙上的壁橱
wall coat 墙壁涂层
wall-coated open tubular column 壁涂开管柱
wall coating 墙面涂料;墙壁涂料;墙壁涂;墙镀层
wall coefficient 墙面吸收系数;墙面吸声系数
wall coil 墙盘管;墙管
wall collision 墙碰撞
wall column 壁柱;墙内柱;墙墩
wall column footing 墙柱脚
wall complex 壁覆植物
wall composition 墙的组成;墙的结构
wall concrete strength 墙用混凝土强度
wall concreting 墙混凝土浇筑
wall conduit 墙内(水电)管道
wall construction 墙体结构

wall construction method 墙施工方法
wall construction type 墙的施工形式
wall cooling 管壁冷却;孔壁冷却;井壁冷却
wall-cooling surface 水冷受热面;水冷壁面
wall coping 墙帽;墙压顶
wall coping plate 墙顶挡板
wall corner 墙角
wall corner guard 墙角护板
wall cornice 墙(头)挑檐;墙壁挑檐
wall correction 壁(影)修正
wall-correction factor 壁(效应)修正系数
wall cover 墙面覆盖
wall covering 墙面覆料;墙面材料;墙面涂料;塑料墙纸;壁面涂料;壁纸;内墙装饰覆盖物
wall-covering fabric 贴墙布
wall covering material 墙面覆盖材料
wall crab 墙装卷扬机;墙上卷扬机;壁装卷扬机
wall crack 墙裂缝
wall cracking 墙体开裂
wall cramp 墙腰箍
wall crane 沿墙起重机;墙装起重机;墙上起重机;墙座起重机;壁装(式)起重机;壁吊
wall crazing 墙体开裂
wall creep 畴壁蠕移
wall crest 墙压顶;墙顶
wall crib 孔壁基;井壁基
wall crossing 墙交叉
wall cross-section 墙横断面
wall crown 墙(压)顶
wall curtain 幕墙
wall curvature 管壁曲率
wall damp-proof course 墙壁防潮层;墙身防潮层
wall damp-proofing 墙身防潮
wall decoration 墙壁装饰;墙面装饰
wall decorative fixture 墙壁装饰件
wall deformation 墙壁变形
wall-derrick 墙装转臂起重机;墙上转臂起重机
wall division 隔墙
wall door catch 墙上门后夹
wall door stop 墙上装的止门器
wall dowel 挡土墙泄水管;墙上预留插铁;墙上预埋木砖;墙上暗榫;墙木栓
wall drag 孔壁阻力;井壁阻力
wall drilling machine 墙装钻床
wall drinking fountain 墙上饮水器;墙面饮水器
wall dryer 墙干燥器
wall duct 墙中空气通道;墙内风道
wall echo 侧壁回波
walled 有围墙的
walled base 壁形基板
walled collector 壁形集电极
walled concrete 移动模板浇注的混凝土
walled enclosure 筑墙围地
walled frame 壁式框架
walled frame structure 壁式框架;立贴架结构;框架剪力墙结构
walled garden 有围墙的花园
walled lake 砾壁湖
walled monastery 有围墙的修道院
walled palace 有围墙的宫殿
walled passage 有围墙的通道
walled town 有城墙的城镇
wall effect 管壁效应;模壁效应;墙效应;堤岸效应;畴壁效应;侧壁效应;边际效应;边壁效应;器壁效应;边界效应
wall element 墙构件
wall emission 壁发射
wall enclosure 围墙
wall-enclosure material 围护墙材料
wall energy 壁能
wall enrichment 墙壁装饰
wall entasis 凸肚墙
wall entrance insulator 穿墙进线绝缘导管
wall erection 墙的建造
wallerian degeneration 次级退变
waller's hammer 墙工用锤
Waller stone 华勒石(美国俄亥俄州产的一种砂岩)
wallet 零星工具袋
wallette 小墙
Walley engine 韦利发动机
wall fabric 墙上挂毯
wall face 工作面前壁;墙面

wall facing 墙壁面层;墙身饰面
wall-facing board 墙面覆盖板
wall-facing material 墙面材料
wall-facing quarry tile 墙面缸砖
wall-facing tile 墙面砖
wall-facing tile factory 墙面瓷砖制造厂
wall fan 隔墙风扇;墙上排风扇;墙上排气窗
wall feed-through sleeve 穿墙套管
wall fence 护墙;墙式栏杆
wall fiberboard sheathing 墙壁纤维盖板;墙衬纤维板
wall filler 墙壁填料
wall finish 墙面终饰;墙面装修;墙面最后涂刷;墙面重饰
wall fire warning device 墙上火警装置
wall fittings 壁灯;墙上固定件
wall-fixed drinking fountain 壁装式饮水龙头
wall-fixed handrail 靠墙扶手
wall flagpole 装在墙上的旗杆
wall flash 挡火墙(倒焰窑火箱部位)
wall flashing 墙面冷水;墙面泛水
wall flow 壁流
wall foil 墙壁用金属箔
wall footing 墙脚;墙(基)脚;墙基础
wall form 墙体模(板)
wall-forming 构成墙的;墙的造型
wall formwork 墙模板;墙模工程
Wallfortis 华尔福蒂斯牌防水液(商品名)
wall foundation 墙式基础;墙基(础)
wall fountain 壁上喷泉;壁上喷泉式饮水器;壁泉
wall frame 壁式框架;立贴架;承托墙框架;墙内电器盒
wall framing 承托墙壁框架
wall fresco 壁画
wall friction 孔壁摩擦;井壁摩擦;墙面摩擦(力);墙背摩擦(力);器壁摩擦(力);壁面摩擦(力)
wall friction angle 墙面摩擦角
wall furnace 墙炉
wall furring 抹墙充筋;墙(上)板条
wall furring base clip 墙面龙骨卡
wall garden 围墙花园;植物生长在墙上的花园;墙面花圃;墙顶花池;墙园
wall gas-fired warm air unit heater 壁装式燃气热风加热器
wall gas geyser 壁装式气体间歇淋浴;壁装式气体间歇喷泉;墙上煤气热水锅炉
wall girder 墙梁
wall glazing 墙上玻璃窗
wall grab bar 墙上抓条
wall greening 墙壁绿化
wall grid 墙盘管
wall grill(e) 墙篦子;墙(洞)栅
wall grinder 墙面磨光机
wall grinding machine 磨墙机
wall grip 孔壁支持器;井壁支持器
wall guard 围墙杆;墙面保护;立墙杆;护墙栏
wall gypsum baseboard 墙石膏踢脚板
wall gypsum sheathing 墙面石膏盖板
wall handrail 靠墙(楼梯)扶手
wall hanger 墙锚栓;墙嵌梁座;墙挂梁座;墙上梁托;墙上承梁件;铁靴
wall hanging 墙上挂毯;墙帷;壁挂
wall hanging in carved lacquer 雕漆挂屏
wall hanging type tapestry 墙壁悬挂式挂毯
wall heater 墙上暖热器;墙面取暖器;墙上取暖器;墙上加热器;墙内取暖器;墙内加热器
wall heat flux 壁面热通量
wall heating 火墙
wall heating panel 火墙护板
wall heel 墙踵
wall height 墙高
wall hitch 墙障
wall hoist 固定在壁上的卷扬机
wall hold (构件的)墙上支点
wall hole 墙孔
wall hollow 墙洞;壁龛
wall hook 孔壁悬挂装置;井壁悬挂装置;墙(托)钩;壁钩
wall hospital luminaire fixture 医疗用照明装置;医疗用壁灯
wall hot water heater 壁装式热水器
wall humidity 墙壁湿度
wall-hung boiler 墙挂式锅炉
wall-hung cabinet 墙面柜

wall-hung closet 壁挂式便桶;墙挂壁橱
wall-hung cupboard 墙挂碗橱;吊柜
wall-hung fixtures 挂墙式器具
wall-hung lavatory 挂式洗脸盆;挂墙洗手盆;墙挂式洗脸器;挂墙式洗脸盆
wall-hung radiator 挂式散热器
wall-hung shape 墙挂式
wall-hung shower 挂墙式淋浴器
wall-hung toilet 壁挂式便橱;墙挂壁橱
wall-hung type 挂墙式;墙挂式
wall-hung urinal 壁挂式小便器;挂式小便器;墙挂式小便器
wall-hung wash down urinal 墙挂式冲落式小便器
wall-hung water closet 壁挂冲水小便器;挂式大便器;墙挂冲水箱
wall hydrant 室内消火栓;墙壁式消火栓;消防龙头;墙上消防栓
wall illumination 墙上照明设备
wall impedance 墙壁声阻抗
wall in advance of arched roof tunneling method 先墙后拱法
walling 筑墙;夹桩;墙体工程;墙体材料;墙砌体
walling board 坑壁板;坑壁横撑板;墙板
walling cast in situ 现场浇筑墙
walling component 墙体建筑单元
walling crib 垛式挡土墙;壁座;矿井圈;井框
walling hammer 墙工锤;筑墙锤
walling mason 砌墙石工
walling masonry 筑墙圬工
walling material 墙体材料
walling of a shaft 砌筑井壁
walling panel 墙板
walling pattern 墙体图案;墙体式样
walling piece 护壁板工件;护壁板板件
walling plank 墙板
walling plant 爬墙植物;壁栽植物
walling slab 预制墙板;搁墙撑头木;墙板
walling system 墙体系统;砌壁方法
walling thread 墙线
walling timber 夹桩木
walling unit 预制墙体砌块;墙体砌块
walling up 炉衬;封墙;砌墙;用墙堵住;用墙堵死;装药卷道封墙;造壁
Walling Wall 饮泣墙(耶路撒冷犹太人会堂的残壁)
walling work 墙体工程
wall insulation 墙壁(热)绝缘;墙身保温
wall insulation lining 墙保温衬
wall insulator 墙上绝缘子;穿墙绝缘子
wall interference 缸壁干涉
wall in trench 沟墙
wall iron 墙铁件
wall ironing machine 罐壁烫压机
Wallis formulas 瓦里斯公式
wallisite 铜红铊铅矿
Wallis product 瓦里斯乘积
Wallis theorem 瓦里斯公式
wall jack scaffold 墙撑脚手架
wall jet 沿墙喷射;壁面射流
wall jib crane 旋臂吊车;旋臂起重机
wall joint 墙面灰浆缝;墙缝
wall junction 墙壁连接
wall key cabinet 可上锁的壁橱
wallkilldellite 水砷钙锰石
wall knot 绳头结
wall lamp 壁灯
wall lamp bracket 壁灯架
wall lampholder 壁灯头
wall lamp socket 壁灯座;墙上灯座
wall lavatory basin 靠墙壁安装的盥洗盆
wall layer 壁层
wall length 工作面长度
wall-less 无墙壁的
wall lift 墙外升降机
wall light fitting 墙壁照明灯具;墙壁照明装置;固定在墙上的照明设备
wall lighting 墙壁照明
wall light with guard 带灯罩壁灯
wall line 墙界线;墙直线
wall lining 炉壁内衬;墙衬;壁衬
wall lining board 墙壁衬板
wall lining material 墙壁衬里材料
wall lino(leum) 墙壁油毡
wall losses 墙壁热损失

wall luminance 墙面亮度
wallman 拆墙工
wall map 挂图
wall marble 墙壁大理石
wall material 墙壁建筑材料
wall mats 墙席
wall method 壁式陷落法
wall mirror 隔断镜;墙镜
wall mixed plaster 墙壁混合灰泥
wall mixer 挂墙式混合机
wall mix(ture) 墙壁灰泥混合物
wall moisture 墙壁潮湿度
wall mosaic 墙壁马赛克
wall mo(u)lding 墙上线脚
wall-mounted 壁装(式)的
wall-mounted closet 装在墙上的便桶
wall-mounted cooker 嵌壁炉灶
wall-mounted fan 挂墙式风扇
wall-mounted fire warning device 嵌壁式火警装置
wall-mounted flagpole 嵌装在墙上的旗杆
wall-mounted gas geyser 壁装式煤气淋浴;壁装式煤气喷泉
wall-mounted grab bar 壁装拉手
wall-mounted grinder 壁装磨机
wall-mounted heater 壁装(式)采暖器;壁装(式)加热器
wall-mounted heating panel 壁装采暖板
wall-mounted hospital luminaire fixture 医院壁灯;医院壁装式照明装置
wall-mounted hot water heater 壁装式热水加热器
wall-mounted lavatory(basin) 壁装式盥洗盆
wall-mounted light fitting 壁装式照明配件
wall-mounted light(ing) 嵌装灯;墙上挂灯;壁灯
wall-mounted lighting fixture 壁装式照明装置
wall-mounted luminaire 壁灯
wall-mounted luminaire fixture 壁装式照明装置
wall-mounted mixing cock for cold and hot water 壁装式冷热水混合开关龙头
wall-mounted oven 壁炉;挂在墙上的烘箱
wall-mounted shower 壁装式淋浴
wall-mounted slop bowl 墙装式倾倒残渣盆
wall-mounted urinal 壁装式小便池
wall-mounted warm air heater 壁装式热风加热器
wall-mounted wash basin 壁装式洗涤室
wall-mounted washbowl 壁装式洗手池;壁装式洗手盆
wall-mounted washdown W. C. pan 墙式式坐便器;墙装冲落坐便器
wall-mounted wash-handbasin 壁装式洗手池;壁装式洗手盆
wall-mounted waste receptacle 墙上污水储[贮]槽;贴壁式垃圾箱
wall-mounted water heater 壁装式热水加热器
wall-mounted water heating appliance 壁装式热水加热装置;壁装式热水加热器
wall-mounted winch 壁装绞车
wall mounting 墙上安置;墙上安装;壁式安装;壁上安装
wall mounting flush 墙面平镶连接
wall mounting-projected 墙面凸起连接
wall mounting telephone 壁挂式电话机
wall mounting vertical electric(al) water heater 垂直墙挂式电热水器
wall movement 岸壁位移
Wallner line 瓦尔纳线
wall niche 壁龛
wall of a basin 水池墙
wall of borehole 钻孔壁
wall of container 容器壁
wall off 套管固井;堵塞墙缝;堵塞壁缝;封闭井壁
wall offset 内收墙身
wall of hole 孔壁
wall of hollow masonry 空心砌体墙
wall of partition 隔墙
wall of patterned glass 压花玻璃墙
wall of pipe 管壁
wall of planks 木板墙
wall of portal 洞门墙
wall of sand 沙壁
wall of tube 管壁
wall of wound 创壁
wall-on type 壁装型
wall opening 墙洞(口)
wall oriel 墙(上)凸(肚)窗

wall ornament 墙装饰;墙上装饰品
wall ornamentation 墙壁装饰术
wall outlet 墙上插座;墙壁插头;壁装(电源)插座
wall outlet connector 穿墙连接器
wall packer 孔壁封隔器;井壁封隔器
wall packing 孔壁泥封;井壁泥封
wall paint 墙上涂料;墙漆;墙壁涂料;墙壁漆
wall painter 墙壁油漆工;漆墙工
wall painting 壁画
wall panel 护墙板;墙面格间;大型壁面板;槽段;壁板;镶板;墙板
wall-panel heating 墙壁辐射采暖;墙板供暖
wall-paneling 镶护墙板;木护墙;护墙板安装法
wall panel in prestressed clay 预应力黏土墙板
wall-panel system 墙构筑法
wall-panel with window 带窗墙板
wall paper 糊墙纸;墙纸;壁纸
wall-paper brush 壁纸裱糊刷
wall-paper cleaning preparation 壁纸清洗剂
wall-paper colo(u)r 壁纸色调的卷筒纸
wall-paper cover mo(u)lding 墙纸覆盖线条
wall-papered 贴墙纸的
wall-paper factory 墙纸制造厂
wall-paper hanging paste 贴墙纸糨糊
wall papering 贴墙纸;裱糊墙纸
wall-papering work 贴墙纸工作
wall-paper printing machine 壁纸印刷机
wall-paper removing preparation 壁纸清除剂
wall-paper roll 墙纸卷
wall-paper sheet(ing) 墙纸裁切
wall-paper trimmer 墙纸修剪工;墙纸修剪器
wall partition 隔(断)墙;间墙
wall passage 穿墙通道;沿墙通道;墙式通道
wall pattern 墙壁图案
wall pedestal fan 壁挂式牛角扇
wall pellitory 墙草
wall piece 承梁板;贴墙立木;搁墙撑头木
wall pier 窗间墙;墙墩
wall pier capital 墙墩柱头
wall pillar 墙支柱;墙(支)墩
wall pipe 穿墙管
wall piping 穿墙管
wall plant 攀墙植物;长在墙上的植物
wall plaster 墙面刷;墙面粉饰;刷墙灰;粉刷石膏
wall plastering 墙面抹灰
wall plate 墙顶垫板;砌入墙内的托梁垫板;承梁板;承椽板;卧梁;墙上承梁板;梁垫;搁墙撑头木;平巷靠壁纵向棚子
wall plug 裸井堵塞;墙插插头;墙上灯座;墙上插头
wall pocket 预留梁洞;墙内座梁;墙内梁座(墙上安装梁的凹穴);墙凹口
wall-pointing machine 墙勾缝机
wall-port system 边墙输水系统(闸门)
wall post 墙柱;壁柱
wall pot 砌墙用空心砌块
wall pouring 墙的浇筑
wall pressure 侧压(力);壁压
wall pressure hole 壁面测压孔
wall primer 墙壁(打)底漆
wall profile 壁面轮廓
wall-projected shelf 壁架
wall-protection kerb 墙裙;墙边保护勒脚
wall protector 缓冲挡;室内护栏杆
wall quartz clock 石英壁钟
wall radial drill 旋臂壁钻
wall radial drilling machine 墙装摇臂钻床
wall radiant tube 墙壁辐射管
wall radiator 挂式散热器;墙装散热器;墙挂(式)散热器;壁挂式散热器
wall rail 墙轨;靠墙扶手
wall rail bracket 靠墙扶手托架
wall ratio 壁厚比
wall recess 壁龛;墙凹槽;岸壁凹槽
wall reef 壁礁
wall reflection 壁壁反射
wall reflection factor 壁反射系数;壁反射率
wall register 墙面风口
wall reinforcement 墙体配筋
wall relief 浮雕墙;墙壁浮雕
wall resistance 管壁热阻
wall rib 墙肋(梁);穹半肋;附墙拱肋
wall ring (支承圆顶的)墙环梁
wall rising through all the stories 所有楼层高的墙
wall rock 孔壁岩石;井壁岩石;围岩

wall-rock alteration 围岩蚀变
wall-rock anomaly 围岩异常
wall-rock aureoles 围岩环
wall-rock halo 围岩晕
wall rosace 园囿浮雕墙
wall roughness 墙粗糙度;边壁糙度;壁面粗糙度;壁面糙率
wall rubber 墙橡胶
wall run 墙(临时性的)上施工通道洞;墙的通道(墙侧的一个临时通道)
wall saddle 墙鞍
wall safe 壁藏保险箱
wall saltpeter 钙硝石
wall sample 墙试验样品;墙模型
wall sampler 薄壁取土(样)器;孔壁取样器;井壁取样器
wall sampling 薄壁采样
walls below grade 地面标高以下的墙;墙脚;基础墙
wall scabbler 墙面粗琢饰面机
wall scone 装饰用壁灯;壁装饰灯
wall scrape 刮洗井壁
wall scraper 孔壁刮洗器;孔壁刮刀;孔壁钢丝刷;井壁刮洗器;井壁刮刀;井壁钢丝刷;墙刮刀
wall scraper bit 孔壁刮刀钻头;井壁刮刀钻头
wall scraper blade 墙面抹灰整平板
wall screw 棘螺栓;锚栓;墙锚栓;墙螺丝;墙螺栓
wall sealer 墙体封底漆
wall-sealing compound 止水添加剂;堵漏添加剂
wall section 墙剖面
wallsend 高级家用煤
wall sensor 壁式传感器
wall set 墙式电话机
wall settee 嵌墙长椅
wall set telephone 壁挂式电话机
wall settle 嵌长靠椅
wall settlement 墙身沉陷
wall settlement fissure 墙身沉陷裂缝
wall shaft 承穹肋小柱;墙面支承穹肋的小柱;墙面小柱
wall shaping 壁形选择
wall shear(ing) stress 壁剪应力;边壁剪应力
wall sheathing 壁外壳;墙体衬板
wall shelf 墙隔板;墙搁板;墙架;吊架
wall shelve (安装在墙上的)吊架;架壁;墙上吊架;壁架
wall shift 壁移
wall shingle 墙面板
wall shower 挂墙式淋浴器;墙式淋浴器;墙挂式莲蓬头
wall shrine 壁龛
wall shuttering 墙模壳;墙模型
wall-sided float 侧壁浮筒
wall siding (木架房屋的)外墙木板
wall sign 墙上标志;墙面广告
wall sill 窗台
wall sized 按室内墙壁尺寸的
wall skin 壁外层
wall slab 壁板;墙板
wall slab in prestressed clay 预应力黏土砖墙板
wall slab structure 板桩结构
wall slate 墙壁石板
wall sleeve 墙套管;穿墙套管
wall sleeving crane 壁行起重机
wall slenderness limit 墙体长细比极限
wall slenderness ratio 墙体长细比
wall slewing crane 墙上旋臂起重机
wall slope 墙的坡度
wall slope bowl 墙倾倒残渣盆
wall-slot lighting 耳光
wall socket 墙装插座;墙式插座;墙上插座;墙壁插座;电器插座;壁装插座;壁式插座;墙上插座
wall socle 墙座
wall space 门窗间墙;墙面;墙壁面积
wall spacer 系模钢条;墙模定位拉伸杆
wall sprinkler 墙上喷水器
wall stability 墙壁稳定性
wall stack 墙内竖管;墙烟囱;供暖墙管;采暖墙管
wall standing by itself 独立墙
wall static pressure 壁面静压
wall static pressure tap 壁式静压龙头
wall sticking 孔壁黏着;井壁黏着
wall stiffness 墙劲性
wall stone 砌墙石料
wall stopping 墙的填充(物);隔墙

wall strength 墙壁强度;壁面强度
wall stress 壁应力
wall stress controller 缸壁应力控制器
wall string(er) 靠墙楼梯斜梁;贴墙楼梯小梁;墙梁;墙楼梯斜梁
wall stucco 墙壁拉毛粉刷
wall stud 墙筋;管线支架;壁柱
wall stuff 刷墙粉;涂墙用的灰泥
wall support 墙(支)座;墙支承
wall surface 墙(表)面
wall surface finish 墙面最后饰;墙面修饰
wall surface line 墙面位置线
wall surfacing 墙表面处理;墙表面加工
wall surfacing board 墙面铺板
wall surfacing compound 墙面混合材料
wall surfacing material 墙面材料
wall system 墙(体)系
wall system installation 墙体系安装
wall tap 墙壁龙头;靠墙龙头
wall tapestry 壁毯
wall telephone set 墙式电话机;墙机
wall temperature 墙壁温度;壁温(度)
wall tent 有墙帐篷
wall thickness 墙厚;管壁厚度;膜壁厚度;墙壁厚度;壁厚
wall thickness after reinforcement 补强后壁厚
wall thickness ga(u)ge 壁厚计
wall thickness micrometer ga(u)ge 壁厚千分尺
wall thickness of collars 钻铤壁厚
wall thickness of pipe 管壁厚度
wall thickness reduction 壁厚减薄
wall-through type air conditioner 窗墙式空调机;穿墙式空调机
wall tie 空心墙连系件;饰面砖;系墙铁;系墙索;炉墙拉杆
wall tie closer 镶墙(边)的砖石
wall tile 瓷面砖;墙砖;墙面(贴)砖;贴墙面砖;面砖
wall tile adhesive 墙面砖黏结剂
wall tile body 墙面砖块
wall tile bonding adhesive 墙面砖黏结剂
wall tile cementing agent 墙面砖胶黏剂
wall tile clay body 墙面黏土砖块
wall tile galzed on both sides 双面釉墙面砖
wall tile press 面砖压机;瓷砖(压)机
wall tile setting 面砖铺贴
wall tile stone 贴墙面用石板
wall tilework 墙面砖镶铺工作
wall tiling 墙面贴砖;饰面砖
wall toe 墙趾
wall toilet 墙装式马桶
wall top 墙顶
wall-to-wall carpeting 全室地毯;满铺地毯
wall-to-wall inventory 全面盘存
wall tower 墙塔
wall tracery 墙花格
wall trimming 壁面修整
wall tube 穿墙管;墙上进线导管
wall tube insulator 穿墙套管;穿墙绝缘管
wall tubing 穿墙管
wall turbulence 贴壁紊流;墙面紊流
wall tying 墙体加固;墙箍扎
wall type 岸壁式;墙壁式
wall-type breakwater 直墙式防波堤
wall-type caisson 壁式沉箱
wall-type door stop 墙面门碰头
wall-type flagpole 嵌墙式旗杆
wall-type foundation 墙式基础
wall-type heat exchange 间壁式换热
wall-type heat exchanger 间壁式换热器
wall-type hydrant 壁墙式消火栓;壁墙式消防栓
wall-type pyrometer 壁式高温计
wall-type switchboard 墙挂式交换机;墙式配电盘
wall-type thickness 壁墙式厚度
wall-type underground diaphragm wall 墙式地下连续墙
wall unevenness 壁厚不均
wall unit 墙构件;(单元墙板;镶(墙)板单元;壁柜
wall unit with recess 有壁龛墙单元
wall unit with yielding to blast 抗暴墙体砌块
wall up 用墙堵住;造墙
wall urinal 壁装式小便池
wall urn 壁盒
wall vapor barrier 墙壁防蒸汽层;墙壁防湿层;墙壁防潮层

wall veneering 镶板饰墙面
wall vent (管道、管沟、房顶等的)墙通风孔
wall ventilator 墙上通风孔;墙上通风机;穿墙式通风机;壁上通风器
wall voltage 墙电压
wall warm air heater 壁墙式热风加热器
wall wash basin 墙上洗手盆
wall washbowl 壁墙式盥洗盆
wall-washer 撑墙支架(垫)板
wall-washing 墙面冲洗;墙面泛光照明;用贴墙光源照亮墙面
wall-wash luminaire 墙面泛光灯具;墙面照明装置
wall water heater 壁装式热水器
wall waterproofing 墙身防水
wall weight 墙重
wall white 墙上碱霜
wall winch 壁装绞车;墙(装)绞车
wall wiring 墙内暗线;墙内装设电线
wall wiring conduit 墙壁内电线管
wall wooden mo(u)lding 挂镜线(条)
walnut 胡桃(木);核桃木
walnut comb 胡桃冠
walnut juglans regia 核桃木
walnut oil 胡桃油
walnut shells 果壳粉(堵漏材料)
Walpole colo(u)rimeter 沃波尔(目视)比色器
walpurgite 砷铀铋矿;砷铋铀矿
walrus 海象
Walsh transformation 沃尔什变换
walstromite 瓦硅钙钡石
walter pine 裸松(产于美国南部)
waltherite 砷铀铋矿
Walton process 华尔顿油毛毡制作法
Walton's cement 华尔顿水泥
Walton's lino(leum) 华尔顿亚麻油毡
Walvis ridge 沃尔维斯海岭
wamoscope 行波示波管;调波示波器;波调示波管
wampum 贝壳
wander 迁移;漂移
wander correction 加速误差改正;漂移校正
wandering 游动;游荡;迁移;飘移;漂移;漂动
wandering albatross 漂泊信天翁
wandering belt 河道蜿蜒带
wandering bird 漂鸟
wandering block method 连续块焊法
wandering block sequence 连续块焊工序;分段跳焊工序;跳焊法;电弧漂移多层焊次序
wandering block welding 游块块焊接
wandering channel 游荡性水道
wandering coal 不规则煤层
wandering dune 游移沙丘;游动沙丘;移动沙丘;流动沙丘
wandering effect (汽车行驶或飞机起降时的)游动效应
wandering fiber 游离纤维
wandering heart 髓部不直(的木材);髓心偏离
wandering lake 游移湖
wandering lead 游动引线
wandering mark 浮动测标
wandering of an arc 电弧漂移
wandering of belt 皮带走偏;皮带跑偏
wandering of river 河道摆动
wandering of river channel 河道摆动
wandering push button 流动按钮开关
wandering reach 游荡(性)河段
wandering river 游荡性河流;游荡型河流;蜿蜒(性)河流
wandering sequence 游荡焊接工序
wandering stream 游荡性河流;游荡型河流;蜿蜒(性)河流
wandering water 悬着水;重力水;通气层水;渗流水;包气带水
wandering welding sequence 跳焊法
wanding 读条形码;读磁卡过程
wand reader 磁卡片信息读出器
wane 缺损;翘棱;(锯材的)钝棱
wane cloud 卷层云
wane-edged wood 缺角镶边方木
wane end off 尖灭
wane lumber 圆角木材
waney 高低不平的;宽窄不齐的;缺角的;缺棱;不等径的
waney edge 板的不齐边皮;缺损边缘;缺角方木边
waney edged 缺一边角的(木材)

waney log 不等径圆木;有缺损圆木
wangle accounts 造假账
Wang Rui classification 王锐分类
wanigan 木筏上的陋室;小橱房;小寝室;储[贮]物柜;储[贮]物箱
waning development 凹坡发育
waning moon 下弦月;亏月;渐亏月
waning slope 凹坡
waning stage 衰退期;衰减阶段
Wankel engine 汪克尔转子发动机;三角转子发动机
Wankel engine's displacement 汪克尔转子发动机排量
Wankel rotary compressor 汪克尔旋转压缩机
Wankel rotating engine 汪克尔转子发动机
Wankel's engine 汪克尔发动机
Wankel's rotating piston 汪克尔转子活塞
Wankel's sealing grid 汪克尔密封系统
Wankel's simple rotating engine 汪克尔单纯旋转式发动机
Wankel type compressor 汪克尔式压气机
Wanner optic(al) phrometer 汪纳光测高温计
Wanner's pyrometer 汪纳高温计
Wannier function 沃尼埃函数
want 材棱凹陷
wantage rod 液面高度计;测量标尺
wanton profusion 浪费
want slips 缺货单
wany log 不等径圆木
wapping off 抽头
war 战争
war anchorage 战时锚地
warbler 电抗管调制器
warble rate 调频深度
warble tone 啭声;低昂音
Warburg respirometer 瓦渤格呼吸计
Warbury model 瓦尔堡模型
war cancellation clause 战争险注销条款
war clause 战争险条款
war communications 战时通信
ward 医院病房;监护;大病房;病室;病房
war-derastated area 战争破坏地区
warded 储[贮]藏室
warded block 房房大楼
warded building 病房大楼
warded key 带齿槽钥匙
warded key hole 暗锁眼
warded lock 弹子锁;撞锁;暗锁
warded locking mechanism 暗锁机构
warded mortise lock 暗插锁
warded rim lock 碰撞弹簧锁
warder house 储藏室;守卫室
war-devastated area 战争破坏地区
war deviation clause 战争绕航条款
ward floor 病房楼层
warding broker 运输经纪人
warding file 平锉;薄片锉;锁孔锉
wardite 水磷铝钠石
Ward-Leonard control 沃特－勒奥那多控制
Ward-Leonard system 沃特－勒奥那多控制系统
ward luminaire light fitting 病房照明灯光设备
ward office 区公所
ward patient's room 病人房间
wardrobe 壁柜;藏衣室;(皇族和贵宅的)保管库;衣柜;立柜
wardrobe closet 衣帽壁橱
wardrobe lock 橱锁
wardrobe type closet bank 衣帽壁橱墙板
wardship 监护
wardsmithite 瓦钙镁硼石
ward stor(e)y 病房楼层
ward unit 病房的护理单元
ware 制品;器皿
ware crack 干裂(木材)
ware glass 器皿玻璃
warehouse 栈房;港区仓库;货栈;货仓;后方货场;后方仓库;堆栈;储存室;成品仓库;仓库
warehouse account 仓库科目
warehouse allocation 仓库分配;仓库分派
warehouse bond 入库保税品
warehouse book 仓库物资账;仓库(账)簿
warehouse booking 仓库货品登记
warehouse building 仓库;货栈
warehouse charges 仓库费用;仓储费(用);仓储成本

warehouse charge rate 储存费率
warehouse cost 仓库费用;仓储成本
warehouse crane 仓库(用)起重机
warehouse delivery 仓库交货
warehouse district 仓库区
warehouse entry 进仓;入库(存货);存仓报单;仓库入库收据
warehouse entry inspection 入库验收
warehouse expenses 仓储费(用)
warehouse fee 仓储费
warehouse fork lift truck 仓库叉车
warehouse goods 库存物资;库存货物
warehouse keeper 货栈主;仓库管理员;仓库管理人
warehouse-keeper's certificate 仓库管理员受货证书
warehouse-keeper's receipt 仓库管理员受货凭单
warehouse-keeper's warrant 仓库保管员栈单
warehouse-keeping method 仓库管理法
warehouse loans 仓单贷款
warehouseman 货栈主;仓库保管员;仓库业者;仓库管理员;仓库管理人;仓库工作人员
warehouse management 仓库管理
warehouse management by specific norms 仓库定额管理
warehouse mark 仓库标志
warehouse market 仓库市场
warehouse network 仓库网
warehouse-out inspection 出库检验
warehouse owner 仓库业者
warehouse problem 库存问题;仓库(合理使用)问题
warehouse purchase 仓买
warehouse receipt 栈单;码头收货单;仓库提单;仓库收据;仓单
warehouse receipt clause 仓单贷款
warehouse rental rate 仓租率
warehouse service charges 仓库服务费
warehouse set 库内结硬(水泥);库凝;库存硬化;储存凝结
warehouse set cement 在储存过程中部分结块的水泥;库凝水泥;库存硬化水泥;货栈内结硬的水泥;仓库内硬化水泥
warehouse setting (水泥的)存储[贮]硬化
warehouse ship 仓库船
warehouse space 仓库容积;仓库面积
warehouse stock 仓库存货
warehouse storage charges 仓储费(用)
warehouse supply 仓库供应
warehouse tally 仓库理货
warehouse-to-warehouse 仓库到仓库;仓至仓
warehouse-to-warehouse clause 仓库至仓库条款
warehouse track 仓库线
warehouse warrant 货栈证券;货仓保付证;仓单
warehousing 存款放贷;仓库房屋;堆栈房屋;匿名控制股票;入仓;仓库作业;仓库业务;仓储
warehousing and carrying charges 仓储入库费
warehousing and storage 仓库和储存
warehousing and storage activities 仓储作业
warehousing and storage margins 库存及储[贮]藏费用
warehousing district 仓储区
warehousing entry 入库单;入仓申请书
warehousing management 仓储管理
warehousing of grain 谷物入仓;谷物仓库
warehousing problem 库存问题
warehousing shelter 仓储[贮]用棚
warehousing storage 仓库储存
warehousing system 仓储系统
war emergency pipe line 战时应急管线
ware pipe 排水管;釉陶管
Warerite 沃拉莱积(一种塑料贴面胶合板)
wareroom 货物陈列室;商品储藏室;商品陈列室
warfare 冲突
warfare agent 军用毒剂
warfare gas 军用毒气
warfarin 灭鼠灵
war geology 军事地质学
war housing 战时住房
warikahnite 瓦水砷锌石
war industry 军事工业
Waring blender 韦林混合器;韦林掺和器
warking structure of population 人口劳动构成
warm air 热空气;热风
warm-air blower 热风鼓风机

warm-air central heating 热风集中加热
warm-air circulation 热风循环
warm-air circulator 热风装置;热风循环器
warm-air curtain 火墙通气(小门);热风幕
warm-air distribution 热风分配
warm-air distribution duct 热风分配通道
warm-air door curtain 门上热风幕
warm-air duct(ing) 热风通道;管道暖气;热风管(道)
warm-air duct riser 暖气竖管;暖气管竖井
warm-air fan system 热风扇供暖系统;热风供暖系统
warm-air feed 暖气供给
warm-air front 暖锋【气】
warm-air furnace 空气减流器;空气加热器;热风炉
warm-air furnace heating 高炉热风采暖;高炉热气设备;热风炉供暖
warm-air heating 热风供暖;热风采暖
warm-air heating installation 热风设备;空气加热设备
warm-air heating plant 暖气设备;热风供暖装置
warm-air heating system 暖气加热系统;热风供暖系统;热风采暖系统
warm-air installation 热风采暖装置
warm-air mass 暖气团;温暖气团
warm-air outlet 热风排出口
warm-air pipe 热空气管(道);暖气管道;热风管(道)
warm-air pipeless heating 无管热风采暖
warm-air register 热风调节器
warm-air rising duct 热风上升管;升气管
warm-air stove 暖气(火)炉
warm-air with duct 管送热气;管送暖气
warm anticyclone 暖高压【气】;暖性反气旋
war map 作战图
warm area 温热区
war material 军事物资
warm avalanche 融雪崩
warm belt 暖水带(黑潮的高温水带)
warm bending test 热弯试验
warm bituminous mix 温铺沥青混合料
warm-blooded animal 定温动物;温血动物
warm-blooded animal pollution 温血动物污染
warm boost 热启动
warm cave 暖洞
warm climate with dry summer 夏干温暖气候
warm cloud 暖云
warm-cold seed treatment 变温浸种
warm colo(u)r 暖色(有暖感的颜色,指红黄橙等)
warm concrete 加温混凝土;热混凝土
warm core 暖核
warm-cored disturbance 暖心扰动
warm curing 采暖养护
warm current 暖流
warm current zone 暖流带
warm cutting production line for bar steel 棒钢温剪生产线
warm cyclone 暖性气旋
warm desert 热荒漠
warm-die forging 热模锻造
warmed air 热风
warmed water 热水
warm enough 温度充足
warm epoch 温暖期
warmer 取暖器;加温器;加热器;加热辊
warmer candle 保温蜡烛
warm fertilizer treatment 热性肥料处理
warm-film anemometer 暖膜风速表
warm floor cover(ing) 加温的地板面层;加温的楼板面层
warm floor(ing) 加温的地板;加温的楼板
warm flow 热流;暖流
warm fog 暖雾
warm forging 降温锻造法;温锻
warm-front fog 暖锋雾
warm-front precipitation 暖锋降水
warm-front rain 暖锋雨
warm-front surface 暖锋面
warm-front thunderstorm 暖锋雷暴
warm-front wave 暖锋波动
warm glacier 暖冰川
warm glue 热胶
warm ground 温热地面

warm hardening 加温人工硬化;加热硬化;热沉淀硬化;温时效;温人工硬化;人工硬化
warm heating 供暖
warm heating installation 供暖设备
warm high 暖性反气旋;暖高压【气】
warmhouse 温室;暖房;暖室
warmhouse effect 温室作用;温室效应
warm humid subzone 暖湿亚带
war mineral 战略矿物
warming 加温;加暖
warming appliance 加温装置
warming calender 预热压延机
warming chamber 保温箱
warming compartment 保温箱
warming device 加温设备;加温装置
warming-house 暖房;温室
warming-in 重热;加温
warming machine 暖气机
warming oven 食物保温箱
warming pan 暖床器
warming period 加热周期;加温周期
warming pipe 加热管;暖管
warming plate 暖盘
warming rate 加热速度;加温速度
warming roll 加热辊
warming time 加热时间;加温时间
warming unit 加热器;加温器
warming up 烤窑;加热;预热;升温
warming-up device 加温设备
warming-up mill 加热辊磨机
warming-up orifice 加热管路限流孔板
warming-up period 加热期
warming-up process 暖机过程
warming-up time 烤窑时间
warm-laid 温铺的(沥青混合料等)
warm-laid surfacing 热铺表面
warm lake 暖湖;热湖;温水湖
warm lime softening 热石灰软化法
warm loess 暖黄土
warm low 暖性气旋;暖低压【气】
warm manure 热性肥料
warm mastic wax 热用接蜡
warm memorial 战争纪念碑
warm mineral spring 温矿泉
warm moist air 暖湿空气
warm monomictic lake 热单对流混合湖
warm mortar 热性灰浆
warm neeedle-leaf forest 暖性针叶林
warm-occluded front 暖型锢囚锋
warm-operation 暖机运转
warm period 暖期
warm pool 暖(水)池;温水塘
warm premixing 热预混
warm-pressing 温压
warm rain 暖雨
warm register 热通风装置
warm ridge 暖脊;暖高压脊
warm rising duct 加热管道
warm rod 暖棒
warm rolling 温卷;温轧
warm roof 保暖屋房;保温屋面;保温屋顶
warm room 暖房
warm saw 湿热锯;热锯
warm sea 暖水区(指南北极圈以外的海洋)
warm season 暖季;温降
warm-season crop 暖季作物
warm-season grass 暖季牧草
warm-season plant 暖季植物;温季植物
warm section 暖区
warm sector 暖区
warm-setting adhesive 中温硬化黏合剂;热固化黏合剂
warm-setting glue 中温硬化胶
warm sludge 热泥
warm spot 热斑
warm spraying 温热喷雾;热喷雾
warm spring 暖泉;温泉
warm start 热启动;半热态起动
warm temperate district 暖带植物区;温带植物区
warm temperate forest 暖带林;温带林
warm temperate rain forest 暖带雨林
warm temperate subzone 暖温带亚带
warm temperate to subtropical subzone 暖温—亚热带亚带

warm temperate zone 暖温带
warm temperate zone mix forest yellow cinnamon soil zone 暖温带混交林黄褐土带
warm-temperature zone 温暖带
warm thunderstorm 热成雷暴雨
warm tone 暖色调
warm-toned 暖色调的
warm tongue 暖舌
warm to the tread 全部加热的;全部加温的;彻底加热的;彻底加温的
warm trough 暖流低压槽
warm-up 准备运动;活跃
warm-up characteristics 暖机特性
warm-up factor 暖机因数
warm-up mill 加热辊
warm-up pad 加热垫
warm-up period 预热期;增暖期;暖机时间
warm-up phenomenon 加温现象
warm-up temperature control 升温控制
warm-up time 预热时间;增温时间;加热时间
warm-up valve 暖机阀
warm upwelling (深层的)暖水上涌
warm wash 热洗净;热清洗
warm water 温水
warm water area 暖水区(指南北极圈以外的海洋)
warm water curing 热水养护
warm water drainage 温排(电站)
warm water environment 暖环境
warm water irrigation 温水灌溉
warm water lagoon 热水池
warm water mass 暖水团
warm water meter 暖水表
warm water method 热水供热方法
warm water port 不冻港
warm waters 暖水区(指南北极圈以外的海洋)
warm water sphere 暖水层
warm water stream 暖流
warm water thawing 温水融化解冻
warm water wetting 热水浸解
warm wave 暖浪;热浪
warm wear 防寒服装
warm wet climate zone 温湿气候带
warm white 暖调白色;暖白光
warm white light 热白光
warm winds 暖风
warm working 中温加工;温加工
warm zone 暖区;温带
warn 警戒;警报
warner 报警器
Warne's metal 白色装饰用合金
warning 警告;警报
warning agent 警戒剂
warning and alarm system 警戒和警报系统
warning apparatus 报警信号器
warning approach 接近预告牌
warning area 警戒区
warning beacon 报警无线电信号台
warning bell 警钟
warning blinker 警告闪(光警戒)灯
warning board 警告牌;警报牌;危险标志牌;危险标示牌
warning buzzer 蜂鸣警报器;警报蜂鸣器
warning center 警戒中心
warning character 报警符(号)
warning circuit 告警电路;报警电路
warning cock 警报龙头
warning colo(u)r 警告(颜)色
warning colo(u)ration 警戒色
warning communication 警报通信
warning condition 警告条件;报警条件
warning conduit 溢流管;溢流管
warning device 警告装置;警报装置;示警装置;报警装置
warning dye 警戒色
warning flag 警告旗
warning float 报警浮标
warning gas 警戒气
warning horn 警报汽笛
warning indicator 报警指示器
warning information 警报信息
warning lamp 警号灯;警告灯
warning laser radar 报警激光雷达
warning lead time 从发出警报到实际发生事情的时间
warning level 警戒级(别)
warning level for spectrochemical smog 光化学烟雾警报标准
warning light 保安警灯;险滩警灯;告警信号灯;警号灯;警告灯;警报灯;示警灯;报警信号灯;报警灯光
warning lighting 警报灯光
warning limit 预警界限;警报限
warning line 警戒线
warning line of water probing 探水警戒线
warning mark 警戒标志;报警标志
warning message 警告信息;报警报文
warning net 预警无线电网;报警网
warning notice 警告通知;警告牌
warning of air pollution 空气污染警报
warning of imminent movement (滑坡)临滑报警
warning paint 报警信号漆
warning period 准备信号时间
warning piece 报警部件
warning pipe (抽水马桶)水箱溢流管;露天溢流管;溢水管
warning post 警报标;示警桩
warning radio beacon 无线电警告航标
warning radio signal station 报警无线电信号台
warning receiver 警报接收机;监视接收机;报警接收机
warning region 警戒区域
warning response 报警响应
warning seismonet 地震预报网
warning service 报警站
warning ship 警戒船
warning sign 预告标(志);警号;警告标(志)
warning signal 警戒信号;警告信号;报警信号;报警信号
warning signal alarm device 报警装置
warning signal station 报警信号台
warning sonar 警戒声呐
warning stage 预警阶段;警戒水位;警报阶段;报警阶段
warning stage water level 警戒水位
warning station 警报站
warning system 警告系统;警告装置;警报装置;警报系统;报警系统
warning time 净预জ期
warning trap 报警陷阱
warning tube 溢水管
warning-up time 升温时间
warning value 警戒值
warning valve 警阀;报警阀
warning valve spring 警告阀簧
warning water depth 警戒水深
warning water level 警戒水位;警报水位;报警水位
warning wheel 警轮
warn-ocean fish 暖水性海鲜鱼类
warn start emissions 热启动排放物
war office map 军用地形图
warp 三角坑;扭曲;淤积物;光学扭曲;绞船索;挠曲;翘曲;反卷;凹凸;卷曲;弯翘
warpage 折曲;翘曲;弯曲度;弯翘;卷曲
warpage analysis 翘曲分析
war paint 战时伪装油漆
warp bands 分条痕
warp clay 冲积土;淤积黏土
warped basin 翘曲盆地
warped bollard 牵船柱
warped coordinates 歪曲坐标
warped roof tile 翘曲瓦
warped sheet 翘曲的窗玻璃
warped surface 挠曲面;弯翘面;翘曲面;扭曲面
warped tile 翘曲瓦
warped timber 翘曲木材
warped transition 曲面超高缓和段
warped transition approach 翘曲渐变进口(段);翘曲过渡进口(段)
warped wing wall 扭曲形翼墙;弯翘翼墙
warper 整经机
warper drum 整经机辊筒
war perils 战争(风)险
warper's beam 经轴
warp faulty lift 糙经
warping 翘曲;淤灌;折曲;绞缆;绞缆绞盘;绞缆绞车;绞船;扭屈;扭曲;挠曲;翘面;放淤;放灌;弯曲
warping bank 围淤地;放淤围堤

warping barrel 系缆卷筒
warping bollard 牵船柱;系船柱;绞缆柱;挽缆桩
warping bridge 尾桥【港】;船尾桥楼
warping buoy 绞滩浮筒;绞缆浮筒
warping capstan 卷缆绞盘;绞缆机
warping chock 翘曲导缆器
warping coefficient (混凝土路面)翘曲系数
warping constraint 翘曲约束
warping crack 翘曲裂缝
warping creel 经轴架
warping dolphin 绞缆筒;绞缆墩
warping drum 绞缆筒
warping due to temperature difference 温差扭曲
warping earthquake 挠曲地震;翘曲地震
warping effect 翘曲作用;弯翘作用
warping guide 导缆器
warping hawser 系缆;拖索
warping head 绞缆筒
warping joint 铰接缝;翘曲缝;弯翘缝;翘变接缝;翘变缝
warping lease 分经绞
warping line 绞船索
warping machine 整经机
warping machine with bobbin creel 带筒子架的整经机
warping moment 翘曲力矩
warping movement 翘曲运动
warping of cross-section 断面翘曲
warping of geoid 大地水准面起伏
warping of model 模型扭曲
warping of rail 钢轨上翘
warping of slab 板翘曲
warping post 绞滩岸桩
warping rigidity 挠曲刚度;翘曲刚度
warping speed 铰缆速度
warping stress 挠曲应力;翘曲应力;弯翘应力
warping test 翘曲试验
warping vessel 绞滩船
warping weir 放淤围堰
warping winch 卷绕式绞车;绞缆机;绞车;牵曳(式)绞车
warping works 放淤工程
warp in wood 木材翘曲
warp knitting 经编
warp laid rope 软搓绳
warp land 淤积地;淤灌地;洪漫地;潮汛积土;放淤地
warp line 翘曲线
warp-rebeaming machine 并轴机;倒轴机
war profiteer 发战争财的人
warp satin 经面缎纹织物
warp separator 分经器
warp soil 淤积土;沉积土壤;放淤土
warp strength 扭曲强度
warp surface 扭曲面;翘曲面
warp tape 包带
warp thread 帆布经线
warp-transmitting boat 递缆船;接头船
warp-transmitting ship 接头船
warp tying machine 结经机
warp-way twist 反手捻
warp winder 络经机
warrant 选购权证;支付命令;委托书;付款凭单
warranted clause 栏外条款
warranted efficiency 保证效率
warranted free 特约不赔
warranted output 保证出力
warranted unemployment rate 保证失业率
warranted value 评估价值;保证价值
warrantee 被保证人;被担保人;被保人
warranter 保证人
warrant fuel consumption 保证燃料消耗量
warranties 特约条款
warrant in bankruptcy 破产财产扣押令
warrant money 保证金
warrant of attorney 委托书;委任状
warrantor 担保人;保证人
warrant payable 应付收款凭单
warranty 质量担保;担保(书);保证;保单
warranty clause 保证条款
warranty cost 保证成本
warranty deed 全权证书
warranty for goods purchased 产品保证
warranty liability 保单负债
warranty life 保证使用期限;保用寿命

warranty of merchandise purchased 购得货品的保证
warranty of neutrality 中立保证
warranty of seaworthiness 海值保证
warranty of title 权利担保
warranty period 保用期
warranty test 验收试验;认可试验;保证性试验;证试验;保用性(鉴定)试验
warranty to sail 开航保证
war rate 战争险费率
war-ravaged area 战争破坏地区
warren 拥挤的地区;养兔场;拥挤房屋
Warren girder 华伦斜腹杆大梁;华伦(式)桁架;华伦大梁
Warren grinder 华伦磨木机
Warren hill series 华伦山统【地】
Warrenite-bitulithic pavement 双层沥青混凝土路面
Warren kiln 华伦砖窑
Warren motor 华伦式同步马达
Warren truss 华伦式屋架;华伦(式)桁架;斜腹杆桁架
Warren truss bridge 斜腹杆桁架桥
war reserves 战备物资;军需储备物资;军需储备物品
war reserve stock 战争储备物资
Warrington hammer 细木工用锤;华灵顿式铁锤
warringtonite 水胆矾
warrior 立射俑
warrior figures 武士俑
war risk 战争(风险);兵险
war risk agreement 战争险协定
war risk clause 兵险条款
war risk insurance 战争险保险;战时保险
war risk policy 战争险保险单
war risk rate 战争险费率
war risks only 仅保战争险
war ship 军舰;军用船;兵舰
Warsop 沃尔索牌工具(一些机动、风动工具)
war, strike and riot clause 战争、罢工及暴动条款
wart 树瘤;木节;小疙瘩
war term 军用术语
war theater 战区;战场
Warthe stade 沃尔特冰阶
wartime air supply duct 战时送风口
wartime blockage 战时封堵
wartime passageway 战时出入口
warty orebody 瘤状矿体
warty projection 窟状突起
warty structure 瘤状构造
war vessel 军舰
warwickite 硼钛镁石
was 金刚石晶体磨光面
wash 洗流;洗积;稀薄液体;海岸被淹地;漂流;排出流;涂覆;冲积砂矿;防止沾模的稀液;水洗;冲洗
washability 可选性;可洗(涤)性;可擦洗性;耐洗性;耐冲洗性
washability curves 可选性曲线
washability graduation 煤可选性分级
washability of coal 煤可选性
washability of ore 矿石可选性
washability sample of coal petrology 煤岩可选样
washable 耐洗的
washable distemper 可洗颜色胶料;可洗性色浆;可洗水浆涂料;可洗涤的刷墙粉;色浆
washable wallpaper 可洗的墙纸;可擦洗壁纸
washable water paint 可用水洗掉的水溶液颜料;可洗性水涂料;可洗掉水溶性颜料;耐洗的水性漆;水溶性颜料
washable white 耐洗涤的白漆
wash and brush of passenger train stock 车底洗刷
wash-and-drive method 边打入套管边冲洗岩粉钻进法
wash and strain ice foot 冲挤冰脚
wash ashore 冲到岸上
wash away 冲毁;冲走;冲洗掉
wash back 反冲洗
washbanding 刷彩
wash bare 冲洗暴露
wash basin 厕所;盥洗间;洗面器;洗脸盆;手手盆;盥洗槽
wash basket 洗衣篓
wash bath 洗槽

wash bead 盖面焊道
wash board 防波板;搓板;踢脚板;洗衣板;舷侧导水板;制荡板;墙脚板;(放在船头)防浪板;壁脚板
washboard course 洗衣板(式道路)
washboard effect (路面上的)搓板现象
washboard erosion 梯形蚀痕;搓板蚀痕
washboard glass 洗衣板玻璃
washboard groove 踢脚板企口
washboard heater 踢脚板加热器
washboard heating 踢脚板加热
washboarding 瓦楞纹;(路面上的)搓板现象;(路面上的)凸波纹状
washboard-like wave 搓板状波纹
washboard member 踢脚板部件;踢脚板构件
washboard pan 分级抖动板
washboard radiator 散热踢脚板;踢脚板散热器
washboard road 搓板状道路
washboard unit 踢脚板部件;踢脚板构件
wash boiler 洗衣锅炉;洗涤用锅炉
wash boring 冲钻法;喷射钻进;水冲钻进(法);水冲钻探;冲洗钻孔;冲水钻探
wash boring drill 洗水式凿岩机;水力钻(机);湿式凿岩机
wash-boring method 水冲钻进法;冲水钻孔法
wash boring rig 洗水式凿岩设备;水力钻进设备
wash bottle 洗涤瓶
wash-bottle for gases 气体洗涤瓶
wash bowl 脸盆;厕所;洗面器;洗手盆
washbowl willow 打土机
wash box 砂洗箱;洗涤箱;洗矿槽;跳汰机
washbox air cycle 跳汰机空气进出周期
washbox cell 跳汰机筛下气室
washbox centre sill 跳汰机中部拦板(在中部排渣室上)
washbox compartment 跳汰机筛上分室
washbox discharge sill 跳汰机溢流拦板
washbox feed sill 跳汰机给料拦板
washbox piston valve 跳汰机滑阀
washbox screen plate 跳汰机筛板
washbox slide valve 跳汰机滑阀
wash bulkhead 挡水舱壁
washburn core 隔片泥芯
washburn riser 易割冒口
Washburn-William's viscometer 瓦西布恩—威廉黏度计
wash cement 最后涂的水泥层;表面水泥
wash channel 洗涤槽;冲沙渠道
wash clay 黏土浆;黏土漂洗;黏土纯化
wash cloth 毛巾
wash coat 罩面层;木材封闭底漆
wash day 洗衣日
wash deck gear 洗甲板器
wash deck hose 洗甲板软管
wash deck pipe 洗甲板软管
wash deck pump 洗甲板水泵
wash deposit 冲积砂矿(床)
wash dirt 含金泥
wash dissolve 波纹叠化
wash down 洗清;冲刷掉;冲洗掉;洗掉
wash-down bowl 冲水马桶
wash-down closet 冲洗式便器;抽水马桶;冲洗厕所
wash-down pan 抽水大便盆;冲水马桶
wash-down type water closet 抽水马桶;冲水马桶
wash down water 洗孔水;洗井水;冲洗水
wash-down yard 冲洗车辆的场地
wash-down yard for cars 洗车场
wash drawing 水彩画;淡水彩画
wash drill 冲水钻孔装置
wash drilling 冲钻法;回水钻探;水冲(式)钻探;冲洗式钻进;(软基中的)冲水钻井法
washed air duct 洗净空气通道
washed by the sea 受海水冲刷的
washed chip(ping)s 洗净石屑
washed clay 洗黏土;淘洗(的)黏土;淘出的黏土;澄洗的黏土
washed coal 洗(选)煤;精煤
washed concrete 冲洗混凝土
washed concrete slab paper 洗过的混凝土平板纸
washed concrete surface 洗刷过的混凝土表面
washed drained weight 洗涤沥干物重
washed drift 成层冰碛;层状冰碛;侵蚀堆积物
washed drown 清洗过的;冲刷过的;冲洗过的
washed emitter technology 泡发射极工艺
washed finish 汰石子子面;洗石子面;水刷石(面

层);粉刷面
washed granolithic plaster 水刷石
washed graphite 精制石墨
washed gravel 洗净砾石
washed-gravel ballast 洗砾石道砟
washed kaolin 淘洗高岭土;纯高岭土
washed kaoline 洗陶土
washed material 洗净材料
washed metal 洗铁;精炼生铁
washed off soil 洗土壤
washed ore 洗矿;精选矿
washed-out diamond 被冲刷掉的金刚石
washed-out picture 模糊图像;淡白图像;冲溃图像
washed-out sand 水洗砂
washed-out section 冲蚀井段
washed precipitate 洗过的沉淀
washed pumice (gravel) 冲洗过的浮石
washed-residue 洗渣
washed sand 净砂;洗净砂;精造矿砂;湿法再生砂
washed sieve analysis 水冲筛分析
washed stock 低档砖
washer 洗涤塔;垫板;垫片;洗衣机;洗涤器;洗机;皮圈;垫圈
washer bearing 自洗轴承
washer boot 洗涤槽下料筒
washer bottle 洗气瓶
washer box 洗涤箱
washer breaks 洗涤折痕(整理疵点)
washer cutter 垫圈切割机
washer-dryer 洗衣脱水机;清洗干燥机
washer element 垫圈式滤清元件
washer for balance piston 平衡柱塞垫圈
washer-grader 清洗分级机
washer-locating pin 垫圈定位销
washer mark 洗呢斑
washer pump 冲洗泵
washer punch 垫圈冲床
washer roller 垫圈辊
washer shim 锚垫圈
washer solvent 洗涤液
washer-stoner 清洗碎石机
washer thermistor 垫圈形热敏电阻
washer wastewater 洗气器废水;洗涤器(废水)
washery 洗选厂;洗涤
washery effluent 洗涤厂泄出水
washery product 洗矿厂产品
washery refuse 洗涤废渣;尾矿
washery slag 洗渣
washery waste 洗涤废物
washery wastewater 洗煤厂废水
washery water 洗矿厂用水
washes 洗涤废水
washeteria 自助洗汽车店;自助洗衣店
wash-fast 耐洗的
wash filtrate 洗涤滤液
wash filtrate vacuum receiver 洗涤滤液真空接受器
wash fixture 洗涤设备
wash fountain 公用洗涤池;公共洗手池;喷水式洗手器
wash goods 耐洗织物
wash gravel 洗过的砾石;含金砾石
wash gun 洗涤用喷枪;冲洗用喷枪
wash gutter 冲洗水槽
wash-hand basin 洗面器;脸盆
wash-hand stand 脸盆架
wash header 洗涤集管
wash heat 渣洗;清洗熔炼
wash heating 熔化清除(氧化皮);渣洗法
wash heating furnace 清洗炉
washholder 防松垫圈
wash hollander 洗浆机
wash house 洗衣房
wash house effluence 洗衣房废水
wash house waste 洗衣房废水
washine out 淋溶
washiness 水分多
washing 洗选;洗矿;洗涤;金属涂覆;淘洗;冲刷浪蚀;冲刷;水洗;冲洗
washing action 冲洗作用
washing agent 洗涤剂;清洗剂
washing analysis 冲洗分析
washing analysis test 冲洗分析试验
washing and conditioning drum 洗涤和预加工转鼓

washing and screening machine 洗筛机
washing-and-sizing tower 冲洗分级设备
washing apparatus 洗矿机;洗涤器
washing appliance 洗选设备
washing apron 冲沙护坦
washing arm 吹洗枪
washing assistant 洗涤助剂
washing away 冲刷
washing away of bank 边坡冲刷
washing bank 岸滩冲刷层;冲刷(河)岸;冲刷段
washing barrel 清洗滚筒
washing basin 洗盆
washing beater 洗涤打浆机
washing beck 洗涤桶
washing blue 耐洗蓝;沉淀蓝
washing bottle 洗瓶;洗涤瓶
washing bowl 洗涤盆
washing box 冲洗箱
washing by jet 喷水洗涤
washing capability 冲洗功能
washing capacity 冲洗能力;冲洗功能
washing cell 洗选格
washing chamber 盥洗室;洗濯室;洗气室;洗涤室
washing circle 冲洗周期
washing circuit 洗选系统
washing classifier 洗涤分级器;洗涤分级机
washing column 洗涤柱
washing content 冲洗容量
washing cost 洗选费
washing cycle 洗涤周期;冲洗周期
washing cyclone 旋流洗矿机;洗选旋流器
washing cylinder 洗筒
washing cylinder lubrication 洗筒的润滑
washing dedusting 洗涤除尘(法)
washing device 洗涤设施;洗涤器
washing device motor 洗涤装置电动机
washing dolly 洗呢机
washing down 冲掉;冲刷;洗清;冲洗
washing drum 洗鼓
washing effect 冲洗效果
washing efficiency 洗选效率
washing elevator 洗矿提升机
washing engine 洗浆机
washing equipment 洗涤设备
washing erosion 冲洗侵蚀
washing fastness 耐洗性;耐洗坚牢度
washing filter 洗滤器
washing filtrate 洗滤液
washing flow 冲洗流量
washing fluid 冲洗液
washing fluid circulation 冲洗液循环
washing gates 转桨;(淘泥机的)洗耙
washing hand 洗手间;洗脸间
washing head 冲洗水头
washing-in method 冲洗油层法
washing installation 清洗装置
washing installation for vehicle 车辆清洗装置
washing jar 冲洗瓶
washing land 河漫滩
washing line 洗车线
washing liquid 洗(涤)液;清洗液
washing liquor 洗涤液
washing liquor exhaust pipe 洗涤液出口管
washing liquor tank 洗涤液储[贮]槽
washing-loss 洗矿损耗
washing lye 洗用碱水
washing machine 洗涤机;洗濯机;洗衣机;家用洗衣机
washing machine for gravel and sand 砂石料冲洗机
washing machinery 洗选设备;洗矿机械
washing marks 冲刷痕
washing medium 冲洗介质
washing method 洗涤法;冲洗法
washing method deodorizing 水洗法脱臭
washing mill 洗涤碾磨机
washing nozzle 洗涤液喷嘴;清洗喷嘴
washing of coal 洗煤
washing off 洗去;冲去;掀起;洗净
washing-off property 洗涤性能
washing of pipe 水管冲洗
washing of precipitate 沉淀(物的)洗涤
washing of tanks 洗舱
washing of water pipe 水管冲洗

washing oil 洗(涤)油
washing-out 冲坍;洗去;冲去;洗净
washing out boiler 洗刷锅炉
washing (out) loss 洗涤损失
washing out plug and drain cock 洗净放水开关
washing overboard 浪击落水
washing pan 淘洗盘
washing performance 洗选效果
washing period 水洗时间;冲洗周期
washing pipe 冲洗(水)管;洗涤水管
washing piping 冲洗管
washing place 洗衣间
washing plant 选矿厂;洗矿机;冲洗设备;冲洗场
washing power 洗涤本领
washing precipitation 洗气式沉淀
washing press 洗气式压滤机;洗涤压滤机;洗涤式压滤机
washing pressure 冲洗压力
washing process 洗选过程;洗选法;冲洗过程
washing pump 冲洗泵
washing rate 淘洗率
washing recess 洗涤凹槽;冲刷槽
washing refuse 洗选尾矿
washing roller 洗涤机
washing room 洗涤间
washings 洗涤物
washing salinity by irrigation 灌水洗盐
washing salt 洗盐
washing sand out 冲砂;冲沙
washing screen 洗石筛分机;洗矿筛;冲洗筛
washing screw 螺旋分选机;弹簧垫圈
washing section 洗涤段
washing-settling tank 洗涤沉降罐
washing siding for passenger vehicle 客车洗车线
washing sink 洗涤盆
washing soda 洗涤(用)碱;碳酸钠
washing solution 洗液;洗涤溶液;洗涤(混合)液
washing spray 洗矿喷水
washing stage 洗涤阶段
washing system 冲洗系统
washing table 洗矿台
washing tank 冲洗水箱;洗涤罐;洗涤槽;水洗槽;冲洗桶;冲洗槽
washing test 洗涤试验;冲刷试验
washing the well 洗井
washing thickener 洗涤浓密机
washing time 水洗时间;冲洗时间
Washington Bridge 华盛顿桥
Washington fir 美国松;洋松
washingtonite 钛铁矿
Washington rule 华盛顿法则(设计路拱的一种方法)
washing tower 洗涤塔
washing track 洗刷线
washing tray 冲洗盘
washing treatment 洗涤处理
washing trommel 洗涤转筒筛
washing-trough 洗涤槽;冲洗槽
washing truck 清洗卡车
washing tub 洗涤盆
washing tube 冲洗管
washing tubing 冲洗管
washing unit 冲洗单元
washing-up 清洗
washing-up sink 洗碟池;洗碟盆
washing vat 冲洗桶
washing velocity 冲洗速度
washing water 洗(涤)水;冲洗水
washing water 淋洗水
washing water inlet 洗涤水入口
washing water pump 洗涤水泵
washing water system 冲洗水系统
wash inside out 反面洗涤
Washita series 瓦希塔统【地】
Washita stone 瓦希塔(油)石
washland 河中滩地;河漫滩;泛滥地
wash leather (擦拭用的)鹿皮;油鞣革;软皮
wash leather glazing 软皮镶玻璃;嵌玻璃皮条
wash leather strip 软皮条;软皮带
wash light 贴墙光源照射墙面
wash line 清洗管路
wash load concentration 冲泻质浓度
wash load discharge 冲泻质输移率
wash load discharge per unit width 单宽冲泻质

输沙量
wash load(ing) 冲刷泥沙量;悬移质;细悬移质;河道冲泻质;泥沙冲泻质;冲刷质;冲刷量;冲刷负载;冲刷搬运(物);冲刷搬运泥沙
wash load material (泥沙)冲泻质
wash marking 冲刷痕;水痕
wash metal 洗炉金属液;清洗金属
wash method 冲洗法
wash mill 洗涤装置;洗泥机;搅浆池;搅拌池;淘洗机;淘泥机;冲洗机
wash off 冲毁;冲掉
wash-off coefficient 冲洗系数
wash-off relief 晕渣地貌
wash-off relief map 地貌晕渣图
wash-off soil 侵蚀土(壤)
wash of relief 水洗浮雕
wash of sea 海浪冲刷
wash oil 清洗油舱用的轻汽油
wash oil still 洗油蒸馏釜;洗涤蒸馏釜
washout(by flood) 水毁
washout closet 旧式抽水马桶(其中的一种)
washout developer 冲洗显影液
washout emitter 冲蚀发射极
washout flow rate 冲洗水流量
washout gate 冲洗口;冲沙闸(门)
washout hole 清洗孔
washout ink 容易洗去的织物油墨
washout machine 冲洗机
washout of earthwork 路基冲刷
washout of subcloud layer 云下洗脱
washout of the way 道路冲毁
washout of trickling filter 滴滤池冲洗
washout pipe 冲洗管
wash out pipet(te) 洗出移液管
wash out plug 洗炉塞
washout pump 冲洗泵
washout rate 清除速率
washout structure 冲蚀构造
washout test 冲刷试验
washout thread 不完整螺纹
washout type 冲洗式
washout type water closet 冲出式坐便器
washout urinal stall 冲洗型立式小便器
washout valve 泄水阀(门);泄水截门;清渣阀;排空阀;冲洗阀;冲刷阀;冲砂阀
washout W.C. 高盘抽水马桶(粪便承盘在存水弯之上)
washover (波浪冲成的)小三角洲;越浪堆积;冲溢三角洲;冲溢沉积;冲溢;冲积物;波成三角洲
washover bit 冲洗钻头
washover crescent 新月形波痕
washover delta 溢流三角洲
washover fan 冲溢扇
washover fan deposit 冲溢扇沉积
washover fan trap 浪积扇圈闭
washover fishing operations 洗鞋解卡法(用装有管鞋钻头的冲洗管柱钻透被卡钻具)
washover pipe 冲洗管柱
washover run 带管鞋钻头的冲洗管柱钻进回次
washover safety joint 冲洗安全接头
washover shoe 套靴式冲洗钻头;冲洗管鞋钻头
washover spear 冲洗捞管器
washover string 洗涤塔;冲洗液管线;冲洗管柱
washover velocity 冲洗液流返速度
wash painting 水墨画
wash pipe 冲(洗)管;洗井管柱
wash plain 冲积平原;冰水砂砾平原;冰水沉积平原
wash plugs and rings 清洗旋塞阀(过滤器)
wash-point penetrometer 水冲式贯入仪;带冲头触探仪;冲洗式贯入仪;贯入度仪
wash-point soil penetrometer 冲洗式贯入仪;洗点式土壤触探计
wash port 舷墙排水口;排水孔
wash port of ship 船舷排水孔
wash prime coat 可洗擦层;可洗擦涂层;防冲涂层;防冲层
wash primer 洗涤底漆;磷化底漆;金属底层处理用漆;蚀洗用涂料;蚀洗涂料;打底涂料;防蚀底层涂料;反应性底漆
wash primer process 涂料蚀洗处理
washproof 耐洗的
wash rack 冲洗格栅
wash rag 毛巾
wash ratio 洗涤比

wash river run rock 洗净的河卵石
wash rods 冲洗钻杆
washroom 洗涤间;洗衣房;厕所;盥洗室;洗手间;盥洗间
wash room equipment 洗涤间设备;盥洗室设备
wash room tray 盥洗室灰缸
wash room waste 洗衣房废水
wash sale 虚抛;诈欺交易
wash sample 冲洗样
wash season 冲洗季节
wash sink 洗涤盆
wash siphon 冲洗虹吸管;冲砂虹吸管
wash slope 崖脚缓坡;重力坡(脚);剥蚀基坡
wash solution 洗出溶液
wash stand 盥洗台;盥洗架;立柱式盥洗盆;洗车处;脸盆架
wash station 洗选间
wash strake 船首舷墙;防浪板(放在船头);舷侧导水板
wash tank 冲洗(水)箱
wash temperature 洗涤温度
wash thickener 洗矿浓密机;洗涤浓缩器
wash thinner 洗涤用稀释剂
wash throughly 充分冲洗
wash time 洗涤时间
wash transaction 冲销交易
wash trap 拦沙阱;拦沙槛;截沙阱;截沙池;沉沙坑
wash tray 洗涤塔盘
wash trommel 洗涤转筒筛
wash trough 冲洗水槽;洗槽
wash tub 洗槽;洗濯盆;洗澡盆;洗涤桶;洗盆
wash tub pipe 冲洗水管
wash up 洗矿精矿
wash-up sink(unit) 洗盆
wash waste 洗涤废水
wash-water 冲洗水;洗涤水;洗余水;洗桶水;洗矿水;冲洗用水;冲孔水
wash-water channel 冲洗水沟
wash-water gutter 冲洗排水槽
wash-water head 冲洗水头
wash-water pipe 冲洗水管
wash-water piping 冲洗水管道
wash-water pump 冲洗水泵;洗涤水泵
wash-water rate 冲洗水强度;冲洗水率
wash-water recirculation 冲洗水循环
wash-water reclamation 洗水回收
wash-water recovery system 洗涤水回收系统
wash-water settling tank 洗水澄清槽
wash-water speed 冲洗水流速
wash-water sump 洗水坑
wash-water system 洗涤水系统
wash-water tank 冲洗(水)箱
wash-water tower 冲洗水塔
wash-water trough 冲洗(水)排水槽
wash-water velocity 冲洗水流速
wash-wear 耐洗的
wash welding 熔面焊接
wash wheel 洗皮转鼓
wash zone 冲刷带
wasout 洗除;洗净;侵蚀处;淘汰;水害冲毁;冲掉;冲刷(坑);冲刷沟;冲刷地段;冲蚀;冲溃;冲毁处;冲坏
Wassermann reaction 乏色曼反应
Wasserman test 乏色曼试验
wastage 干缩;漏失量;破损;损耗(量);废料;剥蚀
wastage allowance 材料许可损耗率
wastage expenses 损耗费(用)
wastage in bulk or weight 总消耗;体积或重量总损耗
wastage moraine 消融冰碛
wastage zone 消融区
waste 下脚料;浪费;垃圾;弃渣;损耗;风化物;废弃;废的
waste account 废料账
waste acid 废酸
waste acid liquid 废酸液
waste acid tank 废酸箱;废酸桶;废酸罐;废酸容器
waste activated carbon 废活性炭
waste activated sludge 活性废污泥;废(弃)活性污泥
waste active sludge 废活性污泥
waste aggregate concrete 废集料混凝土;废骨料混凝土;废石集料混凝土;废石骨料混凝土;废品集料混凝土;废品骨料混凝土

waste air 余风;废(空)气
waste alkali 废碱液
waste alkali liquor 废碱液
waste ammonia liquor 废氨液
waste and old materials 废旧物资
waste and repair 消耗与补充
waste and scrap 废旧材料
waste and shower trap 下水口及存水弯
waste and stormwater collection section 排水管管理区
waste and vent 排渣及放气
waste appliance 剩余设备;无用设备
waste area 弃土场;弃土区;废料场
waste assets 损耗资产
waste assimilation 废水同化
waste assimilation capacity 废物同化能力
waste-assimilative capacity 废水同化能力;吸收废物容量
waste bank 弃土堆;弃土堤;废土堆;废料堆
waste basket 废纸篓;字纸篓
waste bin 废物箱;垃圾箱
waste blasting 凿岩爆破
waste boiler 废物锅炉
waste box 废物箱;废料箱;废件箱
waste bunker 矸石堆;矸石仓;废料槽
waste calcination facility 放射性废液煅烧设施
waste canal (水库的)溢水渠;溢水沟;溢流渠(道);弃水渠;退水渠;废水渠;废水沟
waste carbon 废碳
waste card 废纺梳棉机
waste cartage 废物搬运费;搬运废物;搬运废料
waste casting 废铸件
waste caustic 废(烧)碱;腐蚀废水
waste cell melt 废电解质
waste chamber 废料室
waste channel 溢流渠(道);弃水渠;弃水槽;退水渠
waste characteristic 废水特性
waste chemical reagent 废化学试剂
waste chemical substances 废化学物质
waste chemistry 废物化学
waste chrome liquor 废铬液
waste chute 弃水陡槽;废物溜槽;废物滑槽
waste classification 废物分类
waste coal 废煤
waste cock 排出开关
waste collecting chamber 废物收集室;废物储藏室
waste collecting station 废物收集站
waste collecting system 废水汇集系统
waste collection 废物储运;废物储(贮)藏
waste compactor 废物压实机
waste component 废物成分
waste concrete 弃置混凝土;废弃混凝土
waste concrete aggregate 废混凝土骨料
waste concrete pile 废渣混凝土桩
waste conditioning 废水调质
waste container 盛废物容器
waste containment 废料污染
waste contamination 废料污染
waste control 废物管理;废水控制;废料控制
waste conversion equipment 废物转化装置
waste conversion technique 废物转化技术
waste copper 废铜
waste cotton 废棉
waste courses 分离横列
waste cover 废料覆盖层
waste crude oil 废原油
waste cut 废方
waste cyanide 废氰化物
waste dam 溢流坝;弃水坝;废石堆场
waste database 废物数据库
waste data sheet 废物数据卡
wasted energy 损失的能量
waste-derived fuel 废物回收利用再生燃料
waste destructor 废物焚化灶;废物焚化炉
waste detection sounder 污水检测探针
wasted filtrate 废滤液
wasted industrial oil 工业废油
waste discharge 排废;废物排除
waste discharge point 排污点
waste disposal 垃圾处理;弃土处理;废液处理;废物弃置;废物处置;废(弃)物处理;废料处理;放射性废物处置
waste disposal basin 废物处置池
waste disposal chute 废物处理槽

waste disposal facility 废料处理设备
waste disposal in the depths of borehole 钻孔深部排污
waste disposal in the ground 废物土地处置
waste disposal plant 污水处理装置；废物处置厂；废物处理设备；废物处理场
waste disposal problem 废物处理问题
waste disposal shaft 废物处理竖井
waste disposal tank 废物处置池
waste disposal taxes 废物处置税
waste disposal unit 垃圾研磨机；废物处理装置；废物处理设施；废物处理设备
waste disposal works 废物处理厂
waste disposer 废物清除器；废物处置者
waste dissolver 回料搅拌机
waste district 漏损水分区
waste ditch 废水沟
wasted lime 废石灰
wasted power 损失功率
waste drainage pipe in basement 地下室清除垃圾管道；地下室清除废物管道
waste drainage point 排污点
waste drying 废热干燥
waste dump 矸石堆；废物堆；废石料；废料堆
waste-dump plan 废料堆平面图
waste-dump sampling 废石堆采样
wasted water 用过的水；污水；废水
wasted work 耗功
waste effluence 废水
waste effluent flume 废水沟
waste elbow 废弯管；清除垃圾弯管；清除废物弯管
waste electric(al) and electronic equipment 废弃电气电子设备
waste elimination 消除浪费
waste embankment 废石堆；废石坝
waste emulsion 废乳化液
waste enamel glaze 回收釉
waste energy 无用功；损失能量
waste exchange 废物交换处
waste fiberdisposal 废丝处理
waste-fill 废石充填
waste-filled stope 废石充填回采法
waste-fill mining 废石充填开采
waste firing 凿岩爆破
waste fish 杂鱼
waste fitting 清除垃圾装置；(设在隐蔽处的)清除废物装置
waste flow 废物流
waste fluid 废液
waste fluid burning plant 废液燃烧装置
waste food 食物垃圾
waste form stock-raising production 畜牧饲养业废水
waste from atomic energy industry 原子能工业废物
waste from coal chemical industry 煤炭化学工业废料
waste from coal mine 煤矿废物
waste from corn starch processing 谷物淀粉加工废水
waste from manufacturing processing 制造加工废物
waste from petroleum refinery 石油精炼废水
waste from stockfarming 畜牧废物；牧场废物
waste fuel 可燃废料；燃料渣；废燃料；燃料废渣
wastefull use of natural resources 自然资源的浪费使用
wasteful production 亏本的生产
waste gas 废气
waste gas absorption tower 废气吸收塔
waste gas analysis 废气分析
waste gas burner 废气燃烧器
waste gas burning 废气燃烧
waste gas burning furnace 废气燃烧炉
waste gas cleaning 废气净化；废气除尘
waste gas cleaning plant 废气净化设备
waste gas cleaning system 废气净化系统
waste gas collection 废气收集
waste gas combustion 废气燃烧
waste gas desulfurization 废气脱硫
waste gas emission 废气排放
waste gas emission control 废气排放控制
waste gas emission standard 废气排放标准
waste gas feed heater 废气送料加热器

waste gas filter 废气过滤器
waste gas header 废气总管
waste gas heat 废气热
waste gas heater 废(煤)气加热器
waste gas inlet 烟道气入口
waste gas loss 废气损失
waste gas pipe 废气管
waste gas pollution 废气污染
waste gas pollution control 废气治理；废气污染控制
waste gas purification 废气净化
waste gas purifying equipment 废气净化装置；废气净化设备
waste gas recovery 废气回收
waste gas recuperator 废气换热器
waste gas superheater 废气过热器
waste gas treatment 废气处理
waste gas treatment equipment 废气处理设备
waste gas treatment method 废气治理方法
waste gate 闸门；弃井闸门；弃气门；排放闸门；泄气门；废料排出门；放水闸门；放水门
waste geotechnics 废料岩土力学
waste glass 多条绞玻璃；废玻璃
waste grassland 荒芜草地
waste grating 地漏箅子
waste graveyard 废物填埋场
waste grease 下脚油；废油脂
waste grinder 垃圾破碎机；废胶磨
waste grinding machine 碎胶机
waste gypsum 废石膏
waste handling modification 废物处理装置的改建
waste heap 废料堆；废石堆场
waste heat 余热；废热
waste heat boiler 废热锅炉；余热锅炉
waste heat boiler survey 废气锅炉检验
waste heat box 废气箱
waste heat discharge 废热排放
waste heat disposal 废热处置
waste heat dryer 废热干燥机
waste heat engine 废热蒸汽机
waste heat evapo(u)rator 废热蒸发器
waste heat exchanger 余热交换器；废热交换器
waste heat flue 废气烟道
waste heating 废热加热；废热供暖；废气供暖；废煤气加热法
waste heat loss 余热损失；废热损失；废热损耗
waste heat management 余热处理；废热处理
waste heat oven 余热炉；废热炉
waste heat power generation 废热发电
waste heat recoverer 废热回收机
waste heat recovering equipment 余热回收设备
waste heat recovery 废热回收；余热回收
waste heat rejection 排气热损失
waste heat rotary dryer 废热干燥筒；余热回转烘干机
waste heat superheater 废热过热
waste heat treatment 余热处理；废热处理
waste heat utilization 废热利用
waste heat utilization of rotary kiln 回转窑余热利用
waste incineration 废物焚烧
waste incineration installation 废物焚烧装置
waste incinerator 废物焚烧炉；废物焚化炉；废料焚化炉
waste incinerator plant 废物焚化厂
waste-injection well 污水喷射井
waste inspector 检漏员
waste instruction 空指令
waste iron 废铁
waste land 荒芜地；荒地(地带)；废墟；废的
waste-land reclamation 荒地开垦
waste-land reclamation index 垦殖指数
waste leachate 废物渗滤液
waste leaf pulp 废叶渣
waste legislation 废物管理法规；废物管理立法
wasteless production 无废物生产
waste liquid 废液
waste liquid containing silver ion 含银离子废液
waste liquid from ammonia still 氨蒸馏塔废液
waste liquid solidification 废液固化
waste liquid solidify technology 废液固化技术
waste liquid tank 废液罐
waste liquor 余液；废液
waste liquor for pulping mill 纸板厂废水

waste liquor recovery 废液回收
waste liquor storage tank 废液储[贮]罐
waste load 废(物)负荷；废物含量
waste load allocation 废负荷分配
waste load control 废负荷控制
waste lot 荒地
waste lubrication 废油润滑
waste lye 废碱(液)
waste management 废物管理
waste management program(me) 废物管理规划
waste mantle 土的表层；风化层；废料覆盖层
waste material 矸石料；废弃物；废料
waste material dropped by farm animals 牲畜粪便
waste matter 废物
waste mercury salt 废汞盐
waste metal 金属废料
waste meter 漏损水表；漏水计
waste methanization 废物甲烷化
waste minimization 少废；废物最小量化；废物产生减到最低程度
waste molasses 废糖蜜
waste monitoring 废物监测
waste mo(u)ld 废模；一次性模
waste mo(u)ld gypsum 模型石膏废料
waste mo(u)ld process 废模法
waste mountain 废山
waste mud 废泥浆
waste of energy 能量消耗；能的消耗
waste of fuel 燃料渣
waste of stockfarming 饲养场废物
waste oil 废油
waste oil clot 废油凝块
waste oil collecting tank 废油收集箱
waste oil collecting vessel 废油收集船
waste oil collecting tank 废油收集箱
waste oil collection vessel 废油收集船
waste oil disposal 废油处置
waste oil disposal facility 废油处置设施
waste oil pollution 废油污染
waste oil proof 防废油(的)
waste oil-refining catalyst 废炼油催化剂
waste oil screw 废油桶塞
waste ore 废矿石；矿山废石堆
waste outlet 下水排水口；弃水出口；泄水口
waste packaging 废物包装
waste pad 绒垫
waste pad lubrication 绒垫润滑法
waste paper 废纸
waste paper basket 字纸篓；废纸篓
waste paper deinking agent 废纸脱墨剂
waste paper duster 废纸除尘器
waste paper-making wastewater 废纸造纸废水
waste paper recovery plant 废纸回收装置
waste pesticide 废农药
waste pickle 酸洗废液
waste pickle liquor 废酸(浸)液
waste pile 弃土堆；弃石堆；废石坝；废料堆
waste pipe 排泄管；排水管；浴洗废水管；污水管；废水管(道)
waste pipe elbow 废水弯管
waste piping 废水管
waste pit 废物坑
waste plain 冲积平原
waste plant 废物处理装置
waste plastics 废塑料
waste-plus-waste method 废物加废水法；废水混用法
waste pollution 废料污染
waste pollution strength 废物污染度
waste preventer (厕所用)冲洗水箱
waste prevention 废物预防
waste processing 废物处置；废物处理
waste product 工业废物；废(产)物；废(产)品
waste prohibited against dumping 禁止倾倒的物质
waste puller 打回丝机
waste pulping 废物研磨
waste pulp sludge 废纸浆污泥
waste purification system 污水净化系统
waster 浪费者；等外品；废物；废品；石工凿；次砖
waste range 荒地草场
waste receptacle 废物箱；废物储存器；垃圾箱
waste receptacle with pedal lift 脚踏污物桶

waste reception area 废物接收处
waste-reception facility of port 港口废物接收设施
waste reclamation 资源回收(利用);废物回收(利用);废料回收(利用)
waste recovering salvage of waste materials 废料回收
waste recovery 废物回收;废品回收
waste recovery by physical process 废物物理法回收
waste recovery power plant 燃烧废料发电厂
waste recycle system 废水再循环系统
waste recycling 废物再循环;废物循环使用;废物回收利用;废水再循环
waste reduction 减少废物
waste reduction at the source 废物源头削减
waste removal 污水排放
waste residual 废弃残余
waste residue 废渣
waste residue analysis 废渣分析
waste residue treatment 废渣处理
waste river 荒废河流
waste rock 弃渣;矸石;废石(头)
waste-rock yard 废石场
waster product 废品率
waster storehouse 废品库
waste rubber 废橡胶(料)
waster-waster 废镀锡薄钢板
wastes 废弃物
waste safety disposal 废物安全处治
waste safety treatment 废物安全处理
waste salvage 废水利用
waste sand 废砂
waste sand and gravel 废砂石
waste scavenging 废物清除;废气清除
waste shredder 废物撕碎机
waste silk 废丝(筋)
waste site 荒地;废弃场地
waste slag 废渣
waste sludge 废(弃)污泥
waste sludge flow rate 废弃污泥流量
waste sludge management 废渣管理
waste sluice 弃水闸;排放闸门
waste sluice gate 退水闸门
waste solid 废渣
waste solidification 废物固化
waste solidification equipment 废物固化装置
waste solids 废渣;废弃固体
waste solvent recovery unit 废溶液回收器
waste stabilization 废水的稳定
waste stabilization lagoon 废水稳定池
waste stabilization pond 废水稳定塘
waste stack 废水立管;废气烟囱
waste startup engineer 废水起动工程师
waste steam 废(蒸)汽
waste steam boiler 废汽锅炉
waste steam connecting branch 废汽支管
waste steam heat 废汽热量
waste steam heating 废汽取暖
waste steam heating facility 废汽采暖设备;废蒸汽供暖设备
waste steam heating installation 废汽采暖装置;废汽采暖设备
waste steam heating system 废汽采暖系统
waste steam injector 废汽喷嘴
waste steam network 废汽管网
waste steam pipe 废蒸汽管
waste steam(pipe) line 废蒸汽管道
waste steam pipe network 废汽管网
waste steam turbine 废汽汽轮机;废汽涡轮机
waste steam type winding engine 废汽卷扬机
waste steam utilization 余汽利用;废汽利用
waste stone 废石(头)
waste stopper 废水塞;废气塞
waste storage 废物储[贮]存;废料场
waste storage chamber 废料存放室
waste storage farm 废物储[贮]存场
waste storage tank 废物储存罐;废物储存箱
waste stream 荒废河流
waste stuff 废纸
waste substance 废物
waste sulfate pickle liquor 废硫酸盐浸液
waste sulfate solution 废硫酸盐溶液
waste sulfite 亚硫酸盐废液
waste sulfite liquor 亚硫酸盐(纸浆)废液

waste tank 漏水箱;废物储存槽
waste thermal energy 废热能量
waste time and energy 费时费力
waste tip 垃圾场;垃圾堆;废物堆
waste toiler 余热容器
waste traffic 空载交通;空驶车辆;无效交通
waste trap 污水管弯头;废物收集器
waste treatment 污染处理;废液处理;废物处理;废料处理
waste treatment biology 废水处理生物学
waste treatment plant 废物处理装置;废水处理厂
waste treatment requirement 废物处理要求
waste treatment station 废水处理厂
waste treatment system 废物处理系统
waste treatment technique 废物处理技术
waste truck 弃土车
waste tube 废水管
waste tubing 废水管
waste uranium 废铀
waste utilization 尾矿综合利用;废物(综合)利用;废料利用
waste valve 废料排出阀;放泄阀
waste vapo(u)r treatment 废汽处理
waste vault 废物库
waste vitrification 废物玻璃化
wastewater 损耗水;废水
wastewater advanced treatment 废水改善处理
wastewater aerating 污水曝气
wastewater agricultural reuse 废水农业回用
wastewater analysis 废水分析
wastewater analyzer 废水分析器
wastewater aquaculture 污水养殖
wastewater assimilation 污水同化
wastewater assimilative capacity 污水同化容量;污水同化能力
wastewater biological phosphorus removal 污水生物除磷
wastewater bio-treatment 污水生物处理
wastewater blending 废水掺和
wastewater bypass 废水岔道
wastewater characteristic 污水特征;污水特性
wastewater charges 污水征收费;废水收费
wastewater collection 污水收集;废水收集
wastewater collection and management system 污水收集管理系统
wastewater collection and treatment system 污水收集处理系统
wastewater collection facility 污水收集装置
wastewater collection system 污水收集系统;废水收集系统
wastewater composition 污水组分;污水组成;废水组成;废水成分
wastewater concentration 废水浓度
wastewater condition 废水状况;废水状态
wastewater containing aluminum 含铝废水
wastewater containing aniline 含苯胺废水
wastewater containing bio-refractory compound 含难生物降解化合物废水
wastewater containing formaldehyde 含甲醛废水
wastewater containing heavy metal 重金属废水
wastewater containing heavy metal ions 含重金属离子废水
wastewater containing high molecular weight compound 含高分子量化合物废水
wastewater containing high strength particulate organics 含高浓度颗粒有机物废水
wastewater containing low phosphorus 含低磷废水
wastewater containing methanol 含甲醇废水
wastewater containing naphthenic acid 含环烷酸废水
wastewater containing nickel 含镍废水
wastewater containing nitrobenzene 含硝基苯废水
wastewater containing organic pollutant 含有机污染物废水
wastewater containing phenol 含苯酚废水
wastewater containing phenol compound 含苯酚化合物废水
wastewater containing zinc 含锌废水
wastewater control 污水控制
wastewater conveyance 污水输送
wastewater customer 废水用户
wastewater decay rate 废水衰减率
wastewater decomposition 废水分解

wastewater dilution 污水稀释;废水稀释
wastewater discharge 污水排放;废水排放
wastewater discharge hydraulics 污水排放水力学
wastewater discharge location 污水排放位置
wastewater discharger 污水排放装置
wastewater discharge standard 排污标准;废水排放标准
wastewater disinfectant 污水消毒剂;废水消毒剂
wastewater disinfection 污水消毒;废水消毒
wastewater disposal 废水处理;废水处置
wastewater disposal basin 废水池
wastewater disposal engineer 废水治理工程师
wastewater disposal facility 废水处置设施;废水处理设施
wastewater disposal pump 废水泵
wastewater disposal system 污水处理系统
wastewater effluent 污水处理后出水;废水出水
wastewater effluent quality 污水处理后出水水质
wastewater ejector 污水射流泵;废水喷射器
wastewater engineering 排水工程(学);废水工程
wastewater environmental science 污水环境科学
wastewater equalization pond 污水调节池;废水调节池
wastewater facility 废水设备
wastewater farming 废水农用
wastewater fees 污水税
wastewater field 污水场;污染(地)带;废水带
wastewater flow 污水量;废水流量
wastewater flow equalization basin 污水径流调节池
wastewater flow norm 排水定额
wastewater flow quota 排水定额
wastewater flow rate 废水率
wastewater from acetylene slag 电压渣废水;电石渣废水
wastewater from alcoholic fermentation 酒精发酵废水
wastewater from ashpit 灰坑污水
wastewater from blast furnace 高炉煤气洗涤废水
wastewater from car washer 洗车废水
wastewater from cement mill 水泥厂废水
wastewater from cement plant 水泥厂废水
wastewater from cement product 水泥厂废水
wastewater from chemical industry 化学工业废水;化工废水
wastewater from coal chemical 煤炭化学加工废水
wastewater from coal chemical industry 煤炭化学工业废水
wastewater from coal gas plant 煤气厂废水
wastewater from coal mine 煤矿废水
wastewater from coking gas plant 炼焦煤气废水
wastewater from cold rolling 冷轧厂废水
wastewater from debarking operations 卸货场废水
wastewater from dyestuff industry 染料废水
wastewater from fermentation industry 发酵工业废水
wastewater from foundry 铸造厂废水
wastewater from gas generation plant 煤气发生站废水
wastewater from grapewine factory 葡萄酒厂废水
wastewater from hot rolling 热轧钢废水
wastewater from living 生活废水
wastewater from manufacturing process(ing) 制造加工废水
wastewater from mechanical industry 机械工业废水
wastewater from non-ferrous metallurgical industry 有色冶金工业废水
wastewater from ore dressing 选矿废水
wastewater from paper machine 造纸机器废水
wastewater from paper mill 造纸厂废水
wastewater from pesticide factory 农药厂废水
wastewater from petrochemical 石化废水
wastewater from petrochemical industry 石化工业废水
wastewater from petrochemical industry 石油化工废水
wastewater from petroleum refinery 炼油厂废水
wastewater from pharmaceutical industry 制药工业废水
wastewater from plating 电镀废水
wastewater from potato processing 马铃薯加工废水

wastewater from potato starch processing 马铃薯淀粉加工废水
wastewater from poultry farm 饲养场废水
wastewater from processing of coffee 咖啡加工废水
wastewater from pulp and paper mills 造纸(工业)废水
wastewater from rayon mill 人造丝(工)厂废水
wastewater from starch mill 淀粉工厂废水
wastewater from starch processing 淀粉厂废水
wastewater from sulphuric acid plant 硫酸厂废水
wastewater from textile mill 纺织废水
wastewater from town gas plant 城市煤气废水
wastewater from wet scrubber 烟道排气净化废水
wastewater from wood preparation 木材厂废水
wastewater groundwater recharge 废水地下水回灌
wastewater incinerating 废水焚烧
wastewater industrial reuse 废水工业回用
wastewater infiltration 废水渗透
wastewater influent 废水进水
wastewater infrastructure 排污系统基础设施
wastewater in receiving water 纳污水体污水
wastewater irrigation 废水灌溉;污水灌溉;污灌
wastewater irrigation norm 废水灌溉定额
wastewater irrigation system 废水灌溉系统
wastewater isolation flow 污水离析水流
wastewater lagoon 污水氧化塘;废水氧化塘;废水塘
wastewater levy act 废水征税条例
wastewater liquid residual 废水液态残余物
wastewater loading 废负荷
wastewater management 污水管理;废水管理
wastewater management facility 废水管理设施
wastewater management system 污水管理系统
wastewater mud 废水泥渣
wastewater municipal reuse 废水城市回用
wastewater nutrient removal 从废水中除去营养物;除废水营养物
wastewater of deposit 矿坑废水
wastewater of domestic use 生活污水
wastewater of heating and power stations 热电站废水
wastewater of paper mill 造纸厂废水
wastewater of spa 矿泉疗养池废水
wastewater of spring 温泉废水
wastewater of steel industry 钢铁工业废水
wastewater of textile industry 纺织工业废水
wastewater organism 污水生物
wastewater outfall 污水流出口;废水排出口
wastewater outlet 废水出口
wastewater oxidation 废水氧化
wastewater oxidation basin 废水氧化池
wastewater parameter 废水参数;废水特性参数
wastewater pickling 酸浸废液
wastewater pit 废水坑
wastewater pollutant 污水污染物
wastewater pollutant discharge standard 污水污染物排放标准
wastewater pretreatment 污水预处理
wastewater preventer 冲洗水箱
wastewater processing 废水处理工艺
wastewater processing station 废水处理场;废水处理站
wastewater property 废水特征
wastewater pump 废水泵
wastewater pumping 废水提升
wastewater pumping room 废水泵房
wastewater purification 废水净化
wastewater purification by algae 污水藻类净化
wastewater purification system 污水净化系统
wastewater quality 污水水质
wastewater quality index 污水水质指数;污水水质指标
wastewater quality standard 污水水质标准
wastewater rate 废水收费率;废水排放率;废水费率
wastewater recirculating pump 废水循环泵
wastewater reclamation 废水回收
wastewater reclamation facilities 废水回收装置
wastewater reclamation plant 废水回收装置;废水回收(工)厂
wastewater recreational reuse 废水的改善利用
wastewater recycle 废水循环
wastewater recycle system 废水再循环系统
wastewater recycling 废水(再)循环
wastewater recycling system 污水再循环系统
wastewater regeneration engineering 污水再生技术
waste water re-injection 污水回注
wastewater renovation 废水复用;废水重复利用;废水再生;废水再净;废水净化回
wastewater renovation plant 污水净化装置
wastewater renovation system 废水净化系统
wastewater renovation unit 污水净化装置;废水净化装置
wastewater requirement 废水处理要求
wastewater resources 污水资源
wastewater retention reservoir 污水滞留池
wastewater reuse 废水复用;废水重复利用;废水再用;废水回用
wastewater runoff 污水径流
wastewater sampling 污水采样;废水取样
wastewater seasonal variation 废水季节变化
wastewater segregation 废水离析
wastewater segregation system 废水分流系统
wastewater separator 废水分离器
wastewater sewer 污水沟道
wastewater sludge 污水污泥;废水污泥
wastewater soil purification 污水土壤净化法
wastewater solid 废水固渣
wastewater sources 污水源
wastewater spreading 废水蔓延
wastewater stabilization 废水稳定
wastewater stabilization pond 污水稳定塘
wastewater strength 废水浓度
wastewater stripping 污水汽提
wastewater sulfide 硫化物废水
wastewater survey 废水调查
wastewater system 废水管系统
wastewater tax law 废水征收法令
wastewater technology 废水技术
wastewater toxicity 污水毒性
wastewater treatability 废水可处理性
wastewater treatability study 废水处理可行性研究
wastewater treating process 污水处理工艺
wastewater treatment 废水处理
wastewater treatment by activated carbon process 废水活性炭处理法
wastewater treatment by algal growth 污水藻类生长处理
wastewater treatment by chemical precipitation process 废水化学沉淀处理法
wastewater treatment by high-gradient magnetic separation process 废水高梯度磁分离处理法
wastewater treatment device 污水处理装置
wastewater treatment efficiency 污水处理效率
wastewater treatment equipment 污水处理装置;废水处理设备
wastewater treatment evaluation facility 污水处理测定中心
wastewater treatment facility 污水处理装置
wastewater treatment in chemical industry 化工工业废水处理
wastewater treatment in textile-mills 印染工业废水处理
wastewater treatment method 污水处理法;废水处理方法
wastewater treatment objective 废水处理目的
wastewater treatment of dyestuff industry 染料生产废水处理
wastewater treatment of food processing 食品工业废水处理
wastewater treatment of ore-dressing 选矿废水处理
wastewater treatment of paper mills 造纸工业废水处理
wastewater treatment of wastewater 污水生物处理
wastewater treatment operation 污水处理运作
wastewater treatment optimization model 污水处理最优化模型
wastewater treatment plant 废水处理场;污水处理厂;废水处理工厂
wastewater treatment plant effluent 污水处理厂出水
wastewater treatment process 废水处理方法;污水处理工艺;废水处理过程
wastewater treatment scheme 污水处理方案
wastewater treatment station 废水处理站;废水处理厂
wastewater treatment system 污水处理系统;废水处理系统
wastewater treatment tank 废水处理池
wastewater treatment unit 废水处理设备;废水处理装置
wastewater treatment using highly concentrated ozone 高浓度臭氧处理废水系统
wastewater treatment works 废水处理工程
wastewater underground disposal 污水地下处置
wastewater use 污水利用;废水利用
wastewater utility 污水利用
wastewater utilization 污水利用;废水利用
wasteway 溢流道;弃水渠;弃水道;污水道;退水道【给】;废水道;废路
wasteway channel 排沙槽
wasteway section 退水段;废路段
waste weir 溢洪道;泄洪堰;溢流堰;量水溢流堰;弃水堰;退水堰;超量水溢流堰;废水溢汽堰
waste well 渗漏井;渗水井;泻水井;污水渗井;废水渗井
waste wood 废木料;废材
waste wood chipper 废木材削片机
waste wool 废羊毛
waste yard 垃圾场;弃土堆;废物场
waste yardage 不计价土方量;废土(土)方;废弃土方
waste yarn 废纺纱
wasting 石料次品;凿平;砌石整形;铲平
wasting assets 老旧的产业;消耗资产;减耗资产
wasting mineral resources 资源浪费
wasting of fund 基金的浪费
wasting of resources 资源的浪费
wasting rate of fixed assets 固定资产损耗率
wastometer (厕所)便器冲水计量装置
wastrel 废品
wat 大桶;大缸
watch 值班;观看;观察;警戒;监视;手表;表
watch and duty list 值勤名单
watch and watch 双班制
watch bell 报时钟声
watch bill 值班表
watch boat 巡逻艇;巡逻船;警戒艇
watch-bottom thread-ing lathe 表后盖螺纹车床
watch box 岗亭;守望亭
watch buoy 注意浮标;临时信号浮标;标位浮标
watch cap 烟囱罩;导烟帽
watch case (钟表)表壳
watch-case mo(u)ld 表壳式压模
watch chain 表链
watch compass 罗盘表;怀式罗盘
watch correction 船钟改正量;船表改正量
watch crystal 表面皿;表面玻璃
watchdog 监控设备;监控器;看门狗;监视器
watchdog subsystem 监视子系统
watchdog timer 监视(计)时器;监控时钟
watched crossing 有人看守道口
watched light 有人管理灯标
watch engineer 值班轮机员
watch engineer room 值班轮机室
watcher 值班员;监视器
watch error 手表误差;船钟误差
watch glass 表面皿;表(面)玻璃
watch glass clamps 表面皿夹
watch glass method 表玻璃法
watch glass size 表玻璃大小
watch-guard 表链带
watch holder 手表调整台
watch house 岗房;哨所;值班室;守望所
watching chamber 看炉
watching loft 屋顶瞭望所;顶层瞭望所
watching stick 探水杆
watching timer 监视计时器
watchkeeper 值班(人)员
watch loft 阁楼瞭望所
watchlog 监控设备
watch lubricant 钟表油
watch-maker's lathe 钟表车床
watch-maker's oil 钟表油
watch-maker's shop 钟表店
watchman 了望员;看守员;看守人;看守工
watchman's expenses 看守费
watchman's house 看守房
watchman's indicator 便嚣冲水阀的漏水指示器
watchman shed (古罗马)守夜人的职位

watchman's room 守望员室
watch officer 值班员;值班驾驶员
watch oil 钟表油
watch on deck 舱面值班
watch-quench tank 水淬槽
watch rate 日差率
watch rate recorder 校表仪
watch room 警卫室;值班室
watch section (测斜仪)钟表机构
watch spring 表的发条
watch system 值班制度
watch tackle 预备副滑车;轻便绞辘
watch test 表声试验
watch tower 瞭望台;瞭望塔;岗楼;望楼;守望台
watch turret 看守角塔
watch winding stem (测斜仪)钟表上弦手把
Watco 瓦特杀剂(一种木材杀虫剂)
WAT curves 重高温曲线
Wate 瓦特牌装配式房屋(商品名)
water 浇灌
water above frozen layer 冻结层上水
water above oil reservoir (油藏上的)上层水
water abrasion 水磨蚀;水(力)冲蚀
water-abrasive jet cutting 水磨料喷射切割
water-absorbed 吸水的
water-absorbed capacity 吸水能力
water absorbent 吸水剂
water-absorbent capacity 吸水容量
water-absorbent insole 吸水内底
water-absorbing 吸水的
water-absorbing capacity 水分吸收能力;吸水度;吸水能力
water-absorbing capacity of wood 木材吸水性
water-absorbing force 吸水力
water-absorbing form 吸水模板
water-absorbing natural stone 天然吸水石
water-absorbing polymer 吸水性聚合物
water-absorbing quality 吸水性;吸湿性
water-absorbing stone 吸水石
water-absorption 吸水性;吸水率;水分吸收
water-absorption character 吸水特性
water-absorption of malthoid 油毡吸水性
water absorption quality 吸水度
water absorption rate of rock test 岩石吸水率试验
water absorption test 吸水试验
water-absorptive capacity 吸水能力
water absorptivity 吸水性;吸水率
water absorptivity test 吸水率试验
water abstraction (从江河中)引水
water abundance 富水性
water accumulation 积水
water acidity 水的酸度
water act 水法
water action 水力作用
water-activated battery 水激活电池组
water activity 水分活性
water activity meter 水放射性计
water addition 加水
water adjustment 水分调节
water administration 水政
water administration supervision 水政监察
water admixing device 混合装置
water adsorption 吸水(性)
water adsorption test 吸水试验
water adsorption tube 吸水管
water aeration kinetics 水曝气动力学
water aesthetic quality 水的感官质量
water affairs dispute 水事纠纷
water affinity 润湿性;亲水性;水亲和力
water after-treatment 水的后处理;湿治
waterage 水运(运)费;水运
water aggressive to concrete 对混凝土有侵蚀作用的水
water-air-cement ratio 水气灰比
water-air-cooled machine 水外冷空气循环式电机
water-air mix(ture) 水气混合物
water-air ratio 水气比
water-air regime 水气状况;水气体系
water alarm 水位警报器
water albedo 水反射率
water allocation 水量分配
water alone 清水
water alumin(i)um chloride 水氯化铝
water ampo(u)e 炮眼用塑料水袋

water analysis 水的分析;水质检验;水分(分)析
water analysis index 水质分析指标
water analysis method 水分析法
water analysis of bacteria 水的细菌分析
water anchor 流锚;海锚;浮锚
water and effluent treatment plant 水与污水处理厂
water and electricity sector expenses 水电部门支出
water and gas connection 水及煤气接头
water-and-ink rendering 水墨渲染
water and land transportation 水陆运输
water and oil separator 油水分离器;水油分离器
water and power development authority 水利水力开发机构
water and power supply section 水电段
water and sediment content 水和沉渣含量
water and sediment test 水和沉渣含量测定
water and sewage pipeline 上下水道
water and sewer line 给排水线
water and sewer tunnel 供水隧洞;供水隧道
water and silt carrying capacity 过水输沙能力
water and soil conservation 水土保持
water-and-soil-conservation engineering 水土保持工程
water-and-soil-conservation measure 水土保持措施
water and steam distillation 水上蒸馏
water and wastewater industry 水工业
water and wastewater piping engineering 供水与排污管道工程;给排水管道工程
water annealing 水冷退火
water annulus 环形水道
water anomaly 天然水异常
water anti-pollution 防止水污染
water anti-pollution legislation 防止水污染法规;防水污染立法
water application efficiency 供水效率;施水效率
water application rate 用水率;供水率;施水率
water applicator 灌溉装置
water appropriation 水的专用权
water appropriation right 专用水权
water apron 护坦;消力池;净水池;消水池
water aqua 水剂
water aqueduct 渡槽
water arch 水拱
water area 水的面积;水域(面积);水区
water area environmental sanitary administration 水域环境卫生管理
water area environmental sanitary level 水域环境卫生水平
water area environmental sanitary management 水域环境卫生管理
water area environmental sanitation 水域环境卫生
water area facilities of ports 港口水域设施
water area for braking ship 制动水域【船】
water area for stopping ship 制动水域【船】
water area method 水域法
water arrester 水收集器
water as fuel 水作燃料
water-asphalt affinity test 地沥青亲水试验
water-asphalt preferential test 地沥青亲水试验
water as refrigerant 冷剂水
water assessment 水质评定
water-associated disease 与水有关的疾病
water atmosphere 水汽圈
water atomizer 水雾化器
water atomization 水雾化
water atomizing nozzle 雾化水喷头
water-attracting 亲水的;亲水物
water attraction 水分引力
water audit 水的审计
water authority 供水管理局;水管理局
water avid surface 不能浮起的面;亲水面
water back [贮]水器;加热水箱;加热管道;壁炉后热水箱
water backflow 回流水
water backwash 水反洗
water bacteria 水细菌;水生细菌
water bacteriology 水细菌学
water bag 蒸汽滚筒;水袋
water bailiff 船舶检查员;灌渠管水员
water balance 水(量)平衡;水文计算;水文过程量平衡;水量计算;水均衡

water-balance accounting 水量平衡核算
water balance method 水(量)平衡法
water balance observation method in closed watershed 闭合流域水均衡观测法
water balance plat 水平衡场
water balancing main 水平衡干管
water ballast 压载水;压舱水;水压载水衡重(镇船);水(压)载;水压舱;水衡重
water ballast chamber 底水舱;平衡水舱
water-ballasted tire 充水轮胎
water ballast hand roller 载水手推路碾
water ballast in caisson 沉箱注水;沉箱压载水
water ballasting 水荷载
water-ballast roller 水镇重压路机;水镇重路碾
water ballast space 压载水舱
water ballast system 水压载系统
water ballast tank 压载水舱
water ballast type roller 载水式压路机
water band 水带
water banking 水压坡度
water bar 止水带;阻水栅栏;拦木坎;止水条;拦水梗;水冷炉条;水冷炉排片;挡水条;路肩排水沟
water barge 运水驳船;供水驳船
water bark 供水船
waterbar material 止水材料;水密封材料
water barrel 水桶;提水桶
water barrier 防水层;拦水栅;拦水建筑(物);截水墙;防水材料
water bars 阻水栅栏
water-base 水封
water-base adhesive 水基黏结剂
water-base bore fluids 水基洗井液
water-base cleaner 水基清洁剂
water-base coating 水性涂料
water-base compound 水基封口胶
water-based 水基的
water-based epoxy anti-rust coating 环氧水性防锈涂料
water-based paint 水基涂料;水溶性涂料
water-base dressing 水基涂饰剂
water-based spray 兑水喷洒
water-based vehicle 水基载色剂;水基混合物
water-base finish 水基涂饰剂
water-base fluid 水基冲洗液
water-base foam 水基泡沫
water-base ink 水(溶)性油墨
water-base latex dispersion 水基胶乳分散
water-base latex paint 水基胶乳漆
water-base lubricant 水基润滑剂
water-base mechanical generator 水上的机械发烟器
water-base mud 水基钻泥;水基泥浆(钻泥)
water-base oil emulsion mud 水基油性井乳化泥浆
water-base paint 水溶性材料;水性涂料;水漆;水基涂料
water-base photosensitizer 水溶剂增感剂
water-base rotary drilling fluid 回转钻进用水基泥浆
water-base system 水基质系统
water-base transport 水上运输机
water basin 水盆;水池
water basin pollution prevention and control planning 水源区防污控制规划
water basin protection 水源区保护
water batcher 量水箱;量水槽
water batching 分批用水量;量水
water-bath 水浴(器);水浴锅
water-bath cooling 水浴冷却
water-bath evapo(u)rator 水浴蒸发器
water-bearing 含水(的);水垫层
water-bearing bed 含水(地)层;水层
water-bearing belt 含水带
water-bearing capacity 容水量
water-bearing concrete 含水混凝土
water-bearing contact zone 接触含水带
water-bearing deposit 含水(地)层;含水沉积层;含水沉积(物)
water-bearing fault zone 断裂含水带
water-bearing fissure zone in bedrock 基岩含水带
water-bearing fissure zone of interlayer 层间裂隙含水带
water-bearing formation 含水(地)层;含水岩层;含水结构;含水建造;含水层组织
water-bearing fracture zone 断裂含水带

water-bearing gas bed 含水气层
water-bearing ground 含水地层
water-bearing layer 含水层;蓄水层
water-bearing material 含水材料
water-bearing medium 含水层;含水介质
water-bearing oil bed 含水油层
water-bearing ratio curve method 含水率曲线法
water-bearing rock 含水岩(石);水岩
water-bearing sand 含水砂层
water-bearing section number 含水段编号
water-bearing sediment 含水沉积(物)
water-bearing soil 含水土(壤)
water-bearing stone 含水石
water-bearing strata formation 含水层建造
water-bearing stratum 含水岩层;含水(地)层;含地下水层;渗水地层
water-bearing structure 含水构造
water-bearing subsoil 含水底土
water-bearing vein zone 岩脉含水带
water-bearing zone 含水带;含水(砂)层
water-bearing zone of anticline axis 背斜轴部含水带
water-bearing zone of anticline pitching 背斜倾没端含水带
water-bearing zone of syncline 向斜含水带
water-bearing zone of weathering fissure 风化裂隙含水带
water bears 缓步类
water bed 透水层;含水层;水褥;水垫床
water behavio(u)r observation for mine 矿井水动态观测
water behavio(u)r observation for well 井水动态观测
water below oil reservoir 下层水
water belt 含水带
water benefit use 水的有效利用
water between frozen layer 冻结层间水
water-binder ratio 水与黏结料之比;水与结合料之比;水胶比;水和胶结料之比
water bind(ing) 水结(合力)
water-binding agent 水结合剂
water biology 水生物学
water birch 水桦
water bird 水鸟
water blanket conversion current regeneration 水顶压逆流再生
water-blast 集水爆炸;水流鼓风器;射水处理
water blasting face 水力爆破工作面
water bleeding from grease 润滑脂水分
water block 水箱;水封
water blocking 形成水柱
water blocking wall 挡水墙
water block seal 堵水密封膏
water bloom 湖靛;水华
water bloom control 水华控制
water blue 水溶蓝
water blush(ing) 水致发白(漆病)
water board 放水闸板;管水机构;水闸板;水利管理机构;防浪板(放在船头);水利局;舷侧导水板
water boat 供水船;给水船
water body 水体
water body assimilative capacity 水体同化容量;水体同化能力
water body biological pollution control 水体生物污染控制
water body classification 水体类别
water body environmental capacity 水体环境容量
water body environmental state 水体环境条件
water body environmental state assessment 水体环境条件评价
water body heat pollution 水体热污染
water body pesticide pollution 水体农药污染
water body pollution 水体污染
water body pollution control 水体污染控制
water body pollution control standard 水体污染控制标准
water body pollution source 水体污染源
water body protection 水域保护
water body self-purification 水体自净(化作用)
water body thermal pollution 水体热污染
water body thermal pollution source 水体热污染源
water body turbidity 水体浊度
water body water quality standard 水体水质标准

water boiler 热水锅炉;水锅炉反应堆;水锅炉
water-boiler reactor 水锅炉反应堆
water book 流水账
water booster system 供水增压系统
water boot 雨靴
water bore 水井
water-boring method 水冲钻孔法
water-borne 可干燥的;含水的;漂流的;水运的
water-borne adhesive product 耐水胶合产品
water-borne agreement 战争险水面协定
water-borne and soil-borne disease 水土病
water-borne architectural coating 水性建筑涂料
water-borne bacteria 水生细菌
water-borne bacterial disease 水媒细菌性疾病;水传染细菌性疾病
water-borne body 漂流物;水载物;水传物;水生物体
water-borne carrier 水上运输工具;船舶
water-borne clause 陆上危险不保条款;水运条款(指对陆上危险不保)
water-borne coating 水基涂料;水性涂料
water-borne commerce 水运量;水运业
water-borne contaminant 水载污染物;水媒污染物;水传播的污染物
water-borne deposit 水成沉积层;水生沉积层;水生沉积层;水成沉积(物)
water-borne detritus 漂流岩屑;漂流碎屑
water-borne disease 经水传播的疾病;水传疾病;水源病;水源性疾病;水媒疾病;水媒传染病;水传播病
water-borne displacement 水上排水量
water-borne dumper 倾卸船
water-borne electrode 水成电极
water-borne epidemic 水致流行病;水媒流行病;水媒传染病;水传时疫
water-borne floor paint 水性地面涂料
water-borne glue 水中胶
water-borne infection 介水传染病;水源传染;水媒传染;饮水传染
water-borne material 水生物质;水成物质
water-borne only 限于水上范围;只限于水运
water-borne pathogen 水传病原体
water-borne pathogenic organism 水传播病原生物
water-borne patrol service 水上巡逻工作
water-borne pesticide 水溶农药
water-borne policy 水运政策
water-borne preservative 水(溶)性防腐剂
water-borne radioactivity 水中放射性;水传播放射性
water-borne salt 水成盐
water-borne sediment 冲积层;水流挟带泥沙;水生沉积层;水生沉积层;水成沉积(物)
water-borne sewage 沟渠合流制
water-borne sewerage 沟渠合流制;雨污水合流制下水;水运系统管道;水流沟渠;水冲污水;水冲式沟渠
water-borne sound 水声;水传声音
water-borne throughput 水运通过能力
water-borne trade 海外贸易
water-borne traffic 水上交通;水运(运量)
water-borne transmission 经水传播
water-borne transport 航运;水运;水路运输
water-borne-type preservation 水溶性防腐剂
water-borne typochemical compound 水迁移标志化合物
water-borne typomorphic element 水迁移标志元素
water-borne vessel 水上运输工具;船舶
water-borne waste 水载废物
water bosh 冷却水槽;水桶;水封
water bottle 水瓶;采水器;采水瓶
water bottom 水垫;水底;双层底【船】
water-bottom gas producer 水封煤气发生炉;水底煤气发生炉
water-bound 水结的
water boundary 水界
water-bound base 水结路基
water-bound broken stone 水结碎石
water-bound broken stone road 水结碎石路
water-bound broken stone road surface 水结碎石路面
water-bound hoggin 水结碎石
water-bound macadam 水结碎石;水结碎石路面;马克当路面
water-bound macadam road 水结碎石路

water-bound macadam surface 水结碎石路面
water-bound surfacing 水结碎石面层
waterbowl 饮水器
water bowl paddle 饮水器阀门
water box 水箱
water brake 液压刹车;水压制动器;水力制动器;水力闸式测功器;水力闸
water-break 破坏堤;破浪堤;水膜破裂;水膜残迹;防冲结构;缓冲设施;断水;断流;减冲设备;紊流段
water breaker 消波器;淡水桶
water break-free 无水膜残迹的
water break-free surface 水膜不破表面
water-breathing animal 水呼吸动物
water-bubble 水泡
water budget 水量收支平衡;水文计算;水平衡;水量平衡(预算);水量计算
water budget method 水量平衡法
water buoyancy 水浮力;浮托力
water bursting coefficient 突水系数
water bursting yield from apical bed 顶板突水量
water bursting yield from bottom bed 底板突水量
water-bus 小型客船轮渡
water butt 储水桶;储[贮]水桶;接雨水桶;水桶
water by-laws 水法规
water by-laws advisory 水法规咨询
water cadaster 河流志
water calculation 水文计算
water calorimeter 水卡计;水量热器;水量热计
water caltrop 菱角
water-caltrop flower pattern 菱花图案
water can 浇水壶
water-cannon rock-breaker 水冲碎石机
water capacity 含水量;容水量;水量;水容度;土壤持水量;水容度;水容量
water capacity of bottle 水瓶容量
water capacity of the boiler 锅炉水容积
water car 运水车;洒水车;(运水用的)水车
water-carriage 水运(输);水上运输
water-carriage of garbage 水运垃圾
water-carriage pipe system 水力输送管系
water-carriage system 沟道系统;沟道制;水送系统;(粪便等)水送制;水输系统;水冲系统;输水系统;水运系统
water-carriage tunnel 输水隧道
water-carried paint 水基涂料
water-carried sewerage system 水输污泥系统
water-carried wastes 水运废物
water carrier 蓄水层;运水船;供水船;水上运输工具
water-carrier method 输水方法
water carrying 含水的
water-carrying agent 载水剂
water-carrying bridge 高架输水道
water-carrying capacity 水道容量
water-carrying capacity of channel 河槽过水能力
water carrying section 过水断面
water-carrying sewerage 水输污泥道
water carrying tunnel 输水隧洞;输水隧道
water cart 洒水车;运水车
water cascade 小瀑布;水梯级
water catch 集水工作
water catcher 水分分离器
water catchment 流域;取水口(取水头部)
water catchment area 集水(地)区;集水面积;水湖流域;受水区
water catchment range 集水范围
water catchment region 集水区
water catchment zone 集水区
water cavity 水流空蚀;水冷用空腔
water cell 滤水器
water cellar 水窑;地下水池
water cellulose 充水纤维素
water cement 水硬性胶凝材料;水硬(性)水泥;水凝水泥
water cement paint 水凝水泥涂料
water-cement ratio 水灰比(例)
water-cement ratio law 水灰比定律
water-cement ratio of concrete 混凝土的水灰比
water-cement ratio strength law 水灰比强度定律
water-cement ratio theory 水灰比理论
water cement slurry 水泥浆
water cement value 水灰比值
water chamber 水套;冷却水套;水箱;水池;水舱
water channel 水渠;水道;水槽

water channel drainage 水沟排水
water charges 水费
water charge line 加水管线
water charger (指泵的)充水装置;充水器
water charges of irrigation 灌溉水费
water charter 水事宪章
water check 挡水;阻水活栓;逆流截门
water checked 具有滴水槽的
water checked casement 阻水窗框
water check groove 阻水槽;滴水线
water chemical component 水化学组分;水化学成分
water chemical formation 水化学形成作用
water chemical property 水化学特征
water chemical type 水化学类型
water chemistry system 水化学系统
water chickweed 水繁缕
water chiller 冷水器;水冷装置;水冷却器
water chiller equipment 水冷却设备
water chiller package 水冷却成套设备
water chilling unit 冷水机组
water chlorination 加氯水处理;水加氯(处理);水的氯化处理
water circuit 循环水路;水流循环系统;水流流程
water circuit breaker 水中开关;断水器
water circulating pipe 水循环管
water circulating pump 循环水泵;水循环泵
water circulation 水文循环;水分循环;水(的)循环
water circulation coefficient 水循环系数
water circulation detector 水循环检测器
water circulation gravity system 水(的)重力循环系统
water circulation in the ground 地下水循环;地下水流通
water circulation pipe 水循环管线
water circulation pump 水循环泵
water circulation pumping station 循环水泵站
water circulation system 循环水系统;水(流)循环系统
water circulation velocity 水循环速度
water circulator 环流器;水循环器
water cistern 蓄水槽;蓄水池;储[贮]水器;水柜
water city 水都
water clarification 水的净化(作用);洗水澄清;水的澄清
water clarifier 滤水器;净水器(滤水器);水澄清器;水澄清剂;水净化器
water clarity 水(的)透明度;水(的)澄清度
water clarity meter 水明晰度计;水澄明度仪;测水澄清浓度
water classification 水的分类;水澄清(作用)
waterclear 清澈如水的
water clearance 隔水距离
water clock 水钟
water-clogged 水阻塞的
water closet 抽水马桶;冲洗式便器;盥洗室;卫生水厕;水冲马桶;水冲厕座;水厕;厕所
water closet bowl 抽水马桶桶身;便盘;便盆;大便器陶瓷盆
water closet compartment 厕所间
water closet flush tank 厕所冲水箱
water closet pan 抽水马桶(盆式桶身);冲洗水大便器;抽水便桶
water closet seat 马桶座圈
water closet suite 厕所套间
water closing 水封
water cloud 水(态)云
water cluster 水分子族
water coagulation 水混凝
water coating colo(u)r 水溶性涂饰着色剂
water cock 水旋塞;水龙头
water cock body 水塞体
water code 水利规范;水利法规;水的法规
water code for water rights 水权法规
water-coil 水蟠管
water collar 水冷套
water collecting 集水的
water collecting area 集水区;汇水面积;集水面积;汇水区
water collecting basin 集水池
water collecting gallery 集水廊道
water collecting header 高位集水箱;集水总管;集水联箱;集水横管;集水仓
water collecting pipe 集水管

water collecting range 集水区
water collecting region 集水区
water collecting ridge 集水土埂
water collecting zone 集水区
water collection area 汇水面积;流域面积
water collection work 集水工程
water collector 集水器;水收集器
water collect pump 集水泵
water colo(u)r 水色;水溶性色素;水合颜料;水的颜色;水彩颜料;水彩画颜料;水彩画(色);水彩
water colo(u)r ink 水性油墨
water colo(u)r meter 水色计
water colo(u)r painting 水彩画;水粉画
water colo(u)r paper 水彩画纸
water colo(u)r pigment 水合颜料
water colo(u)r scale 水色分级
water column 水位计;量水管;水柱;水面计;充水塔;水柱高度
water column arrester 水柱避雷器
water column in the well 井内水柱
water column oxygen demand 水柱需氧量
water column pressure 水柱压力
water combining power 水分结合力
water commissioner 水利专员;水利官员
water compaction 土壤压实排出的水
water companies association 给水公司协会
water company 自来水公司
water compartment 水舱
water compatibility 水相溶性
water compensation 补偿水
water complex 水络物
water composition 水的组分
water compressibility 水压缩性
water comsumption 水分耗损
water concentration 水选;湿选
water condenser 水冷凝器
water conditioner 净水器
water conditioning 调质处理【给】;水质调整;水质调理;水质处理;水的调制
water conditioning boiler 锅炉水软化
water conditioning device 水质净化设备
water conditioning plant 水处理设备;水净化厂;水处理厂
water conductivity 水分的传导度
water conduit 水渠;水管道;输水道;导水管
water-conduit bridge 水渠桥;渡槽桥;水管桥
water cone 水锥
water cone method 灌水法
water coning 形成水锥
water coning in oil reservoirs 油锥漏斗
water connection 水管结合配件;水管接头
water connection fitting 水管结合配件
water conservancy 水资源管理;水利(管理)局;水利(工程测量)
water conservancy annals 水利志
water conservancy commission 水利委员会
water conservancy conference 水利会议
water conservancy construction 水利建设
water conservancy facility 水利设施
water conservancy history 水利史
water conservancy law and regulation 水利法规
water conservancy measures 水利措施
water conservancy of agriculture land 农田水利
water conservancy project 水利工程
water conservancy residue sluicing method 水力冲渣法
water conservancy works 水利工程
water conservation 节水;水源保护;水利资源利用;水利;水库蓄水;水分保持;水的保护;保护
water conservation engineering 水源保护工程
water conservation forest 水源涵养林;水源保护林
water conservation law 水源保护法
water conservation measures 水源保护措施
water conservation measures in agriculture 农田水利工程;农业保水措施
water conservation method 节水方法
water conservation plan 节水计划
water conservation program(me) 节水计划
water conservation reservoir 蓄水水库
water conservation water closet 节水型坐便器
water consistency 标准稠度
water-consolidated 水(固)结的
water constant 水(分)常数

water constructional works 水工结构物;水工筑物;水工构造物
water consumed in water works 水厂自用水
water consumer 自来水消费者;自来水用户;用水户
water consumption 用水量;耗水量;水消耗(量);水(的)耗量
water consumption curve 耗水量曲线
water consumption figure 耗水量图
water consumption graph 耗水量图
water consumption in the field 田间耗水量
water consumption in water-works 水厂自用水;自用水量
water consumption norm 用水定额
water consumption of steam locomotive 机车用水
water consumption per day 日用水量
water consumption per unit output value 单位产值耗水量
water consumption per unit product 单位产品耗水量
water consumption quota 用水定额
water consumption rate 用水定额;耗水率
water contact 水流接点
water contact angle 水接触角
water contained in aggregate 骨料含水集料含水;骨料中的固有水分;(混凝土)集料中的固有水分
water-containing ratio 含水率
water-containing soda 碱水
water contaminant 水污染物
water contaminant level 水的污染含量
water contamination 水(的)污染
water contamination level 水污染水平
water content 含水率;含水重量;含水量;水(汽)含量
water content after swelling 膨胀含水量
water content analysis in inclusion 包体中水含量分析
water content as limit test 界限含水量试验
water content by weight 重量含水率
water content coefficient of mine 矿井富水系数
water content in capillary zone 毛细带含水量
water content in percent of dry weight 以(土壤)干重百分比计的含水率;以(土壤)干重百分比计的含水量
water content(in percent of total weight) 含水率
water content of frozen soil 冻土含水量
water content of mud 泥浆含水量
water content of particles 粉尘含水率
water content of soil 土的含水量
water content of timber 木材含水率;木材含水量
water content per unit volume of concrete 单位体积立方体混凝土用水量
water content profile 含水量纵剖面图
water content ratio 含水比
water content test 含水量试验
water contour 给水等值线
water contour percolation 水分水平渗漏
water control 治水;水利措施;水分控制;水的控制运用
Water Control Act 水管理法令
water control agent 止水化学(处理)剂
water control and delivery 治水和输水
Water Control Board 水管制局
water control device 水控装置
water-control facility 水流控制设施
water-controlled field 水驱油田
water controlled reservoir 水驱油层
water-controlling structure 壅水建筑物;控水工程
water control valve 水控阀
water control work 治水工程
water conveyance 水的输送;输水(设备)
water conveyance facility 水的输送设备;输水设施
water-conveying facility 输水设施
water cooked reclaim 水煮回收
water coolant 冷却水;水冷却剂
water-cooled 水散热;水冷(却)的
water-cooled absorptive shielding 隔热水箱
water-cooled after-condenser 水冷再冷凝器
water-cooled air compressor 水冷式空气压缩机
water-cooled air conditioner 水冷式空调器
water-cooled and water-moderated reactor 水冷却慢化堆
water-cooled automatic electrode holder 水冷式自动焊枪

water-cooled base 水冷底板;水冷炉底
water-cooled bearing 水冷轴承
water-cooled bottom plate 水冷底板
water-cooled brake 水冷式制动器
water-cooled brake drum 水冷式刹车卷筒
water-cooled brass connection 水冷黄铜接头
water-cooled capstan 水冷拉丝卷筒
water-cooled cathode 水冷阴极
water-cooled clamp 水冷夹头
water-cooled coil 水冷旋val
water-cooled compressor 水冷却压缩机
water-cooled condenser 水(冷式)冷凝器
water-cooled contact 水冷电接头
water-cooled copper electrode ram 水冷铜电极顶杆
water-cooled copper melting platform 水冷铜坩埚熔化台
water-cooled copper shoe 水冷铜极靴
water-cooled copper tube mo(u)ld 水冷管式铜坩埚
water-cooled cover 水冷炉盖
water-cooled crucible 水冷坩埚
water-cooled cylinder 水冷汽缸;水冷式汽缸
water-cooled diesel engine 水冷柴油机
water-cooled drum 水冷式转鼓
water-cooled electrode 水冷电极
water-cooled electrode bar 水冷电极
water-cooled electrode clamp 水冷电极夹
water-cooled electro-magnet 水冷电磁铁
water-cooled engine 水冷式发动机
water-cooled exhaust manifold 水冷式排气歧管
water-cooled footstep 水冷却轴承
water-cooled forced-oil transformer 水冷强制油循环式变压器
water-cooled former 水冷定型模
water-cooled furnace 水冷炉膛
water-cooled furnace body 水冷炉壳
water-cooled furnace wall 水冷炉壁;水冷炉墙
water-cooled generator 水冷发电机
water-cooled grate 水冷(却)炉箅
water-cooled grate incinerator 水冷炉箅焚化器
water-cooled grating 水冷炉栅
water-cooled head plate 水冷顶板
water-cooled heater 水冷加热器
water-cooled high frequency lead 水冷高频引线
water-cooled hoist 水冷式绞车
water-cooled housing 水冷式壳体
water-cooled ignition fuse 水冷式引燃管
water-cooled jacket 水冷式套筒;水冷(却)套;水套
water-cooled jacket for stuffing box 填料箱水冷水夹套
water-cooled lattice 水冷却栅格
water-cooled launder 水冷流槽
water-cooled load resistance 水冷式负载电阻
water-cooled machine 水冷(式)电机
water-cooled magnetron 水冷式磁控管
water-cooled mill 水冷磨
water-cooled motor 水冷电动机
water-cooled mo(u)ld 水冷模
water-cooled needle valve 水冷针形阀
water-cooled oil-immersed transformer 水冷油浸变压器
water-cooled oil-insulated transformer 水冷油绝缘变压器
water-cooled piston 水冷活塞
water-cooled probe 水冷式测针
water-cooled reactor 水冷反应堆;水冷堆
water-cooled refrigerant condenser 水冷式制冷剂冷凝器
water-cooled resistance 水冷电阻
water-cooled rolls 水冷式开炼机;水冷辊
water-cooled rotor 水冷转子
water-cooled scintillation counter 水冷闪烁计数器
water-cooled seal 水冷密封
water-cooled shell 水冷炉壳
water-cooled silencer 水冷式消音器
water-cooled source 水冷辐射源
water-cooled spacer 水冷分隔器
water-cooled stopper 水冷注塞
water-cooled suit 水冷服装
water-cooled transformer 水冷变压器
water-cooled tungstentipped rod 水冷钨尖电极
water-cooled turbine 水冷涡轮机;水冷汽轮机
water-cooled turbogenerator 水冷汽轮发电机

water-cooled tuyere 水冷风口
water-cooled type 水冷式
water-cooled valve 水冷式阀门;水冷阀
water-cooled wall 水冷壁;水冷却内壁
water-cooled welding torch 水冷焊炬
water-cooler 冷水器;水冷(却)器;船用水冷压缩机
water cooling 水冷却(法);水冷(法)
water-cooling cabinet 冷饮水器
water-cooling chamber 水冷室
water-cooling coil 水冷却盘管
water-cooling engine 水冷发动机
water-cooling groove 水冷沟
water-cooling induced draft tower 水冷却诱导通风塔
water-cooling jacket 水冷却套
water-cooling plant 水冷装置
water-cooling stabilizer panel 水冷稳定板
water-cooling system 水冷(却)系统;水散热系统
water-cooling tank 水冷却槽
water-cooling tower 凉水塔;水冷却塔
water-cooling tube 水冷管
water correction 改正处理;水质改正;水质改善;水的软化
water correction plant 水净化设备
water corrosion 水腐蚀
water cost 水费
water-cote method 洗出图形法【测】
water coulometer 水解电量计
water coupling 通水接头
water course 水道;水系;水路;水流;河流;渠道
water-course bed 常水河床;水道(基)床
water course line 山谷线
water course modification 河流整治
water courses pollution 水道污染
water course survey 水道勘测
water coverage test 水面遮盖力试验
water-covered area 水体覆盖区
water covering capacity 水面覆盖力
water-covering factor 水覆盖因子
water crack 抹面水纹;水纹;水裂;水淬裂纹
watercraft 艇;水运工具;船舶;船(只)
water crane 液力起重机;水柱;水压起重机;水力起重机;水鹤
water crane arm 水鹤臂
water crane column 水鹤柱
water crane indicator 水鹤表示器;水鹤标志
water crane jib 水鹤臂
water crane sign 水鹤标志
water crane stand 水鹤柱
water crane well 水鹤室
water creep 管涌;基底渗流;潜蚀;水的蠕升;渗水
water creeper 潜水蟒
water creep prevention 防止渗水
water crisis 水危机
water crop 出水量
water culture 溶液培养;溶液培养;水栽培;水培(养);水耕法
water cure 热水熟化;热水硫化;水养护;水疗;水处治
water cured concrete 水养护混凝土;湿养护混凝土;洒水养护混凝土;湿治混凝土
water curing 湿治;热水硫化;水养护;湿治水养护;湿养生;(混凝土)湿养护;(保护层)水养法
water-curing room (混凝土)湿养护室
water curing tank (混凝土)养护水箱
water current 水流
water current meter 水流表
water curtain 水幕
water curtain cave 水帘洞
water curtain collector 水幕除尘器
water curtain spray booth 水幕喷漆橱
water cushion 消力池;水垫
water cushion movable launch platform 水垫式活动发射台
water cushion pressure 水垫压力
water-cushion principle 水垫原理
water cut 冲刷坑
water cutoff 截водostena
water cutoff core barrel 断流阀式岩芯管(岩芯自卡时隔断内洗液)
water cutoff head 断流阀式岩芯管接头
water cutoff mastic 截水玛琋脂
water cut oil 含水原油

water cutting 水插
water cutting grafting 水插接
water cycle 水(文)循环;水分循环
water cycle area 水循环区
water-cycling system 水循环系统
water dam 水坝;防水墙
water damage(d) 水损(货物受水渍);水渍
water damaged cargo 水渍货
water damage insurance policy 水渍保险单
water damping effect 消波作用
water data unit 水资料机构;水资料单位
water deaerating equipment 水除气设备
water deck pipe 甲板水管
water decolo(u)ring 水除色
water decontamination 水的净化
water defence 防洪
water deficiency 水分不足;供水不足;缺水;水分不定
water deficient 水分不足
water deficient area 缺水(地)区;水量不足(地)区;缺水地
water deficient dehydration 缺水性脱水
water deficit 亏水量;水分欠缺;水分亏损;水分不足
water deficit region 缺水(地)区;水量不足(地)区
water deflector 导水器
water defluorination 水除氟
water defrosting 水融霜
water degassing 水除气
water deionized 除去离子的水
water delay line 水延迟线
water delivery 供水;输水
water delivery assessment 供水评议
water delivery by pipeline 管道输水
water delivery contract 供水合同
water delivery facility 输水设备
water delivery head 输水水头
water delivery lift 输水水头;输水高度
water delivery loss 输水损失
water delivery manifold 供水管线
water delivery tunnel 输水隧洞;输水隧道
water delivery valve 输水阀(门)
water demand 需水量
water-demand curve 需水量曲线
water demand of canal 运河需水量
water demand of output of ten-thousand-Yuan 万元产值取水量
water demineralization 水质软化;水脱矿物质;除去水中矿物质
water demineralizer 水软化器
water demineralizing 水的脱盐;水的软化
water deodo(u)ring 水除臭
water department 给排水管理局
water dependent well 水压驱动的井
water-deposited bank gravel 水积岸砾石
water-deposited material 水沉积物;水成沉积(物)
water-deposited matter 水成沉积(物)
water-deposited soil 水积(泥)土
water deprivation 缺水;脱水
water depth 水深
water depth above sill 闸室槛水深;门槛水深
water depth before and after dredging 施挖前后水深
water depth device 测深仪;测深设备
water depth diagram 水深图
water depth figure 水深图
water depth in front of bank 岸前水深
water depth in front of wharf 码头前水深
water depth line 水深线
water depth map 水深图
water depth marker 水深标志
water depth measurement 水深测量
water depth of an underground river 暗河水深度
water depth of ground river 暗河水深
water depth of port 港口水深
water depth of river 河床水流深度
water depth of shallow shoal 浅滩水深
water depth of submergence 淹没水深
water depth on front of wharf 码头前沿水深
water depth on sill 闸槛水深
water depth on the end launching way 滑道末端水深
water depth rod 测深竿;测深杆
water depth-ship draft ratio 水深吃水比;吃水系数

water depth signal 水深信号
water depth-to-ship draft ratio 水深吃水比
water depth under ocean bottom seismometer 洋底测震仪下水深
water depth under shot 炮点下水深度
water-derived fulvic acid 从水中取得的富里酸
water desalination 水的脱盐(作用);给水除盐处理;水脱盐
water desalting 除去水的盐分;海水淡化;脱水盐;水(的)脱盐;除去水中盐分
water descaling sprayer 水力除鳞喷嘴
water descaling unit 水力除鳞装置
water-detecting in advance 超前探水
water detection 水侦查;水鉴定法;水检测;水的检定
water determination apparatus 水分测定仪
water development 水显影;水利开发;水的开发;水资源开发
water-development policy 水开发政策
water diaphragm 水压膜片
water different pressure transducer 水差压变送器
water digestion reclaim 水煮回收
water dilutable 水稀释性
water dip 水负峰
water dip lacquer 湿表面浸绘漆;带水(工件)浸涂漆
water-dipper 船舶
water discharge 流量;排水量;水流(流)量
water discharge cock 放水旋塞
water discharge coefficient 排水量系数
water discharged output of ten-thousand-Yuan 万元产值排水量
water discharge pipe 泄水管(道);排水管(道)
water discharge piping 泄流管;泄水管(道);排水管(道)
water discharge relation curve 泄水关系曲线
water disinfection 水质消毒;给水消毒;水(的)消毒
water disinfection method 水消毒法
water disintegrating 水解法
water dispense 水分配站
water dispenser 饮水器
water dispersal 水散布的
water dispersible drier 水分散性催干剂
water dispersible epoxide 水分散性环氧
water dispersible liquid 水分散液;水分散剂
water dispersible oil 水分散性油
water dispersible paint 水分散性涂料
water dispersible pigment 水性颜料
water dispersible zinc stearate 水分散性硬脂酸锌
water dispersion 水分散体;水分散法
water dispersion baking paint 水分散性烘漆
water dispersion halo types 水分散晕类型
water displacement 排水量
water displacement agent 水置换剂(用于防锈剂中)
water displacement contact type generator 排水接触式乙炔发生器
water displacement method 换水法
water displacement test 水置换试验
water displacement vessel 排水容器
water-displacing liquid 排水液
water disposal observation 排污量观测
water disposal system 灌水装置系统;灌水配置系统;水处理系统
water disputes 水(纠)纷
water disputes tribunal 水纠纷法庭
water dissociation reaction 水离解反应
water dissolvable organic acid 溶水性有机酸
water distilling apparatus 水蒸馏器
water distilling barge 水蒸馏驳
water-distributing pipe 配水管
water distribution 配水;水(量)分布;水分配
water distribution and transmission 输配水
water distribution capacity 配水量
water distribution installation 配水设备;配水装置
water distribution network 配水(管)网
water distribution pipe 配水管(网)
water distribution piping 配水管道
water distribution policy 水分配政策
water distribution system 配水系统;配水网
water distribution to buildings 房屋供水
water distributor 布水管
water district 治水管理区;供水区;供水管理区;给水管理区
water disturbance 水体扰动;水的扰动
water ditch 水沟

water diversion 排水;引水;调水;导流;分水
water-diversion ditch 排水沟;分水沟
water-diversion gallery 分水坑道;分水隧道;分水廊道
water-diversion hole 分流孔
water-diversion project 引水枢纽;取水工程
water-diversion structure 分水建筑物
water-diversion tunnel 引水隧洞;引水隧道;导流隧洞;导流隧道;分水隧洞;分水隧道
water-diverting ridge 分水土埂
water-diverting structure 引水建筑物;引水构筑物;引水结构;分水建筑物
water diviner 试水器
water diving 用叉杆探水
water diving rod 探水杆;测水杆
water diving work 潜水作业
water division box 分水箱
water drag 静水阻力
water drain 排水沟
water drainage 排水
water drainage conduit 排水管渠
water drainage pipe 泄水管(道);排水管(道)
water drainage piping 泄水管(道);排水管(道)
water drainage system for yard 站场排水系统
water drainage tube 排水管(道)
water drainage tubing 排水管(道)
water drainage works 排水工程
water drain apparatus 排水器;排水阀(门)
water drain cock 放水旋塞
water drain funnel 排水漏斗
water drain valve 放水阀
water draught ga(u)ge 吃水深度计
water-drench system 水喷淋系统
water dresser 水理机
water drip 滴水槽;拔水屋檐;水滴;巢水井;集水井
water drip cooler 喷淋式冷却器;水淋冷却器
water drip test 滴水(腐蚀)试验
water drive 水驱
water-driven pool 水驱油藏
water-driven reservoir 水驱油层;水驱油藏
water-driven stamp mill 水力捣矿机;水碓
water drop 水滴;跌水
water drop facility 跌水设施
water drop glass 水滴花纹玻璃
water droplet 小滴;微水滴;水点;水滴
water dropper 滴水器;水滴集电器
water dropping 滴水
water-dropping aerating-bio-contact oxidation process 跌水曝气—生物接触氧化工艺
water-dropping aeration 跌水曝气
water drop scavenging 水滴冲刷
water drop step 跌水台阶
water drop test 水滴试验
water drum 下汽包;水包
water-drying 水干
water dumping dam 水中倒土坝
water dust scrubber 水膜除尘器
water duty 用水率;用水量;用水定额;灌溉率;灌溉定额
water duty per unit area 单位面积用水量
water ecologic model 水生态模型
water economic analysis 水利经济分析
water economics 水(利)经济学
water economizer 节水器;省水器
water economy 水收支;水利经济
watered 加水的
watered asset 实值低于账面的资产
watered capital 虚增资本;掺水资本
watered cup 加水漏斗
water eddy force 水漩涡力
water edge 河畔;水滨;水边
water-edge line 水边线
watered ground 含水土(壤)
watered oil 含水的油
watered out 出水开采
watered plug 加水旋塞
watered plug packing 加水旋塞填圈
water eductor 喷水器
water ejection 喷水
water ejector 喷水装置;水喷射器;水力喷射器;水喷射
water ejector pump 喷水抽气泵
water-electricity nuclear power plant 海水淡化发电两用核电站

water electric resources 水电资源
water electrode 水极
water-electrolyte imbalance 水电解质代谢紊乱
water-electrolytic hydrogen making equipment 电解水制氢设备
water elevation 水位高程
water elevator 扬水器;扬水机;提水机
water elimination 放水
water elutriation 水流冲涤;水淘选;水淘洗
water emptying 放空水
water emptying system 放空水系统
water emulsifiable disinfectant 水乳消毒剂
water emulsifiable oils 水溶性乳化油
water emulsifiable paste 水溶性乳化脂膏
water emulsion 水乳胶
water emulsion inhibitor 水乳化抑制剂
water emulsion paint 乳液涂料;乳胶漆;水乳化漆
water emulsion sludge 水乳化淤渣
water encroachment (开采石油时)水浸
water encroachment coefficient 水浸系数
water end 水端
water energy 水流能量
water engine 扬水机;水压机;水力机;水力发动机
water engineer 自来水工程师;给水工程师;水利工程师
water engineering 水道工程;给水工程;水利工程(学)
water enrichment 水的养分富化措施
water entering condition in wells and canals 井渠进水条件
water entering from the bottom and wall of well 井底井壁进水
water entering from the bottom of well 井底进水
water entering from the wall of well 井壁进水
water entering on one side 一侧进水
water entering on two side 两侧进水
water entering well with semi-spheric bottom 半球形井底进水井
water entrained by steam 蒸汽带水
water entrance 进水
water environment 水圈环境;水环境
water environmental assessment 水环境评价
water environmental carrying capacity 水环境承载容量;水环境承载能力
water environmental characteristic 水环境特征
water environmental chemistry 水环境化学
water environmental evaluation 水环境评价
water environmental information 水环境信息
water environmental information system 水环境信息系统
water environmental management model 水环境管理模型
water environmental manager 水环境管理工作者
water environmental parameter 水环境参数
water environmental planning 水环境规划
water environmental pollution 水环境污染
water environmental pollution forecasting 水环境污染预测
water environmental pressure 水环境压力
water environmental pressure index 水环境压力指数
water environmental protection 水环境保护
water environmental quality 水环境质量
water environmental quality assessment 水环境质量评价
water environmental quality index 水环境质量指数
water environmental quality standards 水环境质量标准
water environmental restoration 水环境修复
water environmental sample 水环境样本
water environmental simulation 水环境模拟
water environmental system 水环境系统
water environmental system engineering 水环境系统工程
water environmental treatment 水环境治理
water environment capacity 水环境容量
water environment limit 水环境极限
water environment management 水环境管理
water environment quality 地下水污染指数
water environment residual capacity 剩余水环境容量
water environment risk 水环境风险
water environment security 水环境安全
water environment security evaluation 水环境安

全评价
water environment worsening 水环境恶化
water equilibrium 水分平衡
water equivalent 含水当量;水当量(雪融后的水深)
water equivalent of snow 雪(的)水当量
waterer 饮水器
water erosion 水蚀;水侵蚀;水流冲刷;水冲刷;水冲蚀(作用)
water eruption 喷水
water escape 弃水道;泄水道
water-escape structure 泄水构造
water escaping 漏水(钻杆)
water evaluation 水质评价
water-evapo(u)rating furnace 水冷壁炉膛
water evapo(u)ration 水分蒸发
Waterex 水勒斯(一种防水液)
water examination 水质试验;水质检验;水质检查
water exchange 水量交换;水交换
water exclusion 止水;堵水
water excretion 水(的)排泄
water exhaust 排水沟
water exhaust conditioner 水洗式废气净化器;水洗废气净化装置
water exit interval 出水层段
water expansion 吸水膨胀
water-extended polyester 水扩张树脂;充水聚酯
water-extended resin 水扩张树脂
water extract 水浸提液;水抽出物
water extraction 水浸出物;抽水
water extraction method 水提取法;水萃取法
water extract method 水浸提法
water extractor 脱水器;脱水机
water eye 水眼(钻头)
water face 迎水面
water facility 水域设施
water factor 油水比;水因数
water factory 给水厂
water fade 水衰减
water failure 停水事故
waterfall 瀑布;跌水
waterfall aeration 跌水曝气
waterfall aerator 跌水曝气器
waterfall erosion 瀑布侵蚀;跌水侵蚀
waterfall model 瀑布模型
waterfall process 淋釉法;浇釉法
waterfall sequence theory 瀑布学说
waterfall siphon 虹吸式落水溢流管;落水虹管
waterfall syphon 落水虹管
waterfall theory 瀑布学说
water fascine 水下沉排;沉捆
water fast 不溶于水中
water fastness 耐水性;耐水度
water faucet 放水嘴;水嘴;水龙头
water-fearing 抗水的;憎水的
water-fed rock drilling 湿式凿岩
water fee 水费
water-feed cock 给水龙头
water feeder 供水器;供水管;给水设备;给水器;给水管
water feeder for boiler 锅炉给水器
water-feeding conduit 供水管道
water-feeding pump 给水泵
water-feeding system 供水系统
water-feed pump 供水泵
water-feed rock drill 湿式凿岩机
water-feed selector 进水调整器
water-feed sluice 给水闸
water-feed tank for drill 钻眼用供水筒
water fence 水中篱笆
water field 出水量
water filled capillary 充水毛细孔
water-filled pressure hull 充水耐压壳体
water-filled radiator 充水散热器;水暖气片
water-filled roller 充水镇压器
water-filled roof 充水屋面;水密封屋面;水封屋面
water-filled rubber membrane method 充水橡胶膜法
water-filled stemming bag 水封袋(代替炮泥用)
water filler 木器水性填孔剂;水溶性填料
water filling 灌水;充水
water filling plastic dam 充水塑料坝
water filling plug 注水塞
water filling system 注水系统;(船闸的)充水系统

water filling test 盛水试漏;充水试验
water-film 水膜
water-film bearing 水膜垫
water-film coefficient 水膜系数
water-film cyclone 水膜除尘器
water-film deduster 水膜除尘器
water-film dust collector 水膜集尘器
water-film electrostatic precipitator 水膜静电吸尘器
water-film separator 水膜除尘器
water-film theory 水膜理论
water filter 滤水器;滤水池;水滤器
water filtering glass wire 玻璃纤维滤水丝
water filtration 水的渗透;水(的)过滤
water filtration plant 水过滤装置;水过滤厂
water finder 探水者;地下水脉探测人员;油舱底部水量测定器;找水仪;试水器;测水器
water finish 水纹面饰;湿润处理
water finishing 湿修坯体
water fire-extinguishing system 水灭火系统
water fits 抽水工具
water fitting 水安装
water flea 鱼虫;水蚤
water flo(a)tation paint 水线漆
water flocculation 水絮凝
water flood 洪水
water-flooding 泛滥;注水;洪汛;水攻法(采油);人工水驱
water-flooding additives 水驱添加剂
water-flooding agent 水驱剂
water-flooding operation 注水作业
water flood line 注水管线;注水管路
water flood-retaining structure 挡水结构
water flow 水流(动)
water-flow-alarm system 水流量警报系统
water flow calorimeter 水流量热器
water flow chart 水流图表
water flow circuit 水流循环回路
water flow circulation system 水流循环系统
water flow control measures 水流控制措施
water flow deposit 水流堆积物
water flow direction 水流方向
water flowers 水花
water flow formula 流量(计算)公式
water flow ga(u)ge 水表
water flow impact pressure 水流冲击力
water flow indicator light 水流指示灯
water flow in river bend 弯道水流
water flow into tunnel 隧道涌水
water flow measurement 水流量测量
water flowmeter 水量(计)(水)表;流量表
water flow monitor 水流监测器
water flow net 水流场
water flow process 水流过程
water flow pyrheliometer 流水式太阳热量计;水流日射强度表
water flow recorder 自记水表
water flow regulating valve 水流调节阀
water flow regulator 水流量调节器
water flow relay 水流继电器
water flow retardation 拦滞水流
water flow retarding structure 缓流建筑物;滞洪建筑物
water flow simulation 水流模拟
water fluid 水基液体
water fluorination 水氟化
water flush 水冲;清水洗孔;清水洗井
water flush at well mouth 钻孔口涌水
water flush boring 清水洗孔钻进;清水洗井钻进;水冲钻探
water flush drilling 清水(洗孔)钻进;清水洗井钻进
water flushing 水冲;用水冲洗;清水洗孔;清水洗井
water flushing expulsion mechanism 水冲刷排驱机理
water flushing system 冲水系统
water flush system 清水洗孔法;清水洗井法
water flux 水通量
water foaming method (熔渣的)水淬泡沫法
water fog 水(态)雾
water fog spray method 水雾喷洒法
water fog sprinkler 水雾灭火系统
water foliage 水叶
water for agriculture 农业用水
water force main 给水压力干管

water for concrete 混凝土用水
water for consumption 消耗用水
water for curing 养护水
Waterford glass 蓝色燧石玻璃;华特福德玻璃
water for domestic usage 生活用水;家庭用水
water for domestic use 生活用水
water forecooler 水预冷器
water for fire-fighting 消防用水
water for fire-fighting purposes 消防用水;防火用水
water for human consumption 生活用水
water for injection 注射用水
water for irrigation 灌溉用水
water for living 生活用水
water-formed deposit 水生沉积物;水生沉积层;水成沉积层;水成沉积(物)
water-formed sediment 水生沉积物;水生沉积层;水成沉积(物)
water for road-washing 浇洒道路用水
water for use 循环用水
water foul 水禽
water fountain 饮水喷头;喷水池;引水喷头;团体洗手池
waterfowl 水禽;水鸟
waterfrac treatment 水力压裂处理
water frame 水力纺纱机
water-free 无水的
water-freedom arrangement 排水设备
water-free material 无水材料
water-free oil 不含水原油
water-free plankton 脱水浮游生物
water-free product 不含油品
water-free well 无水井
water freezes from the surface downwards 水从表面向下冻结
water freezing point 水的冰点
water front 滨水区;海水锋面;水缘;湿锋;岸线;岸边线;江边;水滨地;水边
water frontage (船厂、码头的)面临水域
waterfront area 港区;沿河地区;江岸地区;海港区
waterfront building 沿岸建筑
waterfront construction 沿岸建筑(物);驳岸工程;堤岸工程;水边建筑(物)
waterfront container complex 集装箱码头区
waterfront container crane 岸边集装箱装卸桥
waterfront factory 沿岸工厂
waterfront green 水边绿化;水边绿地
waterfront hydrostructure 沿岸水工建筑物
waterfront industrial area 沿海工业区
waterfront line 水沫线
waterfront of wharf 码头岸线
waterfront park 河滨公园;水边公园;水滨公园;滨水公园;沿岸公园
waterfront planning 港区规划
waterfront plant 沿岸工厂
waterfront property 沿河产业
waterfront region 沿岸地区
waterfront retaining wall 岸边挡土墙
waterfront station 港湾站;港口站
waterfront structure 堤岸结构;临水结构物;水滨建筑物;沿岸水工建筑物;沿岸结构(物);沿岸建筑(物);码头建筑(物);海岸建筑(物);水边建筑(物);驳岸结构(物)
waterfront utilization planning 岸线使用计划
waterfront zone 水旁带
water funnel 漏斗;水漏斗
water furnishing ability 供水能力
water furrow 明沟;排水灌毛沟;排灌犁沟;无盖排水沟;水沟
water gain 混凝土渗出水;(混凝土)泌水;泛浆(指混凝土表面泛出水泥浆);泌水(混凝土);增加水量
water gain of concrete 混凝土渗出水
water gallery 水廊;引水隧洞;引水隧道;输水廊道
water gang 河槽;人工水道
water gap 水峡;河流狭窄处;峡谷;水口
water garden 水生植物园;(二十世纪初美国庭园的)流水庭园;水景(庭)园;流水庭院;水上花园
water gardening 水生园艺
water gas 水煤气
water-gas catalyst 水煤气催化剂
water-gas cell 水煤气电池
water-gas coke 水煤气焦炭

water-gas condenser 水煤气冷凝器
water-gas generator 水煤气发生器;水煤气发生炉
water-gas lapweld process 水煤气搭焊钢管法
water-gas machine 水煤气发生器
water-gas pipe 水煤气管
water-gas pipeline 水煤气管线
water-gas pitch 水煤气沥青
water-gas process 水煤气过程
water-gas ratio 气水比
water-gas reaction 水煤气反应
water-gas set 水煤气装置;水煤气成套设备
water-gas shift 水煤气转换
water-gas tar 水煤气焦油
water-gas tar creosote 水煤气焦油杂酚油
water-gas tar emulsion 水煤气焦油乳状液
water-gas tar pitch 水煤气硬沥青;水煤气(焦油)沥青;水煤气焦油脂
water-gas weld 水煤气焊
water-gas welding 液压焊;水煤气焊接
water gate 水门;闸门;水闸(闸门);沉箱式坞门
water gate bridge 闸(门)桥
Watergate City 水门城(美国华盛顿的俚称)
water gate valve 自来水闸阀
water ga(u)ge 水位计;水表;量水表;水位(标)尺;水流量计;水尺;水标(尺);水压计
water ga(u)ge bell 测水钟
water ga(u)ge cock 水量表旋塞;水表旋塞
water ga(u)ge glass 水位计玻璃;水位表玻璃(管);测水管
water ga(u)ge lamp 水表灯
water ga(u)ge level 水尺高程;水尺水位
water ga(u)ge shield 水表护套
water ga(u)ge steam pipe 水位表汽管
water gel 水胶
water gel explosive 水胶炸药
water generating plant 水力发电厂
water generating station 水力发电厂;水电站
water geologist 水文地质学家
water geology 水文地质学
water giant nozzle 水枪喷嘴
water gillyflower 水生紫罗兰
water gland 水封压盖
water glass 古代的滴漏;硅酸钠;可溶性玻璃;水位玻璃管;水玻璃;水平表;玻璃水标尺
water-glass acid-proof concrete 水玻璃耐酸混凝土
water-glass-bonded refractory castables 水玻璃结合耐火浇注料
water-glass bonded sand 水玻璃砂
water-glass coat 水玻璃涂层;水玻璃胶
water-glass colo(u)r 水玻璃颜料
water-glass enamel 硅酸盐颜料;水玻璃搪瓷
water-glass grout 水玻璃注浆材料
water-glass mastic 水玻璃玛琋脂;水玻璃黏结剂;水玻璃胶
water-glass mix(ture) 水玻璃混合物
water-glass paint 硅酸盐油漆;硅酸盐涂料;水玻璃涂料;水玻璃油漆;水玻璃颜料;水玻璃漆
water-glass paint coat 水玻璃油漆涂层
water-glass putty 水玻璃油灰;水玻璃嵌料
water-glass refractory concrete 水玻璃耐火混凝土
water-glass slag mortar 水玻璃矿渣砂浆
water-glass solution 水玻璃溶液
water-glass with cover 带盖的玻璃器皿
water glaze 水面光泽;水光
water grade 水力坡度
water gradient 水力坡度
water grain 水粒纹
water-granulated slag 水淬粒状熔渣;水淬粒状矿渣
water granulation 水淬成粒(法)
water granulator 水碎槽
water grass 水草;湿地草原
water grate 水冷炉箅
water gravity selection 水选
water-greenhouse feedback 水气温室效应反馈
water grid 给水管网
water groove 水槽
water ground 含水层;湿法粉碎;富水土壤
water grown 水生的
water gruel 薄糊
water guard 海关稽查员;税务缉私人员;水上警察
waterguard branch 海关水上巡逻队
water guide 导水
water gun 喷水枪;水枪

water gutter 排水沟
water gypsum ratio 水膏比
water hammer 水击(作用);水锤;水冲(作用);压力冲击
water hammer absorber 水击吸收器
water hammer action 水锤作用
water hammer arrester 水锤制止器;水锤消除器;水锤防护装置
water hammer detector 水击吸收器
water hammer drill 湿式凿岩机;水冲钻;水冲钻孔(机)
water hammering 水击作用;水锤作用
water hammer in pipeline 管道水锤
water hammer noise 水锤噪声
water hammer of heat-supply system 供热系统水击
water hammer press(ing) 水击压力
water hammer pressure 水击压力;水锤压力
water hammer pulse 水冲状脉
water hammer pump 水锤泵
water hammer shock absorber 水锤击吸收器
water hammer tester 水锤试验机
water hammer wave 水击波;水锤波
water handling ability 保水能力
water handling system 水处理系统
water-hardened steel 水淬钢
water-hardened wire rod 水淬硬盘条
water hardening 水淬硬(化);水淬(火)
water hardness 水(的)硬度
water hardness indicator 水硬度计
water hardness recording instrument 水硬度记录仪器
water harvesting 回收用过的水
water-hating 疏水的
water haul 失败试验;打捞失败
water head 水柱高度;河源;水源;水位差;水头
water head above dam top 坝顶水头
water head canalized step 渠化梯级水头
water header 水总管
water head gradient 水头梯度;水头坡降
water head height of front dam 坝前水头高度
water head loss by filtration 过滤水头损失
water head loss of pumps 水泵水头损失
water head of step 梯级水头
water head site 水源地
water head value of grid point 格点水头值
water head value of node 节点水头值
water head value on boundary 边界水头值
water-heated calorifier 热水加热器
water heater 开水器;热水器;水火炉;温水器;水加热器;热水锅炉;煤气热水器
water-heater used for central heating 集中采暖热水器
water-heat exchanger 水热交换器
water heat(ing) 热水供暖;水加热;水暖
water heating appliance 热水加热器
water heating coil 水加热盘管
water heating system 水加热系统;水暖系统
water heating wash boiler 温水洗炉
water hemisphere 水半球
water hold down of conversion current regeneration 水顶压逆流再生
water holding 挡水;持水;保水
water-holding capacity 吸水量;积水量;保水(能)力;蓄水能力;含水容量;持水能力;持水(容)量
water-holding material 含水材料
water-holding power 持水力
water-holding rate 保水率
water-holding structure 挡水建筑物
water hole 水潭;(钻头的)水眼;水坑;水洞;沙漠中的水泉;水洼
water-hole peaking 水隙中的峰值
water horizon 含水层
water horning 喷水清理
water horsepower 有效功率;水马力;水功率
water hose 灌水软管;浇水软管;皮带管;水(龙)带;输水胶管
water hose inner diameter 水管内径
water hose resistance 水带比阻
water house 水房
water household 水量平衡;水分收支
water hull 水壳
water humus 水溶性腐殖质
water hyacinth 土葫芦;水风信子;风信子
water hydrant 消防龙头;消防栓;自来水消防栓;

自来水救火龙头;路边取水管
water hydraulic concrete pump 水压混凝土泵
water hydraulic-feed mechanism 水介质液压给进机构
water hydraulics 水压技术
water hydraulic valve 水压阀
water hygiene 水(质)卫生
water hygiene control 水质卫生控制
water ice 流下水冻成冰;人造冰;水冰
water immersion 浸水;水浸(渍);水浸装置
water-immersion objective 水浸物镜
water-immersion stability 浸水稳定性试验
water-immersion system 水浸系统
water-immersion test 浸水试验
water-immiscible medium 不混水培养基
water-immiscible solvent 不溶于水的溶剂
water impact detector 水冲击探测器
water impeller 排水叶轮
water impermeability 抗渗透性;水密性
water impermeability of malthoid 油毡不透水性
water impermeability test 不透水性试验;不透水性实验
water-impermeable compound 不透水化合物
water-impermeable zone 不透水带
water impervious 不透水带的
water imperviousness 水的不渗透性
water impinger (测定空气中灰尘含量的)湿式碰撞器
water importing from transwater shed 跨流域引水;跨流域调水
water impoundment 蓄水
water-impregnated 水饱和的;水充满的
water improvement 改水
water-impulse separator 水冲式分选机
water-inch 一英寸直径管孔24小时的放水量(约14立方米);水英寸
water inclination 水力坡度
water indicator 水位标志
water industry 水工业
wateriness 稀薄;多水
water infiltration and redistribution 水分渗透作用和再分配
water infiltration test method on large area 大面积灌水入渗试验法
water inflow 涌水;流入水量;进水口
water-in-flow of foundation pit 基坑涌水量
water influx 水注入;水侵量;水流入
water influx rate 水注入速度;水流入速度
water in frozen layer 冻结层水
water infusion 浸渍;浸泡;水浸;注水;回灌
water infusion blasting 注水爆破;分封爆破
water infusion method 灌水法;煤工作面注水法
water infusion tube (岩层注水用)注水管
watering 下河;灌水;加水;施水;洒水;超过资产实质的估价;掺水
watering barrow 洒水车
watering bucket 水桶
watering can 浇水壶;喷(水)壶;洒水器
watering car(t) 洒水车
watering device 饮水器;洒水装置
watering drilling 有冲洗液钻进
watering flower 浇花
watering frequency 灌水频率
watering groove 泻水槽
watering hole (夜总会酒吧间)卖酒处;社交场所
watering hygiene 给水卫生
watering machine 喷灌机
watering of farmland 农田灌水
watering period 灌水(周)期;浇水时期
watering place 饮水池;温泉疗养地;海滨浴场;饮水处;加水站;海水浴场;水疗场
watering pot 喷壶
watering rate 灌水标准
watering rating 灌水标准
watering station 加水站
watering trough (畜生的)饮水槽;浇灌槽
watering truck 洒水车
watering wagon 洒水车
water inhabitating insect 水生昆虫
water in intermediate water 浅水波
water injecting method 注水法
water injecting test 注水试验
water injection 地层注水;注水(钻井);喷水
water injection cooled engine 喷水冷却式发动机

water injection dredge(r) 喷水挖泥船
water injection drill 湿式凿岩机
water injection nozzle 水喷嘴
water injection of well 水井注入法
water injection pattern 注水方式
water injection pump 喷水泵
water injection rate 注水系数;注水速度
water injection servo 喷水伺服机构
water injection system 喷水系统
water injection test 注水试验
water injection test hole 注水试验孔
water injection-triggering earthquake 注水诱发地震
water injection valve 注水阀;水阀
water injection well 注水井
water injector 喷水口
water injector for coal train 煤注水机
water inlet 取水口;进水口
water inlet gallery 进水廊道
water inlet hose 进水胶管
water inlet line 进水管线
water inlet main 进水干管
water inlet pipe 进水管
water inlet piping 进水管
water inlet tube 进水管
water inlet tubing 进水管
water inlet tunnel 地下进水道;进水渠道
water-in-oil 水中水;油包水;水混油
water-in-oil emulsion 油包水乳剂;乳化液;油中水型乳液;油包水(型)乳液
water-in-oil test 油中含水量测定
water-in-oil type 油包水型
water in ore body 矿体水
water in percent of dry weight 含水比
water inrush 水的流入;水涌入
water inrushing of foundation pit 基坑涌水
water-in-sand estimator 砂内含水量测定仪
water insectivorous plant 水生食虫植物
water insolubility 不溶性水
water insoluble 不溶于水
water-insoluble cutting oil 水不溶性切削油
water-insoluble gum inhibitor 水不溶性防胶剂
water-insoluble matter 不溶于水的物质;水不溶物(质)
water-insoluble polyeletrolyte 水不可溶聚电解质
water insulation course 隔水层
water intake 进水口;取水量;取水口(取水头部)
water intake capacity of a well 钻孔的吸水能力
water intake gallery 取水道
water intake in maintain river 山溪进水结构
water intake on pontoon 浮船式取水口;浮船进水口
water intake on scaffold bridge 栈桥(式)进水口
water intake rate 吸水率
water intake sluice 引水闸;进水闸(门);渠首闸
water intake structure 引水建筑物;引水构筑物;取水建筑物;取水构筑物
water intake with filter 渗滤进水口
water intake with filter gallery 廊道式进水口;渗渠水口;渗滤式进水口;渗滤廊道进水口
water intake works 取水构筑物
water-intensive industry 耗水工业
water interception 断流
water interception cover 隔水盖
water internal cooling hydraulic generator 水内冷水轮发电机
water intoxication 水中毒
water intrusion 浸入水;水的渗透
water inundation 洪水泛滥
water invasion 水侵
water investigation 水文调查
water ionization 水离子化
water isotope 水的同位素
water-jacked metal ignition 水套冷却金属点燃管
water jacker 水冷套
water jacket 冷却水套;水衣;水冷套;水(夹)套
water jacket cooling ring 冷却水套
water jacket cover plate 水套盖板
water jacket cupola 水冷冲天炉
water jacket drain cock 水套放水旋塞
water jacket drain pipe 水套放水管
water-jacketed 水套的
water-jacketed carburettor 水套式化油器
water-jacketed chamber 水套冷却室

water-jacketed condenser 水套冷凝器
water-jacketed cooler 水套冷却器
water-jacketed furnace 水套冷却炉
water-jacketed oven 水套加热烘箱
water-jacketed producer 水套气体发生器;水套煤气发生炉
water-jacketed screw conveyer 水套冷却的螺旋输送机
water-jacketing 水套装置
water jacket outer wall 水套外壁
water jacket plug 水套塞
water jacket safety valve 水套安全阀
water jacket space 水套腔
water jacket top cover 水套顶盖板
water-jet 喷水口;喷射水;喷汽水;水注射;水射流;射水器;射水;射流;冲水;水射;水力喷射;水冲
water-jet actinometer 喷水日光能量测定仪;喷水日光计
water-jet aerator 水射曝气器
water-jet air ejector 喷水空气泵;射水抽气器
water-jet air pump 喷水空气泵;射水抽气泵
water-jet arrester 水柱避雷器
water-jet aspirator 水射抽气泵
water-jet blast 水喷射
water-jet blower 水喷射器
water-jet boat 喷水船
water-jet condenser 喷水冷凝器
water-jet cutting 水射流切割
water-jet descaling 喷水除鳞
water-jet drilling 射流钻进
water-jet driver 冲水打桩机;水射打桩机
water-jet driving 冲水打桩法;水冲打桩(法);射水打桩(法);射水打桩
water-jet dust absorber 水射吸尘器;射水除尘器
water-jet injector 水射流喷射器;水喷射器
water-jet-lift pump 水力扬射泵
water-jet method 喷射洗石法;射流打桩法;冲水法
water-jet method of dust-control 喷水抑尘法
water-jet method of pile-driving 水冲打桩法;冲射打桩法;水冲打桩;射水打桩法
water-jet pile driver 射水打桩机;冲水打桩机;射式沉桩机;水冲打桩机
water-jet pile-driving 冲水打桩法
water-jet piling 水冲沉桩
water-jet process 冲水法;水冲法
water-jet propelled boat 喷水船
water-jet pump 射流泵;喷水泵;喷射(水)泵;喷汽泵;水射泵;水射(真空)泵;射水泵
water-jet scrubber 喷水洗涤器
water jetting 注水;射水法;水冲法
water-jetting at high-pressure 高压射水
water-jetting method 水冲法
water-jetting process 水力射填法;水力充填法
water-jetting vehicle 街道喷洒水车
water-jet vacuum 水喷真空(用喷水泵能达到的真空)
water-jet ventilator 水射通风器
water joint 密接;水密接合;防水接头;防水接缝;水密接头;水接头
water jointing 水接头
water jump 水跃
water knockout drum 分水器
water knockout trap 分水器
water knockout valve 放水阀
water knockout vessel 分水器
water-laid 沉积的
water-laid deposit 水力冲填
water-laid material 水沉材料
water lancing cleaning 水枪清洗
water landing gear 水上起落架
water-land transshipment 水陆联运
water lane 水道;水上起落跑道
water law 水法
water law of state 国家水法
Water Law of the People's Republic of China 中华人民共和国水法
water layer 含水层
water-layered clay 饱水黏土
water-leach 水浸出
water-leach liquor 水浸出液
water lead 流送距离;水运起讫点
water-leaf 幌菊(饰);水叶装饰(柱头的树叶状装饰)
water leak(age) 漏水

water leakage zone 渗漏带
water leaking paddy field 漏水田
water leaves 水生植物叶(形饰)
water leg 水涨落速度装置;(锅炉下部)水夹套;水腿
waterless gasholder 活塞式储气罐;无水储气器;干式储气柜
waterless mo(u)lding sand 无水型砂
waterless technology 无水工艺
waterless toilet 干厕
water lettuce 水浮莲
water-level 水准器;水准(面);水位(高程);水平(面);水面
water-level alarm 水位警报装置;水位报警(器)
water-level altitude mining funnel 开采漏斗中心水位标高
water-level altitude of aquifer 含水层水位高程
water-level amplitude 水位变幅
water-level anomaly 水位异常
water-level at spring tide 春汛水位;大潮水位
water-level automatic regulating device 水位自动调节装置
water-level automatic regulator 水位自动调节器
water-level capacity curve 水位一水量曲线;水位一容积曲线;水位一库容曲线
water-level contour 水位等高线
water-level contour map 潜水等水位线图
water-level control 水位控制;水位调节
water-level controller 水位控制器
water-level control network 水位控制网
water-level control system 水位控制系统
water-level curve 水位曲线
water-level decline of recharge layer 补给层水位下降
water-level depression 水位下降
water-level depressor 水位波动缓冲器
water-level depth in soil sample cylinder 土样筒中的水位深度
water-level depth of aquifer from surface 含水层水位埋深
water-level descent curve 水位下降曲线
water-level diagram 水位图
water-level discharge relation curve 水位一流量关系曲线
water-level drawdown at the center of mining area 开采区中心水位降
water-level duration curve 水位延时曲线;水位历时(关系)曲线
water-level elevation 水面高程
water-level float 水位指示浮子;水位指示浮标;水位浮子;水平面浮子
water-level fluctuation 水位涨落;水位波动;水面波动
water-level frequency curve 水位频率曲线
water-level ga(u)ge 水位表;水位指示器;水位计;水位(标)尺
water-level gradient field 水位梯度场
water-level in common season 平水期水位
water-level indicator 水准指示器;水位指示器;水位表;水位标(示仪)
water-level in drought season 枯水期水位
water-level in rainy season 丰水期水位
water-level in survey area 测量区水位
water-level in well after casing 下管后的钻孔水位
water-level in well before casing 下管前的钻孔水位
water-level-duration curve 水位持续曲线
water-level line 水位线
water-level lowering 水位下降
water-level lowering curve 水位下降曲线
water-level mark 水位痕迹;水位标记
water-level measurement 水位量测;水位测量;水位测定
water-level measure post 水位尺
water-level measuring 水位测量;水位测定
water-level measuring post 水位站;水位尺;水位标示仪
water-level meter 自记水位仪;自记水位计
water-level needle ga(u)ge 水位针尺
water-level observation 地下水位观察;水位观测
water-level of lake 湖泊水位
water-level of reservoir 水库水位
water-level probe 水位探测头
water-level raised by wind 风激水位
water-level reactivity coefficient 水位反应性系数

water-level receiver 水位传感器
water-level record 水位记录
water-level recorder 水位记录仪;水位记录器;水位计
water-level recovery method 水位恢复法
water-level regulator 水准调节器;水位调节器;水平面调节器
water-level rising or falling at same velocity in rivers and canals 河渠水位等速升降
water-level scale 水位尺
water-level signal 水位信号
water-level staff 水位尺;水位标尺
water-level tell-tale 水位指示器
water-level transmitter 水位发送器;水位传送器
water-level variation 水位变化
water-level width 水位宽(度)
water lift 扬水机;提水工具;抽水工具
water lifting conduit 升水管
water-lifting device 提水设备
water lifting implement 抽水工具
water lifting machine 抽水机;提水机;扬水机
water lifting machinery 扬水机械;提水机械;抽水机械
water lifting pipe 升水管
water lifting piping 升水管
water lifting structure 升水沟筑物
water lifting tube 升水管
water lifting tubing 升水管
water lifts 提水器
waterlike solvent 类水溶剂
water-lily pool 睡莲池
water lime 水硬(性)石灰;熟石灰
water lime mortar 水凝石灰砂浆
waterline 水管线;水线附近;水(位)线;水迹印;水平线;水管线(路);吃水线;分水岭;上水道
waterline attack 吃水线处的浸蚀;水管腐蚀
waterline belt 船舶水线带
waterline bucket capacity 水线斗容
waterline coefficient 水下线面饱满系数
waterline construction equipment 水管线路施工设备
waterline control dam (控制水位的)拦河坝
waterline girth 水线周长
waterline length 水线长
waterline length of tanker 油轮的水线长度
waterline model 水线模型
waterline network 水管网
waterline paint 水线(带)漆;水线涂漆;水线涂料
waterline target 水位标志
waterline topping 水线带
water lining 波状水线
water liquid 水基液体
water load(ing) 水荷载;水负载;水负荷;水载;水压力
water loading of air 空气中水分含量
water loading test 水压加载试验
water load power meter 水负载功率计
water locating device 水位指示装置
water lock 水厕;水封;存水弯;(内河的)船闸;水闸
water-locked 被水包围的
water lodge 集水坑;水舱
waterlodged 地下水位过高的;积水的;浸透的
waterlodged ground 积水地
waterlodging field 水涝地
waterlodging land 水涝地
water log 积水;水涝
water-logged 浸透水的;浸泡水的;涝的;浸满水的
water-logged area 积水地区;水渍地
water-logged clay 饱和黏土
water-logged compost 沤肥;塘泥
water-logged depression 积水洼地
water-logged farmland 水涝地;水浸地
water-logged ground 含水饱和土;水涝地
water-logged land 水涝地;湿地
water-logged lowland 涝洼地
water-logged marsh deposit 覆水沼泽沉积
water-logged reed marsh deposit 开阔沼泽沉积
water-logged soil 积水土;浸透水的土壤;水浸土;渍水土(壤);浸润土
water-logged stratum 积水地层;含水层
water-logged wood 积水森林
water-logging 积水现象;浸透;浸满水的;水渗;水(浸)渍;渍涝(水);浸透水的;内涝;水浸(透)
water-logging control 除涝;防涝

water-logging disaster 涝灾;水涝灾
water-logging from autumnal rains 秋涝
water-logging resistance 抗淹性
water loop 水回路
water loss 消耗水量;损失水量;水量损失(量);水分流失;水的损失;失水量;失水
water loss and soil erosion 水土流失
water-loss control agent (泥浆)失水量调节剂
water-loss in drilling mud 钻井泥浆失水量
water-loss meter of mud 泥浆失水量计
water-loss shrinkage 干缩;失水收缩
water-loss shrinkage behavio(u)r 失水收缩情况;失水收缩性能
water-loss shrinkage crack 失水收缩裂纹;失水收缩裂缝
water-loss shrinkage curve 失水收缩曲线
water-loss shrinkage limit 失水收缩极限
water-loss shrinkage stress 失水收缩应力
water-loss shrinkage value 失水收缩值
water-loving plant 喜水植物;亲水植物
water lowering 降低(地下)水位;水位降低
water-lubricated 水润滑的
water-lubricated bearing 水润滑轴承
water-lubricated feed pump 水润滑给水泵
water-lubricated fluid-film bearing 水润滑液膜轴承
water lute 水封
water luted producer 水封发生器
water machine 抽水机
water magic 流水魔术
water magnetization 水(的)磁化
water magnetization treatment 水磁化处理
water magnetizer 磁水器
water main 总水管;供水总管;给水总管;输水主管;输水管;输水干管
water main booster pump 总水管增压泵
water main fittings 给水干管配件
water main(pipeline) 给水干管
water-make 涌水量
water-make in day and night 昼夜涌水量
water make-up 补给水
water make-up tank 补给水箱;配水箱
water management 用水管理;水源管理;水量水质管理;水分管理;水(的)管理
Water Management Act 水管理法令
water management activity 水管理活动
water management agency 水管理机构
water management agreement 水管理协议
water management and conservation 水分管理和保持
water management area 水管理区
water management arena 水管理区界
water management complex 水管理综合体
water management criteria 水管理准则
water management decision 水管理决策
water management decision-making process 水管理决策法
water management district 水管理区
water management index 水管理指标
water management institution 水资源管理机构;水资源管理部门
water management measure 水管理措施
water management method 水管理方法
water management model 水管理模型
water management objective 水管理目标
water management option 水管理方案
water management outline planning 水管理规划
water management plan 水管理计划
water management policy 水管理政策
water management practice 水管理实践
water management program(me) 水管理计划
water management regulation 水管理法规
water management scenario 水管理情况
water management strategy 水管理对策
water management system 水管理系统
water management unit 水管理单位
water manager 水管理人员
water mangle 压力脱水机
water manometer 水柱压力计
watermanship 水手技术
water mark 最高高潮痕;水渍;水准面;水浸;水(迹)印;水斑;吃水标志;潮汐(标)尺;水位标志
watermark 透明水印;水位标
watermark board 水位揭示牌
watermark disease 水纹病

water marking 水市场
water marks 船舶吃水标尺
water maser 水微波激射
water mass 水团;水体
water mass analysis 水团分析
water mass continuity 水体连续性
water mass density 水的密度
water mass identification 水质辨识
water meadow 淹水草地;渍水草甸;水草滩;水草地;低水草地;漫洪草原
water measuring 量水(的)
water measuring and monitoring system 水质监测系统
water measuring structure 量水建筑物
water measuring tank 计量水箱;水计量罐
water-melon green 西瓜皮绿
water membrane 水膜
water-membrane scrubber 水膜吸尘器
water metabolism 水分代谢
water meter 量水器;水量计;水(量)表
water meter box 水表箱
water meter chamber 水表井
water metering 量水;水表计量
water meter lid 水表盖
water-meter load factor 水表(面)负荷系数
water meter metering 水表计量
water meter pit 流量计水坑;流量计水槽
water meter system 水表系统
water microbiology 水(域)微生物学
water microorganism 水微生物
water mill(ing) 水磨;水车
water mines 水资源
water mining 地下水开采
water mining and use 地下水超量开采
water mint 水生薄荷
water miscibility 水混溶性
water miscible 水溶性的
water miscible formation 水混性剂
water miscible formulation 水溶性制剂
water miscible liquid 水乳液
water mist drilling 水雾凿岩
water mite 水螨
water-mixing 水混合(的)
water-mixing direct connection 水混连接
water-mixing nozzle 水混合喷嘴
water-mixing valve 水混合阀
water model 水模型
water model(l)ed stone 太湖石
water-moderated reactor 水慢化反应堆;水减速反应堆
water moderator 水慢化剂
water modifier 水调节剂
water molecule 水分子
water molten by de-icing agents 解冻剂融冰
water monitor 水位控制器;水放射性记录器;水质监测仪;水质监测器;水位监控器;水放射性监测器
water monitoring 水(质)监测
water monitoring equipment 水质监测仪
water monitoring system 水监测系统
water moss 水藓
water motion 水流运动
water motor 水动机;水力发动机
water movement 水流运动;水分动移;水的运动
water mud 清水泥浆
water-mud bursting 突水突泥
water-mud management 水浆管理
water mug with cover (有把的)大矮盖杯
water need 需水
water needle 水针
water net 水网
water notch dam 溢洪坝
water nozzle 喷水嘴
water number 水值
water oak 水栎
water object 供水目标
water obstacle 水上障碍物
water odo(u)r 水的臭味
water of capacity water 毛细管水
water of capillarity 毛细(管)水
water of capillary 毛细管渗透作用
water of cistonnage 超量水(年汲水量与补给量之差);地下水过量开采的量
water of combination 结合水;化合水

water of compaction 碾压水;压实(土壤时挤出的)水;压出水
water of composition 含水量;组成水分
water of condensation 冷凝水;凝结水
water of constitution 结合水;结构水;化合水
water of coordination 配位水
water of crystallization 结晶水
water of decomposition 分解水
water of dehydration 脱结合水;脱出水
water of dilatation (沉积层)过饱和水;膨胀水
water of dilation 沉积层过饱和水
water of gelation 胶质水
water of hot decomposition 热解水
water of hydration 残留水;结合水;水合水;水化水
water of hydrolysis 水解水
water of imbibition 吸收水;渗吸水
water of infiltration 渗透水;渗入水;浸透水
water of mixing 拌和水;(混凝土的)拌和用水
water of percolation 渗透水;渗入水
water of plasticity 可塑水量;塑性水
water of productive us(ag)e 生产用水
water of saturation 饱和水
water of setting (水泥的)硬化水;凝结水
water of supersaturation 过饱和水
water-oil displacement 水驱油
water-oil emulsion 油水乳化液;水油乳浊液;水油乳剂
water-oil factor 水油比
water-oil interface 油水界面
water-oil ratio 水油比
water-oil ratio curve method 水油比曲线法
water-oil ratio in flooding 注水油水比
water-oil separator 水油分离器;油水分离器
water-oil stain 水—油污渍
water-only cyclone 水介质旋流器
water oozing 渗水
water opal 玻璃蛋白石
water opening 冰间水面;冰缝
water-operated vacuum pump 水力真空泵
water operating rules 用水调度规则
water operation 水厂运作
water outlet 泄水结构;出水口
water outlet branch 出水支管
water output 供水量
water output on maximum day 最高日供水量
water ouzel 河鸟
water-oven 热水式干燥炉
water overflow settlement ga(u)ge 溢水式沉降仪;溢水式沉降计
water overtopping 漫水;水漫顶
water oxidation reaction 水氧化反应
water ozonization 水的臭氧化(作用)
water packer 水封隔器
water packing 水封
water paint 可洗去的涂料;水性涂料;水(性)漆;水稀释漆;水溶性颜料;水溶漆;水粉涂料
water particle 水质点
water particle motion 水质点运动
water particle movement 水质点运动
water part(ing) 分水界;分水岭;分水线
water parting line 分水线
water passage (钻头)水眼;水道
water passage surface 过流部件表面;流道表面
water-passing caisson 过水沉箱
water passway 水道
water path 水程
water-pearl-ash glass 可溶钾玻璃;水珠灰玻璃;水钾玻璃;硅酸钾玻璃
water penetration 水渗透;建筑漏水;进水;水体穿透能力;渗水(深度)
water penetration test 水浸透测定
water percentage of soil 土壤含水率
water percolating capacity 水分渗透能力;水道渗透能力;渗水量
water-percolating of apical bed 底板透水
water percolation 透水;水渗漏;渗水
water-percolation of apical bed 顶板透水
water performance (水陆两用车的)水中特性
water perfusion experiment 充水实验
water permeability 透水性;水渗透率;渗水性
water permeability method 渗水法
water permeability test 水渗透性试验;透水(性)试验
water permeation 水的渗透

water pervious materials 透水材料
water phase 水相
water-phased emulsion 水包油型乳液
water phase modifier 水相调节剂
water phone 水音测漏装置
water pier 码头;河流突堤
water pigment colo(u)r 水溶性颜料着色剂
water pigment finishes 水溶性颜料涂饰剂
water-pillow 水垫
water pine 水松
water pipe 隔水套管;水管
water pipe head 水管接头
water pipe hose 水管软接管
water pipe hose clip 水管软管夹
water pipeline 给水管道;供水管线;水管线(路)
water pipe manhole 水管(窨)井;水管进孔;水管检查井
water pipe network 供水管网
water piping 管涌;水管线(路);水管;敷设水管
water piping manhole 水管检查井
water piping right 水管敷设权
water piston 水泵活塞
water pit 水坑
water pitcher 供液壶;大水瓶
water plan 给水计划;水利规划;水的规划
water plane 水位;地下水面;(指地下水)潜水面;(水工建筑物等的)水线平面;水线面;水上飞机;水面;吃水面
water plane area (水工建筑物等的)水线平面面积;水线面面积
water plane coefficient 满载水线面面积系数;水线面系数
water planning region 水规划区
water plant 自来水厂;水厂;水草;水生植物
water planting 水中栽植
water plant of railway 铁路给水厂
water plash 过水公路;积水坑
water-plasticity ratio 水塑比
water-plastic ratio 水塑比
water plate 水系版
water plug 消防栓;止水塞;水塞;抽水装置
water plug and chain 水塞及链
water pocket 集水窝;水窝;水囊;袋形水团
water point 引水点
water poisoning 水中毒
water policy 水政策;水利政策
water policy review 水政策审查
water polishing unit 水精处理装置
water pollutant 水污染物
water pollution 水域污染;水体污染;水(的)污染
water pollution accident 水污染事故
water pollution bioassay 水污染生物测定
water pollution by dyeing 印染水污染
water pollution by inorganic substance 无机物水污染
water pollution by organic substance 有机物水污染
water pollution by organism 生物水污染
water pollution charge refund 返还排污费
water pollution chemistry 水污染化学
water pollution control 水质污染防治;水质污染保护管理;水污染控制
Water Pollution Control Act 水污染控制法令;水污染控制条例
Water Pollution Control Administration 水污染控制管理局
Water Pollution Control Board 水污染控制局
water pollution control compact 水污染控制协定
water pollution control engineering 水污染控制工程;水污染防治工程
Water Pollution Control Federation 水污染控制联合会
water pollution control goal 水污染控制目标
water pollution control(ing) 水污染防治
water pollution control law 水污染控制法规
water pollution control legislation 水污染控制立法
water pollution control planning 水污染控制规划
water pollution control plant 水质污染处理厂;水污染控制装置;水污染控制厂
water pollution control policy 水污染控制政策;水污染防治政策
water pollution control regulations 水污染控制法规
water pollution control station 水污染控制站;水污染防治站
water pollution control strategy 水污染对策
water pollution control system 水污染控制系统
water pollution control system planning 水污染控制系统设计;水污染控制系统规划
water pollution detection system 水污染监测系统
water pollution discharge permit 水污染排放许可证
water pollution discharge permitting system 水污染排放许可证制度
water pollution dispute 水污染纠纷
water pollution from petroleum refinery 炼油厂水污染
water pollution from strip-mining 露天矿矿水污染
water pollution hazard 水污染危害
water pollution index 水污染指数;水污染指标
water pollution indicating organism 水污染指示生物
water pollution load 水污染负荷
water pollution load index 水污染负荷指数
water pollution map 水污染图
water pollution mode 水污染模式
water pollution model 水污染模型
water pollution monitor 水污染监测器;水污染监测仪
water pollution monitoring 水污染监测
water pollution monitoring system 水污染监测系统
water pollution potential 水污染潜在能力
water pollution prevention 水质污染防止;水污染预防
water pollution prevention and control law 水污染预防控制法;水污染防治(控制)法
water pollution problem 水污染问题
water-pollution-prone agricultural chemicals 易使水污染的农药
water-pollution-prone pesticide 易于水污染的农药;水污染型农药
water pollution research 水污染研究
Water Pollution Research Board 水污染研究厅
water pollution research laboratory 水污染研究实验室
water pollution risk 水污染风险
water pollution sources 水污染源
water pollution survey 水污染调查;水体污染调查
water pollution system 水污染系统
water pollution tracing 水污染追踪
water pollution treatment 水污染处理
water polo 水球
water polo timer 水球计时器
water pond 蓄水池;水盘
water ponding 围水养生;围水养护
water pool 水池
water pore 水孔隙
water post 水位标尺
water-post reference point 水柱参考点
water pot 水池;水桶
water-potash glass 可溶钾玻璃;硅酸钾玻璃;水钾玻璃
water potential 水势
water potential gradient 水势梯度
water power 水能;水力
water power dam 水力发电坝
water power development 水力开发;水电开发
water power development authority 水利水电发展管理局
water power engine 水力发动机
water power engineering 水力发电工程;水电工程
water power machinery 水力机械
water-power map 水力资源分布(图)
water power plant 水力发电厂;水电站
water power potential 水力蕴藏量
water power project 水力发电工程;水力发电计划
water power reserves 水力资源
water power resources 水能资源;水利资源;水力资源
water power rights 水能权
water power scheme 水力发电计划;水力发电工程
water power set 水力机组
water power site 水电站站址;水力开发地点;水电站地址;水力发电厂址
water power station 水力发电站;水电站
water power station in river channel 河床式水电站

water power survey 水力开发调查
water power utilization 水能利用
water power works 水力发电工程
water precooler 水预冷器
water preheater 水预热器
water press 水压机
water pressing 水压
water pressure 水压(力);水的压力
water pressure distribution 水压分布
water pressure drive 水压驱动
water pressure drop 水压降
water pressure engine 水力发动机
water pressure force 水压力
water pressure ga(u)ge 水压表;水压计
water pressure grid 水压网格
water pressure line 压力输水管线
water pressure load 水压荷载
water pressure main 压力送水总管;压力输水干管
water pressure method 高压水法
water pressure pattern 水压图形
water pressure pipe 压力输水管
water pressure piping 压力输水管
water pressure reducing valve 降压阀
water pressure regulating device 水压调整装置
water pressure regulation 水压变动率
water pressure regulator 水压调整器
water pressure relay 水压继电器
water pressure relief 减压(阀)
water pressure test 压水试验;水压试验;水压法
water pressure test for strength 水压强度试验
water pressure test for strength and tightness 水压强度和密封性试验
water pressure test for tightness 水压紧密性试验
water pressure test hole 压水试验孔
water pressure test of strength and tightness 强度和气密性水压试验
water pressure tests in borings 钻孔水压测试
water pressure tube 压力输水管
water pressure tubing 压力输水管
water pressure type leak detector 水压式渗透仪
water pressure type sounding 水压测深
water pressurizing test 压水试验
water pressurizing test packer 压水试验封(闭)隔器
water pretreatment 水预处理
water price 水费;水价
water privilege 用水权;富能水体;富能河流
water probing in well 井下探水
water problem 水问题
water processing 水处理
water processing element 水处理单元
water produced 产水量
water producer 含水钻孔;出水孔
water producing area 汇水面积
water producing formation 涌水层;含水层
water producing zone 含水层;出水层;产水带
water production 出水量;产水量
water production measuring 测水产量
water productivity 出水能力;产水能力
water project 给水工程;水利工程;水厂工程
waterproof 不透水(的);隔水的;绝湿;耐水;防水(的)
waterproof abrasive cloth 防水砂布
waterproof abrasive paper 水砂纸;水磨砂纸
waterproof adhesive 耐水性黏结剂;防水黏胶剂;防水性胶黏剂
waterproof and drainage system plan 防水排水系统图
waterproof and dustproof lamp 防水防尘灯
waterproof barrier 防水层
waterproof belt 防水皮带
waterproof blanket 不透水盖面;防水毡;不透水铺盖;防渗铺盖
waterproof board 防水纸板
waterproof bulkhead 防洪护岸
waterproof cable 防水光缆
waterproof case 防水外壳
waterproof case pipe 防水套管
waterproof casing(pipe) 防水套管;隔水套管
waterproof cement 不透水水泥;防水胶
waterproof chart 防水海图
waterproof chart container 水密海图筒
waterproof checker 防水性检查仪
waterproof clay concrete 防水黏土混凝土

waterproof cloth 防雨布;防水布
waterproof coating 外防水层
waterproof concrete 不透水混凝土;防渗混凝土
waterproof construction 防水结构
waterproof corrugated board 防水瓦楞纸板;防水波纹纸板
waterproof course 防水层
waterproof cover 防水盖
waterproof cuff 水密袖口
waterproof dam 防水坝
waterproof detector 防水探测器
waterproof device 防水设施
waterproofed background concrete 防水的基体混凝土
waterproofed cement 防水水泥;防潮水泥
waterproofed concrete 防水混凝土
waterproofed mortar 防水砂浆
waterproof electric(al) blasting cap 防水电雷管
waterproof electric(al) equipment 防水(式)电气设备
waterproof electric(al) torch 防水手电筒
waterproof emulsion 防水乳液
waterproof encasing 防水密封层
waterproofer 隔水层;防水层;抗渗外加剂;防水布;防水物;防水工;防水材料
waterproof expansion cement 防水膨胀水泥
waterproof expansive cement 憎水膨胀水泥;防水膨胀水泥
waterproof explosive 防水炸药
waterproof fabric 防雨布;防水布
waterproof felt 防水油毛毡
waterproof finish 防水涂层装修
waterproof finishing 防水防层
waterproof fuse 防水引信;防水导火线
waterproof garment 防水服装
waterproof glass 耐水玻璃
waterproof glue 防水(性)胶;耐水胶
waterproof-glued 上防水胶的;涂防水胶的
waterproof grease 防水脂;防水润滑脂
waterproof hollow concrete block 防水空心混凝土块
waterproofing 防水处理;防水做法;防水(层);防渗
waterproofing additive 防水剂
waterproofing adhesive 防水胶黏剂
waterproofing admixture 防水外加剂;防水添加剂;防水混合物
waterproofing against ground water 防止地下水(的)
waterproofing agent 防水剂;耐水剂
waterproofing agent in powder form 防水粉
waterproofing and dampproofing of masonry (work) 砖石建筑的防水防潮处理
waterproofing asphalt 防水沥青
waterproofing basement 防水地下室
waterproofing building paper 建筑用油毛毡;建筑用防水纸
waterproofing capacity 防水能力
waterproofing cement 防水胶;防水水泥
waterproofing cement mortar 防水水泥砂浆
waterproofing cement plaster coat 防水水泥抹面
waterproofing coat 防水涂层
waterproofing coating 防水涂层
waterproofing coating material 防水涂料
waterproofing coating rubber 防水涂层橡胶
waterproofing compound 防水剂;防水化合物
waterproofing concrete 防水混凝土
waterproofing course 防水层
waterproofing door 防水门
waterproofing emulsion 防水乳胶
waterproofing facing 防水面
waterproofing felt 防水油毛毡
waterproofing finish 防水终饰;防水修整
waterproofing floor 防水地面;防水地板
waterproofing for toilet 厕所防水
waterproofing grout 防水灌浆
waterproofing grouting compound 防水薄浆化合物
waterproofing gum mortar 防水胶浆
waterproofing highly polymerized compound 高分子材料防水
waterproofing in powder form 粉状防水剂
waterproofing layer 防水层
waterproofing liquid 防水液体
waterproofing material 防水材料;抗渗外加剂

waterproofing membrane 防水(薄)膜
waterproofing mixed plaster 防水混合灰膏;防水混合灰泥
waterproofing mortar 防水灰泥;防水灰浆;防水砂浆
waterproofing of basement 地下室防水
waterproofing of pipes 管道防水
waterproofing of segment 管片防水
waterproofing of tube 管段防水
waterproofing paint 防水涂料
waterproofing paper 防水纸
waterproofing paste 防水浆;防水油膏
waterproofing plaster 防水粉刷;防水灰膏;防水灰泥;防水涂料
waterproofing plaster coat 防水涂层;防水抹面
waterproofing powder 防水粉
waterproofing pressure 防水压力
waterproofing process 防水做法
waterproofing product 防水成品
waterproofing property 防水性能
waterproofing protection 防水保护层
waterproofing protection course 防水涂层
waterproofing quality 水密性;抗水性;防水性
waterproofing roll-roofing 防水卷材
waterproofing screed 防水用(砂浆)刮板
waterproofing sheet 防水板
waterproofing solution 防水溶液
waterproofing stabilizer 防水稳定剂
waterproofing system 防水系统
waterproofing wall 防水墙
waterproofing with asphalt 沥青防水层
waterproofing works 防水工程;防水工作
waterproof ink 防水墨
waterproof integrity 防水完好性
waterproof joint 防水接头;防水接缝
waterproof kraft paper 防水牛皮纸
waterproof layer 防水层;防潮层
waterproof lining 防水衬砌
waterproof lining of basement 地下室防水层
waterproof machine 防水式电机
waterproof map 防水地图
waterproof material 防水材料
waterproof measure 防水措施
waterproof membrane 防水隔墙;防水(隔)膜
waterproof mortar 防水砂浆
waterproof mortar plaster 防水砂浆粉刷
waterproof mortar with ferric chloride 氯化铁防水砂浆
waterproofness 防水性;水密性
waterproofness test 水密性试验
waterproof optic(al) fiber cable 防水光缆
waterproof packaging 防水包装
waterproof packing 不透水包;防水包装
waterproof paint 抗水性涂料;防水涂料;防水漆
waterproof painting film 涂层防水
waterproof panel roofing 防水屋面建筑
waterproof paper 防潮纸;防水纸
waterproof paper for curing 养护防水纸
waterproof paper packing 防水包装纸;防潮纸包装
waterproof pavement 不透水路面
waterproof plaster 防水粉刷;防水抹灰;防水灰泥
waterproof plastic tape 防水塑料卷尺
waterproof plywood 防水胶合板
waterproof pulverization 湿法粉碎
waterproof quality 防水性
waterproof retaining wall 截水挡土墙
waterproof rubber strip 橡皮防水条;橡皮防水带
waterproof sand paper 耐水砂纸
waterproof seal 防水封面
waterproof sealed camera 防水密封摄影机
waterproof sheathing paper (建筑用)防水柏油纸
waterproof socket 防水插座
waterproof strain ga(u)ge 防水应变计
waterproof stratum 不透水层;防水层
waterproof structure 防水结构
waterproof switch 风雨开关
waterproof tarpaulin 防水柏油布
waterproof test 水密试验;防水(性)试验
waterproof type 防水式
waterproof type induction motor 防水式感应电动机
waterproof varnish 耐水清漆
waterproof wax 防水蜡
waterproof wire 防水线;防潮线

water propeller 水推进器
water property 水的性质
water proportioning 水量配比;配水
water prospecting 水源勘探
water prospect map 水源勘探图
water protection 水体保护
water protection activity 水源保护活动
water protection area 水源保护区
water protection management 水源保护管理
water protective complex 水源保护体系
water protective measure 水源保护措施
water pump 水泵;抽水机;抽水泵
water pumpage 抽水量
water pump bearing 水泵轴承
water pump body 水泵体
water pump bushing packing 水泵套垫
water pump by-pass elbow 水泵旁通弯头
water pump capacity 水泵出力能力
water pump casing 水泵体
water pump cast housing 水泵铸铁外壳
water pump casting 水泵铸铁外壳
water pump chain driving gear 水泵链动齿轮
water pump combiner 水泵结合器
water pump coupling 水泵连接器
water pump cover 水泵壳体
water pump cover plate 水泵盖板
water pump drive oil thrower 水泵传动挡油圈
water pump drive spindle 水泵传动轴
water pump drive sprocket 水泵传动链轮
water pump driving chain 水泵传动链
water pump driving gear 水泵主动装置
water pump driving shaft 水泵主动轴
water pump felt 水泵衬毡
water pump fixing screw 水泵固定螺钉
water pump front gasket 水泵前垫密片
water pump gasket 水泵垫密片
water pump gland 水泵封垫
water pump gland flange 水泵水封压盖凸缘
water pump gland nut 水泵密封圈螺帽
water pump gland ring 水泵水封环
water pump grease 水泵油膏
water pump house 水泵房
water pump housing 水泵壳体
water pump hub puller 水泵轮毂拆卸器
water pump impeller 水泵叶轮
water pump impeller spring 水泵叶轮弹簧
water pumping 抽水
water-pumping device 抽水设备
water-pumping equipment 抽水设备
water-pumping instrument 抽水设备
water-pumping plant 抽水设备
water-pumping set 抽水机(组)
water-pumping station 抽水站
water-pumping stop time 停止抽水时间
water-pumping test 抽水试验
water-pumping test hole 抽水试验孔
water-pumping unit 抽水设备
water pump inlet hose clamp 水泵进水软管夹
water pump lock ring 水泵护圈锁环
water pump lubricant 水泵润滑剂
water pump oiler 水泵油杯
water pump outlet fitting 水泵排水连接管
water pump outlet pipe 水泵出水管
water pump packing 水泵盘根
water pump plier 水泵用手钳
water pump pulley 水泵皮带轮
water pump pulley and fan 水泵滑车及风扇
water pump seal 水泵护圈
water pump seal cup 水泵水封皮碗
water pump seal retainer 水泵水封座环
water pump seal retainer washer 水泵水封垫圈
water pump seal spring 水泵护圈弹簧
water pump seal spring guide 水泵密封弹簧导套
water pump seal thrust spring 水泵封圈推力弹簧
water pump shaft 水泵轴
water pump shaft bearing 水泵轴轴承
water pump shaft gear 水泵轴齿轮
water pump shaft slinger 水泵轴抛油环
water pump spindle 水泵轴
water pump spindle distance piece 水泵轴隔套
water pump spring 水泵弹簧
water pump suction pipe 水泵吸水管
water pump thrust washer 水泵止推环
water pump vane 水泵叶轮

water pump works 水泵厂
water purification 净水法;水净化(作用);水的净化
water purification agent 净水剂;水净化剂
water purification and reuse 水的净化与再利用
water purification bio-pretreatment 水净化生物预处理
water purification efficiency 水净化效率
water purification plant 净水(工)厂;水净化装置;水净化厂;水处理厂;水净化设备
water purification process 水净化过程
water purification project 水净化工程
water purification station 净水站;水净化站
water purification structure 净水构筑物
water purification system 净水系统;净化系统;水净化系统
water purification unit 净水设备
water purification vehicle 流动净化水站;净化水汽车
water purification works 净水工程
water purifier 净水器;净化器;水净化器;纯水器
water purifying apparatus 净水设备;净水器
water purifying box 净水箱
water purifying performance 净水性能
water purifying revetment technology 净水型护岸技术
water purifying stone-cage 净水石笼
water purifying tank 净水池;净水槽;净化池;水净化池
water purveyor 供水商
water putty 水(调)腻子
water quality 水质
water quality acidization 水质酸化
water quality act 水质条例;水质法令;水质法案
Water Quality Administration 水质管理局
water quality after back-pumping 回扬后的水质
water quality analogy 水质模拟
water quality analyzer 水质计
water quality analysis 水质分析
water quality analysis index 水质分析指标
water quality analysis method 水质分析法
water quality analysis simulation program(me) 水质分析模拟程序
water quality analysis system 水质分析系统
water quality appraisal 水质鉴定
water quality assessment 水质评价;水质评定
water quality assessment index 水质评价指标
water quality assessment model 水质评价模型
water quality assessment standard 水质评价标准
water quality association 水质协会
water quality background 水质本底
water quality balance 水质平衡
water quality based effluent limitation 限度排污水质;基于水质排污极限
water quality before back-pumping 回扬前的水质
water quality canal network monitoring 渠网水质监测
water quality change 水质变化;水质变动
water quality characteristic analysis 水质特征分析
water quality chart 水质图
water quality chemical model 水质化学模型
water quality chemical oxygen demand 水质化学需氧量
water quality chemicals 水质化学成分
water quality class 水质类别
water quality colo(u)rity 水质色度
Water Quality Commission 水质委员会
water quality concentration 水质浓度
water quality condition 水质状况
water quality conditioning 水质调理
water quality conservation 水质保护;水质保持
water quality constituent 水质组成;水质成分
water quality constituent mass 水质组成物质
water quality control 水质控制
water quality control area 水质控制区
water quality control cross-section 水质控制断面
water quality control(river) reach 水质控制河段
water quality control stream reach 水质控制河段
water quality correction 水质改善
water quality criterion 水质基准;水质准则;水质判据;水质标准
water quality data 水质资料;水质数据
water quality decline 水质恶化
water quality degradation 水质恶化
water quality deterioration 水质劣化;水质恶化

water quality determination 水质测定
water quality diagnosis 水质诊断
water quality economic index 实用水质指标
water quality element 水质要素
water quality engineering 水质工程
Water Quality Engineering Division 水质工程局
water quality evaluation 水质评价
water quality event 水质事件
water quality examination 水质检验;水质检查
water quality factor 水质因子
water quality field 水质场
water quality fluctuation 水质变化;水质变动
water quality forecast 水质预报
water quality for river-reservoir system model 河流—水库系统水质模型
water quality ga(u)ge 水质测定仪
water quality goal 水质目标
water quality grade 水质等级
water quality guideline 水质指标
water quality hardening 水质硬化
water quality health risk assessment 水质健康风险评价
water quality hydrograph 水质过程线
water quality hydrology 水质水文学
water quality identification index 水质标识指数法
water quality improvement 水质改善
water quality improvement act 水质改善条例;水质改善法令;水质改善法案
water quality indentification 水质标识指数
water quality index 水质指数;水质指标;水质模数
water quality indicator 水质指示器;水质指示剂
water quality instrument 水质分析仪(器);水质测定仪
water quality insurance 水质保险
Water Quality Insurance Syndicate 水质保险企业联合组织
water quality investigation 水质调查
water quality legislation 水质立法
water quality level 水质优度
water quality limit 水质限制;水质容限
water quality management 水质管理
water quality management and protection 水质管理及保护
water quality management center 水质管理中心
water quality management facility 水质管理设施
water quality management goal 水质管理目标
water quality management planning 水质管理规划
water quality map 水质图
water quality mathematic(al) model 水质数学模型
water quality measurement 水质测量
water quality measurement station 水质测站
water quality measuring point 水质测站;水质测点
water quality measuring station 水质测站
water quality measuring system 水质测定系统
water quality mineralization 水质矿化
water quality mode 水质模式
water quality model 水质模型
water quality model calibration 水质模型校正
water quality model evaluation 水质模型评价
water quality model of bay 海湾水质模型
water quality model of estuary 河口水质模型
water quality model of lake and reservoir 湖泊水库水质模型
water quality model of reservoir 水库水质模型
water quality model of river mouth 河口水质模型
water quality model of river network 河网水质模型
water quality monitor 水质监测仪;水质监测器
water quality monitoring 水质监视;水质监测
water quality monitoring in pipe network 管网水质监测
water quality monitoring network 水质监测网
water quality monitoring of contaminated water 污水水质监测
water quality monitoring planning 水质监测计划
water quality monitoring program(me) 水质监测计划
water quality monitoring ship 水质监测船
water quality monitoring station 水质监测站
water quality monitoring system 水质监测系统
water quality monitoring van 水质监测车
water quality network 水质站网
water quality objective 水质目标

water quality observation 水质观测
Water Quality Office of USEPA 美国环境保护局水质处
water quality of polluted waters 污染水体水质
water quality of sewage 污水水质
water quality optimization 水质最优化
water quality parameter 水质参数
water quality parameter group 水质参数组
water quality physicochemical parameter 水质理化参数
water quality planning 水质规划
water quality planning model 水质规划模型
water quality pollutant 水质污染物
water quality pollution 水质污染
water quality pollution control 水质污染控制;水质污染防治;水质污染保护管理
water quality pollution control law 水质污染管制法(规);水质污染管理法(规)
water quality pollution control legislation 水质污染管制立法;水质污染管理立法
water quality pollution index 水质污染指数;水质污染标志
water quality pollution monitor 水质污染监测仪
water quality pollution monitoring 水质污染监测
water quality pollution research laboratory 水质污染研究实验室
water(quality) pollution source 水质污染源
water quality prediction 水质预测;水质预报
water quality prediction of river 河流水质预测;河流水质预报
water quality processing 水质处理
water quality program(me) 水质规划
water quality programming 水质规划
water quality protection 水质保护
water quality recoverability 水质可复性
water quality regionalization 水质区划
water quality regulation 水质调节
water quality requirement 水质要求
water quality requirement for agricultural water 农业用水水质要求
water quality requirement for fishery 渔业用水水质要求
water quality research 水质调查
water quality research program(me) 水质调查计划
water quality response curve 水质响应曲线
water quality restoration 水质修复
water quality risk judgment 水质风险评价
water quality sample 水质样本
water quality sampling 水质采样
water quality sampling network 水质采样网
water quality sanitary index 水质卫生指标
water quality sanitary standard 水质卫生标准
water quality security 水质安全
water quality sensor 水质传感器
water quality ship 水质监测船
water quality simulation and prediction 水质模拟预测
water quality simulation model 水质模拟模型
water quality situation 水质状况
water quality stabilizer 水质稳定剂
water quality standard 水质量标准;水质(标准)
water quality standards for drinking water 饮用水水质标准
water quality standards for fishery 渔业(用水)水质标准
water quality standards framework 水质标准框架
water quality state 水质状态
water quality stratitification 水质分层
water quality stress 水质应力
water quality substance 水质物质
water quality surveillance 水质监视;水质监测
water quality surveillance network 水质监视网
water quality synthetic evaluation 水质综合评价
water quality target 水质目标
water quality target management 水质目标管理
water quality testing agent 水质测试试剂
water quality testing method 饮用水水质检验法
water quality tolerance 水质容限
water quality transport module 水质输送模数
water quality treatment district 水质处理区
water quality value 水质值
water quality variable 水质变量
water quality variation 水质变化;水质变动

water quality volume 水质优度
water quality vulnerability zone 水质脆弱带
water quality vulnerability zone map 水质脆弱带图
water quanlity model of lake 湖泊水质模型
water quantity analogy 水量模拟
water quantity calculation 水量计算
water quantity of back-pumping 回扬水量
water quantity-quality simulation model 水量水质模拟模型
water quantity security 水量安全
water quantity system 水量系统
water quartz 泡石英
water quench blast furnace slag 水淬高炉渣
water-quenched cinder sintered product 水淬煤渣烧结制品
water-quencher 水急冷器
water quench(ing) 水中急冷;水淬硬(化);水淬(火)
water quench tank 水淬火槽
water race 河段;水道;(工矿的)引水渠
water radiator 水散热器
water radioactivity meter 水放射性测量计
water rail 船舷步道
water-rail terminal 水-铁联运码头
water raiser 扬水机;提水机
water-raising capacity 扬水高程;水扬程;扬水能力
water-raising current wheel 汲水车
water-raising engine 升水机;提水机;抽水机
water-raising machine 抽水机;扬水机
water ram 水力夯锤;水锤扬水机;水锤(泵)
water range 水上靶场
water-rat 水老鼠
water rate (混凝土)渗水率;用水率;自来水费;灌水率;流水率;耗水率;水消费率;水速;水气率;水费(率)
water-rating 用水率
water ratio 冷却水与排气量之比;含水率;水灰比(例)
water ratio of concrete 混凝土水灰比
water ratio theory 水比理论
water reactor 水反应器;水反应堆
water reading 水界读数
water re-aeration 水的再曝气
water receptacle 水容器
water receptive 亲水表面;吸水的
water recession 退水
water recharging 水源补给
water reclamation 水回收
water reclamation plant 水回收装置
water reclamation system 水回收系统
water recooling system 水二次冷却系统
water recovery 排水利用;水蒸气凝结;水回收
water recovery apparatus 回水器;水回收装置;水回收设备
water recovery condenser 回水冷凝器
water recovery system 水回收系统
water recreation area 水上康乐活动
water recycle 水循环
water recycled 循环回用水
water recycle system 循环回用水系统;用水再循环系统
water recycling 水的重复利用;(已处理的)水的循环使用
water reducer 减水剂
water reducible 含水的;可干燥的
water-reducible coating 水稀释性涂料
water-reducible finish 水稀释性面漆
water-reducible water-resin 水稀释性树脂
water-reducing accelerator 减水早强剂
water-reducing action 减水作用
water-reducing admixture 减水(外加)剂
water-reducing agent 减水(添加)剂;水还原剂
water-reducing and accelerating admixture 减水促凝外加剂
water-reducing and retarding admixture 减水及缓凝外加剂
water-reducing and set retarding admixture 减水缓凝(外加)剂
water-reducing effect 减水效果
water-reducing plasticizer 水泥塑化剂
water-reducing retarder 减水缓凝(外加)剂
water-reduction reaction 水还原反应
water reed fiber 芦苇纤维

water reference point 水参考点
water-reflected 水面反射
water-reflected system 水反射系统
water refreshing 水再生
water regain 吸水量;吸水
water regime 水文状况;水文特性;水文情势;水文情况;水位状况;水情;水分状况
water regime in mine 矿坑水动态
water regime in well 井水动态
water region 水域
water regulating function 水分调节作用
water regulating valve 水(量)调节阀;水流调节阀
water regulation 水量调节;给水法规
water regulator 控水器;水量调节器
water-rejecting 疏水的;防水的;防水性
water-related disease 与水有关的疾病
water relation 水文关系;水分关系
water relative permeability 水相对渗透率
water release 排水;泄水;放水
water release from weakly permeable layer 弱透水层水量释放
water release gate 泄水闸;放水旋塞
water release pattern of aquifer 含水层释水形式
water release structure 泄水建筑(物)
water release system 泄水系统;排水系统
water release tunnel 放水隧洞
water release works 放水建筑物
water relief winged cap 带翼放水塞
water removal 去水;脱水;除水
water renovation 水质改进;水再生
water renovation and reuse 水的再生和再用
water renovation process 污水再生法;水质修复工艺
water-repellency 防水性;抗水性
water-repellent 憎水的;驱水剂;防水的
water-repellent additive 防水剂;防水掺合剂
water-repellent admixture 抗水剂;憎水剂;憎水混合剂
water-repellent agent 防水剂
water-repellent aggregate 憎水骨料
water-repellent cement 憎水水泥;防水水泥
water-repellent compound 防水剂;憎水化合物
water-repellent concrete 憎水(性)混凝土;防水混凝土
water-repellent facing 防水覆盖层;防水面层
water-repellent film 防水(隔)膜
water-repellent grease 防水脂
water-repellent material 憎水材料
water repellent membrane 防水膜
water-repellent mortar 防水砂浆;憎水砂浆
water-repellent pad 防水垫
water-repellent paper 防水纸
water-repellent preservation 防水性防腐
water-repellent preservative 憎水(性)防腐剂
water-repellent property 憎水性(能)
water-repellent quality 防水性能;防水特征
water-repellent soils 非亲水性土壤灌溉
water-repellent surface 拒水面;能浮起面(矿物)
water-repellent treatment 表面防水;憎水处理;防水处理;抗水处理
water-repellent treatment for lien 亚麻布的防水处理
water-repellent treatment of masonry 砖石砌筑防水处理
water-repeller 防水剂;疏水剂
water-repelling 防水的
water-repelling admixture 防水掺和物
water-repelling agent 憎水剂
water-repelling material 防水材料
water replenishing 补水
water representation 水系表示法
water requirement 需水(量);水需要量;水的需求量
water requirement for normal consistency for cement paste 水泥净浆标准稠度需水量
water requirement of canal 运河需水量
water requirement of cement mortar 水泥砂浆需水量
water requirement of sailors 舰员需水量
water requirement of slurry 料浆需水量
water requirement ratio of cement mortar 水泥砂浆需水比
water research 水研究
water research abstract 水研究文摘
water research association 水研究协会

water research centre 水研究中心
water reserves 蓄水保水；蓄水区
water reservoir 蓄水池；水库
water reservoir area 水库面积
water reservoir system 水库系统
water resistance 防水性；抗水阻力；抗水性；耐水性；水电阻；水的阻力
water resistance corrugated board 抗水波纹纸板
water-resistance glass 抗水玻璃
water resistance measurement 抗水性测定
water-resistance of malthoid 油毡抗水性
water resistance of rock 岩石的抗水性
water-resistant 隔水的；防水的；抗水的；耐水的
water-resistant adhesive 耐水性胶
water-resistant admixture 防水混合物
water-resistant commercial explosive 防水工业炸药
water-resistant concrete 防水混凝土
water-resistant core 耐水芯
water-resistant course 防水层
water-resistant electric(al) blasting cap 防水电雷管
water-resistant explosive 防水炸药
water-resistant fuse 防水导火线
water-resistant glue 隔水胶；耐水胶；防水黏合剂
water-resistant grease 防水脂
water-resistant gypsum backing board 耐水石膏衬板
water-resistant gypsum board 防水石膏板；耐水石膏板
water-resistant gypsum plaster board 耐水纸面石膏板
water-resistant layer 防水层；隔水层
water-resistant paint 抗水油漆；防水(油)漆；耐水涂料；耐水漆
water-resistant red enamel 红色耐水瓷漆
water-resistant varnish 耐水清漆
water-resistant watch 防水表
water-resisting 耐水的；隔水的；防水的
water-resisting admixture 防水外加剂；防水(添加)剂
water-resisting enamel 防水瓷漆
water-resisting fault 阻水断层
water-resisting layer 隔水层
water-resisting of apical bed 顶板隔水
water-resisting of bottom bed 底板隔水
water-resisting paint 抗水(油)漆；防水油漆
water-resisting property 隔水性；抗水性
water-resisting red enamel 红色耐水瓷漆
water-resisting rock vein 阻水岩脉
water-resisting spar varnish 耐水桅杆清漆
water-resisting structure 阻水构造
water-resisting test 耐水试验
water-resisting varnish 防水清漆
water resistor 水阻器
water resonance 水共振
water resource 水源
Water Resource Act 水资源法令
water resource balance method 水量均衡法
water resource carrying capacity 水资源承载力
water resource conservation 水源保护
water resource facility 水利设施
water resource investigation report 水资源调查报告
water resource planning 水资源规划
water resource pollution 水资源污染
water resource project 水利工程
water resource protection 水源保护
water resource quality 水资源质量
water resources 水资源；水利资源
water resources act 水资源法
water resources and hydropower technique 水利水电技术
water resources and water and soil conservation 水资源和水土保持
water resources assessment methodology 水资源评价法
Water Resources Branch of United States Geologic(al) Survey 美国地质调查局水资源处
Water Resources Bulletin 水资源通报(杂志)
water resources charges 水资源费
water resources conservation zone 水资源保护区
Water Resources Council 水资源委员会
water resources development 水(利)资源开发

water resources development center 水资源发展中心
water resources engineering 水(利)资源工程
water resources exploration 水源勘探
water resources for artificial recharge 人工回灌的水源
water resources input-output table 水资源投入产出表
water resources management 水资源管理
water resources ownership 水资源所有权
water resources planning 水(利)资源规划
Water Resources Planning Act 水资源规划法令
water resources planning and management 水资源规划与管理
water resources policy 水资源政策；水资源利用方针
water resources problem 水资源问题
water resources project 水(利)资源工程
water resources protection 水资源保护
water resources research 水(利)资源研究
water resources study 水资源研究
water resources system 水(利)资源系统；水源系统
water resource survey 水源测量
water-restricting equipment 限流设备
water-retaining 保水(性)的；挡水
water-retaining admixture 保水剂
water-retaining agent 保水剂
water-retaining capacity 水分保持能力；持水量；持水度；保水能力；保水量
water-retaining dam 挡水坝
water-retaining dike[dyke] 挡水堤
water-retaining facing 挡水面板
water-retaining plasticizer 保水塑化剂
water-retaining structure 蓄水建筑物；蓄水结构；挡水建筑物；挡水构筑物；储水结构
water-retaining wall 地下挡水墙
water-retardant mixed plaster 防水混合灰泥
water-retention 水分保持；保水性
water-retention ability 持水能力；保水能力
water-retention agent 保水剂
water-retention capacity 保水效率
water-retention curve 滞水曲线
water-retention effectiveness 保水效率
water-retention efficiency 保水效率
water-retention property 保水(性)；持水性
water-retention test 保水性试验
water-retention value 水分保持值
water-retentive 水分保持的
water-retentive agent 保水剂
water-retentive cement 保水(性)水泥
water retentiveness 保水性
water-retentive Portland cement 保水性硅酸盐水泥；保水性波特兰水泥
water retentivity 保水率；保水性；保水量
water reticulation network 河网站网
water reticulation system 水系；水网系统；水网
water retting 沤麻渍；水浸渍
water return 回归水
water return head 集水器
water return pipe 回水管
water return system 回水系统
water return valve 回水阀
water reuse 水再用；水再生；水的再利用；水的回用；废水利用
water reuse system 水再用系统；复用水系统
water rheostat 水电阻器；水变阻器；电解变阻器
water-rib tile 波纹瓦；突肋瓦
water rights 用水权；水权；地面水使用权
water rights acquired by prescription 时效用水权
water rights system 水权体系
water rights value 水权价值
water ring 敛水圈(井筒的)；井筒集水槽；水圈；(防尘用)水幕；(混凝土喷枪中的)水环；井口壁外推排水沟
water ring air pump 水环空气泵
water ring of bridge pier 桥墩翼墙
water ring pump 水环泵
water ring vacuum pump 水环式真空泵
water riser pipe 给水立管
water riser tube 给水立管
water road 水道
water-rock ratio 水岩比值
water roll 水辊
water-rolled 水冲圆的；被水磨圆的

water root 水生根
water rot 湿腐
water route 水上运输线
water rudder (水上飞机的)水舵
water runoff 径流
waters 水体
water saddle 给水管鞍座
Water Safety Plans of World Health Organization 世界卫生组织水安全计划
water salination 水盐碱化
water salinization 水盐碱化
water salt 可溶性盐
water salvage 废水利用
water sample 水质试样；水样(本)；水分试样
water sample bottle 水样瓶
water sample filter 滤水样器
water sample for measuring the absolute age 绝对年龄测定水样
water sample kind 水样类别
water sample of comprehensive analysis 全分析水样
water sample of isotope measurement 同位素测定水样
water sample of simple analysis 简分析水样
water sample of well or boring 井孔水样
water sample processing 水样处理
water sampler 取水样器；水取样器；水采样器；采(水样)器
water sample sampling device 水样采集装置
water sample stabilization 水样稳定(化)
water sample storage 水样储存
water sampling 取水样；水样采集；水采样；采水样
water sampling bottle 水样取样瓶；采水样瓶
water sampling condenser 锅炉水取样器
water sampling device 采水装置
water sampling equipment 水采样装置
water sampling point 水样采样点；采水点；采集水样点
water-sand resources 水砂资源
water-saturated 水饱和的；饱水的
water-saturated bed 水饱和(岩)层；水饱和(地)层；饱含水的岩层；饱含水的地层
water-saturated coefficient 饱和系数
water-saturated curve 水饱和曲线
water-saturated formation 水饱和(岩)层；水饱和(地)层；饱含水的岩层；饱含水的地层
water-saturated layer 饱和水层
water-saturated sample 水饱和样品
water-saturated soil 水饱和土壤
water saturated state 饱水状态
water-saturated stratum 饱和水层
water-saturated test 水饱和试验
water saturation 含水饱和率；含水饱和度；水分饱和(水)(的)饱和；饱水的
water saturation coefficient 饱水率系数；水饱和系数
water saturation coefficient of rock 岩石的饱水率系数
water saturation deficit 水分饱和亏欠度
water saturation of rock 岩石的饱水率
water saturation rock layer 饱水岩层
water saturation test 水饱和试验
water saturation value 水饱和值
water saving 节约用水
water-saving basin 省水池
water-saving cistern 节水型水箱
water-saving device 节水装置
water-saving hardware 节水硬件
water-saving installation 省水装置
water-saving lock 省水船闸
water-saving method 节水方法
water-saving plan 节水计划
water-saving policy 节水政策
water-saving pressure regulator 回复式减压阀
water-saving software 节水软件
water-saving technique 节水技术
water-saving valve 节水阀(门)
water scalping tank 箱形水力分级机；水选箱；水力分选槽
water-scarce area 缺水(地)区
water scarcity 水荒
water scavengine agent 水净化剂
water scheme 供水方案

water scooping machine 吸水机
water scour 水力冲刷
water scouring 水冲刷;冲刷
water screen 拦污栅;水幕;水滤网;水帘;水冷壁
water screen header 水冷壁联箱
water screening 水拦污
water screen tube 水冷壁管
water screw 螺旋推进器
water scrubber 水洗塔;水洗(涤)器;水洗除尘器
water scrubbing 水洗
water scrubbing tower 水洗塔
water seal 止水;水封;水闸水
water sealant 密封防水油膏
water seal arrangement 水封装置
water seal chamber of dredge pump 泥泵水封腔
water seal conveyer 水封运输机
water-sealed annealing furnace 水封式退火炉
water-sealed basin 水封池
water-sealed bearing 水封轴承;防水轴承
water-sealed drainage 闭式引流排气法
water-sealed equipment 水封设备
water-sealed explosion door 水封防爆门
water-sealed gas holder 湿式储气罐;水封储气罐
water-sealed gas meter 水封气表
water-sealed gland 水封套
water-sealed holder 湿式储气罐
water-sealed joint 止水接头;止水(接)缝;存水弯头;存水接头
water-sealed oil storage 水封油库
water-sealed packing 水封装置
water-sealed rotary gas meter 潮湿气体流量计
water-sealed tank 水封池;水封罐
water-sealed well 水封井
water sealing 水(密)封
water sealing gas holder 水封式气柜
water sealing in different diameter 异径止水
water sealing in same diameter 同径止水
water sealing outside the casing 管外止水
water sealing pipe 水封管
water sealing ring 水封环
water sealing with asphalt 沥青止水
water sealing with casing 套管止水
water sealing with cattle hide 牛皮止水
water sealing with cement 水泥止水
water sealing with clay 黏土止水
water sealing with gasbag 气囊止水
water sealing within the casing 管内止水
water sealing with kelp 海带止水
water sealing with rubber 橡胶止水
water sealing with soybean 黄豆止水
water seal joint 水封接缝
water seal ring 防水圈
water seal tank 水封槽
water seal wall 止水墙
water seasoning 浸水风干法;木材浸水两周后风干法;木材浸水处理法;水浸法
water section (厨房,浴室等的)用水间
water security 用水安全
water's edge 水岸
water sedimentation 水沉积
water-sediment complex 水砂混合体;挟砂水
water-sediment condition 水沙条件
water-sediment interface 水-底泥界面
water-sediment mixture 水砂混合体;水与泥沙混合物;水沙混合体;水沙混合体
water-sediment motion 水沙运动
water-sediment movement 水沙运动
water-sediment-rating curve 水—泥沙—流量关系曲线
water-sediment regime 水与泥沙状况;水沙状况
water seedbox 水生丁香蓼
water seeded rice 水稻
water seepage 进水;渗入(水);渗出水
water seepage test 渗水试验
water seeping 渗水性
water segregator 分水器
water-sensitive 亲水的;水敏
water sensitive formation 吸水岩层;水敏性地层
water sensitive sandstone 吸水砂岩;水敏性砂岩
water sensitive shale 吸水页岩;水敏性页岩
water sensitive soil 水敏性土;湿陷性土
water sensitivity 水敏感性
water sensitivity of rocks 岩石水敏性
waters environment characteristic 水域环境特征

waters environment pollution 水域环境污染
water separating gas filter 分水滤气器
water separating tank 脱水罐;分水罐
water separation capability 脱水能力
water separation cover 隔水盖
water separation index 水分离指数
water separator 分离水器;脱水器;水(分)分离器;除水器
water service 给水工程;供水设备;供水;给水服务
water service cock 供水旋塞
water service conduit 供水管;接户管
water service installation 给水装置
water service pipe 供水管;接(用)户管;服务管【给】
water service plan 给水管图;配水管系图
water service pump 配水泵
water service schedule 水分一览表;水费一览表
water service tube 供水管
watershed 排水面积;分水岭;流域分界线;披水;遮水檐;流域;汇水区;排水盆地;水源区;窗檐
watershed acidification 流域酸化
watershed acidification model 流域酸化模型
watershed area 流域面积
watershed-based water quality evaluation 流域水质评价
watershed boundary 流域边界
watershed characteristic 流域特征(值)
watershed dam 源头坝
watershed development 流域开发
watershed development factor 流域发展因子;流域发展因素
watershed development planning 流域开发规划
watershed development project 流域开发项目;流域开发工程
water-shedding mo(u)lding 流水线脚
watershed divide 分界线;流域分水线;流域分水岭;流域分界线;分水线;分水岭
watershed ecology 流域生态学
watershed ecosystem 流域生态系统;集水区生态系统;水域生态系(统)
watershed elevation 分水岭高程
watershed environmental hydrology 流域环境水文学
watershed exposure 流域暴露位向
watershed form 流域形状
watershed harnessing 分水岭治理
watershed industry 分水岭工业
watershed lag 流域集水时间滞后;流域滞时
watershed land use 流域土地使用状况
watershed leakage 流域渗流量
watershed level ecological risk assessment methodology 流域生态风险评价方法
watershed line 分水线;分水岭
watershed management 流域管理
watershed model 流域模型
watershed morphology 流域地貌
watershed non-homogeneity factor 流域不均匀因子
watershed outlet 流域出水口
watershed planning 流域规划;水资源规划
watershed plot 流域小区;径流实验小区
watershed project 流域开发项目;流域开发工程
watershed protected engineering 流域保护工程
watershed protected project 流域保护工程
watershed protection 流域保护;水域保护
watershed protection and flood control act 流域保护和洪水控制法规
Watershed Protection and Flood Prevention Act 流域保护及防洪法案
watershed routing 流域(汇流)演算
watershed sanitation 流域(环境)卫生
watershed satellite 流域卫星
watershed sediment yield 流域产沙量
watershed shape factor 流域形状因子
watershed size 流域大小
watershed slope 流域坡降;流域坡度
watershed storage capacity 流域蓄水量
watershed tunnel 越岭隧道
watershed unsymmetry factor 流域不对称因子
watershed water balance 流域水量平衡
water sheet 倾盆状流水;成片状流水;水帘
water shield 水屏(蔽);水防护屏
water shielding 水屏蔽
water shielding tank 屏蔽水箱

water ship 供水船
water shock 冲激;水流冲击
water shoot 排水管;排水沟;(屋檐)排水槽;滴水石;放水槽
water shooting 水中爆炸;水下爆破
water-short 缺水
water-shortage 缺水;水量不足;水荒
water-short area 缺水(地)区;水量不足(地)区
watershot (乱石墙中的)贯墙石
watershot walling 甩水墙
water shutoff 截流;堵水;止水;断水;封水
water shutoff chemicals 止水化学(处理)剂
water shutoff test 挡水试验;堵水试验
water shutter 水隔器;水封阀;水闸门
waterside 河畔;水边
waterside bank 迎水坡;外坡
waterside banquette 临水面护堤;前护堤;外戗堤
waterside construction 水边建筑(物)
waterside corrosion 水侧腐蚀
waterside development 湖滨开发;海滨开发;水边开发
waterside face 迎水面;上游面
waterside facility 港口水域设施
waterside foreland 堤外滩地;堤外(土)地
waterside fouling factor 水侧污垢系数
waterside fouling resistance 水侧污垢热阻
waterside gravel 水边砾石
waterside land 水边地;堤外滩地;堤外(土)地
waterside park 水滨公园
waterside pavilion 水榭
waterside plant 港口水域设施
waterside planting 水畔栽植
water sider 码头工人
waterside rail 码头起重机前轨;前轨
waterside slope 堤岸外坡;临水坡(度);河岸外坡;堤外边坡;迎水坡
waterside station 港区车站;水畔车站
waterside step 岸坡踏步
waterside structure 水边建筑(物);岸边建筑物;水滨建筑物
waterside transfer 外档换装;外档过驳
waterside worker 港口工人
water sieve 水筛;水过滤网
water signal 水深信号
water sink 灰岩坑;落水洞;水洞;水漩穴;渗水坑
water ski 滑水橇;滑浪板
water skiing 滑水
water skip 水斗
water sky 水照云光;水映光
water slab 井式楼盖
water slaked lime 水解石灰
water slide valve 水动滑阀
water slip 流水滑动面
water slope 水坡升船机;水坡水升船设施;水力坡度;外坡;迎水坡(坝的)
water slope shiplift 水坡式升船设施;水坡式升船机
water slot 环向流水槽;水口;冲洗水槽
water sludge 水淤泥
water slug 水冲击
water sluice 水排泄
water sluice gate 闸门
water sluicing system 水力除渣系统
water slurry 水悬浮物;水淤泥;水悬浮液
water slurry of pigment 水浆状颜料
water-slurry transportation 稀浆传运法;水和稀浆传运法
water smoke 海上蒸汽雾;水雾
water smoking 水雾;脱水阶段;冒水汽;水熏作用
water smoking period 水分蒸发期(烧容初期)
water smothering system in machinery space 机舱喷水灭火装置
water snake 水蛇
water snow 水雪;湿雪
water-soaked 浸水的;水浸透的
water-soaked till 水浸冰碛
water soaking 吸水现象;(混凝土)浸水现象;水力清扫;水力除灰;(混凝土)渗水现象
water soaking process 水浸法
water soaking test 水浸试验
water soak timber 高含水材
water-soap solution 肥皂水溶液
water soda ash glass 钠水玻璃;水玻璃
water soda glass 水(苏打)玻璃
waters of oligotrophic lake type 寡营养湖型水体

water softener 水(质)软化剂;软水剂;硬水软化器;软水器;软水罐;软化剂;水质软化器
water softening 水净化;硬水软化(法);给水软化;软化水法;球化水冷退火;水质软化;水(的)软化
water softening agent 硬水软化剂;软水剂;水软化剂
water softening apparatus 水软化装置
water softening by heat 水的加热软化法
water softening by heating 水的加热软化法
water softening equipment 水质软化设备;硬水软化装置
water softening facility 软水设施
water softening in boiler 炉内软水
water softening method 硬水软化法;软水法
water softening out of boiler 炉外软水
water softening plant 硬水软化设备;硬水软化厂;软水所;软水设备;软化水厂;水软化设备;水软化(工)厂
water softening room 软水室
water softening tank 软水罐
water softening treatment 给水软化法处理
water softening unit 硬水软化装置;硬水软化设备;软水装置
water-soil-plant relations 水土植物关系
water soldier 水兵草
water-solid ratio 水料比;水固比
water solube paint 水溶性涂料
water solubility 水溶性;水溶度
water solubility of rocks 岩石水溶性
water-soluble 可溶于水的;水溶性的
water-soluble acid and alkali 水溶性酸和碱
water-soluble acrylic resin 水溶性丙烯酸树脂
water-soluble active dye 水溶性活性染料
water-soluble air 水溶态空气
water-soluble air drying paint 水溶性自干涂料
water-soluble alkali 可溶性碱
water-soluble alkyd resin 水溶性醇酸树脂
water-soluble amino resin 水溶性氨基树脂
water-soluble baking paint 水溶性烘漆
water-soluble boron 水溶性硼
water-soluble carbon 水溶性碳
water-soluble carbonhydrate 水溶性碳水化合物
water-soluble carbon test 水溶性碳试验
water-soluble catalyst 水溶性催化剂
water-soluble cationic polymer 水溶性阳离子型聚合物
water-soluble drying paint 水溶性气干型漆
water-soluble dye(stuff) 水溶性染料;水溶性颜料
water-soluble emulsion 水溶性乳剂
water soluble fraction 水溶度
water-soluble gasoline fraction 水溶性汽油分数
water-soluble glue 水溶性胶
water-soluble granule 水溶性颗粒
water-soluble grease 水溶性(油)脂
water-soluble group 水溶性基团
water-soluble heavy metal ion 水溶性重金属离子
water-soluble humus 水溶性腐殖质
water-soluble inorganic substance 水溶性无机物
water-soluble lubricant 水溶性润滑剂
water-soluble material 水溶性物质
water-soluble matter 水溶性物
water-soluble metal ion 水溶性金属离子
water-soluble metalworking liquid 水溶性切削液
water-soluble oil 水溶性油
water-soluble organic substance 水溶性有机物
water-soluble oxidized starch 水溶性氧化淀粉
water-soluble paint 水溶性颜料;水(性)漆
water-soluble pesticide 水溶性农药
water-soluble petroleum sulfonic acid 水溶性石油磺酸
water-soluble pharmaceutics 水溶性药物
water-soluble phenol resin 水溶性酚醛树脂
water-soluble phosphate 水溶性磷
water-soluble phosphate fertilizer 水溶性磷肥
water-soluble phosphoric acid 水溶性磷酸
water-soluble polymer 水溶性聚合物;水溶性高分子聚合物
water-soluble polyvinylalcohol fiber 水溶性聚乙烯醇纤维
water-soluble powder 水溶性粉(剂)
water-soluble preservative 水溶性防腐剂
water-soluble product 水溶性产物
water-soluble resin 水溶性树脂
water-soluble resin coating 水溶性树脂涂料
water-soluble salt 水溶(性)盐
water-soluble salts 水溶性盐类
water-soluble state pollutant 水溶态污染物
water-soluble urea resin coating 水溶性脲醛树脂涂料
water-soluble vitamin 水溶性维生素
water-soluble wood preservative 水溶性木材防腐剂
water solution 水剂
water solvent 水溶剂
water sorption curve 水吸附曲线
water-sorted material 水选材料
water sound absorber 水声吸收器
water source 给水水源
water source classify of spring 温泉水的来源分类
water source conservation 水源保护
water source control 水源控制
water source of mineral deposit inundation 矿床充水水源
water source pollution 水源污染
water source project 水源工程
water source protection 水源保护
water source quality protection zone 水源水质保护区
water source selection 水源选择
water source survey 水源调查
water source utilization 水源利用
water space 水域(面积);水溶积
water spar 提水吊杆
water specimen 水质试样
water speed 水速;水面行驶速度
watersphere 生活圈;生物圈;球形容水器;水圈
water splash 试车水潭;溅水
water splashing 溅水现象;漂水现象
water splash mat 滴水层
water-splitting 水裂解
waters pollution ecology 水域污染生态学
water sport area 水上运动场
water sports 水上运动
water spots 水斑(病)
water spotting 水沾污白斑;水溃试验;水斑
water spout 海龙卷;水落管;水口;水柱;喷水口;水龙卷
water spouting 喷水
water-spout prominence 龙卷日珥
water spray 喷水雾;喷淋
water spray cone 水雾锥
water spray cooling 喷射冷却;洒水降温
water spray defrost system 水喷雾除霜系统
water sprayer 洒水机;喷洒机;洒水器;喷水器
water spray extinguishing system 水喷雾灭火系统
water spray for damping dust 喷雾消尘
water spray granulated slag sand 喷水形成的粒状矿渣砂
water spraying 喷水
water spray injector system 水雾喷射系统
water spray kiln 喷水干燥窑
water spray nozzle 喷水嘴;水幕喷嘴
water spray pipe 水喷洒管
water spray piping 喷水管系
water spray separator 喷水式除尘器;喷雾式除尘器
water spray system 淋水系统;水系统;水雾系统;洒水系统
water-spray treatment 喷水处理
water spray treatment method 喷水处理法
water spray treatment of odo(u)r 臭气喷水处理法
water spread 水流分布;水面
water spreading 漫流;水流扩展;水流分散;布水;引流
water spring 涌泉
water spring tail 水跃尾部
water-sprinkled bosh 喷水式炉腹
water sprinkler 人工降雨装置;洒水车;喷淋器;喷水装置;喷灌机器;洒水机
water sprinkler car 洒水车
water sprinkler tank 喷洒机;洒水机;带水罐洒水机
water sprinkling 喷雾洒水;喷水;洒水;泼水
water-sprinkling protection system 水幕保护系统
waters protection 水域保护
water stability 水稳(定)性;水上稳定性
water stabilization 水质稳定(化)
water stabilized plasma thrower 水稳定等离子弹射器
water stabilizer 水稳定剂
water-stable aggregate 水稳性骨料;水稳性团聚(体)
water-stable area 水稳定场
water stage 水位
water-stage data 水位资料
water-stage fluctuation 水期变化
water-stage ga(u)ge 水位计
water-stage record 水位记录
water-stage recorder 自记水位站;自记水位仪;自记水位计;水位记录器
water-stage recorder well 水位计井
water-stage register 自记水位仪;自记水位计;水位记录表
water-stage transmitter 水位记录传递器;水位自动记录仪;水位传感器;水位传递器
water stain 水溶色料;湿斑;水渍;水性着色剂;水污斑;水色;水染料;水痕(陶瓷缺陷)
water stair(case) 岸坡踏步
water stake 水尺
water stand 停潮
water standard 水质(标准)
water standard at its normal level 正常水位水面
water standard substance 水标准物质
water station 供水站;给水站;给水所
water-stemming bag 水封堵塞袋
water sterilization 消毒水;水(的)消毒
water still 蒸馏水器
water-stilling device 消能设备;静水设备
water stock 蓄水量
waterstone 水磨石;水石层
water-stone flow 水石流
water stop 关闭阀门;防水密封条;止水条;止水塞;止水剂;止水(带);截水墙;阻水片;关闭活门;水密封
water-stop board 隔水板
water-stop key 止水键
water-stop material 止水材料;水密封材料
water stopper 止水剂
water-stop plate 止水板;截水板
water-stop strip 止水条;止水片
water storage 蓄水(量);缺水;水中养护;水上储[贮]木;水储存
water storage capacity 储水量;蓄水量;库容
water storage dam 蓄水坝
water storage facility 蓄水设备;储水设备
water storage in a reach 河段蓄水(量);河段储水
water storage in channel network 河网蓄水
water storage of specimen 水中养护试件
water storage project 蓄水工程
water storage reservoir 蓄水池;水库
water storage tank 蓄水箱;蓄水罐;蓄水槽;储[贮]水箱;储[贮]水池;储水槽
water storage time 蓄水时间
water strainer 滤水管;滤水器;花管(进水管有滤网的一段)
water stratification 湖水分层
water stratum 含水层;层水
water streak 夹层水
water stream injection pump 水喷射泵
water stress 水应力
water string 防水管柱;止水管柱;防水套管
water stripping 遇水剥落
water-struck brick 水脱坯砖;水湿模砖;湿模砖
water structure 水的化学结构
water structure theory 水的化学结构理论
water stuffing box 水阀帽
water substance 水体
water sub-type 水的亚型
water sucking 吸水
water suction curve 吸水曲线
water suction hose 吸水软管
water suction pump 吸水泵;水力抽气泵
water supplies for railroad 铁路给水
water supply 给水(工程);水库;蓄水池;蓄水与供给系统;供水(系统);水供给
water supply allocation 给水分配
water supply and demand balance 水资源供需平衡;水量供需平衡
water supply and drainage 给排水
water supply and drainage and fire fighting system 给排水及消防系统
water supply and drainage drawing 给排水图
water supply and drainage plant 给排水设备

water supply and drainage works 给(水)排水工程
water supply and sewage plant 给排水设备
water supply and sewerage 给排水
water supply and sewerage discipline 给排水专业
water supply and sewerage drawing 给排水图纸
water supply and sewerage works 给(水)排水工程;上下水道工程
water supply area 给水区;供水区(域)
water supply board 供水局;供水部门
water supply boat 供水船
water supply canal 给水渠
water supply capacity 给水度
water supply company 自来水公司
water supply conduit 给水管
water supply connector 进水接头
water supply curve 供水曲线
water supply discipline 给水专业
water supply distance 供水里程
water supply district 给水区(域)
water supply engineer 供水工程师;给水工程师
water supply engineering 供水工程;给水工程;给排水工程(学)
water supply engineering 给水工程学
water supply equipment 给水设备
water supply facility 供水设备;给水设施
water supply fittings 水管接件;供水配件;供水部件
water supply flume 供水槽
water supply for canal 运河供水
water supply forecasting 供水预报;给水预测
water supply hose 进水软管
water supply hygiene 给水卫生
water supply industry 给水工业
water supply installation 给水装置;给水装置
water supply law 供水法规
water supply line 给水管线;上水道
water supply mains 给水干管
water supply net 给水管网
water supply network 供水网;给水网
water supply of the site 工地供水
water supply output 供水量
water supply pipe 供水管(道);给水管(道);上水管
water supply(pipe) line 给水管线;上水(管)道;给排水管线
water supply pipe network 供水管网
water supply pipe well 供水管井
water supply piping 给水管
water supply planning 给水规划
water supply plant 自来水厂;供水设备;给水装置;给水所
water supply point 给水所;给水站
water supply programming 给水规划
water supply protection 供水保护
water supply pump 冷水泵
water supply pump station 给水泵站
water supply quality 供水水质
water supply quanlity turbidity index 供水水质浊度指标
water supply quay 供水码头
water supply rate 供水速率
water supply regulator 给水调节器
water supply riser 上水立管
water supply rubber hose 供水橡皮软管
water supply scheme 供水系统;供水规划;供水方式;给水计划;给水方案;给水布置
water supply section 给水段
water supply selector 进水选择器
water supply sludge treatment 供水污泥处理
water supply sourc 供水(水)源;给水水源;给水管源
water supply station 供水站;给水站
water supply station for passenger locomotive 客机给水站【铁】
water supply station for passenger train 客车给水站【铁】
water supply survey 给水勘测
water supply system 供水系统;给水系统;上水系统
water supply system appurtenances 给水系统附属设备
water supply system rehabilitation 给水系统恢复能力
water supply tank 供水柜
water supply terminal 供水码头
water supply to navigation canals 通航河道的供水
water supply treatment 供水处理;给水处理

water supply treatment plant 给水处理站;给水处理厂
water supply treatment system 给水处理系统
water supply treatment works 给水处理厂
water supply tube 给水管
water supply tubing 给水管
water supply tunnel 给水隧洞;给水隧道;给排水隧道
water supply valve 给水阀
water supply valve well 给水阀门井
water supply well 供水钻孔;供水井;给水井
water supply wharf 供水码头
water supply works 供水工程;给水设备;给水工程;水厂
water supporting deck 挡水面板
water surface 过流(表)面;水(表)面
water surface area 水面面积
water surface burst 水面爆炸
water surface curve 水面线
water surface disguise 水面伪装
water surface elevation 水面高程
water surface float 水面浮子;水面浮标
water surface fluctuation 水面升降(波动);水面波动
water surface gradient 水面比降;水面坡降;水面坡度
water surface gradient 水面比降
water surface level 水面水平
water surface nuclear burst 水面核爆炸
water surface profile 水面(曲)线;水面剖面图;水流纵剖面(图)
water surface profile method 水面曲线法
water surface response 水面反应
water surface roller 水面大浪
water surface slope 水面坡降;水面坡度;水面比率;水面比降
water surface temperature 海面温度;水面温度
water surface temperature distribution 水面温度分布
water surface temperature measuring system 水面温度测定系统
water surface width 水面宽(度)
water surge tower 水塔
water surplus 水剩余物
water surplus region 水量盈余区
water surveillance network 水监视网
water survey 水文测量;水文勘测;水文调查;水文查勘;水利勘测
water survival gear 水上救生设备
water survival training 水上救生训练
water suspension 水悬浮体
water suspension paste 水糊剂
water suspernsion 水悬浮液
water swallow 落水洞;地面水流失处
water swelling 湿胀;水胀
water-swelling fluoromica 水胀氟云母
water switch 水压开关
water swivel 水旋转接头;水龙头【岩】;水接头
water swivel with bail 提引水接头
water syphoning 用虹吸管供水
water syringe 水枪
water system 给水系统;河系;水系(统);水网
water system deterioration 水系退化;水系恶化
water system map 水系图
water system of circulation 循环用水系统
water system of closed-circuit circulation 闭路循环用水系统
water system planning 水系规划
water system survey 水系调查
water system telemetry 给水系统遥测术
water table 泻水台;(地下水)潜水面;披水面;水位;地下水位;承雨水脚
water-table after water blocking 堵水后的水位
water-table aquifer 潜水(含水)层;地下水面含水层;非承压含水层
water-table beams 天车梁
water-table before water blocking 堵水前的水位
water-table contour 地下水等水位线;地下水等高线;潜水等高线
water-table(contour) plan 地下水(位)等高线图
water-table decline 水面下降;地下水下降;地下水面降落;潜水面下降
water-table depression cone 地下水面下陷锥;潜水位漏斗;水位降落锥;潜水面降落锥

water-table divide 分水界
water-table fluctuation 地下水位升降;潜水(面)波动;水位波动;水位变化;地下水位波动
water-table gradient 水头梯度;地下水位坡降;地下水面坡降;地下水面坡度;潜水面梯度
water-table in picking up 水位回升值
water-table isobath 地下水位等深线;地下水面等深线;潜水(位)等深线
water-table isohypse 地下水等高线
water-table level 地下水位
water-table map 地下水位图;地下水等水位线图;潜水等高线图
water-table observation 地下水位观察
water-table opening 天车台检视口
water-table outcrop 地下水面出露;地下水露头(处);潜水面露头【地】
water-table profile 地下水剖面;地下水面剖面(图);潜水剖面
water-table ratio 水位比;地下水水位比
water-table rise 地下水面上升;地下水位上升;潜水面上升
water-table slope 水面坡度;地下水位坡度;地下水面斜率;潜水坡度
water-table spring 地下水面出露泉;地下水露泉;潜水泉
water-table stream 地下水面集中水流;地下水流;潜水流
water-table wave 地下水面波;潜水面波
water-table well 地下水供水井;潜水井
water tabling 飞檐
water tagging 示踪测流(法)
water tail piece 进水部尾管
water taking bottle 泵水瓶
water tank 水塔;水箱;水柜;水池;水舱;储水箱;蓄水池
water tank area 水罐区面积
water tank car 水柜车;水罐车
water tank for heat insulation 隔热水箱
water tank for thermal insulation 隔热水箱
water-tank indicator 水池指示器
water tank lorry 水罐车
water tank model(l)ing 水槽模拟
water tank model(l)ing test 水槽模型实验
water tank trailer 载水箱拖车
water tap 水龙头
water tap noise 给水龙头噪声
water taste control 水除味
water taxi 出租快艇;水上出租船
water technology 水工艺学
water teleindicator 水位传递器
water telescope 水下望远镜
water temperature 水温
water temperature anomaly 水温异常
water temperature control unit 水温控制装置
water temperature ga(u)ge 水温计;水温表
water temperature indicator 水温表
water temperature in well 孔内水温
water temperature measurement 水温测量
water temperature model 水温模型
water temperature of mine 矿井水温
water temperature regulator 水温调节器
water temperature relay 水温继电器
water temperature sensor 水温传感器
water temperature shut-off 水温自动控制器
water temperature stratification 水温成层
water tempering 水回火
water tender 煤水车【铁】;消防水车;供水船;供水车;给水船
water tension 水张力
water terminal 港埠;码头(区);水码头
water termination 水界面
water test 液压试验;止水试验;通水试验;水质试验;水质检验;水密性试验;水的试验
water test evaluation 水测定评估
water testing 水检验
water-test performance of the pump 泵的试验性能
water test pressure 水试验压力
water test unit 液压试验装置;水质试验装置;水压试验装置;水密性试验装置
water tex 水性(可塑)拉毛漆
water tex finishing 水性(可塑)拉毛涂装法
water thermometer 水温计;水温表
water-thermometer dash unit 水温表指示器

water thickness in water collecting pipe 集水管水层厚度
water thickness in well 井中水层厚度
water thief 多接头泄放阀
water thinnable 能用水稀释的
water thinnable paint 水稀释性涂料;水稀释性漆
water thinned 水稀释的
water thinned paint 水稀释性涂料;水稀释性漆;水释涂料
water thread 水纹
water-through wool scouring 水透式洗毛法
water tidal flat 潮浸洼地
water-tight 隔水的;耐水(压的);防水;水密的;不透水(带)的;不漏水的
water-tight air case 水密空气箱
water-tight barrier 不透水栅栏;防水墙;水密堤;防渗墙
water-tight basement 不透水地下室;防水地下室;水密性地下室;不透水基地
water-tight bed 不透水基床
water-tight bedrock 不透水基岩
water-tight blanket 不透水铺盖
water-tight box 不渗水箱;防渗水箱
water-tight branch box 水密分线盒
water-tight building 防水建筑物
water-tight bulkhead 防水围堰;防水舱壁;水密隔堵;水密舱;水密舱壁
water-tight cable 水密电缆
water-tight case 水密外壳
water-tight cesspool 不透水污水坑
water-tight closure 水密关闭
water-tight closure log 水密关闭记录
water-tight cofferdam 防水围堰
water-tight compartment 水密室;水密分舱;水密(分)舱;不透水间隔
water-tight concrete 防水混凝土;防渗混凝土;抗渗混凝土;水密性混凝土;不透水混凝土
water-tight concrete flat roof 水密性的混凝土平屋顶;不漏水的混凝土平屋顶
water-tight connector 水密接头
water-tight construction joint 水密施工缝;不透水施工缝;不透水工作缝
water-tight core 防水心墙;防渗心墙
water-tight core wall (坝的)防水心墙;防渗心墙
water-tight cover 水密盖
water-tight curtain 防渗帷幕;防渗地下连续墙
water-tight deck 水密甲板
water-tight diaphragm 防水(薄)层;阻水层;防水(隔)膜;(刚性的)防渗心墙;防渗斜墙;防渗护面;防渗地下连续墙
water-tight diaphragm dam 有防渗心墙的坝
water-tight door 防水门;水密门
water-tight door test 水密门试验
water-tight electric(al) torch 防水手电筒
water-tight facing 阻水面层;防渗面层
water-tight facing arch (堤坝的)不透水面层拱;防渗护面拱
water-tight fitting 水密配件
water-tight floor 水密肋板
water-tight formation 不透水岩层
water-tight frame 水密肋骨;水密隔舱骨架
water-tight gasket 止水胶垫
water-tight hatch 水密舱口(盖)
water-tight hatch cover 水密舱口盖
water-tight housing 不透水外壳
water-tighting of segment 管片防水
water-tight integrity 水密完整性
water-tight joint 防水接头;防水接缝;水密接头;水密接合;水密接缝;不透水接头;不透水(接)缝
water-tight lamp 水密灯
water-tight layer 不透水层;隔水层
water-tight lobe 水密灯罩
water-tight machine 防水水型电机
water-tight material 水密性材料;不透水材料
water-tight membrane 防水膜
water-tightness 闭水性;不透水(性);不漏水;防水性;水密性;不适水性
water-tightness grouting 防水灌浆
water-tightness of joint 接缝防水
water-tightness test 水密性试验
water-tight packing 不透水密封;不漏水填密
water-tight paper packing 水密纸张包装
water-tight pitch 水密铆距
water-tight riveting 水密铆接;水密铆合

water-tight rock stratum 不透水岩层
water-tight rubber membrane (三轴试验用)防水橡皮膜
water-tight screen 防渗帷幕;防渗墙;阻水帷幕;止水帷幕;防渗板桩;阻水板桩
water-tight seal 止水;水密封(口)
water-tight shaft timbering 导坑防水支撑
water-tight sluice door 水密闸门
water-tight socket 不透水插座;防水插座
water-tight stratum 不透水地层;隔水层;土壤不透水层
water-tight structure 防水结构;水密结构;不透水结构
water-tight subdivision 水密分段;水密分舱;水密(分)舱
water-tight switch 水密开关
water-tight tank 水密柜
water-tight tarpaulin 不透水柏油布
water-tight test 水密试验
water-tight vessel 水密容器;不透水容器;不漏水容器
water-tight wall 防水墙;水密墙;不漏水墙
water-tight watch case 水密性表壳
water-tight work 水密工作;水密工程
water-tight zone 不透水带
water tillage 水耕法
water-timed sampler 水定时式取样机;水定时取样器
water tissue 储水组织
water titration(method) 水滴定法
water to air cooler 水冷空气冷却器
water to be filtered 待滤水
water to be rolled and foamed 翻花冒泡
water to be turbid 水质变浑
water to carbide acetylene generator 电石加水式乙炔发生器
water to carbide(type) generator 注水式乙炔发生器
water-to-cement ratio 水灰比(例)
water-to-earth ratio 蓄水量与挖方量之比;库容土方比
water tolerance 耐水性;耐水度
water tolerance test 耐水性试验
water tongue 水舌
water-to-oil area 水油过渡带
water-to-oil cooler 水冷油冷却器
water-to-pelt ratio 水与裸皮重量比
water-to-rail 水路转铁路运输
water torch 水炬
water toughening 水冷韧化处理
water-to-water heat exchanger 水—水换热器;水换热器
water tower 水塔;高喷水塔;救火用高喷水塔
watertown process 炮筒离心铸造
water tracing 流水跟踪;示踪测流(法)
water tractor 小型拖轮;小拖轮
water trailer 水罐拖车
water trajectories 水流轨迹
water transfer 输水;调水
water translocation 水分迁移
water transmissibility 导水性
water transmissibility coefficient 导水系数
water transmission 水的传输
water transmitting ability 导水能力
water transparency 水(的)透明度
water transparency measurement 水透明度测量
water transparency measurement device 水透明度测量装置
water transparency recorder 水透明度表
water transport(ation) 水运;水上货运;水的输送;航运;水力输送;水上运输
water transport engineering 水运工程(学)
water transport line 水上运输线
water trap 水封;水坑;虹吸管;存水弯;聚水器;脱水器;除水器
water trap vent 水封通气管
water travel(time) curve 汇流曲线;水流传播时间曲线
water treating 水处理
water treating chemicals 水处理药品
water treating equipment 水处理装置;水处理设备
water treatment 水再生;水净化(作用);水的净化处理;水的处理;水处理;净水处理
water treatment agent 水处理剂

water treatment biocide 水处理杀菌剂
water treatment chemicals 水处理剂
water treatment cleaning agent 水处理清洗剂
water treatment corrosion inhibitor 水处理缓蚀剂
water treatment entity 水处理单元
water treatment equipment 水处理装置;水处理设备
water treatment facility 水处理装置;水处理设施
water treatment flocculant 水处理絮凝剂
water treatment in chemical indsutry 化学工业水处理
water treatment in garden 园林理水
water treatment medium and reagent 水处理料剂
water treatment operation 水处理运作
water treatment plant 给水处理厂;净水所;净水厂;水处理厂;水处理装置
water treatment plant sludge 水处理厂污泥
water treatment plant wastewater 给水厂废水;水处理厂废水
water treatment process 水处理工艺
water treatment system 水处理系统
water treatment system by recirculation 水再循环处理系统
water treatment system by recirculation within a factory 厂内水再循环处理系统
water treatment system of closed loop 封闭式循环水处理系统
water treatment technological process 给水处理工艺流程;水处理工艺流程
water treatment technology 水处理技术;水处理工艺
water treatment unit 水处理装置;水处理单元
water treatment works 水处理工程;净水处理工程;净水处理厂
water treeing resistant 抗水树形击穿
water trenching density 淋水密度
water troubles 涌水事故;水侵事故
water trough 饮水槽;排水沟;水槽
water truck 洒水车;水罐车;(运水用的)水车;水槽车
water tub 水盆
water tube 隔水套管;水管
water-tube boiler 水管锅炉
water-tube boiler with bent tubes 弯管式水管锅炉
water-tube boiler with straight tubes 直管式水管锅炉
water-tube circuit 水管环路
water-tube cleaner 水管清垢器
water-tube condenser 水管冷凝器
water-tube gas condenser 水管气体冷凝器
water-tube hose 水软管
water-tube plate 水管板
water-tube waste-heated boiler 水冷式余热锅炉
water tubing 水管
water tunnel 有水隧洞;供水隧洞;供水隧道;水洞;输水隧洞;输水隧道
water tunnel portal 输水隧洞入口;输水隧道入口
water tunnel test 水洞试验
water tupelo gum 水紫树
water turbidity 水的混浊度
water turbine 水轮机;水力涡轮机
water turbine driven generator 水轮驱动发电机
water turbine generator set 水轮发电机组
water turbine oil 水轮机油
water turbine regulation system 水轮机调速系统
water turbo-pump 水涡轮泵
water twist 水扭
water type 水型;水团类型
water type diagram 水型图
water type dust collector 水型集尘器
water type extinguisher 湿式灭火机
water type fire extinguishing system 水型灭火系统
water type sludge 水型油泥
water undertaker 自来水供应服务人
water undertaking 供水企业;自来水公司;给水事业
water uptake 水的吸收;吸水
water usage 耗水量
water usage graph 耗水量图
water use 水的利用
water use coefficient 用水系数
water used for emergency case 事故水
water used in lockage 过闸用水
water use efficiency 用水效率;水的利用效率

water use for boiler 锅炉用水
water-use planning system 计划用水制度
water user 自来水用户;用水户
water-use ratio 用水比;蒸腾比;蒸腾率
water-using industry 用水量大的工业
water utility 水公用事业
Water Utility Council 水公用事业协会
water utility industry 给水(公用)事业工业
water utility management 水公用事业管理
water utilization 水的利用
water utilization project 水利工程
water value 水值
water valve 水(压)阀;水门;水管阀门;水调节阀
water valve bonnet 水阀帽
water valve bonnet nut 水阀帽螺母
water valve gland 水阀压盖
water valve packing nut 水阀填密螺母
water valve seat 水阀座
water valve stem 水阀杆
water vapo(u)r 水蒸气;水汽
water vapo(u)r absorption 水蒸气吸收;水汽吸收
water vapo(u)r arc welding 水蒸气保护电弧焊
water vapo(u)r barrier 水蒸气隔离层
water vapo(u)r barring 水蒸气阻塞
water vapo(u)r bubble 水汽泡
water vapo(u)r capacity 水汽容量
water vapo(u)r chimney 蒸气烟囱
water vapo(u)r concentration 水蒸气浓度;水蒸气提浓法
water vapo(u)r condensation 水蒸气凝结
water vapo(u)r content 水蒸气含量;水汽含量
water vapo(u)r continuum 水汽连续吸收
water vapo(u)r density 水蒸气密度;水汽密度
water vapo(u)r diffusance 水蒸气扩散
water vapo(u)r diffusion 水蒸气扩散
water vapo(u)r impermeability 水蒸气不渗透性
water vapo(u)r imperviousness 水蒸气不渗透性
water vapo(u)r laser 水蒸气激光器
water vapo(u)r layer 水汽层
water vapo(u)r maser 水气微波激射
water vapo(u)r migration 水蒸气迁移
water vapo(u)r mist 水蒸气雾
water vapo(u)r penetration coefficient 水蒸气渗透系数
water vapo(u)r permeability 透湿性;透湿度;水蒸气渗透性;水蒸气透过性;水蒸气渗透率
water vapo(u)r-permeable 可渗透水蒸气的
water vapo(u)r permeance 水蒸气渗透(速率)
water vapo(u)r pipe 水蒸气管
water vapo(u)r plasma 水蒸气等离子体
water vapo(u)r pressure 蒸气压力;水蒸气压;水汽压(力)
water vapo(u)r pressure drop 水蒸气压力降落
water vapo(u)r pressure graph 水蒸气压力图
water vapo(u)r-proof packaging 防潮包装
water vapo(u)r quantity 水蒸气量
water vapo(u)r release 水蒸气释放
water vapo(u)r resistance 蒸气渗透阻力;耐水蒸气;水蒸气不渗透性能
water vapo(u)r resistant 耐水蒸气的;抗水蒸气的
water vapo(u)r retarder 隔气层;水蒸气阻滞层
water vapo(u)r saturation 水汽饱和(率);水汽饱和(度)
water vapo(u)r seal 水汽密封
water vapo(u)r spectroscope 水汽分光镜
water vapo(u)r tension 水汽张力
water vapo(u)r test 水蒸气试验
water vapo(u)r transfer 水汽输送
water vapo(u)r transfer coefficient 水蒸气传递系数
water vapo(u)r transmission 水蒸气传递;水蒸气透湿性;水蒸气渗透;水蒸气透过(性)
water vapo(u)r transmission rate 水蒸气渗透率;水蒸气传递速率;水蒸气透过(速)率
water vapo(u)r transport 水汽输送
water vapo(u)r valve 蒸气活门;蒸气阀
water vapo(u)r welding 水蒸气保护焊
water-varnish 水性清漆
water vascular system 水管系统
water vehicle 船舶
water vein 裂隙地下水;水脉
water velocity 流速;水速;水流速度
water vesicle 储[贮]水水泡;储水泡
water vessel 喷水器;浇水器

water vine 水藤
water violet 水生紫罗兰
water viscosity 水的黏度
water void 水孔隙
water void ratio 水隙比
water vole 河鼠
water voltmeter 水解电压计
water volume 水(容)量
water vortex brake 水涡流测功器
water vulcanization 热水硫化
water wagon 洒水车;运水车;(运水用的)水车
water wall 水幕(即水冷壁)【港】;防水墙;水墙
water-wall air-cushion vehicle 水壁式气垫艇
water-wall circuit 水冷壁回路
water-wall craft 侧壁式气垫船;水壁式气垫艇
water-wall furnace 水壁炉子;水壁炉膛;水壁炉
water-wall header 水冷壁联箱
water-wall panel heater 水屏式加热器
water-wall refuse incinerator 水夹套式垃圾焚化炉;水夹带式垃圾焚化炉;水壁式垃圾焚化炉
water-wall surface 水冷壁受热面
water-wall tube 水冷壁管
water wash 水冲洗
water-washed disease 缺乏卫生用水带来的疾病
water washing 水(冲)洗
water washing device 水洗装置
water wash spray booth 喷水式喷漆室;水洗式喷漆室
water waste preventer (厕所用)冲洗水箱
water waste survey 漏水检查
water-wastewater system 上下水道系统
water-wasting pressure regulator 弃水式压力调节器
water water energetic reactor 水—水动力堆
water-water heat exchanger 水—水换热器
water-water reactor 水—水堆
water-water type heater 水—水式加热器
water-water type heat exchanger 水—水换热器
water wave 水纹;水波
water-wave analog(ue) 流体动力学模拟
water-wave arrivals 水波到达时间
water-wave energy 水波能
water wax 水乳胶蜡
waterway 出水道;排水沟渠;(木船的)梁压材;航道;水路;水冷腔;水道;水口;水槽
waterway administration 航道管理
waterway bed 水道床
waterway capacity 航道通过能力;水道通过能力
waterway carrier 船舶
waterway development 水道开发
waterway dredging 航道疏浚
waterway engineer 航道工程师
waterway engineering 航道工程;水道工程
waterway engineering survey 航道工程勘测
Waterway Experiment Station 美国水道实验站
waterway experiment station 水道实验站
waterway facility 航道设施
waterway habo(u)r 内河港(埠)
water way in front of mountain 山前区河段
waterway installation 航道设施
waterway mileage 航道里程
waterway milepost 航道里程标
waterway milestone 航道里程
waterway net(work) 航道系统;水道网;排水沟网
waterway opening (桥、涵的)出水孔径;过水孔径
waterway planning 航道规划
waterway project 航道工程;水道建设工程
waterway protection 航道保护
waterway regulation 航道整治
waterways and harbo(u)r engineering 航运水工工程
waterway section 河段
waterway size 水路尺寸
waterway stabilization structure 稳定航行水路结构;水道稳定措施结构
waterway station 航道站
waterways transport 水路运输
waterway survey 航道测量;水道勘测
waterway traffic 河运运量;河运
waterway transit capacity 航道通过能力
waterway transport 水路运输
waterway transportation 河运
waterway tunnel 过水隧洞;水道隧道;水道隧洞;水道隧道;输水隧洞;输水隧道

waterway works 水道工程
water wedge shiplift 水坡式升船设施;水坡式升船机
water weed 水生杂草;水草
water weed control 水生杂草防除
water weed cutter 水草切割机
water weed growth 水槽生长
water weed removal 水槽去除
water weighing device 水梁衡器
water well 水井
water-well driller 水井钻机
water-well drilling 水井钻探;水井钻进
water-well drill(ing machine) 水井钻机;冲击式打井机
water-well drilling rig 水井钻架;水井钻机
water-well formula in effective zone 有效带的水井公式
water-well maintenance 水井维修
water-well mechanics 水井力学
water-well rig 水井钻机
water-well technology 水井工艺
water-well yield 水井出水量
water-well yield test equipment 水井出水量测定设备
water wet 水润湿(的)
water wet core 水浸岩芯
water wet sand 水润湿砂
water-wettable 水可湿性
water-wetted 水润湿的
water-wetted sharpening stone 水润湿的磨石
water-wheel 水车;轮船的叶轮;水轮
water-wheel driven distributor 水轮驱动布水器
water-wheel generator 水轮发电机
water-wheel meter 水轮式水表
water-wheel pit 水轮槽;水轮坑
water-white 无色透明的;水白色(的)
water-white acid 水白酸
water-white high-grade burning oil 水白色高级燃料油
water-white kerosene 无色煤油;水白色煤油
water-white mineral oil 水白色矿物油
water-white oil 水白色油
water-white paraffin(e) wax 水白色石蜡
water willow 水柳
water winding 汲水;水力提升
water wing 桥墩翼墙;水翼;挡水翼墙
water witch 找水仪
Waterwite 瓦特怀特牌玻璃(一种高度漫光压制玻璃)
water withdrawal 取水量
water with less precipitate 沉淀物较少的水
water with little precipitate 沉淀物很少的水
water with more precipitate 沉淀物较多的水
water with most precipitate 沉淀物很多的水
water works 给水设备;自来水厂;给水装置;水事工程;水工建筑物;水厂工程;水厂
waterworks alumin(i)um 水厂铝
waterworks and sewerage system 供水与排水系统
Water Works Authority 给水工程管理局
waterworks code 给水厂法规
waterworks construction 自来水厂建设
waterworks design 自来水厂设计
waterworks engineer 水厂工程师
waterworks engineering 给水工程
waterworks operation 水厂的运转;自来水厂运作;给水系统管理
waterworks planning 水力开发区划;地区给水区划;给水厂规划
waterworks plant 给水厂
waterworks project 给水工程计划
waterworks sludge 给水厂污泥
water worn 给水磨损;水成磨损;流水冲刷的;流水冲蚀的;水蚀的;被水磨损的;被水磨光的
water-worn gravel 水磨砾石
water-worn pebble 小圆石
watery 稀薄的;富水性的
watery blue 流青(色料);浅蓝色
watery city 水上城市;水都
watery discharge 水性排出物
watery distillate 含水馏分
water year 水文年(度);水年
water year book 水文年鉴
watery fusion 结晶熔化
water yield 出水量;产水量

water yield after well cleaning 洗井后的水井出水量
water yield before well cleaning 洗井前的水井出水量
water yield formation 产水层
water-yielding layer 出水层
water-yielding stratum 含水层;供水(地)层;出水层
water yield in peak runoff period 峰期涌水量
water yield of development opening 开拓井巷涌水量
water yield of mine 矿井涌水量
water yield of well after explosion 爆破后的水井出水量
water yield of well before explosion 爆破前的水井出水量
water yield of wells 水井出水量
water yield per well 单井出水量
water yield prediction 流量预报
water yield property grade of aquifer 含水层富水性等级
water yield property index of aquifer 含水层富水性指标
water(y) solution 水溶液
watery stratum 含水层
Wates 韦特斯式建筑(英国创造的一种建筑体系)
wathlingite 镁硬石膏
Watkin's pyroscope 韦氏高温计
Watkin's recorder 韦氏测温记录仪
Watling shales 瓦特林页岩
Watson equation 瓦特生方程
Watson four electric system method 瓦特生四电极系统法
Watson-Sommerfeld transformation 瓦特生—索末菲变换
watt 瓦特(电功率单位);瓦【电】;瓦特单位
wattage 瓦特数;瓦数
wattage consumption 功率消耗
wattage dissipation 损耗瓦数
wattage input 输入瓦数
wattage output 输出瓦数
wattage rating 额定瓦特数
watt component 有功分量;有功部分
watt current 有功电流
watt density 功率密度
watterillite 灰芒硝
wattful loss 欧姆损耗
wattful power 有功功率
watt governor 瓦特调速器;飞球式调速器
watt-hour 瓦特小时;瓦时
watt-hour capacity 瓦特小时容量;瓦时容量
watt-hour consumption 瓦时消耗量
watt-hour demand meter 最大需量瓦时计
watt hour efficiency 瓦时效率
watt-hour meter 瓦特小时计(通称电表);瓦时计;瓦时计;电(度)表
wattle 枝条;编条
wattle and da(u)b 灰板墙;泥笆墙;荆笆墙;抹灰篱笆墙;外表涂泥的篱笆墙
wattled 篱笆的;用枝条编织的
wattled fence 篱棚
wattled wall 篱笆墙
wattle fencing 篱笆围墙;篱笆围栏
wattle gum 澳洲产水溶性或水分散性(树)胶;澳洲胶
wattle house 篱笆房屋
wattle hurdle 篱笆栅栏
wattless component 虚数部分;无功分量;无功部分
wattless current 无功电流;电抗性电流
wattless kilowatt 无功千伏安
wattless load 无功负载;无功负荷
wattless output 寄生功率
wattless power 无功功率
watt level 功率大小
wattle work 枝条编制品;河岸防冲拦淤栅;防冲拦淤栅;编篱工
wattling 编篱;柴排;柴捆
watt loss 功率损耗
watt-meter 功率表;瓦特表;电力表;瓦特计
watt[meter-Kelvin] 瓦特(米开)
wattmeter method 功率表法;瓦特计法
wattmetric relay 瓦特计式继电器
watt-minute 瓦分
watt relay 瓦特继电器

watt-second 瓦(特)秒
Watt's governor 飞球调速器
watt's horsepower 英制马力
Watt's law 瓦特定律
Wattson zoom lens 瓦特逊变焦距镜头
watts-out 发出瓦数
watt[steradian] 瓦特(球面度)
watt transducer 电功率传感器
Wauchope(type) starter 沃科普式起动器
Waucobian 沃科巴统
waughammer(drill) 架(柱)式凿岩机
waughoist 带架绞车;柱装绞车;架柱式绞车
wave 浪;波形物;波浪(形);波筋;凹印
wave aberration 波像差
wave abrasion 波蚀
wave absorber 消波装置;消波设备;消波器;吸浪器;波浪吸收器
wave absorbing beach 消波滩;消波岸坡
wave absorbing construction 消波设施
wave absorbing revetment 消波护岸
wave absorbing structure 消波结构;消波建筑物
wave acoustics 波动声学
wave action 波浪作用;波动作用
wave activated generator 波浪发电机
wave activated generator group 波浪发电机组
wave activated power generation 波浪发电
wave advance 波浪推进;波浪前进
wave age 波龄
wave agitation 波扰动;波浪激动
wave allowance 波高留量
wave amplitude 波浪振幅;波幅;波(的)振幅
wave amplitude ratio 波幅比
wave analyzer 波形分析仪;波形分析器;波浪分析器
wave analysis 波(形)分析
wave and tide monitoring system 波浪潮汐监测装置
wave and tide recording system 波浪潮汐记录仪
wave angle 波(程)角
wave angle filtering 波角度滤波
wave antenna 行波天线;波天线
wave anti-node 波腹
wave arch 波形拱
wave attack 波浪袭击;波浪冲蚀;波浪冲击
wave attenuation 波(浪)衰减
wave band 频段;波段;波带;频带
wave band changing switch 波段转换开关
wave band selector 波段开关
wave band switch 波段开关
wave barrier 波状沙洲
wave base 浪蚀基面(深度);浪基面;波浪基部;波基面;波底
wave basin 消波池;造波水池;波浪水池;波浪试验港池;波浪试验槽;波浪盆地;波浪港池;波浪(试验)池
wavebeam guide 光波管;波导管
wave beam width 波速宽度
wave blade 造波板;阻波板
wave bottom 波谷
wave bottom interaction 波浪与海底相互作用
wave braid 波纹编带
wave break 挡浪墙;防浪堤
wave break forest 防浪林
wave breaker 破浪墩;破浪堤;破坏型;防波堤
wave breaker block 消波块体;防波块体
wave breaking 波浪破碎
wave breaking block 消波块体;消波块体
wave breaking coefficient 波浪破碎系数
wave breaking condition 波浪破碎条件
wave breaking facility 消波设施
wave breaking measure 消波措施
wave breaking point 碎波点;波浪破碎点
wave breaking works 消波建筑物;消波工程;防波构筑物;防波(堤)构筑物
wave breaking zone 波浪破碎带
wave-built 浪成的
wave-built beach 浪成滩地;浪成海滩
wave-built crest 浪积波峰;浪成波峰
wave-built crest line 浪成波峰线
wave-built cross bedding 波浪交错层理
wave-built current 浪成水流
wave-built facies 浪成岩相
wave-built formation 浪成地层
wave-built lamination 浪成纹理
wave-built platform 波积台地;浪成台地;浪成平台;浪成阶地
wave-built ripple 浪成波痕
wave-built structure 波成构造
wave-built terrace 浪成阶地;波成台地;波成阶地
wave-built terrace sand trap 波成阶地砂体圈闭
wave bulge 波状隆起
wave buoy 波浪浮标
wave calming oil 镇浪油
wavecanal 波道
wave celerity 波速;波浪传播速度
wave chamber 消波室
wave changer 波段变换器
wave changing switch 波段变换开关;波长转换开关
wave channel 造波槽;波浪(试验)槽
wave character 波动性
wave characteristic curve 波浪特征曲线;波浪特性曲线
wave characteristics 波(浪)特征;波浪特性;波浪特征值
wave chart 波浪图;波浪记录图
wave clearance 波峰间隙
wave climate 波候
wave clipper 削波器
wave cloud 波状云
wave cluster 波族
wave clutter 海面杂乱回波;海浪干扰;波浪回波干扰
wave coda 尾波;波尾
wave collector 波收集器
wave compensation 波浪补偿
wave component 波组成;波(动)分量
wave compression 压缩波;波(的)压缩;波的压力
wave concept of light 光的波动概念
wave condition 波浪状况;波浪情况;波况
wave configuration 波形
wave constant 波常数
wave contour 波轮廓
wave contouring raft 波面筏
wave convergence 波浪辐聚;波浪辐合
wave conversion 波形变换
wave converter 波(形)变换器
wave cord 波纹(玻璃缺陷)
wave corpuscle duality 波粒二象性
wave corpuscular duality 波动粒子二象性
wave-covered water 起波水面
wave crack 等压裂隙
wave crest 波脊;波峰;波顶
wave crest combination 波峰组合
wave crest height 波顶高度
wave crest length 波峰长度
wave crest line 波峰线;波脊线
wave crest tectonic belts 波峰构造带
wave crest touchdown point 波峰拍击点
wave crest width 波峰宽度
wave cross ripple(mark) 浪成交错波痕
wave current 波流;波动电流
wave current interaction 波流相互作用
wave curtain 防波帷幕;波帘
wave curvature buoy 波形曲率浮标
wave curve 波形曲线
wave cut 波浪冲刷;波浪切割;波蚀的
wave-cut base 波蚀基面
wave cut bench 浪蚀台地;浪蚀平台;波蚀台地;海蚀台
wave-cut chasm 浪蚀裂隙
wave-cut cliff 浪蚀崖
wave-cut delta 浪切三角洲
wave-cut notch 浪蚀龛;浪蚀洞
wave-cut plain 浪蚀平原
wave-cut platform 波蚀台地;海蚀台地
wave-cut scarp 浪蚀崖
wave-cut shore 浪蚀岸
wave-cut terrace 浪蚀台地;浪蚀阶地;浪成阶地;波蚀阶地
wave cut(ting) 浪蚀;波蚀
wave-cut topography 浪蚀地形学;浪蚀(底)地形
wave cyclone 波动性气旋
wave damper 消波装置;消波器;波浪消能装置
wave damping 波浪衰减
wave damping device 波浪消能装置
wave damping effect 消波作用
wave data 波浪资料
wave decay 波浪消衰;波浪衰减
wave decay distance 波浪消衰延程;波浪衰减距离
wave decay for deep-water area 深水波浪衰减

wave defence works 防浪工程
wave deflector 挡浪墙
wave deflector on the quay 码头胸墙
wave deformation 波浪变形
wave defraction 波浪绕射;波浪衍射
wave delay distortion 波导时延失真
wave delta 越浪堆积;波成三角洲;冲溢三角洲
wave depression 锋面低压;波动性气旋
wave detector 检波器
wave detonation 爆炸波
wave differential equation 波动微分方程
wave diffraction 波浪衍射;波(浪)绕射
wave diffraction diagram 波浪绕射图
wave diffraction intensity factor 波绕射强度系数
wave diffraction theory 波浪绕射理论
wave diffusion analysis 波扩散分析
wave diffusion spectrometer 波谱仪
wave dipole 波偶极子
wave direction 浪向;波向
wave directional spectrum 波向谱;波浪方向谱
wave direction analyzer 波向分析仪
wave direction meter 波向仪;波向计
wave director 导波体;波导
wave dispersion 波浪消散;分散波;波浪弥散;波速分散;波色散;波浪散射;波(的)弥散
wave dispersion effect 波浪弥散效应
wave dissipating concrete block 消波混凝土块体
wave dissipating wall 波浪消能墙
wave dissipation revetment 消波护岸
wave distortion 波形失真;波形畸变
wave disturbance 波(状)扰动;波浪扰动
wave divergence 波浪辐散;波扩散
wave division multiplexing 分波多路
wave-dominated delta 浪控三角洲
wave-dominated facies 波浪控制相
waved pique 波纹凹凸组织
wave drag 波浪拖曳;兴波阻力;造波抵抗;波阻(力);波阻力;波浪曳力
wave drag coefficient 波阻系数
wave driven generator 波力发电机
waved roof 瓦楞屋面;波状屋面;波状屋顶
waved shell roof 波形薄壳屋顶
waved stitch 波状线迹
waved switch 波动开关
waved tee bar 钢筋混凝土用型钢(凸缘平直,腹板呈S形)
waved tube 波形管
wave duct 波管;波道
wave duration 波动历时;波持续时间
waved wire 波形钢丝
wave dynamics 波浪动力学;波动动力学
wave dynamometer 波(动)力计
wave edge 波形边
wave effect 波浪影响
wave efficiency 波动效率;波动高峰
wave electric(al) power system 波浪发电系统
wave electronics 波动电子学
wave element 波浪要素
wave elevation 波面升降
wave energy 波(浪)能;波动能
wave energy absorber 波能吸收器
wave energy absorption 波能吸收
wave energy air turbine 波能空气透平
wave energy coefficient 波能系数
wave energy conversion 波能转换
wave energy conversion buoy 波能转换浮筒
wave energy conversion system 波能转换系统
wave energy conversion technique 波能转换技术
wave energy converter 波能转换器
wave energy density 波能密度
wave energy device 波能装置
wave energy dissipation 波能耗散
wave energy distribution 波能分布
wave energy extractor 波能发电站;波能萃取器
wave energy flux 波能量通量
wave energy input 波能输入
wave energy of sea floor 波浪冲击能
wave energy partition 波能劈分
wave energy power generating system 波浪能发电系统
wave energy propagation 波能传播
wave energy recovery device 波能吸取装置
wave energy recovery system 波能吸取系统
wave energy rectifier 波能检波器

wave energy research 波能研究
wave energy resource 波能资源
wave energy rose diagram 波能玫瑰图
wave energy spectrum 波能谱
wave energy technology development 波能技术开发
wave energy transmission 波能传递
wave envelope 波包络;波包(迹)
wave equation 波动方程
wave equation analysis 波动方程分析
wave equation migration 波动方程偏移
wave erosion 波浪冲蚀;浪蚀;波蚀;波浪侵蚀;浪冲刷
wave erosion coast 波浪侵蚀海岸
wave erosion shoreline 浪蚀岸线
wave estimation 波浪推算;波浪估算
wave exciter 波激励器
wave exciting device 波激发器
wave face 波阵面;波前波
wave fence 挡浪栅
wave fetch 波浪行程
wave field 波(浪)场
wave field decomposition 波场分解
wave field equation 波场方程
wave field extrapolation 波场延拓
wave field separation 波场分离
wave filter 滤波设备;陷波器;滤波器;波形滤波器
wave filtering 滤波
wave flume 造波水槽;波浪(试验)槽
wave foam 船首白浪
wave focusing 波浪聚焦
wave foot 波脚
wave force 波(浪)力
wave forced device 波力发电装置
wave forced station 波力发电站
wave forecast(ing) 波浪预报
wave forecasting method 风浪预报法
wave form 波形
waveform analysis 波形分析
waveform analyzer 波形分析仪;波形分析器
waveform calculator 波形计算器
waveform characteristic 波形特性
waveform converter 波形变换器
waveform deconvolution 波形反褶积
waveform design 波形设计
waveform distortion 波形失真;波形畸变
waveform distortion factor 波形畸变系数
waveformed mouth 波状齿
waveformed ripple(mark) 浪成波痕
waveform error 波形误差
waveform estimation 波形估计
waveform factor 波形因素;波形因数;波形系数
waveform generator 定波形发生器;波形(信号)发生器
waveform height 波高
waveform influence 波形影响
wave-forming resistance 兴波阻力
waveform interference 波干扰
waveform measurement 波形测量;波形测定
waveform meter 波形测量仪
waveform modelling 波形模拟
waveform monitor 监视示波器;波形监视器;波形监控器
waveform oscilloscope 波形显示器
waveform parameter 波形参数
waveform processing of determine stratigraphic(al) lithology 确定地层岩性的波形处理
waveform pulse 波形脉冲
waveform record distribution 波形记录分布
waveform recorder 波形记录器
waveform recording 波形录制;波形记录
waveform recording method 波形记录法
waveform reference 波形依据
waveform segment 波形段
waveform separation 波形区分;波形分离
waveform shaping 波形整形
waveform survey 波形测量
waveform synthesizer 波形综合器;波形合成器
waveform test 波形试验
waveform translation(al) shell 波形侧移壳体(结构)
waveform transmission 波形传输
waveform undulation 波形起伏
waveform viewer 波形观测仪

waveform washer 波浪式垫圈
wave frequency 电波频率;波频(率);波的频率
wavefront 波(阵)面;波前
wavefront aberration 波阵面像差;波前像差
wavefront aberration function 波像差函数
wavefront advance 波前行进
wavefront advance in two layers 双层波阵面推进
wavefront angle 波前(倾)角
wavefront chart 波前方程
wavefront chart method 波前图法
wavefront chromatic aberration 波色差
wavefront compensation 波前补偿
wavefront control 波前控制
wavefront curvature 波前曲率
wavefront distortion 波前畸变
wavefront distortion of light wave 光波波面变形
wavefront divergence compensation 波前扩散补偿
wavefront folding interferometer 波阵面折叠干涉仪
wavefront interferogram 波阵面干涉图
wavefront length 波前宽度
wavefront matching 波阵面匹配;波前匹配
wavefront method 波阵法;波前(沿)法
wavefront plotter 波前标绘器
wavefront reconstructing record 波阵面再现记录
wavefront reconstruction 波阵面再现
wavefront reconstruction imaging 波阵面再现成像
wavefront reconstruction photography 波阵面再现照相术;波前重现照相(术)
wavefront reproduction 波前再现
wavefront reversal 波前反转
wavefront shearing interferometer 波阵面剪切干涉仪;波阵面错位干涉仪;波前错位干涉仪
wavefront slope 波准面斜度
wavefront splitting 波阵面分裂
wavefront-splitting interference 分波前干涉
wavefront steepness 波前陡度
wavefront tilt errors 波前倾斜误差
wavefront tilt warping 波阵面倾斜变形
wavefront travel 波前行程
wave function 波(动)函数
wave function modulation 波函数调制
wave furrow 波蚀沟
wave ga(u)ge 测波仪;测波计;波浪计;波高计
wave-generated current 波生流;波成流
wave-generated delta 波成三角洲
wave-generated flow 波生流
wave-generated ripple 浪成波痕
wave generating apparatus 造波器;波浪发生器
wave generating area 波发形成区;波浪发生区
wave generating equipment 造波设备
wave generating machine 波发生器
wave generating region 波浪形成区
wave generation 造波;波浪形成;波浪生成;波浪生长;波浪发生
wave generator 波发生机;造波器;造波机;生波器;生波机;波浪发生器
wave glass 瓦楞玻璃;波纹玻璃
wave group 波群
wave group celerity 波群速度
wave group correlation 波组对比
wave group effect 波浪群集效应
wave grouping effect 波群影响;波群效应
wave group spectrum 波群谱
wave group velocity 波群速度
wave group velocity in deep water 深水波群速度
wave growth 风浪成长;波浪成长
waveguide 导波体;导波器;导波管;波导(管)
waveguide admittance 波导纳
waveguide antenna 波导式天线;波导管天线
waveguide aperture 波导孔
waveguide arc detector 波导弧型探测器
waveguide arm 波导支路;波导分路
waveguide assembly 波导装置
waveguide attenuation 波导衰减
waveguide attenuator 波导衰减器
waveguide below-outoff attenuator 波导活塞衰减器
waveguide bend 波导弯头
waveguide bolometer 波导测辐射热计
waveguide bolometer-coupler unit 波导测辐射热器—耦合器组合
waveguide bolometer unit 波导测辐射器组合
waveguide branching filter 波导管分波器;波导

分支滤波器
waveguide bridge 波导电桥
waveguide bundle 波导束
waveguide cavity 波导空腔
waveguide cell 波导(管)节
waveguide choke flange 波导管扼流突缘
waveguide choke plunger 波导管阻波突缘
waveguide CO_2 laser 波导二氧化碳激光器
waveguide communication 波导通信
waveguide component 波导元件;波导节;波导管部件
waveguide conical taper 波导管锥形接续器
waveguide connector 波导连接器;波导接头
waveguide corner 波导弯角
waveguide coupler 波导接头;波导(管)耦合器
waveguide coupling 波导(管)耦合
waveguide critical dimension 波导临界尺寸
waveguide cutoff frequency 波导管截止频率
waveguide delay line 波导延迟线
waveguide diode 波导二极管
waveguide directional coupler 波导型定向耦合器
waveguide dispersion 波导色散
waveguide dispersive line 波导色散线
waveguide distortion 波导畸变
waveguide dummy load 波导管假负载
waveguide effective thickness 波导有效厚度
waveguide elbow 波导肘管;波导弯头;波导弯管
waveguide electro-optic prism 波导光电棱镜
waveguide element 波导元件
waveguide far infrared laser 波导远红外激光器
waveguide feed 抛物镜面天线;波导(管)馈电
waveguide fiber 波导纤维
waveguide film bolometer 波导薄膜测辐射热器
waveguide filter 波导(管)滤波器
waveguide fixed attenuator 波导式固定衰减器
waveguide flange 波导管凸缘
waveguide flow calorimeter 波导流体量热计
waveguide gas laser 波导(式)气体激光器
waveguide hybrid 波导桥接岔路
waveguide impedance 波导阻抗
waveguide impedance bridge 波导阻抗电桥
waveguide impedance transformer 波导阻抗变换器
waveguide iris 波导可变光栏;波导隔膜
waveguide isolator 波导隔离器
waveguide junction 波导管接头
waveguide junction circulator 波导环器
waveguide klystron 波导管(型)速调管
waveguide laser 波导激光器
waveguide laser resonator 波导激光谐振器
waveguide lens 波导(管式)透镜
waveguide linear accelarator 波导直线加速器
waveguide lowpass 波导低通滤波器
waveguide magnetron 波导磁控管
waveguide matching plate 波导管匹配膜片
waveguide match load 波导匹配负载
waveguide meter 波导测定计
waveguide mode 波导振荡模;波导管模式;波导方式
waveguide modulator 波导调制器
waveguide mount 波导(管)支架
waveguide network 波导管端网络
waveguide output 波导输出
waveguide output circuit 波导型输出器
waveguide plunger 波导(管)活塞;波导管短路器
waveguide post 波导柱
waveguide power divider 波导管功率分配器
waveguide prism 波导(管)棱镜
waveguide propagation 波导传播
waveguide pump 波导管增压器
waveguide radiator 波导(管)辐射器
waveguide reactance attenuator 波导管电抗衰减器
waveguide receiver 波导管接收机
waveguide reflector 波导反射器
waveguide resonator 空腔共振器;波导管(谐)振器;波导共振器
waveguide-resonator laser 波导谐振激光器
waveguide ring network 波导环形网络
waveguide rotary joint 波导管旋转接头
waveguide scattering 波导散射
waveguide segment 波导管节
waveguide shim 波导填隙片
waveguide shutter 波导管开闭器;波导保护器
waveguide simulation 波导模拟

waveguide simulator 波导模拟器
waveguide slot 波导槽
waveguide spectrum analyzer 波导光谱分析仪
waveguide storage 波导存储
waveguide stub 波导管短截线;波导短截
waveguide stub tuner 波导短截线调谐器
waveguide sweep reflectometer 波导扫描反射计
waveguide switch 波导转换开关;波导管(转换)开关
waveguide synchrotron 波导同步加速器
waveguide taper 波导管锥形联接器
waveguide tee 波导三通;波导管的T形接头
waveguide-to-coaxial adapter 波导同轴转接器
waveguide transformer 波导变换器
waveguide transition 波导过渡
waveguide transmission line 波导传输线
waveguide tuner 波导式调谐器;波导管调谐器
waveguide twist 波导(管)扭转接头
waveguide type phase meter 波导式相位计
waveguide type resistant attenuator 波导式抗衰减器
waveguide-type resonator 波导式谐振器
waveguide type standing wave detector 波导式驻波检测器
waveguide wall 波导壁
waveguide wall current detector 波导壁电流检波器
waveguide wire 波导线
wave hammer 波浪引起的水锤作用
wave height 浪高;波高
wave height analyzer 波高分析器
wave height coefficient 波高系数
wave height contour 等波高线
wave height correction 六分仪的波高校正
wave height distribution 波高分布
wave height forecast 波高预报
wave height ga(u)ge 波高仪
wave height in deep water 深水波高
wave height inside harbo(u)r 港内波高
wave height-length ratio 波高—波长比;波陡比
wave height measuring equipment 波高测量设备
wave height meter 浪高仪;波高仪
wave height of accumulation rate of wave train 累积率波高
wave height on breaking 破波波高;碎波波高
wave height parameter 波高参数
wave height record 波高记录
wave height reduction factor 波高折减系数
wave height sensor 波高传感器
wave height spectrum 波高谱
wave hindcast(ing) 波浪后报;波浪追算
wave hole 波孔
wave hollow 波谷
wave horsepower 兴波马力
wave impact 波冲击
wave impedance 波(动)阻抗
wave impedance section 波阻抗剖面
wave impulse 波脉冲
wave indicator 波指示器
wave induced 波浪引起的
wave-induced circuit 波浪环流
wave-induced current 波生流;波浪流
wave-induced electromagnetic field 浪感电磁场
wave-induced flow 波生流
wave-induced oscillation 波导振荡
wave-induced pressure 波压(力)
wave-induced saline intrusion 波浪引起盐水入侵
wave-induced scouring 波浪冲刷
wave-induced source 波浪冲刷
wave in easterlies 东风波
wave inside harbo(u)r 港内波浪
wave intensity 波浪强度;波(的)强度
wave interference 波浪干涉;波(的)干涉
wave interference error 波干扰误差
wave interference ripple mark 摆动干扰波痕
wave in westerlies 西风波
wave kinetic energy 波浪动能
wavelength 参考波长;波长
wavelength accuracy 波长精度
wavelength adjustment 波长调节
wavelength allocation 波长分配
wavelength blocking filter 波长闭塞滤波器
wavelength calibration 波长校准
wavelength centroid 波长重心

wavelength change 波长改变
wavelength choice 波长选择
wavelength coefficient 波长系数
wavelength comparator 波长比长仪
wavelength constant 波长常数
wavelength conversion 波长转换
wavelength correction 波长校正
wavelength corresponding to peak in turbulent level 湍流谱极大尺度
wavelength counter 波长计数器
wavelength coverage 波长(分布)范围
wavelength demultiplexer 波长去复用器
wavelength dispersion 波长色散
wavelength dispersion method 波长色散法
wavelength dispersion spectrometer 波长离散分光计
wavelength distribution 波长分布
wavelength division demultiplexer 波分去复用器
wavelength division multiplexer 波分复用器
wavelength division multiplex(ing) 波分复用;波段复用
wavelength effect 波长效应
wavelength exchange cam 波长变换凸轮
wavelength flattened coupler 波长平坦化耦合器
wavelength in deep water 深水波长
wavelength independent singlemode coupler 非波长依赖型单模耦合器
wavelength interferometer 波长干涉仪
wavelength interval 波长间隔
wavelength limit 波长极限
wavelength loss 波长损失
wavelength measurement 波长测量
wavelength minimum 最短波长;波长最小值
wavelength modulation spectroscopy 波长调制光谱学
wavelength multiplexer 波长多路复用器
wavelength of light 光的波长
wavelength of radiation 辐射波长
wave length of sound 声波长度
wavelength of vessel 船舶的水线长度
wavelength plate 波长盘
wavelength programmer 波长程序推进器
wavelength range 波长(分布)范围
wavelength ratio 波长比
wavelength reading spectrometer 波长读数分光计
wavelength region 某一波长区域
wavelength repeatability 波长重复性
wavelength resolved solar flux 按波长测定的太阳辐射流
wavelength response 波长响应
wavelength scale 波长标度
wavelength scan midpoint 波长扫描中点
wavelength-scanning photometer 波长扫描光度计
wavelength scpectroscopy 波长分光镜
wavelength selective star coupler 波长可选择的星形耦合器
wavelength selector 波长选择器
wavelength sensitivity 波长灵敏性
wavelength separation 波长分离
wavelength shift 波长移动;波长偏移
wavelength shifter 波长转换器;波长移动器;波长变换器
wavelength shortening 波长减短
wavelength spectrometer 波长分光仪;波长分光计
wavelength spectroscope 波长分光仪
wavelength spectrum 波长谱
wavelength spread 波长展开度;波长分布范围
wavelength stabilization 波长稳定化
wavelength standard 波长基准;波长标准
wavelength transducer 波长转换器
wavelength unit 波长单位
wavelength wedge filter 波长渐变滤光片
waveless 无波浪
wavelet 波列;小波;子波;弱波;扰动线;成分波;二级浪
wavelet analysis 小波分析
wavelet deconvolution 子波反褶积
wavelet processing 子波处理
wavelet processing section 子波处理剖面
wavelet process time section 子波处理时间剖面
wave level 瞬时水面;波动平面
wave lift 波浪浮托力
wavelike 波状的;波形的
wavelike bond 波状结合

wavelike corridor 波形廊
wavelike cross striation 波状横纹
wavelike extinction 波状消光
wavelike lamination 波状纹理
wavelike line 波状线
wavelike motion 波状运动
wavelike uplift 波形上升
wavelike variable 波形变量
wavelike wall 波形墙
wave line 冲痕;波锋线
wave line of propagation 波的(传播)方向
wave lining 波浪形内衬
wavellite 银星石
wave load(ing) 波浪荷载;波浪作用力
wave loop 波腹
wave machine (船模试验池的)造波机
wave maker 造波器;造波机;波浪发生器
wave-making apparatus 造波设备
wave-making drag 兴波阻力
wave-making resistance 波动阻力;兴波阻力;成波阻力
wave manometer 波压(力)计
wave mark 冲痕;波痕
wave mass 波质量
wave master buoy 波浪主人测波浮标
wave measurement instrument 波浪量测仪;波浪量测器;波浪测量仪器
wave-measuring buoy 波浪测量浮筒;波浪测量浮标;测波浮标;测波浮标
wave-measuring instrument 波浪量测器
wave-measuring point ga(u)ge 点波测波长计
wave mechanical perturbation theory 波动力学微扰论
wave mechanics 波动力学
wave memorizer 波形存储器
wave meter 测波计;测波仪;波频计;波长仪表;波长计
wave microphone 波型传声器
wave mode 波型
wave mode conversion 波型转换
wave model 波浪模型
wave mode selector 波型选择器;波模选择器
wave mode transducer 波模变换器
wave mode transformation 波型变换
wave-modulated oscilloscope 行波示波管;调波示波器
wave-modulated oscilloscope tube 波调示波管
wave momentum 波动(动)量
wave momentum density 波动量密度
wave monitor 波形监视器;波形监控器
wave motion 波浪运动;波动
wave motion compensator 升沉补偿器(海上钻探时,钻探船随波浪升沉而钻压不变)
wave motor 波浪发动机
wave mo(u)lding 波形线条;波形线脚;浪花形脚;波形饰
wave movement 波浪运动
Wavene 瓦应尼牌玻璃(一种供隔墙等用的压制玻璃)
wave node 波节
wave noise 波动噪声
wave normal 波(面)法线
wave normal ellipse 波法向椭圆
wave normal surface 波法向面
wave number 波数
wave number calibration curve 波数校准曲线
wave number counter 波数计数器
wave number difference 波数差
wave number domain 波数域
wave number filter 波数滤波器
wave number frequency spectrum 波数频谱
wave number marker lamp 波数标识灯
wave number migration 波数偏移
wave number operator 波数算子
wave number range 波数范围
wave number registration 波数记录
wave number scale 波数标度
wave number space 波数空间
wave number spectrum 波数谱
wave number unit 波数单位
wave number vector 波数矢量
wave object 波动物体
wave observation 波浪观测
wave observation radar 测波雷达

wave observation tower 波浪观测塔
wave of blocks 地块波浪
wave of condensation 凝聚波;密波
wave of continental shelf 陆架波
wave of detonation 爆震波
wave of displacement 船首波
wave of distortion 畸变波
wave of earthquake 地震波
wave of infinitely small amplitude 无限小振幅波
wave of infinitesimal amplitude 微振幅波
wave of island shelf 岛架波
wave of low amplitude 小振幅波
wave of oscillation 振荡波;往复波;摆动波
wave of rarefaction 疏波
wave of sinusoidal form 正弦波(形)
wave of small amplitude 微振幅波
wave of translation 推移波;移动(水)波;平移波;推进波
wave ogive 波脊
wave operator 波算子
wave optical modulation transfering function 波动光学调制传递函数
wave optics 波动光子;波动光学
wave orbit 波浪轨迹
wave origin 波源
wave ornament 波纹装饰;波形装饰(品)
wave oscillation 波浪振荡
wave overlapped 重叠波
wave overtopping 越浪;波涛越顶(防波堤);波浪越顶
wave packet 波束;波群;波包
wave packet reduction 波包收缩
wave packet theory 波包理论
wave paddle 造波板;生波板
wave parameter 海浪要素;波浪要素;波(浪)参数
wave parenthesis 波形括弧
wave particle duality 波粒二重性;波动粒子二象性
wave particle interaction 波粒子相互作用
wave particle resonant effect 波粒共振效应
wave particle velocity 波浪水分子速度
wave path 波径;波程;电波路径
wave path-difference 波程差
wave path equation 波径方程
wave pattern 波型;波形;波纹图案;波模型
wave peak 波峰
wave period 波周(期);波浪周期;波动周期;波的周期
wave period distribution 波周期分布
wave perturbation 波状扰动;波浪小扰动
wave phase 波相
wave phase velocity 波相速度
wave phenomenon 波浪现象;波动现象
wave physics 波(动)物理学
wave planation 波浪均夷作用
wave plate 波片
wave plate analyser 波片检偏器
wave plate modulator 波片调制器
wave platform 浪蚀台地;浪成台地
wave point 波点
wave pole 测波(标)杆;波高杆
wave potential 波势
wave potential energy 波浪势能
wave power 波力
wave power buoy project 波能浮标工程
wave powered air-turbine generator 波能汽轮发电机
wave powered boat 波力驱动船
wave powered buoy 波能供电浮筒
wave powered device 波力发电装置
wave powered generator 波能发电机
wave powered lightbuoy 波力发电灯标
wave powered station 波力发电站
wave power generation 海浪发电;波力发电;波浪发电
wave power generator 波浪发电机
wave power machine 波能发电机
wave power recovery device 波能回收装置
wave power recovery system 波能回收系统
wave power statistics 波功率统计
wave power system 波浪发电系统
wave prediction 波浪预报;波浪推算
wave prediction model 波浪预报模式
wave prediction parameter 波浪推算参数
wave pressure 浪压力;波压(力)

wave pressure diagram 波压图
wave pressure distribution diagram 波压力分布图
wave pressure equation 波压公式
wave pressure formula 波压(力分布)公式
wave pressure meter 波压(力)计;波压表
wave producer 波发生器
wave profile 波形;波剖面
wave program(me) 波浪研究计划
wave progressing line 波浪进行曲线
wave propagation 波浪传播;波(的)传播
wave propagation angle 波传播角
wave propagation array 波传播台阵
wave propagation line 波浪传播线
wave propagation velocity 波浪传播速度;波的传播速度
wave property 波性(质);波特性
wave protection pile 防波障;防波栅
wave protection works 防浪工程;防波工程
wave quarrying 湖岸浪蚀;河岸浪蚀;海岸浪蚀;湖岸侵蚀;河岸侵蚀;海岸侵蚀
wave quelling oil 镇浪油
waver 动摇;波形转换器;波形转换开关
wave radiation 波辐射
wave range 波段
wave ray 波向线;波射线
wave ray separation 波浪射线间距
wave rear 波尾;波后
wave reckoning 波浪推算
wave record 波浪记录
wave recorder 自记测波仪;波浪记录仪
wave reduction 消波;波浪减弱
wave reflecting spur wall 消波丁坝
wave reflection 波反射;电波反射;波(浪)反射
wave refraction 水波折射;波浪折射;波(的)折射
wave refraction coefficient 波浪折射系数
wave refraction diagram 波浪折射图;波浪曲折图
wave refractor wall 波浪消能墙
wave regime 波浪动态
wave resistance 抗浪能力;波阻(力);波浪阻力
wave resonator 波浪共振器
waverider 浮标测波器
waverider buoy 波浪测波浮标
waverider measuring buoy 波浪测量浮筒
wave ridge 波脊;波峰
wave ripple 波痕
wave ripple bedding 波痕层理
wave ripple mark 波涟痕;摆动波痕
wave rise 波浪升高
Waverly group 瓦味利岩群【地】
wave roller 波形碾;波形辊;波纹面路碾
wave rose 波向频率图;波浪玫瑰图;波像图
wave rose diagram 波浪玫瑰图
wavers 匀墨辊
wave run-up 波浪上(升)高(度);波浪上爬;波浪上溅;波浪上冲
wave run-up height 波浪爬高
wave sampler 波取样器
wave scale 浪级(表);波浪(分)级;波浪等级;波级
wave scan buoy 波浪扫描浮标
wave scan float 波浪扫描浮子
wave scattering 波的散射
wave scattering effect 波散效应
wave scatterometer 波浪散射计
wave science 波科学
wave scrambler 波形加扰器
wave screen 挡浪墙;防浪(胸)墙;波浪障;防波栅
wave scroll 波形涡卷;波形装饰品;波形卷涡饰
wave selector 波(型)选择器
wave separator 电波分离器;分波器
wave series correlation 波系对比
wave setdown 波(浪)减水
wave setup 波(浪)增水;波浪壅水;波浪壅高
wave shape 波形
wave shape distortion 波形失真;波形畸变;波浪变形
wave shaped line 波形曲线
wave shape equalization 波形均衡
wave shape kit 简化波形空气枪
wave shape line 波形曲线
wave shape parameter analysis along layer 沿层波形参数分析
wave shaper 波形控制物;波形形成器;波形成形器
wave shape selector 波形选择器
wave shaping 波形整形;波形形成

wave-shaping electronics 波整形电子学;波成形电子学
wave shed 波形开口
wave shed loom 波形开口织机
wave simulation 波浪模拟
waves in plasma 等离子体中的波
wave size 波浪尺度
wave slam force 波浪砰击力
wave slap 波浪拍击
wave slope 波坦;波面斜率;波面斜度;波浪陡度;波陡
wave slowness 波慢度
waves of compression 压缩波;纵波
waves of distortion 横波;平移波
waves of pollution 污染波
wave soldering 波焊;波峰(钎)焊;波动钎焊;波动焊接
wave soldering unit 波峰焊接机
wave sound 波形声音
wave sources 波源
wave spacing 波距
wave spectra analysis 波谱分析
wave spectral reflectivity 波谱反射率
wave spectral resolution 波谱分辨率
wave spectroscopy 波谱学
wave spectrum 波(浪)谱
wave spectrum analysis 海浪谱分析
wave spectrum analyzer 波谱分析仪;波谱分析器
wave spectrum anomaly 波谱异常
wave spectrum method 波谱方法
wave spectrum method of forecasting sea state 波谱法预报海况
wave speed 波速
wave splitter 碎浪器;防波堤
waves propagation vector 波传播矢量
waves superposition 波叠加
wave staff 测波(标)杆
wave station 波浪站
wave statistics 波浪统计
wave steepness 波浪陡度;波陡;波动坡度;波(的)陡度
wave steepness ratio 波陡比
wave-straightened coast 浪成平直海岸
wave strength 波浪强度
wave stress 波应力
wavestrip 波带
wave stroke 波浪袭击;波浪冲蚀;波浪冲击
wave structure 波结构
wave subducer 消波器;防波装置
wave summit 波峰;波顶
wave superposition 波的叠加
wave suppressor 消波器;防波装置;波浪抑制器
wave surface 波(状)面;波阵面;波动曲面
wave swell compensator 波浪补偿器
wave swell compensator pressure vessel 波浪补偿器压力罐
wave swell compensator ram 波浪补偿器柱塞
wave-swept 受波浪袭击的
wave system 波族;波系
wave tail 信号尾部;波尾
wave tank 造波水池;造波槽;波槽
wave telemetering system 船用遥测波浪仪;波浪遥测系统
wave testing tank 波浪试验槽;波浪试验池
wave theoretical tomographic method 波动理论层析成像法
wave theory 波浪理论;波动(学)说;波动理论
wave theory of light 光的波动理论;光波动说
wave theory of the propagation of stress 应力传递的波浪理论;应力传播波浪理论
wave tilt 波倾斜;波前倾斜
wave tolerance 波浪富裕深度;波浪富余深度
wave top 波峰
wave trace 波迹
wave train 进行波;同向波簇;波序;波列
wave trajectory 波轨道;波的(传播)路径;波传播轨迹
wave transformation 波浪变形;波浪变化
wave transmission 波浪透入
wave trap (海口处导堤形成的)减波池;消能阱;消波阱;陷波器;除波区;防波阱;波谷
wave trapping 波谷
wave travel 波浪行程;波浪推进;波浪前进
wave travel(l)ing direction 波浪推进方向

wave trough 波谷;波槽
wave trough tectonic belts 波谷构造带
wave twill 波形斜纹
wave-type deformation of rail head 轨头波形磨损
wave-type dynamometer 波式测力计
wave-type seal tape 波形花纹带
wave-type sprinkler 喷头间歇转位式喷灌器
wave variability 波的可变性;波的变化性
wave vector 波矢(量);波浪矢量
wave vector filtering 波向量滤波
wave vector space 波矢(量)空间
wave vector surface 波矢面
wave velocity 相位速度;波速(度)
wave velocity anomaly 波速异常
wave velocity logging 波速测井
wave velocity measurement 波速测定
wave velocity meter 波速计
wave velocity method 波速法
wave velocity ratio section 波速比剖面
wave velocity surface 波速面
wave velocity survey 波速勘探法
wave volume 波量
wave wake 兴波伴流
wave wall 迎水坡墙;挡浪墙;防浪墙;防波墙;坝顶防浪墙
wave wash 波浪冲刷
wave wash levee 防浪堤;波浪冲积堤;波冲堤
wave wave interaction 波波相互作用
wave wave resonant effect 波波共振效应
wave width 波幅;波宽
wave winding 波(状)绕组;波状绕法;波形绕组(线圈)
wave winding machine 波状绕组电机;波状绕线机;波形绕法绕线机
wave with horizontal crest 平顶波
wave-worn 浪蚀的
wave-worn material 被波浪冲蚀的材料
wave wound coil 波绕组线圈;波绕线圈
wave zone 波浪区;波动区;波动带;波带区;波带
wavicle 波粒子
waviness 起浪;成波浪形;波纹度;波纹;波浪;波度
waviness curve 波纹曲线;波度曲线
waviness-meter 波纹度测量仪
waviness of a surface 表面波度
waviness of metal body 金属坯胎皱痕
waviness phenomenon 波纹现象
waving groin 拱顶相交出的肋;拱顶交线;波状穹顶肋
waving inspection for piles 波动法测桩
waving of track 轨道波形弯曲
waving rod level(l)ing method 摇尺水准测量法
wavy 波状的;波动的
wavy bedding hummocky bedding 波状层理
wavy bedding structure 波状层理构造
wavy boundary 波形边界
wavy cord 波形绳索
wavy corridor 波形廊
wavy curve 波形曲线
wavy discontinuity 波浪状结构面
wavy disk colter 波形圆犁刀
wavy dressing 波形装饰;波纹表面抛光
wavy edge 波状边;波纹纸边;(薄钢带的)波浪形边
wavy-edge 波浪边
wavy-edge blade 波形刃锯条
wavy-edge disk 波纹刃口圆盘
wavy extinction 波状消光
wavy fiber 波状纤维;波纹硬纸板
wavy fibred growth 波状乱纹
wavy finish 波状釉面;波纹釉面
wavy gneiss 波状片麻岩
wavy grain 波状纹(理);波形纹理;波浪纹(理);织纹;卷纹
wavy-grained 波状木纹
wavy-grained wood 波浪纹木材;皱纹木材;波状纹理木
wavy grown timber 扭曲木材;波纹木材
wavy lamination 波状纹理
wavy leaf margin 波状叶缘
wavy line 波状线;波状条纹;波形线;波纹线道;波筋
wavy-line screen 波线网屏;波纹网版
wavy mosaic structure 波浪状镶嵌构造
wavy mosaic structure hypothesis 波浪状镶嵌构造说
wavy mosaic structure of the earth's crust 地板壳波浪状镶嵌构造
wavy mosaic tectonic network of earth crust 地壳波浪状镶嵌构造网
wavy paper 卷边纸
wavy penciling 波状细条纹
wavy pipe 波纹管
wavy rule 波形铅线
wavy scroll 波形涡卷
wavy structure 波浪状构造
wavy surface 波形面
wavy terrain 起伏地形;起伏不平的地形
wavy texture 波状结构
wavy trace 起伏路线
wavy tube 波纹管
wavy upheaval 波状隆起
wavy vein 波形石纹
wavy wall 波形墙
wavy water surface 波状水面
wawa 非洲轻木
wax 蜡;石蜡
wax-absorbing furnace 吸蜡炉
wax agent (混凝土)蜡养护剂
wax ball 蜡球
wax bandage 蜡布
wax based granulation 蜡基质成粒法
wax bath 蜡浴;石蜡浴
wax-bearing crude 含蜡原油
wax-bearing oil 含蜡油
waxberry 白浆果
wax block photometer 蜡块光度计
wax-bonded yarn 浸蜡纱
wax candle 蜡烛
wax carver 蜡雕刻刀;雕蜡刀
wax carvers and spatulas 蜡雕刻刀与蜡匙
wax casting 蜡型铸件
wax cleaner 除蜡剂
wax cloth 油布;蜡布;涂蜡漆布(铺地板用);上蜡防雨布
wax coal 土褐煤;蜡煤
wax coat 蜡涂层
wax-coating machine 上蜡机
wax collar 蜡垫焊瘤
wax concrete agent 蜡混凝土养护剂
wax concrete curing compound (混凝土)养护蜡涂料;蜡混凝土养护剂;混凝土养护涂料
wax-containing asphalt 含蜡沥青
wax-containing crude 含蜡原油
wax content 含蜡量
wax content of asphalt 沥青含蜡量
wax crayon 蜡笔
wax curing compound 蜡混凝土养护剂
wax disc 蜡片
wax distillate 蜡馏出物
waxed cotton covered wire 涂蜡纱包线
waxed impregnation 蜡浸透
waxed rubber wire 上蜡橡皮线
waxed tissue 薄蜡纸
wax emulsion 蜡乳(化)液;乳状石蜡
waxen 蜡制的
wax engraved plate 蜡刻版
waxer 涂蜡机
wax extractor 制蜡器;蜜蜡提取器
wax figure 蜡像
wax finishing 打蜡(擦光);擦蜡
wax fractionation 石蜡分馏
wax-free 无蜡的
wax-free crude 无蜡原油
wax-free(crude) oil 无蜡石油
wax gland 蜡腺
wax impregnated concrete 石蜡浸渍混凝土
wax-impregnated graphite electrode 浸蜡石墨电极
wax-impregnated paper 石蜡浸纸
wax-impregnated yarn 浸蜡纱
wax impression 蜡版
waxing 蜡补;用有色材料对大理石的天然孔进行补蜡;涂蜡;上蜡;打蜡
waxing machine 上蜡机
waxing moon 盈月;渐盈月;上弦月
wax injection machine 压蜡机
wax injection press 压蜡机
wax injector 压蜡机
wax-like 似蜡的
wax manufacturing 制蜡过程
wax master 原始蜡盘

wax melter 熔蜡炉
wax modifier 蜡改性器;蜡改性剂
wax mo(u)ld(ing) 蜡(铸)模
wax museum 蜡像陈列馆
wax ocher 黄赭石
wax oil 蜡油;石蜡蒸馏液
wax original 原始蜡盘
wax painting 蜡画(法)
wax palm 蜡棕
wax paper 蜡纸
wax paste polish 蜡状糨糊上光;蜡状糨糊抛光
wax pattern 蜡型
wax pattern preparation 蜡型制作
wax pattern wetter 蜡型表面湿润剂
wax pocket 蜡囊
wax polish 打蜡;上光蜡
wax polisher 蜡抛光模
wax-polishing 打蜡
wax-polishing machine 打蜡机
wax printing production wastewater 蜡染印花生产废水
wax-proofing agent 防蜡剂
wax release agent 脱模蜡
wax removal 排蜡
wax residue 蜡油残渣
wax resinite 蜡质体
wax resist 防止黏釉的蜡;防蚀蜡
wax resist painting 涂蜡装饰法
wax scraper 刮蜡器
wax-sealed 用蜡密封的;蜡封的
wax-sealed sample 蜡封土样;蜡封试样
wax-sealing 蜡封
wax-sealing method 蜡封法
wax sequestering agent 蜡分离剂
wax-set ink 蜡凝固油墨;浸蜡固着油墨
wax shale 油页岩
wax sheets and plates 蜡片
wax solubility 蜡溶解性
wax solution 蜡溶液
wax spatula 蜡刀
wax spoon 蜡匙
wax stain 蜡污点;蜡染剂;蜡溶颜料
wax tailings 蜡渣
wax-tree 野漆树;蜡树;木蜡树
wax vent 蜡线出气孔
wax wall 黏土密封墙
wax wire 蜡线;打蜡线
wax work 蜡制品;蜡像陈列馆;蜡像
waxy 蜡状的;蜡质的;蜡样的
waxy bitumen 蜡质沥青
waxy cast 蜡样管型
waxy crude 含蜡原油
waxy degeneration 蜡样变性
waxy distillate fraction 含蜡馏分
wax yellow 淡黄色;蜡黄色
waxy flexibility 蜡样屈曲
waxy fuel 含蜡燃料
waxy hydrocarbon 含蜡烃
waxy infiltration 蜡样浸润
waxy luster 蜡光泽
waxy necrosis 蜡样坏死
waxy oil 含蜡(石)油
waxy petroleum distillate 含蜡石油馏分
waxy production inhibitor 对蜡质产生的抑制
waxy stain 蜡性着色料
way 路线;路径;道路;方式;方法;办法
way aloft 升到顶
way and mean 岁入财源
way and structure sector expenses 工务部门支出
way beam 纵梁;轨道(纵)梁
waybill 货(运)单;货票;运(货)单;海运单;乘客单
way block 支承块;导轨块
way-board (两厚层中的)薄隔板
way bracket 巷道托座;巷道隔撑
way capacity 道路容量
way car 拘留所;监狱;(列车尾部的)守车
way-cover 导轨罩
way end 船台末端
way end pressure 滑道末端压力;滑道端压
wayfarer 走路人
way for measurement self-potential method 自然电场测量方法
way for measuring resistivity of rock and mineral 岩矿石电阻率测定方法

way-freight car 沿途零担车
way grinding machine 导轨磨床
way headway 道路车间时距
way in 入口
waylandite 磷铝铋矿
wayleave 林内修筑林;通行费;通行道;通行权;道路通行权;道路使用权
way lubricant 导轨润滑剂
way mark 路标
way of anchoring 锚泊方式
way of correction to electromagnetic coupling 电磁耦合校正方法
way of digital interpretation of electric(al) sounding curve 电测深曲线数字解释方法
way of direct development of electric(al) prospecting 电法勘探正演计算方法
way of infection 传染途径
way of life 生活方式
way of loading 加载方式;加荷方式
way of measurement in mise-a-la-masse 充电法测量方式
way of metamorphism 变质作用方式
way of payment 支付方式
way of qualitative analysis 定性分析法
way of quantitive interpretation of electric(al) sounding curve 电测深曲线定量解释方法
way of reciprocating motion 往复摆动式
way of recording 记录方式
way of settlement 解决办法
way of specific activity 比放射性法
way of straight going 一线前进式
way of transportation 运输方法
way-operated circuit 分路工作线路;方向转换接线路
way out 出路;(水、气、汽、油等)出口
way point 中途港;航线中间站;航路点
ways and means 经费来源
ways and means advances 财政短期借款;赋税方式贷款
Ways and Means Committee 财政立法委员会
ways and means of finance fund 财政资金筹措办法
ways and means of taxation 赋税方法
ways-end 滑道末端
wayshaft 摇臂轴
wayside 路旁;线旁;沿线;道边
wayside contactor 地面接触器
wayside cross 路旁十字形标记;路旁十字形标牌
wayside equipment 线路地面设备;地面设备
wayside inductor 地面感应器
wayside inn 路边客栈
wayside marker 线路标志
wayside pavilion 凉亭
wayside shrine 路边神龛
wayside sign 线路标志
wayside signal(ling)区间信号;地面信号
wayside station 中间车站
wayside trees 林荫道
ways of airplane's magnetization compensation 飞机磁干扰场补偿方式
ways of exploration 勘探手段
ways of fault terminating 断层消失方式
ways of going about tax evasion 偷税漏税行为
ways of investment 出资方式
way sort 通路分类
way station 小站;中途站;中间停车站;路站
ways to apply 施工方法
ways to cure 固化方法
way train 慢车;普通客车
way-type drilling unit (由稳定器、钻铤、扩孔器组成)组合钻具
W.C. and urinal 大小便器
W coefficient 拉卡系数
W.C. pan 大便器
W-cut 双楔掏槽
weak 弱的;软弱的;轻震;微弱的;薄弱
weak abrasive formation 弱研磨性地层
weak absorption 弱吸收
weak-acid 弱酸
weak acid/base 弱酸/碱
weak acid/base pair 弱酸/碱偶
weak acid/base system 弱酸/碱系统
weak-acid cation exchanger 弱酸型阳离子交换剂
weak-acid cation exchange resin 弱酸阳离子交换树脂
weak acid concentration 弱酸浓度
weak acidic resin 弱酸性树脂
weak acidic water 弱酸性水
weak acidity 弱酸度
weak acid solution 弱酸溶液
weak adsorption 弱吸附
weak alkaline water 弱碱性水
weak anomaly 弱异常
weak aquifer influence 不活跃水驱
weak axis 弱轴
weak base 弱碱
weak-base anion exchanger 弱碱阴离子交换器
weak-base anion-exchange resin 弱碱性阴离子交换树脂
weak base concentration 弱碱浓度
weak base type ion exchanger 弱碱型离子交换剂
weak basic ion exchange resin 弱碱性离子交换树脂
weak basicity 弱碱性
weak basic medium 弱碱性介质
weak basic type ion exchanger 弱碱型离子交换树脂
weak battery 电压不足的电池
weak beam 弱梁
weak bind 松软黏土
weak birefringence 弱双折射
weak black liquor 淡黑液
weak blasting crater 减弱爆破漏斗
weak-bonded molecules 弱结合的分子
weak boundary condition 弱边界条件
weak bound water bond 弱结合水联结
weak brake spring 弱制动弹簧
weak bridge 不能承受重载的桥梁;不能重载的桥梁
weak caustic solution 弱苛性碱溶液
weak cement grout 水泥薄浆
weak-cementing matrix 弱黏结母岩
weak children's facility 幼儿抚育设施
weak closure 弱闭包
weak coal 易碎煤;脆煤
weak colo(u)r 淡颜色
weak compactness 弱紧性
weak complexing agent 弱络合剂
weak component 弱分量
weak compression 弱压缩
weak concrete 贫灰混凝土;低强度(等级)混凝土
weak conjugate base 弱共轭根碱
weak convergence 弱收敛
weak converse theorem 弱逆定理
weak coupling 弱耦合;疏耦合
weak course 软层
weak current 弱电流
weak-current cable 弱电流电缆
weak-current control 弱电控制
weak-current engineering 弱电工程学
weak-current installation 弱电流装置
weak-current insulator 弱电绝缘子
weak-current line 弱电流线路
weak-current paper-insulated lead-covered cable 弱电流纸绝缘铅包皮电缆
weak-current relay 弱电流继电器
weak-current switch 弱电流开关
weak decrepitation 爆裂弱
weak deformation retract 弱变形收缩核
weak depression 弱低压
weak derivative 弱导数
weak diagonal dominance 弱对角线优势
weak differentiability 弱可微性
weak digraph 弱双图
weak discontinuity 弱间断
weak discontinuity surface 弱间断面
weak dispersion 弱分散
weak-double band 弱双潜带
weak earthquake 弱(地)震
weak earthquake zone 弱地震区
weak echo region 弱回波区
weak echo wave 弱回波
weak electrolyte 弱电解质
weak electromagnetic mineral 弱电磁性矿物
weaken 削弱(的);减弱(的);变弱的;变稀薄
weakened cement 强度等级降低的水泥
weakened plane contraction joint (混凝土路面的)假缝;弱面缩缝
weakened plane joint 槽式接合;凹缝;槽缝;半

缝;假缝;弱面缝;路面弱面缝;成形干燥后易分离的接口
weakened plate 强度减弱的板
weakened profile 弱发育剖面
weakener 减光板
weakening 消振;削弱;衰减
weakening factor of axial holes in pattern 排孔轴向削弱系数
weakening factor of holes in pattern 排孔削弱系数
weakening in demand 需求疲乏
weakening of mo(u)lding sand 型砂减泥
weakening of the cross-section 断面缩小
weak equivalence principle 弱等效原理
weaker oxidant 较弱氧化剂
weakest 极弱
weakest shear plane 最弱剪切面
weak extremum 弱极值
weak-eyed 视力差
weak field 弱场
weak-field condition 弱场条件
weak-field method 弱场法
weak fix 弱定位【航海】
weak flocculation 弱絮凝作用
weak-flutter accelerator 弱颤加速器
weak-flutter field accelerator 弱颤场加速器
weak-flutter field cyclotron 弱颤场回旋加速器
weak focusing 弱聚焦
weak-focusing accelerator 弱聚焦加速器
weak-focusing betatron 弱聚焦电子感应加速器
weak-focusing cyclotron 弱聚焦回旋加速器
weak-focusing magnet 弱聚焦磁铁
weak-focusing storage ring 弱聚焦存储环
weak-focusing synchrotron 弱聚焦同步加速器
weak formation 软弱构造;软弱地层
weak foundation 软(弱)地基
weak framework 弱框架
weak frost heaving 弱冻胀
weak ground 软土层;软弱地基;软地基
weak hydraulic jump 弱水跃
weak hydrophilicity 弱亲水的
weak hypothesis test 弱假设检验
weak identifiability 弱能识性
weak image 浅淡影像
weak integrability 弱可积性
weak intensity 弱强度
weak interaction 弱相互作用
weak-interaction experiment 弱相互作用实验
weak-interaction physics 弱相互作用物理学
weak interbed 软弱夹层
weak intercalated layer 软弱夹层
weak interference 弱干扰
weak jamming 弱干扰
weak large number law 弱大数定律
weak law of large numbers 弱大数定律
weak light 弱光灯标
weak light flux 弱光通量
weak light source 弱光源
weak lime 弱灰池
weak link 薄弱环节
weak linkage 弱键
weak low-pressure 弱低压
weakly absorbing material 弱吸收物质
weakly absorbing object 弱吸目标
weakly adhering oil film 弱黏附油膜;附着油膜(金属表面)
weakly alkaline 弱碱性土;弱碱性的
weakly alkaline soil 弱碱性土
weakly anisotropic(al) resonator 弱各向异性共振腔
weakly anisotropic(al) waveguide 弱各向异性波导管
weakly basic 弱碱性的
weakly basic anion exchange fiber 弱碱性阴离子交换纤维
weakly basic species 弱碱物种
weakly bound water 弱结合水
weakly caking 弱黏结
weakly caking coal 弱黏结(性)煤
weakly cemented 弱胶合的;弱黏合的;弱黏结的;弱胶结的;微弱胶结的
weakly closed set 弱闭集
weakly coercive mapping 弱强制映射
weakly coking coal 弱焦性煤

weakly compact linear mapping 弱紧线性映射
weakly compact set 弱紧集
weakly connected 弱连通的;弱连接
weakly connected digraph 弱连通有向图
weakly continuous representation 弱连续表示
weakly continuous stochastic process 弱连续随机过程
weakly convergent 微弱收敛的
weakly coupled system 弱耦合系统
weak lye 弱碱液
weakly equivalent transformations 弱等价变换
weakly ergodic measure space 弱遍历测度空间
weakly guiding optic(al) waveguide 弱导光波导
weakly Hermitian scalar product 弱埃尔米特纯量积
weakly hyperbolic differential operator 弱双曲型微分算子
weakly hyperbolic equation 弱双曲型方程
weakly hyperbolic operator 弱双曲线算子
weakly inaccessible ordinal number 弱不可达序数
weakly integrable function 弱可积函数
weakly integrable operator 弱可积算子
weakly ionized gas 弱电离气体
weakly leached 轻度淋洗
weakly leached brown soil 轻度淋溶棕壤
weakly leaching 轻度淋溶
weakly magnetic 弱磁性物料
weakly magnetic material 弱磁性物料
weakly magnetization 弱磁化
weakly measurable 不易测的
weakly measurable function 弱可测函数
weakly melt-settlement 弱融沉
weakly mixing 弱混合
weakly mixing automorphism 弱混合自同构
weakly mixing hypothesis 弱混合假设
weakly monotonic function 弱单调函数
weakly non-linear differential equation 弱非线性微分方程
weakly non-linear system 弱非线性系统
weakly oriented 弱定向的
weakly pervious boundary 弱透水边界
weakly podzolic soil 弱灰化土
weakly quasi-invariant measure space 弱拟不变测度空间
weakly space-like surface 弱类空曲面
weakly stable species 弱稳物种
weakly stationary 弱平稳
weakly stationary moving average 弱平稳移动平均数
weakly stationary process 弱平稳过程
weakly stationary random distribution 弱平稳随机分布
weakly stationary sequence 弱平稳序列
weakly stationary stochastic process 弱平稳随机过程
weakly symmetric(al) Riemannian space 弱黎曼对称空间
weakly weathered layer 弱风化层
weakly weathered zone 弱风化带
weakly welded agglomerate 弱熔结集块岩
weakly welded agglomeratic texture 弱熔结集块结构
weakly welded breccia texture 弱熔结角砾结构
weakly welded tuff 弱熔结凝灰岩
weakly welded tuffaceous texture 弱熔结凝灰结构
weakly welded volcanic breccia 弱熔结火山角砾岩
weak machine 功率小的机器
weak magnetic mineral 弱磁性矿物
weak magnetic substance 弱磁性体
weak market 市场疲软
weak metamorphic rock 弱变质岩
weak method 弱方法
weak mineralized water 弱矿化水
weak minimum 弱极小
weak mixing property 弱混合性
weak mixture 贫混凝土混合料
weak monotonic property 弱单调性
weak multiplication 弱乘法
weakness 弱点;缺点;无力;薄弱环节
weakness of rocks 岩体弱面
weakness plane 最小抵抗面;薄弱面
weak nitric acid 稀硝酸
weak non-cohesive soil 弱非黏性土
weak non-isentropic(al) flow 弱非等熵流

weak nonlinearity 弱非线性
weak object 弱对比度景物
weak optimal solution 弱最优解
weak oxidizing environment 弱氧化环境
weak plane 弱面;软弱结构面【地】
weak plastic soil 弱塑性土体
weak point 弱点
weak point method 弱点法
weak pressure gradient 弱气压梯度
weak pulse 脉微
weak reciprocity 弱互易
weak reducing environment 弱还原环境
weak relative minimum 弱相对极小
weak rock 软质岩石;软岩(石)
weak roof 不稳定顶板
weak runoff zone 弱径流带
weak salinization 弱度盐渍化
weak sand 瘦(型)砂;低强砂
weak section 危险断面;薄弱部分
weak sewage 稀污水;淡(水)污水
weak shock 弱激波;弱(地)震
weak shock tectonic environment 微震构造环境
weak shock zone 弱地震区
weak signal 弱信号
weak signal detection 弱信号检测
weak soil 软土;松水土壤
weak solution 稀溶液;弱解
weak solvent 弱溶剂
weak spark 弱火花
weak-spiral accelerator 弱螺旋(扇)加速器
weak spiral cyclotron 弱螺旋回旋加速器
weak spiral geometry 弱螺旋几何形状
weak spot 弱光点;疵点;薄弱点
weak spring 弱弹簧
weak stability 弱稳定性
weak stor(e)y 薄弱楼层
weak structure 脆性弱结构
weak subgrade 软弱地基;软(弱)路基;软(地)基;松软路基
weak substitution 弱代入
weak surrounding rock 软弱围岩
weak swelling-shrinkage soil 弱胀缩性
weak throw blasting charge 减弱抛掷药包
weak tidal environment 弱潮汐环境
weak tide river mouth 弱潮河口
weak unit 弱单位
weak variation 弱变分
weak viscidity 弱黏性
weak-walled 围岩不稳固的
weak-wash slope 弱洗刷坡
weak watery overflow zone 弱富水溢出带
weak wave 弱波
weak weathering 微风化
weak weld 不强固焊缝
weak wind 微风
weal and flare response 风团及潮红应答
weald 荒漠旷野;密林地区;山林地带;森林地带
Wealden (产于萨塞克斯的有褐色条纹的)淡黄色砂石
Wealden house 韦尔登房屋
Wealden series 韦尔登统
Wealden stage 韦尔登阶(早白垩世)
wealhy villa 豪华别墅
wealth 财富;财帛
wealth accumulation 财富积累
wealth and fame 名利
wealth constraint 财富约束
wealth distribution 财富分配;财富分布
wealth effect 财富效应
wealth holder 财富所有者
wealth-holding 财富持有
wealth hyperplane 财富超平面
wealth inequality 财富分配不均
wealth of nation 国家财富
wealth of the oceans 海洋资源
wealth of views 景色优美
wealth sharing policy 财富分享政策
wealth tax 财富税;财产税
wealth transfer taxation 财富转移课税
weapon 凶器;军械
weapon debris 武器碎片
weapon delivery run 投射武器的进入
weapon monitoring center 武器监控中心
weaponry 兵器

weapons 兵器
weapon system stock control point 武器系统库存控制站
wear 磨耗;磨;顺风掉抢
wearability 耐磨(损)性;耐腐性;磨损性
wearable 经磨的,耐磨的
wear across the edge 刃口磨损
wear across the ga(u)ge (钻头、钎头)直径磨损
wear alarm transmitter 磨损报警传感器
wear allowance 磨损容差;磨损留量;磨损公差;容许磨损;容许磨耗;磨损限值
wear allowance zone 磨损公差带
wear and tear 消损;磨损;磨耗(及损伤);耗损;损耗
wear and tear allowance of capital 资本损耗设备抵
wear and tear ga(u)ge 磨损量测仪;磨耗规
wear and tear in use 使用中磨损
wear and tear of fixed assets 固定资产损耗
wear and tear part 损耗件;磨损零件
wear and tear volume 磨耗量
wear away 磨损
wear block 磨损量块
wear bushing 防磨补心
wear-compensating 磨损补偿
wear course 磨耗层;磨损层;磨损面
wear curve 磨损曲线
wear cycles in Taber test 泰伯尔法中的循环次数
wear debris 磨损产物
wear down 磨平;磨薄;冲刷
wear extent 磨损量
wear flat 磨损面;磨平
wear-free rotation 无磨损旋转
wear hardness 抗磨硬度;抗磨力;耐磨性;磨损硬度
wear hardness testing apparatus 耐磨硬度试验仪
wear history 磨损曲线
wear hollow 耗损成空壳
wear-in 磨合
wear index 磨耗指数
wear index analyzer 磨损指数分析仪
wear indicator 磨耗指示器
wearing 磨损
wearing action 磨损作用
wearing area 磨损面积;耐磨房屋;磨损范围
wearing away 磨蚀
wearing away of bank 河岸磨蚀
wearing blade 研磨刃;磨损刀片;切削刃
wearing bush 防磨衬套
wearing capacity 磨损量;耐磨性
wearing carpet 磨损层;磨耗层
wearing characteristic (路面)抗磨特性
wearing coat 耐磨层;磨损表面;磨损层;磨耗层
wearing coat of bridge deck 桥面磨耗层
wearing coefficient 耐磨系数
wearing control 磨损控制
wearing course 路面表层;耐磨层;磨损(面)层;磨耗面;磨耗层
wearing course aggregate 磨耗层集料;磨耗层骨料
wearing depth 磨损深度
wearing floor 楼面磨耗层;耐磨地板;桥面磨损层
wearing ga(u)ge 磨损量规
wearing hardness 磨损硬度;抗磨硬度
wearing inhibitor 抗磨损添加物;抗磨损添加剂
wearing inspection of discharge pipeline 排泥管线磨损检查
wearing layer 磨损层;磨耗层
wearing life 磨损期限;抗损寿命;耐磨寿命
wearing limit 磨损极限;磨损限度
wearing loss 研磨耗损
wearing measurement 磨损量测
wearing parts 受磨部件;磨损部分;磨损部件;易损件;磨损零件;磨耗件
wearing piece 耐磨件;磨擦片;摩擦片
wearing plate 防磨板;耐磨板;磨擦片;摩擦片;衬砌板
wearing plate adjustment 磨擦片调节
wearing power 抗磨损能力
wearing-proof alloy 耐磨合金
wearing property 耐磨性;耐用性
wearing quality 耐磨性;磨损性;磨损度
wearing resistance 磨损阻力
wearing ring 抗磨环;耐磨环
wearing ring clearance 止漏环间隙
wearing strength 抗磨强度;耐磨度
wearing strip 调整板;耐磨条;接触板

wearing surface 道路面层;耐磨表面;磨损面;磨耗面;磨耗层;防蚀面
wearing terrain 磨损部位;磨损部件
wearing test(ing) 耐磨性试验;磨耗试验;磨损试验
wearing thickness 磨损厚度
wearing time 磨耗时间
wearing value 磨耗值;耐穿性
wear inhibitor 抗磨损物;防磨剂
wear intensity 磨损度
wear iron 锻铁
wear land 磨损带
wear layer 磨损层;耐磨层
wearlessness 抗磨性;耐磨性
wear-life 抗磨寿命
wear lining 工作衬
wear loss 磨耗减量
wear lug 防磨凸耳
wear mechanics 磨损力学
wear model 磨损模型
wear of brush 电刷磨损
wear off 消失;擦掉;磨掉;磨耗
wear of fixed asset 固定资产磨损
wear of liner plate 衬板磨损
wear of rail on curves 曲线上的钢轨磨耗
wear of slick line 滑线耗损
wear of work 工件磨损
wearometer 磨损计;磨耗器
wear out 磨损
wear-out failure 消耗失效;磨损破坏;磨损故障;磨耗故障;疲劳故障;损耗故障
wear-out-failure period 磨损失效周期
wear-out of ga(u)ge 磨损矢径
wear-out part 磨损件
wear-out period 终结期;磨损期限
wear oxidation 磨损氧化;摩擦腐蚀
wear particle 磨损(的)颗粒
wear parts 耐磨部件
wear pattern 磨耗图纹
wear plate 抗磨板;磨耗增强板;磨耗板;耐磨铁板
wear plate leakage 侧板泄漏
wear prevention 防止磨损
wear prevention agent 抗磨剂
wear-preventive additive 防磨添加剂
wear process 磨损过程
wear products 磨损生成物
wear profile 磨耗型
wear profile of rail 钢轨磨耗型断面
wear-proof 抗磨的;耐磨的
wear-proof coating 耐磨面层
wear pump 耐磨泵
wear rate 耐磨强度;磨损率
wear rate of sand grains 砂粒磨损率;沙粒磨损率
wear rating 相对磨耗率;磨耗等级
wear ratio 磨耗比
wear-reducing value of lubricant 润滑剂减磨值
wear reduction factor 磨耗减低因数
wear reference point 磨损测量的基本尺寸
wear resistance 耐磨(性);耐磨强度;耐磨能力;抗磨力
wear resistance concrete 耐磨混凝土
wear-resistant 耐磨的;抗磨损的
wear-resistant alloy 耐磨合金
wear-resistant cast iron 耐磨铸铁
wear-resistant coat 耐磨涂层
wear-resistant coating 抗磨镀层;耐磨涂层;耐磨镀层
wear-resistant material 抗磨材料;耐磨材料
wear-resistant metal 耐磨金属
wear-resistant paint(ing) 耐磨涂料;高度耐机械磨损漆
wear-resistant parts 耐磨件
wear-resistant rail 耐磨(钢)轨
wear-resistant steel 耐磨钢
wear-resistant surface 耐磨面
wear-resisting 耐磨性的
wear-resisting alloy 耐磨合金;耐磨钢
wear-resisting alloy steel 耐磨合金钢
wear-resisting casting iron 耐磨铸铁
wear-resisting coating 抗磨涂层
wear-resisting composite 抗磨复合材料
wear-resisting floor 耐磨地板
wear-resisting lining board 耐磨衬板
wear-resisting material 耐磨材料
wear-resisting metal 耐磨金属

wear-resisting plate 耐磨板
wear-resisting property 耐磨性能
wear-resisting rail 耐磨(钢)轨
wear-resisting road cement 耐磨道路水泥
wear-resisting rubber 抗磨(损)橡胶
wear rib 耐磨条;耐磨肋骨
wear ring 磨损圈;磨损环;承磨环
wear segment 耐磨扇形衬板;耐磨部件;磨损部件
wear strength 抗磨损强度
wear strip 防磨条
wear surface 磨耗面;磨损面;磨损表面
wear tables 磨损表
wear tester 磨损试验机
wear test(ing) 耐磨试验;抗磨试验;磨损试验
wear-testing machine 耐磨试验机
wear trace 磨损痕迹
wear track 磨损痕迹
wear value 磨耗值
wear washer 抗磨垫片;耐磨垫圈
wear well 耐用
weasel 水陆两用自动车
weather 气象;天气;吹干
weatherability 经得住风吹雨打;耐气候性;耐候性;耐风化性
weather actual 天气报告
weather adjustment 天气调整装置
weather adjustment device 气温调节装置
weather advisory 气象通报
weather advisory warning 天气预报
weather ag(e)ing 气候老化;天然老化;气温老化
weather analysis 天气分析
weather analysis and climate monitoring 天气分析和气候监测
weather analysis center 气象分析中心
weather analysis in abbreviation form 天气分析电码
weather anchor 力锚;抗风锚;上风锚
weather and climate monitoring 天气及气候监测;天气和气候监测
weather and technology trend 天气和技术趋势
weather anomaly 天气异常;天气反常
weather a storm 度过风暴
weather attenuation 气候衰减
weather avoidance 气象回避
weather avoidance radar 气象回避雷达
weather away 使不受风化
weather awning 风雨篷
weather back 设于墙内侧的抗风化材料
weather bar 气象气压计;止水带;防水条;止水条;挡水条;挡风雨条
weather-beaten 经受风雨损耗;经风雨吹打的;经风雨剥蚀的;风雨损耗的;风雨侵蚀的;风雨磨蚀的
weather-bitt 风暴系船柱
weather-board 披水条;棚面板;舷墙;舷侧导水板;迎风舷;檐板;后风雨板;挡风板;风雨板;防浪板(放在船头);墙面板;门窗披水板;护壁板;封檐板
weather-boarding 披叠板;外墙抗候护板;风雨墙面板;风雨护壁板;防雨板;安装风雨板;围护板;外墙板;横钉外墙板;风雨板
weather-bound 天气阻挠;被天气阻隔的
weather boundary 露天部分
weather bound vessel 避风暴船
weather bow 上风首舷
weather box 晴雨指示箱
weather briefing 气象简语
weather broadcast 天气广播
weather bulletin 气象通报;天气公报
weather buoy 气象浮标
weather bureau 气象局
weather cap 伞形风帽
weather cast 天气预报
weather category 气象类型
weather cement shack 水泥防潮存放处
weather center 气象中心
weather central 天气中心
weather change 天气变化
weather chart 气象(要素)图;天气图;地面天气图
weather chart analysis 天气图分析
weather chart facsimile apparatus 气象图传真机
weather check 滴水槽;防水接头;防水接缝;屋檐;企口
weather checked tie 风化辐射裂纹枕木

weather classification 气象分类
weather closure 挡风雨罩
weather cloth 防雨(帆)布
weather coat 耐风雨涂层;耐候涂层
weather-coated 耐风雨涂层的
weather-coating material 耐风雨涂料
weather-coating renewal 耐风雨涂层的更新
weather cock 定风计;风向(标);风标志;风信计
weather cocking 受风浪冲击而转向的;风标作用
weathercock stability 方向稳定性
weather code 天气电码;标准气候条件
weather communication 天象勤务通信
weather communication center 气象通信中心
weather computer 气象计算机
weather condition 天气状态;气候条件;天气状况;天气条件;天气情况
weather condition factor 气候条件因素
weather control 气象控制;天气控制
weather controlled message 气象控制电报
weather cross 受潮线间漏电
weather cycle 天气循环
weather-damp 矿井瓦斯
weather data 气象资料;天气资料
weather data communication network 天气数据通讯网
weather data processing 天气资料整编;天气资料加工
weather deck 露天甲板
weather delay factor 气象延迟系数
weather detector 天气预报器;天平预报器
weather development 天气演变;天气发展
weather disturbance 天气扰动
weather divide 天气区划分
weather door 坑门通气门;外重门;外层门;风雨门;防风门;防暴风门
weather drying 干燥天气;阳光干燥
weathered 倾斜的;泻水的;晾干的;老化的;风化的
weathered clay 风化黏土
weathered crack 风化裂纹
weathered crevasse 风化裂隙
weathered crude 风化原油;风化的原生矿物
weathered crust 风化壳
weathered effect 天气造成的效果;风化效应
weathered escarpment 风化崖
weathered exposure 天然曝晒;暴露于露天;暴露于大气中
weathered face 风化表层
weathered feldspar 风化长石
weathered glass 蒙上水气的玻璃
weathered granite 风化花岗岩
weathered hay 经雨淋日晒的干草
weathered horizon 风化层
weathered ice 风化冰
weathered icebery 风化冰山
weathered ilmenite 蚀变钛铁矿
weathered joint 泻水勾缝;斜(勾)缝
weathered layer 风化层
weathered matrix 风化母岩
weathered mica 风化云母
weathered outcrop 风化露头
weathered pointing 防水刮缝;斜(水)沟缝
weathered ridge 风化冰脊
weathered rock 风化岩石;风化岩(层)
weathered rock-soil 风化物;风化(土石)料
weathered rock surface 风化岩面
weathered rocky soil 风化岩性土
weathered sample 风化样品
weathered sand-gravel fills 风化砂砾填料
weathered sandstone 风化砂岩
weathered shale 风化页岩
weathered siltpelite 风化粉砂泥岩
weathered silt rock 风化粉砂岩
weathered slope (窗台等)泻水坡度
weathered soil 风化土(壤)
weathered stone 风化石料
weathered sycamore 经蒸汽处理过的梧桐木
weathered zone 风化层;风化(地)带
weather effect 天气影响
weather event 天气时件
weather exposure 日晒夜露;暴露于大气中;风化
weather exposure test 耐候性试验;天然曝晒试验;风蚀曝露试验;风化曝露试验
weather face 上风面;迎风面
weather facility 气象站

weather facsimile 气象传真
weather facsimile network 气象传真通信网
weather facsimile recorder 气象传真记录器
weather factor 天气因素
weather fast 耐风雨的;抗风蚀的;天气阻挠
weather fastness 耐天气性;耐气候性
weather fillet 水泥砂浆防水条;水泥填角;水泥八字角;挡风雨贴角压条
weather flag 气象预报信号;气象旗号
weather flight 气象飞行
weather forecast(ing) 天气预报;气象预报
weather forecasting center 气象预报站
weather freezing 天然冻结
weather ga(u)ge (测大气)气压表
weather glass 晴雨计;晴雨表;气压计
weather(-going) tide 气象潮(气压变化所引起的潮升变化)
weather groove 泻水槽
weather hood 防风罩
weather house 晴雨指示箱
weather indicator 气象指示器
weather information 气象情报;天气资料;天气情报
weather information network and display 气象情报接收网;天气情报网与显示;天气情报接收网
weathering 自然时效;披水;天然时效;风雨侵蚀;风化(作用);发霉
weathering age 自然老化
weathering ag(e)ing 大气老化
weathering agency 风化营力
weathering availability (土壤)风化效率
weathering back (烟囱或桥墩的)泻水坡
weathering belt 风化带
weathering bridge-steel 高强耐锈蚀桥梁钢
weathering capacity 风化(能)力
weathering cave 风化洞
weathering characteristic 风化特性
weathering condition 风化条件;风化情况
weathering corrosion 风化侵蚀
weathering crust 风化壳
weathering crust mineral deposit 风化壳矿床
weathering-crust-type bauxite deposit 风化壳型铝土矿矿床
weathering cycle 风化循环;老化周期
weathering degree of rock mass 岩体风化程度
weathering deposit 风化矿床
weathering depth 风化深度
weathering disintegration 风化分解;风化崩解
weathering effect 风化效应
weathering element 天气要素
weathering eluvial type 风化残积型
weathering (exposure) test 风化试验
weathering face 风化面
weathering-fastness 抗风化性;防风化度
weathering fissure 风化裂隙
weathering fissure water 风化裂隙水
weathering fracture 风化裂隙
weathering front 风化面
weathering index 风化指数
weathering intensity 风化强度
weathering intercalated layer 风化夹层
weathering joint 风化缝;风化节理
weathering layer 风化层
weathering machine 大气老化试验机
weathering map 气象图;风化厚度图
weathering mo(u)lding 泻水线条
weathering of building stones 房屋石块的风化
weathering of glass 玻璃发霉
weathering ore-forming process 风化成矿作用
weathering phenomenon 风化现象
weathering plane landslide 风化面滑坡
weathering plate 防风雨板
weathering pointing 泻水勾缝;斜勾缝
weathering potential index 风化势指数
weathering process 风化过程
weathering product trap 风化产物圈闭
weathering-proof 抗风化性;抗风化的
weathering quality 抗风化性;抗风雨性;耐风蚀性(能)
weathering rate 风化速率
weathering residual texture 风化残余结构
weathering residue 风化残积物
weathering resistance 抗风化(能力);耐气候性;耐老化(性);风化抗力;抗风化性
weathering resistant adhesive 耐老化黏结剂

weathering rock 风化岩
weathering series 风蚀系列;风化系列
weathering service 天气服务
weathering shooting 风化带测定地震勘探
weathering shot 风化层爆破
weathering slope 泻水斜度
weathering soil 风化土(壤)
weathering stage 风化作用阶段
weathering stain 雨水污迹
weathering steel 耐大气腐蚀钢;耐老化钢;耐蚀钢
weathering test 老化试验;耐气候性试验;耐气候性检验;耐候试验;发霉实验
weathering thickness at receiver point 检波点低速带厚度
weathering thickness at shot point 炮点低速带厚度
weathering trough 泻水槽
weathering velocity 风化速度
weathering warning 气象警报;气象警报
weathering zone 风化区;风化层;风化(地)带
weather instrument 气象仪(器)
weather intelligence 气象情报
weather interference 气候干扰
weatherizing 使适应气候条件
weather joint(ing) 泻水缝;斜勾缝;防水刮缝
weather ladder 露天梯子
weather lay days 晴天停驳时间
weather life 保漏时间;防蚀寿命
weather limitation 气候限度
weather log 气象记录簿
weather lore 天气谚语
weatherly qualities 适浪性
weather map 气象(地)图;天气图
weather map analysis 天气图分析
weather map facsimile equipment 气象图传真机
weather map type 天气型;天气图类型
weather maxim 天气谚语
weather message 气象电文;气象电报
weather meter 老化仪;大气老化试验机;风蚀仪
weather minimum 最低气象条件
weather model 天气模式
weather modification 人工影响天气;天气演变;天气控制
weather modification process 人工改造气候法
weather modification program(me) 人工影响天气方案
weather monitoring 天气监测
weather mo(u)lding 挡水条;门窗披水板;滴水槽;泻水垂脚;外墙腰线;滴水线脚
weather notation 天气符号
weather observation 气象观测;天气观测
weather observation flight 气象观测飞行
weather observation ship 气象观测船
weather observatory 气象台
weather observer 气象观测员
weather-observing station 天气观测站
weatherometer 人工曝晒机;气象计;耐风蚀测试机;老化测试机;老化试验计;老化试验机;耐候试验机;人工老化机;风雨侵蚀测量仪;风蚀计
weatherometer test 人工老化机试验
weather operator 气象操纵员
weather orbiter 气象卫星
weather outlook 天气展望;天气趋势
weather parameters 天气参数
weather parting 天气区界限
weather parts 露天部分
weather patrol 气象侦察飞行
weather patrol ship 气象侦察船
weather pattern 天气模式
weather penetration 气象穿越
weather penetration investigation 风雨侵入试验
weather permitting 天气允许
weather permitting clause 雨天顺延条款
weather permitting days 天气条件下可能工作天数
weather phenomenon 天气现象
weather plane 天气侦察飞机
weather plotter receiver 气象标图接收机
weather portions 露天部分
weather prediction 天气预告;天气预测
weather process 天气过程
weather prognosis 气象预测;气象预报;天气预报
weather prognostics 天气形势预报
weatherproof 防暴风(雨)的;抗风蚀的;防日晒防风雨的;耐风蚀;全天候(的);不受天气影响

的;不受气候影响的
weatherproof adhesive 防水性胶黏剂;耐候性胶黏剂
weatherproof cabinet 防风雨箱
weatherproof circuit breaker 防雨断路器
weatherproof clothing 风衣
weatherproof dressing 防潮包装;耐风雨的包装
weatherproof enclosure 耐风雨的外壳
weatherproof engine 耐风雨发动机
weatherproof finish 耐气候整理
weatherproof fused switch 防雨熔丝开关;防雨熔断开关
weatherproof glue 抗风化黏结剂;耐候胶;抗老化胶;防水性胶
weatherproofing 不受气候影响的;抗风化的;耐候的
weatherproofing compound 耐风雨的化合物;抗风化的化合物
weatherproofing motor 气候防护电动机
weatherproofing sealant 耐候密封膏
weatherproofness 全天候性;耐风雨性;耐候性;风雨密性;耐风雨侵袭能力
weatherproofness test 风雨密性试验
weatherproof oil paint 防霉油漆
weatherproof plywood 耐风化胶合板;耐候(性)胶合板
weatherproof receptacle 水密插头
weatherproof rope 耐风雨绳索
weatherproof sealant 耐候胶
weatherproof steel 耐气候钢
weatherproof switch 防雨开关
weatherproof telephone set 全天候电话机
weatherproof type motor 防风雨式电动机
weatherproof wire 风雨线;橡皮绝缘风雨线;耐风雨线;防风雨绝缘电线
weather prospect 天气展望;天气预兆;天气形势;天气趋势
weather protected 不受气候影响的
weather protected machine 防风雨机器;风雨防护式电机
weather protected motor 防潮电动机
weather proverb 天气谚语
weather quarter 尾舷上风
weather radar 气象雷达;天气雷达
weather radar data processor and analyzer 天气雷达资料处理分析器
weather radar display 气象雷达显示
weather radar map 气象雷达图
weather radar scanner 气象雷达天线
weather radar system 气象雷达系统
weather reconnaissance 气象侦察
weather reconnaissance aircraft 气象侦察机
weather reconnaissance code 气象侦察报告格式
weather reconnaissance flight 天气侦察飞行
weather reconnaissance mission 气象侦察任务
weather reconnaissance system 气象侦察设备
weather recording 天气记录
weather recurrences 天气重现
weather-reflective paint 耐老化涂料
weather regime 天气状况
weather-related error 气候所造成的误差
weather report 气象报告;天气预报;天气报告
weather resistance 耐天气性;耐候性;耐风雨侵蚀能力;耐风化(性)
weather-resistant 防风雨的;抗风化的;不受天气影响的;抗大气腐蚀;耐风化;防日晒;不易风化的
weather-resistant insulation 耐风雨绝缘
weather-resistant steel 耐气候钢
weather-resistant tar 抗老化焦油
weather resisting 抗老化的;耐风化的
weather-resisting barrier 防雨条
weather-resisting material 抗风化性材料
weather-resisting property 抗老化性能;耐候性
weather routing 气象选线;按气候定航线
weather satellite 气象卫星;气候探测卫星;天气卫星
weather satellite data 气象卫星数据
weather-scouting mission 气象侦察飞行任务
weather seal 不受天气影响的密封
weather sealant 耐风化密封剂;耐风化密封胶;耐候密封膏
weather seal channel 门上漏水;(门上的)滴水
weather search radar 气象探测雷达
weather sequence 天气演变序列
weather service 气象预报广播;气象业务;气象服务
Weather Service Offices 天气预报局

weather service station 气象服务站
weather shake 干裂;干冻裂隙;风蚀裂纹
weather shed 雨棚
weather shingling 墙面木板;盖墙木瓦;竖铺木瓦;防风薄板
weather ship 海洋天气船;气象(观测)船;天气船
weather shore 迎风岸;天气之岸;上风岸
weather side 上风侧;迎风侧;迎风面;上风的
weather sign 气象符号;天气预兆;天气符号;天气标志
weather signal 气象信号;天气信号
weather situation 天气形势;天气情况
weather skin 耐风雨外层;耐风化外层
weather slating 挂墙石板;(防天气侵蚀的)墙顶石盖板;外挂石瓦;石板瓦墙面;墙挂石板瓦;铺石板瓦
weather slating tile 墙石板瓦
weather stain 因经受风雨日晒而褪色;老化变色
weather stained 因风雨褪色的
weather station 测候所;气象站;气象台;天气站
weather strip 挡风雨条;密封条;门槛嵌缝条;门窗挡风雨条;挡风条
weather stripped door 加防风雨条的门
weather stripping 挡缝止水条;挡风雨条
weather-struck 勾凹水缝;(防水勾缝的;防水接头;防水接缝
weather-struck joint 泻水勾缝;斜(勾)缝;防水刮缝;防雨刮缝
weather summary 气象报告
weather surveillance radar 天气监视雷达
weather symbol 气象符号;天气(字码)符号
weather system 气象系统;气候系统;天气系统
weather teletype and facsimile system 气象电传与传真系统
weather television system 气象电视系统
weather tester 耐候试验机
weather tide 迎风潮;逆风潮
weather-tight 不透风雨;防日晒;不透风雨的
weathertight deck 防风雨甲板
weathertight door 防风雨门
weather tightness 防风雨性;不透风雨性;水气不透性
weather tile 盖墙瓦;盖顶瓦;墙顶盖瓦
weather tiling 盖墙顶瓦片;墙面挂瓦;挂式板瓦
weather track 气象侦察飞行
weather trend 天气趋势
weather type 天气分类;天气类型
weather typing 天气分类
weather vane 定风针;风信标;风向(标);风标
weather vaning 风标作用
weather variation 天气变化
weather warning 天气警报
weather window 防风雨窗;(浇筑混凝土)最佳气候
weather withstand steel 耐候钢
weather working consecutive hours 连续24小时晴天工作日
weather working days 晴天装卸作业天数;晴天工作日;气候条件许可日数;天气良好工作日
weather-worn 风雨损耗的;风化损坏的;风雨侵蚀的
weather yaw adjustment 天气偏航调整
weave 交织;波状失真
weave bead 摆动焊道
weave beading 摆动焊
weave design 织纹设计
weave formation 织纹组合
weave screen 织成的网
weave setting 定纹
weaving 交织(焊接);横摆运动
weaving area 交织段(交通);交织区(交通)
weaving distance 交织距离
weaving industry 编织业
weaving length 交织距离;交织长度
weaving mattress 柴排
weaving mill 织造厂;纺织厂
weaving movement 摇摆运动
weaving of traffic 车流交织
weaving point 交织点
weaving-pole 木排
weaving section 交织路段;交织段;行车交织段
weaving sight distance 交织视距
weaving space 交织区段(车辆)
weaving speed of electrode 焊线横摆速度;焊条横摆速度
weaving traffic 交织交通

weaving vehicles 交织车辆
weaving weld 横向摆动焊缝
weaving welding 摆动焊
weaving wire 编织用钢丝
weavy grain 织纹;卷纹
web 钻trick;织网;联结板;金属薄板;桁材;万维网;丝网;腹板;薄毡(美国称法);箔材轧料;箔材坯料
web address 网址
web bar 横向钢筋;箍筋;腹杆;抗剪钢筋;腹筋
web beam 宽板横梁;桁板梁;强梁
web-beam action 腹板—梁作用
webbed chain 杆式升运链
webbed dried film 干燥的网膜
webbed panel 节间腹板
Webb effect 威布效应(泥浆解凝时体积变化)
webbing 织物带;桁架腹杆构件;起皱或污气网纹;喷丝涂装法;网纹涂装(法);网纹;吊带
webbing finish paint 喷丝涂料;网纹涂料
webbing lacquer 网纹漆
webbite 炼钢合金剂
web buckling 腹板压屈;腹板屈曲
web clamp 薄板夹具
web cleat 腹板锚固夹板
web compressor 毛网压紧装置
web connection 腹板连接
web connection of beams to columns 梁柱腹板联结
web-conveyor digger 输送链式挖掘机
web core sandwich panel 腹板夹芯板
web cover 钢板梁腹板接合缝盖板;梁腹连接盖板
web-crack (混凝土梁的)腹板裂缝
web-cracking 腹板裂缝
web crippling 腹板折曲;腹板断裂;腹板损伤;腹板压屈;腹板屈曲
web distortion 腹板变形
Weber 韦伯(磁通单位,等于108麦克斯伟);成网机
Weber cavity 岩层分离形成的空洞;韦伯空洞
Weber Deep 韦伯海渊
Weber differential equation 韦伯微分方程
Weber dynamometer 韦伯功率计
Weber evapo(u)rator 韦伯蒸发器
weber fraction 韦伯分数
weberite 氟铝镁钠石
Weber joint 韦伯接合
Weber law 韦伯定律;韦伯定理
Weber number 韦伯数
Weber number 1 第一韦伯数
Weber number 2 第二韦伯数
Weber number 3 第三韦伯数
Weber photometer 韦伯光度计
Weber process 韦伯炼铁法
Weber ratio 韦伯比
Weber similarity criterion 韦伯相似准则
web failure 梁腹破裂
web former 成网机
web forming zone 浆幅形成区
web form(work) 腹板模壳
web frame 宽板肋骨;加强肋骨
web gear 薄片齿轮
web hay loader 杆条式装干草机
web joint 腹板接合
web-like system 蛛网式(道路)系统;放射环式道路网
web longitudinal stringer 纵桁
web member 腹件;腹杆;腹板
web-member system 腹杆系
web of beam 梁腹板
web of crank 曲柄连接板
web of dry felt 干毡毛织网;干毡毛织物
web of fabric 织物经纱;织物纤维网
web of felted fabric 缩绒织物布匹
web offset ink 卷筒胶印油墨;胶印轮转油
web of girder 大梁腹板
web of hollow tile 空心砖腹隔
web of rail 轨腹
web of saw 锯条
web of tile 空心砖腹板;空心砖腹隔
webpage 网页
web plate 桁板;腹板
web plate angle 腹板角铁
web plate connection 腹板接头;腹板连接
web plate depth 腹板深度;腹板高度
web plate joint 腹板接合;腹板接头;腹板接缝
web plate length 腹板长度

web plate longitudinal connection 腹板纵向连接
web plate moment 腹板力矩
web plate of I-beam 工字梁腹板
web plate stay 腹板撑条;腹板加劲缀条
web plate stiffener 腹板加劲角钢
web plate stiffening 腹板加劲件
web plate stress 腹板应力
web plate thickness 腹板厚度
web plate transverse joint 腹板横向接头;腹板横向连接
web plate width 腹板宽度
web press 卷筒纸印刷机
web printing 卷筒纸
web reinforcement 箍筋;梁腹钢筋;抗剪钢筋;横向钢筋;腹筋
web rod 梁腹钢筋
web roll 斜辊
web saw 框锯;架锯;片锯
web server 网络服务器
website 网址;网站
web sling 吊货索网兜
web splice 梁腹搭接;梁腹镶接;梁腹拼接;腹板拼接;腹板搭接
web splice plate 梁腹镶板
web squeezer 毛网压紧装置
web stay 梁板撑条;腹板撑杆
websterite 矾石;二辉(辉石)岩
web stiffener 梁腹加劲杆;梁腹加劲材;梁腹加劲板;梁腹加筋角铁;腹板加劲肋;腹板加劲件;梁腹加劲角铁
web stiffening 腹板加固件
web system 腹杆系
web thickness 腹板厚
web thinning 修磨横刃
web thinning gashing grinder 横刃修磨机床
web transverse connection 腹板横向连接
web TV 网际电视
web-type flywheel 盘式飞轮
web-type spar 腹板式翼梁
web wheel 板轮;盘式轮
weddellite 草酸钙石
Weddell Sea 威德尔海
wedding agent 乳化剂
Wedding Tower 结婚纪念塔(1907年由奥别列希设计的建于德国的达姆施塔)
Weddle's rule 韦德尔法则
wedge 楔(子);楔形物;楔形图;楔形体;楔入;光楔;高压脊;尖劈;木楔;劈;不平行度;波导楔
wedge absorber 楔形吸收器;吸声尘楔
wedge action 楔(劈)作用;楔紧作用;尖劈作用
wedge-action die 侧楔模
wedge action valve 楔形闸(门)阀
wedge adjuster 调整楔
wedge adjustment 楔块调整
wedge aerofoil 楔形翼型
wedge analysis 土楔分析法
wedge anchor(age) 楔入锚固;楔形锚扎;楔形锚固;楔式锚具;楔块锚;预应力筋楔锚器
wedge anchoring system 楔形锚固系统
wedge and shims 插楔开石工具
wedge-and-sleeve bolt 楔壳式锚杆;插楔和套筒插销
wedge angle 楔角;研磨角;尖角
wedge antenna 楔形天线
wedge arch 楔体拱
wedge attachment 光楔附件
wedge bar 楔形筛条
wedge-bar screen 楔形棒条筛
wedge-bar screen deck 楔形条状筛板
wedge bar(shell) liner 加楔条的衬板
wedge-base lamp 楔形底板指示灯
wedge beam 楔形梁
wedge bed 楔状岩层
wedge belt 楔形带
wedge bit 楔形钻头
wedge block 楔座;楔块;卡盘
wedge block ga(u)ge 楔形角度块规;楔形块规
wedge block tunnel 楔石砌隧洞
wedge boarding 楔镶板;楔接板
wedge bolt 楔形螺栓
wedge bonder 楔式接合器
wedge bond(ing) 楔形接合;楔形压焊;楔形焊接;楔焊
wedge brake 楔形制动器;制动楔

wedge brick 楔(形)砖
wedge buckle 楔形系杆;楔形手杆
wedge busbar 插接母线
wedge-caulking 楔紧
wedge channel grooved piston ring 楔形槽活塞环
wedge chuck 楔形夹盘
wedge clamp 楔形压板;楔形线夹;楔形夹板
wedge clamp joint 楔嵌接
wedge clip 楔形扣板;楔形夹
wedge clothing 楔闭
wedge-colorimeter 楔形比色计
wedge column 楔形柱
wedge compression coupler 楔形压力联结装置
wedge constant 光劈常数
wedge contact 插接触点
wedge coping 楔形墙压顶;楔形盖顶石;单坡屋顶
wedge core lifter 楔形岩芯提断器;楔式提芯器
wedge cotter 楔形扁销
wedge coupler 楔形联结装置
wedge coupling 楔形联轴器
wedge crystal 楔形晶
wedge cuneatic arch 楔形砌拱
wedge curb 楔形壁座
wedge cut 楔形切割;楔形开挖;楔式切割;楔式开挖
wedge cutter 楔形掏槽孔
wedge cutting 楔形插条
wedged 楔形的
wedge data 楔形块数据
wedge design 楔形设计;楔体结构;尖劈设计
wedge director 楔形导向器
wedged joint(ing) 楔接(合)
wedged leading edge 楔形进气边
wedge drive 楔式传动;槽轮传动
wedge driving 打楔子
wedged-shaped lot 楔形土地
wedged-shaped paper 楔形纸
wedged-shaped technique 楔形技术
wedged slot 楔形狭槽
wedged tenon 楔榫
wedge effect 楔效应
wedge electrode 楔形电极
wedge error 楔形误差
wedge factor 楔滤因子
wedge failure 楔形破坏;楔体破坏
wedge film 劈形膜
wedge filter 楔形过滤器
wedge flow 楔(形)流
wedge focusing 楔聚焦
wedge-form 楔形的
wedge formation 楔式队形
wedge-form reef 楔形礁
wedge friction gearing 楔形轮摩擦传动
wedge friction wheel 楔形摩擦轮
wedge gap 楔形槽沟
wedge gasket closure 楔垫密封
wedge gate 楔形闸门;楔形浇口;楔形浇道;三角闸门
wedge gate flooding valve 楔形通海阀
wedge gate valve 楔形闸阀;楔形活门
wedge ga(u)ge 楔(形)规;楔形角度块规
wedge gear 楔形摩擦轮
wedge grafting 楔接(合)
wedge grip 楔形尖;楔形夹
wedge grip gear 楔形安全装置
wedge guide 楔形罐道
wedge hammer 楔锤
wedge in plastic state 塑性状态时楔体
wedge insert 楔形嵌入件
wedge isobars 楔形等高线
wedge joint(ing) 楔形接头;楔形接缝;楔形接合
wedge key 楔(形)键
wedge lens condenser 楔形透镜聚光镜
wedge lewis bolt 楔形地脚螺栓
wedge-like delta 楔形三角洲
wedge line 楔形高压线;楔线
wedge method 楔体法;楔取法
wedge method of attachment 楔形装置法
wedge micrometer 楔形千分尺
wedge mount 楔形座
wedge nozzle 楔形喷管
wedge-nut bolt 楔壳式锚栓
wedge-nut expansion shell bolt 楔螺母胀壳锚杆
wedge of buoyancy 浮楔
wedge of earth 土压力楔体

wedge of emersion 露出楔
wedge off 钻孔偏斜
wedge of failure 滑动楔状岩层;滑楔
wedge of high-pressure 高压楔
wedge-opening-loading 楔形开裂加荷
wedge out 劈成;尖灭
wedge pattern 楔形图
wedge photometer 光楔光度计;光劈光度计;尖劈刀光度计;劈片光度计
wedge piece 楔形零件
wedge pile 楔形桩
wedge pile chuck 楔式桩夹
wedge pin 锥销
wedge plan 楔形设计
wedge planting 楔形栽植
wedge plate 楔板
wedge plug 楔形塞
wedge point 楔形点
wedge post 楔形柱
wedge press 楔式热模锻压力机
wedge pressure 楔形压;楔嵌压
wedge probe 楔形测针
wedge product 楔积
wedge prop 楔锁式支柱
wedge pyrometer 楔(光)片测温计
wedge reaming bit 楔形扩孔钻头
wedge relaxation function 楔形松弛函数
wedge reservoir 楔形储池
wedge ring 卡盘
wedge-riser flange 楔形法兰
wedge roller gate 楔形滚动闸门
wedge rolling 楔形轧制
wedge rolling tool 楔楞轧模
wedge rose bit 偏斜楔玫瑰形钻头
wedge sampling technique 楔形取样器技术
wedge scale 楔形比例尺
wedge seal 楔状密封片
wedge section 楔形断面
wedge-section wire 楔形钢丝
wedge segment 楔状扇形块
wedge-set bit 楔形底唇金刚石钻头
wedge setting gear 楔形定位装置
wedge sewer brick 阴沟用楔形砖;楔形阴沟砖
wedge shape 楔形
wedge-shape cut 楔形陶槽
wedge-shaped 劈形的;楔状的;楔形的
wedge-shaped absorber 劈形吸声体
wedge-shaped armor 楔形装甲
wedge-shaped belt 楔形传动带;V形传动皮带
wedge-shaped block 楔形砌块;楔状地块[地];楔形地块
wedge-shaped brick 楔形砖
wedge-shaped brush 楔形电刷
wedge-shaped clip 楔形扣板
wedge-shaped coil 楔形线圈
wedge-shaped cross-bedding 楔形斜层理
wedge-shaped cross bedding structure 楔形交错层理构造
wedge-shaped curb 楔形路缘石
wedge-shaped cusp 楔形牙尖
wedge-shaped cut 楔形掏槽
wedge-shaped dam 楔形防水墙
wedge-shaped deflector 楔形偏导器
wedge-shaped form 楔形定型模板
wedge-shaped fracture 楔形裂缝;楔形断裂
wedge-shaped gap 楔形隙
wedge-shaped groove 楔形槽
wedge-shaped hair 楔形丝
wedge-shaped high 楔形高气压
wedge-shaped joint 楔形接头
wedge-shaped lightning arrester 楔形隙避雷器
wedge-shaped load 楔形荷载
wedge-shaped mica 楔形云母
wedge-shaped mortar joint 楔形砂浆接头
wedge-shaped oil film 楔形油膜
wedge-shaped opening 楔形孔口
wedge-shaped plate 楔形板
wedge-shaped point 楔形刻针尖
wedge-shaped relaxation spectrum 楔形松弛谱
wedge-shaped rock body 楔状岩体
wedge-shaped runner 楔形滑刀式开沟器
wedge-shaped share 凿形犁铧
wedge-shaped shelter-wood system 楔形伞伐作业方式

wedge-shaped slot 楔形槽
wedge-shaped sound absorber 尖劈吸声器
wedge-shaped step 楔形楼梯踏步
wedge-shaped stone 拱石;楔形石
wedge-shaped template (混凝土路面的)接缝楔形样板
wedge-shaped tile 楔形砖
wedge-shaped tread 楔形踏步;楔形楼梯踏步板
wedge sheet pile 楔形板桩
wedge shell 楔形贝
wedge shock 楔形物激波
wedge shot 楔形掏槽爆破
wedge sleeve for anchor 下锚楔套
wedge slide 楔形导板
wedge slope 锥角
wedge slot 楔形孔;尾栓孔
wedge sluice valve 楔形闸(门)阀
wedge socket 楔形联结套筒
wedge socket fitting 楔形承口装置
wedge spectrograph 楔形摄谱仪;劈片摄谱仪
wedge spring 楔弹簧
wedge sprue 锥形直浇道
wedge stacking 楔形体堆垛
wedge-stilt 楔形坯架
wedge stone 楔形石;拱楔石
wedge storage 楔形体储存;(仓库中)楔形堆垛;楔形蓄水体;楔形体储[贮]存;楔形储池
wedge-strip chromatography 楔形条色谱
wedge structure 楔形结构
wedge stud 楔形销钉
wedge surface 楔面
wedge system 楔状渐伐作业法;楔形伞伐作业
wedge tensile strength test (螺钉等的)楔垫抗拉强度试验
wedge test 楔形块检验;锤击韧性试验
wedge test piece 楔形试片;三角试样
wedge theory 楔体理论;楔块理论;土楔理论
wedge thickness 楔厚
wedge thrusts 楔冲式冲断层
wedge tile 楔形砖
wedge-tip bar 楔尖沙坝
wedge type 楔形闭锁型
wedge type anchor anti-creeper 穿销式防爬器
wedge-type anti-creeper 楔形防爬器
wedge-type breaking piece (轧机)楔形安全臼
wedge-type connector 楔形连接器
wedge-type failure 楔型破坏
wedge-type foundation of yielding soil 流动土楔形基础;软土楔形基础
wedge-type fracture 楔形晶间断裂
wedge-type gate valve 楔形闸(门)阀;楔形活门
wedge-type rock bolt 楔缝式锚杆;前端黏接型锚杆
wedge-type seal 楔形密封
wedge-type sewer brick 下水道楔形砖
wedge type slide valve 楔形滑阀
wedge-type split ring connector 楔形开口环接件
wedge-type ultrasonic delay line 楔型超声延迟线
wedge-type weir 楔形堰
wedge up 楔起
wedge valve 楔形阀;楔形闸(门)阀
wedge valve closet 楔杆阀冲水厕所
wedge wave 楔形波
wedge weir panel 楔形堰板
wedge-wire 楔形金属丝
wedge wire cloth 楔形丝织筛布
wedge-wire deck 楔形丝筛面
wedge-wire drying bed 楔形条干化床
wedge-wire mesh 楔形丝织网
wedge-wire screen 楔形条网;楔形丝筛网
wedge-wire screening 楔形金属丝筛网
wedge wise 成楔型
wedge yoke 楔锁
wedging 楔状岩芯自卡;楔入;楔裂;楔块劈裂作用;楔紧;楔固;扩张楔;人工揉练泥料;劈开;打楔子;产品厚薄不匀
wedging action 楔(入)作用;楔紧作用;楔合作用
wedging agent 楔入剂
wedging bit 楔形钻头
wedging block 楔块;楔合块
wedging compound 楔入化合物
wedging crib 楔形井框
wedging curb 楔形壁座
wedging cut 楔形掏槽
wedging down 楔落;劈碎

wedging effect 楔入效应
wedging folds 楔入褶皱
wedging-in action of setting cement 凝结水泥的楔紧作用
wedging inscribing 楔形内接
wedging method 楔缝(堵漏)法
wedging mill 捏泥机
wedging of arch top 砌成拱顶
wedging of drilling rod 夹钻
wedging of drill tools 卡钻
wedging plate 楔入板;楔紧钢板
wedging reamer 楔形扩孔钻头;楔形扩孔器
wedging separation method 打楔分离法
wedging shot 楔形掏槽爆破
wedging table 揉泥台
wedging thrust 楔冲断层
wedging up 楔起
Wedgwood pyrometer 威吉伍德高温计
wed up 填焊
weed and loose soil machine 除草松土机
weed attachment 除草附加装置
weed boom 除草喷杆
weed breaker 压草机
weed buffer 水缓冲器
weed burner 火焰除草机;烧草机
weed cleaner 杂草清除器
weed clearing 除莠
weed control 杂草控制;杂草防治
weed-control agent 除草剂
weed-control equipment 除草机具
weed-control in rights-of-way 路旁杂草控制
weed-control method 杂草防除法
weed-covered 杂草丛生的
weed cutter 割草机;杂草切除机;除草机
weed-cutting blade 除草锄铲
weed-cutting launch 割草船
weed-cutting machine 除草机
weed destroyer 灭草器;灭草机;除莠器
weed encroachment 杂草为害
weeder 除草工具;割草工;除草器;除草机
weed eradication 灭草;除草
weeder-mulcher attachment 除草覆盖附加装置
weeder-mulcher broadcast planter 除草覆盖撒播机
weed extractor 除草器
weed-filled lake 杂草丛生湖
weed-filled river 杂草丛生河流
weed fouling 藻污
weed-free 无杂草的;无水藻的
weed-free fallow 无草休闲地
weedicide 除草剂
weediness of waters 水草丛生的水体
weeding fork 除草叉
weeding harrow 除草耙
weeding hoe 除草锄;草锄
weeding machine 除草机
weeding of grease 除去脂
weed injury 杂草为害
weed killer 除莠剂;喷洒除草机;除草喷雾车;除草剂
weed killer spray 除草剂喷洒
weed killing 灭草;除草
weed-killing equipment 除草机具
weed-killing machine 喷洒除草机
weed-killing solution 除草溶液
weed-killing train 除草列车
weed mower 除草机;割草机
weed oil 除草油
weed prevention 杂草防治
weed re-growth 再生杂草
weed removal 除草
weed removal from surface waters 水域除藻
weed remover 割草机;杂草筛
weeds 野草;杂草
Weed Science Society of America 美国杂草科学协会
weed-seed eliminator 草籽排除器
weed separator 杂草分离器
weed shovel 除草铲
weed sprayer 灭草喷雾机
weed spraying gear 灭草喷雾器
weed sprays 喷雾除草
weedtree 杂木
weed vine 蔓草
weedy 杂草似的;多杂草的

weedy grassland 杂草型草地
weedy river 水草丛生的河流
weedy water 多藻海水
weedy waters 水草丛生的水域;多藻水域
weekday 星期日以外的任何一天;工作日
week-day chapel 工作日礼拜堂;平日小教堂
week-day load curve 平日负荷曲线
weekend house 周末休假住宅
weekend house zone 周末休假住宅区
weekend traffic 周末交通
weekly average traffic volume 周平均交通量
weekly change of SO$_2$ concentration 二氧化硫浓度周变化
weekly check 周检制
weekly condition report 情况周报
weekly cycle 周期环
weekly measurement 每周检测
weekly payroll 每周工薪单
weekly premium 每周保险费
weekly report 周报
weekly returns 周报表
weekly run 每周运转
weekly service 星期服务
weekly storage 周调节水库
weekly storage plant 周调节电站
weekly time distribution 每周工时分配
weekly traffic pattern 周交通形式;交通量周变(化)图
weekly(volume)variation 交通量周变(化)图
weekly wage 周工资
weeksite 多硅钾铀矿
week-supply reservoir 周供量调节水库;周供量蓄水池
week-supply tank 周供量油罐;周供量水罐
weel 鱼笼;旋涡;深池;静水潭
weep 泄漏处;流挂;渗漏处;滴漏
weepage 渗水;泌水;水分的渗流;水分的分泌
weep drain 分水管;排水管;渗水管
weeper 排水孔;小排水管;滴水孔;哀像
weeper drain 泄水沟;集水盲沟;排水盲沟;渗水管沟;渗水沟管
weep hole 女儿墙上排水孔;挡土墙排水孔;出水孔;泄(流)水孔;流水孔;泪孔;排水孔;排气预留孔;渗水孔;残液放出孔
weep holes in lining 衬砌渗漏洞隙
weeping 漏液(现象);流泪现象(水泥浆在混凝土表面渗水);渗出的;分泌;下多雨的;微微下垂的;渗漏;泌油;泌水;滴落
weeping beech 垂枝山毛榉
weeping channel 泄水槽
weeping conduit 排水孔管
weeping core 油浸岩芯;透水岩芯
weeping fungus 腐木菌
weeping pipe 泄水管;排水(孔)管(混凝土砌体背后排水用);滴水管
weeping piping 排水孔管
weeping rivet 铆钉渗水
weeping rock 渗水岩石
weeping rocks 渗水岩类
weeping spring 渗漏泉
weeping tree 垂枝树
weeping tube 排水孔管
weeping tubing 排水孔管
weeping willow 水柳;垂(杨)柳
weep point 泄漏液点
weep trough 流水槽
weevil 牙轮钻头中的油壶润滑器
weevil-proof 防止钻具部件装错的
weevils got him eat up 因工人无经验造成故障
weft 纬线;织物
weft backed cloth 纬二重织物
weft bar 纬档
weft bobbin 纬管
weft corkscrew twill 纬向螺旋斜纹组织
weft disalignment 纬向排列错乱
weft divider 分纬格子板
weft faced fabric 纬面织物
weft hairline 纬向细线条
wefting 投纬
weft knitting 纬编
weft loop 纬编线圈
weft pirn winder 卷纬机
weft protector 纬停装置
weft replenishing mechanism 换纬机构

weft sense 纬向
weft shiner 糙纬
weft stripe 纬向条纹
weft winder 卷纬机
weft winding 卷纬
weft wire (织物的)纬丝;(织物的)纬线
weft yarn 帆布纬线
Wegener hypothesis 魏格纳大陆漂移说
wegscheiderite 碳氢钠石
Wehnelt cathode 氧化物阴极;控制阴极
Wehnelt cylinder 调制极
Wehnelt electrode 调制极
wehrlite 异剥辉石橄榄岩;针碲铋矿
Weibel probability density function 韦布尔概率密度函数
Weibull coefficient 韦布尔系数
Weibull distribution 韦布尔分布
weibullite 硒硫铋铅矿
Weibull's modulus 韦布尔模数
Weibull plot 韦布尔图
Weibull's statistics 韦布尔统计学
Weibull's theory 韦布尔理论
Weichselian glacial epoch 魏奇塞尔冰期
Weichselian glacial stage 魏奇塞尔冰期
Weierstrass approximation theorem 维尔斯特拉斯逼近定理
Weierstrass canonical form 维尔斯特拉斯标准型
Weierstrass curve 维尔斯特拉斯曲线
Weierstrass elliptic(al) integral 维氏椭圆积分
Weierstrass function 维尔斯特拉斯函数
Weierstrass M test 维尔斯特拉斯 M 判别法
Weierstrass's canonical product 维尔斯特拉斯典范积
Weierstrass's elliptic(al) functions 维尔斯特拉斯椭圆函数
Weierstrass's point 维尔斯特拉斯点
Weierstrass transform 维尔斯特拉斯变换
Weigert effect 魏格特效应
weigh 重压;衡(量);推敲;称量进料皮带机;权衡;称重;量
weighable 可称量的
weigh-a-day-a-day 每日测产一天记录
weighage 过秤费
weigh anchor 起锚;启航
weigh appliance 衡器
weigh bar shaft 枢轴
weigh batcher 重量配料机;(工地用)配料秤;重量配料器;分批称料机
weigh batching 重量配料(法);按重量配料(法)
weigh-batching device 按重量配料装置
weigh-batching equipment 按重量配料装置;按重量配料设备
weigh-batching hopper 重量配料斗;分批配料斗;(按重量的)配料斗;称量配料斗
weigh-batching plant 分配称料设备;按重量配料设备;称重配料搅拌设备;按重量配料装置;按重量配料设备;按重量配料厂
weigh batch plant 称重配料搅拌设备;分批称料设备
weigh-beam 秤杆;平衡梁;天平杆
weigh-beam weight beam 大杆秤
weigh belt bin 皮带秤卸料仓
weigh belt feeder 胶带秤喂料机;皮带秤喂料机;配料皮带秤
weigh bill 称量清单
weigh bin 称料斗;称重箱;计量仓
weigh box 称料斗;称重器;计重箱
weigh-bridge 轨道衡;竖旋桥;地(中)衡;地秤;地磅(台);台秤;桥秤
weigh-bridge track 轨道衡线
weigh bucket (混凝土拌和)称料斗
weigh by volume 按体积计重;按体积比称重;按体积称量
weigh dial 磅秤标盘
weighed 权衡的;衡量的;称量的
weighed additive (泥浆)加重剂
weighed amount 称出重量;称出试样
weighed average 加权平均数
weighed average porosity 加权平均孔隙度
weighed average value 加权平均值
weighed drilling mud 加重泥浆
weighed integral 估量积分
weighed means 加权平均值
weighed mean value 加权平均值

weighed sample 称出试样
weighed transmission loss 计权隔声器
weighed volumetrically 体积加权的
weighed yearly mean precipitation 加权平均年降水量
weigher 秤;自动秤;秤量器;验秤员;司磅(员);过秤员;计量器;称重器;称量机
weigher feeder 秤料—给器;称量给料器
weigh error 权衡误差
weigh feeder 称量送料器;定量进给装置;称量给料器
weigh feeding and blending 定量供料与混合
weigh-gross 毛重
weighhouse 计量所;过磅处;过秤房;过磅房
Weigh House at Haarlem 哈勒姆计量所(荷兰)
weigh-in buggy 称量小推车
weighing 秤;加权;称重(量);权衡;称量
weighing accuracy 秤重准确度;秤量精度;称量准确度;称量精度
weighing and drying test 称量法干燥试验
weighing-and-recording gear 称重与记装置
weighing apparatus 称量装置;衡器
weighing appliance 秤重设备;称量设备;称具
weighing area 配料间
weighing batch box 重量配料器;按重量配料箱
weighing batcher 秤重配料器;称量配料器;按重量配料斗
weighing batcher 分批量料器
weighing batching mixer 称量配料拌和机
weighing batching plant 称料分批设备
weighing beam 秤(杆);杆式秤
weighing bed 衡重床
weighing belt 称量(皮)带
weighing belt bin 称重胶带卸料仓
weighing bin 称量箱;称量筐;称量斗
weighing bottle 称重瓶;称(量)瓶
weighing box 秤料箱;秤桶;称量器
weighing bridge 称量台
weighing bucket 称料斗
weighing burette 称重滴管;称量滴定管
weighing by difference 减差称量法
weighing by swings 摆动称法
weighing capacity 秤量限度;秤量能力
weighing car 称量车;计量小车
weighing charges 过磅费
weighing controller 称量控制器;重量定量器
weighing cycle 称量周期
weighing data recorder 称量数据记录器
weighing device 称量设备;计量设备
weighing dial 称料计;磅秤针盘;磅秤刻度盘
weighing disk 称量皿
weighing equipment 称量装置;称量设备
weighing equipment 衡器;分批配料装置;分批配料设备
weighing error 称量误差;过秤误差
weighing evapotranspirometer 称重式蒸发仪
weighing factor 权衡因素
weighing feeder 称量喂料机;称量加重器;喂料皮带秤
weighing float-type lysimeter 漂浮称重式土壤蒸发器;漂浮称重式测渗仪
weighing floor 磅秤台
weighing gear 衡器;称重设备
weighing hopper 称料漏斗;称料斗;称量计;称量(给料)斗
weighing hopper with loadcell 设有重力传感器的计量斗
weighing house 磅房;过秤房
weighing installation 称量设备;称量设施
weighing larry car 称量车
weighing level ga(u)ge 重量位面计
weighing lever 秤杆
weighing lock 秤船闸;称船闸
weighing lorry 过秤车
weighing lysimeter 秤重式土壤蒸发计;称重式测渗仪
weighing machine 称桥;称量器;称量机;过秤机;计重器;计量机;机械杆;衡器;磅秤;台秤;体重计
weighing machine for iron and steel materials 钢材秤
weighing machine standard 体重秤
weighing machine track 轨道衡线
weighing machine with tipping hopper 倾斗式秤
weighing material (泥浆)加重剂

weighing material of mud 泥浆加重剂
weighing matrix 权矩阵
weighing measuring range 称重量程
weighing method 称重法;称量法;重量法(加气混凝土配料)
weighing pipette 称量移液管;称量吸移管
weighing plant 称量设备;分批配料装置;分批配料设备
weighing platform 称重平台;称(料)台;地磅台;磅秤台
weighing rain-ga(u)ge 秤重雨量器;衡重式雨量计
weighing ring 加重环
weighing room 称料间;称量室;称量间
weighing scale 秤盘;秤;称量秤
weighing scale quantifying instrument 磅秤计量计
weighing scoop 称量(量)勺
weighing sensibility 称量灵敏度
weighing sensitiveness 称量敏感性
weighing sensitivity 称量灵敏度
weighing shaft 秤轴
weighing station 过秤台
weighing tank 称重筒;称量桶;称量槽
weighing technique 称量方法
weighing tower 秤量塔
weighing truck 过秤车
weighing tube 称量管
weighing weigher 分批配料装置;分批配料设备
weigh-in-motion scale 运用车轨道衡
weigh-length 秤杆长度
weighlock 衡闸;船舶称重闸
weigh-machine 地秤;秤桥
weighmaster 矿场量产品重量者
weigh mixing 重量配合法
weigh more than one thousand pounds 体重超过一千磅
weigh-out method 减量法
weigh plant 衡器厂
weigh process in water 水中称重法
weigh shaft 摇臂轴;秤轴;摇臂钻
weight 重物;重量;载重;荷重;权重;权数;平衡块;秤砣;砝码
weight a brake 平衡锤调节绳索下降
weight according to the manifest 货单内的重量
weight account 重量报表
weight-actuated filler 重量定量式加料器
weight aid 体重辅助
weight analysis 重量分析
weight and balance sheet 重量平衡表
weight and measurement mark 重量及尺码标志
weight application 施加重力
weight applied to the bit 钻头(轴向)压力
weight assumption 重量假定
weight average 加权平均数
weight-average degree of polymerization 重量平均聚合度
weight-average molecular weight 重量平均分子量
weight balance 平衡配重
weight-balanced abutment 衡重式桥台
weight-balanced spindle 重锤平衡式联结轴
weight balance measurement 重量平衡测定
weight-barograph 重力气压计;动气压计
weight barometer 重力气压表;称气压计
weight basis 重量本位;重量基位;按重量
weight basket 压载箱;配重箱
weight batcher 称重台;按重量配料斗;按重量分配投配设备
weight batching 按重量配料(法)
weight beam 秤杆;杠杆秤;弹簧秤;弹簧测力计
weight bearing capacity 承载能力
weight-bearing mechanism 重量支撑机构
weight bearing power 承重能力;荷载能力
weight bearing property 承重性质
weight bearing quality 承重性能
weight bearing wheel 支承轮
weight belt feeder 喂料皮带秤
weight bias error 权偏差
weight bin 重量喂料仓
weight box 压载箱;重锤箱;平衡锤盒;配重箱;窗平衡锤箱;窗锤箱;称重器
weight bracket 重锤架
weight brake 重力闸
weight break 自重破裂
weight buckle 重锤扣
weight by volume 容积重量;按体积比称重;按体

积比称量
weight by volume percent 容积重量百分数
weight capacity 钻进给进力;抓取重量;承重能力
weight cargo 重量货(物);重货;计重货物
weight carrier 载重机
weight-carrying 载重的
weight-carrying ability 载重性能
weight-carrying brick 载重砖
weight-carrying clay brick cross-wall 黏土砖承重横墙
weight-carrying construction 承重结构
weight-carrying face work 承重砖石饰面工作;承重砖石镶面工作;承重砖石砌面工作
weight-carrying facing masonry (work) 承重砖石墙面圬工
weight-carrying floor block 承重楼板空心砖
weight-carrying floor brick 承重楼板砖
weight-carrying floor clay block 承重楼板黏土砖
weight-carrying floor clay brick 承重楼板黏土砖
weight-carrying frame 载重构架;载重桁架;承重构架;承重框架
weight-carrying in longitudinal direction 纵向载重;纵向承重
weight-carrying in transverse direction 横向载重;横向承重
weight-carrying masonry wall 载重圬工墙;承重圬工墙
weight-carrying masonry work 载重圬工砌筑工作;承重圬工砌筑工作
weight-carrying mechanism 载重机制;载重机构
weight-carrying partition wall 承重隔墙
weight-carrying plane 载重面
weight-carrying power 载重能力
weight-carrying rib 载重肋
weight-carrying skeleton 载重骨架;承重骨架
weight-carrying skeleton construction 载重骨架结构;承重骨架结构
weight-carrying skeleton member 载重骨架构件
weight-carrying skeleton structure 载重骨架结构
weight-carrying structure 载重结构
weight-carrying structure of plain web girders 载重普通腹板梁结构
weight-carrying structure of solid web girders 载重实心腹板梁结构
weight-carrying system 载重系统
weight-carrying wall 载重墙
weight-carrying wall construction 载重墙结构
weight-carrying wall structure 载重墙结构
weight case 重量箱
weight case conversion factor 重量箱折算系数
weight certificate 重量证(明)书
weight change determination 重量变化测定
weight change indicator 重量变化指示器
weight charges 称重费
weight chart 重量图
weight class 重量级
weight coating 防浮包裹层(包裹在水下管线上的水泥或钢筋水泥层);加重包裹层
weight coefficient 加权系数;重量系数;权重因数
weight coefficient matrix 权系数矩阵
weight component 重力分量
weight concentrated in the back 重量集中在背部
weight concentration 重量浓度
weight concentration method 重量浓度法
weight conductivity 重电导率
weight constant 加权常数
weight control 重量控制
weight controller 重量控制器
weight control of tube 沉管重量控制
weight conversion 重量转化
weight correlation bias 权相关偏倚
weight coulometer 重量电量计
weight crack 冰隙
weight curve 重量曲线;重量分布曲线
weight cylinder 重锤导向筒
weight density 重量密度
weight detector 重力检测器
weight determination 重量测定
weight differential 重量差异
weight disc 重锤圆盘
weight discharge 重力排放
weight displacement 重量排水量
weight distributed ratio 分配重量比
weight distribution 重量分配;重量分布

weight distribution coefficient 重量分配系数
weight distribution curve 重量分布曲线
weight distribution effect (汽车的)荷载分配效应
weight distribution factor 重量分配系数
weight distribution function 重量分布函数
weight distribution pattern 重量分布方式
weight distribution ratio 重量分配比
weight drilling mud 高比重泥浆
weight-driven 重锤装置
weight-drop 落重法
weight dropping method 落锤法
weighted 权重的;加权的;载重的;负荷的;受力的;加重的;计权
weighted 4 bit code 加权的四位代码
weighted accumulator 重锤式蓄能器
weighted adjustment 加权平差
weighted aggregate index 加权综合指数
weighted amplitude coefficient 加权振幅系数
weighted applications 加权作用次数
weighted approximation 加权近似(值);加权逼近
weighted approximation method 加权近似法
weighted area masks 加权区域分离字
weighted arithmetic average 加权算术平均数
weighted arithmetic index number 加权算术指数
weighted arithmetic mean 加权(算术)平均数
weighted array 加权组合;加权数量
weighted at one side 做成偏心的
weighted average 加权平均(值);加权中数
weighted average earings per share 每股加权平均收益
weighted average effective unit weight 加权有效重度
weighted average efficiency 加权平均效率
weighted average grade 加权平均品位
weighted average head 加权平均水头
weighted average inventory method 加权平均存货法
weighted average method 加权平均法
weighted average of post adjustments 调整后的加权平均数
weighted average of skin temperature 加权平均皮温
weighted average porosity 加权平均孔隙度
weighted average pricing 加权平均计价法
weighted average thickness 加权平均厚度
weighted average traffic 加权平均车辆通过数
weighted average utility 加权平均效用
weighted average volume weight 加权平均体重
weighted cement 高密度水泥
weighted check 加权校验
weighted code 加权(编)码
weighted coefficient 加权系数
weighted concept 加权概念
weighted constraint 加权条件
weighted correlation coefficient 加权相关系数
weighted cost of capital 加权资本成本
weighted creep length 加权渗流长度
weighted creep ratio 加权蠕变比
weighted curve for noise measurement 噪声计权曲线
weighted deviational variable 加权离差变量
weighted diffusion constant 加权扩散常数
weighted digraph 加权有向图
weighted directed graph 加权有向图
weighted effective continuous perceived noise level 计权有效连续感觉噪声级
weighted entropy 加权熵
weighted equation 加权方程
weighted error 加权误差;加权误差差;权差
weighted factor 加权因子;权重因数
weighted factor method 加权因子法
weighted factor semiquantitative analysis method 加权因子半定量分析法
weighted filter 重力过滤器;盖重滤水体;盖重反滤层;加重反滤层
weighted filtering 加权滤波
weighted geometric(al) index number 加权几何指数
weighted geometric(al) mean 加权几何平均值
weighted graph 加权图;权图
weighted high-duty 载重高载重
weighted index number 加权指数
weighted-in-points method 加权计分法
weighted least square 加权最小二乘方

weighted lever 均重杆
weighted lever lubricator 均重杆润滑器
weighted linear regression 加权直线回归;加权线性回归
weighted linear regression method 加权直线回归法
weighted market capitalization method 加权资本市值法
weighted mean 加权平均(数);计权平均
weighted mean ash per cent 加权平均灰分百分数
weighted mean ash production rate by thickness 厚度加权平均灰分产率
weighted mean ash production rate by volume 体积加权平均灰分产率
weighted mean ash production rate by weight 重量加权平均灰分产率
weighted mean of unit weight 平均重度
weighted mean square 均值加权平方;加权均方
weighted mean temperature 加权平均温度
weighted mean(value) 加权平均值
weighted method 加权法
weighted mirror 重力镜
weighted monthly mean precipitation 加权平均月降水量
weighted moving average 加权移动平均值;加权移动平均数
weighted mud 加重泥浆
weighted net head 加权净水头
weighted noise 计权噪声;权重噪声
weighted observation 加权观测
weighted ordinate method 等间隔波长法
weighted path-length 加权路径长
weighted P-digraph 加权 P 偶图
weighted platform 加载平台
weighted power-sum symmetric(al) function 加权幂和对称函数
weighted probit method 概率单位加权法
weighted property 加权性质
weighted regression analysis 加权回归分析
weighted regression line method 加权直线回归法
weighted relative value 加权相对值
weighted residual method 加权残值法;加权残数法
weighted residuals vector 加权剩余向量
weighted-resistor digital-to-analog converter 加权电阻数—模转换器
weighted safety valve 重量安全阀
weighted sensor data 加权传感数据
weighted shear modulus of soil 土加权剪切模量
weighted silk 增重的丝绸;加重的丝绸
weighted sound level 计权声级
weighted spectral density 加权谱密度
weighted stack 加权叠加
weighted sum 加权总数;加权和
weighted superposition 权重叠加
weighted switch-stand 衡重式转辙器
weighted value 加权值
weighted valve 重量阀;重锤阀;加重阀
weighted velocity head 加权速头
weighted wire precipitator 挂重锤电晕线的集尘器
weight efficiency 重量效率
weight empty 空载
weight-empty of a car 空车重量
weight enumerator 权重计数子
weight equalizer 平衡重物;平衡器;平衡磁极
weight equation 权方程
weight equipment 重型装备
weight equivalent 重量相当
weight equivalent continuous perceived noise level 加权等连续听见噪声级
weighter 衡器
weight estimation 重量估计
weight factor 重量因数;重量系数;权重因子;权重因数
weight feeder 重量送料器
weight feeding 按重量喂料
weight fineness 重量细度
weight floor block 承载地面砖
weight flow 重量消耗量;重量流量;重力流
weight flow rate 重量流率
weight force 重力
weight for erection 建造装配重量;安装配件重量
weight fraction 重量分数
weight frame 承载刚架
weight gain 重量增加;重量的增加;增重

weight gaining agent 增重剂
weight ga(u)ge 指量表
weight governor 荷重调节器
weight grader 重量分级计;重量分级机
weight graduation(mark) 称量分度
weight-growth factor 重量增长因数
weight handling equipment 起重设备
weight hanger 重锤吊架
weight hauled 牵引重量
weigh the advantages and disadvantages 权衡利弊
weigh the pros and cons 权衡利弊
weight holder 重锤架
weight hook 重锤扎钩
weight hourly space velocity 重量时空速(度)
weight house 过磅处
weight hypothesis 重量架设
weightimeter 重量计
weight in average 平均重(量)
weight increase 重量增加;增重
weight index of engine 发动机重量指标
weight indicator 示重计;指重表
weight indicator chart 示重图表
weight indicator tie down 固定指重表
weight in excess 超重重量
weighting 衡量;加重;加权;取权
weighting agent 填充剂
weighting bar 加重杆
weighting coefficient 加权系数;权系数
weighting equipment 测重设备
weighting factor 加权因子;加权因数;加权系数;取权因子
weighting filter 加权滤波器;加权过滤器
weighting function 加权函数;权函数;重量函数;衡重函数;权重函数
weighting function method 权函数法
weighting function of filtration 滤波权函数
weighting material 加重材料;填充物
weighting method 加权法
weighting method of physical quantity 实体数量加权法
weighting network 加权网络;计权网络
weighting of hydrometer ;波减水比重计称重
weighting of slope 土坝边坡压重
weighting out 计价
weighting platform 秤台
weighting rail 测重机
weighting recorder 重量记录器;重量记录计
weighting rectification 加压矫正法
weighting statistical data 加权统计数据
weight in least square 最小二乘方的权
weight in vacuo 真空重量
weight in vacuum 真空重量
weight in water method 水中称重法
weight in wet base 含水重量
weight item 重量成分
weight jaw 重锤
weight kit 加重箱
weightless 失重的
weightless cohesive soil 失重黏性土(壤)
weightless fluid 无重量流体
weightlessness 失重状态;失重
weightlessness curve 失重曲线
weightlessness effect 失重效应
weightless spring 无重弹簧
weight lever 重锤杆
weight-lever regulator 秤杆调节器;重锤杠杆调节器;杠杆调节器;权杆调节器
weight-lift 杠铃
weight lifting room 举重室
weight liftings 举重器
weight limit 重量限度
weight limit sign 限重标志;重量限制标志
weight list 重量单
weight-loaded 加载重量
weight-loaded accumulator 重物锤式储能器;重力蓄力器;重锤式蓄能器
weight-loaded governor 载重调节器
weight-loaded regulator 重锤调节阀
weight loss 重量减轻;重量(的)损失;干耗;失重
weight loss by ignition 烧灼减量
weight loss ratio 重量损失率;重量损耗率
weight machine 重量计
weight mark 重量标志

weight masonry wall 承重圬工墙
weight matrix 加权矩阵;权矩阵
weight mean diameter 重量平均直径;重量平均粒径
weight measurement 重量量测
weight measuring device 测重装置
weight memo 重量单
weight method 重量法(加气混凝土配料);重量比配料法
weight mixing 重量配合;按重量混合
weight mix(ture) method 重量配料法
weight molality 重量摩尔浓度
weight molar concentration 重量摩尔浓度
weight moment of inertia 重量转动惯量;重量惯性矩
weightness 可称量性
weight normality 重量当量浓度;重量规度
weight number 加权系数;权数
weight-number ratio 重量数量比
weight of a measurement 测量的权
weight of a ship 船重
weight of attachment 附件重量
weight of blow 冲击重量;冲击力
weight of casing section 套管段重
weight of charge 装药重量
weight of compressor 压缩机重
weight of dam 水坝重量
weight of drilling mud 钻探泥浆重量
weight of drilling rig 钻机重量
weight of drill tool 钻具重量
weight of equipment 设备重量
weight of function 函数的权
weight of hammer 落锤重;锤重
weight of instrument 仪表重量
weight of lathe 车床重量
weight of leveling line 水准路线的权
weight of lump with unit volume 块比重
weight of machine 机重
weight of main machine 主机重量
weight of monkey 打桩锤的重量;打桩锤重(量)
weight of observation 观测值的权;观测权数
weight of one measurement 一次测权
weight of one section 单组重量
weight of pile 桩的重量
weight of pile hammer 打桩锤重(量)
weight of precipitate 沉淀重量
weight of rock and soil 岩土的重量
weight of rock in underlayer 垫层块石重
weight of rock mass 岩体重
weight of soil mass 土体重
weight of soil wedge 土楔体的重量
weight of tensor 张量的权
weight of the atmosphere 大气重量
weight of the bit 钻压;钻头压力
weight of train 列车重量
weight of unit volume 单位容重
weightograph 自动记录衡纵器;自动记录式称重仪;自记称重仪;自动秤
weightometer 自动称量记录装置;自动秤;自动称重仪;重量计;皮带秤;称量计
weight on bit 钻压
weight on drivers 主动轮重量
weight or measurement 重量吨或容积吨
weigh tower 过磅塔
weight parameter 重量参数
weight partition wall 承重隔墙
weight per axle 轴负重
weight per batch 每批重量;每盘重量
weight percent(age) 重量百分数;重量百分比(率);重量比
weight percentage content 重量百分含量
weight percent solids 固体重量百分率
weight per epoxy equivalent 单位环氧当量
weight per foot 每英尺重量(1 英尺 = 0.3048 米)
weight per gallon 每加仑重(1 加仑 = 3.785 升)
weight per horsepower 每马力重量(1 马力 = 735.5 瓦)
weight per package 每包重量
weight per pint 每品脱重量
weight per unit area 单位面积重量
weight per unit kelly 方钻杆单位重量
weight per unit of collars 钻铤单位重量
weight per unit of power 单元功率的质量
weight pit 重锤井

weight pocket 平衡锤盒;双悬窗的锤箱
weight polynomial 重量多项式
weight potential 重力势
weight-power ratio 重量动力比
weight proportional retarder 重敏缓行器
weight rate 重量流量
weight rate of flow 重量流率
weight rating 重量定额
weight ratio 重量比;体重比
weight ratio compensator 重量配比补偿器
weight ratio of constituent 成分之重量比
weight ratio of hydrogen to carbon 氢碳重量比
weight reciprocal 权逆;权倒数
weight recovery 重量回收率
weight reduction 重量减轻
weight regulator 重量调节器
weight-relation 重量关系
weight replacing tank 重量代换罐
weight residual approach 加权余法法;加权余量法
weight residual method 加权余法法;加权余量法
weight responsive car retarder 重敏缓行器
weight restriction 重量限制
weight-retained curve 筛余重量曲线
weight ring 重量环
weight rod 重锤杆
weight running average 加权移动平均值;加权移动平均数
weights and measures 计量制;度量衡
weights and measures act 重量及计量条例
weights and measures department 度量衡部门
weights and measures factory 度量衡厂
weights and measures law 权度法
weight-saturation method 岩芯饱和率测定法;岩芯饱和度测定法
weight saving 减轻重量;重量节省;重量减轻
weight-scale table 称量台
weight schedule 重量一览表
weight sensing 测重
weight sensitive retarder 重敏缓行器
weight sensor 重量传感器
weight setting plate 平衡锤固定板
weight shift 重量转移
weight shipped 装船重量
weight shortage 重量短少;减秤
weight sizer 重量分级机
weight snow-ga(u)ge 称雪器
weight sorter 重量分级机
weight sounding apparatus 重力触探仪
weight-space velocity 重量空间速度
weight specific heat 重量比热
weight spring 重锤弹簧
weight standard of dots and lines 点线表
weight station 秤站
weight strap 坠砣夹板
weight strength 重量强度
weight strength ratio 重量强度比
weight stripping ratio 重量剥采比
weight supporting capacity (钻塔)起重能力
weight-supporting roll 支重滚筒
weight tension balance 坠砣补偿器
weight thermometer 重量温度计
weight-to-capacity ratio 重量对容量的比
weight-to-horsepower ratio 重量马力比(1 马力 = 735.5 瓦)
weight ton 重量吨
weight tonnage 重量吨位
weight-to-power ratio 重量功率比
weight-to-volume ratio 重量体积比;重量容量比
weight-to-yield ratio 重量威力比
weight training 负重训练
weight training room 体力重量锻炼房
weight transfer adjuster 重量
weight transfer attachment 重量转移附属装置
weight transfer device 重量转移装置
weight transference 重量转移
weight transfer unit 重量转移装置
weight tray 配重箱;承载底板
weight type pressure retaining valve 重量保压阀
weight type regulator 重力型调节器
weight unevenness 重量不匀率
weight uniformity 重量均匀度
weight unit 重量单位;权单位;单位权
weight unknown 重量不明

weight utilization factor 重量利用系数
weight valve 重力阀
weight variation 重量差异
weight variation test 重量差异试验
weight variation tolerances 重量变异耐受性
weight vector 权向量
weight vehicle 重型潜水服
weight voltameter 重量电量仑计;重力(型)电量计
weight volume 重量体积
weight-volume relationship 重量体积关系
weight wide line stack section 加权宽线叠加剖面
Weigner method 威格纳法
Weigner sedimentation tube 威格纳沉降分析管
weigt balance 重量平衡
weilerite 砷钡铝矾
weilite 三斜砷钙石
Weinberg angle 温伯格角
Weinberg-Salam theory 温伯格—萨拉姆理论
Weining flo(a)tation cell 万尼格型浮选机
weinschenkite 高铁褐闪石
weir 堰;挡坎
weirage 水库
weir and lock 堰闸
weir board 堰板
weir box 堰板量水建筑物;溢流堰箱;堰箱;堰板量水箱
Weirchrome 镀铬薄板
weir cill 堰下槛
weir coefficient 堰流系数
weir crest 堰顶
weir crest control device 堰顶控制装置
weir crest control unit 堰顶控制装置
weir crest curve 堰顶曲线
weir crest depth 堰顶水深
weir dam 溢流坝;堰(流)坝
weir discharge 溢流堰排料
weir discharge coefficient 溢流堰流量系数
weir divider 溢流隔板
weir edge 堰缘
weir floor 堰底
weir flow 溢流量;堰流(量);过堰流量
weir flow meter 堰式流量计
weir formula 堰流公式
weir gate 溢流堰闸门;堰闸门;堰门
weir ga(u)ge 流量计;堰式流量计;堰流计;堰顶水位计
weir head 堰上水头;堰顶水头
weir height 堰高
weir inlet 溢流堰进水口
weir inlet spillway 溢流引水
weir jetty 潜丁坝
weir level 溢流堰水平
weir lip 堰前缘;堰唇
weirload 越流负荷
weir loading 溢流堰负荷;溢流率;溢流负荷量;堰(口)负荷
weir loading in sedimentation tank 沉淀池堰口负荷
weir loading rate 溢流堰负荷率
weir lock 堰闸
weir meter 溢流计
Weir method 韦尔溢流测定法;堰口测流法;堰测(定)法
weir nappe 堰顶(流下的)水帘;贴附水舌
weir notch 堰孔;堰槽;堰上缺口
weir of dry stone and timber work 木笼填石堰
weir of fascines 梢捆堰坝;柴捆堰坝
weir on sand 砂基堰
weir overflow rate 堰溢流率
weir penstock 堰式压力管道
weir pier 堰墩
weir plate 堰板
weir pond 堰(旁水)池
Weir process 韦尔脱蜡过程
Weir pump 蒸汽直接联动式活塞泵;韦尔氏泵
weir rate 堰口率
weir root 堰根
Weir's azimuth diagram 韦尔氏方位图
weir section 溢流堰段;堰口断面;堰段
weir shutter crest 堰闸顶
weir sill 堰底;堰(下)槛
weir skimmer 溢流撇油装置
Weir stabilizer 韦尔稳频器
weir station 堰址测站

weir tank 堰式罐;溢流罐
weir trough 溢流堰槽
weir-type applicator 堰式施胶器
weir-type baffle 堰式挡板
weir-type fish passway 堰式鱼道
weir washer 冲碱装置
weir waste 溢水道;溢弃水;堰溢流
weir with a free fall 自由溢流堰
weir with contraction 束缩堰;收缩堰
weir with free fall 自由溢流堰;自流式溢流堰
weir with inclined crest 斜顶堰
weir with linear discharge relation 有线性流量关系的堰坝
weir with lock 拦河坝;船闸堰;有船闸的堰
weir with overhead bridge 桥跨堰
weir with thin-wall 薄壁堰
Weirzin 镀锌钢
Weisbach-Darcy equation 韦斯巴契—达西方程
weishanite 围山矿
weissbergite 维硫锑铊矿
Weiss constant 魏斯常数;魏斯常量
Weissenberg effect 爬杆效应;魏森堡效应
Weissenberg method 魏森堡法
Weissenberg photograph 魏森堡图
Weissenberg camera 魏森堡照相机
Weissenhof siedlung 魏森霍夫居住住宅方案展览会(1927年在德国举行)
Weiss quadrilateral 韦斯四边形
weisstein 白粒岩;粒变岩
Weiss theory 外斯理论
Weisz ring oven 魏斯环炉
weize 敷设
Weizsacker's theory 魏茨泽克学说
Welch plug 油孔塞;气孔塞;孔塞
welcome mat 门口擦鞋垫
weld 焊缝;(金属的)平锻脚;煅接
weldability 可焊性;焊接性
weldability test 可焊性试验;焊接性试验
weldable 可焊的
weldable copper 可焊复合铜板
weldable steel 可焊钢;焊接钢
weldable structural steel 可焊的结构钢
weldable test 可焊试验
weld all around 围焊
weld-all-over 全周焊接
weld appearance 焊缝外观
weld assembly 焊接组件
weld backing 焊缝衬垫
weld backup 焊缝背负
weld bead 焊缝;熔接层;熔敷焊道;堆焊层
weld bead height 焊道高度
weld beading 焊瘤
weld bead length 焊缝长度
weld bond 熔合线
weld-bonding 黏接点焊;点焊黏结
weld bridge seam 桥接焊缝;搭桥焊缝
weld buildup 堆焊
weld cap 焊帽
weld centerline 焊缝中心线
weld centre line 焊缝中心线
weld collar 加强焊缝焊瘤
weld crack 焊接裂纹;焊缝裂纹
weld crater 焊口;熔池
weld crosswise 交叉焊接
weld decay (焊接后)敏化区腐蚀;焊接区晶间腐蚀;焊接侵蚀;焊接接头晶间腐蚀;焊缝腐蚀
weld defect 焊缝缺陷
weld delay time 焊接延迟时间
weld deposits cleaning 焊接熔敷金属清理
weld dimension 熔接尺寸;焊接尺寸
welded 焊接的
welded agglomerate 熔结集块岩
welded all round 沿周边焊接的
welded and expansion joint 焊接加焊接法
welded and high strength bolted steep bridge 栓焊钢桥
welded and rolled tube joint 焊胀
welded arch 焊接拱
welded area grating 焊接的井盖
welded armor 焊接甲板
welded article 焊接件
welded asphaltic-bitumen sheet(ing) 焊接沥青板
welded attachment 焊接附件
welded bar-mesh reinforcement 焊接钢筋网

welded base 焊接基座
welded beam 全焊梁
welded body 焊接车身
welded bond 焊接式钢轨接续线
welded bonnet 焊接阀盖
welded brass nipple 焊接黄铜嘴
welded breccia 熔结角砾岩
welded bridge 焊接桥梁
welded butt joint 对接焊
welded butt splice 对接焊拼接板;对焊接头
welded cast steel node 焊接铸钢接头
welded cathode 焊接阴极
welded chain 焊接链
welded chain cable 焊接锚链
welded clastic texture 熔结碎屑结构
welded column-girder connection 柱梁之间的焊接
welded connection 焊接连接;焊接联结;焊合
welded connection sheet pile 焊接板桩
welded construction 焊接结构
welded-contact rectifier 熔接整流器
welded corner joint 角焊接
welded cover plate 焊接盖板
welded defect 焊接缺陷
welded edge 搭接缝
welded edgewise 边缘焊接
welded element 焊接构件
welded eye 焊眼
welded fabric mat 焊制金属网
welded ferrous pipe 焊接铁管
welded fin tube 焊片式翅片管
welded fissure 焊接裂纹;焊接裂缝
welded fitting 焊接式管接头
welded flange 焊制法兰;焊接翼缘;焊接法兰
welded flat head 焊接式平盖
welded flat roof 焊接的平屋顶
welded flaw 焊接缺陷
welded frame 焊接框架;焊接骨架
welded frog 焊接辙叉
welded gasket 焊接垫
weld edge 焊缝表面边缘;焊缝边缘
weld edgewise 沿边焊接
welded girder 焊接大梁
welded grating 焊制格栅
welded head 焊接点饰
welded-hollow spherical node 焊接(空心)球节点
welded heat 焊接热量
welded hull 焊接船体
welded impeller 焊制叶轮
welded installation 焊接装配
welded joint 焊接缝;焊接接头;焊接连接;焊接联结;焊接节点;焊接缝;熔接头
welded joint coefficient 焊接接头系数
welded joints seal 焊接连接密封
welded joint strength test (钢筋)焊接接头强度试验
welded knee 焊接肘板
welded latticed construction 焊接花格结构
welded latticed girder 焊接的格构大梁
welded length 焊接长度
welded line 焊接线
welded liquefied petroleum gas cylinder 焊接式液化石油气瓶
welded long rail 焊接长钢轨
welded metal mesh 焊接金属网
welded module 焊接微型组件
welded neck flange 管子上焊法兰
welded nose 焊接节点
welded nose connector 焊接节点接头
welded obsidian 熔结黑曜岩
welded oil-storage tank 焊接油类储[贮]槽
welded-on 焊上的
welded on cast carbide 堆焊的铸造碳化钨
welded on end 焊制管封
welded-on flange 焊上的翼缘;焊上的翼板
welded on head 焊制封头
welded(on) joint 焊接接头;焊接接合
welded on the bottom of vessels 焊在容器底上的
welded pipe 焊制管;焊接管
welded pipe connect(ion) 焊管连接
welded pipe construction 焊管结构
welded pipe system 焊接管道系统
welded pipe tube 焊缝接管
welded piping 焊接管
welded pitchstone 熔结松脂岩

welded plate 焊合板
welded plate construction 焊接板结构
welded plate girder 焊接板梁
welded plate girder bridge 焊接板梁桥
welded plate-steel foot 焊接的钢板底座
welded process 焊接方式；焊接法
welded proclastic rock 熔结火山碎屑岩
welded product 焊接的产品
welded rail 焊接铁轨
welded rail joint 钢轨焊接头
welded reinforcement 焊接钢筋
welded reinforcement of nozzle opening 接管开孔的焊接补强
welded rigid frame 焊接刚性构架；焊接钢架
welded rigidity 焊接刚度
welded roof truss 焊接屋架
welded sample 焊接试样
welded scoria 熔结岩渣
welded seam 焊缝
welded section 焊接部分
welded sheet condenser 焊接板式冷凝器
welded shell course 焊接锅炉筒
welded shell ring 焊接锅炉筒
welded ship 焊接船
welded specimen 焊接试样
welded spherical node 焊接球节点
welded splice 焊接接头；焊接接合
welded stayed construction 焊接的支撑结构
welded steel 焊接钢
welded steel angle seating 焊接角钢支座
welded steel base 焊接钢底座
welded steel box-girder 焊接箱线大梁；焊接箱形钢梁
welded steel bridge 焊接钢桥
welded steel fabric 焊接钢筋网
welded steel fences 钢材焊接围栏
welded steel forging 熔焊钢锻件
welded steel frame(work) 焊接钢构架
welded steel pipe 焊接钢管
welded steel plate construction 钢板焊接结构
welded steel reinforcement 焊接钢筋骨架
welded steel space frame 钢焊接空间框架
welded steel structural frame 焊接钢构架
welded steel structure 焊接钢结构
welded steel tank 金属焊接油罐
welded steel tube 焊接钢管
welded structural shape 焊接结构型钢
welded structural steel 焊接结构钢
welded structure 焊接(的)结构
welded system 焊接(管道)系统
welded tank 焊接的箱柜；焊接储[贮]油罐
welded tensile reinforcement 焊接受力钢筋
welded together 焊合
welded truss 焊接桁架；焊接材架
welded truss bridge 焊接桁架桥；全焊桁架桥
welded tube 焊接管；焊缝管；电焊管
welded tube connection 焊接管连接；焊接管接头
welded tube galvanized 镀锌焊接管
welded tuff 熔凝灰岩；熔结凝灰岩；溶接灰岩
welded tuffaceous texture 熔结凝灰结构
welded turning rolls 焊接滚轮架
welded turret 焊合转塔
welded unit 焊接构件
welded vessel 焊接容器
welded wire 焊接线
welded wire fabric 焊接钢丝网
welded wire fabric reinforcement 焊接钢丝网配筋
welded wire lath 焊接钢丝网
welded wire lathing 焊接的金属网板
welded wire mesh 焊接钢丝网；电焊网
weld-end fittings 带斜面焊接端的配件；带斜面焊接端的零件
weld-end parts 带斜面焊接端的配件；带斜面焊接端的零件
welder 焊接装置；焊接工(人)；焊机；焊工；电焊机
welder cable 焊接电缆
welder performance qualification test 焊工技术鉴定试验
welder rectifier 电焊机整流器
welder's clipping hammer 尖锤；电焊工的整平锤；焊工锤
welder's equipment 电焊设备
welder's face mask 电焊工的面罩
welder's gauntlets 电焊(工的)长手套

welder's generator 电焊发电机
welder's gloves 焊工手套；电焊工手套
welder's goggles 焊工护目镜
welder's hand screen 焊工手持式面罩
welder's hand shield 焊工手持护目镜
welder's head screen 焊工头戴式面罩
welder's head shield 焊工护目(帽)罩
welder's health 焊工保健
welder's helmet 焊工面罩；焊工帽(罩)
welder's helper 焊工助手
welder's lifting platform 焊工升降台
welder's mark No. 焊工钢印号码
welder's metal fume fever 焊工金属烟尘热
welder's pneumoconiosis 焊工尘肺
welder's protective head mask 焊工的保护头罩
welder's spats 焊工护脚
welders spatter ease 焊接易飞溅液
welder test 焊工考核
welder with condenser 带电容焊机
welder with taps 抽头式焊机
weld flange connection 焊接法兰连接
weld flash 焊瘤
weld flush 焊缝隆起
weldforged 锻焊
weld-form factor 焊缝形状系数
weld-fusion line 焊缝熔合线
weld ga(u)ge 焊缝量规；焊缝卡规；焊缝检查规
weld heat input 焊接线能量；焊接热输入
weld holder 焊接架；焊接夹持架
welding 定位焊接；压实作用；固化；焊接(的)；熔接；熔化极惰性气体保护焊
welding allowance 焊接留量
welding alloy 焊条合金；焊接合金
welding and cutting combine 割焊两用机
welding and cutting torch 焊割两用气焊枪；焊割两用炬
welding apparatus 焊接工具；焊接装置；焊接设备
welding apron 焊工围裙；焊工的围裙
welding arc 焊接电弧；焊弧
welding arc analyzer 焊接电弧分析仪
welding arc viewer 焊弧观察器
welding arc voltage 焊接电弧电压；弧电压
welding area 焊接部面积
welding assembly 焊接部件；焊接构件
welding atmosphere 焊接气氛
welding backing 焊接衬垫
welding base metal 焊条金属
welding bead 焊缝
welding bead height 焊缝高度
welding bead length 焊缝长度
welding bell (焊管用)碗模
welding bellows 焊接波纹管
welding bench 焊接(工作)台
welding between pipe connect 管口和筒
welding blow lamp 焊炬
welding blowpipe 焊炬；焊接吹管；喷焊器
welding boat 电焊船
welding booth 焊工室；焊接工作间
welding bottle 焊接用气瓶
welding burner 焊枪；焊接喷灯；焊炬(喷嘴)
welding by both sides 双面焊(接)
welding by one side 单面焊
welding cable 电焊电缆；焊接(设备)电缆
welding cam (接触对焊机的)加压凸轮
welding capacity 可焊性
welding carbon 焊接碳棒
welding characteristic 焊接特性
welding circuit 焊接回路
welding clamp 电焊钳
welding code 焊接规程
welding coefficient 焊接系数
welding composition hopper 焊料斗
welding compound 焊药
welding condition 焊接条件；焊接规范；焊接工艺条件；焊接工艺参数
welding condition 焊接规范
welding connection 焊接连接；焊接联结；焊接合
welding connector 电缆夹头
welding contactor 电焊接触器
welding contractor 焊接承包商
welding control 焊接接触电焊断流器
welding crack 焊接裂缝；焊接裂纹；焊接裂缝
welding crew 焊接人员；焊工组
welding crystalline corrosion 焊缝晶间腐蚀

welding current 焊接电流
welding curtain 焊接幕帘
welding cut bit 焊割工具
welding-cutting tool 焊割具
welding cycle 焊接周期；焊接循环
welding defect 焊接缺陷
welding deformation 焊接变形
welding deposit 焊缝熔敷金属；焊敷金属
welding device 焊接装置；焊接设备
welding direction 焊接方向
welding drawing 焊接图
welding dribble (焊接时间的)金属熔滴或飞溅
welding dynamo 焊接发电机
welding electrode 焊条；焊接电极；电焊条
welding electrode hold 焊条夹
welding electrode production line 电焊条生产线
welding ends 焊接端
welding engineer 焊接工程师
welding engineering 焊接工程
welding equipment 电焊设备；电焊机；焊接设备
welding experiment 焊接试验
welding fabrication 焊接装配
welding factor 焊接系数
welding fixture 焊接夹具
welding flame 焊接焰；焊接火焰
welding flash 弧光灼伤
welding flux 助焊剂；焊药；焊接熔剂；焊剂
welding flux backing 焊剂垫
welding force 阻焊加压；电极(压)力
welding from one side 单面焊
welding fume 焊接烟尘
welding furnace 焊接炉；烧结炉
welding gang 焊接人员；焊接工(作)班；焊工组；焊工班
welding gap 焊件装配间隙
welding ga(u)ge 确定焊缝尺寸用样板
welding generator 焊接用发电机；焊接发电机；电焊发电机
welding generator with differential excitation 差激电焊发电机
welding generator with splitting pole 裂极式电焊发电机
welding glass 护目镜玻璃；黑玻璃；焊接保护玻璃
welding gloves 焊接(用)手套；焊工手套
welding goggles 焊接护目镜；焊工护目镜；焊工保护眼镜
welding grade argon 焊接级用氩
welding grade shield gas 焊接级保护气体
welding ground 工件引线；焊接地线；电焊地线
welding gun 焊烙铁；焊枪；焊接喷枪
welding gutter 焊管拉模
welding handle 焊条柄；焊条夹
welding head 焊头；烙铁头；焊(接)机头；烧焊枪
welding heat 焊接温度；焊热；焊接热(量)
welding heat input 焊接输入热量
welding helmet 焊接用安全帽；焊工帽(罩)；电焊帽
welding holder 焊接握把
welding hood 焊工罩面
welding hose 焊接(用)软管
welding impulse 电焊脉冲
welding inspection 焊接检查
welding inspection ruler 焊接检验尺
welding installation 焊接设施；焊接设备
welding interval 焊接周期；焊接时间(间隔)
welding in water 水中焊接
welding jig 焊接夹具；焊条夹钳
welding joint 焊接连接；焊接联结；焊接(接)头；焊接合；焊(接)缝
welding laboratory 焊接试验室
welding lamp 焊接灯
welding layer 焊层
welding lead 电焊引线
welding length 焊缝长度
welding line 焊缝
welding machine 自动焊机；焊(接)机；电焊机
welding machine with power factor correction 功率因数改善了的焊机
welding manipulator 焊接机械手；焊接操作者；焊接操作机；焊件支架
welding material 焊接材料
welding mesh 焊接金属网
welding mesh fabric 焊接网
welding metal 焊接金属；焊着金属
welding metal cracking 焊缝金属裂纹

welding metallurgy 焊接冶金
welding metal zone 焊接区
welding method 焊接方法
welding micrometer 检验卡规
welding mill 焊管机
welding motor generator 旋转式直流焊机;电动旋转式电焊机;电动焊接发电机
welding neck 对焊
welding neck fitting 对焊件
welding neck flange 对焊法兰;高颈法兰;焊颈法兰
welding nipple 焊接丝头;焊接短管
welding nozzle 焊嘴
welding of pipe in fixed position 定位焊管法
welding of tubes 炉管焊接
welding of tube-tube sheet 管子与管板氩弧焊接
welding on 焊上;焊接;镶嵌;焊合
welding on bottom 底焊
welding operation 焊接操作
welding operator 自动焊工;焊接操作者;焊接(操作)机;焊工
welding outfit 焊接设备;焊接配备;焊接机组;焊接工具
welding overhead position 仰焊(接)
welding parameter 焊接工艺参数
welding party 焊接人员
welding pass 焊道
welding paste 焊接药膏;焊接涂料;焊糊;焊膏
welding performance 焊接性能
welding period 焊接时间;焊接循环周期
welding pipe 焊接管
welding pistol 焊枪
welding plant 电焊车间;焊接装置;焊接车间;焊接厂
welding platform 焊接平台
welding pool 焊接熔池
welding pore 焊点
welding portion 焊接部分
welding position 焊位;焊接姿势;焊接位置;焊接部位
welding positioner 可动焊接工作台;焊接胎架;焊接翻转台
welding positioner roller 焊接定位滚轮
welding powder 焊药;焊熔剂;焊粉
welding power lead 焊接用输电线
welding power source 焊接电源
welding press 焊接压力机
welding pressure 焊接压力
welding primer 焊接打底涂料
welding procedure 焊接施工;焊接规程;焊接工艺(过程);焊接次序
welding procedure drawing 焊接施工图
welding procedure qualification 焊接程序评定
welding procedure qualification test record 焊接工艺鉴定试验记录
welding procedure qualification test specimen 焊接工艺资格考核焊样件
welding procedure sheet 焊接工艺卡
welding procedure specification 焊接工艺规程
welding process 焊接过程;焊接工艺;焊接(方)法;焊接程序
welding puddle 焊接熔池
welding quality 焊接质量
welding rate 焊接速度
welding rectifier 整流式焊机;焊接整流器;电焊整流器
welding regulator 焊接调节器;电焊调节器
welding requirements 焊接要求
Welding Research Council 焊接研究委员会
welding residual deformation 焊接残余变形
welding residual stress 焊接残余应力
welding resistor 焊接电阻器
welding robot 焊接机器人
welding rod 焊条;焊棒;电焊条
welding rod coating 焊条上涂药;焊条上的焊药;焊棒上的焊药
welding rod core 焊条芯
welding rod extrusion press 焊条自动涂抹机;焊条压涂机;焊条挤压机
welding rod holder 焊条(夹)钳;焊把
welding rod steel 焊条钢
welding rod with acidic coating 带酸性涂层的焊条
welding rod with alkaline coating 带碱性涂层的焊条
welding roll 焊接滚轮;焊滚;焊辊

welding rules 焊接规则
welding runner 焊接转枪
welding schedule 焊接工艺过程;焊接规范;焊接计划表;焊接工艺程序
welding screen 电焊遮光罩
welding seam 焊缝
welding seam area 焊接范围
welding-seam cleaner 焊缝清理机
welding seam number 焊缝编号
welding sequence 焊接顺序;焊接次序;焊接程序
welding set 电焊机组;焊接装置;焊接设备;焊接机(组);电焊装置
welding set with a surge injector 脉冲稳弧焊机
welding shop 焊接工场;焊接车间;电焊车间
welding simulator 焊接模拟器
welding slag 焊渣;焊接熔渣
welding sleeves 焊接用套袖
welding socket 焊接套管;焊接管套
welding solvent 焊接用溶剂
welding source 焊接电源
welding spats 焊接用护脚
welding spatter 焊渣
welding speed 焊接速度
welding spot 焊接部位
welding stainless electrode 不锈电焊条
welding station carrier 焊接机组运载车
welding steel 焊钢
welding steel tube 焊接钢管
welding stick 焊棒
welding strain 焊接应变
welding strength 焊接强度
welding stress 焊接应力
welding structure member 焊接构件
welding sub assembly 焊接部件
welding supplies 焊接工具用料
welding symbol 焊接符号;焊接代号;焊缝代号
welding table 焊接工作台
welding team 焊工组
welding technique 焊接技巧;焊接技术
welding technology 焊接工艺
welding temperature 焊接温度
welding temperature field 焊接温度场
welding terminal 焊机输出端
welding thermal cycle 焊接热循环
welding thermit 焊接(用铝)热剂
welding tilter 焊接翻转机
welding time 焊接时间
welding timer 焊接定时器
welding tin billet 焊锡棒料
welding tip 焊接喷嘴;电极工作端
welding together 焊接在一起
welding tongs 焊钳;焊接夹钳
welding tool 焊接工具;电焊工具
welding torch 焊枪;焊接喷灯;焊(接)炬;焊接吹管;熔接炬;气焊枪
welding torch pipe 焊枪管
welding tractor 焊接小车;焊车
welding transformer 焊接变压器;电焊变压器
welding tube 焊接管
welding unit 焊接装置;焊接机
welding variable 焊接规范参数
welding voltage 焊接电压;电弧电压
welding wheel 盘状电极
welding wire 焊条(钢丝);焊丝
welding wire machine 焊丝机
welding-wire reel 焊丝卷圈
welding with backing 垫板焊;衬垫焊
welding with flux backing 焊剂垫焊
welding with weaving 横向摆动焊接
welding work 焊工工作
welding worker 电焊工
welding workshop 焊接车间
weld-in nozzle 焊入式喷嘴
weld inspection 焊接检查
weld interface 焊接界面;焊缝界面
weld-interval timer 焊缝间隔计时器
weld is accepted 焊缝合格
weldite 白硅铝钠石
weld junction 焊接接头;熔合线;熔焊线
weld layer 焊接层
weld leg 焊脚
weld length 焊接长度
weldless 无焊缝的;非焊接的
weldless connection 非焊接式接头

weldless derrick 无缝吊杆
weldless drawn pipe 冷却无缝管;拉制无缝(钢)管
weldless drawn piping 冷却无缝管
weldless drawn tube 冷却无缝管
weldless drawn tubing 冷却无缝管
weldless fitting 非焊接式接头
weldless rolled steel 无缝型钢
weldless steel conduit 无缝钢管
weldless steel pipe 无(焊)缝钢管
weldless steel tube 无(焊)缝钢管
weldless tooled steel 无缝型钢
weld line 熔接线;熔合缝;熔接线;接缝线;焊线;焊缝(轴)线
weld machined flash 闪光焊
weld machined flush 削平焊缝;削平补强的焊缝
weld mark 焊接痕;焊痕;焊缝记号;焊缝符号;熔缝
weldment 焊成件;焊接装配;焊(接)件
weld mesh crate 焊接网格
weldmesh reinforcement 网状钢筋
weld metal 焊缝金属
weld metal area 焊接金属截面(积);焊接金属截面面积
weld metal buildup surface 金属堆焊表面
weld metal composition 焊接金属组成
weld metal cracking 焊缝裂纹
weld metal zone 焊接金属熔化区;焊接(金属)区
weld neck flange 焊接颈圈凸板;焊接颈圈法兰
weld neck valve 气焊用氧气瓶颈阀
weld nugget 焊点熔核
weld nut 焊装螺母
weldolet 韦都拉特焊接法(一种支管焊接法)
weld on 焊上;焊接
weld on bottom 底焊
weld-on tool joint 对焊钻杆
weldor 焊工
weld overlap (焊接的)重叠区
weld pass 焊接通道
weld penetration 焊透深度;焊接穿透度;焊缝熔深;焊穿;浸透;焊接深度;贯穿
weld pitch 焊距
weld plate 焊接板
weld porosity 焊缝气孔
weld positioner 焊接定位器
weld preheating 焊前预热;焊接预加热
weld preparation 焊缝坡口加工
weld puddleh 熔池
weld reinforcement 焊接补强;焊缝加强;焊缝补强
weld repairs 焊缝修补
weld ripple 焊缝波纹
weld root 焊根;焊接根(部);焊件剖口
weld root opening gap 焊缝根部间隙
weld root reinforcement 焊缝反面加强
weld rotation 焊缝转角
weld screw 焊接螺丝
weld sealing 焊封
weld seam 焊缝
weld seam pipe 焊缝管
weld shape 焊缝形状
weld shielding 焊接熔池保护;焊接区保护
weld-shrunk cylinder 包扎式圆筒
weld-shrunk multilayered cylinder 多层包扎式圆筒;包扎式多层圆筒
weld shut 焊合
weld size 焊接尺寸
weld slope 焊缝倾角
weld spacing 焊缝间距;焊点距
weld spatter 焊接飞溅;焊渣
weld strain 焊接应变
welds types 焊接类型
weld surface 焊缝表面
weld swell 焊缝隆起
weld symbol 焊缝符号
weld thermal cycle 焊接热循环
weld thickness 焊缝厚度
weld time 接(触)焊通电时间;焊接通电时间
weld to 焊上
weld toe 焊趾
weld toxic gases 焊接有害气体
weld trimmer 焊缝清理机
weld up 焊补;补焊
weld wheel conditioner 焊轮调节器
weld width 焊缝宽度
weld wire 包铜钢丝
weld with backing strip 用条状垫板的焊接

weldwood 特制胶合板
weld zone 焊缝区
welfare 福利
welfare agency 福利机关;公益机关
welfare analysis 福利分析
welfare building 生活间
welfare center 福利中心
welfare economics 福利经济学
welfare effects 福利效应
welfare expenses 福利费用;福利保健费
welfare facility 生活福利设施;生活福利建筑;福利设施;卫生福利设施
welfare function 福利职能;福利函数
welfare fund 福利基金
welfare hospital 福利医院
welfare office 福利办事处
welfare payable audit 应付福利费审计
welfare payments 福利性支出
welfare policy 福利政策
welfare premises 福利机构;福利单位
welfare program(me) 福利计划
welfare projects 福利项目
welfare quarter 生活间
welfare rent 福利租金
welfare service 福利服务
welfare standard 福利标准
welfare state 福利国家
welfare statism 福利国家论
welfare subsidiary fund 福利补助金
welfare subsidies 福利补助金
welfare system 福利体系
welfare work 福利事业
welfarism 福利主义
welinite 硅钨锰矿
well 炉缸;楼梯井;矿井;井孔,井(坑);浇口窝;水井
well-advised 深思熟虑的
well-aligned reach 平顺河段
well-aligned river 线形良好的河流
well-appointed 装备齐全的;配备齐全的;装备好了的;设备完善的
well-appointed dock 设备齐全的船坞
well-arranged 良好排列的
well arrangement 井的布置
well array 井排;井的位置
well assorted 配搭极佳
well atmosphere pollution 水井污染
well-atomized 雾化良好的
well auger 土钻;凿岩钻;开井钻;手推钻
well-balanced 平衡的;匀称的;良好衡的
well barge 敞口驳(船)
well base rim 凹形轮缘;凹陷式轮缘
well base tyre 凹缘轮胎;凹凸轮胎
well beam floor 井梁楼板
well-bedded 成层良好的;层理明显的;层次分明的
well-behaved 性能良好的;优质的
well-being 机器保持良好;福利;保持良好状态
well blowing out incident 井喷事件
well boat 带活鱼舱的渔船
well-bonded 黏结良好的;充分黏结(的);充分黏结
well bore 孔眼;井眼;井筒;井身
well-bore damage 井眼的损害
well-bore damage ratio 井筒损坏比
well borer 凿井工;钻井工;深眼钻机;钻井机
well-bore storage coefficient 井筒储存效应系数
well boring 钻井
well bottom 基土坚实;井底
well-bottomed road 底层坚固的路面;基层坚实的道路;基层坚固的道路
well brick 砌井砖
well-bucket 吊桶
well builder 筑井工人
well-burned 充分燃烧的
well-burned brick 烧透的砖
well-burned clinker 烧透的熟料;烧成熟料
well-burnt brick 硬砖;炼砖
well-burnt tile 烧透的瓦
well cap 井帽
well capacity 井的出水率;井最大产水率;井的出水能力;井的出水量
well car 元宝车;凹型车
well casing 井壁;井筒;井(口)管;竖井井管;孔壁管
well casing pipe 井壁管
well casing starter 钻井套管鞋
well cavity 井腔;井穴

well cellar 圆井;方井
well-cemented rock 胶结良好的岩石
well cementing 胶结良好的;井台胶结凝合
well chamber 泵室;井室;检查井(室)
well characteristic 井孔动态;井孔特性
well-characterized carboxylic acid 完全特性化羧酸
well checker 井穴检验人;井穴检验器
well cleaning 井穴清洗;洗井
well cleaning trough the agency of pumping and compressing of pump 水泵抽压洗井
well cleaning with acid 酸洗井
well cleaning with air compressor 空压机洗井
well cleaning with liquid carbon dioxide 液态二氧化碳洗井
well cleaning with machine 机械洗井
well cleaning with perforator 冲孔器洗井
well cleaning with piston 活塞洗井
well cleaning with poly sodium phosphate 多磷酸钠洗井
well cleaning with solid carbon dioxide 干冰洗井
well-closed container 密闭容器
well-collimated beam 高准直光束;直线状注流
well column 钻井支柱
well completion 完井;成井【岩】
well completion method 成井方法
well completion process 完井过程
well completion system 完井系统
well completion technique 完井技术
well completion technology 完井工艺;成井工艺
well conditioned 情况良好的
well-conditioned triangle 图形很好三角形
well conditioning 修井;油井准备
well cone of influence 井水影响漏斗;井水降落漏斗
well-considered 考虑周到的
well construction 井(身)结构;井的建造;水井结构
well construction plant 筑井设备;打井设备
well construction rig 筑井机具;筑井设备;打井设备
well construction tool 筑井工具;打井工具
well construction work 筑井工作;打井工作
well-control equipment 自喷装置
well-control manifold 井口控制管汇
well core 钻井岩芯;岩芯;钻孔岩芯
well-core analysis 钻孔岩芯分析
well cost 成井费
well counter 检测波射性流体的计数器
well crib 井框
well crib pattern 井字形花纹
well-crystallized 结晶良好
well curb 井圈;井栏
well curbing 井壁支护板;基坑支托板;井壁挡土板
well-cuts 油中杂质
well cuttings 钻孔岩屑
well deck 井型甲板;凹甲板
well deck barge 凹甲板驳
well decked vessel 凹甲板船
well decker 井甲板船;凹甲板船
well deck ship 井甲板船;凹甲板船
well deck vessel 凹甲板船
well-defined 轮廓分明的;良定义的;界线分明的;明确定义的
well-defined beam 带明显边缘的线束
well-defined crystal 完好晶体
well defined feature 明置地物;明显地物
well-defined function 严格定义的函数
well-defined horizons 土层界限
well-defined notion 严格定义的概念
well-defined peak 轮廓分明的峰态
well-defined population 界限分明的总体
well-defined set 良定集合
well deflection 钻孔弯曲;井斜
well deflect standard 井斜标准
well-deformed wire 规律变形钢丝
well demarcated margin 边缘清楚
well depth 井深
well depth of water yielding at well mouth 孔口涌水时的孔深
well-designed 精心设计的;设计完美的;设计妥善的;设计良好的
well-developed 发展良好的;发育良好的
well-developed cyclone 成熟气旋
well diameter 钻孔直径;井径;井的直径
well-differentiated 高度分化的;分化良好的
well digger 掘井人;钻工;凿井机

well-digging clamshell 圆井抓斗
well distinguish 井别
well-distributed 分布得很好的
well documented model 已获得良好记录的模型
well drain 井排(水);井式排水;井沟;井点排水;排水
well drainage 井排水
well drainage area 井的排水面积
well-drained 排水良好的
well-drained soil 畅水土
well-drained stream basin 排水良好的流域;水系发育的流域
well-drain point 排水井点
well drawdown curve 井的抽降曲线
well drill 凿井钻;钻(井)机;井钻;打井机
well driller 挖井工;挖井机;打井钻头
well drilling 凿井;钻井
well drilling contractor 钻井承包商
well drilling sewage 钻井污水
well drilling technique 钻井技术
well drilling technology 成井工艺
well-drilling with bentonite 泥浆钻进法
well-drill method 钢绳冲击钻进(方)法
well-drill sample 未破坏岩芯;完整钻样
well dry 井水枯竭
well ear 凹底平车
well edge 井口;井边
well effluents 油气井产物
well electrode 井管
well equation 钻井公式
well-equipped 设备完善的;设备齐全的
well-equipped dock 设备齐全的船坞
well eruption 井喷
well-established 大家公认的;非常明确的
well face 油井底
well-fed basin 补偿盆地
well-fed graben 补偿性地堑
well field 井群;井田;井区;井场
well field name 井田名
well field structure 井田构造
well figures 井下(测量)数据
well filter 滤井;井过滤器
well-fired 烧透的
well-fired brick 烧透的砖
well-fitting 正合适的
well flooding 井水漫顶;井漫顶;分水井
well flow 井流量
well flow index 油井生产率
well flushing 洗井
well flushing by piston 活塞沉井
well-focused beam 强聚焦注流;高聚焦束
well for drinking 饮用水井
well for industrial use 工业用水井
well for irrigating 灌溉水井
well-formed 结构良好的;合适
well-formed formula 命题逻辑式
well-formed program(me) 构造良好的程序
well-formed sequent 成形的相继式
well-formed valley 成形河谷
well formula 钻井公式
well formula in step flow rate 阶梯流量水井公式
well formula of leaky recharge 有越流补给的水井公式
well formula of steady state flow 稳定流水井公式
well formula of unsteady flow 非稳定流水井公式
well formula of water-level recovery test 水位恢复试验的水井公式
well formwork 沉井模板
well for passenger train hydrant 客车给水栓室
well for waste disposal 排污井
well foundation 井筒基础;沉井基础
well-founded 基础坚固的
well-founded function 整序函数
well-founded set 整序集;良基集
well-founded-set method 整序集法
well fracturing 油层水力压裂;钻井水力压裂;井围扩碎法
well function 井函数
well function of leaky system 越流系统井函数
well function of matching point 配合点的井函数值
well ga(u)ging 油井产量测定
well gone to water 井出水
well grab 井内抓具

well-graded 颗粒级配良好的;良好级配的;级配良好的;分选好的
well-graded aggregate 良好级配的集料;良好级配的骨料;级配良好的集料;级配良好的骨料
well-graded gravel 良好级配的砾石;级配良好的砾石
well-graded grinding media 级配良好的研磨体
well-graded sand 良好级配的砂;级配良好的砂
well-graded soil 级配良好的土
well grading 良级配
well grain size distribution 良好级配
well-grate 炉格
well grounded 有基础的
well group pumping 井组取水
well head 源泉;井(水)源;井口;吸油井管
well-head assembly 井口装置
well-head cellar 水下井口装置
well-head connector 井头连接器
well-head control device 井口控制
well-head control unit 井口控制
well-head control valve 井口控制阀
well-head disassembling 折井口
well-head elevation 井口高程;井口标高;地面海拔
well-head equipment 井口设备;井口装置
well-head fittings 井口装置配件
well-head flowing pressure 井口喷流压力
well-head fracturing manifold 井口水力压裂管线
well-head gas 井口气(体);套管头气体
well-head installing 装井口
well-head plant 井口设备
well-head pressure 关井井口压力;井口压力
well-head price 井头价格
well-head protection 井水源保护
well-head protection area 井水源保护区
well-head protection area boundary 井水源保护区边界
well-head reentry apparatus 重返井口装置(海洋钻井水下钻具)
well-head re-entry system 井口再入系统
well-head retrieval apparatus 重返井口装置(海洋钻井水下钻具)
well-head sampling 取井口样
well-head setup 井口装置
well-head system 井头系统
well-head tap 井口短节
well-head temperature 井口温度
well-head tool 井口工具
well hole (升降机)井道;井底大仓;深入孔;升降机井道;泉井;楼梯井;垂直大炮眼
well hole car 凹型车
well hole drilling 井孔钻进
well hook 井内打捞钩
well hook up 井口装置
well horizontal departure 钻井水平位移
well house 井亭;井房
well hydraulics 管井水力学
well hydrograph 井水位曲线;井过程线
well-identifiable 容易辨认的
well inactive 静止井
well in aquifer with free water surface 井中自由水面无压
wellin davit 弧齿式吊艇柱
well infiltration area 井渗水面积
Wellington boots 安全靴
well injection 深井注入法
well-injection method 注井法
well in operation 生产井
well intake 井进水口;井的进水孔口
well interference 井群干扰;井(的)干扰
wellin type 弧齿式吊艇柱
well irrigation 井灌
well-jointed 节理发育的
well-keyed structure 锁结良好的结构
well kick 轻微井喷
well kicked off natural 天然喷油井
well killing technique 压井技术
well-knit surface 结实的路面;密实的表面;很好成型的路面
well-known 著名的;公认的
well-known architecture 良好的体系结构
well lining 井壁封套;井衬套
well location 井位
well log 钻井剖面;钻井记录;测井曲线;测井记录;测井

well log columnar section 钻井柱状剖面图
well-log geophone 测井检波器
well logging 测井记录;测井
well-logging probe 测井探头
well loss 井流水头损失;井的水头损失
well lowering 钻井;凿井;井筒下沉;井的下沉
well-lubricated 滑润的
well-made 做工好的
well-manage 经营良好
Wellman-Lord process 韦尔曼—劳德法;亚硫酸钠法烟气脱硫
Wellman's method 韦尔曼法
well-marked 明显标出的
well masonry 井垡工
well-measure line 井口深测绳
well-measurement 井下测量
well-measure wire 井口深测绳
well measuring device 测井装置
well-mixed estuary 混渗型河口
well-moderated 充分慢化的
well mouth 井口
well name 井名
well naming 钻井定名
well number 水井数目;井号
well of complete penetration 完整井
well-off 停井;供应充足的;经济宽裕;富裕的;富有的
well-off stand pipe pressure 关井立管压力
well of one diameter 上下等径的井
well of partial penetration 非完整井
well of shore 充分离岸
well-ordered 安排很好的
well-ordered principle 良序原则
well-ordered set 正序集
well-ordering principle 正序原理
well out 涌出(水、油)
well out of control 无控制井;敞喷井
well-paid 高工资的
well pattern 井网
well pattern dead area 井网死区
well pavilion 井亭;井上亭子
well perforator 射孔器
well performance 油井动态
well performance testing 油井性能试验
well permit 钻井许可证
well pipe 井筒;井管;井壁管
well piping 井管
well-planned 周密计划的;计划得好的
well platform 油井平台;钻井平台
well point 井点
well-point back pumping 井点回灌
well-point dewatering 井点排水;井点降水
well-point dewatering method 井点降水(方)法
well-point dewatering system 降低水位的井点系统;井点排水系统
well-point drainage 井点排水;井点抽水
well-point drainage installation 井点排水装置
well-point dry 井点排水
well-point excavation 井点开挖法
well-point in series 成列井点;井点串联
well-point method 井点(排水)法;井点降低地下水位法;井底法
well-point method of dewatering 井点排水法
well-point method of excavation 井点排水开挖法
well-point method of pumping 井点排水法;井点降水(方)法
well-point pump 井点泵
well-point pumping system 井点排水系统;井点抽水系统
well-point system 井点(排水)系统;轻型井点
well-point unwatering 井点排水
well-point works 井点工程
well-preserved 保养良好的;保养很好的
well pressure 井压力
well production 井的产量;生产井
well program(me) 钻井方案
well-proportioned 相当均匀的;良好配比的;配合良好的
well-proportioned concrete 配合良好的混凝土
well-protected harbo(u)r 掩护良好的港口;掩蔽良好的港口
well-proven 充分证明了的
well pulling 油井起套管
well pulling machine 修井设备

well pump 井泵
well pumped by suction 吸力泵井
well pumping 抽油井
well pumping test 井抽水试验;抽水试验
well radioactivity logging 放射性测井
well radius 井身半径;井眼半径;井筒半径;水井半径
well rating 油井额定生产能力
well reconditioning 井的重建修整
well record 钻井记录;井(的)记录
well reef 壁伏礁
well-refined 精炼良好
well-regulated 管理良好的
well rejuvenating 恢复钻井产量工作
well repair 修井
well resistance 水井损失
well-rig 打井设备;凿井机(器);打井机器;打井机具;钻井机(架)
well rig(ging) 钻井设备
well ring 井圈
well-roaded 筑有良好道路的;路况好的;道路设施完善的
well roads (建筑物的)井道
well rock mass 好的岩体
well-rounded 滚圆的;极圆的
well rounded shape 浑圆状
well running 运转正常
well sample 井样
well sampling 钻井取(土)样;井下取样
well scintillation counter 井式闪烁计数器
well screen 过滤器;井筒滤管;井(管)滤网;井管防砂罩
well seepage area 井渗流面积
well set on packer 装封隔器的井
well shaft 竖井;井筒;井身
well-shaped 很好修整过的
well-shaped counter 流体孔道式计数管;井式计数管
well-shaped sink hole 井状落水洞
well-sheltered harbo(u)r 掩护良好的港口;掩蔽良好的港口
well-sheltered place 掩护良好的地方;掩蔽良好的地方
well shooting 井下爆炸;井喷;井底爆炸;井喷破;地震测井
well shooting record 地震测井记录
well shuttering 筑井模架
well sinker 油井钻工;钻井工;凿井工
well sinking 井的下沉;沉井;井筒下沉
well site 井位
wellsite calibration 现场刻度
wellsite calibrator 现场刻度器
well size standard 井径标准
well slot 船井(海洋钻探船)
well-sorted 级配良好的;分选好的
well-sorted grains 分选良好的颗粒
well spacing 井的布置;井距
well spacing chart 井距图
well spring 井泉;泉源
wells spacing 钻井间距
well stimulation 井的增产
well-stocked 存货丰富的
well straightening 井身弯曲矫正
well strainer 钻井过滤器;井滤网;网式(过)滤器
well stream 流入井内液体
well structure 管井结构;井身结构;优良结构
well-sump type pumping plant 深井式抽水站
well surveyed chart 精测海图
well surveying 测井;井测
well surveying company 测井公司
well system 井系系统
well temperature 井温(度)
well-tempered 拌和得好的
well-tempered enterprise 经过考验的企业
well test 试井
well thermometer 井温计
well-thinned 稀薄得很的;稀释好的
well-timbered 用木材撑牢的;用木材加固的
well-to-do 经济宽裕;富有的
well top 井口
well-trained 良好训练的
well-trained crew 训练有素的班组人员
well-trained gang 训练有素的班组人员
well-trained party 训练有素的班组人员
well trap 存水弯

well-traveled 交通量大的;大交通量的
well treatment 井的处理
well tried 经过反复试验后证明
well-tried export quota technique 久经试验的出口限额技术
well tube 孔壁管;井(壁)管
well tube filter 油井管滤器;滤管
well tubing 井管
well turned 车削得好的
well type 钻井类别
well-type counter 井型计数器
well-type furnace 井式炉膛
well-type manometer 杯式压力表
well-type scintillation counter 井型闪烁计数器
well-type scintillation crystal 井型闪烁晶体
well under control 自喷油井管
well up 涌出(水、油)
well vent 井式排气道
well-ventilated 通风良好的
well-verified 充分证明了的
well vertical section 井的铅直剖面
well wagon 元宝车;凹底平车
well wall 竖井墙
well-wall potential function 井壁势函
well washing 洗井
well water 井水
well water level 井水水位
well water pollution 井水污染
well water quality 井水水质
well water supply 井水供应
well-weighed 经慎重考虑的
well winch 钻井绞车;矿井绞车
well with different diameters 吊管井
well-wooded 多森林的;森林资源丰富的
well workover 修井
well works 钻井工程
well yield 井流量;井的涌水量;井的出水量;井的产量;地下水水源地
weloganite 水碳锆锶石
well-point drain 井点降水
Welsh arch 有楔形拱心石的手拱;威尔士拱
Welsh brick 威尔士硬砖
Welsh cell 威尔士型充气式浮选机
Welsh groin 直交高低穹式交叉拱;交插穹棱;威尔士穹顶肋
welshite 锑钙镁非石
Welsh vault 交插穹顶;威尔士穹顶
welt 盖缝棕;盖缝条;金属板咬口;衬板;贴边;铅板(间)接口;盖缝木条
welt and seam rubbing machine 沿条和沿条缝擦平机
welt beating machine 沿条敲平机
welt beveller 沿条片边机
welt bevelling machine 沿条片边机
welt brush 沿条清洁刷
welt butter 沿条片茬机
welt butting 沿条片茬
welt coil 沿条卷
welt cutter 沿条裁断器
welt cutting 沿条切割
welt-cutting machine 沿条裁断器
welted drip 折叠滴水线;卷边滴水;摺边;搭接边
welted edge 折边;搭接缝;贴边;搭接边
welted insole 沿条内底
welted insole channel 沿条内底槽
welted insole channeller 沿条内底开槽机
welted insole mo(u)lding 沿条内底成型
welted joint 咬口接缝
welted last 沿条楦
welted nosing 贴边;装饰边(用于室内装饰和家具套);(金属薄板屋面的)咬口;折叠突棱;滚边
welted seam 卷边接缝;嵌条接缝
welted seam roof cladding 折边屋顶盖面
welted seam spacing (贴边的)折缝间距
welted standing seam (贴边的)垂直接缝
welt end skiving 沿条片茬
welter 镶边机
welter screen 翻滚筛
welt feather bending machine 沿条边敲平擦亮机
welt feathering 沿条片边
welt groove 沿条槽
welt groover 沿条刻槽机
welt-grooving machine 沿条刻槽机
welt hammer 沿条敲平锤

welt holder (缝沿条机的)沿条夹持器
welting 滚边;装饰边(用于室内装饰和家具套);贴边
welting strip 咬口金属带;折叠金属带;盖缝条;(屋顶与墙的)压缝条
welt inseaming machine 沿条内缝机
welt pattern 沿条样
welts and furrows 陡壁沟槽型
welt-skiving machine 沿条片边机
welt-splitting machine 沿条修平机
welt straightener 沿条矫直机
welt substance 沿条厚度
welt-tempering machine 沿条加湿和阴干装置
welt trimming 沿条修边
welt wetting 沿条浸湿
welt wheel 沿条花纹压轮
welt wheeling machine 沿条压道机
welt width 沿条宽
Welwyn Garden City 魏尔温田城市(英国伦敦)
Wemco cyclone 威姆科型旋流器
Wemco flo(a)tation cell 威姆科型浮选机
wend as course 保持航向
wenkite 钡钙霞石
Wenner arrangement 温纳排列
Wenner method 温纳(电阻测量)法
Wentworth classification 温氏分类
Wentworth scale 温氏泥沙粒度分级标准;温氏分级表【地】
Wenzel's blue 文泽尔蓝
Werfenian stage 赛特阶
wermlandite 羟铝钙镁石
Werner band 魏纳谱带
Werner complex 配位化合物
wernerite 中柱石;方柱石
Werner line 维纳谱线
Werner mixer 维纳混合机
Werner's projection 维纳地图投影
Werner's theory 维纳理论
Werner system crusher 维纳式粉碎机
Werner System earth resistivity method 维纳土地电阻探测法
werssite 黑碲铜矿
Wertheim effect 威德曼效应
Wesco pump 黏性泵;摩擦泵
West African bass 西非洲酒椰纤维
west African Copal 西非(硬)树脂
West African fever 黑水热
West African jute 西非黄麻
West African padauk 西非紫檀
West Asian architecture 西亚建筑
west Australian basin 西澳大利亚海盆
west Australian current 西澳大利亚海流
westblock 西端(多层)廊屋(德国与荷兰的塔楼教堂)
west bound 向西航行
west by north 西偏北
west by south 西偏南
west cardinal mark 西界标;方位标
west coast climate 西岸气候
west-coast marine climate 西岸海洋性气候
West Coast of Africa 非洲西海岸
west country whipping 反手结编扎法
west elevation 西立面
west elongation 西距角
west end (中世纪教堂的)西端
west end tower 西端塔楼
wester 西风;转向西面
Westergaard theory (设计混凝土路面的)威士卡德理论
westerlies 西(风)带
westerlies rain belt 西风多雨带
westerly 西风
westerly belt 西风带
westerly current 西风气流
westerly drift current 西风吹流
westerly limit 西图廓
westerly trough 西风槽
westerly type 西风型
westerly wave 西风波
western 西风的
western amplitude of the sun 日没时的幅角
western architecture 西方建筑
western balsam 巨冷杉
western Byzantine 西方拜占庭(建筑)格式
western Byzantine style 西方拜占庭(建筑)风格

western cargo hook 开口吊货钩
Western-Cathay tectonic zone 华西构造带
western cedar 西方桧柏
Western Central Atlantic Fishing COmmission 西中太平洋渔业委员会
western choir 西方教堂唱诗班席位
western classical garden 西方古典园林
western coralbean 西部刺桐
Western East China Sea depression region 东海西部拗陷地带
Western Equatorial countercurrent 西赤道逆流
western equatorial Pacific 赤道西太平洋
Western European Union 西欧联盟(北大西洋公约组织)
western facade 西方式(房屋)正面;西部(房屋)正面
western food restaurant 西餐馆
western frame 平台式木框架
western framing 平台式框架;西部式框架
western gallery 西部走廊;西部外廊
western hemisphere 西半球
western hemlock 异叶铁杉;西部铁杉
Western Himalaya tectonic segment 西喜马拉雅构造段
western interior bivalve endemic center 西内双亮类繁盛中心
western jackpine 扭叶松(北美西部)
western juniper 西方桧柏
western larch 西部落叶松(美国)
western lubricating oil 西部润滑油(美国加利福尼亚润滑油)
western method 铺浆砌砖法;西部砌砖法
western ocean plant 绳网吊索
Western Pacific geosyncline system 西太平洋地槽系
western pediment 西部山墙;西部柱廊上三角形或弧形檐饰
western pine (美国和加拿大的)部黄松
western porch 西部门廊
western portal 西侧入口;西侧门座
western quire 西端唱诗班席位
western red cedar 乔柏属;落基山柏木材;西部红(雪)杉
western side 西面;西侧
western spruce 西部云杉
western style 西方风格
western suburbs 西郊
western tamarack 西部落叶松(美国)
western tower 西部塔楼
western transept 西部建筑翼部
western white pine 西部山地白松(美国)
westerveldite 砷钴镍铁矿
West-European inland waterway classification scale 西欧内河流分级标准;西欧内河航道分级标准
west-facing wall 西(面)墙;西侧墙
west-facing window 面西的窗;西窗
Westfalian(stage) 威期特发利阶【地】
West Fork terrane 威斯特福克地体
west Gobi basin 西戈壁盆地
west hemibeam 西半球波束
west hemispheric beams 西半球波束
West Indian ebony 西印度乌木
West Indian mahogany 西印度红木
west Indian ocean 西印度洋
West Indies architecture 西印度群岛建筑
westing 向西航行;西行航程;西进;偏西距离
Westing-arc welding 西屋电弧焊(惰性气体保护金属极弧焊)
west inland basin 西内部盆地
west lake 莲花
Westland-Irving balance 内封补偿
Westleton beds 威斯特莱顿层
west longitude 西经
west Melanesian trench 西美拉尼西亚海沟
Westmorland slate 威斯特摩兰石板(瓦)
west northwest 西西北;西北西;北西西
Weston pulley block 差别滑轮组
Weston's differential pulley 韦斯顿差动滑轮
Weston standard cell 韦斯顿标准电池
west orientation 西向
west Pacific molluscan realm 西太平洋软体动物地理区系
west pacific ocean 西太平洋

Westphal and Mohr balance 韦斯特法尔及莫尔比重天平
Westphal balance 韦斯特法尔比重秤
Westphalian(stage) 韦斯特法尔亚阶
westphalt 用沥青粉拌和的冷铺沥青混合料
Westphal(type) balance 韦斯特法尔比重天平
West pile 威斯德桩
west point(of the horizon) 西点
west Siberian basin 西西伯利亚盆地
west southwest 西西南;西南方
West Virginia Spruce 西弗吉尼亚红云杉
west wall 西墙
westward 向西(的);西方;西部
westward course 西行航线
westward drift of geomagnetic field 地磁场西向飘移动
westward drift rate 西向漂移速率
west wind belt 西风带
west wind drift 南极绕极(海)流;西风漂流;西风(吹)流
west winds 西风
westwork (仿罗马式教堂的)西屋
West Wovaya Zemlya trough 西新地岛海槽
wet 润湿的;湿的;潮湿的
wet abrasion resistance 耐湿磨性
wet abrasion tester 耐洗刷试验机
wet abrasive blasting 湿喷砂
wet abrasive blasting machine 湿喷砂机
wet abrasive paper 水砂纸
wet adhesion 湿黏性
wet adhesion test 湿附着试验
wet adhesive bonding 湿黏合
wet adiabat 湿绝热(线);饱和绝热线
wet adiabatic 湿绝热的
wet adiabatic change 湿绝热变化
wet adiabatic instability 湿绝热不稳定(性)
wet adiabatic lapse rate 湿绝热直减率
wet adiabatic temperature difference 湿绝热温度差
wet-aggregate process 湿骨料搅拌;湿集料搅拌
wet air 高湿度空气;湿空气
wet air compressor 湿空气压缩机
wet air duct 湿气管
wet air oxidation 湿空气氧化(法)
wet air oxidation of sludge 污泥的湿空气氧化
wet air oxidation process 湿式空气氧化法
wet air oxidation system 湿(式)空气氧化系统
wet air oxidation technology 湿空气氧化技术
wet air oxidation unit 湿空气氧化装置
wet air pump 湿气泵
wet analysis 水下选法;(湿)(法)分析
wet-and-dry batch concrete plant 干湿配料混凝土拌和机
wet-and-dry bulb hygrometer 干湿球湿度表
wet-and-dry bulb recording hygrometer 干湿球自记湿度计
wet-and-dry bulb thermometry 干湿球温度测定法
wet-and-dry bulb hygrometer 干湿球湿度计
wet-and-dry bulb thermometer 干湿球温度计
wet-and-dry sand paper 干湿砂纸
wet-and-dry screening 湿干联合筛分法
wet-and-dry test 干湿试验
wet arcing length 湿弧距离
wet area 潮湿地区
wet ashing 湿(法)灰化
wet ash removal 湿法除灰
wet assay(ing) 湿法分析;湿鉴定
wet-autoclaved 湿法压炼处理的
wet autogenous mill 湿式无介质磨;湿法自磨
wet automatic sprinkler system 湿式自动喷水灭火系统
wet avalanche 湿土堆雪崩;湿(土)崩坍;湿雪崩
wet-back boiler 湿背式锅炉
wet baffle chamber 湿挡板室
wet bag isostatic pressing 湿袋法等静压成型
wet ball grinding mill 湿式球磨机;湿球研磨机
wet ball mill 湿式球磨机;湿球磨
wet barrel tumbling 湿法转筒抛光
wet base 湿基
wet basin 湿(船)坞;封闭式港池;有闸港池;港池
wet basis 连湿计算;湿基准;按湿量计算
wet batch 湿(配)料
wet batched rating 混凝土湿料计盘
wet batch hopper 混凝土出料斗

wet batch method 湿料运送法;湿料计盘法;(混凝土)湿料分批法
wet batch rating 湿料计盘;湿法搅拌批量
wet batch sample device 混凝土取样设施(拌和楼)
wet battery 湿电池组
wet beating 湿打浆
wet blacking 黑(色)涂料;碳素涂料
wet blast 湿式喷砂;湿喷砂;湿法喷砂;(用含有磨蚀剂的水)湿喷
wet blast cleaning 湿法喷丸清理
wet blasting 蒸汽空气鼓风;水砂抛光(处理);湿鼓风;湿法清理;(用含有磨蚀剂的水)湿喷
wet blasting process 湿喷砂法
wet blast mo(u)ld cleaning 喷湿砂洗模
wet blending 湿搅拌;湿混;湿掺和
wet blue 湿铬鞣革
wet board former 湿抄机
wet board machine 湿式纸板机
wet bonding 湿法黏合
wet bond strength 湿黏合强度
wet boring 湿式凿岩;湿法钻进
wet bottom boiler 液态排渣锅炉
wet bottom boiler slag 湿底炉渣
wet bottom furnace 液态排渣炉膛;湿底燃烧炉
wet bottom gas producer 水封式煤气发生炉
wet bottom precipitator 湿底电集尘器
wet bottom producer 湿底发生炉
wet box 湿气箱
wet braking distance 湿跑道刹车距离
wet brick 湿泡砖
wet broke 湿损纸
wet buckling 湿曲率
wet bulb 湿球
wet-bulb depression 干湿球温度差;湿球温降;湿球温差
wet-bulb globe temperature 湿一黑一干球温度指标;湿球温度
wet-bulb potential temperature 湿球位温
wet-bulb temperature 湿球温度
wet-bulb temperature measuring method 湿球温度测量法
wet-bulb thermometer 湿泡温度计;湿球温度计
wet-bulb thermometry 湿球温度法
wet bulk 湿(态)容积
wet bulk density 湿重度
wet bulk depression 湿球温差
wet bursting strength 湿纸耐破强度
wet caisson 湿沉箱
wet cake 湿泥饼
wet calender stack 湿式光泽机
wet calorific power 湿热值
wet calorific value 湿热值
wet cap 喷水火花捕集器
wet cap collector 湿式除尘器;湿法吸尘器
wet carbonization 湿碳化
wet cargo 液体货(物);湿货
wet cast pipe 湿浇注管
wet cast process 湿浇制法
wet cell 湿电池
wet-cell caplight 湿式矿灯
wet-cell type barge carrier 湿舱型载驳货船
wet-cemented 湿板胶接;湿胶合的
wet cement grout 水泥薄浆
wet cement paste 稀水泥浆
wet centrifugal dust scrubber 湿式离心除尘器
wet cheek 滞颊
wet chemical analysis 湿(法)化学分析
wet chemical etching 湿化学腐蚀
wet chemical method 化学湿选法
wet chemical process 湿化学法
wet chemical separation method 湿式化学分离法
wet chemical technique 湿化学技术
wet chemistry 湿法化学
wet classification 湿式分级;湿法分级
wet classifier 湿法分级器
wet cleaner 湿式水洗涤器;湿式滤清器;湿式除尘器;湿法洗涤器;湿法洗涤器;湿法清刷机
wet cleaning 水选;湿洗;湿(法)净化;湿法精选
wet cleaning method 湿选法
wet cleaning process 水选法;湿法
wet climate 多雨气候;潮湿气候;常湿气候
wet closed-circuit fine grinding 湿法闭路系统细粉磨

wet clutch 油浸离合器;浸油离合器
wet coating procedure 湿法包膜法
wet cobbing 湿法磁选
wet cold 湿冷
wet collector 湿式收集器;湿式集尘器
wet collector of particulate matter 湿式除尘器
wet collodion plate 棉胶湿板
wet colo(u)r 未干色;湿色(料);湿润色
wet combining 湿贴
wet combustion 湿式燃烧(法);湿式焚烧;湿(法)燃烧
wet combustion chamber 湿燃室
wet combustion chamber boiler 湿燃室锅炉
wet combustion method 湿烧法
wet combustion process 湿式燃烧过程;湿燃烧法
wet comminuter 湿磨机
wet comminution 碎湿;湿磨
wet compacted weight 湿密度;湿压实重量;湿重度;湿容量
wet compaction 偏(压)湿压实
wet compass 液体罗盘;液体罗经
wet compost 沤肥
wet compression 湿压缩
wet compression refrigeration cycle 湿式压缩制冷循环
wet-compressive strength 显压强度;湿压强度
wet concentrate 湿精矿
wet concentrate bin 湿精矿仓
wet concentration 湿选;湿法提浓;湿法精选
wet concrete 未干的湿混凝土;未结硬混凝土;塑性混凝土;水灰比高的混凝土
wet concrete mix 塑性混凝土混合料;湿混凝土混合料
wet condense pipe 湿式冷凝管
wet condenser 湿式冷凝器
wet condition 湿润状态
wet connection 湿连接;湿接缝
wet consistency 塑性稠度
wet consistency of concrete 混凝土塑性稠度;湿混凝土(混合料的)塑性稠度
wet construction 浆砌施工;湿作业施工;湿法施工
wet construction partition(wall) 湿构造分隔墙;湿作业隔断
wet contact 湿触点
wet conversion process 湿法转化工艺
wet cooling 湿冷却
wet cooling thermometer 湿式冷却温度表
wet cooling tower 蒸发冷却塔;湿式冷却塔;湿空气氧化塔
wet corrosion 液体腐蚀;湿(式)腐蚀;湿蚀
wet cottrell 湿式静电除尘器
wet cover 潮湿物覆盖
wet cover curing 潮湿物覆盖养护
wet crepe 湿法起皱
wet critical 湿临界的
wet criticality 湿临界
wet crop 水田作物
wet cross-linking process 湿交联法
wet cross-section 水下断面
wet cross-sectional area 水下断面面积
wet crude 含水原油
wet crushed 湿法破碎的
wet crushing 湿碎法;湿法破碎
wet crushing mill 湿碎磨;湿粉碎机;湿法破碎机;湿法粉碎磨
wet cup method 潮湿度扩撒测量法
wet-cured 湿养护的
wet-cured concrete 湿养护混凝土
wet cured polurethane paint 湿固化聚氨酯涂料
wet curing 湿治;对(混凝土)湿养护;湿(润)养生;湿润养护
wet cut 水下开挖;(基坑的)湿挖(方);湿开挖
wet cutting 水下开挖
wet cyclone 旋液分离器;水力旋流器;湿式旋风分离器;湿式旋风除尘器
wet cylinder liner 湿(式)汽缸套
wet damage 湿损(货被水损)
wet damage accident 湿损事故
wet dash 粗涂
wet day 下雨天;雨日
wet decating 湿蒸呢
wet decatizer 湿蒸呢机
wet decatizing 湿蒸呢
wet dedusting 湿法除尘

wet denitrification 湿法脱氮
wet density 湿重度;湿密度
wet deposition 湿沉积(物)
wet desulphurization 湿法脱硫
wet development 湿显色
wet digestion method 湿消化法
wet disc brake 湿式盘式制动器
wet disintegrator 湿浆离解机
wet distillation 水蒸气蒸馏
wet distillation process 水蒸气蒸馏过程
wet dividing rod 湿分绞棒
wet diving 湿式潜水
wet diving bell 湿式潜水钟
wet diving suit 湿式潜水服
wet dock 系船船坞;有闸港池;闸控港池;港池;两突堤码头间的港池;湿(船)坞;封闭(式)港池;泊船坞;泊船池
wet docking 不进坞检修;船壳水上检修
wet-docking facility 不进坞检修设备
wet doubling 湿捻
wet draw 湿部牵引力
wet drawing 湿法拉拔;湿拔
wet-drawn wire 湿(冷)拉钢丝
wet dressing 湿敷裹;湿法选矿
wet drifter 湿架式凿岩机
wet drill 湿式凿岩机
wet drill cutting 湿法凿岩
wet drill hole 湿钻孔
wet drilling 湿钻孔;水下钻探;水下钻井;水下钻进;湿钻探;湿式凿岩;湿法钻法;湿法打眼儿
wet drill method 湿钻法
wet drum cobber 湿式圆筒选矿机;湿法鼓式磁选机
wet drum magnetic separator 湿式圆筒磁选机
wet-dry bulk depression 干湿球温差
wet-dry sieving technique 湿法干法混合筛分技术
wet-dry strength ratio (混凝土的)干湿强度比
wet-dry test 干湿试验
wet-dry tropical climate 干湿热带气候
wet dust arrester 湿法集尘器
wet dust collection 湿法除尘
wet dust collection device 湿式集尘装置;湿式除尘装置
wet dust collection equipment 湿式集尘装置
wet dust collector 湿式除尘器;湿式收尘器;湿式集尘器
wet dust extraction 湿法除尘
wet dusting 湿式喷粉
wet dust removal 湿式除尘;湿法除尘
wet dust separator 湿式除尘器;湿式尘埃分离器
wet edge 湿边
wet-edge agent 湿边剂
wet-edge extender 湿边增充剂
wet-edge retention 湿边保持性
wet-edge runner mill 湿式轮碾(粉碎)机
wet-edge time 湿边时间
wet electric(al) dust precipitator 湿式电气除尘器
wet electrochemical oxidation 湿空气电化学氧化
wet-electrolytic capacitor 湿电解电容器;电解液电容器;液体电解质电容器
wet-electrolytic condenser 电解质电容器
wet elutriator 水析器
wet emery mill 金刚砂湿磨机
wet emplacement 水冷发射台
wet-emulsion technique 湿乳胶技术
wet end 湿部
wet end additive 湿部添加剂
wet end chemistry 湿部化学
wet end efficiency 湿部效率
wet end finish 湿部装饰
wet end furnish 湿部配比
wet engine 湿发动机
wet equatorial climate 潮湿赤道气候
wet excavation 水下挖土;水下挖方;水下开挖;湿挖(方);湿开挖
wet exhaust manifolds 湿排气总管
wet-expand 湿膨胀
wet expansion 湿伸长率;湿膨胀
wet extrusion molding 湿式挤压成型
wet fading 湿褪色
wet fastness 耐湿性
wet fastness property 湿牢度性能
wet fat-liquoring 湿态加油
wet-fed cement kiln 湿式给料水泥回转窑
wet feed 湿料(输送)

wet feed device 湿投装置
wet feed equipment 湿投装置
wet feeder 湿料加料器
wet feed facility 湿投设备
wet feeding 湿喂法;湿进料
wet feed kiln 湿式喂料窑
wet feed material 湿原料
wet feed method 湿送料法
wet feed mixer 湿进料混合器
wet feed unit 湿投装置
wet-felting 湿法铺装;湿法铺毡
wet fibre filter 湿式纤维过滤器
wet field 机耕船
wet field gas 伴生富气
wet-filling 湿法填装
wet filling filter 湿式填充过滤器;湿法装填
wet film 湿膜
wet film ga(u)ge 湿膜测厚仪
wet film hanger 洗片挂
wet film thickness 湿膜厚度
wet film thickness ga(u)ge 湿膜测厚仪
wet filter 湿式过滤器
wet filtration 湿法过滤
wet finish 湿部装饰
wet finishing 湿磨精加工
wet flashover distance 湿闪弧距离
wet flashover potential 湿闪络电压
wet flashover test 湿闪络试验
wet flashover voltage 湿飞弧电压
wet flash test 湿闪试验
wet flex test 湿弯曲试验
wet float 湿浮筒
wet flong 湿式纸型纸
wet flue gas desulfurization system 湿式烟道气脱硫系统
wet flue gas treatment 湿式烟气处理
wet flushing 湿法冲洗
wet fog 湿雾
wet formation 含水地层
wet foundation 水下地基;湿地基地
wet Fourdrinier board machine 长网湿抄机
wet front 湿润界面
wet fuel 液体燃料;湿燃料
wet galvanizing 湿法熔剂镀锌;湿法镀锌
wet gas 含油天然气;湿天然气;湿气
wet gas cleaning 煤气湿法除尘;湿气洗气
wet gas flow meter 湿式气体流量计;湿球气流计
wet gas holder 水封储气罐
wet gas meter 湿式气体流量计;湿式量气表;湿球气体流量计
wet gas phase-out zone 湿气消失带
wet gas purification 湿煤气净化法
wet gas purifier 湿法煤气净化器
wet gas scrubber 湿气气体洗涤器
wet gas zone 湿气带
wet gate oxidation process 液栅氧化过程
wet geothermal field 湿地热田
wet geothermal steam 地热湿蒸汽
wet glazing 湿法安装玻璃
wet glued 湿板胶接
wet-glued plywood 湿黏胶合板;湿(式)胶合板
wet gluing 湿性胶合法
wet gluten 湿面筋
wet goods 液体货(物)
wet-grained 湿搓纹的
wet gravel 湿砾石
wet-gravity concentration method 湿式重选方法
wet grind 湿研磨
wet grinder 湿磨机
wet grinding 湿研;湿磨(法);湿法研磨;湿法细碎;湿法碾磨;湿法粉磨
wet grinding and dry process burning 湿磨干烧
wet grinding attachment 湿磨削装置
wet grinding machine 湿磨机;湿式磨矿机
wet grinding mill 湿法磨
wet grinding operation 湿磨操作
wet grinding pebble mill 湿式砾磨机
wet grinding plant 湿磨设备;湿磨车间
wet grinding tube mill 湿法管磨
wet grip 湿(路)附着力
wet ground 含水岩石;湿地
wet ground mica 水磨云母
wet ground muscovite mica 湿磨白云母
wet ground slag 湿渣粉;湿研矿渣;湿碾矿渣

wet grout 液状灰浆;灰浆
wet gun 甲板炮
wet gypsum process 湿式石膏法
wet handling 湿法处理
wet head 顶阀消防栓
wet heat 蒸汽或热水取暖系统
wet heating system 热水暖气系统;湿供暖系统
wet heat setting 湿热定形
wet heat sterilization 湿热杀菌
wet hole 积水钻孔;湿孔
wet hookup 湿接合
wet hopper 湿料储箱
wet hub rotor 湿毂式转子
wet hydrogen 湿氢
wet hydrogen technique 湿氢技术
wet impinger 湿式撞击式滤尘器;(测定空气中灰尘含量的)湿式碰撞器;湿式冲击(取样)器
wet inflation 湿式膨胀法
wet injury 湿害
wet instability 湿润不稳定
wet installation 湿装配;湿砌
wet job 是抽泥筒作业
wet job-site after-treatment 现场湿法事后处理
wet joint 湿接头;湿缝
wet katathermometer 湿球冷却温度表
wet laid deposit 水力冲填
wet-laid mat 湿法薄毡
wet-laid non-woven fabric 湿法成网非织造织物
wet-laid tissue machine 湿法制毡机
wet laminate process 湿层合法
wet-laminating process 湿的层压法
wetland 塘地;潮湿地面;沼泽地
wetland ecosystem 湿地生态系统
wetland habitat 湿地生境
wetland restoration 湿地修复
wet landslide 湿土坍坡;湿地土崩;湿地坍方;湿地坍坡
wetland soil 湿地土壤
wetland treatment of wastewater 废水沼泽地处理
wetland system 湿地系统
wetland treatment system 湿地处理系统
wet lapping 湿式研磨;湿式抛光法
wet-laying 湿法成网
wet laying process 湿法成网工艺
wet lay system 湿法制造
wet layup 湿法铺放;湿法敷(涂)层;树脂浸渍湿增强材料
wet layup material 湿态层贴材料
wet layup method of bonding 湿态层贴胶合法
wet layup technique 湿铺工艺
wet lead 粗铅
wet-lifting 湿运
wet lime 湿石灰
wet lime mortar 塑性石灰砂浆
wet limestone scrubber 湿石灰涤汽器
wet limestone scrubbing 湿石灰涤汽法
wet liming 湿澄清法
wet-line correction 湿绳改正;湿线校正;(水文测验的)湿绳校正
wet liner 湿式汽缸套
wet location 潮湿区
wet look 湿视
wet look gloss film 湿视有光膜
wet loss 湿滴损失
wet lubricant 湿式润滑剂
wet lump 湿块
wet machine 浆纸机;湿抄机
wet machine tender 湿抄机操作工
wet magnetic cobber 湿式(粗选)磁选机
wet magnetic dressing 湿法磁选
wet magnetic particle(powder) 湿式磁粉
wet magnetic separation 湿法磁选;湿式磁选
wet magnetic separator 湿法磁选机;湿式磁选机
wet manufacture 湿法制造
wet mash 湿粉料
wet masher 湿料碾碎机
wet masher with tap 湿料搅碎机
wet masonry 浆砌
wet mat 湿木浆
wet material 湿料
wet meadow 湿草甸
wet meadow soil 湿草甸土
wet mechanical analysis 机械湿选分析;机械湿选法;湿化粒径分析;湿分析;湿法粒径分析;比重

（计）分析
wet mechanical method 机械湿选法
wet membrane curing agent （混凝土的）湿膜养护剂
wet-memory effect 湿定形效应
wet metallurgical processing 水冶处理
wet meter 湿式煤气表
wet meter of gas 湿式煤气表
wet method 湿（磨）法
wet method operation 湿法作业
wet migration process 湿泳移法
wet mill 洗涤机；湿磨机；湿法磨
wet-mill concentration 湿法选矿
wet-milling 湿磨；湿粉；湿法精选；湿法粉磨
wet-milling plant 湿法选矿厂
wet-milling timber pulp 湿磨木浆
wet mining 湿法开采
wet mist collector 湿式雾滴收集器
wet-mix aggregate 湿拌（混凝土）集料；湿拌（混凝土）骨料
wet-mix base 湿拌级配碎石基层
wet-mix concrete 湿拌（喷射）混凝土
wet mixer 湿（式）搅拌机；湿法混料器；湿拌和机
wet-mix gun 湿料喷枪
wet-mix gunite 湿料喷浆
wet-mixing 湿式混合；湿拌；湿法混合（带有液态燃油的混合气）
wet-mixing process 湿法混合（带有液态燃油的混合气）
wet-mixing time 湿混合时间
wet-mix macadam 湿拌碎石混合料
wet-mix method 湿式喷射方法
wet-mix process 湿拌法；湿拌步骤；湿拌作业
wet-mix shotcrete 湿拌喷射混凝土
wet mix(ture) 湿拌；含过多水的混凝土混合料；湿料；湿拌混合料；湿拌和料
wet-mix weight 湿拌重量
wet module 厨房，厕所等部分
wet monsoon （夏季西南季风）湿季风；潮湿季风
wet mortar 灰浆；塑性砂浆；湿灰浆
wet motor 潜水电动机
wet mo(u)lding 湿法成型
wet mud brick 湿泥砖；湿法压制砖
wet mullen 湿纸耐破度
wet natural gas 含大量石油天然气；富天然气；湿天然气
wetness 湿度；潮湿
wetness index 湿度指数
wetness loss 湿度损失
wet net 湿绳梯
wet-net training 湿绳梯训练
wet niche 湿龛
wet nodule kiln 湿法球窑
wet nonwoven fabrics 湿法无纺布
wet of optimum 最优湿度
wet oil 含水原油
wet-on-wet 湿压湿喷漆法（前一道漆未干时就喷后一道漆）；湿压湿印刷；湿碰湿（工艺）
wet-on-wet coating 湿碰湿涂装
wet-on-wet painting 湿压湿漆法；湿态喷漆法
wet opener 湿式开松机
wet operation 湿法作业
wet out 打湿
wet-out agent 湿润剂
wet-out rate 浸透率
wet-out speed 浸透速度
wet oxidation 湿（式）氧化
wet oxidation deodorizing 湿法氧化脱臭
wet oxidation method 湿式氧化法
wet oxidation process 湿（式）氧化法
wet packing 湿法装填；湿法填装
wet pad 湿发射场
wet paint 油漆未干（警告用语）
wet pan 碾矿机；湿式轮碾（粉碎）机；湿润容器；湿磨碎机；湿辊磨机
wet pan grinder 湿法磨床
wet pan grinding 湿法盘磨
wet pan mill 湿研盘
wet pan milling 湿辊磨机研磨
wet paper atmometer 湿纸蒸发表
wet parallel flow low lying condenser 湿式并流低位冷凝器
wet paste 水泥浆
wet pelt 湿皮

wet pen 墨水笔
wet perimeter 浸润周边
wet periphery 浸润周边
wet peroxide oxidation 湿式过氧化物法；湿空气过氧化物氧化
wet phosphoric acid 湿法制备的磷酸
wet physical processing 湿法物理处理
wet pick 湿式风镐
wet pick resistance 防潮性（能）
wet pick-up 湿涂层量
wet piece 湿料
wet-pipe sprinkler system 湿管喷水灭火系统；湿管喷水器系统；湿式喷水系统
wet-pipe system 湿管系统
wet pit 沉污池
wet pit pump 排水泵
wet plaster 粉刷；抹灰
wet-plate 湿板
wet plate efficiency 湿板效率
wet-plate photography 湿版摄影术；湿版照相
wet-plate process 湿版法
wet point 湿点
wet polishing 湿抛光；湿法抛光
wet powder grinder 湿式粉磨机；湿粉研磨机
wet precipitator 水力除尘器；湿式收尘器
wet preliminary splitting 湿法预裂
wet preparation 湿制剂；湿选法
wet press 纸板机；湿压机
wet pressed hardboard 湿压硬质纤维板
wet pressed pulp 湿压纸浆
wet pressing 湿压榨；湿压成型
wet pressing loss 湿压损失
wet press loss 湿压损失
wet pressure loss 湿压损失
wet process 湿处理；（湿法生产水泥的）；湿法加工；湿法（工艺）；湿版洗印
wet-process cement production 湿法水泥生产
wet-process development 湿过程显影
wet-process enameling 湿法搪瓷；湿法涂搪
wet-process for making mineral wool slab 湿法制矿棉板
wet-process hardboard 湿法生产的硬质纤维板
wet-processing 湿式作业；湿加工；湿（法）处理
wet-process installation 湿法生产装置；湿法加工装置
wet-process kiln 湿法工艺窑；湿料炉；湿法窑
wet-process kiln with slurry filter 带料浆过滤机的湿法窑
wet-process mat 湿法薄毡
wet-process metallurgy 湿法冶金
wet-process of cement manufacture 湿法水泥生产工艺
wet-process of fume gas desulfurization 湿法烟气脱硫
wet-process porcelain 湿法制瓷
wet-process rotary cement kiln 湿法水泥回转窑
wet-process rotary kiln 湿法（作业）回转窑；湿法烧制转窑
wet-process-type rotary kiln 湿法转窑
wet-process with filter and boiler 带过滤机和余热锅炉的湿法工艺
wet-proof 防潮的；防湿的
wet property 湿润性
wet puddling 湿精炼
wet pulp 湿纸浆
wet pulverizer 湿磨机
wet purification 湿法提纯；湿法净化
wet purification of combusting gas 燃烧气的湿式净化；湿法燃气净化
wet purifying 湿法净化
wet quenching 湿法熄焦
wet raising 湿起绒；湿起毛
wet rating 湿料计盘
wet reaction 湿反应
wet reagent feeder 湿式给药机；湿式给药器
wet reclamation 湿法再生
wet reduction 湿法还原；湿碾磨细；湿碎；用水磨细
wet-reed relay 湿式笛簧继电器
wet reeling 湿式摇纱
wet relaxation 湿松弛复原
wet removal rate 雨水清除污染速率
wet-rendered 蒸气炼油厂
wet rendering 湿法炼油

wet return 湿（式）回水
wet return line 湿回水管
wet return pipe 湿式凝结水管
wet return system 湿式回水系统
wet rock 含水岩石
wet rock cuttings 湿凿（碎）石
wet-rolled granulated slag concrete 湿碾矿渣混凝土
wet room 湿室；湿房间
wet room dampproofing 湿室防潮
wet room fixture 湿室电气设备
wet room light fitting 湿室照明装置
wet room light(ing) fixture 湿室照明装置
wet room luminaire(fixture) 湿室照明装置
wet room partition(wall) 湿室隔墙
wet room service 湿室事务
wet rot 湿朽；（木材的）湿腐朽；湿腐
wet rotary drilling 湿法回转钻进
wet rotary grinder 旋转湿磨机
wet rotoclone collector 湿式旋风集尘器；湿式旋风除尘器
wet rub 湿润摩擦阻抗
wet rubbing 湿擦
wet rub quality 湿耐磨性能
wet rub strength 湿耐磨强度
wet runway 湿跑道；湿的跑道
wet sacks curing 湿麻袋养护
wet salt （含镁钙等杂质的）湿盐
wet sample 湿样品；湿试样
wet sampling 湿法取样
wet sand-binder construction 湿集料法施工；湿砂黏结施工
wet sand blast 湿式喷砂
wet-sand blaster 湿砂喷射器
wet-sand blasting 喷湿砂法；湿砂喷射
wet-sand blasting process 喷湿砂法
wet sand cure （混凝土的）湿砂养护
wet sand cure of concrete 混凝土湿砂养护；铺湿砂养护混凝土
wet sand curing （混凝土的）湿砂养护
wet sanding 湿砂打磨；水打磨；湿打磨
wet sand mix 湿砂拌和
wet sand process （混凝土）湿砂养护法
wet-saw closed-circuit grinding 湿法原料闭路系统粉磨
wet-scale disposal 湿法除氧皮
wet screed 湿态刮平
wet screen analysis 湿法筛分分析
wet-screened 湿筛（分）的；湿（法）过筛的
wet-screened mortar 湿筛砂浆
wet screening 湿选法；湿筛析；湿筛（分）；湿法筛分；带水筛分
wet screening factor 湿筛系数
wet screening(method) 湿筛法（筛分的）
wet scrubber 喷水除尘装置；湿洗器；湿式洗涤器；湿式涤气器；湿式除尘器
wet scrubbing 湿法洗涤
wet season 雨季；湿季
wet section 水下断面
wet separation 湿法分离
wet separator 湿式分离机；湿式除尘器；湿法分选机；湿法磁选机
wet setting 湿定形
wet-settling losses 湿陷性黄土
wet ship lift 湿式升船机；湿运升船机
wet shoots 湿枝
wet shot 湿喷（喷火器不经点燃而喷射胶状燃料）
wet shot rattle test 湿球磨耗试验
wet sieve 湿筛析；湿筛（分）；湿格筛
wet sieve analysis 湿法筛分分析
wet sieve method （筛分的）湿筛法
wet sieving 湿筛析；湿筛（分）；湿法筛分
wet sinker drill 湿式下向凿岩机
wet-skid 炉底滑道水管
wet-skidding test 湿滑试验
wet-skid resistance 湿滑阻抗；湿滑抗力
wet slag 湿渣
wet slaked lime 湿熟石灰
wet slaking process 石灰水化法
wet-sleeve construction 湿汽缸套结构
wet-sleeve cylinder 带湿式套筒汽缸；湿式缸套汽缸
wet slide （运木材用的）水滑道
wet slip 舣装港池；试车船坞
wet slip coating （两突岸码头间的）港池覆盖

wet slope 迎水坡;临水坡
wet sludge 湿污泥
wet sludge tank 湿污泥池
wet smear 湿擦拭
wet snow 湿雪
wet soil 湿土;湿地
wet soiling 湿污染
wet solonchak 湿盐土
wet sorting 湿式拣选
wet spell 雨期;雨季;闷湿的阴雨天;湿期
wet sphere device 湿球装置
wet spinning 湿纺;湿法纺丝
wet-spinning frame 湿法纺丝机
wet-spinning machine 湿法纺丝机
wet split 湿分离;湿法分选
wet spraying 湿法喷补
wet spreader and extractor 给湿脱湿机
wet sprinkler 洒水器
wet stable consistency 砂浆的湿稳定稠度
wet-stage cyclone 湿段旋风分离器
wet stamp 湿捣碎
wet stamping 湿捣碎法
wet stamping mill 湿式捣磨
wet stamp mill 湿式捣矿机
wet standpipe system 充水立管系统;湿式立管系统
wet starch 湿淀粉
wet start 湿起动
wet state 湿润状态
wet state strength 湿(态)强度
wet steam 湿(蒸)汽
wet steam cooled reactor 湿蒸汽冷却堆
wet steam cure 湿蒸硫化
wet steam curing 湿蒸汽养护
wet steamer 饱和蒸汽锅炉
wet steam field 湿蒸汽田
wet steam geothermal power generator 利用湿蒸汽的地热电站
wet steaming 湿汽蒸
wet steam pack apparatus 蒸汽湿敷器
wet steam turbine 湿蒸汽汽轮机
wet sticking process 湿接成型
wet stock 湿物料;湿材料
wet stock dryer 湿物料干燥机
wet storage 湿(储)藏法
wet storage battery 湿式蓄电池
wet storage holder(s) 湿式气柜
wet storage stain 湿法储[贮]存污染;镀锌件上的白锈;(镀锌层表面的)白膜
wet stoving 湿硫熏漂白
wet stowage 湿储法
wet stream sampling 湿流取样
wet strength 耐湿强度;(型砂的)湿态强度;湿强度;湿法胶接强
wet strength agent 湿强度剂
wet strengthening 增湿强度
wet-strengthening impregnant 防湿浸渍剂
wet strength of adhesion 湿润黏合强度
wet strength paper 抗湿纸
wet strength paper sack 湿强纸袋
wet strength retention 湿强度保留率
wet string 充液钻杆柱
wet submersible 湿式潜水器
wet subsea X-mass tree 湿式海底采油树
wet-subsidence due to overburden 计算自重湿陷量
wet-subsidence for classification 分级湿陷量
wet suction fan 湿式吸风机
wet suit 水中服;湿(式)潜水服
wet suit for water skiing 滑水连衣裤
wet sump 湿式油底壳;湿槽
wet sump lubricating system 湿槽润滑系统
wet sump lubrication 湿槽润滑
wet surface 湿润面
wet surface hardenable adhesive 潮湿面可固化胶黏剂
wet surface profilometer 湿面轮廓测定仪
wet system of evacuating 湿体系抽空(用喷水空气泵抽空的装置)
wettability 吸湿度;可湿性;浸润性;润湿性;受湿性;湿润性;湿润度
wettable 可润湿的;可湿性的
wettable powder 可湿性粉剂
wet tabling 摇床湿法洗选
wet takeoff power 湿起飞力(以水喷射作为助力的起飞力)
wet tan 湿废鞣料
wet tandem drawing machine 串列式湿拉机
wet tank 湿箱(用以储[贮]存液体);湿塔
wetted area 过水面积;受潮面积;湿润面积
wetted cross-section 水下断面;湿润断面;过水断面;过流断面;浸水断面
wetted diameter 浇灌直径(喷灌机)
wetted graded stone base 湿拌级配碎石基层
wetted land 湿地
wetted midship and canal area ratio 断面比(指船中水下面积与运河断面面积的比值)
wetted nappe 淹没水舌
wetted parts 润湿部件;沾湿部件
wetted perimeter 浸润周界;润周;湿周(边);湿润周边;湿润范围
wetted perimeter of cross-section area 过水断面湿周
wetted saline soil 潮湿盐土
wetted section 过水断面
wetted surface 浸水面;湿润(表)面;湿面;潮湿表面;潮片
wetted surface area 浸水面积
wetted surface coefficient 湿面系数
wetted surface column 沾湿表面分馏塔;黏湿表面分馏塔;湿面分馏塔
wetted wall column 湿壁塔
wetted wall tower 湿壁塔
wet tempering 湿式回火
wettenschlick 潮泥
wet tensile(strength) 湿抗张强度;湿拉力
wet tensile test 湿抗张强度试验
wet tensile tester 湿抗张强度仪
wet test 湿度试验
wet test meter 湿式试验气表;测湿用表;湿式仪表;湿式气表
wettest stable consistency 最湿稳定稠度
Wetthauer test 魏特豪尔测试
wet thermometer 湿式温度计
wet time 湿润时间;雨天工资
wetting 浸湿;加湿;润湿(作用);受潮;湿润
wetting ability 润湿能力;湿润能力
wetting action 润湿作用
wetting additive 湿润剂
wetting agent 湿润剂;浸润剂;润湿剂
wetting and dispersing agent 湿润分散剂
wetting-and-drying 干湿
wetting-and-drying cycle 干湿循环
wetting-and-drying test 干湿试验
wetting and weighting rectification 浸水与加压矫正法
wetting angle 浸润角;润湿角;湿润角
wetting angle of molten glass 玻璃液润湿角;玻璃液接触角
wetting apparatus 给湿装置;润湿装置
wetting balance 润湿天平
wetting brick 浸湿砖;砖的湿润;砖的加湿
wetting capacity 润湿能力;湿润能力
wetting characteristics 润湿特性
wetting colo(u)r 易润湿颜料
wetting device 加湿装置;润湿装置
wetting-drying test 干湿循环试验
wetting efficiency 润湿效率
wetting error 润湿误差
wetting film 湿润薄膜
wetting force 润湿力;濡湿力
wetting front 湿润线;湿润界面;湿润锋
wetting heat 润湿热;湿润热
wetting index 润湿指数
wetting irrigation 润湿灌溉
wetting line 浸湿线
wetting liquid 润湿液体
wetting mechanism 湿润装置
wetting of packing 充填物的沾湿
wetting-out deck 润湿槽
wetting perimeter 润湿周边
wetting period 润湿时期
wetting phase 湿相
wetting power 润湿(能)力;濡湿力;湿润能力
wetting property 润湿性(能)
wetting rate 湿润率
wetting rectification 浸水矫正法
wetting system 洒水系统;洒水设备
wetting tension 润湿张力
wetting time 润湿时间
wetting unit 湿润装置
wetting zone 湿润带
wet to dry process conversion 湿法改干法
wet to modification 湿法改干法
wet tower 湿塔
wet trade 湿工序;湿作业
wet transfer rescue 湿式转运救生
wet treatment 湿(法)处理;湿选(矿石)
wet trim 湿纸边
wet tube grinding mill 筒式湿磨机
wet tube(-type) mill 管式湿磨机
wet tumbler 湿式清理滚筒
wet tumbling 湿式转鼓抛光;湿(法)抛光
wet tumbling barrel 湿式清理滚筒
wet tundra soil 湿水沼土;湿冻原土
wet twisting 湿捻
wet-type air cell 湿式空气电池
wet-type air cleaner 湿式空气净化设备
wet-type air cooler 湿式空气冷却器
wet-type alarm system 湿式报警
wet-type bolting machine 湿式锚杆安装机
wet-type cleaner 湿式空气滤清器
wet-type clutch 湿离合器
wet-type cooling system 湿式冷却系统
wet-type dust collector 水(力)除尘器;湿式除尘器
wet-type electric precipitator 湿式电集尘器
wet-type gasholder 湿式气柜
wet-type hyperbolic cooling tower 湿式双曲线面冷却塔
wet-type machine 湿式凿岩机
wet-type meter 湿式水表
wet-type photocell 湿式光电池
wet-type refuse channel 湿式垃圾管道
wet-type rock drill 湿式凿岩机
wet-type servo-valve 湿式伺服阀
wet-type system of dust control 湿法集尘器系统
wet-type tower abrasion mill 湿式塔式磨粉机
wet-type water ga(u)ge 湿式水表
wet-type water meter 湿式水表
wet-type wire drawing machine 湿式拉丝机
wet unit weight 湿重度;湿单位重;潮湿状态重度
wet unstable 湿润不稳定
wet-use adhesive 湿用胶;耐水胶;湿用胶黏剂
wet vacuum distillation 真空蒸汽蒸馏
wet vacuum pump 湿(式)真空泵;湿气泵
wet vapo(u)r 湿蒸气
wet vehicle system 湿式潜水器系统
wet vent 湿通风管道;排湿气孔;排湿气管
wet ventilator 湿式通风器
wet volume 湿容积
wet volume capacity 湿容量
wet volume weight 湿体重
wet wall 抹灰墙
wet wash allowance 湿法选矿后的水分容差
wet washer 洗泥机;湿洗机;湿法洗选机
wet washing 湿洗法
wet washing plant 湿式集尘器设备
wet water 润湿水(含有强润湿剂的水,灭火用)
wet way 湿法
wet weather flow 雨天流量
wet weather peak flow 雨天高峰流量
wet weather rill 下雨形成的小河
wet weathers 雨衣雨裤;潮湿天气;雨天
wet weather spell 潮湿季节
wet weather suit 防湿衣
wet web 湿纸幅
wet-web former 湿法成网机
wet weight 湿重
wet well 吸水井;湿井;出水井
wet well cuttings 湿钻屑;湿法钻井
wet well pumping house 湿室泵房
wet wheel mill 湿轮碾机
wet-wick static dissipator 湿芯天电耗散器
wet winding 湿法缠绕
wet wire drawing 湿法拉丝
wet without-dry recovery 湿折皱恢复
wet without rain 湿
wet wood 湿材;湿芯材;发潮材
wet work 湿(法)作业
wet-worked cursher 湿式操作破碎机
wet-wound 湿绕(涂有湿树脂的线圈)
wet wrapper damping machine 卷布给湿机

wet wrinkle 湿皱纹
wet year 湿年;多雨年;多水年;丰水年
Weyl unified field theory 韦尔统一场论
Weymouth formula 韦茅斯公式
Weymouth pine 韦茅斯松
Weymouth seam-face granite 韦茅斯裂石面
Weyroc 韦洛克牌木料(一种木花树脂做成的材料)
whacker 捣实器;巨物
whale 鲸鱼
whale and fish processing at sea 海上鲸和鱼加工
whaleback 船尾防浪损的拱形架;鲸背石
whaleback deck 鲸背型甲板
whaleback roof 拱形坡顶;凸弧形坡屋顶;鲸背形坡屋顶
whaleback vessel 鲸背船
whale boat 捕鲸船;船载划艇
whalebone brush (清沟渠用)鲸须刷
whale deck 鲸背型甲板
whale in kiln colo(u)r 窑彩鲸鱼
whale meal 鲸鱼粉
whale mother boat 捕鲸母船
whale mother ship 捕鲸母船
whale quenching oil 淬火鲸油
whaler (用于开挖支撑的)横枋木;捕鲸船;模板横撑;横档
whales 鲸类
whale ship 捕鲸船
whale shot 粗鲸蜡
whale stocks 鲸资源;鲸群
whale-tide cofferdam 满潮围堰
whale-towing craft 拖鲸艇
whaling 捕鲸
whaling craft 捕鲸船
whaling ground 鲸鱼场;捕鲸场
whaling industry 捕鲸加工工业
whaling mother craft 捕鲸母船
whaling winch 捕鲸绞机
whammel 鲑鱼漂网
whare 棚座;(毛利人的)住房
wharf 码头;顺岸码头
wharfage 港务费;码头业务;码头(使用)费;停靠码头费
wharf apron 码头前沿
wharf apron space 码头前沿区
wharf apron track 码头前沿区轨道
wharf boat 码头趸船
wharf borer 码头蛀虫
wharf breasting structural member 码头靠船构件
wharf breast wall 码头胸墙
wharf conduit 码头管沟
wharf crane 码头起重机;岸边集装箱装卸桥
wharf demurrage 码头(卸货)延误费
wharf dues 码头税;码头费
wharf equipment 码头设施;码头设备
wharf extension 码头扩建
wharf face structure 码头岸面构造
wharf facility 码头设施;码头设备
wharf floating pontoon 浮船码头
wharf frontage 码头前沿
wharf-frontage depth 码头前沿水深
wharfhand 码头工人
wharf house 船员临时办公室
wharfie 码头工人
wharfinger 码头管理人;码头管理(人)员
wharfinger's receipt 码头管理人收据
wharfinger's warrant 码头栈单;码头仓库;码头仓单
wharf landing account 码头卸货账单
wharf line 码头边线
wharfman 码头装卸工(人);码头(带缆)工人
wharf master 码头管理(人)员;码头老板;码头管理人
wharf of precast hollow slab 空心大板码头
wharf of precast reinforced concrete slab without beam 无梁面板高桩码头
wharf of skeleton construction 桁架式高桩码头
wharf price 仓库交货价(格)
wharf quay 码头岸壁
wharf shed 码头前方货栈;码头(前方)仓库
wharfside 里档;码头边
wharfside crane 码头沿岸起重机;码头前沿起重机;码头边起重机
wharfside lightering 里档过驳
wharfside transfer 里档过驳
wharf siding 码头线

wharf structure 码头建筑(物)
wharf structure of high level platform supported on pile 高桩码头结构
wharf superintendent 码头监督员
wharf surface 码头(地)面;码头用地
wharf surface level 码头面高程
wharf throughput 码头通过能力
wharf wall 码头岸墙;码头岸壁
what economic law is to regulate 经济法调整对象
Whatman paper 瓦特曼纸(一种适于水彩画的优质图画纸)
what-not 陈列书籍的架子
wheal 锡矿;风团
wheat aggregated 麦秸集料;麦秸骨料
wheat awn cap 麦芒帽
wheaten 淡黄色的
wheaten chaff 铡碎小麦干草
wheatgrass 冰草
wheatland harrow plow 直圆盘犁
wheatland plow 垂直圆盘犁
wheat paddock 小麦田
wheat paste 小麦糊浆
wheat pit 现期小麦交易部(美交易所)
wheat producing area 产麦区
wheat producing countries 产麦国
wheat screenings 小麦筛渣
wheat stalk ornament 麦秸装饰品
wheat stalk patchwork 麦秸贴画
wheat starch 小麦淀粉
Wheatstone bridge 惠斯通电桥;单臂电桥
Wheatstone perforator 惠斯通凿孔机
wheat straw 麦秸
wheat-straw tar 麦秸焦油
wheel 轮状花纹;轮形图;辘轳机;舵轮;车轮;操纵轮
wheel abrader 转轮式喷砂机;转轮式耐磨性试验机
wheel abrating 喷丸清理
wheel abrator 抛丸清理装置;喷丸清理装置;喷丸清理机;喷丸器;打毛刺机;带式喷丸清理机
wheel abrator cabinet 喷丸室
wheel abrator machine 转轮喷砂机
wheel aligner 车轮对准器
wheel aligner indicator 车轮校准指示器
wheel alignment 车轮定位;轮位对准
wheel alignment ga(u)ge 车轮对准器;车轮对准规
wheel aliner 车轮对准器
wheel and axle parallelism 轮轴平行度
wheel and bearing roller system 轮与轴承式滚轮系统
wheel and bucket ditch dredger 轮斗式挖沟机
wheel-and-disc integrator 摩擦轮(式)积分器
wheel and rack 齿轮与齿条
wheel angle 轮偏转角;导向轮偏转角
wheel angle indicator 舵角指示器
wheel arch 轮罩拱
wheel arm 轮弯臂;轮辐
wheel arrangement 轴式
wheel axle 轮轴
wheel back-toback diameter dimension 轮对内侧距离
wheel balancer 车辆平衡器
wheel band 轮箍;车轮轮箍
wheel-barometer 轮形气压表
wheel barrow 手推小车;独轮车;单轮手推车;小车;独轮手推车
wheel-barrow broadcast seeder 手推车式撒播机
wheel-barrow man 推车工
wheel-barrow powered 动力小车
wheel-barrow run 手推车狭桥;手推车小路
wheel-barrow scale 手推车过磅秤
wheel-barrow sprayer 手推车式喷雾机
wheel-barrow with inflatable tire 独轮小车;有充气车胎的手推车
wheel-base (指车)轴距;轮轴距;前后轮距;车(轴)距;车轮底面积;车辆轴距
wheel-base of a bogie 转向架轴距
wheel-base of motor vehicle (指车)轴距
wheel batch method 按车计量法;按车分批法
wheel bearing 车轴轴承;车轮轴承
wheel-bearing outer cup 轮轴承外杯
wheel-bearing retaining ring 轮轴承扣环
wheel beckets 防动绳制
wheel blade 轮叶
wheel blank 轮坯
wheel block 制动块;轮挡;轮坯

wheel body 轮体
wheel boss 轮心;轮毂
wheel bound theory 车轮跳动理论
wheel box 齿轮箱;变速箱
wheel brace drill 手摇轮钻
wheel bracket 导轮支架
wheel brake 轮闸;车轮制动器
wheel brake cylinder 轮闸储气筒
wheel brake support 轮闸支座
wheel brake system 轮制动系统
wheel-bucket trencher 轮斗式挖沟机
wheel bull dozer 轮式推土机
wheel burn 钢轨擦伤;车轮灼伤
wheel camber 轮曲面
wheel cap 轮盖
wheel carriage 磨床床头
wheel carrier 轮架
wheel carrier bracket 轮托架
wheel carrier flange 轮架凸缘
wheel carrier support 轮架支柱
wheel case 转轮外壳;蜗壳
wheel casing 轮罩
wheel center 轮心;车轮心
wheel center distance 轮距
wheel chain 舵链;齿轮链系
wheel chair 轮椅;车椅;椅车
wheelchair housing 轮椅住宅
wheelchair housing unit 轮椅住房单元
wheelchair store 轮椅存放处
wheelchair tractor 轮椅牵引机
wheel chamber 叶轮腔室
wheel checker 车轮检验器
wheel chock 止轮块
wheel circumference 轮周
wheel circumference measure 轮周量尺
wheel clamp nut 车轮固定螺帽
wheel cleaner 轮子刮泥板
wheel cleaning station 车轮清洗站
wheel column control 轮杆控制
wheel concentration 轮集重
wheel concentration loading 集中轮载;车轮集中荷载
wheel configuration 轮胎形状;轮的构型;车轮配制图
wheel contact 轮压面;轮接面
wheel contact area 车轮与道路的接触面积;车轮接触面积
wheel container 带轮集装箱;备轮储箱
wheel control linkage 车辆转向联接装置
wheel conveyer 辊道;轮式输送机
wheel cover 轮罩
wheel cowl 轮罩
wheel crane 轮式起重机;汽车起重机;汽车吊
wheel crank 盘形曲柄
wheel cutter 铣捣;轮式铰;齿轮刀具
wheel cutting machine 切齿机
wheel cylinder 制动轮泵;制动分泵;轮闸储气筒;车轮(液压)制动分泵缸;车轮刹车泵
wheel cylinder adjusting pinion shaft 轴闸储气筒调准小齿轮轴
wheel cylinder cover 轮缸盖
wheel cylinder piston 轮闸储汽缸活塞
wheel davit 车体千斤顶
wheel deflection force 车辆偏转力
wheel detachable side ring 轮胎压圈
wheel detector 轮式探测器
wheel diameter 轮(直)径;车轮直径
wheel digging arm loader 轮式立爪扒渣机
wheel discharge capacity 水轮(机)泄流量
wheel disk 轮盘
wheel disk mill 车轮轧机
wheel ditcher 轮式挖沟机;轮式挖泥机
wheel dobby 齿轮式多臂机
wheel dozer 轮式推土机
wheel drag 轮阻力;轮闸
wheel drawer 车轮拆卸器
wheel dredge(r) 轮式挖泥船;轮式挖泥机;斗轮式采砂船
wheel dresser 砂轮修整器
wheel drive 车轮传动装置
wheeled 带行走轮的
wheeled cable drum car riage 放线车
wheeled carrier 轮架
wheeled carrier-and-stacker 轮式堆垛车

wheeled crane 装轮的起重机;轮(胎)式起重机
wheeled ditching machine 轮胎式挖壕机
wheeled dry powder fire extinguisher 轮式干粉灭火机
wheeled fire extinguisher 车载灭火机;轮式灭火机
wheeled foam extinguisher 轮式泡沫灭火器
wheeled gate 定轮闸门
wheeled litter 床车
wheeled plough 轮式犁
wheeled polishing machine 轮式抛光机
wheeled pusher 装载车;手推车
wheeled rig 带轮的装备
wheeled scraper 轮(胎)式铲运机;轮(胎)式铲土机;轮式刮土机
wheeled stack 轮式码垛机
wheeled stretcher 手推床
wheeled tractor 轮式拖拉机;轮式牵引车
wheeled trailer 拖车
wheeled trolley 拖车
wheeled vehicle 车辆
wheel efficiency (水轮机)转轮效率
wheel electrode 滚极;电极滚轮
wheeler 轮车;手车工
Wheeler bottom 惠勒底
wheeler filter 轮式过滤器
wheel excavator 轮式挖土机;轮胎式挖掘机
wheel face 轮缘外表面;轮面
wheel facewidth 大齿轮齿宽
wheel finder 仿形轮
wheel fire extinguisher 轮式灭火机
wheel fit 轮座配合
wheel flange 轮缘
wheel flange baseline 轮缘高度测定线
wheel flange height 轮缘高度
wheel flange lubrication equipment 轮缘润滑装置
wheel flange lubrication system 轮缘润滑系统
wheel flange thickness 轮缘厚度
wheel forging press 车轮坯成型模锻压力机
wheel forming tool 砂轮成形工具
wheel frame 导轮架
wheel friction loss 叶轮摩擦损失
wheel front support 车轮前悬挂支架
wheel ga(u)ge 轮距;车轮量规
wheel gear 齿轮
wheel gearing 齿轮传动装置
wheel girdles 轮上防滑套爪
wheel grader 自行式平地机
wheel grating 操омог站台
wheel grease retainer 轮护脂圈
wheel guard 护轮板;汽车挡泥板;车轮挡板
wheel guards for timber bridge floor 木桥上的护轮木
wheel hanger 导轮吊架
wheel-head 陶轮的顶部表面;砂轮头
wheel-hoe cultivator 轮式中耕锄
wheel hook 制动钩
wheelhorse 架辕马;操舵室;轮箱;驾驶室;明轮罩
wheelhouse console 驾驶室操纵台
wheelhouse control 驾驶室操纵
wheelhouse flat 驾驶室平台
wheel hub 轮毂
wheel hub bearing adjusting nut 车轮毂轴承调整螺母
wheel hub cap 轮毂帽
wheel hub drive 轮毂驱动
wheeling 旋转;运行;道路行车鉴定;车运;薄板滚压法
wheeling machine 滚压机;薄板压延机
wheeling steps 转向踏步;(楼梯的)斜踏步;扇形踏步
wheel key way 轮键槽
wheel lathe 车轮镟床;车轮车床
wheel lean 车胎充气未满
wheel lean crank 车轮倾斜曲柄
wheel lean cylinder 车轮倾斜液压缸
wheel lean tie bar 车轮倾斜连杆
wheelless plow 无轮犁
wheel lifter 轮式挖掘铲
wheel load 轮重;轮荷载;车轮载重;轮载(荷)压力
wheel load capacity 路面承载能力;承载能力【道】;车轮承载能力
wheel-load distribution 轮载分布
wheel loader 轮胎装载机;轮式装载机;单斗车;铲斗车

wheel loading 车轮荷载
wheel loading shovel 轮式铲斗装载机
wheel load strength 车轮荷载强度
wheel-load stress 轮载应力
wheel locking 车轮抱闸
wheel lug 车式防滑钉;车轮防滑板
wheelman 汽车驾驶员;舵工
wheelmark 轮辙;轮迹
wheelmark eliminator 轮迹消除装置;轮迹消除耙
wheelmark eraser 轮迹消除装置;轮迹消除器
wheel meter 轮式计量器
wheel mill 轮碾机
wheel mill edger 车轮轧机的立辊
wheel miter 等径伞齿轮
wheel model 轮状模型
wheel mo(u)lding machine 齿轮制型机
wheel-mounted 装有轮胎的;装有车轮的
wheel-mounted carrier 装轮通用机架
wheel-mounted crane 轮胎(式)起重机;汽车式吊车
wheel mounted device 轮胎行走装置
wheel-mounted dragline 轮胎式拖拉挖土机;轮式拉铲挖土机
wheel-mounted drill 轮式钻机;车钻
wheel-mounted excavator 轮式挖掘机
wheel-mounted gate 带轮的门
wheel-mounted main frame 轮载主车架
wheel-mounted mobile crusher 轮胎式移动破碎机
wheel-mounted mobile plant 轮胎式移动破碎设备
wheel-mounted mucker 车轮行走式装岩机
wheel-mounted power shovel 轮胎式机动铲;轮胎式动力铲
wheel-mounted sprayer 装轮式喷雾机
wheel mounting 车轮压装
wheel mounting press 装轮机
wheel nave 轮毂
wheel nut 车轮螺母
wheel nut wrench 车轮螺母扳手
wheel oil deflector 轮导油器
Wheelon forming process 橡皮模成形法
wheel order 轮位
wheel over point 用舵点
wheel pair repair of vehicles 车辆轮对修理
wheel pants 机轮整流罩
wheel path 行车轨迹
wheel path area 轮迹地带
wheel pattern 轮形图式【道】
wheel pit 轮坑;水轮坑
wheel plate 轮板;车轮辐板
wheel play 车轮轴毂间隙
wheel plough 轮式犁
wheel power unit 轮动力头
wheel press 车轮压机;车轮装配压力机;车轮安装压力机
wheel press ga(u)ge 车轮压装机压力表
wheel-pressure 轮压
wheel printer 轮式打印机
wheel pull 车轮拉力
wheel puller 卸轮器;拆轮器
wheel pump 轮式泵;水轮泵;齿轮泵
wheel-quartering machine 卧式双轴对钻床
wheel race 水轮(机)尾水道
wheel-rail 轮轨【铁】
wheel-rail adhesion 轮轨间黏着力
wheel-rail contact force 轮轨接触力
wheel-rail contact stress 轮轨接触应力
wheel-rail force 轮轨力
wheel-rail interaction 轮轨关系;轮轨相互作用
wheel-rail interface 轮轨接触面
wheel-rail reaction force 轮轨反力
wheel-rail system 轮轨系统
wheel ratio 转轮比值
wheel resistance 车轮阻力
wheel reversing gear 齿轮回动装置
wheel rib 轮辐
wheel rim 轮缘;轮辋
wheel rim clamp 轮环夹
wheel rim wedge 制动铁鞋;轮周楔块
wheel rim with tyre 整体车轮
wheel rods 舵链连杆
wheel roller 轮式碾轧机;轮式压路机
wheel rolling 车轮轧制
wheel rolling mill 轮子轧机;车轮轧机
wheel runner 飞轮;转子

wheel rutting test 轮辙试验
wheel scraper 轮式刮土机;碾轮用刮板;轮胎式铲运机
wheel screw 手轮锁紧螺钉;车轮螺钉
wheel seat 轮座
wheel set 轮副;轮组;轮对
wheel set factor 轮组系数
wheel set load 轴重
wheel set measuring device 轮对测量装置
wheel set press 压轮机
wheel setting ga(u)ge 轮位规
wheel set workshop 轮对间
wheel shaft 轮轴
wheel shake 轮裂
wheel shaped 轮状的
wheel shimmy 轮摆动
wheel shop 车轮厂
wheel shroud 轮盖
wheel-ski 轮橇
wheel skid 止轮铁鞋
wheel skidder 轮式集材拱架机
wheel sleeve 轮套筒;砂轮套筒
wheel slide 磨床床头滑架
wheel-slide protection 车轮打滑保护
wheel slide release valve 车辆防滑缓解阀
wheel sliding 车轮打滑
wheel-slip 车轮打滑
wheel slippage 轮滑转动
wheel slip protection 车轮防滑保护
wheels-locked testing 刹车试验;车轮刹住(滑行)试验
wheel slot 轮槽
wheel spacer 轮隔片
wheel spad 轮罩
wheelspan 轮距
wheel spider 针齿盘;轮辐条盘
wheel spike 轮滑转
wheel spin 车轮空转;滑转
wheel spindle 轮轴
wheel spinner 转向盘
wheel spoke 轮辐
wheel stand 轮架;磨架;舵工
wheel station 车轮静电干扰
wheel steel 车轮钢
wheel steps 扇形踏步;(楼梯的)斜踏步
wheelstop 止轮器
wheel stretcher 轮压展机;扩轮器;齿轮展宽器
wheel stretching machine 轮压展机
wheel stud 车轮双头螺栓
wheel swing arm 行走轮摇动臂
wheel tension assembly 补偿棘轮
wheel-tire 车轮轮箍
wheel-tire lathe 轮箍车床;车轮车床
wheel tooth 轮齿
wheel tracer 轮式定向器
wheel tracery 轮幅式花格窗;轮形花格窗;轮形窗花格
wheel track 轮距;存轮线;车辙;轮轧【铁】
wheel tracking depth 车辙深度
wheel tracking test 车辙试验
wheel tractor 轮胎式拖拉机
wheel tractor crane 轮胎式拖拉起重机;轮式牵引起重机
wheel tractor front-end loader 轮式拖拉机前端装载机
wheel tractor scraper 轮胎式铲土机;轮式自行铲运机;轮胎式(拖拉机)铲运机
wheel trailer 双胶轮运输车
wheel train 轮系;轮列
wheel transducer 车轮传感器
wheel traverse 横向进磨;砂轮横向进给
wheel tread 轮胎花纹;轮距;车轮踏面
wheel trencher 轮式挖沟机
wheel trench excavator 轮(胎)式挖沟机
wheel trenching machine 轮式挖沟机
wheel truing 砂轮修正
wheel turn failure 车轮空转损失
wheel-type chassis 轮式底盘;轮式车架
wheel-type continuous extrusion machine 轮形连续挤出机
wheel-type excavator 轮式挖沟机;轮式挖土机
wheel-type factor 轮形系数
wheel-type hanging roof 圆形悬索屋顶
wheel-type lift 轮盘式启闭机

wheel-type pickup 轮式捡拾器
wheel-type rake 轮式搂草机
wheel-type sampler 轮式取样器
wheel-type tractor 轮式牵引车；轮胎式拖拉机
wheel-type trencher 轮式挖沟机
wheel-type trenching machine 轮式挖槽机；轮式挖沟机
wheel type truck crane 轮式起重机；汽车起重机
wheel undercarriage 轮式起落架
wheel unloading test 轮重减载试验
wheel washer 洗轮机
wheel wear 轮对磨损
wheel web 轮辐条；车轮腹板
wheel-web-rolling mill 轮辐轧机
wheel weight 轮上配重
wheel window 玫瑰形窗；圆花窗；轮形窗；车轮窗；轮形窗扇
wheel wobble 车轮摇摆；车轮摆振
wheelwork 齿轮装置；传动装置
wheel-wright 制轮业者
wheelwright's wood 车辆木材
wheel yoke 轮轭
whelp 链轮扣链齿；扣链齿；绞盘筒筋；绞缆筒筋条；长方条耐火砖
when filling with substance 装料时
when issued 假定发行
when not required 当不要求时
when permitted 当被允许时
when radiographed 射时规照相时
when viewed from the driver end 从传动机侧看
whereabouts search 下落搜索
whereabouts strategy 下落搜索策略
whereas clause 鉴于条款
wherryite 氯碳铜铅矾
wherryman 小船桨手
whet 研磨；磨
whetstone 砂轮；刺激品；油石；磨(刀)石；砥石
whetstone holder 磨刀石架
whetter 磨锐工具
whetting 磨块
whet wastewater 研磨废水
whewellite 水草酸钙石
whey wastewater 乳清废液
whiff 小舟桨手
whiffletree switch 横杠开关
whiffling breeze 无定向微风
while schema 当型模式
while symbol 当符号
whim 绞盘；辘轳；后盘辘轳；绕绳滚筒
whine 晃动
whine-rock 玄武土；暗色岩；杂砂岩；硬砂岩
whin setts 暗色岩小方石
whin sill 暗色岩床
whinstone 闪绿岩；闪长岩；玄武岩类岩石；暗色岩；粗玄岩
whinstone setts 暗色岩小方石
whip 滑轮吊车；钢丝绳吊索；钢丝绑绳；急速动作机件；拍打；定单滑车组；垂曲；鞭子树；鞭(抽)
whip action 鞭梢作用
whip and derry 简易滑轮起重机；起重用绳子和滑轮
whip-and-tongue graft 舌接
whip antenna 鞭状天线
whip apparatus 救生索投射器
whip a rope 缠扎绳端
whip bridge 止摆杆夹板
whip cover 止摆杆压片
whip crane 简易起重机；动臂起重机
whip cutting 斜削插条
whip gin 定单铁绞辘
whip grafting 镶接；舌接
whip guide 防振控制
whip hammer 止摆心杆
whip-hammer spring 止动心杆簧
whip-hammer stud 止动心杆柱
whip hoist 动臂起重机；单轮起重装置
whip lashing response 鞭梢效应
whiplash style 鞭舞艺术风格
whip lever 止动杠杆
whiplike 鞭样的
whip line 转臂起重机；滑轮吊车(辅助卷扬机)；火箭抛绳；火箭救生绳；撒绳；单绞辘绳
whip lock 止动锁片
whip on whip 定单与定单复绞辘
whip operating lever 止摆操作杆

whip operating lever pin 止动操作杆销
whip operating lever spring 止摆杆簧
whip pan 快速摇摄
whipped toppings 人造稠黄油
whipper 净毛机；搅打器
whipper deseeding machine 亚麻脱籽机
whipping 搅打；锁缝；甩尾；甩击；甩动；抖动；缠扎绳头
whipping crane 回转式起重机；摇臂起重机
whipping during spraying 喷涂时喷枪发生飞弧
whipping effect 鞭梢效应
whipping gear 悬臂摆动机构
whipping machine 搅打器；打包机
whipping method 挑弧焊；挑焊运条法；抖焊运条法
whipping welding 抖动焊
whipple hammer 摆动锤
Whipple-Murphy truss 惠莆—马非(式)桁架；多腹杆桁架
Whipple truss 惠莆桁架；多腹杆桁架
whip-poor-will storm 春暴
whippy 有弹性的；易弯曲的
whip roll 织机后梁
whips 鞭策者
whip saw 双人横切锯；狭边(钩齿)粗木锯；大锯
whip-shaped 鞭状的
whip-shaped lamp post 鞭形灯杆；鞭形灯柱
whipstock 楔尖式钻孔定向器；造斜器；偏斜器；槽式变向器
whipstock grab 钻孔造斜器打捞器
whipstocking 人工造斜
whipstock point 造斜点
whip stud 止摆杆柱
whip support 止摆杆座
whiptail 尾鞭病
whip tube 止摆杆套
whirl 急旋；涡流；涡转
whirl aggregate spreader 旋转式骨料撒布机
whirl blast 旋风
whirl burner 旋流式烧嘴；旋流式喷燃器；涡流式燃烧器
whirl coating 旋转涂覆法；甩胶
whirlcone 旋流器
whirl core 旋涡核(心)
whirl crane 旋转式起重机
whirl current 涡流
whirled lamination 旋涡纹理
whirled sub 旋流短节
whirler 旋转台；离心式滤气器；回转机；盘碟底部不稳定检测器；底部不稳
whirler crane 旋转式起重机；回转式起重机
whirlering 旋转注浆成型
whirler machine 拷版机
whirler shoe 旋流套管鞋；斜排泄眼注水泥管鞋
whirler-type collector 离心式集尘器
whirler(y) crane 旋臂吊车；回转式起重机
whirl gate 旋涡侧滤渣浇口；旋涡侧浇口；离心渣浇口
whirlgate dirt trap 离心集渣器
whirlgate feeder 离心集渣冒口
whirlgate head 离心集渣冒口
whirligig 旋转运动
whirling 旋转作用；离心甩残漆；回转；涡转
whirling action 旋转(动作)
whirling arm 旋转臂；旋转器
whirling current 涡流；旋涡流
whirling flower decorative painting 旋子彩画
whirling flower pattern 旋子团花图案
whirling flywheel 端面摆动飞轮
whirling horizontal drum 旋流卧式滚筒
whirling hygrometer 旋转(式)湿度计；旋转(式)干湿表
whirling mixer 旋转搅拌机；旋流搅拌器
whirling mixing 旋流混合
whirling motion 旋转运动；涡流
whirling of fusing charge 熔料串炉
whirling pool 旋涡
whirling psychrometer 旋转(式)湿度计；旋转式干湿表；旋动式干湿球湿度计
whirling reaction 涡流反应
whirling reaction chamber 涡流反应室
whirling runner 旋流除渣机；回旋除渣器
whirling speed 旋涡失稳速度；涡流速度；共振回转速度；临界转速；切向分速度
whirling sprinkler 旋转式洒水器

whirling test 旋转试验；离心力试验
whirling test stand 旋转试验台
whirling thermometer 旋转湿度计；旋转式温度表
whirl insert 涡流插芯
whirl jet nozzle 旋涡喷嘴
whirllike arrangement 旋涡状排列
whirl mix 摆轮式混砂机
whirl-on 旋转式涂布
whirlpool 旋涡(水)；旋流
whirlpool bath tub 旋涡浴槽
whirlpool effect 旋涡效应
whirlpools and eddies 泡漩
whirlpool-type hopper 旋涡型仓斗
whirl riser 离心渣冒口
whirl screen 回旋筛
whirl spreader (松散材料)旋涡式摊铺机
whirl structure 旋涡状构造
whirl tide 旋转潮
whirl tube 空气动力管；风洞
whirl velocity 切向分速度；涡流速度
whirl vortex 旋转涡流
whirlwind 涡流；旋风
whirlwind air separator 旋风式气选机；旋风式空气分离器
whirlwind clarifier 旋风分级机
whirlwind classifier 旋风分级机；离心式空气分级器；离心式空气分级机
whirlwind computer 旋风型计算机；旋风式计算机
whirly 旋转的；塔吊；塔式筑坝起重机；小旋风
whirly-bird 直升(飞)机
whirtle plate 拉模板
whisk 搅拌(液体使之产生气泡)
whisker 晶须；纤维状结晶；螺旋触簧；金属须
whisker column 晶须柱
whisker contact 点接触
whisker crystal 须状晶体
whiskering 墙面风化起霜；毛绒疵点；泛碱(砖或混凝土表面上泛起的白色盐基类粉末)
whiskering cupola 回音圆屋顶；耳语屋顶
whiskering dome 回音穹顶；耳语穹顶
whiskering gallery 回音廊；传声廊；耳语廊
whiskerizing 晶须化
whisker reinforced metal 线体加强金属
whisker technology 晶须工艺学
whisking machine 搅打器；打盘机
whispering cupola 回音廊
whispering dome 回音穹顶；低音廊；耳语穹顶；混响隆
whispering gallery 回音廊
whistle 振鸣声；号哨
whistle board 鸣汽笛
whistle budding 管状芽接
whistle buoy 响哨浮标；汽笛浮标；哨浮标
whistle control 汽笛按钮
whistle cord 拉绳
whistle interference 啸声干扰
whistle lever 汽笛杆
whistle lever shaft arm 汽笛杆轴臂
whistle lever shaft arm key 汽笛杆轴臂键
whistle lever shaft bracket 汽笛杆轴托
whistle lever shaft key 汽笛杆轴键
whistle mark 鸣笛标
whistle-pipe sampler 笛管取样机
whistle post 鸣笛牌
whistle pull 汽笛拉索
whistler 冒口；啸声干扰；机车笛出气孔；排气道；通气孔；出气帽口
whistler-type noise 啸音
whistler wave 电子回旋波
whistle signal 汽笛信号
whistle valve 汽笛开关阀；气笛阀
whistling buoy 鸣笛浮标；响哨浮标；号笛浮标；汽笛浮标；哨音浮标
whistling mark 鸣笛标
whistling swan 小天鹅
Whitbed 惠特石(一种优质建筑用石灰石)
Whitby cement 惠特比水泥(一种天然水泥)
white 橄榄树；白色(的)；白的
white acid 白酸(一种氟化氢铵溶液)
white-acid embossing 陶瓷腐蚀浮雕
white acid-free tissue paper 白拷贝纸
white afrodite 白泡石
white agate 白玛瑙
white aggregate 白色集料；白色骨科

white alder 白桤木
white alkali 白苏打
white alkali soil 盐土；白碱土
white alkyd paint 白色醇酸漆
white alkyd resin enamel 白色醇酸瓷漆
white alloy 白色合金；假银
white alloy plate 白合金片
white alum 明矾；白矾
white alumina 白色氧化铝；白电刚玉
white alumina grain 白色氧化铝质颗粒
white aluminate cement 白(色)高铝水泥
white aluminium oxide 白色氧化铝
white alundum 白刚玉
white amino bakingcan coating 白氨基烘干罐头漆
white-and-black 墨地白花
white and greasy tongue coating 白腻苔
white annealing 二次退火
white ant 白蚁
white antimony 锑华
white architecture 白色建筑(风格)
white arsenic 白砷；白砒
white as a jade 白如玉
white asbestos 白石棉
white asbestos-cement board 白水泥石棉板
white asbestos flour 白石棉粉
white ash 美国白蜡树
white asparagus 白芦笋
white balance 白(色)平衡
white balsam 冷杉
white bark pine 白皮松
white base 白色基层
white basswood 白椴
white beeswax 白蜂蜡
white beryl 白柱石
white bina 页岩土
white birch 欧洲白桦；白桦
white blast 喷砂除锈到出白
white board 白纸板
white body 反射体；白粘土坯体；白体
white bole 高岭土；白(陶)土
white bombway 榄仁木(产于印度的一种褐色硬木)
white bond paper 白证券纸
white bone fan 白骨扇
white bone folding fan 白骨折扇
white book 白皮书
white book on science and technology 科学技术白皮书
white bowl lamp bulb 白碗罩灯泡
white-box model 白箱模型
white brass 高锌黄铜；白(黄)铜
white brick 白垩黏土砖；白砖
white brightness 白场亮度
white broken rice 白碎米
white bronze 淡色青铜；白(青)铜
white buckskin shoes 白鹿皮鞋
white buoy 白色浮标
white cable 白色电缆
white camphor oil 白樟油
white cap 浪沫；浪花；白浪(花)；白涌头
white carbon black 白炭黑
white cardboard 白卡纸
white cast iron 白(口)铸铁；白口铁
white caustic 白色灼剂
white cedar 兴叶扁柏；白杉
white cement 白(色)水泥
white cement asbestos board 白水泥石棉板
white cement plaster 白水泥抹灰；白水泥粉刷
white ceramic tile 白色釉面砖；白瓷砖
white chalk 白粉笔
white chalk lime 熟石灰
white characters on blue background 蓝底白字
white charcoal 白炭
white chip(ping)s 白碎块；白碎片
white chop 白浪(花)
white chromium cast iron 白铬铸铁
white chuglum 印度银灰木
white cigarette paper 白色卷烟纸
white city 娱乐城
white clay 白泥
white cliff sandstone 白崖砂岩
white clip circuit 白色电平限制电路
white clipper 白电平削波器
white coal 塔斯曼油页岩；白煤

white coat(ing) 软管用白墨；白涂料；白底漆；刷白层；白灰罩面；白灰涂面
white cobalt 辉砷钴矿
white cold 白冷
white cold blast pig iron 冷风白口铁
white colony 白色菌落
white colo(u)r 白色
white colo(u)r temperature 白场色温
white communicating branch 白交通支
white compression 白色区域压缩
white concrete 白(色)混凝土
white concrete carriageway markings 白混凝土路面上行车道白色标志
white concrete frame 白混凝土构架
white concrete panel 白混凝土板
white content 白色成分
white copal 白色硬树脂
white copper 德银(一种镍铜锌合金)；白铜；铜锌镍合金
white copperas 皓矾；针绿矾；硫酸锌
white coral 白珊瑚
white cornet 白色球节
white correction fluid 白色修正液
white corundum aggregate 白色刚玉集料；白色刚玉骨料
white crane 白鹤
white crystal 白色晶体
white cycle 白周期
white damar 白硬胶树；白达马树脂
white damp 一氧化碳
white deal 挪威杉木；白枞木
white deer 白鹿
white deposit (混凝土表面的)渗斑；(混凝土表面的)霜斑
white detail 白细节
white dew 冻露
whited face 刷白面
white dhup 白色桃花心木(印度产)
white diamond 白色钻石
white Ding glaze 白定釉
white dirt 素烧坯黏附的垫砂；白斑点
white discharging agent 拔白剂
white double glossy paper 白色厚光纸
white dragon design on blue ground vessel 青地白龙纹尊
white drying 全干燥
white dune 白砂丘
white duplex board 白纸板
white earth 白硅质土
white edge 白框
white embroidery 白色刺绣
white exposed concrete 表面白色的混凝土
white factice 白油膏；浅胶；白胶
white fair-faced concrete 白色磨光面混凝土
white feld-spar 钠长石
white fiber 白纤维
white fibrocartilage 白纤维软骨
white field balance 白色场平衡
white field balance current ratio (彩色管的)白色场平衡电流比
white field equipollence 白色场均匀性
white finger 白手指
white finish coat 罩面白灰；白灰罩面
white finished sheet 酸洗薄板
white fir 白冷杉；白枞(树)
white fish 白鲑；白鱼
white flake 白点
white flame heat 白热
white-flowered varieties 白花品种
white flush switch 白色入墙式开关
white fog 白雾
white fraction 乳白
white French polish 白色法国罩光漆
white frequency 白频率
white frequency noise 白频率噪声
white frost 白霜
white functionalism 白色功能主义
white fused alumina 白刚玉
white gale warning 狂风警报
white garnet 白榴石
white gasoline 无铅燃料
white Gauss noise 白高斯噪声
white glassine paper 半透明玻璃纸
white glaze 填白；白釉

white glazed coat(ing) 白釉涂层；上白釉
white glazed finish 白釉终饰；白釉饰面
white glazed latex cup 白釉胶杯
white glazed wall tile 白釉壁砖
white glaze vase with carved Taotie design 白釉刻花饕餮纹瓶
white glaze with brown mottles 白釉褐斑
white glaze with green mottles 白釉绿斑
white glaze with underglaze black colo(u)rs 白釉釉下黑彩
white glazing 上白釉
white globe lamp 白球灯
white glue 白胶
white gold 白金
white gold alloy 白色金银合金
white gold solder 白金焊料
white goods 日用漂白织物
white grained 白粒的
white granite 白麻石
white granite chip(ping)s 白花岗石碎片
white granite-ware 白花岗岩质器皿
white graphite 白石墨
white ground coat(enamel) 白底釉
Whitehall 白厅(英国)
white-hard 白陶土块；干透；白硬的
white-hard clay 白硬土
white-hard dryer 脱模干燥器
white head 白穗
white heat 白热
white hemlock 白铁杉
white hemp rope 白麻绳索
white hole 白洞
white hole effect 白洞效应
white horses 白涌头；白浪(花)
white hot 白热(的)
white-hot filament 白热丝
White House 白宫
white hydraulic cement 白色水硬性水泥
white ice 白色冰；白冰
white India rubber substitute 白色代胶
white iron 白铸铁；白(皮)；白口(铸)铁
white iron pyrite 白铁矿
white Italian marble 白色意大利大理石
whiteite 磷铝镁铁钙石
white jade 白玉
white jade carving 白玉雕刻品
white joint mortar 白色接缝灰浆；白色接缝砂浆
white Jun glaze 白钧釉
White Jura 白侏罗统
White Jura period 白侏罗纪
white jute 白黄麻
white lac 漂白紫胶；漂白虫胶；脱色紫胶；白漆片；白虫胶
white lake 矾土白
white lamp 白灯
white latex adhesive 白乳胶
white lauan 白柳安
white layer 白色亮层
white lead 碱式碳酸铅；铅白；白铅(粉)
white lead cement 白铅水泥
white lead flakes 铅白片
white lead in oil paste 油质白铅
white lead paint 白铅(油)漆
white lead paste 白铅浆
white lead plant 白铅工厂
white lead putty 铅白油灰；白铅油灰
white leather 白革
white-leaved oak shales 白叶栎页岩
white lens 白透镜
white level 白色信号电平；白(色)电平
White Lias 白里阿斯层
white light 白(色)光
white-light corona 白光日冕
white-light coronameter 白光日冕光度计
white-light coronograph 白光日冕观察仪
white-light event 白光事件
white-light exposure 白光曝光
white-light flare 白光耀斑
white-light fluorescent lamp tube 白光荧光灯管
white-light fringe 零次干涉条纹；白光干涉条纹
white-light hologram 白光全息图
white-light holography 白光全息术
white-light image processing 白光图像处理
white-light laser 白光激光器

white-light processing system 白光处理系统
white lime 消石灰;熟石灰;白灰
white lime mortar 白石灰砂浆
white lime paste 白色灰浆
white lime putty 白色灰膏
white limestone powder 白石灰石粉
white line 空行;板件装配线;白线;白色标线【道】；路面白色标线;车道线
white linear scar 白色线状瘢痕
white line machine 路面标白线机;路面划线机
white lines in background area 粉纹路
white lines in decoration area 花纹路
white line slag 白渣
white liquid 白色液体
white liquor 白液
white lotion 白色洗液
white lump lime 白块石灰;白生石灰;块石灰
white magnesium oxide plate 白色氧化镁板
white mahogany 白桃花心木
white maize 白玉米
white malleable casting 白口可锻铸件
white Manila board 灰底白板纸
white marble 汉白玉(石);白大理石
White marble balustrade 汉白玉栏杆
white marble carving 汉白玉石刻
white marble sitting buddha 白石坐释迦(工艺品)
white marble table screen 白云石台屏
white masonry mortar 圬工白灰浆
white mastic 白腻子;白色玛琋脂
white matter 白质
white matt paper 白色无光纸
white metal 轴承合金;喷砂除锈到出白级的金属；铜硫;白(色)金属;白合金;白冰铜;巴氏合金
white metal alloy 白金属合金
white metal babbit bearing 巴比合金轴承
white metal bearing 巴比合金轴承
white metal bearing alloy 铅基巴氏合金
white metal lining 白合金衬层
white mica 白云母
white mineral oil 石蜡油;白矿物油
white mink 白水貂
white mixture alba 白色合剂
white mortar 白砂浆;白色灰泥
white mo(u)lding plaster 造型白石膏
white mulberry 桑树
white mushroom 白蘑菇
white muskmelon slice 白香瓜片
white mustard(seed) oil 白芥子油
whiten 漂白;刷白
white narcissus 白水仙
whitened surface 刷白(表)面
whiteness 洁白色;白色的性质或状态;白度
whiteness index 白度指数
whiteness meter 白度计
whiteness of paper 纸张洁白度
white nickel ore 白镍矿
white night 白夜
whitening 增白;加白;刷白;泛白(现象);变白;白粉
whitening coat 白色涂层
whitening filter 白化滤波器
whitening in the grain 粗纹木材涂清漆后呈现的难看白条;木纹发白
whitening mark 白涂料标
whitening power 消色力
white noise 混合噪声;白噪声
white-noise generator 白噪声发生器
white-noise level 白噪声水平;白噪声大小
white-noise measuring set 白噪声测量仪
white-noise process 白噪声作用过程
white-noise signal 白噪声信号
white-noise test set 白噪声测试仪
white non-opaque pigment 白色体质颜料
white nucleus 白核
white oak 白橡木;白栎木
white object 反射体;白体;白目标
white oil 白油
white olivine 镁橄榄石
white-on-black 墨地白花
white opacifier 白色乳浊剂
white opaque pigment 白色不透明颜料
white-out 乳白现象;疏排;乳白天空
white page directory 白页目录
white paint 白色涂料;白漆

white-painted 涂过白漆的;刷过白漆的
white paper 白票
white paper on the environment 环境白皮书
white papers 白皮书
white paraffin wax 白石蜡
white Parian marble 伯利安白大理石
white paste 铅油;铅漆
white patch 白色片痕(色彩缺陷)
white peach flower 白桃花
white peak clipping 白峰削波
white peak point 白色峰点
white pearl lamb skin 白珍珠羔皮
white pearl lamb skin plate 白珍珠羔皮褥子
white peat 白泥炭
white pecan 白山核桃木
white Peony scented tea 白牡丹花茶
white pepper 白胡椒
white perch 白蛉(美国)
white perilla seed 白紫苏籽
white petrolatum 白矿脂
white petroleum jelly 白凡士林
white phosphate opal 磷酸盐乳白玻璃
white phosphorus 白磷
white phosphorus grenade 白磷榴弹
white phosphorus match 白磷火柴
white picking 二次酸洗
white pig iron 白生铁;白口铁
white pigment 白(色)颜料
white pigmented 涂白色颜料的
white pigment powder 白色颜料粉
white pine 白松
white plastering 白色涂料
white plastic 塑性白漆
white pocket 白囊;囊状腐朽
white point 白(色)点
white pointing mortar 白(色)勾缝砂浆
white pollution 白色污染
white poplar 银白杨;白杨
white porcelain tile 白瓷砖
white Portland cement 白水泥;白(色)硅酸盐水泥;白色波特兰水泥
white pottery 白陶
white powdery filling 白色粉末状填充物
white precipitate 白淀汞
white print 白色图;白背景图片
white products 轻油
white putty 白油灰
white pyrite 白铁矿
white quebracho 白坚木
white quicklime 白生石灰;浓生石灰
white radiation 白光;白辐射
white rainbow 雾虹
white rami 白支
white rattan products 白藤制品
white reference 基准白
white reference level 白色参考电平
white reflectance standard 白反射率标准
white rendziness 浅黑色碳土
white residue 硅渣
white resin 白松脂;白树脂
white rigid polyvinyl chloride exterior-profile extrusion 白色硬质聚氯乙烯外用挤出型材
white rigid polyvinyl chloride insulating storm window 白色硬质聚氯乙烯绝热防风雨窗
white room 绝尘室;洁净室
white rot 白黏土;白斑污;白(斑)腐木
white rot fungus 白腐真菌
white rouge 微细二氧化硅;白铁丹;白抛光粉
white Russia architecture 白俄罗斯建筑
white rust 白锈
white sand 白砂
white sandalwood 白檀
white sapphire 刚玉;白(色)蓝宝石
white scale 白粗石蜡片
white schorl 钠长石
white sea 白海;白浪(花)
white sea-elephant 废船(体)
white seal 白色密封料;白色填缝料
white sesameseed 白芝麻
white shade 白场色调
white shellac 白色虫胶片;脱色虫胶
white shellac varnish 白虫胶清漆
white signal 白色信号
white silky paper 白色绸纹纸

white single glossy paper 白色薄光纸
white slag 白渣
white slate pencil 白滑石笔
white slime 白泥;硅渣
white smear 白色拖尾
whitesmith 锡匠;白铁工
white smog 白雾
white smoke 白色烟雾
White softening 白色软化
white soil 白土
white soot 白炭黑
white space 空白
white speck 白斑;白麻点
white specks in decoration 花斑点
white spindle oil 白锭子油
white spirit 漆溶剂;松香水;石油溶剂(油)
white spots (涂层水作用的)白斑
white spotted finish 涂层发白
white spruce 白云杉
white squall 无形的飑风;无形飑
white stain 镀锌件上的白锈
White's test 怀特试验
white stone 白石(美国);白宝石
white stone chip(ping)s 白石屑
white stone-made mill 白石磨
white stone powder 白石粉
white stoneware 白炻器
white suds 白色泡沫
white sulfate of alumina 纯硫酸铝
white synthetic(al) baking enamel 白色合成烘漆
white synthetic(al) flat enamel 白色合成平光瓷漆
white talc 白滑石
white teacup 白茶盏
white tellurium 针碲金银矿;白蹄金银矿
white thread blight 白线疫病
white tin 白锡
white-to-black frequency swing 黑白最大频偏
white transmission 白信号传输;白色传输
white tungsten ore 白钨砂
white varnish 达玛琋脂清漆;白瓷漆用漆基
white Vaseline 白凡士林
white velvety paper 白色绒面纸
white vinegar 白醋
white vitriol 锆矾;皓矾;七水合硫酸锌;白矾
white wall tile 白瓷砖
white wall tire 白边轮胎
white war 经济竞争
whiteware 白色陶瓷(器皿);白瓷;卫生陶瓷
whitewash 刷石灰水;刷墙粉;刷白;白色涂料;白灰粉饰;白刷;白涂料;白灰水;白垩灰浆;白灰粉饰;盐泡;涂白;石灰水;大白浆;粉饰
whitewash brush 粉白涂料的刷子;白灰刷帚
whitewash coat 白灰粉刷涂层
whitewashed wall 白粉墙
whitewash finish 大白浆罩面
whitewashing 白灰粉刷涂层;喷大白;刷大白浆;刷白
white water 碎波水花;白水(即水中掺气);白浪(花)
white water clarifier 白液澄清器
white waters 白色水域
white water silo 白水槽
white water storage 白水池
white water treatment by air flotation 白水气浮处理
white wave 白波
white wax 漂白蜡;白蜡
white wax plant basket 白蜡杆篮
white wax plant products 白蜡杆制品
white way 白道;白路
white wedge 白楔
white wine wastewater 白酒废水
white wing 街道清洁工
whitewood 白木;鹅掌楸木
white woodfree writing paper 白色书写纸
white wool 白色纤维毡;白棉
white yolk 白卵黄
white zinc 锌白;白锌(粉)
white zinc ore 白锌矿
white zinc paint 锌白漆;白锌(油)漆
whiting 细白垩;研细的白垩;磨细白垩粉;铅粉;刷白;大白粉;粉末状碳酸钙;白涂料;白(铅)粉
whiting putty 铅白油灰
whitish 带白色的;白相;微白色的

whitish or brownish bloom 白色或褐色的粉衣
whitish soil 白土
whitleyite 杂顽火无球粒陨石
whitlockite 白磷钙石;白磷钙矿
whitmoreite 褐磷铁矿
Whitney key 半月销;半圆键
Whittaker criterion 惠特克准则
Whittaker differential equation 惠特克微分方程
Whittery shales 维特里页岩
whittle 在木头上破坏性刻划;石板整边工具;石斧
Whitworth's die 惠氏螺纹扳牙
Whitworth's form(of thread)惠氏螺纹牙样
Whitworth's ga(u)ge 惠氏规号
Whitworth's quick return motion 惠氏急回运动
Whitworth's screw pitch ga(u)ge 惠氏螺距规
Whitworth's screw(thread)惠氏螺纹
Whitworth's standard screw thread 惠氏标准螺纹
Whitworth's thread 惠氏螺纹
Whitworth's variable speed gear (带有回转连杆的)惠氏变速机构
Whitworth's wire ga(u)ge 惠氏线规
Whitworth thread 英制标准螺纹
whiz 旋离;离心分离;水气提取
whizzer 离心机;分离机
whizzer air separation 离心空气分离
whizzer air separator 离心空气分离器;离心空气除离器
whizzer blade 离心机大风叶
whizzer classifier 离心式选粉机;离心式空气分级器;离心式空气分级机
whizzer separation 离心空气分离
whole 全体;全部的
whole-and-half compasses 比例规
whole area 全面面积
whole article autoclaving test 整件试验
whole back 全背式
whole beam 滚子;圆木
whole block 整个街坊;整个区段;整块块材
whole body 整体
whole-body autoradiography 整体放射自显影术
whole-body counter 全身计数器
whole-body dose 全身剂量
whole-body exposure 全身辐照(量)
whole-body irradiation 整身辐照(量)
whole-body monitor 全身监测仪
whole-body outrigger hydraulic excavator 整体支腿式液压挖掘机
whole-body radiation 全身辐射
whole-body radiation dose 全身照射剂量
whole-body radiation meter 全身辐射剂量计
whole-body radiometer 全身辐射剂量计
whole brick 整砖
whole-brick wall 一砖墙;整砖墙
whole bridge cast in site 就地浇筑桥梁
whole-circle bearing 全圆(周)方位角
whole circle protractor 全圆分度器
whole coil 全节(距)线圈
whole coiled winding 全(线)圈绕组
whole-colo(u)red 纯色的;同一色的
whole column cryotrapping method 全柱冷阱法
whole course tracing 全程追踪
whole coverage 全赔保险
whole current 全流
whole current classifier 全流(式)分级机
whole current method 全流方法
whole current settler 全流式分级机
whole current theory 全流理论
whole day grazing 昼夜放牧
whole deals 整块松板(1.25英寸3.175cm厚)
whole depth 全齿高;齿全高
whole depth coefficient 全齿高系数
whole diamond 整粒金刚石
whole diamond bit 整粒金刚石钻头
whole differential spectrophotometry 全差示分光光度法
whole dislocation 全位错
whole-dollar accounting 整元会计
whole drawn tube 拔管
whole earth movement 全球性运动
whole event 全事件
whole fill 全部充填
whole gale 狂风(暴雨);十级风
whole gale warning 狂风警报
whole grain 全粒

whole hip 整个战脊;四坡屋顶
whole-house equipment 全家用设备
whole-house treatment unit 全家用处理装置
whole-house unit 全家用装置
whole job ranking 全部作业评级法
whole latex 全胶乳
whole latex rubber 全胶橡胶;全浆橡胶
whole length 船舶全长
whole-line width 全线宽
whole log barker 长木剥皮机
wholeness 完整性
whole number 整数【数】;全数
whole paper 全张
whole pipet(te) 全容吸移管;全节吸量管
whole plot 完全小区
whole process 全部过程
whole range 全距
whole range angle 全射程角
whole range distance 全程距离
whole range point 全靶场点
whole record spectra 全记录频谱
whole reflecting zone 全反射带
whole resistance of cars 车辆总阻力
whole rock 全岩
whole rock ages 全岩年龄
whole rock isochron 全岩等时线
wholesale 一揽子;总批发;批量;批发
wholesale account 批发账户
wholesale business 批发业务
wholesale center 批发站
wholesale commerce 批发贸易
wholesale commerce expenses 批发商业费用
wholesale cost 批发成本
wholesale cut-off 连续截弯;全面截弯
wholesale dealer 批发商
wholesale department 批发部
wholesale distributor 总经销人
wholesale establishment 批发业务
wholesale financing 大额资金融通
wholesale firm 批发店
wholesale food 批发食品市场
wholesale grain store 粮栈
wholesale house 批发店
wholesale insurance 整批保险
wholesale inventory 批发商业库存
wholesale market 批发市场
wholesale market value 批发市场价值
wholesale middleman 批发中间商
wholesale money 巨额借款;批发款项
wholesale network 批发商业网
wholesale of small articles 小商品批发
wholesale price 批发价(格)
wholesale price curve 批发价格曲线
wholesale price index 批发(物)价指数
wholesale quantity 批发量
wholesaler 批发商
wholesale receiver 批发承受商
wholesaler trade 批发交易
wholesale sale 批发销售
wholesale shop 批发商店
wholesale trade 批发业务;批发贸易
wholesale warehouse(building)批发零售商店;批发货栈
wholesaling middlemen 批发中间商
wholescale agreement 大规模协议
whole scene 全景
whole section member 整体式构件;非拼接式构件
whole service life 全部使用寿命(固定资产)
whole set 全套装置
whole set of products 整套产品
whole shift zone 整体移动带
whole size 原尺寸;实际尺寸;足尺
whole-sky camera 全天空摄影机
whole-society-productivity method 全社会劳动生产率法
wholesome 耐巨浪的船体
wholesome fountain 卫生饮水(喷)泉
wholesome water 卫生的水;卫生饮用水
whole staff safety management 全员安全管理
whole stage tendering 全过程招标
whole-stem logging 全干集材
whole stone 整粒金刚石
whole stone bit 整粒金刚石钻头
whole-structure model 整体结构模型

whole-sun sensor 太阳全光谱传感器
whole tide cofferdam 高潮围堰
whole tile 整砖
whole timber 大木;末锯木材;整方料;整材;非拼接材
whole tine grab 带全齿的抓斗;全齿式抓斗;带齿抓斗
whole-tree logging 伐倒木集材
whole tunnel method 全巷法
whole tyne grab 全齿式抓斗
whole-tyre reclaim 全轮胎再生胶
whole vehicle population 车辆总保有量
whole vision mask 全视面罩
wholly 完全
wholly-owned ownership 百分之百所有权
wholly-owned subsidiary 独家开办的子公司;附属公司
wholly rough zone 完全粗糙区
wholly saturation 完全饱和
whorl 螺纹;斗形纹
whorl of roots 根环
wibit 锯石成板(美国约克郡方言)
wiborgite 奥环状花岗岩
Wichart truss 芝铰桁架
wichtisite 玄武玻璃;玻基粒玄岩
wick 灯芯
wick drain 排水板法;塑料排水管;灯芯式排水井;灯芯绳排水孔;袋装砂井
wickenburgite 铝硅铅石
wicker 柳条
wicker dam 柳枝坝;柴束坝
wicker mat 柳条垫
wicker protection works 柳枝护岸设施
Wickersham quoin 威克歇姆版楔
wicker wall 柳条墙
wicker work 柳条编制品;编制品;柳枝织物;柴束工作;编枝工
wicker work fence 柳条篱笆
wicket 小门;水闸门;售票口;大门上的附门;船闸闸门上的放水孔;封窗墙;放水门;边门;导叶
wicket dam 旋转桁架木板坝;有旋转闸门的活动坝;门坝;撑板坝;旋转桁架木板活动坝
wicket door 便门;边门
wicket gate 导叶;旋闸;小门;导叶;导水门;边门
wicket gate-actuating rod 导叶操纵杆
wicket gate-adjusting servomotor 导叶调节接力器
wicket gate circle 导叶(中心)分布圆
wicket gate servomotor 导叶接力器
wicket gate stem 导叶转动轴;导叶(叶)柄
wicket girder 闸板立墩
wicket hole 平炉观察孔
wicket screen (纱窗上的)小纱窗
wicket-type gate 旋转式闸门
wicket type heater 门式炉
wicket weir 撑板堰;撑板坝
wick feed 虹吸油см
wick-feed oil cup 虹吸油芯注油杯
wick-feed oiler 虹吸油芯注油器
wick filtration 纱布过滤
Wick flagstone group 威克板层砂岩群
wicking 芯吸;浸吸;灯芯作用
wicking height 毛细升高值
wicking property 毛细管特性
wickiup 简陋的临时住处;(美国印第安人的)小屋子;窝棚
wick lubrication 芯绳润滑;油绳润滑
wick lubricator 吸油绳润滑器;油绳润滑器
Wickman ga(u)ge 威克曼螺纹量规;凹口螺纹量规
wickmanite 羟锡锰石
wick oiler 油绳润滑器
wick screw 灯芯螺旋
wick siphon 油绳虹吸
wicksite 魏磷石
wick trimmer 灯芯剪
wicky 帐篷屋子
Widal test 维达尔试验
wide 广阔的
wide address bus 宽地址总线
wide angle 宽角度(的);广角(的);大角度(的)
wide-angle aerial camera 广角航摄仪
wide-angle anastigmat 广角消像散透镜
wide-angle camera 广角照相机;广角摄影机
wide-angle collimator 广角准直仪
wide-angle converter 宽角前辅透镜

wide-angle coordinator 广角坐标方位仪
wide-angle coverage 广角视界
wide-angle deflection 大角度偏转
wide-angle diffusion 广角漫射
wide-angle direction 广角方向
wide-angle distortion 广角失真;广角畸变
wide-angled object glass 广角物镜
wide-angle eyepiece 广角目镜
wide-angle field 广角视场
wide-angle hologram 广角全息图
wide-angle horizon sensor 广角地平仪
wide-angle image 广角像
wide-angle lens 广角透镜;广角镜头;大视角镜头
wide-angle lighting 广角照明
wide-angle lighting fittings 广角型照明器
wide-angle luminaire 广角型灯具;广角型照明器
wide-angle magnifier 广角放大镜
wide-angle mapping objective 广角制图物镜
wide-angle multiplex projector 广角多倍投影(测图)仪
wide-angle nozzle 广角喷嘴
wide-angle objective 广角物镜
wide-angle(object) lens 广角物镜
wide-angle optics 广角光学系统
wide-angle panoramic photo 广角全景照片
wide-angle periscope 广角潜望镜
wide-angle perspective 广角透视
wide-angle photogrammetric camera 广角摄影测量照相机
wide-angle photographic system 广角照相系统
wide-angle photography 广角照相术;广角摄影
wide-angle radiometer 广角辐射计;宽视角辐射计
wide-angle reflection 广角反射
wide-angle reflection seismic survey 广角反射地震调查
wide-angle scanner 广角扫描器
wide-angle scanning 广角扫描
wide-angle shower 广角簇射
wide-angle sight 广角瞄准具
wide-angle spectroradiometer 广角光谱辐射计
wide-angle system 广视角系统
wide-angle telescope 广角望远镜
wide-angle viewing device 广角观察装置
wide-angular field 广角视场
wide anode 宽阳极
wide apart 间距很大
wide-aperture antenna 大孔径天线
wide-aperture lens 大孔径物镜;大孔径透镜
wide-aperture photo objective 特大开口率摄影物镜
wide-aperture system 大孔径系统
wide approach 大航线进场
wide area 广域
wide area comprehensive water pollution survey 广域水污染综合调查;宽频带综合水污染测量;大面积水污染综合调查
wide area data service 广域数据服务
wide area detector 广域检测器
wide area information servers 广域信息服务器
wide area network 广域网(络);广布网络
wideband 宽(频)带;宽波段
wideband acoustic(al) emitter 宽带声发射计
wideband air diffusion 宽带空气扩散
wideband amplifier 宽带放大器
wideband antenna 宽频带天线
wideband cable television 宽频带有线电视
wideband channel 宽(频)带信道
wideband coating 宽频带膜层
wideband code division multiple access 宽带码分多址
wideband coherent light modulator 宽带相干光调限制器
wideband communication system 宽带通信系统
wideband cophased antenna array 宽带同相天线阵
wideband data set 宽带数传机
wideband data transmission 宽带数据传输
wideband data transmission system 宽带数据传输方式
wideband discriminator 宽带鉴别器
wideband earthquake loading 宽频带地震荷载
wideband earthquake seismometer 宽频带反馈地震计
widebanded coal 宽条带状煤

widebanded structure 宽条带状结构
wideband equipment 宽带设备
wideband exchange unit 宽带交换机
wideband filter 宽带带滤波器;宽带滤光片
wideband instrument 宽频带仪
wideband modem 宽带调制解调器
wideband multichannel electron multiplier 宽频带多通道电子倍增器
wideband optic(al) frequency translation 宽带光频变换
wideband optic(al) modulator 宽带光调制器
wideband oscilloscope 宽频带示波器
wideband packet technology 宽带分组技术
wideband process 宽频带过程
wideband pulse 宽带脉冲
wideband pulse amplifier 宽带脉冲放大器
wideband ratio 宽带比
wideband repeater 宽带转发器
wideband seismograph 宽频带地震仪
wideband seismometer 宽频带地震计
wideband switch 宽带交换机
wideband switching 宽频开关
wideband transformer 宽带变压器
wideband transmission system 宽带传输系统
wide band video detector 宽频带视频检波器
wideband voltage tunable oscillator 宽带电压调谐振荡器
wide base 宽基极
wide base rim 阔口轮辋
wide base rims for tires 宽底胶胎轮辋
wide base tire 宽轮辋轮胎
wide beam coma 宽束彗形像差
wide beam echo sounder 宽声束回声测深仪
wide beam ship 肥大型船舶;浅吃水肥大型船舶
wide beam vessel 肥大型船舶
wide bearing 宽阔接触区
wide belt sander 宽带打磨机
wide berth 保持安全距离
wide-bore 开大孔;大口径的
wide-bore conduit 大口径管
wide-bore glass capillary column 大孔径玻璃毛细管柱
wide-bore pipe 大口径管
wide-bore piping 大口径管
wide-bore return bend 大口径回转弯头
wide-bore return bending 大口径回转弯头
wide-bore tee 大口径三通
wide-bore tube 大口径管
wide-bore tubing 大口径管
wide-bore well 大口井
wide bottom flange rail 宽底钢轨
wide bucket excavator 宽斗挖土机
wide chamber lock 广室船闸
wide chamber mill 宽室研磨机
wide column 宽柱
wide concrete curb 宽的混凝土侧石;宽的混凝土路缘石
wide corridor 宽檐廊
wide crack 宽裂缝
wide-crested measuring weir 宽顶量水堰
wide-crested weir 宽顶堰
wide curb 宽侧石;宽路缘石;宽路边石
wide cut fuel 宽馏燃料
wide cut gasoline type fuel 宽馏分汽油型燃料
wide cutting knife 宽幅平切铲
wide display 宽显示器
wide distribution 广泛分布
wide drum winch 宽筒绞车;宽筒卷扬机
wide dynamic(al) range 宽动态范围
wide energy gap semiconductor 宽禁带半导体
wide face drivage 宽工作面掘进
wide face wheel 宽面轮
wide field 广视场;宽场
wide-field eyepiece 广视野目镜
wide-field ocular 广视野目镜
wide field-of-view eyepiece 大视野广角目镜
wide film 宽胶片
wide fire box 下涨式火箱
wide flange 宽翼缘
wide flange beam H 型梁;宽缘工字钢;宽缘梁;宽翼(缘)梁
wide flange beam mill 宽缘钢梁轧机
wide flanged 宽(翼)缘的
wide flanged section 宽翼缘型材;宽翼缘断面;宽

翼工字钢
wide flanged steel beam 宽翼缘工字钢梁
wide flange girder 宽缘大梁;宽翼缘梁
wide flange H-beam 宽翼工字钢
wide flange I-beam 宽缘工字钢
wide flange steel H 型钢
wide flange steel section 宽翼工字钢
wide flat steel 宽带钢
wide flood plains 宽阔漫水地带
wide foil 宽幅薄膜;宽幅箔
wide frequency filtering 宽频滤波
wide gap junction 宽禁带结
wide gate 宽电闸
wide ga(u)ge 宽轨(距)
wide ga(u)ge railroad 宽轨铁路;宽轨铁道
wide ga(u)ge railway 宽轨铁路;宽轨铁道
wide grained wood 横纹木材
wide handrail 宽扶手
wide hearth kiln 宽体窑
wide-heel London 宽而短的泥刀
wide jet spray nozzle 宽幅喷嘴
wide joint 宽接头;宽缝
wide kerb 宽路边石
wide ledge tub 宽边浴盆
wide level disk 垂直圆盘犁
wide line 宽线
wide line array 宽线组合
wide line profile method 宽线剖面法
wide liner 宽衬垫
wide line stack section 宽线叠加剖面
widely dispersed 分布范围很广
widely grown 广泛栽培
widely-pitched 宽螺距的
widely spaced anticline 隔档式褶皱
widely spaced syncline 隔槽式褶皱
wide-meshed 粗筛孔的;宽网孔(的)
wide-meshed screen 大孔筛;粗筛网
wide-meshed sieve 大孔筛
wide-mesh screen 粗矿筛
wide meter 宽刻度仪表
wide-mounted collector (复合冲压发动机的)宽口收集器
wide-mouthed flat-bottom extraction flask 广口平底抽提瓶
wide mouth socket 漏斗形打捞筒;宽口打捞筒;钟形打捞筒
wide mouth vacuum bottle 大口保温瓶
wide mouth ware 广口制品
widen 加宽;放宽
widenable to 可拓展为
wide natural stone curb 宽天然侧石
widen differences in personal income 拉开收入差距
wide neck container 大口瓶
wide neck flask 广口(烧)瓶
widened channel 拓宽的渠道;拓宽的河段;拓宽的河道;拓宽的河槽
widened intersection 加宽式交叉口
widened pier 加宽的突堤码头
widened planer 宽式刨床
widened planing machine 宽式刨床
widener 扩孔器;展宽机;加宽工具
widen import and export trade 扩大进出口贸易
widening 加宽;横轧展;拓宽;变宽
widening angle 扩张角;扩大角
widening at turns 弯段加宽
widening banks 加宽路堤
widening cuts 加宽路堑
widening line 拓宽线【道】
widening loss 扩展损失
widening of curve 曲线(段)加宽
widening of embankment 路堤加宽
widening of fill 填方加宽
widening of line 线间间距加宽
widening of roadbed 道床加宽
widening of subgrade 路基加宽
widening of the band 扩大变动幅度
widening on curve 曲线加宽;弯道加宽
widening rate 放宽率
widening ribbon 放边
widening road 拓宽道路
widenmannite 炭铅铀矿
widen the road 加宽路面
wide open 全开的

wide open flow 敞喷
wide open hatch 大舱口
wide opening 宽面巷道
wide open space 开敞空间
wide pair 远距双星
wide passband infrared system 宽通带红外系统
wide period range 宽周期范围
wide planting 疏植
wide plate mill 宽板轧机
wide projective bundle 宽投射光束
wide range 宽量程;宽范围的;宽波段;大量程
wide range crystal video receiver 宽带晶体电视接收机
wide range exposure meter 大量程曝光表
wide range meter 宽量程仪表
wide range monitor system 宽范围监视器系统
wide range of habitat 大范围生境
wide range orifice meter 宽压差锐孔流量计;宽压差孔板流量计
wide range oscillator 宽频带振荡器;宽波段振荡器
wide range regulation 宽范围调整
wide range strain 大幅度应变
wide range temperature controller 宽幅度调温器
wide range tunable resonator 宽带可调谐振器
wide-ranging 广泛的
wider currency band 扩大变动幅度
wide rebate 宽裁口
wide return bend 宽径(回转)弯头;宽回转弯头
wide ring 宽年轮
wide-ringed timber 宽年轮木材;粗纹木材;疏纹木材;疏年轮木材
wider strip 宽管带
wide scope 宽频带示波器
wide screen 宽银幕
wide screen cinema 宽银幕电影院
wide-screen film 宽银幕电影
wide seam 宽裂缝
wide selvage 宽留边
wide-selvage asphalt roll roofing 宽面沥青屋面卷材
wide-sense stationary process 广义平稳过程
wide shallow channel 宽浅水道
wide sizing 加大网眼
wide slate 宽石板瓦
wide-spaced 疏柱式的;宽间距的
wide-spaced building 疏柱式建筑物(柱距约等于柱径的四～五倍)
wide-spaced temple 疏柱式庙
wide-spaced wells 按宽井距布置的井
wide space electric(al) precipitator 宽间距电除尘器
wide space electrostatic precipitator 宽间距静电除尘器
wide spacing 大节距
wide span 大跨度
wide-span beam 大跨度梁
wide-span compressed component 大跨度受压构件
wide-span floor slab 大跨度楼板
wide-span frame 大跨度框架
wide-span gantry crane 大跨度装卸桥
wide-span latticed beam 大跨度格构梁
wide-span latticed girder 大跨度格构大梁
wide-span load-bearing system 大跨度承重体系
wide-span precast concrete beam 大跨度预制混凝土梁
wide-span prestressed concrete beam 大跨度预应力混凝土梁
wide-span rib 大跨度肋
wide-span roof 大跨度屋顶
wide-span shell 大跨度薄壳
wide-span shell vault 大跨度薄壳拱
wide-span trussed beam 大跨度桁构架
wide-span trussed girder 大跨度桁构大梁
wide-span wall arch 大跨度墙拱
wide spectral range 宽光谱区;宽光谱段
wide spray nozzle 宽幅喷嘴
widespread 广泛的;宽布的
widespread availability 广泛的利用
widespread corrosion 大面积腐蚀
widespread distributor 宽幅撒肥机
widespreader 撒布装置
widespread grab 大开度抓斗
widespread metastasis 广泛转移

widespread occurance 广泛分布的矿藏
widespread period of basin 盛盆期
widespread rain 大面积降雨;大面积降水
widespread shower 大面积降雨;大面积降水
widespread spiral 螺旋形抛撒轮
widest boiling point range 最宽沸程范围
wide strip 宽带钢;宽带材
wide strip foundation 宽条形基础
wide strip mill 宽带材轧机
wide strip train 宽钢皮列车
wide tape without selvage 无织边宽带
wide tee 大口径三通
wide-throw hinge 宽叶铰链
wide-to-tele zoom lens 广角远摄变焦镜
wide tracked 宽轨距的
wide-tracked caterpillar bulldozer 宽履带推土机
wide-wale corduroy 粗条灯芯绒
wide wheel grinding 宽磨
widget 小机械;小机具
Widia bit 韦地亚钻头
Widmanstatten figure 韦德曼司特顿花纹
Widmanstatten pattern 韦德曼司特顿花纹
Widmanstatten structure 韦德曼司特顿结构
widow maker 喷孔器
widow's walk 屋顶走道;带栏杆屋面;面对大海的屋顶阳台
width 宽(度);面阔
width adjustment 宽度调整
width and depth ratio in aeration tank 曝气池宽深比
width and format of standard 标准的幅面及格式
width angle 齿宽角
width at dam crest 坝顶宽度
width at dam top 坝顶宽度
width at the rubble mound breakwater crest 斜坡堤堤顶宽度
width at top of roadbed 道床顶宽
width at top of subgrade 路基顶(面)宽(度)
widthband decoupling method 宽带去耦法
width basis 路宽分级标准
width between centers 中心距(离)
width choke 调宽扼流圈
width coding 信号宽度编码;宽度编码
width coil 调宽线圈
width constriction 河宽束狭(段);河宽收缩(段)
width contraction 河宽束狭(段);河宽收缩(段)
width control 宽度控制;宽度调整
width control of dredge-cut 挖槽宽度控制
width cutting 纵切
width depth ratio 宽深比
width diameter ratio 宽径比
width flexure vibration mode 宽度弯曲振动模式
width ga(u)ge 测宽仪
width-integrated method 积宽法
width of abrasion shoal patch 磨蚀浅滩宽度
width of a canal 运河宽度
width of active fault 活断层宽度
width of aeration tank 曝气池池宽
width of anomaly 异常宽度
width of anomaly body 异常体宽度
width of an underground river 暗河宽度
width of apron 码头前沿地带宽度
width of area to be covered 覆盖面积宽度
width of a traffic lane 每一车道宽度
width of ballast bed 道床宽度
width of brake drum 刹车鼓宽度
width of bridge carriage-way 桥面净宽
width of butt strap 平接盖板宽度
width of car body 车体宽度
width of carriageway 车行道宽度
width of channel 河宽;渠宽
width of checker channel 格孔间距
width of coast zone 海岸带宽度
width of contribution 补给宽度
width of court 庭院宽度
width of coverage 总面积宽度;视觉宽度
width of cracks 裂缝宽(度);地裂缝宽度
width of cushion 垫层宽度
width of cut 挖泥宽度
width of dam 坝宽
width of depth domain 深度域宽度
width of discharge opening 出料口宽度
width of dispersion belt 弥散带宽度
width of dock barrel 坞室宽度

width of dock entrance 坞口宽度
width of dock land 港口陆域纵深
width of dredge-cut 挖槽宽度
width of dry dock barrel 干坞墙宽度
width of echelon array 雁列带宽度
width of entrance 入口宽度
width of fault scarp 断层崖宽度
width of fault trace 断层迹宽度;断层痕迹宽度
width of fault zone 断裂带宽度
width of feed opening 进料口宽度
width of fetch 风区宽度
width of fissure 裂缝的宽度
width of fold 褶皱宽度
width of footing 基础宽度
width of foundation 基础宽度
width of frequency 频率域宽度
width of fringe 条纹宽度
width of glide mass 滑体宽度
width of groove 轧槽宽度
width of hole 孔径
width of immersion zone 浸没带宽度
width of influence zone 影响带宽度
width of joint 缝宽
width of kink band 膝折带宽度
width of lagoon channel 泄湖水道宽
width of landslide crack 滑坡裂隙宽度
width of lane 车道宽度
width of leakage zone 渗漏带宽度
width of lifting bridge 吊桥宽度
width of mark 标牌宽度
width of navigable passage 通航水道宽度
width of opening 孔隙宽度
width of orebody 矿体宽度
width of outcrop 露头宽度
width of oxidized coal seam 煤层氧化带宽度
width of oxidized coal zone 煤层氧化带宽度
width of parasitic amplitude modulation 寄生调幅宽度
width of parking stall 车位宽度
width of pavement 路面宽度;路基宽度
width of planed surface 刨面宽度
width of port land area 港口陆域纵深
width of recharge zone 补给带宽度
width of regulating line 整治线宽度
width of regulation line 整治线宽度
width of relaxed zone 卸荷带宽度
width of reverse drag structure 逆牵引构造宽度
width of right-of-way 用地宽度【道】
width of river 河宽
width of riverbed 河床宽度
width of road 路幅宽;道路宽度
width of roadbed 路基面宽度;道床宽度
width of road property line 道路红线宽度
width of road reservation 路幅保留宽度
width of roadway 车行道宽度
width of root face 钝边高度
width of rootzone 限带宽度
width of row 行宽
width of ruled area 格网宽度
width of ruling 格网线宽度
width of runoff zone 径流带宽度
width of saddle support 鞍座宽度
width of sailing path 航迹(带)宽度【航海】
width of sand-driving wind 风沙流宽度
width of sanitary protection zone 卫生防护距离
width of seedling shoot 苗宽
width of separation area on acopy 分离边宽度
width of shakes (木材的)环裂宽度
width of shuttles 货叉宽度
width of sidewalk 人行道宽度
width of site 工地宽度;场地宽度
width of slope surface 坡面宽度
width of spectral lines 谱线宽(度)
width of standard section 标准段宽度
width of step 踏步宽度;阶宽
width of stone fall 石瀑宽度
width of strip 航线宽度
width of strip hole 条孔宽度
width of structural broken zone 构造破碎带宽度
width of subgrade 路基(面)宽度
width of swath 铲雪宽度;犁幅宽度;锄幅宽度
width of tectonic region 构造区的宽度
width of tension zone 斜坡张应力带宽度
width of terrace surface 阶地面宽度

width of the high reflectance zone 反射膜的带宽
width of the machined surface 被加工表面宽度
width of the magnetic body 磁性体宽度
width of the workpiece 工件宽度
width of thread 螺纹宽度
width of throat 辙叉喉宽度
width of time domain 时间域宽度
width of tooth 齿宽
width of transitional belt salt concentration 盐分浓度过渡带宽度
width of transition steepness 前沿陡度
width of trap 圈闭宽度
width of tree 树冠宽度；冠径
width of velocity anomaly band 速度异常带宽度
width of vertical strip 竖条宽度
width of water cross-section 过水断面宽度
width of water-level 水面宽(度)
width of watershed 分水岭宽度
width of weathered coal zone 煤层风化带宽度
width of wheel tread 车轮踏面宽度
width or diameter of submerged structure or member 水下结构或构件的宽度或直径
width span ratio 宽跨比
width specification 宽度说明
width surface classification 路面宽度分级
width thickness ratio (集料的)宽厚比
width-to-depth ratio (截面的)宽高比
width-to-span ratio 宽跨比
widthwise 横向地
Wiechert method 维谢尔法接地电阻测定法
Wiechert seismograph 维谢尔地震仪
Wiedemann effect 威德曼效应
Wiedemann-Franz constant 威德曼—弗蓝兹常数
Wiedemann-Franz law 威德曼—弗朗兹定律
Wiedemann-Franz ratio 威德曼—弗朗兹比值
Wiedemann's additivity law 威德曼加和律
Wiegmann roof truss 维格曼屋架
Wiegold 齿黄铜
wiemet 维梅特硬质合金
Wien constant 维恩常数
Wien displacement law 维恩位移定律
Wien effect 维恩效应
Wien equation 维恩公式
Wiener best filter 维纳最佳滤光片
Wiener deconvolution 维纳反褶积
Wiener ergodic theorem 维纳遍历定理
Wiener experiment 维纳实验
Wiener filter 维纳滤波器
Wiener filtering 维纳滤波
Wiener-Hoof equation 维纳—霍普夫方程
Wiener-Hoof technique 维纳—霍普夫技术
Wiener information theory 维纳信息论
Wiener-Khinchine relation 维纳—钦辛关系式
Wiener-Khinchine theorem 维纳—钦辛定理
Wiener-Levinson algorithm 维纳—雷文森算法
Wiener optimization 维纳最佳化
Wiener process 维纳过程
Wiener reaction 维氏反应
Wiener's optimum system 维纳最佳系统
Wien frequency bridge 维氏频率电桥
Wien's displacement law 维恩位移定律
Wien's distribution law 维恩分布律
Wien's law 维恩定律
Wien's rediation law 维恩辐射定律
Wieri equation 威里方程
Wiese formula 威斯公式
Wiesen 草甸
wifatite 酚醛氨基酸塑料
wigan 帆布似的平纹棉布
wiggle-beat method 摆动拍频方法
wiggled cleavage 波状劈理
wiggle effect 失常效应
wiggle nail 波形钉
wiggle-spring casing hook 螺旋形弹簧提引钩
wiggle tail 肘节机具；铰链式钻具
wiggle trace 波型曲线；波形曲线；波形记录
wiggle trace superimposed on variable area 波形加变面积记录
wiggle trace superimposed on variable density 波形加变密度记录
wiggling-in 移站定线法【测】
wightmanite 伟硼镁石
Wigner approximation 维格纳近似
Wigner-Eckart theorem 维格纳—埃卡特定理
Wigner effect 维格纳效应
Wigner energy 维格纳能
Wigner force 维格纳力
Wigner gap 维格纳间隙
Wigner growth 维格纳增长
Wigner isobar 维格纳同量异位素
Wigner release 维格纳释放
Wigner-Seitz cell 维格纳—赛茨晶胞
Wigner-Seitz method 维格纳—赛茨法
Wigner's rational approximation 维格纳合理近似
Wigner supermultiplet 维格纳超多重态
Wigner theorem 维格纳定理
wig-wag 摆动器；信号旗；打信号
wig-wag motion 汽车摆头
wig-wag signal 摇摆(式)信号(表示列车接近)；灯光信号；(信号兵、火车站等)打信号
wig-wag tool 摆动抛光机
wigwam 圆柱形焚化炉；棚屋
wigwam burner 工厂废料焚化炉
wigwam-type incinerator 圆柱形焚化炉；棚型焚化炉
wig-way signal 信号
wiilfingite 正羟锌石
Wijs method 韦伊斯碘值测定法
wilcoxite 水氟铝镁矾
wild 操舵不稳
wildacat 空头的
wild animal of national priority protection 国家重点保护野生动物
wild animals protection area 野生物保护区
wild barley 芒麦草
wild boar 野猪
wild bore 失控井孔
wild card 通配符
wildcard character 通配符
wildcat 冒险性企业；冒牌的；起锚(机)绞筒；初探井；持链轮；非法经营的；无计划钻探
wildcat area 盲目钻井区；初探区
wildcat drilling 盲目钻探；钻野猫探井；普查钻探
wildcat sprocket 绞盘链轮
wildcat stage 初步勘探阶段
wildcatter 投机分子
wildcatting 盲目开采的；石油试钻井；钻野猫探井；盲目钻探；(石油)探测井
wildcat well 野猫井；初探井
wild cat wheel 锚机持链轮
wild coal 野煤
wild effect 不稳定现象
Wild elevator 魏尔德起重机
wilderness 荒野；荒漠；荒地
wilderness area 原生环境保护区；荒芜地；保留自然环境地区；保留自然面貌地区
wilderness preservation area 原始自然环境保护区；保留自然面貌地区
Wilderness(stage) 魏尔德尼期阶
wild fauna 野生动物区系
Wild fence 魏尔德栅栏；栅栏(魏尔德雨量器四周围用)
wildfire 荒地火
wild flooding 失控洪水；浸灌；暴风雨泛滥
wild-flooding irrigation 漫灌；大片漫灌；暴雨泛滥
wild-flooding method 大水漫灌法；漫灌法
wild flowing 敞喷猛喷
wildflysch 野复理石
wild folds 无序褶皱
wild garden 野趣园；野生园
wild gasoline 不稳汽油
wild gas well 敞喷气井
wild ginger 杜衡
wild grain 乱纹理
wild grass ground 荒草地
wild grassland 野生草地
wild heat 强烈沸腾熔炼
wild hole 乱开钻孔；乱开炮眼
wild interest 难以追索的产权变迁
wild land 未开垦的土地；野地；荒地
wildlife 野生动物；野生生物
wildlife conservation 野生生物保护
wildlife fauna 野生生物区系
wildlife habitat 野生生物栖息地
wildlife park 新式动物园
wildlife preserve 野生生物保护
wildlife protection area 野生生物保护区；野生动物保护区
wildlife protection system 野生动物保护系统
wildlife refuge 野生动物保护区
wildlife reserves 禁猎区
wildlife resources 野生生物资源
wildlife state 野生状态
wild metal 猛沸腾金属；冒涨金属
wild-mouth bottle 广口(烧)瓶
wildness 猛烈沸腾
wild night 暴风雨之夜
wild observation 不切实际的观测值
wild park 天然公园
wild pasture 野生草场；天然牧场
wild plants botanical garden 野趣园
wild plants garden 野生植物园
wild rose 蔷薇
wild rubber 天然橡胶
wild sea 狂风大浪；大风浪海洋；大风浪的海
wild seeding 天然下种
wild snow 原始雪
wild snow avalanche 狂暴雪崩
wild species 野生种
wild steel 冒涨钢；强烈沸腾钢
wild trade 不法贸易
wild trajectory 任意轨道
wild turmeric 姜黄
wild well 野喷井；敞喷井
wild wind 狂风
wild wood 原始林；天然林
Wilfley table 维尔弗莱型摇床
wilful act 蓄意行为
wilful inflation of prices 随意涨价
wilfully caused loss 蓄意造成的损失
wilful misconduct 蓄意的违法行为
wilkeite 氧硅磷灰石；硅硫磷灰石
wilkinite 皂土；胶膨润土
Wilkinson oven 威尔金逊窑
Wilkinson's blue 威尔金森蓝
wilkmanite 斜硒镍矿
will 遗嘱
Willans line 总耗汽量曲线；威兰斯线
willemite 天然硅酸矿；硅锌矿；大硅锌矿
willemite crystalline glaze 硅锌矿结晶釉
willemseiye 镍滑石
Willesden paper 铜铵纸；威尔斯登式纸材
Willett forceps 威勒特氏钳
willey 开松机；打土机
willful arrangement 任意排列；任意布置
willhendersonite 板沸石
William Janney pump 威廉珍尼泵
William's core 通气芯；大气压冒口泥芯
Williams-Landel-Ferry relation 威一兰一菲三氏关系式(液体黏度与温度关系)
Williamson amplifier 威廉逊型放大器
Williamson kiln 威廉逊窑
Williamson's test 威廉逊氏试验
Williamson synthesis 威廉逊合成
Williamson turn 威廉逊旋回
William's plastometer 威廉氏塑度计；威廉斯塑性仪(平板式)
William's refractometer 威廉斯折射计；威廉士折射计
William's riser 大气压力冒口
William stone 含镍蛇纹石
willingness to pay 支付意愿
willingness to pay princple 愿付原则
willingness to sell 销售意愿
Williot deflection diagram 威洛(变位)图
Williot diagram 威洛图
Williot displacement diagram 威洛位移图
Williot graphic(al) method 威洛图解法
Williot-Mohr diagram 威洛—摩尔图
williwaw 威利瓦飑
will not investigate 不予追究
willow 柳属；柳木；帚料
willow blue 柳叶青
willow box groyne 柳梢捆丁坝
willow brush 柳梢
willow check dam 柳枝谷坊
willow fascine 柴排
willow fascine mattress 柳排褥；编柳排工
willow fascine revetment 柳(柴)排护坡
willow green 柳绿
willow groin 柳梢(透水)丁坝
willow groyne 柳梢(透水)丁坝

willow-herb 柳兰
willowing machine rag duster 破布除尘机
willow mattress 柳枝排;柳条沉排
willow oak 北美柳栎
willow pattern ware 柳叶图案陶瓷器
willow revetment 柳枝护岸
willow sapling 摇摆;柳条
willow tree 柳树
willow twig 柳条
willyamite 辉锑钴矿
wilshite 威尔什石
Wilson cycle 威尔逊旋回
Wilson depression 威尔逊凹陷
Wilson effect 威尔逊效应
Wilson experiment 威尔逊实验
Wilson gear 威尔逊齿轮
Wilson line 威尔逊线
Wilson seal 威尔逊密封
Wilson's theorem 威尔逊定理
Wilson tracks 威尔逊射线径迹
Wilson-θ method 威尔逊—θ 法
wilt 枯萎病
wilting coefficient 枯萎系数;凋萎系数
wilting moisture content 凋萎含水量
wilting percentage 枯萎百分数;凋萎(含水)率
wilting phenomenon 凋萎现象
wilting point 植物萎蔫点;凋萎点;凋萎点
wilting range moisture 凋萎含水量范围
Wilton carpet 威尔顿提花地毯
Wilton rug 威尔顿机织绒头地毯
Wilton still 威尔顿管式炉
wiluite 硼符山石
wimble 清孔器;钻孔清渣器;螺杆钻;手摇钻
win a bid 承包中选;中标
win a tender 中标;得标
winch 有柄曲拐;辘轳;卷扬机;绞盘;绞车;起货机
winch attachment 卷扬机附属设备;绞车附设备;绞车附件
winch barrel 绞车卷(缆)筒;绞车滚筒
winch bed 绞车座
winch brake 绞车刹车
winch brake band spring 绞车闸带弹簧
winch brake case 绞车闸箱
winch braking valve 绞车制动阀
winch cable 卷扬(缆)绳;绞盘缆索;绞车缆(索);绞车钢丝绳
winch cable clamp 绞车缆头
winch cable drum 绞盘鼓筒
winch cable guide 绞盘钢索导向夹
winch capacity 绞车提升能力;绞车牵引力
winch capstan 起锚机
winch chain 绞车链
winch chain and sprocket 绞车链及链轮
winch clutch 绞车离合器
winch control room 绞车操纵室
winch cover 绞车罩
winch deck 绞车甲板
winch drag brake 绞车闸
winch drawn dung channel scraper 绞盘牵引式粪沟清理铲
winch drive 卷扬驱动;绞车驱动
winch drive gear 绞车传动机构
winch drive shaft 绞车主动轴
winch drive shaft universal joint 绞车驱动轴万向节
winch driving gear 绞车驱动装置
winch driving sprocket 绞车驱动链轮
winch drum 卷筒;绞车卷(缆)筒;绞车滚筒
winch drum clutch shift fork 绞盘鼓筒离合器拨叉
winch drum shaft 绞车卷筒轴
winch drum sliding clutch 绞盘卷筒合套;绞盘鼓筒滑动离合器;绞盘卷筒离合器
winchester 标准细长玻瓶
Winchester cut method 温切尔特切割法
Winchester cutting 温切尔特切割;瓦片切割
winch for electric(al) logging 电测绞车
winch for pod 控制盒绞车
winch frame 绞车架
winch guard 绞车护板
winch handle 绞车操纵手柄
winch-hauled plastic pipe layer 绞盘拖曳式塑料管铺设机
winch-hauled trenchless pipe layer 绞盘拖曳式塑孔铺管机

winch head 绞车副卷筒;辘轳头
winch hoist 起重绞车;卷扬吊起;绞车起重机;绞车卷扬机
winch house 绞车房
winching 卷扬
winching station 绞滩站
winchite 蓝透闪石
winch launch 绞盘车牵引起飞
winch line 绞车钢索;卷扬绳;绞车索;绞车绳
winch locking pin 绞车制动销
winch machine 绞盘机;绳状洗布机
winchman 绞车司机;绞车手;绞车操纵者
winch motor 绞车马达
winch operator 卷扬机操作工;绞车操作工;绞车工
winch pawl 绞车爪
winch pawl wheel 绞车棘轮
winch pinion 绞车小齿轮
winch platform 绞车操纵台;起货机台
winch power takeoff 绞车功率输出
winch pull 绞车拉力
winch ratchet wheel 绞车棘轮
winch reductor 绞盘减速器
winch reductor worm 绞盘减速蜗轮
winch rope 卷扬绳;绞车索;绞车绳
winch runner 绞车手;绞车操纵者
winch shear pin 绞车安全销
winch shift lever 绞车安速杆
winch sliding clutch 绞车滑动离合器
winch speed 绞车速度
winch spool 绞筒
winch support 绞车支架
winch suspension 绞盘悬置
winch table 绞车操纵台;起货机台
winch truck 卷扬机车;绞盘车
winch type cold-drawing machine 卷扬机式冷拉机
winch with double barrel 双股绞车
winch worm 绞车蜗杆
winch worm bearing cover 绞车蜗杆轴承盖
winch worm gear 绞车蜗轮
winch worm gear case bushing 绞车蜗轮壳衬套
winch worm housing cover 绞车蜗杆壳盖
winch worm shaft 绞车蜗杆轴
winch worm shaft bearing 绞车蜗杆轴承
winch worm shaft bearing cap 绞车蜗杆轴轴承盖
winch worm shaft bushing 绞车蜗杆衬套
wind 开发去;卷绕;缠
wind abaft the beam 后侧风
wind abeam 正横风;侧风
wind abrasion 风蚀(作用);风磨蚀;风力侵蚀
wind abreast 横风
wind accelerating dike[dyke] 风力加速堤
wind action 风力作用;风的作用
wind after 尾后风
windage 风力影响;风力修正量;风吹流
windage area 受风面积
windage loss 风阻损失;风吹损失
windage resistance 风阻(力)
windage scale 风力计
windage set-up 风扬起
windage stress 风应力
windage suction 风吸力
windage vane 风向标
windage yaw 风偏
wind ahead 迎风;顶头风
wind aloft 高空风
wind aloft observation 高空风观测
wind and drift chart 风力偏流修正表
wind and dust protection planting 风尘防护林
wind and fire kiln 风火窑
wind and flood damage 风洪损害
wind and snow loads 风雪荷载
wind area 受风面积;风压面积
wind arm 操纵杆
wind around 旋绕
wind arrow 风向标;风矢
wind axes 风轴系
wind axis 风轴
wind backs 风向逆转
wind backwater 风壅水
wind-baffle door 风门
wind bar 大风浪
wind-barometer table 风速—气压换算表
wind barrier 防风设施(防风林、挡风墙等)
wind basin 风蚀盆地

wind beam 抗风系杆;抗风梁
wind belt 风带;防风林(带)
wind bench 风箱台
windbend 风弯
wind bending moment 风弯矩
wind bent 抗风木架;风排架
wind blast 矿内暴风;风害
wind blocking coefficient 风阻系数
wind blower 通风机;吹送器
wind blow in 倒灌风
wind-blown deposit 风积物
wind-blown dirt 飞灰
wind-blown dust 风扬尘土;风积尘;风成尘
wind-blown sand 风(积)沙;风沙磨蚀
wind-blown silt 风化粉土
wind-blown soil 风积土(壤);飞土
wind bore 进气管(泵)
wind-borne 风送的;风成的
wind-borne anomaly 风积物异常
wind-borne debris 风积碎屑
wind-borne deposit 风积物;风积土(壤);风成沉积(地)层
wind-borne deposition 风成沉积作用
wind-borne dust 风载尘
wind-borne load 风携质
wind-borne sand deposit 风成沉积砂层
wind-borne sediment 风积物;风积土(壤);风成堆积物;风成沉积(物);风成沉积(地)层
wind-borne sedimentation 风成沉积作用
wind-borne soil 风积土(壤)
wind-borne succession 风成序列
wind box 空气室;风箱;风室
wind brace 抗风支撑;抗风联结系;抗风联杆;牵条;风撑;屋架间抗风力剪刀撑
wind-braced boom 风撑杆;有抗风支撑的吊杆;抗风支撑吊杆;置有抗风支撑的吊杆
wind bracing 抗风支撑;抗风联结系;风撑
wind bracing connection 抗风支撑连接
wind bracing system 风撑系统
windbreak 挡风墙;防风墙;防风带;防风设备;防风障;防风设施(防风林、挡风墙等);防风林;风障;风折
wind breakage 风折
windbreak and sand-fixation forest 防风固沙林
windbreak equipment 防风设备
windbreaker 防风设施(防风林、挡风墙等);风雪衣;防风衣
windbreaker cloth 防风布
windbreak fence 防风篱笆;防风围栏
windbreak forest 防风林
windbreak planting 防风栽植
windburn 风害;风吹性皮肤伤
wind by day 昼风
wind cable 抗风(缆)索;抗风钢索;风缆
wind cable tie 抗风拉条;抗风索系杆
wind calculator 风速计算器
wind cap 防风(雨)罩;防风帽
wind-carried pebble 风蚀砾岩
wind carried sand 风蚀沙
wind-carved 风蚀的
wind-carved conglomerate 风蚀砾岩
wind-carved gravel 风蚀砾石
wind-carved pebble 风蚀砾石(卵石)
wind catcher 舷窗风斗;风斗
wind cave 风穴
wind chamber 抽风箱;抽风室
wind channel 空气风洞;风洞;风道
wind charger 风力充电机
wind chart 风力图
windcheater 防风衣
wind chill 吹风冷却;风力降温;风寒
windchill blower cooling 吹风冷却
wind chill index 气流冷却指数;风冷指标;风寒指数
wind chop 白浪(花)
wind cirrus 风卷云
wind class 风力等级;风级
wind classification 风的分类
wind clock 钟面式风向指示器
wind cock 风(向)标
wind coefficient 风系数
wind condition 风条件;风况
wind cone 风向器;圆锥形风标;风向锥;风向袋
wind conveyer 风力输送机;气力(式)输送机
wind-cooled 风冷的

wind cooling 风冷
wind cork 风标
wind correction 风力修正
wind corrosion 风蚀
wind cowl 烟囱风帽;喇叭式风斗;风帽
wind crack 风裂
wind cross 抗风剪刀撑
wind crust 风成雪壳
wind current 吹(送)流;风(送)流;风吹流;风成流;气流;风驱流
wind-cut stone 风棱石
wind damage 风害
wind deflation 风力吹蚀作用
wind deflation pavement 风蚀覆盖层
wind deflection 风向偏转;风力偏移
wind deflector 挡风板
wind denivellation 增水倾斜度;减水倾斜度;风浪倾斜度
wind deposit 风积层;风成沉积层;风积物
wind-deposited 风积的
wind-deposited coast 风积海岸
wind-deposited sediment 风积泥砂
wind-deposited soil 风积土(壤)
wind deposition 风积;风成沉积
wind-deposition coast 风积海岸
wind desert 风成沙漠
wind desiccation 风干
wind-designed 抗风设计的
wind-detecting radar 测风雷达
wind dew 风(生)露;风力露
wind diagram 风向频率图;风玫瑰(图);风(力)图
wind dial 测风仪;风速表度盘;风力表
wind direction 风向
wind directional diagram 风玫瑰图
wind directional effect coefficient 风向影响系数
wind directional indicator 风向标;风向指示器
wind directional recorder 风向自记器;自记风向仪
wind directional rose 风向玫瑰图
wind directional shaft 风矢杆
wind directional spectrum 风向谱
wind disk 挡风盘
wind dispersal 风散播
wind disturbance 风扰动
wind divide 风向界线
winddorn 风刺
wind drift 漂流;吹流;风漂移;风吹浪;风成流;风积;飞沙;表层漂流
wind-drift convection 风力混合作用
wind-drift current 风生(海)流
wind-drift line 风向突变线
wind-drift of ice 冰漂浮
wind-drift sand 风(积)沙
wind-drift surface movement 表层吹流运动
wind-driven 风驱动的
wind-driven atomizer 风轮传动式喷雾器
wind-driven circulation 风生环流
wind-driven current 风海流;漂流;风生(海)流;风驱(动)流;风吹流;风成流
wind-driven dynamo 风力发电机
wind-driven generator 风动发电机;风力发电机
wind-driven mixed layer 风成混合层
wind-driven ocean-surface current 风生洋面流
wind-driven rotor ship 旋筒风力推进船
wind-driven turn indicator 风动转弯指示器
wind-driven water pump 风力提水机;风动水泵
wind-driven wave 风驱(动)波;风成浪;风成波
wind duration 吹送延时;风时;风的吹送延时;风吹持续时间
winded 印刷版的分离通风
wind eddy 涡动气流;风涡
wind edge 风蚀边
wind effect 风力作用;风力影响;风力效应
wind electric(al) power generation 风力发电
wind-enduring plant 抗风树木;耐风植物
wind-enduring tree 抗风树木;耐风树木
wind energy 风能;风力能源
wind energy utilization 风能利用
wind engineering 抗风工程;风力工程(学);风工程
wind engineering test 风工程试验
winder 斜踏步;卷绕机;卷片机;卷簧器;绕线机;盘梯;扇形踏步;缠绕机;缠线机;拨禾轮
winder injury control 冬害防治
winder motor 提升机电动机
wind-eroded castle 风蚀城堡

wind-eroded pillar 风蚀柱
wind-eroded soil 风蚀土壤
wind erosion 风蚀(作用);风侵蚀作用;风力侵蚀
wind erosion cave 风蚀洞
wind erosion class 风蚀等级
wind erosion formula 风蚀公式
wind erosion lake 风蚀湖
wind erosion of soil 土壤风蚀
winder reel 卷取机
winder stair(case) 斜踏步楼梯
winder with reciprocating collet 机头往复移动拉丝机
W-index W 指数;平均渗透率
wind-excited oscillation 风激振荡
wind-excited vibration 风激振动
wind-faced pebble 风棱卵石
wind factor 风力系数
wind fairing 风嘴
wind fall 风折(树)林;吹落的果实;风折枝;风落物;意外收获;风倒区;风倒木
windfall profit 暴利;意外利益;超额利润
windfall profits tax 暴利税
wind farm 风电场
wind farming 风力农场
wind fence 防风栅栏
wind fetch length 风区宽度
wind field 风区;风场
wind field chart 风场图
wind-field interaction 风场互作用
wind filling 梁间填充的砖石砌体
wind-finding radar 测风雷达
wind-fire angle 射向与风后夹角
wind fire-extinguisher 风力灭火机
wind flow 风力量
wind flower 风银莲花属
wind force 风力
wind-force coefficient 风力系数;风荷载系数
wind-force diagram 风荷载图;风力图
wind-force direction ga(u)ge 风力方向计
wind-force distribution 风力分布
wind-force scale 风级(表);风力级
wind-formed bar 风成沙洲
wind frame 抗风构架
wind-free plummet 防风垂球
wind frequency 风频率
wind friction 风摩擦
wind frost 风霜
wind furnace 自然通风(式)炉;通风炉
wind gap 风隙;风口;风害造成的空隙;风谷;风坳
wind ga(u)ge 风速计;风力计;风力表;风压计;测风速器;测风速计
wind ga(u)ge sight 风速计观测器
wind-generated current 风生(海)流
wind-generated gravity wave 风生重力波
wind-generated noise 风生噪声
wind-generated sea wave 海上风浪;风成海浪
wind-generated turbulence 风致湍流
wind-generated wave 风生波;风(成)浪;风成波
wind-generated wave spectrum 风浪谱
wind generating set 风力发电机组
wind girder 抗风大梁;储[贮]罐防风梁;防风梁
wind gorge 风峡
wind-grooved stone 风棱石
wind guard 防风构造;抗风构造
wind guard wall 挡风墙
wind gust 阵风
wind gust loading factor 风振系数
wind gust speed 阵风速度
wind header 风罩
wind hield wiper equipment 挡风玻璃雨刷
wind hole 风穴;风眼
wind in 绞进
wind in Beaufort scale 布伦福特风级
wind indicator 风向仪;风向风速指示器;风力指示器;风力器
wind-induced 风成的,风引起的
wind-induced current 吹流;风致海流;风生(海)流
wind-induced current gyro 风生环流
wind-induced mixing 风致混合
wind-induced motion 风致运动;风生运动
wind-induced oscillation 风致振荡;风致振荡;风激振荡
wind-induced random wave 风致随机波;风引起的随机波

wind-induced surge 风成涌浪
wind-induced turbulence 风致湍流;风生湍流
wind-induced vibration 风致振动;风力引起的振动
wind-induced wave 风生波;风浪;风成波
winding 绕组;绕线;绕圈;蜿蜒;提升;缠绕(的)
winding angle 绕纱角度;缠绕角
winding apparatus 绕线机;打轴机
winding arrangement 绕组布置
winding availability 绕组利用率
winding barrel 卷筒;绞车卷(缆)筒;提升绞筒
winding beam 绞盘棒
winding bridge 曲桥
winding by hand 手绕
winding cable 提升钢丝绳
winding cage 提升罐笼
winding circuit 绕组电路
winding coastline 弯曲海岸线
winding coefficient 绕组因数;绕组系数
winding coil 绕纱圈层
winding constant 绕组常数
winding construction 绕组结构
winding core type transformer 芯式变压器
winding corridor 曲廊;曲廊
winding current 绕组电流
winding cutter 卷切机
winding cycle 提升循环
winding data 绕组数据
winding department 绕线间
winding depth 提升高度
winding diagram 绕组图;绕线图
winding displacement 绕线位移;排线
winding distribution 绕组分布
winding distribution ratio 绕组分布系数
winding down frame 卷取机构
winding drum 卷筒;提升卷筒;提升绞筒;缠索轮
winding-drum machine 卷扬机;滚筒式卷扬机
winding eddy current loss 绕组涡流损失
winding element 绕组元件;绕组单元
winding engine 卷扬机;提升机
winding engine drum 卷扬机卷筒
winding engine indicator 提升机深度指示器
winding face 滚筒面;(卷扬机的)转筒长度
winding factor 绕组因数;绕线系数;缠绕速比
winding film 卷片
winding former 绕组模;绕线模
winding gear 拖拉设备;卷扬装置;卷扬机齿轮;卷绕机构;升降装置
winding guide 绕线导架
winding handle 旋转手柄;摇手柄
winding head 绕组端部
winding induction motor 三相绕线式感应电动机
winding installation 卷绕装置
winding insulation 绕组绝缘
winding interval 提升间歇时间
winding knob 卷钮
winding layout 绕组布置
winding length 绕组长度
winding lever 卷拔杆
winding line 盘山线
winding loss 绕组损耗
winding machine 卷绕机;卷切机;绕线机;提升机;缠绕机;缠绕机
winding material 绕组用导线
winding mechanism 绕线机构
winding motor 提升电动机
winding number 绕数
winding-off 退卷;退绕
winding-off device 卷取装置
winding-off installation 退绕装置
winding-off length 退绕长度
winding-off machine 放轴机
winding of river 河流曲折;河道蜿蜒;河道蜿曲;河道曲折;河道的摆动
winding operation 绕线操作
winding over-hang 绕组端部
winding overhang support 绕组端部支撑
winding path 曲径
winding pattern 缠绕线型
winding pipe 风管;弯管
winding pitch 绳索绕组节距;线圈节距;绕组节距;缠索节距;缠绕间距
winding-point 分枝点
winding positioning machine 绕线定位机
winding power 绕组功率

winding protection 绕组保护
winding pulley 绕线滑轮
winding ratio 卷绕比
winding reel 辊式卷线机;卷线筒;绕线筒
winding resistance 线组电阻
winding river 蜿蜒河流;弯曲河流
winding river winding stream 蛇曲河(流)
winding road 盘陀道路;弯曲道路;弯弯曲曲的道路
winding roll 卷线盘
winding rope 绳索;绞车;卷绳;起重索;提升(绳)索;提升钢丝绳
winding section 绕组单元;弯曲河段
winding shaft 卷轴
winding shield 绕组护罩
winding space 绕组空间;绕组间隙
winding space factor 绕组占空系数
winding speed 卷绕速度
winding square 上条方柄头
winding stair(case) 盘旋式楼梯;圆或椭圆形盘旋楼梯;旋梯;螺旋(式)楼梯;盘旋柱
winding stem 攀缘茎
winding stick 直规平面检测器
winding stream 蜿蜒河流;弯曲河流
winding support 绕组支撑
winding tackle 起重滑车组
winding temperature indicator 线圈温度指示器;绕组温度指示器
winding temperature trip 绕组温度跳闸
winding tension 缠绕张力
winding terminal 绕组出线端
winding tester 绕组试验器
winding-to-earth capacitor 绕组对地电容
winding tower 卷扬塔;井塔;井架
winding tube 卷绕管;绕线管
winding up 清盘;清理(场地)
winding up drum 卷绕筒
winding up roller 卷辊
winding up sale 结业出售
winding utilization 绕组利用率
winding voltage 绕组电压
winding volume 绕组体积
winding weight drum 绕锤辊
winding wire 绕组线
wind injury 风害
wind instrument 气流计;风速计;风力计
wind intensity 风的强度
wind-laid deposit 风力堆积;风成沉积(物);风成堆积物
wind-laid soil 风积土
wind lake 风成湖
windlass 小绞车;辘轳;链式绞盘;卷扬机;绞盘;绞车卷(缆)筒;起锚机;提升;手摇(式)绞盘
windlass bitts 起锚机(系)柱
windlass bucket 小吊桶;绞盘吊桶
windlass capstan 起锚绞盆
windlass gate lifting device 绞盘式闸门启闭机
windlass heaver 起锚机推杆
windlass machine 吹风机
windlass main 抽风总管
windlass meter 风速表
windlass mill 风车
windlass room 起锚机舱
windlean 风歪
windless coast 无风海岸
windless region 无风区
wind load area 风载面积
wind load assumption 风荷载假设
wind load capacity 风载能力
wind load(ing) 风荷载;风压力
wind loading rating 抗风能力
wind load moment 风荷载力矩
wind load regulation 风荷载规范;风荷载规则
wind load resistance 耐风载性
wind load stress 风荷载应力
wind load test 风压试验
wind lock 风向标
wind lop 短(涌)浪
windlown tree 风扭木
wind lull 风暂停;风速暂减
wind map 风图
wind measurement 风(的)测定
wind measurement from satellite 卫星测风
wind meter 风速计;风力计
windmill 风力站;风力发动机;风车;风磨(坊)
windmill anemometer 风车式风速计;风车式风速表
windmill brake 风车式制动装置
windmill-braking condition 风车制动状态
windmill for drainage 风力排水机;排水风车
windmill generator 风力发电机
windmill governor 风力调速器
windmill palm 棕榈
windmill pump 风力泵;风(车)式泵;风车驱动泵
windmill sail 风力发动机工作轮叶
wind-mixed isothermal layer 风成等温层
wind mixing 风混合
wind model 风模型
wind moment 风载力矩;风力矩
wind moment coefficient 风力力矩系数
wind motor 风力马达;风力发动机
wind movement 风的运动
wind noise 风噪声
wind of 120 days 十二旬风
wind of Beaufort force eight 八级风
wind of Beaufort force eleven 十一级风
wind of Beaufort force five 五级风
wind of Beaufort force four 四级风
wind of Beaufort force nine 九级风
wind of Beaufort force seven 七级风
wind of Beaufort force six 六级风
wind of Beaufort force ten 十级风
wind of Beaufort force three 三级风
wind of Beaufort force twelve 十二级风
wind of Beaufort force two 二级风
wind off 退卷;退绕
wind onset 风增水;风壅水
window 观察窗;构造窗;干扰雷达用的金属带;金属干扰带;人工雷达干扰;窗口;车窗;冰窗口
window absorption 窗吸收
window accessories 窗的附件
window air conditioner 窗式空调器;窗式空气调节器
window air conditioning unit 窗式空调(机)
window alignment error 窗口定位出错
window amplifier 上下限幅放大器;窗放大器
window and frame packaged unit 窗与窗框成套件
window aperture 窗孔;窗洞;窗口
window application 窗口应用程序
window apron 窗间墙板;窗台下的板;窗台板;窗盖口条
window arch 窗拱;窗的拱弧
window area 窗面积
window area sign 窗面招牌
window assembly 窗的装配
window awning 窗凉篷;窗的遮阳篷
window awning blind 窗遮阳篷
window back 窗后板;窗台线;胸高护墙线;窗腰
window band 带形窗;窗带;带状窗
window bar 窗闩;窗棂;窗栅;窗芯
window bay 凸出墙外的凸窗
window bead 窗内侧挡条;吊窗压条;吊窗压条;吊窗(挡钩);窗框压条;窗压玻璃条
window blind 百叶窗;板窗;遮光(窗)帘;窗帘
window-blind fastener 百叶窗扣
window blind Holland 窗帘布
window block 窗口块
window board 窗台板
window bole 采光通风孔
window bolt 窗插销
window bossing 窗腰
window bottom rail 窗扇下横档;窗扇下冒头
window boundary 窗口边界
window box 窗台花箱;窗锤箱;平衡锤盒;吊窗锤箱;窗框两侧空槽;窗槛空槽
window box garden 窗前花池
window breast rail 窗台栏杆;窗腰栏
window brightness 窗亮度
window budding 窗形芽接
window builder 窗的施工人员
window burst transfer 窗口分页转换
window butt 窗铰链
window button 窗开关旋钮
window cardfile 窗口卡片文件
window case 窗(扇)框;窗樘
window casement 平开窗;(平开的)窗扇
window casement frame 窗樘;窗框
window casement hardware 窗五金
window casement stay 窗撑
window casing 窗套;装修的窗框
window catch 窗栓;窗插销;窗钩;风钩
window cellar 酒窖
window center line 中腰窗
window check 窗铲口;窗半槽边;窗边槽口
window cill 窗台;窗底
window cill block 窗台块
window cill brick 窗台砖
window cill cover 窗台盖板
window cill head height 窗台楣高
window cill rail 窗台栏杆
window cill slab 窗台板
window cill tile 窗台砖
window class structure 窗口分类结构
window cleaner 擦窗器;玻璃刮水器
window cleaner's anchor 擦窗人系安全带用锚钩
window cleaner's platform 擦窗吊篮;擦窗活动平台
window cleaning 窗的清洗;擦窗
window cleaning apparatus 擦窗器
window cleaning balcony 清洗窗的平台
window cleaning cradle 窗的清洗台架
window cleaning equipment 窗的清洗设备
window cleaning facility 擦窗设施
window clipping 窗口剪取
window clipping plane 窗口剪取平面图
window coating 窗户涂层
window column 窗间柱
window comparator 窗口比较器
window component 窗的部件
window conditioner 窗式空气调节器
window conversion 窗口转换【计】
window cooler 窗式冷却器
window coordinate 窗口坐标
window coordinate system 窗口坐标系统
window cornice 窗上线脚;窗口檐板
window counter tube 窗口计数管
window cradle machine 修窗吊篮;擦窗吊篮;洗窗机
window cross 窗横撑
window crown 开面冠
window curtain 窗帘
window cut of baffle 折流板弓形缺口
window data structure 窗口数据结构
window decoration fitting 窗装饰配件
window definition block 窗口定义块
window definition block status 窗口定义块状态
window defroster (汽车玻璃的)防霜装置
window deposition 波浪形堆料
window design 窗的设计
window dial 窗梯
window dimension 窗的尺寸
window directory 窗口目录
window display 橱窗陈设;窗展
window display mode 窗口显示模式
window division 窗格
window drape paper 窗帘纸
window draughtproofing 窗的防气流性
window dressing 橱窗设计;窗橱布置;商店橱窗装饰;橱窗装饰;浮夸经营成果;粉饰结算;粉饰(橱窗);布置商店橱窗;装门面措施
windowed 装窗的
windowed bay 装窗节间;装窗间距
windowed frieze 装窗壁缘
windowed pattern primitive 窗口模式基元
windowed veranda 轩
window efficiency ratio 窗面采光率;窗面采光效率比
window end time 时窗终了时间
window engineering 窗工程
window erection 窗户安装
window extension 窗口扩充
window fan 窗通风扇;窗户扇
window fastener 窗闩销;掣柄
window fastener of metal 窗用金属钩扣
window film 窗户薄膜
window finishing 窗户装修
window fittings 窗配件
window fixing 窗户固定
window flanning 窗框两侧斜边
window flaring 八字形窗洞
window form 窗形
window frame 窗(扇)框
window frame aerial 车窗天线
window frame antenna 车窗天线

window function 窗(口)函数
window furniture 窗五金;窗附件
window garden 窗园
window gasket 窗的密封垫
window gearing 窗开关联动装置;气窗联动开关;气窗开关联动装置
window glass 窗玻璃
window glass bumper 窗玻璃垫条
window glass pane 窗玻璃
window glass run channel 窗玻璃滑道
window glazing 窗玻璃;窗配玻璃
window glazing bar 窗芯横木;窗棂
window grate 窗楣
window grating 窗栅
window grillage 窗格栅
window grille 窗栅(栏);窗护栏;窗格栅
window groove 窗槽口
window guard 护窗网;窗(铁)栅;窗护栏
window guard rail 护窗栏
window guide rail 窗玻璃导轨
window handle 窗户手柄
window hanging 窗帘;窗帷
window hardware 窗(金属)附件;窗五金
window hatch 镶玻璃的舱口
window head 窗樘的顶横木;窗头;窗框上槛
window head mo(u)lding 窗头线
window hinge 窗铰链;窗合页
window Holland 窗帘布
window hood 窗兜;窗罩
window icon editor 窗口图标编辑程序
window identifier 窗口标识符
windowing 窗口式选择框;开窗(口);开窗技术
windowing facility 开窗装置;开窗条件
windowing scheme 开窗口方案
windowing technique 开窗法
window insulator 引入绝缘子;穿心绝缘子
window in work stone 料石窗
window jack(scaffold) 窗槛上挑出的脚手架;窗口挑脚手架;脚手架活动平台;窗挑脚手架;从窗洞伸出的脚手架
window jamb 窗(口)侧墙;窗樘边梃;窗框边梃;窗侧壁;窗子边梃
window konb 窗握手
window lead 窗镶玻璃铅条;镶玻璃铅条
window ledge 窗台(板);外窗台
window lend-in 穿窗引(入)线
windowless 无窗的
windowless building 无窗建筑;无窗结构物;无窗房屋
windowless house 无窗房屋
windowless panel 无窗墙板
windowless story 无窗楼层
window lift 提窗把手;提拉窗扣手;吊窗提手
windowlight (玻璃)窗扇
window-lighting 窗采光
window lining 窗衬套;窗框镶板
window lintel 窗过梁
window lintel in prestressed clay 预应力黏土窗过梁
window location 窗的位置
window lock 窗键;窗插销
window lowering crank 开窗摇柄
window machine 端末装置;终端机
window management software 窗口管理软件
window manufacture 窗的制造;窗门生产
window masonry wall 带窗的砖砌墙
window material 窗口材料;窗片材料
window measurement 窗口测量
window menu button 窗口菜单按钮
window menu handle 窗口菜单处理
window module 窗的模数;门窗模数
window mullion 窗框的中梃;窗中梃;窗间直梃;窗间小柱
window multimedia 窗口多媒体
window niche 床框或窗帮凹槽;窗(形)壁龛;窗龛
window niche arch 窗(壁)龛拱
window object 窗口目标
window of a detector 探测器窗
window of the slope 斜率窗口
window of tube 荧光屏;管窗口
window of water vapo(u)r 水蒸气窗口
window opener 开窗装置;开窗器
window opening 窗孔;钻窗大门;窗口;窗洞
window opening device 开窗设备

window opening size 窗口大小;窗口尺寸;窗洞(口)尺寸
window openwork gablet 窗上透雕细工花山头;花格窗小山墙
window operation 窗口操作
window ornament 窗饰
window oscillation 窗的振荡
window pan 窗玻璃腔
window pane 窗玻璃;窗格(子)玻璃;玻璃窗格
window panel 窗墙板;带窗口的墙板;窗镶板
window pane separation 窗玻璃分隔
window parapet 窗下墙;窗栏杆;窗槛墙;窗心墙
window parts 窗部件;窗配件
window paster 糊窗纸
window pier 窗墩
window pillar 窗口中柱;窗棂
window pipe (疏浚机的)带孔送泥管;孔管
window placement 窗户设置
window plane 窗口平面图
window position 窗口(波门)位置
window post 窗樘子梃;窗间柱;窗肋柱
window preformed gasket 预制型窗密封垫
window profile 窗型材;窗(纵)剖面图
window projecting inwards with sliding pivot 带滑动枢轴的内开窗
window pull 扯窗把手;窗拉手
window pulse 窗(口)脉冲
window rabate 窗铲口
window rabbet 窗铲口;窗边半槽;窗边槽口
window radiator 窗形散热器
window rail 窗横档;窗轨
window-range 频критерный范围;窗频范围
window rebate 窗半边槽
window recess 窗龛;窗帮凹槽;窗框或窗帮的凹槽;窗框凹槽;窗壁凹
window recess arch 窗龛拱
window recess with slipping walls 有外层壁的窗框
window regulator 车窗开闭调节器;窗开关
window reveal 窗筒子板;窗帮
window reveal brick 窗沿侧墙砖
windows 视窗操作系统
window sandwich 夹层玻璃窗
window sash 钢窗框;窗扇;水下开关的窗扇;窗框
window sash stripping 窗框挡风条
window scale factor 窗口比例系数
window schedule 窗一览表
window screen 窗栅;窗格;纱窗;窗纱
window sealing 窗的填缝;窗的密封
window sealing fillet 窗的压缝条
window sealing gasket 窗框衬垫
window sealing rope 窗的填缝绳
window seat 窗座
window section 窗型材;窗形式
window section with thermal break 热穿透窗断面
windows environment 视窗环境
window shade 遮光帘;窗口遮阳篷;窗户遮阳棚;遮阳窗帘
window shade blind 遮光(窗)帘
window shape 窗的形状;窗形
window shelter 窗户掩蔽体
window-shop 浏览窗橱
window shutter 百叶窗
window sill 窗台(板);窗盘;窗(下)槛;窗扇下冒头
window sill brick 窗台砖;砌窗台砖
window sill unit 窗台板单元
window size 窗口大小;窗口尺度;窗口尺寸
windows-method 波浪形堆栽法
window space 窗面积
window spandrel 窗台墙;上下层窗之间的空间
window spandrel panel 窗台墙板
window splay 窗口斜角;八字形窗洞
window spring bolt 窗扇弹簧插销;窗门弹簧螺栓
window start time 时窗起始时间
window stay 窗撑;窗支撑
window stile 窗扇边梃;窗竖梃;窗扇中梃;窗扇边框;窗(边)梃
window stool 抹头;内窗台;窗台板
window stop 窗内侧挡条;窗止条
window strap 窗子开关带
window strip 窗玻璃密封条
window structural gasket 窗的定型衬垫
window stud 窗樘;窗间柱;窗板墙筋
window style 窗(边)框

window sub 侧眼短节
window subdivision 窗分格
window sunblind 窗百叶;窗遮阳
window surround 窗框
window system 窗口系统
window tax 窗税(按窗计税)
window template 窗垫木;窗框垫块
window test signal 窗口测试信号
window tier 窗列;窗层;分层窗
window title 窗口标题
window top rail 窗扇上横档;窗扇上冒头;窗扇冒头
window tracery 窗饰;花格窗;窗格
window transom 腰窗;窗框中槛
window trim 窗姿饰;窗贴脸(又称窗头线);窗饰;修理窗扇;饰窗花格;窗线脚
window type 窗的形式;窗的型号
window type air conditioner 窗式空调器;窗装型温度调节器
window type cage 框型保持架
window type current transformer 贯通式变流器
window type panel 带窗幕墙板;窗式墙板
window type restraint weld cracking test 窗形拘束抗裂试验;窗口拘束焊缝抗裂试验
window type sampler 侧壁开口取样器
window unit 窗单元;窗组合体;窗式空气调节器
window van 修理车;工具车
window ventilation 窗的通风
window ventilator 车窗通风器;通风窗;气窗;通气窗
window vibration 窗的振动
window visor 遮阳板
window wall 窗间墙;窗墙;玻璃幕墙
window washing 窗的清洗
window-washing cradle 洗窗台架
window-washing equipment 窗清洗设备
window weight 窗扇上下扯窗平衡锤;吊窗平衡锤;窗的平衡锤
window winder 车窗卷绕器
window wiper 风挡雨刮子;风挡刮水器;车窗刮水器;擦窗器
window with sashes opening outwards 外向窗口窗
window with wings opening inwards 内向侧翼窗
window yoke 双旋窗的顶樘子
wind pack 风成硬水泥
wind path 风路
wind penetration 透风
wind pipe 风管
wind plumber 操作气管的水暖子
wind-polished stone 风棱石
wind-polish rock 风磨(光岩)石
wind pollinated plant 风媒植物
wind porch 风廊
wind portal 抗风框架;风门
wind power 风力;风功率
wind power accumulation type 风力堆积型
wind power charger 风力充电机
wind-powered lifting 风力提水
wind power engineering 风力工程(学)
wind power generation 风力发电
wind power generator 风力发电机
wind power machine 气动机械;气力机械
wind power plant 风力发电厂
wind power station 风力发电站
wind pressure 风压(力);风压荷载
wind pressure coefficient 风压力系数
wind pressure diagram 风压图
wind pressure distribution 风压分布
wind pressure distribution graph 风压分布图
wind pressure effect 风压作用
wind pressure force 风压力
wind pressure ga(u)ge 风压计
wind pressure test 风压试验
wind pressure value 风压值
wind profile 风速分布图
wind prone tree 风害树
wind proof 不透风(的);防风的
wind-proof construction 防风构造
wind-proofed monitor 避风天窗
wind-protecting plantation 防护林;防风(植)林
wind protection 防风
wind protection cover 防风罩
wind-protection plantation 防风林
wind-protection screen 防风遮板;挡风板

wind pulse-vibration factor 风振系数
wind pump 风力提水机;风动泵
wind race 屋架间抗风剪刀撑
wind-raised level 风力吹起的水平线
wind-raised water level 风激水位;风增水
wind rate 风量
wind ratio 缠绕圈数;缠绕比
wind ratio of cake 成型速比
wind recorder 自记测风器
wind regime 风源;风态;风情;风况;风的状况
wind resistance 风阻力;抗风;受风率
wind resistance loss 风阻损失
wind resistance truss 抗风桁架
wind resistant 抗风能力;抗风性能
wind resistant construction 抗风构造
wind resistant design 抗风设计
wind resistant designed 抗风设计的
wind resistant plant 抗风植物
wind resistant stability 抗风稳定性;空气动力稳定性
wind-resisting glazing 耐风压玻璃
wind-resisting truss 抗风桁架
wind riddle 风选筛
wind ridge 卷曲脊瓦;风成波脊
wind ripple 风成雪波;风成波痕
wind ripple mark 风成波纹;风成波痕
wind road 风巷
wind rock 风成岩
wind rocking 风摇
wind rode 顶风锚泊
wind rose 风玫瑰;风向(频率)图;风向玫瑰图;风向风花图;风图;风徽;风花
wind rose diagram 风向玫瑰图;风向风花图
wind rose map 风向玫瑰图;风向风向图;风向风花图
windrow 风干处理物料堆;行堆;料堆;条形料堆;堆行;堆料;成列物体堆积;长形料堆;(风干的)长料堆;草条;按长堆放料
windrow baller 草条(捡拾)压捆机
windrow chute 铺条滑槽
windrow composting 野外堆积
windrow equalizer 分(料)堆器
windrower 割晒机;料堆整形机;料堆摊平机
windrower evener 平堆机
windrow evener 料堆摊平机
windrowing 堆场风干;废物堆积风干处理
windrowing attachment for motormower 动力割草机的铺条装置
windrow loader 料堆装载机;草条捡拾装载机
windrow method 成堆法
windrow pickup 禾铺捡拾器;草条捡拾器
windrow pickup baller 草条捡拾压捆机
windrow planting 堆间栽植
windrow sizer 料堆断面整型机;料堆断面测定样板
windrow travel plant 堆料式移动拌和设备
windrow-type 堆料式的
windrow-type mechanical cement spreader 堆料式水泥撒布机
wind run 风程;迎风航行
wind sail 帆布通风筒
wind sand investigation 风沙调查
wind sand ripple 风成砂纹
wind sandstone 风成砂岩
wind scale 风力(等)级;风级(表)
wind scale of Beaufort 蒲福风级
wind scoop 招风斗;集气器吸气口;风穴;风斗
wind-scoured basin 风蚀盆地
windscoured stone 风棱石
windscreen 挡风玻璃;挡风板;挡风屏;风挡;防风屏;防风(雨)罩
windscreen cleaning liquid 挡风玻璃清洗剂
windscreen defogging unit 挡风玻璃除雾器
windscreen glass 风挡玻璃
windscreening 挡风板;风挡
windscreen wiper 挡风玻璃刮水器
wind sea 波浪;风浪
wind-sensitive structure 风敏结构
wind separator 风力选粉机
wind set-down 风减水
wind set-up 风增水;风涌水;风壅水;风扬起;风潮;海面倾斜;气象潮
wind shaft 风矢杆
wind shake 木材辐裂;木材裂;环裂;顺年轮干裂;树干风裂;风裂
wind shallow 风影
wind-shaped stone 风棱石
wind shear 风的切力;风(的)切变
wind shelf 烟道凸缘
wind shield 雨量计风挡;遮风屏;挡风罩;挡风玻璃;挡风装置;挡风屏;挡风板;防风(雨)罩;风挡(玻璃)
windshield adjusting arm 风挡调整臂
windshield anti-icer 风挡防冰器
windshield apron 玻璃风挡
windshield card method 挡风玻璃贴通知单法(汽车违章)
windshield clamp 风挡夹
windshield cleaner 挡风罩清洁剂
windshield cover 风挡罩
windshield defroster 风挡除霜器;挡风玻璃除霜器;风挡去霜器
windshield equipment and signal horn 挡风玻璃设备和信号喇叭
windshield glass 挡风玻璃;风挡玻璃
windshield glass against dazzle 防眩风挡玻璃
windshield glass channel 挡风玻璃槽
windshield heating equipment 挡风玻璃加热设备
windshield pillar 风挡柱
windshield polish 挡风罩上光剂
windshield strap 风挡嵌带
windshield washer 挡风玻璃冲洗器;风挡玻璃喷洗器
windshield wiper 挡风板雨刷;挡风板刮水器;风挡刮水器;风窗刮水器
windshield wiper arm 风挡刮水臂;风雨刮臂
windshield wiper blade 刮水片
windshield wiper vacuum governor 风挡水刮真空调节器
windshield wiping paper 擦风挡纸
wind shift line 风向转变线
wind shifts 风变向
wind shoe 风斗
wind sign 风标
wind signal 风汛信号;风力风向信号
wind signal pole 风汛信号杆
wind slab 风成雪堆;风成雪板
wind slash 风倒落物;风害迹地
wind sleeve 布风袋;袋形风向标;风向袋
wind sock 布风袋;圆锥形风标;风向袋
wind soil erosion 土壤风蚀;土壤吹失
Windsor chair 温莎椅
Windsor probe 温莎探测器
wind-sorted material 经风分选的材料
wind-sorted soil 经风分选土壤
wind spectrum 风谱
wind speed 风速
wind speed alarm 风速警报;风速报警
wind speed and direction 风速与风向
wind speed counter 风速计数器
wind speed indicator 风速(指示)仪;风速指示器
wind speed meter 风速表
wind speed power law 幂指数风廓线;风廓线幂指数律
wind speed recorder 风速记录仪;风速记录器
wind speed scale 风速等级;风速刻度盘
wind speed schedule 风速刻度盘
wind split 风裂伤口
wind spout 旋风;龙卷风
wind spread sterility (土壤)风沙瘠化
wind spun vortex 风动涡旋
wind squall 风飑
wind stability 风稳定性
wind stack 气力集草车
wind star 等边三角形测风法
windstau 风增水
wind stick 飞机的螺旋桨
wind stop 挡风(雨)条;压缝条;盖缝条
wind storm 暴风;风飑;风灾
windstorm insurance 风灾险;风害保险;风暴险
windstorm insurance policy 风灾险保单;风暴险保单
wind strata 风层
wind street 弯曲街道
wind strength 风力
wind stress 风(压)应力
wind-stress coefficient 风应力系数
wind stress formula 风压公式
wind stripping 防风条带
wind structure 抗风结构;风(区)结构;风的构造
wind suction 风(的)吸力
windsurfer 风帆冲浪船
wind-swept 风吹过的
wind-swept area 风吹区;受风区;风灾区
wind system 风系
wind tape 缠线带
wind technology 风能技术
wind tee 丁形灯光风标;丁字风向标
windthroat 鼓风机排气口;风扇排气口
windthrow 风倒木
windthrow tree 风倒树
wind tide 风增水;风涌水;风壅水;风(暴)潮
wind tie 抗风斜撑;抗风斜杆;抗风联杆
windtight 不透风的;不通风的
windtight material 不透风的材料
wind-tipped stereo-comparator 偏轴立体坐标仪
wind-to-jet difference 风—射差
wind transducer 风向传感器
wind transport(ation) 气动运输
wind triangle 风速三角形
wind trunk 大风道
wind truss 抗风桁架
wind tunnel 风洞;风道
wind tunnel balance 风洞天平
wind tunnel computer 风洞(型)计算机
wind tunnel cooling tower 风筒式冷却塔
wind tunnel experiment 风洞试验
wind tunnel fan 风洞鼓风机
wind tunnel installation 风洞设备
wind tunnel instrument 风洞测试仪器
wind tunnel instrumentation 风洞测试设备
wind tunnel investigation 风洞研究
wind tunnel laboratory 风洞实验室
wind tunnel model 风洞模型
wind tunnel modelling 风洞(式)模拟
wind tunnel plant 风洞
wind tunnel propeller 风洞螺旋桨
wind tunnel silencer 风洞消声器
wind tunnel simulation 风洞(式)模拟
wind tunnel study 风洞研究
wind tunnel technique 风洞实验技术
wind tunnel test 风洞试验
wind tunnel turbulence 风洞湍流;风道湍流
wind turbine 风轮机;风力涡轮机
wind up 卷起;卷紧;卷取;绞起;清算;缠绕;终结;结局
wind-up a case 结案
wind-up an account 结算
wind uplift 风吸力;风上举力;风揭性
wind uplift resistance 耐风揭性
wind-up of suspense 停业
wind-up one's business 闭业
wind-up roll 收卷辊
wind-up station 卷取工段
wind valley 风口;风谷
wind vane 风标翼;风向(标);风标
wind variability 风变度
wind vector 风矢量
wind vector diagram 风矢量图
wind velocity 风速
wind velocity alarm 风速报警
wind velocity counter 风速计数器
wind velocity fluctuation 风速涨落
wind velocity gradient 风速梯度
wind velocity indicator 风速指针;风速(指示)仪;风速指示器
wind velocity pressure 基本风压值;风压
wind velocity profile 风速廓线
wind velocity profiler 风速廓线绘图仪
wind velocity rose 风速玫瑰图;风玫瑰(图)
wind vibration 风振
wind volume of air compressor 空气压缩机风量
windward 向上风;向风;迎风时;迎风的;在上风;上风(面)
windward area 受风面积
windward bank 向风岸;上风岸;迎风岸
windward breakwater 上风防波堤
windward chord 上风弦
windward coast 向风海岸
windward drift 向风飘流
windward end of fetch 风区迎风端
windward face 迎风面

windward flood 迎风涨潮
windward sailing 抢风航法
windward set 迎风流
windward shore 迎风(海)岸;上风岸
windward side 向风面;迎风面;上风面
windward slope 迎风坡
windward tide 迎风潮
windward truss 上风桁架;迎风(面)桁架
wind wave 风(成)浪;风(成)波
wind-wave flume 风成玻璃水力试验槽;风浪试验槽;风成波试验(水)槽
wind-wave frequency spectrum 风波频率谱
wind-wave hydraulics 风浪水力学
wind-wave interaction 风浪相互作用
wind-wave spectrum 风浪谱
wind way 通风道;风巷
windway test 风道试验
wind wheel 风(动)轮;风车
wind wheel anemometer 风轮风速计
wind-worn 风蚀
windworn pebble 风蚀砾石(卵石)
windy downpour 狂风暴雨
windy hammer 气动铆钉锤
windy shot 空炮
wind zone 风区;风带
wine 紫红色;深红色
wine cellar 酒窖;酒窨
wine gallon 液加仑;酒加仑
wine jar 酒罐
wine-pot in the shape of a calabash 扁蒲壶
wine producing factory 制酒厂
Winer spectrum 维纳频谱
winert wastewater 制酒厂废水
winery 酒厂;酿酒厂
wineshop 酒店
wine stillage 酒槽
wine vault 酒吧间;酒窖
win fame both at home and abroad 驰名中外
Winford red 温福特红颜料
wing 翼瓣;机翼;翼
wing abutment 大字形桥台;翼墙,翼座;翼形桥台;翼式桥台;翼墩;八字翼形桥台
wing and body combination 机身与机翼的组合
wing antenna 翼式天线
wing area 边角地带
wing assembly 弹翼组
wing axis 翼轴线
wing balcony 剧院的侧翼楼座
wing bar 翼杆;覆尾羽
wing barricade 路障;路侧栅栏
wing base 侧底座
wing beam 翼梁;翼杆
wing bearing 铧翼
wing bit 翼状钻头
wing boat 水翼船
wing bolt 翼形螺栓;蝶形螺栓
wing bracket 双层底翼板;舭肘板
wing brake 叶象制动条
wing bulkhead 翼舱壁
wing bunker 边煤舱
wing car 边车
wing carry through structure 机翼支承贯穿梁结构
wing cascade 翼形叶栅
wing cellule 翼组结构
wingceltis 青檀
wing chair 翼形高背软垫椅;高背椅子
wing chord 翼弦
wing clearance angle 翼隙角
wing compartment 舷舱
wing compasses 弓形两脚规;翼罗盘;象限圆规
wing contour 翼轮廓
wing coping (桥墩的)出缘石
wing core 楔形芯头泥芯;下落芯;下落式顶填泥芯;爬芯
wing core bit 肋骨式取芯钻头
wing cover 翼形培土器
wing covering 翼罩
wing covering fabric 翼蒙布
wing curve 翼轮廓线
wing dam 挑水坝;侧坝;翼(形)堤;翼坝;丁坝
wing dyke 翼形堤;翼堤;丁坝
wing disc (古建筑)翼托盘装饰
wing dividers 翼形量规
wing door 侧(面)门

wing drag 翼阻力
wing duster 翼式除尘器
winged bull 双扇小圆窗
winged cork 翼状木栓
winged disc 翼圆盘;古建筑中带翅膀形的托盘装饰;古建筑翅膀托盘装饰
winged ga(u)ge taper 钎刃锥度
winged headland 有翼岬角;双翼岬
winged hollow reamer 肋骨式空心扩孔器
winged insect 飞虫
winged mo(u)ld board 装翼模板
winged scraping bit 翼状钻头;刮刀钻头
winged screw 翼形螺钉
winged vehicle 气翼船;冲翼艇
wing embankment 侧堤;侧岸
winger 舱边板
Winget 温伽特牌搅拌机(一种混凝土搅拌机)
wing feeder 舱侧灌补器
wing fence 翼形栅栏
wing flap control valve 翼瓣控制阀
wing flap relief valve 翼瓣安全阀
wing float 翼梢浮筒
wing flutter 翼震动
wing footing 翼形基础
wing form 翼形;翼状
wing-front 翼前羽
wing furnace 侧炉
wing gided valve 翼状导管阀
wing girder 旁纵桁;侧桁材
wing groin 翼状穹顶肋
wing guided 有导向肋骨的
wing guided valve 导翼阀
wingheaded bolt 翼形螺栓;双叶螺栓
wing heating tube 翼式散热管
wingheaviness 翼重
wing heavy 翼重的
wing hinge 翼绞链
wing horse 有翼飞马
wing incidence 翼倾角
wing keel 边内龙骨
wing key 翼状钥匙
wing length 翼长;刃长
wingless abutment 无翼墙桥台
wingless aircraft 无翼飞行器
winglet 小翼
wing levee 翼形堤;翼堤;丁坝
wing light 侧投灯光;侧身灯光;翼窗
wing lion 有翼飞狮
wing load 翼荷载
wing masonry wall 翼墙
wing mast 翼柱
wing member 翼构件
wing mixer 翼形混砂机
wing navigational light 翼尖航行灯
wing nut 元宝螺母;元宝螺帽;翼形螺母;翼螺帽;蝶形螺母;蝶形螺帽
wing of a dormer 天窗的侧翼
wing of a lucarne 天窗的侧翼
wing of door 门侧翼;门侧扇
wing oil bunker 边燃油舱
wing pallet 翼式托盘
wing pallet with projecting deck 有突板的翼式托盘
wing panel 翼片;翼段;翼板
wing passage 侧通道
wing pavilion 角亭
wing photograph 偏斜像片
wing pile 端部加宽的桩;翼柱
wing plane 翼面
wing plate 翼板;望板
wing plough 翼犁
wing point 旁交点
wing power 翼力
wing profile 翼型
wing pump 翼式泵;叶片泵;叶轮泵;轮叶泵
wing radiator 翼形散热器
wing rail 翼轨
wing reaming bit 翼状扩孔钻头
wing resistance 翼阻力
wing room 厢房
wings 舞台工作区
wingsail 翼帆
wing scattering 翼形散射
wing screen 翼式拦污栅;叶轮筛

wing screw 蝶形螺钉;翼形螺钉;元宝螺丝;翼状螺钉;翼形螺丝
wing screw propeller 外侧螺旋推进器
wing section 翼型(指托盘两端吊翼);翼剖面
wing sets 边幕
wing shaft 侧轴
wing shape 翼形
wing shape float 袖网浮子
wing shape water ga(u)ge 叶轮式水表
wing shelter 驾驶室侧棚
wing-side car 边车
wing skid 翼梢滑橇
wing slip 翼滑
wing slot 翼缝
wing snow plough 翼式雪犁
wing socket 麻绳卡
wings of anticline 背斜的两翼
wing space 舷舱
wing span (of aircraft) 翼长;翼展
wing spar 翼梁
wing spot 侧投点照灯光
wing spot generator 翼点发生器
wing spread 翼展
wing stay 翼子板托架
wing stick 飞机螺旋桨
wing structure 翼结构
wing strut 翼支柱
wing stull 翼状横撑支柱
wing-supporting float 翼梢浮筒
wing surface 翼面
wing surface radiator 翼面散热器
wing-tail arrangement 翼尾排列
wing tank 舷侧水柜;翼舱;上边舱
wing taper 翼斜削度;翼弦斜度
wing thickness 钎刃厚度;刀翼厚度
wing tip 翼梢;翼端;翼尖
wing-tip aileron 翼梢副翼
wing-tip antenna 翼梢天线
wing-tip chord 翼梢弦
wing-tip drag rudder 翼梢阻流方向舵
wing-tip flare 翼梢着陆闪光灯
wing-tip float 翼尖浮筒
wing-tip landing flare 翼梢着陆闪光灯
wing-tip launcher 翼梢发射架
wing-tip light 翼梢灯
wing-tip loss 翼梢损失
wing-tip nozzle 翼端喷嘴
wing-tip parachute 翼梢伞
wing-tip plate 翼梢板
wing-tip pontoon 翼梢浮筒
wing-tip radius 翼梢半径
wing-tip rake 翼尖斜削
wing-tip rigid boom air-borne electromagnetic system 翼尖硬架航电仪
wing-tip rudder 翼梢方向舵
wing-tip slot 翼梢缝
wing-tip store 翼梢外挂物
wing-tip truss 翼梢构架
wing-tip turret 翼梢炮塔
wing-tip vortex 翼梢涡流;翼梢旋涡
wing toe 翼形坡脚
wing track 八字线【铁】
wing transom 船尾横梁材
wing trenches 侧翼截水槽
wing type anemometer 翼式风速表
wing type hydraulic motor 叶片式液压马达
wing type meter 翼式水表
wing type pallet 翼型集装架
wing type pulley 翼形滚筒
wing type water meter 叶轮式水表
wing valve 圆盘导翼阀
wing-walk 翼上走道
wing wall 侧墙;边墙;翼墙;浮坞墙;耳墙;八字墙
wingwall 浮船坞边墙;屏式凝渣管
wingwall abutment 翼墙式桥台
wingwall at quay end 码头端部翼墙
wingwall flare 翼墙的张开部分
wingwall of bridge pier 桥墩翼墙
wingwall of lock (船)闸翼墙
wing-warp 翼挠曲
wing wheel 翼轮
wing wheel riser 翼轨升高
wing width 刀翼宽度
wing window 侧窗

wink 瞬息;瞬间
wink bending flexure strength 瞬时抗挠曲强度
wink-dry 瞬间干燥
winker (汽车的)方向指示灯;信号灯;汽车用闪光灯
winker beacon 闪光灯标
Winkler's assumption 温克尔假设
Winkler's burette 温克尔量管;温克尔气体分析管
Winkler's coefficient 温克尔系数
Winkler's foundation 温克尔地基
Winkler's gasification method 温克尔气化法
Winkler's generator 温克尔发生炉
Winkler's hypothesis 温克尔假设
Winkler's method 温克尔法
Winkler's model 温克尔模型
Winkler's theory 温克尔理论
Winkler's titration 温克尔滴定
Winkler's triangle 温克尔三角形(图)
Winkler's value 温克尔值
winning 提炼;开采与搬运;采运
winning and opening 采掘
winning bid 得标;中标;获得投标
winning bidder 中标人;中标公司;中标商;中标厂家;得标人
winning cell 电积槽
winning design 获胜设计
winning equipment 采矿设备
winning of natural stone 石料开采
winning of sand 采砂
winning-post 终点
winnowed sediment 漂流沉积
winnower 风选机
winnowing 风(力)选;簸选
winnowing conveyer 风选输送机
winnowing gold 风选金矿
winnowing machine 风扬机;风选机
winnow mill 风选机
Winslow effect 温斯洛效应
winter 冬季
winter air-conditioning 冬季空气调节
winter air-conditioning installation 冬季空调设备
winter annual 冬性一年生植物
winter antarctic stratospheric circumpolar vortex 南极冬季平流层绕极涡旋
winter atmosphere 冬季大气
winter axle oil 冬季轴油
winter balancing reservoir 冬季调节水库
winter bed 冬季河床
winter bourne 间歇河(流);季节性小河
winter breed 冬季品种
winter bud 冬芽
winter budding 冬季芽接
winter bud growth 冬芽生长
winter building 冬季用房屋;冬季施工
winter building construction 冬季建筑施工;冬季房屋建造;建筑(物)冬季施工
winter building technique 建筑物冬季施工技术
winter buoy 冬季浮标
winter capacity 冬季容量
winter cereals 越冬谷物;冬性谷类物;冬季谷物
winter compound oil 冬季齿轮油
winter concreting 混凝土冬季施工;冬季浇注混凝土;冬季浇灌混凝土;冬季混凝土施工;冬季混凝土浇筑;冬季(混凝土)浇注;冬季混凝土浇灌
winter condition 冬态
winter construction 冬季施工
winter construction equipment 冬季施工设备
winter cover crops 冬季覆盖作物
winter damage 冬害
winter design temperature 冬季设计温度;设计冬季温度
winter dyke 冬季堤;冬堤
winter discharge 冬季流量
winter dormancy 冬季休眠
winter draft 冬季吃水
winter draught 冬季吃水
winter drought injury 冬季旱害
winter drying 冬季干燥;冬干梢
winter emulsion 冬季用乳化液
winter energy 冬季用能量
winter fallow 冬季休耕
winter flood 冬汛;冬季洪水
winter flow 冬季径流
winter flower 腊梅
winter fluctuation 冬季水位变化

winter fog 冬雾
winter freeboard 冬季干舷
winter front 散热器保温罩
winter garden 冬园;冬景花园
winter-grade gasoline 冬季级汽油
winter green oil 冬青油
wintergreen plants 冬绿植物群
winter grit spreader 冬季铺砂机
winter gritting machine 冬季铺砂机
winter half year 冬半年
winter harbo(u)r 过冬港;卧冬港;冬(季)港
winter hardiness 越冬性;抗寒性;耐寒性;耐冬性
winter hump 冬季驼峰
winter hydrant 冬季用消防栓;冬季用水龙头
winter ice 冬冰
wintering 越冬;冰冻风化
wintering ground 越冬地;越冬场所
winter injury 冻害;冬害
winter irrigation 冬季灌溉;冬灌
winterization 准备过冬;过冬准备;安装防寒装置
winterized concrete plant 冬季混凝土拌和厂;冬季(制造)混凝土工厂;防寒混凝土拌和厂
winterized oil 冻凝油;冬化油
winterized pavement 模拟冰雪条的路面
winterized unit 过冬装置
winter jasmine 迎春【植】
winter jet 冬季飞行用喷嘴
winter kill 使植物冻死
winter-killing 冻害;寒害;冻死
winter light 冬季灯标
winter load line 冬季载货吃水线
winter load line mark 冬季载重线标志
winter marker 冬季浮标
winter mode 冬季模式
winter monsoon 冬季季(节)风
winter monsoon climate 冬季季风气候
winter monsoon current 冬季季风海流
winter months 冬季
winterness crop 冬性作物
winterness form 冬种性类型
winterness plant 冬性作物;冬播作物
winter North Atlantic load line mark 冬季北大西洋载重线标志
winter nurseries 冬季场圃
winter oil 冬季用润滑油;冷冻油;耐(冻润)滑油;冬(用)油
winter operation 冬季运行
winter oxygen depletion rate 冬季耗氧速率
winter patching 冬季修补工作;冬季修理工作
winter-placed concrete 冬季浇捣的混凝土;冬季灌筑的混凝土
winter plankton 冬季浮游生物;冬大麦
winter port 过冬港;冬港
winter power plant 冬季用发电厂
winter precipitation 冬季降水(量)
winter profile 冬季剖面
winter-proofing 防寒
winter protection 防寒
winter pruning 冬季整枝
winter quarters 冬季藏艇坞
winter range 冬季草场
winter refuse 冬季垃圾
winter reservoir 冬季调节水库
winter resident 冬留鸟;冬候鸟
winter resistance 抗寒性;防寒性
winter resistant plant 耐冬植物
winter rest 冬眠
winter runoff 冬季径流
winter runoff low 冬季枯水径流(量)
winter seamark 冬季航海标志
winter season greenhouse 冬季温室
winter service 冬季防雪设施;道
winter shutdown 冬季关闭;冬季停工
winter solstice 冬至日;冬(点)
winter solstitial point 冬至点
winter sowing 冬播
winter spreading for road-ice control 冬季撒沙
winter stagnation 冬季停滞
winter stagnation period 越冬停滞期
winter standstill 冬季停航
winter stop 冬季停航
winter storage 冬季储[贮]藏;冬季调节水库
winter sunscald 冬季日灼病
wintersweet 腊梅

winter-talus ridge 冬岩屑堆脊
winter timber load line 冬季木材载重线
winter time 冬天;冬季时
winter train working diagram and timetable 冬季运行图和时刻表
winter variety 冬季品种
winter wheat 冬小麦
winter window 冬季双层窗;双层窗
winter work 冬季工作;冬季施工
winter work cost 冬季施工费
winter working 冬季工程;冬季施工
winyl acetic ester 醋酸乙烯酯
winze 盲井;(由上而下开凿的)小暗井;开暗井;(连接矿内各层的)斜通道;地井
wipe 挤锌辊;抹
wipe circuit 消除电路
wiped film evapo(u)rator 转膜蒸发器;扫壁蒸发器
wiped joint 包裹式嵌铅锡合金接头;裸接(头);焊接点;拭接;试接;软焊接接头;热焊接;铅管接头
wipe-down trim die 切边成形模
wipe-dry 擦干
wiped wall still 转膜蒸馏器;扫壁蒸馏器
wipeless windscreen 无雨刷玻璃
wipe-off 擦去;划变;擦拭
wipe-off type 擦拭型(磷化清洗剂)
wipe on plate 即涂感光板
wipe on release agent 擦涂脱模剂
wipe out 擦掉
wipe out an account 抹去账目
wipe out area 抹除区
wiper 刮水器;(汽车风挡的)刮水片;刮水器;揩布;接点;接触刷;滑壁;弧刷;清洁工(人);擦刷工具;擦拭器;擦具;擦净器
wiper arm 擦拭器;刮油器;刮水器;雨刷臂杆;接寻臂;弧刷臂
wiper assembly 弧刷组
wiper blade 雨刷刮片;刮水片
wiper chatter 弧刷振动
wiper cup 油杯
wipe resistance 抗擦拭性
wiper hammer 杆锤
wiper handle 刮水器柄
wipe ring 擦油圈
wiper lubricator 擦拭式润滑器
wiper motor 刮水器电动机
wiper plug (注水泥)上木塞
wiper redistributor 挡板式再分配器
wiper ring 刮垢环;清洁环
wiper rod 刮水杆
wiper roll 擦拭辊
wiper scraper seal 防尘圈
wiper seal 刮油密封环;接触密封;弹性密封
wiper shaft 弧刷轴;擦拭器轴
wiper-type contact 瞬动接点
wiper vacuum booster 刮水器真空助力器
wiper vacuum governor 刮水器真空调节器
wiping 擦拭;挤干
wiping action 接寻作用;擦拭作用
wiping arrangement 挤干装置;擦拭装置
wiping berth 消磁泊位
wiping blade 密封的框架
wiping cloth 裹布;拭布
wiping contact 滑触接点;扫描接点;扫接触点;扫动接点;擦拭触点
wiping effect 拭擦效应;擦洗效果;擦拭效应;擦浆效果
wiping off glaze 除釉
wiping paper 擦拭纸
wiping rag 破布;拭布
wiping seal 柔性密封(条)
wiping solder 擦拭焊料;铅锡合金焊料
wiping solvent 擦拭溶剂
wiping solver 脱漆(溶)剂
wiping stain 擦拭着色
wiping waste 擦拭用回丝
wipped fascine 柴笼;柴捆
wippets 石板的匠称(石工按尺寸命名的石板)
wipple hammer 摆动锤
wire 线;金属线;金属丝;电线
wire adjuster 导线调整器
wire-adjusting screw 导线调整螺丝;导线反正扣
wire agency 新闻电讯社
wire anchor 钢丝锚固
wire anchorage 钢丝锚固位

wire and bar sweeping 拖杆式扫测
wire and cable marker 电线电缆印标机
wire and pipe precipitator 圆管式电集尘器
wire and plate counter 线绕和板极计数管
wire and tube condenser 钢丝盘管式冷凝器
wire angle indicator 倾角仪(吊索)
wire annealing 钢丝退火
wire antenna 线状天线；导线天线
wire-armo(u)red 钢丝铠装的；钢丝紧缠的
wire-armo(u)red cable 钢丝铠装电缆
wire article 钢丝制品；金属丝制品
wire back-tie 模板的反拉钢丝；反拉钢丝
wire bar 钢锭；线材坯；条锭
wirebar applicator 绕线棒刮涂器
wirebar copper 条锭铜
wire basket 金属丝滤网；网篮；金属丝篮；钢丝笼；铁丝笼；尼丝笼；铁丝筐
wire basket strainer 铁丝下水口箅子
wire bead 钢丝橡皮撑轮圈
wire bead core 钢丝撑轮圈
wire becket splice 钢索环插接
wire belt 丝传送带
wire belting 线带装置
wire bending and forming machines 线材弯曲和成形机
wire binders 铁丝扎捆机；(预应力混凝土用)钢丝；火烧丝；绑扎(用)铁丝
wire bolster 钢丝石笼；钢丝笼枕垫；填石钢丝笼
wire bonder 引线接合器；丝焊器
wire bonding 电缆接合
wire bonding method 金属线接合法
wire bound box 钢丝板箱
wire box 铅丝笼；铁丝笼；铁丝笼；铁丝盒
wire bracing 线拉条
wire brad 铁丝钉；角钉；曲头钉；铁丝制的无头钉
wire brad bead 无头钉(固定)圆线脚
wire braid 金属丝编织物
wire braid hydraulic hose 钢丝编织液压胶管
wire braiding machine 金属丝编带机
wire breakage detector 断丝探测器
wire breakage lock 导线断线锁闭器
wire-break relay 断线继电器
wire bridge 缆式悬桥；悬索桥；钢索吊桥；铁索桥
wire broadcasting 有线广播
wire broom 钢丝(路)刷
wire brush 金属丝刷；线刷；钢丝刷
wire brush finish 用钢丝刷(进行)表面抛光；钢丝刷刷面；钢丝刷出面
wire brushing 钢丝刷清理表面；钢丝刷打光；钢丝刷除锈；刷钢丝刷
wire brush scraper 钢丝刷刮蜡器
wire brush tail seal 金属刷盾尾密封
wire buff 钢丝刷磨光轮；金属丝刷抛光轮
wire burn-off rate 焊丝熔化率
wire cable 钢(丝)索；钢丝绳；钢缆(索)；多股(钢)缆；电缆
wire cable anchorage 钢丝绳锚固；钢缆锚定
wire cable clamp 缆索夹
wire cable clip 钢缆夹具
wire cable construction 钢缆绳构造
wire cable grease 钢缆润滑油
wire cable guard 钢丝绳防护装置
wire cable lubricant 缆索润滑剂
wire cable spacer 钢丝缆分隔器；钢丝束定位板
wire cableway 钢(丝)缆索道；钢(丝)绳索道
wire cage 铅丝笼
wire cage packed with stone 铁丝石笼
wire calorimeter 金属丝量热计
wire capacitor 线绕电容器
wire cargo net 吊货铁丝网
wire carrier 导线轮；便携式线盘
wire casing 钢丝套管
wire center system 中心布线方式
wire changer 换绳装置
wire chisel 钢(丝)网截断凿；钢丝凿
wire clamp 引线夹头；线夹；钢丝钳
wire-cleaning line 线材清净作业线
wire clip 线夹；钢索眼夹；钢丝绳扎头；钢丝绳剪；钢丝夹；钢缆夹；绳头夹
wire clippers 断线钳
wire cloth 筛网；钢丝网布；金属丝织物；金属网筛；金属丝布；铁丝(纱)
wire cloth area 筛布面积
wire cloth galvanizing 布辊擦拭镀锌

wire cloth lath 钢丝网条板
wire cloth standards 金属丝布规格
wire coarse sieve 金属丝粗网筛
wire coating 线缆涂料；漆包线漆涂装；漆包线漆涂层
wire coating enamel 漆包线漆
wire code 接线标志
wire coil 线盘；细丝盘条；盘条；盘丝
wire-coiling and winding machine 线材卷取和绕卷机
wire-coiling machine 绕线机；绕丝机
wire-coil stripper 吊具勾架
wire-coil wrapping machine 电线包装机
wire cold-drawn test 钢丝冷拔试验
wire comb 钢齿耙；钢丝梳
wire communication 有线通信
wire communication facility 有线通信设备
wire communication line 有线通信线路
wire comparison 线尺比较
wire compensator 导线调整器
wire conductor 导线【电】
wire conduit 导线(导)管
wire connector 接线柱【电】；电线连接器；线夹
wire-control 导线操纵
wirecord fabric 钢丝帘布
wire core (钢丝绳的)钢丝芯；焊条芯；线芯
wire-core coil 线芯线圈
wire counter 线绕计数器；线绕电极计数器
wire covering machine 金属丝包编机
wire cradle 金属网兜；金属丝篮
wire-cut brick 钢丝切割砖；钢丝切砖法；丝切砖
wire-cut brickmaking 钢丝切割制砖法
wire-cut cake machine 模烤压切机
wire-cut process 钢丝切(砖)坯法(用金属线切割泥段的方法)
wire cutter 剪线钳；钢丝钳；钢丝截断器；钢丝剪；铁丝剪
wire cutting pliers 克丝钳
wire dam 钢丝网屏障；钢丝网坝；石笼坝
wired back 布线背面
wired broadcast station 有线广播站
wired call 有线电话
wired decorative glass 嵌丝装饰玻璃
wired edge tyre 直边式胎
wired figured glass 嵌丝图案玻璃
wired glass 钢丝网玻璃；夹丝玻璃；嵌丝玻璃；铅丝玻璃；embedded网玻璃；铁丝网玻璃
wired glass caisson 嵌丝玻璃格子顶棚；嵌丝玻璃藻井
wired glass cassette 嵌丝玻璃格子顶棚
wired glass coffer 嵌丝玻璃格子顶棚
wired glass domed rooflight 嵌丝玻璃穹隆形天窗
wired glass dome light 嵌丝玻璃穹隆形天窗
wired glass light cupola 嵌丝玻璃采光圆顶；嵌丝玻璃采光穹隆
wired glass pan 嵌丝玻璃板
wired glass roof light 嵌丝玻璃天窗
wired glass saucer dome 嵌丝玻璃穹隆形天窗
wired glass waffle 嵌丝玻璃格子顶棚
wired hose 钢丝软管
wire diagram 布线图【电】
wire diameter 线径
wire die (金刚石)拉绳模
wired-in 固定的
wired-in memory 线定存储器
wire distance 线距
wire distance measuring machine 索式测距仪
wired laminated glass 嵌丝叠层玻璃
wired logic 布线逻辑
wired mat 钢丝网底板；钢丝网席垫
wired-opaque white glad 嵌丝乳白玻璃
wired ornamental glass 嵌丝装饰玻璃
wiredot printer 点线打印机
wire dowel 穿钉
wired patterned glass 嵌丝图案玻璃；压花夹丝玻璃
wired plate glass 嵌丝平板玻璃；夹丝玻璃板
wired polished plate glass 嵌丝抛光平板玻璃
wired program(me) computer 配线程序计算机；插接程序计算机；插线程序计算机
wire drag 钢丝拖海；钢丝网索；扫海
wire-drag survey 拖索扫测
wire-drag sweeping 拖索扫测
wire-draw 拔丝
wire drawer 拉丝机；拔丝机
wire-drawer's plate 拔丝模板

wire drawing 拔丝；张拉钢丝法；拉丝
wire-drawing bench 拔丝机
wire-drawing block 拉丝模；拉模板
wire-drawing compound 拉拔用乳剂
wire-drawing die 拔丝模
wire-drawing die orifice 拉丝模孔口
wire-drawing die profilometer 拉丝模模孔检查仪
wire-drawing machine 拔丝机
wire-drawing reel 拔丝工字轮
wire-draw tongs 拉线钳
wire drive unit 送丝装置
wired roofing glass 夹丝玻璃瓦
wire drum 钢丝绳卷筒
wired safety glass 夹丝安全玻璃
wired sheet glass 夹丝玻璃板
wired system 有线系统
wired teletext system 有线电视文字传输系统
wired television 闭路电视
wired television receiver 有线电视接收机
wired television system 有线传输的电视系统
wired tire 钢丝轮胎
wired trunking 有线中继
wired tube 编织套
wired TV 有线电视
wired TV amplifier 有线电视放大器
wire duct 缆槽；线槽；电线管道
wire enamel 线材漆；电缆涂料；导线涂料
wire end cap 导线帽式接头；导线管端盖帽
wire entanglement 茨篱网；钢丝网障碍物；铁丝网障碍物；带刺钢丝网
wire extruding machine 钢筋拉丝机
wire eye splice 钢索环插接
wire fabric 钢丝布；钢丝围栅；钢丝网；钢丝编织品；金属网；金属挂网；铜丝布；铁纱
wire fabric reinforcement 钢丝网(配筋)；钢筋网配筋
wire-fabric reinforcing 钢丝网加强
wire fabric sheet 铁纱片；金属丝布
wire fabric standards 金属丝布规格
wire facsimile system 有线传真系统
wire failure 断线事故
wire fate item 要求电告已否收妥的托收款项
wire feed 金属丝送料；送丝
wire feeder 送丝器；送丝装置；送丝器；送丝机构
wire feeding device 送丝机构
wire feed rate 送丝速度
wire feed rolls 送丝轮
wire feed system 送丝系统
wire fence 铁丝网；钢丝围栏；钢丝栅栏；铁丝栅栏；铁丝围栏
wire fence groin 铁丝拦污栅
wire fence groyne 铁丝拦污栅
wire fence picket 铁丝网栅栏桩
wire ferrule 线箍
wire file 成串文件
wire filter 金属丝过滤器；绕线过滤器
wire filter cloth 金属丝滤布
wire fix 无线电船位
wire flame spray gun 线材火焰喷枪
wire flame spraying 线材火焰喷涂
wire flattening mill 线材压扁机
wire flexible shaft 钢丝软轴
wire-floored coop 铁丝网底笼
wire-floored house system 铁丝网地面养鸡法
wire flue brush 铁丝烟筒刷子
wire former 成丝机
wire forming machinery 线材成形机
wire for needle products 制针钢丝
wire for prestressed concrete 预应力混凝土用钢丝
wire for welding 焊丝
wire frame 线圈架；接线框；网架
wire-frame image 轮廓像
wire-frame representation 结构线表示图
wire fuse 保险丝；熔丝；线状保险丝
wire fusing current 电线熔融电流
wire gabion 钢丝石笼
wire galvanizing 钢丝涂锌
wire galvanization 钢丝镀锌
wire ga(u)ge 线卡；线材规格；线规；钢丝号码；金属丝规；电阻应变计
wire-ga(u)ge number 线规编号
wire-ga(u)ge plate 线规板
wire ga(u)ging equipment 线径测量装置
wire gauze 细金属丝网；钢丝网；金属(丝)网；金属

丝布;铁丝网;铁纱网
wire gauze diaphragm 线网隔层
wire gauze filter 线网过滤器
wire gauze filter element 钢丝滤芯
wire gauze packing 线网填充料
wire gauze sieve 编织筛
wire glass 格网玻璃;夹丝玻璃;钢丝网玻璃
wire glass mesh 嵌丝玻璃用钢丝网
wire grade 钢丝号
wire grafting 铁丝嫁接
wire-grate concave 钢丝网筛式凹板
wire grating 线栅
wire grid 线栅;终栅
wire grill(e) 金属丝网格
wire grip 金属丝扎紧;拉线器;紧线器
wire gripping device 钢丝锚固设备;钢丝夹具;钢丝夹紧设备;夹线装置
wire guard 钢丝护网;保护网罩;铁丝防护罩
wire guide 线材导板;针导管;钢丝绳(导)道;捆丝导向器;焊丝导向装置;铜网校正辊;导丝孔;导丝装置;导electric铜管
wire guillotine 断线钳
wire gun 钢丝枪;金属丝喷涂枪
wire gutter top 金属丝滤网;天沟网罩
wire guy 缆索;钢丝稳索;钢丝牵索
wire guy line 绑绳
wire halyard 升降钢索
wire hanger 钢绳挂钩;电缆挂钩;钢丝吊筋
wire harness 束线
wire heald 钢丝综
wire height 绒高
wire holder 绝缘子;线夹;接线座;夹绳器
wire holding forceps 持钢丝钳
wire hook 钢丝摘钩
wire image penetrameter 线成像穿透计
wire-in check 硬件检验
wire index line 钢丝标记线;钢丝标度线
wire industry 线材工业;金属线工业
wire insert(ion) 钢丝嵌入层
wire insertion gasket 金属丝基衬垫
wire installation 无线电装置;导线装置
wire installation curve 导线安装曲线
wire insulation 导线绝缘
wire interval 丝距
wire introducer 钢丝导引器
wire joining 导线连接;导线接点
wire joint 线连接;钢丝绳接头;钢丝绳铰接器;金属线缝接
wire jumper 跳线
wire knotter 铁丝扭结器
wire knot tightener 钢丝缚扎器
wire lacquer 线材漆;钢丝喷漆
wire lacquering machine 钢丝刷涂漆机
wire laminated glass 金属丝叠层玻璃
wire lapping 导线重叠(法)装配
wire lath 抹灰用钢丝网;钢丝网;金属网;铁丝网
wire lath and paster suspended ceiling 钢丝网灰泥吊顶棚
wire lath and paster wall 钢丝网灰泥墙
wire lathing 钢丝板条;钢丝肋网
wire lattice wall 网格墙
wire lawn 钢丝细筛
wire layer 钢丝网层
wire laying vehicle 布线车
wire lead drop out 线头脱落
wireless 无线电
wireless access 无线接入
wireless aerial 无线电天线
wireless apparatus 无线电设备
wireless application protocol 无线应用协议
wireless automatic block 无线电自动闭塞系统
wireless beacon 无线电航标
wireless bonding 无线电接合
wireless chart 无线电(信标)导航图
wireless communication 无线电通讯;无线电通信
wireless communication line 无线通讯电路
wireless compass 无线电罗盘;无线电罗经;无线电测向仪;无线电测向器;无线电测向计
wireless compass error 无线电测向仪误差
wireless contact 无线电联络
wireless control vessel 无线电操纵船
wireless data link 无线数据线路
wireless direction(al) finder 无线电定向仪;无线电测向仪

wireless direction finding 无线电定向;无线电探向
wireless engineer 无线电工程师
wireless engineering 无线电工程
wireless fix 无线电船位
wireless fog signal 无线电雾号
wireless gateway 无线网关
wireless house 无线电室
wireless installation inspection certificate 无线电安全证书
wireless in the loop 无线环路系统
wireless jammer 无线电通信干扰机
wireless link 无线电联系
wireless mast 无线电天线杆
wireless message 无线电报
wireless mic 无线传声器
wireless microphone receiver 无线传声器接收机
wireless monitoring 无线电监听
wireless navigational warning 无线电导航警报
wireless officer 无线电通信主任
wireless order 无线电指令
wireless panel 无线电控制盘
wireless power 无线电功率
wireless receiver 无线电接收机
wireless remote control 无线电遥控
wireless reply 无线回复
wireless room 无线电(报)室
wireless section 无线电组
wireless serial number 无线电报流水号数
wireless set 无线电台;无线电收发机
wireless signal 无线电信号
wireless silence 无线电静默
wireless station 无线电台
wireless team 无线电小组
wireless telegram 无线电报
wireless telegraph direction finding 无线电报测向
wireless telegraph fog signal 无线电雾号
wireless telegraph station 无线电报台
wireless telegraphy 无线电报学;无线电报术
wireless telegraphy board 无线电报交换台
wireless telegraphy direction finder 无线电报测向器
wireless telephone 无线电话
wireless telephone exhanger 无线电话交换机
wireless telephonic communication 无线电话通信
wireless time signal 无线电(报)时信号
wireless tower 无线电发射塔;无线电天线塔;无线电天线杆
wireless transformer 无线电变压器
wire ligature 金属缝线
wire line 线道;细铁丝绳;有线线路;钢(丝)绳
wire-line bit 投入式钻头;绳索式钻头
wire-line boring method 钢绳岩心管钻探法
wire-line breakout 钢绳折断
wire-line bridge plug 钢绳下套管定点塞
wire-line cable 绳索取芯内管
wire-line clamp 钢丝绳卡
wire-line connection 钢丝接头
wire-line constant tensioning system 钢绳恒张力装置
wire-line core barrel 绳索取芯管;钢索岩心钻管
wire-line core bit 绳索取芯钻头;钢索取芯钻头
wire-line core drilling 钢绳岩芯钻进;绳索心钻进
wire-line core reel 绳索取芯绞车
wire-line coring 绳索取芯
wire-line coring apparatus 绳索取芯装置
wire-line coring bit 绳索取芯钻头
wire-line coring system 绳索取芯钻具
wire-line diamond drilling 绳索取芯金刚石钻进
wire-line downhole guidance tool 钢丝绳井底导向工具
wire-line drill pipe 绳索取芯钻杆
wire-line drill rod 绳索取芯钻杆
wire-line drill rod coupling 绳索取芯钻杆接头
wire-line drum 绳索取芯绞车
wire-line electrodrill 绳绳电钻
wire-line formation tester 电缆式地层测试器
wire-line grab 钢缆抓钩
wire-line guide 绳索导向装置
wire-line hammer sampler 钢绳锤击取土器
wire-line hoist 绳索取芯绞车
wire-line jar 钢丝振击器
wire-line log 电缆测井
wire-line measurement 钢丝绳测井
wire-line MODEM 有线调制解调器

wire-line oil pump 用绳索起下油泵
wire-line preventer 钢丝绳防喷盒
wire-line reel 绳索取芯绞车
wire-line shoe 钢丝绳卡;钢绳冒头
wire-line sidewall corer 电缆式井壁取芯器
wire-line slippage 钢(丝)绳滑程
wire-line socket 钢绳绳卡
wire-line sub 绳索接头
wire-line thimble 钢(丝)绳环
wire-line tool 绳索取芯工具
wire-line truck 绳索导轮
wire-line unit 录井钢丝绞车
wire-line winch 绳索取芯绞车
wire-line wiper 钢丝绳刮子
wire-line work 绳索取芯作业
wire-line workover unit 通井机
wire-link 有线通信(线路);有线联系;连接导线【电】
wire-link guidance 有线制导
wire-link telemetry 有线遥测术
wire list 连线表;连线表
wire lock 钢绳条锁
wire lock equipment 线锁设备
wire locking ring 线锁环
wire long splice 钢丝绳长插接
wire loop 线环;钢丝(套)圈;钢丝绳套眼
wire loop method 金属环法
wire lying machine 铜网接头机
wire man 电路检修工;线务员;接线工;架设或修理电线的工人;电气装配工;线路工;架线工;装线工;电工
wire matrix 线矩阵
wire mattress 钢丝垫(子)
wire memory 磁线存储器
wire memory matrix 磁线存储器矩阵
wire mesh 钢丝网;金属网(格);金属丝网;线网
wire-mesh belt 钢丝网运输带
wire-mesh belt conveyer 钢丝网带运输机;钢丝网带输送机
wire-mesh bulkhead 钢丝网舱壁;铁丝网隔堵
wire-mesh cement 钢丝网水泥
wire-mesh concrete plate 钢丝网混凝土板
wire-mesh demister 线网除雾器
wire-mesh door 钢丝网门
wire-mesh facing 金属网覆面层
wire-mesh fence 钢丝网栅栏;钢丝网围栏
wire-mesh frame 钢丝网框架
wire-mesh hurdle 铁丝网栅
wire-mesh lathing 钢丝板条;钢丝网网
wire-mesh mattress 金属网棉毡
wire-mesh oven 线网炉
wire-mesh pallet 网眼集装箱
wire-mesh partition 钢丝网隔墙;钢丝网隔断
wire-mesh plotting device 线网绘图仪
wire-mesh rail 钢丝网护栏
wire-mesh-reinforced asbestos cement corrugated sheet 钢丝网石棉水泥波瓦
wire-mesh-reinforced concrete 钢丝网混凝土
wire-mesh-reinforced shotcrete 钢丝网加固喷射混凝土
wire-mesh-reinforced slab 钢丝网混凝土板
wire-mesh reinforcement 钢丝网配筋;钢丝网;金属网加固;网状钢筋;钢丝网加强
wire-mesh screen 金属丝方孔筛;金属丝细网筛;金属(大孔)网筛;线网屏(蔽);铅丝筛;丝网
wire-mesh section 线网分隔层
wire-mesh shearing device 丝网剪断机
wiremesh-shotcrete-rock bolt support 钢筋喷射混凝土锚杆支护
wiremesh-shotcrete-rock support 钢筋喷射混凝土支护
wire-mesh steam washer 钢丝网洗汽器
wire-mesh water screen 铁丝网水幕
wire-mesh welding machine 金属网焊机
wire-mesh window 钢丝网窗
wire-mesh with brickbats 钢丝网加筋块
wire-mesh with close grid 细格钢丝网
wire-mesh with fine grid 细格钢丝网
wire metallizing 金属丝喷涂
wire mile 线英里
wire mill 线材轧机
wire milling 线材轧制
wire modem 导线调制解调器
wire motor 线材用电动机

wire mounting 线固定式
wire nail 圆形截面钉;销钉;线钉;细铁丝钉;圆(头)钉;圆铁钉;圆钢钉
wire needle 钢丝针
wire net 粗制金属网
wire net fencing 钢丝网围墙;铁丝网栅
wire net sling 钢丝网络;吊货铁丝网
wire-netted colloidal pipe assembly 钢丝编织胶管总成
wire netting 金属丝网;钢丝网;金属栅栏;铁丝网;粗制金属网
wire netting fence 钢丝网围篱
wire netting for construction(al) purposes 建筑用钢丝网
wire netting lathing 钢丝网肋网;钢丝网板条
wire netting letter basket 钢丝编信件篮
wire network 导线网络
wire nippers 剪线钳;剪镊钳;钢丝剪;剪丝钳;电工钳
wire nonpayment 如遭拒付请电告
wire nut 金属线帽;接线柱【电】
wire of high tensile strength 高强钢丝
wire of irregular shape 齿纹钢丝;有刺钢丝
wire of stainless steel 不锈钢丝
wire pair 线对
wire parametron 线式参变管
wire pattern glass 嵌丝图案玻璃
wire peg 导线桩
wire photo 有线传真
wire pistol 金属丝喷涂枪;喷丝枪
wire pitch 钢丝捻距;导线间距
wire plotting 线路测绘
wire pointer 钢丝压尖机
wire pole 电线杆
wire printer 型板印刷机;点阵式打印机
wire product 金属丝制品;钢丝制品
wire pull 拉索
wire-puller 预留在线管中拉电线的铁丝
wire race ball-bearing 钢丝滚道球轴承
wire radio 有线载波通信
wire raising machine 钢丝起绒机
wire rake 铁丝耙
wire reading 基线尺读数
wire recording 钢丝记录
wire reel 绕线盘;钢丝盘
wire reinforced 钢丝配筋的
wire-reinforced concrete 钢丝加劲混凝土;钢纤维混凝土
wire-reinforced rubber hose 金属丝加强橡皮管
wire reinforcement 钢丝网配筋
wire relay broadcasting 有线转播
wire relaying 有线中继
wire release 线条快门开关;尺度索松放器
wire-remote control 导线遥控
wire resistance strain ga(u)ge 线阻应变仪
wire retention 保丝网性
wire rod 线材;钢丝棒;金属丝标尺;盘圆;盘条
wire-rod applicator 绕线棒刮涂器
wire-rod concave 钢丝网筛式凹板
wire rod for screws 螺丝用线材
wire-rod mill 线材轧机
wire-rod milling 线材轧制
wire-rod pass 线材孔型
wire-rod reel 线材卷架
wire-rod rolls 线材卷盘条
wire roll 线材轧辊;钢筋盘条;导网辊
wire rope 钢缆;钢绞线;钢索;铁丝绳;钢丝绳;钢丝缆(索);金属线绳
wire rope anchor 钢丝绳锚栓
wire rope anchorage 钢丝绳锚碇
wire rope back-tie 钢丝绳夹头
wire rope block 钢索滑石;钢丝滑车
wire rope bolt 钢丝绳锚杆
wire rope chopper 钢绳切断器
wire rope clamp 钢绳卡子;网索夹;钢丝绳固定(卡)
wire rope clip 钢索夹(子);钢丝绳卡;钢丝绳夹
wire rope compound 钢丝绳润滑脂
wire rope cutter 钢丝绳割刀;钢丝绳剪
wire rope detector 钢丝绳探伤器
wire rope diameter 钢绳直径
wire rope efficiency 钢绳绳效率
wire rope extensimeter 钢丝绳伸长计
wire rope fall 滑车钢索

wire rope fittings 钢丝绳配件
wire rope for mine 矿山用钢丝绳
wire rope for shipping and fishery 航海和渔业用钢丝绳
wire rope gearing 钢丝绳传动(装置);钢绳传动装置
wire rope grease 钢丝绳润滑脂;涂钢丝的油
wire rope gripping device 钢丝绳锚固设备
wire rope guiding support 防钢丝绳脱落架
wire rope guy 钢丝绳牵索
wire rope handling 钢绳装卸
wire rope hempwicking grease 钢丝绳麻芯脂
wire rope hoist block 钢丝绳葫芦
wire rope interchanger 换绳装置
wire rope knife 钢绳割刀
wire rope lay 钢丝绳捻法
wire rope measuring 井深测绳
wire rope nippers 钢丝绳钳
wire rope oil 钢丝绳润滑油
wire rope pulley 钢索滑轮;钢丝绳轮;钢丝绳滑轮
wire rope rolling 线材轧制
wire rope screen 钢索筛网;金属绳筛
wire rope sling 钢丝绳吊具;钢丝绳吊环
wire rope socket 钢丝绳套环;钢丝绳卡
wire rope splicer 钢丝绳接头;钢丝绳铰接器
wire rope splicing 接长钢丝绳
wire rope suspended conveyer 钢丝绳吊挂式带输送机
wire rope suspension 悬索;吊索
wire rope tester 钢丝绳试验机
wire rope thimble 钢绳绳环
wire rope tramway 架空索道
wire rope tricing lines 钢索拉紧装置
wire ropeway 钢缆索道;缆索道
wire ropeway of double rope 复线铁索道
wire saw 钢丝绳锯岩机;钢丝锯
wire saw guide 线锯导子;线锯导引器
wire scissors 钢丝剪
wire scrap 废线材
wire scratcher 带钉镊板;刮痕抹子;钢丝刷;钢丝梳;钉镊板
wire screen 金属细筛;金属网格;金属丝(网)筛;铅丝纱;铁丝网
wire screen cloth 金属丝网布
wire screen floor pen 铁丝网地面鸡圈
wire screen ladle 网勺;筛网漏勺
wire section 钢丝截面
wire sensitivity 透度计灵敏度;钢丝针入度计的灵敏度
wire separator 钢丝分隔板
wire service 新闻电讯社
wire setting machine 钢丝针布置针机
wire shackle 钢索钩环
wire sharing 线路共享
wire-sharing system 线路共享系统;分线传输系统
wire shears 线材剪切机
wire sheath 钢丝绳外皮
wire sheathing 金属编织层
wire short splice 钢丝绳短插接
wire sieve 金属细筛;金属丝(细)网筛
wire size 线号;线规;金属丝尺寸
wire skinner 刮除线末端绝缘包皮的工具
wire sling 钢丝绳套;钢丝绳扣;钢丝吊索;吊货钢丝索环
wire sling for salvage pontoon 打捞浮筒钢丝索环
wire sling package 绳扣成组货件
wire slip 钢丝滑动
wire snips 线剪
wire solder 焊剑
wire sonde 系留探空仪;有线探空仪
wire sounding 钢索测深法;测深索测深
wire sounding line 钢索测深索
wire spacer 钢丝隔离物;钢丝衬垫
wire span 顶跨钢索
wire spinner 铁丝扭结器
wire spiral 盘旋箍筋
wire splice detector 接线检查器
wire splicer 钢丝连接器
wire spoke 线辐;钢丝轮辐;钢丝辐条
wire spoke wheel 线辐轮
wire spool 引线卷轴;绕线盘
wire-spray gun 金属喷涂枪
wire spraying 金属丝喷涂
wire spring relay 线簧继电器

wire spring ring 铁丝弹簧圈
wire srapped screen 绕丝胶管;缠丝过滤器
wire staple 钢丝卡板;订书钉;U形钉;马蹄钉
wire-steel 钢丝
wire stirrer 线机网搅拌器
wire stirrup 钢箍
wire-stitching machine 钢丝订书机;缝线机
wire-stock 线材
wire stopper 钢缆制动器
wire straightener 线材整直器
wire-straightening and cutting machine 线材拉直切割机
wire-straightening machine 线材矫直机;线材矫直机
wire strainer 钢丝拉紧器;紧线机
wire strain ga(u)ge 线张应变计;线式变形测定仪;电阻丝应变仪
wire strand 钢丝索;钢丝股
wire stranding machine 绞绳机
wire stress 钢丝应力
wire stress variation 钢丝应力变化
wire stretcher 钢索拉伸机;钢丝拉伸机;拉线机;紧线机;紧线机;拔丝机
wire stretching 张拉钢丝法;拉丝;拔丝
wire stripper 紧线钳;导线绝缘剥除钳;剥线钳;剥线器;剥皮钳
wire stripping pliers 剥线钳
wire strop 钢丝套索;钢环索
wire-supported model 张线式悬挂模型
wire surface 针表面
wire suspension 悬索;吊索
wire suspension bridge 钢索吊桥;缆式悬桥;悬索桥
wire sweep 用钢丝扫床【疏】
wire sweep survey 钢丝扫床测量
wire synchronization system 有线同步方式
wire table 导线表
wire tach 铁钉
wire tach(e)ometer 蛛丝视距仪
wire tack 钢丝平头钉
wire tail 线尾
wiretap 秘密接头;窃听装置;窃取情报
wire telecommunication 有线通信
wire telephone 有线电话
wire telephone installation section 有线电话设备分排
wire teletype 有线电传机
wire television 有线电视
wire tendon 钢丝束
wire tension 紧线台
wire tensioner 钢丝绳拉紧器
wire terminal 线接头;钢丝接头;电线接头
wire terminal clamp 电线接头
wire thickness 钢丝厚度
wire thimble 钢索套环
wire thread brush 钢丝刷
wire tie 钢条系件;扎钢丝;扎钢筋;钢丝系件;拉结钢筋
wire tightener 钢丝紧张器;钢丝缠扎钳;紧丝器
wire-to-earth capacity 线地间电容
wire torque tester 线材扭转试验机
wire-to-water efficiency 水泵总效率;水泵机械效率
wire-to-wire capacitance 线间电容
wire-to-wire capacity 线间电容
wire train 涂线序列
wire tramway 轻便铁索路;索道
wire transposition 导线换位;导线编花
wire triangle 线三角
wiretron 线型变感元件
wire tube 电线导管
wire turn 线匝
wire turn-buckle 导线正反螺丝扣
wire twined filter 缠丝过滤器
wire twisting machine 绞线机
wire-tying baler 铁丝捆扎式拾检压捆机
wire-tying device 铁丝打捆装置
wire-tying machine 捆线盘机;包线机
wire-tying mechanism 铁丝捆扎装置
wire-type acoustic(al) delay-line 线声波延迟线
wire-type borehole extensometer 弦式钻孔伸长计
wire-type penetrometer 金属丝透度计
wire under voltage 火线
wire vehicle detector 钢丝车辆检测器
wire watcher 值班报务员

wireway 电线槽;线槽;钢丝绳道;金属线导管;提升绳道;电缆槽;布线槽
wire weaver 钢丝织机
wire weaving machine 金属网织机
wire weeder 钢丝除草耙
wire weight ga(u)ge 悬锤水标尺;悬锤式水位计;悬锤式水(标)尺
Wireweld 歪尔韦德牌钢筋(商品名)
wire wheel 钢丝(刷)轮;钢丝辐轮;轮形钢丝刷
wire wheel brush 圆钢丝刷;钢丝轮刷
wire winch 卷车
wire winding 绕丝
wire-winding machine 绕钢丝机;卷线机;缠绕钢丝机
wire/wireless switcher 有线/无线转接器
wire work 金属丝制品;金属丝织物;金属丝厂;线制品;制线厂
wire working 钢丝加工;金属线加工
wire-wound 绕线(的);线绕式
wire-wound armature 线绕电枢
wire-wound card 绕线纸板
wire-wound component 线绕部件
wire-wound concrete core-pipe process 混凝土管芯缠丝工艺;三阶段预应力制管工艺
wire-wound cryotron 线绕冷子管
wire-wound doctor 绕线棒刮涂器
wire-wound filter 绕线式滤油器
wire-wound frame 绕丝框架
wire-wound furnace 线绕电炉;绕丝炉
wire-wound high pressure vessel 绕线高压容器
wire-wound hydraulic press 绕带式液压机
wire-wound induction motor 线绕式感应电动机
wire-wound isostatic press 绕丝等静压机
wire-wound joint 线绕接头
wire-wound pipe joint 线绕管接头
wire-wound piping joint 线绕管接头
wire-wound pole 线绕磁板
wire-wound potential meter 线绕电位器
wire-wound potentiometer 线绕电位器;线绕电位计
wire-wound resistance 线绕电阻
wire-wound resistor 绕线电阻器;绕射电阻器;线绕电阻(器)
wire-wound rheostat 线绕变阻器
wire-wound rod 绕线棒刮涂器
wire-wound rubber hose 铠装胶管
wire-wound strainer 绕线式滤油器
wire-wound tube joint 线绕管接头
wire-wound tubing 线绕管接头
wire-wound vessel 绕丝式容器
wire woven 钢丝绕织
wire wrap connection 绕线连接;绕接
wire wrap machine 绕线机
wire wrappable assembly 绕接装置
wire wrappable module 绕接模件
wire wrappable module board 绕接模件板
wire wrapped dam (防洪用)钢丝网防护的堆石坝;铁网包坝
wire wrapped hose 高压胶管
wire wrapped socket board 绕接插孔板
wire wrapped tool 绕接工具
wire wrapper 绕接机
wire wrapping 绕接;箍筋;绕线
wire wrapping board 绕接板
wire-wrapping connection 绞接
wire-wrapping machine 包线机,绕线机,绕接机;钢丝缠绕机
wire-wrapping method 包线法;绕线法
wire wrapping system 绕接方式
wire wrap spacer 绕丝定位件
wire wrap tool 绕线工具
wiring 线路【电】;装线;配线;敷线;布线
wiring and sealing 钢丝绑扎与封铅
wiring board 接线板;控制板;插接(线)板;布线底板
wiring box 接线盒
wiring capacitance 接线电容;分布电容;布线电容
wiring capacity 布线电容
wiring card 接线卡
wiring case 接线箱;接线盒
wiring clip 线夹;钢丝剪
wiring coach 架线车
wiring code 布线规程
wiring condition 配线情况
wiring conduit 电线管道;布线管道
wiring conduit line 电力管线
wiring conduit network 电力管线网
wiring connection 接线
wiring design 电路设计;布线设计
wiring diagram 配(接)线图;线路(平面)图;线路布置图;接线图;电路图;布线图【电】;安装图
wiring diagram of contactor switch gear of control panel 控制盘接触器接线图
wiring error 布线误差
wiring grommet 索眼;电线索眼;电线环状接头;布线环;布线套
wiring harness 装配电路;成行导线
wiring installation 布线工程
wiring layout 线路敷设;线路设计;接线图
wiring list 布线表
wiring material 布线材料
wiring method 布线法
wiring pattern 布线图【电】
wiring pen 布线笔
wiring plan 线路敷设图
wiring plate 布线板
wiring plug 线路插头
wiring point 联结点;接线点
wiring press 钢丝弯边压力机;卷边压力机;嵌线卷边机
wiring problem 布线问题
wiring process 线路接法
wiring regulation 布线规程;接线规程
wiring scheme 布线图;接线图;安装图
wiring strategy 布线策略
wiring switch 线路开关
wiring system 导线系统;布线系统
wiring terminal 接线柱【电】;接线端子
wiring test 导测
wiring topography 接线布局
wiring trough 导线槽
wiring unit 布线器
wiring up 接线
wiring work 配线工工作
Wirsung's duct 维尔松管
Wisconsian glacial stage 威斯康辛冰期
Wisconsin 威斯康辛期
Wisconsinan(stage) 威期康星阶【地】
Wisconsin general testing apparatus 威斯康辛通用测验仪
Wisconsin glacial epoch 威斯康辛冰期
wiserite 羟硼锰石
wishbone 叉形杆
wishbone arm 叉形定位臂
wishing line 愿望线
wisp 亮条
wisper wind 威兹珀风
wispy annular flow 卷束环状流
wistaria trellis 藤萝花架
wistaria violet 紫藤花色
witch grass 茅草
witch mirror 魔镜
with 10's carry 逢十进位
with all risks 保一切险
withamite 黄红帘石
with an elevation of 2000 metres 海拔高达两千米
with approval of authority 经有关当局批准;经上级认可
with average irrespective of percentage 不受免赔额限制的水渍险
with certain qualifications 附带声明;附带某些条件;保留意见
with code symbol 带有规范标记的
withdraw 退回;收回;抽取;抽出
withdraw a bid 撤消投标
withdrawable 能撤回的;能提取的;能移开的
withdrawable axle 可卸卸车轴;可拆卸的轮轴
withdrawable beating 可卸轴承;可卸支承;活络轴承
withdrawable bit 可拆装钻头;活络钻头
withdrawable shoe 可拆桩靴;可移动座脚
withdraw a claim 放弃要求
withdrawal 回采;取款;排水量;排出量;提取;提款;抽水量;撤消;撤退;放出
withdrawal and reducing roll 拉轧辊
withdrawal bias 失访偏性
withdrawal casing 拔去套管
withdrawal crystal 拉晶
withdrawal current 生产流
withdrawal device 取出装置;退卸工具;抽取器
withdrawal feeder 取料喂料机
withdrawal load 拔出阻力;拔拉荷载
withdrawal mechanism 取锭机构
withdrawal of aids-to-navigation 航标撤除
withdrawal of assets 提取资产
withdrawal of bill of lading 收回提单
withdrawal of currency from circulation 货币回笼
withdrawal of international standard 取消国际标准
withdrawal of oil 油出;抽去油分;去油;除油
withdrawal of water 脱水
withdrawal receipt 提款收据
withdrawal resistance 抗拉力;拔出阻力
withdrawal resistance test of pile 拔桩试验
withdrawal roll 拉辊;夹送辊
withdrawal shares 退股
withdrawal sleeve 退卸套
withdrawal slip 取款单;提款单
withdrawal space 拆卸空间
withdrawal straightening stands 拉矫机组
withdrawal test (桩的)拔出试验
withdrawal test of pile 拔桩试验
withdrawal ticket 提款单
withdrawal tool 拆卸工具
withdrawal tube 吸管;抽吸管
withdrawal well 出油井
withdraw an offer 取消报价;收回报价;撤回原发价;撤回发盘;撤回承诺
withdraw collar 分离环
withdraw deposit 提取存款;提款
withdrawer 拉轮器;回收器
withdraw from partnership 退伙
withdrawing buoy 撤回浮标
withdrawing device 拉轮器;抽出装置
withdrawing pattern 拔模
withdrawing pile 拔桩
withdrawing room 坐起室;休息室
withdraw money 提款
withdraw share 退股
withdraw the drill pipes 抽出套管
withe 烟道隔墙;半砖隔墙;枝条;柳条;空斗墙的立砌隔砖
wither 干枯;凋萎
withered 干枯
withering point 枯萎点
witherite 碳酸钡矿;碳钡矿;毒重石
withers height 体高
with feather bolt 带鼻螺栓
with fixed base collector 固定式接电器
with full and without full 全有全无
with good geologic(al) conditions of ore deposition 成矿地质条件有利
withhold 扣留;扣缴;拒绝放弃
withhold business 停止交易
with holder rule 带柄尺
withhold from the market 停止售货
withholding 扣款;预提;扣缴;扣交
withholding detain 扣款
withholding income tax 代扣所得税;预提所得税;扣缴所得税
withholding income tax return 扣缴所得税报告表
withholding of tax 扣缴税款
withholding or impounding appropriation 拨款扣留
withholding period 停止期
withholding statement 扣缴所得税凭单;扣缴(税款)清单;扣缴报表
withholding system 扣缴制
withholding tax 预提税;预扣赋税;代扣所得税
withholding tax credit 预提税抵免
withholding tax rate 预提税率
withhold payment 拒绝支付;不予支付
with horizontal surface 带水平面的
within-basin depth-area curve 流域内雨深一面积曲线
within battery limit 界区内
within block information 区组内信息
within close tolerance 在高精度范围内
within inside journals axle 内预轴
within large water boiler 多水锅炉
within site 厂区内
within the next two decades 在今后的二十年内
within the tole-rance specified in the drawing 在图纸规定公差范围

within the trade 同业间
with lubricator 用润滑剂
with microfold 具微褶皱
with oblique surface 带倾斜面的
with off a score 付账
with order 按程序处理
without articulations 固定的;无铰的
without benefit of salvage 无救助利益
without boron 没有硼
without charges 免费;不计价
without consideration 不考虑
without consideration of salary 不计报酬
without cost 不计价
without cracks 无裂缝
without delay 不得延误
without exception 无例外
without fraudulent intent 无欺骗意图
without harming crop seeds 不影响作物种子
without interference to navigation 不碍航行
without interruption 连续供应;不中断
without obligation 不受约束;不计价
without our responsibility 我方无责任;我方概不负责
without page citation 不注页数
without particular average 单独海损
without prejudice 不使权利受到损害
without previous notice 不预先通知
without project 无项目
without putrefaction 不腐烂
without qualification 没有严格限制条件;无条件的;不受限制地
without recourse 免予追索;无(受)追索权;无权追索;无偿还义务
without recourse letter of credit 无追索权信用证
without regard for 不考虑
without return 不能退货;不可退货
without second-pollution 无二次污染
without stock removal 非切削成形
without striations 无擦痕(条纹)
without top man 塔上无人
without wear 抗磨损
with particular average 水渍险
with practical experience 有实践经验的
with preexisting foliation 具先存面理
with project 有项目
with proviso 附有条件
with recourse 有追索权;追索权;保留追索
with reference to 比照
with rights 附认股权
with sight flow 带有可视流量
with slickenside 有镜面
withstand current 耐电流
withstand electric(al) strength 耐电强度
withstand electric(al) strength test 耐电强度试验
withstand fire 耐火
withstanding capability 承受能力
withstand test 耐压试验;容许电压试验
withstand test voltage 耐压试验电压
withstand voltage 耐(电)压
withstand voltage test 耐(电)压试验【电】
with striations 有擦痕
with tension pulley 带有张紧轮
with the current 顺潮
with the grain 顺纹
with the priority to be selected 优先选用的
with the sun 顺时针转动
with through drive gear 直接传动变速装置
with tide 顺潮
with traps 带汽水阀
with two arms hanging on both sides 两臂自然下垂
with two spring draw gear 双簧牵引装置;双簧车钩
with well 带套管的
with wind 顺风(航行)
withy 烟道纵隔;柳条绳
witness angle 联测角
witness corner 地界角标;联测点;土地测量联测点;测量中的联系点;参考角
witnessed performance test 证实性性能试验
witnessing monument 参考标石
witness inspection 见证检查
witness line 证示线
witness mark 联系点;(测量中的)参证点;验证记号;联测点;测量(中)参考(点);参考点

witness point 见证点;参考点
witness post 参考桩【测】
witness sample 检验样品;检验样板;核查用试样
witness stake 参考桩【测】
witness test 目击试验;订货人在场的试验;法人在场作证的试验
Witte-Margules equation 维特—马古列斯方程
wittichenite 硫铋铜矿
wittite 威硒硫铋铅矿
Witt's colo(u)r theory 威特色理论
Witt theory 威特理论
Witt vector 威特向量
wivet 石板的匠称
wivot 石板的匠称
wizard 精灵
wizened 干枯
wlolframite 锰铁钨矿
woad 菘蓝;大青
Wobbe index 华白(指)数;沃布指数
Wobbe number 华白(指)数;沃布值
wobble 摇摆;摆动
wobble coefficient 变异系数;摆动系数;(预应力混凝土)后张钢筋摩阻损失系数
wobble control signal 抖动控制信号
wobble crank 摆动曲柄
wobble-disk meter 摆盘式水表
wobble effect 摆动效应
wobble friction 变异摩擦;后张预应力筋套管摩擦;摆动摩擦
wobble gear 摆动齿轮
wobble hypothesis 变偶假说;摆动假说
wobble joint 振动连接
wobble of wheel 转向轮摆头
wobble plate 摇摆板
wobble plate engine 摇摆板多缸发动机
wobble plate feeder 摇板式进料器;摇摆式进料器;摆动式平板喂料机;摆动式平板给料机
wobble plate fuel pump 摇摆板燃料泵
wobble plate mixer 摆动板式搅拌机
wobble plate pump 摇摆盘式泵
wobble pump 摇摆泵;手摇泵
wobbler 摇摆机构;偏心装置;偏心轮
wobbler action 偏心作用
wobbler machine 摇摆机(构)
wobbler mechanism 摇摆机构
wobbler milling machine 梅花头铣床
wobble rolling machine 摆动辗压机
wobbler seal 振动压实(法)
wobbler shaft 偏心轴;凸轮轴
wobbler spindle 梅花连接轴
wobble saw 开槽锯;摇摆锯
wobble seal 振动压实
wobble shaft 滚转凸轮轴;滚转(偏心)轴
wobble wheel 摆(动)轮
wobble wheel roller 活动式轮胎压路机;摆轮式轮胎碾压机;摆动式压路机;摆动式轮胎碾压机;摆动式轮胎压路机;振动式压路机
wobbling 摇摆
wobbling action 甩摆动作;摆动动作
wobbling action of train 列车甩摆动作
wobbling bit 可调整的摆动铣床
wobbling drum 转鼓
wobbling effect 贴纹效应;颤动效应
wobbling system 摇频制
wobbulation 频率摆动
wobbulator 摆频振荡器;摆频信号发生器
wodanite 钛云母
wodginite 锡锰钽矿
woehlerite 硅铌锆钙钠石
woelsendorfite 红铅铀矿
woggle joint 挠性连接;挠性接头
Wohl block 伍尔黏土空心块
Wohler curve 疲劳强度曲线;沃勒曲线
wold 荒原;山地;高原
wolf 经验丰富的投机商
wolfachite 辉砷锑镍矿
Wolf diagram 沃尔夫图
wolfeite 羟磷铁石
Wolf-Lundmark system 沃尔夫—伦德马克系统
Wolf number 沃尔夫数
wolf-proof fence 防狼围栏
wolfram 钨
wolframate 钨酸盐
wolfram brass 钨黄铜

wolfram bronze 钨青铜
wolfram diboride 二硼化钨
wolfram filament 钨丝
wolframic acid 钨酸
wolframine 钨华
wolframite 黑钨矿;钨锰铁矿
wolframite ore 黑钨矿矿石
wolframite-quartz vein deposit 黑钨矿石英脉矿床
wolframium 钨
wolframium oxide 氧化钨
wolfram ocher 钨华
wolfram ore 钨砂
wolfram steel 钨钢
wolf's claw 卷柏;石松
Wolf's curtain 沃尔夫式浮坝
Wolfsholz bored pile 沃尔夫斯霍茨钻孔(灌注混凝土)桩
wolf tooth gearing 棘齿式啮合
wolf tree 老狼木
wollastonite 硅灰石;钙硅石
wollastonite ceramics 硅灰石陶瓷
wollastonite content 硅灰石含量
wollastonite deposit 硅灰石矿床
wollastonite diopside grossular hornfels 硅灰石透辉石钙铝榴石角岩
wollastonite-gneiss 硅灰石片麻岩
wollastonite in lump 硅灰石块
wollastonite in powder 硅灰石粉
wollastonite ore 硅灰石矿石
wollastonite porcelain 硅灰石瓷
Wollaston polarizing prism 沃拉斯通偏光棱镜
Wollaston prism 渥拉斯顿棱镜;沃拉斯顿棱镜
Wollaston's reflecting goniometer 吴尔拉斯顿反射测角仪
wollen fabrics 毛织品
Wolman salt 防腐盐(剂);沃尔曼盐(木材防腐剂);氟化钠—砷酸二硝基酚溶液
Wolman salt agent 沃尔曼盐剂(木材防腐用)
Woltman current meter 沃尔特曼流速仪
women drawing room 妇女休息室
women hairdressing shop 妇女美发厅
women quarter 闺房;妇女室
women's changing room 妇女更衣室
Women's Council of Realtors 女房地产经纪人协会
women's shop 妇女用品商店
women toilet 女厕所;女盥洗室
Wommer safety clamp 沃梅尔安全夹持器
wonder gun 水泥喷射枪
wonderland 仙境
wonderstone 奇异石(叶蜡石的俗称)
wonesite 翁钠金云母
wood 木料;木材;密集小林
wood-stone carving 木石刻品
wood adhesive 木材胶黏剂
wood aggregate 木质集料;木质骨料
wood alcohol 木精;木醇
wood alcohol oil 木醇油
wood allowance 木窗框裕量
wood and brick-clad home 砖木(结构)住宅
Wood-Anderson seismic instrument 伍德—安德逊地震仪
wood and metal construction 钢木混合结构
wood-and-steel composite truss 钢木屋架
wood angle cutter 木料斜角铣刀
wood apron 木护坦
wood arch 木制大拱;木拱
wood area 林区
wood ash 木灰;草木灰
wood ash black 木炭黑
wood ash glaze 灰釉;草木灰釉
wood ball 木球
wood ball turning lathe 木球车床
wood bandle cross peen hammer 木柄十字尖头锤
wood bark tar 树皮焦油
wood base 木踢脚板
wood base broad 木踢脚板
wood-based product 木基制品
wood base fiber board 木基纤维板
wood base fiber panel materials 木基纤维板料
wood base fibre and particle panel material 木质人造板
wood base materials 木基材料(指纤维板和碎料板等人造板)
wood base panel material 木质人造板材

wood base particle panel 木基刨花板
wood base plastics 木基塑料
wood batten 木挂瓦条;挂瓦条;板条;压缝条;木窄板
wood beam-base 木梁基
wood beetle 木蠹蛾
wood bending machine 弯木机
wood block 地板木条;木砖;木块;木版画
wood block filler 木嵌块;木砌块;木垫块
wood block floor 镶木楼板;镶木地板;木辘砖地面;木块铺楼面;木块地面
wood block flooring 拼花地板;拼木地板;木块铺屋面;木块拼花地板;木块(楼)地面;木块铺楼面
wood block floor-installation 木地板铺装
wood block humidity controller 木块湿度控制器
wood blocking 木枕
wood block parquetry flooring 木块拼花地板
wood block pavement 木块铺面
wood block paving 木块路面;木块铺砌;木块铺面
wood block printing 木刻水印;木刻版印刷(术)
wood blocks core door 拼块芯木门
wood board 木侧板;木墙板;木板壁
wood boat bottom paint 木船船底漆
wood bolt 木锚钎;木锚杆
wood borer 木钻床;木钻;木质部蛀虫
wood boring machine 木(工)钻床
wood boring organism 钻木生物
wood box 木砖
wood break pin 木制安全销
wood brick 木(落)砖;木块
wood buffer 木缓冲器
wood building panel 木制建筑嵌板
wood built-in unit 固定的木装置;固定的木构件
wood bundle 桦树条
wood-burning kiln 柴窑器
wood butcher 鳖脚木工(俚语)
wood caisson 木沉箱
wood carbohydrate 木材碳水化合物
wood carbonization plant 木料炭厂
wood card 木纹板
wood carrier 运木船
woodcarver's file 木工雕刻锉
woodcarver's rasp 木工雕刻锉
wood carving 木刻(工艺);木雕(刻品)
wood carving articles 木刻制品
wood carving tool 木雕刻刀;木刻工具
wood carving wares 木雕制品
wood casing system 木槽板条布线法
wood ca(u)lker 捻缝工人
wood ceiling blocks 天花板木块
wood cell 木垛;木笼
wood cellulose 木(材)纤维素
wood cement 木质(感)水泥;仿木水泥
wood cement block 水泥木屑砌块
wood cement board 木丝水泥板;水泥木丝板;水泥木橡板
wood cement concrete 木屑水泥混凝土;木质水泥混凝土
wood cement concrete slab 木屑砂浆板
wood cement particle board 水泥刨花板
wood cement roof 木材水泥屋顶
wood centering 木拱鹰架;木拱架
wood char 木炭
wood charcoal 木炭
wood chemistry 木材化学
wood chip 木屑;木片;刨花
wood chip absorbent ceiling 碎木片吸音天花板
wood chip absorbent ceiling board 木屑吸声顶棚板
wood chipboard 碎木电板;碎木胶合板
wood chip concrete 木质感混凝土;仿木混凝土
wood chip grab 木屑抓斗
wood chip paper 木浆粗纤维纸
wood chipper 木片切削机
wood chip products 木屑制品;木花制品
wood chip recovery 木屑回收
wood chip storage 木屑储存
wood chip wallpaper 木花墙纸;木屑墙纸
wood chisel 木(工)凿
wood chock 轨撑;钢轨支撑
wood chopper 柴刀;伐木刀
wood chopper's maul 伐木木锤
wood chuck 木夹盘
wood clamp 木夹

wood clapper (木的)踏脚板
wood clear finish 木材的清漆罩面
wood cleaver 劈木器
wood coal 木炭
wood collapsible box pallet 木制折叠箱形托盘
wood column 木柱
wood combined beam 木组合梁
wood compressometer 木材压缩测定仪
wood concrete 木(质感)混凝土;仿木混凝土
wood concrete block 木混凝土块
wood concrete composite beam 木混凝土混合梁;木混凝土合成梁;木一混凝土组合梁
wood concrete filler 木混凝土填料
wood concrete roofing slab 木混凝土屋面板
wood concrete roof slab 木丝混凝土屋顶板
wood concrete slab 木混凝土板
wood concrete soffit floor filler block 木混凝土地板垫块
wood concrete tile 木混凝土块
wood conduit 木制管;木导管
wood connection 木结点;木接头
wood connector 木连接器;木接头
wood construction 木建筑;木结构;木构造
wood construction set 木质建筑玩具
wood consumption 木材消耗量
wood container 木制集装箱
wood-containing paper 含磨木浆纸张
wood contractor 木材包商
wood copper 橄榄铜矿
wood core box 木花盆
wood cotton tree 木棉树
wood cover fillet 木压缝条
wood covering 木地板;木覆面;木覆板
wood cover strip 木压缝条
woodcraft 木工(技)(艺);木材加工
wood creosote 木质酚油;木(焦)杂酚油
woodcrete 木质感混凝土;仿木混凝土
wood crisis 木材危机
wood culvert 木质涵洞
wood curing waste 木材加工废物;木材加工废料
wood curing wastewater 木材加工废水
woodcut 木刻(画);木版画;木材锯解
woodcut knife 木刻刀
woodcutter 版画家;木刻家;伐木工人
wood cutter's axe 伐木斧
wood cutting 木材锯解;木刻
wood-cutting blade 木刻刀
wood-cutting mechanist 锯木技师
wood-cutting saw 伐木锯
wood dado 木墙裙
wood decaying fungus 木材腐朽菌
wood deck 木平台;木质屋面
wood decomposing fungi 木材腐朽菌
wood depot 储(贮)木场;集材场
wood derrick 木钻塔;木起重臂
wood-destroying fungus 腐木菌;木腐菌
wood-destroying organisms 木腐生物
wood destruction 木材破坏
wood deterioration 木材腐;木材变质
wood distillation 木材蒸馏(法);木材干馏
wood distillation methanol 木馏甲醇
wood distillation plant 木材干馏厂
wood distilling plant 木材蒸馏设备;木材蒸馏厂
wood door 木门
wood dough 木纤油灰;木纤腻子
wood downspout 木水落管
wood drain pipe 木质排水管
wood drawing 木画
wood drill 木工钻;麻花钻
wood drill bit chuck 木钻钻头夹头
wood driller 木钻床
wood drilling machine 钻木机;木钻床
wood dry distillation 木材炭化
wood drying 木材干燥(法)
wood dust 锯木屑;木屑;木尘
wood dye 植物染料
wooded 有森林的;树木繁茂的;多树木的
wooded area 森林面积;产木地区
wooded bridge 木桥
wooded coaming 木围板
wooded concrete composite beam bridge 木混凝土组合梁桥
wooded form 木架
wooded lath stud partition 灰板条隔墙;灰板条隔断

wooded meadow 林中草地
wooded mo(u)ld(ing) 木线脚(饰)
wooded patch 木制堵漏板
wooded rammer 木夯
wooded site 有树木的场地
wooded stair(case) 木楼梯
wooded stave flume 木板水槽;木板渡槽
wooded washboard 木踢脚板
wood element 木构件;木质素;木材纤维细胞
wooden 木质的;木制的;木模
wooden airplane 木质飞机
wooden and steel furniture 钢木家具
wooden arch 木拱
wooden arch bridge 木拱桥
wooden articles 木制品;木器
wooden astragal 木窗棂;木窗玻璃镶条;木制玻璃格条;木制窗芯子
wooden backing 木支撑
wooden baluster 木栏杆;木栏杆(小)柱
wooden banister 栏杆小木柱
wooden bar 檩条;木挡;木条
wooden barrack (临时的)木板工房
wooden barrel principle 木桶箍紧原理
wooden base 木模基;木底板
wooden base plate 木踢脚板
wooden bathtub 木浴桶;木浴(澡)盆
wooden batten 木压条;木(板)条
wooden beam 木梁
wooden beam bridge 木梁桥
wooden beam floor 木梁楼板
wooden bearing 木制轴承
wooden bearing structure 承重木结构
wooden bed plate 木垫板
wooden bench 板凳
wooden bevel wood siding 鱼鳞板墙;披叠木板壁
wooden bin 木储仓;木箱
wooden binder 木联系(小)梁;木联结梁
wooden binding beam 木系梁
wooden binding joist 木联系格栅
wooden bitt 木质木船柱
wooden blade construction 木制叶片结构
wooden block 木制砌块;木(砌)块;木滑车
wooden-block pavement 木块铺面;木块路面
wooden board 木板
wooden board covering 木板覆盖;木板盖面
wooden board lining 木衬板;木侧板;木板衬里
wooden board sheathing 木望板;木屋面板
wooden boat 木船
wooden bolt 木锚杆
wooden box culvert 木制箱形涵洞
wooden box drain 木箱式暗沟
wooden box headgate 木箱式进水闸
wooden box package 木箱包装
wooden bracket 木托座;木牛腿;木托座
wooden brad 木曲角钉;木曲头钉
wooden brake block 木闸瓦;木闸块
wooden brake shoe 木闸瓦
wooden brick 木砖
wooden brick floor 木砖地面
wooden bridge 木桥
wooden bridge sleeper 木桥枕
wooden broach 木钻头;木尖塔
wooden building 木材建筑物
wooden building block module 木建筑块材模数
wooden bulkhead 木舱壁
wooden bumper 木质缓冲器
wooden car 木制车
wooden carved furniture 木雕家具
wooden carving lantern 木雕灯
wooden case 木箱
wooden casement window 木平开窗;木制竖铰链窗
wooden casing saw 木柄套锯
wooden casks 木桶
wooden centering 建拱木模架
wooden centring 建拱木模架
wooden chipboard 木屑板;碎木胶合板
wooden chock 木楔;钢轨楔形垫木;木塞块;木垛
wooden church 木教堂
wooden ciborium 木龛室;木华盖
wooden cistern 木蓄水池;木水槽
wooden clamp 夹固板;夹骨齿夹固板
wooden coal 木炭
wooden column 木柱

wooden column order 木柱式
wooden comb 木梳
wooden composite pile 木混合桩
wooden concrete composite beam 木材混凝土组合梁
wooden concrete form 木制混凝土模板
wooden connection 木结构联结
wooden construction 木结构；木材建筑
wooden construction type 木构造类型
wooden container 木制容器
wooden core 木芯
wooden counter ceiling 木吊顶棚
wooden crate 木箱；木条箱
wooden crib 木围栏；木笼；木(井)框；木垛
wooden cross arm 木横担；木横臂
wooden crusher block 木垫块
wooden cushion pad 木垫板；木垫块
wooden dam 木(堰)堤；木坝
wooden dark slide 木制蝉滑板
wooden deck 木甲板
wooden decoration 木装修
wooden derrick 木质钻塔
wooden dome 木穹隆；木制圆屋顶
wooden domino 木质骨牌
wooden door frame 木门樘；木门框
wooden door threshold 木(门)槛；木制门槛
wooden double beam 木叠(合)梁
wooden double block 木制二拼滑车
wooden dovetail 木销钉
wooden dowel 木销钉
wooden drain 木质排水管
wooden drainage 木制排水槽
wooden drilling derrick 木钻架
wooden dunnage 木衬垫；垫舱木板
wooden dust shield 木质防尘罩
wooden eave(s) gutter 木檐沟
wooden engraving 木刻
wooden engraving picture 木雕画
wooden envelope 木质蒙皮
wood energy 木柴含能量
wooden facing 木贴面；木(板)饰面
wooden falsework 木脚手架
wooden fastener 木紧固件；木扣件
wooden fence 木栅栏；木围栏
wooden fence post 木护栏柱
wooden fender 挡木；木挡板；木栅；木围墙；木碰垫；木防护板；护舷木
wooden filler 嵌缝木；填木；木填缝料
wooden fillet 木嵌条；木(楞)条
wooden finger plate 木推手板
wooden firring 木灰板条
wooden fish 木鱼
wooden flat roof 木平屋顶
wooden float 刮板；镘刀；木镘板；木蟹；木抹子
wooden floating floor 架空木地板
wooden floor 木地板；木楼板
wooden floor chip 木地板皮
wooden floor cover(ing) 木地板面层；木楼板面层
wooden flower-vase 木花瓶
wooden flume 导水木管；木水槽
wooden folded plate roof 木折板屋盖；木折板屋顶
wooden folding chair 木折椅
wooden folding door 木折(叠)门
wooden folding rule 木折尺
wooden foot bridge 木行人桥
wooden form 木质模板
wooden former 木模
wooden formwork 木模(板)
wooden frame 木制框板；木框(架)
wooden frame construction 木框架结构
wooden framed building 木构架建筑(物)
wooden framed house 木框架房屋
wooden framed mirror 木框镜
wooden-framed spiked harrow 木架钉齿耙
wooden framed wall 木框架墙
wooden frame with dougong 大式
wooden frame without dougong 小式
wooden framework building 木框架建筑
wooden furniture 木制家具
wooden furring 钉木板条
wooden fuselage 木质机身
wooden garage 木结构车库
wooden gate 木大(闸)门；木制闸门
wooden gatehouse 木过街楼；木门房

wooden girder 木大梁
wooden glazing bar 镶玻璃木窗棂条；(镶嵌玻璃)木窗芯子；玻璃窗木嵌条
wood engraving 木板画；木刻术；木(雕)刻画；木版刻
wood engraving material 木刻材料
wooden grid 木格子；木格栅；木窗格
wooden grid footing 木格子基础
wooden grid packing 木栅填料
wooden grillage 木格框；木格排
wooden grillage footing:wooden grillage foundation 木格排基础；木格床基础
wooden grillage foundation 木格床基础
wooden ground 木基底
wooden ground bracing 木地撑
wooden ground floor 木(底层)地板
wooden ground slipway 木滑道
wooden gutter 木天沟；屋顶木雨水槽；屋顶木檐槽
wooden hammer 短锤；手锤；大锤；木锤；木槌
wooden handle 木执手；木柄；木把手
wooden handle nest of saws 木柄套锯
wooden handle ratchet screwdriver 木柄棘轮式螺丝起子
wooden handle screwdriver 木柄螺丝起子
wooden handle with ferrule 木柄套箍木凿
wooden hand rail 木扶手
wooden hatch board 木舱口盖
wooden header 木横梁；木顶梁；露头木砖
wooden hipped-plate roof 木饺脊折板屋顶；四坡木板屋顶
wooden hoist tower 木起重(机)塔；木吊扣塔
wooden hollow floor filler 架空木地板填料
wooden horse 木马
wooden house 木屋
wooden hull 木质船体；木船壳
wooden hut 木小舍；木制临时营房；木棚屋
wooden image 木偶
wooden inlay 木镶嵌饰
wooden insert 木衬垫；木制嵌件；木楔；木垫板
wooden Ionic column 木爱奥尼亚式柱
wooden jack plane 木壳粗刨
wooden joist 木托梁；木格栅
wooden joist floor 木格栅楼板；木格栅地板
wooden joist roof floor 木格栅屋顶地板
wooden joist upper floor 上层木格栅楼板；木格栅上层盖板
wooden key 木键；木楔
wooden knocker 木鱼
wooden ladder 木爬梯；木梯(子)
wooden lagging 木隔板
wooden lamella cupola 木板条拼成的圆顶；薄板条叠层圆屋顶
wooden lamella dome 木板条拼成的穹隆；薄板条叠层穹顶
wooden latch 木(门)闩
wooden lath 抹灰木条；灰板条
wooden lathing 抹灰木板条；粉刷木板条
wooden latticed beam 木格(式)梁
wooden latticed cupola 木格构穹隆；木格构圆顶
wooden level 木水平尺
wooden lifecraft 木质救生筏
wooden lining 木贴面；木隔板；木衬板
wooden lintel 木过梁
wooden louvres[louvres] 木百叶
wooden mallet 木锤；木槌
wooden mast 木桅(杆)
wooden matting 木垫排
wooden maul 木锤；木槌
wooden mitering gate 木制人字形闸门
wooden mitring gate 木制人字形闸门
wooden model 木模型
wooden mopboard 木踢脚板
wooden mosaic 镶嵌细木工
wooden mo(u)ld 木模(型)
wooden mo(u)lding 木制线脚装饰；木饰线条
wooden mo(u)lding wiring 槽板布线
wooden nail 木钉
wooden nailing plug 木钉块；木钉条；(墙上)受钉木塞
wooden order 木柱式
wooden packing 木包装；木衬垫
wooden-packing case 木箱
wooden palisade fence 木栅栏；木围篱
wooden pallet 木托板

wooden panel 木镶板
wooden panel(l)ing 木护墙板；木壁板；木镶板
wooden particle board 木屑板
wooden partition(wall) 木隔墙；木隔断；板壁
wooden patent glazing bar 木玻璃条
wooden pattern 木模
wooden pavement 木材铺面；木路面
wooden paving 木铺面
wooden paving block 木铺砌块
wooden paving work 木铺砌工作
wooden peat 木质泥炭
wooden peg 小木桩；木柱橛；木栓；木钉
wooden pier 柁墩
wooden pile 木桩
wooden pile joint 木桩接头
wooden pile stage 木桩工作台
wooden pile staging 木桩工作平台
wooden pillar 木柱；木支墩
wooden pin 木钉
wooden pipe 木管
wooden pipe penstock 木制管道闸门；木制压力水管
wooden plain webbed beam 实腹木梁
wooden plain webbed girder 实腹木大梁
wooden plain(web) girder 木制实腹梁；木板梁
wooden plane 木刨
wooden plate girder 木板梁；木板桁架
wooden plug 木栓；木塞；木楔
wooden pole 木杆
wooden pole line 木杆线路
wooden post 木立柱；木桩；木杆
wooden post and rail fences 木杆围栏
wooden prefabricated construction 预制木建筑；预制(装配)木结构
wooden principal post 木主柱
wooden prismatic shell roof 棱形薄壳木屋顶；木折板屋盖；木折板屋顶；木棱柱壳屋顶
wooden product 木制品
wooden propeller 木螺旋桨
wooden pulpit 木讲堂；木制讲坛
wooden pump logs 木制溢流导管
wooden punner 木夯；木捣
wooden purlin(e) 木檩(条)
wooden puttyless glazing bar 无油灰玻璃木条木
wooden quoin 木楔
wooden raceway 木制穿线板；木制槽板
wooden raceway work 木制线槽敷设工程
wooden raft 木筏
wooden rafter 木椽子
wooden rail 木扶手；木横挡；木护栏
wooden rail ga(u)ge 木道尺
wooden railroad tie 铁路枕木；铁轨枕木
wooden rainwater gutter 木雨水檐沟；木天沟
wooden rake 木耙
wooden rammer 木夯
wooden reciprocating potato sorter 木制往复式马铃薯分选机
wooden revolving door 木(旋)转门
wooden rigid frame 木刚架
wooden road 木路面
wooden rock bolt 木锚杆
wooden rod 木制标尺
wooden roller bearing-block 导链滚筒木质轴承座
wooden roller shutter 木卷百叶窗；木制卷帘百叶窗
wooden roof 木屋顶
wooden roof bar 坑道木梁
wooden roof cladding 木屋面(盖)板
wooden roofed 木屋顶
wooden-roofed basilica 有木屋顶的会议厅
wooden roof floor 木平屋顶
wooden roof gutter 木檐沟
wooden roofing 木屋面
wooden roof shake 木鱼鳞板
wooden roof sheathing 木屋面盖板
wooden roof spire 木屋顶尖塔；木尖塔屋顶
wooden roof truss 木屋(顶)架
wooden round 木弹
wooden ruler 木尺
wooden runner 木轨；木碾轮
wooden runway 木路面
wooden sanitary cove 木槽板；木踢脚板
wooden sash 木窗扇
wooden sash bar 木窗棂；(镶嵌玻璃的)木窗芯子
wooden scraping straight edge 木刮板

wooden screed 木样板;木刮板;木准条
wooden screen 木隔板;木挡板
wooden screw 木螺旋桨
wooden screw pile 木螺旋桩
wooden scrub board 木踢脚板;木擦洗板
wooden sculpture 木雕刻
wooden seat 木座位
wooden separator 木刮板;木隔板
wooden shake 木片瓦;木鱼鳞板
wooden shed 木棚
wooden sheet pile 木板桩
wooden sheet pile cofferdam 木板桩围堰
wooden sheet piling 木板桩
wooden sheet piling weir 木板桩堰
wooden shell 木薄壳
wooden shell cupola 木薄壳圆顶
wooden shell roof 木薄壳屋顶
wooden shingle 鱼鳞板;木片瓦;木板瓦
wooden shingle roof 木铺片瓦屋顶
wooden ship 运木船
wooden shipyard 木(壳船)造船厂
wooden shutter 木百叶窗
wooden shuttering 木模板
wooden siding shake 木墙雨淋板;木墙鱼鳞板
wooden skirting shake 木踢脚板
wooden slab 木板
wooden slat 木板条
wooden slatted drum 木格子转鼓
wooden slatted roller blind 木板卷百叶窗
wooden slatted Venetian blind 木板软百叶帘
wooden sleeper 枕木;木枕
wooden snatch block 木开口滑车
wooden soffit 木拱底面;木梁底面
wooden sole 木底
wooden space load-bearing structure 框架承重木结构
wooden spacer 木隔离物
wooden spar 木质翼梁
wooden spike 道钉口木塞
wooden spire 木尖塔
wooden spirit level 木水准仪
wooden splash-dam 木堰
wooden splintage 木夹板固定术
wooden spoke 木制轮辐
wooden square 木制直角尺
wooden staff 木标尺
wooden stage 木吊盘
wooden stair(case) 木楼梯
wooden stair(case) builder 木楼梯施工人员
wooden stanchion 木(支)柱
wooden stand 木架
wooden stave 木制排气管;木板条
wooden stave flume 木板槽
wooden stave pipe 木板条管
wooden steeple 木尖塔
wooden step 木梯级;木踏步;木阶梯
wooden string(er) 木梯梁;木楼梯斜梁
wooden structural system 木结构体系
wooden structure 木结构
wooden structure factory 木材构件厂
wooden strut 木支撑;木压杆;木斜杆;木支柱
wooden stud 木墙筋;木立筋
wooden stud partition(wall) 木筋隔墙;木筋隔断
wooden sub-floor 木毛地板;底层木地板
wooden support 木支撑;木(支)架;木支护
wooden surfacing 木饰面
wooden surround 木镶框;木镶边
wooden suspended floor 木浮筑地板
wooden swing door 木弹簧门;木制双开式弹簧门
wooden system 木体系
wooden system construction 木体系构造
wooden tamper 木夯
wooden target drone 木质靶机
wooden three-dimensional load-bearing structure 三向承重木结构
wooden three-dimensional weight-carrying structure 三向承重木结构
wooden thresher 木制脱粒机
wooden tie 木枕;枕木
wooden tilted-slab roof 木斜板屋顶;木倾板屋顶;斜木板屋顶
wooden timbering 木支架
wooden tower 木塔;木井架;木舰标
wooden toy 木制玩具

wooden track 木轨道;木路面
wooden transom 木腰窗;(门窗的)木横挡
wooden tread 木踏步;木楼梯踏步板
wooden trestle 木栈道;木栈桥
wooden trestle bridge 木排架桥;木桥
wooden triangle 木三角板
wooden tripod 木(制)三角架
wooden truss 木屋架;木桁架
wooden truss bridge 木桁架桥
wooden truss button 木窗搭闩
wooden trussed girder 木桁构大梁
wooden truss frame 木桁构桁架
wooden turbine 木制水轮机
wooden turned baluster 车制木栏杆柱
wooden type of construction 木类构造
wooden underdrain(age) pipe 木制地下排水管
wooden under frame 木底框
wooden unitized unit 统一化木构件
wooden upper roof framing 屋面木基层
wooden valley gutter 木天沟;木斜沟槽
wooden vault 木拱顶
wooden Venetian blind 木制威尼斯软百叶窗
wooden ventilator 木制通风筒
wooden vessel 木船
wooden wainscoting 木护墙板;木镶板
wooden wall 木墙
wooden wall plug 墙中木栓;墙中木楔;可受钉木塞;墙上木质摇头
woodenware 木器;木制品
wooden water tower 木水塔
wooden wedge 木楔
wooden weir 木堰
wooden wicket 木质小门;木质放水门
wooden window 木窗
wooden window shutter 木窗百页
wood ether 二甲醚
wood excelsior absorbent board 木丝吸声板
wood excelsior absorbent sheet 木丝吸音板
wood excelsior acoustic(al) sheet 木刨花吸声板
wood excelsior building slab 木刨花建筑板
wood excelsior concrete 木丝混凝土
wood excelsior concrete slab 木丝混凝土板;木屑混凝土板
wood excelsior hollow filler 木丝空心垫衬
wood excelsior insulation 木丝绝缘
wood excelsior partition 木丝板隔墙
wood excelsior slab 木丝板
wood exterior 外装修木材
wood extractive 木材抽出物
wood-faced fire door 防火处理木门
wood failure 木破损;木破(率);木材破坏
wood failure ratio 木质破坏率;木材破坏率
wood-falling machine 植树机;伐木机
wood fender 木护舷;护木;防撞木
wood fender pile 木护舷桩;(码头)木栏桩;防冲桩
wood fiber 木(质)纤维;木材纤维
wood fiber aggregate 木纤维骨料;木纤维集料;木质骨料
wood fiber board 木(材)纤维板
wood fiber board absorbent ceiling 木纤维板吸声顶棚
wood fiber board acoustic(al) tiled ceiling 铺木纤维板吸声顶棚;木纤维吸音天花板
wood fiber board(sound absorptive) ceiling 木纤维吸音天花板
wood fiber concrete 木纤维混凝土
wood fibered plaster 木纤维(灰)泥
wood fiber inorganic composite 无机胶凝木纤维材料
wood fiber insulation 木纤维保温材料;木纤维隔热材料
wood fiber plaster 木纤维灰泥;木纤维灰浆(粉刷);木纤维粉刷
wood fiber plaster baseboard 木纤维抹灰底板;抹灰木纤维护壁板
wood fiber silicification 木纤维硅化
wood fiber slab 木丝板;木质纤维板
wood fiber sound-deadening board 木纤维隔声板
wood fibre board 木材纤维板;木纤维板
wood fibreboard ceiling 木纤维板顶棚
wood fibre concrete 木纤维混凝土;木丝混凝土
wood fibre(d) plaster 木纤维灰泥;木纤维灰浆;木纤维粉刷

wood fibre pane 木纤维板
wood fibre plaster baseboard 木纤维抹灰底板
wood fibre saturation point 木材纤维饱和点
wood figurine 木雕像
wood file 木锉(刀)
wood filler 木垫梁;木材填料缝;木材填孔剂;木材封闭底漆;木质填料;木材嵌填料;木材腻子;嵌木缝泥脂
wood filler block 木填块
wood filling 油灰;木填料;木材填孔用油灰;(木材的)木缝油灰
wood finish 木材涂装
wood finishing 木制品装饰;木材油饰工序;木制品装修
wood finishing lacquer 木饰面蜡克;木饰面清漆
wood-fired boiler 烧木锅炉
wood fire-retardant treatment 木材阻燃处理;木材滞燃处理
wood flake core panel 木质刨花芯板
wood flask 木砂箱
wood flask paper 木砂纸
wood flight 木刮板
wood float 木镘刀;木抹子;木镘板;木浮子
wood floated 木抹子粉平的
wood float finish 木抹修整;(混凝土路面的)木镘修整;(混凝土路面的)木镘抹光面
wood flock 木绒
wood floor 木(条)地板
wood flooring 铺设木地板;木地面
wood flooring contractor 木地板承包者
wood floor strip 木地板条
wood flour 锯末;细木屑;木屑;木粉
wood flume 木渡槽
wood flush door 木平面门
wood fly back 梭箱扬起背板
wood folding chair 木折椅
wood for fittings 门窗木料
wood for fixture 门窗木料
wood for furniture 家具用材
wood form 木模板
wood form drawing 木模图
wood formwork 木模工程
wood for piles 木桩
wood for shingles 木瓦用材
wood frame construction 木(结构)建筑;木框架结构;木架构造;木框架结构
wood-framed facade 木构造正面
wood-framed floor 木构架地板
wood-framed in-built units 木构造嵌入构件;木构造固定构件
wood-framed mirror 有木框的镜子
wood-framed plywood 木框胶合板
wood frame house 木构架房屋;木框架房屋
wood frame panel wall 木框镶板墙
wood framing 木框架
woodfree paper 不含木浆的纸张
woodfree white laid paper 白罗纹花纹书写纸
wood fret 木回纹花饰;木格子细工
wood fretter 蛀木动物;蛀木虫
wood fungus 腐木菌;木菌
wood furring 钉木板条
wood gas 木瓦斯;木炭气;木煤气
wood girder 木大梁
wood glue 木胶;细木工胶
wood grab 圆木夹具
wood grabber 抓木工具
wood grain 木纹;木材纹理
wood grain design 木理纹
wood grained 涂成木纹的;成木纹的
wood-grained board 仿木纹板
wood grain finish 木纹加工
wood grain paint 木纹漆
wood grain print 木纹印花
wood grain printing 木纹印刷
wood grain synthetic(al) decorative board 木纹合成装饰板
wood grapple 圆木抓具
wood grating 木制格子板;(浴室地面用)木栅板;木踏板
wood gravel pavement 木块砂砾路面;木块砾石路面
wood grid packing 木格填料
wood grinder 磨木机;碎木机
wood ground 衬条;受钉的墙角小木条

wood guide shoe 木制引鞋
wood gum 木胶;树脂;树胶
wood gutter 木檐;木天沟;木檐沟
wood head frame 木质井架
wood house 木房(屋)
woodhouseite 磷钙铝矾
wood hurdle scrubber 木棚洗涤器
wood hydrolysate effluent 木材水解产生的污水
wood hydrolysis 木材水解
wood hydrolyzate 木材水解物
wood in building sizes 建筑用木材尺寸
wood industry 木材工业
woodiness 木质;多树林
wood interior 内装修木材
wood jetty 木码头
wood joiner 木工接合器
wood joining method 木接合方法
wood joint 木榫头;木接合;拼缝
wood joint filler 木质接缝填料
wood joist 木格栅;木龙骨
wood joist floor 木龙骨楼板
wood lacquer 木器涂料;木器漆
wood laminated arch 层叠胶合木拱
woodland 林区;林地;小块林;温带疏林;树林;森林区;森林地;矮疏林
woodland board 森林版
woodland drawing 森林原图
woodland ecosystems 林地生态系统
woodland hawthorn 山楂树林
woodland information 林区资料
woodland pasture 林间牧地
woodland plate 森林版
woodland reserves 森林保留地
woodlands irrigation with wastewater 废水森林灌溉
woodland soil 森林土
wood lath 抹灰板条;木灰条;木板条
wood lathe 木旋床
wood lath facing 木条墙面
wood lathing 钉抹灰板条;钉灰板条
wood letters 木刻活字
wood line 森林分布线
wood-lined passage 木贴面通道
wood lined pipe 木衬管
wood-lined shaft 木材支护的井筒
wood lining 木墙面
wood lintel 木过梁
wood loader 木材装载机
woodlock 摩擦力止动的窗撑
woodlot 小块林地;小片林;植林地
wood louver window 木百叶窗
wood machine planing knife 木工机器刨铁;木工机器刨刀
wood making 制材
wood mallet 木槌;小木槌
woodman 伐木工人;护林工人
woodman's axe 木工斧
wood meal 木屑;木粉填料;木粉
wood meal filler 木粉填料
wood measurement 木材量积
wood meat 细木屑
wood member 木构件
wood metal window 钢木窗
wood milling-slotting machine 木工刨榫机
wood mill works 木材加工厂
wood-modelling shop 木模车间
wood moisture content 木材含水率;木材含水量
wood mosaic 木质镶嵌工艺品;镶嵌细木工;镶木块;镶木马赛克;木砖;木墙块;木镶板;拼(花)木地板块
wood mo(u)lding 木线脚;木模
wood mo(u)lding machine 制木线(脚)机;木饰线条机
wood nailing strip 受钉木条
wood naphtha 木石脑油
wood naval stores 木材树脂产品
wood nog 嵌入墙体的木砖;木钉;木榫;木栓
wood of commerce 商品木材
wood of coniferous tree 针叶树木材
wood of deciduous tree 落叶树木材
wood of ripe age 可用材;成熟材
wood oil 木油;桐油;中国桐油;油树脂
wood oil reactive resin 桐油活性树脂
wood oil stand oil 熟桐油

wood oil tree 桐油树;木油树
wood oil varnish 桐油清漆
wood opal 木化石;木蛋白石
wood overcast 木制风桥
wood packing 木填块;垫木
wood pad 垫木
wood pagoda 木塔
wood pallet 木制托盘
wood panel 木镶板
wood paneling of interior wall 木护墙
wood panelled partition 木镶板隔墙;木格板隔墙
wood panel(l)ing 镶木板;木护墙板;木镶板
wood paper 木制纸
wood particle material 木屑材料;碎木材料
wood partition 木隔墙;木隔断
wood paste filler 木材填缝腻子;木材填缝膏
wood patio door 庭院木门
wood pattern 木模
wood pattern colo(u)r 木模涂色
wood pattern glue 木模胶
wood pattern maker 木模工
wood pattern shop 木模车间
wood pattern store 木模仓库
wood pavement 木块路面;木块铺地;木块铺砌层
wood paving 铺砌木块;木块路面;木块铺砌;木块铺面;木材铺面
wood pavior 木块路面铺设工人;木铺砌工人
wood peat 木泥煤
wood pecker hole 绳卡中穿绳的孔
wood-peeling machine 薄板锯床
wood-pegged boot 木钉结合靴
wood pegging 木钉结合
wood pendulum rod 木摆杆
wood pile 木桩
wood pile cluster 木桩群;木群桩
wood pile dolphin 木桩靠船墩;木簇桩
wood pipe 木(制)管
wood piping 木(制)管
wood pitch 木沥青
wood plane 木刨
wood planer 木(工)刨床;刨木机
wood plank 木板
wood planking 模板;木(铺)板
wood planning machine 木刨床;木工(平面)刨床;刨木机
wood plastic article 木屑电木制品;木屑塑料制品
wood plastic composite 木塑料复合材料
wood plastic paequetry 木屑塑料拼花地板
wood plastic product 木塑制品
wood plastics composite 木材复合塑料板
wood plastic window 木屑塑料窗;木塑窗
wood-plate lintel 平板枋
wood plug 木栓;木钉;木榫
wood post 木柱
wood preparation 备木
wood preservation 木料防腐;木材防腐处理;木材保存
wood preservative(agent) 木材防腐剂;木料防腐剂
wood preservative coating 木材防腐涂料
wood preservative creosote 木材防腐油
wood preservative salt 木材防腐盐
wood preserving 木材防腐;木材保存
wood preserving industry 木材防腐工业
wood preserving oil 木材防腐油
wood preserving plant 木材防腐蒸炼厂;木材防腐厂
wood preserving process 木材防腐处理
wood preserving waste 木材防腐废液
wood processing 木材加工
wood processing industry 木材加工业
wood product 木器品
wood product industry 木材产品工业
wood product processing 木材产品加工
wood protection 木护坦
wood pulp 绝缘用木材浆粕;纸浆;木(纸)浆
wood pulp board 木浆(纸)板
wood pulp cellulose 木浆纤维
wood pulp chemistry 木浆化学
wood pulp hydrolysis plant 木浆水解厂
wood pulp material 木浆浆料
wood pulp paper 木浆纸
wood pulp wadding paper 木浆卫生纸
wood pulp waste 木浆废料;木纸浆厂废水
wood putty 木材油灰
wood pyrolysis 木材热解;木材干馏

wood ranger 森林巡视员;森林管理员
wood rasp 木锉(刀)
wood ray 木心髓线;木射线
wood ray parenchyma 木射线薄壁组织
wood receptacle 木插座
wood refuse 碎木;废材
wood resin 木材树脂;天然树脂
wood roll (薄金属屋面咬口)加固木条;木制漩涡形柱头装饰;木滚筒
wood roof 木屋顶
wood roof truss 木屋架
wood rosin 木(蒸)松香;松香
wood rot 木腐病
wood rotting fungi 木腐菌;木腐朽菌
wood rotting-fungi preventative 木材菌霉防腐剂
wood rubbing strip 木擦条
woodruff drill 枪孔钻;半圆钻
woodruffite 纤锌锰矿
woodruff key 半月锁;半圆键
woodruff key seat cutter 半圆键槽铣刀
woodruff keyway milling cutter 半圆键槽铣刀
wood rule 木尺;木制隔空条
wood-run 林场
wood rush 地杨梅
woods 树林;森林
Wood's alloy 伍氏(低熔)合金;伍德合金
wood sanding machine 木料砂处理机
wood sash 木窗扇
wood sash jamb block 装嵌木门窗樘的混凝土砌块
wood sash putty 木窗框油灰
wood saw 木锯;横切锯
wood sawer 锯木工人
wood's butt end section 木材截面
wood scraps separator 木屑分离器
wood screw 木螺旋;木螺丝;木螺钉
wood screw drill 木麻花钻
wood screw lathe 木螺丝车床;木螺钉车床
wood screw pump 木螺旋泵
wood screw thread 木螺钉螺纹
wood sculpture 木雕
wood sealer 木材填孔剂;木材底涂料
Wood's effect 伍德效应
wood separator 木隔板
Wood's glass 伍德玻璃
wood shake 木材环裂
wood shaper 木牛头刨
wood shaping machine 木牛头刨;木工刨床
wood shaving 木刨花;刨花
wood shaving cement plate 水泥刨花板
wood shaving filter 木刨花过滤器
wood shaving tar 木刨花焦油
wood sheathing 木屋面板;木护墙板;木覆板
wood shed 柴薪棚;木棚
wood shingle 木瓦板
wood shirting 木踢脚
wood shirting rough ground 木踢脚板龙骨
Woods Hole sediment analyzer 伍兹·霍尔沉降分析(仪)器
wood shop 木工车间
wood shrinkage 木材干缩(性);木材收缩
wood siding 木板壁;木拔叠板;木墙板
wood siding wall 木板墙
wood sill 窗下槛;木门槛
wood skip 木制翻斗车
wood skirting board 木踢脚板
wood slat 木板条
Wood's light 伍德光线
wood slip 木滑道;嵌缝木条
Wood's metal 伍德(低熔)合金;铋基低熔点合金
wood soaking vat wastewater 木材浸渍瓮废水
wood soil 林地土壤
wood spatula 木刮刀
wood species 木材品种
wood spirit 甲醇;木精;木醇
wood splitter 木材分割机
wood splitting machine 劈木机
wood splittings tar 木刨花焦油
wood splitting tar 碎木焦油
wood spring-type fender 木弹簧式护舷;木弹簧式护木
Wood's process 伍德法(垂直拉玻璃管法)
wood stain 木染料;木材着色料;木材着色剂
wood staining 木材染色
wood staining fungi 木材变色菌

wood stair(case) 木楼梯
wood stand 木台
wood starch 木材淀粉
wood-stave flume 板条渡槽;木板过水渡槽;木板(渡)槽
wood-stave pipe 木板条管;狭木条拼成的水管
wood steaming 木材蒸干
wood stick break 木制绝缘子
wood stilt 木垫板
woodstone 石化木;硅化木;木化石
wood stone interface 木石界面
wood stopper 木塞
wood storage 储[贮]木场
wood storage basin 水上储木场
wood strength 木材强度
wood strip 木板条
wood strip flooring 条木楼板;条木地板;木楼板;木条地板
wood strip parquet 拼花地板条
wood strips-solid core door 板条实心木门
wood structural member 木结构构件;木结构杆件
wood structural system 木结构体系
wood structural technique 木结构技术
wood structure 木结构;木材构造
wood strut 木撑
wood stud 木壁柱
wood stud anchor 木立筋锚件
wood stud construction 木立筋结构;木壁柱构造
wood studding 木间柱
wood stud partition(wall) 木立筋隔断;木立筋隔墙
wood studs construction 木立筋构造;木柱结构
wood-stud wall 木筋墙
wood substructure 木制下部结构
wood superficial treatment 木材表面处理
wood support 木材支架
wood tannin 木材丹宁
wood tar 木溚;柏油;木焦油;木柏油
wood tar creosote 木焦油杂酚油;木焦油防腐油
wood tar fraction 木焦油馏分
wood tar pitch 木焦油沥青;松焦油沥青
wood taste 木桶味
wood technology 木材工艺学
wood terebene oil 木材松节油
wood terrace door 平台木门
wood tile 木瓦
wood tin 纤锡矿;木锡矿
wood tip(ping) wagon 木制翻斗车
wood tool grinder 木工磨床
wood transport by water 木材水运
wood treating plant 木材处理厂
wood treatment 木材处理
wood trestle(bridge) 木栈桥
wood trim 木压条;木门窗镶边;木包角
wood tripod dam 杩杈坝
wood trowel 木镘板
wood tube 木制管
wood tubing 木制管
wood turner 木车工
wood turning lathe 木旋床;木(工)车床
wood turning tool 木工车刀
wood turnover elevator 木格板回转式升运器
wood turpentine(oil) 木松节油;松节油
wood used for bridges 桥梁木材
wood used for foundation's buildings 基础工程用材
wood varnish 木材清漆
wood vault 木拱顶;木拱形圆顶
wood veneer 木饰面板;木皮;薄木片;镶板;饰面板;制胶合板用木单板
wood veneered 单板贴面材
wood veneer laminate 木质层压板
wood ventilation pipe 木质风筒
wood vessel 木导管
wood volume 木材体积;木材材积
wood waling 木围图;木横挡
Woodward governor 伍德华德液压调速器
woodwardite 水铜铝矾
Woodward speed governor 伍德华德调速器
Woodward visco(si) meter 伍氏黏度计;伍德华德黏度计
woodware 木制家具;木质器皿;木制品;木器
wood warm 木材虫
wood washer 木垫圈
wood waste 木材废料制品;木材废料;废木(材)

wood waste concrete 木屑混凝土;碎木混凝土
wood waste processing 废木料加工
wood waste tar 废木焦油
wood weather boarding 木鱼鳞板
wood wedge 木楔
wood wedge cutter 制木屑器;制木屑机
wood weir block 木闸板
wood wind instruments 木管乐器
wood window 木窗
wood winds 木管乐器
wood with crooked fibre 纹理扭曲的木材;弯曲纤维木材
wood with large annual rings 宽年轮木材
wood with radial pore band 辐射孔状木材
wood wool 木刨花;细刨花;木渣;木丝
woodwool board 木丝板;万利板
woodwool building slab 木丝板
woodwool cement 木丝水泥
woodwool cement plate 水泥木丝板
woodwool cement slab 刨花水泥板;木丝水泥板
woodwool concrete 木丝混凝土
woodwool covering rope 木丝木屑绳;木丝刨花绳
woodwool filler 木丝填料
woodwool filter 刨花滤器
woodwool hollow filler 木丝填空料
woodwool insulation 木丝绝缘
woodwool making machine 刨丝机;木丝制造机
woodwool nailing slab 受钉刨花板;受钉木丝板
woodwool permanent formwork 木丝永久性模板
woodwool plate soaked with bitumen 浸沥青木丝板
woodwool slab 木丝板;木棉板;万利板;木丝板皮料;刨花板
woodwool slab ceiling 木丝板平顶;木丝板吊顶
woodwool slab partition(wall) 木丝板隔墙
woodwork 木工工程;细木工;木作;木制品;木工活;木工工作;木工(指工作);细木作;(房屋内部的)木建部分
woodwork coating 木器涂装
woodwork construction 木结构;细木工程
wood worker 木工人;木工
wood worker's clamp 木工夹
woodwork industry 木制品工业;木工工业
wood working 木工的;制造木商品的;木器加工;木材加工
wood working anvil vice 木工台虎钳
wood working band saw 木工带锯床;木工带锯
wood working band sawing machine 木工带锯机
wood working blade 木工锯条
wood working boring machine 木工钻床
wood working cutting head 木工刀头
wood working drill 木工钻
wood working driller 木工打眼机
wood working drilling machine 木工钻孔机;木工钻床
wood working factory 木材加工厂
wood working grinder 木工磨床
wood working industry 木制品加工业;木制品工业;木材加工业
wood working instrument 木工工具
wood working lathe 木工车床
wood working machine 木工(加工)机械;木工机床;木工加工机
wood working machinery 木材加工机械;木工机械
wood working machine shop 木工车间
wood working machine tool 木工机床
wood working marking machine 木工画线机
wood working mechanist 木工机械师
wood working milling machine 木工铣床
wood working plane edge 木工刨刀
wood working planer 木材加工刨床
wood working planing machine 木工刨床
wood working rip saw 解木工锯
wood working round sawing machine 木工圆锯机
wood working saw frame 木工锯架
wood working sawing machine 木工锯机
wood working shaper 木工成形机
wood working shop 木工车间
wood working surface planer 木工平面刨床
wood working tool 木工工具
wood working vice 木工虎钳
wood work shop 木工车间
woodwork's vise 木工台钳
woodwork with metal facing 金属面木作

woodworm 蚀木虫;木蠹虫
woody 木质的;木制的
wood yard 储[贮]木场;木材堆(置)场;堆木场
woody brown coal 木质褐煤
woody fiber 木丝;木(质)纤维
woody fiber aggregate 木纤维集料;木纤维骨料
woody fiber-cement composite 木纤维增强水泥复合材料
woody fiber insulation 木纤维绝热材料;木纤维隔热材料
woody fiber sound-deadening board 木纤维隔声板
woody flowering plant 木本花卉
woody fracture 木纹状断口
woody headframe 木质井架
woody lignite 木质褐煤
woody ornamentais 木本观赏植物
woody peat 木质泥炭
woody perennial stems 多年生木质茎
woody plant 木本植物;树木
woody stem 木质梗
woody structure 木质状结构;木状结构
woody terminal cleft grafting 木质顶劈接
woody type 木质型
woofer 低频扬声器
woofer-and-tweeter 高低音两用喇叭
woof threads 帆布纬线
wool 毛纤维
Woolaway 伍拉维牌混凝土(一种轻质混凝土)
wool carpet 羊毛地毯
woolder 绞扎绳
wool-drag 颜色流痕
wool dust 羊毛尘
woolen blanket 毛毯
woolen fabrics 毛织品
woolen-knitting yarn 毛线
woolen mill 毛纺厂;毛制厂
woolen mill waste 羊毛厂废物
woolen mill wastewater 毛纺厂废水;羊毛厂废水
woolen oil 粗纺油
woolens 毛织品
woolen yard 粗纺毛线
woolen yarn 毛线
wool fabric 毛织物;毛织品
wool fat 羊毛脂;无水羊毛脂
wool felt 建筑用粗制毡;羊毛毡
wool felt board 羊毛毡板
wool-felt floor cover(ing) 毛毡地面
wool fiber 羊毛纤维
woolfil yarn 长丝芯毛纱
wool grease 羊毛脂;无水羊毛脂
wool grease pitch 羊毛脂沥青;脂肪酸沥青
wooliness 混响过度
wool labeling 毛织物质量标签
wooll-carding mill 梳毛厂
woollen jersey 毛针织物
woollen knitwear 毛针织品
wool-like handle 毛状手感
wool linen 毛衬布
woollybutt 澳洲桉树
woolly grain 模糊的木纹
woolly-type engine 低转速发动机
wool of fine staple 优质羊毛
wool oil 无水羊毛脂
woolpack 卷毛
woolpack cloud 卷毛云
wool pitch 羊毛脂沥青
wool processing wastewater 羊毛加工废水
wool rinsing machine 洗毛机
wool scouring effluent 洗毛污水
wool scouring industry 洗毛工业
wool scouring machine 洗毛机
wool scouring manufacturing wastewater 洗毛生产废水;洗毛工业废水
wool scouring waste liquor 洗毛废液
wool scouring wastewater 洗毛废水
wool shavings 羊毛屑
Woolton 伍尔顿砂岩(一种暗红色砂岩建筑材料,英国兰开郡产)
wool tricot lining 长筒线里
wool tuft technique 贴线法
wool washing industry 洗毛工业
wool washing machine 洗毛机
wool-washing plant 羊毛洗涤厂

wool waste 毛回丝;废羊毛
wool wax 无水羊毛脂
Woolwich beds 伍尔维奇层
wool yarn 毛纱
wool yield 净毛量;净毛率
word address format 字地址形式
word arithmetic 字运算
word by word 按词序排列
word capacity 字长【计】
word cell 字单元
word code 字代码
word combination 词组(合)
word control list 词控表
word count 字(记录)计数
word counter 字计数器
word delimiter 字定义符
word direction 字方向
word element 词素
Worden gravimeter 渥登重力仪;沃登重力仪
Worden gravity meter 沃登重力仪
word for word translation 逐字翻译
word gap 字间间隔
word group 词组
word indexing 字索引
wording 措辞
wording area 选词分区
word length 字码长度;字长【计】
word list 词汇表;词表
word management 作业管理
word mark 字标志;字标记
word marking 路面文字标记(交通标志)
word order 词序
word-organized memory 字选存储器
word problem of semi-Thue system 半图厄系统的判字问题
word processing 文字处理
word processing system 字处理系统
word queue entry 作业排队入口
word separator 字分隔符
word serial 字串行
word sign 词符
word space 字间距;字间间隔
words per minute 每分(钟)字数
word stock 词库
word table 字表
word time 小周期
work 著作;功;工作;工程;机件
workability 易加工性;作业性;工作度;可使用性;可加工性;和易性;施工性能;操作性能
workability admixture 增塑剂;增和易性剂;改善和易性的添加剂
workability agent 增塑剂;工作性改善剂;塑化剂
workability aid 增塑剂
workability annealing 改善加工性的退火
workability for cutting 切削加工性
workability index 可加工性指数
workability loss 和易性损失;可加工性损失
workability of coal 煤可采性
workability of concrete 混凝土(的)和易度;混凝土(的)和易性
workability of soil 土壤适耕性
workability period 和易性时间;可加工性时间
workability range 工作水分范围
workability test 工作度试验;加工性试验;混凝土和易性试验
workability two-point test 和易性二点试验法
workable 可开采的;可加工的
workable boundary line of ore body 矿体可采边界线
workable by open-cast mining 可露天采矿的
workable by underground mining 可地下采矿的
workable coal seam 可采煤层
workable coal seam sample 可采煤样
workable competition 行得通的竞争
workable concrete 有和易性混凝土;可加工性好的混凝土混合料;和易性(好)的混凝土;塑性混凝土
workable group of gypsum seams 可采膏组
workable hatch 可使用舱口
workable machine-hours 装卸机械完好台时
workable metal 可加工的金属
workable metal pigment 可加工的金属颜料
workable mixture 流动混凝土混合物
workable moisture 适合湿度

workable nail 可加工的钉
workable ore 可加工的矿石
workable oxide 可加工的氧化物
workable oxide pigment 可加工的氧化物颜料
workable pigment 可加工的颜料
workable pipe 可加工的管子
workable plan 切实可行的计划
workable poor primer 可施工的劣质底漆
workable powder 可加工的粉末
workable rainwater gutter 可使用的雨水天沟
workable raw lead 含银的铅;可加工的生铅
workable reserves 可采矿量;可采储量;开采储量
workable resinate 可加工的树脂酸盐
workable sludge 可铲污泥
workable soil 适耕土壤
workable system 可工作的系统
workable thickness 可采厚度
workable time 可工作时间
work abroad 海外工程
work accident 现场事故
work against time 抢时间作业
work ahead 延深工作
work allocation 工作分配
work amount 工程量
work analysis 工作分析;工程分析
work analysis program(me) 工作分析程序
work analysis report 工程分析报告
work anchor 工作锚
work and power 功与能
work and recreational therapy 工娱治疗
work angle 操作角度
work arbor 工作心轴
work area 工作存储区;作业地带;工(作)区;工作面积;施工区
work arm 工作臂
work arrangement diagram of sounding 电测深工作布置图
work arrangement figure in induced polarization method 激电法工作布置图
work as executed drawing 按已做的进行制图
work assignment plan 工作分配计划
work attachment 工作设备附件
work bag 工具袋
work being held up for lack of materials 停工待料
work bench 工作台;工作架;成型台
work bin 零件箱;零件盒;料箱
workblank 毛坯
work-board 承坯板
work boat 作业船;施工吊篮;工作吊篮;工作船
workbook 工作手册
work box 用具箱;工具箱;机修工具箱;工具袋
work breakdown structure 把最终目标按统筹方法分解为分支的作业分解结构;工作分解结构;工程项目分解结构;任务分解结构
work breaking structure 工作分解结构;任务分解结构
work breaking structure chart 工作分解结构图
work brittleness 加工脆性
work capacity 作业能力;加工容量;切削容量
work card 工作卡;派工单
work center 工作中心
work center line 工件中心线
work chamber 工作室
work classification 工程细目
work clearance 加工余隙
work coil 工作线圈;加热电感器;热感应线圈
work compensation 劳工赔偿
work content 功函
work contract 承包合同
work contracted to specialized group 包工到作业组
work contracting 工程承包
work cost 生产成本
work cycle 工作循环;工作周期
work cylinder 工作同位标磁道组
work data file 工作数据文件
work data set 工作数据组;工作数据集
workday 工作日;劳动日
workday per draught animal 畜工
work-day quota 工期定额
work deck 施工平台;工作平台
work decrement 工作下降
work defect 工程缺陷
work derricks 钻塔安装工作;井架安装工作

work design 业务设计;工作设计
work diagram 工作图
work-distribution chart 任务分配表;工作关系分布图
work documents 工作资料
work done 已完成工作量;工作量
work done by hand 人工
work done by impact 冲击功
work done by steam 蒸汽操作的工作
work done during compression 压缩功
work done factor 做功系数;作功因素;减功系数;耗功系数
work done in crushing 粉碎功
work done in grinding 粉碎功
work drawing 工作图;施工图
work drive motor 工作电机
work driver 自动偏
work driving arm 拨杆
work duty separation control room 职务分离控制室
worked consistency 混合稠度
worked copper 加工铜
worked crystal 加工晶体
worked efficiency 工作效率
worked edge 木工件的基准边;加工面;加工边
worked grease 用过的润滑脂;工作过的润滑脂
worked grease penetration 用过润滑脂的针入度
worked lumber 已切割木材;已加工木料;加工过的木材
worked material 自制材料;加工材料
worked penetration 用过的针入度
worked rubber 反复处理过的橡胶
work efficiency 工效;劳动效率
work element construction program(me) 施工作业计划
work end 木工件的基准端
work-energy principle 功能原理
work-energy relation 功能关系
work-engineer 工人工程师
work equipment 施工设备;工具
worker 工作者;工人;劳动者;定点者
worker accounting clerk in shift-group 班组核算员
worker doctor 工人医生
worker entrance 工人入口
worker estate 工人宿舍区
worker occupancy rate 工人工作面积率
worker officially employed 正式工人
worker participation in management 工人参与管理
worker's accommodation 工人宿舍
workers and staff members 职工
worker's barrack 临时性(木板)工房;单身职工宿舍
worker's club 工人俱乐部
worker's compensation 工人病假补助
workers' compensation insurance 劳动保险
worker's cultural palace 工人文化宫
worker's dwelling unit 工人居住处
worker's gymnasium 工人体育馆
worker's hospital 职工医院
worker's housing 工人住房
worker's insurance 工人保险
workers insurance expenses 劳动保险费
workers labo(u)r productivity 工人劳动生产率
worker's living quarters 工人宿舍区
worker's main camp 工人生活区
worker's quarters 工人寓所;工人宿舍
worker's rate of attendance 工人出勤率
worker's sanatorium 工人疗养院
worker's shed 工棚
worker's stadium 工人体育场
workers-staff contracting system 职工集体承包制
workers with difference technical classifications 不同技术种类的工人
work-exchange team 变工队
work extension 工厂扩建
work face 木工件的基准面;木作表面;工作面;规准面
work factor 功系数;工作因数
work factor system 工作因子法
workfare 劳动福利
work farm 劳教农场
work feasibility 工程可行性
work feed 工作福利;工作进给;工作进程
work feeder 进给装置;进刀装置
work first 头道操作

work fixture 工件夹具
work flow 业务流程;作业流程;加工流程;工作流程
workflow analysis 工作流分析
workflow diagram 工作流程图
workflow language 工作流语言
work for berthing 靠船建筑物
work force 工人总数;劳动力;施工人员;施工力量;施工队伍
work for deflecting avalanche 疏导坍塌工程
work for hire 做雇工
workfront labo(u)r 港口工人
work function 功函数
workhand 雇工
work-handling 工件装卸
work-hardened bronze 加工硬化青铜
work-hardened concrete 硬化混凝土
work-hardened steel 冷加工钢;加工硬化钢
work-hardened threads 工作时冷作硬化的螺纹
work hardening 加工硬化;冷作硬化;冷作加工;机械硬化
work-hardening capacity 加工硬化合金
work-hardening index 加工硬化指数
work-hardness 加工硬度
work hard together 齐心协力
work head 工作台;绘图头;工作水头
work head bridge 工作台挡板
work head slide 工作台滑板
workhead transformer 工件变流器
work high above the ground 高空作业
work history 工作历史
workholder 工件夹持装置;工件夹具;工件夹具
workholding 工件夹紧
workholding device 工件夹具
workholding fixture 工件夹具
workholding jaw 工件夹爪
workholding table 工件夹紧工作台
work horn 工件套承
work horse 设备;工具;广为应用的工具;重要机具
work-hour rate 人工小时率
work hours 上班时间
work-house 习艺所;工场
work in cooperation with a due division of labo(u)r 分工合作
work in coordination 互相配合
work index 功指数;工作指标;工作定额;做功指数
work in dusty environment 粉尘作业
work in echelon 阶梯操作
work in four shift 分四班工作
working 作业;运转;工作的;加工;维护;操作
working ability 工作能力;作业能力;加工能力
working above the water 水上开采
working access 工作通道
working accident 工伤事故;操作事故
working account 活期借贷账户;往来账户
working accuracy 精加工精确度;加工精度
working adit 掘进平洞
working against regulations 违章作业
working age 工龄
working agreement 合作协议
working air pressure 工作风压;工作大气压
working aisle 工作通道
working allowable stress 工作许可应力
working aloft 高空作业
working amount accomplishment rate 工作量完成率
working anchor 受力锚
working angle 工作角度;切削过程角
working apartment 工作室
working aperture 工作孔径
working approach angle 余偏角
working area 作业地带;工作面积;施工区;有效面积;作业区;作业面积;中间结果存储区;暂存储区;工作(地)区;加工面积
working arrangement 工作安排
working assets 流动资产;营运资产
working at full capacity 满负荷工作
working atmosphere 工作环境
working attachment 工作装置
working back clearance angle 纵向工作后角
working back plane 车刀的工作纵向剖面
working back wedge angle 纵向工作楔角
working band 工作频率
working bank 工作台阶
working barge 工作舱;工作篮;工作趸船;工作驳

working barrel 工作筒;泵桶;泵缸
working barrel cups 工作筒皮碗
working barrel seat 工作筒座;工作筒衬套
working barrel valve 深井泵泵阀
working barrel valve seat 深井泵泵阀座
working base 工作基地
working basic stand 工作基座
working battery 工作电池
working beam 平衡杆;钻臂;操作杆;摇臂;制动杠(杆);工作杆
working bench 工作台;作业台;工作站;工作平盘;工作架
working bill 作业票
working boat 工作艇;工作船
working body 办事机构
working branch 作用分路
working budget 运用预算
working buoyancy 载货浮力
working cam curve 工作凸轮曲线
working canvas 常用帆
working capacity 工作能力;工作量;劳动能力
working capital 运用资本;营运资金;营运资本;周转资金;流动资金;流动资本;经营流动资金;生产资金;生产资本
working capital and reserve fund 周转准备金
working capital cycle 流动资本循环
working capital fund 周转资金;周转基金
working capital loan 流动资金放款
working capital ratio 流动资金比率
working capital reserves 流动资金准备
working capital suspense 未定流动资本
working capital turnover 流动资本周转(率)
working capital turnover time 流动资本周转时间
working carriage 移动式运料车;移动式吊车
working cash 周转现金
working cell 工作元件;工作单元
working chain 工作链
working chamber 工作室;工作间;熔炼室;成型部;工作池;成形室;沉箱工作室
working chamber of caisson 沉箱工作室
working chamber of pneumatic caisson 气压沉箱作业室
working channel 加工导槽;施工用航道
working characteristic 操作特性
working chart 常用海图
working circle 施业区
working class 工人阶级
working-class dwelling 工人住居
working clearance 工作间隙
working clothes 工作服
working committee 执行委员会;工作委员会
working concrete in frosty weather 冰冻天气使用的水泥
working condition 工作状况;工作条件;工作情况;工况;劳动条件;使用状况;使用条件;施工条件
working condition of dredge pump 泥泵工况
working connection 主油路
working consistency 工作稠度
working content 工作范围
working control 实际控制
working cost 经营费;工作成本;工作费(用);加工使用费用
working craft 工作船
working crew 作业队;工作队;工作(班)组
working critical temperature 实用临界温度
working current 工作电流
working curve 工作曲线
working cutting edge angle 工作主偏角
working cutting edge inclination 工作刃倾角
working cycle 工作循环;工作周期;作业周期
working cylinder 工作汽缸
working date file 工作数据文件
working datum 施工基准面
working day 工作日;劳动日
working day norm 工期定额
working day of a machine 机器的运转日;机器的工作日
working days 工作日数
working deadline 工作期限
working deck 工作(平)台;工作甲板;轻甲板
working dedendum 节圆齿根高;节圆齿顶高
working depth 扎根深度;工作面深度;加工深度
working depth of tooth 工作齿高

working design 施工图设计
working details 施工详图
working device 工作装置
working diagram 工作图;作业图;加工图;施工(详)图
working diameter 工作直径
working diamond 菱形工作稳定区
working dimension 施工尺寸;工作尺寸
working direction 前进方向;使用须知
working directory 工作目录
working distance 工作距离
working distance of microscope 显微镜的工作距离
working district 作业区;采区
working district boundary 采区边界
working division 施业区划
working document working file 工作文件
working draft 工作草案
working drawing 作业图;制作图;加工图;图纸;施工(详)图;工作图
working drawing estimate 施工图预算
working drawings based on estimate 施工图预算
working durability 切削寿命
working duration of locomotive crew 乘员连续工作时间
working earthing 工作接地
working edge 工作边;加工边
working efficiency 工作效率;加工效率
working electrode 工作电极
working element 工作部件;工作构件;施工构件
working end 杠杆的作用端;杠杆的工作端;工作池;成型部
working engagement 端铣时工件被切部分宽度
working environment 工作环境;作业环境
working equations 运算式
working equipment 工作设备;施工设备;施工机具
working exchange capacity 工作交换容量
working expenditures 工作费(用);经营费
working expenses 工作费(用);经营(消)费
working extent 工作范围
working face 工作面;作用面;作业面;掌子面;规准面;生产工作面;采煤工作面;采掘面
working face efficiency 工作面效率
working face locomotive 采掘面机车
working factor 安全系数
working fastening 暂钉;临时固定
working field 工作现场
working fit 转动配合;间隙配合;松动配合
working floor level 井架台板平面
working flow 工作水流
working fluid 工作液体;工作流体
working-fluid characteristic 工作流体特性
working-fluid level 工作液面
working flux 有效磁通
working force 劳动力
working form 工作制式
working frequency 工作频率
working from the flat 就工选料工作
working function 工作函数;生产功能
working fund 周转(资)金
working fund for capital construction 基本建设周转金
working gallery 施工通路;施工通道
working gang 作业队;工班
working gangway 常用通道
working gas 工作气体
working gate 工作闸门
working ga(u)ge 工作量规;工作测规
working graphics 施工图
working group 工作组
working group documents 工作组文件
working group of experts on environment(al) law 环境法专家工作组
working group on climatic data management 气候资料管理问题工作组
working group on effects 效应问题工作组
working group on impacts 影响问题工作组
working group on policy 政策问题工作组
working group on science 科学问题工作组
working group on sea ice 海冰问题工作组
working group on volatile organic compounds 挥发性有机化合物问题工作组
working head 工作水头;工作压头;工作台;作用水头
working height 有效高度;工作高度;加工高度

working hoist 绞车
working hold 工作舱
working hole 工作口;挑料口
working hour meter 工时计
working hour method 工时法
working hours 运转时间;工作时间;作业时间;工作小时;工作时数;劳动时间;运行时间;工时
working hours depreciation method 工作时间折旧法
working hours quota 定额工时
working hypothesis 可使用假设
working imperfection 操作缺陷
working index 工作指数;工作指标
working in on line 双倒锤定线法
working input 指示功
working in shifts 工作换班
working instruction 工作命令;操作说明书;操纵规程;操作规程
working intensity 工作强度
working interest 石油业的开采利益
working interval 工作行程
working item 工作项目
working jig 工作架
working joint 施工缝;工作缝
working knowledge 作业知识
work(ing) lamp 作业灯;工作灯
working language 工作语言
working laser material 激光激活材料;激光工作物质
working length 有效长(度);资用长度;工作长度;使用长度
working level 施工层面;工作层面;工作水平
working lever 操纵杆
working licence 工作证
working life 工作寿命;工作期限;工作年限;使用寿命;使用年限;操作期
working life of furnace 炉龄
working light 工作灯;作业(照明)灯
working lighting 工作照明
working limit 工作极限
working line 运行线路;作业线;工作线(路)
working lining 工作炉衬;工作衬(里);工作层
working list 工程明细表
working load 资用荷载;工作荷载;工作负载;工作负荷;许用荷载;活荷载;使用荷载;使用负载;施工荷载
working load design 工作荷载设计;使用荷载设计(法)
working load limit 安全荷载限度;工作负荷极限值
working location 工作位置
working loss 工作损耗
working machine 工作机
working major cutting edge 工作主切削刃
working management 工作管理
working map 作业图;工作(蓝)图;工作地图;加工图;施工图;底图
working mask 工作掩蔽
working material 工作材料
working mean 假定平均数
working measuring instrument 工作测量器具
working medium 工作介质;工质;使用介质
working medium pressure 工作介质压力
working memory 暂存器;工作存储器
working method 工作方法;操作(方)法;加工方法
working minimum blast 最低有效鼓风
working minor cutting edge 工作副切削刃
working minor cutting edge angle 工作副偏角
working mixture 工作混合气
working mode 工作模式
working model 工作模型;可使用模型;实用模型
working mode of locomotive crew 乘务方式
working modulus 加工率
working moisture 工作水分
working motion 加工运动
working mo(u)ld 工作模具;工作模型
working movement 加工运动
working normal clearance 工作法向后角
working normal rake 工作法向前角
working normal wedge angle 工作法向楔角
working ocular 观察目镜
working of an invention 发明的实施
working of clay 黏土加工
working-off process 销除法
working offshore platform 活动海洋(钻探)平台;活动海洋钻井平台

working of kiln 窑作业
working of mesh 筛孔有效面积
working of metal 金属加工
working on live ultra-high tension power lines 超高压线路带电作业
working operation 工序
working operations 操作规程
working order 运转状态;运转情况;工作状态;工作命令;工作订单;工作程序;操作顺序;正常运转(状态)
working orthogonal clearance 工作后角
working orthogonal plane 工作主剖面
working orthogonal wedge angle 工作楔角
working-out 规划;制订;计算
working outer size 工作时外形尺寸
working over 加工处理
working paper 工作文件;工作报告;结账计算表
working parameter 运行参数;工作参数
working partner 执行业务股东;积极合伙人
working parts 工作部件
working party 工作组;作业队;工班
working party on air pollution problem 空气污染问题工作队
working pass 工作孔型
working path 工作冲程
working pendulum 工作摆
working performance 工作性能
working period 工作周期
working period shortening rate 工作周期缩短率
working period target 工作周期指标
working personnel 作业人员
working pile 工作桩【疏】
working pit 工作坑
working place 工作区;工作空间;施工地点;工作场地;掌子面;工作地点
working place illumination 工地照明
working plan 施工图;工作计划;工作规划;作业计划;工程程序图;施工计划
working plan area 工作案面积
working plane 工作面;工作横向剖面
working planning 工作规划
working plan period 施业期
working plans circle 施业案管理区
working plans conservator 施业案总监
working plans officers 施业案编制员
working plate 工作板
working platform 工作(平)台;操作台;操作平台
working platform of scaffold 脚手架上的工作平台
working point 作用点;作业温度;加工点;受力点;成型温度
working point of amplifier 放大器工作点
working pontoon 工作趸船
working pool 可航水域
working population 工作人口;就业人口
working port 工作油口
working portion 工作部分
working portion of die 模具工作部分
working position 工作状态;工作位置
working post 作业标
working power 使用动力
working power supply 工作电源
working practice 工作实践
working press 工作加压
working pressure 作用压力;资用压力;工作压力;使用压力
working pressure head 工作压力水头
working principle 工作原理
working procedure 工作程序;运行程序
working procedure of environment(al) impact assessment 环境影响评价工作程序
working process 工作过程;操作过程
working program(me) 工作程序;工作计划;工作规划
working progress 工作进程
working property 加工性质
working property of paint 涂料的作业性
working quality 工作质量;加工性能;加工性;施工质量
working quality qualified rate 工作质量合格率
working quantity of investigation 勘察工作量
working quantity of locomotive affairs 机务工作量
working quota for one stratum 每层中的工作定额
working quota of investigation 勘察工作定额;调查工作定额

working quota of specific area 单位面积的工作定额
working Q(value) 负载时的Q值
working radius 作用半径;作业半径;工作幅度;工作半径
working radius of grab machine 抓斗机工作半径
working raft 工作筏
working rail 脚手栏杆;舞台天桥栏杆
working range 工作范围;仪表工作范围;作业范围;工作区域;热加工范围;操作范围
working range of temperature 作业温度范围
working rate 加工率
working reactance 工作电抗
working redesign 工作再设计
working reduction ratio 工作破碎比
working reference plane 工作基面
working region 工作区
working register 工作寄存器
working register group 工作寄存器组
working regulation 工作条例;工作规程
working relief 工作后角
working report 工作报告
working resistance 工作阻力
working roll 工作辊
working room 工作室;工作舱;回采工作空间
working rope 传动索
working routine 工作程序;工作顺序;操作程序
working rule 经营法则;操作规则;操作惯例
workings 矿内巷道
working safety 运转安全度
working sails 常用帆
working scaffolding 工作架;作业脚手架;操作用的脚手架
working scale 工作比例尺;操作规模
working schedule 工作进度(表)
working scheme 工作计划;工作规划
working season 工作季节;施工季节
working section 工作剖面
working section area 工作段横截面面积
working section recovery 采区回采率
working section window 工作段观察窗
working sequence 作业程序;操作程序;施工程序
working session 工作会议
working set 工作组;工作(页面)区;工作集
working set algorithm 工作组算法
working set swapper 工作集交换程序
working shaft 工作轴;工作井;施工竖井
working sheave 天车或滑车的工作轮
working shift 工作班(制);工组;作业班制;工班
working ship-shift 工作艘班
working side clearance 横向工作后角
working side rake 横向工作后角
working sign 作业标
working signal 作业标志
working site 现场;工作场地;工地;工作地点
working situation 工作情况
working slag 初渣
working slope 工作帮
working slope face 工作帮坡面
working sludge 可铲污泥
working space 工作区;工作空间;工作地区;工作场所
working specifications 操作规程;作业规程
working speed 工作转速;工作速率;工作速度
working speed of mill 磨机工作转速
working spud 工作(钢)桩【疏】
working stage 舞台工作间
working stand 工作台;工作梯
working standard 工作标准;工作基准;现行标准;通用标准;操作标准
working state 运转状态
working statement 工作报告(书)
working station 工作站
working steam 工作(新)蒸汽;新鲜蒸汽;新汽;活汽
working steam pressure 工作汽压
working stem 工作杆
working stock 运行储备器
working storage 暂存器;工作库容;工作存储区;工作存储器
working storage in reservoir 水库工作库容
working storage section 工作存储段;暂存区;暂存节;暂存段
working strain 工作应变;资用应变
working strength 工作强度;资用强度;许用强度

working stress 工作应力;许用应力;资用应力;安全应力
working stress design 工作应力设计;资用应力设计;容许应力设计
working stress method 工作应力法;许用应力法
working stroke 工作行程;工作冲程
working study 工作研究
working submergence 活动沉没
working substance 作用物质;工作(用)物;工作介质;工质;可用物质
working surface 工作(表)面;施工路面;工作活动面
working surface of rail 钢轨工作边
working surface of table 工作台台面
working surface of the pivot 轴颈工作面
working system 作业法;生产法
working system in management 管理工作制度
working system of locomotive crew 机车乘务制
working table 工作台
working tank 工作池
working tape 工作带
working team 工作组;工班
working temperature 加工温度;工作温度;使用温度;操作温度
working tension 作用张力;工作张力
working thickness 加工板厚
working time 运行时间;工作时间;作业时间
working time between trains 列车空隙作业时间
working time of closed section 列车空隙作业时间;封闭线路作业时间
working torque 工作扭矩
working tour 工作巡视;工作行程
working to windward 逆风驶帆
working trailer with jib crane 悬臂吊拖车
working train 工作列车
working transmission 工作传量
working trial balance 核数试算表
working trip 工作出行
working type of investigation 勘察工作类型
working under capacity 开工不足
working under the water 水下开采
working unit 工作装置
working up 修整;计算工程用料;咬底;量方;结算;计算工程量
working value 工作值
working valve 工作阀(门);工作调节阀
working vehicle-days 工作车日
working vehicle for optic(al)-fibre cable 光缆敷设设备
working vehicle hour 工作车时
working vehicle rate 工作车率
working vehicles 工作车辆数
working velocity 工作速度
working vest 工作救生衣
working viscosity test 涂刷黏度试验
working voltage 工作电压
working volume 工作体积;工作容积
working wastewater 作业废水
working water-jet capacity 喷水量
working water-level 施工水位
working water-stage 施工水位
working wave 工作波;符号波
working-week 工作周
working weight 资用重量;工作重量;使用重量
working width 埋管时占地宽度;工作幅度;耕幅;加工宽度
working winch 绞车
working with a basic slag 碱性炉渣冶炼
working with naked fire 明火作业
working yard 工作场地
working years 工龄
working zone 施工区;工作区(段)
work in hot environment 高温作业
work in mid-air 高空作业
work-in-process 阶段产品;在制品;在加工中
work-in-process inspection 生产过程的检验
work-in-process inventory 在制品盘存
work-in products 在产品
work in progress 进行中的工程;施工中的工程
work input 机器总输入;机器的输入功
work in shifts 倒班
work in succession 顺序施工
work intensity 加工强度
work interchange system 交叉作业制
work in three shifts 分三班工作

work in two shift 分两班工作
work item 工程项目;单项工程
work joint 施工缝
work justification 工程论证
work-kinetic energy theorem 功动能定理
work lead 工件引线;生铅;粗铅
work-leisure time ratio 工作空闲时间比
work letter 房地产主与租户为准备的房屋租赁所作的分工协议书
work level 动液面
work-life boat 工作救生艇
work light 工作照明;工作灯
work load 有用负荷;劳动负荷量;任务;作用荷载;资用荷载;活荷载;工作荷载;工作负载;工作负荷;作业量;工作量
workload analysis 工作量分析
workload calculation audit 工程量计算的审计
workload estimate sheet 工作量估计表
workload manager 工作负荷管理人;工作负荷管理程序
workload of biochemical oxygen demand 有效生化需氧量
workload of suspended solid 有效悬浮固体
work location 工作位置;工作地点;运算位置
workman 职工;工作者;工人;劳动者
workmanlike 精巧的;熟练的
workman's compensation 工人伤残赔偿
workmanship 施工质量;制造工艺;工作质量;工作法;手艺;作品;完工工艺;技能;工程质量
workman's protective face-shield 工人防护面罩
workmaster 监督者;工长;监工
work measurement 工作计量;工作测定;工效计量
workmen's compensation 对工人的补偿(金)
workmen's compensation insurance 工人补偿保险;劳工保险
workmen shelter 工棚
work metal 被加工金属
workmeter 工作表
work method 操作法
workmistress 女职工
work motor push button 工作电机按钮
work noise 工作噪声
work norm 劳动定额
work notice 作业通知单;工作通知书;工作通知(单);派工单
work of adhesion 附着功
work of art 艺术品;工艺品
work of basic building geology 基建地质工作
work of basic exploration 基建勘探
work of cultivating soil 土壤耕作
work of defence [defense] 防护设施;防护建筑(物)
work of deformation 形变功;变形功
work of elastic strain 弹性应变力
work of environmental protection 环保工程
work of erosion 风化作用;腐蚀作用;侵蚀作用
work off 除去
work off the debt 清偿债务;偿清债务
work off arrears 陆续补还欠款
work of fracture 断裂功
work of protection 防护设施;防护建筑(物)
work of resistance 阻力功
work of screening pseudo-collectives 清理假集体工作
work of separation 分离功
work on rotating-basis 轮流工作制
work order 工作订单;工作程序;工作指令;工作通知书;工作通知单;工单;工程任务书;开工令;派工单;正常运转(状态)
work order sheet 作业通知单
work order sheet of construction task 建筑施工任务单
work organization 作业组织
work organization plan 施工组织计划;施工部署计划
work orifice 工作口
work out 制定;拟订;编制
work out a development planning 编制发展规划
work out a production plan 编制生产计划
workout loan 延期贷款
work output 工时研究;劳动量
work output queue 工作输出排队;工作输出队列;输出排队
work out the cost of production 核算生产成本

workover 详细研究;修理工作;修整
workover and downhole operation 井下作业
workover derrick 修井井架
workover equipment 修理设备
workover platform 修井平台
workover rig 修井机
workover the beam 架上操作
workover well 延深井身
work overtime 加班加点
work passenger flow 工作客流
work period 工期
work permit 工作(许可)证
work physiology 劳动生理学
workpiece 分部工程;轧件;工作部件;工件
workpiece arrangement device 理件装置
workpiece center 工件中心
workpiece holder 工件夹具
workpiece process 加工过程
workpiece ratio 效率
workpiece revolution 工件旋转
workpiece temperature 工件温度
work place 作业地带;工区;工作场所;车间
workplace air 厂房空气
workplace layout 工作场地布局;工地地点布置
workplace layout chart 工作地点布置图
workplace population 工作地点人口
workplace study 现场研究
work plan 业务计划;工程计划图
work plane 工作面
work-point credited on norm 定额记工
work point register 工作点寄存器
work position 运算位置
work principle 工作原理
work procedure 工序
work process scheduler 工作处理调度程序
work productivity 工作量;劳动生产率
work programme specification 工艺规程说明书
work progress 工作进度;工作进程
work project 工程计划图
work propeller 活车
work protective gloves 劳保手套
work queue 工作队列
work queue entry 工作队列目录
work quota 劳动定额
work range 仪表的量程;工作范围
work-related activity 同工作有关的管理活动
work-related injury 公伤;因公负伤
work relief 以工代账
work report form 工程报表
work request 加工申请书
work required of size reduction 破碎需要功
work rest 工件支架;工件架
work rest blade 导向尺;刀形支承
work retainer 工件定位器
work roll 工作轧辊
work roll cage 工作辊的轴承套
work roll change 工作辊的换辊装置
work roller supporting apparatus 工作辊支撑装置
workroom 工作室
workroom atmosphere 车间气氛
workroom man-lock 工作室
work-rotating internal grinder 工件旋转内圆磨床
work routine 工作例行程序
work routinization 工作常规
work rule 工作细则;就业规则
works abstract 工程摘要表
work sampling 工作取样;工作的抽样检验;工作抽样;工作抽查法
works annealed sheet 金属薄板
works approval 工厂认可
work saving device 助力装置;省力装置
works beam test 现场梁试验;工地梁试验
work's berth 工厂泊位
works bottling 工厂装瓶
works bus 上班公共汽车
works canteen 工厂食堂;工地食堂;工地小卖部
work-schedule 工作日程;工程进度;工作进度表;施工进度表
work science 实践科学
works code 工厂代号
works contract 工厂合同
works council 员工福利咨询委员会;劳资协议会
works cubes 工厂混凝土试块
works cube strength 工厂混凝土试验强度;工地

混凝土的立方体强度
works cube test 现场试块试验
works depot 工程器械库
works drawing 施工图
work sea area 工作海区
workseat 工件座
works equipment 工厂设备
work session 工作约会;工作会话;工作对话(期)
works fire brigade 工地消防队
works for berthing 靠船构筑物
works for centrifugally cast concrete 离心法浇筑混凝土
works for dissipating wave 消波构筑物
works for dissipating wave energy 波浪消能设施
works for doors and windows 门窗工程
work's harbo(u)r 工厂港口
work shed 工棚
work sheet 记工单;作业票;工作卡;工作单;加工单;任务命令单;工作周
work sheet for distribution 分配计算表
work sheet of construction task 建筑施工任务单
worksheet to prorate service department cost 服务部门成本分配计算表
work shift 工作班
workship requirement 工艺要求
workshop 修理间;车间;工作间;研究会;专题研究会;专题讨论会;工场;工厂;学术方面讨论会;车间
workshop appliance 车间用具;车间设备
workshop assembly 工场装配;工厂装配;车间装配
workshop barge 修理驳;水上工厂
workshop building 厂房
workshop coat 车间涂装面层;车间涂刷面层
workshop conduit system 车间管网
workshop crane 车间吊车;机修起重机;车间起重机
workshop director 车间主任
workshop drawing 车间用加工图
workshop equipment 车间设备
workshop expenses 车间经费
workshop fitter 车间装配工
workshop for lead pipes 铅管工场
workshop for plumbing 管子工场
workshop for timetable editor 时刻表编辑工作间
workshop ga(u)ge 工作量规;工作测规
workshop manufacture 车间制造
workshop microscope 车间用显微镜
workshop outfit 车间设备
workshop pipe system 车间管网
workshop piping system 车间管网
workshop power source 车间电源
workshop power supply 车间供电
workshop practice 现场实习
workshop section 工务(分)段;工段
workshop storage area 修理间堆场
workshop trial assembly 车间试装配
workshop truss 锯齿形桁架;锯齿形厂房屋架;厂房桁架
workshop tube system 车间管网
workshop tubing system 车间管网
workshop wagon 车辆修理厂
workshop-welded 车间焊接的
workshop welding 车间焊接
work simplification 工作简化;工艺简化
works initial 厂名缩写
works inspection 工厂检查
work's inspection certificate 工厂检查证明书
work site 工区;施工现场;工地
worksite allowance 工地补助费
worksite management 现场管理
worksite mixer 现场搅拌机
worksite of bridge construction 桥梁工地
worksite premium 工地奖金
work slice 工作时间片
works load 自用负载
works management 工程管理;工厂管理;施工管理
work's manager 工厂经理
works occupation time 列车空隙作业时间;封闭线路作业时间
works of intercepting underground flow 截潜流工程
works of man 人工建筑物
works of portal 洞口工程
work-softening 加工软化
work-softening material 加工软化材料

work space 工作区域;工作空间
work space address 工作区地址
work space architecture 工作区结构
work space file register 工作区文件寄存器
work space location 工作区单元
work space pointer 工作区指示字;工作区指示器;工作空间指针;工作空间指示字
work space pointer address 工作区指示器地址
work space register 工作区寄存器;工作空间寄存器
work space register address 工作区寄存器地址
work space register addressing 工作区寄存器寻址
work space register content 工作区寄存器内容
work space structure 工作区结构
work specification 工作规范
work speed 工件速度;操作速度
work spindle 工作主轴
work's port 工厂港口
works process specification 典型工艺规程
works progress 工程进展
works progress committee 工务进度委员会
work's quay 工厂码头
works railway 工厂(专用)铁路;工厂铁道
works scope 工程范围
works siding 工厂专用线
works staff 低级职员
works sub-contracting 工程分包
works subletting 工程转包
works suspension 工程暂停;工程停工;工程停顿
work stack 工作栈
work standard 作业标准
work standardization 工作标准化
work statement 工作报表;加工语句
workstation 工作站
workstation category 工作站类型
workstation display 工作站显示
workstation for service operation 维护操作工作站
workstation for simulation and demonstration 模拟演示工作站
workstation hard-copy unit 工作站硬拷贝设备
workstation interface 工作站接口
workstation printer 工作站打印
workstation process 工作站过程
workstation state table 工作站状态表
workstation transformation 工作站变换
workstation viewpoint 工作站视区
workstation window 工作站窗口
work statistics 工程统计
work status 就业状况
work steady 工件架
works test 工程检修
works to continue 工程的继续
work stone 料石
work stoppage 停工
work stoppage loss 停工损失
work study 作业研究;工作效率研究;操作研究
work-study system 半工半读制
work supervision company 工程监理公司
work support 工件支架
work supporting device 工件支架
work surface 工作(表)面
works visit 工厂参观;工厂巡视
work's wharf 工厂码头
worktable 工作项目表;工作台
worktable rotation 工作台旋转
worktable traverse 工作台横动
work test 工作试验
work the pipe to free 上下提动被卡钻杆使之解卡
work the pipe up and down 上下活动钻具
work therapy 工作疗法
work the string up and down 上下活动遇阻钻具
work through 掘通(巷道)
work ticket 工作通知书;工作通知单;派工单;工票
work time 有效工时
work-to-home peak hour 下班高峰时间
work to maximum load 最大荷载工作量
work to pattern 按样板加工
work to proportional limit 比例极限功
work to rule (迫使资方让步的)变相怠工
work tour 工作行程
work track 工作磁道;施工便线
work train 工作列车;工程列车
worktrain management manual 工程车运输管理手册
worktrain manual 工程车手册

worktrain notice 工程车通告
worktrain scheduler 工程车计划员
work trestle 工件支架
work tugboat 工件拖轮
work underconstruction 施工中的工程
work unit 作业单位;工作单位
work-ut period 工作周期
work volume 工作(文)卷
work water depth 工作水深
workweek 工作周;一周的总工时
work-yard 施工场地;工作场地
world advanced level 世界先进水平
world aeronautical chart 世界航空图
world air route 世界航空线
world area forecast system 世界区域预报系统
World Association of Industrial and Technological Research Organization (WAITRO) 世界工业和技术研究组织协会
world atlas 世界地图集
World Atmosphere Fund 世界大气基金
World Bank 世界银行
World Bank Environmental Ministry 世界银行环境部
World Bank Group 世界银行集团
world calendar 世界历
world-class 世界第一流水平的
world climate 世界气候
world climate research programme 世界气候研究计划
world clock 世界钟
world commerce 世界商业;世界贸易
World Commision on Environment and Development 世界环境与发展委员会
World Computer Citizen 世界计算机公民
world concordant time 世界协调时间
World Confederation of Labo(u)r 世界劳工联合会
World Conservation Monitoring Center 世界养护监视中心;世界自然保护中心;世界保护监测中心
world conservation strategy 世界自然资源保护大纲
world coordinates 世界坐标;通用坐标;完全坐标;全局坐标
world coordinate system 世界坐标系;通用坐标系
world currency 世界通货;世界货币
world data center 世界数据中心
world data center on microorganism 世界微生物资料中心
World Data Centers for Meteorology 世界气象资料中心
world data referral system 世界数据查询系统
world data transmission service 世界数据传输业务
world day 世界日
world days of International Geophysical Year 国际地球物理日
world development budget 世界开发预算
world distribution 全球性分布;世界分布
World Dredging Association 世界疏浚协会
World Dredging Conference 世界疏浚会议
World Dredging Organization 世界疏浚组织
world dynamics 世界动力学
world economic survey 世界经济概览
world economy 世界经济
World Energy Conference 世界能源会议
world environment 世界环境
World Environment and Resources Council 世界环境和资源委员会
world environment day 世界环境日
World Environment Institute 世界环境研究所
world export prices indices 世界出口价格指数
World Federation for the Protection of Animals 世界动物保护联合会
world fertilizer economy 世界肥料经济
world financial market 世界金融市场
World Food Day 世界粮食日
World Food Program(me) 世界粮食计划署
world forest inventory 世界森林目录
world forest resources 世界森林资源
world geodetic coordinate system 世界大地测量坐标系(统)
world geodetic network 世界大地网
world geodetic system 全球大地坐标系统;世界测地系统
world geographic(al) reference system 全球地

理(坐标)参考系统;世界地理基准系
world giant oil and gas field 世界巨型油气田
world gravimetric system 世界重力测量系统
world grid 世界方格坐标
World Health Organization 世界卫生组织
world heritage 世界遗产
World Heritage Convention 世界人类遗产公约
world heritage site 世界遗产保护地
world inflation 世界性通货膨胀
World Intellectual Property Organization 世界知识产权组织
world island 世界岛(即欧亚非大陆)
world knowledge 世界知识
world level 世界水平
world-line 世界线
world map 世界地图
world map of soil degradation and hazards 世界土壤退化和危害图
world map series 世界图组
World Maritime University 世界海事大学
world market 世界市场
world market price 国际市场价格;世界(市场)价格
world merchant fleet 全世界商船总数;世界商船队
world merchant tonnage 世界商船总吨位
World Meteorological Center 世界气象中心
world meteorological day 世界气象日
World Meteorological Organization 世界气象组织
World Meteorological Organization Code 世界气象组织天气码
world meteorologic (al) organization ozone project 气象组织臭氧计划
world model 世界模型
world money 世界货币
world network 全球组网
world ocean 世界海洋;世界大洋
world ocean circulation experiment 世界海洋环流实验
World Ocean Organization 世界海洋组织
world oceans day 世界海洋日
world oil 世界石油
World Organization of Intellectual Property 世界知识产权组织
world outlook 世界观
world ozone data center 世界臭氧资料中心
world ozone network 世界臭氧监测网
World Packaging Organization 世界包装组织
world patents index 世界专利索引
world plan of action on the ozone layer 世界臭氧层行动计划
world point 世界点
world polyconic(al) grid 全球多圆锥投影坐标网;世界多圆锥投影坐标方格
world population year 世界人口年
world port 州际大港
World Port Index 世界港口指南
world price 世界市场价格
world production pattern 世界生产模式
world quota 世界配额
world radiation map 世界辐射地图
world radiation network 世界辐射监测网
world rainfall record 世界降雨记录
world resources 世界储量
world resources institute 世界资源研究所
world's atmosphere 世界大气
world scale 世界运价表;世界运费率
World's Fair 万国博览会;世界博览会
world's major basin name 世界主要盆地名称
world soil map 世界土壤地图
world speed record 世界车速记录
world's record 世界记录
world standard 国际水平;世界水平
World Standard Day 世界标准日
world's three-major cereals 世界三大谷类作物
world's total trade volume 世界出口贸易总额
world strategy 世界战略
world's water resources 世界水资源
world system model 世界系统模型
world time 世界时
world time clock 世界钟
world time zone 世界时区
world-touring liner 环great球客轮
world trade 世界贸易
World Trade Center 世界贸易中心
World Trade Centers Association 世界贸易中心联合会
World Trade Center Towers 世界贸易中心塔楼
world trade matrix 世界贸易矩阵
World Trade Organization 世界贸易组织
world use of fertilizer 世界肥料使用量
world view 世界观
Worldwatch Institute 世界环境研究所
world water day 世界水日
world weather 世界天气
world weather extremes 世界气候极值
world weather watch 世界天气监视网;世界天气监测网
world-wide communication network 世界交通图
world-wide disturbance 全球扰动
world-wide dose commitment 全球性剂量分担
world-wide dynamic(al) system 全球动力系
world-wide earthquake data 世界地震资料
world-wide fallout 全球微粒散落;全球(放射)性沉降
World Wide Fund for Nature 世界大自然基金
world-wide geodetic net 全球大地控制网
world-wide gravimetric basic point 世界重力基点
world-wide gravity base station network 全球重力基本点网
world-wide income 全球范围收入
world-wide natural disaster warning system 全世界自然灾害警报系统;全球自然灾害报警系统
world wide navigation warning service 全球航行警告业务
world wide production 世界产量
world wide sales 世界销售量
world-wide standard seismograph network 世界标准地震仪台网
world-wide structural pattern 全球构造图
world-wide synchronization 全球同步
world wide telecommunication network 世界性的电信网络
world-wide time 世界时
world-wide topographic(al) readiness concept 全球测绘保障方案
world wide treaty 全球性分保合同
world wide web[www] 全球信息网;万维网;环球信息网
World Wild Fund for Nature 世界自然保护基金会
World Wildlife Fund 世界野生动物基金会
worm 蛀虫;螺杆;蠕虫形;爬行;蜗杆;填平绳纹;虫状金属珠;虫
worm adjustment 蜗杆调节;用蜗杆调节
worm-and-double-roller steering gear 蜗杆双滚轮式转向机构
worm and gear 蜗轮蜗杆
worm and gear reducer 蜗杆涡轮传动减速器
worm-and-nut type steering 蜗杆螺母式转向
worm-and-roller 蜗杆与滚轮
worm and roller-lever steering gear 蜗杆滚轮式转向器
worm-and-roller steering gear 蜗杆滚轮式转向机构
worm-and-sector 蜗杆与扇形轮;扇形蜗轮蜗杆
worm-and-sector steering device 蜗杆齿弧式转向机
worm-and-single-roller steering gear 蜗杆单滚轮式转向机构
worm-and-spiral-teeth sector 蜗杆与螺旋齿扇齿
worm-and-three-teeth sector 蜗杆与三齿式齿扇
worm-and-twin-lever steering gear 蜗杆曲柄双销式转向机构
worm-and-wheel axle 蜗杆蜗轮式驱动桥
worm-and-wheel gear 蜗杆蜗轮机构
worm-and-wheel steering gear 蜗杆蜗轮转向机构
worm and worm gear 蜗杆与蜗轮
worm and wormwheel 蜗杆蜗轮
worm and wormwheel pulley block 蜗杆蜗轮滑车组
worm auger 麻花钻;螺旋钻
worm ball 蜗杆球
worm ball return guide 蜗杆球回行导架
worm ball return guide clamp 蜗杆球导管夹
worm bearing 蜗杆轴承
worm bearing adjusting nut 蜗杆轴承调整螺母
worm bearing adjustment 蜗杆轴承的调整
worm bearing ball race 蜗杆轴承滚珠座圈
worm bearing cup 蜗杆轴承杯
worm bearing thrust screw 蜗杆轴承的止推螺钉
worm bending strength 蜗轮抗弯强度
worm bit 螺旋钻头
worm box 蜗杆传动箱
worm bracket 蜗杆架
worm brake 蜗杆制动装置;蜗杆制动器
worm brake cone 蜗杆制动器内制动锥体
worm brake spring 蜗杆制动器弹簧
worm-cam roller feeder 蜗杆凸轮辊式送料装置
worm case 蜗杆罩
worm casing 蜗杆罩
worm chain block 蜗杆差动滑车
worm channel 蛀眼;蛀孔
worm clip 蜗杆式夹紧机构
worm coil 蛇管
worm compressor 蜗杆式压缩机
worm condenser 旋管冷凝器
worm conduit 盘管
worm conveyer 螺旋输送机;螺旋式输煤机;蜗杆输送器;螺旋输送器;绞刀
worm couple 蜗轮蜗杆副
worm crusher 螺旋式破碎机
worm diagram 蠕虫图
worm distributor 螺旋给料机
worm divider 螺旋式分配器
worm drive 蜗杆传动
worm-driven rear axle 蜗杆传动后轴
worm driving pinion 蜗杆传动主动齿轮
worm dust 虫蛀粉
worm-eaten 虫蚀的;虫蛀的
worm eccentric adjusting sleeve 蜗杆偏心调整套
worm elevator 螺旋提升机
worm eye 蛀眼
worm feed 螺旋喂送器;蜗杆钻进;蜗杆推进器
worm feeder 螺旋喂料机;蜗杆加料器;螺旋加料机;螺旋给料器;蜗杆给矿机
worm feeding core extrusion machine 螺旋挤芯机
worm fence 曲折栅栏;弯弯曲曲的栅栏
worm gear 蜗轮蜗杆装置;蜗轮
worm gear cam drive 蜗轮凸轮传动
worm gear case 蜗轮箱;蜗杆箱
worm gear conjugation tester 蜗轮副检查仪
worm gear differential 蜗轮蜗杆式差速器
worm gear drive 蜗轮传动
worm gear drive mechanism 蜗轮传动装置
worm gear drive ratio 蜗轮传动比
worm-geared hand counter 手摇蜗轮传动计数器
worm-geared hoist 蜗轮卷扬机
worm-geared machine 蜗轮蜗杆传动机械;蜗轮式卷扬机
worm-geared winch 蜗杆传动绞车;蜗杆传动卷扬机;蜗轮传动绞车
worm gear feed 蜗轮传动进给
worm gear grease 蜗轮润滑脂
worm gear hob 蜗轮滚(铣)刀
worm-gearing 蜗轮传动装置;蜗杆传动装置;蜗杆传动;蜗杆传动装置
worm gear milling machine 蜗轮铣床
worm gear oil 蜗轮油
worm gear reducer 蜗轮减速器
worm gear screw jack 蜗流螺旋起重器
worm gnawed 虫蛀的
worm grinder 蜗杆磨床
worm grinding machine 蜗轮磨床
worm hob 蜗杆滚刀
worm hobbing machine 蜗杆滚齿机
worm hoist 蜗杆起重机
worm hole 蛀眼;蛀孔;虫孔;毛虫状气孔;条命状气孔;(木材的)虫眼
worm housing 蜗轮箱壳体
worming 填卷
wormkalk 竹叶状灰岩
worm knotter 螺旋式结筛
worm lift 垂直螺旋提升机;垂直螺旋输送机;螺旋提升机
worm lock 蜗杆止动
worm mesh 蜗杆啮合
worm milling cutter 蜗杆铣刀
worm milling machine 蜗杆铣床
worm motor 蜗杆传动马达
worm-operated bar bender 螺旋式弯钢筋机
worm pipe 螺形管;盘管;螺形管;蜗轮管;蛇管
worm piping 盘管
worm press 螺旋压力机;螺旋压出机
worm pulley block 螺旋滑车;蜗杆滑车

worm rack 蜗杆(传动)齿条
worm rack type 蜗杆齿条式
worm rack type driving device 蜗杆齿条式驱动装置
worm recirculating ball nut 蜗杆往复球螺母
worm reduction box 蜗轮蜗杆减速箱
worm reduction gear 蜗轮减速机;蜗杆减速传动装置
worm reduction gear unit 蜗轮减速齿轮箱
worm reduction unit 蜗轮减速器
worm reverse gear 蜗轮换向齿轮
worm reversing gear 蜗轮换向齿轮
worm roll 麻花辊
worm roller 螺旋滚筒
worm rose shell 超前式螺旋泵筒
worm screw 蜗钉(螺钉)
worm screw feed 螺旋供料
worm screw feeder 螺旋给料机
worm sector 扇形蜗轮
worm's eye map 不整合面底板图
worm's-eye view 仰视图
worm shaft 蜗杆轴
worm shaft and gear type 蜗杆轴齿轮式
worm shaft and nut type 蜗杆轴螺母式
worm shaft and sector type 蜗杆轴扇轮式
worm shaft end play 蜗杆轴向间隙
worm shaft roller conical bearing 蜗杆锥棍轴承
worm shaft spacer tube 蜗杆轴隔套
worm-shaped 虫状的
worm speed reducer 蜗轮减速器;蜗轮减速机
worm steer(ing) 蜗杆转向
worm steering gear 蜗轮转向装置;蜗杆转向装置
worm tester 蜗杆检查仪
worm thread 蜗杆螺纹
worm-thread tool ga(u)ge 齿轮滚刀用样板
worm thrust bearing 蜗杆止推轴承
worm thrust screw 蜗杆推力螺钉
worm trace 虫孔
worm tube 盘管
worm tubing 盘管
worm-type cooler 螺旋式冷却器
worm-type lubricant 蜗轮润滑油
worm-type mechanical hand feed 蜗杆式机械手动给进
worm-type propeller 蜗杆式推进器
worm-type rotary compressor 螺杆回转式压气机
worm-type steering gear 蜗杆式转向机构
worm-up apron 跑道导路(飞机场跑道前段)
worm waist 蜗杆腰部
worm wheel 蜗轮
worm-wheel bracket 蜗轮支架
worm-wheel dial 蜗轮(标)度盘
worm-wheel dredge(r) 蜗轮式挖泥船
worm-wheel friction brake 蜗轮摩擦制动器;蜗轮摩擦闸
worm-wheel gearing 蜗轮传动装置
worm-wheel hobbing machine 蜗轮滚齿机
worm-wheel hob thread 蜗轮滚刀螺纹
worm-wheel hub 蜗轮轮毂
worm-wheel shaft 蜗轮轴
worm with disengaging motion 分离蜗杆
wormwood 蒿属
worm work dressing 虫蚀状琢面;虫蚀状錾痕
wormy 虫蛀的
wormy gravel clastic texture 竹叶状砾屑结构
worn 磨损的
worn bit 磨钝的钻头;磨钝的钎头
worn bushing 磨损(的)导套
worn coin 残币
worn flange 磨耗轮缘
worn flat 磨平
worn guide pin 磨损(的)导柱
worn-in journal 磨合轴颈
worn into ruts 磨成凹痕;陷损成辙
worn on a taper 磨成锥形
worn-out 磨损的;磨穿的;用旧的
worn-out article 磨损物件
worn-out fixed assets 磨损的固定资产
worn-out pavement 破损路面
worn-out rail 磨损轨
worn-out soil 贫瘠土壤
worn-out surface 磨损表面
worn parts 磨损件
worn pavement 磨损路面;磨损地面

worn rail with overlapping fin 有飞边的旧钢轨
worn wheel 磨耗车轮
worn wheel profile 磨耗型车轮踏面
worry box 折断钻具(焊)接点
worseness 劣等品位
worsening 恶化
Worstall heat test method 沃斯塔试验法
worst-case 最坏条件;最坏情况
worst-case analysis 最坏情况分析
worst-case bound 最坏情况界线
worst-case condition 最不利情况;最坏条件
worst-case design 最坏情况设计;按最坏情况设计
worst-case device 最劣情况下应用的器件
worst-case in motion 动态值
worst-case loop 最坏情况循环
worst-case method 最坏情况法
worst-case noise 最坏情况噪声;最大的噪声
worst-case noise pattern 最坏情况噪声模式
worst-case test 最不利情况试验
worst-case value 最不利值
worst-case water 最劣质水
worst dynamic(al) load 危险动负载
worsted yard 精梳毛纱;精纺毛纱
worst face 最坏的材面
worst pattern of stored information 存储信息的最坏图样
worst position of load 荷载最不利位置
worst rolling car 最难行车【铁】
worst status check 最坏状态检验;最坏布局检查
wort 野草;草本植物
wortgagee 受抵押人
worth debt ratio 资产负债比(率);资本负债比率
worthiness of wrecking 打捞价值
Worthington pump 蒸汽往复泵
worthless 无价值的
worth to current debt ratio 资本与流动负债比率
worth to fixed assets ratio 资本与固定资产比率
worth to fixed debt ratio 资本与固定负债比率
wortle plate 拉丝模板
Wotton 沃屯牌窗(一种专利天窗)
Wouldham hod 沃尔特哈姆牌运斗(装运砖块、灰浆等用)
wound 创伤
wound armature 绕线式电枢
wound clip 创缘夹
wound clip forceps 创缘夹缝合镊
wound clip removing forceps 创缘夹拆除钳
wound edge scissors 创缘剪
wound field motor 绕线磁极式电动机
wound gasket 缠绕垫
wound gum 伤胶
wound heart wood 假心材
wound in both directions 双向缠绕
wound irrigation syringe 创口冲洗器
wound margin 创缘
Woundplast 创可贴
wound poisoning 铅中毒
wound probe 铂探针
wound roller-type bearing 螺旋滚柱轴承
wound rotor 线绕式转子;绕线转子
wound-rotor induction motor 线绕式感应电动机;绕线式转子感应电动机
wound-rotor type motor 绕线式(转子)电动机;线绕转子式电动机
wound-type die extrusion press 缠绕式模挤压机
wound-type induction motor 线绕式感应电动机
wound-wound tube furnace 铂丝管式炉
woven asbestos brake lining 石棉制动衬片
woven belt 帆布皮带
woven board 木板条编织的板;编成(的)板;编织板
woven bone 编织骨
woven canvas tyre 帆布轮胎
woven carcase(type) belting 多层帘布胶带
woven carpet 机织地毯
woven cloth for asphalt felt 油毡基布
woven cloth for electric(al) laminate 层压布
woven cloth for fishing rod 钓鱼竿基布
woven cloth for grinding wheel 增强砂轮布
woven cloth for reinforcing plastics 玻璃钢纺布
woven cloth for reinforcing rubber 增强橡胶基布
woven design 布纹
woven fabric 织物;机织物;纺成(土工)织物
woven-fiber optics 编织纤维光学
woven filter medium 滤网;筛网

woven geotextile 有纺型土工织物;纺成土工织物;编织型土工织物
woven glass fabric 玻璃织物;玻璃布;机织玻璃布
woven glass roving fabric 玻璃纤维无捻粗纱
woven glass tube 玻璃织物管
woven jacket 帆布套
woven lining 织物衬里
woven lumber 重型柴排;柴排
woven mattress 编织沉排;柴排;编织成的沉排衬垫;编梢沉排;编成的沉排
woven membrane 织物滤层
woven memory 编织存储器
woven-oak fencing 栎木条编成的篱笆;橡木条编成的篱笆
woven pattern 编织纹
woven pile carpet 编织绒毛地毯
woven polypropylene filter cloth 聚丙烯滤布;编织的聚丙烯滤布
woven roving fabric reinforcing 机织粗纱织物增强材料
woven screen 编织筛
woven-screen storage 编线网存储器
woven scrim 网格布
woven staple fiber[fibre] fabric 定长纤维布
woven steel fabric 编织(成的)钢网;编的钢丝网
woven tape 布卷尺
woven valley 搭瓦天沟
woven willow 柳枝排;柳条编的柴排
woven willow mat 柳条(编成的)衬垫
woven wire 编钢丝网
woven-wire belt 网带
woven-wire cloth 金属丝布
woven-wire cloth standards 金属丝布规格
woven-wire cloth with cropped up edges 裁边金属丝布
woven-wire cloth with folded up edges 折边金属丝布
woven-wire cloth with turned up edges 卷边金属丝布
woven-wire dam 钢丝网坝
woven-wire fabric 钢丝网织物;编织成的钢丝网;编织钢丝网
woven-wire fence 钢丝网栅栏;铁丝网围栏
woven-wire gate 铁丝网院门
woven-wiremesh filter 编织线网滤油器
woven-wire reinforcement 织网钢筋;编织钢丝网;钢丝网配筋
woven-wire screen 金属线筛;金属丝细网筛;编织金属筛网
woven-wire screen cloth 金属丝网布
woven-wire squaremesh screen 织丝方眼筛
woven-wire window guard 铁丝网窗护栏
woven wood fences 木板条围栏
wove paper 布纹纸;横纹纸
wow 频率颤动
wow and flutter 速度不均匀性
wow-flutter 晃抖度
wow fluttermeter 抖晃仪
W placer ore 砂钨
wrack 次质木材;最低级的板材;劣等材;交叉扎绳;草袋;失事船只;残骸
wracking 船体横断面结构变形
Wrangellia terrane 兰格利亚地体
wrap-and-solder splice 缠绕焊接
wrap angle 包角
wrap-around 返回抵押;卷绕;卷包;返转
wrap-around angle 包角
wrap-around arrangement 环配置
wrap-around astragal 门扇护条;门扇盖缝条;搭接的半圆饰
wrap-around band brake 环绕式刹车带
wrap-around frame 包墙的门窗框
wrap-around hanger 半合式油管挂;半合成油管挂
wrap-around hologram 曲面全图息
wrap-around label 绕贴标签
wrap-around lease 返回租赁
wrap-around loan 返回贷款
wrap-around mortgage 二次抵押
wrap-around oven 壳芯电热炉
wrap-around tie bars 加固围框
wrap around type solar cell 卷包式太阳能电池
wrap-around windscreen 曲面挡风玻璃
wrap attachment 吊线装置
wrap bending test 卷弯试验

wrap forming 拉伸成形(法);卷缠成形
wrap of spiral case 蜗壳包角
wrappage 包皮;包袋
wrapped acid-alkali hose 包扎输酸碱胶管
wrapped angle 包角
wrapped bearing bush 卷制轴套
wrapped booster 侧置助推器
wrapped cable 绕扎电缆
wrapped connection 绕接
wrapped electrode 绕丝焊条
wrapped fascine 沉捆
wrapped hose 夹布胶管;布卷软管
wrapped in teflon 外包聚四氟乙烯
wrapped joint 缠绕接线头
wrapped pipe 包裹好的管道
wrapped underdrain system 包裹地下排水系统
wrapped water hose 夹布输水管;包布输水胶管
wrapper 包皮;包装纸;助卷板;封套;包装板;包裹料
wrapper rolls 助卷机辊
wrapper sheet 包板
wrapping 包装(材料);包扎;包封;包边
wrapping and package 贴身包装
wrapping angle of blade 叶片包角
wrapping choice model 包裹选择模型
wrapping connection 缠线接线法;缠绕接法
wrapping connector driving 柔性传动
wrapping counter 包装计数器
wrapping head 缠绕器
wrapping-in machine 包裹机
wrapping machine 绕线机;皮带助卷机;打包机;包装机
wrapping material 包装物料;包装材料
wrapping paper 包装纸;包皮纸
wrapping paper and cardboard 包装纸及纸板
wrapping plane 包装机
wrapping plate 包装板
wrapping post 绕线柱
wrapping test 缠绕试验
wrapping tool 包装工具
wrapping wiring 绕线连接
wrap-round booster 侧置助推器
wrap-round evapo(u)rator 封套式蒸发器
wrap-round motor 侧置助推器
wrap-round plate 薄卷筒凸印版
wrap-round tee 现焊三通
wrap system 包裹系统
wrap-up 简要报告;收卷装置
Wratten filter 拉登滤光片
Wray's rule 雷氏法则(用于圬工设计)
wreath 转弯扶手;扭弯;花圈;扶手弯子
wreath at turn (楼梯扶手的)环状饰;花环
wreathe 编成环
wreathed column 旋卷柱;绞花柱;扭形柱;麻花形柱;蟠花柱;盘绕柱身的饰带;螺旋柱
wreathed handrail 曲线形楼梯外斜梁;楼梯扶手弯头;转弯扶手;扭弯扶手;扭曲形扶手
wreathed stair(case) 几何形楼梯;弯曲楼梯;盘绕楼梯
wreathed string 曲线形楼梯外斜梁;转弯扶手;楼梯扶手弯头
wreathed stringer 盘旋楼梯斜梁
wreath filament 环形灯丝;屈折螺旋灯丝
wreathing 注件内部浆痕;螺旋状纹
wreath piece 扭曲梁的直段;(曲线形楼梯的)外斜梁
wreath string(er) (楼梯井弯曲处)外斜梁
wreck 失事船(舶);船舶失事;沉船
wreckage 遇险船残骸;遇难船残骸;折断;(船)残骸
wreckage value 残骸价值
wreck ball 撞锤;(拆房屋用)撞击锤;破碎锤
wreck buoy 沉船浮标
wreck car 救援车
wreck crane 救险起重机;救险吊车;救援吊车
wrecked concrete 拆毁的混凝土
wrecked ship 失事船(舶)
wrecked submarine 失事潜艇
wrecked vessel 失事船(舶)
wrecker 救援船;救援车;救险车;拆卸旧建筑物者;营救船;工程救险车;救险人员;救难者;急救船;急救车;失事抢救车;打捞人员;救险承包人
wrecker track 救援线路;救援列车停留线
wrecking 拆毁房屋;海难救助作业;打捞;(船舶)拆毁
wrecking anchor 救援用锚
wrecking and breaking hammer 拆毁用锤;拆毁破碎锤
wrecking and drill hammer 拆毁钻锤
wrecking ball 拆房铁球;落锤破碎机;破碎球;拆房重锤
wrecking ball demolition 重锤拆除法
wrecking bar 铁棒;起钉撬棍;铁锯;拔钉撬棍
wrecking bit 拆捣棒头
wrecking block 特大滑车(用来架设临时吊重货专用吊杆的大型滑车)
wrecking bond 拆房房屋承包人的具结保证
wrecking building 拆除房屋
wrecking by government authorities 由政府机关拆屋
wrecking cable 特大缆绳
wrecking car 救援车;抢险车
wrecking company 救援队;旧屋拆除公司;打捞公司;拆除公司
wrecking contract 拆除合同
wrecking crane 救援起重机;救险起重机;轨道起重机
wrecking crew 抢修队;救援队;打捞队;拆除房屋工程队
wrecking firm 打捞公司;拆船机构;拆屋机构;打捞机构
wrecking frog 救险复轨器
wrecking gang 打捞队;拆屋队;拆除房屋工作队
wrecking hose 救难用软管
wrecking master 打捞监察
wrecking of buildings 房屋拆毁
wrecking operation 海难救助作业
wrecking party 拆屋组;打捞组
wrecking permit 拆船许可证;打捞许可证;拆屋许可证;允许拆除
wrecking pick 拆屋镐
wrecking pump 打捞泵
wrecking raising 沉船起捞
wrecking screw 抢救队
wrecking site 拆屋现场;拆除现场
wrecking strip 模板的活络搭扣;拆条
wrecking team 拆屋队;打捞队
wrecking tool 拆捣工具;救险船;救险车;打捞者;拆除工具
wrecking train 救援列车
wrecking truck 救险车
wrecking tug 海难救助拖轮;打捞拖轮
wrecking workboat 打捞船
wreck light buoy 沉船浮标
wreck mark(ing) 沉船标(志)
wreck-marking buoy 沉船浮标
wreck-marking light 沉船灯标
wreck-marking vessel 沉船标志船;沉船位置设标船
wreck obstacles removal 清除沉船障碍
wreck railway crane 救援用铁路起重机
wreck raising 沉船打捞
wreck-raising and shifting 沉船移位
wreck-raising camel 打捞(沉船)浮筒
wreck-raising lighter 打捞沉船浮筒
wreck reflo(a)tation 沉船起浮
wreck remains 沉船残骸
wreck salvage 沉船打捞
wreck signal 失事信号
wreck sunk vessel 沉没船
wreck surveying 沉船勘测
wreck train 救援列车
wrench 旋钳;扳子;扳手;扳钳;扳紧器
wrench circle 半圆形齿条
wrench clearance 扳手空位
wrencher 拧螺栓工人;扳钳工(人);钻工
wrench fault 走向(滑动)断层;扭(转)断层;平移断层
wrench-fault and associated structures 平移断层和伴生构造
wrench faulting 走向断层作用
wrench flats 扳手切口
wrench fork 叉形扳手
wrench hammer 锤柄开口扳手
wrench-head bolt 扳头螺栓
wrenching 扳紧
wrenching bar 拧螺钉杆;撬棍;起钉杆
wrench jaw 钳子叉头;钳子牙口;钳子牙板;扳钳叉头
wrench motion 扭转运动
wrench movement 扭转运动
wrench opening 扳子开度;扳手开度
wrench set 成套扳手
wrench shaft for pneumatic wrench 风扳机扳轴
wrench-shear fracture 扭剪(切)破裂;扭剪(切)破坏
wrench socket 套筒扳手;扳手套筒
wrench squares 方切头
wrench up 拧上
wrest 绞具
wrestle 操作加重管
wrestling ring 摔跤场
wrest of saw 锯齿修整
wriggle 泄水槽;遮水楣;蠕动
wriggle rigol 窗楣
wriggling instability 扭曲不稳定性
wriggling perturbation 弯曲型扰动
wright 木工;制造者;匠;工人
Wright's phenomenon 赖特现象
Wright system 赖特系统
Wright telescope 赖特望远镜
wring 绞出;收缩量
wring a lashing 绞紧绑索
wringer 绞衣机;榨水机;绞拧器
wringer injury 机器绞轧伤
wringer neck 钻杆折断;钻杆断裂
wringing 绞接
wringing fit 转入配合;紧动配合
wringing machine 挤水机
wring out 绞漆;拧干
wrinkle 皱折;折叠;起皱
wrinkle a joint of pipe 管子螺纹缠棉纱
wrinkle bend 管弯(同平面内弯成的弯管)
wrinkle chasing machine 除皱机
wrinkle crease 皱
wrinkled 皱纹的
wrinkled enamel 皱纹瓷漆
wrinkled leather 皱纹革
wrinkle finish 波纹装修;皱褶饰面;皱纹罩面(漆);皱纹饰面;皱纹漆;波纹面饰
wrinkle formation 皱纹的形成
wrinkle lacquer 皱纹清喷漆
wrinkle laminar flame 皱纹层流火焰
wrinkle mark 皱痕
wrinkle pattern 折皱标样
wrinkle pipe 车管子螺纹
wrinkle proofing 防皱整理
wrinkle rating 皱折度
wrinkle recovery 折皱回复度
wrinkle-recovery tester 折皱回复度试验仪
wrinkle resistance 抗皱性(能)
wrinkle ridge 皱脊;纹脊
wrinkle varnish 皱纹清漆;波纹清漆
wrinkling 皱纹(饰面);局部扭曲;起皱;皱褶
wrist 机械腕;锚腕
wrist action 肘节运转;肘节作用;肘节动作
wrist-action drive 摆杆式传动机构
wristing head 记录头
wrist movement 机械腕动作
wrist pin 肘节销;活塞销;曲柄销
wrist-pin bearing 十字头销衬套
wrist-pin brass wedge block 肘销铜楔垫块
wrist-pin brass wedge bolt 肘销铜楔螺栓
wrist-pin brass wedge set screw 肘节销铜楔止动螺钉
wrist-pin brass wedge washer 肘节销铜楔垫
wrist-pin bush 连杆小端衬套
wrist-pin collar 十字头销衬套
wrist-pin dearing lubrication 活塞销承润滑
wrist-pin end of connecting rod 连杆小端
wrist-pin hole 连杆小端孔
wrist-pin nut 活塞销螺帽
wrist-pin nut wahser 肘节销螺母垫圈
wrist-pin oil cup 肘节销油杯
wrist-pin spool 曲柄销猫头
wrist plate 肘板
wrist watch 手表
wrist-watch dial locating pin welding machine 表盘焊钉机
writability 可写性
writable control storage 可控制存储器
writable information 可写信息
write addressing 写寻址
write amplifier 写放大器
write and compute 读写与计算

write a receipt 开发票
write bus 写入总线
write data 写数据
write data out signal 写入数据输出信号
write direct 直接写入
write down 降低账面价值;减少账面资产;冲转;折旧;减记资产账面值;减记
write-down cost 减记成本
write driver 写数驱动器
write drive winding 写驱动绕组
write enable line 写命令线
write gating 写控制
write head 写头
write in 存入
write input 记录输入信号
write-in time 写入时间
write key 记录键
write lock-out 写入锁定
write off 销账;注销;冲销;删除账面值
write off and exempt the circulating funds 核销流动资金
write off an entry 抹去账项
write off method 冲销法
write off un-collectible account 冲去坏账;冲销坏账
write only 只写
write out a receipt 开收据
write pen register 光笔记录寄存器
write pen unit 光笔记录装置
write protect 记入保护
writer driver 写数策动器
writer dynamometer 记录的保管
write signal 记录信号
write time 写时间
write-to-operator 给操作员送信息;记录信息通知操作员
write up 提高账面价值;增记
write validity check 写入确实性检查
writing 书面的
writing apparatus 写字仪
writing beam 记录电子束
writing brush 毛笔
writing byte rate 写入字节速率
writing-chair 写字椅;扶手椅
writing circuit 记录电路
writing density 记录密度
writing desk 写字台;记录板;书桌
writing down value 冲销后账面价值
writing field 写区
writing galvanometer 自记电流计
writing gun 记入电子枪
writing head 写头
writing ink 墨水
writing lamp 记事灯
writing lens 记录透镜
writing lever 描记杠杆
writing materials 文具
writing-method 记录法
writing-off process 销帐法
writing pen recorder 笔记录器
writing plate 记录板
writing pointer 自动记录器;自动记录计
writing position 记录位置
writing pulse 记录脉冲
writing room 写信室;写作室
writing scan 录储扫描;记入扫描;记录扫描
writing short 表格填错;单据填错
writing speed 写入速度;记录速度
writing station 写入装置;记录站
writing system 书写体
writing table 写字台;书桌
writing task 写任务
writing tip 记录笔尖
writing tissue 薄书写纸
writing velocity 写入速度
writ of ejectment 收回出租的房地产的诉讼
writ of execution 执行令
written agreement 协议书;书面协议
written application 书面申请;申请书

written approval 书面同意;书面批准;书面批复
written authorization 核准书
written calculation 笔算
written catalogue 稿本目录
written confirmation 书面确认;书面证明;书面凭证
written contract 书面契约;书面合同
written declaration 书面声明
written discharge 书面结清单
written documents 书面单据;书面说明
written down value 减记的价值
written estimate 预算书;概算书;施工预算书
written examination 笔试
written expert testimony 专家鉴定
written form of order 指令的书写形式
written inquiry 书面质询;书面提问
written liaison method 书面联络法
written log 记录钻井剖面【岩】
written mark 手写标志
written material 书面材料
written notice 书面通知
written off 摊提
written of equipment 设备报废
written order 书面订货
written permission 书面许可;书面同意
written pledge 保证书
written record 书写记录
written reply 书面答复
written report 书面报告
written request 书面要求
written statement 书面声明;书面记录
written statement of claim 索赔清单
written translation 笔译
written type 书写法
written undertaking 书面保证
written-up 已提账面价值
written verdict 裁定书
wroewolfeite 斜蓝铜矾
wrong 错误的
wrong channel 虚假信道
wrong claim 不当的索赔
wrong clearance 未达标准规定的间隙;非标准间隙;不标准间隙
wrong clearing of a signal 错误开放信号
wrong colo(u)r 颜色不正
wrong delivery 交货错误
wrong file type 错位文件类型
wrongful dealing 不公平的交易
wrongful dismissal 非法解雇
wrong handling 误装卸;错误办理
wrong indication 错误显示
wrong line 反方向运行线路
wrongly calculated 算错
wrong-reading negative 反阴片
wrong-reading positive 反阳片
wrong side 反面
wrong stacking method (集装箱)错误堆码方式
wrong track 反方向运行线路
wrot lumber 刨光的木料
wrought 锻(制)的
wrought alloy 可锻合金;锻制合金;锻造合金
wrought alloy of alumin(i)um 轧制铝合金;锻铝合金
wrought alumin(i)um 熟铝;锻铝
wrought alumin(i)um alloy 轧制铝合金;熟铝合金;锻压铝合金;变形铝合金
wrought alumin(i)um plate 锻压铝板
wrought board 装修用板;精制板
wrought brass 熟铜
wrought concrete formwork 加工混凝土模板
wrought copper 熟铜
wrought Everdur 埃弗托尔形变硅青铜
wrought full crown 锤造全冠
wrought grounds 钉在墙的刨光木砖;(钉在墙的)刨光木砖
wrought iron 熟铁;锻铁
wrought iron bend 熟铁弯头
wrought iron bolt 熟铁插销
wrought iron conduit 熟铁管

wrought iron finishing turning tool 熟铁光车刀
wrought iron pipe 熟铁管
wrought iron piping 熟铁管
wrought iron plate 熟铁板
wrought iron scrap 熟铁废料
wrought iron sectional boiler 熟铁片式锅炉
wrought iron shield 锻铁保护板
wrought iron tube 熟铁管
wrought iron tubing 熟铁管
wrought iron window grille 熟铁窗栅
wrought iron wye 熟铁斜三通
wrought lumber 刨光木料;刨光木材
wrought metal 锻制金属;轧制金属
wrought nail 锻(制)钉
wrought pipe 熟铁管;锻制管
wrought seamless copper tube 轧制无缝紫铜管
wrought shuttering 平滑模壳;平滑模板
wrought stainless steel wire 不锈钢丝
wrought steel 搅炼钢;焊接钢;熟钢;锻钢
wrought-steel fittings 锻钢配件;锻钢零件
wrought-steel pipe 轧制钢管;熟钢管;锻钢管
wrought stuff 已加工件;抛光的木材
wrought superalloy 变形高温合金
wrought timber 精致木材;精制木材;刨光木材;刨光木材
wrought tool steel 锻造工具钢
W-shaped valley W 形谷
W-S ratio 水固比
W-truss W 形桁架
W-type expansion joint W 形膨胀水泥
W-type guardrail W 型护栏
W-type superposed folds W 型叠加褶皱
Wuchen loess 午城黄土
wudian roof 庑殿屋顶
Wufoshan group 五佛山群【地】
wu-jin glaze 乌金釉
Wulf electrometer 伍尔夫静电计
wulfenite 黄铅矿石;钼铅矿
Wulff net 经纬网;伍氏网
wulff pyrolysis furnace 伍尔夫裂解炉
Wulf string electrometer 伍尔夫弦线静电计
wulstchialas 海水细碧岩
Wurm 武木期【地】
Wurster's red 沃斯特氏红
wurtite 纤锌矿
wurtzilite 沥青焦沥青;韧沥青
wurtzite 纤(维)锌矿
wurtzite BN 纤锌矿型氮化硼
wurtzite structure 纤锌矿结构
Wurtz reaction 维尔茨反应
Wusong horizontal zero 吴淞零点
Wu-song zero of elevation 吴淞零点
wustite 方铁体;方铁矿
wustite iron 方铁体铁
Wutai system 五台系
ww-dichloroacetophenone 二氯乙酰苯
wyamin 甲苯丁胺
wyartite 黑碳钙铀矿;水碳钙铀矿
wychelm 榆木
wye 星形;二叉
wye bearing Y 形支架
wye branch Y 形支管;Y 形叉管;叉管
wye connection Y 形接法;星形连接
wye-fitting Y 形管接头;叉形管
wye level 回转水准仪;华氏水准仪;Y 型水准仪
wye pipe Y 形管
wye rectifier 星形全波整流器
wyethe 单层砖墙厚
wyethern 矿脉
wye track 折返三角线;三角(回车)线【铁】;三岔轨道;转向三角线
wye tube 叉形管
wyllieite 磷铝铁锰钠石
Wyoming valley stone 怀俄明州谷岩(美国)
Wyring adsorption coefficient 艾润吸声系数
wythe 烟囱隔板;吊杆端箍;烟囱纵向分隔板;墙的垂直断面;立砌砖墙;空心墙的一面墙;=withe
WZ alloy 碳化钛烧结合金

X

X-alloy 铜铝合金
xalostocite 蔷薇榴石
xalsonte 砾石堆积;粗粒砂
X-antenna X 形天线;双 V 形天线
xanthan gel 黄原凝胶
xanthan gum 黄原胶
xanthate 黄原酸酯;黄原酸盐
xanthation 黄原酸化作用
xanthic 黄色的
xanthic acid 乙氧基二硫代甲酸;黄原酸
xanthic amide 黄原酰胺
xanthine 黄嘌呤
xanthiosite 砷镍石
xanthochroism 黄色现象
xanthoconite 黄银矿
xanthogenate 黄原酸酯;黄原酸盐
xanthogenate sulfur 黄原酸酯硫
xanthogenic acid 黄原酸
xanthonate 黄原酸酯
xanthonation 黄原酸化作用
xanthone 氧杂蒽酮;咕吨酮
xanthoxenite 黄磷铁钙矿
xanthsiderite 针铁矿
X-arm 横臂
x-arm machine 多臂成形机
xarque 纪念柱
xat (印第安人的)雕刻纪念柱;物像柱
X-axis 横轴【测】;X 轴;横坐标轴
X-axis length of markers 标志体 X 轴长度
X-back 防静电背面层
X-band diode phase shifter X 波段二极管移相器
X-band ferrite modulator X 波段铁氧体调制器
X-band limiter attenuator X 波段限幅衰减器
X-band microwave source X 波段微波源
X-band parametric amplifier X 波段参量放大器
X-band passive array X 波段无源阵
X-band phase shifter X 波段移相器
X-band planar array X 波段平面阵
X-band power amplifier X 波段功率放大器
X-band pulse transmitter X 波段脉冲发射机
X-band radar beacon X 波段雷达信标
X-band travelling wave amplifier X 波段行波放大器
X-band triode oscillator X 波段三极管振荡器
X-bar structure 交叉棒结构
X-bit 十字形钻(头)
X-brace X 形条柱;交叉支条;交叉支撑;剪刀撑;十字撑;X 形支撑
X-bracing 交叉支条;交叉支撑;交叉联接;剪刀撑(系统);X 形支撑
X-bridge 电抗电桥;X 形电桥
X-chair 交叉折椅
X-chisel 十字形钻(头)
X-component X 分量
X-conn 十字接头
X-coordinate 横轴坐标;X 坐标
X-correction X 轴方向改正
X-crossing member X 形横梁
X-datum line 穿孔卡片基准线
X-direction X(轴)方向
X-displacement X 轴方向移动
X eliminator 静电消除器
xenene 联苯
xenidium 胶合板
xenobiotics 外来化合物;宾主共栖生物
xenoblast 他形变晶
xenoblastic texture 他形变晶结构
xenocryst 捕房晶;捕获晶【地】
xenodocheum 古希腊的客栈或旅店;(古建筑)客厅
xenodochium (中世纪的)救济院;收容所;旅店;客栈
xenolith 俘虏岩;捕房体【地】;异晶体
xenolithic enclave 捕房岩包体
xenology method 氙法
xenomorphic 外来形的;他形的
xenomorphic crystal 他形晶
xenomorphic grain 他形晶粒
xenomorphic-granular 他形晶粒状(的)

xenomorphic granular texture 似花岗岩状结构;他形晶粒状结构
xenon aid lamp 氙航标灯
xenon arc lamp 氙气灯;氙弧光灯
xenon arc weatherometer 氙弧老化机;氙灯老化机
xenon-filled flashtube 充氙闪光管
xenon-filled linear flash lamp 充氙线性闪光灯
xenon filled quartz discharge tube 充氙气石英放电管
xenon flash lamp 脉冲氙灯
xenon flash light source 脉冲氙灯光源
xenon fluoride 氟化氙
xenon high-pressure lamp 高压氙灯
xenon-iodine method 氙—碘法
xenon lamp 氙灯管;氙(气)灯
xenon-mercury lamp 汞氙灯
xenon oxide 氧化氙
xenon repeated pulse flash light 脉冲氙灯
xenon standard white light source 氙气标准白(色)光源
xenon weatherometer 氙灯老化(试验)机
xenoparasite 宿主寄生物
xenopathic experience 外界体验
xenotest apparatus 氙气灯(耐候)试验机;氙灯老化机
xenothermal 浅成高温热液的
xenothermal deposit 浅成高温热液矿床
xenothermic period 干温期(冰后期)
xenotime 磷钇矿(含量)
xenotime ore 磷钇矿矿石
xenotopic fabric 他形组构
xenyl 联苯基
xenylamine 联苯基胺;对联苯基胺;苯基苯胺
xenysalate 苯柳胺酯
xeponent 幂
xerad 旱生植被
xeralf 干热淋溶土
xeransis 干化
xeraphium 干燥粉
xerarch succession 旱生演替(系列)
xerasium 干燥演替;旱涝演替
xerert 季节性干旱变性土
xeric 干旱的
xerium 干燥粉
xerochasy 干裂
xerochore 无水沙漠(区)
xerocole 旱生动物
xerogel 干凝胶
xerogram 静电复印副本
xerograph 干板摄片
xerographic(al) print 静电复印图;静电复印件
xerographic(al) printer 静电复印机;电子照相印刷机
xerographic(al) printing paper 静电复印纸
xerographic(al) technique 静电复印技术
xerography 干印法;干板摄影;静电印刷(术);静电复印(法)
xeroll 干热软土
xeromorphic vegetation 旱生植物;旱生植被
xeromorphism 旱生形态
xeromorphosis 适旱变态
xeromorphy 旱性形态
xerophile 喜旱植物;喜旱性;喜旱的
xerophilization 旱生化
xerophilous 适旱的
xerophilous animal 旱生动物
xerophilous crop 喜旱植物
xerophilous plant 喜旱植物;旱生植物;适旱植物
xerophious plant 旱生植物
xerophobous 嫌旱的;避旱的
xerophyte 旱生植物;干地植物;旱生植被
xerophytia 旱生植物群落
xerophytic vegetation 旱生植被
xeropolum 荒野群落;干草原
xeroprinting 静电复印(法)
xeroradiogram 干板 X 线片
xeroradiograph 干板 X 线照片;干板 X 线片
xeroradiographic(al) plate 干式射线照相干板

xeroradiography 静电放射线照相术;干板 X 线照相术
xeroradiography apparatus 干放射照相设备
xerosere 旱生演替系列
xerosol 干旱土
xerothermal index 干热指标
xerothermal period 干热气候期
xerothermic period 干热期
xerothermic theory 干热学论
xerotripsis 干擦
xerox 硒静电复印;静电复制
Xerox copy 静电复印图;静电复印件;复印件;硒静电复印
xerox-graphic(al) printer 硒静电复印
xeroxing 复印
xerox machine 硒静电复印
xerult 干热石成土
Xestobium 窃蠹属
Xestobium rufovillosum 盗窃虫
X-fluorescence survey in borehole X 荧光井中测量
X-fluorescence survey in field X 荧光现场测量
X-fluorescence survey in laboratory X 荧光实验室分析
X-fluorscence logging meter X 荧光测井仪
X-frame X 形构架;交叉形框架;交叉形架;十字形框架;十字形机架
X-frame brace 框架对角撑木
X-gear 径向变位齿轮
Xiashu clay 下蜀黏土
xifengite 喜嶂矿
xilingolite 锡林格勒矿
ximenic acid 山梅酸
xinanite 新安石
xing pedestrian 人行横道
X-intercept X 截距
xiphoid process 鸠尼
Xite 耐热镍铬铁合金
xitieshanite 锡铁山石
Xi-type 多字型
Xi-type structural system 多字型构造体系
Xi-type structure 多字型构造
X-joint X 形接合
X-leg 叉形腿
X-line X 轴(线)
X-linked polyvinyl chloride insulated power cable 交联聚氯乙烯绝缘电力电缆
xlonzonitic graphic(al) granite 二长花岗斑岩
X-member 叉形杆件;交叉形构件;X 形梁 X 形构件
X-motion 沿 X 轴方向运动;X 轴方向运动
X-movement X 轴方向移动
X-network X 形网络
X normal polarity subzone X 正向极性亚带
xoanon 希腊原始木雕神像
xocoecatlite 绿碲铜矿
X-offset X 轴方向偏移
xonotlite 硬硅钙石
xonotlite model insulating board 无石棉硅酸钙板
X-parallax X 视差
X-plate X 水平偏转板
X-quadripole 形四端网络;斜格形端网络
X-radiation 伦琴辐射;X 光辐射
X-radiography X 射线照相术
X-ray X 射线;X 光线
X-ray absorbing glass 防伦琴射线玻璃
X-ray absorption edge spectrometry X 射线吸收限光谱法
X-ray absorption method X 射线测量密度法
X-ray absorption spectrometry X 射线吸收光谱法
X-ray adsorption X 射线吸收
X-ray analysis X 射线分析;X 光分析
X-ray analysis method X 射线分析法
X-ray analyzer X 射线分析仪
X-ray apparatus X 光机
X-ray application X 射线应用
X-ray beam X 射线束
X-ray core X 射线型岩芯
X-ray crystal density X 射线晶体密度
X-ray crystal diagram method X 射线晶体照相法

X-ray density probe X 射线密度探测仪
X-ray department 放射科;X 光部分
X-ray detectoscope X 射线探伤法
X-ray detectoscopy X 射线探伤法
X-ray detector X 射线探伤仪
X-ray diagnosis X 光诊断
X-ray difference method X 射线组织分析
X-ray diffraction X 射线衍射;X 光衍射
X-ray diffraction analysis X 射线衍射分析
X-ray diffraction camera 伦琴射线绕射照相室;X 射线照相机
X-ray diffraction distribution image X 射线衍射分布象
X-ray diffraction in crystals X 射线在晶体中的衍
X-ray diffraction pattern X 射线衍射图案
X-ray diffraction system X 射线衍射系统
X-ray diffractogram X 射线衍射图
X-ray diffractometer X 射线衍射仪
X-ray dispersion X 射线色散
X-ray drill 轻型钻探机;X 射线型金刚石钻机
X-ray emission X 射线发射
X-ray emission analysis X 射线发射分析
X-ray emission ga(u)ge X 射线发射表
X-ray emission spectrum X 射线发射谱
X-ray energy spectrum analysis X 射线能谱分析
X-ray examination X 射线检查;X 光透视;X 光检查
X-ray film X 光片
x-ray film drying cabinet 胶片干燥箱
X-ray flaw detection for pressure vessel 受压容器 X 光探伤
X-ray flaw detector X 射线探伤器
X-ray fluoremetry log X 射线荧光灯测井
X-ray fluoremetry log curve X 射线荧光灯测井曲线
X-ray fluoremetry logger X 射线荧光灯测井仪
X-ray fluorescence X 射线荧光
X-ray fluorescence absorption X 射线荧光吸收
X-ray fluorescence analysis X 射线荧光分析
X-ray fluorescence spectrometer X 射线荧光光谱仪
X-ray fluorescence spectrometry X 射线荧光光谱法
X-ray fluorescent spectrometer X 射线荧光光谱仪
X-ray goniometer X 射线测角仪
X-ray hazard meter X 危险性测量器
X-ray high temperature camera X 射线高温照相机
X-ray holography X 射线全息摄影
X-ray identification X 射线鉴定
X-ray illumination X 射线照射
X-ray image X 射线图像
X-raying 伦琴射线照射
X-ray inspection X 射线检查;X 光检查
X-ray installation X 射线设备
X-ray instrumentation X 射线检测仪
X-ray intensity X 射线辐射强度
X-ray intensity meter X 射线强度计
X-ray interferometer X 射线干涉仪
X-ray laser X 射线激光器
X-ray machine X 光机
X-ray method X 射线法
X-ray microanalysis method X 射线显微分析法
X-ray microanalyzer X 射线微量分析仪
X-ray mineral phase identification X 射线物相鉴定
X-ray monochromator X 射线单色仪
X-ray pattern X 射线图案
X-rayphase quantitative analysis X 射线相定量分析
X–ray photo detection X 射线照相检验
X-ray photoelectron spectroscopy X 射线光电子光谱学
X-ray photogrammetry X 射线摄影测量
X-ray photograph X 射线照片
X-ray powder crystal data X 射线粉晶数据
X-ray powder diffraction X 射线粉末衍射
X-ray powder diffractometer 粉末衍射照相机
X-ray powdered crystal diffraction method X 射线粉晶衍射法
X-ray powder method 粉末法;X 射线粉末法

X-ray protective glass 防伦琴射线玻璃;X 射线防护玻璃
X-ray quantitative analysis X 射线定量分析
X-ray radiation X 射线辐射
X-ray rockfabric analysis X 射线岩组分析
X-ray sedimentometer X 射线沉降分析仪
X-ray sensor X 射线传感器
X-ray shield door X 光防护门
X-ray sorting of diamond 金刚石的 X 射线拣选
X-ray spectrochemical analysis X 射线光谱化学分析
X-ray spectrograph X 射线摄谱仪
X-ray spectrometer X 射线荧光谱仪
X-ray spectroscopy X 射线频谱学
X-ray spectrum X 射线谱
X-ray stress analysis X 射线应力分析
X-ray structural analysis X 射线结构分析
X-ray structural analysis instrument X 射线结构分析仪器
X-ray structuring goniometer X 射线组构测角仪;X 射线应力仪
X-ray take out angle X 射线取出角
X-ray technic-film 工业 X 射线胶片;工业 X 光胶片
X-ray test X 射线检验
X-ray testing X 射线分析法
X-ray texture goniometry method X 射线结构测角法
X-ray thickness ga(u)ge X 射线测厚仪
X-ray topography X 射线地形测量术;X 光形貌法
X-ray tube X 射线管;X 光管
X-ray turbidimetry X 射线比浊法
X-rayunit X 光机
X-ray use X 射线应用
X-ray waste X 射线废物
X-R control chart 平均值极差控制图
X-roads 交叉道路
X-rotation 绕 X 轴旋转
X-scale 水平线比例尺
X-section 横截面;交叉截面
X-shape 交叉形
X-shaped frame 交叉型架
X-spring X 形弹簧
X-stage X 期
X-stool 小折凳
X-stretcher 家具的十字形横档;(家具的)十字形横档
Xtra-dull 埃克斯特拉无光纤维
X-tube X 形管
X-type 交叉形;X 型
X-type frame 交叉形框架
X-type groove (熔缝的)X 形坡口
X-type member X 形梁
Xuande ware 宣德窑
xulovitro-vitrite 微木质镜煤—微镜煤质煤
X-unit X 单位
X-wave X 波
x-y chromaticity diagram 麦克斯韦三角形
X-Y digitizer X-Y 数字化器
X-Y function recording apparatus X-Y 函数记录仪
xylamon 氯化萘防腐杀虫剂
xylan(e) 木聚糖;木糖胶
xylan polysulfuric acid 多硫酸木聚糖
xylanthrax 木炭
xylary 木(质)的
xylem 木质部
xylem ray 木射线
xylene 二甲苯
xylene dichloride 二氯二甲苯
xylene disulfonic acid 二甲苯二磺酸
xylene equivalent 二甲苯当量(测定沥青材料均匀性用)
xylene halide 卤代二甲苯
xylene monosulfonic acid 二甲苯磺酸
Xylene resin modified phenolic mo(u)lding compound 二甲苯树脂改性酚醛压塑粉
xylenesulfonic acid 二甲苯磺酸
xylenesulfonyl chloride 二甲苯磺酰氯

xylenol 混合二甲酚;二甲苯酚
xylenol blue 二甲苯酚蓝
xylenol carboxylic acid 二甲酚酸
xylenol-formaldehyde resin 二甲苯酚甲醛树脂
xylenol orange 二甲酚橙
xylenol resin 二甲苯(酚)树脂
xylic acid 二甲苯甲酸
xylidine 二甲苯胺
xylidyl blue 二甲苯胺蓝
xylinoid group 木质组
xylitol 木糖醇
xylium 树木群落
xylochrome 木质光敏素
xylofusinite 木质丝煤
xylogen 木素
xyloglyphy 艺术木雕;木刻术
xylograph 木刻;木版画;木板画
xylographer 木刻师;木刻版工
xylography 木刻术;木版印画
xyloid 木质的;木制的
xyloid coal 木质煤
xyloidin 淀粉炸药
xylol glass cloth plate 二甲苯玻璃布板
xylolite 菱苦土木屑板;木屑板;木花板
xylolite slab 刨花板;菱苦土桐板;菱苦土木屑板;木屑板
xylolith 木屑板;木花板
xylology 木材构造学
xyloma 木质瘤
xylometer 木材测容器;木材比重计;测容器(木材比重计)
xylometer drag 木刮板
xylon 木质
xylonamide 木质酰胺
xylonic acid 木质酸
xylonite 假象牙;赛璐珞
xylonite-solution 硝基纤维清漆
xylophagan 蚀木虫
xylophagous 毁木的;食木的
xylophone 八管发射机
xylophyta 木本植物
xylopropamine 二甲苯丙胺
xylopyranose 吡喃木糖
xylorcinol 木间二酚
xyloritrosemifusinite 木质镜煤半丝炭体
xylose 木糖;吡喃木糖
xylose and xylitol pharmaceutical wastewater 木糖和木糖醇制药废水
xylotelinite 木质结构镜质体
xylotile 木纹石
xylotomous 能钻木的;能蛀木的
xylotomy 木材截片术;木材解剖术
xylotya 蛀木海虫
xylovitrain 镜煤
xylovitrofusinite 木质镜质丝质体
xyltile 木材化石
xylulose 木酮糖
xylyene carbinol 二甲苯基乙醇
xylylalcohol 甲苄醇
xylyl amine 甲苄胺
xylylene cyanides 二氰甲基苯
xylylene diamine 亚二甲苯基二胺
xylylene diisocyanate 亚二甲苯基二异氰酸酯
xylyl-mercaptan 二甲苯基硫醇
xylylol 二甲苯基二醇
X-Y plotter X-Y 绘图仪;X-Y 绘图机
xypohyta 木本植物
xyptal 醇酸塑料
X-Y recorder X-Y 记录器;双变量记录器
X-Y scaler X-Y 坐标测定器
X-Y scheme 二维图
xyst(us) 花园周围柱廊;绿廊;林荫道
X-zero gear 零变位齿轮
X-zone X 带
X-zoom scale factor X 方向缩放比例系数

Y

Y2K problem 两千年问题
yabunikkei seed oil 天竺桂(籽)油
yacca 浅黄色细纹木材(做家具用);草树树脂
yacca gum 禾木树胶
yacht 游艇;快艇
yacht basin 游艇港池;快艇碇泊地
yacht building yard 游艇制造厂
yacht chair 折叠帆布椅
yacht chart 游艇用图
yacht club 游艇俱乐部
yachter 游艇驾驶人员
yacht harbo(u)r 游艇港;快艇港
yachting 游艇驾驶术
yachting chart 快艇航海图
yachting fleet 快艇舰队
yacht insurance 游艇保险
yacht landing area 游艇停泊区
yacht landing stage 游艇码头
yacht marline 小油绳
yacht regatta 赛艇会
yacht rope 游艇系缆绳
yachtsmanship 游艇驾驶术
yacht wharf 游艇码头
yafsoanite 雅碲锌石
Yagi antenna 引向反射天线;波道式天线;巴木天线;八木天线
yagi antenna 波渠天线
yagiite 隙钠镁大隅石
yalca 耶尔卡雪暴
yale brass 低锡黄铜
yale bronze 低锡青铜
Yale lock 耶尔锁;普通圆柱式门锁
Y-alloy Y 合金
yamase 山背
Yamato metal 亚马托铅锡锑轴承合金
Yangshao culture pottery 仰韶文化陶器
Yangshao painted pottery 仰韶彩陶
Yangtze alligator 扬子鳄
Yangtze River 长江
Yangtze valley 长江流域
Yanisei polarity superchronzone 叶尼塞极性超时间带
Yankee dryer 杨基干燥箱
Yankee gutter 美国式天沟;英国大水槽;美国式檐沟
yankee paper machine 单光造纸机
Yankee ratchet screw driver 棘爪自动旋凿
Yankee screwdriver 美国式螺丝刀;美国式攻锥
Yanshan conjunct arc structure 燕山联合弧形构造
Yanshan cycle 燕山旋回
Yanshanian geosyncline 燕山期地槽
Yanshanian subcycle 燕山亚旋回
Yanshan mountain 燕山
Yanshan movement 燕山运动
Yanshan Mt platform fold belt 燕山台褶带
Yanshan orogeny 1st episode 燕山运动 1 幕
Yanshan orogeny 2nd episode 燕山运动 2 幕
Yanshan orogeny 3rd episode 燕山运动 3 幕
Yanshan orogeny 4th episode 燕山运动 4 幕
Yanshan orogeny 5th episode 燕山运动 5 幕
Yanshan orogeny episode A 燕山运动 A 幕
Yanshan orogeny episode B 燕山运动 B 幕
Yanshan period 燕山期
Yanshan platform folded belt 燕山台褶带
Yaozhou type 耀州窑系
Yaozhou ware 耀州窑器
yapp 卷边装订
yard 院子;英码;码;桁;前方堆场;桅横杆;围场;庭院;堆(置)场;车场;场院
yardage 立方码数;码数;平方码数;土方;尺码;按码计算
yardage clock 码分表
yardage distribution 土方分配
yardage indicator 航程指示器
yardage meter 土方计(英制)
yardage recorder 码数表
yardage table 土方分配;土方表
yard-and-stay system 联杆吊货法
yard-and-stay tackle 联杆吊滑车组
yardang 风蚀土脊
yardarm 横桅杆臂;横杆端;桁端;帆桁端;桅横杆
yardarm earing 横桅杆端耳索
yardarm group 货舱移动丛灯
yardarm iron 横端铁箍;桁端铁箍
yardarm to yard-arm 两船并列靠拢
yard bank 储纱库
yard becket 横桅安全环索
yard blue 储[贮]材蓝变
yard boom 舷外吊杆;大关【船】
yard capacity 调车场作业能力
yard carpenter 船厂木工
yard catch basin 庭院集水井;场地截沙雨水口;宅院渗水井;庭院雨水井;庭院截流井;场地落底雨水口
yard clerk 场地员;场地管理员
yard controller's tower 站调楼【铁】
yard craft 船厂工作艇
yard crane 移动油车;移动吊车;堆场起重机;堆场吊机;场内起重机;场地起重机
yard crane system 场地起重机装卸方式
yard crew 调车场人员
yard derrick 舷外吊杆;大关【船】
yard development cost 堆场建设费;场地建设费
yard drain 庭院排水沟;场地排水(沟)
yard drainage 庭院排水;场站排水;场院排水
yard drainage system 场地排水系统
yard drainage works 场地排水工程
yard dried lumber 场干木材;风干木材
yard engine 调车机车
yard equipment 堆场设备
yarder 木场曳引机;蒸汽集材机;集材绞盘机;码垛机
yard floating dock 船厂用浮船坞
yard floating drydock 船厂用浮船坞
yard gull(e)y 庭院带格栅下水道进口;场院排水口;有格栅的下水道进口;庭院雨水口
yard gutter 场院篦栅进水口;排水沟
yard handling 堆场作业
yard harbo(u)r 船厂码头
yard heart common 普通红心木
yarding 集居;堆置;库藏;进场作业;集材;堆放;动力集材
yard layout 车场布置
yard limit 调车场范围
yard limit board 调车场站界标
yard limit sign 车场地界标;车场地界线
yard locomotive 场用机车
yard lumber 商品木材;场堆木材;场存气干材;料场分类堆放的木材;锯剩木材;风干木材
yard machinery 堆场机械
yardman 车场工作人员;场地工作人员;调度员【铁】
yard management 堆场管理
yard manure 圈粪
yard masonry wall 宅院工墙
yardmaster 车场场长;调度长【铁】
yard measure 码尺;码
yard number 造船编号;调车场股道号;船舶建造序号;仓库数
yard of materials 材料场
yard oiler 船厂油船
yard operation 站场作业;堆场作业
yard pavement 宅院铺面
yard piping 场地管路
yard plan 地面建筑平面图;地面图;场地计划
yard planning 堆场调度计划
yard pound method 码磅度量衡法
yard repair ship 基地修理船
yard rope 吊桁索;帆桁索;桁索
yard rubbish 庭院垃圾;庭园垃圾;庭园废物
yard-service lighter 船厂驳船
yard sling 横桁吊链
yard slip 下横桁索
yard space 庭院空地
yard speed 调车场速度
yards per minute 码分
yard standing capacity 调车场容量
yard station 土方站
yardstick 丈尺;码尺;尺度(标准);标尺
yardstick compass 码尺圆规
yard system 舍饲制
yard tackle 底轮滑车组;桁端绞辘;底轮滑轮组
yard to yard 场到场
yard track 编组轨道;站线【铁】;调车轨道;调车场线路;车场线;编组站调车线
yard tractor 堆置场牵引车;堆场索引车;堆场牵引车
yard transportation expenses 厂内运输费
yard trap 进水口回水井;下水道进口截污井;进水口防臭设备;集水井存水隔间;场地水封井
yard truck 堆场搬运车
yard truss 桁架连接环
yard tug 船厂拖轮
yard vehicle 站场车辆【铁】
yard wagon-handling capacity 调车场办理车数能力
yard wand 码尺;直的码尺杆
yard waste 庭院废物;庭园垃圾;庭院废物
yard whip 吊货杆牵索
yard work 场内作业;站场作业
yard working 堆场作业
Yarkant block 叶尔羌地块
Yarkant-Eastern Dian block-zone 叶尔羌—滇东地块带
yarn 绳条
yarn and cloth quadrant 纱布扇形天平
yarn ballistic test 纱线冲击强力试验
yarn break detector 断丝检测器
yarn bundle cohesion 纱束包合性
yarn bundling machine 纱线打包机
yarn count balance 纱线支数秤
yarning iron 填隙捻缝凿
yarn meter 纱线长度测定器
yarn package dyeing machine 筒子纱染色机
yarn size 纱线支数
yarn solder 线状焊料
yarn speed meter 纱线测速仪
yarn storage device 储纱器
yarn swellant 麻纱胀堵漏剂(用于铅麻子管)
yarn tester 缆线试验机
yarn trap 捕纱器
yaroslavite 水氟铝钙矿
yarovization 春化处理
yarrow boiler 船用水管锅炉
yarrowite 雅硫铜矿
Yates' correction for continuity 耶茨连续性校正
yate tree 桉(木材具有弹韧性)
yavapaiite 斜钾铁矾
yaw 摇首;起泡沫
yaw acceleration 偏航加速表
yaw amplifier 偏荡波道放大器
yaw and pitch 偏航与俯仰
yaw angle 船头摆角
yaw-angle gyro 偏航摆首摇摆角陀螺仪
yaw-angle pickoff 偏航角传感器
yaw-angle sensor 偏航角传感器
yaw calibrator 偏航校正仪
yaw-checking anchor 止荡锚
yaw control 偏航操纵;船首偏控制
yaw control channel 偏控制通道
yaw damper 减摆器;偏阻尼器;偏航阻尼器
yaw detector 偏航传感器
yaw direction stability 航向稳定性
yawed body 偏航物体
yawer 偏航控制器;偏航操纵机构
yaw error 偏航误差
yaw freedom 偏航自由度
yaw frequency 偏航频率
yaw guy 系塔索;正向系索
yaw gyroscope 航向陀螺仪;偏航陀螺仪
yawhead 偏航传感器
yaw heel 摇首倾侧
yaw in bore 腔内偏转角
yaw indicator 偏航(角)指示器
yawing 偏转;偏航;偏荡;左右摇摆;偏航力矩;船摇;船头摆摇
yawing amplitude 偏荡幅度;首偏荡幅度

yawing angle 偏航角;偏荡角;船首摇摆角;摆艏角
yawing axis 偏航轴;偏转轴线
yawing balance 偏荡
yawing couple 船首摇摆力偶
yawing force 侧向(偏航)力
yawing instability 偏航不稳定性
yawing maneuverability 偏航机动性
yawing moment 偏航力矩;偏转力矩;偏荡力矩;盘旋力矩
yawing motion 偏航
yawing oscillation 扭转振荡
yawing rotation 绕垂直轴的转动;偏航运动
yaw instability 航向不稳定性;航向不稳定度
yawl 杂用船;舰载小艇
yaw line 系塔索
yaw loop 偏航回路
yaw maneuver 偏航动作
yawmeter 偏航指示器;偏航计;偏转仪;偏流计;偏荡计;测向计
yawn 呵欠;未填补裂缝
yawning moment 方向力矩
yaw-position 偏航姿态
yaw rate 偏航角速度
yaw rate control 偏航角速度控制;船首摇摆率控制
yaw rate gyroscope 偏航率陀螺仪
yaw-roll gyroscope 偏航滚转陀螺仪
yaw-roll maneuver 偏航滚转动作
yaw rudder indicator 方向舵指示器
yaw-sensing accelerograph 偏航加速度自记仪
yaw-sensing accelerometer 偏航(运动)加速表;船首摇摆加速度计
yaw sensitivity 偏航灵敏度
yaw simulator 偏航模拟机
yaw stability 航向稳定性;偏航安定性
yaw stiffness 方向稳定性
yaw synchro 偏航同步机
yaw synchronizer 偏航同步器
yaw vane 偏航翼
Y-axis Y 轴;纵坐标轴
Y-axis amplifier 垂直信号放大器
Y-axis length of markers 标志体 Y 轴长度
Y-azimuth 基准方向角;Y 方位角
Yazoo clay 亚祖黏土(美国,为一种膨胀性黏土)
Yazoo stream (支干流平行的)亚祖式河川
Y-bend 二叉;Y 形弯头;三通弯矩;叉形弯头;分叉弯头;分叉接头;分叉接管
Y-block 楔型试块
Y-box 星形连接电阻箱
Y-branch 三通支管;叉管;分叉(支)管;Y 形支管
Y-branch fitting Y 形三通;Y 形支管
Y-channel Y 信道;分叉河槽
Y-clean-out Y 形清扫设备;分叉清污口;分叉清扫;分叉清理
Y-connected 星形连接的;Y 形接法的
Y-connection 星形网络;星形连接;星形接法;叉形接头;分叉接头;分叉接管;Y 形管接(头);Y 形连接;Y 形接头;Y 形接法;星状连接
Y-coordinate Y 坐标;纵坐标
Y-crossing Y 形交叉口
Y-curve 叉形曲线;Y 形曲线
Y-datum line 穿孔卡片基准线
Y-delta starter 星形三角(形)起动器
Y-delta starting method 星形三角形起动法
Y-delta transformation 星形三角变换
Y-direction Y 轴方向
Y-displacement Y 轴方向位移
Y-divider 分水岔
year acquired 购置年份
year-around 全年的
year-bearer 年承
year book 年历;年鉴;年报;年刊
year-book of waterway work 航道工作年鉴
year built 制造年份
year class 年龄级(动物出生年)
year-climate 年气候
year clock 年钟
year displacement 年位移量
year-end 年终
year-end adjustment 年终调整
year-end audit 年终审计
year-end balance 年终结余
year-end balance sheet 年终资产负债表
year-end bonus 年终奖金
year-end capacity 年终生产能力

year-end closing 年终结账
year-end dividend 年终股息;年终股利
year ending 终了年度
year-end report 年终报告
year-end stock 年终库存
year-end summarization 年终汇总
year footage 年进尺
year increment 年增长量;年生长量
year installed 安装年代
yearling face 一周年割面
year-long navigable waterway 常年通航航道
year-long range 全年放牧地
year-long river 常年不涸河川;常年河流
year-long stream 常年不涸河川;常年河流
yearly 全年
yearly accumulated air temperature 年累积气温
yearly budget 年度预算
yearly capacity 年生产能力
yearly change 年变化
yearly consumption 年度消耗量
yearly correlation 年相关
yearly depreciation 年折旧率
yearly efficiency 年效率
yearly evapo(u)ration discharge 年蒸发量
yearly fish 当年鱼
yearly fluctuation 年涨落;年(升降)变化;年起伏;年波动
yearly goods transport plan 年度货物运输计划
yearly heat load 常年性热负荷
yearly hydrologic balance 年水文平衡
yearly income 年底收入
yearly increment 年度增加额
yearly inspection 年度检查
yearly installment 按年摊付
yearly load curve 年荷载曲线
yearly loaded curve 年负载曲线;年负荷曲线
yearly load factor 年荷载率;年负荷因子
yearly load variation 年荷载变化
yearly maintenance 年度养护;年度维修;年度维护;年度检修;年度保养;全年维修;岁修
yearly maintenance charges 年度维修费(用)
yearly maintenance cost 年度养护费
yearly man sunshine percentage 年均日照百分率
yearly maximum 年最大(值)
yearly maximum load 年最高负载
yearly mean air temperature 年平均气温
yearly mean runoff 年平均径流量
yearly mean sea-level 年平均海平面
yearly mean sediment concentration 年平均含沙量
yearly mean sediment discharge 年平均输沙量
yearly mean sunshine percentage 年总辐射量
yearly minimum 年最小(值)
yearly operating cost 年经营费
yearly output 年产(电)量;年产出
yearly overhaul 岁修
yearly payment 年费
yearly plan 年度计划
yearly plant 一年生植物
yearly pollution load 年污染负荷
yearly precipitation 年(降)雨量;年降水量
yearly progress report 进展情况年度报告
yearly rainfall 年降雨量;年降水量
yearly rate of aquifer dewatering 含水层年疏干率
yearly regulation 年调节
yearly renewable term 每年续保制;每年更新期
yearly ring (树木的)年轮
yearly runoff 年径流(量)
yearly sea-level 年平均海平面
yearly sediment discharge 年输沙量
yearly sediment transport 年输沙量
yearly silt discharge of river 河流年输沙量
yearly storage 年蓄水量;年蓄能量
yearly sunshine time 年日照时数
yearly taxation 按年计税
yearly test 年度检验
yearly transportation capacity 年运输能力
yearly trench 年度性趋向
yearly variation 年变量;年变化
yearly variation of runoff 径流年变化
yearly variation of shoal channel 浅滩河槽的年变化
yearly working hours 年工作小时数
yearly working volume 年度工作量

year of abundance 丰水年
year of account 结算年(度)
year of assessment 征收年度
year of average rainfall 平均雨量年份
year of completion 竣工年份
year of construction 建造年份
year of delivery 交付年代
year of dried spring 泉水干枯年份
year of drought 旱年
year of famine 灾年;荒年
year of grace 宽限年限
year of grace survey 长期检验年份
year of operation 运转年份
year of publication 出版年份
year purchase method 按年采购法
year ring (树木的)年轮
year-round 全年的
year round employment 常年雇佣
year-round fishery 全年渔业
year-round grazing 终年放牧
year-round heating load 常年性热负荷
year-round pours (混凝土)全年浇筑量;整年浇筑
year-round surface 全天候通车路面
year-round type 全年通用式
years 年代
years-digit depreciation method 使用年限积点折旧法
years in return of capital investment 投资回收年限
years of observation data 观测资料积累年数
years of operation 运转年份
years of schooling completed 学习年数
years of successive observation 连续观测的年数
year temperature difference 年温(度)差
year-to-year 多年的
year-to-year change 逐年变化
year-to-year comparison of financial statement 财务报表逐年比较
year-to-year pressure difference 年际气压差
year-to-year reservoir 多年调节水库
year-to-year storage 多年调节库容
year-to-year tenancy 逐年租赁
year-to-year variation 年年变化
year with abundance of water 丰水年
year with low water 枯水年
yeast 发酵
yeast-containing wastewater 含酵母废水
yeast-extractor agar 酵母抽提琼脂
yeast industry 酵母工业
yeatmanite 硅锑锌锰矿
yeddo raphiolepis 伞状石班木
yedinite 氯铅铬矿
yeild 捕获量
yelloe lamp 黄灯
yellow 黄色的
yellow acaroides 黄禾木胶
yellow book 黄皮书
yellow brass 黄铜;铜锌合金
yellow brass alloy 黄铜合金
yellow brick 黄砖
yellow-brown 黄棕(色)
yellow-brown earth 黄棕壤
yellow-brown forest earth 黄棕色森林土
yellow-brown forest soil 黄棕色森林土
yellow buoy 检疫浮筒;检疫浮标
yellow cadmium fringe on teeth 牙黄色环;镉环
yellow cake 黄饼(铀浓缩物的一种)
yellow cedar 黄杉木;黄桧【植】;杉木材
yellow chrome 铬黄
yellow cinnamon soil area 黄褐土区
yellow clay 黄(色)黏土
yellow clunamon soil 黄褐土
yellow coal 塔斯曼油页岩
yellow concrete 黄色混凝土
yellow croaker 黄鱼
yellow cypness 阿拉斯加扁柏
yellow deal 波罗的海红木;黄松板;松板
yellow dog contract 不入工会契约
yellow dwarfism 黄矮病
yellow dyes 黄色染料
yellow earth 褚黄土;黄壤
yellow earth area 黄壤区
yellowed rice toxication 黄变米中毒
yellow enamel 黄瓷漆
yellow ferralsol 黄色铁钴土

yellow ferric oxide 铁黄
yellow fever 黄热病
yellow fever vaccine 黄热病疫苗
yellow filter 黄色滤色镜;黄色滤光器
yellow fir 黄枞木
yellow flag 检疫旗
yellow flame 黄焰
yellow forest soil 黄色森林土
yellow glass bulb 黄色玻璃灯泡
yellow glaze 黄釉
yellow glaze vase with white floral design 黄釉白花瓶
yellow gold 金银铜合金
yellow grease 黄牛油
yellow-green 黄绿色
yellow-green algae 黄藻类;黄绿藻
yellow-green filter 黄绿滤光镜
yellow-green fluorescence 黄绿荧光
yellow-grey 黄灰色
yellow ground 黄泥带;黄地
yellow gum 桉树(澳大利亚)
yellow heat 黄热
yellow index 黄度指数
yellow indicator lamp 黄色指示灯
yellowing 黄变;泛黄;返黄;发黄;变黄
yellowing on ag(e)ing 老化变黄
yellow iron oxide 黄色氧化铁;氧化铁黄;铁黄
yellowish 浅黄色的;淡黄色的
yellowish-brown 黄棕(色);黄褐色
yellowish eosin 黄色曙红
yellowish green 黄相绿;黄光绿
yellowish litharge 黄相黄丹
yellowish-orange 黄橙色
yellowish pea green glaze 豆青釉
yellowish pink 肉色;肉红色
yellow jack 检疫旗
yellow jewel tea set 宝石黄茶具
yellow lake 黄色色淀
yellow layer 黄色层
yellow lead ore 钼铅矿
yellow lead(oxide) 黄丹;一氧化铅;黄色氧化铅;氧化铅;铅黄
yellow light 黄色灯光
yellow limonite 天然氧化铁黄
yellow line 黄色线
yellow lithopone 黄立德粉
yellow mercuric oxide 黄色氧化汞
yellow mercury oxide 黄色氧化汞
yellow metal 黄金;黄铜
yellow mineral oil 黄矿油
yellow-mud 黄泥(浆)
yellowness 黄色
yellowness index 泛黄指数
yellow oak 栎树;黄栎
yellow ocher 赭土;黄赭石;铁黄颜料;赭石黄;赭石;赫黄(土)
yellow oil silk 黄油绸
yellow organic pigment 黄色有机颜料
yellow orthoclas 黄色正长石
yellow oxide 铁黄
yellow paint 黄色油漆;黄漆
yellow phenolic enamel 黄色酚醛瓷漆
yellow phosphorus 黄磷;白磷
yellow phosphorus match 黄磷火柴
yellow pigment 黄色颜料;黄色素
yellow pine 黄松;长叶松(木)
yellow pine oil 黄松油
yellow podzolic soil 黄色灰化土
yellow poplar 鹅掌楸木(美国)
yellow porcelain 黄瓷
yellow precipitate 氧化汞
yellow prussiate 亚铁氰化钾;黄血盐
yellow prussiate of soda 亚铁氰化钠
yellow prussiate soda 黄血盐钠
yellow pulp 稻草纸浆
yellow pyrite 黄铜矿
yellow quartz 黄(水)晶
yellow-red 黄红色
yellow resin 黄树脂
Yellow River 黄河
yellow sand 黄砂;黄沙
yellow sandal 黄檀木
yellow sandalwood 黄檀
yellow sapphire 金黄宝石

Yellow Sea datum 黄海基面
Yellow Sea mean sea level 黄海平均海平面
Yellow Sea warn current 黄海暖流
yellow shellac 黄虫胶片
yellow slate 黄色板岩
yellow slow board 慢行牌
yellow snow 黄雪
yellow soda ash 火碱;碳酸钠;苏打
yellow soil 黄色土(壤);黄壤
yellow soil relief 黄土地貌
yellow spot 黄斑
yellow spotting 黄色斑点
yellow-stain 黄变
yellow stone 黄石
Yellow Stone National Park 黄石公园(美国)
yellow straw pulp 稻草纸浆
yellow stringbark 黄棕色桉木
yellow tellurium 针碲金银矿
yellow tubercle 黄色结核结节
yellow ultramarine 佛青
yellow varnished cambric cloth 黄漆布
yellow varnished cambric tape 黄漆细麻布
yellow varnished insulating cloth 黄蜡布
yellow ware 淡黄色陶瓷器(皿)
yellow wash 黄色洗剂
yellow wax 黄蜡
yellow wind 黄土风
yellow wood 非洲罗汉松;美洲香槐(木);黄色木材;黄桑;黄木
yellowy 带黄色的;淡黄色的;黄色的
yelm 茅草捆;茅草屋顶
yelven 茅草捆;茅草屋顶
yen 日元
Yen base 日元基准
Yenbond 日元债券
Y-engine 三缸星形发动机
Yerkes classification 耶基斯光谱分类法
Yerkes objective 耶克物镜;耶基斯物镜
Yerkes refractor 耶基斯折射望远镜
yermosol 漠境土
yeso 耶索石膏(南美洲刷墙用)
yew 紫杉;水松
yew podocarpus 罗汉松
yezo spruce 鱼鳞松
Y-fitting Y形管接头;叉形管件
Y-frame 叉形架
yftisite 氟硅钛钇石
Y-grade separation Y形立体交叉;叉形立体交叉
Y-grid Y格网线
yield 屈服;让渡;收率;收获率;收获量;实际获利;得率;出成率;成品率;产量
yieldability 可屈服性;沉陷性
yieldable arch 可缩性拱形支架;可变式拱架
yieldable arch steel set 可缩性拱形钢支架
yieldable arch support 可缩性支架
yieldable lining 可缩性衬砌
yieldable steel arch 镶压性金属地架
yieldable support 伸缩式支护
yield ahead sign 前面让路标志
yield band 屈服带
yield behavio(u)r 屈服性能
yield capacity 生产力
yield class indication 产量等级指标
yield coefficient 产率系数
yield components 产量成分
yield compression strength 压缩时屈服强度;屈服压力强度
yield condition 屈服状况;屈服状态;屈服条件(试验材料的弹性变度等);塑性能量条件
yield criterion 屈服准则
yield curve 屈服曲线;屈服界线;收益曲线
yield density effect 产量密度效应
yield elongation in tension 屈服伸长
yield factor 出水率;屈服极限安全系数
yield factor of a catchment 回水面积出水率
yield factor of safety 屈服安全系数
yield failure 屈服破坏
yield flow 塑性流动
yield function 屈服函数
yield gap 普通股与优先股收益差额
yield good economic returns 经济效益好
yield guard 吊挂卫板
yield hinge 屈服铰
yield index 产量指标

yielding 屈服;易变形的;塑性变形;产生
yielding ability 丰产性
yielding acceleration 屈服加速度
yielding and inelastic behavior 可压缩性和非弹性
yielding arch 镶压性拱形支架;可缩性拱形支架;让压性拱形支架
yielding arch steel set 可缩性拱形钢支架
yielding attachment 发条外钩
yielding capacity of a well 井的出水量
yielding content of a well 井的出水量
yielding curvature 屈服曲率
yielding deflection 屈服挠度
yielding deformation 屈服变形
yielding flow 总径流(量);塑流
yielding foundation 易沉陷基础
yielding ground 易沉陷地面;易沉陷(的软)土;松散土;松软土地;松软土(壤)
yielding intensity 屈服强度
yielding level coefficient 屈服程度系数
yielding locus 屈服位置
yielding material 可缩性材料
yielding mechanism 屈服机构
yielding of crystals 晶体形成
yielding of foundation 基础沉陷;基础沉降
yielding of reinforcement 钢筋屈服
yielding of springing support 拱脚支撑松弛
yielding of structure 结构屈服
yielding of supports 支座下沉;支座沉陷;支点沉陷
yielding pattern 屈服类型
yielding point 屈服点
yielding point elongation 屈服伸长点
yielding point of reinforcement 钢筋屈服点
yielding roadway arch 可塑性拱;柔性巷道拱构件;柔性拱
yielding rubber 减振橡皮;缓冲橡皮
yielding seat 下沉支座;弹性支座;沉陷支座
yielding soil 流动土;软土;松软土(壤)
yielding spreading 屈服扩散
yielding steel arch 镶压性金属支架;让压性金属支架
yielding steel prop 让压可缩性钢支柱
yielding stiffness ratio 屈服刚度比
yielding strength 屈服强度
yielding strength ratio 屈服强度比
yielding structure 屈服结构
yielding support 可变位的支承;下沉支座;可缩性支撑
yielding-type support 变形式支承
yielding under pressure 压力下出水量;压力下产出量
yielding water 出水
yield intensity 屈服强度
yield limit 流限;流动性范围;屈服(极)限;屈服点(流限);产量极限
yield line 屈服线;塑性变形线
yield line method 屈服线法
yield line theory 破坏阶段理论;屈服线理论
yield load(ing) 屈服荷载
yield locus 屈服轨迹
yield loss 收得率损失
yield method 内部收益率计算法;(运输经济)产出法
yield modulus 屈服模量
yield moment 屈服弯矩;屈服(力)矩
yield net 净产出
yield of base 基底屈服点
yield of capital 资本收益
yield of clay 黏土造浆率
yield of clinker 熟料产量
yield of coal 煤产量
yield of concrete 混凝土(出)产量
yield of counter 计数器效率
yield of deposit 矿床产量;存款收益
yield of drainage basin 流域产水量
yield of glass sizing 玻璃切裁率
yield of grasses 产草量
yield of groundwater 地下水供水量;地下水出水量
yield of lime 石灰产浆量
yield of metal 金属屈服
yield of real capital 实际资本收益率
yield of spring 弹簧的屈服点;泉的涌水量;泉水流量
yield of steel 钢的屈服点
yield of water 出水量
yield of well 井的出水量

yield of well group 井组出水量
yield of well-point 井点(稳定)出水量
yield performance 产量表现
yield per ha 每公顷产量
yield per pass 每次收率
yield pick-up swap 收益掉期法
yield-pillar system 让压矿柱法
yield point 拐点;流动点;降伏点;击穿点;软化点;屈服点
yield point elongation 屈服平台;屈服点延伸;屈服点伸长
yield point force 屈服点力
yield point jog 屈服平台
yield point load 屈服点荷载
yield point of steel bar 钢筋屈服点
yield point strain 屈服点应变;屈服点变形
yield point strength 屈服点强度
yield point stress 屈服点应力
yield point value 屈服值;流动值;塑变值
yield point value of displacement 位移屈服点值
yield polymer films 成膜率
yield possession 让与所有权;让出所有权
yield power 生产力;(电厂)产电量
yield prediction 产量预测
yield pressure 屈服压力
yield printed board 产量印刷电路板
yield range 流限
yield rate 屈服率;投资实得率;收益率;实际获利率
yield ratio 屈强比;屈服比
yield region 屈服区
yield regulation 收获量调节
yield response 产量反应
yield returns 利润收益
yield safety of solid-web composite structures 实腹复合结构的屈服点安全
yield sample plot 收获表标准地
yield sign strain 屈服点应变
yields point 软化点;屈服极限;屈服点;流限;击穿点;拐点
yield strain 屈服应变
yield strength 屈服强度;软化强度;屈变力
yield stress 屈曲应力;屈服应力
yield stress controlled bonding 屈服应力黏结 (在高于屈服点应力的压力作用下进)
yield stress model 屈服应力模型
yield stress of flow 流动屈服应力
yield-stress ratio 屈服应力比
yield support 屈服支承
yield surface 屈服(曲)面
yield-survey 产量调查
yield table 收获表
yield temperature 流动温度;屈服温度;安全塞温度;易熔塞变形温度
yield test 屈服试验
yield theory of small scope 小范围屈服理论
yield timbering 让压支柱
yield-time diagram 塑性变形时间图;下沉时间图
yield to adjusted minimum maturity 调整后最早到期日收益率
yield to average life 债券平均期限收益率;平均年限收益率
yield to call 通知收益率
yield to crash 破产收益率
yield to maturity 全期收益率;全期获利率
yield to put 出售收益率
yield to worst 最差收益率
yield trial 产量试验
yield unit 屈服单位
yield value 屈服极限;降伏值;屈服值;起始值;塑变值
yield value of stress 屈服应力值
yield valve 屈服阀
yield variance 收益率差异
yield zone 屈服区
yig filter 钇铁石榴石滤波器
Yilgran nucleus 伊尔加恩陆核
Yilgran old land 伊尔加因古陆
yimengite 沂蒙矿
Ying bowl 英碗
Yingch'uang 膺窗
yinshanite 阴山石
Y-intercept Y 形截距
Y-intersection Y 形交叉口
yip-top table 折叠桌

Y-joint 叉形接头;分叉管接(头);Y 形接头;Y 形接法
Y-junction 三通管接头;Y 形交叉;Y 形道路枢纽;Y 形连接点;星形连接;星形交叉
ylan-ylan oil 衣兰油
Y-level Y 型水准仪;活镜水准仪
Y-line Y 轴线
Y-matching Y 形匹配
Y-member 叉形杆件
Y-motion 沿 Y 轴方向运动
Y-movement Y 轴方向移动
yoderioite 紫硅镁铝石
yodowall 搪瓷面冷轧钢板
Y-offset Y 轴方向偏移
yoforterite 锰坡缕石
yogoite 等辉正长岩
Yohen spot 曜斑
yohen Tenmoku 曜变天目釉
yoke 门上槛;横舵柄;偏转线圈;铁芯;舵柄横木;磁头组;磁轭;窗头板;叉臂;扼铁
yoke assembly 换挡叉;离合器分叉叉
yoke bolt 系铁螺钉
yoke cam 定幅凸轮;等直径径向凸轮
yoke cap 托架上盖
yoke clamping frame 固定架
yoke connect 轭连接
yoke cup nut 杯形压紧螺母
yoke current 偏转线圈电流
yoked basin 配合地槽
yoked connecting rod 叉形连杆
yoke deflection coil 扼形偏转线圈
yoke end 杠杆叉
yoke for steady arm 定位器联板
yoke joint 叉形连接
yoke lanyard 舵柄绳
yoke lever 叉形杆件;叉杆
yoke line 舵柄绳;舵柄操舵索
yoke magnetizing method (磁粉探伤的)极间法;磁轭法
yoke method 磁轭法
yoke of mast 下桅帽
yoke of the magnet 磁轭
yoke pass 平顶山口
yoke permeameter 框式磁导计
yoke piece 轭架
yoke pin 叉头销;轭销
yoke plate 联板
yoke ring 磁轭圈;轭环
yoke riveter 叉架铆钉机
yoke rope 操舵索;舵柄操舵索
yoke shifter 齿轮拨叉
yoke trunnion 叉形十字头
yoke vent 轭状气管;叉形排泄口;结合通气管;轭式通气管
yoke venting 轭管通气
yoke vent pipe 轭管通气管;结合通气管
yokkaichi asthma 四日市哮喘
Yokohama fender 横滨式充气橡胶护舷
Yokohama Port 横滨港(日本)
yoldia clay 夹有小贝壳黏土
yolk 蛋黄
yolk coal 松软煤;不结焦煤
Yoloy 铜镍低合金高强度钢
Yorcalbro 约卡尔布劳牌合金(一种制管用铝铜合金)
Yorcalnic 铝镍青铜
Yorcwyte 约克怀特牌金属(一种制管用白色金属)
York-Scheibel column 赛贝尔萃取器
Yorkshire bond 每层两顺一丁砌法;跳丁砖砌合;跳丁砖砌法
Yorkshire light 横扯窗;约窗
York stone 约克郡砂岩
yorky 曲裂石板;曲线劈裂的石板
yosemitite 淡色花岗岩
Yosenmite 约森米克牌屋面卷材(商品名)
yoshimuraite 硅钛锰钡石
youg 尤格风
young alluvial soil 新冲积土
young blow 欠吹;吹炼不足
young clay 新(近)沉积黏土
young clearing virgin soil 新垦荒地
young coast 幼年海岸
young coastal ice 初期岸冰
young crops 青苗
young crops compensating fee 青苗补偿费

Young Crops Law 青苗法
young ebb 潮水初落
younger generation 下一代
youngest Dryas stade 晚得利亚斯冰阶
young fault scarp 年轻断层崖
young flood 涨潮初期;初(涨)潮;初汛
young folded belt 幼褶皱带
young folding zone 年青褶皱带
young growth 幼龄林
young ice 新冰;初期冰;初冰
young industrial country 新兴工业国家;发展中国家
young iron 糊状搅炼铁
young karst 幼年期喀斯特;幼年(期)岩溶
young land 幼年地形
young loess 新黄土
young marine soil 幼年海成土
young mountain 幼年山;年青山地
young of the year 当年鱼
young peat 年青泥炭
young period 青年期
young plain 幼年平原
Young plan 扬格计划
young plant 秧苗;树苗
young platform 年青地台
young platform stage 年青地台阶段
young rift 年青裂谷
young river 初成河道;幼年河(流)
Young's coefficient 杨氏系数
Young's construction 杨氏作图法
Young's diffraction theory 杨氏衍射理论
Young's double slit 杨氏双狭缝
Young's double slit experiment 杨氏双缝实验
young sea 初生波;初浪
Young's eriometer 杨氏测微径计
Young's experiment 杨氏实验
young shore ice 初期岸冰
young shoreline 幼年海岸线;幼年滨线
Young's interfenrece fringes 杨氏干涉条纹
Young's modulus 杨氏模数;杨氏模量;杨氏棱镜;弹性模数;弹性模量
Young's modulus in flexure 杨氏挠曲模量
Young's modulus of elasticity 杨氏弹性模量
Young's modulus of rail 钢轨弹性模量
young snow 新降雪;初雪
young soil 幼年土
Young's prism 杨氏棱镜
Young's projection with total area true 杨氏全区域等积投影
Young's spectrograph 杨氏摄谱仪
young stage 幼年期
young stage of erosion cycle 幼年期侵蚀循环
young star 年轻恒星
young stream 幼年河(流)
Young's two-slit interference 杨氏双缝隙干涉
Young's twoslit interferometer 杨氏双缝干涉仪
young tide 初潮
young topography 幼年地形
young valley 幼年河谷;幼年谷
young wave 初生波
your cable 贵方电报
your telex 贵方电传
youth center 青年中心
youthful stream 青年河
youthful topography 幼年地形
youthful valley 青年期河谷;青春期河谷
youth hostel 青年招待所
youth hotel 青年旅店
youth stage 幼年期
youth-stage landform 幼年期地貌
yo-yo 双联推土机(俚语);前后索连推土机(俚语);运输机装车溜槽
Y-parallax 上下视差;Y 视差
Y-pattern design Y 形设计
yperite 芥子气
Y-piece Y 形肘管;叉形肘管;叉形件
Y-pipe 斜叉三通;三通管;三叉管;叉形管;分叉管;Y 形管
Y-plan Y 形平面
yrast state 晕转态
yrneh 利亨
Y-scale 主纵线比例尺
Y-section 三通管接头;三角分线杆;Y 形截面;Y 形接头;Y 形接法

Y-shaped 分叉形;Y 形的
Y-shaped bend Y 形支管
Y-shaped building 星形建筑;Y 形砌块
Y-shaped design Y 型设计
Y-shaped network 星形电路
Y-shaped valley Y 形谷
Y-shaped valve Y 形阀
Y signal 亮度信号
Y-strut Y 形支柱
Y-tile 三通瓦管;三通陶管;叉形陶管
Y-tilt 纵向倾斜
ytong 轻质混凝土;多孔混凝土
Y-tong block Y 通轻质多孔砌块
Y-track 折返三角线;转向三角线;三叉形轨道;分叉线;Y 形轨;Y 形岔线
ytterbium gallium garnet 镱镓石榴石
ytterbium glass laser 镱玻璃激光器
ytterbium nitride 氮化镱
ytterbium ores 镱矿
ytterbium pentetate 放射性喷替酸镱
ytterbium silicide 硅化镱
yttrialite 硅钍钇矿
yttrium aluminitum garnet 钇铝石榴石
yttrium boride 硼化钇
yttrium carbide 碳化钇
yttrium ferrite 钇铁氧体

yttrium gallium garnet 钇镓石榴石
yttrium garnet 钇石榴石
yttrium iron garnet 钇铁石榴石
yttrium iron garnet material 钇铁石榴石材料
yttrium nitride 氮化钇
yttrium nodular iron 钇球墨铸铁
yttrium ores 钇矿
yttrium oxide 氧化钇
yttrium silicate 硅酸钇
yttrium silicide 硅化钇
yttrium vanadate crystal 钒酸钇晶体
yttroalumite 钇铝石
yttrocerite 铈钇矿
yttrocolumbite 钇铌铁矿
yttrocrasite 钛钇钍矿
yttrogarnet 钇榴石
yttropyrochlore 钇铀烧绿石
yttrotantalite 钇钽铁矿;钽钇矿
yttrotungstite 钇钨华
Y-tube 斜角支管;三通管;叉形管;Y 形管
Y-type 叉形
Y-type arch 分叉碹
Y-type branch Y 形分叉管
Y-type gasket Y 形密封条;Y 形衬垫;Y 形密封垫
Y-type pipe 叉管
Yucatan Current 尤卡坦海流

yucca 丝兰花
Yudatan channel 尤卡坦海峡
yugawaralite 条沸石;汤河原石
Yugoslav architecture 南斯拉夫建筑
Yugoslavian architecture 南斯拉夫建筑
Yugoslavian Ship Classification Society 南斯拉夫船级社
yuju oil 油树油
Yukin-Tanana upland terrane 育空—坦那纳高地地体
yukonite 英闪细晶岩;水砷钙铁石;水砷钙铁矿
yuksporite 针碱钙石
yulan magnolia 玉兰
yuloh 摇橹
Yunnan epsilon tectonic system 云南山字形构造体系
Yunnan hemlock 云南铁杉
yurt(a) 圆顶帐篷;毡包;蒙古包
yurt-type earth house 蒙古包式土房
Y-voltage 相电压;Y 电压
Yvon photometer 伊冯光度计
Y-wing 叉形翼;Y 形翼
Y-Y connection 双星形接法;Y-Y 形接线;双星型接法
Y-Z interchange knob Y-Z 变换钮
Y-zone Y 带

Z

Z-absolute counter Z 绝缘计数器
ZAB suspension preheater ZAB 悬浮预热器
zaccab 白泥石灰浆
zacco 柱石;座石
zaffer 钴蓝釉;钴蓝色料;钴焙砂
zaguan (西班牙建筑的)门厅;(西班牙建筑的)前厅
zaherite 水羟铝矾
Zahn cup 察恩杯
Zahnup 察恩黏度杯
Zahn viscosimeter 锥盘黏度计
zaibatsu 财阀
zaibatsu combine 财阀联合企业
zakharovite 札哈罗夫石
zala 硼砂
Zaldecide 扎尔德赛德牌杀虫剂(一种木材杀虫剂)
Z-alloy 铝轴承合金
Zamak 扎马克压铸锌合金
Zambia nucleus 赞比亚陆核
Zamium 扎密阿姆镍铬合金
Zam metal 电动机电枢用合金
Z-angle Z 形钢;Z 形角铁
Zang ware 藏器
zantac 甲胺呋硫
Zanzibar gum 桑给巴尔树胶
zapatalite 水磷铝铜石
Zapata scorpion structure 扎帕塔式蝎尾型结构(钻井台架)
Zapata tripod structure 扎帕塔式三脚结构
zapon 硝化纤维清漆;硝基清漆
zapon enamel 硝化纤维素瓷漆
zapon lacquer 硝酸纤维素漆;硝基清漆;硝化纤维清漆;透明漆
zapon varnish 硝化纤维清漆;硝酸纤维素清漆;硝基清漆
Zapotec architecture 萨波特克建筑(中美洲)
zaratite 翠镍矿
Zarchariasen's rule 查哈里阿森规则
Zarchariasen's theory 查哈里阿森学说
Zarden sprinkler 庭园洒水器
zare(e)ba 防御栅(苏丹人用于防御);栅塞;木栅;篱笆;防御工事营地;刺栅
Zargros geosyncline 扎格罗斯地槽
Zariba 宰里拜圈地(位于东北非)
zastruga 雪面波纹;雪波
Zat 扎特牌防水乳状液(商品名)
zavaritskite 氟氧铋矿
Zavaritsky method 扎瓦里茨基法
zawn 崖上砂洞
zax 石工用斧子;石斧
Z-axis Z 轴
Z-axis length of markers 标志体 Z 轴长度
Z axis modulation Z 轴调制
Z-axis propeller 全回转螺旋桨
Z-bar Z 形格钢;Z 字钢;Z 形钢
Z-bar column Z 形钢柱
Z-bar column with covers 覆盖有翼缘板的 Z 形型钢柱
Z-beam Z 字钢;Z 形梁
Z-bit Z 形钻头
Z-blade mixer 曲拐式搅拌机;Z 形刮板式混料机
Z-buffer geometry Z 缓冲区几何
Z-buffer projection Z 缓冲区投影
Z-buffer rendering Z 缓冲区绘制
Z-connection 曲折接法
Z-corbel Z 形托臂
Z-crank Z 形曲柄
Z-direction Z 轴方向
Z-drill bit Z 形钻头
zeal 积极性
Zea moys 玉蜀黍
zebra 有斑马纹的
zebra crossing 斑马线人行横道;马路人行横道;人行横道;斑马线;斑马横道
zebra(crossing) method 斑马法
zebra dolomite 条带状白云岩
zebra instrument 条纹仪;斑马仪
zebra layering 韵律层
zebra lines 斑马线

zebra marking 斑马线;斑马横道
zebrano 斑纹木
zebra roof 联合式屋顶;斑马状炉顶
zebras 斑马纹
zebra tube 齐勃拉管;斑纹彩色显像管
zebrawood 斑马纹木(产于西非);斑纹木
Zechstein 蔡希斯坦统【地】;镁灰岩
zed Z 形钢;Z 形铁钎
Zedoary oil 蓬莪术油
zed purlin(e)Z 形檩条
zee Z 形截面构件;Z 形钢
zee bar Z 形钢
zee iron rod Z 形铁钎
Zeeman broadening 塞曼(谱线)加宽
Zeeman coherence 塞曼相干性
Zeeman component 塞曼子线
Zeeman displacement 塞曼位移
Zeeman effect 塞曼效应
Zeeman effect correction 塞曼效应校正
Zeeman effect frequency stabilization 塞曼效应稳频
Zeeman energy 塞曼能量
Zeeman laser 塞曼效应激光器
Zeeman level 塞曼能级
Zeeman phenomenon 塞曼现象
Zeeman shift 塞曼效应位移
Zeeman slit 塞曼缝隙
Zeeman spectrogram 塞曼图谱;塞曼光谱片
Zeeman split of magnetic 磁的塞曼分裂
Zeeman-splitted level 塞曼效应分裂能级
Zeeman splitting 塞曼分裂;塞曼分离
Zeeman state 塞曼态
Zeeman-tuned laser 塞曼调谐激光器
zee purlin(e)Z 形檩条
zee-type piling Z 形板桩
zein fiber 玉米纤维;玉米蛋白
Zeisel's method 蔡泽尔法
Zeiss-Abbe apertometer 蔡斯—阿贝数值孔径计
Zeiss aspheric grinding machine 蔡斯非球面磨床
Zeiss convertible Protar 蔡斯可换普劳夫特镜头
Zeiss-Endter particle-size analyser 蔡斯—恩特粒度分析仪
Zeiss ga(u)ge block interferometer 蔡斯块规干涉仪
Zeissig green 蔡齐绿
Zeiss indicator 蔡斯杠杆式测微头;蔡斯导程检查仪
Zeiss konimeter 蔡斯灰尘计
Zeiss lead tester 蔡斯螺距(导程)检查仪
Zeiss lens 蔡斯透镜
Zeiss macrohardness tester 蔡斯显微硬度试验计
Zeiss objective 蔡斯物镜
Zeiss optimeter 蔡斯光学比较仪
Zeiss orthometer lens 蔡斯奥杜尔美泰镜头
Zeiss orthotest 蔡斯奥托比较仪
Zeiss parallelogram 蔡斯平行四边形
Zeiss particle-size analyzer 蔡斯粒径分析仪
Zeiss planetarium 蔡斯天象仪
Zeiss sonnar 蔡斯索那镜头
Zeiss universal theodolite 蔡斯全能经纬仪
zeitgeber 给时者
zektzerite 硅锆钠锂石
Zelco 泽尔科锌铝合金
zelder's gauntlets 焊工的长手套
zellerite 碳钙铀矿
zelling 零长导轨发射
zemannite 水碲锌矿
Zener barrier 齐纳阻挡层;齐纳防爆栅
Zener breakdown 齐纳击穿
Zener diode 齐纳二极管;稳压二极管
Zener effect 齐纳效应
Zener efficiency 齐纳效应
Zener impedance 齐纳阻抗
Zener model 齐纳模型
Zener region 雪崩区;齐纳区
Zener tunnel(l)ing 齐纳隧道效应
zenith 天顶;顶点
zenith absorption 天顶大气吸收
zenithal 天顶的

zenithal angle diagram 天顶角图
zenithal angle measurement 顶角测量
zenithal chart 天顶投影海图;方位投影(地)图
zenithal equal-area projection 天顶等积投影
zenithal equidistant projection 天顶等距投影;等距(离)天顶投影;方位等距投影
zenithal hourly rate 天顶流星出现率
zenithal map 方位投影(地)图
zenithal projection 方位投影(法);天顶投影(法)
zenithal rain 天顶季雨
zenithal refraction 天顶大气折射
zenith angle 天顶角
zenith-angle distribution 天顶角分布
zenith-angle effect 天顶角效应
zenith-angle variation (宇宙射线强度的)天顶角变化
zenith astrograph 天顶照相仪
zenith attraction 天顶引力;天顶吸引
zenith blue 天顶蓝(淡紫光蓝色)
zenith camera 天顶摄影机
zenith discontinuity 天顶间断
zenith distance 天顶距;顶距
zenith distance measurement 天顶距测量
zenith eyepiece 天顶目镜
zenith hourly rate 天顶每时出现率
zenith instrument 天顶仪
zenith magnitude 天顶星等
zenith measurement 天顶距观测
zenith photography 天顶摄影
zenith plummet 顶点对点器
zenith point 天顶点
zenith position 天顶位置
zenith prism 天顶棱镜
zenith pump 天顶泵
zenith reading 天顶读数
zenith sector 天顶扇形仪;地平纬仪
zenith star 天顶星
zenith sun 天顶太阳
zenith telescope 天顶仪;天顶望远镜
zenith tube 天顶仪;天顶筒
zenocentric coordinate 木心坐标
zenographic coordinate 木面坐标
zenographic latitude 木面纬度
zenographic longitude 木面经度
zenography 木面学
zenolite 硅线石
Zeo-Dur 硅酸盐阳离子交换剂
Zeo-Karb 磺化煤阳离子交换剂
zeolite 泡沸石;沸石
zeolite biological aerated filter 沸石曝气生物滤池
zeolite catalysis 沸石催化
zeolite catalyst 沸石催化剂
zeolite cement 沸石水泥;沸石胶结物
zeolite content 沸石含量
zeolite ester 沸石酯
zeolite exchanger 沸石软水交换器
zeolite facies 沸石相
zeolite filter 沸石(软水)滤池
zeolite filtration 沸石过滤
zeolite filtration column 沸石滤柱
zeolite-metal catalyst 沸石—金属催化剂
zeolite molecular sieve 沸石分子筛
zeolite molecular sieve membrane separating technology 沸石分子筛膜分离技术
zeolite process 沸石软水法;沸石化作用;沸石(处理)法
zeolites 沸石类
zeolite softener 沸石软水剂;沸石软化器
zeolite softening 沸石软水法;沸石软化
zeolite sorption pump 沸石吸附泵
zeolite tuff 沸石凝灰岩
zeolite-vermiculite biological aerated filter 沸石—蛭石曝气生物滤池
zeolite water 沸石水
zeolite water softener 沸石软水剂
zeolite water softening 沸石软水法;沸石软化法
zeolite water softening tank 沸石软水池
zeolitic clay 沸石质黏土

zeolitic ore deposit 含沸石矿床
zeolitic rock 沸石质岩
zeolitic water 沸石水
zeolitiform 沸石状的
zeolitization 沸石化(作用)
zeolitized fly-ash 粉煤灰沸石
Zeolox 分析纯极分子筛
zeophyllite 叶羟硅钙石;叶沸石
zeotropic system 非共沸体系
zeotropy 非共性
zepharovichite 银星石
zephiran 苯甲烃铵
zephyr 和风
Zepp antenna 泽普天线
Zeppelin alloy 齐柏林合金
Zeppelin antenna 齐柏林天线
Zerener process 间接作用碳弧焊
Zerener welding 间接作用碳弧焊
zerk 加油嘴;加油咀
zermattite 叶蛇纹石
Zernike's image-forming process 查涅克成像法
Zernike's phase contrast method 查涅克相衬法
Zernike's test 查涅克检验
Zernike's theorem 查涅克定理
zero 原点;零值;零位;零度
zero acceptance level 零点验收标准
zero access 高速存取;立即访问;立即存取
zero access addition 立即取数加;立即存取加法;超高速存取加法
zero access instruction 立即存取指令
zero access memory 立即访问存储器;立即存取存储器
zero access storage 零存储时间存储器;立即存取存储器
zero activity 零活度
zero address 无地址
zero address code 零地址码;无地址码
zero address computer 无地址计算机
zero address instruction 零地址指令
zero address instruction format 无地址指令方式
zero adjust 零调节
zero adjuster 零(位)调整器;零位调节器;零点校正器
zero adjust(ing) 调零
zero adjusting lever 零度调整杆
zero adjusting screw 零位调节螺钉;调零旋钮
zero adjustment 归零;零(位)调整;调整零点;对准零点
zero adjustment knob 指零调置钮
zero air void 零空(气孔)隙;无空隙
zero air void curve 零空气孔隙曲线;饱和曲线
zero air void density 零空气孔隙密度;无空隙密实度;饱和密度
zero air void ratio 饱和孔隙比
zero alignment 调零
zero allowance 零容差;无容差;无穷差;无公差
zero altitude 零高度;超低空
zero angle 零角
zero angle cut 零度截断法
zero argument 零变元;零变数
zeroaxial 过坐标原点的;通过零点
zero background technique 零背景技术
zero balance 零位调整;零数余额;零点平衡
zero balancing 零平衡;零差法;零比较
zero base approach 零基(方)法
zero base budget 零基预算
zero base budgeting 零基预算(编制)法
zero base budgeting system 零基预算制度
zero-based budget 从零开始的预算
zero-based budgeting 零基预算法
zero base line 零基线
zero base planning and budgeting 零基计划与预算编制法;以零为基数的计划编制核预算
zero beat 零拍;零差
zero beat method 零拍法
zero beat reception 零拍接收法;零差接收
zero bed load transport 零底质输移
zero Bessel function 零阶贝塞尔函数;零次贝塞尔函数
zero bevel gear 零度弧齿锥齿轮
zero bias 零偏压
zero bias tube 零偏管
zero bit 零位
zero bit detection 零位检测

zero blanking 消零
zero blanking circuit 消零电路
zero bleed 零泄漏
zero blowdown 零排污
zero blowdown system 零排污系统
zero boundary 零点边界线
zero boundary line of ore body 矿体零点边界线
zero bracket amount 免税额;免征额
zero branch 零转移;零支
zero calibration 零点校准
zero carbon steel 低碳薄钢板
zero carrier 零(振幅)载波
zero center instrument 中心零位仪表
zero centre meter 中心指零式仪表
zero charge 零电荷
zero check 零检查
zero checker 零位校验器;定零位装置
zero check routine 零检验程序
zero circle 零周;零圆;基圆;点圆
zero circuit 零电路
zero clearance 零间隙;无隙;无间隙
zero clearing 清零
zero compensation 零位补偿
zero complement 补码
zero compression 零压缩
zero condition 零状态
zero constant 零点常量;冰点常量
zero contact angle 零接触角
zero control 指零旋钮;零位控制;零位调整
zero correction 零位修正;零(位)校正;零点(差)修正
zero correlation 零相关;无相关
zero creep 仪器零点漂移;零蠕变;零点漂移
zero cross comparator 零交比较器
zero cross detection 过零检查;零交叉检波
zero cross(ing) 零交叉;零交点;零跨越;零穿越
zero crossing carrier 零交波
zero crossing detector 零交点检测器;过零点检测器
zero crossing distribution 非零分配
zero crossing method 零交法
zero crossing rate 零交率
zero crossing time 零相交时间
zero crossing wave 零交波
zero crossing wave period 跨零波周期
zero cross level 零交叉电平
zero cross pulse 零交叉脉冲
zero cross signal 零交叉信号
zero current 零压电流;零位电流;零点电流;超导电流
zero curtain 零度面;土壤零温度层
zero curve 零线
zero cutout 无电自动断路器
zero damage 无损伤;无损害
zero damping 零阻尼
zero date 起算日
zero datum 零点(高程)
zero dead stop 停在零位
zero decrement 零衰减量;无衰减
zero defect 无缺陷;无事故;无故障;无疵残
zero defect casting 无缺陷铸件
zero defect management 无缺陷管理;无缺点管理
zero defect program(me) 无缺陷计划;无缺点计划
zero defect team 无次品小组
zero deflection 零偏转;零度偏料;无偏差
zero deflection method 零偏法
zero degree 零度;零次
zero degree signal 零度信号
zero delivery 零流量
zero-detection 检零
zero-detection circuit 检零电路
zero deviation 零偏差
zero deviation situation 零漂移位置
zero device-miles 汽车试用期
zero difference contour line 零差点廓线
zero difference detection 零差探测
zero diffraction spot 零点衍射斑点
zero dilatancy 无膨胀
zero dimension 零维数;零维;零次;无因次
zero dimensional 零维的;无因次的
zero dimensional water quality model of river 零维河流水质模型
zero dimension of river water quality model 零维河流水质模型
zero diopter 零屈光度

zero direction 初始零方向;原方向;零方向
zero discharge 零排放;无出料;零污染排放;空转
zero discharge design 零排放设计
zero discharge of pollutant 污染零排放
zero discharge system 无出料系统
zero displacement 零位移;零点漂移
zero dissolved oxygen 零溶解氧
zero division 零划分;零分划
zero divisor 零因子
zero down-crossing 下跨零点
zero draining 零排放
zero drift 零位偏移;零点漂移
zero drift compensation 零漂移补偿
zero drift error 零位漂移误差;零点漂移误差
zero droop governor 无差调速器
zero-duty binding 免税待遇的承诺
zero dynamic(al) pressure 零动压
zero economic growth 经济增长停滞
zero economic zone 经济无增长
zero electrode 零电极
zero element 零元(素);零成分
zero elevation 零高度;零(点)高程
zero elevation of tidal station 验潮站零点高程
zero elimination 消零(法)
zero elongation 零伸长
zero end 零点
zero end of survey 路线测量起点桩;(路线)测量的起点
zero energy 零功率
zero energy coast 零能海岸
zero energy reactor 零功率反应堆
zero energy state 零能量状态
zero energy thermonnuclear apparatus 零功率热核装置
zero energy uranium system[zeus] 零功率铀系统
zero error 零(位)误差;零点误差
zero error capacity 零误差容量
zero error of level 水准仪零点误差
zero error of microwave distance measuring instrument 微波测距仪零点差
zero error of staff 标尺零点差
zero error reference 零误差基准线
zero expansion glass 零膨胀系数玻璃
zero expansion glass ceramics 零膨胀微晶陶瓷
zero exponent 零指数
zero-extended field 零扩展域;补零字段
zero fill 填零;补零
zero five brick wall 半砖墙
zero five contour 半距等高线
zero flag 零标记
zero fluctuation 零点漂移
zero form 零形式
zero forward velocity 零向前速度
zero frame 零框架
zero frequency 零位频率;零频
zero frequency current 零频电流
zero frequency normalization 零频归一化
zero friction 无摩擦
zero galvanometer 零值电流计
zero game 零博弈
zero gas 零气
zero gate 置零门
zero gear 弧齿伞齿轮;零度弧齿锥齿轮;零度弧齿伞齿轮
zero gel 零凝胶
zero geodesic 零短程线
zero governor 零点调节器
zero grade line 零点标高线;零点高程线;施工基准线;(路基的)不填不挖线
zero gradient 零梯度
zero graduation 零刻度;零分度;零(点)分划
zerographic copy 复印相片
zero gravity 零重力;无重力;失重
zero-gravity condition 失重状态
zero-gravity effect 失重效应
zero-gravity environment 失重环境
zero-gravity facility 失重装置
zero-gravity period 失重阶段
zero-gravity switch 无引力开关
zero-gravity test 失重试验
zero-gravity time 失重时间
zero-gravity tower 失重试验塔
zero growth 零增长
zero growth rate 零增长率

zero-guy 杆间水平拉线
zero heat current 无热交换流动
zero heat transfer 无换热效应
zero heel stress 坝踵零应力;坝趾零应力
zero height 零高度
zero height of burst 零高度爆炸
zero hold 零保持
zero hoop stress head contour 零周向应力封头曲面
zero hour 零时;开始时刻;行动开始时刻;紧急关头;预定行动开始时间
zero-hour mixing 无间歇搅拌(法)
zero impedance 零阻抗
zero-in 瞄准具校正;归零校正
zero incidence 零攻角;零冲角
zero incremental cost 无增支成本
zero indicator 零位指示器;零点指示器
zero inflow curve 进水零值基线
zeroing 零位调整;零点调整;调整零点;调整到零;调零;定零点
zeroing circuit 调零电路
zeroing of instrument 仪器调零
zero initial 零起点偏压
zero initial bias 零始偏压
zero initial data 原始数据;零初始值
zero insertion 零插入
zero insertion and deletion 零插入和零删除
zero insertion force 无插拔力
zero insertion force socket 无插拔力插座
zero instrument 零位仪表
zero intensity 零强度
zero interfacial tension 零界面张力
zero interference 零点干扰
zero investment supply 零投资供给;无投资款项
zero isochron(e)零等时线;零时水波线
zero isotherm 零度等温线
zeroize 填零;补零
zero kill 零消失;零位抑制;零分类
zero label 零标号
zero lag 零位延迟
zero-lag filter 无滞后过滤器
zero-lag point 零滞后点
zero lap 零遮盖;零开口
zero lapped 零叠合
zero lap valve 零遮盖阀;零叠量阀
zero lash valve lifter 无间隙气门挺杆
zero lateral strain test 无侧限应变试验
zero layer 零面
zero leakage 零泄漏;零排泄
zero length 零长
zero length buffer 零长度缓冲器
zero length launcher 零长发射架
zero length launching 垂直发射
zero length rail 零长导轨;超短型导轨
zero length spring 零长度弹簧
zero length spring gravimeter 零长弹簧重力仪
zero length spring seismograph 零长度弹簧地震仪
zero level 零水平;零级;零电平;零标高;起点级
zero level address 零级地址;立即地址
zero level addressing 直接选址;零级定址
zero level binding 零级结合
zero level channel 零阶通路
zero level stability 零电平稳定性
zero-level transmission reference point 零电平传输参考点
zero lift 零升力
zero lift angle 零升力迎角
zero lift chord 零升力翼弦
zero line 零(位)线;基准(水位)线;基丝;起点线
zero line mirror 零线镜
zerolling 零下温度压延加工
zero load 零(位)负荷;无载(荷);空负荷
zero load method 零(荷)载法
zero load test 零荷载试验;空载试验;零载实验;无载试验
zero-loss circuit 无(损)耗电路
zero lot line 地界线
zero lot line district 零地块线地区
zero luminance 零亮度
zero luminance plane 零亮度平面
zero-man engine room 无人机舱
zero mark 零位刻度;零位记号;零米标高;零度;零点刻度;零点标志;零标;基准记号
zero mask 零屏蔽

zero material dispersion wavelength 零材料色散波长
zero matrix 零矩阵
zero mean 零平均;同度盘位置测回中数
zero mean excitation 零平均激振
zero mean process 零平均过程
zero mean value 零平均值
zero member compensation solution 零补偿溶液
zero meridian 高程零点;零(经度子)午线;起点子午线
zero meter 指零表
zero method 指零法;零值方法;零位法;零点法
zero modified Bessel function of the 1st kind 第一类零阶虚宗贝塞尔函数
zero modulation 零调制
zero moisture index 零值湿度指数;零水分指数
zero moment 零弯矩;零力矩
zero moment point 零弯矩点
zero momentum 零动量
zero motor 低速马达(0~300转/分)
zero norm 最低限度
zero of coordinate system 坐标原点
zero of elevation 高程标准原点;高程标准零点
zero offset 零(位)偏移;零偏置;零点偏移;零不均匀性
zero offset correction 零点漂移改正
zero offset data processing 零偏移距资料处理
zero offset optic(al) square 零距直角头
zero offset point 零偏置点
zero offset voltage 零补偿电压
zero of ga(u)ge 水尺零点;表计零点
zero of maximum concentration 零位最大浓度
zero of vernier 游标零分划
zero ohm adjustment 零欧姆调整
zero-one distribution 零一分布;一点分布;单点分布
zero one integer programming 零一整(数)规划
zero one law 零一律
zero one process 零一过程
zero operator 零算符
zero order 零序;零阶;零级;零次
zero order aberration coefficient 零级像差系数
zero order biotransformation 零级反应生物转变
zero order component 零阶分量
zero order correlation 零次相关
zero order elimination kinetics 零级消除动力学
zero order fringe 零阶条纹
zero order hold 零阶保持器
zero order image 零级像
zero order reaction 零级反应
zero order spectrum 零级光谱
zero order term 零级项
zero order transformation 零级转化
zero-order traversing 特级导线测量
zero order wave 零级波
zero output 零输出
zero output signal 零输出信号
zero padding 补零
zero passage 过零
zero path difference 零程差
zero pause 零值停止
zero peak amplitude 零峰值振幅
zero percent mortgage 零点抵押
zero period 零周期
zero period acceleration 零周期加速度
zero permeability isolator 零导磁率隔离器
zero phase deconvolution 零相位反褶积
zero phase effect 零相位效应
zero phase filter 零相位滤波器
zero phase lines 零相位线
zero phase-sequence 零相序
zero phase-sequence component 零相序分量;零相序部分
zero phase sequence protection 零相序保护(装置)
zero phasesequence relay 零相序继电器
zero phase shift 零相移
zero phase-shift filter 零相(位)移滤波器
zero phase window 零相窗
zerophyte 旱766旱植物;旱地植物
zero place function 零位函数
zero plane 零平面
zero point 起点;零位;零度;零点;致死临界温度
zero point adjustment 调零位;零点校正;对零
zero point constancy 零点稳定性

zero point correction 零点校正
zero point energy 零点能量
zero point error 零点误差
zero point motion 零点运动
zero point mutation 起点突变
zero point of charge 零点电荷
zero point of rod 水尺零点
zero point of tidal station 验潮站零点
zero point shift 零点偏移
zero point signal 零点信号
zero point variation 零点变化
zero point vibration 零点振动
zero-pollution goal 零污染目标
zero population growth 人口零增长
zero-porosity composite 无孔隙复合材料
zero position 零位(置);零点位置;零始位置
zero position of shaft plummet 竖井垂球静止位置
zero position signal of echo sounder 回声测深仪零位信号
zero potential 零电位;零电势
zero potential surface 零位面
zero potentiometer 调零电位计
zero pour 凝固点
zero power 零(值)功率;零功会聚度;零屈光度
zero power aberration 零屈光度像差
zero power air spaced doublet 零屈光度空气隙双(分)透镜
zero power factor 零值功率因数
zero power-factor characteristic 零功率因数特性曲线
zero power factor method 零功率因数法
zero power-gain 零功率增益
zero power lens 零屈光度透镜
zero power level 零功率电平
zero power reactor 零功率反应堆
zero power space 零屈光度空间
zero pressure 零压(力);无压力
zero-pressure gradient flow 无压力梯度流动
zero pressure position 零压位置
zero pressure resin 无压(固化)树脂;常压(固化)树脂
zero probability 零概率
zero probability of failure 零概率破坏
zero proof 零证明;零数验证;零检查
zero pulse 基准脉冲
zero punch 零穿孔(位)
zero range mark 零距离标记
zero range trigger 零距触发脉冲
zero-rated goods 零税率货物;免税货物
zero rate of duty 免税
zero rate of penetration (桩的)零值贯入度
zero reactance line 零电抗线
zero reader 零位读出器
zero reading 零点调整;零(位)读数;零点读数;起点读数
zero reading correction 零位读数修正
zero redundancy array 零重复阵
zero reference point 零参考点
zero release 零排放;不排放(废物)
zero release plant 不排放废物的工厂
zero reset 零位复位;零点重调;回零;重新对零;复零
zero resetting device 零位恢复装置;复零装置
zero resistance ammeter 零内阻安培计
zero resistivity 零值电阻率
zero rigidity 零劲度
zero risk goal 零风险目标
zero root 零根
zerorotation main sequence 无自转主序
zero salvage value 无残值
zero-saturation colo(u)r 白色
zero scan speed 零位扫描速度
zero section 零断面
zero sediment discharge 零输沙率
zero sediment supply 零来沙量
zero sequence 零序
zero sequence component 零序分量
zero sequence current 零序电流
zero sequence impedance protection 零序阻抗保护(装置)
zero sequence overcurrent protection 零序过流保护
zero sequence protection 零序保护(装置)
zero sequence protective relay 零序保护继电器
zero sequence reactance 零序电抗

zero set	零点调置;对准零位;零位调整;调零
zero set control	置零控制装置;过零控制;零位修整调节
zero setter	零调整器
zero setting	置零;零(点)调整;零位调整;对准零位;对零;调零
zero setting knob	置零旋扭
zero sharpening control	平衡器旋钮
zero shear	零剪力
zero shear point	零剪力点
zero shear viscosity	零剪切黏度
zero shift	零位(漂)移;零点误差;零点漂移
zero shifting	零点移位法
zero signal	零(位)信号;无信号
zero signal zone	无信号区
zero-size image	零尺寸图像
zero slip	零位滑移
zero slip point	无滑动点
zero slope	零斜率;零坡度;平坡
zero-slump concrete	干硬性混凝土;零坍落度混凝土;无坍落度混凝土
zero solids treatment	全挥发水处理
zero span	零档
zero speed position	零速度位置
zero speed switch	零速断路器
zero speed torque	零速力矩
zero stability	零点稳定性
zero state	零状态;零态
zero status	零状态
zero status bit	零状态位
zero status flag	零状态标志
zero step-up voltage	零阶跃电压
zero stereo	零立体
zero stiffness	零劲度;零刚度
zero stoichiometric method	零当量法
zero strain	零点应变
zero strain velocity	零应变速度;无应变速度
zero strength temperature	零强温度;失强温度
zero strength time	零强时间;失强时间
zero stress	零应力
zero stress-optic(al) constant	零应力光学常数
zero sum	零和
zero sum two-person game	两人零和对策
zero suppression	消零;零(点)抑制
zero suppression character	消零字符
zero suppression picture character	消零图像字符
zero surface	零面
zero temperature	基准点温度
zero temperature coefficient	零温度系数
zero temperature level	零温度层
zero terminal	中心线端子;零线端子
zero test	零位试验
zeroth diffraction order	零级衍射
zero thickness	零厚度
zeroth mode	零次波型
zeroth order	零级;零阶
zeroth-order approximation	零阶近似
zeroth-order differential rectification	零级微分纠正
zeroth period	第零周期
zero thrust pitch	零推力螺距;无推力螺距
zero tide	基准潮位;潮水零点
zero tillage	免耕法;不耕地种植
zero time	零时(间);时间零点;计时起点;起始瞬间
zero time reference	零时间基准(点);零时参考点
zero time zone	零时区
zero to-peak amplitude	零至峰值振幅
zero-to-peak record motion	零峰记录运动
zero torque pitch	无扭力螺距
zero transmission level reference point	零电平传输参考点
zero trimmer	调零电位计
zero twist	零缠度
zero twist fabric	零捻织物
zero-type dynamometer	零式电测功率计;归零式电测力计
zero up-crossing	上跨零点
zero valent compound	零价化合物
zero valent iron	零价铁
zero valent metals	零价金属类
zero valent oxidation state	零价氧化态
zero valent zinc	零价锌
zero value	零值
zero value conversion method	零值换算法
zero variance problem	零方差问题
zero variation	零点偏差
zero vector	零向量
zero velocity	零速度
zero velocity surface	零速度面
zero velocity update	零速度修改正
zero visibility	零能见度;极为不良的能见度
zero voltage	零电压
zero volt switch	零伏开关
zero wander	零点漂移
zero water level	零(点)水位;水位零点
zero water pollution	零水污染;无水质污染
zero welding	冷焊接
zero working	零下温度加工
zero-zero	能见度为零的
zero-zero fog	能见度等于零的浓雾
zero-zero gel	(起始和十分钟均为零的)双零值静切力胶体
zero zero transition	零零迁
zero zone	零时区
zestocautery	蒸汽烙管
zeta	房间(教堂门廊上的);密闭小间
zero Energy Thermonuclear Assembly [ZETA]	零功率热核装置
zeta function	黎曼 ξ 函数
zeta-potential	电动电位
zeta-type structure	歹字形构造【地】
Zet meter	拉丝模圆柱孔长度测量表
Zetonia	泽托尼阿铅锑锡合金
zeugenberg	外露层
zeugogeosyncline	联隆地槽;联合地槽;配合地槽
zeunerite	铜砷铀云母;翠砷铜铀矿
zeyssatite	硅藻石
Z-factor	玻璃棉空气阻力系数
Z flashing	Z 形泛水
Zhayier geodome series	扎依尔地穹列
zheltozem	黄壤
zhemchuzhnikovite	草酸铝钠石
zhonghuacerite	中华铈矿
Zhonghua old land	中华古陆
Zhoukoudian deposit	周口店沉积
Zhoukoudian limestone	周口店灰岩
Zhoukoudian (stage)	周口店阶;周口店期
zianite	蓝晶石
Zibell anchoring system	支撑设备;(用于大理石薄墙板的)紧固设备
Ziehl's carbol-fuchsin stain	齐尔氏石炭酸品红染剂
Ziehl's stain	齐尔染色剂
zig	锯齿形转角;急转
zig bench terrace	折坡梯田
ziggurat	古代亚述及巴比伦之宝塔式建筑;锥形塔;塔庙;叠级方尖塔
zigzag	曲折(的);曲折图案;曲折(的)线条;锯齿状;锯齿形的
zigzag air gap leakage	曲折气隙泄漏
zigzag anchor bar	曲折形锚杆;交错锚杆;交错锚杆
zigzag and oblique angle earth pushing process	之字斜角推土法
zigzag antenna	曲折天线
zigzag arch	曲折拱;之字形拱;锯齿拱
zigzag baffle chamber	曲折挡板室
zigzag bond	人字砌合;锯齿形砌合
zigzag bridge	锯齿形开合桥;九曲桥;之字形桥
zigzag bulkhead	曲折舱壁
zigzag chain	曲折链
zigzag clock	折航指示钟
zigzag closing	曲折缝合
zigzag coast	曲折航线
zigzag coefficient	曲折系数
zigzag configuration	曲折构型
zigzag connection	交错连接;曲折连接;曲折接线;曲折接法
zigzag continual non-regular variation	跳跃连续不规则
zigzag continual regular variation	跳跃连续有规则
zigzag continuous blender	曲折连续掺和机
zigzag contour reflection	锯齿状边缘反射
zigzag corridor	曲廊
zigzag coupling	曲折连接
zigzag course	曲折航线;曲径
zigzag cracks	锯齿状裂缝;曲折裂缝;不规则裂纹;不规则裂缝;不规则开裂
zigzag crevasse	锯齿形裂缝
zigzag crossing-bedding	锯齿状交错层理
zigzag curve	之字形曲线
zigzag curve sign	之字形曲线标志
zigzag-delta connection	曲折三角形连线
zigzag development	之字形展线(法);盘旋展线法
zigzag dislocation	之字形位错
zigzag divide	锯齿状分水岭
zigzag fastening	交错连接;参叉紧固
zigzag fault	锯齿状断裂;锯齿断口;锯齿状断层;锯齿形层错;曲折断层
zigzag feed	锯齿形送料;交错送料;曲折送料
zigzag filter	锯齿形滤波器;曲折法滤波器
zigzag fluting	曲折柱槽
zigzag fold	来回折叠;锯齿状褶皱;曲折褶皱
zigzag form	Z 形线;锯齿形
zigzag frieze	曲折雕带
zigzagging	锯齿形运动;曲折航行
zigzag girder	三角(孔)梁
zigzag harrow	之字形耙;曲形耙
zigzaging	之字形运动
zigzag intermittent fillet weld	锯齿形断续角焊缝
zigzag kiln	折线窑;锯齿形窑;曲折窑;Z 形窑
zigzag laying	交错排列
zigzag lead	曲折水道
zigzag leakage	曲折漏磁
zigzag leakage flux	曲折漏磁通
zigzag line	之字形线(路);之字形路线;之字线;锯齿形曲线;曲折线(路);Z 形线
zigzag mill	之字形轧机
zigzag mode	锯齿模
zigzag motion	之字形运动
zigzag mo(u)lding	波浪文饰;人字文饰;波纹线脚;人字形线脚;锯齿形线脚;曲折线脚;曲折的回纹形装饰;曲折的回纹形线条
zigzag-mountain type	之字山型
zigzag ornament	人字纹装饰
zigzag paper filter	曲折纸过滤器
zigzag partition	曲折隔断;曲折隔墙;错列龙骨隔墙
zigzag path	曲折路线
zigzag path of wave	曲折波线
zigzag pattern	Z 形图式;Z 形图像;Z 形图案;曲折图形;曲折图案
zigzag power transformer	曲折接线电力变压器
zigzag ramp	折返坑线
zigzag ray model	锯齿形射线模型
zigzag reactance	曲折漏抗
zigzag reflection	曲折反射;多次反射
zigzag resistance unit	曲折状电阻器;弯曲形电阻器
zigzag ridge	锯状(状)山脊
zigzag rift	锯齿状裂谷
zigzag riveted joint	之字形铆接;交错铆接;错列铆接
zigzag riveting	交错铆接;错列铆钉
zigzag roadway	曲折车道
zigzag route	之字路;之字(形线)路;回头线
zigzag rule	花杆;测杆;折尺;曲尺
zigzag sampling	交错抽样;盘旋抽样
zigzag scan	锯齿形扫描
zigzag scanning	折线扫描
zigzag separator	折板形分离器
zigzag-shaped bar	波形钢筋
zigzag sill	锯齿槛;曲折槛
zigzag slip surface	折线型滑动面
zigzag solid laser	锯齿形固体激光器
zigzag-star connection	曲折星形连接
zigzag stitching	曲线缝制
zigzag stone slab bridge	多曲石板桥
zigzag superheater	蛇形管过热器
zigzag tectonic belt	锯齿状构造带
zigzag test	之字形航行试验
zigzag tooth	交错齿
zigzag truss	锯齿形桁架;华伦式屋架
zigzag type shear wall	交错型剪切墙;交错布置的剪力墙
zigzag vernada	曲廊
zigzag watershed	锯齿状分水岭
zigzag wave	锯齿状波形;锯齿(形)波
zigzag (wave type)	之字形
zigzag weld(ing)	交错焊(接)
zikustal	围谷
zillion	千千万
Zilva	齐瓦牌钢材(一种店面橱窗用的不锈钢)
Zimbabwe nucleus	津巴布韦陆核
Zimmerman process	齐默尔曼过程;湿式焚化法;齐默尔曼法
Zimmerman roller bascule bridge	齐默尔曼滚柱竖旋桥;齐默尔曼辊转跳开桥;齐默尔曼辊轴平衡桥

zinalsite 硅锌铝石
zinc acetate 乙酸锌;醋酸锌
zinc acetylide 乙炔锌
zinc-air battery 锌空气电池
zinc-air cell 锌空气电池
zincalism 慢性锌中毒
zinc alkyl 烷基锌
zinc alloy 锌合金
zinc alloy coated sheet steel 锌合金涂层钢板
zinc alloy for anti-friction metal 电动机电枢用合金
zinc alloy for stamping 冲模锌合金
zinc alloy lining plate 锌合金衬板
zinc aluminate 锌尖晶石
zinc aluminite 锌明矾;锌矾石
zinc alumin(i)um 锌铝合金
zinc amalgam 锌汞齐;锌汞合金
zincamide 氨基锌
zinc ammonium chloride 锌氯化铵
zinc anode 锌阳极
zinc anode cathodic protection system 带锌阳极防蚀系统
zinc anode protection equipment 防电化学腐蚀装置
zinc antimonide 锑化锌
zinc arsenate 砷酸锌
zinc arsenite 亚砷酸锌
zinc ash 锌浮渣;锌粉;锌灰;镀锌浴面的氧化锌
zincate 锌酸盐
zincate treatment 锌酸盐处理
zinc balance 锌摆轮
zinc bar 锌棒
zinc-barium-lead glass 锌钡铅玻璃
zinc-base alloy 锌基合金
zinc-base bearing metal 锌基轴承合金
zinc-based foam activator/stabilizer 锌基泡沫活化稳定剂
zinc base grease 锌基润滑脂
zinc-base slush casting alloy 锌基糊状铸造合金
zinc bath 镀锌槽
zinc batten roof 锌板条屋顶
zinc battery 锌蓄电池
zinc-bearing 含锌的
zinc-bearing alloy 含锌合金
zinc benzoate 苯甲酸锌
zinc bias 负偏压
zinc black 锌黑
zinc blasting cap 锌引爆管
zinc blende 闪锌矿
zinc bloom 氧化锌(矿);水锌矿;锌华
zinc boiler plate 锌锅炉板
zinc borate 硼酸锌
zinc-borate glass 锌硼酸盐玻璃
zinc box 锌箱
zinc box rainwater gutter (匣形的)锌皮雨水槽沟
zinc bromide 溴化锌
zinc bronze 锌青铜
zinc butyrate 丁酸锌
zinc-cadmium sulfide 硫化锌镉
zinc can 锌罐
zinc carbonate 碳酸锌
zinc casting 锌铸件
zinc chlorinate primer 氯化锌底漆
zinc-chlorine battery 锌氯电池
zinc chlorite 氯化锌
zinc chlorite solution 氯化锌溶液
zinc chromate 锌黄颜料;锌铬黄;铬酸锌
zinc chromate anti-rusting paint 锌铬黄防锈漆
zinc chromate anti-rust paint 锌黄防锈漆
zinc chromate coated pigment 铬酸锌包核颜料
zinc chromate-coated silica 铬酸锌包二氧化硅;包核铬酸锌
zinc chromate hydroxide 锌黄
zinc chromate primer 锌铬黄底漆;铬酸锌涂底料;铬酸钭底漆;铬化锌底漆
zinc chromate rust-proofing paint 铬酸锌防锈涂料
zinc chrome 锌(铬)黄
zinc-clad steel strand wire 镀锌钢绞线
zinc coat 镀锌(层)
zinc-coated barbed wire 镀锌刺丝
zinc-coated bolt 镀锌螺栓
zinc-coated iron nail 镀锌铁钉
zinc-coated material 镀锌材料
zinc-coated nut 镀锌螺母
zinc-coated screw 镀锌螺钉

zinc-coated sheet steel 锌涂层钢板
zinc-coated steel 镀锌钢板
zinc-coated steel sheet 镀锌薄钢板
zinc-coated steel tile 镀锌钢瓦
zinc-coated washer 镀锌垫圈
zinc-coated wire 镀锌钢丝(绳);镀锌金属线
zinc coating 锌面层;锌镀(层);涂锌;包锌
zinc coat thickness ga(u)ge 镀锌层测厚仪
zinc compound 锌化合物
zinc concentrate 精炼锌
zinc concentration 锌的浓度
zinc condenser 锌冷凝器
zinc conduit 锌管渠
zinc-copper melanterite 锌铜水绿矾
zinc-copper-titanium alloy 锌铜钛合金
zinc corrosion 锌腐蚀
zinc corrosion test 锌腐蚀试验
zinc couple 锌电偶;锌半电池
zinc covered flat roof 锌皮平屋顶
zinc covering 锌皮披水;包锌
zinc creosote 锌化杂酚油
zinc crown 锌冕玻璃;上等锌玻璃
zinc crown glass 锌冕玻璃
zinc crust 锌壳;银锌壳
zinc cure 锌板防蚀;防蚀锌板
zinc current 锌电流
zinc cyanide 氰化锌
zinc-desilverization plant 加锌除银设备
zinc detonator 锌引爆管;锌起爆雷管
zinc dibutyl dithiocarbamate 二丁基二硫代氨基甲酸锌
zinc dichromate 重铬酸锌
zinc die casting 锌压铸;压铸锌
zinc die casting alloys 压铸锌合金
zinc diffusion 扩散渗锌
zinc dihydrogen phosphate 磷酸二氢锌
zinc diisobutyl 二异丁锌
zinc dimethyl 二甲锌
zinc drier 锌干料;锌催干剂
zinc dross 锌渣
zinc dust 锌粉
zinc dust anticorrosive paint 锌粉防锈涂料;锌粉防锈漆
zinc dust distillation 锌粉蒸馏
zinc dust paint 锌粉漆
zinc dust paste 锌粉膏
zinc dust pigment 锌粉颜料
zinc dust primer 锌粉底漆
zinc dust purification 锌粉精制
zinced sheet 镀锌铁皮;镀锌钢板
zinc electrode 锌电极
zinc electrolyzing plant 锌电解设备
zinc engraving 锌凸版
zinc-enriched 含锌多的
zinc enriched paint 富锌涂料;富锌漆
zinc etching 锌(凸)版
zinc ethylenebis-dithiocarbamate 代森锌
zinc fastening 浇锌固定件
zinc-fauserite 锌锰泻盐
zinc ferrite 铁酸锌颜料
zinc ferrocyanide 氰亚铁酸锌
zinc fertilizer 锌肥
zinc filled socket 灌锌锚杯
zinc finish 锌辊轧花整理;镀锌
zinc flash 锌着色;锌闪烁
zinc flashing 锌铁披水;锌皮泛水
zinc fluo 氟砷酸锌
zinc fluoride 氟化锌
zinc fluoroatsenate 氟砷酸锌
zinc fluoroborate 氟硼酸锌
zinc fluosilicate 氟硅酸锌
zinc foil 锌箔
zinc fume 锌尘
zinc furnace 熔锌炉
zinc galvanizing 镀锌
zinc ga(u)ge 锌皮标号;锌板厚度规
zinc gelatin 氧化锌明胶
zinc glaze 锌釉
zinc glaze tile 锌釉砖;锌玻璃砖
zinc granule 锌粒
zinc gray 锌灰油(漆);锌灰(漆)
zinc green 锌绿
zinc grey 锌灰
zinc half-cell 锌电偶

zinc halide 卤化锌
zinc hausmannite 锌黑锰矿
zinc hydroxide 氢氧化锌
zinc hydroxide precipitation 氢氧化锌沉淀物
zinc hydroxycarbonate 碱式碳酸锌
zinc hydroxyl carbonate 碳酸羟锌
zinc hypophosphite 次磷酸锌
zincian paratacamite 含锌三方氯铜矿
zincic acid 锌酸
zinciferous 含锌的
zincification 锌饱和;加锌;包锌
zincify 包锌
zinc impregnation 锌化
zinc impurity P-type detector 掺锌P型探测器
zincing 镀锌
zinc ingot metal 锌锭
zinc iodide 碘化锌
zinc ion 锌离子
zinc iron 锌铁
zinc iron cell 锌铁电池
zinc iron oxide 氧化铁锌
zincite 锌石;红锌矿;红色氧化锌
zincite crystal 氧化锌晶体
zincity 镀锌
zinck 含似锌的
zincked sheet 镀锌钢板
zinckenite 辉锑铅矿
zincky 锌的
zinclavendulan 锌水砷铜矿
zinc-lead accumulator 锌铅蓄电池
zinc liner 锌衬里
zinc-magnesium chalcanthite 锌镁胆矾
zinc-magnesium galvanizing alloy 锌镁电镀合金
zinc manganate 锰酸锌
zinc-manganese battery 锌锰电池
zinc manganic cobaltic drier 铅锰钴催干剂
zinc melanterite 锌水绿矾
zinc-mercuric oxide cell 锌汞电池
zinc mercury battery 锌汞电池
zinc mercury cell 锌汞电池
zinc meta-aresnite 偏亚砷酸锌
zinc metallurgy 炼锌
zinc metal sheet 锌片
zinc metasilicate 偏硅酸锌
zinc mine 锌矿
zinc-modifier system 锌盐变性剂工艺
zinc molybdate 钼酸锌
zinc nail 锌钉
zinc naphthenate 环烷酸锌
zinc-nickel coated sheet 锌镍涂层薄钢板
zinc-nickel storage battery 锌镍蓄电池
zinc niobate 铌酸锌
zinc nitrate 硝酸锌
zinc nitride 氮化锌;二氮化三锌
zinco 锌凸版
zincobotryogen 锌赤铁矾
zincocalcite 锌方解石
zincocopiapite 锌叶绿矾
zincode (电池的)锌极
zincograph 锌凸版印件;制锌版
zincography 锌版(印刷)术;制锌版术
zinc oleate 油酸锌
zincolith 锌白;白色颜料
zincon 锌试剂
zincorange 锌橙色
zinc ore 锌矿(石)
zinc orthoarsenate 砷酸锌
zinc orthophosphate 磷酸锌
zincous 锌的;含锌的;阳极的
zinc oxalate 草酸锌
zinc oxide 锌白;氧化锌
zinc oxide bath 氧化锌槽
zinc oxide calcine 氧化锌焙砂
zinc oxide cement 氧化锌接合剂
zinc oxide ceramics 氧化锌陶瓷
zinc oxide fume 氧化锌尘雾
zinc oxide ointment 氧化锌软膏
zinc oxide paint 氧化锌油漆
zinc oxide paper 氧化锌纸
zinc oxide pigment 氧化锌颜料
zinc oxide poisoning 氧化锌中毒
zinc oxide powder 锌氧粉;锌白粉
zinc oxide process desulfurization 氧化锌法脱硫
zinc oxide thickening value 氧化锌增稠试验值

zinc oxide viscosity test 氧化锌黏性试验
zinc-oxygen battery 锌氧电池
zinc oxysulfide 氧硫化锌
zinc paint 锌漆;锌(粉)涂料;含锌油漆
zinc perchlorate 高氯酸锌
zinc perhydrol 过氧化锌
zinc permanganate 高锰酸锌
zinc peroxide 过氧化锌
zinc phosphate 磷酸锌
zinc phosphate binder 磷酸锌胶结料
zinc phosphate cement 磷酸锌黏结固粉
zinc phosphate coating 磷酸锌处理
zinc phosphide 磷化锌
zinc phosphide poisoning 磷化锌中毒
zinc phosphide wastewater 磷化锌废水
zinc pig 锌锭
zinc pigment 锌(系)颜料
zinc pipe 锌管
zinc piping 锌管
zinc plate 锌版;锌板;镀锌钢板
zinc-plated heavy iron hinge 镀锌厚铁铰链
zinc-plated iron pad bolt 镀锌铁挂插
zinc-plated light iron hinge 镀锌薄铁铰链
zinc plate finish 锌版装饰
zinc plate process 镀锌法
zinc plating 镀锌
zinc-plating additive 镀锌添加剂
zinc-plating brightener 镀锌光亮剂
zinc-plating electrowelding net 镀锌电焊网
zinc-plating material 镀锌材料
zinc-plating metal plastics-coated hose 镀锌金属包塑软管
zinc-plating pipe 镀锌铁管
zinc point 锌点
zinc poisoning 锌中毒
zinc pollutant 锌污染物
zinc pollution 锌污染
zinc polycarboxylate cement 聚羧酸锌黏固粉
zinc polysilicate 聚合硅酸锌
zinc potassium chromate 铬酸钾锌
zinc potassium cyanide 氰化锌钾
zinc powder 锌粉
zinc protector 锌护板;防蚀锌板
zinc putty 锌皮油灰
zinc pyrophosphate 焦磷酸锌
zinc rebate 锌凹凸槽
zinc reducing method 锌还原法
zinc refiner 锌精炼厂
zinc resin 含锌(松香)钙脂
zinc resinate 松香酸锌;树脂酸锌
zinc retort 锌蒸馏罐
zinc-rich anti-rust paint 富锌防锈漆
zinc-rich crust (冷凝器内壁上的)富锌壳
zinc rich epoxy 多锌环氧树脂
zinc-rich epoxy paint 富锌环氧树脂漆
zinc rich epoxy primer 环氧富锌底漆;富环氧底漆
zinc rich paint 多锌油漆;富锌涂料;富锌漆;白色底漆
zinc rich primer 富锌底漆
zinc-rockbridgeite 锌绿铁矿
zinc roof 锌皮屋面;锌皮屋顶;马口铁屋顶
zinc roof cladding 镀锌铁皮屋面
zinc roof gutter 锌皮屋顶天沟
zinc roof sheathing 镀锌铁皮屋面
zinc roof surround 镀锌锌皮屋顶
zinc rule 锌质嵌线
zinc salt 锌盐
zinc-saponite 锌皂石
zinc-schefferite 锌锰红辉石
zinc sheet 锌薄板;锌片;锌皮
zinc sheet cover 镀锌锌皮盖层;镀锌铁皮面层
zinc sheet roofing 锌皮屋面;锌皮屋顶
zinc shingle 锌盖板
zinc silicate 硅酸锌
zinc silicate coating 锌硅酸盐涂层
zinc silicate glass 硅酸锌玻璃;锌钠硅玻璃;锌冕玻璃
zinc silicate glaze 硅酸锌釉
zinc silicate primer 硅酸盐富锌底漆
zinc silicate type crystalline glaze 硅酸锌型结晶釉
zinc silicofluoride 氟硅酸锌
zincsilite 无铝锌皂石
zinc-silver accumulator 锌银蓄电池
zinc-silver alloy 锌银合金
zinc-silver-chloride primary cell 锌氯化银原电池
zinc slab 锌片
zinc slag 锌渣
zinc sludge 锌矿泥
zinc smelter 锌熔炼炉
zinc smelting 炼锌
zinc smoke candle 锌烟罐
zinc soap 锌皂
zinc solder 锌焊料
zinc spar 菱锌矿
zinc spelter 锌块
zinc spinel 锌尖晶石
zinc sponge 锌棉
zinc spraying 喷镀锌;喷涂锌
zinc stability time test 锌稳定性时间试验
zinc stearate 硬脂酸锌
zinc storage battery 锌蓄电池
zinc strip 条形镀锌板;锌(板)条
zinc subcarbonate 碱式碳酸锌
zinc succinate 琥珀酸锌
zinc sulfanilate 氨基苯磺酸锌
zinc sulfate 硫酸锌
zinc sulfate flo(a)tation method 硫酸锌浮选法
zinc sulfide 硫化锌
zinc sulfide crystal 硫化锌晶体
zinc sulfide precipitation 硫化锌沉淀物
zinc sulfite 亚硫酸锌
zinc sulphate 硫酸锌
zinc sulphide 硫化锌
zinc tallate 妥尔油酸锌
zinc-tannin 锌化单宁酸
zinc templet 锌模
zinc tetroxy chromate 碱式铬酸锌;四盐基铬酸锌
zinc thickness 锌板厚度
zinc thymotate 百里酸锌
zinc tile 镀锌铁皮瓦
zinc titanate 钛酸锌
zinc titanate ceramics 钛酸锌陶瓷
zinc toxicity 锌毒性
zinc tube 锌管
zinc tubing 锌管
zinc valley gutter 镀锌铁皮天沟
zinc valley strip 镀锌铁皮天沟条
zinc vertical retorting 竖罐炼锌
zinc vitriol 锌矾;七水合硫酸锌
zinc white 白锌粉;锌皮漆;锌白(粉);氧化锌;中国白
zinc white stand oil enamel 锌白熟油瓷漆
zinc wire 镀锌铁丝
zinc yellow 锌铬黄;锌黄(粉)
zincy lead 含锌铅
zinc zippeite 水铀矾
zinc zirconium silicate 硅酸锌锆
zine alloy die 锌基合金模
zine matte 含锌锍
zine sulfide atmosphere tracer 硫化锌大气示踪物
zinethyl 二乙锌
zing bench terrace 隔坡梯田
zingiberene 姜烯
zingiberone 姜酮
Zinkalium 津卡锌铝合金
zink balance 锌制摆轮
zink covering 锌皮屋面
zink dural 含锌硬铝
zinkenite 辉锑铅矿
zinkify 包锌
zinking 包锌
zinkite 红锌矿
zinkosite 锌矾
zink rod 锌条
Zinn 齐恩合金
Zinnal 双面包锡双金属轧制耐蚀铝板
zinnia 百日菊属
zinnober 辰砂;朱砂
zinnwald mica 铁锂云母
zinoc 双代森锌
Z-intercept Z 截距
zinwaldite 铁锂云母
zip code 邮区编号(美国);邮政编码
zip-fastener 拉链
Zipf's Law 齐波夫定律
Zipf's law method 齐波夫法
zippeite 水铀矾
zipper 拉链;闪光环
zipper clause 拉链条款;不能重开谈判条款
zipper conveyer 拉链式输送机;密闭式运输带
zippered 拉链式的
zippered pencil case 拉链文具盒
zipper glazing 拉链式装配玻璃
zipper strip 拉链条;锁紧条;嵌固玻璃橡皮条
zip pouch 透镜遮光罩
zips 铜夹丝外露
zircal 锌镁铜铬锰铁硅合金
zircaloy 锆铝合金
zircite 天然斜锆石矿
zircon 锆石(含量);风信子石;风信子矿
zircon alba 提纯氧化锆
zircon-alumina brick 锆英石—氧化铝砖
zircon arc lamp 锆石弧(光)灯
zirconate 锆酸盐
zirconate ceramics 锆酸盐瓷
zircon brick 锆(英)砖
zircon cement 锆镁耐火水泥;锆石耐火水泥
zircon ceramics 锆英石陶瓷
zircon-corundum refractory 锆刚玉耐火材料
zircon hot repair patch 锆英石热补料
zirconia 氧化锆;锆氧
zirconia alumina 锆刚玉
zirconia brick 氧化锆砖;锆砖
zirconia ceramics 氧化锆陶瓷
zirconia corundum block 锆刚玉砖
zirconia enamel 氧化锆搪瓷;锆白釉
zirconia refractory 氧化锆耐火材料;锆耐火材料
zirconiated tungsten 锆钨电极
zirconiated tungsten electrode 锆钨电极
zirconia titanate 钛酸锆
zirconia whiteware 氧化锆瓷;锆质瓷
zirconic 锆的
zirconic acid 锆酸
zirconin 锆试剂
zirconite 锆(英)石;锆土
zirconium alloy 锆合金
zirconium analyzer 锆含量分析仪
zirconium anhydride 二氧化锆瓷
zirconium antimonide 锑化锆
zirconium-bearing ceramic products 含锆陶瓷制品
zirconium boride 硼化锆
zirconium boride ceramic 硼化锆陶瓷
zirconium carbide 一碳化锆
zirconium chloride 四氯化锆
zirconium copper 锆青铜
zirconium deposit 锆矿床
zirconium diboride 硼化锆;二硼化锆
zirconium dioxide 二氧化锆
zirconium dioxide refractory 二氧化锆瓷耐火材料
zirconium drier 锆干料;锆催干剂
zirconium electrode 锆电极
zirconium electrolysis cell 锆电解槽
zirconium ferrosilicon 锆硅铁
zirconium filled flash lamp 锆闪光灯
zirconium glass 锆玻璃
zirconium halide 卤化锆
zirconium ingot 锆锭
zirconium iron pink 锆铁红
zirconium lamp 锆(弧)灯
zirconium naphthenate 环烷酸锆
zirconium nitride 氮化锆;二氮化锆
zirconium ore 锆矿(石)
zirconium oxide 氧化锆
zirconium oxychloride 氧氯化锆;二氯氧化锆
zirconium phosphate 磷酸锆
zirconium plant 锆厂
zirconium powder 锆粉
zirconium-rich mineral 富锆矿
zirconium salt 锆盐
zirconium silicate gel 锆硅胶
zirconium silicide 硅化锆
zirconium spinel 锆尖晶石
zirconium starter plug 锆始锭
zirconium steel 锆钢
zirconium tanning agent 锆鞣剂
zirconium tetrachloride 四氯化锆
zirconium titanate 钛酸锆
zirconium ultrafiltration dynamic membrane 锆超滤动态膜
zirconium vanadium blue 锆钒蓝
zirconium white 锆白

zircon lamp 锆石灯
zircon mortar 锆石火泥
zircon mullite block 锆莫来石砖
zircon ore 锆英石矿石
zircon porcelain 锆英石(质)瓷
zircon-pyrophyllite brick 锆英石—叶蜡石砖
zircon refractory 锆英石耐火材料
zircon salt 锆盐
zircon sand 锆(英石)砂
zircon-SiC brick 锆英石—碳化硅砖
zircon slurry 锆砂粉浆料
zircon-taltalite-rutile maturity index 锆石—电气石—金红石成熟度指数
zircon web offset printing machine 锆英石转轮版印刷机
zircon whiteware 锆英石瓷
zirconyl chloride 二氯氧化锆
Zircopax 齐尔可帕克斯(锌、锡、硅质乳浊剂的商品名)
zircophyllite 锆星叶石
zirkelite 钛锆钍矿
zirkite 斜锆石砾;天然斜锆石矿
zirklerite 氯铁铝石
Zirkonal 泽科纳尔铝合金
zirkophyllite 锆叶石
Z-iron Z字钢;Z形铁
zirsinalite 硅锆钙钠石
Zirtan 泽坦碳化锆烧结合金
zirten 锆碳烧结合金
Zisman apparatus 蔡斯门测定仪
zitterbewegung 狄拉克颤动
ZJ1031 type container ZJ1031型集装箱
Z-lay 右捻
Z-line 间线
Z-motion 沿乙方向的运动;沿Z轴方向运动;Z轴方向运动
Zn-bearing copper ore 锌铜矿石
Zn-bearing Mo ore 锌钼矿石
zneith column 大气气柱
Zn-Pb ore 锌铅矿石
ZnS-film 硫化锌膜
Zobel filter 佐贝尔滤波器
zocco 柱脚;柱墩最下层;柱子的方石座
zoccola 柱墩最下层;柱底座部分;墩底部分;座石
zocle (雕像、石柱等的)座仓;柱脚
Zodiac 左迪阿克铜镍锌合金
zodiac 黄道带
zodiacal band 黄道带
zodiacal belt 黄道区
zodiacal circle 黄道圈
zodiacal cloud 黄道云
zodiacal cone 黄道角锥
zodiacal constellation 黄道星座
zodiacal counterglow 对日照
zodiacal diagram 黄道带星图
zodiacal dust cloud 黄道尘云
zodiacal light 黄道光
zodiacal light bridge 黄道光桥
zodiacal pyramid 黄道角锥
zodiacal signs 黄道十二宫(符号)
zodiacal star 黄道带恒星
zodiakos kyklos 黄道带
zoelly turbine 多级压力式汽轮机
Zoiga block 若尔盖地块
zoisite 黝帘石
zolertine 苯哌乙四唑
zolestin 天青石
zonal 区域(性)的;纬向;地区性的;带状的;成带的;分区的;分带的
zonal aberration 域像差;带域像差;带像差
zonal aeration 分区充气
zonal air flow 纬向气流
zonal anisotropy 带状异向性
zonal approach 分区(办)法
zonal biostratigraphy 分带生物地层学
zonal boundary 带型边界
zonal catalogue of stars 分区星表
zonal centrifugation 区带离心(法)
zonal centrifuge 分带离心机
zonal charges 分区收费
zonal circulation 纬向气流;纬向环流;地区性环流;带状环流
zonal coefficient 区域系数
zonal combustion 区域燃烧

zonal crenulation cleavage 带型褶劈理
zonal current 纬向流
zonal curve 晶带曲线
zonal dam 分区坝
zonal deposition 带状沉积
zonal deposition of ore 矿石带状沉积
zonal discrete boundary 带型分离边界
zonal distribution 带状分布
zonal distribution of mineral deposits 矿床分带
zonal electrophoresis 分段电泳
zonal embankment 分区填筑
zonal esterlies 纬向西风带
zonal factor 地带性因素
zonal filter 区域滤光器
zonal flow 纬向(气)流;带状环流
zonal geologic(al) profile 综合地质剖面
zonal growth 带状生长;带状结晶;(晶体的)带状成长
zonal guide fossil 分带标准化石
zonal harmonic 带调和函数;球带调和数;带谐函数
zonal hyper-spheric(al) function 超球带函数
zonal index 纬向指数;纬向指数;地带性指标
zonal injection 分层注水
zonality 地区性;地带性
zonality of the factors of regional engineering geology 区域工程地质因素分带性
zonal kinetic energy 纬向动能
zonal light flux 球面带光通量
zonallysin 透明带溶素
zonal mineral 分带矿物
zonal ozone transport 臭氧纬向输送
zonal plankton 分区浮游生物
zonal planting 地带性种植
zonal production 分层开采
zonal profile 综合剖面;分带剖面
zonal ray 区域射线
zonal rectification 分带纠正
zonal reflux 熔区回流
zonal rotation 分带自转
zonal sampling 区域抽样(法);分区抽样(法)
zonal scan 区域搜索
zonal settling velocity of sludge 污泥成区沉降速度
zonal soil 显域土;区域性土;地带性土(壤)
zonal spheric(al) function 球带函数;带球函数
zonal spheric(al) harmonics 球面带谐函数;球带调和
zonal structure 环带构造;区域构造;多壳结构;带状组织;带状构造;层状组织
zonal temperature 线圈温度
zonal texture 环带结构;带状组织;带状结构
zonal theory 分带理论
zonal tide 带潮
zonal vegetation 地带性植被;分区植被
zonal ventilation 区域通风
zonal westerlies 西风带
zonal wind 纬向风;带状气流;带状风;西风带
zonal withdrawal 分层开采
zonary 带状的
zonate(d) 带状的;成带的
zonation 地带性;带状排列;成带(现象);分(布)带
zonation criterion 区划准则
zonation map 区划图
zonation of basin 盆地分带
zonation of coral reef 珊瑚礁分带
zonation pattern 成带型
zone 区域;区(段);三行区;地区;地带
zone and overtime registration 按区域和时间记录
zone-area method 分区面积法
zone axis 晶带轴;带轴线
zone beam 区域波束
zone bit 区域元;区域位;区位;区段位;分区位;标志位
zone blanking 区域遮没;地区消隐
zone boundary 区域(办)界;区界;时区范围;分区界
zone broadening 区域宽化
zone bundle 晶带束
zone center 中心台;区长途电话局;分区中心
zone centroid 小区中心
zone charge 装药号
zone chart 分区图板
zone circle 晶带圈;区划圈;带圈
zone cleared of buildings 建筑物的成区拆除;建筑物的成区清理
zone coefficient 地区系数

zone condemnation 地带征用;地带收用
zone control 分区(温度)控制;恒温器控制;区域控制
zone control system 分区控制系统
zone coverage 区域波束覆盖范围
zone coverage beam 区域覆盖波束
zone curve 距平距限曲线
zoned (高层建筑分区缩迭的)各层楼层;分区
zoned-air 分区送风
zoned air-conditioning 分区空气调节
zoned-air control 分区通风调节
zoned-air distribution 分段送风
zoned decimal 区段十进制
zoned decimal number format 区域式十进数格式
zoned earth dam 分区式土坝;分段式土坝;非匀质土坝;非均质土坝
zoned earth fill dam 分区填土;分段填土
zoned embankment 分区填坝;分块填筑
zone description 区号;时区(标)号
zoned format 区位格式;区标格式;分区形式;分区格式;分段格式
zoned housing code 区划住房法规
zone diagram 区域图;分区图表
zone digit 区段数
zone distance 区间距离
zone dividing meridian 分带子午线
zone-diving of Gaussian projection 高斯投影分带
zoned lens 区域透镜
zoned lens antenna 分区透镜天线
zoned lining 均衡炉衬;特殊部位砖衬;特殊部位炉衬
zoned magnetic field chart 磁场分区图
zoned scheme 划区方案;分区布置
zoned tempered glass 区域钢化玻璃
zoned type of construction 分区式施工
zoned unit 有单独采暖设备的房间;分区采暖单元
zone electrophoresis 区带电泳;层(状)电泳法
zone embankment 分段填筑
zone extinction angle method 晶带消光角法
zone face 晶带面;带面
zone fare 分区计价
zone-fire two-zone continuous furnace 分段加热双段连续炉
zone focus 区域调焦
zone focusing 区域聚焦;区间调焦
zone fo mild pollution 轻污染带
zone for economic activities 经济建设地区
zone formet 区段格式
zone fossil 分带化石
zone freezing 区域凝固
zone freight rate 地区统一运费率
zone fusion 区域熔融
zone grouting 分区灌浆
zone-hardened alloy 区域硬化合金
zone heating 暖气分区;分区供暖
zone height compliance 符合区划规定高度
zone identification 区识别
zone law 晶带定律
zone law of Weiss 外斯晶带定律
zone length 熔区长度;区域长度
zone letter 区字号
zone-leveled 区域匀化
zone leveled single crystal 区域均化单晶
zone leveler 区域匀化器
zone level(l)ing 区域平均法;区域均化;区域致匀
zone line 带线
zone location 区域定位
zone map 区域图案
zone marker 区域标志器
zone marker beacon 区域指点标
zone marking paint 标区涂料
zone melted metal 区域精炼金属
zone melt(ing) 区域熔融;区域熔化(法);区域精炼法;区熔
zone melting analog computer 区熔模拟计算机
zone melting apparatus 区熔法装置
zone melting by electric(al) discharge 放电加热的区域熔炼
zone melting chromatography 区熔色谱法
zone melting crystallization 区域熔融结晶
zone melting growth 区域熔融法成长;区熔生长法
zone melting process 分区熔化法
zone melting purification 区域熔融提纯
zone melting technique 区域熔化技术;区熔技术
zone melting with focused radiation 辐射聚焦的

区域熔炼
zone melting with laser 激光区域熔化
zone meridian 分区子午圈
zone metering 分区统计；分区记录；分区计量；按区统计；按区登录
zone metering system 分区计量方式
zone migration 区域迁移
zone movement 区的移动
zone noon 分区正午
zone observation 分区观测
zone obstruction 区域遮没
zone of ablation (冰川)消融带；消融区
zone of absolute fatality 绝对致死带
zone of accidental distribution 偶然分布区；偶然分布带
zone of accumulation 堆积带
zone of action 作用区(域)
zone of aeration 曝气层；通气带；通气层；渗水曝气区；渗水触气区；充气带；掺气层；饱气带
zone of aerodynamic(al) shadow 空气动力阴影区
zone of affected overburden 上部沉积影响带；地表影响带
zone of ambiguity 不定区
zone of audibility 听阈；耳闻阈；能听区
zone of avoidance 隐带
zone of bank storage 河岸调蓄带
zone of belt 褶皱层
zone of bending tension 弯曲张力区
zone of biologic(al) effect 生物作用带
zone of boat mooring 船只停泊区
zone of capillarity 毛细管(水作用)带
zone of capillary 毛细管作用带；毛细管水带
zone of capillary saturation 毛细管水饱和带；毛管水饱和层
zone of capillary water 毛细水带
zone of cementation 胶结带
zone of circulation 环流层；流动层
zone of combustion 燃烧带；燃烧层
zone of comfort 舒适区(域)
zone of competition 竞争地带
zone of concentration 淀积层
zone of conductivity 导电区
zone of connection 胶接带
zone of consolidation 凝固带；凝固区
zone of constant temperature 恒温带；等温带
zone of contact 接触区；接触面
zone of contact strain 接触应变
zone of convergence 聚合区域；交汇带；辐合区(域)
zone of cracking 裂化区
zone of crush 挤压带；破碎带
zone of crustal weakness 地壳薄弱带
zone of crystalline rocks 结晶岩带
zone of deformation 变形带
zone of deformation around pile 桩周围变形区
zone of departure 出发区
zone of deposition 沉积带
zone of depression 沉降带
zone of different permeability 不同渗透带
zone of diffuse(d) scattering 漫射带；散射带
zone of diffusion 扩散带
zone of discharge 泄水带
zone of discontinuity 不连续(能)带
zone of discontinuous soil moisture 不连续土壤水分带
zone of disease 发病带
zone of dispersion 散布面积
zone of dissipation 消融区
zone of effective temperature 有效温度(地)带
zone of eluviation 淋滤层
zone of enrichment 富集带
zone of erosion 侵蚀带
zone of eventual death 最后致死(地)带
zone of faces 晶带
zone of fault 断层带
zone office 地区办事处
zone of fire 射界
zone of flame 火焰层
zone of flowage 岩流带；软流带；泛溢带；过流带；流动带；流变带
zone of fold 褶皱带
zone of fossil 化石层
zone of fracture 破碎带；破裂带；断裂带
zone of freezing 冻结带；冰冻带

zone of frozen 冻结区
zone of function 功能区
zone of growth 生长带
zone of heating 加热区；加热带；加热层
zone of height limitation 净高控制；建筑高度限制区
zone of highest breakers 最高碎波带；最高破波带
zone of highest breaks 最高碎波带
zone of high fatal temperature 致死高温带；高致死温度带；高温致死地带
zone of illuviation 淀积带
zone of immediate death 直接致死(地)带；立即致死带
zone of incandescence 白热带
zone of increasingly favorable temperature 逐渐适应温度带
zone of indifference 两可地带
zone of influence 势力范围；影响区；影响范围
zone of instability 不稳定地带
zone of intense fracture 严重破碎地带
zone of intense fracturing 剧烈断裂带
zone of interfingering 交叉层
zone of interior 后方地带；内部区域
zone of intermediate tide 中潮带
zone of isolation 生产层隔离法
zone of latitude 纬度带
zone of leaching 淋滤带
zone of least resistance 最小阻力区
zone of littoral drift 沿岸漂沙区；沿岸漂沙带；海岸漂沙区；海岸漂流带
zone of loose rock 坍塌带
zone of loss 漏失带
zone of low critical temperature 临界低温范围
zone of low fatal temperature 致死低温带
zone of low-temperature carbonization 低温碳化层
zone of magma 岩浆带
zone of maturation 成熟区
zone of maximum crystal formation 最大结晶区
zone of maximum pollution concentration 最大污染浓度带
zone of maximum precipitation 最大降水带
zone of mean tide 中潮带
zone of merchandise 商品销售区
zone of metamorphism 变质带
zone of middle and small industry 中小工业区
zone of migration 洄游区；迁移区；变迁范围
zone of mild pollution 轻(度)污染区；轻(度)污染带
zone of mineralization 矿化带
zone of mixing 混合带
zone of mobility 活动带
zone of natural foci 自然疫源地带
zone of negative pressure 负压区；负压带；负压层
zone of normal distribution 正态分布区；正态分布带
zone of observation 观察区；观测区
zone of opacity 不透明区(域)；不透明带；暗区；暗带
zone of optic(al) visibility 光学通视区
zone of optimal proportion 最适比例带
zone of optimum condition 最适条件区；最适条件带；适合条件带
zone of ore rank 矿石品级带
zone of ore type 矿石类型带
zone of oxidation 氧化带；氧化层
zone of parking 停车区
zone of percolation 浸透带；渗滤带；渗漏带
zone of phreatic fluctuation 潜水波动带
zone of pivoting pressure action 尾浮加强区(船)
zone of plastic flow 塑性流动带；塑流区
zone of plastic flow of soil 土的塑性流动区
zone of pollutant groundwater 地下污染水扩散带范围
zone of pollution 污染(地)带
zone of positive pressure 正压区
zone of possible distribution 可能分布区；可能分布带
zone of potential heave 潜伏隆起带；可能隆起带
zone of preference for rejection 不合格域
zone of preparation 预热带
zone of pressure 气压带
zone of processing 加工区
zone of production 产油带

zone of protection 保护区；保护范围
zone of protective gummosis 流胶保护圈
zone of provisional calcification 临时钙化区
zone of radar coverage 雷达覆盖空域
zone of radial shear(ing) 辐向剪力区
zone of recent pollution 新污染区
zone of recompression 再压缩带
zone of reduction 还原带
zone of reflections 反射带
zone of relaxation of rock 围岩松弛范围
zone of removal 表土层
zone of residential 住宅区
zone of residual spheric(al) aberration 剩余球面像差带
zone of reverse flow 逆流区；倒流区
zone of rip current 裂流带
zone of rock flowage 岩石流变带；石隙滤层；石隙流层；软流层
zone of rock fracture 岩石裂隙带；岩石破碎带；石隙层
zone of saturation 饱水带；饱和区(域)；饱和带；饱和层
zone of seasonal change 季节性变化地带；季节变化层
zone of seasonal variation 季节性变化地带
zone of secondary enrichment 次生富集带
zone of self-purification 自净区
zone of semi-coking 半焦化层
zone of sharp focus 锐聚焦区；调焦清晰区域
zone of sharpness 清晰区域
zone of silence 静风区；无声区；无风带；跳跃区；声盲区；平静区；静水区
zone of slippage 滑移区
zone of slippage on the entry side 后滑区
zone of soil water 土壤水带
zone of specific effect 特异作用带
zone of sphere 球带
zone of stability 稳定地带
zone of stagnation 停滞区
zone of subsidence 下沉区；沉降带
zone of suspended water 悬着水带；重力水层；土中悬着水；通气层；包气带
zone of swelling 膨胀区
zone of synchronization 同步区
zone of tectonic disturbance 构造变动带
zone of time 时区
zone of trade-wind 信风带
zone of transition 过渡区；转变区；过渡(地)带；渐变区；缓和区；变化层
zone of transverse force 横向力区
zone of undesigned land use (城市规划中)未指定土地用途区
zone of vadose water 流动水带；浅潜水带
zone of variable phreatic level 潜水位变化带【地】
zone of vegetation 植被(地)带
zone of vision 视区
zone of wake 尾流区
zone of wastage (冰川)消融带；(冰川)消融带
zone of waste 消融区
zone of wave breaking 波浪破碎带
zone of wave convergence 波浪汇聚区；波浪辐聚区
zone of wave divergence 波浪收敛区；波浪扩散区；波浪辐散区
zone of weakness 弱点
zone of weathering 风化层；风化(地)带
zone of weld 焊缝表面
zone of welding current 焊接电流调节范围
zone of worker's homes 工人居住区
zone of workingmen's homes 工人住宅区
zone of zero gravity 无重力区
zone of zero slip 零位滑移区；黏着区；无滑移区
zone of zero stress 零应力区；零应力带
zone og folding 褶皱区
zone passage 区域通过
zone patterns of sewerage system 分区式下水道系统
zone pen 区划掩蔽所
zone perturbation 区域扰动
zone phenomenon 区域现象；带状现象
zone plate 环板；波片区；波带片；同心圆绕射板
zone-plate coded imaging 波带片编码成像
zone-plate interferometer 波带片干涉仪
zone plate telescope 波带片望远镜；波域片望远镜

zone position indicator 分区位置显示器;区域位置指示器
zone precipitation 区域沉淀法
zone price 区域(性)价格
zone pricing 区域(性)定价;分区定价法
zone pricing system 分区定价制度
zone-pumping 局部汲水
zone punch 区段穿孔;三行区穿孔;分位穿孔;分区穿孔
zone purification 区域提纯;区域精制;区域精炼;区熔提纯;纯化区
zone purification device 区域提纯装置
zone-purified metal 区域提纯金属
zone rate 地区价格
zone receiver 区域波束接收机
zone recrystallization method 区域再结晶法
zone-refine 区域精炼;区熔提纯
zone-refined material 区域提纯材料
zone refiner 区域提纯器;区域熔炼器;区域精炼炉
zone refining 区域提纯法;区域熔炼;区域精制;区域精炼(法);区域熔化;纯化区
zone refining of chemicals 化学试剂区域精炼
zone refining unit and equipment 区域提纯设备和装置
zone reflector 区域波束反射器
zone registration 按区记录
zone sampler 区层取样器;分层取样js;分层取样管
zone scanning 区域扫描
zone segregation 熔区偏析
zone service 区段间运输
zone-setting 区域设置;存储区建立
zone settlement 区域沉降
zone settlement rate 区域沉降速率
zone settling 分区沉降;分区沉淀;区域沉降;成层沉淀;层状沉降
zone settling rate 区域沉降速率
zone settling velocity 区域沉降速度
zones for sand grading 砂子级配区
zone shape 熔区形状
zone sintering 区域烧结
zones of deformation around piles 桩周围变形区
zones of different 不同区域
zones of environment(al) disaster 环境灾害区
zones of groundwater quality suitable 适用地下水水质带
zone soil 区域性土
zone spreading 区域扩展
zone standard time 区标准时;当地标准时(间);标准时区
zone star 分区星
zonesthesia 束带状感觉
zone switch(ing) center 区域交换中心;区转换中心;电话区域交换中心
zone symbol 晶带符号
zone system 分区式(供暖)系统;区域制;城市分区系统
zone system of pricing 分(地)区定价制度
zone television 分片电视
zone theory 能带理论
zone tillage 分层耕作
zone time 区域时间;区时;时区时间;标准时(间)
zone time system 区时制
zone-to-zone traffic 区间交通
zone-to-zone travel 区间交通
zone transfer technique 层递技术
zone transport method 区域输运法;区域传输法
zone transport refiner 熔区传输精炼炉;区域输运提纯器
zone travel mechanism 熔区移动机构
zone velocity 区域速度
zone-void method 熔区空段法;区域空段法
zone-void process 区熔空段法
zone-void refiner 熔区空段精炼炉;区域空段提纯器
zone yard 地区编组场【铁】;分区车场【铁】
zoning 机场区域注记;环带生长;区域制;区域划分;区域规定;区划;区分;地带划分;带现象;成区;分区(制);分区取样;分区规划;分期制(美国);分局布局;分带
zoning act 分区法令
zoning administrator 区划行政官
zoning a drawing 将图纸分区
zoning amendment 区划修正
zoning board 分区规划委员会
zoning by-law 地方性用地分区管理条例

zoning by right 区划授权
zoning classification 分区等级
zoning coefficient 分带系数
zoning commission 城市规划委员会
zoning control 区域(规划)控制
zoning district 区划市区
zoning district dividing property 分区地产
zoning factor 区域系数
zoning gradient 分带梯度
zoning if intercalated soft layer 泥化夹层分带
zoning index 分带指数
zoning intensity 基本烈度
zoning in urban area 城市区域的分区
zoning law 分区法规;分区管制;分区法
zoning map 分区图;分带图
zoning mechanism 分区机理
zoning of concrete 混凝土分区
zoning of gas 瓦斯分带
zoning of ore deposit 矿分带
zoning of phreatic water 潜水的分带性
zoning of sedimentary mineral deposits 沉积矿床分带
zoning of yard 调车场分区
zoning order 分带等级
zoning ordinance 分区条例;分区规则;分区法规
zoning permit 区域规划执照;区划许可证
zoning plan 区域划分图;区间图;区划(平面)图;分区规划图
zoning planning 分区规划
zoning price 分区定价
zoning process 分区过程
zoning regulation 分区规则;地区法规;分区条例;分区法规;分区规章
zoning scheme 划区方案;分区布置
zoning seismic activity 分区地震活动
zoning sequence 分带序列
Zono 稠诺牌灰浆(一种吸声灰浆)
zonobiome 地带生物群系
zonochlorite 绿纤石
zonoecotone 地带生物群落过渡带
zonolite 钛水蛭石;金蛭石;烧蛭石(可做建筑材料)
zonule 小区域;小环;小带;小城区
zonuls occludens 闭锁小带
zoo 动物园
zoobenthos 底栖动物
zoobiocenose 动物群落
zooecium 虫室
zoo-ecology 动物生态学
zoogenic argillaceous texture 动物泥质结构
zoogenic rock 动物岩
zoogenous rock 动物岩
zoogeographic(al) barrier 动物地理阻限
zoogeographic(al) region 动物地理区
zoogeography 动物地理学
zooglea 菌胶团
zooglea mass 菌胶团块
zoogloea ramigera 分支菌胶团
zooid 个员
zoolgeal matrix 菌胶团基质
zoolite 动物化石
zoolith 动物岩;动物化石
zoological garden 动物园
zoology 动物学
zoom 影像放大;可变焦距;图像电子放大;缩放;变焦距
zoom adapter 焦距变换器
zoomar 可变焦距透镜系统;可变焦距镜头系统;变焦距物镜系统
zoomaric acid 棕榈油烯酸
Zoomar lens 摇臂变焦距镜头
zoom axis 缩放轴
zoom back 移后
zoom binocular 变焦距双筒望远镜
zoom converter 变焦倍率镜
zoom copier 变焦复印机
zoomed-out window 放大窗口
zoom factor 缩放因子
zoom fast Fourier transformation 变频距快速傅立叶变换
zoom finder 变焦距寻像器;变焦探视器;变焦距取景器;可变焦距寻像器
zoom graph 缩放图
zoom image intensifier 变焦距图像增强器

zoom image intensifier tube 变倍率像增强管
zoom image tube 变倍像管
zoom in 移前
zooming 图形变比;缩放;跃升;移向目标;电算;变焦
zooming in 缩小
zooming out 放大
zooming ratio 变焦比
zoom lens 可变焦距透镜;可变焦距镜头;变焦距透镜;变焦(距)镜头
zoom lens optical system 变焦距透镜光学系统
zoom macroscope 变焦放大观测器
zoom magnification 变焦放大
zoom microscope 变焦(距)显微镜
zoom optics 变焦距光学装置;变焦光学器件
zoomorphic 动物形象的;兽形的
zoomorphic column 兽形柱
zoomorphic ornament 兽形装饰
zoomorphism 动物图案
zoomorphism in ornament 兽形装饰
zoom pick-up tube 变倍率摄像管
zoom ratio 焦距比;变焦距倍率;变焦比
zoom ring 变焦距圈
zoom scale factor 缩放比例系数
zoom stereo interpretoscope 变倍立体判读仪
zoom stereoscope 可变焦立体镜
zoom stereoscopic microscope 变焦距立体显微镜
zoom system 变焦距系统
zoom table 速查表
zoom transferscope 自动调焦转绘仪;变焦转绘仪
zoom type 可变焦距式
zoom viewfinder 变焦距取景器
zoop 调制噪声
zooparasite 寄生动物
zoophilous 人兽饰盘壁
zoophoric column 兽形柱
zoophorism in ornament 装饰小兽
zoophorus 古建筑挑檐装饰;古建筑挑檐腰线上的人或兽形装饰;屋柱的中楣
zoophyte 植物性动物;植虫
zooplankton 浮游动物
zooplankton production 浮游动物产量
zoosphere 动物圈
zoostratigraphy 动物地层学
zootechnical science 畜牧学
zootope 动物生境
zooturbation structure 动物扰动构造
zooxanthellae 虫黄藻
zores bar 瓦垅铁;波纹钢板
zores beams 波纹钢板
zorite 佐硅钛钠石
zorsite 糟绿帘石
zotepine 苯噻庚乙胺
zotheca 起居室外的壁龛;起居凹室;套间
Z-parameter 开路阻抗参数
Z-piling Z 形钢板桩
Z-piling bar Z 字形钢板桩
Z plot Z 值图
Z-profile wire Z 形剖面钢丝
Z-rib metal lath Z 形钢丝网;Z 肋钢丝网
Z-rib pattern Z 形形式;Z 形肋布置
ZrO2-bearing magnesia brick 含锆镁砖
Z-section Z 形截面;Z 形断面
Z-section steel sheet pile Z 形钢板桩
Z-shaped double nail puller Z 形双拔钉机
Z-shaped fold Z 形褶皱
Z-shape splitting Z 形分岔
ZSM quartz-spring gravimeter ZSM 型石英弹簧重力仪
Z-steel Z 字型钢;Z 形钢
Z-steel section Z 形型钢
Z-stool Z 形钢筋垫
Z-tie Z 形埋铁;Z 形系墙铁
Z time 格林威治平时
Z-transform Z 形变换
Z-transform value Z 变换值
Z-truss Z 形桁架
Z-twist 右捻;反手捻;Z 捻度
Z-type piling bar Z 字板桩;Z 形板桩
Zublin bit 独牙轮钻头;单牙轮钻头
Zublin differential bit 差异式独牙轮钻头;差动式单牙轮钻头
zulu time 格林威治平时
zungenbecken 盘谷;冰川舌状盆地
zunyite 氯黄晶

Zurich classification 苏黎世分类
Zurich number 苏黎世数
zussmanite 硅钾镁铁矿;菱硅钾铁石
Z variometer 竖直强度磁力仪
zvyagintesevite 等轴铅钯矿
Zwenger's test 茨文格试验
zwieslite 氟磷铁石
zwinger 附廊要塞;城堡外廊;保卫城市的要塞;外栅;城镇的防御要塞
zwitter 锡石云英岩
zwitterion 阴阳离子;偶极离子

zwitterionic compound 两性离子化合物
zwitterionic detergent 两性离子洗涤剂
zwitterionic surfactant 两性离子表面活性剂
zwitterions 两性离子
zyglo 荧光探伤器;荧光探伤法
Zyglo penetrant method 杰格罗油浸探伤法;荧光探伤法
zygodactyl 对趾
zygodactylous 对趾的
zygodont 对尖齿
zygo-pachynema 偶粗线

zygospore 接合孢子
zygotene stage 合线期
zykaite 水硫砷铁石
Zylex 齐肋斯牌油毛毡(一种加强的屋面油毛毡)
zymogeneous bacteria 发酵性细菌
zymogeneous flora in soil 土壤中发酵性微生物区系
zymolysis 发酵
zymosis zone 发酵带
Zyrianka glacial stage 兹良冰期
Zyrianka stade 赞卡冰阶

数 字

1-aminocyclopropane-1-carboxylic acid 氨基环丙烷羧酸
1st titanium cooler 一段钛冷器
210Po method in borehole 钻孔中钋法
230Th-231Po deficiency method 钍 230—镤 231 亏损法
230Th-231Po excess method 钍 230—镤 231 过剩法
232Th-208Pb isochron 钍—铅 208 等时线
234U method 铀 234 法
234U/238U method 铀 234—铀 238 法
235U-207Pb isochron 铀 235—铅 207 等时线
238U-206Pb isochron 铀 238—铅 206 等时线
2-butanone 丁酮
2-cyanopropyl silicone 2—氰丙基硅酮
2D drafting system 二维制图系统
2-methylbenzoic acid 邻甲苯甲酸
2-ply tar(red) paper 二毡一油柏油纸
39Ar-40Ar isochron 氩—氩等时线
3-butyn-2-one 丁炔酮
3D model(l)ing system 三维模型系统
3D terrain modeling 三维地貌模型
3-phenyl-1-propanol 苯基丙醇
3-phenylpropionaldehyde 苯基丙醛
3-wing bit 三翼钻头
40Ar/39Ar ages 氩 39—氩 39 年龄
40Ar/39Ar method 氩—氩法
40K-40Ar isochron 钾 40—氩 40 等时线
4-aminophenetole 氨基苯乙醚
4-bolt splice bar 四栓孔鱼尾板
4-nitrophenyl phosphate 磷酸对硝基苯酯

参 考 文 献

[1] 罗新华.汉英港湾工程大词典[M].北京:人民交通出版社,2000.
[2] 中国社会科学院语言研究院.新华字典[M].11版.北京:商务印书馆,2011.
[3] 方中天.英汉港口工程词典[M].北京:人民交通出版社,2006.
[4] 本词典编写组编.日英汉土木建筑词典[M].北京:中国建筑工业出版社,1996.
[5] 邱民.英汉航海·航运·船舶大词典[M].北京:中国人民大学出版社,1995.
[6] 洪庆余.现代英汉水利水电科技词典[M].武汉:武汉出版社,1990.
[7] 谢凯成.英汉·汉英涂料技术词汇[M].北京:化学工业出版社,1999.
[8] BENGT B. BROMS[瑞典].英汉对照图示基础工程学[M].史佩栋,编译.北京:人民交通出版社,2005.
[9] 宁滨.英汉汉英铁路词典[M].北京:中国铁道出版社,2005.
[10] 张文健.汉英英汉地铁轻轨词汇[M].成都:西南交通大学出版社,2003.
[11] 本词汇编辑组.英汉测绘词汇[M].北京:测绘出版社,1985.
[12] 王业俊,许保玖.英汉给排水辞典[M].北京:中国建筑工业出版社,1989.
[13] 陈名扬.英汉船舶机电词典[M].北京:人民交通出版社,1996.
[14] 李若珊,柳克惜.新编英汉科技常用词汇[M].北京:宇航出版社,1994.
[15] 陆谷孙.英汉大词典(缩印本)[M].上海:上海译文出版社,1993.
[16] 夏行时.新英汉建筑工程词典[M].北京:中国建筑工业出版社,1999.
[17] 李开运.英汉建筑工程大辞典[M].南京:河海大学出版社,1989.
[18] 王同亿.英汉科技词天[M].北京:中国环境科学出版社,1987.
[19] 本词汇编订组.新英汉机械工程词汇[M].北京:科学出版社,2008.
[20] 马家驹,李骝.英汉铁路综合词典[M].北京:中国铁道出版社,1992.
[21] 潘钟林.英汉起重装卸机械词典[M].北京:人民交通出版社,1983.
[22] 李轸,朱和.英汉石油化工词典[M].北京:化学工业出版社,1997.
[23] 交通部基建管理司.英汉水运工程词典[M].北京:人民交通出版社,1997.
[24] 余昌菊,于渤,李学丹,等.英汉电工词汇[M].2版.北京:科学出版社,1987.
[25] 麦世基,陈宝春,刘惠,等.英汉工程机械词汇[M].哈尔滨:黑龙江科学技术出版社,1992.
[26] 本词典编辑组.英汉地质词典[M].北京:地质出版社,1983.
[27] 马怀平,邵伯岐.最新汉英审计·会计·金融大辞典[M].北京:中国审计出版社,1993.
[28] 周湘寅.实用汉英机电词典[M].北京:人民交通出版社,1994.
[29] 戎培康.英汉-汉英建材工业大词典[M].北京:中国建材工业出版社,2000.
[30] 张人琦.汉英建筑工程词典[M].北京:中国建筑工业出版社,1993.
[31] 赵祖康,徐以枋.汉英土木建筑工程词典[M].北京:人民交通出版社,1997.
[32] 赵祖康,黄兴安.英汉道路工程词汇[M].4版.北京:人民交通出版社,2001.
[33] 李育才.英汉建筑装饰工程词典[M].北京:中国建筑工业出版社,1997.
[34] 本词典编委会.新英汉建筑工程词典[M].北京:中国建筑工业出版社,1991.
[35] 科学出版社名词室.汉英生物学词汇[M].北京:科学出版社,1998.
[36] 科学出版社名词室.汉英化学化工词汇[M].北京:科学出版社,2007.
[37] 吴钰.英汉铁路工务工程词汇[M].北京:中国铁道出版社,2003.
[38] 辞海编辑委员会.辞海(缩印本)[M].6版.上海:上海辞书出版社,2010.
[39] 林鸿慈.英汉港口航道工程词典[M].2版.北京:人民交通出版社,1997.
[40] 白英彩.英汉计算机技术大词典[M].上海:上海交通大学出版社,1997.
[41] 中国土木工程学会,土力学及基础工程学会.土力学及基础工程名词(汉英及英汉对照)[M].北京:中国建筑工业出版社,1991.
[42] 李浑成,王逢辰,王建平,等.英汉/汉英航海词典[M].北京:人民交通出版社,1998.
[43] 林宗元.岩土工程勘察设计手册[M].沈阳:辽宁科学技术出版社,1996.

[44] 本词汇编辑组.英汉现代科学技术词汇[M].上海:上海科技出版社,1982.
[45] *Ground Engineer's Reference Book*, Edited by FG Bell with specialist contributors, Butterworths and Co (publishers) Ltd ,1987.
[46] British Standard(BS).
[47] American Society for Testing and Materials(ASTM).
[48] 中华人民共和国国家标准.GB/T 16803—97 采暖、通风、空调、净化设备术语[S].北京:中国标准出版社,1997.
[49] 中华人民共和国国家标准.GB 50155—92 采暖通风与空气调节术语标准[S].北京:中国标准出版社,1999.
[50] 中华人民共和国国家标准.GB/T 14911—2008 测绘基本术语[S].北京:中国标准出版社,2008.
[51] 中华人民共和国国家标准.GB/T 50280—98 城市规划基本术语标准[S].北京:中国标准出版社,1999.
[52] 中华人民共和国国家标准.GB/T 50833—2012 城市轨道交通工程基本术语标准[S].北京:中国建筑工业出版社,2012.
[53] 中华人民共和国国家标准.GB/T 17159—2009 大地测量术语[S].北京:中国标准出版社,2009.
[54] 中华人民共和国国家标准.GBJ 124—88 道路工程术语标准[S].北京:中国计划出版社,1989.
[55] 中华人民共和国国家标准.GB/T 16820—2009 地图学术语[S].北京:中国标准出版社,2009.
[56] 中华人民共和国国家标准.GB/T 17228—98 地质矿产勘查测绘术语[S].北京:中国标准出版社,1998.
[57] 中华人民共和国国家标准.GB/T 50297—2006 电力工程基本术语标准[S].北京:中国计划出版社,2006.
[58] 中华人民共和国行业标准.JGJ/T 30—2003 房地产业基本术语标准[S].北京:中国建筑工业出版社,2003.
[59] 中华人民共和国国家标准.GB 50186—93 港口工程基本术语标准[S].北京:中国标准出版社,1994.
[60] 中华人民共和国行业标准.JT/T 392—2013 港口装卸工属具术语[S].北京:人民交通出版社,2014.
[61] 中华人民共和国国家标准.GB/T 50125—2010 给水排水工程基本术语标准[S].北京:中国计划出版社,2010.
[62] 中华人民共和国国家标准.GB/T 50228—2011 工程测量基本术语标准[S].北京:中国计划出版社,2012.
[63] 中华人民共和国国家标准.GB/T 8843—2002 工程船术语[S].北京:中国标准出版社,2003.
[64] 中华人民共和国国家标准.GB/T 7391—2002 海洋调查船术语[S].北京:中国标准出版社,2003.
[65] 中华人民共和国行业标准.JTJ/T 204—96 航道工程基本术语标准[S].北京:人民交通出版社,1997.
[66] 中华人民共和国国家标准.GB/T 50731—2011 建材工程术语标准[S].北京:中国计划出版社,2012.
[67] 中华人民共和国行业标准.JGJ/T 191—2009 建筑材料术语标准[S].北京:中国建筑工业出版社,2010.
[68] 中华人民共和国国家标准.GB/T 50083—97 建筑结构设计术语和符号标准[S].北京:中国建筑工业出版社,1998.
[69] 中华人民共和国行业标准.JGJ 84—92 建筑岩土工程勘察基本术语标准[S].北京:中国建筑工业出版社,1995.
[70] 中华人民共和国行业标准.JGJ/T 119—2008 建筑照明术语标准[S].北京:中国建筑工业出版社,2009.
[71] 陈宽基.科技标准术语词典 第一卷 综合[M].北京:中国标准出版,1995.
[72] 中华人民共和国国家标准.GB/T 50504—2009 民用建筑设计术语标准[S].北京:中国计划出版社,2009.